Comprehensive Organic Functional Group Transformations

Comprehensive Organic Functional Group Transformations

Editors-in-Chief

Alan R. Katritzky, FRS
University of Florida, Gainesville, FL, USA

Otto Meth-Cohn
University of Sunderland, UK

Charles W. Rees, FRS
Imperial College of Science, Technology and Medicine, London, UK

Volume 4
SYNTHESIS: CARBON WITH TWO HETEROATOMS,
EACH ATTACHED BY A SINGLE BOND

Volume Editor
Gordon W. Kirby
University of Glasgow, UK

PERGAMON

UK	Elsevier Science Ltd., The Boulevard, Langford Lane, Kidlington, Oxford OX5 1GB, UK
USA	Elsevier Science Inc., 660 White Plains Road, Tarrytown, New York 10591-5153, USA
JAPAN	Elsevier Science Japan, Tsunashima Building Annex, 3-20-12 Yushima, Bunkyo-ku, Tokyo 113, Japan

Copyright © 1995 Elsevier Science Ltd.

All rights reserved. No part of this publication may be reproduced, stored in any retrieval system or transmitted in any form or by any means: electronic, electrostatic, magnetic tape, mechanical, photocopying, recording or otherwise, without permission in writing from the publishers.

First edition 1995

Library of Congress Cataloging in Publication Data
Comprehensive organic functional group transformations / editors-in-chief, Alan R. Katritzky, Otto Meth-Cohn, Charles W. Rees. — 1st ed.
   p. cm.
   Includes indexes.
   Contents: v. 1. Synthesis: carbon with no attached heteroatoms / edited by Stanley M. Roberts — v. 2. Synthesis: carbon with one heteroatom attached by a single bond / edited by Steven V. Ley — v. 3. Synthesis: carbon with one heteroatom attached by a multiple bond / edited by Gerald Pattenden — v. 4. Synthesis: carbon with two heteroatoms, each attached by a single bond / edited by Gordon W. Kirby — v. 5. Synthesis: carbon with two attached heteroatoms with at least one carbon-to-heteroatom multiple link / edited by Christopher J. Moody — v. 6. Synthesis: carbon with three or four attached heteroatoms / edited by Thomas L. Gilchrist — v. 7. Indexes.
   1. Organic compounds—Synthesis.   I. Katritzky, Alan R.   II. Meth-Cohn, Otto.   III.   Rees, Charles W.
QD262.C534   1995
547'.2—dc20   95–31088

British Library Cataloguing in Publication Data
A catalogue record for this book is available from the British Library.

ISBN 0–08–040604–1 (set : alk. paper)
ISBN 0–08–042325–6 (Volume 4)

∞™ The paper used in this publication meets the minimum requirements of the American National Standard for Information Sciences—Permanence of Paper for Printed Library Materials, ANSI Z39.48–1984.

Printed and bound in Great Britain by Cambridge University Press, Cambridge, UK.

Contents

	Preface	vii
	Contributors to Volume 4	ix
	Contents of All Volumes	xi
	Introduction	xv
4.01	Dihalo Alkanes, $R_2C(Hal)_2$ R. A. HILL, *University of Glasgow, UK*	1
4.02	Functions Incorporating a Halogen and a Chalcogen N. W. A. GERAGHTY, *University College Galway, Republic of Ireland*	41
4.03	Functions Incorporating a Halogen and Another Heteroatom Group Other Than a Chalcogen A. C. CAMPBELL and D. R. JAAP, *Organon Laboratories Ltd, Newhouse, UK*	95
4.04	Functions Bearing Two Oxygens, $R^1_2C(OR^2)_2$ D. T. MACPHERSON and H. K. RAMI, *SmithKline Beecham Pharmaceuticals, Epsom, UK*	159
4.05	Functions Incorporating Oxygen and Another Chalcogen R. H. WIGHTMAN, *Heriot-Watt University, Edinburgh, UK*	215
4.06	Functions Incorporating Two Chalcogens Other Than Oxygen Y. VALLÉE, *Université Joseph Fourier, Grenoble, France*, and A. BULPIN, *Institut Jacques Boy, Reims, France*	243
4.07	Functions Incorporating a Chalcogen and a Group 15 Element C. D. GABBUTT and J. D. HEPWORTH, *University of Central Lancashire, Preston, UK*	293
4.08	Functions Incorporating a Chalcogen and a Silicon, Germanium, Boron or Metal M. J. GOUGH, *Technical Typesetters UK, Ashford, UK*, and J. STEELE, *Pfizer Central Research, Sandwich, UK*	351
4.09	Functions Bearing Two Nitrogens D. R. BUCKLE and I. L. PINTO, *SmithKline Beecham Pharmaceuticals, Epsom, UK*	403
4.10	Functions Containing a Nitrogen and Another Group 15 Element F. HEANEY, *University College Galway, Republic of Ireland*	451
4.11	Functions Incorporating a Nitrogen and a Silicon, Germanium, Boron or a Metal J. STEELE, *Pfizer Central Research, Sandwich, UK*, and M. J. GOUGH, *Technical Typesetters UK, Ashford, UK*	505
4.12	Functions Containing One Phosphorus and Either Another Phosphorus or As, Sb, Bi, Si, Ge, B or a Metal R. A. AITKEN, *University of St Andrews, UK*	543
4.13	Functions Containing at Least One As, Sb or Bi with or without a Metalloid (Si or Ge) or a Metal W. M. HORSPOOL, *University of Dundee, UK*	591
4.14	Functions Containing at Least One Metalloid (Si, Ge or B) Together with Another Metalloid or Metal C. G. BARBER, *Pfizer Central Research, Sandwich, UK*	601
4.15	Functions Containing Two Atoms of the Same Metallic Element W. J. KERR and P. L. PAUSON, *University of Strathclyde, Glasgow, UK*	667

4.16	Functions Containing Two Atoms of Different Metallic Elements W. J. KERR and P. L. PAUSON, *University of Strathclyde, Glasgow, UK*	705
4.17	Functions Incorporating Two Halogens or a Halogen and a Chalcogen P. D. KENNEWELL, *Hoechst Roussel Ltd, Swindon, UK*, R. WESTWOOD, *Roussel-Uclaf, Romainville, France*, and N. J. WESTWOOD, *University of Oxford, UK*	729
4.18	Functions Incorporating a Halogen or Another Group Other Than a Halogen or a Chalcogen D. I. SMITH, *Sanofi Research Division, Alnwick, UK*	789
4.19	Functions Bearing Two Chalcogens G. N. SHELDRAKE, *The Queen's University of Belfast, UK*	823
4.20	Functions Containing a Chalcogen and Any Group Other Than a Halogen or a Chalcogen P. D. KENNEWELL, *Hoechst Roussel Ltd, Swindon, UK*, R. WESTWOOD, *Roussel-Uclaf, Romainville, France*, and N. J. WESTWOOD, *University of Oxford, UK*	879
4.21	Functions Containing at Least One Nitrogen and No Halogen or Chalcogen G. L. PATRICK, *University of Paisley, UK*	967
4.22	Functions Containing at Least One Phosphorus, Arsenic, Antimony or Bismuth and No Halogen, Chalcogen or Nitrogen J. M. BERGE, *SmithKline Beecham Pharmaceuticals, Epsom, UK*	1021
4.23	Functions Containing at Least One Metalloid (Si, Ge or B) and No Halogen, Chalcogen or Group 15 Element; also Functions Containing Two Metals R. B. WEBSTER, *Pfizer Central Research, Sandwich, UK*	1043
4.24	Tri- and Dicoordinated Ions, Radicals and Carbenes Bearing Two Heteroatoms $(RC^+X^1R^2, RC^-X^1X^2, RC \cdot X^1X^2, :CX^1X^2)$ W. M. HORSPOOL, *University of Dundee, UK*	1071
	References	1085
	Subject Index	1231

Preface

Some years ago the three of us met in a London club reviewing an ongoing publishing venture in Organic Synthesis. The conversation drifted to a consideration of volumes on the synthesis of key functional groups. No doubt the good wine helped since we actually broached the idea of a work on the synthesis of *all* functional groups. Would it be useful? Definitely. Would it be feasible? How would it be organized? Where do you start? We recognized that functionality was based on the coordination and heteroatom attachment of a carbon atom. But putting together a complete framework seemed particularly daunting. Two of us became very interested in the fascinating bouquet of the Muscat de Beaumes de Venise.

At our next dinner together Alan announced that he had solved the problems posed last time—problems that Charles and I hoped he had forgotten! He brought out a remarkable matrix analysis of *all* functional groups, analysed rigorously and logically. Even unknown functions were covered. Although we were all very impressed, the practicalities of the idea still seemed daunting. Those who know Alan's terrier instincts will appreciate that he would not give up such a challenge so easily. Our twice yearly club get-togethers, occasionally with friends from Pergamon, refined our thinking. Alan's cosmic vision was tempered by Charles's intuitive realism and fully supported by the publishers.

Another major problem remained: how to reduce our thinking into a practical handbook for authors—a dismaying task for three busy chemists. We settled on a seven-volume work and the indomitable ARK produced a rough breakdown to fit such a format. Putting flesh on these bones became feasible during a fortuitous three-month break between jobs by myself, and the largest handbook ever assembled by Pergamon (120 pages) was written and page allocations agreed—even for little or unknown functional groups. Sample chapters were commissioned and finally proved very encouraging, despite our first chosen topic uncovering virtually no known examples!

Contracts were defined and agreed, volume editors approached, and potential authors considered during a pleasant preconference stay in Grasmere. Following the sale of Pergamon to Elsevier Science Ltd there was a lull in the project but soon *Comprehensive Organic Functional Group Transformations* was back on track, and everyone adhered to a very businesslike timetable.

OTTO METH-COHN
Sunderland

CHARLES W. REES
London

ALAN R. KATRITZKY
Florida

Contributors to Volume 4

Dr R. A. Aitken
Department of Chemistry, University of St Andrews, St Andrews, Fife KY16 9ST, UK

Dr C. G. Barber
Pfizer Central Research, Sandwich, Kent CT13 9NJ, UK

Dr J. M. Berge
SmithKline Beecham Pharmaceuticals, Great Burgh, Yew Tree Bottom Road, Epsom, Surrey KT18 5XQ, UK

Dr D. R. Buckle
SmithKline Beecham Pharmaceuticals, Great Burgh, Yew Tree Bottom Road, Epsom, Surrey KT18 5XQ, UK

Dr A. Bulpin
Institut Jacques Boy, 45 Rue Cognacq-Jay, BP 1430, F-51061 Reims Cedex, France

Dr A. C. Campbell
Medicinal Chemistry Department, Organon Laboratories Ltd, Newhouse, Lanarkshire ML1 55H, UK

Dr C. D. Gabbutt
Department of Chemistry, University of Central Lancashire, Preston PR1 2HE, UK

Dr N. W. A. Geraghty
Department of Chemistry, University College Galway, Galway, Republic of Ireland

Dr M. J. Gough
Technical Typesetters UK, 61 Queens Road, Ashford, Kent TN24 8HL, UK

Dr F. Heaney
Department of Chemistry, University College Galway, Galway, Republic of Ireland

Professor J. D. Hepworth
Department of Chemistry, University of Central Lancashire, Preston PR1 2HE, UK

Dr R. A. Hill
Department of Chemistry, University of Glasgow, Glasgow G12 8QQ, UK

Dr W. M. Horspool
Department of Chemistry, University of Dundee, Dundee DD1 4HN, UK

Dr D. R. Jaap
Medicinal Chemistry Department, Organon Laboratories Ltd, Newhouse, Lanarkshire ML1 55H, UK

Dr P. D. Kennewell
Hoechst Roussel Ltd, Kingfisher Drive, Covingham, Swindon, Wiltshire SN3 5BZ, UK

Dr W. J. Kerr
Department of Pure and Applied Chemistry, University of Strathclyde, Thomas Graham Building, 295 Cathedral Street, Glasgow G1 1XL, UK

Dr D. T. MacPherson
SmithKline Beecham Pharmaceuticals, Great Burgh, Yew Tree Bottom Road, Epsom, Surrey KT18 5XQ, UK

Dr G. L. Patrick
Department of Chemistry, University of Paisley, High Street, Paisley PA1 2BE, UK

Dr P. L. Pauson
Department of Pure and Applied Chemistry, University of Strathclyde, Thomas Graham Building, 295 Cathedral Street, Glasgow G1 1XL, UK

Dr I. L. Pinto
SmithKline Beecham Pharmaceuticals, Great Burgh, Yew Tree Bottom Road, Epsom, Surrey KT18 5XQ, UK

Dr H. K. Rami
SmithKline Beecham Pharmaceuticals, Great Burgh, Yew Tree Bottom Road, Epsom, Surrey KT18 5XQ, UK

Dr G. N. Sheldrake
School of Chemistry, The Queen's University of Belfast, David Keir Building, Belfast BT9 5AG, UK

Dr D. I. Smith
Sanofi Research Division, Alnwick Research Centre, Willowburn Avenue, Alnwick, Northumberland NE66 2JH, UK

Dr J. Steele
Pfizer Central Research, Sandwich, Kent CT13 9NJ, UK

Professor Y. Vallée
LEDSS, Université Joseph Fourier, BP 53X, F-38041 Grenoble Cedex, France

Dr R. B. Webster
Pfizer Central Research, Sandwich, Kent CT13 9NJ, UK

Mr N. J. Westwood
The Dyson Perrins Laboratory, University of Oxford, South Parks Road, Oxford OX1 3QY, UK

Dr R. Westwood
Roussel-Uclaf, 102 Route de Noisy, F-93235 Romainville Cedex, France

Dr R. H. Wightman
Department of Chemistry, Heriot-Watt University, Riccarton, Edinburgh EH14 4AS, UK

Contents of All Volumes

Volume 1 Synthesis: Carbon with No Attached Heteroatoms

(Part I Tetracoordinated Carbon with No Attached Heteroatoms)

1.01   One or More CH Bond(s) Formed by Substitution: Reduction of C–Halogen and C–Chalcogen Bonds
1.02   One or More CH Bond(s) Formed by Substitution: Reduction of Carbon–Nitrogen, –Phosphorus, –Arsenic, –Antimony, –Bismuth, –Carbon, –Boron, and –Metal Bonds
1.03   Two or More CH Bond(s) Formed by Addition to CC Multiple Bonds
1.04   One or More CC Bond(s) Formed by Substitution: Substitution of Halogen
1.05   One or More CC Bond(s) Formed by Substitution: Substitution of Chalcogen
1.06   One or More CC Bond(s) Formed by Substitution: Substitution of Carbon–Nitrogen, –Phosphorus, –Arsenic, –Antimony, –Boron, –Silicon, –Germanium, and –Metal Functions
1.07   One or More CC Bond(s) Formed by Addition: Addition of Carbon Electrophiles and Nucleophiles to CC Multiple Bonds
1.08   One or More CC Bond(s) Formed by Addition: Addition of Carbon Radicals and Electrocyclic Additions to CC Multiple Bonds
1.09   One or More CH and/or CC Bond(s) Formed by Rearrangement

(Part II Tricoordinated Carbon with No Attached Heteroatoms)

1.10   One or More =CH Bond(s) Formed by Substitution or Addition
1.11   One or More =CC Bond(s) Formed by Substitution or Addition
1.12   One or More C=C Bond(s) Formed by Addition
1.13   One or More C=C Bond(s) by Elimination of Hydrogen, Carbon, Halogen, or Oxygen Functions
1.14   One or More C=C Bond(s) by Elimination of S, Se, Te, N, P, As, Sb, Bi, Si, Ge, B, or Metal Functions
1.15   One or More C=C Bond(s) Formed by Condensation: Condensation of Nonheteroatom Linked Functions, Halides, Chalcogen, or Nitrogen Functions
1.16   One or More C=C Bond(s) Formed by Condensation: Condensation of P, As, Sb, Bi, Si, Ge, B, or Metal Functions
1.17   One or More C=C Bond(s) by Pericyclic Processes
1.18   One or More =CH, =CC and/or C=C Bond(s) Formed by Rearrangement
1.19   Tricoordinate Anions, Cations, and Radicals

(Part III Dicoordinate and Monocoordinate Carbon with No Attached Heteroatoms)

1.20   Allenes and Cumulenes
1.21   Alkynes
1.22   Ions, Radicals, Carbenes and Other Monocoordinated Systems

Volume 2 Synthesis: Carbon with One Heteroatom Attached by a Single Bond

(Part I Functions Linked by a Single Bond to an sp^3 Carbon Atom)

2.01   Alkyl Halides
2.02   Alkyl Chalcogenides: Oxygen-based Functional Groups
2.03   Alkyl Chalcogenides: Sulfur-based Functional Groups
2.04   Alkyl Chalcogenides: Selenium- and Tellurium-based Functional Groups
2.05   Alkylnitrogen Compounds: Amines and Their Salts
2.06   Alkylnitrogen Compounds: Compounds with N—Halogen, N—O, N—S, N—Se, and N—Te Functional Groups
2.07   Alkylnitrogen Compounds: Compounds with N—N, N—P, N—As, N—Sb, N—Bi, N—Si, N—Ge, N—B, and N—Metal Functional Groups

2.08 Alkylphosphorus Compounds
2.09 Alkylarsenic, -antimony, and -bismuth Compounds
2.10 Alkylboron and -silicon Compounds
2.11 Alkyl Metals

(Part II Functions Linked by a Single Bond to an sp^2 Carbon Atom)

2.12 Vinyl and Aryl Halides
2.13 Alkenyl and Aryl Chalcogenides: Oxygen-based Functional Groups
2.14 Vinyl and Aryl Chalcogenides: Sulfur-, Selenium-, and Tellurium-based Functional Groups
2.15 Vinyl- and Arylnitrogen Compounds
2.16 Vinyl- and Arylphosphorus Derivatives
2.17 Vinyl- and Arylarsenic, -antimony, and -bismuth Compounds
2.18 Vinyl- and Arylsilicon, -germanium, and boron Compounds
2.19 Vinyl- and Arylmetals
2.20 Stabilized Substituted Ions and Radicals Bearing One Heteroatom ($R^1R^2C^-X$, $R^1R^2C^+X$, $R^1R^2C^{\cdot}X$)

(Part III Functions Linked by a Single Bond to an sp Carbon Atom)

2.21 Alkynyl Halides and Chalcogenides
2.22 Alkynylnitrogen and -phosphorus Compounds
2.23 Alkynylarsenic, -antimony, -bismuth, -boron, -silicon, -germanium, and -metal Compounds

Volume 3 Synthesis: Carbon with One Heteroatom Attached by a Multiple Bond

(Part I Tricoordinated Carbon Functions, $R_2C{=}Y$)

3.01 Aldehydes: Alkyl Aldehydes
3.02 Aldehydes: α,β-Unsaturated Aldehydes
3.03 Aldehydes: Aryl and Heteroaryl Aldehydes
3.04 Ketones: Dialkyl Ketones
3.05 Ketones: α,β-Unsaturated Ketones
3.06 Ketones Bearing an α,β-Aryl or -Hetaryl Substituent
3.07 Aldehyde and Ketone Functions Further Substituted on Oxygen
3.08 Thioaldehydes and Thioketones
3.09 Seleno- and Telluroaldehydes and -ketones
3.10 Imines and Their N-Substituted Derivatives: NH, NR, and N-Haloimines
3.11 Imines and Their N-Substituted Derivatives: Oximes and Their O-R Substituted Analogues
3.12 Imines and Their N-Substituted Derivatives: Hydrazones and Other ${=}$NN Derivatives Including Diazo Compounds
3.13 Synthesis of P, As, Sb and Bi Ylides ($R_3P{=}CR_2$, etc.)
3.14 Doubly Bonded Metalloid Functions (Si, Ge, B)
3.15 Doubly Bonded Metal Functions

(Part II Dicoordinated Carbon Functions, $R_2C{=}C{=}Y$)

3.16 Ketenes, Their Cumulene Analogues and Their S, Se, and Te Analogues
3.17 Ketenimines and Their P, As, Sb and Bi Analogues

(Part III Dicoordinated Carbon Functions, $R{-}C{\equiv}Z$)

3.18 Nitriles: General Methods and Aliphatic Nitriles
3.19 α,β-Unsaturated and Aryl Nitriles
3.20 N-Substituted Nitriles and Other Heteroanalogues of Nitriles of the Type RCZ

(Part IV Monocoordinated Carbon Functions)

3.21 Isocyanides and Their Heteroanalogues (RZC)

Volume 4 Synthesis: Carbon with Two Heteroatoms, Each Attached by a Single Bond

(Part I Tetracoordinated Carbon Functions Bearing Two Heteroatoms, R_2CXX')

4.01 Dihalo Alkanes, $R_2C(Hal)_2$
4.02 Functions Incorporating a Halogen and a Chalcogen
4.03 Functions Incorporating a Halogen and Another Heteroatom Group Other Than a Chalcogen
4.04 Functions Bearing Two Oxygens, $R^1{}_2C(OR^2)_2$

4.05 Functions Incorporating Oxygen and Another Chalcogen
4.06 Functions Incorporating Two Chalcogens Other Than Oxygen
4.07 Functions Incorporating a Chalcogen and a Group 15 Element
4.08 Functions Incorporating a Chalcogen and a Silicon, Germanium, Boron, or Metal
4.09 Functions Bearing Two Nitrogens
4.10 Functions Incorporating a Nitrogen and Another Group 15 Element
4.11 Functions Incorporating a Nitrogen and a Silicon, Germanium, Boron, or a Metal
4.12 Functions Containing One Phosphorus and Either Another Phosphorus or As, Sb, Bi, Si, Ge, B, or a Metal
4.13 Functions Containing at Least One As, Sb, or Bi with or without a Metalloid (Si or Ge) or a Metal
4.14 Functions Containing at Least One Metalloid (Si, Ge, or B) Together with Another Metalloid or Metal
4.15 Functions Containing Two Atoms of the Same Metallic Element
4.16 Functions Containing Two Atoms of Different Metallic Elements

(Part II Tricoordinated Carbon Functions Bearing Two Heteroatoms, $R_2C\!=\!CXX'$)

4.17 Functions Incorporating Two Halogens or a Halogen and a Chalcogen
4.18 Functions Incorporating a Halogen or Another Group Other Than a Halogen or a Chalcogen
4.19 Functions Bearing Two Chalcogens
4.20 Functions Containing a Chalcogen and Any Group Other Than a Halogen or a Chalcogen
4.21 Functions Containing at Least One Nitrogen and No Halogen or Chalcogen
4.22 Functions Containing at Least One Phosphorus, Arsenic, Antimony or Bismuth and No Halogen, Chalcogen or Nitrogen
4.23 Functions Containing at Least One Metalloid (Si, Ge, or B) and No Halogen, Chalcogen or Group 15 Element; Also Functions Containing Two Metals

(Part III)

4.24 Tri- and Dicoordinated Ions, Radicals and Carbenes Bearing Two Heteroatoms ($RC^+X^1X^2$, $RC^-X^1X^2$, $RC^{\cdot}X^1X^2$, $:CX^1X^2$)

Volume 5 Synthesis: Carbon with Two Attached Heteroatoms with at Least One Carbon-to-Heteroatom Multiple Link

(Part I Tricoordinate Carbon Functions with One Doubly Bonded and One Singly Bonded Heteroatom, $RC\!=\!YX$)

5.01 Acyl Halides
5.02 Carboxylic Acids
5.03 Carboxylic Esters and Lactones
5.04 Other Acyloxy Compounds
5.05 Acylsulfur, -selenium, or -tellurium Functions
5.06 Amides
5.07 N-Heterosubstituted Amides
5.08 Acylphosphorus, -arsenic, -antimony, or -bismuth Functions
5.09 Acylsilicon, -germanium, or -boron Functions
5.10 Acyl Metal Functions
5.11 Thio-, Seleno-, and Telluroacyl Halides
5.12 Thio, Seleno, and Telluro Acyloxy Functions, $R^1C(S)OR^2$, $R^1C(Se)OR^2$, $R^1C(Te)OR^2$, etc
5.13 Functions with Two Chalcogens Other Than Oxygen
5.14 Thionoamides and Their Se and Te Analogues
5.15 N–Substituted Thionoamides and Their Se and Te Analogues
5.16 Thioacyl Functions linked to a Metalloid (Si, Ge, or B) or Metal; and Their Seleno and Telluro Analogues
5.17 Iminoacyl Halides and Oxy Functions
5.18 Iminoacyl Functions Linked to Chalcogens Other Than Oxygen
5.19 Amidines and N-Substituted Amidines
5.20 Iminoacyl Functions Linked to Any Heteroatom Other Than Halogen, Chalcogen or Nitrogen
5.21 N-Heterosubstituted Iminoacyl Functions
5.22 Diazo Functions with an α-Heteroatom ($RC(X)N_2$)

5.23 Phosphoacyl Functions and Their As, Sb, and Bi Analogues
5.24 Doubly Bonded Metalloid Functions, $R^1C(X)=SiR^2_2$, $R^1C(X)=BR^2$, $R^1C(X)=GeR^2_2$
5.25 Functions Doubly Bonded to a Metal

(Part II Dicoordinate Carbon Functions with Two Doubly Bonded Heteroatoms, Y=C=Y′)

5.26 Functions with at Least One Oxygen, Y=C=O
5.27 Functions with at Least One Chalcogen Other Than Oxygen
5.28 Functions with at Least One Nitrogen and No Chalcogens
5.29 Functions with Heteroatoms Other Than Chalcogen or Nitrogen (Y=C=Y′)

(Part III Dicoordinate Carbon Functions with One Singly Bonded and One Triply Bonded Heteroatom, X—C≡Z)

5.30 Nitriles with a Heteroatom Attached to the Cyanocarbon
5.31 Triply Bonded Heteroatom Derivatives Other Than Nitriles with Another Heteroatom Attached to the *sp*-Carbon Atom

Volume 6 Synthesis: Carbon with Three or Four Attached Heteroatoms

(Part I Tetracoordinated Carbon with Three Attached Heteroatoms, RCXX′X″)

6.01 Trihalides
6.02 Functions Containing Halogens and Any Other Elements
6.03 Functions Containing Three Chalcogens (and No Halogens)
6.04 Functions Containing a Chalcogen and Any Other Heteroatoms Other Than a Halogen
6.05 Functions Containing at Least One Group 15 Element (and No Halogen or Chalcogen)
6.06 Functions Containing at Least One Metalloid (Si, Ge, or B) and No Halogen, Chalcogen, or Group 15 Element; Also Functions Containing Three Metals

(Part II Tetracoordinated Carbon with Four Attached Heteroatoms, CXX′X″X‴)

6.07 Functions Containing Four Halogens or Three Halogens and One Other Heteroatom Substituent
6.08 Functions Containing Two Halogens and Two Other Heteroatom Substituents
6.09 Functions Containing One Halogen and Three Other Heteroatom Substituents
6.10 Functions Containing Four or Three Chalcogens (and No Halogens)
6.11 Functions Containing Two or One Chalcogens (and No Halogens)
6.12 Functions Containing at Least One Group 15 Element (and No Halogen or Chalcogen)
6.13 Functions Containing at Least One Metalloid (Si, Ge, or B) and No Halogen, Chalcogen, or Group 15 Element; Also Functions Containing Four Metals

(Part III Tricoordinated Carbon with Three Attached Heteroatoms, Y=CXX′)

6.14 Functions Containing a Carbonyl Group and at Least One Halogen
6.15 Functions Containing a Carbonyl Group and at Least One Chalcogen (but No Halogen)
6.16 Functions Containing a Carbonyl Group and Two Heteroatoms Other Than a Halogen or Chalcogen
6.17 Functions Containing a Thiocarbonyl Group and at Least One Halogen; Also at Least One Chalcogen and No Halogen
6.18 Functions Containing a Thiocarbonyl Group Bearing Two Heteroatoms Other Than a Halogen or Chalcogen
6.19 Functions Containing a Selenocarbonyl or Tellurocarbonyl Group—Se(X)X′ and TeC(X)X′
6.20 Functions Containing an Iminocarbonyl Group and at Least One Halogen; Also One Chalcogen and No Halogen
6.21 Functions Containing an Iminocarbonyl Group and Any Elements Other Than a Halogen or Chalcogen
6.22 Functions Containing Doubly Bonded P, As, Sb, Bi, Si, Ge, B, or a Metal

(Part IV)

6.23 Tricoordinated Stabilized Cations and Radicals, $^+$CXYZ and $^{\cdot}$CXYZ

Volume 7 Indexes

Author Index
Cumulative Subject Index

Introduction

OBJECTIVES, SCOPE, AND COVERAGE

Comprehensive Organic Functional Group Transformations (COFGT) aims to present the vast subject of organic synthesis in terms of the introduction and interconversion of functional groups. All organic structures can be considered as skeletal frameworks of carbon atoms to which functional groups are attached[a]; it is the latter which are mainly responsible for chemical reactivity and which are highlighted in COFGT. All known functional groups fit a logical and comprehensive pattern and this forms the basis for the detailed list of contents. The format of the present work was designed with the intention to cover systematically all the possible arrangements of atoms around a carbon, including those which are quite unfamiliar. The work also considers the possibility of as yet unknown functional groups which may be constructed in the future and prove to be important; thus COFGT also indicates what is not known and so points the way to new research areas.

The philosophy of the present work has been to rationalize this enormous subject within as logical and formal a framework as possible, in a scholarly and critical fashion. COFGT is designed to provide the first point of entry to the literature for synthetic organic chemists, together with an unrivalled source for anyone interested in less common, obscure, or unknown functional groups.

All functional groups are viewed as being carbon based (even if the group contains no carbon). Thus, a nitro compound is considered from the standpoint of the immediately attached carbon atom, whether di- (sp), tri- (sp^2), or tetracoordinated(sp^3). The work is organized on the basis of formation or rupture of bonds to a carbon atom and it is the nature of the carbon atom left after the transformation that determines the classification of the overall sequence. Several key criteria have been used to organize the work and to minimize overlap. These are, in order of priority:

1. the number of attached heteroatoms;
2. the coordination of the carbon atom involved in the functional group;
3. the nature of the immediately attached heteroatom(s); and
4. the Latest Placement Principle.

These four key principles have been used to determine the content of each volume, and to develop the detailed chapter breakdown within each volume.

Thus, according to the number of attached heteroatoms:

Volume 1 deals with synthetic reactions which result in the alteration of bonding at carbon atoms which are left with *no* attached heteroatoms.

Volume 2 deals with syntheses which result in carbon atoms attached to *one* heteroatom by a single bond.

Volume 3 deals with syntheses which result in carbon atoms attached to *one* heteroatom by a double or by a triple bond.

Volume 4 deals with syntheses which result in carbon atoms attached to *two* heteroatoms, each by a single bond.

Volume 5 deals with syntheses which result in carbon atoms attached to *two* heteroatoms by one single and one double bond, or by two double bonds, or by one single and one triple bond.

Volume 6 deals with syntheses which result in carbon atoms attached to *three* or *four* heteroatoms.

Volume 7 comprises the author and subject indexes.

Certain key principles apply to all the volumes because all functional groups are viewed as carbon based (e.g. a nitro group is either alkyl-, vinyl-, aryl-, or alkynyl-); these are:

(a) Volumes are subdivided according to the *coordination* of the carbon atom which is the product of the reaction, i.e., tetra- coming before tri- before di- before monocoordinated carbon functions.

---

[a] The major exception to this lies in heterocyclic compounds, where the cyclic heteroatoms are more logically considered as part of the framework. The subject of heterocycles has been treated elsewhere in the companion work *Comprehensive Heterocyclic Chemistry* published in 1984 with a second edition to be published in 1996.

In Volumes 1 and 6, reactions producing four-coordinated carbon are considered first, followed by three- and then two-coordinated carbon. The other volumes contain a more limited range of coordination types (Volumes 2 and 4 only four-coordinated, Volumes 3 and 5 only two- or three-coordinated). Each type of coordination is allocated a separate section in each volume.

(b) Attached *heteroatoms* are discussed in the following order of priority:

Halogens—F, Cl, Br, I
Chalcogens—O, S, Se, Te
Nitrogen—N
Other group 15 elements—P, As, Sb, Bi
Metalloids—B, Si, Ge
Main group metals—Sn, Pb, Al, Ga, In, Tl, Be, Mg, Ca, Sr, Ba, Li, Na, K, Rb, Cs
Transition metals—Cu, Ag, Au, Zn, Cd, Hg, Ti, Zr, Hf, Cr, Mo, W, Mn, Fe, Co, Ni, Pd, Pt, and others.

Higher coordination of heteroatoms is treated after lower. Thus, in sections dealing with iodo compounds, monocoordinate (e.g. iodides) are discussed before dicoordinate (e.g. iodoxyls) and tricoordinate functions.

(c) The *Latest Placement Principle* (or Last Position Principle) is used to avoid undue overlap in the work. Thus, the carbon attached to the heteroatom is discussed at the last possible position in the above prioritizing of heteroatoms. Examples of its application are noted later. On this basis, for example, when both C—C and C—H bonds are formed the reaction will appear in the latest chapter (i.e. Chapters 1.04–1.10 rather than in the earlier Chapters 1.01–1.03), and when both C—C and C=C bonds are formed, this will be found in the later Chapter 1.17. The Latest Placement Principle is particularly important in determining where to find electrocyclic reactions in Volume 1. If C—H, =C—C, and C=C bonds are all formed in a reaction, then the latest appropriate chapter will deal with the reaction. Only if a change in heterofunction occurs is the reaction left to a later volume.

Exceptions to the above principles are rare. However the reactions of heteroarenes are mentioned along with those of arenes. If on reduction no change in the heterofunction occurs (e.g. in going from thiophene to tetrahydrothiophene or from pyridine to 2,3,4,5-tetrahydropyridine) the reaction is found in Volume 1. However, when the function changes (e.g. pyridine to piperidine), the conversion is considered in Volume 2. Conversion of methyl phenyl sulfone into methyl cyclohexyl sulfone appears in Volume 2, whereas the formation of cyclohexyl methyl ketone is treated in Volume 1, since the coordination of the carbon atom to the heteroatoms is changed in the first but not in the second hydrogenation.

Some further exceptions to the rigorous ordering of the work have been made for the purpose of easy reference. Thus, in Volume 1, a special chapter on ions, radicals, and carbenes is added: this chapter is limited to the treatment of species capable of more than a transitory existence. Throughout the work aspects of the Latest Placement Principle are occasionally ignored for reasons of clarity. Thus metal ligands that are incidental to the chemistry under discussion are not considered when prioritizing. Also, references to aromatic substituents, some of which involve a heteroatom (e.g. pyridyl, thienyl, etc.) but are incidental to the chemistry being described, are not viewed as changing the priority.

Within each section, we have endeavored to explain the influence of important secondary effects such as inclusion in a ring, degree of strain, degree of substitution, various types of activation, influence of stereochemistry, and so on, on the transformation under consideration. General synthetic methods are treated before specific methods.

Transient intermediates, as such, do not fall within the scope of this work. Although there is clearly no sharp division, we have attempted to restrict coverage of radicals, etc., to more stable, longer lived species. It is the aim of this work to consider all organic functional groups provided that the molecules which incorporate them, though they may be unstable, can have a finite lifetime and chemistry. The whole work deals with the generation and transformation of functional groups, *not* of molecules such as CO_2, COS, CS_2, $ClCN$, etc. Such simple carbon derivatives are not treated unless a further carbon is attached (e.g. RN=C=O).

VOLUME 1 SYNTHESIS: CARBON WITH NO ATTACHED HETEROATOMS

Volume 1 deals solely with the formation of nonheteroatom functional groups and as such is different in style to the remaining volumes.

Introduction xvii

In addition to the general principles, Volume 1 is further organized as follows:

1. By the type of bond formed (i.e. C—H before C—C).
2. By the type of reaction involved (i.e. substitution, then addition, then rearrangement). With C=C bond formation the order is addition, elimination, condensation, then electrocyclic and other methods. One rearrangement chapter only is devoted to each of the Parts I and II.
3. In Parts II and III the treatment of formation of ions, radicals, and carbenes is added at the end of the section dealing solely with those species with a significant rather than a transient lifetime.

In Volume 1, the heteroatom sequence is a secondary feature since only remote heteroatom functions are involved in the products: but the standard order pertains in reactants that contain heteroatoms (see, e.g. Chapters 1.01 and 1.02).

All the major structural influences that are treated throughout this work apply equally (or perhaps more importantly) in Volume 1. Thus the effects of conjugation, remote substituents, rings, stereochemistry, strain, kinetic or thermodynamic factors, solvation, primary, secondary and tertiary nature, etc., are mentioned whenever relevant.

VOLUME 2   SYNTHESIS: CARBON WITH ONE HETEROATOM ATTACHED BY A SINGLE BOND

Volume 2 is arranged in three parts: I, II and III, dealing respectively with sp^3, sp^2, and sp carbon linked to the heteroatom. In each chapter we have endeavored to explain important effects due to such features as the primary, secondary, tertiary nature, ring effects, strain activation, effect of beta, gamma, and more remote functionality, stereochemical effects, and so on. Methods that are common to a larger group are dealt with at their first appearance and suitably cross-referenced.

Volumes 2–6 all deal with the synthesis of functions involving at least one heteroatom. To avoid major overlap we have applied the Latest Placement Principle; that is, the chemistry is discussed at the last possible position based on the prioritization of the carbon attached to the heteroatom. Thus the compound CH_3ONH_2 is treated under "Alkyl Chalcogenides" in the subsection "Functions Based on the RON-Unit" (i.e. 2.02.6). However, CH_3ONHCH_3 appears under "Alkyl Nitrogen Compounds" (2.06.2.3) since the Latest Placement Principle prevails. Also, dialkyl ethers appear in Part I of Volume 2 (Functions Linked by a Single Bond to an sp^3 Carbon Atom), while alkyl aryl ethers appear in Part II of Volume 2 (Functions Linked by a Single Bond to an sp^2 Carbon Atom). Exceptions to the rule are:

(a) When a fully unsaturated heterocyclic substituent (e.g. thienyl, pyridyl, etc.) is used as an example of an aryl group, the ring heteroatom(s) is (are) not taken into account (e.g. 2-methoxypyridine should strictly appear in Volume 6, but is covered in Volume 2 along with 3- and 4-methoxypyridine).

(b) Carbon-based metal ligands that are incidental to the synthesis under discussion (e.g. carbonyls, cyclopentadienyls, etc.) are not taken into consideration.

VOLUME 3   SYNTHESIS: CARBON WITH ONE HETEROATOM ATTACHED BY A MULTIPLE BOND

Volume 3 follows the logical development indicated in Volume 2. Thus, according to the Last Placement Principle, the imines, RCH=N—R, appear in Volume 3 rather than in Volume 2 (where functions singly bonded to carbon are treated). Furthermore, acetophenone, $PhCOCH_3$, is treated under α,β-unsaturated ketones (3.05) rather than saturated ketones (3.04). Chloronitroacrylonitriles would appear under the section "α,β-Vinylic Nitriles with Nitrogen-based Substituents" (3.19.2.7), not under the related earlier section dealing with halo-substituents (3.19.2.3).

VOLUME 4   SYNTHESIS: CARBON WITH TWO HETEROATOMS, EACH ATTACHED BY A SINGLE BOND

Volume 4 is in three parts. Part I deals with tetracoordinated carbon bearing two heteroatoms, Part II with tricoordinated carbon bearing two heteroatoms, and Part III (a brief chapter) with stabilized radicals, ions, and the like bearing two heteroatoms. The material is arranged according to the Latest Placement Principle: thus, the synthesis of $CHBr_2CHI_2$ would appear in the section dealing with diiodo, not dibromo functions (i.e. in 4.01.5, not 4.01.4), and the synthesis of $CF_3CHBrCl$ is discussed in Volume 6 (carbons bearing three heteroatoms), rather than in Volume 4.

VOLUME 5 SYNTHESIS: CARBON WITH TWO ATTACHED HETEROATOMS WITH AT LEAST ONE CARBON-TO-HETEROATOM MULTIPLE BOND

Volume 5 is in three parts. Part I deals with functions with one doubly bonded and one singly bonded heteroatom, Part II with functions containing two doubly bonded heteroatoms and Part III with one triply bonded and one singly bonded heteroatom. Part I constitutes the bulk of Volume 5.

The arrangement of the chemistry in each part follows the same logical sequence. The multiply bonded heteroatom is focused on first and then the other heteroatom in a secondary classification, both following the priority rules already described. Each section excludes the coverage of the previous sections. Thus, all carbonyl derivatives will appear in Chapters 5.01–5.10 but not in Chapters 5.11, *et seq*.

According to the Latest Placement Principle structure RC(O)OC(S)R is discussed in the chapter dealing with carbons bearing a doubly bonded sulfur and singly bonded oxygen (5.12.3), *not* in that dealing with doubly and singly bonded oxygen (5.04.1). Another effect of the Latest Placement Principle is that the amides RCONMePh are discussed under *N*-arylalkanoamides (5.06.2.4), rather than *N*-alkylalkanoamides (5.06.2.2). Again, exceptions are made to the latest placement rules for: (a) hetaryl rings used as examples of aryl substituents which are not viewed as functional groups. Thus, 2-methylimidazole is not considered as an example of an amidine function and 2-methoxypyridine is not an example of a doubly bonded nitrogen, singly bonded oxygen function; (b) metal ligands that are incidental to the organic chemistry under discussion are not viewed as functions in priority considerations.

VOLUME 6 SYNTHESIS: CARBON WITH THREE OR FOUR ATTACHED HETEROATOMS

Volume 6 is in four parts. Part I deals with tetracoordinate carbons bearing three heteroatoms. Part II covers tetracoordinate compounds bearing four heteroatoms, i.e. substituted methanes, and Part III deals with tricoordinate systems bearing three heteroatoms, i.e. where one heteroatom is attached by a double bond. Part IV is brief and deals with stabilized radicals and ions. Not surprisingly, the coverage of Volume 6 is very large—and also shows that many gaps in the development of organic chemistry still exist.

The organization within the three sections not only follows the same broad logic developed in the previous volumes, but also has a structure unique to the multiheteroatom volume. According to the Latest Placement Principle $CF_3C(NR_2)_3$ appears in the section dealing with carbons bearing three nitrogens (6.05.1.1), not that dealing with carbons bearing three halogens (6.01.2), while $(CF_3CH_2O)_2CO$ appears in Part III, not in Part I.

In the chapter dealing with iminocarbonyl functions in Part III, the substituents on nitrogen are discussed in each appropriate subsection in the order outlined above. Thus, the RN= group would be first considered with R = H, then alkyl, alkenyl, aryl and hetaryl, alkynyl and then heteroatom substituents in the usual order.

In each relevant section, we have endeavored to explain the influence of important secondary effects on the synthesis such as structure (primary, secondary, etc.), ring effects, strain, activation, stereochemistry, remote substituent effects, etc.

The arrangement of the chemistry in each of Parts I–III follows a similar pattern. Thus, each section commences with functions containing at least one halogen. This section deals with all combinations of halogen with other heteroatoms in the described order. The next section deals with functions containing at least one chalcogen in combination with any other heteroatoms except halogens. Subsequent sections each exclude the previous title heteroatom functions.

VOLUME 7 INDEXES

Subject Indexes are included in each of Volumes 1–6 and Cumulative Subject and Author Indexes appear in Volume 7. Most entries in the Subject Index consist of two or three lines: the first line is the entry itself (e.g. Lactones) and the second line is descriptive of that entry (e.g. reduction); in many cases more detail is given (e.g. with 9-BBN).

REFERENCES

The references are handled by the system previously used successfully in *Comprehensive Heterocyclic Chemistry*. In this system reference numbers appear neither in the text, nor as footnotes, nor at the end of chapters. Instead, each time a reference is cited in the text there appears (in parentheses) a two-letter code assigned to the journal being cited, which is preceded by the year (tens and units only for twentieth-century references) and followed by the page number. For example: "It was shown ⟨80TL1327⟩ that . . .". In this phrase, "80" refers to 1980, "TL" to *Tetrahedron Letters*, and "1327" to the page number. For those journals which are published in parts, or which have more than one volume number per year, the appropriate part of the volume is indicated, e.g. as in ⟨73JCS(P2)1594⟩ or ⟨78JOM(162)611⟩, where the first example refers to *J. Chem. Soc., Perkin Trans 2*, 1973, page 1594; and the second to *J. Organomet. Chem.*, 1978, volume 162, page 611.

This reference system is adopted because it is far more useful to the reader than the conventional "superscript number" system. It enables readers to go directly to the literature reference cited, without first having to consult the bibliography at the end of each chapter.

References to the last century quote the year in full. Books have a prefix "B-" and if they are commonly quoted (e.g. *Organic Reactions*) they will have a code. Otherwise, as with uncommon journals, they are given a miscellaneous code (MI) and numbered arbitrarily *abb1*, *abb2*, etc., where *abb* refers to the volume and chapter number and *1*, *2*, etc., are assigned sequentially. Patents are assigned appropriate three-letter codes.

The references are given in full at the end of each volume. They include *Chemical Abstract* references when these are likely to help; in particular, they are given for all patents, and for less accessible sources such as journals in languages other than English, French, or German, company reports, obscure books, and theses.

4.01
Dihalo Alkanes, $R_2C(Hal)_2$

ROBERT A. HILL
University of Glasgow, UK

4.01.1	GENERAL METHODS	2
4.01.2	DIFLUORO ALKANES—R_2CF_2	2
4.01.2.1	*Difluoro Alkanes from Alkanes*	2
4.01.2.2	*Difluoro Alkanes from Dihalo Alkanes*	3
4.01.2.3	*Difluoro Alkanes from Trihalo Alkanes*	5
4.01.2.4	*Difluoro Alkanes from Alkenes*	5
4.01.2.5	*Difluoro Alkanes from Alkynes*	6
4.01.2.6	*Difluoro Alkanes from Difluorocarbene*	7
4.01.2.7	*Difluoro Alkanes from Aldehydes and Ketones*	8
4.01.2.8	*Difluoro Alkanes from Imines*	10
4.01.3	DICHLORO ALKANES—R_2CCl_2	11
4.01.3.1	*Dichloro Alkanes from Alkanes*	11
4.01.3.2	*Dichloro Alkanes from Dihalo Alkanes*	13
4.01.3.3	*Dichloro Alkanes from Trihalo Alkanes*	13
4.01.3.4	*Dichloro Alkanes from Alkenes*	14
4.01.3.5	*Dichloro Alkanes from Alkynes*	15
4.01.3.6	*Dichloro Alkanes from Dichlorocarbene*	16
4.01.3.7	*Dichloro Alkanes from Aldehydes and Ketones*	18
4.01.3.8	*Dichloro Alkanes from Imines*	19
4.01.4	DIBROMO ALKANES—R_2CBr_2	19
4.01.4.1	*Dibromo Alkanes from Alkanes*	19
4.01.4.2	*Dibromo Alkanes from Dihalo Alkanes*	22
4.01.4.3	*Dibromo Alkanes from Trihalo Alkanes*	23
4.01.4.4	*Dibromo Alkanes from Alkenes*	23
4.01.4.5	*Dibromo Alkanes from Alkynes*	24
4.01.4.6	*Dibromo Alkanes from Dibromocarbene*	24
4.01.4.7	*Dibromo Alkanes from Aldehydes and Ketones*	25
4.01.4.8	*Dibromo Alkanes from Imines*	27
4.01.4.9	*Dibromo Alkanes from Carboxylic Acids*	27
4.01.5	DIIODO ALKANES—R_2CI_2	28
4.01.5.1	*Diiodo Alkanes from Alkanes*	28
4.01.5.2	*Diiodo Alkanes from Halo Alkanes*	28
4.01.5.3	*Diiodo Alkanes from Alkynes*	29
4.01.5.4	*Diiodo Alkanes from Diiodocarbene*	29
4.01.5.5	*Diiodo Alkanes from Imines*	29
4.01.6	FLUOROHALO ALKANES—R_2CFHal	30
4.01.6.1	*Chlorofluoro Alkanes—R_2CClF*	30
4.01.6.1.1	*Chlorofluoro alkanes from halo alkanes*	30
4.01.6.1.2	*Chlorofluoro alkanes from halo alkenes*	30
4.01.6.1.3	*Chlorofluoro alkanes from chlorofluorocarbene*	32
4.01.6.1.4	*Chlorofluoro alkanes from imines*	32
4.01.6.1.5	*Chlorofluoro alkanes from carboxylic acids*	33
4.01.6.2	*Bromofluoro Alkanes—R_2CBrF*	33
4.01.6.2.1	*Bromofluoro alkanes from halo alkanes*	33
4.01.6.2.2	*Bromofluoro alkanes from halo alkenes*	34

4.01.6.2.3 *Bromofluoro alkanes from bromofluorocarbene*	35
4.01.6.2.4 *Bromofluoro alkanes from carboxylic acids*	35
4.01.6.3 *Fluoroiodo Alkanes—R$_2$CFI*	36
4.01.6.3.1 *Fluoroiodo alkanes from halo alkanes*	36
4.01.6.3.2 *Fluoroiodo alkanes from halo alkenes*	36
4.01.6.3.3 *Fluoroiodo alkanes from fluoroiodocarbene*	36
4.01.6.3.4 *Fluoroiodo alkanes from carboxylic acids*	37
4.01.7 CHLOROHALO ALKANES—R$_2$CCl Hal(not F)	37
4.01.7.1 *Bromochloro Alkanes—R$_2$CBrCl*	37
4.01.7.1.1 *Bromochloro alkanes from halo alkanes*	37
4.01.7.1.2 *Bromochloro alkanes from halo alkenes*	38
4.01.7.1.3 *Bromochloro alkanes from bromochlorocarbene*	38
4.01.7.1.4 *Bromochloro alkanes from ketones*	38
4.01.7.1.5 *Bromochloro alkanes from carboxylic acids*	38
4.01.7.2 *Chloroiodo alkanes—R$_2$CClI*	39
4.01.7.2.1 *Chloroiodo alkanes from halo alkanes*	39
4.01.7.2.2 *Chloroiodo alkanes from halo alkenes*	39
4.01.7.2.3 *Chloroiodo alkanes from ketones*	40
4.01.7.2.4 *Chloroiodo alkanes from carboxylic acids*	40
4.01.8 BROMOIODO ALKANES—R$_2$CBrI	40

4.01.1 GENERAL METHODS

There are many general methods for the preparation of *gem*-difluoro, *gem*-dichloro and *gem*-dibromo alkanes. These are given in detail in the following sections. Direct halogenation of alkanes is of limited use as there is generally little control of the site of halogenation. The method can be useful, however, when there is some control such as halogenation of benzylic positions or α to a carbonyl group. Replacement of one halogen for another can be useful for diiodo and mixed *gem*-dihalo alkanes, but it is often very difficult to control the degree of exchange. One of the major problems in the generation of *gem*-dihalo alkanes by this method is the possibility of elimination of hydrogen halide under the reaction conditions. This is a particular problem for dihalo alkanes where one of the halides is bromine or iodine.

Addition of hydrogen halides or halogens to halo alkenes has been used extensively for the production of dihalo alkanes. Radical addition of hydrogen halides often leads to 1,2-dihalo compounds and care must be taken to reduce the possibility of radical formation. Other problems of direction of addition occur when interhalogen compounds are added across halo alkenes; mixtures of products are often obtained.

Dihalocarbenes have been used extensively in addition reactions to double bonds to form dihalocyclopropane derivatives. There are many methods for the generation of carbenes or effecting a carbene transfer, particularly for difluoro-, dichloro- and dibromocarbene. The other dihalocarbenes have been studied less extensively.

The conversion of an aldehyde or ketone into a dihalo alkane works well with fluoro and chloro alkanes, but bromo and iodo alkanes are easily hydrolysed back to the aldehyde and ketone. Many preparations of dibromo and diiodo alkanes result in carbonyl compounds as side products.

4.01.2 DIFLUORO ALKANES—R$_2$CF$_2$

The preparation of *gem*-difluoro alkanes is included in a general review by Henne on the synthesis of aliphatic fluorine compounds ⟨44OR(2)49⟩.

4.01.2.1 Difluoro Alkanes from Alkanes

Direct fluorination of saturated compounds has been used since 1900 to replace hydrogen by fluorine ⟨44OR(2)49⟩. However, the reaction is not easy to control; most organic compounds react violently with fluorine. The reaction of elemental carbon with fluorine has been reported to give a mixture of products from which perfluoropropane, perfluorobutane and perfluoropentane have been isolated ⟨37JA1407⟩. This method is clearly not of general application. More-controlled fluorination of ethane using fluorine diluted with nitrogen yielded partially fluorinated ethanes from which CHF_2CHF_2 and CHF_2CH_2F could be isolated ⟨40JA1171⟩. Electrochemical fluorination of ethane with a solution in hydrogen fluoride is a more controllable method but again mixtures were obtained, however, CH_3CHF_2 could be obtained in usable amounts ⟨66BCJ219⟩. Cobalt trifluoride is a useful reagent for the perfluorination of unsaturated compounds. For example, cyclopentane can be perfluorinated (Equation (1)), however the substitution of the last few hydrogens in a compound requires higher temperatures ⟨51JA4241⟩. Perfluorocyclohexane has been prepared from benzene with fluorine and a catalyst (Equation (2)) ⟨50JCS2689⟩. Gold was found to be the best catalyst. Perfluorocyclohexane has also been made from methyl benzoate by the action of potassium tetrafluorocobaltate at high temperatures (Equation (3)) ⟨73JFC(3)329⟩. Active methylene compounds have been reported to be fluorinated efficiently with two equivalents of sodium ethoxide in ethanol followed by perchloryl fluoride (Equations (4)–(6)) ⟨58JA6533⟩; however, a later report suggests that the reaction is quite complex ⟨66JOC916⟩.

4.01.2.2 Difluoro Alkanes from Dihalo Alkanes

The substitution of halide in dihalo alkanes using metal fluorides is of general use for the preparation of difluoro alkanes as the corresponding dichloro and dibromo alkanes are generally more accessible. The ease of substitution is I > Br > Cl; the substitution of chlorine frequently requires very high temperatures. Potassium fluoride will displace the chlorine in the relatively reactive α-keto alkyl chlorides (for example Equation (7)) ⟨86JA7739⟩, whereas the chlorine of N,N-diethylchlorofluoroacetamide can only be displaced at high temperatures (Equation (8)) ⟨77CCC2537⟩. Substitution of unreactive chlorines such as in dichloromethane requires harsher conditions, for example a melt of potassium hydrogen difluoride, KHF_2 (Equation (9)) ⟨66AG(E)314⟩. KHF_2 has also been used to prepare 1,1-difluoroacetone from 1,1-dichloroacetone (Equation (10)) ⟨71JCS(C)279⟩. Mercuric fluoride has been extensively used for the preparation of fluoro alkanes by displacement ⟨44OR(2)49⟩. Bromine is substituted at low temperature with good yields (Equation (11)) whereas chlorine requires high temperatures and results in low yields (Equation (12)) ⟨36JA889⟩.

$$\text{PhCOCHCl}_2 \xrightarrow[28\%]{\text{KF}} \text{PhCOCHF}_2 \quad (7)$$

$$\text{ClCHFCONEt}_2 \xrightarrow[75\%]{\text{KF, 140 °C}} \text{CF}_2\text{HCONEt}_2 \quad (8)$$

$$\text{CH}_2\text{Cl}_2 \xrightarrow[82\%]{\text{KHF}_2} \text{CH}_2\text{F}_2 \quad (9)$$

$$\text{CH}_3\text{COCHCl}_2 \xrightarrow[50\%]{\text{KHF}_2} \text{CH}_3\text{COCHF}_2 \quad (10)$$

$$\text{CH}_3\text{CHBr}_2 \xrightarrow{\text{HgF}_2,\ 0\ °C} \text{CH}_3\text{CHF}_2 \quad (11)$$

$$\text{CH}_2\text{ClCHCl}_2 \xrightarrow[10\%]{\text{HgF}_2,\ 140\ °C} \text{CH}_2\text{ClCHF}_2 \quad (12)$$

Dibromo alkanes are generally smoothly substituted by mercuric fluoride (Equation (13)) but 3,3-dibromobutan-2-one gives side reactions including the production of diacetyl (Equation (14)) ⟨77JOC3527⟩. Silver fluoride has been used in these reactions; however, it is difficult to prepare in anhydrous form and it forms insoluble, complex silver halides ⟨44OR(2)49⟩. Antimony trifluoride with a catalytic amount of bromine converts dichloro(diphenyl)methane into difluoro-(diphenyl)methane in high yield (Equation (15)) ⟨38JA864⟩. Antimony pentafluoride is very effective at substituting alkyl bromides (Equation (16)) and alkyl chlorides (Equation (17)) but it does not exchange vinyl halides ⟨66JA2481⟩. A mixture of antimony trifluoride, antimony pentachloride and hydrogen chloride has been used to convert 2,2-dichlorobutane into 2,2-difluorobutane (Equation (18)) but many side reactions occurred ⟨79JFC(13)325⟩

$$\text{PhCOCBr}_2\text{CH}_3 \xrightarrow[89\%]{\text{HgF}_2} \text{PhCOCF}_2\text{CH}_3 \quad (13)$$

$$\text{CH}_3\text{COCBr}_2\text{CH}_3 \xrightarrow{\text{HgF}_2} \text{CH}_3\text{COCOCH}_3 \quad (14)$$

$$\text{Ph}_2\text{CCl}_2 \xrightarrow{\text{SbF}_3,\ \text{Br}_2\ (\text{cat.}),\ 140\ °C} \text{Ph}_2\text{CF}_2 \quad (15)$$

$$\text{C}_3\text{Br}_4 \xrightarrow[51\%]{\text{SbF}_5,\ 109\ °C} \text{C}_3\text{Br}_2\text{F}_2 \quad (16)$$

$$\text{C}_3\text{Cl}_4 \xrightarrow{\text{SbF}_5,\ 110\ °C} \text{C}_3\text{Cl}_2\text{F}_2 \quad (17)$$

$$\text{CH}_3\text{CCl}_2\text{CH}_2\text{CH}_3 \xrightarrow{\text{SbF}_3,\ \text{SbCl}_5,\ \text{HCl}} \text{CH}_3\text{CF}_2\text{CH}_2\text{CH}_3 \quad (18)$$

4.01.2.3 Difluoro Alkanes from Trihalo Alkanes

Reduction of the bromodifluoromethyl group with sodium borohydride in DMSO seems an attractive method of preparation of compounds containing the difluoromethyl group as long as the starting material is readily available as (Equation (19)) ⟨91JOC4322⟩.

$$\text{(cyclohexyl-CHBr-CF}_2\text{Br)} \xrightarrow[51\%]{\text{NaBH}_4,\ \text{DMSO}} \text{cyclohexyl-CH}_2\text{-CHF}_2 \quad (19)$$

4.01.2.4 Difluoro Alkanes from Alkenes

Addition of an acid to a 1,1-difluoro alkene will lead to a difluoromethyl group. The high electronegativity of fluorine ensures that hydrogen adds to the carbon bearing the fluorines. Thus hydrogen bromide (Equation (20)) and hydrogen iodide (Equation (21)) add efficiently to 1,1-difluoroethene ⟨56JCS61⟩. Methanol will add across tetrafluoroethene in the presence of a catalytic amount of sodium methoxide (Equation (22)) ⟨51JA1329⟩. The addition to the electron-deficient tetrafluoroethene is initially by nucleophilic attack. Cyanide will add to chlorotrifluoroethene to give, after acid hydrolysis, 3-chloro-2,2,3-trifluoropropanoic acid (Equation (23)) ⟨60OSC(5)239⟩. Tetrafluoroethene can be alkylated using aluminum trichloride as a catalyst, for example, dichlorofluoromethane can be effectively added across the double bond as (Equation (24)) ⟨71CCC1867⟩.

$$CF_2=CH_2 \xrightarrow[100\%]{\text{HBr}} CHF_2-CH_2Br \quad (20)$$

$$CF_2=CH_2 \xrightarrow[100\%]{\text{HI}} CHF_2-CH_2I \quad (21)$$

$$CF_2=CF_2 \xrightarrow[81\%]{\text{MeOH, MeONa (cat.), 35 °C, 5 h}} CHF_2-CF_2OMe \quad (22)$$

$$CClF=CF_2 \xrightarrow[76-79\%]{\text{i, KCN; ii, H}^+} CClF_2-CF_2CO_2H \quad (23)$$

$$CF_2=CF_2 \xrightarrow[58\%]{\text{CHFCl}_2,\ \text{AlCl}_3,\ 10\ ^\circ\text{C},\ 5\ \text{h}} F_3C-CFCl-CHCl_2 \quad (24)$$

The [2 + 2] adducts of fluoro alkenes can be prepared at high temperatures, probably involving a radical mechanism. Tetrafluoroethene can be dimerised at 600 °C to give perfluorobutane (Equation (25)); temperatures above 600 °C give various side reactions including polymerisation ⟨53JCS2083⟩. Mixed cycloaddition reactions such as tetrafluoroethene with ethene as in (Equation (26)), with butadiene (Equation (27)) and with acrylonitrile (Equation (28)) are possible, as they occur much more readily than the dimerisation of tetrafluoroethene ⟨49JA490⟩. Tetrafluoroethene will also add to acetylene to give 3,3,4,4-tetrafluorocyclobutene (Equation (29)) ⟨61JA382⟩. A variety of other fluorinated ethenes will cyclodimerise (Equations (30) and (31)) at lower temperatures than tetrafluoroethene ⟨47JA279⟩. Intramolecular [2 + 2] cycloaddition of 1,1-difluorobutadiene takes place under UV irradiation (Equation (32)) ⟨87JOC1872⟩.

$$C_2F_4 \xrightarrow[42\%]{600\ °C} \text{octafluorocyclobutane} \qquad (25)$$

$$C_2F_4 + H_2C=CH_2 \xrightarrow[40\%]{150\ °C,\ 8\ h} \text{1,1,2,2-tetrafluorocyclobutane} \qquad (26)$$

$$C_2F_4 + \text{butadiene} \xrightarrow[90\%]{125\ °C,\ 8\ h} \text{1,1,2,2-tetrafluoro-3-vinylcyclobutane} \qquad (27)$$

$$C_2F_4 + CH_2=CHCN \xrightarrow[84\%]{150\ °C,\ 8\ h} \text{1,1,2,2-tetrafluoro-3-cyanocyclobutane} \qquad (28)$$

$$C_2F_4 + HC\equiv CH \xrightarrow[35\%]{225\ °C,\ 12\ h} \text{3,3,4,4-tetrafluorocyclobutene} \qquad (29)$$

$$\text{CFCl=CFCl} \xrightarrow[80\%]{200\ °C,\ 12\ h} \text{1,2,3,4-tetrachloro-1,2,3,4-tetrafluorocyclobutane} \qquad (30)$$

$$\text{CFCl=CF}_2 \xrightarrow[80\%]{200\ °C,\ 8\ h} \text{1,3-dichloro-1,2,2,3,4,4-hexafluorocyclobutane} \qquad (31)$$

$$\text{2,3-difluoro-1,3-butadiene} \xrightarrow{h\nu,\ 4\ \text{days}} \text{1,2-difluorocyclobutene} \qquad (32)$$

4.01.2.5 Difluoro Alkanes from Alkynes

The addition of two equivalents of hydrogen fluoride across a triple bond is a general method of preparing difluoro alkanes (Equation (33)) ⟨47JA281⟩. Fluorination of alkynes by fluorine in methanol leads to the formation of a *gem*-difluoro dimethyl acetal (Equation (34)) ⟨86JA7739⟩.

$$HC\equiv C-CH_2CH_2CH_2-Cl \xrightarrow[50\%]{HF} F_2C(CH_3)CH_2CH_2CH_2Cl \qquad (33)$$

$$Ph-C\equiv CH \xrightarrow{F_2,\ MeOH} Ph-C(OMe)_2-CHF_2 \qquad (34)$$

4.01.2.6 Difluoro Alkanes from Difluorocarbene

The generation of difluorocarbene has been extensively reviewed ⟨63OR(13)55, B-69MI 401-01, B-71MI 401-01, 77FCR119, B-85MI 401-01⟩. Difluorocarbene transfer is most commonly achieved by decomposition of a trifluoromethyl–metal complex. Pyrolysis of trimethyltrifluoromethyl tin generates perfluorocyclopropane (Equation (35)), formed by difluorocarbene dimerisation to tetrafluoroethene, which undergoes a difluorocarbene addition ⟨60JA1888⟩. Pyrolysis of potassium trifluoromethylfluoroborate also gives perfluorocyclopropane together with perfluorocyclobutane (Equation (36)) ⟨60JA5298⟩. The complex of bis(trifluoromethyl)cadmium and DIGLYME reacts with acetyl chloride to produce acetyl fluoride and difluorocarbene, which can be trapped with 2,3-dimethylbut-2-ene in high yield (Equation (37)) ⟨81JA2995⟩. Metallic lead and dibromodifluoromethane have been used to produce difluorocarbene and its capture by several alkenes studied ⟨81ZN(B)1375⟩. Tetrabutylammonium bromide was added to form a complex with the $PbBr_2$ produced in the reaction. Excellent yields were achieved with 2,3-dimethylbut-2-ene (Equation (38)) but the yields decrease with less substituted alkenes (Equations (39) and (40)).

$$Me_3SnCF_3 \xrightarrow{150\ °C,\ 20\ h} \text{perfluorocyclopropane} \quad (35)$$

$$KCF_3BF_3 \xrightarrow{300\ °C} \text{perfluorocyclopropane} + \text{perfluorocyclobutane} \quad (36)$$

$$\text{2,3-dimethylbut-2-ene} \xrightarrow[70\%]{(CF_3)_2Cd,\ DIGLYME,\ AcCl,\ -27\ °C} \text{1,1-difluoro-2,2,3,3-tetramethylcyclopropane} \quad (37)$$

$$\text{2,3-dimethylbut-2-ene} \xrightarrow[80–90\%]{CBr_2F_2,\ Pb,\ Bu_4NBr} \text{1,1-difluoro-2,2,3,3-tetramethylcyclopropane} \quad (38)$$

$$\text{α-methylstyrene} \xrightarrow[55\%]{CBr_2F_2,\ Pb,\ Bu_4NBr} \text{1,1-difluoro-2-methyl-2-phenylcyclopropane} \quad (39)$$

$$\text{styrene} \xrightarrow[17\%]{CBr_2F_2,\ Pb,\ Bu_4NBr} \text{1,1-difluoro-2-phenylcyclopropane} \quad (40)$$

Bromodifluoromethylphosphonium salts, prepared *in situ*, are good sources of difluorocarbene. Treatment with caesium fluoride formed difluorocarbene, which added to 2,3-dimethylbut-2-ene (Equation (41)) ⟨73JA8467⟩, whereas potassium fluoride was used likewise with butadiene (Equation (42)) ⟨82JA2494⟩. Difluorotris(trifluoromethyl)phosphorane has been used to transfer difluorocarbene to a variety of halogenated alkenes (Equation (43)) ⟨70JCS(C)178⟩.

$$\text{2,3-dimethylbut-2-ene} \xrightarrow[79\%]{CBr_2F_2,\ PPh_3,\ CsF,\ RT,\ 24\ h} \text{1,1-difluoro-2,2,3,3-tetramethylcyclopropane} \quad (41)$$

$$\text{butadiene} \xrightarrow[55\%]{CBr_2F_2,\ PPh_3,\ KF} \text{1,1-difluoro-2-vinylcyclopropane} \quad (42)$$

One of the most useful reagents for generating difluorocarbene is phenyltrifluoromethylmercury ⟨72ACR65⟩; an example of its use is the addition of difluorocarbene to benzobarrelene (Equation (44)) ⟨79TL1913⟩. One of the earliest methods used to generate difluorocarbene was pyrolysis of the sodium chlorodifluoroacetate ⟨60PCS81, 64TL1461⟩; it has been used to add to a double bond (Equation (45)) ⟨73TL1319⟩. The hindered base, sodium bis(trimethylsilyl)amide, has been used to generate difluorocarbene from chlorodifluoromethane. The difluorocarbene reacted with a malonate anion to give an addition product (Equation (46)) ⟨85TL2445⟩.

4.01.2.7 Difluoro Alkanes from Aldehydes and Ketones

Sulfur tetrafluoride was the first reagent used to convert aldehydes and ketones into *gem*-difluoro alkanes. Two excellent reviews cover the use of sulfur tetrafluoride ⟨74OR(21)1, 85OR(34)319⟩; a few examples will be given here to highlight the advantages and disadvantages. Aldehydes and ketones with α-hydrogen atoms need to be treated at low temperatures for long periods to prevent decomposition as shown in Equations (47) and (48) ⟨71JOC818⟩. Aromatic aldehydes (Equation (49)) ⟨71T945⟩ and higher temperatures, generally 150–200 °C, give much higher yields. Formaldehyde (in the form of paraformaldehyde) at a high temperature (150 °C) gave only a modest yield (Equation (50)) ⟨60JA543⟩.

$$H_2C=O \xrightarrow[49\%]{SF_4,\ 150\ °C,\ 6\ h} F\text{-}CH_2\text{-}F \qquad (50)$$

Hindered ketones require an acid catalyst; hydrogen fluoride, boron trifluoride, arsenic trifluoride and titanium tetrafluoride have been used. Hydrogen fluoride is conveniently produced *in situ* by the hydrolysis of sulfur tetrafluoride. This reagent has been used extensively in the production of fluoro steroids (Equation (51)) ⟨71JOC575⟩. Large-ring ketones give particularly low yields (Equation (52)) ⟨71JOC818⟩. A tosyloxy group has been introduced adjacent to the carbonyl group to increase the latter's reactivity, but the yields are disappointingly low (Equation (53)) ⟨72JA2020⟩. One way round this problem is to convert the ketone (or aldehyde) into a 1,3-dithiolane derivative followed by treatment with the hydrogen fluoride–pyridine complex and 1,3-dibromo-5,5-dimethylhydantoin (DBH). This gives high yields of difluoro alkanes as in Equations (54) and (55) ⟨86JOC3508⟩.

One of the problems with sulfur tetrafluoride as a reagent is that it is a gas. Liquid fluorinating reagents have been developed to overcome this problem. Phenyl sulfur trifluoride, $PhSF_3$, ⟨73OSC(5)959⟩ has been used to convert cyclooctanone into 1,1-difluorooctane in 9% yield (Equation (56)) ⟨69JA1386⟩. This same transformation has been reported with sulfur tetrafluoride to give only 1.6% yield ⟨71JOC818⟩. Phenyl sulfur trifluoride has been used to convert other ketones (Equation (57)) ⟨71JOC818⟩ and aldehydes (Equation (58)) ⟨73OSC(5)390⟩ into the corresponding difluoro compounds.

$$\text{cyclohexanone} \xrightarrow[9\%]{\text{PhSF}_3} \text{1,1-difluorocyclohexane} \quad (56)$$

$$\text{1,4-cyclohexanedione} \xrightarrow[36\%]{\text{PhSF}_3} \text{1,1,4,4-tetrafluorocyclohexane} \quad (57)$$

$$\text{PhCHO} \xrightarrow[71-80\%]{\text{PhSF}_3} \text{PhCHF}_2 \quad (58)$$

Diethylaminosulfur trifluoride (DAST) and related aminofluorosulfuranes have become very useful fluorinating agents ⟨88OR(35)513⟩. They also have the advantage of being liquids. DAST is about equivalent to SF_4 for the preparation of geminal difluorides from aldehydes and ketones, however DAST has the advantage of not reacting with carboxylic acids and esters. N,N-Dialkyl-aminosulfur trifluorides have been used to prepare a wide range of geminal difluoro compounds in good yield ⟨88OR(35)513⟩. Aromatic and aliphatic aldehydes and ketones are converted into the corresponding difluoro derivatives in good-to-excellent yields in the presence of several other functional groups apart from hydroxy groups ⟨73S787⟩.

Two other reagents that have seen limited use for the conversion of aldehydes and ketones into *gem*-difluoro derivatives are selenium tetrafluoride and molybdenum hexafluoride. Molybdenum hexafluoride has been used in dichloromethane with boron trifluoride as catalyst at room temperature and gives moderate-to-good yields ⟨71T3965⟩. Selenium tetrafluoride is also used under mild conditions and gives excellent yields of *gem*-difluoro products (Table 1) ⟨74JA925⟩.

Table 1 Yields of *gem*-difluoro compounds prepared from ketones and aldehydes with selenium tetrafluoride, SeF_4.

$$R^1COR^2 \rightarrow R^1CF_2R^2$$

R^1	R^2	Yield (%)
Me	Me	78
Et	Me	80
Et	Et	75
$(CH_2)_5$	$(CH_2)_5$	70
Ph	Me	75
Ph	Ph	90
Ph	H	70

4.01.2.8 Difluoro Alkanes from Imines

Diazo alkanes can be converted into *gem*-difluoro alkanes in excellent yield using fluorine dissolved in Freon-11 at $-70\,°C$ (Equations (59)–(62)) ⟨81JOC3917⟩. An interesting method for the preparation of *gem*-difluoro alkanes with adjacent amino groups is by the treatment of azirines with the hydrogen fluoride–pyridine complex (Equations (63) and (64)) ⟨80JOC5333⟩.

$$Ph_2C=N_2 \xrightarrow[71\%]{F_2,\ \text{Freon-11},\ -70\,°C} Ph_2CF_2 \quad (59)$$

$$\text{9-diazofluorene} \xrightarrow[88\%]{F_2,\ \text{Freon-11},\ -70\,°C} \text{9,9-difluorofluorene} \quad (60)$$

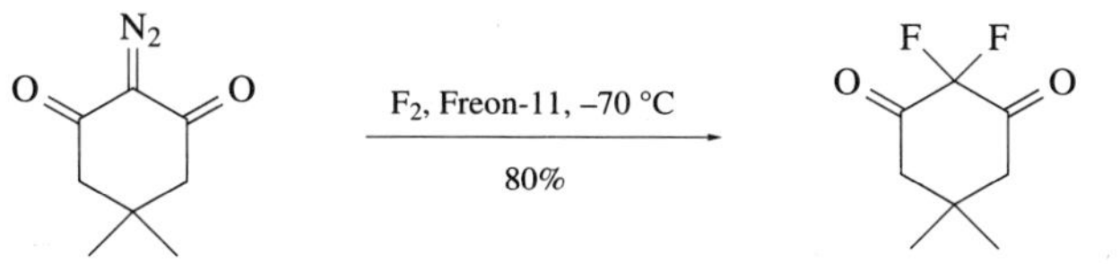

Dichloro Alkanes

(61)

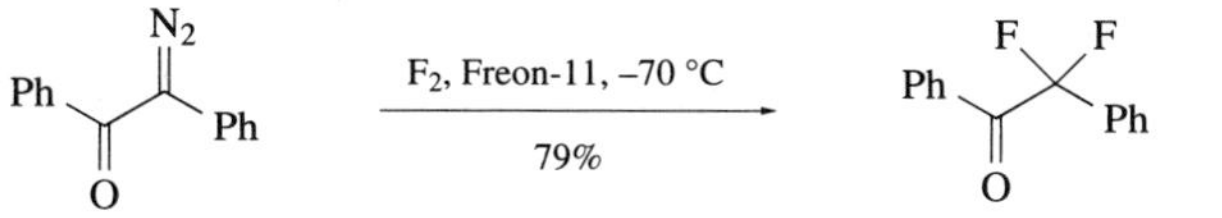

(62)

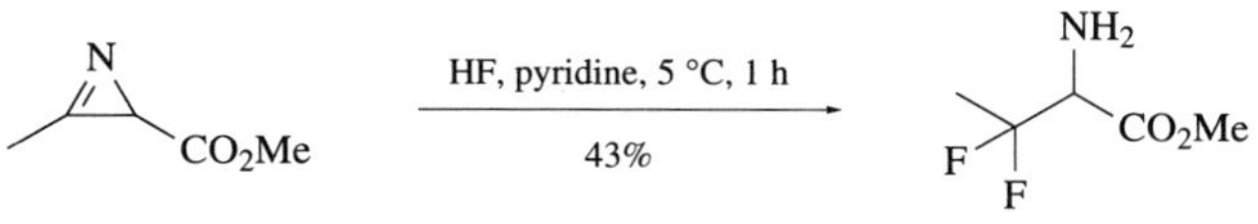

(63)

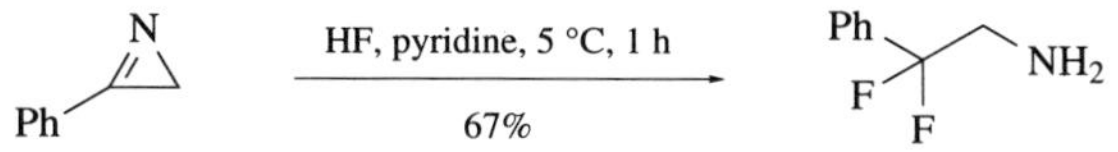

(64)

4.01.3 DICHLORO ALKANES—R_2CCl_2

4.01.3.1 Dichloro Alkanes from Alkanes

Direct chlorination of alkanes generally gives a mixture of chlorinated products. For example, direct chlorination of methane at 485–510 °C gives a mixture of products from which dichloromethane can be isolated, although better yields can be obtained by chlorination of methyl chloride at 340–350 °C in the presence of a catalyst ⟨78KO686⟩. Direct chlorination is only of synthetic use if the rest of the molecule either has no hydrogens or is an unreactive aromatic system. Direct chlorination of pivalic acid yields 3,3-dichloro-2,2-dimethylpropanoic acid under irradiation conditions (Equation (65)) ⟨64T1567⟩. Chlorination of aromatic methyl groups to give dichloromethyl groups can be achieved relatively easily as the replacement of the last hydrogen is much harder; examples include the chlorination of methylbenzoyl chlorides (Equation (66)) ⟨22JCS2202⟩, fluorotoluenes (Equation (67)) ⟨44JIC112⟩ and hexamethylbenzene (Equation (68)) ⟨87JOC3713⟩. Toluene when treated with sulfuryl chloride in the presence of dibenzoyl peroxide gives an excellent yield of dichloromethylbenzene (Equation (69)) ⟨39JA2142⟩.

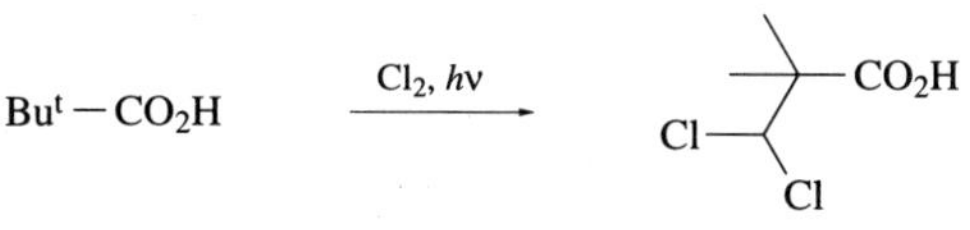

(65)

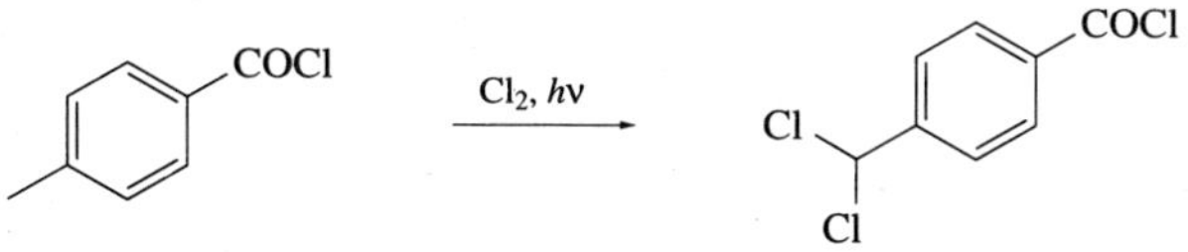

(66)

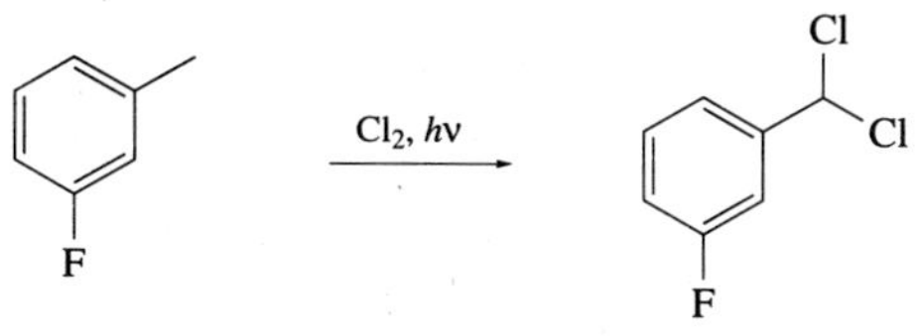

(67)

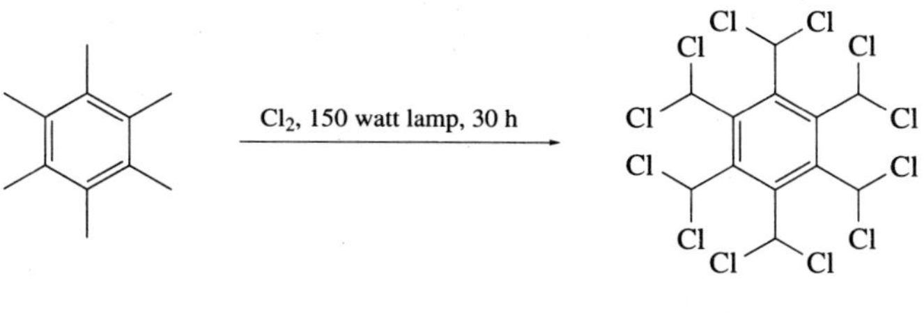

(68)

α,α-Dichlorination of aldehydes is relatively straightforward using acid catalysis. Propanal has been chlorinated using chlorine in hydrochloric acid in good yield (Equation (70)) ⟨62JOC272⟩. The chlorination of a series of aldehydes has been reported with chlorine in DMF and HCl (Table 2) ⟨85T4057⟩. Ketones with only one α-methylene group can also be α,α-dichlorinated without too many problems (Equation (71)) ⟨55JA3278⟩. Several phenyl ketones have been α,α-dichlorinated in excellent yield with chlorine in DMF at 100 °C (Table 3) ⟨79SC575⟩. Sulfuryl chloride has also been used to α,α-dichlorinate phenyl ketones ⟨77JOC3527⟩. Caprolactam can be α,α-dichlorinated with chlorine and phosphorus pentachloride (Equation (72)) ⟨58JA6233⟩.

Table 2 Yields of 2,2-dichloro aldehydes from aldehydes with Cl_2/DMF/HCl.

$RCH_2CHO \rightarrow RCCl_2CHO$

R	Yield (%)
Me	83
Et	76
Pr^n	84
Pr^i	83
Bu^n	81
n-C$_5$H$_{11}$	46

Table 3 Yields for the α,α-dichlorination of phenyl ketones with Cl_2 in DMF at 100 °C for 30–45 min.

$Ph\,COCH_2R \rightarrow PhCOCCl_2R$

R	Yield (%)
Me	87
Et	94
Pr^n	96
Bu^t	96
Ph	86

Trifluoromethanesulfonyl chloride is a mild chlorinating agent when used with an equivalent of triethylamine or 1,5-diazabicyclo[5.4.0]undec-5-ene (dbu). Dimethyl malonate and ethyl acetoacetate have been dichlorinated under these conditions (Equations (73) and (74)) ⟨79TL3643⟩ and β-dicarbonyl compounds containing double bonds have been similarly dichlorinated without affecting the double bonds ⟨88JA5533⟩. Ketones with two α-methylene groups cause more problems and mixtures of products are normally obtained. The best reagent for α,α-dichlorination in this case appears to be sulfuryl chloride (Table 4).

$$MeO_2C\text{---}CH_2\text{---}CO_2Me \xrightarrow[100\%]{CF_3SO_2Cl, Et_3N, RT} MeO_2C\text{---}CCl_2\text{---}CO_2Me \quad (73)$$

$$CH_3COCH_2CO_2Et \xrightarrow[100\%]{CF_3SO_2Cl, Et_3N, RT} CH_3COCCl_2CO_2Et \quad (74)$$

Table 4 Yields for the α,α-dichlorination of ketones with sulfuryl chloride, SO_2Cl_2.

$$R^1COCH_2R^2 \rightarrow R^1COCCl_2R^2$$

R^1	R^2	Yield (%)	Ref.
Me	H	61	64JOC1956
Me	Me	48	64JOC1956
Me	Me	80	77JOC3527
Et	Me	58	64JOC1956

4.01.3.2 Dichloro Alkanes from Dihalo Alkanes

The benzylic proton of dichloromethylbenzene can be removed with butyllithium and the resulting carbanion alkylated with methyl iodide to give 1,1-dichlorophenylethane (Equation (75)) ⟨65JA4147⟩. A similar reaction occurs with allyl dichloride; the carbanion can be treated with carbon dioxide to give 2,2-dichlorobut-3-enoic acid (Equation (76)) or with ketones to give α,α-dichloro alcohols (Equation (77)) ⟨77JA5317⟩.

$$Ph\text{---}CHCl_2 \xrightarrow[\text{ii, MeI}]{\text{i, BuLi}} Ph\text{---}CCl_2Me \quad (75)$$

$$CH_2=CH\text{---}CHCl_2 \xrightarrow[91\%]{\substack{\text{i, BuLi} \\ \text{ii, CO}_2}} CH_2=CH\text{---}CCl_2\text{---}CO_2H \quad (76)$$

$$CH_2=CH\text{---}CHCl_2 \xrightarrow[\text{ii, RCOR}]{\text{i, BuLi}} CH_2=CH\text{---}CCl_2\text{---}C(OH)R_2 \quad (77)$$

4.01.3.3 Dichloro Alkanes from Trihalo Alkanes

The reduction of trichloromethyl compounds to the corresponding dichloromethyl derivatives has been achieved with a wide variety of reagents. Tin and hydrochloric acid have been used to reduce trichloro-α-picoline to its dichloro analogue (Equation (78)) ⟨51JCS1145⟩. Tin(II) chloride has also been used for this transformation ⟨70JOC508⟩. General reagents for the conversion of trichloromethyl groups into dichloromethyl groups include diethyl phosphite ⟨83BCJ1881⟩, tributyltin hydride ⟨68ACR299⟩ and a hydrogen donor such as triethylsilane or isopropanol together with a transition metal complex, for example $Fe(CO)_5$ ⟨83S773⟩. Electrochemical reduction is also effective ⟨75TL997⟩. Trichloroacetic acid can be reduced to dichloroacetic acid by a metal such as zinc, cadmium, iron or copper. Copper has been reported to give an 80% yield of dichloroacetic acid

⟨31JA1594⟩. Chloral hydrate can be converted into the calcium salt of dichloroacetic acid with calcium carbonate and a cyanide catalyst (Equation (79)) ⟨43OSC(2)181⟩.

$$\text{2-PyCCl}_3 \xrightarrow[49\%]{\text{Sn, HCl, 100 °C, 1 h}} \text{2-PyCHCl}_2 \quad (78)$$

$$\text{Cl}_3\text{C-CH(OH)}_2 \xrightarrow{\text{CaCO}_3, \text{KCN (cat.)}} (\text{Cl}_2\text{HCCO}_2)_2\text{Ca} \quad (79)$$

Trichloromethyl groups undergo metal–halogen exchange with butyllithium to give carbanions that can be alkylated (Equation (80)) ⟨80S644⟩. A similar metal–halogen exchange with allyl trichloride gives the allyl carbanion intermediate involved in Equations (76) and (77). This time it was alkylated with methyl iodide (Equation (81)) ⟨77JOM(141)71⟩. The allyl radical is an intermediate in the reaction of chloroform or 1,1,1-trichloroethane with allyltributyltin to give 4,4-dichlorobut-1-ene and 4,4-dichloropent-1-ene, respectively (Equations (82) and (83)) ⟨83BCJ2480⟩. Zinc has also been used to remove a chlorine from a trichloromethyl group, the resulting carbanion reacting readily with formaldehyde (Equation (84)) ⟨86HCA881⟩. Copper(I) catalysis has been used to add a trichloromethyl group across a double bond (Equation (85)) ⟨80HCA1947⟩.

$$\text{PhCCl}_3 \xrightarrow[81\%]{\text{i, BuLi; ii, HCO}_2\text{Me}} \text{PhCCl}_2\text{CHO} \quad (80)$$

$$\text{CH}_2=\text{CHCH}_2\text{CCl}_3 \xrightarrow{\text{i, BuLi; ii, MeI}} \text{CH}_2=\text{CHCH}_2\text{CCl}_2\text{Me} \quad (81)$$

$$\text{CHCl}_3 + \text{CH}_2=\text{CHCH}_2\text{SnBu}_3 \xrightarrow[40\%]{\text{AIBN}} \text{CH}_2=\text{CHCH}_2\text{CHCl}_2 \quad (82)$$

$$\text{MeCCl}_3 + \text{CH}_2=\text{CHCH}_2\text{SnBu}_3 \xrightarrow[42\%]{\text{AIBN}} \text{CH}_2=\text{CHCH}_2\text{CCl}_2\text{Me} \quad (83)$$

$$\text{CF}_3\text{CCl}_3 + \text{HCHO} \xrightarrow[76\%]{\text{i, Zn, DMF, RT, 3 d; ii, HCl}} \text{F}_3\text{C-CCl}_2\text{CH}_2\text{OH} \quad (84)$$

$$\text{CF}_3\text{CCl}_3 + \text{CH}_2=\text{CHCO}_2\text{H} \xrightarrow[76\%]{\text{CuCl, MeCN, 140 °C, 4 h}} \text{F}_3\text{C-CCl}_2\text{CH}_2\text{CHClCO}_2\text{H} \quad (85)$$

4.01.3.4 Dichloro Alkanes from Alkenes

Addition of chlorine to a chloro alkene is a straightforward method for producing a *gem*-dihalo alkane (Equations (86) ⟨48JA2813⟩ and (87) ⟨51JA4393⟩). Nitrogen trichloride is an effective reagent for the addition of chlorine to a chloro alkene (Equation (88)) ⟨71JOC3566⟩. Many other chlorine-containing reagents can be added to chloro alkenes. These include iodine monochloride, effectively generated from iodine and $CuCl_2$ (Equation (89)) ⟨71JOC3324⟩, acetyl chloride catalysed by aluminum trichloride (Equation (90)) ⟨90JCS(P1)3317⟩, chloroform, also catalysed by aluminum chloride (Equation (91)) ⟨43OSC(2)312⟩ and phosgene, effectively generated from $PdCl_2$ and carbon monoxide (Equation (92)) ⟨64JA4851⟩.

The addition of chlorine to an enamine leads to an α,α-dichloro aldehyde (Equation (93)) ⟨73TL4237⟩. A range of Grignard reagents were added to the chloro enamine (**1**) to give, after acid hydrolysis, dichloromethyl ketones (Table 5) ⟨74BSF1533⟩.

Table 5 Yields for the preparation of dichloromethyl ketones.

R	Yield (%)
Me	40
Et	78
Pr^i	25
$n\text{-}C_5H_{11}$	77
Ph	90
Vinyl	25
BuC≡C	60

4.01.3.5 Dichloro Alkanes from Alkynes

The addition of two molecules of HCl to an alkyne, in principle, should give a *gem*-dichloro alkane; however, in practice, the yields are poor (Equation (95)) ⟨65JA3151, 67JOC2651⟩. Chlorination of terminal alkynes with chlorine in water or acetic acid gives dichloromethyl ketones (Equation (96)); however, again the yields are poor ⟨39JA1460⟩. A better method for the preparation of dichloromethyl ketones from terminal alkynes is to use N-chlorosuccinimide in methanol followed by hydrolysis of the intermediate acetal (Table 6) ⟨65JOC2195⟩. This method may also be used for symmetrical alkynes (Equation (97)).

$$\text{HC} \equiv \text{CH} \xrightarrow{\text{HCl}} \text{Cl}_2\text{C}(\text{CH}_3) \quad (95)$$

$$R-\equiv \xrightarrow{\text{Cl}_2, \text{H}_2\text{O or AcOH}} R-\text{CO-CHCl}_2 \quad (96)$$

Table 6 Yields for the preparation of dichloromethyl ketones from terminal alkynes.

$$R-\equiv \xrightarrow{\text{NCS, MeOH}} R-\text{C(OMe)}_2\text{CHCl}_2 \xrightarrow{\text{H}^+, \text{H}_2\text{O}} R-\text{CO-CHCl}_2$$

R	Yield (%)
Me	68
Et	73
Pr^n	66
Bu^n	68
Ph	68

$$\text{MeC}\equiv\text{CMe} \xrightarrow[70\%]{\text{i, NCS, MeOH; ii, H}^+} \text{MeC(O)C(Cl)_2Me} \quad (97)$$

4.01.3.6 Dichloro Alkanes from Dichlorocarbene

Dichlorocarbene production, properties and uses have been covered in several reviews ⟨63OR(13)55, B-64MI 401-01, B-71MI 401-01, B-85MI 401-01⟩. There are many ways to generate dichlorocarbene efficiently. One of the earlier methods was to use chloroform and potassium t-butoxide (Equation (98)) ⟨54JA6162⟩. The same reaction can be performed in excellent yield (91%) using bromotrichloromethane and butyllithium to generate the dichlorocarbene ⟨59JA5008⟩. Phase-transfer conditions have been used extensively to generate dichlorocarbene, for example chloroform, 50% aqueous sodium hydroxide and a quaternary ammonium salt such as benzyltriethylammonium chloride (TEBA). These conditions have been used for the production of a range of dichlorocyclopropanes (Equations (99) ⟨71LA(744)42⟩ and (100) ⟨90CB583⟩). The addition of dichlorocarbene to alkenes that contain base-sensitive groups can be achieved using the thermal decomposition of sodium trichloroacetate, for example, for the reaction with cyclohexene (Equation (98)) in 65% yield ⟨59PCS229⟩. However side reactions tend to occur ⟨62RTC925⟩. The best method for base-sensitive compounds appears to be the use of phenyl(bromodichloromethyl)mercury (PhHgCBrCl$_2$) in an inert solvent at 80 °C ⟨65JA4259⟩. For example 7,7-dichloronorcarane can be prepared from cyclohexene (Equation (98)) in 89% yield using this method. A large range of dichlorocyclopropanes has been produced in high yield from the corresponding alkenes (Equation (101)). *cis*-Alkenes and *trans*-alkenes react with retention of configuration.

$$\text{cyclohexene} \xrightarrow[\text{or PhHgCBrCl}_2]{\text{CHCl}_3\text{-KOBu}^t \text{ or CBrCl}_3\text{-Bu}^n\text{Li}} \text{7,7-dichloronorcarane} \quad (98)$$

$$Ph\diagup\!\!\!\diagdown Ph \xrightarrow{CHCl_3,\ 50\%\ NaOH,\ BnEt_3NCl} \text{Ph-cyclopropane(Cl,Cl)-Ph} \qquad (99)$$

$$\text{MeCH=CHCO}_2Bu^t \xrightarrow{CHCl_3,\ 50\%\ NaOH,\ Me_4NCl} \text{Me-cyclopropane(Cl,Cl)-CO}_2Bu^t \qquad (100)$$

$$\text{MeCH=CHCO}_2Me \xrightarrow[76\%]{PhHgCBrCl_2} \text{Me-cyclopropane(Cl,Cl)-CO}_2Me \qquad (101)$$

Dichlorocarbene can also be inserted into C–H bonds, particularly when there is an α-oxygen anion (Table 7) ⟨83JA2771⟩. The carbene C–H insertion competes favourably with addition to an alkene except for more highly substituted alkenes (Equation (102)). Chloroform may be added across the double bond of styrene using copper metal and 1,10-phenanthroline (as a Cu(I) complex) (Equation (103)) in high yield. The reaction is less efficient with aliphatic alkenes ⟨83CCC1710⟩.

Table 7 Yields for the insertion of carbenes into the C–H bonds of alkoxides.

$$R\text{-CH}_2\text{-OLi} \xrightarrow{Bu^tOLi,\ CHCl_3} R\text{-CH(OH)-CHCl}_2$$

R	Yield (%)
Ph	87
CH$_2$=CH	55
CH$_3$CH=CH	76
PhCH=CH	83

$$(CH_3)_2C=CHCH_2OLi \xrightarrow{Bu^tOLi,\ CHCl_3} (CH_3)_2C=CH\text{-CH(OH)-CHCl}_2 + \text{dimethylcyclopropane(Cl,Cl)-CH}_2OH \qquad (102)$$

$$Ph\text{-CH=CH}_2 \xrightarrow[95\%]{CHCl_3,\ Cu,\ 1,10\text{-phenanthroline},\ 66\ °C,\ 15\ h} Ph\text{-CH}_2\text{-CHCl-CHCl}_2 \qquad (103)$$

Dichlorocarbene is involved in the Reimer–Tiemann reaction for the production of aromatic aldehydes from phenols with chloroform and aqueous sodium hydroxide ⟨82OR(28)1⟩. The intermediate dichloromethylated compounds are normally hydrolysed during the reaction, however if the *ortho*- and *para*-positions are blocked the abnormal Reimer–Tiemann product, a chloromethylcyclohexadienone, is formed (Equation (104)) ⟨74T2661⟩. Chloromethylation of aromatic compounds can be achieved with aluminum trichloride and chloroform (Equation (105)) ⟨91JOC5445⟩.

$$\text{2,6-di-Bu}^t\text{-4-Me-phenol} \xrightarrow[78\%]{CHCl_3,\ NaOH,\ 50\ °C} \text{2,6-di-Bu}^t\text{-4-Me-4-(CHCl}_2)\text{-cyclohexadienone} \qquad (104)$$

4.01.3.7 Dichloro Alkanes from Aldehydes and Ketones

Aldehydes and ketones can be readily converted into dichloro alkanes by phosphorus pentachloride, which is by far the most widely used reagent for this conversion ⟨80LA1⟩. Aliphatic ketones do not normally give the corresponding dichloro alkane in particularly high yield since elimination products are frequently formed (Equation (106)) ⟨34JA2730, 39JA938⟩. Aliphatic aldehydes generally give better yields (Equation (107)) ⟨63BSF1868⟩ and if the conditions are carefully controlled, spectacular yields can be attained (Equation (108)) ⟨69JOC2618⟩.

Aromatic aldehydes can be easily converted in high yield into the corresponding dichloromethyl compound (Equation (109)) ⟨33JCS496⟩; there are many examples in the literature ⟨24LA(435)219, 60JA6115⟩. α,β-Unsaturated aldehydes are also converted smoothly into the corresponding dichloromethyl compounds with phosphorus pentachloride (Equation (110)) ⟨79S425⟩.

Ketones that lack α-hydrogens can be converted into the dichloro alkane but particularly high temperatures are needed as in Equations (111) and (112) ⟨43JA389, 49JA3439, 65JOC1241⟩.

Only a few other reagents have been used for the transformation of aldehydes and ketones into dichloro alkanes. Thionyl chloride in DMF has been used for this transformation (Equations (113) and (114)). Ketones require a higher reaction temperature than aldehydes ⟨78JOC4367⟩. Oxaloyl chloride has also been used for the preparation of benzal chloride from benzaldehyde ⟨09CB3966⟩ and antimony pentachloride with a trace of iodine as catalyst has been used in a similar transformation (Equation (115)) ⟨31RTC753⟩.

Dibromo Alkanes

Ph-CO-Ph $\xrightarrow[\text{85%}]{\text{SOCl}_2,\text{ DMF, reflux, 16 h}}$ Ph-CCl$_2$-Ph (113)

PhCHO $\xrightarrow[\text{89%}]{\text{SOCl}_2,\text{ DMF, RT, 4 h}}$ PhCHCl$_2$ (114)

2,5-dichlorobenzaldehyde $\xrightarrow{\text{SbCl}_5,\text{ I}_2\text{ (cat.)}}$ 2,5-dichloro-α,α-dichloromethylbenzene (115)

Dichloromethyllithium, generated from dichloromethane and lithium dicyclohexylamide, adds smoothly to carbonyl compounds (Equation (116)) ⟨74JA3010, 77BCJ1588⟩.

cyclohexanone $\xrightarrow[\text{89%}]{\text{(c-C}_6\text{H}_{11}\text{)}_2\text{N-Li, CH}_2\text{Cl}_2,\text{ 0 °C}}$ 1-(dichloromethyl)cyclohexan-1-ol (116)

4.01.3.8 Dichloro Alkanes from Imines

Diazoketones can be converted into the corresponding dichloromethyl ketones by chlorine (Equation (117)); however, the yields tend to be poor ⟨66ACS253⟩. In a similar reaction with triphenylphosphoranylidenehydrazones, sterically hindered dichloro alkanes can be produced (Equation (118)), again with a low yield ⟨83OMR(21)64⟩. In an interesting reaction of chlorine on 1,2,3-triazolo[1,5-a]pyridine, 2-dichloromethylpyridine is produced in high yield (Equation (119)) ⟨81JCS(P1)78⟩.

Br$_2$C(Me)-CO-CH=N$_2$ $\xrightarrow{\text{Cl}_2,\text{ CCl}_4}$ Br$_2$C(Me)-CO-CHCl$_2$ (117)

Bu^t-C(=N-N=PPh$_3$)-Bu^t $\xrightarrow[\text{31%}]{\text{Cl}_2,\text{ CCl}_4}$ Bu^t-CCl$_2$-Bu^t (118)

1,2,3-triazolo[1,5-a]pyridine $\xrightarrow[\text{67%}]{\text{Cl}_2,\text{ CaCO}_3,\text{ CCl}_4,\text{ 0 °C}}$ 2-(dichloromethyl)pyridine (119)

Oxidative deamination of a variety of primary amines using copper(II) chloride and isopentyl nitrite gives *gem*-dichloro compounds in moderate to good yields (Table 8) ⟨76CC433, 76JA1627⟩.

4.01.4 DIBROMO ALKANES—R$_2$CBr$_2$

4.01.4.1 Dibromo Alkanes from Alkanes

Bromination of alkanes under radical conditions has been used since the mid 1920s. Sunlight has been used to produce dibromomethyl aromatic compounds from toluene derivatives (Equation

Table 8 Yields for the oxidative deamination of primary amines by copper(II) chloride and isopentyl nitrite to give *gem*-dichloro alkanes.

$$R-CH_2-NH_2 \rightarrow R-CHCl_2$$

R	Yield (%)
Ph	14
PhCH$_2$	58
PhCH$_2$CH$_2$	34
PhCH$_2$CH$_2$CH$_2$	32
Ph(CH$_2$)$_5$	34
CH$_3$(CH$_2$)$_8$	26
EtO$_2$C(CH$_2$)$_4$	30
HO(CH$_2$)$_5$	39
Cyclohexyl	26

(120)); however, the yields tend to be low ⟨26JA1093⟩. One of the reasons is that, unless there is a constraint on the number of bromines that can be added to the molecule, there is generally a mixture of brominated products. Steric crowding of *ortho*-substituents results in better yields of dibromomethyl products (Equation (121)) ⟨54OSC(4)807⟩. Extreme crowding can completely inhibit the production of tribromomethyl products (Equation (122)) ⟨88JCS(P1)961⟩. A 200 or 300 W sunlamp is found to be best for these benzylic brominations, the lamp providing both light and heat for the reaction. The concentration of bromine should be low during the reaction and for highly hindered cases the concentration of the precursor should also be low.

Bromination α to ketones is generally performed with bromine without irradiation (Equations (123) ⟨75JCS(P1)251⟩ and (124) ⟨66JCS(C)533⟩). Bromination of ketones with two equivalents of bromine under acid conditions generally gives an α,α'-dibromo derivative (Equation (125)) rather than a *gem*-dibromo derivative; however, four equivalents of bromine gives a reasonable yield of the bis(*gem*-dibromo) derivative (Equation (126)) ⟨62ACS2467⟩. α,α-Dibromination of aldehydes is generally poor due to the variety of side reactions that occur (Equation (127)) ⟨79LA278⟩. Bromination between β-dicarbonyl groups can be carried out under basic conditions (Equations (128) ⟨58JA1942⟩ and (129) ⟨76TL4577⟩).

$$\text{CH}_3\text{COCO}_2\text{H} \xrightarrow[50\%]{\text{Br}_2,\text{ CHCl}_3,\text{ reflux, 30 h}} \text{Br}_2\text{CHCOCO}_2\text{H} \quad (124)$$

$$\text{CH}_3\text{COCH}_3 \xrightarrow{\text{Br}_2\text{ (2 equiv.), H}^+,\text{ 0 °C}} \text{BrCH}_2\text{COCH}_2\text{Br} \quad (125)$$

$$\text{CH}_3\text{COCH}_3 \xrightarrow{\text{Br}_2\text{ (4 equiv.), H}^+,\text{ 0 °C}} \text{Br}_2\text{CHCOCHBr}_2 \quad (126)$$

$$\text{CH}_3\text{CH}_2\text{CHO} \xrightarrow[36\%]{\text{Br}_2,\text{ CH}_2\text{Cl}_2} (\text{CH}_3)(\text{CBr}_2)\text{CHO} \quad (127)$$

(128) Meldrum's acid → 5,5-dibromo-Meldrum's acid, Br$_2$, NaOH, 72%

(129) malonaldehyde → 2,2-dibromomalonaldehyde, Br$_2$, NaOH

N-Bromosuccinimide (NBS) is often the reagent of choice for bromination ⟨B-74MI 401-01⟩. Bromination of aromatic methyl groups under radical conditions normally leads to a mixture of mono-, di- and tribrominated products together with products due to radical dimerisation, although reasonable amounts of dibromomethyl compounds can generally be obtained (Equation (130)) ⟨84JHC1157⟩. As observed for bromination with bromine, steric crowding increases the yields of dibromination of methyl groups (Equation (131)) ⟨86JCS(P1)1495⟩. A range of ketones have been α,α-dibrominated in very good yields by NBS in carbon tetrachloride with a 300 W sunlamp (Table 9) ⟨77JOC3527⟩. The yield for the reaction of pentan-3-one is low as the main product is 2,4-dibromopentan-3-one. Radical initiators, such as dibenzoyl peroxide and 2,2′-azobisisobutyronitrile (AIBN), have been used extensively in NBS reactions (Equations (132) ⟨54HCA90⟩ and (133) ⟨62JOC757⟩). NBS can also be used under nonradical conditions (Equation (134)) ⟨89JCS(P1)2009⟩. Bromotrichloromethane in the presence of dbu has been shown to be effective at dibrominating active methylene compounds (Equation (135)) but diethyl malonate gives monobromination and dimerisation products ⟨78CL73⟩.

(130) 4-methyl-1,2,3-thiadiazole → 4-(dibromomethyl)-1,2,3-thiadiazole, NBS, CCl$_4$, $h\nu$

(131) 2,3-dimethylpyridine → 2,3-bis(dibromomethyl)pyridine, NBS, CCl$_4$, $h\nu$, 7 h, 78%

(132) 8-methylquinoline → 8-(dibromomethyl)quinoline, NBS, CCl$_4$, dibenzoyl peroxide, 60%

Table 9 Yields of α,α-dibromo ketones formed with NBS in CCl$_4$ and a 300 W sunlamp.

$$R^1COCH_2R^2 \rightarrow R^1COCBr_2R^2$$

R^1	R^2	Yield (%)
Ph	Me	95
Me	Me	95
Me	Ph	97
Et	Me	15

benzocyclobutene-1,2-dibromide → benzocyclobutene-1,1,2,2-tetrabromide, NBS, CCl$_4$, dibenzoyl peroxide, 65% (133)

3-bromooxindole → 3,3-dibromooxindole, NBS, Bu^{t}OH (aq.), 98% (134)

fluorene → 9,9-dibromofluorene, BrCCl$_3$, dbu, 65% (135)

4.01.4.2 Dibromo Alkanes from Dihalo Alkanes

Dichloro alkanes can be converted into dibromo alkanes by calcium bromide under phase-transfer conditions with tetrahexylammonium bromide (Equation (136)) ⟨84S34⟩, boron tribromide (Equation (137)) ⟨66JA2481⟩ or boron tribromide with aluminum tribromide and bromine (Equation (138)) ⟨73JOC153⟩.

CH$_2$Cl$_2$ → CH$_2$Br$_2$, CaBr$_2$, (n-C$_6$H$_{11}$)$_4$NBr, 63% (136)

tetrachlorocyclopropene → tetrabromocyclopropene, BBr$_3$ (137)

hexachlorocyclopentadiene → hexabromocyclopentadiene, BBr$_3$, AlBr$_3$, Br$_2$, 78% (138)

Dibromomethyllithium, prepared from dibromomethane and lithium diisopropylamide (LDA) can be alkylated (Equation (139)) ⟨90JOC5719⟩. α,α-Dibromo alkyllithium compounds can in general be prepared from the corresponding dibromo alkanes and LDA, and they can be used in a variety of reactions such as alkylations and reactions with esters to give α,α-dibromo ketones (Equation (140)), with carbon dioxide to give α,α-dibromo carboxylic acids (Equation (141)) ⟨75BSF1797⟩ and with methyl formate to give α,α-dibromo aldehydes (Equation (142)) ⟨80S644⟩.

$$Br\diagdown Br \xrightarrow[92\%]{\text{i, LDA} \\ \text{ii, Cl}\diagup\diagdown Br} Cl\diagdown\diagup\diagdown\text{CHBr}_2 \qquad (139)$$

$$Br\diagdown Br \xrightarrow[90\%]{\text{i, LDA} \\ \text{ii, }\diagup\text{CO}_2\text{Et}} \text{(iPr)C(O)CHBr}_2 \qquad (140)$$

$$\text{BuCHBr}_2 \xrightarrow[48\%]{\text{i, LDA} \\ \text{ii, CO}_2} \text{BuCBr}_2\text{CO}_2\text{H} \qquad (141)$$

$$\text{PhCHBr}_2 \xrightarrow[73\%]{\text{i, LDA} \\ \text{ii, HCO}_2\text{Me, }-110\,°\text{C}} \text{PhCBr}_2\text{CHO} \qquad (142)$$

4.01.4.3 Dibromo Alkanes from Trihalo Alkanes

Bromoform can be reduced to dibromomethane by sodium arsenite in good yield (Equation (143)) ⟨32OSC(1)357⟩. Tribromoquinaldine has been reduced to dibromoquinaldine (Equation (144)) with tin and hydrobromic acid ⟨51JCS1145⟩ but also with ethanol in concentrated sulfuric acid producing a 98% yield ⟨46JOC55⟩.

$$\text{HCBr}_3 \xrightarrow[88–90\%]{\text{Na}_3\text{AsO}_3,\text{ NaOH}} \text{CH}_2\text{Br}_2 \qquad (143)$$

$$\text{quinoline-2-CBr}_3 \longrightarrow \text{quinoline-2-CHBr}_2 \qquad (144)$$

4.01.4.4 Dibromo Alkanes from Alkenes

Addition of hydrogen bromide to bromo alkenes under ionic conditions gives good yields of *gem*-dibromo alkanes (Equation (145)) whereas addition of hydrogen bromide to 2-bromobut-2-ene under UV irradiation gives 3,4-dibromobutane ⟨59JA5937⟩. The addition of hydrogen bromide to bromo alkenes under radical and ionic conditions has been reviewed ⟨40CRV351⟩. Ionic addition of hydrogen bromide can be assisted by a small amount of a Lewis acid catalyst such as $FeCl_3$ (Equation (146)) ⟨55JA3465⟩; however an excess of $FeCl_3$ causes elimination and halogen exchange reactions ⟨57JA6270⟩.

$$\text{2-bromobut-2-ene} \xrightarrow{\text{HBr}} \text{2,2-dibromobutane} \qquad (145)$$

$$\text{1-bromocyclohexene} \xrightarrow{\text{HBr, FeCl}_3} \text{1,1-dibromocyclohexane} \qquad (146)$$

Addition of bromine to bromo alkenes also gives *gem*-dibromo alkanes in good yields (Equation (147)) ⟨35JA1088, 67BCJ594⟩ and (Equation (148)) ⟨54JA479⟩. Addition of bromine to dibromo alkenes often requires radical conditions (Equation (149)) ⟨66JA2481⟩. Bromination of 2,4,6-tribromophenol produces a tetrabromocyclohexadienone ⟨76OS(55)20⟩ which is itself a mild brominating agent (Equation (150)).

$$\text{CH}_2=\text{CHBr} \xrightarrow{\text{Br}_2} \text{BrCH}_2\text{CHBr}_2 \quad (147)$$

$$\text{CH}_2=\text{C(Br)CO}_2\text{Me} \xrightarrow[56\%]{\text{Br}_2} \text{BrCH}_2\text{C(Br)}_2\text{CO}_2\text{Me} \quad (148)$$

$$\text{(tribromocyclopropene)} \xrightarrow{\text{Br}_2, h\nu} \text{(pentabromocyclopropane)} \quad (149)$$

$$\text{2,4,6-tribromophenol} \xrightarrow[61-67\%]{\text{Br}_2, \text{AcOH}, \text{AcONa}, 70\,°\text{C}} \text{2,4,4,6-tetrabromocyclohexa-2,5-dienone} \quad (150)$$

An interesting rearrangement occurs when *trans*-2,3-dibromobut-2-ene is treated with trifluoroperacetic acid and boron trifluoride giving 3,3-dibromobutan-2-one (Equation (151)) ⟨67JOC2669⟩.

$$\text{MeC(Br)=C(Br)Me} \xrightarrow{\text{CF}_3\text{CO}_3\text{H}, \text{CH}_2\text{Cl}_2, \text{BF}_3} \text{MeC(O)C(Br)}_2\text{Me} \quad (151)$$

4.01.4.5 Dibromo Alkanes from Alkynes

Addition of two equivalents of hydrogen bromide to propyne under ionic conditions gives good yields of 2,2-dibromopropane (Equation (152)) ⟨35JA2463⟩. However, under radical conditions the yields are poor and several side reactions occur ⟨65JA3151⟩. A variety of other terminal alkynes give good yields in addition reactions with hydrogen bromide (Equations (153) ⟨36JA1806⟩ and (154) ⟨07JCS816⟩).

$$\text{MeC}\equiv\text{CH} \xrightarrow{\text{HBr}} \text{MeCBr}_2\text{CH}_3 \quad (152)$$

$$\text{CH}_3(\text{CH}_2)_3\text{C}\equiv\text{CH} \xrightarrow{\text{HBr}} \text{CH}_3(\text{CH}_2)_3\text{CBr}_2\text{CH}_3 \quad (153)$$

$$\text{HC}\equiv\text{CCH}_2\text{CH}_2\text{CO}_2\text{H} \xrightarrow{\text{HBr}} \text{Br}_2\text{CHC(Br)CH}_2\text{CH}_2\text{CO}_2\text{H} \quad (154)$$

4.01.4.6 Dibromo Alkanes from Dibromocarbene

The formation and reaction of dibromocarbene has been extensively reviewed ⟨63OR(13)55, B-64MI 401-01, B-71MI 401-01, B-85MI 401-01⟩. The classic production of this carbene using bromoform and potassium *t*-butoxide ⟨64JOC2951⟩ gives good yields in addition reactions to a variety of substituted alkenes, for example Equation (155) ⟨62JOC748, 70JA5469⟩, Equation (156) ⟨56JA5430, 83JA3411⟩ and Equation (157) ⟨56JA5430, 77OS(56)32⟩. Phase-transfer conditions using benzyltriethylammonium chloride (TEBA), 50% sodium hydroxide solution and bromoform, have been used to add dibromocarbene to a variety of alkenes (Equations (158) ⟨90CB583⟩ (159) ⟨71LA(744)42⟩ and (160) ⟨73TL1367⟩) again with good yields. In many cases the traditional method using potassium *t*-butoxide gives better yields than the phase-transfer method, for example with cyclohexa-1,3-diene (Equation (161)) KOBu^t gives 70% yield ⟨59JA992⟩ whereas phase-transfer conditions give only 50% ⟨89S188⟩.

$$\text{cyclopentene} \xrightarrow[85\%]{\text{CHBr}_3, \text{Bu}^t\text{OK, BuOH}} \text{6,6-dibromobicyclo[3.1.0]hexane} \qquad (155)$$

$$\text{isobutylene} \xrightarrow[93\%]{\text{CHBr}_3, \text{Bu}^t\text{OK, BuOH}} \text{1,1-dibromo-2,2-dimethylcyclopropane} \qquad (156)$$

$$\text{1,1-diphenylethylene} \xrightarrow[63-78\%]{\text{CHBr}_3, \text{Bu}^t\text{OK, BuOH}} \text{1,1-dibromo-2,2-diphenylcyclopropane} \qquad (157)$$

$$\text{acrylate-CO}_2\text{Bu}^t \xrightarrow{\text{CHBr}_3, 50\% \text{NaOH, BnEt}_3\text{NCl}} \text{dibromocyclopropane-CO}_2\text{Bu}^t \qquad (158)$$

$$\text{stilbene (Ph-CH=CH-Ph)} \xrightarrow{\text{CHBr}_3, 50\% \text{NaOH, BnEt}_3\text{NCl}} \text{1,1-dibromo-2,3-diphenylcyclopropane} \qquad (159)$$

$$\text{2-methyl-2-butene} \xrightarrow[73\%]{\text{CHBr}_3, 50\% \text{NaOH, BnEt}_3\text{NCl}} \text{1,1-dibromo-2,2,3-trimethylcyclopropane} \qquad (160)$$

$$\text{cyclohexa-1,3-diene} \longrightarrow \text{7,7-dibromobicyclo[4.1.0]hept-2-ene} \qquad (161)$$

Phenyl(tribromomethyl)mercury has been reported to produce a range of dibromocarbene addition reactions ⟨72ACR65⟩ and, as for the chloro analogue (Section 4.01.3.6), it is useful for base-sensitive compounds (Equation (162)) ⟨90CB583⟩.

$$\text{crotonate-CO}_2\text{Bu}^t \xrightarrow[30\%]{\text{PhHgCBr}_3} \text{dibromocyclopropane-CO}_2\text{Bu}^t \qquad (162)$$

4.01.4.7 Dibromo Alkanes from Aldehydes and Ketones

Compared with the formation of *gem*-dichloro alkanes from aldehydes and ketones, there are fewer general methods for the conversion of aldehydes and ketones into *gem*-dibromo alkanes. Phosphorus pentabromide has been used to convert benzaldehyde derivatives into the corresponding dibromomethylbenzenes (Equation (163) ⟨23LA(431)270⟩). PBr$_2$Cl$_3$, formed from phosphorus trichloride and bromide *in situ* is an alternative reagent (Equation (164)) ⟨71JA4472; see also 83JOC2084⟩. Phosphorus tribromide has been reported to give a high yield of dibromo alkane (Equation (165)) ⟨84JA8174⟩. An interesting new reagent, (PhO)$_3$PBr$_2$, formed from a 1:1 mixture of triphenyl phosphite and bromine, seems to have general application for the production of *gem*-dibromo alkanes from aldehydes (Table 10) ⟨90S657⟩.

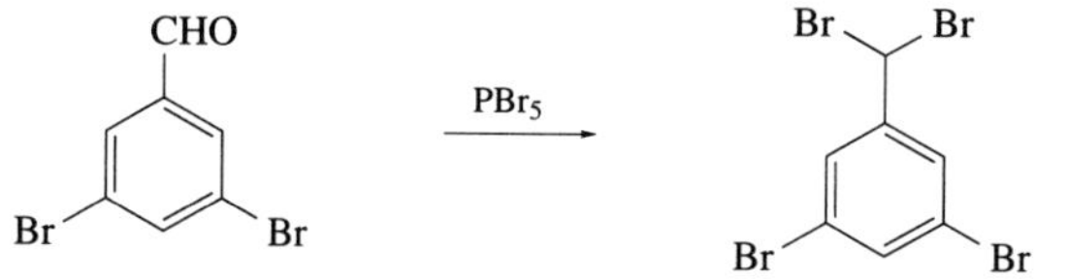

Dihalo Alkanes

(163) 3,5-dibromobenzaldehyde + PBr$_5$ → 1-(dibromomethyl)-3,5-dibromobenzene

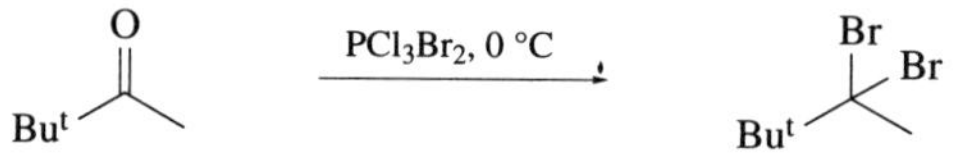

(164) Bu^{t}COCH$_3$ + PCl$_3$Br$_2$, 0 °C → Bu^{t}C(Br)$_2$CH$_3$

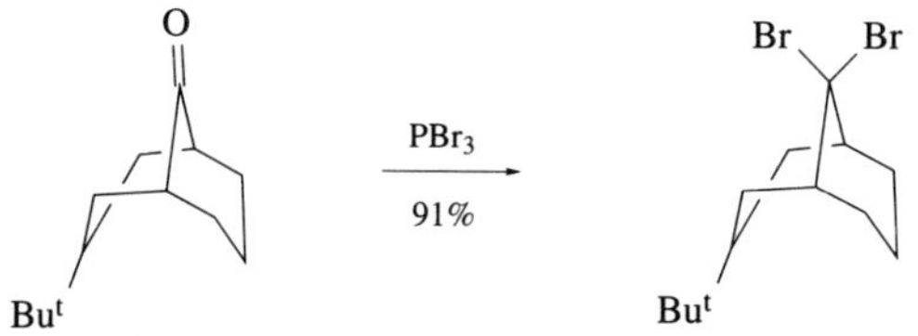

(165) 4-tert-butylcyclohexanone + PBr$_3$ → 4-tert-butyl-1,1-dibromocyclohexane, 91%

Table 10 Yields for the production of dibromo alkanes from aldehydes and (PhO$_3$)PBr$_2$ in CH$_2$Cl$_2$ at −15 °C.

R—CHO → R—CHBr$_2$

R	Yield (%)
n-C$_6$H$_{13}$	70
2-pentyl	56
cyclohexyl	55
Bu^t	50
BnOCH$_2$	64
Ph	76
3-chloro-4-nitrophenyl	91

Boron tribromide was found to convert an aromatic aldehyde into the corresponding dibromomethylbenzene derivative during the demethylation of a methoxyl group. The reaction has been shown to be generally applicable (Table 11) ⟨79SC341⟩. Ketones, when converted into their catechol acetal derivatives, can be converted into the corresponding dihalo alkane in good yield using boron tribromide (Scheme 1) ⟨86S122⟩.

Table 11 Yields for the preparation of benzal bromides from aldehydes with BBr$_3$.

X—C$_6$H$_4$—CHO → X—C$_6$H$_4$—CHBr$_2$

X	Yield (%)
H	83
4-Me	92
4-Cl	87
4-NO$_2$	53
3-NO$_2$	70
3-OMe	90 (of the 3-OH derivative)

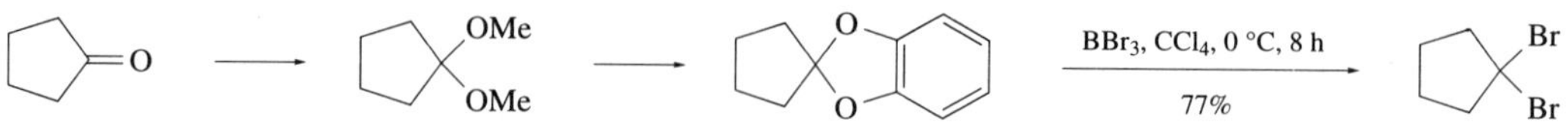

cyclopentanone → 1,1-dimethoxycyclopentane → cyclopentanone catechol acetal → (BBr$_3$, CCl$_4$, 0 °C, 8 h, 77%) → 1,1-dibromocyclopentane

Scheme 1

An interesting route for the preparation of dibromo ketones from α-diketones utilises the reaction of bromine with the adduct of the α-diketone and triphenyl phosphite (Scheme 2) ⟨68JOC25⟩.

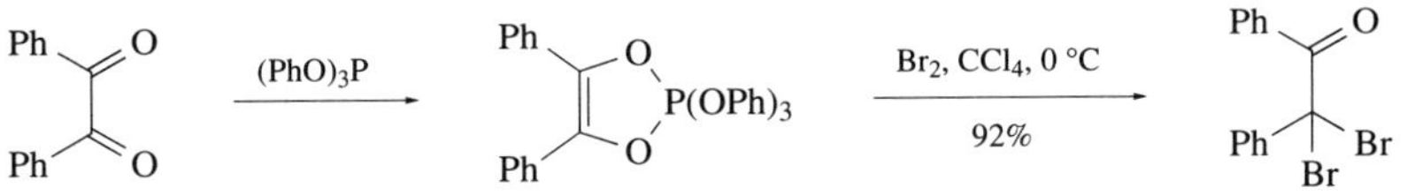

Scheme 2

4.01.4.8 Dibromo Alkanes from Imines

A general method for the preparation of sterically hindered dibromo alkanes from hydrazones (Equation (166)) with bromine and triethylamine has been described ⟨88S547⟩. They are also available from triphenylphosphoranylidene hydrazones (Equation (167)) ⟨83OMR(21)64⟩. In a reaction analogous to that with chlorine, 1,2,3-triazolo[1,5-a]pyridine when treated with bromine yields 2-dibromomethylpyridine (Equation (168)) ⟨81JCS(P1)78⟩. A range of primary amines have been converted into *gem*-dibromo alkanes (Table 12) ⟨76JA1627⟩ with copper(II) bromide and isopentyl nitrite.

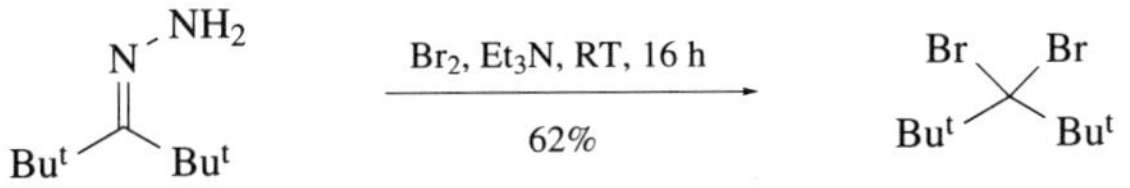

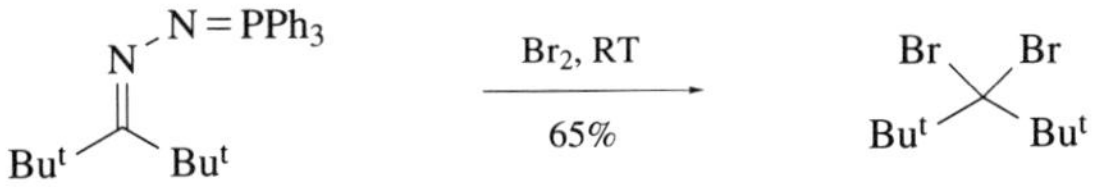

Table 12 Yields for the oxidative deamination of primary amines by copper(II) bromide and isopentyl nitrite to give *gem*-dibromo alkanes.

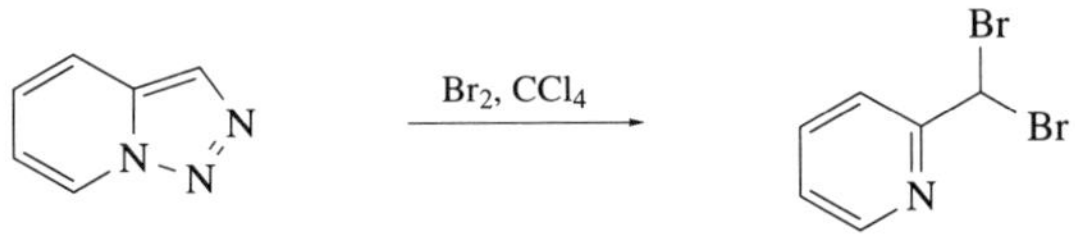

$$RCH_2NH_2 \rightarrow RCHBr_2$$

R	Yield (%)
Ph	30
PhCH$_2$	52
Ph(CH$_2$)$_5$	39
CH$_3$(CH$_2$)$_8$	37

4.01.4.9 Dibromo Alkanes from Carboxylic Acids

The Hunsdiecker reaction of the silver salt of α-bromo carboxylic acids with bromine has been used to prepare dibromo alkanes (Equation (169)) ⟨53JA1148⟩. Apart from the reaction of silver salts ⟨56CRV219, 57OR(9)332⟩, there are several other ways to bromodecarboxylate acids, including the use of lead tetraacetate and lithium bromide ⟨72OR(19)279⟩ and of thallium(I) salts and bromine ⟨81JCS(P1)2608⟩. Silver acrylate, when treated with bromine, undergoes addition across the double bond as well as the Hunsdiecker reaction (Equation (170)) ⟨67BCJ594⟩.

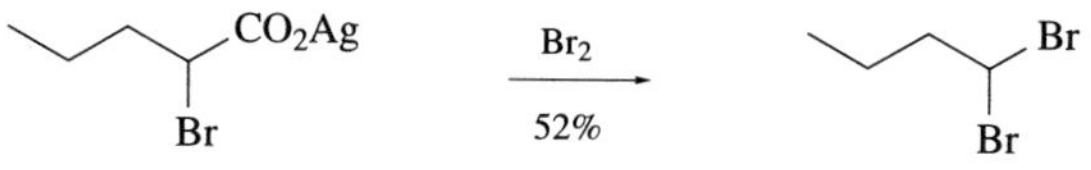

(169) [structure: 2-bromobutyl silver carboxylate → 1,1-dibromobutane derivative], Br₂, 52%

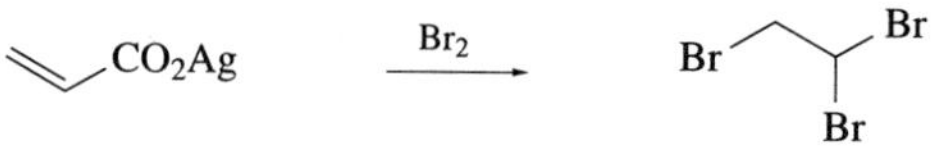

(170) [CH₂=C(Br)CO₂Ag → BrCH₂CHBr₂], Br₂

Other methods of bromodecarboxylation are exemplified by treatment of an α-keto acid with bromine and sodium acetate (Equation (171)) ⟨53JA3297⟩ and of cyanoacetic acid with NBS (Equation (172)) ⟨63OSC(4)254⟩ and by the Hofmann degradation of α-bromo amides with sodium hydroxide and bromine (Equation (173)) ⟨56JA2264⟩.

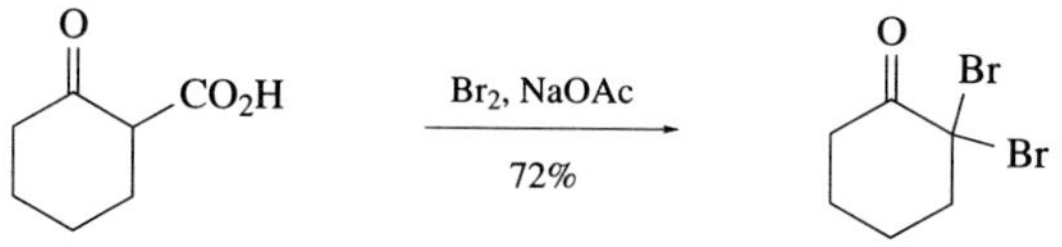

(171) 2-oxocyclohexanecarboxylic acid → 2,2-dibromocyclohexanone, Br₂, NaOAc, 72%

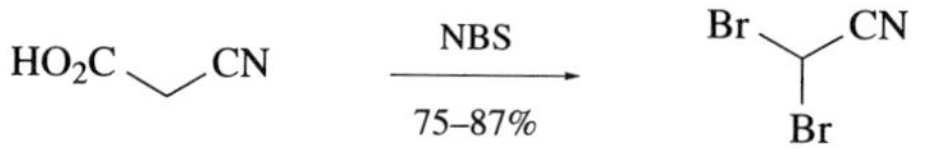

(172) HO₂C-CH₂-CN → Br₂CH-CN, NBS, 75–87%

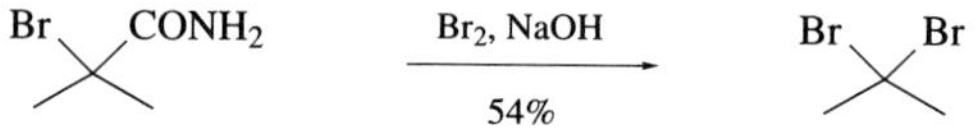

(173) BrC(CH₃)₂CONH₂ → Br₂C(CH₃)₂, Br₂, NaOH, 54%

4.01.5 DIIODO ALKANES—R₂CI₂

4.01.5.1 Diiodo Alkanes from Alkanes

Iodination of alkanes is restricted to the iodination of active methylene groups. Malonic acid can be diiodinated with iodine and potassium iodide ⟨51JIC675⟩ but better yields are obtained with iodine and potassium iodate (see Scheme 3) ⟨14JA1899, 58JOC1368⟩. Heating the diiodomalonic acid causes decarboxylation to give diiodoacetic acid. Decarboxylation also readily occurs when acetone dicarboxylic acid is treated under the same conditions as Equation (174) ⟨72ACS1735⟩.

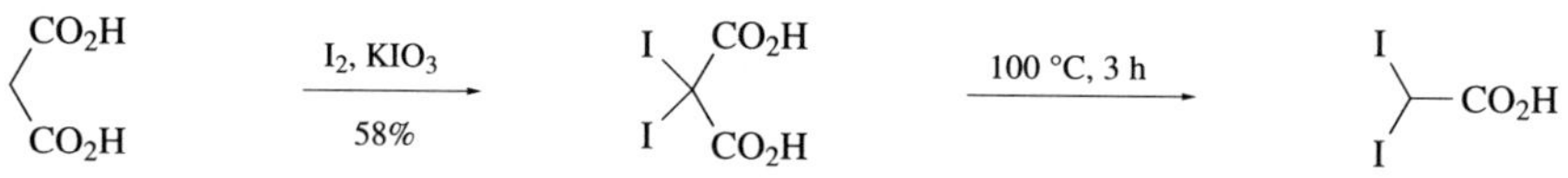

CH₂(CO₂H)₂ →[I₂, KIO₃, 58%] CI₂(CO₂H)₂ →[100 °C, 3 h] CHI₂CO₂H

Scheme 3

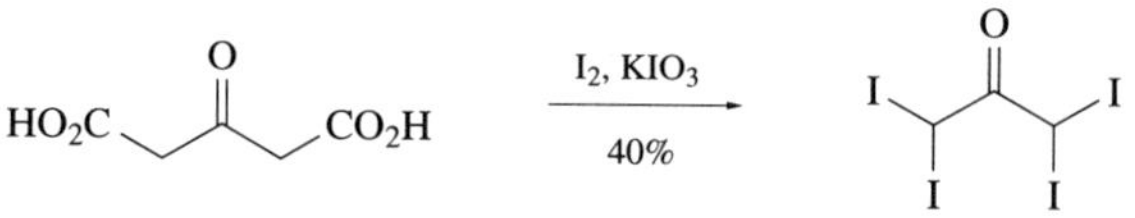

(174) HO₂C-CH₂-CO-CH₂-CO₂H →[I₂, KIO₃, 40%] I₂CH-CO-CHI₂

4.01.5.2 Diiodo Alkanes from Halo Alkanes

There are only a few reports of replacement of dihalides by iodides, for example dichloromethane when treated with sodium iodide in DMF gives diiodomethane (Equation (175)) ⟨73CI(L)331⟩ and benzal bromide is converted into benzal iodide by sodium iodide in carbon disulfide in the presence of silver nitrate (Equation (176)) ⟨76JCS(P1)416⟩. An interesting exchange reaction is used in the conversion of 1,1-dichloroethane into 1,1-dibromoethane where ethyl iodide is the iodine source with aluminum trichloride as a catalyst as Equation (177) ⟨51JA4476⟩.

$$\text{Cl} \diagdown \text{Cl} \xrightarrow{\text{NaI, DMF}} \text{I} \diagdown \text{I} \quad (175)$$

$$\underset{\text{Br}}{\overset{\text{Br}}{\text{Ph}}}\!\!\!\diagup \xrightarrow[\text{82\%}]{\text{NaI, AgNO}_3,\,\text{CS}_2,\,20\,°\text{C},\,48\,\text{h}} \underset{\text{I}}{\overset{\text{I}}{\text{Ph}}}\!\!\!\diagup \quad (176)$$

$$\underset{\text{Cl}}{\overset{\text{Cl}}{\diagup}}\!\!\!\diagdown \xrightarrow[60\%]{\text{EtI, AlCl}_3} \underset{\text{I}}{\overset{\text{I}}{\diagup}}\!\!\!\diagdown \quad (177)$$

Iodoform was reduced to diiodomethane by sodium arsenite (Equation (178)) ⟨32OSC(1)358⟩.

$$\text{CHI}_3 \xrightarrow{\text{Na}_3\text{AsO}_3,\,\text{NaOH}} \text{I}\diagdown\diagup\text{I} \quad (178)$$

4.01.5.3 Diiodo Alkanes from Alkynes

Two equivalents of hydrogen iodide add readily to terminal alkynes to give diiodo alkanes (Equations (179) ⟨65JA3151⟩ and (180) ⟨07JCS816⟩).

$$\equiv \xrightarrow{\text{HI}} \overset{\text{I}\quad \text{I}}{\diagup\!\!\!\diagdown} \quad (179)$$

$$\equiv\!\!\!\diagdown\text{CO}_2\text{H} \xrightarrow{\text{HI}} \overset{\text{I}\quad \text{I}}{\diagup\!\!\!\diagdown}\!\!\!\diagdown\text{CO}_2\text{H} \quad (180)$$

4.01.5.4 Diiodo Alkanes from Diiodocarbene

The generation of diiodocarbene has received relatively little attention compared to that of other dihalocarbenes, largely because the adducts with alkenes are too unstable. However the diiodocyclopropanes can be isolated when formed from potassium *t*-butoxide and iodoform at low temperatures ⟨76JCS(P1)54, 76S313⟩. Diiodocarbene can also be generated under phase-transfer conditions ⟨81T1215⟩.

4.01.5.5 Diiodo Alkanes from Imines

The production of diiodoalkanes from imino compounds is the most generally applicable strategy, even though the yields can be low. Diazo alkanes react with iodine to give *gem*-diiodo alkanes in yields in the range 27–37% (Equation (181) ⟨66JOC1857, 83OMR(21)64⟩). Higher yields can be achieved from hydrazones with iodine and triethylamine; a range of diiodo alkanes has been prepared by this method (Equation (182)) ⟨70AJC989, 75SC33⟩.

$$\diagup\!\!\!\diagdown\!\!=\!\!\text{N}_2 \xrightarrow[30\%]{\text{I}_2} \diagup\!\!\!\diagdown\!\!\overset{\text{I}}{\diagdown}\text{I} \quad (181)$$

$$\text{Ph}\diagdown\!\!=\!\!\text{N}\diagdown\text{NH}_2 \xrightarrow{\text{I}_2,\,\text{Et}_3\text{N}} \text{Ph}\diagdown\!\!\overset{\text{I}}{\diagdown}\text{I} \quad (182)$$

4.01.6 FLUOROHALO ALKANES—R₂CFHal

4.01.6.1 Chlorofluoro Alkanes—R₂CClF

4.01.6.1.1 Chlorofluoro alkanes from halo alkanes

Chlorination of fluoro alkanes (Equation (183)) is not a good method for the preparation of chlorofluoro alkanes as a mixture of chlorinated by-products is obtained ⟨71JCS(B)1723⟩. The best method for the preparation of chlorofluoro alkanes is by exchange of a halogen for fluorine. Several reagents are available for this transformation. Heating dichloromethane with a mixture of sodium and potassium fluoride at high temperatures is reported to give high yields of chlorofluoromethane (Equation (184)) ⟨86CI(L)490⟩. The same transformation has been achieved using antimony trifluoride ⟨37JA1400⟩. Antimony trifluoride has been used to produce a variety of chlorofluoro alkanes (Equation (185)) but care must be taken not to replace both *gem*-chlorines ⟨66JA2481⟩.

Silver fluoride has also been used to effect the last reaction in 40% yield ⟨78HCA2482⟩. Potassium fluoride in the polar solvent, *N*-methyl-2-pyrrolidone, has been reported as an effective reagent for the preparation of chlorofluoro alkanes (Equation (186)) ⟨63JOC112⟩. One of the most common reagents for replacement of a halogen by fluorine is mercury difluoride (Equations (187) ⟨77JOC3527⟩ and (188) ⟨36JA889⟩). In a different approach, iodine has been replaced by chlorine under radical conditions in very high yield (Equation (189)) ⟨78JFC(11)527⟩.

4.01.6.1.2 Chlorofluoro alkanes from halo alkenes

Addition of hydrogen fluoride to a chloro alkene such as 1-chlorocyclohexene (Equation (190)) ⟨63HCA1818⟩, 2-chloropropene (Equation (191)) ⟨73JOC2091⟩ and 2,3-dichloropropene (Equation (192)) ⟨46JA496⟩ gives very good yields of chlorofluoro alkanes.

[Equation (190): 1-chlorocyclohexene + HF (96%) → 1-chloro-1-fluorocyclohexane]

[Equation (191): 2-chloropropene + HF (75%) → 2-chloro-2-fluoropropane]

[Equation (192): 3-chloro-2-chloromethylpropene + HF (70%) → corresponding chlorofluoro product]

Addition of fluorine to chloro alkenes (Equation (193)) ⟨63JOC494⟩, chlorine to fluoro alkenes (Equation (194)) ⟨51JA711⟩, FCl to chloro alkenes (Equation (195)) ⟨76JFC(7)569⟩, FBr (effectively formed from Br_2 and AgF) to chloro alkenes (Equation (196)) ⟨73JA182⟩, BrCl to fluoro alkenes (Equation (197)) ⟨73JA182⟩ and hydrogen bromide to chlorofluoro alkenes (Equation (198)) ⟨54JCS3747⟩ have all been reported; however in many cases mixtures of products are obtained due to elimination or halogen exchange reactions ⟨70JOC4201⟩ or by interhalogen compounds adding to an unsymmetrical alkene both ways round ⟨73JA182⟩.

[Equation (193): ClCH=CHCl + F_2 (79%) → ClCHF-CHFCl]

[Equation (194): CF_2=CF_2 + Cl_2 (100%) → $CClF_2$-$CClF_2$]

[Equation (195): CH_2=CHCl + FCl (90%) → CH_2Cl-CHFCl]

[Equation (196): PhCF=CFCl + Br_2, AgF (57%) → PhCF(F)-CF(Br)Cl]

[Equation (197): PhCF=CF_2 + BrCl (70%) → PhCF(F)-CF_2Br with Cl]

[Equation (198): CF_2=CFCl + HBr, hν (66%) → CHF_2-CFBrCl (approx)]

Other reagents for the addition of fluorine to chloro alkenes such as cobalt trifluoride (Equation (199)) ⟨63JOC494⟩ and sulfur tetrafluoride and lead dioxide (Equation (200)) ⟨64JOC1591⟩ have been reported.

[Equation (199): CH_2=CHCl + CoF_3 (43%) → CH_2F-CHFCl]

[Equation (200): ClCH=CHCl + PbO_2, SF_4 → ClCHF-CHFCl]

A useful method for the preparation of chlorofluoro alkanes is by the addition of HX to a suitable chlorofluoro alkene. The alkene needs to be polarised so that the H of the HX adds to the correct carbon. This can be achieved by using chlorotrifluoroethene in addition reactions such as the addition of ethanol ⟨48JA1550, 52JCS4259, 63OSC(4)184⟩, diethylamine ⟨74CCC2616, 77CCC2537⟩, hydrogen cyanide (the nitrile is hydrolysed to a carboxylic acid under the reaction conditions) ⟨73OSC(5)239⟩, trichlorosilane ⟨60JCS4503⟩ and hydrogen fluoride (using formamide as a hydrogen donor) (Scheme 4) ⟨61JCS3825, 84JFC(26)29⟩. Cyclodimerisation of chlorotrifluoroethene gives a good yield of a 1,2-dichloroperfluorocyclobutane (Equation (201)) ⟨47JA279⟩.

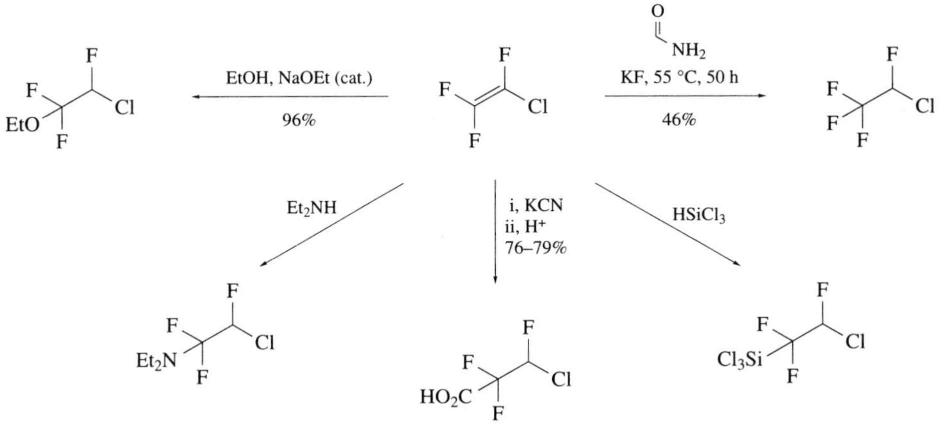

Scheme 4

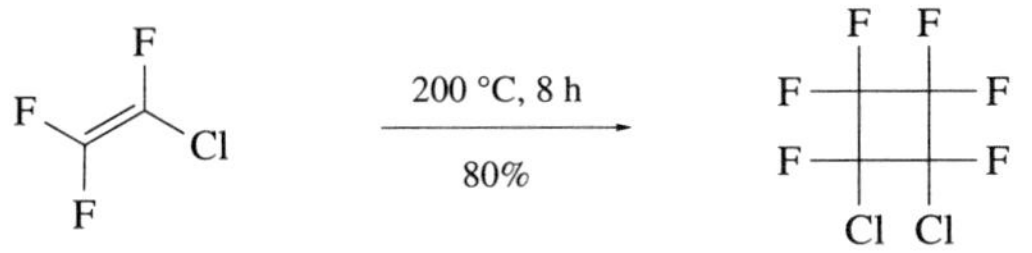

(201)

4.01.6.1.3 Chlorofluoro alkanes from chlorofluorocarbene

Chlorofluorocarbene ⟨77FCR119⟩ has been generated by a variety of methods and added to a range of alkenes to form chlorofluorocyclopropanes (Equation (202)). The methods for the generation of chlorofluorocarbene are listed in Table 13.

(202)

Table 13 Reagents used to generate chlorofluorocarbene.

Reagents	Ref.
$CHFCl_2$ and $KOBu^t$	57JOC730
$CHFCl_2$ and MeLi	71CB1921
$CFCl_3$ and $TiCl_4$, $LiAlH_4$	90JOC589
$(CFCl_2)_2CO$ and $KOBu^t$	63JOC2494, 67T2549
$PhHgCFCl_2$	70JOC1297

4.01.6.1.4 Chlorofluoro alkanes from imines

Chlorofluoromethyl ketones can be readily prepared from the corresponding diazo ketone using N-chlorosuccinimide and pyridinium poly(hydrogen fluoride) (Equation (203)) ⟨79JOC3842⟩.

$$R-C(=O)-CH=N_2 \xrightarrow{\text{NCS, pyridine, HF}} R-C(=O)-CHFCl \quad (203)$$

4.01.6.1.5 Chlorofluoro alkanes from carboxylic acids

The Hunsdiecker reaction of silver fluoroacetate with chlorine gives a good yield of chlorofluoromethane (Equation (204)) ⟨52JCS4259⟩. The Barton generation of a cyclopropyl radical using sodium mercaptopyridine *N*-oxide, followed by abstraction of a chlorine atom from carbon tetrachloride lead to a chlorofluoroalkane in poor yield (Equation (205)) ⟨91JOC2193⟩.

$$AgO_2C\text{-}CH_2\text{-}F \xrightarrow[52\%]{Cl_2} Cl\text{-}CH_2\text{-}F \quad (204)$$

(205) — cyclopropane with F, Cl, 2 Ph groups reacting via i, pyridine-N-oxide-S⁻/dmap; ii, CCl₄ (5%) to give fluorochloro-diphenylcyclopropane.

4.01.6.2 Bromofluoro Alkanes—R₂CBrF

4.01.6.2.1 Bromofluoro alkanes from halo alkanes

Bromination of a benzylic or α-keto monofluoro alkane leads to a bromofluoro alkane in good yield. Benzylic fluoro alkanes require radical conditions for bromination (Equation (206)) ⟨79T2661⟩, whereas bromination of a α-keto monofluoro alkane can be performed with bromine (Equations (207) ⟨61JCS3452⟩ and (208) ⟨53JA4091⟩) or NBS (Table 14) ⟨77JOC3527⟩. Replacement of one bromine of a dibromo alkane can be achieved with one equivalent of mercury difluoride (Equations (209) ⟨36JA889⟩ and (210) ⟨54JA479⟩). The Grignard reagent made from a dibromofluoro alkane will react with a series of aldehydes and ketones to give α-hydroxy bromofluoro alkanes (Equation (211)) ⟨84JFC(26)467⟩.

$$Ph\text{-}CHF\text{-}CF_2Cl \xrightarrow[95\%]{Br_2,\ CCl_4,\ h\nu,\ 24\ h} Ph\text{-}CFBr\text{-}CF_2Cl \quad (206)$$

$$CH_3C(=O)CH_2F \xrightarrow[45\%]{Br_2,\ CHCl_3,\ 10\ °C,\ 90\ min} CH_3C(=O)CHFBr \quad (207)$$

$$F_3C\text{-}C(=O)\text{-}CHF\text{-}CO_2Et \xrightarrow[95\%]{Br_2,\ H_2SO_4,\ 80\text{-}90\ °C,\ 2\ h} F_3C\text{-}C(=O)\text{-}CFBr \quad (208)$$

$$BrCH_2\text{-}CHBr_2 \xrightarrow[100\%]{HgF_2} BrCH_2\text{-}CHFBr \quad (209)$$

Table 14 Yields for the synthesis of α,α-bromofluoro ketones from fluoro ketones using *N*-bromosuccinimide.

$$R^1COCHFR^2 \rightarrow R^1COCBrFR^2$$

R^1	R^2	Yield (%)
Me	Me	95
Ph	Me	95
CH$_2$Cl	Me	98
Me	Ph	97

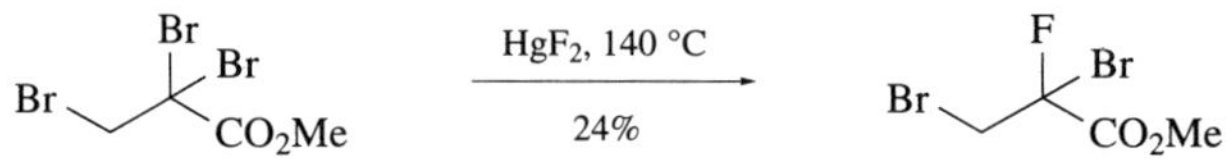

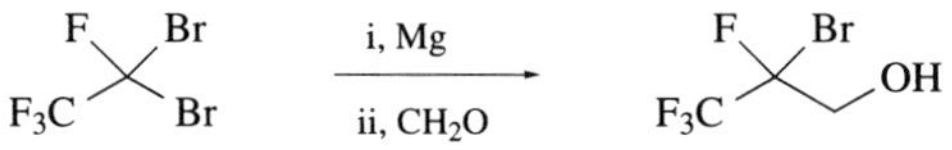

4.01.6.2.2 *Bromofluoro alkanes from halo alkenes*

As with the other classes of *gem*-dihalo alkanes, a good way to generate them is by addition to halo alkenes. The various ways to make bromofluoro alkanes are exemplified by the following reactions: bromination of fluoro alkenes by bromine either with irradiation (Equations (212) ⟨51JA711⟩ and (213) ⟨57JA4170⟩) or under ionic conditions with bromine in the dark (Equation (214)) ⟨76JCS(P1)2349⟩ or with a catalytic amount of sodium iodide (Equation (215)) ⟨79JOC1394⟩; fluorination of bromo alkenes, (Equation (216)) with fluorine or cobalt trifluoride ⟨63JOC494⟩; addition of hydrogen bromide to a fluoro alkene (Equation (217)) ⟨76JCS(P1)2349⟩ or hydrogen fluoride (Equation (218)) ⟨75JOM(92)7⟩ or hydrogen chloride (Equation (219)) ⟨79T2661⟩ to a bromo-fluoro alkene and of BrF (Equation (220)) ⟨61JCS3779⟩ or BrCl (Equation (221)) ⟨73JA182⟩ to a fluoro alkene.

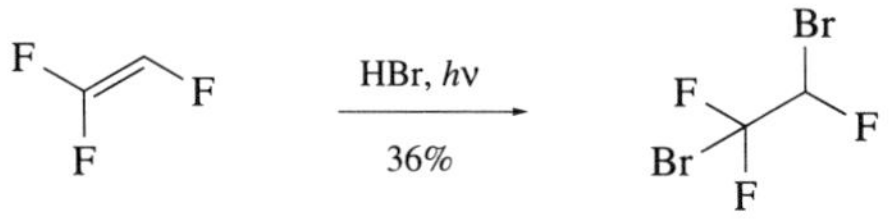

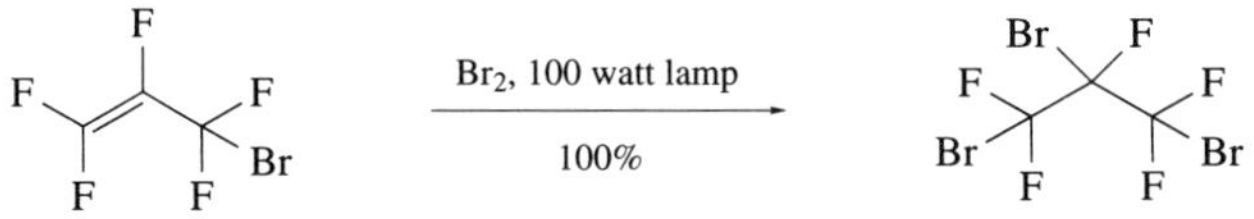

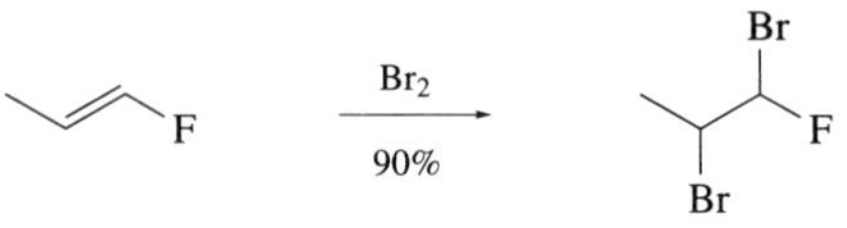

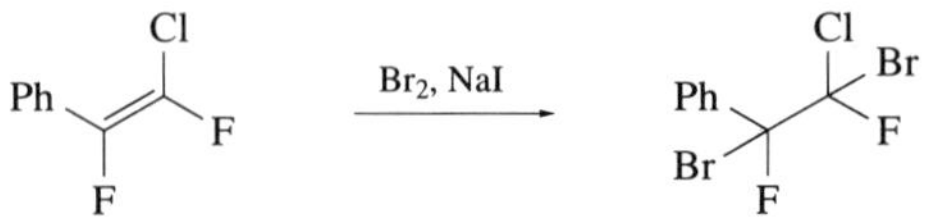

(217)

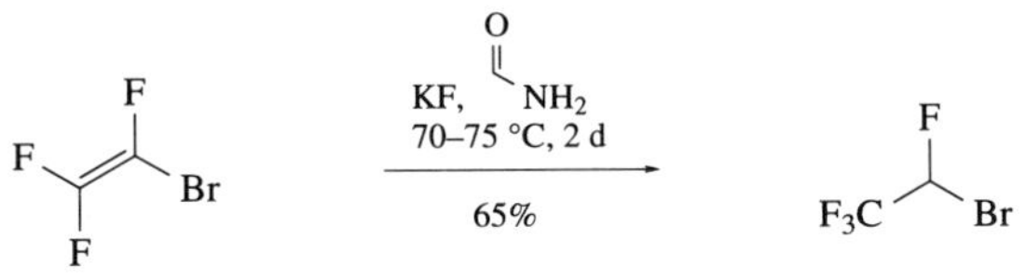

(218)

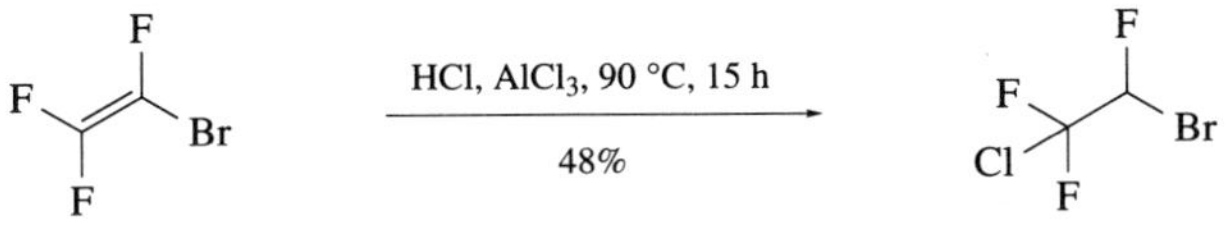

(219)

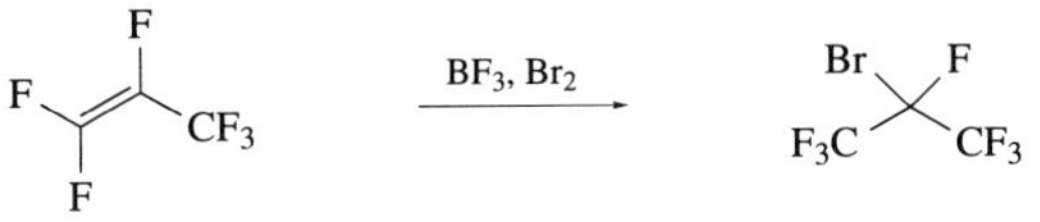

(220)

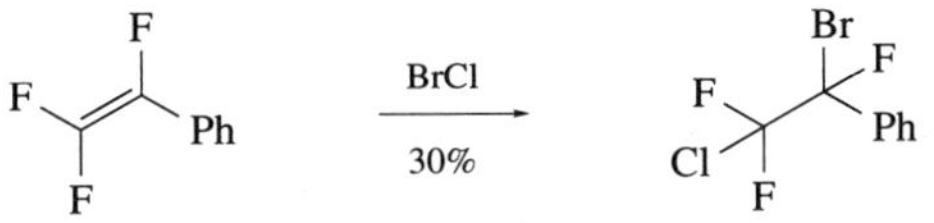

(221)

4.01.6.2.3 Bromofluoro alkanes from bromofluorocarbene

The generation of bromofluorocarbene has been reviewed ⟨77FCR119⟩; see Table 15 for the methods of generation. The reported yields for the adducts formed with phenyl(dibromofluoromethyl)mercury are better than those from other methods (Equation (222)) ⟨73JOM(51)77⟩.

Table 15 Reagents used to generate bromofluorocarbene.

Reagents	*Ref.*
$CHFBr_2$ and $KOBu^t$	69TL1957
$CHFBr_2$, 50% NaOH and $BnEt_3NCl$	71TL3869, 83ZOR1625
$PhHgCFBr_2$	73JOM(51)77

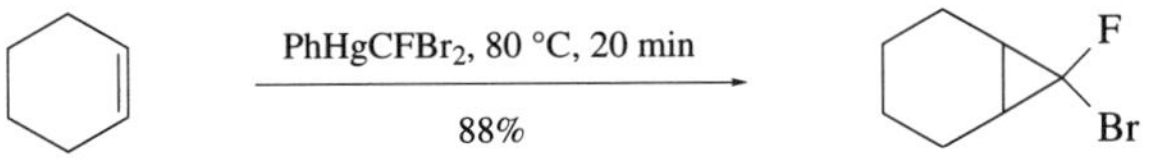

(222)

4.01.6.2.4 Bromofluoro alkanes from carboxylic acids

The Hunsdiecker reaction ⟨56CRV219⟩ gives good yields of bromofluoro alkanes from the corresponding silver α-fluoro carboxylate and bromine (Equation (223)) ⟨52JCS4259⟩. The radical formed in the Barton method from an α-fluoro acid chloride will abstract a bromine atom from bromotrichloromethane (Equation (224)) ⟨91JOC2193⟩.

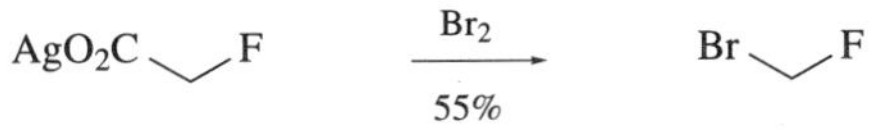

(223)

4.01.6.3 Fluoroiodo Alkanes—R₂CFI

4.01.6.3.1 Fluoroiodo alkanes from halo alkanes

Exchange of one iodine of a diiodo alkane using mercury difluoride is possible, however the yields are low (Equation (225)) ⟨75JOC2796⟩.

$$I-CH_2-I \xrightarrow[20\%]{HgF_2} F-CH_2-I \qquad (225)$$

4.01.6.3.2 Fluoroiodo alkanes from halo alkenes

Addition of IF to a fluoro alkene is a useful way to generate a fluoroiodo alkane. For example, IF, generated by various methods, has been added across the double bond of hexafluoropropene (Table 16). ICl has also been added to a fluoro alkene (Equation (226)) ⟨60JCS1398⟩. The overall addition of an iodo alkene across the double bond of fluoroethene has been achieved with UV irradiation initiating a radical mechanism (Equation (227)) ⟨56JA59⟩.

Table 16 Reagents for the generation of IF for the addition to hexafluoropropene.

Reagents	Ref.
I₂, KF	62JOC1813
IF₅, I₂	61JCS3779
i, HgF₂, HF; ii, I₂	61JA4105

4.01.6.3.3 Fluoroiodo alkanes from fluoroiodocarbene

Fluoroiodocarbene has been generated under phase-transfer conditions and added to a small number of alkenes (Equation (228)) ⟨73TL611⟩.

4.01.6.3.4 Fluoroiodo alkanes from carboxylic acids

The Hunsdiecker reaction ⟨56CRV219⟩ gives poor yields of iodofluoro alkanes from the corresponding silver α-fluoro carboxylate and iodine (Equation (229)) ⟨52JCS4259⟩. The radical formed in the Barton method from an α-fluoro acid chloride will abstract an iodine atom from trichloroiodomethane (Equation (230)) ⟨91JOC2193⟩; however, the yield is low.

4.01.7 CHLOROHALO ALKANES—R$_2$CCl Hal(not F)

4.01.7.1 Bromochloro Alkanes—R$_2$CBrCl

4.01.7.1.1 Bromochloro alkanes from halo alkanes

Replacement of one chlorine in dichloromethane by a bromine with calcium bromide under phase-transfer conditions has been achieved (Equation (231)). However, the major product is dibromomethane ⟨84S34⟩. Better methods to produce bromochloro alkanes are chlorination of α-bromo carbonyl compounds (Equation (232)) ⟨27LA(453)113⟩, bromination of α-chloro carbonyl compounds (Equation (233)) ⟨71JCS(C)279⟩ and radical chlorination of benzyl bromides (Equation (234)) ⟨91JOC1663⟩.

4.01.7.1.2 Bromochloro alkanes from halo alkenes

As with the other mixed *gem*-dihalo alkanes, there are a variety of ways to produce bromochloro alkanes by addition to halo alkenes. Some of the methods are illustrated by the following reactions: addition of hydrogen bromide (Equations (235) ⟨55JA3465⟩ and (236) ⟨34JA712⟩) and addition of bromine to a chloro alkene as in Equations (237) ⟨52JA3895, 70JA7359⟩, (238) ⟨28JCS2125⟩ and (239) ⟨52JA3895⟩ and chlorine to a bromo alkene (Equations (240) ⟨79T2661⟩ and (241) ⟨35JA1088⟩).

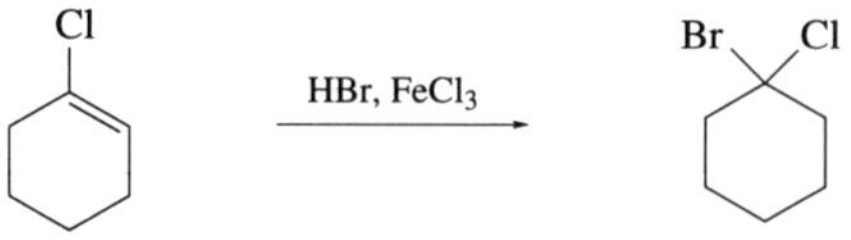

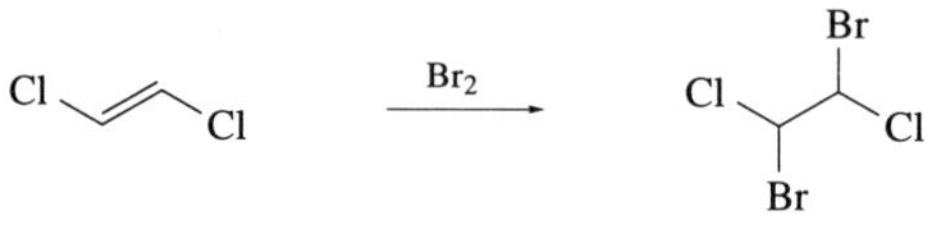

4.01.7.1.3 Bromochloro alkanes from bromochlorocarbene

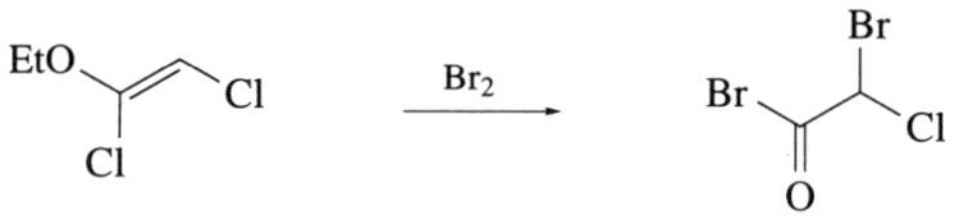

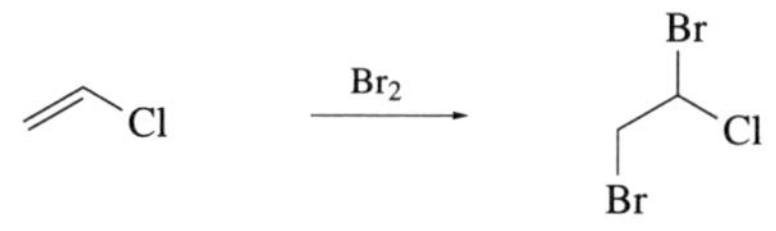

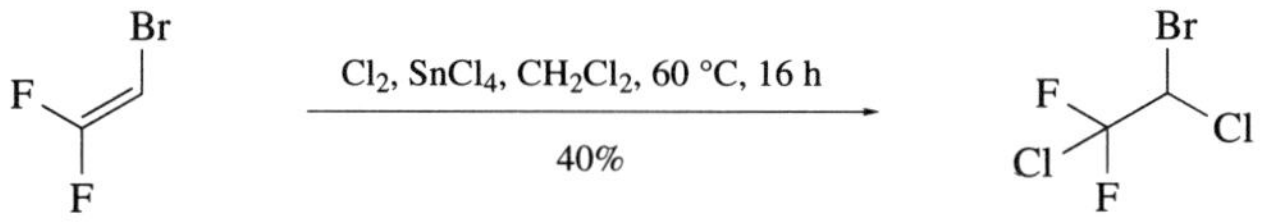

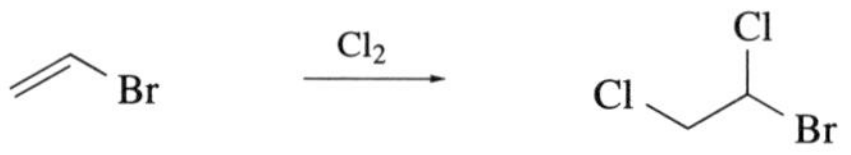

Bromochlorocarbene has been generated by electroreduction of tetrabromomethane in the presence of chloride; however, dichlorocarbene is also produced by this method ⟨90IZV1802⟩. The best method for generating bromochlorocarbene for the addition to alkenes is from dibromochloromethane and potassium *t*-butoxide at 25 °C for 18 h ⟨85AG(E)585⟩.

4.01.7.1.4 Bromochloro alkanes from ketones

Bromochloro alkanes can be conveniently synthesised from ketones via oximes and *gem*-chloronitroso compounds in high yield (Table 17) ⟨76TL943⟩.

4.01.7.1.5 Bromochloro alkanes from carboxylic acids

The Hofmann degradation of α-chloro amides is a useful method for the generation of bromochloro alkanes (Equation (242)) ⟨56JA2264⟩.

Table 17 Yields for the preparation of *gem*-bromochloro alkanes from ketones.

$$R^1COR^2 \xrightarrow{NH_2OH} R^1R^2C=NOH \xrightarrow{Cl_2} R^1R^2C(Cl)(NO) \xrightarrow{Br_2, h\nu, 10-30 \text{ min}} R^1R^2C(Cl)(Br)$$

R^1	R^2	Yield (%)
Me	Me	75
Me	Et	85
Et	Et	90
Me	Bu^t	70

$$\text{Me}_2\text{C(Cl)(CONH}_2) \xrightarrow[95\%]{Br_2, \text{ NaOH}} \text{Me}_2\text{C(Cl)(Br)} \quad (242)$$

4.01.7.2 Chloroiodo alkanes—R₂CClI

4.01.7.2.1 Chloroiodo alkanes from halo alkanes

Replacement of one chlorine of a *gem*-dichloro alkane by iodine can be efficiently performed using sodium iodide in DMF as the solvent of choice (Equation (243)) ⟨71BCJ2864, 73CI(L)331⟩. The substitution of the bromine of a bromochloro alkane by an iodine has been accomplished with the iodide form of Amberlyst A-26 resin (prepared from the chloride form and methyl iodide) in excellent yield (Equation (244)) ⟨88JOC1331⟩.

$$\text{Cl-CH}_2\text{-CH}_2\text{-Cl} \xrightarrow[83\%]{\text{NaI, DMF}} \text{I-CH}_2\text{-CH}_2\text{-Cl} \quad (243)$$

$$\text{Br-CH}_2\text{-CH}_2\text{-Cl} \xrightarrow[92\%]{\text{Amberlyst A-26 (iodide form)}} \text{I-CH}_2\text{-CH}_2\text{-Cl} \quad (244)$$

4.01.7.2.2 Chloroiodo alkanes from halo alkenes

As with other mixed *gem*-dihalo alkanes additions to halo alkenes is a useful method of preparation of chloroiodo alkanes. The addition to chloro alkenes of hydrogen iodide is exemplified by Equation (245) ⟨34JA712⟩, of ICl by Equation (246) ⟨23JCS576, 27JCS538⟩ and of IF by Equation (247) ⟨61JA2383⟩.

$$CH_2=CHCl \xrightarrow{HI} CH_3CH(I)(Cl) \quad (245)$$

$$\text{(EtO)(Cl)C=CHCl} \xrightarrow{ICl} \text{ClCH(Cl)C(O)I} \quad (246)$$

$$CF_2=CFCl \xrightarrow{IF_5, I_2} F_3C-CH(I)(Cl) \quad (247)$$

4.01.7.2.3 Chloroiodo alkanes from ketones

The reaction of *gem*-chloronitroso alkanes (prepared from ketones as shown in Table 17) and iodine with UV irradiation has the potential to be a useful route to chloroiodo alkanes (Equation (248)) ⟨76TL943⟩.

$$R^1R^2C(Cl)(NO) \xrightarrow{I_2, h\nu} R^1R^2C(Cl)(I) \quad (248)$$

4.01.7.2.4 Chloroiodo alkanes from carboxylic acids

Treatment of α-chloro carboxylic acids with lead tetraacetate and iodine gives a poor yield of iodo decarboxylated products (Equation (249) ⟨66JOC1857⟩).

$$CH_3CH(Cl)CO_2H \xrightarrow[12\%]{Pb(OAc)_4, I_2} CH_3CH(Cl)I \quad (249)$$

4.01.8 BROMOIODO ALKANES—R₂CBrI

There are only a few methods for the preparation of bromoiodo alkanes. These include the substitution of one bromine of a *gem*-dibromo alkane for an iodine (Equation (250)) ⟨71BCJ2864, 73CI(L)331⟩ and iodo decarboxylation of an α-bromo carboxylic acid (Equation (251)) ⟨66JOC1857⟩.

$$CH_2Br_2 \xrightarrow[84\%]{NaI, DMF} CH_2BrI \quad (250)$$

$$CH_3CH_2CH(Br)CO_2H \xrightarrow[18\%]{Pb(OAc)_4, I_2} CH_3CH_2CH(Br)I \quad (251)$$

4.02
Functions Incorporating a Halogen and a Chalcogen

NIALL W. A. GERAGHTY
University College Galway, Republic of Ireland

4.02.1 HALOGEN AND OXYGEN DERIVATIVES, $R^1_2CHal(OR^2)$	41
4.02.1.1 α-Halo Alcohols (Geminal Halohydrins), $R_2CHal(OH)$	41
4.02.1.2 α-Halo Ethers, $R^1_2CHal(OR^2)$	43
4.02.1.2.1 α-Fluoro ethers, $R^1_2CF(OR^2)$	44
4.02.1.2.2 α-Chloro ethers, $R^1_2CCl(OR^2)$	46
4.02.1.2.3 α-Bromo ethers, $R^1_2CBr(OR^2)$	50
4.02.1.2.4 α-Iodo ethers, $R^1_2CI(OR^2)$	51
4.02.1.3 Other Derivatives of α-Halo Alcohols (Geminal Halohydrins), $R^1_2CHal(OR^2)$ and $R_2CHal(OX)$	51
4.02.1.3.1 α-Haloalkyl esters, $R^1_2CHalOCOR^2$	51
4.02.1.3.2 α-Haloalkyl haloformates ($R_2CHalOCOHal$) and carbonate derivatives ($R^1_2CHalOCOOR^2$) etc.	58
4.02.2 HALOGEN AND SULFUR DERIVATIVES, $R^1_2CHal(SR^2)$, etc.	61
4.02.2.1 Dicoordinate α-Halo Sulfur Derivatives, $R^1_2CHal(SR^2)$, etc.	61
4.02.2.1.1 α-Halo sulfides, $R^1_2CHal(SR^2)$	61
4.02.2.1.2 Other dicoordinate α-halo sulfur derivatives, $R_2CHal(SX)$	71
4.02.2.2 Tricoordinate α-Halo Sulfur Derivatives, $R^1_2CHalS(O)R^2$, etc.	73
4.02.2.2.1 α-Halo sulfoxides, $R^1_2CHalS(O)R^2$	73
4.02.2.2.2 Other tricoordinate α-halo sulfur derivatives, $R_2CHalS(O)X$	78
4.02.2.3 Tetracoordinate α-Halo Sulfur Derivatives, $R^1_2CHalS(O)_2R^2$, etc.	79
4.02.2.3.1 α-Halo sulfones, $R^1_2CHalS(O)_2R^2$	79
4.02.2.3.2 Other tetracoordinate α-halo sulfur derivatives, $R_2CHalS(O)_2X$	85
4.02.3 HALOGEN AND SELENIUM AND TELLURIUM DERIVATIVES, $R^1_2CHal(SeR^2)$ AND $R^1_2CHal(TeR^2)$, etc.	87
4.02.3.1 α-Halo Selenium Derivatives, $R^1_2CHal(SeR^2)$, etc.	87
4.02.3.1.1 Dicoordinate α-halo selenium derivatives, $R^1_2CHal(SeR^2)$	87
4.02.3.1.2 Tri- and tetracoordinate α-halo selenium derivatives, $R^1_2CHalSe(O)R^2$, $R^1_2CHalSe(O)_2R^2$, etc.	92
4.02.3.2 α-Halo Tellurium Derivatives, $R^1_2CHal(TeR^2)$, etc.	93
4.02.3.2.1 Dicoordinate α-halo tellurium derivatives, $R^1_2CHal(TeR^2)$	93
4.02.3.2.2 Tri- and tetracoordinate α-halo tellurium derivatives, $R^1_2CHalTe(O)R^2$ and $R^1_2CHalTe(O)_2R^2$, etc.	93

4.02.1 HALOGEN AND OXYGEN DERIVATIVES, $R^1_2CHal(OR^2)$

4.02.1.1 α-Halo Alcohols (Geminal Halohydrins), $R_2CHal(OH)$

α-Halo alcohols in general are unstable relative to mixtures of the appropriate hydrogen halide and aldehyde or ketone (Equation (1)); despite this, the simpler α-halo alcohols in particular have been the subject of considerable interest, much of it theoretical. The early work in this area has been reviewed ⟨B-64MI 402-01⟩ and the status of fluoro- and chloromethanol has been summarised

⟨81JOC571⟩. Some α-fluoro alcohols are relatively stable in ionic form: thus the protonated form of fluoromethanol has been obtained in HF/SbF$_5$/SOCl$_2$F at −78 °C (Scheme 1) ⟨71JA781⟩, although it was concluded that the free alcohol is unstable relative to HF and formaldehyde, and the α-fluoroalkoxy anion (**1**) is stable at room temperature as its tris(dimethylamino)sulfonium salt ⟨85JA4565⟩.

$$R\!\!-\!\!X(OH)R(H) \longrightarrow R(H)C\!\!=\!\!O + HX \quad (1)$$

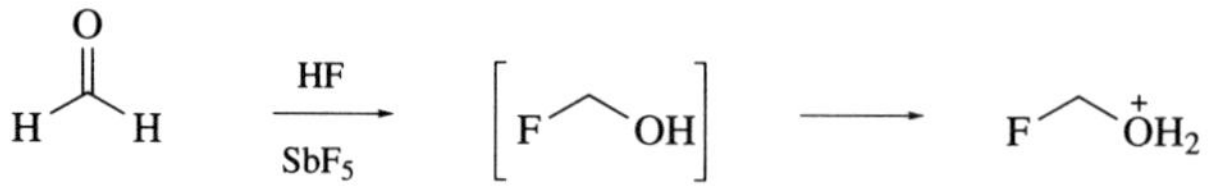

Scheme 1

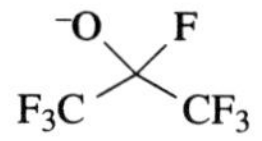

(**1**)

Evidence for the protonated form of chloromethanol has been obtained at −80 °C in FSO$_3$H/SbF$_5$/SO$_2$ (Scheme 2); attempts to obtain the protonated form of higher homologues failed ⟨75JA2293⟩. More recently evidence has been obtained for the photochemical, gas-phase chlorination of methanol, under which conditions the chloromethanol formed undergoes a rapid decay to HCl and formaldehyde, its lifetime being only some minutes ⟨93JPC1576⟩.

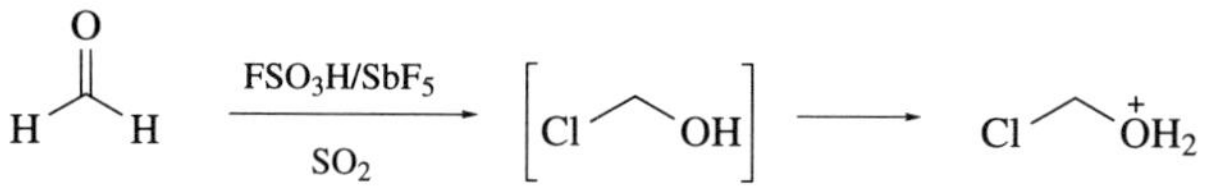

Scheme 2

Substantial NMR evidence confirms the formation of the α-bromo alcohol (**2**) by the reaction of HBr with 1,1,1-trifluoropentane-2,4-dione in dibromodifluoromethane at temperatures below −23 °C (Equation (2)); the reaction of 1,1,1-trifluoroacetone with HBr under the same conditions leads to the formation of the α-bromo alcohol (**3**) ⟨88JCS(P2)1107⟩. Similar but somewhat less substantial evidence has also been provided for the formation of α-bromo alcohols in the reactions of acetaldehyde, 2-methylpropanal and 2,2-dimethylpropanal with HBr in dibromodifluoromethane ⟨82JCS(P2)881⟩. All of these bromo alcohols are unstable at room temperature, decomposing to a mixture of HBr and aldehyde or ketone.

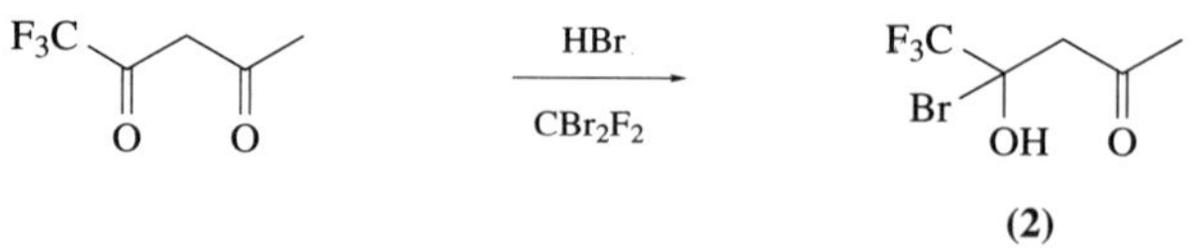

(2)

(**2**)

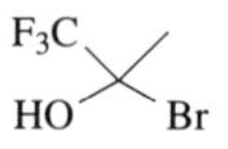

(**3**)

IR evidence alone has been adduced to support the formation of the α-bromo alcohol (**5**) from levoglucosenone (**4**) (Scheme 3) ⟨81CAR284⟩, and no evidence is provided to support the formation of the α-bromocyclopropanol (**6**) (Equation (3)) ⟨74TL909⟩. The formation of iodomethanol in the photochemical reaction between ozone and iodomethane, in an argon matrix at 17 K, has been reported ⟨85IC3285⟩, as has ^{1}H NMR evidence for the formation under basic conditions of the α-iodo alcohols (**8**) and (**9**) from the 4,4-diiodo-1,1-dimethyl-1,4-dihydroquinolinium cation (**7**) (Equation (4)) ⟨74CJC951⟩.

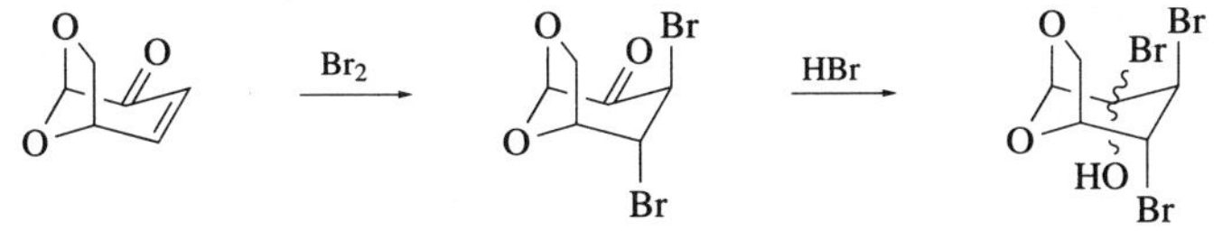

(4) **(5)**

Scheme 3

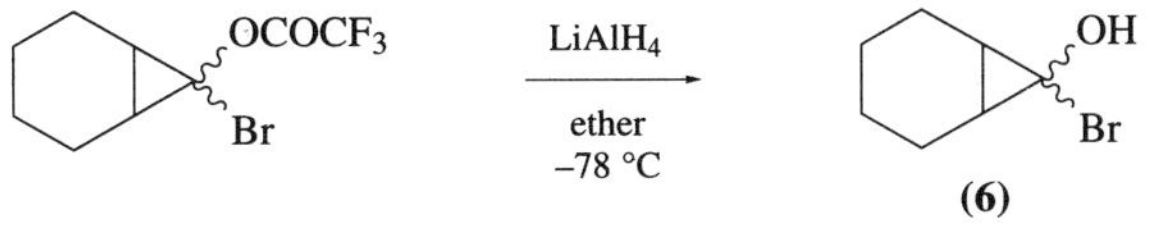

(6) (3)

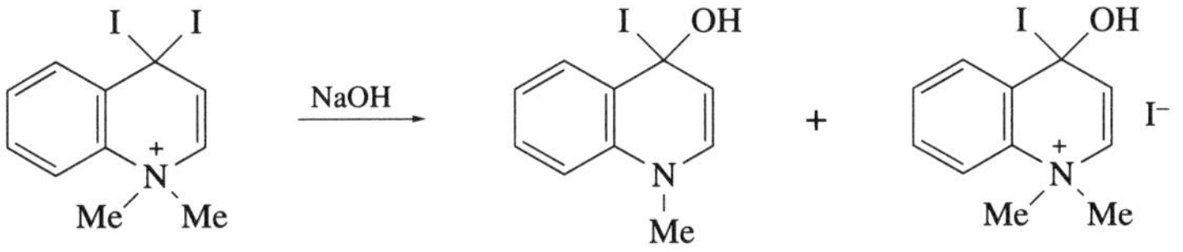

(7) **(8)** **(9)** (4)

There are, however, a few α-halo alcohols which are stable. The chlorination of the 2-hydroxy-1,3-diketones (**10**) and (**11**) with sulfuryl chloride gives the α-chloro alcohols (**12**) and (**13**), respectively, which function as masked cyclohexane-1,2,3-triones (Equation (5)) ⟨81CB1951⟩. α-Halocyclobutanols (**14**) are formed in good yield when perfluorocyclobutanone reacts with hydrogen halides (Equation (6)) ⟨61JA4670⟩; these compounds are stable in the absence of water but revert to a mixture of the ketone and hydrogen halide when heated. Neither ring strain nor perfluorination alone is sufficient to account for the stability of these materials, as neither cyclobutanone nor perfluoroacetone gives isolable α-halo alcohols under these conditions.

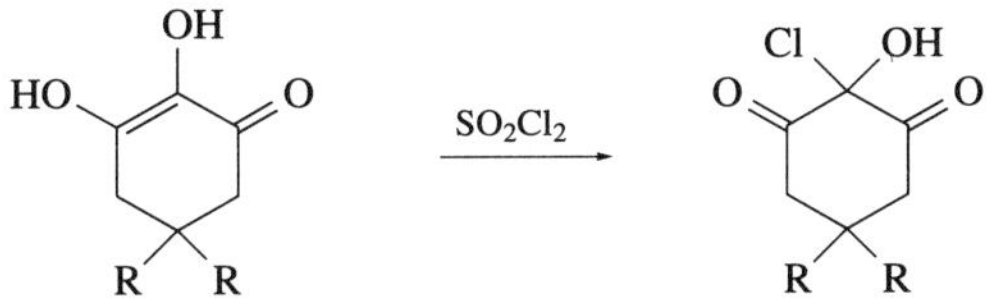

(10) R = Me
(11) R, R = $(CH_2)_5$

(12) R = Me, 66%
(13) R, R = $(CH_2)_5$, 68%

(5)

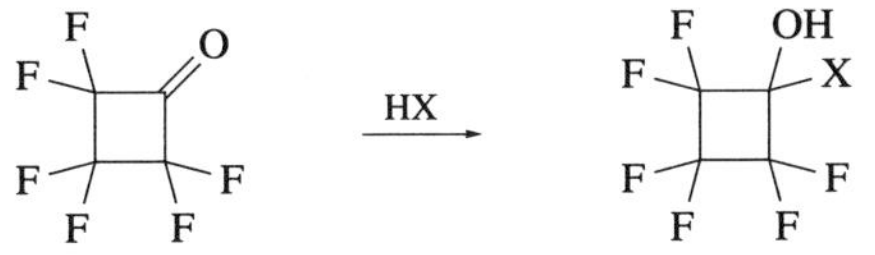

(14)
X = F, 75%
X = Cl, 73–90%
X = Br, 73–90%
X = I, 73–90%

(6)

4.02.1.2 α-Halo Ethers, $R^1_2CHal(OR^2)$

The synthetic utility of α-halo ethers (geminal halohydrin ethers) is partly due to their reactivity; however, this property in some cases results in limited thermal stability, a susceptibility to hydrolysis and poor storage properties. A number of α-halo ethers are lachrymatory.

4.02.1.2.1 α-Fluoro ethers, $R^1{}_2CF(OR^2)$

An early report of the synthesis and properties of α-fluoro ethers concluded that they are inherently unstable, being very susceptible to hydrolysis, unless they contain a CF_3 or a CF_2 group in the β position ⟨50JA4378⟩. Although a number of α-fluoro ethers have been characterised which do not possess such a structural feature, the conclusion does reflect the situation in general. Spectroscopic evidence for the formation of bis(fluoromethyl) ether in HF/SO_2ClF at $-78\,°C$ has been obtained ⟨71JA781⟩, although its preparation by the reaction of formaldehyde with hydrofluoric acid in a two-phase system has also recently been claimed ⟨94MIP9322265⟩. The potentially useful physical properties of perfluoro ethers ⟨81JCS(P1)1321⟩ has led to considerable interest in their synthesis and, of those prepared, a number contain the $(R_F)_2CFOR_F$ grouping, where R_F is a perfluoroalkyl group; the work has also led, incidentally, to the isolation of some hydrofluoro ethers whose physical properties, however, are less interesting. The reactions of polypropylene oxide (Equation (7)) and paraformaldehyde (Equation (8)) with fluorine lead to the formation of a number of these compounds, which were isolated by preparative GC of the volatile products in relatively small amounts ⟨81JCS(P1)1321⟩. Similar perfluorinated ethers have been obtained by the nucleophilic displacement of fluorosulfate from perfluoroallyl fluorosulfate by perfluoroalkoxide anions (Scheme 4) ⟨81JA5598⟩. The direct fluorination of a number of monomeric α-s-alkyl and phenyl ethers also leads to the formation of monomeric α-fluoro ethers, the products again being isolated by preparative GC (Scheme 5) ⟨88JOC78⟩.

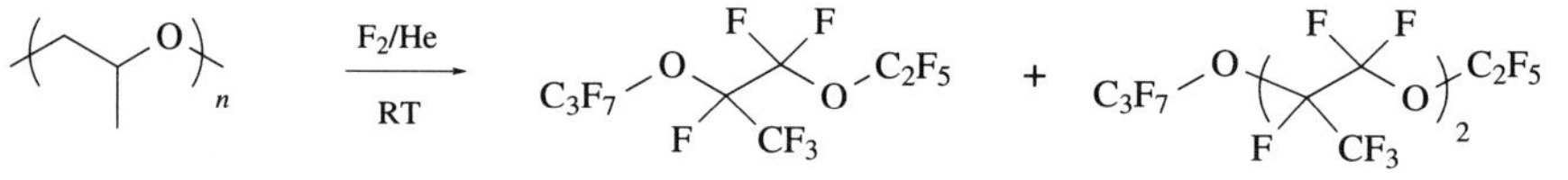

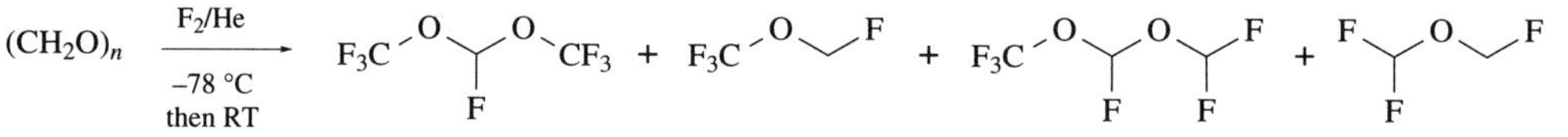

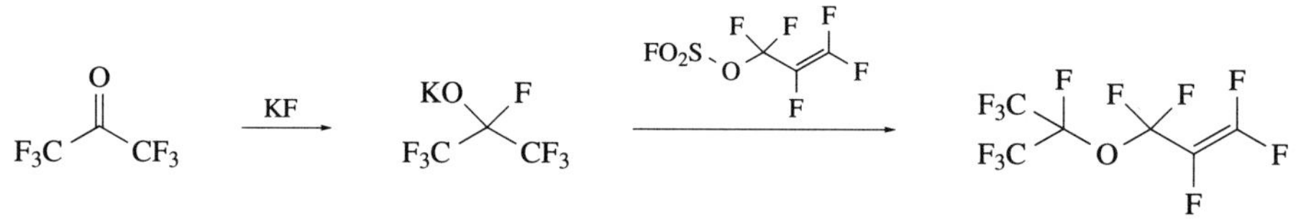

Scheme 4

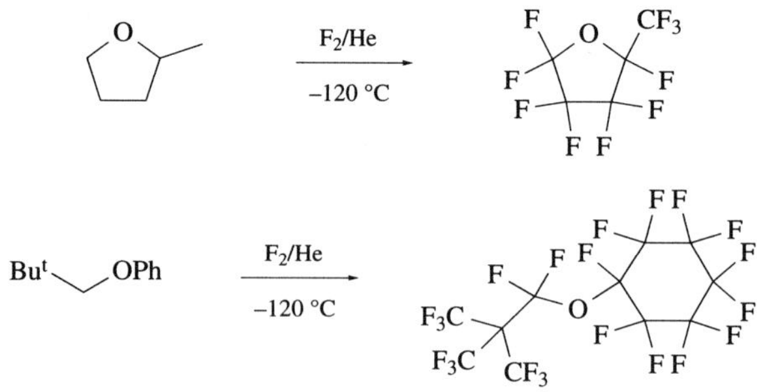

Scheme 5

Addition reactions in which one of the addends contains fluorine have been used to synthesise α-fluoro ethers. Thus, acetyl hypofluorite undergoes 1,2-addition to the aromatic ring of piperonal and related molecules, over-reaction being controlled by restricting the conversion to low levels (Equation (9)) ⟨84JOC806⟩. The Prins addition of formaldehyde to tetrafluoroethene in hydrogen fluoride gives (16) as the major product, together with some of the alcohol (17) which is formed from (16) by hydrolysis; the reaction is believed to involve bis(fluoromethyl) ether (15) as an intermediate (Scheme 6) ⟨63JOC494⟩.

$$\text{(Equation 9)}$$

$$\text{(Scheme 6)}$$

Although the selective introduction of a fluorine atom into an organic molecule is difficult, a number of α-fluoro ethers have been synthesised in this way. Glycosyl fluorides, for example, which are useful in building fluorine-containing carbohydrates, have been prepared from phenyl thioglycosides using DAST (diethylaminosulfur trifluoride)/NBS or HF·pyridine/NBS (Equation (10)) ⟨84JA4189⟩, the reactions proceeding with retention of configuration; the use of the HF·pyridine/NBS system is reported to be compatible with most of the functional groups found in carbohydrates ⟨84JA4189⟩. Glycosyl fluorides have also been prepared from substrates which contain a free ⟨84CL1751⟩ or acetylated ⟨84CL1747⟩ hydroxyl in the 1-position with HF·pyridine complex (Equation (11)).

$$\text{(Equation 10)}$$

$$\text{(Equation 11)}$$

Xenon difluoride converts benzyl alcohol, and benzyl alcohols containing electron-withdrawing groups, into fluoromethyl aryl ethers (Equation (12)) ⟨93TL4355⟩; the reaction becomes less controlled when electron-donating groups are introduced, producing a complex mixture from which only 20% of the fluoromethyl ether could be isolated in the case of a methyl substituent, and no α-fluoro ether with substituents such as OH, OR or NHR. The reaction also failed with diphenylmethanol and 2-phenyl-2-propanol, both of which give largely dimeric products. Selective monofluorination of the ether (**18**) was achieved using bromine trifluoride in a synthesis of the partially deuteriated anaesthetic sevoflurane (**19**) (Equation (13)) ⟨93MI 402-01⟩.

$$\text{(Equation 12)}$$

X = H, o-, m- or p-NO$_2$, p-CF$_3$, m-F

$$\text{(Equation 13)}$$

The chemistry of fluorinated epoxides has been reviewed ⟨71FCR77⟩. α-Fluoro epoxides have been prepared in the usual way by the reaction of fluorinated alkenes with sodium hypobromite (Equation (14)) ⟨B-92MI 402-01⟩; the vanadyl acetylacetonate-catalysed reaction of *t*-butyl hydroperoxide (tbhp) with 2,3-difluoroallylic alcohols is reported to proceed with good diastereoselectivity for the (Z) isomer (**20**) (Equation (15)) ⟨93JFC(63)157⟩. The development of the chemistry of hexafluorobenzene has led to the isolation and characterisation of a number of structurally unusual α-fluoro epoxides.

Thus, the diene (**21**), obtained from hexafluorobenzene, reacts with trifluoroperacetic acid to give the epoxide (**22**), which was subsequently converted into hexafluorobenzene oxide (**23**) (Scheme 7) ⟨90JA6715⟩. The diene (**24**) was converted into the monoepoxide (**25**) with trifluoroperacetic acid, the stereochemistry of the epoxide being assigned on steric grounds (Equation (16)) ⟨89JOC5520⟩. Dewar hexafluorobenzene reacts with bis(fluoroxy)difluoromethane under photochemical conditions to give small amounts of the α-fluoro epoxides (**26**) and (**27**), the latter rearranging thermally to the α-fluoro acetal (**28**) (Scheme 8) ⟨79JOC2813⟩. Hexafluorobenzene also reacts photochemically in the vapour phase with oxygen to give a low yield of the unsaturated α-fluoro epoxide (**29**), which undergoes addition reactions with halogens, and cycloaddition reactions with dienes and 1,3-dipoles, to give a series of other α-fluoro epoxides (Equation (17)) ⟨80CC158⟩.

4.02.1.2.2 α-Chloro ethers, $R^1_2CCl(OR^2)$

Methods for the synthesis of α-chloro ethers have been reviewed ⟨55CRV301⟩ and a general account of their synthetic utility is available ⟨64ZC401⟩. Many synthetic routes to one of the most potentially useful α-chloro ethers, chloromethyl methyl ether, involve the potent carcinogen bis(chloromethyl) ether ⟨72CEN55, 72CEN62, 73OSC(5)218⟩ as a by-product. Early methods for the preparation of both these ethers ⟨63OSC(4)101⟩ have been listed, as have the chemical reactions of chloromethyl aryl ethers ⟨73OSC(5)221⟩.

(i) Haloalkylation of aldehydes and ketones

The most general method used for the synthesis of these compounds involves the haloalkylation of a carbonyl group with an alcohol and hydrogen chloride (Equation (18)) ⟨55CRV301⟩; the reaction is believed to involve the formation of a hemiacetal which is then chlorinated by the hydrogen chloride. The reaction is reasonably general, being applicable to aldehydes and ketones and primary and secondary alcohols; aromatic aldehydes and ketones, however, give poor yields (Table 1). A key element in securing an acceptable yield is the removal of excess hydrogen chloride prior to distillation of the crude product; even then, the limited thermal stability of a number of α-chloroethers precludes their purification by distillation (e.g., ⟨63OSC(4)748⟩).

$$R^1\text{COR}^2 + R^3\text{OH} + \text{HCl} \longrightarrow R^1R^2\text{C(Cl)(OR}^3) + H_2O \quad (18)$$

Table 1 Preparation of α-chloro ethers by chloroalkylation of ketones and aldehydes with hydrogen chloride.

Product	Ketone/aldehyde	Alcohol	Yield (%)	Ref.
MeO–CH₂–Cl	HCHO(aq.)	MeOH	87	32OSC(1)369
CH₃CH(Cl)(OEt)	Paraldehyde	EtOH	90	63OSC(4)748
EtCH(Cl)(OMe) related	EtCHO	MeOH	70	59CB1818
iPr-CH(Cl)(OMe)	Me₂CHCHO	MeOH	70	59CB1818
1-chloro-1-methoxycyclohexane	cyclohexanone	MeOH	a	59CB1818
PhCH(Cl)CH(OEt)	PhCHO	EtOH	94	49JA4007
Cl(CH₂OCH₂)₁₀Cl	HCHO	HO(CH₂)₁₀OH	a	49JOC754

a Yield not given.

(ii) Chlorination of ethers

The direct α-chlorination of ethers has also been used as a route to certain α-chloro ethers (Table 2), and chlorine itself ⟨35JA2364, 50JOC715, 77CI(L)127⟩, (dichloroiodo)benzene under photochemical conditions ⟨69LA(728)12⟩ and phosphorus pentachloride ⟨73OSC(5)221⟩ have been successfully used for this purpose. The claim that treating anisole with sulfuryl chloride affords chloromethyl phenyl ether ⟨67OS(47)23⟩ has, however, been refuted ⟨68JOC3335⟩, the product being in fact *p*-chloroanisole;

this problem underlines the complications that can arise when using this method with aromatic substrates which are susceptible to electrophilic, nuclear halogenation (Equation (19)).

Table 2 Preparation of α-chloro ethers by chlorination of ethers.

Product	Substrate	Conditions	Yield (%)	Ref.
CH(Cl)(OEt)(Me)	Et$_2$O	Cl$_2$ (1 equiv.)	42	50JOC715
(iPrO with Cl on each iPr)	Et$_2$O	Cl$_2$ (2 equiv.)	51	50JOC715
PhOCH$_2$Cl	PhOMe	Cl$_2$, hν	73	77CI(L)127
CH(Cl)(OEt)(Me)	Et$_2$O	PhICl$_2$, hν	a	69LA(728)12
4-Cl-C$_6$H$_4$-O-CH$_2$Cl	4-Cl-C$_6$H$_4$-OMe	PCl$_5$, Δ	68–80	73OSC(5)221
2,3-dichloro-1,4-dioxane	1,4-dioxane	Cl$_2$	61	35JA2364

[a] Yield not given.

$$\text{1,4-(MeO)}_2\text{C}_6\text{H}_4 \xrightarrow[\text{CCl}_4, 10\%]{\text{Cl}_2,\ h\nu} \text{1,4-(ClCH}_2\text{O)}_2\text{-2,5-Cl}_2\text{-C}_6\text{H}_2 \quad (19)$$

(iii) α-Chloro ethers from acetals

Treatment of acetals with an acid chloride produces α-chloro ethers in fair yields (Equation (20)) (Table 3). The addition of a small amount of thionyl chloride ⟨32LA(493)191⟩ or copper bronze ⟨32LA(498)101⟩ to the reaction mixture has been found to improve the yield, the latter possibly making the chloride ion more available through complex formation ⟨B-70MI 402-01⟩. A potential advantage of this method is that it allows chloromethyl methyl ether to be prepared free (GC) of the carcinogenic bis(chloromethyl) ether ⟨79S970⟩, although it contains methyl acetate and a small amount of acetyl chloride; the product has been used directly in this form for the synthesis of cephalosporin esters ⟨75TL3979⟩. Although aldehyde acetals are most frequently used, the dimethyl acetal of acetone is reported to react with phosphorus trichloride to form the thermally unstable

Table 3 Preparation of α-chloro ethers from acetals.

Product	Substrate	Conditions	Yield (%)	Ref.
PhCHClOMe	PhCH(OMe)$_2$	AcCl/SOCl$_2$	80	32LA(493)191
PhCH=CHCHClOMe	PhCH=CHCH(OMe)$_2$	AcCl/SOCl$_2$	a	32LA(493)191
Me(CH$_2$)$_2$CHClOEt	Me(CH$_2$)$_2$CH(OEt)$_2$	AcCl/Cu bronze	a	32LA(498)101
Me$_2$CClOMe[b]	Me$_2$C(OMe)$_2$	PCl$_3$	a	32LA(498)101
ClCH$_2$CHMeOCH$_2$Cl	(ClCH$_2$CHMeO)$_2$CH$_2$	BzCl	66	39JOC234
MeOCH$_2$Cl[c]	CH$_2$(OMe)$_2$	AcCl	100	79S970

[a] Yield not given. [b] Decomposes at $T > 40\ °C$. [c] Obtained as a solution containing methyl acetate and some acetyl chloride.

1-chloro-1-methylethyl methyl ether ⟨32LA(498)101⟩ and that of cyclopropanone reacts with thionyl chloride to give 1-chlorocyclopropyl methyl ether in poor yield ⟨81RTC194⟩.

$$R^1\text{-}C(OR^2)_2 + R^3\text{-}COCl \longrightarrow R^1\text{-}C(OR^2)(Cl) + R^3\text{-}C(O)OR^2 \quad (20)$$

(iv) α-Chloro ethers from α-alkoxy and α-aryloxy acid chlorides

The decarbonylation of α-alkoxy ⟨85S490⟩ and α-aryloxy ⟨71S150⟩ acid chlorides leads to the formation of α-chloro ethers in good yield (Table 4). Although the reaction proceeds quite well in refluxing thionyl chloride ⟨84S727⟩, aluminum chloride has been used as a catalyst both under those conditions ⟨85S490⟩ and on its own ⟨71S150⟩; the α-chloro ether can be produced directly from the alkoxy acid in a one-pot procedure, using thionyl chloride on its own ⟨84S727⟩ or together with aluminum chloride ⟨85S490⟩. The reaction has been used to prepare chloromethyl methyl ether, producing not only a bis(chloromethyl) ether-free product (GC, NMR), but one which, unlike that obtained from dimethoxymethane (see Section 4.02.1.2.2(iii)), contains neither methyl acetate nor acetyl chloride ⟨84S727⟩.

Table 4 Preparation of α-chloro ethers by decarbonylation of α-alkoxy and α-aryloxy acid chlorides.

Product	Substrate	Conditions	Yield (%)	Ref.
X–C₆H₄–O–Cl	X–C₆H₄–O–COCl	AlCl₃	a	71S150
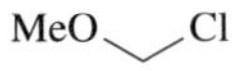 MeO–CH₂–Cl	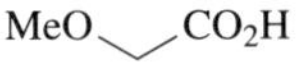 MeO–CH₂–CO₂H	SOCl₂, reflux	81	84S727
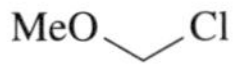 MeO–CH₂–Cl	 MeO–CH₂–CO₂H	i, SOCl₂; ii, AlCl₃	63	85S490
MeO–CH₂–Cl	MeO–CH₂–CO₂H	SOCl₂, AlCl₃, DMF	79	85S490

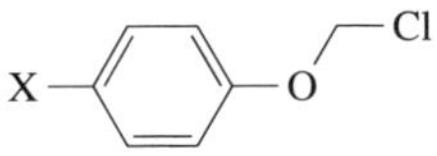

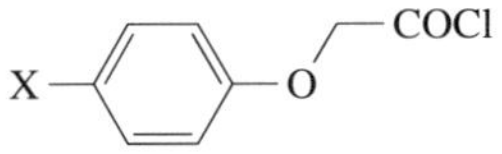

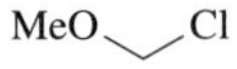

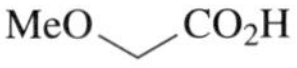

[a] Yield not given

(v) Miscellaneous methods

Chloromethyl phenyl ether has been prepared from sodium chloromethanesulfonate (Scheme 9) ⟨63CB2266⟩. The addition of chlorine and hydrogen chloride to vinyl ethers has been used to synthesise a number of cyclic α-chloro ethers; the stable 2,3-dichloro-1,4-dioxane is formed on addition of chlorine to dihydro-1,4-dioxine, whereas addition of dry hydrogen chloride gives the unstable mono-α-chloro ether, which continuously evolves hydrogen chloride (Scheme 10) ⟨35JA2364⟩. 2-Chlorotetrahydropyrans have been prepared in the same way and appear to be more stable ⟨67JOC607⟩. The inhalation anaesthetic agent isoflurane (**30**) is an α-chloro ether, the individual enantiomers of which have been synthesised by a process in which an α,α-dichloro ether is photochemically reduced to an α-monochloro ether (Scheme 11) ⟨93JOC7382⟩.

$$Na_2SO_3 + CH_2Cl_2 \xrightarrow[30\text{ atm, }100\,°C]{H_2O/EtOH,\ CuCl_2} ClCH_2SO_3Na \xrightarrow[\text{ii, PCl}_5]{\text{i, PhO}^-\text{Na}^+} PhOCH_2Cl$$

Scheme 9

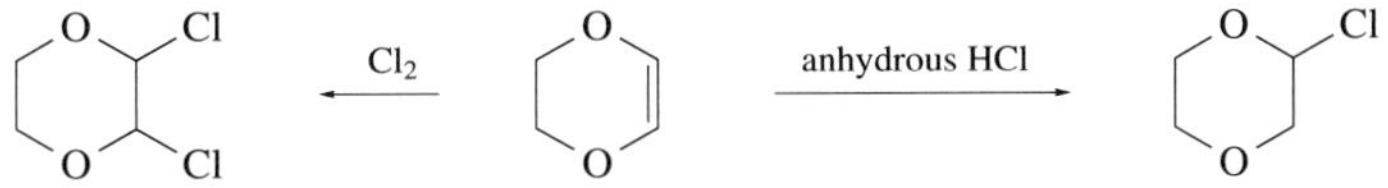

2,3-dichloro-1,4-dioxane ⟵(Cl₂)— 1,4-dioxine —(anhydrous HCl)⟶ 2-chloro-1,4-dioxane

Scheme 10

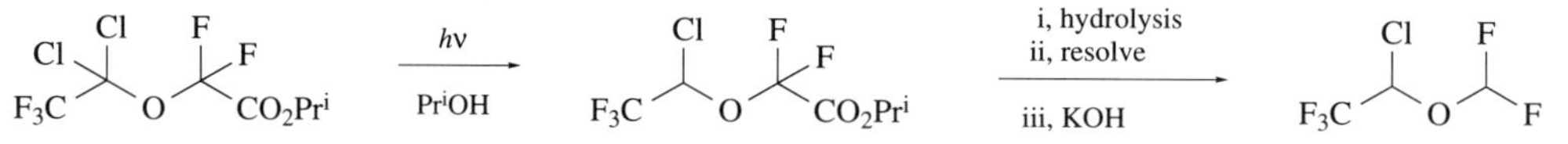

A Halogen and a Chalcogen

Scheme 11

The early literature relating to α-chloro epoxides has been reviewed ⟨63JA4004⟩; they have been prepared by the reaction of peracids with chloro alkenes (Equation (21)) and in general are very reactive materials. 1-Chloroepoxycyclohexane undergoes a rapid rearrangement to 2-chloro- and 2-hydroxycyclohexanone on exposure to moist air ⟨63JA4004⟩, although the corresponding 4-methyl derivative is more stable and can be distilled ⟨67JA6573⟩. *exo*-2-Chloro-2,3-epoxynorbornane (**31**), the optically active form of which has also been prepared ⟨70JA5664⟩, is stable at dry-ice temperature but on being left at room temperature undergoes a violent exothermic reaction with evolution of hydrogen chloride (Scheme 12) ⟨68JOC2934⟩.

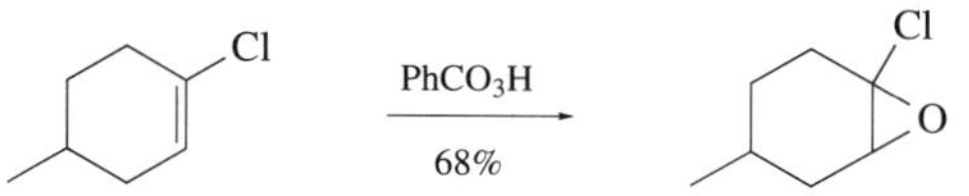

(21)

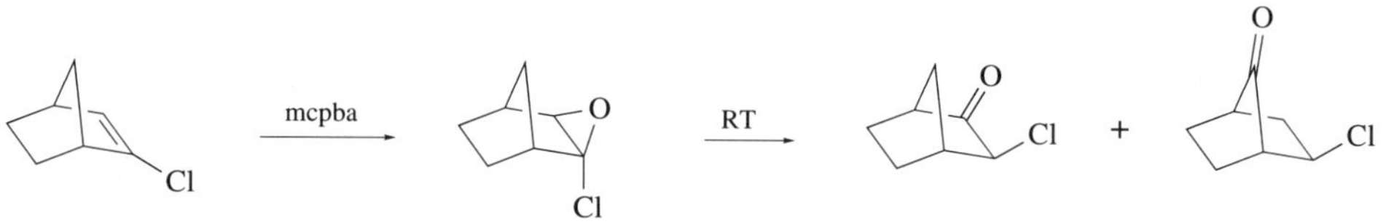

Scheme 12

4.02.1.2.3 α-Bromo ethers, $R^1{}_2CBr(OR^2)$

α-Bromo ethers have not been as widely used synthetically as the analogous chloro compounds, presumably because their even greater reactivity is offset by a corresponding decrease in thermal and general stability.

A number of the methods for the preparation of α-chloro ethers (Table 5) (Section 4.02.1.2.2) have also been used for the α-bromo ethers: thus the haloalkylation procedure using hydrogen bromide is reported to be effective for all except tertiary alcohols, with again the inclusion of a drying agent such as calcium chloride being of value in improving the yield ⟨49JA258⟩. The reaction of acetals with acetyl halides and related reactions also appears to be equally effective for the preparation of α-chloro and α-bromo ethers ⟨32LA(498)101⟩; the reaction of 1-ethoxy-1-(trimethylsiloxy)cyclopropane, for example, with phosphorus tribromide gives 1-bromo-1-ethoxycyclopropane (**32**) (Equation (22)) ⟨85JOC3255⟩, although the previously reported ⟨79JA7617⟩ reaction of the corresponding methyl compound could not be repeated. Although (**32**) is unstable at room temperature, it can be stored for two months at −20 °C without appreciable decomposition. The

Table 5 The preparation of α-bromo ethers.

Product	Reactants	Conditions	Yield (%)	Ref.
PhCHBrOMe	PhCH(OMe)$_2$	AcBr	a	32LA(498)101
Me$_2$CHCH$_2$OCH$_2$Br	HCHO(aq.),Me$_2$CHCH$_2$OH	HBr(2 equiv.), −5 °C	79	49JA258
Me$_2$CHCH$_2$OCH$_2$Br	HCHO(aq.),Me$_2$CHCH$_2$OH	HBr(1 equiv.), CaCl$_2$	96	49JA258
Me$_2$CH(CH$_2$)$_2$OCH$_2$Br	HCHO(aq.),Me$_2$CH(CH$_2$)$_2$OH	HBr(1 equiv.), CaCl$_2$	96	49JA258
PhCHBrCHBrOEt	PhCH$_2$CHClOEt	Br$_2$, no solvent	a	49JA4007
BrCH$_2$CHBrOEt	MeCHClOEt	Br$_2$, no solvent	70	63OSC(4)748

[a] Yield not given.

addition of bromine or hydrogen bromide to cyclic vinyl ethers has been used to prepare α-bromo ethers (Equation (23)); 2,3-dibromo-1,4-dioxane (**33**) is unstable, hydrolysing in moist air ⟨35JA2364⟩, but α-bromotetrahydropyrans, like the corresponding chloro derivatives, are reported to be stable ⟨67JOC607⟩. The regiospecific addition of benzene selenenyl bromide to vinyl ethers gives α-bromo-β-selenenyl ethers, which were not isolated but made to react inter- and intramolecularly with alcohols to give α-benzeneselenenyl acetals (Scheme 13) ⟨78HCA2286, 78HCA3075⟩. α-Bromo ethers have also been prepared from α-chloro ethers and bromine, through a process of dehydrochlorination and subsequent bromination (Equation (24)) ⟨30JA651, 49JA4007, 63OSC(4)101⟩.

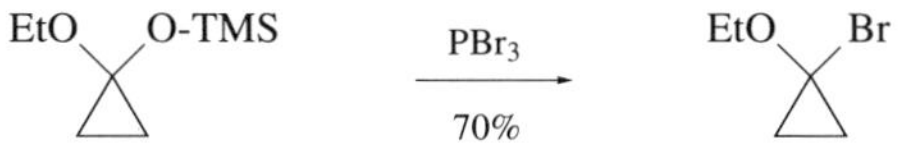

$$\text{EtO-C(O-TMS)} \xrightarrow[70\%]{\text{PBr}_3} \text{EtO-C(Br)} \quad (22)$$

(**32**)

$$\text{1,4-dioxene} \xrightarrow[92\%]{\text{Br}_2} \text{2,3-dibromo-1,4-dioxane} \quad (23)$$

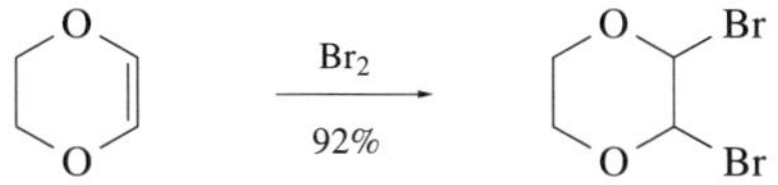

(**33**)

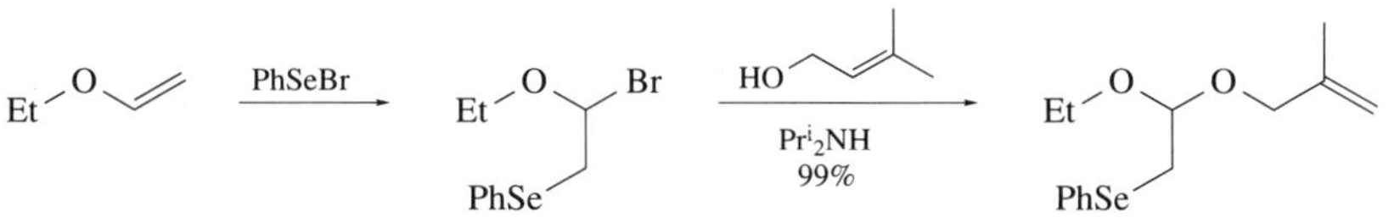

Scheme 13

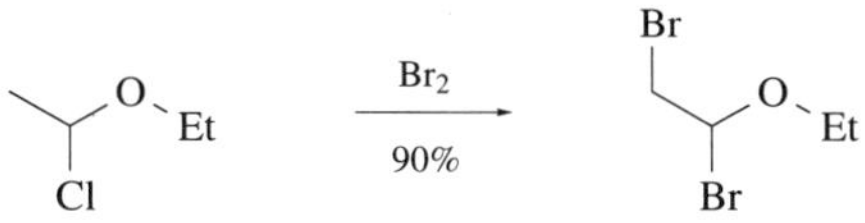

$$\text{CH}_3\text{CHCl(OEt)} \xrightarrow[90\%]{\text{Br}_2} \text{CHBr}_2\text{CH(Br)OEt} \quad (24)$$

4.02.1.2.4 α-Iodo ethers, $R^1{}_2CI(OR^2)$

The room temperature reaction of trimethylsilyl iodide with dimethoxymethane has been used to prepare iodomethyl methyl ether ⟨78S588⟩. The standard, sodium iodide-in-acetone procedure forms the unstable α-iodo ether (**34a**) and (**34b**) (Equation (25)) from the corresponding α-chloro compounds, and although they could not be isolated in a pure form their structures were confirmed by NMR spectroscopy ⟨67JOC607⟩.

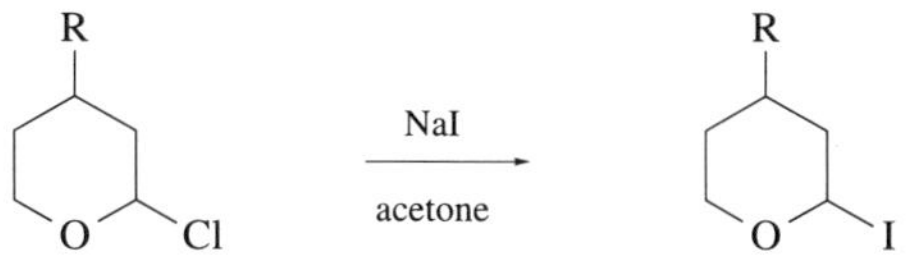

$$\xrightarrow[\text{acetone}]{\text{NaI}} \quad (25)$$

(**34a**) R = H
(**34b**) R = Me

4.02.1.3 Other Derivatives of α-Halo Alcohols (Geminal Halohydrins), $R^1{}_2CHal(OR^2)$ and $R_2CHal(OX)$

4.02.1.3.1 α-Haloalkyl esters, $R^1{}_2CHalOCOR^2$

(i) α-Fluoroalkyl esters, $R^1{}_2CFOCOR^2$

Compounds of this type have been prepared by the addition of fluorine-containing reagents to vinyl acetates and the addition of acetate derivatives to fluorinated alkenes. Thus, methyl hypofluorite which is prepared by the reaction of fluorine with methanol and is a source of the electrophilic methoxylium ion, adds to the enol acetate of acetophenone to give the α-fluoroalkyl acetate (**35**) (Equation (26)) ⟨92JA7643⟩. The fluoro lactone (**36**) is also formed by the addition of a reagent containing fluorine to a vinyl acetate; in this case the reagent, benzeneselenenyl fluoride, is formed *in situ* by the reaction of silver fluoride with benzeneselenenyl chloride (Equation (27)) ⟨90TL973⟩. An example of the opposite approach is the formation in reportedly good yield of the α-fluoro ester (**38**) by the phase transfer-catalysed addition of carboxylic acids to the perfluoro alkene (**37**) (Equation (28)) ⟨77TL2893, 81BCJ1151⟩. Acetyl hypobromite, also formed *in situ*, reacts with the difluoro alkene (**39**) giving the adduct (**40**) (Equation (29)) ⟨80JOC1394⟩.

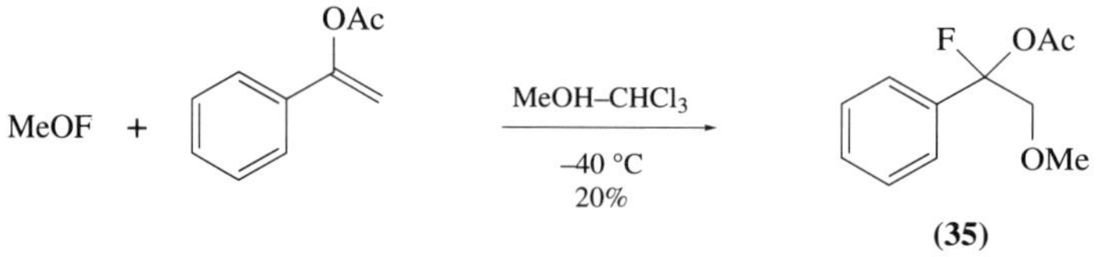

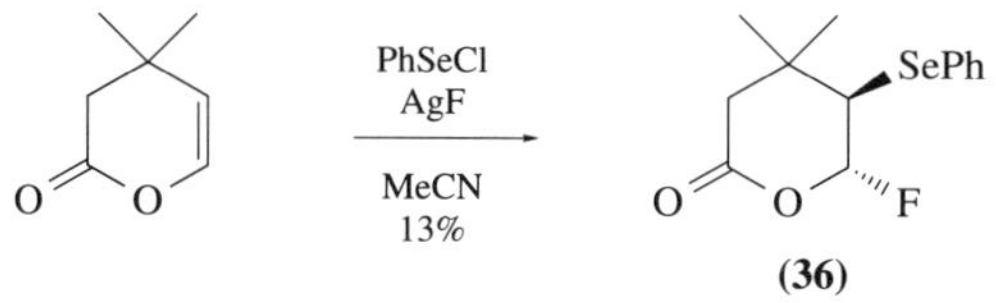

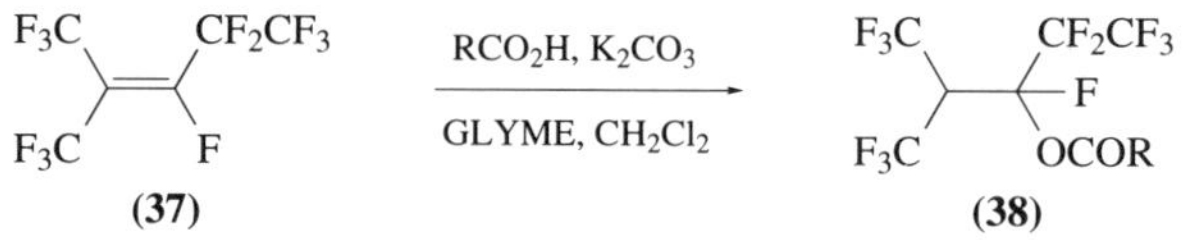

α-Fluoroalkyl esters have also been prepared by substitution reactions, which include that of perfluoroacyl fluorides with the perfluoro alkoxide (**41**) at −183 °C to give the perfluoro esters (**42**), which are said to be stable at 25 °C and above when pure ⟨72JOC3332⟩, and the reaction of ethyl 1-bromo-1-fluoroacetate (**43**) with potassium benzoate in DMF which gives ethyl benzoylfluoro acetate (**44**) (Scheme 14) ⟨88JCS(P1)1149⟩.

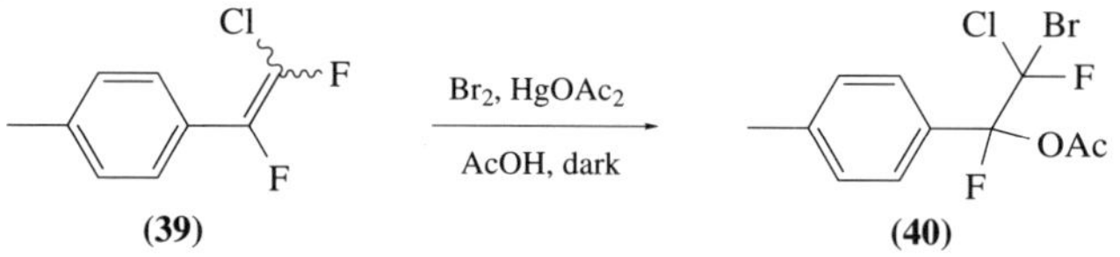

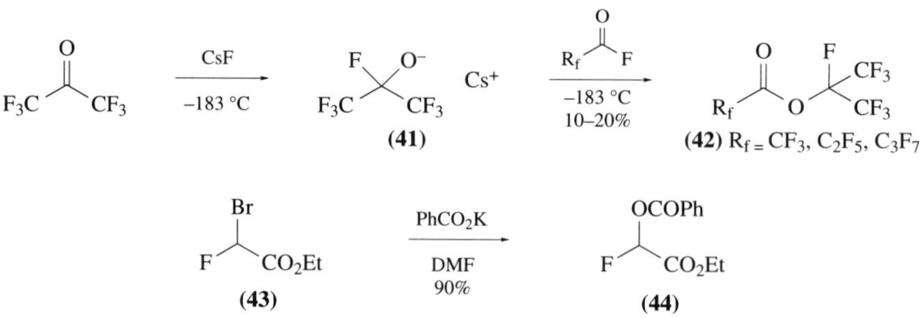

Scheme 14

The reaction of ethyl acetate with fluorine at −100 °C gives as the major product a moisture-sensitive ester in which one of the methylene protons remains unsubstituted (Equation (30)) ⟨74JA7588⟩. The DAST-promoted cyclization of 3-oxo carboxylic acids provides a general synthesis of γ-fluoro γ-lactones and 3-fluorophthalides (Scheme 15) ⟨84TL1019⟩; the intermolecular variation of the reaction has been used to prepare the α-fluoroalkyl ester (**45**). The cyclisation of the

3-fluoromuconate (**46**) in concentrated hydrochloric acid results in the formation of a mixture of butenolides in which the compound containing the α-fluoroalkyl ester unit (**47**) is the major component (Equation (31)); surprisingly cyclisation of the corresponding bromo and chloro compounds under the same conditions results in the formation of the isomeric lactones (**48**) ⟨90JOC3029⟩.

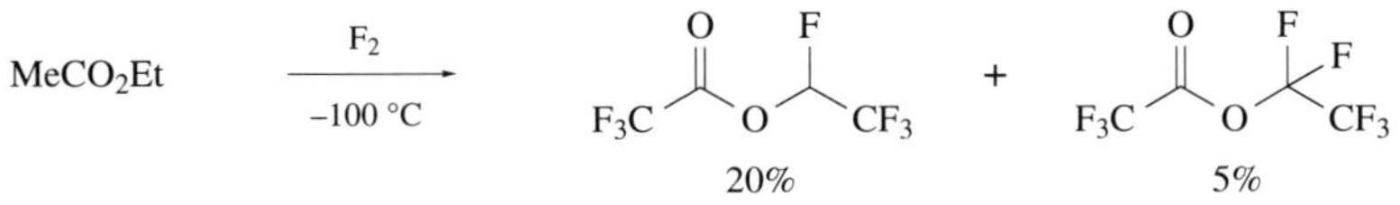

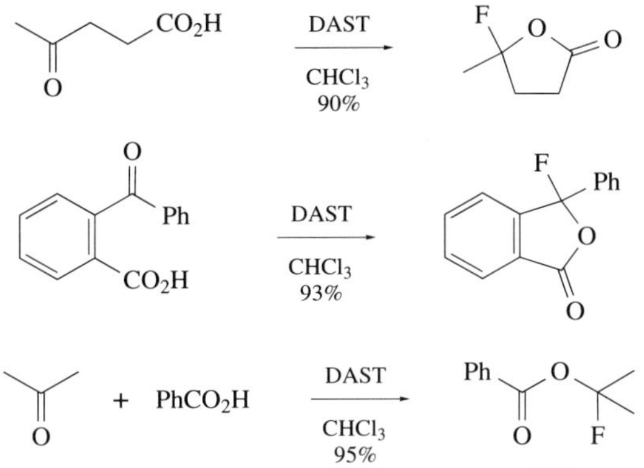

DAST = diethylaminosulfur trifluoride

Scheme 15

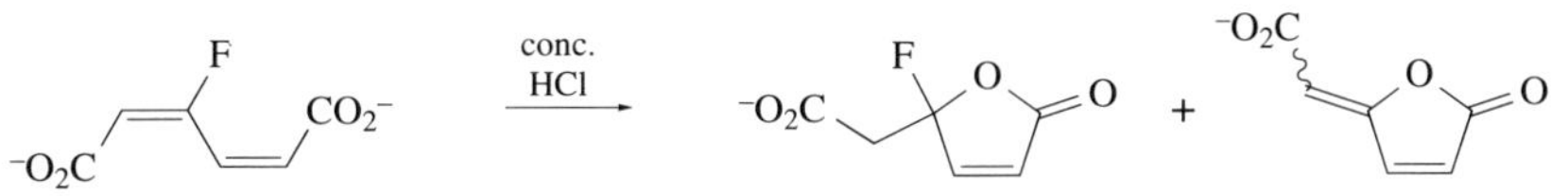

(ii) α-Chloroalkyl esters, $R^1_2CClOCOR^2$

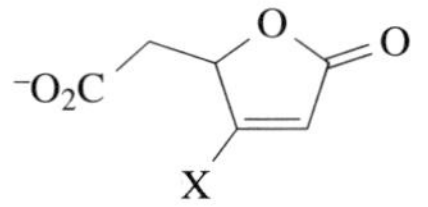

The chloroacylation of ketones and aldehydes is the most general method for the synthesis of α-chloroalkyl esters (Equation (32)) ⟨21JA651, 21JA660⟩. The acetylation of aldehydes in chloroform or dichloromethane, with zinc chloride as catalyst, proceeds in high yield, the conditions being compatible with a wide variety of other functional groups in the aldehyde (Table 6) ⟨77HCA1061, 78S593, 93JOC588⟩. Aromatic aldehydes are sufficiently reactive to give good yields of adduct in the absence of a catalyst (Table 6, entry 5) ⟨81JCS(P1)785⟩. Aromatic acid chlorides have been used, although the yields are significantly lower (Table 6, entry 7) ⟨51JA5870, 67CB2120⟩ and in at least one case the inherent reversibility of the reaction results in the product's reversion to starting materials on standing (Table 6, entry 7) ⟨67JOM(8)361⟩. Ketones react equally well under these conditions (Table 6, entries 8–11) ⟨93JOC588⟩, although the use of aluminum chloride has been preferred in some cases (Table 6, entries 10 and 11) ⟨78HCA2047⟩.

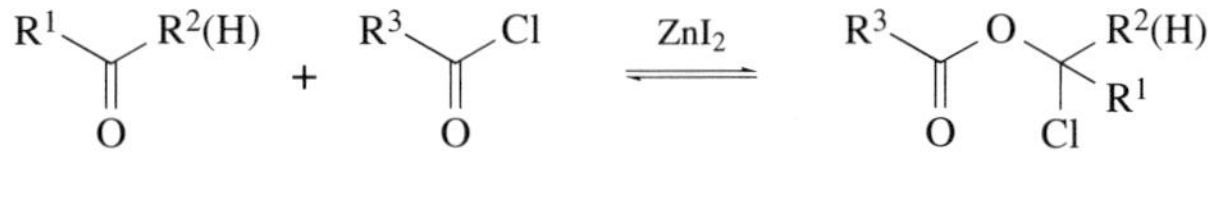

Table 6 Synthesis of α-chloroalkyl esters. Chloroacylation of aldehydes and ketones.

Entry	Product	Carbonyl compound	Acyl chloride	Yield (%)	Ref.
1	(α-chloropropyl acetate)	EtCHO	AcCl	91	71HCA1037
2	(α-chloro furfuryl acetate)	furfural (2-furaldehyde)	AcCl	75	77HCA1061
3	(1-chloro-3-acetoxyallyl acetate)	OHC-CH=CH-OAc	AcCl	80	77HCA1061
4	(chloromethyl crotonate)	HCHO	CH₃CH=CH-COCl	b	93JOC588
5[b]	(α-chlorobenzyl acetate)	PhCHO	AcCl	62	81JCS(P1)785
6	(1-chloro-2-propynyl acetate)	HC≡C-CHO	AcCl	94	71HCA1037
7	(α-chlorocinnamyl benzoate, Ph)	PhCH=CH-CHO	BzCl	c	67JOM(8)361
8	(1-acetoxy-1-chlorocyclobutane)	cyclobutanone	AcCl	93	71HCA1037
9	(2-chloro-2-methylpropyl acetate, t-Bu)	Me₂CO	AcCl	50	71HCA1037
10[d]	(2-chloro-2-methylpropyl acetate)	Me₂CO	AcCl	85	78HCA2047
11[d]	(1-acetoxy-1-chlorocyclohexane)	cyclohexanone	AcCl	77	78HCA2047

[a] ZnCl₂ used as catalyst unless otherwise indicated. [b] No catalyst used. [c] Yield not given. [d] AlCl₃ used as catalyst.

Thionyl chloride has been used to prepare α-chloroalkyl esters from the corresponding alcohols in phthalide systems (Equation (33)) ⟨79JOC4722, 83JOC635⟩; the same interconversion has been achieved in butenolides with titanium tetrachloride (Equation (34)) ⟨73JOC3878⟩. α-Phenylthioalkyl esters also react with thionyl chloride to give α-chloroalkyl esters in a process which allows a wide range of functional groups to be incorporated into the acyl alkyl group (Equation (35)) ⟨89ACS74⟩. The cyclisation of benzoic acids with o-acyl groups to give 3-chlorophthalides (Scheme 16) ⟨71JCS(C)1772, 75JCS(P1)2048, 81CJC3055⟩ and the formation of a γ-chloro γ-lactone from levulinic acid (Scheme 16) ⟨61BSB77⟩ in thionyl chloride are further examples of the importance of the reagent in this area. The potential of the liquid-phase chlorination of esters, although it produces a mixture of products, has also been investigated ⟨82ACS(B)721⟩.

$$\text{3-methyl-3-hydroxyphthalide} \xrightarrow[\text{FeCl}_3 \text{ 85\%}]{\text{SOCl}_2} \text{3-methyl-3-chlorophthalide} \tag{33}$$

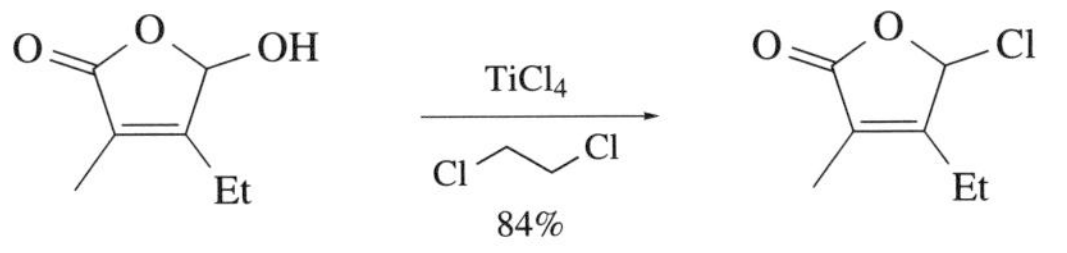

[Scheme 16 reactions shown as equation (34) and (35) and additional unnumbered reactions]

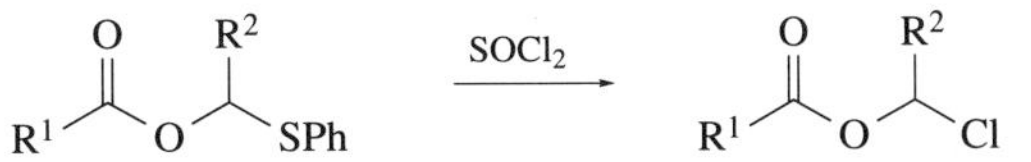

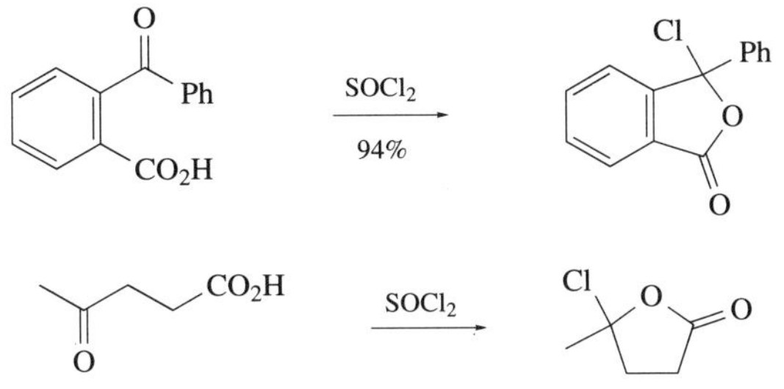

Scheme 16

2-Diazo-1,3-dicarbonyl compounds react to form 2-acyloxy-2-chloro-1,3-dicarbonyl compounds in good to excellent yield when they are treated with *t*-butyl hypochlorite in formic or acetic acid, the reaction being formally the result of a carbene insertion into the O—Cl bond of formyl or acetyl hypochlorite which is formed *in situ* (Equation (36)) ⟨68CB3604, 81CB1958⟩. These 2-acyloxy-2-chloro-1,3-dicarbonyl compounds have also been prepared by the acylation of the corresponding α-halo alcohols and the chlorination of 2-acyloxy-1,3-dicarbonyl compounds (Scheme 17) ⟨81CB1951⟩. The acylation of ninhydrin also leads to this type of compound (Equation (37)) ⟨80LA1919⟩.

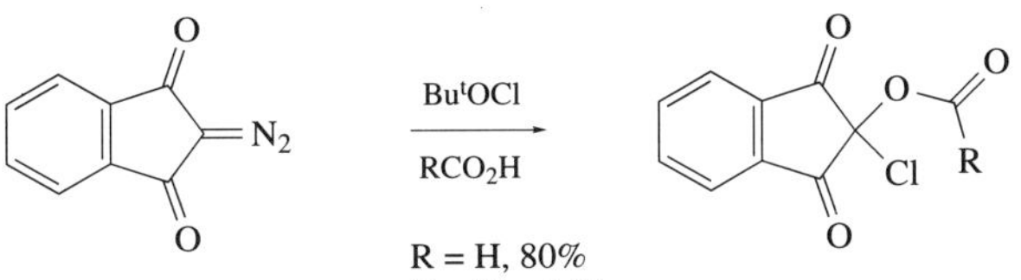

[Equation (36): indanedione diazo + Bu^{t}OCl / RCO$_2$H → 2-acyloxy-2-chloro indanedione; R = H, 80%; R = Me, 74%]

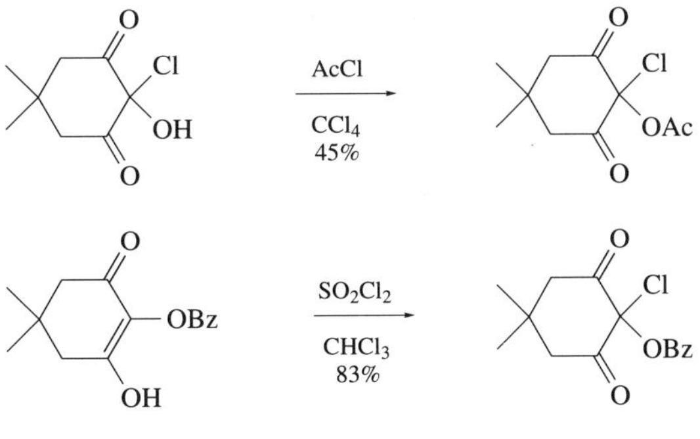

Scheme 17

[Equation (37): ninhydrin + AcCl → 2-chloro-2-acetoxy indanedione]

A number of miscellaneous methods have been used for the preparation of α-chloroalkyl esters. The reaction of arenesulfenyl halides with vinyl acetate is reported to be regioselective, giving only the product of Markovnikov addition (Scheme 18) ⟨88ZOR1945⟩; the ruthenium-catalysed addition of trifluoromethanesulfonyl chloride to vinyl benzoate gives the chloroalkyl ester (**49**) (Scheme 18) ⟨89CC1559⟩. An α-chloroalkyl ester is also formed when the isocrotonic acid derivative (**50**) undergoes an unusual dimerisation reaction in acetyl chloride at room temperature ⟨82JOC2582⟩ and when benzyl alcohol, among others, is oxidised with benzyl(triethyl)ammonium permanganate in dichloromethane (Scheme 19) ⟨81AG(E)104⟩.

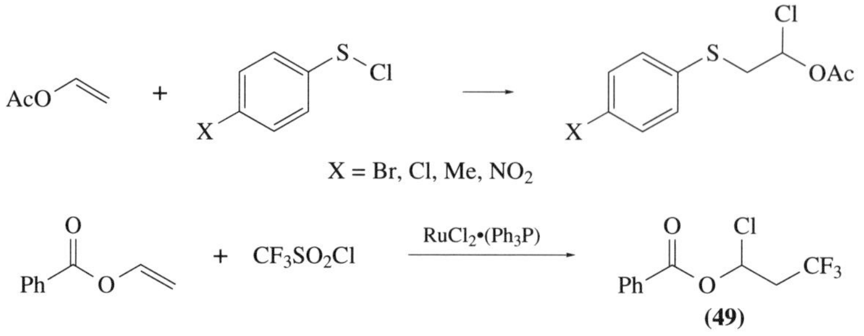

Scheme 18

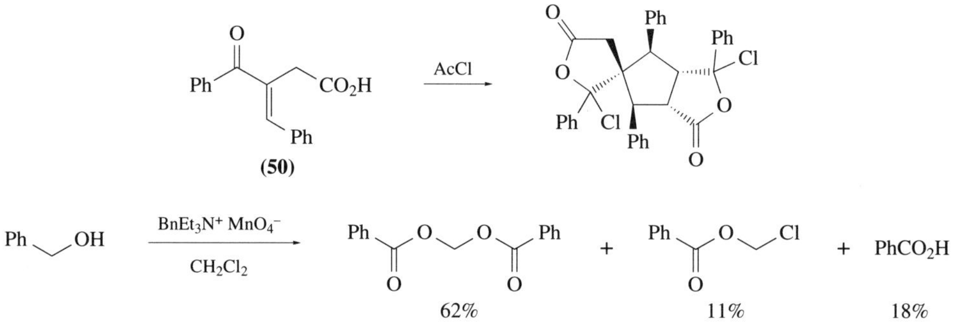

Scheme 19

(iii) α-Bromoalkyl esters, $R^1_2CBrOCOR^2$

The Lewis acid-catalysed addition of acyl bromides to aldehydes and ketones has been used extensively to prepare α-bromoalkyl esters (Scheme 20) ⟨67CB2120, 78HCA2047⟩; as for the chloro compounds, it is the most general method, being compatible with thioether, cyano and ester groups ⟨71HCA1037⟩. Aromatic aldehydes are again sufficiently reactive not to require a catalyst ⟨74AG(E)676⟩.

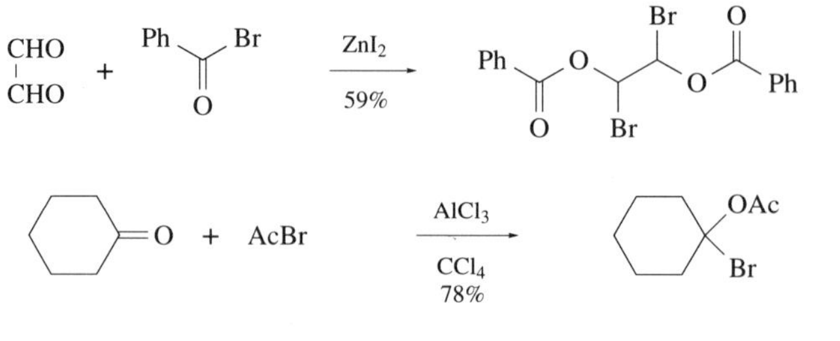

Scheme 20

Free radical, allylic bromination with NBS, together with light ⟨73OSC(5)145⟩, benzoyl peroxide ⟨64JCS766, 81CJC3055⟩ or 2,2′-azobisisobutyronitrile (AIBN) ⟨65JCS3075, 85JCS(P1)1567⟩, has been used extensively with butenolide (Equation (38)) and phthalide (Equation (39)) molecules. Although the products are often used directly without purification ⟨93JCS(P1)1493⟩, distillation, chromatography and crystallisation have, on occasion, been employed to obtain the materials in a pure form ⟨64JCS766⟩. Bromo butenolides have been prepared by the bromination of the corresponding hydroxy compounds with hydrogen bromide and 'pyrocatechol phosphorus tribromide' (Scheme 21) ⟨73JOC3878⟩, and a range of α-bromoalkyl esters have been prepared by the bromination of phenylthioalkyl esters in a process analogous to that used for the corresponding chloro compounds (Section 4.02.1.3.1(ii)) (Equation (35)) ⟨89ACS74⟩.

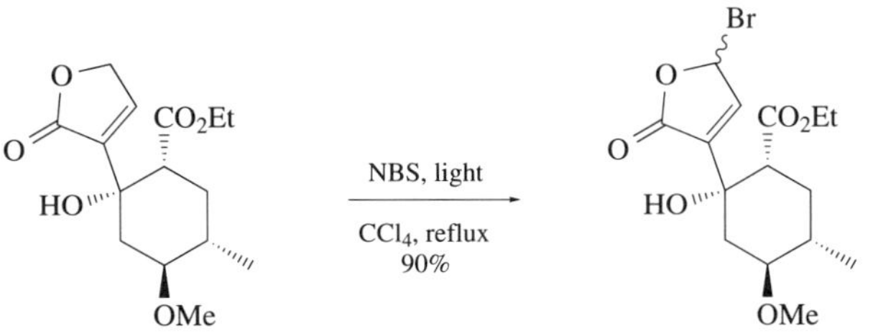

(38)

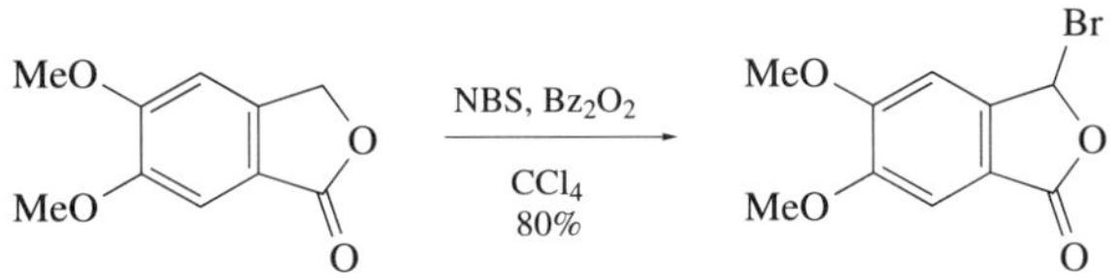

(39)

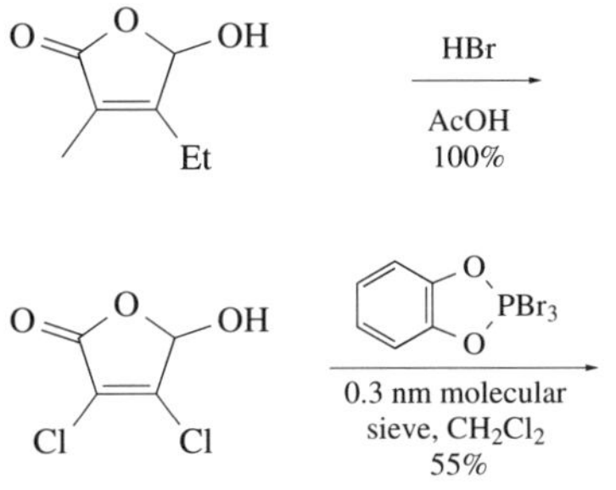

Scheme 21

The addition of bromine reagents to vinyl acetates has also been used to prepare this functional group. Thus the addition of bromine to the vinyl acetate (**51**) ⟨86ZOR2327⟩, and of arenesulfenyl bromides to vinyl acetate itself ⟨88ZOR1945⟩, both give α-bromoalkyl esters (Scheme 22). 2-Diazo-1,3-dicarbonyl compounds react with formyl and acetyl hypobromite to give 2-bromo-2-acyl derivatives ⟨68CB3604⟩, the acetate being also formed by the bromoacetylation of ninhydrin (Scheme 23) ⟨80LA1919⟩. 1-Bromocyclopropyl acetates are available through the reaction of *gem*-dibromocyclopropanes with silver acetate ⟨76JA6752⟩ and trifluoroacetate (Equation (40)) ⟨74TL909⟩.

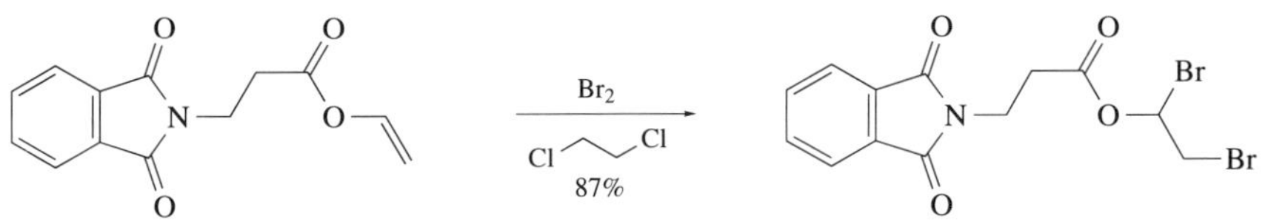

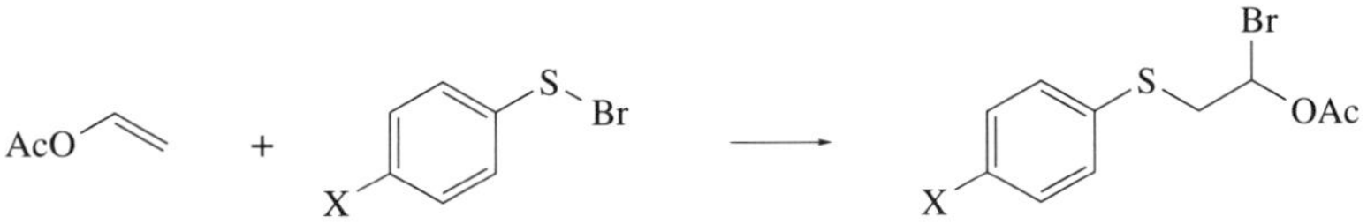

Scheme 22

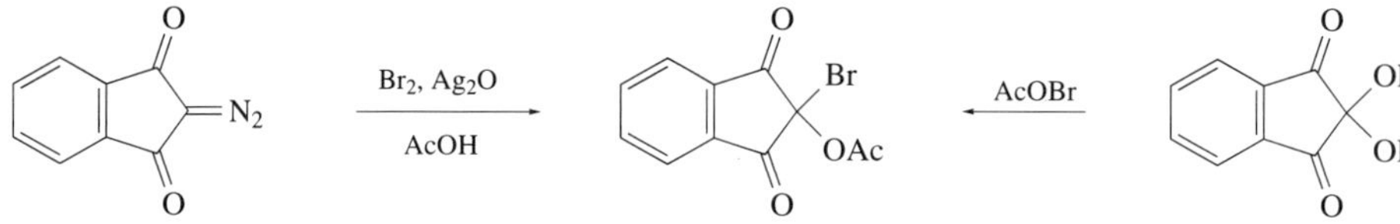

Scheme 23

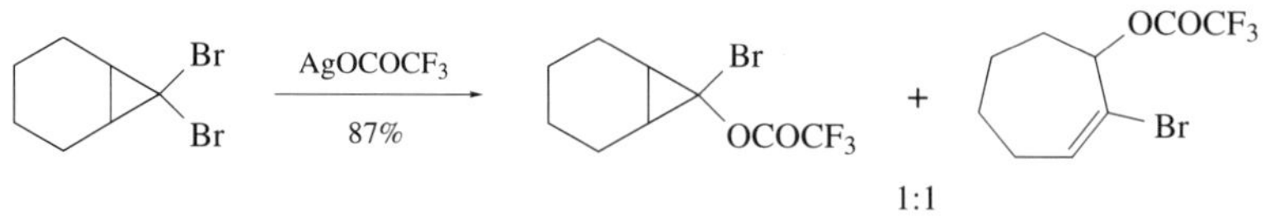

(40)

(iv) α-Iodoalkyl esters, $R^1_2ClOCOR^2$

The Finkelstein reaction is the only method which has been employed for the synthesis of α-iodoalkyl esters (Scheme 24). Although most of the compounds thus prepared appear to have limited stability ⟨83JOC635, 83JOC5280, 93JOC588⟩, (benzoyloxy)iodomethanes (**52**) ⟨93JCS(P1)2303⟩ appear to be reasonably stable. In one case the iodide undergoes homolytic cleavage, giving radicals which subsequently dimerise (Scheme 25) ⟨71JCS(C)3344⟩.

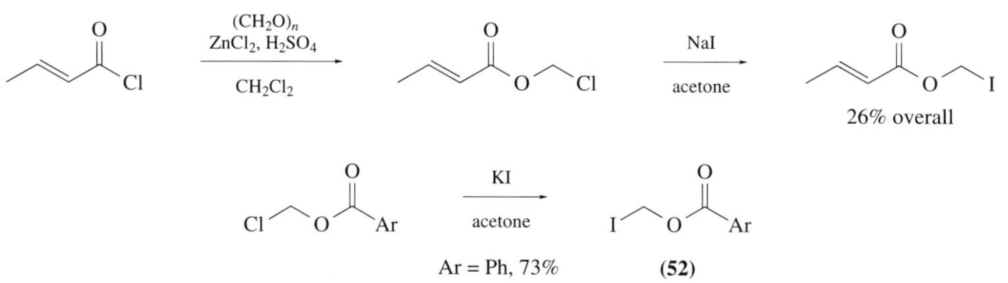

Scheme 24

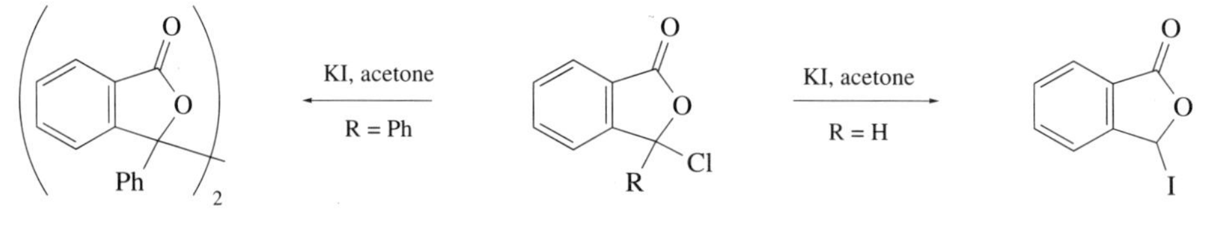

Scheme 25

4.02.1.3.2 α-Haloalkyl haloformates ($R_2CHalOCOHal$) and carbonate derivatives ($R^1_2CHalOCOOR^2$) etc.

(i) α-Haloalkyl haloformates, $R_2CHalOCOHal$

The most commonly used method of synthesising α-chloroalkyl chloroformates (α-chloroalkyl carbonochloridates) is a special case of the carbonyl chloroacylation reaction used to synthesise α-chloroalkyl esters and involves the reaction of liquid phosgene with an aldehyde in the presence of a quaternary ammonium salt (Equation (41)) ⟨81EUP40153, 84JOC2081, 86S627, 90JOC2240⟩; the reaction is applicable to alkyl, aromatic and heterocyclic aldehydes ⟨83GEP3241568⟩. The uncatalysed reaction between chloral and phosgene in dry benzene also gives a reasonable yield of the adduct (**53**) ⟨57JCS618⟩. Trichloromethyl carbonate can be used as an *in situ* source of phosgene and thus constitutes an attractive alternative to the liquid form of this reagent (Equation (42)) ⟨89TL2033⟩. Although the photochemical chlorination of methyl chloroformate gives chloromethyl chloroformate in only poor yield, the method does allow the compound to be prepared without the involvement of phosgene in any form (Equation (43)) ⟨90S1159⟩.

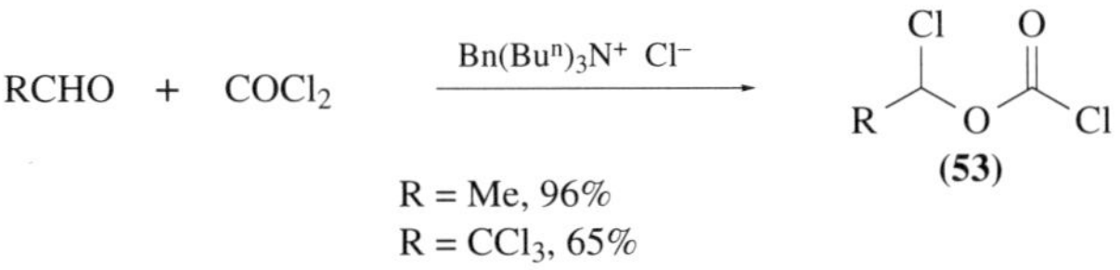

$$RCHO + COCl_2 \xrightarrow{Bn(Bu^n)_3N^+ Cl^-} \underset{(53)}{R\text{-CHCl-O-C(O)Cl}} \quad (41)$$

R = Me, 96%
R = CCl$_3$, 65%

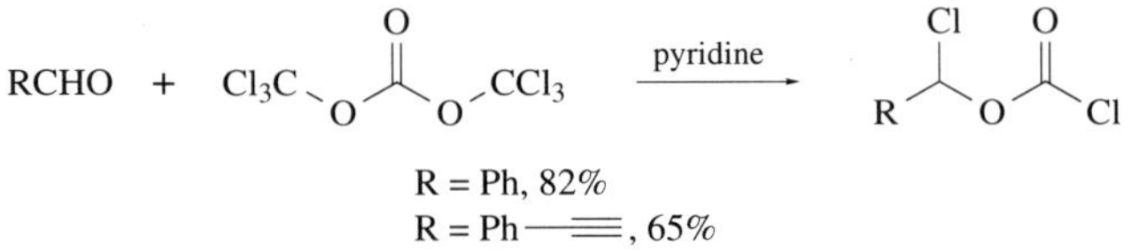

$$RCHO + Cl_3C\text{-O-C(O)-O-}CCl_3 \xrightarrow{pyridine} R\text{-CHCl-O-C(O)Cl} \quad (42)$$

R = Ph, 82%
R = Ph–≡–, 65%

$$Me\text{-O-C(O)Cl} \xrightarrow[22\%]{Cl_2, h\nu} Cl\text{-CH}_2\text{-O-C(O)Cl} \quad (43)$$

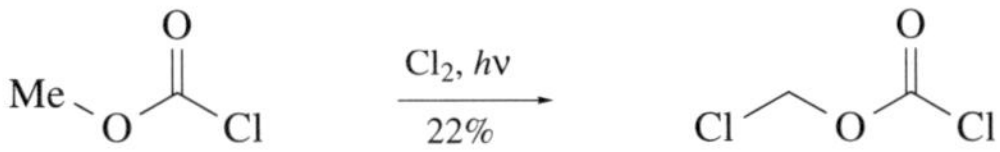

The α-fluoroalkyl fluoroformate (**55**) has been prepared from the cyclobutanone (**54**) and carbonyl fluoride (Scheme 26) ⟨71JA2481⟩; the reaction of the perfluoroalkoxide (**41**) with formyl fluoride also gives a 1-fluoroalkyl fluoroformate (Scheme 26) ⟨72JOC3332⟩.

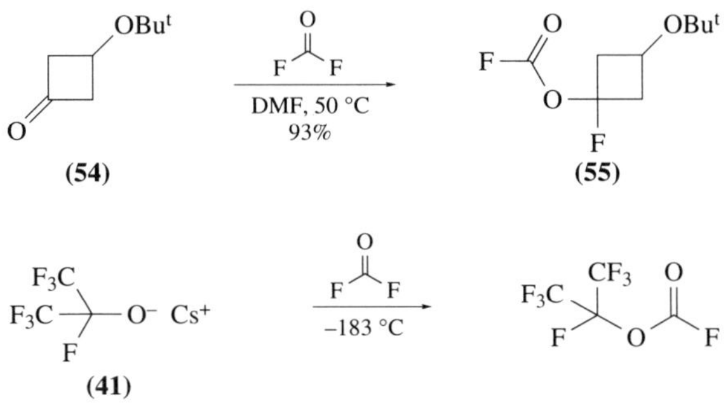

(**54**) + F-C(O)-F $\xrightarrow[93\%]{DMF, 50\ °C}$ (**55**)

(**41**) + F-C(O)-F $\xrightarrow{-183\ °C}$ F$_3$C-C(CF$_3$)(F)-O-C(O)-F

Scheme 26

(ii) α-Haloalkyl alkyl and aryl carbonate derivatives, $R^1_2CHalOCOOR^2$, etc.

α-Haloalkyl phenyl carbonates can be prepared by the addition of phenyl haloformates to aldehydes ⟨88S407⟩, and α-chloro-, α-bromo- and α-iodoalkyl compounds have been prepared in this way using pyridine in dichloroethane (Equation (44)); the synthesis of the corresponding fluoro compound requires the use of 4-dimethylaminopyridine (dmap), or potassium fluoride and 18-crown-6 ⟨88S407⟩. Chloromethyl chloroformates react with alcohols and phenols in dichloromethane or THF containing pyridine to give α-chloroalkyl carbonates (Equation (45)) ⟨84JA1809, 86S627, 90JOC1847⟩; 1,2,2,2-tetrachloroethyl *t*-butyl carbonate, which has been used to introduce the *t*-butoxycarbonyl group into amino acids, has been prepared by this method ⟨85JOC3953⟩. The chloroalkyl alkyl carbonates (**56**) and (**57**) have been prepared by the photochemical chlorination of methyl carbonate ⟨19CR(169)1143⟩ and the reaction of the enolate of mesityl oxide with 1-chloroethyl chloroformate ⟨90TL1405⟩, respectively (Scheme 27).

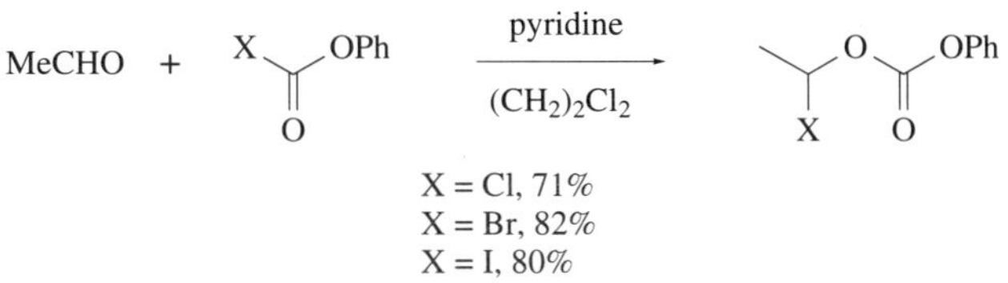

$$MeCHO + X\text{-C(O)-OPh} \xrightarrow[(CH_2)_2Cl_2]{pyridine} \text{CH}_3\text{CH(X)-O-C(O)-OPh} \quad (44)$$

X = Cl, 71%
X = Br, 82%
X = I, 80%

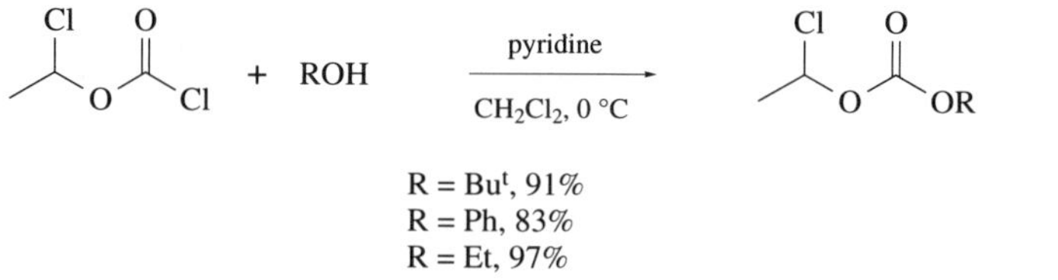

Scheme 27

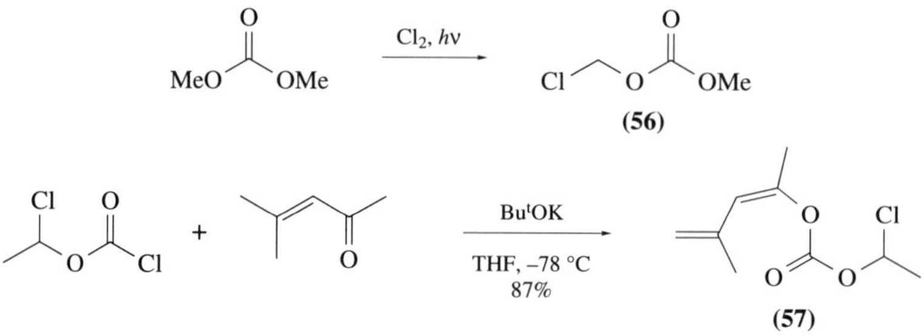

(45)

R = Bu^t, 91%
R = Ph, 83%
R = Et, 97%

(56)

(57)

The reaction of α-chloroalkyl chloroformates with amines gives α-chloroalkyl carbamates ⟨76TL3381, 93SL195⟩, which in some cases have been elaborated further (Scheme 28) ⟨87S1027⟩; a bis(carbamate) is formed when benzimidazole reacts with 1-chloroethyl chloroformate and allyltributyltin (Equation (46)) ⟨92TL5399⟩. Hydrazines react in an analogous manner to give carbazates (Equation (47)) ⟨86TL6319⟩.

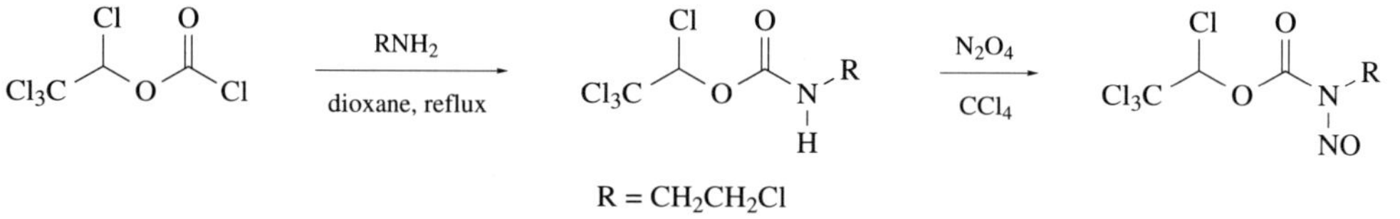

R = CH$_2$CH$_2$Cl

Scheme 28

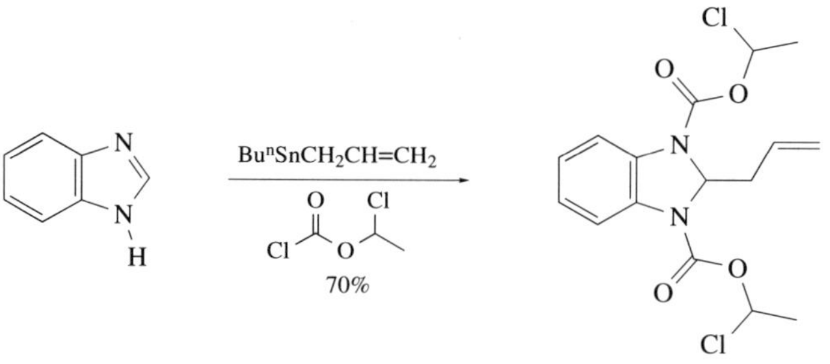

(46)

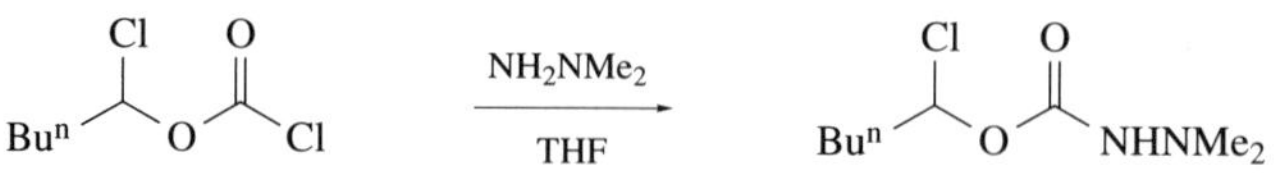

(47)

Chloromethyl thiocarbonates are formed by the reaction of chloromethyl chloroformate with thiols; this results in an inversion of the reactivity of the two electrophilic centres and allows iodomethyl thiocarbonates to be prepared therefrom by treatment with sodium iodide in acetone (Scheme 29) ⟨90S1159⟩.

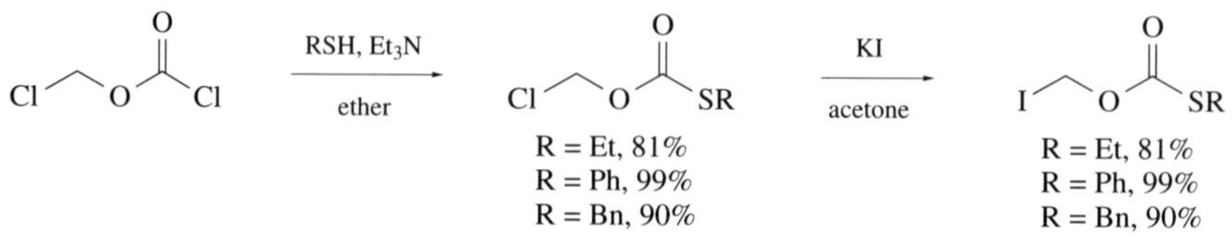

R = Et, 81%
R = Ph, 99%
R = Bn, 90%

R = Et, 81%
R = Ph, 99%
R = Bn, 90%

Scheme 29

4.02.2 HALOGEN AND SULFUR DERIVATIVES, $R^1_2CHal(SR^2)$, etc.

4.02.2.1 Dicoordinate α-Halo Sulfur Derivatives, $R^1_2CHal(SR^2)$, etc.

4.02.2.1.1 α-Halo sulfides, $R^1_2CHal(SR^2)$

(i) α-Fluoro sulfides, $R^1_2CF(SR^2)$

Although phenyl fluoromethyl sulfides can be prepared from the corresponding chloro compounds by the nucleophilic displacement of chloride with potassium fluoride and 18-crown-6 ⟨77S791⟩, the first general method to be introduced for the synthesis of α-fluoro sulfides involved the direct fluorination of sulfides with xenon difluoride. This reagent is available commercially but is expensive and requires the use of specialised handling techniques ⟨76JFC(8)305, 77CJC3031, 83JFC(22)557⟩; however despite these disadvantages, which have seen it replaced by more experimentally convenient reagents, xenon difluoride has been used for the selective fluorination of sulfide groups in nucleosides (Table 7) ⟨91JOC6878⟩.

Table 7 Synthesis of nucleosides containing α-fluoro sulfide groups.

Product	Reagent	Substrate	Yield (%)	Ref.
X = OMe	DAST[a](SbCl$_3$)	X = OMe, Y = S(O)	96	91JOC6878
X = OMe	DAST(SbCl$_3$)	X = OMe, Y = S(O)	85	88TL5729
X = OMe	XeF$_2$	X = OMe, Y = S	91	91JOC6878
X = OAc	DAST(SbCl$_3$)	X = OAc, Y = S(O)	89	91JOC6878
X = OAc	XeF$_2$	X = OAc, Y = S	83	91JOC6878

[a] DAST = diethylaminosulfur trifluoride.

The increasing interest in fluorinated, biologically active molecules has led to the development of a number of other general methods, the earliest of which was the introduction of diethylaminosulfur trifluoride (DAST) for converting sulfoxides into α-fluoro sulfides in a process which was reported to be catalysed by zinc iodide ⟨85JA735, 88TL5729⟩. However, this catalyst was subsequently found, in some cases, to give mixtures of fluorinated and unfluorinated sulfides of varying composition ⟨88TL3365, 88TL5729, 89JMC997⟩, thus leading to the introduction of an alternative catalyst, antimony trichloride (Table 8) ⟨88TL5729, 90JOC4757, 91JOC6878⟩.

The DAST/SbCl$_3$ combination has since been used to fluorinate sulfides directly, thus eliminating

Table 8 Synthesis of α-fluoro sulfides by fluorination of sulfoxides.

Product	Reagent	Substrate	Yield (%)	Ref.
PhSCHFR		PhS(O)CH$_2$R		
R = H	DAST[a]	R = H	85	85JA735
R = H	DAST(SbCl$_3$)	R = H	88	90JOC4757
R = (CH$_2$)$_3$Me	DAST(SbCl$_3$)	R = (CH$_2$)$_3$Me	81	90JOC4757
R = CH$_2$CH$_2$CO$_2$Et	DAST(SbCl$_3$)	R = CH$_2$CH$_2$CO$_2$Et	79[b]	90JOC4757
R = CH$_2$Ph	DAST(SbCl$_3$)	R = CH$_2$Ph	88[b]	90JOC4757
R = Ph	DAST(SbCl$_3$)	R = Ph	94[c]	90TL5449

[a] DAST = diethylaminosulfur trifluoride. [b] Yield of the fluoro sulfoxide obtained by direct oxidation of the fluoro sulfide. [c] Yield of the fluoro sulfone obtained by direct oxidation of the fluoro sulfide.

the need to convert them to sulfoxides before fluorination ⟨93JOC3800⟩; in addition the *p*-methoxy group, which was considered necessary to activate phenyl sulfides and sulfoxides for fluorination by DAST, is no longer required when DAST/SbCl₃ is used, as the unsubstituted compounds are equally reactive ⟨93JOC3800⟩. Sulfides have also been fluorinated directly using *N*-fluorotrimethylpyridinium triflate (NFPT), a reagent which, although less frequently used than DAST, is reported to be more easily handled and to have a greater thermal stability; the latter property may be of importance if high reaction temperatures are required ⟨86BCJ3625⟩. An electrochemical technique, which involves $Et_3N \cdot 3HF$ as the fluorinating agent, has also been developed for the direct fluorination of sulfides ⟨90JOC6074, 90TL2287⟩. Although initially it was reported that the method was only synthetically useful for sulfides with an electron-withdrawing group, a change of reaction solvent from MeCN to THF has been shown to increase the yield significantly for sulfides such as ethyl phenyl sulfide ⟨91CC1027⟩. This electrochemical procedure can be carried out in standard laboratory glassware and thus is competitive with the chemical methods discussed above; experiments suggest that it can be successful where NFPT fails ⟨92JOC3755, 93JOC4200⟩, but comparisons with DAST are not valid as they were based on experiments which do not involve the use of antimony trichloride.

Tables 9–11 illustrate the application of these fluorinating systems to various classes of sulfide.

Table 9 Synthesis of α-fluoroalkyl alkyl sulfides by fluorination of sulfides.

Product	Reagent	Substrate	Yield (%)	Ref.
$MeSCH_2F$	DAST[b]	MeSMe	83[a]	85JA735
$MeSCH_2F$	XeF_2	MeSMe	100	77CJC3031
$EtSCH_2F$	DAST(SbCl₃)	MeSEt	[a]	93JOC3800
$MeSCHFCO_2Et$	NFPT[c]	$MeSCH_2CO_2Et$	46[a]	86BCJ3625
$FCH_2SCH_2CH_2C[NHCOCF_3]HCO_2Me$	NFPT	$MeSCH_2CH_2C[NHCOCF_3]HCO_2Me$	41[a]	86BCJ3625
$EtO_2CCH_2SCHFCO_2Et$	Electrochemical MeCN, $Et_3N \cdot 3HF$	$S(CH_2CO_2Et)_2$	50	90TL2287

[a] Product not purified. [b] DAST = diethylaminosulfur trifluoride. [c] NFPT = *N*-fluorotrimethylpyridinium triflate.

Table 10 Synthesis of α-fluoro sulfides by fluorination of benzyl sulfides.

Product	Reagent	Substrate	Yield (%)	Ref.
$PhCHFSMe + PhCH_2SCH_2F$ (4:3)	NFPT[b]	$PhCH_2SMe$	77[a]	86BCJ3625
$PhCHFSMe + PhCH_2SCH_2F$ (1:1; PhCHFSMe unstable)	DAST[c] (SbCl₃)	$PhCH_2SMe$	55	93JOC3800
$PhCH_2SCHFCF_3$	Electrochemical MeCN, $Et_3N \cdot 3HF$	$PhCH_2SCH_2CF_3$	25	90JOC6074
$PhCH_2SCHFCO_2Et$	Electrochemical MeCN, $Et_3N \cdot 3HF$	$PhCH_2SCH_2CO_2Et$	44	90JOC6074
F-substituted thiazolidinone (F on C adjacent to S; N–Ph, C=O)	Electrochemical MeCN, $Et_3N \cdot 3HF$	thiazolidinone (N–Ph, C=O)	84	92JOC3755
$PhCH_2SCHFPh$	XeF_2	$PhCH_2SCH_2Ph$	100	77CJC3031

[a] Product not purified. [b] NFPT = *N*-fluorotrimethylpyridinium triflate. [c] DAST = diethylaminosulfur trifluoride.

(ii) α-Chloro sulfides, $R^1_2CCl(SR^2)$

The considerable synthetic importance of α-chloro sulfides has led to the development of a range of methods for their synthesis; these, and the synthetic uses of α-chloro sulfides, have been reviewed ⟨86T3731, 91COS(7)206⟩. The reactivity which contributes to the synthetic utility of these compounds

Table 11 Synthesis of α-fluoroalkyl phenyl sulfides by fluorination of sulfides.

Product	Reagent	Substrate	Yield (%)	Ref.
PhSCH$_2$F	DAST[a] (SbCl$_3$)	PhSMe	94	93JOC3800
PhSCH$_2$F	NFPT[b]	PhSMe	49	86BCJ3625
PhSCH$_2$F	Electrochemical THF, Et$_3$N·3HF	PhSMe	50	91CC1027
PhSCHFMe	Electrochemical MeCN, Et$_3$N·3HF	PhSEt	18	91CC1027
PhSCHFMe	Electrochemical THF, Et$_3$N·3HF	PhSEt	45	91CC1027
PhSCHF(CH$_2$)$_6$Me	DAST(SbCl$_3$)	PhSCH$_2$(CH$_2$)$_6$Me	88	93JOC3800
PhSCFMe$_2$	XeF$_2$	PhSCHMe$_2$	>90	77CJC3031
PhSCHFCF$_3$	Electrochemical MeCN, Et$_3$N·3HF	PhSCH$_2$CF$_3$	35	90JOC6074
PhSCHFCF$_3$	Electrochemical MeCN, Et$_3$N·3HF	PhSCH$_2$CF$_3$	62	91CC1027
PhSCHFCO$_2$Me	NFPT	PhSCH$_2$CO$_2$Me	38	86BCJ3625
PhSCHF(CH$_2$)$_2$CO$_2$Et	DAST(SbCl$_3$)	PhSCH$_2$(CH$_2$)$_2$CO$_2$Et	85	93JOC3800
PhS, F-substituted γ-butyrolactone	Electrochemical MeCN, Et$_3$N·3HF	PhS-γ-butyrolactone	84	90JOC6074
PhSCHFCOMe	Electrochemical MeCN, Et$_3$N·3HF	PhSCH(COMe)$_2$	55	90JOC6074
PhS, F-substituted N-Bu^n β-lactam	Electrochemical MeCN, Et$_3$N·3HF	PhS-substituted N-Bu^n β-lactam	92	93JOC4200
X-C$_6$H$_4$-SCH$_2$F; X = H	18-crown-6, KF	X = H, Y = Cl	83	77S791
X = Cl	DAST(SbCl$_3$)	X = Cl, Y = H	92	93JOC3800
X = Cl	NFPT	X = Cl, Y = H	76	86BCJ3625
X = NO$_2$	18-crown-6, KF	X = NO$_2$, Y = Cl	98	77S791

[a] DAST = diethylaminosulfur trifluoride. [b] NFPT = N-fluorotrimethylpyridinium triflate.

does limit their stability, particularly if the molecule contains a β hydrogen, and so many are best used directly after preparation. Indeed, the often modest yields reported for the preparation of these substances may reflect losses during attempted purification rather than any inherent inefficiency of the reaction. In addition a number of α-chloro sulfides have been reported to be irritating to the skin ⟨57JA376, 67JOC204⟩ and so appropriate precautions should be taken in their use.

The ready availability of alkyl sulfides, and of a range of chlorinating agents of different reactivities, makes the synthesis of α-chloro sulfides by direct chlorination particularly attractive. Although chlorine has occasionally been employed for this purpose ⟨49LA(563)54⟩ since it was first used for the preparation of chloromethyl methyl sulfide from dimethyl sulfide ⟨1855MI 402-01⟩, a relatively recent example being the conversion of 1,3,5-trithiane (**58**) into di(chloromethyl) sulfide (Scheme 30) ⟨73CC49⟩, more convenient reagents such as sulfuryl chloride and N-chlorosuccinimide are now favoured; it is generally possible with these to find a system which provides the required combination of selectivity and reactivity to monochlorinate the majority of dialkyl and alkyl aryl sulfides. Sulfuryl chloride is the most reactive reagent currently employed for this purpose and has been used at temperatures as low as −40 °C and −75 °C to monochlorinate 1,3-dithiane (**59**)

(Scheme 30) ⟨77TL885⟩ and dimethyl sulfide ⟨52JA3594⟩, respectively. The reagent has been used successfully in a range of solvents including pentane, dichloromethane, chloroform and carbon tetrachloride to chlorinate a wide range of alkyl methyl, aryl methyl, benzyl and cyclic sulfides in good to moderate yields (Table 12).

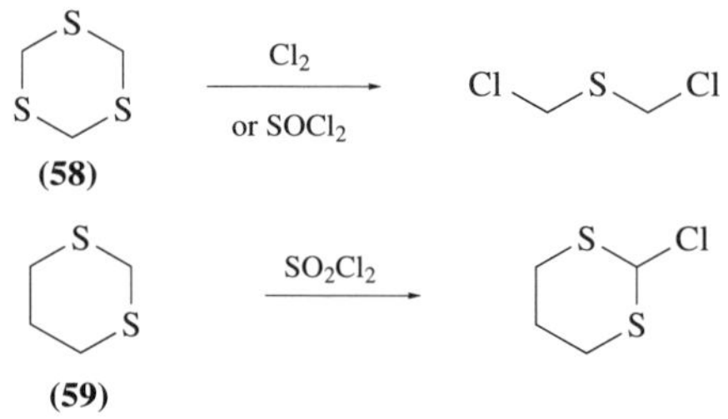

Scheme 30

Table 12 Synthesis of α-chloro sulfides by direct chlorination of sulfides with sulfuryl chloride.

Product	Reagent	Substrate	Yield (%)	Ref.
ClCH$_2$SMe	SO$_2$Cl$_2$ (1.05 equiv.)	Me$_2$S	40	55JA572
ClCH$_2$SCHCl$_2$	SO$_2$Cl$_2$ (1.2 equiv.)	ClCH$_2$SCH$_2$Cl	78	52JA3594
ClCH$_2$SCCl$_3$	SO$_2$Cl$_2$ (4.4 equiv.)	Me$_2$S	72	52JA3594
ClCH$_2$SCCl$_3$	SO$_2$Cl$_2$ (3.2 equiv.)	ClCH$_2$SCH$_2$Cl	81	52JA3594
EtO$_2$CCClHSCH$_2$CO$_2$Et	SO$_2$Cl$_2$ (1.05 equiv.)	EtO$_2$CCH$_2$SCH$_2$CO$_2$Et	70	55JA572
PhSCH$_2$Cl	SO$_2$Cl$_2$	PhSMe	99	74JOC2648
PhCHClSPh	SO$_2$Cl$_2$ (1.05 equiv.)	PhCH$_2$SPh	65	55JA572
PhCHClSMe	SO$_2$Cl$_2$ (1.02 equiv.)	PhCH$_2$SMe	73	55JA572
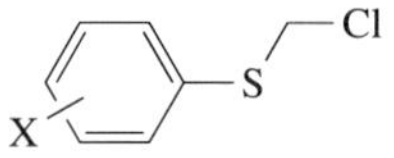 X-C$_6$H$_4$-SCH$_2$Cl	SO$_2$Cl$_2$ (1.02 equiv.)	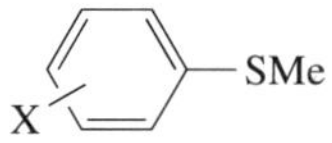 X-C$_6$H$_4$-SMe		57JA376
p-Me		p-Me	83	57JA376
m-Cl		m-Cl	75	57JA376
p-NO$_2$		p-NO$_2$	91	57JA376
p-OMe		p-OMe	79	55JA572
PhCH$_2$SCHClPh	SO$_2$Cl$_2$ (1 equiv.)	PhCH$_2$SCH$_2$Ph	a	51JA5187
PhCH$_2$SCHClPh	SO$_2$Cl$_2$	PhCH$_2$SCH$_2$Ph	90	73TL4395
PhCHClSCHClPh	SO$_2$Cl$_2$	PhCH$_2$SCH$_2$Ph	a	70JOC3002

a Yield not given.

Polychlorination is not generally a problem when one equivalent, or slightly more, of sulfuryl chloride is used, however the use of a higher ratio results in further chlorination of the carbon already carrying a chlorine atom; it is only when this position is completely chlorinated that attack occurs at the other carbon atom ⟨52JA3594⟩. The behaviour of the dibenzyl sulfide (**60**) is atypical in that on treatment with 2 equiv. of sulfuryl chloride appreciable amounts of the α,α'-dichlorinated product is obtained (Scheme 31) ⟨64JA4089⟩; significant amounts of the α,α'-dichlorinated product (**61**) were also obtained when chloromethyl p-nitrobenzyl sulfide was chlorinated, whereas the sulfides with p-F, p-Cl and m-CF$_3$ substituents gave only the α,α-dichlorinated product ⟨68JOC1080⟩.

Although the mono-α-chlorination in tetrachloromethane of dibenzyl sulfide itself ⟨51JA5187, 53LA(581)133⟩ and the cyclic dibenzyl sulfide (**60**) ⟨64JA4085⟩ in a conventional manner has been reported, it was found that running together dilute streams of the sulfide and sulfuryl chloride in tetrachloromethane ⟨73TL4395⟩ was the most effective method of minimising dichlorination. The reaction is not successful for sulfides containing β hydrogen atoms as elimination of hydrogen chloride from the initially formed α-chloro sulfides leads to the formation of a complex mixture of

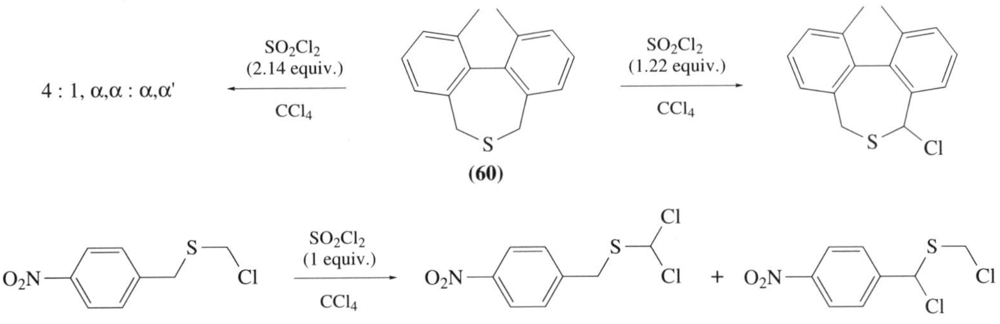

Scheme 31

products. Thus phenyl ethyl sulfide, methyl isopropyl sulfide, diisopropyl sulfide, thiacyclopentane, 2-methylthiacyclopentane, thiacyclohexane and diallyl sulfide all failed to give the required α-chloro sulfide when treated with sulfuryl chloride ⟨55JA572⟩. The formation of 3,4-dihydro-2*H*-thiin (**62**) and 2,3-dichlorothiacyclopentane from thiacyclohexane and thiacyclopentane (Scheme 32), respectively, provides evidence for the importance of this elimination process ⟨55JA572⟩, as does the observation that good yields of 2-bromothiacyclopentane can be obtained when thiacyclopentane is treated with bromine in the presence of 1 equiv. of triethylamine, an effect which was attributed to the suppression of HBr elimination ⟨73JOC2160⟩.

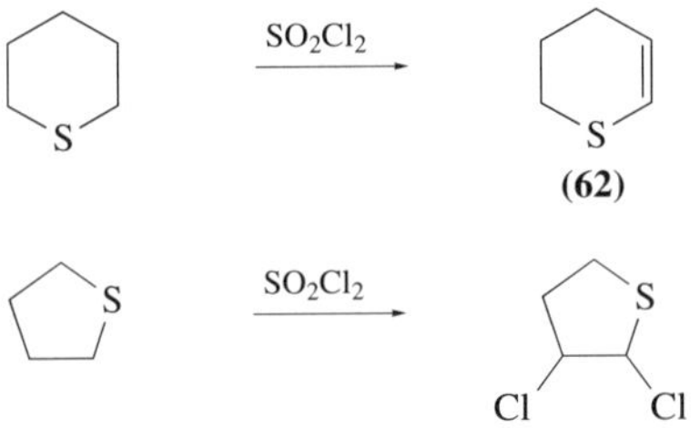

Scheme 32

The use of pyridine was also found to be an effective way of selectively introducing two chlorine atoms into phenyl alkyl sulfides when the reaction was carried out at −5 °C in tetrachloromethane; elimination of hydrogen chloride to form an α-chloro α,β-unsaturated sulfide occurred when the reaction was carried out in refluxing solvent ⟨82CC857⟩. The preparation in tetrachloromethane of the α-chloro sulfide (**63**) (Equation (48)) demonstrates that although the use of sulfuryl chloride has generally been restricted to relatively simple sulfides, it can also be used with more highly functionalised sulfides ⟨83CC1349⟩.

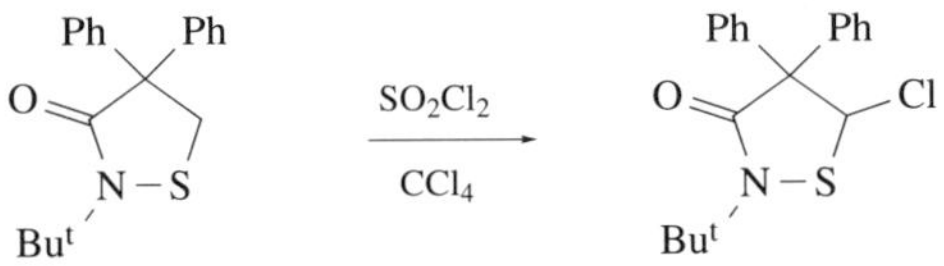

(48)

Although thionyl chloride has not been used extensively as a chlorinating agent, its lower reactivity compared to that of sulfuryl chloride has occasionally been exploited. Thus di(chloromethyl) sulfide was obtained in 75% yield from methyl sulfide and a slight excess of thionyl chloride at 25–90 °C, conditions which avoid the low temperatures required in the sulfuryl chloride reaction; if a larger excess is used, the reagent demonstrates a regioselectivity similar to that of sulfuryl chloride and 1,1,1-trichloromethyl methyl sulfide is obtained in good yield ⟨52JA3594⟩. The reaction of thionyl chloride with 1,3,5-trithiane also gave di(chloromethyl) sulfide in 98% yield (Scheme 30) ⟨52JA3594⟩;

this is not available directly from dimethyl sulfide because, as explained above, complete chlorination of one α-carbon is favoured. *t*-Butyl chloromethyl sulfide has also been prepared using this reagent ⟨55JA572⟩.

Currently, the most commonly used reagent for the direct chlorination of sulfides is *N*-chlorosuccinimide (NCS) (**64**) (Equation (49)) ⟨66CI(L)1555⟩. This solid reagent is more easily handled than sulfuryl chloride, is less reactive and does not produce hydrogen chloride during its reaction with sulfides (Equation (49)). Thus NCS is compatible with a wider range of functional groups, including alkenes, esters, imides, acyl chlorides, anhydrides, amides, trimethylsilyls, acetals, β-lactams, ethers and *N*-BOC groups, many of which are acid-sensitive, and in addition rarely leads to the formation of polychlorinated products (Table 13).

The fact that NCS is soluble in tetrachloromethane whereas its reaction product, succinimide, is not, permits the isolation of the α-chloro sulfide by simple filtration and evaporation of the solvent. The purity of the product obtained in this way is such that α-chloro sulfides required for synthetic purposes are not generally subjected to further purification by chromatography or distillation, but are used directly in tetrachloromethane solution. The preparation of α-chloro sulfides with β hydrogens such as (**65**) ⟨66CI(L)1555⟩, (**66**) ⟨69JHC115⟩ and (**67**) ⟨82TL2399⟩ and reactive α-chloro sulfides such as (**68**) ⟨76BCJ553⟩, without the use of low temperatures, exemplifies the advantages of NCS over sulfuryl chloride.

Allylic sulfides, which can undergo an allylic rearrangement (Equation (50)) ⟨75TL4433⟩, and certain small ring compounds which can give ring-opened products (Equation (51)) ⟨82CL587⟩ are among the few sulfides which do not undergo simple α-chlorination with NCS. Although there are some differences in reactivity ⟨69JOC31⟩, the mechanism by which NCS functions is believed to be the same as that involving sulfuryl chloride ⟨86T3731, 91COS(7)206⟩, and thus the factors controlling the regiochemistry of α-chlorination should be effectively the same in both cases. Although many of these factors became clear from experiments involving sulfuryl chloride (see above), the wider range of sulfides which could be chlorinated with NCS allowed them to be defined more precisely. Thus it is found that the major, or only, product obtained when unsymmetrical sulfides are chlorinated using NCS is that obtained by replacement of the more acidic α hydrogen with chlorine (Scheme 33) ⟨69JOC31⟩.

Table 13 Synthesis of α-chloro sulfides by direct chlorination of sulfides with *N*-chlorosuccinimide (NCS).

Product	Substrate	Yield (%)	Ref.
TMS-O–CHCl–SPh	TMS-O–CH2–SPh	>95	82TL5083
PhS–CHCl–CO2Me	PhS–CH2–CO2Me	a	83TL327
(bicyclic thiepane with Cl α to S)	(bicyclic thiepane)	100	83JA667
3-PhS-3-Cl-succinic anhydride	3-PhS-succinic anhydride	95	83JOC1096
MeS–CCl2–C(O)Cl (α-Cl)	MeS–CH2–C(O)Cl	82	81S534
Ph-N(CH2C≡C-TMS)-C(O)-CHCl-SMe	Ph-N(CH2C≡C-TMS)-C(O)-CH2-SMe	100	89JCS(P1)879
MeO-cyclohexenyl-(methylenedioxyphenyl)-N(Me)-C(O)-CHCl-SPh	MeO-cyclohexenyl-(methylenedioxyphenyl)-N(Me)-C(O)-CH2-SPh	100	93JOC2360
β-lactam with CHCl–SPh side chain, MeO, N-CO2CH2CH2TMS	β-lactam with CH2–SPh side chain	a	83TL139
methylenedioxyphenyl-N(Ts)-CH2CH2-S-CHCl-CO2Et	methylenedioxyphenyl-N(Ts)-CH2CH2-S-CH2-CO2Et	a	86JHC1163

^a Yield not given.

Studies of the regioselectivity of chlorination with sulfides such as methyl ethyl sulfide and methyl isopropyl sulfide (Scheme 34) reveal a preference for reaction at the internal position (Scheme 34) ⟨69JOC31⟩. As there is no marked difference in the acidities of the respective α-hydrogens, it has been suggested that the capacity to stabilise a carbocation may also be a factor in determining the regioselectivity of chlorination in these cases ⟨69JOC31⟩. The strong directing effect of a chlorine atom, which has already been noted for sulfuryl and thionyl chlorides, operates for NCS as well

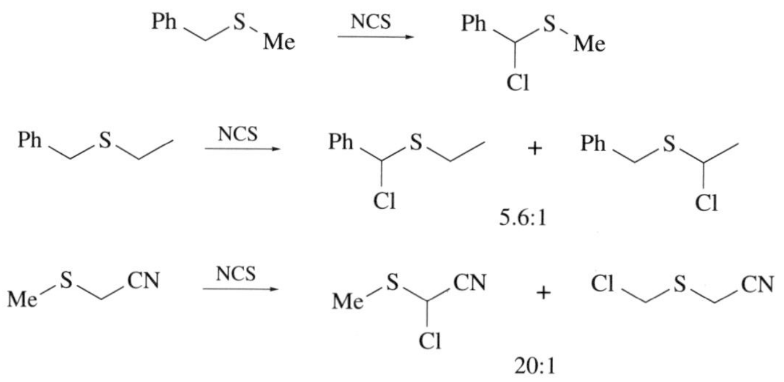

Scheme 33

(Equation (52)) ⟨69JOC31⟩; as with sulfuryl chloride there are exceptions to this (Equation (53)) ⟨86TL757⟩. N-Chlorosuccinimide has also been used to carry out the chlorinative decarboxylation of phenylthioacetic acid giving chloromethyl phenyl sulfide in good yield ⟨74JOC2516⟩.

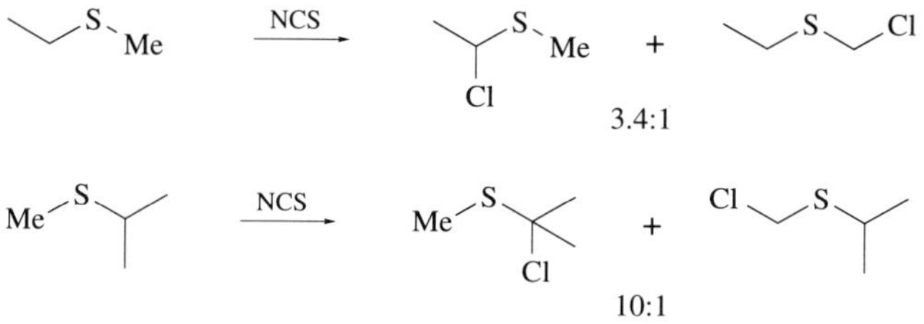

Scheme 34

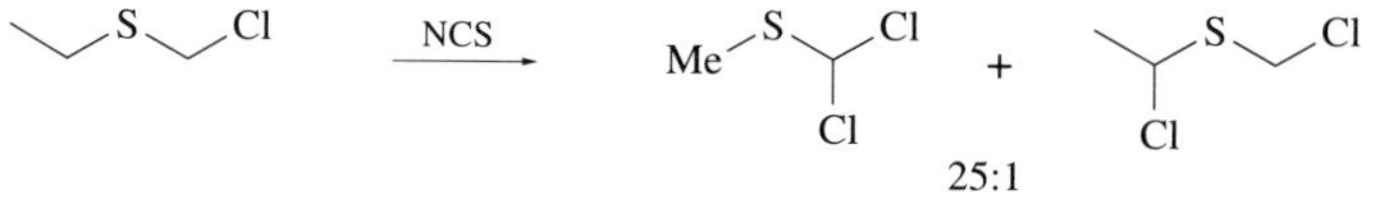

$$\text{(52)}$$

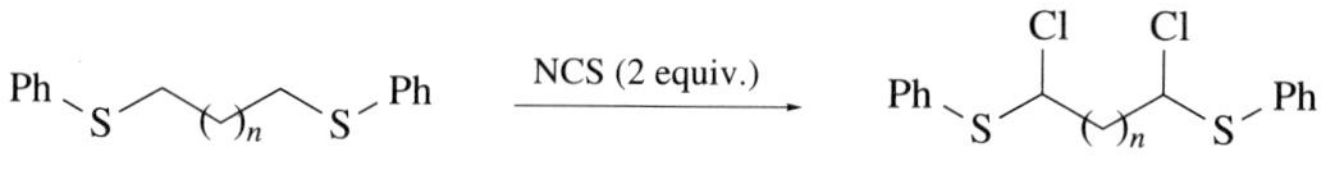

$$\text{(53)}$$

$n = 0, 1, 2$

The direct α-chlorination of sulfides has also been achieved with trichloroisocyanuric acid (Chloreal) (**70**) (Scheme 35) ⟨75TL4433⟩, (dichloroiodo)benzene ⟨61JOC2478⟩ and benzenesulfenyl chloride ⟨70BCJ1223⟩. It would appear that of these only Chloreal has general synthetic utility: it was found to chlorinate allylic sulfides such as phenyl crotyl sulfide (**69**) more efficiently though less stereoselectively than NCS ⟨75TL4433⟩ and was also used to prepare the chlorocyclopropyl sulfide (**71**) (Scheme 35) ⟨81TL2455⟩. N-Chlorosuccinimide is reported to be a more efficient reagent than Chloreal for simple alkyl phenyl sulfides ⟨86T3731⟩.

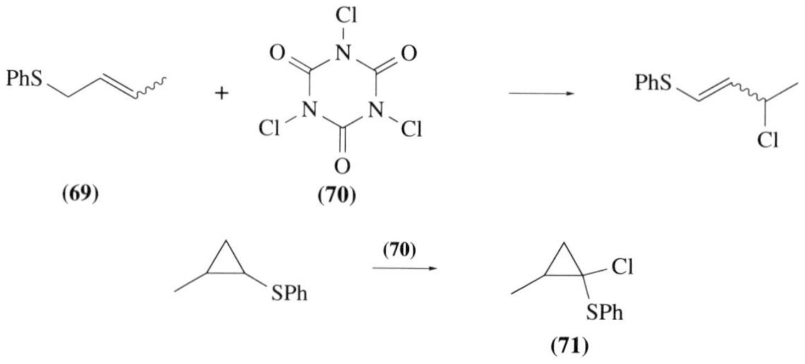

Scheme 35

Thionyl chloride, benzoyl chloride and acetyl chloride have been used to convert simple alkyl and alkyl aryl sulfoxides into α-chloro sulfides (Scheme 36) ⟨55JA572⟩; hydrogen chloride in the presence of molecular sieves has been used for the same purpose ⟨71TL3553⟩, but in general, as many sulfoxides are prepared from the corresponding sulfide, the method is less important than direct sulfide chlorination.

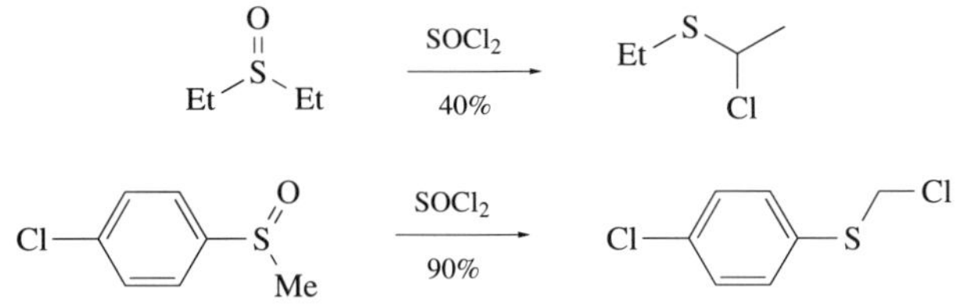

Scheme 36

The reaction of thiols with aldehydes in the presence of dry hydrogen chloride is a versatile way to prepare relatively simple α-chloro sulfides (Table 14) ⟨36CB1610, 49LA(563)54, 64CB179, 67JA4483⟩. The use of paraformaldehyde leads to the formation of primary chloro sulfides (Equation (54)) ⟨45JA655⟩, whereas the use of paraldehyde or higher aldehydes gives secondary chlorosulfides (Equation (55)) ⟨45JA657⟩. A number of improvements in the original procedure have been suggested ⟨45JA655, 45JA657⟩. The method is particularly useful as there is no possibility of a mixture of regioisomers being formed, the chlorine invariably appearing on what was the carbonyl carbon of the aldehyde. Thus, although NCS gives a mixture of the α-chloro sulfides (**72**) and (**73**), the thiol–aldehyde reaction gives the sulfide (**72**) exclusively (Scheme 37) ⟨67JOC204⟩. A mechanistically related process is the use of hydrogen chloride in dichloromethane to convert the benzoate (**74**) into the α-chloro sulfide (**75**) ⟨75JA5957⟩; the reaction occurs with retention of configuration and appears to be general for benzoates of this type ⟨78JA3933⟩.

Table 14 Synthesis of α-chloro sulfides by the reaction of aldehydes and thiols in the presence of hydrogen chloride.

Product	Reagent	Substrate	Yield (%)	Ref.
EtSCH$_2$Cl	Paraformaldehyde	EtSH	50	69JOC31
Pr^{i}SCH$_2$Cl	Paraformaldehyde	Pr^{i}SH	63	69JOC31
EtSCHClPr^n	Pr^{n}CHO	EtSH	'Excellent'	45JA657
EtSCHClCHEt$_2$	Et$_2$CHCHO	EtSH	'Excellent'	45JA657
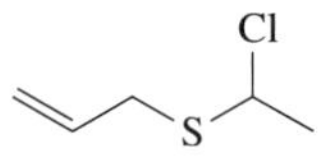 allyl-S-CHClMe	Paraldehyde	 allyl-SH	'Excellent'	45JA657
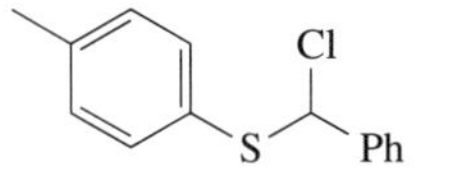 4-MeC$_6$H$_4$-S-CHClPh	PhCHO	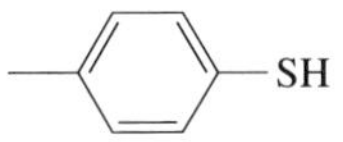 4-MeC$_6$H$_4$-SH	a	67JA4483
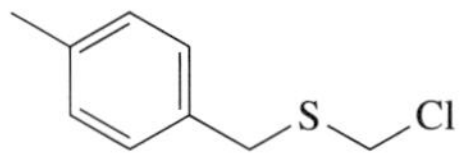 4-MeC$_6$H$_4$-CH$_2$-S-CH$_2$Cl	Paraformaldehyde	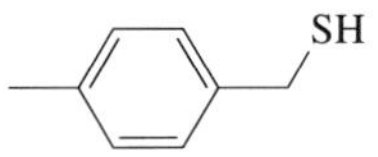 4-MeC$_6$H$_4$-CH$_2$SH	77	68JOC1080

^{a} Yield not given.

$$\text{Pr}^n\text{SH} + (\text{CH}_2\text{O})_n + \text{HCl} \longrightarrow \text{Pr}^n\text{SCH}_2\text{Cl} + \text{H}_2\text{O} \qquad (54)$$

$$\text{Bu}^n\text{SH} + \text{EtCHO} + \text{HCl} \longrightarrow \text{Bu}^n\text{SCHClEt} + \text{H}_2\text{O} \qquad (55)$$

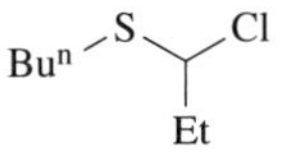

The reaction of various sulfenyl chlorides, particularly benzenesulfenyl chloride, with α-diazo ketones ⟨55ZN296, 83TL117⟩, α-diazo esters ⟨60CB2340⟩ and 2-diazo-1,3-dicarbonyl compounds (Scheme 38) ⟨60CB2340⟩ allows α-chloro sulfides to be prepared by the simultaneous introduction

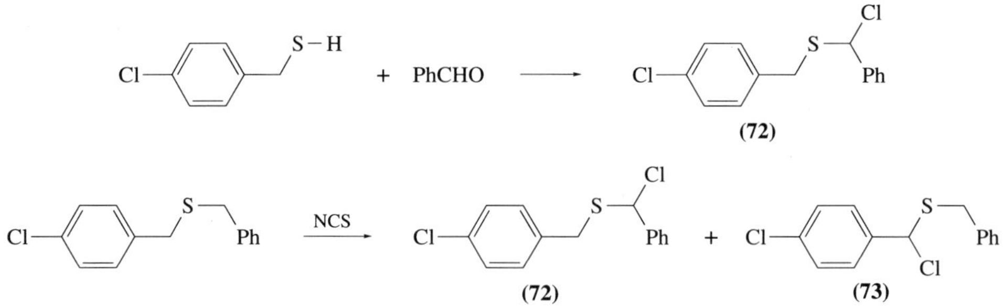

Scheme 37

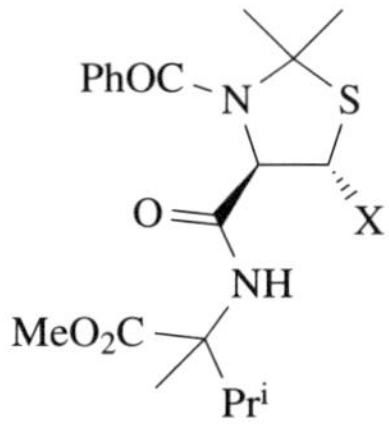

(74) X = OCOPh
(75) X = Cl

of the chlorine and sulfur atoms; the reaction has also been used with α-diazo lactams (Equation (56)) ⟨78JOC2203⟩. The 1,3-diketone dimedone reacts directly with benzenesulfenyl chloride to give the α-chloro sulfide shown in Equation (57). A similar reaction has been reported for methanesulfenyl chloride ⟨81JOC4911⟩.

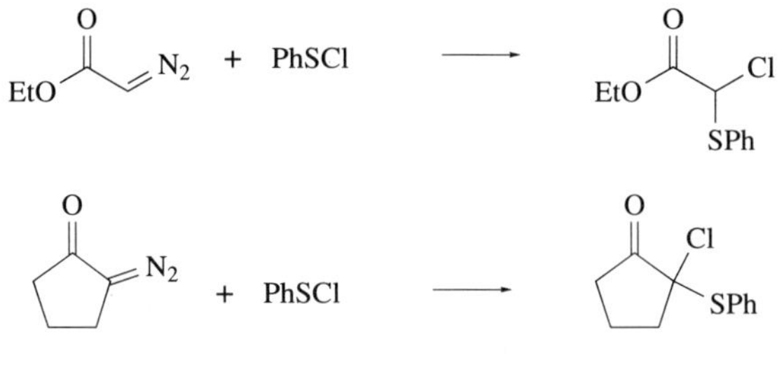

Scheme 38

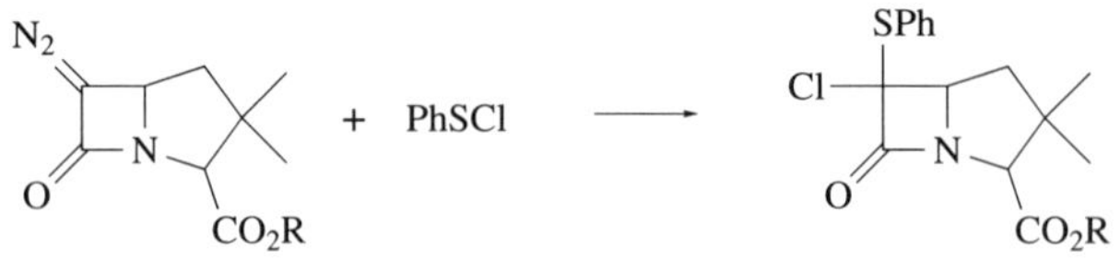

(56)

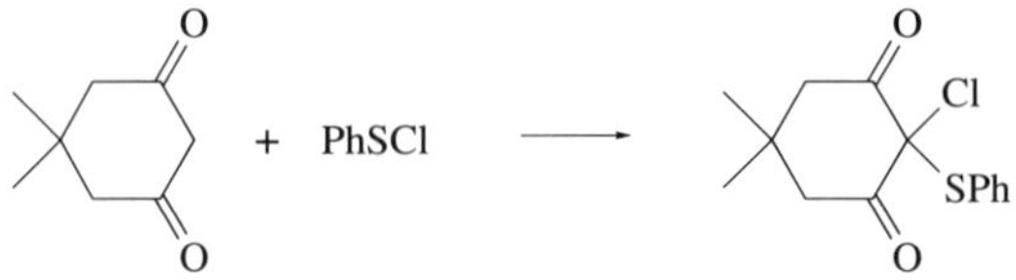

(57)

The chlorine attached to the sulfur in α-chloro-α-(chlorosulfenyl)carboxylic esters is susceptible to substitution by malonate-type anions giving α-chloro sulfides ⟨80TL3579, 81JA2757⟩.

(iii) α-Bromo sulfides, $R^1_2CBr(SR^2)$

α-Bromo and α-iodo sulfides are more reactive but less stable than the corresponding α-chloro compounds and thus have been used less frequently as synthetic intermediates. However most of the methods used for α-chloro sulfides have been applied with varying degrees of success to the synthesis of α-bromo sulfides (Table 15). Thus the direct bromination of sulfides using bromine ⟨70JOC3002, 81AG(E)585⟩ and NBS ⟨70JOC3002⟩ and the reaction of aldehydes with thiols in the presence of hydrogen bromide ⟨86JOC2981⟩ have all been used. N-Bromosuccinimide has also been used to carry out the brominative decarboxylation of α-phenylthio carboxylic acids ⟨74JOC2516⟩.

Table 15 Synthesis of α-bromo sulfides.

Product	Reagent	Substrate	Yield (%)	Ref.
MeSCHBrCO₂Me	Br₂, hv	MeSCH₂CO₂Me	>40	81AG(E)585
MeCH₂SCHBrCO₂Me	Br₂, hv	MeCH₂SCH₂CO₂Me	>71	81AG(E)585
PhCH₂SCH₂Br	HBr, (HCHO)ₙ	PhCH₂SH	72	86JOC2981
PhCH₂SCHBrPh	Br₂, CCl₄	PhCH₂SCH₂Ph	a	70JA3002
PhCH₂SCHBrPh	NBS	PhCH₂SCH₂Ph	a	70JA3002
PhCH₂SCHBrCO₂Me	Br₂, hv	PhCH₂SCH₂CO₂Me	>79	81AG(E)585

[a] Yield not given.

(iv) α-Iodo sulfides, $R^1_2CI(SR^2)$

Iodomethyl phenyl sulfide has been used as an alkylating agent, being prepared *in situ* by the reaction of the commercially available chloromethyl methyl sulfide with lithium iodide in THF/hexamethylphosphoramide ⟨73TL3831⟩ or THF ⟨80JOC752⟩, or immediately before use by the reaction with sodium iodide in acetone ⟨74JOC2648⟩.

4.02.2.1.2 Other dicoordinate α-halo sulfur derivatives, $R_2CHal(SX)$

(i) α-Chloroalkanesulfenyl chlorides, $R_2CCl(SCl)$

α-Chloroalkanesulfenyl chlorides, which are highly coloured, have been prepared from trithianes and chlorine at low temperature ⟨50JOC795⟩ or from disulfides and chlorine or sulfuryl chloride also at low temperature (Scheme 39) ⟨50CB87, 54CB300, 79JOC1177⟩.

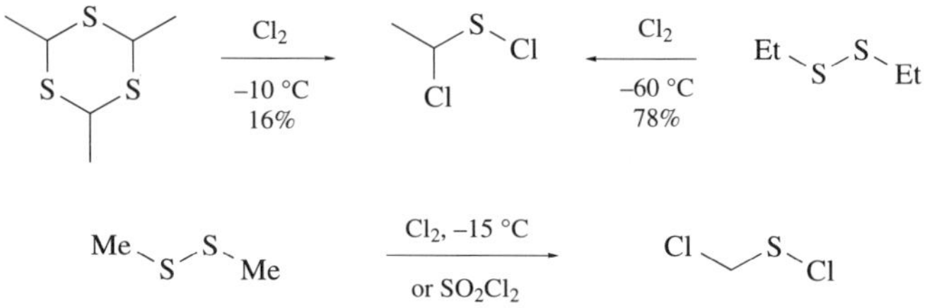

Scheme 39

An alternative approach involves the reaction of active methylene compounds with thionyl chloride under basic conditions (Scheme 40) ⟨77TL695⟩. Arylacetic and β,γ-unsaturated acids react under the same conditions to give α-chloro-α-(chlorosulfenyl)acyl chlorides (Scheme 40) ⟨67JA5838, 75JOC3037⟩, whereas acetophenone and methyl *t*-butyl ketone give an equimolar mixture of the expected sulfenyl chloride (**76**) and a thioformyl chloride (**77**) (Scheme 40) ⟨76TL2783⟩. The α-thiolacetanilide (**78**) also gives an α-chloroalkanesulfenyl chloride on being treated with thionyl chloride; however, no base is required (Equation (58)) ⟨72JOC1526⟩. The reaction is believed to involve the formation of an intermediate sulfinyl chloride (**79**) which then undergoes a Pummerer-type process giving the α-chloroalkanesulfenyl chloride (Scheme 41). Phenylacetonitrile and other nitriles having two α-hydrogens give a similar reaction in the presence of anhydrous hydrogen chloride but less cleanly, the product having to be separated from a number of minor by-products (Equation (59)) ⟨75JOC3540⟩.

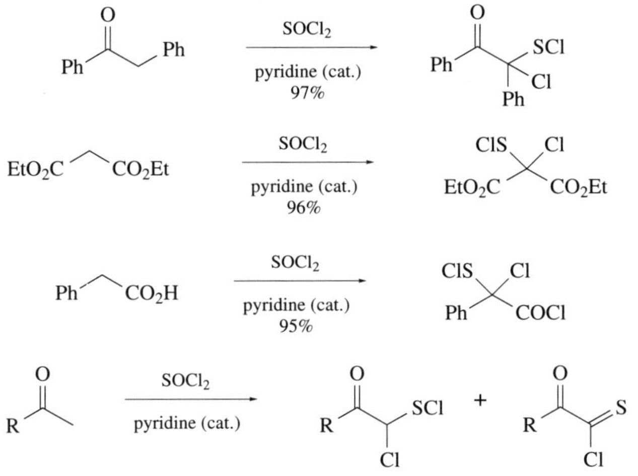

R = Ph, Bu^t **(76)** **(77)**

Scheme 40

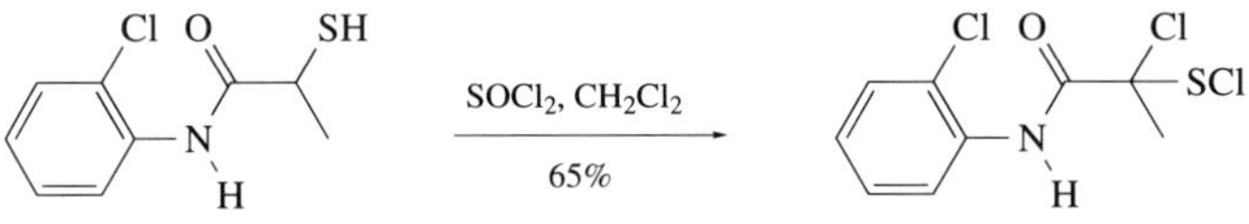 (58)

(78)

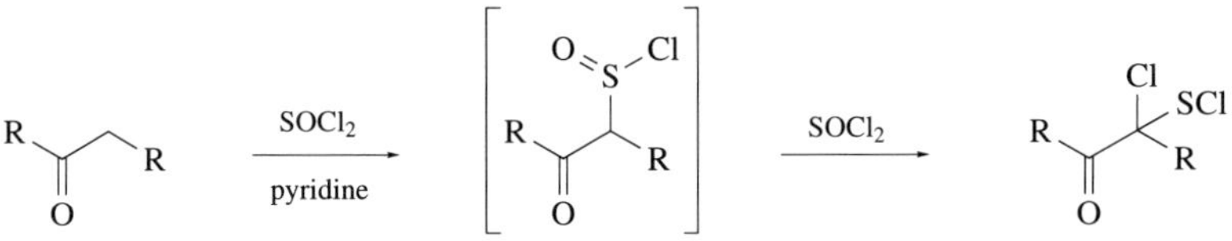

(79)

Scheme 41

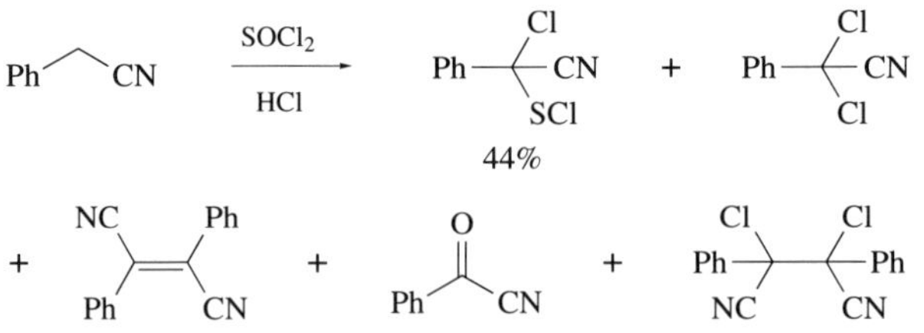 (59)

(ii) Miscellaneous dicoordinate α-halo sulfur derivatives, $R_2CHal(SX)$

α-Chloroalkanesulfenyl chlorides react with primary and secondary amines to give the corresponding sulfenamides (Equation (60)) ⟨77TL695⟩. 1-Chloro-1-(chlorosulfenyl)propanoyl chloride reacts with sodium ethoxide to give a 1-chloro-1-(chlorosulfenyl) ester which can be converted into a thiol phosphate (**80**) and a disulfide (**81**) by reactions with triethyl phosphite and triphenylphosphine, respectively (Scheme 42) ⟨79JOC1736⟩.

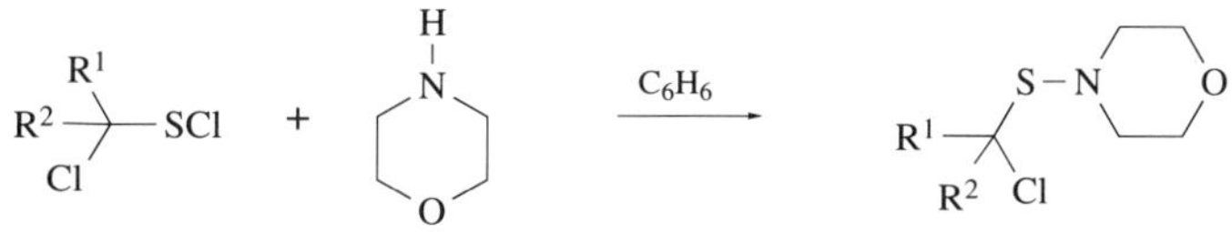

$$R^2 = Ph, R^1 = PhCO, 97\%$$
$$R^2 = Me, R^1 = PhCO, 98\%$$
$$R^2 = R^1 = CO_2Et, 97\%$$

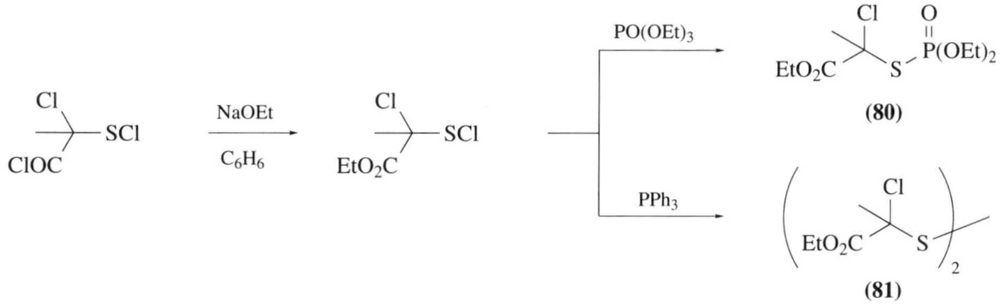

Scheme 42

4.02.2.2 Tricoordinate α-Halo Sulfur Derivatives, $R^1_2CHalS(O)R^2$, etc.

4.02.2.2.1 α-Halo sulfoxides, $R^1_2CHalS(O)R^2$

The early work on the synthesis and reactions of α-halo sulfoxides has been reviewed ⟨74OPP77⟩. Other developments, particularly relating to the use of 1-haloalkyl aryl sulfoxides in organic synthesis, have also been reviewed ⟨92SL455⟩. The most widely used procedure for the synthesis of these compounds begins with the appropriate sulfide and involves a chlorination–oxidation sequence or the reverse.

(i) α-Fluoro sulfoxides, $R^1_2CHalS(O)R^2$

The reaction of DAST (diethylaminosulfur trifluoride) with sulfoxides results in the formation of α-fluoro sulfides and so the usual route to α-fluoro sulfoxides involves the fluorination of a sulfide followed by oxidation, the most commonly used reagent for the latter step being mcpba ⟨92SL455⟩. Aryl fluoromethyl sulfoxides ⟨85JA735, 88TL3365, 90JOC4757⟩ and α-fluoroalkyl aryl sulfoxides ⟨85JA735, 89JA1127, 90JOC4757, 91JOC6878, 91TL1463⟩ (Scheme 43) have been prepared with this reagent in good to excellent yield. α-Fluoro sulfide groups in nucleosides have also been successfully oxidised using mcpba ⟨89JA1127⟩, although in one case it is reported that some over-oxidation to the α-fluoro sulfone occurred (Equation (61)) ⟨91JOC6878⟩. N-Bromosuccinimide in aqueous dioxan, methanol or THF ⟨77S791⟩, and tbhp in the presence of vanadyl acetylacetonate (Scheme 44) ⟨91TL1463⟩, have also been used to oxidise α-fluoro sulfides, although the latter reagents give consistently lower yields than mcpba.

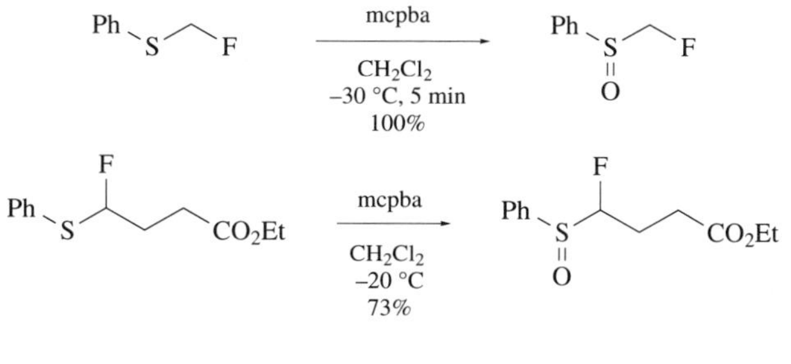

Scheme 43

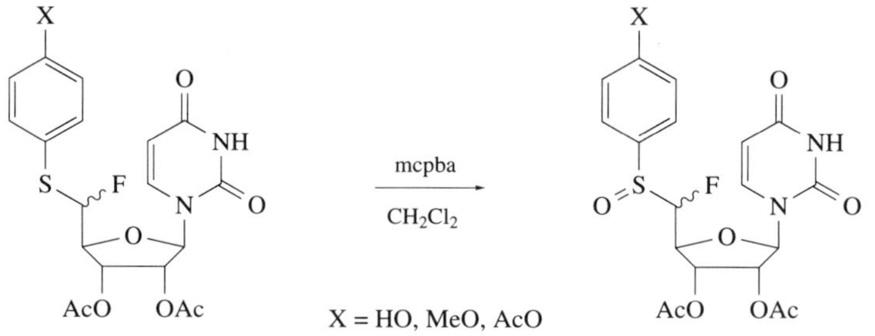

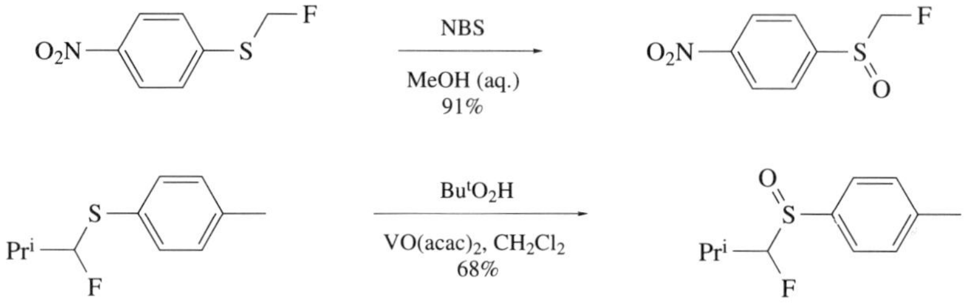

Scheme 44

The lithiation of phenyl fluoromethyl sulfoxides gives α-sulfinyl carbanions which, as described below for the corresponding chloro derivatives (Section 4.02.2.2.1(ii)), can be alkylated with alkyl halides or added to ketones and aldehydes, giving more highly substituted and functionalised derivatives ⟨83TL725⟩. The two diastereomers produced in addition reactions with aldehydes and unsymmetrical ketones can be separated chromatographically ⟨92TL1483⟩.

(ii) α-Chloro sulfoxides, $R^1{}_2CClS(O)R^2$

The older oxidation methods used to convert α-chloro sulfides into the corresponding sulfoxides have been reviewed ⟨73JOC17⟩. More modern methods (Table 16) include the use of ozone, which was employed to oxidise chloromethyl phenyl sulfide to the sulfoxide (Table 16, entry 1) ⟨64JA4645⟩, but more particularly mcpba (Table 16, entry 2) ⟨69JA1034⟩, which has become one of the most widely used reagents for this purpose ⟨85BCJ1983⟩. A general problem with this route to α-chloro sulfoxides can be the formation, due to over-oxidation, of sulfones, which can be difficult to remove from the product. A number of methods have been promoted on the basis that a sulfone-free product is obtained, including the vanadium pentoxide-catalysed reaction with hydrogen peroxide in *t*-butanol (Table 16, entry 3) ⟨69JCS(C)2334⟩, the use of chlorine in aqueous acetic acid (Table 16, entry 4) ⟨77CJC421⟩ and the use of sulfuryl chloride in the presence of wet silica (Table 16, entry 5) ⟨76TL613⟩. The use of chlorine does result in some C—S bond cleavage, a process which, like over-oxidation, can result in the formation of by-products; in this case, however, the sulfonyl chlorides which are formed can easily be removed. The direct conversion of a sulfide into an α-chloro sulfoxide has been achieved with sulfuryl chloride in the presence of a metal salt (Table 16, entries 6 and 7) ⟨93S209⟩, the initial step being chlorination. Polychlorination can be a problem with this approach if more than 1 equiv. of sulfuryl chloride is used.

Although the chlorination of sulfoxides avoids the need to use unpleasant-smelling sulfides and the possibility of sulfone formation, C—S bond cleavage can be a problem if a stable carbocation can be formed as a result. A very wide range of chlorinating systems has been used: nitrosyl chloride ⟨68TL5415⟩, toluene-*p*-sulfonyl chloride in pyridine ⟨68JA4496⟩, *t*-butyl hypochlorite ⟨69TL5259⟩, chlorine ⟨70BCJ2271, 70JPR683⟩ sulfuryl chloride in the presence of calcium oxide ⟨70TL4643⟩, sulfuryl chloride in the presence of pyridine ⟨71S89⟩, NCS ⟨71BCJ1726⟩, *N*-chlorobenzotriazole ⟨72S259⟩, (dichloroiodo)benzene in the presence of pyridine ⟨72JCS(P2)296⟩ and NCS–silica ⟨86S831⟩. *N*-Chlorosuccinimide and sulfuryl chloride are currently the most commonly used methods ⟨92SL455⟩. The cleavage of a C—S bond has been observed with some of these reagents for molecules which can give a stable carbocation on cleavage. Thus, benzhydryl benzyl sulfoxide reacts with sulfuryl

Table 16 Oxidation of α-chloro sulfides to α-chloro sulfoxides.

Entry	Product	Reagent	Substrate	Yield (%)	Ref.
1	PhS(O)CH$_2$Cl	O$_3$, −78 °C, CH$_2$Cl$_2$	PhSCH$_2$Cl	74	64JA4645
2	PhS(O)CH$_2$Cl	mcpba	PhSCH$_2$Cl	70	69JA1034
3	C$_{12}$H$_{25}$S(O)CH$_2$Cl	V$_2$O$_5$, H$_2$O$_2$, Bu^{t}OH	C$_{12}$H$_{25}$SCH$_2$Cl	69	69JCS(C)2334
4	ClCH$_2$S(O)CH$_2$Cl	Cl$_2$, AcOH, H$_2$O	ClCH$_2$SCH$_2$Cl	77	77CJC421
5	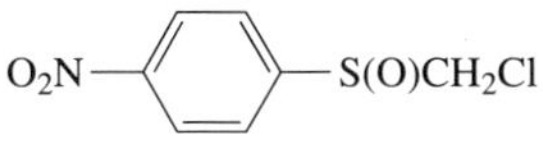 O$_2$N—C$_6$H$_4$—S(O)CH$_2$Cl	SO$_2$Cl$_2$, SiO$_2$	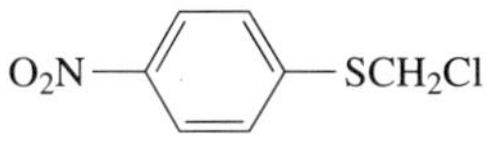 O$_2$N—C$_6$H$_4$—SCH$_2$Cl	100	76TL613
6	MeS(O)CH$_2$Cl	SO$_2$Cl$_2$, AgNO$_3$, MeCN	Me$_2$S	61	93S209
7	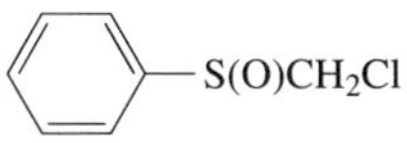 C$_6$H$_5$—S(O)CH$_2$Cl	SO$_2$Cl$_2$, AgNO$_3$, MeCN	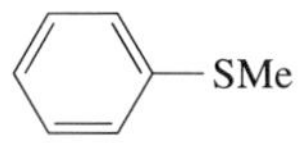 C$_6$H$_5$—SMe	88	93S209

chloride in the presence of calcium oxide or pyridine to give products resulting from such a cleavage (Scheme 45) ⟨73TL289⟩. A similar pattern of behaviour was observed for NCS (Scheme 45) ⟨73CC4⟩.

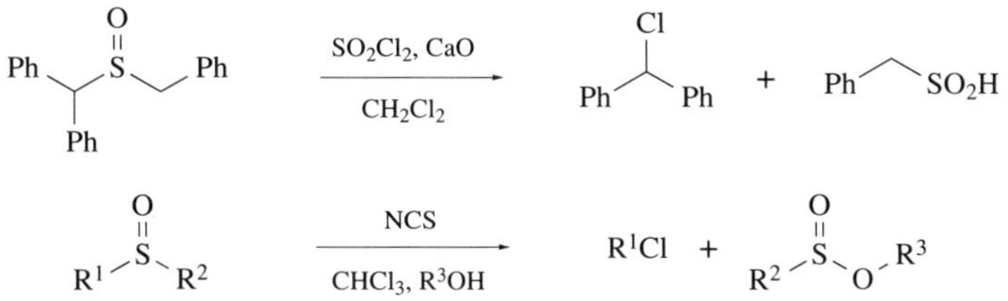

$$Ph_2CH-S(O)-CH_2Ph \xrightarrow{SO_2Cl_2, CaO, CH_2Cl_2} Ph_2CHCl + PhCH_2SO_2H$$

$$R^1-S(O)-R^2 \xrightarrow{NCS, CHCl_3, R^3OH} R^1Cl + R^2-S(O)-O-R^3$$

R^1 = Bu^t, Bn, PhCHMe
R^2 = 2° or 1° alkyl

Scheme 45

The chlorination of unsymmetrical sulfoxides results in many cases in the formation of a mixture of regioisomers (Table 17), although a number of reagents do regioselectively chlorinate the benzylic position of benzyl methyl sulfoxide. The reaction of diazo compounds with sulfinyl halides (see below) may be the method of choice if control of regioselectivity is a problem. Although chlorination of sulfoxides which have a prochiral α-carbon may lead to a mixture of diastereomers with many of these reagents (Equation (62)) ⟨92SL455⟩, halogenation with a number of other reagents occurs with high diastereospecificity ⟨74JOC643⟩. Thus, halogenation of sulfoxides such as ethyl phenyl sulfoxide with (dichloroiodo)benzene, a controlled source of chlorine, gives a single diastereomer, a similar result being obtained with N-chlorobenzotriazole and also bromine in pyridine ⟨72JCS(P1)1883⟩. In addition, when an optically active sulfoxide such as (+)-methyl p-tolyl sulfoxide was halogenated with (dichloroiodo)benzene, N-chlorobenzotriazole, or bromine in pyridine, the stereochemical outcome of the reaction at the sulfur atom was altered by the addition of silver nitrate, proceeding with retention in the absence of the salt and with inversion in its presence ⟨72JCS(P1)1886⟩.

Table 17 Regioselectivity of α-chloro sulfoxide formation.

Method	Substrate	PhCHClS(O)Me[a]	PhCH$_2$S(O)CH$_2$Cl[a]	Ref.
Bu^{t}OCl, Pyridine	PhCH$_2$S(O)Me	15	45	69TL5259
N-Chlorobenzotriazole	PhCH$_2$S(O)Me	37	44	72S259
PhICl$_2$, pyridine	PhCH$_2$S(O)Me	30	32	72JCS(P2)296
SO$_2$Cl$_2$, CaO	PhCH$_2$S(O)Me	25	0	70TL4643
Bu^{t}OCl, KOAc	PhCH$_2$S(O)Me	40	0	69TL5259
CH$_2$N$_2$	PhCH$_2$S(O)Cl	0	100	73JOC17
PhCHN$_2$	MeS(O)Cl	49	0	73JOC17

[a] Yield (%).

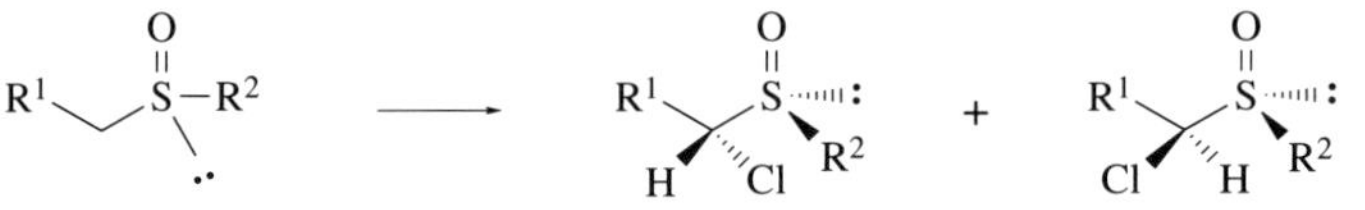

$$R^1\underset{..}{\overset{O}{\underset{\|}{S}}}{-}R^2 \longrightarrow R^1\underset{H}{\overset{O}{\underset{\|}{\underset{Cl}{S}}}}\!\!\!\!: \; R^2 \;+\; R^1\underset{Cl}{\overset{O}{\underset{\|}{\underset{H}{S}}}}\!\!\!\!: \; R^2 \quad (62)$$

This sensitivity to the conditions of halogenation is general, and other reagents give products with varying degrees of retention, inversion and racemisation ⟨73JA7431, 86S831⟩. Optically active 1-chloromethyl *p*-tolyl sulfoxide has been obtained in high chemical yield and enantiomeric excess by the chlorination of the optically pure sulfoxide (**82**) with NCS in dichloromethane containing potassium carbonate (Equation (63)) ⟨88TL313⟩.

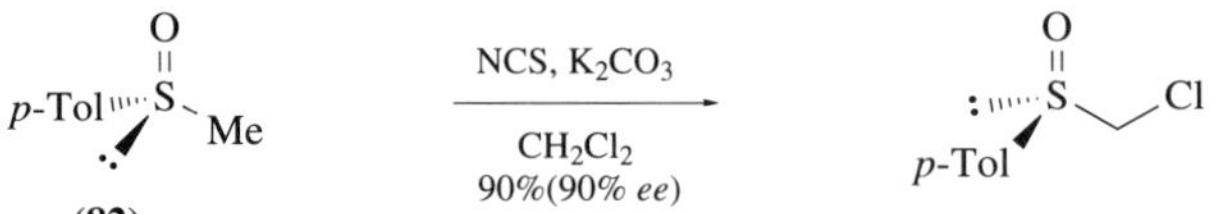

$$p\text{-Tol}\overset{O}{\underset{..}{S}}{-}\text{Me} \quad\xrightarrow[\substack{\text{CH}_2\text{Cl}_2 \\ 90\%(90\%\ ee)}]{\text{NCS, K}_2\text{CO}_3}\quad :\overset{O}{\underset{p\text{-Tol}}{S}}\!\!\!\diagdown\!\!\text{Cl} \quad (63)$$
(**82**)

Diazo compounds insert into the S—Cl bond of sulfinyl chlorides to give α-chloro sulfoxides, the first reported example being the reaction of diazomethane with thionyl chloride (Scheme 46) ⟨48JCS699⟩. More recently the reaction of diazomethane with alkane- and arenesulfinyl chlorides has been used to synthesise α-chloromethyl sulfoxides in high yield; other diazo compounds have also been used, but in these cases the yields are significantly lower (Scheme 46) ⟨73JOC17⟩. A major advantage of this method is that it gives unambiguously a single α-chloromethyl sulfoxide, whereas chlorination of methyl alkyl sulfoxides with many reagents gives a mixture of regioisomers (Table 17); the regioselectivity of the insertion reaction may in some cases complement that of sulfoxide chlorination (Table 17; Scheme 47 ⟨70TL4643, 73JOC17⟩). Thus, the reaction of diazomethane with sulfinyl halides is the method of choice for the preparation of α-chloromethyl sulfoxides, and, even though the yields are lower than with diazomethane, the insertion reactions of other diazo compounds may constitute the best method of preparing, regioselectively, a particular α-chloro sulfoxide.

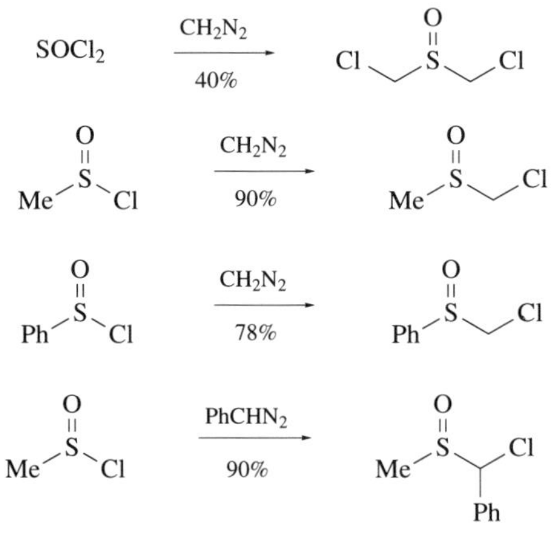

Scheme 46

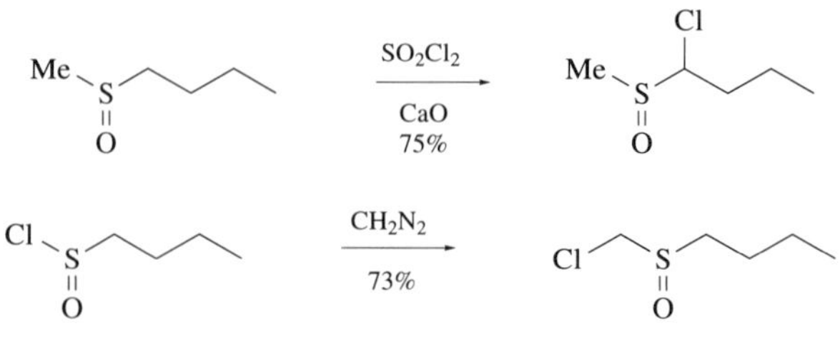

Scheme 47

The acidic nature of the α hydrogens of α-chloro sulfoxides facilitates the formation of α-sulfinyl carbanions, which can undergo substitution and addition reactions, forming a wide range of other α-chloro sulfoxides ⟨92SL455⟩. Thus deprotonation of chloromethyl phenyl sulfoxide gives an α-sulfinyl carbanion (**83**), which reacts with ketones and aldehydes to give adducts in good to excellent yield (Scheme 48) ⟨69JA1034, 79TL617, 92TL1483, 93TL2331⟩. The reaction is stereospecific, being completely under the control of the stereochemistry of the sulfur atom, and gives a single diastereomer with symmetrical ketones and only two with aldehydes or unsymmetrical ketones. In a similar fashion the α-sulfinyl carbanion (**83**) can be alkylated with alkyl halides (Scheme 48) ⟨77TL1225⟩.

Chlorine has been added to divinyl sulfoxide to give a tetrachloro adduct ⟨31JCS1913⟩.

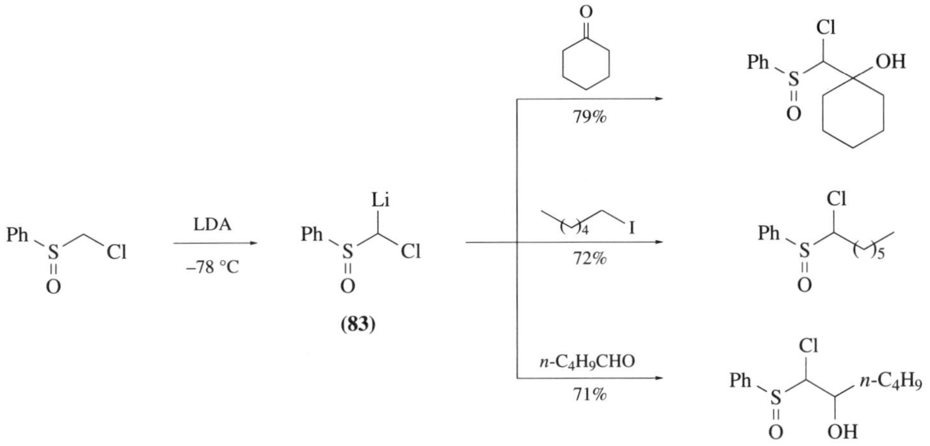

LDA = lithium diisopropylamide

Scheme 48

(iii) α-Bromo sulfoxides, $R^1{}_2CBrS(O)R^2$

Most of the methods used for the synthesis of α-chloro sulfoxides have, appropriately modified, been applied to the preparation of the α-bromo compounds. Thus, a mixture of bromine and NBS in pyridine ⟨70S588, 71BCJ1726⟩, and bromine alone in pyridine ⟨72JCS(P1)1883⟩, have been used to brominate sulfoxides. The insertion of diazomethane into the S—Br bond of trichloromethanesulfinyl bromide gives the α-bromo sulfoxide (**84**) (Equation (64)) ⟨68ACS3256⟩. The bromination reactions are subject to the same constraints as applied to chlorination: they are not regioselective (Equation (65)) ⟨72JCS(P1)1883⟩ and can lead to C—S bond cleavage if this results in a stable carbocation ⟨73CC4⟩. The facility to generate α-sulfinyl carbanions from these compounds has been used, as for the chloro analogues, to synthesise other α-bromo sulfoxides by substitution and addition reactions ⟨79CL209⟩.

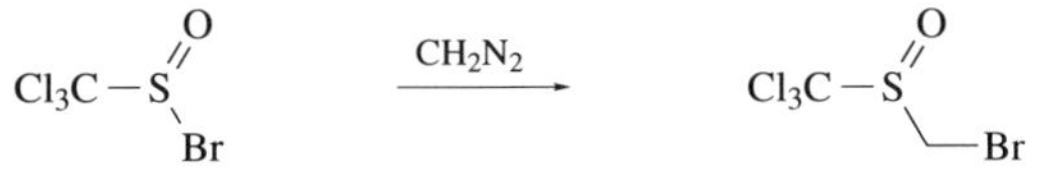

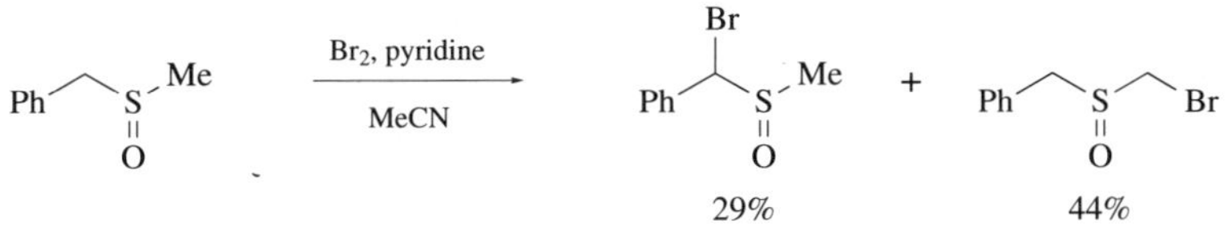

(iv) α-Iodo sulfoxides, $R^1{}_2CIS(O)R^2$

Despite the inertness of α-halo sulfoxides to nucleophilic attack, α-iodo sulfoxides have been synthesised by the standard halogen-interchange reaction in acetone ⟨64JA4645⟩; a more effective procedure involves DMSO as solvent and this has been used to prepare a range of iodomethyl aryl sulfoxides ⟨76S697⟩ (Equation (66)). Alkyl and aryl iodomethyl sulfoxides have been prepared from the appropriate sulfinyl chloride and diazomethane in THF, modified by the inclusion of an alkali-metal iodide salt ⟨73CC319⟩ (Equation (67)).

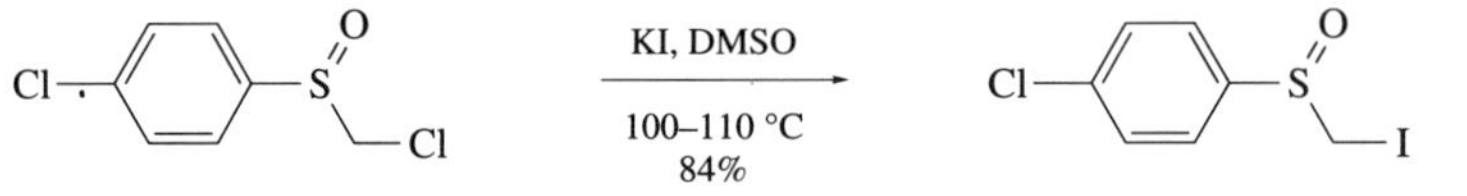

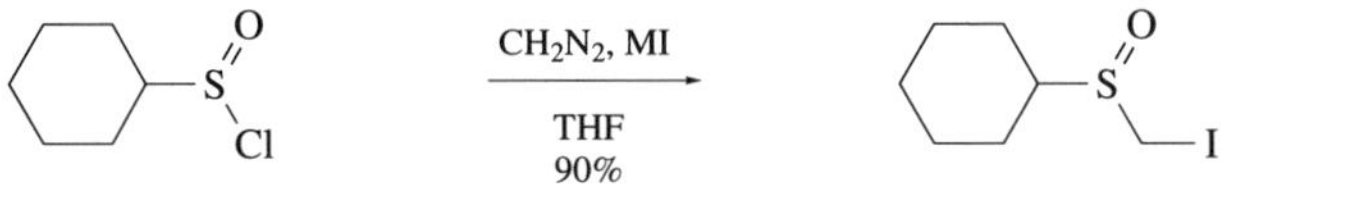

4.02.2.2.2 Other tricoordinate α-halo sulfur derivatives, R_2CHalS(O)X

The general method for the synthesis of sulfinyl chlorides involves the oxidation of the appropriate sulfenyl chloride ⟨70OPP235⟩; α-chloroalkanesulfinyl chlorides have been prepared by this method (Equation (68)) ⟨79JOC1177⟩. The reaction of 1-chloroethanesulfinic acid with thionyl chloride gives the corresponding sulfinyl chloride and the chlorination of α-chloro-α′-polychloro sulfoxides has been used to prepare chloromethanesulfinyl chloride (Scheme 49) ⟨77CJC421⟩. The chlorination of symmetrical trithianes has also been used to prepare these compounds (Equation (69)) ⟨73JOC17⟩. Sulfinyl chlorides have limited stability, decomposing to give gaseous products, and thus should not be stored in closed containers for prolonged periods ⟨68JOC2104⟩.

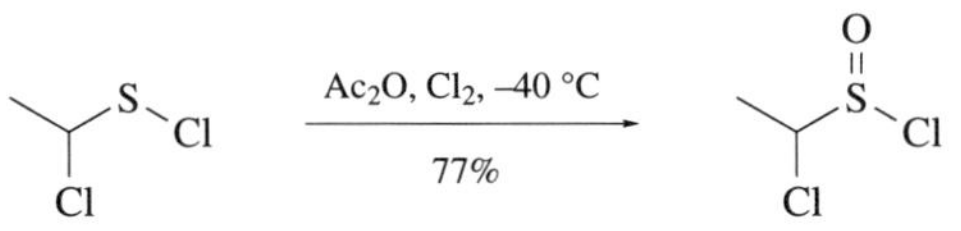

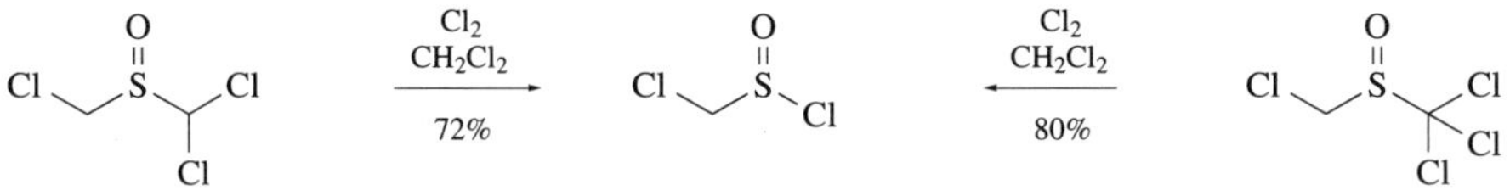

Scheme 49

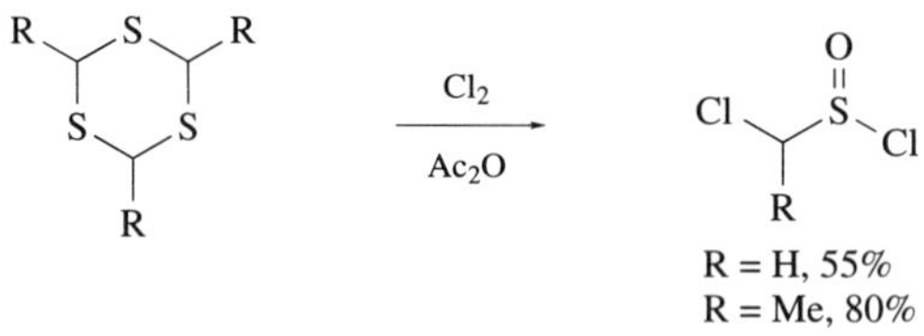

The reaction of water with chloromethanesulfinyl chloride is reported to give chloromethanesulfinic acid (**85**) ⟨73CC49⟩; sulfinic acids in general are unstable in air ⟨67JA4099⟩. 1-Chloroethanesulfinic acid (**86**) has been prepared by the sulfite reduction of the corresponding α-chloro sulfonyl chloride ⟨77CJC2323⟩.

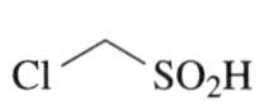

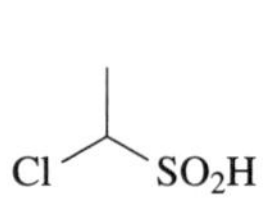

4.02.2.3 Tetracoordinate α-Halo Sulfur Derivatives, $R^1_2CHalS(O)_2R^2$, etc.

4.02.2.3.1 α-Halo sulfones, $R^1_2CHalS(O)_2R^2$

The preparation and chemistry of α-halo sulfones has been reviewed in the context of the Ramberg–Backlund reaction ⟨77OR(25)1⟩; a more general account of the preparation and synthetic applications of sulfones, including α-halo sulfones, is also available ⟨B-93MI 402-02⟩.

(i) α-Fluoro sulfones, $R^1_2CFS(O)_2R^2$

The standard method of producing α-fluoro sulfones involves the mcpba oxidation of α-fluoro sulfides (Scheme 50) ⟨85JA735, 90JOC4757, 90TL5449⟩ formed by the fluorination of sulfides or sulfoxides. This procedure has been used in the synthesis of fluoro nucleosides and is thus compatible with a wide range of functional groups (Scheme 50) ⟨91JOC6878⟩. The addition of α-oxy radicals to 1-fluoro-1-(benzenesulfonyl)ethene (**87**), formed by the base-promoted dehydrochlorination of 1-fluoro-2-chloroethyl phenyl sulfone, has been used to prepare an extensive range of α-fluoro β-substituted ethyl phenyl sulfones (Scheme 51) ⟨90JOC2973⟩. The products are obtained as a mixture of diastereomers which can be separated. The α-sulfonyl carbanion (**88**), generated from α-fluoromethyl phenyl sulfone by *n*-butyllithium, gives adducts with ketones and aldehydes in high yield (Scheme 52) ⟨85CC678⟩. The selective monofluorination of diethyl 1-(benzenesulfonyl)methyl phosphonate (**89**) has been achieved with perchloryl fluoride ($FClO_3$) (Equation (70)), and the carbanion from the resulting α-fluoro sulfone has been alkylated to produce a series of substituted derivatives ⟨87CPB3959⟩.

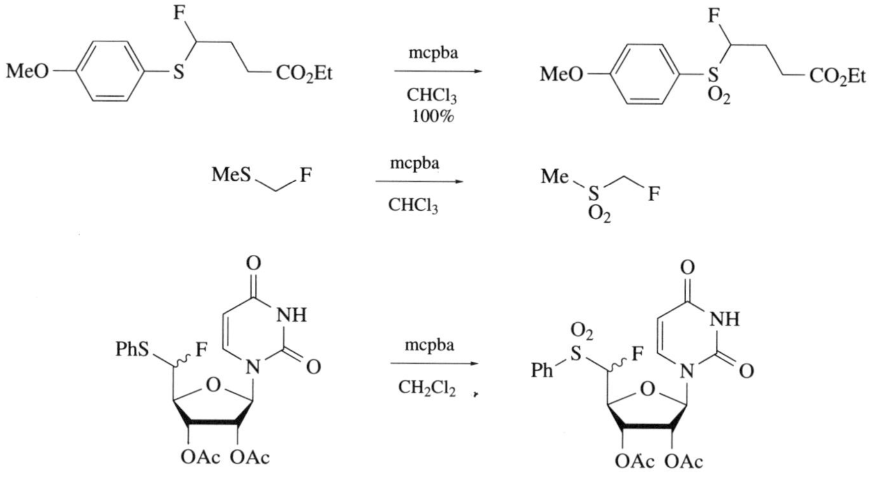

Scheme 50

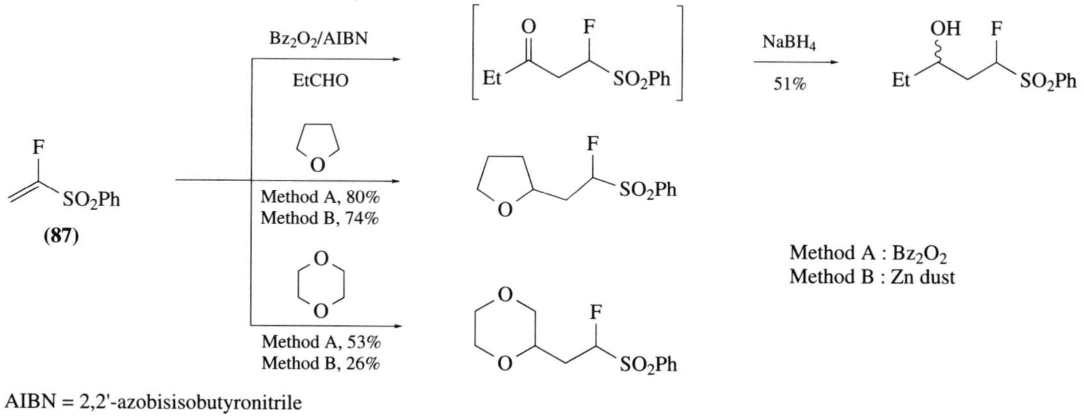

Scheme 51

AIBN = 2,2'-azobisisobutyronitrile

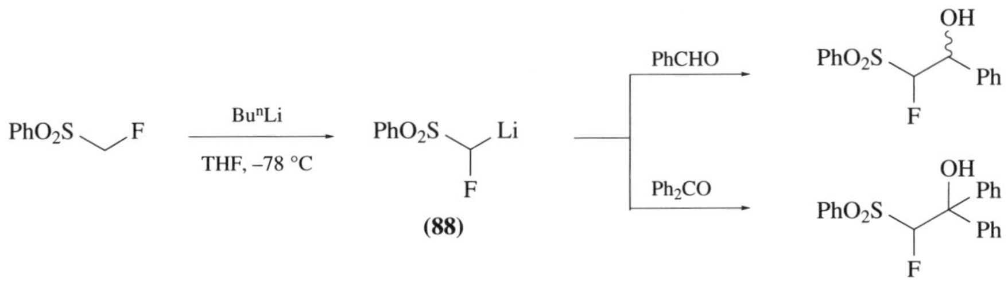

Scheme 52

(89) → (70)

(ii) α-Chloro sulfones, $R^1{}_2CClS(O)_2R^2$

The ready availability of α-chloro sulfides (Section 4.02.2.1.1(ii)) and the ease with which they can be oxidised contribute to the importance of this method of preparing α-chloro sulfones. The older literature has been reviewed ⟨52JA3594⟩ and although an extensive range of oxidising systems have been employed—peracetic acid ⟨51JA5184⟩, chromic anhydride in glacial acetic acid ⟨52JA3594⟩, potassium permanganate ⟨68JA4496⟩ and monoperphthalic acid ⟨49LA(563)54, 70JOC3002⟩—the reagent of choice is undoubtedly mcpba ⟨64JA4383, 68JOC1080⟩, which has been used in a range of solvents including diethyl ether, dichloromethane and chloroform. α-Chloro sulfides, prepared from sulfuryl chloride and sulfides in tetrachloromethane, can be oxidised *in situ* by addition of a solution of mcpba in dichloromethane directly to the crude reaction mixture (Scheme 53) ⟨74JOC2521⟩. A particularly direct method of preparing 2-chloro-2-(benzenesulfonyl)cycloalkanones involves α-sulfenylation of the ketone, chlorination with NCS and finally mcpba oxidation (Scheme 54) ⟨93BCJ2339⟩.

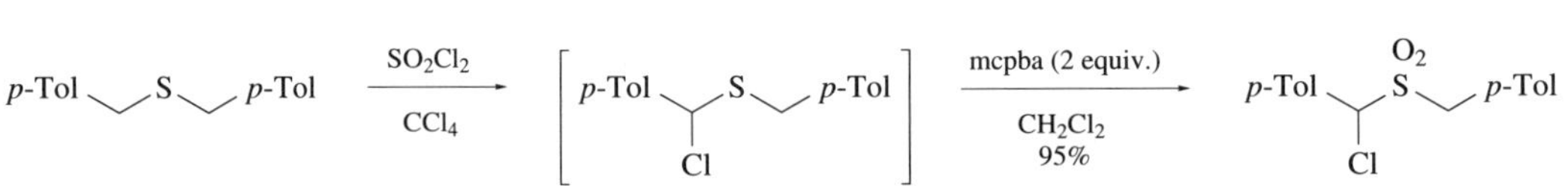

Scheme 53

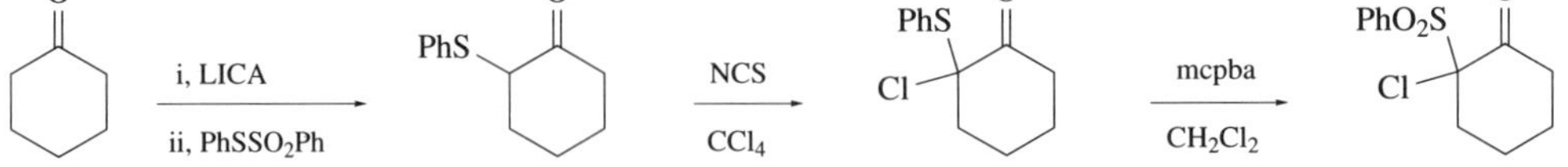

LICA = lithium isopropylcyclohexylamide

Scheme 54

α-Chlorination of sulfones is considerably more difficult than of sulfoxides, requiring the generation of a carbanion which is subsequently chlorinated. Chlorine has been introduced into the bridgehead position of bicyclic sulfones by the use of strong bases such as *t*-butyl- and *n*-butyllithium to generate carbanions which are subsequently chlorinated (Scheme 55) ⟨69JA3870, 69JOC1233⟩; 1-chlorocyclopropyl sulfones have been prepared in a similar fashion (Equation (71)) ⟨71JOC1015⟩.

The use of sodium hydroxide in a polar aprotic solvent, DMF, with tetrachloromethane as chlorinating agent, was also found to be effective (Equation (72)) ⟨82TL2539⟩.

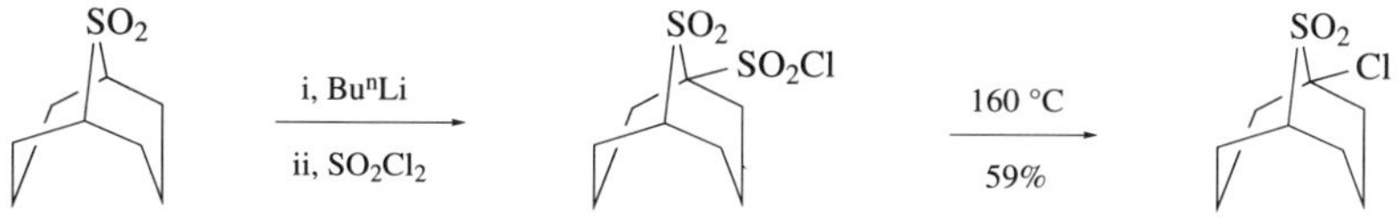

Scheme 55

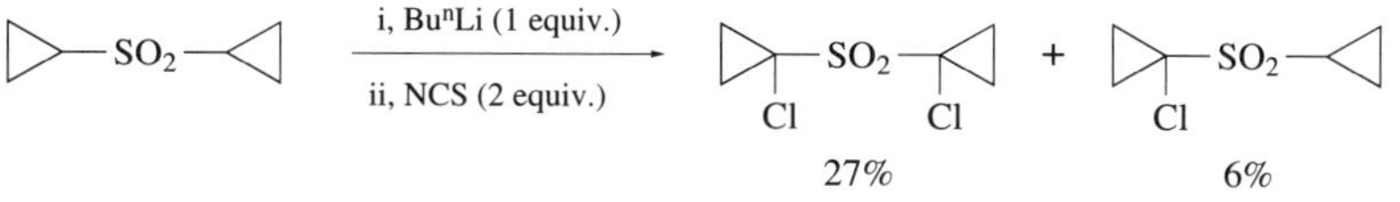

(71)

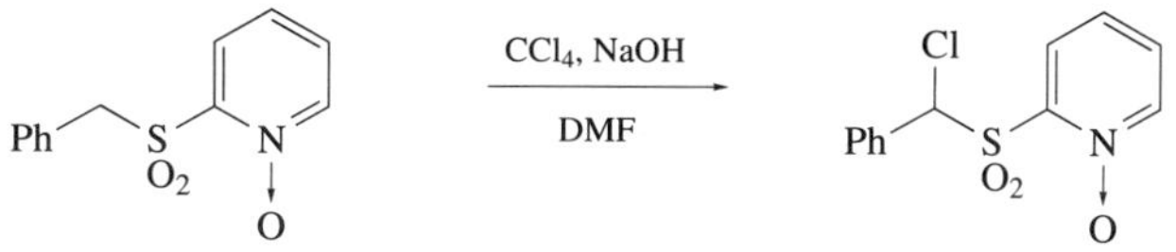

(72)

The basic conditions required to chlorinate sulfones are similar to those required to convert the product α-chloro sulfones into alkenes by the Ramberg–Backlund reaction. Thus under phase-transfer conditions the initially formed α-chloro sulfone (**90**) is not isolated but is converted directly into *trans*-stilbene (Scheme 56) ⟨82S504⟩, and under similar conditions it was found that although aryl alkyl sulfones, which cannot undergo the Ramberg–Backlund reaction, were quantitatively chlorinated (Equation (73)), dibenzyl and di-*s*-alkyl sulfones gave alkenes ⟨69JA7510⟩. It was also found that di- and tri-chlorination proceeded at increasing rates, thus making it difficult to achieve selective monochlorination of certain sulfones ⟨69JA7510⟩.

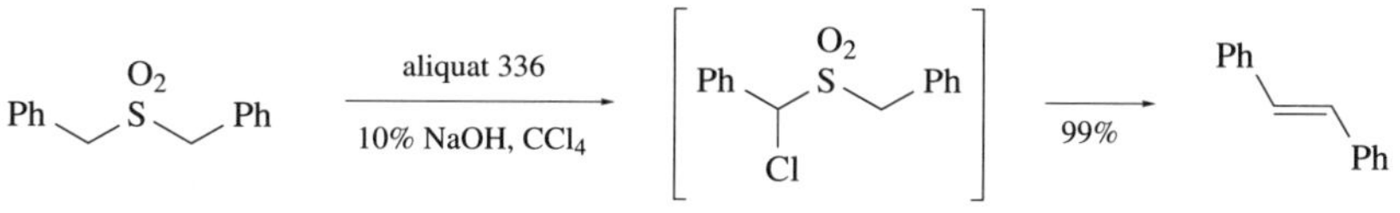

Scheme 56

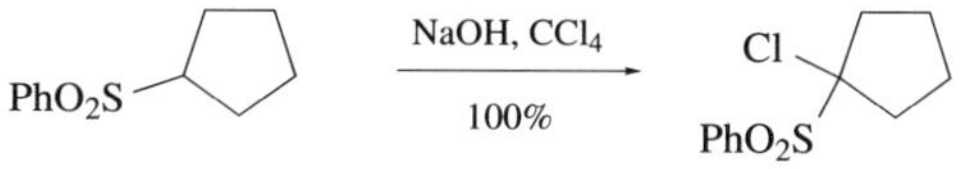

(73)

α-Sulfonyl carboxylic acids undergo a process of halogenative decarboxylation on treatment with a halogen or *N*-halosuccinimide, giving α-halo sulfones (Equation (74)); an essential requirement for this reaction is the presence of an enolisable α hydrogen in the acid and overall it can be considered to be an example of a general reaction of carboxylic acids which contain a strongly electron-withdrawing group, such as ArSO, RSO, CN, NO, COR or COR, in the α position ⟨74JOC2516⟩. Aryl and alkyl halomethyl sulfones, and aryl and alkyl haloalkyl sulfones, can all be synthesised by this method ⟨74JOC2516⟩; the early work relating to the reaction has been reviewed ⟨B-40MI 402-01⟩. Aryl and alkyl chloromethyl sulfones are prepared by a sequence which begins with chloroacetic acid and the appropriate thiolate anion and gives a dichloromethyl sulfone following halogenative decarboxylation of the intermediate α-sulfonyl acetic acid (**91**) (Scheme 57); the monochloro derivative is obtained by subsequent sulfite reduction. The alkyl compounds can also be prepared beginning with an alkyl halide and sodium mercaptoacetate (Scheme 57). α-Aryl and α-alkyl sulfonylacetic acids can be prepared in an analogous manner and on halogenative decarboxylation give α-haloalkyl sulfones directly (Equation (75)). Cyclopropyl 1-chloro-1-phenyl-methyl sulfone (**92**), for example, can be prepared in this way, being formed together with some of the dichloro compound (**93**), the result of a secondary chlorination of the activated benzylic position in (**92**) (Equation (76)) ⟨74JOC2516⟩.

α-Sulfonyl carbanions can be generated from α-chlorosulfones by strong bases; they give substitution and addition products with alkyl halides and carbonyl compounds, respectively (Equation (77)) ⟨69JA1034⟩. These products are identical with those obtained from the corresponding sulfoxides by α-lithiation, addition or substitution and finally oxidation. Phase-transfer catalysis has also been used to produce α-sulfonyl carbanions which were then alkylated (Equation (78)) ⟨75JOC266⟩. The reaction of diazomethane with the chloro sulfene (**94**), generated *in situ* from triethylamine and chloromethanesulfonyl chloride, has been used to prepare the chloro episulfone (**95**) (Scheme 58) ⟨67JA4487⟩. The addition of chloromethanesulfonyl halides to unsaturated systems also produces chlorosulfones. Thus chloromethanesulfonyl chloride, in acetonitrile containing a copper(I) salt, adds to reactive alkenes such as styrene to give β-chlorosulfonyl adducts in reasonable yield (Equation (79)) ⟨64JCS4962⟩, and chloromethanesulfonyl bromide adds photochemically to alkenes and alkynes to give a wide range of chloromethyl sulfones (Section 4.02.2.3.1(iii)) ⟨86JA4568⟩.

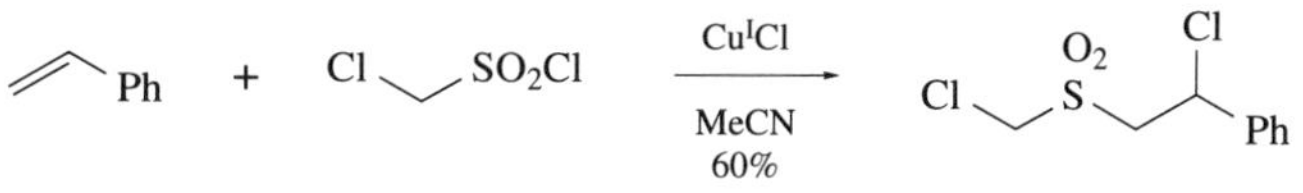

$$CH_2=CHPh + Cl-CH_2-SO_2Cl \xrightarrow[\text{MeCN}]{\text{Cu}^I\text{Cl}} Cl-CH_2-S(O_2)-CH_2-CHCl-Ph \quad 60\% \quad (79)$$

(iii) α-Bromo sulfones, $R^1_2CBrS(O)_2R^2$

m-Chloroperbenzoic acid (mcpba) has been used to oxidise α-bromo sulfides in good yield ⟨71JA476⟩. In some cases α-bromo sulfides, obtained by the bromination of sulfides (Section 4.02.2.1.1(iii)), have been oxidised *in situ* by addition of mcpba to the bromination mixture ⟨68JA435⟩.

Although α-bromo sulfones have been prepared by the bromination of α-sulfonyl carbanions, following the pattern established for the chloro compounds (Equation (80)) ⟨40JA2596, 65JOC1313, 69JOC1233⟩, free-radical bromination involving NBS and a catalytic quantity of benzoyl peroxide is also possible for sulfones which have an α-position that is both tertiary and benzylic (Equation (81)) ⟨74JOC2526⟩; simple benzylic and allylic sulfones do not react under comparable conditions ⟨48RTC451⟩.

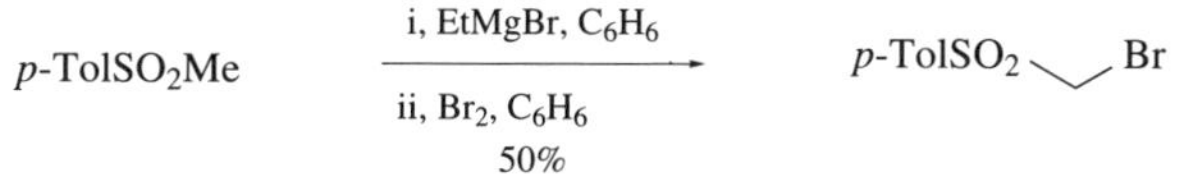

$$p\text{-TolSO}_2\text{Me} \xrightarrow[\text{ii, Br}_2, \text{C}_6\text{H}_6]{\text{i, EtMgBr, C}_6\text{H}_6} p\text{-TolSO}_2\text{-CH}_2\text{Br} \quad 50\% \quad (80)$$

$$\text{Ph-CH(Me)-SO}_2\text{-CH}_2\text{-Ph} \xrightarrow[\text{Bz}_2\text{O}_2 \text{(cat.), CCl}_4]{\text{NBS (1-2 equiv.)}} \text{Ph-C(Me)(Br)-SO}_2\text{-CH}_2\text{-Ph} \quad 75\% \quad (81)$$

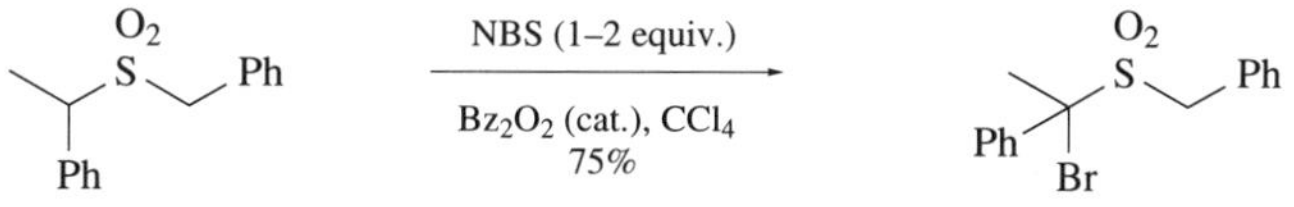

α-Bromo sulfones can be prepared by halogenative decarboxylation with the same synthetic flexibility which is possible for the α-chloro compounds (Section 4.02.2.3.1(ii)) (Scheme 57, Equation (75)) ⟨40JA2596, 74JOC2516⟩. A typical application of this approach is the preparation of di(bromobenzyl) sulfone from phenylacetic acid (Scheme 59) ⟨88OSC(6)403⟩.

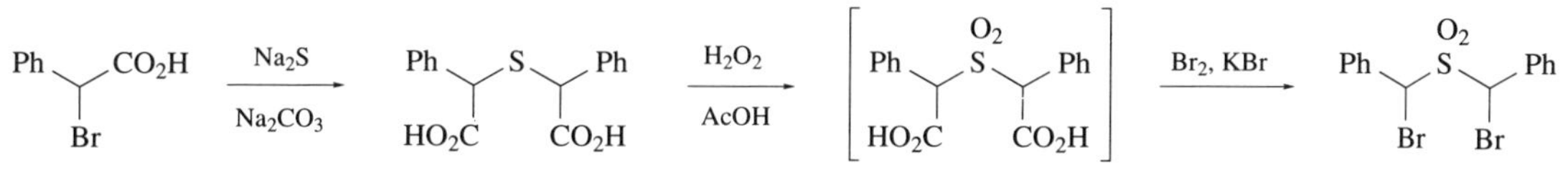

$$\text{Ph-CHBr-CO}_2\text{H} \xrightarrow[\text{Na}_2\text{CO}_3]{\text{Na}_2\text{S}} \text{Ph-CH(CO}_2\text{H)-S-CH(CO}_2\text{H)-Ph} \xrightarrow{\text{H}_2\text{O}_2, \text{AcOH}} [\text{Ph-CH(CO}_2\text{H)-SO}_2\text{-CH(CO}_2\text{H)-Ph}] \xrightarrow{\text{Br}_2, \text{KBr}} \text{Ph-CHBr-SO}_2\text{-CHBr-Ph}$$

Scheme 59

The reaction of sulfinate anions with dibromomethane in ethanol gives a modest yield of bromomethyl sulfones (Equation (82)) ⟨40JA2596⟩; the reaction gives lower yields with aromatic sulfinates as these are less reactive, but in common with the halogenative decarboxylation route it constitutes a regioselectively unambiguous route to bromomethyl alkyl sulfones. α-Sulfonyl carbanions have been generated from α-bromo sulfones under phase-transfer conditions and alkylated to give α-bromoalkyl sulfones which are more highly substituted in the α-position (Equation (83)) ⟨75JOC266⟩. 1-Bromo-1-methylethene episulfone has been prepared in 64% yield from α-bromoethanesulfonyl chloride by the same method used to synthesise related chloro compounds (Scheme 58) ⟨66JA5682⟩.

$$\text{Bu}^n\text{SO}_2^-\text{Na}^+ + \text{CH}_2\text{Br}_2 \longrightarrow \text{Bu}^n\text{O}_2\text{S-CH}_2\text{Br} + \text{NaBr} \quad 44\% \quad (82)$$

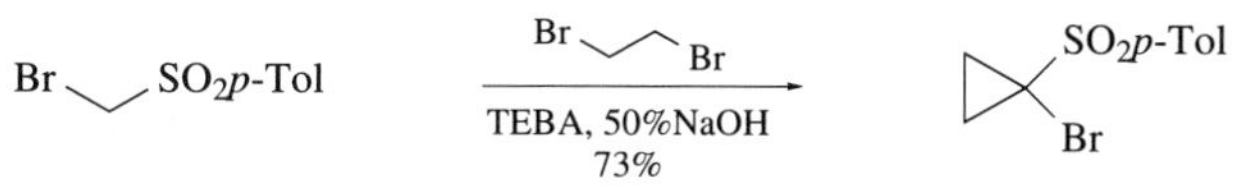

$$\text{Br-CH}_2\text{-SO}_2 p\text{-Tol} \xrightarrow[\text{TEBA, 50\%NaOH}]{\text{CH}_2\text{Br}_2} \text{(episulfone with SO}_2 p\text{-Tol and Br)} \quad 73\% \quad (83)$$

TEBA = Et$_3$BnN$^+$ Cl$^-$

An extremely wide range of alkyl bromomethyl sulfones is available from the photochemical addition of bromomethanesulfonyl bromide to alkenes ⟨86JA4568⟩. The reaction involves the homolytic cleavage of the S—Br bond, a process that is facilitated by its light-sensitive nature and by the fact that it is weaker than the corresponding S—Cl bond in halomethanesulfonyl chlorides, compounds which undergo photochemical addition with difficulty and copper(I)-catalysed addition only to more reactive alkenes. The reaction is regioselective, the bromine adding to the more substituted end of the alkene, and the reagent to the less substituted alkene in a diene (Scheme 60). α-Bromomethyl vinyl sulfones can be obtained through the dehydrohalogenation of the initially formed β-bromoalkyl sulfones (Equation (84)), a procedure which can be extended to bromomethyl dienyl sulfones giving dienes stereospecifically. Bromomethanesulfonyl bromide adds to alkynes stereospecifically to give *trans* adducts (Equation (85)). The reaction of bromomethanesulfonyl bromide with silyl enol ethers, in the presence of ethylene oxide as acid scavenger, gives good yields of α-bromomethanesulfonyl ketones (Equation (86)).

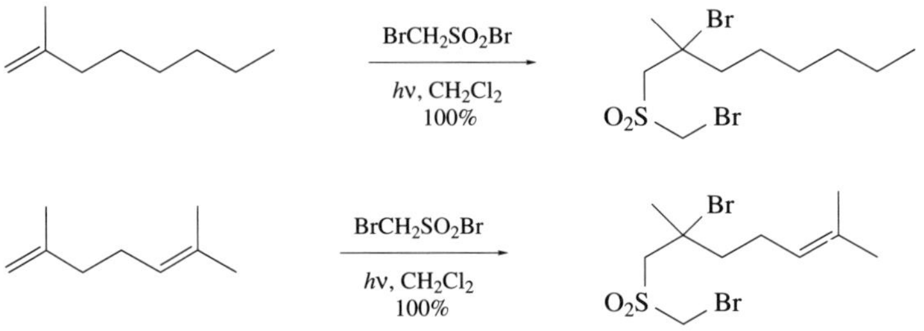

Scheme 60

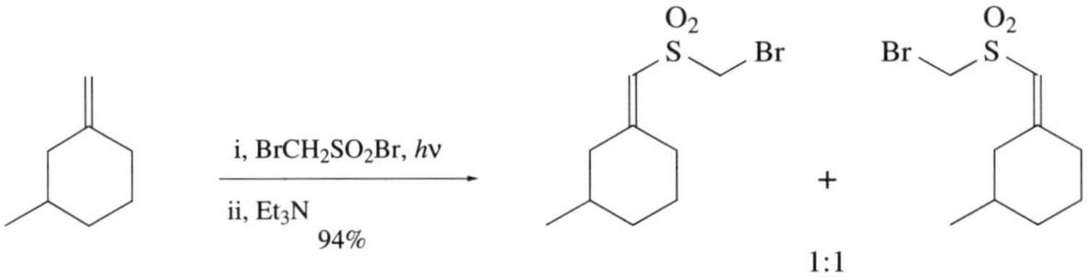

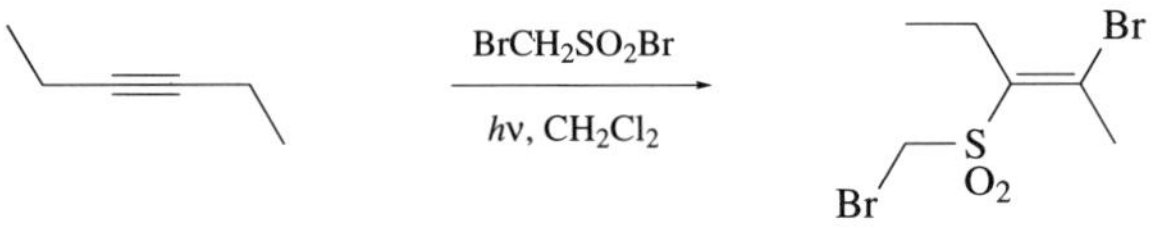

(iv) α-Iodo sulfones, $R^1_2CIS(O)_2R^2$

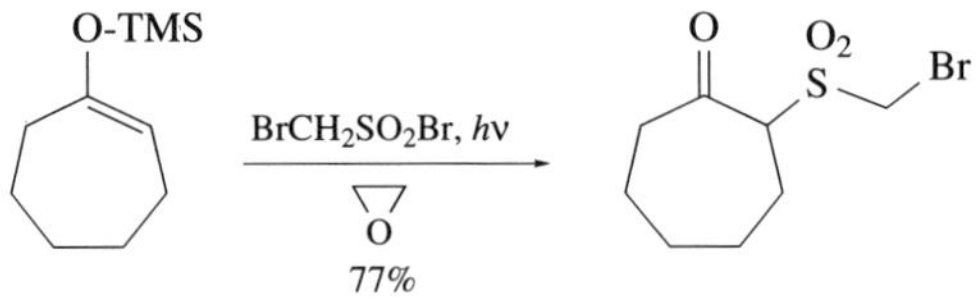

α-Sulfonyl carbanion chemistry has been used to introduce iodine into sulfones and to make structurally more complicated molecules from sulfones already containing iodine. Thus α-iodo sulfones have been prepared by the iodination of α-sulfonyl carbanions with iodine (Scheme 61) ⟨68JA435, 69JOC1233⟩ and the α-sulfonyl carbanion generated from iodomethyl phenyl sulfone under phase-transfer conditions has been alkylated to give a range of α-iodoalkyl sulfones (Equation (87)) ⟨78S883⟩. The substitution reaction of sodium benzenesulfinate with diiodomethane gave iodomethyl phenyl sulfone ⟨40JA2596⟩. Halogenative decarboxylation of α-sulfonyl carboxylic acids with *N*-iodosuccinimide (NIS) is a generally applicable method for the synthesis of these compounds (Equation (88)) ⟨74JOC2516⟩. The iodo sulfone (**97**) is formed by the addition of trimethylsilyl iodide to the vinyl sulfone (**96**) and subsequent reaction with 1,2-ethanediol (Equation (89)) ⟨89TL3267⟩. Iodomethanesulfonyl bromide behaves in the same way as chloro- and bromomethanesulfonyl bromide (Section 4.02.2.3.1(iii)), adding photochemically to alkenes and alkynes to give a wide range of iodomethyl sulfones ⟨86JA4568⟩.

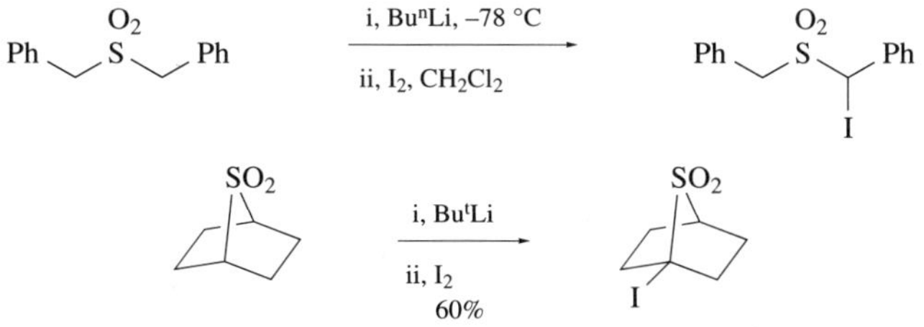

Scheme 61

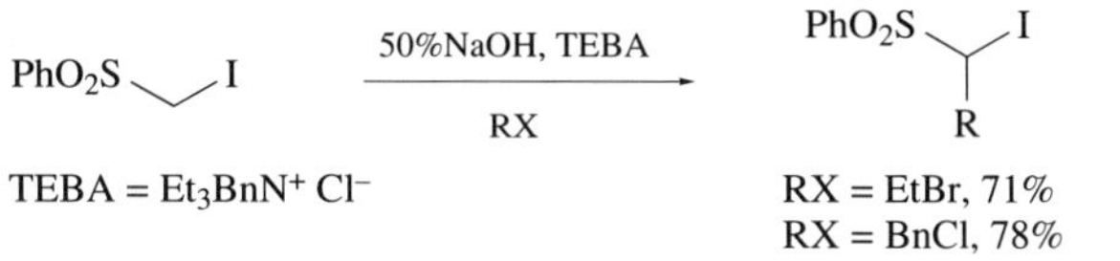

(87)

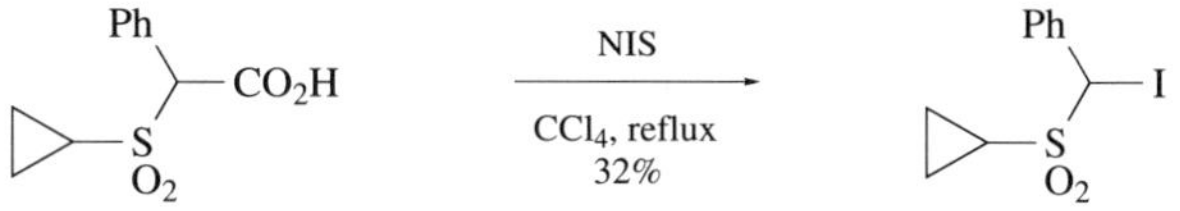

RX = EtBr, 71%
RX = BnCl, 78%

TEBA = Et₃BnN⁺ Cl⁻

(88)

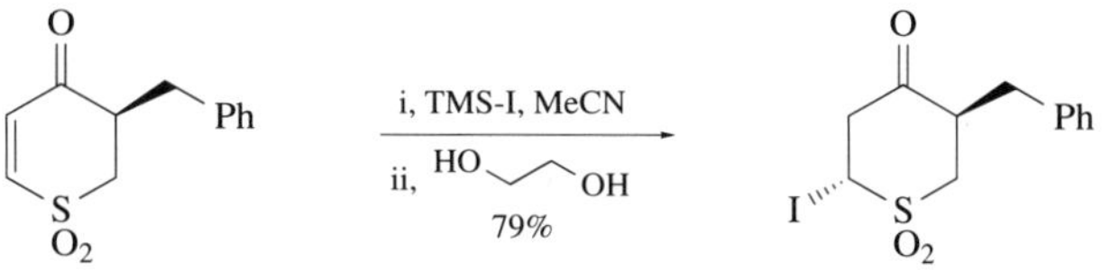

(89)

(96)   (97) major isomer

4.02.2.3.2 Other tetracoordinate α-halo sulfur derivatives, $R_2CHalS(O)_2X$

Apart from α-halo sulfones, the most important α-halogenated tetracoordinate sulfur derivatives are the α-halo sulfonyl halides. Chloromethanesulfonyl chloride, a lachrymatory liquid, has been prepared in low yield by the chlorination of 1,3,5-trithiane (Scheme 62) in water ⟨41JA1571, 60JCS3058⟩ or aqueous acetic acid ⟨73OSC(5)231⟩; the reaction is somewhat unpredictable and gives variable yields ⟨60JCS3058⟩. The chlorination of chloromethyl trichloromethyl sulfide in water gives a better yield of this sulfonyl chloride, but from a less readily available starting material (Scheme 62) ⟨73CC50⟩; it has also been prepared from phosphorus pentachloride and sodium chloromethanesulfonate ⟨60JCS3058⟩. Chloromethanesulfonyl bromide is formed when chloromethanesulfonyl chloride is treated with aqueous sodium sulfite and bromine ⟨86JA4568⟩.

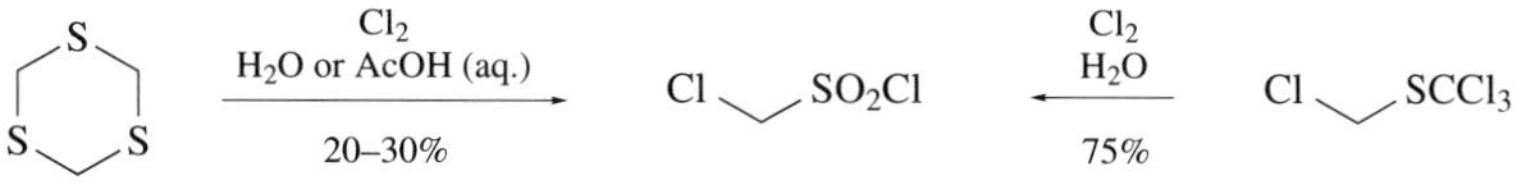

Scheme 62

Bromoalkanesulfonyl halides have been prepared by the chlorination of sodium bromoethanesulfonate ⟨71JA476⟩, by the bromination on a mole scale, but in modest yield, of symmetrical trithianes (Equation (90)) ⟨35MI 402-01, 86JA4568⟩ and by the reaction of bromoethanesulfonyl chloride with aqueous sodium sulfite and bromine ⟨86JA4568⟩. The reaction of sodium iodomethanesulfonate with phosphorus pentabromide ⟨86JA4568⟩ and phosphorus pentachloride ⟨60JCS3058⟩ gives iodomethanesulfonyl bromide and chloride, respectively.

$$\underset{H_2O}{\overset{Br_2}{\longrightarrow}} \quad Br\diagup SO_2Br \qquad (90)$$

α-Halo alkanesulfonamides (**98**) have been prepared in the usual way by the reaction of an amine with the appropriate α-halosulfonyl halide; if the reaction is carried out in benzene, 2 moles of amine are required, but only 1 mole if pyridine is used as solvent; chloro-, bromo- and iodoalkanesulfonamides have been prepared in this way (Equation (91)) ⟨60JCS3058⟩. The reaction of bromo- and iodomethanesulfonyl chloride with anhydrous ammonia has also been used to prepare the respective sulfonamides ⟨69JCS(C)652, 71JA476⟩. Fluoromethanesulfonanilides have been prepared by the reaction of the amine with a mixture of sodium fluoromethanesulfonate and chlorofluoromethanesulfonate, the required anilide being easily separated by crystallization ⟨60JCS3058⟩. The reaction of iodomethanesulfonyl chloride with benzamidine under Schotten–Baumann conditions gives *N*-(iodomethanesulfonyl)benzamidine in modest yield ⟨69JCS(C)652⟩.

$$R^1\underset{X}{\diagup}SO_2Y + R^2R^3NH_2 \longrightarrow R^1\underset{X}{\diagup}\overset{O_2}{\underset{}{S}}\underset{R^3}{\diagup}N-R^2 \qquad (91)$$

$R^1, R^2, R^3 =$ H, alkyl, aryl (**98**)

α-Halo alkanesulfonyl halides react with alcohols to give the corresponding sulfonic acid esters (**99**) (Equation (92)), an example being the formation of 2-methylpropyl bromomethanesulfonate from bromomethanesulfonyl bromide and 2-methylpropanol ⟨74JOC1449⟩. The unusual α-fluoro alkanesulfonate (**100**) is formed from sulfur trioxide and perfluoropropene at 100 °C and under pressure ⟨81JA5598⟩. The salts of α-halo alkanesulfonic acids are readily available by the alkylation of the sulfite anion with *gem*-dihalo alkanes ⟨35JA2360, 60JCS3058, 63CB2266, 71JA476⟩.

$$R^1\underset{X}{\diagup}SO_2Y + R^2OH \longrightarrow R^1\underset{X}{\diagup}SO_3R^2 \qquad (92)$$

$R^1 =$ H, alkyl, aryl
$R^2 =$ alkyl, aryl   (**99**)

$$F_3C\underset{O_2S-O}{\overset{F\quad F}{\diagup\diagdown F}}$$

(**100**)

α-Chloroalkyl *N*-alkylsulfoximides can be prepared by the chlorination of *N*-alkylsulfoximides using *t*-butyl hypochlorite in the presence of potassium carbonate (Equation (93)), the product being obtained as a single diastereomer if the α carbon is prochiral; *N*-chlorosulfoximides can be chlorinated in the same way (Equation (93)) ⟨78JOC4136⟩. A mixture of diastereomers is obtained on bromination of the sulfoximide (**101**) with sodium hydride and bromine in DMF (Equation (94)) ⟨78JOC4140⟩. The reaction of α-chloro and α-bromo sulfoxides with mesitylenesulfonyloxyamine (**102**) can also be used to produce α-haloalkylsulfoximides, diastereoselectively where possible (Equation (95)) ⟨78JOC4136⟩. Although the direct fluorination of the *N*-methylsulfoximide (**103**) gives an unsatisfactory yield of the fluoromethylsulfoximide (**104**), it can be obtained in better yield by the reaction of fluoromethyl phenyl sulfoxide with sodium azide, which gives the free sulfoximine which is subsequently methylated ⟨88TL3365⟩ (Scheme 63). Chloromethyl phenyl *N*-chlorosulfoximide has been reduced to the free imine and tosylated, and the α carbanions from the fluoromethyl sulfoximine (**104**) have been added to ketones and aldehydes (Equation (96)), giving overall a wide range of tetracoordinated sulfur derivatives.

$$\underset{NR}{\overset{O}{Ph-S-Me}} \xrightarrow[CH_2Cl_2]{Bu^tOCl, K_2CO_3} \underset{NR}{\overset{O}{Ph-S}}\diagup Cl \qquad (93)$$

R = Me, 75%
R = Cl, 76%

4.02.3 HALOGEN AND SELENIUM AND TELLURIUM DERIVATIVES, $R^1_2CHal(SeR^2)$ AND $R^1_2CHal(TeR^2)$, etc.

4.02.3.1 α-Halo Selenium Derivatives, $R^1_2CHal(SeR^2)$, etc.

4.02.3.1.1 Dicoordinate α-halo selenium derivatives, $R^1_2CHal(SeR^2)$

α-Fluoro selenides (**105**) have been obtained by an electrochemical monofluorination process (Equation (97)). The presence of the electron-withdrawing group is essential and the yield is somewhat dependent on the precise type of electrochemical cell used; selenides which contain two electron-withdrawing groups have also been successfully fluorinated by this technique (Equation (98)) ⟨92TL3161⟩. Deprotonation of the α-fluoro selenide (**106**) gives a selenium-stabilised carbanion which has been alkylated with benzyl bromide giving (**107**) (Equation (99)) ⟨92TL3161⟩.

$$\text{Ph}\diagdown\text{Se}\diagup\text{C}(\text{CO}_2\text{Et})_2\text{H} \xrightarrow[\text{Et}_3\text{N}\cdot 3\text{HF, MeCN}]{\text{electrolysis}} \text{Ph}\diagdown\text{Se}\diagup\text{CF}(\text{CO}_2\text{Et})_2 \quad 55\% \qquad (98)$$

$$\underset{(\textbf{106})}{\text{Ph-Se-CF(CO}_2\text{Et})_2} \xrightarrow[\text{ii, PhCH}_2\text{Br}]{\text{i, LICA}} \underset{(\textbf{107})}{\text{Ph-Se-CF(CH}_2\text{Ph})(\text{CO}_2\text{Et})} \quad 70\% \qquad (99)$$

LICA = lithium isopropylcyclohexylamide

Many of the methods used for the synthesis of α-bromo and α-chloro sulfides can be successfully adapted to the synthesis of the corresponding selenium compounds. Thus, for example, the reaction of selenols (**108**) or selenophenols with ketones or aldehydes, in the presence of anhydrous hydrogen halide, has been used to prepare α-chloro and α-bromo selenides (**109**) (Equation (100)) (Table 18). The reaction is quite general but the reaction product may contain quantities of the selenoacetal (**110**) ⟨77AG(E)541⟩.

$$\underset{(\textbf{108})}{R^2R^3C{=}O} + R^1\text{SeH} \xrightarrow{\text{anhydrous HX}} \underset{(\textbf{109})}{X{-}CR^2R^3{-}SeR^1} + \underset{(\textbf{110})}{R^1\text{Se}{-}CR^2R^3{-}SeR^1} \qquad (100)$$

Table 18 Synthesis of α-halo selenides by the reaction of selenols and selenophenols with carbonyl compounds in the presence of hydrogen halides.

Product	RSeH	HX	Carbonyl compound	Yield (%)	Ref.
PhCH$_2$SeCH$_2$Cl	PhCH$_2$SeH	HCl	(CH$_2$O)$_n$	a	48JA1244
PhCH$_2$SeCH$_2$Br	PhCH$_2$SeH	HBr	(CH$_2$O)$_n$	95	86JOC2981
PhSeCH$_2$Br	PhSeH	HBr	(CH$_2$O)$_n$	70	86JOC2981
PhSeCHBrMe	PhSeH	HBr	MeCHO	93	77AG(E)541
PhSeCHClMe	PhSeH	HCl	MeCHO	40 (23% seleno-acetal)	77AG(E)541
MeSCHCl | n-C$_{10}$H$_{21}$	MeSeH	HCl	n-C$_{10}$H$_{21}$CHO	45 (20% seleno-acetal)	77AG(E)541
MeSCHBr | n-C$_{10}$H$_{21}$	MeSeH	HBr	n-C$_{10}$H$_{21}$CHO	54 (17% seleno-acetal)	77AG(E)541

[a] Yield not given.

The Pummerer-type processes by which reagents such as NCS, NBS and sulfuryl chloride produce α-halo sulfides also operate for selenides ⟨87T4309⟩. The reaction of selenium(IV) dichlorides, which are readily formed from methyl ketones and phenylselenium trichloride or selenium tetrachloride, with pyridine in dichloromethane (Equation (101)), constitutes one of the most extensively studied reactions of this type, and produces α-chloro α-phenylselenenyl ketones in good to excellent yield ⟨89TL2665⟩ (Table 19, entries 1–4). The reaction has also been carried out with triethylamine or thermally ⟨88CL1317⟩ (Table 19, entries 5 and 6). A number of other similar reactions have been described (Table 19, entries 7–10), although the products have not, in some cases, been isolated in a pure form (entries 7–9) or are unstable (entry 10).

$$\text{RCH}_2\text{-Se}(\text{Cl})_2\text{-R}^1 \xrightarrow{\text{base}} \text{RCHCl-Se-R}^1 \qquad (101)$$

α-Halo selenides can also be synthesised by a variety of substitution reactions. Thus the reaction of selenide anions with dihalomethanes has been extensively used to prepare halomethyl selenides. The anions can be generated from the appropriate diselenide by a reducing agent (Scheme 64) ⟨91CB1315⟩ or under alkaline, phase-transfer catalysis conditions (Equation (102)) ⟨82JCR(S)212⟩. The anions can also be formed from a selenophenol with sodium hydride ⟨76JOM(114)281⟩, a stoichiometric amount of triethylamine or, again, a phase-transfer catalyst ⟨77AG(E)541⟩; however, in these cases a selenoacetal is produced as a by-product, or when 1,1-dibromoethane is used, as the

Table 19 Synthesis of α-halo selenides using seleno-Pummerer reactions.

Entry	Product	Substrate	Reagent	Yield (%)	Ref.
1	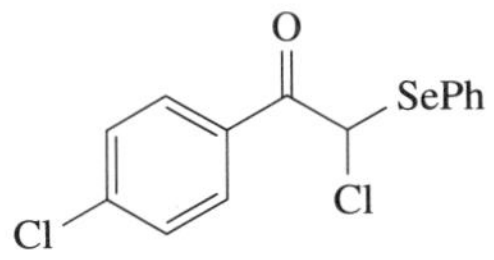 4-Cl-C₆H₄-C(O)-CH(Cl)-SePh	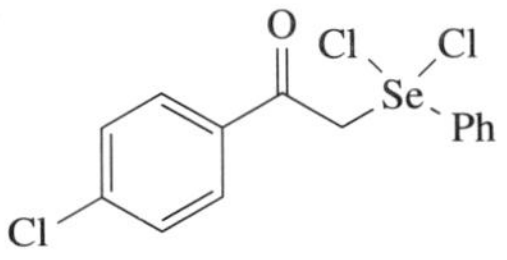 4-Cl-C₆H₄-C(O)-CH₂-SeCl₂Ph	Pyridine, CH₂Cl₂	89	89TL2665
2	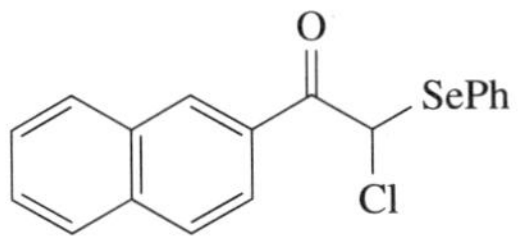 2-Naphthyl-C(O)-CH(Cl)-SePh	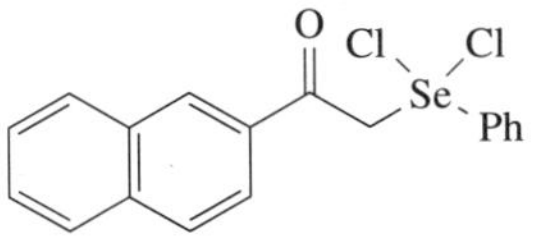 2-Naphthyl-C(O)-CH₂-SeCl₂Ph	Pyridine, CH₂Cl₂	92	89TL2665
3	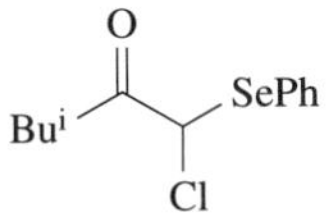 Bu^i-C(O)-CH(Cl)-SePh	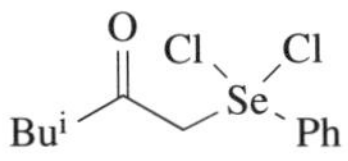 Bu^i-C(O)-CH₂-SeCl₂Ph	Pyridine, CH₂Cl₂	87	89TL2665
4	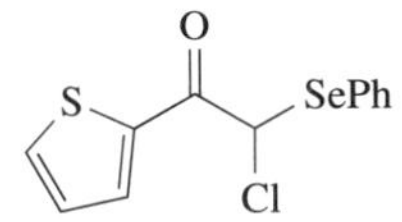 2-Thienyl-C(O)-CH(Cl)-SePh	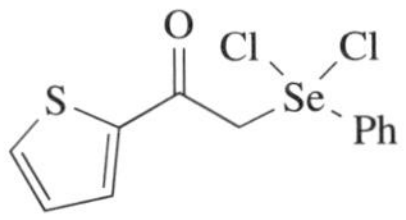 2-Thienyl-C(O)-CH₂-SeCl₂Ph	Pyridine, CH₂Cl₂	79	89TL2665
5	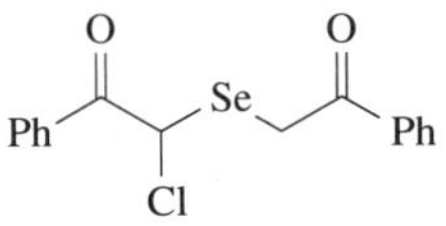 Ph-C(O)-CH(Cl)-Se-CH₂-C(O)-Ph	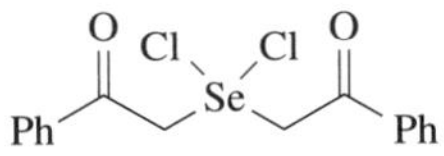 Ph-C(O)-CH₂-SeCl₂-CH₂-C(O)-Ph	Et₃N/THF	86	88CL1317
6	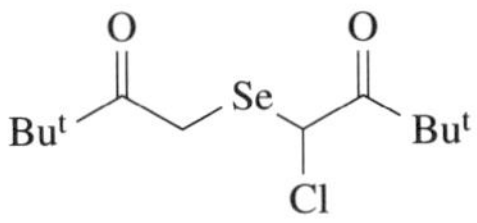 Bu^t-C(O)-CH₂-Se-CH(Cl)-C(O)-Bu^t	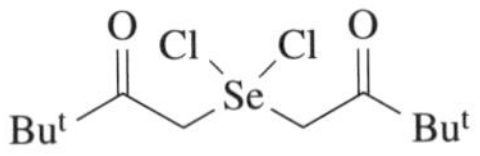 Bu^t-C(O)-CH₂-SeCl₂-CH₂-C(O)-Bu^t	Et₃N/THF	90	88CL1317
7	 Ph-C(Cl)(CHO)-SePh	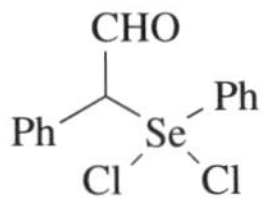 Ph-CH(CHO)-SeCl₂Ph	CCl₄, 60 °C	a	88TL5893
8	 Et-C(Cl)(CHO)-SePh	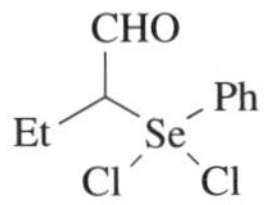 Et-CH(CHO)-SeCl₂Ph	CCl₄, 60 °C	a	88TL5893
9	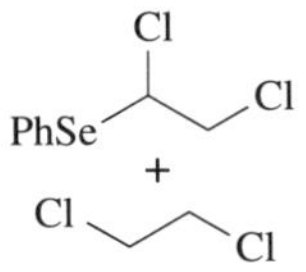 PhSe-CH(Cl)-CH₂Cl + Cl-CH₂-CH₂-Cl	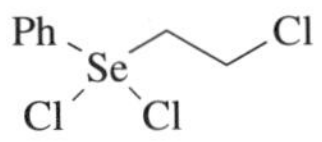 Ph-SeCl₂-CH₂-CH₂-Cl	Δ, 115 °C	a	75CJC1922
10	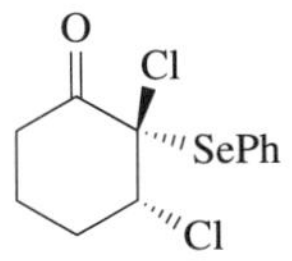 2-Cl-2-SePh-3-Cl-cyclohexanone	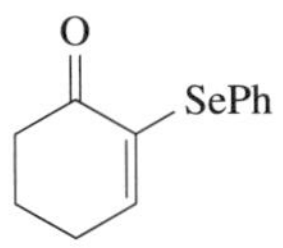 2-SePh-cyclohex-2-enone	SO₂Cl₂	98	90JOM(391)165

^{a} Yield not given.

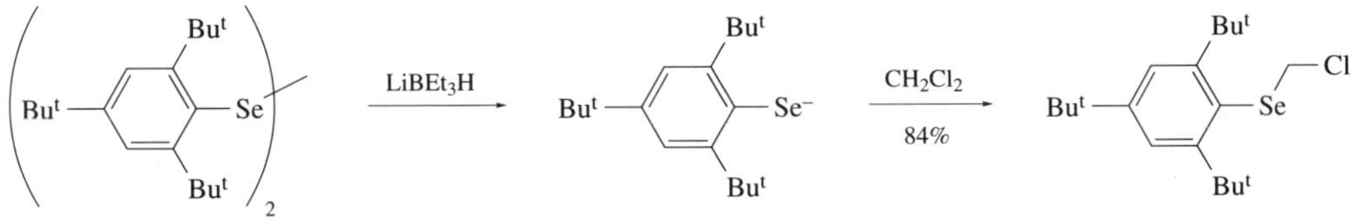

(2,4,6-Bu$^t_3$-C₆H₂-Se)₂ $\xrightarrow{\text{LiBEt}_3\text{H}}$ 2,4,6-Bu$^t_3$-C₆H₂-Se$^-$ $\xrightarrow[84\%]{\text{CH}_2\text{Cl}_2}$ 2,4,6-Bu$^t_3$-C₆H₂-Se-CH₂Cl

Scheme 64

only product of the reaction (Equation (103)) ⟨76JOM(114)281⟩. The reaction of benzeneselenenyl bromide with the carbanion formed from methyl 2-chloropropionate produces an α-chloro selenide (**111**) ⟨78JOC1607⟩, and in a similar fashion tetrachloromethane reacts with the carbanion formed from (**112**), producing diethyl chloro(phenylseleno)methylphosphonate (Scheme 65) ⟨87S169⟩.

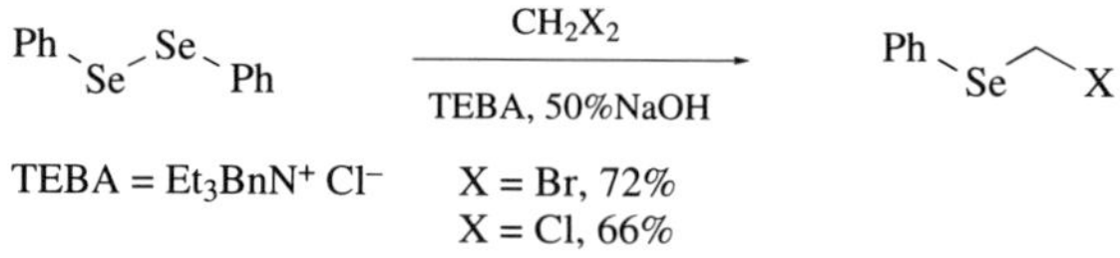

Ph Se Se Ph
CH_2X_2
TEBA, 50%NaOH
Ph Se X
TEBA = $Et_3BnN^+ Cl^-$
X = Br, 72%
X = Cl, 66%

(102)

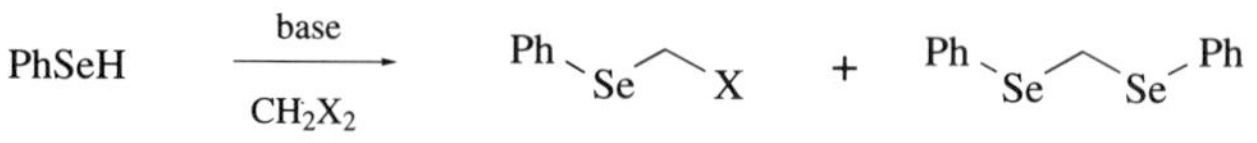

PhSeH
base
CH_2X_2
Ph Se X
+
Ph Se Se Ph

(103)

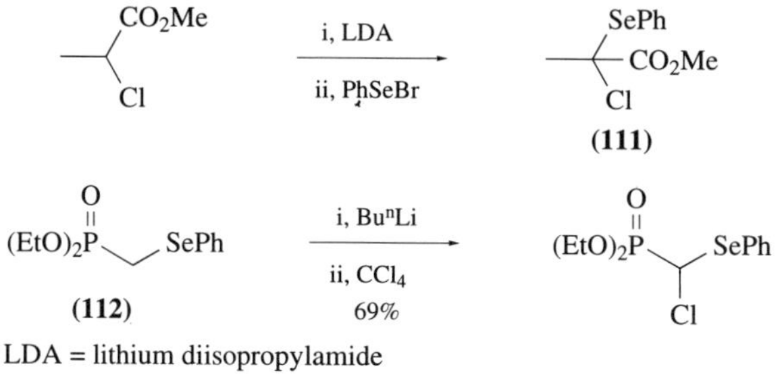

CO_2Me
Cl
i, LDA
ii, PhSeBr
SePh
CO_2Me
Cl
(111)
O
$(EtO)_2P$
SePh
i, Bu^nLi
ii, CCl_4
69%
(112)
O
$(EtO)_2P$
SePh
Cl
LDA = lithium diisopropylamide

Scheme 65

Diazo compounds react with benzeneselenenyl chloride and bromide in a process which is formally the insertion of a carbene into the selenium–chlorine bond; the reaction does not require the metal catalysts usually associated with the insertion reactions of diazo compounds and is more typical of a nucleophilic than a carbene-based process. The reaction was initially applied in the synthesis of α-bromo selenides and was found to be an attractive alternative to the alkylation of selenide anions as selenoacetal formation did not occur (Scheme 66) or could be eliminated by the use of low temperatures (Scheme 67) ⟨76JOM(114)281, 81JOM(216)287⟩.

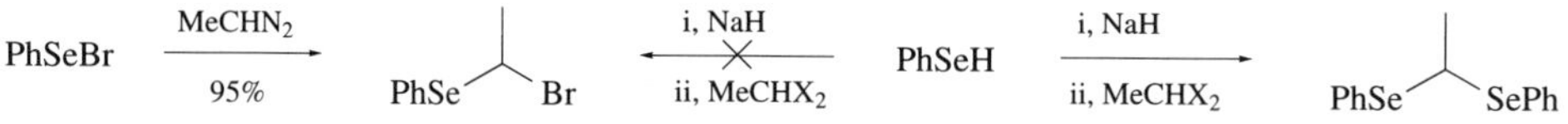

PhSeBr
$MeCHN_2$
95%
PhSe Br
i, NaH
ii, $MeCHX_2$
PhSeH
i, NaH
ii, $MeCHX_2$
PhSe SePh

Scheme 66

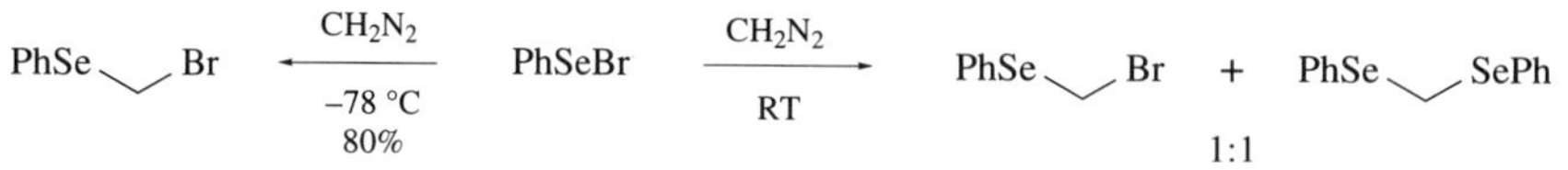

PhSe Br
CH_2N_2
−78 °C
80%
PhSeBr
CH_2N_2
RT
PhSe Br
+
PhSe SePh
1:1

Scheme 67

The reaction of α-diazocycloalkanones with benzeneselenenyl chloride or bromide gives the α-halo-α-benzeneselenenyl ketones; such compounds have considerable synthetic utility as they can efficiently be converted into either α-halo enones or α-benzeneselenenyl enones (Scheme 68) ⟨80CC506, 85JCS(P1)2193⟩. The procedure has also been applied to 6-diazopenicillanate esters (Equation (104)) ⟨80TL395, 86JCS(P1)2207⟩ such as the 2,2,2-trichloroethyl compound (**113**) which gives the α-chloro selenide (**114**) in reasonable yield; the stereochemistry of the 6-position was not determined but was assigned on mechanistic grounds ⟨82JCS(P1)2757⟩. An alternative procedure to this diazoactivation of the α position of ketones involves the use of boron trifluoride to catalyse their reaction with benzeneselenenyl chloride ⟨85CPB1745⟩; the alkaloidal derivative (**115**) was synthesised in this way as part of a 1,2-carbonyl transposition procedure ⟨91CPB1365⟩.

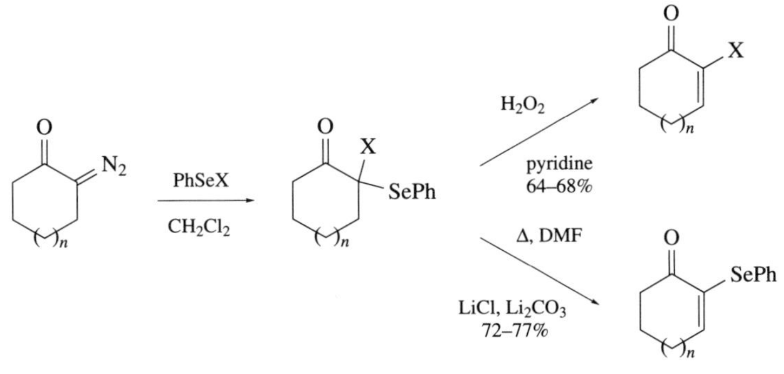

Scheme 68

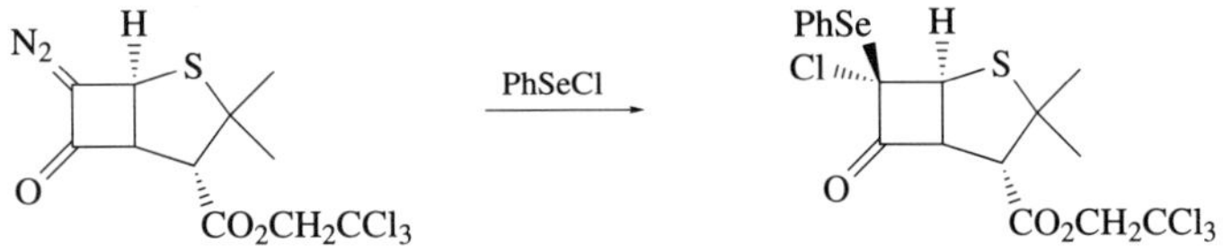

(113) → (114)   (104)

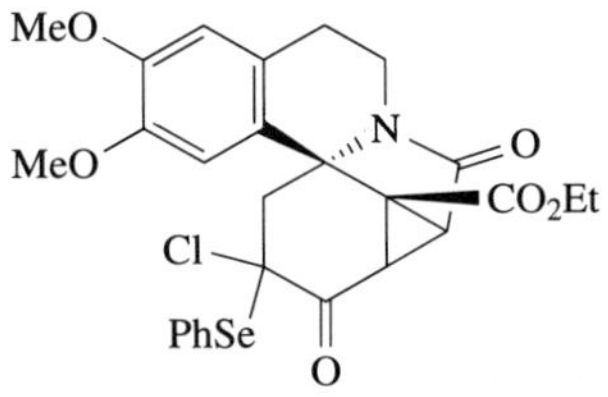

(115)

A range of addition reactions has also been employed for the synthesis of α-halo selenides. One of the most general involves the addition of dry hydrogen chloride or bromide to alkyl or aryl vinyl selenides, the yields generally being about 95% (Equation (105)) ⟨77AG(E)541⟩. Areneselenenyl and arenesulfenyl chlorides are also reported to react with aryl vinyl selenides giving products of Markovnikov addition (Equation (106)) ⟨88ZOR1945⟩. However, the addition of benzeneselenenyl chloride to alkenes containing electron-withdrawing groups is reported to give both regioisomers, which can be interconverted in refluxing acetonitrile ⟨85T2527⟩; its addition to the vinyl chloride (**116**) is also reported to give a mixture of products (Equation (107)) ⟨81JOC3721⟩.

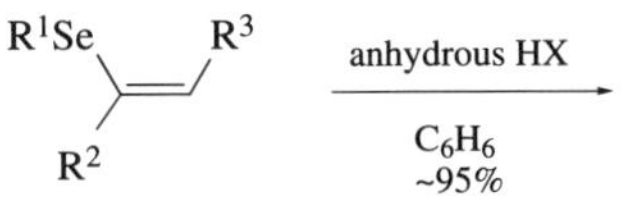

R^1 = Ph, Me
R^2 = H, Me
R^3 = H, n-C_9H_{19}, n-C_5H_{11}

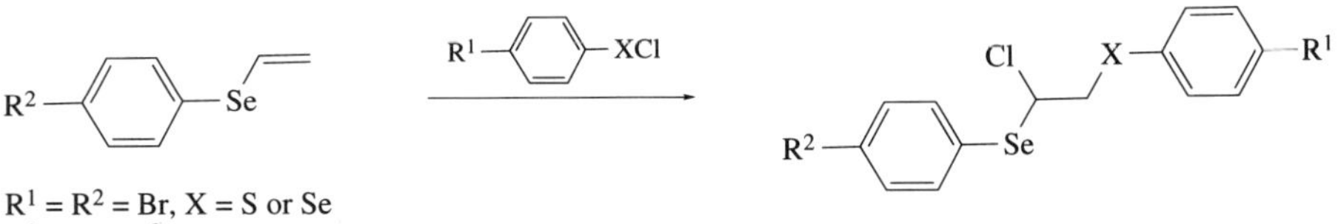

(105)

$R^1 = R^2 = Br$, X = S or Se
$R^1 = Me$, $R^2 = Br$, X = S

(106)

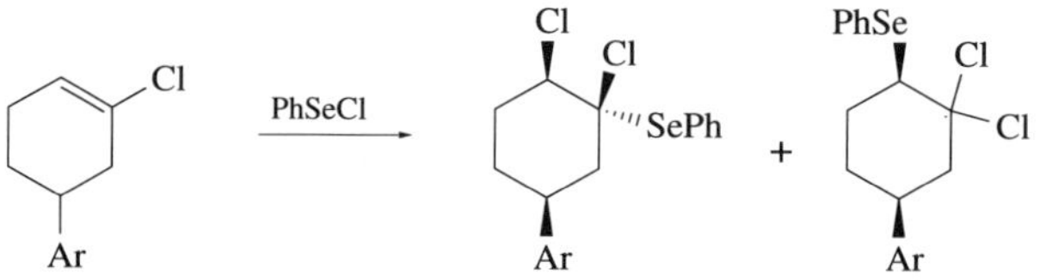

(116) Ar = 2,5-$Cl_2C_6H_3$ 60:40 (107)

The α-chloro selenide (**117**) was obtained in low yield from a reaction which involves the initial addition of benzeneselenenyl chloride to an alkyne (Scheme 69) ⟨91JOC4529⟩. α-Halovinyl selenides undergo thermal [2 + 2] cycloaddition to alkenes with captodative substituents ⟨85T4183⟩; thus (**118**) adds to the alkene (**119**) giving the cyclobutane (**120**) as a mixture of stereoisomers (Equation (108)).

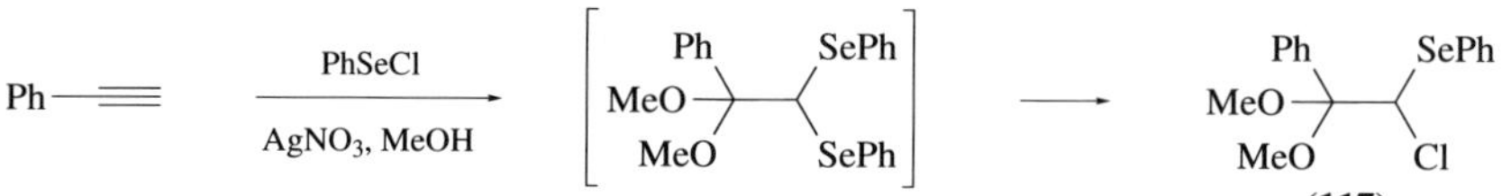

(**117**)

Scheme 69

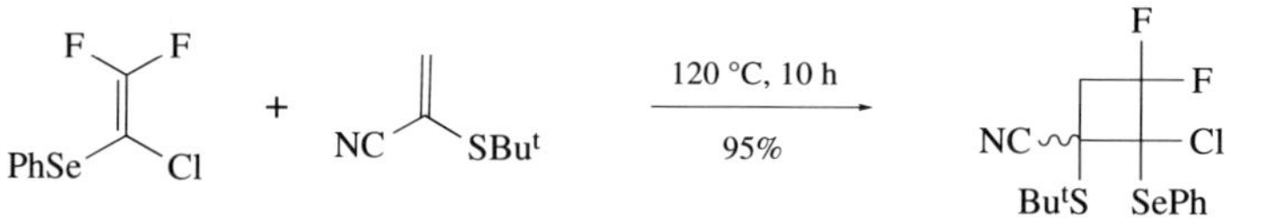

(**118**) (**119**) (**120**) (108)

The electrochemical monofluorination of ethyl phenylselenoacetate has been described, and the α-anion of the resulting α-fluoro α-phenylselenoacetate has been alkylated, e.g., with benzyl bromide; the fluorination of arylselenomethanes with the electron-withdrawing substituents CN and $CONH_2$ in place of CO_2Et has been effected similarly ⟨92TL3161⟩.

4.02.3.1.2 *Tri- and tetracoordinate α-halo selenium derivatives, $R^1{}_2CHalSe(O)R^2$, $R^1{}_2CHalSe(O)_2R^2$, etc.*

The oxidative deselenenylation of selenides constitutes an important method of synthesising alkenes and in general occurs at room temperature; thus in general reports of the isolation of α-halo selenoxides and α-halo selenones are uncommon. The α-halo selenoxide (**121**) (Equation (109)) has been isolated and is reported to be reasonably stable at room temperature ⟨81JOC3721⟩, *syn* elimination to give 2,3-dichlorocyclohexene requiring heating for 20 minutes in refluxing tetrachloromethane. In general however α-halo selenoxides should be less stable than the corresponding unhalogenated compounds as the presence of a halogen atom in the α position has been found to accelerate *syn* elimination; thus 1-chloroethyl phenyl selenoxide reacts 12 times faster than ethyl phenyl selenoxide and has a half-life of only 30 minutes at 38 °C in $CDCl_3$ ⟨78JOC1697⟩. Although the monofluoro compound is unknown, 1,1-difluoro- and 1,1,1-trifluoromethyl phenyl selenoxide have been prepared by the hydrogen peroxide oxidation of the selenides in dichloromethane ⟨93TL1311⟩.

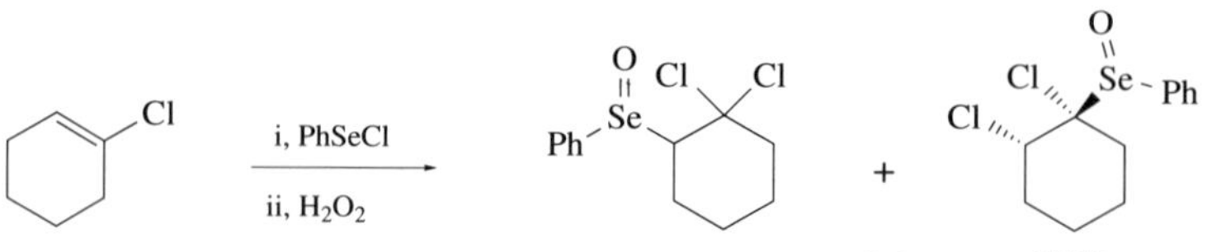

3:1 (**121**) (109)

There are no references in *Chemical Abstracts* to simple α-halo selenones of the type $R^1{}_2CHal\,Se(O)_2R^2$ (R^1,R^2 = H, alkyl or aryl). The α-halo selenium(IV) dichlorides (**122**) were obtained by treating the appropriate selenide with a stoichiometric amount of sulfuryl chloride in chloroform (Equation (110)) ⟨89TL2665⟩.

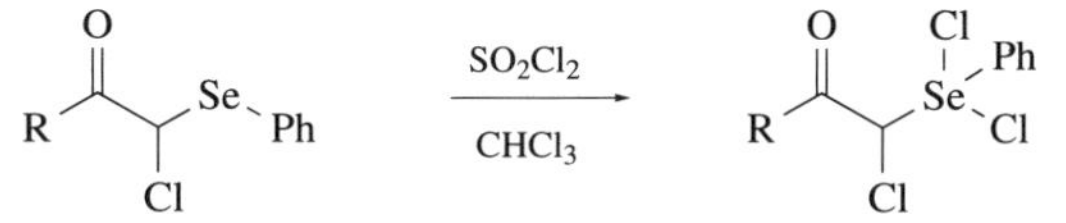

$$R \underset{Cl}{\overset{O}{\|}} Se{-}Ph \xrightarrow[CHCl_3]{SO_2Cl_2} R \underset{Cl}{\overset{O}{\|}} \underset{Cl}{\overset{Cl}{|}} Se{-}Ph \quad (110)$$

R = Ph, 95%
R = Me, 83%
(122)

4.02.3.2 α-Halo Tellurium Derivatives, $R^1_2CHal(TeR^2)$, etc.

4.02.3.2.1 Dicoordinate α-halo tellurium derivatives, $R^1_2CHal(TeR^2)$

Although there are few examples in the literature, it would appear that the methods used to prepare the corresponding selenium and sulfur compounds are applicable to the synthesis of molecules of the type $R^1_2CHal(TeR^2)$ as well. Thus diphenyl ditelluride, paralleling the behaviour of diphenyl diselenide, reacts with dichloromethane under alkaline, phase-transfer conditions to give chloromethyl phenyl telluride, which is reported to be unstable (Equation (111)) ⟨84JOM(277)261⟩. The reaction of acetylide anions with tellurium powder generates telluride anions which can be alkylated with chloroiodomethane (Scheme 70) ⟨82TL1531⟩. The procedure has been used to prepare the ^{125}Te-labelled compound (**123**; R = TMS) (Scheme 71) in 63% yield ⟨85JA6298⟩ and is modelled on a method used to synthesise the corresponding sulfides ⟨81RTC10⟩, with the exception that tellurium powder is used because sodium telluride, unlike sodium sulfide, does not react with acetylide anions.

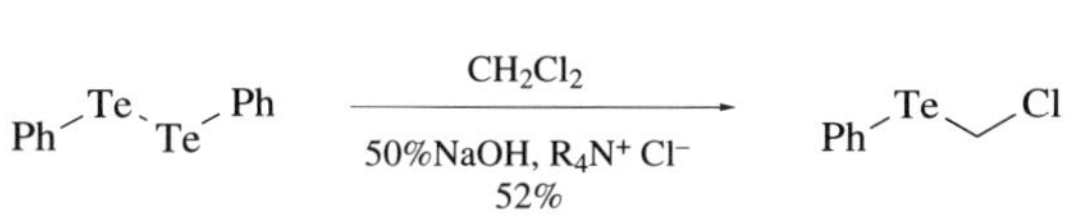

$$Ph{-}Te{-}Te{-}Ph \xrightarrow[50\%NaOH, R_4N^+Cl^-]{CH_2Cl_2} Ph{-}Te{\frown}Cl \quad (111)$$
52%

$$R{\equiv\equiv} \xrightarrow{Bu^nLi} R{\equiv}C^-Li^+ \xrightarrow{Te\ powder} R{\equiv\equiv}Te^-Li^+ \xrightarrow[-40\ °C\ to\ -60\ °C]{ICH_2Cl} R{\equiv\equiv}Te{\frown}Cl$$

R = TMS, Me, Bu^n, Ph

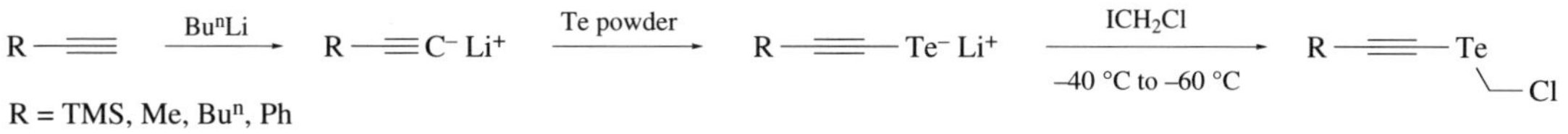

Scheme 70

4.02.3.2.2 Tri- and tetracoordinate α-halo tellurium derivatives, $R^1_2CHalTe(O)R^2$ and $R^1_2CHalTe(O)_2R^2$, etc.

Ethyl diazoacetate is reported to react with benzenetellurenyl bromide to give an unstable α-bromo telluride which was oxidized to the tellurone (**123**) (Scheme 71) ⟨93JOM(460)31⟩.

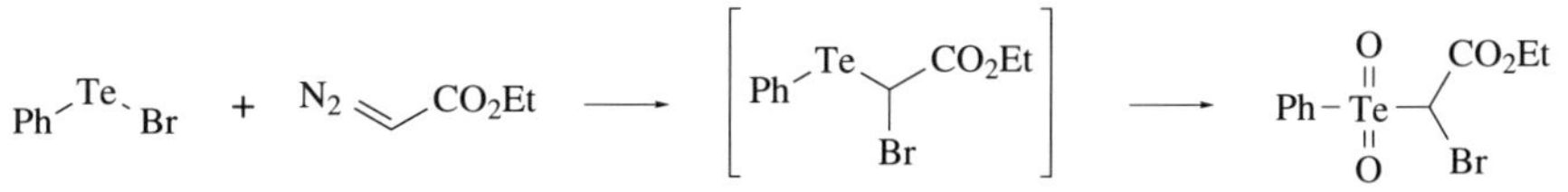

$$Ph{-}Te{-}Br + N_2{=}CHCO_2Et \longrightarrow \left[Ph{-}Te{\frown}\underset{Br}{CHCO_2Et} \right] \longrightarrow Ph{-}\underset{\underset{O}{\|}}{\overset{\overset{O}{\|}}{Te}}{-}\underset{Br}{CHCO_2Et}$$

(123)

Scheme 71

4.03
Functions Incorporating a Halogen and Another Heteroatom Group Other Than a Chalcogen

ALEX. C. CAMPBELL and DAVID R. JAAP
Organon Laboratories Ltd, Newhouse, UK

4.03.1 HALOGEN AND NITROGEN DERIVATIVES—$R^1{}_2CHal(NR^2{}_2)$, $R^1{}_2CHal(NR^2X)$, $R_2CHal(NX_2)$, $R_2CHal(NY)$	96
4.03.1.1 α-Halo Amines—$R^1{}_2CHalNR^2{}_2$ (Where R^1, R^2 = H, Alkyl or Aryl)	97
4.03.1.2 N-Substituted α-Halo Amines—$R^1{}_2CHal(NR^2X)$, $R_2CHal(NX_2)$, $R_2CHal(NY)$	98
4.03.1.2.1 α-Halo amine derivatives—$R^1{}_2CHal(NR^2X)$, $R_2CHal(NX_2)$ (singly bonded nitrogen)	98
4.03.1.2.2 Other α-halo amines—$R_2CHal(NY)$ (doubly bonded nitrogen)	107
4.03.2 HALOGEN AND PHOSPHORUS DERIVATIVES—$R^1{}_2CHalPR^2$, $R^1{}_2CHalPR^2{}_2$, $R^1{}_2CHalPO(OR^2)_2$, etc.	110
4.03.2.1 Dicoordinate Phosphorus Derivatives—$R^1{}_2CHalPR^2$	110
4.03.2.2 Tricoordinate Phosphorus Derivatives—$R^1{}_2CHalPR^2{}_2$, etc.	110
4.03.2.2.1 Primary and secondary α-halophosphines—$R_2CHalPH_2$, $R^1{}_2CHalPH(R^2)$	111
4.03.2.2.2 Tertiary α-halophosphines—$R^1{}_2CHalPR^2{}_2$	111
4.03.2.3 Tetracoordinate Phosphorus Derivatives—$[R^1{}_2CHalPR^2{}_3]^+X^-$, $R^1{}_2CHalP(O)R^2{}_2$, $R^1{}_2CHalPO(OH)R^2$, $R^1{}_2CHalPO(Hal)R^2$, $R_2CHalPO(OH)_2$, etc.	117
4.03.2.3.1 α-Halophosphonium salts—$[R^1{}_2CHalPR^2{}_3]^+X^-$	117
4.03.2.3.2 α-Halophosphine oxides and sulfides—$R^1{}_2CHalP(Y)R^2{}_2$	118
4.03.2.3.3 α-Halo oxo acids of phosphorus	123
4.03.2.4 Penta- and Hexacoordinate Phosphorus Derivatives—$R^1{}_2CHalPR^2{}_4$, $[A]^+[R_2CHalPX_5]^-$, etc.	133
4.03.3 α-HALO ARSENIC, ANTIMONY AND BISMUTH DERIVATIVES—$R^1{}_2CHalAsR^2{}_2$, $R_2CHalAsO(OH)_2$, etc.	134
4.03.3.1 α-Halo Arsenic Derivatives	134
4.03.3.1.1 Tricoordinate α-halo arsenic derivatives	134
4.03.3.1.2 Tetracoordinate α-halo arsenic derivatives	135
4.03.3.1.3 Pentacoordinate α-halo arsenic derivatives	135
4.03.3.2 α-Halo Antimony and Bismuth Derivatives	136
4.03.4 α-HALO ALKYLMETALLOIDS—$R_2CHalMETALLOID$	136
4.03.4.1 α-Halo Silicon Derivatives—$R^1{}_2CHalSiR^2{}_3$	136
4.03.4.1.1 Alkyl- and aryl(α-haloalkylsilanes)	136
4.03.4.1.2 Halo(α-haloalkyl)silanes	139
4.03.4.1.3 (α-Haloalkyl)oxysilanes	140
4.03.4.1.4 Miscellaneous α-halo silicon derivatives	140
4.03.4.2 α-Halo Germanium Derivatives—$R^1{}_2CHalGeR^2{}_3$, etc.	141
4.03.4.2.1 Alkyl- and aryl(α-haloalkyl)germanes	141
4.03.4.2.2 Halo(α-haloalkyl)germanes	142
4.03.4.2.3 Miscellaneous α-halogermanes	142
4.03.4.3 α-Halo Boron Derivatives—$R^1{}_2CHalBR^2{}_2$, etc.	142

4.03.4.3.1 (α-Haloalkyl)boron hydrides	142
4.03.4.3.2 Alkyl- and aryl(α-haloalkyl)boranes	142
4.03.4.3.3 Halo(α-haloalkyl)boranes	143
4.03.4.3.4 (α-Haloalkyl)oxyboranes	144
4.03.4.3.5 Miscellaneous (α-haloalkyl)boranes	145
4.03.5 α-HALO METAL DERIVATIVES—R_2CHalM, etc.	145
4.03.5.1 Group 1 and Group 2 Derivatives—R_2CHalLi, etc.	145
4.03.5.1.1 α-Haloalkyllithium derivatives	145
4.03.5.1.2 α-Haloalkylmagnesium derivatives	147
4.03.5.2 Transition Metal Derivatives—R_2CHalFeX_n, etc.	147
4.03.5.2.1 Derivatives of chromium, molybdenum and tungsten	147
4.03.5.2.2 Derivatives of manganese, iron and cobalt	148
4.03.5.2.3 Derivatives of ruthenium, rhodium and palladium	149
4.03.5.2.4 Derivatives of rhenium, osmium, iridium and platinum	150
4.03.5.2.5 Derivatives of copper, silver and gold	151
4.03.5.2.6 Derivatives of zinc, cadmium and mercury	152
4.03.5.3 Group 3 and Group 4 Derivatives—R_2CHalSnX_3, etc.	153
4.03.5.3.1 Derivatives of aluminum, gallium, indium and thallium	153
4.03.5.3.2 Derivatives of tin	153
4.03.5.3.3 Derivatives of lead	156

4.03.1 HALOGEN AND NITROGEN DERIVATIVES—R^1_2CHal(NR^2_2), R^1_2CHal(NR^2X), R_2CHal(NX_2), R_2CHal(NY)

α-Halo amines are probably best represented as an equilibrium between the covalent form (**1a**) and the iminium salt (**1b**) (Equation (1)).

$$\text{Hal} \overset{R^1}{\underset{R^2}{\rightarrow}} N \overset{R^3}{\underset{R^4}{\diagdown}} \quad \rightleftharpoons \quad \overset{R^1}{\underset{R^2}{=}} \overset{+}{N} \overset{R^3}{\underset{R^4}{\diagdown}} \quad \text{Hal}^- \qquad (1)$$

(**1a**)   (**1b**)

Simple α-halo amines, R^1R^2CHal(NR^3R^4) where R^1, R^2, R^3 and R^4 are hydrogen or small alkyl groups and the halogen is other than fluorine, exist predominantly as iminium halides, for example the commercially available N,N-dimethylmethyleneammonium iodide (Eschenmoser's salt) (**2**) is a crystalline solid, insoluble in nonpolar solvents. Derivatives containing atoms or groups capable of limiting iminium ion formation, however, may have considerable covalent character. Thus, the fluoromethyl dialkyl amines (**3**) and (**4**) ⟨70CB104⟩ and the N-α-chlorobenzyl amide (**5**) ⟨63CB600⟩ are distillable liquids. The highly covalent nature of the fluoro compounds is due to the strong carbon–fluorine bond and the stability of the amide is attributed to the reluctance of the N atom to participate in iminium ion formation. However, the products obtained from the characteristically covalent N-α-haloalkyl amides and nucleophiles are of the same chemical class as those obtained in related reactions involving iminium halides ⟨B-60MI 403-01, 65OR(14)52, 70S49, 75ZN(B)245⟩ and it is likely that transient iminium ions are involved in reactions of these α-halo amides.

$$H_2C=\overset{+}{N}\overset{Me}{\underset{Me}{\diagdown}} \quad I^- \qquad F\diagdown\overset{R^1}{\underset{R^2}{N}} \qquad \overset{Cl}{\underset{Ph}{\diagdown}}\overset{}{\diagup}\overset{Me}{\underset{COPh}{N}}$$

(**2**)   (**3**) $R^1 = R^2 = $ Me   (**5**)
          (**4**) $R^1R^2 = (CH_2)_2O(CH_2)_2$

For the purpose of this chapter, the synthesis of α-halo amines will be confined to compounds which have been described as predominantly covalent in the literature. Compounds with essentially ionic character are discussed in Chapter 3.10. The literature does not always clearly distinguish between covalent and ionic forms, however, and some compounds represented by covalent structures may in fact be more ionic than covalent.

The preparation, physical properties and reactions of α-halo amines and the corresponding iminium halides have been the subject of a number of reviews ⟨B-65MI 403-01, B-69MI 403-01, B-70MI 403-01, B-76MI 403-01, 79COC61, 84H585, 90T1791⟩. Early work on the preparation and reactions of N-α-haloalkyl amides and imides are dealt with in three reviews ⟨B-60MI 403-01, 65OR(14)52, 70S49⟩.

General methods are available for the preparation of covalent α-halo amines and their derivatives. These include direct treatment of amines with halogenating agents; addition of halogens, hydrogen halides and alkyl halides to imines and enamines; halogenation of N-α-hydroxyalkyl amines and their derivatives and cleavage of *gem*-diheteroatomic species, for example methylenediamines, methylene aminoacetals and methylene aminothioacetals. These and other miscellaneous methods are exemplified in Sections 4.03.1.1 and 4.03.1.2, below.

4.03.1.1 α-Halo Amines—$R^1_2CHalNR^2_2$ (Where R^1, R^2 = H, Alkyl or Aryl)

Reports on the preparation of covalent primary and secondary α-halo amines are rare. The fluoro amine $(CF_3)_2CFNH_2$ was prepared by addition of hydrogen fluoride to $(CF_3)_2C=NH$ ⟨65JOC1398⟩. Similar addition of anhydrous HCl to $ClCF_2(CF_3)C=NH$ afforded $ClCF_2(CF_3)CClNH_2$ in 72% yield ⟨73JOC3924⟩.

The compounds (**8**) and (**10**), which are analogues of the neuroprotective agent MK-801 [(+)-(**6**)], are notable examples of stable secondary α-halo amines. The fluoro derivative (**8**) was obtained from the corresponding alcohol (**7**) with (diethylamino)sulfur trifluoride ⟨88EUP264183⟩ and the chloro analogue (**10**) was prepared in 88% yield by treatment of the alcohol (**9**) with thionyl chloride ⟨79JOC3117⟩.

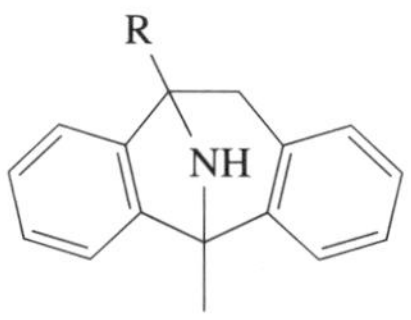

(**6**) R = H
(**7**) R = OH
(**8**) R = F

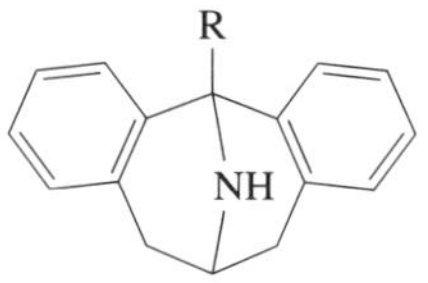

(**9**) R = OH
(**10**) R = Cl

There are a few notable examples of the preparation of covalent tertiary α-halo amines. The addition of HCl to the carbazole derivative (**11**) furnished the α-chloro amine (**12**) (Equation (2)) ⟨66MI 403-01⟩. The 3-acetyl-*N*-chloromethylindole (**14**) has been prepared in good yield by treatment of the *N*-dimethylaminomethyl amine (**13**) with AcCl (Equation (3)) ⟨89HCA93⟩. Similarly the *N*-methoxymethyl derivative (**15**) reacted with AcCl to give the *N*-chloromethyltriazacyclohexane (**16**) (Equation (4)) ⟨50JCS2925⟩. The latter reaction was accompanied by the formation of methyl acetate. Cleavage of the methylenediamine and the aminomethyl ether moieties therefore proceeds by attack of the acetyl group on the exocyclic heteroatom.

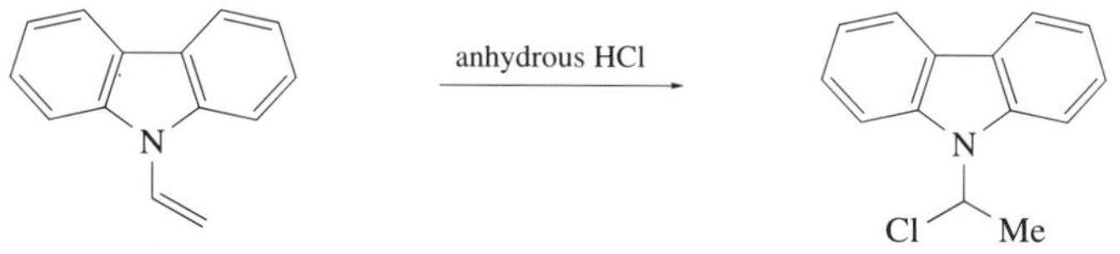

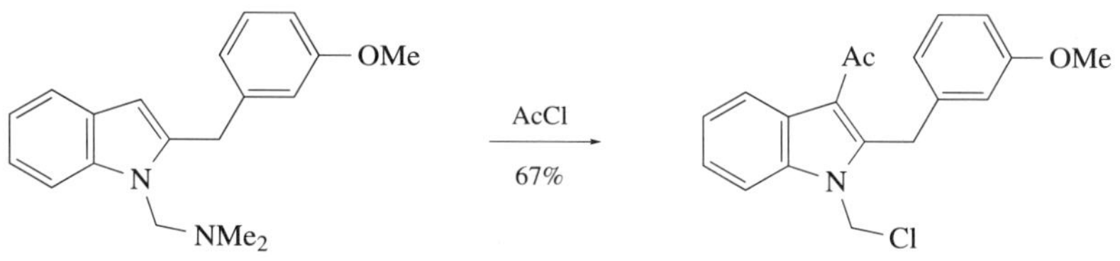

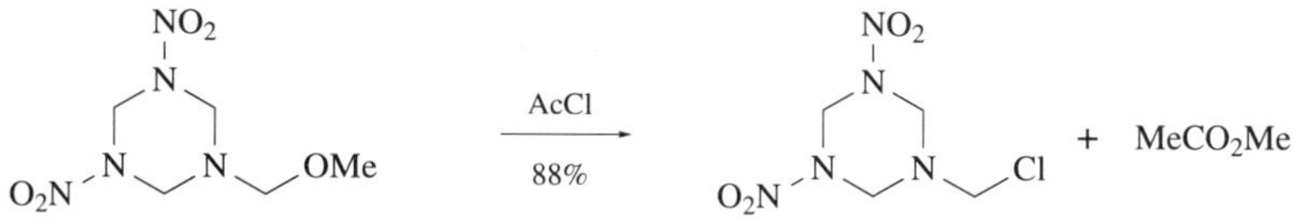

Davies *et al.* have described the chlorotropic shift and the accompanying skeletal rearrangement which resulted in the conversion of the *N*-chloro amine (**17**) into the α-chloro amine (**18**) on treatment with alumina (Equation (5)) ⟨85CC686⟩. The authors report that this rearrangement appears to have some generality for the preparation of strained bicyclic systems having bridgehead nitrogen.

$$(17) \xrightarrow[CH_2Cl_2]{Al_2O_3} (18) \qquad (5)$$

Bakker and Speckamp ⟨75TL4065⟩ irradiated quinuclidine (**19**) in C_2Cl_6 to furnish the 2-chloro derivative (**20**) (Equation (6)); use of CCl_4 in place of C_2Cl_6 gave poor yields.

$$(19) \xrightarrow[66\%]{C_2Cl_6} (20) \qquad (6)$$

4.03.1.2 *N*-Substituted α-Halo Amines—$R^1{}_2CHal(NR^2X)$, $R_2CHal(NX_2)$, $R_2CHal(NY)$

This section is divided into two parts. The first (4.03.1.2.1) covers derivatives of α-halo amines having atoms or groups attached to nitrogen by single bonds and the second (4.03.1.2.2) those having atoms or groups doubly bonded to nitrogen.

4.03.1.2.1 α-Halo amine derivatives—$R^1{}_2CHal(NR^2X)$, $R_2CHal(NX_2)$ (singly bonded nitrogen)

The most important derivatives of this type are the α-halo amides and imides, nearly all of which are essentially covalent in character. The preparations and properties of those compounds which were reported before 1970 have been extensively reviewed ⟨B-60MI 203-01, 65OR(14)52, 70S49⟩. Although there have been no up-to-date surveys, the occurrence of the compounds in the literature has been widespread. Other important *N*-substituted α-halo amine derivatives in this category are carbamic acid derivatives [$R_2CHal(NRCO_2R)$] and those formed with N—N and N—O bonds, etc.

(i) α-Halo amides and imides

Most α-halo derivatives of amides and imides have been prepared by the replacement of α-hydroxy groups or by direct halogenation. Other useful routes include the fragmentation of *gem*-diheteroatom groups of the type $R^1CON(R^2)CR^3R^4Y$ where $Y = NR^5R^6$, OR^5 and SR^5, and the addition of electrophiles to imines and enamines. Examples of these and other miscellaneous methods are described below. Many of these compounds are unstable, reactive, alkylating agents and are often used immediately after their preparation.

(a) By replacement of an α-hydroxy group. α-Chloro amides and imides are typically prepared by treatment of the α-hydroxy analogue with $SOCl_2$, often in the presence of pyridine or a more basic analogue, or a catalytic amount of DMF, for example the α-chloro amides (**21**) ⟨73JOC2251⟩, (**22**) ⟨87CZ247⟩ and (**23**) ⟨87H(25)221⟩ have been prepared in 93%, 88% and 63% yields respectively by this method. PCl_5 ⟨86AP(319)954⟩, Me_2SiCl_2 ⟨87JGU334⟩ and TMS-Cl ⟨91JGU1875⟩ have also been used to good effect for this replacement of OH by Cl.

Amides containing an unsubstituted chloromethyl group on N are often prepared from a primary or secondary amide by treatment firstly with HCHO to give the *N*-hydroxymethyl derivative and then with the chlorinating agent *in situ* ⟨81CB3421, 87JGU334, 88JGU2192, 91JGU1875⟩, for example $CH_3C(O)NHCH_2Cl$ is formed in 73% overall yield by this process ⟨81CB3421⟩. Many of the homologous *N*-α-chloroalkyl compounds derived from primary amides, however, cannot be prepared in this way because of the tendency of the higher aldehydes to react with primary amides to give bisamides $R^1CH(NHCOR^2)_2$ ⟨65OR(14)52⟩.

Methyl *N*-acetyl-α-hydroxyglycinate has been converted into the bromo derivative ($AcNHCHBrCO_2Me$) in 59% yield by treatment with TMS-Br ⟨89SC1479⟩. Similarly, the hydroxy group in $PhCONHCH_2OH$ was smoothly substituted by Br on treatment with PBr_3 ⟨73JA7813⟩. The commercially available *N*-bromomethylphthalimide has been prepared in 95% yield by the action of a mixture of HBr and H_2SO_4 on the *N*-α-hydroxy precursor ⟨87OS(65)119⟩.

(b) By direct halogenation. As a method for the synthesis of *N*-haloalkyl amides and imides, direct halogenation has been essentially limited to substrates bearing an activating function on the α-carbon atom. NBS and Br_2 have been the most frequently used brominating agents, the preferred solvent is CCl_4 and radical initiation is often employed; for example, the α-bromo lactam **(24)** ⟨89T6113⟩, the α-bromoacetamide **(25)** ⟨88T5403⟩ and the α-bromophthalimide **(26)** ⟨90T5263⟩ have been prepared in excellent yields from their precursors by the action of NBS in CCl_4 containing AIBN, in MeOH under photolysis and in MeOH in the absence of a radical initiator, respectively.

(24)   **(25)**   **(26)**

phth = phthaloyl

(c) By cleavage of gem-*diheteroatomic compounds.* There is a large number of reports on the preparation of α-halo derivatives from compounds containing the —$CON(R^1)CR^2R^3Y$ moiety (where Y = NR^4R^5, SR^4 or OR^4 and R^4, R^5 = H, alkyl, aryl, COR, SiR_3, etc.) by cleavage of the C—Y bond with halogens, acid chlorides, Lewis acids, etc. The action of acid halides on *N*-aminomethyl amines also affords α-halo amides. In contrast to the reactions of many of the corresponding amines, electrophilic attack on the amides is highly selective for the heteroatom Y and the reaction usually furnishes good yields.

As a method of preparation of *N*-α-halo amides and imides, cleavage of *gem*-diheteroatomic systems has been applied mainly to the preparation of chloro derivatives, although the synthesis of a few fluoro, bromo and iodo compounds have been described. *N*-(1-Piperidinylmethyl)benzamide **(27)** reacted with acetyl chloride to give the *N*-chloromethyl amide **(28)** and the expected *N*-acetylpiperidine (Equation (7)) ⟨63CB595⟩. Acyl chlorides react with the readily prepared, highly reactive 1,3,5-hexahydrotriazines to give the corresponding *N*-alkyl-*N*-chloromethyl amides in excellent yields ⟨79S810⟩. This method has been applied to the formation of chiral amides. Thus 1,3,5-tris[(*S*)-phenylethyl]hexahydrotriazine **(29)**, obtained simply and in quantitative yield by the addition of (*S*)-1-phenylethylamine to formalin, reacted with αβ-unsaturated acyl chlorides (3 molar equivalents) to give the corresponding *N*-[(*S*)-1-phenylethyl]-*N*-chloromethyl amides **(30)** in good yield (Equation (8)) ⟨92JOC1082⟩.

(27)   **(28)**   (7)

(29)   **(30)**   (8)

The *N*-trimethylsilyloxymethyl lactam (**31**) has been converted in good yield into the chloro derivative (**32**) by treatment with TMS-Cl ⟨89JGU200⟩.

(**31**) R = O-TMS
(**32**) R = Cl

There are many reports on the preparation of α-halo amides and imides from *N*-alkylthiomethyl amides, particularly in the chemistry of β-lactams (see Table 1 for examples).

Table 1 Preparation of *N*-α-halo amides from *N*-alkylthiomethyl amides.

Reactant	Product	Reagent	Yield (%)	Ref.
[phthN β-lactam with SCH₂CH₂CO₂Me] [a]	[phthN β-lactam with F]	i, Me₃OBF₄ ii, dbu	40	86TL3199
[TsO β-lactam with SMe, N-Ph]	[TsO β-lactam with Cl, N-Ph] [b]	Cl_2	80	90JOC3244
[Br penicillinate with S, CO₂Pom] [a]	[Br β-lactam with Cl, CO₂Pom]	Bu^{t}OCl	80	88T7007
[EtS, MeO₂C, NHZ, Ph amide] [c]	[Br, MeO₂C, NHZ, Ph amide] [c]	Br_2	94	91TL3163
MeO₂C-CH(SMe)-NHCOPh	MeO₂C-CH(Cl)-NHCOPh	Cl_2	92	75T863

Z = benzyloxycarbonyl, Pom = pivaloyloxymethyl.
[a] Penicillinate derivative. [b] Mixture of *cis* and *trans* isomers. [c] Diastereomeric mixture.

(d) By addition of electrophiles to imines and enamines. The imines (**33**) and (**35**) combined with acetyl chloride and chloroacetyl chloride respectively to form in good yields the α-chloro amides (**34**) (Equation (9)) ⟨88AG(E)304⟩ and (**36**) (Equation (10)) ⟨87CZ149⟩.

(**33**) →[AcCl, 66%] (**34**)   (9)

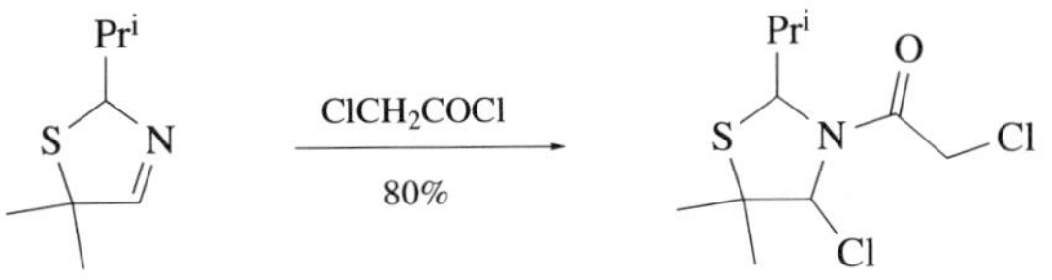

(35) (36)

(10)

N-Vinylphthalimide (**37**) and *N*-vinylisatin (**39**) reacted with HCl and HBr to give in nearly quantitative yields the respective adducts (**38**) (Equation (11)) ⟨86CB2387⟩ and (**40**) (Equation (12)) ⟨87CB1897⟩. The phthalimide (**39**) also reacted smoothly with HI to give the iodo derivative (**40**; X = I).

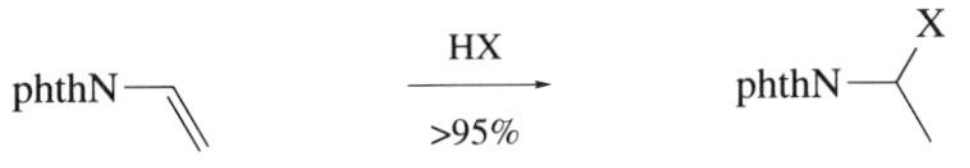

(37) X = Cl, Br, I (38)

(11)

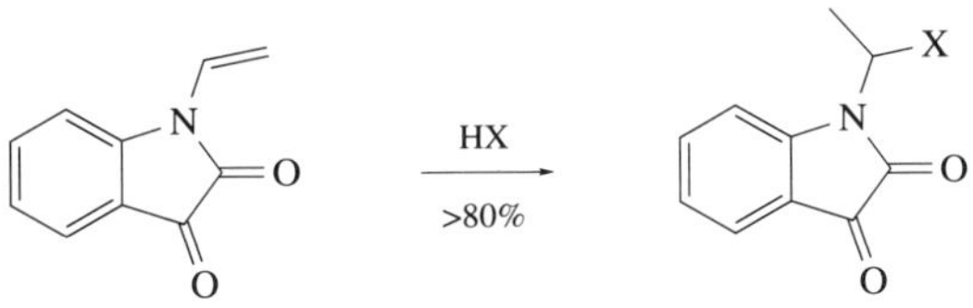

(39) X = Cl, Br, I (40)

(12)

(e) By miscellaneous methods. In a potentially versatile procedure for the synthesis of the elusive α-fluoro amino acids, potassium phthalimide reacted with $XFCHCO_2Et$ (where X = Cl and Br) to give the *N*-protected ethyl fluoroglycinate (**41a**) ⟨87JCS(P1)2203, 88JCS(P1)1149⟩.

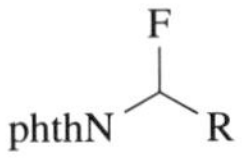

(**41**)
a; R = CO_2Et
b; R = H

The phthalimidomethyl fluoride (**41b**) has been prepared from the corresponding chloride in 76% yield by treatment with CsF and 18-crown-6 in THF ⟨87JFC(35)677⟩.

Unfortunately, no asymmetric induction was observed when the exocyclic double bond in the (*S*)-2′-thiazinylideneglycinate (**42**) was attacked by NCS to give the α-chloro glycinate (**43**) (Equation (13)) ⟨91TA157⟩.

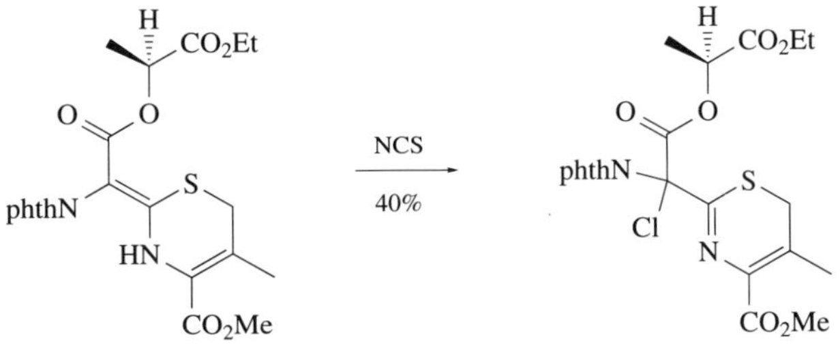

(42) (43)

(13)

(ii) α-Halo carbamoyl derivatives

(a) α-Halo carbamates. These compounds are well documented in the literature. They are generally prepared by the methods used for the synthesis of the structurally related amides.

N-Alkylation of dibenzyl iminodicarboxylate potassium salt (**44**) with *t*-butyl bromofluoroacetate furnished *t*-butyl *N,N*-bis(benzyloxycarbonyl)-α-fluoroglycinate (**45**) in 38% yield (Equation (14)) ⟨91JCS(P1)49⟩. Chlorination of the benzyloxycarbonyl-α-hydroxy-glycinate (**46**) with thionyl chloride and pyridine gave crude α-chlorocarbamate (**47**) in good yield ⟨85JHC957⟩. Using the method involving cleavage of *N*-alkoxymethyl derivatives, Williams *et al.* showed that this compound (**47**) (67% yield), and its bromo analogue (**49**) (88% yield), could also be readily obtained by treatment of the methoxy carbamate (**48**) with PCl_5 and PBr_3, respectively ⟨90JOC4657⟩.

$$\begin{array}{c} BnO_2C \\ N-K \\ BnO_2C \end{array} \quad \xrightarrow{\begin{array}{c} F \\ Br-\overset{|}{C}H \\ CO_2Bu^t \end{array}} \quad \begin{array}{c} BnO_2C F \\ N-\overset{|}{C}H \\ BnO_2C CO_2Bu^t \end{array} \quad (14)$$

(**44**)   (**45**)

(**46**) R = OH   (**48**) R = OMe
(**47**) R = Cl   (**49**) R = Br

Photolytic halogenation of *t*-butyl *N-t*-BOC-glycinate (*t*-BOC = *t*-butoxycarbonyl) (**50**) with NBS in CCl_4 afforded in 97% yield the crystalline α-bromo carbamate (**51**) ⟨87S223⟩. This compound, which can be stored at 4 °C for several months without decomposition, is a versatile synthon for α-amino acids. Hamon *et al.* used such a radical-induced halogenation in the asymmetric synthesis of α-amino acids. Thus 8-phenylmenthyl *N*-BOC-glycinate (**52**) underwent free radical bromination to give the unstable bromo derivative (**53**) obtained as one diastereoisomer in 95% yield ⟨92T5163⟩. This reacted with a number of Grignard reagents to give the corresponding 2-alkyl glycinates with high diastereoselectivity. The benzyloxycarbonylmorpholinone (**54**) and its enantiomer also underwent regiospecific and stereospecific bromination when treated with NBS to give respectively the 3-bromo derivative (**55**) and its enantiomer in nearly quantitative yields ⟨88JA1547⟩. The *trans* configuration of the bromine atom to the phenyl groups was based on spectroscopic and chemical evidence. The chloro derivative (**56**) was similarly prepared by treating the morpholinone (**54**) with *t*-BuOCl ⟨88JA1547⟩. A similar *anti* selectivity was observed when the protected (*R*)-imidazolidinone (**57**) furnished the bromo derivative (**58**) (90% yield) on treatment with NBS in the presence of AIBN ⟨91LA655⟩.

(**50**) R = H   (**52**) R = H   (**54**) R = H   (**57**) R = H
(**51**) R = Br   (**53**) R = Br   (**55**) R = Br   (**58**) R = Br
                                                  (**56**) R = Cl

(b) Miscellaneous α-halo carbamoyl derivatives. *N*-Methylene-2,6-diethylbenzenamine (**59**) was added to phosgene to give the carbamoyl chloride (**60**) in good yield (Equation (15)) ⟨87JHC945⟩ and the photohalogenation of ethyl isocyanate (**61**) with Cl_2 afforded the analogous adduct (**62**) in 83% yield (Equation (16)) ⟨90T7729⟩. The parent chloromethylcarbamoyl chloride has been prepared in good yield by the addition of monomeric HCHO to HNCO and then treatment of the resultant hydroxymethyl isocyanate $HOCH_2NCO$ with $SOCl_2$ ⟨63JOC1825⟩. The intermediate $HOCH_2NCO$ polymerizes explosively above 0 °C.

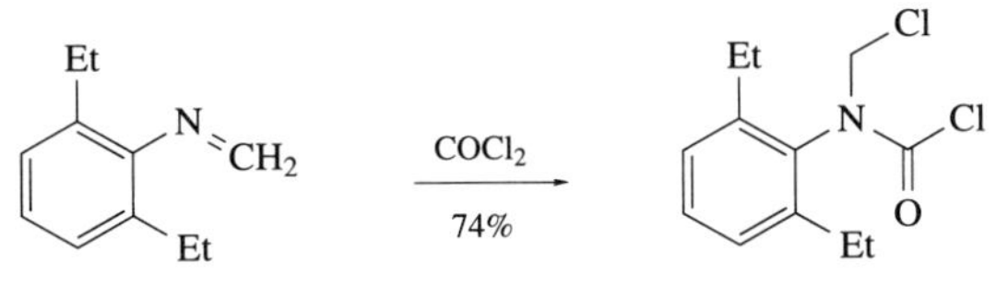

5-Chlorohydantoin (**64**) has been prepared by reduction of imidazolidinetrione with KBH_4 and treatment of the resultant 5-hydroxy derivative (**63**) with $SOCl_2$ in 50% overall yield ⟨71BSF942⟩. Cleavage of the tolylthio group of the uracil (**65**) with SO_2Cl_2 gave the corresponding chloro compound (**66**) in 45% yield ⟨85JCS(P1)93⟩. Similarly, cleavage of the benzyloxymethylamino function in the uracil (**67a**) and the dihydro analogue (**67b**) with TMS-I afforded the N-iodomethyl derivatives (**68a**) and (**68b**) respectively (Equation (17)) ⟨85S323⟩.

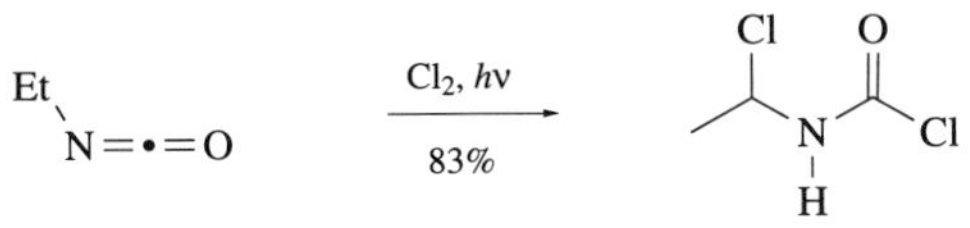

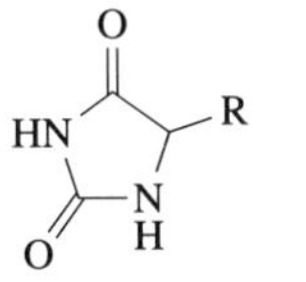

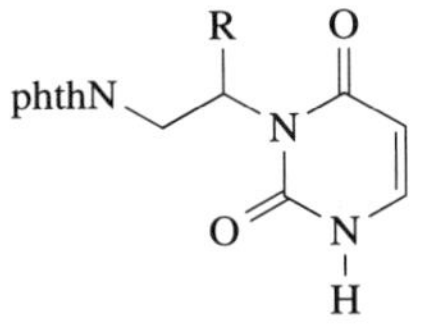

The triazines (**69a–e**) have been converted into the carbamoylsulfenyl chlorides (**70a–e**) by treatment with 3 molar equivalents of chlorocarbonylsulfenyl chloride. The intermediates, which were not isolated, were then converted into the 3-chloromethyl-2(3H)-benzothiazolones (**71a–e**) by a Friedel–Crafts process employing aluminum chloride (Scheme 1) ⟨88S482⟩. The overall yields ranged from 35–65%.

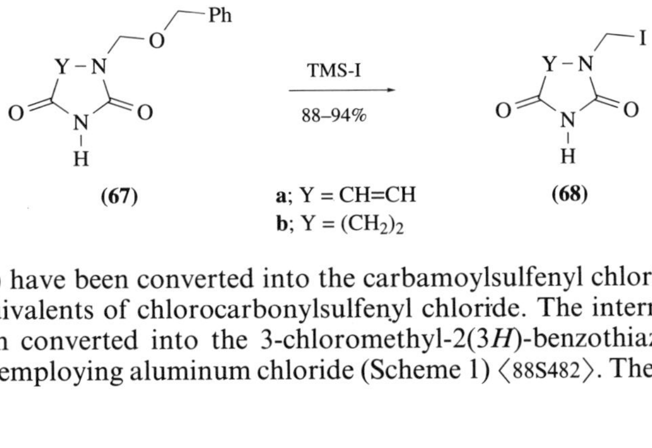

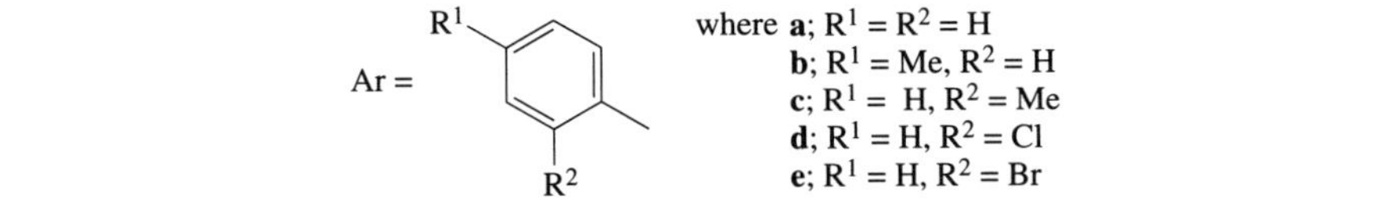

where a; $R^1 = R^2 = H$
b; $R^1 = Me, R^2 = H$
c; $R^1 = H, R^2 = Me$
d; $R^1 = H, R^2 = Cl$
e; $R^1 = H, R^2 = Br$

Scheme 1

(iii) N,N-Difluoro α-halo amines

N,N-Difluoro α-halo amines have been prepared in good yields by the free radical addition of tetrafluorohydrazine to the appropriate halo alkene ⟨67JOC4034, 68JOC1861, 68JOC2330⟩, for example the α-fluoro amine (**73**) was obtained in 87% yield by the addition of N_2F_4 to the dicyanodifluoroethene (**72**) (Equation (18)) ⟨68JOC2330⟩. Analogous α-chloro and α-bromo compounds were also prepared in good yields by this method ⟨67JOC4034⟩.

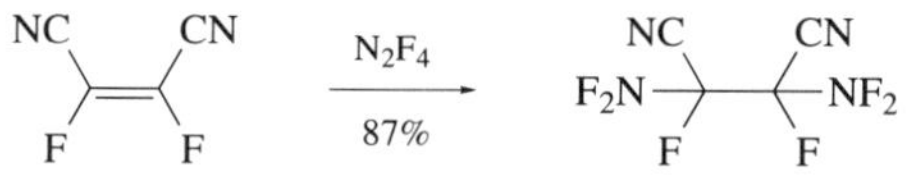

$$\underset{(72)}{\underset{F\quad F}{NC\diagup\hspace{-0.3em}=\hspace{-0.3em}\diagdown CN}} \xrightarrow[87\%]{N_2F_4} \underset{(73)}{\underset{F\quad F}{F_2N-\overset{NC}{\underset{}{C}}-\overset{CN}{\underset{}{C}}-NF_2}} \qquad (18)$$

(iv) N-Oxy α-halo amines

A few stable compounds of this type have been reported, for example the perfluorooxaziridine (**75**) was prepared in 39% yield by treatment of the imine (**74**) with mcpba (Equation (19)) ⟨92EUP496413⟩ and the (2-nitrosotrifluoroethyl)oxysulfonyl fluoride (**77**) was obtained in 49% yield by treatment of trifluoroethene with nitrosonium fluorosulfate (**76**) (Equation (20)) ⟨86IZV810⟩. The perfluoro analogues of these nitroso compounds were shown to be capable of [2 + 4] addition reactions with dienes.

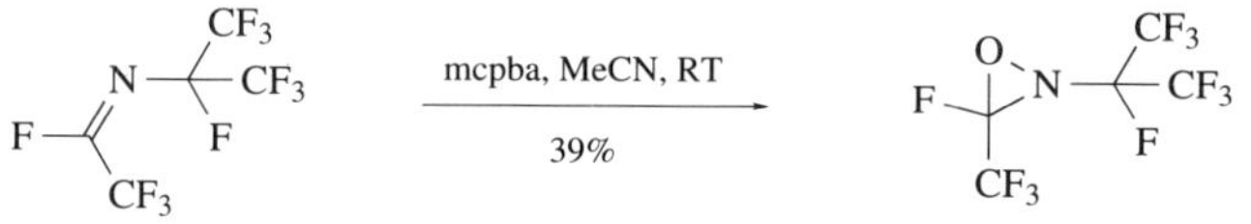

$$\underset{(74)}{\text{imine}} \xrightarrow[39\%]{\text{mcpba, MeCN, RT}} \underset{(75)}{\text{oxaziridine}} \qquad (19)$$

$$NOSO_3F \;+\; \underset{(76)}{CF_2=CHF} \xrightarrow{49\%} \underset{(77)}{ON-CHF-CF_2-OSO_2F} \qquad (20)$$

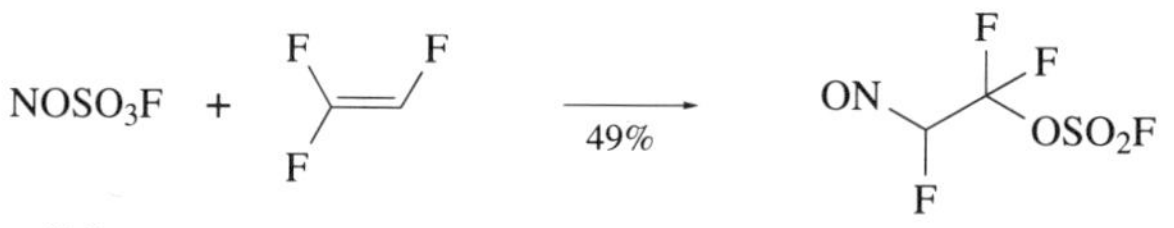

(v) N-Thio α-halo amines

This substantial category of α-halo amines is comprised essentially of sulfonamides and, because of their resemblance to the imides, it is not unexpected that the methods of transformation for the two series are similar.

Thus the sulfonamide (**78**) afforded the α-chloro derivative (**79**) in nearly quantitative yield when treated with $SOCl_2$ ⟨73JOU418⟩. The N-methyl alkanesulfonamides (**80a**) and (**80b**), on concomitant treatment with HCHO and TMS-Cl, gave the corresponding N-chloromethyl derivatives (**81a**) and (**81b**) in good yields (Equation (21)) ⟨89JGU1625⟩.

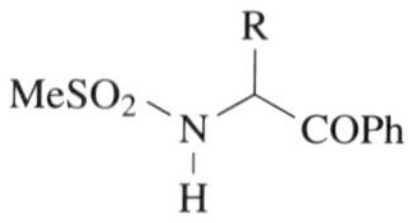

$$MeSO_2-\underset{H}{N}-\underset{R}{CH}-COPh$$

(**78**) R = OH
(**79**) R = Cl

Treatment of dimethyl benzenesulfonamidomalonate (**82**) with Br$_2$ under the action of ultraviolet light afforded the α-bromo sulfonamide (**83**) in 77% yield ⟨92AP(325)411⟩. Similarly, the *N*-tosylbenzazepine (**84**) has been halogenated with NBS in the presence of benzoyl peroxide ⟨67JCS(C)58⟩ and with Br$_2$ ⟨81JCS(P1)2435⟩ to give the bromo derivative (**85**) in excellent yields. Unlike isoquinoline, the 1-cyano-2-tosyl derivative (**86**) added bromine readily to give the isolable *cis*-dibromo derivative (**87**) (Equation (22)) ⟨74JOC1965⟩.

(vi) N-Amino α-halo amines

The compounds most frequently reported within this series are 1,2-diazoles and 1,2-diazines. Ethyl 1-hydroxymethylpyrazole-4-carboxylate (**88**) has been transformed into the fluoro derivative (**89**) in 76% yield by treatment with CsF and MsF in the presence of 18-crown-6 in THF ⟨87JFC(35)677⟩. Replacement of the hydroxy group by chlorine in the triazole (**90**) prepared in 99% yield by treatment of 1*H*-1,2,4-triazole with (HCHO)$_n$ and Et$_3$N was accomplished with SOCl$_2$ to give the *N*-chloromethyltriazole (**91**) in 93% yield ⟨87PS(33)41⟩. This inherently unstable compound (**91**) can be stored for months in CH$_3$CN. Thionyl chloride converted the *N*-hydroxymethylthiadiazolethione (**92**) into the chloro derivative (**93**) in 79% yield ⟨90JHC139⟩. Treatment of the pyrazoline (**94**) with Br$_2$ afforded the surprisingly stable bromo derivative (**95**) ⟨71MI 403-01⟩.

As a part of their studies on bridgehead hydrazines, Sheradsky and Moshenberg irradiated the *cis*-2,5-diphenyltetrahydropyridazines (**96**) and (**98**) in CCl$_4$ containing NBS and benzoyl peroxide to give in 62–68% yields the monobromo derivatives (**97**) ⟨87JOC101⟩ and (**99**) ⟨86JOC3123⟩ respectively. The *N*-methoxymethylhydrazines (**100**; R^1 = alkenyl or alkynyl) were converted into the chloro derivatives (**101**) by interaction with PCl_5 in moderate to good yields ⟨88TL6975⟩. By treatment with Lewis acids, these products were readily cyclized to 1,2-dinitrogen heterocycles via intramolecular nucleophilic attack at the multiple bond.

(**96**) R = H
(**97**) R = Br

(**98**) R = H
(**99**) R = Br

(**100**) R^1 = OMe
(**101**) R^1 = Cl

(vii) N-Nitro α-halo amines

N-Acetoxymethyl-*N*-nitromethylamine (**102**) has been cleaved with $SOCl_2$ in the presence of CH_3COOH and H_2SO_4 to give the corresponding chloro derivative (**103**) in 90% yield ⟨88JOU2003⟩. Forceful conditions were required to effect the addition of Br_2 to the acylvinyl nitro amine (**104**) to give the dibromo adduct (**105**) in 73% yield (Equation (23)) ⟨85JOU806⟩.

(**102**) R = OAc
(**103**) R = Cl

$$\text{(104)} \xrightarrow{Br_2, PhH} \text{(105)} \quad (23)$$

(viii) N-Phosphoryl α-halo amines

Of the possible nitrogen–phosphorus α-halo amines, only the phosphorus(V) acids and their derivatives have been reported. These are related to the carbamic acids and similar methods have been used for their preparation, for example photochemical chlorination of the phosphoramidic dichlorides (**106**) and (**108**) with Cl_2 and SO_2Cl_2 respectively gave in 45–55% yields the corresponding chloro derivatives (**107**) (Equation (24)) and (**109**) ⟨68JGU1276⟩. The α-hydroxy phosphoramidates and thiophosphoramidates (**110**; R^1 = Et,Pr^n,Pr^i,Bu^n,Bu^i;Y = O,S) were converted into the corresponding chloro derivatives (**111**) with $SOCl_2$ ⟨79EGP137839, 87JPR871⟩. Anti-Markovnikov addition of benzenesulfenyl chloride to the alkene (**112**) readily afforded the *N*-(α-chloro-β-phenylthio)phosphoramidate (**113**) (Equation (25)) which is stable in solution in hexane ⟨90JGU2208⟩.

$$\text{(106)} \xrightarrow[55\%]{Cl_2, h\nu} \text{(107)} \quad (24)$$

(**108**) R = Me
(**109**) R = CH_2Cl

(**110**) R^2 = OH
(**111**) R^2 = Cl

(ix) Miscellaneous α-halo amines

There are a few α-halo amines having groups attached to nitrogen by single bonds which do not fall into the above categories.

Peterman and Shreeve reported the preparation of the N-α-chloroalkyl aminosilyl derivative (**115**) in good yield by the action of HCl on the silyl diimine (**114**) (Equation (26)) ⟨76IC743⟩. Treatment of the imine (**116**) with BCl$_3$ gave the versatile synthon (**117**) in nearly quantitative yield (Equation (27)) ⟨70IC975⟩. While BBr$_3$ in place of BCl$_3$ gave a similar result, BF$_3$ gave the dipolar adduct (**120**). During attempts to obtain analogues of the neuroprotective agent MK-801 [(+)-(**6**)], Monn and Rice inadvertently obtained the remarkably stable bridgehead iodide (**119**) in 81% yield by sequential treatment of the amidine (**118**) with s-butyllithium and CF$_3$CH$_2$I ⟨89TL911⟩.

4.03.1.2.2 Other α-halo amines—R$_2$CHal(NY) (doubly bonded nitrogen)

The compounds within this subsection are those in which Y of the general formula R$_2$CHal(NY) is doubly bonded to nitrogen, for example N=O, N=N, N=P and N=S. Although there are no reviews on the synthesis of these compounds, they are featured in many publications.

(i) α-Halo nitroso derivatives

There are a few reports on the synthesis of stable α-halo nitroso compounds. A notable method is the chlorination of oximes. For example, Oppolzer et al. ⟨92JA5900⟩, in a search for chiral gem-chloro nitroso reagents capable of aminating prochiral ketones, formed the chloro nitroso compound (**122**) from the 2-bornanone (**121**) by treating the derived oxime with t-BuOCl (Equation (28)). Molecular Cl$_2$ has also been employed for this transformation ⟨85AJC1505⟩.

(ii) α-Halo nitro derivatives

Most of the many reported α-halo nitro compounds have been prepared by one of two methods, namely from nitro compounds by direct halogenation and from oximes by oxidation of the foregoing N-α-halo nitroso intermediates. The halogenation–oxidation procedure as applied to oximes can be carried out simultaneously by a single reagent, for example NBS (see ⟨91JOC316⟩ and references cited therein).

(a) Halogenation of nitro compounds. Takeuchi *et al.* reported the efficient α-fluorination of a number of α-nitro carboxylic acid derivatives (**123**; $R^2 = R^3O_2C$) and α-nitro sulfones (**123**; $R^2 = SO_2Ph$) by sequential treatment with KF (acting as a base) and freshly prepared $FClO_3$ to give the corresponding versatile intermediates (**124**) ⟨87JOC5061, 89JOC5453, 91CPB3120⟩. This is a particularly useful method of fluorination because it avoids the use of special equipment. The removal of the nitro group and/or the carboxylate function was readily achieved. Banks *et al.* were able to convert the lithium salt of 2-nitropropane into the *gem*-fluoro nitro derivative (**125**) in good yield by treatment with N-fluoroquinuclidinium fluoride ⟨86JFC(32)461⟩. Aliphatic nitro compounds have also been readily converted into the corresponding chloro and bromo derivatives by treatment with NCS and NBS in the presence of KOH ⟨86S826, 86S828⟩. A reducing agent was employed to prevent overbromination in the conversion of CH_3NO_2 into $BrCH_2NO_2$ with Br_2 ⟨91USP5043489⟩.

(**123**) $R^1 = H$
(**124**) $R^1 = F$
(**125**) $R^1 = Me, R^2 = F$

(**126**) $R = H$
(**127**) $R = Cl$

On reaction with dbu and TMS-Cl, the nitro carboxylate (**126**) selectively halogenated α to the nitro group to give the chloro derivative (**127**) in 54% yield ⟨88LA1169⟩.

Wade *et al.* ⟨91JA8807⟩ found that *trans*-1,2-dinitrospiropentane (**128**) was converted in 72% yield into a mixture (1:1) of the *cis* and *trans* isomers of the diiodo derivatives (**129**) by treatment with CH_3ONa and I_2 (Equation (29)). α-Halo nitro compounds can also be prepared by the addition of halogens to nitro alkenes, for example treatment of the nitroethene (**130**) with Cl_2 gave the dichloro derivative (**131**) in 81% yield (Equation (30)) ⟨89IZV1107⟩.

(29)

(30)

(b) Halogenation—oxidation of oximes. Sequential treatment of the cyclobutanone oximes (**132**) with Cl_2 and then NaOCl under phase-transfer catalysis furnished the *gem*-chloro nitro derivatives (**133**) in good yields (Equation (31)) ⟨89JOC2869⟩.

(31)

a; R = H
b; R = CO_2Et

Both stages in the conversion of an oxime into the α-halo nitro compound can be accomplished by hypobromite at a pH greater than 7. Thus, 2-adamantanone oxime was converted into the bromo nitro derivative (**134**) in 80% yield by treatment with NBS in aqueous sodium carbonate ⟨88JOC4645⟩. Walters *et al.* used, *inter alia*, the commercially available triazine (**135**) to prepare this compound in similar yield ⟨91JOC316⟩.

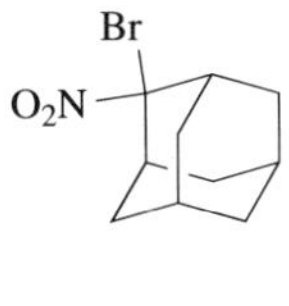

(**134**) 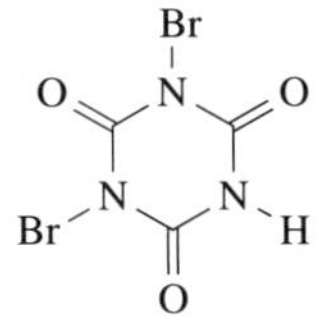 (**135**)

(iii) α-Halo azo alkanes and α-halo azides

The two principal methods for the preparation of α-halo azo alkanes are the electrophilic halogenation of hydrazones and the [2 + 3] cycloaddition of halo alkenes to diazo alkanes.

Aryl and alkyl hydrazones (**136**) of alkyl ketones and propanal were smoothly converted into the α-chloro azo alkanes (**137**) by the action of Bu^{t}OCl (Equation (32)) ⟨92S710⟩. Treatment of the pyrazoline (**138**) with NBS afforded the bromo derivative (**139**) with equal facility (Equation (33)) ⟨89CJC1125⟩.

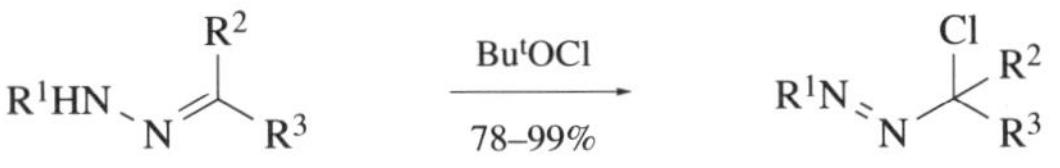

(**136**) → (**137**) (32)

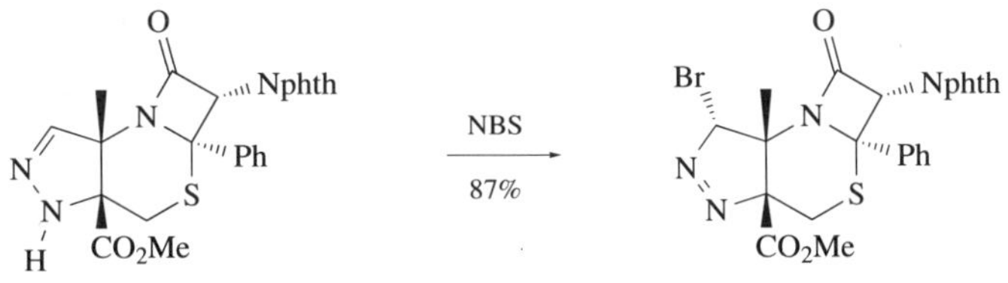

(**138**) → (**139**) (33)

The addition of diazomethane to methyl 2-chloro-2-cyclopropylideneacetate (**140**) occurred readily and regiospecifically to afford the 3-chloropyrazoline (**141**) (Equation (34)) ⟨89T2957⟩. The more bulky diphenyldiazomethane also added with high regiospecificity but in the reverse sense. Dimethyl chlorofumarate (**142**) and diphenyldiazomethane formed an adduct by regiospecific and stereospecific attack of nitrogen on the carbon bearing the chlorine atom to give the unstable *trans*-3,4-bis(methoxycarbonyl) derivative (**143**) which spontaneously decomposed to give the *trans*-cyclopropane (**144**) (Scheme 2) ⟨89JOC1135⟩.

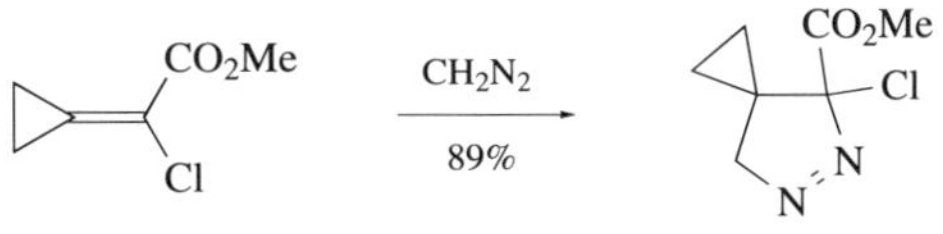

(**140**) → (**141**) (34)

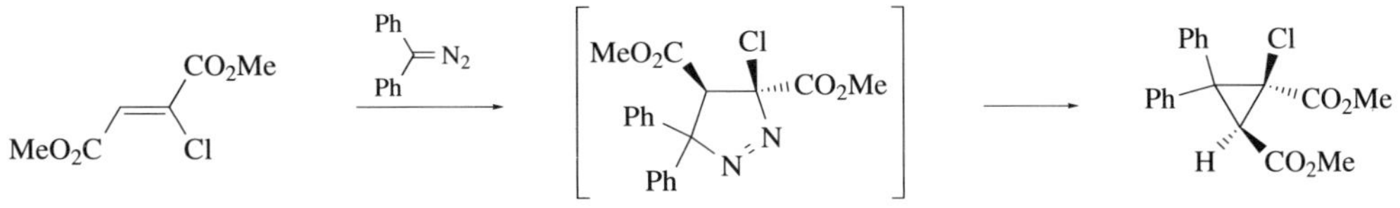

(**142**) → (**143**) → (**144**)

Scheme 2

Azido fluoro acetates have been prepared by treatment of the corresponding bromo compounds with sodium azide ⟨88JCS(P1)1149⟩, for example this method furnished EtO_2CCHFN_3 in 68% yield. Note that an attempted distillation of this azide resulted in an explosion.

(iv) Miscellaneous α-halo amines

α-Halo amines where the N atom is doubly bonded to an atom other than N or O are relatively few. Of these, compounds containing N=P and N=S are the most common. These have been formed by the reaction of Lewis acids with *N*-silyl imines and *N*-chloro imines, for example the silyl imine (**145**) furnished the phosphine imide (**146**) and the sulfinylamine (**147**) in good yields when heated with PCl_5 and $SOCl_2$ respectively (Scheme 3) ⟨79S747⟩. Shermolovich *et al.* have reported that sulfur dichloride and arenesulfenyl chlorides react with the *N*-chloro imine (**148**) to give the *N*-α-chloroalkyl sulfinimidoyl chlorides (**149**) and (**150**), generally in good yields (Scheme 4) ⟨82JOU2240⟩. The presence of an electron-withdrawing substituent at the α-carbon of the imine impedes the reaction.

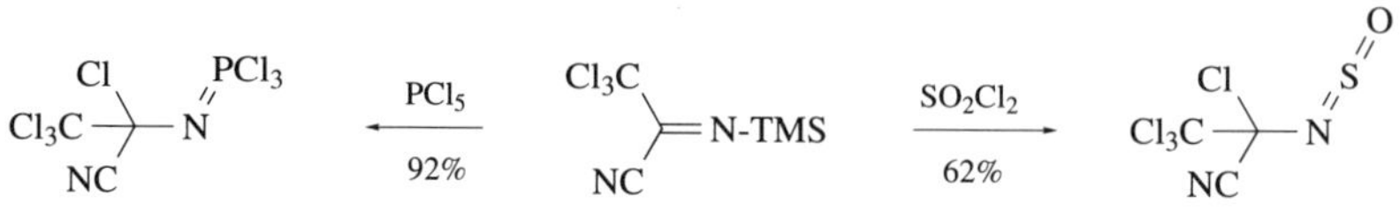

Scheme 3

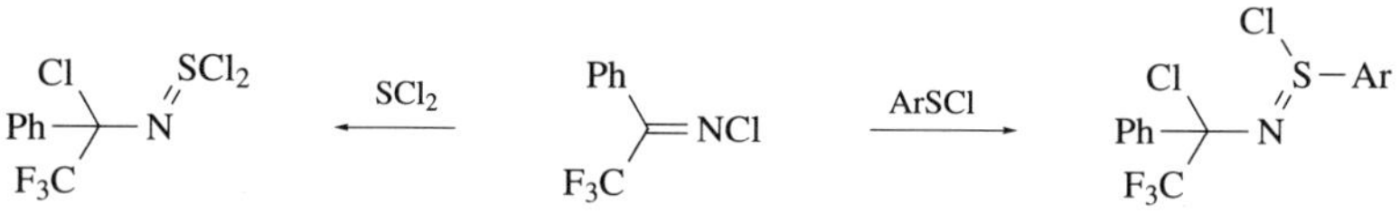

Scheme 4

4.03.2 HALOGEN AND PHOSPHORUS DERIVATIVES—$R^1_2CHalPR^2$, $R^1_2CHalPR^2_2$, $R^1_2CHalPO(OR^2)_2$, etc.

4.03.2.1 Dicoordinate Phosphorus Derivatives—$R^1_2CHalPR^2$

There are no reports in the literature of stable compounds which contain an sp^3-hybridized carbon atom bearing a halogen atom and a dicoordinate phosphorus atom. Reference to their existence as unstable intermediates in reactions first appeared in 1989 when the 2-chloro-2*H*-phosphirenes (**153a**) were postulated as transient species in the preparation of 1-chloro-1*H*-phosphirenes (**154a**) by the reaction of phosphaalkynes (**151**) with substituted chlorocarbenes (**152a**) (Scheme 5) ⟨89AG(E)225⟩. In 1991, the bromo and fluoro analogues (**154b**) and (**154c**) were similarly prepared ⟨91CB1207⟩.

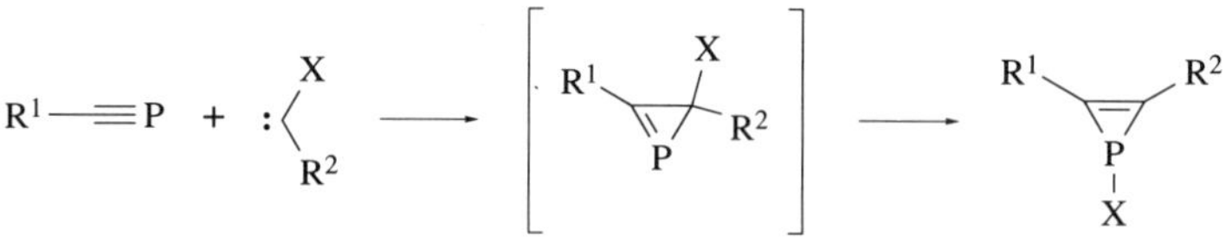

(a) X = Cl; (b) X = Br; (c) X = F

Scheme 5

4.03.2.2 Tricoordinate Phosphorus Derivatives—$R^1_2CHalPR^2_2$, etc.

Due to the instability of α-haloalkylphosphines, there are few general reactions for their synthesis. Examples of the routes employed, some of which indicate the inherent difficulties, are described below. Rigorous exclusion of moisture and oxygen is mandatory when manipulating these compounds and low temperatures and pressures are often required. There are no α-halophosphines commercially available.

4.03.2.2.1 Primary and secondary α-halophosphines—$R_2CHalPH_2$, $R^1_2CHalPH(R^2)$

Primary and secondary α-halophosphines are of particular importance as precursors of phosphaalkenes ($R_2C{=}PR$), a highly reactive series which has attracted considerable attention as intermediates in synthesis ⟨89T6019, B-89MI 403-01⟩. Due to the low stability of these α-halophosphines only a few of those prepared have been characterized. Some have been stored for several months in the presence of catalytic amounts of hydroquinone ⟨89MI 403-01⟩.

Irradiation (UV) of a mixture of PH_3 and $FHC{=}CF_2$ afforded $F_2HCCHFPH_2$ contaminated with the 1,1,2-trifluoro isomer (ca. 13%) (75% combined yield based on the alkene consumed ⟨66JCS(C)2075⟩). The secondary α-fluorophosphine $FCH_2(CF_3)PH$ has been obtained in nearly quantitative yield by the reaction of $FCH_2(CF_3)PCl$ with HI and Hg ⟨81IC2739⟩.

Cabrioch et al. have synthesized the primary α-chloroalkyl phosphines (**156a–c**) by reduction of the corresponding phosphonates (**155a–c**) with AlH_3 (Equation (35)) ⟨89MI 403-02⟩.

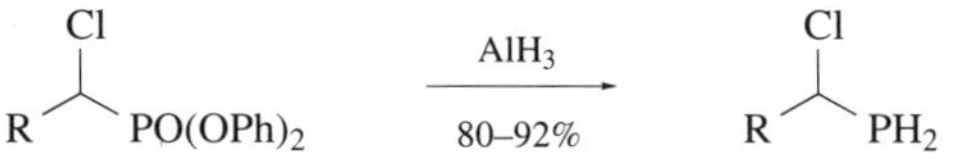

$$\underset{(155)}{R\overset{Cl}{\underset{}{\diagup}}PO(OPh)_2} \xrightarrow[80-92\%]{AlH_3} \underset{(156)}{R\overset{Cl}{\underset{}{\diagup}}PH_2} \quad \text{a; R = H} \quad \text{b; R = Me} \quad \text{c; R = Et} \tag{35}$$

Treatment of (dichloromethyl)phosphine (**157**) with pyridine and then 2,3-dimethylbutadiene afforded a mixture of the *cis-* and *trans-*chlorotetrahydrophosphorines (**159**) via the chloro phosphaalkene (**158**). The existence of the phosphaalkene as an intermediate is supported by the formation of the adduct (**160**) when ethanethiol was used as a trapping agent (Scheme 6) ⟨89CC988⟩.

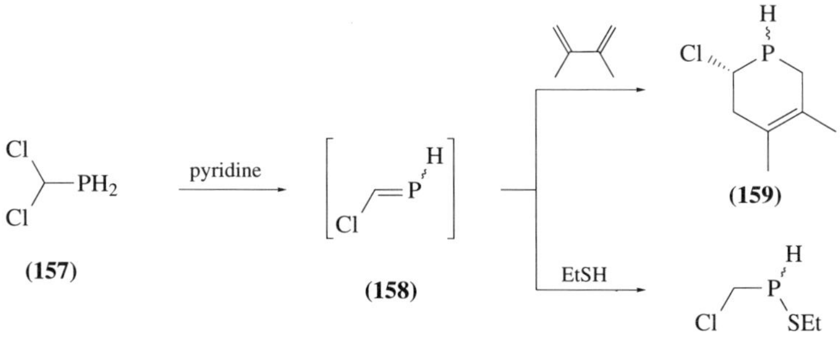

Scheme 6

4.03.2.2.2 Tertiary α-halophosphines—$R^2_2CHalPR^1_2$

(i) Dialkyl (or aryl) (α-haloalkyl)phosphines

Haszeldine and coworkers have prepared the tertiary α-fluoroalkylphosphines (**161a**) and (**161b**) in good yields by treating $CF_2{=}CFH$ with Me_2PH and $(CF_3)_2PH$ respectively under UV irradiation ⟨70JCS(C)744⟩. The bisphosphines (**162a**) and (**162b**) were obtained similarly by the action of $CH_2{=}CFH$ on Me_2PPMe_2 and $(CF_3)_2PP(CF_3)_2$ ⟨71JCS(C)3031⟩.

The reaction of fluoro(trifluoromethyl)carbene $CF_3CF{:}$ (generated by pyrolytic decomposition of

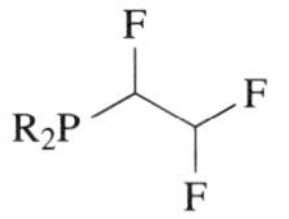

(**161**) a; R = Me
b; R = CF_3

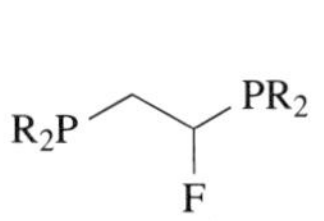

(**162**) a; R = Me
b; R = CF_3

$C_2F_5SiF_3$) with $(CF_3)_3P$ gave the perfluorotrialkylphosphine (**163**) (44% yield), formally by insertion of the carbene into a P—C bond 〈76IC1697〉.

(**163**)

The dienophilicity of the phosphaalkenes was exemplified when Grobe *et al.* demonstrated that the perfluoro phosphaalkene (**164**), prepared as a single isomer (probably (Z)) by thermal decomposition of $Me_3SnP(C_2F_5)_2$, reacted smoothly with cyclic and acyclic dienes to give Diels–Alder adducts, for example cyclopentadiene afforded the 2-phosphabicyclo[2.2.1]heptene (**165**) in 80% yield (Equation (36)) 〈86ZN(B)974, 88JOM(344)61〉. Likewise, the phosphaalkenes $F_3CP{=}CHF$ 〈89ZN(B)175〉, and $F_3CP{=}CFCF_3$ and $F_5C_2P{=}CF_2$ 〈90ZN(B)148〉 were effective dienophiles for [4 + 2] addition reactions. The phosphaalkene (**164**) also combined with the elements of HX (where X = OMe, Br, NMe_2, NEt_2) to give the adducts $C_2F_5P(X)CFH(CF_3)$ 〈86ZN(B)974〉.

(**164**) (**165**) (36)

Langhans *et al.* have reported that interaction of commercially available Ph_2PH with CH_2Cl_2 using phase-transfer catalysis afforded Ph_2PCH_2Cl in 99% yield 〈90CB995〉. Other efficient methods previously employed for the preparation of this compound involved the treatment of the phosphide Ph_2PNa with CH_2Cl_2 in liquid NH_3 〈75JOM(94)327, 86OM2030〉 and reduction of $Ph_2P(O)CH_2Cl$ with $HSiCl_3$ 〈78TL2407〉.

Treatment of the alkyldiphenylphosphines (**166a–e**) with CCl_4 in aprotic solvents gave a mixture of the (dichloromethyl)phosphonium salts (**167a–e**) and the (α-chloroalkyl)diphenylphosphines (**168a–e**) in good yields (Scheme 7). In protic solvents, interaction of the phosphines (**166a–e**) with CCl_4 formed mixtures of the phosphonium salts (**167a–e**) and the alkyl(dichloro)diphenylphosphoranes (**169a–e**) (Scheme 7). These dichlorophosphoranes, on reaction with Et_3N, were converted in nearly quantitative yield into the corresponding α-chloroalkylphosphines (**168**) via a 1,2-chlorotropic rearrangement of the intermediate (alkylidene)chlorophosphoranes (**170a–e**) (Scheme 7) 〈79ZAAC(459)7, 83CB114〉. The dichlorophosphoranes (**169a–e**), *inter alia*, were also readily obtained in ca. 80% yields by treatment of the appropriate alkyldiphenylphosphines with C_2Cl_6 in MeCN 〈77CB2382〉 and this procedure and subsequent treatment of the formed dichlorophosphoranes with Et_3N therefore constitutes a convenient synthesis of α-chloroalkyldiphenylphosphines.

Pentaphenylphosphole, on treatment with lithium, gave the lithium phospholide (**171**) which interacted with CH_2Cl_2 to furnish the chloromethyl derivative (**172**) in 35% overall yield 〈71T5523〉. Chloromethylation of $PhPCN^-$ with CH_2Cl_2 gave $PhP(CN)CH_2Cl$ as a distillable liquid 〈84CB1695〉. The unstable (chloromethyl)dimethylphosphine (**174**) has been prepared albeit in 11% yield by treatment of chloro(chloromethyl)methylphosphine (**173**) with LiMe 〈82CB823〉.

The alkylbis(chloromethyl)phosphines (**176a**) and (**176b**) have been prepared from the chlorophosphine (**175**) in 23–29% isolated yields by treatment with Grignard reagents (Equation (37)) 〈71HCA1651〉. A similar procedure afforded the *t*-butyl analogue (**176c**) in 81% yield 〈84JGU2250〉. An alternative route to the ethyl derivative (**176b**) involved treatment of ethyltris(hydroxymethyl)phosphonium fluoroborate (**177**) with $SOCl_2$ and then interaction of the resultant tris(chloromethyl) derivative (**178**) with NaOH (Scheme 8) 〈62LA(659)49〉.

(**171**) R = Li
(**172**) R = CH_2Cl

(**173**) R = Cl
(**174**) R = Me

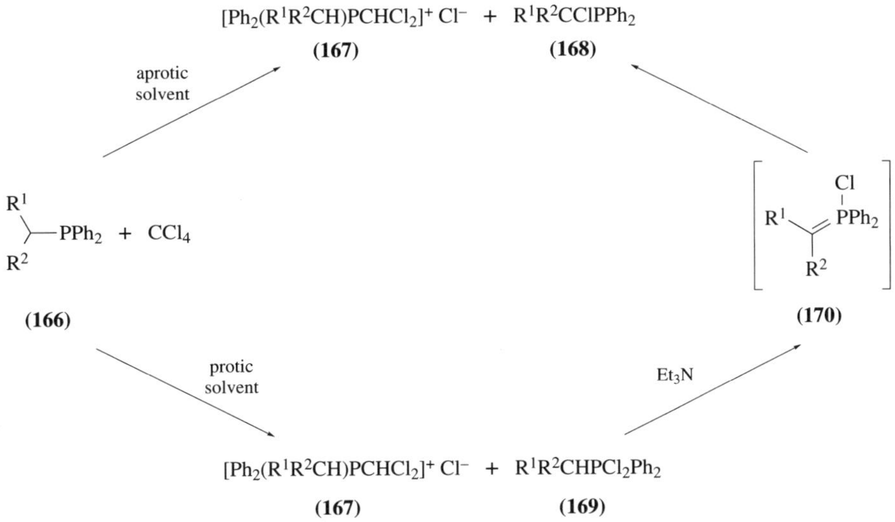

[Ph₂(R¹R²CH)PCHCl₂]⁺ Cl⁻ + R¹R²CClPPh₂
(167)   (168)

(166) R¹R²CH—PPh₂ + CCl₄

aprotic solvent → (167) + (168)

protic solvent → [Ph₂(R¹R²CH)PCHCl₂]⁺ Cl⁻ + R¹R²CHPCl₂Ph₂
(167)   (169)

(170) [R¹R²C=PPh₂Cl]

Et₃N → (170)

(a) R¹ = Me, R² = H; (b) R¹ = Et, R² = H; (c) R¹ = R² = Me; (d) R¹ = Pr^n, R² = H; (e) R¹ = Pr^i, R² = H

Scheme 7

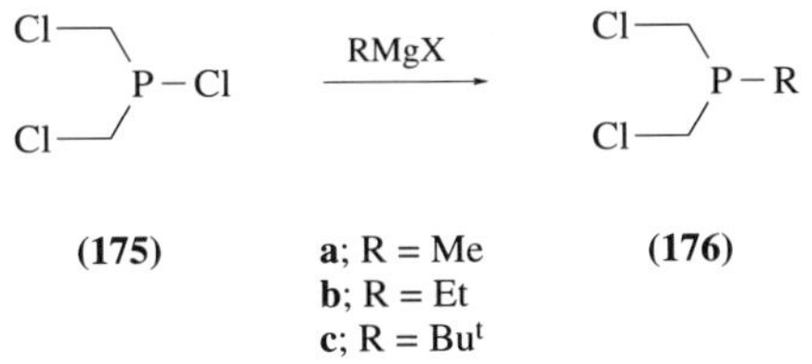

(175) Cl₃P—Cl $\xrightarrow{\text{RMgX}}$ (176) Cl₂P—R   (37)

a; R = Me
b; R = Et
c; R = Bu^t

(177) (HOCH₂)₃P⁺—Et BF₄⁻ $\xrightarrow[90\%]{\text{SOCl}_2}$ (178) (ClCH₂)₃P⁺—Et BF₄⁻ $\xrightarrow[72\%]{\text{NaOH}}$ (176b) (ClCH₂)₂P—Et

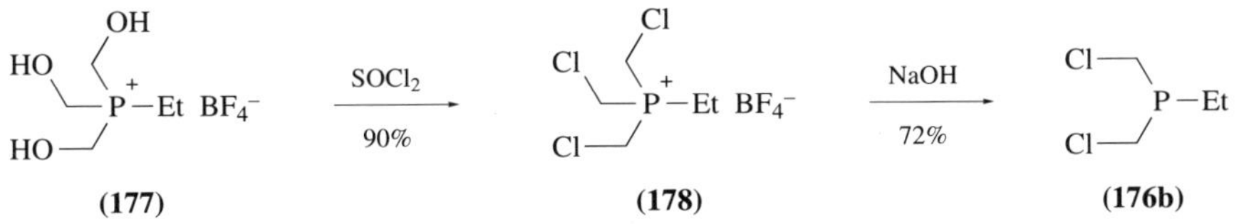

Scheme 8

Tris(chloromethyl)phosphine (**181a**) has been obtained in 78% yield from tetrakis(hydroxymethyl)phosphonium chloride (**179**) by chlorination with PCl₅ and then treatment of the resultant tetrakis(chloromethyl) derivative (**180a**) with aqueous NaHCO₃ ⟨30JA2995, 69HCA858⟩ or aqueous sodium hydroxide (Scheme 9) ⟨70JGU255⟩. Similarly, (BrCH₂)₃P (**181b**) has been prepared from the phosphonium chloride (**179**) by treatment with PBr₃ and then NaHCO₃ (Scheme 9) ⟨67JCED282⟩.

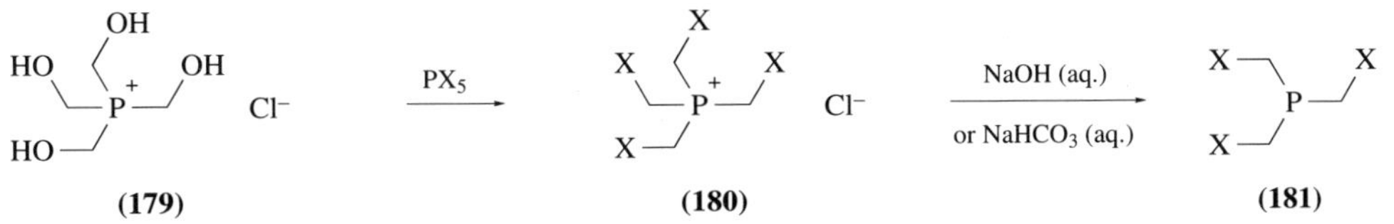

(179) (HOCH₂)₄P⁺ Cl⁻ $\xrightarrow{\text{PX}_5}$ (180) (XCH₂)₄P⁺ Cl⁻ $\xrightarrow{\text{NaOH (aq.) or NaHCO}_3\text{ (aq.)}}$ (181) (XCH₂)₃P

(a) X = Cl; (b) X = Br

Scheme 9

The phosphetan-1-oxide (**182**) has been reduced with trichlorosilane to give the phosphetane (**183**) with retention of configuration (Equation (38)) ⟨81JCS(P2)1138⟩.

$$\text{(182)} \xrightarrow[>78\%]{\text{HSiCl}_3} \text{(183)} \tag{38}$$

(ii) Halo(α-haloalkyl)phosphines

Photoinduced addition of fluoroethene to P_2F_4 afforded the adduct (**184**) in 52% yield. The yield decreased rapidly with increasing fluorine content of the alkene ⟨76JFC(7)153⟩. The α-fluoroalkyl(trifluoromethyl)iodophosphines (**186a**) and (**186b**) have been prepared in 77% and 65% yields respectively by treating $(CF_3P)_4$ with the appropriate alkyl iodide (**185**) (Equation (39)) ⟨75JGU1668⟩. Heptafluoroisopropyl iodide (**185b**), when treated with red phosphorus in a steel bomb for 40 h at 220 °C yielded the diiodophosphine (**187**) in 10% yield. When this reacted with Hg, a surprisingly stable cyclotriphosphine (**188**) was formed in quantitative yield ⟨78ICA(30)(L331)⟩. Typically, halogen bonded to phosphorus readily exchanges with an excess of another nucleophilic halogen. Thus the diiodide (**187**) has been converted into the corresponding dichlorophosphine $(CF_3)_2CFPCl_2$ in quantitative yield by treatment with AgCl ⟨78JIC1254⟩. Perfluorovinyl-substituted phosphines have been shown to add HCl, for example the (trifluorovinyl)phosphine (**189**) furnished the adduct (**190**) (Equation (40)) ⟨69JA1929⟩.

$$(CF_3P)_4 \xrightarrow[\text{(185)}]{RI} \text{(186)} \tag{39}$$

(a) R = CF_2ClCHF; (b) R = $(CF_3)_2CF$

(**187**)   (**188**)

$$\text{(189)} \xrightarrow[56\%]{\text{HCl}} \text{(190)} \tag{40}$$

Treatment of PCl_3 and of $MePCl_2$ with CH_2Cl_2 and $AlCl_3$ and then reduction of the resultant complexes with Sb provided the chloromethyl derivatives (**191**) and (**192**) respectively in 60% yields ⟨73CB2733⟩. Dichloro(methoxy)phosphine $(MeO)PCl_2$ has also been used to good effect for the reduction of the complex ⟨84JGU2250⟩. The dichloro(chloromethyl)phosphine (**191**) together with the α-chloroethyl analogue (**193**) had been prepared earlier in 40% yields by the interaction of PCl_3 with CH_2N_2 and $MeCHN_2$ respectively ⟨52JGU1575⟩. Likewise, $ClCH_2PClMe$ (**192**) was prepared by treating $MePCl_2$ with CH_2N_2 ⟨64USP3161607⟩. A high yield of $ClCH_2PCl_2$ (**191**) has been obtained through a sulfur exchange reaction between the commercially available phosphine sulfide (**194**) and dichloro(phenyl)phosphine (**195**) (Equation (41)) ⟨61JA2299⟩. Treatment of ClCH$_2$PCl$_2$ with Bu^{t}MgCl gave (ClCH$_2$)Bu^{t}PCl in 82% yield but alkylations with Grignard reagents containing α-hydrogen in the alkyl group resulted in dichloromethyl derivatives Cl$_2$CHPClR ⟨84JGU2250⟩. Diethyl chloromalonate has been shown to react with PCl$_3$ in the presence of Et$_3$N to give the α-chlorophosphine (**196**) in 55% yield ⟨74JGU1386⟩. The dichlorophosphiranes (**199a–d**) have been obtained in 47–89% yields by interaction of the appropriate phosphaalkenes (**197**) and chlorocarbenes (**198**) (Equation (42)) ⟨89TL3951⟩.

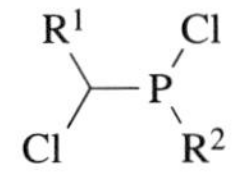

(**191**) R^1 = H; R^2 = Cl
(**192**) R^1 = H; R^2 = Me
(**193**) R^1 = Me; R^2 = Cl

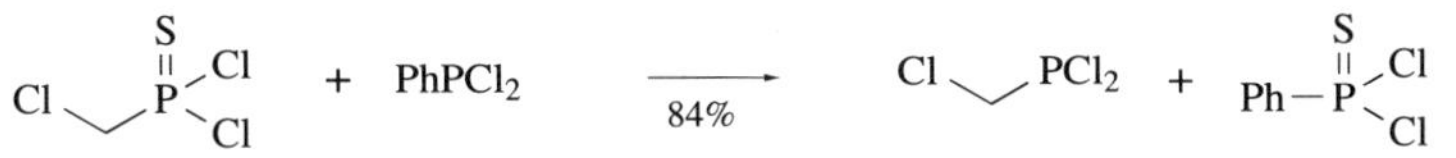

$$(194) + (195) \xrightarrow{84\%} (191) + \text{Ph–P(S)Cl}_2 \quad (41)$$

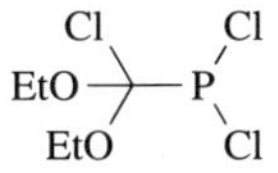

(**196**)

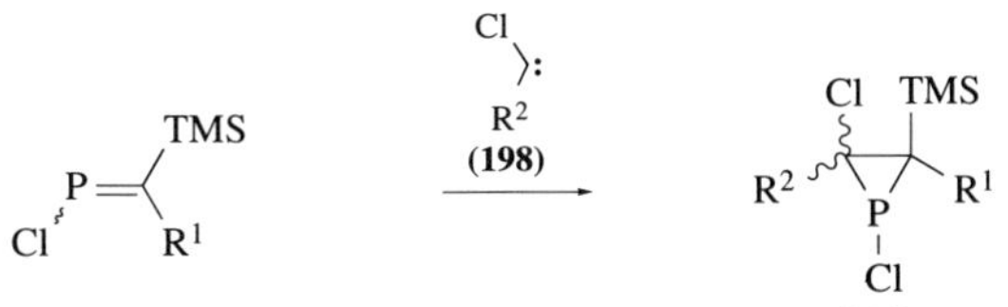

$$(197) + (198) \longrightarrow (199) \quad (42)$$

(a) R^1 = R^2 = Ph; (b) R^1 = TMS, R^2 = Ph
(c) R^1 = Ph, R^2 = OPh; (d) R^1 = TMS, R^2 = OPh

The chloromethyldihalophosphines ClCH$_2$PF$_2$ and ClCH$_2$PBr$_2$ have been prepared from ClCH$_2$PCl$_2$ (**191**) by halogen exchanges using SbF$_3$ ⟨63MI 403-01⟩ and HBr ⟨83PS(15)93⟩. The difluoride ignites spontaneously on contact with air.

Reduction of bis(chloromethyl)thiophosphinic chloride (**200**) with (PhO)$_3$P furnished the bis(chloromethyl)phosphine (**201**) in good yield (Equation (43)) ⟨71HCA1651⟩.

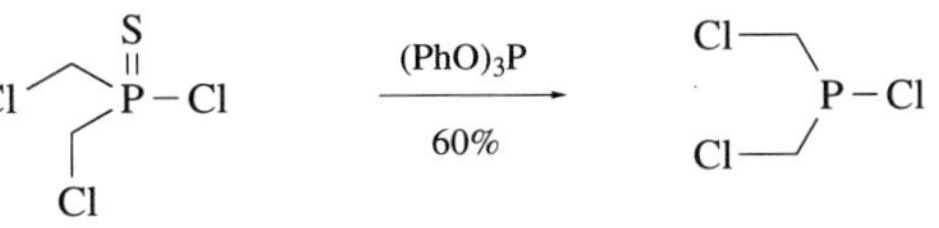

$$(200) \xrightarrow[60\%]{(PhO)_3P} (201) \quad (43)$$

Dibromo(bromomethyl)phosphine (**202**) has been synthesized by a method similar to that employed for the trichloro analogue (**191**), that is treatment of PBr$_3$ with CH$_2$Br$_2$ and AlBr$_3$ and the reduction of the resultant complex with (MeO)PCl$_2$ ⟨84JGU1354⟩.

Treatment of F$_2$PI with commercially available iodo(iodomethyl)mercury gave difluoro(iodomethyl)phosphine (**203**) in 50% yield ⟨76JINC55⟩.

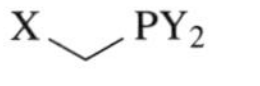

X–CH$_2$–PY$_2$

(**202**) X = Y = Br
(**203**) X = I, Y = F

(iii) Amino (α-haloalkyl)phosphines

The secondary phosphine (**204**) reacted with an excess of Me$_2$NH to afford a mixture of the (dimethylamino)phosphines (**206**) and (**207**) (61% and 12% yields respectively) (Scheme 10). Replacement of an α-fluorine atom by hydrogen in the formation of the former product lends support to (but does not differentiate between) the two proposed mechanisms which involve (a) formation of the phosphaalkene (**205**) as an intermediate by base-induced elimination of HF (Scheme 10) and (b) hydride shift and expulsion of the α-fluorine atom ⟨78JFC(11)441⟩. Later, Grobe and co-workers carried out related studies on the addition of amines to preformed perfluoroalkyl phosphaalkenes ⟨86ZN(B)974, 90ZN(B)148, 91ZN(B)978⟩. Tris(dimethylamino)phosphine (Me$_2$N)$_3$P reacted with (CF$_3$)$_2$CFI to give (CF$_3$)$_2$CFP(NMe)$_2$ (68% yield) which ignites on exposure to air ⟨68JINC1715⟩.

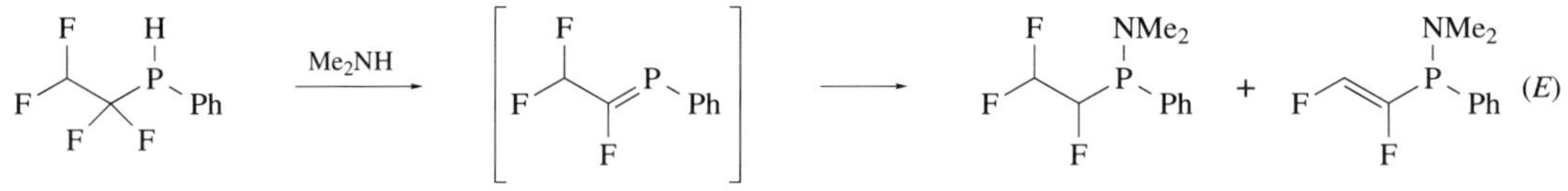

Scheme 10

The α-chloro- and α-bromo-alkylbis(diisopropylamino)phosphines (**210a–d**) have been obtained in good yields by interaction of the diamino(alkyl)phosphines (**208a–d**) with CCl$_4$ (1 molar equivalent) and with CBrCl$_3$ (1 molar equivalent) in polar solvents (Scheme 11) ⟨89JGU2194, 89TL2445, 90JGU1536, 90JGU1541⟩. The equilibrium (**209**)–(**210**) (Scheme 11) is dependent on temperature, the nature of substituents on the P atom and the ylide C atom, and on the polarity and basicity of the medium. Solvents with high dissociating powers favour the C—P chlorotropic rearrangement while nonpolar media favour the formation of the (α-halomethyl)phosphine ⟨91JGU627⟩.

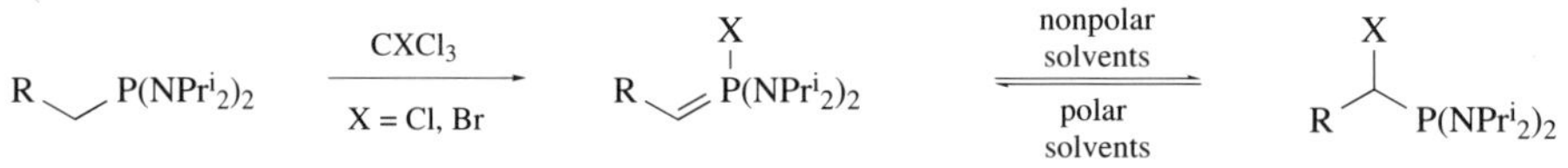

(a) R = H; (b) R = Me; (c) R = Pr^n; (d) R = Pr^i

Scheme 11

Amino (α-haloalkyl)phosphines are also readily formed by the action of primary and secondary amines on the corresponding chlorophosphines in nonpolar solvents, for example the (di-*t*-butylamino)phosphine (**212**) has been prepared from the chlorophosphine (**211**) and Bu$^t_2$NH (2 molar equivalents) (Equation (44)) ⟨91JGU627⟩. Similarly prepared were the amino(chloromethyl)phosphines (**213a–b**) ⟨85JGU1065⟩, (**213c**) ⟨65JCS5630⟩ and (**213d**) ⟨69IZV181⟩.

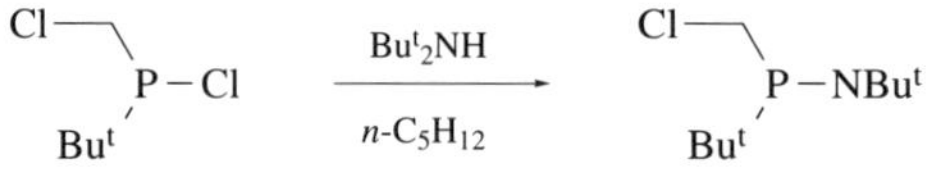

(44)

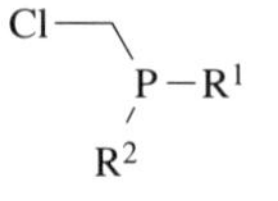

(**213**) **a**; R^1 = Bu^t, R^2 = Et$_2$N
b; R^1 = R^2 = Et$_2$N
c; R^1 = Cl, R^2 = Me$_2$N
d; R^1 = CH$_2$Cl, R^2 = 4-MeOC$_6$H$_4$NH

The fluorophosphine ClCH$_2$PF(NMe$_2$) has been prepared by the replacement of the phosphorus-bound chlorine atom in the aminophosphine (**213c**) using SbF$_3$ at room temperature and also by treatment of ClCH$_2$PF$_2$ with the stoichiometric amount of Me$_2$NH ⟨70OMR(2)81⟩.

(iv) Alkoxy (α-haloalkyl)phosphines

The unstable diethoxy(heptafluoroisopropyl)phosphine (**215**) has been prepared in moderate yield by the action of $(CF_3)_2CFI$ on tetraethyl pyrophosphite (**214**) in the presence of di-*t*-butyl peroxide (Equation (45)) ⟨81CC1173⟩. Addition of MeOH to $C_2F_5P{=}CFCF_3$ gave the adduct (**216**) as a mixture of diastereoisomers ⟨86ZN(B)974⟩. Dropwise addition of NaOMe (1 molar equivalent) in MeOH to the secondary phosphine (**217**) furnished a mixture of the phosphinites (**218**) (38% yield) and (**219**) (24% yield) (Equation (46)) ⟨77JFC(10)27⟩.

$$[(EtO)_2P]_2O \xrightarrow[40\%]{(Bu^tO)_2 \atop (CF_3)_2CFI} (F_3C)_2CF\text{-}P(OEt)_2 \qquad (45)$$

(**214**)   (**215**)

(**216**): F_5C_2–P(F)(OMe)(CF_3)

(**217**) CF_2H–CF_2–P(H)(Ph) $\xrightarrow{MeO^-}$ (**218**) CF_2H–CHF–P(OMe)(Ph) 34% + (**219**) CHF=CF–P(OMe)(Ph) 28%   (46)

Treatment of $ClCH_2PCl(Me)$ with MeOH and pyridine gave the methoxyphosphine (**220a**) in 55% yield ⟨64USP3161607⟩. The chloromethylbis(octyloxy)phosphine (**220b**) has been obtained in 96% yield from $ClCH_2PCl_2$ and *n*-octanol in benzene containing pyridine. The bis(alkyloxy) derivatives from the lower alcohols, for example MeOH and EtOH are unstable and could not be isolated ⟨67USP3314900⟩.

(**220**) $Cl\text{-}CH_2\text{-}P(R^1)(R^2)$
    a; $R^1 = Me, R^2 = OMe$
    b; $R^1 = R^2 = O\text{-}n\text{-}C_8H_{17}$

4.03.2.3 Tetracoordinate Phosphorus Derivatives—$[R^1_2CHalPR^2_3]^+X^-$, $R^1_2CHalP(O)R^2_2$, $R^1_2CHalPO(OH)R^2$, $R^1_2CHalPO(Hal)R^2$, $R_2CHalPO(OH)_2$, etc.

4.03.2.3.1 α-Halophosphonium salts—$[R^1_2CHalPR^2_3]^+X^-$

Most phosphonium salts exist as an equilibrium mixture of the covalent phosphorane form $R^1_2CHalPXR^2_3$ and the ionized form $[R^1_2CHalPR^2_3]^+X^-$. Two factors appear to influence the equilibrium: (a) the higher the electron withdrawing power of the group R^2, the greater the tendency towards the covalent form; and (b) when the P atom is part of a ring, the phosphorane structure is stabilized ⟨79COC1233⟩. The discrete α-halophosphoranes are discussed in Section 4.03.2.4.

α-Halophosphonium salts are of particular interest as precursors of Wittig reagents for the conversion of carbonyl compounds into vinyl halides ⟨87TL6317⟩. Several methods have been employed for their synthesis.

Treatment of (hydroxymethyl)triphenylphosphonium tetrafluoroborate (**221**) with diethylaminosulfur trifluoride (DAST) afforded the fluoro derivative (**222**) in 88% yield. A less expensive, albeit less direct route to this fluoride was accomplished by interaction of Ph_3P and $CFBr_3$ to give the bromide (**223**), exchange of the anion to give the stable crystalline tetrafluoroborate and then hydrolysis of this salt ⟨85JFC(27)85⟩. The latter strategy was also employed for the preparation of the analogous tri-*n*-butyl(fluoromethyl)phosphonium salts (**224**) ⟨88JOC366⟩. It is noteworthy that salts of the type (**223**) are not formed from trialkylphosphines in which the alkyl groups are branched at the α-carbon atoms and contain α-hydrogen atoms ⟨88JOC366⟩. The (ethoxycarbonylfluoromethyl)phosphonium salts (**225**; R^2 = Ph or Bu^n) in homogeneous solutions have been prepared in 90% conversions from $BrFCR^1CO_2Et$ (where R^1 = H, alkyl) and the required phosphine R^2_3P. Deprotonation of the tri-*n*-butylphosphonium salt (**225**; $R^2 = Bu^n$) with Bu^nLi gave the ylide as a mixture of geometric isomers in almost quantitative yield. This reacted with primary alkyl iodides and activated bromides to furnish the α-alkylated derivatives (**226**; $R^2 = Bu^n$) ⟨90JOC2311⟩.

$\left[R \diagup PPh_3 \right]^+ BF_4^-$ $\left[\begin{array}{c} Ph_3\overset{+}{P} \diagdown \diagup \overset{+}{PPh_3} \\ F \end{array} \right] Br^-$ $\left[F \diagup PBu^n_3 \right]^+ X^-$ $\left[\begin{array}{c} EtO_2C \\ R^1 \diagdown \diagup PR^2_3 \\ F \end{array} \right]^+ X^-$

(**221**) R = OH (**223**) (**224**) (**225**) R^1 = H
(**222**) R = F   (**226**) R^1 = alkyl

Tris(2,4,6-trimethoxyphenyl)phosphine, which is one of the most basic and nucleophilic tertiary phosphines known, and the 2,6-dimethoxy analogue reacted readily with CH_2Cl_2 in the presence of $HClO_4$ to give the corresponding (chloromethyl)phosphonium salts (**227a–b**) in 70–96% yields ⟨85JCR(S)38⟩. The commercially available tetrakis(chloromethyl)phosphonium chloride (**228a**) has been prepared from the tetrahydroxy precursor $[(HOCH_2)_4P]^+Cl^-$ in nearly quantitative yield by the action of PCl_5 ⟨30JA2995⟩. Reduction of $[(ClCH_2)_4P]^+Cl^-$ with Ph_3P (1 molar equivalent) in concentrated aqueous HCl gave $[(ClCH_2)_3MeP]^+Cl^-$; two mole equivalents of Ph_3P gave $[(ClCH_2)_2Me_2P]^+Cl^-$ ⟨69JGU1490⟩.

$\left[R_3P \diagup Cl \right]^+ ClO_4^-$ $[(XCH_2)_4P]^+ X^-$

(**227**) a; R = 2,4,6-$(MeO)_3C_6H_2$ (**228**) a; X = Cl
       b; R = 2,6-$(MeO)_2C_6H_3$ b; X = Br

The commercially available (bromomethyl)triphenylphosphonium bromide (**229**) has been conveniently prepared in 52% yield by heating a mixture of Ph_3P and CH_2Br_2 ⟨87OM2489, 90SC1671⟩.

$\left[Br \diagup PPh_3 \right]^+ Br^-$

(**229**)

Tetrakis(bromomethyl)phosphonium bromide (**228b**) has been obtained in 93% yield by the portionwise addition of Br_2 to a mixture of $[(HOCH_2)_4P]^+Cl^-$ and PBr_3 ⟨67JCED282⟩.

A general route to α-halophosphonium bromides (Hal = Cl, Br, I) has been reported by Li and Hu whereby the ylides Ph_3P=CHR, where R = H, Me, Et and Pr^n, reacted readily with perhalofluoro alkanes in the presence of LiBr to give α-halophosphonium bromides $[Ph_3PCH(Hal)R]^+Br^-$ in good yields, for example treatment of the ylide (**230**) with ICF_2CF_2Cl and LiBr gave the (α-iodopropyl)phosphonium bromide (**231**) in good yield (Equation (47)) ⟨87TL6317⟩.

$Ph_3P=\!\!\!\diagup^{Et} \quad \xrightarrow[88\%]{LiBr,\ ICF_2CF_2Cl} \quad Ph_3\overset{+}{P}\!\!\diagup\!\!\diagdown^{I}_{Et}\ Br^-$ (47)

(**230**)   (**231**)

4.03.2.3.2 α-Halophosphine oxides and sulfides—$R^1_2CHalP(Y)R^2_2$

(i) α-Halophosphine oxides

(a) α-Fluorophosphine oxides. Few (α-fluoroalkyl)phosphine oxides are known. (α-Fluorobenzyl)diphenylphosphine oxide (**233**; R^1 = Ph) has been prepared in 99% yield by the action of

diethylaminosulfur trifluoride (DAST) on the hydroxy compound (232; $R^1 = Ph$) ⟨86JCS(P1)913⟩. The unstable fluoromethyl analogue (233; $R^1 = H$) was obtained similarly ⟨90TL5571⟩.

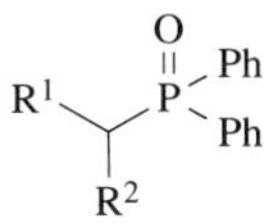

(232) $R^2 = OH$
(233) $R^2 = F$

(b) α-Chlorophosphine oxides. (α-Chloroalkyl)phosphine oxides are useful as phosphinylmethylating agents. The reactivity of the chlorine atom is essentially a function of the nature of the substituents at the phosphorus atom ⟨90JGU1351⟩.

General methods for the preparation of (α-chloroalkyl)phosphine oxides have been developed. The procedure used most frequently involves chlorination of the corresponding hydroxy analogues with PCl_5, $SOCl_2$, $COCl_2$, $(COCl)_2$ and HCl. Other methods include direct and indirect replacement of acetoxy groups by Cl, treatment of chlorophosphines R_2PCl with aldehydes and ketones and the alkylation of phosphinic chlorides $R_2P(O)Cl$. Examples of these and other methods are described below.

In 1953, Kabachnik and Shepeleva reported that treatment of chlorodiphenylphosphine (234) with formaldehyde furnished (chloromethyl)diphenylphosphine oxide (235) in moderate yield (Scheme 12) ⟨53IZV763⟩. Gallagher modified the conditions and improved the yield to 72% ⟨68AJC1197⟩. This oxide (235) has also been prepared by treating the complex (236), formed by the addition of $AlCl_3$ and PCl_3 to benzene, with $POCl_3$ and then HCHO (Scheme 12) ⟨86JGU2394⟩.

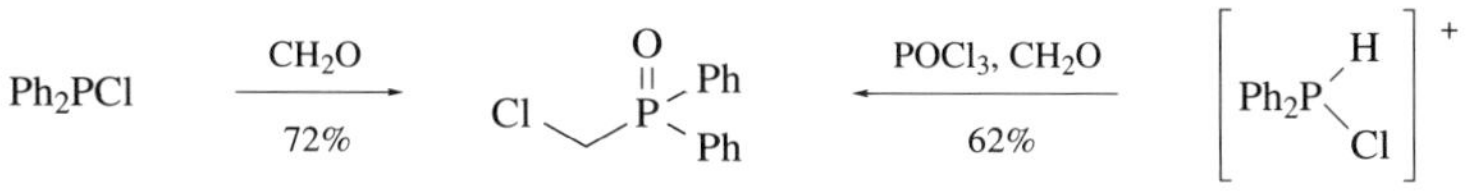

Scheme 12

A number of dialkyl-, diaryl- and alkyl(aryl)(chloromethyl)phosphine oxides (239) have been prepared in good overall yields by treatment of dialkyl(or aryl)phosphinous acids (secondary phosphine oxides) (237) with HCHO in the presence of base to give the hydroxymethyl derivatives (238) and then with PCl_5 (Scheme 13) ⟨86JGU501, 86JGU1430, 90JGU1351⟩. The syntheses of the dialkyl (chloromethyl)phosphine oxides (239; $R^1 = R^2 = Me,Et,Pr^n,Bu^n$) by treatment of the hydroxy analogues (238) with $COCl_2$, ClCOCOCl or HCl (80–96% yields) (Scheme 13) have been reported in a patent ⟨72GEP(O)2060217⟩. Petrov et al. used $SOCl_2$ as the chlorinating agent in the preparation of the di-*t*-butyl derivative (239; $R^1 = R^2 = Bu^t$) ⟨65JGU2053⟩.

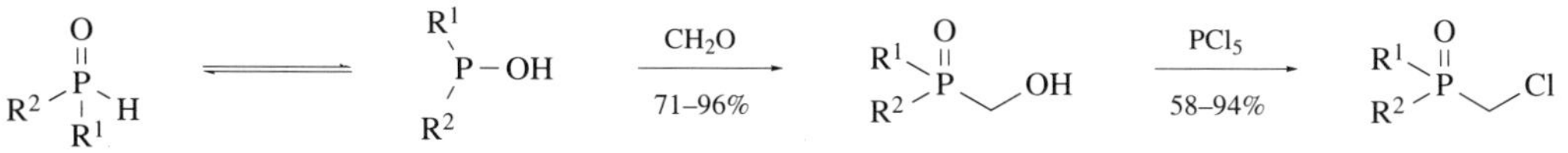

R^1 and R^2 = alkyl or aryl

Scheme 13

(Acetoxymethyl)phosphine oxides, which are readily prepared and more easily purified than the hydroxy analogues, have been converted into the (chloromethyl)phosphine oxides both directly by treatment with anhydrous HCl and also by chlorination of the crude acetoxy derivatives with PCl_5. For example, $ClCH_2PO(Me)_2$ was obtained in 83% yield from $AcOCH_2PO(Me)_2$ by the former method ⟨80IZV491⟩ and $ClCH_2PO(Pr^n)_2$ from $AcOCH_2PO(Pr^n)_2$ in 69% yield by the latter ⟨90JGU1351⟩. $Ph_2PO(CH_2Cl)$ could not be formed by the direct displacement method ⟨80IZV491⟩.

Good yields of $ClCH_2PO(Me)_2$ (92%) and the cyclohexyl analogue $ClCH_2PO(Me)(c-C_6H_{11})$ (83%) were obtained by treatment of the methyl- and cyclohexyl-tris(hydroxymethyl)phosphonium chlorides $[(HOCH_2)_3RP]^+Cl^-$ (R = Me, $c-C_6H_{11}$) with HCl at ca. 200 °C ⟨75GEP(O)2412800⟩. Elaboration of commercially available tetrakis(hydroxymethyl)phosphonium chloride by the route shown in Scheme 14 afforded (chloromethyl)dimethylphosphine oxide in 55% overall yield ⟨80IZV491⟩.

$$[(HOCH_2)_4P]^+Cl^- \xrightarrow{NaOH} (HOCH_2)_3P \xrightarrow{Me_2SO_4} [(HOCH_2)_3MeP]^+MeSO_4^- \xrightarrow{HCl} ClCH_2PO(Me)_2$$

Scheme 14

Chlorination of bis(2-hydroxyethyl) (hydroxymethyl)phosphine oxide (**240**) with PCl$_5$ gave the trichloro derivative (**241**) ⟨70HCA2069⟩. This readily formed the divinylphosphine oxide (**242**) when treated with Et$_3$N (Scheme 15) ⟨72MI 403-01⟩. Use was made of the different reactivities of the chlorine substituents in the phosphine oxides (**241**) to prepare, *inter alia*, the diethoxy derivative (**243**) which was hydrolysed to the dihydroxy compound (**244**) (Scheme 15) ⟨72MI 403-01⟩.

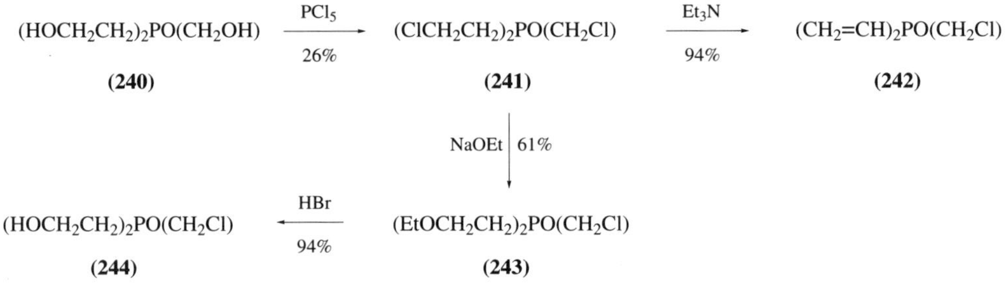

$$(HOCH_2CH_2)_2PO(CH_2OH) \xrightarrow[26\%]{PCl_5} (ClCH_2CH_2)_2PO(CH_2Cl) \xrightarrow[94\%]{Et_3N} (CH_2=CH)_2PO(CH_2Cl)$$

(**240**)   (**241**)   (**242**)

NaOEt | 61%

$$(HOCH_2CH_2)_2PO(CH_2Cl) \xleftarrow[94\%]{HBr} (EtOCH_2CH_2)_2PO(CH_2Cl)$$

(**244**)   (**243**)

Scheme 15

The (chloromethyl)phosphine oxides (**246a–c**) have been prepared in 26–37% yields from the phosphinic chloride (**245**) by treatment with the appropriate Grignard reagents (Equation (48)). Likewise, treatment of chloromethylphosphonic dichloride ClCH$_2$PO(Cl)$_2$ with PhMgBr afforded ClCH$_2$PO(Ph)$_2$ ⟨86ACH145⟩.

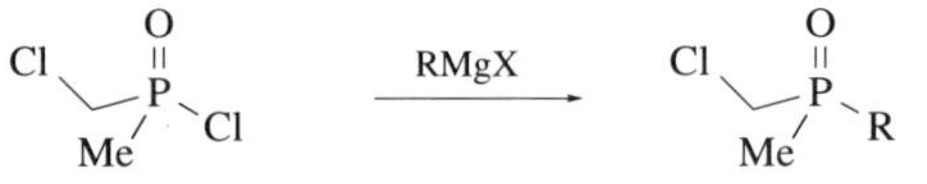

$$\begin{array}{c} Cl\diagdown\;\;\;O \\ \;\;\;\diagdown\;\|\;\;\; \\ \;\;\;\;\;\;P\diagdown \\ Me\diagup\;\;\;\;Cl \end{array} \xrightarrow{RMgX} \begin{array}{c} Cl\diagdown\;\;\;O \\ \;\;\;\diagdown\;\|\;\;\; \\ \;\;\;\;\;\;P\diagdown \\ Me\diagup\;\;\;\;R \end{array} \quad (48)$$

(**245**)   **a**; R = Ph   (**246**)
          b; R = 4-ClC$_6$H$_4$
          c; R = 4-MeOC$_6$H$_4$
          d; R = CH$_2$Ph

Chloro(diethyl)phosphine has been shown to react with the ketones (**247a–d**) to form the (α-chloroalkyl)phosphine oxides (**248a–d**) in good yields (Equation (49)) ⟨71JGU2183⟩.

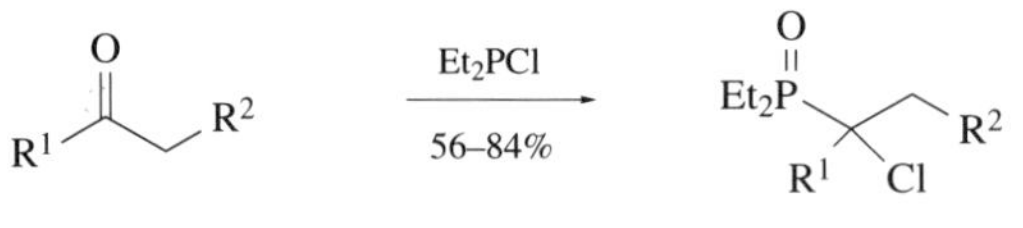

$$\begin{array}{c} O \\ \| \\ R^1\;\;\;\;\;\;R^2 \end{array} \xrightarrow[56-84\%]{Et_2PCl} \begin{array}{c} O \\ \| \\ Et_2P\;\;\;\;\;\;R^2 \\ \;\;\;R^1\;\;Cl \end{array} \quad (49)$$

(**247**)   **a**; R^1 = Me, R^2 = H   (**248**)
          b; R^1 = R^2 = Me
          c; R^1 = Me, R^2 = Pr^n
          d; R^1 = Ph, R^2 = H

Treatment of the chlorovinylphosphine oxide (**249**) with thiophenol furnished the adduct (**250a**) (95% yield) which was oxidized smoothly to the corresponding sulfoxide (**250b**) and sulfone (**250c**) with H$_2$O$_2$ and a mixture of SeO$_2$ and H$_2$O$_2$, respectively ⟨89T337⟩.

Bis(chloromethyl)methylphosphine oxide (**251a**) has been prepared from (ClCH$_2$)$_3$P (**181a**) in good yield by treatment with boiling water. The conversion apparently proceeds by way of a pseudoallylic rearrangement (Scheme 16) ⟨70JGU255⟩. Conversions into the phosphine oxides (**251a**) also occurred when the phosphine (**181a**) was heated with hydrochloric acid ⟨62DOK(143)211, 80IZV491⟩ and with aqueous ammonia ⟨67JCED282⟩. Bis(chloromethyl)methylphosphine oxide (**251a**) has also been prepared by passing anhydrous HCl through (i) (AcOCH$_2$)$_2$POMe (72% yield),

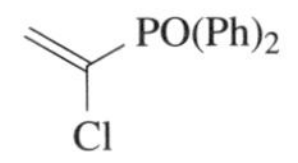

(249) *(structure: CH2=C(Cl)PO(Ph)2)*

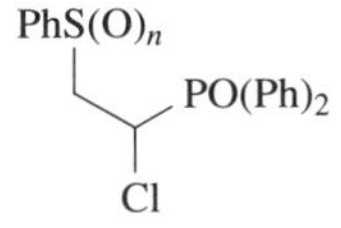

(250) *(structure: PhS(O)n-CH2-CH(Cl)-PO(Ph)2)* a; n = 0 b; n = 1 c; n = 2

(ii) (AcOCH$_2$)$_3$P (61% yield) and (iii) molten [(HOCH$_2$)$_4$]P$^+$Cl$^-$ (45% yield) ⟨80IZV491⟩. Oxidation of (ClCH$_2$)$_2$PEt with NaOBr gave the oxide (**251b**) in 76% yield ⟨71HCA1651⟩. Bis(chloromethyl)*n*-propylphosphine oxide (**251c**) has been prepared by chlorination of the bis(hydroxymethyl) analogue with SOCl$_2$ ⟨61JGU3189⟩.

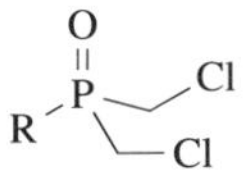

(251) a; R = Me b; R = Et c; R = Pr^n

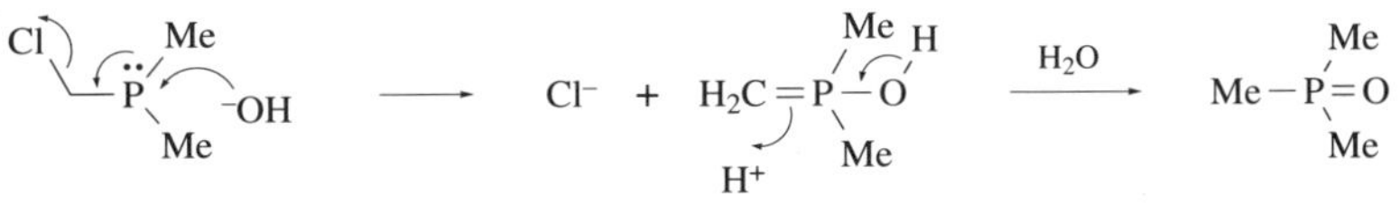

Scheme 16

The bis(chloromethyl) (1-hydroxyalkyl)phosphine oxides (**253a–f**) have been prepared in good yields by the action of (ClCH$_2$)$_2$PCl on the aldehydes and ketones (**252a–f**) followed by the addition of H$_2$O (Equation (50)) ⟨71JGU2228⟩.

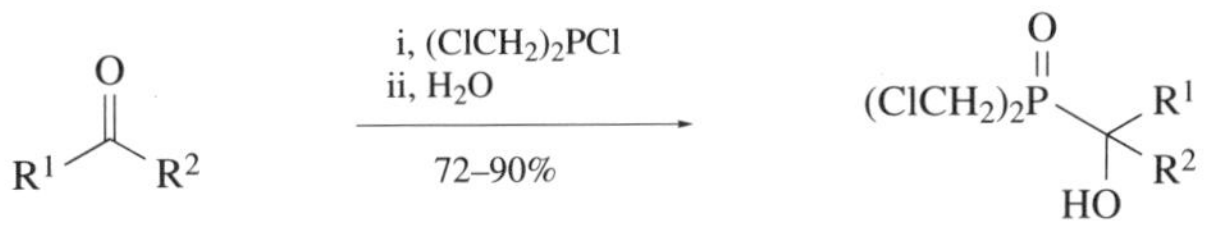

(**252**) → (**253**)   (50)

i, (ClCH$_2$)$_2$PCl
ii, H$_2$O
72–90%

a; R^1 = R^2 = H
b; R^1 = H, R^2 = Me
c; R^1 = H, R^2 = Ph
d; R^1 = R^2 = Me
e; R^{1}R^2 = –(CH$_2$)$_5$–
f; R^1 = H, R^2 = CCl$_3$

Tris(chloromethyl)phosphine oxide (**254**), which was first prepared by oxidation of (ClCH$_2$)$_3$P with nitric acid ⟨30JA2995⟩ and later with NaOBr (59% yield) ⟨69HCA858⟩, has been obtained in 45–75% yield by chlorination of (HOCH$_2$)$_3$PO with PCl$_5$ or Ph$_3$PCl$_2$ ⟨69HCA858⟩. It has also been prepared by the action of anhydrous HCl on (AcOCH$_2$)$_3$PO (79% yield) ⟨80IZV491⟩ and in quantitative yield by treatment of (TMS-OCH$_2$)$_3$PO with PCl$_5$ ⟨80JGU802⟩.

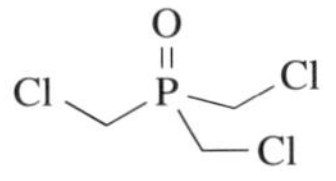

(254)

(c) α-Bromophosphine oxides. (Bromomethyl)dimethylphosphine oxide (**255a**) was prepared in 96% yield by treatment of (HOCH$_2$)PO(Me)$_2$ with oxalyl bromide ⟨74LA751⟩ and the corresponding diethyl compound (**255b**) has been obtained similarly ⟨72GEP(O)2060217⟩. Displacement of the tosyloxy group in (TsOCH$_2$)PO(Ph)$_2$ by Br was accomplished with KBr in DMF to give the diphenylphosphine oxide (**255c**) ⟨88JGU465⟩. The triacetate (**256a**) was converted into the tribromo derivative

(**256b**) with hydrogen bromide (21% yield) ⟨72MI 403-01⟩. Bromination of bisbenzyl(phenyl)phosphine oxide (**257a**) with Br_2 furnished the dibromo derivative (**257b**) in 67% yield ⟨82CB3384⟩.

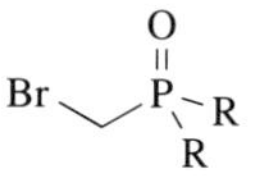

(**255**) a; R = Me
b; R = Et
c; R = Ph

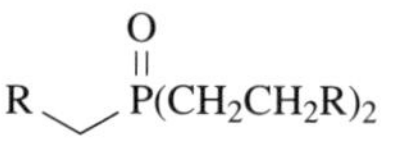

(**256**) a; R = OAc
b; R = Br

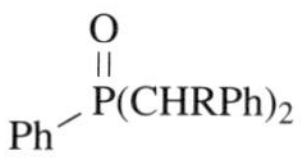

(**257**) a; R = H
b; R = Br

1-Phenyl-2-phospholene 1-oxide (**258**) reacted with *N*-bromoacetamide or Br_2 to give the *threo* (**259**) and *erythro* (**260**) bromohydrins in 43% and 24% isolated yields respectively (Equation (51)) ⟨92CL407⟩.

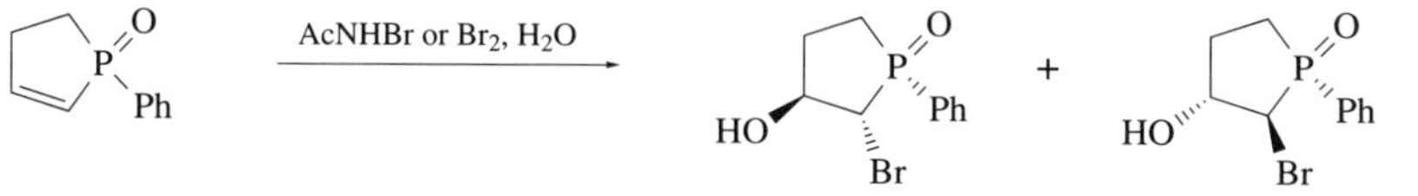

(**258**) → *threo* (43%) (**259**) + *erythro* (24%) (**260**)   (51)

Tris(bromomethyl)phosphine oxide $(BrCH_2)_3PO$ has been prepared from (a) $(BrCH_2)_3P$ (**181b**) and H_2O_2 ⟨67JCED282⟩, (b) $(TMS-OCH_2)_3PO$ and PBr_5 (69% yield) ⟨80JGU802⟩ and (c) $(AcO-CH_2)_3PO$ and HBr ⟨67USP3306937⟩.

(d) α-Iodophosphine oxides. Treatment of $ClCH_2PO(Ph)_2$ with KI in boiling DMF gave $ICH_2PO(Ph)_2$ in 71% yield ⟨88JGU465⟩. Sequential treatment of the diaryl(methyl)phosphine oxides (**261**) with Bu^nLi and $AlEt_3$ gave the aluminates (**262**) which were readily iodinated to give the diaryl(iodomethyl)phosphine oxides (**263**) in good yields (Scheme 17) ⟨85S982⟩.

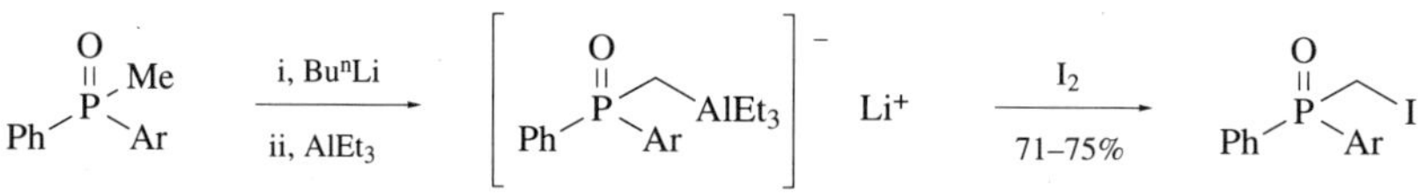

(**261**) →[i, Bu^{n}Li; ii, AlEt$_3$] (**262**) →[I_2, 71–75%] (**263**)

Scheme 17

Bis(iodomethyl)methylphosphine oxide (**264b**) was prepared in 47% yield from the corresponding bis(chloromethyl) analogue (**264a**) and NaI in boiling acetone ⟨67JCED282⟩.

$MePO(CH_2R)_2$

(**264**) a; R = Cl
b; R = I

Bis(chloromethyl)phenylphosphine oxide (**265**) and TMS-I in Et_2O gave the (trimethylsilyloxy)phosphonium iodide (**266**) in excellent yield (Scheme 18). When this was treated with a further 2 mole equivalents of TMS-I, a good yield of bis(iodomethyl)phenylphosphine oxide (**267**) was obtained (Scheme 18) ⟨80S823⟩.

$PhPO(CH_2Cl)_2$ →[TMS-I, 98%] $[PhP(O-TMS)(CH_2Cl)_2]^+ I^-$ →[i, TMS-I; ii, $Na_2S_2O_3$ (aq.), 60%] $PhPO(CH_2I)_2$

(**265**)   (**266**)   (**267**)

Scheme 18

Tris(iodomethyl)phosphine oxide $(ICH_2)_3PO$ was readily prepared by treatment of the corresponding chloro, bromo or tosyloxy derivatives $(XCH_2)_3PO$ with TMS-I (4 mole equivalents) in 60–70% yields ⟨80S823⟩.

(ii) α-Halophosphine sulfides

(Chloromethyl)phosphine sulfides ClCH$_2$PS(R)$_2$ have been prepared by treatment of dialkyl(halo)phosphines with thioformaldehyde or its cyclic trimer S-trithiane, for example Me$_2$PCl gave the sulfide ClCH$_2$PS(Me)$_2$ when treated with S-trithiane ⟨67USP3360556⟩. (α-Chloroalkyl)phosphine sulfides have been obtained by direct oxidation of the corresponding trialkylphosphines with sulfur. Thus (ClCH$_2$)$_2$PR, where R = Me,Et, gave the sulfides (ClCH$_2$)$_2$PS(R) ⟨71HCA1651⟩.

4.03.2.3.3 α-Halo oxo acids of phosphorus

(i) α-Halophosphinic acids and derivatives

(a) α-Halophosphinic acids. Few α-halophosphinic acids containing a P—H bond have been reported. The parent compound FCH$_2$PO(OH)H of the fluoro series is unknown. The heptafluoroisopropyl derivative (**268a**) was prepared as the sodium salt in nearly quantitative yield by treatment of (CF$_3$)$_2$CFP(CF$_3$)$_2$ with NaOH ⟨78JIC1254⟩. α-Chloromethylphosphinic acid (**268b**) has been prepared in almost quantitative yield by acid hydrolysis of ClCH$_2$PCl$_2$ ⟨61JA2299, 63BRP934090⟩. It has also been synthesized from the same starting material in 70% yield by interaction with MeOH to produce the methyl phosphinate ClCH$_2$PO(OMe)H and then treatment of this with NaOH ⟨66JA3572⟩.

(**268**) a; R = (CF$_3$)$_2$CF
b; R = ClCH$_2$

(Fluoromethyl)phenylphosphinic acid (**269**) was obtained from phenyltetrafluorophosphorane PhPF$_4$ by interaction with CH$_2$N$_2$ and then H$_2$O (Equation (52)) ⟨60JGU3986⟩. (2-Cyano-1-fluoroethenyl)n-butylphosphinic acid (**270**) has been hydrogenated over a rhodium catalyst to give the phosphinic acid (**271**) (Equation (53)) ⟨90EUP402312⟩.

$$\text{PhPF}_4 \xrightarrow[37\%]{\text{i, CH}_2\text{N}_2 \\ \text{ii, H}_2\text{O}} \text{(269)} \quad (52)$$

$$\text{(270)} \xrightarrow{\text{H}_2\text{, Rh–C}} \text{(271)} \quad (53)$$

The reaction of aldehydes and halophosphines is a general one for the preparation of α-halophosphoryl compounds (Equation (54)). (Chloromethyl)phenylphosphinic acid (**272a**) has been prepared in 22% yield by alkylation of PhPCl$_2$ with paraformaldehyde and then treatment with water (Equation (55)) ⟨88IC1787⟩. Similarly, alkyl- and aryldichlorophosphines in the presence of acids reacted with benzaldehydes bearing electron donating groups and with 4-methoxyacetophenone to afford α-chlorobenzylphosphinic acids in good yields, for example PhPCl$_2$ reacted with 4-methoxybenzaldehyde to give the phosphinic acid (**272b**) (84% yield) (Equation (55)) ⟨86DOK(286)19⟩.

$$R^1\text{CHO} + \begin{array}{c} R^2 \\ | \\ :P-\text{Hal} \\ | \\ R^3 \end{array} \longrightarrow R^1 \underset{\text{Hal}}{\overset{O}{\underset{|}{\stackrel{\|}{P}}}} \begin{array}{c} R^2 \\ R^3 \end{array} \quad (54)$$

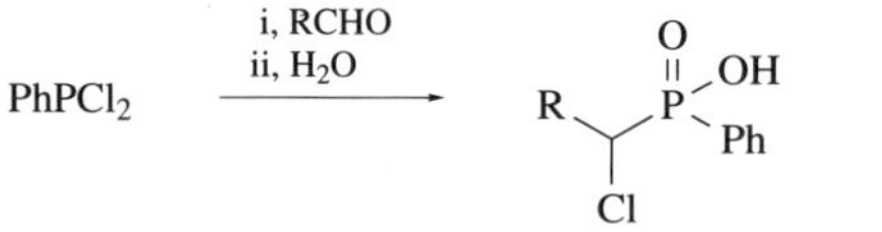

$$PhPCl_2 \xrightarrow[\text{ii, H}_2\text{O}]{\text{i, RCHO}} \underset{\substack{(272) \ \mathbf{a}; R = H \\ \mathbf{b}; R = 4\text{-MeOC}_6H_4}}{R-\underset{Cl}{\overset{\underset{\displaystyle\|}{O}}{\underset{|}{P}}}(OH)(Ph)} \qquad (55)$$

α-Chloroalkylphosphinic acids are readily obtained by hydrolysis of the corresponding phosphinic chlorides, for example, the acid chloride (**273**), prepared by addition of ClCH$_2$PCl$_2$ to acrylic acid, was converted into the phosphinicopropionic acid (**274**) by treatment with ice (Equation (56)) ⟨69JGU577⟩. Elaboration of the unsaturated phosphinic chloride (**275**), obtained by treatment of ClCH$_2$PCl$_2$ with propiolic acid, gave by the reaction sequence shown in Scheme 19, a nearly quantitative yield of the phosphinicoacrylic acid (**276**) ⟨71IZV1159⟩.

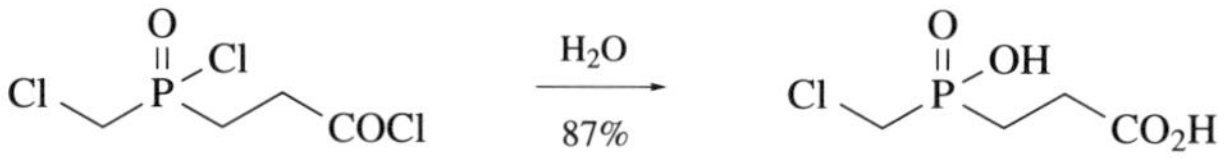

(273) → (274) H$_2$O, 87% (56)

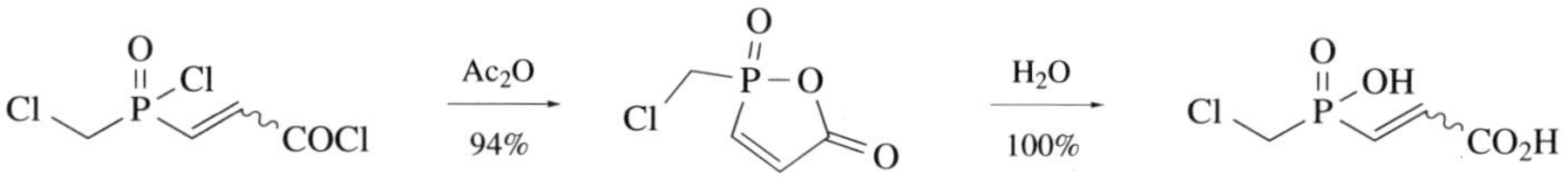

(275) → Ac$_2$O, 94% → (intermediate) → H$_2$O, 100% → (276)

Scheme 19

(1-Aminomethyl)(chloromethyl)phosphinic acid (**277a**) has been synthesized in good yield by amidoalkylation of ClCH$_2$PCl$_2$ with *N*-hydroxymethylbenzamide and then hydrolysis with strong acid (Scheme 20). The α-aminoalkyl analogues (**277b–f**) were obtained by amidoalkylation of ClCH$_2$PCl$_2$ with benzyl carbamate and the appropriate aldehydes (**278**) (Scheme 20) ⟨87S477⟩.

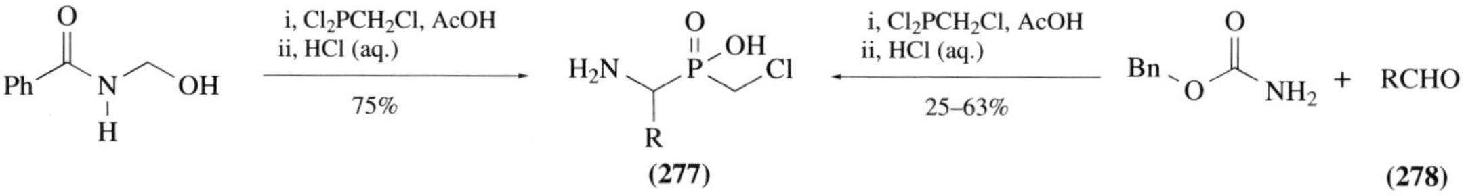

(**277**) ← i, Cl$_2$PCH$_2$Cl, AcOH; ii, HCl (aq.), 75% ← PhC(O)NHCH$_2$OH
(**277**) ← i, Cl$_2$PCH$_2$Cl, AcOH; ii, HCl (aq.), 25–63% ← BnOC(O)NH$_2$ + RCHO (**278**)

(a) R = H; (b) R = Me; (c) R = Pr^i; (d) R = Bu^i; (e) R = Ph; (f) R = Bn

Scheme 20

Chloromethyl(1-hydroxyalkyl)phosphinic acids ClCH$_2$PO(OH)[C(OH)R$_2$] have been obtained by alkylation of ClCH$_2$PCl$_2$ with carbonyl compounds, for example ClCH$_2$PCl$_2$ with aqueous acetone furnished the α-hydroxyalkyl derivative (**279**) in 70% yield (Equation (57)) ⟨75JGU1950⟩.

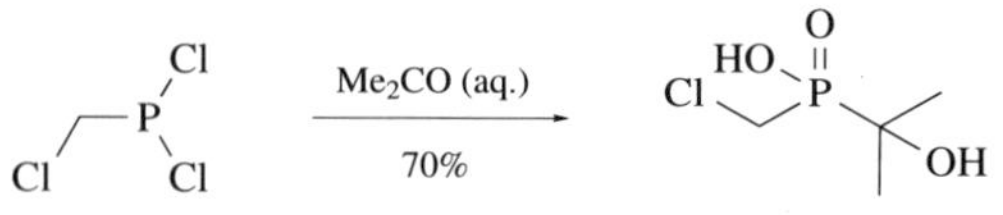

ClCH$_2$PCl$_2$ $\xrightarrow[70\%]{\text{Me}_2\text{CO (aq.)}}$ (**279**) (57)

Hydrolysis of the 1,4,2-benzooxazaphosphorinane (**280a**) afforded the (chloromethyl)phosphinic acid (**281a**) in 86% yield (Equation (58)) ⟨92JGU39⟩. Similarly, the phosphinic acid (**281b**) was obtained from the phosphorinane (**280b**) by reaction with H$_2$O (91% yield) (Equation (58)) ⟨89JGU1778⟩.

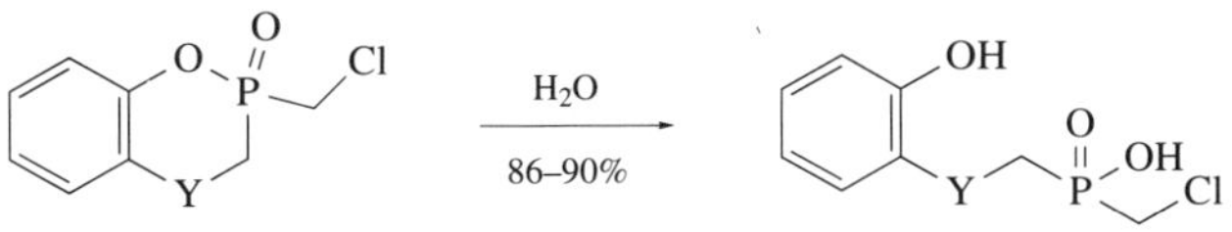

$$\text{(280)} \xrightarrow[86-90\%]{H_2O} \text{(281)} \qquad (58)$$

a; Y = NH
b; Y = O

Bis(chloromethyl)phosphinic acid $(ClCH_2)_2PO(OH)$ has been obtained by hydrolysis of $(ClCH_2)_2PO(Cl)$ in 95% yield ⟨79JOM(178)157⟩.

(b) α-Halophosphinic halides. Alkyl- and arylchloromethylphosphinic chlorides $R(ClCH_2)PO(Cl)$ have been prepared in good yields from alkyl- and aryldichlorophosphines and anhydrous formaldehyde ⟨53IZV763, 61JA1811, 86JGU2246⟩, for example treatment of the perfluoroalkyldichlorophosphines (282) with paraformaldehyde furnished the chloromethylphosphinic chlorides (283) (Equation (59)) ⟨86JGU2246⟩. It has been claimed that better yields are obtained when trioxane in conjunction with an acid catalyst is used in place of paraformaldehyde ⟨80USP4224241⟩. The phosphinic acid anhydrides $[R^1R^2PO]_2O$, formed as by-products in these reactions, have been converted into the desired phosphinic chlorides with PCl_5 ⟨61JA4381⟩.

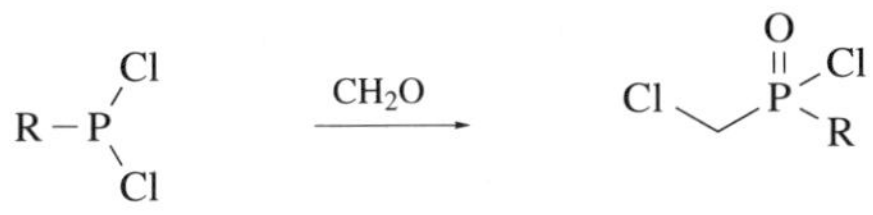

$$\text{(282)} \xrightarrow{CH_2O} \text{(283)} \qquad (59)$$

a; R = CF$_3$
b; R = C$_2$F$_5$

(Chloromethyl)methylphosphinic chloride (285) has been prepared from the phosphinic acid salt (284) in good yield by chlorination with PCl_5 (Equation (60)) ⟨85JGU2497⟩.

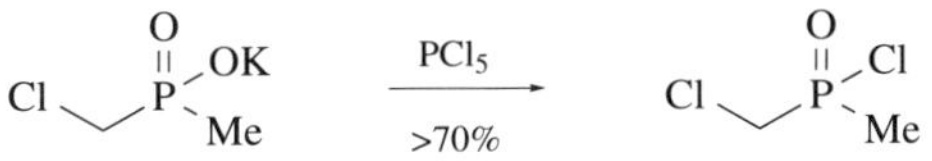

$$\text{(284)} \xrightarrow[>70\%]{PCl_5} \text{(285)} \qquad (60)$$

Electrophilic attack on $ClCH_2PCl_2$ by acrylamide gave the 2-cyanoethylphosphinic chloride (286) in 20% yield ⟨68JGU2006⟩. The foregoing, analagous propanoyl chloride (273) ⟨69JGU577⟩ and acryloyl chloride (275) ⟨71IZV1159⟩ were prepared similarly.

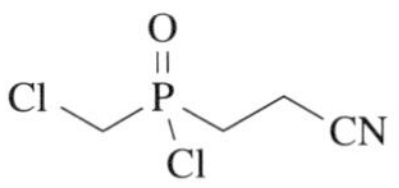

(286)

Phosphinic acid esters are readily converted into acid chlorides by halide donors, for example the ethyl phosphinate (287a) with PCl_5 afforded the phosphinic chloride (287b) in 77% yield ⟨71IZV1159⟩.

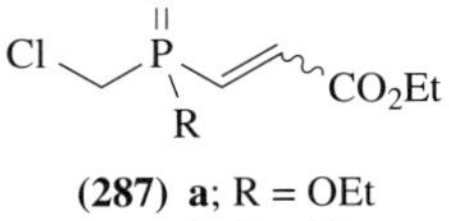

(287) **a**; R = OEt
b; R = Cl

Bis(chloromethyl)phosphinic chloride (289) has been prepared in good yield by treatment of hypophosphorous acid with paraformaldehyde and hydrochloric acid and then chlorination of the bis(hydroxymethyl) derivative (288) with $SOCl_2$ (Scheme 21) ⟨79JOM(178)157⟩. It has also been synthesized from $ClCH_2PCl_2$ and paraformaldehyde under anhydrous conditions in 54% yield and from bis(chloromethyl)phosphinic anhydride $[(ClCH_2)_2PO]_2O$ and PCl_5 in 85% yield ⟨61JA4381⟩.

(c) α-Halophosphinic acid esters. Methyl chloromethylphosphinate (290) was synthesized by treatment of $ClCH_2PCl_2$ with MeOH (Equation (61)) ⟨66JA3572⟩.

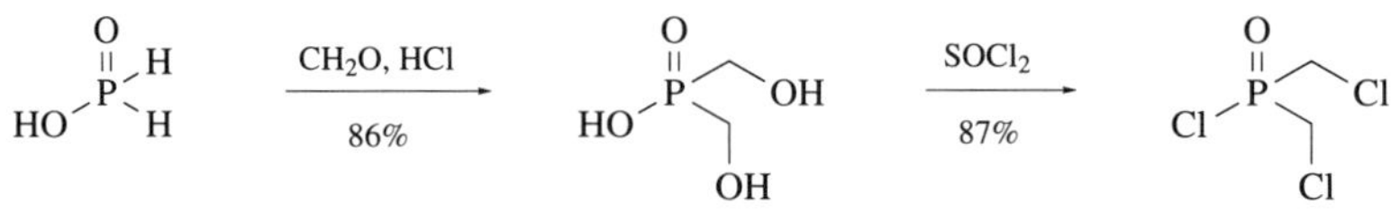

$$\text{HO-P(=O)H}_2 \xrightarrow[86\%]{\text{CH}_2\text{O, HCl}} \text{HOP(=O)(CH}_2\text{OH)}_2 \xrightarrow[87\%]{\text{SOCl}_2} \text{ClP(=O)(CH}_2\text{Cl)}_2$$

(288)   (289)

Scheme 21

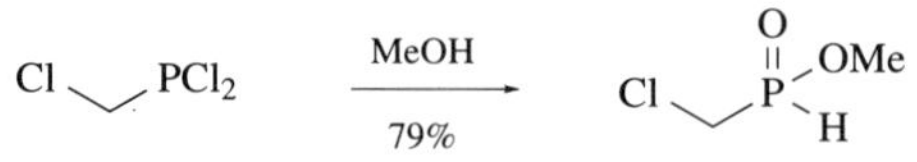

$$\text{ClCH}_2\text{PCl}_2 \xrightarrow[79\%]{\text{MeOH}} \text{ClCH}_2\text{P(=O)(OMe)H} \quad (290) \tag{61}$$

Alkyl- and arylchloromethylphosphinic acid esters have been readily prepared by alcoholysis of the corresponding phosphinic chlorides, with and without an acid scavenger. Thus, the phosphinic chlorides (**291**; $R^1 = \text{Et,Ph}$) were converted into the phosphinates (**292a–c**) by treatment with the appropriate alcohol (Equation (62)) ⟨53IZV763⟩. Ethyl (chloromethyl)methylphosphinate (**292d**) was similarly prepared ⟨86JGU2144⟩. Methyl (chloromethyl)methylphosphinate (**292**; $R^1 = R^2 = \text{Me}$) was prepared by methanolysis of ClCH$_2$PO(Cl)Me in the presence of Me$_3$N ⟨79IZV2073⟩. When the R^1 group in the chloride (**292**) was CF$_3$ or C$_2$F$_5$, the conditions for the alcoholysis had to be rigorously controlled ⟨86JGU2242⟩.

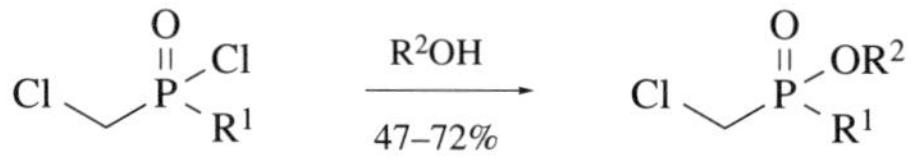

$$\text{ClCH}_2\text{P(=O)(Cl)R}^1 \xrightarrow[47-72\%]{R^2\text{OH}} \text{ClCH}_2\text{P(=O)(OR}^2)R^1 \tag{62}$$

(291)   (292)

(**a**) $R^1 = R^2 = \text{Et}$; (**b**) $R^1 = \text{Ph}, R^2 = \text{Me}$; (**c**) $R^1 = \text{Ph}, R^2 = \text{Et}$; (**d**) $R^1 = \text{Me}, R^2 = \text{Et}$

Interaction of ClCH$_2$PCl$_2$ with triethyl *ortho*-formate (2 mole equivalents) afforded ethyl chloromethyl(diethoxymethyl)phosphinate (**293**) in excellent yield (Scheme 22) ⟨86JGU1258⟩.

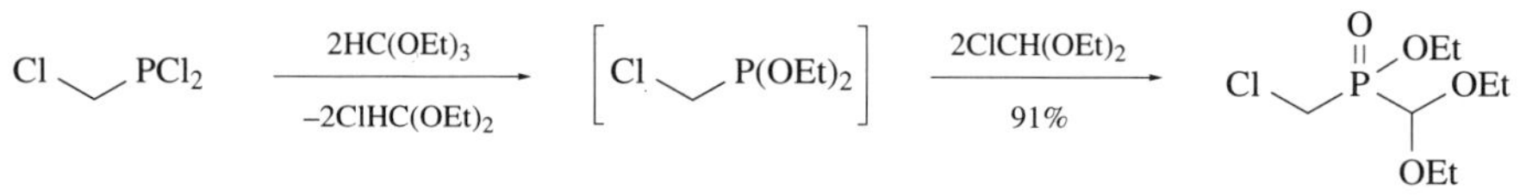

$$\text{ClCH}_2\text{PCl}_2 \xrightarrow[-2\text{ClHC(OEt)}_2]{2\text{HC(OEt)}_3} [\text{ClCH}_2\text{P(OEt)}_2] \xrightarrow[91\%]{2\text{ClCH(OEt)}_2} \text{ClCH}_2\text{P(=O)(OEt)CH(OEt)}_2$$

(293)

Scheme 22

The methyl chloromethyl (1-hydroxyalkyl)phosphinates (**294a–b**) have been prepared in 52–72% yields, respectively, from the appropriate aldehyde and methyl chloromethylphosphinate (**290**) ⟨71JGU2228⟩. The aryloxymethyl derivative (**295**) was obtained by ethanolysis of the benzodioxaphosphorinane (**280b**) in nearly quantitative yield; cleavage of the P—O bond in the ester (**280b**) by nucleophilic reagents is extremely facile ⟨89JGU1778⟩. The benzo-1,3,2-dioxaphosphole (**296**) on treatment with the required Grignard reagents afforded the phosphinates (**297a–e**) in 57–73% yields (Equation (63)) ⟨89IZV850⟩.

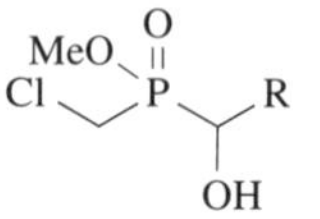

(**294**) **a**; R = Me
b; R = CCl$_3$

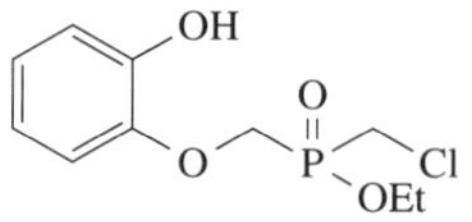

(**295**)

(296) → (297)

(a) R = Et; **(b)** R = Bu^n; **(c)** n-C$_8$H$_{17}$; **(d)** R = Ph; **(e)** 4-MeC$_6$H$_4$

Bis(chloromethyl)phosphinates (**298**; R = alkyl or aryl) have been prepared by the action of alcohols and phenols on (ClCH$_2$)$_2$PO(Cl), generally in the presence of a tertiary amine ⟨67JGU1768, 69HCA827, 71HCA1651⟩. Their conformations have been studied ⟨79IZV2073⟩.

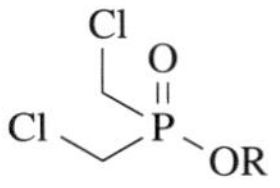

(**298**)

(Diethoxy)phenylphosphine (**299**) reacted with CH$_2$Br$_2$ to give the bromomethylphosphinate (**300**) in good yield (Equation (64)) ⟨88BSF699⟩. Treatment of phenyl-bis(chloromethyl)phosphinate (**301a**), *inter alia*, with KI gave the bis(iodomethyl) analogue (**301b**) in 95% yield ⟨67IZV1535⟩. This was converted into diphenyl [(iodomethyl)phenoxyphosphinyl]methylphosphonate (**302**) by the action of diphenyl methyl phosphite (MeO)P(OPh)$_2$ ⟨88PS(39)27⟩.

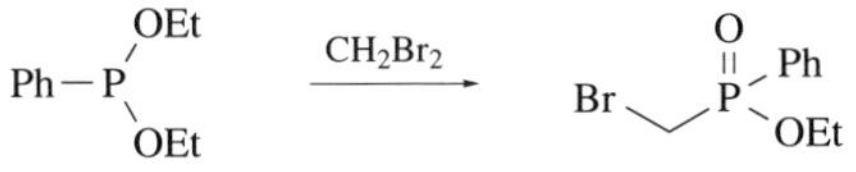

(**299**) → (**300**) (64)

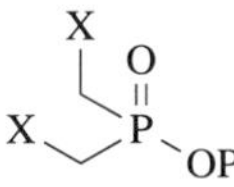

(**301**) **a**; X = Cl
       b; X = I

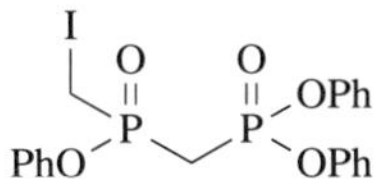

(**302**)

(d) α-Halophosphinic amides. α-Halophosphinic amides (HalCR$_2$)RPO(NR$_2$) are generally prepared by the action of primary and secondary amines on phosphinic chlorides, for example treatment of the (chloromethyl)phosphinic chloride (**303a**) with Et$_2$NH gave the amide (**303b**) in 60% yield ⟨68JGU2006⟩. Bis(chloromethyl)phosphinic chloride (**304a**) reacted with MeNH$_2$ in the presence of Et$_3$N to give the amide (**304b**) in 49% yield ⟨67JGU1768⟩ and with di-n-propylamine to afford the amide (**304c**) in 89% yield ⟨87PS(29)287⟩. Phosphinic amides have also been prepared by interaction of phosphinic anhydrides with amines, for example the diethylamide (**306**) was obtained from the anhydride (**305**) in 93% yield by this method (Equation (65)) ⟨61JA4381⟩.

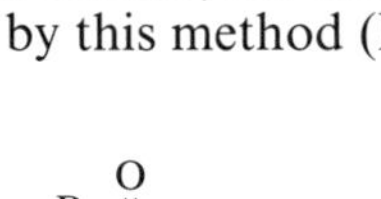

(**303**) **a**; R = Cl
       b; R = NEt$_2$

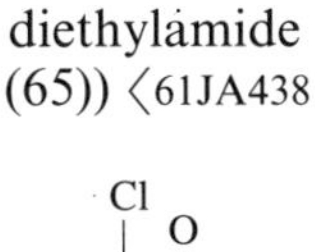

(**304**) **a**; R = Cl
       b; R = MeNH
       c; R = Pr^n$_2$N

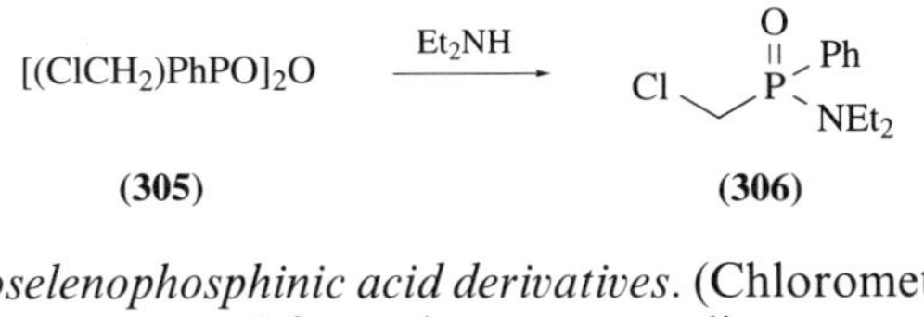

[(ClCH$_2$)PhPO]$_2$O $\xrightarrow{\text{Et}_2\text{NH}}$ (**306**) (65)

(**305**)

(e) α-Halothio- and α-haloselenophosphinic acid derivatives. (Chloromethyl)thiophosphinic chlorides (ClCH$_2$)RPS(Cl) have been prepared from the corresponding oxo acid chlorides by treatment with P$_2$S$_5$ ⟨68IZV1557, 71HCA1651⟩, PSCl$_3$ in the presence of catalytic amounts of PCl$_3$ and AlCl$_3$

⟨61JA1811⟩, and a mixture of red phosphorus and sulfur ⟨78JGU1813⟩, and also by treatment of α-halophosphines with sulfur ⟨73CB2733⟩. For example, (chloromethyl)methylthiophosphinic chloride (**307b**) was synthesized from the phosphinic chloride (**307a**) in 39% yield by treatment with PSCl$_3$, PCl$_3$ and AlCl$_3$ ⟨61JA1811⟩ and the bis(chloromethyl) compound (**308a**) (87% yield) from (ClCH$_2$)$_2$PO(Cl) with P$_2$S$_5$ ⟨68IZV1557⟩. The chlorophosphine ClCH$_2$(Me)PCl was oxidized by sulfur and by selenium to give the thiophosphinic chloride (**307b**) and the selenophosphinic chloride (**307c**) in 82% yields ⟨73CB2733⟩.

(**307**) a; Y = O
b; Y = S
c; Y = Se

(**308**) a; R = Cl
b; R = OEt
c; R = OPh
d; R = NEt$_2$
e; R = NHPh

O-Alkyl and O-aryl thiophosphinates (ClCH$_2$)RPS(OR) and thiophosphinic amides (ClCH$_2$)RPS(NR$_2$) have been obtained by treatment of thiophosphinic chlorides with alcohols and amines ⟨68IZV1557, 71HCA1651⟩. Thus, the thiophosphinic chloride (**308a**) and EtOH and PhOH, or Et$_2$NH and PhNH$_2$, in the presence of Et$_3$N, afforded the O-thioesters (**308b–c**), or the amides (**308d–e**), respectively in 72–95% yields. The O-phenyl thiophosphinate (**308c**) was converted into the bis(iodomethyl) analogue (ICH$_2$)$_2$PS(OPh) with KI ⟨68IZV1557⟩.

S-Alkyl and S-aryl thiophosphinates have been synthesized by the action of thiols on phosphinic chlorides. In this way, the S-n-butyl (chloromethyl)thiophosphinates (**309a**) ⟨68JGU2006⟩ and (**309b**) ⟨71IZV1159⟩ were prepared in 67% yields.

(**309**) a; R = (CH$_2$)$_2$CN
b; R = CH=CHCO$_2$Et

(f) α-Halophosphinic anhydrides. Treatment of the phosphinic acid chlorides (**310a–c**) with paraformaldehyde gave the phosphinic anhydrides (**311a–c**) in 81–86% yields (Equation (66)) ⟨61JA4381⟩. The methyl compound (**311a**) has also been prepared from the commercially available (HOCH$_2$)$_2$PO(OH) in 92% yield by treatment with SOCl$_2$ ⟨71HCA1651⟩. The (chloromethyl)-perfluoroalkylphosphinic anhydrides (**313a–b**) were prepared from the acid chlorides (**312a–b**) in acetic acid (Equation (67)) ⟨87JGU622⟩.

(**310**) a; R = Me
b; R = Ph
c; R = CH$_2$Cl

[(ClCH$_2$)RPO]$_2$O (**311**)   (66)

(**312**) a; R = CF$_3$
b; R = C$_2$F$_5$

[(ClCH$_2$)RPO]$_2$O (**313**)   (67)

(g) Miscellaneous α-halophosphinic acid derivatives. The chloromethylbenzo-1,4,2-dioxaphosphorinane (**314a**) has been obtained in 87% yield by the action of pyrocatechol on (ClCH$_2$)$_2$PO(Cl) ⟨89JGU1778⟩. Similarly, the analogous benzoxazaphosphorinanes (**314b**) and (**314c**) were isolated in about 22% yields by interaction of (ClCH$_2$)$_2$PO(Cl) and o-aminophenol. The yield of the former analogue (**314b**) was improved (51%) when the N,O-di(trimethylsilyl) derivative of o-aminophenol was employed for the condensation ⟨92JGU39⟩. The reaction of ClCH$_2$PCl$_2$ with 3-methyl-3-buten-

2-one in the presence of Ac$_2$O furnished the 2-oxo-1,2-oxaphospholene (**315**) (Equation (68)) ⟨78JGU1168⟩.

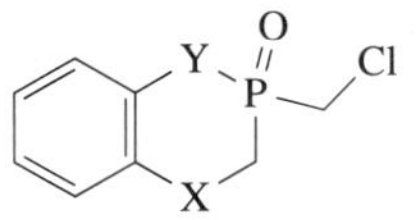

(**314**) **a**; X = Y = O
b; X = NH, Y = O
c; X = O, Y = NH

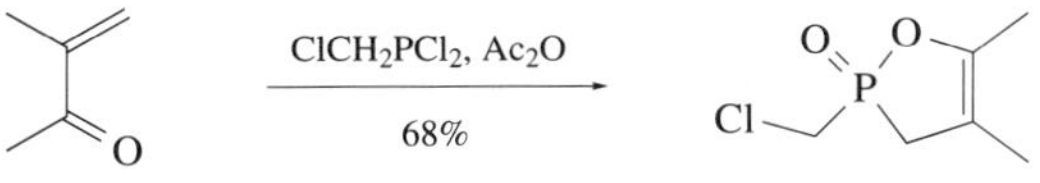

$$\text{(68)}$$

(**315**)

The (bromomethyl)imidophosphinate (**316c**) was prepared from the dimethyl compound (**316a**) by bromination of the lithio derivative (**316b**) ⟨87PS(33)147⟩.

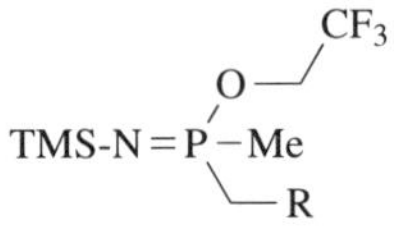

(**316**) **a**; R = H
b; R = Li
c; R = Br

(ii) α-Halophosphonic acids and derivatives

(a) α-Halophosphonic acids. Fluoromethylphosphonic acid (**317b**) has been prepared in 50% yield by treatment of (Pr^{i}O)$_3$P with Na and FCH$_2$Cl to give the diisopropyl fluoromethylphosphonate (**317a**) which was then hydrolysed with hydrochloric acid ⟨67BSF4289⟩. With diethylaminosulfur trifluoride (DAST), α-hydroxybenzylphosphonate esters (**318a**; Ar = Ph,p-ClC$_6$H$_4$,p-MeC$_6$H$_4$) afforded the corresponding α-fluoro derivatives (**318b**). Treatment of these with TMS-I afforded the trimethylsilyl esters which on methanolysis gave the phosphonic acids (**318c**) isolated as their cyclohexylammonium salts ⟨86JCS(P1)913⟩. Lithiation of the phosphonate (**317a**) and treatment of the resultant carbanion with CO$_2$ and with COS gave the carboxylic acid (**319a**) and the thio-carboxylic acid (**319b**) (ca. 60% overall yields). These were converted into the acids (**319c**) and (**319d**) with TMS-Br and then MeOH ⟨85JCR(S)92⟩. The 2-halopropionic acids (**320a–c**), isolated as their pyridinium salts, were prepared by hydrolysis of their triethyl esters with HCl (see ⟨89B8270⟩ and references cited therein).

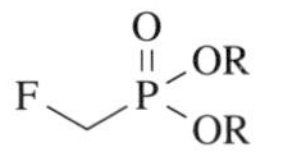

(**317**) **a**; R = Pr^i
b; R = H

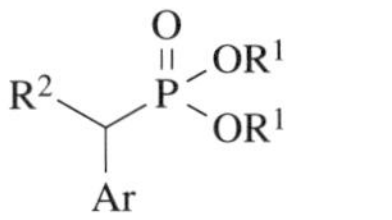

(**318**) **a**; R^1 = Et, R^2 = OH
b; R^1 = Et, R^2 = F
c; R^1 = H, R^2 = F

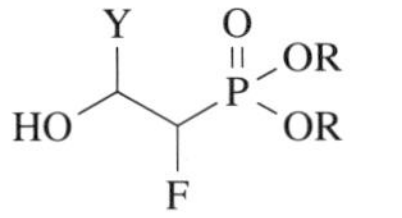

(**319**) **a**; R = Pr^i, Y = O
b; R = Pr^i, Y = S
c; R = H, Y = O
d; R = H, Y = S

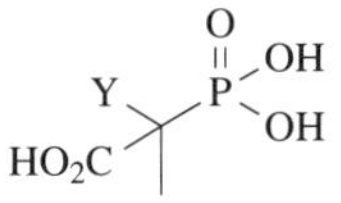

(**320**) **a**; Y = F
b; Y = Cl
c; Y = Br

α-Chloroalkylphosphonic acids are generally prepared by hydrolysis of the corresponding dihalides or diamides, for example the commercially available ClCH$_2$PO(OH)$_2$ has been conveniently prepared by hydrolysis of ClCH$_2$PO(Cl)$_2$ with H$_2$O ⟨49HCA1175, 88PS(40)183⟩. Hanessian *et al.* have described the asymmetric synthesis of the (*R*)-α-chloroalkylphosphonic acids (**322a–g**) from the (*R,R*)-diaminocyclohexane (**321**) in 76–80% yields by the route shown in Scheme 23. The enantiomers of the (*R*)-acids (**322b**) and (**322e**) were obtained by starting with the corresponding

(S,S)-diamine ⟨90TL6461⟩. Treatment of (Z)-1-propenylphosphonic acid with aqueous NaOCl afforded the *threo* derivative (**323**) in 85% yield ⟨69TL4647⟩.

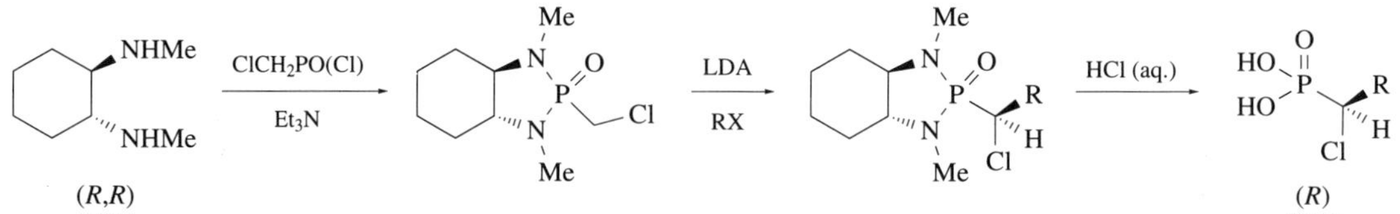

(a) R = Me; (b) R = Et; (c) R = Pr^n; (d) R = CH$_2$CH=CH$_2$; (e) R = Bn; (f) R = TBDPSiO(CH$_2$)$_2$; (g) R = TBDPSiO(CH$_2$)$_3$

TBDP = *t*-butyldiphenyl

Scheme 23

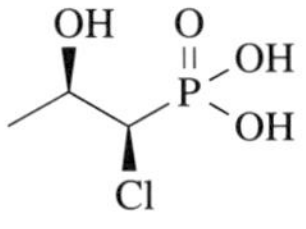

(**323**)

Diethyl α-bromomethylphosphonate (**324a**) hydrolysed on treatment with 48% hydrobromic acid to give the bromomethylphosphonic acid (**324b**) and the iodomethyl analogue (**325b**) was prepared similarly ⟨53JA5738⟩. The transformation of the diethyl iodomethylphosphonate (**325a**) to the acid (**325b**) was nearly quantitative when the diethyl ester was first converted into the di(trimethylsilyl) ester and this was then hydrolysed at neutral pH ⟨79CC739⟩. The 1-bromo-2-aminoethylphosphonic acid (**327**) was readily prepared in 47% yield by sequential treatment of the vinylphosphonate (**326**) with Br$_2$, NH$_4$OH and HCl in a one-pot procedure (Equation (69)) ⟨85T4979⟩.

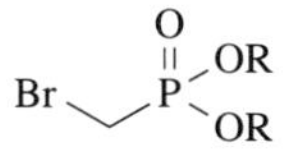

(**324**) a; R = Et
b; R = H

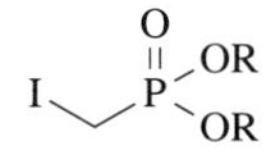

(**325**) a; R = Et
b; R = H

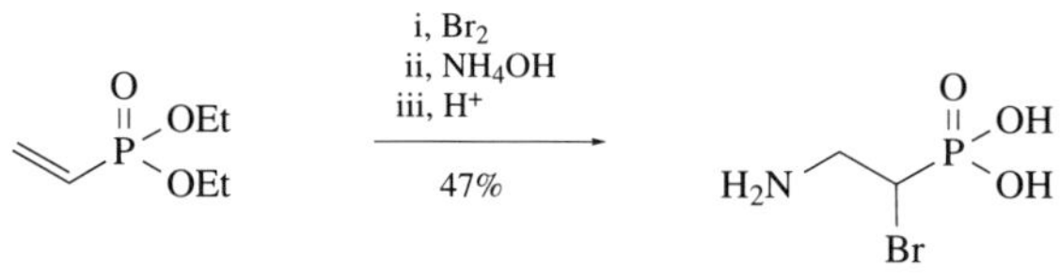

(**326**)   (**327**)   (69)

(b) α-Halophosphonic dihalides. Fluoromethylphosphonic dichloride FCH$_2$PO(Cl)$_2$ has been prepared from the acid (**317b**) in 60% yield by treatment with PCl$_5$ ⟨67BSF4289⟩.

α-Chlorophosphonic dihalides are formed by the reaction of PCl$_3$ and AlCl$_3$ on *gem*-dihalo alkanes and hydrolysis of the resultant complexes. By this method the dichloro alkanes (**328a–d**) afforded the α-chlorophosphonic dichlorides (**329a–d**) in 40–85% yields (Scheme 24) ⟨52JCS3437, 86JCS(P1)1681⟩. Chlorination of HOCH$_2$PO(OH)$_2$ with SOCl$_2$ in the presence of pyridine gave ClCH$_2$PO(Cl)$_2$ in 57% yield ⟨66CJC2593⟩. The α-chlorobenzyl derivative (**329c**) has also been prepared in 55% yield by heating PhCHO with PCl$_3$ ⟨89JCS(P1)563⟩. Treatment of aldehydes with PCl$_3$ to give α-haloalkylphosphonic dihalides is a general procedure first reported by Kabachnik and Shepeleva ⟨50DOK(75)219, 50IZV39⟩. Sodium fluoride converted chloromethylphosphonic dichloride (**329a**) into the difluoride (**330a**) in 76% yield ⟨60JOC2016⟩. Antimony trifluoride has also been used for this transformation ⟨85JCP1517⟩. Treatment of CHCl$_2$PO(F)(OR) with chlorine gave the mixed dihalide ClCH$_2$PO(F)(Cl) (83% yield) of which only the chlorine atom of the dihalophosphoryl group was replaceable by alkoxy and amino groups ⟨68JGU1284⟩. Treatment of ClCH$_2$PO(Cl)$_2$ in pyridine with H$_2$O (1 mole equivalent) gave the selective monophosphonoylating agent (**330b**) ⟨85CCC1507⟩.

Iodomethylphosphonic dichloride $ICH_2PO(Cl)_2$ has been prepared in 85% yield by treatment of $ICH_2PO(OEt)_2$ [from $P(OEt)_3 + I_2CH_2$] with PCl_5 ⟨86JCS(PI)1681⟩.

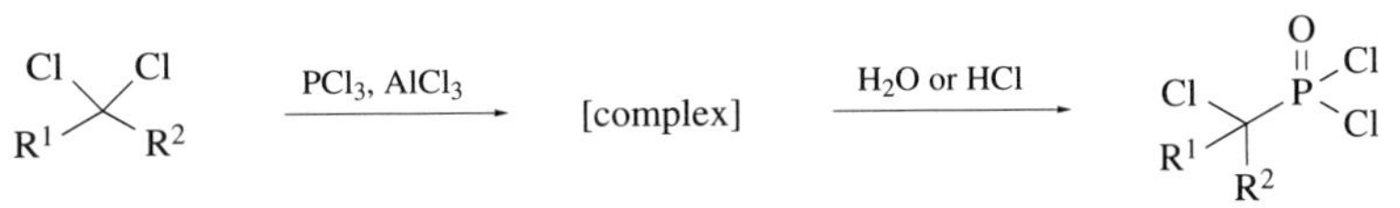

$$\begin{array}{c} Cl \\ \diagdown \\ R^1 \end{array} \begin{array}{c} Cl \\ \diagup \\ R^2 \end{array} \xrightarrow{PCl_3, AlCl_3} [complex] \xrightarrow{H_2O \text{ or } HCl} \begin{array}{c} Cl \\ R^1 \end{array} \begin{array}{c} O \\ \parallel \\ P \\ R^2 \end{array} \begin{array}{c} Cl \\ \diagdown \\ Cl \end{array}$$

(328)
a; $R^1 = R^2 = H$
b; $R^1 = H, R^2 = Me$
c; $R^1 = H, R^2 = Ph$
d; $R^1 = R^2 = Me$

(329)

Scheme 24

$$Cl \diagdown \overset{O}{\underset{\parallel}{P}} \diagup \overset{R^1}{\underset{R^2}{}}$$

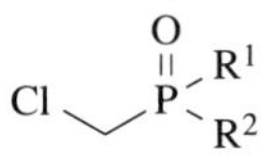

(330) **a**; $R^1 = R^2 = F$
b; $R^1 = Cl, R^2 = OH$

(c) α-Halophosphonochloridates. The compounds, $R_2CHalPO(Cl)(OR)$, are usually prepared by monoesterification of phosphonic dichlorides or by replacement of one of the alkoxy groups of phosphonate esters with a chlorine atom, for example $ClCH_2PO(Cl)(OEt)$ has been prepared in 72% yield by treating $ClCH_2PO(Cl)_2$ with ethanol in the presence of pyridine ⟨67BRP1087066⟩ and in 85% yield from $ClCH_2PO(OEt)_2$ and $POCl_3$ ⟨91SC793⟩.

(d) α-Halophosphonamidic halides. N,N-Dialkyl α-haloalkylphosphonamidic fluorides and chlorides $R_2CHalPO(Hal)(NR_2)$ are prepared by allowing the dichlorides and difluorides to react with the stoichiometric amount of amine; for example, $ClCH_2PO(F)(NMe_2)$ ⟨67JSP32⟩, $ClCH_2PO(Cl)(NMe_2)$ ⟨68JGU2563⟩ and $ClMe_2CPO(Cl)(NHMe)$ ⟨89JCS(P1)563⟩ have been prepared by this procedure.

(e) α-Halophosphonic acid esters. α-Fluorophosphonates are generally prepared by (i) treatment of the salts of dialkyl phosphites with fluoroalkyl halides, (ii) alkylation of α-fluoroalkylphosphinate anions, (iii) electrophilic fluorination of alkylphosphonate anions and (iv) replacement of the OH group in α-hydroxyalkylphosphonates using diethylaminosulfur trifluoride (DAST). For example, diisopropyl fluoromethylphosphonate (332) has been prepared by treatment of sodium diisopropylphosphite (331) with FCH_2Cl (Scheme 25) ⟨67BSF4289⟩. Conversion of this into the lithium salt (333a) and then addition of $ClCO_2Et$ gave the phosphonofluoroacetate (333b) (Scheme 25) ⟨87JCS(P1)181⟩. The salt has been alkylated with a number of other carbon electrophiles ⟨83CC886⟩. α-Deprotonation of the phosphonate (334a) with potassium diisopropylamide and then treatment with the selective fluorinating agent N-fluorobis(benzenesulfonyl)imide $(PhSO_2)_2NF$ gave the α-fluoro derivative (334b) in 54% yield ⟨91SL395⟩. Likewise, $(F_3CSO_2)_2NF$ has been used to good effect as the source of electrophilic fluorine in the preparation of $NCFHCPO(OEt)_2$ ⟨92JCS(PI)313⟩. α-Fluorobenzylphosphonate esters have been conveniently prepared in good yield by treating the α-hydroxy analogues with DAST. Dehydration rather than substitution occurred when the secondary α-hydroxyalkylphosphonates were treated with this reagent ⟨86JCS(PI)913⟩.

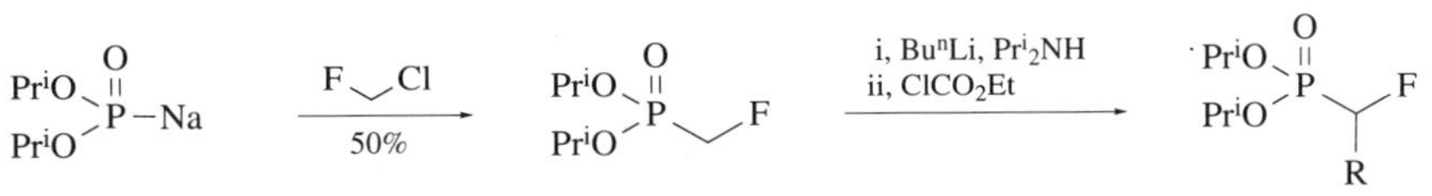

$$\begin{array}{c} Pr^iO \\ \diagdown \\ Pr^iO \end{array} \overset{O}{\underset{\parallel}{P}} -Na \xrightarrow[50\%]{F\diagdown Cl} \begin{array}{c} Pr^iO \\ \diagdown \\ Pr^iO \end{array} \overset{O}{\underset{\parallel}{P}} \diagdown F \xrightarrow[\text{ii, } ClCO_2Et]{\text{i, } Bu^nLi, Pr^i_2NH} \begin{array}{c} Pr^iO \\ \diagdown \\ Pr^iO \end{array} \overset{O}{\underset{\parallel}{P}} \diagdown \underset{R}{\overset{F}{\diagup}}$$

(331)   (332)   (333)

(**a**) R = Li; (**b**) R = CO_2Et

Scheme 25

α-Chlorophosphonic acid esters have been synthesized by (i) alkylation of diethyl chlorolithio(trimethylsilyl)methylphosphonates and then removal of the silyl group with sodium ethoxide (Scheme 26), (ii) replacement of the OH group in α-hydroxyalkylphosphonates and (iii) alcoholysis of phosphonic acid dichlorides. Thus, α-chloroalkylphosphonates (**338a–f**) have been prepared from the trichloromethylphosphonate (335) in 80–90% overall yields by the reaction sequence shown in

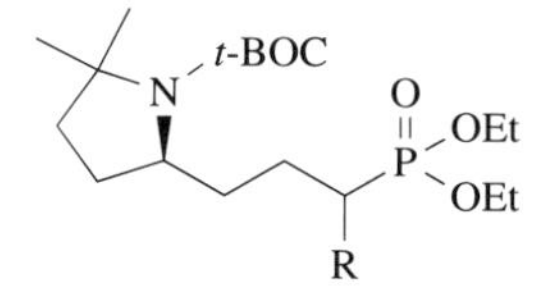

(334) a; R = H
b; R = F

Scheme 26 ⟨88JOM(338)295⟩. The diethyl α-chloroalkylphosphonates (**339a–e**) have been prepared in 81–90% yields from the corresponding α-hydroxy analogues by chlorination with $Ph_3P\text{-}CCl_4$ ⟨90S717⟩. Condensation of α-chloroethylphosphonic dichloride (**340a**) with EtOH in the presence of Et_3N gave the diethyl ester (**340b**) in 88% yield ⟨69JGU1490⟩. The α-halo aldehydes (**342a–b**) have been obtained in excellent yield by the action of Cl_2 or Br_2 and then H_2O on the enol ether (**341**) (Equation (70)) ⟨91JGU1459⟩.

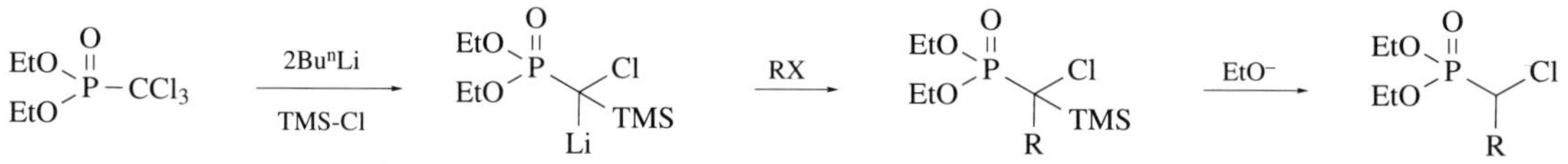

(335) (336) (337) (338)

(a) R = Me; (b) R = Et; (c) R = $CH_2CH=CH_2$; (d) R = Bu^n; (e) R = $CH_2CH=CHMe$; (f) R = CH_2Ph

Scheme 26

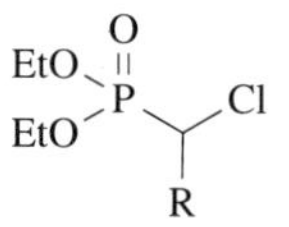

(339) a; R = H
b; R = Me
c; R = Et
d; R = Pr^n
e; R = Ph

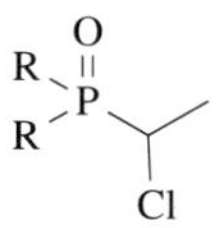

(340) a; R = Cl
b; R = OEt

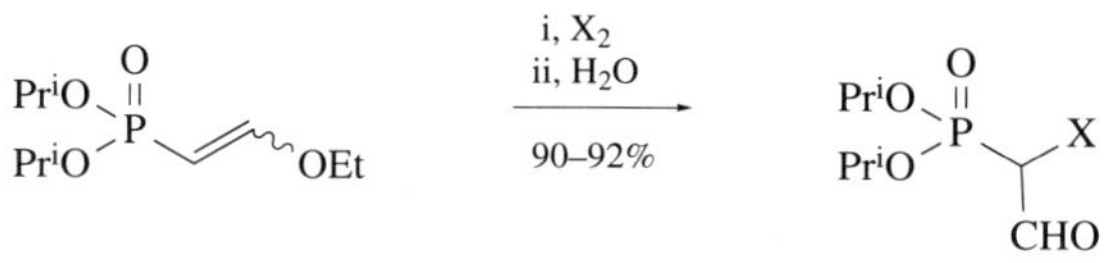

(341) (342) a; X = Cl
b; X = Br

$$\tag{70}$$

Treatment of the α-chlorophosphonates (**337a–c**) with Bu^nLi and then with $Br(CH_2)_2Br$ gave the bromo derivatives (**343a–c**), the silyl groups of which could be removed with sodium ethoxide to afford the α-bromoalkylphosphonates (**346b**), (**346c**) and (**346f**) respectively in 76–88% yields from the chloro compounds (**337a–c**) ⟨88JOM(338)295⟩. The diisopropyl α-bromoalkylphosphonates (**345a–e**) have been prepared in good yields by bromination of the corresponding alkylphosphonates (**344a–e**) with NBS in the presence of a free radical initiator (Equation (71)) ⟨91SC1039⟩.

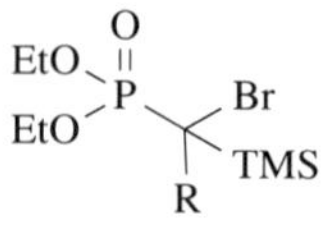

(343)

(a) R = Me; (b) R = Et; (c) R = $CH_2CH=CH_2$

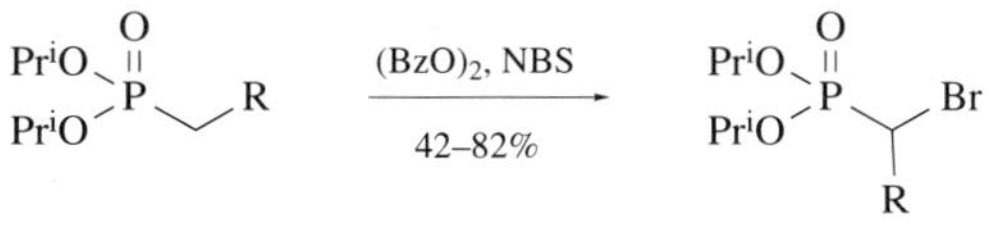

$$(344) \xrightarrow[42-82\%]{(BzO)_2, NBS} (345) \quad (71)$$

(344) Pr^iO-P(=O)(OPr^i)-CH_2R
(345) Pr^iO-P(=O)(OPr^i)-CHBrR

(a) R = H; (b) R = Me; (c) R = Pr^i; (d) R = Bu^i; (e) R = Ph

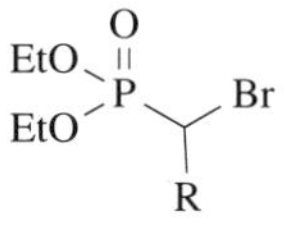

(346) EtO-P(=O)(OEt)-CHBrR

(a) R = H; (b) R = Me; (c) R = Et; (d) R = Pr^n; (e) R = Ph; (f) R = CH_2CH=CH_2

The diethyl α-bromoalkylphosphonates (**346a–e**) have been synthesized in 39–65% yields by replacing the OH group of the corresponding α-hydroxyphosphonates by Br with Ph$_3$P and CBr$_4$ (or Ph$_3$PBr$_2$) ⟨90MI 403-01⟩.

Diethyl iodomethylphosphonate ICH$_2$PO(OEt)$_2$ has been prepared in 60% yield by the action of I$_2$CH$_2$ on (EtO)$_3$P ⟨36JGU283⟩. The mixed ester ICH$_2$PO(OEt)OPh has been obtained similarly from I$_2$CH$_2$ and (EtO)$_2$P(OPh) ⟨86ZAAC(536)187⟩.

(f) α-Halophosphonic diamides. These compounds are generally formed by treatment of the phosphonic dichlorides with an excess of an amine, for example ClCH$_2$PO(Cl)$_2$ reacted with Et$_2$NH to give the diamide ClCH$_2$PO(NEt$_2$)$_2$ in 73% yield ⟨69JGU1490⟩. The phosphonic diamides (**349a–d**; R^1 = Et, Pr^i) have been prepared in good yield from the corresponding alkyl(diamino)phosphines (**347**) by the action of an excess of CCl$_4$ or CBrCl$_3$ and then treatment of the resultant dihalo derivatives (**348a–b**) with MeOH (Scheme 27) ⟨89JGU285, 90JGU1536, 91JGU627⟩. Treatment of ClCH$_2$PO(Cl)$_2$ with Et$_2$NH (2 mole equivalents) gave ClCH$_2$PO(Cl)(NEt$_2$) (60% yield) which reacted further with PhNH$_2$ (1 mole equivalent) and Et$_3$N to give the 'mixed' diamide ClCH$_2$PO(NEt$_2$)(NHPh) in 65% yield ⟨76JGU2501⟩.

$$(R^1{}_2N)_2PCH_2R^2 \xrightarrow[40-70\%]{CXCl_3} (R^1{}_2N)_2PCR^2X_2 \text{ (348)} \xrightarrow[70-90\%]{MeOH} (R^1{}_2N)_2PO(CHXR^2)$$

(347)   X = Cl, Br   (348)   (349)

(a) R^2 = H; (b) R^2 = Me; (c) R^2 = Pr^n; (d) R^2 = Pr^i

Scheme 27

(g) α-Halophosphonamidates. N-t-Butyl α-chlorophosphonamidates RCHClPO(OMe)(NHBu^t) were prepared from the appropriate phosphonic dichlorides by sequential treatment with Bu^{t}NH$_2$ (2 mole equivalent) and then NaOMe ⟨89JCS(PI)563⟩.

(h) α-Halothiophosphonic acid derivatives. The phosphonous diamides (Pr$_2^i$N)$_2$PCH$_2$X where X = Cl, Br were oxidized with S to give (Pr$_2^i$N)$_2$PS(CH$_2$X) in 85–90% yields ⟨89TL2445⟩. Treatment of ClCH$_2$PO(Cl)$_2$ with P$_2$S$_5$ afforded the commercially available ClCH$_2$PS(Cl)$_2$ in 65% yield ⟨61JA2299, 66OS(46)21⟩.

4.03.2.4 Penta- and Hexacoordinate Phosphorus Derivatives—R$^1{}_2$CHalPR$^2{}_4$, [A]$^+$[R$_2$CHalPX$_5$]$^-$, etc.

α-Halophosphoranes existing predominantly in the ionized form as phosphonium salts are described in Section 4.03.2.3.1.

Treatment of ClCH$_2$PCl$_2$ with SbF$_5$ effected halogen exchange (confined to the P—X bonds) and concomitant oxidation to give ClCH$_2$PF$_4$ in 83% yield ⟨64IC410, 67IS63⟩. The analogue ClCH$_2$PCl$_4$, prepared by careful chlorination of ClCH$_2$PCl$_2$ with Cl$_2$, is molecular above room temperature. Below room temperature, the salt form [ClCH$_2$PCl$_3$]$^+$[ClCH$_2$PCl$_5$]$^-$ predominates ⟨91CC234⟩. Interaction of ClCH$_2$PF$_4$ with trimethylsilyl(trimethylsilyloxy)acetate TMS-OCH$_2$CO$_2$-TMS (2 mole equivalents) afforded the spirophosphorane (**350**) in 82% yield ⟨92CB801⟩. The dimeric benzo-1,3,2-oxazaphospholes (**351a–c**), of which (**b**) and (**c**) were obtained as a mixture of diastereoisomers, were synthesized by the route shown in Scheme 28 ⟨90TL5361⟩.

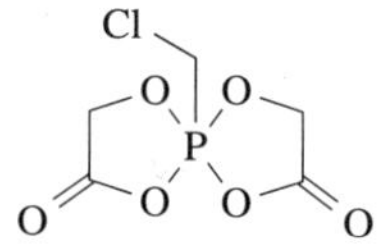

(**350**)

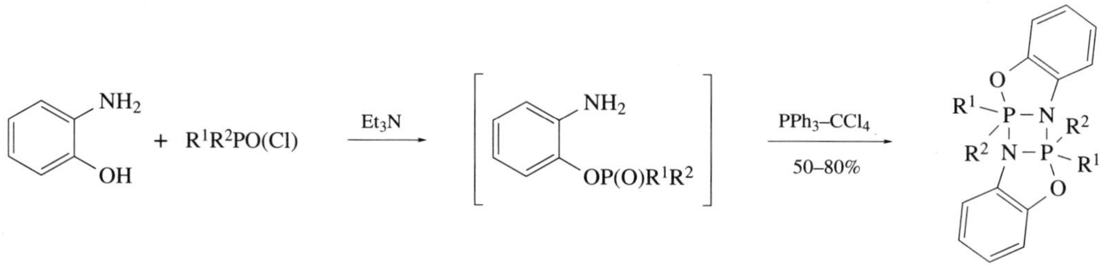

(**351**) **a**; $R^1 = R^2 = CH_2Cl$
b; $R^1 = CH_2Cl, R^2 = Ph$
c; $R^1 = CH_2Cl, R^2 = OPh$

Scheme 28

4.03.3 α-HALO ARSENIC, ANTIMONY AND BISMUTH DERIVATIVES—$R^1_2CHalAsR^2_2$, $R_2CHalAsO(OH)_2$, etc.

4.03.3.1 α-Halo Arsenic Derivatives

4.03.3.1.1 Tricoordinate α-halo arsenic derivatives

Cullen *et al.* have reported the synthesis of the bis(dimethylarsines) (**352a–c**) in 82–94% yields by the addition of fluoroethene, trifluoroethene and chloroethene to tetramethyldiarsine, $Me_2AsAsMe_2$, with cleavage of the weak As—As bond ⟨72JA5702⟩. Addition of $(C_2F_5)_2AsF$ to $CHF=CF_2$ in the presence of SbF_5 afforded $(C_2F_5)_2AsCHFCF_3$ in 76% yield ⟨72IZV215⟩.

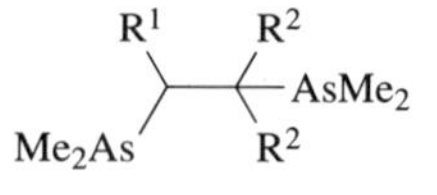

(**352**) **a**; $R^1 = F, R^2 = H$
b; $R^1 = R^2 = F$
c; $R^1 = Cl, R^2 = H$

(Chloromethyl)arsine $ClCH_2AsH_2$ was prepared by the reduction of (chloromethyl)arsonic acid $ClCH_2AsO(OH)_2$ with a Zn–H_2SO_4 system. The corresponding bromo and iodo analogues, however, could not be prepared by this procedure ⟨75JOM(102)437⟩. Markl *et al.* showed that the metal arsenide (**353**; M = Li or K), formed from 1,2,5-triphenylarsole and two mole equivalents of Li or K, was alkylated with CH_2Cl_2 and with Cl_2CHCO_2Na to give the 1-(1-chloroalkyl)arsoles (**354a**) (28% yield) ⟨83JOM(249)335⟩ and (**354b**) (39%) respectively (Equation (72)) ⟨74TL303⟩.

(72)

(353) a; R = H
(354) b; R = CO₂H

Meyer *et al.* reported that dibromotris(trimethylsilylmethyl)arsorane (**355**) fragments under reduced pressure to give the (bromomethyl)arsine (**357**) and TMS-Br. The ylide (**356**) was proposed as an intermediate in the reaction (Scheme 29) ⟨83CB348⟩.

$$(TMS\text{-}CH_2)_3AsBr_2 \longrightarrow [Br(TMS\text{-}CH_2)_2As=CH_2] \longrightarrow (TMS\text{-}CH_2)_2AsCH_2Br$$

(**355**)   (**356**)   (**357**)

Scheme 29

(Iodomethyl)diphenylarsine ICH₂AsPh₂ was obtained in 76% yield by the addition of phenyllithium to a mixture of Ph₂AsCl and CH₂I₂ ⟨85CB2353⟩.

The first attempts to prepare dihalo(halomethyl)arsines by the reactions of AsCl₃ and AsBr₃ with CH₂N₂ resulted in low yields (<10%) ⟨41JGU41⟩. Rheingold and Bellama improved the yields in these methylene insertion reactions by the use of copper catalysts. Using this modification, mixtures of the mono- and bis(α-halomethyl)arsines XCH₂AsX₂ and (XCH₂)₂AsX where X = Cl or Br were formed in nearly quantitative yield. No tris(halomethyl)arsines were obtained by this method ⟨75SRI199⟩. The α-chloroarsines (ClCH₂)MeAsCl and (ClCH₂)₂MeAs were prepared in 39% and 10% yields respectively by the interaction of MeAsCl₂ and CH₂N₂ ⟨73CB2742⟩. Moderate yields of (MeCHCl)₂AsCl and (MeCHCl)₃As were obtained by treating AsCl₃ with 2.5 and 4.0 mole equivalents respectively of CH₃CHN₂ ⟨52JGU1569⟩. Sommer prepared ClCH₂AsCl₂ by treatment of AsCl₃ with (ClCH₂)₂Hg (ca. 65% yield) and of AsCl₃ with ClCH₂AsPh₂ (54%) ⟨70ZAAC(377)128⟩.

Reduction of ClCH₂AsO(OH)₂ with hypophosphorous acid H₂PO(OH) resulted in the formation of pentakis(chloromethyl)pentaarsacyclopentane (**358**) ⟨75JOM(102)445⟩.

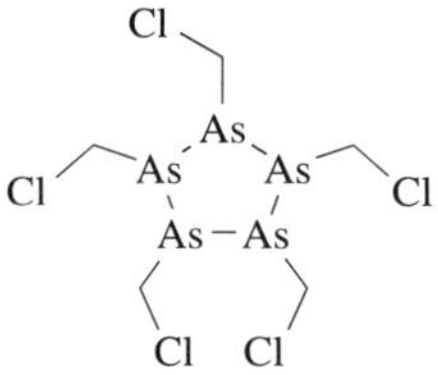

(**358**)

4.03.3.1.2 Tetracoordinate α-halo arsenic derivatives

α-Haloarsonium salts [(XR¹₂C)R²₃As]⁺X⁻ are formed by treatment of trialkyl(or aryl)arsines with alkylidene dihalides, for example the (iodomethyl)trimethylarsonium iodide [(ICH₂)Me₃As]⁺I⁻ was formed from trimethylarsine and CH₂I₂ ⟨25MI 403-01⟩.

No α-haloalkylarsine oxides have been reported.

Of the α-haloarsinic acid and α-haloarsonic acid series, only (ClCH₂)₂AsO(OH) and ClCH₂AsO(OH)₂ are known. They were prepared from (ClCH₂)₂AsCl and ClCH₂AsCl₂ and H₂O₂ (15%) ⟨41JGU41, 75JOM(102)437⟩.

4.03.3.1.3 Pentacoordinate α-halo arsenic derivatives

The α-fluoroarsorane Ph₃AsF(CHFCO₂Et) was formed by the addition of N₂F₄ to Ph₃As=CHCO₂Et and subsequent elimination of N₂ ⟨71JGU481⟩.

4.03.3.2 α-Halo Antimony and Bismuth Derivatives

There are only a few α-halo antimony derivatives known. Depending on the conditions used, dichloro(chloromethyl)stibine $ClCH_2SbCl_2$ and tris(chloromethyl)stibine $(ClCH_2)_3Sb$ were formed by interaction of $SbCl_3$ with CH_2N_2. Likewise, $(MeCHCl)_2SbCl$ was produced from $SbCl_3$ and diazoethane. Treatment of $(ClCH_2)_3Sb$ with bromine readily afforded $(ClCH_2)_3SbBr_2$ ⟨52JGU1569⟩. Unexpectedly, $Me_3Sb(CH_2I)CHI_2$ was obtained in 31% yield when a mixture of Me_3Sb and MeI in acetone was heated under reflux ⟨68MI 403-01⟩.

No reports of rigorously characterized α-halo bismuth compounds are evident.

4.03.4 α-HALO ALKYLMETALLOIDS—R_2CHalMETALLOID

4.03.4.1 α-Halo Silicon Derivatives—$R^1{}_2CHalSiR^2{}_3$

α-Halosilanes have widespread application in the preparation of organosilicon compounds ⟨91OM1960⟩, and a number of methods have been developed for their synthesis. These include direct halogenation of alkylsilanes, selective reduction of halo(α-haloalkyl)silanes with LAH, reduction of α,α-dihalosilanes with tin hydrides, addition of silanes to halo alkenes, halogen exchange reactions, silylation of haloiodomethyl compounds, treatment of halosilanes with carbenoids, addition of halogen-containing electrophiles to vinylsilanes, nucleophilic substitution of the OH group in α-hydroxyalkylsilanes and the alkylation of carbanions from halomethylsilanes. Examples of these and other methods for the preparation of α-halosilanes are described below.

4.03.4.1.1 Alkyl- and aryl(α-haloalkylsilanes)

(i) Compounds with Si—H bonds

Due to their instability, few α-fluorosilicon compounds have been fully characterized. Treatment of trichloro(dichlorofluoromethyl)silane (**359a**) with $Bu^n{}_3SnH$ gave the α-fluoromethylsilane (**359b**) isolated in 26% yield ⟨92JOM(427)293⟩. The mercury-photosensitized addition of an excess of SiH_4 to $CF_3CF=CF_2$ furnished a mixture (3:2) of the silanes $CF_3CHFCF_2SiH_3$ and $CF_3CF(SiH_3)CHF_2$ in 85% yield ⟨76JCS(D)694⟩.

(**359**) **a**; R = Cl
b; R = H

α-Chloromethyl- and α-bromomethylsilanes containing Si—H bonds have been prepared in good yields by reduction of the corresponding chlorosilanes with LAH. In this way, $ClCH_2SiCl(Me)_2$ was converted into $ClCH_2SiH(Me)_2$ (53% yield) ⟨90JOM(388)57⟩, $ClCH_2SiCl(Ph)_2$ gave $ClCH_2SiH(Ph)_2$ (82%) ⟨82LA1946⟩ and $BrCH_2SiCl_3$ gave $BrCH_2SiH_3$ (82%) ⟨86ZN(B)1527⟩. Likewise, (chloromethyl)cyclohexyl(phenyl)silane (**360a**) was obtained in good yield by reduction of the methoxysilane (**360b**) with LAH ⟨87LA51⟩. Chloro(methyl)phenylsilane Me(Ph)SiH(Cl) reacted with chloromethyllithium to produce $ClCH_2SiH(Me)(Ph)$ in 61% yield ⟨91OM1960⟩.

The (iodomethyl)silane $ICH_2SiH(Ph)_2$ was obtained in 89% yield by treatment of $ClCH_2SiH(Ph)_2$ with NaI ⟨82LA1946⟩.

(**360**) **a**; R = H
b; R = OMe

(ii) Compounds with Si—alkyl and Si—aryl bonds

(a) α-Fluorosilanes. Reduction of Et$_3$SiCHFBr with Bu$^n_3$SnH gave Et$_3$SiCH$_2$F in 74% yield ⟨73JOM(51)77⟩. TMS-CH$_2$F was prepared in moderate yields by treatment of TMS-CH$_2$OTs with KF ⟨76CCC386⟩ and by the reduction of TMS-CFCl$_2$ with Bu$^n_3$SnH ⟨90JOM(381)315⟩. Addition of TMS-H to the perfluorocyclobutene (**361**) produced the adduct (**362**) in 60% yield (Equation (73)) ⟨66JOM(6)633⟩. TMS-H and (chlorodifluoromethyl)fluorocarbene ClF$_2$C(F)C:, generated by pyrolysis of CF$_2$ClCF$_2$SiF$_3$, furnished CF$_2$ClCHF-TMS in 92% yield ⟨76JCS(P1)513⟩. Likewise, addition of the carbenes (**364a–c**), formed by thermolysis of the organomercurials (**363a–c**) where X = Cl,Br, to Et$_3$SiH produced the (α-fluoroalkyl)silanes (**365a–c**) in 53–72% yields (Scheme 30) ⟨73JOC4031, 75JOM(92)7⟩. The carbene Ph$_3$Si(Ph)C: (derived from the diazoalkane Ph$_3$SiCN$_2$Ph) and HF produced Ph$_3$SiCFHPh in 82% yield ⟨75CJC332⟩. Sequential treatment of the α-fluoro-α-iodo esters (**366**) with *n*-BuLi in pentane followed by TMS-Cl readily afforded the α-fluoro-α-trimethylsilyl esters (**367a–c**) (Equation (74)) ⟨90JOC4782⟩. Treatment of the αβ-epoxysilane (**368**) with SiF$_4$ in the presence of Pr$^i_2$NEt and H$_2$O produced the α-fluoro-β-hydroxysilane (**369**) in 80% yield (Equation (75)) ⟨89TL967⟩.

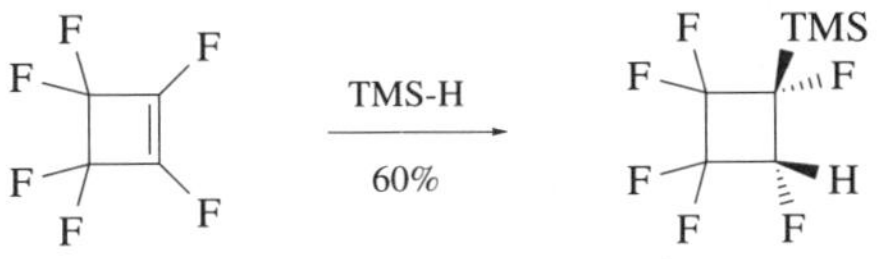

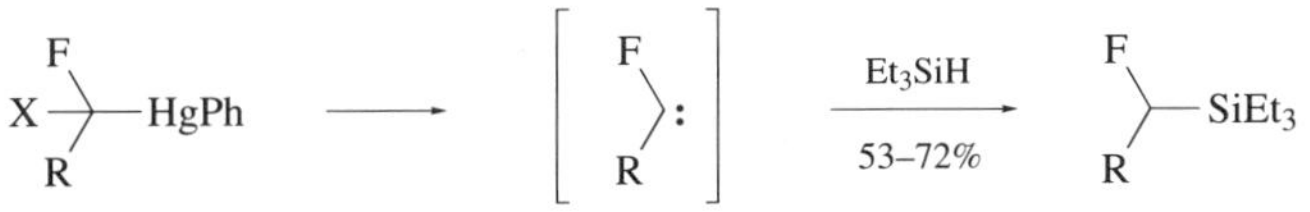

Scheme 30

(a) R = CF$_3$; (b) R = CO$_2$Me; (c) R = CO$_2$Et

a; R = Et
b; R = 2,4,6-Me$_3$C$_6$H$_2$
c; R = 2,4,6-Bu$^t_3$C$_6$H$_2$

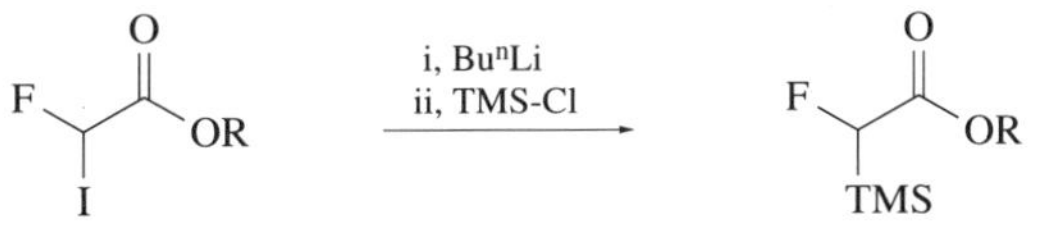

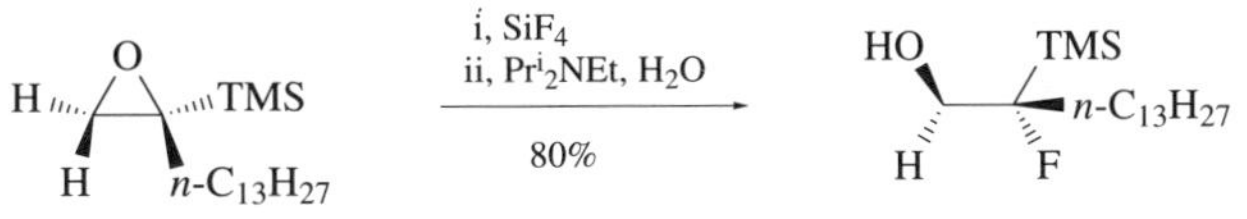

(b) α-Chlorosilanes. Vapour phase photochemical chlorination of TMS-Me afforded ClCH$_2$-TMS in 76% yield ⟨51JA1879⟩. (α-Chloroethyl)triethylsilane MeCHClSiEt$_3$ (75% yield) contaminated with ClCH$_2$CH$_2$SiEt$_3$ (7%) was obtained when Et$_4$Si was treated with SO$_2$Cl$_2$ and PCl$_5$ ⟨37JGU2495⟩.

A number of alkyl- and aryl(chloromethyl)silanes have been formed from chloro(methyl)silanes by preferential nucleophilic substitution at the silicon atom, for example ClCH$_2$SiCl(Me)$_2$ produced ClCH$_2$Si(CH$_2$Ph)Me$_2$ in 95% yield when treated with PhCH$_2$MgCl ⟨89JA8737⟩ and ClCH$_2$-TMS (90%) was obtained from ClCH$_2$SiCl(Me)$_2$ and MeMgBr ⟨47JA1976⟩. α-Chlorosilanes have also been prepared by the addition of electrophiles to vinylsilanes, for example Cl$_2$ and Ph$_3$SiCH=CH$_2$ furnished Ph$_3$SiCHClCH$_2$Cl in 84% yield ⟨76JOM(122)31⟩. Treatment of TMS-CH=CH$_2$ with 1,1,1-trichloroethane (2 mole equivalents) in the presence of Fe(CO)$_5$ and a 'nucleophilic coinitiator' (e.g. PPh$_3$, DMF, HMPA, etc.) produced MeCl$_2$CCH$_2$CHCl-TMS in 70–92% yields ⟨87IZV1087⟩. A number of arenesulfenyl chlorides ArSCl and $R_3SiCH=CH_2$ (R=EtO,Ph) afforded the adducts $R_3SiCH(SAr)CH_2Cl$ preferentially ⟨92JOM(437)111⟩.

Barrett et al. showed that the α-hydroxysilanes (**370a–d**), on treatment with the PPh_3—CCl_4 system, were converted into the corresponding α-chlorosilanes (**371**) in 65–87% yields (Equation (76)) ⟨92JOC386⟩.

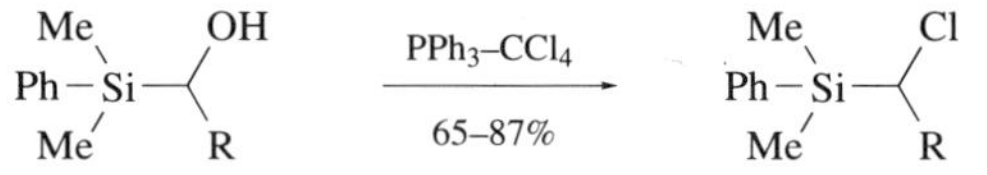

$$\text{Me}\diagdown\text{Ph}-\underset{\text{Me}}{\overset{}{\text{Si}}}-\underset{\text{R}}{\overset{\text{OH}}{\diagup}} \xrightarrow[65-87\%]{PPh_3-CCl_4} \text{Me}\diagdown\text{Ph}-\underset{\text{Me}}{\overset{}{\text{Si}}}-\underset{\text{R}}{\overset{\text{Cl}}{\diagup}} \quad (76)$$

(**370**) (**371**)

(**a**) R = n-C_5H_{11}; (**b**) R = n-C_7H_{15}; (**c**) R = c-C_6H_{11}; (**d**) R = Ph

Reduction of $R_3SiCHClBr$ (R = Et, Ph), and $R_3SiCHBr_2$ (R = Et,Ph) with Bu^n_3SnH afforded the corresponding monohalo derivatives R_3SiCH_2Cl and R_3SiCH_2Br in good yields and $Me_2Si(CH_2Cl)_2$ was similarly prepared from $Me_2Si(CH_2Cl)CHClBr$ ⟨70JOM(23)99⟩. The silanes R_3SiCCl_2Ph (R = Me, Et) were reduced to TMS-CHClPh (46% yield) and $Et_3SiCHClPh$ (40% yield) ⟨92OM859⟩.

Addition of TMS-CHClBr to s-butyllithium in THF followed by the addition of n-butyl iodide gave TMS-CH(Bu^n)Cl in 72% yield ⟨85BSF825⟩. Treatment of TMS-CH_2Cl with s-butyllithium followed by alkylation with EtBr afforded TMS-CHClEt in 58% yield ⟨87JOC809⟩. Deprotonation of allyl chloride with LDA followed by quenching of the resultant carbanion with TMS-Cl furnished CH_2=CHCHCl-TMS in 93% yield ⟨90SL769⟩ and addition of benzyl chloride and TMS-Cl to LDA produced PhCHCl-TMS in 91% yield ⟨87JOM(336)C41⟩. Triphenylsilyl chloride $ClSiPh_3$ and $ClSiMe_2(CH=CH_2)$ reacted with chloromethyllithium to give $ClCH_2SiPh_3$ (79% yield) and $ClCH_2SiMe_2(CH=CH_2)$ (71% yield) ⟨91OM1960⟩.

On thermolysis, 3-chloro-3-phenyldiazirine was converted into chlorophenylcarbene Ph(Cl)C: which reacted with Et_3SiH to form $Et_3SiCHClPh$ in 91% yield ⟨88TL5863⟩.

(c) α-Bromosilanes. Photoinduced bromination of benzyl(trimethyl)silane in CCl_4 or AcOH afforded (α-bromobenzyl)trimethylsilane. The selectivity of this bromination is largely dependent on the substrate and the solvent ⟨90AG(E)658⟩. Treatment of the amide (**372**) with NBS in the presence of benzoyl peroxide with UV irradiation gave the (α-bromobenzyl)trimethylsilane (**373**) in 65% yield (Equation (77)) ⟨86JOC3325⟩. Bromination of Ph_3SiEt with NBS (2 mole equivalents) gave Ph_3SiCBr_2Me in 73% yield which on sequential treatment with Bu^nLi and then HBr afforded $Ph_3SiCHBrMe$ in 83% yield. This compound was also obtained in 24% yield by treatment of $Ph_3SiCH(OH)Me$ with PBr_3 ⟨70CJC561⟩.

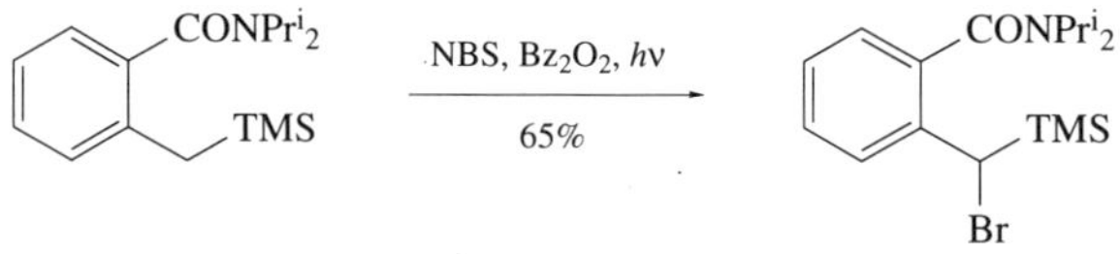

$$\underset{(372)}{\text{Ar-CH}_2\text{TMS}} \xrightarrow[65\%]{NBS, Bz_2O_2, h\nu} \underset{(373)}{\text{Ar-CHBr-TMS}} \quad (77)$$

(with Ar = 2-$CONPr^i_2$-C_6H_4)

A number of α-bromosilanes have been prepared by addition of electrophiles to vinylsilanes, for example addition of Br_2 to $Ph_3SiCH=CH_2$ and to $Ph_3SiC(Me)=CH_2$ produced $Ph_3SiCHBrCH_2Br$ (95% yield) and $Ph_3SiC(Me)BrCH_2Br$ (92% yield) respectively ⟨76JOM(122)31⟩. Brook et al. showed that addition of Br_2 to the trimethylsilyl(vinyl)silane (**374**) was exclusively *cis* giving the αβ-dibromosilane (**376**) (Scheme 31). The greater the electronegativity of the groups on silicon, the greater is the extent of *trans* addition. Thus, addition of Br_2 to the trifluorosilyl(vinyl)silane (**375**) gave the αβ-dibromosilane (**377**) as the major product (Scheme 31) ⟨89CC957⟩. Irradiation of a mixture of TMS-CH=CH_2 and $PhSO_2CBr_3$ furnished the adduct $PhSO_2CBr_2CH_2CHBr$-TMS in 81% yield ⟨86JOC3369⟩.

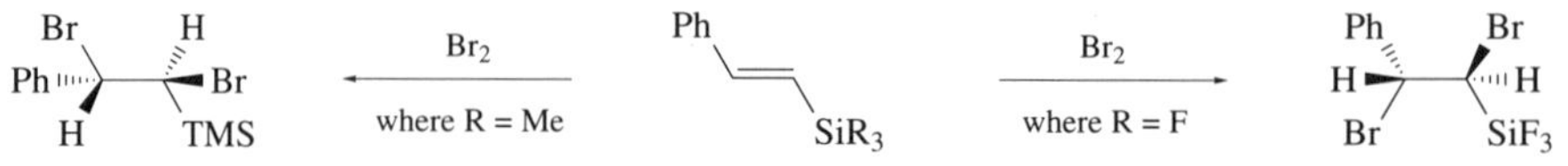

$$\underset{\substack{(376)\\cis\,(100\%)}}{\text{Ph-CHBr-CHBr-TMS}} \xleftarrow[\text{where R = Me}]{Br_2} \underset{\substack{(374)\,R=Me\\(375)\,R=F}}{\text{Ph-CH=CH-SiR}_3} \xrightarrow[\text{where R = F}]{Br_2} \underset{\substack{(377)\\trans\,(100\%)}}{\text{Ph-CHBr-CHBr-SiF}_3}$$

Scheme 31

Addition of TMSCHBr$_2$ to *s*-butyllithium in HMPA and alkylation of the resultant carbanion with *n*-butyl iodide afforded TMS-CHBrBu^n in 95% yield ⟨86BSF470⟩. Interaction of 3-bromo-3-phenylaziridine at ambient temperature with Ph(Me)$_2$SiH and Ph$_2$(Me)SiH resulted in insertion of the intermediate carbene Br(Ph)C: into the Si—H bonds to produce PhCHBrSiMe$_2$(Ph) (70% yield) and PhCHBrSiMe(Ph)$_2$ (60% yield) respectively ⟨92OM859⟩.

In the presence of an excess of MgBr$_2$ in diethyl ether the $\alpha\beta$-epoxysilane (**378**) was converted into the bromohydrin (**379**) in 97% yield (Equation (78)) ⟨92MI 403-01⟩.

(d) α-Iodosilanes. Addition of reagents containing electrophilic iodine to cyclic vinylsilanes has been shown to occur with high regio- and stereospecificity, for example the trimethylsilylcyclohexene (**380**) with iodine and methanol in sulfolane furnished the *trans* iodo-methoxy derivative (**381**) in high yield (Equation (79)) ⟨88T4087⟩. Copper catalysed addition of the iododifluoroacetates (**382a–c**) to trimethyl(vinyl)silane afforded the adducts (**383a–c**) (Equation (80)) ⟨89JFC(45)435⟩. Radical induced addition of freshly prepared tosyl iodide to the trimethyl(vinyl)silane (**384**) produced the (α-iodoalkyl)silane (**385**) in quantitative yield (Equation (81)) ⟨88JCS(P1)2585⟩.

Treatment of the (α-hydroxyalkyl)trimethylsilanes (**386a–b**) with methyl(triphenoxy)phosphonium iodide afforded the stable α-iodosilanes (**387a–b**) in good yields (Equation (82)) ⟨91JOC638⟩.

(Iodomethyl)trimethylsilane ICH$_2$-TMS was formed in 77% yield by treatment of ClCH$_2$-TMS with NaI ⟨89JA7199⟩.

Iodomethylation of Ph$_3$SiBr with ICH$_2$Li produced ICH$_2$SiPh$_3$ in 55% yield ⟨85CB391⟩.

4.03.4.1.2 *Halo(α-haloalkyl)silanes*

Although the halogenation of silicon hydrides under mild conditions is a standard procedure for the synthesis of halosilanes, the chlorination of FCH$_2$SiH$_3$ could not be controlled (explosions occurred). Bromination of FCH_2SiH_3 and FCH_2SiH_2Me, however, gave the bromosilanes FCH_2SiBr_3 (75% yield) and FCH_2SiBr_2Me (63% yield) respectively ⟨93OM4930⟩. Monochloro(α-fluoro)silanes were prepared by treatment of silicon hydrides with $SnCl_4$, for example FCH_2SiH_3, FCH_2SiH_2Me and FCH_2SiHMe_2 afforded FCH_2SiH_2Cl, $FCH_2SiHClMe$ and $FCH_2SiClMe_2$ in 98%, 40% and 98% yields respectively by this method ⟨93OM4930⟩. Chloro(α-fluoro)silanes have also been prepared by the photochemical addition of chlorosilanes to fluoro alkenes, for example Haszeldine *et al.* prepared $FCH_2CHFSiCl_3$ in 85% yield by the reaction of *cis*-1,2-difluoroethene with an excess of $HSiCl_3$ ⟨75JCS(D)2177⟩.

The photochemical, gas-phase reaction of the methylchlorosilanes Me_nSiCl_{4-n} ($n = 1,2,3$) with chlorine gave the chloromethyl derivatives $(ClCH_2)Me_nSiCl_{3-n}$ ($n = 1,2$) in high yields ⟨51JA824, 86ZN(B)1527⟩. The bromo analogues $(BrCH_2)Me_nSiCl_{3-n}$ ($n = 1,2$) have been prepared similarly ⟨51JA826⟩. Addition of trifluoromethyl hypochlorite CF_3OCl to the vinylsilane (**388**) provided a mixture (87:13) (90% yield) of the α-chlorosilane (**389**) and the β-chlorosilane (**390**) (Equation (83)) ⟨86IC376⟩. Chloro(chloromethyl)silanes and (bromomethyl)chlorosilanes undergo preferential substitution at Si with nucleophiles, for example PhMgBr (1 mole equivalent) with $ClCH_2SiMeCl_2$ and with $BrCH_2SiMeCl_2$ gave $ClCH_2SiClMePh$ ⟨71JOM(30)349⟩ and $BrCH_2SiClMePh$ ⟨73JOM(57)261⟩ in good yields. Tetrachlorosilane $SiCl_4$ interacted with CH_2N_2 to furnish $ClCH_2SiCl_3$ in 60% yield ⟨60IS37⟩. The reaction of $ClCH_2SiH_3$ with $SnBr_4$ provided $ClCH_2SiH_2Br$ in 72% yield ⟨69DOK(185)311⟩. In the presence of $Fe(CO)_5$—PPh_3 (1:1) as a catalyst, $CHBr_3$ and $R_3SiCH=CH_2$ (R = Me,Cl) formed the corresponding adducts $R_3SiCHBrCH_2CHBr_2$ in ca. 52% yields based on the vinylsilane consumed ⟨83JGU109⟩. Photoinduced bromination of the chlorosilanes $MeSiCl(R)_2$ (R = alkyl, Cl) with a mixture of Cl_2 and Br_2 (BrCl) gave the bromomethyl derivatives $BrCH_2SiCl(R)_2$, for example TMS-Cl furnished $BrCH_2SiClMe_2$ in 62% yield based on Br_2 ⟨51JA826⟩. $(Pr^i)Me_2SiCl$ was brominated with Br_2 to give $BrMe_2CSiClMe_2$ ⟨78JOM(154)353⟩. The trichloro(halomethyl)silanes $BrCH_2SiCl_3$ and ICH_2SiCl_3 have been prepared in ca. 70% yields by the reaction of $ClCH_2SiCl_3$ with $AlBr_3$ ⟨75IZV1576⟩ and AlI_3 ⟨76JGU2076⟩ respectively. Treatment of $ClCH_2SiCl_3$ with NaI gave ICH_2SiCl_3 in 51% yield ⟨76IZV2200⟩. The tribromosilane $BrCH_2SiBr_3$ was prepared in 55% yield from $SiBr_4$ and CH_2N_2. Further sequential treatments with CH_2N_2 (2 × 1 mole equivalent) afforded $(BrCH_2)_2SiBr_2$ (55% yield) and then $(BrCH_2)_3SiBr$ (40% yield). $SiCl_4$ and CH_2N_2 reacted similarly ⟨52ZOB1783⟩.

$$\overset{}{\underset{SiCl_3}{\diagup\!\!\!\diagdown}} \xrightarrow[90\%]{CF_3OCl} F_3CO\overset{Cl}{\underset{SiCl_3}{\diagup\!\!\!\diagdown}} + Cl\overset{OCF_3}{\underset{SiCl_3}{\diagup\!\!\!\diagdown}} \quad (83)$$

(**388**)   (**389**)   (**390**)

4.03.4.1.3 (α-Haloalkyl)oxysilanes

(Chloromethyl)- and (bromomethyl)oxysilanes have been formed from the corresponding halo(halomethyl)silanes by preferential nucleophilic substitution at the silicon atom, for example methanolysis of $ClCH_2SiCl_2Me$ furnished $ClCH_2Si(OMe)_2Me$ in 84% yield ⟨91HCA1477⟩. Likewise, $BrCH_2SiClMe_2$ and $HC≡CCH_2OH$ produced $BrCH_2SiMe_2(OCH_2C≡CH)$ in 91% yield ⟨89S687⟩. The reactions of $ClCH_2SiClMe_2$ with NaOAc and $HOSO_2Cl$ afforded $ClCH_2Si(OAc)Me_2$ (40–60% yield) ⟨85SRI321⟩ and $ClCH_2Si(OSO_2Cl)Me_2$ (85%) ⟨86BSF413⟩ respectively.

4.03.4.1.4 Miscellaneous α-halo silicon derivatives

Treatment of $ClCH_2SiCl_2Me$ with hexamethyldisilazane $(TMS)_2NH$ afforded $ClCH_2SiClMeNH$-TMS in 98% yield ⟨91IZV1039⟩. The monochlorosilane $ClCH_2SiClMe_2$ reacted with MeCON-Me(TMS) and TMS-$NHCO_2$-TMS to form the *trans*-silylation products $MeCON(Me)SiMe_2CH_2Cl$ (100% yield) ⟨89JOM(361)147⟩ and $ClCH_2SiMe_2NHCO_2SiMe_2CH_2Cl$ (64% yield) ⟨88JGU81⟩ respectively.

Interaction of hexachlorodisilane Si_2Cl_6 with CH_2N_2 afforded the mono(chloromethyl) derivative $ClCH_2Si_2Cl_5$. This was reduced with LAH to give the disilane $ClCH_2Si_2H_5$ which was converted into the iodomethyl analogue by halogen exchange with sodium iodide ⟨75IC1614⟩.

4.03.4.2 α-Halo Germanium Derivatives—$R^1_2CHalGeR^2_3$, etc.

4.03.4.2.1 Alkyl- and aryl(α-haloalkyl)germanes

α-Haloalkylgermanes, $R^1_2CHalGeR^2_3$, have been prepared by methods similar to those employed for the corresponding silanes, for example alkylation and reduction of halo(α-haloalkyl)germanes, addition of electrophilic halogen to vinylgermanes, addition of germane hydrides to haloalkenes, replacement of the hydroxy group in (α-hydroxyalkyl)germanes, and halogen exchange in the α-haloalkyl group. Examples of these methods are shown below.

Addition of Me_3GeH to the perfluorocyclobutene (**361**) gave the adduct (**391**) (86% yield, Equation (84)) ⟨66JOM(6)633⟩.

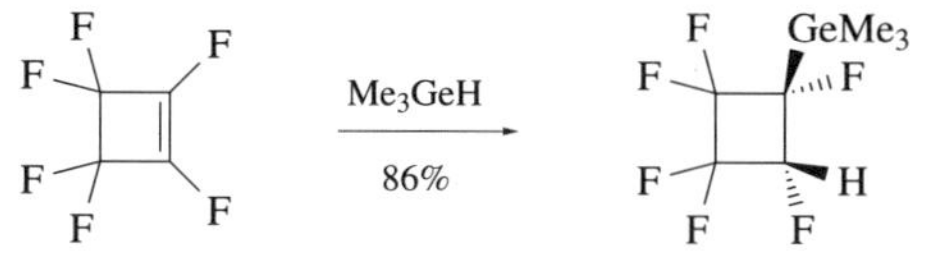

(84)

Reduction of $ClCH_2GeCl_3$ with LAH afforded the chloromethylgermane $ClCH_2GeH_3$, the halogen of which was exchanged for iodine by treatment with NaI to give ICH_2GeH_3 ⟨71MI 403-02⟩. Reduction of $ClCH_2GeClMe_2$ with LAH afforded $ClCH_2GeHMe_2$ in moderate yield ⟨78JOM(154)353⟩. Selective reduction of $Cl_2HCGeH_2Bu^n$ was achieved with Bu^n_3SnH to give $ClCH_2GeH_2Bu^n$ in 75% yield ⟨81JOM(205)311⟩.

Chlorination of tetravinylgermane $(CH_2=CH)_4Ge$ with Cl_2 produced $(ClCH_2CHCl)_4Ge$ but addition of Br_2 to this precursor gave bromo(trivinyl)germane $(CH_2=CH)_3GeBr$ in 80% yield ⟨74JCS(D)2537⟩. Chloroform reacted with $(CH_2=CH)_4Ge$ in the presence of $Fe(CO)_5$ and PPh_3 to give $(CH_2=CH)_3GeCHClCH_2CHCl_2$ in 60% yield ⟨81DOK(261)474⟩.

Treatment of Ph_3GeBr with $ClCH_2Li$ afforded $ClCH_2GePh_3$ in 76% yield ⟨91OM1960⟩. The alkylation of $ClCH_2GeCl_3$ with EtMgBr furnished $ClCH_2GeEt_3$ in 86% yield ⟨72BSF1361⟩ and the bis(chloromethyl)germane $(ClCH_2)_2GeMe_2$ was prepared similarly in 69% yield from $(ClCH_2)_2GeCl_2$ and MeMgI ⟨84SRI21⟩. Stepwise addition of alkyl (or aryl) groups has been accomplished, for example treatment of $ClCH_2GeCl_3$ with p-FC_6H_4MgBr (2 mole equivalents) gave the diaryl derivative $ClCH_2GeCl(p$-$FC_6H_4)_2$ (40% yield) which was alkylated with MeLi to provide $ClCH_2GeMe(p$-$FC_6H_4)_2$ (85%) ⟨92JOM(438)45⟩. Chlorination of bis(hydroxymethyl)dimethylgermane $(HOCH_2)_2GeMe_2$ with the stoichiometric amount of $SOCl_2$ furnished the mono(chloromethyl) derivative $ClCH_2(HOCH_2)GeMe_2$ in 90% yield ⟨88SRI317⟩.

Free radical bromination of $CH_2=CHGePh_3$ with Br_2 gave $BrCH_2CHBrGePh_3$ in 78% yield ⟨76JOM(122)31⟩. The addition of Br_2 to the (Z)-vinylgermane (**392**) proceeded stereospecifically anti to give the dibromide (**393**) (Equation (85)) ⟨84TL3221⟩. Bromination of $Ph_2CHGeMe_3$ (in the absence of a free radical initiator) and $(Ph_2CH)_2GeMe_2$ (in the presence of AIBN) with NBS afforded in 75–82% yields the α-bromobenzyl derivatives $Ph_2CBrGeMe_3$ ⟨80JOM(201)197⟩ and $(Ph_2CBr)_2GeMe_2$ ⟨84SRI21⟩. Ph_3GeBr was bromomethylated and iodomethylated with XCH_2Li (X = Br,I) to give $BrCH_2GePh_3$ (49% yield) and ICH_2GePh_3 (74% yield) ⟨85CB391⟩. The α,α-dibromoethylgermane $MeBr_2CGePh_3$ was reduced to the monobromo derivative $MeBrHCGePh_3$ in 83% yield by the addition of Bu^nLi and then HBr ⟨70CJC561⟩.

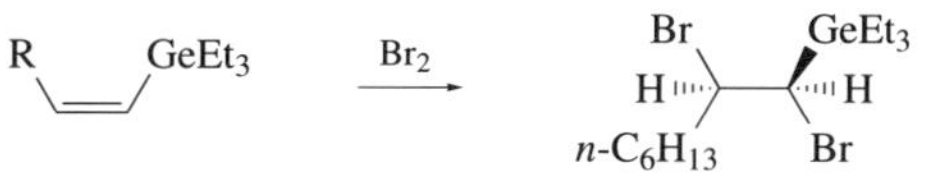

(85)

Bis(chloromethyl)dimethylgermane $(ClCH_2)_2GeMe_2$ was converted into the bis(iodomethyl) derivative $(ICH_2)_2GeMe_2$ in 76% yield by treatment with KI ⟨84SRI21⟩. (Iodomethyl)trimethylgermane ICH_2GeMe_3 has been synthesized from ICH_2GeCl_3 and an excess of MeMgBr ⟨87MI 403-01⟩.

4.03.4.2.2 Halo(α-haloalkyl)germanes

Laser-induced photochemical chlorination of MeGeCl$_3$ gave ClCH$_2$GeCl$_3$ in 48% yield based on the starting material consumed ⟨88IZV1946⟩. The same product was obtained in 36% yield by treating the stable dichlorogermane–dioxane complex GeCl$_2$·C$_4$H$_8$O$_2$ with CH$_2$Cl$_2$ ⟨71DOK(196)85⟩ and in 48–63% yields by the copper-catalysed attack of CH$_2$N$_2$ on GeCl$_4$ ⟨88JOM(354)147, 91CL97⟩. This chloromethyl compound ClCH$_2$GeCl$_3$ was more conveniently prepared, however, by the interaction of trichlorogermane-etherate with paraformaldehyde and treatment of the resultant hydroxymethyl derivative HOCH$_2$GeCl$_3$ with PCl$_5$ (ca. 65% overall yield) ⟨92MI 403-02⟩. Sequential treatment of ClCH$_2$GeCl$_3$ with stoichiometric amounts of PhMgCl and then MeLi afforded ClCH$_2$GeClMePh as the major product ⟨91ZN(B)275⟩.

Tribromo(bromomethyl)germane BrCH$_2$GeBr$_3$ (92% yield) and triiodo(iodomethyl)germane ICH$_2$GeI$_3$ (67%) were synthesized from the dibromogermane-dioxane complex Br$_2$Ge·C$_4$H$_8$O$_2$ and Br$_2$CH$_2$, and the corresponding diiodo complex I$_2$GeC$_4$H$_8$O$_2$ and I$_2$CH$_2$ ⟨85JGU1079⟩. The former product BrCH$_2$GeBr$_3$ and the analogous chloride ClCH$_2$GeCl$_3$ have also been obtained from GeX$_4$ (X = Cl,Br) and CH$_2$N$_2$ ⟨71MI 403-02⟩.

Trichloro(iodomethyl)germane ICH$_2$GeCl$_3$ has been prepared by the action of ICH$_2$ZnI on GeCl$_4$ ⟨87MI 403-01⟩.

4.03.4.2.3 Miscellaneous α-halogermanes

Treatment of ClCH$_2$GeCl$_3$ with Pr^{i}OH in the presence of Et$_3$N afforded ClCH$_2$Ge(OPr^i)$_3$ (78% yield) ⟨91MI 403-02⟩. The chloromethylgermanes ClCH$_2$GeMe$_2$R, where R = NMe$_2$, PMe$_2$ and AsMe$_2$, have been prepared by treating ClCH$_2$GeMe$_2$Cl with Me$_2$NH, Me$_2$PLi and Me$_2$AsLi respectively ⟨77JOM(132)77⟩.

4.03.4.3 α-Halo Boron Derivatives—R^1$_2$CHalBR^2$_2$, etc.

4.03.4.3.1 (α-Haloalkyl)boron hydrides

These have been formed in high yields as extremely labile reaction intermediates by the addition of boron hydrides to vinyl halides. Predominantly, the boron adds to the carbon bearing the halogen atom ⟨66JOC2773, 68JA2915⟩. Thus, the action of diborane on *trans*-β-bromostyrene (**394**) gave the α-bromoalkylborane (**395**) as an intermediate in greater than 95% yield (Equation (86)). This reacted with aqueous NaOH and then with H$_2$O$_2$ to give PhCH$_2$CH$_2$OH in 70% yield ⟨66JOC2773⟩. Similarly, hydroboration of Me$_2$C=CHCl provided the (α-chloroalkyl)borane Me$_2$CHCHClBH$_2$) which was oxidized with a mixture of NaOAc and H$_2$O$_2$ to isobutyraldehyde in 84% overall yield ⟨68JA2915⟩.

$$\underset{\underset{(394)}{(E)}}{\text{Ph}\diagup\hspace{-0.5em}=\hspace{-0.5em}\diagdown\text{Br}} \xrightarrow{B_2H_6} \left[\underset{(395)}{\text{Ph-CHBr-BH}_2}\right] \qquad (86)$$

4.03.4.3.2 Alkyl- and aryl(α-haloalkyl)boranes

There are few reports in the literature which describe preparations of fully characterized alkyl (or aryl) (α-haloalkyl)boranes. The simple dialkyl(α-haloalkyl)boranes show high reactivity undergoing rearrangement in the presence of heat, electrophiles and weak nucleophiles such as water and THF ⟨71JA2796, 72IC1150⟩.

(Chloromethyl)dimethylborane ClCH$_2$BMe$_2$ was prepared in 31% yield by treating trimethylborane with Cl$_2$ (1 molar equivalent) ⟨65JA488⟩. Interaction of Ph$_3$B with dichloromethyllithium gave (α-chlorobenzyl)diphenylborane PhCHClBPh$_2$ ⟨67AG(E)74⟩.

(1-Bromoethyl)diethylborane MeBrHCBEt$_2$ was prepared in 70% yield from Et$_3$B and Br$_2$ in *n*-pentane. It was important to remove the HBr produced as it formed to avoid subsequent C—B bond cleavage ⟨71JA2796⟩. The highly activating effect of the boron moiety was evident in the bromination of *B*-isopropyl-9-borabicyclo[3.3.1]nonane (**396**) which was accomplished in dry CH$_2$Cl$_2$ with removal of HBr, to provide the unstable bromo derivative (**397**) (Equation (87)) ⟨74JA311⟩. Photochemical addition of Br$_2$ to the vinyl-1,2-dicarbadodecaborane (**398**) furnished the dibromide (**399**) in 61% yield (Equation (88)) ⟨85IZV809⟩. Treatment of the 1,2-bis(hydroxymethyl) dicarbaborane (**400a**) with an excess of Br$_2$ and Ph$_3$P produced either the 1-bromomethyl-2-hydroxymethyl derivative (**400b**) or the 1,2-bis(bromomethyl) compound (**400c**) depending on the conditions used; with approximately a three-fold excess of reagents, the monobromide (**400b**) was obtained in 27% yield whereas a four-fold excess gave the dibromide (**400c**) in 36% yield. While the mono(hydroxymethyl) analogue (**400d**) readily gave the chloromethyl derivative (**400e**) with SOCl$_2$, the diol (**400a**) gave only the stable condensation product (**401**) ⟨82IZV1425⟩.

(396) →[Br$_2$] (397) (87)

B$_{10}$H$_9$—9-(CH=CH$_2$) →[Br$_2$] B$_{10}$H$_9$—9-(CHBrCH$_2$Br) (88)
(398)   (399)

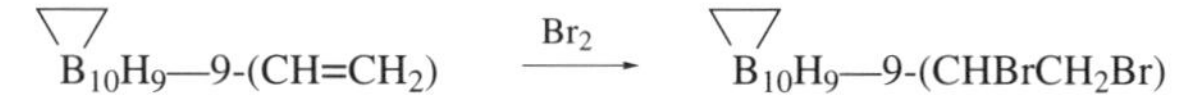

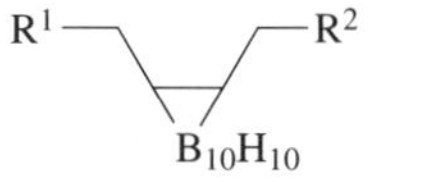

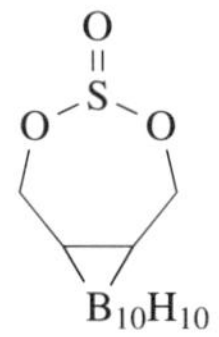

(**400**) a; R^1 = R^2 = OH
b; R^1 = OH, R^2 = Br
c; R^1 = R^2 = Br
d; R^1 = OH, R^2 = H
e; R^1 = Cl, R^2 = H

(401)

Treatment of ClCH$_2$BMe$_2$ with KI afforded ICH$_2$BMe$_2$ in 40% yield ⟨72IC1150⟩. The reaction of 1,12-bis(hydroxymethyl)decahydrododecaborate caesium salt (**402**) with HI readily furnished the corresponding bis(iodomethyl) derivative (**403**) ⟨86IC4309⟩.

1,12-B$_{12}$H$_{10}$(CH$_2$R)$_2$Cs$_2$

(**402**) R = OH
(**403**) R = I

4.03.4.3.3 *Halo(α-haloalkyl)boranes*

(Fluoromethyl)difluoroborane FCH$_2$BF$_2$, the first α-haloalkylboron compound to be prepared, was obtained by interaction of BF$_3$ with CH$_2$N$_2$ ⟨57LA(604)168⟩. Since then, there have been few halo(α-haloalkyl)boranes reported. Brown and De Lue obtained evidence for the formation of the unstable, bridgehead α-bromoborane (**405**) during the bromination of *B*-bromo-9-BBN (**404**) ⟨77TL3007⟩.

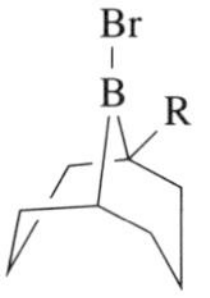

(404) R = H
(405) R = Br

4.03.4.3.4 (α-Haloalkyl)oxyboranes

The aryl(α-bromoalkyl)(n-butoxy)boranes (**407**) (n-butyl aryl(α-bromoalkyl)borinates is an alternative nomenclature often used) were prepared in good yields by photochemical addition of CCl_3Br to the vinylboranes (**406**) (Equation (89)) ⟨63JOC2171⟩.

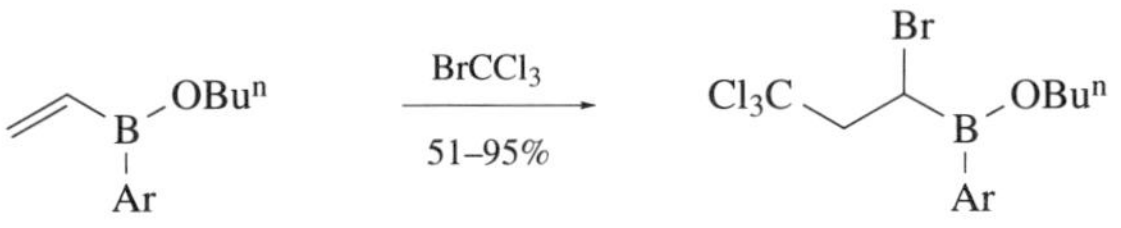

$$\text{(406)} \xrightarrow[51-95\%]{BrCCl_3} \text{(407)} \quad (89)$$

a; Ar = Ph
b; Ar = 2,5-$Me_2C_6H_3$
c; 2,4,6-$Me_3C_6H_2$

(α-Haloalkyl)bis(alkoxy)boranes (α-haloalkylboronic esters) $R_2CHalB(OR)_2$, particularly where $B(OR)_2$ is part of a five- or six-membered ring, have important synthetic applications among which is their ability to provide high stereocontrol in asymmetric synthesis. They have been prepared by addition of Br_2 and hydrogen halides to alkenylboronic esters, bromination of boronic esters with activated α-hydrogens, reduction of α,α-dihaloalkylboronic esters, insertion reactions, etc. The preparation and properties of α-halo boronic esters form the subject of a comprehensive review by Matteson ⟨89CRV1535⟩. In 1992, the (α-chloroalkyl)dioxyborane (**409**) was obtained in 89% yield by the reaction of the corresponding α,α-dichloromethyl analogue (**408**) with m-methoxy-benzylmagnesium bromide ⟨92JOM(431)255⟩. The (1S)-(1-chlorobutyl)boron derivatives (**411**) were prepared by addition of Bu^nLi to the substrates (**410a–b**) in the presence of CH_2Cl_2 followed by rearrangement of the addition product with a Lewis acid (Equation (90)) ⟨90S200⟩. $BrCH_2B(OPr^i)_2$ was synthesized in 89% yield by sequential treatment of a mixture of $(Pr^iO)_3B$ and CH_2Br_2 with Bu^nLi and $MeSO_3H$ ⟨91SL631⟩.

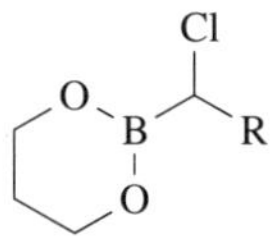

(408) R = Cl
(409) R = 3-MeC_6H_4

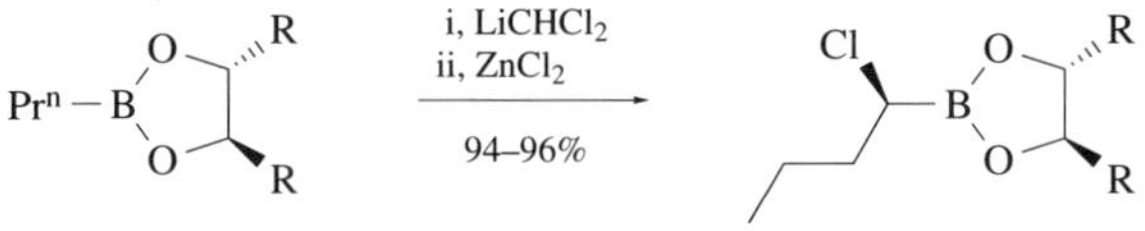

$$\text{(410)} \xrightarrow[94-96\%]{\text{i, LiCHCl}_2 \text{; ii, ZnCl}_2} \text{(411)} \quad (90)$$

a; R = Me
b; R = Pr^i

Hydroboration of 1-chloro-2-methylpropene produced the (chloromethyl)borane (**412**) which on addition of H_2O and careful removal of the solvent gave 1-chloro-2-methylpropylboronic acid (**413**) (Equation (91)) ⟨68JA6259⟩. Exposure of the (iodomethyl)boronic ester (**414**) to atmospheric moisture provided the boronic acid (**415**) ⟨68JOC3055⟩.

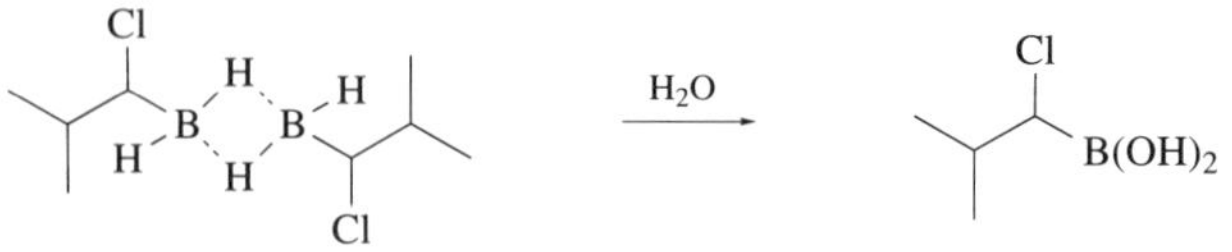

$$\text{(412)} \xrightarrow{H_2O} \text{(413)} \quad (91)$$

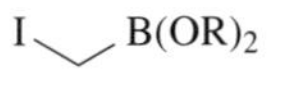

(414) R = Bu^n
(415) R = H

4.03.4.3.5 Miscellaneous (α-haloalkyl)boranes

Addition of Br$_2$ to the vinylborane (416) afforded the α,β-dibromoborane (417) in 90% yield (Equation (92)) ⟨72BSF811⟩. The iminoboranes (418a–b) and diphenyldiazomethane (2 molar equivalents) gave the (α-bromomethyl)iminoboranes (419a–b) (Equation (93)) ⟨71M118⟩. The B-(hydroxymethyl)diazadiboracyclohexane (420a) was converted into the iodomethyl analogue (420c) in 54% yield by treating the trimethylsilyl derivative (420b) with TMS-I ⟨91IC2228⟩.

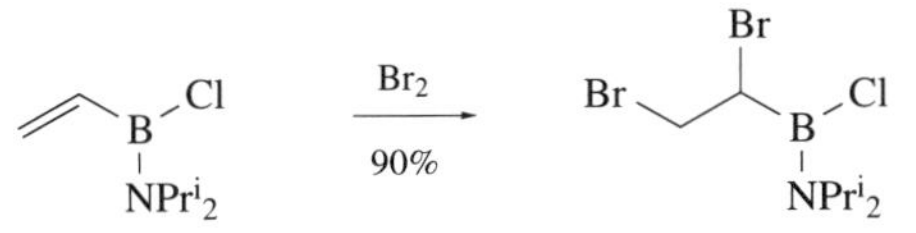

(416)   (417)   (92)

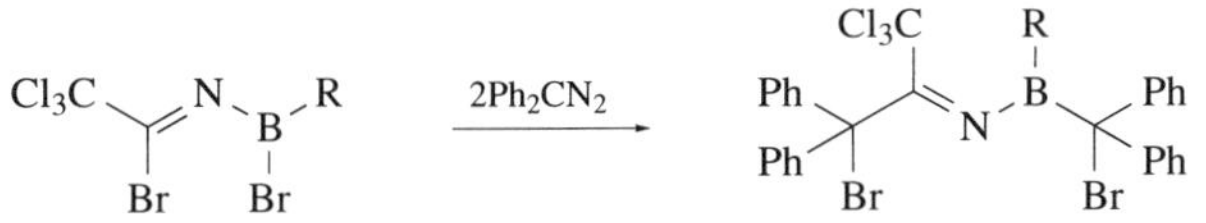

(418) a; R = Me
      b; R = Ph

(419)   (93)

(420) a; R = OH
      b; R = O-TMS
      c; R = I

4.03.5 α-HALO METAL DERIVATIVES—R$_2$CHalM, etc.

4.03.5.1 Group 1 and Group 2 Derivatives—R$_2$CHalLi, etc.

Of the α-haloalkyl derivatives of group 1 and group 2 elements in which the α-carbon is sp^3-hybridized and bears one halogen atom, only the preparations of lithium and magnesium derivatives have been reported.

4.03.5.1.1 α-Haloalkyllithium derivatives

The existence of the reactive species R$_2$CHalLi (carbenoids) was first demonstrated in the early 1960s by Kobrich and Trapp using trapping experiments ⟨63ZN(B)1125⟩. Reports on the preparation and properties of other α-haloorganolithium compounds followed soon afterwards.

Two reviews by Kobrich illustrate the extent of the progress made during this period ⟨67AG(E)41, 72AG(E)473⟩. Since then, improvements in the methods of preparation of these compounds (stable in the temperature range between ca. −130 °C and −70 °C) ensured their place as important synthetic intermediates. A review by Siegel details some of the synthetic applications of these carbenoids ⟨82TCC55⟩.

(i) α-Halomethyllithium compounds

Early attempts to prepare and utilize chloromethyllithium by treatment of ClCH$_2$Br with *n*-butyllithium, typically at temperatures below $-115\,^\circ$C, were hampered by its extreme thermal instability. Villieras and co-workers overcame this difficulty by performing this reaction in the presence of the electron acceptor LiBr in a mixture of THF, diethyl ether and pentane at $-115\,^\circ$C. In this way the destabilizing metal–halogen interaction was eliminated. Thus ClCH$_2$Li and BrCH$_2$Li were prepared from CH$_2$ClBr and CH$_2$Br$_2$ respectively and yielded 60–80% of halohydrins or oxiranes on their addition to aldehydes and ketones, and α-haloketones on their addition to esters ⟨84TL835⟩. Later, ClCH$_2$Li was prepared and captured in nearly quantitative yields by addition of Bu^{n}Li or MeLi to mixtures of ClCH$_2$I and aldehydes and ketones in THF at $-78\,^\circ$C. Immediate acidification gave the chlorohydrins while delayed workup afforded epoxides ⟨86TL795⟩. Einhorn *et al.* showed that either CH$_2$ClBr or CH$_2$Br$_2$, carbonyl compounds and lithium metal in THF under sonochemical conditions reacted to provide the corresponding α-halohydrin, and subsequently the oxiranes, in excellent yields ⟨88CC333⟩. Cainelli and co-workers had earlier reported the generation and capture (in moderate yields) of BrCH$_2$Li from CH$_2$Br$_2$ and lithium dispersion or amalgam in THF at $-78\,^\circ$C ⟨71T6109⟩. In 1991, BrCH$_2$Li was prepared by the addition of Bu^{n}Li to CH$_2$Br$_2$ in the presence of aldehydes and ketones (*inter alia*) in THF at $-78\,^\circ$C to give the corresponding oxiranes in excellent yields ⟨91SL631⟩. Iodomethyllithium has been generated by the action of PhLi on CH$_2$I$_2$ ⟨85CB391⟩.

(ii) Other α-haloalkyllithium compounds

The higher α-haloalkyllithium carbenoids are generally synthesized from *gem*-bromohalo precursors by lithium–halogen exchange at low temperatures. Heptafluoroisopropyllithium (CF$_3$)$_2$CFLi was prepared by the reaction of (CF$_3$)$_2$CFI with MeLi or Bu^{n}Li ⟨62JCS1993⟩. Villieras and co-workers prepared a range of α-haloalkyllithium carbenoids RHCHalLi (R = Me, Bu^n, Me$_3$Si; Hal = Cl,Br) employing similar conditions to those developed for the preparation of the parent α-halomethyllithium compounds, that is using *s*-butyllithium in a mixture of THF, diethyl ether and pentane at temperatures below $-115\,^\circ$C. As with the parent halomethyl compounds, one mole equivalent of LiBr was added to stabilize some of these carbenoids ⟨85BSF825, 88JOM(346)C1⟩. Similarly, chlorobis(trimethylsilyl)methyllithium (TMS)$_2$ClCLi and chlorodiphenylmethyllithium Ph$_2$ClCLi were prepared from the corresponding *gem*-bromochloro precursors using Bu^{n}Li ⟨86CB1977⟩.

In studies to determine the configurational stability of α-haloalkyllithium carbenoids, the diastereomeric cyclopropanecarboxylates (**422**) and (**423**) were prepared by treatment of the *gem*-dibromocyclopropanes (**421**) with alkyllithiums at $-125\,^\circ$C (Equation (94)). Since these carbenoids (**422**) and (**423**) were shown to equilibrate by halogen–metal exchange with the *gem*-dibromo precursor, the product ratios were strongly dependent on the reaction conditions used. Addition of Bu^{n}Li to the precursor (**421**) gave a mixture of carbenoids in which the thermodynamic product (**422**) was the major component. On the other hand, reverse addition of the precursor to Bu^{t}Li produced the kinetic product (**423**) predominantly. The configurations of these carbenoids were found to be stable at $-125\,^\circ$C, but underwent rapid equilibration at temperatures above $-78\,^\circ$C. Decomposition set in at $-60\,^\circ$C ⟨91CB1253⟩.

$$\text{(421)} \xrightarrow{\text{RLi}, -125\,^\circ\text{C}} \text{(422)} + \text{(423)} \quad (94)$$

Hoffman and co-workers observed high asymmetric induction in the formation of epoxides obtained by the addition of acetone to the carbenoids (**425**) and (**426**). The ratio of epoxides reflected the substantial 1,3-asymmetric induction which resulted from the diastereoselective lithium–bromine exchange of the bromine atoms when the substrate (**424**) was treated with Bu^{n}Li (Equation (95)) ⟨88JOM(353)C30, 91CB1259⟩. A similar 1,2-asymmetric induction was observed when the dibromide (**427**) was treated with Bu^{n}Li (Equation (96)) ⟨91LA811⟩.

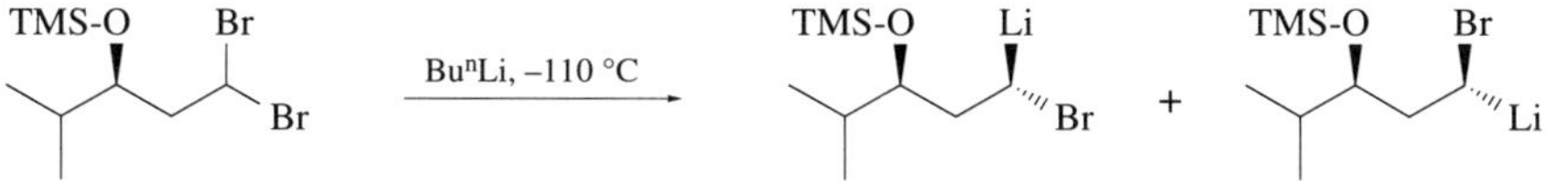

$$\text{(424)} \xrightarrow{\text{Bu}^n\text{Li},\ -110\ °\text{C}} \text{(425)} + \text{(426)} \quad 85:15 \tag{95}$$

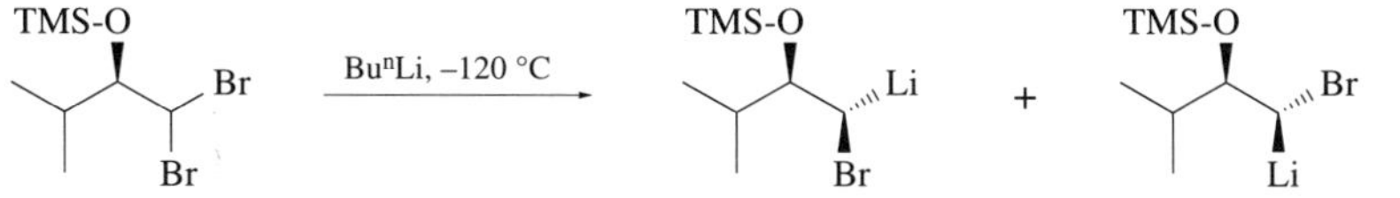

$$\text{(427)} \xrightarrow{\text{Bu}^n\text{Li},\ -120\ °\text{C}} \text{(428)} + \text{(429)} \quad 84:16 \tag{96}$$

4.03.5.1.2 α-Haloalkylmagnesium derivatives

A review by Villieras describes the preparation and properties of α-haloalkyl Grignard reagents synthesized prior to 1971 ⟨71MI 403-03⟩. Therein are procedures described for the preparation of BrCH$_2$MgCl (70% yield) and ICH$_2$MgCl (50% yield) by treatment of isopropylmagnesium chloride with CH$_2$Br$_2$ in THF–diethyl ether and with CH$_2$I$_2$ in THF respectively. Since the magnesium–halogen exchange is known to be an equilibrium process, judicious choice of the solvent system allowed the isolation of the insoluble halomethylmagnesium compounds in high yields. In 1992, the derivatives RCHClMgCl (R = H, Bu^n) were obtained in high yields by the reactions of RCHICl (R = H, Bu^n) with isopropylmagnesium chloride in THF ⟨92SL133⟩.

α-Chlorophenylacetic acid (**430**) was converted by *halogen*–metal exchange into the Grignard reagent (**431**) in 50% yield by treatment with isopropylmagnesium chloride (2 mole equivalents) (Equation (97)) ⟨80MI 403-01⟩.

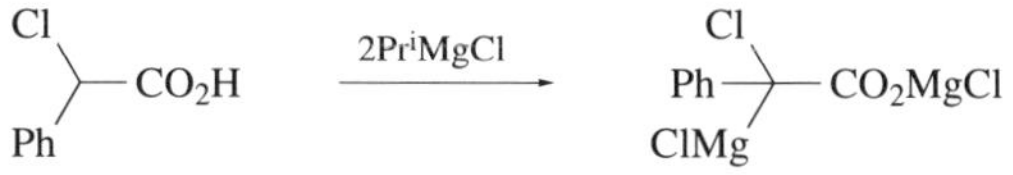

$$\text{(430)} \xrightarrow{2\text{Pr}^i\text{MgCl}} \text{(431)} \tag{97}$$

4.03.5.2 Transition Metal Derivatives—R$_2$CHalFeX$_n$, etc.

α-Haloalkyl derivatives of transition metals have been prepared by a number of routes, for example treatment of alkali metal salts of complexed transition metal cations with dihalomethanes; oxidative addition of dihalomethanes to coordinatively unsaturated metal complexes; reactions of halometal complexes with diazoalkanes; reactions of (alkyloxymethyl)metal complexes with hydrogen halides; reactions of formylmetal complexes with hydrogen halides; direct halogenation of the analogous methyl complex. These and other methods are exemplified below.

4.03.5.2.1 Derivatives of chromium, molybdenum and tungsten

Chloromethyl- and bromomethylchromium complexes (**432c**) and (**432d**), where R^2 = H and Me, were obtained in high yields by the slow addition of CH$_2$N$_2$ to the corresponding halometal analogues (**432a**) and (**432b**) in the presence of copper. Treatment of the chloromethyl compound (**432c**; R^2 = H) with NaI in THF provided the iodide (**432e**; R^2 = H) in 95% yield ⟨90OM2683⟩.

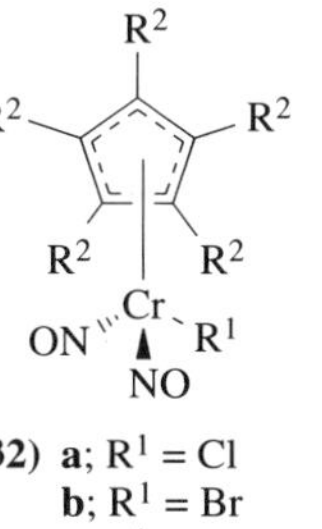

(**432**) a; R^1 = Cl
   b; R^1 = Br
   c; R^1 = CH$_2$Cl
   d; R^1 = CH$_2$Br
   e; R^1 = CH$_2$I

The formylmolybdenum complex *trans*-(η-C_5H_5)Mo(CO)$_2$[P(OPh)$_3$]CHO (**433a**) was converted into the corresponding halomethyl complexes (**433b–d**) in 46–75% yields by the action of the appropriate hydrogen halide ⟨89OM1114⟩. Addition of [(η-C_5Me_5)Mo(CO)$_3$]Li to solutions of CH_2ClX (X = Br or I) and CH_2Br_2 in THF afforded the chloromethyl- and bromomethylmolybdenum derivatives (**434b**) and (**434c**) in 40% and 30% yields respectively. The latter compound was also obtained in 58% yield by treatment of the (methoxymethyl)molybdenum complex (**434a**) with dry HBr. The related iodomethyl derivative (**434d**) could not be prepared by either of these routes ⟨86ICA(119)177⟩. Similarly, the halomethyltungsten complexes (**435b–d**) were obtained from [(η-C_5Me_5)W(CO)$_3$]Li and CH_2Br_2 or CH_2IX (X = Cl, I) in 17–45% yield and from (η-C_5Me_5)W(CO)$_3CH_2OMe$ (**435a**) and HX (X = Cl,Br,I) in 73–94% yields ⟨86ICA(119)177⟩.

(**433**) a; R = CHO
b; R = CH_2Cl
c; R = CH_2Br
d; R = CH_2I

(**434**) a; R = OMe
b; R = Cl
c; R = Br
d; R = I

(**435**) a; R = OMe
b; R = Cl
c; R = Br
d; R = I

4.03.5.2.2 *Derivatives of manganese, iron and cobalt*

The (halomethyl)manganese complexes *cis*-Mn(CO)$_4$(PPh$_3$)CH_2X (X = Cl,Br,I) (**436b–d**) were prepared in 58–78% yields by treatment of the formyl precursor (**436a**) with the appropriate hydrogen halide. The related compounds *mer*, *trans*-Mn(CO)$_3$(PR$^2_3$)$_2$CH$_2$R^1 (**437b–d**; R^2 = Ph or OPh) were obtained in 65–81% yields from the formyl precursors (**437a**; R$_2$ = Ph or OPh) and methyl triflate and then treatment of the resultant methoxymethyl complex with the appropriate hydrogen halide ⟨89OM1114⟩. The (iodomethyl)manganese complex *cis*-Mn(CO)$_4$(PPh$_3$)CH_2I (**438b**) was prepared in 83% yield from the corresponding (methoxymethyl) precursor (**438a**) and TMS-I ⟨90JOM(397)313⟩.

(**436**) a; R = CHO
b; R = CH_2Cl
c; R = CH_2Br
d; R = CH_2I

(**437**) a; R^1 = CHO
b; R^1 = CH_2Cl
c; R^1 = CH_2Br
d; R^1 = CH_2I

(**438**) a; R = OMe
b; R = I

The (halomethyl)iron complexes (**439c,d**) were obtained in 92–94% yield on treatment of the corresponding chloro complex (**439a**) and bromo complex (**439b**) with CH_2N_2 in the presence of copper. The related iodomethyl derivative (**439e**) was obtained in 80–88% yield from the bromomethyl analogue and NaI in acetone–diethyl ether. Similarly obtained in good yields were the corresponding pentamethylcyclopentadienyl analogues ⟨92JOM(429)369⟩. The chloromethyl and iodomethyl compounds (**439c**) and (**439e**) were also obtained by cleavage of the corresponding (methoxymethyl)iron complex (**439f**) with HCl ⟨66JA5044⟩ and HI (60% yield) ⟨89JOM(366)175⟩.

(Chloroacetyl)cobalt tetracarbonyl ClCH$_2$C(O)Co(CO)$_4$, prepared from chloroacetyl chloride and Na[Co(CO)$_4$], underwent decarbonylation to afford (chloromethyl)cobalt tetracarbonyl ClCH$_2$Co(CO)$_4$ in 88–95% yield. With PPh$_3$ this produced ClCH$_2$C(O)Co(CO)$_3$PPh$_3$, which at 50 °C decarbonylated to give the chloromethyl analogue ClCH$_2$Co(CO)$_3$PPh$_3$ in 51% yield (based on PPh$_3$ consumed) ⟨87OM861⟩. The chloromethylcobalt compound (**440**) was prepared in ca. 60% yield by a photolytic reaction of (η-C_5Me_5)Co(CO)$_2$ with CH_2Cl_2 ⟨86OM630⟩. Likewise, the chloromethyl derivative (**441**) was obtained (28% yield) from (η-C_5H_5)Co(CO)(PMe$_3$) and ClCH$_2$I ⟨85JOM(289)141⟩.

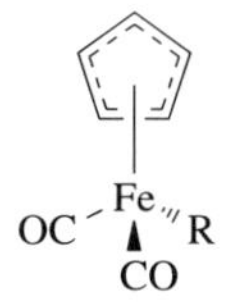

(439) a; R = Cl
b; R = Br
c; R = CH$_2$Cl
d; R = CH$_2$Br
e; R = CH$_2$I
f; R = CH$_2$OMe

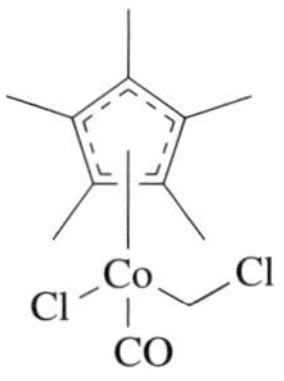

(440)

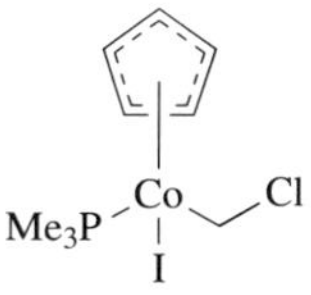

(441)

4.03.5.2.3 Derivatives of ruthenium, rhodium and palladium

In a similar way to the iron complexes (**439**), the (halomethyl)ruthenium complexes (**442**; R^1 = H, Me and R^2 = CH$_2$Cl, CH$_2$Br) were obtained in excellent yields by treatment of the corresponding halo complexes (**442**; R^1 = H, Me and R^2 = Cl, Br) with CH$_2$N$_2$ in the presence of copper. Treatment of the so formed bromomethyl derivatives (**442**; R^1 = H, Me and R^2 = CH$_2$Br) with NaI in acetone–diethyl ether furnished the analogous iodomethyl compounds in 87–90% yields ⟨92JOM(429)369⟩.

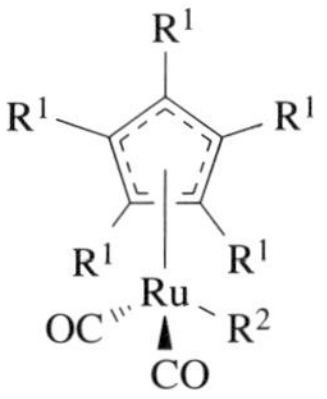

(442)

Treatment of the rhodium(I) complex (**443a**) with CH$_2$ClI or CH$_2$I$_2$ resulted in displacement of the cyclooctene ligand with concomitant oxidation to afford the carbenoid (chloromethyl)- and (iodomethyl)rhodium (III) complexes (**444a,b**) in 93–95% yields ⟨91JOM(417)149⟩. The chelated rhodium complex (**443b**) on treatment with CH$_2$Br$_2$ and CH$_2$I$_2$ furnished the (bromomethyl)- and (iodomethyl)rhodium salts (**444c**) and (**444d**) in 45% and 65% yields ⟨88OM1106⟩. The (α-chloroethyl)rhodium complexes (**445c**) and (**445d**) were obtained in 70% and 62% yields on addition of HCl to the vinylrhodium complexes (**445a**) and (**445b**) respectively ⟨87JOM(336)413⟩.

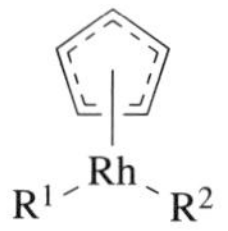

(443) a; R^1 = c-C$_8$H$_{14}$, R^2 = CH$_2$PPr$^i_3$
b; R^1R^2 = Ph$_2$P(CH$_2$)$_2$PPh$_2$

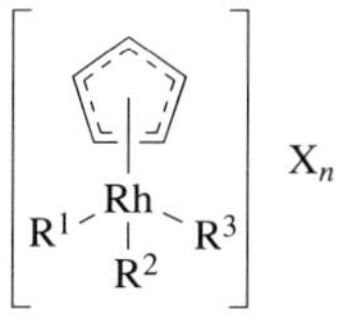

(444) a; R^1 = CH$_2$Cl, R^2 = I, R^3 = CH$_2$PPr$^i_3$, n = 0
b; R^1 = CH$_2$I, R^2 = I, R^3 = CH$_2$PPr$^i_3$, n = 0
c; R^1R^2 = Ph$_2$P(CH$_2$)$_2$PPh$_2$, R^3 = CH$_2$Br, X = Br, n = 1
d; R^1R^2 = Ph$_2$P(CH$_2$)$_2$PPh$_2$, R^3 = CH$_2$I, X = I, n = 1

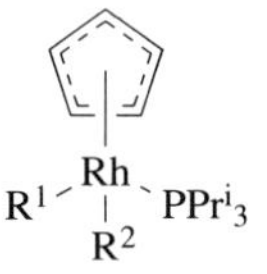

(445) a; R^1 = Cl, R^2 = CH=CH$_2$
b; R^1 = I, R^2 = CH=CH$_2$
c; R^1 = Cl, R^2 = CHClMe
d; R^1 = I, R^2 = CHClMe

A number of palladium dichloride and dibromide complexes containing chelating ligands reacted with CH_2N_2 to give the corresponding mono(chloromethyl) and mono(bromomethyl) derivatives, the former being the more readily formed, for example treatment of the palladium dichloride complex (**446a**) with an excess of CH_2N_2 in acetone containing LiBr produced the related chloromethyl derivative (**446b**) in quantitative yield ⟨89JCS(D)761⟩.

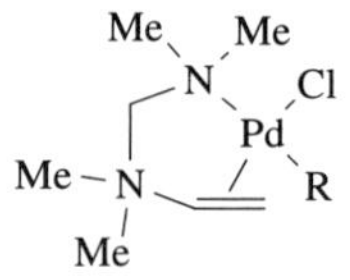

(**446**) **a**; R = Cl
 b; R = CH_2Cl

4.03.5.2.4 Derivatives of rhenium, osmium, iridium and platinum

The (α-chlorobenzyl)rhenium complex (**447b**) was prepared by cleavage of the corresponding trimethylsilyl ether (**447a**) with aqueous HCl ⟨86JA1455⟩. The iodomethyl complex cis-$Re(CO)_4$ $(PPh_3)CH_2I$ was synthesized in 90% yield by the action of TMS-I on the corresponding methoxymethyl complex cis-$Re(CO)_4(PPh_3)CH_2OMe$ ⟨90JOM(397)313⟩. The rhenium complex (**448a**) reacted with I_2 to afford the corresponding iodo derivative (**448b**) in 97% yield ⟨88IC3796⟩.

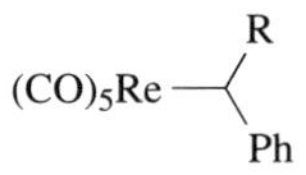

(**447**) **a**; R = O-TMS
 b; R = Cl

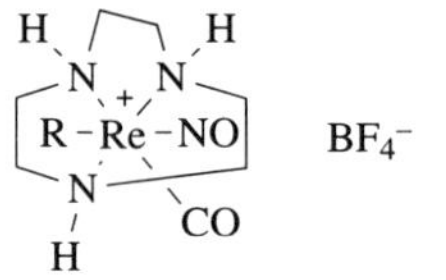

(**448**) **a**; R = Me
 b; R = CH_2I

Interaction of the osmium complex (**449**; R = Cl) with CH_2N_2 did not produce the chloromethyl derivative (**449**; R = CH_2Cl). Instead, the pentacoordinate osmium complex (**450**) was formed which was converted readily into the (chloromethyl)osmium complex (**451**) with Cl_2 (Scheme 32) ⟨83JA5939⟩. The osmium(0) complex (**452**), on sequential treatment with 40% aqueous formaldehyde solution (1.5 mole equivalents) and HCl afforded the chloromethyl derivative (**454**) via the osmium(0) complex (**453**) (Scheme 33) ⟨79JA503⟩.

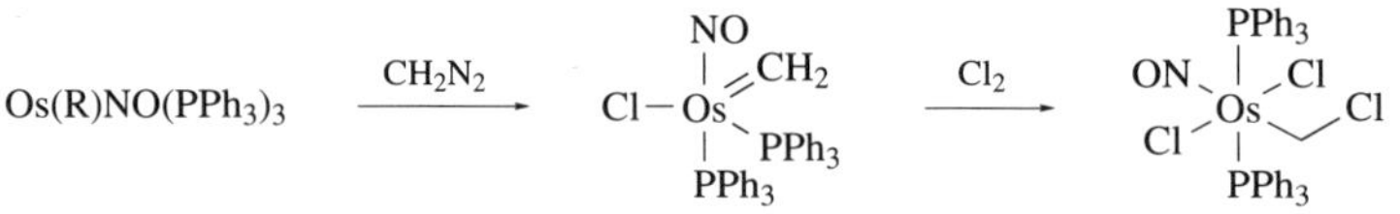

Scheme 32

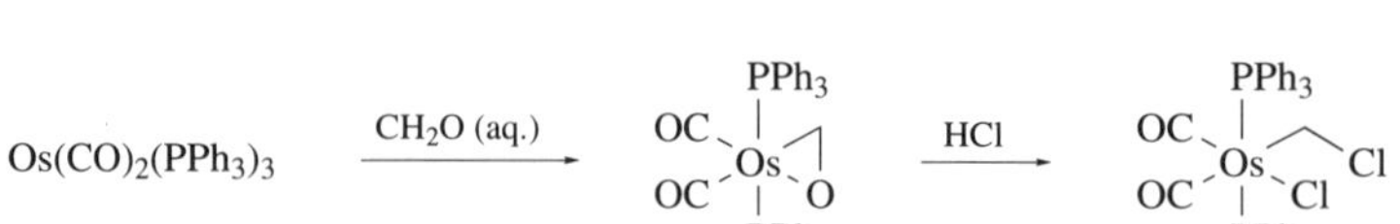

Scheme 33

The addition of CH_2ClI and CH_2I_2 to the iridium complexes (**455a–c**) afforded the octahedral halomethyliridium(1,1,1) complexes (**456a–c**; X = Cl,I) (Equation (98)) in which the phosphorus ligands have the *trans* configurations. The corresponding *cis* isomers of the complexes (**456c**; X = Cl,I) were also found to be present ⟨86AJC1363⟩.

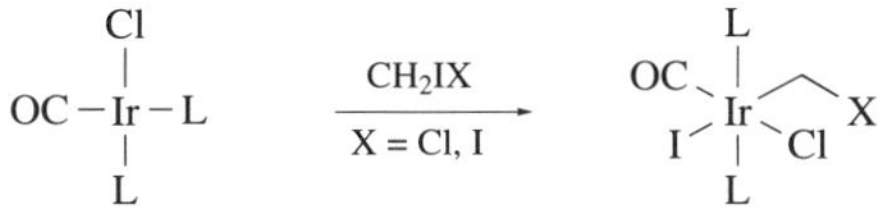

$$\underset{(455)}{\overset{\text{Cl}}{\underset{\text{L}}{\text{OC}-\text{Ir}-\text{L}}}} \xrightarrow[X = \text{Cl, I}]{\text{CH}_2\text{IX}} \underset{(456)}{\overset{\text{L}}{\underset{\text{L}}{\text{OC}\underset{|}{\overset{|}{\underset{\text{I}}{\text{Ir}}}}\text{X}}{\text{Cl}}}} \qquad (98)$$

a; L = PMe$_3$
b; L = PMePh$_2$
c; L = PMe$_2$Ph

Ethyl diazoacetate reacted with a wide range of dihaloplatinum complexes PtX$_2$L$_2$ (X = Cl,Br,I) to give the corresponding mono-α-halo esters PtX(CHXCO$_2$Et)L$_2$ as racemic mixtures in high yields, for example the complex (**457a**) and EtO$_2$CCHN$_2$ furnished the derivative (**457b**) (60% yield). Optically active complexes underwent diasteroselective carbene insertion ⟨92OM3879⟩. Earlier, it had been shown that haloplatinum complexes reacted with CH$_2$N$_2$ to give the corresponding (halomethyl)platinum derivatives ⟨88JCS(D)1773, 90JCS(D)1553⟩. Oxidative addition of CH$_2$Cl$_2$ and CH$_2$Br$_2$ to the 2,2′-bithiazoline (**458**) gave predominantly the *trans*-octahedral platinum(IV) compounds (**459a**) and (**459b**) in 60% and 70% yields respectively (Equation (99)) ⟨90JOM(396)115⟩. The reaction of the oxaplatinacyclobutane complex (**460**) with acetyl chloride (1 mole equivalent) furnished the chloromethyl derivative (**461**) in 83% yield (Scheme 34). With an excess of AcCl, ring cleavage was accompanied by oxidative addition to provide the platinum(IV) complex (**462**) in 69% yield (Scheme 34). The (chloromethyl)platinum complex (**464**) was obtained when the oxaplatina(IV)cyclobutane (**463**) was treated with AcCl (Equation (100)) ⟨89OM2973⟩.

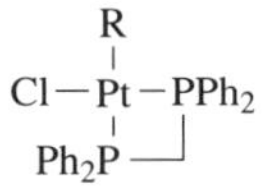

(**457**) **a**; R = Cl
b; R = CH(CO$_2$Et)Cl

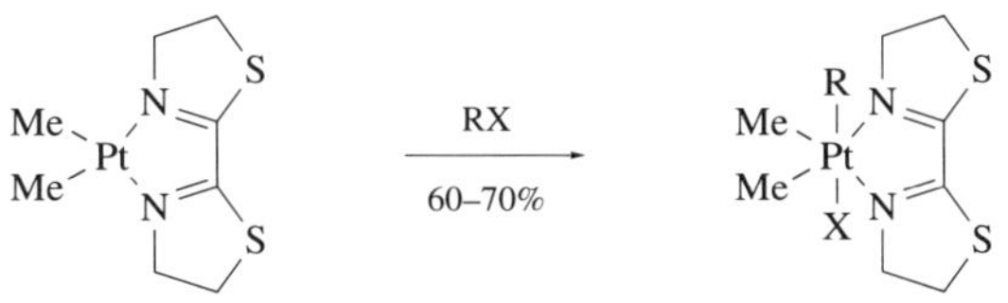

(**458**) (**459**)
a; R = CH$_2$Cl, X = Cl
b; R = CH$_2$Br, X = Br

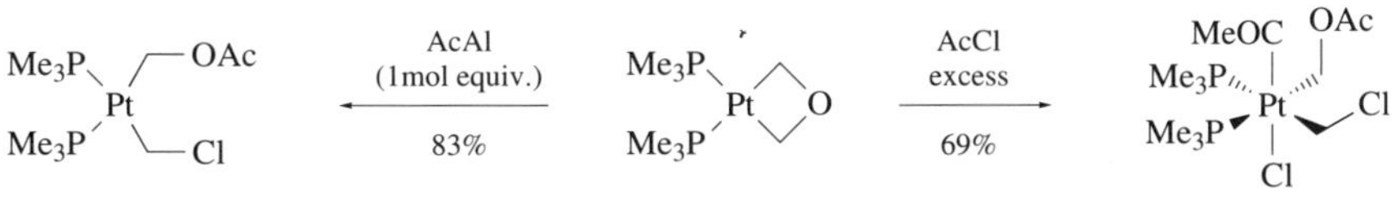

(99)

(**461**) (**460**) (**462**)

Scheme 34

(**463**) (**464**)

(100)

4.03.5.2.5 *Derivatives of copper, silver and gold*

Metathesis of perfluoroisopropylcadmium with the salts CuX, where X = Cl,Br,I, has provided two types of perfluoroisopropyl copper species (CF$_3$)$_2$CFCu and [(CF$_3$)$_2$CF]$_2$Cu$^-$ in quantitative yields. The distribution of the two types depended on which copper(I) salt was used for the reaction and also on the ratio of the copper(I) salt to the cadmium reagent ⟨92JFC(56)341⟩.

Perfluoroisopropyl silver was formed by codeposition of silver vapour with $(CF_3)_2CFI$ at $-196\,°C$ followed by matrix warm-up ⟨76JFC(7)95⟩. Treatment of $CF_2{=}CFCF_3$ with silver trifluoroacetate and CsF (2 mole equivalents) produced $(CF_3)_2CFAg$ in good yield ⟨73JOM(57)423⟩. This procedure was claimed to be more satisfactory than the direct addition of AgF to $CF_2{=}CFCF_3$ ⟨68JA7367⟩. $(CF_3)_2CFAg(MeCN)$ was shown to be heterolytically labile in solution existing in equilibrium with solvated Ag^+ and the complex $Ag[CF(CF_3)_2]^-{}_2$ ⟨86JA5359⟩.

The gold dimer $[Au(CH_2)_2PPh_2]_2(CH_2Cl)Br$ was prepared in good yield by the action of CH_2ClBr on $[Au(CH_2)_2PPh_2]_2$ ⟨85CC1278⟩. Interaction of Ph_3PAuCl with CH_2N_2 afforded Ph_3PAuCH_2Cl in 80% yield ⟨77IZV2417⟩.

4.03.5.2.6 Derivatives of zinc, cadmium and mercury

The classical Simmons–Smith reaction employing a zinc–copper couple and CH_2I_2 provides a useful method for the cyclopropanation of alkenes. The structures of the reactive species are not fully understood, however, and they are often written as the (iodomethyl)zinc compounds ICH_2ZnI, $(ICH_2)_2Zn$, etc. Methods used to prepare Simmons–Smith reagents fall into three general categories, that is treatment of activated zinc with a dihalomethane, a zinc(II) salt with CH_2N_2 and a 1,1-dihaloalkane with an alkylzinc compound. A publication by Denmark and co-workers (on the structure of halomethylzinc reagents) gives an account of these methods ⟨92JA2592⟩. Bromomethylzinc bromide, an alternative to the classical Simmons–Smith reagent, has been prepared by the reaction of zinc metal with CH_2Br_2 ⟨84JOM(269)219⟩.

Bis(perfluoroisopropyl)cadmium was prepared in quantitative yield from $(CF_3)_2CFI$ and dimethyl- or diethylcadmium in the presence of a Lewis base ⟨84JFC(26)1⟩ and by treatment with activated cadmium powder ⟨92JFC(56)341⟩.

The perfluoroisopropylmercury compounds (**465a–c**) were obtained in 42–71% yield from the appropriate organomercury trifluoroacetates and $CF_2{=}CFCF_3$ in the presence of CsF (Equation (101)) ⟨91MI 403-01⟩. Styrene also combined with $(CF_3)_2CFHg(O_2CCF_3)$ in methanol to furnish the adduct (**466**) in 56% yield ⟨89MI 403-03⟩.

$$RHg(O_2CCF_3) \xrightarrow[42-71\%]{CF_3CF=CF_2,\ CsF} (CF_3)_2CFHgR \qquad (101)$$

(**465**) **a**; R = Ph
b; R = $Ph_2C{=}CH$
c; R = CH(Me)Et

Ph-CH(OMe)-CH$_2$-HgCF(CF$_3$)$_2$

(**466**)

A range of alkyl- and aryl (α-chloroalkyl)mercury derivatives have been prepared by treatment of the corresponding alkyl- and aryl(chloro)mercury precursors with diazoalkanes, for example EtCHClHgCl and $MeCHN_2$ afforded EtCHClHgCH(Me)Cl, formally by alkylidene insertion, in 99% yield ⟨79S893⟩. Similarly, (chloromethyl)phenylmercury was prepared by the action of CH_2N_2 on PhHgCl ⟨92RRC393⟩. Nucleophilic cleavage of the Si—C bond of the trifluorosilane $ClCH_2SiF_3$ with phenylmercuric acetate and with mercuric oxide gave $ClCH_2(Ph)Hg$ and $(ClCH_2)_2Hg$ respectively, albeit in 15–18% yields ⟨88JOM(341)225⟩.

A number of alkyl (bromomethyl)mercury compounds were obtained by treatment of the corresponding alkyl(bromo)mercury precursors with CH_2N_2, for example the action of CH_2N_2 on Pr^nHgBr provided $BrCH_2(Pr^n)Hg$ in 85% yield ⟨79S893⟩. Similarly, $(BrCH_2)_2Hg$ and $BrCH_2(Br)Hg$ were prepared from CH_2N_2 and $HgBr_2$ ⟨69JA5027⟩ and an excess of $HgBr_2$ respectively ⟨86JOM(314)13⟩. Treatment of the α-bromolithium compound (**467**) with mercuric chloride gave the α-bromomercury derivative (**468**) (Equation (102)) ⟨75JOM(88)255⟩.

$$\text{(norcaryl)}\begin{smallmatrix}Br\\Li\end{smallmatrix} \xrightarrow[62\%]{HgCl_2} \left(\text{(norcaryl)}\begin{smallmatrix}Br\\\ \end{smallmatrix}\right)_2 Hg \qquad (102)$$

(**467**)   (**468**)

The (iodomethyl)mercury compound (**470**) was prepared in 95% yield from the iodomercury precursor (**469**) and CH_2N_2 (Equation (103)) ⟨79S893⟩. Bis(iodomethyl)mercury $(ICH_2)_2Hg$, an effective transfer agent of one CH_2 group in Simmons–Smith type reactions, was synthesized in 80% yield by sequential treatment of diethylzinc with CH_2I_2 and $HgBr_2$ ⟨92JOM(438)11⟩ and in 75% yield by the reaction of ethyl(iodo)zinc with CH_2I_2 followed by $HgCl_2$ ⟨82JIC111⟩.

$$Ph\underset{OMe}{\overset{}{\diagdown}}HgI \quad \xrightarrow[95\%]{CH_2N_2} \quad Ph\underset{OMe}{\overset{}{\diagdown}}HgCH_2I \quad\quad (103)$$

(**469**) (**470**)

4.03.5.3 Group 3 and Group 4 Derivatives—$R_2CHalSnX_3$, etc.

4.03.5.3.1 *Derivatives of aluminum, gallium, indium and thallium*

Interaction of diethylaluminum iodide Et_2AlI with CH_2N_2 provided (iodomethyl)-diethylaluminum Et_2AlCH_2I in 98% yield; with oxygen this gave (ethoxy)ethyl(iodomethyl)-aluminum $(EtO)EtAlCH_2I$ in 97% yield ⟨67LA(703)1⟩. (Iodomethyl)diethylgallium was similarly obtained by treatment of diethylgallium iodide with CH_2N_2 ⟨80ZN(B)1376⟩.

Indium(I) chloride and CH_2Cl_2 in the presence of TMEDA furnished a complex of (chloro-methyl)dichloroindium $ClCH_2InCl_2$ and TMEDA (2:3) in 84% yield. Likewise, a (bromo-methyl)dibromoindium–DMSO complex was produced in 99% yield from indium(I) bromide, CH_2Br_2 and DMSO ⟨86OM525⟩. Indium(I) bromide and indium(I) iodide, prepared by electrochemical oxidation of indium in CH_2Br_2—MeCN and CH_2I_2—MeCN respectively, reacted further in these media to give (bromomethyl)dibromoindium and (iodomethyl)diiodoindium, isolated as the tetraethylammonium bromide and iodide salts $Et_4N[Br_3InCH_2Br]$ and $Et_4N[I_3InCH_2I]$ in 84% and 79% yields respectively. Indium(I) chloride when prepared electrochemically did not react further with CH_2Cl_2 due to its disproportionation to give $In°$ and ICl_3 ⟨91OM2159⟩.

(Chloromethyl)chloro(phenyl)thallium, $(ClCH_2)PhTlCl$, prepared in 72% yield from dichloro-(phenyl)thallium and CH_2N_2 was treated with mercuric isobutyrate to give $(ClCH_2)Tl(O_2CPr^i)_2$ (61%). This was converted into $Me(ClCH_2)Tl(O_2CPr^i)$ in 70% yield by treatment with tetramethylstannane Me_4Sn ⟨72JOM(43)117⟩.

4.03.5.3.2 *Derivatives of tin*

(i) Trialkyl (or aryl) (α-haloalkyl)stannanes

(a) α-Fluorostannanes. Direct fluorination of Me_4Sn was nonselective and led to a mixture of products which included Me_3SnCH_2F and $Me_2Sn(CH_2F)_2$ (together comprising about 25% of the products) ⟨78IC618⟩. α-Fluorostannanes have been formed by the addition of trialkylstannane hydrides to fluoroalkenes, for example Akhtar and Clark showed that the photoinduced addition of Me_3SnH to $Me_3SnCF=CF_2$ gave a mixture (1:1) of the unstable adducts $Me_3SnCHFCF_2SnMe_3$ and $(Me_3Sn)_2CFCF_2H$ ⟨68CJC2165⟩. Similarly, the unstable, cyclic α-fluorostannane (**472**) was obtained by the *cis*-addition of trimethylstannane hydride to hexafluorocyclobutene (**471**) in ca. 90% yield (Equation (104)) ⟨66JOM(6)633⟩. The α-fluorostannane (**474**) was synthesized in 75% yield from $Me_3Sn(O_2CCF_3)$ and the perfluoroorganocadmium compound (**473**) (Equation (105)) ⟨85JFC(27)309⟩.

$$\underset{(\mathbf{471})}{\text{hexafluorocyclobutene}} \quad \xrightarrow[92\%]{Me_3SnH} \quad \underset{(\mathbf{472})}{\text{Me}_3\text{Sn-adduct}} \quad\quad (104)$$

$$Cd[C(CF_3)_2F]_2 \xrightarrow[75\%]{Me_3Sn(O_2CCF_3)} Me_3SnC(CF_3)_2F \quad (105)$$
(473) **(474)**

(b) α-Chlorostannanes. Generally halomethyl compounds of tin cannot be prepared by direct halogenation because halogens cleave heavy metal–carbon bonds ⟨78IC618⟩. Hillgartner *et al.*, however, have shown that $R_3SnCClPh_2$ (R = Me, Ph) can be prepared in good yields by the reaction of $R_3SnCHPh_2$ with Bu^tOCl ⟨80JOM(201)197⟩.

Thermal decomposition of phenylchlorodiazirine **(475)** in the presence of Bu^n_3SnH provided α-chlorobenzyltri-*n*-butylstannane **(476)** (Equation (106)) ⟨69JOC2728, 88TL5863⟩. In aprotic solvents, the arenesulfenyl chlorides **(478**; R^2 = H,Me,NO_2**)** reacted with the vinylstannanes **(477**; R^1 = Me,Ph**)** to give the adducts **(479)** (Equation (107)) ⟨74JOM(78)395, 75JCS(D)1786⟩.

The α,α-dichlorostannane **(480)** was reduced with Ph_3SnH in the presence of 2,2′-azobisisobutyronitrile (AIBN) to give a quantitative yield of a stannane tentatively assigned the monochloro structure **(481)** (Equation (108)) ⟨79JOM(166)339⟩.

Seitz *et al.* have described the preparation of $Bu^n_3SnCH_2Cl$ from Bu^n_3SnH in 78% yield by sequential treatment with LDA, paraformaldehyde and methanesulfonyl chloride (MsCl) ⟨83SC129⟩. The unstable carbinol **(482)**, prepared by treatment of $Ph(CH_2)_2CHO$ with Bu^n_3SnLi, was converted into the stable chloro derivative **(483)** in 65% yield with either TsCl in the presence of pyridine, or Ph_3P and diethyl azodicarboxylate (dead) in CH_2Cl_2 ⟨81TL2397⟩.

(482) R = OH
(483) R = Cl

Me_3SnCH_2Cl, $Me_2Sn(CH_2Cl)_2$ and $Me_2(Ph)SnCH_2Cl$ were conveniently prepared from the corresponding iodides by treatment with AgCl in MeCN ⟨71JOM(30)151⟩.

Bis(α-chloroethyl)diethylstannane $(MeCHCl)_2SnEt_2$ has been prepared from $(MeCHCl)_2SnCl_2$ by treatment with EtMgBr ⟨52JGU1827⟩.

(c) α-Bromostannanes. Treatment of chlorotrimethylstannane and dichlorodimethylstannane with the Simmons–Smith reagent $BrCH_2ZnBr$ (see Section 4.03.5.2.6) produced (bromomethyl)trimethylstannane and bis(bromomethyl)dimethylstannane respectively in 65% and 31% yields ⟨71JOM(30)151⟩. Nemoto *et al.* showed that treatment of the aldehyde **(484)** with Bu^n_3SnLi and then bromination of the resultant hydroxystannane **(485)** with PPh_3 and CBr_4 in CH_2Cl_2 furnished the α-bromostannane **(486)** in 36% yield (Scheme 35) ⟨85JCS(P1)1185⟩. Similarly, the

α-bromostannane (**488**) was formed in good yield from the αβ-unsaturated aldehyde (**487**), exclusive 1,2-addition to the aldehyde being observed (Equation (109)) ⟨83CL1303⟩.

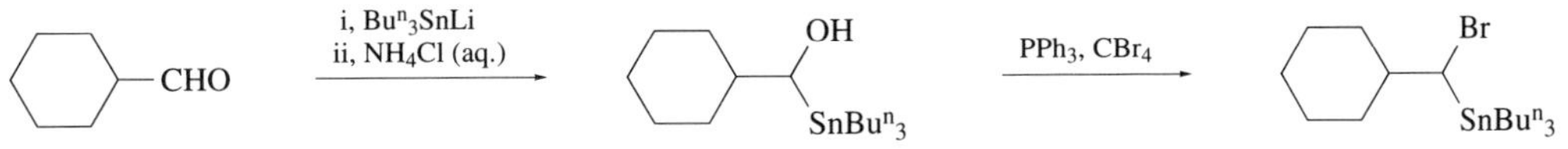

Scheme 35

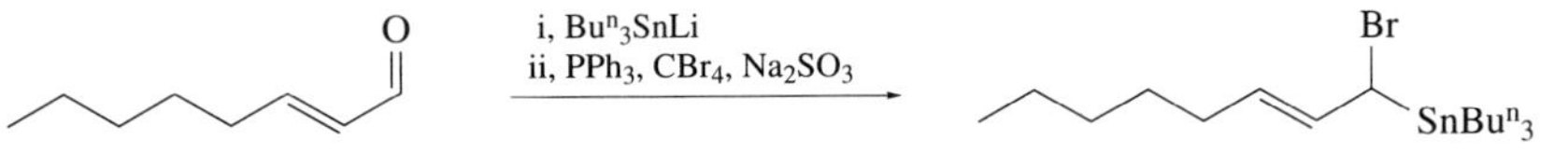

(109)

Treatment of 7,7-dibromo-8,9,10-trinorcarane (**489a**) with Pr^{i}MgCl (1 molar equivalent) at −65 °C gave a mixture (ca. 8:1) of the *anti-* and *syn*-MgCl derivatives (**489b**) and (**489c**) which reacted with Me$_3$SnCl to give the *anti*-7-trimethylstannyl derivative (**489d**) in 49% overall yield ⟨75JOM(88)287⟩. Conversely, when the dibromotrinorcarane (**489a**) was treated with Bu^{n}Li, the *syn*-lithio isomer (**489e**) predominated over the *anti-* isomer (**489f**). Consequently, when Me$_3$SnCl was added, a mixture (4:1) of *syn-* (**489g**) and *anti-* (**489d**) trimethylstannyl derivatives was obtained in 71% overall yield from the dibromotrinorcarane (**489a**) ⟨75JOM(88)255⟩.

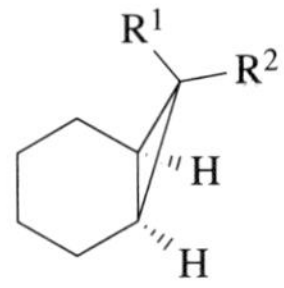

(**489**) a; $R^1 = R^2 = Br$
b; $R^1 = Br, R^2 = MgCl$
c; $R^1 = MgCl, R^2 = Br$
d; $R^1 = Br, R^2 = SnMe_3$
e; $R^1 = Li, R^2 = Br$
f; $R^1 = Br, R^2 = Li$
g; $R^1 = SnMe_3, R^2 = Br$

Ph(CH$_2$)$_2$CHBrSnBu$^n_3$ was prepared in quantitative yield from the hydroxy analogue (**482**) by treatment with Ph$_3$P and CBr$_4$ in CH$_2$Cl$_2$ ⟨81TL2397⟩.

Me$_3$SnCH$_2$Br and Me$_2$Sn(CH$_2$Br)$_2$ were synthesized from the corresponding α-iodomethyl analogues by treatment with AgBr in MeCN ⟨71JOM(30)151⟩.

(d) α-Iodostannanes. Trialkyl (or aryl) (iodomethyl)stannanes have been prepared by the reaction of iodostannanes with iodo(iodomethyl)zinc (see Section 4.03.5.2.6); examples are given in Table 2.

The iodide (**492a**) was prepared in 35% overall yield by treatment of the aldehyde (**490a**) with Bu$^n_3$SnLi and then treatment of the resulting α-hydroxystannane (**491a**) with Ts$_2$O and then NaI (Scheme 36). Improved yields (50–85% from the aldehydes (**490b–g**)) were obtained when the α-hydroxystannane intermediates (**491b–g**) were converted into the iodides with Ph$_3$P, imidazole and I$_2$ in CH$_2$Cl$_2$ (Scheme 36) ⟨93JOC523⟩. The α-iodopropylstannane Me$_2$ClSnBu$^n_3$ was obtained similarly from propanal except that the α-hydroxymethyl group was transformed into the α-iodomethyl group with [(PhO)$_3$PMe]$^+$I$^-$ ⟨92SL891⟩. The system Ph$_3$P–dead–MeI has also been used for converting the α-hydroxy group in stannanes into the α-iodomethyl group ⟨81TL2397⟩.

(ii) Halo(α-haloalkyl)stannanes

Depending on the conditions used and the quantities of reactants employed, SnCl$_4$ and CH$_2$N$_2$ reacted to give ClCH$_2$SnCl$_3$ (24% yield), (ClCH$_2$)$_2$SnCl$_2$ (22%), (ClCH$_2$)$_3$SnCl (95%) and (ClCH$_2$)$_4$Sn (71%) ⟨68IZV270⟩. Similarly, the corresponding bromo(bromomethyl)stannanes BrCH$_2$SnBr$_3$, (BrCH$_2$)$_2$SnBr$_2$ and (BrCH$_2$)$_4$Sn have been prepared from SnBr$_4$ and CH$_2$N$_2$ ⟨52JGU1827⟩. The photochemical reaction of the polymeric organotin species (R$_2$Sn)$_n$, where R = Bu^n

Table 2 Reactions of ICH$_2$ZnI with tin halides.

Halide	Product	Yield (%)	Ref.
Me$_3$SnCl	Me$_3$SnCH$_2$I	82	71JOM(30)151
Me$_2$SnCl$_2$	Me$_2$Sn(CH$_2$I)$_2$	79	71JOM(30)151
Me$_2$PhSnI	Me$_2$PhSnCH$_2$I	94	71JOM(30)151
Ph$_3$SnCl	Ph$_3$SnCH$_2$I	66	71JOM(30)151
SnCl$_4$	Sn(CH$_2$I)$_4$	17	71JOM(30)151
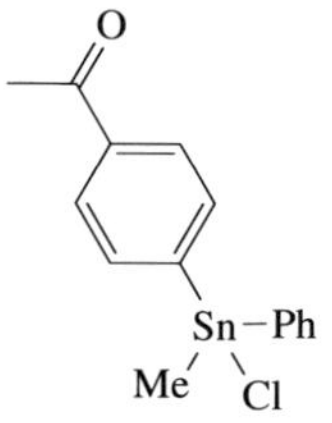 [4-acetylphenyl-Sn(Me)(Ph)Cl]	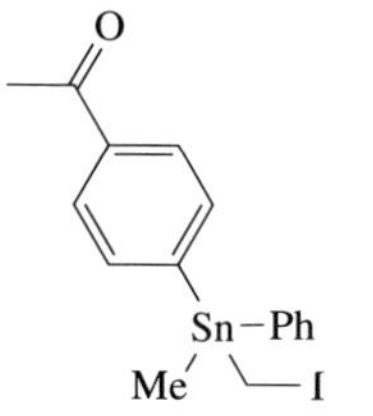 [4-acetylphenyl-Sn(Me)(Ph)CH$_2$I]		81JOM(214)191
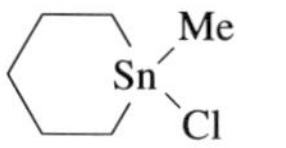 [cyclohexyl-Sn(Me)Cl piperidine]	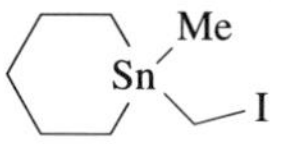 [cyclohexyl-Sn(Me)CH$_2$I piperidine]	70	78JOM(153)305

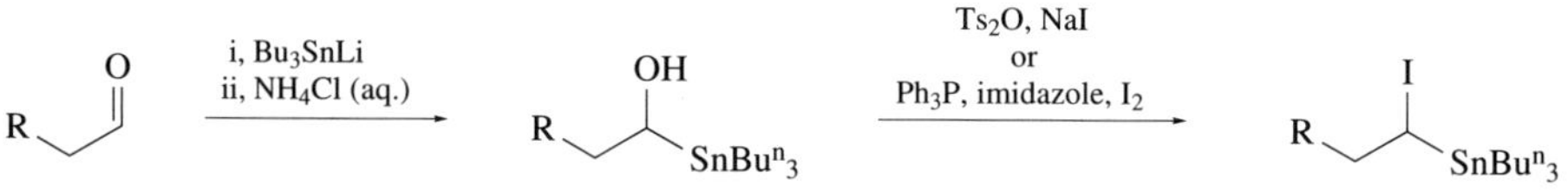

Scheme 36

(a) R = Bu^n; (b) R = TBDMS-O(CH$_2$)$_3$; (c) R = PhCH$_2$; (d) R = TMS—≡C(CH$_2$)$_2$;
(e) R = Me$_2$CH; (f) R = n-C$_5$H$_{11}$CHO-TBDMS; (g) R = n-C$_5$H$_{11}$CHO-MOM

and Ph, with CH$_2$Cl$_2$ and with CH$_2$Br$_2$ produced the corresponding halo(halomethyl)stannanes R$_2$Sn(Cl)CH$_2$Cl and R$_2$Sn(Br)CH$_2$Br in moderate to good yields ⟨76BCJ2837, 77BCJ1353⟩.

Treatment of (BrCH$_2$)$_4$Sn with Br$_2$ (2 mole equivalents) conveniently afforded (BrCH$_2$)$_2$SnBr$_2$ in 76% yield ⟨72JOM(40)115⟩. The reaction of SnBr$_2$ with CH$_2$Br$_2$ in the presence of a catalytic amount of Et$_3$Bu^{n}NCl or SbI$_3$ afforded (BrCH$_2$)SnBr$_3$ in 23–35% yields ⟨78JGU769⟩.

4.03.5.3.3 Derivatives of lead

Addition of Me$_3$PbBr to the mixture (4:1) of syn-7-lithio- (**489e**) and $anti$-7-lithio- (**489f**) bromotrinorcaranes previously described (Section 4.03.5.3.2) gave a mixture (2.5:1) of the α-bromoplumbanes (**493a**) and (**493b**) in 75% yield ⟨75JOM(88)255⟩.

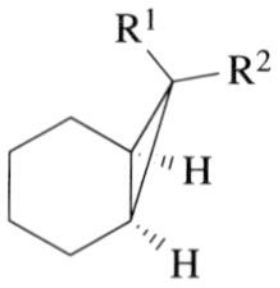

(**493**) a; R^1 = Me$_3$Pb, R^2 = Br
b; R^1 = Br, R^2 = Me$_3$Pb

(Iodomethyl)triphenylplumbane (**495**) was prepared from chlorotriphenylplumbane (**494**) in 31% yield using iodo(iodomethyl)zinc and in 62% yield with (iodomethyl)lithium (Equation (110)) ⟨85CB370, 85CB391⟩.

$$\text{Ph}_3\text{PbCl} \xrightarrow{\text{ICH}_2\text{ZnI or LiCH}_2\text{I}} \text{Ph}_3\text{PbCH}_2\text{I} \qquad (110)$$
(494) **(495)**

The authors could find no reports on the preparation of α-haloalkyl derivatives of the members of the lanthanide or actinide series.

4.04
Functions Bearing Two Oxygens, $R^1_2C(OR^2)_2$

DAVID T. MACPHERSON and HARSHAD K. RAMI
SmithKline Beecham Pharmaceuticals, Epsom, UK

4.04.1 α-DIOLS—$R_2C(OH)_2$	160
4.04.1.1 α-Diols by Hydration of Carbonyl Compounds Bearing Electron-Withdrawing Groups	160
4.04.1.2 α-Diols by Oxidation of Active Methyl or Methylene Groups	161
4.04.1.3 α-Diols from Strained Cyclic Ketones	162
4.04.2 HEMIACETALS—$R^1_2C(OH)OR^2$	163
4.04.2.1 Hemiacetals from Aldehydes and Ketones by Addition of Alcohols	164
4.04.2.1.1 Acyclic hemiacetals	164
4.04.2.1.2 Cyclic hemiacetals from hydroxy carbonyl derivatives	164
4.04.2.2 Hemiacetals from Esters	169
4.04.2.2.1 By reduction	169
4.04.2.2.2 By addition of carbanions	171
4.04.2.3 Hemiacetals from Enol Ethers	171
4.04.2.4 Hemiacetals from Acetals by Deprotection	175
4.04.2.4.1 From noncarbohydrate substrates	175
4.04.2.4.2 From carbohydrate substrates	176
4.04.3 ACETALS—$R^1_2C(OR^2)_2$	176
4.04.3.1 General Methods	177
4.04.3.1.1 From aldehydes and ketones	177
4.04.3.1.2 From acetals	184
4.04.3.1.3 From enol ethers and alcohols	186
4.04.3.1.4 From silyl enol ethers and enol acetates	188
4.04.3.1.5 From gem-dihalides by alkylation	189
4.04.3.2 Symmetrical Acetals	190
4.04.3.2.1 From aldehydes and ketones	190
4.04.3.2.2 From ortho esters and nucleophiles	192
4.04.3.2.3 From alkenes and alkynes	193
4.04.3.2.4 From dithioacetals and O,S-acetals	196
4.04.3.2.5 Electrochemical methods	197
4.04.3.2.6 Miscellaneous methods	198
4.04.3.3 Unsymmetrical Acetals	198
4.04.3.3.1 From α-substituted ethers and alcohols	198
4.04.3.3.2 From ethers	201
4.04.3.3.3 From cyclic hemiacetals	201
4.04.3.3.4 By hetero Diels–Alder reactions	202
4.04.3.3.5 Mixed alkyl silyl acetals	202
4.04.4 OTHER DIOXYGEN DERIVATIVES	203
4.04.4.1 Synthesis of $R^1_2C(OCOR^2)_2$	203
4.04.4.1.1 From aldehydes and ketones	203
4.04.4.1.2 From carboxylic acids	205
4.04.4.1.3 By oxidation of aromatic methyl and methylene groups	206
4.04.4.1.4 By oxidation of furan derivatives	206
4.04.4.1.5 Miscellaneous methods	206
4.04.4.2 Synthesis of $R^1_2C(OCOR^2)OR^3$	207

4.04.4.2.1 From ethers	207
4.04.4.2.2 From enol ethers and enol esters	208
4.04.4.2.3 From carboxylic acids by alkylation with α-halo alkyl ethers	209
4.04.4.2.4 From alcohols and α-haloalkyl carboxylate derivatives	209
4.04.4.2.5 From hemiacetals	209
4.04.4.2.6 From acetals	211
4.04.4.2.7 From aldehydes and ketones	212
4.04.4.3 Other Derivatives	214
4.04.4.3.1 1,2,4-Trioxolanes (ozonides)	214
4.04.4.3.2 1,2,4-Trioxane	214
4.04.4.3.3 $R^1_2C(OR^2)O_2R^3$	214
4.04.4.3.4 $R^1_2C(OR^2)OX$ and $R_2C(OX)_2$ (X = heteroatom)	214

4.04.1 α-DIOLS—$R_2C(OH)_2$

With the exception of strained ring systems, stable α-diols are essentially hydrates of carbonyl compounds bearing electron-withdrawing substituents such as polyhalogenated alkyl or polycarbonyl groups. The syntheses of specific α-diols such as ninhydrin (**1**) and its analogues have appeared in the literature ⟨91SC1055, 91SC2231⟩. Other specific α-diols such as chloral hydrate, additions to polyhalogenated aldehydes and ketones, and hydrates of cyclopropanone have been documented ⟨B-92MI 404-01⟩.

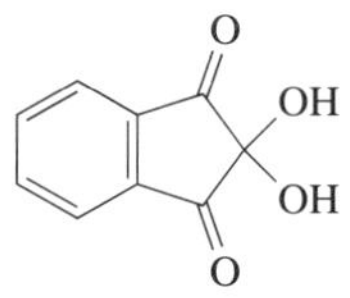

(**1**)

4.04.1.1 α-Diols by Hydration of Carbonyl Compounds Bearing Electron-Withdrawing Groups

Polyhalogenated carbonyl compounds add water readily to form hydrates due to the strongly polarizing nature of the halogens, and the equilibrium lies in favour of the diol. Hydrates of cyclic perfluorinated ketones form readily upon addition of water to give 1:1 adducts which are well-defined distillable compounds obtained in high yields (Equation (1)) ⟨68JOC2692⟩. Highly fluorinated bicyclo[2.2.0]hex-5-en-2-ones have been shown to form hydrates readily (Equation (2)). The carbonyl group has an IR stretching band at 1850 cm^{-1}, a very high frequency because of ring strain and fluorine substitution in its vicinity. These combinations cause the carbonyl group to hydrate voraciously, transforming a mobile volatile liquid ketone into a nonvolatile crystalline α-diol ⟨89JOC5502⟩. Similarly for the above reasons, hexafluorocyclobutanone also forms a very stable hydrate ⟨61JA2205⟩.

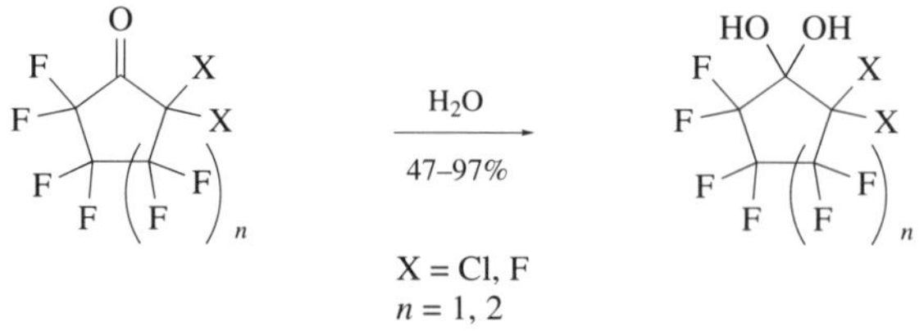

X = Cl, F
n = 1, 2

(1)

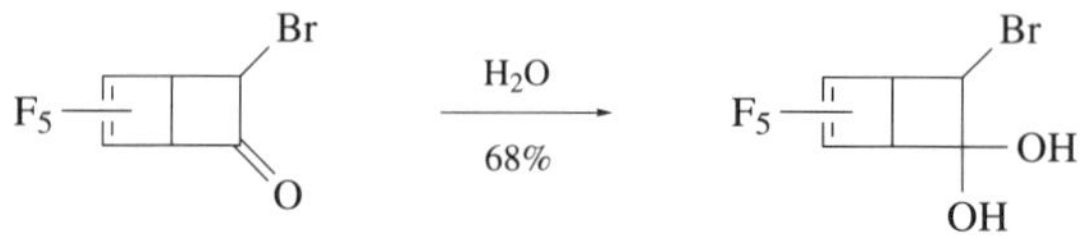

(2)

Acyclic carbonyl compounds bearing perhalo groups also from hydrates. Treatment of the silyl ether (**2**) with hydrochloric acid in dichloromethane followed by the addition of water quantitatively affords the trifluoromethyl hydrate (Equation (3)) ⟨89TL5243⟩. The hydrates (**3**) and (**4**) of trifluoropyruvaldehyde are obtained when 3,3-dibromo-1,1,1-trifluoroacetone is treated with aqueous sodium acetate (Scheme 1). The results from ^{19}F and ^{1}H NMR studies indicate that the major species in aqueous solution is the dihydrate (**4**) ⟨88JOC5088⟩.

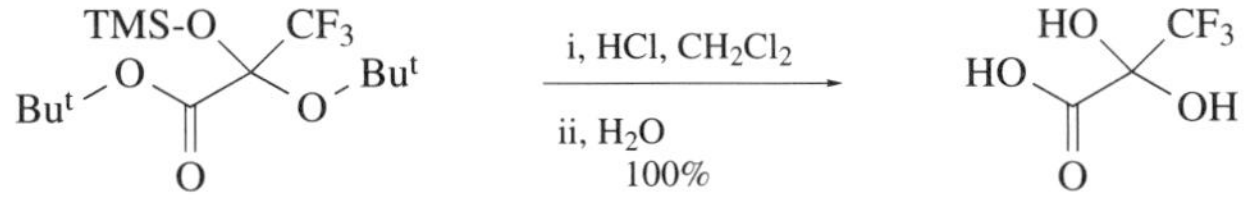

Scheme 1

Perfluorinated sulfonic acid resin (Nafion-H) has been used to catalyse cleavage of α-keto acetals to either α-acetoxy α-methoxy ketones or α-chloro α-methoxy ketones in 90–100% yield (Scheme 2). The α-acetoxy and α-chloro compounds readily hydrolyse to afford the α-keto aldehyde 1,1-diols, the former with potassium hydrogen carbonate in dimethoxyethane–water, the latter with aqueous tetrahydrofuran ⟨83S891, 86S513⟩.

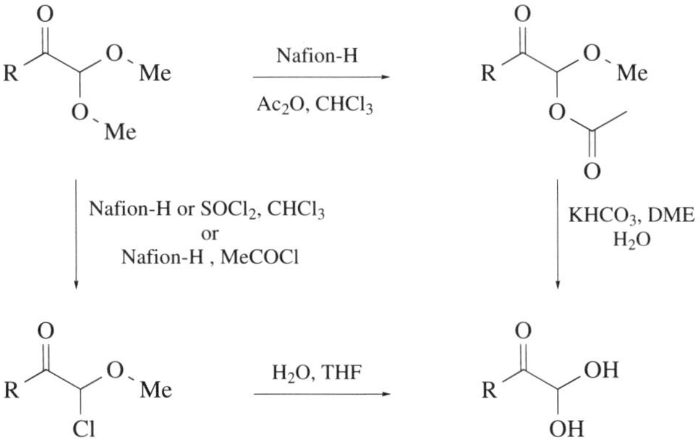

Scheme 2

4.04.1.2 α-Diols by Oxidation of Active Methyl or Methylene Groups

Oxidation of a methylene group bearing electron-withdrawing substituents, that is, aryl or carbonyl groups, leads to α-diols. Dimethyl sulfoxide–iodine has been used to oxidize active methylene compounds in high yield (Equation (4)). Indane-1,3-dione has also been oxidized to ninhydrin (**1**) (33%) using similar conditions ⟨77JCS(P1)372⟩. Phenacyl bromides have been converted into arylglyoxal hydrate derivatives with a variety of reagents, for example DMSO ⟨57JA6562⟩, silver nitrate, sodium acetate and DMSO ⟨66JA865⟩, and N,N-dialkylhydroxylamines ⟨77JOC754⟩. Acetophenone derivatives can also be oxidized to arylglyoxal hydrates via intermediate phenacyl bromides with aqueous hydrobromic acid in DMSO ⟨85JOC5022⟩.

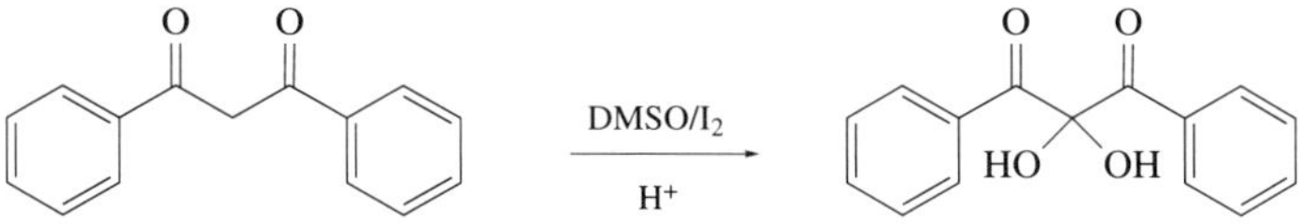

Acetoacetic esters can be oxidized to 2,2-dihydroxy-3-oxobutanoate derivatives in moderate yields by treatment with nitrous acid followed by dinitrogen tetroxide (Equation (5)) (R = ethyl, 30% ⟨54JCS520⟩, R = *t*-butyl, 37% ⟨70HCA1598⟩. A tricarbonyl derivative such as (**5**) can be converted into the bis-diazo compound (**6**) by treatment with toluene-*p*-sulfonyl azide; subsequent treatment with *t*-butyl hypochlorite in formic acid leads directly to the hydrated pentaketones (**7**) (Scheme 3) ⟨86AG(E)999⟩.

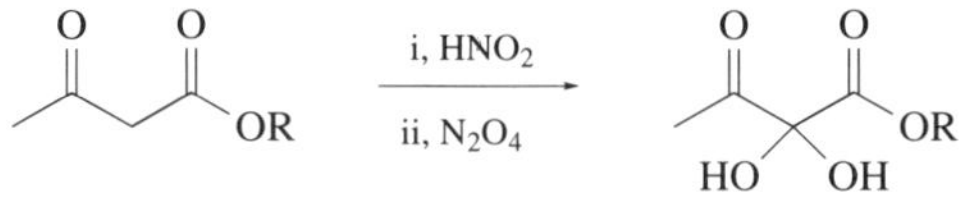

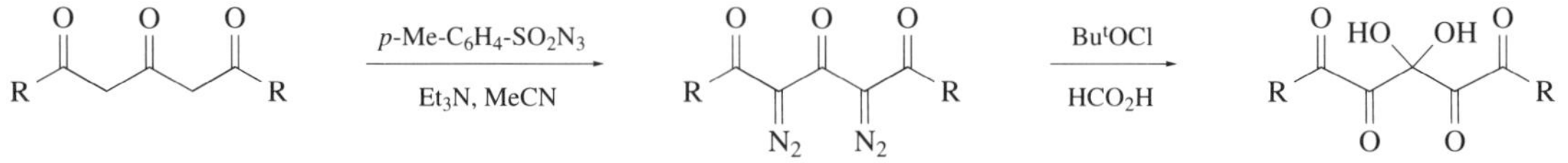

Scheme 3

Dimethyldioxirane has been used as an oxidizing agent for the preparation of α-oxo aldehydes which exist as hydrates. Under neutral conditions, a diazo compound and dimethyldioxirane react virtually quantitatively without any involatile by-products (Equation (6)). The starting diazo compounds were prepared from the corresponding acid chlorides and diazomethane ⟨91TL6215⟩.

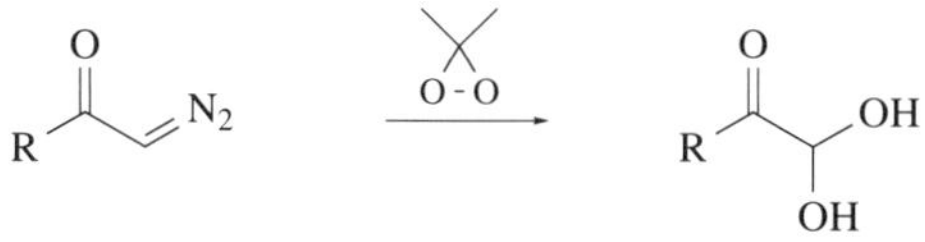

R	Yield (%)
Ph	100
C_6H_{11}	96
2-Furyl	98
2-Thienyl	85
2-Pyridyl	100
3-Pyridyl	100
Ethoxycarbonyl	100

4.04.1.3 α-Diols from Strained Cyclic Ketones

Highly strained ring systems such as cyclopropanone are amongst the most reactive carbonyl compounds, and readily undergo nucleophilic attack at the carbonyl carbon to relieve strain by converting the sp^2-hybridized carbonyl carbon to sp^3. Cyclopropanone can be prepared from ketene and diazomethane, and is extremely susceptible to hydration (for reviews on cyclopropanone chemistry, see ⟨74ACR85, 74FCF73, 75AG(E)473⟩ and ⟨80JOC2874⟩). Its IR spectrum shows a highly strained carbonyl stretching band at 1813 cm^{-1} ⟨66JA3672⟩.

Treatment of dibromocarbene with the ketene acetal (**8**) followed by reduction with tri-*n*-butyltin hydride affords (**9**). The derivative (**9**) lacks the very high reactivity of cyclopropanone itself and serves as a useful substrate for synthesizing cyclopropanone hydrate derivatives by lithiation and subsequent alkylation (Scheme 4) ⟨85TL2279, 85TL2283⟩. The free cyclopropanone hydrate derivative is liberated by hydrogenolytic removal of the *o*-xylyl protecting group using either palladium oxide in ethyl acetate or palladium hydroxide in methanol with potassium carbonate.

The 2,3-diazabicyclo[2.2.1]heptane derivative (**10**), prepared by hydrogenation of the Diels–Alder adduct of dimethylfulvene and diethyl azodicarboxylate, readily underwent ozonolysis to the ozonide (**11**), which on hydrogenolysis gave the stable ketone hydrate (**12**) in moderate yield (Scheme 5) ⟨68JOC2368⟩.

The rigid 10-membered ring ketonic groups incorporated in [2.2]metacyclophanes also exhibit a pronounced tendency towards adduct formation with nucleophiles (Equation (7)). When a solution of (**13**) in dioxane–water is kept at room temperature, the dihydrate (**14**) is obtained as a crystalline solid in good yield. Replacement of the phenyl rings in (**13**) with cyclohexane rings prevents the derivative (**15**) from affording a similar hydrate, due to reduced ring strain ⟨80T1345⟩. The tendency of (**13**) for adduct formation is reminiscent of the corresponding behaviour of cyclopropanone derivatives, in which the driving force for rehybridization has been accounted for by the I-strain hypothesis ⟨51JA212⟩.

4.04.2 HEMIACETALS—$R^1_2C(OH)OR^2$

Hemiacetals serve as useful intermediates in organic synthesis, and a variety of methods for their preparation have been described. A few limited reviews have appeared ⟨66JCE527, B-67MI 404-01⟩ and a review published in 1991 ⟨91HOU(E14a/1)600⟩ gives a good account as part of an overall review on

164 Two Oxygens

acetals. Most hemiacetals are cyclic, and acyclic hemiacetals require electron-withdrawing groups to stabilize them. Many examples of cyclic hemiacetals have been prepared as intermediates in carbohydrate and natural product syntheses. The discussion below sets out important synthetic routes to hemiacetals, and the reader should refer to the original references for greater detail.

4.04.2.1 Hemiacetals from Aldehydes and Ketones by Addition of Alcohols

4.04.2.1.1 Acyclic hemiacetals

In α-halo ketones the presence of the electronegative halogen atoms enhances the electrophilicity of the carbonyl group and facilitates the addition of oxygen nucleophiles (cf. the formation of α-diols). The equilibrium lies far to the right (Equation (8)), in contrast to that of aliphatic ketones, and the hemiacetals are stabilized by the electron-withdrawing groups. Dichlorotetrafluoroacetone ($R^1 = R^2 = CClF_2$) reacts with ethanol at room temperature to give the hemiacetal *in situ*, and subsequent alkylation affords unsymmetrical acetals ⟨65JOC3834⟩. Similarly, the reaction of 2-chloroethanol with 1,1,1-trifluoroacetone in the presence of potassium carbonate affords the corresponding hemiacetal ⟨60JA2288⟩. Self-condensation of cyclic ketones such as hexafluoro-cyclobutanone is observed when two equivalents are combined with one equivalent of hydrogen fluoride at low temperature, resulting in the formation of hemiacetals in high yield ⟨61JA4670⟩.

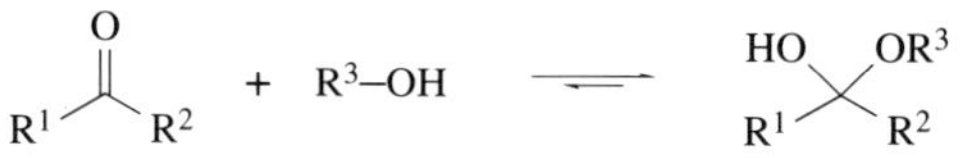

$$\underset{R^1\quad R^2}{\overset{O}{\|}} + R^3\text{-OH} \rightleftharpoons \underset{R^1\quad R^2}{\overset{HO\quad OR^3}{\diagdown\diagup}} \qquad (8)$$

R^1, R^2 = electron-withdrawing groups

4.04.2.1.2 Cyclic hemiacetals from hydroxy carbonyl derivatives

A well-established route for generating the hemiacetal group involves cyclization of hydroxy carbonyl derivatives. Various methods have been employed to generate the requisite hydroxy carbonyl compounds, the most obvious being oxidation of diols or addition of a nucleophile to dicarbonyl compounds. In both cases two different hydroxy carbonyl derivatives are possible if R^1 and R^2 are different, and thus selective formation of one of these is necessary to make this a useful procedure for hemiacetal synthesis (Scheme 6).

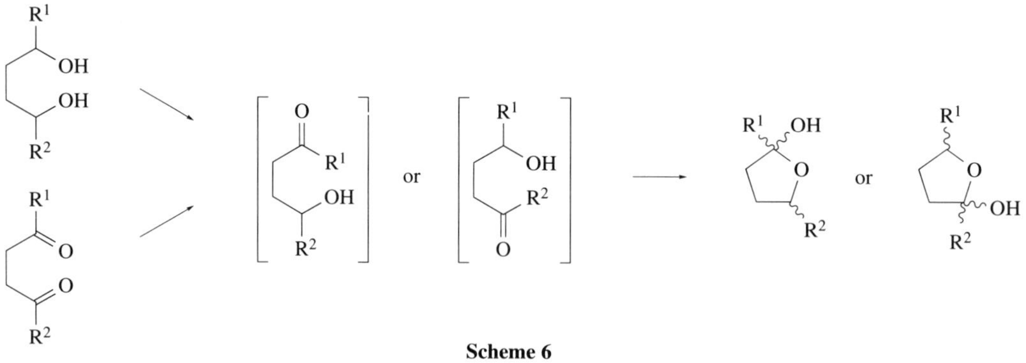

Scheme 6

(i) By oxidation of diols

Manganese dioxide is a mild and selective oxidant used for this purpose, although the reaction conditions may be critical to prevent over-oxidation. For example, short exposure of pyridoxol to activated manganese dioxide under acidic conditions affords pyridoxal in good yield whilst 4-pyridoxic acid is formed under basic conditions (Scheme 7) ⟨67JHC625⟩. In a strategy directed towards the preparation of ellipticine analogues, the diol (**16**) was selectively oxidized to the lactol

Hemiacetals

(17) with activated manganese dioxide (Equation (9)) ⟨84JOC4518⟩. The allylic alcohol in the diol (18) was oxidized with barium manganate to afford the lactol (19), whereas treatment with a ruthenium(VI) complex resulted in oxidation of the other primary alcohol to give the alternative lactol (20) (Scheme 8) ⟨83JCS(P1)1579, 84JCS(P1)681⟩.

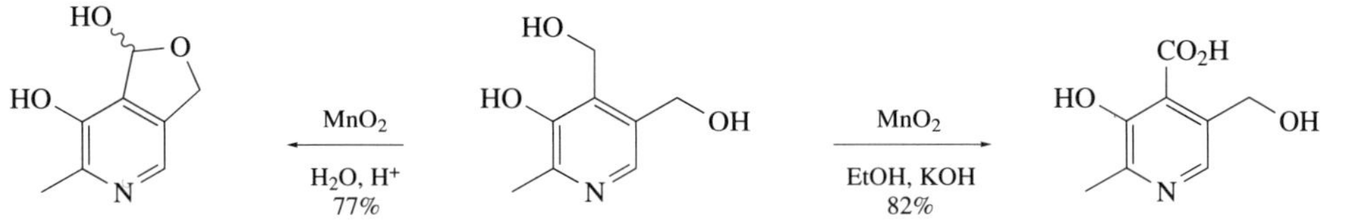

Scheme 7

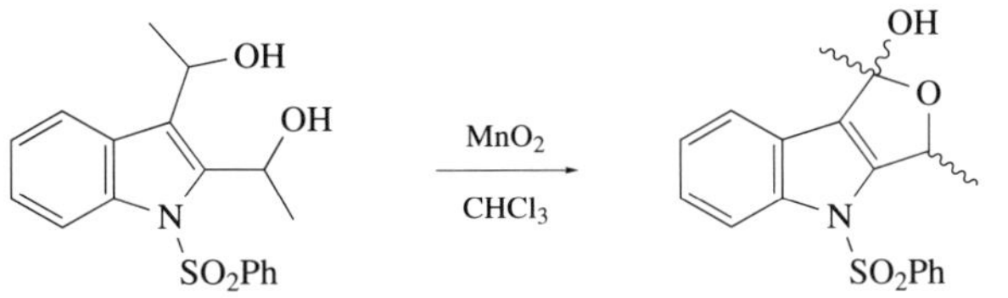

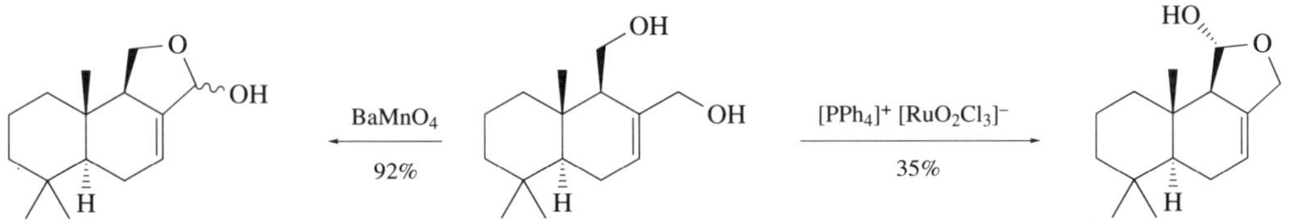

Scheme 8

Selective oxidation of a secondary hydroxy group in the presence of a primary hydroxy group with ceric ammonium nitrate (can) or pyridinium dichromate (pdc) and subsequent cyclization to the hemiacetal is exemplified in Equation (10) ⟨84SC147⟩. Selective oxidation of a primary hydroxy group in the presence of a secondary hydroxy group leading to hemiacetal formation is also possible ⟨84JA1148⟩.

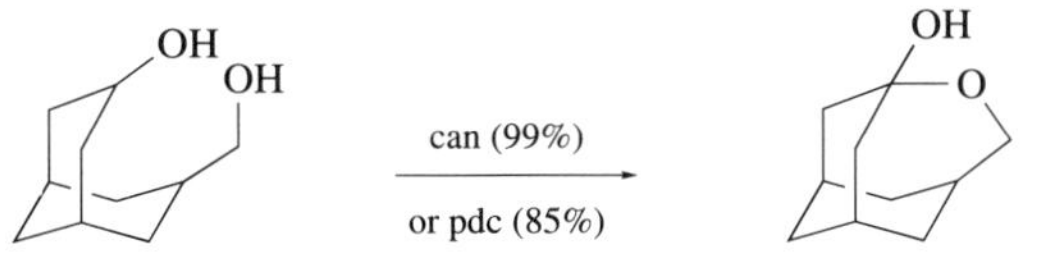

can = ceric ammonium nitrate
pdc = pyridinium dichromate

(ii) By reduction of dicarbonyl derivatives

Sodium borohydride, a common reagent for this purpose, was used to reduce the symmetrical dicarbonyl compound (21) to afford the hemiacetal in a high yield (Equation (11)) ⟨64T2181⟩. In a similar manner the hemiacetals (22) and (23) were prepared by reduction of dicarbonyl precursors with sodium borohydride ⟨89JOC91⟩ and aluminum triisopropoxide ⟨51JA1668⟩, respectively. An attempt to use LAH for this type of reduction resulted in the formation of diols rather than hemiacetals ⟨74H(2)177⟩. Enzymatic reduction of dicarbonyl compounds was used to prepare a hemiacetal from the dialdehyde (Equation (12)). Besides selective reduction of the enal carbonyl, stereospecific hydroxylation also occurred in the saturated ring ⟨80JCS(P1)2535, 83JCS(P1)1579⟩.

(iii) By addition of carbanions to dicarbonyl derivatives

The mono addition of organometallic reagents or other carbanions to dicarbonyl compounds can also generate an alkoxide intermediate which can subsequently cyclize to give hemiacetals. For example, Grignard reagents add to glutaraldehyde to provide δ-lactols in acceptable yields (Equation (13)) ⟨81HCA1247⟩. Mono addition of alkyl Grignard reagents predominates, provided that the reaction is conducted at −70 °C, but yields were reduced by addition to both aldehydes ⟨72HCA249⟩. As expected, 3-methoxyphenylmagnesium bromide selectively adds to the aldehyde carbonyl group of the keto aldehyde (**24**) to provide the hemiacetal after cyclization (Equation (14)) ⟨70JOC468⟩. A highly enantioselective synthesis of lactols involves the addition of dialkylzinc reagents to *o*-phthalaldehyde catalysed by chiral 1,2-substituted ferrocenylamino alcohols (Equation (15)) ⟨92JOC742⟩. The addition of nitromethane under mild basic conditions to the diketone (**25**) affords the tricyclic hemiacetal in high yield (Equation (16)) ⟨77LA1807⟩.

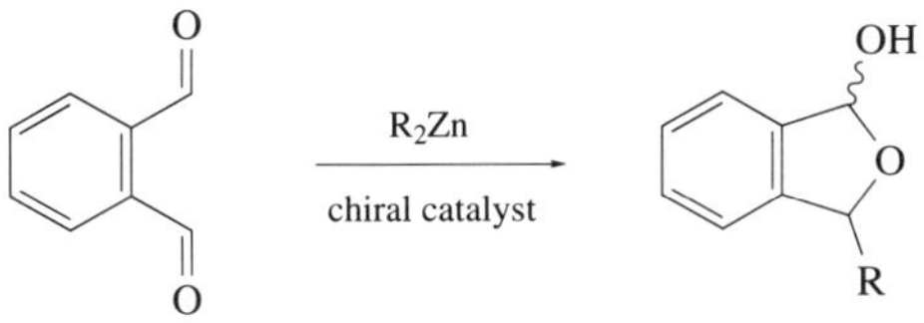

$$\text{(15)}$$

$$\text{R = Et, 98\%, Bu}^n\text{, 94\%}$$

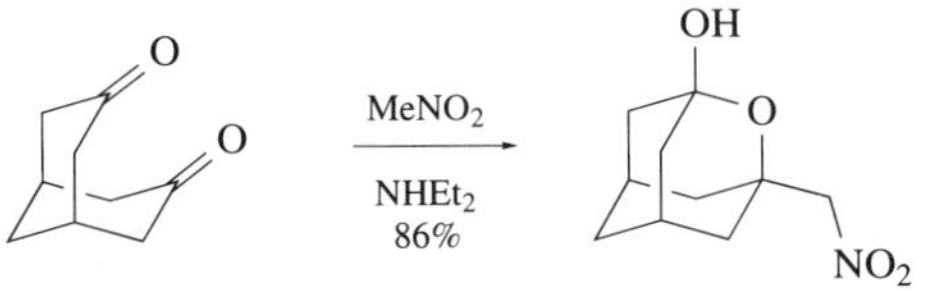

$$\text{(16)}$$

(25)

(iv) Miscellaneous methods for the generation of hydroxy carbonyl derivatives

When selectivity between alcohols in a dihydroxy compound or between carbonyl groups in a dicarbonyl system is a problem, then masked forms of these groups are required. Several examples of hemiacetal synthesis involve unmasking of a hydroxy or phenol group in the presence of a carbonyl group, with subsequent cyclization to hemiacetals. Phenols serve as excellent nucleophiles towards ketones, for example treatment of the aryloxy ketone **(26)** with zinc metal in acetic acid reveals the phenol, which cyclizes in quantitative yield (Scheme 9) ⟨75JOC1371⟩. The benzofuran **(27)** ⟨77JOC1045⟩ and 2-chromanol **(28)** ⟨77JA1631, 81JOC2260⟩ were similarly prepared in high yields by the unmasking of benzyl ethers of the corresponding phenols with hydrogen bromide and catalytic hydrogenation, respectively. Another example of the unmasking of a carbonyl group leading to a hemiacetal is shown in Equation (17) ⟨81JCS(P1)1015⟩.

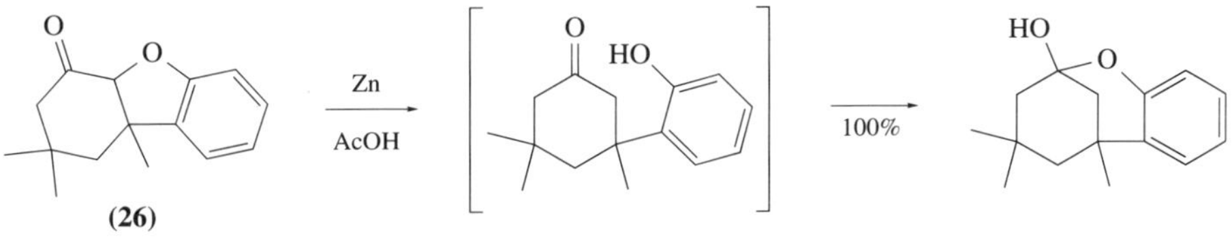

(26)

Scheme 9

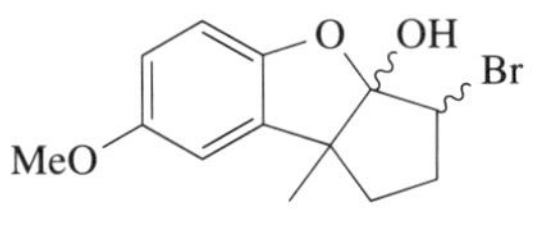

(27) 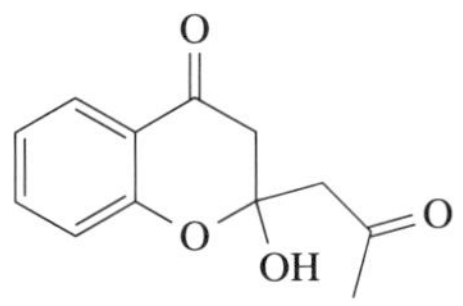   **(28)**

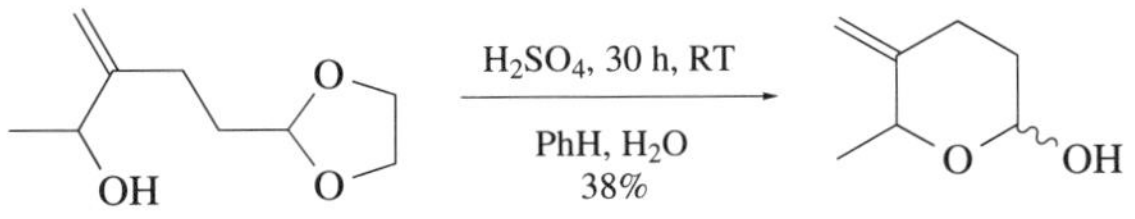

$$\text{(17)}$$

Alkenes serve as convenient precursors of either alcohols or carbonyl derivatives which can subsequently cyclize to hemiacetals. In the synthesis of mevalonolactone the terminal double bond in the diol **(29)** was initially ozonized and then reduced using polymer-supported triphenylphosphine (PTPP) to afford the hydroxy aldehyde, which cyclized *in situ* to the lactol (Scheme 10)

⟨87JCS(P1)2301⟩. Osmium tetroxide and sodium periodate have been used in combination to oxidatively cleave alkenes *en route* to hemiacetals ⟨89JOC2751⟩.

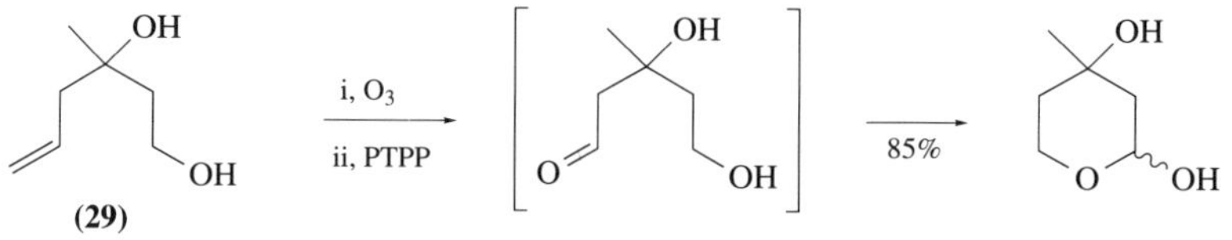

Scheme 10

Potassium permanganate is capable of both *vicinal* dihydroxylation of an alkene and selective oxidation of primary alcohols in the presence of secondary alcohols ⟨87S85⟩. Thus, treatment of the homoallylic alcohol (**30**) with potassium permanganate under carefully defined, carbon dioxide-buffered conditions initially gives the dihydroxy aldehyde (**31**), which spontaneously cyclizes to give a mixture of hemiacetals (Scheme 11), and further oxidation is precluded ⟨88JOC2979⟩. The osmylation of double bonds has also been employed for hemiacetal formation ⟨80JA6816⟩.

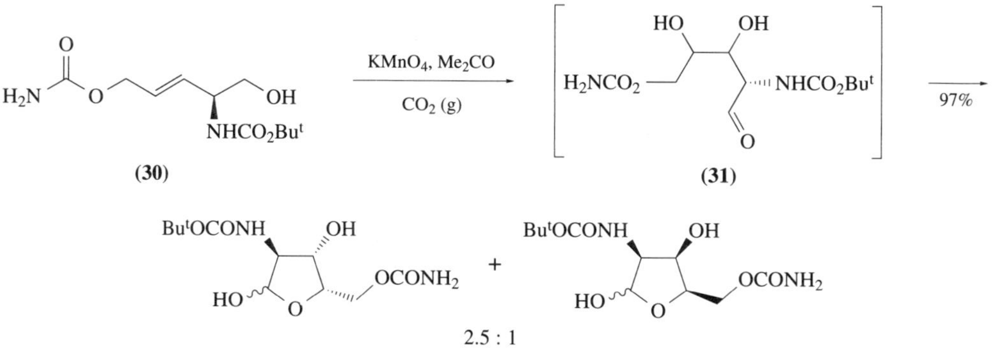

Scheme 11

Periodate degradation is commonly used for conversion of 1,2-diols into dicarbonyl compounds, which may subsequently cyclize to hemiacetals. For example, in an approach to the synthesis of elenolic acid, the 1,2-diol (**32**) was cleaved to the dialdehyde (**33**), which spontaneously cyclized via the enol (**34**) to give the hemiacetal (Scheme 12) ⟨73JA7156⟩. Further examples of hemiacetals prepared from 1,2-diols by periodate cleavage have been described ⟨68JCS(C)7, 83JA3661, 84TL3127, 86CC156⟩.

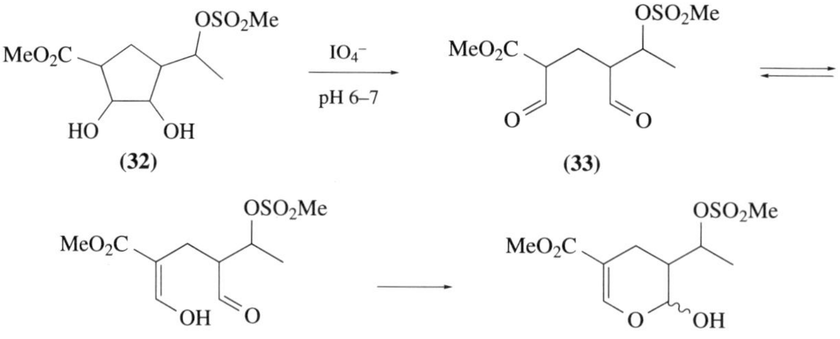

Scheme 12

2-Chromanols have been synthesized in a one-pot, addition–cyclization sequence from (2-hydroxyaryl)mercury chlorides and α,β-unsaturated ketones (Scheme 13). The yields are generally higher when R^1 is other than hydrogen, and R^2 is bonded to the carbonyl group through an sp^3-hybridized carbon atom ⟨82JOC2995⟩.

The Heck-type reaction of aryl halides and (Z)-2-buten-1,4-diol to provide β-aryl-γ-butyrolactols has been reported to be an excellent method for the synthesis of substituted tetrahydrofurans

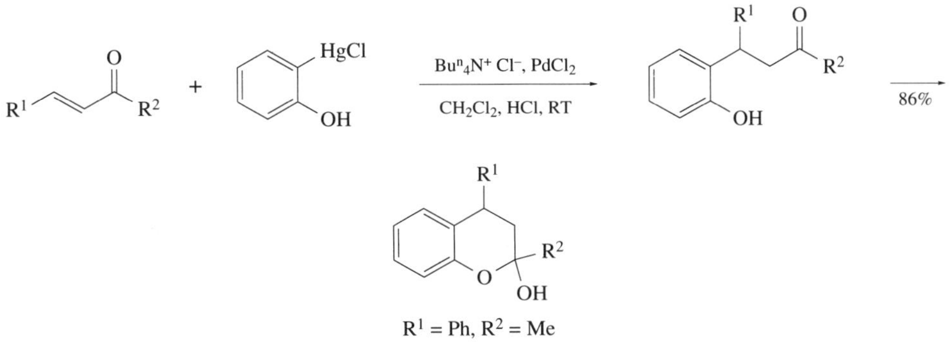

[Scheme 13]

(Equation (18)). The procedure is applicable to electron-donating as well as electron-withdrawing substituents on the aromatic nucleus, with aryl iodides giving higher yields than bromides ⟨91T1525⟩.

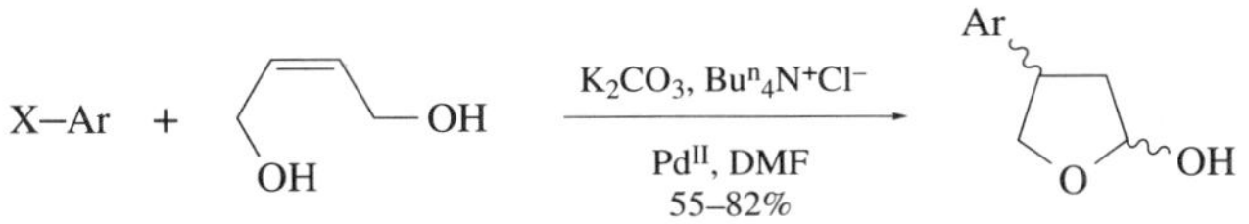

$$X-Ar + \text{(allyl diol)} \xrightarrow[\text{Pd}^{II}, \text{DMF} \atop 55-82\%]{K_2CO_3, Bu^n_4N^+Cl^-} \text{(furanol)} \quad (18)$$

X = Br, I

Many other methods for the generation of hydroxy carbonyl derivatives have been described, but they are, in general, compound-specific, for example fragmentation of the norbornane epoxide (**35**) with base gave the lactol (**36**) in quantitative yield via the hydroxy aldehyde intermediate (Scheme 14) ⟨84TL4455⟩.

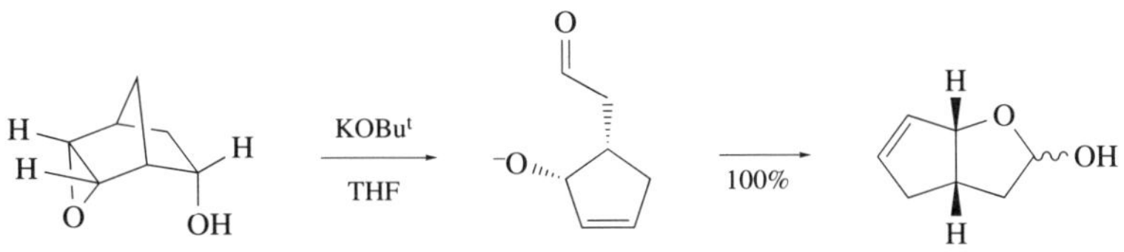

[Scheme 14: (35) → hydroxy aldehyde intermediate → (36), KOBu^t, THF, 100%]

4.04.2.2 Hemiacetals from Esters

The preparation of hemiacetals from esters requires either the reduction or addition of a nucleophile (Scheme 15). Rarely are hemiacetals prepared from acyclic esters, as such substrates require electron-withdrawing groups to stabilize the product. However, many lactones can be reduced to hemiacetals.

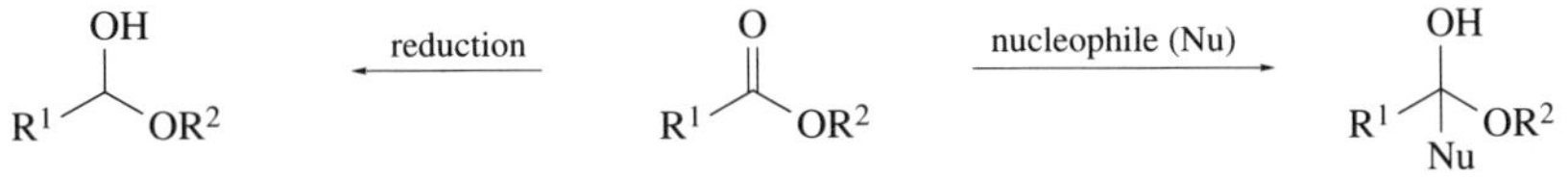

[Scheme 15]

4.04.2.2.1 By reduction

The most widely used reducing agent used for reduction is diisobutylaluminum hydride (dibaL-H) because of its selectivity for the ester group and the mild reduction conditions. As a result, dibal-H has been widely used in the formation of hemiacetals, particularly in natural product synthesis. The general reaction conditions involve treating a solution of the lactone in either tetrahydrofuran, toluene or dichloromethane with dibal-H at low temperature. Representative examples from the many in the literature are listed in Table 1. Several other functional groups are unaffected, including epoxide, acetal and nitro groups, and yields of hemiacetals are good.

Table 1 Hemiacetal formation by the reduction of esters with dibal-H.

Substrate	Product	Reduction Conditions with dibal-H	Yield (%)	Ref.
		Toluene, −70 °C, 2 h	77	83JOC5315
		CH_2Cl_2, −78 °C, 3 h	90	83JA3661
		THF, −30 °C	98 (X = F) 60 (X = Cl)	86JOC1704
		Toluene, −60 °C, 0.33 h	94	87S497
		THF, −50 °C, 10 h	67	89JOC5171

Other reagents have been employed for the reduction of lactones to hemiacetals, for example lithium tri-*t*-butoxyaluminum hydride (LITBAL) reduction of dihydrocoumarin ⟨62JA813⟩ (Equation (19)). The same reagent has also been used to reduce maleic anhydride derivatives to hemiacetals ⟨79JCS(P1)62⟩. Sodium bis(2-methoxyethoxy)aluminum hydride reduced the lactone (**37**) to the lactol (**38**) with the minor side product (**39**) of hemiacetal reduction also being isolated (Equation (20)) ⟨87LA607⟩. Dissolving-metal reductions have also been used for the conversion of lactones into hemiacetals ⟨87T4433⟩.

Reduction of acyclic esters to afford the corresponding hemiacetal requires an electron-withdrawing group to stabilize the hemiacetal. Thus, the ester (**40**) was reduced to the corresponding hemiacetal in moderate yield with sodium borohydride in GLYME at room temperature (Equation (21)) ⟨67JOC2595⟩.

4.04.2.2.2 By addition of carbanions

Addition of carbanions to lactones has been applied widely to generate hemiacetals. There are no common procedures, and reaction conditions have been adapted to suit the substrate and maximize yields. An intramolecular example of this reaction was the preparation of cyclopropanone ethyl hemiacetal by the reduction of ethyl 3-chloropropionate with metallic sodium in the presence of trimethylsilyl chloride followed by methanolysis (Scheme 16) ⟨85OS(63)147⟩. There are many examples of carbanion addition to lactones affording lactols, and Table 2 summarizes examples from the literature. Various nucleophiles have been employed, including organolithium reagents, Grignard reagents, enolates and α-heteroatom-substituted species. A common strategy for the synthesis of spiroacetals involves the addition of an organometallic reagent to a lactone to form a hemiacetal intermediate ⟨89CRV1617, 91CSR211⟩. For example, pretreatment of the functionalized lithium acetylide derived from (**41**) with boron trifluoride etherate in THF at low temperature followed by the addition of a lactone affords the hemiacetal in excellent yield (Equation (22)) ⟨83JOC4427⟩. Other examples of acetylide addition to lactones affording hemiacetals have also been reported ⟨82JOC615, 82JOC3140⟩.

4.04.2.3 Hemiacetals from Enol Ethers

Enol ethers have been converted into hemiacetals by a variety of different reagents, and the transformation has been applied widely in natural product synthesis, particularly in carbohydrate chemistry. Acid-catalysed hydration of enol ethers provides hemiacetals, generally in high yield. For example, treatment of (**42**) with camphorsulfonic acid (CSA) in aqueous THF gave the hemiacetal (Equation (23)) ⟨86TL947⟩. Hydroxymercuration of enol ethers followed by reductive demercuration is an alternative method for achieving the same transformation ⟨B-86MI 404-01⟩, and was used to convert the enol ether (**43**), containing acid-sensitive protecting groups, into the hemiacetal (Equation (24)) ⟨84JOC3994⟩. This procedure has also been applied in the synthesis of leukotriene C and D intermediates ⟨80TL3463⟩.

Table 2 Hemiacetal formation by the addition of carbanions to lactones.

Substrate	Nucleophilic substrate	Conditions	Product	Yield (%)	Ref.
(isobenzofuranone)	(isobenzofuranone)	KOBu^t, DMSO RT, 1 h		75	64JOC3070
(bicyclic lactone)	PhS–CH(OMe)	Bu^{n}Li, THF –30 °C		100 (R^1 = H) 88 (R^1 = Me)	75JA7182
δ-valerolactone	OLi / OLi dienediolate	THF –78 °C to 0 °C		95	81CC556
phthalic anhydride	2,4,5-trimethoxyphenyl-MgBr	THF, TMEDA –78 °C, 12 h		63	81LA2247
(tetra-O-benzyl gluconolactone)	BnO–CH$_2$CH$_2$–C≡CH	Bu^{n}Li, THF –78 °C, 1.5 h		98	83TL4833

Table 2 (continued)

Substrate	Nucleophilic substrate	Conditions	Product	Yield (%)	Ref.
	CH_2Br_2	Bu^nLi, c-$C_6H_{11}NH_2$ THF, −78 °C		68	84TL5009
	$R^2\!\!-\!\!P(OEt_2)$	THF, −78 °C		58–77 (n = 1, 2; R^1 = R^2 = H or alkyl)	85TL6329
	PhLi	THF, −78 °C, 2 h		86	89JOC610
		Bu^nLi, THF −95 °C, 5 min		87	91JCS(P1)897

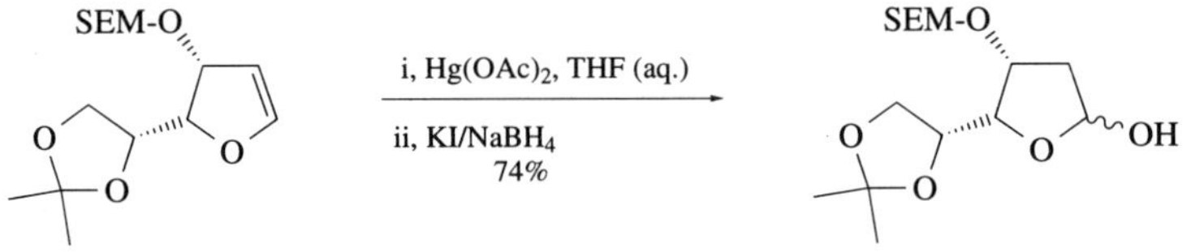

(24)

SEM = β-trimethylsilylethoxymethyl

Hydration of enol ethers in the presence of electrophiles affords functionalized hemiacetals. The benzopyranone (**44**) undergoes bromohydrin formation (Equation (25)) with *N*-bromosuccinimide in wet dimethyl sulfoxide ⟨75JHC981⟩. Perphthalic acid in moist ether converts the tetrahydrochroman (**45**) into the glycol (Equation (26)) ⟨66JOC3032⟩. Azidonitration of glycals promoted by CAN has been reported to produce 2-azido hemiacetals ⟨87TL1981⟩.

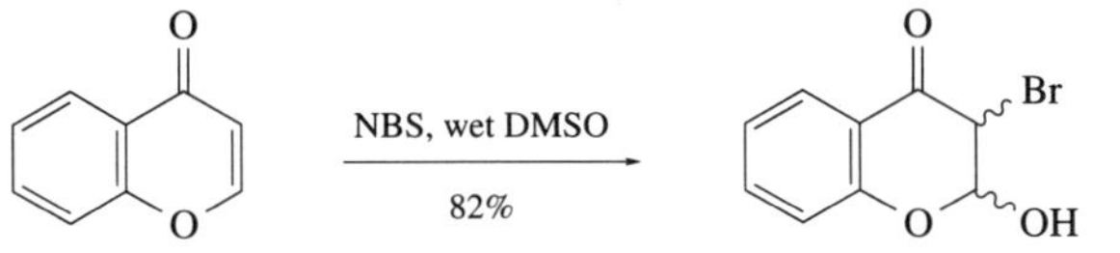

(25)

(44)

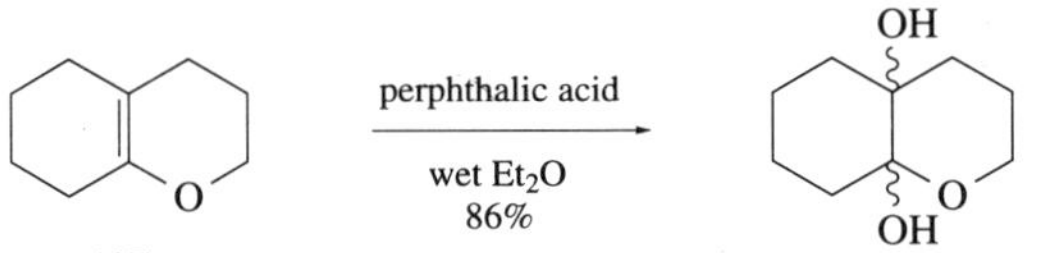

(26)

(45)

Osmylation of cyclic enol ethers also provides hemiacetals with an adjacent hydroxy group (Equation (27)) ⟨91TL2565⟩ (for further examples see ⟨86JOC902⟩ and ⟨87JOC622⟩). *Vicinal* dihydroxylation of enol ethers has also been carried out with molybdenum oxide–hydrogen peroxide ⟨82JA358⟩.

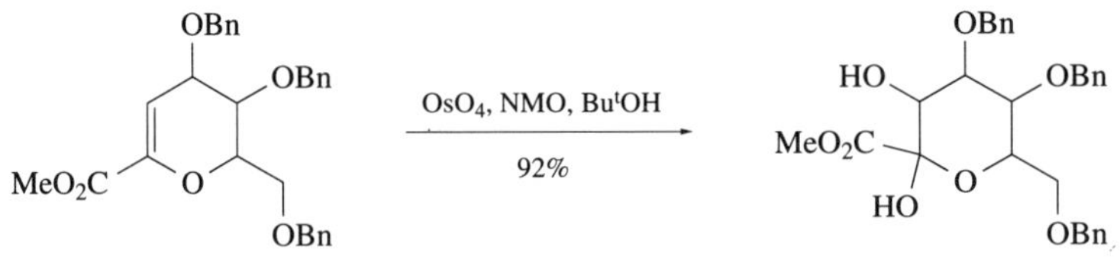

(27)

A recent procedure which promises to be quite general for the formation of six-membered ring hemiacetals, involves the *in situ* preparation of enol ethers such as (**46**) from cyclic acetals with an excess of triisobutylaluminum at low temperature (Scheme 17). The intermediate (**46**) undergoes cyclization to the hemiacetal (**47**) in high yield (>95%) when exposed to triflic anhydride and diisopropylethylamine. Several examples were reported, and high yields were obtained starting from both substituted and unsubstituted cyclic acetals ⟨85JOC5444, 90JOC5814⟩.

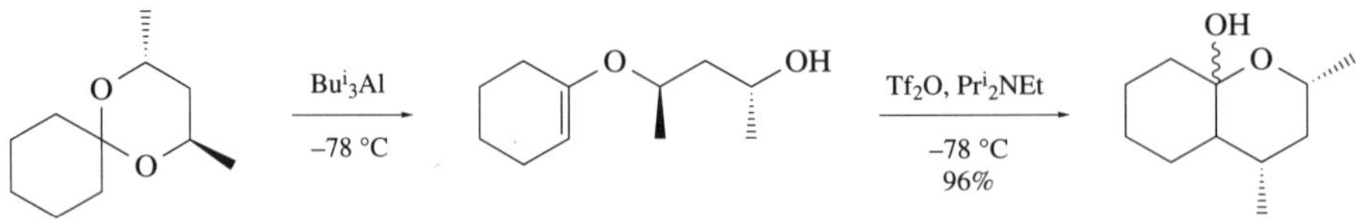

(46) **(47)**

Scheme 17

Lead tetraacetate has been reported to convert cyclic enol ethers into allylic hemiacetals in moderate yield ⟨80T1763⟩. Other methods for the conversion of enol ethers into hemiacetals have been described but have been aimed at specific targets or are useful for only certain types of substrates (see, for example, ⟨84TL4797⟩ and ⟨91TL3313⟩).

4.04.2.4 Hemiacetals from Acetals by Deprotection

4.04.2.4.1 From noncarbohydrate substrates

Hemiacetals can be prepared by the deprotection of their corresponding acetals, and this is a widely used reaction in natural product synthesis. Most methods involve deprotection of acetals under aqueous acidic conditions, and such reactions proceed via an intermediate carbonium ion with the hydroxy group of the hemiacetal arising from water in the reaction mixture. Alternatively, protecting groups have been developed which allow deprotection under nonacidic conditions, and in these cases the hydroxy group is liberated directly in the deprotection step. Acidic deprotection simply involves stirring the substrate in an aqueous acidic solution to afford the hemiacetal, as illustrated in Equation (28) ⟨82JOC946, 94TL1825⟩.

Several protecting groups and variations in reaction conditions which allow deprotection under nonacidic conditions have been developed (see also the following section on carbohydrate derivatives). Cyclic hemiacetals protected as methyl ethers can be selectively deprotected using boron trihalides ⟨87TL5595⟩. An example is shown in Equation (29) for the synthesis of 2-hydroxy-2H-benzoxazin-3-ones where the precursor acetals were found to be resistant to preparative acid hydrolysis. Yields are moderate to good, and electron-withdrawing groups in the aromatic ring slow down or, in some cases (e.g., R = CN, CF$_3$), prevent the reaction ⟨91JOC1788⟩.

The chloroethyl group in the furan derivative (**48**) was used because of the sensitivity of this substrate towards acid hydrolysis conditions. Removal of this protecting group with sodium in tetrahydrofuran–liquid ammonia, however, is extremely efficient and provides the hemiacetal without affecting the double bond (Equation (30)) ⟨83JA3720⟩. Unsymmetrical acetals can sometimes be cleaved selectively to afford hemiacetals. In a synthesis of prostaglandin analogues, treatment of the acetal intermediate (**49**) with acetyl chloride and titanium tetrachloride gave the hemiacetal in high yield (Equation (31)). The same reaction using classical acidic hydrolysis (acetic acid) gave only a moderate yield (38%) of the corresponding hemiacetal ⟨82JOC824⟩.

4.04.2.4.2 From carbohydrate substrates

Protection and deprotection of the anomeric hydroxy group plays a significant role in the manipulation of these important organic molecules. Several of the protecting groups described in the previous section are useful. Glycoside bonds can be cleaved under acidic conditions to afford sugar hemiacetals in good yield as exemplified in Equation (32) ⟨85TL2065⟩. Similarly, anomeric *t*-butyl ethers can be cleaved under relatively mild acidic conditions which do not affect most hydroxy protecting groups utilized in oligosaccharide synthesis ⟨88CAR(181)246⟩. Alkyl glycosides bearing adjacent electron-withdrawing groups or atoms such as fluorine are more difficult to hydrolyse, and require strong mineral acids such as hydrochloric acid to form the sugar hemiacetal (see, for example, ⟨88CJC187⟩. 1-*O*-Acylated sugar derivatives have been deprotected to hemiacetals under a variety of basic as well as some acidic conditions ⟨85CAR(144)342, 87CAR(162)145, 87TL3569, 89JCR(S)152⟩.

$$\underset{EtO}{\overset{OH}{\bigcirc}}\overset{OH}{\underset{O}{\bigcirc}} \xrightarrow[83\%]{HCO_2H} \underset{HO}{\overset{OH}{\bigcirc}}\overset{OH}{\underset{O}{\bigcirc}} \qquad (32)$$

The allyl group has been employed widely for the protection of the anomeric hydroxy group. Deprotection to the hemiacetal generally involves initial isomerization of the allyl group to the more labile 1-propenyl group using potassium *t*-butoxide ⟨77JCS(P1)2513⟩ or transition metal catalysts ⟨91TL7369⟩. Subsequent cleavage of the propenyl glycoside to the sugar hemiacetal can be carried out in good yield under mild acidic conditions ⟨80MI 404-01⟩, under neutral conditions in the presence of mercury(II) salts ⟨82CAR(102)99⟩, with iodine in the presence of 1,5-diazabicyclo[5.4.0]undec-5-ene ⟨82CC1274⟩ or by catalytic osmylation ⟨91TL7369⟩. The *n*-pentenyl glycoside protecting group allows chemospecific liberation of the anomeric hydroxy group under nonacidic conditions (NBS, aqueous acetonitrile) ⟨88JA2662⟩.

The silylethoxymethyl (SEM) group is useful for protecting the anomeric hydroxy group and can be removed in high yield under mild conditions with lithium tetrafluoroborate ⟨81TL4603⟩. Boron trifluoride etherate has also been used to deprotect SEM derivatives of carbohydrates in good yield ⟨86TL753⟩. Phenyldimethylsilylmethanol has recently been introduced for the protection of anomeric hydroxy groups, and deprotection to the hemiacetal sugar is achieved in high yield by peracetic acid and potassium bromide in acetic acid–sodium acetate ⟨90TL2197⟩.

4.04.3 ACETALS—$R^1_2C(OR^2)_2$

Several general reviews covering the synthesis and chemistry of acetals have appeared ⟨B-67MI 404-01, B-70MI 404-01, B-80MI 404-02, 81S501, B-89MI 404-01⟩, and the synthesis of acetals of all types has been covered extensively in a monograph ⟨91HOU(E14a/1)1⟩. Reviews on the preparation and use of acetals as protecting groups ⟨B-91MI 404-01, B-91MI 404-02, 91COS(6)631⟩, chiral reagents ⟨88PAC49, 90TA477⟩ and versatile intermediates in synthesis ⟨87S1043, 92SL97⟩ have also appeared. The synthesis of certain specific types of acetals will not be covered here unless methods have more general applicability. The reader should therefore consult reviews for more details on the preparation of spiroacetals ⟨89CRV1617, 91CSR211, 91CSR271⟩ and glycoside coupling methodology ⟨91COS(6)33, 93CRV1503⟩.

The following survey is divided into three sections: general methods (suitable for symmetrical and unsymmetrical acetals), methods for symmetrical acetals (**50**) ($R^1 = R^2$), and those for unsymmetrical acetals (**50**) ($R^1 \neq R^2$). The situation is confused by the large array of cyclic acetals described in the literature. All of the methods described in the general section are suitable for the preparation of symmetrical acetals; however, some of these methods are restricted to the preparation of cyclic unsymmetrical acetals from an unsymmetrical diol, and do not allow the use of two different monohydric alcohols. Such restrictions will be obvious or will be indicated at appropriate points in the discussion. Methods covered in the sections on symmetrical and unsymmetrical acetals represent the general situation found in the literature, but an appropriate choice of reaction components may result in the crossing of boundaries between sections. The reader should therefore consult all sections to avoid overlooking a potentially useful method. Intramolecular variants of many of the methods are also feasible.

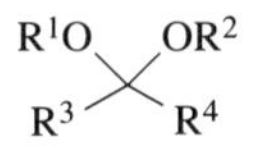

$$\begin{array}{c} R^1O \quad OR^2 \\ \diagdown\!\!\diagup \\ R^3 \quad R^4 \end{array}$$
(50)

4.04.3.1 General Methods

4.04.3.1.1 From aldehydes and ketones

(i) With alcohols and protic or Lewis acid catalysts

Aldehydes and ketones react with alcohols under acidic conditions to form acetals and water in an equilibrium process which proceeds via an oxonium ion (**51**), a common intermediate in many acid-catalysed acetal syntheses (Scheme 18) ⟨81S501⟩. The following order of carbonyl reactivity is generally observed: aliphatic aldehydes > aromatic aldehydes > acyclic ketones and cyclohexanones > cyclopentanones > α,β-unsaturated ketones, and α,α-disubstituted ketones ≫ aromatic ketones. Thus, chemoselective reactions in polycarbonyl systems may be possible. Many monohydric alcohols, including functionalized alcohols, may be used, and cyclic acetals (formed by the addition of 1,2- or 1,3-diols, such as ethylene glycol) are generally formed more easily than acyclic acetals. Steric hindrance in the alcohol slows down the rate of acetal formation. In some cases, particularly for saturated aliphatic aldehydes and primary alcohols, the equilibrium conversion to the acetal may be good. However, in general it is necessary to shift the equilibrium in favour of the product by removing the water by-product, either by physical or chemical methods.

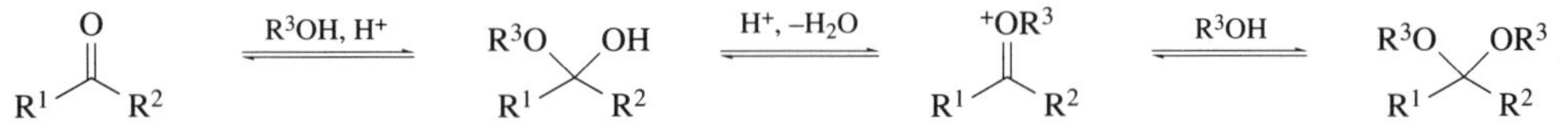

Scheme 18

The most commonly employed method for acetal formation involves heating an aldehyde or ketone with an alcohol or diol in an inert solvent such as benzene, toluene or xylene, which allows removal of water by continuous azeotropic distillation with a Dean–Stark or similar water separatory head ⟨38CB1803⟩. Some examples are shown in Table 3, and numerous others have been reported.

Alternatively, water may be removed with dehydrating reagents such as molecular sieves ⟨71RTC1141, 72S419, 77RTC44⟩, calcium sulfate ⟨74JOC2815⟩, copper sulfate ⟨78JOC438⟩ and alumina ⟨79CB3603⟩. Although this approach often allows acetalization to be carried out at room temperature or below, it appears to give high yields with only more reactive aldehydes or ketones.

Where there is a very unfavourable equilibrium for acetal formation, it is necessary to remove the water completely by a reaction with a suitable reagent. Ortho esters are the most widely used reagents for this purpose, and react with the water to form an ester and an alcohol (Equation (33)) ⟨55JOC1695, B-70MI 404-02⟩. This is a general procedure which is particularly suitable for the preparation of ketone acetals and because of the mild conditions it is frequently employed as an alternative to the azeotropic removal of water. Some examples are illustrated in Table 4. The mechanism of this reaction has been studied ⟨55JOC1695, 69CC1175⟩. The orthoformate is usually chosen to match the alcohol, but with higher-boiling alcohols trimethyl or triethyl orthoformate can be used, and the ethanol or methanol and alkyl formate distilled out of the reaction mixture to displace the equilibrium (Table 4, entries 4 and 7). This may not be necessary for cyclic acetals if an excess of diol is used.

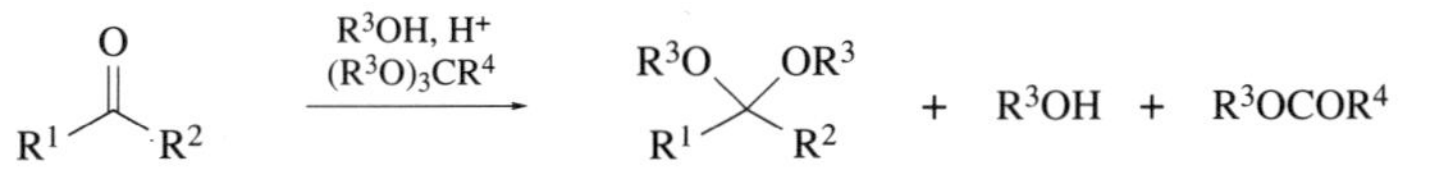

$$R^1COR^2 \xrightarrow[(R^3O)_3CR^4]{R^3OH,\ H^+} R^1R^2C(OR^3)_2 + R^3OH + R^3OCOR^4 \quad (33)$$

Dialkyl sulfites displace the equilibrium by reacting with water to form the alcohol and sulfur dioxide (Equation (34)) ⟨71JCS(C)1213⟩. Cyclic sulfites have been employed in cyclic acetal formation ⟨60CB1249⟩. Water can be removed by reaction with another reactive acetal, usually

Table 3 Acetals prepared by the azeotropic removal of water.

Entry	Substrate	Conditions	Acetal	Yield (%)	Ref.
1	PhC(O)CH₃	TsOH, benzene, 3.5 h; HOCH₂CH₂OH	2-methyl-2-phenyl-1,3-dioxolane	85	48JA2827
2	n-C$_{15}$H$_{31}$CHO	MeOH, sulfosalicylic acid, xylene, 3 h	n-C$_{15}$H$_{31}$CH(OMe)$_2$	88	58JA6613
3	acetone	TsOH, 35–55 °C, petroleum ether, 21–36 h; HOCH$_2$CH(OH)CH$_2$OH	2,2-dimethyl-4-(hydroxymethyl)-1,3-dioxolane	90	55OSC(3)502
4	CH$_3$C(O)CH$_2$CO$_2$Et	TsOH, benzene; HOCH$_2$CH$_2$OH	2-methyl-2-(ethoxycarbonylmethyl)-1,3-dioxolane	87	38CB1803
5	3-(ethoxycarbonylmethyl)cyclopentanone	ppts, benzene, 1 h; HOCH$_2$CH$_2$OH	corresponding 1,3-dioxolane	95	79S724
6	(hydroxymethyl)cyclobutyl methyl ketone	ppts, benzene, 1 h; HOCH$_2$CH$_2$OH	corresponding 1,3-dioxolane	91	79S724

ppts = pyridinium p-toluenesulfonate.

2,2-di-methoxypropane to produce acetone and methanol, which is often removed by distillation to displace the equilibrium if a different alcohol is being employed for the acetalization (Equation (35)), ⟨60JOC521⟩. Other less widely used reagents for water removal are tetraalkoxysilanes (RO)$_4$Si ⟨24CB795, 70JOC3375⟩ and ethylene carbonate ⟨60CB1249⟩, and in some cases the acid catalyst may serve this purpose, for example sulfuric acid ⟨66JOC853⟩ and selenium dioxide ⟨54JA6113⟩.

$$\text{3-Cl-C}_6\text{H}_4\text{-O-CH}_2\text{-C(O)-CH}_3 \xrightarrow[\text{HCl, reflux, }-\text{SO}_2]{\text{ROH, (RO)}_2\text{SO}} \text{3-Cl-C}_6\text{H}_4\text{-O-CH}_2\text{-C(OR)}_2\text{-CH}_3 \quad (34)$$

R = Et, 67%
R = Pr^i, 9%

$$\text{PhC(O)CH}_3 + \text{Pr}^n\text{OH} \xrightarrow[\text{TsOH, }n\text{-hexane}]{\text{MeO OMe; }63\%} \text{PhC(OPr}^n\text{)}_2\text{CH}_3 \quad (35)$$

A wide range of protic acids, Lewis acids and heterogeneous catalysts have been employed for the acetalization of aldehydes and ketones, and these are displayed in Table 5. Some of these are exemplified in Tables 3 and 4 and more references are available in a monograph ⟨91HOU(E14a/1)136⟩. The choice of catalyst is governed by the reactivity of the carbonyl group, the thermal and chemical stability of the substrate and alcohol, and the reaction conditions. Aldehydes in general can be acetalized in the presence of weaker acids such as ammonium chloride, ammonium nitrate, calcium chloride, alumina and lanthanide halides, although for convenience other acids are often used (see

Table 4 Acetal formation using ortho esters as dehydrating agents.

Entry	Substrate	Conditions	Acetal	Yield (%)	Ref.
1	3,4,5-trimethoxybenzaldehyde	MeOH, (MeO)$_3$CH, NH$_4$Cl, reflux, 3 h	dimethyl acetal	83	90S313
2	CH$_2$=CHCHO	EtOH, (EtO)$_3$CH, NH$_4$NO$_3$, RT, 6–8 h	CH$_2$=CHCH(OEt)$_2$	72–80	63OSC(4)21
3	2-(hydroxymethylene)-4-ethoxycyclohex-3-enone	MeOH, (MeO)$_3$CH, TsOH, reflux, 2 h	2-(dimethoxymethyl)-4-ethoxycyclohex-3-enone	—	77SC409
4	pentan-3-one	Bu^{n}OH, (EtO)$_3$CH, TsOH	3,3-bis(butoxy)pentane	84	55JOC1695
5	3-oxo-N,N-dimethylcyclobutanecarboxamide	EtOH, (EtO)$_3$CH, HCl, RT, overnight	3,3-diethoxy-N,N-dimethylcyclobutanecarboxamide	98	58JA5837
6	dienynone	trimethyl orthoformate / 1,3-propanediol, OMe, TsOH, THF, RT, 48 h	1,3-dioxane acetal	84	80JOC4283
7	tetralone CO$_2$Et ester	HOCH$_2$CH$_2$OH, (MeO)$_3$CH, TsOH, 40–50 °C, 4 h	1,3-dioxolane acetal	87	79JA2171

Tables 3 and 4). Ketones require stronger acids and mineral acids such as HCl or H$_2$SO$_4$, or sulfonic acids such as toluene-p-sulfonic acid, are used most often. Some catalysts display potentially useful characteristics and others have been developed to meet specific needs. Pyridinium p-toluenesulfonate (ppts) is a particularly mild catalyst which has been used on acid-sensitive substrates and is often employed when toluene-p-sulfonic acid is unsuitable (see Table 3, entry 6), ⟨79S724⟩. In general, saturated carbonyl groups are more reactive than α,β-unsaturated carbonyl groups, but this reactivity can be reversed when the bulky collidinium toluene-p-sulfonate is employed as catalyst, although isomerization of the double bond usually accompanies acetalization ⟨84TL3047⟩. The extent of double bond isomerization during acetalization of enones is dependent on the pK_a of the acid catalyst, and acetalization of cyclic enones can be carried out without double bond migration with fumaric acid (Equation (36), ⟨73RTC1047⟩). Adipic acid has been used for the same purpose with steroidal enones ⟨64JA2183⟩. Diaryl ketones are particularly difficult to acetalize under standard conditions, but triflic acid in nitromethane appears to overcome this long-standing problem (Equation (37), ⟨88S233⟩). Sulfuric acid has been suggested as being suitable for acetal formation from aldehydes or ketones and/or alcohols containing electron-withdrawing groups ⟨66JOC853⟩.

Table 5 Catalysts for the acetalization of aldehydes and ketones with alcohols.

Protic acids

(i) Inorganic acids: HCl, HBr, HI, HClO$_4$, H$_2$SO$_4$, HNO$_3$, H$_3$PO$_4$

(ii) Carboxylic acids: MeCO$_2$H, ClCH$_2$CO$_2$H, Cl$_3$CCO$_2$H, F$_3$CCO$_2$H, (CO$_2$H)$_2$, HO$_2$C(CH$_2$)$_6$CO$_2$H, fumaric acid

(iii) Sulfonic acids: benzenesulfonic acid, toluene-*p*-sulfonic acid, sulfosalicylic acid, camphorsulfonic acid

(iv) Pyridine salts: pyridinium chloride, pyridinium toluene-*p*-sulfonate (PPTS), 2,4,6-collidinium toluene-*p*-sulfonate

(v) Ammonium salts: NH$_4$Cl, NH$_4$NO$_3$

(vi) Others: electrochemically generated acid, PhSO$_2$NHOH

Lewis acids

(i) Metal salts: AlCl$_3$, BF$_3$·OEt$_2$, CaCl$_2$, CuSO$_4$, FeCl$_3$, SnCl$_4$, ZnCl$_2$, ZrO$_2$, LnCl$_3$ (Ln = Ce, Er, Yb, Nd), SeO$_2$

(ii) Silyl halides: TMS-Cl, TMS-I, Me$_2$SiCl$_2$

(iii) Tin reagents: R$_n$SnCl$_{4-n}$, [tin cluster structure] ⟨92T1449⟩

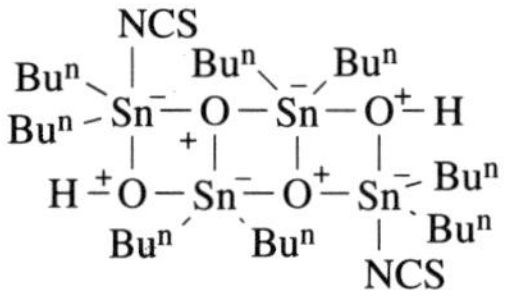

(iv) Transition metal complexes: Rh^{III} complexes, Pd^{II} complexes

(v) Others: I$_2$, [pyridinium SbF$_6^-$ structure] ⟨91S368⟩

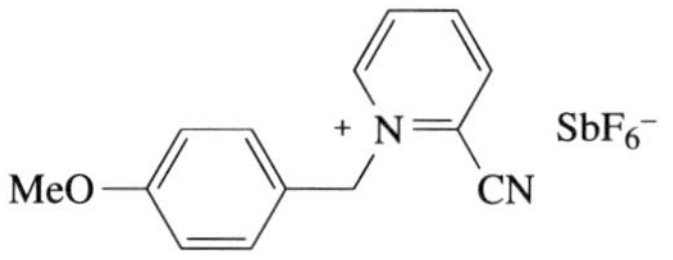

Heterogeneous catalysts

Nafion-H, Amberlyst 15, Dowex 50, polyvinylpyridinium hydrochloride, montmorillonite K10, graphite bisulfate, alumina, acidic zeolites, polystyryldiphenylphosphine–iodine complex, aminopropylated silica gel·hydrochloride (—Si—O—Si—(CH$_2$)$_3$NH$_3^+$ Cl$^-$)

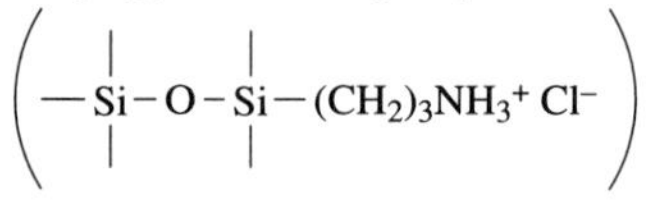

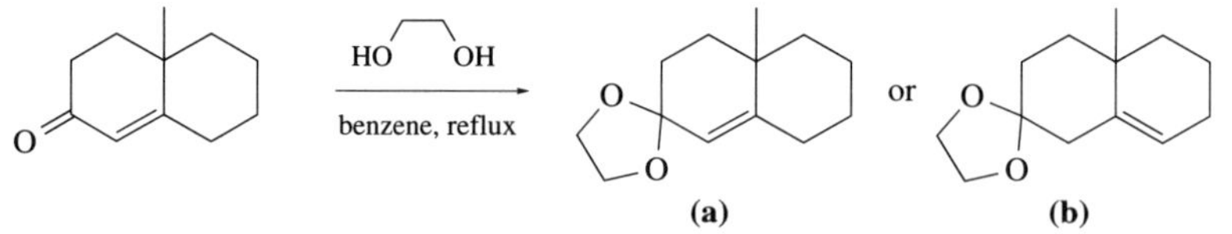

Catalyst	pK$_a$	Yield (**a**) (%)	Yield (**b**) (%)	Conversion (%)
p-toluenesulfonic acid	~1.00	0	100	100
fumaric acid	3.03	100	0	90

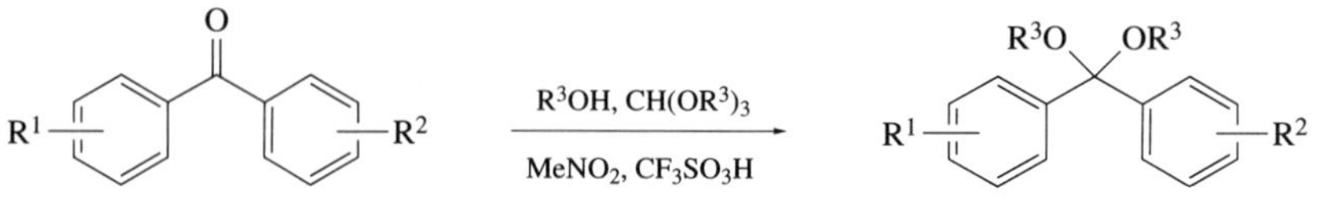

(36)

(37)

Both 1,2- and 1,3-diols in natural-product synthesis and in carbohydrate chemistry are frequently protected as acetals by condensation with simple aldehydes, such as benzaldehyde, or ketones, such as acetone (for reviews see ⟨79CRV491, 91COS(6)631⟩ and ⟨B-91MI 404-01⟩).

(ii) With silyl ethers

The reaction of silyl ethers with carbonyl compounds, catalysed by trimethylsilyl triflate, is a particularly useful method for acetal synthesis, which occurs under mild aprotic conditions (Equation (38)) ⟨80TL1357⟩. Formation of stable hexamethylsiloxane (**52**) drives the equilibrium towards the acetal. Some examples are shown in Table 6; several features are worthy of note—the reaction occurs under mild conditions ($-78\,°C$), double bond isomerization in enones does not occur (entries 3 and 4) and acid-sensitive functionalities may be tolerated (entry 5). The bulkiness of the trimethylsilyl group has been exploited to allow chemoselective acetalization of dicarbonyl systems at the less hindered carbonyl group (entry 4), and the mildness of the reaction conditions allows facile acetalization of α,β-unsaturated aldehydes (entry 5). Functionalized diols which decompose with acid catalysts have been used to prepare cyclic acetal protecting groups from their silyl ethers (entries 7 and 8). Trimethylsilyl iodide ⟨84JOC2808⟩ (Table 2, entry 7) and electrochemically generated acid ⟨83CL1349⟩ have also been used as catalysts with alkoxysilanes. High pressure may be advantageous for the acetalization of very hindered substrates ⟨86JOC4964⟩.

$$R^1\text{COR}^2 + 2\,R^3\text{O-TMS} \xrightarrow[\text{CH}_2\text{Cl}_2,\,-78\,°C]{\text{TMS-OTf (cat.)}} R^1R^2\text{C(OR}^3)_2 + \text{TMS-O-TMS} \quad (38)$$
$$(52)$$

(iii) With epoxides

Hydrolysis of an epoxide under acidic conditions produces a diol which can react with aldehydes and ketones to form cyclic acetals as described in Section 4.04.3.1.1.i ⟨73JOC834⟩. Alternatively, cyclic acetals (1,3-dioxolanes) may be formed by direct condensation of epoxides with aldehydes and ketones in the presence of Lewis acids (Equation (39)) ⟨84MI 404-01⟩. As no water is formed, the reaction conditions are often milder than those for the diol. Early reactions using stannic chloride as catalyst with simple epoxides gave low yields of acetals due to side reactions ⟨33JA3741⟩, although better yields were obtained when epihalohydrins were used as the epoxide components ⟨41CB145, 81LA1105⟩. Stannic chloride is still a commonly employed catalyst for this reaction, and others include aluminum chloride ⟨88S854⟩, copper sulfate ⟨78JOC438⟩ and montmorillonite K10 clay ⟨80BSB759⟩. Tetraethylammonium bromide has been used as catalyst to prepare simple acetals, but the reaction required an autoclave ⟨67LA85⟩. A systematic study of boron trifluoride etherate as catalyst, with a view to optimizing the acetalization conditions, indicated that best yields were obtained with cyclic ketones, catalytic $BF_3\cdot OEt_2$, and simple epoxides such as ethylene oxide or propylene oxide (Equation (40)) ⟨93JOC7274⟩. The air and moisture stable pyridinium salt (**53**) promotes acetalization with epoxides in high yield and is the only reported catalyst that gives a useful yield of an acetal from an aromatic ketone ⟨90CL2019⟩.

$$\text{epoxide}(R^1,R^2) + R^3\text{COR}^4 \xrightarrow{\text{Lewis acid}} \text{1,3-dioxolane} \quad (39)$$

$$\text{cyclopentanone} \xrightarrow[\text{BF}_3\cdot\text{OEt}_2,\,\text{CH}_2\text{Cl}_2,\,\text{RT, 2 min}]{\text{ethylene oxide (10 equiv.)}} \text{spiro-dioxolane} \quad 79\% \quad (40)$$

(**53**) — 4-methoxybenzyl-(2-cyanopyridinium) SbF_6^-

Table 6 Acetal formation with silyl ethers.

Entry	Substrate	Silyl ether[a]	Product	Yield (%)	Ref.
1	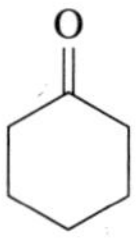 cyclohexanone	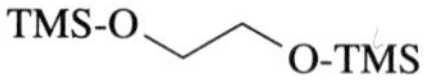 TMS-O~~O-TMS	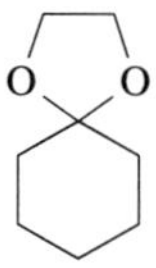 cyclohexanone 1,3-dioxolane	96	80TL1357
2		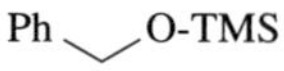 Ph-CH$_2$-O-TMS	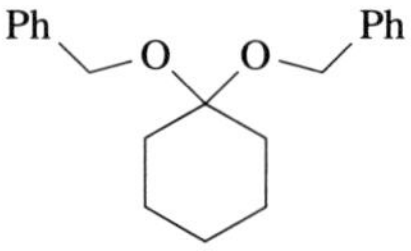 cyclohexanone bis(benzyloxy) acetal	99	
3	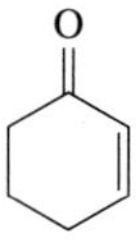 cyclohex-2-enone	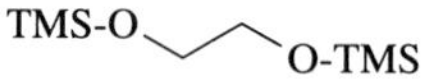 TMS-O~~O-TMS	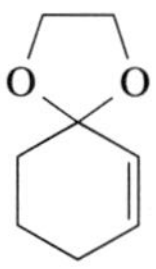 cyclohex-2-enone 1,3-dioxolane	92	
4	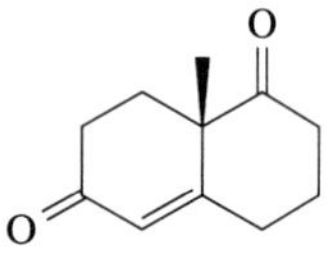 Wieland–Miescher-type dione	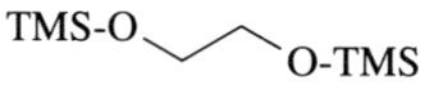 TMS-O~~O-TMS	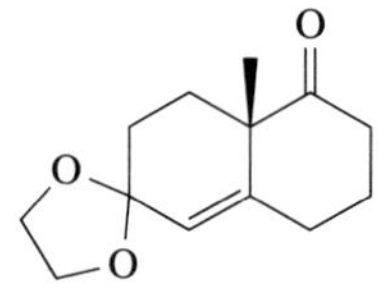 mono-dioxolane product	65	85JOC3946
5	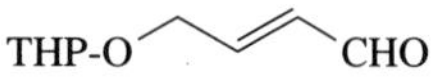 THP-O-CH=CH-CHO	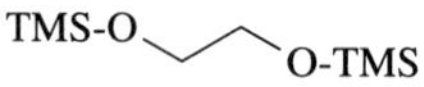 TMS-O~~O-TMS	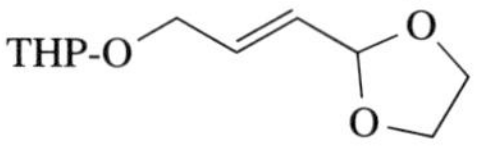 THP-O-CH=CH-CH(OCH$_2$CH$_2$O)	78	87JOC188
6	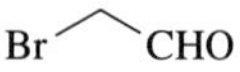 Br-CH$_2$-CHO	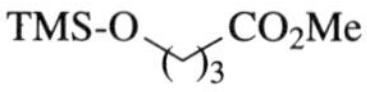 TMS-O-(CH$_2$)$_3$-CO$_2$Me	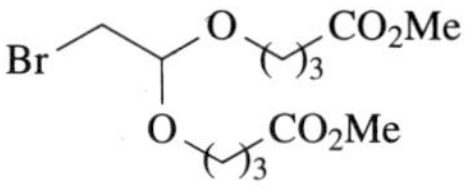 Br-CH$_2$-CH(O(CH$_2$)$_3$CO$_2$Me)$_2$	96	88LA559
7[b]	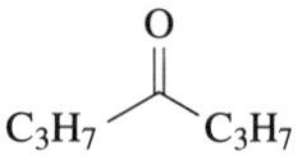 C$_3$H$_7$-CO-C$_3$H$_7$	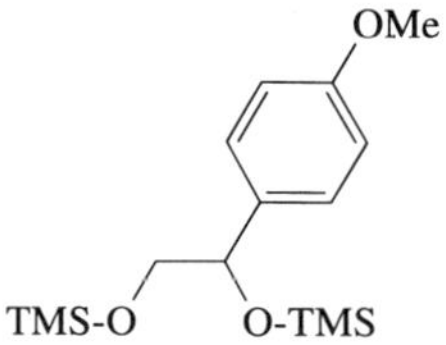 4-MeO-C$_6$H$_4$-CH(OTMS)-CH$_2$-OTMS	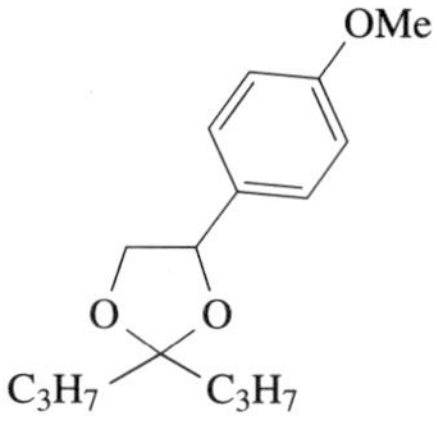 dioxolane with 4-MeOC$_6$H$_4$ and two C$_3$H$_7$ groups	76	94TL57
8	PhCHO	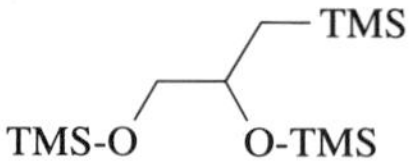 TMS-O-CH$_2$-CH(CH$_2$TMS)-O-TMS	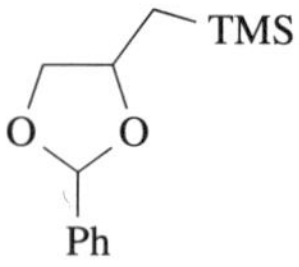 dioxolane with CH$_2$TMS and Ph	100	94TL969

[a] Reactions were carried out in CH$_2$Cl$_2$ at −78 °C with TMS-OSO$_2$CF$_3$ as the catalyst. [b] TMS-I was used as the catalyst.

The mechanism of acetal formation has been studied with the epoxides of *cis*- and *trans*-but-2-ene and involves attack on the epoxide by the carbonyl oxygen with inversion of configuration, followed by bond rotation and ring closure (Scheme 19) ⟨70T1311, 74AJC679⟩. The utility of this procedure in stereoselective synthesis is exemplified in Equation (41), where, because the epoxide is valuable, the carbonyl component (acetone) was used as the solvent ⟨88S854⟩.

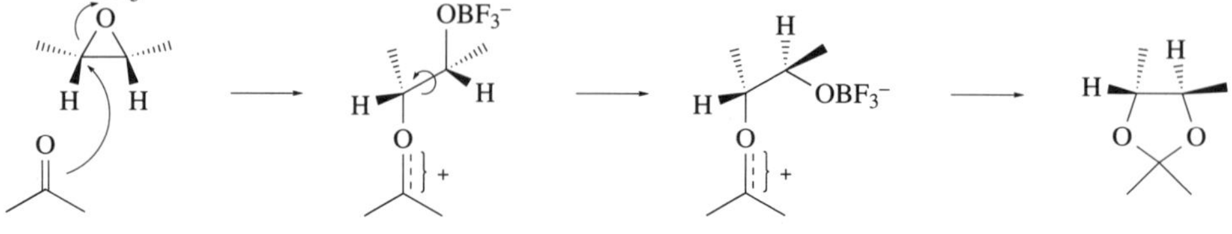

Scheme 19

$$\text{[vinyl epoxide]-OTs} \xrightarrow[\text{RT, 24 h} \atop 94\%]{\text{acetone, AlCl}_3} \text{[acetonide]-OTs} \quad (41)$$

Acetals have also been formed from epoxides in the presence of the one-electron oxidant tetracyanoethylene ⟨93CL17⟩ and from a vinyl epoxide and benzaldehyde with palladium(0) catalysis ⟨86TL69⟩, but the scope of these processes has not been explored. Certain activated epoxides, for example (**54**), react with carbonyl compounds under thermal or photochemical conditions, via the 1,3-dipole arising from C—C bond cleavage (Equation (42)) ⟨71TL231, 72TL5133, 74JOC3145⟩.

$$\underset{(\mathbf{54})}{\text{Ph-epoxide-CO}_2\text{Et}} \xrightarrow[75\%]{\text{HCHO} \atop h\nu} \text{[1,3-dioxolane with Ph and CO}_2\text{Et]} \quad (42)$$

(iv) Under basic or neutral conditions (addition–alkylation)

Carbonyl derivatives whose carbonyl group is flanked by electron-withdrawing groups readily form hemiacetals (see Section 4.04.2.1.1) which are usually difficult to convert into acetals under acidic conditions owing to the destabilizing effect of the electron-withdrawing group towards subsequent carbonium ion formation. Such aldehydes and ketones may form acetals by alkylation of the intermediate hemiacetals formed *in situ* under basic conditions. Compounds of this type include highly halogenated ketones (Equation (43)) ⟨60JA2288⟩, nitrobenzaldehydes (Equation (44)) ⟨58CB410⟩ and 1,2-dicarbonyl derivatives ⟨61CB2258⟩. The dipyridyl ketone (**55**) reacted with 2-chloroethanol under basic conditions to form the acetal (**56**) in 45% yield, compared to a best yield of 20% under a variety of acidic reaction conditions (Equation (45)) ⟨73TL1599⟩.

$$\text{ClCF}_2\text{C(O)CF}_2\text{Cl} \xrightarrow[\text{Me}_2\text{SO}_4 \atop 46\%]{\text{EtOH, K}_2\text{CO}_3} \text{ClCF}_2\text{C(OMe)(OEt)CF}_2\text{Cl} \quad (43)$$

$$\text{NO}_2\text{-C}_6\text{H}_4\text{-CHO} \xrightarrow[\text{Me}_2\text{SO}_4 \atop 84-85\%]{\text{MeOH, NaOH}} \text{NO}_2\text{-C}_6\text{H}_4\text{-CH(OMe)}_2 \quad (44)$$

$$\underset{(\mathbf{55})}{(\text{2-py})_2\text{C=O}} \xrightarrow[45\%]{\text{ClCH}_2\text{CH}_2\text{OH} \atop \text{Li}_2\text{CO}_3, \text{ reflux}} \underset{(\mathbf{56})}{(\text{2-py})_2\text{C(OCH}_2\text{CH}_2\text{O})} \quad (45)$$

A special case of this type of reaction is the addition of an alkoxide to an α-halo ketone. In highly substituted examples the reaction can stop at the intermediate epoxy ether (**57**) (Scheme 20) ⟨58JA4072, 83CB3631⟩ but in other cases this opens to form the α-hydroxy acetal (**58**). The generality of this reaction for the preparation of α-hydroxy acetals is limited by the many possible side reactions, but the use of α-tosyloxy ⟨93JCR(S)430⟩ or α-(4-nitrobenzenesulfonyloxy)(α-nosyloxy) ketones ⟨86JOC130⟩ results in high yields of this type of acetal. A more direct approach to α-hydroxy acetals involves the treatment of ketones with hypervalent iodine reagents, by a reaction which proceeds in part via a similar mechanism (Equation (46)) ⟨86ACR244, 86OS(64)138⟩.

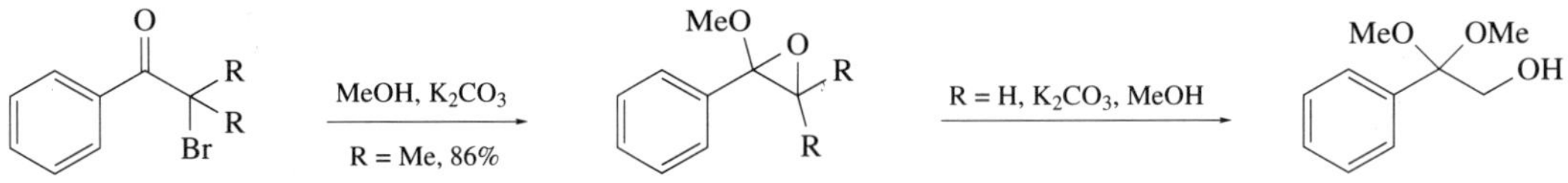

Scheme 20

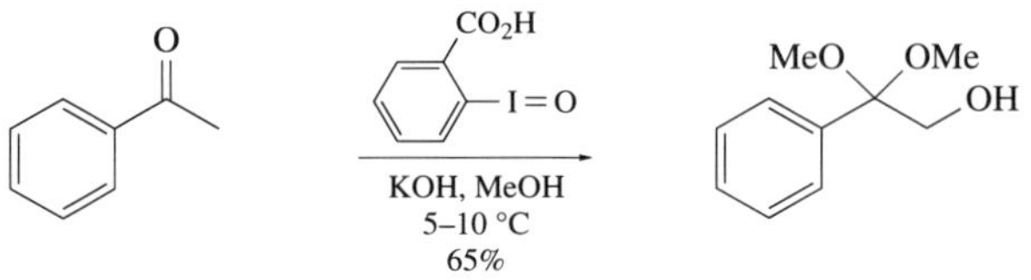

The adducts of DMF and dialkyl sulfates (**59**) also form acetals from aldehydes and some ketones with alcohols in a reaction which proceeds under neutral conditions by alkylation of an intermediate hemiacetal (Equation (47)) ⟨79LA522⟩. Aldehydes and more reactive ketones such as cyclohexanone give good yields of dimethyl and diethyl acetals. A variation of this reaction involves alkylation with (**59**) of the hemiacetal formed from the addition of a diol, followed by cyclization to provide a cyclic acetal ⟨79LA1362⟩.

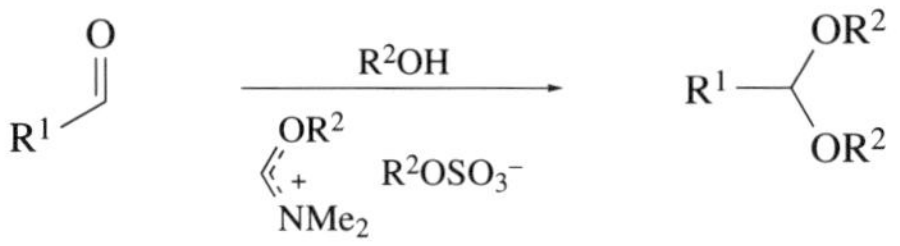

4.04.3.1.2 From acetals

(i) With alcohols

The alkoxy groups of acetals formed from low boiling alcohols such as methanol can be exchanged under acidic conditions with higher boiling alcohols, the equilibrium being displaced by the removal of the lower boiling alcohol by distillation (Equation (48)). Control of the reaction conditions enables the mixed acetal (**60**) to be formed, but is often necessary to separate this from the symmetrical acetal (**61**) which is formed exclusively in the presence of an excess of the alcohol R^3OH ⟨60JOC521, 60JOC525, 64USP3127450⟩. The use of an excess of the starting acetal allows the unsymmetrical acetal (**60**) to be isolated in good yield at room temperature ⟨87RTC545⟩. Unsymmetrical acetals were obtained in good yield when 2,2,2-trichloroethanol reacted with dimethyl or diethyl acetals (Equation (49)). The electron-withdrawing effect of the trichloroethyl group stabilizes the unsymmetrical acetal towards subsequent carbonium ion formation, although more forcing conditions enable bis(2,2,2-trichloroethyl) acetals to be produced in good yield ⟨73JOC554⟩. Cyclic acetals are formed in good yield from methyl acetals with diols ⟨58JA6613, 66JMC127⟩, and this method gave a wide range of acetals of the type (**62**), including some aromatic acetals which could not be prepared from the corresponding ketones (Equation (50)) ⟨66JMC127⟩. 2,2-Dimethoxypropane ⟨73CAR(29)209⟩ and analogous acetals ⟨B-91MI 404-01⟩ are very widely used in synthesis for the protection of diols as cyclic acetals (Equation (51)) ⟨81JOC2419⟩. Dipent-4-enyl acetals have recently been introduced for this purpose ⟨94CC749⟩.

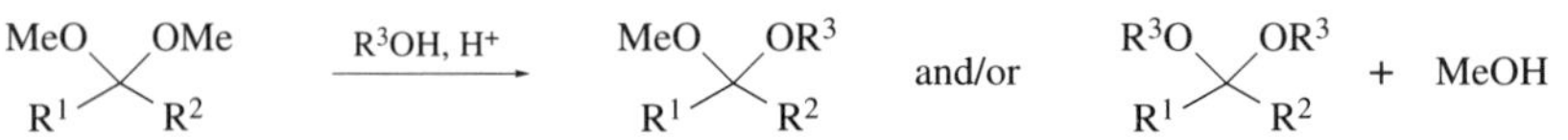

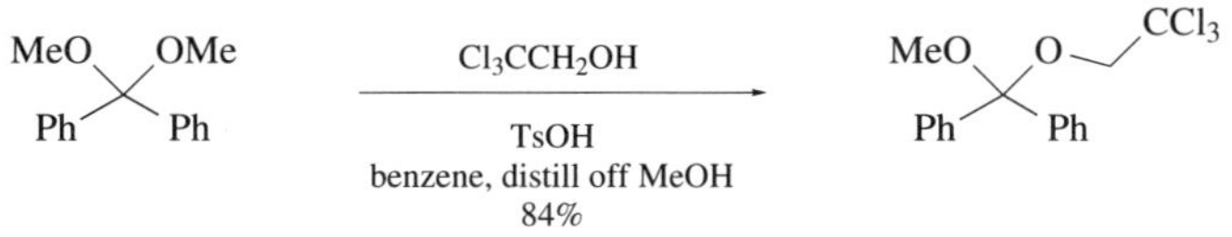

Acetals

(49) [Scheme showing dimethyl acetal of benzophenone + Cl₃CCH₂OH → mixed acetal, TsOH, benzene, distill off MeOH, 84%]

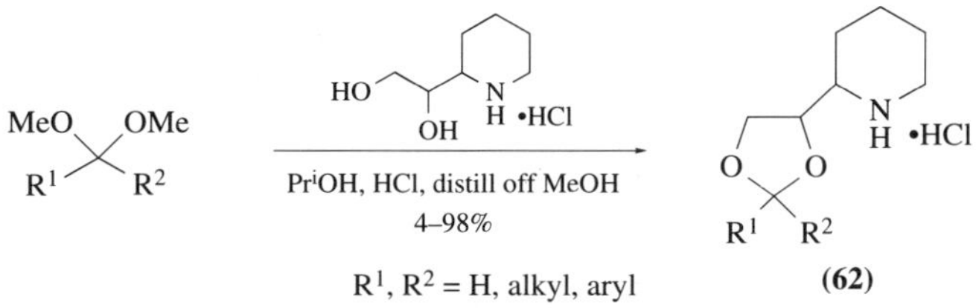

(50) [Scheme showing R¹R²C(OMe)₂ + diol·HCl → cyclic acetal (62), Pr^iOH, HCl, distill off MeOH, 4–98%, R¹, R² = H, alkyl, aryl]

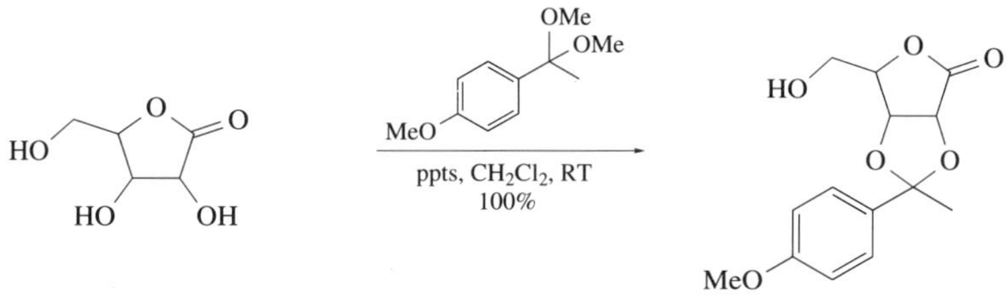

(51) [Scheme showing lactone triol + p-MeO-C₆H₄-C(OMe)₂Me → cyclic acetal, ppts, CH₂Cl₂, RT, 100%]

ppts = pyridinium p-toluenesulfonate

Methoxymethyl (MOM) ethers are formed from alcohols and dimethoxymethane (methylal) under acidic conditions. The use of an excess of the volatile methylal, often as the solvent, enables the unsymmetrical acetal to be isolated in high yield (Equation (52)). As this is a common method for the protection of alcohols and phenols (cf. MeOCH₂Cl, Section 4.04.3.3.1.i), many variations in reaction conditions have been developed ⟨B-91MI 404-01, 93SL429⟩. Polyols react with methylal to form cyclic acetals, and with triols there is high selectivity for the formation of six-membered ring acetals ⟨87TL6601⟩.

$$ROH + MeO\frown OMe \xrightarrow{H^+ \text{ or Lewis acid}} RO\frown OMe \quad (52)$$

(ii) By acetal interchange

MOM ethers or ethoxyethyl (EE) ethers undergo acetal interchange when heated in the presence of acid to afford formaldehyde or acetaldehyde acetals, respectively, with the equilibrium being displaced by evaporation of the volatile acetal coproduct (**63**) (Equation (53)) ⟨81JOC2981⟩. A similar reaction, which is particularly suitable for the preparation of strained cyclic acetals owing to its irreversible nature, involves heating mixed acetals, formed from diols and volatile carbonyl and alcohol components, under acidic conditions (Scheme 21) ⟨74S23⟩.

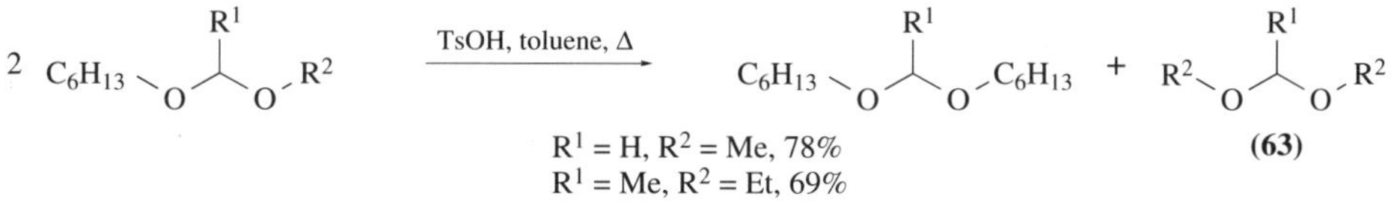

(53) [2 C₆H₁₃-O-CHR¹-O-R² → C₆H₁₃-O-CHR¹-O-C₆H₁₃ + R²-O-CHR¹-O-R² (63); TsOH, toluene, Δ; R¹ = H, R² = Me, 78%; R¹ = Me, R² = Et, 69%]

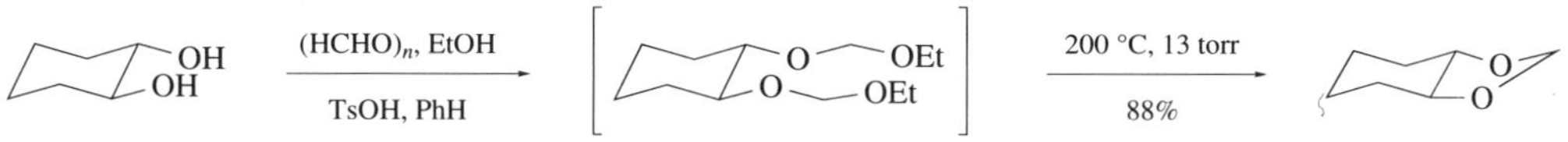

[Scheme 21: cyclohexane-1,2-diol + (HCHO)ₙ, EtOH, TsOH, PhH → [mixed acetal with OEt groups] → bicyclic acetal, 200 °C, 13 torr, 88%]

Scheme 21

4.04.3.1.3 From enol ethers and alcohols

(i) By acid catalysis

The acid-catalysed addition of alcohols to enol ethers proceeds via the intermediacy of the oxonium ion (**64**) (cf. Scheme 18) to form acetals (Scheme 22) ⟨35USP2000252, 83MI 404-01⟩. In general, an excess of the enol ether enables the unsymmetrical acetal (**65**) to be isolated in good yield, whereas an excess of the alcohol generally leads to the formation of the symmetrical acetal (**66**) by replacement of the original alkoxy group of the enol ether ⟨62USP3024284⟩. As enol ethers themselves are commonly prepared from acetals by elimination ⟨88JOC5574⟩, this method is generally more useful for the synthesis of unsymmetrical acetals of the type (**65**). Some examples illustrating the generality of the reaction are given in Table 7. Numerous protic and a few Lewis acids catalyse the reaction, the most commonly employed catalysts being HCl, toluene-*p*-sulfonic acid and PPTS. Other catalysts have been summarized ⟨91HOU(E14a/1)323⟩, and some examples appear in Table 7.

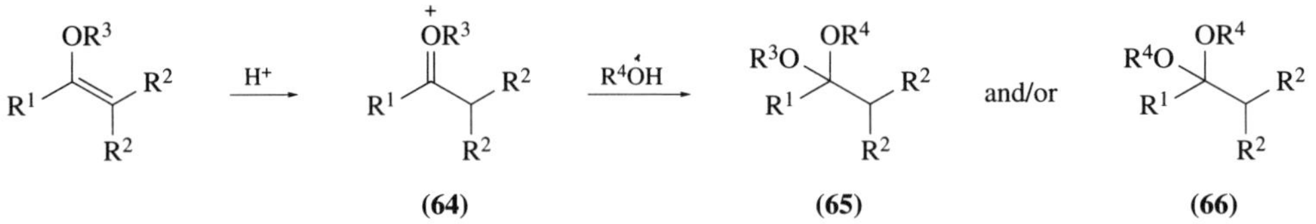

Scheme 22

Much of the literature deals with the use of this reaction for the protection of alcohols (Table 7), and cyclic enol ethers such as dihydrofuran and particularly dihydropyran ⟨47JA2246⟩, which readily form acetals with alcohols, have been widely used for this purpose (Equation (54)). Toluene-*p*-sulfonic acid is the most common catalyst for the tetrahydropyranylation of alcohols ⟨73S169⟩, but the popularity of tetrahydropyranyl (THP) ethers has resulted in the development of many sets of mild reaction conditions for their formation from a variety of alcohols ⟨B-91MI 404-01, 93S1069, 93TL5269⟩. For example, tertiary alcohols (Equation (55)) ⟨88TL4583⟩ and an alcohol containing an acetal group (Equation (56)) ⟨92BCJ304⟩ have been converted into THP ethers in high yield. Formation of THP ethers introduces a stereogenic centre, but enol ethers such as (**67**) have been developed to overcome this complication ⟨67JA3366, 73S169⟩. Protection of diols is also commonly achieved by their reaction with enol ethers such as 2-methoxypropene to form cyclic acetals (Table 7, entry 6). This method is widely used in natural product synthesis and offers particular advantages in carbohydrate chemistry as it allows access to products of kinetic acetonation (Equation (57)) ⟨81H(16)1587⟩.

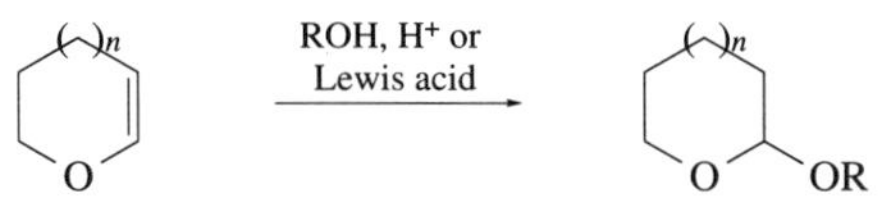

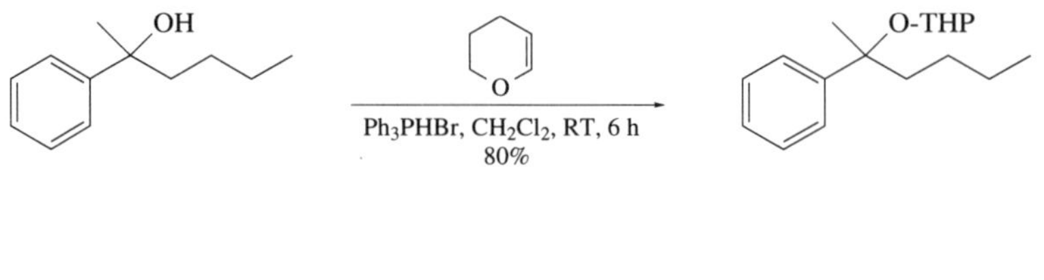

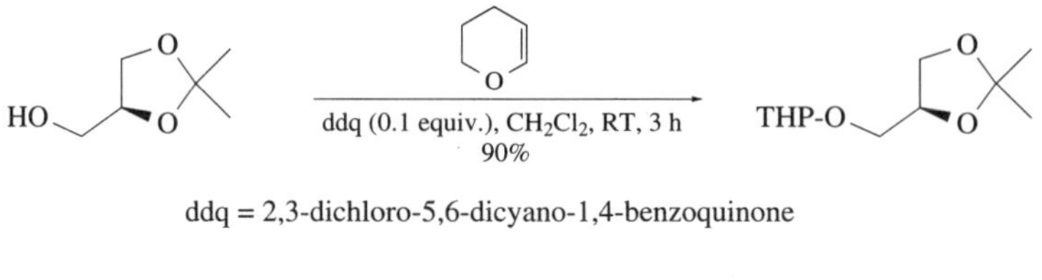

ddq = 2,3-dichloro-5,6-dicyano-1,4-benzoquinone

Table 7 Acetal formation from enol ethers, alcohols and phenol under acid catalysis.

Entry	Enol ether	Alcohol/conditions	Acetal	Yield (%)	Ref.
1	EtO-CH=CH₂	PhOH, Et₂O, 30 °C; CH₃CH(Cl)OEt (catalyst)	CH₃CH(OEt)(OPh)	80–85	35USP2000252
2	EtO-CH=CH₂	HOCH₂CH(CO₂Me)- (S), ppts, CH₂Cl₂, Et₂O, RT	EtO-CH(CH₃)-O-CH₂CH(CO₂Me)-	98	88S36
3	EtO-CH=CH₂	furfuryl-CH₂OH, CoCl₂, RT	furfuryl-CH₂-O-CH(CH₃)OEt	72	89SC901
4	MeO-CH=CH-CH₂Ph	Ph(CH₂)₂CH(OH)CH₂C(=CH₂)CH₃, ppts, CH₂Cl₂, 0 °C to RT	Ph(CH₂)₂CH(OCH(OMe)CH₂Ph)CH₂C(=CH₂)CH₃	44	89JOC5695
5	CH₂=C(OMe)CH₃	I-CH=CH-CH(OH)C₅H₁₁, POCl₃ (trace)	I-CH=CH-CH(O-C(OMe)(CH₃)₂)C₅H₁₁	100	72JA7827
6		CH₃CH(OH)CH(Bu)CH=CHCH₃ with adjacent OH, picric acid	acetonide of diol	93	84TL5155
7		Pr^iOH, CuBr₂, Et₂O, 15–20 °C	Pr^iO-C(OMe)(CH₃)₂	80	83SC629
8	CH₂=C(OCH₂Ph)CH₃	PhCO₂(CH₂)₃OH, PdCl₂(COD), C₆H₆, RT, 24 h	PhCO₂(CH₂)₃-O-C(CH₃)₂-O-CH₂Ph	98	84CL265

(67) 4-methoxy-3,6-dihydro-2H-pyran

glucose + CH₂=C(OMe)CH₃ →(TsOH, DMF, 0 °C, 95%) 4,6-O-isopropylidene glucose (57)

(ii) With electrophiles

Alcohols can also add to enol ethers in the presence of a range of electrophiles to form α-functionalized acetals (Equation (58)). Electrophiles which introduce chloro ⟨63CB77⟩, bromo ⟨72TL4055, 86JCS(P1)1351⟩, iodo ⟨89SC21⟩, benzeneselenyl ⟨80CC477⟩ and benzenesulfenyl ⟨87TL2723, 92JOC2084⟩ substituents have been employed, and the halogen or sulfur residues are sometimes removed to afford the unfunctionalized acetal. Oxidation of enol ethers with mcpba in alcoholic solvents gives α-hydroxy acetals (Equation (59)) ⟨77S578, 80SC83⟩. Dimethyldioxirane has been employed in a similar reaction for glycoside bond formation from glycals ⟨89JA6661⟩. In the absence of alcohols, enol ethers and dimethyldioxirane form epoxides, which can be isolated in some cases (Equation (60)) ⟨89TL257, 89TL6497⟩. Alkoxymercuration of enol ethers, followed by reduction of the intermediate organomercury derivative, is a mild, versatile and predictable alternative to acid catalysis for the preparation of unsymmetrical acetals ⟨83TL4923⟩. In unfunctionalized derivatives, the demercuration is carried with sodium borohydride (Equation (61)), whereas sodium trithiocarbonate is preferred for substrates bearing electron-withdrawing groups (Equation (62)). Alkenyl and alkynyl groups in the alcohol component are generally unaffected by the reaction conditions. Other reactions of the organomercury intermediates are also possible ⟨B-86MI 404-01⟩. The Lewis acid-catalysed addition of acetals to enol ethers provides acetals that are usually intended as intermediates in the synthesis of enals (Equation (63)) ⟨81S137, 88CL1101, 88JOC2920⟩. Addition of allyl alcohols to enol ethers in the presence of palladium acetate leads to tetrahydrofuranyl ethers (Scheme 23) by alkene insertion into the C—Pd bond in the intermediate (**68**) ⟨89BCJ2050⟩.

$$\ce{CH2=CHOR^1} \xrightarrow{E^+, R^2OH} \ce{E-CH2-CH(OR^2)(OR^1)} \qquad (58)$$

E = Cl, Br, I, PhSe, PhS, OH, etc.

$$(59)$$

$$(60)$$

$$(61)$$

$$(62)$$

$$(63)$$

4.04.3.1.4 From silyl enol ethers and enol acetates

Silyl enol ethers react with alcohols in a similar manner to alkyl enol ethers to form acetals ⟨73JOC3935, 82S1089⟩. Generation of silyl enol ethers from enones enables additional functionality to be introduced (Scheme 24) ⟨84TL3805, 85JOC3627⟩.

Vinyl acetate and isopropenyl acetate react with alcohols to form symmetrical acetals of acetaldehyde and acetone, respectively (Equation (64)). Best yields are obtained when the reaction is catalysed by mercuric oxide or mercuric acetate in combination with boron trifluoride etherate ⟨48JA2805, 58JA1083⟩. Although phenol was reported not to react ⟨48JA2805⟩, catechol does form an acetal with vinyl acetate in 58% yield ⟨79JMC1264⟩.

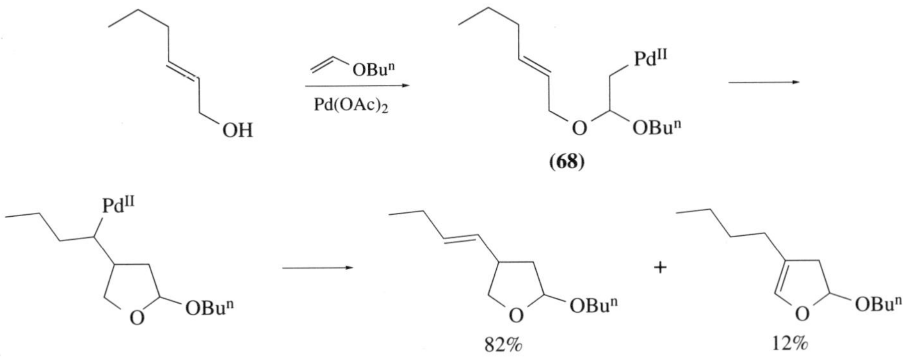

Scheme 23

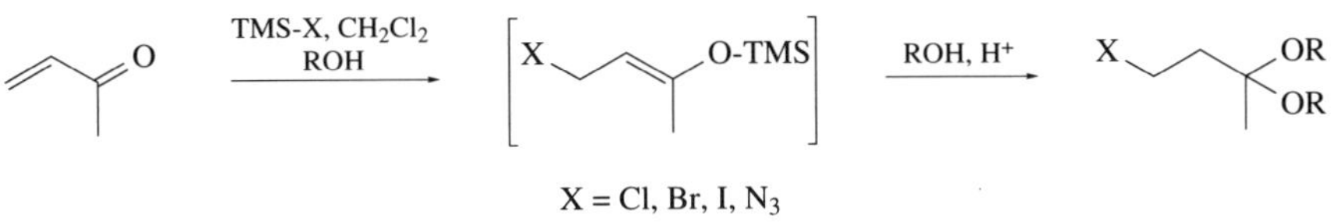

Scheme 24

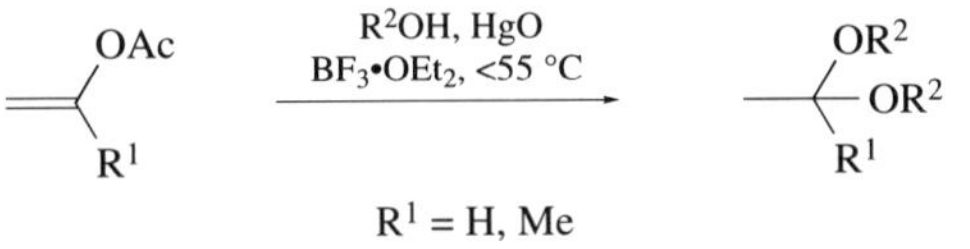

$$R^1 = H, Me \tag{64}$$

4.04.3.1.5 From gem-dihalides by alkylation

Displacement of both halides from geminal dihalides with alkoxides affords acetals (Equation (65)). The method is particularly useful when the dihalide is activated by a neighbouring aryl or carbonyl group (R^1 and/or R^2 = Ar, COR), and is often the best method for the preparation of acetals of this type. Several acetals of substituted benzophenones which were not available by other methods were prepared by conversion of the benzophenones into the *gem*-dichlorides, followed by displacement with alkoxides (Scheme 25) ⟨52JA6189, 66JMC127, 71JOC2357⟩. A notable example of this type of reaction is the preparation of the catechol acetal of benzophenone (Equation (66)) ⟨89JOC4549⟩. Di-*t*-butyl acetals of substituted benzaldehydes were prepared in a similar fashion, although no yields were reported ⟨60CI(L)656, 71JA1701⟩. The diethyl acetal of ethyl glyoxylate (**69**) was prepared from dichloroacetic acid by displacement with ethoxide followed by esterification (Equation (67)) ⟨63OSC(4)427⟩. Displacement of the halogens in α-bromo α-chloro ketones with alcohols in the presence of silver(I) carbonate gave α-keto acetals ⟨94S427⟩.

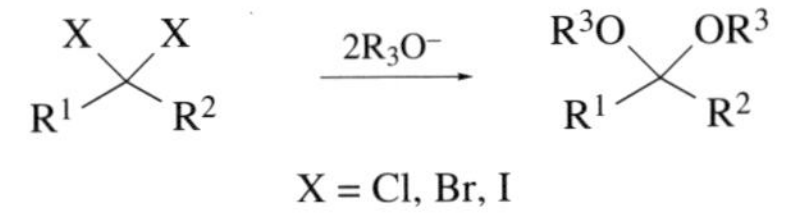

$$X = Cl, Br, I \tag{65}$$

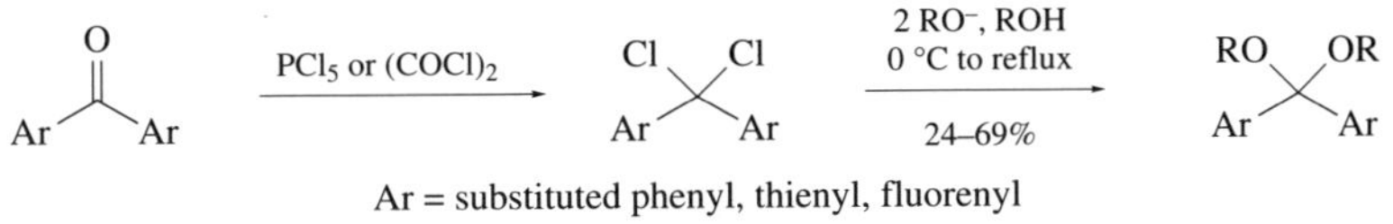

Scheme 25

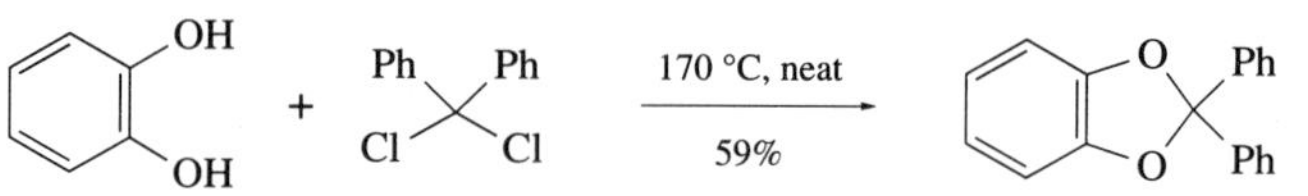

Dihalomethanes undergo displacement with alcohols, diols (symmetrical and unsymmetrical) and phenols to provide acetals of formaldehyde. Phase transfer catalysis is often employed and generally gives best yields (Equation (68)) ⟨76TL95, 82S162⟩. The use of equimolar amounts of two different monohydric alcohols leads to product mixtures ⟨90SC2527⟩. The method is commonly employed for the methylenation of catechols ⟨69JCS(C)1202, 75TL3489⟩, and the use of the fluoride ion as a base is particularly effective for this purpose (Equation (69)) ⟨76TL3361⟩. Diols in carbohydrate derivatives have been converted into methylene acetals by alkylation ⟨78S48, 82S421, 85S751⟩. Phenyl halodiazirines, e.g. (**70**) ⟨93HCA211⟩, and acrolein hydrate diacetate (**71**) ⟨91H(32)1445⟩ are alternative reagents which have been developed for the acetalization of diols, by alkylation then carbene insertion and by palladium(0)-catalysed alkylation, respectively.

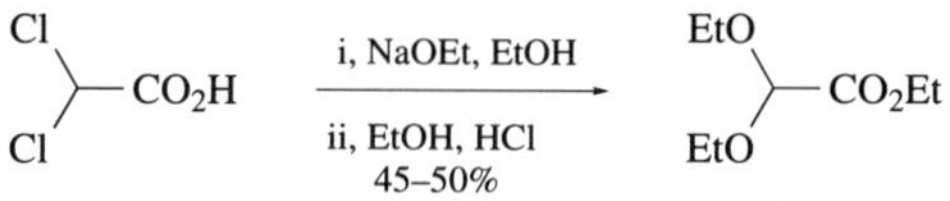

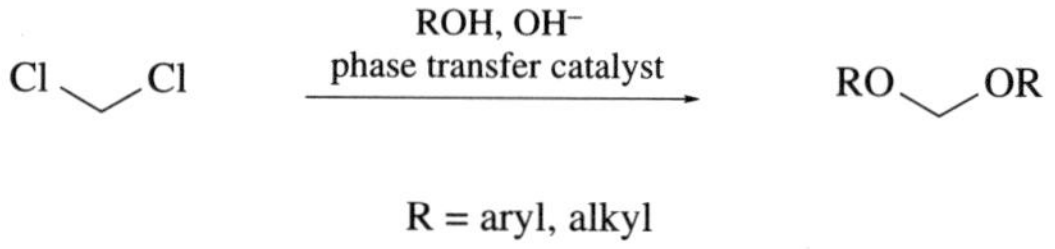

4.04.3.2 Symmetrical Acetals

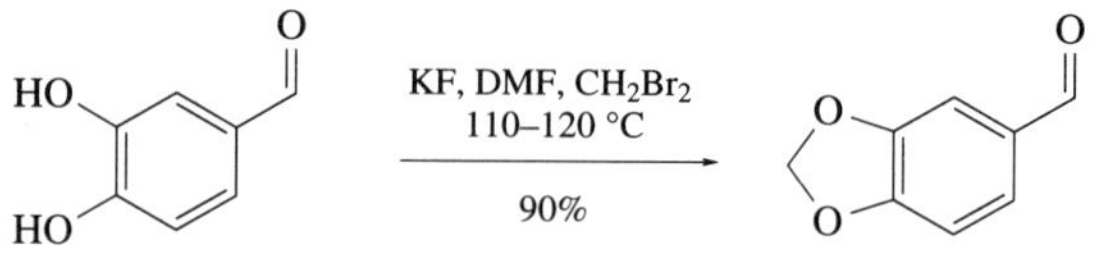

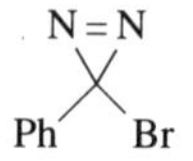

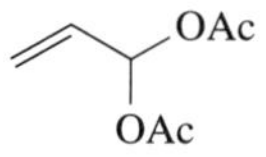

4.04.3.2.1 *From aldehydes and ketones*

(i) With orthoformates

Ortho esters, in addition to acting as dehydrating agents in acetalization reactions with alcohols (see Section 4.04.3.1.i), can themselves undergo acid-catalysed reactions with aldehydes and ketones to form acetals (Table 8), although the presence of the alcohol accelerates the reaction ⟨69CC1175⟩. Several of the acids listed in Table 5 are useful catalysts, and heterogeneous catalysts are particularly efficient (Table 8, entries 4 and 5). Diethylene orthocarbonate (**72**) is an effective reagent for the formation of ethylene acetals and has been recommended for the preparation of *o*-hydroxy acetals of aromatic aldehydes (Equation (70)) ⟨75CC432⟩.

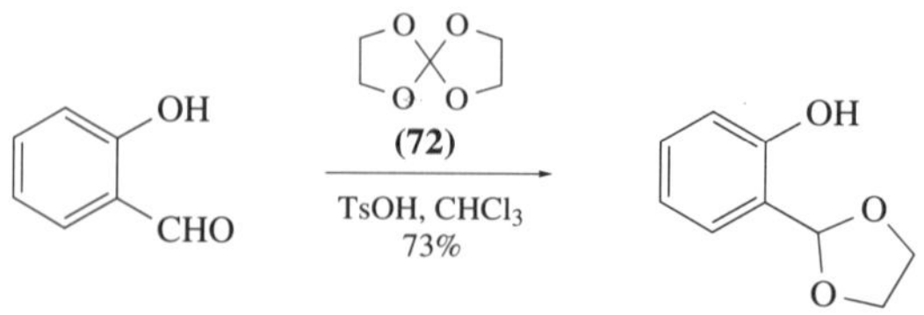

(ii) By acetal exchange

In a similar fashion, acid-catalysed transfer of an alcohol, or more usually a diol, from a simple acetal, for example (**73**)–(**75**), to form a more complex acetal is a useful method of acetal synthesis;

Acetals 191

Table 8 Acetalization with ortho esters.

Entry	Substrate	Conditions	Acetal	Yield (%)	Ref.
1	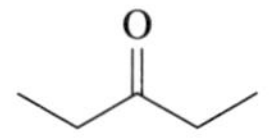 pentan-3-one	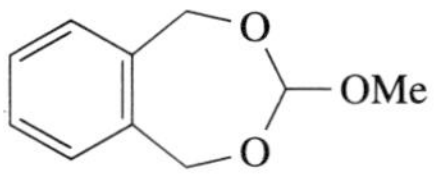 benzo-fused cyclic orthoester with OMe; TsOH, dimethoxyethane, RT, 0.5 h	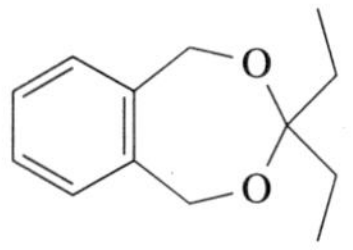 benzo-fused cyclic diethyl acetal	91	89TL4165
2	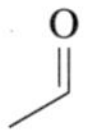 acetaldehyde	(EtO)$_3$CH, conc. H$_2$SO$_4$, RT, 24 h	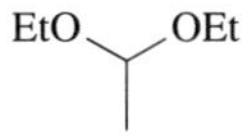 CH$_3$CH(OEt)$_2$	58	40JOC244
3	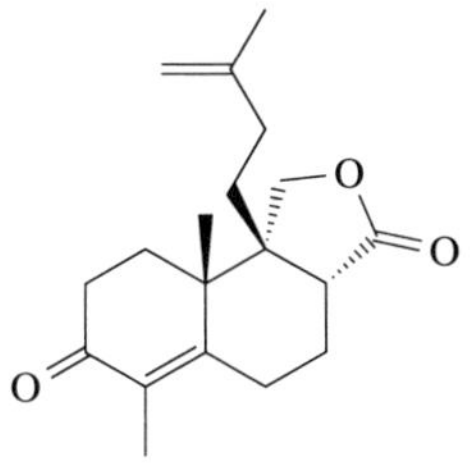 steroid ketone/lactone	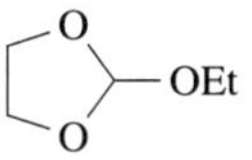 2-ethoxy-1,3-dioxolane; ppts, benzene, reflux, 8 h	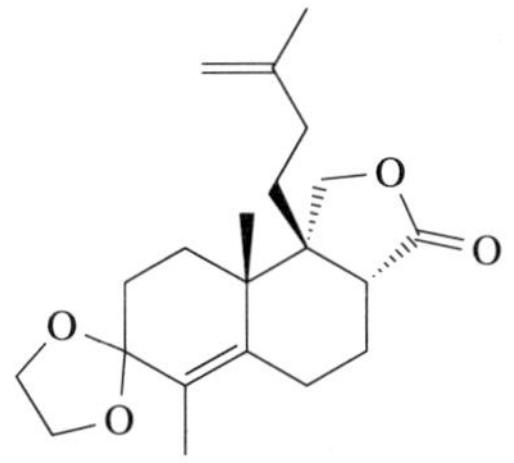 dioxolane-protected steroid	89	87JA1597
4	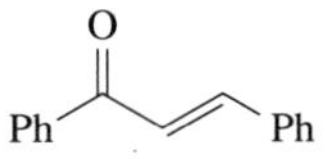 PhCOCH=CHPh (chalcone)	Montmorillonite clay K-10, (MeO)$_3$CH, CCl$_4$ or n-hexane, 15 h	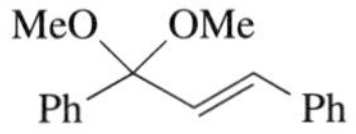 PhC(OMe)$_2$CH=CHPh	92	77S467
5	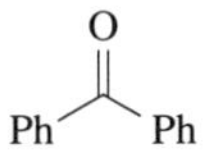 Ph$_2$CO (benzophenone)	Nafion-H, (MeO)$_3$CH, CCl$_4$, RT	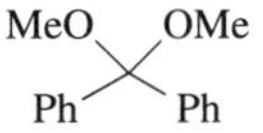 Ph$_2$C(OMe)$_2$	91	81S282

examples are shown in Equations (71) and (72). This reaction has been widely employed in steroid chemistry and is sometimes more selective than conventional procedures ⟨59T269, B-63MI 404-01⟩. It is usually necessary to distill off the liberated ketone to displace the equilibrium, but this may be unnecessary if a favourable equilibrium exists. As with orthoformates, acceleration of the reaction by traces of alcohol may also occur ⟨54JA1359, 73T4225⟩. The relative stability of various ketones has been determined to aid the prediction of exchange acetalization in polycarbonyl systems ⟨77T3105⟩. A particularly mild exchange dioxolanation was demonstrated with the ethylene acetal of DMF ⟨63MI 404-02⟩.

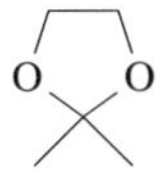

(73) 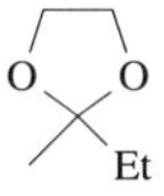   (74) 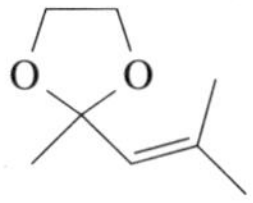   (75)

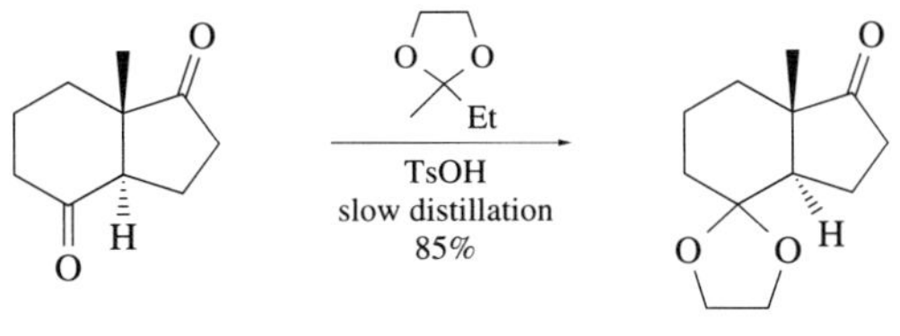

(71) bicyclic diketone + 2-ethyl-2-methyl-1,3-dioxolane, TsOH, slow distillation, 85% → mono-dioxolane product

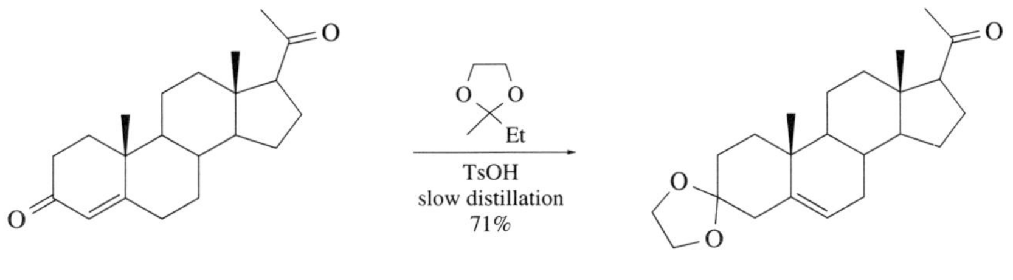

(72) progesterone-type dienone + 2-ethyl-1,3-dioxolane, TsOH, slow distillation, 71% → 3-dioxolane diene product

4.04.3.2.2 From ortho esters and nucleophiles

The displacement of an alkoxy group from an ortho ester with a carbon nucleophile has become a well-established method for the synthesis of a wide range of acetals ⟨B-70MI 404-03⟩.

(i) With organometallic reagents

Grignard reagents react with orthoformates to provide aldehyde acetals by displacement of an alkoxy group (Equation (73)). Alkyl, aryl, alkynyl, vinyl and heterocyclic Grignard reagents have all been employed, and the reaction is generally carried out in refluxing diethyl ether with yields ranging from fair to good ⟨B-70MI 404-03⟩. The mixed ortho ester (**76**) prepared from triethyl orthoformate and phenol ⟨70CB643⟩ reacts with a wide variety of Grignard reagents at room temperature and in higher yields than standard ortho esters ⟨81JCR(M)4016⟩. Treatment of (**76**) with acetylenic Grignard reagents in dichloromethane allowed the preparation of some sensitive acetals (Equation (74)), ⟨84JOC2031⟩. Allyl and propargyl aluminum reagents react with orthoformates or trimethyl orthoacetate at $-80\,°C$ to form unsaturated acetals ⟨86CB1725⟩. Organocuprates ⟨84TL3075⟩ and allylsilanes ⟨89S128⟩ also react readily with ortho esters in the presence of Lewis acids. The reaction of alkynes with ortho esters catalysed by zinc salts ⟨58JA4607, 63OSC(4)801⟩ is an alternative to Equations (73) and (74) for the preparation of acetylenic acetals (Equation (75)). Although this method allows access to ketone acetals, it requires a pressure vessel for the reaction of volatile alkynes.

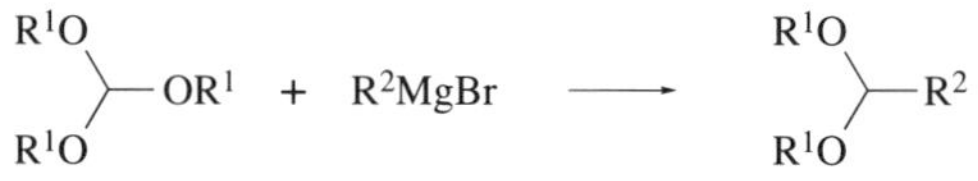

$$R^1O\text{-CH}(OR^1)_2 + R^2MgBr \longrightarrow R^1O\text{-CHR}^2(OR^1) \quad (73)$$

R^1 = Me, Et; R^2 = alkyl, aryl, vinyl, alkynyl

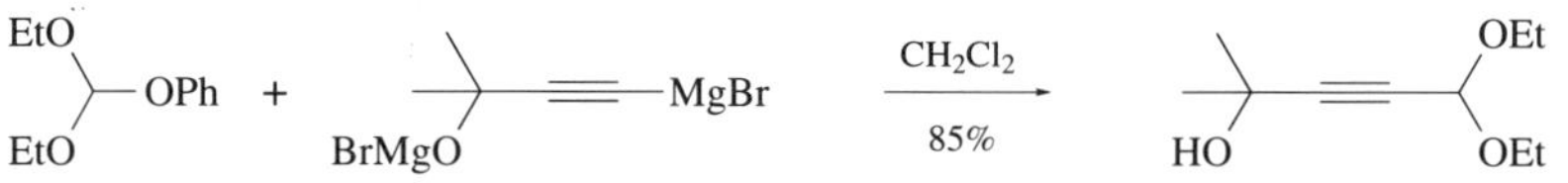

$$(EtO)_2CH\text{-OPh} + BrMgO\text{-C≡C-MgBr} \xrightarrow[85\%]{CH_2Cl_2} HO\text{-C≡C-CH}(OEt)_2 \quad (74)$$

(**76**)

$$(R^1O)_3CR^2 + R^3\text{-C≡CH} \xrightarrow[15-80\%]{ZnX_2} R^3\text{-C≡C-CR}^2(OR^1)_2 \quad (75)$$

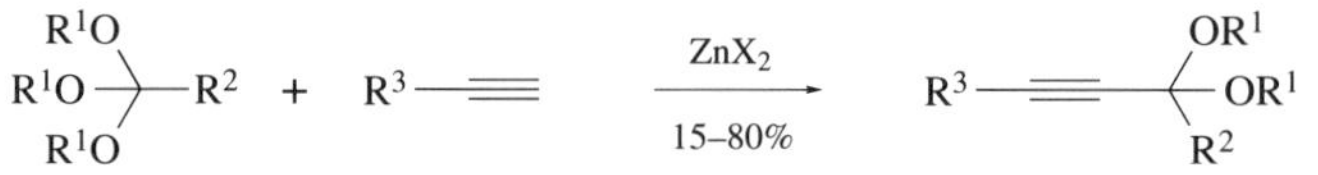

R^1 = Me, Et; R^2 = H, Me, Bu^n; R^3 = H, alkyl, phenyl

(ii) With enolate derivatives

Lithium enolates ⟨82TL3595⟩, silyl enol ethers ⟨74CL15, 80JA3248, 93TL7335⟩ and enamines ⟨84BCJ1876⟩ react with orthoformates in the presence of a Lewis acid in a useful procedure for the preparation of monoprotected 1,3-dicarbonyl derivatives (Scheme 26). The reaction has been

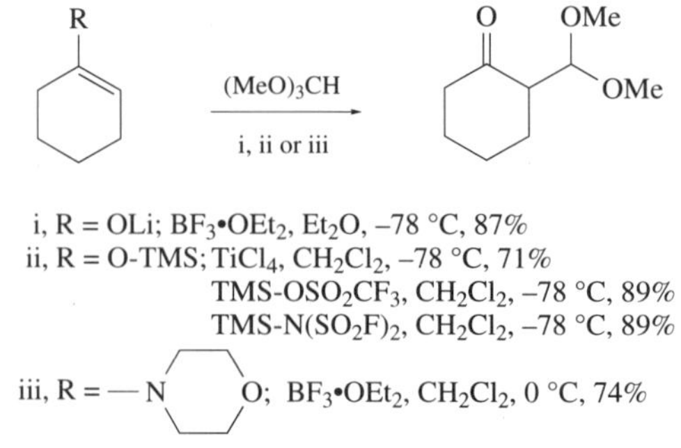

cyclohexene-R + (MeO)₃CH →(i, ii or iii)→ 2-(dimethoxymethyl)cyclohexanone

i, R = OLi; BF₃·OEt₂, Et₂O, –78 °C, 87%
ii, R = O-TMS; TiCl₄, CH₂Cl₂, –78 °C, 71%
          TMS-OSO₂CF₃, CH₂Cl₂, –78 °C, 89%
          TMS-N(SO₂F)₂, CH₂Cl₂, –78 °C, 89%
iii, R = —N(morpholino); BF₃·OEt₂, CH₂Cl₂, 0 °C, 74%

Scheme 26

extended to prepare cyclic acetals of ketones ⟨84M587, 85LA2472⟩ (Equation (76)). Dienol silyl ethers react to give a monoprotected 1,5-dicarbonyl derivative ⟨84S227⟩, whereas dienamines form α-dimethoxymethyl β,γ-unsaturated ketones ⟨84BCJ1876⟩. α-Dialkoxyalkylation of enones can be achieved by the one-pot process of Equation (77) ⟨81TL1809⟩. Diethoxycarbenium tetrafluoroborate (77) generated *in situ* from triethyl orthoformate and $BF_3 \cdot OEt_2$ reacts with ketones in the presence of diisopropylethylamine to afford α-diethoxymethyl ketones ⟨81JOC2557, 85TL1581⟩.

A dialkoxymethyl group can be attached to the α-carbon of an ester by the reaction of an orthoformate with Reformatski reagents ⟨47JA2233⟩ or ketene silyl acetals ⟨76CL769⟩.

$(EtO)_2\overset{+}{C}HBF_4^-$ $(EtO)_2\overset{+}{C}HNR^1R^2X^-$

(77)   (78)

(iii) With other nucleophiles

Other nucleophiles have been reported to react with ortho esters to form acetals, including nitroalkanes ⟨81S878⟩, acidic methine derivatives ⟨79JOC4825⟩ and the hydride ion ⟨51JA5005⟩. Dialkoxymethylammonium salts, for example (78), generated from ortho esters react with a range of acidic methylene derivatives such as malonates to form acetals ⟨71S312⟩.

4.04.3.2.3 From alkenes and alkynes

(i) From alkenes and alcohols with electrophilic metal derivatives

The palladium(II)-catalysed oxidation of terminal alkenes in the presence of water, the Wacker oxidation, is a well-established procedure for the preparation of methyl ketones ⟨84S369⟩. If anhydrous alcohols are substituted for water, acetals are formed (Equation (78)), ⟨69JOC3949, 84S369⟩. Although terminal alkenes are most commonly employed as substrates, internal alkenes may also react, albeit in lower yield. A similar reaction may be carried out after initial alkoxymercuration of the alkene, followed by transmetallation with palladium, although stoichiometric amounts of palladium are required for optimum yield ⟨72TL3595⟩. Such an approach may be advantageous for the acetalization of internal alkenes, as potential palladium-catalysed isomerization of the alkene is avoided. An intramolecular variant of the palladium-catalysed reaction has been employed in natural-product synthesis; an example is shown in Equation (79) ⟨84JCS(P1)1643⟩. The scope of the palladium-catalysed acetalization of alkenes was extended with the discovery that terminal alkenes bearing electron-withdrawing groups are regioselectively acetalized with diols at the terminal carbon atom (Equations (80) ⟨90ACR49, 90BCJ166⟩ and (81) ⟨89JHC1405, 90JHC1419⟩). This promises to be a more useful and reliable reaction than the original variant. Alkyl nitrites have been employed in the reaction to provide acetals of monohydric alcohols ⟨92JHC1625⟩. Acetalization of the terminal carbon may also occur if the substrate contains an appropriately positioned group which can coordinate the palladium (Scheme 27) ⟨89CL737⟩.

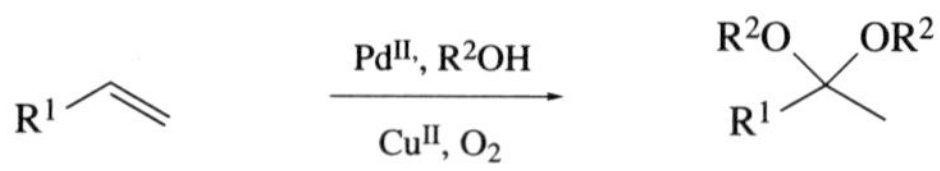

(78)

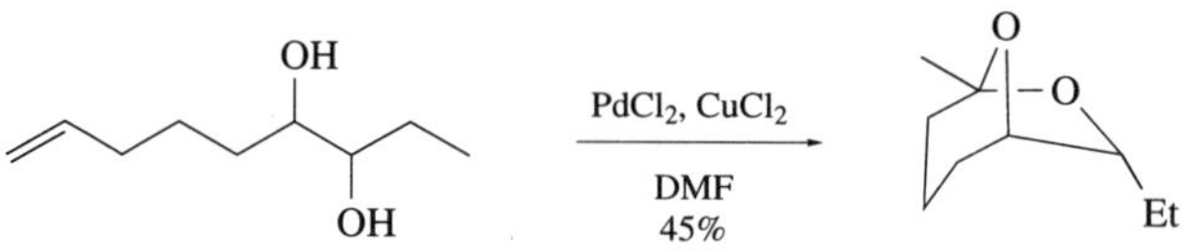

(79)

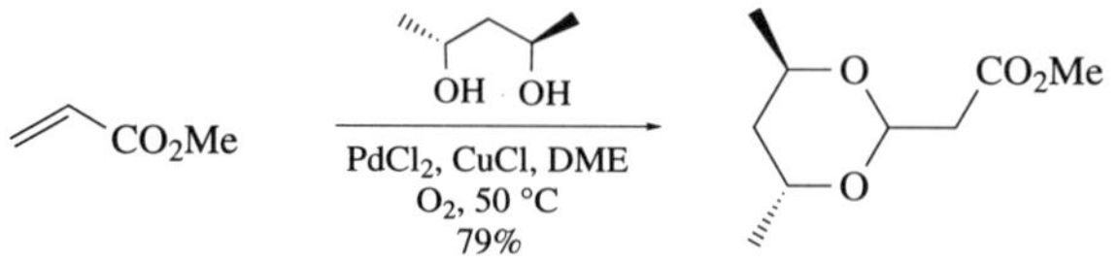

(80)

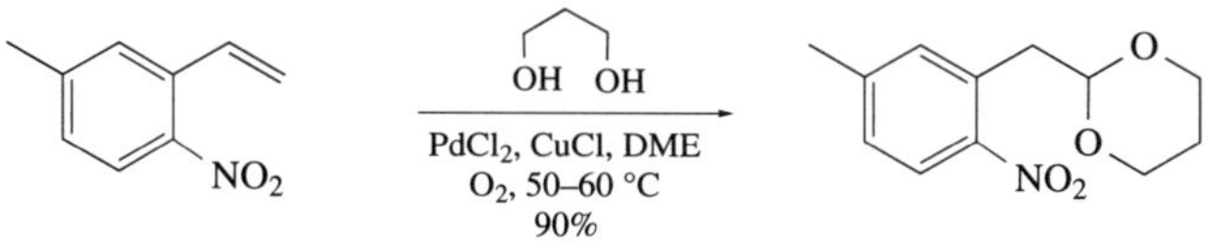

(81)

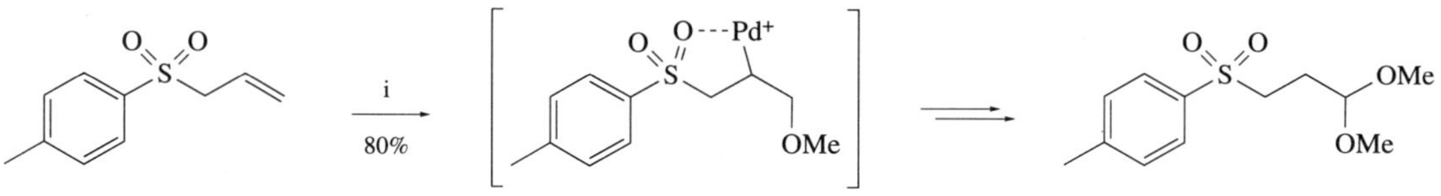

i, PdCl$_2$, CuCl$_2$, N, N, N', N'-tetramethylurea, ethyl *ortho*-acetate, MeOH, reflux

Scheme 27

Treatment of alkenes with thallium(III) nitrate (ttn) in methanol gives acetals via an oxidative rearrangement (Scheme 28) ⟨73JA3635, 82COMC-I(7)465⟩. Disubstituted alkenes react more readily than trisubstituted alkenes, and tetrasubstituted double bonds are unreactive or form mixtures of products. The reaction is successful only when the migrating bond can adopt a *trans* relationship to the departing thallium (Scheme 28). The use of trimethyl orthoformate alone or as a 1:1 mixture with methanol as the solvent leads to improved yields (Equation (82)) ⟨77TL1827⟩ and even allows rearrangement of deactivated alkenes ⟨76JA3037⟩. Further improvements in efficiency may be achieved by supporting the reagents (methanol, trimethyl orthoformate and ttn) on montmorillonite K10 clay ⟨76JA6750⟩.

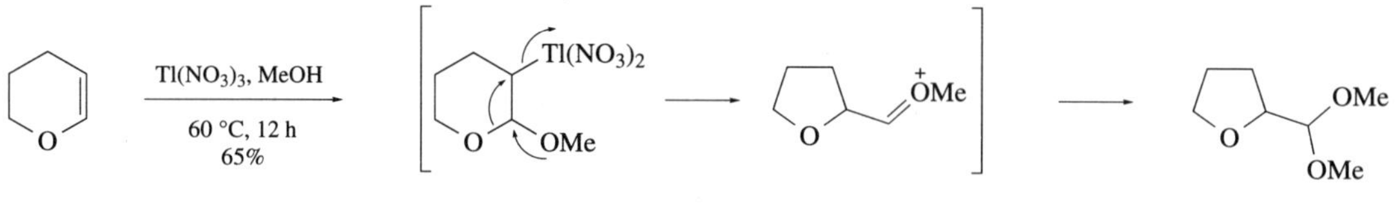

Scheme 28

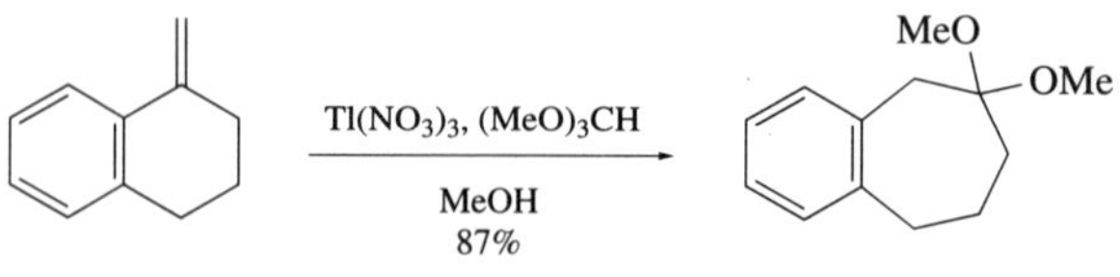

(82)

(ii) From alkynes and alcohols

Alkynes react with alcohols in the presence of mercury(II) salts to form acetals, the reaction presumably occurring via an enol ether intermediate (Scheme 29) ⟨B-86MI 404-01⟩. A mixture of mercuric oxide and boron trifluoride appears to be the most effective catalyst, and the reaction is generally carried out in the alcohol as the solvent. Ethyne forms acetals of acetaldehyde, and monosubstituted alkynes form acetals of methyl ketones by the addition of alcohol to the most substituted carbon atom. Increasing the acidity of the reaction mixture by the addition of trichloroacetic acid is recommended for the addition of alcohols higher than methanol to substituted alkynes (Equation (83)), ⟨36JA80⟩. Disubstituted alkynes also undergo the reaction, but this is only useful when the substrate is biased to avoid formation of regioisomers (Equation (84)) ⟨36JA892⟩. A wide range of alcohols and diols have been employed in the reaction, and functionalized alcohols such as glycolates ⟨48JOC223⟩ may also be used.

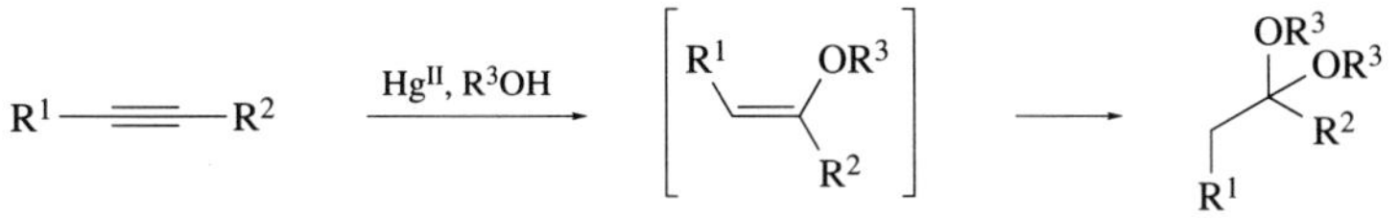

Scheme 29

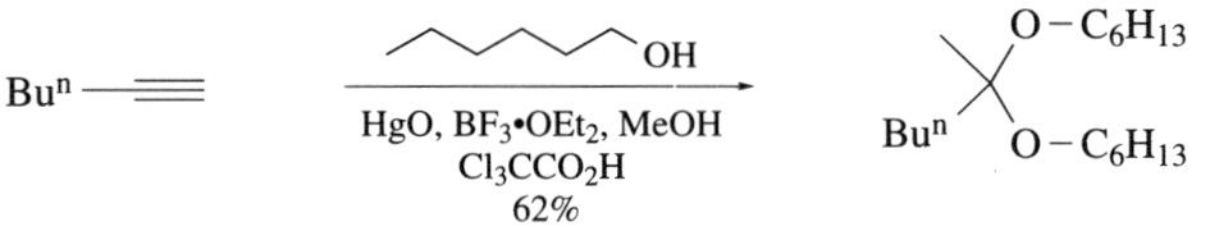

(83)

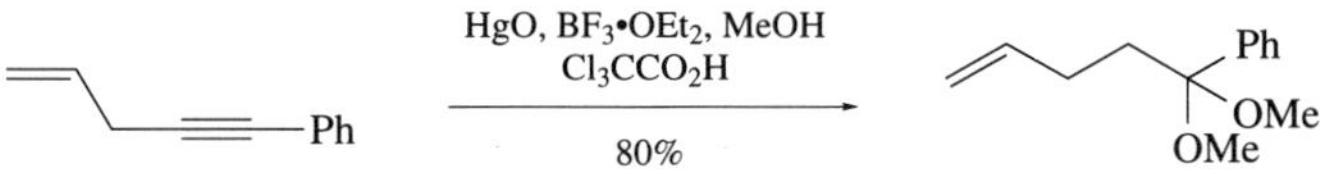

(84)

(iii) From electron-deficient alkynes and alkenes by conjugate addition of alcohols

Conjugate addition of alcohols to electron-deficient alkynes or β-functionalized alkenes affords acetals (Equation (85)). Most examples in the literature cover the synthesis of aldehyde acetals, but the synthesis of ketone acetals is also possible (Equation (86)) ⟨76JOC3765⟩. Several variants of this methodology have been used to prepare ethyl 3,3-diethoxypropionate (**79**), a useful synthon (Scheme 30). Copper(I) triflate catalyses the addition of ethanol to ethyl propiolate to provide (**79**) in high yield ⟨82JOC2216⟩. The reaction presumably proceeds via the intermediate (**80**), which itself has been converted into (**79**) in refluxing ethanol containing sodium bisulfate ⟨49JA2736⟩. The trichloromethyl ketone (**81**), prepared from ethyl vinyl ether and trichloroacetyl chloride, readily undergoes a haloform-type reaction and conjugate addition to form (**79**) when treated with ethanol and a catalytic amount of potassium carbonate ⟨88S274⟩. Also, addition of a different alcohol to (**81**) provides unsymmetrical acetals. The ethoxy groups in (**79**) can be exchanged for higher-boiling alcohols or diols, in a reaction which proceeds through (**80**) by elimination then addition (Scheme 31) ⟨49JA2736, 49JA2741⟩. Conjugate addition of one equivalent of an alcohol to β-alkoxyacrylates of the type (**80**) gives unsymmetrical acetals.

(85)

X = Br, Cl, OR; EWG = electron-withdrawing group

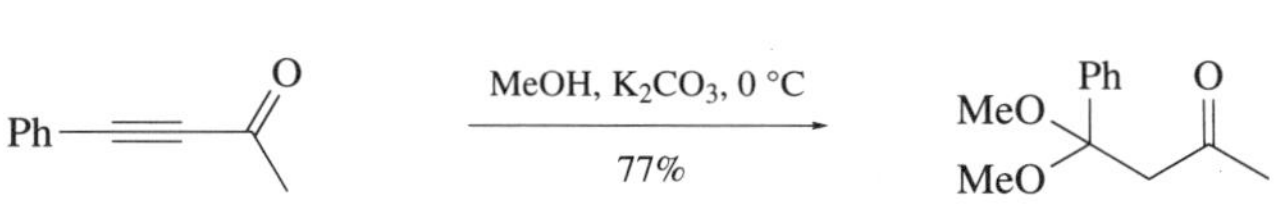

(86)

Alkynes activated by an electron-deficient aryl ring or by heteroaryl groups also undergo addition of alkoxide to form acetals. Ethynyl aza-arenes with the ethynyl group at an active position, for example the quinoline (**82**), form acetals in refluxing methanolic sodium methoxide (Equation (87)) ⟨84S245⟩. When the ethynyl group is at an inactive position, as in the quinoline (**83**), prolonged reaction times are required, and only one equivalent of methanol adds to form an enol ether. Similarly, in an approach to indole synthesis, alcohols were added to ethynylnitroarenes to form acetals in good yield (Equation (88)) ⟨86TL1653, 86CPB2362⟩. The ready availability of β-chlorovinyl ketones from the addition of acid chlorides to alkynes makes them useful precursors for acetal synthesis by treatment with alkoxides (Scheme 32) ⟨37USP2091373, 50JA2613, 63OSC(4)558⟩. The treatment of ketones with the Vilsmeier reagent produces β-chlorovinyl aldehydes, which react with the monosodium salt of ethylene glycol to form acetals in moderate yield ⟨85S496⟩. Other Michael acceptors which add alcohols to form acetals include perfluorinated alkynes ⟨82JOC2251⟩ and nitroalkenes ⟨88S707⟩.

4.04.3.2.4 From dithioacetals and O,S-acetals

The alcoholysis of dithioacetals induced by thiophilic metals, for example mercury(II), copper(II) or silver(I), or by S-alkylation or oxidation is a potentially nonacidic procedure for acetal formation

⟨77S357⟩. Some examples are shown in Table 9. Intramolecular variants of the reaction have been used in spiroacetal and other natural-product syntheses (see, for example, ⟨78CJC2700⟩ and ⟨80HCA1960⟩). In a similar manner, the methyl methylthiomethyl sulfoxides (**84**) ⟨71TL3151⟩, dithiocarbamates (**85**) ⟨74S705⟩, *O,S*-acetals (**86**) ⟨83TL4993, 88S95⟩ and α-TMS sulfides (**87**) ⟨91T615⟩ have all been developed as acyl anion equivalents and have served as precursors to acetals.

Table 9 Acetals from dithioacetals.

Substrate	Conditions	Acetal	Yield (%)	Ref.
[naphthyl-CH₂CH₂C(SPh)₂CO₂Me, MeO-naphthyl]	I₂, MeOH, reflux	[naphthyl-CH₂CH₂C(OMe)₂CO₂Me, MeO-naphthyl]	87	75JOC148
n-C₆H₁₃-CH(1,3-dithiolane)	i, MeOSO₂F; ii, MeOH, CH₂Cl₂	n-C₆H₁₃-CH(OMe)₂	87	75TL3267, 75TL4543
PhS-C(O)-(CH₂)₃-CH(SPh)₂	(CF₃CO₂)₂IPh, ethylene glycol	PhS-C(O)-(CH₂)₃-CH(1,3-dioxolane)	91	89TL287
2-furyl-C(1,3-dithiane)-CH₂CH₂OTs	Chloramine-T, THF, ethylene glycol	2-furyl-C(1,3-dioxolane)-CH₂CH₂OTs	95	81JA3112

R—C(SMe)(SMe)(S(O)Me)   R—C(SMe)(NMe₂)(S=S)   R—C(S(O)ₙPh)(OMe) n = 0, 2   R—C(SPh)(TMS)

(**84**)   (**85**)   (**86**)   (**87**)

4.04.3.2.5 *Electrochemical methods*

A few electrochemical methods have been developed for acetal formation, and, although they require specific types of substrate, they may be useful for the preparation of certain acetals. Electrolysis of α-sulfenyl esters in methanol gave methyl acetals in high yield (Equation (89)) ⟨80CL617⟩. Similar types of acetals are also formed by the anodic oxidation of 2-arylacetic acid esters ⟨93GEP4122315⟩. Electrolysis of α-alkoxyphenyl- and α-alkoxy- or α-(arylthio)diphenyl-acetic acids provides acetals of benzaldehyde and benzophenone by decarboxylation (Equation (90)) ⟨62CI(L)1868, 62JOC281⟩. Conditions have been developed to convert styrenes into α-bromo acetals ⟨88TL1603⟩ or arylacetaldehyde acetals ⟨89TL5309⟩ (Scheme 33). Electrochemical oxidation of α-methoxy silanes in methanol provides dimethyl acetals ⟨92JOC1321⟩.

$$R-C(S-BT)(CO_2Me) \xrightarrow{-3e, CuCl_2, MeOH} R-C(OMe)_2(CO_2Me) \qquad (89)$$

BT = 2-benzothiazolyl

$$Ph-C(R^1)(R^2)(CO_2H) \xrightarrow{R^3OH, electrolysis} Ph-C(R^1)(OR^3)_2 \qquad (90)$$

R¹ = Ph, H; R² = SPh, OR³; R³ = Me, Et

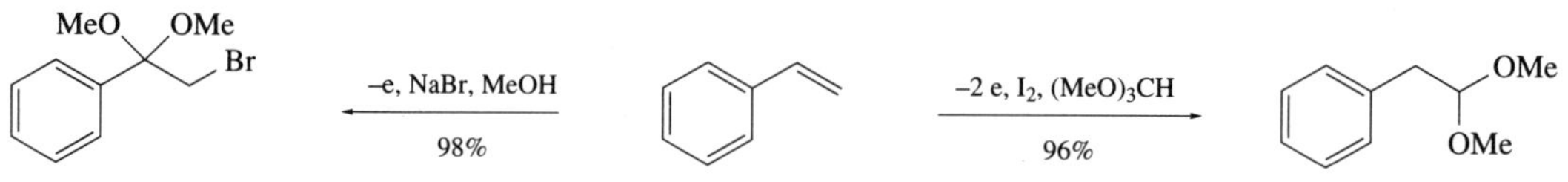

Scheme 33

4.04.3.2.6 Miscellaneous methods

The acetalization of formaldehyde catalysed by acid is generally inferior to alternative procedures. Methods have already been described which provide formaldehyde acetals by alkylation or acetal exchange. Two other similar methods using activated DMSO have been developed which produce symmetrical formaldehyde acetals in high yield (Equation (91)) ⟨72JA8929, 79JOC3727⟩. Two methods which are suitable for the preparation of acetals starting from aldehydes only are the treatment of aldazines with hypervalent iodine reagents and alkoxides ⟨82TL1537⟩ and the reaction of aldehydes with organoantimony alkoxides ⟨92ACR182⟩. Diazoalkanes react with alcohols in the presence of *t*-butyl hypochlorite to form acetals in moderate to good yield ⟨66AG(E)420⟩. The treatment of commercially available diphenyldiazomethane (**88**) with alcohols in the presence of 2,3-dichloro-5,6-dicyano-1,4-benzoquinone (ddq) provides a welcome addition to the few methods available for the high-yield formation of benzophenone acetals (Equation (92)) ⟨80TL3919⟩. α-Sulfenyl carbonyl derivatives are converted into acetals with ttn ⟨78TL4115⟩.

i, DMSO, NBS (2 equiv.), 50 °C, 86%; ii, DMSO (1 equiv.), TMS-Cl (1 equiv.), PhH, reflux, 84%

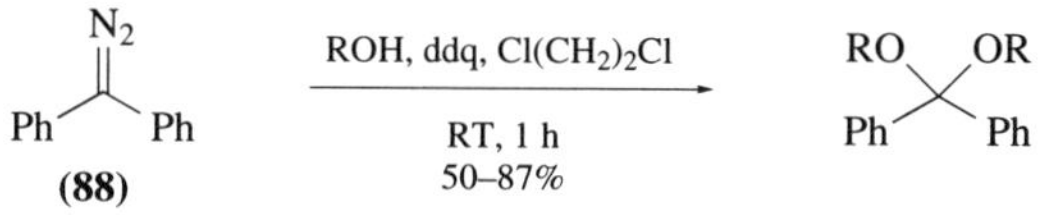

4.04.3.3 Unsymmetrical Acetals

4.04.3.3.1 From α-substituted ethers and alcohols

The displacement of a leaving group from the α carbon of an ether is the basis of the vast majority of glycoside-coupling methods. It is also one of the most general methods for the synthesis of unsymmetrical acetals. Many leaving groups have been developed for glycoside coupling that can be activated under a range of reaction conditions and also allow the control of stereochemistry at the anomeric centre in a variety of carbohydrate systems. This topic has been comprehensively reviewed ⟨86AG(E)212, 91COS(6)33, 93CRV1503⟩ and will not be covered in detail in the following discussion.

(i) From α-halo ethers

Alkylation of an alcohol with an α-halo ether is a general method for the formation of unsymmetrical acetals and is one of the oldest methods for the formation of a glycoside bond (Koenigs–Knorr reaction). This has been developed into a commonly employed strategy for the protection of alcohols as unsymmetrical acetals, usually of formaldehyde (Equation (93)) ⟨88CLY402, B-91MI 404-01⟩. A variety of α-chloro ethers have been employed for this purpose to form acetals which possess different stability characteristics; these are listed together with their adopted abbreviations in Table 10. The alkylation is usually carried out either by the treatment of the lithium or sodium alkoxide with the α-chloro ether in an ethereal solvent, or by treatment of the alcohol and a tertiary amine base such as diisopropylethylamine with the α-chloro ether in dichloromethane. The crystalline salt (**89**) was prepared from methoxyethoxymethyl (MEM) chloride to allow base-sensitive alcohols to form acetals (MEM ethers), simply by reacting with (**89**) in hot acetonitrile ⟨76TL809⟩.

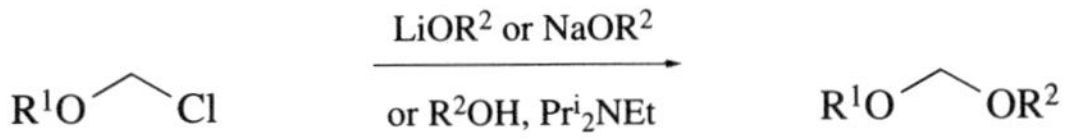

$$R^1O\frown Cl \xrightarrow[\text{or } R^2OH, Pr^i_2NEt]{LiOR^2 \text{ or } NaOR^2} R^1O\frown OR^2 \tag{93}$$

Table 10 α-Chloro ethers for the protection of alcohols as unsymmetrical acetals.

α-Chloro ether	Name of product ether (acetal)	Abbreviation	Ref.
MeOCH$_2$Cl	Methoxymethyl	MOM	72JA7827
MeOCH$_2$CH$_2$OCH$_2$Cl	Methoxyethoxymethyl	MEM	76TL809
TMS-CH$_2$CH$_2$OCH$_2$Cl	β-(Trimethylsilyl)ethoxymethyl	SEM	80TL3343
Cl$_3$CCH$_2$OCH$_2$Cl	Trichloroethoxymethyl	TCEM	79SC57
PhCH$_2$OCH$_2$Cl	Benzyloxymethyl	BOM	75JA6260
Bu^{t}OCH$_2$Cl	*t*-Butoxymethyl		78JOC3964
Me$_2$PhSiCH$_2$OCH$_2$Cl	(Dimethylphenylsilyl)methoxymethyl	SMOM	90TL2197
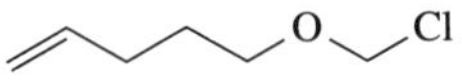 CH$_2$=CHCH$_2$CH$_2$CH$_2$OCH$_2$Cl	4-Pentenyloxymethyl	POM	88TL6549
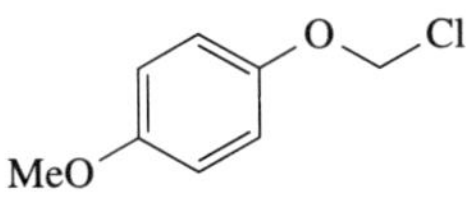 4-MeO-C$_6$H$_4$-OCH$_2$Cl	*p*-Anisyloxymethyl	*p*-AOM	89CL659
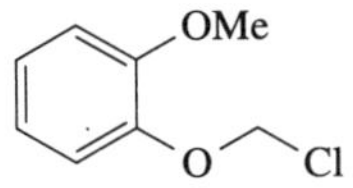 2-MeO-C$_6$H$_4$-OCH$_2$Cl	2-Methoxyphenoxymethyl or guaiacylmethyl	GuM	81TL1973
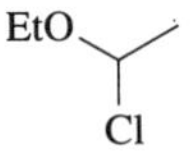 EtOCH(Cl)CH$_3$	Ethoxyethyl	EE	78JA1481

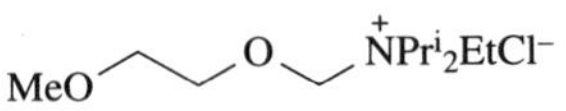

$$\text{MeO}\frown O\frown \overset{+}{N}Pr^i_2EtCl^-$$
(**89**)

MOM and MEM ethers and other simple acetals are converted into α-halo ethers by boron trichloride ⟨86JOC4711⟩ or dimethylboron bromide ⟨85JOC5379, 87TL2225⟩, and this has been used to prepare more-complex, unsymmetrical acetals (Scheme 34) ⟨88JA2248, 89JOC5695⟩. Chlorination of THF with sulfuryl chloride provides the 2-chloro derivative which reacts with alcohols to form 2-tetrahydrofuryl ethers ⟨79RTC371⟩. Alkylation of phenoxides with the α-bromo ester (**90**) gave diaryloxy acetals (Scheme 35) ⟨83S568⟩. Few unsymmetrical acetals of ketones have been prepared from α-halo ethers, presumably due to the instability of such halo ethers, but an example is shown in Scheme 36 ⟨72JOC521⟩. A vinylogous chloromethyl methyl ether, 3-chloro-1-methoxypropene (**91**), undergoes clean S$_N$2′ displacement of chloride when treated with alcohols to provide unsymmetrical acetals of acrolein (Equation (94)) ⟨87JOC782⟩.

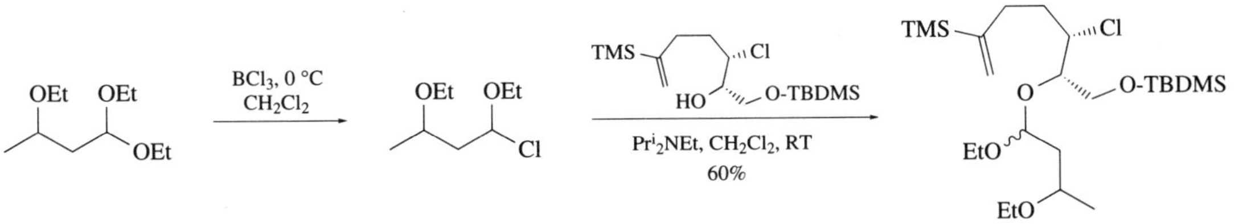

Scheme 34

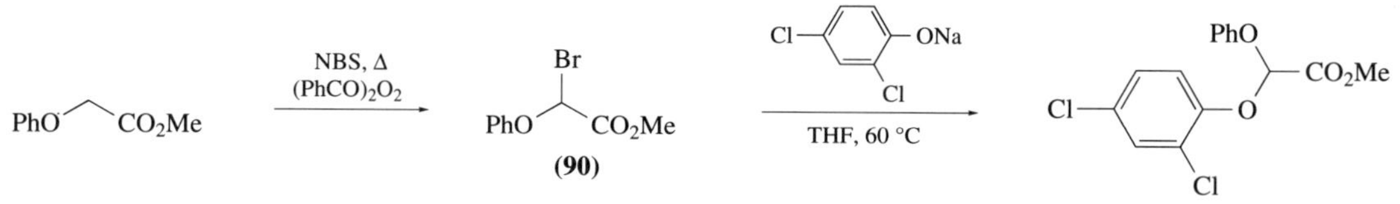

PhO-CH₂-CO₂Me $\xrightarrow{\text{NBS, }\Delta\text{ (PhCO)}_2\text{O}_2}$ PhO-CHBr-CO₂Me **(90)** $\xrightarrow[\text{THF, 60 °C}]{\text{2,4-Cl}_2\text{C}_6\text{H}_3\text{ONa}}$ 2,4-Cl₂C₆H₃-O-CH(OPh)-CO₂Me

Scheme 35

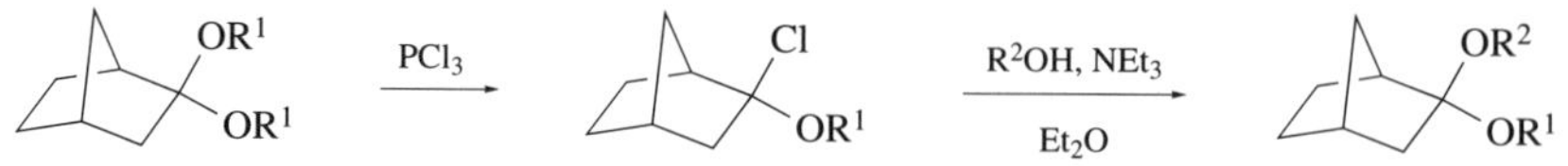

norbornyl-C(OR¹)₂ $\xrightarrow{\text{PCl}_3}$ norbornyl-C(Cl)(OR¹) $\xrightarrow[\text{Et}_2\text{O}]{\text{R}^2\text{OH, NEt}_3}$ norbornyl-C(OR²)(OR¹)

R¹ = Me, Et, Pr^i
R² = Me, Et, Pr^i, Bu^t

Scheme 36

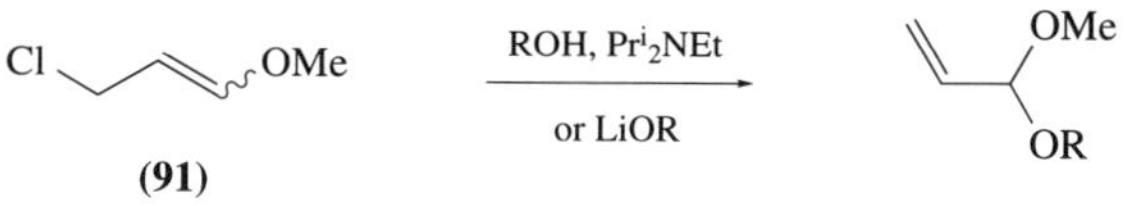

Cl-CH₂-CH=CH-OMe **(91)** $\xrightarrow{\text{ROH, Pr}^i_2\text{NEt or LiOR}}$ CH₂=CH-CH(OMe)(OR) (94)

(ii) From α-acyloxy ethers

Displacement of an acyloxy group from an α-acyloxy ether has been employed for the preparation of cyclic ether acetals only, including glycosides. The 2-diphenylacetoxytetrahydrofuranyl **(92)** and 2-diphenylacetoxytetrahydropyranyl **(93)** derivatives undergo displacement with primary and secondary alcohols under mild conditions (Equation (95)) ⟨78JOC3548⟩. Cyclic hemiacetals are converted in a mild, one-pot procedure via the trifluoroacetate derivative in a reaction that tolerates a range of sensitive functionality (Equation (96)) ⟨93SL111⟩. Other analogous leaving groups which have been developed for glycoside coupling but which may have broader applicability are trichloro-acetimidate ⟨89PAC1257⟩ and 1-imidazolylcarbonyl ⟨90SL255⟩.

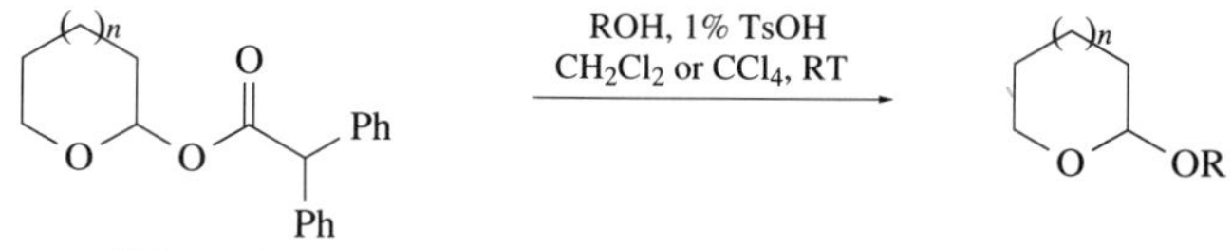

(cyclic ether with OC(O)CHPh₂) $\xrightarrow[\text{CH}_2\text{Cl}_2\text{ or CCl}_4\text{, RT}]{\text{ROH, 1% TsOH}}$ (cyclic ether with OR) (95)

(92) n = 0
(93) n = 1

HO-tetrahydrofuranyl-Bu^n $\xrightarrow[\text{ii, thiazolyl-CH}_2\text{CH}_2\text{OH}]{\text{i, (CF}_3\text{CO)}_2\text{O, NEt}_3\text{, dmap (cat.)}}$ thiazolyl-CH₂CH₂O-tetrahydrofuranyl-Bu^n 98% (96)

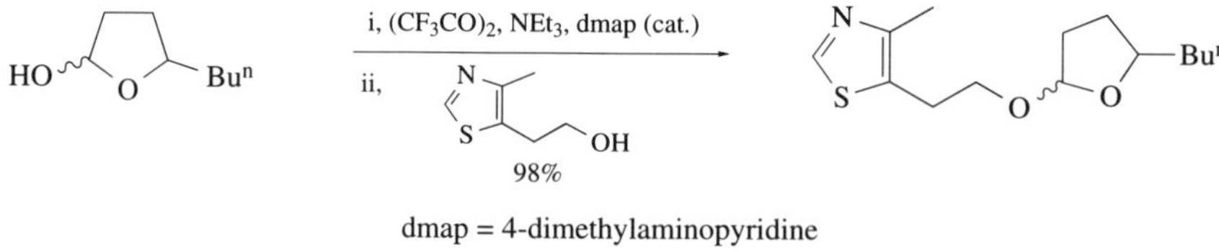

dmap = 4-dimethylaminopyridine

(iii) From O,S-acetals and derivatives

Activation of anomeric sulfides with thiophilic metal salts or by S-alkylation or oxidation is a common strategy for glycoside coupling. α-Sulfonyl ethers have also been developed as useful precursors for the preparation of other unsymmetrical acetals. Lactols, lactol ethers, dihydropyrans and dihydrofurans can all be converted into α-sulfonyl cyclic ethers which undergo easy displacement of the sulfonyl group with alcohols in the presence of magnesium bromide etherate and sodium bicarbonate to provide the corresponding acetals in high yield (Equation (97)) ⟨91T1329⟩. The readily prepared *t*-butoxymethyl phenyl sulfone **(94)**, can be alkylated and then converted into methyl acetals with acidic methanol (Scheme 37) ⟨91SL501⟩. However, of potentially greater utility is the displacement of the sulfonyl group only, to form mixed acetals (**95**) with alcohols. The sulfone (**94**) is also a useful alternative to *t*-butyl chloromethyl ether (Table 10) for the preparation of *t*-butoxymethyl ethers (Scheme 37) ⟨91SL503⟩. Dimethyl sulfide reacts with α,β-unsaturated acetals in the presence of trimethylsilyl triflate to form, for example, the sulfonium salt (**96**), which undergoes S_N2' displacement of dimethyl sulfide when treated with alkoxytributylstannanes to provide α,β-unsaturated mixed acetals (Scheme 38) ⟨93TL5769⟩.

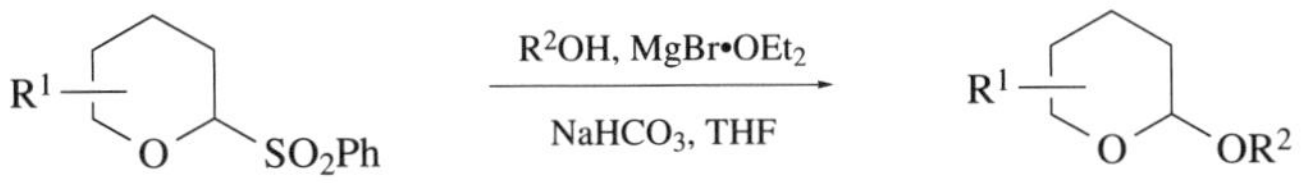

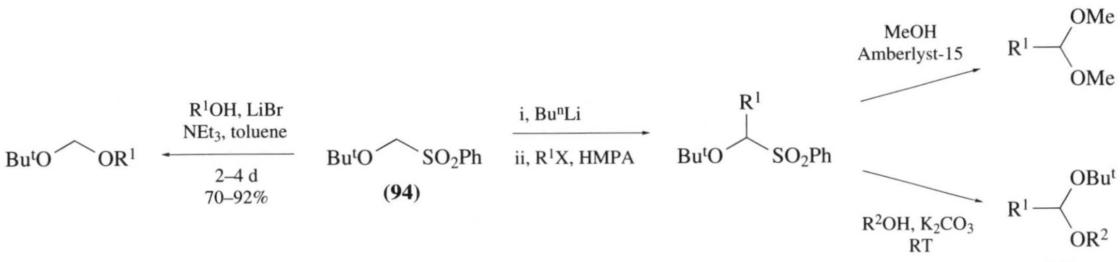

Scheme 37

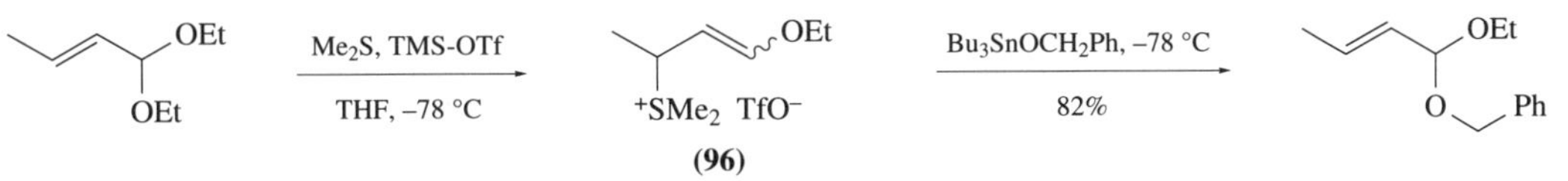

Scheme 38

4.04.3.3.2 From ethers

A few methods have been reported for acetal formation by the α-alkoxylation of ethers, and these are generally more useful for the preparation of acetals from cyclic ethers. Anodic oxidation of ethers in methanol introduces an α-methoxy group in low yield ⟨69JA2803⟩, but improved yields are obtained when the electrolysis is carried out in the mixed solvent methanol–acetic acid, although mixtures of acetals are formed from unsymmetrical ethers (Equation (98)) ⟨87S1099⟩. 2-Tetrahydrofuranyl ethers are formed in good yield either by the oxidation of THF with ceric triethylammonium nitrate (CTAN) in the presence of alcohols ⟨87S250⟩ or by radical coupling of THF with alcohols mediated by tetra-*n*-butylammonium peroxydisulfate (Equation (99)) ⟨93TL3581⟩. Methoxybenzyl ethers containing additional hydroxy groups at the α or β positions undergo intramolecular formation of methoxybenzylidene acetals when treated with ddq by attack of the hydroxy group on an intermediate benzyl cation ⟨82TL889⟩.

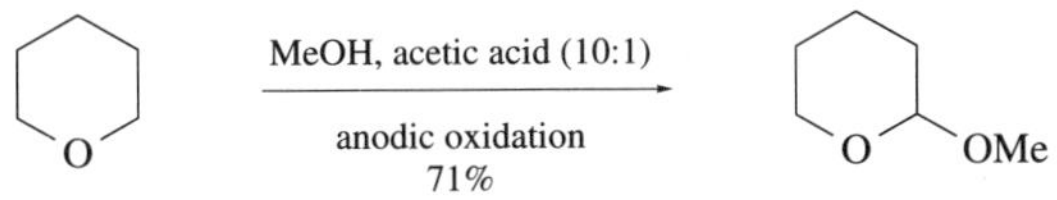

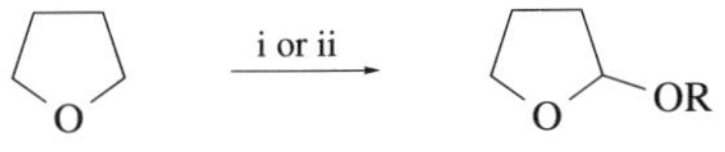

i, ROH, ceric triethylammonium nitrate, 50–100 °C, 30–98%; ii, ROH, $(Bu^n_4NOSO_2O)_2$, reflux, 81–97%

4.04.3.3.3 From cyclic hemiacetals

Cyclic hemiacetals can be converted into acetals under either acidic or basic conditions, and acetals often serve to protect the hemiacetal group during complex syntheses. The most common method involves acid-catalysed exchange of water for the alcohol (cf. Scheme 18), but as the reaction begins with the hemiacetal, milder reaction conditions are usually employed. This method and several variants have been employed for glycoside bond formation ⟨93CRV1503⟩. Alkylation of the hemiacetal hydroxy group is generally carried out with a strong base such as sodium hydride and an alkyl halide, although the method appears to be limited to primary alkylating agents. Again this reaction has been exploited for glycoside bond formation ⟨91COS(6)33⟩. Phase transfer catalysis has also been employed for anomeric O-alkylation ⟨90SC687⟩.

4.04.3.3.4 By hetero Diels–Alder reactions

The hetero Diels–Alder cycloaddition is a useful method for the preparation of 2-alkoxypyran derivatives which has been widely exploited in natural-product synthesis. Two modes of reaction are possible and have been studied. Cycloaddition of an alkoxy-substituted diene with an aldehyde as the dienophile has been most widely studied (Equation (100)). Alternatively, an inverse electron demand, Diels–Alder reaction between an enol ether and an α,β-unsaturated carbonyl derivative provides pyran derivatives (Equation (101)). Both types of reactions may occur under thermal conditions, but they are generally carried out with Lewis acid catalysis or under high pressure. Numerous examples have appeared in the literature, and reviews should be consulted for further details ⟨86ACR250, 91COS(2)661, 91HOU(E14a/1)412, 91HOU(E14a/1)431⟩.

4.04.3.3.5 Mixed alkyl silyl acetals

A few methods for the preparation of mixed alkyl silyl acetals such as (**97**) have been reported. Silylation of hemiacetals is an obvious method but is generally limited to cyclic hemiacetals ⟨93SL349⟩. The most general method depends on the low temperature trapping of the intermediate from dibal-H reduction of esters with a silylating reagent (Scheme 39) ⟨93TL1491⟩. The intermediates from organolithium addition to esters have also been trapped with TMS-Cl to form mixed acetals of ketones, but no attempt was made to isolate these derivatives ⟨86JOC951⟩. Several chloromethoxysilanes (**98**) have been used to protect alcohols as mixed alkyl silyl acetals (Equation (102)) ⟨89ACS706⟩. Silyl ketene acetals serve as precursors to alkyl silyl acetals by [3,3]-sigmatropic rearrangement ⟨86TL1557⟩ or aldol condensation ⟨84TL511⟩. An interesting example is shown in Scheme 40, where it is proposed that the intermediate (**99**) in the aldol reaction is reduced by hydride transfer from the borane promoter ⟨91JOC2276⟩.

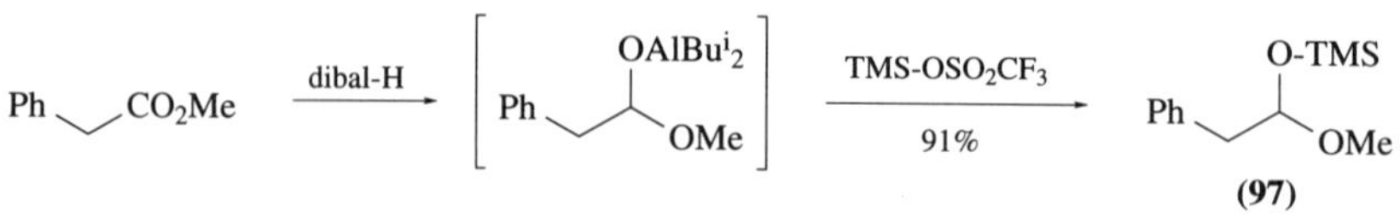

Scheme 39

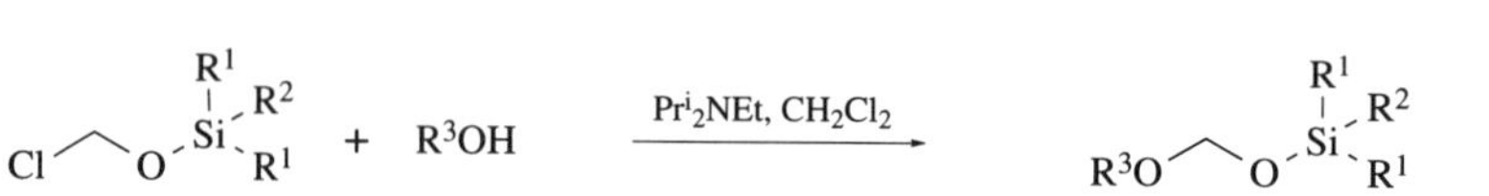

R^1 = Me, Ph; R^2 = Bu^t, thexyl

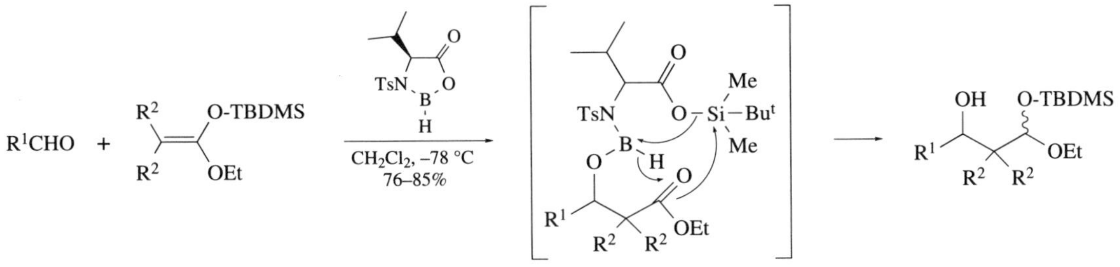

R^1CHO + [alkene with R^2, R^2, O-TBDMS, OEt] → (scheme reagents: TsN-B(H)-O-iPr, CH_2Cl_2, −78 °C, 76–85%) → [**(99)** transition state] → R^1-CH(OH)-CR^2_2-CH(O-TBDMS)(OEt)

TBDMS = *t*-butyldimethylsilyl

Scheme 40

4.04.4 OTHER DIOXYGEN DERIVATIVES

4.04.4.1 Synthesis of $R^1_2C(OCOR^2)_2$

A large number of methods have been described for the synthesis of *gem*-diacyloxy derivatives from different substrates (see ⟨B-91MI 404-03⟩ and ⟨91HOU(E14a/1)683⟩). These types of derivatives serve as useful protecting groups for aldehydes in particular because of their relative stability to acidic conditions and lability to mild basic conditions.

4.04.4.1.1 From aldehydes and ketones

(i) From aldehydes

The acid-catalysed reaction of aldehydes and acetic anhydride is the most common procedure for the preparation of 1,1-diacetoxy alkanes (Equation (103)). Sulfuric acid was the original catalyst, and is still the most widely used ⟨63OSC(4)489⟩. Other acid catalysts include methanesulfonic acid or phosphoric acid ⟨77JCED355⟩, perchloric acid ⟨77JOC1794⟩, boric and oxalic acids ⟨60JOC1699⟩, and nitric acid ⟨86JOC3811⟩. Nafion-H, a perfluorinated sulfonic acid resin, may be advantageous for some substrates due to the short reaction times and simple nonaqueous workup, and yields are good with aromatic aldehydes (including those having electron-withdrawing substituents), alkanals and 2-alkenals ⟨82S962⟩. Phosphorus trichloride is also a good catalyst for aromatic aldehydes (except those with electron-withdrawing substituents) and α,β-unsaturated aldehydes ⟨81S824⟩.

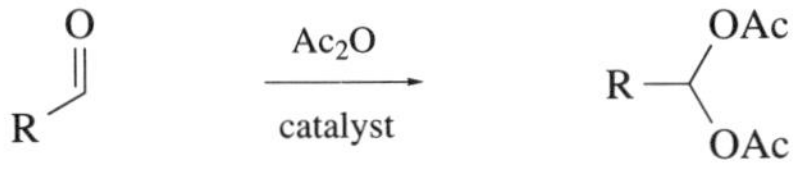

$$R-CHO \xrightarrow[\text{catalyst}]{Ac_2O} R-CH(OAc)_2 \quad (103)$$

Lewis acids have also been used to catalyse the formation of 1,1-diacetates when protic acids afforded poor yields or were not compatible with the substrate. Iron(III) chloride was reported to be a good catalyst for the preparation of geminal diacetates from aromatic and aliphatic aldehydes ⟨83JOC1765⟩. This method was used selectively to convert aldehydes into diacetates in the presence of ketones ⟨86SC833⟩. Several polycyclic aromatic aldehydes react with acetic anhydride in the presence of an excess of cobalt(II) chloride to afford the diacetates in good yield (49–74%) ⟨91JOC3283⟩. Cobalt(II) chloride is a much weaker Lewis acid for diacetate formation than iron(III) chloride, zinc chloride or boron trifluoride etherate, all of which have been used for this purpose ⟨60LA355, 81S824⟩. It reacts only with the most reactive aromatic aldehydes and therefore can be useful for selective functionalization of polycarbonyl compounds. Mixed diacyl derivatives can be prepared from aldehydes and formic acid and an anhydride with phosphorus pentoxide as the catalyst (Equation (104)) ⟨73S151⟩. Aldehydes such as chloral can be converted to the acylal in quantitative yield with acetic anhydride in pyridine, presumably by acylation of the hydrate ⟨70HCA1330⟩.

(ii) From ketones

The acid-catalysed reaction with anhydrides (see above) cannot generally be applied to ketones, but when trichloroacetic anhydride is employed, acylals are formed from ketones without a catalyst ⟨69T1679⟩. Most acylals derived from ketones depend on the formation of cyclic structures. Meldrum's acid (2,2-dimethyl-4,6-dioxo-1,3-dioxan) is readily prepared from malonic acid and acetone in acetic anhydride with sulfuric acid as the catalyst (Equation (105)) ⟨48JA3426⟩. Other ketones can also be employed to produce a large variety of 1,3-dioxan derivatives ⟨61CB929⟩. A similar method which has been widely employed involves the reaction of isopropenyl acetate with malonic acid derivatives to afford 5-substituted 2,2-dimethyl-4,6-dioxo-1,3-dioxans (Equation (106)). Sulfuric acid is generally used as the catalyst, and the products with the exception of the dimethyl derivative ($R^1 = R^2 = Me$) are isolated in moderate to good yields ⟨48JA3426, 75JOC3807, 76JOC1668, 81OS(60)66⟩.

Keto diacids such as (**100**) can be dehydrated by dissolution in neat acetyl chloride at room temperature to afford spirolactones (Equation (107)). Acid catalysis and azeotropic removal of water may also be used in some cases ⟨81JA5618⟩.

4- or 5-Oxo carboxylic acids exist as equilibria between acyclic and cyclic forms, and acylation of the cyclic form affords diacylal derivatives (Equation (108)). Many different acylating agents have been used, including acid chlorides and anhydrides, and isocyanates; examples selected from the literature are shown in Table 11.

Table 11 Diacyl acetals formation by the acylation of alcohols.

Substrate	Acylating reagent	Conditions	Product	Yield (%)	Ref.
3,4-dichloro-5-hydroxy-2(5H)-furanone	PhCOCl	110 °C, 3 h	3,4-dichloro-5-(benzoyloxy)-2(5H)-furanone	95	50JA2535
3,4-dichloro-5-hydroxy-2(5H)-furanone	PhNCO	PhH, reflux, 2 d	3,4-dichloro-5-(phenylcarbamoyloxy)-2(5H)-furanone	82	50JA2535
4-ethoxy-6-hydroxy-5,5-dimethyl-5,6-dihydro-2H-pyran-2-one	(MeCO)$_2$O	Pyridine, 20 °C, 12 h	4-ethoxy-6-acetoxy-5,5-dimethyl-5,6-dihydro-2H-pyran-2-one	70	89JOC4866
4,4-dimethyl-1-hydroxy-1-phenyl-isochroman-3-one	(MeCO)$_2$O	Reflux, 1 h	4,4-dimethyl-1-acetoxy-1-phenyl-isochroman-3-one	40	68BSB379
5-hydroxy-5-methyl-γ-butyrolactone (levulinic acid cyclic form)	(MeCO)$_2$O	Amberlyst-15, 40 °C, 15 min	5-acetoxy-5-methyl-γ-butyrolactone	70	75JHC749

4.04.4.1.2 *From carboxylic acids*

The alkylation of carboxylic acids with α-haloalkyl esters or lactones provides *gem*-diacyloxy derivatives (Equation (109)). The procedure has been used to prepare a number of important ester prodrugs of β-lactam antibiotics from their corresponding carboxylates in high yield (e.g., (**101**) ⟨76CPB102⟩ and (**102**) ⟨76TL3739⟩). The carbonate derivative (**103**) ⟨79CB148⟩ and the phenol (**104**) ⟨47JA2358⟩ were prepared similarly from the corresponding acetate and benzoate salts, respectively, although the yields were moderate in these cases. Alkylation involves the treatment of the sodium or potassium salt of the carboxylic acid with the α-haloalkyl ester. For unreactive substrates, the use of the iodoalkyl ester or its *in situ* formation from a less reactive derivative by addition of potassium iodide increases yields and shortens reaction times.

$$R^1COOH + X-CH_2-O-CO-R^2 \longrightarrow R^1-CO-O-CH_2-O-CO-R^2 \qquad (109)$$

(**101**) (**102**) (**103**) (**104**)

4.04.4.1.3 By oxidation of aromatic methyl and methylene groups

Aromatic methyl groups can be oxidized to diacyloxy derivatives with chromium trioxide, sulfuric acid and acetic anhydride (Equation (110)). Several functional groups on the aromatic ring are unaffected by the reaction conditions, including nitro ⟨63OSC(4)713⟩, chloro ⟨55RTC1429⟩, carboxylic acid or ester ⟨50JPP764⟩ and alkyl ⟨1900LA353⟩ moieties, but yields in most cases are only moderate.

$$\text{Ar-Me} \xrightarrow[\text{Ac}_2\text{O}]{\text{CrO}_3, \text{H}_2\text{SO}_4} \text{Ar-CH(OAc)}_2 \quad (110)$$

Methyl groups attached to heteroarenes such as 1,2-benzoxazoles can also be oxidized to the diacyloxy derivatives under similar conditions, again in moderate yields ⟨71JCS2166⟩. 4-Picoline was oxidized electrochemically with acetic anhydride, potassium acetate and a cobalt(III) species generated *in situ* from cobalt(II) acetate, to afford pyridine-4-carbaldehyde hydrate diacetate (Equation (111)) ⟨90H(31)1959⟩. Arylmethyl acetates such as benzyl acetate can be oxidized to diacyloxy derivatives with diacyl peroxides and copper(I) bromide ⟨60JOC899⟩. Similarly, 6-cyano-2-picolyl-*N*-oxide acetate was converted into the diacetate under the same conditions ⟨81JMC1181⟩.

$$\text{4-methylpyridine} \xrightarrow[\substack{\text{Ac}_2\text{O, KOAc, 80 °C} \\ 55\%}]{\substack{\text{carbon anode, 4 F mol}^{-1} \\ \text{Co(OAc)}_2 \text{ (10 mol\%)}}} \text{4-CH(OAc)}_2\text{-pyridine} \quad (111)$$

4.04.4.1.4 By oxidation of furan derivatives

Oxidation of 2-silyloxyfurans with lead(IV) acetate affords acetoxyfuranone derivatives (Equation (112)) ⟨80BCJ1061, 82TL353⟩. Similarly, 2-(trialkylstannyl)furans can also be oxidized to acetoxyfuranones by lead(IV) acetate, although the yields are generally lower ⟨92BCJ2366⟩. Other oxidants such as iodosobenzene and boron trifluoride etherate in combination with acetic acid generated furanone derivatives from 2-(trimethylsilyloxy)furan ⟨89TL3019⟩.

$$\xrightarrow[\substack{\text{PhH, 4 °C} \\ 83\%}]{\text{Pb(OAc)}_4} \quad (112)$$

2-Tri-*n*-butylstannylfuran derivatives also undergo oxidation by lead(IV) acetate to acetoxyfuranones. Aryl, alkyl and alkoxy groups survive the oxidation conditions, but yields are variable ⟨88CC560, 92BCJ2366⟩.

4.04.4.1.5 Miscellaneous methods

Alkynes afford diacylal adducts when treated with carboxylic acids and using mercuric salts or ruthenium complexes as catalysts ⟨82JOC3707, 83OM1689⟩. Malonic acids undergo bisdecarboxylation with lead(IV) acetate to afford diacetoxy derivatives ⟨66TL6145⟩. A palladium-catalysed intramolecular redox reaction has been described leading to novel allylic *gem*-diacetates in high yields. The procedure involves the treatment of propargylic acetates with acetic acid and palladium(0), giving exclusively the (*E*) isomer of the product (Equation (113)). The procedure is also applicable to the preparation of mixed geminal carboxylates ⟨92AG(E)1335⟩. 1,3-Dicarbonyl compounds such as 5,5-dimethyl-1,3-dioxocyclohexane can be oxidized to 2,2-bis-(4-methoxybenzoyloxy) derivatives using bis-(4-methoxybenzoyl)peroxide and sodium hydride in acetonitrile (Equation (114)) ⟨81CB1938⟩.

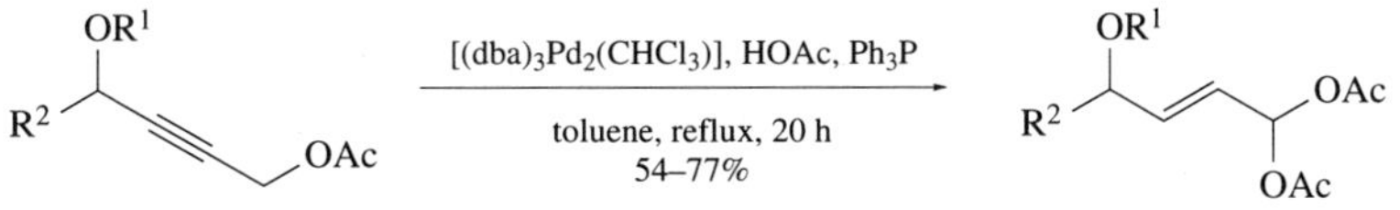

Other Derivatives   207

$$R^1 = \text{Me, TBDMS}$$
$$R^2 = \text{alkyl, aryl}$$

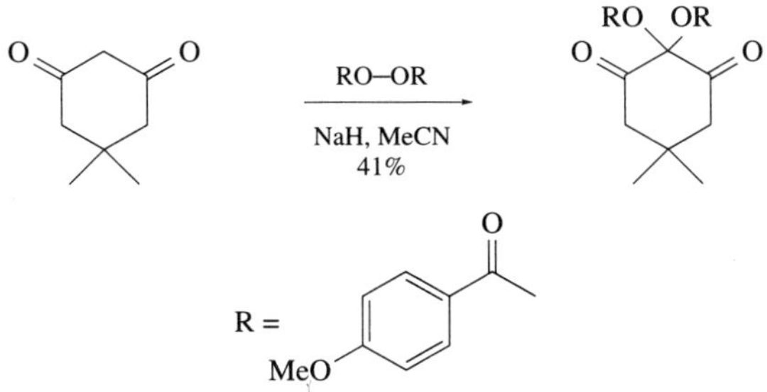

(113) Reaction of HC≡C–CH(OR¹)R² with CH₂OAc using [(dba)₃Pd₂(CHCl₃)], HOAc, Ph₃P, toluene, reflux, 20 h, 54–77%, giving R²(R¹O)CH–CH=CH–CH(OAc)₂.

dba = dibenzylideneacetone

(114) 5,5-Dimethyl-1,3-cyclohexanedione + RO–OR, NaH, MeCN, 41%, giving 2,2-bis(RO)-5,5-dimethyl-1,3-cyclohexanedione.

$$R = \text{4-MeO-C}_6\text{H}_4\text{-C(O)-CH}_2\text{-}$$

Phenyl vinyl sulfoxides derived from aromatic aldehydes or aliphatic ketones and methyl phenyl sulfoxide in the presence of triflic anhydride and sodium acetate in acetic anhydride undergo Pummerer reactions with migration of the phenylthio group to afford 2-phenylthio acylals (Equation (115)). The yields are generally high except with a methoxy phenyl group ⟨91TL6973⟩. Further methods which are specific for certain substrates are described in the literature ⟨91HOU(E14a/1)683⟩.

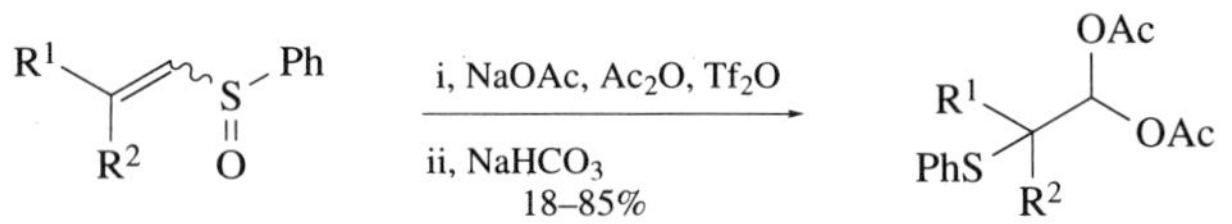

(115) R¹R²C=CH–S(O)Ph → (i, NaOAc, Ac₂O, Tf₂O; ii, NaHCO₃, 18–85%) → R¹R²C(SPh)–CH(OAc)₂

$$R^1 = \text{aryl, alkyl}$$
$$R^2 = \text{H, Me}$$

4.04.4.2 Synthesis of $R^1_2C(OCOR^2)OR^3$

A number of functional groups serve for the preparation of acyl alkyl acetals, and the discussion below sets out the major synthetic approaches.

4.04.4.2.1 From ethers

The oxidation of aliphatic ethers with diacyl peroxides is a commonly employed procedure, and was used to convert diethyl ether into 1-benzoyloxy-1-ethoxyethane (Equation (116)) ⟨47JA500⟩. *t*-Butyl methyl ether was converted into the corresponding benzoyloxy derivative under similar conditions ⟨64JCS1217⟩. The addition of copper(I) bromide or chloride shortens the reaction time, as exemplified in Equation (117) for the preparation of a 1,3-dioxolane derivative ⟨76AG(E)688⟩. Further examples starting from tetrahydrofuran ⟨65T871⟩, 1,4-dioxane ⟨60JOC899⟩ and an isochroman derivative ⟨63AK(20)225⟩ have also been described.

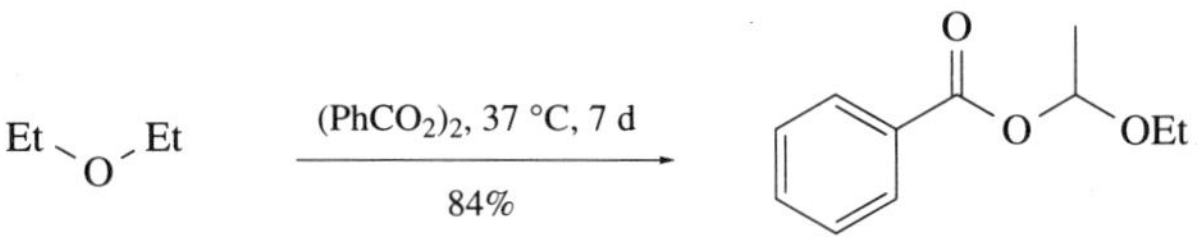

(116) Et–O–Et → (PhCO₂)₂, 37 °C, 7 d, 84% → PhC(O)–O–CH(CH₃)–OEt

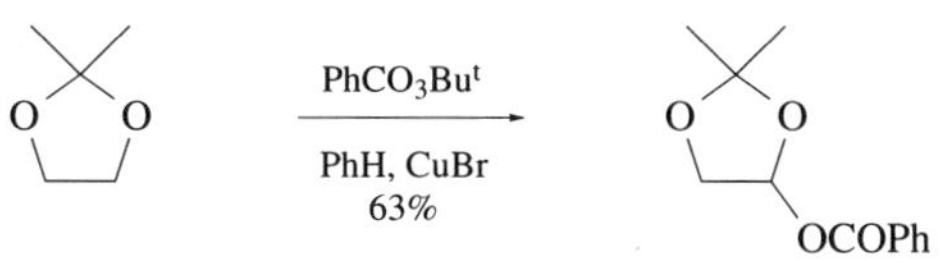

(117) 2,2-dimethyl-1,3-dioxolane → PhCO₃Bu^t, PhH, CuBr, 63% → 2,2-dimethyl-4-(OCOPh)-1,3-dioxolane

Allyl ethers with pendant alkyltin residues can be oxidized to acyl alkyl acetals with lead(IV) acetate in good yield (Equation (118)) ⟨89CL221⟩. A variety of (acyloxy)boranes have been shown to cleave oxiranes and tetrahydrofurans to generate unsymmetrical ethers. With a substrate containing an oxirane as well as a pyran ring (Equation (119)), oxybis(diacetoxyborane) affords the ring-enlarged product in a moderate yield. The acetoxy derivative (**105**) was similarly prepared ⟨92SL565⟩.

4.04.4.2.2 *From enol ethers and enol esters*

(i) Enol ethers

The acid-catalysed addition of carboxylic acids to both cyclic and acyclic enol ethers is generally an efficient method for the preparation of acyl alkyl acetals. Several catalysts have been employed including TFA (Equation (120)) ⟨82JOC3517⟩, sulfuric acid ⟨48JA2805⟩ and 2,4-dimethylbenzenesulfonic acid ⟨78JOC3548⟩. Intramolecular addition of a carboxylic acid to a dihydropyran ring has been demonstrated in the synthesis of cephem spiroacetal lactones ⟨94SL152⟩. Carboxylic acids also add to ethyl vinyl ether in the presence of copper(II) bromide to give 1-ethoxyethyl esters (Equation (121)) ⟨83SC629⟩.

The nitration of ethyl vinyl ether with acetyl nitrate (caution—acetyl nitrate can be hazardous, and reactions using this reagent should be conducted behind a safety screen) at $-33\,°C$ affords the acyl alkyl acetal (**106**) in 75% yield ⟨88S706⟩. 3-Chlorobenzoyl hypobromite, prepared from potassium bromide and mcpba in the presence of 18-crown-6, reacts with 3,4-dihydro-2*H*-pyran to afford the bromopyran derivative (**107**) in 75% yield ⟨89SC197⟩. Allyl enol ethers can be cyclized to give acetoxytetrahydrofuran derivatives with palladium(II) acetate (Equation (122)). The yields are higher for phenyl derivatives, and no Claisen rearrangement is observed under the reaction conditions ⟨85TL5411⟩.

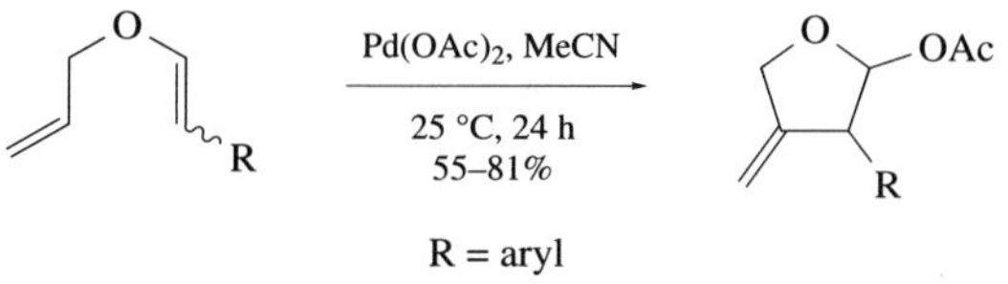

Other Derivatives

(122)

(ii) Enol esters

Epoxidation of enol esters with peroxy acids such as perbenzoic acid ⟨54JA2943⟩, peracetic acid ⟨54JA743⟩ and mcpba ⟨74JOC77⟩ provides the corresponding epoxy esters. Dimethyldioxirane is also a very efficient reagent for this reaction and has been used to prepare the epoxy esters in high yields from the corresponding enol esters ⟨89TL4223⟩. This reagent was also suitable for the epoxidation of γ-methylene-γ-butyrolactones in high yield (Equation (123)) whereas mcpba failed to afford the spiro epoxy lactones ⟨89TL4223⟩.

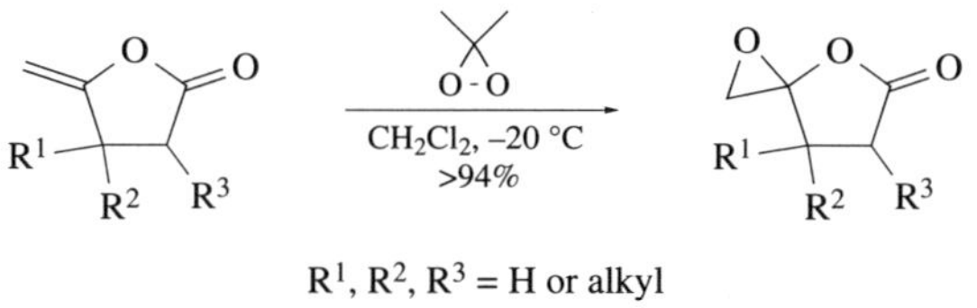

(123)

$R^1, R^2, R^3 = H$ or alkyl

4.04.4.2.3 From carboxylic acids by alkylation with α-halo alkyl ethers

The esterification of carboxylic acids with α-halo alkyl ethers is a widely used procedure for the preparation of acyl alkyl acetals, and is a common tactic for protection of the carboxyl group (Equation (124)) ⟨B-91MI 404-04⟩. Some representative examples are given in Table 12, and numerous others have been reported. Most examples involve the use of chloromethyl ethers, which lead to acyl alkyl acetals of formaldehyde, but cyclic chloro alkyl ethers (Table 12) and α-bromo ethers ⟨88SC2337⟩ have also been used.

$$R^1-CO_2H + X\frown OR^2 \longrightarrow R^1\frown O\frown OR^2 \quad (124)$$

4.04.4.2.4 From alcohols and α-haloalkyl carboxylate derivatives

Alkylation of alcohols with α-haloalkyl carboxylate derivatives under basic conditions (Equation (125)) provides acyl alkyl acetals. The reaction is applicable to both cyclic and acyclic esters, and carbamates also react to afford high yields of the products (Table 13).

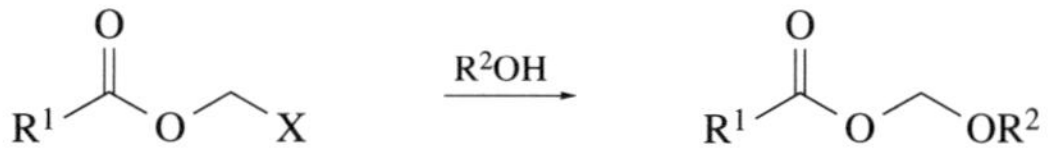

(125)

4.04.4.2.5 From hemiacetals

Acylation of hemiacetals with anhydrides or acid chlorides is effective in producing the acyl alkyl acetals (Equation (126)) ⟨69T4257⟩ (for further high yielding examples see ⟨51JA5252, 62JA813⟩ and ⟨71JA746⟩. The acyclic hemiacetal (**108**) was acylated with ethyl chloroformate or acetyl chloride to afford the corresponding acyl alkyl acetals (Equation (127)) ⟨65JOC3834⟩.

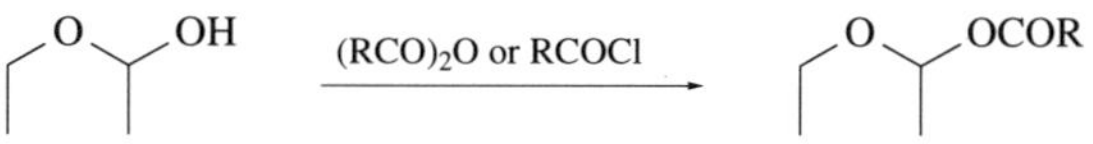

(126)

Table 12 Acyl alkyl acetals by the alkylation of carboxylic acids.

Carboxylic acid	α-Halo ether	Product	Conditions	Yield (%)	Ref.
cyclopropane with HO and CO$_2$H	TMS–CH$_2$CH$_2$–O–CH$_2$Cl	cyclopropane(HO)(C(O)OCH$_2$OCH$_2$CH$_2$TMS)	Et$_3$N, THF, 0 °C	80	84TL4195
pentyl–CO$_2$Na	Ph–CH$_2$–O–CH$_2$Cl	pentyl–C(O)–O–CH$_2$–O–CH$_2$Ph	HMPA, 20 °C, 48 h	73	75IOC2962
3-MeO-C$_6$H$_4$–CO$_2$H	MeO–CH$_2$Cl	3-MeO-C$_6$H$_4$–C(O)–O–CH$_2$–OMe	Ag$_2$O, Et$_2$O, reflux, 1 h	80	72SC361
CH$_3$CH(OH)CO$_2$H	MeO–CH$_2$Cl	CH$_3$CH(OH)C(O)–O–CH$_2$–OMe	Pr$^i_2$NEt, CH$_2$Cl$_2$, 20 °C, 22 h	91	84TL5409
Ph–CH$_2$CH$_2$–CO$_2$H	MeO–CH$_2$Cl	Ph–CH$_2$CH$_2$–C(O)–O–CH$_2$–OMe	DMF, NMP, Et$_3$N, 20 °C, 1 h	99	86JOC546
EtCO$_2$H	MeO–CH$_2$CH$_2$–O–CH$_2$Cl	EtC(O)–O–CH$_2$–O–CH$_2$CH$_2$OMe	KOBu^t, Et$_2$O	82	79JA2501
Ph–CH$_2$–CO$_2$H	2-chloro-tetrahydrofuran	Ph–CH$_2$–C(O)–O–(tetrahydrofuran-2-yl)	Et$_3$N, THF, 20 °C, 0.5 h	90	76TL1725
MeCO$_2$Na	chloro-epoxide	acetoxy-epoxide	Bu^{n}OH, EtOH, 0 °C, 20 h	66	79CB148
MeCO$_2$Na	2,5-dichloro-1,4-dioxane	2,5-diacetoxy-1,4-dioxane	AcOH, 20 °C, 2 h	33	79CB148

Table 13 Acyl alkyl acetal by the alkylation of alcohols.

Alcohol	α-Halo-alkyl carboxylate derivative	Conditions	Product	Yield (%)	Ref.
MeOH	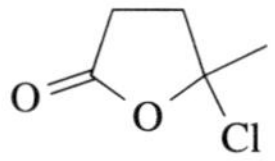 [chloro-methyl-γ-butyrolactone]	Na$_2$CO$_3$, 20 °C, 0.5 h	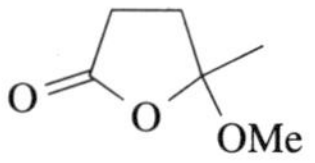 [methoxy-methyl-γ-butyrolactone]	62	48JA2624
MeOH	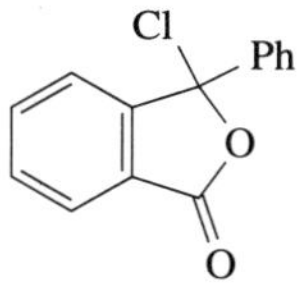 [3-chloro-3-phenylphthalide]	Pyridine, 20 °C	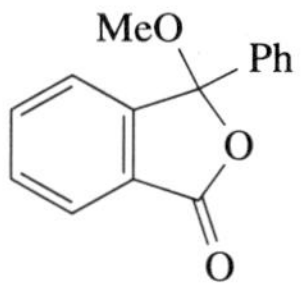 [3-methoxy-3-phenylphthalide]	95	41JA1537
MeOH	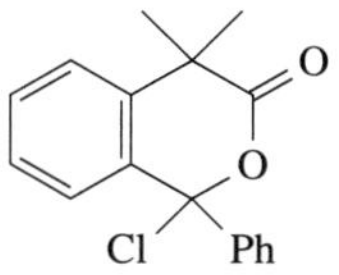 [1-chloro-1-phenyl-4,4-dimethylisochroman-3-one]	Pyridine, 20 °C, 1 h	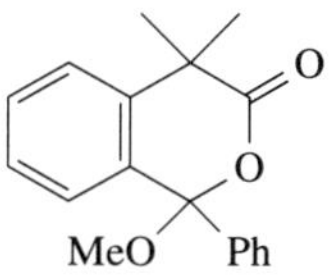 [1-methoxy-1-phenyl-4,4-dimethylisochroman-3-one]	45	62BSB379
Bu^{t}OH	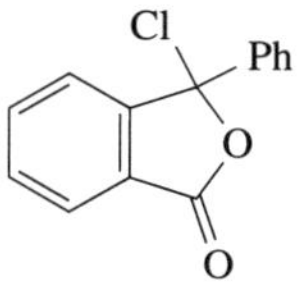 [3-chloro-3-phenylphthalide]	Pyridine, 20 °C	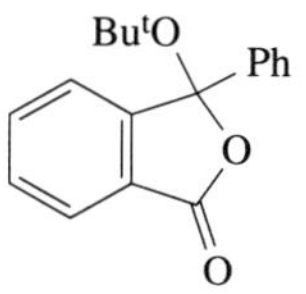 [3-t-butoxy-3-phenylphthalide]	59	50JA514
PhOTl	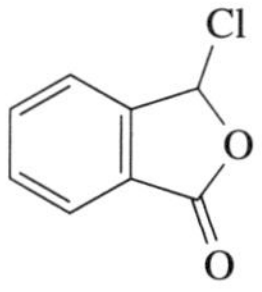 [3-chlorophthalide]	DMF, 20 °C, 1.6 h	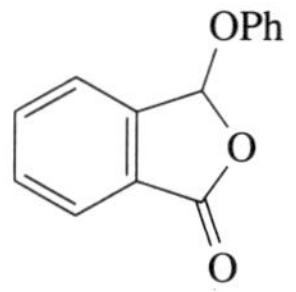 [3-phenoxyphthalide]	85	83JOC5280
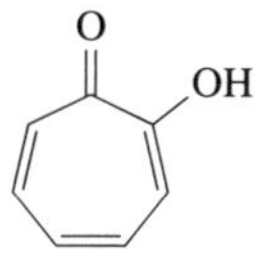 [2-hydroxytropone]	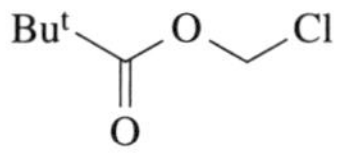 Bu^{t}C(O)OCH$_2$Cl	K$_2$CO$_3$, 18-crown-6, MeCN, reflux, 10 h	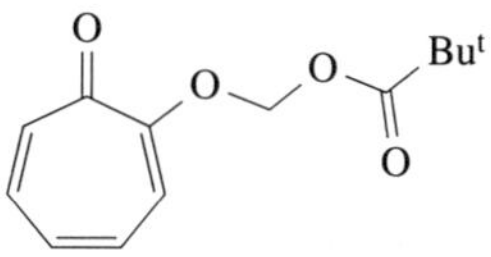 [tropone-O-CH$_2$-O-C(O)Bu^t]	65	85SC225
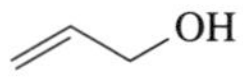 allyl-OH	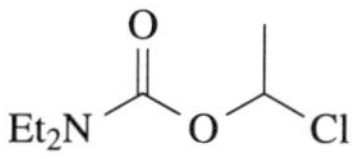 Et$_2$N-C(O)-O-CH(Me)-Cl	NaHCO$_3$, 50 °C, 1.75 h	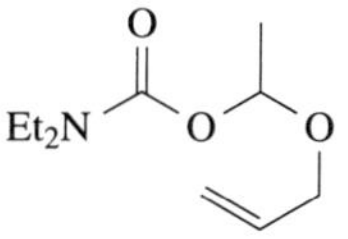 [Et$_2$N-C(O)-O-CH(Me)-O-allyl]	90	87SC1467

$$\text{ClF}_2\text{C}-\underset{\text{OH}}{\overset{\text{OEt}}{\text{C}}}-\text{CF}_2\text{Cl} \quad \xrightarrow[\text{Et}_2\text{O}]{\text{ClCOR, pyridine}} \quad \text{ClF}_2\text{C}-\underset{\text{O-C(O)R}}{\overset{\text{OEt}}{\text{C}}}-\text{CF}_2\text{Cl} \qquad (127)$$

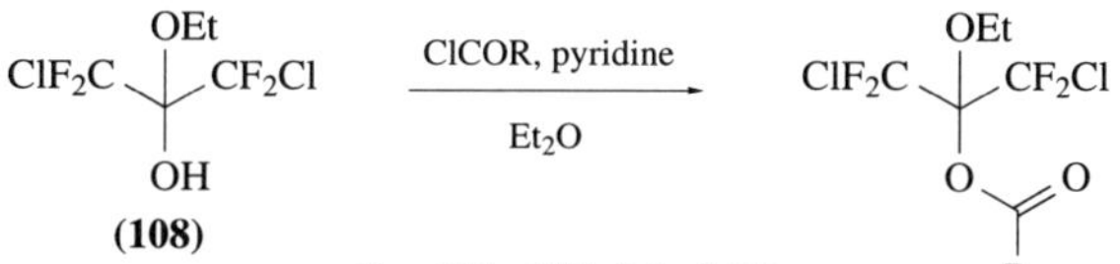

(108)

R = OEt, 67%; Me, 84%

The acid-catalysed addition of alcohols to hydroxyfuranone derivatives leads to alkoxyfuranones (Equation (128)). Sulfuric acid is the catalyst of choice, and the product is isolated in good yield ⟨64JOC1371⟩. Lewis acids such as zinc chloride have also been used to prepare similar derivatives ⟨72MI 404-01⟩, and in some examples simply heating the alcohol and hydroxyfuranone without a catalyst is sufficient to form the product ⟨57JCS158, 72CPB2123⟩.

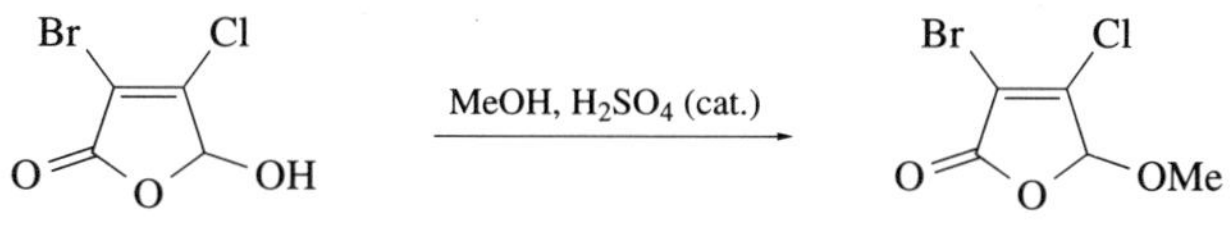

$$[\text{3-bromo-4-chloro-5-hydroxy-2(5H)-furanone}] \xrightarrow{\text{MeOH, H}_2\text{SO}_4 \text{ (cat.)}} [\text{3-bromo-4-chloro-5-methoxy-2(5H)-furanone}] \qquad (128)$$

4.04.4.2.6 *From acetals*

Acetals react with anhydrides with or without an added acid catalyst to afford acyl alkyl acetals by the replacement of an alkoxy group. Dimethoxymethane is converted into methoxymethyl acetate simply by heating with acetic anhydride (Equation (129)) ⟨54JA5161⟩. 1-Ethoxyisochroman ⟨56CB1254⟩ and 1-methoxy-2-benzopyran-4-one ⟨87JCS(P1)195⟩ undergo similar reactions with acetic anhydride to afford the corresponding 1-acetoxy derivatives in 87% and 63% yields, respectively. The addition of sulfuric acid as a catalyst also leads to high yields of product ⟨54JA693⟩.

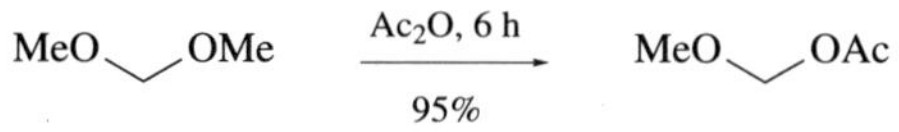

$$\text{MeO} \diagdown \text{OMe} \xrightarrow[95\%]{\text{Ac}_2\text{O, 6 h}} \text{MeO} \diagdown \text{OAc} \quad (129)$$

Cyclic acetals undergo ring cleavage when treated with anhydrides under acidic conditions to afford acyclic derivatives (Equation (130)). The reaction is applicable to 1,3-dioxolanes ⟨46JA734⟩ and 1,3-dioxanes ⟨59JOC1768, 67BSF4172, 84JOC4958⟩. Nafion-H and acetic anhydride cleave α-keto acetals to the α-acetoxy α-methoxy analogues ⟨83S891⟩ (see Scheme 2 ⟨86S513⟩).

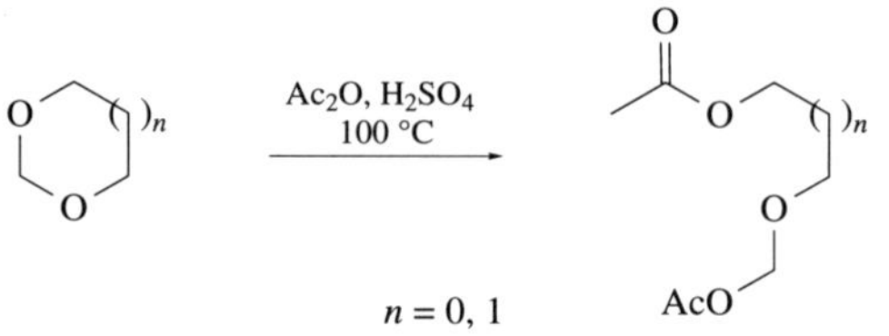

(130)

$n = 0, 1$

4.04.4.2.7 From aldehydes and ketones

(i) By Baeyer–Villiger oxidation

Baeyer–Villiger oxidation of α-alkoxy ketones is a high-yielding route to acyl alkyl acetals (Equation (131)). The oxidant of choice is often mcpba, and the reaction is usually conducted at room temperature in dichloromethane. The tricyclic derivative (**109**) and related analogues were prepared in high yield under these conditions (Equation (132)) ⟨80JA1198⟩. Similarly prepared in high yield, from the corresponding cyclic ketones, were the derivatives (**110**) ⟨89CC178⟩, (**111**) ⟨77JOC3458⟩ and (**112**) ⟨85JOC5167⟩. Ethyl 3-oxo-4-alkoxybutyrates can be oxidized with mcpba in 70–80% yields (Equation (133)). The presence of the 4-alkoxy group has a profound effect on the course of the reaction as no hydroxylation at the central methylene group is observed ⟨91TL2413⟩.

(131)

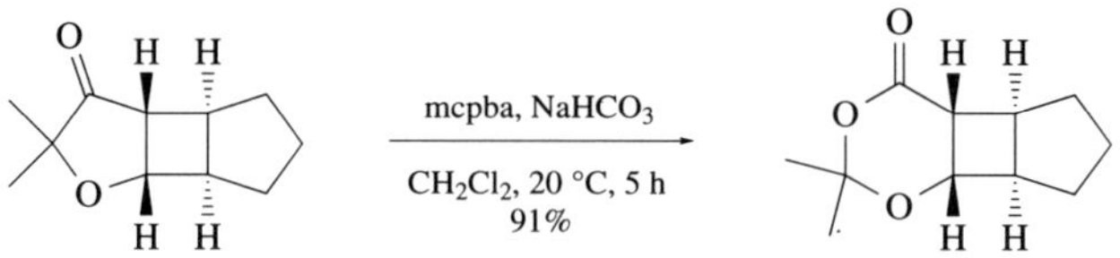

(132)

(**109**)

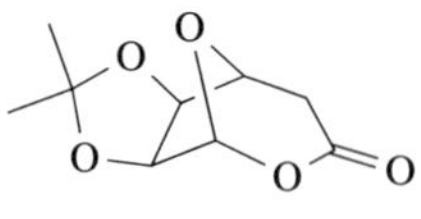

(**110**) 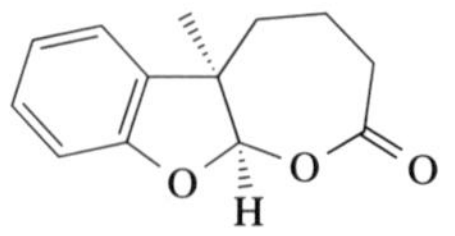   (**111**) 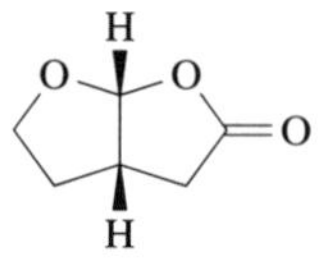   (**112**)

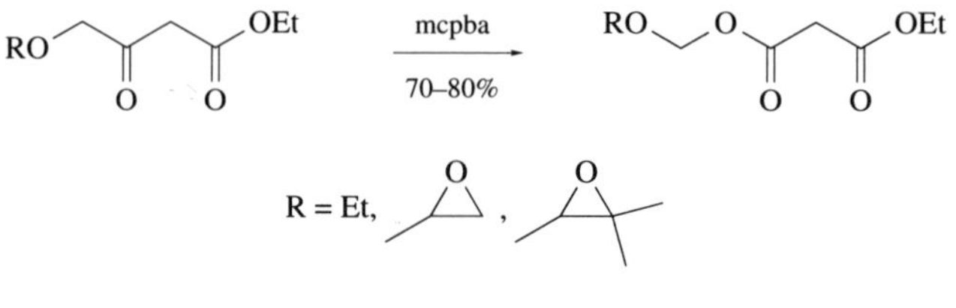

(133)

$R = \text{Et,}$

The epoxidation and Baeyer–Villiger oxidation of α,β-unsaturated ketones occur under similar mild conditions with peroxy acids (cf. epoxidation of enol acetates) to afford acyl alkyl acetals (Equation (134)) ⟨59JOC284, 74JOC77⟩. Peroxyphthalic acid has also been employed in reactions of this type ⟨57JA456⟩.

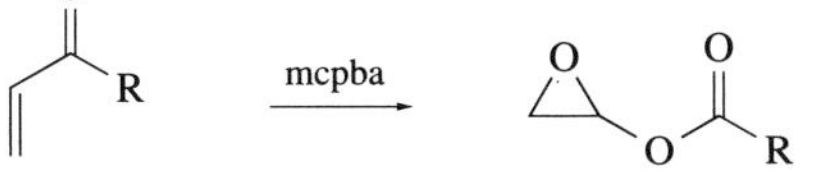

$$\text{(134)}$$

(ii) With hydroxy acid derivatives

(a) Preparation of 4-oxo-1,3-dioxolanes with α-hydroxy acids. 4-Oxo-1,3-dioxolanes can be prepared from glycolic acids and an aldehyde or a ketone (Equation (135)) with acid catalysis (sulfuric acid) in benzene or toluene. The yields are generally low for simple aldehydes whereas cyclic ketone derivatives can be prepared in high yields ⟨70BSF332⟩. Other similar methods have been described including the preparation of spiro derivatives ⟨84JCS(P1)1531, 84T1313, 94TL2537⟩. An improved procedure involves trimethylsilyl trifluoromethanesulfonate or trimethylsilyl iodide catalysed addition of trimethylsilyl (trimethylsilyloxy)acetate to carbonyl compounds. The yields by this method are superior and the procedure is also applicable to acetal substrates ⟨87JOC1353⟩. A similar method has also been reported for the preparation of 5-alkyl-2-*t*-butyl-1,3-dioxolan-4-ones by the trimethylsilyl triflate-catalysed reaction between bis(trimethylsilyl) derivatives of α-hydroxy carboxylic acids and pivaldehyde ⟨87JOC1351⟩.

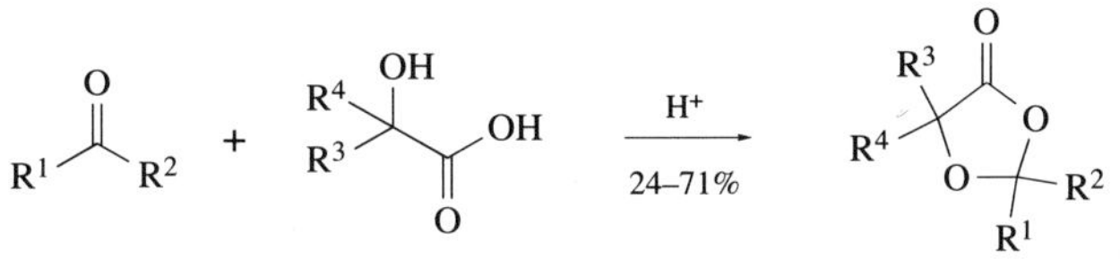

$$\text{(135)}$$

(b) Preparation of 4-oxo-1,3-dioxane derivatives from β-hydroxy acids. The acid-catalysed addition of β-hydroxy carboxylic acids to aldehydes or acetals ⟨86S649, 87HCA448⟩ provides 4-oxo-1,3-dioxane derivatives (Equation (136)). The substrates are heated with an acid catalyst such as toluene-*p*-sulfonic acid ⟨87HCA448, 89HCA690⟩, acetic acid or sulfuric acid ⟨78JOC1248⟩ or DOWEX 50 ⟨70TL4095, 87HCA448⟩ in benzene, toluene or dichloromethane. The reaction is quite versatile, and derivatives have been prepared from formaldehyde ($R^1 = R^2 = H$) ⟨70TL4095⟩ and aliphatic aldehydes ($R^1 = $ alkyl, $R^2 = H$) ⟨86AG(E)178, 87HCA448⟩. β-Hydroxy acids which provide spiro (R^5, $R^6 = (CH_2)_5$) ⟨78JOC1248⟩ and bicyclo compounds (R^4, $R^5 = (CH_2)n$, $n = 3, 4$) ⟨87TL3791, 89HCA690⟩ have also been used. The addition of salicylic acids to vinyl acetate affords similar derivatives (4-oxo-2*H*,4*H*-1,3-benzodioxin), but yields are moderate (Scheme 41) ⟨47JA2358⟩.

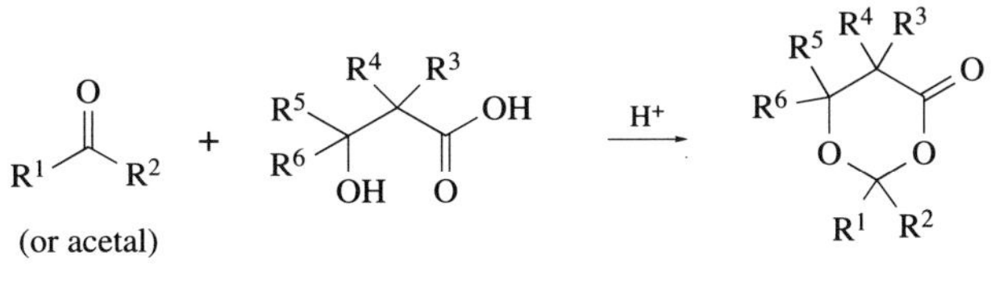

$$\text{(136)}$$

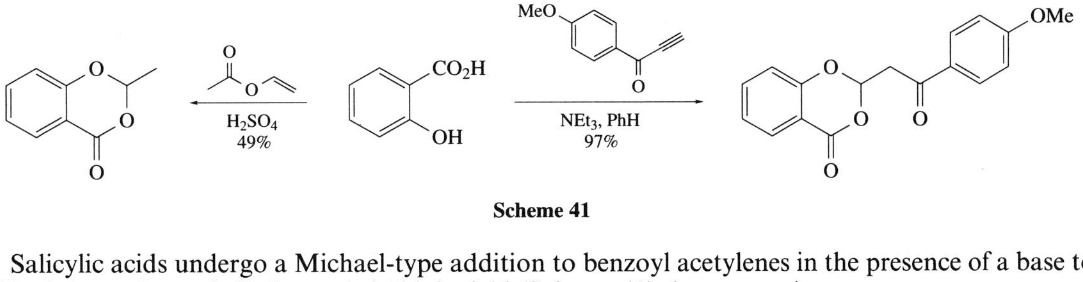

Scheme 41

Salicylic acids undergo a Michael-type addition to benzoyl acetylenes in the presence of a base to afford the cyclic acyl alkyl acetals in high yield (Scheme 41) ⟨79JCS(P1)36⟩.

4.04.4.3 Other Derivatives

4.04.4.3.1 1,2,4-Trioxolanes (ozonides)

No discussion is offered here, and adequate references are cited in the literature ⟨B-92MI 404-02⟩.

4.04.4.3.2 1,2,4-Trioxane

No discussion is offered here, but see ⟨83CC1064, 91JA8168, 91CC947, 94BMC931⟩ for further information.

4.04.4.3.3 $R^1_2C(OR^2)O_2R^3$

No discussion is offered here, but compounds containing this group ($R^2 = H$, $R^3 = C$ residue or $R^2 = C$ residue, $R^3 = H$) have been reported in the literature.

4.04.4.3.4 $R^1_2C(OR^2)OX$ and $R_2C(OX)_2$ (X = heteroatom)

Other dioxygen derivatives of the type $R^1_2C(OR^2)OX$, where X is a heteroatom such as sulfur, nitrogen or phosphorus, are also known (for X = Si, see Section 4.04.3.3.5). Many variations of this type of derivative have been described in the literature, and some of these, for example OX = sulfonate or phosphate, have been used in glycoside bond formation ⟨93CRV1503⟩. Symmetrical variants such as $R^1_2C(OSO_2R^2)_2$ have also been reported, but are less common (see, for example, ⟨87S49⟩ and ⟨88JOC5783⟩).

4.05
Functions Incorporating Oxygen and Another Chalcogen

RICHARD H. WIGHTMAN
Heriot-Watt University, Edinburgh, UK

4.05.1 FUNCTIONS CONTAINING OXYGEN AND SULFUR	215
4.05.1.1 Monothioacetals and Other Derivatives with Dicoordinate Sulfur	215
4.05.1.1.1 Acyclic compounds	215
4.05.1.1.2 Compounds with oxygen in a ring	223
4.05.1.1.3 Compounds with sulfur in a ring	226
4.05.1.1.4 Cyclic monothioacetals	229
4.05.1.2 Derivatives with Tricoordinate Sulfur	232
4.05.1.2.1 α-Alkoxy sulfoxides	232
4.05.1.2.2 α,β-Epoxy sulfoxides	233
4.05.1.2.3 α-Hydroxy sulfinates	234
4.05.1.3 Derivatives with Tetracoordinate Sulfur	234
4.05.1.3.1 α-Hydroxy sulfones	234
4.05.1.3.2 α-Alkoxy sulfones, α-acyloxy sulfones and related compounds	235
4.05.1.3.3 α,β-Epoxy sulfones	236
4.05.1.3.4 α-Hydroxy sulfonic acids	237
4.05.2 FUNCTIONS CONTAINING OXYGEN AND EITHER SELENIUM OR TELLURIUM	237
4.05.2.1 Dicoordinate Selenium and Tellurium Derivatives	237
4.05.2.1.1 From carbonyl compounds, other acetals and enol ethers	238
4.05.2.1.2 From α-halo ethers and related compounds	239
4.05.2.1.3 By seleno-Pummerer reactions	240
4.05.2.2 Tricoordinate Selenium Derivatives	241

4.05.1 FUNCTIONS CONTAINING OXYGEN AND SULFUR

4.05.1.1 Monothioacetals and Other Derivatives with Dicoordinate Sulfur

4.05.1.1.1 Acyclic compounds

(i) From carbonyl compounds and other acetals

Adducts of hydrogen sulfide and carbonyl compounds are of low stability, except those of aldehydes and ketones with highly electron-deficient carbonyl groups. Thus, H_2S will react under pressure with trifluoroacetaldehyde and hexafluoroacetone to give the 1:1 adducts in 57% and 84% yields respectively ⟨65JOC2190⟩.

Similarly, acyclic adducts of thiols with aldehydes and ketones are generally of low stability ⟨75T809⟩, although benzyl thiol and formaldehyde interact to form $PhCH_2SCH_2OH$ in 65% yield ⟨59LA(620)1⟩. Again, electron-deficient carbonyl groups form more stable adducts, such as those isolated by addition of various thiols to trichloroacetaldehyde ⟨74JPR304⟩ and to α-ketoaldehydes (glyoxal derivatives) ⟨52JA1068, 69JOC1799⟩, although the author feels there is no evidence which excludes these latter adducts being cyclic dimers (1,4-dioxan derivatives).

However, hemithioacetals can be trapped as their *O*-silylated derivatives [α-(trimethylsilyloxy)sulfides] when the thiol is added to the carbonyl compound and TMS-Cl in the presence of pyridine ⟨76TL319⟩, and the same class of compounds can also be made by treatment of aldehydes and ketones with TMS sulfides, either preformed ⟨77JA5009⟩ or prepared *in situ* from the thiol and TMS-imidazole ⟨90S104⟩.

Monothioacetals in which neither heteroatom forms part of a ring can be prepared directly from the thiol and either the carbonyl compound or its *O*,*O*-acetal, but careful attention to conditions is necessary to avoid the formation of significant amounts of the dithioacetal. The monothioacetal (**1**) is obtained by mixing equimolar amounts of benzaldehyde and thiophenol in partially aqueous methanol containing HCl, the product separating out as an oil. Other similar compounds such as (**2**) were obtained from the *O*,*O*-acetal and the thiol in CCl_4 containing HCl, the reaction being monitored by NMR spectroscopy ⟨79JA1476⟩. However, good yields of monothioacetals can be obtained from *O*,*O*-acetals by treatment with one equivalent of the thiol at low temperatures in the presence of $BF_3 \cdot Et_2O$, as in the formation of $PhSCH_2OMe$ from dimethoxymethane ⟨86JOC879⟩, and in the synthesis of (**3**) (61%) ⟨80JA6900⟩ and (**4**), obtained (73%) as a single isomer of undetermined stereochemistry ⟨77JA4835⟩.

The use of diethylaluminum thiophenoxide can give better control in acetal exchange procedures, as in the example in Equation (1), although again use of larger excesses of the reagent can give significant amounts of dithioacetals ⟨85CL1933⟩. In the late 1980s, reagents of the type $Bu_nSn(SPh)_{4-n}$ have been used, in the presence of BF_3, to effect acetal exchange, Equations (2) and (3) being typical examples ⟨89T1209⟩, whilst the simple reagent combination of a thiol and $MgBr_2$ in ether can effect the formation of *O*,*S*-acetals from acetals in good yield ⟨89TL6697⟩.

Activation of a dithioacetal by methylation can also give rise to a monothioacetal (Equation (4)) ⟨75TL3267⟩, and similar activation by bromine was used to convert the methylthiomethyl (MTM) ether (**5**) into the dinucleotide analogue (**7**) (Equation (5)) which was subsequently incorporated into oligodeoxynucleotides ⟨91JA7767⟩.

i, (**5**), Br$_2$, 2,6-diethylpyridine, mol. sieves, C$_6$H$_6$; ii, add (**6**)
R = Ph(p-MeOC$_6$H$_4$)$_2$C

Sequential use of dimethylboron bromide and a thiol can be used to convert O,O-acetals into monothioacetals, as in the examples of Equations (6) and (7). This latter case represents the conversion of a methoxymethyl (MOM) protecting group into the MTM group, and methoxyethoxymethyl (MEM) groups could be similarly transformed ⟨85JOC5379⟩. The same reagent combination was used to open cyclic benzylidene acetals to give, for example, the monothioacetal-alcohol (**8**), and the combination Ph$_2$BBr/PhSH gave the diastereomeric mixture (**9**), where the 4,6-O-benzylidene acetal was cleaved with 4:1 regioselectivity in favour of (**9**) ⟨90CJC897⟩. Corey *et al.* have also employed the reagent BrB(SPr^i)$_2$ to convert MEM ethers into monothioacetals of the type ROCH$_2$SPr^i ⟨84TL3⟩.

Aldehydes can be converted directly into monothioacetals by treatment with equimolar amounts of a silyl ether and phenyl(trimethylsilyl)sulfide in the presence of catalytic quantities of TMS-OTf, as exemplified by Equation (8) ⟨91TL467⟩. An interesting sequence of two allylic displacements can be used to convert acetals of α,β-unsaturated aldehydes into their monothioacetals; an example is given in Scheme 1, and some more structurally-complex cases have also been reported ⟨93TL5769⟩.

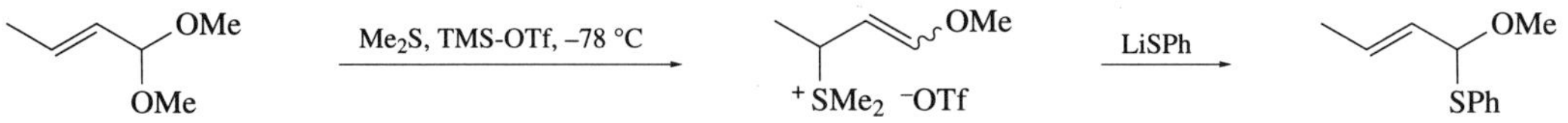

Scheme 1

[Scheme 1: CH₃CH=CHCH(OMe)₂ → (via Me₂S, TMS-OTf, –78 °C) → intermediate with ⁺SMe₂ ⁻OTf and OMe → (via LiSPh) → CH₃CH=CHCH(OMe)(SPh)]

(ii) From α-halo ethers and thiols

It is likely that the boron-based methods noted above involve α-bromo ethers as intermediates, and the interaction of an α-halo ether with a thiol is a fairly general route to *O,S*-acetals, used for the preparation of PhSCH₂OMe ⟨12BSB323⟩ (see above for a route to this compound avoiding the use of the carcinogenic ClCH₂OMe), and various other methoxy (arylthio)methanes ⟨62JCS3686, 64TL543, 67TL3057⟩. The functionalised diketopiperazine (**10**) was made in 90% yield by treatment of the bisthiol with ClCH₂OMe and KOBuᵗ ⟨74TL1549⟩, and compounds of type (**11**; X = OMe, Me, H, Cl, NO₂) were prepared from the thiolate and the benzylic chloro ether ⟨70JA5464⟩. More complex monothioacetals can be made from PhSCH₂OMe via its lithio derivative ⟨92JCS(P1)2303⟩; interaction of this organolithium species with 3-ethoxycyclohex-2-enone gave the product (**12**) in 93% yield ⟨86JOC879⟩.

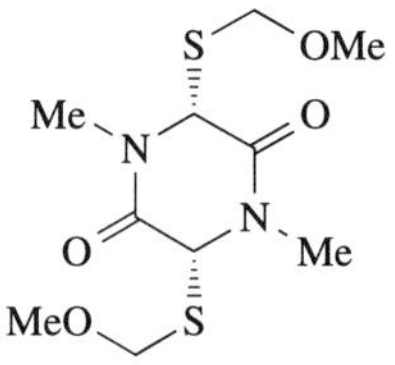

[Structures (10), (11), (12)]

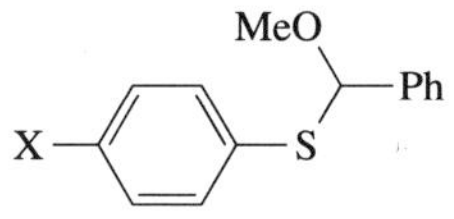

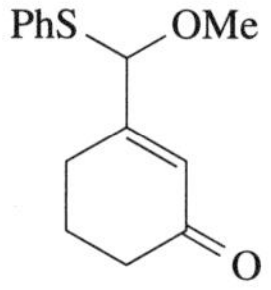

Monothioacetal functions have been used as protecting groups for the sulfur of cysteine. Thus, the derivative (**13**) was obtained in high yield from reaction of the thiolate, made by reductive cleavage of cystine, with the α-chloro ether ⟨64JCS3832, 70JOC215⟩, and the benzyloxymethyl protecting group has been similarly introduced and used to protect the thiol during peptide synthesis ⟨89CPB526⟩.

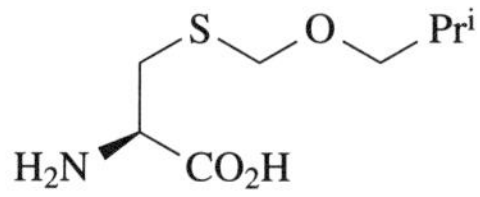

[Structure (13)]

Allylic rearrangement can be used to make acyclic monothioacetals of acrolein (Equation (9)) ⟨87JOC782⟩.

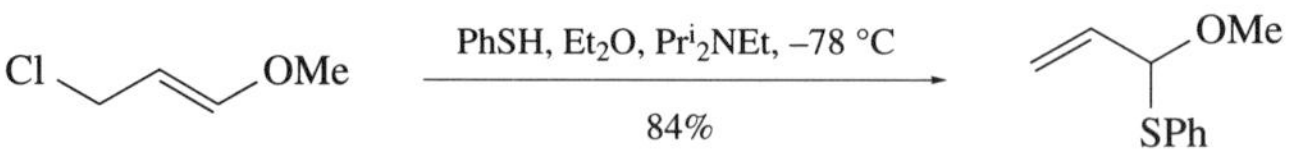

[Equation (9): Cl-CH₂-CH=CH-OMe → (PhSH, Et₂O, Pri₂NEt, –78 °C, 84%) → CH₂=CH-CH(OMe)(SPh)]

(iii) From α-halo sulfides and oxygen nucleophiles

The interaction of alcohols with α-halo sulfides represents a fairly general route to monothioacetals. Examples include those in Equations (10) and (11) ⟨80JA6900⟩, where in each case the α-halo sulfide was made by halogenation of the sulfide. A combination of a Lewis acid and a silver(I) salt can also be used to facilitate reactions of this type, as in the case of Equation (12), where the diastereomer indicated predominates in a ratio of 9:1 over the epimer ⟨82CL1555⟩. In 1994, it was reported that the α-fluoro sulfide (**14**), in which the fluorine was introduced using XeF₂, undergoes substitution on chromatography in the presence of methanol (Equation (13)) ⟨94JOC544⟩.

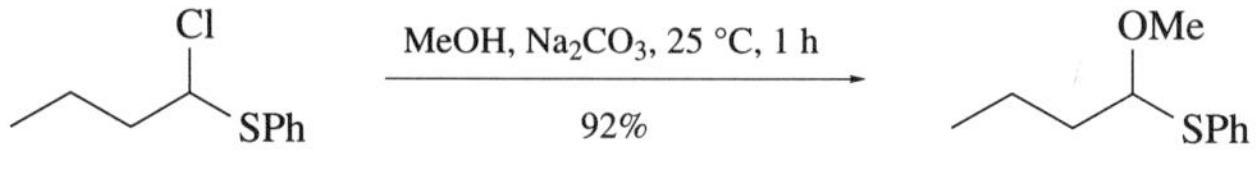

[Equation (10): CH₃CH₂CH(Cl)(SPh) → (MeOH, Na₂CO₃, 25 °C, 1 h, 92%) → CH₃CH₂CH(OMe)(SPh)]

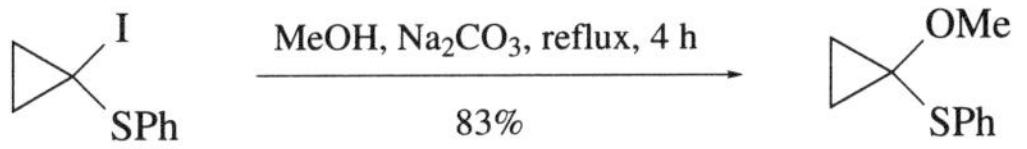

Oxygen and Sulfur

$$\text{(cyclopropyl)(I)(SPh)} \xrightarrow[83\%]{\text{MeOH, Na}_2\text{CO}_3\text{, reflux, 4 h}} \text{(cyclopropyl)(OMe)(SPh)} \quad (11)$$

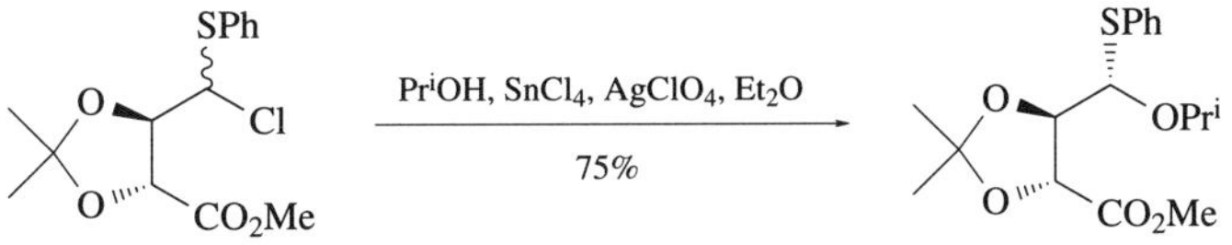

$$\text{(12)}$$

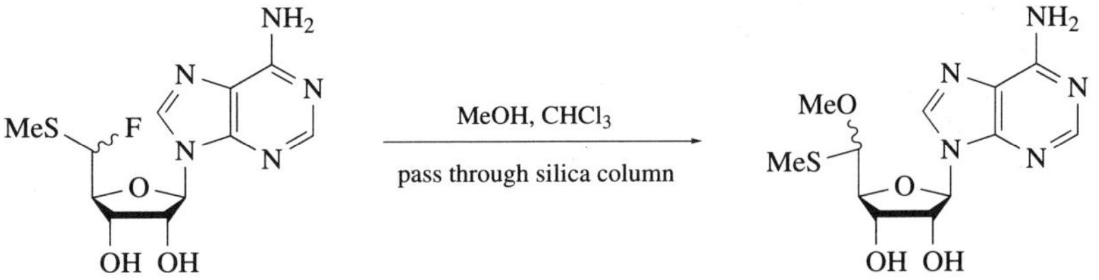

$$\text{(13)}$$

(14)

However, probably the most common use of this approach to *O,S*-acetals is for the protection of alcohols as methylthiomethyl (MTM) ethers. As originally employed by Corey and Bock, the alkoxide of a primary alcohol was treated with ClCH$_2$SMe in DME in the presence of NaI to give the MTM ether in good yield ⟨75TL3269⟩; poorer yields were reported for secondary alcohols, but the MTM ether (**15**) has been obtained in 80% yield by essentially this procedure ⟨90TL2385⟩. Primary, secondary and allylic alcohols react smoothly with ClCH$_2$SMe in the presence of AgNO$_3$ and triethylamine in benzene or cyclohexane; the MTM ether (**16**) was obtained in 82% yield by this method ⟨79CL1277⟩. MTM ethers of phenols are also easily made ⟨77TL533⟩, and the related phenylthiomethyl ethers of phenols have also been employed ⟨80SC911⟩. The ((2-(methylthio)phenyl)thio)methyl (MPTM) protecting group has been developed for use in nucleoside chemistry ⟨89JOC5998⟩, and used for protection of the 2′-hydroxy group in oligoribonucleoside synthesis; Scheme 2 indicates the introduction of the protecting group for use in this context ⟨91CL121⟩.

The MTM group has also been used for the protection of carboxylic acids, the MTM ester being

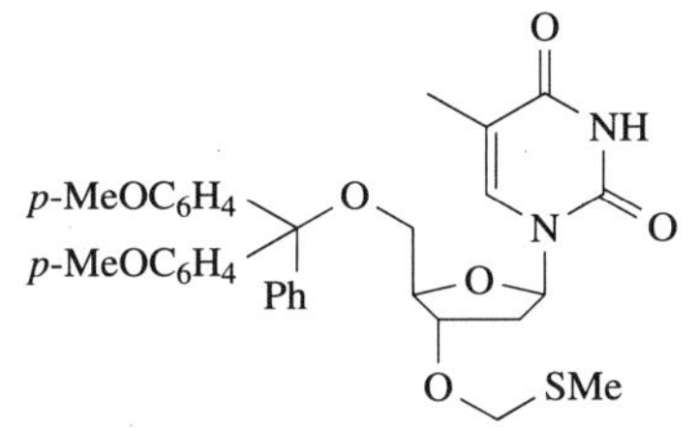

(15) 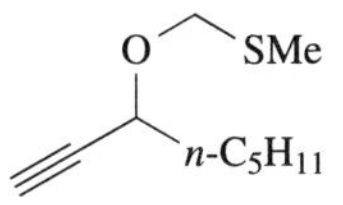   (16)

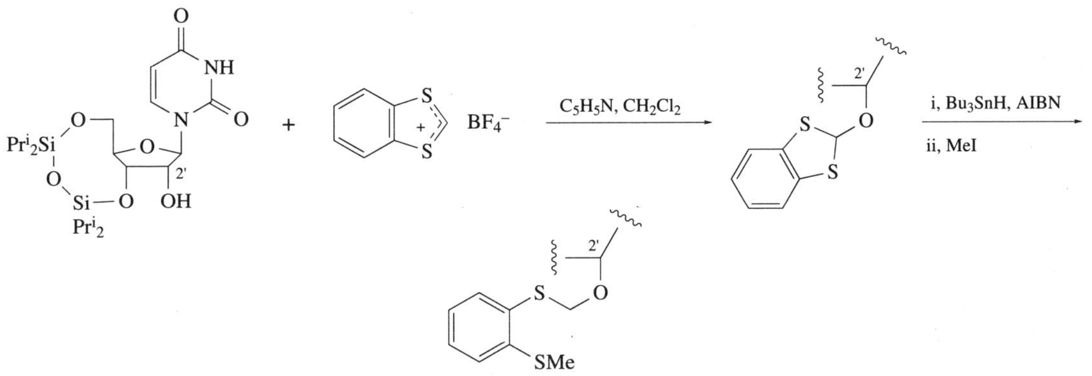

AIBN = 2,2′-azobisisobutyronitrile

Scheme 2

made from the acid and ClCH$_2$SMe in the presence of a tertiary amine 〈73CC224, 78BCJ2401〉, or from the potassium salt and ClCH$_2$SMe in presence of NaI and 18-crown-6 〈78TL731〉.

Some other mechanistically distinct routes to MTM ethers and esters are mentioned below (see 4.05.1.1.1(iv) and 4.05.1.1.1(v)).

(iv) By direct oxidation of sulfides

Clearly, given the ease by which sulfides undergo oxidation to sulfoxides and sulfones, specific oxidants must be employed in order to achieve α-oxidation. Lead tetraacetate (LTA) was shown in work from the 1940s to effect α-acetoxylation of dibenzyl sulfide in high yield in nonpolar solvents, but other sulfides gave lesser amounts of this type of product 〈49LA(563)54〉, and this process does not seem to be generally applicable in cases where an α-radical does not have an adjacent stabilising group. In the late 1970s, Trost et al. have shown that α-(phenylthio)ketones, made by sulfenylation of the enolate, undergo α-acetoxylation readily on treatment with LTA in benzene at reflux 〈77JA4405〉; Equation (14) provides an example from work directed towards the taxane skeleton 〈86T3323〉.

The same group have also shown that β-hydroxy sulfides, which can be prepared by an interesting oxidative addition of the elements of benzenesulfenic acid to the alkene, can be cleaved oxidatively to give α-acetoxy sulfides; the procedure is illustrated by Scheme 3 and Equation (15) provides another example from the synthesis of a natural product 〈78JA7103〉.

The interaction of symmetrical acyclic sulfides with t-butyl peracetate or perbenzoate in the presence of CuBr leads to α-acyloxy sulfides in moderate to good yields (Equation (16)) 〈62T15〉, and diacyl peroxides have also been used to convert sulfides into their α-acyloxy derivatives 〈57LA(602)135, 72JOC2885〉. This latter oxidation has been developed into an oxidative route for the formation of MTM ethers, involving the treatment of a mixture of the alcohol and dimethyl sulfide in acetonitrile with dibenzoyl peroxide; primary, secondary and tertiary alcohols all react well, the MTM ethers of geraniol and menthol, for example, being obtained in 88% and 96% yield respectively 〈88TL3773〉. This method has also been used, in the presence of 2,6-lutidine, to prepare MTM ether (**15**) in 75% yield 〈91T1547〉.

Electrochemical methods have also been used for oxidation of sulfides at the α-position. Acetoxylation was achieved by electrolysis of a solution of the sulfide in acetic acid containing sodium acetate, with platinum electrodes and a current density of 0.05 A cm^2, as in the formation of the α-acetoxy sulfide (**17**) in 71% yield from octyl phenyl sulfide. Unsymmetrical dialkyl sulfides not surprisingly gave a mixture of products, but the method was applied successfully to a number of α-(phenylthio)ketones 〈80TL2557〉. α-Acetoxyphenylthiomethane has been similarly prepared (70%)

from methyl phenyl sulfide ⟨84SUL1⟩. Anodic oxidation has also been used to make the α-acetoxy sulfide (**18**), and electrolysis in methanol containing tetraethylammonium tosylate gave the *O*,*S*-acetal (**19**) in high yield; this methoxylation was unsuccessful with the nonfluorinated ethyl phenyl sulfide ⟨86TL3869⟩. Phenylthiomethyl ethers such as (**20**), obtained in 52% yield, can be prepared by anodic oxidation of phenyl trimethylsilylmethyl sulfide in the presence of alcohols (graphite anode, current density 1.0 A dm^{-2}, CH$_3$CN containing Et$_4$NOTs and the alcohol) and α-acyloxy compounds (phenylthiomethyl esters) can be made similarly ⟨87CL1095⟩. Electrolysis of cyanomethyl sulfides in methanol containing TsOH gives products (**21**; R = Me, Et, Bu^t, Ph) in yields of over 70% ⟨87CC122⟩.

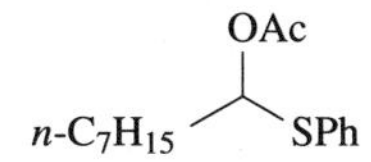

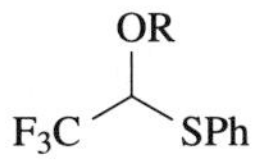

(v) From sulfoxides

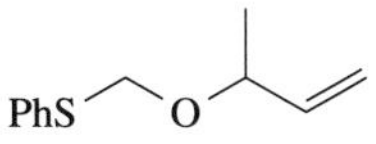

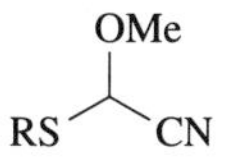

The formation of an α-functionalised sulfide from a sulfoxide bearing at least one α-hydrogen atom is referred to as the Pummerer reaction. A review and comprehensive tabulation of examples of Pummerer reactions is available ⟨91OR(40)157⟩, and only selected examples will be given here. A generalised mechanism for the reaction is shown in Scheme 4, and it is likely that some of the procedures mentioned in 4.05.1.1.1(iv) above operate by the direct oxidative formation of intermediates of types (**22**) and (**23**) from the sulfide. Most commonly, the electrophilic activating agent is a carboxylic anhydride, and very often the incoming nucleophile Nu$^-$ is identical with EO$^-$, leading, if the activator is an anhydride, to the formation of an α-acyloxy sulfide. However, other nucleophiles, including oxygen nucleophiles, can intercept the sulfonium cation. Typical examples of the formation of α-acetoxy sulfides by Pummerer reactions are given in Equations (17) ⟨74JA4280⟩, (18) ⟨72JHC175⟩ and (19) ⟨78JCS(P2)1302⟩, whilst Equation (20) shows a reaction catalysed by aqueous mineral acid, leading to a hemithioacetal of ninhydrin; the same product could also be obtained by direct addition of MeSH to ninhydrin ⟨64JOC1358⟩.

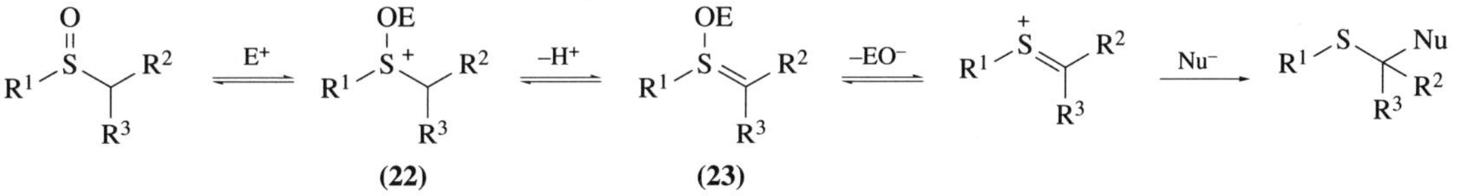

Scheme 4

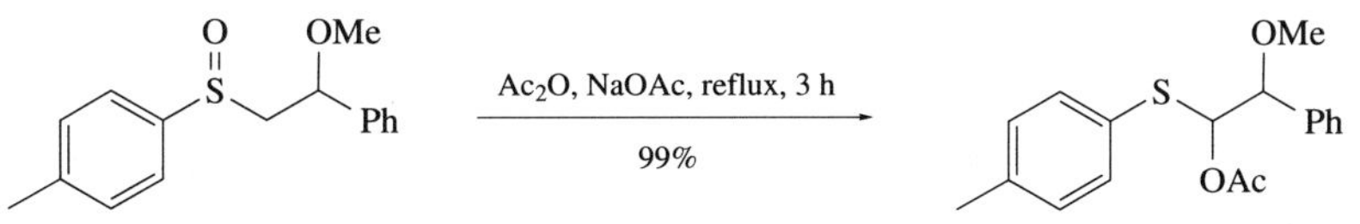

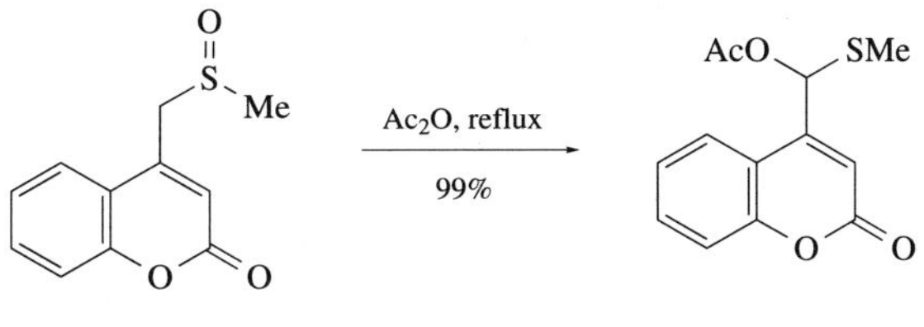

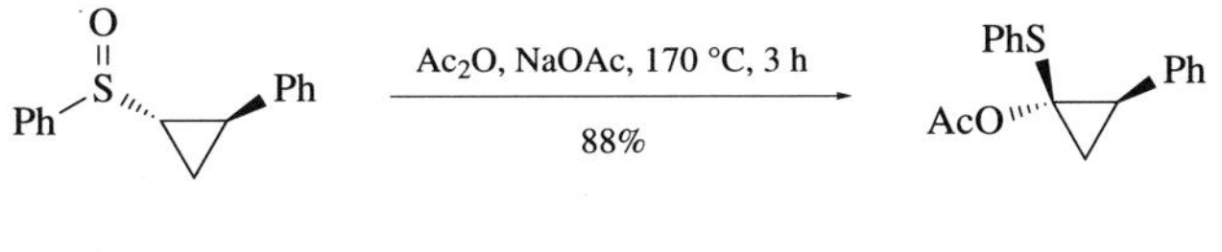

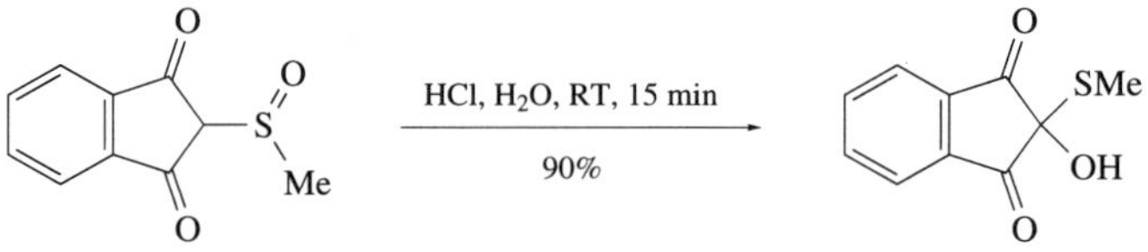

The activated sulfoxide (**22**) of Scheme 4 is of the same type as the intermediates formed during DMSO-dependent oxidations of primary and secondary alcohols (the Moffatt, Swern and related oxidations), where the oxidation proceeds by attack of the alcohol on the sulfur of the activated DMSO. It is well known that MTM ethers are often formed as by-products in such oxidations, particularly with DMSO-Ac$_2$O, by the competing mechanism of Scheme 4 (R^1 = Me; R^2 = R^3 = H) ⟨90OR(39)297⟩. Indeed, the reagent combination DMSO-Ac$_2$O has been employed for the formation of MTM ethers of nonoxidisable tertiary alcohols in high yield ⟨76TL65⟩. Pojer and Angyal have shown that, if acetic acid is also present, DMSO-Ac$_2$O will give good yields of MTM ethers from primary and secondary alcohols. These workers suggest that, in the reactions of Scheme 5, the formation of the alkoxysulfonium ion (**24**) is a reversible process, so that the oxidative reaction is suppressed in the presence of AcOH. Thus, for example, the fructopyranose derivative (**25**) was formed from the di-*O*-isopropylidene compound in 82% yield ⟨78AJC1031⟩; a more recent example is the formation of 3′-*O*-MTM nucleoside (**26**) using DMSO, Ac$_2$O and AcOH ⟨91TL7593⟩.

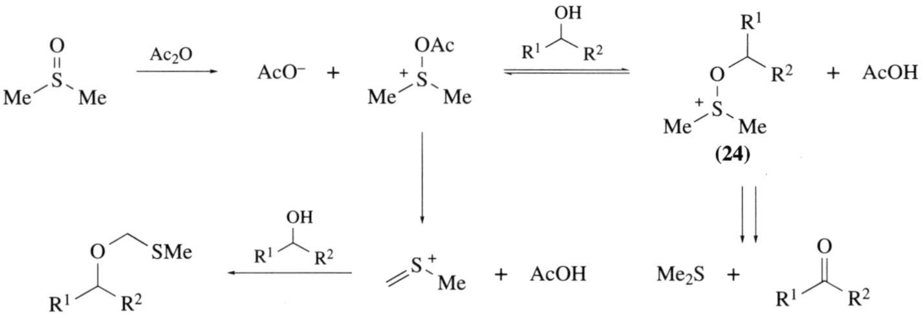

Scheme 5

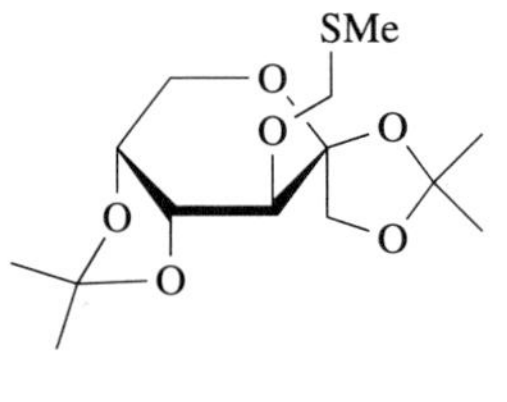

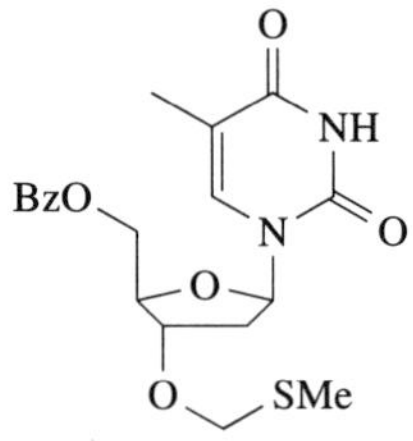

In 1968, Brook and Anderson reported that trimethylsilylmethyl phenyl sulfoxide underwent thermal rearrangement (60 °C, 1 h) to give the Pummerer-type product PhSCH$_2$OTMS in 79% yield ⟨68CJC2115⟩. This is now known to be a general reaction of α-silylsulfoxides, referred to as the sila-Pummerer reaction ⟨91OR(40)157⟩, and a further example is given in Equation (21); when this reaction was carried out in methanol, the analogous α-methoxysulfide was the product ⟨87TL4793⟩. The mechanism of the sila-Pummerer reaction is thought to involve rearrangement of the α-silylsulfoxide through a 4-centre transition state to give an intermediate of type (**23**; E = TMS) (Scheme 4). This contention is suggested by the observations of Scheme 6, diastereomer (**27**) undergoing rearrangement at a higher temperature due to the steric strain present during the process ⟨75TL2017⟩.

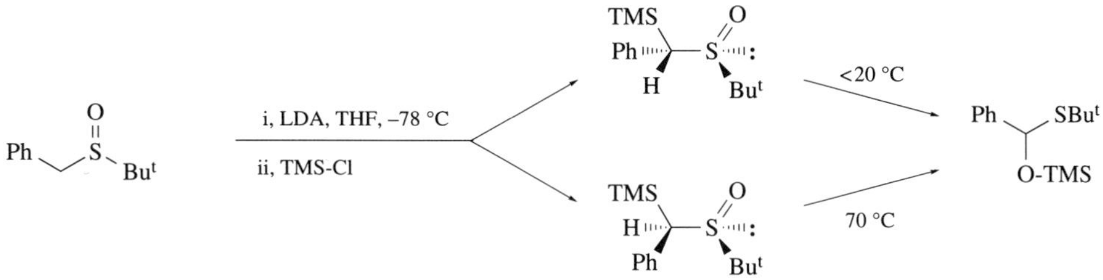

Scheme 6

There has been interest in trying to achieve chirality transfer from sulfur to carbon during Pummerer reactions, but with varying degrees of success in the case of acyclic sulfoxides ⟨83BCJ257, 83BCJ266⟩. For success it would seem to be necessary to ensure that the rearrangement of the ylide of type (**23**) is substantially intramolecular, and to guard against the formation of a sulfurane by attack of EO⁻ at the sulfur of (**22**). In the early 1990s, however, a silicon-induced Pummerer reaction has been reported which yields an α-silyloxy sulfide with 87% *ee* and in 75% yield (Scheme 7) ⟨93TL4063⟩.

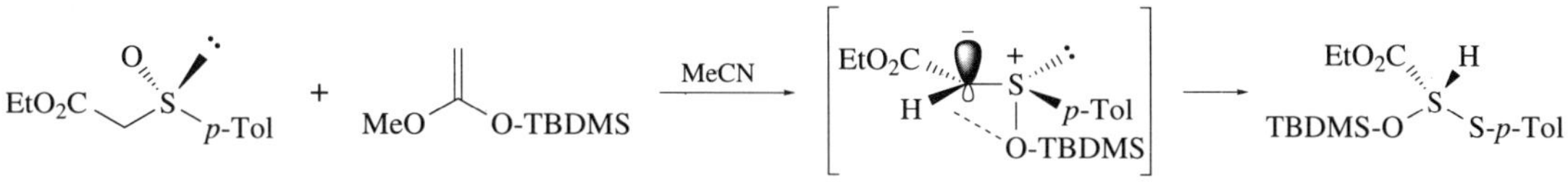

Scheme 7

4.05.1.1.2 Compounds with oxygen in a ring

Although a number of synthetic methods in this area are similar to those used for acyclic compounds, there are some significant differences. These reflect both the stability of cyclic hemiacetals and the much reduced tendency of monothioacetals in which the oxygen is *endo*-cyclic to undergo further acid-catalysed reaction with thiols to give acyclic dithioacetals. Much of the interest in this area has concerned carbohydrate monothioacetals (thioglycosides), and this activity has been maintained by the prominence such compounds have gained as intermediates in oligosaccharide synthesis ⟨87MI 405-01⟩.

The hemithioacetal (**28**) can be prepared by addition of H₂S to dihydropyran (Equation (22)) ⟨74JOC2010⟩. 1-Thiosugars can be synthesised by treatment of peracetylated glycosyl halides with sulfur nucleophiles such as potassium ethyl xanthate ⟨43JA1477⟩, potassium thioacetate ⟨31CB2696⟩, or thiourea; this last nucleophile is advantageous, since the thiol can be liberated under milder conditions which leave *O*-acetyl groups unchanged, as for the D-glucose derivative (**29**) (Scheme 8) ⟨59CCC64⟩. The thiouronium salt can also be cleaved reductively, with the per-*O*-acetyl-1-thioaldose being extracted as formed into an organic layer, thus minimising side reactions; this procedure was applied for the synthesis of 2,3,4,6-tetra-*O*-acetyl-1-thio-D-galactose ⟨63M290⟩.

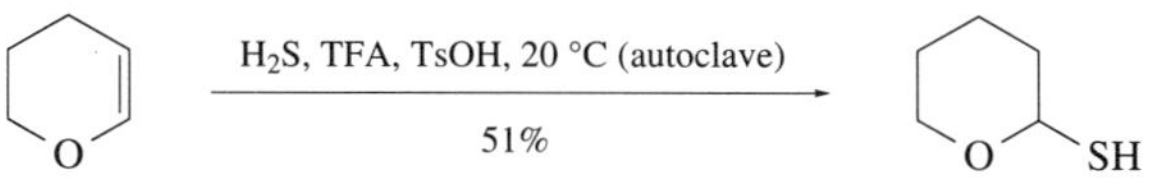

Monothioacetals of type (**30**; R = various alkyl groups) ⟨51JA822⟩ and (**30**; R = Ph) ⟨54JA4962⟩ have been made from dihydropyran and the appropriate thiol in the presence of HCl. In the late 1980s, the use of pyridinium *p*-toluenesulfonate in CH₂Cl₂ in such reactions has led to the synthesis of (**30**; R = CH₂TMS) ⟨86JOC3428⟩ and (**30**; R = Ph) ⟨89JA658⟩ in high yields. The THP group has also been used for protection of the thiol unit in cysteine and its derivatives, but has the disadvantage of introducing an extra chiral centre ⟨58JA3765, 62JA4789⟩.

The first synthesis of a 1-thioglycoside was the preparation of the *S*-phenyl compound (**31**) from the α-glycosyl bromide and thiophenolate anion ⟨09CB1476⟩, and this approach has been used many

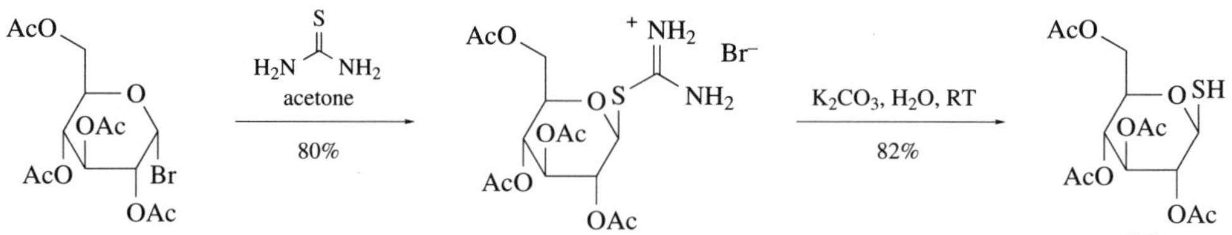

Scheme 8

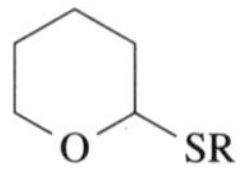

(30)

times since then, with earlier work in this area being reviewed in 1963 ⟨63MI 405-01⟩. An example of this base-catalysed process is the formation of the pyridylthioglycoside (**32**) ⟨80CAR(80)C17⟩. Another much-used procedure for making 1-thioglycosides involves the interaction of a 1-*O*-acetyl glycopyranoside with a thiol in the presence of a Lewis acid. Thus, Lemieux obtained a 71% yield of the *S*-ethyl thioglucoside (**33**) from penta-*O*-acetyl-β-D-glucopyranose and neat ethanethiol in the presence of anhydrous $ZnCl_2$ ⟨51CJC1079⟩, and the same product (**33**) has been obtained from near stoichiometric amounts of EtSH with either BF_3 ⟨76CAR(52)63⟩ or $SnCl_4$ ⟨93CAR(248)377⟩ as catalysts, in yields of 83% and 89% respectively. The glucosamine derivative (**34**) was similarly made in 72% yield with $TiCl_4$ as catalyst in CH_2Cl_2 at 0 °C ⟨85CAR(139)115⟩.

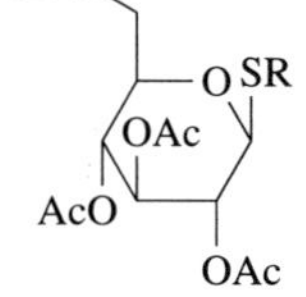

(**31**) R = Ph
(**32**) R = 2-pyridyl
(**33**) R = Et

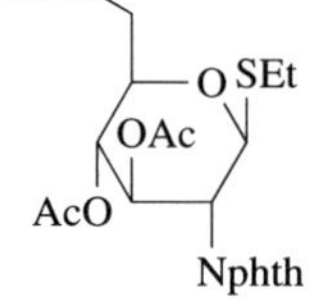

(**34**)

O,O-Acetals can also be converted into *O,S*-acetals of the type under discussion by treatment with a thiol and BF_3, Equation (23) ⟨78JA1938⟩ and Equation (24) ⟨88JOC2953⟩ being illustrative. The lack of an oxygen function α to the acetal probably facilitates these reactions as compared with similar reactions in most sugar systems. In sugar systems, *O*-glycosides have been converted into thioglycosides by PhS-TMS and ZnI_2 in refluxing dichloroethane ⟨80CAR(86)C3⟩, or, under milder conditions, by PhS-TMS and TMS-OTf (Equation (25)) where the *t*-butyldimethylsilyl (TBDMS) protecting group is lost under the conditions of thioglycoside formation ⟨83JA2430⟩. As might be expected these same conditions are effective in converting 1-*O*-acetylglycopyranoses into 1-thioglycosides ⟨89TL2179⟩. Reagents of the type RS-TMS can also be used to convert glycals into unsaturated thioglycosides with Ferrier-type rearrangement (Equation (26)) ⟨88JOC845⟩.

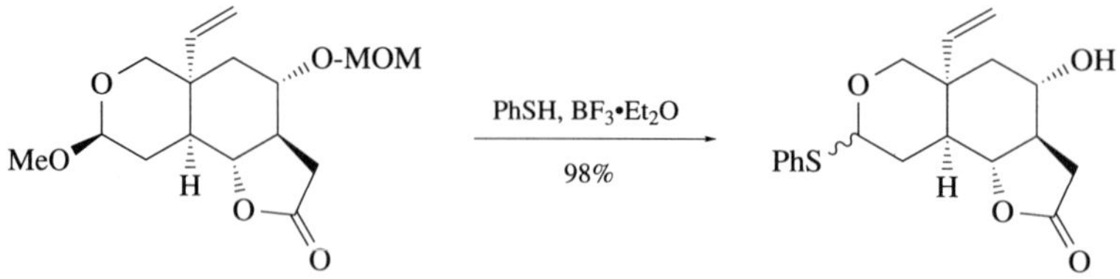

(23)

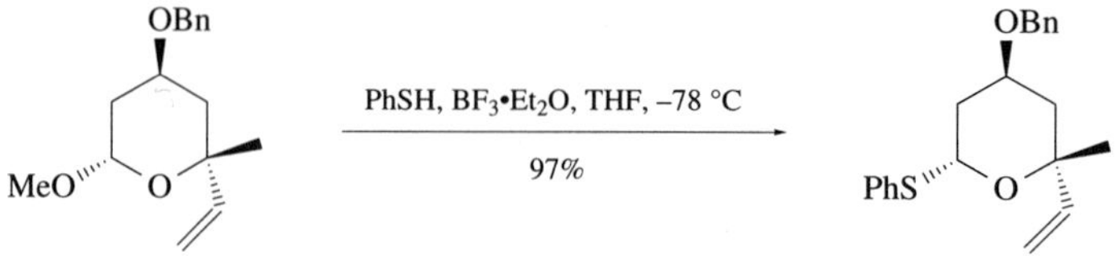

(24)

[Equation (25): reaction with i, PhS-TMS, TMS-OTf, CH₂Cl₂, 25 °C; ii, TBDMS-Cl, imidazole; 74%]

[Equation (26): reaction with 2-(TMS-thio)pyridine, BF₃·Et₂O, C₆H₆, 5 °C]

Thiostannanes have also been used as reagents for thioglycoside formation. Treatment of glycosyl halides or acetates with reagents of the type $n\text{-}Bu_3SnSR$ in the presence of $SnCl_4$ gives good yields of the thioglycosides ⟨77CAR(54)C17, 81CAR(95)308⟩, as in Equation (27), whilst more recently reagents of the type $Bu^n_2Sn(SR)_2$ have been recommended for conversion of O-glycosides and glycosyl acetates into thioglycosides, when catalytic amounts of $Bu^n_2Sn(OTf)_2$ are present (Equation (28)) ⟨92TL239⟩.

[Equation (27): Bu₃SnSMe, SnCl₄, ClCH₂CH₂Cl, 20 °C, 15 h; 85%]

[Equation (28): Bu₂Sn(SPh)₂, Bu₂Sn(OTf)₂ (0.3 equiv.), ClCH₂CH₂Cl, 50 °C, 2 h; 100%; α:β 70:30]

O-Glycosides have also been converted into S-glycosides by sequential treatment with Me_2BBr in CH_2Cl_2, which is presumed to form the glycosyl bromide *in situ*, followed by treatment with a thiol and Pr^i_2NEt ⟨85JOC5379⟩. Glycosyl fluorides can also be converted into thioglycosides under Lewis acidic conditions (Equation (29)) ⟨93CAR(249)197⟩, and Mitsunobu-type reactions have been used to make thioglycosides (Equation (30)) ⟨81JA3215, 83JA2430⟩.

[Equation (29): Pr^iSH, TiF₄, MeCN, 0 °C, 15 min; 82%]

[Equation (30): 2-mercaptopyrimidine, dead, PBu^n_3, toluene; then ClCO₂Me, NaHCO₃; 63%]

dead = diethyl azodicarboxylate

S-Alkylation can also be used to make O,S-acetals of this type, as in the synthesis of compounds of type (**30**; R = alkyl) from the thiol (**28**) with alkyl bromides and NaOMe ⟨74JOC2010⟩. Similarly, the 1-thioglucopyranose derivative (**29**) (Scheme 8) can be S-alkylated without change in the anomeric configuration ⟨59CCC2566⟩, as can the equivalent compound of D-galacto-configuration ⟨63M290⟩. It has been reported that such S-alkylations can be carried out under free-radical conditions (Equation (31)) ⟨88TL4293⟩. Treatment of the thiol (**29**) with aryldiazonium salts, followed by thermolysis of the resultant *S*-diazocompounds, can be used as a route to aryl 1-thio-β-D-glucopyranosides ⟨61CCC2206⟩.

Intramolecular trapping of Pummerer intermediates can be used to prepare γ-phenylthio-γ-butyrolactones (Equation (32)) ⟨81BCJ817⟩, and similar trapping by phenols has been used to make bicyclic systems such as (**35**) ⟨81JHC587⟩. Heating of methyl vinyl ketone (MVK) with phenyl vinyl sulfide in the presence of a radical trap gives a moderate yield of the cycloadduct (**36**) ⟨80JA6900⟩. When caprothionolactone is treated with organolithium reagents, followed by methylation with MeI, monothioacetals of type (**37**) are obtained. This method can be used for larger ring thionolactones and in cases where R = H (from LiEt₃BH); cases were not reported for ring sizes smaller than seven ⟨90JA6263, 93JOC506⟩, but the *O*,*S*-acetal (**38**) has been made in moderate yield by treatment of the thionolactone with Me₂CuLi followed by MeI, the predominant (4:1) formation of the *endo*-adduct being rationalised in terms of a single-electron transfer mechanism ⟨93HCA995⟩.

4.05.1.1.3 Compounds with sulfur in a ring

(i) By functionalisation of preformed sulfur heterocycles

Treatment of tetrahydrothiophene with *t*-butyl peracetate and Cu₂Br₂, as described above for acyclic sulfides ⟨62T15⟩, gave the α-acetoxy derivative (**39**) in 73% yield. This could be hydrolysed to the hemithioacetal (**40**), and similar chemistry was also carried out for the 6- and 7-membered rings ⟨67JCS(C)1130⟩. Use of dibenzoyl peroxide in refluxing chloroform led to the α-benzoyloxy derivative (**41**) (95%) from the 1,4-thiazin-3-one, in a reaction that was accelerated by light ⟨82S312⟩, whilst the α-hydroxy compound (**42**) was made in 88% yield from the thiazoline by oxidation with singlet oxygen followed by reduction of the intermediate hydroperoxide with Me₂S ⟨85T2133⟩.

Clearly compounds of the type under discussion can, in principle, be obtained from Pummerer reactions on cyclic sulfoxides. Many examples can be found in a compilation ⟨91OR(40)157⟩, and the cases in Equations (33) ⟨85CC1286⟩, (34) ⟨82JCR(S)116⟩ and (35) ⟨83CJC2103⟩ are typical. In examples involving unsymmetrical sulfoxides, the direction and extent of any regioselectivity is not always easy to predict; in the case of Equation (34), the regioselectivity in favour of migration towards the oxygen substituent was based on precedent, whilst the example of Equation (35) illustrates that, in the absence of strong electronic effects, migration is likely to occur preferentially towards the least hindered carbon. This is also illustrated by Equation (36), which also indicates a strong preference (19:1) in favour of the product with an equatorially oriented acetoxy group, a result independent of the initial stereochemistry at sulfur. Labelling experiments indicated the intermolecular nature of the rearrangement ⟨83BCJ270⟩. Cyclic α-chloro sulfides, also accessible by Pummerer-type chlorinations, can be used to prepare monothioacetals, as in the formation of the 2α-methoxycephem (Equation (37)) ⟨78JOC79⟩.

(ii) By cyclisation of thiol carbonyl compounds and thiol acetals

In this approach to 5- and 6-membered cyclic *O,S*-acetals, a sulfhydryl group, usually protected, is introduced into a molecule which also contains a protected carbonyl group, usually an acetal, and cyclisation then ensues when the functionalities are released. This approach has been widely adopted for the preparation of sulfur-in-ring analogues of sugars and glycosides, where some early applications have been reviewed ⟨63MI 405-01⟩. A more recent illustration is in the synthesis of 5-thio-D-allose (Equation (38)) ⟨86CAR(148)25⟩, and two other ways of generating the appropriate functional groups for cyclisation are shown in Equations (39) ⟨91CC1421⟩ and (40) ⟨92JMC533⟩, where the products shown were used as precursors for sulfur-containing nucleoside analogues.

(iii) By cyclisation of dithioacetals

Acyclic dithioacetals with a suitably disposed leaving group can undergo dealkylative cyclisation to give a cyclic dithioacetal, which in turn can be converted into a cyclic α-acetoxy sulfide by treatment with mercuric acetate. An application of this approach is seen in the formation of the 5-thio-L-idose derivative (**43**) (Scheme 9) ⟨79CAR(70)217⟩ and another example (Scheme 10) is in the synthesis of the D-xylo-configured thionucleoside precursor (**44**) ⟨90TL2759⟩. The use of S,S-dibenzyl dithioacetals and the presence of iodide ions seem to be important in such reactions, and alternative modes of cyclisation involving alkoxy groups and leading to anhydrosugars (tetrahydrofuran derivatives) can easily intervene ⟨76CAR(46)237, 79CAR(70)217⟩.

(iv) By cycloadditions to thionoesters and thionolactones

There have been two isolated reports of [4 + 2] cycloadditions to thionoesters, adduct (**45**) being obtained from 2,3-dimethylbutadiene and dimethyl dithionooxalate ⟨80LA1665⟩, and the thiadiazine (**46**) from cycloaddition of ethyl thionoformate and the appropriate tetrazine ⟨84AG(E)890⟩. Vasella and co-workers have reported various cycloadditions to a carbohydrate-derived thionolactone to produce spirocyclic systems including (**47**) from the use of a Danishefsky-type diene ⟨93HCA1779⟩.

4.05.1.1.4 Cyclic monothioacetals

Most compounds in this category are substituted 1,3-oxathiolanes and 1,3-oxathianes; the chemistry of these systems, and of 1,3,5-oxadithianes and -dioxathianes, was reviewed in 1980, with particular emphasis on structural and conformational aspects ⟨B-80MI 405-01⟩. Synthetic methods for cyclic monothioacetals and related compounds can be classified under three categories.

(i) From carbonyl compounds or their O,O-acetals

The acid-catalysed reaction between an aldehyde or ketone and a 1,2- or 1,3-thiol alcohol is certainly the most general method for making cyclic monothioacetals with 5- and 6-membered rings, and a tabulation of some 25 examples, involving various protonic and Lewis acids as catalysts has been given ⟨91HOU(E14a/1)785⟩. Typical cases of this procedure are the formation of the oxathiolane (**48**) (92%) from cycloheptanone and thiolethanol using $BF_3 \cdot Et_2O$ as a catalyst ⟨68JOC2133⟩, and the synthesis of the 1,3-oxathiane (**49**) in 80% yield with TsOH in refluxing dichloroethane ⟨87JMC24⟩. An interesting example (Scheme 11) is the formation of the two epimers (**50**) and (**51**) in a ratio of 1:2, (**51**) being subsequently used to make the oxathiolanyl nucleoside analogue (**52**), which possesses high anti-HIV activity ⟨92TL4625⟩. Another case of interest involves the interaction of a chiral 1,3-hydroxy thiol derived from pulegone with racemic dimethyl *trans*-cyclopentanone-3,4-dicarboxylate (TsOH, benzene, reflux) to produce a mixture of diastereomers from which the isomer (**53**) was obtained in 43% yield by crystallisation; the mother liquors, rich in the other diastereomer, could be re-equilibrated to a 1:1 mixture of diasteromers by base treatment, and after three cycles, 80% of (**53**) was obtained ⟨93TA1547⟩. For many aldehydes and ketones, the reaction with 2-thiolethanol and 3-thiolpropanol will lead to a mixture of two epimeric heterocycles, and it was observed that, when thiolethanol condenses with cyclohexanones there is a preference at equilibrium for the epimer with sulfur in the axial orientation ⟨62TL103, 65JOC855⟩. Thus, for example, isomer (**54**) is present in a 79:21 mixture with its *S*-equatorial epimer at equilibrium ⟨69JOC2080⟩.

It has been observed that the reaction of 1-thioglycerol with acetone in the presence of P_2O_5 gives a mixture of the oxathiolane (**55**) and the dioxolane in a ratio of 65:35 ⟨42CB13⟩. It is noteworthy that the *trans*-fused bicyclic oxathiolane (**56**) can be formed easily from the *trans*-hydroxy thiol (acetone, $ZnCl_2$, 69%) ⟨64JOC724⟩, and a similar *O,S*-isopropylidene derivative could also be made from *trans*-2-thiolcyclopentanol ⟨65JCS1298⟩.

Acetal exchange processes involving *O,O*-acetals and 1,2- or 1,3-hydroxy thiols can also be used to prepare 1,3-oxathiolanes and 1,3-oxathianes, as for example in the synthesis of (**57**) (65%), as a mixture of diastereomers, from 1,1,3,3-tetramethoxypropane and 2-thiolethanol ⟨88JOC5179⟩. Condensation between α-thiolcarboxylic acids and aldehydes or ketones gives 1,3-oxathiolan-5-ones; examples are the formation of (**58**) from thiolsuccinic acid and cyclohexanone (TsOH, toluene, reflux with water separation, 79%) ⟨85JCS(P1)587⟩, and the uncatalysed condensation in refluxing toluene to produce (**59**), used to prepare oxathiolane nucleosides of type (**52**) ⟨91JA9377⟩. From (+)-camphor, the epimer (**60**) was produced with good stereoselectivity, and alkylation of the enolate of (**60**) gave 4-alkyl derivatives with very high diastereoselectivity ⟨90TL257⟩. 1,3-Oxathian-4-ones of type (**61**) can be obtained by treatment of β-hydroxy thioamides with formaldehyde in DME at reflux ⟨84TL5797⟩.

More complex structures can be prepared from simpler 1,3-oxathianes by lithiation at C-2 and subsequent reaction with an electrophile, as for example in the formation of (**62**) (66%) by alkylation of the compound unsubstituted at C-2 ⟨85TL1927⟩. An important application of this principle has been in work, predominantly by Eliel, on the use of 1,3-oxathianes as chiral auxiliaries. Thus, the bicycle (**63**) derived from pulegone ⟨87OS(65)215⟩, was converted into 2-acyl-derivatives of type (**64**) as outlined in Scheme 12, an 80% overall yield being achieved where $R^1 = Me$ ⟨85JOC3402⟩. Subsequent reactions with Grignard reagents gave adducts (**65**) with high diastereoselectivity, explicable in terms of a chelated transition state involving the oxygen of the heterocycle, and further manipulation involving removal of the chiral auxiliary gave α-hydroxy acids, α-hydroxy aldehydes and 1,2-diols with high optical purity ⟨84JA2943⟩. Hydride reduction of the 2-acyl-1,3-oxathianes (**64**) was also studied, with the extent and sense of the diastereoselectivity dependent on the reducing agent ⟨84T1333⟩. Utimoto *et al.* have shown that, when alkynyllithium and -magnesium reagents are used, the diastereoselectivity indicated in Scheme 12 can be reversed in the presence of YbCl$_3$ ⟨90JA8189⟩, and a similar effect has been found for borohydride reductions, using a simpler monocyclic oxathiane ⟨92CL2173⟩. 1,3-Oxathianes derived from 10-thiol-*exo*-borneol have also been used as chiral auxiliaries in various reactions (⟨88TL4773⟩, and refs. therein).

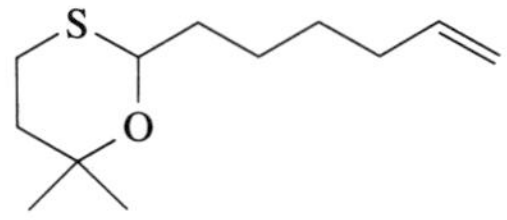

(**62**)

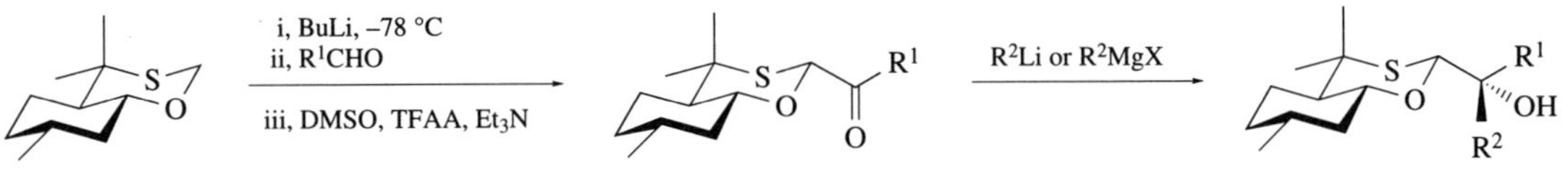

(**63**) → (**64**) → (**65**)

i, BuLi, −78 °C
ii, R^{1}CHO
iii, DMSO, TFAA, Et$_3$N

R^{2}Li or R^{2}MgX

Scheme 12

(ii) By intramolecular alkylation

The *cis*-disubstituted oxathiane (**66**) can be prepared (53%) by treatment of the hydroxy thiol with CH$_2$Br$_2$ and KOH under phase-transfer conditions, and the *trans*-isomer was similarly made from the *threo*-precursor ⟨93CB1227⟩. Rather similar phase-transfer conditions had been employed earlier to make 1,3-benzoxathiole (**67**), and various derivatives with substituents in the benzene ring, from the appropriate 2-mercaptophenol and CH$_2$Br$_2$ ⟨76S797⟩.

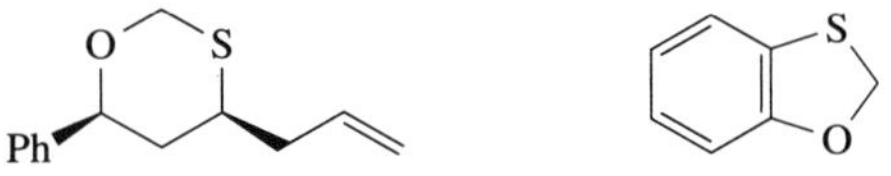

(**66**) (**67**)

An interesting intramolecular *S*-alkylation was used to form the bicyclic oxathiolane (**68**) (Equation (41)), similar chemistry having previously been reported for sugars of different configurations. Oxidative cleavage of the pyranose ring of (**68**) led ultimately to a further synthesis of the oxathiolane nucleoside (**52**) as a pure enantiomer ⟨92JOC2217⟩. An intramolecular *O*-alkylation (Equation (42)) led to the *trans*-disubstituted oxathiane (**69**) from the *threo*-precursor shown ⟨75T327⟩.

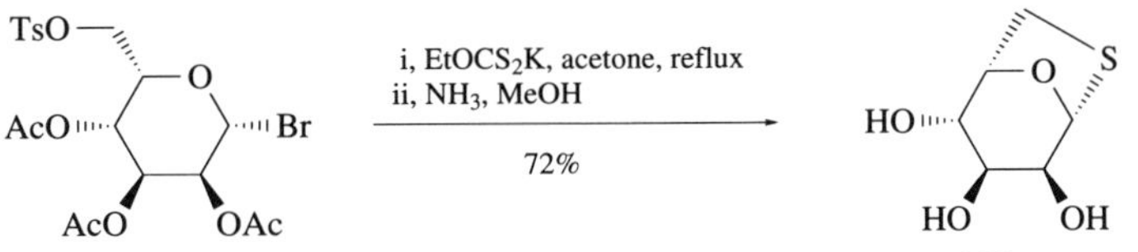

i, EtOCS$_2$K, acetone, reflux
ii, NH$_3$, MeOH

72%

(**68**) (41)

CH$_2$O, MeOH, H$^+$

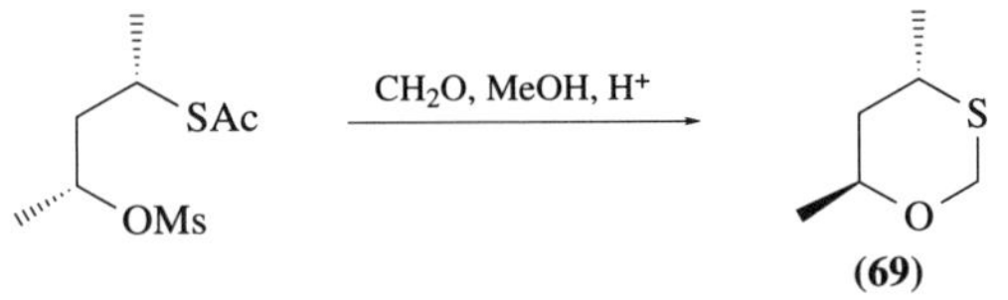

(**69**) (42)

(iii) By cycloaddition reactions

1,3-Dipolar cycloaddition of bis(methylene)sulfurane to aldehydes has been used in a novel approach to 1,3-oxathiolanes (Equation (43)) ⟨87CC1442⟩. Various other cycloadditions have been described which lead to heterocycles which include a cyclic monothioacetal unit as a substructure, although many of these routes may well be of limited generality. These procedures include the formation of 1,4,2-oxathiazolidines from thioketones and nitrones or nitronates ⟨73AJC2491, 86JOC117⟩, and of 1,4,2-oxathiazolines such as (**70**) from thiocarbonyl compounds and nitrile oxides ⟨84JCS(P1)2641, 86JOC117⟩. Alternatively, nitrile sulfides, generated *in situ*, will cycloadd to electron-deficient aldehydes and ketones to give 1,3,4-oxathiazolines such as (**71**), obtained in 76% yield ⟨81JCS(P1)2991⟩. 1,3-Oxathioles of type (**72**) can be obtained from certain α-oxo thiocarbonyl compounds and diazo alkanes ⟨80JHC1655, 90T1783⟩. [4 + 2]-Cycloaddition of thiobenzophenone and α-nitrosostyrene gave (76%) the 4*H*-1,5,2-oxathiazine (**72**) ⟨85TL2131⟩, whilst trimethylsilyl vinyl ketone and thiocarbonyl compounds gave 4*H*-1,3-oxathiins (**73**) ⟨91TL2971⟩. 4*H*-3,1-Benzoxathiines (**74**) can be prepared by thermolysis of benzothiete in the presence of electron-deficient carbonyl compounds ⟨90CB1143⟩.

4.05.1.2 Derivatives with Tricoordinate Sulfur

4.05.1.2.1 α-Alkoxy sulfoxides

(i) By oxidation of monothioacetals

The oxidation of monothioacetals is by far the most common approach to the synthesis of α-alkoxy sulfoxides, and can be carried out on monothioacetals of various types using the reagents that are most frequently employed for the oxidation of simple sulfides. Thus, for example, mcpba has been used in the preparation (86%) of the acyclic sulfoxide (**75**) ⟨88ACS(B)515⟩, and for the oxidation of thioglycosides to sulfoxides such as (**76**), obtained in 85% yield ⟨89JA6881⟩, such sulfoxides being the basis of an important new procedure for glycosylation ⟨93JA1580⟩. The oxidant in the formation of (**77**) was also mcpba ⟨84TL757⟩ and in the oxidation of the camphor derivative (**60**) to give a single sulfoxide (91%), with the oxygen *anti* to the bridgehead methyl group ⟨90TA143⟩. Hydrogen peroxide in acetic acid was used as oxidant to form the muscarine analogue (**78**), the major epimer formed (4:1 ratio) being shown ⟨87JMC1934⟩, whilst sodium periodate was used, for example, in the preparation of (**79**), formed in a 3:1 ratio with the axial sulfoxide ⟨87LA451⟩.

Acyclic α-alkoxy sulfoxides are of limited thermal stability. Thus (Scheme 13) methoxymethyl phenyl sulfoxide, made by oxidation with mcpba, underwent rearrangement to the sulfenate and this in turn was more slowly converted into *S*-phenyl benzenethiosulfinate and bis(methoxymethyl)ether ⟨72JA5115⟩.

(ii) From α-chloro sulfoxides

Chloromethyl and bromomethyl sulfoxides react with alkoxide and phenoxide ions to give alkoxymethyl and aryloxymethyl sulfoxides in moderate to good yields, as in the formation of ethoxymethyl methyl sulfoxide (77%) from chloromethyl methyl sulfoxide ⟨70CC1689⟩, and of methoxymethyl *p*-tolyl sulfoxide in 80% yield from the bromomethyl precursor ⟨70CC1441, 72BCJ2794⟩. The kinetics of such substitutions have been studied, and the rates of reaction found to be comparable with those of a typical straight-chain chloro alkane and thus much slower than those of, for example, phenacyl halides ⟨72CC734⟩. This synthetic approach seems to be restricted to alkoxymethyl sulfoxides, since treatment of α-haloethyl sulfoxides under similar conditions leads to β-alkoxyethyl sulfoxides, presumably by an elimination–addition mechanism (Equation (44)) ⟨72JCS(P1)1883, 76JCS(P2)996⟩.

4.05.1.2.2 α,β-Epoxy sulfoxides

It was first reported by Durst in 1969 that deprotonation of chloromethyl phenyl sulfoxide, followed by addition of a ketone and subsequent cyclisation of the resultant halohydrin, gives α,β-epoxy sulfoxides (sulfinyloxiranes) by a Darzens-type process ⟨69JA1034⟩. This reaction has subsequently been studied by other workers, and can be carried out in a one-pot procedure as illustrated by Equation (45), where the predominant formation of the isomer shown (3:1 ratio with the *E*-isomer) could be rationalised in terms of an irreversible initial reaction between the sulfoxide anion and pinacolone ⟨72BCJ2023⟩. Alternatively, the procedure can be carried out with the isolation of the chlorohydrin (Scheme 14) ⟨86BCJ2463⟩, and in many examples involving unsymmetrical ketones or aldehydes no effective stereocontrol is obtained. However, when the carbonyl component is an aldehyde, oxidation of the mixed chlorohydrins to the ketone can be followed by stereocontrolled reduction with dibal-H to give, after cyclisation, the pure (E)-α,β-epoxy sulfoxide ⟨89CPB184⟩. An alternative approach to α,β-epoxy sulfoxides, applied to cyclohexenyl systems, involves the oxidation of thioenol ethers with two molar equivalents of ozone ⟨86JOC2276⟩. α,β-Epoxy sulfoxides have been used synthetically in a variety of ways, mostly involving nucleophilic attack at the β-position, followed by spontaneous decomposition of the resultant α-hydroxy sulfoxide to generate a carboxyl group (⟨86BCJ2463⟩, and refs. therein).

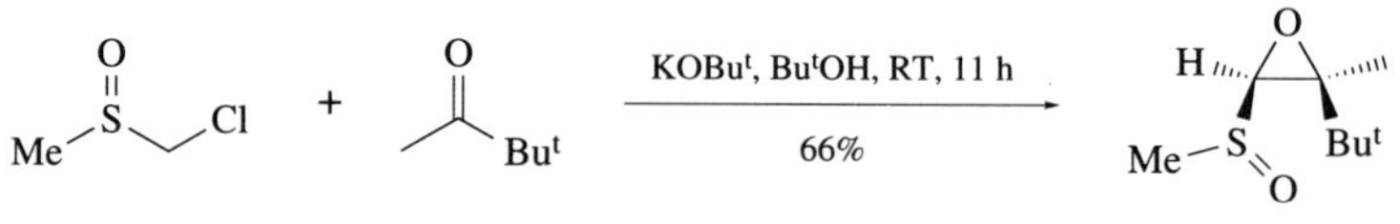

[Scheme with reaction (45): MeS(O)CH2Cl + acetone tBu, KOBut, ButOH, RT, 11 h, 66% → epoxide with S(O)Me and But]

[Scheme 14: PhS(O)CHCl-CH2-C6H4-OMe + i, LDA, -78 °C; ii, acetone → PhS(O)C(Cl)(CH2Ar)C(Me)2OH → KOH, H2O, RT, 2 h, 49% → epoxide product]

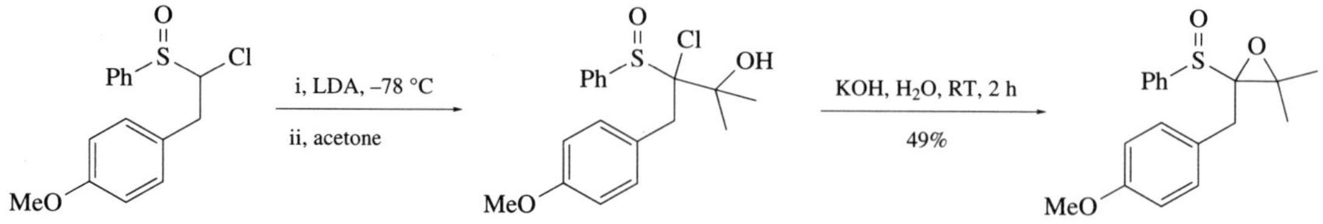

Scheme 14

4.05.1.2.3 α-Hydroxy sulfinates

It was first established at the beginning of the twentieth century that the reaction of sodium dithionite with certain aldehydes under basic conditions and at room temperature or below can give α-hydroxy sulfinates as isolable salts ⟨05CB1057, 09CB4634⟩. At higher temperatures, these α-hydroxy sulfinates decompose to give the corresponding alcohol (Scheme 15), and thus dithionite can be used as a reducing agent for aldehydes and ketones ⟨80JOC4126⟩. The intermediacy of α-hydroxy sulfinates in such reductions has been demonstrated by ^{1}H-NMR studies ⟨81JOC5457⟩. The commercially available sodium hydroxymethanesulfinate ('Rongalite') can also reduce organic halides in a variety of ways depending on the structure of the substrate ⟨88JOC5750⟩.

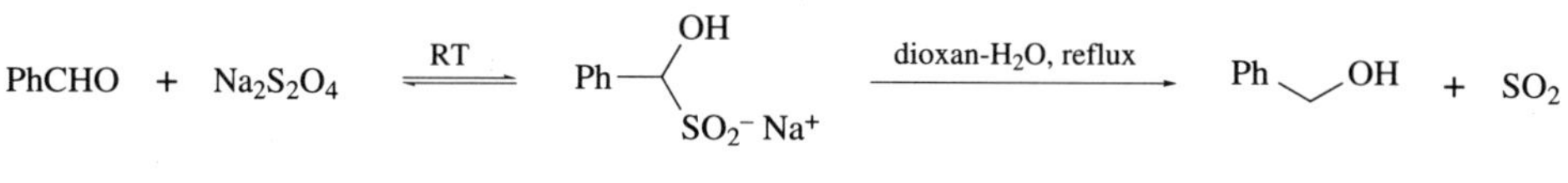

[Scheme 15: PhCHO + Na2S2O4 ⇌ (RT) Ph-CH(OH)-SO2⁻Na⁺ → (dioxan-H2O, reflux) Ph-CH2-OH + SO2]

Scheme 15

4.05.1.3 Derivatives with Tetracoordinate Sulfur

4.05.1.3.1 α-Hydroxy sulfones

α-Hydroxy sulfones are formed reversibly from sulfinic acids and aldehydes, but usually not from ketones ⟨54CB129⟩. The adducts from formaldehyde (hydroxymethyl sulfones) are not surprisingly among the more stable examples (e.g., ⟨84AP(317)15⟩), as are those from α-keto aldehydes, like the adduct (**80**), obtained in 83% yield from phenylglyoxal and toluene-*p*-sulfinic acid ⟨66CB48⟩. An alternative route to hydroxymethyl sulfones is via the acid-catalysed hydrolysis of α-diazo sulfones ⟨64TL547⟩.

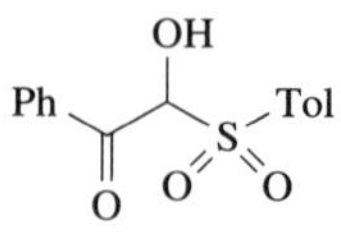

[Structure 80: PhC(O)-CH(OH)-SO2-Tol]

(**80**)

Presumably α-hydroxy sulfones are intermediates in procedures which have been reported for the oxidative desulfonylation of sulfones to carbonyl compounds by formation of the α-anion and treatment with a hydroxylating agent such as the $MoO_5 \cdot$ pyridine $\cdot$ HMPA complex ⟨80TL3339⟩ or bis(trimethylsilyl)peroxide ⟨83JOC4432⟩.

4.05.1.3.2 α-Alkoxy sulfones, α-acyloxy sulfones and related compounds

(i) By oxidation of O,S-acetals and related compounds

Many of the oxidants commonly employed for the oxidation of sulfides to sulfones have been used to make α-alkoxy sulfones from monothioacetals. Hydrogen peroxide–acetic acid has been used in the preparation of methoxymethyl *p*-tolyl sulfone ⟨62JCS3686⟩ and in early syntheses of glycosyl sulfones ⟨48JA2435⟩, a more recent example being the preparation of the sulfone (**81**) in 83% yield from the thioglycoside ⟨66CAR(2)461⟩. A two-phase system using $KMnO_4$ has been studied as an oxidant for various hemithioacetals, with the oxathiolane dioxide (**82**) being obtained in 80% yield ⟨80JOC3634⟩, and $KMnO_4$ and HOAc have also been used to make glycosyl sulfones ⟨48JA2435⟩ such as the L-rhamnose derivative (**83**) ⟨86TL4355⟩. Monoperphthalic acid has been used to oxidise aryl methoxymethyl sulfides to their sulfones ⟨64TL543, 67TL3057⟩, and mcpba to make glycosyl sulfones in high yield ⟨77CAR(58)397⟩. Similarly high yields of glycosyl sulfones have been achieved using $NaIO_4$ and catalytic quantities of $RuCl_3$ under Sharpless' conditions ⟨89CAR(188)81⟩, whilst the 1,3-oxathiane dioxide (**84**) was formed in 81% yield using *t*-butyl hydroperoxide (tbhp) and molybdenyl acetylacetonate as oxidant ⟨86CL1655⟩.

α-Acetoxy sulfides can be oxidised to their sulfones using H_2O_2 ⟨54CB784⟩, and similar oxidation of MTM esters to α-acyloxy sulfones converts the protecting group into a reasonable leaving group, so that subsequent hydrolysis with base regenerates the carboxylic acid ⟨79TL689⟩. Alternatively, *trans*-esterification can occur when the sulfone is treated with lithium alkoxides, and this approach was used in a macrolactonisation reaction ⟨84JA2954⟩.

(ii) From α-diazo sulfones

It was reported some time ago that photolysis of arenesulfonyl diazomethanes in methanol gives aryl methoxymethyl sulfones (76% in the case of tosyl diazomethane) as a result of carbene insertion ⟨67TL3057⟩; other work on this type of insertion has been reviewed ⟨70QRS67⟩. In the late 1980s it was shown that similar products can be obtained from tosyl diazomethane and stoichiometric amounts of alcohols in the presence of HBF_4, with the alkoxy sulfone (**85**), for example, being formed in 73% yield ⟨88TL5233⟩. Similarly, treatment of tosyl diazomethane with sulfonic acids gives α-sulfonyloxy sulfones (**86**; R = Me, Bu^t, *p*-tolyl, CF_3) ⟨72TL2477, 77JOC2792⟩.

(iii) From sulfinic acids and their salts

The reaction of alkyl halides with sulfinate anions is a standard route for the synthesis of sulfones, and has been applied to some α-halo ethers. Thus methoxymethyl *p*-tolyl sulfone is formed from $MeOCH_2Cl$ and either sodium *p*-toluenesulfinate ⟨67TL3061⟩ or preferably tetrabutylammonium *p*-toluenesulfinate (THF, 40 °C, 59%) ⟨75S519⟩, and a further example is the formation of the isochromanyl sulfone (**87**) from the corresponding bromide and sodium benzenesulfinate ⟨78CB2859⟩.

It has been noted, however, that this approach is not successful for the formation of glycosyl sulfones from glycosyl halides ⟨86TL4355⟩.

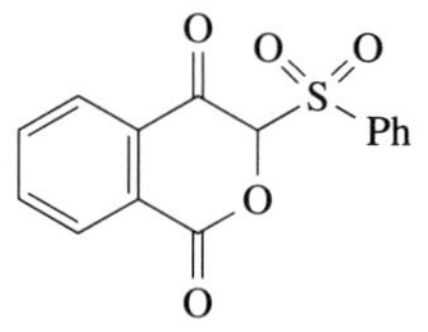

(87)

The interaction of acetals with sulfinic acids in the presence of a Lewis acid also gives α-alkoxy sulfones, as in the formation of (**88**) (48%) from 1,1-dimethoxyethane and *p*-chlorobenzenesulfinic acid in the presence of $BF_3 \cdot Et_2O$ ⟨77CB3235⟩. In a similar type of reaction, the sulfone (**89**) can be prepared from benzenesulfinic acid and either dihydropyran or 2-methoxytetrahydropyran (CH_2Cl_2, RT, 2 h), the product being obtained in ca. 80% yield in both cases ⟨86T4333⟩. This approach to 2-(benzenesulfonyl)tetrahydropyrans has been applied to the transformation in Equation (46) ⟨93TL5649⟩.

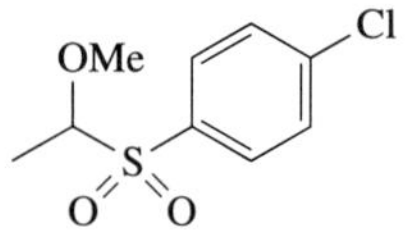

(88) 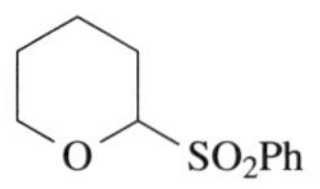   (89)

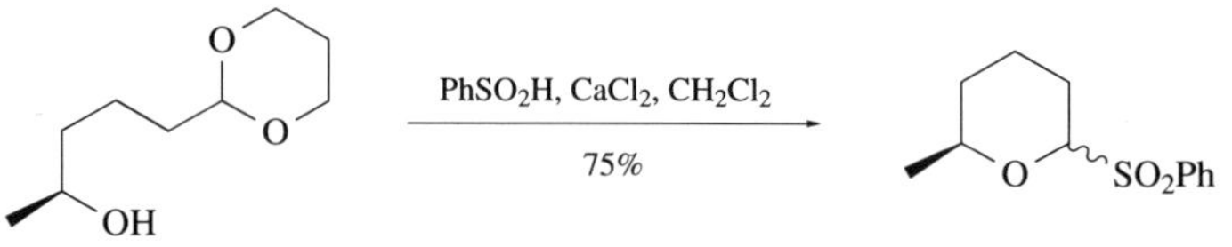

(46)

(iv) By C—C bond formation

Simpler α-alkoxy sulfones can in principle be converted into more complex systems by formation of the α-sulfonyl carbanion and treatment with an appropriate electrophile. With simple acyclic alkoxymethyl phenyl sulfones, such reactions do not proceed well except for deuteration and methylation ⟨77LA1116⟩, but the chemistry of Equation (47) has proved much more successful, presumably due to the extra chelation sites available for the lithium. The ethoxyethyl compounds (**90**; R^1 = H, Me) were prepared conventionally from the α-hydroxy sulfones, and a range of primary alkyl halides R^2X were employed. With (**90**; R^1 = H), a dianion could be formed by use of 2.2 equivalents of lithium diisopropylanide (LDA), giving products (**91**; $R^1 = R^2$) directly. The products of type (**91**) were converted by successive acid and base treatments to ketones R^1COR^2, thus illustrating the use of the α-alkoxy sulfone unit as an acyl anion equivalent ⟨80BCJ3619⟩. The oxathiolane dioxide (**82**) can also be metallated effectively, and either alkylated or hydroxyalkylated in good to excellent yields. The alkylated products fragment to an aldehyde, SO_2 and isobutene on heating ⟨79TL3375⟩. The *gem*-dimethyl group in (**82**) is important in directing metallation to the correct site; it is noteworthy that the 1,3-oxathiane dioxide (**84**) metallates at C-4 rather than at C-2 ⟨86CL1655⟩.

(90)   (91)   (47)

HMPA = hexamethylphosphoramide

4.05.1.3.3 α,β-Epoxy sulfones

One principal route to α,β-epoxy sulfones involves Darzens-type reactions between α-chloro sulfones and carbonyl compounds under the influence of $KOBu^t$ ⟨69CJC2875⟩ or under phase-transfer conditions ⟨75JOC266⟩. An example is given in Equation (48), *trans*-oxiranes being obtained from aldehydes, and mixtures of stereoisomers from unsymmetrical ketones ⟨91JCS(P1)3103⟩.

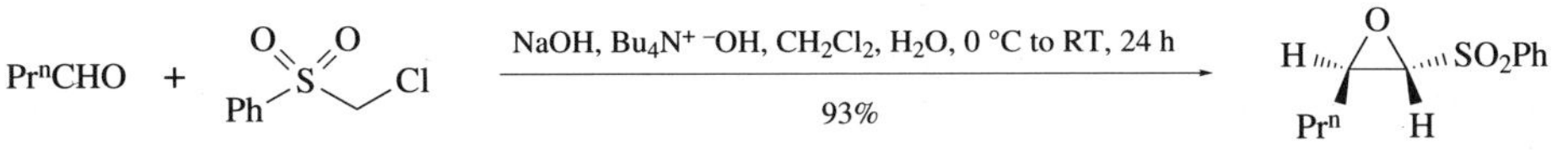

$$Pr^nCHO + PhS(O)_2CH_2Cl \xrightarrow[93\%]{NaOH, Bu_4N^+\ ^-OH, CH_2Cl_2, H_2O, 0\ ^\circ C\ to\ RT,\ 24\ h} \underset{Pr^n\ \ \ H}{H\ \ \ \ SO_2Ph} \quad (48)$$

Alternatively, α,β-epoxy sulfones can be made by addition of peroxides to α,β-unsaturated sulfones. Hydrogen peroxide ⟨70TL935⟩ or *t*-butylhydroperoxide ⟨74TL4085⟩ under aqueous conditions gave nonstereospecific results, but lithium *t*-butyl hydroperoxide under nonaqueous conditions gave a stereospecific reaction (Scheme 16) ⟨88JCS(P1)2663⟩ which also illustrates that epoxy sulfones can be stereoselectively functionalised with reactive electrophiles ⟨91JCS(P1)897⟩.

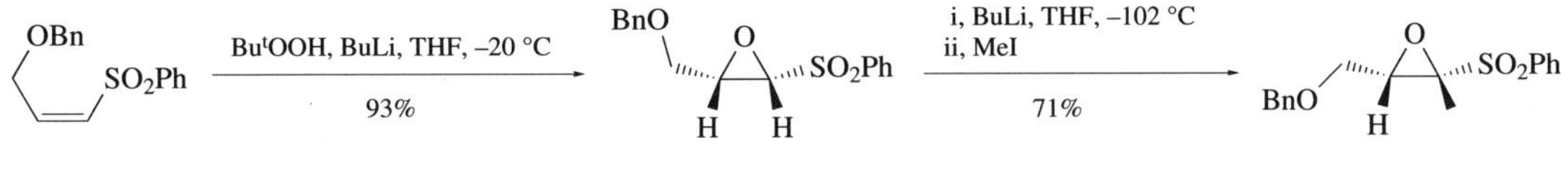

Scheme 16

The oxidation of epoxy sulfides, by either mcpba or monoperphthalic acid, has also been used to make α,β-epoxy sulfoxides ⟨90SUL157⟩.

4.05.1.3.4 α-Hydroxy sulfonic acids

The interaction of aldehydes and some ketones with aqueous sodium bisulfite to form bisulfite addition compounds has been known for very many years, and was reviewed some time ago ⟨B-66MI 405-01⟩. After some initial controversy, it became clear that these adducts were α-hydroxy sulfonic acids or their salts ⟨41JOC888⟩. The reaction is sensitive to steric effects; most aldehydes react readily, but the reaction with ketones is generally limited to methyl ketones or cycloalkanones. A typical example is the reaction of an aqueous solution of pyridine 2-carboxaldehyde with SO_2 to give the α-hydroxy sulfonic acid (**92**) in high yield ⟨51CB648⟩. With α,β-unsaturated ketones, bisulfite reacts mainly by conjugate addition, whilst with α,β-unsaturated aldehydes either 1,2- or 1,4-addition can occur depending on the conditions. Acrolein, for example, gives the kinetic product (**93**) with sodium bisulfite below pH5, whilst the bis adduct (**94**) forms more slowly ⟨62JOC649⟩. Sugars react readily with potassium metabisulfite to give α-hydroxy sulfonates derived from the open-chain form of the sugar, such as the adduct (**95**) obtained in 65% yield from D-glucose ⟨59AJC97⟩.

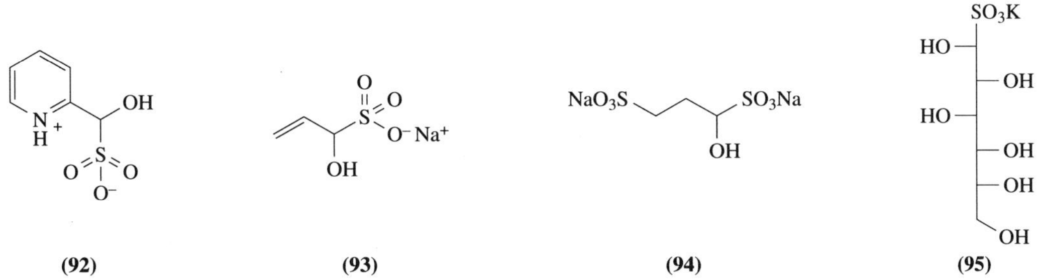

(**92**)   (**93**)   (**94**)   (**95**)

4.05.2 FUNCTIONS CONTAINING OXYGEN AND EITHER SELENIUM OR TELLURIUM

4.05.2.1 Dicoordinate Selenium and Tellurium Derivatives

The general methods that have been reported for compounds of this type can be classified under three main headings, which, perhaps unsurprisingly, have good parallels in the main methods used for the analogous sulfur compounds.

4.05.2.1.1 *From carbonyl compounds, other acetals and enol ethers*

The reaction between benzeneselenol and dimethoxymethane in the presence of BF_3 leads to acetal exchange to give $PhSeCH_2OMe$ ⟨77LA846⟩. A rather more general procedure for such acetal exchanges involves the treatment of an *O,O*-acetal with tris(phenylseleno)borane (typically 0.35 mol. equiv., toluene, RT), as in the formation (87%) of the *O,Se*-acetal (**96**) from the dimethoxycompound ⟨79JOC1883⟩. The reagent diisobutylaluminum benzeneselenolate, $Bu^i_2AlSePh$, made *in situ* by treatment of $(PhSe)_2$ with dibal-H, has been used to effect the same type of transformation, the monoselenoacetal (**97**) being obtained in 58% yield ⟨91CL1775⟩. Directly analogous chemistry can be used to make monotelluroacetals such as (**98**) using $Bu^i_2AlTePh$, but an excess of the reagent can lead to formation of the ditelluroacetal as well ⟨91CL415⟩.

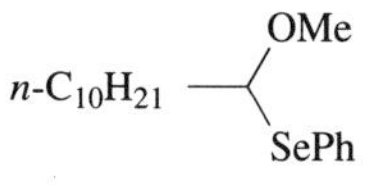

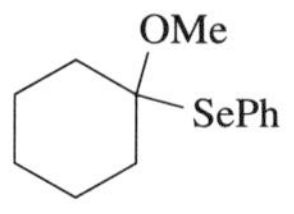

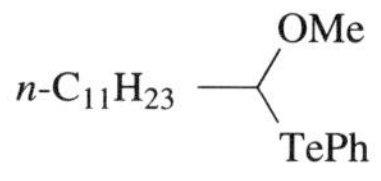

Treatment of cyclic hemiacetals with PhSeH and $BF_3 \cdot Et_2O$ can give monoselenoacetals with the oxygen within a ring, as in the formation of (**99**) ⟨88TL2179⟩, and (**100**) with $MeSO_3H$ as catalyst (Equation (49)) ⟨87T4875⟩. Somewhat similar conditions can be used for the synthesis of selenoglycosides (for a review of earlier work on carbohydrates containing selenium, see ⟨82H(19)1719⟩; thus the L-rhamnopyranosyl selenoglycoside (**101**) is formed from the α-1-*O*-acetyl compound on treatment with PhSeH and $BF_3 \cdot Et_2O$ ⟨91TL4435⟩, and the D-*manno*-compound (**102**) is formed in 90% yield from the methyl 1,2-*ortho*-acetate and PhSeH in the presence of a catalytic quantity of $HgBr_2$ ⟨93SL522⟩. Selenoglycosides can also be made from 1-*O*-acetyl sugars with the reagent combination $Me_2Sn(SePh)_2$–$Bu_2Sn(OTf)_2$ ⟨92TL239⟩. The simple analogue (**103**) can be prepared in 90% yield by acid-catalysed addition of PhSeH to dihydropyran ⟨75TL1613⟩. A sugar analogue with selenium in the ring has been made (Scheme 17) by a procedure based on those described above (Schemes 9 and 10) for sulfur-in-ring compounds ⟨77CAR(59)351⟩.

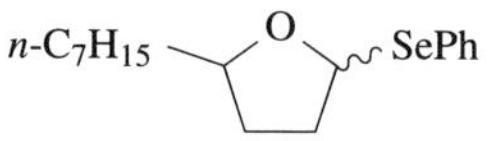

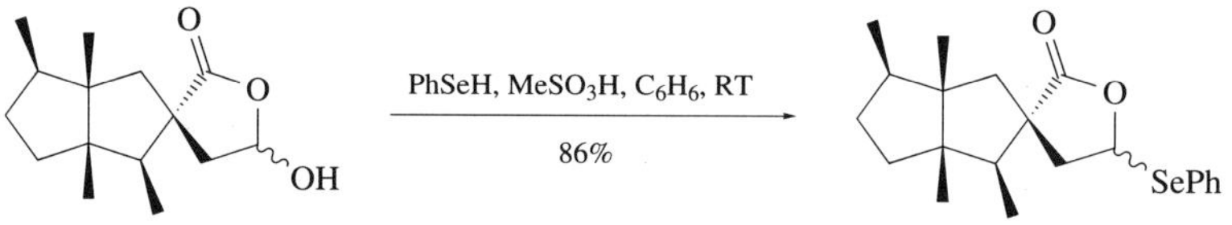

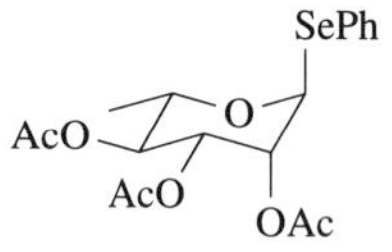

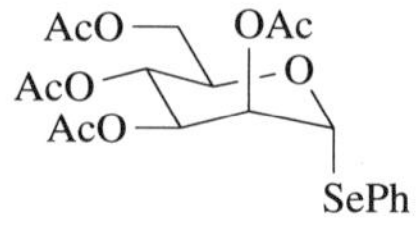

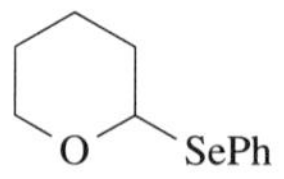

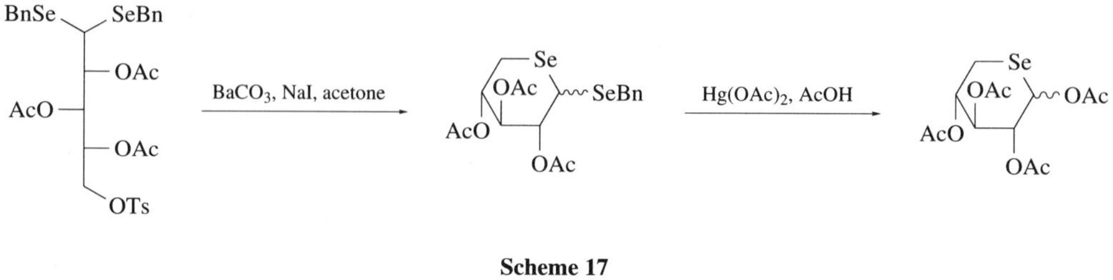

Scheme 17

When aldehydes, but not ketones, are treated with PhSeTMS in the presence of a Lewis acid, *O*-silylated monoselenoacetals (α-(trimethylsilyloxy) selenides) are obtained, *n*-butanal, for example, giving the adduct (**104**) quantitatively with ZnCl$_2$ as catalyst; α,β-unsaturated aldehydes and ketones give 1,4-adducts instead ⟨78TL5091⟩. It has also been reported that the same type of adduct can be obtained from both aldehydes and ketones by treatment of the carbonyl compound with a selenol and TMS-Cl in the presence of pyridine, *n*-heptanal giving an 80% yield with PhSeH, and acetone giving the product (**105**) in 45% yield ⟨77AG(E)540⟩. In 1994, in a variant of this procedure, successive treatment of dihydrocinnamaldehyde in pyridine with PhSeH, Cl$_2$SiMe$_2$ and allyl alcohol gave a high yield of the silylated hemiselenoacetal (**106**). The phenylselenyl group in (**106**) was used to effect free radical coupling between the two groups tethered by the silicon, and this principle was used in an elegant approach to the tunicamycin antibiotics ⟨94JA4697⟩.

4.05.2.1.2 *From α-halo ethers and related compounds*

An 80% yield of PhSeCH$_2$OMe has been reported from the reaction between bromomethyl methyl ether and PhSeK ⟨75TL1613⟩. Reich *et al.* obtained an even higher yield (94%) of the same product from PhSeNa, generated *in situ* by borohydride reduction of diphenyl diselenide, and ClCH$_2$OMe. These workers also described the lithiation of the analogous *m*-trifluoromethylphenyl methoxymethyl selenide and subsequent methylation to give (**107**); treatment of the anion with aldehydes and ketones gave hydroxyalkylated products ⟨79JA6638⟩. A range of compounds of the type R^{1}SeCH$_2$OR^2, with R^1 = C$_6$H$_{13}$ or longer, or an aryl group and R^2 = Me, Et, Pr^i, has been prepared from α-chloro ethers and alkylseleno- or arylselenomagnesium halides ⟨80JOU1377⟩. Similarly, reagents of the type ArSeMgBr and di(chloromethyl) ether give products ArSeCH$_2$OCH$_2$SeAr ⟨82JOU876⟩.

Similar chemistry has been carried out using telluride anions, the compounds Bu^{n}TeCH$_2$OCH$_2$CH$_2$OMe and Bu^{n}TeCH$_2$OBn being prepared by the action of Bu^{n}TeLi on the appropriate α-chloro ether ⟨90OM1355⟩.

Displacements of halides under basic conditions have also been used to make selenoglycosides. The β-D-glucopyranosyl phenyl selenide (**108**) can be produced in 87% yield from the α-bromide by treatment with PhSeH and KOH in ethanol–chloroform at reflux ⟨50JA354⟩, and a high-yielding reaction on a glycosyl chloride has been reported in which Et$_3$N was used as base (CH$_3$CN, RT) ⟨89TL6311⟩. In a similar way, the *Se*-benzoyl compound (**109**) was prepared from acetobromoglucose and potassium selenobenzoate, and subsequent deacylation gave 1-seleno-D-glucose, isolated as its sodium salt ⟨63CI(L)1397⟩. Alternatively, the *O*-acylated derivative (**110**) of 1-selenoglucose (potassium salt) can be obtained by a reaction sequence directly analogous with Scheme 8, but with use of selenourea in place of thiourea. The salt (**110**) was condensed with another equivalent of either the same or a different per-*O*-acetyl glycosyl bromide to give, after deacetylation, bis-sugar selenides ⟨64AP(297)461⟩.

Equation (50) outlines a procedure to make anisyl glycosyl tellurides, which were used as precursors of glycosyl radicals ⟨90JA891⟩.

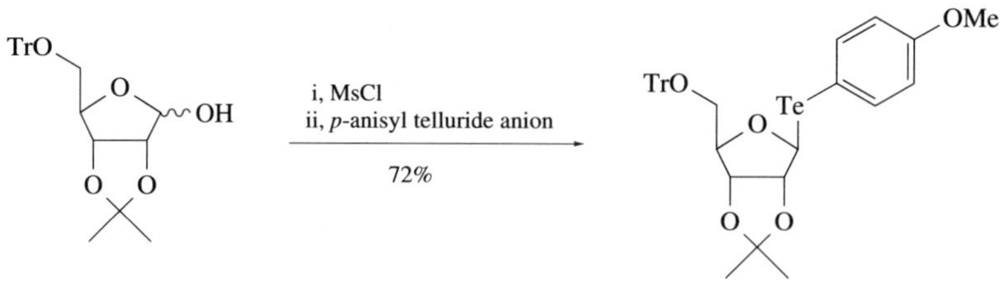

4.05.2.1.3 By seleno-Pummerer reactions

Due to the instability of selenoxides which have β-hydrogen atoms, the seleno-Pummerer reaction has not been as well developed as the sulfur equivalent (see 4.05.1.1.1(v) above), but nonetheless in suitable cases the desired transformation has been achieved (for a tabulation of examples, see ⟨91OR(40)157⟩). The reaction sequence in Scheme 18 has been carried out starting from dimethyl selenoxide, where clearly elimination of a selenenic acid cannot occur ⟨77TL851⟩, and oxidation of cyanomethyl phenyl selenide with peracids leads directly to α-acyloxy selenides ⟨82TL4371⟩. Treatment of ethyl phenyl selenide with dibenzoyl peroxide forms an isolable selenurane (Scheme 19), which underwent rearrangement as shown on heating; two other similar cases were reported ⟨73JOC3172⟩. In Equation (51) a case is shown where both generation of the selenoxide and its activation are carried out at low temperatures, with smooth rearrangement ensuing ⟨82JOC693⟩, whilst the selenoxide (**111**) in Equation (52) was stable at room temperature, presumably due to the presence of electron-withdrawing substituents at both β-positions, and also underwent rearrangement stereoselectively in good yield on treatment with acetic anhydride, or with pivalic anhydride (73%) ⟨92MI 405-01⟩. Scheme 20 outlines an interesting reaction sequence in which an intermediate of the type proposed in seleno-Pummerer reactions is generated by a [3,3]-sigmatropic rearrangement ⟨87TL4925⟩.

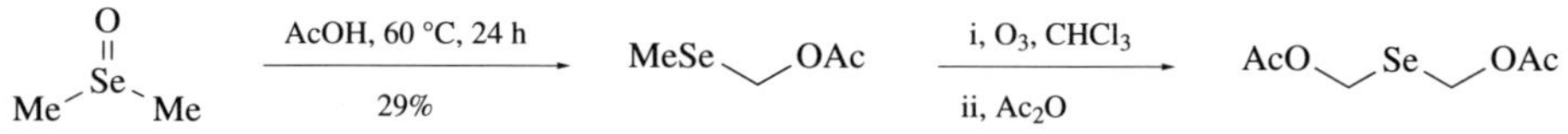

Scheme 18

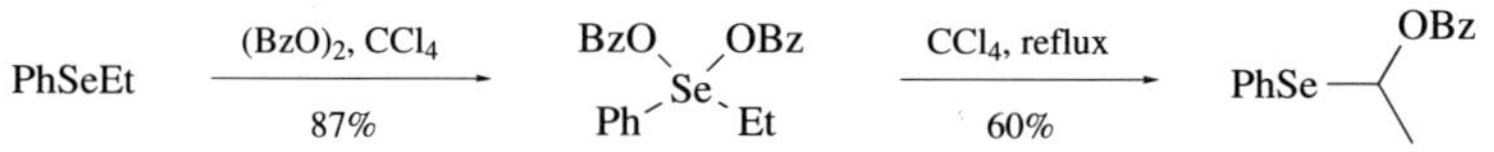

Scheme 19

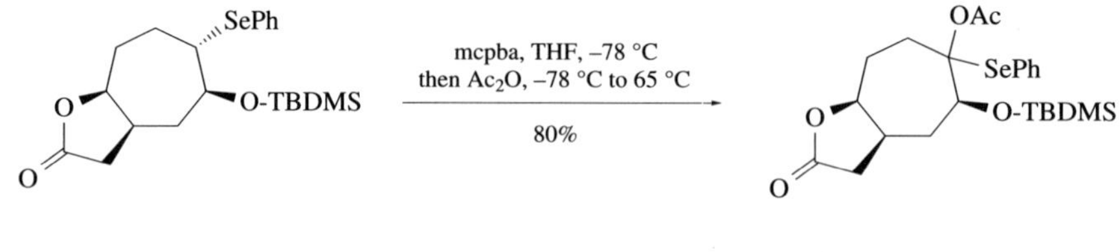

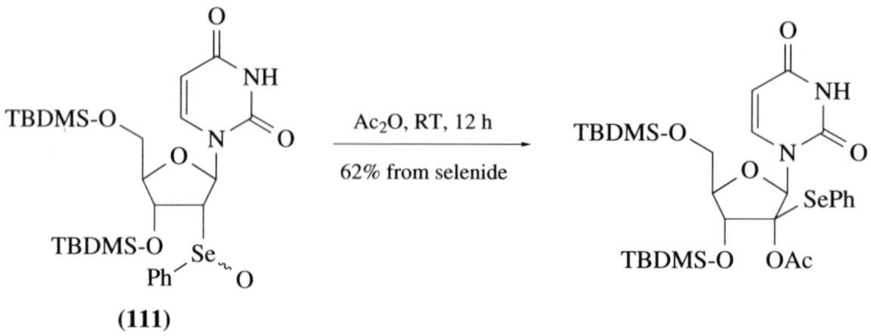

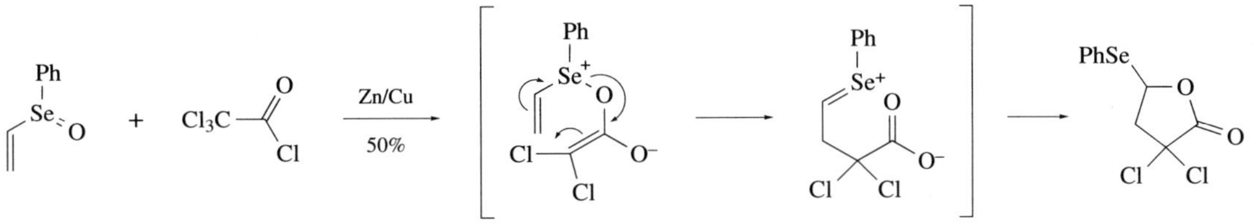

Scheme 20

4.05.2.2 Tricoordinate Selenium Derivatives

Selenoxides with oxygen substituents in the α-position seem to have the normal instability towards elimination shown by other selenoxides which have a β-hydrogen atom, the products of elimination being enol ethers or enol esters. Thus, for example, when the selenide (**100**) was treated with ozone at −75 °C, it was assumed that the selenoxide had been formed. Addition of diisopropylamine and warming to RT then gave the enol lactone in high yield ⟨87T4875⟩. Similarly, oxidation of selenide (**99**) with tbhp and Ti(OPr^i)$_4$ in the presence of EtNPr$^i_2$ at 0 °C gave the corresponding enol ether in good yield ⟨88TL2179⟩. In both these cases it was important that the selenoxide fragmentation occurred in the presence of a base to prevent formation of substantial amounts of by-products in which benzeneselenenate had been displaced by other nucleophiles present.

4.06
Functions Incorporating Two Chalcogens Other Than Oxygen

YANNICK VALLÉE
Université Joseph Fourier, Grenoble, France

and

ANDREW BULPIN
Institut Jacques Boy, Reims, France

4.06.1	FUNCTIONS CONTAINING TWO SULFURS—$R^1_2C(SR^2)SO_2R^3$, etc.	244
	4.06.1.1 Introduction	244
	4.06.1.2 Two Dicoordinated Sulfurs—$R^1_2C(SR^2)_2$	244
	4.06.1.2.1 gem-Dithiols	244
	4.06.1.2.2 Hemidithioacetals	245
	4.06.1.2.3 Dithioacetals	245
	4.06.1.3 One Dicoordinated Sulfur and One Higher Coordinated Sulfur—$R^2_1C(SR^2)SO_2R^3$, etc.	258
	4.06.1.3.1 α-Thio sulfoxides	258
	4.06.1.3.2 α-Thio sulfones	265
	4.06.1.3.3 Other derivatives	267
	4.06.1.4 Two Tricoordinated Sulfurs—$R^1_2C[S(O)R^2]_2$	267
	4.06.1.4.1 Bis(sulfoxides)	267
	4.06.1.4.2 Other compounds	270
	4.06.1.5 One Tricoordinated and One Higher Coordinated Sulfur—$R^1_2CS(O)OR^2S(O)_2R^3$, etc.	271
	4.06.1.6 Two Tetracoordinated Sulfurs—$R^1_2C[S(O)_2R^2]_2$	272
	4.06.1.6.1 Bis(sulfones)	272
	4.06.1.6.2 Bis(sulfonic)acids and their derivatives	281
	4.06.1.6.3 Other compounds	282
4.06.2	FUNCTIONS CONTAINING ONE SULFUR AND ONE SELENIUM OR TELLURIUM—$R^1_2CSR^2SeR^3$, etc.	283
	4.06.2.1 Dicoordinated Sulfur Derivatives	283
	4.06.2.2 Tricoordinated Sulfur Derivatives	285
	4.06.2.3 Tetracoordinated Sulfur Derivatives	286
4.06.3	FUNCTIONS CONTAINING SELENIUM AND/OR TELLURIUM—$R^1_2C(SeR^2)_2$, $R^1_2C(SeR^2)TeR^3$, etc.	287
	4.06.3.1 Diselenium Derivatives	287
	4.06.3.2 Ditellurium Derivatives	290
	4.06.3.3 Other Derivatives	291

4.06.1 FUNCTIONS CONTAINING TWO SULFURS—$R^1_2C(SR^2)SO_2R^3$, etc.

4.06.1.1 Introduction

In 1885 the condensation of thiols with aldehydes and ketones to form dithioacetals was reported for the first time (Equation (1)) ⟨1885CB883⟩. This reaction was applied to sugar chemistry before the turn of the century (Equation (2)) ⟨1894CB673⟩. However, it was not until 70 years later, in 1965, with the innovative work of Corey and Seebach, that dithioacetals took on any major importance in the arsenal of tools available to the organic chemist.

$$\underset{CO_2H}{\overset{O}{\|}} \xrightarrow{PhSH, H^+} \underset{CO_2H}{\overset{PhS \quad SPh}{\diagup\diagdown}} \tag{1}$$

$$(C_5H_{11}O_5)\underset{}{\overset{O}{\|}} \xrightarrow{EtSH, H^+} (C_5H_{11}O_5)\underset{SEt}{\overset{SEt}{\diagup\diagdown}} \tag{2}$$

In particular, dithianes and the carbanions derived from them are now widely used and are featured in numerous syntheses of complex natural products. Combined with the concept of umpolung, a novel field of chemistry employing reversed polarity was born. For the formation of C—C bonds the chemist now has—depending on the requirements of the synthesis or his or her predilections—the choice between two polarities.

The applications of dithioacetals are now so diverse and the methods of preparing them so numerous that it is impossible to cover all aspects of their chemistry comprehensively in this chapter. However, several review articles dealing more or less exclusively with dithioacetals, their carbanions, and their synthetic applications have already been published ⟨69S17, 77S357, 78CRV363, 89T7643, 91ACR257⟩, and the attention of the more avid reader is drawn to them.

Sections 4.06.1.3 to 4.06.1.6 inclusive describe mainly the syntheses of variously oxidized dithioacetals (Equation (3)). The most convenient route to all these compounds is the direct oxidation of the corresponding dithioacetals. This chapter concentrates particularly on the problems associated with the selectivity of these oxidations and the questions of diastereo- and enantioselectivity, which in several cases remain unresolved.

$$\tag{3}$$

4.06.1.2 Two Dicoordinated Sulfurs—$R^1_2C(SR^2)_2$

There are effectively three types of compounds which contain two dicoordinated sulfurs: *gem*-dithiols, hemidithioacetals, and dithioacetals. Although dithioacetals are by far the most important, for completeness the first two will initially be considered, before looking in more detail at dithioacetals.

4.06.1.2.1 gem-Dithiols

gem-Dithiols—compounds renowned for being particularly malodorous and frequently unstable—may be obtained by an acid-catalyzed reaction between hydrogen sulfide and a ketone

⟨62CB1764⟩. This was how the dithiol **(1)** (Equation (4)) was prepared, which has the peculiarity of being a stable crystalline solid ⟨59JA3148⟩.

$$\text{Ph} \overset{O}{\underset{}{\bigtriangleup}} \text{Ph} \xrightarrow[\text{HCl}]{\text{H}_2\text{S}} \text{Ph} \overset{\text{HS} \quad \text{SH}}{\underset{}{\bigtriangleup}} \text{Ph} \qquad (4)$$

$$(1)$$

4.06.1.2.2 Hemidithioacetals

Hemidithioacetals have received very little investigation, although they have been obtained by the reaction of hydrogen sulfide and a thiol with an aldehyde at pH 5 (Equation (5)) ⟨71TL2321⟩. The reduction of dithioesters with sodium borohydride is another convenient way to produce hemidithioacetals ⟨74RTC242⟩.

$$R^1 \overset{O}{\underset{}{\bigtriangleup}} \xrightarrow[\text{pH 5}]{\text{H}_2\text{S}, \text{R}^2\text{SH}} R^1 \overset{\text{SH}}{\underset{\text{SR}^2}{\bigtriangleup}} \qquad (5)$$

4.06.1.2.3 Dithioacetals

The accessibility of dithioacetals has been explored extensively. Their most common precursors are aldehydes and ketones, or occasionally carbonyl derivatives such as acetals and *gem*-dichlorides. These precursors have the similarity in that they all give dithioacetals by functional transformations, with the creation of two C—S bonds but without C—C bond formation. The most common way to obtain a new dithioacetal with C—C bond formation is to treat the carbanion derived from a simple dithioacetal such as dithiane with an electrophilic reagent. Ketene dithioacetals, although less frequently employed, are other potential precursors in which the double bond may be hydrogenated or alkylated. Certain thiocarbonyl compounds can also be used to prepare dithioacetals, for example in the thiophilic addition of alkyl Grignard reagents to dithioesters, which is a fairly general method.

(i) From aldehydes, ketones, acetals, and gem-dichlorides

The condensation of thiols with aldehydes or ketones is catalyzed by Brönsted and Lewis acids. Toluene-*p*-sulfonic acid, for example, has found particular use with solvents which allow for the azeotropic distillation of the water formed, but the most favored catalyst currently employed in synthesis is boron trifluoride etherate ⟨75JOC231⟩. Several representative examples of dithioacetal formation are given in Scheme 1.

The dithioacetal **(2)** is an intermediate in the synthesis of gingerols, the active ingredients of ginger ⟨93JOC2181⟩. The amide **(3)** is a precursor of dithyreanitrile, an insect antifeedant ⟨93TL1085⟩. Other functionalities may also be present in the molecule. For example, the bromo dithioacetal **(4)** was obtained from the corresponding aldehyde ⟨93S149⟩. Likewise, conjugated aldehydes also give the expected dithioacetals, as shown for the acetylenic aldehyde **(5)**, which contains a C—Si bond ⟨82JOC1145⟩. Bisdithioacetals such as compound **(6)** may also be prepared from the corresponding dicarbonyl compounds ⟨93T2151⟩.

In certain cases it has been found preferable to carry out the reaction in the absence of a solvent ⟨91SC1369⟩. The keto diester **(7)** was transformed in this way into the dithiolane **(8)** in 98% yield after only 4 h (Equation (6)). When the same reaction was carried out in dichloromethane, a standard solvent for such reactions, the yield dropped to 55% even after 100 h.

$$\text{EtO} \overset{O}{\underset{}{\diagup}} \overset{O}{\underset{}{\diagup}} \overset{O}{\underset{}{\diagup}} \text{OEt} \xrightarrow[98\%]{\text{HS(CH}_2)_2\text{SH} \atop \text{BF}_3 \cdot \text{OEt}_2, \text{ no solvent}} \text{EtO} \overset{O}{\underset{}{\diagup}} \overset{\text{S} \quad \text{S}}{\underset{}{\diagup}} \overset{O}{\underset{}{\diagup}} \text{OEt} \qquad (6)$$

(7)   **(8)**

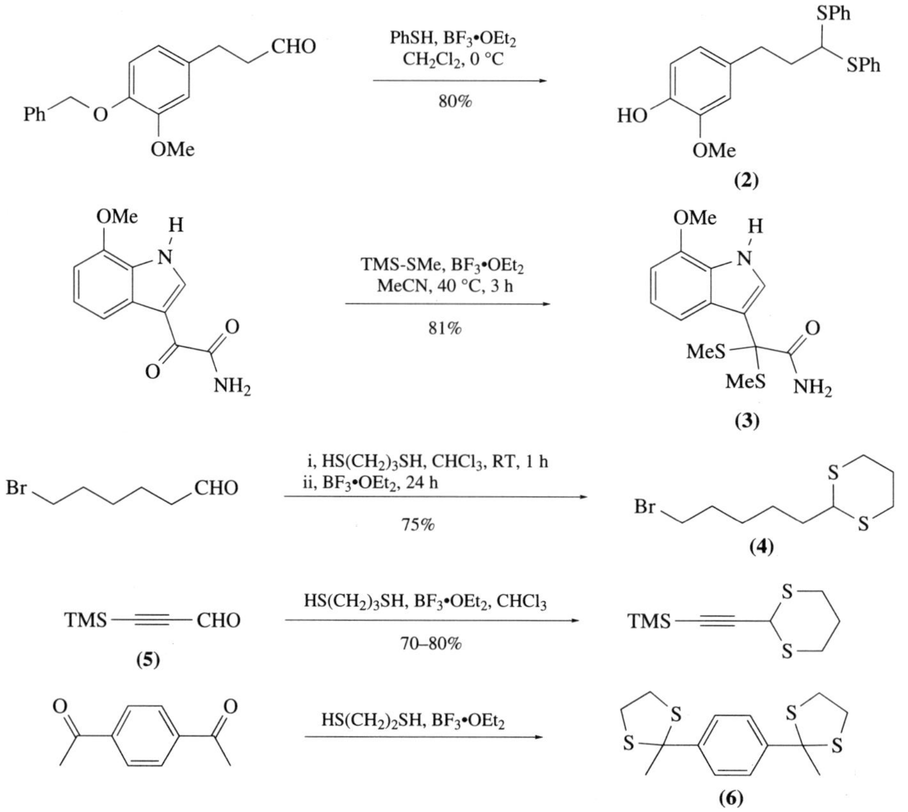

Scheme 1

Numerous other catalysts have also been proposed. However, it is generally difficult to perceive the improvements, if any, that these catalysts offer over $BF_3 \cdot Et_2O$.

Anhydrous lanthanum chloride has been employed to catalyze the reaction between ethanedithiol and cyclohexanone, giving the dithiolane derivative in 90% yield after only 1 h ⟨90TL5815⟩. By contrast, when used with acetophenone, the corresponding dithiolane was obtained in only 25% yield, even after 100 h. This disparity, although somewhat reducing the general application of this catalyst, could be advantageous for the chemoselective reaction of aliphatic ketones in the presence of aromatic ketones. The use of tetrachlorosilane has also been proposed for this type of reaction ⟨89SC433⟩. In 1993, bis(trimethylsilyl) sulfate in the presence of silica gel was reported to be a useful catalyst for the synthesis of dithiolanes ⟨93TL7127⟩.

When ethanedithiol reacted with cyclohexanone in the presence of bentonitic earth, the corresponding dithiolane was formed quantitatively ⟨90SC153⟩. However, with the same catalyst, the dithiolane (**8**) was isolated in only 60% yield (compared with 98% with $BF_3 \cdot Et_2O$—see Equation (6)). Other zeolites have been tested for catalytic activity ⟨93S67⟩. With the zeolite H–Y, the dithiolanes of cyclopentanone and acetophenone were each obtained in 95% yield. Selected clays may also be used for the preparation of dithioacetals ⟨89SC31⟩. Montmorillonite KSF, a solid Brönsted acid, catalyzes the condensation of benzaldehyde with thiophenol (Equation (7)). Dihydropyran, an enol ether, has been used as the starting material to give a hydroxy dithioacetal (Equation (8)).

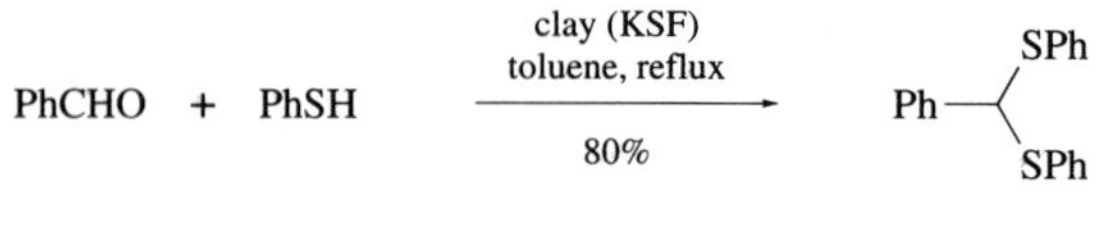

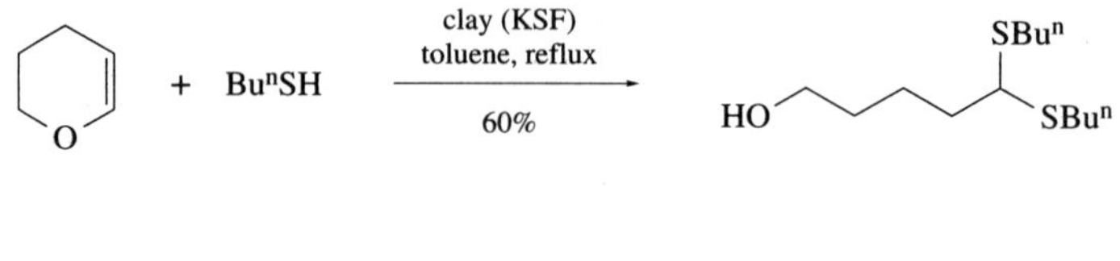

Polyphosphoric acid trimethylsilyl ester (PPSE) has been employed as a condensation agent ⟨87S164⟩. It intervenes both as a Lewis acid and as a dehydrating agent giving, on the whole, good yields (62–96%). However, with *t*-butyl thiol and cyclohexanone, the major product is the enethiol ether (**9**) (Equation (9)). This alternative reactivity is rather common for enolizable ketones.

The results obtained with three relatively new catalysts and certain dicarbonyl compounds are listed in Table 1. The selectivity is influenced by steric and electronic factors which favor the dithioacetalization of an aldehyde over a ketone. A thorough comparison, however, remains to be made before it is known which of these three new catalysts, or which of the more established catalysts, are really the most useful, and for which substrates and under which conditions.

Table 1 Catalysts for the regioselective preparation of cyclic dithioacetals.

Starting materials	Catalyst	Conditions	Product	Yield (%)	Ref.
p-acetylbenzaldehyde + 1,2-ethanedithiol	Amberlyst 15	25 °C, CHCl$_3$ overnight	acetyl-aryl-1,3-dithiolane	83 90	89SC2383 86JOC1427
5-oxohexanal + 1,3-propanedithiol	SiO$_2$–SOCl$_2$	C$_6$H$_6$, reflux 24 h	2-(4-oxopentyl)-1,3-dithiane	98	86JOC1427
6-oxoheptanal + 1,3-propanedithiol	TeCl$_4$	C$_2$H$_4$Cl$_2$, RT 2–3 h	2-(5-oxohexyl)-1,3-dithiane	76	91TL2039
Wieland–Miescher-type diketone + 1,3-propanedithiol	TeCl$_4$	C$_2$H$_4$Cl$_2$, RT 2–3 h	mono-dithiane product	68	91TL2039

All the methods above relate to the preparation of symmetrical dithioacetals, $R^1_2C(SR^2)_2$, and they cannot be efficiently extended to asymmetrical dithioacetals, $R^1_2C(SR^2)(SR^3)$ (Equation (10)). The reaction of two different thiols and a ketone will lead *a priori* to a mixture of three dithioacetals, the proportions varying somewhat with the steric constraints of the system and the relative nucleophilicities of the two thiols.

Therefore, when the aldehyde (**10**) (Scheme 2) was treated with equimolar quantities of the thiols (**11**) and (**12**), a mixture was obtained from which the asymmetrical dithioacetal (**13**) was isolated in 49% yield ⟨89JOC3718⟩. An alternative synthesis of compound (**13**) was effected by treatment of the thiol (**12**) with the silylated thioacetal (**14**). However, the yield for this step alone was only 61%.

Alternatively using a thiol and a thiol acid it is possible to prepare selectively and isolate an acyl dithioacetal which, as shown in Equation (11), can then be treated with a nucleophile and an alkylating agent to produce an asymmetrical dithioacetal ⟨88TL6729⟩. This method gives good yields

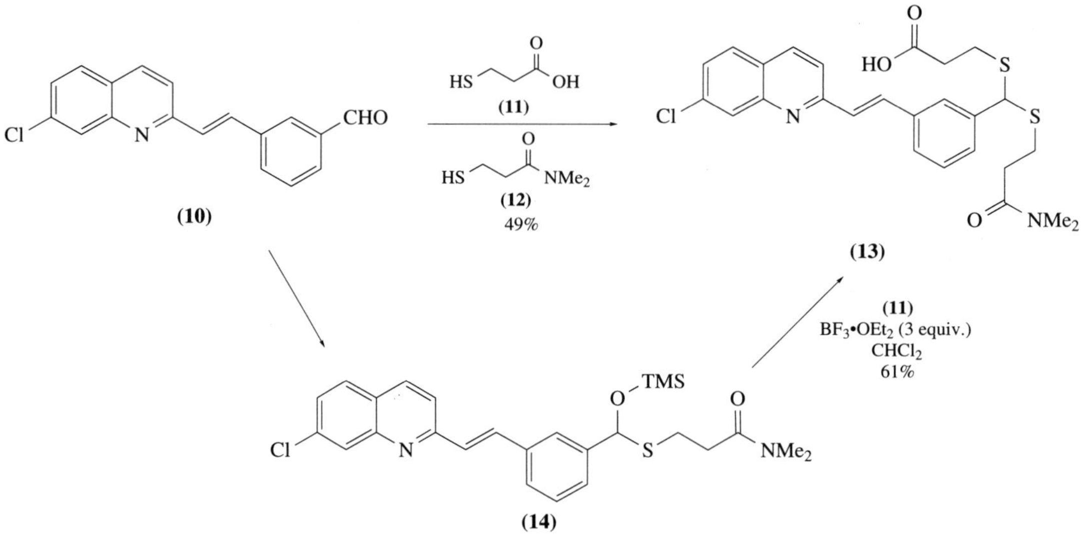

Scheme 2

with aldehydes. Any traces of the symmetrical dithioacetal obtained after the first step are eliminated using GLC. This strategy has been used to prepare enantiomerically enriched dithioacetals ⟨88TL6733⟩.

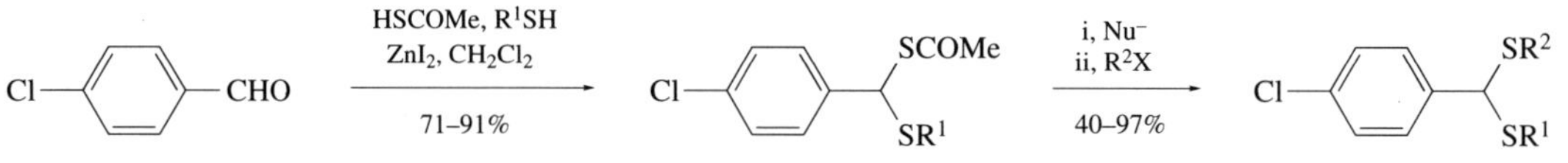

(11)

Returning to symmetrical dithioacetals, rather than simply using thiols as the thiolating agents, a slightly different strategy employed certain thiol derivatives containing S—Si ⟨86TL6305⟩, S—Sn ⟨88TL3971⟩, and S—B bonds ⟨79JOC656⟩. Formation of the dithioacetal is then promoted by a favorable energy difference between the sulfur–heteroatom bond and the corresponding oxygen–heteroatom bond. Several typical examples are given in Table 2. The yields are generally good. However, it should be remembered that, unlike the direct preparation ($R^1SH + R^2R^3CHO$), these methods transform thiols into dithioacetals in two steps via relatively unstable intermediate compounds which undergo rapid hydrolysis in the presence of only traces of water vapor.

Table 2 Uses of silicon, tin, and boron derivatives for the synthesis of dithioacetals.

Starting materials	Product	Yield (%)	Ref.
PhCHO 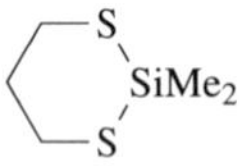 + S-SiMe₂-S (1,3-dithiane)	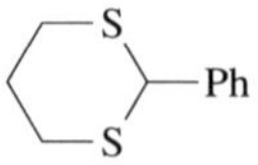 S-Ph-S (2-phenyl-1,3-dithiane)	98 (GLC)	86TL6305
PhCHO 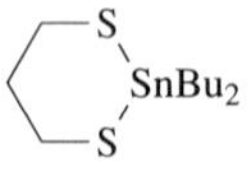 + S-SnBu₂-S	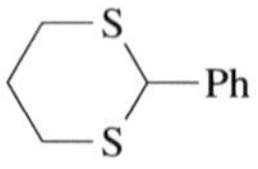 S-Ph-S	73 (isolated) 100 (GLC)	88TL3971
PhCHO 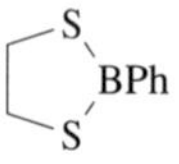 + S-BPh-S (1,3-dithiolane)	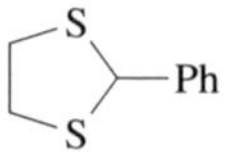 S-Ph-S	98 (isolated)	79JOC656
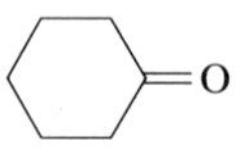 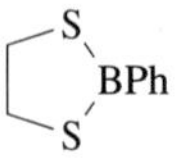 cyclohexanone + S-BPh-S	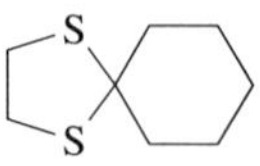 dithiolane spiro cyclohexane	98 (isolated)	79JOC656

Dithioacetals can also be prepared from the acetal derivatives of aldehydes and ketones. Although this generally adds another synthetic step compared with their direct preparation from carbonyl compounds, in certain circumstances this may be necessary or at least advantageous. One example ⟨93TL1141⟩, presented in Equation (12), is the transformation of the acetal (**15**), obtained from a quinone derivative, into the dithioacetal (**16**). Notably, transacetalization of the aldehyde acetal occured in preference to reaction at either of the two ketone groups. In a similar way, the dithiol (**17**) was transformed into the corresponding formaldehyde dithioacetal by a BF_3-catalyzed reaction with formaldehyde dimethyl acetal (Equation (13)) ⟨90BSF734⟩.

(Haloalkyl)-1,3-dithiolanes ⟨93T199⟩ have been prepared from ethanedithiol and chloro acetals in the presence of cobalt(II) chloride and chlorotrimethylsilane, thus avoiding the formation of any rearrangement products (Equation (14)).

gem-Dichlorides have also been used as dithioacetal precursors but, like acetals, they are used less frequently than carbonyl compounds, with notable and valid exceptions. One exception (Equation (15)) is a simple and inexpensive route to deuterium-labelled dithioacetals from dithiols and the relatively cheap dideuteriodichloromethane ⟨81TL1821⟩. The reaction takes place in a basic medium. Ono *et al.* proposed the use of 1,5-diazabicyclo[5.4.0]undec-5-ene (dbu) to mediate this type of reaction ⟨80S952⟩. Formaldehyde dithioacetals were obtained in up to 91% yield. In the early 1990s, bis(diphenylphosphino)methane complexes of platinum(II) were proposed as catalysts ⟨92T5933⟩. In such cases, however, no dithioacetals could be isolated from dichloro- or dibromomethane, and diiodomethane has to be used giving a (70–73% yield). When comparing these two methods, note that one equivalent of dbu was used whereas the quantity of platinum complex used was truly catalytic. Finally, Le Floc'h reported that the treatment of *gem*-diamines with thiols in an acidic medium also gives dithioacetals ⟨78BSF(2)595⟩.

(ii) From other dithioacetals and related compounds

The production of dithioacetals from other dithioacetals or related compounds is, of course, the archetypal synthetic reaction of the dithioacetals. Although in the strictest sense of the term such reactions are not really functional group transformations, they permit the conversion of a dithioacetal derived from an aldehyde into the dithioacetal (dithioketal) of a ketone. Several examples concerned with dithianes, including those reported by Corey and Seebach in their original papers, are presented in Scheme 3 ⟨65AG1134, 75JOC231, 81JOC1513, 92T7265⟩. The review by Bulman Page *et al.* is recommended for a more thorough account of the possibilities of such reactions ⟨89T7643⟩.

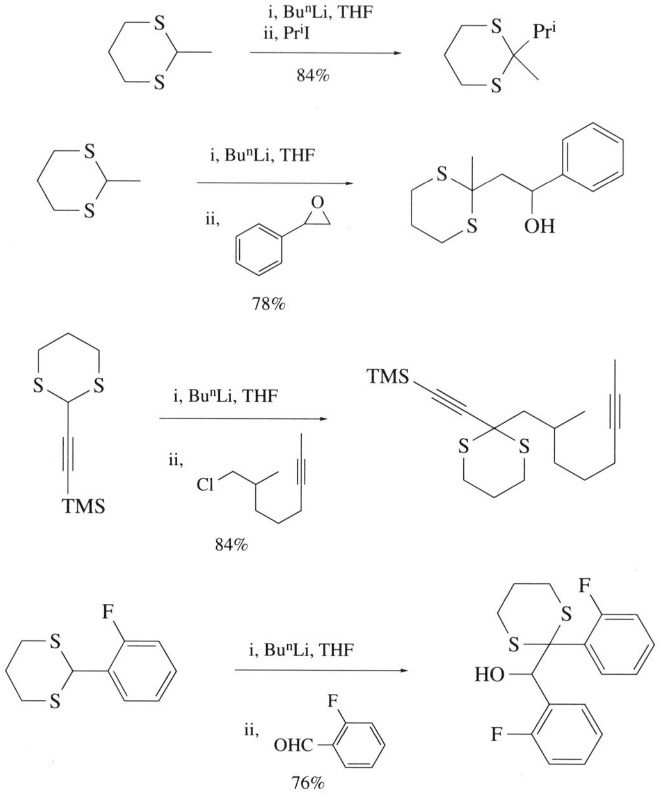

Scheme 3

Apart from the dithianes, other types of dithioacetal group can be introduced into a molecule by this method (Scheme 4). Of particular note amongst the cyclic dithioacetals employed are the trithiane (**18**) ⟨71OS(51)39⟩, the benzodithioles such as (**19**) ⟨78TL2345⟩, the benzodithiepine (**20**) ⟨75S720⟩, and the chiral dithiepine (**21**) ⟨89SL28⟩; and amongst the acyclic dithioacetals, di(phenylthio)methane (**22**) ⟨66JOC4097, 80TL4763⟩ and di(methylthio)methane (**23**) ⟨82TL1047⟩.

All of the examples above make use of a carbanion stabilized by two adjacent sulfur atoms. Although much less frequently, carbenes, carbocations and radicals have all been investigated as possible intermediates. The trimethylenedithiocarbene (**24**) (Scheme 5), obtained by decomposition of the dianion (**25**) at −20 °C, was trapped by an organolithium reagent present in the reaction mixture to give the corresponding carbanion, which subsequently reacted with an electrophile ⟨69TL173⟩. This method, which transforms a trithioorthoformate into a dithioacetal, has not been further exploited as of 1994.

The dithio-*ortho*-formate (**26**) has been treated with various silyl enol ethers in the presence of zinc chloride to give the keto dithioacetals (**27**) (Equation (16)). This route permits easy access to the synthetically useful, half-protected 1,3-dialdehydes and 1,3-keto aldehydes ⟨81TL3243⟩. Contrary to the classic reaction of Corey and Seebach, where a sulfur-stabilized carbanion is employed as the nucleophile, the dithioacetal functionality was introduced here by the electrophilic reagent. This was also true for the syntheses of the deuterated dithioacetals shown in Equation (17), where the electrophile was the 2-deuterio-1,3-benzodithiolium perchlorate (**28**) ⟨81TL1821⟩. In these cases, the polarity of the dithioacetal is the same as that of the parent carbonyl compounds; there is no umpolung under these circumstances. When considering the perchlorates for potential synthetic applications, the danger of explosions with these treacherous reagents should be considered.

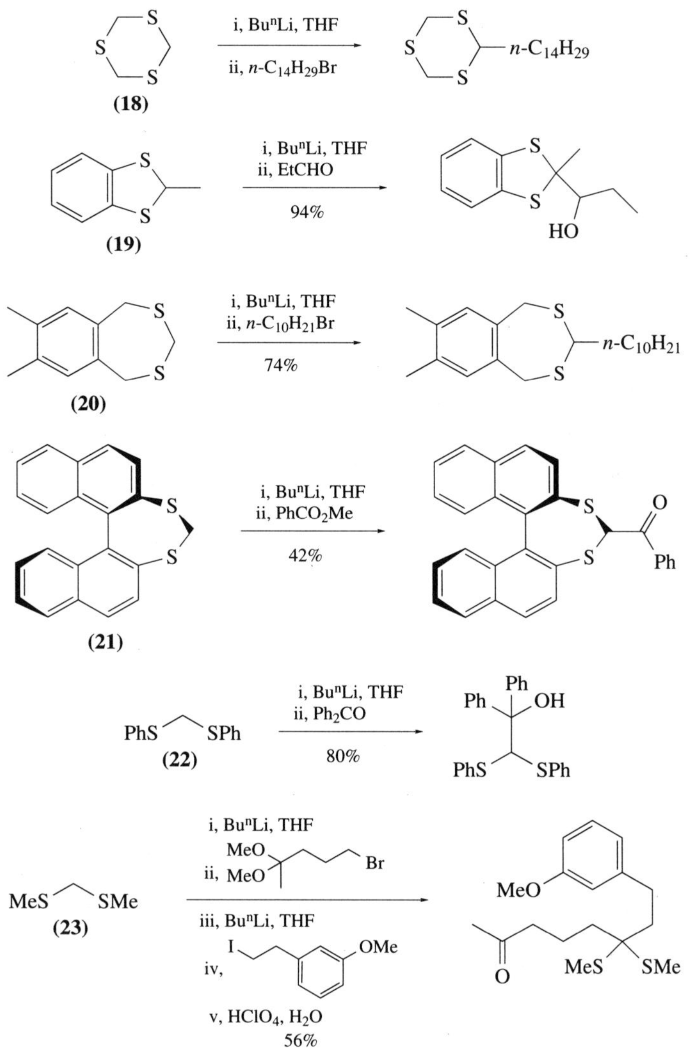

Scheme 4

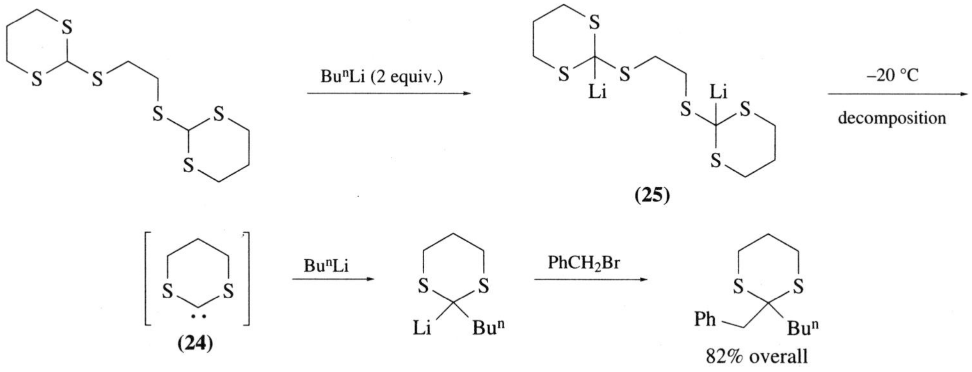

Scheme 5

Alternatively, when the organotin compound (**29**) was treated with ceric(IV) ammonium nitrate (CAN) in acetonitrile, the radical (**30**) was obtained (Equation (18)). This electrophilic radical was then trapped *in situ* with silyl enol ethers or other alkenes to give the desired dithioacetals in 52–92% yield ⟨92CL1229⟩.

(iii) From ketene dithioacetals

The transformation of a ketene dithioacetal into a saturated dithioacetal can proceed either by a simple reduction, without C—C bond formation, or by a reductive alkylation, with the creation of a C—C bond. The latter process is, of course, more interesting from a synthetic viewpoint.

(a) Without C—C bond formation. The formal addition of H_2 to the double bond of ketene dithioacetals leads to dithioacetals. However, such a 'one-step' reduction is unusual and the transformation is generally effected in two ionic steps. Moreover, given the ability of sulfur to stabilize both adjacent carbocations and carbanions, the transformation may be started either by proton addition to the double bond to give a carbocation, or by hydride addition to give a carbanion.

The former method (H^+ addition) has the broadest scope. Thus, treatment of the alkylidene-1,3-dithianes (**31**) with trifluoroacetic acid, as shown in Equation (19), in the presence of triethylsilane gave the corresponding dithianes in 48–85% yield ⟨71JOC2731⟩. This method has been extended to the fluoro compounds (**32**) (Equation (20)). In these cases, however, the more strongly acidic trifluoromethanesulfonic acid had to be used as the proton source, probably because the electron-withdrawing effect of the CF_3 group reduced the basicity of the substrate ⟨92S965⟩.

Rearrangement products have sometimes been observed following the formation of the carbocation. An example is shown in Equation (21). When the dithiane (**33**) reacted with methanesulfonic acid and triethylsilane, the cyclized compound (**34**) was isolated from a mixture of products ⟨75TL4547⟩.

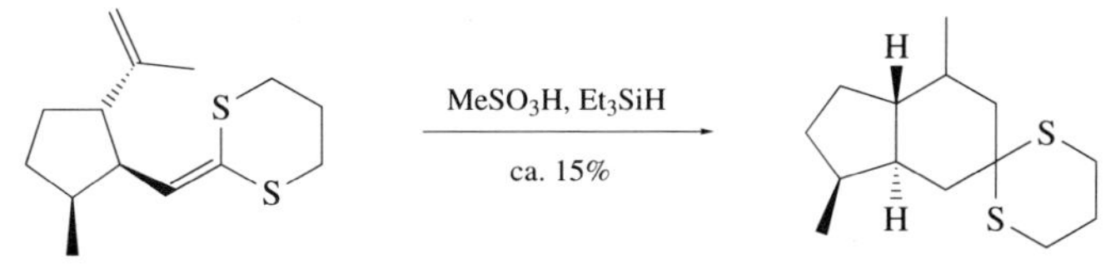

α-Oxoketene dithioacetals can be selectively reduced to β-oxo dithioacetals using acetic acid as the proton source and sodium borohydride as the hydride source (Scheme 6) ⟨90T2195, 92S1075⟩. Alternatively, a formally analogous [1,4]-reduction can be effected by first forming the carbanion. The reagent of choice is then a diisobutylaluminum hydride (dibal-H)–triethylamine complex ⟨81JOC3555⟩. However, dibal-H alone or catecholborane have also been used. When LAH is used the carbonyl group is reduced and β-hydroxy dithioacetals are isolated ⟨80JA3095⟩.

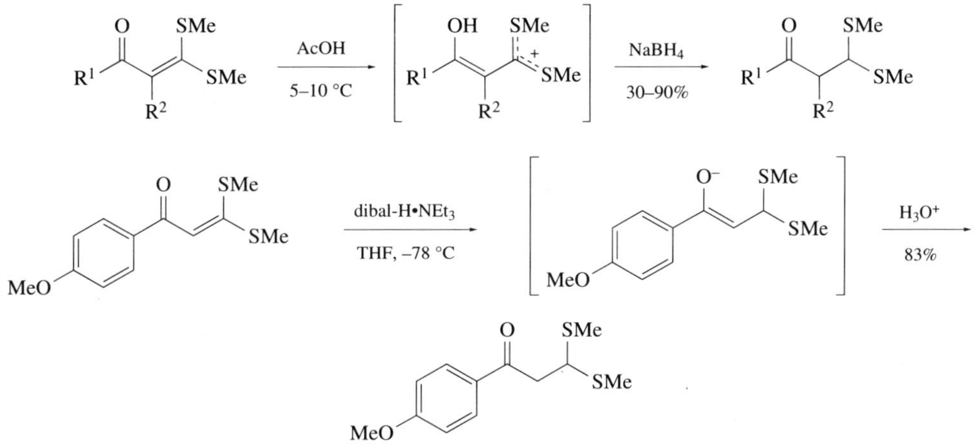

Scheme 6

LAH is also an effective reducing agent for sugar-derived ketene dithioacetals (Scheme 7) ⟨77TL1617, 78JA3548⟩. Lithium aluminum deuteride (LAD) gave the 2-deoxy-2-deuteriopentose derivative (35). This, in conjunction with the fact that the methoxy compound (36) did not react with LAH, was purported to show that the alkoxy aluminum hydride anion (37) was an intermediate in the reaction. Such an intermediate probably also intervenes in the reduction of the α-oxoketene dithioacetals by LAH ⟨78JA3548⟩.

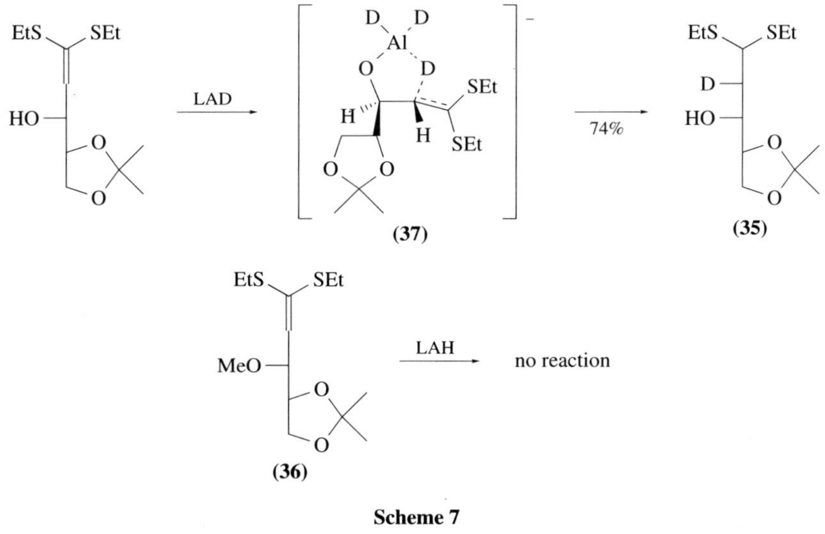

Scheme 7

Marchand and Rajapaksa ⟨93TL1463⟩ have demonstrated that hydrogen itself can be made to add to oxoketene dithioacetals. The reactions were carried out as indicated in Equation (22) in the presence of palladized charcoal at 3.4×10^5 Pa. In some cases cyclization products were isolated, but the scope of the reaction remains to be explored.

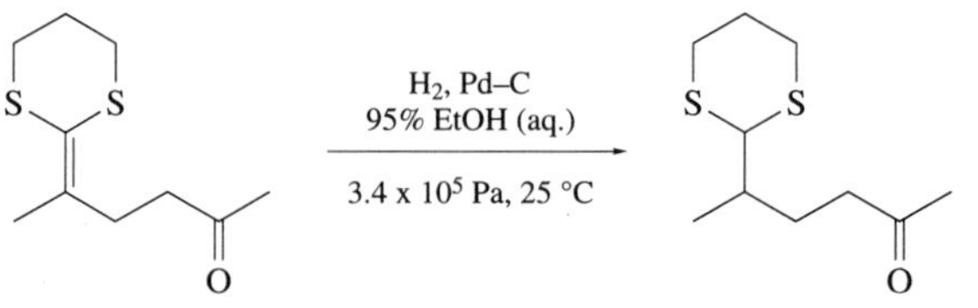

$$\text{dithiane-C(CH}_3\text{)=CH-CH}_2\text{-C(=O)-CH}_3 \xrightarrow[3.4 \times 10^5 \text{ Pa, 25 °C}]{\text{H}_2, \text{Pd-C}, 95\% \text{ EtOH (aq.)}} \text{dithiane-CH(CH}_3\text{)-CH}_2\text{-C(=O)-CH}_3 \tag{22}$$

(b) With C—C bond formation. The addition of alkyllithium reagents to 2-methylene-1,3-dithiane leads to the formation of a carbanion which can then react with selected electrophiles to give a variety of dithioacetals (Scheme 8) ⟨77LA811, 77LA830⟩. Although a classic example of umpolung, as its stands this reaction unfortunately has limited scope; when an alkylidenedithiane possessing an allylic proton is used, the organolithium reagent behaves preferentially as a base and deprotonation occurs. The resulting allylic carbanion may, however, be hydrolyzed or alkylated as shown in Scheme 9. Some representative results are listed in Table 3. Whether the organolithium acts as a nucleophile or as a base, dithianes have been prepared in good yields ⟨74TL3171, 75TL925⟩.

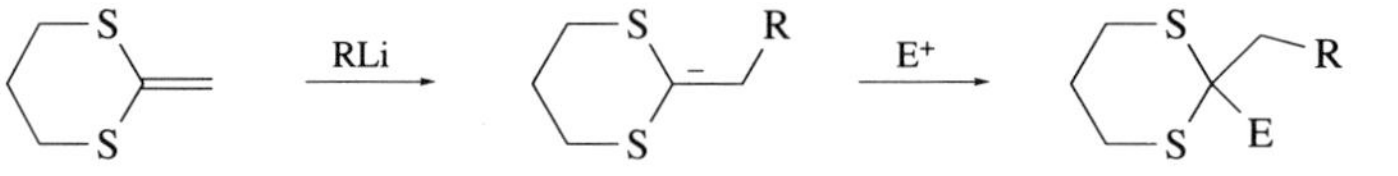

Scheme 8

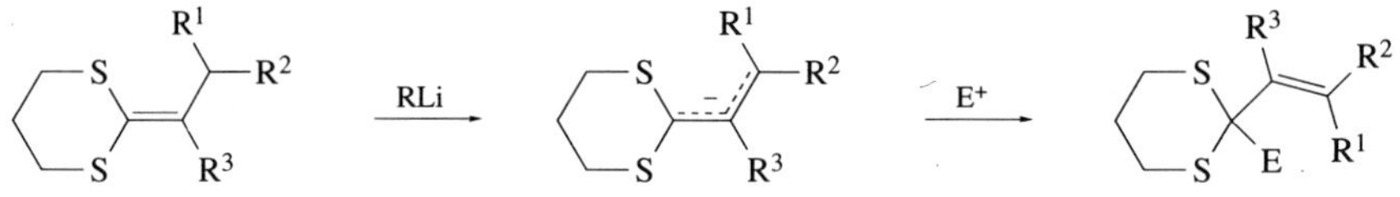

Scheme 9

Table 3 Dithianes from alkylidene dithianes (Scheme 9).

Starting materials	Electrophile	Product	Yield (%)	Ref.
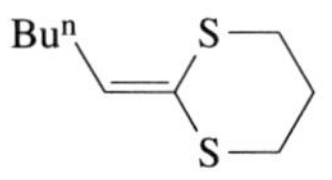 Bu^n-dithiane	MeI	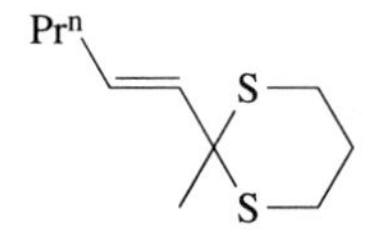 Pr^n-substituted dithiane	90	74TL3171
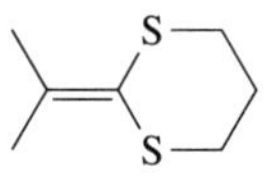 (CH$_3$)$_2$C=dithiane	Pr^{i}Cl	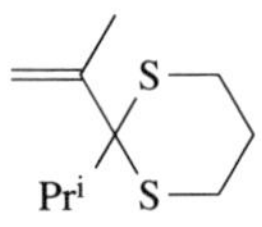 Pr^i-substituted product	92	74TL3171
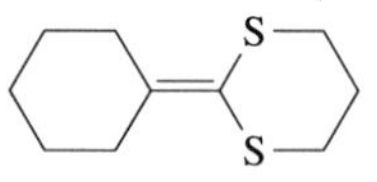 cyclohexylidene dithiane	MeI	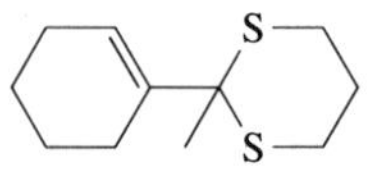 methylated product	74	75TL925
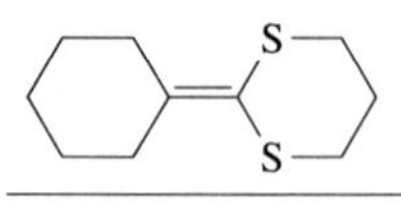 cyclohexylidene dithiane	MeOH	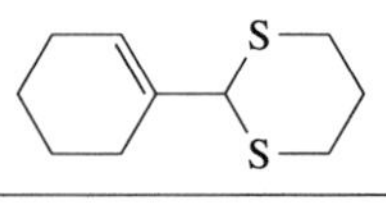 protonated product	93	75TL925

In contrast to these last results, the formation of dithioacetals by the nucleophilic addition of organolithium reagents to 1,1-dithio 1,3-dienes, including those which possess allylic protons, has been carried out with some success. For example, when the dithioacetal **(38)** (Equation (23)) was treated between $-78\,°\text{C}$ and $-20\,°\text{C}$ in THF with the organolithium reagent **(39)**, the addition product **(40)** was isolated in 53% yield. Compound **(40)** is an intermediate in Cazes and Julia's synthesis of lanceol ⟨78TL4065⟩. No products derived from the potentially competitive deprotonation of the dithioacetal (**38**) were reported in this case.

On the other hand, when Thuillier and co-workers treated the 1,1-dithio 1,3-diene (**41**) shown in Equation (24) with ethyllithium followed by iodomethane, they obtained a mixture of three products. The addition compound, the dithioacetal of manicone (**42**), was isolated along with the other dithioacetal (**43**) and the ketene dithioacetal (**44**), both of which result from an initial deprotonation of the substrate (**41**) ⟨80CR(C)183⟩.

(iv) From thiocarbonyl compounds

The addition reaction between a dithioester and an organometallic reagent—allylic reagents excluded—occurs most frequently with sulfur addition (thiophilic addition) ⟨75BSF1439⟩. The first example of this reaction was reported by Beak and Worley who described the reaction between phenyllithium and phenyl dithiobenzoate to give (diphenylthio)methylbenzene (Equation (25)) ⟨72JA597⟩.

Given the numerous and convenient methods that now exist for the preparation of dithioesters, this reaction can unquestionably be considered as another general synthetic approach to dithioacetals. Furthermore, it represents an efficient and practical route to asymmetrically substituted dithioacetals without any modification to the methodology. However, due to the basic nature of organolithium reagents, their application is limited to non-enethiolizable dithioesters only, for a hydrogen α to the C=S group deprotonates preferentially.

Grignard reagents, however, are less basic than their lithiated analogues and no deprotonation is observed with them. Their reaction with dithioesters thus represents a "clean" source of dithioacetals ⟨74CR(C)695, 75BSF657⟩. To avoid the formation of secondary products, it generally suffices to work at −20 °C. Several examples are shown in Scheme 10, outlining the synthetic utility of this method. Compound (**45**) is the synthetic equivalent of a 1,5-diketone masked by two different protecting groups ⟨82CC335, 85BSF881⟩. The transformation of the dithioester (**46**) into the dithioacetal (**47**) was used by Meyers et al. during their synthesis of 4,5-deoxymaysine ⟨79JA4732⟩, whereas the dithioacetal (**48**) was an intermediate in the synthesis of maysine ⟨79JA7104⟩. It should be noted that, by suitable choice of experimental conditions for the second reaction, the desired chemoselectivity was obtained and the epoxide was not cleaved by the organomagnesium reagent. The addition of ethylmagnesium bromide to the furfuryl dithioate (**49**) leads to the allylic dithioacetal (**50**), which was used in a synthesis of perillane ⟨79JOC2807⟩.

Cyclopropyl dithioacetals have been prepared from β-oxo dithioesters ⟨76TL4775⟩. A [1,4]-homoaddition has been proposed as a possible explanation for the selectivity observed during the reaction (Scheme 11). In effect, the only cyclopropanols formed exhibit a *cis* relationship between the hydroxy group and the SR^2 substituent.

Allylic Grignard reagents, as opposed to their saturated analogues, undergo carbon addition (carbophilic attack) to dithioesters; furthermore, the allylic chain is inverted during the reaction (Scheme 12). This difference in reactivity, however, can still be exploited for the preparation of dithioacetals in average to good yields ⟨77T2949⟩. Moreover, by acid quenching of the intermediate thiolate ion rather than alkylation, hemidithioacetals can also be prepared using this methodology. Judicious application of the carbophilic addition process resulted in an efficient preparation of isoartemisia ketone ⟨78TL2717⟩. A general problem, however, is the need to use at least 2 to 3 molar equivalents of the Grignard reagent. This is obviously a major drawback if the organomagnesium reagent is prepared from an expensive or commercially unavailable halide.

Several cases of thiophilic addition have been reported with nucleophiles other than organometallic reagents. For example, a special group of dithioacetals, α-alkylthio disulfides ⟨78HCA2351⟩, can be obtained from the thiophilic addition of certain thiols to electron-poor dithioesters. Hence, due to the inductive effect of the fluorine atoms, CF_3SH reacts thiophilically with the hexafluoro dithioate (51) (Equation (26)) to give the disulfide (52) ⟨65JOC1384⟩. More markedly, dithioesters containing an electron-withdrawing substituent, such as a carbonyl group, α to the thiono group undergo thiophilic addition with even nonhalogenated thiols (Equation (27)) ⟨88ZC269⟩.

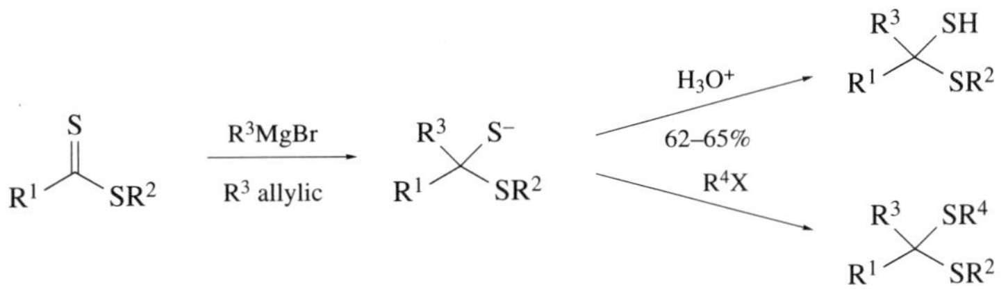

$R^1 = R^2 = R^4 = Me, R^3 = $ allyl, 65%
$R^1 = R^2 = R^4 = Me, R^3 = $ propargyl, 80%

Scheme 12

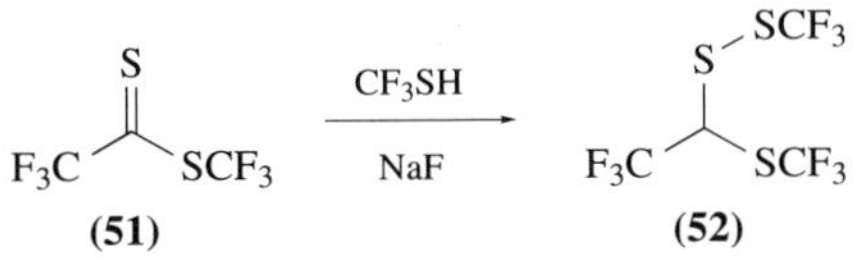

Another reaction which has been used to convert dithioesters into dithioacetals is the Diels–Alder reaction ⟨80TL4657⟩. The dithioester behaves as the dienophile, producing compounds which contain one intracyclic and one extracyclic sulfur. As for other Diels–Alder reactions, the stronger dienophiles possess an electron-withdrawing substituent attached to the reacting double bond, that is, the thiocarbonyl group. This therefore limits the general utility of the method since simple dithioesters are poor dienophiles. However, when an electron-withdrawing group is present, such as in the cyanodithioformate (**53**) shown in Equation (28), excellent results have been obtained ⟨87JOC2442⟩.

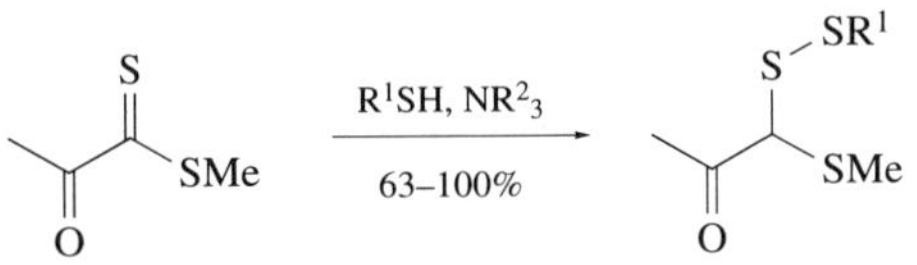

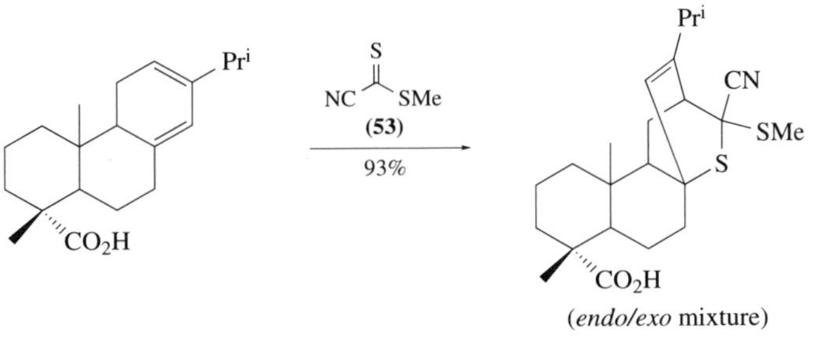

(v) From various precursors

Compounds bearing an acidic methylene group may be disulfenylated by the action of a base and a sulfenylating agent, such as a disulfide. This type of reaction has already been reviewed ⟨78CRV363⟩. In the early 1990s, the use of potassium-fluoride-impregnated alumina as the base was suggested ⟨92SC1359⟩. The reaction is then effected by microwave irradiation; an example is shown in Equation (29). The yields reported, the short reaction times (2 min), and the ease with which the products are isolated make this a very attractive technique.

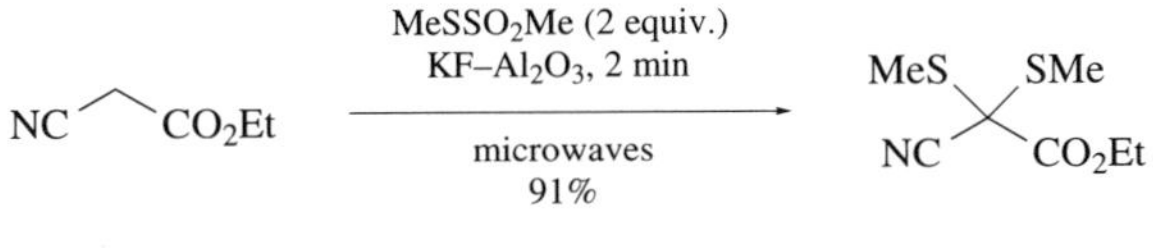

Other functional groups can also be transformed into dithioacetals, for example carboxylic acids as indicated in Equation (30). They are directly reduced to masked aldehydes by the action of 1,3,2-dithiaborinane-dimethylsulfide (**54**) in the presence of tin(II) chloride ⟨87JOC2114⟩. The reaction is carried out at ambient temperature in THF. Given that carboxylic acids are commercially more numerous and frequently less expensive than the corresponding aldehydes, this method represents an excellent alternative for the synthesis of aldehyde-derived dithianes.

$$RCO_2H \xrightarrow[59-90\%]{(54) \; THF, \; 3.5-20 \; h} R-\text{(dithiane)} \quad (30)$$

α-Chloro-sulfides are convenient precursors for dithioacetals. The conditions used are similar to those described for the conversion of simple halides into sulfides (see Chapter 2.03.2). 2,4,6-Trithiaheptane, $MeSCH_2SCH_2SMe$, has been obtained in this way ⟨78CJC1183⟩. Elsewhere, Ranu *et al.* have reported that conjugated acetylenes may be converted into dithianes ⟨92JOC7349⟩. As indicated in Equation (31), the reaction of propanedithiol with acetylenic ketones in the presence of alumina leads, via a double Michael addition, to monoprotected 1,3-diketones. The reaction can be effected without solvent or in dichloromethane. Given that acetylenic ketones are relatively accessible, in certain cases this method represents a practical alternative to the more classical syntheses of β-keto-1,3-dithianes, such as the reduction of α-oxoketene dithioacetals (cf. Section 4.06.1.2.3(iii)) or the oxidation of the alcohols formed by the action of a dithiane anion on an epoxide.

$$R^1-C(=O)-C \equiv C-R^2 \xrightarrow[70-85\%]{HS(CH_2)_3SH \; Al_2O_3, \; 4-6 \; h} R^1-C(=O)-CH_2-C(S)(S)-R^2 \quad (31)$$

Finally, although it is normal to synthesize the oxidized analogues from the parent dithioacetals, the seemingly illogical reverse process of reduction is also possible. This strategy is useful when, for example, the oxidized compound is more accessible than the parent compound. Thus, 1,3-dithietane-1-oxide was obtained from a dichloride precursor as indicated in Equation (32), and was subsequently reduced to dithietane with an excess of borane–THF complex at room temperature ⟨82JA3119⟩.

$$Cl-CH_2-S(=O)-CH_2-Cl \xrightarrow[34\%]{Na_2S} \text{(dithietane-oxide)} \xrightarrow[44\%]{THF \cdot BH_3} \text{(dithietane)} \quad (32)$$

4.06.1.3 One Dicoordinated Sulfur and One Higher Coordinated Sulfur—$R^2_1C(SR^2)SO_2R^3$, etc.

4.06.1.3.1 α-Thio sulfoxides

(i) Oxidation of dithioacetals

Without doubt, the most practical method for preparing α-alkylthio sulfoxides remains the direct oxidation of the corresponding dithioacetals. The essential problems are those of oxidation selectivity: chemoselectivity—avoiding formation of a bis-sulfoxide or a sulfone; regioselectivity—for asymmetrical dithioacetals; diastereoselectivity—particularly with 1,3-dithioacetals already possessing an asymmetric or prochiral carbon between the two sulfurs; and enantioselectivity—as in the more general case of sulfoxides.

In theory, if a dithioacetal needs to be mono-oxidized, any of the oxidizing agents used to transform sulfides into sulfoxides could be employed. However, the most frequently used agent is mcpba, which has the advantage of being soluble in organic solvents. Three typical examples are shown in Scheme 13. The reaction is generally effected in dichloromethane, and temperatures around 0 °C prevent overoxidation. If these conditions are respected, the yields are generally high, as in the case of the anthracene derivative (**55**) ⟨90T5093⟩. Oxidation of the heterocycle (**56**) represents an application of this method in the presence of a thiolester ⟨92LA1039⟩. Elsewhere, when one of the dithioacetal sulfur atoms was also part of a thiolester, as in compound (**57**), regioselective oxidation of the other, more nucleophilic sulfur occurred ⟨90JCS(P1)2035⟩.

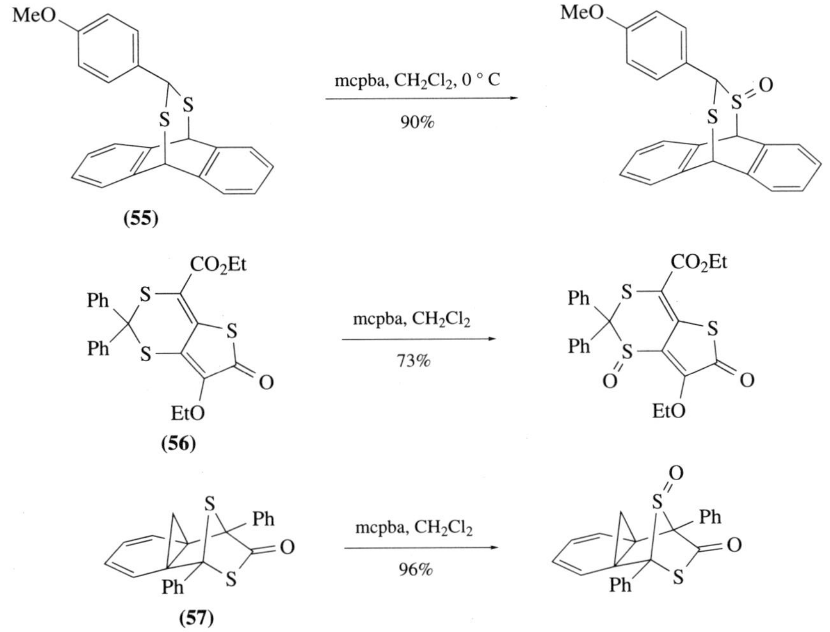

Scheme 13

Sodium periodate has also been used. The reaction is carried out by the addition of an aqueous solution of the periodate to a solution of the dithioacetal in a water-miscible solvent, such as methanol. This represents an obvious disadvantage with nonpolar dithioacetals. However, the easy elimination by simple filtration of the sodium iodate by-product allows this method to remain attractive. It would appear equivalent to the mcpba method, providing that the dithioacetal is suitably soluble in the water–methanol solvent, as was the butyrate (**58**) (Equation (33)) ⟨73RTC117⟩.

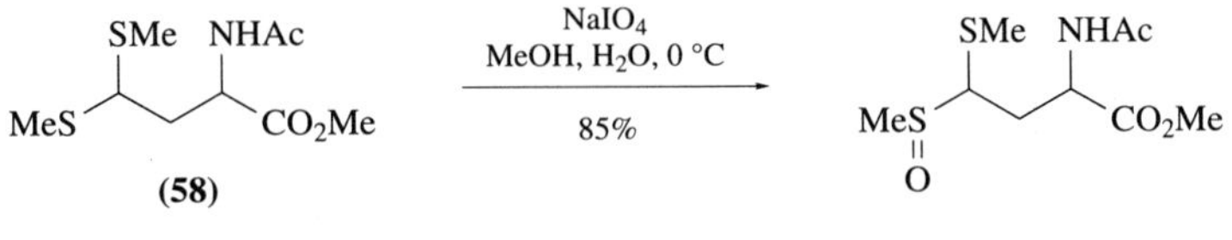

Hydrogen peroxide, on the other hand, suffers from a lack of chemoselectivity, as demonstrated by its reaction with simple sulfides to give significant quantities of the corresponding sulfones. However, given its cheapness, it still represents a tempting alternative, especially if the reactions need to be carried out on a large scale. This reagent was used with acceptable results for the theophylline dithioacetal (**59**) (Equation (34)) ⟨89EJM635⟩ and for the chlorinated compound (**60**) (Equation (35)) ⟨90S271⟩. In the latter case, vanadium(V) oxide, present in a catalytic quantity, is the actual oxidizing agent and the hydrogen peroxide serves to regenerate it *in situ*.

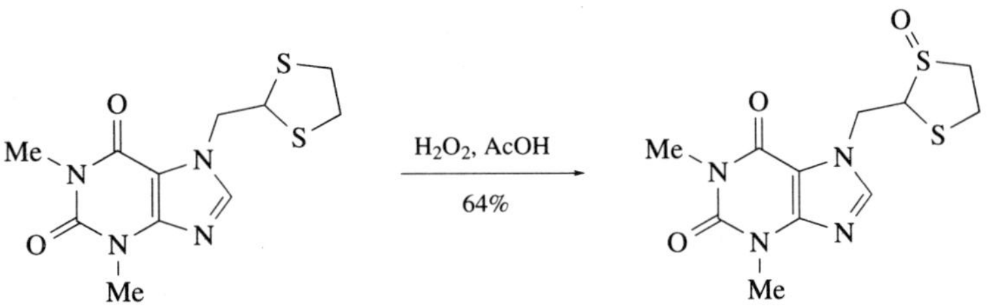

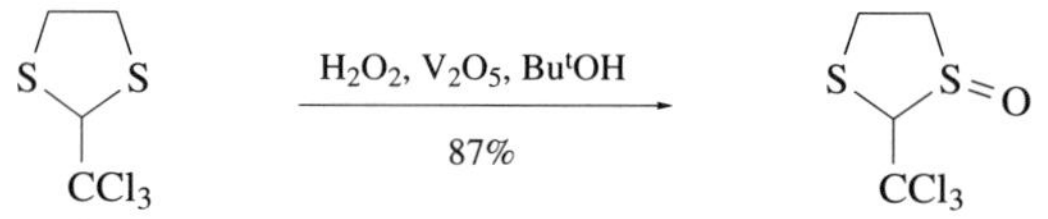

$$\text{(60)} \xrightarrow[87\%]{H_2O_2,\ V_2O_5,\ Bu^tOH} \text{product} \qquad (35)$$

Nitric acid has been proposed as an oxidizing agent in this type of reaction ⟨90JOC1323⟩. Gasparrini *et al.* have obtained good results with biphasic conditions (nitromethane–water) in the presence of tetraethylammonium tetrabromoaurate(III) ($Et_4N^+AuBr_4^-$), as shown in Equation (36). The oxidizing agent is in fact the $AuBr_4^-$ anion, which is reduced to $AuBr_2^-$ during the dithioacetal oxidation and then reoxidised by the nitric acid.

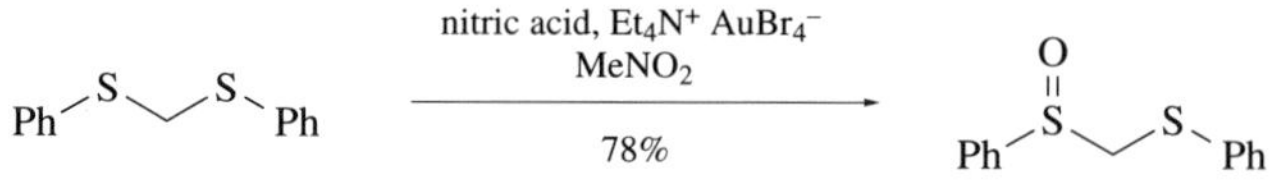

$$Ph-S-CH_2-S-Ph \xrightarrow[78\%]{\text{nitric acid, }Et_4N^+\ AuBr_4^-,\ MeNO_2} Ph-S(O)-CH_2-S-Ph \qquad (36)$$

Various photochemical reactions have been studied (Scheme 14). Oxidation occurs either by singlet oxygen ⟨90JCS(P1)3217⟩ or by single electron transfer ⟨89TL4007⟩. Whatever the mechanism, which is still being debated ⟨92TL5085⟩, the formation of by-products such as disulfoxides, sulfones, and even carbonyl compounds limits the scope of this method.

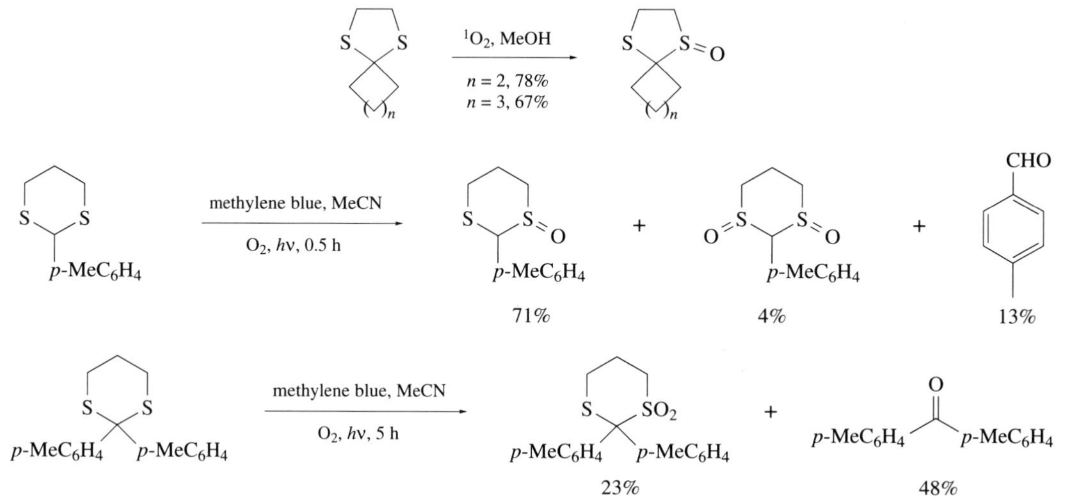

Scheme 14

The oxidation of one of the sulfur atoms of a dithioacetal almost always leads to the formation of an asymmetric center. Moreover, if the dithioacetal carbon is prochiral, a mixture of diastereoisomers is generally formed. This problem has been widely studied, especially with cyclic dithioacetals. Carey *et al.* ⟨76JOC3975⟩ have shown that the oxidation of 2-methyl-1,3-dithiane with either mcpba or sodium periodate gives, due to the steric interaction, mainly the *trans* isomer shown in Equation (37). Furthermore, this selectivity is general, such that synthesis of the *cis* isomer is only possible after strategic modifications including choice of a different precursor. Hence, from 2-methyl-2-trimethylsilyl-1,3-dithiane (Scheme 15), in which the trimethylsilyl substituent is the more sterically demanding, the *cis* isomer was successfully prepared by oxidation and subsequent desilylation.

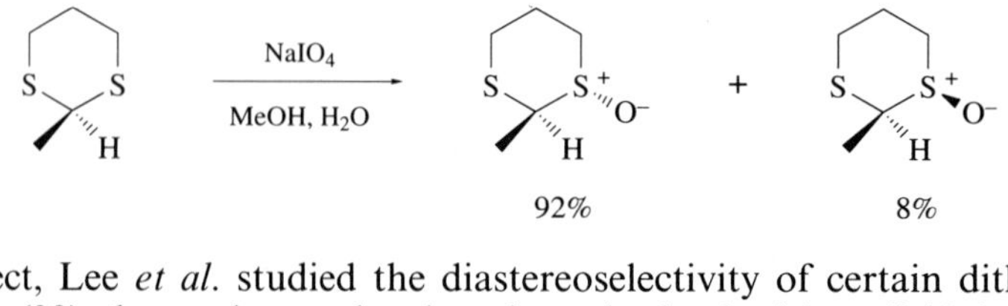

$$\text{2-methyl-1,3-dithiane} \xrightarrow[MeOH,\ H_2O]{NaIO_4} \text{trans (92\%)} + \text{cis (8\%)} \qquad (37)$$

In a similar project, Lee *et al.* studied the diastereoselectivity of certain dithiolane oxidations ⟨91T8091⟩. Equation (38) shows the results that they obtained with a dithiolane-2-carboxamide. Both mcpba and sodium periodate gave mainly the *cis* isomer. Conversely, iodobenzene dichloride

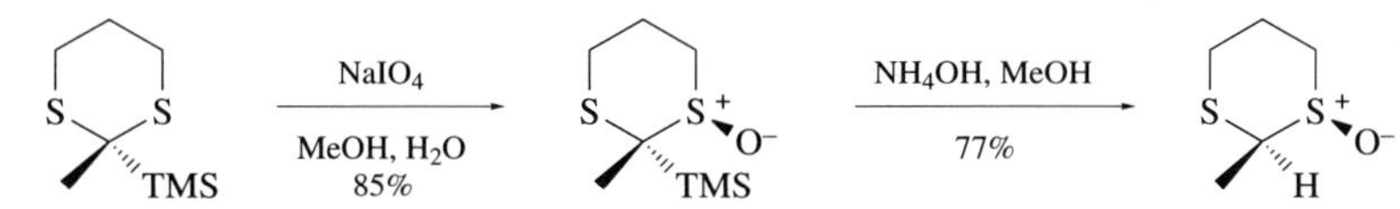

Scheme 15

in pyridine gave the *trans* isomer selectively. Furthermore, unlike the first two oxidizing agents which gave small amounts of disulfoxide, none whatsoever was detected with the dichloride. Iodobenzene dichloride might well prove to be the reagent of choice for the oxidation of dithioacetals.

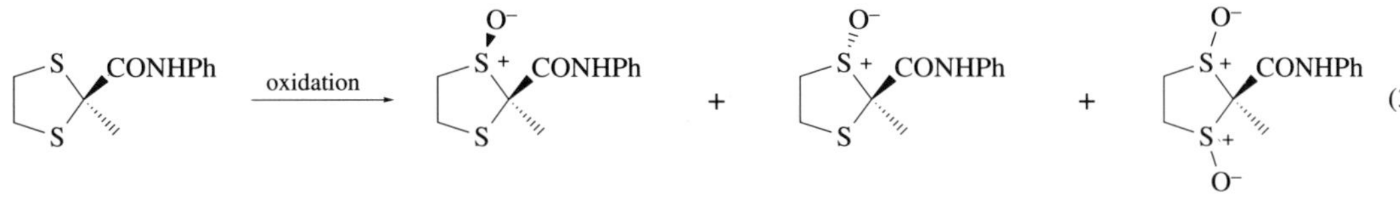

(38)

H_2O_2, AcOH, 0 °C	64	28	8
mcpba, $CHCl_3$, 0 °C	82	8	8
$PhICl_2$, C_5H_5N, H_2O, 15–25 °C	30	70	

The most challenging problem still remains, as for simple sulfoxides (see Chapter 2.03.2): the synthesis of optically pure *S*-oxides of dithioacetals. Various solutions have been proposed. Kagan, Di Furia, Bulman Page and co-workers have shown that a modified Sharpless reagent gave optically active, cyclic *S*-oxide dithioacetals. The reaction is effected either with a Kagan-modified Sharpless reagent (dithioacetal : *t*-butyl hydroperoxide : titanium tetraisopropoxide : (+)-diethyl tartrate : water in the ratio 1 : 1.1 : 1 : 2 : 1) in dichloromethane ⟨90SL457, 90SL643⟩ or a Di-Furia-modified reagent (1 : 0.5 : 0.25 : 1 : 0) in dichloroethane ⟨86TL6257, 89TL2575, 92TL3043⟩. Kagan himself preferred to use $Me_2C(Ph)OOH$ in place of Bu^tOOH ⟨89JOM(370)43⟩. Several representative results are shown in Table 4. They clearly show that in the absence of polar substituents the enantiomeric excesses are low; however, with polar substituents favoring chelation, high enantiomeric excesses result. The chemical yields are generally good, however, since the separate research teams adopted different premises upon which to calculate the yields; as a direct comparison would be invalid, the yields are not included here. Nevertheless, dithiolanes generally give better enantiomeric excesses than do dithianes ⟨90G165⟩, whilst the results remain very much substrate-dependent as demonstrated by the two isomers of the dithiane (**61**).

As mentioned, these methods do not give satisfactory results with dithioacetals lacking a polar substituent. For such compounds, a multistep route has been proposed whereby the necessary polarity is temporarily introduced into the molecule ⟨91SL80⟩. It is illustrated in Scheme 16 with 2-ethyl-1,3-dithiane as an example. The starting dithiane is deprotonated and then acetylated with a large excess of ethyl acetate. The resultant acetyl dithiane thus contains the prerequisite polar substituent and is a good candidate for Sharpless–Kagan oxidation. Following oxidation, the acetyl group is removed by treatment with 10% sodium hydroxide at room temperature for 1–3 days. The *cis*-*S*-oxides of dithianes are isolated with a 77–94% *ee*. The sulfoxide group of the major enantiomer

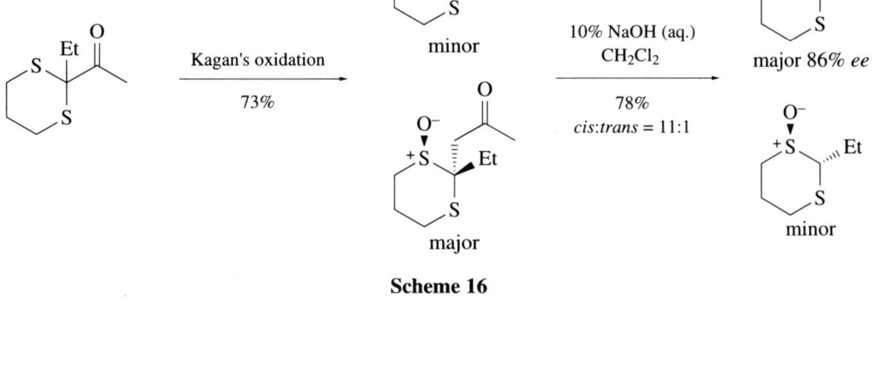

Scheme 16

Table 4 Representative examples of enantioselective oxidations of cyclic dithioacetals to monosulfoxides.

Substrate	Method	Diastereoisomer	ee (%)
2-Ph-1,3-dithiane	A[a]	trans	10
	B[b]	trans	14
2-Bu^t-1,3-dithiane	A	trans	17
2-Bu^t-1,3-dithiolane	B	trans[c]	70
2-CONEt$_2$-1,3-dithiolane	B	trans[d]	94
2-Ph-2-COMe-1,3-dithiane	A	trans (65%)	99
		cis (6%)	99
2-Ph-2-COEt-1,3-dithiane (61)	A	trans (42%)	97
		cis (15%)	57

[a] Method A: Kagan–Bulman Page. [b] Method B: Di Furia. [c] trans:cis, 85:15. [d] trans:cis, 99:1.

has been assigned the (R) configuration and the cis:trans ratios vary from 4:1 to 30:1. This methodology has recently been improved ⟨93TA2139⟩.

Although this circuitous route requires two extra steps, it still seems to be the most reliable method currently available for nonpolar dithioacetals. However, endeavors in the mid 1990s suggest that an alternative, direct method may replace it. For example, the chiral oxaziridine shown in Equation (39) has been proposed as a stereoselective oxidizing agent, with a promising 80% ee being obtained with the nonpolar 2,2-dimethyl-1,3-dithiane ⟨92JA1428⟩.

$$\text{2,2-dimethyl-1,3-dithiane} \xrightarrow[\text{CCl}_4,\ 3\ \text{h}]{\text{chiral oxaziridine}} \text{monosulfoxide} \quad (39)$$

60%, 80% ee

One can equally envisage the use of enzymes for these reactions. Colonna et al. have studied the oxidation of 1,3-dithiane with sodium periodate in the presence of bovine serum albumin, but the enantiomeric excess obtained was very disappointing ⟨86JOC891⟩. However, with bis(p-tolylthio)methane the result was somewhat better ⟨80TL2233⟩. Cashman et al. ⟨92JA8772⟩ have compared the oxidation of 2-methyl-1,3-benzodithiole using chemical and enzymatic methods, concluding that enzymes gave the better enantioselectivity but also the thermodynamically less stable cis isomer (see also ⟨92JCS(P1)1105⟩). The best result was obtained with rabbit-lung, flavin-containing monooxygenase which gave exclusively the (−)-1(S),2(R) compound. One hundred percent diastereo and enantioselectivity were observed with 2-(p-methoxyphenyl)-1,3-dithiolane ⟨90JA3191, 90MI 406-01⟩. However, these reactions were only carried out on very small scales and it remains to be seen whether they can be successfully scaled up for preparative purposes. In conclusion, the enantioselective oxidation of dithioacetals, in particular nonpolar dithioacetals, remains a problem to be mastered.

(ii) From α-thio sulfoxide carbanions

The same proviso which applied to dithioacetals is also valid here. That is, although this method is frequently used for the introduction of the α-thio sulfoxide group, the reaction of an α-thio sulfoxide anion with an electrophile to give a substituted α-thio sulfoxide is not, in the strictest sense of the term, a functional group transformation. However, given the synthetic importance of the reaction, several leading references illustrating the necessary experimental conditions will be given.

Carey *et al.* have studied the stereoselectivity of the reaction of 2-lithio-1,3-dithiane-1-oxides with various electrophiles ⟨76JOC3979⟩. Deprotonation of the mono-oxidized dithianes can be achieved by treatment with *n*-butyllithium or lithium diisopropylamide (LDA) in THF. The chemical yields are good but the stereoselectivities, of which three examples are given in Scheme 17, are mediocre. More recently, a similar study with 2-crotyl-1,3-dithiane-1-oxide has been published ⟨90JOC5515⟩. High yields were reported when aldehydes were used as the electrophiles at −78 °C. Products resulting from an attack at the γ-carbon were generally not observed (Equation (40)). Examination of the stereochemical implications of such reactions continues ⟨89JCS(P1)2441, 91SL84⟩. An example of notable interest has been reported by Tanaka *et al.* ⟨90BCJ466⟩. As indicated in Scheme 18, the carbanion (**62**) undergoes Michael addition to the unsaturated ester (**63**), leading to the enolate (**64**) which is subsequently treated with an aldehyde. This one-pot procedure gives, after Peterson alkenation, the final products (**65**) in good isolated yields.

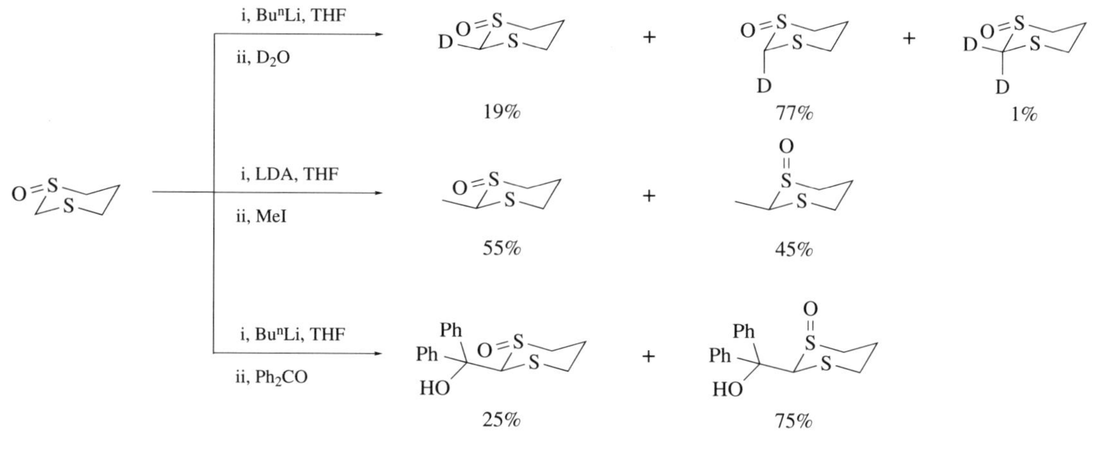

Scheme 17

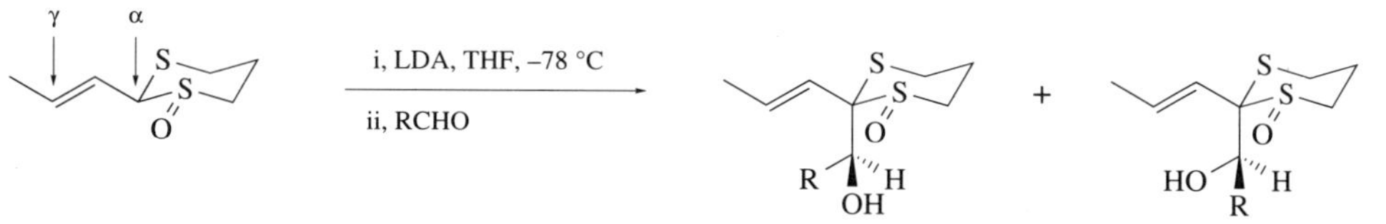

(40)

R = Et	61%	26%
R = Pr^i	89%	
R = CH=CH$_2$	49%	37%

(iii) By various methods

Although much less frequently, other methods have also been used for the synthesis of α-thio sulfoxides. This discussion will start with two methods which complement each other; the first uses an electrophile, equivalent to $RSOCH_2^+$, and the second uses the carbanion $RSOCH_2^-$. Both

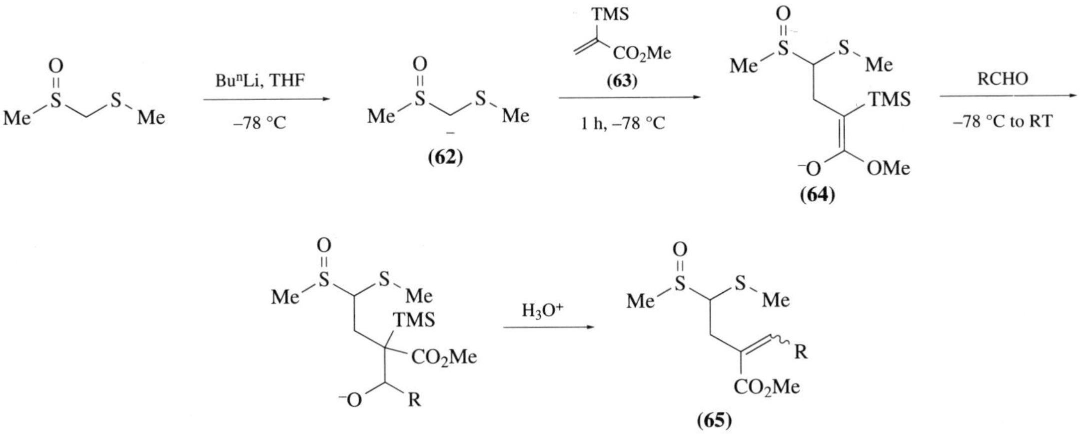

Scheme 18

R = Et, 72% overall

methods were investigated for application to the synthesis of the antibiotic sparsomicin (**66**) ⟨92JA5946⟩. Ottenheijm *et al.* have shown that it is possible to use a chloromethyl sulfoxide efficiently ⟨81JOC3273⟩; an example is shown in Equation (41). Treatment of compound (**67**) by sodium methanethiolate leads to the expected product (**68**) in quantitative yield. The method had previously been used for the preparation of optically active α-thio sulfoxides ⟨76BCJ256⟩ and has now been expanded to encompass other thiolates for the preparation of sparsomicin analogues (Equation (42)) ⟨92RTC163⟩.

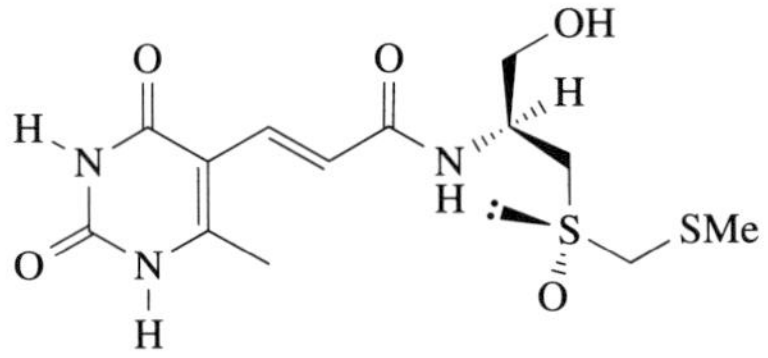

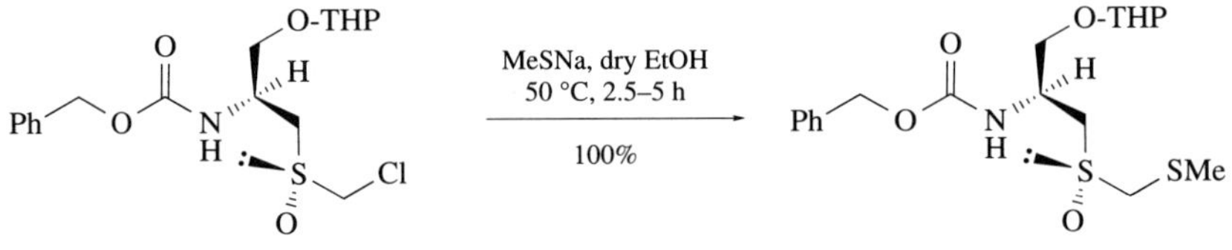

In the method above, the alkanethiolating agent was the nucleophile. Conversely, however, one can use an electrophilic alkanethiolating agent, such as a disulfide, which reacts with the carbanion of a sulfoxide. Two examples, one from a sparsomicin synthesis ⟨79JA1057⟩ and the other which includes the use of a chiral base to deprotonate the dicyclohexyl sulfoxide ⟨91CB2489⟩, are illustrated in Equations (43) and (44). The choice between these two strategies depends very much on the individual case and the constraints of the synthesis envisaged.

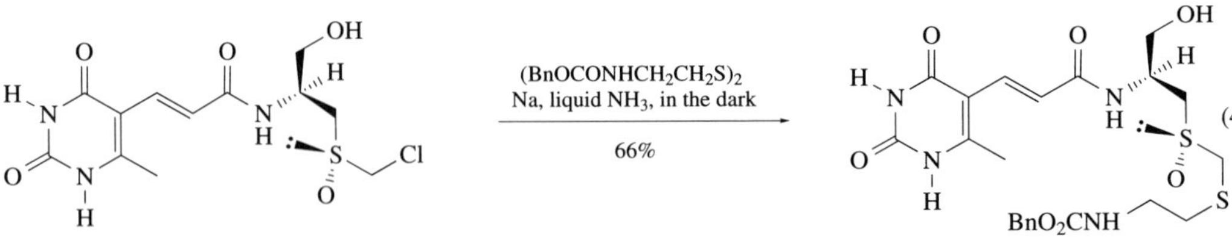

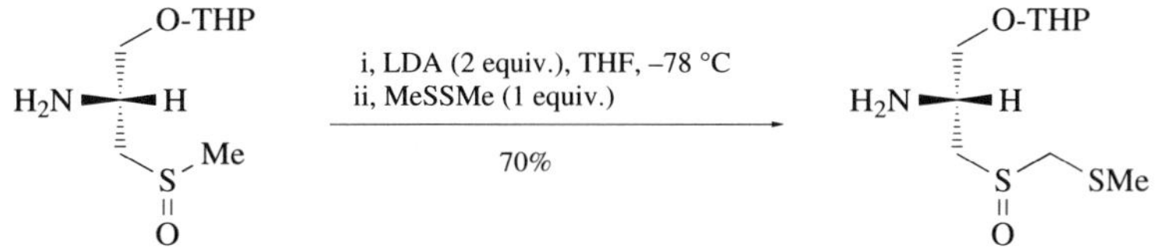

(43) [Scheme showing reaction with LDA (2 equiv.), THF, −78 °C; MeSSMe (1 equiv.); 70%]

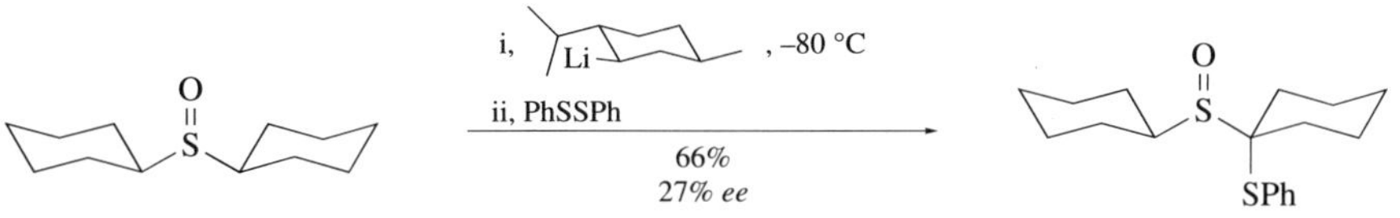

(44) [Scheme showing reaction with Li-cyclohexyl, −80 °C; PhSSPh; 66%, 27% ee]

Instead of a carbanion α to a sulfoxide, one can employ a carbanion α to a sulfide. It is then necessary to use an electrophile containing the sulfoxide group. Equation (45) shows such an example, where dabco represents 1,4-diazabicyclo[2.2.2]octane ⟨78TL3861⟩. The optically active menthyl toluene-p-sulfinate reacts with p-tolylthiomethyllithium to give, after complete inversion of chirality at the sulfur atom, the expected mono-oxidized dithioacetal. The optical purity of this product was estimated to be 100%. It seems, therefore, that this method is an excellent route to optically pure α-thio sulfoxides and that, for similar cases, it is competitive with the oxidations described in Section 4.06.1.3.1(i).

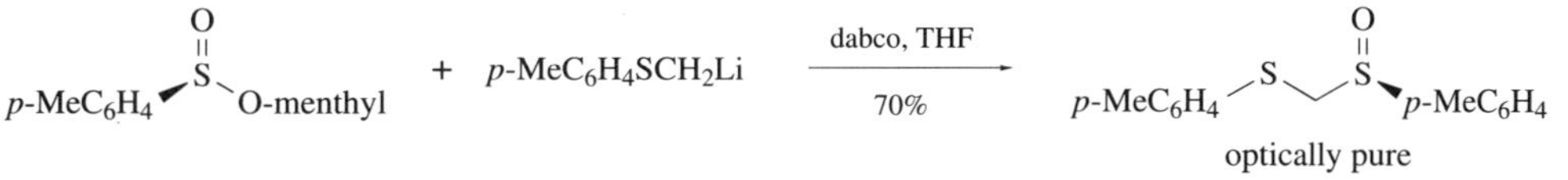

(45) $p\text{-MeC}_6\text{H}_4\text{S(O)-O-menthyl} + p\text{-MeC}_6\text{H}_4\text{SCH}_2\text{Li} \xrightarrow{\text{dabco, THF}, 70\%} p\text{-MeC}_6\text{H}_4\text{-S(O)-CH}_2\text{-S-}p\text{-MeC}_6\text{H}_4$ (optically pure)

Finally, the Diels–Alder reaction has been used for the preparation of two specific types of α-thio sulfoxides. The [4π + 2π]-cycloaddition of the dithioester sulfine (**69**) with cyclopentadiene or 2,3-dimethyl-1,3-butadiene (Scheme 19) gave α-thio sulfoxides, with the oxidized sulfur in the endocyclic position ⟨89CB1757⟩. The general application of this method should be possible, providing that the dithioester sulfines are sufficiently stable. Alternatively, the presence of the electron-withdrawing SO group in mono-oxidized ketene dithioacetals will render these compounds more dienophilic than their nonoxidized analogues. An example of the application of these alkenes in the Diels–Alder reaction is given in Equation (46) ⟨90CPB3242⟩.

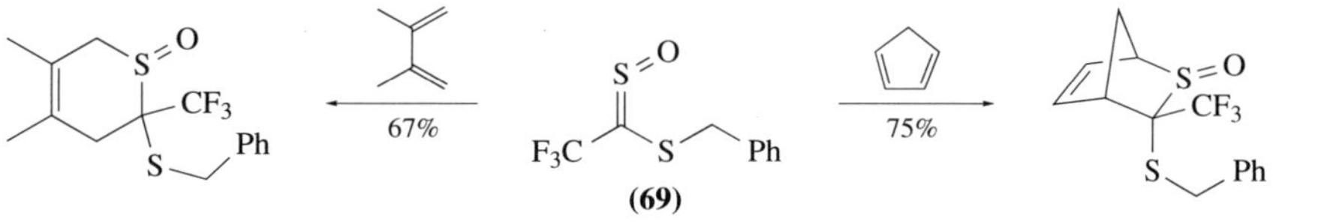

Scheme 19

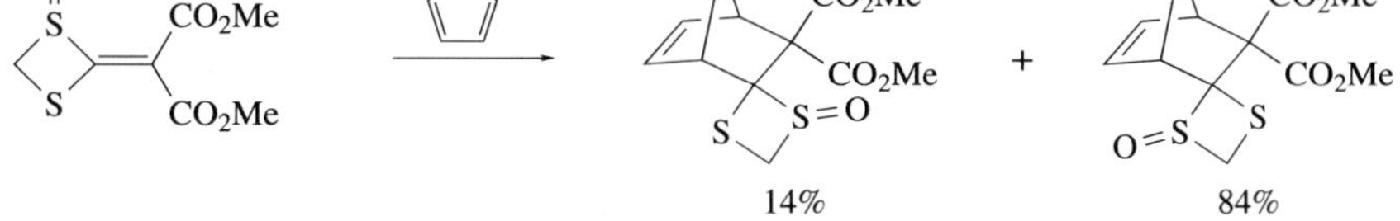

(46) [Scheme showing Diels-Alder reaction with cyclopentadiene giving two products 14% and 84%]

4.06.1.3.2 α-Thio sulfones

Although they have been the subject of much less investigation than α-thio sulfoxides, sufficient work has been done on α-thio sulfones to warrant a review of these compounds ⟨91PS(58)207⟩. They can be prepared by the oxidation of α-thio sulfoxides, although the yields are sometimes modest. The major problem lies in the regiocontrol of the oxidation to minimize bis(sulfoxide)formation. The reagent of choice for this regioselectivity is potassium permanganate. Thus, oxidation of the sulfoxide (**70**) in acetone gave the sulfone (**71**) in 44% yield (Equation (47)) ⟨92MI 406-01⟩. Moreover, it has been proposed that these reactions are facilitated by an 18-crown-6 ether or methyltrioctylammonium chloride, proposed as a phase-transfer agent ⟨89BAP117, B-90MI 406-02⟩. However given that α-thio sulfoxides are themselves often prepared by the oxidation of dithioacetals, a more direct route to α-thio sulfones would be the one-pot, double oxidation of dithioacetals. Scheme 20 indicates that the one-pot, double oxidation of the fluorinated dithiolane (**72**) not only works but is more convenient and more efficient than the two-step approach ⟨90CB177⟩.

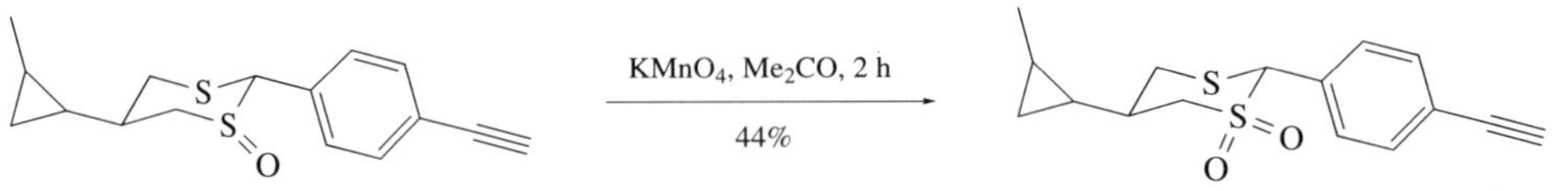

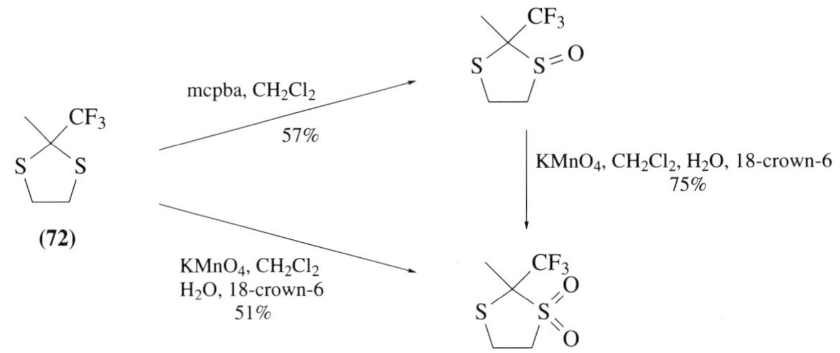

Scheme 20

α-Thio sulfones can be easily deprotonated by *n*-butyllithium in THF. Subsequent treatment of the anions formed with various electrophiles gives access to a wide range of substituted α-thio sulfones. Two examples are depicted in Scheme 21, where the electrophiles are an alkyl halide ⟨81PS(10)169⟩ and an epoxide ⟨90CAR(202)1⟩.

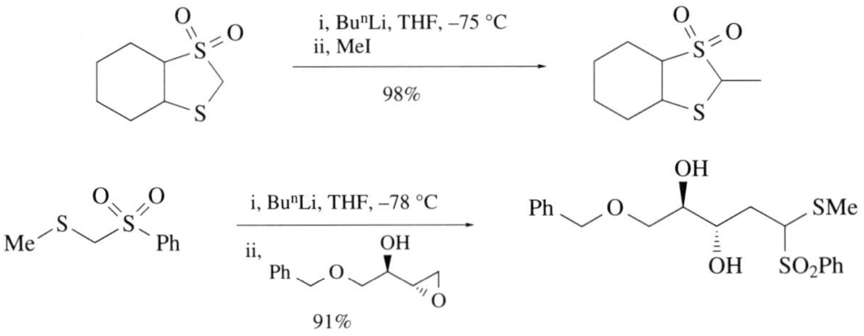

Scheme 21

Several other methods have also been employed, such as the action of a sulfenylating agent on a carbanion α to a sulfone. When the carbanion formed by deprotonation of the corresponding sulfolane was treated with diphenyl disulfide, 2-phenylthiosulfolane was formed in good yield (Equation (48)) ⟨89AQ22⟩. Alternatively, the treatment of certain activated methylene compounds with a sulfinyl chloride also led to α-thio sulfones and not to the expected bis(sulfoxides), as indicated in Equation (49) ⟨82SUL63⟩. Finally, Barton *et al.* have effected the free radical addition of an alkyl group and a pyridylthio group onto phenyl vinyl sulfone ⟨91T7091⟩. The radical R· is formed by the reaction of *N*-hydroxy-2-thiopyridone with a carboxylic acid. As indicated in Scheme 22, the alkyl group attacks the β-position and the pyridylthio group the α-position of the sulfonylethene.

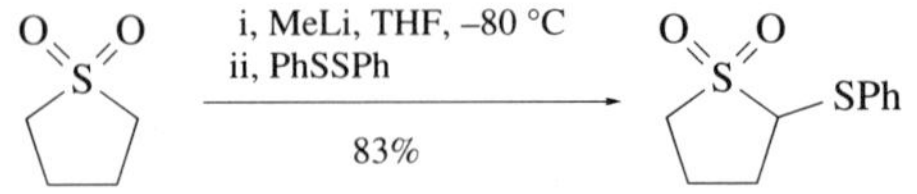

4.06.1.3.3 Other derivatives

Compounds containing both a divalent sulfur and a sulfinic or sulfonic acid derivative on the same carbon have rarely been reported in the literature. Equation (50), however, shows one such example ⟨73JA6962⟩. The attention of the reader is therefore drawn to Chapters 2.03.4 (sulfinic and sulfonic acids), 2.03.5 (sulfinates and sulfonates), and 2.03.6 (their halogen derivatives).

4.06.1.4 Two Tricoordinated Sulfurs—$R^1_2C[S(O)R^2]_2$

4.06.1.4.1 Bis(sulfoxides)

(i) Oxidation

Bis(sulfoxides) can be prepared by oxidation of α-thio sulfoxides or by double oxidation of dithioacetals. A variety of reagents including peroxides, ozone, and the ozonide of triphenyl phosphite have been employed (Scheme 23) ⟨80CJC878, 90JCS(P2)1987, 90ZOR1259⟩. However, the most commonly used reagents are mcpba or sodium periodate. Potassium permanganate is not suitable, giving preferentially α-thio sulfones (see Section 4.06.1.3.2).

The reagent mcpba has been used extensively to prepare bis(sulfoxides) for fundamental research, as synthetic intermediates, and they are of potential pharmacological interest. For example, treatment of 2,2-bis(methylthio)-1,3-diphenylpropane with two equivalents of mcpba gave uniquely the *meso*-S^1,S^2-dioxide in quantitative yield (Scheme 24) ⟨80TL3089⟩. During a synthesis of the insecticide pluridone, the intermediate bis(sulfoxide) (**74**) was obtained by mcpba oxidation of the dithioacetal (**73**) ⟨83TL5563⟩. No oxidation of the thiolester sulfur was observed. Likewise, mcpba was the reagent of choice for the preparation of the steroidal dithiolane S^1,S^2-dioxide (**75**) from its dithiolane precursor (**76**) ⟨91JIC368⟩. Sodium metaperiodate has been used to oxidize 1,3-dithiolane-1-oxide (Scheme 25) ⟨81PS(10)163⟩. The *trans*-1,3-dioxide (**77**) is formed quantitatively. Similarly, when 1,3-dithiane is treated with two equivalents of $NaIO_4$ the 1,3-dioxides (**78** *trans*) and (**78** *cis*) are produced in the ratio 86:14, from which 58% of the *trans*-1,3-dioxide was isolated and crystallized ⟨91JCS(P1)662⟩. Alternatively, the reaction of 5,5-dimethyl-1,3-dithiane with two equivalents of $NaIO_4$ gives the 1,3-dioxides (**79** *trans*) and (**79** *cis*) in the ratio 20:80, from which 60% of the *cis*-1,3-dioxide was isolated.

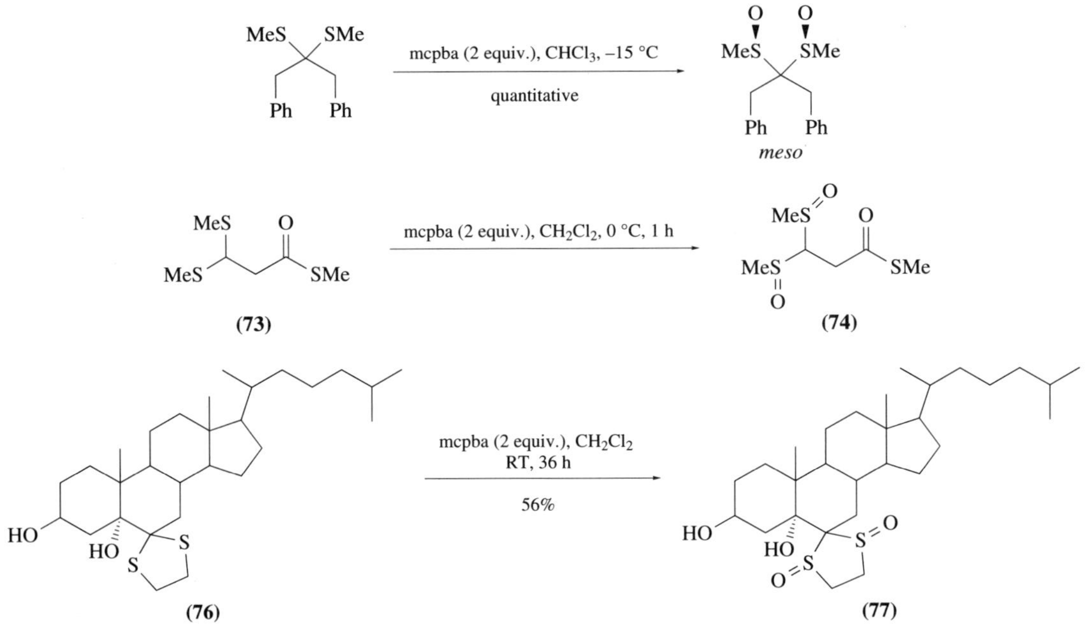

Scheme 24

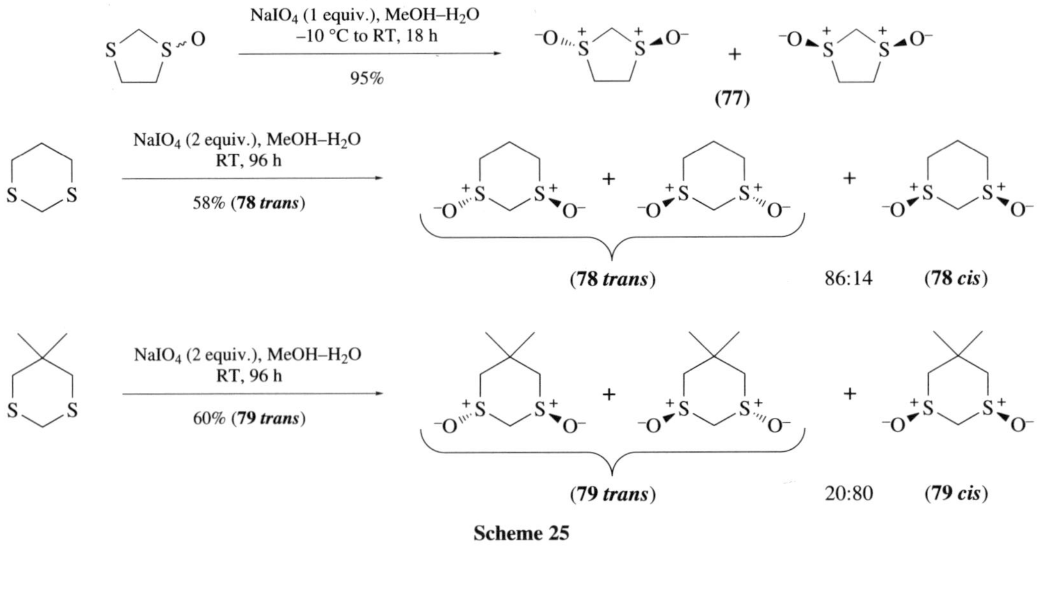

Scheme 25

(ii) From methylene bis(sulfoxides)

Dithioacetals have proved themselves to be very useful reagents in organic synthesis, representing masked carbonyl groups with umpolung activity. Their oxidation to bis(sulfoxides) increases the acidity of the C-2 hydrogens, opening the way to milder metallation processes with a greater variety of bases. Moreover, the alternative reactivity of the alkanesulfinyl moiety in elimination reactions makes bis(sulfoxides) truly versatile synthetic reagents. For example, the allylic alcohol (**80**) was derived from chalcone by way of the bis(sulfoxide) (**81**) (Scheme 26) ⟨92TL4913⟩.

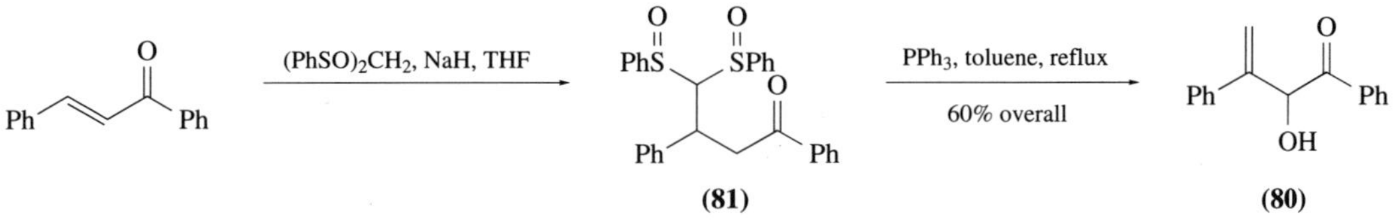

Scheme 26

In contemporary asymmetric synthesis, chiral acyl anion equivalents have become highly sought-after reagents. One such reagent is the C-2 symmetrical bis(sulfoxide) *trans*-1,3-dithiane-S^1,S^2-dioxide (**78** *trans*). The advantages of C-2 symmetry during addition reactions with trigonal electrophiles include a reduction in the number of competing diastereomeric transition states and the formation of only one new stereocenter. This simplification therefore automatically increases the stereoselectivity of these reactions ⟨90TL135⟩. Metallation of the bis(sulfoxide) (**78** *trans*) and treatment of the resulting carbanion with both aromatic and heteroaromatic aldehydes immediately gave a 1:1 mixture of isomeric adducts ⟨91TL7743⟩. Equilibration of this mixture over 30 min at 0 °C subsequently led to excellent diastereoisomeric excess (ca. 90% *de*) (Scheme 27). The major diastereoisomers were then isolated and their absolute stereochemistries determined by x-ray crystallography. By comparison, the stereoselectivity obtained with aliphatic aldehydes was only modest.

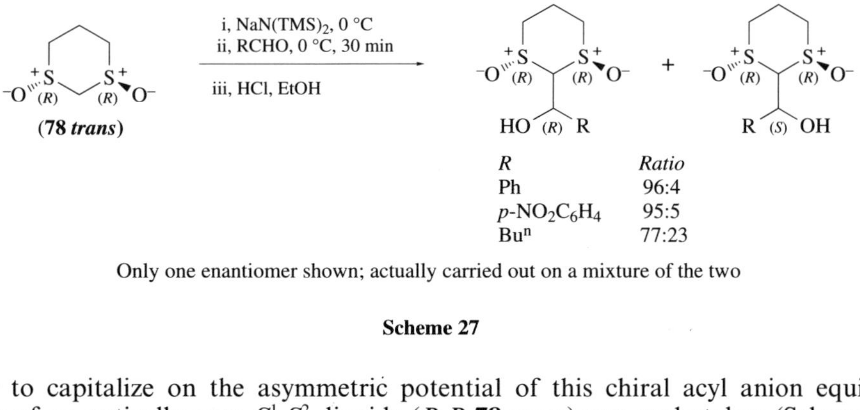

R	Ratio
Ph	96:4
p-NO$_2$C$_6$H$_4$	95:5
Bu^n	77:23

Only one enantiomer shown; actually carried out on a mixture of the two

Scheme 27

In order to capitalize on the asymmetric potential of this chiral acyl anion equivalent, the preparation of an optically pure S^1,S^2-dioxide (*R,R*-**78** *trans*) was undertaken (Scheme 28, where det represents diethyl tartrate). It has been shown that highly stereoselective oxidations only occur with compounds possessing polar substituents. 2-Carbethoxy-1,3-dithiane was therefore chosen for this reason, and because of its commercial availability and the ease with which the control substituent could be removed at the end of the synthesis ⟨92JOC6390⟩. A Modena-modified Sharpless oxidation of the 2-carbethoxy substrate proceeded with excellent enantioselectivity (>97% *ee*). Subsequent hydrolysis and decarboxylation gave the enantiomerically pure 1,3-dithiane-S^1,S^2-dioxide (*R,R*-**78** *trans*). The application of this enantiomerically pure reagent and the final hydrolysis of the chiral auxillary are under investigation.

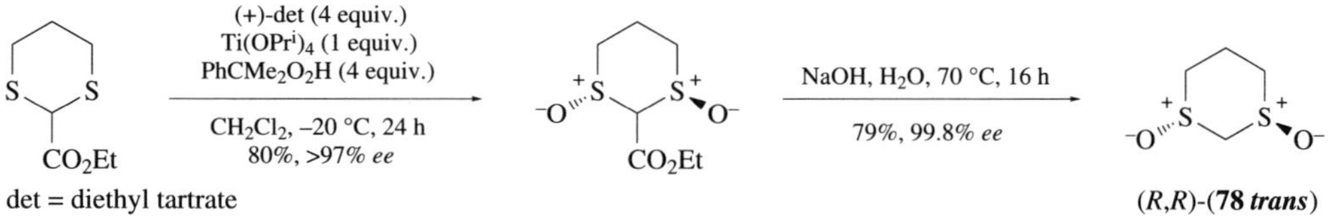

Scheme 28

A similar C-2 symmetrical reagent under investigation is (S,S)-bis-(toluene-p-sulfinyl)methane (**82**) prepared from optically pure precursors (Scheme 29) ⟨91TL3695⟩. Reactions of its carbanion gave good stereoselectivities with a variety of aromatic aldehydes (ca. 70% de), but disappointing results with aliphatic aldehydes. The stereoselectivities of both reagents (**78**) and (**82**) appear to be dependent on the choice of the metal counterion employed. In contrast, the reaction of lithiated (**82**) with α,β-unsaturated aldehydes follows a different sequence, giving in one step the corresponding (E)-elimination product (**83**). The utility of these optically pure bis(toluene-p-sulfinyl)butadienes in asymmetric synthesis is being explored ⟨91TL3695⟩.

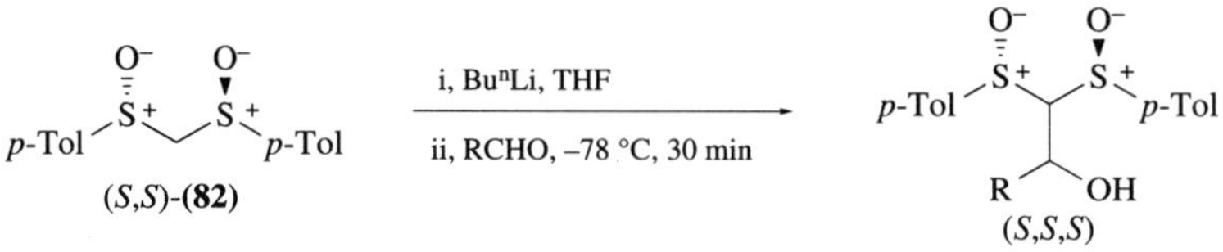

R	de(%)	yield(%)
Ph	80	70
p-O$_2$NC$_6$H$_4$	70	93
p-MeOC$_6$H$_4$	68	37
n-C$_5$H$_{11}$	20	60

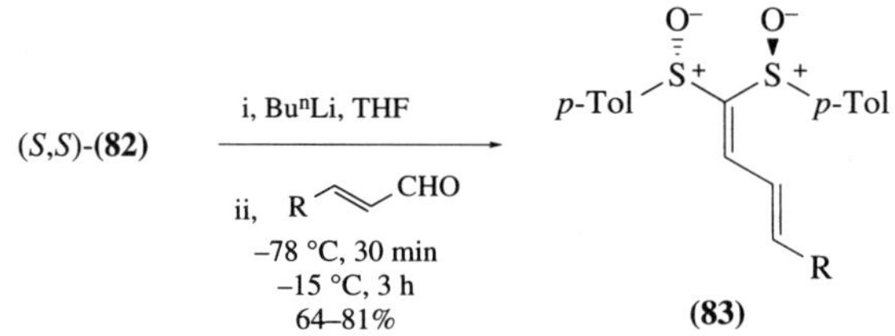

Scheme 29

(iii) From various precursors

Optically active 1,1-bis-(toluene-p-sulfinyl)ethene has been used as a chiral dienophile in an asymmetric Diels–Alder reaction with cyclopentadiene (Equation (51)) ⟨86SC233⟩. The unoptimized selectivity (ca. 60% de) is a promising debut for this type of reagent. Finally, metallated DMSO reacts at the sulfine sulfur atom (thiophilic addition) of di-toluene-p-sulfine to give the bis-sulfoxide (**84**) in 80% yield (Equation (52)) ⟨80RTC39⟩.

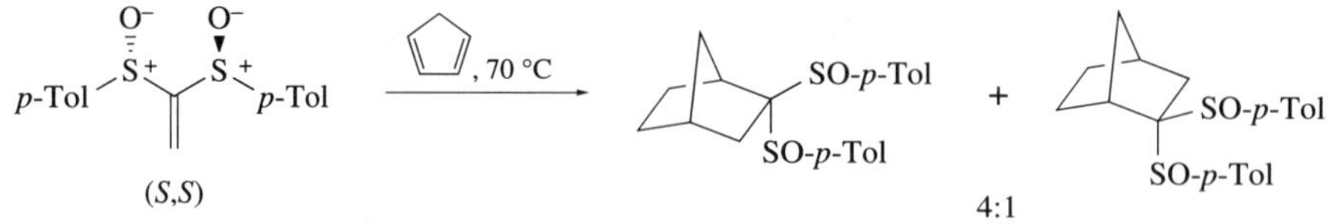

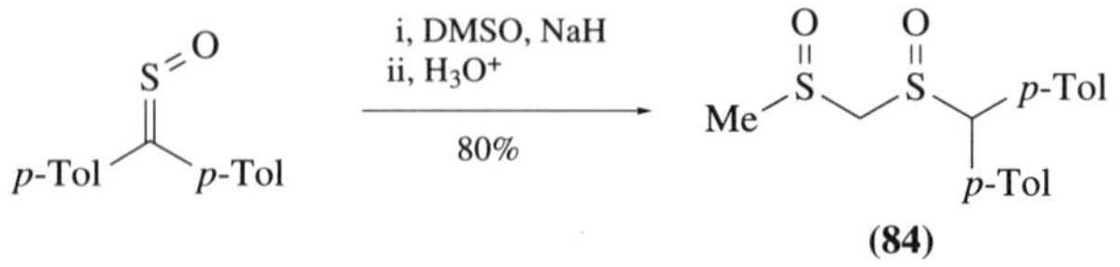

4.06.1.4.2 *Other compounds*

This section should deal with compounds which are either sulfinic acid derivatives containing a sulfoxide moiety on the α-carbon or bis-sulfinic acids. However, such compounds are very rare! One example is an anion derived from MeS(O)CH$_2$S(O)OMe which was obtained as an intermediate in a gas-phase reaction ⟨90JA607⟩. The attention of the reader is, therefore, drawn to Chapter 2.03.4, which deals with the synthesis of sulfinic acids.

4.06.1.5 One Tricoordinated and One Higher Coordinated Sulfur—R$^1_2$CS(O)OR^{2}S(O)$_2$R^3, etc.

Amongst the possible oxides of dithioacetals, S^1,S^1,S^2-trioxides have been, without doubt, the least studied. They are nevertheless easily accessible, especially by oxidation. Hydrogen peroxide and mcpba are the most suitable reagents. For example, Ogura et al. prepared two gem-sulfoxide-sulfones with hydrogen peroxide in acetic acid (Equation (53)) ⟨81TL4499⟩. It is often more practical to start directly from the dithioacetal. Peroxide oxidation of di(phenylthio)methane led to the corresponding trioxide (Equation (54)) ⟨62JA684⟩. Alternatively, the use of three molar equivalents of mcpba on the dithianes (**85**) led to the trioxides (**86**) in high yields (Equation (55)) ⟨92MI 406-02⟩.

Oxidation of the dithioacetal (**87**) containing both an endocyclic sulfur and an exocyclic sulfur led to the formation of the trioxide (**88**) which is doubly oxidized on the endocyclic sulfur. There were no indications, however, as to whether this selectivity was general. Of interest, however, is the chemoselectivity of the oxidation: the alkene double bond remains unchanged (Equation (56)) ⟨75JCS(P1)180⟩.

As would be expected, dithioacetal trioxides can be deprotonated. Condensation of the resulting anions with various electrophiles gives access to novel trioxides. For example, Böhme and Clément used a Mannich-type reaction to obtain some aminomethylated products ⟨79TL1737⟩. Elsewhere, treatment of the trioxide (**89**) with sodium hydride and acetal-protected 4-bromobutanal gave the alkylated product (**90**) (Equation (57)) ⟨90T7197⟩.

Another method of preparing dithioacetal trioxides is to treat an anion α to a sulfone with a sulfinylating agent. Furthermore, if the sulfinylating agent is chiral, optically active trioxides may be prepared by this method. For example, deprotonation of methyl phenyl sulfone with n-butyllithium leads to the corresponding anion. When this anion was treated with (−)-menthyl (S)-toluene-p-sulfinate, the expected trioxide was obtained in 75% chemical yield and >98% ee (Equation (58))

⟨79S535, 90T7197⟩. This method has also been employed for the preparation of the (S)-(sulfinylmethyl)-sulfoximine (**91**) (Equation (59)) ⟨82S767⟩.

$$\text{PhSO}_2\text{Me} \xrightarrow[\text{ii, } p\text{-Tol}\overset{-O}{\underset{+}{S}}\text{Omenthyl}]{\text{i, Bu}^n\text{Li}} \text{PhSO}_2\overset{-O}{\underset{+}{S}}p\text{-Tol} \quad 75\%$$ (58)

$$p\text{-Tol}\overset{-O}{\underset{+}{S}}\text{Omenthyl} + \text{Ph}\overset{O}{\underset{NMe}{\overset{\|}{S}}}\text{Me} \xrightarrow[67\%]{\text{Bu}^n\text{Li}, -78\,°\text{C, THF}} \overset{O}{\underset{MeN}{\overset{\|}{Ph\,S}}}\overset{-O}{\underset{+}{S}}p\text{-Tol} \quad (\mathbf{91})$$ (59)

The trioxides of dithioesters, sulfonyl sulfines (**92**), already contain both the SO and the SO$_2$ moieties. A thiophilic addition of an organometallic reagent onto the sulfine function would therefore lead to the formation of a dithioacetal trioxide. Such a thiophilic addition, enhanced by the adjacent electron-withdrawing SO$_2$R group, has been demonstrated by Veenstra and Zwanenburg ⟨78T1585, 80RTC39⟩. Treatment of the compounds (**92**) with methyllithium followed by acid-mediated hydrolysis led to the isolation of dithioacetal trioxides in good yields (Equation (60)). This method, however, is limited to those dithioester trioxides which are stable, essentially those where R^2 is aromatic. Finally, certain *gem*-sulfoxide-sulfones can be prepared by the Diels–Alder reaction of a ketene dithioacetal trioxide and a diene. An example is given in Equation (61) ⟨91SL565⟩.

$$\underset{(\mathbf{92})}{\overset{O}{\underset{R^1}{\overset{\|}{S}}}\underset{SO_2R^2}{}} \xrightarrow[80-84\%]{\text{i, MeLi}\atop\text{ii, NH}_4\text{Cl, H}_2\text{O}} \overset{O}{\underset{R^1}{\overset{\|}{S}}}\overset{Me}{\underset{SO_2R^2}{}}$$ (60)

(Equation 61 structural scheme, CDCl$_3$, RT, 12 h, 96%)

4.06.1.6 Two Tetracoordinated Sulfurs—R$^1{}_2$C[S(O)$_2$R^2]$_2$

4.06.1.6.1 Bis(sulfones)

(i) From dithioacetals and their derivatives

The accessibility of dithioacetals has been extensively investigated. The direct oxidation of these readily available precursors represents the most common method employed to prepare methylene disulfones. Furthermore, since the target tetraoxides represent the highest possible oxidation state of the parent compounds, the problems of selectivity and mixed products are minimized. In consequence, the yields are frequently high. The most favored reagents currently employed are hydrogen peroxide, mcpba, oxone (KHSO$_5$) and potassium permanganate.

For example, bis(arenesulfonyl)methanes have been obtained in excellent yields from their corresponding dithioacetals with a 30% solution of H$_2$O$_2$ in glacial acetic acid, as shown in Equation (62) ⟨82BSF43⟩. Likewise, the dithioacetals derived from thioglycolic acid, a chemical of ever-growing industrial importance, underwent successful peroxide oxidation (Scheme 30) ⟨81S995, 89SUL79⟩. Moreover, subsequent decarboxylation of methylenebis(sulfonylacetic acid) gave access to bis(methanesulfonyl)methane in good yield ⟨91ZOR216⟩.

Elsewhere, H$_2$O$_2$ was employed for the selective preparation of the camphor-derived bis(sulfone) imine shown in Equation (63). No formation of the potential oxaziridine product was observed with this reagent ⟨90JCS(P1)2919⟩. The partially oxidized dithioacetal derivative, trifluoromethane-sulfonyl(phenylthio)methane (**93**) (Equation (64)), also gave a dithioacetal tetraoxide upon oxidation of the phenylthio group ⟨92TL745⟩. For the oxidation of cyclic sulfur atoms, particularly in larger rings, peroxycarboxylic acids seem to be the reagents of choice. The yields are generally good, as indicated in Scheme 31, although a large excess of reagent (four equivalents) is often employed ⟨89MRC760, 90JA8084, 92JCS(P1)1179⟩. The reagent mcpba has also been used for the one-pot conversion of the alkenic dithioacetal shown in Equation (65) into the corresponding epoxy bis(sulfone) in 77% yield ⟨91JOC3530⟩. However, attempts at a one-pot conversion of a dithioacetal imine resulted in a mixture of products of which only 10% was the desired, fully oxidized oxaziridine bis(sulfone) (Equation (66)) ⟨90JCS(P1)2919⟩.

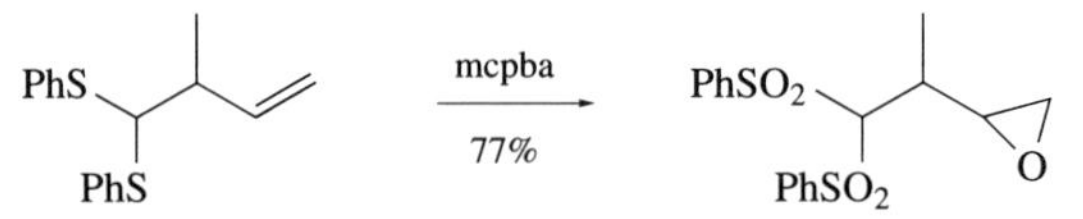

(65)

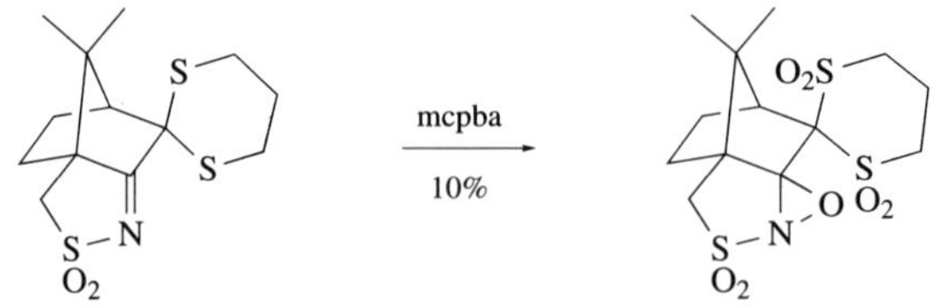

(66)

The convenient and inexpensive oxidizing agent oxone was used to prepare bis(methanesulfonyl)methane, shown in Equation (67), from the commercially available methyl methanesulfinylmethyl sulfide ⟨92JOC3496⟩. Of potential interest, this reagent selectively oxidizes sulfides to sulfones in the presence of isolated double bonds ⟨81TL1287⟩. So far this selectivity has not been used for the preparation of alkenic disulfones. Finally, various thiacepham derivatives were oxidized using $KMnO_4$ for the preparation of β-lactam derivatives of potential pharmaceutical interest (Equation (68)) ⟨90JCS(P1)773⟩.

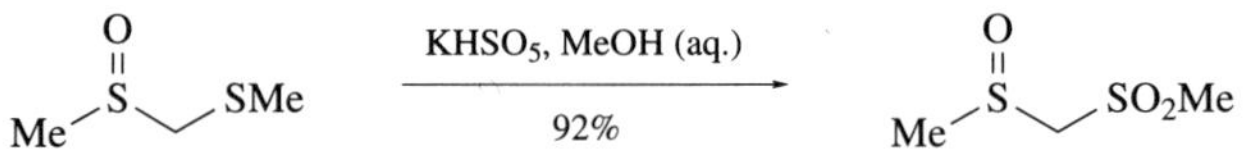

(67)

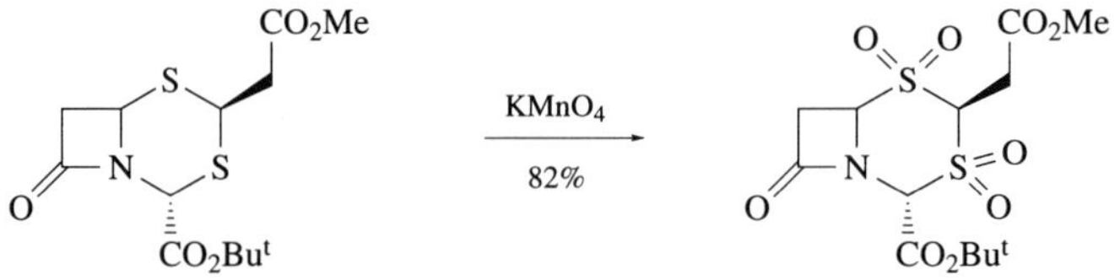

(68)

(ii) From methylene and alkylidene disulfones

Metallation of bis(benzenesulfonyl)methane with sodium hydride followed by alkylation with a bromo ester led to the substituted bis(sulfone) in excellent yield, as reported in Scheme 32 (where TBDMS represents the *t*-butyldimethylsilyl group). Subsequent remetallation and alkylation with a different primary bromide also proceeded in excellent yield ⟨88BSF989⟩. However, the formation of dialkylated derivatives from bis(benzenesulfonyl)alkylanes has been shown, as in the fruitless reaction with 2-iodopropane presented in Scheme 33, to be limited by steric factors. If the disulfone is being used as a C^{2-} synthon, a suitable alternative to overcome this problem is the 1,3-benzodithiole tetraoxide, wherein the bulky sulfur groups, now part of a fused ring system, are both held back in the same plane. The steric interactions are therefore reduced and the reaction proceeds smoothly ⟨88T6855⟩.

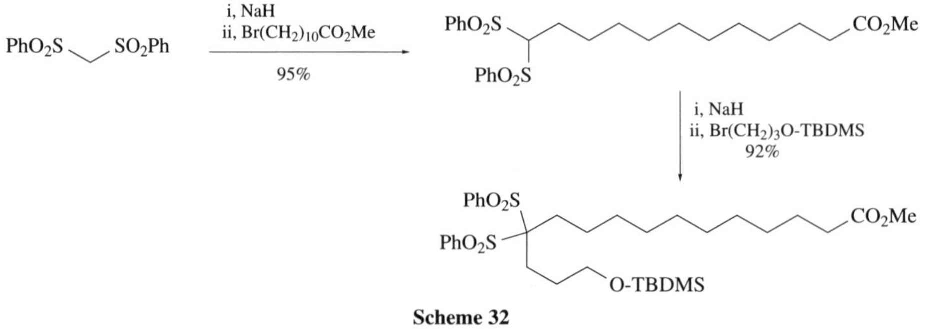

Scheme 32

Other electrophiles may be used. For example, the treatment of certain 3,4-epoxy 1,1-bis(sulfonyl)alkanes with ethanolic sodium ethoxide resulted in a concerted ring-opening/ring-closure reaction to form hydroxymethylcyclopropanes (Equation (69)) ⟨91JOC3530⟩. Whilst exploring its

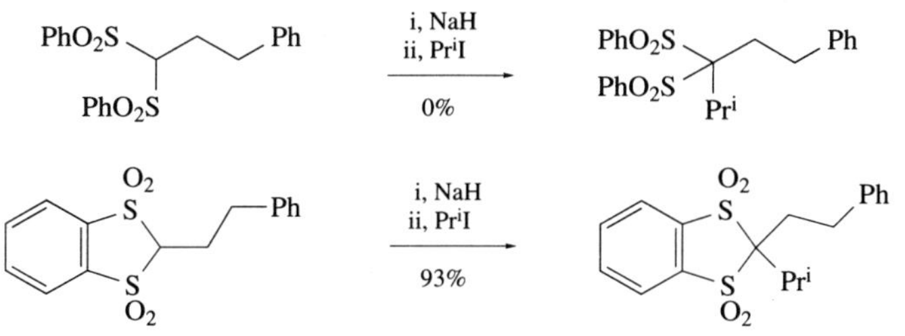

Scheme 33

reactivity as a potential multicoupling reagent, 2,3-bis(benzenesulfonyl)-1-propene (Equation (70)) was shown to undergo clean S_N2' displacement when treated with the lithium salt of bis(benzenesulfonyl)methane. The resulting novel tris(sulfone) should also find use as an interesting synthetic building block ⟨91JOC6386, 92JOC298⟩.

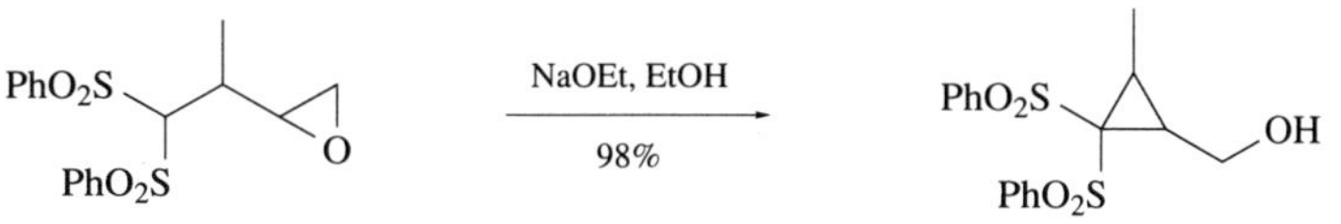

(69)

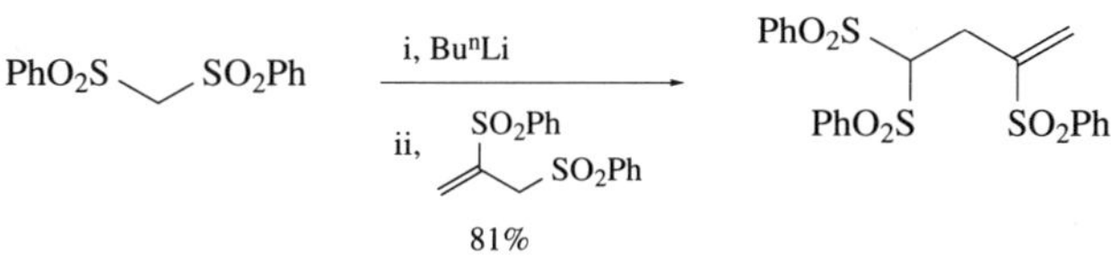

(70)

Bis(sulfones) have also been used successfully for the stereoselective preparation of alkenes by base-catalyzed condensation with aldehydes. The reaction is carried out with piperidinium acetate and molecular sieves at room temperature to give uniquely the (E) isomers in high yields, as shown in Scheme 34. Subsequent metallation and alkylation of the alkenyl bis-sulfones occurs cleanly without compromising the stereointegrity of the alkenes ⟨82BSF43⟩. A similar Knoevenagel condensation was observed between arene- and phenylmethanesulfonylmethane sulfonylacetic acid (**94**) and aromatic aldehydes ⟨89SUL79⟩. However, with these sulfones, the methylene hydrogens between the carboxyl and sulfone groups are more acidic and reaction therefore occurs here. After decarboxylation, (E)-styryl sulfones (**95**) resulted.

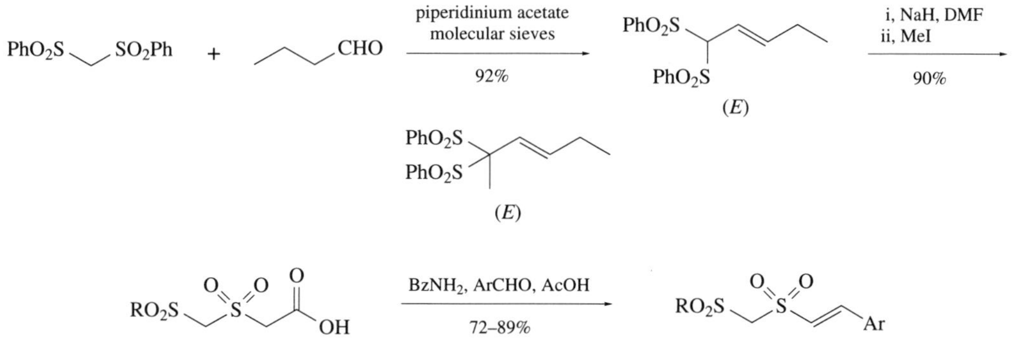

R = aryl or benzyl

Scheme 34

With two equivalents or more of alkylating agent in a suitable basic medium, the corresponding bis-alkylated disulfones may be obtained in good yields by a one-pot procedure (Scheme 35) ⟨89UKZ1216, 92TL745⟩. It then follows that treatment of disulfones with dihalides will lead to the formation of cyclic compounds ⟨90JOC247, 92TL745⟩. Accordingly, the reaction of cyclic disulfones with dihalides will lead to spiro-disulfones ⟨88T6855, 89UKZ1216⟩. Elsewhere, dialkylated bis(sulfones) resulting from the combination of two different reactions can be found. For example, the tricyclic diene (**96**) shown in Equation (71) was formed via an initial allylic substitution followed by an intramolecular S_N2' reaction ⟨91CB1827⟩, and the uridine nucleoside derivative (**97**) (Equation (72)), where MMTr represents the monomethoxytrityl group) via a Michael addition/substitution sequence ⟨90T2587, 91T3431⟩.

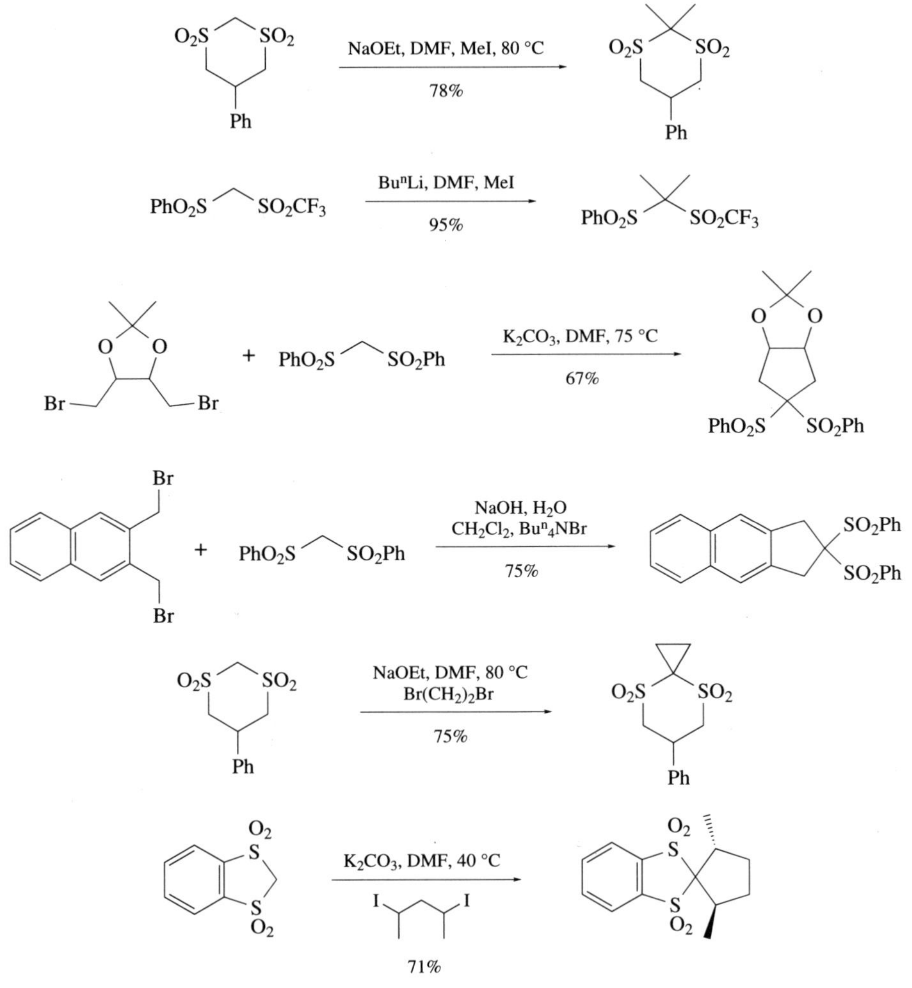

Scheme 35

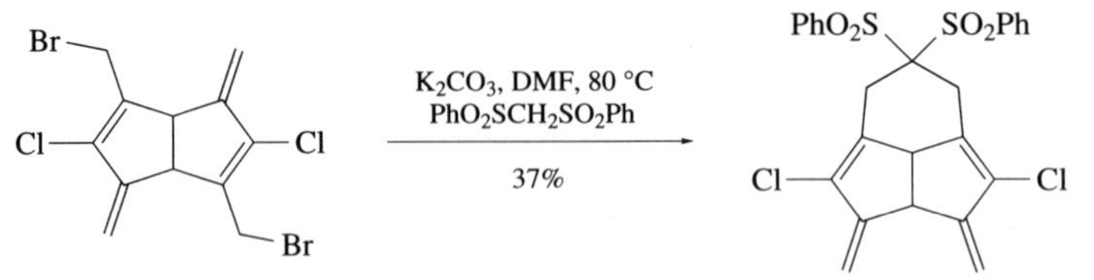

(**96**) (71)

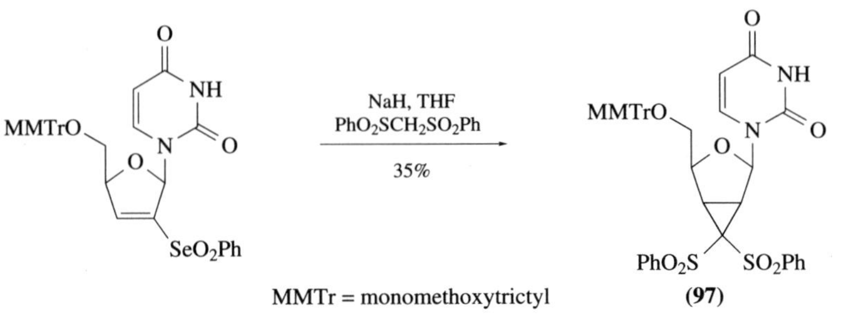

MMTr = monomethoxytrictyl (**97**) (72)

When dimethyl methylenebis(sulfonylacetate) was condensed with benzaldehyde in the presence of ammonium acetate, a 1,3-dithiane-1,1,3,3-tetraoxide was formed, albeit in low yield (Scheme 36) ⟨81S995⟩. Moreover, subsequent metallation and alkylation of this compound occurs uniquely at the C-2 position. No alkylation at C-4 and C-6 was observed ⟨89UKZ1216⟩.

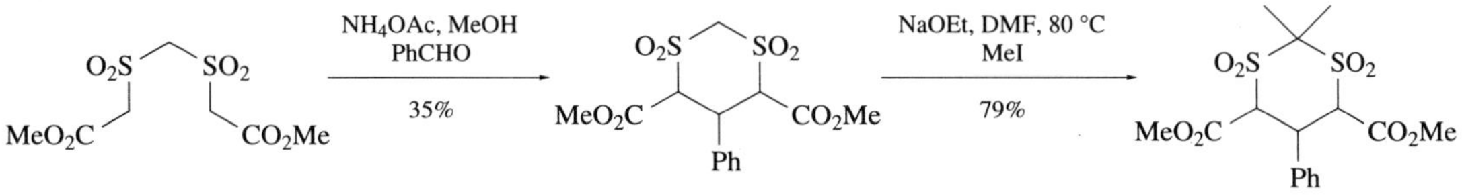

Scheme 36

A study of the relative acidities of the methylene (α^1) and methyl (α^2) hydrogens of bis(methanesulfonyl)methane has been reported ⟨92JOC3496⟩. Treatment with one, two, or three molar equivalents of Bu^{n}Li gave the α^1-anion, α^1,α^1-dianion and the $\alpha^1,\alpha^1,\alpha^2$-trianion, respectively. Treatment of the trianion with prenyl bromide then led to a regioselective α^2-alkylation, although in somewhat modest yield, as shown in Equation (73).

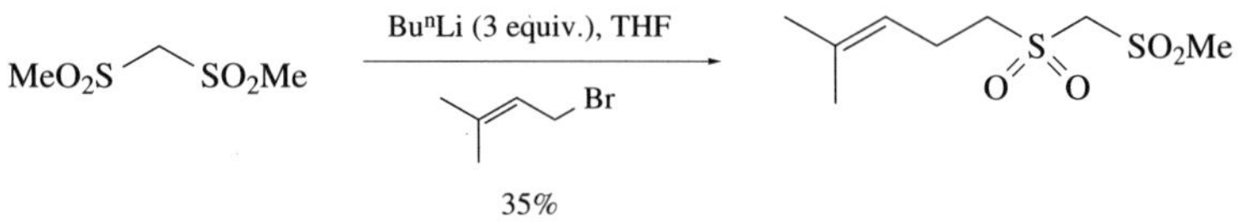

Finally, as shown in Equation (74), Le Guillanton and Simonet have reported that the electrochemical oxidation of bis(sulfone) anions may be used to prepare tetrasulfones by anodic coupling. Moreover, these interesting polysulfones represent a potentially controllable radical source due to their facile radical cleavage ⟨90TL3149⟩.

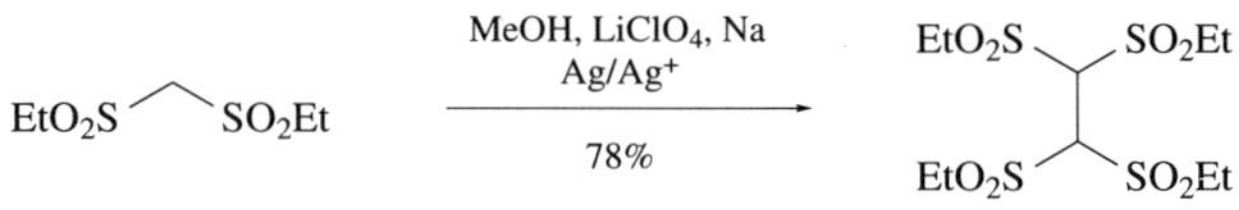

(iii) From ketene dithioacetal tetraoxides

(a) By cycloaddition. Ketene dithioacetal tetraoxides readily undergo Diels–Alder reactions with dienes to give [4π + 2π]-cycloaddition products in good yields (Scheme 37) ⟨84S757⟩. When the reaction is effected with an asymmetrical diene, a mixture of regioisomers may be formed. In general, 1-substituted dienes favor the *ortho*-substituted bis(sulfone) and 2-substituted dienes favor the *para*-substituted bis(sulfone) ⟨84S757, 92T1485⟩. This selectivity has been explained by molecular orbital considerations. However, the reaction of N-benzoylindole-2,3-quinodimethane, a 2,3-disubstituted diene, gave equal amounts of both regioisomers (Scheme 38) ⟨91T1925⟩. In an attempt to control the stereochemical outcome of the reactions, biphenyl and binaphthyl substituted precursors have been employed. The best results were obtained with the binaphthyl compounds shown in Scheme 38, which although giving total facial selectivity only gave an *endo*:*exo* ratio of 4:1 ⟨92T1485⟩.

Two other cycloaddition reactions with ketene dithioacetal tetraoxides have also been described. The reaction of 1,1-bis(benzenesulfonyl)ethene with the 3H-indolium ylide (**98**) gave a [3 + 2]-addition product as a single regioisomer (Scheme 39) ⟨90SL359⟩, and the addition of diazomethane to the allylic selenide (**99**) gave a 3,3-bis(benzenesulfonyl)pyrazoline (**100**). Selenoxide-mediated alkenation of this pyrazoline followed by thermal decomposition led to a hydroxycyclopropyl bis(sulfone) in 92% yield ⟨92TL1321⟩.

(b) Ketene dithioacetal tetraoxides as electrophiles. Ketene dithioacetal tetraoxides are also powerful Michael acceptors and have been treated with a variety of nucleophiles. In particular, 1,1-bis(sulfonyl)ethene has been the subject of much synthetic interest, representing a two-carbon electrophile with easily transformed hetero-substituents. For example, its reaction with a 2,3-dihydropyran-4-one (Equation (75)) gave the C-3-substituted derivative in high yield ⟨91HCA27⟩.

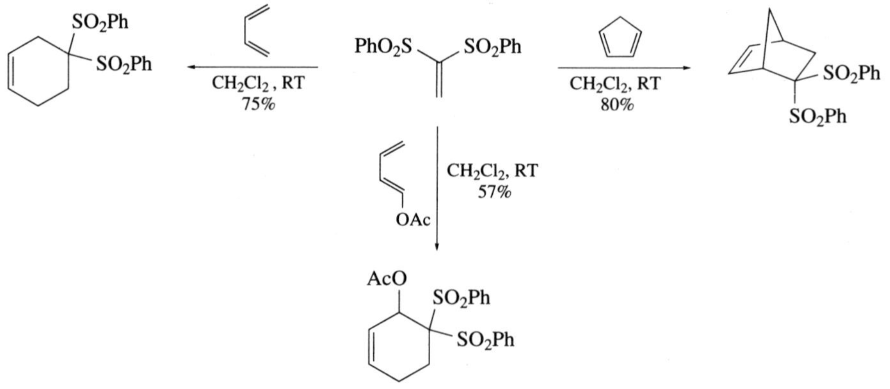

Scheme 37

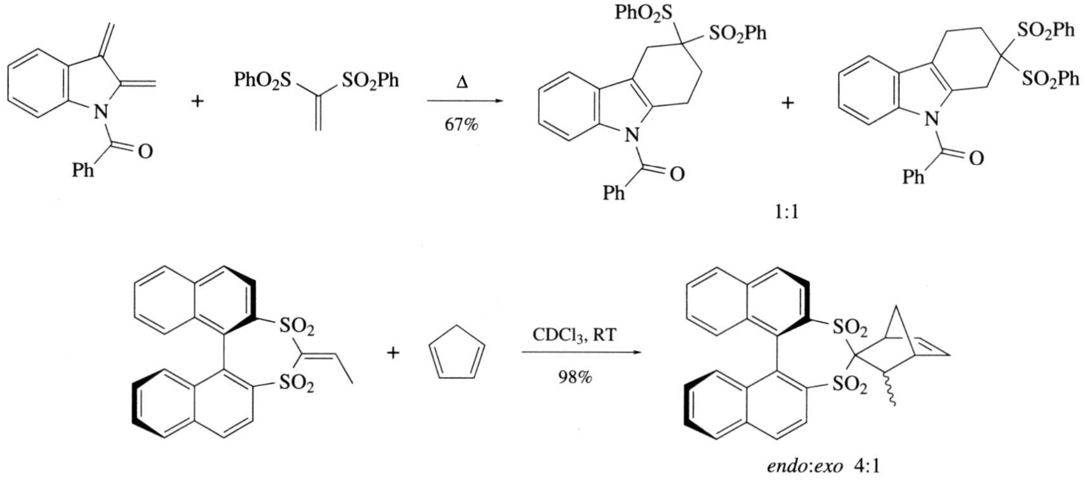

Scheme 38

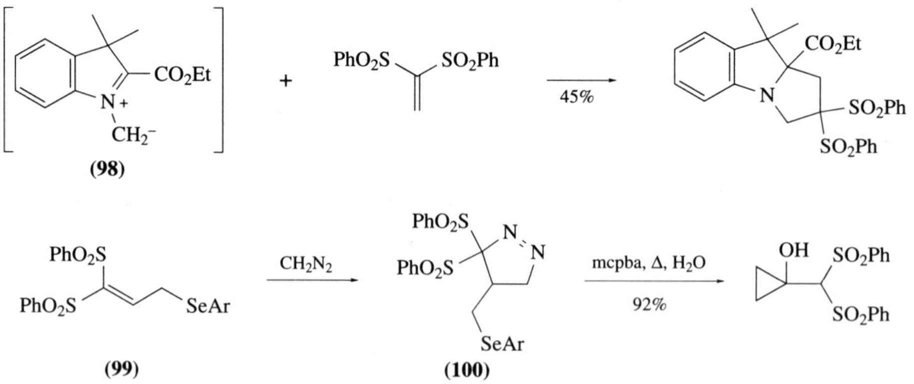

Scheme 39

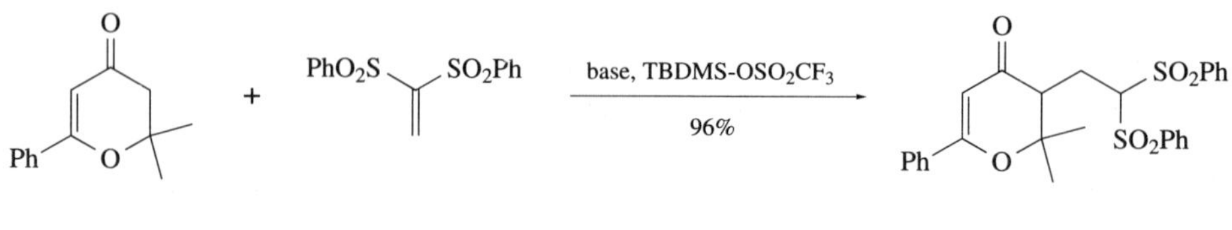 (75)

Unsurprisingly, its reaction with chiral nucleophiles has been studied. Two examples are presented in Scheme 40. Addition to an enamino ester, followed by hydrolysis, gave the corresponding 1,3-keto diester **(101)** (50% *ee*) ⟨92TA1003⟩. Alternatively, alkylation of a 5-benzyl-1,4-oxazine-2-one gave the expected *trans* product (98% *de*) ⟨92TL1573⟩.

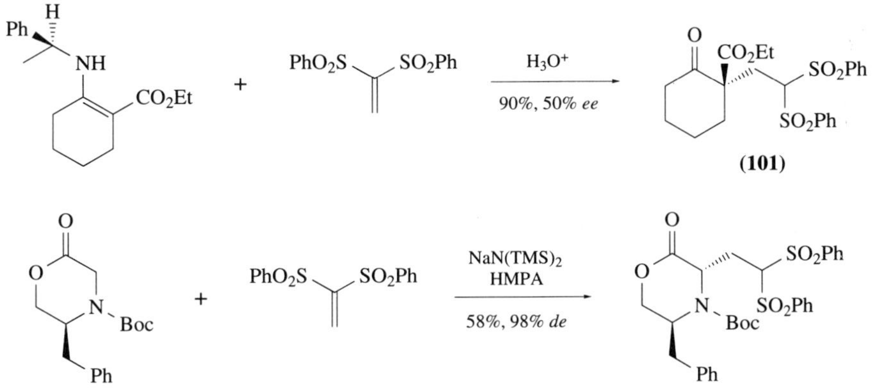

Scheme 40

HMPA = hexamethyl phosphoramide

(iv) Palladium(0) catalysis

The importance of (π-allyl) Pd⁰-mediated chemistry in the synthesis of carbon–carbon bonds is now well established. The development of this field must be due, in part at least, to the facility and frequent stereoselectivity of these reactions. For example, when the nucleoside model compound shown in Equation (76) was treated with the anion of bis(benzenesulfonyl)methane, a Pd⁰-catalyzed displacement of the allylic acetate occurred in 88% yield. Furthermore, starting with a racemic precursor, a single diastereoisomer was formed ⟨90TL1043⟩. During the synthesis of prostaglandin I₁ analogues, a similar nucleophilic substitution of an allylic carbonate also yielded the corresponding regiospecifically γ-alkylated compound (Equation (77)) ⟨91MI 406-01⟩. However, steric demands of the bulky bis(benzenesulfonyl)methane may favor α-alkylation, depending on the system involved. Thus, the urethane-protected allylic amine shown in Equation (78) (where bsa represents *N,O*-bis(trimethylsilyl)acetamide) was obtained by terminal coupling of the bis(sulfone) anion with the corresponding allylic acetate precursor. Furthermore, the single *trans* isomer was obtained in 98% *ee* ⟨90TL6819⟩.

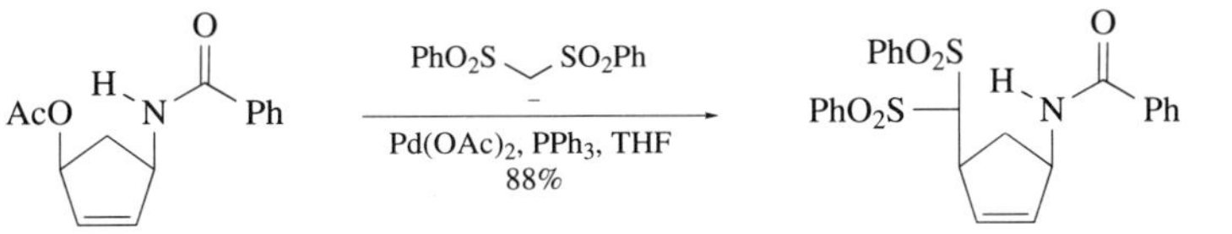

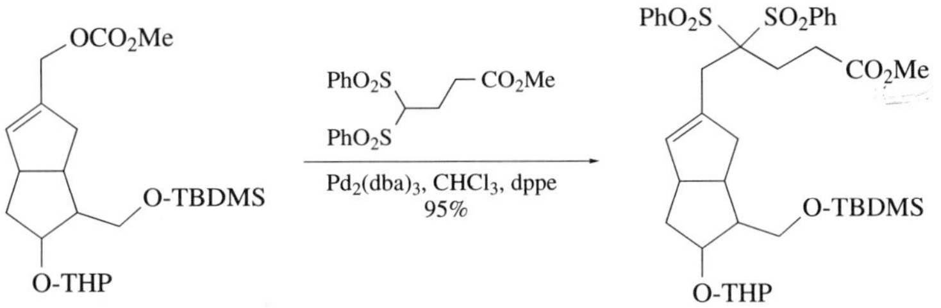

dba = dibenzylideneacetone

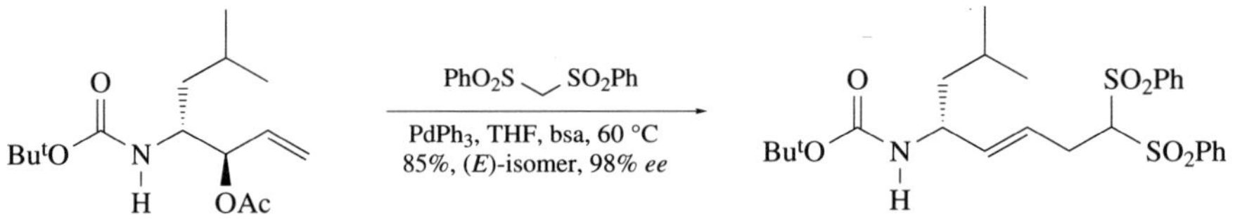

(78)

bsa = *N*, *O*-bis(trimethylsilyl)acetamide

Likewise, in the preparation of certain insect pheromones, twofold, regioselective α-coupling of a vinyl epoxide with bis(benzenesulfonyl)methane was required. This was achieved, as shown in Equation (79), in 91% yield, of which 74% was recrystallized as the desired (*E*,*E*) isomer ⟨91TL2193⟩. Indeed, Trost *et al.* capitalized on these steric constraints, enhanced by a suitable choice of catalyst, to favor nine-membered ring formation by α-alkylation over the possible seven-membered ring formation by γ-alkylation (Scheme 41, where dba represents dibenzylideneacetone). This did not apply in the case of homologous six–eight-membered ring systems, where, despite the same steric demands of the bis(sulfone) nucleophile, the six-membered carbocycle was formed by γ-alkylation, suggesting that the regioselectivity results from a fine balance between many comparable energies ⟨92TL717⟩.

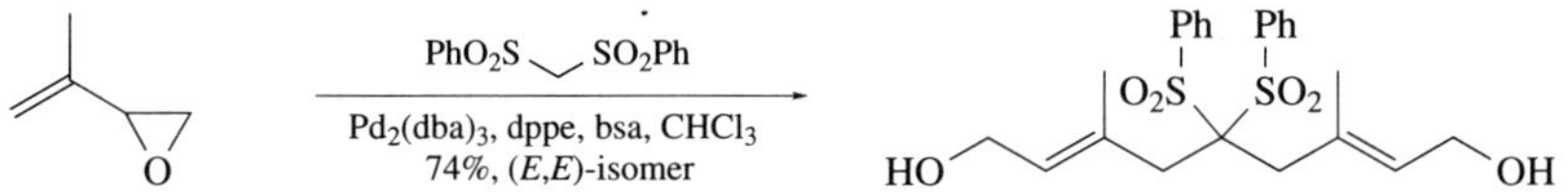

(79)

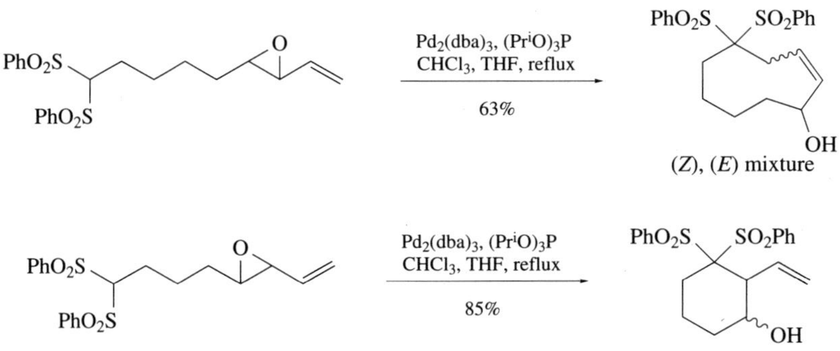

Scheme 41

The Pd^0-catalyzed addition of dienes to bis(sulfones) has also been studied and some examples are presented in Scheme 42. Appropriate choice of catalyst and conditions make this reaction a valuable preparative route. However, a lack of genuine regioselectivity limits synthetic utility to symmetrical dienes ⟨92TL1831⟩.

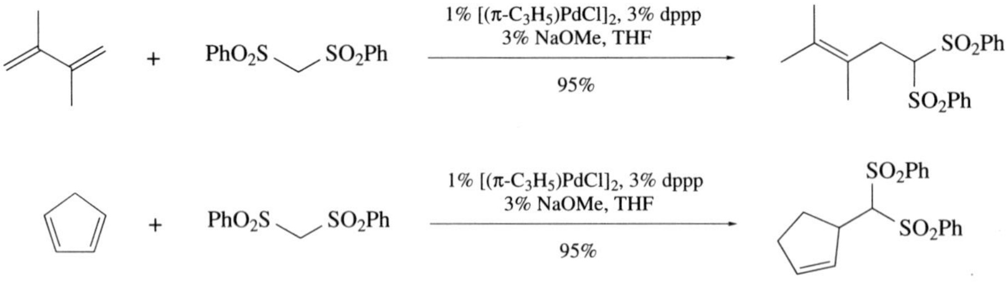

Scheme 42

(v) Various precursors

One of the oldest preparative methods known for bis(sulfones) (Equation (80)) is the coupling of a sulfonyl fluoride with a Grignard reagent. The overall conversion involves formation of the monosulfone, α-metallation, and finally bis(sulfone) production ⟨63JOC1420⟩. More recently, a bis(sulfone) was formed by absorbing diiodomethane onto sodium benzenesulfinate-impregnated alumina followed by sonification (Equation (81)) ⟨90SC925⟩.

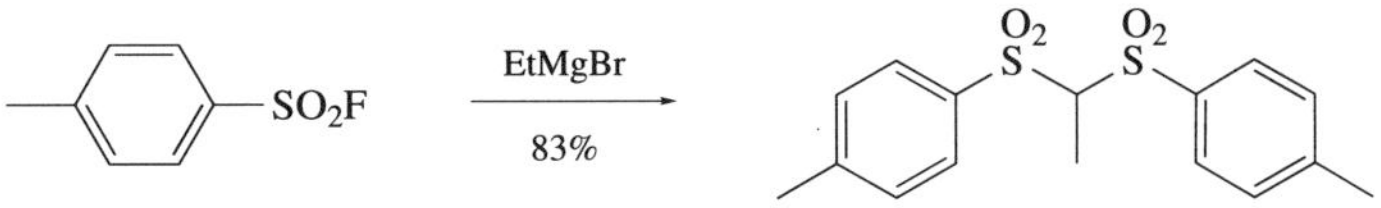

$$\text{Tol-SO}_2\text{F} \xrightarrow[83\%]{\text{EtMgBr}} \text{Tol-SO}_2\text{-CH(CH}_3\text{)-SO}_2\text{-Tol} \quad (80)$$

$$\text{NaSO}_2\text{Ph} + \text{CH}_2\text{I}_2 \xrightarrow[88\%]{\text{Al}_2\text{O}_3,\text{ sonification}} \text{PhO}_2\text{S-CH}_2\text{-SO}_2\text{Ph} \quad (81)$$

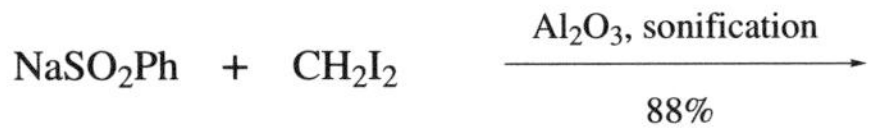

The reaction of quinoline-1-oxide with bis(benzenesulfonyl)methane in the presence of acetic anhydride gave access to the corresponding quinolyl bis(sulfone) in 74% yield (Equation (82)) ⟨90JHC1433⟩. Elsewhere, protonation of a tetrakis(benzenesulfonyl)cyclopentadienide led, after migration of the benzenesulfonyl groups, to a tetra(benzenesulfonyl)cyclopentadiene (Equation (83)) ⟨91LA243⟩.

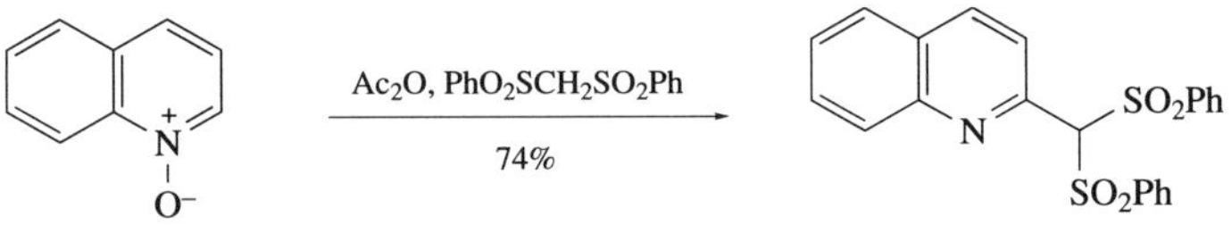

$$\text{quinoline-N-oxide} \xrightarrow[74\%]{\text{Ac}_2\text{O, PhO}_2\text{SCH}_2\text{SO}_2\text{Ph}} \text{2-quinolyl-CH(SO}_2\text{Ph)}_2 \quad (82)$$

$$\text{tetrakis(PhO}_2\text{S)cyclopentadienide Et}_4\text{N}^+ \xrightarrow[94\%]{\text{FSO}_3\text{H}} \text{tetrakis(PhO}_2\text{S)cyclopentadiene} \quad (83)$$

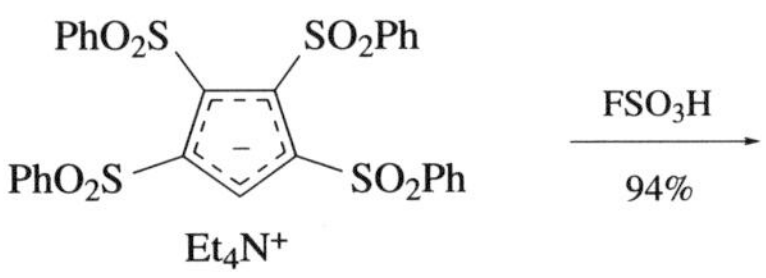

A stable, crystalline iodinium ylide (Scheme 43) has been developed as a convenient bis(perfluoroalkanesulfonyl)carbene source. Photolysis of the ylide in the presence of an alkene or toluene led to the formation of addition and insertion products, respectively ⟨90CC1459⟩. Finally, Stoodley and coworkers have investigated the application of methanesulfonylsulfene (**102**) (Scheme 44) in synthesis. Reactions of this species with enamines and enol ethers gave thietane dioxides, and with tropone gave a novel cycloaddition product ⟨92JCS(P1)2371⟩.

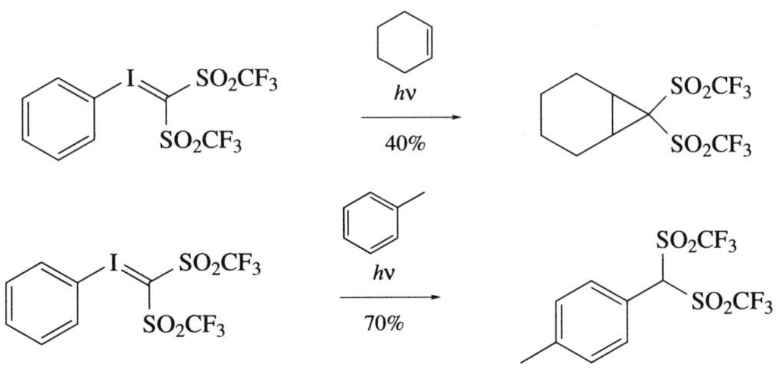

Scheme 43

4.06.1.6.2 Bis(sulfonic) acids and their derivatives

Bis(sulfonic) acids and their derivatives have most frequently been prepared as model compounds for spectroscopic studies ⟨81SA(A)819⟩. Hydrated methanebis(sulfonic) acid, methionic acid, is obtained upon acidification of its sodium salt prepared from iodoform and sodium sulfite. The sodium salt is converted into the hydrated acid via the barium salt. Treatment of the hydrated acid with thionyl chloride–DMF does not give methanedisulfonyl chloride as expected, but leads to the formation of the anhydrous acid (Scheme 45) ⟨65T2743⟩. Methanedisulfonyl chloride, however, can be obtained either by treatment of the anhydrous acid with phosphorus pentachloride or by

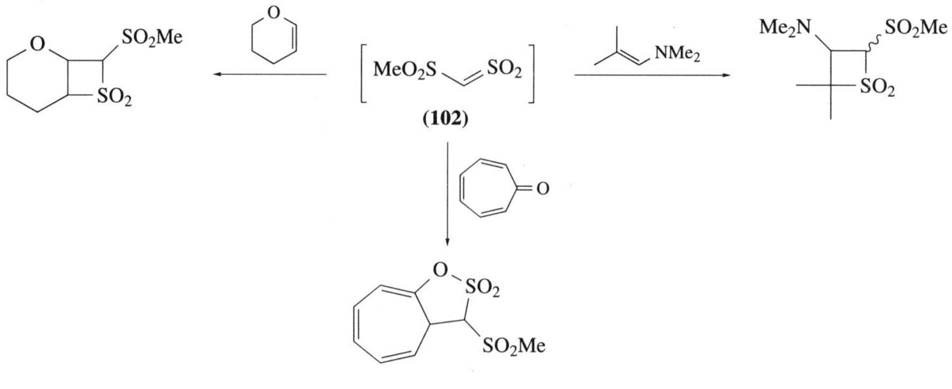

Scheme 44

treatment of the hydrated acid with phosphorus oxychloride. The former method is reported to be more reliable (Scheme 46). Alternatively, the chloride can be prepared directly in good yield by the reaction of acetic acid with chlorosulfuric acid and phosphorus pentachloride. The dideuterated analogue was successfully prepared from the corresponding deuterated reagents ⟨65T2743⟩.

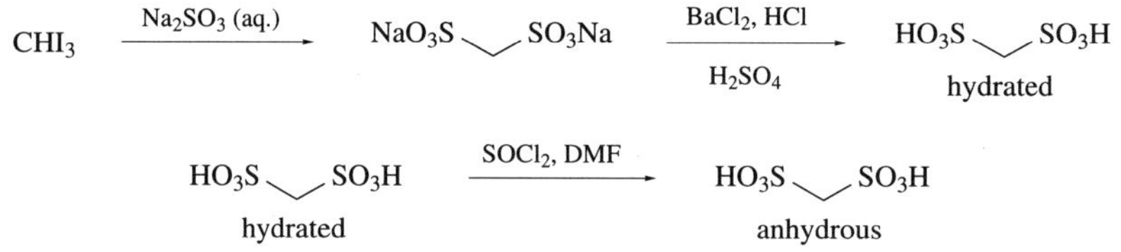

Scheme 45

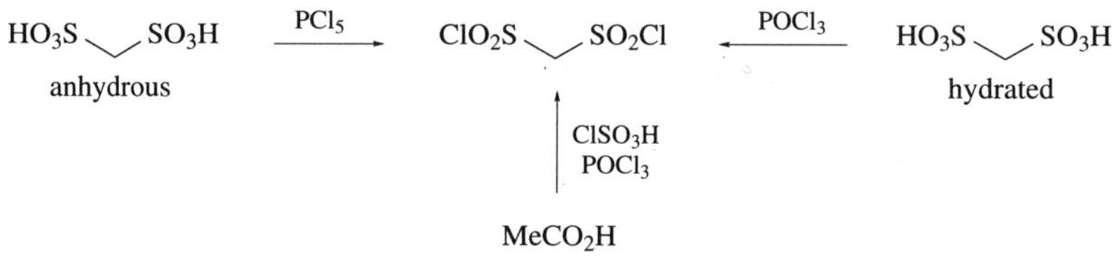

Scheme 46

Methanedisulfonyl chloride is without doubt the most versatile intermediate for the preparation of bis(sulfonic) acid derivatives. It has been esterified with ethanol to form the bis-sulfonic ester (Scheme 47) ⟨76ZN(B)153⟩, aminated with aniline to form the bis-sulfonamide ⟨65T2743⟩, and fluorinated with antimony trifluoride to form methanedisulfonyl fluoride ⟨73ZN(B)98⟩. Methanedisulfonyl chloride can also be hydrolyzed back to methionic acid ⟨76ZN(B)153⟩. The bis(anilide) has also been prepared by treatment of phenoxyfluoroxonium di(fluorosulfonyl)methylide (**103**) with aniline ⟨83TL87⟩.

4.06.1.6.3 Other compounds

Compounds containing both a sulfone moiety and a sulfonic acid moiety (or a derivative thereof) on the same carbon have rarely been reported in the literature. Senning has described an efficient synthesis of the sodium salt of methanesulfonylmethanesulfonic acid and of two esters of this acid ⟨73S211⟩. Opitz *et al.* have obtained certain sulfone-sulfonamides by trapping the relatively stable sulfene (**104**) with amines at −40 °C (Equation (84)) ⟨66AG(E)594⟩. In a subsequent study, the same group successfully employed $CH_3SO_2CH_2SO_2Cl$ in a modified preparation ⟨66TL5263⟩. Finally, a European patent has described the synthesis of sulfone-chlorinated sulfonic esters. One was obtained from methanesulfonyl chloride and 2-chloroethanol (Equation (85)) ⟨85EUP146838⟩.

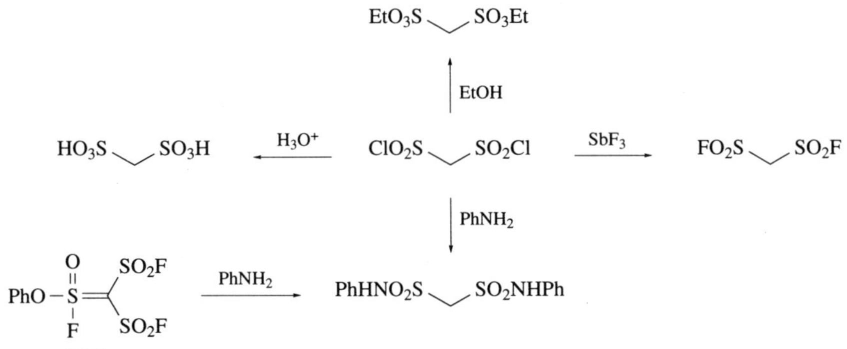

4.06.2 FUNCTIONS CONTAINING ONE SULFUR AND ONE SELENIUM OR TELLURIUM—$R^1{}_2CSR^2SeR^3$, etc.

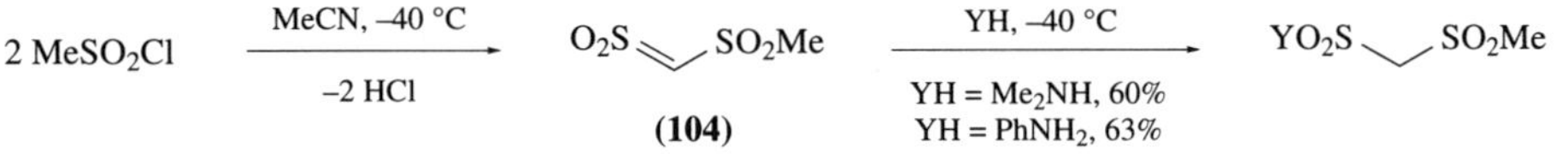

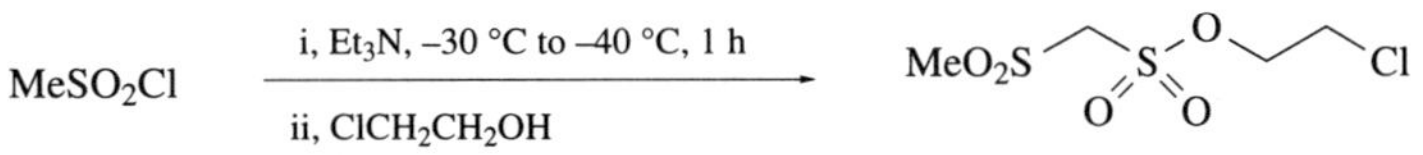

4.06.2.1 Dicoordinated Sulfur Derivatives

For the preparation of thioselenoacetals, four major ionic pathways are theoretically possible (Scheme 48). There is the choice of coupling a selenenyl anion or cation onto a sulfur derivative, or inversely coupling a sulfenyl anion or cation onto a selenium derivative. All four pathways have been attempted successfully.

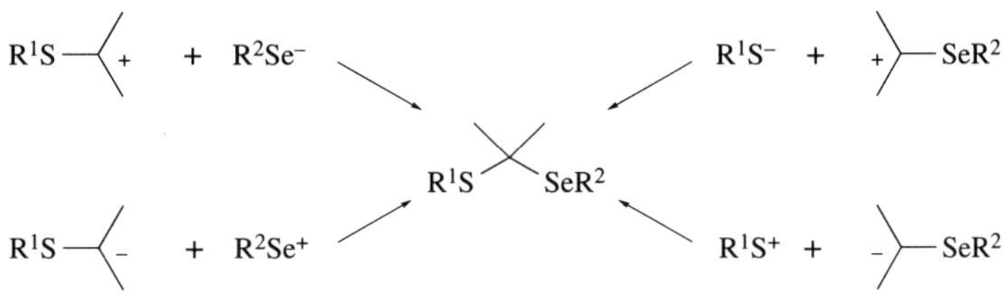

Scheme 48

For the first route ($R^1SC^+ + R^2Se^-$) it is necessary to find a suitable synthetic equivalent for an alkylthiocarbocation. This can be accomplished by the reaction of N-chlorosuccinimide with a sulfide. The resulting chlorinated compound can then be treated with potassium selenophenoxide to give the desired thioselenoacetal (Equation (86)) ⟨75TL1613⟩. The chlorination is carried out in carbon tetrachloride and the substitution in DMF. Sodium and magnesium selenides have also been successfully applied to this reaction. For example, sodium methyl selenide can be generated from dimethyl diselenide and sodium metal in liquid ammonia (Equation (87)) ⟨87JCS(D)757⟩. Reaction of this nucleophile with 1-chloro-2-thiapropane gave 2-thia-4-selenapentane in 64% overall yield.

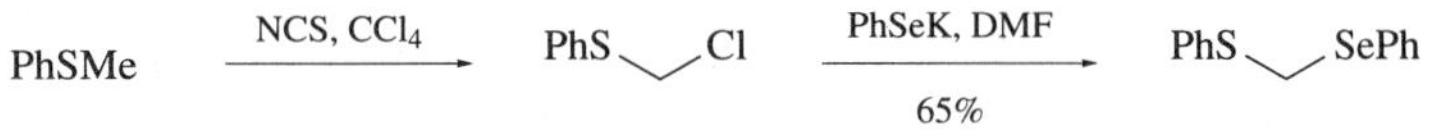

MeSeSeMe $\xrightarrow{\text{Na, NH}_3}$ MeSeNa $\xrightarrow[\text{64\% overall}]{\text{ClCH}_2\text{SMe}}$ MeSe⌒SMe　　(87)

Ferreira *et al.* have used the reaction of a sulfide anion with the equivalent of an α-selenated carbocation ($R^1S^- + R^2SeC^+$). They initially prepared sodium thiophenoxide by treating diphenyl disulfide with sodium hydroxide under phase-transfer conditions (Equation (88)) ⟨82SC595⟩. The resulting sodium salt was then treated with an α-bromoselenide. The yield of the thioselenoacetal was in the region of 80%. A second example of this method is given in Equation (89). This time, however, the sodium salt was obtained by the action of sodium hydroxide in DMF and the selenide electrophile was an iodo compound ⟨91JA9864⟩.

PhSSPh $\xrightarrow[\substack{\text{PhSeCH}_2\text{Br, 2 h, reflux} \\ 81\%}]{\substack{50\% \text{ NaOH, THF} \\ \text{phase-transfer conditions}}}$ PhS⌒SePh　　(88)

(89)

The treatment of α-seleno carbanions with sulfur nucleophiles ($R^2SeC^- + R^1S^+$) has been attempted by Seebach *et al.* This group obtained the carbanions by butyllithium cleavage of diselenoacetals in THF at −78 °C (Equation (90)) ⟨77LA846⟩. Reaction of the anions with benzenesulfenyl chlorides gave the required thioselenoacetals in good yields. In an analogous reaction, Krief *et al.* obtained 2-phenylthio-2-phenylselenopropane in 72% yield ⟨89T2023⟩. Overall, this synthetic method represents the transformation of a diselenoacetal (see Section 4.06.3.1) into a thioselenoacetal. The last route employs a carbanion adjacent to a sulfur with a selenated electrophile ($R^1SC^- + R^2Se^+$). An example is shown in Equation (91) ⟨92JA3910⟩. The sulfide (**105**) is treated with a base to give, due to the two acidic hydrogens adjacent to the lactone carbonyl and phenylthio groups, a dianion. Subsequent treatment with benzene selenenyl chloride gives a di(phenylseleno) product.

(90)

(91)

(**105**)

Apart from the four ionic methods outlined in Scheme 48, it is also possible to prepare these compounds by free-radical reactions. Hence, when diphenyl diselenide is treated with NaBH$_4$ in dry ethanol in the presence of a perfluoroalkyl iodide, PhSe• radicals are generated. They result from a one-electron transfer between a phenylselenide anion and the perfluorinated iodide (Scheme 49). The two radicals, formed this way, may then add to alkenes. One example is the addition to an enethiol ether leading to the formation of a fluorinated thioselenoacetal ⟨91TL375⟩. Curran and Thoma have studied the radical addition of 2-methyl-2-phenylselenomalonitrile to alkenes (Equation (92), where AIBN represents 2,2′-azobisisobutyronitrile) ⟨92JA4436⟩. Two examples of enethiol ethers were described, the resulting β,β-dicyano thioselenoacetals being formed in good yields. However, too few examples of these radical reactions have been reported to confirm whether they represent a general, preparative method for thioselenoacetals.

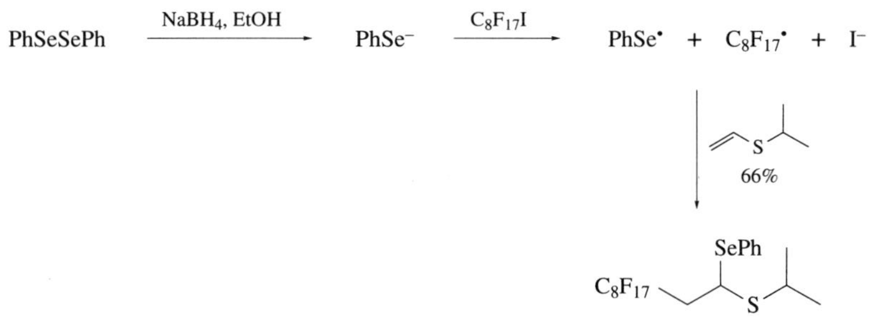

Scheme 49

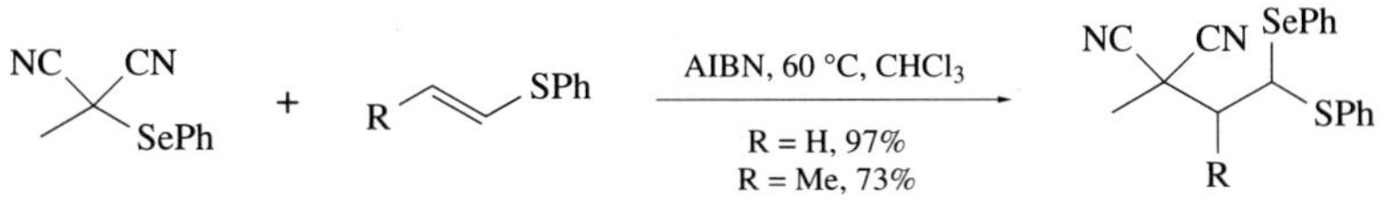

$$PhSeSePh \xrightarrow{NaBH_4, EtOH} PhSe^- \xrightarrow{C_8F_{17}I} PhSe^\bullet + C_8F_{17}^\bullet + I^-$$

(leading, 66%, to C$_8$F$_{17}$CH$_2$CH(SePh)(S-iPr))

$$\text{(NC)}_2\text{C(SePh)(Me)} + R\text{-CH=CH-SPh} \xrightarrow[\text{R = H, 97%; R = Me, 73%}]{\text{AIBN, 60 °C, CHCl}_3} \text{(NC)}_2\text{C(Me)-CH(R)-CH(SPh)(SePh)} \quad (92)$$

AIBN = 2,2'-azobisisobutyronitrile

Finally, like dithioacetals, thioselenoacetals can be deprotonated with LDA. Alkylation of the resulting lithiated anions gives new thioselenoacetal derivatives (Equation (93)) ⟨90JA5609⟩. To the authors' knowledge, no thiotelluroacetals have been reported.

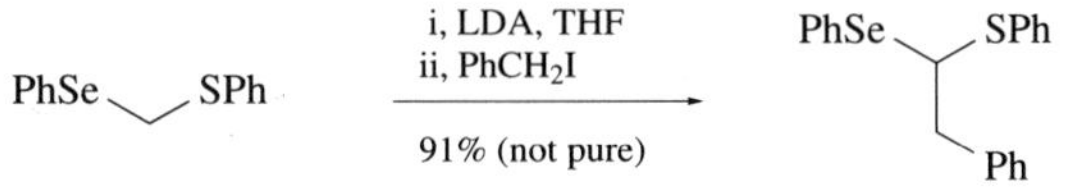

$$\text{PhSe-CH}_2\text{-SPh} \xrightarrow[\text{91% (not pure)}]{\text{i, LDA, THF; ii, PhCH}_2\text{I}} \text{PhSe-CH(CH}_2\text{Ph)-SPh} \quad (93)$$

4.06.2.2 Tricoordinated Sulfur Derivatives

Practically all the α-seleno sulfoxides reported (and there are not very many) have been prepared by the reaction of an electrophilic selenol derivative (R^{1}Se$^+$) with an α-sulfoxide carbanion (R^{2}S(O)C$^-$). In this way, Renaud obtained 2-phenylselenothiopyran S-oxides in yields of 60–75% (Equation (94)) ⟨91HCA1305⟩. Similarly Koizumi and co-workers prepared the α-phenylseleno sulfoxide (**106**) from an optically pure sulfoxide (Equation (95)) ⟨85TL6205⟩. The seleno sulfoxide (**106**) was subsequently deprotonated and treated with CO$_2$, to give the acid (**107**) in 82% yield.

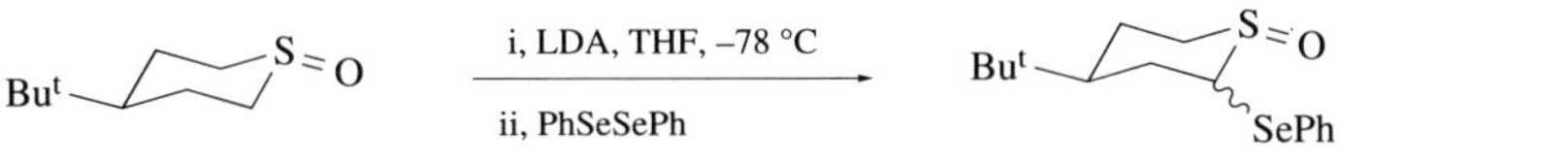

$$\text{Bu}^t\text{-thiane-S=O} \xrightarrow[\text{ii, PhSeSePh}]{\text{i, LDA, THF, }-78\text{ °C}} \text{Bu}^t\text{-thiane(SePh)-S=O} \quad (94)$$

$$p\text{-Tol-S(O)-Et} \xrightarrow[\text{50%}]{\text{i, Bu}^n\text{Li, }-78\text{ °C; ii, PhSeCl}} p\text{-Tol-S(O)-CH(Me)(SePh)} \xrightarrow{\text{i, Bu}^n\text{Li, }-78\text{ °C; ii, CO}_2\text{; iii, H}_3\text{O}^+} p\text{-Tol-S(O)-C(Me)(SePh)(CO}_2\text{H)} \quad (95)$$

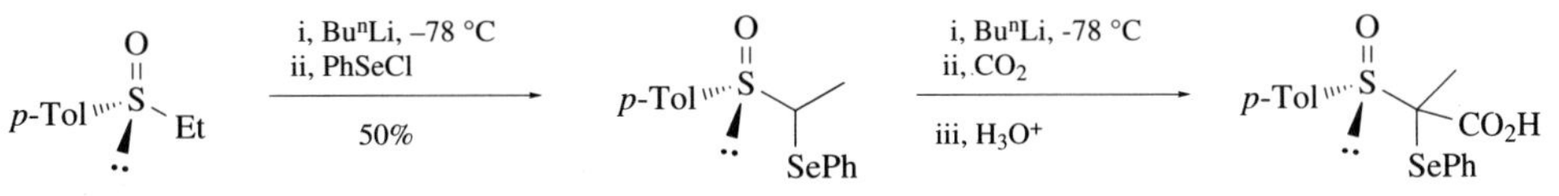

(**106**) (**107**)

Equation (96) outlines a third example of an α-seleno sulfoxide synthesis by selenylation of a carbanion ⟨84JCS(P1)21⟩. The intermediate compound (**108**) was not isolated but *in situ* oxidation led via PhSeOH elimination to an ethylene derivative. Another example of an ethylene derivative prepared via a selenoxide-sulfoxide is given in Equation (97) ⟨75JA5434⟩. The isolation of these *gem*-selenoxide sulfoxides, however, remains unreported.

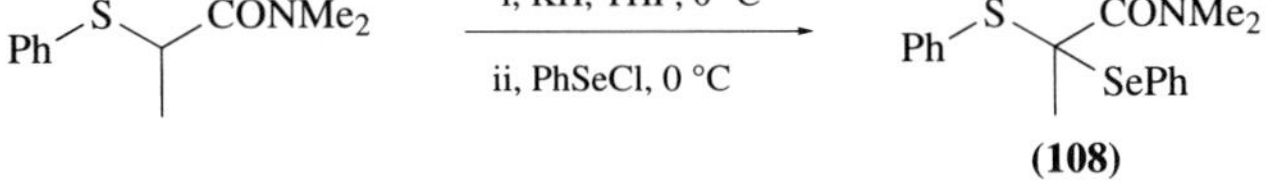

$$\text{Ph-S(O)-CH(Me)-CONMe}_2 \xrightarrow[\text{ii, PhSeCl, 0 °C}]{\text{i, KH, THF, 0 °C}} \text{Ph-S(O)-C(Me)(SePh)-CONMe}_2 \quad (96)$$

(**108**)

4.06.2.3 Tetracoordinated Sulfur Derivatives

As for α-seleno sulfoxides, the most convenient preparative method for the corresponding sulfone compounds comprises the treatment of an α-sulfone carbanion (RSO_2C^-) by a selenyl electrophile (RSe^+). Four examples of this approach are outlined in Scheme 50 ⟨80JOC1486, 81T2547, 89T7161, 91HCA1305⟩. The sulfones are generally deprotonated by an organolithium reagent in THF at −78 °C. Sodium hydride has been employed for the removal of more acidic protons. Diphenyl diselenide is the electrophile which has been most frequently used. The inverse strategy of using a selenyl anion (RSe^-) and an α-halogenated sulfone has also been investigated. It was by this route that Simpkins prepared phenylselenomethyl phenyl sulfone (Scheme 51) ⟨91T323⟩. He used sodium phenyl selenide, prepared by sonification of sodium metal and diphenyl diselenide, with an α-brominated sulfone in THF. A variation of this protocol where the sodium is replaced by Na_2Te has also been reported ⟨89BCJ1358⟩.

α-Seleno sulfones can, of course, be deprotonated by bases—LDA, for example—and subsequently alkylated to give new α-seleno sulfones ⟨91T323⟩. Equation (98) presents the results obtained by Simpkins ⟨88TL6787⟩. The reaction of ketene thioselenoacetal S,S-dioxides with enamines is another route to seleno sulfones ⟨87JOC4943⟩.

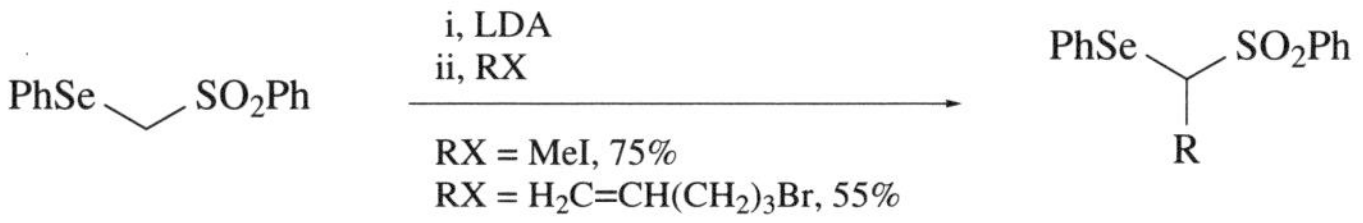

$$\text{PhSe}\diagdown\text{SO}_2\text{Ph} \xrightarrow[\substack{RX = \text{MeI}, 75\% \\ RX = \text{H}_2\text{C=CH(CH}_2)_3\text{Br}, 55\%}]{\substack{i, \text{LDA} \\ ii, RX}} \text{PhSe}\diagdown\underset{R}{\text{C}}\diagup\text{SO}_2\text{Ph} \quad (98)$$

This method, reported by Clive *et al.*, constitutes an interesting synthesis of monoprotected [1,4]-diketones, of which the exposed carbonyl group has been treated with an acetylenic anion (Scheme 52). The addition of PhSeCl to vinyl sulfones also leads to seleno sulfones ⟨85T2527⟩. Further, a cyclobutane containing both an α-seleno sulfone function and an azide group was obtained in about 50% yield by the reaction of PhSeN₃ with a bicyclic sulfone (Equation (99)), where NMP represents *N*-methyl-2-pyrrolidone) ⟨89T2819⟩.

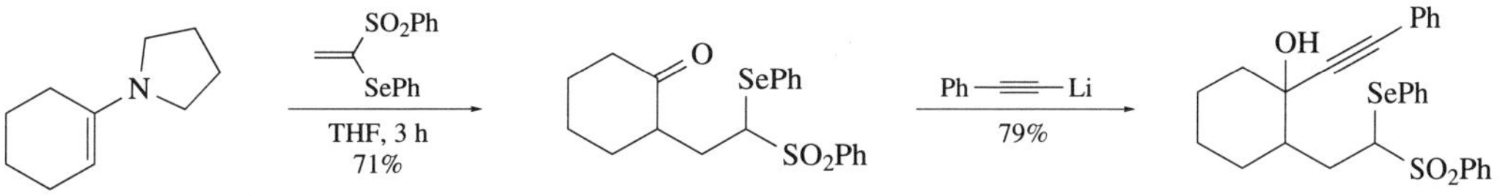

Scheme 52

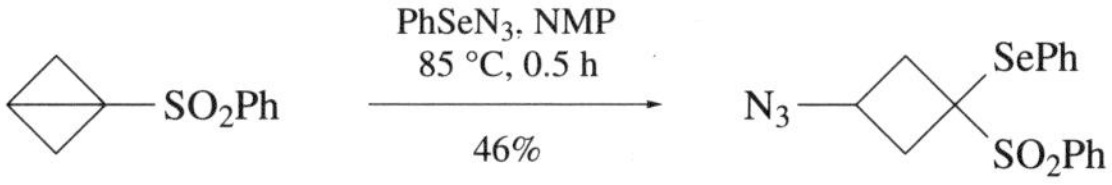

$$\text{Equation (99): bicyclic sulfone} \xrightarrow[46\%]{\substack{\text{PhSeN}_3, \text{NMP} \\ 85\,°\text{C}, 0.5\,\text{h}}} \text{N}_3\text{-cyclobutane-SePh, SO}_2\text{Ph} \quad (99)$$

NMP = *N*-methyl-2-pyrrolidone

Finally, the Diels–Alder reaction of certain selenothioester *S,S*-dioxides, trapped *in situ*, leads to the formation of α-seleno sulfones, in which the selenium atom is a part of a ring. The high instability, however, of selenocarbonyl intermediates may well limit the synthetic utility of this method (Scheme 53) ⟨88JA8671, 88JA8679⟩.

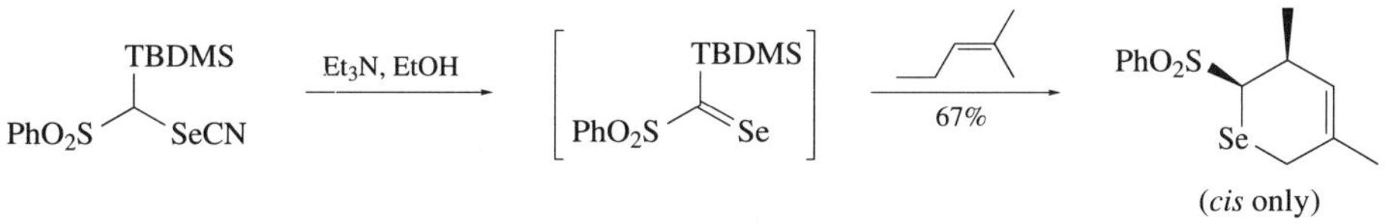

Scheme 53

4.06.3 FUNCTIONS CONTAINING SELENIUM AND/OR TELLURIUM—$R^1{}_2C(SeR^2)_2$, $R^1{}_2C(SeR^2)\,TeR^3$, etc.

4.06.3.1 Diselenium Derivatives

Although diselenoacetals have been known for a long time ⟨26JA520⟩, their chemistry has really only become significant over the last 20 years, particularly through the work of Krief's group. As for dithioacetals, the development of efficient synthetic routes to these diselenated compounds only became important once their utility as umpolung reagents had been established. The parallels do not stop there, since the majority of diselenoacetal syntheses are the direct transpositions to selenium of known sulfur chemistry. Not surprisingly, therefore, the most general preparative method for diselenoacetals is an acid-catalyzed condensation of selenols with aldehydes or ketones. For those planning laboratory preparations, a comprehensive paper by Krief and co-workers which describes the various possible methods, giving numerous experimental details, is highly recommended ⟨85T4793⟩.

Hydrogen chloride gas was the first catalyst employed ⟨26JA520⟩. The selenol and the carbonyl compound were mixed together without a solvent and the acid was bubbled through the mixture at 0 °C ⟨75AG(E)700, 85T4793⟩. Bis(benzylseleno) sugars have been obtained in a similar way ⟨77CAR351⟩. From Equation (100) it can be seen that the results are generally good, excepting the condensation of selenophenol with aldehydes where the yields were modest (34–40%), owing to the competitive formation of a chloro selenide (Equation (101)). This by-product may in some cases be the major product ⟨77AG(E)541⟩. To avoid this problem, other catalysts, including sulfuric acid, have been proposed ⟨77AG(E)540⟩. However, the oxidizing nature of sulfuric acid sometimes leads to the formation of diselenides via selenol oxidation. Although of little significance with PhSeH, it becomes the major reaction with MeSeH.

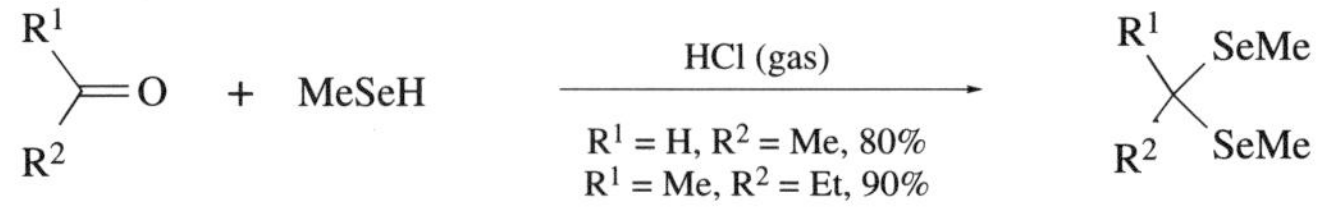

$$\underset{R^2}{\overset{R^1}{>}}=O + MeSeH \xrightarrow[R^1 = Me, R^2 = Et, 90\%]{HCl\ (gas)} \underset{R^2}{\overset{R^1}{>}}\!\!\!<\!\!\!\overset{SeMe}{\underset{SeMe}{}} \quad (100)$$

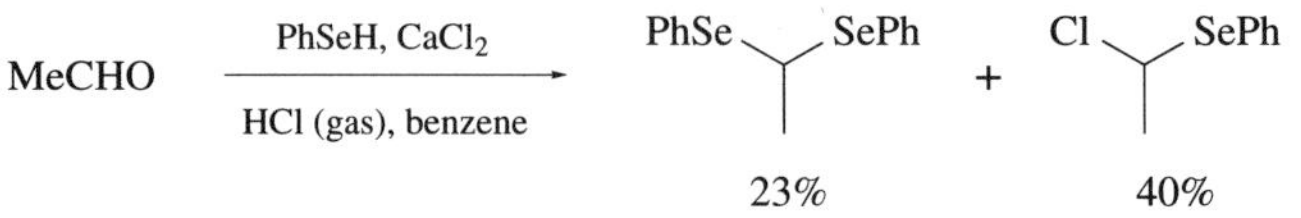

$$MeCHO \xrightarrow[HCl\ (gas),\ benzene]{PhSeH,\ CaCl_2} PhSe\!\!\!\diagdown\!\!\!SePh + Cl\!\!\!\diagdown\!\!\!SePh \quad (101)$$
$$\qquad\qquad\qquad\qquad\qquad\quad 23\% \qquad\quad 40\%$$

Amongst the Lewis acids which have been studied, zinc chloride gives the best results and Krief recommends its use in the majority of cases (Equation (102)) ⟨77AG(E)540⟩. The few limitations which exist concern its use with sterically hindered ketones; however, it still remains the most general catalyst ⟨79S877, 86TL1723⟩. Tin tetrachloride has also been used ⟨92LA643⟩.

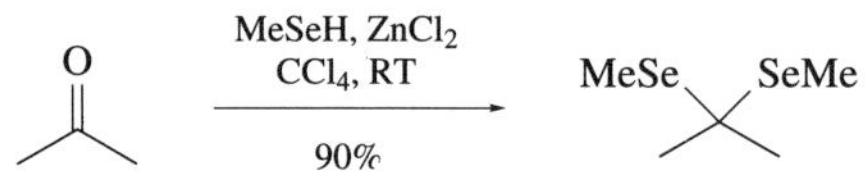

$$\underset{}{\overset{O}{\bigtriangleup}} \xrightarrow[CCl_4,\ RT]{MeSeH,\ ZnCl_2} MeSe\!\!\!\diagdown\!\!\!\diagup\!\!\!SeMe \quad (102)$$
$$\qquad\qquad 90\%$$

As for dithioacetals, it is of course possible to replace the selenols by their boron or silicon derivatives. In such reactions, the use of an acid catalyst is sometimes not necessary. The selenoboranes are generally more efficient than the corresponding selenosilanes ⟨85T4793⟩. Several examples are depicted in Scheme 54 ⟨79JOC1883, 79JOC4279, 85T4793, 91TL105, 92TL269⟩. As shown, the substrates can be aldehydes, ketones, or acetals and the presence of carbon–carbon double or triple bonds is unimportant. Alternatively, the use of selenide salts (RSeLi, RSeNa, etc.) with *gem*-dihalides also leads to the formation of diselenoacetals ⟨84JCS(P2)429, 84S439, 91CB1315⟩. This method has been applied to the preparation of a bis(seleno) ester (Equation (103)) ⟨76CL203⟩. The reaction of a potassium selenocarboxylate with dichloromethane gave the desired product (**109**) in 98% yield. It is of course possible to start from a *gem*-dihalide analogue, wherein one of the halogen atoms is replaced by a selenyl substituent. For example, the diselenoacetal (**110**) was prepared by the reaction of sodium benzylselenide with the bromo selenide (**111**) (Equation (104)) ⟨86JOC2981⟩.

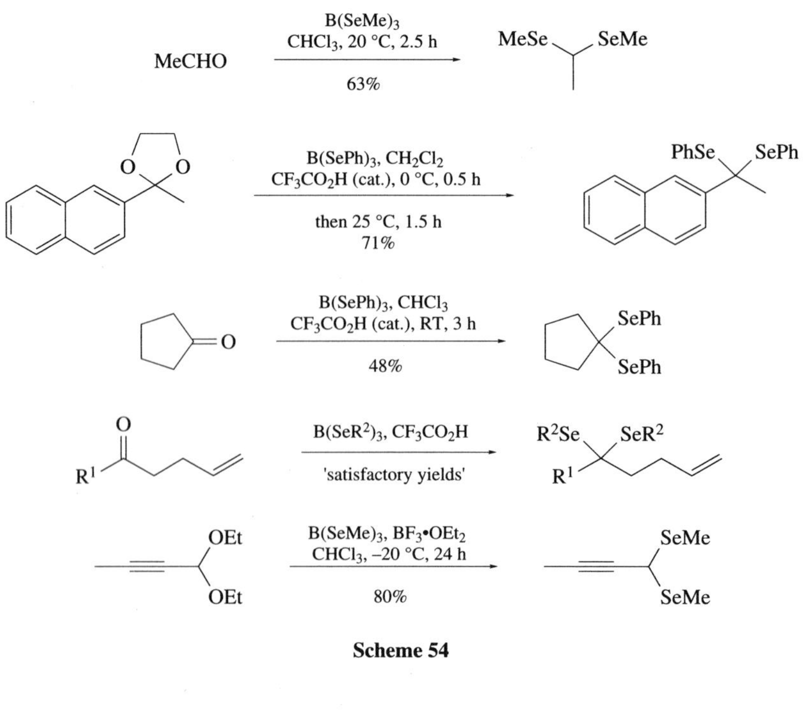

$$MeCHO \xrightarrow[63\%]{B(SeMe)_3,\ CHCl_3,\ 20\,°C,\ 2.5\ h} MeSe\!\!\!\diagdown\!\!\!SeMe$$

Naphthyl-C(Me)(O-CH2-CH2-O) $\xrightarrow[then\ 25\,°C,\ 1.5\ h,\ 71\%]{B(SePh)_3,\ CH_2Cl_2,\ CF_3CO_2H\ (cat.),\ 0\,°C,\ 0.5\ h}$ Naphthyl-C(Me)(SePh)(SePh)

cyclopentanone $\xrightarrow[48\%]{B(SePh)_3,\ CHCl_3,\ CF_3CO_2H\ (cat.),\ RT,\ 3\ h}$ 1,1-bis(SePh)cyclopentane

$R^1\text{-CO-CH}_2\text{CH}_2\text{-CH=CH}_2 \xrightarrow[\text{'satisfactory yields'}]{B(SeR^2)_3,\ CF_3CO_2H} R^1\text{-C}(SeR^2)_2\text{-CH}_2\text{CH}_2\text{-CH=CH}_2$

HC≡C-CH(OEt)(OEt) $\xrightarrow[80\%]{B(SeMe)_3,\ BF_3·OEt_2,\ CHCl_3,\ -20\,°C,\ 24\ h}$ HC≡C-CH(SeMe)(SeMe)

Scheme 54

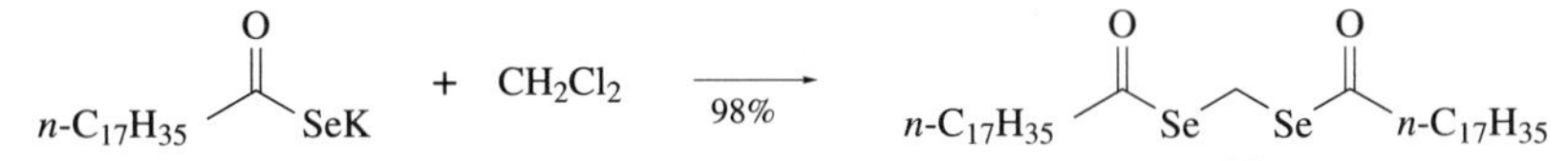

$$n\text{-}C_{17}H_{35}\text{-C(O)-SeK} + CH_2Cl_2 \xrightarrow{98\%} n\text{-}C_{17}H_{35}\text{-C(O)-Se-CH}_2\text{-Se-C(O)-}n\text{-}C_{17}H_{35} \quad (103)$$

(109)

$$Ph\text{-CH}_2\text{-Se-CH}_2\text{-Br} \xrightarrow{PhCH_2Se^-} Ph\text{-CH}_2\text{-Se-CH}_2\text{-SePh} \quad (104)$$

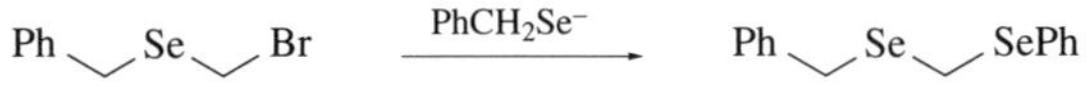

(111) → **(110)**

Gabriel and Seebach have prepared a series of ^{13}C-labelled diselenoacetals ⟨84HCA1070⟩. They used a variety of methods, both those described above and another method which takes advantage of the easy cleavage of the carbon–selenium bond by alkyllithium reagents (Scheme 55). Upon treatment of triphenyl(triseleno)*ortho*-formate with butyllithium, one of the C—Se bonds is cleaved and lithiated bis(phenylseleno)methane is formed. Subsequent protonation with water gave the corresponding diselenoacetal in 70% yield. Given that (triseleno)orthoformates are relatively easy to prepare (see Chapter 6.03.4), this method can be considered as a convenient general preparation of diselenoacetals. A second method based on (triseleno)orthoformate chemistry proceeds by their reaction with silyl enol ethers in the presence of a Lewis acid catalyst (Equation (105)). The products obtained are the synthetic equivalents of monoprotected 1,3-dicarbonyl compounds ⟨85TL6513⟩.

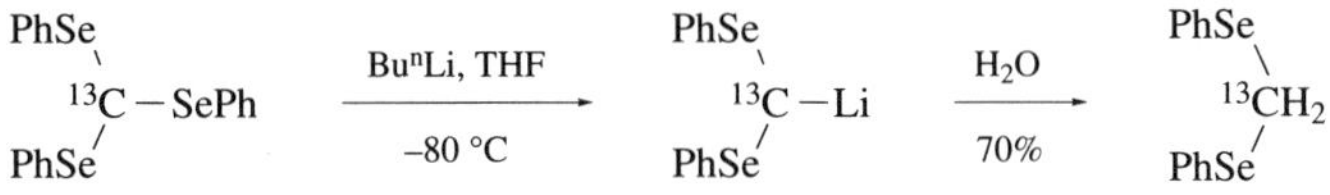

$$(PhSe)_2{}^{13}C\text{-SePh} \xrightarrow[-80\,^\circ C]{Bu^nLi,\,THF} (PhSe)_2{}^{13}C\text{-Li} \xrightarrow[70\%]{H_2O} (PhSe)_2{}^{13}CH_2$$

Scheme 55

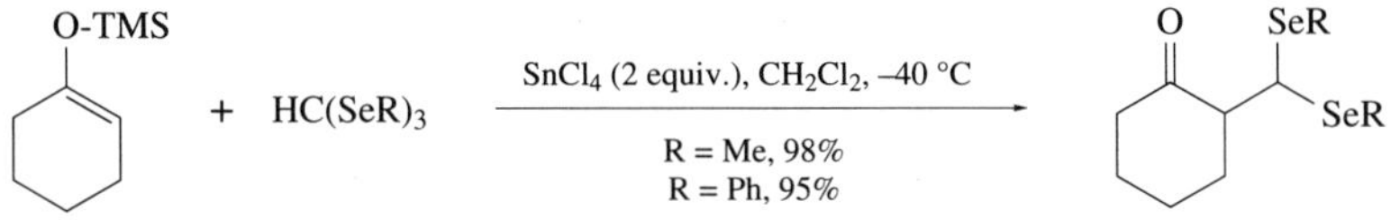

Cyclohexenyl O-TMS + HC(SeR)$_3$ $\xrightarrow[R = Me,\,98\%;\,R = Ph,\,95\%]{SnCl_4\,(2\,equiv.),\,CH_2Cl_2,\,-40\,^\circ C}$ 2-(CH(SeR)$_2$)cyclohexanone $\quad (105)$

Like dithioacetals, diselenoacetals are deprotonated upon treatment with certain bases, for example LDA, and the resulting anions react with a range of electrophiles. A variety of substituted diselenoacetals have been prepared this way (Scheme 56) ⟨69AG(E)450, 78JOC3794, 78TL3971, 81JOC2775⟩. Of particular interest is the formation of bis(phenylseleno)cyclopropane, the diselenoacetal of cyclopropanone, formed by an intramolecular version of the reaction. It should be noted that for those compounds in which the carbon to be deprotonated only possesses one hydrogen (a methine carbon), the base of choice is potassium diisopropylamide (KDA) ⟨78JOC3794⟩.

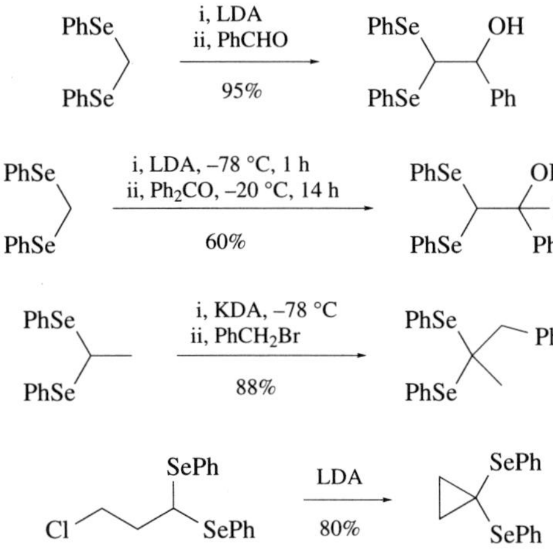

CH$_2$(SePh)$_2$ $\xrightarrow[95\%]{i,\,LDA;\,ii,\,PhCHO}$ (PhSe)$_2$CH-CH(OH)Ph

CH$_2$(SePh)$_2$ $\xrightarrow[60\%]{i,\,LDA,\,-78\,^\circ C,\,1\,h;\,ii,\,Ph_2CO,\,-20\,^\circ C,\,14\,h}$ (PhSe)$_2$CH-C(OH)Ph$_2$

(CH$_3$)CH(SePh)$_2$ $\xrightarrow[88\%]{i,\,KDA,\,-78\,^\circ C;\,ii,\,PhCH_2Br}$ (PhSe)$_2$C(CH$_3$)(CH$_2$Ph)

Cl-CH$_2$CH$_2$-CH(SePh)$_2$ $\xrightarrow[80\%]{LDA}$ cyclopropane-C(SePh)$_2$

Scheme 56

Up to 1994, selenocarbonyl compounds have rarely been used for the preparation of diselenoacetals. However, certain results of Krief *et al.* ⟨92SL638⟩ concerning the possible addition of selenide anions to selenoaldehydes, and of Kirby and Trethewey concerning the formation of certain products during the Diels–Alder reaction of transient diseleno esters ⟨88JCS(P1)1913⟩, suggest that this should soon be possible.

Finally, the insertion of alkylidene groups, generated from diazo compounds, into the selenium–selenium bond of diselenides is a well-established route to diselenides ⟨70CB2271, 85JOM(286)171⟩.

4.06.3.2 Ditellurium Derivatives

The methods generally employed for the preparation of ditellurium derivatives are often similar to those used for the preparation of diselenoacetals. For example, the reaction of diazomethane with a ditelluride leads to the insertion of a methylene moiety between the two tellurium atoms. Petragnani and Schill have reported two examples of this reaction, both of which were effected in quantitative yield ⟨70CB2271⟩. Also, Torres has prepared a series of nine different ditelluroacetals with yields of 74–100% (Equation (106)) ⟨90JOM(381)69⟩. The reactions are carried out in benzene at 0 °C.

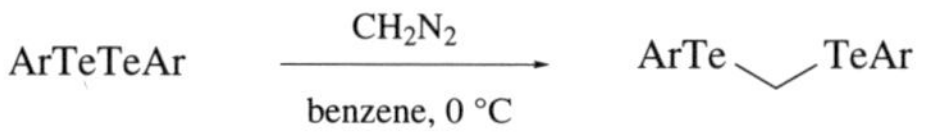
ArTeTeAr
CH_2N_2
benzene, 0 °C
ArTe
TeAr

(106)

An alternative method for the preparation of ditelluroacetals is by the reaction of two equivalents of telluride anion with a *gem*-dihalide. The reaction between lithium phenyl telluride and diiodomethane gives di(phenyltelluro)methane, although the yield is very low (6%) ⟨75CB314⟩. This method has been applied to the synthesis of a compound containing two ferrocene groups (Scheme 57) ⟨87JOM(336)153⟩. The precursor, a lithium telluride, obtained by lithium triethylborohydride reduction of diferrocenyl ditelluride, was treated with diiodomethane in THF. The expected ditelluroacetal was obtained in 12% yield. Although offering an alternative route to these compounds, the poor yields obtained by this method limit its synthetic utility relative to the diazomethane approach.

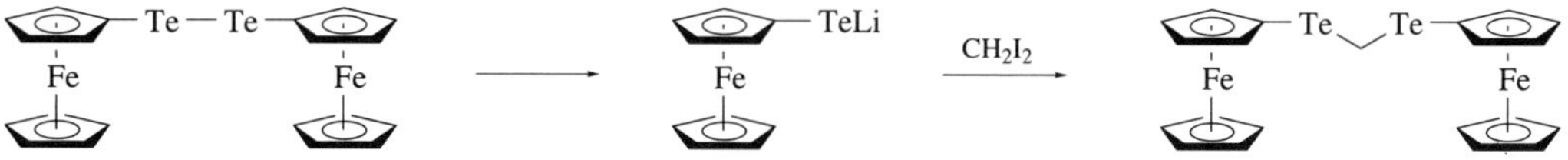
Te — Te
Fe
Fe
TeLi
Fe
CH_2I_2
Te
Te
Fe
Fe

Scheme 57

Nevertheless, the method has been employed for the synthesis of benzo-1,3-ditellurole ⟨88KGS1144⟩. Another means of preparing this heterocycle is by zinc reduction of the tetrachloride derivative (**112**) (Scheme 58), which in turn was prepared from the hexachloride (**113**) and *o*-(bistrimethylsilyl)benzene (**114**). The overall yield of benzoditellurole starting from (**114**) was 19.4% ⟨91HAC307⟩. Other ditelluroles have been prepared from acetylene precursors ⟨82TL1531⟩.

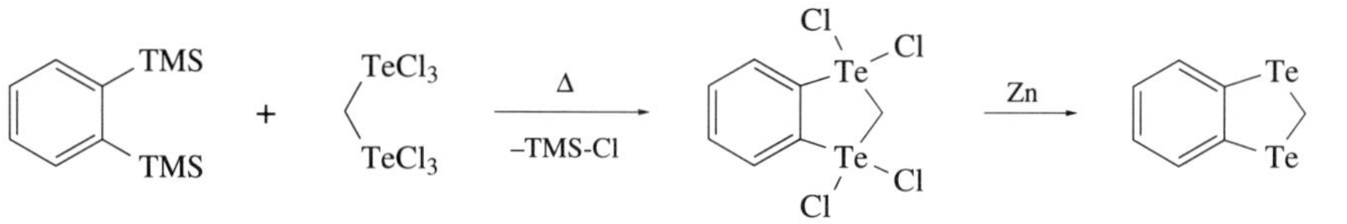
TMS
TMS
+
$TeCl_3$
$TeCl_3$
Δ
–TMS-Cl
Cl
Cl
Te
Te
Cl
Cl
Zn
Te
Te

(**114**) (**113**) (**112**)

Scheme 58

Like dithioacetals, ditelluroacetals can be deprotonated and their corresponding anions treated with electrophiles to give new ditelluroacetals. For example, Seebach and Beck treated di(phenyltelluro)methane with LDA at −78 °C in THF, and alkylated the derived anion with benzyl bromide. The benzylated product was isolated in quantitative yield (Equation (107)) ⟨75CB314⟩. A similar study concerning the deprotonation and alkylation of 1,3-ditelluroles has also been published ⟨83TL237⟩.

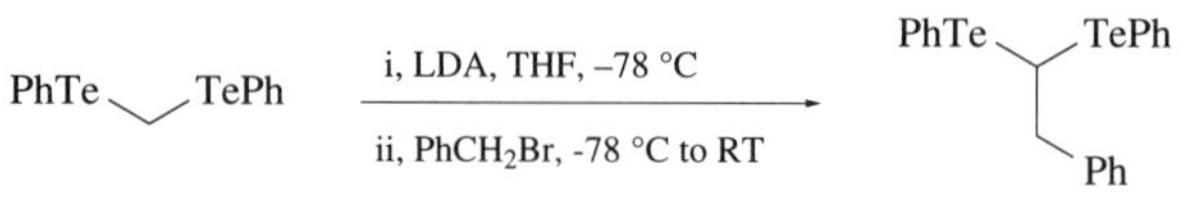
PhTe
TePh
i, LDA, THF, –78 °C
ii, $PhCH_2Br$, -78 °C to RT
PhTe
TePh
Ph

(107)

A previously described precursor of benzo-1,3-ditellurole was a ditelluroacetal derivative possessing two chlorines on each of the tellurium atoms (compound (**112**), Scheme 58). This class of compound is easily prepared by the reaction of chlorine with the corresponding ditelluroacetals ⟨86OM805, 87OM2164⟩. The reactions are effected in carbon tetrachloride and the yields are close to quantitative (Scheme 59). Likewise, the bromination of benzotellurole leads quantitatively to the tetrabrominated derivative (Equation (108)) ⟨88KGS1144⟩.

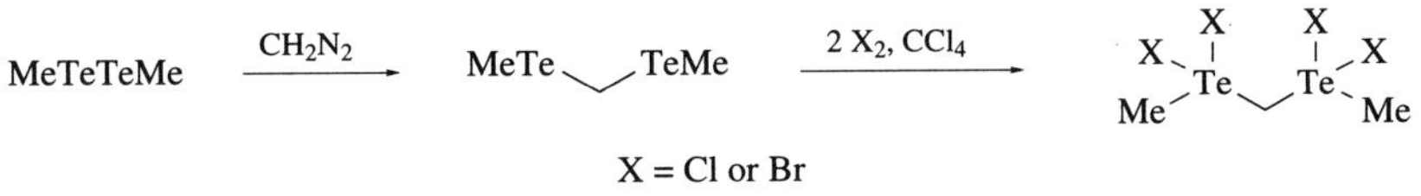

$$\text{MeTeTeMe} \xrightarrow{\text{CH}_2\text{N}_2} \text{MeTe}\frown\text{TeMe} \xrightarrow{2\,X_2,\,\text{CCl}_4} \underset{\text{Me}}{\overset{X}{\underset{|}{\text{Te}}}}\frown\underset{\text{Me}}{\overset{X}{\underset{|}{\text{Te}}}}\!\!\!\!\!\!\!\!\!\!\!\!\!\!\!\!\!\!\!X\;X$$

X = Cl or Br

Scheme 59

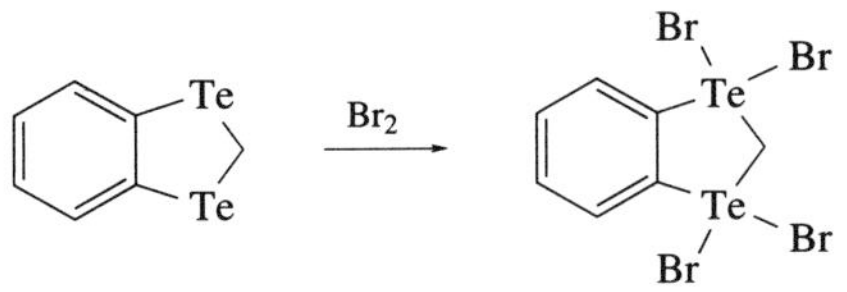

$$\text{benzo-1,3-ditellurole} \xrightarrow{\text{Br}_2} \text{tetrabromo derivative} \tag{108}$$

Another derivative is the hexachlorinated compound (**113**) which is readily obtained from tellurium tetrachloride and acetic anhydride (Equation (109)) ⟨85JA675⟩.

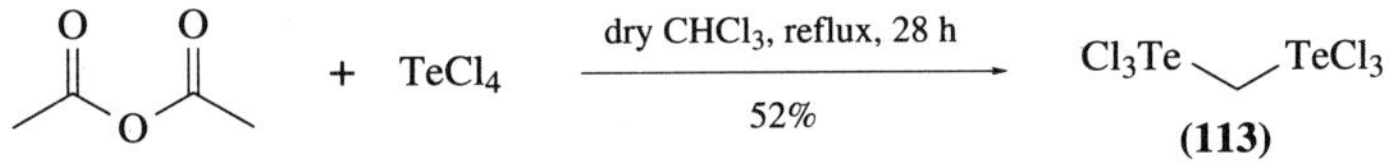

$$(\text{AcO})_2\text{O} + \text{TeCl}_4 \xrightarrow[52\%]{\text{dry CHCl}_3,\,\text{reflux, 28 h}} \text{Cl}_3\text{Te}\frown\text{TeCl}_3 \quad\text{(113)} \tag{109}$$

4.06.3.3 Other Derivatives

Selenotelluroacetals, which contain two different heavy heteroatoms, have been the subject of very little study. These componds are, however, conveniently obtained by the reaction of a telluride anion (ArTe⁻) with a selenated alkylating agent (RSeCH₂X). Scheme 60 depicts a method proposed by Brandt *et al.* ⟨83MI 406-01⟩. A bromomethyl selenide is prepared from diazomethane and a selenenyl bromide; in addition sodium aryl tellurides are obtained by sodium borohydride reduction of the corresponding ditellurides. Subsequent reactions of these two precursors lead to the desired selenotelluroacetals in isolated yields of 62–98%.

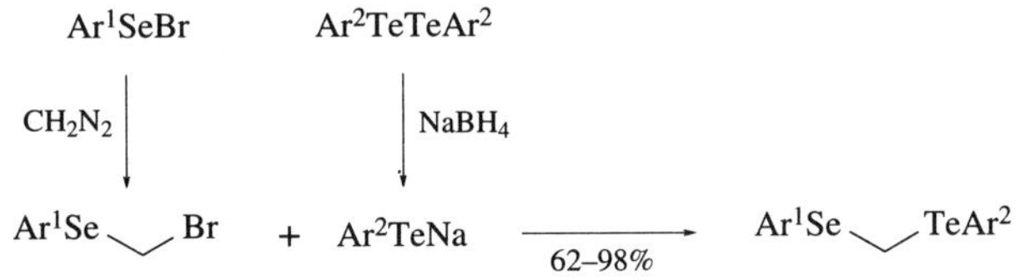

$$\text{Ar}^1\text{SeBr} \xrightarrow{\text{CH}_2\text{N}_2} \text{Ar}^1\text{Se}\frown\text{Br}\quad+\quad \text{Ar}^2\text{TeTeAr}^2 \xrightarrow{\text{NaBH}_4} \text{Ar}^2\text{TeNa} \xrightarrow{62-98\%} \text{Ar}^1\text{Se}\frown\text{TeAr}^2$$

Scheme 60

An alternative method for the preparation of telluride anions is by reaction of sodium hydroxide with diaryl ditellurides under phase-transfer conditions (Scheme 61) ⟨84JOM(277)261⟩. The phase-transfer agent used was a mixture of dialkyldimethylammonium chlorides.

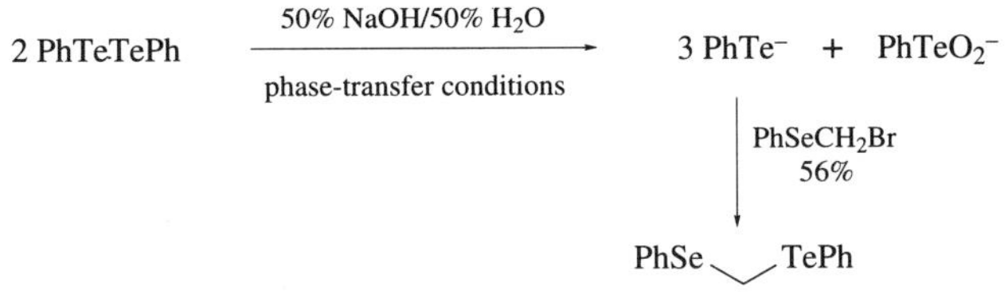

$$2\,\text{PhTeTePh} \xrightarrow[\text{phase-transfer conditions}]{50\%\,\text{NaOH}/50\%\,\text{H}_2\text{O}} 3\,\text{PhTe}^- + \text{PhTeO}_2^- \xrightarrow[56\%]{\text{PhSeCH}_2\text{Br}} \text{PhSe}\frown\text{TePh}$$

Scheme 61

4.07
Functions Incorporating a Chalcogen and a Group 15 Element

CHRISTOPHER D. GABBUTT and JOHN D. HEPWORTH
University of Central Lancashire, Preston, UK

4.07.1	FUNCTIONS CONTAINING A CHALCOGEN AND A NITROGEN FUNCTION	293
4.07.1.1	*Functions Bearing Oxygen and Nitrogen*	293
4.07.1.1.1	*Hemiaminals with tricoordinate nitrogen bearing alkyl, aryl or acyl substituents*	294
4.07.1.1.2	*Functions with tricoordinate nitrogen bearing heteroatom substituents*	311
4.07.1.1.3	*Functions with dicoordinate nitrogen*	313
4.07.1.2	*Functions Bearing Sulfur and Nitrogen*	315
4.07.1.2.1	*Dicoordinate sulfur derivatives*	315
4.07.1.2.2	*Tricoordinate sulfur derivatives*	329
4.07.1.2.3	*Tetra- and higher coordinate sulfur derivatives*	330
4.07.1.3	*Functions Bearing Selenium or Tellurium, Together with Nitrogen*	333
4.07.1.3.1	*From compounds containing a multiply bonded functional group*	333
4.07.1.3.2	*From compounds containing two singly bonded X—C—N groups (X = Li, Na)*	335
4.07.2	FUNCTIONS CONTAINING A CHALCOGEN AND PHOSPHORUS, ARSENIC, ANTIMONY OR BISMUTH	335
4.07.2.1	*Functions Bearing Oxygen*	336
4.07.2.1.1	*Oxygen and phosphorus*	336
4.07.2.1.2	*Oxygen and arsenic, antimony or bismuth*	344
4.07.2.2	*Functions Bearing Sulfur*	345
4.07.2.2.1	*Sulfur and phosphorus*	345
4.07.2.3	*Functions Bearing Selenium or Tellurium*	347
4.07.2.3.1	*Selenium or tellurium with phosphorus*	347
4.07.2.3.2	*Selenium or tellurium with arsenic, antimony or bismuth*	349

4.07.1 FUNCTIONS CONTAINING A CHALCOGEN AND A NITROGEN FUNCTION

4.07.1.1 Functions Bearing Oxygen and Nitrogen

Functional groups possessing an sp^3 hydridized carbon bonded to oxygen and to nitrogen, for example (**1**) and (**2**), are known as *O,N*-acetals or hemiaminals. There is a large number of possible variations of the substituents in (**1**) and (**2**). If the O—C—N function forms part of a ring system, then the potential for interesting structures becomes considerable. For example, either the oxygen or the nitrogen function may be incorporated into a ring to give a number of semicyclic *O,N*-acetals (Scheme 1). Alternatively, both oxygen and nitrogen may form part of a ring structure to generate a vast number of heterocyclic systems. Some of these possibilities are shown in Scheme 1. Even here there remains much scope for variation since the nitrogen function may be di- or tetracoordinate.

Scheme 1

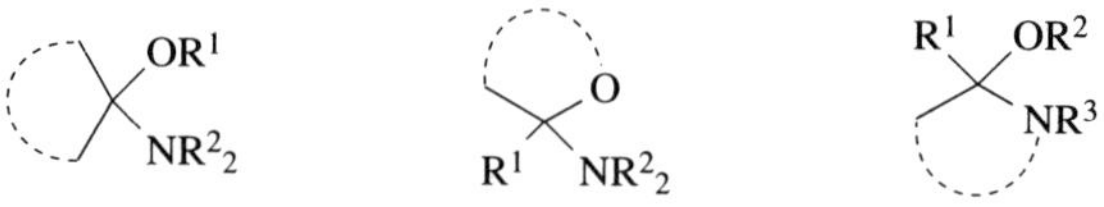

The literature concerning hemiaminals is both immense and widely scattered but fortunately the treatise by Rasshofer ⟨91HOU(E14a/2)1⟩ surveys synthetic routes to almost every permutation on the O—C—N theme and coverage of the heterocyclic examples is particularly impressive.

4.07.1.1.1 Hemiaminals with tricoordinate nitrogen bearing alkyl, aryl or acyl substituents

(i) From compounds containing a multiply bonded functional group

(a) From aldehydes and ketones. In principle, the simplest way to obtain the O—C—N unit is from the condensation of an amine with an aldehyde or ketone. However, the outcome of the reaction is influenced both by the nature of the carbonyl compound and amine and by the reaction conditions.

The behaviour of aliphatic aldehydes towards amines is outlined in Scheme 2. Condensation with a primary amine leads initially to the hemiaminal (**3**), which frequently dehydrates to the imine (**4**) which is tautomeric with the enamine (**5**). Aromatic aldehydes afford imines very readily, especially so with aromatic primary amines to give the Schiff bases ArCH=NAr. Secondary amines react analogously to give (**6**), although if an excess of the amine is employed, the iminium ion may be trapped as the aminal (*N,N*-acetal) (**7**). Aminal formation is particularly efficient from aldehydes which lack an α-CH group such as aromatic aldehydes. Ketones exhibit similar reaction pathways to those in Scheme 2, but unless particularly reactive, for example cyclohexanone, require more stringent conditions.

Mechanistic features of these reactions have been reviewed ⟨72CRV705⟩. Separate reviews also deal with enamine formation ⟨82T1975, B-88MI 407-01⟩. The synthesis of aminals from aldehydes and ketones has also been reviewed ⟨92HOU(E14a/3)545⟩.

It is necessary to operate under carefully controlled conditions in order to obtain *O,N*-acetals and to prevent the formation of by-products. The simplest compounds are the 1-amino-1-alkanols but the parent member of the series, $HOCH_2NH_2$, has not been obtained. The condensation reactions of aldehydes with ammonia or amines have a long history and it is only since the 1970s that many problems in this area have been resolved. The early work in this area has been reviewed ⟨40CRV297⟩. The reaction of aliphatic aldehydes with ammonia is often rationalized as a cyclocondensation involving 3 mol of reactants which generate hexahydro-1,3,5-triazines (**9**), the so-called 'aldehyde ammonias' (Scheme 3). The hemiaminals (**8**) are putative intermediates in this transformation. The first compound prepared was from acetaldehyde (**9**; R = Me) and is relatively stable as the trihydrate. It was later shown that the parent compound (**9**; R = H) can be obtained from formaldehyde ⟨48JA3659⟩, though at high temperatures the only compound obtained is hexamethylenetetraamine (**10**).

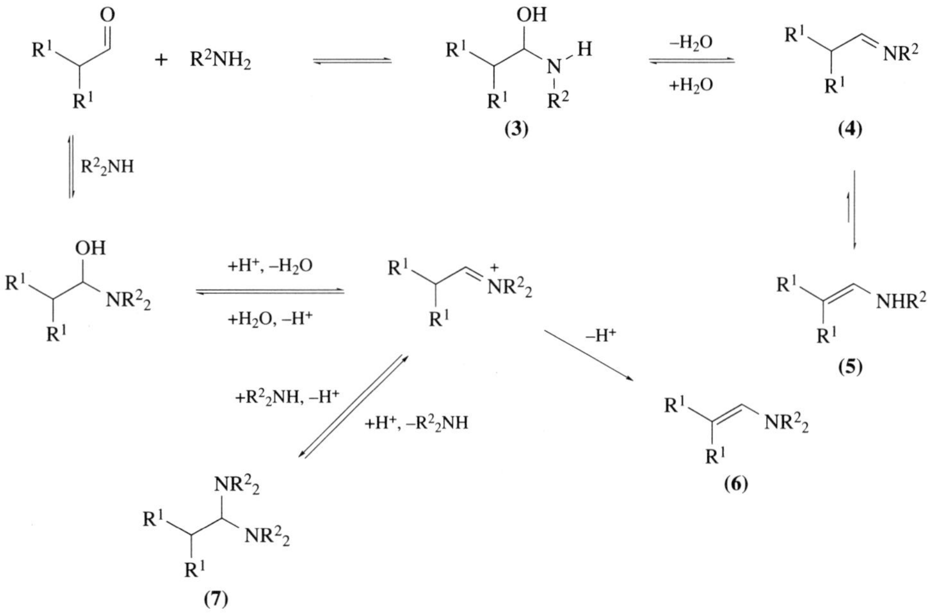

Scheme 2

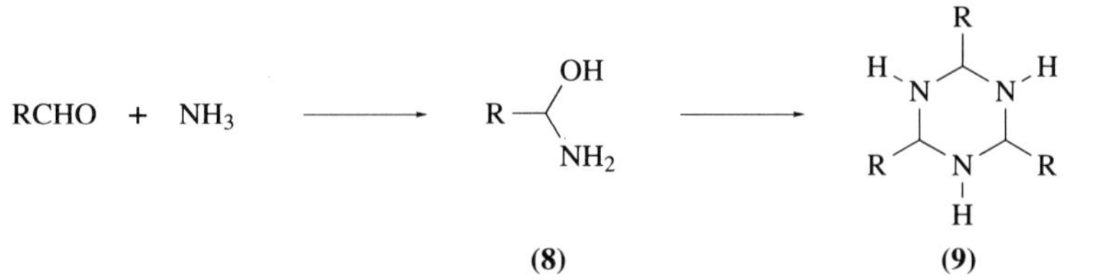

Scheme 3

The reactivity of ammonia towards aliphatic aldehydes has given rise to much confusion since the products have often been formulated as (**8**) with little supporting evidence. However, the situation has now been clarified by the detailed studies of Nielsen *et al.* ⟨73JOC3288⟩. This paper provides an excellent overview of much of the initial work. A wide range of *O*,*N*-acetals (**8**) can be obtained by treatment of aliphatic aldehydes with 15 M aqueous ammonia. The product generally precipitates from the reaction mixture within a few minutes as the trihydrate when R = Et, Pr, Bu^n, Bu^i or Bu^t or as the dihydrate when R = Me. Many of these hemiaminals are unstable, decomposition often being complete within a few hours at ambient temperature. Attempts to obtain *O*,*N*-acetals (**8**) from benzaldehyde or other aromatic aldehydes are unsuccessful, the only products being hydrobenzamides, (ArCH=N)$_2$CHAr ⟨40CRV297⟩.

Secondary amines react readily with formaldehyde. However the reaction does not usually proceed cleanly. For example, piperidine and formaldehyde condense at ice-bath temperature to give (**11**) as an unstable oil contaminated with ca. 12% of the aminal (**12**) ⟨56CB81⟩. Application of a similar protocol to aziridine gave the 1-hydroxymethyl compound (**13**) ⟨60DOK(135)853⟩.

In general, hemiaminals of the type HOCH$_2$NR$_2$ are unstable and decompose on exposure to water, though *O*-silylation confers some stability. Dimethylaminotrimethylsilyloxymethane, TMS-OCH$_2$NMe$_2$, has been obtained from the reaction of gaseous formaldehyde with TMS-NMe$_2$ ⟨81ZOB2382⟩. There is a marked increase in stability if the nitrogen function is incorporated in a

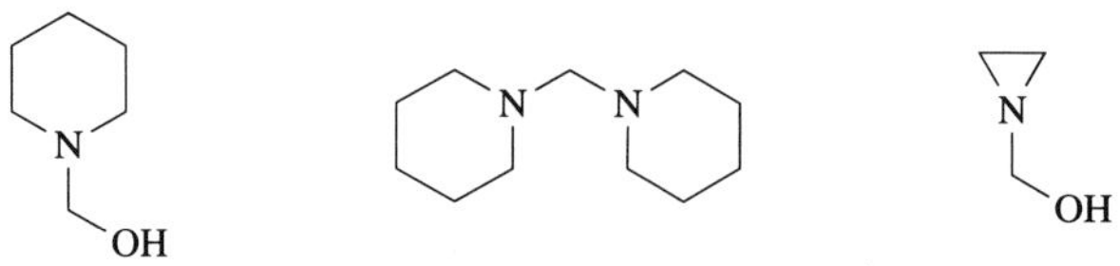

(11)   (12)   (13)

heteroaromatic ring and many of these compounds have been obtained. Their preparation is straightforward, involving the addition of aqueous formaldehyde to a solution of the heterocycle in methanol, ethanol or water and allowing precipitation of the product to occur. The *N*-hydroxymethyl heterocycles are formed cleanly and yields are generally high. A selection of these compounds is shown in Scheme 4.

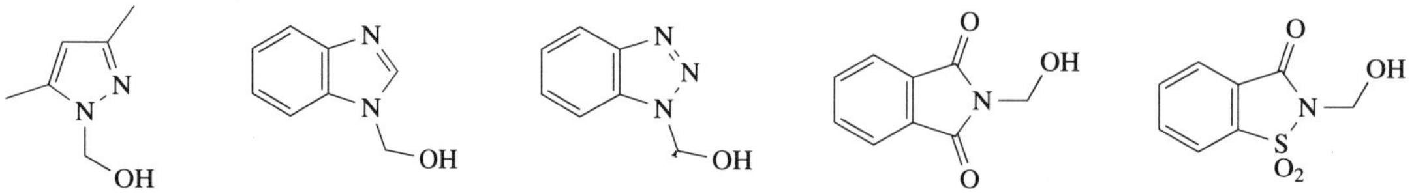

⟨50JOC1285⟩   ⟨59JPR150⟩   ⟨52JA3868⟩   ⟨47JA254⟩   ⟨59JPR150⟩

Scheme 4

The use of aldehydes other than formaldehyde for the preparation of heterocyclic *O,N*-acetals has received little attention. However, the reactivity of benzotriazole is such that simply mixing it with an equimolar amount of an aliphatic aldehyde results in rapid formation of 1-hydroxyalkylbenzotriazoles (**14**). In solution, these 1:1 adducts are ionized and exist in equilibrium with the starting materials (Scheme 5). The position of the equilibrium is dependent on the substituent, solvent and concentration. Compounds in which intramolecular hydrogen bonding between the *N*-1 substituent and *N*-2 is possible exhibit high stability and are formed very efficiently. For example (**15**) is obtained from glyoxylic acid, whereas the diol (**16**) is obtained in quantitative yield from glyoxal. Aromatic aldehydes do not yield these hemiaminals unless they possess an electron-withdrawing group ⟨87JCS(P1)791⟩. A related series of *N*-hydroxyalkylations has been explored with 1,2,4-triazole, which reacts regiospecifically at *N*-1 with aliphatic and aromatic aldehydes ⟨91SL485⟩. In certain cases, bifunctional amines condense with 2 equiv. of formaldehyde to generate heterocyclic *O,N*-acetals, for example (**17**) from 2-aminomethylbenzimidazoles ⟨70CPB1245⟩.

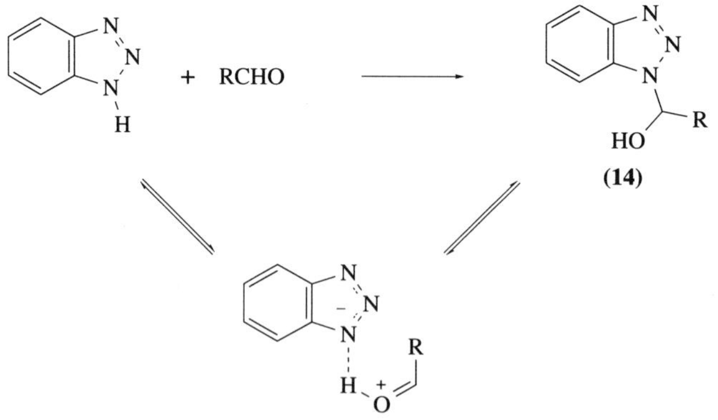

R = Me, Et, Pr^n, Pr^i, Bu^n, Bu^t, 4-pyridyl

Scheme 5

Many amides condense with formaldehyde under a wide range of conditions to give compounds of the type HOCH$_2$NHCOR ⟨05LA(343)207, 08LA(361)113⟩. An *Organic Syntheses* procedure provides large amounts of *N*-(hydroxymethyl)acetamide, HOCH$_2$NHCOMe, from acetamide and aqueous formaldehyde in the presence of potassium carbonate ⟨88OSC(6)5⟩. Analogous amidoalkylations have been carried out using acrylamide, benzamide and formamide ⟨66LA(697)171⟩. *N*-(Hydroxy-

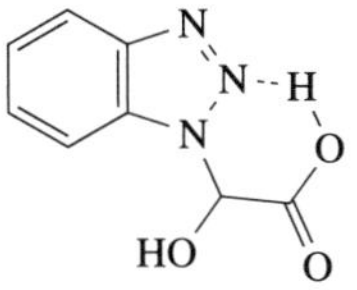

(15) 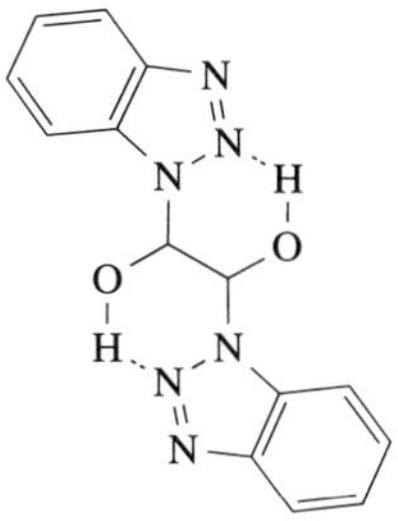   **(16)** 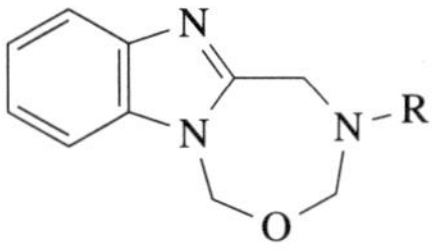   **(17)**

methyl)-thioamides are obtained similarly ⟨67AP(300)241⟩. These compounds are of value as sources of *N*-acyliminium ions, H₂C=NHCOR, which are potent amidoalkylating agents.

Considerable variation of the amide component is possible and with a bifunctional compound such as urea both the mono- and bis(hydroxymethyl) compounds HOCH₂NHCONH₂ and HOCH₂NHCONH₂OH have been obtained ⟨08CB24⟩.

The foregoing discussion has centred on the application of formaldehyde and simple aliphatic aldehydes for the synthesis of *O,N*-acetals. A wide range of other more complex aldehydes and ketones may be employed. Whereas the hemiaminals **(8)** are normally of very low stability when obtained from aliphatic aldehydes, use of some simple analogues can lead to stable, crystalline, high-melting compounds. This is usually the case when the carbonyl group is attached to an electron-withdrawing group. Chloral reacts readily with aqueous ammonia to give a stable hemiaminal and will even condense with amides and thioamides ⟨75S789⟩. When heated with benzyl urethane, chloral gave CCl₃CH(OH)NHCO₂Bn ⟨66CB1944⟩. Hexafluoroacetone reacts even more readily with ammonia and a wide range of amides ⟨74CB1488⟩. The less electrophilic 2,2,2-trifluoroacetophenone also affords a stable *O,N*-acetal (Scheme 6).

⟨61DOK(139)877⟩   ⟨66IZV1108⟩, ⟨88OSC(6)664⟩   R = COMe, CONMe₂ ⟨64JOC3114⟩
R = Pr^i ⟨75JOC2414⟩   ⟨50JA5409⟩

Scheme 6

Aromatic amines normally condense readily to form Schiff bases with aryl aldehydes. However, with electron-deficient amines, dehydration of the initially formed hemiaminal is greatly retarded ⟨61JOC4029⟩ and in some cases the hemiaminal is isolable (Scheme 7) ⟨21JA341⟩.

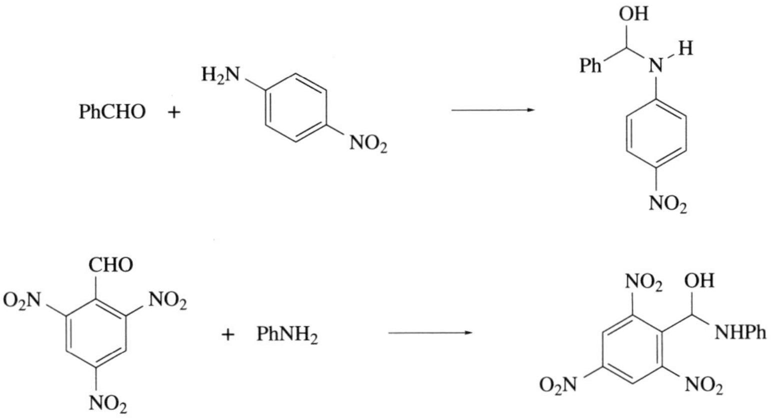

Scheme 7

Whilst aminals are normally obtained from aromatic aldehydes and secondary amines, this is not always the case. Unusual behaviour has been observed in the reactions of 4-fluorobenzaldehyde, which undergoes amine–halogen substitution when heated with piperidine but with morpholine yields the O,N-acetal as a stable crystalline solid (Scheme 8) ⟨94PC 407-01⟩. Hemiaminals have been obtained from a wide range of polyfunctional carbonyl compounds and amines or amides. Thus, PhCOCHO affords (**18**) in good yield by condensation with acetamide ⟨85AP(318)473⟩ whilst HCO-CO$_2$Me condenses with alkenylurethanes to give, for example, the O-acetyl derivative (**19**) after acylation with acetic anhydride–4-dimethylaminopyridine ⟨87TL3285⟩.

Numerous 2-amino-2-hydroxyindan-1,3-diones, for example (**20**) and (**21**) are available in high yield from ninhydrin and primary or secondary amines ⟨78T1285⟩.

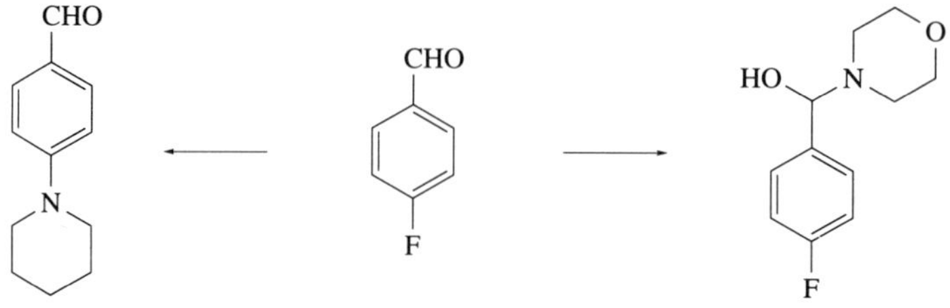

Scheme 8

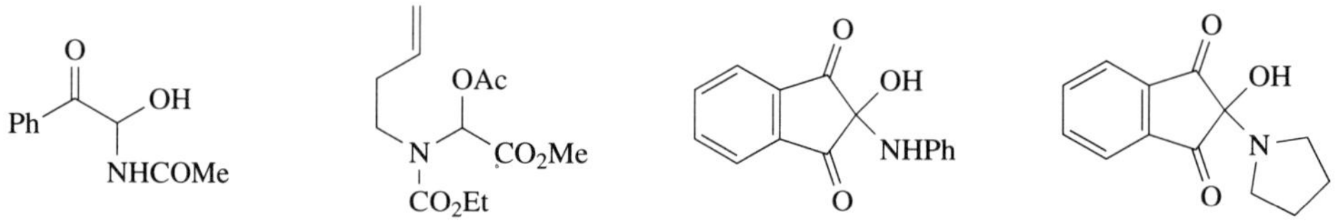

(**18**)   (**19**)   (**20**)   (**21**)

The reactivity of glyoxal and other bisaldehydes towards amines has been investigated. The outcome of the reaction is dependent upon the nature of the amine, solvent and reaction temperature. Glyoxal usually condenses with 2 equiv. of primary amines to give the bisimines (**22**), but with aniline or 2-chloroaniline the bis O,N-acetal (**23**) may be isolated. When either 3- or 4-nitroaniline is used, the hemiaminals are stable and not readily dehydrated to the imines (Scheme 9) ⟨70JOC3140, 70T2555⟩.

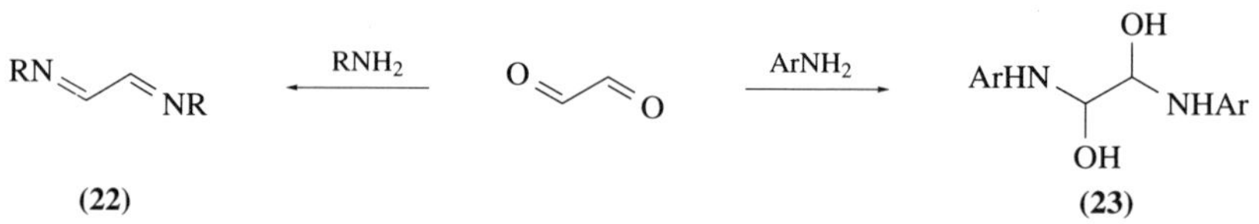

(**22**)   (**23**)

Scheme 9

The complexity of some of these condensations is illustrated by the behaviour of glyoxal towards benzylamines. In aqueous THF, under formic acid catalysis, benzylamine itself affords (**24**) as a crystalline hydrate. However if the reaction is conducted in methanol or acetonitrile with formic acid present, then good yields of the polyazapolycycle (**25**) are obtained, which represents the combination of 3 equiv. of glyoxal with 6 equiv. of the amine. Conversion of (**24**) into (**25**) results from treatment with formic acid in acetonitrile (Scheme 10). This transformation proceeds through a series of self-condensations of the bisimines BnN=CH—CH=NBn ⟨90JOC1459⟩. Analogous bisimines are the main products when either α-methyl- or α,α-dimethylbenzylamine is employed ⟨84CB694⟩.

Phthalaldehyde exhibits complex behaviour towards ammonia and primary amines. The diol (**26**) is formed rapidly on addition of ammonia or a primary amine to a dilute DMSO solution of the aldehyde. Although isolable, these compounds are of low stability and are readily dehydrated to give phthalimidines (**27**). Secondary amines such as diethylamine or morpholine react to give (**28**) as a mixture of diastereomers (Scheme 11) ⟨77JOC4217⟩. Thiourea reacts with phthalaldehyde in dilute aqueous alkali to give (**26**; R = CSNH$_2$) ⟨78JOC3838⟩. Other aromatic *ortho*-dicarbonyl compounds are a source of cyclic hemiaminals. When 2-benzoylbenzoic acid is heated with aniline, **(29)** is obtained, which on heating with thionyl chloride is isomerised to the isoindolone **(30)** ⟨70CB3205⟩.

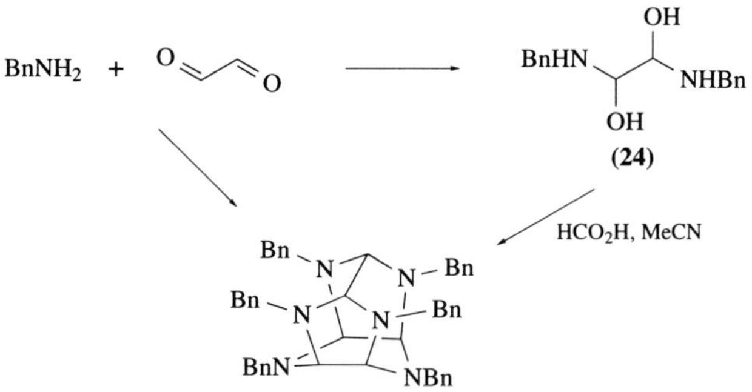

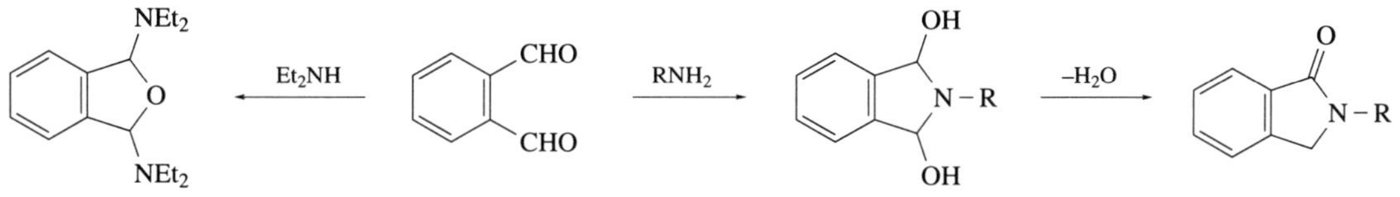

Scheme 11

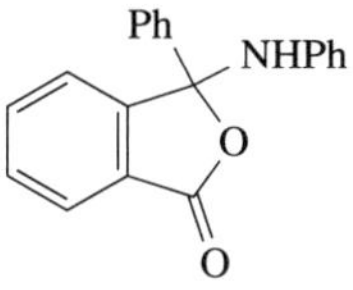

An interesting series of hemiaminals has been obtained from cyclopropanone in which relief of ring strain facilitates the reaction. Ammonia reacts rapidly to give bis(1-hydroxycyclopropyl)amine **(31)** ⟨73SC189, 74RTC294⟩ which may be hydrolysed to give the salt **(32)** ⟨77JCS(P1)684⟩. With secondary amines such as piperidine a rapid reaction occurs at low temperature to give **(33)** ⟨70TL1729⟩.

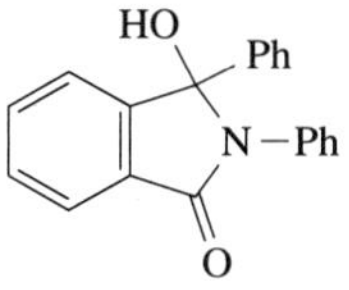

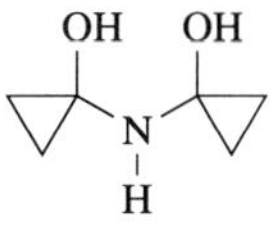

α-Amino alkoxides are formed when aromatic aldehydes react with lithium amides in ether or THF. This provides a useful means of *in situ* protection of the aldehyde function during metal–halogen exchange or directed lithiation reactions. Although the alkoxides are instantly cleaved on aqueous workup, they may be trapped and isolated in some cases (Equation (1)). The usefulness of α-amino alkoxides has been reviewed ⟨92SL615⟩. An extension of this amine–carbonyl condensation provides a wide range of *N*(α-alkoxyalkyl)amines, $R^1_2NCH(R^2)OR^3$. They are obtained from an aldehyde and an amine in the presence of an alcohol. Assembly of the O—C—N function in this manner by aminoalkylation of a hydroxyl group is a variation of the Mannich reaction. Mechanistic details have been reviewed ⟨68JPS715⟩.

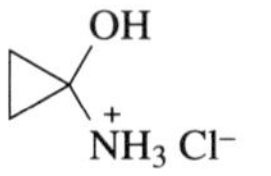

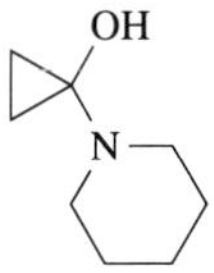

$$\underset{\text{NHMe}}{\text{pyridyl}} \xrightarrow[\text{ii, Bu}^t\text{Me}_2\text{SiCl}]{\text{i, Bu}^n\text{Li, PhCHO}} \underset{\underset{\text{Ph}}{\text{OSiMe}_2\text{Bu}^t}}{\text{pyridyl-N(Me)CH}} \quad (1)$$

The initial application of this reaction was investigated by McLeod and Robinson who demonstrated the generality of the condensation between formaldehyde, a secondary amine and an alcohol. Potassium carbonate was used as a catalyst and equimolar amounts of reactants were employed (Equation (2)) ⟨21JCS1470⟩. Other aliphatic aldehydes did not react and aromatic aldehydes also failed to give any hemiaminals. Later work demonstrated that improved yields result from use of an excess of the alcohol and under these conditions aromatic aldehydes reacted efficiently although aliphatic aldehydes were inert ⟨32JA4172, 55JA1098, 63JCED600, 71JOC3112, 84S495⟩. The yields of hemiaminals obtained by these procedures are generally high.

$$\text{HCHO} + \text{R}^1\text{R}^2\text{NH} + \text{R}^3\text{OH} \xrightarrow{\text{K}_2\text{CO}_3} \underset{\text{R}^2}{\overset{\text{R}^1}{\text{N}}}\diagup\text{OR}^3 \quad (2)$$

A variant of this reaction has been used to obtain the very unstable *N*-(alkoxymethyl)arylamines. Paraformaldehyde, an aromatic amine and a solution of a sodium alkoxide in the appropriate alcohol react to give the hemiaminals (**34**) (Equation (3)). Yields of products were in the range 40–90% and the reaction is compatible with both electron-rich and electron-deficient aryl groups. Although the products decompose rapidly at room temperature, they survive at $-18\,^\circ\text{C}$ ⟨88JCS(P1)1631⟩.

$$(\text{HCHO})_n + \text{ArNH}_2 + \text{RONa} \xrightarrow{\text{ROH}} \text{ArNHCH}_2\text{OR} \quad (3)$$
$$(\mathbf{34})$$

R = Me, Et, Pr^n; Ar = Ph, *o*-Tol, *m*-Tol, *p*-Tol, 4-NO$_2$C$_6$H$_4$

The *N*-methoxymethylation of *N*-methylaniline with paraformaldehyde in methanol at ambient temperature gave a low yield of PhN(Me)CH$_2$OMe, which was contaminated with an equal amount of the corresponding aminal ⟨82JA5753⟩.

A number of acyclic *O*,*N*-acetals have been obtained from the condensation of benzotriazole with aromatic or aliphatic aldehydes in the presence of an alcohol. The reaction, which is performed in boiling carbon tetrachloride, provides excellent yields of 1-(1-alkoxyalkyl)benzotriazoles and is an extension of that outlined in Scheme 5 ⟨87JCS(P1)791⟩.

A considerable number of other variations of this Mannich-type reaction are possible. The bisaminomethylation of diols has been accomplished and a number of unusual ring systems may be obtained by this route. For example, the 1,5,3-dioxazepine (**35**) is obtained in good yield by refluxing a mixture of paraformaldehyde, ethane-1,2-diol and isopropylamine in benzene ⟨80TL2949⟩, whilst (**36**) has been obtained from *trans*-cyclopentane-1,2-diol and cyclohexylamine. Many other examples are collated in the review by Rasshofer ⟨91HOU(E14a/2)1⟩.

(**35**)   (**36**)

The condensation reactions of phenylglyoxal with aromatic amines are complex and the nature of the product correlates to the pK_a of the amine. For example, when phenylglyoxal was refluxed with 4-nitroaniline (pK_a 1.0) in ethanol, the hemiaminal (**37**) was obtained, whilst (**38**) resulted from 2,4-dichloroaniline (pK_a 1.53) in methanol. When stronger bases such as 4-bromo- or 4-methylaniline (pK_a 4.0 and 4.58 respectively) were used, the products were the aminals BzCH(N-HAr)$_2$ ⟨67JCS(C)2696, 81JCS(P1)2435⟩.

Morpholine reacts rapidly with glyoxal to afford the bisaminal 1,1,2,2-tetramorpholinoethane. However if the reaction is carried out in methanol, the hemiaminal (**39**) is obtained. The bifunctional nucleophile *N,N'*-dimethylethylenediamine reacts under similar conditions to give the piperazine (**40**) as a 88:12 mixture of *trans* and *cis* isomers ⟨68JCS(C)2721⟩.

Although the Mannich reaction can provide a considerable range of *N*-(α-alkoxyalkyl)amines, the reaction is not applicable to the preparation of *N*-(α-aryloxyalkyl)amines of the type ArOCH(R)NR$_2$, since phenols undergo ring dialkylaminoalkylation. This aromatic Mannich reaction can however be used for the synthesis of cyclic *O,N*-acetals, or more specifically 1,3-benzoxazine derivatives (Equation (4)). For example, a phenol and 2 equiv. of formaldehyde condense with a primary amine to give the heterocycle ⟨64JOC407⟩. The reaction has been extended to a include a wide range of other phenols, amines and aldehydes ⟨91COS(2)953⟩.

The bishydroxymethylation of urea and thiourea derivatives forms the basis for the preparation of a wide range of cyclic systems. Acid-catalysed condensation of 1,3-dimethylurea with 2 equiv. of formaldehyde at ca. 90 °C is complete within 1 h and affords (**41**) in 96% yield. Many variations of this reaction are possible. For example, urea affords (**42**) with formaldehyde, but when 4 mol are employed in the presence of methanol the *N,N'*-bis(methoxymethyl)compound can be obtained in 80% yield. This adaptation of the Mannich reaction has been termed α-ureidoalkylation. The reader is referred to the excellent comprehensive review by Petersen which provides a wealth of experimental procedures for many of these O—C—N functions. The reaction may be extended to other aldehydes and in some instances also to ketones ⟨73S247⟩.

The Mannich-type condensation between ketones, secondary amines and alcohols cannot be used to obtain compounds of the type R^{1}OCR$^2_2$NR$^3_2$. However benzotriazole condenses with cyclohexanone and an alcohol to afford (**43**; R = Pr^n, Pr^i or Bn) in high yields. Other ketones do not react under these conditions ⟨89JOC6022, 91S279⟩.

Bifunctional compounds such as amino acids, amino alcohols and hydroxy amides are valuable starting materials for the preparation of heterocyclic *O,N*-acetals. The condensation of 2-aminoethanol with aldehydes and ketones is readily accomplished and usually involves heating the reactants with azeotropic distillation of water. The oxazolidines are generally obtained in good

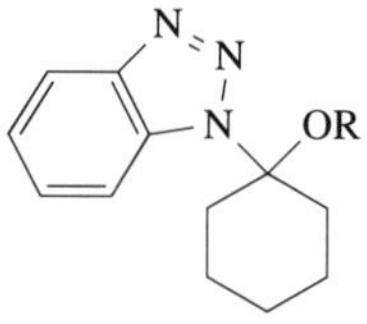

(43)

yields and a wide range of substituents is compatible with the reaction. Some examples are given in Scheme 12. 2-(Methylamino)ethanol condenses with glyoxal to give (**44**) ⟨78BSF(2)83⟩. Fused bicyclic systems based on oxazolidine are readily prepared. Thus the oxazolo[3,4-c]oxazole (**45**) is formed from the condensation of 2-aminopropan-1,3-diol with formaldehyde ⟨51JA2596⟩. Tetrahydro-1,3-oxazines, for example (**46**), are available through an analogous series of condensations using 3-aminopropanol. Extensive reviews chart methods for the preparation of these systems ⟨63AHC(2)311, 78AHC(23)1⟩, whilst the benzologues are available from 2-hydroxybenzylamine and 2-aminobenzyl alcohol ⟨84CHEC(3)995⟩. Hydroxy amides are also useful for the preparation of 1,3-oxazines. For example, 1,3-benzoxazin-4-ones can be prepared simply by treating salicylamide with an aldehyde or ketone in boiling benzene; pyrrolidine is the catalyst of choice, although p-toluenesulfonic acid has been used. Excellent yields of products are normally obtained and a wide array of substituents is tolerated in the reaction ⟨81JOC3340, 89S677⟩.

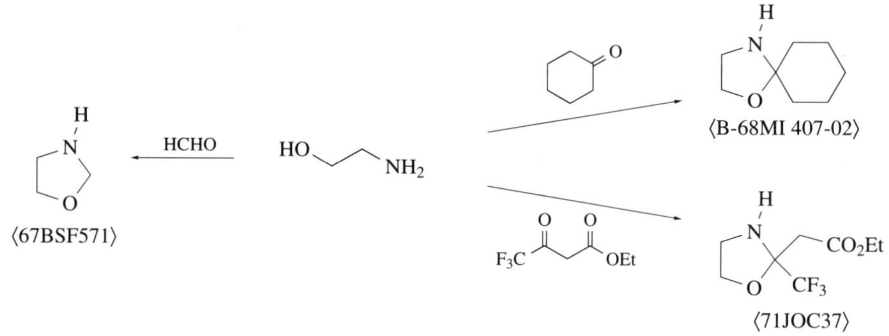

Scheme 12

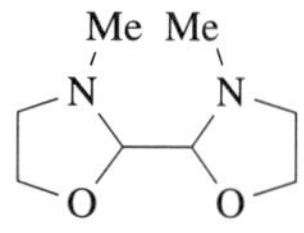

(**44**) 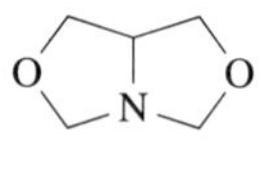   (**45**) 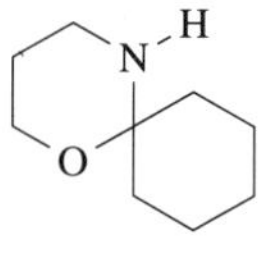   (**46**)

Amino acids are of value for the preparation of some oxazolidine derivatives. Brief heating of N-phenylglycine with paraformaldehyde in toluene gave the thermally labile (**47**) in quantitative yield ⟨87BCJ4079⟩. The fused system (**48**), although moisture sensitive, can be obtained in large quantities from the condensation of pivaldehyde with L-proline in pentane in the presence of TFA as catalyst ⟨83JA5390⟩. This pivaldehyde condensation has also been applied to alanine, methionine, phenylalanine and valine and forms the basis of an efficient method for the asymmetric functionalization of α-amino acids. 2-Aryloxazolidinones have also been exploited in a similar manner ⟨84TL4337, 85HCA1243, 88HCA224, B-89MI 407-01⟩.

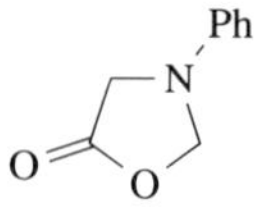

(**47**) 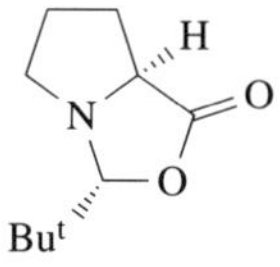   (**48**)

(b) From amides and imides. Nucleophilic attack at an amide or imide carbonyl group is generally of little value for the preparation of *O,N*-acetals, since the intermediate α-amino alkoxide anion **(49)** is easily cleaved giving a carbonyl compound and an amine. However this approach is of value in some instances. Although carbamates are susceptible to reductive cleavage, the carbonyl group of ethyl carbamate, $H_2NCOOEt$, is smoothly reduced by lithium aluminum hydride (LAH) to give H_2NCH_2OEt. The reaction is also successful when the nitrogen function is incorporated into an azepine ring ⟨73CC601⟩.

(49)

Although not involving nucleophilic attack at an amide carbonyl group, a cyclic *O,N*-acetal **(50)** is obtained from the self-condensation of MeCOCONHPh. The dimerization is catalysed by diethylamine and is rapid at ambient temperature ⟨72AJC1737⟩. Both secondary and tertiary amides may be oxidized to hemiaminals, although this approach is not always convenient. For example, the anodic oxidation of *N*-ethylacetamide in methanol affords MeCONHCH(OMe)Me in good yield ⟨75GEP2503114⟩. Considerable structural variation is possible. The carbamate **(51)** was oxidized regiospecifically to the hemiaminal in methanol; tetrabutylammonium tosylate was used as the supporting electrolyte (Equation (5)). The protocol is also applicable to heterocyclic carbamates as illustrated by the C-2 methoxylation of 1-methoxycarbonylpyrrolidine ⟨75JA4264⟩.

(50)

(51) (5)

Of the various metal hydrides that have been used for the reduction of cyclic imides, sodium borohydride has proved to be the most useful. Clean and efficient formation of hydroxy lactams results when either *N*-alkylsuccinimides or -phthalimides are reduced with $NaBH_4$ in ethanol containing a trace of acid. Whilst *N*-arylsuccinimides are reduced efficiently, *N*-arylphthalimides have a strong tendency to give ring cleavage products ⟨61JOC2273, 71SC103, 75T1437, 78T179⟩. The only other addition reaction of value for the preparation of hemiaminals from imides is their condensation with Grignard reagents. An example is illustrated in Equation (6) ⟨90JCS(P1)83⟩. Addition of a second mole of Grignard reagent has been achieved under forcing conditions ⟨90G677⟩.

(6)

Np = 1-naphthyl

Directed metallation has been employed for the synthesis of phthalimidines. Dilithiation of **(52)** gives the C-2 lithio compound regiospecifically. This intermediate may be intercepted with DMF to give the heterocyclic product (Equation (7)) ⟨82JCS(P1)2227⟩. Quenching the dianion with an acid chloride leads to 3-substituted 3-hydroxy-1,3-dihydroisoindole-1-ones.

(c) *From imines and iminium salts.* The addition of oxygen nucleophiles to an azomethine unit provides a useful means of obtaining *O,N*-acetals. The reaction is particularly efficient when the imine unit is attached to one or more electron-withdrawing groups which both facilitate the addition reaction and confer stability on the product. Schiff bases, PhCH=NAr, usually suffer hydrolytic cleavage under acidic conditions, but the hemiaminals PhCH(OH)NHAr have been obtained as their hydrochlorides under carefully controlled conditions ⟨02CB984⟩.

Both *C*- and *N*-acylimines react readily with alcohols to give hemiaminals. Thus PhCOCH=NPh gave PhCOCH(OEt)NHPh in quantitative yield after boiling in ethanol for 0.5 h. Considerable variation is possible here and related compounds with differing *N*-aryl substituents also reacted smoothly with either ethanol or methanol ⟨84G405, 84JCR(M)144⟩.

The isomeric imine, PhCH=NCOPh, affords PhCH(OMe)NHPh on brief treatment with methanol. Again there is scope for structural modification and a number of *N*-aroylimines have been generated and treated in this way ⟨67T2869⟩. The introduction of additional functional groups on the azomethine unit causes few problems, for example (53) was obtained from BnC(CO$_2$Et)=NAc in 75% yield when treated with methanol containing potassium *t*-butoxide ⟨80JOC1880⟩. The imine, Me$_2$CHC(CO$_2$Me)=NH, generated *in situ* from oxidation of valine methyl ester with *t*-butyl hypochlorite, gives (54) when the reaction is performed in methanol ⟨77CB948⟩. Imines derived from perhalo aliphatic aldehydes and ketones react especially readily with oxygen nucleophiles giving high yields of the *O,N*-acetals. Hexafluoroacetone imine, (F$_3$C)$_2$C=NH, reacts with acetic anhydride in the presence of H$_2$SO$_4$ to give (55) in high yield. The reaction involves initial formation of the highly electrophilic *N*-acylimine which is intercepted by acetate. Similar behaviour is exhibited by the *N*-benzoyl imines, (F$_3$C)$_2$C=NCOPh, which afford (56; R = H) when heated with aqueous acetone ⟨66IZV1108⟩. The ethers (56; R = Me or Et) are obtained from the imine and the appropriate alcohol ⟨65IZV2046⟩. Addition of either water or methanol to (F$_3$C)$_2$C=NCO$_2$Bu^t is rapid at room temperature, giving the hemiaminal ⟨75JOC2414⟩. Trichloroacetaldehyde imines, Cl$_3$C—CH=NR (R = CO$_2$Me or Ts), react readily with water, alcohols or phenols to give good yields of *O,N*-acetals ⟨64CB483, 64CB490, 69ZOR2181⟩.

Imines can also function as heterodienophiles and have been used to intercept *o*-quinonemethides generated *in situ* from the thermolysis of phenolic Mannich bases (Equation (8)). Although the yields from this reaction are only modest, it provides a convenient means of obtaining otherwise inaccessible, fused 1,3-oxazines ⟨69JHC429⟩. Certain activated methylene compounds have been shown to undergo cyclocondensation reactions with Schiff bases. When benzylidineaniline and 4-hydroxycoumarin were stirred in acetic acid for a few hours, the benzopyrano[1,3]oxazine (57) was obtained. The reaction involves initial attack at the imine function followed by addition of benzaldehyde, generated by *in situ* hydrolysis, and cyclodehydration ⟨78TL3607⟩. Semicyclic *O,N*-acetals have been obtained from the reaction of imidates with ketenes. Diphenylketene undergoes an efficient cycloaddition under mild conditions (MeCN, 25 °C) with PhCOC(OMe)=NMe to give (58) ⟨81JCS(P1)2443⟩. In a similar vein, ethyl *N*-cyclohexylformimidate and chlorocyanoketene gave (59). The reaction is not concerted but proceeds through a dipolar intermediate ⟨85JOC4231⟩.

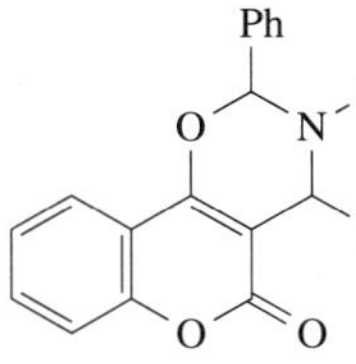

(57) 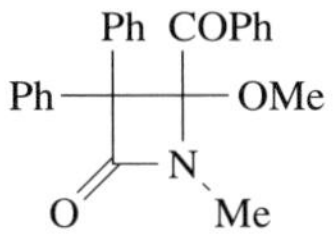 (58) 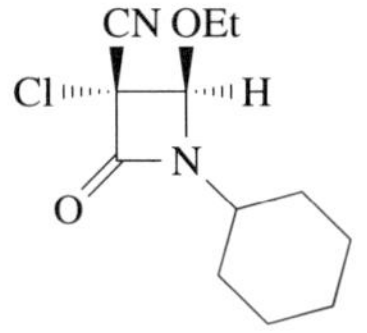 (59)

C-Alkenyl N-acylimines are 1-azabuta-1,3-dienes and represent a useful class of heterodiene which react with vinyl ethers to give 2-alkoxytetrahydropyridines. The example outlined in Equation (9) affords a mixture of cycloadducts but exhibits pronounced *endo* diastereoselectivity ⟨90JOC2999⟩.

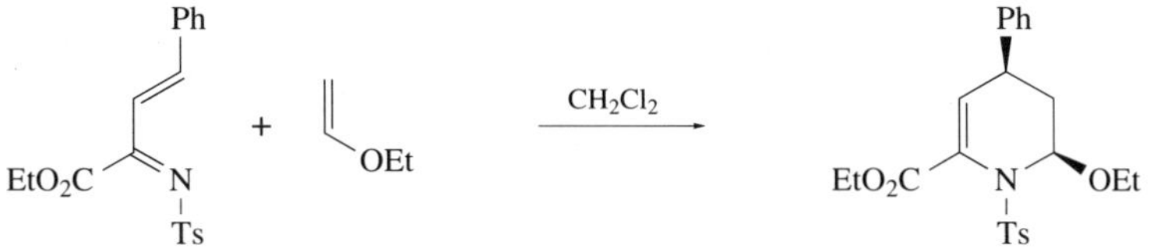

(9)

Alkoxide addition to cyclic imines also provides a route to O,N-acetals as illustrated by the facile formation of 2-alkoxyaziridines from 1-azirines ⟨67JA4456⟩ in which the driving force is the relief of ring strain. A highly reactive 3H-indole derivative is generated when the diester (**60**) is treated with sodium methoxide. The indolone, which cannot be isolated, is spontaneously intercepted giving (**61**) (Scheme 13) ⟨81JCS(P1)2443⟩.

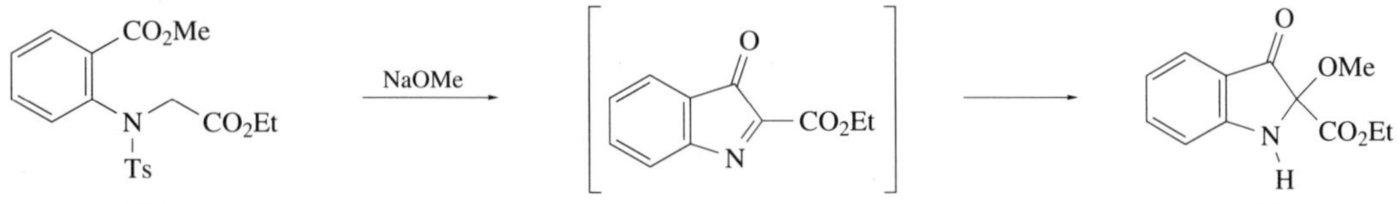

Scheme 13

The azomethine function of 4-methyl-2-trifluoromethyl-5(2H)-oxazolone is susceptible to nucleophilic attack with concomitant ring fracture. Treatment with methanolhydrochloric acid gives 1-methoxy-2,2,2-trifluoroethylamine hydrochloride, $F_3CCH(OMe)NH_2 \cdot HCl$ ⟨65CB487⟩.

Iminium salts are of value for the preparation of both acyclic and cyclic hemiaminals although relatively few examples have been reported. The simplest application involves the addition of dimethylmethyleneammonium chloride, $Me_2N=CH_2{}^+Cl^-$ (**62**), to alcohols. Although good yields of dimethylaminomethyl alkyl ethers, Me_2NCH_2OR, are obtained ⟨63CB604⟩ they are best prepared by the three component condensation described previously. This approach is however applicable to the synthesis of O-functionalized hemiaminals. When a suspension of (**62**) in dichloromethane was treated with an excess of sodium acetate, dimethylaminomethyl acetate, Me_2NCH_2OCOMe, was obtained. Variation of the iminium salt is possible and N-methylenemorpholinium chloride has also been employed ⟨75LA1790⟩.

The 2-chloroethyl ether (**63**) can be obtained from (**62**) either by treatment with 2-chloroethanol or with ethylene oxide. Ring closure of (**63**) is facile and affords the oxazolidinium salt (**64**) (Scheme 14) ⟨69CB2651⟩. Iminium salts are also effective heterodienophiles and the reaction of Eschenmoser's salt ($Me_2N=CH_2{}^+I^-$) with 1-methoxy-3-trimethylsilyloxybuta-1,3-diene (Danishefsky's diene) proceeds at room temperature to give the tetrahydropyridinium salt (**65**) in near quantitative yield ⟨76JA6715⟩.

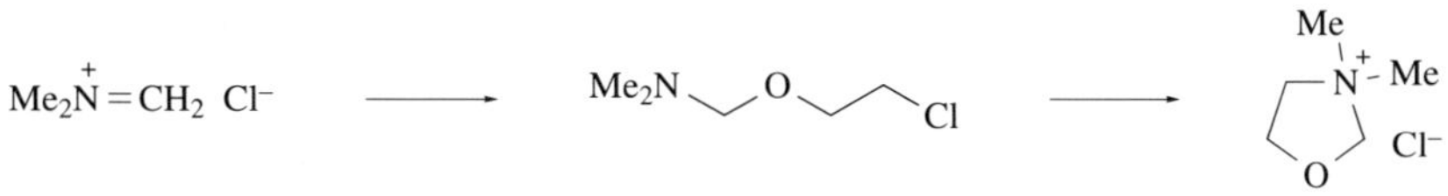

Scheme 14

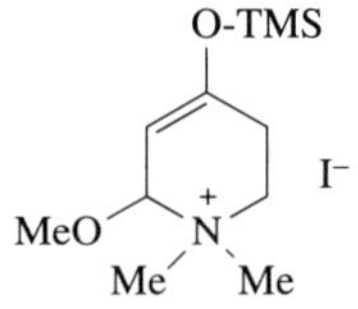

Heterocyclic systems which incorporate an imidate function are useful starting materials for cyclic O,N-acetals, although the approach is best suited to the modification of partially saturated systems. Oxazoles can be reduced either by LAH or sodium in ethanol to give oxazolidines ⟨56JA2167⟩. Benzoxazoles usually suffer reductive cleavage. Reduction of 4,5-dihydrooxazoles is best accomplished by quaternization followed by treatment with sodium borohydride. Oxazolinium salts react readily with aryl Grignard reagents although the process is much less efficient with alkylmagnesium halides (Scheme 15) ⟨66JHC531, 88OSC(6)64⟩. Applications of these reactions of 4,5-dihydrooxazoles have been extensively reviewed by Meyers ⟨76AG(E)270, 85T837, 94T2297⟩. The azomethine group in 5,6-dihydro-4H-1,3-oxazines is readily reduced by sodium borohydride in acidified ethanol at −30 °C, as exemplified by the synthesis of the tetrahydrooxazine (66) ⟨88OSC(6)905⟩.

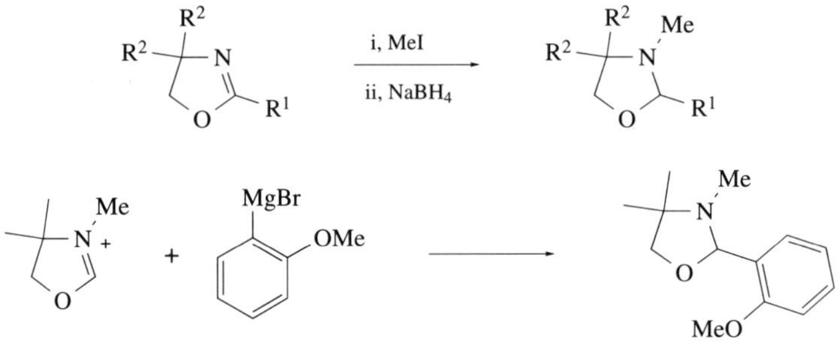

Scheme 15

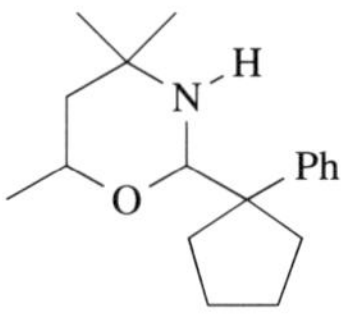

(d) From alkenes. Whilst diethyl 2-nitrobenzylidenemalonate reacts straightforwardly as a Michael acceptor with alkoxides and amines, anomalous reactions supervene when a second nitro group is introduced. Treatment of diethyl 2,4-dinitrobenzylidenemalonate with diethylamine promotes a complex rearrangement leading to the O,N-acetals (67) ⟨83S654⟩. Primary amines do not promote the rearrangement and afford only the corresponding malonamides. With triethylamine in pyridine an alternative pathway operates giving the 3,1-benzoxazine (68) (Scheme 16) ⟨88S111⟩.

In the presence of an acid catalyst, electron-rich vinyl ethers react readily with nucleophiles. Although this approach has been used for the synthesis of O,N-acetals, examples appear to be restricted to the use of nitrogen heterocycles. 3,4-Dihydro-2H-pyran reacts with a variety of six-substituted purines to give the corresponding 9-(tetrahydropyran-2-yl) compounds. Ethyl vinyl ether reacts analogously ⟨61JA2574, 84IJC(B)1286⟩. When the nucleophile is benzotriazole (pK_a 8.2), no catalyst is necessary and quantitative yields of the α-benzotriazolyl ethers are obtained simply by refluxing the reactants in carbon tetrachloride ⟨90JCS(P1)1717⟩.

Other addition reactions to vinyl ethers include the mercury(II)-catalysed addition of 2-aminoalkanols which afford oxazolidines ⟨57JA2833⟩. Tetrahydro-1,3-oxazines are also available by this route ⟨57JA2825⟩. Vinyl acetate undergoes a rapid [2 + 2] cycloaddition to chlorosulfonyl isocyanate at low temperature to give 4-acetoxyazetidine-2-one after *in situ* reductive cleavage of the initially formed N-chlorosulfonyllactam ⟨93OSC(8)3⟩.

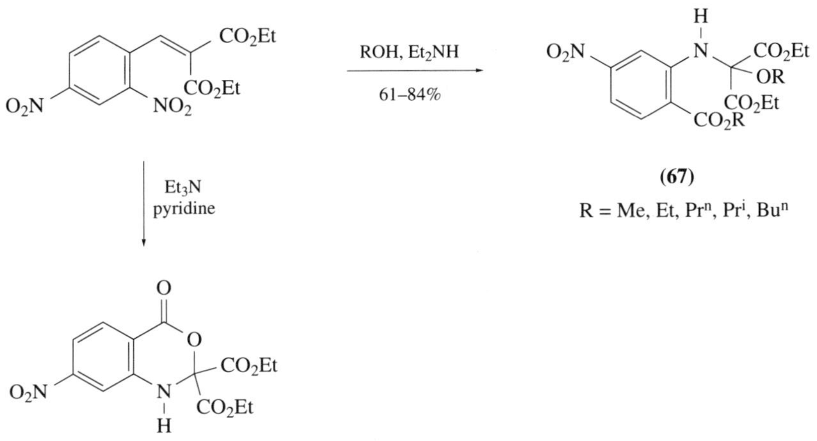

Scheme 16

A number of N-t-BOC allylic amines have been prepared and in the presence of a catalytic amount of [RhH(PPh$_3$)$_4$] isomerize to the corresponding enamines. If the reaction is performed in the presence of an alcohol, then the imine tautomer is trapped to give the hemiaminal. The sequence is outlined in Equation (10) ⟨90CC1304⟩.

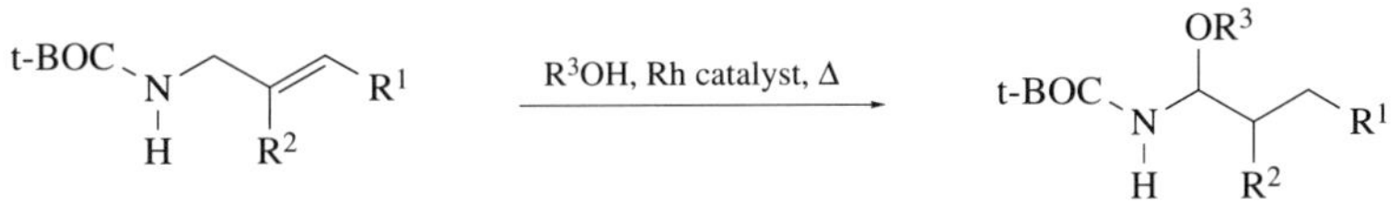

$$R^1, R^2 = H, Me; R^3 = Me, Pr^n \quad (10)$$

The utility of enamines is considerably greater for the synthesis of heterocyclic O,N-acetals than for the open-chain systems. A diverse range of 2-aminobenzopyrans is accessible from the reaction of salicylaldehyde with enamines (Scheme 17). The xanthene (**69**) is generated as a mixture of diastereomers from the condensation with 1-morpholinocyclohexene in benzene. Subsequent oxidation by CrO$_3$ in pyridine also resulted in elimination of the amine with the formation of 1,2,3,4-tetrahydroxanthone ⟨66JOC1232⟩. N-Styrylmorpholine reacted with salicylaldehyde in refluxing benzene providing (**70**) in good yield and here the addition operates with concomitant dehydration to generate the extended conjugated system ⟨82JCS(P1)1193⟩. The enamine may be modified to prevent dehydration. For example Me$_2$C=CHN(Me)Ph and salicylaldehyde gave a chromanol which was oxidized to (**71**) in excellent yield ⟨82JCS(P1)2771⟩. The heterocyclic enamine (**72**) condenses readily to give the spirobenzopyran (**73**) ⟨B-71MI 407-01, B-90MI 407-01⟩. Similarly 1-nitroso-2-naphthol condenses with (**72**) in either ethanol or toluene to give the spirooxazine (**74**) ⟨B-90MI 407-02⟩.

1,2-Benzoquinone reacts with 1-morpholinocyclohexene to give (**75**) ⟨65LA(687)187⟩. A similar reaction with 1,4-benzoquinone affords (**76**) ⟨66JPR144⟩ and in some instances polycyclic systems result from addition of 2 mol of the enamine ⟨66CB930⟩. In a reaction analogous to that in Equation (8), the o-quinonemethide generated from 1-dimethylamino-2-naphthol has been intercepted with 1-pyrrolidinocyclohexene to give (**77**). An extensive range of ring systems has been obtained by this route ⟨70JHC1311⟩. There are numerous other examples which illustrate the cyclophilic nature of enamines. C,N-Diphenylnitrone, PhCH=N(O)Ph, reacts readily with 1-phenyl-1-pyrrolidinoethene to give the isoxazolidine (**78**) ⟨67TL3769⟩. The heterodiene α-nitrosostyrene can be generated in situ and trapped to give the [4 + 2] cycloadduct (**79**) with α-morpholinostyrene ⟨79JCS(P1)249⟩. Heterocyclic hemiaminals obtained by this route have been reviewed by Rasshofer ⟨91HOU(E14a/2)1⟩.

(ii) From compounds containing two singly bonded functional groups

(a) From X—C—N functions (X = Hal, OR, SR, SO$_2$R, NR$_2$). Tris(chloromethyl)amine (ClCH$_2$)$_3$N is a relatively stable α-chloroalkylamine and is available in quantity from hexamethylenetetraamine. When reacted with sodium methoxide in dichloromethane, all three halogens

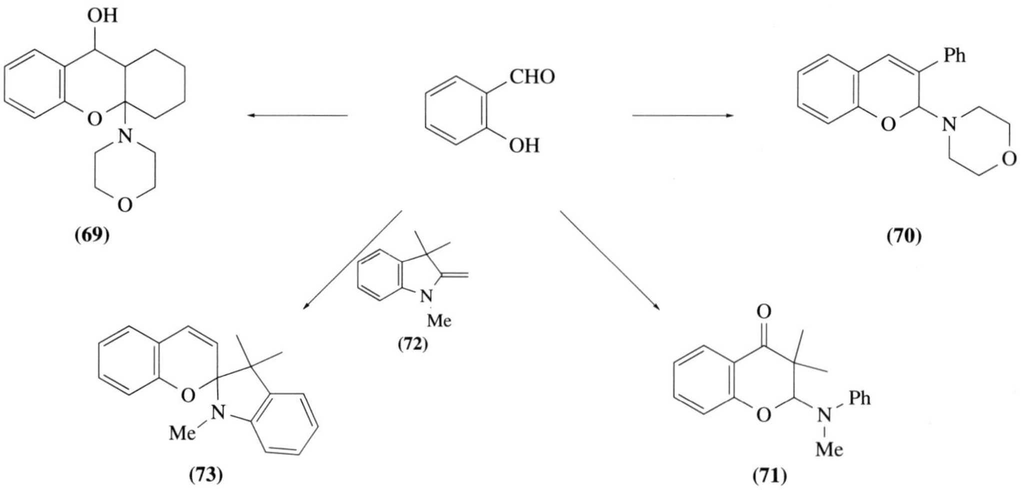

Scheme 17

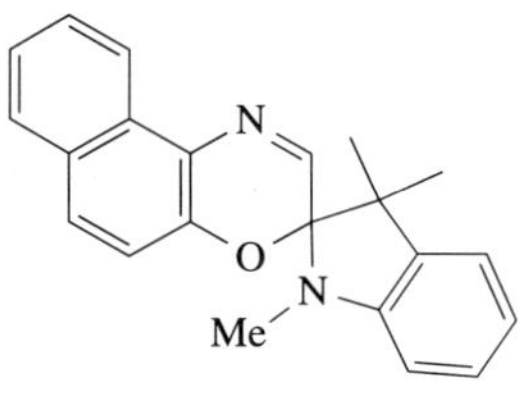

are substituted and (MeOCH$_2$)$_3$N is obtained in high yield ⟨73CB69⟩. Sodium trimethylsilanolate, NaO-TMS, reacts in an analogous way ⟨73HCA1117⟩. *N*-(α-Chloroalkyl)azoles are readily available materials from which the halogen is easily displaced. For example, 1-(chloromethyl)imidazole gave the corresponding 1-(phenoxymethyl) compound when refluxed with phenol in ethanol ⟨75JCS(P1)1670⟩. Nucleophilic displacements from 1-(benzotriazol-1-yl)-1-chloroalkanes have been studied in more detail. These compounds undergo smooth halide displacement with a variety of alkoxide, phenoxide and carboxylate ions ⟨87JCS(P1)811, 89JOC6022⟩.

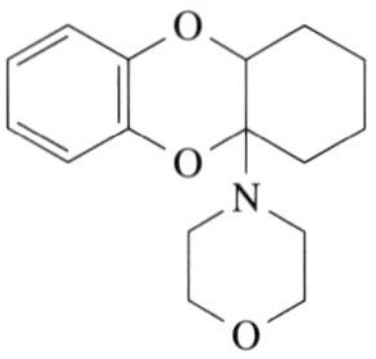

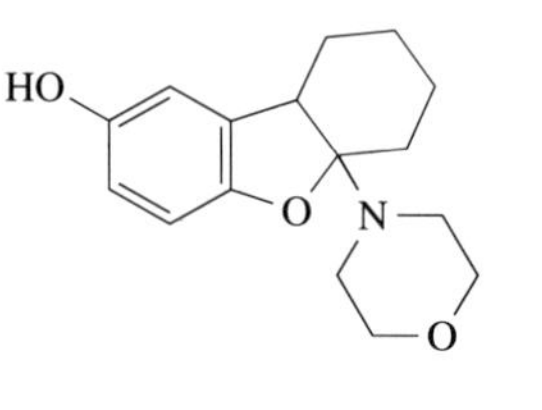

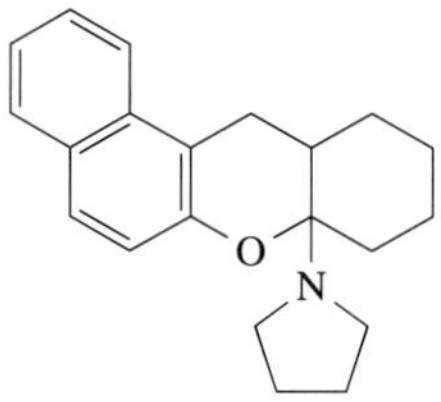

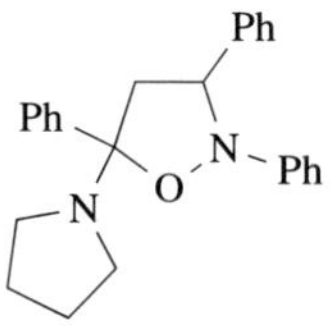

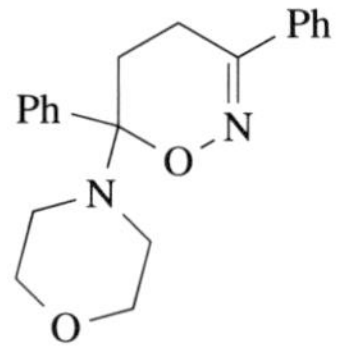

Halide displacement reactions are of considerable value for the preparation of *N*-acyl *O*,*N*-acetals owing to the accessibility of *N*-(α-chloroalkyl)amides and -imides. A large number of compounds have been prepared by this route ⟨91HOU(E14a/2)1⟩. Preparations of *N*-(haloalkyl)amides are tabulated in the review by Zaugg and Martin ⟨65OR(14)52⟩. As a rule, reactions of haloalkylamides with oxygen nucleophiles proceed under mild conditions giving high yields of products. Compounds containing an S—C—N unit have also been used to obtain O,N-acetals. The oxidative cleavage of methyl 2-benzylthiohippurate, BnSCH(NHCOPh)CO$_2$Me, by NBS in methanol gives MeO-CH(NHCOPh)CO$_2$Me in high yield by displacement of the sulfur function ⟨75T863⟩. Although (**80**) possesses two potential leaving groups, treatment with NaOMe results in exclusive displacement of thioacetate. The reaction is complete within minutes at room temperature ⟨75TL3579⟩. Rapid solvolysis occurs with displacement of trifluoromethanesulfinate when (**81**) is treated with water or ethanol giving the O,N-acetals in good yield (Scheme 18) ⟨66CB1932⟩.

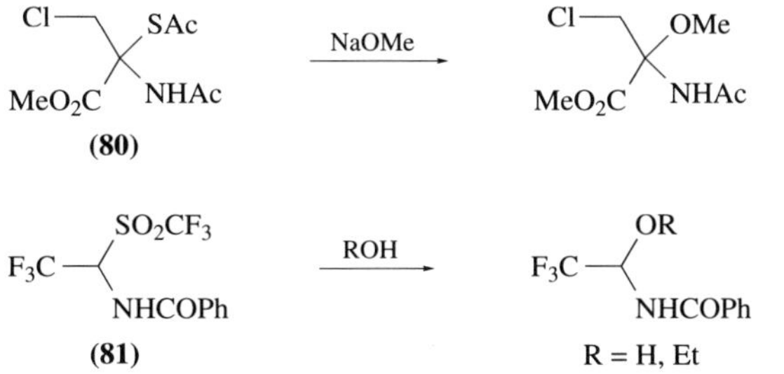

Scheme 18

Bis(dimethylamino)methane reacts with acid anhydrides to give Me$_2$NCH$_2$OCOMe from acetic anhydride or Me$_2$NCH$_2$OCOPh from benzoic anhydride ⟨63LA(664)130⟩. The solvolysis of aminals to hemiaminals is rapid and many examples have been reported. 2,2-Dimorpholinoacetophenone affords (**82**) when treated with hydrochloric acid in ethanol ⟨74CR(C)221⟩. A high yield of the cyclopropane hemiaminal (**83**) was obtained from the hydrolysis of 1,1-bis(dibenzylamino)cyclopropane in HCl-THF ⟨91SL87⟩. Hemiaminal O-alkyl ethers may be similarly prepared. When 1,1,2,2-tetramorpholinoethane was refluxed in methanol for 20 minutes, substitution of two vicinal morpholino groups occurred to give (**39**) ⟨68JCS(C)2721⟩. This procedure offers an alternative, though less efficient, method to the three-component reaction described previously.

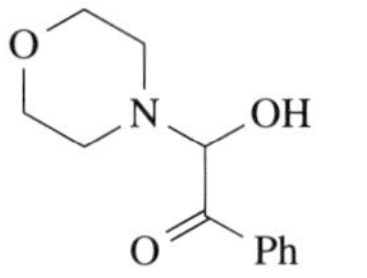

The most useful of the aminal → hemiaminal conversions makes use of heterocyclic aminals obtained from the aminoalkylation of benzotriazole. This aminoalkylation reaction is a very general and efficient process which involves heating benzotriazole with an aliphatic, aromatic or heteroaromatic aldehyde, and a primary or secondary aliphatic or aromatic amine in benzene, toluene or ethanol. The products, N-[1-(benzotriazol-1-yl)alkyl]amines, are obtained in very high yields ⟨87JCS(P1)799, 89JCS(P1)225⟩. Subsequent treatment with an alkoxide effects displacement of benzotriazolate to give the N-(α-alkoxyalkyl)amines (**84**) ⟨93S229⟩. The sequence is illustrated in Scheme 19. The methodology is capable of considerable variation and has been used to obtain N-acyl O,N-acetals (**85**) via aminoalkylation of benzotriazole ⟨91JOC4439⟩ with primary amides followed by treatment with an alkoxide in the appropriate alcohol (Scheme 19) ⟨92JOC547⟩. These procedures represent a very significant development in hemiaminal synthesis and for many compounds it is undoubtedly the method of choice. The full scope of heteroalkylations mediated by benzotriazole has been reviewed ⟨94S445⟩.

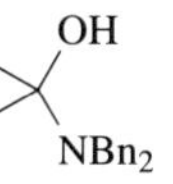

(b) *From O—C—X functions (X = Hal or OR)*. Alkyl chloromethyl ethers, ClCH$_2$OR, are smoothly transformed into dialkylaminomethylalkyl ethers, R$_2$NCH$_2$OR, when treated with 2 equiv. of a secondary amine ⟨48JCS2174⟩. Tertiary amines are quaternized by haloalkyl ethers and many examples of long-chain alkoxyalkylamines have been prepared ⟨72BRP1299180⟩. Reactions with heteroaromatic compounds such as pyridine or quinoline are also known. Numerous heterocyclic amides and imides have been N-alkoxyalkylated using a variety of methods ⟨91HOU(E14a/2)170⟩.

Acetals are of considerable value for the synthesis of hemiaminals, since they react readily with a variety of nitrogen nucleophiles. The reaction is generally promoted by acid catalysis. Benzotriazole when heated with 2,5-dimethoxytetrahydrofuran in the presence of TsOH gave approximately equal amounts of (**86**) and (**87**) together with a small amount of 2-(benzotriazol-1-yl)-5-(benzo-

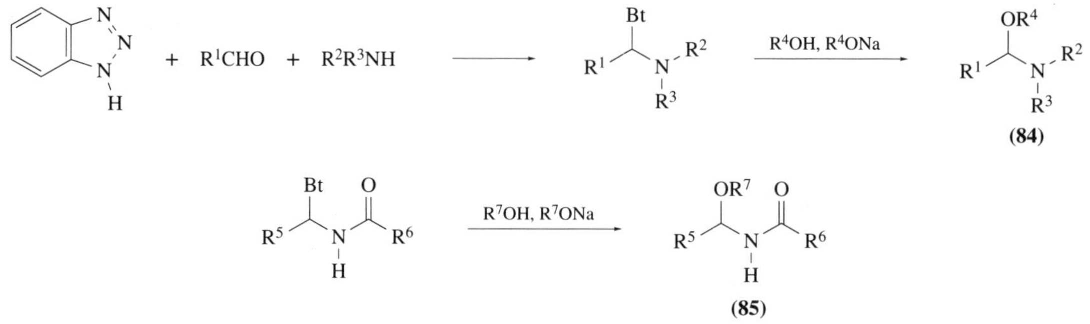

Scheme 19

R^1 = H, Ph, 2-pyridyl; $R^2 = R^3 = (CH_2)_2O$; R^2 = H, Et, Ph, Bn; R^3 = Bn;
R^4 = Me, Et, Pr^i; R^5 = H, aryl; R^6 = Me, Ph; R^7 = Me, Et, Pr^i, Bu^i

triazol-2-yl)tetrahydrofuran ⟨90JCS(P1)1717⟩ 1,1-Dialkoxyalkanes react readily with carbazole to give 9-(1-alkoxyalkyl)carbazoles (**88**) in a reaction promoted by mineral acid. An extensive range of these compounds has been obtained by this procedure although the Mannich alkoxyalkylation of carbazole is more versatile ⟨78ZOR1723⟩. Amides react with acetals in an analogous fashion; the formation of (**89**) from propionaldehyde diethyl acetal and 2-pyrrolidinone is representative. A great number of compounds are accessible by this means ⟨66CB2127⟩. Thioamides and carbamates react in a similar way. Isocyanates undergo an insertion reaction into the C—O bond of acetals. Dimethoxymethane reacts readily with isocyanates to give (**90**; R = alkyl or aryl). The reaction is promoted by a Lewis acid catalyst ⟨62AG872, 65LA(686)102⟩. With chlorosulfonyl isocyanate, formation of $MeO_2CN(SO_2Cl)CH_2OMe$ is rapid ⟨68AG(E)172⟩. Whilst none of the foregoing reactions proceed via an S_N2 mechanism, this pathway is possible in some instances and may operate when the acetal possesses a particularly good nucleofuge. For example, when $BnOCH_2OTs$ is treated with pyridine in dichloromethane, (**91**) is obtained ⟨71S150⟩.

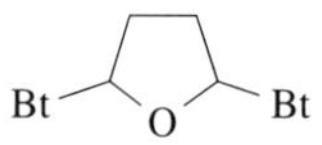

(c) By functionalization of a preformed O,N-acetal. Most of the examples here appear to be limited to the acylation of *N*-(α-hydroxyalkyl)azoles, in which the heterocycle is either benzotriazole ⟨87JCS(P1)791⟩ or 1,2,4-triazole ⟨94H(37)1951⟩. This contrasts with *N*-(α-hydroxyalkyl)amides and imides, which are readily *O*-alkylated and acylated by a wide variety of reagents. Cleavage of an *O*-function is possible in certain cases. Aryl *N*-(benzyloxymethyl)carbamates, $BnOCH_2NHCO_2Ar$, undergo catalytic hydrogenolysis to give $HOCH_2NHCO_2Ar$ although the yields are only moderate. A wide range of substituents in the aryl ring is tolerated ⟨68MI 407-01⟩. Modification of the *O*-substituent in *N*-(*t*-butoxymethyl)piperidine has been accomplished by heating with acetic anhydride. The product and related alkoxyalkylamines are readily *N*-alkylated by iodoalkanes or dialkyl sulfates ⟨63LA(664)130, 64GEP1183513⟩. Similarly, chloromethyl methyl ether effected the *N*-alkylation of $MeOCH_2NEt_2$ to the hygroscopic salt $(MeOCH_2)_2NEt_2{}^+Cl^-$ ⟨70CB3918⟩.

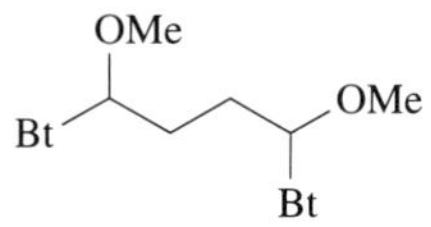

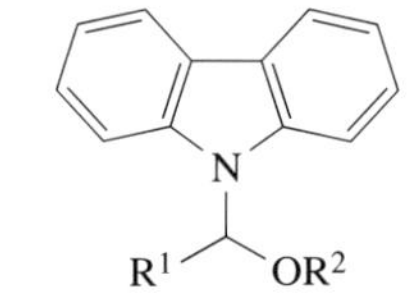

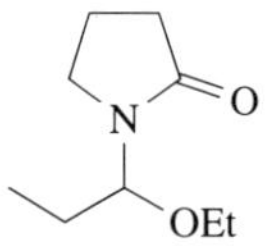

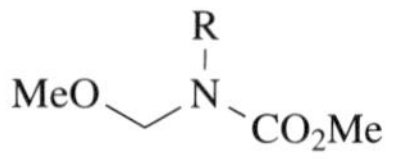

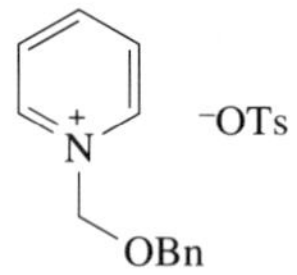

The acetoxy group in the *N*-acyl-2-acetoxyamino acid (**92**) is displaced on treatment with methanoltriethylamine to give (**93**) (Equation (11)) ⟨79JOC391⟩.

$$\underset{(\mathbf{92})}{\text{EtO}_2\text{C}\diagdown\!\!\!\diagup\text{NHCOMe}}^{\text{OAc}} \quad \xrightarrow[98\%]{\text{MeOH, Et}_3\text{N}} \quad \underset{(\mathbf{93})}{\text{EtO}_2\text{C}\diagdown\!\!\!\diagup\text{NHCOMe}}^{\text{OMe}} \qquad (11)$$

(d) Formation by miscellaneous routes. The anodic oxidation of amides (see Section 4.07.1.1.1(i)(b)) is also applicable to *N*-alkylanilines. The reaction is highly regiospecific; thus when *N*-ethyl-*N*-methylaniline was oxidized in methanol the methyl group was methoxylated to give PhN(Et)CH$_2$OMe. *N,N*-Dimethylaniline affords either the mono- or dimethoxylated product depending on the conditions ⟨82JA5753⟩.

Oxidation of amino acids can provide *O,N*-acetals. Lead(IV) acetate promotes oxidative cleavage of *N*-benzoylserine and *N*-benzoylthreonine ethyl esters to give either HOCH(CO$_2$Et)NHCOPh or AcOCH(CO$_2$Et)NHCOPh, depending on the conditions ⟨70CB2314, 91JA1042, 91TL3163⟩. Anodic oxidation of monoethyl acetamidomalonate in methanol results in decarboxylation and methoxylation with the formation of MeOCH(CO$_2$Et)NHAc ⟨79BCJ826⟩.

The Polonovski reaction of trimethylamine *N*-oxide with acetic anhydride give Me$_2$NCH$_2$OCOMe ⟨68CJC385, 75LA1790⟩. Later variants of the reaction have been used to obtain α-silyloxy alkylamines, R$^1_2$NCHR^{2}OSiMe$_2$Bu^t ⟨87BCJ3291⟩ and α-selenenyloxyalkylamines, R$^1_2$NCHR^{2}OSePh ⟨87CL1575⟩. These reactions involve initial reaction of the amine oxide with either TBDMS triflate or benzeneselenenyl triflate followed by treatment with base.

The electrocyclic ring opening of substituted aziridines generates azomethine ylides which afford *O,N*-acetals when intercepted by water or alcohols. However, many of these reactions are complex and mixtures of products frequently result ⟨83CHE(42-1)133⟩.

4.07.1.1.2 *Functions with tricoordinate nitrogen bearing heteroatom substituents*

When compared to those considered in the previous section there are relatively few hemiaminals possessing NCl, NOH, NNH$_2$, NNO, NNO$_2$ or NSR functions. Of these, the last three are probably the most accessible.

(i) From compounds containing a multiply bonded functional group

(a) From aldehydes and ketones. Amines possessing an electron-withdrawing group on nitrogen generally react readily with aldehydes to give relatively stable *O,N*-acetals.

N-Nitroamines condense readily with aqueous formaldehyde to give the corresponding *N*-hydroxymethyl-*N*-nitroamine, RN(NO$_2$)CH$_2$OH, in which the R group can be Me, Et, Pr^i or Bu^n ⟨60CCC2334, 84LA1494⟩. The products are relatively stable and may be *O*-acylated by standard procedures.

p-Toluenesulfonamide and phenylglyoxaldehyde condense when heated in dioxane to give PhCO-CH(OH)NHTs. The reaction is also successful with a variety of substituted phenylglyoxaldehydes ⟨68AP(301)867⟩.

Aldehydes which lack an α-hydrogen atom generally react with *N*-sulfinylsulfonamides to give *N*-sulfonylimines. However the reaction with trifluoroacetaldehyde hydrate affords an *O,N*-acetal ⟨64CB490⟩. Highly electrophilic ketones such as hexafluoroacetone or 1,3-dichloro-1,1,3,3-tetrafluoroacetone react readily with arenesulfinamides to give, for example (**94**) ⟨77LA624⟩. The stable N—N linked hemiaminal (**95**) is obtained from hexafluoroacetone and hydrazine ⟨73AG(E)502⟩. The double addition of hydrazine to both carbonyl functions in α-keto perfluoroalkane carboxylic esters has been described and a number of complex *O,N*-acetals have been obtained by this procedure ⟨83IZV2568⟩.

Successful hemiaminal preparations using hydrazines include the formation of (**96**) from *o*-phthalaldehyde and phenylhydrazine ⟨85CC1183⟩.

(**94**) (**95**) (**96**)

A variation of the Mannich reaction permits access to *N*-nitroso *O*,*N*-acetals. Aldehydes, primary aliphatic amines and an alcohol react in acetic acid. In the presence of sodium nitrite the initially formed aminal is transformed to the nitrosamine (Equation (12)). Both formaldehyde and aromatic aldehydes can be used, as can both primary and secondary alcohols ⟨72LA(765)55⟩.

$$\text{PhCHO} + \text{Bu}^n\text{NH}_2 + \text{MeOH} \xrightarrow[33\%]{\text{NaNO}_2, \text{AcOH}} \text{Ph}-\overset{\text{OMe}}{\underset{\underset{\text{ON}}{\text{N}-\text{Bu}^t}}{|}} \quad (12)$$

3-(Hydroxymethyl)triazenes of the type HOCH$_2$N(Me)N=NAr have been obtained from the reaction of formaldehyde, methylamine and a diazonium salt. The reaction is assisted by electron-withdrawing groups in the diazonium salt ⟨84JCR(S)108⟩.

(b) From imines and nitrones. The addition of water to *N*-sulfonylimines such as Cl$_3$CCH=NTs proceeds under mild conditions to give hemiaminals. The hexafluoroacetone imine (F$_3$C)$_2$C=NSOPh reacts readily with alcohols to give (**97**; R = Me or Et) ⟨77LA624⟩. The addition of alcohols to the C=N bond in imidoyl chlorides such as Cl$_2$CHC(Cl)=NSO$_2$Ar proceeds without halide displacement to give *N*-sulfonylhemiaminals ⟨84ZOR1502⟩. The addition of nitrosyl chloride to imines at low temperature in dichloromethane generates α-chloroalkylnitrosamines which were not isolated but were intercepted by silver acetate. Application of this procedure to EtCH=NPr^n gave (**98**). A considerable range of alkoxyalkyl nitrosamines is accessible by this procedure ⟨84LA1468⟩.

Although possessing two electrophilic sites, only one of the azomethine functions in hexafluoroacetone azine reacts with alcohols. Monoaddition was also observed with ethane-1,2-diol which gave (**99**) ⟨76JFC(7)471⟩.

(**97**) (**98**) (**99**)

The oxidation of imines to oxaziridines may be achieved under a great variety of conditions, the most commonly employed oxidants being hydrogen peroxide or a peracid. The scope of this reaction is immense and a huge number of oxaziridines have been prepared. Two examples which illustrate the usefulness of this reaction are provided by the perbenzoic acid oxidation of *N*-*t*-butylbenzaldimine to (**100**) ⟨73OSC(5)191⟩ and by the preparation of *trans*-2-phenylsulfonyl-3-phenyloxaziridine. The latter was obtained by oxidation of PhCH=NSO$_2$Ph with mcpba under phase transfer catalysis ⟨93OSC(8)546⟩. Nitrones may be oxidized to give hemiaminals. For example, *N*-benzyl-*C*-phenylnitrone has been oxidized to (**101**) by Pb(OAc)$_4$ ⟨73TL1889⟩. The reaction with ketonitrones leads to cleavage of the imine bond.

(**100**) (**101**)

(c) From alkenes. The examples here relate to the preparation of small ring heterocycles. Good yields of substituted nitrooxiranes are obtained from the oxidation of nitroalkenes with dilute alkaline hydrogen peroxide ⟨69CC369⟩. The addition of *N*-sulfinylsulfonamides, ArSO$_2$N=S=O, to enol ethers gives the 3-alkoxy-1,2-thiazetidines 1-oxides (**102**) ⟨66CB3903⟩.

(**102**)

(ii) From compounds containing two singly bonded functional groups

(a) From X—C—N functions (X = Hal). N-Chloromethyl-N-ethylnitramine is readily prepared and when treated with sodium alkoxides affords compounds of the type $ROCH_2N(NO_2)Et$. An alternative approach involves substitution of alkyl chloromethyl ethers by sodium nitramides, although a mixture of both N- and O-alkylated products is obtained which is difficult to separate ⟨73S302⟩.

(b) From O—C—X functions (X = Hal). Chloromethyl methyl ether reacts with lithium bis(trimethylsilyl)amide in THF at $0\,°C$ to give $MeOCH_2N(TMS)_2$ in high yield ⟨84AG(E)53, 84CC794⟩. Sulfamides such as Me_2NSO_2NHMe are alkylated with α-halo ethers under phase transfer conditions ⟨82AP(315)852⟩.

4.07.1.1.3 Functions with dicoordinate nitrogen

A number of hemiaminals possessing dicoordinate nitrogen are known. In many cases these compounds are available by simple variations of procedures outlined in the foregoing sections. The systems obtained include O—C—N=C, O—C—N=C=O, O—C—N=NR, O—C—N=O and O—C—N₃ groups.

(i) From compounds containing a multiply bonded functional group

(a) From aldehydes and ketones. Formaldehyde condenses with isocyanic acid at low temperatures to give the dangerously unstable hydroxymethyl isocyanate. However, if the reaction is performed in the presence of 3,4-dihydro-2H-pyran, the ether (THP)OCH₂N=C=O is obtained ⟨63JOC1825⟩. Trialkylsilyl isocyanates react readily with formaldehyde to give silyloxymethyl isocyanates ⟨81ZOB2382, 82ZOB1386, 83ZOB119⟩.

Although HN_3 will add to highly electrophilic carbonyl compounds such as hexafluoroacetone or perfluorocyclobutanone, a more general procedure for the preparation of α-azido alcohols is available. Trimethylsilyl azide and aliphatic aldehydes react rapidly when heated in the presence of zinc chloride to give 1-trimethylsilyloxyalkyl azides, $TMSOCH(R)N_3$ ⟨67T2781, 75LA266⟩.

Aliphatic or aromatic diazenes (e.g. $Bu^tN=NH$) may be generated from a variety of precursors. In the presence of an aldehyde, O,N-acetals of the type $HOCH(R)N=NBu^t$ are obtained ⟨71CB1104⟩.

The three-component reaction between an aldehyde, hydrogen azide and an alcohol provides an alternative means of obtaining 1-alkoxyalkyl azides. The reaction, which is performed in dichloromethane at ambient temperatures, is catalysed by $TiCl_4$. A wide range of compounds may be obtained by this procedure (Equation (13)). A variation of this method has been applied to ketones, which with HN_3 and trimethyl orthoacetate give 1-azido-1-methoxyalkanes ⟨88JOC22⟩.

$$Ph_2CHCHO + HN_3 + MeOH \xrightarrow[84\%]{TiCl_4} Ph_2C(OMe)(N_3) \quad (13)$$

The condensation of an α-hydroxy ketone, an aliphatic or aromatic aldehyde and ammonia provides 2,5-dihydrooxazoles (Scheme 20). Although the yields are variable, the attraction of the procedure lies in its experimental simplicity. This reaction represents an extension of the Asinger condensation ⟨95T755⟩. 5,6-Dihydro-2H-1,3-oxazines are available by a similar route (Scheme 20) ⟨93T9495⟩.

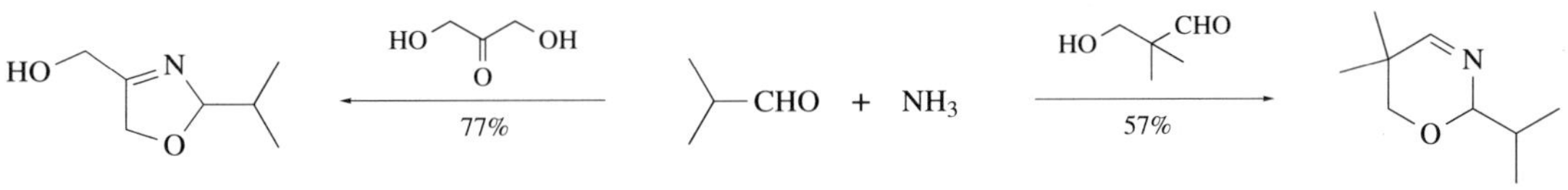

Scheme 20

(b) From hydrazones and oximes. Formaldehyde 2,4-dinitrophenylhydrazone was oxidized to (**103**; R = H) in high yield by $Pb(OAc)_4$ in acetic acid. A small amount (ca. 3%) of the isomeric acylhydrazine (**104**; R = H) was also obtained ⟨75JCS(P1)61, 77JCS(P1)282⟩. The reaction is also effective

with other dinitrophenylhydrazones, but in most cases the yield of (**104**) is considerably greater than (**103**). The procedure is also successful when the 2-nitro-, 4-nitro-, 2-bromo- or 2-methyl-phenylhydrazones are employed.

(**103**)   (**104**)

The oxidation of acetophenone phenylhydrazone with lead(IV) acetate in dichloromethane gave (**105**) ⟨61JA747⟩ and a number of other ketone hydrazones have been similarly oxidized. When benzophenone phenylhydrazone was treated with Pb(OAc)$_4$ in ethanoldichloromethane, two products were formed. The minor compound was (**106**; R = Ac) whilst (**106**; R = Et) was obtained in 80% yield ⟨67JCS(C)735⟩.

The Pb(OAc)$_4$ oxidation of phenylhydrazones with a functionalized side chain leads to heterocyclic systems. Thus (**107**) was obtained from levulinic acid ⟨69CJC3983⟩, whilst Me$_2$C=NNHCONHPh gave the oxadiazole (**108**) ⟨69JOC3230⟩.

The oxidation of ketoximes with Pb(OAc)$_4$ leads to nitroso compounds. Cyclohexanone oxime gave 1-acetoxy-1-nitrosocyclohexane. If the reaction is performed in the presence of a carboxylic acid, a variety of 1-acyloxy-1-nitrosocyclohexanes can be prepared ⟨66LA(700)1⟩.

(**105**)   (**106**)   (**107**)   (**108**)

(c) From alkenes. Enol ethers have proved to be a useful source of 1-alkoxyalkyl isocyanates by the acid-catalysed addition of isocyanic acid (Equation (14)) ⟨63JOC2082⟩. Hydrazoic acid reacts readily with silyl enol ethers, which, in the presence of an alcohol, undergo *trans*-etherification giving alkyl 1-azidoalkyl ethers. The reaction is catalysed by TiCl$_4$ or TFA ⟨84JOC4237, 88JOC22⟩. An alternative approach to these compounds utilizes a cohalogenation reaction of vinyl azides with alcohols (Equation (15)) ⟨71JA5469⟩.

(14)

(15)

(ii) From compounds containing two singly bonded functional groups

(a) From X—C—N functions (X = Hal). Halide displacements from compounds with an adjacent dicoordinate nitrogen function are mostly unremarkable and proceed as expected. However, α-chloroalkyl isocyanates such as Ph$_2$C(Cl)N=C=O show differing behaviour towards alkoxides and phenoxides. The former give the α-alkoxyalkyl isocyanate whilst with the latter the initially formed aryloxy isocyanates, Ph$_2$C(OAr)N=C=O, are unstable and rearrange to the ketimines Ph$_2$C=NCO$_2$Ar ⟨74JOU1565⟩. Interestingly, α-aryloxyalkyl isothiocyanates, obtained by a similar procedure, are stable and show no tendency to rearrange ⟨79CB1956⟩.

(b) From O—C—X functions (X = Hal, OR). Azidoalkyl alkyl ethers, N$_3$CH(R^1)OR^2, are readily obtained from the corresponding halo ethers and sodium azide in either water, acetonitrile, DMF or DMSO as solvent ⟨69JHC921⟩. In some cases phase transfer catalysis is useful ⟨83S568⟩. Acetals undergo substitution of one alkoxy group when treated with trimethylsilyl azide in the presence of

SnCl₄. The method is adaptable to the preparation of a wide range of compounds of the type $R^1R^2C(OR^3)N_3$ ⟨83S500, 84S683⟩. Alkoxyalkyl isocyanates are available from halo ethers and silver cyanate ⟨B-77MI 407-01⟩.

4.07.1.2 Functions Bearing Sulfur and Nitrogen

Functional groups represented by the general structure (**109**) are known as *S,N*-acetals or hemithioaminals. The number of possible permutations of groups based on the S—C—N array is even greater than for the *O,N*-acetals (see Section 4.07.1.1.) since the sulfur may be a di-, tri-, tetra- or higher coordinate. By far the most commonly encountered examples of the last two are represented by (**110**) and (**111**) which possess sulfinyl and sulfonyl groups respectively. The S—C—N fragment is commonly encountered in many heterocyclic structures and a plethora of examples abound in which this unit is either wholly or partially incorporated into partially saturated semicyclic, cyclic or polycyclic ring systems.

Synthetic routes to *S,N*-acetals have been the subject of an excellent review which provides a wealth of experimental procedures for the preparation of a wide range of structural types including many complex fused and spiroannulated heterocycles ⟨92HOU(E14a/3)483⟩.

$$
\begin{array}{ccc}
R^1\!\!\!\diagup\!\!\!\!\diagdown\!SR^4 & R^1\!\!\!\diagup\!\!\!\!\diagdown\!\overset{\overset{O}{\|}}{S}\!-\!R^4 & R^1\!\!\!\diagup\!\!\!\!\diagdown\!\overset{R^4\diagdown\!\!\diagup O}{\underset{\diagup\!\!\diagdown O}{S}} \\
R^2\quad NR^3{}_2 & R^2\quad NR^3{}_2 & R^2\quad NR^3{}_2 \\
(\mathbf{109}) & (\mathbf{110}) & (\mathbf{111})
\end{array}
$$

4.07.1.2.1 Dicoordinate sulfur derivatives

(i) Functions with tricoordinate nitrogen bearing alkyl or aryl substituents

(a) From compounds containing a multiply bonded functional group. (1) From aldehydes and ketones. The most versatile route to *S,N*-acetals is based on a variant of the Mannich reaction ⟨73S703, 91COS(2)893⟩ in which the S—C—N array is formed in a one-pot, three-component reaction between an aldehyde and an amine in the presence of a thiol (Equation (16)). Wide variations of R^1, R^2 and R^3 are compatible with the reaction. The condensation may be performed under a variety of conditions, either in the absence of a solvent or with ethanol, chloroform or toluene as diluent. Reaction times, conditions and temperatures may vary considerably. Although the transformation depicted in Equation (16) resembles the classical Mannich α-aminoalkylation at nitrogen, it may also be regarded as an α-thioalkylation reaction. The full scope of thioalkylations with aldehydes and thiols has been reviewed ⟨87S589⟩.

$$R^1CHO \ + \ R^2SH \ + \ HNR^3{}_2 \quad \longrightarrow \quad R^1\!\!-\!\!\!\diagup\!\!\!\!\diagdown\!\!\!\overset{SR^2}{\underset{NR^3{}_2}{}} \tag{16}$$

Hemithioaminals are readily available from the condensation of formaldehyde and a primary or secondary amine in the presence of a thiol. Initial studies on the synthesis of alkyl dialkylaminomethyl sulfides, $R^1SCH_2NR^2{}_2$, by this route utilized saturated aqueous potassium carbonate as solvent ⟨21JCS1470⟩. Subsequent investigations found that this medium is unnecessary and could even lead to reduced yields. Morpholine condenses with aqueous formaldehyde and propane-2-thiol at 30 °C in the absence of solvent to give (**112**) in high yield. An analogous product is obtained from 2-methylpropane-1-thiol (Scheme 21). Morpholine may be replaced by other secondary amines such as piperidine or diethylamine. Aromatic thiols behave similarly, affording good yields of products ⟨66JCED620⟩, although in some instances higher reaction temperatures may be necessary. The reaction is compatible with a variety of substituents in the aromatic ring. In general, the yields of products are independent of electronic influences since the hemithioaminals (**113**) derived from 4-chloro-, 4-methyl-, 4-methoxy- and 4-nitrothiophenol are obtained in high yields. However, the reaction appears to be more susceptible to steric influences. Thus yields of the

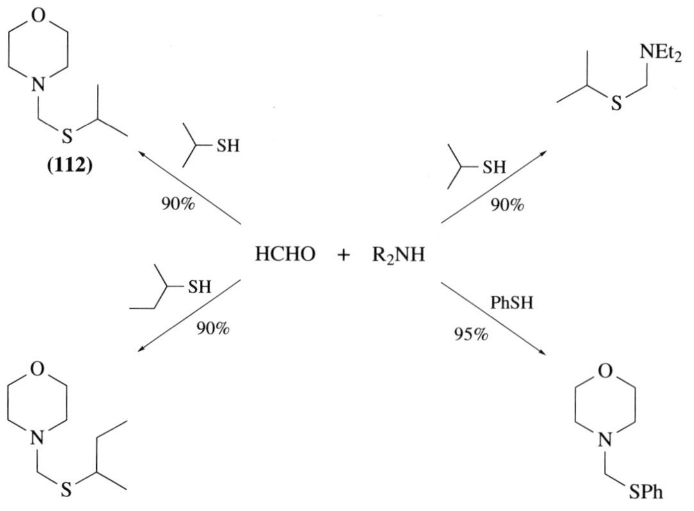

Scheme 21

ortho-substituted compounds (**114**), (**115**) and (**116**) are lower. The hindered *S,N*-acetal (**116**) derived from 2,4,6-trimethylthiophenol and diethylamine is obtained in 40% yield, but with the bulkier morpholine this is reduced to 22% ⟨54JA3969⟩. Thiophenols show no tendency to form the ring aminomethylated products (**117**), a characteristic reaction of phenols ⟨91COS(2)953⟩.

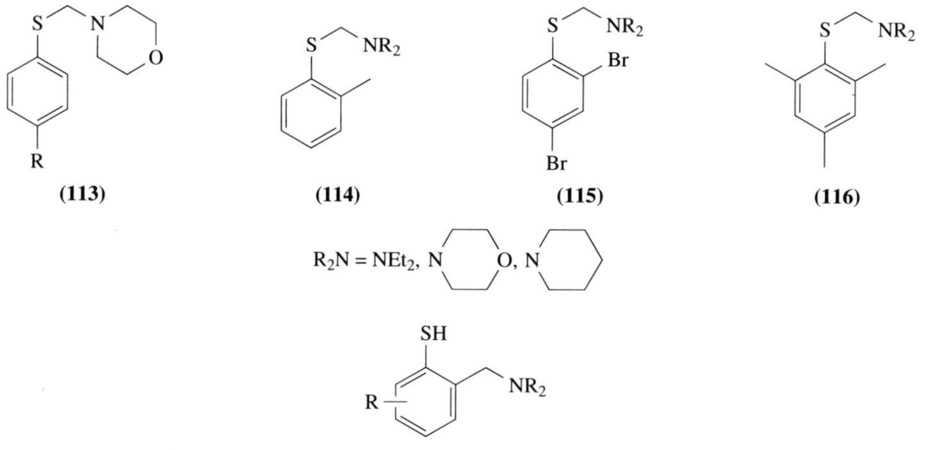

$R_2N = NEt_2$, N⌒O, N⌒ (morpholine, piperidine)

Variation of the amine component is also possible. Whilst aliphatic and alicyclic secondary amines afford hemithioaminals from thiophenols under mild conditions, the corresponding reaction with *N*-alkylanilines proceeds much more sluggishly unless the reaction is conducted in boiling ethanol (Equation (17)). Yields of products are generally high ⟨59JOC1035⟩.

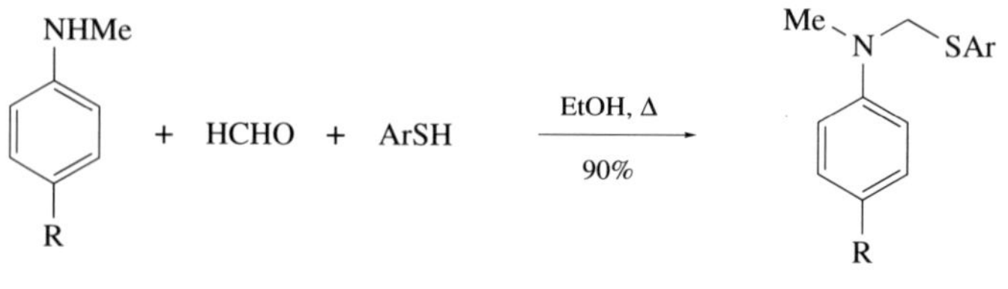

In a similar vein, primary aromatic amines undergo mono-*N*-arylthiomethylation with equimolar quantities of formaldehyde and a thiophenol. Yields of products are only moderate owing to competitive formation of *N,N*-bis(arylthiomethyl)anilines, $Ar^1N(CH_2SAr^2)_2$. The latter are the sole products when the ratio of amine : formaldehyde : thiol is 1 : 2 : 2 ⟨59JOC1035⟩. Optimal conditions to effect the mono-*N*-arylthiomethylation of primary amines have been developed. Use of a twofold excess of the thiophenol appears to suppress the bis-alkylation reaction. This procedure is also applicable to acetaldehyde, propionaldehyde and benzaldehyde. The amine component may be either a substituted aniline or a heteroaromatic amine (Equation (18)) ⟨81JCS(P1)1569⟩. Aliphatic primary amines may also be both mono- and bis-*N*-thioalkylated, although this reaction has been much less widely applied. Methylamine hydrochloride reacts readily with formaldehyde and thiophenol in warm ethanol to give $MeNHCH_2SPh \cdot HCl$ ⟨77CPB2964⟩. The procedure is also applicable to primary aliphatic amines possessing a variety of functional groups. Thus equimolar quantities of alanine, formaldehyde and benzyl thiol condense in boiling ethanol to give $BnSCH_2NHCH(Me)CO_2H$ ⟨77CPB3385⟩. Excess formaldehyde and thiophenol effect a bis-alkylation giving $(PhSCH_2)_2NCH(Me)CO_2H$ ⟨81ZC403⟩.

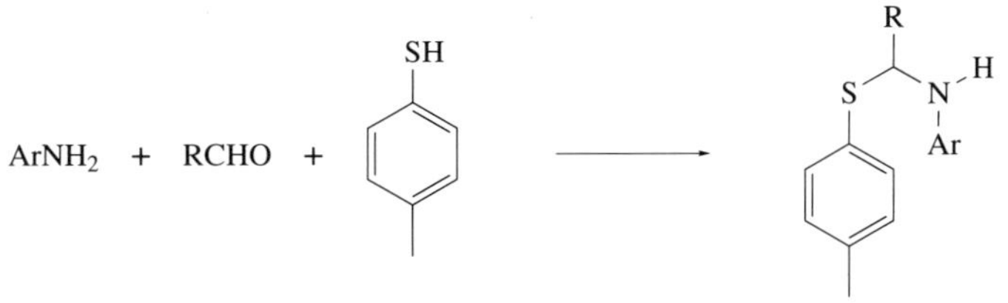

$$ArNH_2 \;+\; RCHO \;+\; \text{(thiocresol)} \longrightarrow \text{(S,N-acetal)} \qquad (18)$$

The *N*-arylthioalkylation reaction has also been applied to formaldehyde, thiophenol and ammonia where the only product, tris(phenylthiomethyl)amine $(PhSCH_2)_3N$, is formed cleanly and efficiently in ethanol at ambient temperature ⟨67JOC2891⟩. Earlier workers reported that analogous compounds are produced when thiophenols are condensed with hexamethylenetetraamine in refluxing anhydrous dioxane for 24 h ⟨33JA4588⟩ or in ethanol–acetic acid for 30 min ⟨47G375⟩. An anomalous trisalkylthiomethylation was observed in the reaction of 2,4-dimethyl-2-thiolpentan-3-one with formaldehyde and gaseous ammonia, from which $(Me_2CHC(O)CMe_2SCH_2)_3N$ was obtained ⟨58LA(619)169⟩. Sterically unencumbered α-thiol ketones normally afford 2,5-dihydrothiazoles under these conditions (see below).

A detailed study of conditions appropriate to the formation of *S*,*N*-acetals from alkyl or aryl aldehydes with thiols and secondary amines has been undertaken by Katritzky *et al.* Wide variations in the reactants are possible. The condensation proceeds efficiently with both straight-chain or branched and hindered aliphatic aldehydes and thiols. Aromatic aldehydes, such as furan- and pyridine-2-carbaldehydes, and aromatic and alkaryl thiols, for example benzyl thiol, can also be employed. The amine component may be pyrrolidine, morpholine, dibenzylamine or *N*-methylaniline. In general, the products are formed simply by allowing equimolar amounts of the reactants to stand for some time in a solvent such as ethanol, benzene, chloroform or water ⟨89CS33⟩. Earlier work by Katritzky demonstrated that *S*,*N*-acetals could be formed from benzaldehyde, thiophenol and either dimethylamine or piperidine under very mild conditions ⟨86S804⟩. The bis-hemithioaminal (**118**) was obtained from terephthalaldehyde together with a small amount of the aminal (**119**), a reaction intermediate. The latter could be readily converted into (**118**) by treatment with thiophenol (Scheme 22). This aminal → hemithioaminal conversion has been observed previously and is especially rapid under acid catalysis ⟨69CR(C)1718⟩.

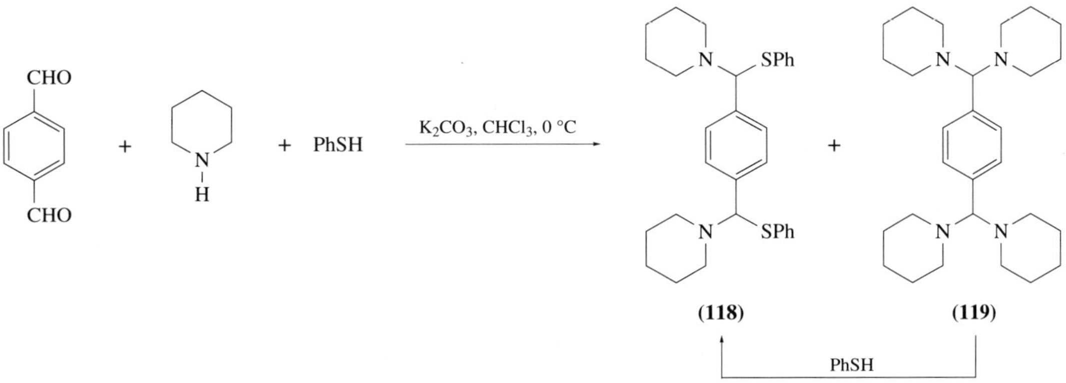

Scheme 22

There are relatively few examples of acyclic *S,N*-acetals in which the nitrogen forms part of a heteroaromatic ring. Benzotriazole reacts as a normal secondary amine and condenses with thiols and carbonyl compounds when refluxed in benzene containing a catalytic amount of *p*-toluenesulfonic acid. Although the procedure works well with aldehydes and cycloalkanones, aliphatic ketones are more troublesome. Acetone affords only low yields of products, whilst acetophenone fails to react. The products from these reactions, *N*-[1-(alkylthio)alkyl]- and *N*-[1-(arylthio)alkyl]-benzotriazoles, are formed as mixtures of 1*H*- and 2*H*-isomers in which the former usually predominate. In some instances these isomers can be separated by flash chromatography 〈91HCA1924, 91HCA1936〉. The reaction with cyclohexanone is noteworthy, since it appears to provide the first example of an acyclic *S,N*-acetal prepared from a ketone (Scheme 23).

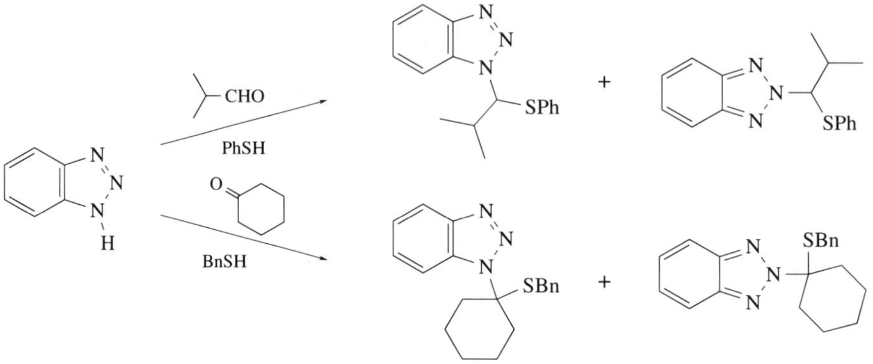

Scheme 23

Another variation of the amine–thiol–carbonyl compound condensation provides heterocyclic *S,N*-acetals. Two approaches are shown in Scheme 24. 2,3-Dihydrothiazoles are obtained 〈59LA(622)83, 59LA(622)94〉 when α-thiol ketones such as (**120**) are treated with an amine and a carbonyl compound. Although mechanistic details are obscure, the reaction may be envisaged to proceed via aminoalkylation at sulfur to generate an acyclic hemithioaminal, which subsequently cyclizes to the product. The reaction is applicable to other thiol ketones and aldehydes or ketones. Rühlmann observed a somewhat similar reaction in the formation of 1,3,4-thiadiazolidines from hydrogen sulfide, a carbonyl compound and hydrazine. Yields of products are generally high although the reaction with formaldehyde is less successful 〈59JPR285〉.

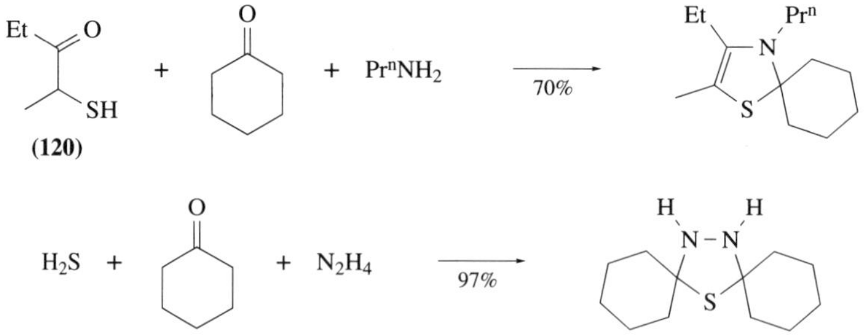

Scheme 24

The reaction of carbonyl compounds with thiols possessing a proximal amino group provides a very useful means of obtaining heterocyclic *S,N*-acetals and a great many compounds have been prepared, exemplified in Equations (19)–(25). Thiazolidines are readily available from derivatives of 2-aminoethanethiol and an aldehyde. The reaction is usually performed in ethanol at room temperature (Equations (19) and (20)) 〈B-49MI 407-01〉. In a modification of this reaction, heating the amino thiol with formaldehyde and formic acid effects both ring closure and methylation in a single step by an Eschweiler–Clarke reaction (Equation (21)) 〈86S139〉. Cysteine methyl ester readily condenses with a variety of aliphatic aldehydes under mild conditions leading to methyl 2-alkyl-thiazolidine-4-carboxylates (Equation (22)) 〈85CL939〉. The outcome of the reaction of 2-amino-thiophenol with aldehydes is complex. In some instances, benzothiazolines are isolated (Equation (23)), 〈84CHEC(6)321〉, although these are often prone to aromatization by air oxidation. *N*-Substituted benzothiazolines are air stable and have been prepared by the reaction of 2-(methylamino)thiophenol with both aldehydes and ketones (Equation (24)) ⟨89BCJ1215⟩ in boiling ethanol. Six-membered rings are accessible by an analogous reaction giving 1,3-benzothiazine derivatives (Equation (25)) ⟨72ACH363, 87H(26)2381⟩.

(2) *From thiocarbonyl compounds.* Thioketones have however found limited application in the preparation of heterocyclic hemithioacetals. Thiobenzophenone is an effective 1,3-dipolarophile and readily adds to the unstabilized azomethine ylide generated *in situ* by fluorodesilylation of (**121**) with lithium fluoride in acetonitrile. The efficiency of the reaction is increased markedly by sonication ⟨87JOC235⟩. A [4 + 2]-cycloaddition between the dithioester (**122**) and 1-piperidinopropene affords a high yield of the 2-aminothiopyran (**123**), a semicyclic *S,N*-acetal (Scheme 25). The orientation of the product suggests that the reaction is not concerted but proceeds through a zwitterionic intermediate ⟨84JCS(P1)865⟩.

(3) *From imines and iminium salts.* The addition of thiols to the azomethine group in Schiff bases occurs under mild conditions to give *S,N*-acetals in high yields. Although some of the initial studies cast doubt on the generality of this reaction ⟨30JA4573, 52JA3885⟩, a careful reinvestigation confirmed the usefulness of this approach. Failures were ascribed to the use of an excess of thiol and the high temperatures reached in boiling dioxane or toluene causing reduction of the imine or to hydrolysis during the reaction by adventitious moisture or as a result of an aqueous alkaline workup. Benzylideneaniline and its derivatives react with both aromatic and aliphatic thiols in benzene at room temperature. Product formation is generally complete within 1 h. Electron-withdrawing substituents in the benzylideneaniline do not appear to influence the rate of the reaction (Equation (26)) ⟨55JA3869⟩. When the reaction is performed with a functionalized thiol such as thiolacetic acid, the initially formed *S,N*-acetal cyclizes to a 4-oxothiazolidine ⟨92HOU(E14a/3)483⟩.

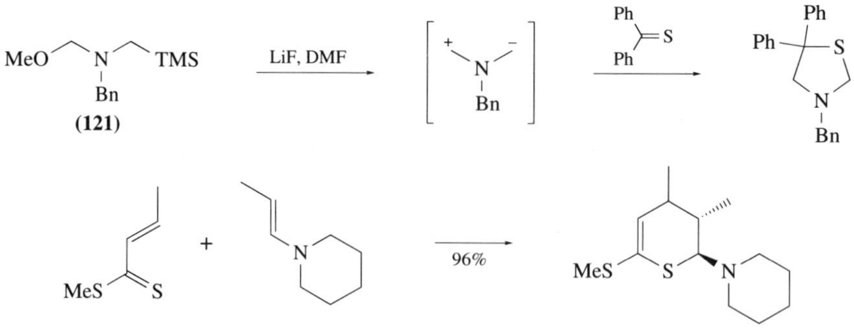

Scheme 25

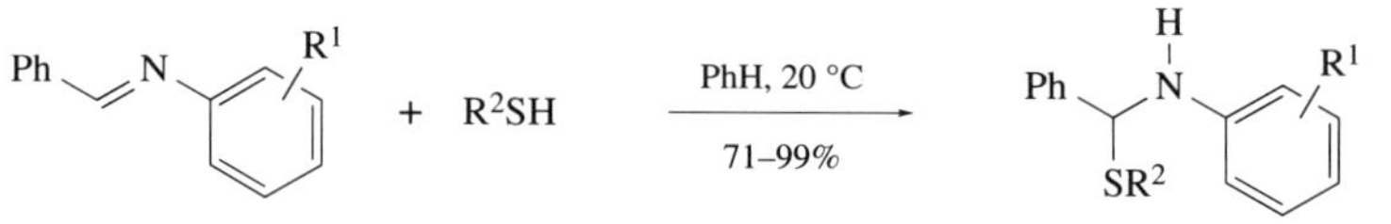

R¹ = H, 2-CO₂H, 4-CO₂H, 2-COMe, 4-COMe; R² = alkyl, aryl (26)

The thionation of benzil monoanils with 2,4-bis(phenylthio)-1,3,2,4-dithiadiphosphetane-2,4-dithione (**124**) produces the hemithioaminals (**125**) in high yields. The product results from initial thionation followed by addition of thiophenol, generated by *in situ* hydrolysis of (**124**), to the azomethine group. This behaviour contrasts with that of benzil dianils which on treatment with (**124**) give a thioimidate produced by a pathway involving homolysis of the dianil to generate an iminyl radical (Scheme 26) ⟨92PS(66)87⟩.

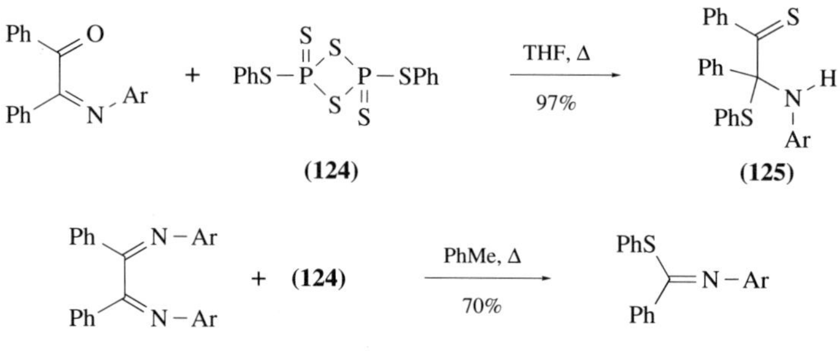

Scheme 26

Further variations on the imine–thiol reaction are possible. The azomethine moiety of 5(4*H*)-oxazolones, which are tautomeric with 5(2*H*)-oxazolones (**126**), is readily attacked by thiols, especially under acidic conditions. Thus, 4-substituted 2-trifluoromethyl-5(4*H*)-oxazolones, available by the trifluoroacetylation of α-amino acids, when treated with 3 equiv. of ethanethiol in acetic acid–HBr afford high yields of the α-keto acid dithioacetals (**127**) and 1-ethylthio-2,2,2-trifluoroethylamine hydrobromide (Scheme 27). The reaction proceeds through an initial attack of EtSH on protonated (**126**) which subsequently fragments to generate highly electrophilic sulfonium and iminium ions. These species are intercepted by excess thiol to give the products ⟨62LA(658)128⟩. Under identical conditions, 2-difluoromethyl-4-methyl-5(2*H*)-oxazolone fragments to give 2,2-difluoro-1-ethylthioethylamine hydrobromide, H₂NCH(CHF₂)SEt·HBr, in 93% yield ⟨93S961⟩. In general, thiols attack C-5 of 5(4*H*)-oxazolones to give thiol esters as the products of ring cleavage ⟨84CHEC(6)117⟩.

The application of preformed iminium salts for the synthesis of *S,N*-acetals has not been investigated to any great extent, despite its considerable potential. However, the iminium salt, Me₂N⁺=CH₂ Cl⁻, has been treated with metal dithiocarbamates to give Me₂NCH₂SCSNR₂

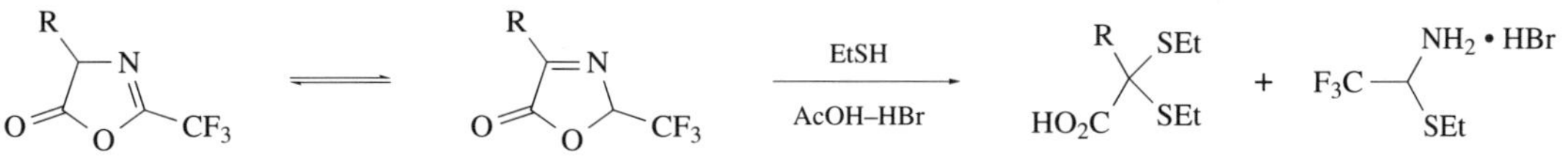

Chalcogen and a Nitrogen

Scheme 27

⟨66AP(299)906⟩. Related compounds were obtained from potassium alkylxanthates and *O*-alkyl-thiocarbonates ⟨67AP(300)647⟩. Prop-2-yniminium triflates show ambident reactivity towards thiols and thiolates. The salt (**128**) is attacked at C-1, which, according to MNDO calculations, possesses the highest charge density, by sodium cyclohexanethiolate or cyclohexanethiol-Hünigs base (EtNPr$_2^i$) in acetonitrile. The product is the hemithioaminal (**129**). Weaker *S*-nucleophiles such as cyclohexanethiol itself or sodium thiophenoxide give products resulting from C-3 attack leading to (**130**) and the allene (**131**), respectively (Scheme 28) ⟨89CB2311⟩. Nucleophilic addition to alkylthiomethyleneiminium salts, for example (**132**), available from the quaternization of *N,N*-dialkylthioamides with iodomethane, permits access to hemithioaminals. When treated with aqueous potassium cyanide at room temperature, the salts (**132**) afford (**133**) (Scheme 29) ⟨77JCS(P1)1811⟩. The iminium salt (**132**; R^1 = H, R^2 = NMe$_2$) affords PhCH(SMe)NMe$_2$ when reacted with phenyl-magnesium bromide ⟨72CI(L)380⟩.

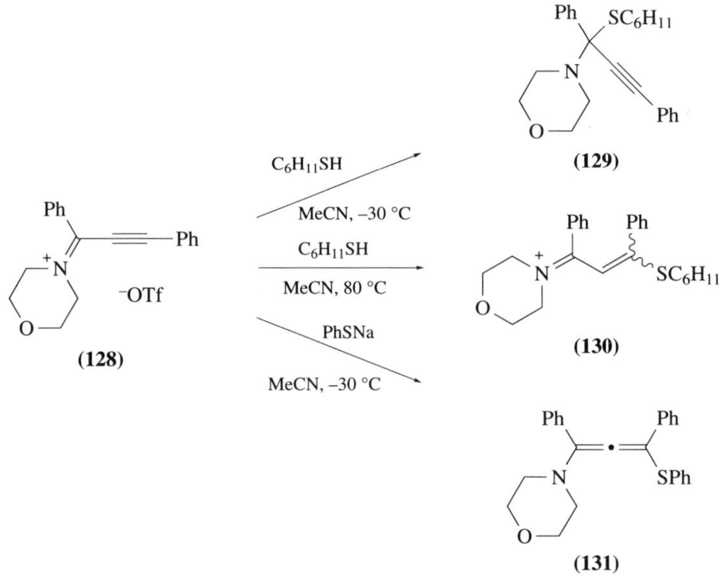

Scheme 28

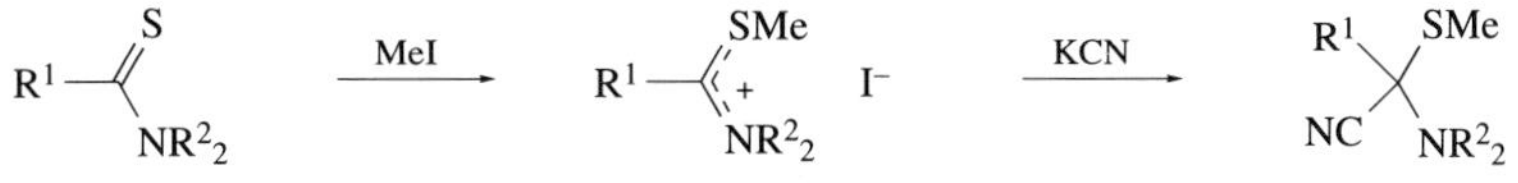

Scheme 29

Systems in which a thioimidate unit is incorporated into a heterocyclic ring are a potential source of cyclic *S,N*-acetals via nucleophilic addition. Examples include the preparation of thiazolidines ⟨67JCS(C)1411⟩, benzothiazolines ⟨78TL5, 85H(23)2509, 86S375⟩ and tetrahydro-1,3-thiazines ⟨66JOC556⟩ by hydride reduction of 4,5-dihydrothiazoles, 3-substituted benzothiazolium salts and 2-substituted 5,6-dihydro-4*H*-1,3-thiazines, respectively. Unsaturated Grignard reagents also add to the azomethine unit in 4,5-dihydrothiazoles ⟨87JOM(335)283⟩ and benzothiazoles ⟨86S638⟩ to give the expected 2-substituted products.

(b) From compounds containing two singly bonded functional groups. (1) From X—C—N functions (X = Hal, CN, OR, NR₂ or a metal). *N*-(Chloroalkyl)azoles are easily prepared shelf-stable materials and are thus excellent substrates for functionalization by halide displacement. 1-(Chloromethyl)benzotriazole undergoes smooth substitution with thiols to give the corresponding 1-(thioalkyl)benzotriazoles ⟨87JCS(P1)781⟩. High yields of products are normally obtained, although the reaction of (**134**) with sodium thiophenoxide failed, presumably as a consequence of steric hindrance (Equation (27)) ⟨91HCA1924⟩. 1-(Phenylthiomethyl)benzotriazole is readily lithiated on the CH₂ group by butyllithium at −78 °C and the resulting carbanion has been quenched with a range of alkyl halides, benzaldehyde, benzophenone and benzonitrile ⟨87JCS(P1)781⟩. 1-(Chloromethyl)-3,5-dimethylpyrazole hydrochloride undergoes an analogous series of reactions ⟨89CJC1144⟩. Further functionalization of these arylthiomethylazoles can be achieved by oxidation with sodium periodate or mcpba to afford the corresponding sulfoxides and sulfones.

$$\text{Bt-CHCl-Et} \xrightarrow[80\%]{\text{NaSPh}} \text{Bt-CH(SPh)-Et} \qquad (27)$$

Bt = benzotriazol-1-yl

(**134**)

Aminals are of value because of the ease with which they ionise to iminium ions, which can be intercepted *in situ* by a nucleophile. The ionization is promoted by either Bronsted or Lewis acids. Depending on the reaction conditions, either one or both of the amino groups may be substituted. An example is the behaviour of the α,α-di(morpholino)xylene (**135**) towards thiophenol. In the absence of solvent but in the presence of anhydrous HCl or sulfuric acid and at temperatures less than 40 °C, the *S,N*-acetal (**136**) is formed. If the reaction is performed in chloroform above 40 °C with 2 equiv. of thiophenol, (**137**) is obtained. A number of hemithioaminals have been obtained by this method ⟨69CR(C)1718, 70CR(C)436, 73BSF3499⟩, although aminals derived from aliphatic aldehydes readily undergo elimination to form enamines ⟨77ZOR1930⟩.

(**135**)   (**136**)   (**137**)

N-Acyl derivatives of aminals are readily available by the Mannich aminomethylation of amides and imides ⟨73S703⟩ and are useful precursors of hemithioaminals. Mannich bases of the type R¹₂NCH₂N(R²)COR³ possess two potential leaving groups and their reaction with thiols could therefore produce either *N*-alkyl or *N*-acyl derivatives of *S,N*-acetals. When *N*-(diethylaminomethyl)benzamide, Et₂NCH₂NHCOPh, was heated with benzyl thiol in boiling toluene in the presence of NaOH, diethylamine was eliminated to give BnSCH₂NHCOPh in 79% yield. A similar result was obtained using thiophenol ⟨57CB444⟩. The outcome of this reaction, an amidomethylation at sulfur, is entirely consistent with earlier observed behaviour of active methylene compounds ⟨54CB1690⟩ and amines ⟨57CB50⟩. A contrasting reactivity has been noted in the analogous reaction with benzyl thiol and *N*-(piperidinomethyl)benzamide (**138**). This substrate reacts by preferential substitution of the benzamido group to give (**139**) in 51% yield, whilst the yield of BnSCH₂NHCOPh was only 23%. Formation of the latter compound was completely suppressed with methanol as solvent, whilst the yield of (**139**) increased to 80%. Benzyl thiol in refluxing methanolic sodium hydroxide effected the substitution of the succinimide unit in (**140**), giving (**139**) in 89% yield. Reactions with thiophenol or ethanethiol follow an identical course although they are lower yielding ⟨73CPB2257⟩. The reaction has been extended to a range of *N*-(α-dialkylaminobenzyl)benzamides (Equation (28)) in which displacement of both amino and amido groups is observed. Treatment of (**141**) with benzyl thiol in boiling methanolic NaOH gave (142) and (143). Relatively minor modifications of the substrate markedly alter the course of events and the *N*-acylhemithioaminal (143) is the sole product of the morpholine analogue of (141) ⟨73CPB2257⟩.

(138)   (139)   (140)

(141) —BnSH, MeOH, Δ→ (142) + (143)   (28)

Other examples of thiolate substitution reactions on aminals result from the extensive work of Katritzky *et al.* on benzotriazole-mediated transformations ⟨91T2683⟩. The aminoalkylation of benzotriazole is accomplished by treating an ethanol solution of the heterocycle with an aldehyde and an amine. The products, 1-(benzotriazol-1-yl)alkylamines, are of value since the benzotriazolyl group is readily displaced by a wide range of nucleophiles. This process is particularly facile when the nucleophile is a thiolate, *S,N*-acetals being obtained in excellent yields. A very wide range of compounds is accessible by this route ⟨93S229⟩.

The *N*-benzoylaminal (144) reacts with 3-methylthiophenol to afford (145) in 84% yield. Analogous compounds have been prepared using formamide, acetamide, or *p*-toluenesulfonamide and other arene or alkane thiols ⟨91S1147⟩. Aminals derived from benzotriazole are available by other methods. 9-Vinylcarbazole affords (146) when heated with benzotriazole in chloroform under TsOH catalysis. Displacement of the triazolyl group by thiophenol, catalysed by zinc bromide, affords (147). An analogous sequence can be applied to enamides, 1-vinyl-2-pyrrolidinone giving (148) ⟨92S1295⟩. All of these routes are depicted in Scheme 30.

(2) *From S—C—X functions (X = Cl, Br, SR)*. The examples here appear to be limited almost entirely to the heterocyclic field. Examples include *N*-(phenylthiomethyl)carbazole ⟨85JOC1351⟩ and benzimidazole derivatives ⟨87JCS(P1)775⟩, which are obtained in high yields by treatment of the appropriate heterocycle with PhSCH$_2$Cl in DMF in the presence of base. Aziridine is similarly alkylated in acetonitrile containing potassium carbonate ⟨87IZV2399⟩.

(c) By functionalization of a preformed S,N-acetal. The sulfur function of hemithioaminals is susceptible to oxidation. Typical examples include the transformation of 1-(phenylthioalkyl)azoles to the corresponding sulfoxides and sulfones. These compounds are readily metallated, allowing functionalization of carbon in the S—C—N group. The *C*-arylation of Me$_2$NCH$_2$SPh has been achieved by alkylation with benzyl chloride and subsequent treatment with potassium *t*-butoxide in 1,2-dimethoxyethane affords a low yield of (149), the product of a Sommelet–Hauser rearrangement ⟨76BSF1482⟩.

(d) Formation by miscellaneous routes. The acyloxysulfonium salts, generated by treatment of dimethyl sulfoxide under Pummerer rearrangement conditions, are readily trapped by added nucleophiles. Amines afford *S,N*-acetals. Thus, (150) is obtained by treating DMSO and ethoxalyl chloride with 2-aminopyridine ⟨79CJC1153⟩, whilst heating 1-trimethylsilylimidazole with DMSO affords (151) and hexamethyldisiloxane ⟨79JHC415⟩. Both amido- and imidoalkylations of DMSO and other alkyl sulfoxides have been achieved. Intramolecular versions of this reaction are also known and an extensive review of this area is available ⟨91OR(40)157⟩.

(ii) Functions with tricoordinate nitrogen bearing acyl or heteroatom substituents

Although there is an immense number of possible *N*-functionalized hemithioaminals, relatively few have been obtained. By far the most commonly encountered examples are those in which the

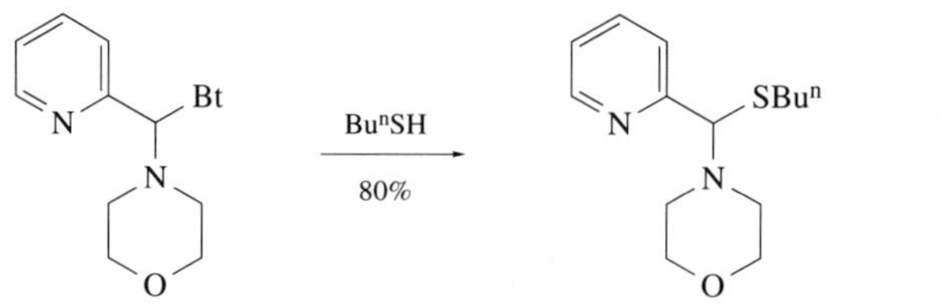

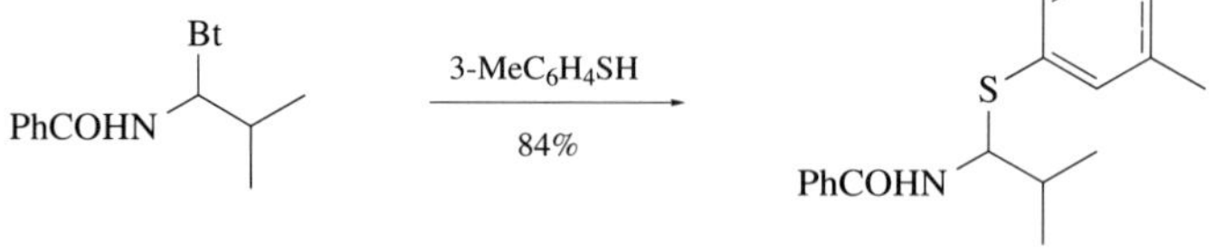

(144) → (145)

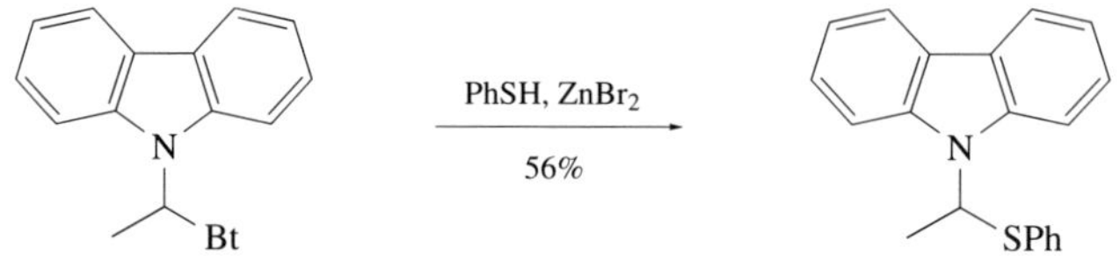

(146) → (147)

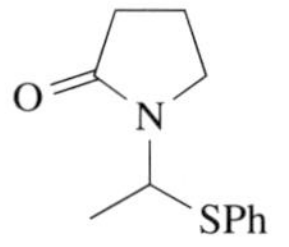

(148)

Scheme 30

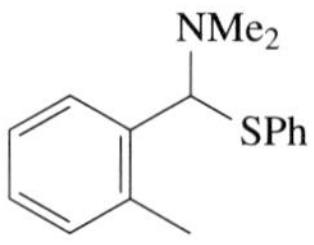

(149)

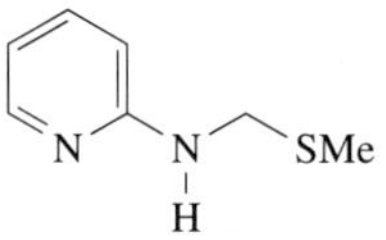

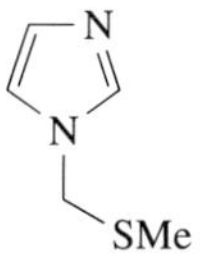

(150)   (151)

nitrogen function is attached to an alkanoyl, aroyl or sulfonyl group. The synthesis of compounds containing NCl, NOH, NNO, NNH$_2$ and NNO$_2$ groups or those possessing an α-nitro sulfide (S—C—NO$_2$) unit is, by comparison, relatively undeveloped. Whilst some of the hemithioaminals considered here, especially the *N*-acyl compounds, are available through adaptations of procedures described in the previous section, many synthetic protocols are sufficiently different to merit separate consideration.

(a) From compounds containing a multiply bonded functional group. (1) From aldehydes and ketones. The Mannich-type condensation between an aldehyde, an amide and a thiol, whilst of some value for the preparation of *N*-acylhemithioaminals, is in general much less efficient than the corresponding *N*-thioalkylation of amines, a consequence of the low nucleophilicity of amides to even the most reactive aldehydes. Frequently, attempts to obtain *N*-(thioalkyl)amides provide *N*-(hydroxyalkyl)amides as the sole products. It is often preferable to obtain these compounds initially and to effect conversion to *N*-acyl *S,N*-acetals in a separate step, usually by treatment with excess thiol.

The *N*-ethylthiomethylation of caprolactam has been achieved in a one-pot process initiated by formation of the *N*-hydroxymethyl derivative from paraformaldehyde in warm ethanol. Subsequent treatment with an excess of ethanethiol gave (**152**) in 65% yield after the mixture was allowed to stand overnight at room temperature. The transformation may also be conducted in the absence of solvent and at elevated temperatures. Under these conditions, (**153**) has been obtained from 2-thiolbenzothiazole after only 2.5 h at ca. 135°C ⟨48JA2115⟩. Thioalkylation of other simple amides such as acetamide or benzamide is best achieved by initial formation of the *N*-hydroxymethyl derivative (RC(O)NHCH$_2$OH; R = Me or Ph) and subsequent condensation with the thiol ⟨87S589⟩. The thioalkylation of urea proceeds in a single step under very mild conditions and (**154**) has been obtained from formaldehyde and 3-thiolpropionic acid in aqueous solution at ambient temperature ⟨87S589⟩. There do not appear to be any reports of the mono-*N*-(thioalkylation) of urea.

(**152**) (**153**) (**154**)

Cyclic *N*-acyl *S*,*N*-acetals are available from the condensation of thiolamides with aldehydes under acid catalysis. For example, the β-thiolamide, HS(CH$_2$)$_2$CONHMe, gave the thiazine (**155**) when condensed with acetaldehyde in boiling xylene. The γ-thiolamide, HS(CH$_2$)$_3$CONHBn, reacted with paraformaldehyde in an analogous fashion to give the thiazepine (**156**) ⟨78H(9)831⟩. There have been few applications of the *N*-thioalkylation reaction for the preparation of *S*,*N*-acetals possessing heteroatom substituents on nitrogen, the only known examples involving hydroxylamine or its derivatives. Both alkane and arene thiols react with formaldehyde and hydroxylamine in aqueous solution to afford bis(*N*-thiomethyl)hydroxylamines, (RSCH$_2$)$_2$NOH, as the sole products in excellent yields ⟨79CPB1691⟩. *N*-Alkyl- or *N*-arylhydroxylamines are *N*-thiomethylated in an analogous manner.

(**155**) (**156**)

(2) *From imines*. The imine (F$_3$C)$_2$C=NCO$_2$Bu^t reacts readily at room temperature with *t*-butyl thiol in methanol to give the acetal (**157**) ⟨75JOC2414⟩. Similarly, *N*-(hexafluoroisopropylidene)benzenesulfinamides condense smoothly with both ethanethiol and thiophenol in benzene at room temperature. The *N*-(benzenesulfinyl) acetals (**158**; R = Et or Ph) are obtained in near quantitative yields ⟨77LA624⟩. Although possessing two reactive imine groups, hexafluoroacetone azine affords only 1 : 1 addition products (**159**) when treated with butane- or hexane-thiol, thiolacetic acid, or thiophenol either neat or in diethyl ether ⟨76JFC(7)471⟩.

The azaallyl anion generated at −78°C from EtO$_2$CCH$_2$N=C(SMe)$_2$ and potassium *t*-butoxide in THF condenses with 1,1,1-trifluoroacetone by a complex pathway to give the *N*-acylhemithioaminal (**160**) ⟨92JCS(P1)291⟩.

(**157**) (**158**) (**159**) (**160**)

Other approaches to *N*-acyl *S*,*N*-acetals have involved halide displacement by thiols from *N*-acyl(α-chloroalkyl)amides. The latter are generated *in situ* by treatment of an imine with an acid chloride. For example, *N*-benzylidenebenzylamine affords (**161**) after acylation and treatment with an alkanethiol in the presence of triethylamine (Scheme 31) ⟨81CPB2496⟩. *N*-Nitrosohemithioaminals have been obtained by an adaptation of this approach. *N*-Methylmethanimine (H$_2$C=NMe) reacts with nitrosyl chloride in dichloromethane to give *N*-methyl-*N*-nitrosochloromethylamine which affords the *S*,*N*-acetals MeN(NO)CH$_2$SR on treatment with an alkane or arene thiol ⟨87LA583⟩.

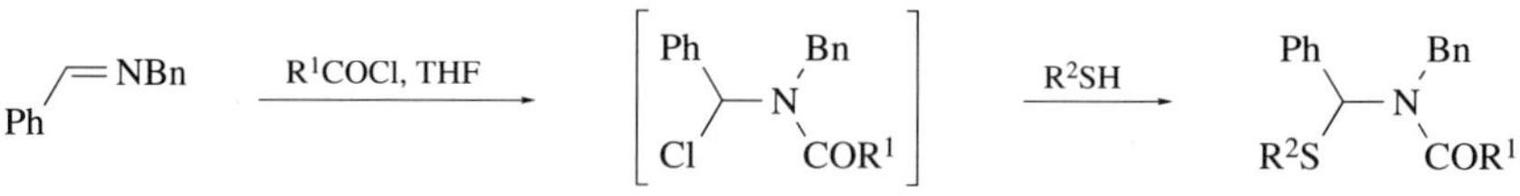

Scheme 31

(3) *From compounds possessing an S=N unit.* There have been relatively few applications of compounds with S=N functions to the synthesis of *S*,*N*-acetals. An example is provided by the behaviour of *N*-sulfinyl-*p*-toluenesulfonamide (TsN=S=O) towards β-keto sulfoxides. When TsN=S=O is heated with (**162**) in ether, the *N*-tosyl *S*,*N*-acetal is obtained directly (Equation (29)) ⟨74JOC3412⟩. Halide displacement from arenesulfinimidoyl chlorides by active methylene compounds such as 1,3-diketones or malononitrile, promoted by triethylamine, leads directly to sulfimides which rearrange under mild conditions (20 °C) to *N*-tosylhemithioaminals. For example, 4-ClC$_6$H$_4$S(Cl)=NTs and acetylacetone generate (**163**) directly ⟨78ZOR1659⟩. The sulfimides obtained by treating (arylthio)dibenzoylmethane with chloramine-T exhibit a marked instability, rearranging in acetone at room temperature to provide the *S*,*N*-acetals (**164**) ⟨82ZOR841⟩.

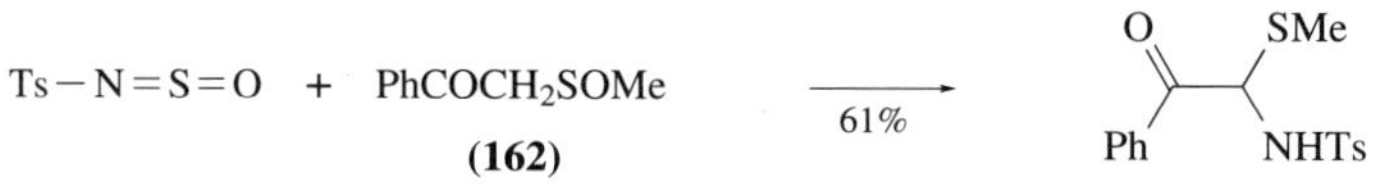

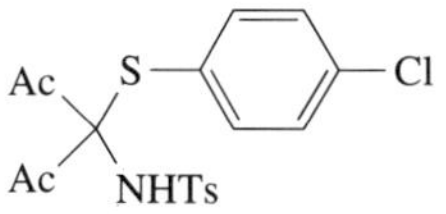

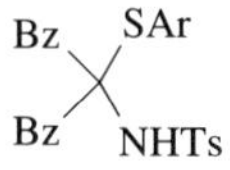

(4) *From alkenes.* When the acid (F$_3$C)$_2$C=C(NHCOPh)CO$_2$H was treated with benzyl thiol in boiling ethanol, none of the anticipated product (**165**) was obtained and instead the isomeric compound (**166**) resulted. The electron-withdrawing effect of two CF$_3$ groups is sufficient to reverse the mode of conjugate addition ⟨80JFC(15)29⟩. An alternative approach to other α-thio α-amino acid derivatives has been developed using the versatile synthon methyl methylthiomethyl sulfoxide, MeSCH$_2$SOMe. The anion from this compound condenses smoothly with both aliphatic and aromatic nitriles in a stereoselective fashion to give enamino sulfoxides ((*Z*):(*E*), ca. 9:1) in high yields. Subsequent treatment with acetic anhydride–pyridine at ambient temperature initiates a Pummerer-type rearrangement to the *S*,*N*-acetals (**167**) (Equation (30)) ⟨74JA1960, 78TL375⟩.

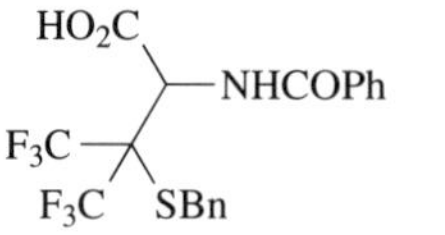

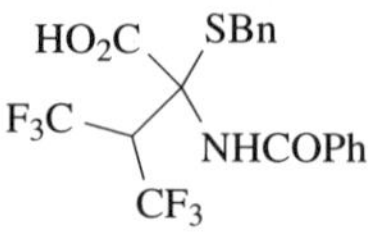

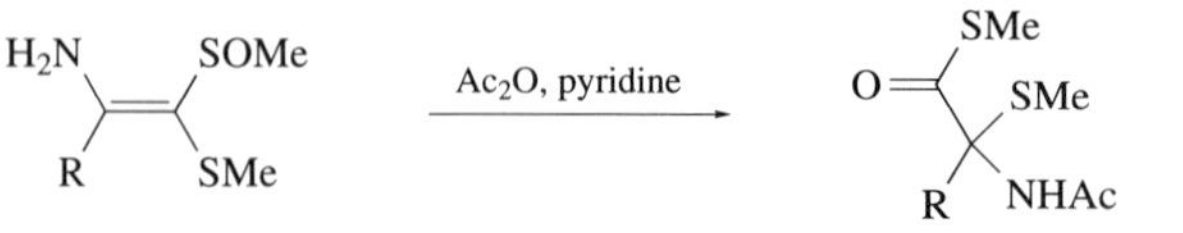

R = Me, Pr^i, Ph

A free radical chain reaction has been developed for the synthesis of α-nitro sulfides. The *O*-acyl thiohydroxamates (**168**), easily prepared from *N*-hydroxypyridine-2-thione, are photolabile, generating alkyl radicals by decarboxylative fragmentation. Nitroalkenes function as efficient radical traps when photolysed with (**168**) at low temperatures, affording good yields of the α-nitro sulfides (Equation (31)) ⟨85T5507⟩.

$$
(168) + \underset{R^2\ R^3}{\overset{NO_2}{\diagup}} \xrightarrow{h\nu,\ PhH,\ CSA} \underset{\text{2-pyridyl-S}}{R^2}\underset{\ }{\overset{R^1\ R^3}{\diagdown\diagup-NO_2}} \quad (31)
$$

(b) From compounds containing two singly bonded functional groups. (1) From X—C—N functions (X = Hal, OR, NR$_2$ or a metal). The accessibility and reactivity of α-chloroalkylamides make these compounds ideal starting materials for the synthesis of *N*-acyl *S,N*-acetals. Halide substitution is accomplished either by direct treatment with a thiol or a thiolate in a solvent such as ethanol or aqueous acetone. In some instances, substitution proceeds satisfactorily under heterogeneous conditions in nonpolar solvents such as dichloromethane. Yields, on the whole, are excellent and the reaction conditions very mild ⟨58CB1432, 61JOC3591, 66CB1944, 74AJC1579, 80S322, 94JOC3721⟩.

Halide displacement reactions constitute a valuable means of obtaining α-nitro sulfides. 2-Bromo-2-nitropropane reacts with thiolates in dipolar aprotic solvents to give the products in good yields (Equation (32)). The reactions are light catalysed and proceed by a radical nucleophilic substitution (S$_{RN}$1) pathway rather than by S$_N$1 or S$_N$2 mechanisms. High yields of products were obtained from both 2- and 4-nitrobenzenethiolates and from benzothiazole-2-thiolate ⟨80JCS(P1)1407, 81TL1551⟩.

$$
Br-C(Me)_2-NO_2 + ArSNa \xrightarrow{h\nu,\ DMF} ArS-C(Me)_2-NO_2 \quad (32)
$$

α-Heteroarylthio nitro sulfides have been prepared from the oxidative addition of thiolates to the anion of 2-nitropropane in the presence of potassium ferricyanide. The highest yields of products result from benzothiazole-2-, 1-methylimidazole-2- and pyrimidine-2-thiolates. When 2- and 4-pyridinethiolates are employed, the efficiency of the process is reduced due to competing side reactions. An alternative protocol to this oxidative addition is the S$_N$2 displacement of thiolate by the nitropropane anion from symmetrical aryl and heteroaryl disulfides ⟨77TL4519, 84JCS(P1)2327⟩.

The large scale preparation of PhSCH$_2$NO$_2$ utilizes chloride displacement from PhSCl by the sodium salt of nitromethane ⟨93OSC(8)550⟩.

N-Alkyl-*N*-chloromethylnitramines undergo smooth chloride displacement by alkane or arene thiols in the presence of sodium ethoxide in ethanol. The reaction time is short and good yields of the products (**169**; R^1 = Me or Et, R^2 = alkyl or aryl) are the norm ⟨73S303⟩.

N-Acylhemithioaminals are accessible from a variety of other X—C—N units. For example, methyl 2-methoxyhippurate, MeOCH(NHCOPh)CO$_2$Me, affords the *N*-benzoyl compounds (**170**; R = Pr^i, Bu^n or Bn) when treated with a thiol in boiling 1,2-dichloroethane in the presence of a sulfonic acid catalyst ⟨75T863⟩. The reactions are greatly accelerated by Lewis acid catalysts ⟨79BCJ826⟩.

Activation of the amino group in *N*-(dimethylaminomethyl)formamide is achieved by quaternization with MeI. Nucleophilic displacement of trimethylamine by both alkane and arene thiols permits access to *N*-formamidohemithioaminals of the type RSCH$_2$NHCHO ⟨70CB2775⟩.

$$
\underset{(169)}{R^1\diagdown N \diagup NO_2 \atop \diagdown SR^2} \qquad \underset{(170)}{MeO_2C\diagdown \diagup SR \atop \diagdown NHCOPh}
$$

(2) *From X—C—S units* (X = Hal). Chloride displacement from chloromethyl methyl sulfide by aromatic amides proceeds smoothly in either trifluoroacetic acid or methanesulfonic acid to give the corresponding *N*-methylthiomethyl amide (RCONHCH$_2$SMe). Good yields of products are obtained from benzamide, nicotinamide or isonicotinamide. Pivalamide also reacts but much less efficiently ⟨75CC320⟩.

(c) By functionalization of a preformed S,N-acetal. The only reaction resulting in modification of a *S*-function of any practicable use involves nucleophilic substitution of S—C—N groups possessing

328   *Chalcogen and a Group 15 Element*

tetracoordinate sulfur, that is displacement of a sulfinate group. Thus, *N*-(arylthiomethyl)-formamides (ArSCH$_2$NHCHO) are obtained in excellent yields from TsCH$_2$NHCHO and arenethiolates ⟨85RTC177⟩. In a similar vein, α-nitro sulfones react with thiolates to give the corresponding α-nitro sulfides ⟨74JA2580⟩.

(iii) Functions with dicoordinate nitrogen

Hemithioaminals possessing a dicoordinate nitrogen function are by far the least common of all the systems with dicoordinate sulfur, although a range of *S,N*-acetals possessing N=CR$_2$, N=C=O, N=C=S, N=C=NR, N=C, N=PR$_3$ N=NR and N$_3$ groups have been obtained.

(a) From compounds containing a multiply bonded functional group. (1) From aldehydes and ketones. Enolizable ketones, sulfur and ammonia in the ratio 2:1:1 condense to give good yields of 2,5-dihydrothiazoles, the initial step involving formation of an α-thiol ketone. A wide range of solvents may be used which include benzene, pyridine, methanol, water or DMF at temperatures between 20–60 °C. The outcome of the reaction is critically dependent on conditions. Mixtures of products frequently result from unsymmetrically substituted ketones. For example, 3-methylbutan-2-one affords a mixture of **(171)** and **(172)**. At 20 °C, the latter is predominant, 94%, but at 80 °C this yield is reduced to 28%. Cyclohexanone reacts straightforwardly to give the spirocycle **(173)** (Scheme 32). Extensive reviews chart the scope and limitations of this reaction ⟨63AG1050, 67AG(E)907, 84CHEC(6)312⟩.

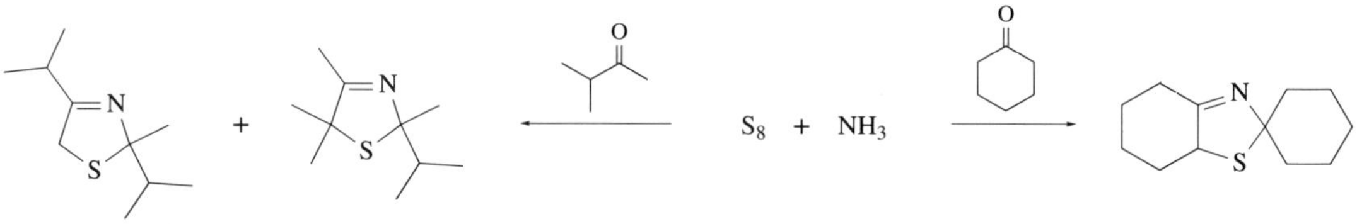

Scheme 32

(2) From acid chlorides. Acyl azides are readily prepared from α-thioalkanoyl chlorides. Curtius rearrangement to the corresponding isocyanate is accomplished in refluxing benzene. Thus phenylthiomethyl isocyanate, PhSCH$_2$N=C=O, has been obtained in 77% yield by this procedure ⟨85S276⟩.

(3) From thiocarbonyl compounds. The cycloaddition of diazoalkanes to thioketones leads to 2,5-dihydro-1,3,4-thiadiazoles, which, unless heavily substituted, fragment with loss of nitrogen below 0 °C. Thus, the thione **(174)** affords a high yield of the relatively stable thiadiazole **(175)** on treatment with diazomethane at −78 °C (Equation (33)) ⟨70JOC1501, 85H(23)2207⟩. Analogous dipolar cycloadditions with thioketenes are possible leading to 2-alkylidene-2,5-dihydro-1,3,4-thiadiazoles ⟨88T1827⟩.

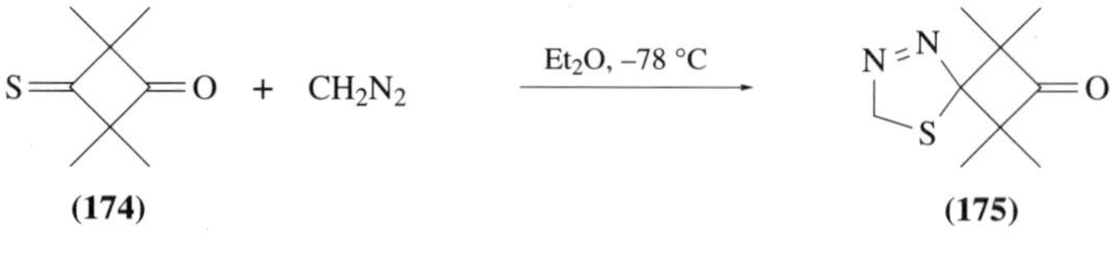

(33)

(4) From imines. *N,N*-Dimethyl-*N*′-thiobenzoylformamidine is *S*-alkylated by iodomethane to afford the iminium salt **(176)**. Although this salt possesses two electrophilic centres, MNDO calculations predict that the highest total charge density resides at C-1. When treated with KCN in aqueous acetonitrile, **(177)** is obtained in quantitative yield ⟨91CR(313)517⟩.

(b) From compounds containing two singly bonded functional groups. (1) From X—C—N units

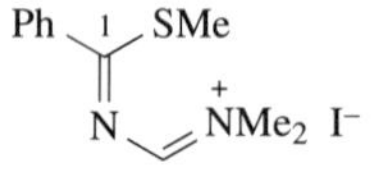

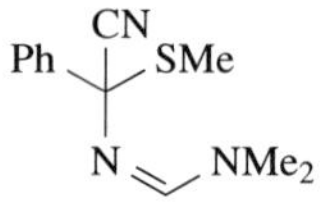

(X = Hal, OR, NR$_2$). α-Chloroalkyl isocyanates are in some instances tautomeric with N-chlorocarbonylketimines [R$_2$C(Cl)N=C=O ⇌ R$_2$C=N—COCl], although it is not possible to distinguish between the two on the basis of reactivity. Reactions with thiols occur rapidly at or below 0°C in the presence of triethylamine giving α-thioalkyl isocyanates which are formally derived by S$_N$2 displacement of halide. Compound (**178**) appears to be relatively stable, although a trace of alkoxide effects rearrangement to the ketimine (**179**) ⟨74JOU1565⟩. The trifluoromethyl compound (**180**) was obtained in an analogous manner, although in this case a significant amount of the N-acylketimine was also formed ⟨74JOU244⟩. Thiophenol and Ph(Bu^t)C(Br)N=C=S react cleanly at −78°C in the presence of triethylamine to afford (**181**) in 90% yield ⟨79CB1956⟩.

α-Azo sulfides constitute a rare class of S,N-acetal; only a single example has been reported. Controlled bromination of acetone 4-chlorophenylhydrazone and subsequent treatment with thiourea gives (**182**). Alkaline hydrolysis of this salt results in extensive decomposition with small amounts of the highly unstable α-thiol azo compound being obtained ⟨85M1329⟩.

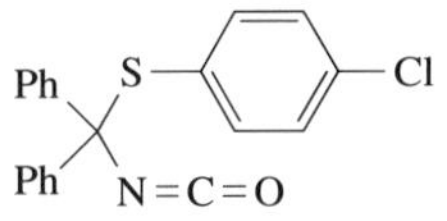

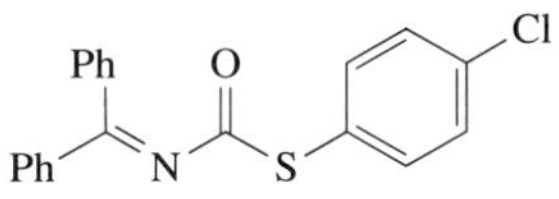

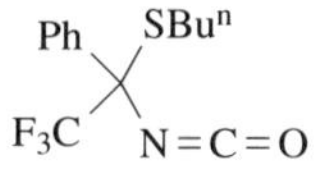

Triphenylphosphine benzotriazol-1-ylmethylimide is readily prepared from 1-(chloromethyl)benzotriazole ⟨90S565⟩ and eliminates the benzotriazole unit when treated with a thiolate. The products, N-(alkylthiomethyl)iminophosphoranes, were not isolated but were quenched with aldehydes and with isocyanates to give (**183**) and (**184**) respectively by aza-Wittig reactions. Acylation of the iminophosphorane with acid chlorides gives N-acyl S,N-acetals (Scheme 33) ⟨94S107⟩.

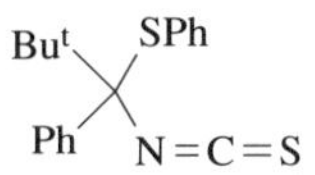

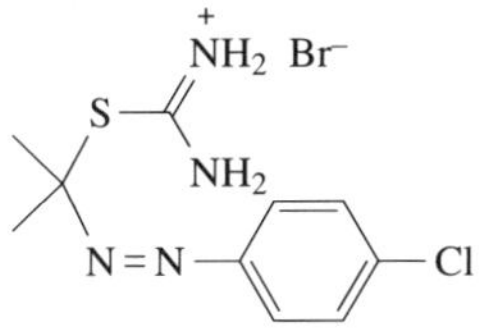

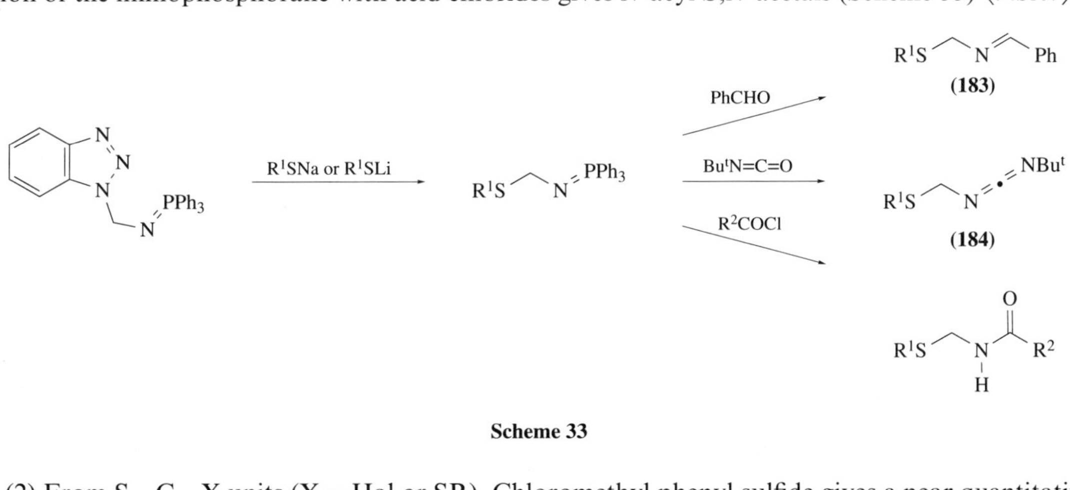

Scheme 33

(2) From S—C—X units (X = Hal or SR). Chloromethyl phenyl sulfide gives a near quantitative yield of PhSCH$_2$N$_3$ when treated with NaN$_3$-NaI. This compound is of interest since it effects the electrophilic amination of aryl Grignard reagents ⟨81JA2483⟩. Dithioacetals have also been used as starting materials. 1,1-Bis(methylthio)cyclohexane reacts exothermically with *in situ* generated IN$_3$ giving 1-azido-1-methylthiocyclohexane in almost quantitative yield ⟨80JA7929⟩.

4.07.1.2.2 Tricoordinate sulfur derivatives

Although in principle a considerable number of hemithioaminals based on tricoordinate sulfur is possible, very few members of this class have been prepared. Examples are confined to α-substituted sulfoxides (**110**). Related *S,N*-acetals such as the sulfimides (**185**) and methylides (**186**) have not been described. Despite the wealth of literature on sulfonium salts ⟨B-81MI 407-01⟩, functionalities exemplified by (**187**) appear to be unknown.

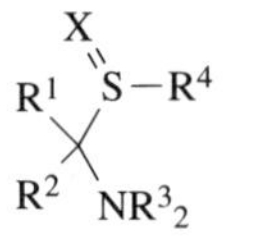

$$R^1R^2C(NR^3_2)S(X)R^4 \quad R^1R^2C(NR^3_2)S^+(R^5)R^4$$

(**185**) X = NR^5
(**186**) X = CR$^5_2$

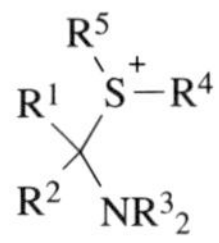

(**187**)

(i) From compounds containing multiply bonded functional groups

The double bond in thiochromone 1-oxide is susceptible to conjugate addition. Treatment with a chloroform solution of hydrogen azide gives 2-azidothiochroman-4-one 1-oxide as a mixture of diastereomers (*trans* : *cis*; 3 : 2) (Equation (34)) ⟨84CJC586⟩.

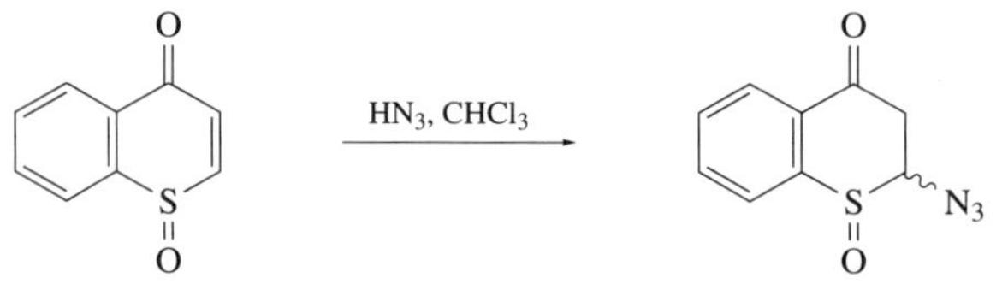

$$\text{thiochromone 1-oxide} \xrightarrow{HN_3, CHCl_3} \text{2-azidothiochroman-4-one 1-oxide} \tag{34}$$

(ii) By functionalization of a preformed S,N-acetal

α-Azido sulfides can be selectively oxidized to sulfoxides, 2-azidotetrahydrothiophene being oxidized to the corresponding *S*-oxide in high yield by sodium periodate ⟨84CJC586⟩.

In view of the ease with which sulfoxides may be obtained, it is rather surprising that so few *S,N*-acetal *S*-oxides have been prepared. Clearly there is much scope for further investigations in this area, since a multitude of reagents for the oxidation of sulfides to sulfoxides is available ⟨91COS(7)762⟩. The formation of sulfimides by oxidative imination of sulfides has been extensively investigated and a review is available ⟨93SR149⟩. However, this reaction has not been applied to *S,N*-acetals.

4.07.1.2.3 Tetra- and higher coordinate sulfur derivatives

The only hemithioaminals possessing a tetracoordinate sulfur are the α-functionalized sulfones (**111**). Here again there is much scope for the preparation of new functional types. For example, whilst the chemistry of sulfoximides has been extensively investigated ⟨75CSR189, 80CSR477⟩, there do not appear to be any examples of α-aminoalkyl sulfoximides (**188**). However, a sulfoximidoylmethyl isocyanide is known. *S,N*-Acetals based on sulfoxonium groups are also unknown. A similar situation exists for sulfur(IV) compounds such as sulfuranes and compounds with five- and six-coordinate sulfur functions.

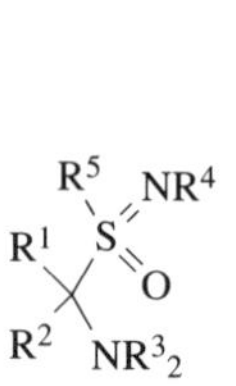

$$R^1R^2C(NR^3_2)S(=NR^4)(=O)R^5$$

(**188**)

(i) From compounds containing multiply bonded functional groups

(a) From aldehydes and ketones. Previous sections have highlighted the value of *S*-amino- and *S*-amido-alkylations using the Mannich reaction. The sulfur function in these instances was a thiol. The analogous *S*-functionalization of sulfinic acids or their salts is particularly facile ⟨55CB41⟩. Sodium *p*-toluenesulfinate, formaldehyde and aniline hydrochloride condense in aqueous solution to give a good yield of the sulfone **(189)**. Related alkylations of primary amides are also successful ⟨72RTC209⟩. Ethyl carbamate reacts readily to give **(190)** in 90% yield and a detailed experimental procedure is available ⟨88OSC(6)981⟩. Undoubtedly the most important of these reactions is the preparation of TsCH$_2$NHCHO, which is readily dehydrated to tosylmethyl isocyanide (TosMIC) itself available in large quantities ⟨88OSC(6)987⟩. Aliphatic aldehydes may be used in place of formaldehyde in these condensations providing a route to α-alkyl TosMIC derivatives ⟨75TL3487⟩. Analogous reactions with *N*-substituted thioureas also proceed smoothly, enabling the carbodiimides **(191)** to be obtained ⟨81JOC2069⟩. These transformations are depicted in Scheme 34.

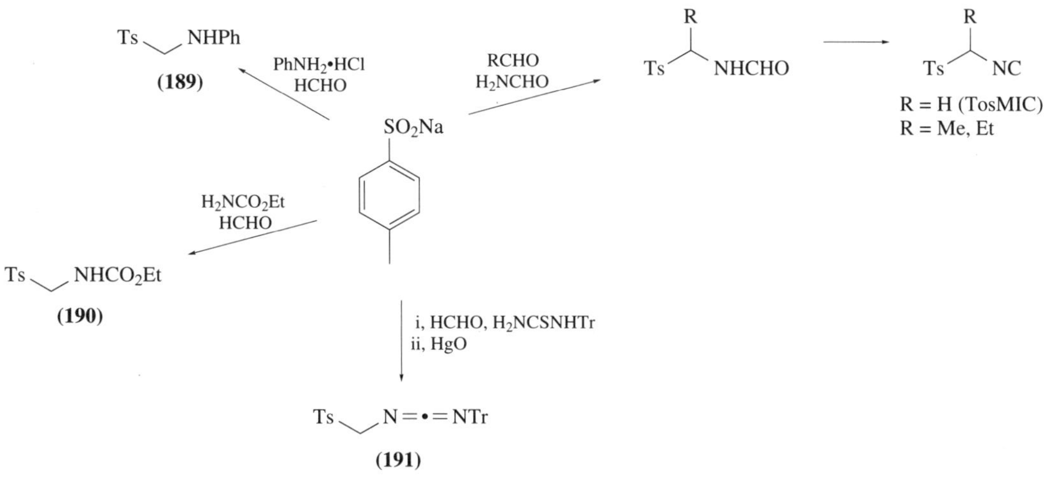

Scheme 34

(b) From sulfonic acid derivatives. The reaction between alkanesulfonyl chlorides and tertiary amines has been investigated in some detail ⟨90CB1563⟩ and provides a route to some unusual *S,N*-acetals. The dehydrochlorination of cyclohexanesulfonyl chloride by trimethylamine proceeds with formation of a sulfene intermediate, which in the presence of an excess of base is intercepted to give the isolable, but unstable adduct **(192)**. Treatment with hydrogen peroxide affords the sulfonate **(193)** as a stable, high-melting solid (Scheme 35). Arylmethanesulfonyl chlorides undergo an analogous series of reactions. *p*-Toluenesulfonyl iodide reacts with potassium nitronates in DMF to give α-nitro sulfones. Tosylnitromethane is obtained in 50% yield, whilst the product from potassium isopropylnitronate, TsC(Me)$_2$NO$_2$ is formed more efficiently (80% yield) ⟨74RTC11⟩.

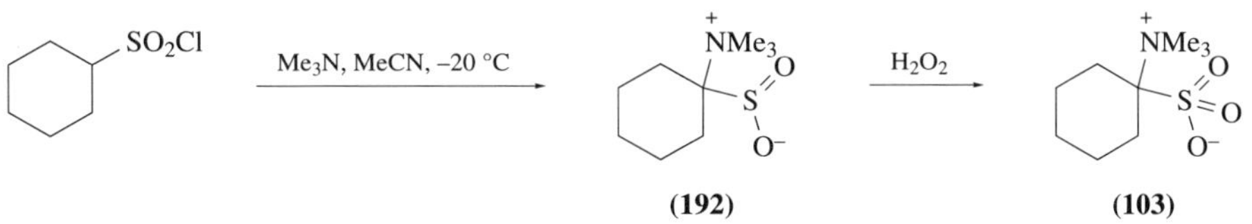

Scheme 35

A highly unusual reaction affords α-azido sulfones when potassium nitronates are treated with tosyl azide in THF. The yields of products are only modest (35–56%) and the reaction with nitromethane is unsuccessful (Equation 35) ⟨87JOC3466⟩.

Although hardly the method of choice, TosMIC has been prepared from LiCH$_2$NC and tosyl fluoride ⟨72TL2367, 72LA(766)130⟩. Under similar conditions, the sulfoximinoyl fluoride, PhS(O)(NTs)F, gave **(194)** ⟨84RTC41⟩.

$$R^1R^2C(NO_2) + TsN_3 \xrightarrow{KH, THF} R^1R^2C(Ts)(N_3) \quad (35)$$

$$Ph-S(O)(NTs)-CH_2-NC$$
(194)

(c) From sulfones. Methylsulfonyldibenzoylmethane, MeSO$_2$CHBz$_2$, couples efficiently with 4-nitrobenzenediazonium tetrafluoroborate in DMF at room temperature to give the azo compound **(195)** in good yield ⟨77JOU2224⟩.

$$Bz_2C(SO_2Me)-N=N-C_6H_4-NO_2$$
(195)

(ii) From compounds containing two singly bonded functional groups

(a) From X—C—N units (X = Hal, OR, NR$_2$ or a metal). α-Halonitroalkanes are useful starting materials for the preparation of α-nitro sulfones. Thus, 2-iodo-2-nitropropane reacts with sodium benzenesulfinate in DMF in the dark at $-20\,°C$ to give **(196)** in very high yield. The reaction is particularly successful with branched iodonitroalkanes. These compounds are conveniently generated *in situ* from the nitroalkane and iodine in the presence of sodium methoxide ⟨73JA3356⟩. A later modification obviates the need to prepare the iodo compound and relies on the ferricyanide-catalysed, oxidative coupling of the nitronate with sodium arenesulfinates ⟨83JOC332⟩. Aminals have also been employed as a source of α-amidoalkyl sulfones. Dimethylamine is readily eliminated from Me$_2$NCH$_2$NHCO$_2$Et on treatment with sodium *p*-toluenesulfinate to give **(190)** in good yield ⟨62RTC966⟩.

$$Me_2C(SO_2Ph)(NO_2)$$
(196)

(b) From S—C—X units (X = Metal). Sulfone-stabilized carbanions are readily generated and have been α-nitrated under a variety of conditions. Dimethyl sulfone was metallated with KNH$_2$ in liquid ammonia and the anion quenched with ethyl nitrate to give MeSO$_2$CH$_2$NO$_2$. The reaction is more efficient with benzylic sulfones, benzyl phenyl sulfone giving PhCH(NO$_2$)SO$_2$Ph in 81% yield ⟨69JOC3104⟩. The reaction has been extended to the α-nitration of alkyl alkanesulfonates, neopentyl 1-butanesulfonate giving **(197)** in good yield. The reaction is less successful with sterically unencumbered sulfonates in which elimination reactions compete ⟨70JOC2551⟩.

$$Pr^n-C(SO_3CH_2Bu^t)(NO_2)$$
(197)

(iii) By functionalization of a preformed S,N-acetal

The acidity of α-nitro sulfones has been exploited to achieve further functionalization. Tosylnitromethane has been both mono- and bis-alkylated by the Mannich bases of ketones in DMF

⟨80S565⟩. Benzenesulfonylalkylation proceeds readily in 90% formic acid with aldehydes and benzenesulfinic acid ⟨74JOC3215⟩. Selective monoalkylation of benzenesulfonylnitromethane has been accomplished using sodium methoxide and an alkyl halide in HMPA ⟨81JOC765⟩.

4.07.1.3 Functions Bearing Selenium or Tellurium, Together with Nitrogen

The 1980s and 1990s have witnessed an explosive growth in the development of organoselenium chemistry. Reviews of organoselenium chemistry ⟨B-86MI 407-01⟩ and its applications in synthesis ⟨B-84MI 407-01, B-84MI 407-02, B-87MI 407-01⟩ are available. Whilst the literature on organotellurium compounds is extensive ⟨90HOU(E12b)1⟩, their use in synthesis has made relatively slow progress but is now gaining in momentum. Developments in this area have been reviewed ⟨91S793, 91S897, B-94MI 407-01⟩. Despite all of this activity, *Se,N*-acetals represent a very rare class of functional group and the *Te,N*-acetals are even more uncommon. In all cases, the chalcogen is dicoordinate and higher coordinate compounds do not appear to have been obtained. Functions possessing a tricoordinate nitrogen are the most common.

4.07.1.3.1 *From compounds containing a multiply bonded functional group*

(i) From aldehydes and ketones

Probably the most straightforward means of obtaining *Se,N*-acetals involves the Mannich aminomethylation of selenols. Selenophenol, aqueous formaldehyde and *N*-methylaniline in the ratio 1:1:1 condense to give (**198**) after brief heating in ethanol. Similarly, aniline undergoes bis-phenylselenomethylation to give an excellent yield of PhN(CH$_2$SePh)$_2$, whilst ammonia is tris-alkylated to afford N(CH$_2$SePh)$_3$. Attempts to prepare *N*-(phenylselenomethyl)aniline, PhNHCH$_2$SePh, were not successful. As with *S,N*-acetal syntheses, the reaction conditions affect the nature of the product. When the alkylation of *N*-methylaniline was conducted in the presence of 1 equiv. of hydrochloric acid, the selenide (**199**) was the only product ⟨66JOC3514⟩. The mechanism of the reaction has been investigated ⟨67JOC3101⟩. Variations of this three component reaction are known which permit access to heterocycles. For example, 2-aminoethaneselenol, H$_2$NCH$_2$CH$_2$SeH, condenses with glyoxylic acid to give the selenazolidine (**200**). Some related routes to these compounds have been described ⟨84CHEC(6)346⟩.

(**198**)   (**199**)   (**200**)

(ii) From selenocarbonyl compounds

Selenocarbonyl compounds, especially the aldehydes and ketones are relatively inaccessible materials exhibiting a marked instability. They are particularly prone to oligomerization and oxidation ⟨B-87MI 407-02⟩. However, routes to sterically stabilized selenones have been developed. The latter are very efficient 1,3-dipolarophiles and readily cycloadd to diazoalkanes. When the diazo compound also possesses bulky substituents, good yields of 2,5-dihydro-1,3,4-selenadiazoles (**202**; M = Se) are obtained. Ketone hydrazones function as precursors to both reactants. The selenone is generated cleanly and efficiently by heating the ylide (**201**) with selenium metal ⟨85JCS(P1)107⟩. An analogous sequence to give the telluradiazole (**202**; X = Te) employed tellurium(IV) chloride as the chalcogen source (Scheme 36) ⟨93CL1047⟩.

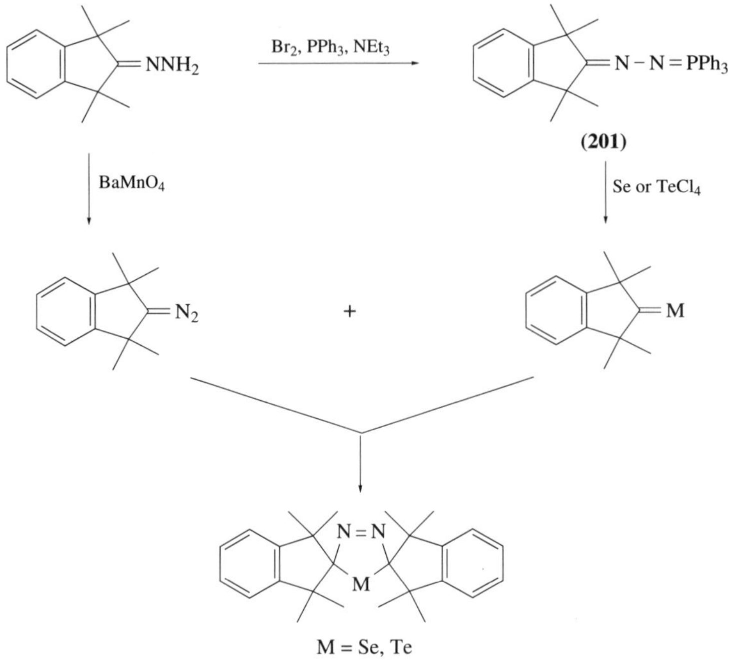

Scheme 36

The selenoamide (**203**) undergoes carbophilic attack when treated with phenyllithium in THF to give (**204**) which, although not isolable, was characterized by its ^{13}C and ^{77}Se NMR spectra. Addition of 1 equiv. of phenyllithium to (**204**) results in deselenation giving 1-benzhydrylpiperidine. This compound is formed instantly when (**203**) is quenched with 2 equiv. of phenyllithium ⟨90CL2053⟩.

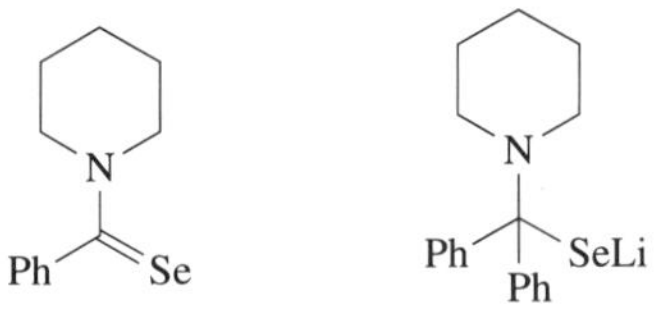

(iii) From iminium salts

N,N-Dimethylmethyleneammonium chloride reacts readily with lithium butanetellurolate in THF to give the Te,N-acetal (**205**). Treatment of this compound with butyllithium promotes an efficient lithium–tellurium exchange. The intermediate aminomethyllithium has been intercepted with benzaldehyde (Scheme 37) ⟨90OM1355⟩.

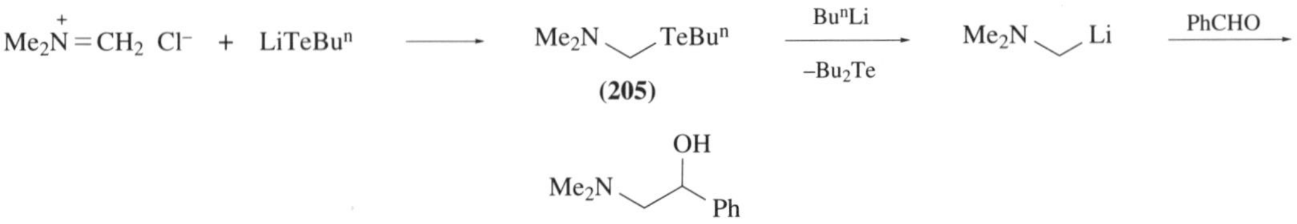

Scheme 37

(iv) From diazoalkanes

Both diazomethane and ethyl diazoacetate react readily with electrophilic organoselenium compounds. With *N*-phenylselenophthalimide, the insertion products (**206**) are obtained. With ethyl diazopropionate, none of the *Se,N*-acetals is produced and an elimination pathway supervenes affording the acrylate (**207**) and phthalimide (Scheme 38) ⟨85JOM(286)171⟩. Some analogous insertion reactions were observed with other selenium electrophiles. Thus, benzeneselenenyl thiocyanate, PhSeSCN, reacts with diazomethane to generate PhSeCH$_2$N=C=S as the principle product together with some of the isomeric *Se,S*-acetal, PhSeCH$_2$SCN ⟨85JOM(286)171⟩.

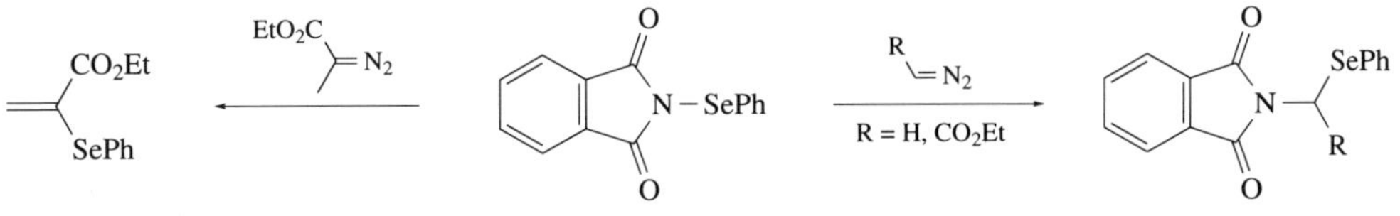

Scheme 38

(v) From alkenes

Conventional selenium electrophiles such as PhSeCl are unreactive towards electron-deficient nitroalkenes. However *in situ* conversion of PhSeCl into PhSeOCOCF$_3$ by treatment with silver(I) trifluoroacetate permits a smooth and efficient trifluoroacetoxy-phenylselenation to proceed. 1-Nitrocyclohexene gave (**208**) in 95% yield ⟨85T4861⟩.

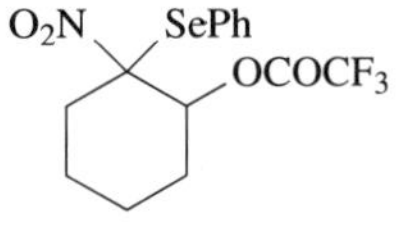

4.07.1.3.2 From compounds containing two singly bonded X—C—N groups (X = Li, Na)

N-Nitrosopiperidine is readily lithiated by LDA in HMPA at low temperatures and addition of PhSeCl gave the α-phenylseleno nitrosamine (**209**) in good yield. Oxidative elimination to (**210**) was accomplished by addition of mcpba ⟨79JOC2326⟩.

Deprotonation of 1-nitroethane by sodium ethoxide followed by treatment with benzeneselenenyl bromide gave (**211**). In the presence of formaldehyde and calcium hydroxide, a good yield of (**212**) was then obtained. This methodology is of value for the preparation of functionalized nitroalkenes, since selenoxide elimination occurs regiospecifically away from oxygen. Thus addition of hydrogen peroxide to 1-hydroxy-2-nitro-2-phenylselenononane gave (**213**) in high yield ⟨82S261⟩. Benzeneselenenylation of *geminal* dinitroalkanes has been achieved by a similar method ⟨82IZV161⟩.

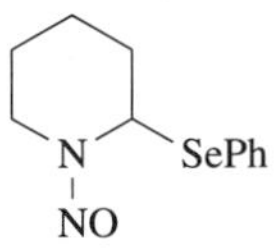

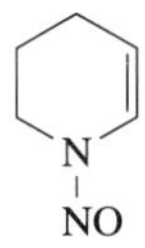

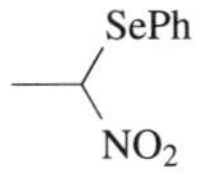

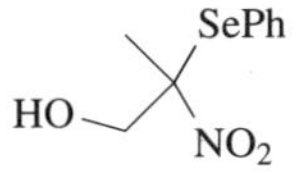

4.07.2 FUNCTIONS CONTAINING A CHALCOGEN AND PHOSPHORUS, ARSENIC, ANTIMONY OR BISMUTH

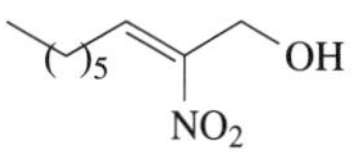

Russian workers have probably carried out the bulk of the studies in this area and their work has been reported widely in the Russian literature. Some of this has been summarized in the major treatises on phosphorus chemistry and the reader is referred to the chapters on the preparation of phosphines and cyclic phosphines and on their nucleophilic reactions in Volume 1 of *The Chemistry of Organophosphorus Compounds* ⟨B-90MI 407-03⟩ and that on the synthesis of phosphine chalcogenides in Volume 2 ⟨B-92MI 407-01⟩. The chapter on phosphorus heterocycles from α-hydroxyalkylphosphines and vinylphosphines is written by one of the foremost Russian workers ⟨94AHC(61)59⟩. Useful information can also be found in *Organic Phosphorus Compounds* by Kosolapoff and Maier ⟨B-72MI 407-01⟩, the *Houben-Weyl* volume on organophosphorus compounds ⟨82HOU(E1/E2)⟩ and in the *Dictionary of Organophosphorus Compounds* ⟨B-87MI 407-03⟩.

4.07.2.1 Functions Bearing Oxygen

4.07.2.1.1 *Oxygen and phosphorus*

This section covers compounds with the general structures (**214**), (**215**) and (**216**) and cyclic derivatives containing these fragments.

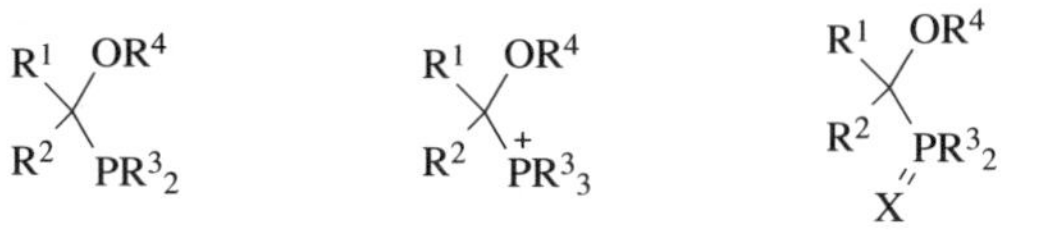

$$
\begin{array}{ccc}
R^1\diagup OR^4 & R^1\diagup OR^4 & R^1\diagup OR^4 \\
R^2\diagdown PR^3{}_2 & R^2\diagdown \overset{+}{P}R^3{}_3 & R^2\diagdown PR^3{}_2 \\
& & \overset{\|}{X} \\
(\mathbf{214}) & (\mathbf{215}) & (\mathbf{216})
\end{array}
$$

(i) From compounds containing multiply bonded functional groups

(a) From aldehydes and ketones. Tricoordinate phosphorus is nucleophilic and can therefore attack the electron-deficient centre of a carbonyl group. However, the reaction is reversible and the equilibrium is generally in favour of the reactants rather than products, presumably because of the dipolar nature of latter. Nevertheless, the reaction is synthetically useful provided that a further reaction such as proton transfer from phosphorus to oxygen is possible. When this cannot occur, as when a tertiary phosphine is a reactant, the initial P—C—O product may rearrange to a P—O—C derivative (Scheme 39) ⟨74PS(4)109⟩. α-Hydroxyalkylphosphonium salts have been isolated when the addition is carried out under anhydrous acidic conditions ⟨90JOC2644⟩. Additionally, examples are known where initial attack by P occurs preferentially at oxygen (Equation (36)), when the betaine is stabilized by resonance.

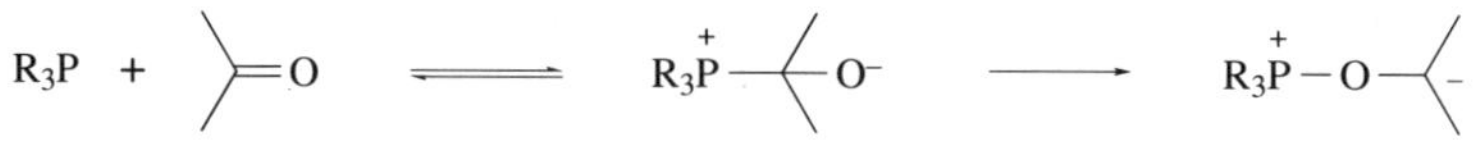

$$R_3P + \underset{}{\diagup}=O \rightleftharpoons R_3\overset{+}{P}\underset{}{\diagup}O^- \longrightarrow R_3\overset{+}{P}-O\underset{}{\diagup}$$

Scheme 39

$$Ph_3P + O=\underset{Ph}{\overset{Ph}{\diagup}}\underset{CN}{\overset{CN}{\diagdown}} \longrightarrow Ph_3\overset{+}{P}-O\underset{Ph}{\overset{Ph}{\diagup}}\underset{CN}{\overset{CN}{\diagdown}} \quad (36)$$

The scope of the reaction is appreciable not only because of the range of carbonyl compounds which can be used but also because of the variation which is acceptable in the phosphorus component. The reaction is typified in Scheme 40 from which it is seen that the product is an α-hydroxymethylphosphine. Acid catalysis of the process has been established, when protonation of the carbonyl oxygen facilitates the attack by phosphorus ⟨61ZOB3417⟩. The presence of an electron-withdrawing group in the carbonyl compound may obviate the need for a catalyst. When the phosphine phosphorus carries more than one hydrogen atom, sequential stages may lead to mono-, bis-, tris- and even tetrahydroxymethyl derivatives. Thus, phosphine and formaldehyde yield tris(hydroxymethyl)phosphine, $P(CH_2OH)_3$ ⟨84MI 407-03⟩ and the phosphonium salt $P^+(CH_2OH)_4Cl^-$ under acidic catalysis ⟨61ZOB3417⟩. Similarly, primary phosphines, RPH_2, can give the bis(hydroxymethyl) derivative, such as (**217**) in which the sterically demanding adamantyl moiety appears to have little effect on the course of the reaction ⟨86JGC425⟩. The reaction of Ph_2PH with benzaldehyde is complex yielding the expected bis(α-hydroxybenzyl)phosphine (**218**) with dilute HCl, but the rearranged phosphine oxide (**219**) in the presence of concentrated acid. Furthermore, when acetonitrile is used as solvent, the cyclic 1,3,5-dioxaphosphorinane is formed by a reaction sequence involving three molecules of benzaldehyde (Scheme 41) ⟨61ZOB3411⟩. When isobutylphosphine reacts with benzaldehyde in dilute HCl, the product is a tertiary phosphine hydrochloride (**220**) ⟨61JCS2813⟩.

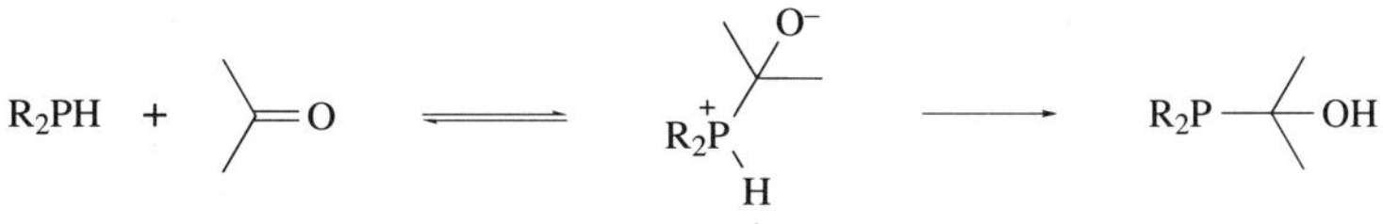

Scheme 40

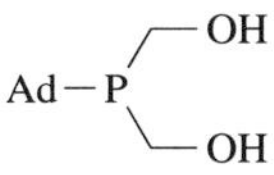

(**217**)
Ad = 1-adamantyl

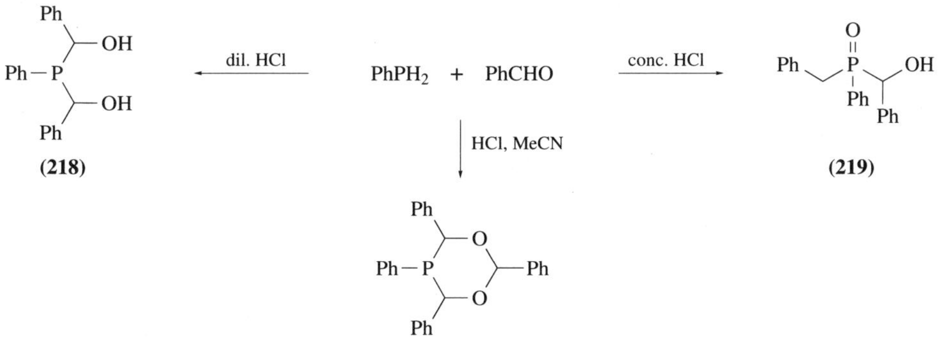

(**218**) ← dil. HCl — $PhPH_2$ + PhCHO — conc. HCl → (**219**)

↓ HCl, MeCN

Scheme 41

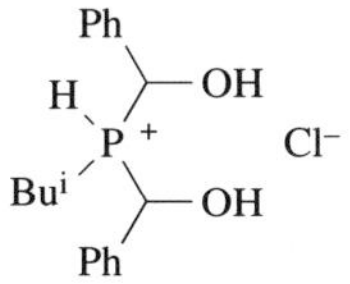

(**220**)

The reaction of phosphines with hexafluoroacetone has been examined in detail. Diphenylphosphine gives the normal α-hydroxyalkyl derivative ⟨73CJC1136⟩, but both phosphine and methylphosphine afford the mono- and bis(hydroxyalkyl)compounds. Dimethylphosphine behaves abnormally, yielding only rearranged products ⟨78ZN(B)131⟩.

Variation in the structure of the phosphine is compatible with the reaction sequence, though the substituents may bring about a further reaction. Thus, a carboxylic acid derivative yields a 1,3-oxaphospholane as a consequence of dehydration of the initial product (Scheme 42) ⟨72JPR66⟩. (3-Hydroxypropyl)phenylphosphine (**221**; $n = 3$), derived by the reaction of $PhPH_2$ with allyl alcohol, reacts readily with aldehydes and ketones to form the hydroxyalkyl derivatives which spontaneously cyclize to 1,3-oxaphosphorinane (Scheme 42) ⟨79ZC57⟩. By the same protocol, phosphines with a 2-hydroxyethyl substituent yield 1,3-oxaphospholanes ⟨72T2587⟩.

Although tertiary phosphines react preferentially at oxygen, providing they incorporate a suitable leaving group attack may occur at a carbonyl carbon atom. For example, trimethylsilylphosphines react with formaldehyde to give, for example the silyl ether (**222**; R = H). The same product arises from the reaction of the phosphine with glyoxal which is accompanied by a decarbonylation (Scheme

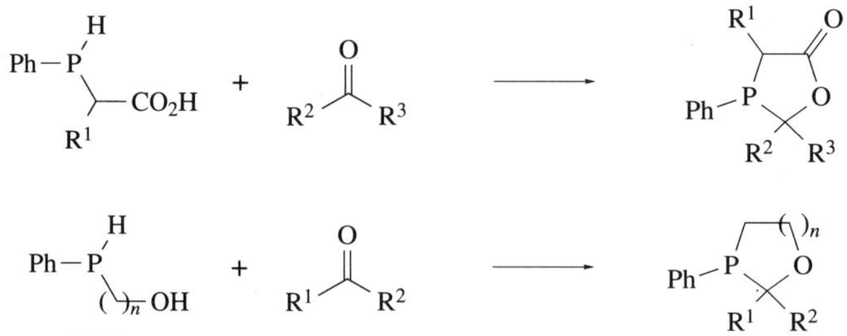

43) ⟨72IC2274⟩. The reaction with ketenes yields (222; R = CR^2_2) ⟨73JOM(47)67⟩. The initial addition of diphenyl(trimethylsilyl)phosphine to trifluoromethyl ketones is also followed by transfer of the silyl moiety to oxygen ⟨88CZ146⟩.

There are many other examples of phosphines and other PH compounds reacting with carbonyl compounds to give α-hydroxyalkylphosphines etc. and a selection chosen to illustrate the diversity of this reaction is given in Table 1.

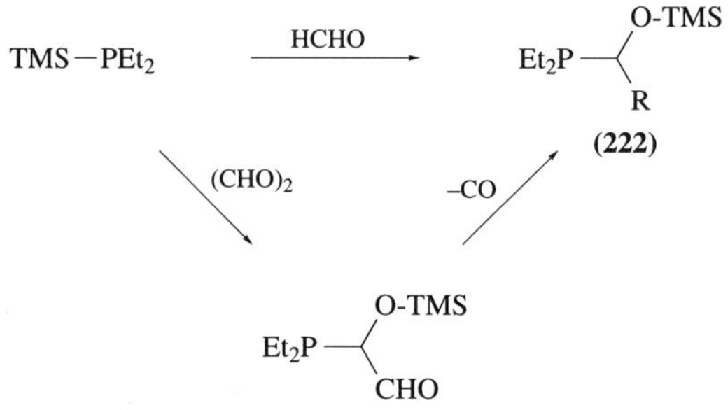

Scheme 43

Table 1 Reactions of P compounds with carbonyl compounds to give α-hydroxyalkylphosphines etc.

P compound	Carbonyl compound	Product	Ref.
PH_3	$MeCOCF_3$	$HP[C(OH)MeCF_3]_2$	89CZ320
$(TMS-O)_2PH$	$MeCOCF_3$	$(TMS-O)_2P[C(OH)MeCF_3]$	89CZ320
$(MeO)_2P(O)H$	$PhCOCH_2CN$	$(MeO)_2P(O)-C(OH)(CN)(Ph)(CH_2?)$	72T6013
$(MeO)_2P(O)H$	$PhCOP(O)(OMe)_2$	$PhC(OH)[P(O)(OMe)_2]_2$	71JOC3843
$Ph_2P(O)H$	$MeCOCO_2H$	$Ph_2P(O)-C(OH)(Me)(CO_2H)$	85PS(25)39
$Me_2P(S)H$	Cl_3CCHO	$Me_2P(S)-CH(OH)(CCl_3)$	92PS(72)171

Tetracoordinate phosphine derivatives also react with carbonyl compounds. The condensation between diethyl phosphite and paraformaldehyde proceeds efficiently under triethylamine catalysis giving good yields of $HOCH_2P(O)(OEt)_2$ which has been converted into its THP ether ⟨90OSC(7)160⟩. Ketones are converted into α-hydroxyalkyl phosphonic esters on treatment with dialkyl phosphites, $(RO)_2P(O)H$, by adsorption onto an alumina–KF mixture ⟨86TL3515⟩ and the bis(phosphine oxides), $(CH_2)_n[P(O)RH]_2$, afford the corresponding bis adducts with carbonyl compounds ⟨79ZC417⟩.

The reaction of phosphines and phosphine oxides with diketones leads to α-hydroxyalkyl derivatives, though a subsequent reaction may occur. Biacetyl is attacked at only one carbonyl group even with an excess of the phosphine oxide ⟨85PS(25)39⟩ and indan-1,3-diones are also attacked only once ⟨76IZV1416⟩. However, after initial reaction at one carbonyl group, 1,4-diketones subsequently yield phosphorylated tetrahydrofurans (223) ⟨91ZOB909⟩. 2-(Acetylmethyl)cyclopentanone (224) and H_3PO_2 give the phosphabicyclooctane (225) (Equation (37)) ⟨91ZOB1263⟩. Similarly, 1,5-diketones may cyclize in a second stage, $PhCOCH_2CHPhCH_2COPh$ giving, for example the acyclic adduct, a 2,6-dihydroxyphosphorinane (226) and the 2,3-dihydropyranyl-2-phosphinic acid (227) in varying amounts depending on the phosphine used ⟨79ZOB1956, 91ZOB1315⟩.

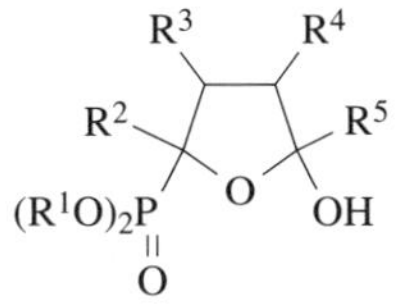

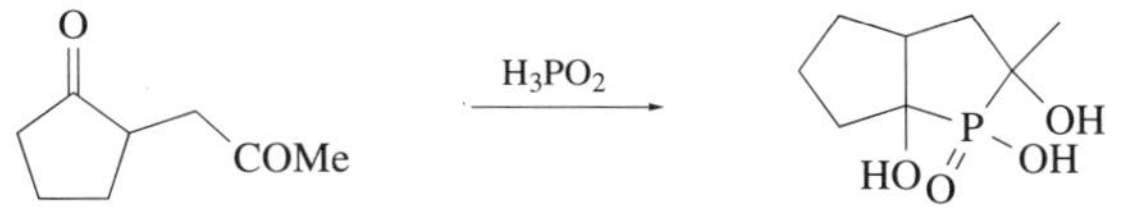

The reaction of phosphines with α,β-unsaturated carbonyl compounds is dependent on the structures of the two reactants and on the reaction conditions. At room temperature, addition of a variety of phosphines and phosphine oxides to the carbonyl function is observed. Thus dibutylphosphine oxide adds in a 1,2-fashion to mesityl oxide, but at 60 °C conjugate addition also occurs ⟨86ZOB711⟩. When 2 equiv. of dimethyl phosphite react with enones $R^1COCH=CHR^2$ both unsaturated groups are attacked leading to (228), which may cyclize to an oxaphospholane or rearrange to a phosphate (Scheme 44) ⟨91CB175, 91LA229⟩. Whereas α,β-unsaturated aldehydes undergo 1,2-addition of tris(trimethylsilyl)phosphite, both α,β-unsaturated ketones and $CH_3CH=CHCO_2Et$ give the 1,4-adducts, for example $(TMS-O)_2P(O)CHRCH=C(R)O-TMS$ ⟨77CL485⟩. Careful choice of reaction conditions and the P reagent enable regioselective 1,2- and 1,4-addition to α,β-unsaturated aldehydes to be achieved ⟨78JA3467⟩.

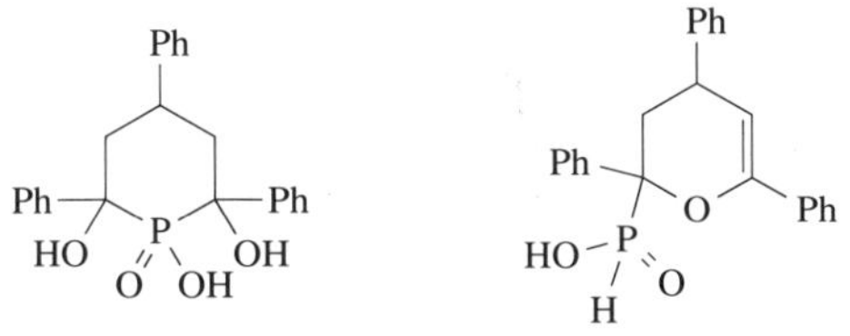

It is clear that the nature of the reactants and the reaction conditions play a significant role in the outcome of the reaction between carbonyl compounds and phosphines and careful control is necessary if α-hydroxyalkyl phosphorus compounds are to be obtained in useful yields.

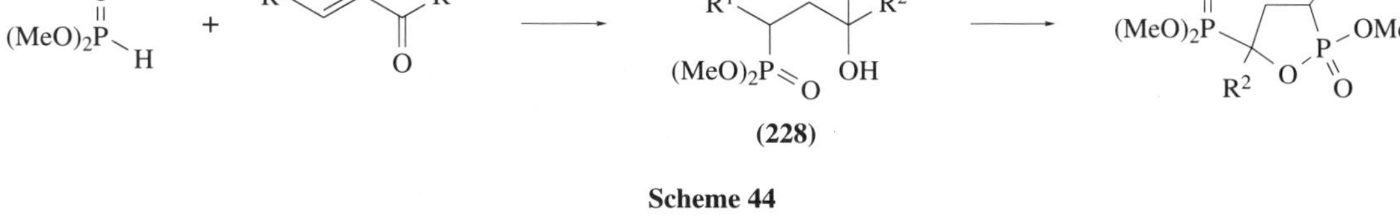

Scheme 44

(b) From compounds containing two singly bonded functional groups. (1) From X—C—P functions (X = Hal). The halogen in an α-chloroalkyl phosphorus compound can be directly replaced by nucleophiles as in the formation of the heterocycle (229) (Equation (38)) ⟨84PS(21)59⟩ and of $MeP(O)(CH_2OR)_2$ from $P(CH_2Cl)_3$ and alkoxides ⟨73ZOB534⟩. The epoxides shown in Scheme 45

⟨72ZOB479, 73TL173⟩ are formed by carbanion generation from the chloromethyl group followed by an intramolecular Williamson ether synthesis.

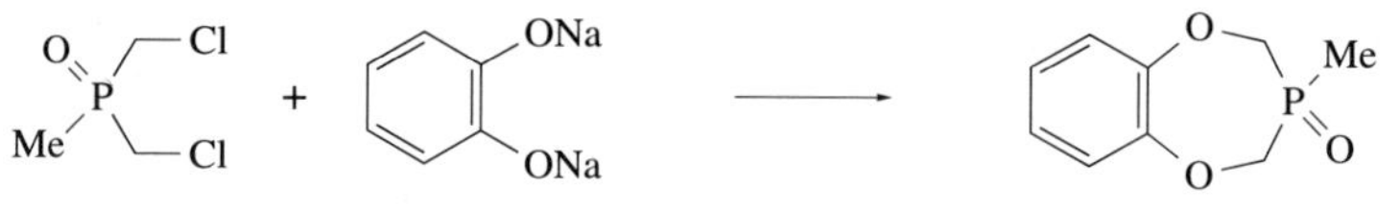

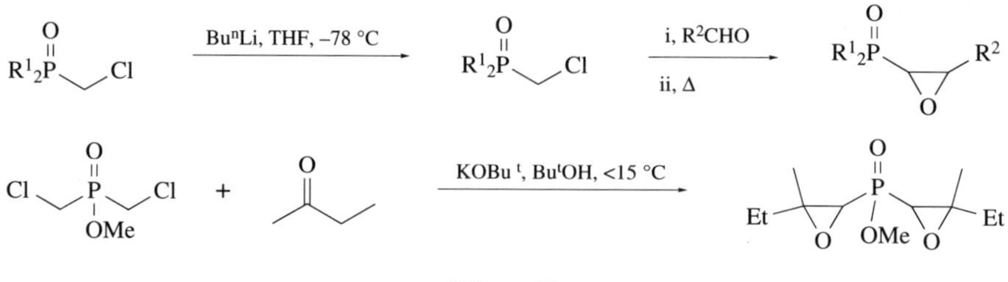

Scheme 45

(2) From O—C—X functions (X = Hal). The chlorine in α-chloro ethers is readily replaced by P nucleophiles yielding α-alkoxyalkylphosphonium salts as illustrated in Scheme 46 ⟨62CB2514, 79JCS(P1)3099, 80JCS(P1)1627⟩. These products may be converted into the corresponding phosphine oxides.

The Arbuzov reaction ⟨81CRV415⟩ provides an easy approach to the formation of a P—C bond, involving the reaction of, usually, an alkyl halide with a phosphite or a phosphinite ester. With the former phosphorus reagent, α-alkoxyalkyl phosphonates are formed from α-alkoxyalkyl halides ⟨77LA88⟩ as in the reaction of MEM-Cl with triethyl phosphite which gives $MeOCH_2CH_2OCH_2P(O)(OEt)_2$ ⟨79JOC4847⟩. In a related manner, P-acetals result when orthoformates react with phosphorus(III) halides, and an aldehyde acetal affords an alkoxyalkylphosphinic acid derivative (Scheme 47) ⟨70ZOB2560, 71ZOB2575, 72ZOB1647⟩.

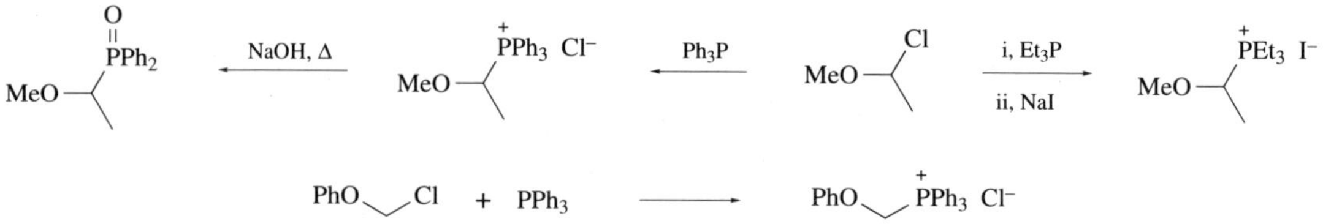

Scheme 46

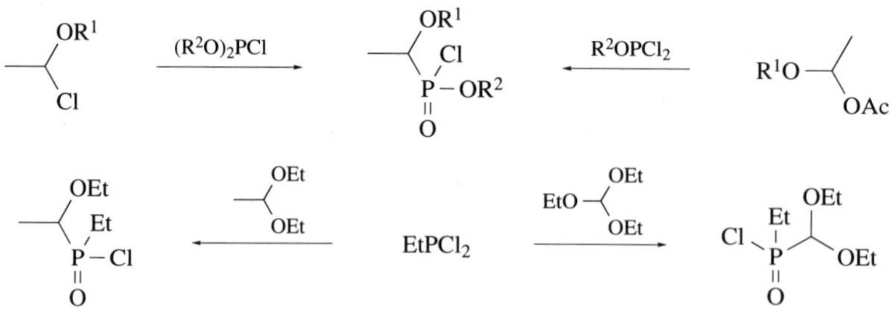

Scheme 47

(c) Formation by miscellaneous routes. Phosphaalkenes undergo a [4 + 2]-cycloaddition with *o*-quinones to yield 1,4,2-dioxaphosphorinanes (**230**) (Equation (39)) ⟨83T3189⟩.

The photochemical rearrangement of the cyclopropenyl phosphaalkenes (231) with methyl propynoate gives the bridged phosphines (232) through Diels–Alder reactions of intermediates, which also undergo 1,3-dipolar cycloadditions with nitrile oxides to give the bicyclic phosphines (233) (Scheme 48) ⟨87JOM(332)C1⟩.

α-Halo ketones react with salts derived from dialkyl phosphonates to give epoxyphosphonates ⟨58QR341⟩.

On treatment with benzoyl chloride, the phosphole (234) yields the acylphosphonium salt (235) which ring expands on hydration (Scheme 49) ⟨73T707⟩. Similar behaviour is shown by dibenzophospholes and phosphole irontricarbonyl complexes ⟨82TL1565, 84JOM(266)285⟩. The reaction of P_2F_4 with CF_3COCF_3 affords $F_2PC(CF_3)_2OPF_2$ ⟨76JFC(7)1⟩.

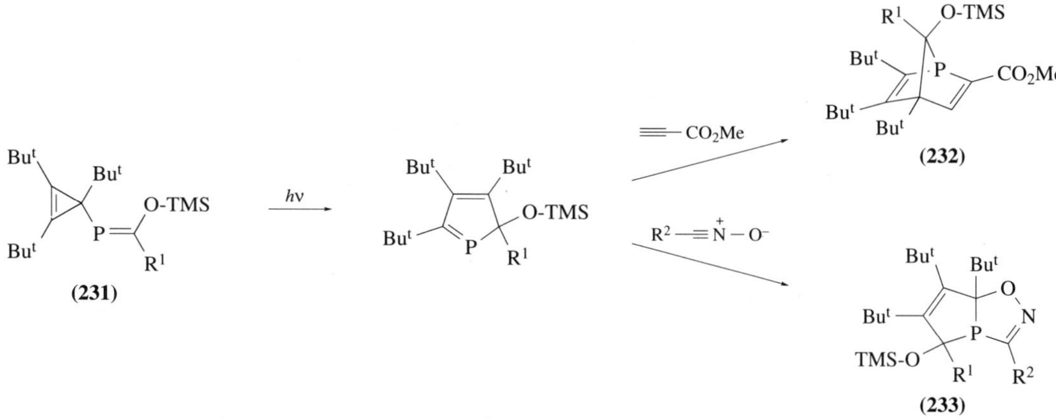

Scheme 48

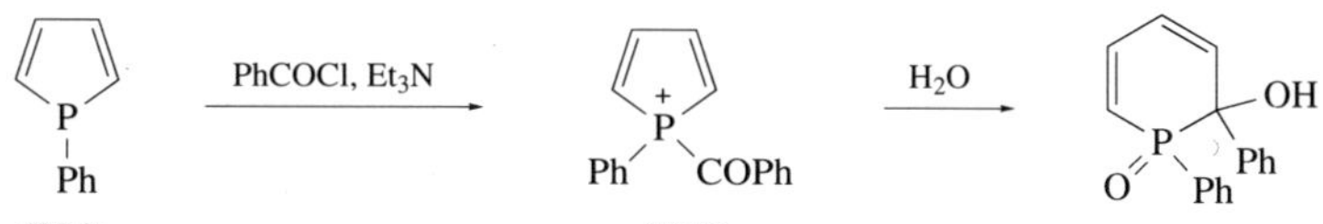

Scheme 49

(d) By functionalization of a preformed O,P-acetal. (1) Functionalization on oxygen. The thermal disproportionation of bis- and tris(α-hydroxyalkyl) phenylphosphines and phosphonium salts yields 1,3,5-dioxaphosphorinanes (Scheme 50) ⟨79IZV866, 80IZV1626⟩. Similar compounds result when bis-α-hydroxyalkylphosphines react with aldehydes, though the product is a mixture of stereoisomers as a consequence of the two chiral centres in the phosphine. The reaction is considered to proceed through a double carbocationic attack at the diol oxygen atoms (Scheme 51) ⟨79DOK(244)610⟩.

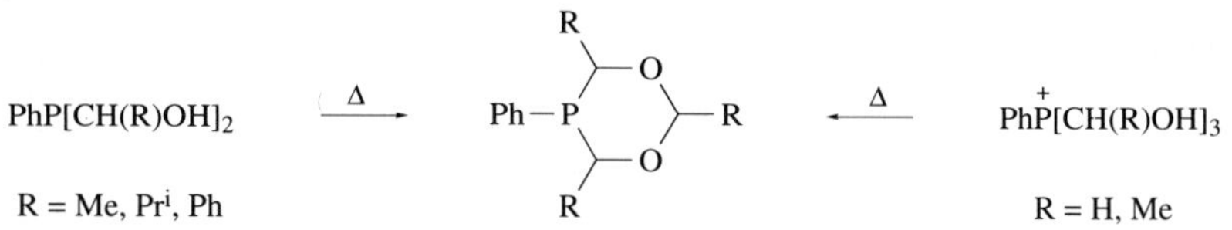

Scheme 50

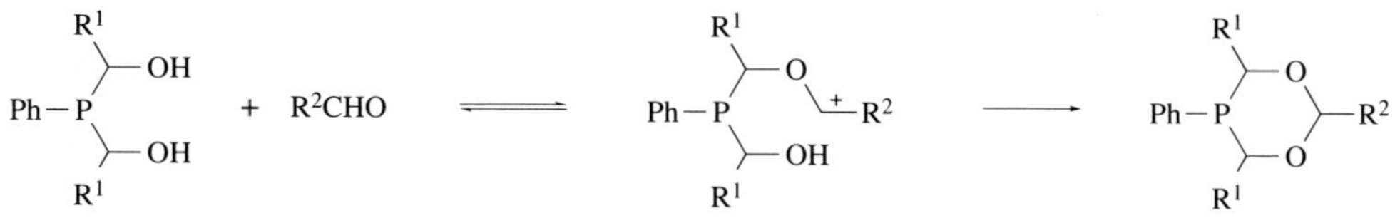

Scheme 51

When the diol reacts with acetals of various aldehydes, moderate yields of dioxaphosphorinanes are obtained ⟨86IZV2502⟩. Parallel behaviour is shown by the phosphine oxides towards benzaldehyde in the presence of an acidic catalyst leading to the 5-oxide ⟨75JOC2056⟩.

The isomerization of tris(α-hydroxyalkyl)phosphines in the presence of formaldehyde also yields a phosphorus(V) dioxaphosphorinane ⟨78ZOB2653⟩, whilst with orthoformates and chlorotrimethylsilane tricyclic species are produced (Equation (40)) ⟨66JA1140, 70JOC2310, 73PS(3)1, 80ZOB2424⟩.

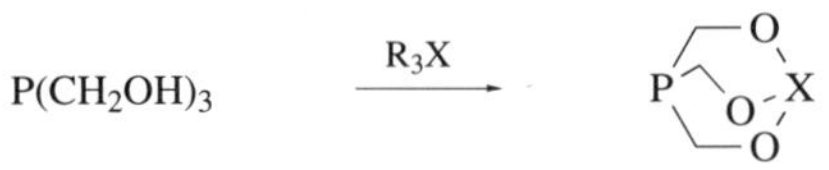

$$P(CH_2OH)_3 \xrightarrow{R_3X} \text{[tricyclic product]} \quad (40)$$

$R_3X = (MeO)_3CH$ or TMS-Cl

Reaction of the diols (**236**) with dichlorosilanes similarly leads to the incorporation of silicon into the heterocyclic product ⟨79DOK(247)609⟩, whilst with boric acid esters 1,3,5,2-dioxaphosphaborinanes result ⟨85IZV1102⟩. The diol phosphine oxide reacts in a similar manner with the anhydride of boric acid ⟨83IZV1374⟩. Borylation in the presence of aldehydes leads to betaines (Scheme 52) ⟨86IZV2502⟩. Prior formation of the diols, several of which are unstable, can be avoided since primary phosphines react with borinic esters and aldehydes in the presence of base to give 1,3,5,2-dioxaphosphaborinane salts ⟨88IZV159⟩. Many variations of this last approach are possible, including the reaction with salicylaldehyde which affords the benzoheterabicyclo[3.3.1]nonane (**237**) ⟨89IZV946⟩ and the formation of the spiro compound (**238**) from tris(hydroxymethyl)phosphine ⟨90IZV1133⟩. When a diphenylphosphinous acid reacts with salicylaldehyde and a boron ester, the product (**239**) contains an exocyclic phosphorus moiety ⟨92IZV196⟩.

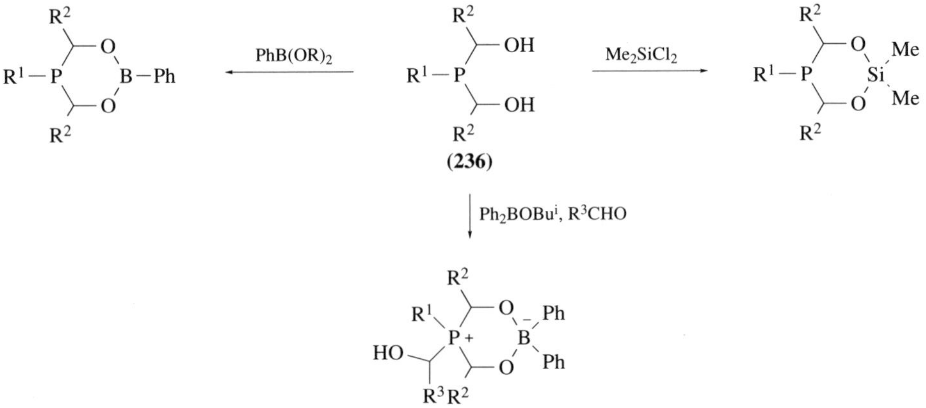

Scheme 52

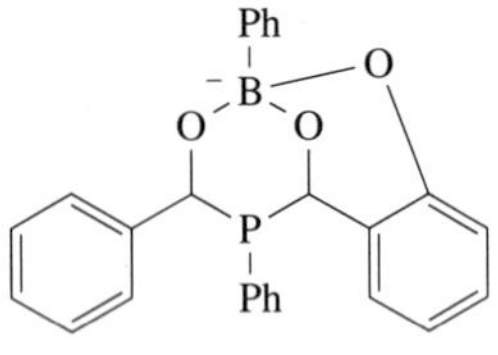

(**237**) 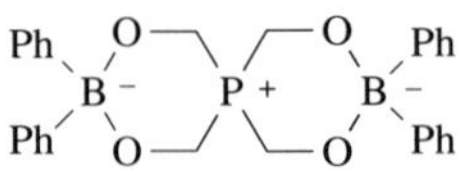 (**238**) 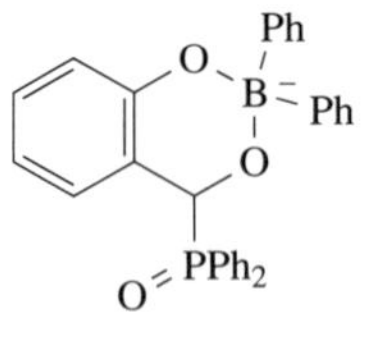 (**239**)

(2) *Functionalization on carbon.* The hydroxy group in hydroxyalkylphosphorus compounds is amenable to alkylation as illustrated by conversion into an orthoester with triethyl orthoformate and into a cyano ether with acrylonitrile (Scheme 53) ⟨71ZOB1964⟩.

The phosphonates derived form acid chlorides by an Arbuzov rearrangement with P(OR)$_3$ are

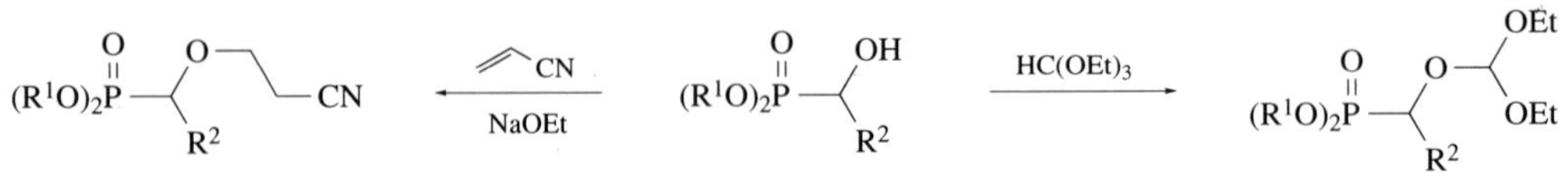

Scheme 53

reduced to the alcohols (**240**) with NaBH$_4$. The significance of this functional group interconversion lies in the ease of hydrolysis of (**240**) to an aldehyde and a dialkyl phosphonate, such that the overall sequence represents the conversion of an acid into an aldehyde (Scheme 54) ⟨70CB2984, 70JOC597⟩.

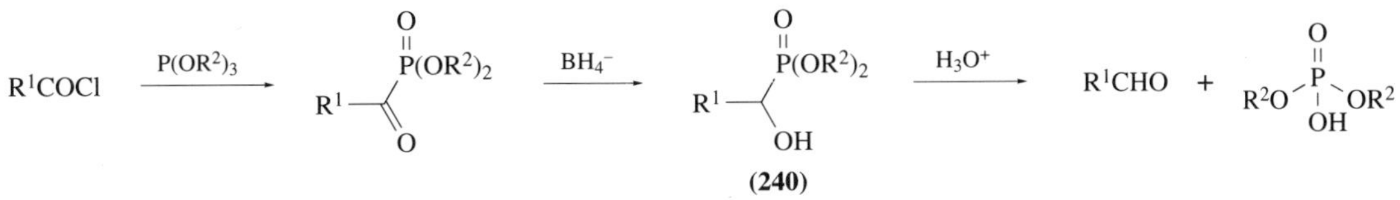

Scheme 54

α-Hydroxyalkylphosphonates result from the reaction of the carbanion derived from nitromethane with acylphosphonates under phase transfer conditions (Equation (41)) ⟨92S258⟩, whilst the chiral acylphosphine oxide (**241**) yields the *erythro* alcohol when treated with methylmagnesium iodide (Equation (42)) ⟨77PS(3)345⟩.

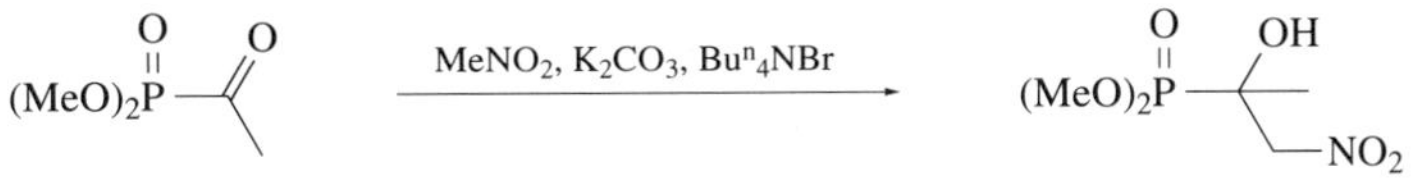

Epoxides with an α-phosphorus-containing substituent have been cleaved by nucleophiles to give α-hydroxyalkyl phosphorus derivatives. Amines ⟨80PJC233⟩ and alcohols in the presence of BF$_3$ etherate ⟨79MI 407-01⟩ attack at the β-carbon, whereas alcohols alone attack at the α-carbon ⟨87ZOB2300⟩ (Scheme 55). An intramolecular Williamson ether synthesis yields the epoxide (**242**) and a similar cyclization spontaneously follows the initial condensation of diethyl phosphonate and CH$_3$COCHClCO$_2$Et (Scheme 56) ⟨72ZOB1861, 76ZOB783⟩. The anions derived from (α-methoxyalkyl)diphenylphosphine oxides with LDA react with aldehydes and ketones to give 2-methoxy alcohols as mixtures of diastereoisomers ⟨77CC314⟩.

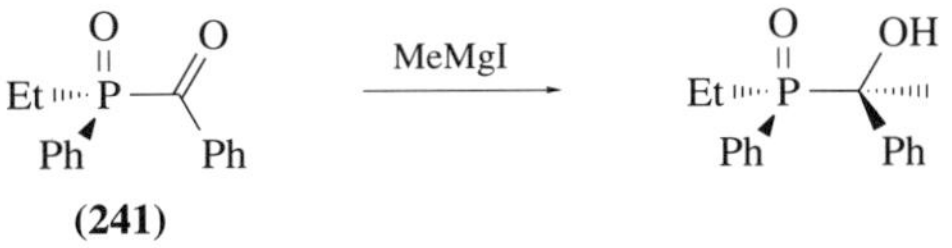

Scheme 55

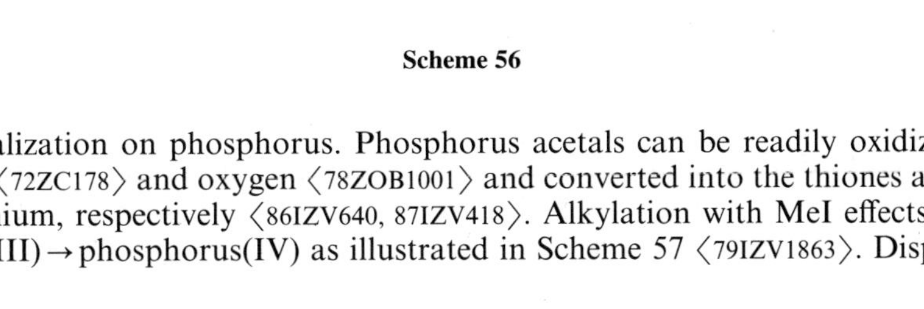

Scheme 56

(3) *Functionalization on phosphorus.* Phosphorus acetals can be readily oxidized at P by, for example, H$_2$O$_2$ ⟨72ZC178⟩ and oxygen ⟨78ZOB1001⟩ and converted into the thiones and selenones by sulfur and selenium, respectively ⟨86IZV640, 87IZV418⟩. Alkylation with MeI effects the conversion of phosphorus(III) → phosphorus(IV) as illustrated in Scheme 57 ⟨79IZV1863⟩. Displacement of an

alkoxy function from either a tri- or tetra-coordinated P can be achieved on treatment with iodomethane ⟨72ZOB2418, 73PS(3)55⟩.

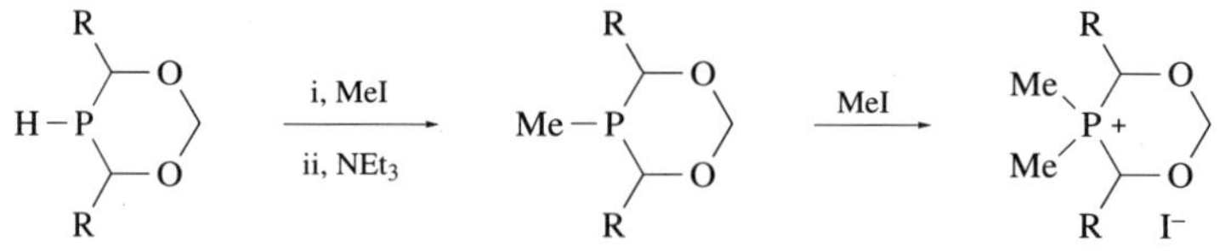

Scheme 57

The reactions of the phosphonium salt (**243**) with chlorine is dependent on the pH of the aqueous reaction mixture. Thus, at pH 5–7, oxidation at P is accompanied by loss of the hydroxymethyl group, but at pH 1–3, a hydroxyethyl group is lost (Scheme 58) ⟨72PS(1)237⟩.

The 1,3-oxapholane oxide (**245**) has been obtained from the bicyclic 1,3,5-dioxaphosphorinane (**244**) through initial functionalization at P as depicted in Scheme 59 ⟨74ZOB1029⟩.

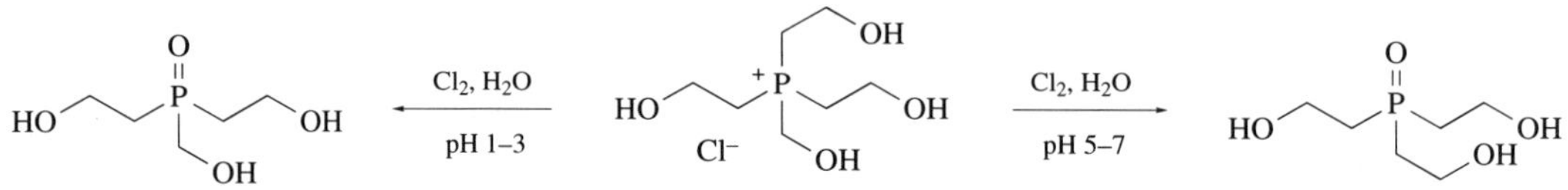

Scheme 58

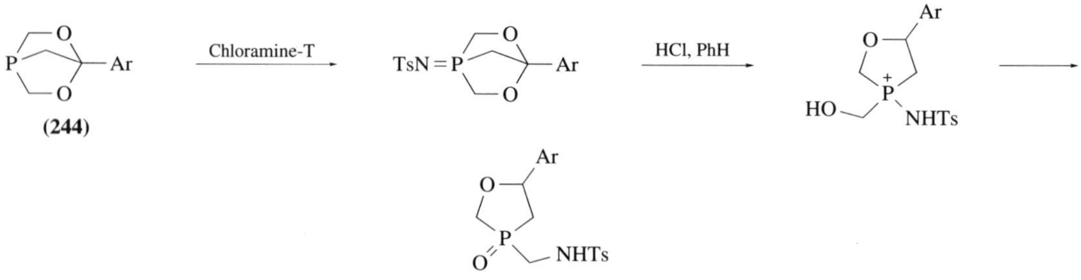

Scheme 59

4.07.2.1.2 Oxygen and arsenic, antimony or bismuth

There is a considerable number of procedures available for the preparation of organo arsenic, antimony and bismuth compounds ⟨82COMC-I(2)681⟩. However few of these have been applied to the synthesis of functions which also contain a chalcogen. Of the systems considered here only the *O,As*-acetals have been described in the literature. The corresponding *O,Sb*- and *O,Bi*-acetals are unknown.

(i) From ketones

Hexafluoroacetone reacts in an similar way with arsines as it does with amines and phosphines. Dimethylarsine gave $(F_3C)_2C(OH)AsMe_2$ ⟨65JOM(4)151⟩. Similarly with diphenylarsine the addition compound $(F_3C)_2C(OH)AsPh_2$ was obtained as a viscous liquid in high yield ⟨73CJC1136⟩. Related products are formed with silylarsines. Thus $(F_3C)_2C(O\text{-}TMS)AsMe_2$ (88%) was obtained when TMS-$AsMe_2$ and hexafluoroacetone were kept, in the absence of solvent, at 20 °C for 12 h. Insertion products were also formed from $Me_2Si(AsMe_2)_2$ and $MeSi(AsMe_2)_3$ and 2 and 3 mols of hexafluoroacetone respectively ⟨69JOM(17)161⟩. A detailed analysis of the ^{19}F NMR spectrum of the product from chloropentafluoroacetone and TMS-$AsMe_2$ has been reported ⟨69TFS1697⟩.

4.07.2.2 Functions Bearing Sulfur

This section comprises compounds containing sulfur and a group 15 element. However, there are no examples involving As, Sb and Bi.

4.07.2.2.1 Sulfur and phosphorus

Both sulfur and phosphorus can exist in a variety of oxidation states, so that in combination a wide variety of structural types can be envisaged. In practice, the large majority of examples involve 4-coordinate phosphorus in association with sulfur in 2- and 3-coordination states. Some 2-coordinate sulfur types are exemplified by (**246**), (**247**) and (**248**). Isolated examples of other structural types are known.

The chemistry of S,P-acetals has not been reviewed, although the chemistry of α-phosphoryl organosulfur compounds has been discussed, including that of α-phosphoryl sulfones ⟨B-92MI 407-02, B-94MI 407-02⟩.

$$R^1 \diagdown SR^4 \qquad R^1 \diagdown SR^4 \qquad R^1 \diagdown SR^4$$
$$R^2 \diagup PR^3{}_2 \qquad R^2 \diagup \overset{+}{P}R^3{}_3 \qquad R^2 \diagup \underset{\underset{X}{\|}}{P}R^3{}_2$$

(**246**) 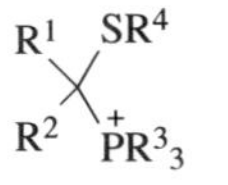   (**247**) 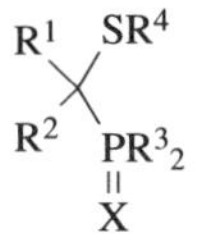   (**248**)

(i) From compounds containing multiply bonded functional groups

(a) From aldehydes and ketones. The simplest and most versatile approach to S,P-acetals possessing dicoordinate sulfur appears to be the reaction between a ketone and a chlorophosphine possessing either alkyl or aryl substituents, in the presence of a thiol. Presumably the reaction proceeds by nucleophilic attack of the phosphine at the carbonyl group with the formation of a phosphonium epoxy salt. Subsequent ring opening by the thiol completes a C to P migration of oxygen and leads to the product (Equation (43)) ⟨87DOK(292)360⟩. In a similar manner, trimethyl thiophosphite reacts with methyl ketones in the presence of PCl_3 to give $MeSC(Me)(R)P(O)Cl_2$. The yields are, however, only moderate ⟨86ZOB216⟩. Benzaldehyde and S-acetyldiphenylthiophosphinous acid react in the presence of HCl to give the dibenzyl sulfide $[Ph_2P(O)CHPh]_2S$ ⟨86ZOB1905⟩.

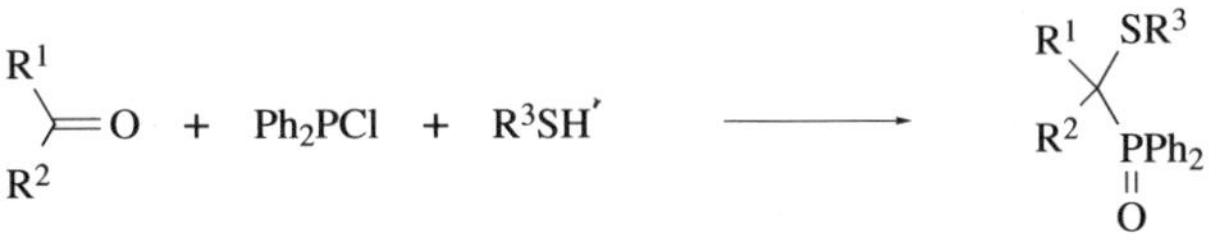

$$R^1{\mathrel{\mathop:}}{=}O + Ph_2PCl + R^3SH' \longrightarrow R^2\underset{\underset{O}{\|}}{\overset{R^1\diagdown SR^3}{\diagup}}PPh_2 \qquad (43)$$

(b) From thiocarbonyl compounds. The reaction between thiocarbonyl compounds and trialkyl phosphites can proceed in two directions following initial nucleophilic attack by P at C. Triisopropyl phosphite leads exclusively to the thiol (**249**) by proton transfer from an isopropyl group. On the other hand, trimethyl phosphite gives the sulfide (**250**) as a result of methyl transfer. Both types of product result when triethyl phosphite is used (Scheme 60) ⟨78JOC1980, 79BCJ3342⟩.

High yields of the S,P-acetals (**251**) result when thioketones react at room temperature with $(EtO)_2P(O\text{-}TMS)$. The free thiols are formed in alcohol solution ⟨81ZOB2140⟩. A similar reaction occurs with sodium dialkylphosphites ⟨83ZOB46⟩.

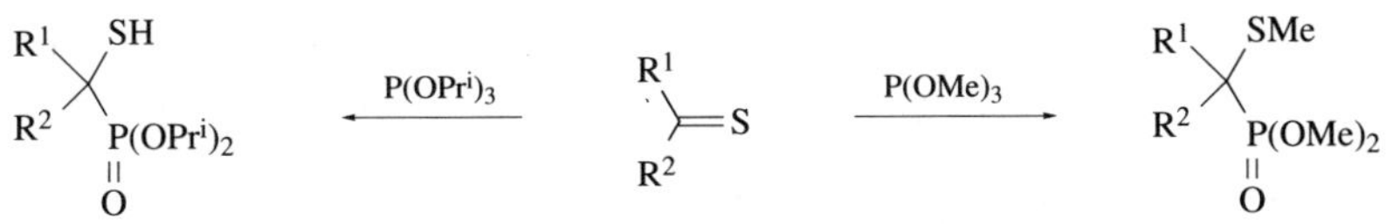

$$R^1 \diagdown SH \qquad\qquad R^1 \qquad\qquad R^1 \diagdown SMe$$
$$R^2 \diagup P(OPr^i)_2 \xleftarrow{P(OPr^i)_3} \; {=\!\!=}S \; \xrightarrow{P(OMe)_3} R^2 \diagup P(OMe)_2$$
$$\underset{O}{\|} \qquad\qquad R^2 \qquad\qquad \underset{O}{\|}$$

(**249**)   (**250**)

Scheme 60

$$\underset{(251)}{\overset{R^1}{\underset{R^2}{>}}\!\!\!\!\!\!<\!\!\!\!\!\!\overset{S\text{-TMS}}{\underset{\overset{\|}{O}}{P(OEt)_2}}}$$

(c) From diazoalkanes. (Diphenylthiophosphinoyl)phenyldiazomethane **(252)** undergoes a Pummerer-like rearrangement with a 1,2-shift of S when kept in acetic acid giving a quantitative yield of **(253)** ⟨79TL2415⟩.

Carbene insertion into the C—S bond in allyl phenyl sulfide and into the S—S bond in dimethyl disulfide yields **(254)** and **(255)**, respectively. The carbene source is diethyl diazomethylphosphonate and the reactions proceed under Rh catalysis ⟨87CL1569, 87PS(30)225⟩. Photolysis of α-diazo phosphine sulfides also involves the intermediacy of a carbene which may be intercepted by alcohols leading to $Ph_2P(O)CHPhSMe$ ⟨79TL2415, 79T81⟩.

(252) $Ph_2P(S)C(N_2)Ph$ **(253)** $Ph_2P(O)C(SAc)Ph$ **(254)** $CH_2=CHCH_2C(SPh)P(O)(OEt)_2$ **(255)** $(EtO)_2P(O)C(SMe)_2$

(d) From phosphorus ylides. (N-Phenylthio)succinimide serves as a S-transfer reagent, reacting with the ylide **(256)** to give the phosphonium salt shown in Equation (44) ⟨75TL4531, 81JOC1828⟩.

$$\underset{(256)}{\triangleright\!\!-\!\!\overset{+}{P}Ph_3} \;+\; \underset{SPh}{\overset{O}{\underset{N}{\bigcirc}}\!\!\!\!\!\!\!\overset{O}{\bigcirc}} \;\xrightarrow{Et_3O^+\,BF_4^-}\; \triangleright\!\!\!\!\!<\!\!\!\overset{SPh}{\underset{\overset{+}{P}Ph_3\;BF_4^-}{}} \qquad (44)$$

(e) From alkenes. Conjugate addition of EtSH to the vinyl sulfoxide **(257)** proceeds straightforwardly to give the ethylthiomethyl derivative **(258)** (Equation (45)) ⟨92TA1515⟩.

$$\underset{P(OEt)_2,\,O}{CH_2=C(S(O)Tol)P(O)(OEt)_2} \;+\; EtSH \;\longrightarrow\; EtSCH_2CH(S(O)Tol)P(O)(OEt)_2 \qquad (45)$$

(ii) From compounds containing two singly bonded functional groups

(a) From X—C—P functions (X = Hal, OTs, Li). Displacement of halide by nucleophilic S species has been used to prepare α-thioalkyl and α-alkylthioalkyl phosphonates. For example, thioacetate reacts with diethyl iodomethylphosphonate to give **(259)**, the free thiol resulting from hydrolysis of the initially formed S-acetate ⟨79JOC2967⟩. A double substitution occurs when Na_2S reacts with diethyl chloromethylphosphonate ⟨57ZOB2360⟩ and a further illustration is provided by the reaction of bis(chloromethyl)phosphinic acid with thiourea to give a bis-thiol after hydrolysis of the isothiouronium salt (Scheme 61) ⟨68IZV1625⟩.

Dialkyl alkylphosphonates are readily metallated with Bu^nLi and subsequent treatment with elemental sulfur gives high yields of α-thioalkylphosphonates, $HSCH(R)P(O)(OEt)_2$ ⟨79JOC2967⟩. Dimethyl sulfide also reacts with α-halomethylphosphonates to give sulfonium salts $(EtO)_2P(O)CH_2S^+Me_2$ ⟨68GEP1924135⟩. Similarly, diethyl ethylphosphonate reacts with diphenyl disulfide, after metallation, to give $(EtO)_2P(O)CH(Me)SPh$ in 87% yield ⟨80S72⟩. Alkyldiphenylphosphine oxides have also been metallated and then treated with some disulfides ⟨77JCS(P1)2263⟩.

Displacement of tosylate from $(EtO)_2P(O)CH(OTs)R$ by thiocyanate gives the α-thiocyanatoalkylphosphonate $(EtO)_2P(O)CH(SCN)R$ in good yield ⟨92ZOB1268⟩.

(b) From S—C—X functions (X = Hal, Li). Nucleophilic displacement of halide by P species is

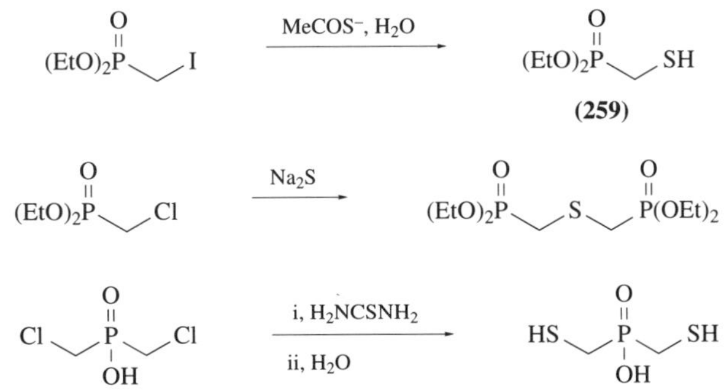

Scheme 61

exemplified by quaternization of Ph$_3$P by α-chloroalkyl sulfides; a number of these phosphonium salts are commercially available.

The initial reaction between P(OEt)$_3$ and MeSCH$_2$Cl is spontaneously followed by an Arbuzov reaction leading to MeSCH$_2$P(O)(OEt)$_2$ ⟨63JCS1324⟩. Oxidation at sulfur gives the corresponding methylsulfinylphosphonate ⟨73S669⟩. Analogous compounds have been synthesized from chloroalkyl sulfides and the sodium salt of diethyl phosphite ⟨76S107⟩. Ethyl diphenylphosphinite behaves similarly towards chloromethyl methyl sulfide giving a high yield of Ph$_2$P(O)CH$_2$SMe ⟨77JCS(P1)2263⟩.

Benzyl methyl sulfide is metallated at the benzylic site by Bu^{n}Li-TMEDA and has been subsequently quenched by chlorodiphenylphosphine to give PhCH(SMe)PPh$_2$ ⟨84PS(19)61⟩. Similarly the carbanion derived from methyl phenyl sulfone reacts with (RO)$_2$P(O)Cl to give PhSO$_2$CH$_2$P(O)(OR)$_2$ ⟨89SC2209, 90SC273⟩.

(iii) By functionalization of a preformed S,P-acetal

Oxidation of sulfur to the sulfoxide and the sulfone has been achieved by the usual variety of reagents and is detailed in ⟨B-94MI 407-02⟩. Alkylation and halogenation at carbon is also covered therein.

The bis-thiol **(260)** yields the 1,3,5-dithiaphosphorinane by condensation with ketones (Equation (46)) ⟨71ZOB1247⟩.

α-Alkylthioalkylphosphonates react with alkyl halides at sulfur in the presence of AgClO$_4$ to give the sulfonium salts and offers an alternative approach to that described in 4.07.2.2.1(ii)(a) ⟨74JCS(P1)1279⟩.

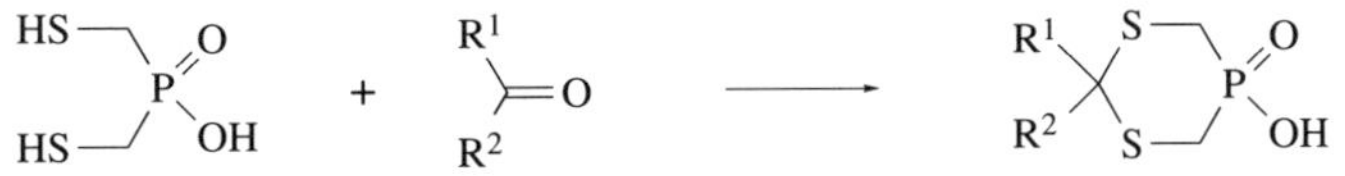

4.07.2.3 Functions Bearing Selenium or Tellurium

4.07.2.3.1 *Selenium or tellurium with phosphorus*

Very few *Se,P*- or *Te,P*-acetals have been prepared. The known compounds appear to be restricted to those containing phosphorus(IV) or phosphorus(V) functions.

(i) From phosphines

Triphenylphosphine and α-haloalkyl aryl selenides react to give the corresponding phosphonium salts ⟨61CB1373⟩. However, the relative inaccessibility of the starting selenides limits the value of this approach. The salts (**261**; R = H or Me) were obtained in high yields from PhSeCH$_2$Br and PhSeCH(Me)Br, respectively, and PPh$_3$ in refluxing benzene. The halides were generated most efficiently from the appropriate diazo alkane and PhSeBr ⟨76JOM(114)281⟩.

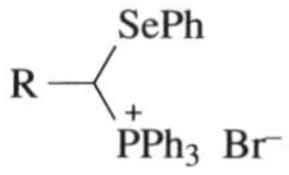

$$R-\underset{\overset{+}{PPh_3}\ Br^-}{\overset{SePh}{|}}$$

(**261**)

(ii) From phosphorus ylides

The stabilized ylide (**262**) reacts readily with areneselenenyl halides. However the initially formed salts (**263**) are not isolable since a spontaneous transylidation reaction ensues giving the phosphoranes (**264**) and the salt (**265**). When (**262**) reacted with TeBr$_4$ *trans*-ylidation was not observed and the only product was the unusual *Te,P*-acetal (**266**) (Scheme 62) ⟨64CI(L)1461⟩. The reaction conditions were not specified in this report. Nonstabilized alkylidenetriphenylphosphoranes have been generated in THF solution and quenched with PhSeBr; the resulting phosphonium salts were obtained as perchlorates following the addition of AgClO$_4$. Compounds such as (**267**) and (**268**) were produced by this method ⟨79CB355⟩. The salt (**263**; Ar = Ph) could be isolated as its perchlorate.

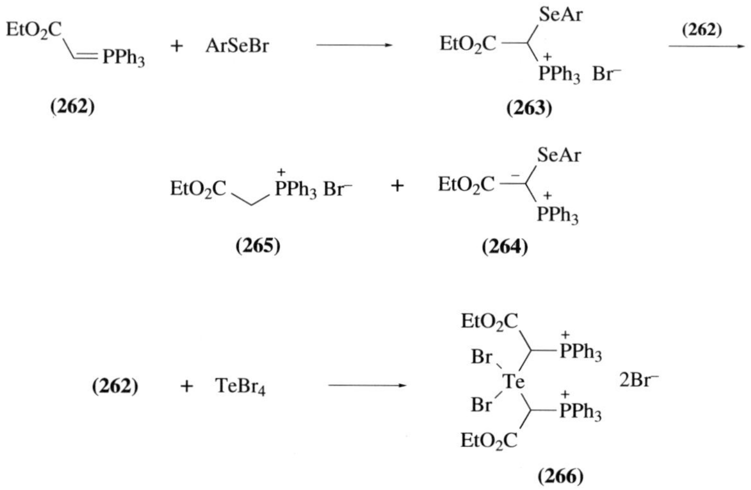

EtO$_2$C—=PPh$_3$ + ArSeBr ⟶ EtO$_2$C—C(SeAr)(PPh$_3$$^+$) Br$^-$ ⟶$^{(262)}$

(**262**)   (**263**)

EtO$_2$C—CH$_2$—$^+$PPh$_3$ Br$^-$ +   EtO$_2$C—C$^-$(SeAr)($^+$PPh$_3$)

(**265**)   (**264**)

(**262**) + TeBr$_4$ ⟶ [Br(EtO$_2$C)C(—$^+$PPh$_3$)—Te—C(Br)(EtO$_2$C)(—$^+$PPh$_3$)] 2Br$^-$

(**266**)

Ar = Ph, 2,4-(NO$_2$)$_2$C$_6$H$_3$

Scheme 62

(CH$_3$)$_2$C(SePh)($^+$PPh$_3$) ClO$_4$$^-$ C$_6$H$_{10}$(SePh)($^+$PPh$_3$) ClO$_4$$^-$

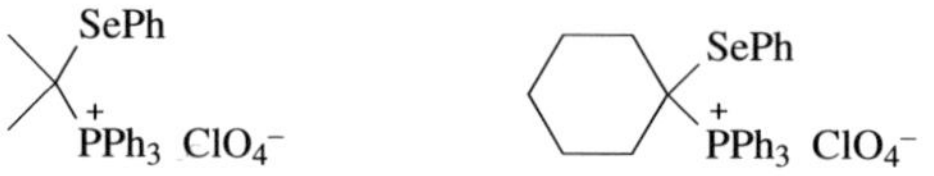

(**267**)   (**268**)

(iii) From phosphates and phosphine oxides

(a) With selenium nucleophiles. Iodide displacement from diethyl iodomethylphosphonate proceeds smoothly with sodium benzeneselenolate giving a high yield of PhSeCH$_2$P(O)(OEt)$_2$. Sodium selenide reacted in a similar way with 2 equiv. of the phosphonate with the formation of Se[CH$_2$P(O)(OEt)$_2$]$_2$. In both cases the reaction proceeds quickly and at ambient temperature ⟨78JOM(152)295⟩.

(b) With selenium electrophiles. This is probably the method of choice for compounds possessing a phosphorus(V) function. Benzyldiphenylphosphine oxide is readily deprotonated by butyllithium at low temperature. The anion has been intercepted by PhSeBr to give PhCH(SePh)P(O)Ph$_2$ ⟨78JOM(152)295⟩.

Phosphonates possessing a β-carbonyl group are easily deprotonated (NaH, THF, 0 °C) and subsequently selenenylated with PhSeBr. Compounds **(269)** ⟨78JOC1256⟩, **(270)** ⟨86JOC1537⟩ and **(271)** ⟨92CC190⟩ were all obtained by this procedure. The value of these materials lies in their facile transformation to vinylphosphonates by selenoxide elimination.

4.07.2.3.2 *Selenium or tellurium with arsenic antimony or bismuth*

Functions with Se—C—As, Se—C—Sb or Se—C—Bi units possessing sp^3 hybridized carbon are unknown, as are the corresponding tellurium analogues. This area therefore offers considerable scope for future research.

4.08
Functions Incorporating a Chalcogen and a Silicon, Germanium, Boron or Metal

MAX J. GOUGH
Technical Typesetters UK, Ashford, UK

and

JOHN STEELE
Pfizer Central Research, Sandwich, UK

4.08.1	FUNCTIONS CONTAINING A CHALCOGEN AND A METALLOID	352
	4.08.1.1 Functions Bearing Oxygen	352
	4.08.1.1.1 Oxygen and silicon—$R^1{}_2C(OR^2)SiR^3{}_3$, etc.	352
	4.08.1.1.2 Oxygen and germanium—$R^1{}_2C(OR^2)GeR^3{}_3$, etc.	360
	4.08.1.1.3 Oxygen and boron—$R^1{}_2C(OR^2)BR^3{}_2$, etc.	362
	4.08.1.2 Functions Bearing Sulfur	364
	4.08.1.2.1 Sulfur and silicon—$R^1{}_2C(SR^2)SiR^3{}_3$, etc.	364
	4.08.1.2.2 Sulfur and germanium—$R^1{}_2C(SR^2)GeR^3{}_3$, etc.	371
	4.08.1.2.3 Sulfur and boron—$R^1{}_2C(SR^2)BR^3{}_2$, etc.	372
	4.08.1.3 Functions Bearing Selenium or Tellurium	374
	4.08.1.3.1 Selenium or tellurium and silicon—$R^1{}_2C(SeR^2)SiR^3{}_3$, etc.	374
	4.08.1.3.2 Selenium or tellurium and germanium—$R^1{}_2C(SeR^2)GeR^3{}_3$, etc.	377
	4.08.1.3.3 Selenium or tellurium and boron—$R^1{}_2C(SeR^2)BR^3{}_2$, etc.	377
4.08.2	FUNCTIONS CONTAINING A CHALCOGEN AND A METAL	377
	4.08.2.1 Functions Bearing Oxygen—$R^1{}_2C(OR^2)M$, etc.	377
	4.08.2.1.1 Lithium, sodium or potassium	378
	4.08.2.1.2 Magnesium	383
	4.08.2.1.3 Titanium or aluminum	383
	4.08.2.1.4 Copper or zinc	384
	4.08.2.1.5 Mercury	385
	4.08.2.1.6 Tin	385
	4.08.2.1.7 Lead	390
	4.08.2.2 Functions Bearing Sulfur—$R^1{}_2C(SR^2)M$, etc.	390
	4.08.2.2.1 Lithium	391
	4.08.2.2.2 Beryllium or magnesium	395
	4.08.2.2.3 Titanium	396
	4.08.2.2.4 Copper, gold or zinc	396
	4.08.2.2.5 Aluminum, indium or gallium	397
	4.08.2.2.6 Tin	397
	4.08.2.3 Functions Bearing Selenium or Tellurium—$R^1{}_2C(SeR^2)M$, etc.	398
	4.08.2.3.1 Lithium	398
	4.08.2.3.2 Copper	401

4.08.1 FUNCTIONS CONTAINING A CHALCOGEN AND A METALLOID

4.08.1.1 Functions Bearing Oxygen

4.08.1.1.1 Oxygen and silicon—$R^1{}_2C(OR^2)SiR^3{}_3$, etc.

Whilst oxygen-stabilized carbanions have been the subject of several reviews (see Section 4.08.2.1), systems containing an oxygen with an α silicon, although quite common synthetic intermediates, have not received the same attention. The major synthetic pathways are reviewed below.

(i) From halomethylsilanes

Displacement of the halide from a halomethylsilane with an oxygen nucleophile is a versatile and convenient synthetic approach to a wide range of α-alkoxy- and α-acyloxysilanes. Chloromethyltrimethylsilane, chloromethyldimethylphenylsilane and chloromethyltrichlorosilane are commonly available commercial precursors. Typical nucleophiles employed have been alkoxides and, more usefully, acetates (Table 1) since subsequent reduction by LAH or hydrolysis yields the parent hydroxymethylsilane (Equation (1)) ⟨88JOM(354)139⟩. It is known, however, that the entire chloromethyl group at the silicon atom can be substituted by the prolonged action of agents nucleophilic towards the silicon itself (e.g., alkoxides or halides ⟨73CCC1522, 91HCA1477⟩. Migration of groups from silicon to the methylene carbon can also reduce overall yields (Equation (2)) ⟨73CCC1522⟩. A systematic study of the influence of reagents and conditions on product distribution has been made for the reaction of halomethylsilanes with methoxide ⟨85TL1115⟩.

Table 1 Reactions of halomethylsilanes with oxygen nucleophiles.

Halomethylsilane	Conditions	Product	Yield (%)	Ref.
TMS–Cl	NaOAc, Bu₄NCl	TMS–OAc	82	80S318
TMS–SiCl₂–Cl	AcOH, TEA	TMS–Si(OAc)–OAc, AcO	63	88JOM(354)139
Ph–Si(Cl)–Cl	KOAc, Bu₄NBr	Ph–Si(OAc)–OAc	80	91TL6325
PhMe₂Si–Cl	NaOAc	PhMe₂Si–OAc	84	90TL2197
PhMe₂Si–Br	MeOH, AgNO₃, TEA	PhMe₂Si–OMe	50	89CC1256
(1-naphthyl)Si(OMe)–CH₂Cl	NaOAc, DMF	(1-naphthyl)Si(OMe)–CH₂OAc	85	92CB591

$$\text{TMS}\diagdown\text{Si}(\text{Cl})(\text{Cl})\diagup\text{OAc} \xrightarrow[80\%]{\text{i, LAH; ii, H}_3\text{O}^+} \text{TMS}\diagdown\text{Si}(\text{H})(\text{H})\diagup\text{OH} \qquad (1)$$

$$\text{PhMe}_2\text{Si}\diagdown\text{Cl} \xrightarrow[\text{EtOH}]{\text{NaOEt}} \text{PhMe}_2\text{Si}\diagdown\text{OEt} + \text{PhMe}_2\text{Si}-\text{OEt} + (\text{EtO})\text{Me}_2\text{Si}\diagdown\text{Ph} \qquad (2)$$

(ii) From acylsilanes

(a) By reduction. Reduction of acylsilanes has been routinely achieved with standard reagents in good yields (Table 2). Asymmetric reduction has also been described (Table 3), with moderate to good *ee* values achieved.

Table 2 Reduction of acylsilanes.

Acylsilane	Conditions	Product	Yield (%)	Ref.
TMS–Si(CHO)(H)–TMS	LAH, then H_3O^+	TMS–Si(CH_2OH)(H)–TMS	82	92CB591
TMS—CHO	BH_3·THF, then H_2O	TMS–CH_2OH	82	78JOM(156)C12
R_3Si—CHO	BH_3·Me$_2$S, then H_2O	R_3Si–CH_2OH	82–94	89JOC4051
$(TMS)_3Si$—CHO	$NaBH_4$	$(TMS)_3Si$–CH_2OH	90	88JA313

Table 3 Asymmetric reduction of acylsilanes.

Acylsilane	Conditions	Product	ee (%)	Yield (%)	Ref.
Ph–C(O)–TMS	Baker's yeast	Ph–CH(OH)–TMS	73–90	10–30	91JOC5213
Ph(Bu^t)Si–C(O)–Me	*C. dioxydans*	Ph(Bu^t)Si–CH(OH)–Me	99	20	91JOM(403)29; 92JOM(424)273
Et–C(O)–TMS	i, (–)-IPC$_2$BCl, (+)-α-pinene; ii, HN(CH$_2$CH$_2$OH)$_2$	Et–CH(OH)–TMS	98	62	90TL4677
Me–C(O)–SiPh$_3$	Itsuno reagent (2:1 borane:(S)(–)-2-amino-3-methyl-1,1-dimethylbutan-1-ol)	Me–CH(OH)–SiPh$_3$	94	71	89CC89

(b) By nucleophilic addition. Nucleophilic addition of Grignard or organolithium reagents to formyl- and other acylsilanes provides access to a range of α-hydroxysilanes (Table 4). Formylsilanes too unstable to be isolated (entries 2–4) can be generated by Swern oxidation of the corresponding alcohol and then treated *in situ* with the organometallic reagent at $-78\,°C$. Ether was shown to be the solvent of choice in these cases. The basic nature of the reaction medium enhances the prospect of carbon to oxygen migration of the silyl group (Brook rearrangement) in the product with concomitant lowering of yield: there is evidence, however, that this is less of a problem following addition of organomagnesium nucleophiles ⟨92S995⟩.

Addition of other carbon nucleophiles, notably the cyanide ion, has been reported; nucleophilic addition of the cyanide to acylsilanes is typically carried out in dichloromethane at $-78\,°C$ (Table

Table 4 Nucleophilic addition to acylsilanes.

Entry	Acylsilane	Reagent	Product	Yield (%)	Ref.
1	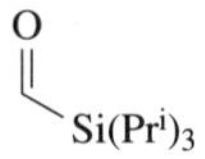 $(Pr^i)_3Si$-CHO	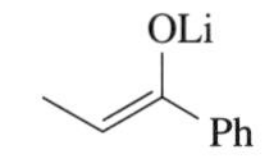 CH$_2$=C(OLi)Ph	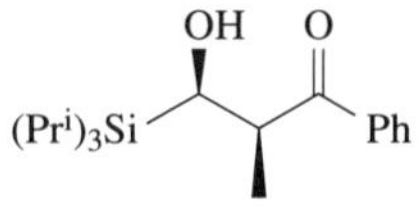 $(Pr^i)_3Si$-CH(OH)-CH(Me)-C(O)Ph	65 (>97% syn)	92JA10078
2	 TMS-CHO	BuLi	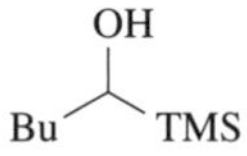 Bu-CH(OH)-TMS	16	88JOC1569
3	 TMS-CHO	BuMgBr	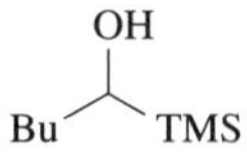 Bu-CH(OH)-TMS	38	88JOC1569
4	 TMS-CHO	PhC≡CLi	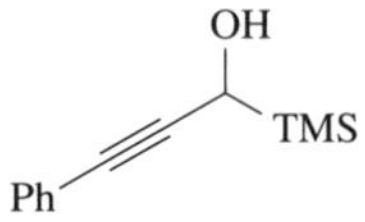 Ph-C≡C-CH(OH)-TMS	76	88JOC1569
5	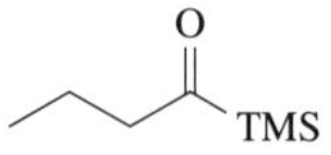 Pr^n-C(O)-TMS	PhC≡CMgBr	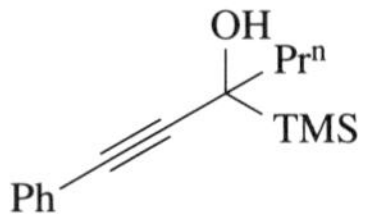 Ph-C≡C-C(OH)(Pr^n)-TMS	96	84BCJ827
6	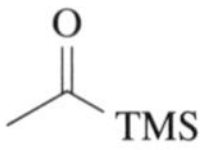 Me-C(O)-TMS	TMS—CN	 NC-C(OH)(Me)-TMS	87	90TL1945

4, entry 6), and the highly functionalized adducts so formed have served as useful intermediates for further synthetic development ⟨90JOC4634, 90TL1945⟩.

Stereoselective additions to acylsilanes have been reported: Panek and Cirillo showed that *syn-α-alkoxy-β-(silyloxy)acylsilanes* undergo diastereoselective, chelation-controlled, Cram-type, nucleophilic additions (Equation (3)) ⟨90JOC6071⟩; Ohno and co-workers reported that acylsilanes with α chiral centres exhibit high diastereofacial selectivity in nucleophilic additions, attributing this, in part, to the large bulk of the trimethylsilyl group exerting strong stereodifferentiation between the proposed transition states (Equation (4)) ⟨88JA4826⟩.

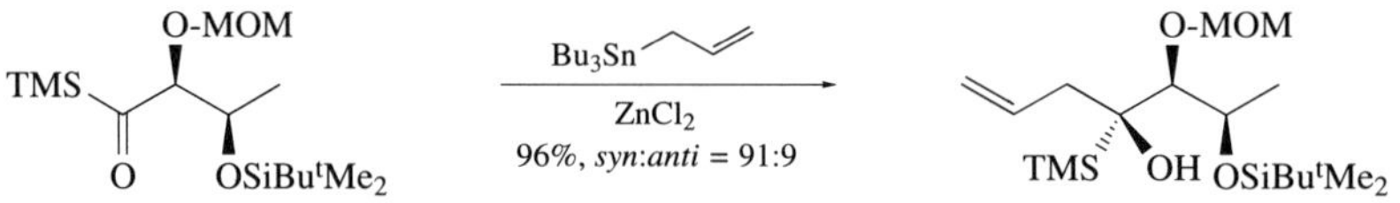

$$\text{TMS-C(O)-CH(Me)-CH(OSiBu}^t\text{Me}_2\text{)-O-MOM} \xrightarrow[\text{ZnCl}_2]{\text{Bu}_3\text{Sn-CH}_2\text{-CH=CH}_2} \text{CH}_2\text{=CH-CH}_2\text{-C(TMS)(OH)-CH(Me)-CH(OSiBu}^t\text{Me}_2\text{)-O-MOM} \quad (3)$$

96%, syn:anti = 91:9

MOM = methoxymethyl

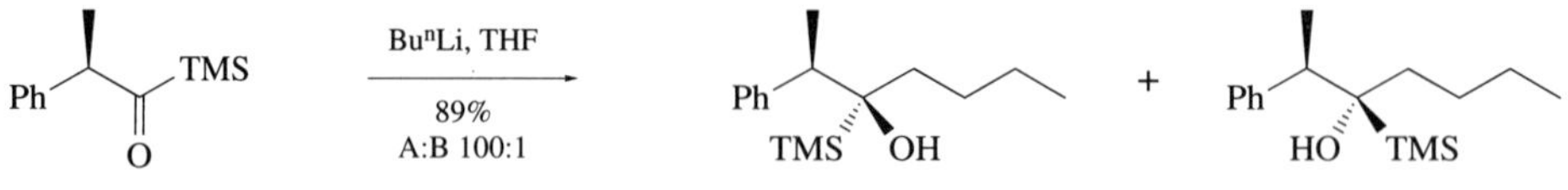

$$\text{Ph-CH(Me)-C(O)-TMS} \xrightarrow[\text{89\%, A:B 100:1}]{\text{Bu}^n\text{Li, THF}} \text{Ph-CH(Me)-C(TMS)(OH)-Bu}^n \text{ (A)} + \text{Ph-CH(Me)-C(OH)(TMS)-Bu}^n \text{ (B)} \quad (4)$$

(iii) From aldehydes and ketones

(a) By reactions with R_3SiLi or R_3SiMgX. In the mid 1950s it was discovered that silyl anions undergo smooth addition to electrophiles, including aldehydes and ketones ⟨54JA2502, 58JA2680⟩. Trimethylsilyl alkali metal reagents were originally prepared by metal exchange between an alkali metal and bis(trimethylsilyl)mercury by a lengthy and unattractive procedure (see ⟨B-81MI 408-01⟩, Chapter 11, for further discussion). The trimethylsilyl group can be more conveniently introduced using chlorotrimethylsilane and either lithium in THF or magnesium in hexamethylphosphoric triamide (HMPT), although with simple alkyl aldehydes and ketones, mixtures of products are frequently encountered, representing not only direct *C*-silylation, but also enolization and reduction

(Equation (5)) ⟨75JOM(87)151⟩. This method does allow direct *C*-silylation of formaldehyde (Table 5, entry 1), and subsequent simple hydrolysis of the product yields trimethylsilylmethanol. The use of hexamethyldisilane and methyllithium in ether–hexamethylphosphoramide (HMPA) appears to promote cleaner reactions, for example with 4-*t*-butylcyclohexanone a single product was formed (entry 2) from direct *C*-silylation, although the course of the reaction was highly solvent-sensitive. Dimethylphenylsilyllithium, prepared from the corresponding chloride and lithium metal in THF, also adds more cleanly to aldehydes (entry 3), ketones (entry 4) and to α,β-unsaturated aldehydes (in a 1,2 sense, entry 5). Magnus and co-workers, in work aimed at producing overall homologation of aldehydes and ketones, have devised two reagents, chloromethyltrimethylsilane and α-chloroethyltrimethylsilane, which when deprotonated undergo clean nucleophilic addition to ketones (entry 6) ⟨77CC513, 77JA4536⟩. They found that *s*-butyllithium was essential for generation of the reactive α-lithio species. The epoxysilanes formed were progressed by nucleophilic ring opening α to silicon (see Section 4.08.1.1.1.v).

$$\text{>=O} \xrightarrow[\text{THF, 0–10 °C}]{\text{TMS-Cl, Li}} \text{>(O-TMS)(TMS)} + \text{>(O-TMS)(O-TMS)} + \text{>=(TMS-O)} \quad (5)$$

42%   21%   14%

Table 5 Nucleophilic silylation of aldehydes and ketones.

Entry	Carbonyl precursor	Conditions	Product	Yield (%)	Ref.
1	CH$_2$O	TMS-Cl, Li, THF, –40 °C to –50 °C	TMS–CH$_2$–O-TMS	30	77JOM(142)C35
2	Bu^t-cyclohexanone	(TMS)$_2$, MeLi, –20 °C, ether, HMPA	Bu^t-cyclohexanol with OH and TMS		76JOC3063
3	*n*-C$_5$H$_{11}$CHO	PhMe$_2$SiCl, Li, THF, –78 °C to 0 °C	*n*-C$_5$H$_{11}$CH(OH)SiMe$_2$Ph	73	92JOC386
4	acetone	PhMe$_2$SiLi	(CH$_3$)$_2$C(OH)SiMe$_2$Ph		92JOC6552
5	BnO–CH=CH–CHO	PhMe$_2$SiLi, THF, –100 °C	BnO–CH=CH–CH(OH)SiMe$_2$Ph	52	90JA4873
6	TMS-O–C(=CH$_2$)–CH$_2$CH$_2$CHO	Cl–CH(TMS)–Li, THF, –78 °C	TMS-O–C(=CH$_2$)–CH$_2$CH$_2$–C(O epoxide)(TMS)	95	78CC297

HMPA = hexamethylphosphoramide.

(b) By reactions with other trialkylsilylmetals.

(i) *Aluminum.* Nucleophilic silylation reactions at carbonyl groups can be initiated using tris(trimethylsilyl)aluminum etherate or lithium tetrakis(trimethylsilyl)aluminate ⟨81AG(E)581⟩, achieving good yields of the corresponding α-hydroxysilanes, for instance the piperonal adduct in Equation (6). Clean 1,2-addition of the trimethylsilyl anion is accomplished to α,β-unsaturated aldehydes with the former reagent in toluene at −78 °C (Equation (7)) ⟨91TL855⟩, and to α,β-unsaturated aldehydes and ketones in ether at room temperature (Equation (8)) ⟨83TL45⟩. The latter paper indicates that 1,4-addition is the preferred pathway in THF or DME at −78 °C.

(ii) *Tantalum.* In contrast to the common route of electrophilic transfer of a trimethylsilyl group from a transition metal to a carbonyl oxygen, η-CpCl$_3$Ta-TMS reacts with aromatic or aliphatic aldehydes and ketones to produce *C*-silylated products by nucleophilic attack, as shown in Table 6 ⟨87JA3318⟩. (Benzophenone gave reductive elimination of TMS-Cl.) The reaction conditions simply require mixing of the carbonyl compound with the silyltantalum reagent in benzene at room temperature under an inert atmosphere: yields of the intermediate tantalum adducts were moderate to good, and subsequent simple aqueous hydrolysis to the corresponding α-hydroxysilanes was quantitative.

Table 6 Nucleophilic silylation with η-CpCl₃TaTMS.

Carbonyl precursor	Product	Yield (%)
acetaldehyde	α-TMS isopropanol	60
acrolein	α-TMS allyl alcohol	71
acetophenone (PhCOCH₃)	Ph-C(OH)(TMS)CH₃	50
cyclohexanone	1-TMS-cyclohexanol	70

(iv) From vinylsilanes by hydroboration

The hydroboration of vinyltrimethylsilane, originally carried out by Seyferth ⟨59JA1844⟩ and subsequently studied by Brown and Soderquist ⟨80JOC3571⟩, gives at best a 60:40 ratio of α- to β-(trimethylsilyl)ethylboranes with borane–THF at 0 °C (Equation (9)) (9-borabicyclo[3.3.1]nonane gives the β product exclusively). Substituents at the β position of the vinylsilane, including the case of cyclic vinylsilanes ⟨77JOM(132)301⟩, promote the formation of the α product (Equations (10)–(12)) ⟨88T4033⟩. In all cases the required α-hydroxysilane was released after alkaline hydrogen peroxide hydrolysis.

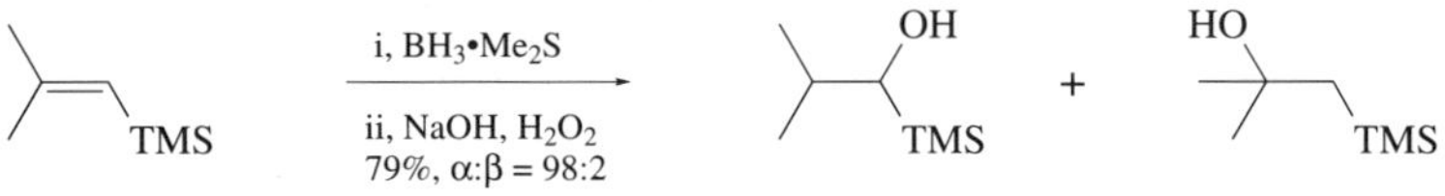

$$\text{(11)}$$

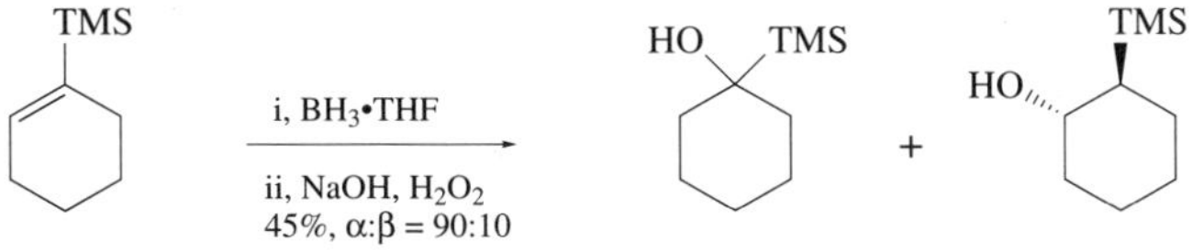

$$\text{(12)}$$

Larson and co-workers have developed an alternative approach to α-hydroxysilanes by the oxidation of α-borylsilane intermediates ⟨78JOM(146)C8, 80JOM(198)15⟩. Reactions of organoboranes with (trimethylsilyl)bromomethyllithium or (phenyldimethylsilyl)chloromethyllithium proceed with transfer of one ligand from boron to the α carbon in the intermediate complex (Scheme 1). The α-silylchloromethyllithium is best generated by deprotonation with lithium tetramethylpiperidide (LITMP): the analogous organomagnesium and organozinc derivatives gave none of the desired products. Generally, good overall yields of α-hydroxysilanes after oxidative work-up were realised, although these diminished with more sterically demanding boranes.

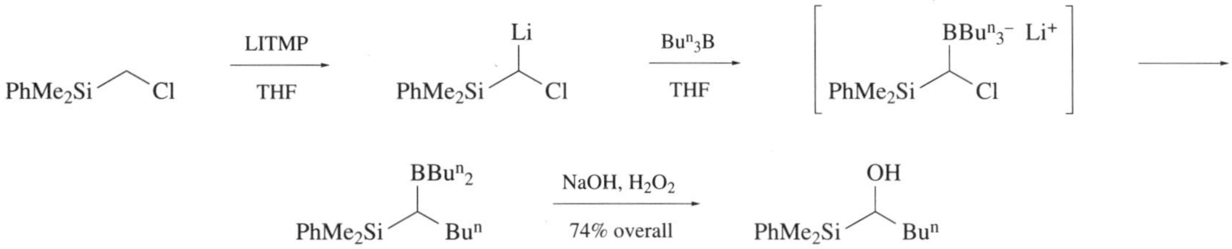

Scheme 1

(v) From vinylsilanes by oxidation

Epoxidation of vinylsilanes under standard conditions (e.g., buffered mcpba in dichloromethane) gives the corresponding epoxysilanes in good to excellent yields (for examples, see ⟨78JOC1620, 80JOC3028, B-81MI 408-01, B-83MI 408-01⟩). This functionality has been exploited as a latent carbonyl group under acidic conditions ⟨75PAC553⟩: Hudrlik and co-workers, however, showed that hydrolysis to a diol is possible in mild acid ⟨77TL591⟩, with ring opening occurring by attack α to the silicon (Equation (13)). (Subsequent elimination to the ketone (via the enol) is facile in the acyclic case, but can be prevented if the required stereochemical relationship is unattainable; see also ⟨76CC697, 85JOM(280)31⟩).

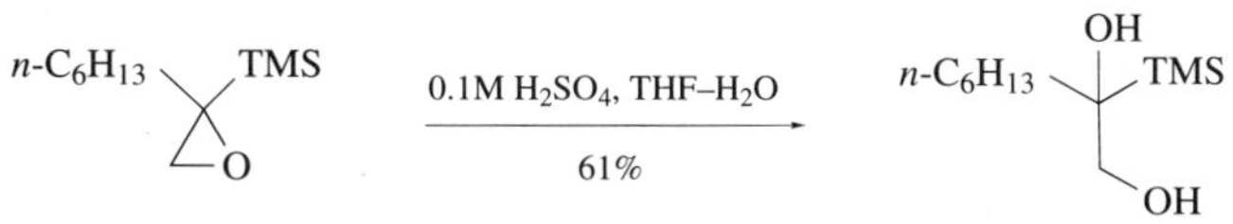

$$\text{(13)}$$

The epoxide ring can be opened by a range of other nucleophiles: in particular, alcohols under acidic or Lewis acidic conditions generate α-alkoxysilanes (Equations (14) and (15)), and acids alone give α-acyloxysilanes (Equation (16)) ⟨81JOC5357, B-83MI 408-01, 88TL6395⟩. Ring opening with organocuprates usually proceeds by attack of the nucleophile on the carbon atom bearing the silicon, as is the case with simple hydrolysis, to give β-hydroxysilanes ⟨79JOC155⟩: Knochel and Normant, however, observed attack β to the silicon, to give an α-hydroxysilane in high yield (Equation (17)) ⟨84TL4383⟩. No reasons for this discrepancy were suggested. Oxidation of vinylsilanes with catalytic osmium tetroxide has been reported to give the expected diols in reasonable yields. Trimethylamine N-oxide (Equation (18)) ⟨85JA4260, 90T2573⟩ or N-methylmorpholine N-oxide (Equation (19)) ⟨86JCS(P1)683⟩ have been used as re-oxidants; it is reported that t-butyl hydroperoxide promotes acid-catalysed decomposition of the sensitive 1,2-diol products.

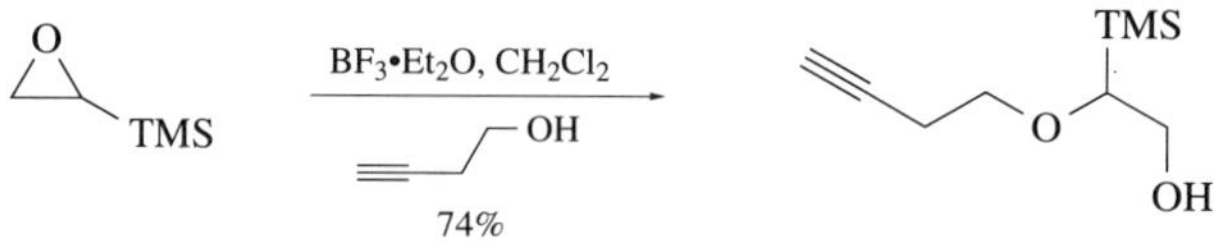

$$\text{(14)}$$
$$\text{(15)}$$ 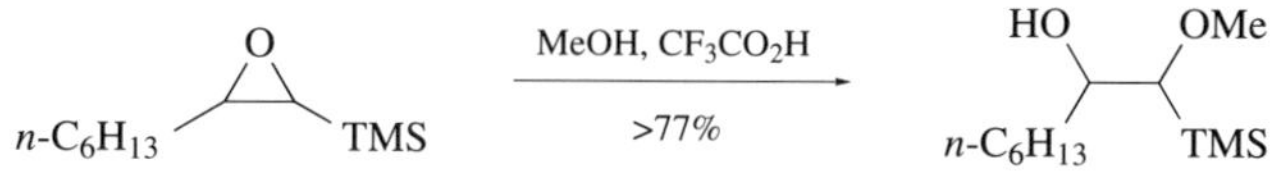
$$\text{(16)}$$ 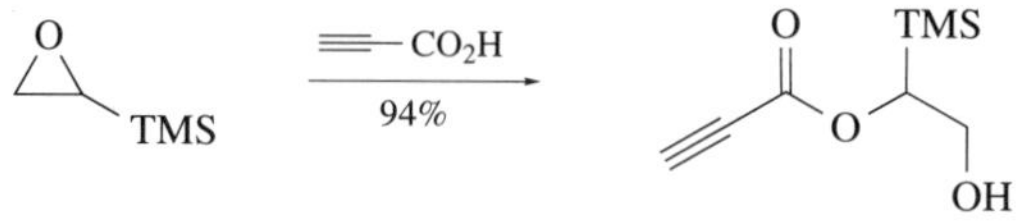
$$\text{(17)}$$ 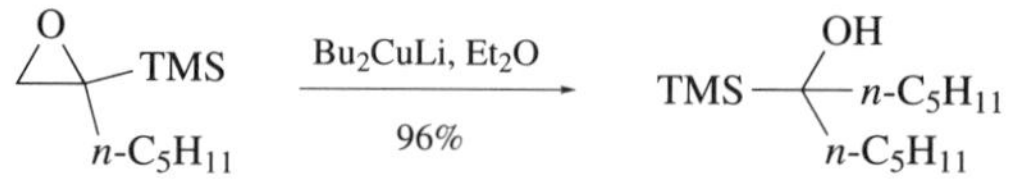
$$\text{(18)}$$ 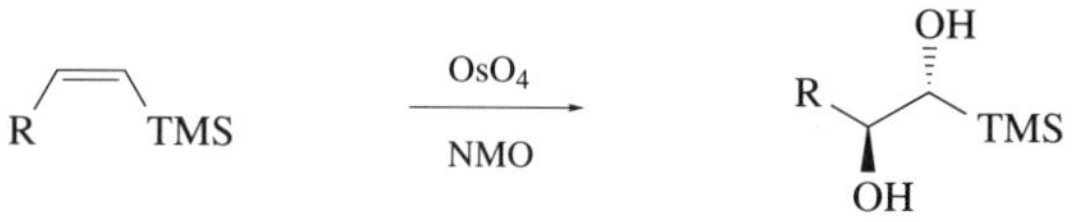
$$\text{(19)}$$ 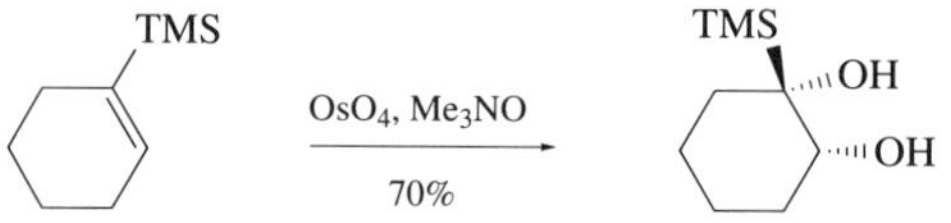

(vi) From alkoxysilanes: reverse Brook rearrangement

Rearrangement of alkoxysilanes to α-hydroxysilyl anions under conditions of strong base is formally analogous to the Wittig rearrangement, although the detailed mechanism appears more closely related to the Brook rearrangement (i.e., α-hydroxysilanes to alkoxysilanes under base catalysis (see ⟨74JA3227⟩)). Thus, α-metallation of alkoxysilanes promotes migration of the silyl group from oxygen to carbon; trapping, for instance by protonation, provides the corresponding silyl carbinols, generally in good to excellent yields (Scheme 2). If formal deprotonation is used, the reaction is limited to alkoxysilanes bearing acidic α protons, typically allylic or benzylic systems (Table 7). Although s-butyllithium is widely used as the deprotonating agent in the literature, Danheiser and co-workers rate t-butyllithium as the reagent of choice ⟨85JOC5393⟩. The alkoxide produced by rearrangement can be trapped by a range of electrophiles (e.g., methyl iodide or chlorotrimethylsilane) as well as a proton source. With allylic substrates, protonation occurs exclusively at C-1 (entry 2). Lindermann and Ghannam have extended the range of substrates for the reverse Brook rearrangement by generating the α-silyl carbanion by tin–lithium exchange ⟨90JA2392⟩. Typically an α-alkoxysilylstannane is treated at −78 °C with n-butyllithium in THF or DME. Reported yields are usually moderate to good (entries 5 and 6). (See Section 4.08.2.1.1.ii for further discussion of tin–lithium exchange processes.) Reverse Brook rearrangements can occur at greater distances: entry 7 shows a [1,4] rearrangement which occurs with retention of configuration at the carbanion centre.

(vii) From ethers, nitriles, esters and amides

Trapping of metallated ethers with trialkylchlorosilanes yields the corresponding α-silyl ethers. The reaction is of course again limited to systems bearing acidic α protons, typically allylic or benzylic ethers (Equation (20)) ⟨88CC503, 90JOM(391)283⟩. Ricci and co-workers have used bis(trimethylsilyl)peroxide as an oxysilylating agent ⟨90T2999⟩ to trap benzylic carbanions additionally stabilized by a sulfenyl or cyano group. Thus, trimethylsilylacetonitrile could be oxysilylated in 48% yield following deprotonation with butyllithium (Equation (21)).

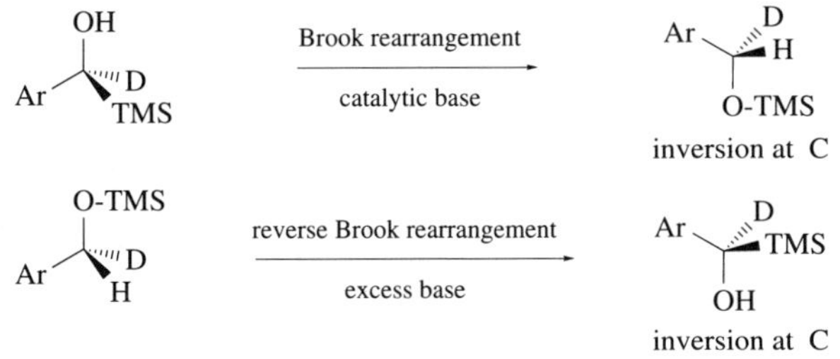

Scheme 2

Table 7 Reverse Brook rearrangements of alkoxysilanes.

Entry	Alkoxysilane	Conditions	Product	Yield (%)	Ref.
1	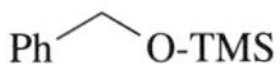 Ph–O-TMS	Bu^{t}Li, pentane, THF, then AcOH	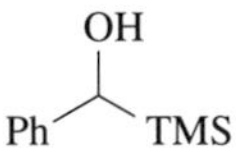 Ph–C(OH)(TMS)	95	74JA3214
2	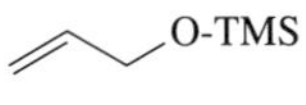 allyl–O-TMS	Bu^{t}Li, hexane, THF, then NH$_4$Cl	 CH$_2$=CH–CH(OH)(TMS)	94	88OS(66)14
3	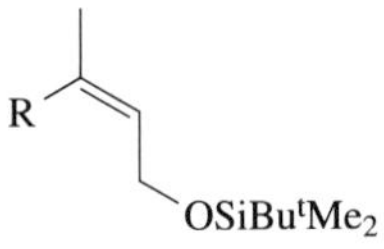 R–C=C–CH$_2$–OSiBu^{t}Me$_2$	Bu^{s}Li, TMEDA, THF, then CF$_3$CO$_2$H	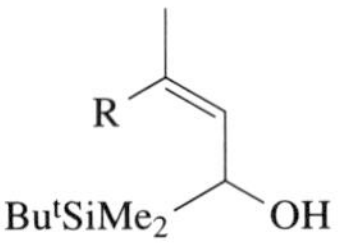 R–C=C–CH(Bu^{t}SiMe$_2$)(OH)	83	85HCA44
4	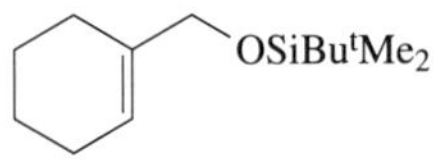 cyclohexenyl-CH$_2$–OSiBu^{t}Me$_2$	Bu^{s}Li, TMEDA, THF, then (EtCO)$_2$O	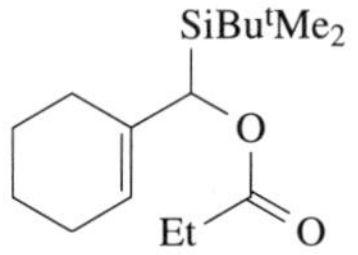 cyclohexenyl–CH(SiBu^{t}Me$_2$)(OC(O)Et)	99	89JOC1789
5	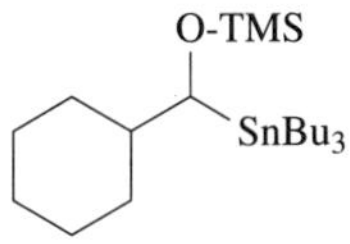 Cy–CH(O-TMS)(SnBu$_3$)	Bu^{n}Li, THF, then H$_2$O	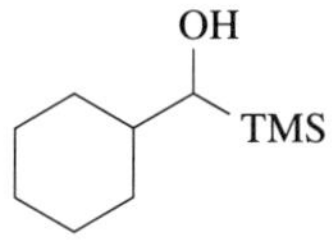 Cy–CH(OH)(TMS)	57	90JA2392
6	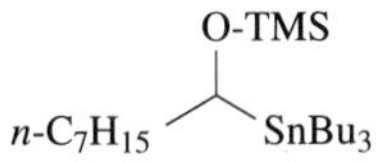 n-C$_7$H$_{15}$–CH(O-TMS)(SnBu$_3$)	Bu^{n}Li, THF, then H$_2$O	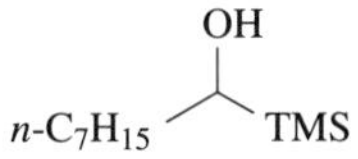 n-C$_7$H$_{15}$–CH(OH)(TMS)	67	90JA2392
7	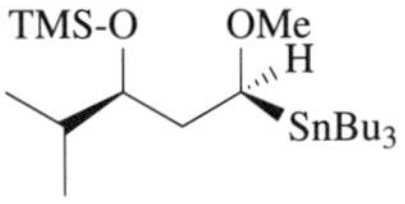 TMS-O, OMe, H, SnBu$_3$	Bu^{n}Li, THF, then NH$_4$Cl	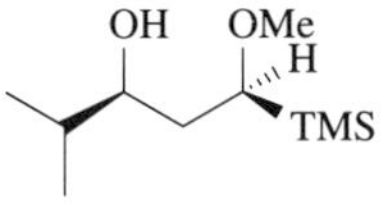 OH, OMe, H, TMS	82	92CB2731

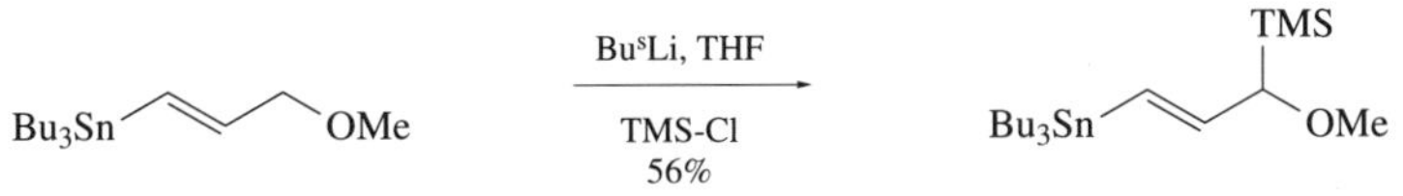

$$\text{Bu}_3\text{Sn–CH=CH–CH}_2\text{OMe} \xrightarrow[\text{TMS-Cl} \atop 56\%]{\text{Bu}^s\text{Li, THF}} \text{Bu}_3\text{Sn–CH=CH–CH(TMS)(OMe)} \quad (20)$$

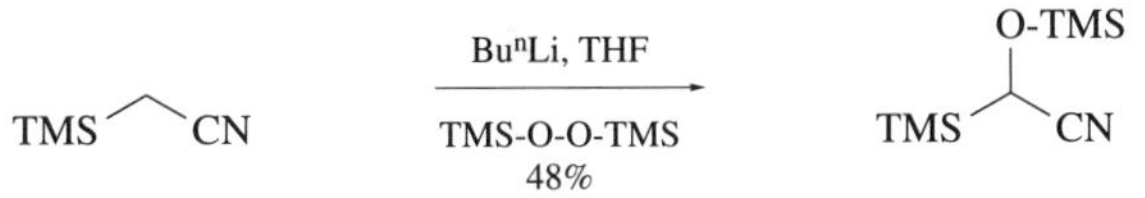

$$\text{TMS–CH}_2\text{–CN} \xrightarrow[\text{TMS-O-O-TMS} \atop 48\%]{\text{Bu}^n\text{Li, THF}} \text{TMS–CH(O-TMS)–CN} \quad (21)$$

Hydroxy-protected cyanohydrins can be deprotonated and alkylated on carbon ⟨92JOC1202, 92JOC6999⟩. A typical procedure to introduce a trimethylsilyl group involves treatment with lithium diisopropylamide (LDA) at low temperature followed by quenching with chlorotrimethylsilane (Equation (22)). More sterically demanding groups attached to the carbon increase the possibility of *N*-alkylation; for instance isopropyl substitution gives only 10% *C*-silylation with 90% of the product *N*-silylated.

$$R-CH(OTMS)(CN) \xrightarrow[\text{TMS-Cl}]{\text{LDA, }-78\,°C} R-C(OTMS)(TMS)(CN) \quad (22)$$

R = H, 77%; n-C$_6$H$_{13}$, 100%; Pr^i, 10%

Direct deprotonation of *O*-alkyl esters, carbonates and carbamates is fully discussed in Section 4.08.2.1.1.v(c). These species have been trapped with a variety of electrophiles including, of course, trialkylchlorosilanes. Examples are given in Equations (23)–(25) ⟨78JOC4255, 90AG(E)1424, 92SL764⟩.

(23) 2,4,6-triisopropylbenzoate ethyl ester → α-TMS ester, Bu^{s}Li, TMEDA, THF, −78 °C then TMS-Cl, 64%

(24) Me$_2$N(CH$_2$)$_3$O-carbamate → α-TMS product, Bu^{s}Li, ether then TMS-Cl, 89%

(25) PhCH(OCONPr$_2$) → Ph-C(TMS)(OCONPr$_2$), i

i, Bu^{s}Li, TMEDA, hexane, −78 °C, then TMS-Cl, 94%, 96%*ee* (absolute stereochemistry of product undetermined)

4.08.1.1.2 Oxygen and germanium—$R^1_2C(OR^2)GeR^3_3$, etc.

(i) From halomethyl ethers, aldehydes and ketones

Early reports of synthetic routes into α-alkoxygermane systems utilize the reaction of alkali metal germanes with electrophiles. For instance, treatment of methoxymethyl chloride with potassium and germane in HMPA gave (**1**) in 41% yield (Equation (26)) ⟨67IC1751⟩ (see also ⟨67IC1989⟩): the same precursor gave (**2**) in 73% yield with phenylgermyllithium (Equation (27)) ⟨92JOM(433)49⟩. In the last-cited paper, addition of arylgermyllithiums to aldehydes, yielding α-hydroxygermanes, is reported. Bis(trimethylgermyl)mercury effects nucleophilic addition of the trimethylgermyl group to (**3**) in 73% yield (Equation (28)) ⟨79CB936⟩; the mechanism in this case involves an intermediate nitrogen to carbon migration of a trimethylgermyl group.

$$\text{MeOCH}_2\text{Cl} \xrightarrow[41\%]{\text{GeH}_4,\ \text{K, HMPA}} \text{MeOCH}_2\text{GeH}_3 \quad (26)$$
(**1**)

$$\text{MeOCH}_2\text{Cl} \xrightarrow[73\%]{\text{PhGeH}_3,\ \text{Bu}^t\text{Li, THF}} \text{MeOCH}_2\text{GePhH}_2 \quad (27)$$
(**2**)

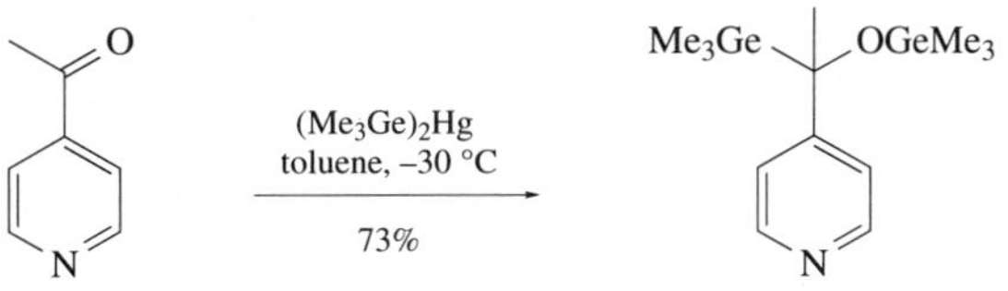

Metal-free germyl anions have been generated from hexaalkylsilylgermanes and mixed alkyl-arylsilylgermanes with catalytic amounts of fluoride ions. These species add usually in largely moderate yields to simple aliphatic or aromatic aldehydes and ketones ⟨90SC3245⟩. Representative is the reaction of trimethylgermyldimethylphenylsilane with the aldehyde (**4**) in THF–HMPA to give (**5**) in 48% yield (Equation (29)).

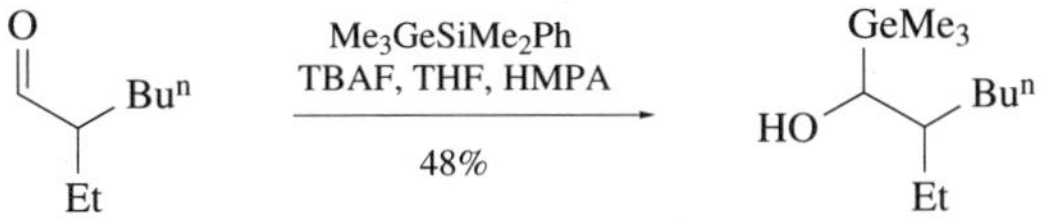

TBAF = tetra-*n*-butylammonium fluoride

Polarization in molecules such as the dichlorogermane (**6**) induced by the combined effects of the two halogens promotes a probably concerted dipolar addition of the germane to aldehydes, for example (**7**); suppression of secondary, radical reactions in nitromethane solvent gives the desired product in high yield (Equation (30)) ⟨71AG(E)267, 82JOM(232)137⟩. Trichlorogermane reacts when stirred with chloromethyl methyl ether at room temperature in high yield ⟨75IC1618⟩. Reductive work-up with LAH gives (**8**) in overall 85% yield (Scheme 3).

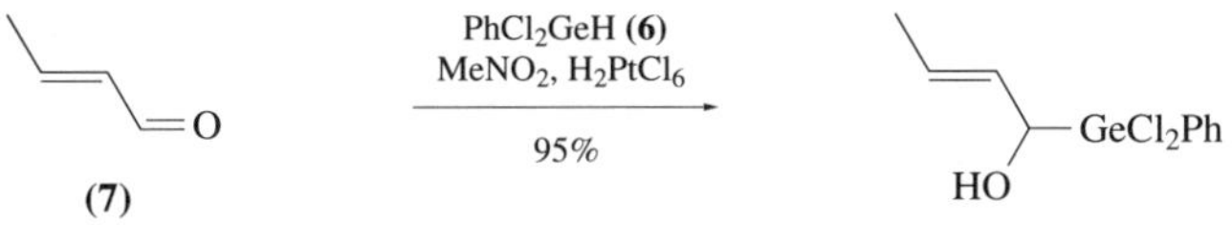

Scheme 3

(ii) Miscellaneous methods

(a) Nucleophilic substitution of halogermanes. There have been two reports in the early 1990s of displacement of halides from germanes: lithiation of (**9**) by transmetallation followed by addition of bromotrimethylgermane (**10**) was reported to give (**11**) in 65% yield (Equation (31)) ⟨92SL843⟩; and Lindermann and Anklekar have used the same procedure to synthesize an α-alkoxygermane from chlorotrimethylgermane ⟨92JOC5078⟩.

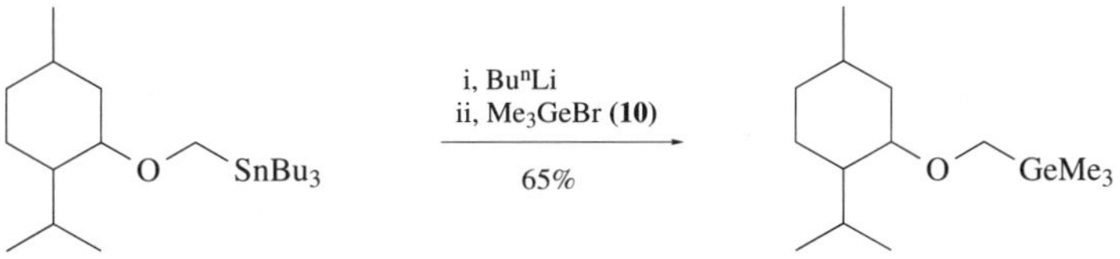

(b) Reverse Brook rearrangement. In contrast to the numerous examples for alkoxysilanes (see Section 4.08.1.1.1.vi), there are few reports of the analogous process of converting alkoxygermanes to α-hydroxygermanes. West and Wright have reported that the trimethylgermyl ether (**12**), upon

treatment with *t*-butyllithium in THF at $-78\,°C$, gave (13) in 70% yield after quenching (Equation (32)) ⟨74JA3214⟩; the other product, isolated in 30% yield, was that of direct cleavage of the trimethylgermyl group.

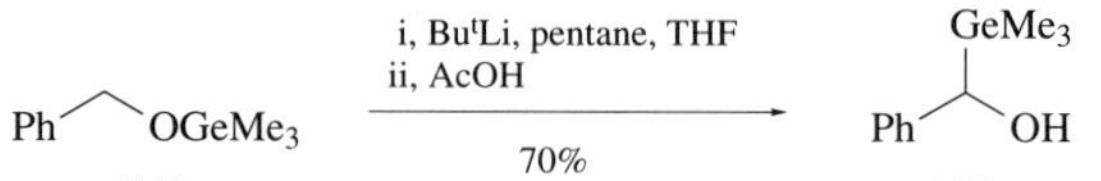

$$\text{Ph}\diagdown\text{OGeMe}_3 \quad\xrightarrow[\text{70\%}]{\text{i, Bu}^t\text{Li, pentane, THF};\ \text{ii, AcOH}}\quad \text{Ph}\diagdown\text{OH}\ (\text{GeMe}_3) \qquad (32)$$

(12) → (13)

(c) From halomethylgermanes. Tacke and Becker synthesized the hydroxymethylgermane (14) from tetrachlorogermane in 32% overall yield ⟨88JOM(354)147⟩ in a four-step sequence that has clear potential for the preparation of a series of variously substituted products (Scheme 4). Yields in the diphenyl series, generated using phenylmagnesium bromide in the second step, were uniformly 15–20% higher.

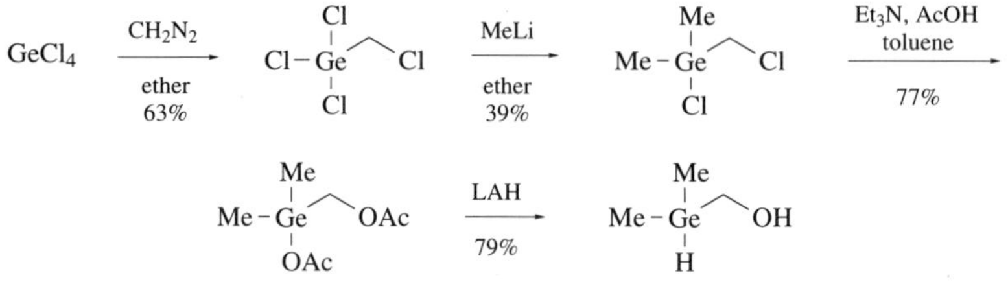

Scheme 4

Czech workers reported that heating (chloromethyl)trimethylgermane (15) with potassium acetate at $200\,°C$ for 18 h gave (acetoxymethyl)trimethylgermane (16) in 73% yield ⟨72CCC1392⟩. Cleavage of the ester was effected with a Grignard reagent and aqueous work-up gave (hydroxymethyl)-trimethylgermane in 77% yield (Scheme 5). In a related study, Barrau *et al.* have looked at the reactions of bis(chloromethyl)dimethylgermane with alkali and with sodium acetate ⟨88SRI317⟩.

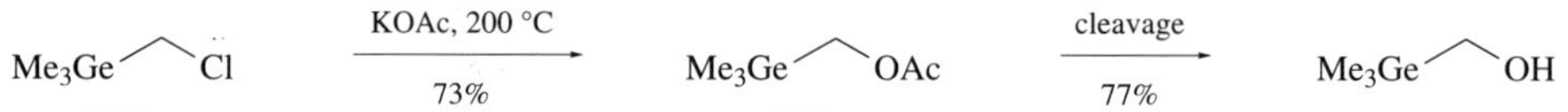

Scheme 5

4.08.1.1.3 Oxygen and boron—$R^1{}_2C(OR^2)BR^3{}_2$, etc.

Matteson provides a good general review on organoboranes ⟨B-87MI 408-01⟩; this also contains further discussion on the reactions discussed in all three sections below.

(i) From α-haloboranes and α-haloboronates

Matteson has covered the chemistry of α-haloboronic esters in two reviews ⟨89CRV1535, 89T1859⟩. The following discussion highlights the general processes applicable to the generation of α-alkoxyboranes and recent developments; for further background information, reference should be made to the latter two reviews. Following the disclosure of a feasible synthetic route to α-chloroboronates by Matteson and Majumdar ⟨80JA7588⟩ and to the corresponding chiral chloroboronates based on pinanediol ⟨80JA7590⟩, Matteson and Sadhu went on to show that the displacement of the α chlorine could be effected by lithium benzyloxide in THF in nonchiral boronates ⟨83OM1529⟩ and in the chiral series with an *ee* of 96% (Equation (33)) ⟨83JA2077⟩. Subsequently, Matteson replaced the pinanediol auxiliary with (*S*,*S*)-2,5-dimethylhexane-3,4-diol (DIPED), which is more easily removed from the boron ⟨86TL3831⟩. Lithium benzyloxide effects displacement of the α chlorine from the corresponding DIPED boronate in 74% yield with a similar *ee*. Using repetitions of this synthetic strategy, Matteson and Peterson achieved a synthesis of L-(+)-ribose from achiral materials ⟨87JOC5116⟩; in the same paper they show that overall displacement with benzyloxide is more efficient on α-bromoboronates. Removal of the benzyl group can be effected by hydrogenolysis over palladium: better is the oxidative removal of the related 4-methoxybenzyl group ⟨87JOC5121⟩; the free alcohol could be mesylated and then inverted by a second benzyloxide displacement (Scheme 6).

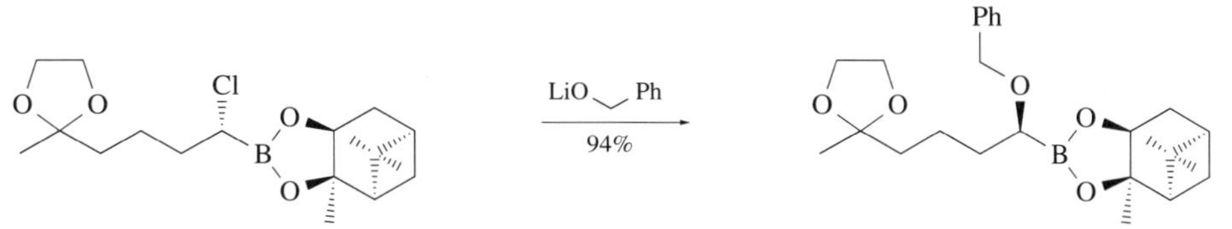

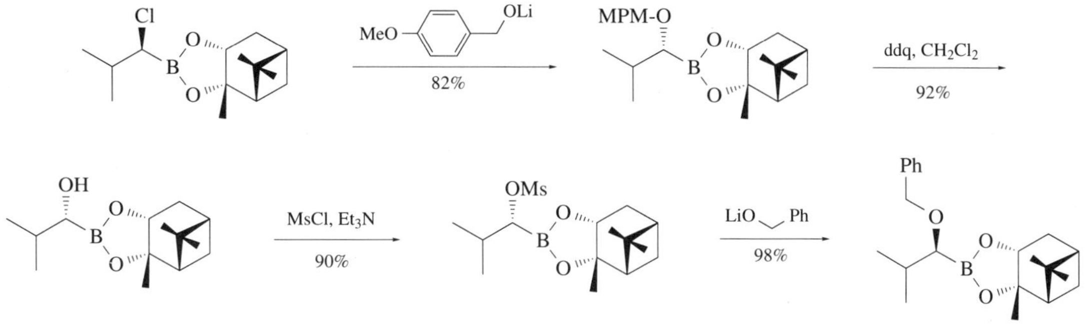

ddq = 2,3-dichloro-5,6-dicyano-1,4-benzoquinone;
MPM = methoxyphenylmethyl;
Ms = methanesulfonyl.

Scheme 6

Hoffmann and Landmann have investigated similar displacements of halides from allyl boronates with alkoxide and thiolate ions (see Section 4.08.1.2.3.ii in the latter case). Thus the reaction of the α-bromoboronate (**17**) with sodium methoxide in dry methanol gave the corresponding α-methoxyboronate in 75% yield ⟨86CB1039⟩ and with lithium phenoxide in dry hexane–THF gave the corresponding α-phenoxyboronate (70% yield) (Equation (34)) ⟨89CB903⟩. Substitution of chlorine by methoxide in the related tertiary α-chloroboronate (**18**) was achieved by Brown and co-workers with sodium methoxide in dry pentane–methanol (Equation (35)) ⟨75JOC813⟩; in these tertiary systems the more usual outcome of attempted solvolysis with aqueous ethanol proceeds via hydride or methide shifts and loss of the elements of chlorodimethoxyborane to give alkenes.

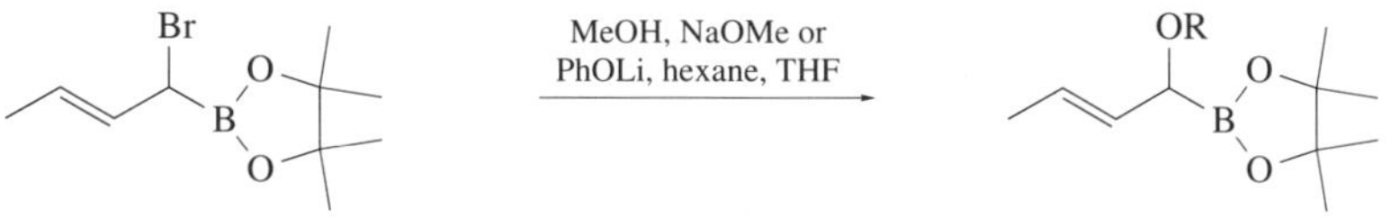

R = Me, 75%
R = Ph, 70%

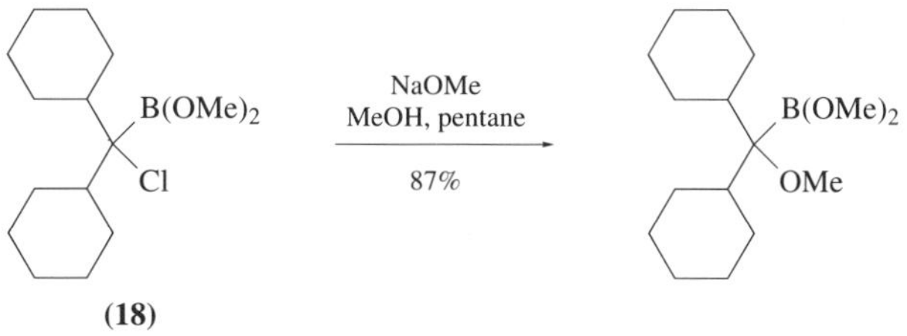

(ii) Miscellaneous methods

(a) Via carbonylation of organoboranes. Treatment of trialkylboranes with carbon monoxide in an excess of water between 25 °C and 75 °C gives excellent yields of the cyclic diboradioxanes (**19**)

(Equation (36)) ⟨62JA4715⟩; higher temperatures cause tertiary alcohol formation. These original conditions called for high carbon monoxide pressures. (Brown and Rathke subsequently showed that the reactions occur under 1 atm of carbon monoxide in DIGLYME at 100–125 °C; the exploitation of this process has been extensively reviewed, notably by Brown himself ⟨B-72MI 408-01⟩).

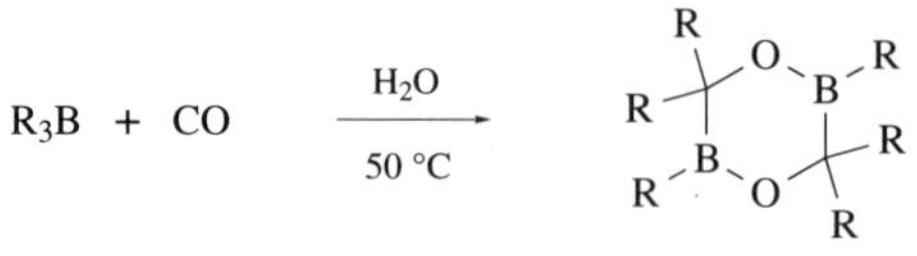

$$R_3B + CO \xrightarrow[50\,°C]{H_2O} \text{(19)} \qquad R = Et, 95\% \qquad (36)$$

Hydride sources such as lithium borohydride and lithium or potassium trialkoxyaluminum hydrides increase the rate of absorption of carbon monoxide; reactions occur at room temperature but stop after the transfer of one alkyl group. The resulting dialkylboryl species (20), valuable synthetic intermediates, are not isolated (Scheme 7); the reaction mixtures are worked up by oxidation or hydrolysis to give the corresponding aldehydes or alcohols in high yields, for example Equation (37) ⟨78S676, 79S701⟩. Carbonylation of alkali metal trialkylborohydrides is catalysed by the presence of traces of free trialkylborane ⟨79JOC467⟩. The true nature of these intermediates was later revealed by Hubbard to be possibly cyclic monomers or dimers ⟨92HAC223⟩.

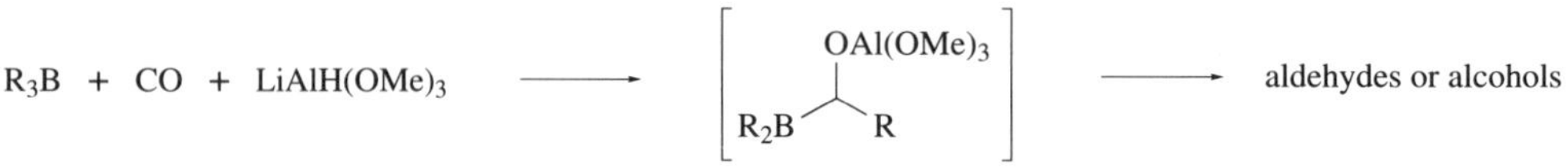

$$R_3B + CO + LiAlH(OMe)_3 \longrightarrow \left[R_2B\overset{OAl(OMe)_3}{\underset{R}{\diagdown\diagup}} \right] \longrightarrow \text{aldehydes or alcohols}$$

Scheme 7

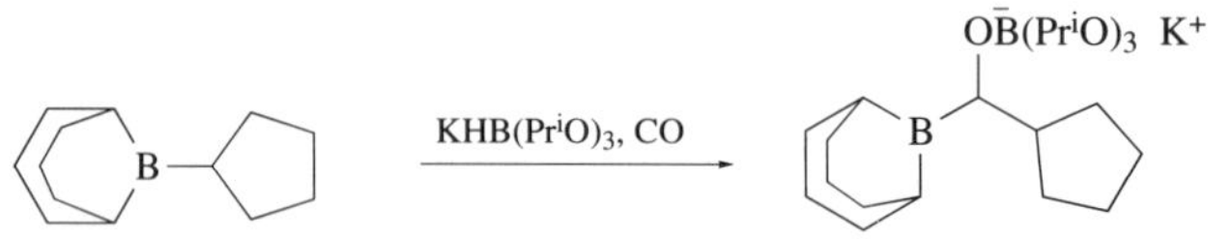

$$(37)$$

(b) Homologation of dioxaborinanes. Treatment of the dioxaborinanes (21) with methoxy(phenylthio)methyllithium (MPML) (22), developed by Brown and Imai ⟨83JA6285⟩, in THF produces an intermediate ate complex. Upon addition of mercuric chloride, migration of the alkyl group from boron to carbon ensues, giving α-methoxyalkyl derivatives in high yields (Equation (38)). The workers only isolated the 2-hexyl derivative (by distillation) since the overriding concern was to effect an overall aldehyde synthesis by the subsequent (*in situ*) oxidation of these materials.

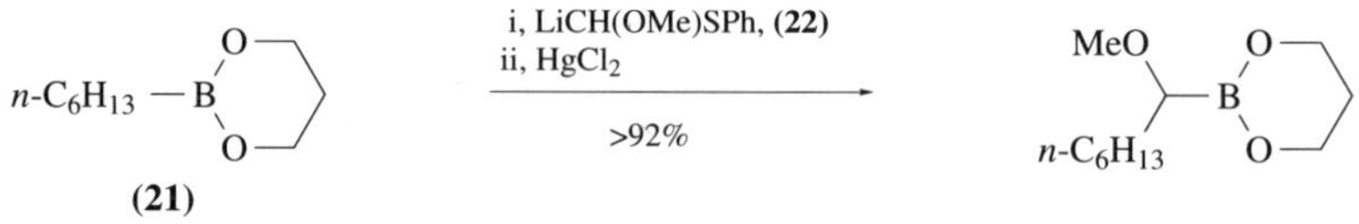

$$(38)$$

4.08.1.2 Functions Bearing Sulfur

4.08.1.2.1 *Sulfur and silicon—$R^1_2C(SR^2)SiR^3_3$, etc.*

The importance of sulfur-stabilized carbanions is reflected in the many reviews on the preparation and reactivity of these species (see Section 4.08.2.2). Although common electrophilic traps include chlorotrimethylsilane, there is only one major review of α-silylated sulfur derivatives to date, in which Block and Aslam discuss mixed organosulfur–silicon compounds containing heteroatoms connected not only to the same but also to neighbouring carbons ⟨88T281⟩. Attention is directed to the above review for further examples on the major synthetic pathways detailed below.

(i) From sulfur-stabilized carbanions

By far and away the commonest route to sulfur systems containing α-silicon functionality involves the C-silylation of anions generated α to sulfur. The direct generation of α-thio anions has been extensively reviewed (for leading references and discussion, see Section 4.08.2.2). Typical procedures involve the low-temperature addition of strong base to the organosulfur substrate in an inert solvent followed by addition of the silylating agent and aqueous work-up. Representative examples of C-silylation are shown in Table 8. Once formed, those silylated compounds still bearing acidic protons α to sulfur can typically be elaborated by further deprotonation and alkylation (for examples, see ⟨82CSR493⟩). (Although (arylthio)- and (alkylthio)allyllithiums intercept electrophiles with a high degree of α selectivity, once silylated the steric bulk of the silyl group directs further alkylation predominantly to the γ site ⟨81JOC5182⟩.) A good example of the combination of deprotonation, silylation followed by further deprotonation and alkylation is shown in Scheme 8. Thioanisole is elaborated to (23), which, after treatment with methyllithium, reacts with the epoxide (24); the resulting alkoxide is caught with benzenesulfonyl chloride, and cyclopropanation ensues after addition of s-butyllithium ⟨85TL2965⟩.

Table 8 Deprotonation/silylation adjacent to sulfur.

Sulfide	Conditions	Product	Yield (%)	Ref.
PhSMe	i, Bu^{n}Li, ether, reflux ii, TMS-Cl	PhS⌒TMS	95	80TL1677
(methallyl SPh)	i, LDA, THF, −78 °C ii, TMS-Cl	(CH$_2$=C(Me)CH(TMS)SPh)	100	81CC377
Me$_2$S	i, Bu^{n}Li, TMEDA, hexane–THF ii, TMS-Cl	MeS⌒TMS	59	67JOC1717 71USP3597463
PhS⌒CHMe$_2$	i, Bu^{t}Li, HMPA, THF ii, TMS-Cl	PhSC(iPr)(TMS)	86	77TL1961
(tetrahydropyranyl SMe)	i, Bu^{t}Li, THF, HMPA ii, Bu^{t}Me$_2$SiCl	(tetrahydropyranyl-S-CH$_2$SiBu^{t}Me$_2$)	87	85JA6729
PhSOMe	i, LDA, THF ii, TMS-Cl	PhSO⌒TMS	90–95	75TL2017
PhSO$_2$Me	i, LDA, THF ii, Bu^{t}Me$_2$SiCl	PhSO$_2$⌒SiBu^{t}Me$_2$	85–95	87TL5121
PhSO$_2$-N-pyrrolidinyl-CH$_2$OMe	i, Bu^{n}Li ii, TMS-Cl	TMS-CH(Ph)SO$_2$-N-pyrrolidinyl-CH$_2$OMe	89	84JOC1691
(cyclohexyl PhSO$_2$)	i, Bu^{n}Li, ether ii, TMS-Cl	(cyclohexyl PhSO$_2$ TMS)	80	91SC(21)1675

Alternatively, α-chlorination and subsequent nucleophilic displacement is a possible continuation ⟨82CSR493, 84JCS(P1)435, 89JCS(P1)115⟩. For example, the phenylthiosilane (25) can be chlorinated quantitatively with N-chlorosuccinimide. The α-chlorosilane then reacts smoothly with the silyl enol ether (26) in the presence of zinc bromide at room temperature to give the propiophenone (27) in 86% yield (Scheme 9).

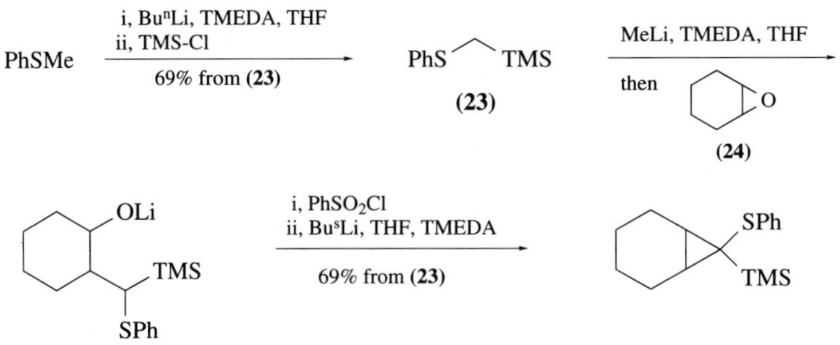

Scheme 8

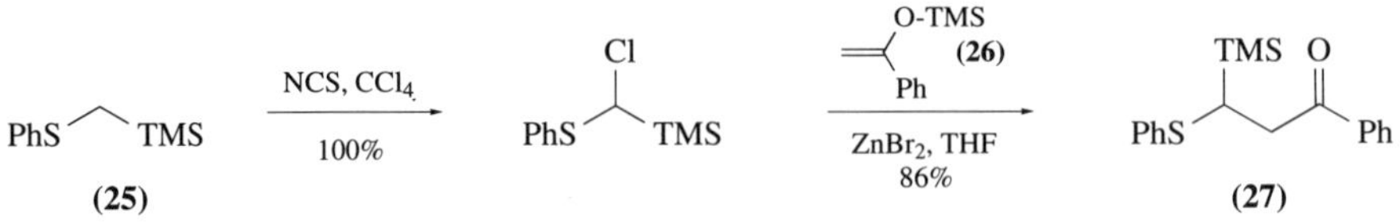

Scheme 9

Carbanions adjacent to sulfur can be produced by other indirect means; these are discussed more fully in Section 4.08.2.2.ii. The following are examples of carbanion generation from reports that specifically mention quenching with electrophilic silicon. Reductive lithiation of diphenyl dithioacetals and dithioketals produces α-carbanions in positions not easily accessible by direct deprotonation. Lithium naphthalenide (LN) ⟨81TL2923⟩ or lithium 1-(dimethylamino)naphthalenide (LDMAN) have been used, although the latter appears to be the reagent of choice ⟨84JA3245⟩. Typical conditions require low-temperature addition of the organosulfur precursor to LDMAN in THF followed by quenching with trimethylsilyl chloride and aqueous work-up (a dilute acid wash removes any 1-(dimethylamino)naphthalene). Excellent yields of the corresponding α-(phenylthio)alkylsilanes are achieved (Equations (39)–(41)).

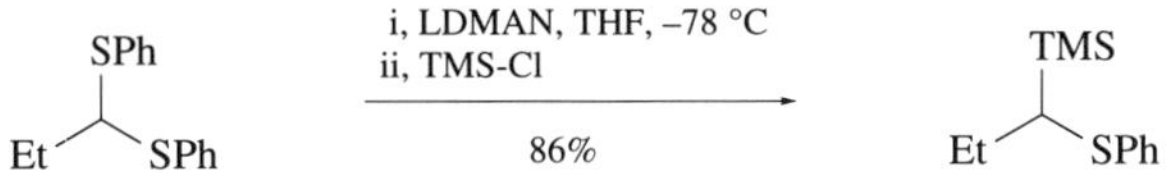

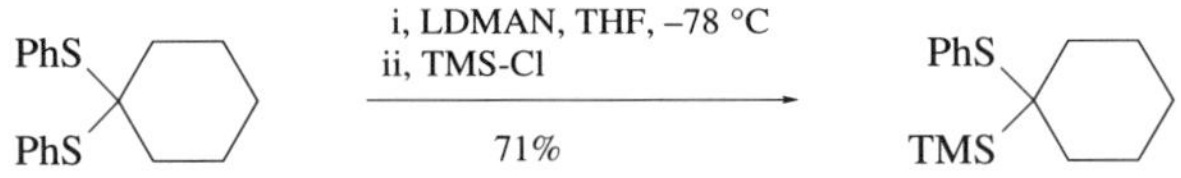

Addition of organolithium nucleophiles to vinyl sulfides results in the formation of intermediate α-lithio sulfides, which can then be caught with chlorotrimethylsilane ⟨81TL587⟩ (see also ⟨82CSR493⟩). Thus, treatment of phenyl vinyl sulfide (**28**) with methyllithium in ether at 0 °C in the presence of TMEDA affords the corresponding α-lithio sulfide (**29**) (Scheme 10); addition of TMS-Cl gives the α-phenylthiopropylsilane (**30**) in high yield. The reaction works well with both aryl- and alkyllithiums. Ringing the changes with this basic sequence provides related strategies: the organolithium can be added to a vinylsilane (e.g., (**31**)) and the adduct then trapped with benzenesulfenyl chloride or diphenyl disulfide (Scheme 11); alternatively, the organolithium can be added to α-(phenylthio)vinylsilane (**32**) and the resulting α-carbanion either protonated or substituted further (Scheme 12). Ager's papers ⟨81TL2923, 83TL95⟩ and review ⟨82CSR493⟩ give excellent summaries of these types of synthetic approach. Cohen and co-workers used a similar strategy to good effect in the preparation of α-(phenylthio)cyclopropylsilanes ⟨85TL2965⟩. Attack of a sulfur-stabilized carbanion (e.g., derived from (**33**)) on α-(phenylthio)vinylsilane (**34**) generates the intermediate carbanion (**35**), which undergoes intramolecular cyclopropanation, in this case in quantitative yield (Scheme 13). Michael addition of organometallic reagents to α-(trimethylsilyl)vinyl sulfoxides and sulfones has also been reported (see ⟨88BCJ3957⟩ and ⟨80CL331⟩, respectively); as have additions to α-(trimethylsilyl)allenyl sulfoxides ⟨90BCJ51⟩.

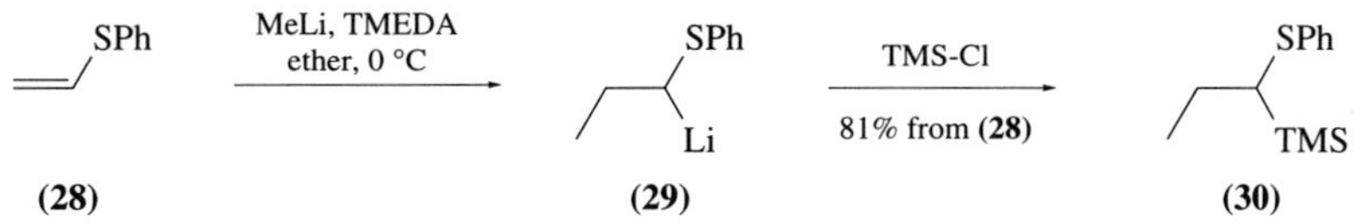

Scheme 10

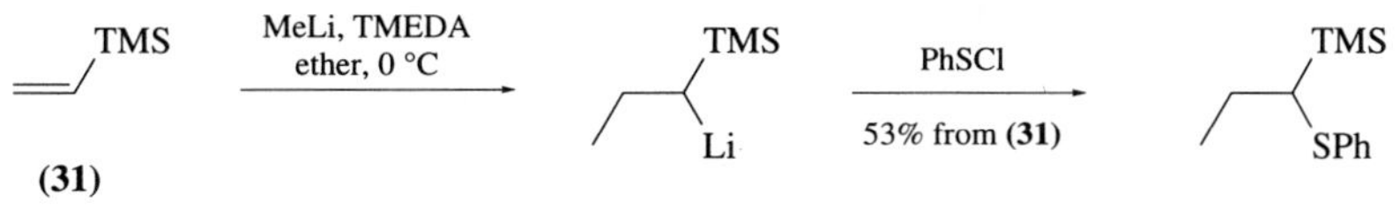

Scheme 11

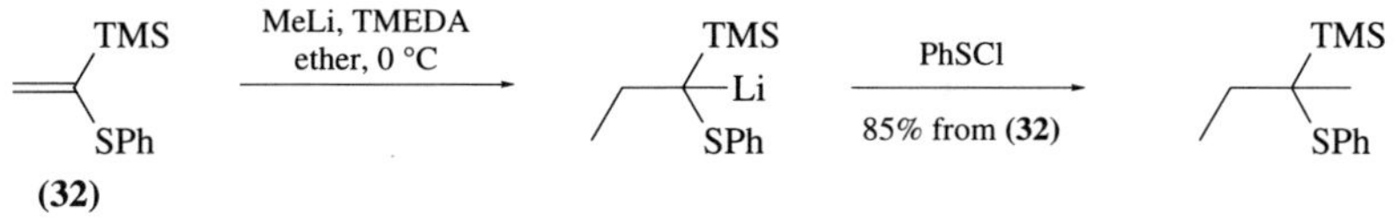

Scheme 12

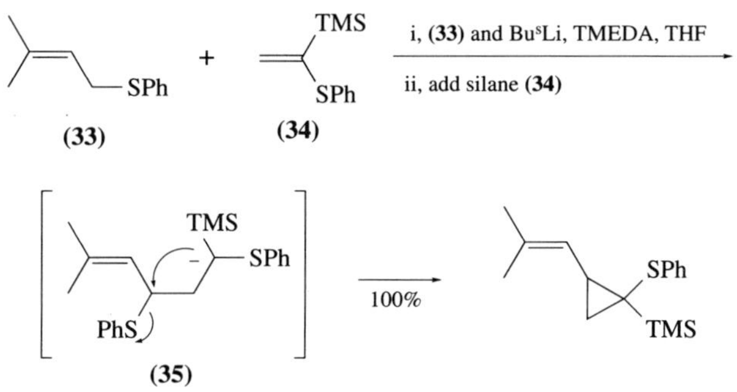

Scheme 13

Seebach has reported on the double deprotonation of benzyl, allyl and 2-methallyl thiols ⟨77CB1833⟩. These species can be trapped successfully with electrophiles (allyl thiols give mixtures by α- and γ-alkylation). Moreover, since the first alkylation occurs on carbon, it is possible to generate specific α-silylated sulfides. For instance, phenylmethanethiol (**36**) can be successfully doubly deprotonated and the dianion quenched with chlorotrimethylsilane followed by an electrophile of choice (Scheme 14).

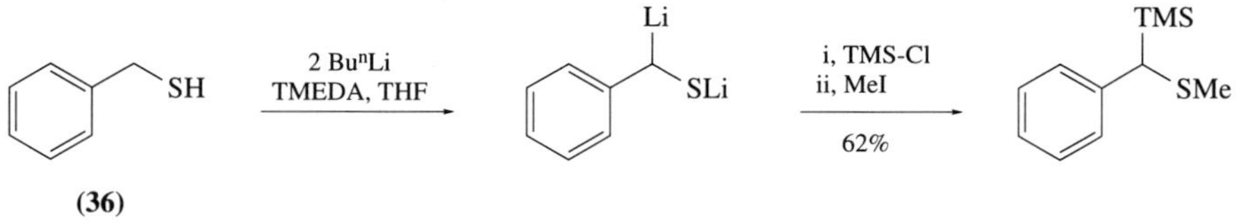

Scheme 14

(ii) From halomethylsilanes

As was the case in the oxygen/silicon system (see Section 4.08.1.1.1.i), displacement of the halide from a halomethylsilane with a sulfur nucleophile is a versatile and convenient synthetic approach to a wide a range of α-thiosilanes. Numerous chloromethylsilanes are available commercially. Typical nucleophiles employed have been metal sulfides, organic thiols and, more usefully, thioacetate (Table 9) since subsequent reduction or hydrolysis yields the parent silylmethanethiol.

Table 9 Reaction of chloromethylsilanes with sulfur nucleophiles.

Sulfur nucleophile	Conditions	Product	Yield (%)	Ref.
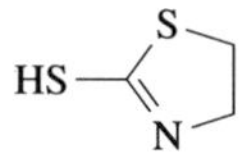 HS−C(=NCH₂CH₂S) (2-mercaptothiazoline)	TMSCH₂Cl, NaI, NaOH, EtOH, 70 °C	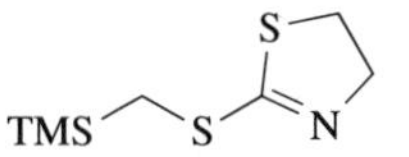 TMS-CH₂-S-C(=NCH₂CH₂S)		92TL85
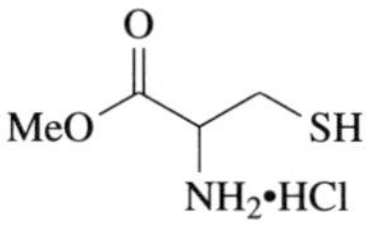 MeO-C(=O)-CH(NH₂·HCl)-CH₂-SH	TMSCH₂Cl, NaI, K₂CO₃, DMF, 25 °C, then H₃O⁺	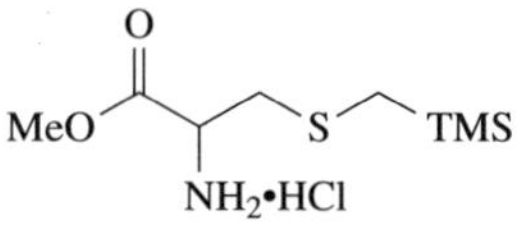 MeO-C(=O)-CH(NH₂·HCl)-CH₂-S-CH₂-TMS	78	90CJC1408
KSCOMe	HMe₂SiCH₂Cl, THF, 25 °C	 HMe₂Si-CH₂-SCOMe	63	90JOM(388)57
KSH	Me₂Si(CH₂Cl)₂, EtOH, 20 °C	 S-CH₂-SiMe₂-CH₂ (ring)	55	81JOM(204)13
Na₂S	PhMe₂SiCH₂Cl, EtOH, reflux	 PhMe₂Si-CH₂-S-CH₂-SiMe₂Ph	72	91SL557

Thiourea has also been employed as the nucleophilic counterpart; the intermediate isothiouronium salt can be either hydrolysed to a thiol ⟨85TL1425⟩ (e.g., Scheme 15) or chlorinated in water to a sulfonyl chloride prior to further elaboration ⟨82TL4203⟩.

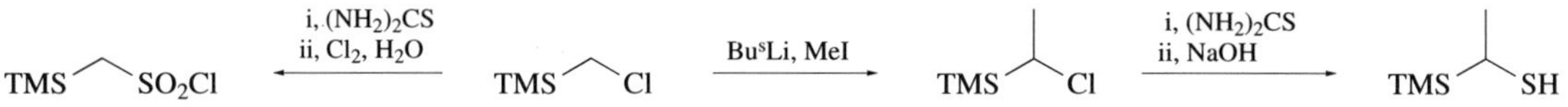

TMS–CH₂–SO₂Cl ⟵ i, (NH₂)₂CS; ii, Cl₂, H₂O ⟵ TMS–CH₂–Cl ⟶ Buˢ Li, MeI ⟶ TMS–CH(Me)–Cl ⟶ i, (NH₂)₂CS; ii, NaOH ⟶ TMS–CH(Me)–SH

no yields specified

Scheme 15

(iii) From silyl thioethers: reverse Brook rearrangement

Benzyl silyl thioethers have been rearranged to the corresponding α-silylphenylmethanethiols in reactions that directly parallel the reverse Brook process described in Section 4.08.1.1.1.vi. The reaction is less well developed than the oxygen analogue, presumably due to the greater instability of the corresponding silyl thioethers. Benzylthiotrimethylsilane (**37**) reacts with *t*-butyllithium at −78 °C. After addition of water, the rearranged product (**38**) is formed in high yield (Equation (42)) ⟨72JA4784, 74JA3222⟩. As was the case with their oxygen counterparts, the thiolates produced by rearrangement can be trapped by a range of electrophiles (e.g., methyl iodide and chlorotrimethylsilane) as well as a proton source.

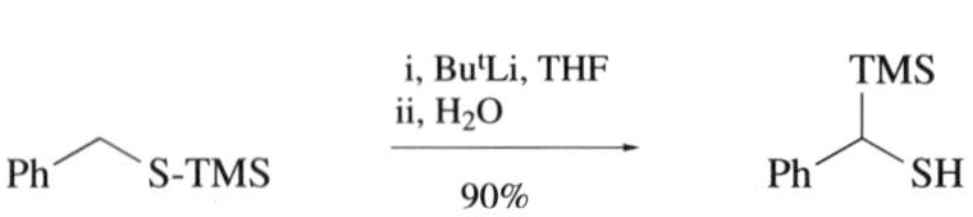

Ph–CH₂–S–TMS → i, Buᵗ Li, THF; ii, H₂O   (90%) → Ph–CH(TMS)–SH   (42)

(**37**)   (**38**)

(iv) From vinylsilanes and α-arylthiovinylsilanes

In contrast to the nucleophilic attack and trapping of the resulting α-carbanions with silicon electrophiles discussed above (see Section 4.08.1.2.1.i), vinyl- and α-arylthiovinylsilanes are also susceptible to direct electrophilic attack; the β effect of silicon usually directs the incoming electrophile to the α position unless steric or stereoelectronic effects dictate otherwise. This process has been exploited in two distinct ways. Magnus and co-workers have shown that vinyltrimethylsilane reacts cleanly with electrophilic arenesulfenyl chlorides at low temperatures in excellent yields with no evidence of loss of HCl or chlorotrimethylsilane under the reaction conditions ⟨80JOC1046⟩. Representative is the coupling of (**39**), to give (**40**) in 94% yield (Equation (43)). The effect of silicon and sulfur ligands on α/β ratios has also been studied ⟨92JOM(437)111⟩; in dichloromethane, the predominant product in all cases is the α-thio adduct shown (i.e., anti-Markovnikov addition).

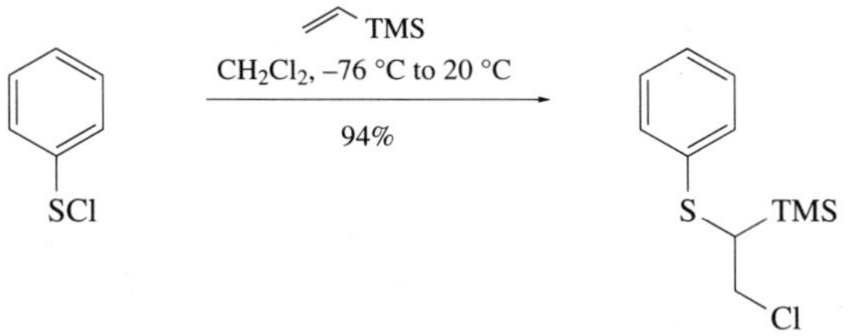

Takeda and co-workers treated the 1-phenylthiovinylsilane (**41**) with titanium tetrachloride/methanol in dichloromethane and obtained an intermediate thionium ion; this species reacts cleanly with nucleophiles to give overall addition products (Scheme 16) ⟨87CL1963⟩. Despite the potential versatility of this reaction (the starting vinylsilanes are readily available from ketene dithioacetals by reductive lithiation and quenching), the only carbon nucleophiles investigated were allyltrimethylsilane and trimethylsilyl cyanide, both of which, however, were captured in good yield.

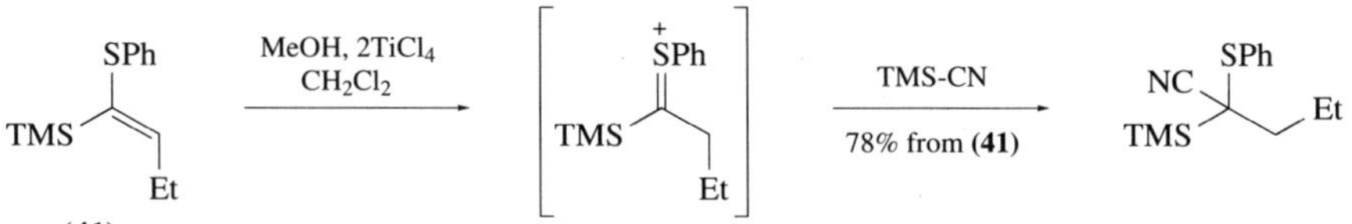

Scheme 16

(v) Rearrangement of ylides from α-thiosilanes

Two groups have reported transformations of α-thiosilanes which provide useful structural modifications whilst retaining the relative heteroatom positions. Kocienski has found that allylic silyl sulfonium ylides can be rearranged to give the corresponding homoallylic α-methylthiosilanes ⟨80CC1096⟩. For instance, treatment of the fluorosulfonate salt (**42**), prepared from the sulfide (**43**) and methyl fluorosulfonate, with *n*-butyllithium gave an ylide which underwent a spontaneous [2,3] shift below −20 °C to give the homoallylic α-methylthiosilane (**44**) in 65–75% yield (Scheme 17). Achiwa and co-workers described the generation of a thiocarbonyl ylide from the bromomethylsilane (**45**), and observed its cycloaddition reactions with reactive dipolarophiles, for example *N*-phenylmaleimide, in excellent yields (Scheme 18) ⟨85TL3011⟩. The intermediate ylide was generated by thermal extrusion of bromotrimethylsilane in DMF. Regiochemical effects with unsymmetrical dipolarophiles were later examined ⟨86H(24)1571, 87CPB1734⟩.

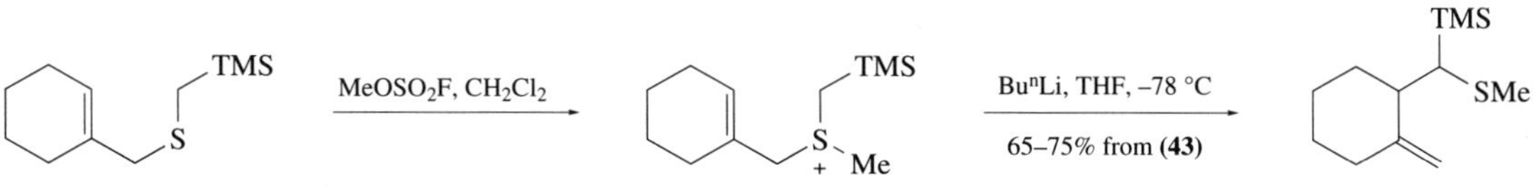

Scheme 17

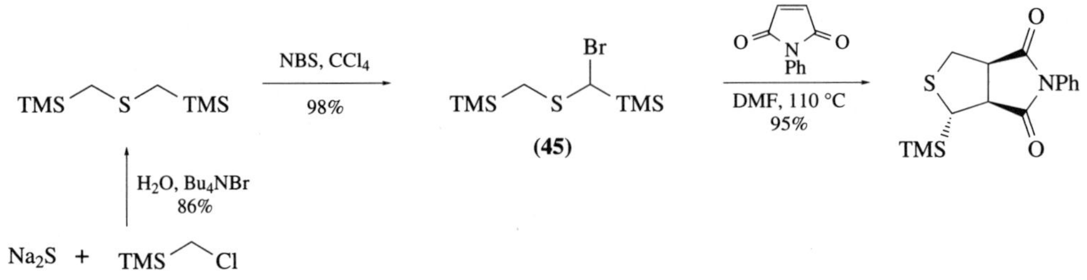

Scheme 18

(vi) Miscellaneous methods

(a) From α-silyl organomagnesium compounds. There are a few examples in the literature of Grignard reagents derived from halomethylsilanes. The Grignard reagent from chlorotrimethylsilane (**46**) reacted with isothiocyanates ⟨91G471⟩ and sulfinate esters ⟨68CJC2115⟩ to give sulfinamides and sulfoxides, respectively. For example, (**46**) reacted with thionylaniline (**47**) in 93% yield and with methyl benzenesulfinate (**48**) in 90% yield (Scheme 19); it also combines with the Diels–Alder adducts of *N*-sulfinyl dienophiles and alkadienes ⟨86T2979⟩ to give ring-opened products. For instance, the dihydrothiazine (**49**) was treated with (**46**) at −40 °C to give the sulfoxide (**50**); subsequent deoxygenation without purification provided the sulfide (**51**) in 77% overall yield (Scheme 20).

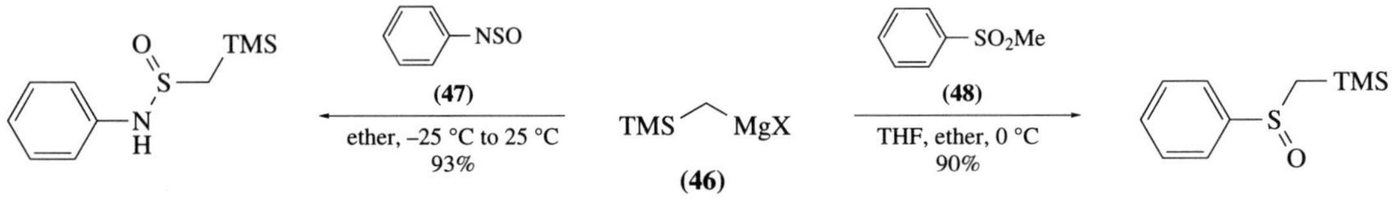

Scheme 19

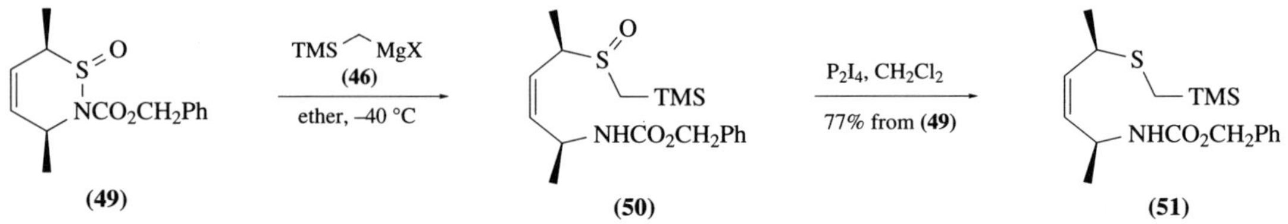

Scheme 20

(b) From α-silylated O,S-acetals. Han and Oh have reported the investigation of *O,S*-acetals as one-carbon homologation reagents. [Methoxy(phenylthio)methyl]trimethylsilane (**52**) reacted with electron-rich arenes, in the presence of Lewis acids, to give Friedel–Crafts products (Equation (44)), and with alkenes, in the presence of tin(IV) chloride, to give ene products (Equation (45)) ⟨89SC2213⟩. The silane (**52**) was synthesized from methoxy(phenylthio)methane by standard deprotonation and silylation.

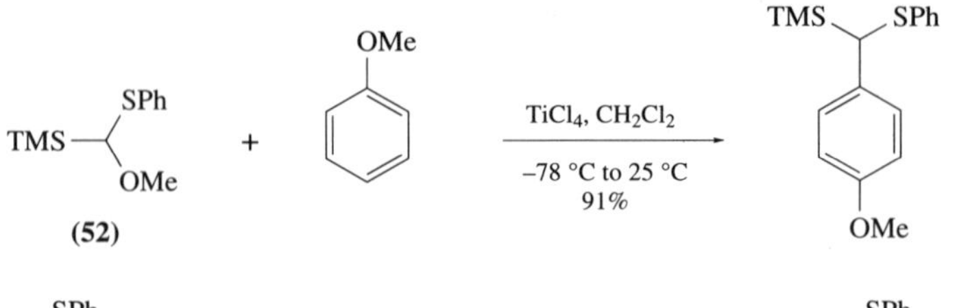

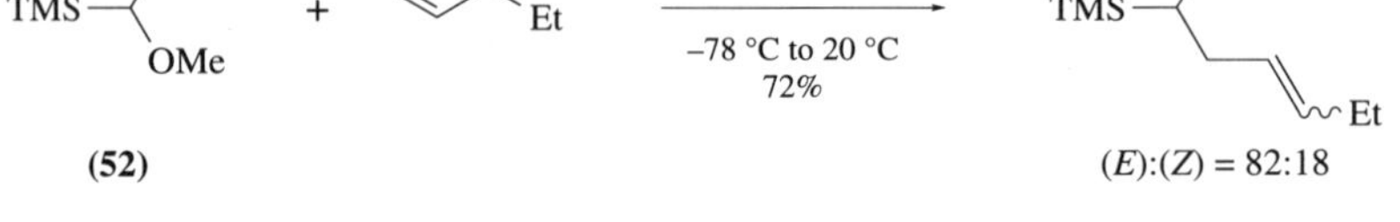

(c) From trimethylsilylthiones. Trimethylsilyl *t*-butyl thione (**53**) can be prepared by the acid-catalysed reaction of the corresponding acylsilane with hydrogen sulfide followed by an alkaline wash of the reaction mixture. Bonini and co-workers showed that (**53**) undergoes nucleophilic attack on sulfur, typically with alkyllithiums, in good yields ⟨89JCS(P1)2083⟩; (**53**) also undergoes cycloaddition with 1,3-dienes in excellent yields after prolonged exposure (Scheme 21).

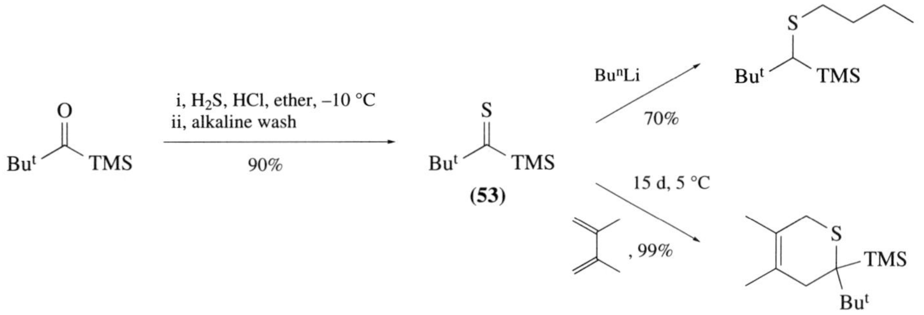

Scheme 21

4.08.1.2.2 Sulfur and germanium—$R^1_2C(SR^2)GeR^3_3$, etc.

(i) From α-germyl organomagnesium compounds

Grignard reagents derived from chloromethylgermanes have been used synthetically in only one paper to generate α-germyl sulfur systems ⟨68CJC2115⟩. Brook and Anderson reported that addition of the Grignard reagent from chloromethyltrimethylgermane (**54**) to methyl benzenesulfinate gave the α-germyl sulfoxide (**55**) in 78% yield (Equation (46)).

(ii) From germyl thioethers: reverse Brook rearrangement

Benzyl germyl thioethers have been rearranged to the corresponding α-germylphenylmethanethiols in reactions that directly parallel the reverse Brook process described in Section 4.08.1.1.1.vi. Thus treatment of benzylthiotrimethylgermane (**56**) with *t*-butyllithium at −78 °C, followed by electrophilic trapping of the resulting thiolate with methyl iodide, gave the rearranged α-germyl thioether (**57**) in good yield (Equation (47)) ⟨74JA3214, 74JA3222, 74JA3227⟩. In contrast to rearrangements involving transfer of a silyl group, the germyl thioether precursors are much more prone to direct cleavage of the sulfur–germanium bond by the metallating reagent; this process accounts for up to 28% of the products with *t*-butyllithium (thioethers were produced after electrophilic trapping). Moreover, *n*-butyllithium gave products derived exclusively from direct S—Ge cleavage.

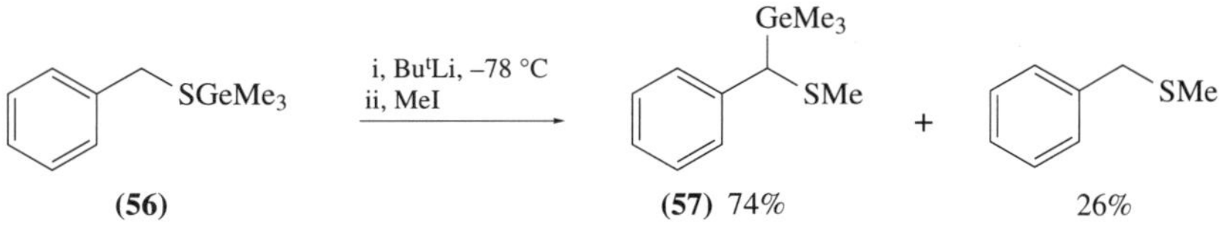

(iii) From halomethylgermanes

The displacement of the halide from a halomethylgermane with a sulfur nucleophile is a less well-documented synthetic approach to α-thiogermanes compared to the corresponding routes to α-alkoxysilanes and α-thiosilanes (see Sections 4.08.1.1.1.i and 4.08.1.2.1.ii). Much of the chemistry in this area is now quite old, although one more recent example is the generation of various 3,3-dialkyl-1,3-thiagermatanes, prepared by treating bis(halomethyl)dialkylgermanes with sodium sulfide or hydrogen sulfide in the presence of triethylamine ⟨84SRI21⟩. Representative is the formation of the thiagermatane (**59**) from the bis(iodomethyl)germane (**58**) in 80% yield (Equation (48)).

$$\text{Me}_2\text{Ge}(CH_2I)_2 \quad \xrightarrow[80\%]{\text{Na}_2\text{S, Et}_3\text{N, THF}} \quad \text{Me}_2\text{Ge}\overset{\diamond}{\underset{}{}}\text{S} \qquad (48)$$

(**58**)   (**59**)

(iv) From vinylgermanes

In a process directly analogous to that discussed in Section 4.08.1.2.1.iv for vinylsilanes, vinyltrimethylgermane reacted with electrophilic arenesulfenyl chlorides at room temperature to give mixtures of the corresponding α and β adducts in unspecified yields ⟨92JOM(437)111⟩. The α/β ratios reported were typically in the range 80:20 to 90:10.

4.08.1.2.3 Sulfur and boron—$R^1{}_2C(SR^2)BR^3{}_2$, etc.

There are relatively few examples of systems containing an α-thioalkylboron unit in the literature, the most important of which are discussed below. This area has not been the subject of comprehensive reviews, although fragments necessarily occur in general organoboron reviews. (The most recent general review on organoboranes is by Matteson ⟨B-87MI 408-01⟩.)

(i) From methyl phenylthio ethers and diphenyl dithioacetals

Some of the most frequently exemplified preparations of α-thioalkylboron compounds involve reactions of a trialkylborane or trialkyl borate with α-thiocarbanions. For example, the reaction of phenylthiomethyllithium with trimethyl borate at −78 °C gives the corresponding phenylthiomethylboronate after aqueous work-up; acidic work-up provides the phenylthiomethylboronic acid in high yield ⟨78JA1325⟩, which can be conveniently isolated by crystallization (Equation (49)). Trialkylboranes have also been treated with the anions derived from methyl phenyl sulfone, dimethyl sulfoxide and dimethyl sulfide ⟨75JOC814⟩. The corresponding organoborate complexes produced, effectively quantitatively, are exemplified in Equation (50). Subsequent addition of methyl iodide effects an overall one-carbon homologation of the starting organoborane.

$$\text{PhS}\frown\text{Li} + \text{B(OMe)}_3 \quad \xrightarrow[77-87\%]{\text{i, THF, } -70\,°\text{C};\ \text{ii, H}_3\text{O}^+} \quad \text{PhS}\frown\text{B(OH)}_2 \qquad (49)$$

$$\text{PhO}_2\text{S}\frown\text{Li} + \text{Bu}_3\text{B} \quad \xrightarrow[100\%]{\text{THF, 0\,°C}} \quad \text{PhO}_2\text{S}\frown\bar{\text{B}}\text{Bu}_3\ \text{Li}^+ \qquad (50)$$

The reactions of anions derived from diphenyl dithioacetals produce α-thioborane intermediates by a [1,2] shift of one ligand from boron to carbon; for example, the initial complex formed between the dithioacetal (**60**) and the trialkylborane (**61**) spontaneously rearranges at 20 °C to (**62**) (Equation

(51)) ⟨74CC863⟩. (For further discussion of the reaction of boranes with α-thioanions, see ⟨B-87MI 408-01⟩.)

$$\text{(61)} \quad (\text{Cy})_3\text{B} \xrightarrow[98\%]{\text{Pr}^n\text{CLi(SPh)}_2 \ (60)} \text{(62)} \quad (\text{Cy})_2\text{B}-\text{C(Pr)(Cy)-SPh} \qquad (51)$$

(ii) From α-haloboronates

Matteson and Hoffmann have independently reported the substitution of α-haloboronates with sulfur nucleophiles. Hoffmann and Landmann treated the unstable allyl bromide (63) with both ethanethiolate and 1,1-dimethylethanethiolate (the boron effectively assists this α-substitution process ⟨68JOC3055⟩). The products, the α-alkylthioboronates (64), were obtained in moderate yields (Equation (52)) ⟨83TL3209, 86CB1039⟩. Matteson and Majumdar have recorded substitution of the closely related α-chloroboronate (65) with sodium thiophenolate. The yields attained were much higher even with these hindered substrates, presumably attesting to the poor stability of (63) (Equation (53)) ⟨83OM1529⟩.

$$\text{(63)} \xrightarrow[\substack{R = \text{Et}, 54\% \\ R = \text{Bu}^t, 44\%}]{\text{RSLi, THF, 0 °C}} \text{(64)} \qquad (52)$$

$$\text{(65)} \xrightarrow[88\%]{\text{PhSNa, THF, } -78\ °\text{C}} \text{SPh derivative} \qquad (53)$$

(iii) Alkylation of α-phenylthioboronates

The combined α-carbanion-stabilizing effects of sulfur and boron provide for clean deprotonation and alkylation of α-phenylthioboronates. Matteson and Arne have lithiated the parent boronate (66) with LDA; subsequent alkylation by a range of reactive electrophiles, mainly primary or benzylic halides, gave the corresponding derivatives in good yields (Scheme 22) ⟨78JA1325, 89CRV1535⟩.

(iv) Miscellaneous methods

(a) Methylthiomethyllithium has been reported to react with the trimethylamine–borane complex to form the lithium salt (67). Treatment of this with trimethylamine hydrochloride gives the trimethylamine–methylthioborane complex (68), which decomposes on heating to form the dithioniadiboratacyclohexane ring system (69) (Scheme 23) ⟨83CB1479⟩.

(b) The reaction of trialkylboranes with the anion from tris(phenylthio)methane (70) occurs with two spontaneous migrations of alkyl groups from boron to carbon, the second being driven by prior attack of a phenylthiolate anion on the boron itself. Thus, when (70) is mixed with the borane (71), the reaction proceeds directly to give (72) (Equation (54)). Yields are generally good unless more sterically demanding trialkylboranes are used ⟨81CC1149⟩.

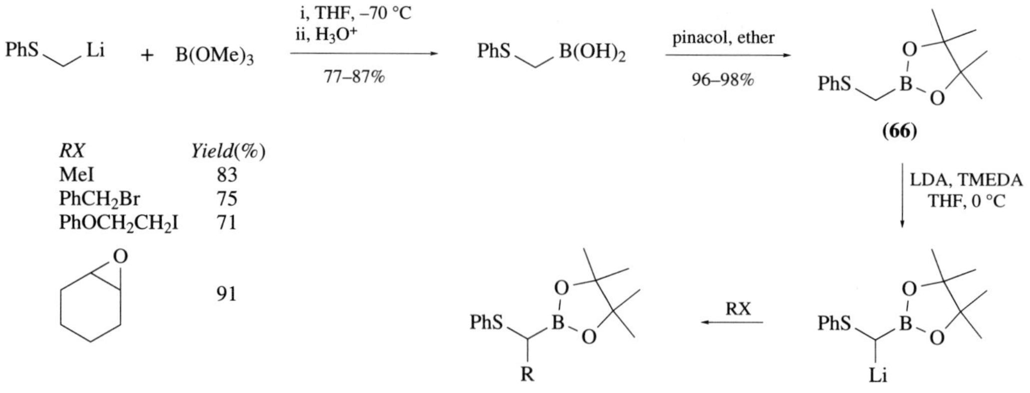

374   *Chalcogen and a Silicon, Germanium, Boron or Metal*

PhS⌒Li + B(OMe)$_3$ →(i, THF, –70 °C; ii, H$_3$O$^+$) PhS⌒B(OH)$_2$ →(pinacol, ether, 96–98%) PhS–CH$_2$–Bpin (**66**)

77–87%

RX	Yield(%)
MeI	83
PhCH$_2$Br	75
PhOCH$_2$CH$_2$I	71
cyclohexene oxide	91

(**66**) →(LDA, TMEDA, THF, 0 °C) PhS–C(Li)(Bpin) →(RX) PhS–C(R)(Bpin)

Scheme 22

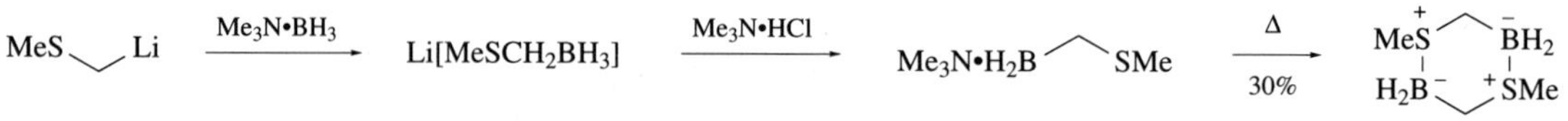

MeS⌒Li →(Me$_3$N·BH$_3$) Li[MeSCH$_2$BH$_3$] (**67**) →(Me$_3$N·HCl) Me$_3$N·H$_2$B⌒SMe (**68**) →(Δ, 30%) [MeS=BH$_2$ / H$_2$B=SMe] (**69**)

Scheme 23

LiC(SPh)$_3$ + BR$_3$ → RB(SPh)CR$_2$SPh   (54)
(**70**)   (**71**)   R = n-C$_6$H$_{13}$ (**72**)

4.08.1.3 Functions Bearing Selenium or Tellurium

4.08.1.3.1 Selenium or tellurium and silicon—$R^1{}_2C(SeR^2)SiR^3{}_3$, etc.

Although generation of carbanions α to selenium is easily achieved experimentally (see Section 4.08.2.3), as is electrophilic capture of the resulting species, for instance by halosilanes, systems containing a selenium with an α silicon have not been extensively reviewed. Appropriate synthetic pathways are outlined below.

(i) From selenium-stabilized carbanions

(a) By direct deprotonation. By far and away the commonest route to selenium systems containing an α silicon functionality involves the *C*-silylation of anions generated α to the selenium. Selenium-stabilized carbanions do not command the same synthetic status as their sulfur cousins, especially in their higher oxidation states, and treatment of aryl and alkyl selenides with alkyllithiums often leads to selenium–lithium exchange rather than deprotonation (see Section 4.08.2.3.ii for further discussion). An additional electron-withdrawing functionality α to a phenylseleno group is, however, sufficient to promote facile α-deprotonation in most cases. Addition of electrophilic halosilanes to solutions of these anions at low temperature gives good to excellent yields of the corresponding α-selenosilanes (Table 10). Deprotonation/silylation of allyl phenyl selenides with either LDA as a base, or lithium diethylamide with more hindered precursors, gives predominantly the corresponding α-selenosilanes in high yields (entries 2 and 3). The proportion of γ-silylation is variable, and α/γ ratios are dependent on the structure of the precursor and the silylating agent. For instance, whilst

chlorotrimethylsilane gives an α:γ ratio of approximately 4:1, ratios with chlorodimethylphenylsilane are frequently nearer 1:1.

Table 10 Deprotonation/silylation adjacent to selenium.

Entry	Selenide	Conditions	Product	Yield (%)	Ref.
1	PhSeMe	i, LiTMP, −55 °C ii, TMS-Cl	PhSe⌒TMS		79JA6638
2	PhSe⌒⌒(Me)	i, LDA, THF, 0 °C ii, Me₂PhSiCl	PhSe–CH(PhMe₂Si)–CH=CMe₂	>74	75JOC2570
3	PhSe⌒⌒	i, LDA, THF, −78 °C ii, TMS-Cl	PhSe–CH(PhMe₂Si)–CH=CH₂		75JOC2570

(b) From selenoacetals and selenoketals. α-Selenoalkyllithiums which cannot be synthesized by direct metallation of selenides can be conveniently prepared by transmetallation of a diselenoacetal, a process which relies upon the ease of selenium–lithium exchange and the availability of these precursors from a range of commercial starting materials (see Section 4.08.2.3.ii for a detailed discussion). Typically, treatment of the acetal with *n*-butyllithium at −78 °C in THF cleaves a C—Se bond; alkylation of the resulting anion with a halosilane gives the α-selenosilane in good to excellent yields (Table 11).

Table 11 Selenium–lithium exchange and silylation.

Entry	Diselenoacetal/ketal	Product	Yield (%)	Ref.
1	(PhSe)₂CMe₂	PhSe–C(Me₂)–TMS	75	75TL1613
2	(MeSe)₂CH–*n*-C₆H₁₃	MeSe–C(*n*-C₆H₁₃)–TMS	85	76AG(E)161
3	cyclopropyl (SePh)₂	cyclopropyl(SePh)(TMS)	74	81TL4737

Once prepared, α-selenosilanes can be elaborated by deprotonation and alkylation. For example, Sachdev and Sachdev have prepared the phenylselenosilane (**73**) in quantitative yield from diphenyl diselenide and sodium borohydride followed by addition of chloromethyl trimethylsilane (Scheme 24) ⟨76TL4223⟩. An alternative route to the same material, utilizing the condensation of benzeneselenic acid and chlorotrimethylsilane, gave a 76% yield of the α-selenosilane. The Sachdevs were able to derivatize this compound by deprotonation at −78 °C and treatment with simple primary alkyl halides in high yield. Krief has similarly derivatized α-selenosilanes by trapping the same carbanion with aldehydes ⟨79TL485, 91COS(1)629⟩.

(c) From vinyl selenides. Raucher and Koolpe have reported the addition of alkyllithiums to vinyl selenides and the trapping of the resulting α-lithio selenides with electrophiles ⟨78JOC4252⟩. Treatment of phenyl vinyl selenide (**74**) with either *n*-butyl- or isopropyllithium in DME or ether followed by addition of chlorotrimethylsilane gave the α-selenosilane (**75**) in high yield (Equation (55)). The choice of the proper reaction conditions was critical since alkyllithiums were also shown to effect α-deprotonation or C—Se bond cleavage in other solvents, resulting in mixtures of products.

$$\text{CH}_2=\text{CH-SePh} \xrightarrow[\text{ii, TMS-Cl}]{\text{i, Bu}^n\text{Li, ether, 0 °C}} \text{Bu}^n\text{-CH}_2\text{-CH(SePh)(TMS)} \quad 90\% \tag{55}$$

(**74**) → (**75**)

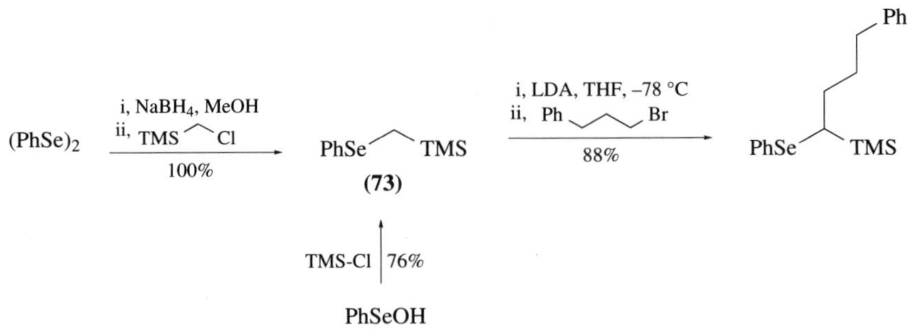

Scheme 24

(PhSe)₂ →[i, NaBH₄, MeOH; ii, TMS-CH₂-Cl; 100%] PhSe-CH₂-TMS **(73)** →[i, LDA, THF, –78 °C; ii, Ph-CH₂-CH₂-Br; 88%] PhSe-CH(CH₂CH₂Ph)-TMS

PhSeOH →[TMS-Cl, 76%] (73)

(ii) From aldehydes and thioaldehydes

Krafft and Meinke have synthesized α-silylalkyl selenocyanates as precursors to selenoaldehydes ⟨87TL5121, 88JA8671⟩. The preparation involves the simple addition of the appropriate alkyl or aryl aldehyde to dimethylphenylsilyllithium in THF at −78 °C and trapping the resultant alkoxide with toluene-*p*-sulfonyl chloride (Scheme 25). Displacement of the tosylate then gives the required α-silylalkyl selenocyanate in good yield.

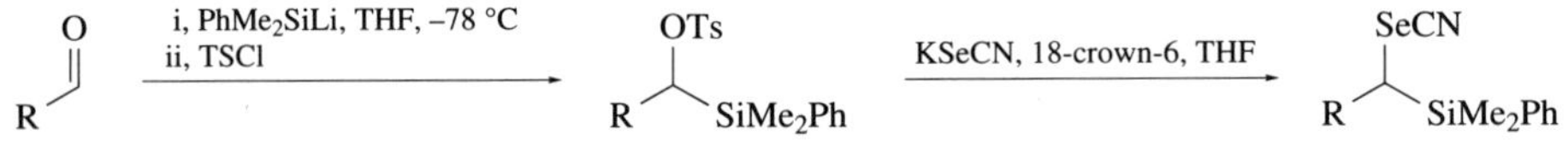

RCHO →[i, PhMe₂SiLi, THF, –78 °C; ii, TSCl] R-CH(OTs)-SiMe₂Ph →[KSeCN, 18-crown-6, THF] R-CH(SeCN)-SiMe₂Ph

R = Prⁿ, 91%; Ph, 97%

Scheme 25

Okazaki and co-workers have described the synthesis of (**76**), a precursor to a stable selenoaldehyde, from aryl thioaldehydes ⟨89JA5949⟩. Treatment of the thiobenzaldehyde (**77**), bearing substituents chosen to confer stability on the overall target molecule, with trimethylsilyllithium followed by sulfur derivatization and reductive lithiation gave the carbanion (**78**), which could be reacted with nucleophilic selenium in 47% overall yield (Scheme 26).

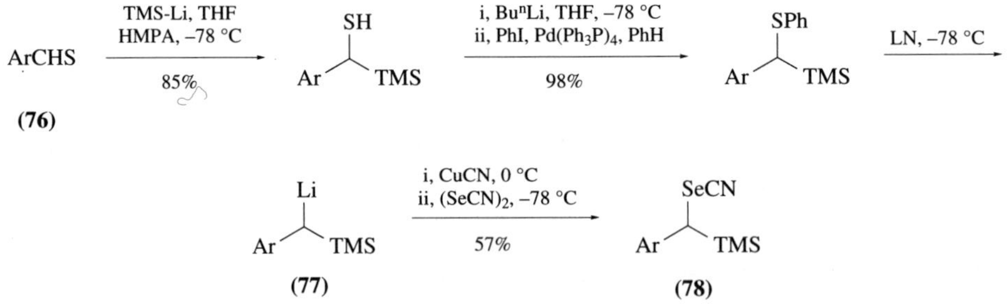

ArCHS →[TMS-Li, THF, HMPA, –78 °C; 85%] Ar-CH(SH)-TMS **(76)** →[i, BuⁿLi, THF, –78 °C; ii, PhI, Pd(Ph₃P)₄, PhH; 98%] Ar-CH(SPh)-TMS →[LN, –78 °C] Ar-CH(Li)-TMS **(77)** →[i, CuCN, 0 °C; ii, (SeCN)₂, –78 °C; 57%] Ar-CH(SeCN)-TMS **(78)**

Ar = 2,4,6-tri-*t*-butylphenyl

Scheme 26

(iii) Miscellaneous methods

(a) From vinylsilanes. Hayama et al. have reported the synthesis of α-selenosilanes from vinylsilanes ⟨84JOC3235⟩ by addition of benzeneselenenyl chloride or bromide to vinylsilanes and subsequent treatment of the β-haloethylsilane intermediates with silver nitrite and mercury(II) chloride. Overall yields of the α-seleno-β-nitrosilanes were moderate, and the range of vinylsilane precursors investigated only included β-alkyl substituents. Representative is the production of (**79**) from (**80**) in 39% yield (Equation (56)).

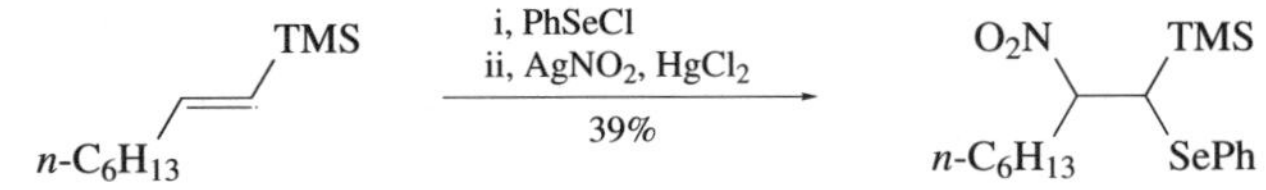

$$\underset{(80)}{\underset{n\text{-}C_6H_{13}}{\diagup}\hspace{-0.5em}\diagdown\text{TMS}} \xrightarrow[39\%]{\text{i, PhSeCl} \atop \text{ii, AgNO}_3\text{, HgCl}_2} \underset{(79)}{\underset{n\text{-}C_6H_{13}}{O_2N}\hspace{-0.5em}\diagdown\hspace{-0.5em}\underset{\text{SePh}}{\diagup}\text{TMS}} \quad (56)$$

(b) From (halomethyl)trimethylsilanes. There are relatively few descriptions of the displacement of the halide from a halomethylsilane with a selenium nucleophile despite the apparent synthetic versatility of such a process. A typical example would be the reaction of (iodomethyl)trimethylsilane with lithium phenylselenide in ether ⟨86ICA51⟩, which produces the α-selenosilane (**81**) in good yield (70–90%) (Equation (57)).

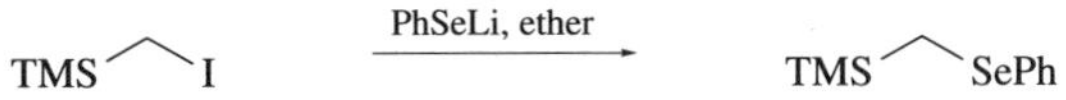

$$\text{TMS}\diagdown\text{I} \xrightarrow{\text{PhSeLi, ether}} \underset{(81)}{\text{TMS}\diagdown\text{SePh}} \quad (57)$$

(c) Preparation of dialkyl tellurides. Dialkyl tellurides can be prepared by alkylation of sodium telluride with alkyl chlorides ⟨79IC2696⟩. When halomethylsilanes are used as electrophiles, α-tellurosilanes can be synthesized. The reported procedure involves initial generation of sodium telluride by reduction of metallic tellurium, followed by addition of (chloromethyl)trimethylsilane and heating under reflux (Scheme 27). Yields were claimed to be quantitative.

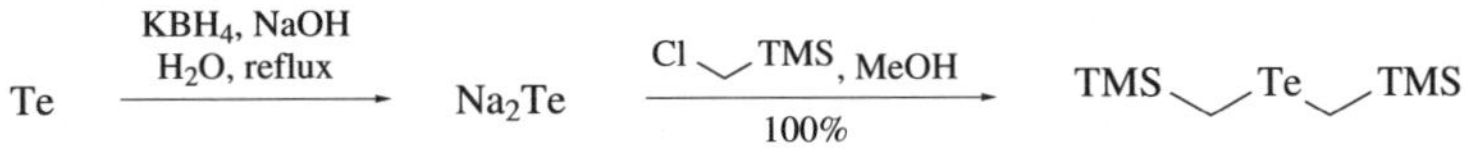

$$\text{Te} \xrightarrow[\text{H}_2\text{O, reflux}]{\text{KBH}_4\text{, NaOH}} \text{Na}_2\text{Te} \xrightarrow[100\%]{\text{Cl}\diagdown\text{TMS, MeOH}} \text{TMS}\diagdown\text{Te}\diagdown\text{TMS}$$

Scheme 27

4.08.1.3.2 Selenium or tellurium and germanium—$R^1{}_2C(SeR^2)GeR^3{}_3$, etc.

There are few examples of compounds containing a germyl group α to a selenium or tellurium. Drake and Chehayber have described iodide displacement from (iodomethyl)trimethylgermane using lithium phenylselenide ⟨86ICA51⟩ in ether; the α-selenogermane (**82**) is obtained in good yield (70–90%) (Equation (58)). These compounds were subsequently reacted with Na_2PdCl_4 and K_2PtCl_4 to produce square planar complexes containing two selenium ligands: $(Me_3GeCH_2SePh)_2PdCl_2$, etc. ⟨86IC611⟩.

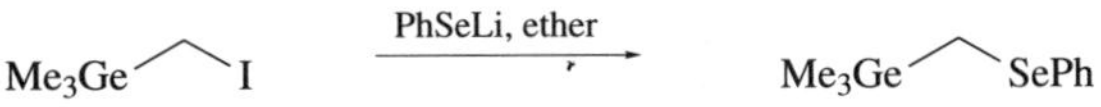

$$\text{Me}_3\text{Ge}\diagdown\text{I} \xrightarrow{\text{PhSeLi, ether}} \underset{(82)}{\text{Me}_3\text{Ge}\diagdown\text{SePh}} \quad (58)$$

4.08.1.3.3 Selenium or tellurium and boron—$R^1{}_2C(SeR^2)BR^3{}_2$, etc.

There are no reports of significant preparative routes to systems containing a boron function α to a selenium or tellurium.

4.08.2 FUNCTIONS CONTAINING A CHALCOGEN AND A METAL

4.08.2.1 Functions Bearing Oxygen—$R^1{}_2C(OR^2)M$, etc.

Cheshire has reviewed the formation of carbanions adjacent to an oxygen function ⟨91COS(3)193⟩. The area has also been included in general reviews of α-heteroatom-substituted organometallics, notably by Peterson ⟨72MI 408-02⟩, Krief ⟨80T2531⟩ and Wardell ⟨B-87MI 408-02⟩. The last source, in particular, contains further discussion about the major synthetic operations covered below.

Despite a higher inductive effect, the prevalence and exploitation of carbanions α to oxygen lag behind the heavier chalcogen, sulfur, since the second-row element can offer other modes of support for an adjacent negative charge, for example its greater polarizability. (The mechanisms and relative importance of stabilizing effects of first- and second-row nonmetals have been discussed by von Rague Schleyer et al. ⟨84JA6467⟩). For oxygen this means that additional carbanion-stabilizing factors are frequently present. The sections below first discuss systems stabilized by oxygen alone, and then systems containing oxygen with an additional functionality.

4.08.2.1.1 Lithium, sodium or potassium

(i) By direct deprotonation

As intimated above, there are few reports of the successful direct deprotonation of a saturated carbon adjacent to oxygen. Attempted lithiation of dimethyl ether with *n*-butyllithium results in decomposition, and similarly ring opening and cleavage of THF. In contrast, Schlosser and Lehmann reported that at temperatures between −100 °C and −40 °C butylpotassium metallates ethers ⟨84TL745⟩; the resulting organopotassium species decompose slowly at −78 °C and more rapidly at −50 °C. At low temperatures, the ether anions could be trapped with chlorotrimethylsilane in good yields (Scheme 28). Corey and Eckrich have also achieved direct ether metallation. They treated *t*-butyl methyl ether with the Schlosser base at −78 °C, subjecting the resulting anion to electrophilic capture (Scheme 29) ⟨83TL3165⟩. Interestingly, in Corey's case as in Schlosser's, the active metallating agent is presumably an organopotassium species. Corey also showed that addition of lithium bromide to the reaction mixture effectively reduces the basicity by precipitation of potassium; the resulting lithiated ether then could be added cleanly to aldehydes and ketones without concomitant α-deprotonation. The *t*-butyl group could be removed with acetic anhydride and anhydrous ferric chloride followed by potassium carbonate hydrolysis, making *t*-butyl methyl ether a hydroxymethyl anion equivalent.

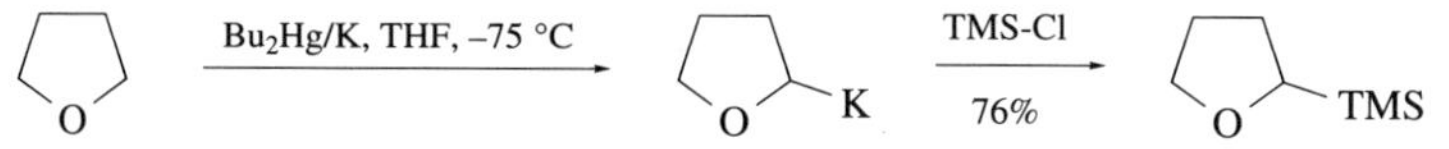

Scheme 28

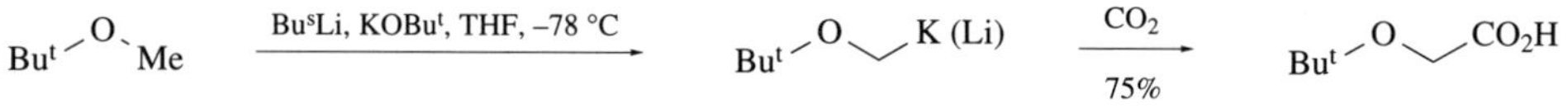

Scheme 29

(ii) By tin–lithium exchange

Given the difficulty of achieving direct deprotonation, one of the most common methods of introducing an alkali metal, most notably lithium, α to an oxygen is by transmetallation of a tin functionality. (The introduction of tin α to oxygen, usually from the corresponding aldehyde, is discussed below in Section 4.08.2.1.6.) One of the earliest reports of tin–lithium exchange was by Seebach, who realized the direct hydroxymethylation of carbonyl compounds by treatment of tributylstannylmethanol (**83**) with two equivalents of *n*-butyllithium, effectively producing the dianion of methanol (**84**) (Scheme 30) ⟨76AG(E)438, 80CB1290⟩. The utility of the method is limited by the instability of the reagent and moderate yields of addition products. Later, Still reported the preparation of α-alkoxyorganolithium reagents ⟨78JA1481, 78JA1927, 79TL593⟩ from α-alkoxytin species, and showed that high yields could be obtained upon addition to carbonyl compounds. The ease of the tin–lithium exchange at low temperatures and the accessibility of the tin precursors has since spurred a widespread exploitation of this route to α-oxygenated organolithiums. The metal exchange is typically carried out at low temperatures in donor solvents such as THF by addition of *n*-butyllithium. The α-alkoxylithiums, in contrast to the α-alkoxystannanes, are only stable at low temperatures and are progressed *in situ*. Substituents reported on the lithiated carbon have included hydrogen, and alkyl, cycloalkyl, aryl and heteroaryl groups (Table 12), although examples of tin–lithium exchange in disubstituted alkoxystannanes are rarer: Still himself reported the failure to transmetallate (**85**) ⟨80JA1201⟩. McGarvey, however, showed that such α-alkoxylithiums can be prepared by transmetallation in DME (Table 12, entry 5) ⟨84JA3376⟩. The oxygen atom is commonly protected (typically as an MOM-type ether) or substituted with either alkyl or aryl groups; frequently an *O*-allyl group has been used to induce the exposed carbanion to undergo [2,3] Wittig rearrangement, resulting in a route to the corresponding homoallylic alcohols (Scheme 31) ⟨92TL5795⟩.

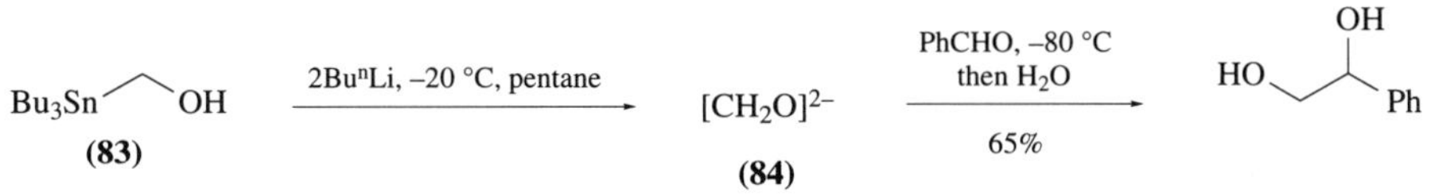

Scheme 30

Table 12 Alkoxylithiums from alkoxystannanes.

Entry	Alkoxystannane	Alkoxylithium	Yield (%)	Ref.
1	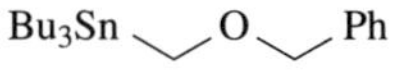 Bu$_3$Sn∼O∼Ph	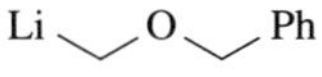 Li∼O∼Ph	>98	78JA1481
2	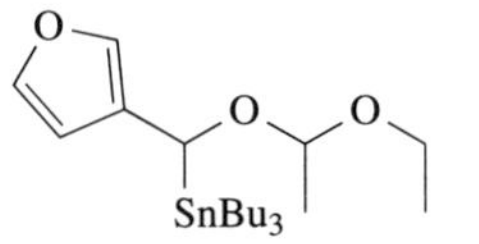 furan-CH(SnBu$_3$)-O-CH(OEt)-CH$_3$	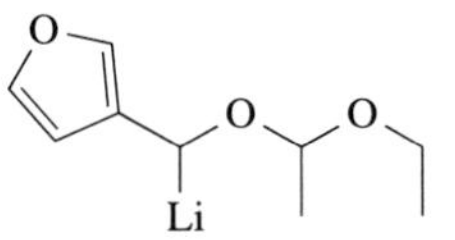 furan-CH(Li)-O-CH(OEt)-CH$_3$	>76	78JA1481
3	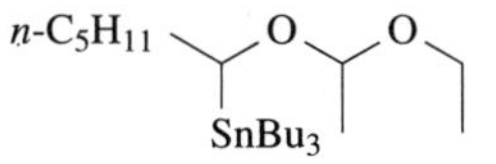 n-C$_5$H$_{11}$-CH(SnBu$_3$)-O-CH(OEt)-CH$_3$	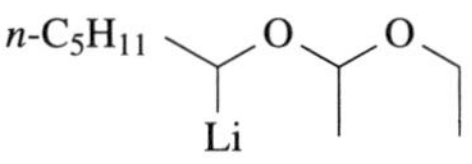 n-C$_5$H$_{11}$-CH(Li)-O-CH(OEt)-CH$_3$		92JOC5078
4	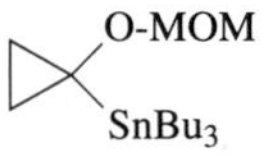 cyclopropyl-C(O-MOM)(SnBu$_3$)	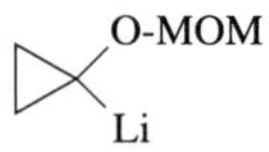 cyclopropyl-C(O-MOM)(Li)	>68	85JOC3255
5	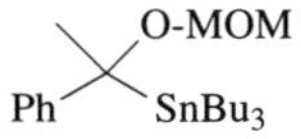 Ph-C(CH$_3$)(O-MOM)(SnBu$_3$)	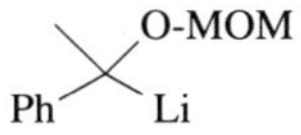 Ph-C(CH$_3$)(O-MOM)(Li)	>92	84JA3376

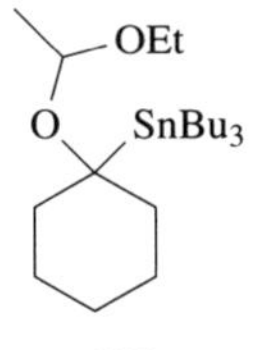

(85)

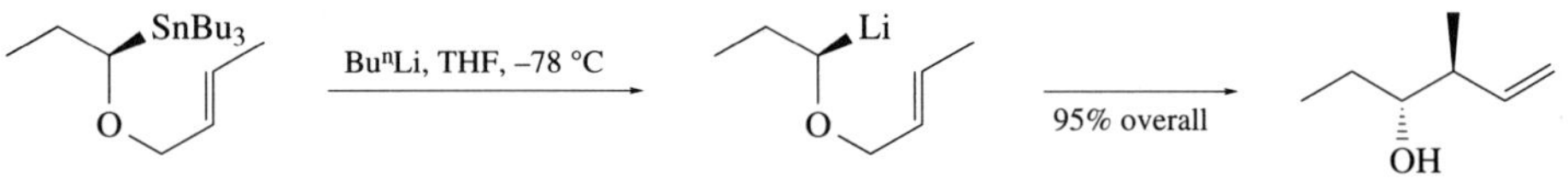

Scheme 31

Sawyer *et al.* have studied the effect of substituents on the thermodynamic stability of α-alkoxyalkyl and alkyllithiums, by examining ligand selectivity in the tin–lithium exchange process ⟨84JA3376, 88JA842⟩. In the same papers they also investigated the effects of solvent and tin ligands on the rate of exchange. Furthermore, Still and Sreekumar first reported that α-alkoxyalkylthiums are effectively configurationally stable carbanions ⟨80JA1201⟩, and can be prepared with retention of configuration from chiral alkoxystannanes (Scheme 32; see also Scheme 31). Although they prepared enantiomerically pure stannanes by resolution, these stannanes can also be conveniently prepared by asymmetric reduction of acylstannanes (see Section 4.08.2.1.6.i) or from α-chloroboronic esters (see Section 4.08.2.1.6.vi (b)).

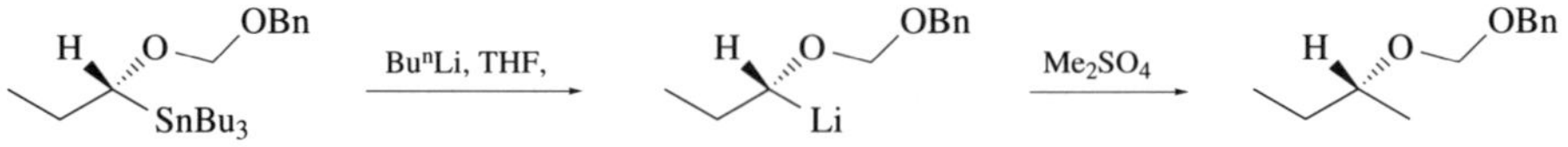

Scheme 32

Corey and Eckrich have demonstrated that transmetallation of trihalostannanes, for example (**86**), prepared *in situ* from chloromethyl ethers (see Section 4.08.2.1.6.vi (c)), with four equivalents of *n*-butyllithium gives α-alkoxylithium reagents ⟨83TL3163⟩ in good yields (Scheme 33). Subsequent reactions with aldehydes produce high yields of the corresponding alcohols, although in all cases there was contamination with traces of butylated products arising from the presence of unreacted butyllithium.

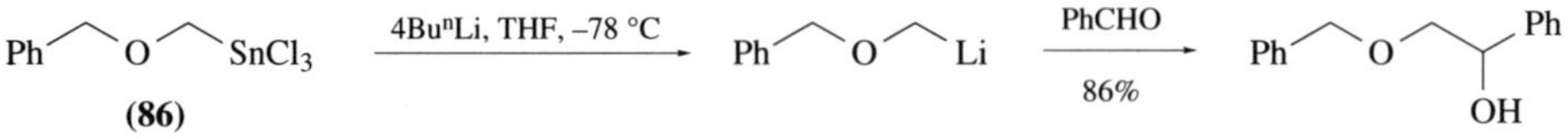

Scheme 33

(iii) By reductive lithiation of α-phenylthioethers

Cohen and co-workers have described a general preparative route to α-alkoxylithium species by the low-temperature reductive lithiation of α-phenylthio ethers in THF with LDMAN or LN ⟨80JA6900⟩. Both reagents have also been used for the reductive lithiation of diphenyl dithioacetals (see Section 4.08.2.2.ii (b)). The advantage of LDMAN is that although it must be prepared and used below $-45\,°C$, the by-product, 1-(dimethylamino)naphthalene, can be easily separated, after subsequent derivatization of the carbanion, by acid extraction during work-up. A variety of stabilized and unstabilized α-alkoxylithiums can be rapidly prepared in THF below $-60\,°C$ (Table 13). The method is particularly attractive since the starting phenylthio ethers are readily accessible, for example from thioethers (by α-chlorination and displacement) or acetals and ketals (by Lewis acid-catalysed exchange with thiophenol).

Table 13 Reductive lithiation of α-phenylthio ethers.

α-Phenylthio ether	Conditions	Product	Yield (%)	Ref.
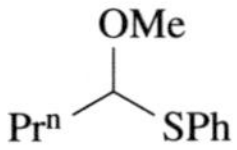 Pr^n–C(OMe)(H)–SPh	LDMAN, THF, $-78\,°C$	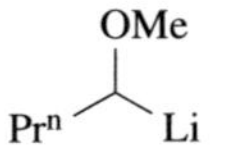 Pr^n–C(OMe)(H)–Li	>85	80JA6900
 PhS–C(OMe)(Me)–Me	LDMAN, THF, $-78\,°C$	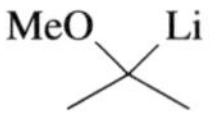 MeO–C(Li)(Me)–Me	>69	80JA6900
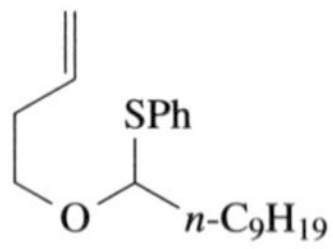 2-(SPh)(n-C$_9$H$_{19}$)-tetrahydrofuran derivative	LN, THF, $0\,°C$	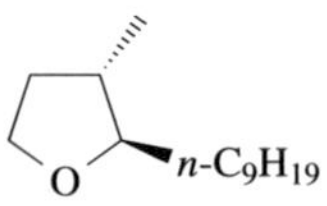 2-Li-tetrahydrofuran with n-C$_9$H$_{19}$	52	89JA2981
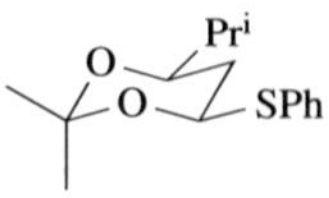 Pr^i-substituted dioxane with SPh	LDBB, THF, $-78\,°C$	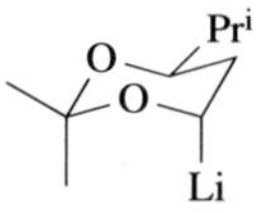 Pr^i-substituted dioxane with Li	>78	89TL3011
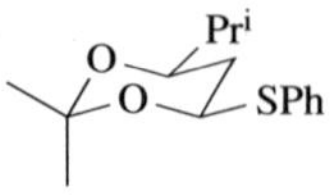 Pr^i-substituted dioxane with SPh	LDBB, THF, $-78\,°C$, then $-20\,°C$	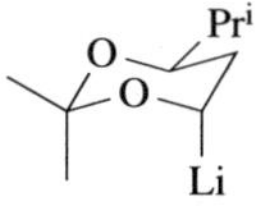 Pr^i-substituted dioxane with Li (epimer)	>54	89TL3011

LDBB = lithium di-*t*-butylbiphenylide.

A paper by Brueckner and Kruse illustrates the related reductive lithiation of arylsulfones ⟨89CB2023⟩. Treatment of the sulfone (**87**) with LN generates the requisite α-alkoxylithium, in this case used as a precursor to a [2,3] shift (Scheme 34).

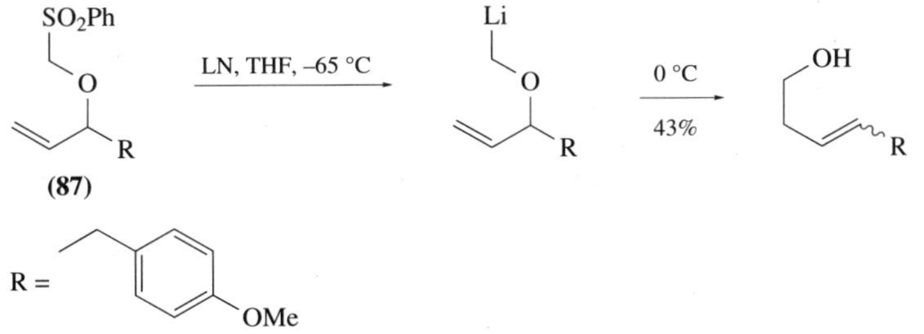

Scheme 34

(iv) By halogen–lithium exchange

There are relatively few examples of the exchange reactions involving α-halo ethers and *t*-butyllithium. Early reports described the generation of α-alkoxymetal species by the reaction of unsubstituted chloromethyl ethers with metals. For example, Schollkopf and co-workers studied the low-temperature metallation of chloromethyl methyl ether with lithium and the subsequent reaction with carbon dioxide and carbonyl compounds (Scheme 35) ⟨64TL1503, 67LA(704)120⟩. Gadwood and co-workers have since shown that exchange with *t*-butyllithium is feasible when the carbanion is stabilized; (**88**) was produced from the corresponding α-alkoxy bromo ether and *t*-butyllithium, and subsequently added to carbonyl compounds (Scheme 36) ⟨85JOC3255⟩.

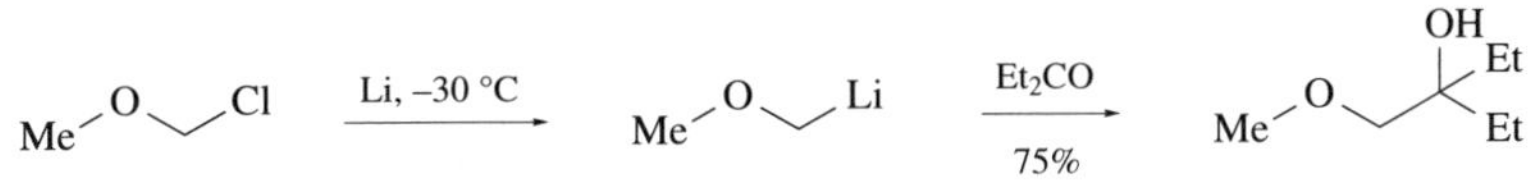

Scheme 35

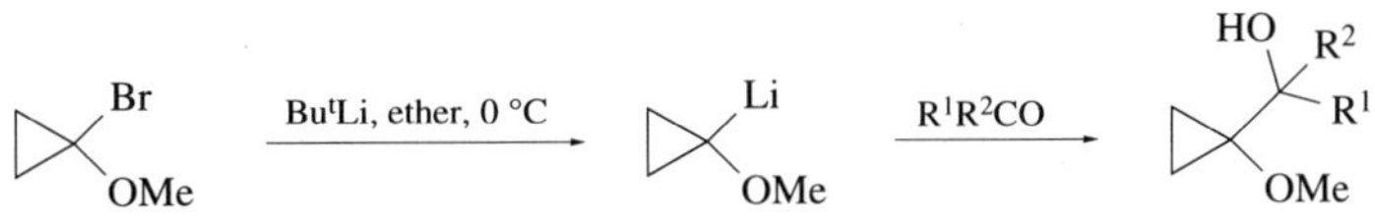

Scheme 36

(v) Carbanions with additional stabilization

(a) Lithiation of allylic and benzylic ethers. The preparation of α-alkoxymetals by direct deprotonation has been reported commonly for allyl and benzyl ethers, in which additional stabilization of the formal carbanion is provided by unsaturation. The former case is covered by general reviews on α-heteroatom-stabilized, allylic carbanions ⟨82OR(27)1, 84AG(E)932⟩. Frequently in these circumstances, carbanion generation is followed by intramolecular rearrangement, for example [1,3] Wittig or reverse Brook migrations (see Section 4.08.1.1.1.vi) in the case of benzyl and *O*-silyl ethers, respectively, and [2,3] Wittig or [3,3] Claisen shifts in allyl ethers. There are fewer references to the generation of anions capable of intermolecular capture; all notable cases have been covered by Cheshire in his review ⟨91COS(3)193⟩, whither further attention is directed.

(b) Lithiation of α-cyano ethers. Deprotonated, protected cyanohydrins are well-known acyl anion equivalents displaying umpolung reactivity at what was the normally electrophilic carbonyl carbon. The development of this chemistry by Stork and co-workers and the scope of recent applications is again covered in Cheshire's review ⟨91COS(3)193⟩, and is not therefore repeated here.

(c) Lithiation of O-alkyl esters, O-alkyl carbonates and O-alkyl carbamates. Direct metallation of *O*-alkyl esters was first reported by Beak and co-workers ⟨77JA5213, 78JOC4255, 81JOC2363⟩ and Seebach and co-workers ⟨78HCA512⟩; stabilization of the carbanion can be in part attributed to the adjacent carbonyl dipole. Alkyl benzoates are typically treated with *s*-butyllithium/TMEDA in THF, and the anions trapped with simple electrophiles to give alkylated products in moderate to good yields (Scheme 37). Efforts to extend the reaction to systems bearing esters derived from secondary alcohols have been unsatisfactory except in favourable cases (Scheme 38) ⟨85JOC3255⟩. The whole area of dipole-stabilized carbanions, including those adjacent to oxygen, has been usefully reviewed by Beak ⟨78CRV275⟩, although the date of this article precluded inclusion of the material below.

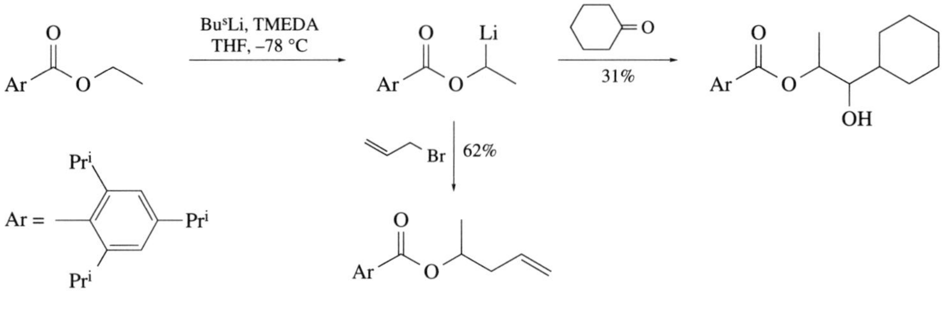

Scheme 37

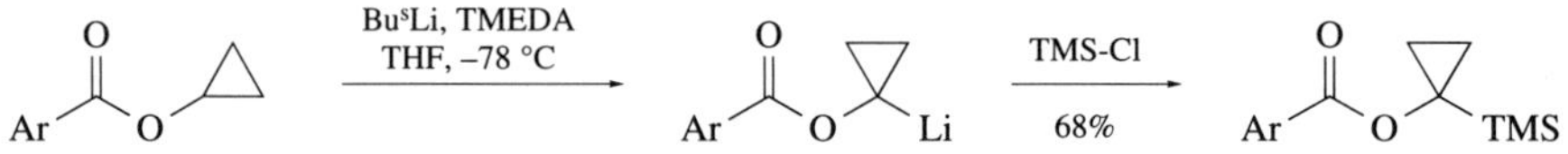

Scheme 38

Hoppe and co-workers have successfully alkylated allyl *N*-alkyl- and allyl *N*-phenylcarbamates ⟨81AG(E)127, 85CB2822⟩. The reaction proceeds via lithiation of the substrate with electrophilic trapping of the carbanion produced. By an analogous process, Hoppe and Broenneke ⟨82S1045⟩ and, separately, Barner and Mani ⟨89TL5413⟩ have also achieved overall alkylation of benzylic carbamates. Interestingly in this latter case, the intermediate carbanions did not undergo any Wittig rearrangement. Both groups achieved high overall yields with allyl precursors, giving products

Table 14 Dipole-stabilized α-oxygenated carbanions.

Precursor	Conditions	Product	Yield (%)	Ref.
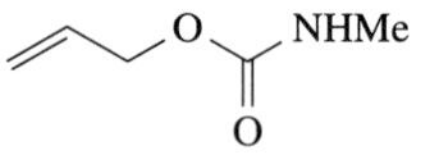 allyl O-C(O)-NHMe	2 Bu^{n}Li, TMEDA THF, −78 °C to −50 °C	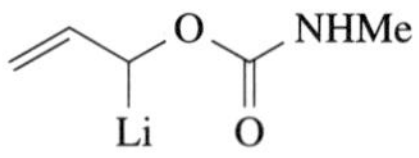 allyl(Li) O-C(O)-NHMe	~75	81AG(E)127
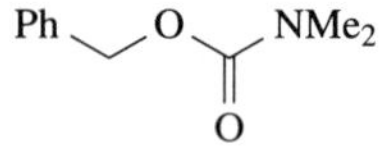 Ph-CH$_2$-O-C(O)-NMe$_2$	Bu^{n}Li, TMEDA THF, −78 °C	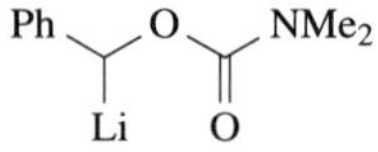 Ph-CH(Li)-O-C(O)-NMe$_2$	>90	82S1045
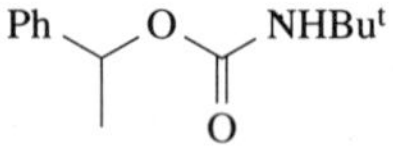 Ph-CH(iPr)-O-C(O)-NHBu^t	2 Bu^{s}Li, THF, −78 °C	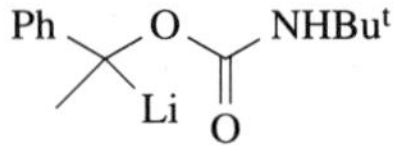 Ph-C(Li)(iPr)-O-C(O)-NHBu^t	~80	89TL5413

arising largely from γ substitution. Table 14 shows examples of the initial deprotonation process.

In later papers, Hoppe and co-workers studied deprotonation of chiral carbamates (derived from chiral alcohols), both benzylic and allylic, discovering that lithiation occurred with retention of configuration and that the derived organolithium species were largely configurationally stable at low temperatures ⟨86AG(E)160, 87TL5149, 90AG(E)1424⟩. Subsequently, Hoppe and Sommerfeld demonstrated that *O*-alkyl carbamates can be enantioselectively deprotonated in the presence of sparteine ⟨92SL764, 92T8377⟩. The carbamate **(89)** was treated with 1.9 equivalents of *s*-butyllithium in ether at −78 °C in the presence of two equivalents of (−)-sparteine. Electrophilic trapping of the carbanion produced gave substituted products in high *ee* (Scheme 39). The presence of an *N,N*-dibenzylamino group was crucial to the production of high selectivity: dimethylamino derivatives gave racemic products, supporting the conclusion that, in the latter case, intramolecular coordination of the lithium by the nitrogen occurs in preference to intermolecular coordination by (−)-sparteine. Hoppe used the sequence for the asymmetric synthesis of 3-hydroxy alkylamines; however, since the dibenzylamino group is apparently only a spectator during the deprotonation, then the reaction might have wider potential. A related process described by Katritzky and Sengupta is the lithiation of the lithium carbonate derivatives of alcohols ⟨87TL1847⟩. Benzyl alcohol and trimethylsilylmethanol were each treated successively with *n*-butyllithium and carbon dioxide. Further deprotonation *in situ* with *s*-butyllithium gave the lithiated species (**90**) and (**91**), which could be trapped with electrophiles (Scheme 40). In the case of the trimethylsilylmethanol series, acidic work-up gave products derived formally from a hydroxymethyl dianion.

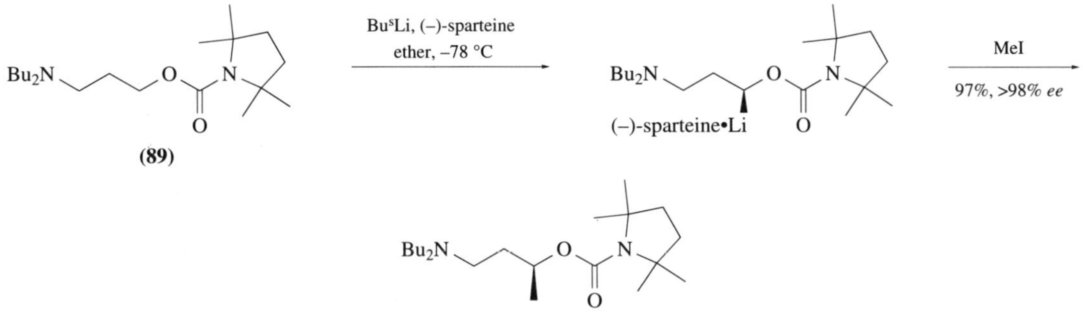

Scheme 39

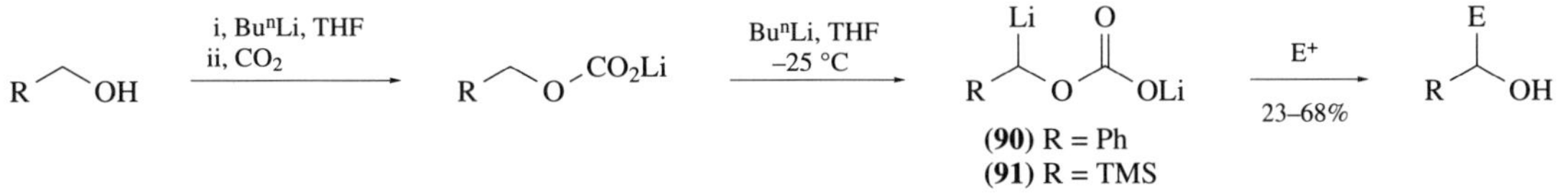

Scheme 40

4.08.2.1.2 Magnesium

Sommlet originally studied the action of magnesium on chloromethyl methyl ether ⟨7BSF394⟩ in diethyl ether as the solvent; subsequent reactions with ketones and esters gave poor yields of alcohols. Normant and Crisan showed much later that in THF the yields of reaction were considerably improved ⟨59BSF459⟩. They noted, however, considerable difficulty in forming Grignard reagents from substituted α-chloroalkyl ethers. Subsequently, Castro demonstrated the preparation of Grignard reagents from a range of chloromethyl ethers at lower temperatures than previously studied (as low as −30 °C) (Equation (59)) ⟨67BSF1533⟩.

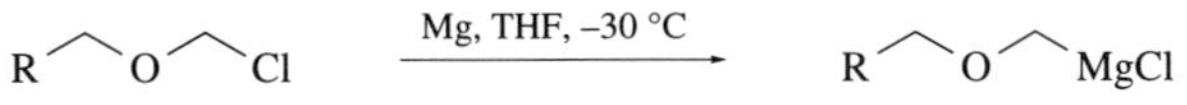

$$R\frown O\frown Cl \xrightarrow{\text{Mg, THF, } -30\,°C} R\frown O\frown MgCl \tag{59}$$

R = H, alkyl, allyl, benzyl

More recently, Still reported the use of benzyloxymethylmagnesium chloride (**92**) in the preparation of the derived stannane, although he gave no preparatory details (Equation (60)) ⟨78JA1481⟩.

(**92**) → (Equation 60)

4.08.2.1.3 Titanium or aluminum

Lithiated carbamates of the type exemplified by (**93**) have been metal exchanged with tris(dimethylamino)titanium chloride or diisobutylaluminum methanesulfonate (Scheme 41); the titanium and aluminum intermediates obtained were used in stereoselective aldol reactions with aldehydes and ketones ⟨84AG(E)239, 85TL411⟩. In the optically active series, it has been shown that titanation can be induced with retention (with $Ti(OPr^i)_4$) or inversion of configuration (with $ClTi(NEt_2)_3$) at the original lithiated carbon centre ⟨87TL5149⟩.

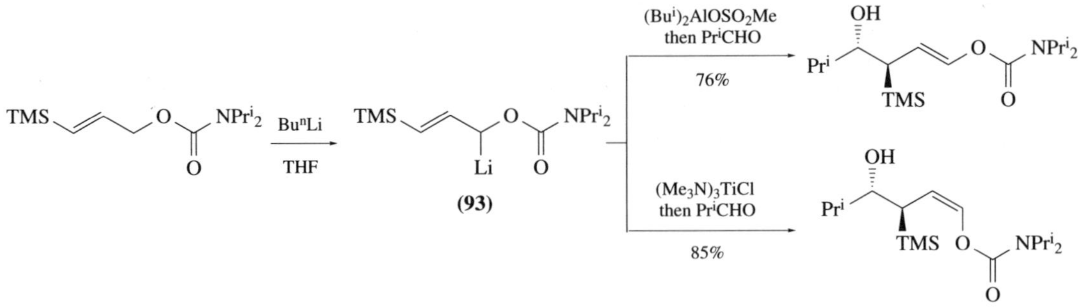

Scheme 41

4.08.2.1.4 Copper or zinc

Linderman and co-workers have reported the high-yielding 1,4-addition of α-alkoxycuprates to α,β-unsaturated ketones, finding that the reaction failed in substrates with additional substitution at the β position. Cuprates were derived from the corresponding lithiated species, by the addition of CuCN, which were in turn obtained from the α-alkoxystannanes by transmetallation with *n*-butyllithium (Scheme 42) ⟨87TL3911, 88JA6249, 89T495⟩. At about the same time, Hutchinson and Fuchs, in a more thorough examination of the copper precursor and reaction additives, showed that α-alkoxycuprates could be synthesized from CuBr/dimethyl sulfide, but that the reagents obtained were relatively unreactive towards enones; improvement in reactivity was seen in some cases upon the addition of chlorotrimethylsilane and, moreover, boron trifluoride etherate ⟨87JA4930⟩. With the latter reagent, addition to β-substituted enones was possible (Equation (61)). Corey and Eckrich transmetallated the *t*-butoxymethyllithium reagent (**94**) (see Section 4.08.2.1.1.i) using CuBr/dimethyl sulfide, and demonstrated a high yield upon addition to 2-cyclohexenone (Scheme 43) ⟨83TL3165⟩.

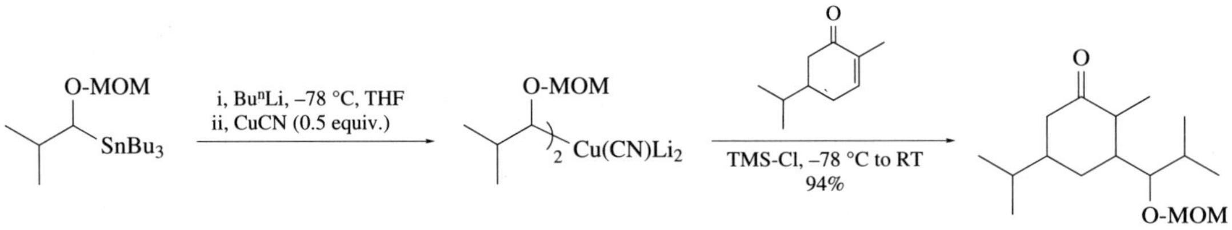

Scheme 42

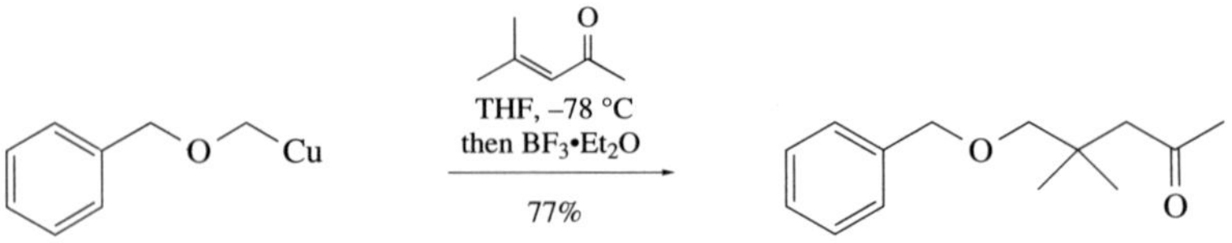

(61)

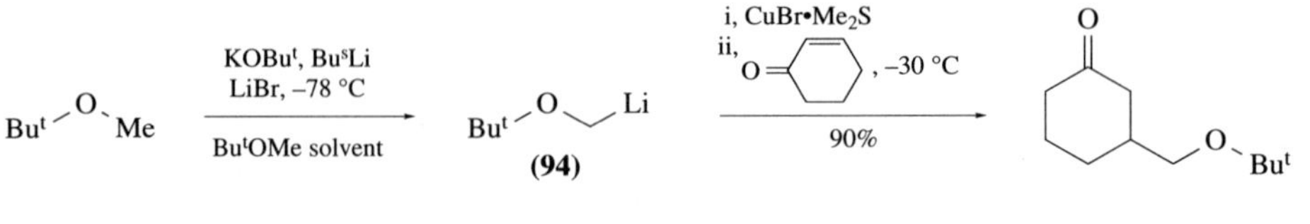

Scheme 43

Linderman and Griedel have studied the generation and addition properties of chiral α-alkoxyorganocuprates from chiral α-alkoxystannanes ⟨90JOC5428, 91JOC5491⟩. They showed that tin–lithium and lithium–copper exchange occur essentially with retention of configuration. Cuprates produced using CuI/TMEDA added to enones with complete retention of configuration whereas cuprates synthesized with CuCN added with only partial retention.

Knochel and co-workers showed that treatment of iodomethyl pivalate (**95**) with activated zinc foil in THF gives a high yield of the zinc reagent (**96**). Addition of this species at $-30\,°C$ to a THF solution of CuCN/LiCl gives the cuprate (**97**). This reagent reacts in good yields with electrophiles (mainly aromatic acid chlorides, aldehydes, enones and allyl halides) (Scheme 44) ⟨89JOC5202⟩: the mixed copper/cadmium species, prepared in an analogous manner, performed significantly better in the corresponding reactions with aliphatic acid chlorides. Wittig and Jautelat have also prepared related ester zincates ⟨67LA(702)24⟩ by treatment of zinc benzoate with diazomethane.

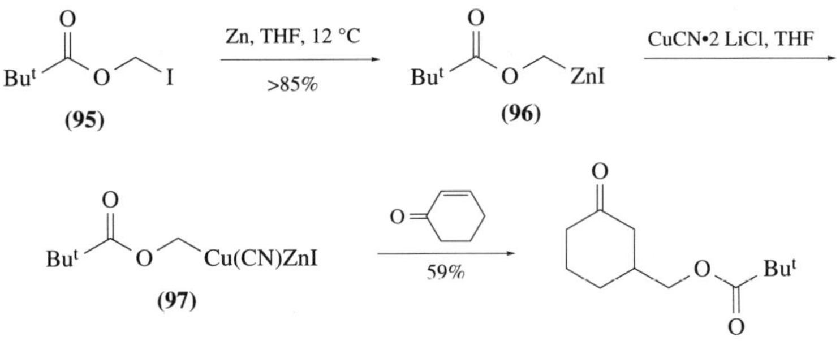

Scheme 44

4.08.2.1.5 Mercury

There are a small number of diverse reports of α-alkoxyorganomercury compounds in the literature, of which two have preparative merit:

(i) Skell and Valenty have reported the synthesis and photochemistry of α-diazomercurials ⟨73JOC3937⟩. Photolysis of the diazoacetate (**98**), prepared from methylmercury chloride and methyl diazoacetate in a one-pot procedure, in methanol gave a virtually quantitative yield of (**99**) (Equation (62)).

(ii) Giese and Erfort obtained the α-acetoxymercury species (**100**) in poor to good yields when they treated a range of dialkyl- and aralkylhydrazones, such as (**101**), with mercuric oxide/mercuric acetate (Equation (63)) ⟨83CB1240⟩.

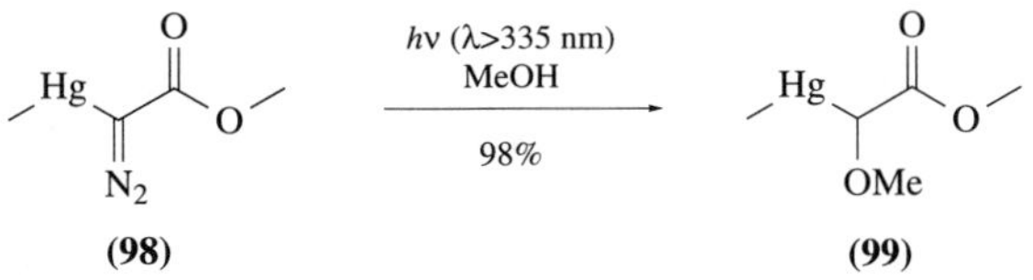

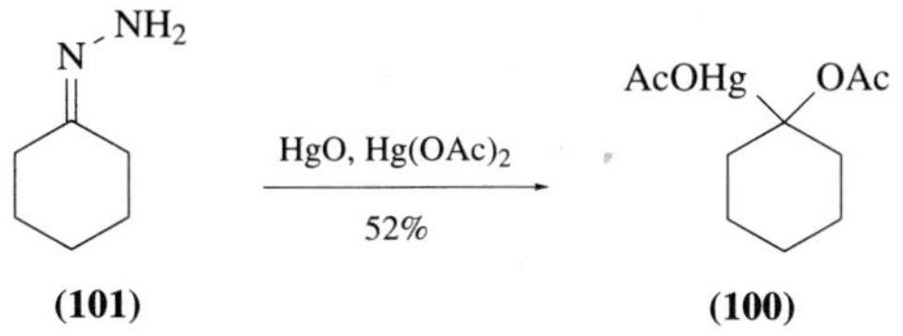

4.08.2.1.6 Tin

Much of the α-alkoxyorganometal chemistry discussed in Section 4.08.2.1 relies on organostannane precursors. The major preparative routes are outlined below.

(i) From acylstannanes

Reduction of acylstannanes can be easily effected by a variety of common reducing agents (e.g., LiAlH$_4$ ⟨66JOM(5)486⟩ and BH$_3$/THF ⟨88JOC5584⟩), although, possibly due to the relative instability of products, this is not a common synthetic operation. In contrast, asymmetric reduction is more widely reported. In the late 1980s, two reports appeared on the enantioselective reduction of acyltributylstannanes with 2,2′-dihydroxy-1,1′-binaphthyl-modified lithium aluminum hydride (BINAL-H) ⟨88JOC5584, 88TL1657⟩. In contrast to reductions of hindered ketones, asymmetric reductions of acylstannanes are complete in a few hours at −78 °C, with work-up usually including the addition of MOM-Cl or benzyloxymethyl chloride (BOM-Cl) to protect the relatively unstable alcohol functionality. The *ee* values are typically 90% or greater. This method of production of chiral α-alkoxystannanes has since proved to be the method of choice (Table 15).

Table 15 Asymmetric reduction of acylstannanes.

Acylstannane	Conditions	Product	ee (%)	Yield (%)	Ref.
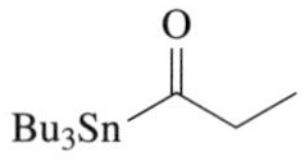 Bu$_3$Sn-C(O)-Et	i, (S)-BINAL-H; ii, BOM-Cl, Pr$^i_2$NEt	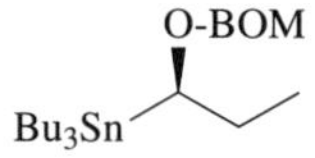 Bu$_3$Sn-CH(OBOM)-Et	96	69	88JOC5584
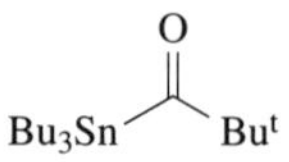 Bu$_3$Sn-C(O)-CH$_2$Bu^t	i, (S)-BINAL-H; ii, BOM-Cl, Pr$^i_2$NEt	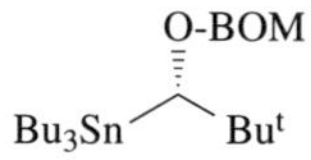 Bu$_3$Sn-CH(OBOM)-CH$_2$Bu^t	80	55	88JOC5584
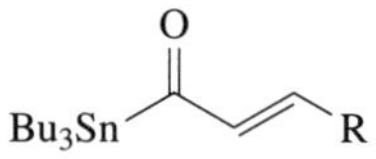 Bu$_3$Sn-C(O)-CH=CH-R	i, (R)-BINAL-H; ii, MOM-Cl, Pr$^i_2$NEt	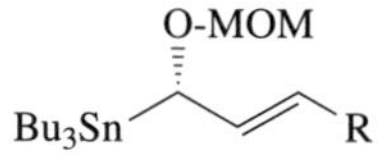 Bu$_3$Sn-CH(OMOM)-CH=CH-R	>55		88TL1657, 89T1043
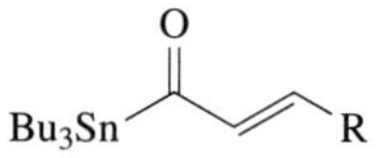 Bu$_3$Sn-C(O)-CH=CH-R	i, LiAlH$_4$, Chirald; ii, BOM-Cl, Pr$^i_2$NEt	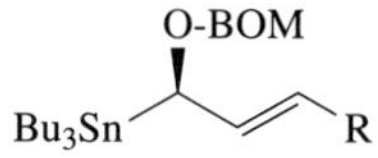 Bu$_3$Sn-CH(OBOM)-CH=CH-R	>60	>75	89TL2183

Chirald = (2S,3R)-4-dimethylamino-1,2-diphenyl-3-methylbutan-2-ol.

(ii) From aldehydes and ketones

The successful addition of tributylstannyllithium to alkyl or aryl aldehydes in high yields was reported by Still in the late 1970s in an improved preparation over existing methodology ⟨78JA1481⟩. The reagent is conveniently prepared at 0 °C by deprotonation of tributyltin hydride with LDA in THF. Addition to aldehydes at −78 °C is rapid and proceeds in high yields. The resulting stannyl alcohols, although purifiable by chromatography, are relatively unstable, and consequently the crude tin adducts are typically protected during work-up (Scheme 45). The less hindered trimethylstannyllithium nucleophile has been used to effect addition to ketones. A representative range of carbonyl precursors is shown in Table 16.

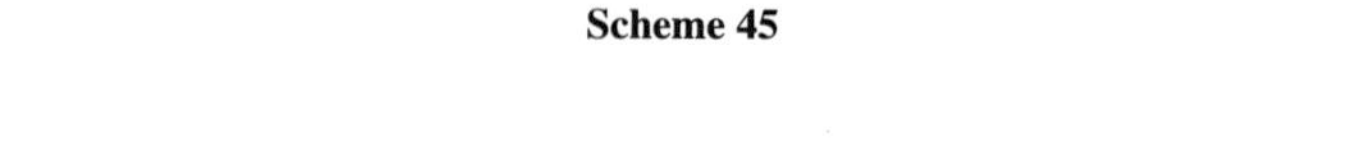

R-CHO $\xrightarrow{Bu_3SnLi}$ R-CH(OH)-SnBu$_3$ $\xrightarrow{protection}$ R-CH(OP)-SnBu$_3$

Scheme 45

There is scant mention of the analogous additions of tin Grignard reagents to carbonyl compounds: the addition of tributylstannylmagnesium chloride to aldehydes and ketones with derivatization of the α-stannyl alcohols as acetates and thiocarbonates prior to thermolysis has been investigated as a route to vinylstannanes (Equation (64)) ⟨87JOM(331)181⟩; *in situ* generation of the same Grignard reagent effected addition to the cyclobutanone (**102**) (Equation (65)) ⟨83JA625⟩.

Table 16 C-Stannylation of aldehydes and ketones.

Carbonyl precursor	Protecting agent	Product	Yield (%)	Ref.
$(CH_2O)_n$	$CH_2(OMe)_2$	$Bu_3Sn\diagup O\diagup OMe$	60	88JOC4131
$n\text{-}C_6H_{13}CHO$	EtO-CH(Me)-Cl	$n\text{-}C_6H_{13}$-CH(O-EE)-SnBu$_3$	97	78JA1481
crotonaldehyde (MeCH=CHCHO)	$ClCH_2OMe$	MeCH=CH-CH(O-MOM)-SnBu$_3$	82	82CC1115
Cyclohexyl-CHO	None	Cyclohexyl-CH(SnBu$_3$)(OH)		83TL4257
5-TMS-furan-3-CHO	EtO-CH(Me)-Cl	5-TMS-furan-3-CH(O-EE)(SnBu$_3$)	95	85TL5827
Cyclohexanone	TMS-CN	1-(TMS-O)-1-(SnMe$_3$)-cyclohexane		88JOC2878

EE = ethoxyethyl.

$$\text{Bu}^t\text{COMe} \xrightarrow[>74\%]{\text{i, Bu}_3\text{SnMgCl then H}_2\text{O} \\ \text{ii, TolOC(S)Cl}} \text{Bu}_3\text{Sn-C(Me)(Bu}^t\text{)-OC(S)OTol} \qquad (64)$$

$$\underset{(\mathbf{102})}{\text{3,3-dimethoxycyclobutanone}} \xrightarrow[68\%]{\text{i, Bu}_3\text{SnLi, MgCl}_2\text{, THF, }-70\,°\text{C} \\ \text{ii, MsCl}} \text{Bu}_3\text{Sn-C(OMs)-cyclobutane-C(OMe)}_2 \qquad (65)$$

(iii) From tributylstannylmethyl iodide

In his original paper ⟨78JA1481⟩, Still discusses briefly two further approaches to α-alkoxystannane construction. In the first of these, deprotonation of benzyl alcohol with sodium or potassium hydride produces a salt which will displace iodide from tributylstannylmethyl iodide (**103**) to give the corresponding α-alkoxystannane in 81% yield (Equation (66)). In general, both primary and secondary alcohols react in high yields; more acidic hydroxy compounds (e.g., phenol) can be coupled in the presence of potassium carbonate alone ⟨82JOC5051⟩.

$$\text{PhCH}_2\text{OH} \xrightarrow[81\%]{\text{i, NaH} \\ \text{ii, Bu}_3\text{SnCH}_2\text{I (103)}} \text{PhCH}_2\text{-O-CH}_2\text{-SnBu}_3 \qquad (66)$$

(iv) From tributyltin chloride

In the second alternative approach to α-alkoxystannane construction, Still added benzyloxymethylmagnesium chloride to tributyltin chloride ⟨78JA1481⟩ in unspecified yield (Equation (67)). There are scarce few further examples of this process in the literature; indeed, there are very few direct references to the required α-alkoxy Grignard reagents (see Section 4.08.2.1.2).

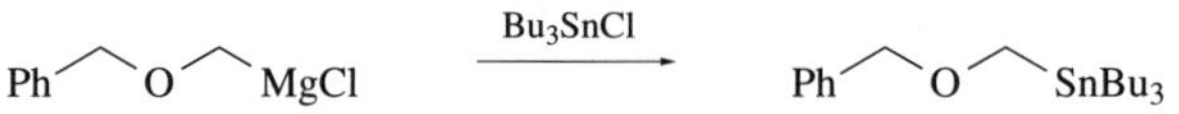

Tributyltin chloride has been used as the electrophilic component in reactions with α-alkoxy-lithium nucleophiles; examples include the quenching of (**104**) ⟨78JOC4255⟩ and (**105**) (Equations (68) and (69), respectively) ⟨83TL3165⟩.

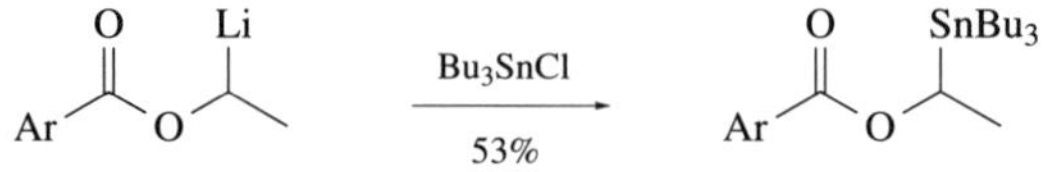

(v) From tributylstannyl acetals

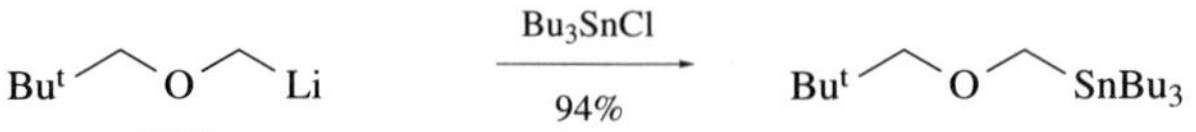

A novel route to substituted α-alkoxystannanes has been pioneered by Quintard and co-workers, who showed that treatment of a stannyl acetal (**106**) with acetyl chloride produced the useful α-chloroalkoxystannane (**107**) ⟨81JOM(212)C31⟩. Although only moderately stable, this highly functionalized material reacts readily with Grignard reagents to give α-substituted alkoxystannanes ⟨83JOC1559⟩ or undergoes reduction of the chlorine to the parent alkoxystannanes (Scheme 46) ⟨85BSF787⟩. (The stannyl acetals were in turn prepared efficiently from the reaction between tri-butylstannylmagnesium chloride and an *ortho*-ester.) An alternative route to the α-substituted alkoxystannanes avoiding the unstable α-chloro intermediate (**107**) involves treatment of the stannyl acetals with an alkenyl- or alkynylaluminum ⟨87CC29⟩. Direct substitution of an alkoxy group occurs in high yield with no propensity for transmetallation of the existing tin. Quintard also showed that chiral cyclic stannyl acetals can be ring opened stereoselectively with organometallics in the presence of a Lewis acid ⟨92JOM(437)C19⟩; the highest selectivities were achieved with organocopper reagents. Representative is the reaction of the acetal (**108**) with dimethylcopperlithium (Equation (70)); optimum yields were achieved at −78 °C, and although selectivity was generally improved at −100 °C, the yields were diminished.

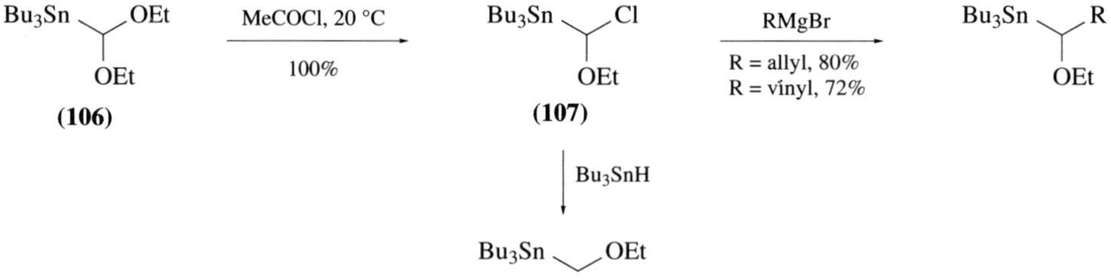

Scheme 46

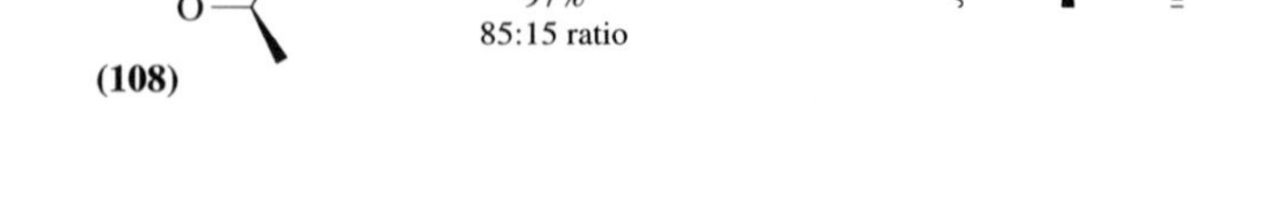

(vi) Miscellaneous methods

(a) From Fischer carbene complexes. Nakamura and co-workers have reported that Fischer-type chromium carbene complexes insert into the Sn—H bond of trialkylstannanes ⟨92JA9715⟩. When the complexes bear an α-stereogenic centre, considerable levels of 1,2-asymmetric induction are

observed. Thus, for example, the complex (**109**) reacts with tributyltin hydride in the presence of pyridine when refluxed in hexane; after filtration and chromatography, the α-alkoxystannane (**110**) is obtained in 81% yield (Equation (71)).

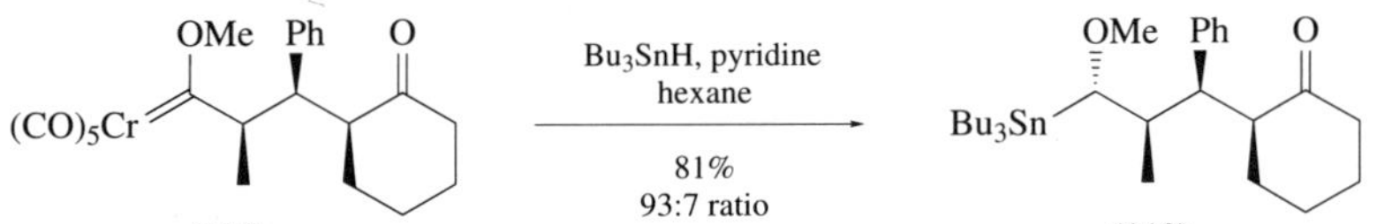

$$\underset{(\mathbf{109})}{(CO)_5Cr\!=\!\!\!\begin{array}{c}OMe\ Ph\ O\\|\ |\ \|\\ \end{array}} \xrightarrow[\substack{81\% \\ 93:7\ ratio}]{Bu_3SnH,\ pyridine\ hexane} \underset{(\mathbf{110})}{Bu_3Sn\!-\!\!\!\begin{array}{c}OMe\ Ph\ O\\|\ |\ \|\\ \end{array}} \quad (71)$$

(b) From α-chloroboronic esters. The conversion of the α-chloroboronic ester (**111**) into the chiral α-alkoxystannane (**112**) can be effected in two high-yielding steps with retention of configuration (Scheme 47) ⟨89JA4399⟩. Overall diastereomeric and enantiomeric excesses are uniformly very high, providing DIPED (optically active 2,5-dimethyl-3,4-hexanediol) is used as the chiral auxiliary; indeed, the authors claim this method gives higher levels of chiral control regardless of structure than BINAL-H-type reductions of acylstannanes (see Section 4.08.2.1.6.i).

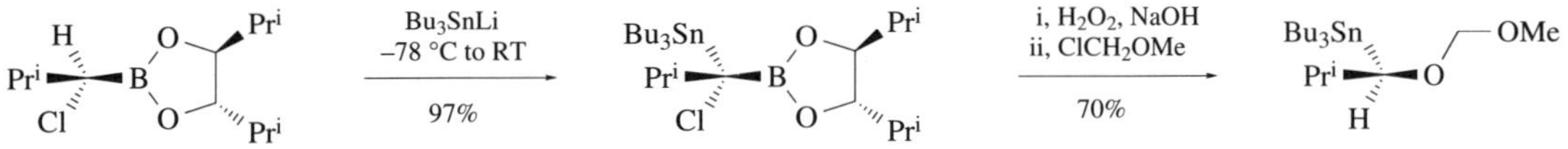

Scheme 47

(c) From chloromethyl ethers. The simple substitution of a chloromethyl ether with a tin nucleophile has little mention in the literature. Clibze and Reist obtained benzyloxymethylstannane (**113**) in 77% yield upon treatment of benzyl chloromethyl ether with tributylstannyllithium (Equation (72)) ⟨92MI 408-01⟩. Quintard and co-workers treated chloromethyl methyl ether with tributylstannylmagnesium chloride on a 0.5 mol scale to give the corresponding ether (**114**) in 70% yield (Equation (73)) ⟨85BSF787⟩. Finally, Corey and Eckrich, in a related process, generated the nucleophilic complex (**115**) from tin(II) chloride and lithium bromide; reaction with alkyl or aryl chloromethyl ethers gave unstable (alkoxymethyl)trihalostannanes, such as (**116**) (Scheme 48) ⟨83TL3163⟩; these intermediates could be transmetallated *in situ* with four equivalents of *n*-butyllithium to the corresponding α-alkoxylithiums (see Section 4.08.2.1.1.ii).

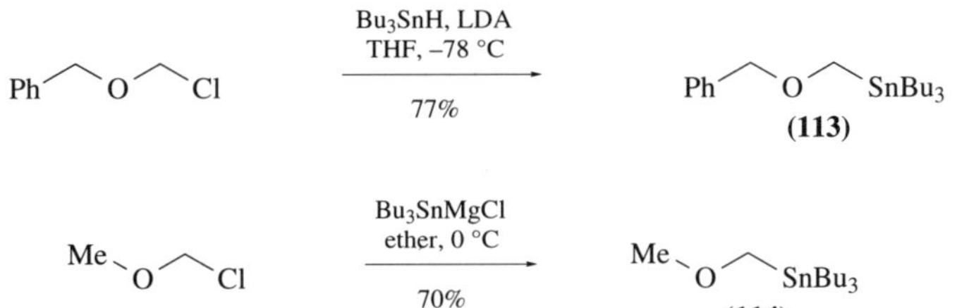

$$Ph\!\frown\!O\!\frown\!Cl \xrightarrow[77\%]{Bu_3SnH,\ LDA\ THF,\ -78\ °C} Ph\!\frown\!O\!\frown\!SnBu_3 \quad (72)$$
(**113**)

$$Me\!\frown\!O\!\frown\!Cl \xrightarrow[70\%]{Bu_3SnMgCl\ ether,\ 0\ °C} Me\!\frown\!O\!\frown\!SnBu_3 \quad (73)$$
(**114**)

$$SnCl_2 + LiBr \xrightarrow{THF} \underset{(\mathbf{115})}{Li^+BrCl_2Sn^-} \xrightarrow{Ph\frown O\frown Cl} \underset{(\mathbf{116})}{Ph\!\frown\!O\!\frown\!SnCl_2Br}$$

Scheme 48

(d) Kinetic resolution of α-hydroxystannanes. Itoh and Ohta reported the hydrolysis of α-acyloxystannanes with lipase P (*Pseudomonas sp.*) in water/acetone. The resulting *ee* values and overall conversions were, however, in general only modest ⟨90TL6407⟩. Subsequently, Chong and Mar showed that it was better to attempt the reverse process of enzymatic esterification in an organic solvent, attributing poor results from hydrolysis to the instability of the α-hydroxystannane under the aqueous reaction conditions ⟨91TL5683⟩. Simple α-alkyl-α-hydroxytrimethylstannanes or triethylstannanes were esterified in the presence of porcine pancreatic lipase (PPL) in ether. The enantiomeric purity of the product ester and recovered alcohol were high, although the extent of

conversion was still modest. Noteworthy is the conversion of the racemate (**117**) into the ester (**118**) and the alcohol (**119**) (Equation (74)). The reaction failed with tributylstannanes.

$$
\begin{array}{c}
\text{OH} \\
\diagup\!\!\diagdown \\
\text{SnMe}_3 \\
(\mathbf{117})
\end{array}
\xrightarrow[\text{BuCO}_2\text{CH}_2\text{CF}_3]{\text{PPL, ether, 25 °C}}
\begin{array}{c}
\text{O-COBu} \\
\diagup\!\!\diagdown \\
\text{SnMe}_3 \\
(\mathbf{118}) \\
38\%,\ ee = 98\%
\end{array}
+
\begin{array}{c}
\text{OH} \\
\diagup\!\!\diagdown \\
\text{SnMe}_3 \\
(\mathbf{119}) \\
41\%,\ ee = 97\%
\end{array}
\quad (74)
$$

4.08.2.1.7 Lead

Following the initial discovery that tetraalkyllead compounds add smoothly to aldehydes, Yamamoto and co-workers showed that α-alkoxylead species react with aldehydes in a stereodivergent fashion dependent on the reaction conditions ⟨90JA6118, 92JOC2981⟩. The lead compounds were prepared by transmetallation of the corresponding organostannanes with *n*-butyllithium followed by tributyllead bromide (Equation (75)). Although relatively unstable at room temperature, purification of the α-alkoxylead derivatives was possible by rapid column chromatography on silica gel. In the later paper ⟨92JOC2981⟩, the displacement illustrated in Equation (76) is cited as a more convenient synthetic approach.

$$
\begin{array}{c}
\text{OMe} \\
R\diagup\!\!\diagdown\text{SnBu}_3
\end{array}
\xrightarrow[\text{ii, Bu}_3\text{PbBr}]{\text{i, Bu}^n\text{Li}}
\begin{array}{c}
\text{OMe} \\
R\diagup\!\!\diagdown\text{PbBu}_3
\end{array}
\quad (75)
$$

R = C$_7$H$_{15}$, 77%; *cyclo*-C$_6$H$_{11}$, 87%

$$
\text{(allyl-O-CH(Cl)-CH}_2\text{CHMe}_2\text{)} \xrightarrow[\text{THF, }-78\text{ °C}]{\text{Bu}_3\text{PbLi}} \text{(allyl-O-CH(PbBu}_3\text{)-CH}_2\text{CHMe}_2\text{)}
\quad (76)
$$

61%

4.08.2.2 Functions Bearing Sulfur—R$^1_2$C(SR^2)M, etc.

The following sections concern the synthesis of systems with a metal formally attached to a carbon bearing a sulfur function. In reality, the actual position of attachment of the metal may be elsewhere; for example, in sulfoxide and sulfone α-anions, the metal may be more closely associated with one of the sulfur oxygens ⟨89AG(E)277⟩. From a synthetic viewpoint, and for the purposes of this review, the metal is regarded as bound to the α carbon atom whence the corresponding hydrogen was removed.

The capacity of sulfur in its various oxidation states to enhance the acidity of adjacent C—H bonds was recognized as early as 1889; exploitation of this effect has since been widespread in organic chemistry. Carbanions adjacent to sulfur have been included in general reviews on sulfur chemistry: Block ⟨B-78MI 408-01⟩, Wolfe ⟨B-85MI 408-01⟩ and, more recently, Ogura ⟨91COS(1)505⟩ cover all possibilities of sulfur-stabilized anions. Barrett has briefly reviewed α-sulfenyl carbanions ⟨79COC(3)33⟩; Magnus ⟨77T2019⟩, Durst ⟨79COC(3)171⟩ and, more recently, Simpkins ⟨B-93MI 408-01⟩ have discussed α-anions in specific reviews on sulfones; Durst ⟨79COC(3)121⟩ has also discussed carbanions adjacent to sulfoxides, and anions associated with both these higher oxidation states appear in a volume on sulfoxides and sulfones ⟨B-88MI 408-01⟩. Carbanions adjacent to sulfur are also included in more general reviews on the metal–carbon bond: since Peterson's now dated review ⟨72MI 408-02⟩, Krief has written extensively on synthetic methods using α-heterosubstituted organometallics ⟨80T2531⟩, with extensive coverage of the preparation and properties of α-thio carbanions, and of alkylations of sulfur- and selenium-containing carbanions ⟨91COS(3)85⟩, again with a comprehensive coverage of α-thio carbanions. A good general review of the metal–carbon bond with numerous references to sulfur systems is by Wardell ⟨B-87MI 408-02⟩. In view of the many reviews in this area, a summary of the main synthetic routes, briefly exemplified, is given below: the reader is directed to the above references for further detail.

4.08.2.2.1 Lithium

(i) By direct deprotonation

(a) Sulfides. Among the heralds of the modern era in organosulfur chemistry is Corey, who published a series of papers in the 1960s on sulfur-stabilized carbanion chemistry. In one of these with Seebach, the successful lithiation of thioanisole with *n*-butyllithium/1,4-diazabicyclo[2.2.2]octane (DABCO) in THF in about 97% yield was achieved (Table 17, entry 1) ⟨66JOC4097⟩. The reaction of the lithio derivative with benzophenone gave a β-hydroxy sulfide in 93% yield. Shortly after, Peterson published the lithiation of dimethyl sulfide with *n*-butyllithium/TMEDA in THF (entry 2) ⟨67JOC1717, 71USP3597463⟩; quenching with benzaldehyde gave the corresponding β-hydroxy sulfide in 84% yield.

These preparations belie the considerable difficulty in achieving selective, clean deprotonation in more complex alkyl and aryl sulfides (e.g., ⟨77TL1961, 86TL4625⟩). (For a general, theoretical discussion of the stability of carbanions α to sulfur, and for comparisons with oxygen, see ⟨84JA6467⟩.) Consequently, instances of deprotonation without the presence of additional sources of carbanion stabilization and/or metal coordination are rare. Table 17 illustrates deprotonation of sulfides in the presence and absence of typical further stabilization.

(b) Sulfoxides. Lithiation of sulfoxides is facilitated relatively to sulfides by the presence of the oxygen; moreover, the lithium atom is likely to be in closer association with it rather than with the α carbon (see opening remarks to this section above). Early pioneers again included Corey, who with Chaykovsky achieved a high-yielding synthesis of a metallated sulfoxide and demonstrated its potential as a synthetic agent ⟨65JA1345⟩. Essentially, dimethyl sulfoxide is heated with sodium hydride at about 70 °C (Table 18, entry 1). Since this report, α-lithio sulfoxides have been routinely prepared by the reaction of the appropriate sulfoxides in an inert solvent with a variety of alkyl-lithiums, usually methyl and *n*-butyl (Table 18). Importantly, Durst has shown that sulfoxides treated with alkyllithiums can undergo two competing reactions: abstraction of an α hydrogen and carbon–sulfur bond cleavage ⟨74CJC761⟩. The relative amounts of products formed by the two pathways are largely determined by the substrate structure and the choice of base: to effect deprotonation over bond cleavage, methyllithium or, better, LDA is the preferred base.

The greater interest in α-sulfinyl carbanions has arisen from the chirality of the $R^1R^2S(O)$ group, and the exploitation of such a chiral species in synthesis (for reviews, see ⟨B-78MI 408-01, 81S185, B-85MI 408-01, 91COS(1)505⟩). Early reports showed that the diastereotopic methylene protons of benzyl methyl sulfoxide underwent hydrogen–deuterium exchange at unequal rates, the relative ratio being 14:1 ⟨65JA5498⟩; in conformationally rigid systems, the rates of base-catalysed hydrogen–deuterium exchange of the diastereotopic hydrogens α to a chiral sulfoxide can differ by up to 10^3 ⟨72JA8795⟩. Solladie's review ⟨81S185⟩ contains a detailed discussion of the factors influencing generation and reaction of carbanions from chiral sulfoxides. A more recent paper applies the Hard/Soft Acid Base (HSAB) principle to the understanding of the stereochemical course of sulfoxide deprotonation and subsequent electrophilic capture, an overall process that is dependent on the reaction conditions (for example, different results can be obtained in protic and aprotic solvents) ⟨87JOC1414⟩.

(c) Sulfones. Hydrogens located α to a sulfone are thermodynamically more acidic than those α to sulfoxides. In addition, it has been demonstrated in the particular case of phenyl sulfones that stabilizing effects operate that are larger than those expected of polar contributions alone ⟨76JOC1883⟩. The result is that removal of hydrogens α to a sulfone is a general and high-yielding process. Truce and Buser reported an early example of the deprotonation of dimethyl sulfone with an alkyllithium in 1954 (Table 19, entry 1) ⟨54JA3577⟩. The generation of α-lithio sulfones is nowadays usually a routine matter involving treatment of the appropriate sulfones with *n*-butyllithium or LDA at low temperatures in an aprotic solvent (Table 19; see ⟨B-93MI 408-01⟩ for many more examples).

Acyclic sulfones having a chiral α carbon atom largely retain their configuration upon hydrogen–deuterium exchange. The debate over α-sulfonyl carbanionic structure is beyond the scope of this review (see ⟨89AG(E)277, 91COS(1)505, B-93MI 408-01⟩ for coverage); however, the impact of the findings has a synthetic bearing, for example in the context of the stereochemistry of carbanion formation involving diastereotopic hydrogens in conformationally restricted systems. The last named review contains an extensive discussion of these factors.

(d) Other oxidized sulfur systems. Deprotonation has been reported adjacent to sulfinamides, sulfonamides, sulfoximines, and sulfonyl esters. Typically, *n*-butyllithium or LDA is used in an aprotic solvent at low temperatures; representative examples are shown in Table 20.

Table 17 α-Sulfenyl carbanions by deprotonation of sulfides.

Entry	Sulfide	Conditions	Product	Ref.
1	PhSMe	Bu^{n}Li, dabco, THF, 0 °C	PhS–CH$_2$–Li	66JOC4097
2	Me$_2$S	Bu^{n}Li, TMEDA, hexane, 0 °C	MeS–CH$_2$–Li	67JOC1717
3	PhS–CH$_2$CH(CH$_3$)$_2$ (PhS-iBu)	Li, HMPA, THF, –78 °C	PhS–CLi(CH(CH$_3$)$_2$)	77TL1961
4	PhSMe	Bu^{n}Li, ether, heat, 15 h	PhS–CH$_2$–Li	83JCS(P1)1131
5	2-(methylthio)tetrahydropyran (O–CH(SMe)– in ring)	Bu^{t}Li, THF, HMPA, –90 °C	tetrahydropyranyl–S–CH$_2$Li	85JA6729
6	Pr$^i_2$N–C(O)–CH(Me)–(CH$_2$)$_n$–SPh	Bu^{s}Li, THF, TMEDA (n = 0 or 1)	Pr$^i_2$N–C(O)–CH(Me)–(CH$_2$)$_n$–CHLi–SPh	87JA5403
7	Ar–C(O)–S–CH$_2$R	Bu^{n}Li, THF, –78 °C (Ar = 2,4,6-triethyl- or 2,4,6-triisopropylphenyl)	Ar–C(O)–S–CHLi–R	78JA5428
8	Het–S–Me	Bu^{n}Li or LDA, –78 °C	Het–S–CH$_2$Li	Het = thiazoline ⟨78CRV275⟩, oxazoline ⟨78CRV275⟩, N-methylimidazole ⟨84CPB1829⟩, pyridine ⟨84CPB1829⟩, benzothiazole ⟨87JOC844⟩
9	2-(methylthio)-3,4,5,6-tetrahydropyridine (N=C–SMe)	Bu^{n}Li, –78 °C, THF	corresponding –S–CH$_2$Li	80JA7929
10	R–CH$_2$–S–C(=S)–X	LDA, –78 °C, THF	R–CHLi–S–C(=S)–X	X = NMe$_2$ ⟨74TL3625⟩, OR ⟨76S413⟩
11	R^1–CH$_2$–S–C(=NR^2)–SMe	LDA, –78 °C, THF	R^1–CH$_2$–S–C(=NR^2)–S–CH$_2$Li	80LA1765
12	PhS–CH$_2$–C(Me)=CH$_2$	LDA, –78 °C, THF	PhS–CHLi–C(Me)=CH$_2$	87CPB1413
13	PhS–CH$_2$–C≡CH	2 Bu^{n}Li, –60 °C, TMEDA, THF	PhS–CLi(–C≡C–Li)	81JOC5041
14	PhCH$_2$SH	2 Bu^{n}Li, THF, TMEDA, –5 °C	Ph–CH(–S$^-$)·2Li$^+$	74AG(E)202, 77CB1833
15	CH$_2$=CH–CH$_2$–SH	2 Bu^{n}Li, THF, TMEDA, 0 °C	CH$_2$=CH–CH(–S$^-$)·2Li$^+$	74AG(E)479, 79CB1420

dabco = 1,4-diazabicyclo[2.2.2]octane.

(ii) By other methods

(a) From vinyl sulfides and thioketones. In work related to Seebach's investigation of nucleophilic addition to ketene dithioacetals, Ager showed that addition of organolithium species to vinyl sulfides yields the corresponding α-lithio sulfides ⟨81TL587, 81TL2923⟩ (see also ⟨82CSR493⟩). (At low temperatures, α-deprotonation is observed.) For instance, phenyl vinyl sulfide (**120**) can be treated with alkyl- or aryllithiums in ether at 0 °C in the presence of TMEDA, affording the corresponding

Table 18 α-Sulfinyl carbanions by deprotonation of sulfoxides.

Entry	Sulfoxide	Conditions	Product	Ref.
1	Me-S(O)-Me	NaH, 70 °C	Me-S(O)-CH$_2$Na	65JA1345
2	Me-S(O)-Me	Bu^{n}Li, THF	Me-S(O)-CH$_2$Li	65JA1345
3	Ph-S(O)-Me	LDA, THF, −78 °C	Ph-S(O)-CH$_2$Li	75TL2017
4	Bu^t-S(O)-CH$_2$R	MeLi, THF, −60 °C	Bu^t-S(O)-CHRLi	73JA3420
5	Ph-S(O)-CH$_2$-C(Me)=CH$_2$	Bu^{n}Li, THF	Ph-S(O)-CH(Li)-C(Me)=CH$_2$	79TL1783

Table 19 α-Sulfonyl Carbanions by deprotonation of sulfones.

Entry	Sulfone	Conditions	Product	Ref.
1	Me-SO$_2$-Me	Bu^{n}Li, benzene	Me-SO$_2$-CH$_2$Li	54JA3577
2	norbornenyl-SO$_2$	Bu^{n}Li, THF	α-lithio norbornenyl sulfone	87JOC809
3	4-MeC$_6$H$_4$-SO$_2$-CH$_2$CH$_2$CO$_2$H	2 Bu^{n}Li, THF	4-MeC$_6$H$_4$-SO$_2$-CH(Li)CH$_2$CO$_2$Li	92TL4065
4	Ph-SO$_2$-Me	LDA, THF	Ph-SO$_2$-CH$_2$Li	88JA8671
5	Ph-SO$_2$-cyclohexyl	Bu^{n}Li, THF	Ph-SO$_2$-C(Li)(cyclohexyl)	91CC297, 91SC(21)1675
6	Me-SO$_2$-Me	2 LiNH$_2$, liquid NH$_3$	LiCH$_2$-SO$_2$-CH$_2$Li	73JOM(59)53

α-lithio sulfides (**121**) in high yields (Equation (77)). Best results were obtained by slowly adding (typically over 1 h) a solution of the alkene to the ethereal organolithium. Organometallic reagents have also been added to β-silyl vinyl sulfones to give related α-thio metal intermediates ⟨85JOM(285)121⟩.

$$CH_2=CH-SPh \xrightarrow[>85\%]{Bu^nLi,\ ether\ TMEDA,\ 0\ °C} Bu^n-CH_2-CH(Li)-SPh \qquad (77)$$

(**120**) → (**121**)

Table 20 Carbanions adjacent to sulfur by deprotonation of miscellaneous precursors.

Sulfur precursor	Conditions	Product	Ref.
R–CH$_2$–S(=O)–NHPh	2 Bu^{n}Li, THF	R–CH(Li)–S(=O)–N(Li)–Ph	84JOC1700, 66JA5656
R–CH$_2$–S(=O)$_2$–NR$^1_2$	LDA, THF, –20 °C	R–CH(Li)–S(=O)$_2$–NR$^1_2$	91MI 408-01
Ph–S(=O)(=NR)–CH$_2$R	Bu^{n}Li, THF, 0 °C	Ph–S(=O)(=NR)–CH(Li)R	85MI 408-02
R–CH$_2$–S(=O)$_2$–OMe	Bu^{n}Li, THF, –78 °C	R–CH(Li)–S(=O)$_2$–OMe	68JA5548

The reaction of phenyllithium with thiobenzophenone (**122**) interestingly gives the α-lithio sulfide (**123**) by thiophilic addition (Equation (78)) ⟨72JA597⟩. A similar pathway is observed with vinyllithium nucleophiles ⟨75JOC3052⟩. In the latter paper, Beak and co-workers discuss the scope and stereochemistry of the process, which extends to aliphatic as well as aromatic thioketones, dithioesters and trithiocarbonates.

$$\text{Ph}_2\text{C=S} \xrightarrow[>70\%]{\text{PhLi, ether, RT}} \text{Ph}_2\text{C(Li)-SPh} \qquad (78)$$

(**122**) → (**123**)

(b) From diphenyl dithioacetals and ketals. Reductive lithiation of diphenyl dithioacetals with LN ⟨81TL2923⟩ or LDMAN offers an alternative approach to α-lithiophenylsulfides. LDMAN is preferred ⟨84JA3245⟩ since, after the reaction of the α-lithio sulfide with some electrophile, the tedious separation of naphthalene from the reaction product is obviated; the 1-(dimethylamino)naphthalene can be simply removed by an acid wash. Typically, the organosulfur precursor is added at low temperature to LDMAN in THF. The method has the advantage that it can provide lithiated species not available by direct deprotonation, for example tertiary α-lithio sulfides (Equation (79)).

$$\text{cyclohexyl(SPh)}_2 \xrightarrow[>71\%]{\text{LDMAN, THF, –78 °C}} \text{cyclohexyl(SPh)(Li)} \qquad (79)$$

(c) By tin–lithium or selenium–lithium exchange. Fuchs and co-workers have assessed the tin–lithium exchange of α-tributylstannyl sulfones as a route to α-lithio sulfones. The tin precursor (**124**) was treated with *n*-butyllithium at –78 °C to give the corresponding α-lithiated derivative in almost quantitative yield (Equation (80)) ⟨91SC(21)1675⟩. A similar reaction with the tributyltin sulfide (**125**) gives the α-lithiated sulfide (**126**) in 85% yield (Equation (81)) ⟨72MI 408-02⟩. The cleavage of the C—Se bond in mixed sulfur/selenium acetals is facile and high-yielding at –78 °C, and allows the synthesis of α-phenylthio and α-methylthio alkyllithiums (Equation (82)) ⟨74AG(E)806, 80T2531⟩. In addition, the carbon bearing the lithium can be mono- or even disubstituted, providing access to secondary or tertiary organolithiums. Experimentally, the reaction is conducted to avoid an excess of the mixed acetal in the presence of the α-thio alkyllithium, since the latter metallates the former, by addition of the precursor to *n*-butyllithium. In some systems, particularly those derived from formaldehyde, this side-reaction is difficult to avoid ⟨80T2531⟩.

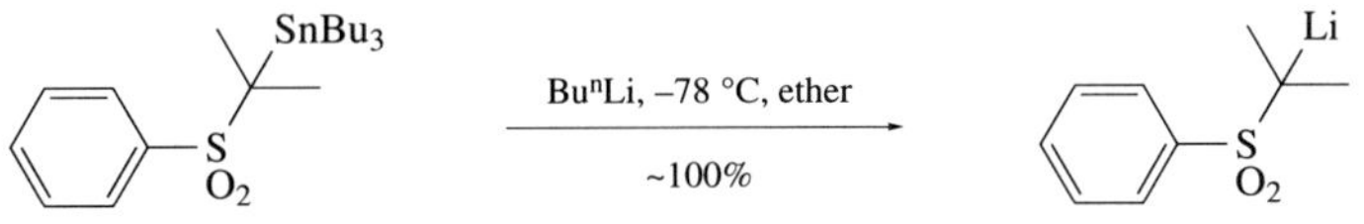

$$\text{(124)} \xrightarrow{\text{Bu}^n\text{Li, } -78\,°\text{C, ether}} \text{product} \quad \sim 100\% \tag{80}$$

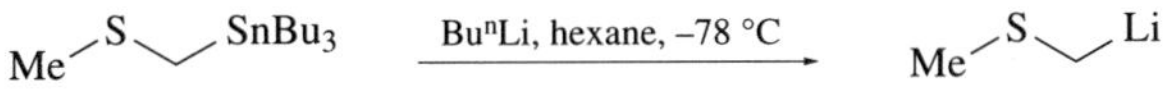

$$\text{Me-S-CH}_2\text{-SnBu}_3 \xrightarrow{\text{Bu}^n\text{Li, hexane, } -78\,°\text{C}} \text{Me-S-CH}_2\text{-Li} \tag{81}$$
$$\text{(125)} \qquad\qquad\qquad \text{(126)}$$

$$\text{PhS-CH(iPr)-SePh} \xrightarrow[96\%]{\text{Bu}^n\text{Li, THF, } -78\,°\text{C}} \text{PhS-CH(iPr)-Li} \tag{82}$$

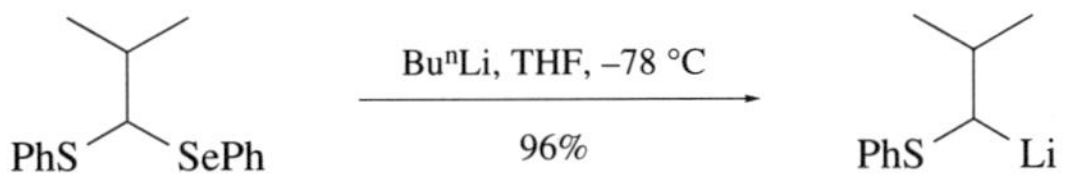

(d) By halogen–lithium exchange. There have been few reports of halogen–metal exchange α to a sulfur function. (The exchange of halogen for magnesium is discussed in the next section.) Krief and co-workers have reported the exchange shown in Equation (83) in moderate yield ⟨80T2531⟩. The presence of additional stabilizing groups, as in Equation (84), leads to increased yields ⟨75CB2368⟩.

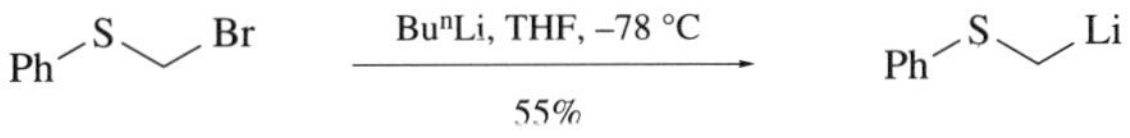

$$\text{Ph-S-CH}_2\text{-Br} \xrightarrow[55\%]{\text{Bu}^n\text{Li, THF, } -78\,°\text{C}} \text{Ph-S-CH}_2\text{-Li} \tag{83}$$

$$\text{(cyclopropane-Br, SPh)} \xrightarrow[92\%]{\text{Bu}^n\text{Li, THF, } -78\,°\text{C}} \text{(cyclopropane-Li, SPh)} \tag{84}$$

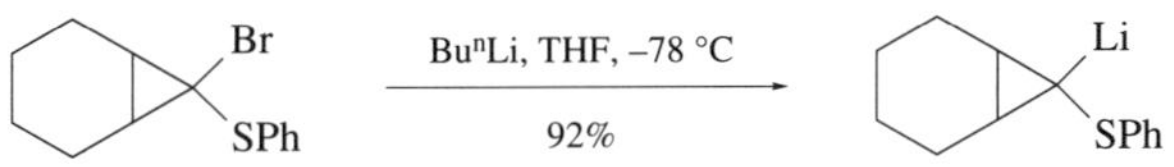

4.08.2.2.2 Beryllium or magnesium

Grignard reagent analogues derived from sulfones have been known since the 1930s, and can be synthesized by the reaction of a sulfone with a Grignard reagent. For example, Field treated methyl phenyl sulfone with ethylmagnesium bromide, and obtained the magnesio derivative in approximately 90% yield (Equation (85)) ⟨52JA3919⟩. Simpkin's book contains a useful summary of this chemistry ⟨B-93MI 408-01⟩. The first report of an α-sulfenyl alkylmagnesium reagent was by Normant and Castro, who prepared benzylthiomethylmagnesium chloride from the corresponding chloromethyl sulfide in THF in 50% yield ⟨64CR(259)830⟩. Later, Sakurai and co-workers obtained a Grignard reagent from chloromethyl methyl sulfide by treatment with magnesium activated with iodine and dibromoethene ⟨67CC889⟩. A black solution was obtained which reacted with TMS-Cl to give (**127**) in 33% yield (Scheme 49). The procedure was optimistically hailed by the authors as a useful synthetic method. Much later, Ogura and co-workers modified the reaction conditions (essentially by controlling the temperature between 10 °C and 20 °C) of Sakurai's experiment, and obtained the same Grignard reagent in yields above 90% (by titration) ⟨82CL1697⟩.

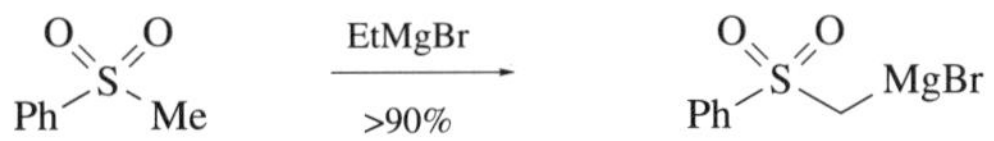

$$\text{Ph-SO}_2\text{-Me} \xrightarrow[>90\%]{\text{EtMgBr}} \text{Ph-SO}_2\text{-CH}_2\text{MgBr} \tag{85}$$

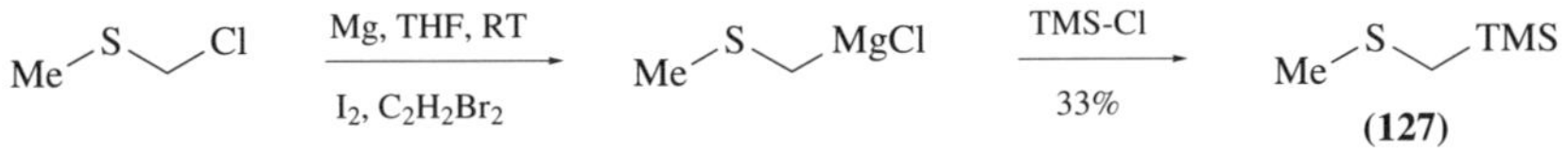

$$\text{Me-S-CH}_2\text{-Cl} \xrightarrow[\text{I}_2, \text{C}_2\text{H}_2\text{Br}_2]{\text{Mg, THF, RT}} \text{Me-S-CH}_2\text{-MgCl} \xrightarrow[33\%]{\text{TMS-Cl}} \text{Me-S-CH}_2\text{-TMS} \quad (\mathbf{127})$$

Scheme 49

Seebach and co-workers have described what amounts formally to an α-thio magnesium species. They transmetallated the thioacrolein dianion, originally prepared from thioacrolein with two equivalents of *n*-butyllithium (see Section 4.08.1.2.1.i), with magnesium bromide, whereupon subsequent reactions with electrophiles were confined to the α over the γ position ⟨76AG(E)437⟩.

The only report of an α-thio beryllium species is by Yamamoto ⟨87BCJ1189⟩, who obtained (**128**) in 89% yield upon mixing the ylide (**129**) with beryllium chloride in THF (Equation (86)).

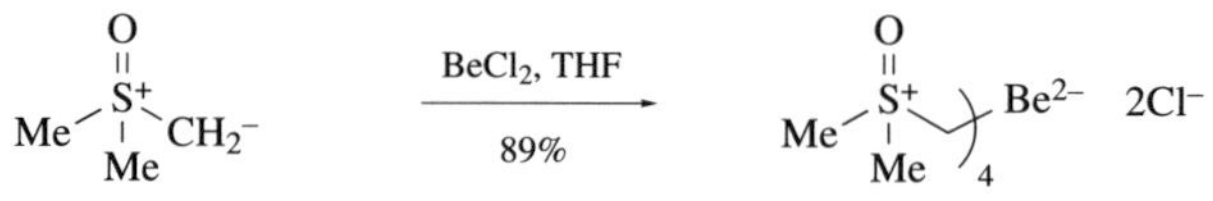

$$\underset{(129)}{\underset{Me}{\overset{O}{\underset{|}{\overset{||}{\underset{Me}{\overset{S^+}{-}}}}}}\text{CH}_2^-} \xrightarrow[89\%]{\text{BeCl}_2,\ \text{THF}} \underset{(128)}{\underset{Me}{\overset{O}{\underset{Me}{\overset{||}{\underset{}{\overset{S^+}{-}}}}}}\underset{4}{\underbrace{\ \ \ }}\text{Be}^{2-}\ \ 2\text{Cl}^-} \quad (86)$$

4.08.2.2.3 Titanium

In an isolated example, Seebach and co-workers describe the titanation of phenylthiomethyl-lithium (130) with trisisopropoxytitanium chloride (Equation (87)) in unspecified yield ⟨81HCA357⟩.

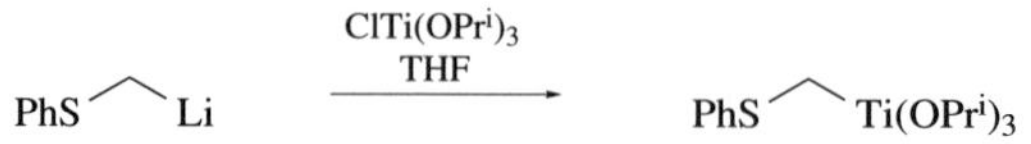

$$\underset{(130)}{\text{PhS}\frown\text{Li}} \xrightarrow[\text{THF}]{\text{ClTi(OPr}^i)_3} \text{PhS}\frown\text{Ti(OPr}^i)_3 \quad (87)$$

More recently, Lin et al. have synthesized a variety of related α-sulfinyl palladium species by treating sulfur ylides with a tetrachloropalladium(II) salt ⟨86JOM(315)135⟩.

4.08.2.2.4 Copper, gold or zinc

Johnson and Dhanoa have reported the development of a class of mixed homocuprates containing sulfinyl-stabilized carbanions as nontransferable ligands ⟨82CC358, 87JOC1885⟩. After lithiation of the corresponding sulfoxides at 0 °C in THF with n-butyllithium, treatment with copper(I) iodide gave the α-thio copper species (131), which could be alkylated between −20 °C and 0 °C by addition of a suitable organolithium, RLi. In subsequent reactions with electrophiles, only the R group was transferred (Scheme 50). Krief and co-workers have elaborated upon work originally detailed by Corey and Jautelat on the copper-promoted allylation of α-phenylthioalkyllithiums ⟨84IJ125⟩. Krief observed elimination of the phenylthio group down to −78 °C and only achieved successful allylation, with presumed cuprate intermediates, at −100 °C.

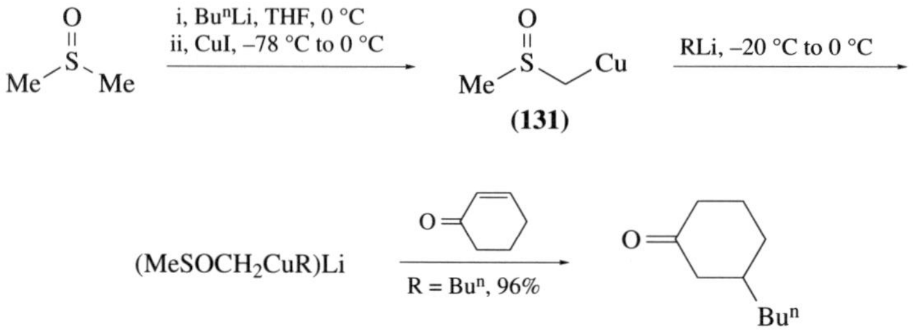

Scheme 50

Knochel and co-workers treated α-chloroalkyl phenyl sulfides (132) and the iodomethylthioester (133) with zinc in THF to give high yields of the corresponding zinc reagents (Schemes 51 and 52, respectively) ⟨92T2025⟩. The presence of the phenylthio function greatly facilitates the insertion of zinc into the carbon–halogen bond: in contrast, α-chloroalkyl alkyl sulfides did not insert zinc under the same conditions (see also Section 4.08.2.1.4 above for the related α-alkoxy zinc species). The zinc reagents themselves are relatively unreactive; transmetallation of these species in THF with CuCN·2LiCl (THF-soluble) gave the more reactive cuprates (134) and (135), which coupled in high yields with common electrophiles. In contrast with α-thio lithiums, these α-thio zinc reagents show high functional group tolerance (e.g., to additional ester and cyano groups) allowing for the preparation and transfer of highly substituted α-thio carbanions.

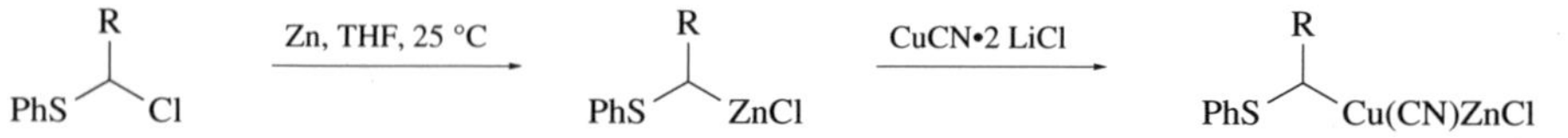

Scheme 51

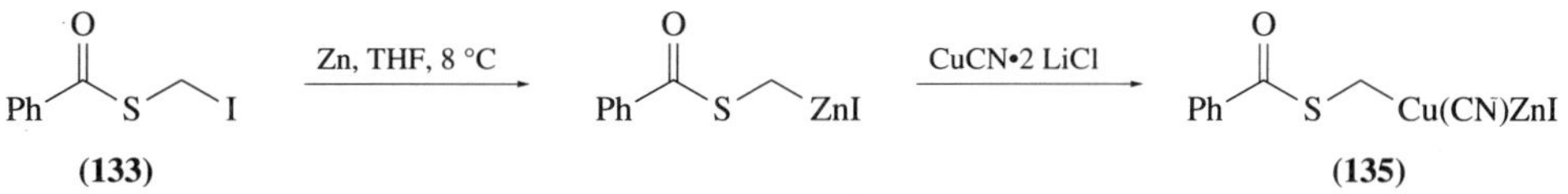

Scheme 52

Braun has transmetallated the α-lithio sulfoxide (**136**) with zinc chloride, and found improved diastereoselectivity over the lithio derivative in its addition to benzaldehyde (Scheme 53) ⟨84CB413⟩.

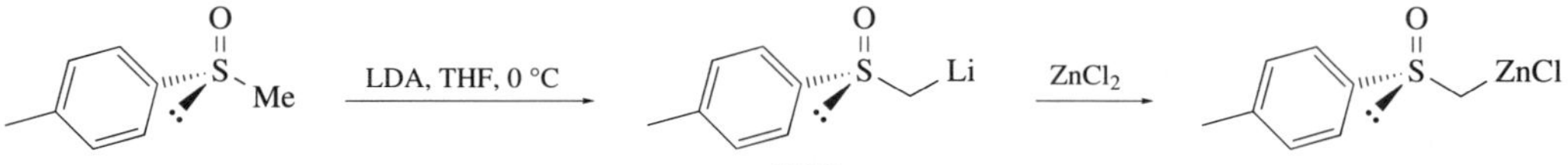

Scheme 53

Yamamoto has prepared a rare example of an α-thio gold species ⟨87BCJ1189⟩, obtaining (**137**) in 86% yield upon mixing the ylide (**138**) with triphenylphosphinegold chloride in THF (Equation (88)).

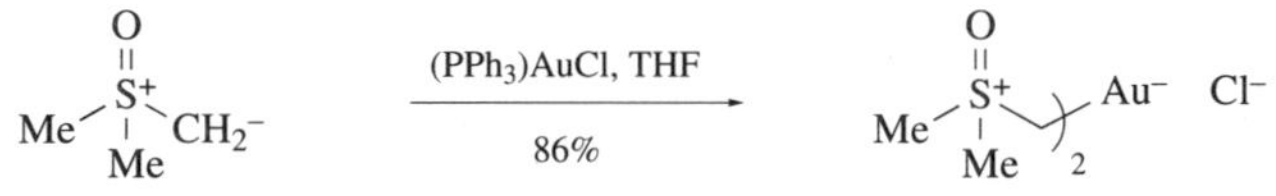

4.08.2.2.5 Aluminum, indium or gallium

Yamamoto has also treated the ylide (**138**) independently with aluminum, gallium and indium trichloride in THF ⟨87BCJ1189⟩ to give the corresponding α-thio metal species (**139**) in moderate to good yields (Equation (89)). Sonnek *et al.* have reported the treatment of tris(silylalkyl)aluminum compounds with sulfur trioxide to give α-thio aluminum intermediates which are hydrolysable to silyl alkanesulfonate derivatives ⟨80JOM(194)9⟩.

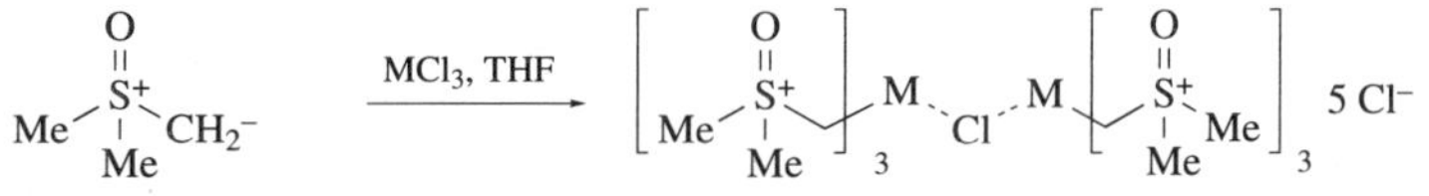

M = Al, 62%; In, 73%; Ga, 57%

4.08.2.2.6 Tin

The synthesis of compounds containing a trialkyltin group α to a sulfur moiety has been exemplified in two patents. In the first, the reaction between alkyl- or arylthiomethyllithiums, prepared from the corresponding sulfides and *n*-butyllithium/TMEDA, and trialkyltin chlorides is reported to give α-stannyl thioethers. For example, exposure of tributyltin chloride in ether to a solution of the methylthiomethyllithium·TMEDA complex (**140**) in hexane yields the α-stannyl thioether (**141**) in 80% yield (Equation (90)) ⟨71GEP2114367⟩. In the second, (sulfinylmethyl)trialkyltins were prepared from the reaction of trialkyltin amines with various dialkyl or alkyl aryl sulfoxides. For example, heating phenyl methyl sulfoxide and (dimethylamino)tributyltin at 100 °C for 20 h yielded (**142**) in 65% yield after distillation (Equation (91)) ⟨76USP3987191⟩.

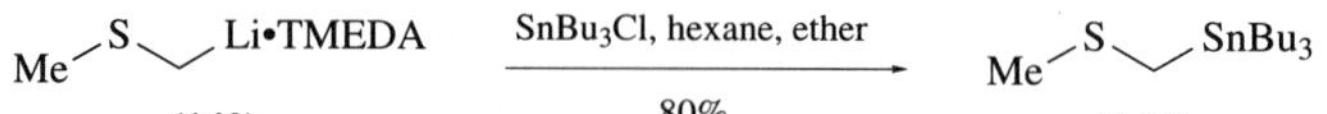

$$\text{Me}^{\diagdown}\text{S}^{\diagdown}\text{Li·TMEDA} \xrightarrow[80\%]{\text{SnBu}_3\text{Cl, hexane, ether}} \text{Me}^{\diagdown}\text{S}^{\diagdown}\text{SnBu}_3 \qquad (90)$$
$$\textbf{(140)} \qquad\qquad\qquad\qquad\qquad\qquad \textbf{(141)}$$

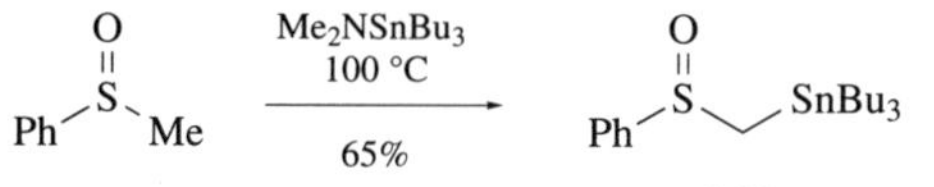

$$\underset{\textbf{}}{\text{Ph}\overset{\overset{\text{O}}{\|}}{\text{S}}\text{Me}} \xrightarrow[65\%]{\substack{\text{Me}_2\text{NSnBu}_3 \\ 100\,°\text{C}}} \underset{\textbf{(142)}}{\text{Ph}\overset{\overset{\text{O}}{\|}}{\text{S}}\diagdown\text{SnBu}_3} \qquad (91)$$

Other occurrences of α-thio tin compounds are rare in the literature; Ando *et al.* have inserted α-stannylcarbenes into C—S bonds ⟨79CC1121⟩, although the yields were low. Knochel and Normant have reported that upon treatment with potassium thiophenoxide, **(143)** reacts smoothly to afford the thioether **(144)** in 81% yield (Equation (92)) ⟨86TL1043⟩. Takeda and co-workers have treated trimethylsilyl enol ethers with α-tributylstannyl thioacetals to give β-phenylthio-β-tributylstannyl ketones ⟨92CL819⟩ in good to excellent yields. These compounds underwent elimination to β-tributylstannyl α,β-unsaturated ketones under basic conditions (Scheme 54).

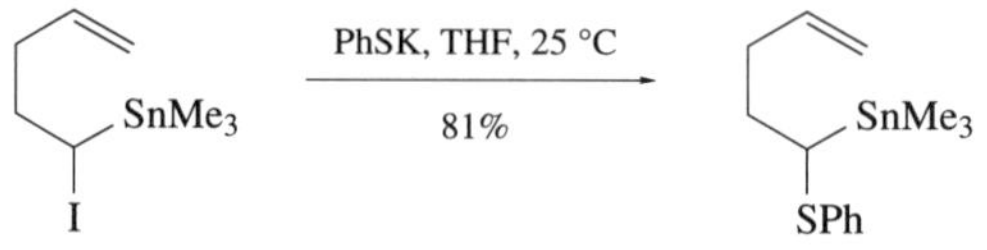

$$(92)$$

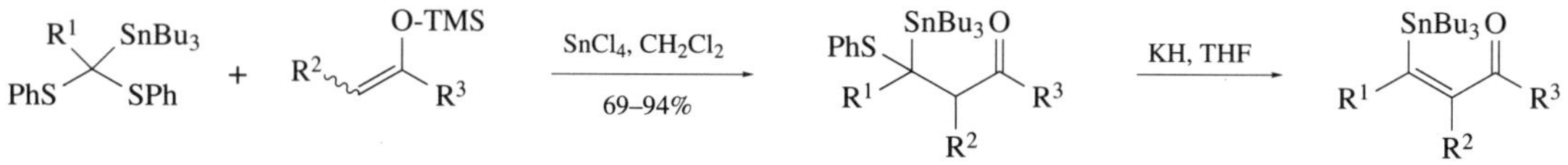

Scheme 54

4.08.2.3 Functions Bearing Selenium or Tellurium—$R^1{}_2C(SeR^2)M$, etc.

Carbanions adjacent to selenium (in all its oxidation states) have been the subject of numerous reviews: in comprehensive books on organoselenium chemistry both Paulmier ⟨B-86MI 408-01⟩ and Reich ⟨B-87MI 408-03⟩ have covered the preparation and reactivity of selenium-stabilized carbanions. Krief has reviewed both the synthesis and reactivity of selenium-stabilized carbanions ⟨B-87MI 408-04, 91COS(1)629⟩ and also specifically the subsequent alkylation of these species ⟨91COS(3)85⟩. He is also the author of an earlier general review concerning synthetic methods using α-heterosubstituted organometallics ⟨80T2531⟩, which contains extensive coverage of the preparation and properties of α-seleno carbanions. Carbanions adjacent to tellurium appear in Krief's review on selenium and tellurium ⟨B-87MI 408-04⟩. (Other good general reviews of the metal–carbon bond with references to selenium and tellurium systems are by Wardell ⟨B-87MI 408-02⟩ and the useful but now dated article by Peterson ⟨72MI 408-02⟩.) In view of the substantial amount of review material covering this area, a summary of the main synthetic routes is briefly exemplified below. The reader is directed to the above references for further details. Most sections below deal with anions adjacent to selenium, with the specific case of α-telluro carbanions being discussed separately in Section 4.08.2.3.4.

4.08.2.3.1 Lithium

(i) By deprotonation

(a) Selenides. Attempts to metallate most aryl and alkyl selenides with alkyllithium reagents result in a competitive (or exclusive) lithium–selenium exchange reaction; thus, a technique applicable to the preparation of α-thio lithium species can only be extended to selenides in but a few cases. For instance, whereas thioanisole is cleanly deprotonated by *n*-butyllithium, selenoanisole is cleaved to phenyllithium and butyl methyl selenide ⟨69AG(E)450⟩. LDA, and related bases such as LITMP, show much less proclivity to exchange and are better suited as deprotonation agents. If an

additional carbanion-stabilizing functionality (e.g., benzylic, allylic or propargylic groups) is present, then deprotonation is usually the exclusive route; however, the presence of additional alkyl groups α to the selenium (i.e., the organolithium would be tertiary) raises the spectre of competing C—Se bond cleavage. Table 21 shows examples of precursors for which direct deprotonation has been a success.

Table 21 α-Selenenyl carbanions by deprotonation of selenides.

Selenide	Conditions	Product	Ref.
PhSeMe	Bu^{n}Li, THF, TMEDA	PhSe-CH$_2$-Li	72CB511
PhSe-CH$_2$-CO$_2$R	LDA, THF, −78 °C	PhSe-CH(Li)-CO$_2$R	78TL2693
R^{1}Se-CH$_2$-C(R^3)=CHR^2	Bu^{n}Li or LiTMP, THF, −78 °C	R^{1}Se-CH(Li)-C(R^3)=CHR^2 (with Li)	75JOC2570, 84TL3629
PhSe-CH$_2$-C≡CH	2 LDA, THF, −78 °C	PhSe-CH(Li)-C≡C-Li	77JA263
PhSe-CH$_2$-Ph	LDA, THF, −78 °C	PhSe-CH(Li)-Ph	79JA6638

(b) Selenoxides and selenones. The inherent reactivity of selenoxides and the corresponding instability of their α-lithio derivatives are the reasons for the marginal amount of work on the preparation and reactions of α-metallo selenoxides compared to their sulfoxide cousins. Even though selenones bear much more acidic α protons than the corresponding sulfones, α-metallo selenones are a rarity. Selenones are strongly oxidizing, and, moreover, phenyl selenones are excellent electrophiles. Attempted deprotonation and derivatization are frequently complicated by reductive processes and alkylation (e.g., see ⟨85CC571⟩). Some of the few examples of direct deprotonation are shown in Table 22.

Table 22 α-Seleninyl and α-selenonyl carbanions by deprotonation of selenoxides and selenones.

Selenoxide/selenone	Conditions	Product	Ref.
Ph-Se(=O)-CH$_2$-R	LDA, THF, −78 °C (selenoxide prepared *in situ*)	Ph-Se(=O)-CH(Li)-R	79JA6648, 83JOC2098
Ph-SeO$_2$-CH$_2$-CH(CH$_3$)$_2$	LDA, THF, −78 °C	Ph-SeO$_2$-C(Li)H-CH(CH$_3$)$_2$	83MI 408-02
Ph-SeO$_2$-CH$_2$-R	Pr$^i_2$NK or KOBu^t, THF, −110 °C	Ph-SeO$_2$-CH(K)-R	88TL3269

(ii) By selenium–metal exchange

In view of the difficulty associated with direct deprotonation of alkyl selenides and the facility of selenium–carbon bond cleavage with alkyllithiums, selenium–lithium exchange of one of the selenyl groups of diselenoacetals has proved to be a reliable alternative route to α-lithio selenides with wide applicability. In general terms, since exchange processes are equilibria, the lithium reagent formed must be more stable that the one chosen to prepare it. Also, with unsymmetrical diselenoacetals, the most stable carbanion will be formed. The rates of these processes depend strongly on the solvent.

Seebach and co-workers first exemplified the cleavage of a C—Se bond using bis(phenylseleno)methane with *n*-butyllithium in THF ⟨69AG(E)450, 72CB511⟩; independently, the Seebach and Krief groups subsequently developed these exchange processes from diselenoacetals and examined the reactivity of the derived α-seleno lithiums with electrophiles ⟨80T2531, 89T2005⟩. Treatment with *n*-, or, preferably, *s*- or *t*-butyllithium at low temperature invokes cleavage of a C—Se bond (methyllithium is unreactive); the corresponding α-lithio selenide is formed in good to excellent yield. Phenyl diselenoacetals exchange more readily than methyl diselenoacetals. Examples of the process are given in Table 23, covering the formation of primary, secondary and tertiary organolithium species.

Table 23   α-Selenenyl carbanions by selenium–lithium exchange.

Selenoacetal	Conditions	Product	Ref.
PhSe⌒SePh	Bu^{n}Li, THF, −78 °C	PhSe⌒Li	69AG(E)450
OMe SePh / SePh (isopropyl)	Bu^{s}Li, ether, −78 °C	OMe SePh / Li	90TL7419
cyclopropyl(SePh)(SePh)	Bu^{n}Li, THF, −78 °C	cyclopropyl(SePh)(Li)	81TL4737
PhSe\\/SePh	Bu^{n}Li, THF, −78 °C	PhSe\\/Li	75AG(E)350
furyl-CH(SePh)(SePh)	Bu^{n}Li, THF, −78 °C	furyl-CH(SePh)(Li)	80TL3209
PhSe, SePh, Bu^t	Bu^{n}Li, THF, −78 °C	PhSe, Li, Bu^t	89T2005

(iii) By halogen–metal exchange

The exchange of a halogen for lithium is much less widely used than the analogous selenium–lithium exchange, due largely to the relative difficulty of obtaining the requisite α-halo selenides compared to diselenoacetals. Exchange is typically initiated at −78 °C; for instance, exposure of the α-bromo selenide in Equation (93) to *n*-butyllithium in THF or ether gives the derived α-lithio selenide ⟨77AG(E)541⟩. The reaction of this intermediate with *n*-butyl bromide is a reported side-reaction, a difficulty which could presumably be alleviated with *t*-butyllithium. The starting α-bromo selenides were prepared by addition of HBr in benzene to a vinyl selenide or by treatment of a mixture of an aldehyde and selenophenol again with HBr in benzene.

$$\text{Br, SePh, R}^1\text{, R}^2 \xrightarrow[\geq 60\%]{\text{Bu}^n\text{Li, THF, }-78\,^\circ\text{C}} \text{Li, SePh, R}^1\text{, R}^2 \qquad (93)$$

R^1, R^2 = H, alkyl

(iv) From vinyl selenides, selenoxides and selenones

In the late 1970s, Raucher and Koolpe reported the addition of alkyllithiums to vinyl selenides and the trapping of the resulting α-lithio selenides with electrophiles ⟨78JOC4252⟩. At about the same time, Krief and co-workers ⟨78AG(E)526⟩ and Woltermann and co-workers ⟨77AG(E)710⟩ described essentially the same process.

For example, treatment of phenyl vinyl selenide (145) with either *n*-butyl- or isopropyllithium in DME or ether at −78 °C gives the α-seleno lithium (146) in high yield (Equation (94)). Methyllithium does not react under the same conditions. The choice of proper reaction conditions is important to avoid C—Se bond cleavage or metallation by deprotonation: reactions in other solvents, for instance THF, give mixtures of products in poor yields.

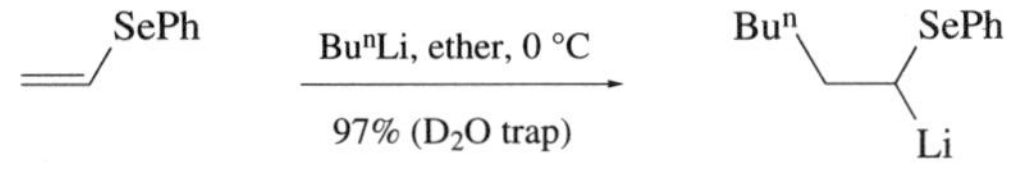

$$\text{(145)} \xrightarrow[\text{97\% (D}_2\text{O trap)}]{\text{Bu}^n\text{Li, ether, 0 °C}} \text{(146)} \tag{94}$$

4.08.2.3.2 Copper

α-Seleno alkyllithiums oxidatively dimerize efficiently upon treatment with copper salts; the reaction is rapid at −40 °C. Below this temperature, the existence of discrete intermediate entities was shown by Krief and co-workers, who studied the reaction of α-phenylselenoheptyllithium and CuI:SMe$_2$ by low-temperature NMR spectroscopy ⟨84IJ125⟩. The exact nature of the low-temperature species (presumed cuprates), which dimerized upon warming and which could be trapped by allyl bromide, was not clarified. Liotta and co-workers produced effectively an α-selenyl cuprate (147) by 1,4-addition of dimethylcopperlithium to the α-phenylseleno α,β-unsaturated ketones (148) (Equation (95)) ⟨80JOC2736, 81JOC4301⟩. Subsequent addition of methyl iodide alkylated the α position.

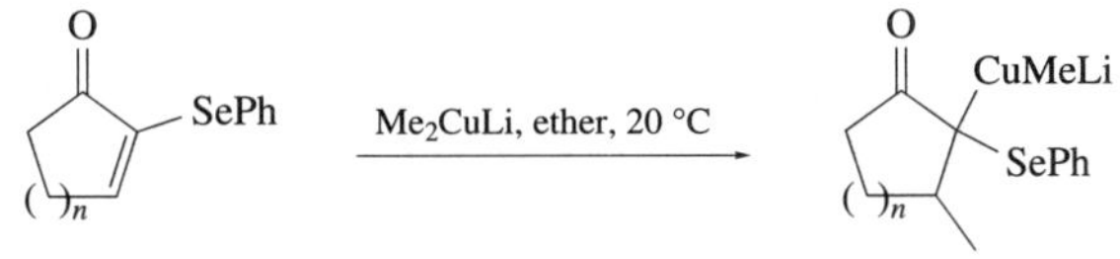

$$\text{(148)} \xrightarrow{\text{Me}_2\text{CuLi, ether, 20 °C}} \text{(147)} \tag{95}$$

4.08.2.3.3 Tin

There are isolated reports of the generation of α-seleno tin species:

(i) Kauffmann and Kriegesmann treated (triphenylstannyl)methyllithium (149) with benzeneselenenyl bromide in ether to give the α-selenenylstannane (150) in 40% yield (Equation (96)) ⟨82CB1810⟩.

(ii) Sarkar and Satapathi treated bis(phenylseleno)methane (151) with lithium diisobutylamide followed by quenching with iodomethyltrimethylstannane and found, to their surprise, none of the desired stannylated acetal: instead, phenyl vinyl selenide (152) and the α-selenostannane (153) were formed in 55% and 35% yields, respectively, presumably after breakdown of the lithiated intermediate (Scheme 55) ⟨89TL3333⟩.

(iii) Standard transmetallation of the diphenyl diselenoacetal (154), followed by quenching with tributyltin chloride, gave two diastereoisomeric α-selenostannanes in 93% yield (Equation (97)) ⟨90TL7419⟩.

(iv) The α-selenostannane (155) can be prepared by the reaction of (iodomethyl)trimethylstannane with phenylselenolithium in ether ⟨86ICA51⟩ in good yield (Equation (98)).

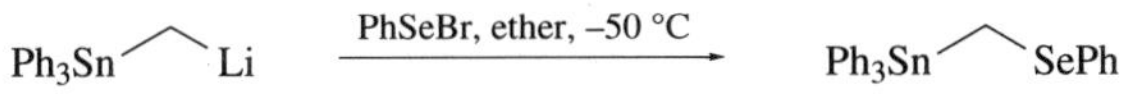

$$\text{Ph}_3\text{Sn}\frown\text{Li} \xrightarrow{\text{PhSeBr, ether, −50 °C}} \text{Ph}_3\text{Sn}\frown\text{SePh} \tag{96}$$
(149) (150)

$$\text{PhSe}\frown\text{SePh} \xrightarrow[\text{THF, −80 °C}]{\text{lithium diisobutylamide}} \text{PhSe}\underset{\text{Li}}{\frown}\text{SePh} \xrightarrow{\text{Me}_3\text{Sn}\frown\text{I}} \text{PhSe}\diagup\!\!\!= \; + \; \text{PhSe}\frown\text{SnMe}_3$$
(151) (152) (153)

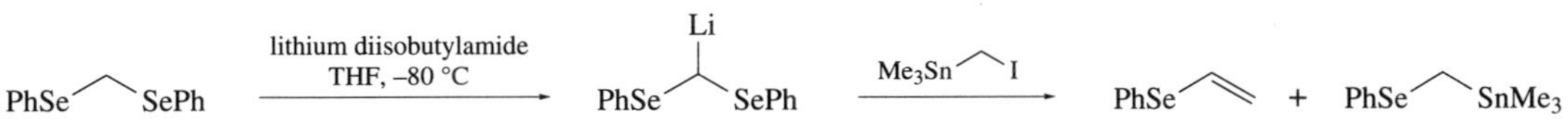

Scheme 55

$$\underset{(154)}{\text{iPr-CH(OMe)-CH}_2\text{-CH(SePh)}_2} \quad \xrightarrow[\text{93\%, 2:1 ratio}]{\substack{\text{i, Bu}^s\text{Li, ether, }-78\,°\text{C} \\ \text{ii, Bu}_3\text{SnCl}}} \quad \text{iPr-CH(OMe)-CH}_2\text{-CH(SePh)(SnBu}_3) \qquad (97)$$

$$\text{Me}_3\text{Sn-CH}_2\text{-CH}_2\text{-I} \quad \xrightarrow[\text{70-90\%}]{\text{PhSeLi, ether}} \quad \underset{(155)}{\text{Me}_3\text{Sn-CH}_2\text{-CH}_2\text{-SePh}} \qquad (98)$$

4.08.2.3.4 α-Telluro lithium species

Tellurium–lithium exchange, by analogy with selenium–lithium exchange, has been successfully used to generate α-telluro lithium species. Thus treatment of bis(phenyltelluro)methane **(156)** with methyl-, *n*-butyl- or *t*-butyllithium in THF at −80 °C effects the metal exchange in excellent yields (Equation (99)) ⟨75CB314⟩. The same exchange has also been effected by Japanese workers with phenyllithium ⟨81CL447⟩. As expected, treatment of the ditelluroacetal with LDA causes deprotonation.

$$\underset{(156)}{\text{PhTe-CH}_2\text{-TePh}} \quad \xrightarrow[>90\%]{\text{RLi, THF, }-80\,°\text{C}} \quad \text{PhTe-CH}_2\text{-Li} \qquad (99)$$

4.08.3 ACKNOWLEDGEMENT

The authors are indebted to Mrs Sandra Wood of the Research Information Department, Pfizer Central Research, for her invaluable assistance in the preparation of this chapter.

4.09
Functions Bearing Two Nitrogens

DEREK R. BUCKLE and IVAN L. PINTO
SmithKline Beecham Pharmaceuticals, Epsom, UK

4.09.1	INTRODUCTION	404
4.09.2	GEMINAL DIAMINO ALKANES—AMINALS	404
	4.09.2.1 Condensation of Aldehydes and Ketones with Amines	405
	4.09.2.1.1 Acyclic aminals	405
	4.09.2.1.2 Cyclic aminals	407
	4.09.2.2 Reaction of Amines with Geminal Dihalo Compounds	410
	4.09.2.3 Amine Addition to Imines and Iminium Salts	411
	4.09.2.3.1 To imines	411
	4.09.2.3.2 To iminium salts	412
	4.09.2.4 Reduction of Amidines, Amidinium Salts and Cyanamides	413
	4.09.2.5 Reduction of Ureas	414
	4.09.2.6 Miscellaneous Procedures	414
4.09.3	GEMINALLY SUBSTITUTED ALKANES BEARING ONE AMINO AND ONE ACYLATED OR SULFONATED AMINO GROUP	415
	4.09.3.1 Acylated Derivatives	415
	4.09.3.1.1 Condensation of amines and amides with carbonyl compounds	416
	4.09.3.1.2 Nucleophilic displacement reactions	417
	4.09.3.1.3 Addition of amines and amides to C—N multiple bonds	418
	4.09.3.1.4 Reductive methods	420
	4.09.3.1.5 Cycloaddition procedures	421
	4.09.3.1.6 Miscellaneous methods	422
	4.09.3.2 Sulfonated Derivatives	423
4.09.4	GEMINALLY SUBSTITUTED ALKANES BEARING TWO ACYLATED OR SULFONATED AMINO GROUPS	423
	4.09.4.1 Acylated Derivatives	423
	4.09.4.1.1 Condensation of amides with carbonyl compounds	424
	4.09.4.1.2 Reductive methods	425
	4.09.4.1.3 Nucleophilic addition to imines, enamines and isocyanates	426
	4.09.4.1.4 Miscellaneous procedures	426
	4.09.4.2 Sulfonated Derivatives	427
4.09.5	GEMINALLY SUBSTITUTED ALKANES BEARING TWO SIMILAR DICOORDINATE OR HETEROSUBSTITUTED NITROGENS	428
	4.09.5.1 gem-Dinitro Alkanes and gem-Dinitroso Alkanes	428
	4.09.5.1.1 Nitration of nitro alkanes	428
	4.09.5.1.2 Oxidative nitration of oximes	429
	4.09.5.1.3 Formation from nitro alkenes	431
	4.09.5.1.4 Formation from halo nitro alkanes	432
	4.09.5.2 gem-Diazido Alkanes	432
	4.09.5.2.1 Formation from gem-dihalo alkanes	432
	4.09.5.2.2 Formation from carbonyl compounds	432
	4.09.5.2.3 Miscellaneous methods	433
	4.09.5.3 gem-Diisocyanates and gem-Diisothiocyanates	433

4.09.5.4	gem-*Dinitrosamines* and gem-*Dinitramines*	434
4.09.5.5	gem-*Dihydroxylamino alkanes*	435
4.09.5.6	gem-*Dicarbodiimides*	436
4.09.5.7	gem-*Diazo alkanes*	436
4.09.5.8	gem-*Dihydrazino alkanes*	436
4.09.5.9	gem-*Diimino Alkanes* and gem-*Diisocyanides*	437
4.09.5.10	gem-*Difluoroamino Alkanes*	439

4.09.6 GEMINALLY SUBSTITUTED ALKANES BEARING TWO DIFFERENT DICOORDINATE OR HETEROSUBSTITUTED NITROGENS   349

4.09.6.1	*Nitro Alkane Derivatives*	439
4.09.6.2	*Nitramino Alkane Derivatives*	440
4.09.6.3	*Azo Alkanes and Hydrazino Alkane Derivatives*	440
4.09.6.4	*Isocyanato Alkane Derivatives*	441
4.09.6.5	*Miscellaneous Derivatives*	441

4.09.7 GEMINALLY SUBSTITUTED ALKANES BEARING ONE AMINO GROUP AND ONE DICOORDINATE OR HETEROSUBSTITUTED NITROGEN   442

4.09.7.1	*Aminomethylhydroxylamines*	442
4.09.7.2	*Aminomethyl Nitramines and Nitrosamines*	442
4.09.7.3	*Aminomethyl Azides and Triazines*	443
4.09.7.4	*Aminomethyl Azo and Hydrazino Compounds*	444
4.09.7.5	*Aminomethyl Imines*	445
4.09.7.6	*Miscellaneous Compounds*	445

4.09.8 GEMINALLY SUBSTITUTED ALKANES BEARING ONE ACYLATED OR SULFONATED AMINO GROUP AND ONE DICOORDINATE OR HETEROSUBSTITUTED NITROGEN   446

4.09.8.1	*Acylaminomethyl Isocyanates and Isothiocyanates*	446
4.09.8.2	*Acylaminomethyl Azides*	447
4.09.8.3	*Acylaminomethyl Nitramines*	447
4.09.8.4	*Acylaminomethyl Hydroxylamines and Hydrazines*	448
4.09.8.5	*Acylaminomethyl Azo Alkanes and Imino Alkanes*	448
4.09.8.6	*Miscellaneous Benzotriazole Derivatives*	449

4.09.1 INTRODUCTION

Alkanes which are substituted geminally by two nitrogen-containing moieties represent a diverse and important chemical class, serving a variety of distinct functions in both nature and synthetic chemistry. Many form the core structures of natural products—particularly the alkaloids—and synthetic compounds of medicinal interest ⟨91JCS(P1)1693⟩. They are also of great interest in the protection of carbonyl and amino groups, as intermediates for further transformations and as reagents in their own right. While this chapter will attempt to highlight the importance of specific representatives of each structural class, it will nevertheless concentrate predominantly on the major routes available for their synthesis. Less generally utilised routes which offer the potential for more extensive application will also be discussed. No attempt has been made to review exhaustively the numerous heterocycles which formally embody the functionality embraced by this chapter, or to give a detailed account of relatively trivial interconversions from one functionality to another. Quaternized amines have not been reviewed.

4.09.2 GEMINAL DIAMINO ALKANES—AMINALS

Although readily hydrolysed into their constituent amine and carbonyl components by aqueous acids (Equation (1)) ⟨53CB1463⟩, the aminals are a relatively stable class of compound which have general synthetic utility. Only those compounds possessing α-hydrogen atoms may need to be handled with care, since such compounds are thermally unstable, eliminating one amino group to form the enamine when heated (Equation (2)) ⟨36CB2106⟩. In addition to their synthetic value, aminals form the framework of many important heterocyclic nuclei, some examples of which occur naturally as members of the quinoline and indoline alkaloid families ⟨B-82MI 409-02⟩. As a consequence, the aminals have received considerable attention, resulting in the availability of a number of versatile synthetic routes. Some of these methods are the subject of a review by Duhamel ⟨B-82MI 409-02⟩.

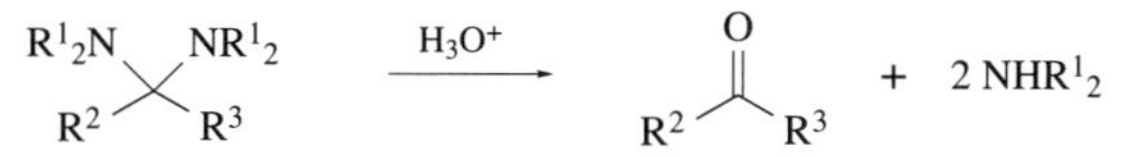

$$\underset{R^2}{\overset{R^1{}_2N}{\diagdown}}\underset{R^3}{\overset{NR^1{}_2}{\diagup}} \xrightarrow{H_3O^+} \underset{R^2}{\overset{O}{\diagdown}}\underset{R^3}{\diagup} + 2\,NHR^1{}_2 \qquad (1)$$

$$\underset{R^3}{\overset{R^2}{\diagdown}}\underset{R^4}{\overset{NR^1{}_2}{\diagup}}\overset{NR^1{}_2}{} \xrightarrow{heat} \underset{R^3}{\overset{R^2}{\diagdown}}=\underset{NR^1{}_2}{\overset{R^4}{\diagup}} + NHR^1{}_2 \qquad (2)$$

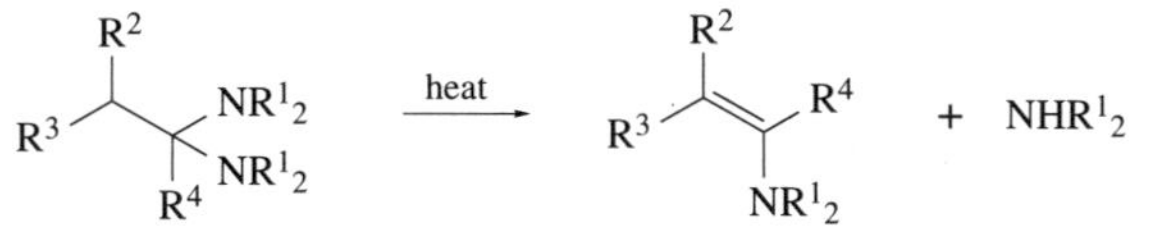

4.09.2.1 Condensation of Aldehydes and Ketones with Amines

The direct reaction of amines with aldehydes or ketones, often under conditions of water removal, constitutes the most commonly used method for the formation of symmetrical aminals. In addition, the ability to discriminate between more reactive aldehyde groups in the presence of ketones, or indeed most other carbonyl functions, is a considerable advantage ⟨53CB1463⟩. Whereas ammonia and aliphatic primary amines do not undergo simple aminal formation with aldehydes (see below), the reaction of secondary amines occurs rapidly on mixing and is considerably less complex ⟨B-82MI 409-02⟩. In the presence of excess amine, the reaction usually proceeds directly to the aminal (1), but the use of stoichiometric quantities of reagents often allows the isolation of the intermediate aminol (2), which can be treated with a second amine to afford the asymmetrical aminal (3) (Scheme 1) ⟨B-82MI 409-02⟩. In general, however, asymmetrical compounds such as (3) are best prepared by alternative routes. Some aromatic primary amines have been shown to form aminals with aromatic aldehydes, but imines are the more usual product ⟨86G229⟩.

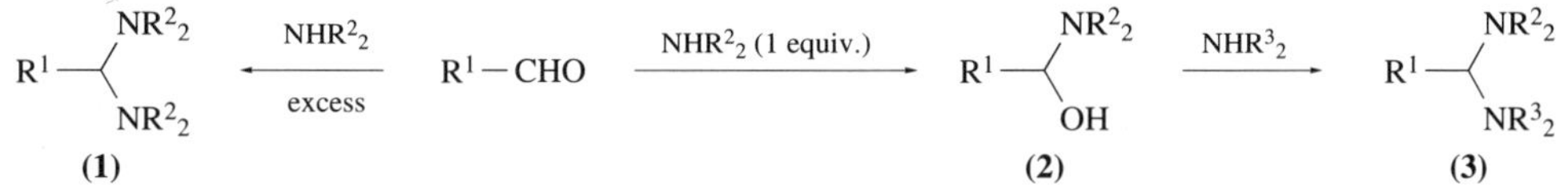

$$R^1\underset{NR^2{}_2}{\overset{NR^2{}_2}{\diagup\diagdown}} \xleftarrow[excess]{NHR^2{}_2} R^1-CHO \xrightarrow{NHR^2{}_2\,(1\,equiv.)} R^1\underset{OH}{\overset{NR^2{}_2}{\diagup\diagdown}} \xrightarrow{NHR^3{}_2} R^1\underset{NR^3{}_2}{\overset{NR^2{}_2}{\diagup\diagdown}}$$
$$(1) \qquad\qquad\qquad\qquad\qquad\qquad (2) \qquad\qquad\qquad (3)$$

Scheme 1

4.09.2.1.1 Acyclic aminals

Methylene aminals are usually prepared by mixing concentrated aqueous solutions of formaldehyde with an appropriate secondary amine. In this manner a large number of derivatives have been made ⟨1895BSF157, B-82MI 409-02⟩, the overall procedure being typified by the formation of bis(dimethylamino)methane in 83–86% yield ⟨88OSC(4)474⟩. Aminals derived from aliphatic, aromatic and many heteroaromatic aldehydes form rapidly on addition of a secondary amine, the reaction usually being carried out in solvents such as alcohols ⟨49JA2271, 53CB1463, 86G229⟩, benzene ⟨55JA1098, 69CPB32, 74BSF331⟩ or pyridine ⟨49JCS2342⟩. Optimal yields, approaching 100%, may be achieved either by the azeotropic removal of water ⟨55JA1098⟩ or by the addition of dehydrating agents. While azeotropic conditions are generally favoured with less volatile amines ⟨55JA1098, 74BSF331, 87CJC687⟩, anhydrous potassium carbonate ⟨36CB2106⟩, boric anhydride ⟨69CPB32⟩, Drierite ⟨75BSF196⟩ and molecular sieves ⟨91SL111⟩ have all been successfully used. In the case of more volatile amines, a useful alternative is treatment of the aldehyde with tris(dimethylamino)arsine ⟨66JOC4041, 67CB1289⟩ or the corresponding stibine ⟨71JCS(C)511⟩ in toluene or diethyl ether at ambient temperature (Equation (3)). Prior formation of the metallo amine may not be essential in such modifications, since a variety of Lewis acids have been effectively used in the presence of secondary amines to generate enamines from ketones ⟨67JOC213⟩. Despite reports to the contrary ⟨B-82MI 409-02⟩, this procedure has not yet been applied to the synthesis of aminals. Asymmetrical aminals may be prepared by the reaction of aldehydes with equimolar proportions of two secondary amines ⟨67CB2515⟩, or by prior formation of the intermediate aminol (Scheme 1) ⟨56CB2873, 65CB4036⟩, but the yields by these methods are generally poor and contamination with symmetrical products is a problem.

$$3\ \text{PhCHO} + 2\ \text{As(NMe}_2)_3 \xrightarrow[86\%]{\text{ether, ambient temperature}} 3\ \text{Ph-CH(NMe}_2)_2 + \text{As}_2\text{O}_3 \qquad (3)$$

As expected, the rate of aminal formation from benzaldehydes and heteroaromatic aldehydes is highly sensitive to the electronic and steric character of the aromatic ring. In a series of substituted benzaldehydes, the reaction rate with morpholine in benzene under conditions of water removal was shown to follow the order 4-Cl ≫ 4-NMe$_2$ ≫ 2,4,6-Me$_3$ ⟨74BSF331⟩, although excellent yields of morpholine aminals have been achieved from a range of benzaldehydes on simply heating for 5 min in ethanol ⟨49JA2271⟩. Aminals of furan- and quinoline-2-carbaldehyde, as well as some nonaromatic heterocyclic carbaldehydes, similarly form in high yield ⟨49JA2271, 66CB868, B-82MI 409-02⟩, but pyrrole-2-carbaldehyde afforded exclusively a 76% yield of the tricyclic system (**4**) from its reaction with piperidine ⟨66CB868⟩. Aromatic compounds such as terephthalaldehyde, having two carboxaldehyde functions, form good yields of the bis aminals under standard conditions ⟨71CB3354⟩.

(**4**)

A great variety of more complex aldehydes have also been shown to form the expected aminals, although problems with multiple amine incorporation have been observed with reactive substituents such as α-halides ⟨B-82MI 409-02⟩.

As well as the standard methods of amine addition to aldehydes, a mild procedure using silylated amines has been developed for aromatic aldehydes which results in excellent yields of the corresponding aminals (Equation (4)), where Tf represents trifluoromethanesulfonyl ⟨88BSF1009⟩. This method appears to be of particular value with volatile amines, and with appropriate caution may be extended to enolisable aliphatic aldehydes by replacing TMS-OTf by tetra-*n*-butylammonium fluoride (tbaf) ⟨88BSF1009⟩.

$$\text{furfural-CHO} + \text{morpholine-N-TMS} \xrightarrow[95\%]{\text{TMS-OTf, CH}_2\text{Cl}_2,\ 22\ °\text{C, 15 min}} \text{furfuryl-CH(morpholino)}_2 \qquad (4)$$

Aminals may also be prepared in a two-step synthesis developed by Katritzky *et al.* ⟨90JCS(P1)541, 91T2683⟩, in which intermediate bis triazoles such as (**5**)—prepared in good yield by a Mannich-type reaction of a primary amine with benzotriazole and formaldehyde—undergo C—C bond formation with the elimination of the triazole on treatment with Grignard reagents (Scheme 2, where BtH represents benzotriazole). While good yields may be achieved in this manner, the method has yet to be fully explored and its potential is currently unknown.

Only reactive ketones such as cyclopropanone undergo rapid aminal formation with secondary amines, and under controlled conditions it is possible to isolate the intermediate aminols (**6**) ⟨67RTC417, 67TL3085, 70TL1729⟩. Subsequent reaction of (**6**) with further amine leads to high yields of the symmetrical aminal (**7**), but attempts to generate asymmetrical aminals by the addition of a second amine lead to mixtures of both the symmetrical (**8**) and asymmetrical (**9**) products (Scheme 3) ⟨67TL3085, 70TL1729⟩.

Aminals of a rather more unusual nature have been prepared from acetophenones and diaryl

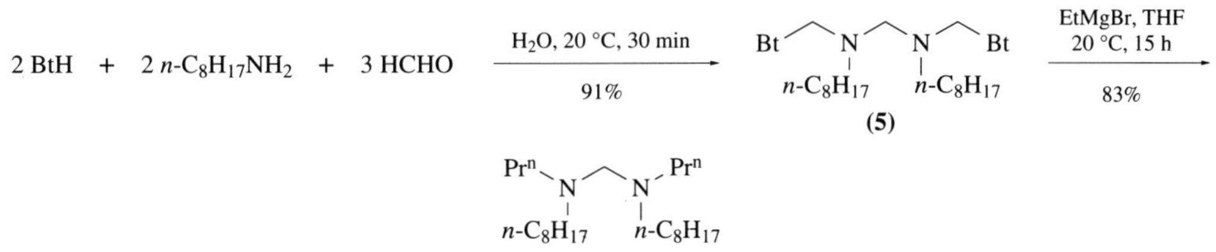

Scheme 2

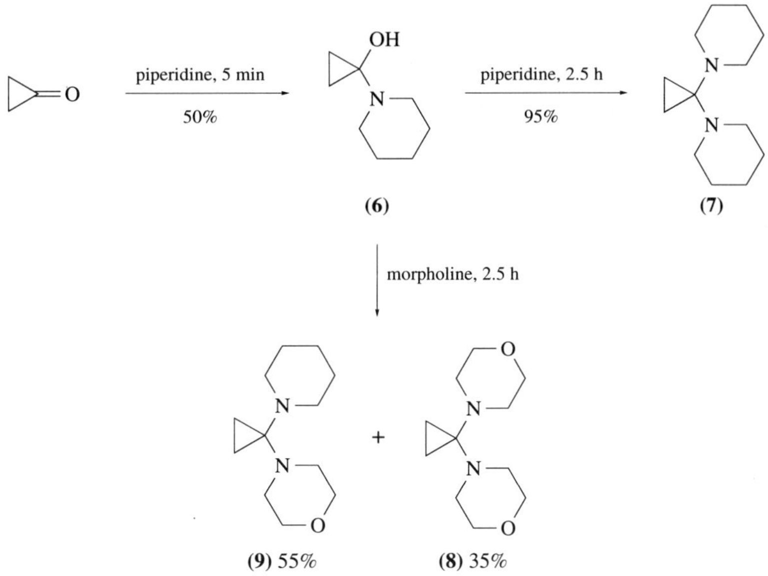

$2\,BtH + 2\,n\text{-}C_8H_{17}NH_2 + 3\,HCHO \xrightarrow[91\%]{H_2O,\ 20\,°C,\ 30\ min}$ **(5)** $\xrightarrow[83\%]{EtMgBr,\ THF\ 20\,°C,\ 15\ h}$

Scheme 3

ketones **(10)** by the addition of carbonyl- or sulfinyldiimidazole in dichloromethane (Equation (5)), but with few exceptions low yields of mixed products are obtained ⟨79TL5011, 80H(14)97⟩.

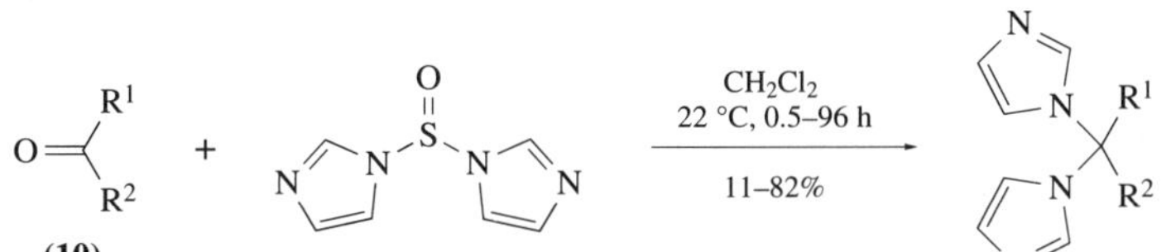

$$\text{(10)} + \text{Im}_2S=O \xrightarrow[11-82\%]{CH_2Cl_2,\ 22\,°C,\ 0.5\text{–}96\ h} \text{product} \quad (5)$$

4.09.2.1.2 Cyclic aminals

Formaldehyde reacts with ammonia to form hexamethylene tetramine **(11)** ⟨42OS(22)65⟩, which has found utility as a methylene transfer reagent in the formation of other aminal derivatives. Primary aliphatic aldehydes also form cyclic products on treatment with saturated aqueous solutions of ammonia, but in this instance the initially formed 'aldehyde ammonias' dehydrate on standing at low temperature with the formation of hexahydro-1,3,5-triazines **(12)** ⟨73JOC3288⟩.

The reaction of 1,2- or 1,3-diamines with aldehydes provides a simple and important synthetic

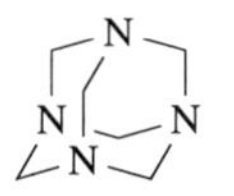

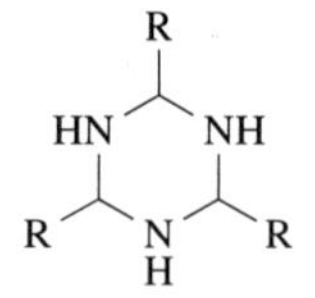

(12)

route to a wide range of heterocyclic amines ⟨B-82MI 409-02⟩, a reaction typified by the high-yielding formation of perhydroimidazopyridines (13) from 2-(methylamino)methylhexahydropyridine ⟨60AP(293)203, 70JHC355⟩. More complex heterocycles such as the diazaadamantane (15) form quantitatively on brief treatment of the diamine (14) with trioxan in methanol (Equation (6)) ⟨58CB598⟩. In a rather unusual reaction for primary amines, the triamine (16) furnishes the triazaadamantane (17) on condensation with trioxan (Equation (7)) ⟨51CB834⟩.

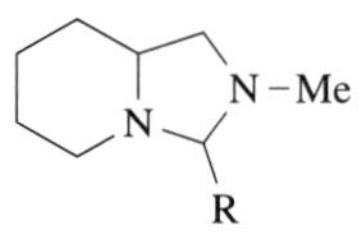

(13)

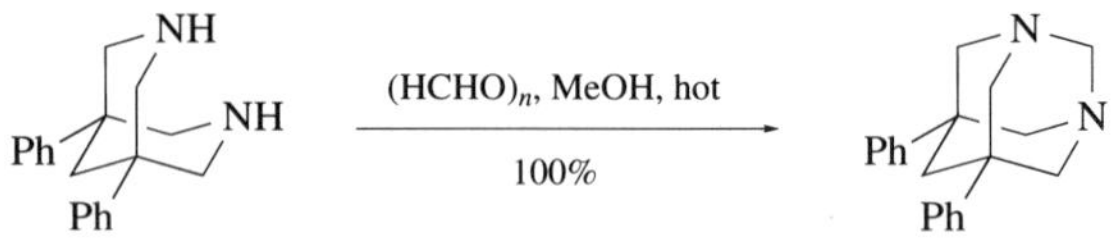

(14) → (15) (HCHO)$_n$, MeOH, hot, 100% (6)

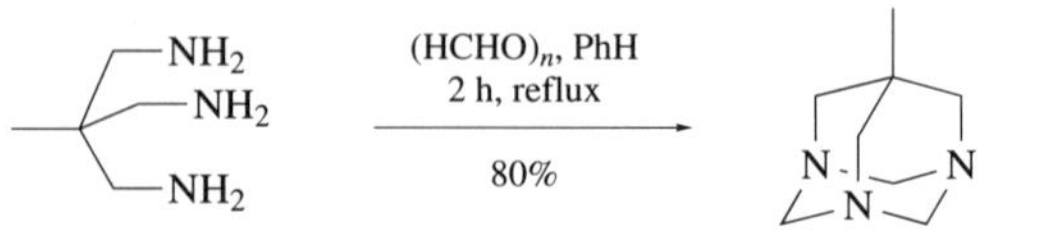

(16) → (17) (HCHO)$_n$, PhH, 2 h, reflux, 80% (7)

The formation of imidazolidines (18) and hexahydropyrimidines (19) from 1,2- and 1,3-diamines and various aldehydes generally proceeds in good yields, although 1,2-diamines are cyclised more readily ⟨53CB1463, 77LA956, B-82MI 409-02⟩. As a result of its propensity to form highly crystalline products in reasonable yields, 1,2-dianilinoethane (Wanzlick's reagent) in particular has been extensively used for the characterisation and purification of aldehydes ⟨53CB1463⟩. Mild acidic hydrolysis readily regenerates the parent aldehyde ⟨53CB1463⟩.

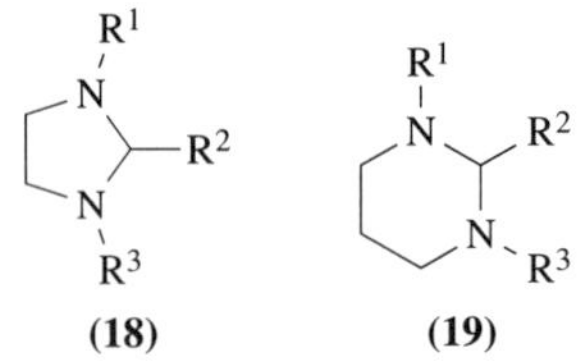

(18)   (19)

Many other 1,2-diamines have been used to prepare imidazolidine derivatives ⟨B-82MI 409-02⟩ and chiral diamines offer particular advantages in the synthesis of chiral heterocycles ⟨78CL1253, 80CL17, 89H865, 91SL111⟩. The (S)-diamine (20) was condensed with 3-methoxybenzaldehyde to give the adduct (21) in 100% de (Equation (8)) ⟨89H(29)865⟩. In a similar manner, the chiral diamine (22) afforded the imidazolidine (23) with retention of chirality (Equation (9)) ⟨91SL111⟩. The formation of chiral dihydropyrimidines (25) in high diastereomeric excess (88–97%) from the amino enimines (24) has been described using (S)-2-benzyloxypropanal as a chiral auxiliary (Scheme 4, where R*CHO represents (S)-2-benzyloxypropanal). Reduction of (25) using sodium borohydride readily leads to the chiral tetrahydropyrimidines (26) ⟨88CC410⟩.

Typically, such reactions may be extended to the synthesis of a large range of structurally diverse heterocycles, as exemplified by formation of the bicyclic compound (**27**) by condensation of glyoxal with N^1,N^2-dimethyl-1,2-diaminoethane (Equation (10)) ⟨76JCS(P2)1564⟩. By contrast, 1,3-diaminopropane reacts with glyoxal to yield the bis(aminal) (**28**) (Scheme 5), a product which may react further to provide perhydro-tetraazafluorenes such as (**29**) ⟨89JOC4771⟩. Similarly, the unmasked aldehyde generated on acid treatment of the diamine (**30**) undergoes an intramolecular cyclisation on basification to afford a 57% yield of the aminal (**31**) (Equation (11)) ⟨72LA(759)84⟩. The intramolecular formation of aminals from suitably orientated aminols (formed as intermediates in the diisobutylaluminum hydride (dibal-H) reduction of amides) has also been demonstrated ⟨78JOC4276⟩.

The efficient formation of aminals in high enantiomeric purity has also been demonstrated in more complex cyclic systems in which existing chiral centres may be used to effect stereocontrol in the aminal-producing step. The sequential reaction of the aldehyde (**32**) with methylamine and LAH has been shown to provide a high yield of the enantiomerically pure physostigmine precursor, esermethole (**33**) (Equation (12)) ⟨93JOC6949⟩. In this instance LAH serves the dual purpose of reducing the intermediate imine after converting the amide into its aminol.

$$(12)$$

Cyclic aminals are also efficiently prepared by the benzimidazole displacement procedure developed by Katritzky *et al.* ⟨90JCS(P1)541, 91T2683⟩. Thus, imidazolidines and hexahydropyrimidines (**34**), prepared in good yield by an extension of the method shown in Scheme 2, gave 68–87% yields of the corresponding aminals (**35**) by reaction with a range of alkyl Grignard reagents (Equation (13), where Bt represents benzotriazole).

$$(13)$$

Ketones tend not to form aminals with 1,2- and 1,3-diamines, and aldehydes can react selectively in their presence ⟨53CB1463⟩. Nevertheless, excellent yields of aminals (**37**) have been obtained by the reaction of various aromatic and alicyclic ketones with diamines of type (**36**), although mild acid catalysis (AcOH) was required to facilitate reaction with the latter (Equation (14)) ⟨82JPR832⟩. A large number of aminals were also prepared by the reaction of (**36**) with aromatic and heteroaromatic aldehydes in the presence of catalytic quantities of AcOH ⟨82JPR832⟩.

$$(14)$$

Optimal methods for the synthesis of aminals formally derived from ketones and other carbonyl compounds are described in Section 4.09.2.2.

4.09.2.2 Reaction of Amines with Geminal Dihalo Compounds

The reaction of secondary amines with geminal dihalo compounds (Equation (15)) is the second-most important procedure for the synthesis of symmetrical aminals, but is unsuitable for dihalo compounds having α-hydrogen atoms due to competitive dehydrohalogenation reactions ⟨B-82MI 409-02⟩. Although in many instances methylene aminals are readily prepared from formaldehyde as described above, they may be synthesised under nonaqueous conditions in 50–71% yields from heterocyclic amines and dichloromethane in methanol at slightly elevated temperature and pressure ⟨84H(22)1417⟩. High yields of methylene aminals have also been obtained from dichloromethane and less basic heterocyclic amines such as pyrrole, but this has generally required the addition of hexamethylphosphoramide (HMPA) ⟨80HCA1190⟩. An alternative, phase-transfer procedure has been shown to afford moderate to good yields of aminals with several substituted pyrroles, but gave only a 20% yield of the parent compound (**38**) ⟨85H(23)1127⟩. Whether diiodomethane, which has been used effectively for the synthesis of bis(2-acetylpyrrol-1-yl)methane ⟨78TL4995⟩, offers any improvement is not known.

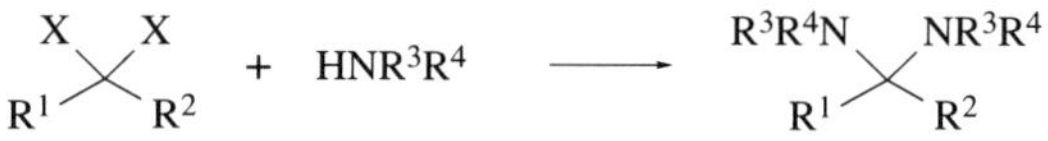

$$\underset{R^1R^2}{XX} + HNR^3R^4 \longrightarrow \underset{R^1R^2}{R^3R^4NNR^3R^4} \qquad (15)$$

X = halogen

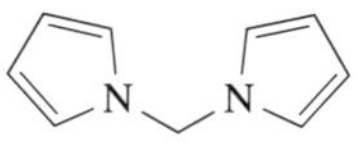

(38)

Efficient aminal synthesis under extremely mild conditions has been demonstrated using bis(chloromethyl) ether rather than dichloromethane (Equation (16)) ⟨84RRC333⟩, although the toxicity associated with this reagent precludes its more general utility.

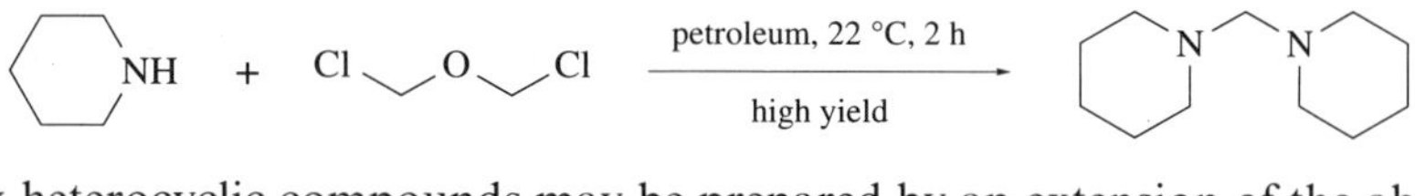

$$\text{piperidine-NH} + Cl\!\!-\!\!CH_2\!\!-\!\!O\!\!-\!\!CH_2\!\!-\!\!Cl \xrightarrow[\text{high yield}]{\text{petroleum, 22 °C, 2 h}} \text{pip-N-CH}_2\text{-N-pip} \qquad (16)$$

More complex heterocyclic compounds may be prepared by an extension of the above reactions, as exemplified by the formation of high yields of the triazepines **(40)** on treatment of 2-(chloromethyl)benzimidazole with primary amines in dichloromethane (Scheme 6) ⟨90AG(E)933⟩. In this instance it is possible to isolate the intermediate aminomethyl derivative **(39)** and subsequently form the triazepine using a second amine ⟨90AG(E)933⟩.

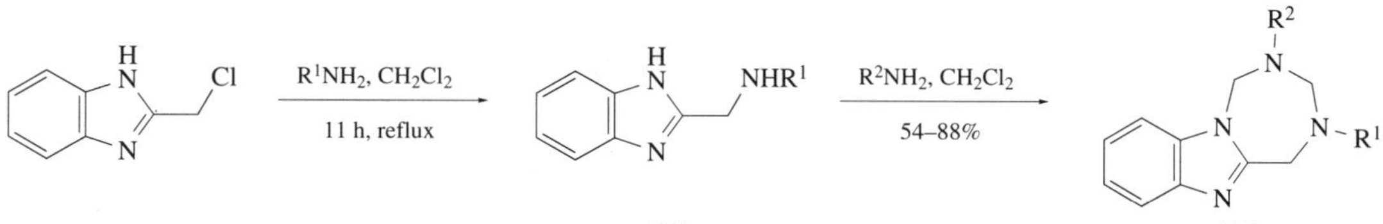

Scheme 6

In addition to methylene derivatives, aminals form readily from dihalomethylbenzenes ⟨75BSF196⟩, α,α-dihalocarbonyl compounds ⟨65CR(261)2232, B-82MI 409-02⟩ and other functionalised α,α-dihalo derivatives ⟨68M990, B-82MI 409-02⟩. Di- and tri-aminals have also been prepared from the corresponding poly(dibromomethyl)benzenes ⟨B-82MI 409-02⟩.

In contrast to the reaction of secondary amines with ketones, amination of dihalo compounds is an efficient process for the preparation of aminals formally derived from ketones ⟨B-82MI 409-02⟩. Only those dihalo compounds which cannot readily undergo dehydrohalogenation reactions, that is, those which lack hydrogen atoms on the carbon atom adjacent to that bearing the halogen substituents, will form aminals in this manner.

Where comparisons have been made, the order of reactivity of haloalkyl compounds towards nitrogen nucleophiles follows the expected order I > Br > Cl ⟨75BSF196⟩, but suitably activated fluorocarbons have been shown to generate aminals under relatively mild conditions (Equation (17)) ⟨90JOC4777⟩.

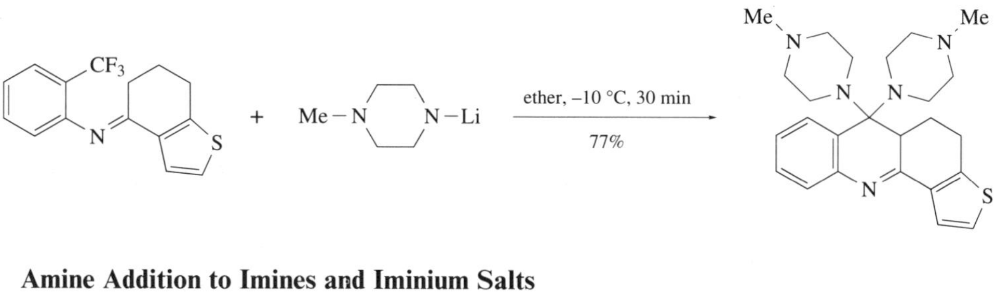

$$(17)$$

4.09.2.3 Amine Addition to Imines and Iminium Salts

4.09.2.3.1 *To imines*

The addition of secondary amines to imines is a potentially versatile procedure for the synthesis of aminals which does not appear to have been fully exploited. Examples of both symmetrical

⟨63CB1630⟩ and asymmetrical ⟨74JOC167⟩ aminals have been produced in excellent yields by this method (Equations (18) and (19), respectively), but only single examples are given.

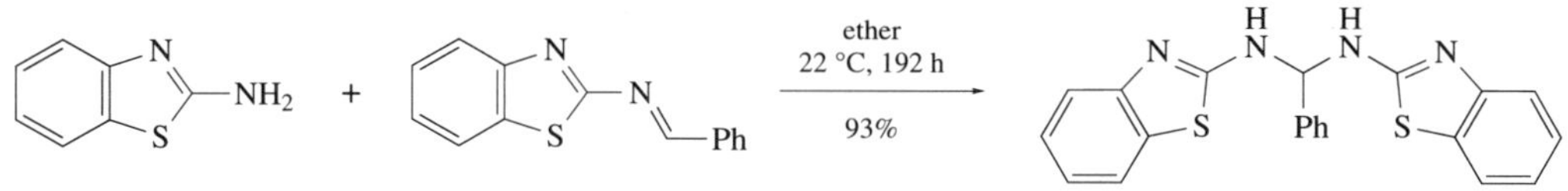

Although imines do not appear to have been isolated during the formation of aminals from amines and aldehydes, they have been proposed as possible precursors in trimerisation reactions from aldehydes with ammonia ⟨73JOC3288, 74JOC1349⟩, and there is kinetic evidence to support their formation (as iminium salts) under neutral aqueous conditions ⟨69JA1860⟩. Moreover, aromatic imines (**42**) are formed as stable intermediates on treatment of *N*-(methoxymethyl)anilines (**41**) with PhLi at a low temperature and may be quenched with a second equivalent of aromatic amine to afford high yields of symmetrical aminals (**43**) (Scheme 7) ⟨88CB1813⟩. Since aminal formation is reversible under such strongly basic conditions, this procedure is not suitable for the synthesis of asymmetrical aminals. However, a simple modification, in which the imine (**42**) is generated under neutral conditions in the presence of a second aromatic amine, allows the formation of excellent yields of asymmetrical products (**44**) (Scheme 7) ⟨88CB1813⟩.

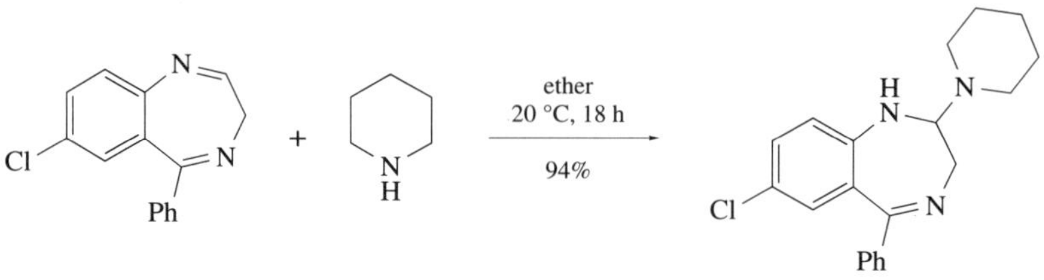

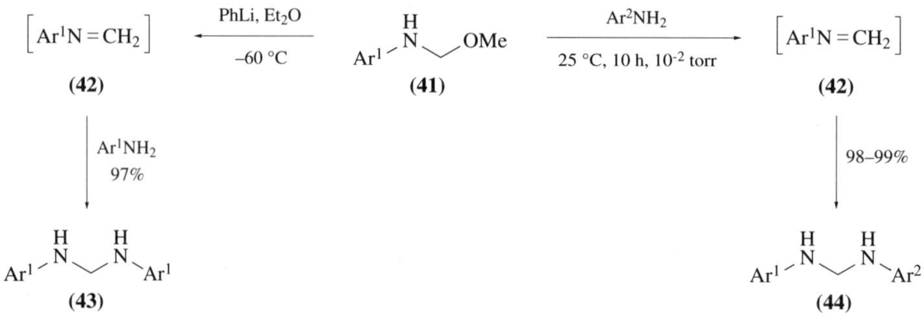

Scheme 7

4.09.2.3.2 *To iminium salts*

Arguably one of the most versatile procedures for the synthesis of asymmetrical aminals involves the reaction of secondary amines with iminium salts, although there is little to suggest that this procedure has found extensive use. *N*-Methylaniline (**45**; R = H) and its two hindered derivatives (**45**; R = Me or Cl) readily form aminals by addition to the iminium salts (**46**) generated from the corresponding *N*-chloromethyl derivatives (Equation (20)), for example, whereas only *N*-methylaniline was able to form an aminal by reaction with either formaldehyde or dichloromethane ⟨67CB2131⟩. Several α-halo iminium salts also form aminals by reaction with secondary amines, although retention of the α-halogen is highly dependent on the steric bulk of the incoming nucleophile ⟨B-82MI 409-02⟩. Intramolecular aminal formation by amine attack at an iminium salt (generated *in situ* from an amide and phosphorus oxychloride) provides a useful procedure for polycyclic aminals ⟨91CC462⟩.

Although not unequivocally proven, it is likely that the formation of the spiroaminals (**48**) through the acid-mediated hydrolysis and decarboxylation of (**47**) also proceeds via an intermediate iminium ion (Scheme 8) ⟨66CB724⟩.

Amination of quaternized quinolines, pyridines and other aromatic nitrogen heterocycles has been shown to occur rapidly and quantitatively below 0 °C in liquid ammonia to give the corresponding aminals (Equation (21)), although the products readily dissociate to the parent compounds on warming ⟨73JOC1949, 76JOC1303⟩. Stable annulated compounds are isolable, however, when the incoming nitrogen nucleophile forms part of the quaternizing side chain (Equation (22)) ⟨73JOC437⟩.

4.09.2.4 Reduction of Amidines, Amidinium Salts and Cyanamides

Dissolving metal reductions proceed readily with simple amidines and provide a versatile route to aldehydes through hydrolytic cleavage of the intermediate aminals (Scheme 9). Such reactions are less efficient with *N*-substituted derivatives ⟨55AJC512⟩. High yields of cyclic aminals may be prepared by either complex metal hydride reduction ⟨71JCS(C)780, 76TL1199⟩ or catalytic hydrogenation ⟨59JA3789⟩ of the amidine moiety of purines and related compounds, although reductions of cyclic amidines leading to less stable aminals may require milder conditions (BH$_3$–THF) to prevent reductive cleavage ⟨93TL6329⟩. Aminals have also been prepared in excellent yield by the reduction of primary cyanamides formed from the corresponding amines and cyanogen bromide (Equation (23)) ⟨82ZN(B)512⟩. Compounds in which the amine is endocyclic or directly attached to a ring system are not readily reduced to aminals.

$$C_8H_{17}NHCN \xrightarrow[90\%]{\substack{Zn, HCl \\ 30 \text{ min, reflux}}} C_8H_{17}NHCH_2NH_2 \qquad (23)$$

In a rather more unusual reaction, formamidinium salts, prepared by the reaction of *ortho*-formamides with mineral acids, furnished aminals on treatment with either NaH or PhLi (Equation (24)) ⟨62JOC3664⟩. A similar reaction with lithium alkyls failed to give the corresponding aminal and the general applicability of the procedure is unknown. In contrast, iminium salts of imidazolines have been shown to form cyclic aminals on addition of lithium alkyls or alkyl Grignard reagents, whereas no addition occurred with the nonquaternized imidazoline ⟨82CC282⟩.

4.09.2.5 Reduction of Ureas

The reaction of LAH with five- and six-membered cyclic ureas results in reasonable yields of the corresponding aminals (Equation (25)), and systematic study of a series of N^1,N^2-disubstituted derivatives indicates that the rate of reduction follows the relative order Ph > Bn > Me > Et ⟨86JOC2228⟩. LAH in ether under reflux has also been shown to effect a clean reduction of hydantoins such as (49) to 1,8-diazabicyclo[4,3,0]nonanes ⟨60JOC2108⟩. While such reactions are of undoubted value, the stability of the resulting aminals will inevitably limit their overall utility.

$$n = 0, 1$$

(49)

4.09.2.6 Miscellaneous Procedures

The stereocontrolled synthesis of more complex aminals such as (51) (Scheme 10), involving the photolysis of *N*-chloro amines (50) prepared *in situ*, has been reported and offers a potentially versatile route to polycyclic natural products ⟨76S201, 88TL1691⟩. In the formation of (51), no trace of the epimeric material was observed. Bicyclic aminals of a similar nature may also be prepared by the cycloaddition reactions of dipolarophiles with zwitterions prepared from imidazolium salts (Equation (26)), and a one-pot modification has given considerably improved yields ⟨93JCS(P1)2391⟩. Inverse electron-demand Diels–Alder cyclisations of the 2-methylene imidazolidines (52) are also known, and lead to diastereomeric mixtures of the unstable spiro aminals (53) (Equation (27)) ⟨87TL2681⟩. High-pressure induced [2 + 2]-cycloaddition reactions of imines with enamines produced 2-aminoazetidines in reasonable purity, but isolation was precluded by their instability ⟨87JOC365⟩.

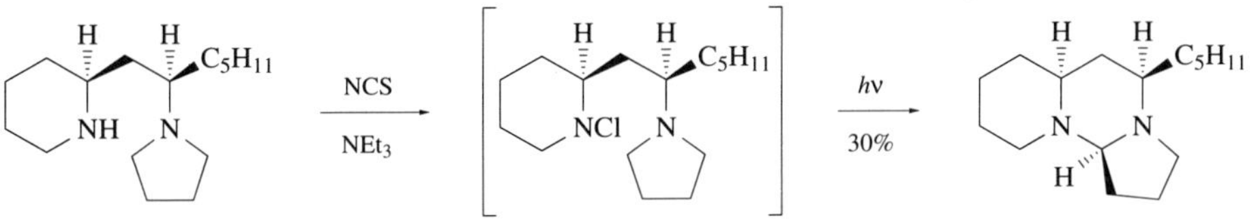

One Amino and One Acylated or Sulfonated Amino Group

Scheme 10

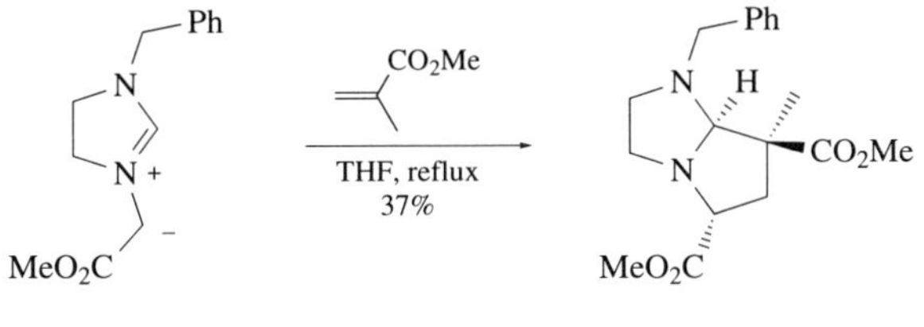

(26)

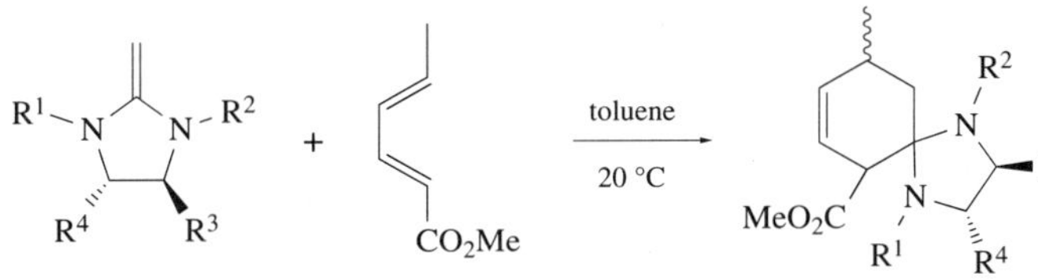

(27)

Like aldehydes, selenoaldehydes generated *in situ* condense with morpholine to produce aminals in ∼70% yield ⟨88CL1145⟩. Thioaldehydes, by contrast, react further under the reaction conditions and lead solely to the formation of thioamides. A moderate yield of the benzimidazoline (**55**) has been prepared on treatment of the 1,3-dithietane (**54**) with 1,2-diaminobenzene (Equation (28)), although this is unlikely to be a general reaction ⟨74BCJ785⟩.

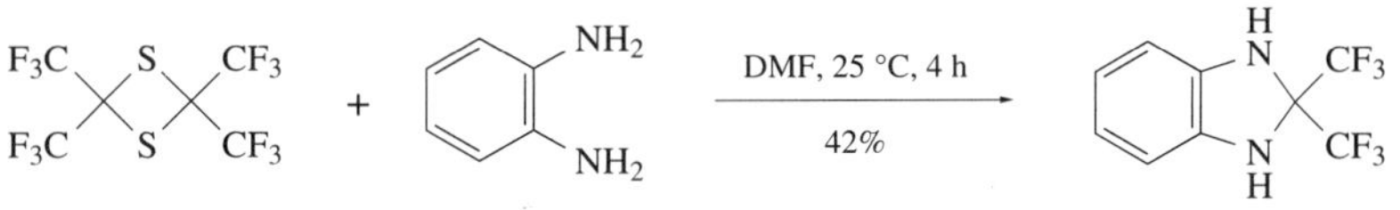

(28)

Highly electron-deficient pyrroles are known to undergo nucleophilic addition and substitution reactions, and high yields of pyrrolines such as (**56**) have been reported from the reaction of *N*-methyl-3,4-dinitropyrrole with secondary amines at room temperature ⟨83JOC162⟩.

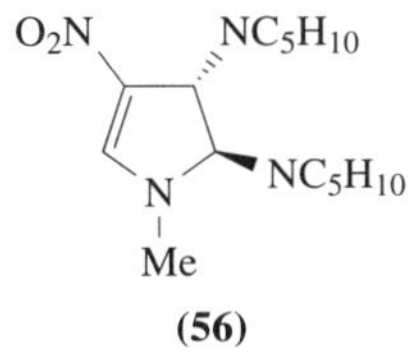

(**56**)

A number of unusual rearrangement reactions resulting in the formation of various aminals have been described, but these are unlikely to be of general application ⟨69JOC2720, 73T4049, 80LA1573, 84TL1023⟩.

4.09.3 GEMINALLY SUBSTITUTED ALKANES BEARING ONE AMINO AND ONE ACYLATED OR SULFONATED AMINO GROUP

4.09.3.1 Acylated Derivatives

While the direct acylation of preformed aminals represents a conceptually straightforward method for the formation of compounds of this type, it has been little used due—primarily—to the sensitivity of many aminals to the usual acylation conditions and/or the availability of alternative, preferential procedures. Nevertheless, intramolecular acylation reactions leading to stable imidazolidinones ⟨86AG(E)345⟩ and the acylation of more stable cyclic aminals, particularly those involving intramolecular reactions (Equation (29)), have provided reasonable yields of the anticipated products ⟨75JMC177, 75JMC182⟩.

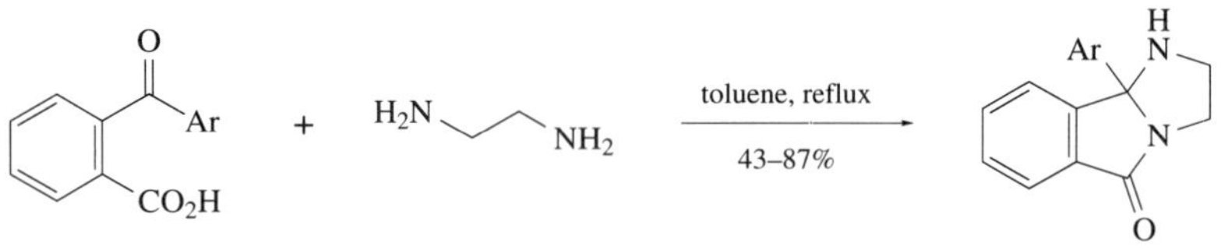

$$\text{(Ar-CO-C}_6\text{H}_4\text{-CO}_2\text{H)} + \text{H}_2\text{N-CH}_2\text{CH}_2\text{-NH}_2 \xrightarrow[43-87\%]{\text{toluene, reflux}} \text{product} \quad (29)$$

4.09.3.1.1 Condensation of amines and amides with carbonyl compounds

One of the most versatile reactions for the formation of *N*-acyl aminals is the Mannich-type Einhorn reaction (Scheme 11) in which stoichiometric mixtures of amide, formaldehyde and amine are heated together at ~70 °C ⟨65OR(14)52, B-66MI 409-01, B-70MI 409-01⟩. Since, as expected, the reaction of amides with formaldehyde requires harsher conditions than those required for amines, it is likely that reaction proceeds via the intermediate aminol (**57**) rather than the *N*-hydroxymethylamide (**58**), even though such products are well known ⟨B-70MI 409-01⟩. The reaction is general for a very large variety of amines and amides, and good yields of acylated aminals have been isolated using both aliphatic and aromatic aldehydes ⟨B-66MI 409-01, 76AP503, 85S1148, 86S804, 88JHC119⟩ in place of formaldehyde. Typically, the reaction proceeds well when the amine and amide are represented on either the same (Equation (30)) ⟨64G595, 69M469, 76AP503, 85S1148, 88JHC119⟩ or different ⟨61AP(294)404, 63JIC777, 86S804⟩ molecules. Imides ⟨75BCJ357, 76H(5)203, 76S748⟩, carbamates ⟨61CB2209, 93SC1467⟩, ureas ⟨47JA2136⟩, thiocarbamates ⟨61JPR72⟩, thioamides ⟨75LA2318⟩ and thioureas ⟨47JA2136⟩ perform equally well as amide alternatives. The extensive scope of this useful reaction has been comprehensively reviewed ⟨B-66MI 409-01⟩. The formation of 1,3,5-triazine derivatives (**60**) by the reaction of bis-*N*-(hydroxyalkyl)ureas (**59**) with primary amines (Scheme 12) is illustrative of the general reaction, although a greater range of products may be obtained in a one-pot reaction by heating 1:2:1 mixtures of the urea (or thiourea), aldehyde and primary amine ⟨47JA2136, 48AG267⟩.

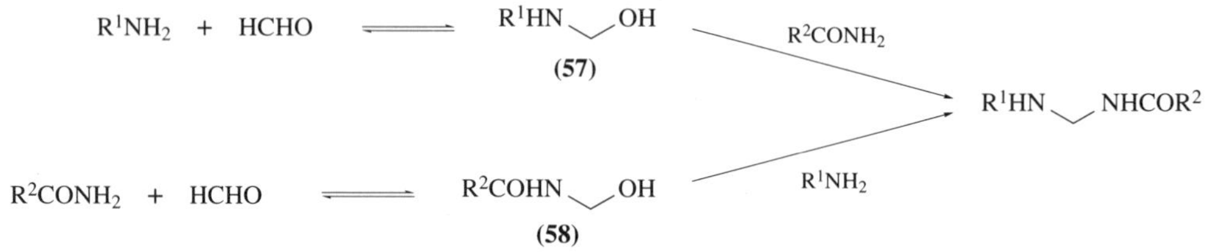

$$R^1NH_2 + HCHO \rightleftharpoons R^1HN\text{-}CH_2\text{-}OH \quad (\mathbf{57})$$

$$R^2CONH_2 + HCHO \rightleftharpoons R^2COHN\text{-}CH_2\text{-}OH \quad (\mathbf{58})$$

$$(\mathbf{57}) + R^2CONH_2 \longrightarrow R^1HN\text{-}CH_2\text{-}NHCOR^2$$
$$(\mathbf{58}) + R^1NH_2 \longrightarrow R^1HN\text{-}CH_2\text{-}NHCOR^2$$

Scheme 11

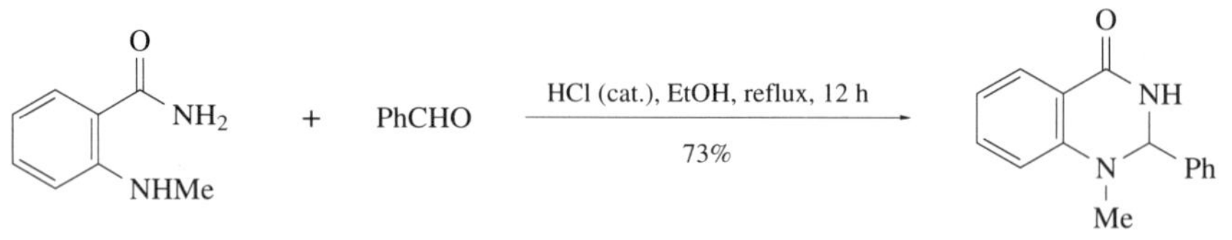

$$\text{2-(MeHN)C}_6\text{H}_4\text{-CONH}_2 + \text{PhCHO} \xrightarrow[73\%]{\text{HCl (cat.), EtOH, reflux, 12 h}} \text{quinazolinone product} \quad (30)$$

$$O=C(NH_2)_2 + R^1CHO \longrightarrow (\mathbf{59}) \xrightarrow{R^2NH_2} (\mathbf{60})$$

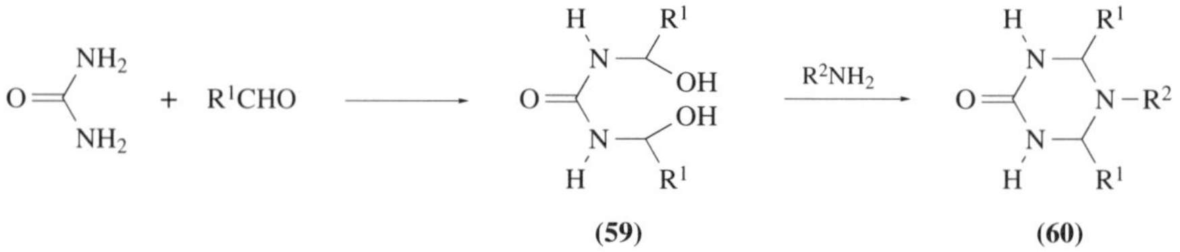

Scheme 12

Under certain circumstances, ammonia has been used effectively as the amine component in the Einhorn reaction, but there are few examples where this is described ⟨85S1148⟩. A useful modification is that developed by Katritzky *et al.* for the synthesis of monoacyl-α-aminoglycines in which the intermediate benzotriazole (**61**) is isolated prior to displacement by ammonia (Scheme 13) ⟨89CC337⟩. While used primarily for the synthesis of amino acid derivatives, it is likely that this methodology could have general application.

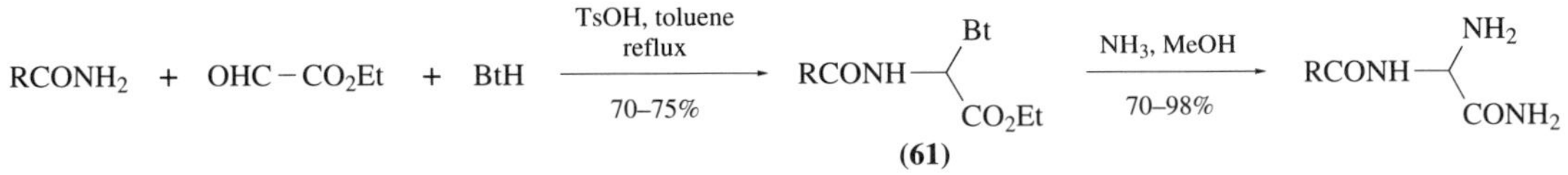

Scheme 13

The condensation of amino amides proceeds most favourably with aldehydes, but reactions with ketones have been reported. Particularly important is the formation of N^1,N^2-isopropylidene-protected dipeptides derived from a variety of amino acids, which occurs in good yield simply on stirring in acetone at room temperature or under reflux (Equation (31)) ⟨77JCS(P1)1954⟩. Neither the formation of the imidazolidinone ring system, nor its subsequent removal by heating in aqueous solution under neutral conditions, destroy the chiral integrity of the constituent amino acids ⟨77JCS(P1)1954⟩. Under comparatively forcing conditions it is also possible to prepare cyclic aminals from more complex ketones (Equation (32)), but the yields in this instance are highly dependent on the nature of the amino substituent ⟨82CPB1036⟩.

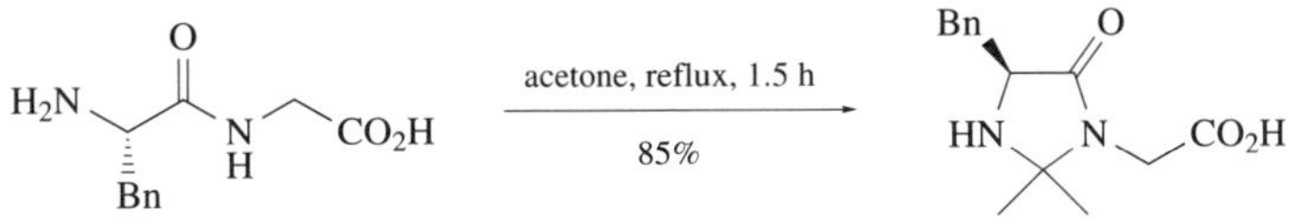

The Einhorn reaction is undoubtedly the most common procedure for the preparation of monoacylated aminals and many modifications involving either the prior formation of the intermediate aminols (**57**) or *N*-hydroxymethylamides (**58**) (Scheme 11) have been described ⟨65OR(14)52, B-66MI 409-01⟩. Like aminols, hydroxymethylamides (**58**) form reversibly from amides and formaldehyde at room temperature ⟨65OR52⟩. Neutral or basic media are generally preferred since acidic conditions often result in further transformation to ethers or methylenebisamides.

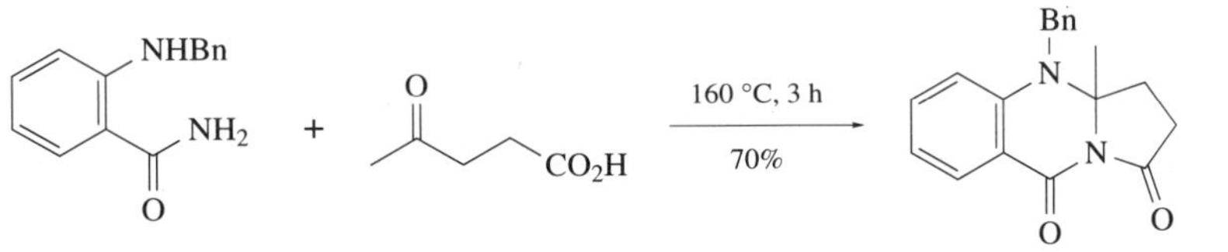

4.09.3.1.2 Nucleophilic displacement reactions

The reaction of amines with amides substituted at the α-carbon atom by an appropriate leaving group offers a potentially versatile route to monoacylaminals (Scheme 14). Halogen atoms ⟨64TL1693, 72CJC2902⟩, *O*-acyl groups ⟨70CB3459⟩, alkanethiols ⟨87JOC3232⟩, arenesulfonamides ⟨58CB2432⟩, dialkylamino groups ⟨57CB50, 57CB53⟩ and benzimidazoles ⟨73MI 409-01⟩ have been displaced in this manner. The benzotriazole displacement methodology discussed above may also be considered as an example of this approach (Section 4.09.3.1.1). The choice of leaving group is highly dependent on both amide accessibility and stability under the transformation conditions, but like the benzotriazole method of Katritzky *et al.* ⟨89CC337⟩, the mercuric ion-assisted thio displacement procedure developed by Bock *et al.* (Equation (33)) is suitable for the formation of amino acid derivatives and is likely to be of general applicability ⟨87JOC3232⟩. A modification of this approach in which differentially protected α,α-diamino acids may be prepared and then selectively monodeprotected offers additional versatility ⟨86JOC3718⟩. The Lewis acid-mediated displacement of benzenesulfonate by *N*-silyl amines (Equation (34), where TBDMS-O represents *t*-butyldimethylsilyl oxide) offers a mild alternative for the formation of 4-(*N*-imidazolo)azetidinones (**62**), although its general applicability has still to be demonstrated ⟨91TL2375⟩.

Scheme 14

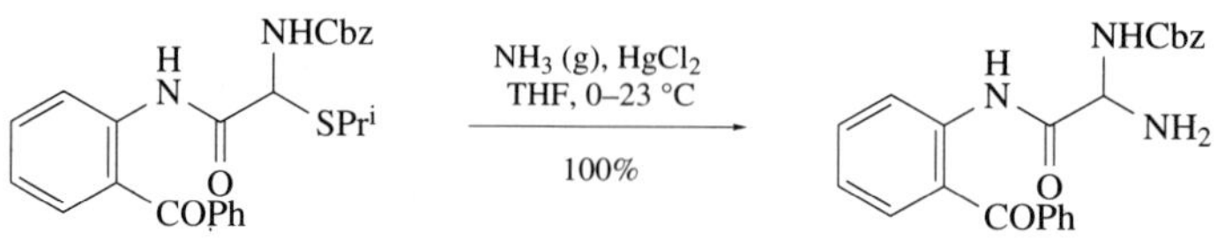

(33)

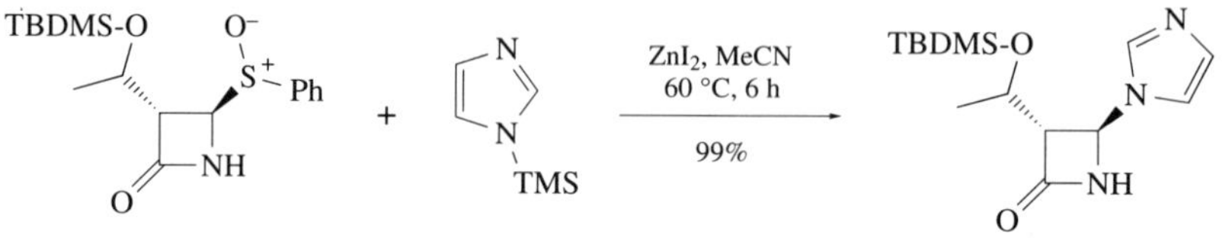

(34)

Displacement by amides of suitable leaving groups on the α-carbon atom of amines also offers a route for the synthesis of monoacylated aminals (Scheme 14), but fewer examples of such transformations exist, presumably due to the instability of the amine precursors. Nevertheless, alkoxy ⟨89ZOB2435⟩, silyloxy ⟨84JGU635⟩ and dialkylamino ⟨69CPB32⟩ groups have been efficiently displaced by amide nucleophiles. In the last method, amine displacement from symmetrical aminals was shown to occur simply on fusing at 85–90 °C for 3.5 h with 0.5 equivalents of a primary amide or an imide.

4.09.3.1.3 Addition of amines and amides to C—N multiple bonds

The nucleophilic addition of amides to unactivated C—N multiple bonds is not a particularly favourable reaction, although such reactions have been reported under relatively mild conditions. Acetamide and other unsubstituted aliphatic carboxamides, for example, add to *N*-methylene-*t*-butylamine (Scheme 15) on gentle heating to afford the bisamides (**64**) in high yield via the presumed intermediate aminals (**63**) ⟨91JGU1072⟩. Formamides were unusual in that hexahydro-1,3,5-triazines were formed as a result of intramolecular condensation ⟨91JGU1072⟩. At slightly higher temperatures an analogous reaction has been demonstrated with urea and its derivatives, which leads exclusively to the formation of perhydro-1,3,5-triazines ⟨90JGU2516⟩. Related reactions involving amide or thioamide addition to unactivated nitriles have been described, but these appear to proceed via a more complex pathway involving initial attack through oxygen or sulfur, respectively (see ⟨87CJC282⟩ and references therein). Good to excellent yields of cyclic thione derivatives (**65**) may be obtained, nevertheless, on reaction of α-amino nitriles with ketones in the presence of hydrogen sulfide (Scheme 16) ⟨87CJC282⟩.

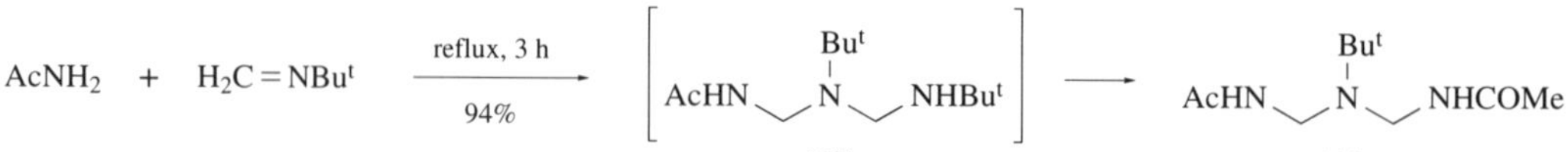

Scheme 15

In contrast to the unactivated systems above, the reaction of amides with activated C—N multiple bonds is comparatively straightforward and well known. The hexahydropyrrolo[2,3-b]indole nucleus, for example, is a common feature of many natural products, and its formation by nucleophilic attack at an indolenium ion by a suitably situated amido function is an important synthetic route (Scheme 17) (see ⟨81T1487, 91JCS(P1)1693⟩ and references therein). Ring closure preferentially

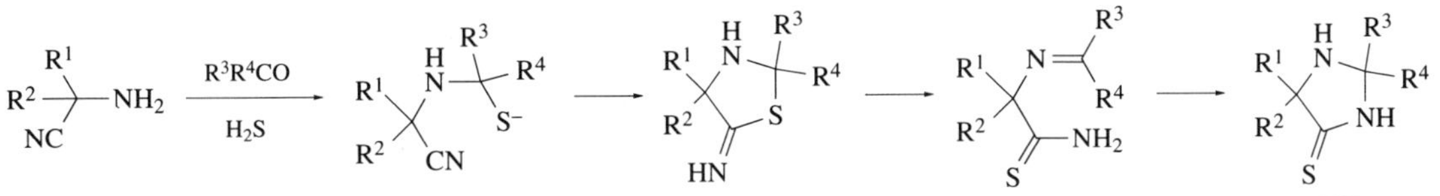

Scheme 16

forms the thermodynamically more stable *trans* isomer (66) and is facilitated by a variety of activators, including protons, positive-halogen or -oxygen donors, singlet oxygen and carbon electrophiles ⟨91JCS(P1)1693⟩. Despite the many options available for effecting ring closure, acidic reagents, particularly phosphoric acid, have been preferred in more recent years ⟨81T1487, 91JCS(P1)1693, 93JMC305⟩. Similar cyclisations of amino acid imines result in the formation of excellent yields of imidazolidinones with high diastereoselectivity (Equation (35)), although for synthetic expedience such products were usually isolated as their acylated derivatives ⟨85HCA135⟩.

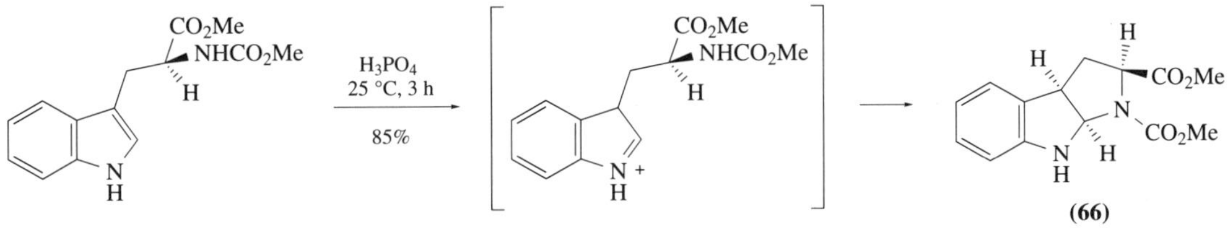

Scheme 17

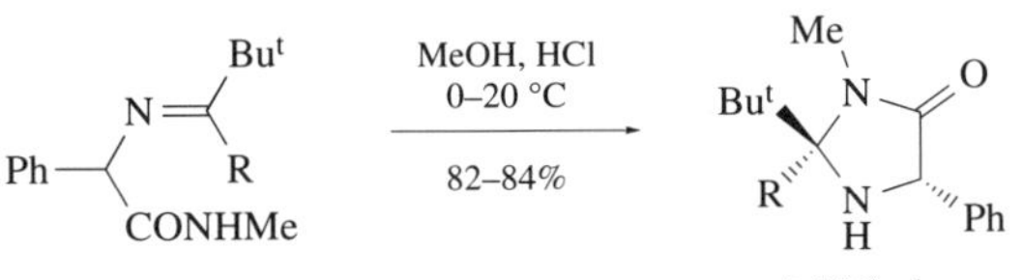

(35)

An interesting variant for the formation of monoacyl aminals has been developed in β-lactam chemistry where the reaction of *N,N*-bis(trimethylsilyl)formamide with the Schiff base (67) produced (69) in 67% overall yield following hydrolysis of the intermediate formamido derivative (68) (Scheme 18, where PMP represents *p*-methoxybenzyl) ⟨91TL2683⟩. Whether this approach is suitable for nuclei other than β-lactams is not known.

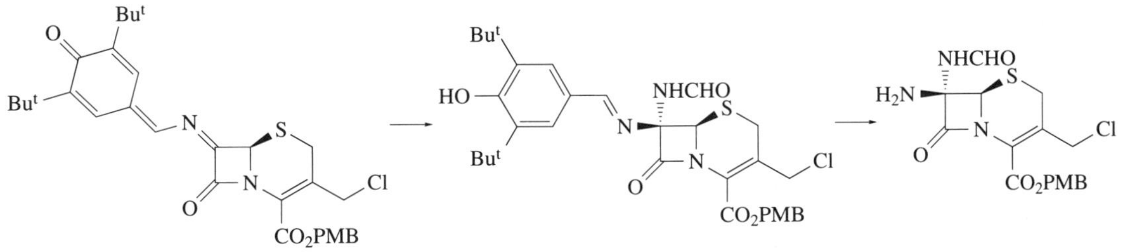

Scheme 18

Under suitable circumstances, amines formally add to C≡N bonds, particularly when such bonds are activated or the reaction is favoured by appropriate geometry. The formation of 3-aminoisoindolinones by the reaction of ammonia, and aliphatic primary or secondary amines with 2-cyanobenzaldehydes (Scheme 19) is a typical reaction which proceeds in high yield through a mechanism involving inversion of an intermediate imidate (70) ⟨84CL1599, 90BCJ1160⟩. Transformations of this type do not proceed well with aromatic amines due to preferential Schiff base formation ⟨84CL1599⟩. Essentially similar reactions have also been demonstrated with cyclic 1- and 2-amino nitriles on treatment with ketones in the presence of strong base, although elevated temperatures may be necessary ⟨71BCJ3445, 84JPR279⟩.

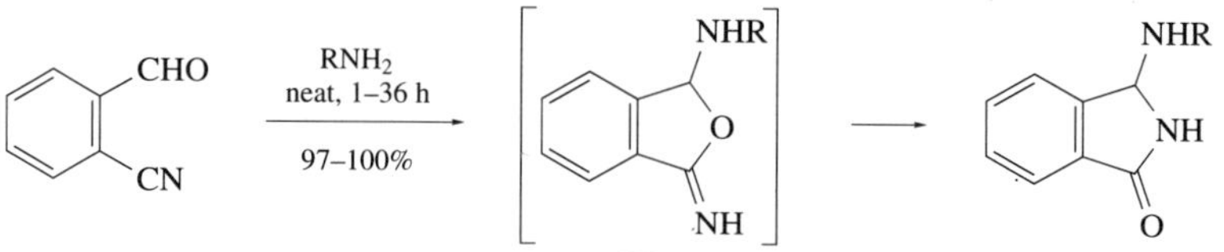

Scheme 19

4.09.3.1.4 Reductive methods

The reduction of imines affords a well-established route to amino compounds which is equally applicable as a route to aminals (Section 4.09.2.4) and their acylated derivatives. Sodium borohydride in particular is an effective reagent, but care may be required to ensure formation of the anticipated product. For example, a detailed study of the *N*-benzylidene compounds (**71**) (Scheme 20) showed that those Schiff bases derived from aromatic aldehydes bearing electron-withdrawing substituents produced the anticipated products (**72**), whereas those derived from aromatic aldehydes bearing electron-donating substituents preferentially formed the tertiary amines (**73**) ⟨64JA1701⟩. It is presumed that the tertiary amines result from nucleophilic attack of the reduced products (**72**) on unchanged imine (**71**), followed by subsequent reductive cleavage. While this procedure does provide a route to acylated aminals, it is less convenient and gives lower yields than the more recently developed methods of Bock and Katritzky (Section 4.09.3.1.2).

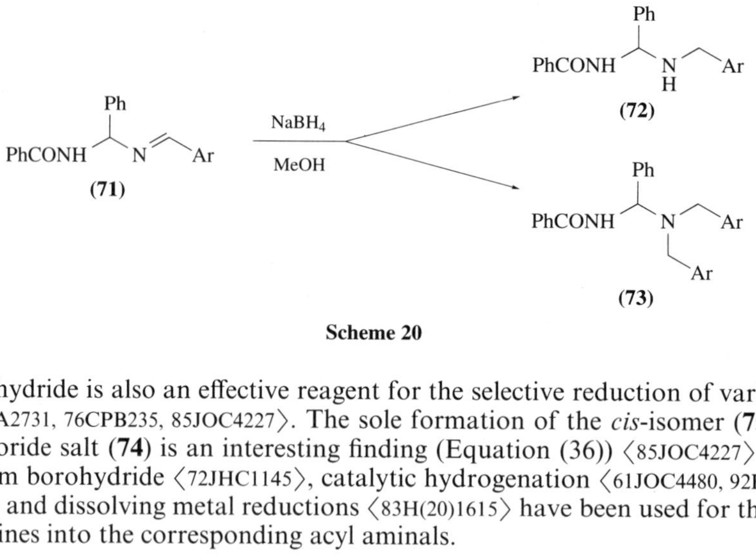

Scheme 20

Sodium borohydride is also an effective reagent for the selective reduction of various pyrimidine derivatives ⟨60JA2731, 76CPB235, 85JOC4227⟩. The sole formation of the *cis*-isomer (**75**) on reduction of the hydrochloride salt (**74**) is an interesting finding (Equation (36)) ⟨85JOC4227⟩. In addition to the use of sodium borohydride ⟨72JHC1145⟩, catalytic hydrogenation ⟨61JOC4480, 92PHA754⟩, Raney nickel ⟨55JA745⟩ and dissolving metal reductions ⟨83H(20)1615⟩ have been used for the conversion of cyclic acyl amidines into the corresponding acyl aminals.

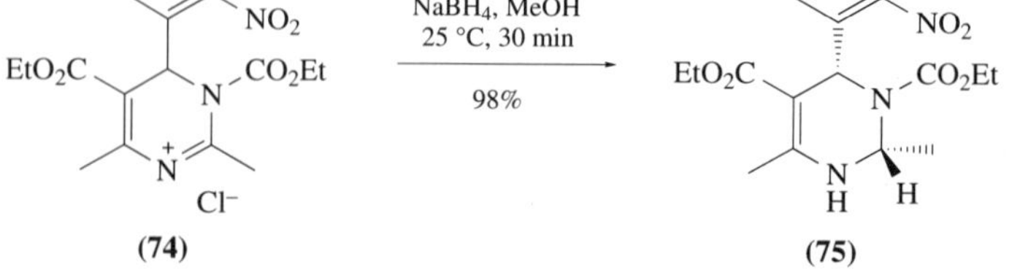

The desulfurisation of thiohydantoins provides a method for the generation of imidazolidinones, although the procedure has not found extensive application. Raney nickel has been used ⟨74JOC1710⟩, but incomplete desulfurisation may result in complex mixtures, requiring the isolation of intermediates and addition of further reductant ⟨55JA745⟩. A more efficient method using sodium in isoamyl alcohol (Equation (37)) is a preferred alternative to Raney nickel in such instances (see ⟨55JA745⟩ and references therein). A somewhat more specialised desulfurisation reaction has been described for 2-aza-1-thiacephems, from which 1-azapenems (Equation (38)) have been isolated in moderate to good yields ⟨81CC1269⟩. It is not known whether similar reactions are applicable to the formation of other acyl aminals.

4.09.3.1.5 Cycloaddition procedures

Monoacyl aminals have been prepared by various cycloaddition reactions, but none appear to have been used to any great extent. Early studies failed to demonstrate any reaction of *N*-*t*-alkyl azomethines with ketene until zinc chloride was added as a catalyst ⟨61JOC949⟩. Under these conditions 1,3-diazinones such as (**76**) were isolated in modest yield (Equation (39)). By contrast, the more reactive ketenes derived from pyrrole-2-carbonyl chloride and its 4-nitro derivative reacted rapidly with benzalaniline and 4-methoxybenzalaniline in the absence of catalysis and afforded 58–67% yields of the respective adducts (**77**) (Equation (40)). Under similar conditions, indole-2-carbonyl chloride gave only a dimeric product ⟨76JOC3050⟩.

As expected from their reaction with other dienophiles, 1,3-diazabutadienes undergo inverse electron demand [4 + 2]-cycloaddition reactions with enamines to give high yields of the corresponding pyrimidines having well-defined relative geometry (Equation (41)) ⟨89TL4573⟩. A similar reaction has been demonstrated with the trimethylsilylthio diene and cyclic enamines, although the full scope of this reaction has yet to be established. Cycloaddition reactions of 2-azabutadienes with trimethylsilyl isothiocyanate also lead to pyrimidine-4-thione derivatives ⟨91CC1704⟩. An interesting extension of this type of reaction involving the cycloaddition of 1,3-diphenyl-2-azaallyllithium to alkyl and aryl isocyanates or isothiocyanates has been shown to furnish moderate yields of imidazolidinones and imidazolidinethiones, respectively (Equation (42)) ⟨77CB651⟩. Imidazolidinones prepared by [6 + 2]-cycloaddition reactions of phenylisocyanate with cumulenes derived by the pyrolytic breakdown of derivatives similar to (**4**) have also been described ⟨75CL607⟩.

In an isolated series of reactions, dimethylketene-*N*-phenylimine was shown to form the bicyclic imidazolidinone (**79**) when heated with the nitrone (**78**) under reflux in benzene (Equation (43)), although a similar reaction with the 2,5,5-trimethyl homologue of (**78**) gave the corresponding adduct in considerably reduced yield ⟨79JOC4543⟩.

4.09.3.1.6 Miscellaneous methods

Hofmann degradation of primary amides proceeds under very mild conditions in the presence of phenyl iodosyl bis(trifluoroacetate) (PIFA), and this rearrangement is particularly suitable for the formation of aminals derived from peptides and acylamino acids (Equation (44)) (⟨84JOC4272, 85JMC769⟩ and references therein). The conditions are such that racemisation does not occur ⟨82CC280, 84JOC4272⟩.

Oxidative cyclisation reactions have proved to be useful for the formation of hexahydropyrrolo[1,2-a]imidazolinones (Equation (45)), although only one-electron oxidants such as potassium hexacyanoferrate(III) provide high product yields ⟨81T4337⟩. While it seems likely that other cyclic acylaminals should be accessible using similar methodology, such reactions do not appear to have been attempted. Derivatives of the same ring system have also been prepared by rhodium acetate-mediated carbene insertion reactions into 1,2-diazetidinones ⟨84JOC113⟩, but this procedure is limited in its generality. Photolytic C—C bond-forming reactions leading ultimately to bicyclic imidazolidinones have also been described ⟨85SC829⟩.

In contrast to the nucleophilic addition of amines and amides to nitriles and imines, addition to unsaturated C—C bonds is disfavoured in the absence of suitable activation. One example is the formation of dihydroquinazolinones by the reaction of anthranilamides with ethyl propiolate

(Equation (46)), which requires a tandem Michael reaction involving first the amino group and subsequently the amido group ⟨80JHC1163⟩.

$$\text{(arene with NHMe and C(O)NHMe)} \xrightarrow[\substack{80\,^\circ\text{C, 18 h} \\ 63\%}]{\substack{\text{ethyl propiolate} \\ \text{NaOEt, EtOH}}} \text{(fused quinazolinone with N-Me, N-Me, CH}_2\text{CO}_2\text{Et)} \qquad (46)$$

4.09.3.2 Sulfonated Derivatives

Sulfonated derivatives are comparatively rare, but when their synthesis has been described it has invariably involved the direct condensation of mixtures of amines and an appropriate sulfonamide with formaldehyde under conditions similar to those used for the Einhorn reaction. Several arenesulfonamides were shown to provide good yields of monosulfonated aminals on reaction with cyclic secondary amines (Equation (47)), although dimethylamine failed to generate the expected product ⟨44JA222, 58CB2432⟩. Whether other acyclic secondary amines would behave in a similar manner is not known. While it is likely that the condensation reaction proceeds via some intermediate aminol, no such intermediates have been reported. Aminols derived from the reaction of sulfonamides alone with certain aldehydes have been described, although these show a strong inclination towards the generation of the corresponding imines through the loss of water ⟨55BSF669⟩. The greater nucleophilicity of amines relative to sulfonamides would suggest that aminols similar to those formed in the Einhorn reaction (cf. Section 4.09.3.1.1) are the probable intermediates.

$$\text{ArSO}_2\text{NHR}^1 + \text{HNR}^2{}_2 + \text{CH}_2\text{O} \xrightarrow{20\,^\circ\text{C}} \text{ArSO}_2\text{NR}^1\text{CH}_2\text{NR}^2{}_2 \qquad (47)$$
$$R^1 = H, Me$$

Under favourable conditions, primary amines undergo related condensation reactions, particularly when stable cyclic compounds result. Typical products are the bicyclic triamine (**80**), formed in 33% yield from benzenesulfonamide, formaldehyde and ethylenediamine ⟨75LA2318⟩, and the cyclic sulfonamides (**81**) formed by the reaction of ammonia or primary aliphatic amines with sulfamide and formaldehyde ⟨73S243⟩.

(**80**)

(**81**)

It is evident that many opportunities exist for the preparation of monosulfonyl aminals using modifications of procedures found to be suitable for their acylated analogues, and that the exploitation of such procedures should lead to a variety of novel compounds.

4.09.4 GEMINALLY SUBSTITUTED ALKANES BEARING TWO ACYLATED OR SULFONATED AMINO GROUPS

4.09.4.1 Acylated Derivatives

In contrast to aminals, N^1,N^2-alkylidene-bis-amides are frequently crystalline solids which are stable to mild alkaline conditions, but which hydrolyse to amides with hot dilute acids. Concentrated acids and hot strongly basic conditions effect cleavage to carboxylic acids ⟨65OR(14)52⟩. This stability enables their synthesis by the further derivatisation of partially acylated aminals to be a viable option. Such reactions are usually carried out by treatment with an acyl halide or anhydride in the presence of triethylamine or pyridine ⟨61JPR72, 70JA343, 77JCS(P1)1954, 85HCA135, 91JCS(P1)1693⟩. Asymmetrical N^1,N^2-alkylidene-bis-amides have also been prepared by the acetylative hydrolysis of the N-[α-(benzylideneamino)benzyl]benzamides (**82**) ⟨64JA1701⟩. These methods, however, require the prior formation of the aminal skeleton and therefore lack the versatility of the procedures discussed below.

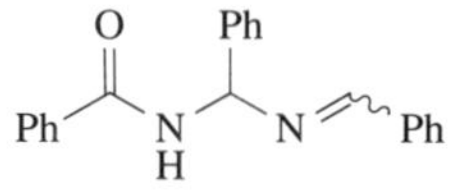

$$Ph\overset{O}{\underset{}{\text{C}}}\underset{H}{N}\overset{Ph}{\underset{}{\text{C}}}N\mathord{=}Ph$$

(**82**)

Thioamide analogues of N^1,N^2-alkylidene-bis-amides are not well reported, but simple derivatives have been prepared by reaction of the parent amide with phosphorus pentasulfide ⟨65OR(14)52⟩. The use of Lawesson's reagent as an alternative to phosphorus pentasulfide does not appear to have been explored.

4.09.4.1.1 *Condensation of amides with carbonyl compounds*

Perhaps the most common and versatile procedure for the synthesis of N^1,N^2-alkylidene-bis-amides utilises the direct condensation of amides with formaldehyde or other aldehydes. Whereas it is possible to isolate the intermediate amido alcohols (**83**) on treatment of formaldehyde with amides (see Section 4.09.3.1.1), aliphatic and aromatic aldehydes generally react further to form the symmetrical bis-amides (Scheme 21) ⟨51RTC269, 65OR(14)52⟩. The reaction, which occurs on heating or under acid catalysis, is particularly suitable for primary amides, but has also been demonstrated for secondary amides, imides, carbamates ⟨65JOC2769, 65OR(14)52⟩ and thioamides ⟨57JOC984⟩. Hexamethylenetetramine may be used as an alternative to formaldehyde for the preparation of N^1,N^2-methylene-bis(amides) ⟨55JA2559⟩, but does not appear to have any marked advantages. Simple ketones fail to undergo similar reactions, whereas reactive ketones such as pyruvic and benzoylformic acids behave in a similar fashion to that of aldehydes ⟨65OR(14)52⟩.

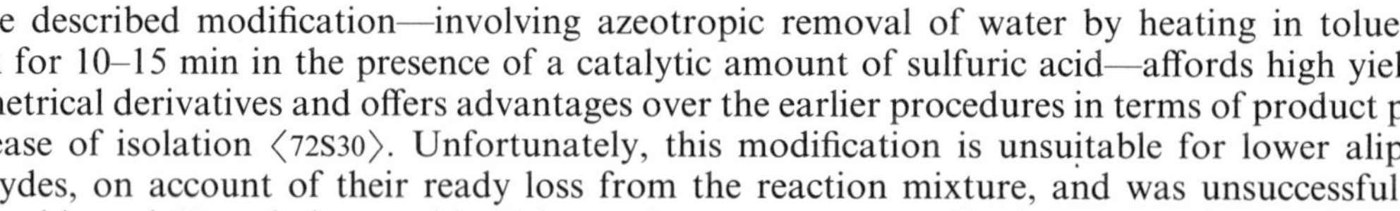

$$R^1CONH_2 + R^2CHO \longrightarrow \left[R^1CONH\overset{R^2}{\underset{OH}{\text{<}}} \right] \overset{R^1CONH_2}{\underset{R^2 = H}{\longrightarrow}} \begin{matrix} R^1CONH \\ R^1CONH \end{matrix} \mathord{>}\!\!R^2$$

(**83**)

Scheme 21

One described modification—involving azeotropic removal of water by heating in toluene at reflux for 10–15 min in the presence of a catalytic amount of sulfuric acid—affords high yields of symmetrical derivatives and offers advantages over the earlier procedures in terms of product purity and ease of isolation ⟨72S30⟩. Unfortunately, this modification is unsuitable for lower aliphatic aldehydes, on account of their ready loss from the reaction mixture, and was unsuccessful with formamide and N-methylacetamide. Trioxan, however, was an effective reagent for the formation of methylene-bis-amides.

In the case of methylene-bis-amides, isolation of the intermediate amido alcohols (**83**) and their reaction with a second amide under acidic conditions provides a high-yielding entry to a wide variety of asymmetrical derivatives (Scheme 21) ⟨65OR(14)52⟩. Addition of the same amide will, of course, result in the formation of symmetrical compounds. Modifications in which the electrofugicity of the leaving group has been changed by conversion to halo, dimethylamino, methoxy or acetoxy substituents have also been described which allow the displacement to occur under basic rather than acidic conditions ⟨57CB50, 65OR(14)52, 74LA539, 87CB1897⟩, but any benefit arising from such changes is likely to be limited to specific transformations. One interesting variant, however, is the Lewis acid-mediated N-alkylation of the TMS-protected pyrimidine (**84**), which furnishes the asymmetrical alkylidene-bis-amide (**85**) (Equation (48)) in high yield ⟨82JOC1706⟩.

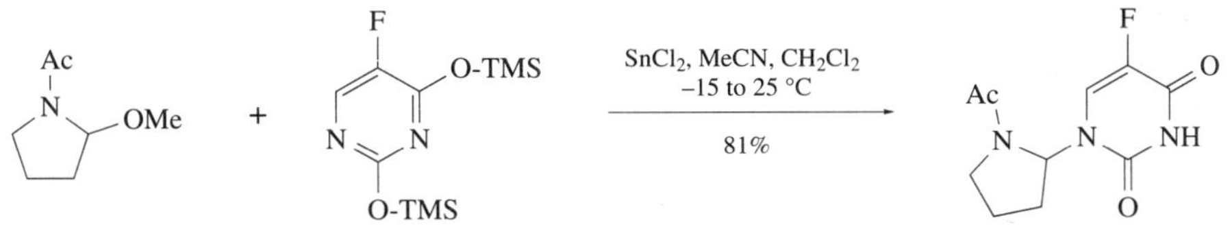

Asymmetrical derivatives may also be prepared in good yield by treatment of nitriles with amido alcohols (**83**) or their derivatives in the presence of concentrated sulfuric acid ⟨74JOC3745⟩. Analogous reactions of nitriles and trioxan result in symmetrical perhydro-1,3,5-triazines (Equation (49)), but sulfonic acid ion-exchange resins (e.g., Amberlyst 15) limit degradation reactions and increase product yield ⟨86S643⟩. The use of ion-exchange resins as substitutes for sulfuric acid in the synthesis of asymmetrical N^1,N^2-alkylidene-bis-amides does not appear to have been investigated.

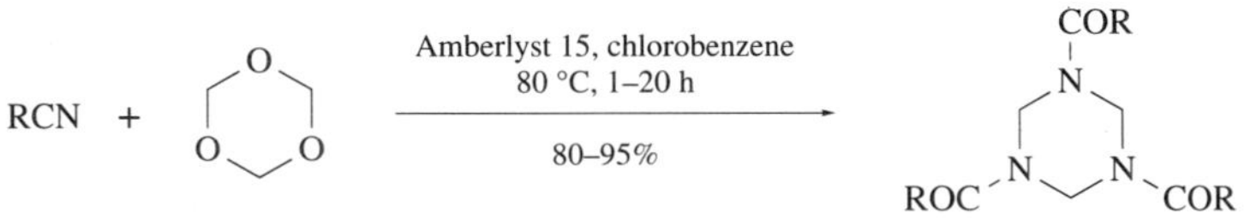

Although the reaction of equimolar quantities of two different amides with a single aldehyde is not an efficient procedure for the formation of asymmetrical alkylidene-bis-amides ⟨51RTC269⟩, intramolecular condensation reactions offer a potential route to both symmetrical and asymmetrical products ⟨65OR(14)52⟩. Moreover, intramolecular reactions have been shown to overcome the relative inertness of simple ketones towards attack by amides, as illustrated by the formation of the tetracyclic derivative (**88**) from (**86**) (Scheme 22, where *p*-TSA represents toluene *p*-sulfonic acid) ⟨78JHC949⟩. In this instance the reaction may proceed by way of the intermediate bis(amide) (**87**), which optimally positions the ketone for attack by both amide groups; alternatively, initial formation of a Schiff base might similarly facilitate closure of the lactam rings.

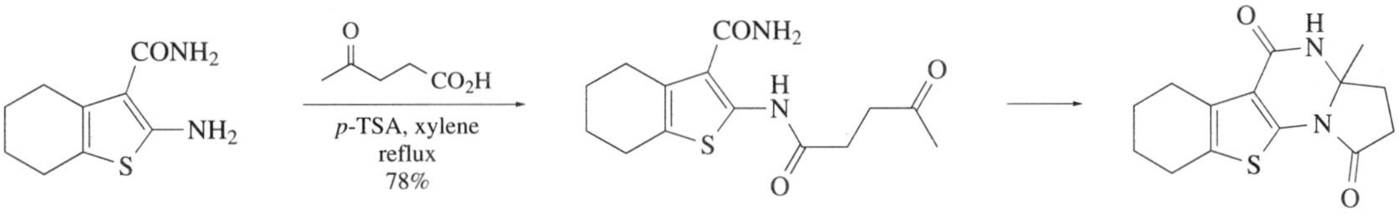

Scheme 22

4.09.4.1.2 Reductive methods

The reductive acylation of amidines is rarely used for the preparation of N^1,N^2-alkylidene-bis(amides), but such reactions do provide an efficient route to the synthesis of N^1,N^2-diacylimidazolidines from imidazole derivatives. Typically, the reduction is carried out hydrogenolytically over platinum oxide in the presence of acetic anhydride (Equation (50)) ⟨61JOC1649, 69LA(729)73⟩, but a greater variety of acylated derivatives are achievable by reduction of the acylium perchlorates with sodium borohydride in acetonitrile ⟨90S951⟩.

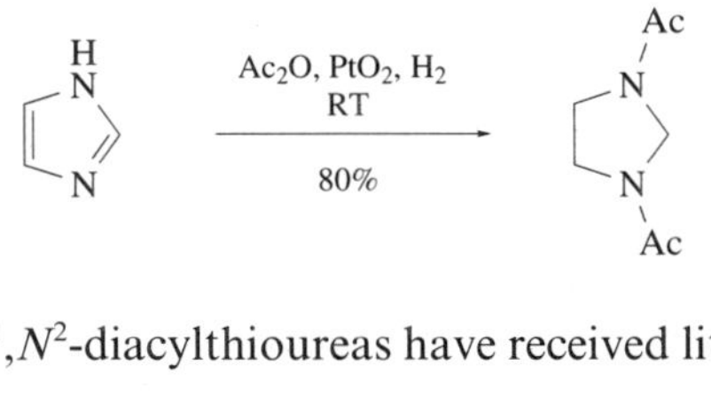

Desulfurisation reactions of N^1,N^2-diacylthioureas have received little attention for the formation of N^1,N^2-methylene-bis(amides), but high yields of hexahydropyrimidinediones (**89**) were obtained from 5,5-disubstituted-2-thiobarbituric acids when heated in refluxing ethanol with Raney nickel for several hours ⟨55JA745⟩.

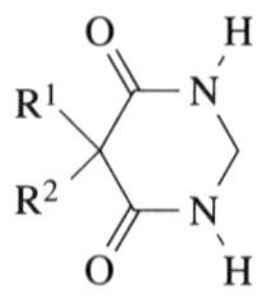

(**89**)

4.09.4.1.3 Nucleophilic addition to imines, enamines and isocyanates

The activation of imidazole to nucleophilic attack at C-2 following acylation at both nitrogen atoms may be exploited for the synthesis of a variety of heterocyclic compounds. In particular, good yields of a large number of 2-aryl and heteroaryl N^1,N^2-diacylimidazolines (Equation (51)) have been prepared by coupling nucleophilic arenes with imidazole activated by alkanoic acid anhydrides ⟨80T2505⟩. Since hydrolysis of the resultant imidazolines leads to the corresponding formylated arenes, the overall conversion resembles the Vilsmeier reaction, the scope of which appears to be similar. A notable advantage of the imidazoline route is the generation of the aldehyde in a protected form. A related addition of trimethylsilyl cyanide to benzimidazole activated by benzoyl chloride gives 2-cyano-1,3-dibenzoyl-2,3-dihydrobenzimidazole in almost quantitative yield ⟨91JOC865⟩. Several analogous nucleophilic additions to imines generated at the C-6 position of 6-amidopenicillins have been shown to give interesting 6,6-diacylaminopenicillin derivatives ⟨78CRV65, 85TL377⟩, but this approach is likely to be limited to closely related compounds.

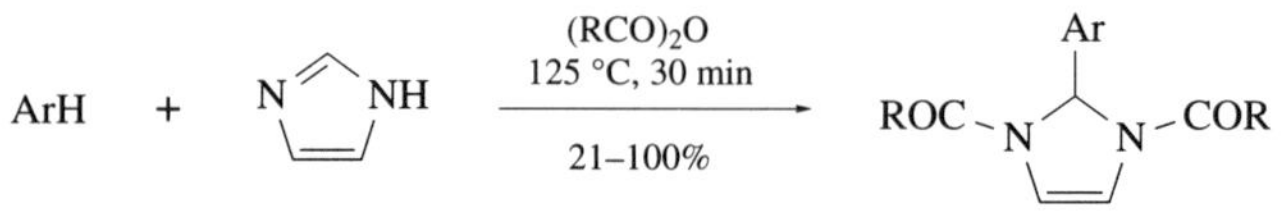

$$\text{ArH} + \underset{\text{N}}{\overset{\text{N}}{\bigvee}}\text{NH} \xrightarrow[21-100\%]{(RCO)_2O, 125\,°C, 30\,\text{min}} \text{ROC}-\text{N}\underset{\text{N}}{\overset{\text{Ar}}{\bigvee}}\text{N}-\text{COR} \qquad (51)$$

The base-catalysed reaction of 4-pyridone or 4-quinolone with isocyanates provides a useful route to 1,3,5-triazine derivatives (Scheme 23) and remains to be fully explored ⟨82TL3181⟩. Preliminary studies suggest that the reaction proceeds through the intermediate urea (**90**), which adds a second molecule of isocyanate followed by Michael addition to the enone. An interesting extension of this type of reaction involves the acid-catalysed reaction of isocyanates with acetals and thioacetals to give near-quantitative yields of alkylidene-bis-carbamic esters and thioesters, respectively ⟨62AG(E)592⟩.

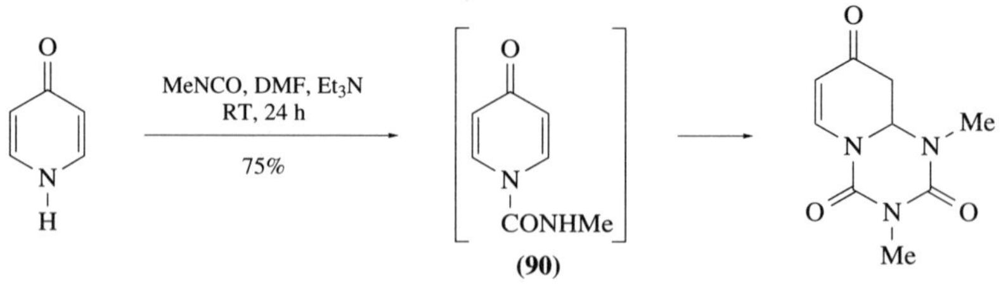

Scheme 23

4.09.4.1.4 Miscellaneous procedures

Several unusual rearrangement reactions have been described which lead to heterocyclic systems embodying the N^1,N^2-alkylidene-bis(amide) moiety ⟨78T2399, 84TL1769, 86H(24)25⟩, but none offers a route of general application. Curtius rearrangements, on N-protected amino acid azides (e.g., by benzyloxycarbonyl, Cbz), however, are likely to have greater versatility since it is possible to trap the intermediate isocyanates when the reaction is conducted in the presence of an appropriate amine

(Equation (52)) ⟨53JA3469⟩. Symmetrical N^1,N^2-alkylidene-bis(carbamates) ⟨55JCS4280⟩ and N-acyl alkylidene carbamates ⟨84JOC821⟩ have been prepared in a similar manner following an alcoholic workup procedure. Cyclic analogues formed as a result of intramolecular trapping have also been observed ⟨75JHC595⟩.

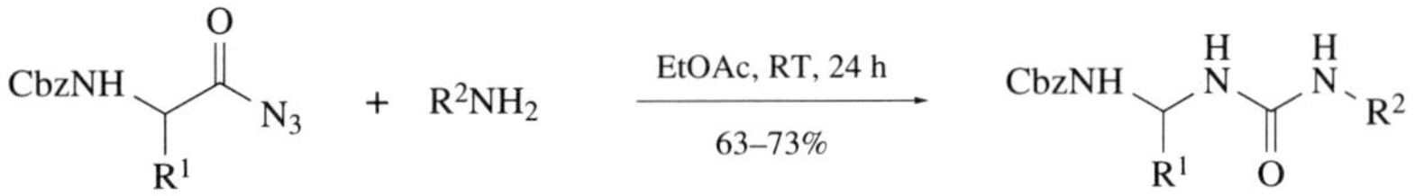

$$\text{CbzNH}\underset{R^1}{\overset{O}{\underset{|}{\text{C}}}}\text{N}_3 + R^2NH_2 \xrightarrow[63-73\%]{\text{EtOAc, RT, 24 h}} \text{CbzNH}\underset{R^1}{\overset{H}{\underset{|}{\text{C}}}}\underset{O}{\overset{H}{\text{N}}}\text{N-}R^2 \quad (52)$$

One N^1,N^2-alkylidene-bis(amide) has also been prepared by the electrolytically induced homolytic fission of an N^1,N^2-diacyldiaziridine ⟨78LA1505⟩, but little is known of the scope of this reaction.

4.09.4.2 Sulfonated Derivatives

In the same way that N^1,N^2-alkylidene-bis(amides) may be prepared by the acylation of monoacyl aminals (see Section 4.09.4.1), sulfonation of the same precursors leads to mixed derivatives in which one nitrogen atom is acylated and the other sulfonated. Reasonable yields of such compounds may usually be isolated following reaction with an alkyl- or arylsulfonyl chloride in pyridine at 0–20 °C ⟨91JCS(P1)1693, 93JMC305⟩. The synthesis of mixed derivatives of this type is also possible by either the reaction of aminols (**91**) with an appropriate amide, or by treatment of the alternative aminol (**92**) with an appropriate sulfonamide (Scheme 24) ⟨65OR(14)52⟩. The latter process gives better yields. The related displacement of dimethylamine from acylated N,N-dimethylamino aminals on treatment with sulfonamides under alkaline conditions similarly affords good yields of mixed derivatives ⟨57CB50⟩.

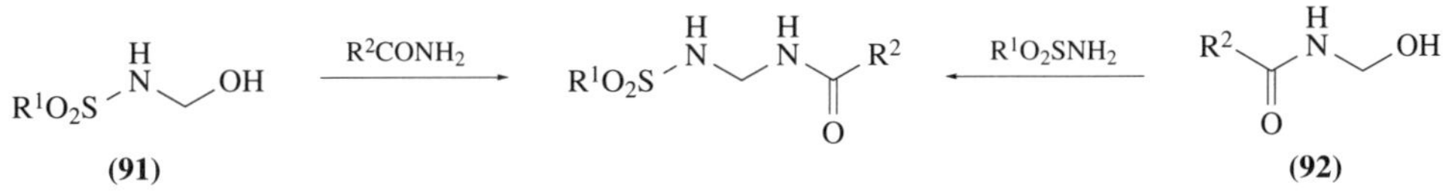

Scheme 24

Unlike amides, sulfonamides do not readily react with substituted aldehydes and in the presence of zinc chloride generally yield N-sulfonyl imines (**94**) rather than the corresponding N^1,N^2-alkylidene-bis(sulfonamides) (**95**) (Scheme 25) ⟨55BSF669⟩. With more reactive aldehydes such as chloral or formaldehyde, however, the intermediate aminols (**93**) may be isolated in quantitative yield ⟨55BSF669⟩. When catalysed by concentrated sulfuric acid, the reaction of sulfonamides with chloral produces 60–65% yields of the bis(sulfonamides) (**95**) ⟨51G80⟩. Symmetrical bis(sulfonamides) (**95**) may also be prepared on heating the aminols (**91**) derived from an arenesulfonamide and formaldehyde with a further quantity of sulfonamide, or by the direct reaction of formaldehyde with two equivalents of the sulfonamide ⟨34BSF990⟩. The cyclic derivatives (**96**) may also be prepared in this manner ⟨73S243⟩.

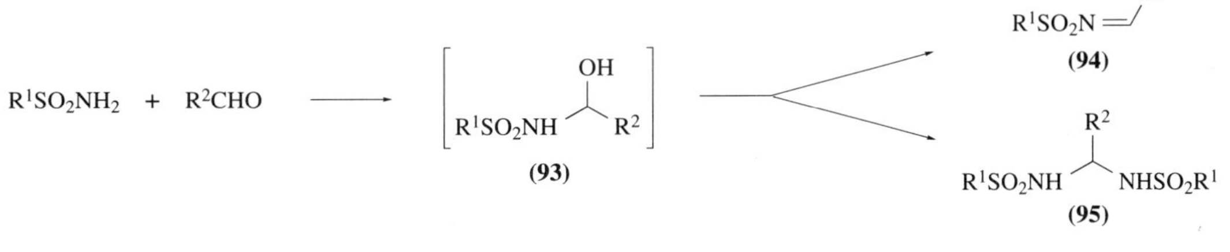

Scheme 25

The reaction of sulfonamides with trioxan is more complex, and a variety of products are possible depending on the conditions used. Yields of 55–91% of the hexahydro-1,3,5-triazines (**96**) are formed with three equivalents of sulfonamide, whereas lower yields (9–15%) of the tetrahydro-1,3,5-oxadiazines (**97**) are isolable when a 2:1 ratio of reactants is used ⟨75JCS(P1)772⟩.

(**96**)   (**97**)

In a rather unusual reaction, *N*-alkyl arenesulfonamides furnished excellent yields of bis(sulfonamides) when heated with dimethyl sulfoxide and phosphoric oxide in refluxing xylene (Equation (53)) ⟨65CR(260)2252⟩.

$$2 \; \text{ArSO}_2\text{NHBu}^t + \text{Me}_2\text{SO} \xrightarrow[\substack{150\,°C,\,3\,h \\ 92\%}]{\substack{P_2O_5 \\ \text{xylene, reflux}}} \text{ArSO}_2\text{N(Bu}^t\text{)CH}_2\text{N(Bu}^t\text{)SO}_2\text{Ar} \quad (53)$$

4.09.5 GEMINALLY SUBSTITUTED ALKANES BEARING TWO SIMILAR DICOORDINATE OR HETEROSUBSTITUTED NITROGENS

4.09.5.1 *gem*-Dinitro Alkanes and *gem*-Dinitroso Alkanes

The synthesis of *gem*-dinitro compounds has achieved significance due to their potential as high-energy explosives, and as such they should be handled with extreme care. Moreover, the observation that aliphatic polynitro compounds undergo nucleophilic substitution reactions has markedly enhanced the synthetic procedures by which such compounds may be prepared ⟨76RCR1052, B-82MI 409-01⟩. No representatives of the *gem*-dinitroso alkane class have been found.

4.09.5.1.1 *Nitration of nitro alkanes*

A versatile and mild method for the preparation of *gem*-dinitro compounds is the Kaplan–Shechter reaction (Equation (54)), in which simple primary or secondary nitro alkanes are further nitrated with mixtures of silver nitrate and sodium nitrite in neutral or alkaline media ⟨61JA3535, 90JOC2920⟩. Although the dinitro compounds may be isolated in good yields, the use of silver nitrate makes the reaction expensive, even when recovery of silver is allowed for. An alternative, electrochemical approach utilising a silver anode has been reported ⟨63T(S)3⟩, but a more attractive procedure, which is particularly suited to large-scale reactions, has been developed in which silver nitrate may be replaced by potassium iron(III) cyanide ⟨79PJC187, 83JOC332, 87JOC4781, 93JOC763⟩. In general, the yields obtained using potassium iron(III) cyanide are comparable to those achieved using silver nitrate (Table 1). Both reactions proceed via a radical anion intermediate derived from the enol form of the mononitro precursor ⟨75AG(E)734⟩.

$$\underset{R^2}{\overset{R^1}{>}}\!\!-\text{NO}_2 \xrightarrow[60-98\%]{\text{AgNO}_2,\,\text{NaNO}_2} \underset{R^2}{\overset{R^1}{>}}\!\!\underset{\text{NO}_2}{\overset{\text{NO}_2}{<}} \quad (54)$$

In large-scale reactions the large excess of potassium iron(III) cyanide (typically fivefold) may cause handling difficulties, although these may be circumvented by the use of catalytic amounts of reagent, provided that an oxidant such as potassium persulfate is included ⟨85JOC1699, 90JOC2920⟩. Thus, while it was not possible to prepare 2,2-dinitropropanediol under the usual conditions, 30–35% yields were isolable following this modified procedure ⟨83JOC332⟩.

Whereas the synthesis of mononitro alkanes by nitro group transfer from alkyl nitrates is well documented ⟨B-82MI 409-01⟩, similar transfer reactions leading to *gem*-dinitro alkanes are poorly exemplified and only a single example, involving an intramolecular transfer reaction (Equation (55)), has been reported ⟨59JOC865⟩. A more usual transfer procedure involves the treatment of an

Table 1 Preparation of *gem*-dinitro alkanes.

Reactant	Product	Conditions	Yield (%)	Ref.
(CH₃)₂CH–NO₂	(CH₃)₂C(NO₂)₂	NaOH, NaNO₂, K₃Fe(CN)₆, MeOH (aq.), 1.5 h	83	83JOC332
cyclopentyl-NO₂	1,1-dinitrocyclopentane	as above	85	83JOC332
2,6-dinitronorbornane	2,2,6,6-tetranitronorbornane	NaOH, NaNO₂, K₃Fe(CN)₆, CH₂Cl₂, H₂O, 0 °C, 18 h	83	93JOC763
2,5-dinitro-cis-bicyclo[3.3.0]octane	2,2,5,5-tetranitro-cis-bicyclo[3.3.0]octane	as above	75	93JOC763
CH₃CH₂–NO₂	CH₃C(NO₂)₂H (2,2-dinitropropane)	C(NO₂)₄, NaOMe, MeOH, 4 days	45	61USP2991315
CH₃CH₂CH₂–NO₂	CH₃CH₂C(NO₂)₂CH₃-like	MeN(NO₂)₂, KOH, MeOH, 0 °C, 2 h	55	77IZV2384

alkaline solution of a primary nitro alkane with tetranitromethane; *gem*-dinitro alkanes may be obtained in 30–80% yields (Table 1) ⟨61USP2991315, 67USP3316311⟩.

$$\text{camphor-derived NO}_2\text{, ONO}_2 \xrightarrow[\text{ii, HCl}]{\text{i, KOH, EtOH, H}_2\text{O, warm}} \text{O}_2\text{N, NO}_2\text{, OH} \quad 90\% \tag{55}$$

An alternative source of NO₂⁺ is methyldinitramine, which provides modest yields of *gem*-dinitro alkanes on reaction with salts derived from primary nitro alkanes (Table 1) ⟨77IZV2384⟩. Secondary nitro alkanes undergo an oxidative dimerisation under similar conditions.

4.09.5.1.2 Oxidative nitration of oximes

The oxidative nitration of oximes provides a route to *gem*-dinitro alkanes, but rarely results in high yields. The Ponzio reaction involving dinitrogen tetroxide, for example, gives 27–38% yields of aryldinitromethanes from aryl aldehyde oximes (Table 2), some representatives of which are highly explosive ⟨06JPR494, 46JA2252⟩. A more unusual oxidation has been demonstrated with the isoxazoline (**98**) (Scheme 26), which undergoes ring opening of the intermediate to afford the *gem*-dinitro compound (**99**) ⟨87JOC3442⟩.

The sequential nitration and oxidation of oximes by nitric acid and hydrogen peroxide is a more convenient substitute for dinitrogen tetroxide, and has been successfully used for the formation of the tetranitronorbornane (**100**) ⟨93JOC759⟩ and *gem*-dinitro steroids ⟨65JCS2601⟩ (Table 2). Nitric acid alone is also suitable in some circumstances, as illustrated by the synthesis of the explosive 1,3,3-trinitroazetidine (Table 2) ⟨93TL6677⟩. Although its general applicability has not been demonstrated, clay-supported iron(III) nitrate (Clayfen) is potentially one of the most useful reagents for the oxidative nitration of oximes, since it can effect conversion under relatively mild conditions ⟨90JOC6198⟩. For example, simply stirring a solution of the oxime (**101**) with Clayfen in dichloromethane at ambient temperature was sufficient to form the intermediate isolable nitro enol (**102**),

Table 2 Preparation of *gem*-dinitro alkanes.

Reactant	Product	Conditions	Yield (%)	Ref.
Ph-CH=N-OH	Ph-CH(NO$_2$)$_2$	N$_2$O$_4$, ether, reflux	38	46JA2252
4-O$_2$N-C$_6$H$_4$-CH=N-OH	4-O$_2$N-C$_6$H$_4$-CH(NO$_2$)$_2$	N$_2$O$_4$, ether, reflux, 1.75 h	27	46JA2252
norbornane with N-OH and geminal NO$_2$, NO$_2$	norbornane with O$_2$N, NO$_2$ and NO$_2$, NO$_2$ (100)	i, HNO$_3$, urea, NH$_4$NO$_3$, reflux, 0-5 h; ii, H$_2$O$_2$, reflux, 15 min	52	93JOC759
decalin oxime derivative	decalin gem-dinitro derivative	i, HNO$_3$, CH$_2$Cl$_2$, reflux, 0.66 h; ii, H$_2$O$_2$, HNO$_3$, CH$_2$Cl$_2$, reflux, 1 h	23	65JCS2601
HO-N=azetidine-N-SO$_2$Tol	(O$_2$N)$_2$C-azetidine-N-NO$_2$	99% HNO$_3$, CH$_2$Cl$_2$, reflux	40–50	93TL6677
Ph-CH=CH-NO$_2$	Ph-CH(OMe)-CH(NO$_2$)$_2$	C(NO$_2$)$_4$, NaOH, MeOH (aq.) 0–10 °C, 1 h	60	78JOC2460
Ph-CH=C(Me)-NO$_2$	Ph-CH(Me)-C(NO$_2$)$_2$-Me	i, MeLi, THF, –40 °C, 1 h; ii, C(NO$_2$)$_4$, 20 °C, 1 h	60	78JOC2460
C(NO$_2$)$_3$-Me	MeO-CH$_2$-C(NO$_2$)$_2$-Me	KOMe, MeOH	80	03CB434
C(NO$_2$)$_3$-Me	NC-CH$_2$-C(NO$_2$)$_2$-Me	KCN, KOMe, MeOH, 6 h	80	06CB2543
C(NO$_2$)$_3$-Me	(EtO$_2$C)$_2$CH-CH(Me)(NO$_2$)$_2$	(EtO$_2$C)$_2$CH$_2$, KOEt, EtOH, reflux	36	57JA4708
Br-CH(NO$_2$)-CH$_2$CH$_2$-CH(Br)(NO$_2$)	O$_2$N-C(NO$_2$)-CH$_2$CH$_2$-C(NO$_2$)-NO$_2$	KNO$_2$, KOH, MeOH (aq.), 0 °C, 2.5 h	28	62JOC3598
HO-CH$_2$-CH(Br)(NO$_2$)	HO-CH$_2$-C(NO$_2$)$_2$	KNO$_2$, KOH, MeOH (aq.), 0 °C	77–87	57JOC1665

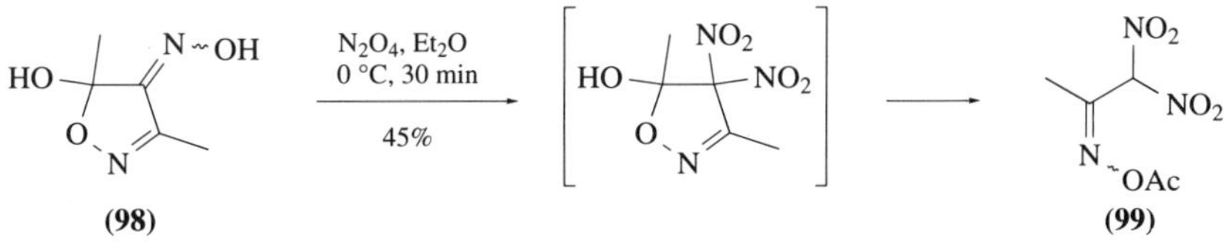

Scheme 26

which generated the dinitro derivative (103) when heated in the same solvent under reflux (Scheme 27) ⟨90JOC6198⟩. The same intermediate was also formed on treatment of the aldehyde (104) with Clayfen. Presumably, montmorillonite-supported copper(II) nitrate (Claycop), which is also a source of NO_2^+ and has a longer shelf-life than Clayfen ⟨85S909⟩, could perform similar transformations, although this does not appear to have been studied.

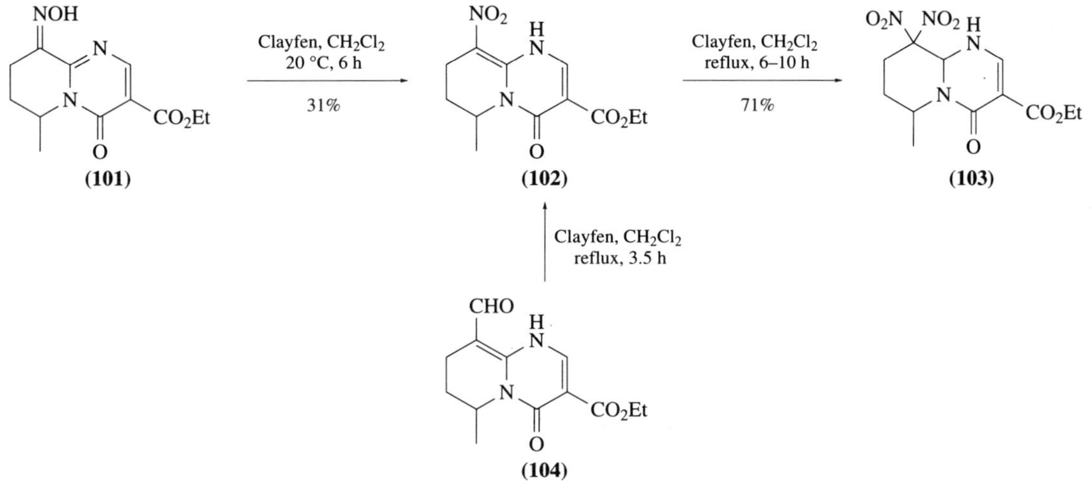

Scheme 27

4.09.5.1.3 Formation from nitro alkenes

As a development from the facile nitration of alkyl nitronates with tetranitromethane ⟨61USP2991315, 67USP3316311⟩ (see Section 4.09.5.1.1), quenching of the resultant anions from Michael-type additions to nitro alkenes has been demonstrated (Scheme 28) and a number of functionalised *gem*-dinitro alkanes have been prepared in this manner ⟨78JOC2460⟩. This reaction is equally effective for the addition of alkoxides and alkyllithium reagents (Table 2) and isolated yields fall in the range 29–60%, although the true yields are generally much higher (70–90%) ⟨78JOC2460⟩.

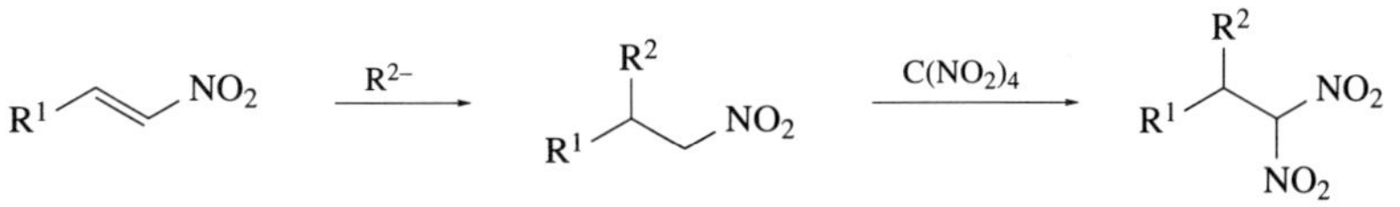

Scheme 28

An alternative Michael approach is also possible from *gem*-trinitro alkanes which eliminate the elements of HNO_2 under some basic conditions to form a vinyl dinitro intermediate, which is readily trapped by subsequent nucleophilic addition (Scheme 29) ⟨57JA4708⟩. A range of α-functionalised *gem*-dinitro compounds derived from 1,1,1-trinitroethane (Table 2) have been prepared by this procedure. With strong bases such as butyllithium or potassium hydroxide, however, nucleophilic attack on one of the nitro groups occurs to generate the anion of 1,1-dinitroethane which can be alkylated or quenched with acid ⟨03CB434, 06CB2543, 57JA4708⟩.

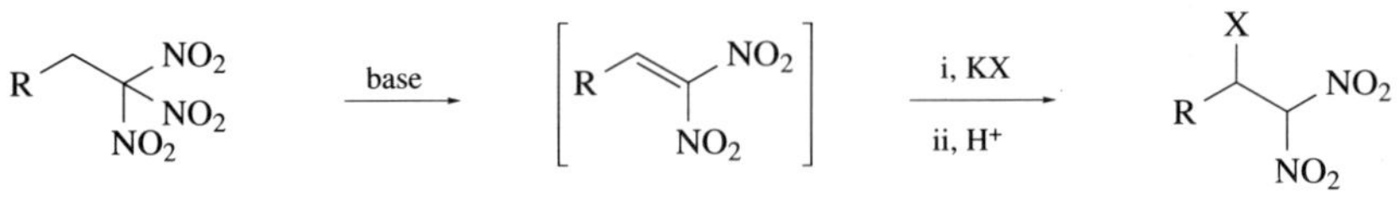

Scheme 29

4.09.5.1.4 Formation from halo nitro alkanes

The formation of *gem*-dinitro alkanes from α-halo nitro alkanes, the ter Mer reaction, is a well-established reaction which proceeds with potassium nitrate in the presence of base ⟨1876LA(181)1, 57JOC1665, 62JOC3598⟩. The yields from this reaction can be low, however (Table 2), and since the halo nitro alkanes are usually prepared from the corresponding nitro alkanes, the methods described above are generally preferable.

4.09.5.2 *gem*-Diazido Alkanes

Primarily as a consequence of their explosive character, this class of compounds has received relatively little attention, although their potential as intermediates in the preparation of tetrazoles and nitriles is recognised ⟨87T693, 93TL5097, 94TL89⟩.

4.09.5.2.1 Formation from gem-dihalo alkanes

Formally the simplest route to *gem*-diazido alkanes is the treatment of the corresponding dihalo compounds with sodium azide (Equation (56)); good yields are possible using this procedure when the halogens are activated to displacement (Table 3) ⟨80ZC437, 87CJC166, 89JHC1555⟩. In situations where the halogen atoms are not activated, displacement using Group 1 metal azides is difficult, but replacement by silver azide in nitromethane ⟨90CC431⟩ or by ammonium azide in liquid sulfur dioxide ⟨70CZ215⟩ has proved successful in specific instances (Table 3).

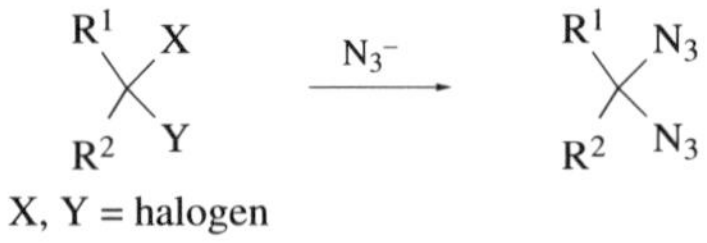

(56)

4.09.5.2.2 Formation from carbonyl compounds

The Lewis acid-catalysed reaction of aldehydes, and to a lesser extent ketones, with trimethylsilyl azide is the most general, and probably the safest, route to *gem*-diazido alkanes ⟨87T693, 88S106, 93S1218⟩. Typically, the reaction has been catalysed by the chlorides of zinc, tin(II) and titanium(IV) and results in good product yields (Table 3). Optimal yields from ketones are obtained in the absence of solvent. Trimethylsilyl azide has also been shown to convert a sugar lactone into the corresponding

Table 3 Preparation of *gem*-diazides.

Reactant	Product	Conditions	Yield (%)	Ref.
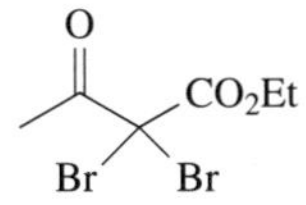 MeC(O)C(Br)(Br)CO₂Et	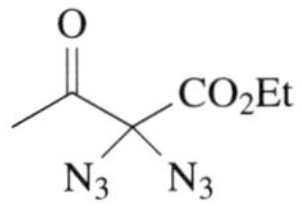 MeC(O)C(N₃)(N₃)CO₂Et	NaN₃, DMF, 0 °C, 1 h	76	87CJC166
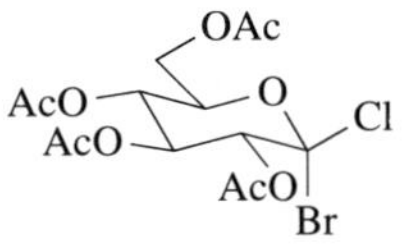 Acetylated sugar with Cl, Br	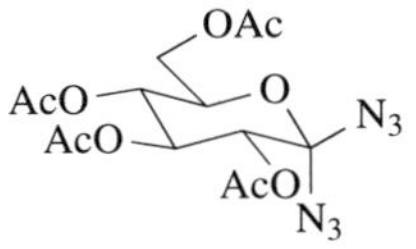 Acetylated sugar with two N₃	AgN₃, MeNO₂, 20 °C, 24 h	70	90CC431
 PrCHO	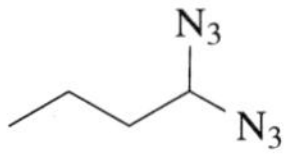 PrCH(N₃)₂	TMS-N₃, SnCl₂, CH₂Cl₂, 0–20 °C, 20 h	78	87T693
PhCHO	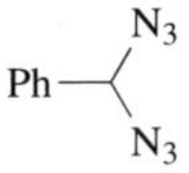 PhCH(N₃)₂	TMS-N₃, SnCl₂, CH₂Cl₂, 0–20 °C, 20 h	87	87T693
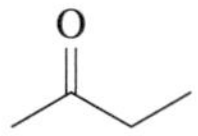 2-butanone	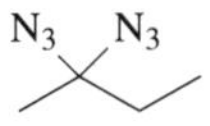 2,2-diazidobutane	TMS-N₃, SnCl₂, 0–20 °C, 20 h	52	88S106
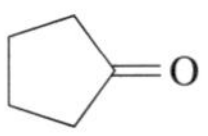 cyclopentanone	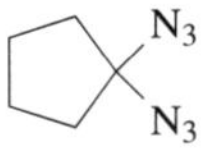 1,1-diazidocyclopentane	TMS-N₃, SnCl₂, 0–20 °C, 20 h	48	88S106

gem-diazide (Equation (57)) in moderate yield, although the general applicability of this reaction is not known ⟨90CC431⟩.

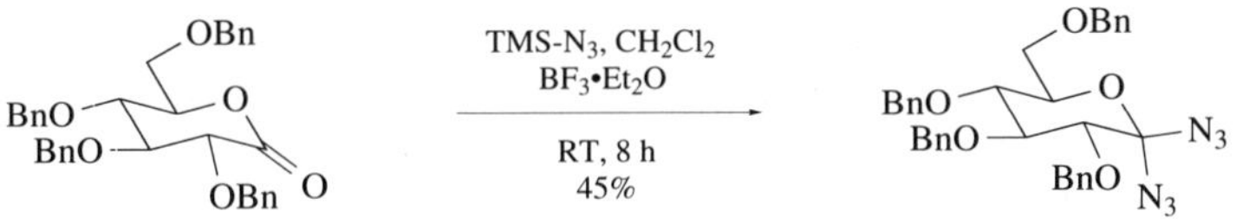

$$\text{tri-OBn sugar lactone} \xrightarrow[\text{RT, 8 h, 45\%}]{\text{TMS-N}_3, \text{CH}_2\text{Cl}_2, \text{BF}_3\cdot\text{Et}_2\text{O}} \text{tri-OBn sugar gem-diazide} \quad (57)$$

4.09.5.2.3 Miscellaneous methods

An isolated example in which bromine azide furnishes the *gem*-diazide (**107**) from the α,β-unsaturated nitroester (**105**) (Scheme 30) has been reported ⟨78BCJ2614⟩. The reaction is believed to proceed through an addition–elimination sequence via the intermediate (**106**).

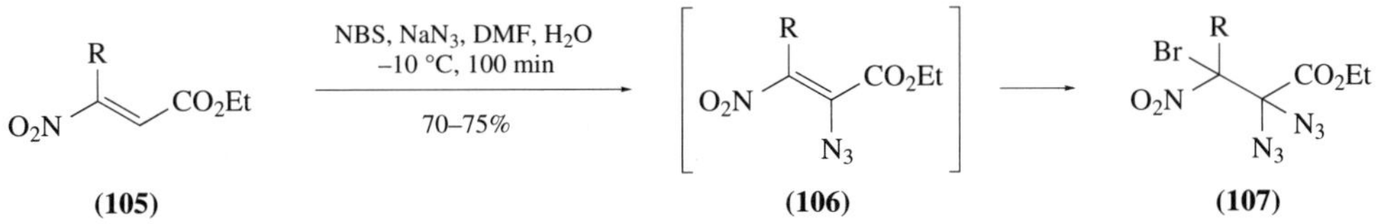

$$\underset{(\mathbf{105})}{\text{O}_2\text{N-C(R)=CH-CO}_2\text{Et}} \xrightarrow[\text{70–75\%}]{\text{NBS, NaN}_3, \text{DMF, H}_2\text{O} \\ -10\,°\text{C, 100 min}} \underset{(\mathbf{106})}{[\text{O}_2\text{N-C(R)=C(N}_3\text{)-CO}_2\text{Et}]} \longrightarrow \underset{(\mathbf{107})}{\text{O}_2\text{N-C(R)(Br)-C(N}_3\text{)-CO}_2\text{Et}}$$

Scheme 30

4.09.5.3 *gem*-Diisocyanates and *gem*-Diisothiocyanates

Each of these classes of compound is virtually unknown. There are two reports describing the synthesis of diisocyanates via a Curtius rearrangement ⟨77JOC4095, 84ACS113⟩, as exemplified by the formation of (**108**) from the corresponding diacyl chloride (Scheme 31). The yields for compound (**108**) and several closely related analogues were generally good, although in most instances the products were transformed further without isolation. The most convenient method for preparing these compounds is by the tetrabutylammonium iodide-catalysed reaction of an α-chloro isocyanate with TMS isocyanate (Equation (58)). Yields by this procedure are typically around 50–70% ⟨77ZOR723⟩.

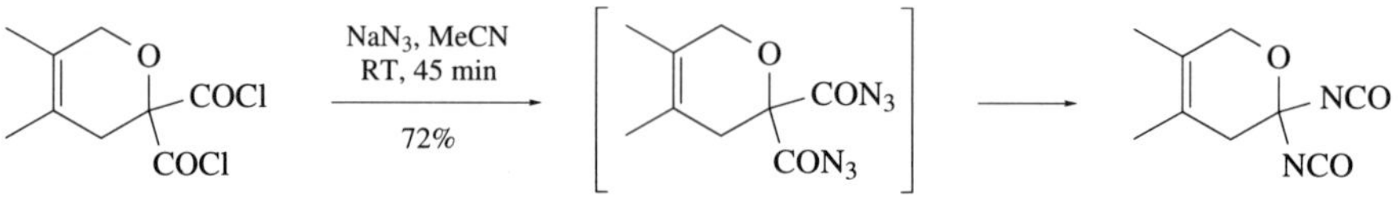

Scheme 31

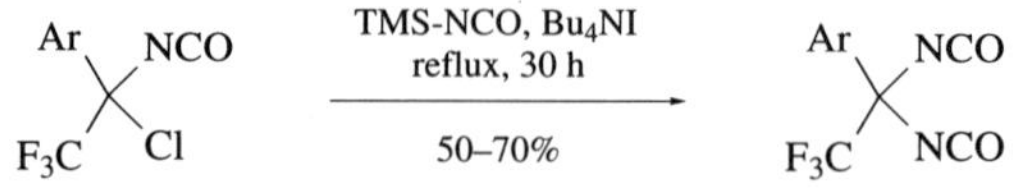

$$\text{Ar}\underset{F_3C}{\overset{NCO}{\diagdown\diagup}}\text{Cl} \xrightarrow[\text{50–70\%}]{\text{TMS-NCO, Bu}_4\text{NI, reflux, 30 h}} \text{Ar}\underset{F_3C}{\overset{NCO}{\diagdown\diagup}}\text{NCO} \quad (58)$$

The first examples of *gem*-diisothiocyanates were described in 1979 following the reaction of a series of α-bromo alkyl isothiocyanates with potassium thiocyanate (Equation (59)) ⟨79CB1956, 87S745⟩. In a similar fashion, α-(isothiocyanato) isocyanates have been prepared in 52–86% yields by the reaction of α-bromo alkyl isocyanates with potassium thiocyanate ⟨82CB860⟩.

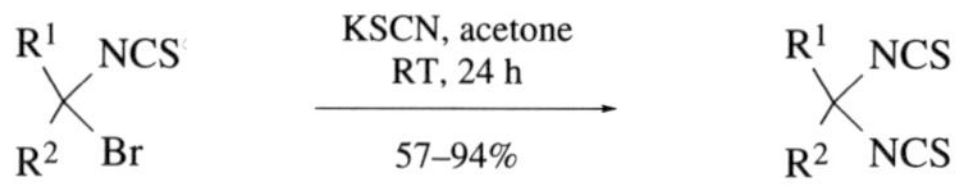

$$\underset{R^2}{\overset{R^1}{\diagdown}}\underset{Br}{\overset{NCS}{\diagup}} \xrightarrow[\text{57–94\%}]{\text{KSCN, acetone, RT, 24 h}} \underset{R^2}{\overset{R^1}{\diagdown}}\underset{NCS}{\overset{NCS}{\diagup}} \quad (59)$$

R^1 = alkyl, R^2 = alkyl, Ph or CO_2Me

4.09.5.4 *gem*-Dinitrosamines and *gem*-Dinitramines

gem-Dinitrosamines are a little known class of compound. However, hexamine (**109**) has been shown to undergo a dealkylative nitrosation on treatment with nitrous acid in which the exact nature of the product isolated is critically dependent on the reaction pH (Scheme 32) ⟨49MI 409-01, 51JA2777⟩. Treatment with nitrous acid generated in hydrochloric acid at pH 1 gave only the trinitroso derivative (**110**), while treatment in acetic acid at pH 3–6 gave predominantly the intermediate dinitrosamine (**111**) ⟨51JA2773⟩. The *gem*-dinitrosamine (**112**) has also been prepared from imidazolidine, prepared *in situ* by the condensation of 1,2-diaminoethane with formaldehyde, on brief treatment (5 min) with nitrous acid at 5 °C ⟨84H(22)2351⟩, as have a number of other examples by variation of the diamine ⟨84JOC5147⟩.

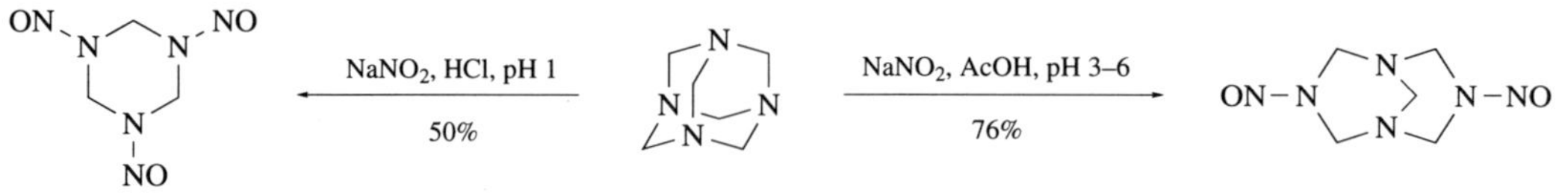

Scheme 32

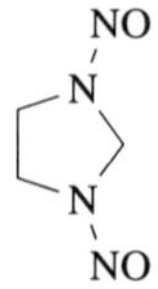

(**112**)

Much of the attention paid to *gem*-dinitramines has been concerned with their potential as high explosives, and as such they should be handled with extreme care. Studies in the 1950s conducted with hexamine (**109**), for example, showed that treatment with nitric acid under a variety of conditions (Scheme 33) afforded a number of products, including the powerful explosive cyclotrimethylenenitramine (RDX or cyclonite, (**113**)) ⟨50JCS2920, 51JA2769, 51JA2773⟩. Oxidation with acetyl nitrate prepared *in situ* at 75 °C afforded RDX (**113**) as the dominant product ⟨51JA2769⟩, whereas with nitric acid alone at a lower temperature (−40 °C) the dinitramine (**114**) was produced ⟨50JCS2920⟩. Hydrolysis of the dinitramine (**114**) gave methylenedinitramine, the simplest member of this unstable family of compounds ⟨50JCS2920⟩. RDX has also been prepared by oxidation of the nitrosamine (**110**) using a mixture of nitric acid and hydrogen peroxide ⟨51JA2777⟩.

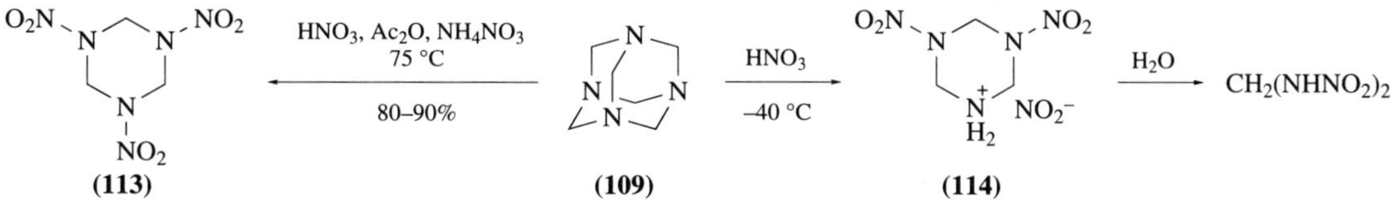

Scheme 33

In addition to their formation by the oxidative cleavage of cyclic amines, *gem*-dinitramines have been prepared by the oxidative removal of *t*-butyl groups from monocyclic amines (Equation (60)), by treatment with a mixture of nitric and sulfuric acids ⟨82JOC2474⟩. A related oxidative nitration of the triacetyl derivative (**115**) under the milder conditions of ammonium nitrate and acetic anhydride (Equation (61)) has also been reported ⟨88S743⟩. An alternative approach relies on the oxidation of *gem*-dinitrosamines using either 100% nitric acid or dinitrogen pentoxide ⟨84JOC5147⟩. This latter process appears optimal for cyclic derivatives, and for five-membered ring compounds in particular.

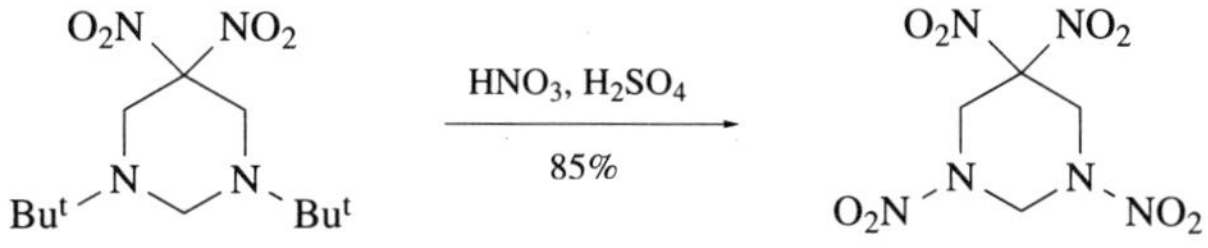

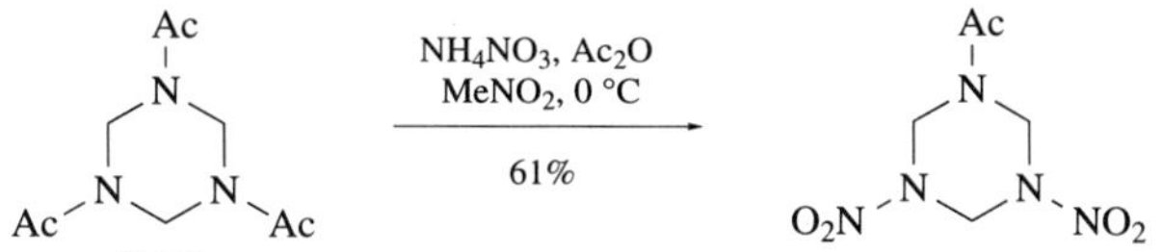

4.09.5.5 *gem*-Dihydroxylamino alkanes

The simplest method for the synthesis of *gem*-dihydroxylamino alkanes is by the condensation of hydroxylamines with aldehydes and ketones. For example, the reaction of several aliphatic hydroxylamines with aqueous formaldehyde under mildly alkaline conditions furnished good yields of the corresponding N^1,N^2-dihydroxy *gem*-diamines (Equation (62)) ⟨66CB2686⟩. A similar reaction of bis-hydroxylamines with a variety of aldehydes was shown to result in excellent yields of 1,3-dihydroxyimidazolidines which were of sufficient stability to allow subsequent acylation with either benzoyl chloride or phenyl isocyanate (Scheme 34) ⟨73AP(306)161⟩. Although this has not been demonstrated, it is likely that many other *gem*-dihydroxylamines could be prepared in this manner.

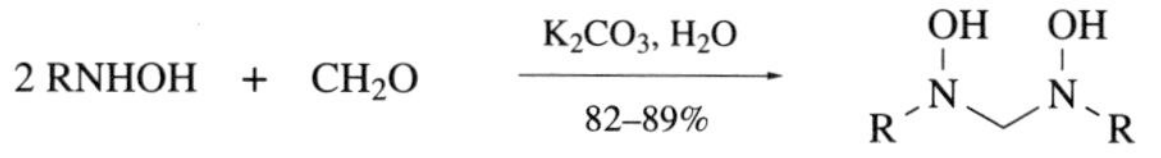

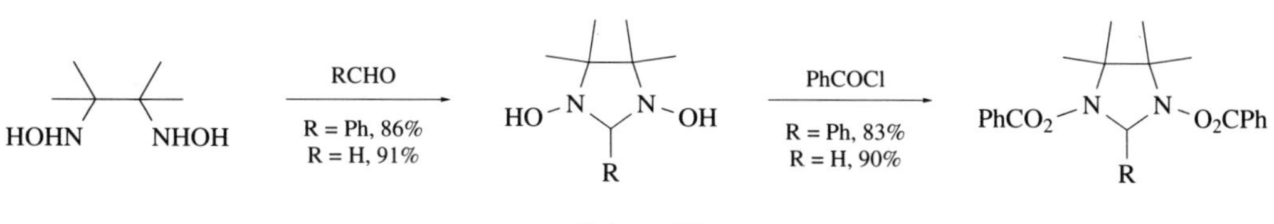

Scheme 34

In specific instances ketones have been shown to react with bis-hydroxylamines, as typified by the trapping of the intermediate (**116**), formed during the reduction of the corresponding dinitro compound, with propanone or butanone (Scheme 35) ⟨72ACS1659, 74ACS539⟩.

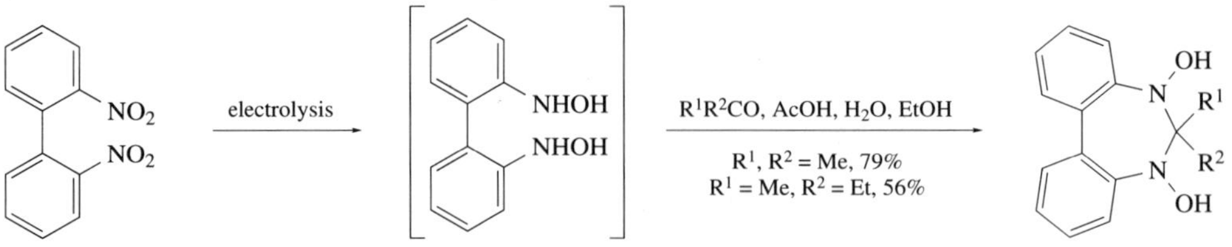

Scheme 35

4.09.5.6 *gem*-Dicarbodiimides

There is little known about this class of compounds, although representative examples have been prepared by the reaction of *gem*-diisocyanates (Section 4.09.5.3) with triphenylphosphine imides (Equation (63)) ⟨77ZOR2449⟩. Those compounds prepared were sufficiently stable to allow purification by distillation under reduced pressure.

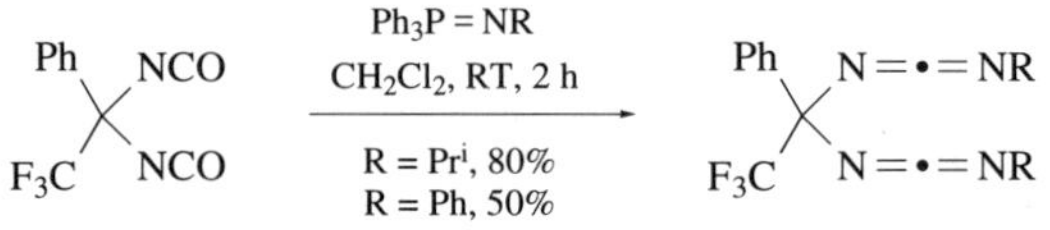

(63)

4.09.5.7 *gem*-Diazo alkanes

The principal method for the synthesis of this class of compound is by the low-temperature coupling of diazonium salts to hydrazones (Equation (64)). Arenediazonium salts have been most commonly used, but since numerous hydrazones are available through the condensation of aldehydes and hydrazines, this procedure offers an entry to a large variety of analogues ⟨67LA(706)107, 81CJC679⟩.

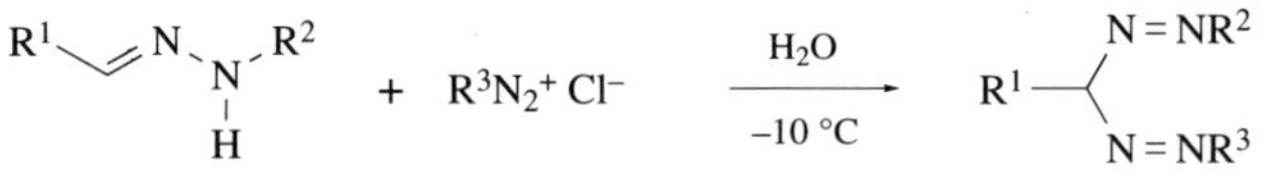

(64)

4.09.5.8 *gem*-Dihydrazino alkanes

In contrast to primary amines, which form aminals on treatment with aldehydes (Section 4.09.2.1), hydrazines having at least one unsubstituted amino group react readily with aldehydes and ketones to form hydrazones. The reaction of monosubstituted hydrazines with formaldehyde is somewhat unusual in that the initially formed hydrazones subsequently dimerize to symmetrical tetrahydro-1,2,4,5-tetrazines under very mild acidic conditions (Scheme 36) ⟨72TL949, 73ACS779⟩.

Scheme 36

The condensation of N^1,N^2-disubstituted hydrazines also leads to tetrahydro-1,2,4,5-tetrazines (Scheme 37) ⟨84CHEC(3)531⟩, although in this instance it has been proposed that the intermediate aminols (**117**) dehydrate to dipolar azomethines (**118**), which then spontaneously dimerize ⟨65JOC74⟩.

This procedure provides a simple and high yielding synthesis for symmetrical tetrahydrotetrazines from aliphatic ⟨70HCA251, 72ACS1258, 72TL949⟩ and aromatic ⟨80AG(E)724⟩ aldehydes.

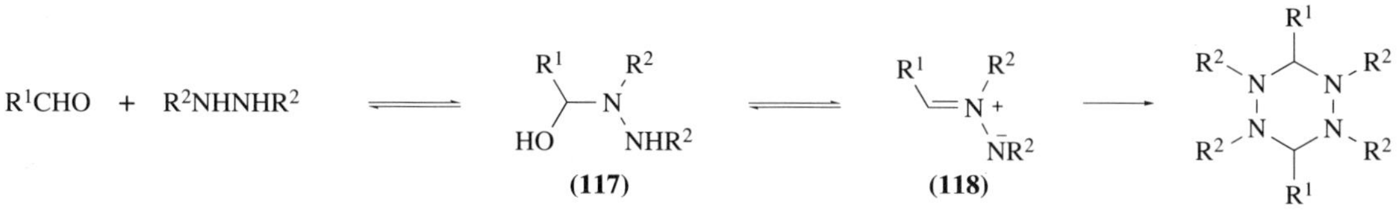

Scheme 37

Trisubstituted hydrazines are unable to form tetrahydrotetrazines and consequently react with aldehydes in a similar manner to that of primary amines ⟨63JOC1144, 66CB1678⟩. The potassium carbonate-catalysed condensation of trimethylhydrazine with formaldehyde to give 80–95% yields of the corresponding bis-hydrazine (Equation (65)) is typical ⟨66CB1678⟩. *N*-Methylhydrazine has also been reported to furnish acyclic intermediates such as (**119**) following reaction with aliphatic aldehydes ⟨69JHC187⟩. Methylene-bis(hydrazine), the simplest member of the series, is best prepared by the reaction of hydrazine with tris(chloromethyl)amine in benzene under reflux, although the yield by this procedure is only 20% ⟨74LA1851⟩.

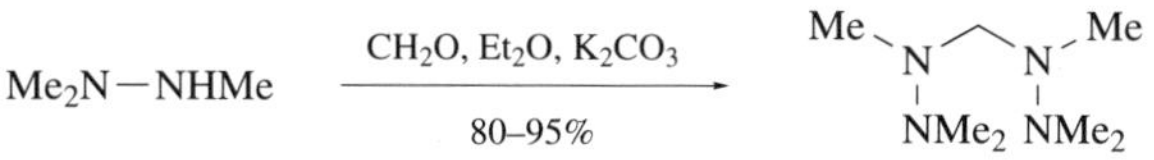

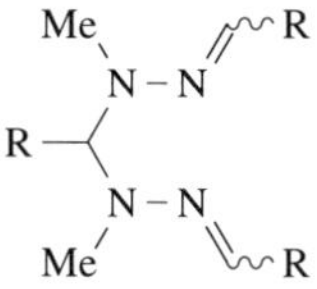

Only in exceptional cases have ketones been shown to form *gem*-dihydrazino derivatives by reaction with hydrazine derivatives ⟨88IJC(B)912⟩, although compounds formally derived from ketones have been prepared by alternative procedures. An unusual reaction involving the transfer of pyrazolyl groups to ketones (Equation (66)) has been described ⟨72CC841, 73CJC2448⟩, but related compounds have been more readily prepared in 34–82% yields by the acid-catalysed reaction of pyrazole with a large variety of acetals and ketals ⟨70JA5118⟩.

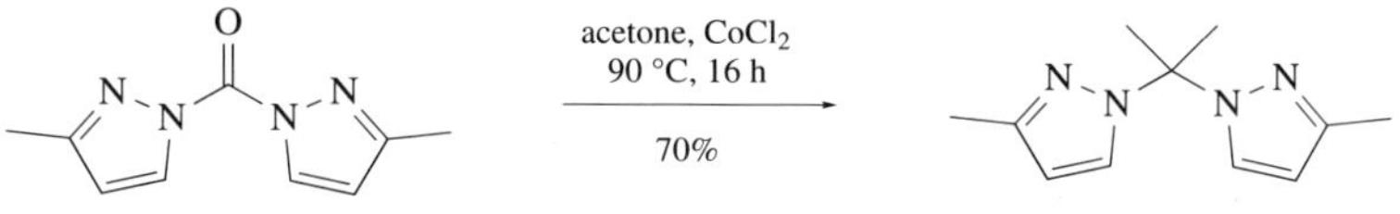

4.09.5.9 *gem*-Diimino Alkanes and *gem*-Diisocyanides

gem-Diimino alkanes have been known for many years following the demonstration by Laurent ⟨1837LA(21)130⟩ that N^1,N^2-bis(benzylidene)phenylmethylenediamine was formed by the reaction of ammonia with benzaldehyde. Since that time, however, members of this class of compound received relatively little attention until it was recognised that they had potential for the synthesis of various unusual heterocyclic compounds ⟨77S647, 82S1080, 84S259⟩. A number of aromatic aldehydes have now been shown to behave in a similar fashion to benzaldehyde, producing reasonable yields of the adducts (**120**) when treated with either concentrated aqueous ammonia ⟨64JOC1985⟩ or ammonium acetate in methanol (Equation (67)) ⟨84S256⟩. In both instances an equilibrium is favoured by precipitation of the product from the reaction medium.

$$\text{ArCHO} + \text{NH}_4\text{OAc} \xrightarrow[44-87\%]{\text{MeOH}} \text{Ar-CH(N=CHAr)}_2 \quad (120) \tag{67}$$

Since the reaction is reversible, it is possible to prepare mixed derivatives by taking advantage of the considerable rate differences between different aldehydes. The adduct (**121**) from salicylaldehyde in particular has been shown to provide good yields of the mixed derivative (**122**), either on treatment with various other bis(imino) alkanes or following treatment with an alternative aromatic aldehyde such as 4-dimethylaminobenzaldehyde (Equation (68)) ⟨84S256⟩. The possible alternative adduct with the 2-hydroxyphenyl group attached to the bridging methylene group was not observed, due to stabilisation of the intermediate salicylaldehyde imine through intramolecular hydrogen bonding. Indeed, mixed products (**122**) derived from both aromatic and aliphatic aldehydes may be prepared in 80–97% yields by simply mixing salicylaldehyde with the appropriate aldehyde and ammonium acetate in concentrated ammonia ⟨88SC2289, 89S307⟩.

$$(\mathbf{121}) \xrightarrow[\substack{\text{i, (120), NH}_4\text{OAc, MeOH, RT, 4 h} \\ 46-85\% \\ \text{or} \\ \text{ii, ArCHO, NH}_4\text{OAc, RT, 1 day} \\ 96\%}]{} (\mathbf{122}) \tag{68}$$

In contrast to primary aliphatic aldehydes, which form polymeric products with ammonia (see Section 4.09.2.1.2), secondary and tertiary aliphatic aldehydes have been shown to behave like aromatic aldehydes and to form 62–85% yields of analogues akin to (**119**) ⟨61JOC1822⟩. These aliphatic adducts, however, are less stable than those derived from aromatic aldehydes and readily decompose on heating ⟨61JOC1822⟩.

Hexamethyldisilazane has proved to be a useful alternative to ammonia and has the advantage of allowing the synthesis of *gem*-diimines under nonprotic conditions ⟨88BCJ609⟩. A number of aromatic and aliphatic aldehydes have been shown to react with this reagent in the presence of anhydrous zinc chloride to provide reasonable yields of the corresponding adducts (Equation (69)).

$$\text{RCHO} + (\text{TMS})_2\text{NH} \xrightarrow[30-79\%]{\substack{\text{ZnCl}_2, \text{CH}_2\text{Cl}_2 \\ \text{RT, 24 h}}} \text{R-CH(N=CHR)}_2 \tag{69}$$

While no similar transformations have been exemplified with ketones, diphenylmethanimine does react with the ketals of aromatic aldehydes in the presence of fluoroboric acid to form excellent yields of the corresponding bis(imines) (Equation (70)) ⟨78LA1928⟩, and it is likely that other diarylmethanimines would form analogous products. Moreover, modest yields of cyclic *gem*-diimino alkanes have been prepared from methyl ketones and diiminosuccinonitrile (Equation (71)), although considerably improved yields are possible using the ketone dimethyl ketals and catalytic amounts of concentrated mineral acid ⟨72JOC4136⟩.

$$\text{Ph}_2\text{C=NH} + \text{ArCH(OEt)}_2 \xrightarrow[95\%]{\text{HBF}_4, 85\,°\text{C, 3 h}} \text{Ar-CH(N=CPh}_2)_2 \tag{70}$$

$$\text{(NC)(HN=)C-C(=NH)(NC)} + \text{RC(=O)Me} \xrightarrow[15-37\%]{\substack{\text{oxalic acid, benzene} \\ 80\,°\text{C, 1 h}}} \text{2H-imidazole product} \tag{71}$$

1,1-Diisonitrilomethane, the only *gem*-diisocyanide described up to early 1995, is an unstable compound (even at $-30\,°C$) that has been prepared by treatment of bis(formylamino)methane in triethylamine at $-60\,°C$ with two equivalents of phosgene ⟨64AG382⟩. Whether more hindered homologues would have greater stability is not known.

4.09.5.10 *gem*-Difluoroamino Alkanes

Due to their highly unstable character, extreme caution should be exercised when handling *gem*-difluoroamino alkanes and the intermediates used in their preparation. Specific reaction vessels have been recommended to minimise the hazards ⟨68JA7083⟩.

Both aldehydes and ketones undergo addition of difluoroamine in strongly acidic conditions (usually at least 92% sulfuric acid) to provide moderate to good yields of *gem*-difluoroamino alkanes (Equation (72)) ⟨68JA7083⟩. Aldehydes generally react with difluoroamine more slowly than ketones, particularly when substituted by electron-withdrawing substituents, and the more unreactive carbonyl compounds may require oleum to facilitate a favourable equilibrium. The reaction of methyl vinyl ketone is an interesting case in which 1,3,3-tris(difluoroamino)butane is formed through an initial, rare, acid-catalysed Michael addition followed by attack at the carbonyl group ⟨68JA7083⟩.

$$\begin{array}{c} R^1 \\ {>}{=}O \\ R^2 \end{array} + HNF_2 \xrightarrow[31-85\%]{\text{conc. } H_2SO_4,\ -23\,°C,\ 4\,h} \begin{array}{c} R^1 \quad NF_2 \\ {>}{<} \\ R^2 \quad NF_2 \end{array} \qquad (72)$$

gem-Difluoroamino alkanes may also be prepared by the nucleophilic displacement of bromo, nitro and nitroso groups by difluoroamine in oleum, as typified by Equation (73), although the yields are highly variable ⟨69JOC2049⟩. The displacement of a third leaving group to generate 1,1,1-tris(difluoroamino)alkanes is not observed. Chlorine is generally an unsuitable replacement for bromine as a leaving group in the formation of difluoroamino alkanes ⟨69JOC2049⟩, but poor yields of difluoroamino alkanes have been obtained by the reaction of 1,1-dichloro-1-alkenes with difluoroamine in oleum ⟨69JOC2046, 72JOC922⟩.

$$\underset{\text{cyclohexane-}O_2N,NO}{\bigcirc} \xrightarrow[31\%]{HNF_2,\ \text{oleum} \atop -23\,°C,\ 5\,\text{min}} \underset{\text{cyclohexane-}F_2N,NF_2}{\bigcirc} \qquad (73)$$

4.09.6 GEMINALLY SUBSTITUTED ALKANES BEARING TWO DIFFERENT DICOORDINATE OR HETEROSUBSTITUTED NITROGENS

4.09.6.1 Nitro Alkane Derivatives

Three members of this general class of compound have been described: those bearing a nitroso, a diazo or an isocyanate group on the carbon atom bearing the nitro group. Nitro nitroso alkanes, commonly described as pseudonitroles, are best prepared by the classic method of Meyer, in which an alkali metal salt of a nitro alkane is treated with nitrous acid generated *in situ* (Equation (74)); the yields from such reactions are generally high ⟨1875LA(175)88, 63T(S)23, 69JOC2049, 73T4195⟩. Oximes have also been oxidized to pseudonitroles by fuming nitric acid ⟨65JCS2601⟩, while cyclohexanone forms the 2-nitro-2-nitroso derivative in 72% yield on treatment with dinitrogen tetroxide and acetic anhydride in tetrachloromethane at $0-10\,°C$ ⟨69OPP5⟩.

$$Pr^nNO_2 \xrightarrow[93\%]{\substack{\text{i, NaOH (aq.)} \\ \text{ii, NaNO}_2,\ \text{EtOH},\ H_2O \\ \text{iii, 18\% HCl},\ -18\,°C\ \text{to}\ 0\,°C}} Et{-}\!\!\!\!<\!\!\!\begin{array}{l}NO_2\\NO\end{array} \qquad (74)$$

Alkali metal salts of secondary nitro alkanes also undergo addition to arenediazonium salts to afford highly coloured diazo nitro alkanes (Equation (75)), a reaction which has found application in the dyestuffs industry ⟨51JOC1507⟩. Similar reactions with primary nitro alkanes also proceed in reasonable yield, although the products in this instance exist predominantly as the thermodynamically more stable nitrohydrazones (123) ⟨54JA3489, 79S380⟩. Cyclic analogues, such as the 3-nitropyrazolines (124), have been prepared in high yields by the cycloaddition of diazomethane to nitro alkenes ⟨50JA3843, 67ZC421⟩.

$$K^+ \underset{SO_3K}{\overset{NO_2}{\diagup\!\!\!\diagdown}} + ArN_2^+ Cl^- \xrightarrow[51-77\%]{pH\ 7-9,\ H_2O,\ 0\ °C,\ 20\ min} \underset{N=N\diagdown Ar}{\overset{O_2N}{\diagup\!\!\!\diagdown}} SO_3K \quad (75)$$

(123) ArNHN=C(NO₂)(R)   (124) 3-R,3-NO₂-pyrazoline

The third class of compounds, the *gem*-nitro isocyanates, may be prepared by Curtius rearrangement of the corresponding nitro acyl azides, themselves produced by diazotisation of nitromethylacylhydrazines (Equation (76)). No yields have been reported for this reaction, presumably because the products are highly unstable and polymerise even on storage at low temperature ⟨57JOC1662⟩.

$$\underset{O_2N}{\overset{R}{\diagup\!\!\!\diagdown}}CONHNH_2 \xrightarrow[ii,\ CHCl_3\ or\ CCl_4,\ reflux]{i,\ NaNO_2,\ HCl,\ 0\ °C} \underset{O_2N}{\overset{R}{\diagup\!\!\!\diagdown}}NCO \quad (76)$$

4.09.6.2 Nitramino Alkane Derivatives

Although not a well-described class of compound, it is known that nitramino alkane derivatives can be formed under the conditions of the Mannich reaction from primary nitramines, which possess an active hydrogen atom, with formaldehyde and a secondary amine ⟨71JOC3846⟩. Furthermore, nitrosation of the cyclic aminomethyl nitramine (125; R = H) provides a useful route to the corresponding nitrosamino derivative (Equation (77)) ⟨66JCS(C)870⟩, and analogues of (125) having labile *N*-substituents (R = CH₂OMe, CH₂OEt, CH₂OAc) form the same product in high yield when treated in a similar manner ⟨66JCS(C)862, 66JCS(C)867⟩. Products of a similar nature are also formed on partial oxidation of cyclic *gem*-dinitrosamino alkanes (see Section 4.09.5.4) with either nitric acid ⟨84JOC5147⟩ or hydrogen peroxide ⟨49MI 409-01⟩. Other oxidants, however, either failed to react (CF₃CO₃H) or resulted in conversion into the *gem*-dinitramine (N₂O₅) ⟨84JOC5147⟩.

$$\text{(125)} \xrightarrow[72-86\%]{NaNO_2,\ H_2SO_4,\ H_2O} \text{product} \quad (77)$$

4.09.6.3 Azo Alkanes and Hydrazino Alkane Derivatives

Azomethylene isocyanates and isothiocyanates may be readily isolated in high yields following the oxidation of 1,2,4-triazolidin-3-ones and thiones, respectively, with potassium permanganate in aqueous ether (Equation (78)), although a subsequent slow rearrangement to dihydro-1,2,4-triazolones has been observed with several compounds on standing at room temperature ⟨84S315, 90S803, 90S1048⟩. Good yields of the bis analogues (126) have also been prepared in a similar fashion from the bicyclic heterocycles (**127**) ⟨85T5525⟩. Bis compounds (**126**) are also available by halide displacement from the corresponding dichloro compounds with potassium cyanate in aqueous isopropanol, although the yields by this procedure are generally inferior ⟨70M568, 80T1753⟩. An alternative route to azomethylene isocyanates involves the oxidation of semicarbazones with chromyl acetate or lead tetraacetate, whereby moderate yields (30–71%) may be isolated following rearrangement of the intermediate nitrene ⟨68AG(E)293⟩.

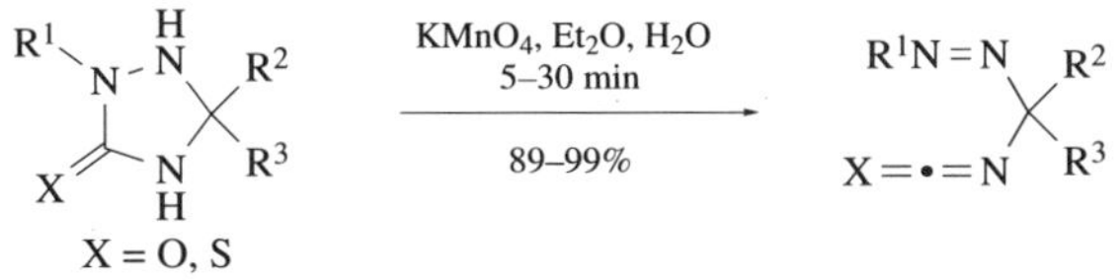

$$\underset{X=O,\ S}{\overset{R^1\underset{H}{\overset{H}{N}}-N\underset{H}{\overset{R^2}{\underset{N}{\longrightarrow}}R^3}}{}} \xrightarrow[89-99\%]{\text{KMnO}_4,\ \text{Et}_2\text{O},\ \text{H}_2\text{O} \quad 5-30\ \text{min}} \quad R^1N=N\underset{X=\bullet=N}{\overset{R^2}{\underset{R^3}{\longrightarrow}}} \tag{78}$$

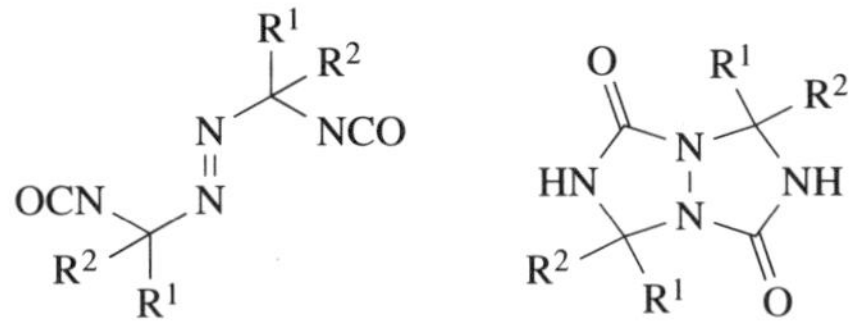

(**126**)   (**127**)

Examples of a number of more unusual heterocyclic compounds containing the azo function linked to a carbon atom bearing a different nitrogen substituent are known ⟨72JHC827, 82TL2103⟩, but the methods used in their synthesis are unlikely to be of a more general synthetic application and will not be discussed here.

Phenylhydrazones derived from both aldehydes and ketones undergo a smooth addition of hydroxylamine in hot ethanol to afford good yields of the corresponding hydroxylaminohydrazines (Equation (79)) ⟨81IJC1003⟩. The subsequent reduction of these products to aminomethylhydrazines is described in Section 4.09.7.4.

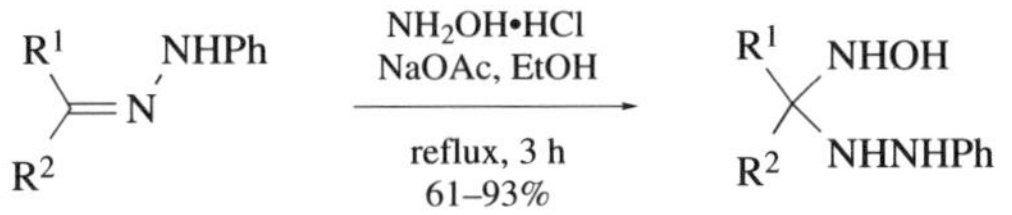

$$\underset{R^2}{\overset{R^1}{\longrightarrow}}=N\overset{\text{NHPh}}{\underset{}{}} \xrightarrow[\substack{\text{reflux, 3 h} \\ 61-93\%}]{\text{NH}_2\text{OH}\cdot\text{HCl} \\ \text{NaOAc, EtOH}} \underset{R^2}{\overset{R^1}{\underset{}{\longrightarrow}}}\overset{\text{NHOH}}{\underset{\text{NHNHPh}}{}} \tag{79}$$

4.09.6.4 Isocyanato Alkane Derivatives

Isocyanatomethyl carbodiimides (**128**) may be prepared in moderate yields by the reaction of mono *N*-silylated carbodiimides with α-chloro alkyl isocyanates in benzene under reflux, but the reaction takes an alternative course when the second carbodiimide nitrogen atom is part of a carbamate moiety, giving isocyanatomethyl cyanamides (Scheme 38) ⟨76ZOR231, 76ZOR2103⟩. Carbodiimides similar to (**128**) are also generated in aza-Wittig-type reactions between triphenylphosphine imides and *gem*-diisocyanato alkanes (Section 4.09.5.3) ⟨77ZOR2449⟩. The yields by this procedure are comparable to those achieved on halogen displacement.

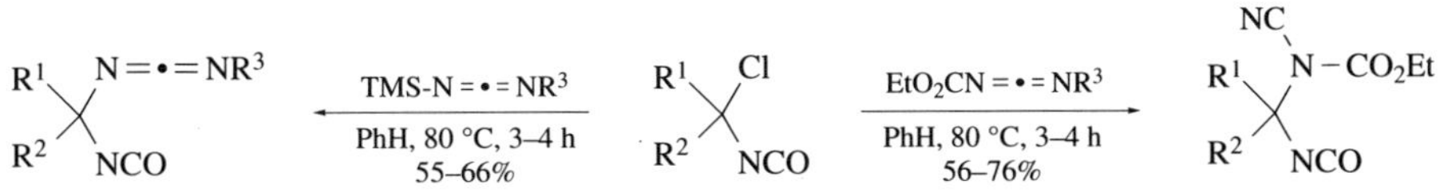

Scheme 38

4.09.6.5 Miscellaneous Derivatives

N-Isocyanidomethyl derivatives of a variety of nitrogen heterocycles have been prepared by dehydration of the corresponding heterocyclic formamido-methyl derivatives with either POCl₃ or

Ph$_3$P/CCl$_4$ ⟨83CPB723⟩. For the various monocyclic and bicyclic heterocycles studied, the latter procedure generally gave superior yields (41–82%) and is the method of choice. Those isocyanidomethyl compounds derived from benzimidazole and benzotriazole were relatively stable, whereas those derived from imidazole or 1,2,4-triazole formed intractable materials on exposure to moisture.

4.09.7 GEMINALLY SUBSTITUTED ALKANES BEARING ONE AMINO GROUP AND ONE DICOORDINATE OR HETEROSUBSTITUTED NITROGEN

4.09.7.1 Aminomethylhydroxylamines

Like many other nucleophiles, *N*-aryl- and *N*-acyl-hydroxylamines undergo Mannich-type reactions with formaldehyde and amines, reactions that give aminomethylhydroxylamines in 40–57% yields ⟨56CB1134⟩. Similar reactions result in good yields of *N*-hydroxy-4-imidazolidinones (Equation (80)) when the amine and hydroxylamine are suitably positioned in the same molecule ⟨58LA(615)34, 70JHC439⟩.

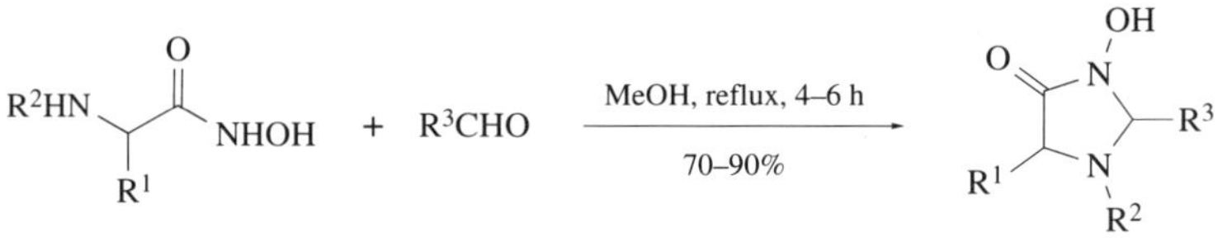

$$R^2HN-\underset{R^1}{\underset{|}{C}}(=O)-NHOH + R^3CHO \xrightarrow[70-90\%]{MeOH, reflux, 4-6\ h} \text{N-hydroxy-4-imidazolidinone} \quad (80)$$

Aminomethylhydroxylamines may also be prepared in good yield by the addition of hydroxylamine to aryl imines (Scheme 39) ⟨83PHA449⟩ and, while not exemplified, an analogous reaction should proceed with aliphatic imines. The reaction of aryl imines with the dilithio derivatives of oximes provides an efficient route to 2-(hydroxylamino)azetidines (Scheme 39) ⟨83S951⟩.

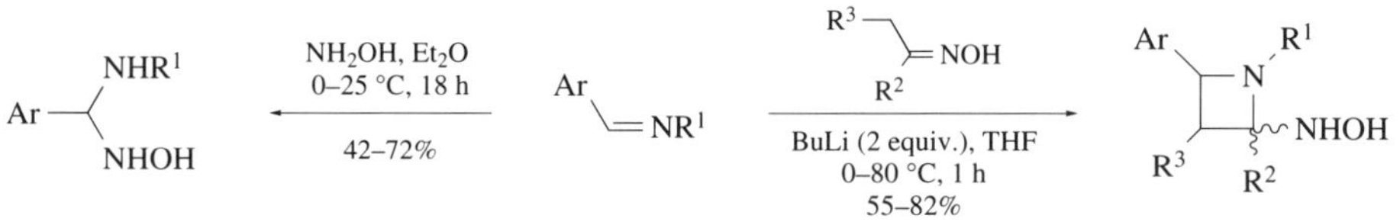

Scheme 39

4.09.7.2 Aminomethyl Nitramines and Nitrosamines

Primary nitramines readily undergo Mannich-type reactions with amines and formaldehyde which generally result in high yields of aminomethyl nitramines (Table 4) ⟨10RTC296, 49JCS1638, 55CJC923, 61JOC4709, 66JCS(C)862, 67JCS(C)562, 71JOC3846⟩. The best results have been obtained by first treating the nitramine with aqueous formaldehyde, usually at 50–60 °C, and then adding the amine component to the cooled solution, from which the product crystallises. Ethyl carbazate has been shown to react in a similar manner (Table 4) ⟨71JOC3846⟩. There is little in the literature regarding the use of aldehydes other than formaldehyde, although there is no evidence to suggest that such reactions would not proceed equally well.

The nucleophilicity of nitramines under alkaline conditions affords an alternative method for the preparation of aminomethyl nitramines from suitably substituted aminomethyl derivatives. In the presence of pyridine a variety of nitramines have been used to effect displacement of the acetoxy group from the triazepine (**129**) (Equation (81)), with the products generally forming in reasonable yield ⟨66JCS(C)862⟩. The partial nitration of cyclic polyamines such as hexamine with nitric acid

Table 4 Preparation of aminomethyl nitramines.

Reactant	Product	Conditions	Yield (%)	Ref.
$O_2N{-}NH{-}CH_2CH_2{-}NH{-}NO_2$	$[\text{cyclic bis-nitramine with N-CH}_2]_2$	CH_2O, NH_3, acetone (aq.)	63	67JCS(C)562
$O_2N{-}NH{-}CH_2CH_2{-}NH{-}NO_2$	7-membered dinitramine ring with $CH_2CH_2C(NO_2)_2$ substituent; also $H_2N{-}CH_2CH_2{-}C(NO_2)_2$	CH_2O, NaOH, H_2O, 15 min	99	61JOC4709
$O_2N{-}NH{-}CH_2CH_2{-}NH{-}NO_2$	7-membered dinitramine ring with CH_2NHCO_2Et substituent	EtO_2CNHNH_2, CH_2O, NaOH, 20 °C, 16 h	80	71JOC3846
$MeNHNO_2$	cyclic bis-nitramine with Me, Me and N–Me	$MeNH_2$, CH_2O, H_2O, 0 °C	50	49JCS1638

offers another route to aminomethyl nitramines, but the method lacks versatility and utilises strongly acidic media ⟨51JA2769⟩. The substitution of nitrous for nitric acid, however, allows the synthesis of cyclic aminomethyl nitrosamines in reasonable yield ⟨51JA2777⟩.

$$\text{(129)} \quad \xrightarrow{\text{RNHNO}_2,\ \text{pyridine}\ \ \text{Me}_2\text{CO or DMF}\ \ 14\text{–}88\%} \quad \text{product} \tag{81}$$

4.09.7.3 Aminomethyl Azides and Triazines

While aminomethyl azides can be isolated and characterised, they are of limited stability, and only in 1993 was a convenient procedure for their synthesis described. *N,N*-dimethylanilines in particular react with iodosylbenzene and TMS-azide to provide excellent yields of the corresponding azidomethyl compounds (Equation (82)) ⟨93JA9347⟩. Alkyl groups other than the methyl group also react under these conditions, but at a reduced rate, such that *N*-ethyl-*N*-methyl-1-naphthylamine gives a 2:1 mixture of the azidomethyl and azidoethyl derivatives, respectively ⟨93JA9347⟩. In addition to aromatic amines, trimethylamine has been shown to produce the corresponding aminomethyl azide under similar conditions.

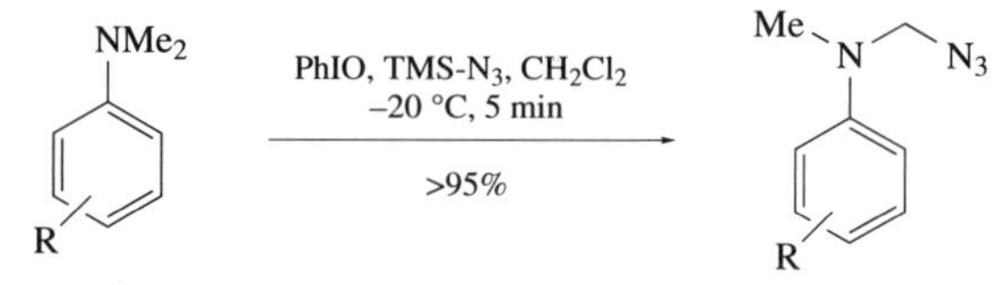

A more conventional approach to the synthesis of aminomethyl azides relies on the nucleophilic displacement of halogen from an aminomethyl halide by azide (Equation (83)); silver azide ⟨58CB660⟩, sodium azide ⟨82JCED94⟩ and acetyl azide ⟨83JOC611⟩ have all been shown to be effective. Although there are only limited published examples of this method, it is likely to be of general applicability and its mildness might make it the procedure of choice. The formation of one aminomethyl azide by the addition of hydrazoic acid to an imine has been described ⟨65JOC1398⟩.

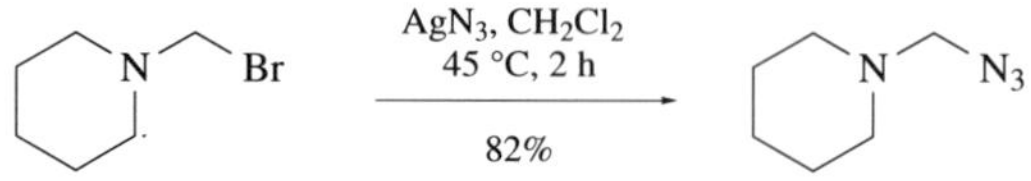

In contrast to aminomethyl azides, the corresponding triazenes form a little-known class of compound that has only been described in the mid 1980s ⟨83CC721, 83JCR(S)108, 84CJC749⟩. Thus, the reaction of arenediazonium salts with aqueous solutions of methylamine and excess formaldehyde (Equation (84)) probably offers the best available route to such compounds, and has been shown to be most efficient with electron-deficient arenes ⟨84CJC749⟩.

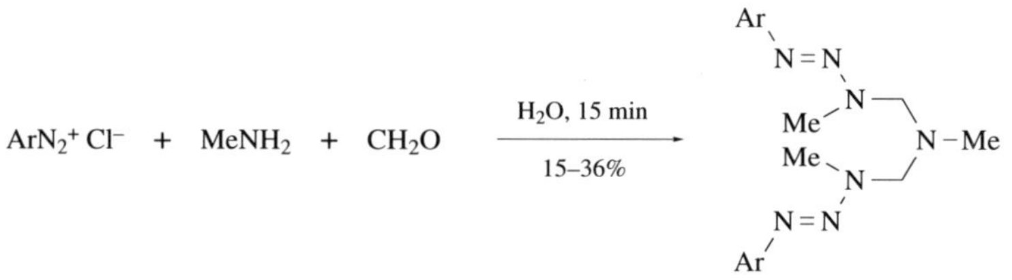

4.09.7.4 Aminomethyl Azo and Hydrazino Compounds

The displacement of halogen from α-halo azo alkanes by primary and secondary amines provides a useful method for the preparation of aminomethyl azo compounds (Equation (85)), although the yields are fairly poor ⟨85M1329⟩. Primary amino compounds may also be prepared by the addition of O-mesitylenesulfonyl hydroxylamine (MSH) to hydrazones followed by subsequent treatment of the intermediate salt with triethylamine (Scheme 40) ⟨82CJC285⟩.

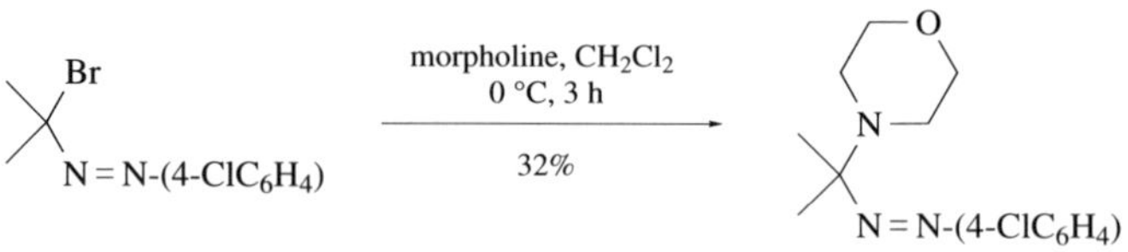

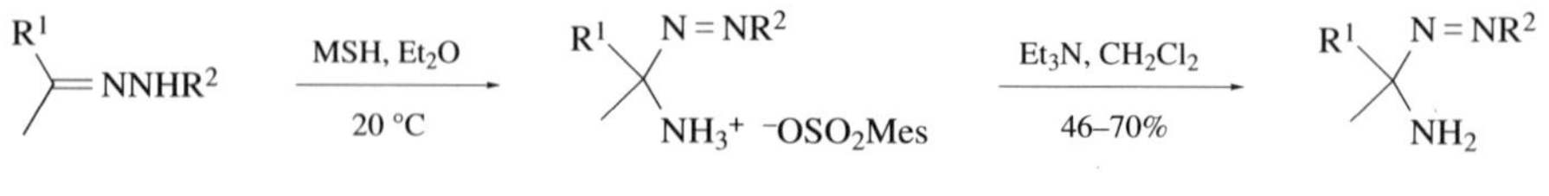

Scheme 40

Representative aminomethyl hydrazines may be prepared by the catalytic reduction of hydroxylaminomethyl hydrazines, in which the N—OH bond is selectively cleaved (Equation (86)) ⟨81IJC1003⟩. Cyclic analogues based on triaza[3.1.0]hexane have also been made via several routes ⟨76JOC3221, 85S100⟩.

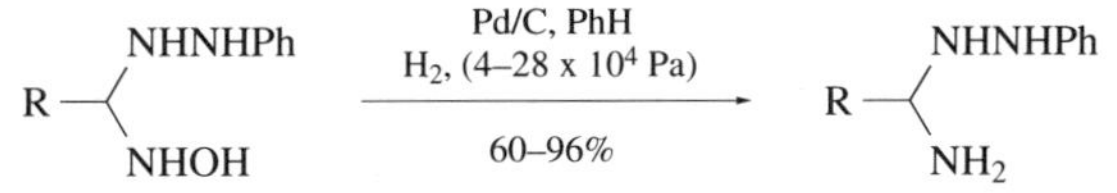

$$R-\underset{NHOH}{\overset{NHNHPh}{\diagdown}} \xrightarrow[60-96\%]{Pd/C, PhH \atop H_2, (4-28 \times 10^4 Pa)} R-\underset{NH_2}{\overset{NHNHPh}{\diagdown}} \quad (86)$$

4.09.7.5 Aminomethyl Imines

The simplest method for the synthesis of this group of compounds is by the Mannich condensation between diaryl methanimines, secondary amines and formaldehyde or an aromatic aldehyde, and the yields generally fall within the range 55–90% ⟨78LA1928⟩. The use of these imines, which are not easily prepared, may be avoided by first forming an aminomethyl aza-Wittig reagent such as (**130**) and constructing the imine in the final step (Scheme 41) ⟨91CJC1153⟩.

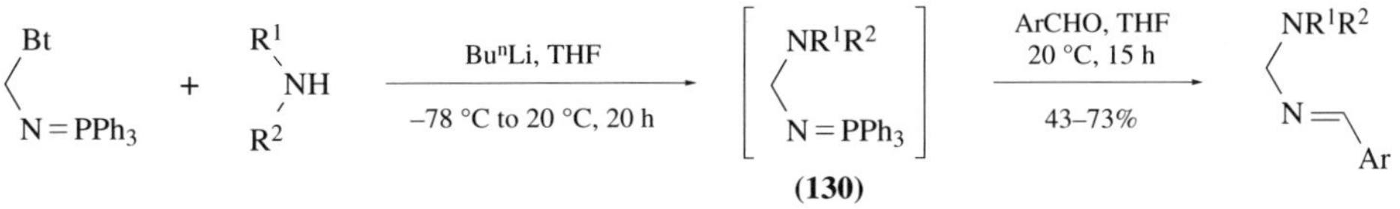

Scheme 41

A route to aminomethyl imines (Equation (87)) involving the displacement of bromine from α-bromo glycine diphenylmethanimine has been described which seems to offer a mild procedure for the formation of these compounds ⟨85TL695⟩. A number of cyclic analogues (**131**) have also been prepared through the reaction of *gem*-diimino alkanes (Section 4.09.5.9) with aralkyl ketones ⟨77S647, 81S151, 82S1080⟩.

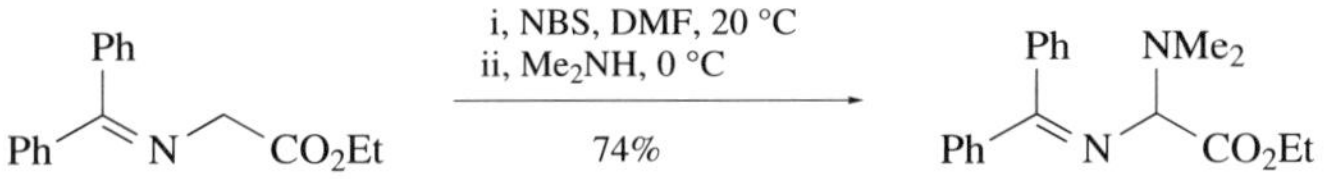

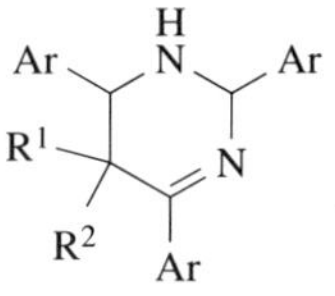

4.09.7.6 Miscellaneous Compounds

Aminomethyl benzotriazoles form a compound class that has been largely developed through the work of Katritzky and co-workers, and some reference to their use has been made in earlier sections (cf. Sections 4.09.2.1.1 and 4.09.2.1.2). A comprehensive review of the synthesis and reactions of this class of compound has been published (⟨91T2683⟩ and references therein). The compounds may be formed directly by the Mannich reaction between benzotriazole, formaldehyde and a dialkylamine or arylamine ⟨75JCS(P1)1181, 87JCS(P1)799⟩ (cf. Scheme 2), or by the reaction of primary aliphatic or aromatic amines with preformed hydroxyalkyl triazoles (**132**) in ethanol at reflux (Equation (88)) ⟨87JCS(P1)799⟩. Chloroalkylbenzotriazoles also reacted with amines to afford similar products ⟨87JCS(P1)811⟩.

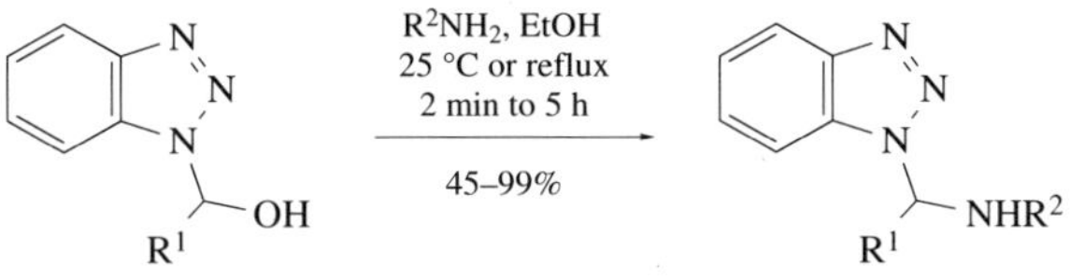

An *N*-(formylaminomethyl)benzotriazole has been used to prepare an *N*-(formylaminomethyl) amine, which was subsequently dehydrated with POCl₃ to give the first reported example of an aminomethyl isocyanide (Equation (89)) ⟨93S45⟩.

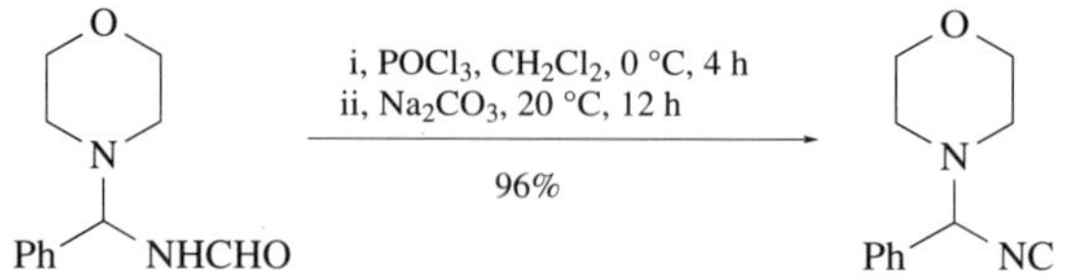

$$\text{morpholine-CH(Ph)-NHCHO} \xrightarrow[96\%]{\substack{\text{i, POCl}_3\text{, CH}_2\text{Cl}_2\text{, 0 °C, 4 h} \\ \text{ii, Na}_2\text{CO}_3\text{, 20 °C, 12 h}}} \text{morpholine-CH(Ph)-NC} \quad (89)$$

4.09.8 GEMINALLY SUBSTITUTED ALKANES BEARING ONE ACYLATED OR SULFONATED AMINO GROUP AND ONE DICOORDINATE OR HETEROSUBSTITUTED NITROGEN

There appear to be few reported examples of alkanes bearing one dicoordinate nitrogen substituent and a sulfonated amino substituent on the same carbon atom, and only a single paper ⟨93JOC2086⟩, discussed below, describes a synthetically useful procedure. Nevertheless, it seems likely that many of the methods described for the synthesis of acylated analogues should be capable of modification to include sulfonated compounds, and this is clearly an area which awaits exploration.

4.09.8.1 Acylaminomethyl Isocyanates and Isothiocyanates

Inversion of one or more of the amide groups of a linear peptide represents an important strategy in drug design, and as such has received a considerable amount of attention. Such modifications have frequently involved trapping intermediate isocyanates formed as a result of Curtius, Lossen or Hofmann rearrangements of *N*-protected amino acid derivatives ⟨83MI 409-01⟩, but the Curtius acyl azide rearrangement is the most commonly used of these three. Typically, various *t*-butoxycarbonyl (t-BOC) or Cbz protected amino acid azides and their analogues readily lose nitrogen on heating to 80 °C in toluene to form the corresponding isocyanates in high yield (Equation (90)) ⟨83JMC129, 83MI 409-01, 83MI 409-02⟩. As a rule, these products are not generally isolated but are treated further with either amines or alcohols. While this approach has been confined largely to amino acids and peptides, it is likely to be of general applicability.

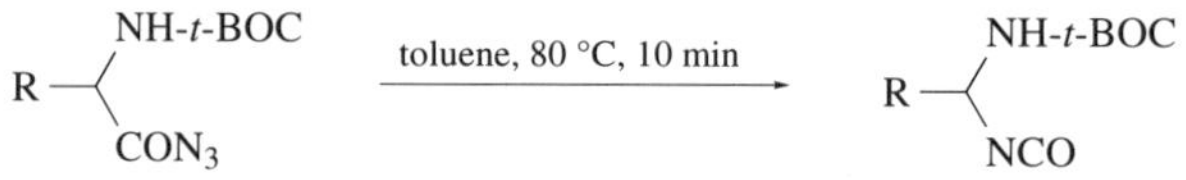

$$\text{R-CH(NH-t-BOC)-CON}_3 \xrightarrow{\text{toluene, 80 °C, 10 min}} \text{R-CH(NH-t-BOC)-NCO} \quad (90)$$

An alternative method involving the displacement of chloride from the aryl isocyanates (**133**) by simply stirring with triethylamine and a secondary amide in benzene for 2 h at room temperature has been reported, and provides 64–75% yields of readily isolable product ⟨92ZOR2119⟩. The general applicability of this reaction remains to be established since it is possible that the trifluoromethyl substituent may be necessary to enhance halogen displacement. The reverse strategy, in which chloride is displaced from an *N*-chloromethyl amide, also offers a route to such compounds and has been demonstrated for the synthesis of isothiocyanates (Scheme 42) ⟨86S817⟩. In this instance the initially formed thiocyanate (**134**) rearranges easily at 80 °C in butanone, although strongly electron-withdrawing groups lower the rate of isomerisation.

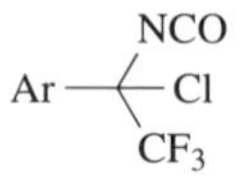

Ar–C(NCO)(Cl)(CF₃)

(**133**)

$$\text{R}^1\text{C(O)N(R}^2\text{)CH}_2\text{Cl} \xrightarrow[56-87\%]{\text{KSCN, Me}_2\text{CO}, 20\,°\text{C, 1 h}} \text{R}^1\text{C(O)N(R}^2\text{)CH}_2\text{SCN} \xrightarrow[40-90\%]{\text{butanone, 80 °C}} \text{R}^1\text{C(O)N(R}^2\text{)CH}_2\text{NCS}$$

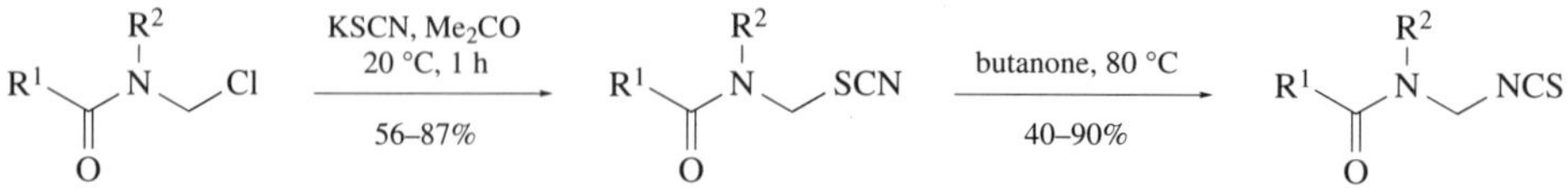

(**134**)

Scheme 42

While the direct displacement of halide by isocyanate appears to be an attractive option, trimerisation on isolation can become a significant problem. One way of circumventing such complications has involved the use of silver nitrocyanamide, which when heated with *N*-bromomethylphthalimide in benzene at reflux for 1 h afforded the isocyanate (135) in 81% yield ⟨88JCS(P1)2137⟩. While this approach was suitable for the synthesis of various isocyanates, only this single example of an acylaminomethyl derivative has been reported.

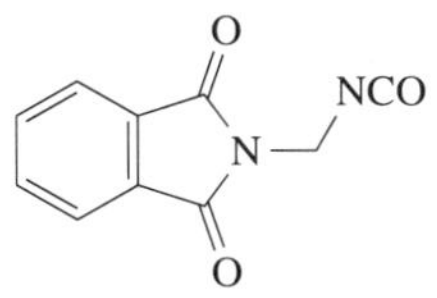

(135)

The formation of good yields of carbamoylmethyl isocyanates by the partial alcoholysis of *gem*-diisocyanates following treatment with one equivalent of alcohol in ether at room temperature overnight has been described, and ureido analogues may be prepared in a similar manner with one equivalent of *N*-ethylaniline ⟨77ZOR723⟩.

4.09.8.2 Acylaminomethyl Azides

These compounds are best known in the monocyclic β-lactam field where they have been particularly useful in the construction of unusual heterocyclic systems ⟨77JCS(P1)189, 86JCS(P1)1077⟩. Typically, the azido group may be introduced into a preconstructed azetidinone such as (136) (Equation (91)) by nucleophilic displacement of chloride using either tetramethylguanidinium azide in chloroform ⟨77JCS(P1)189⟩ or sodium azide in DMF ⟨86JCS(P1)1077⟩. The yields are generally good.

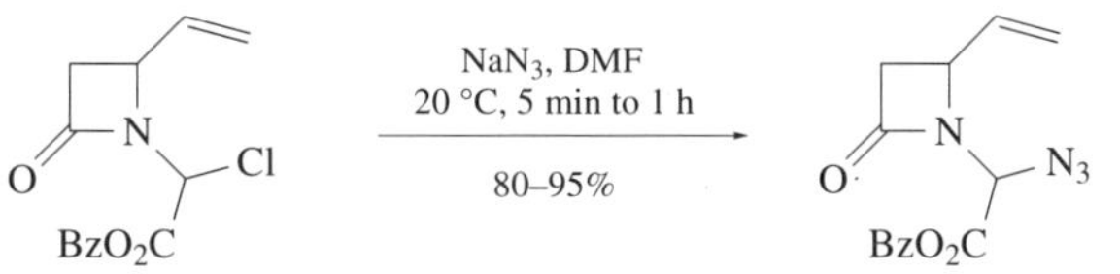

(136) → (91)

NaN$_3$, DMF
20 °C, 5 min to 1 h
80–95%

4.09.8.3 Acylaminomethyl Nitramines

Possibly the best route to this class of compounds involves the nucleophilic displacement of halogen or toluenesulfonyl groups from appropriately substituted acylamines by treatment with primary alkyl nitramines (Equation (92)) ⟨85S973⟩. The reaction is high yielding, appears to tolerate a range of functionality, is suitable for nitraminomethyl carbamates, and occurs rapidly under fairly mild conditions in solvents such as propanone and dichloromethane. By comparison, the synthesis of nitraminomethyl carbamates by Curtius rearrangement and alcoholic quench of the intermediate isocyanates proceeded in only 20–25% yield ⟨85S973⟩.

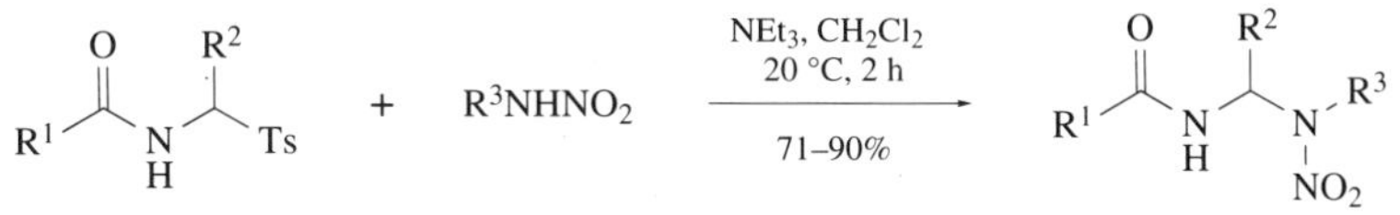

NEt$_3$, CH$_2$Cl$_2$
20 °C, 2 h
71–90%
(92)

Two of the three acetyl groups of 1,3,5-triacetyl-hexahydro-1,3,5-triazine may be replaced to give a 61% yield of the 1,3-dinitro derivative on treatment with ammonium nitrate and trifluoroacetic anhydride (TFAA) at 0 °C ⟨88S743⟩, but the reaction is unlikely to have general applicability.

4.09.8.4 Acylaminomethyl Hydroxylamines and Hydrazines

These two classes of compound are poorly documented, although representative examples of each have been prepared. Thus, ethyl diphenylmethylenecarbamate was shown to give 76% of the adduct (**137**) when briefly heated with hydroxylamine in methanol, but this was the only example reported using this procedure ⟨58JA4921⟩. Nevertheless, it is probable that analogues of (**137**) could be prepared in a similar manner. The nucleophilic displacement of halogen also offers a possible route to such compounds since the hydrazine derivative (**138**) was reported to be formed on treatment of the corresponding α-chloro ester with *t*-butyl carbazate in dichloromethane ⟨84CC1289⟩. Although (**138**) was subsequently converted into bicyclic β-lactams which still formally retained the acylaminohydrazine moiety, neither the yield nor the full experimental conditions for its formation were stated. A number of acylaminomethyl-1,2,4-triazoles (**139**), which may be formally considered as cyclic hydrazino derivatives, have been prepared by the reaction of primary and secondary amides with the corresponding alkoxymethyl triazoles in refluxing dioxan in the presence of TsOH ⟨90H(31)2029⟩. The isolated yields were highly dependent on the amide used.

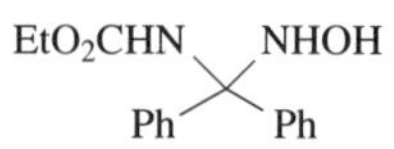

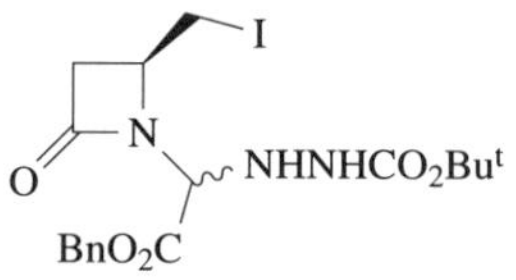

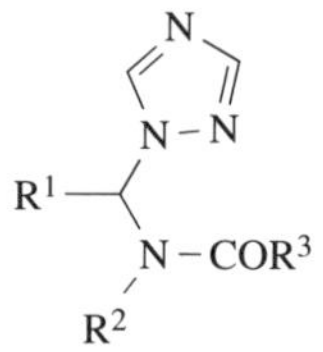

4.09.8.5 Acylaminomethyl Azo Alkanes and Imino Alkanes

Ammonolysis of the pyridinium salts (**141**), formed on treatment of a pyridine solution of the hydrazones (**140**) with bromine (Scheme 43), results in 64–68% yields of the amino compounds (**142**; R = H), which are readily converted by acetyl chloride into 37–49% yields of the acylamino azo alkanes (**142**; R = Ac) ⟨87M851⟩. Acylated derivatives of (**142**; R = H) have also been prepared directly by treatment of the pyridinium salt (**141**) with potassium phthalimide in the presence of 18-crown-6, although the yields by this method were much poorer ⟨87M851⟩. Rather miniscule yields of similar compounds (~4%) have also been reported as intermediates arising during the addition of Grignard reagents to azomethyl isocyanates ⟨85M1051⟩, but this process is unlikely to have any synthetic application.

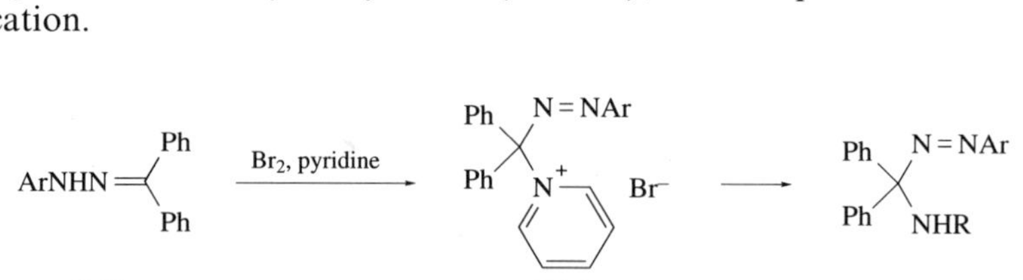

Scheme 43

An alternative route to acylamino azo alkanes has been described involving the addition of acylaminomalonates to arenediazonium salts at pH 6 (Equation (93)) ⟨90HCA1700⟩. While this reaction gives a variety of coupled products in good yields, it is likely to be specific to derivatives of active methylene compounds.

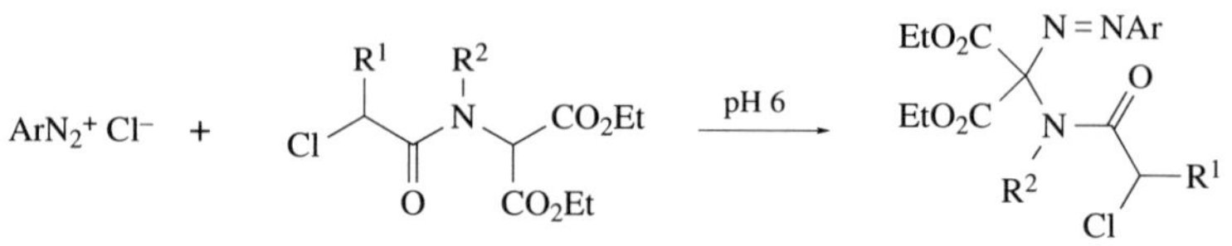

The oxidation of benzylamines with buffered potassium permanganate provides a rapid and general route to acylamino imino alkanes, and probably represents the preferred route to such compounds (Equation (94)) ⟨64JA1701, 82JCED475⟩.

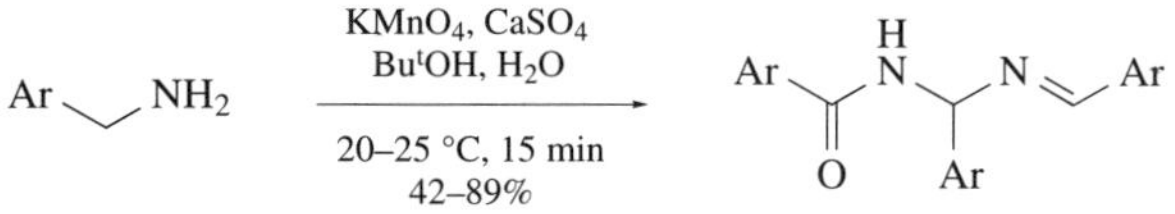

$$Ar\diagup NH_2 \xrightarrow[20-25\,°C,\,15\,min]{\substack{KMnO_4,\,CaSO_4 \\ Bu^tOH,\,H_2O \\ 42-89\%}} Ar-C(=O)-N(H)-C(Ar)=N-Ar \quad (94)$$

4.09.8.6 Miscellaneous Benzotriazole Derivatives

Various primary aliphatic and aromatic amides react well with benzotriazole and aliphatic and aromatic aldehydes in refluxing toluene with azeotropic water removal to provide moderate yields of the corresponding Mannich products (Equation (95)). The versatility of this procedure coupled with the facile removal of the benzotriazole moiety allows the synthesis of a variety of functionalised amides ⟨88JCS(P1)2339⟩. Similar products may also be prepared from the intermediate Mannich product, 1-(hydroxymethyl)benzotriazole, by reaction with primary amides in acetic acid, or toluene containing TsOH, at reflux ⟨93JOC2086⟩. Furthermore, this approach using the Mannich intermediate has been extended to include secondary amides, ureas, thioamides, thioureas and sulfonamides, the yields in all cases falling within the range 43–96% ⟨93JOC2086⟩.

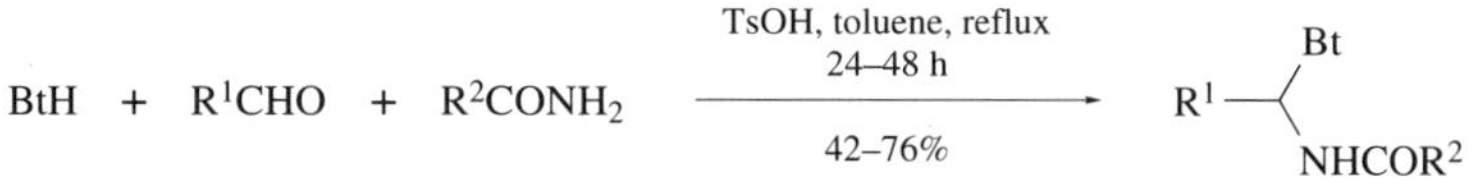

$$BtH + R^1CHO + R^2CONH_2 \xrightarrow[42-76\%]{\substack{TsOH,\,toluene,\,reflux \\ 24-48\,h}} R^1-CH(Bt)(NHCOR^2) \quad (95)$$

In addition to the above reactions, 1-(hydroxymethyl)benzotriazole also serves as a useful intermediate for the formation of the hydroxamic acid derivatives (**143**), which can be isolated in 90–96% yield from arylhydroxamic acids in benzene at reflux in the presence of TsOH ⟨90S663⟩. The versatile 1-(hydroxymethyl)benzotriazole is likely to facilitate the synthesis of further compounds falling within the scope of this review.

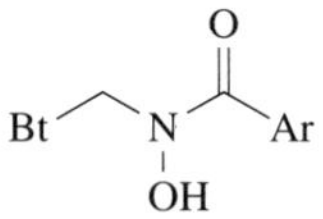

Bt–CH$_2$–N(OH)–C(=O)–Ar

(**143**)

4.10
Functions Containing a Nitrogen and Another Group 15 Element

FRANCES HEANEY
University College Galway, Republic of Ireland

4.10.1 FUNCTIONS CONTAINING ONE NITROGEN AND ONE PHOSPHORUS: $R^1_2C(NR^2_2)PR^3_2$, etc.	451
4.10.1.1 Amino Functions: $R^1_2C(NH_2)PR^2_2$, $R^1_2C(NHR^2)PO(OR^3)_2$, $R^1_2C(NR^2_2)PX$, etc.	452
4.10.1.1.1 Dicoordinate phosphorus functions: $R^1_2C(NR^2_2)PO$, etc.	452
4.10.1.1.2 Tricoordinate phosphorus functions: $R^1_2C(NR^2_2)PR^3_2$, $R^1_2C(NH_2)P(OR^2)_2$, etc.	452
4.10.1.1.3 Tetracoordinate phosphorus functions: $R^1_2C(NR^2_2)P(O)(OR^3)_2$, etc.	460
4.10.1.1.4 Higher coordinate phosphorus functions	478
4.10.1.2 Other Nitrogen Functions: $R^1_2C(NY)PR^2_2$, $R^1_2C(NHX)PR^2_2$, $R^1_2C(NR^2X)PR^3_2$, $R^1_2C(NX_2)PR^2_2$, etc.	479
4.10.1.2.1 Dicoordinate phosphorus functions	479
4.10.1.2.2 Tricoordinate phosphorus functions	480
4.10.1.2.3 Tetracoordinate phosphorus functions	485
4.10.1.2.4 Higher coordinate phosphorus functions	493
4.10.2 FUNCTIONS CONTAINING NITROGEN AND ARSENIC, ANTIMONY OR BISMUTH	495
4.10.2.1 Amino Functions: $R^1_2C(NH_2)ZR^2_2$, $R^1_2C(NHR^2)ZO(OR^3)_2$, $R^1_2C(NR^2_2)ZX$, etc. (Z = As, Sb, Bi)	495
4.10.2.1.1 Dicoordinate Z functions	495
4.10.2.1.2 Tricoordinate Z functions	495
4.10.2.1.3 Tetracoordinate Z functions	501
4.10.2.1.4 Higher coordinate Z functions	502
4.10.2.2 Other Nitrogen Functions	502
4.10.2.2.1 Dicoordinate Z functions	502
4.10.2.2.2 Tricoordinate Z functions	502

4.10.1 FUNCTIONS CONTAINING ONE NITROGEN AND ONE PHOSPHORUS: $R^1_2C(NR^2_2)PR^3_2$, etc.

The literature covering organophosphorus chemistry is vast and the subsection dealing with N—C—P compounds (ca. 29,000 examples) is no small contributor. Major sources of data exist in a number of excellent review series: *The Chemistry of Organophosphorus Compounds* ⟨B-90MI 410-01⟩, *Organophosphorus Chemistry* ⟨B-68MI 410-01⟩, *Topics in Phosphorus Chemistry* ⟨B-64MI 410-01⟩ and *Organic Phosphorus Compounds* ⟨B-72MI 410-01⟩. Additionally, a number of review articles have appeared, many of which are relevant to this topic ⟨68RCR7, 74RCR984, B-76MI 410-01, 84ZC365, 87RCR859, 88AHC(43)2, 88OR(36)176, 92RCR1220⟩. A comprehensive list of known compounds is given in the *Dictionary of Organophosphorus Compounds* ⟨B-88MI 410-01⟩. Inevitably, this chapter has had to be selective.

4.10.1.1 Amino Functions: $R^1_2C(NH_2)PR^2_2$, $R^1_2C(NHR^2)PO(OR^3)_2$, $R^1_2C(NR^2_2)PX$, etc.

4.10.1.1.1 Dicoordinate phosphorus functions: $R^1_2C(NR^2_2)PO$, etc.

Prior to 1958, no phosphorus compound was known where phosphorus had coordination number 2, and up to 1995 there are no examples of molecules of the general structure N—CR^{1}R^2—P with a dicoordinate phosphorus atom.

4.10.1.1.2 Tricoordinate phosphorus functions: $R^1_2C(NR^2_2)PR^3_2$, $R^1_2C(NH_2)P(OR^2)_2$, etc.

An excellent review by Kellner and Tzschach ⟨84ZC365⟩ details the exploitation of the Mannich reaction as a synthetic concept in phosphine chemistry. This theme dominates the synthesis of organophosphorus compounds with tricoordinate phosphorus atoms.

(i) Tricoordinate phosphorus functions by Mannich-type reactions

Modifications of the Mannich reaction constitute by far the most commonly employed route to N—C—P compounds where the phosphorus atom is present in coordination number 3. A nucleophilic phosphorus species (phosphine or a primary, secondary, tertiary or metallo phosphine) reacts with a Mannich base to afford the α-aminomethylphosphine. The Mannich base (**6**) may be preformed or obtained from an amine (**1**) and formaldehyde, an *N*-hydroxy alkylamine (**2**), a diaminomethane (**3**), an imine (**4**) or another $(CH_2)_xN_y$ species, for example (**5**) as shown in Scheme 1.

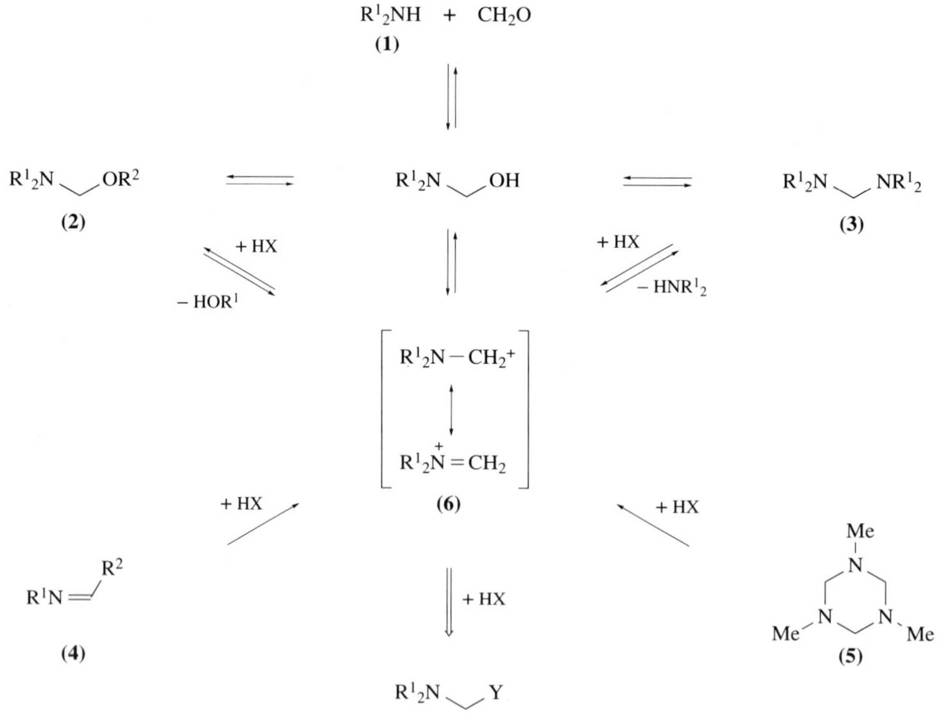

Scheme 1

The direct synthesis of α-aminoalkylphosphines, as one of a number of products, can be realised by the reaction between elemental phosphorus and *N*-hydroxymethyl dialkylamines ⟨65AG(E)527, 67HCA1723⟩. The reaction of diprimary α,ω-bisphosphino alkanes (**7**) and an *N*-hydroxymethyl dialkylamine (formed *in situ* from aqueous formaldehyde and a secondary alkylamine) furnishes the α,ω-bis(bisdialkylaminomethylphosphine)alkane (**8**) (Equation (1)). The reaction decreases in vigour with increasing n ⟨66HCA842⟩. Similar chemistry is observed in the reaction of primary or secondary β-aminoalkylphosphines (9) with N-hydroxymethyl diethylamine (Equation (2)) ⟨68CB3612⟩. The reaction proceeds at room temperature, affording (10) in good yield. By a similar one-pot process, organometallic crown ethers (12) have been prepared by McLain ⟨83JA6355⟩ and by Balch and Rowley ⟨90JA6139⟩. In these cases the amine component was an azacrown species (11) (Equation (3)). This same route to multidonor macrocyclic ligands has been employed by Power and co-workers ⟨84IC2550⟩. *Ortho*-disubstituted aromatic phosphines are also suitable substrates for this reaction; thus, aminoalkylation of o-R^{1}HPC$_6$H$_4$PHR^1 (13) with formaldehyde and a secondary amine leads to mixtures of mono-, di- and tricyclic aminomethylphosphines (14), (15) and (16) in 52–78% yield (Equation (4)) ⟨85ZC172⟩.

$$H_2P(CH_2)_nPH_2 + 4CH_2O + 4HNEt_2 \xrightarrow{62-94\%} (Et_2NCH_2)_2P(CH_2)_nP(CH_2NEt_2)_2 \quad (1)$$
$$(7) \qquad n = 1, 2, 3, 4 \qquad (8)$$

$$H_nP(CH_2CH_2NEt_2)_{3-n} + nEt_2NCH_2OH \xrightarrow[\text{1 h, RT}]{-H_2O} (Et_2NCH_2)_nP(CH_2CH_2NEt_2)_{3-n} \quad (2)$$
$$(9) \qquad n = 1, 76\% \qquad (10)$$
$$\qquad n = 2, 72\%$$

(11) + Ph$_2$PH + CH$_2$O $\xrightarrow[\text{60 °C}]{\text{toluene}}$ (12)   (3)
86%

(13) R^1 = H, Pr^i, Et, Bu
R^2 = H; R^3 = Ph, Pr^i, Bu^t, Et, CH$_2$CO$_2$Et
R^2 = Et; R^3 = Et, Ph

+ H$_2$CO + R^{2}R^{3}NH $\xrightarrow{52-78\%}$ (14) (15) (16)   (4)

Barluenga *et al.* have applied a variation of this methodology to the synthesis of N-(phosphinomethyl)arylamines (18) ⟨90SL261⟩. The N-(alkoxymethyl)-N-arylamines (17) react directly with a secondary phosphine to afford the product in high yield (Scheme 2). Alternatively, the aromatic methyleneamine (19), prepared *in situ*, reacts with the lithium salt of the same phosphine (BuLi, −60 °C) to furnish analogous products—the yield by this route is slightly lower. Another route to N-(phosphinomethyl)arylamines involves the thermal displacement (160–170 °C) of dimethylamine from tris(dimethylaminomethyl)phosphines by aniline ⟨72JOC2752, 80IZV2417⟩.

The synthesis of cyclic secondary and tertiary α-aminoalkylphosphines (22) has been carried out successfully using a parallel approach. Secondary (ω-aminoalkyl)phenylphosphines (20) react with a range of aldehydes and ketones (21) ⟨67CB2685, 67TL1489, 68CB3619, 68CB4032, 73JPR526, 73ZC139⟩ in a cyclocondensation reaction to furnish the 1,3-aza-phospholanes ($n = 1$), -phosphanes ($n = 2$) and -phosphepanes ($n = 3$) (22) (Equation (5)). It is believed the relatively acidic P–H group undergoes intramolecular addition to the Schiff's base formed between the amino group of (20) and the added carbonyl component. The reaction proceeds in good yield although more forcing conditions are required when the carbonyl component is changed from an aldehyde to a ketone. The scope of this reaction includes primary aminoalkylphosphines ⟨68CB3619, 88JOM(355)71⟩, the use of formaldehyde

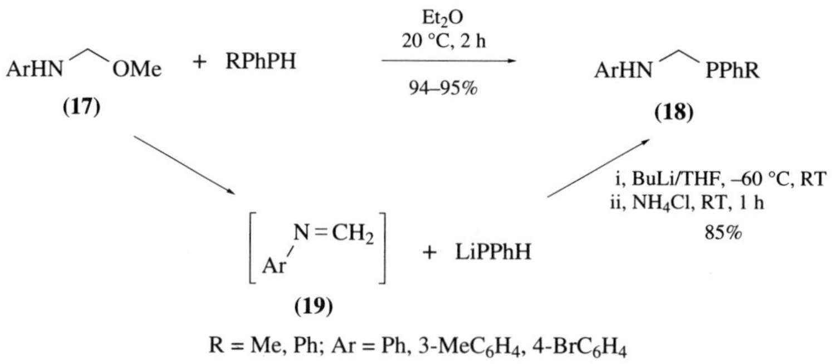

Scheme 2

ArHN–OMe (17) + RPhPH → (Et$_2$O, 20 °C, 2 h, 94–95%) → ArHN–PPhR (18)

(17) → [Ar–N=CH$_2$] (19) + LiPPhH

(19) → (18)
i, BuLi/THF, –60 °C, RT
ii, NH$_4$Cl, RT, 1 h
85%

R = Me, Ph; Ar = Ph, 3-MeC$_6$H$_4$, 4-BrC$_6$H$_4$

and aromatic acyl halides as the carbonyl component. Additionally, several examples are reported where the amino group is secondary in nature ⟨68CB3619, 73JPR526⟩. Some representative examples are given in Table 1.

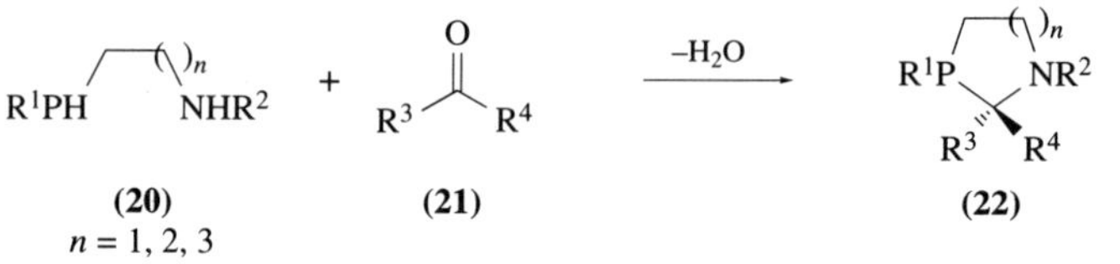

$$R^1PH\text{–}(CH_2)_n\text{–}NHR^2 + R^3COR^4 \xrightarrow{-H_2O} R^1P\text{–}(CH_2)_n\text{–}NR^2\text{–}CR^3R^4 \quad (5)$$

(20) n = 1, 2, 3   (21)   (22)

Table 1 Preparation of α-aminoalkylphosphines (22) by reactions between α,ω-aminoalkylphosphines and carbonyl compounds.

No.	R^1	R^2	R^3	R^4	Yield (%)	Ref.
n = 1						
1	Ph	H	H	H	71	67CB2685
2	Ph	H	Et	H	65	67CB2685
3	Ph	H	Me	Me	79	67CB2685
4	Ph	H	a	a	86	67CB2685
5	H	H	Ph	H	47	68CB3619
6	H	H	b	b	78	68CB3619
7	Bu	H	H	H	68	68CB3619
8	H	Et	H	H	51	68CB3619
9	H	Et	Ph	H	86	68CB3619
10	Et	Et	H	H	56	68CB3619
11	Ph	Et	Ph	H	62	68CB3619
12	H	H	Ar	Cl	76	74JOM(81)187
n = 2						
13	Bu	H	Ph	H	67	68CB4032
14	Bu	H	b	b	73	68CB4032
15	Ph	H	H	H	67	68CB4032
16	Ph	Et	H	H	84	73JPR526
17	Pr^i	Bu^n	H	Et	68	73JPR526
18	Ph	Bu^n	Me	Et	47	73JPR526
n = 3						
19	Ph	H	Ph	H	30	73ZC139
20	Ph	H	b	b	30	73ZC139

^{a} $R^3 + R^4 = (CH_2)_4$. ^{b} $R^3 + R^4 = (CH_2)_5$.

The reaction of the (3-oxoalkyl)phenylphosphines (23) either with aldehydes and primary amines or with aldimines gave the tetrahydro-1,3-azaphosphorines (24) (Equation (6)) ⟨73MI 410-01⟩. Similarly, the benzo-1,3-aza-phospholanes, -phosphanes and -phosphetanes (25) have been prepared in modest yield by Issleib et al. (Equation (7)) ⟨71MI 410-01 (n = 0); 74SRI191 (n = 1); 76ZAAC(424)97, 83AJC2095 (n = 2)⟩. A wide range of carbonyl compounds have been used and in some cases the reaction was carried out in the presence of TsOH. Optically pure α-aminoalkylphosphines and bisphosphine derivatives may be obtained if the amino component in the reaction is itself optically pure. Kellner and co-workers have employed valine, alanine or serine—in the form of the free acid or sodium salt—as the amino segment; the yields were fair ⟨80JOM(193)307, 80TL1845, 87JOM(326)c9⟩.

$$(23) + R^4CHO + NH_2R^5 \text{ or } R^4\text{—}NR^5 \longrightarrow (24) \quad (6)$$

$$\text{(o-H}_2\text{N-C}_6\text{H}_4\text{-(CH}_2\text{)}_n\text{PH}_2\text{)} + R^1COR^2 \xrightarrow[-H_2O]{\text{heat, 2–4 h}} (25) \quad (7)$$

$n = 0$; $R^1 = $ Me, Ph, Et; $R^2 = $ CO$_2$H, Me, Et; $R^1R^2 = $ (CH$_2$)$_5$
$n = 1$; $R^1 = $ Ph, Et, H, Me; $R^2 = $ H, Et
$n = 2$; $R^1 = $ Me, H; $R^2 = $ Et, n-C$_3$H$_7$F, H; $R^1R^2 = $ (CH$_2$)$_5$

The direct addition of a nucleophilic phosphorus species to imines or iminium ions as a route to N—C—P compounds with 3-coordinate phosphorus is a less common route than the analogous preparation of α-amino phosphonic acids and their derivatives. Primary, secondary and metallo phosphines react with variously substituted imine substrates to furnish the products, a range of which are shown in Table 2. The 1,3,5-oxazaphosphane (27) is the product of a tandem reaction; a Mannich reaction between the imine (26) and phenylphosphine precedes cyclocondensation (Scheme 3) ⟨74SRI453⟩. The iminium ion (29), formed *in situ* by treatment of the perhydrotriazine (28) with HCl in a protic solvent, reacts with secondary phosphines to furnish ammonium salts (30), from which free aminomethylphosphines can be obtained with NaOMe at −40 °C (Scheme 4) ⟨52DOK(83)865⟩. Other examples of the preparation of α-aminoalkyl phosphorus compounds from iminium halides generated *in situ* include the reaction of lithium dialkyl(aryl)phosphines with α-chloro tertiary amines ⟨67JOC2383, 69IC1336⟩ and the preparation of (33) by attack of phosphines on the imine (32), generated from acetyl chloride and phenylbis(diethyldiamino)methane (31) (Scheme 5) ⟨67JOC2383, 69IC1336, 78JOM(149)167, 84ZC365⟩. The preparation of α-aminomethyl tertiary phosphines in moderate yield by the slightly exothermic reaction between lithium diphenylphosphide and α-chloro tertiary amines was discovered in 1967 ⟨67JOC2383⟩.

Table 2 Addition of nucleophilic phosphine derivatives to imines.

Phosphine	Imine	Product	Comment	Yield (%)	Ref.
PhPH$_2$	PhCH=NPh	PhP[(CH(Ph)NHPh)]$_2$	CO$_2$, 6 h, 130–135 °C	65	63ZOB3353
PhPH$_2$	CH$_2$=NCMe$_3$	PhHPCH$_2$NHCMe$_3$	catalytic Na	7.5	74ZC243
Et$_2$PH	CH$_2$=NCMe$_3$	Et$_2$PCH$_2$NHCMe$_3$		82	
HOC(CF$_3$)$_2$PH$_2$	(CF$_3$)$_2$C=NH	HOC(CF$_3$)$_2$PHC(CF$_3$)$_2$NH$_2$			86MI 410-01
R$_2$PH (R = PhNC(CH$_2$)$_2$)	(CF$_3$)$_2$C=NH	R$_2$PC(CF$_3$)$_2$NH$_2$	80 °C, 4 days	92	81JINC629
R^{1}R^{2}PH (R^1, R^2 = H, Me)	(CF$_3$)$_2$C=NH	R^{1}R^{2}PC(CF$_3$)$_2$NH$_2$			84ZN(B)356
R^{2}R^{1}PH (R^1,R^2 = alkyl, aryl)	PhCH=NR^3 (R^3 = alkyl, aryl)	R^{1}R^{2}PCHPhNHR^3		76–96	78ZOB1008
MP(TMS)$_2$ (M = Na, Li)	R^{2}CH=NR^1	(TMS)$_2$PCHR^{2}NHR^1			65ZAAC(336)234, 75HCA1316
TMS-PEt$_2$	R^{2}CH=NR^1	Et$_2$PCHR^{2}NHR^1	120 °C	80	75HCA1316
PhHP-TMS	RCH=NPh	Ph(TMS)PCHRNHPh		43	90ZOB1718

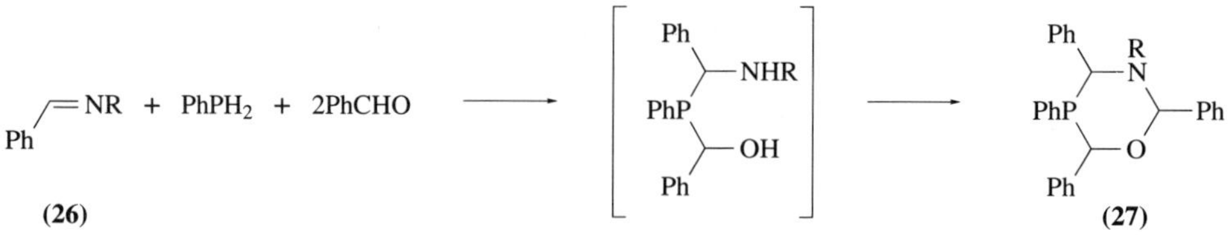

Scheme 3

R = Me, Et, Bu, Bu^s

Scheme 4

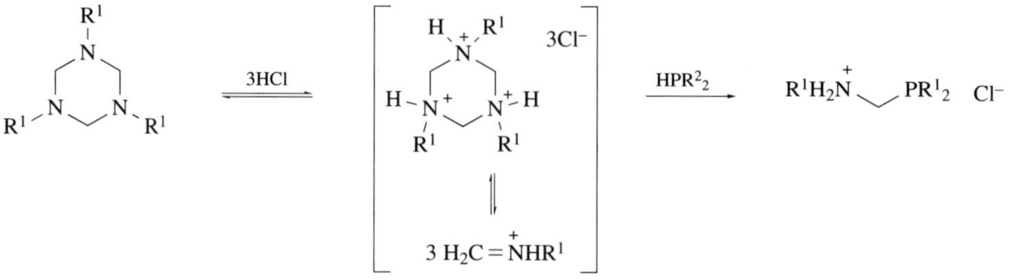

Scheme 5

(ii) Tricoordinate phosphorus functions by reaction with α-hydroxyalkyl phosphorus compounds

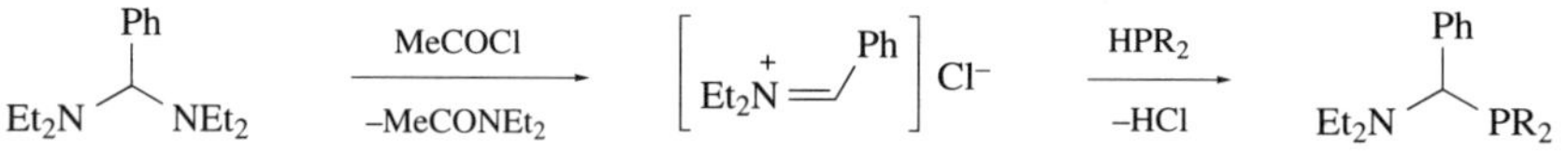

The development of flame-retardant finishes for cotton fabrics has led to the synthesis of a variety of cyclic and acyclic N—C—P compounds. Tetrakis(hydroxymethyl)phosphonium salts (**34**) react with primary amines, for example, aniline, which displace all the hydroxyl groups to furnish the aminomethyl analogues (**35**), from which the phosphine (**39**) is obtained upon treatment with ammonia. Secondary amines react similarly, furnishing tris(dialkylaminomethyl)phosphines; protected amines such as methyl carbamate and various ureas may also participate ⟨81PS(10)147⟩. Historically, halide salts have been employed, although reports in the early 1970s and 1980s describe the use of sulfonium salts as a superior route to these compounds ⟨72JOC2752, 81CJC27⟩. Treatment with a secondary or tertiary amine effects elimination of CH_2O and HCl from (**34**; X = Cl) to afford the tris(aminomethyl)phosphine (**38**). Compounds such as (**39**) and their derivatives may also be furnished by the exothermic reaction between PH_3, CH_2O and NHR_2, and nucleophilic attack by a secondary amine on tris(hydroxymethyl)phosphine (**38**) or tris(dimethylaminomethyl)phosphine (**37**) (Scheme 6) ⟨72JOC2752⟩. Arbuzov *et al.* have demonstrated other examples of such exchange of amino and hydroxymethyl groups in hot benzene with aniline ⟨80IZV2417⟩. Similarly, aryldi(hydroxymethyl)phosphines, arylalkylhydroxymethylphosphines and diarylhydroxymethylphosphines react with two mole equivalents of an amine or an equimolar amount of an α,ω-diamino component; for example, the phosphines (**40**) react as shown in Equation (8) to furnish the 1,5-diaza-3-phosphacycloheptanes (**42**) ⟨80TL3467, 81TL229⟩.

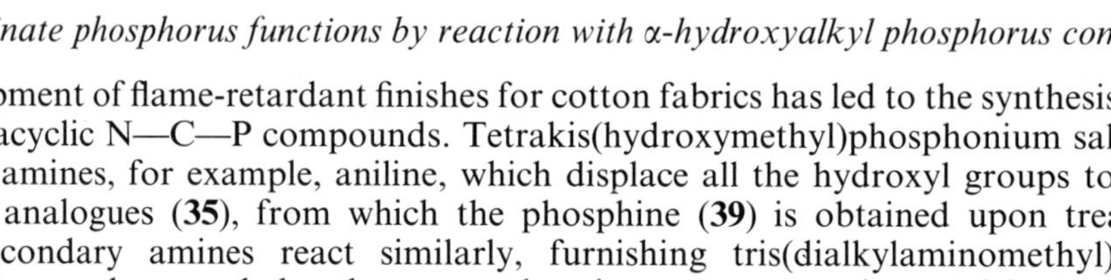

R^1 = Ph, c-C_6H_{11}, Et, Me
R^2 = CMePhH

(8)

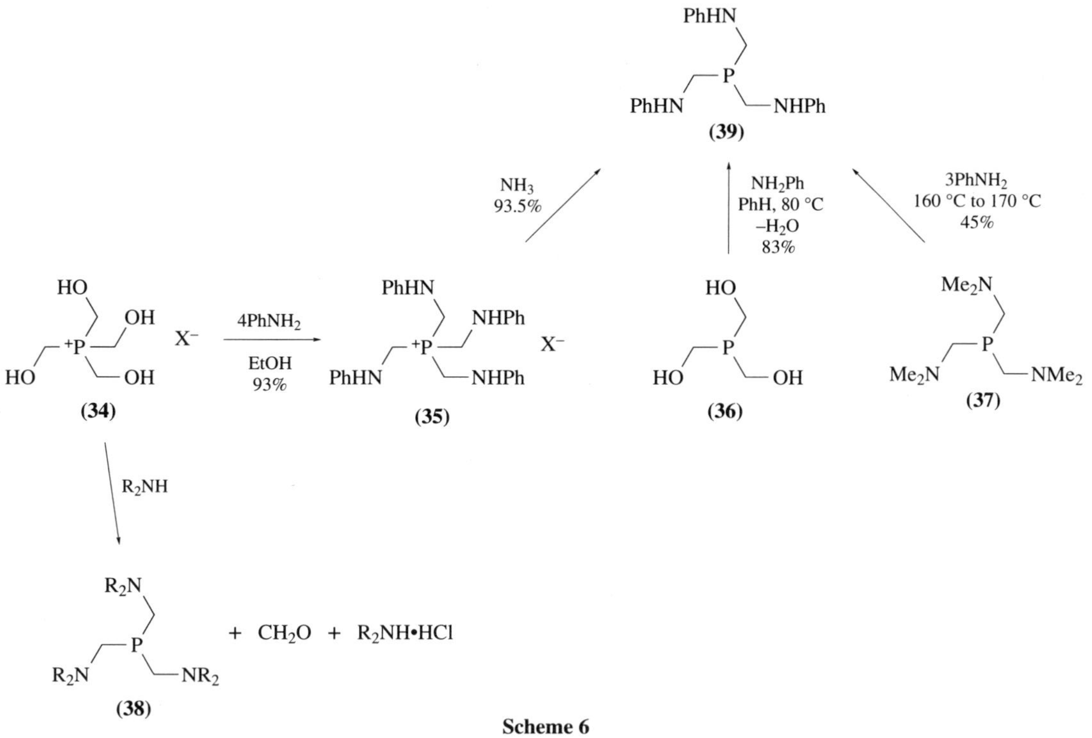

Scheme 6

The adduct (**45**) can be prepared similarly from (**43**); it may also be obtained by a unique heteroatom exchange of the cyclic boron ester (**44**) (Scheme 7) ⟨80IZV952⟩. A further example of this exchange is illustrated by preparation of the 1,5-diaza-3,7-diphosphacyclooctane (**46**) by treatment of (**44**) with 1,3,5-triphenyl-1,3,5-triazine (Scheme 8). The product (**46**) may also be obtained from the phosphonium salt (**47**). The 1,3-azaphosphetanes (**48**) complete this monocyclic series and may be prepared as shown in Scheme 9 ⟨80IZV1438⟩.

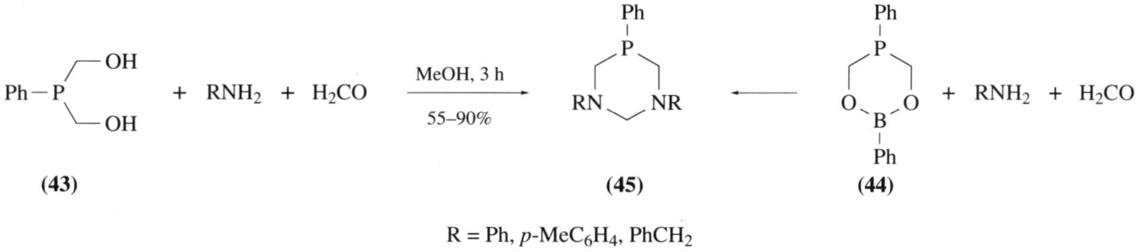

Scheme 7

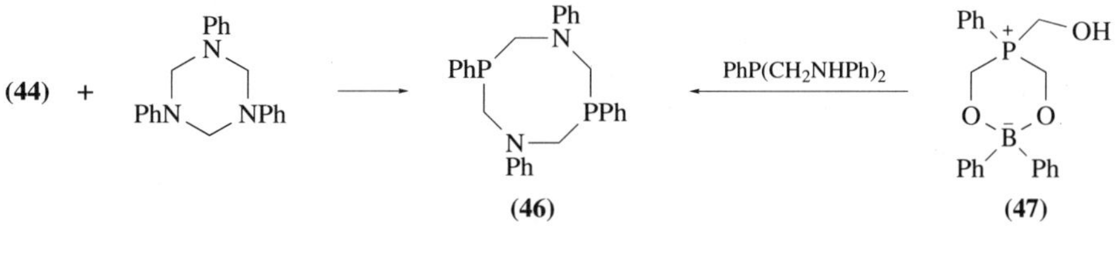

Scheme 8

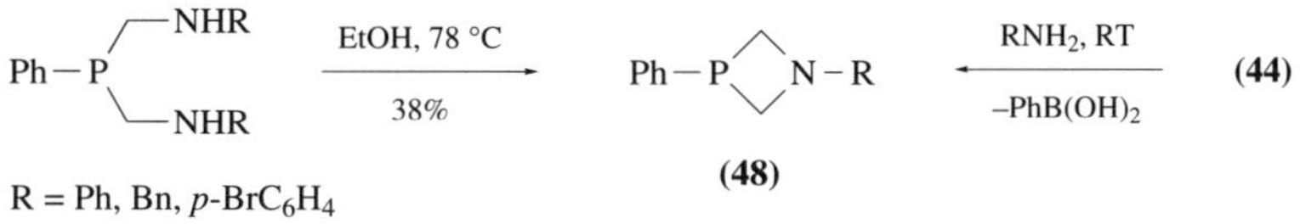

Scheme 9

The 1,3,5-diazaphosphane (**49**) may be obtained by the action of triethylamine on the appropriate tetrakis(aminomethyl)phosphonium salt (84% yield), or sequential treatment of the tetrakis(hydroxymethyl) salt with sodium ethoxide and aniline (77% yield) (Scheme 10) ⟨72JOC2752⟩. The bicyclic product (**50**) represents a further example of this methodology where the amine component is cyanamide (Equation (9)) ⟨72JHC715, 72JHC1295⟩. The reaction has also been extended to include the adamantane derivative (**51**) (Equation (10)) ⟨74JHC407, 75JHC579⟩, and with a mixture of ammonia and ethane-1,2-diamine the ring-expanded analogue (**52**) was obtained.

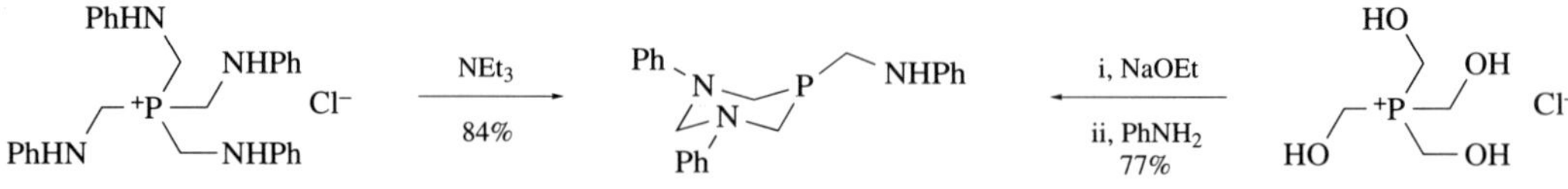

Scheme 10

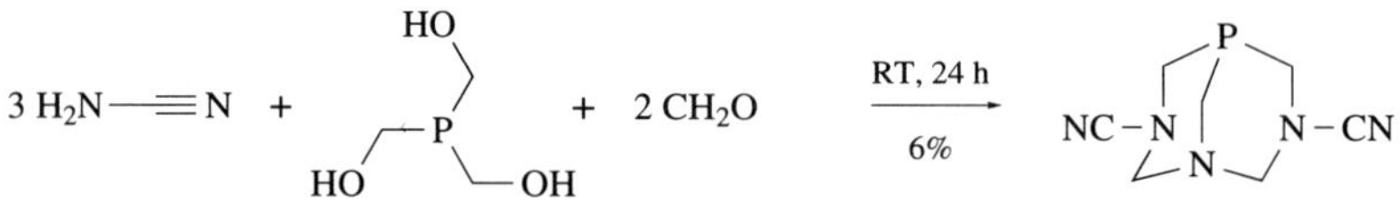

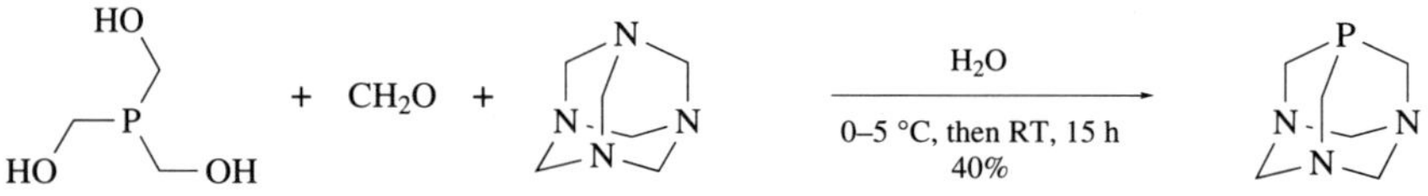

(iii) Tricoordinate phosphorus functions by miscellaneous reactions

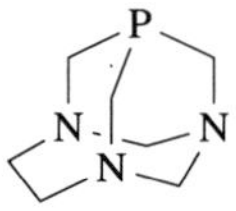

The di(trimethylsilyl)ester of phosphonic acid reacts with the enamines (**53**) in a free-radical reaction to furnish (**54**) in 45–81% yield at 100–120 °C (Equation (11)), where AIBN represents 2,2′-azobisisobutyronitrile ⟨92ZOB946⟩. Secondary and tertiary 1,3-azaphosphanes ((**56**), (**58**) and (**59**)) can also be obtained by free-radical cyclisation of diallylaminomethylphosphines (**55**) (Scheme 11) ⟨83PS(17)73⟩ and from the reaction between formaldehyde and β-allylaminoalkylphosphines (**60**) ⟨85ZAAC(523)7⟩. In the latter case the product(s) formed—the azaphosphane (**61**) or the phosphepane (**62**)—are dependent on the choice of reaction conditions (Scheme 12). The formation of (**61**) is believed to proceed via (**63**) and (**64**), as shown in Scheme 13. The α-oxo-phosphonic acid diethyl ester (**65**) reacts similarly with *N*-ethoxymethyl dialkylamines to afford (**66**) (Equation (12)). A catalyst (BF$_3$·OEt$_2$) may be required, depending on the nature of the R group; for example, no catalyst is necessary when R = Et ⟨91ZOB1478⟩.

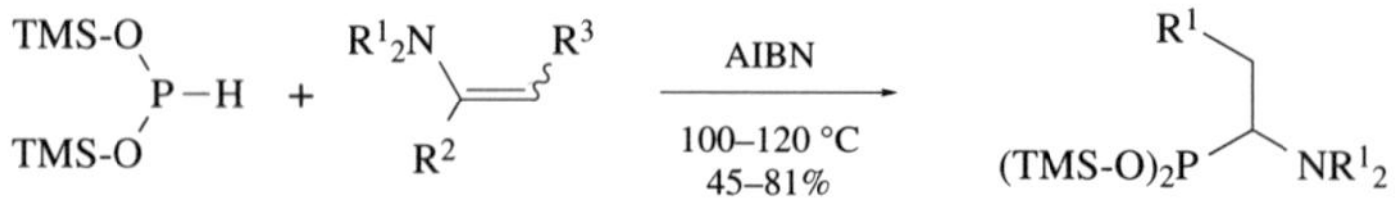

R^1_2N = NEt$_2$, piperidino, morpholino
R^2 = H; R^3 = Me, Et
R^2R^3 = (CH$_2$)$_4$

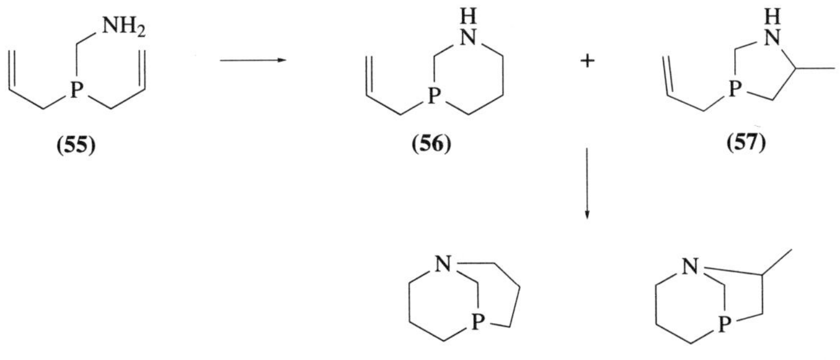

Scheme 11

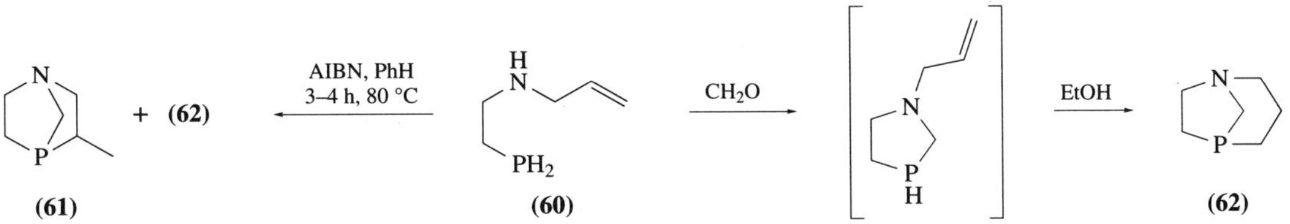

Scheme 12

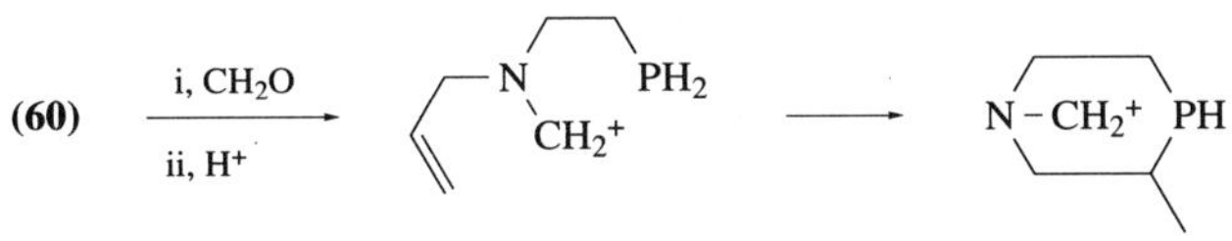

Scheme 13

$$(EtO)_2P(O)Bu^t + R_2N\text{-}CH_2\text{-}OEt \xrightarrow[73-88\%]{EtOH} (EtO)_2P\text{-}CH_2\text{-}NR_2 \quad (12)$$

(65) R = Et, Pr^i, (CH$_2$)$_5$, piperidino, morpholino (66)

Photolysis of a methanolic solution of the methylenephosphine (**67**) leads to (**68**) in addition to a number of other products (Equation (13)) ⟨83TL1975⟩. The methylenephosphine (**67**) also takes part in Diels–Alder reactions with suitable dienes, forming tetrahydrophosphorines.

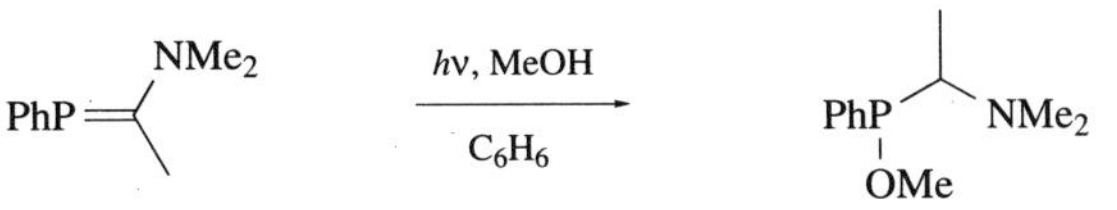

$$PhP=C(NMe_2)(Me) \xrightarrow[C_6H_6]{h\nu, MeOH} PhP(H)\text{-}C(Me)(OMe)(NMe_2) \quad (13)$$

(67) (68)

Tristrimethylsilylphosphine reacts with an excess of aminals or dialkylaminomethanols to yield the trisamino(methyl) derivatives (**69**); the reaction may be carried out in the presence of a zinc chloride catalyst at 80–110 °C (Equation (14)) ⟨91URP1618747⟩ or in CH$_2$Cl$_2$ ⟨91ZOB1016⟩.

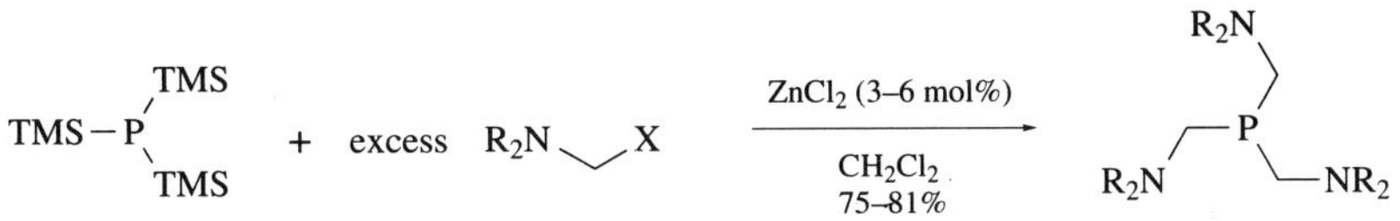

$$TMS\text{-}P(TMS)_2 + \text{excess } R_2N\text{-}CH_2\text{-}X \xrightarrow[\substack{CH_2Cl_2 \\ 75-81\%}]{ZnCl_2\ (3-6\ mol\%)} P(CH_2NR_2)_3 \quad (14)$$

R = Et, Pr, Bu^s
NR$_2$ = piperidino
X = OH, NR$_2$

(69)

4.10.1.1.3 Tetracoordinate phosphorus functions: $R^1{}_2C(NR^2{}_2)P(O)(OR^3)_2$, etc.

The observed biological activity of the phosphorus analogues of α-amino acids has been responsible for an explosion of research activity in the area of synthesis; many analogues of the naturally occurring compounds have been prepared in both racemic and enantiomerically pure form. Much work has also been reported in the area of phosphorus analogues of nucleic acids; however, this material will not be discussed here, and metal complexes of 1-aminoalkyl phosphorus compounds will not be considered.

(i) Tetracoordinate phosphorus functions by reaction of phosphorus nucleophiles at sp^2 centres

(a) Conjugate addition to suitably substituted acceptors. The addition of alkyl phosphites or phosphonites to β-amino-substituted α,β-unsaturated esters furnishes aminophosphonic and -phosphinic acid esters respectively. Free acids may be obtained upon hydrolysis ⟨79PJC2327⟩; yields are poor, about 28–34%. Much better yields are obtained in the corresponding reaction with the β-amido substrate, for example, addition of ethyl phosphinates or diethyl phosphite to (70) in refluxing HCl followed by treatment with aqueous NH_3 furnishes (71) (Equation (15)) ⟨74RZC1119, 76RZC661⟩.

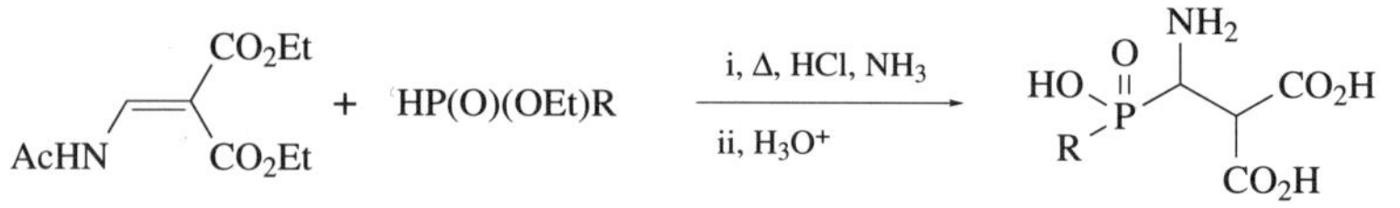

(b) Addition to azines. α-Aminobenzylphosphinic acids (73a) may be obtained (73–81% yield) by addition of ethyl phosphonites to aromatic aldazines followed by reductive N—N bond cleavage (Equation (16)) ⟨79ZC253⟩. Corresponding derivatives of phosphonic acid can be formed by analogous reaction with dialkyl phosphites; thus the aryl aldazine (72) reacts with an excess of the sodium salt of diethyl phosphite to furnish, after hydrolysis, (73b). Yields are moderate to good (56–93%) ⟨73ZC254, 75RZC397, 78TL1609⟩. Reaction with aliphatic aldazines gives the monoadducts (74) (73–93% yield), which can be treated with acid to obtain α-hydrazinophosphonic acids (75) (36–85% yield) or reduced (H_2, Raney nickel (RaNi)) to afford the α-aminophosphonic acids (76) (71–92% yield) (Scheme 14) ⟨76C187, 76RZC477⟩. A parallel reaction with alkyl phosphonites furnishes α-aminophosphinic acids (65–75% yield) ⟨78PJC1315⟩. Ketazines may also be suitable substrates for this type of reaction ⟨78C253⟩.

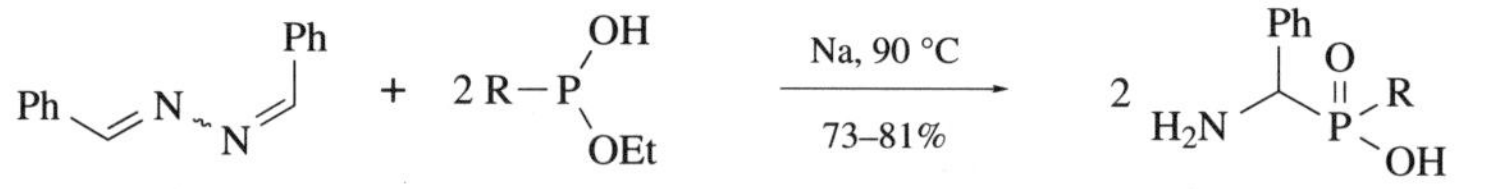

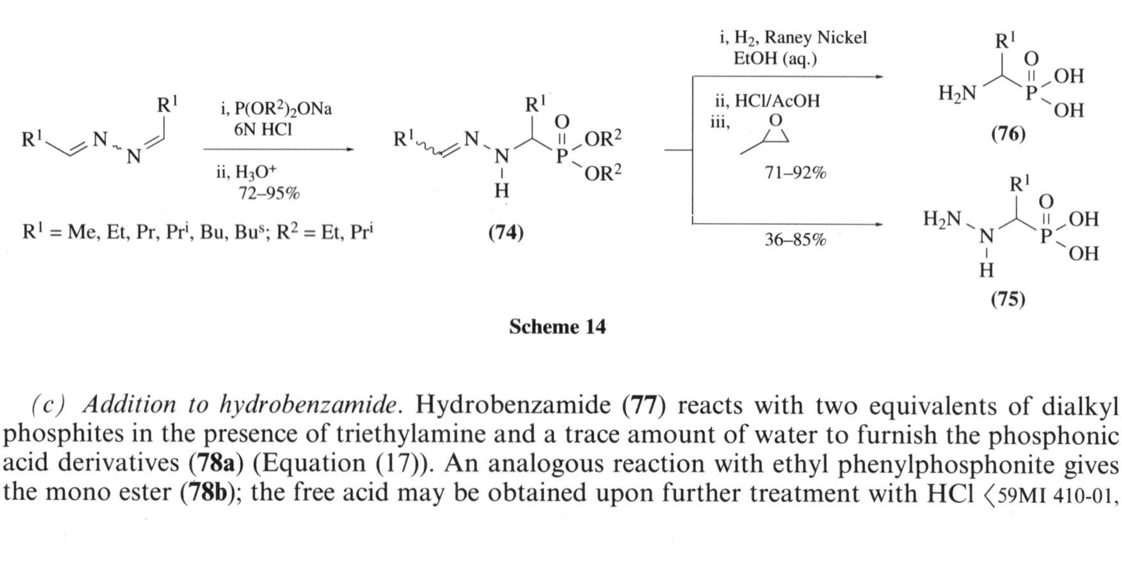

Scheme 14

(c) Addition to hydrobenzamide. Hydrobenzamide (77) reacts with two equivalents of dialkyl phosphites in the presence of triethylamine and a trace amount of water to furnish the phosphonic acid derivatives (78a) (Equation (17)). An analogous reaction with ethyl phenylphosphonite gives the mono ester (78b); the free acid may be obtained upon further treatment with HCl ⟨59MI 410-01,

73IZV955, 80IZV1125⟩. Similar results are observed from the one-pot reaction involving the addition of dialkyl phosphites to N^1,N^2-dialkylidene 1,1-diamino alkanes; the adducts (**79**) are formed in essentially quantitative yield (Scheme 15) ⟨78ZAAC(444)249⟩.

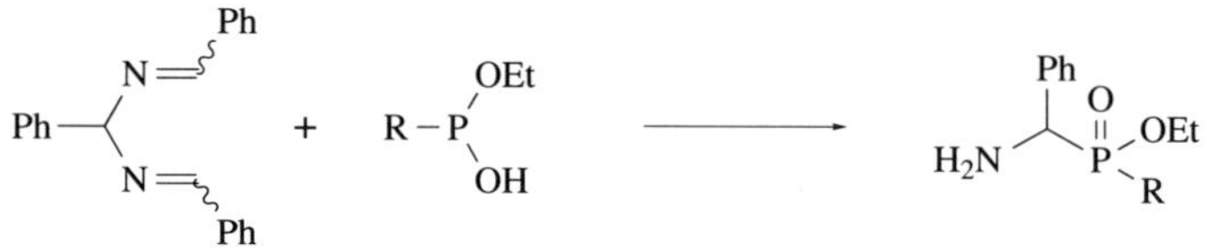

(17)

(**77**)   (**78**) **a**; R = OEt; NEt$_3$, H$_2$O, 100 °C, 3 h, 98%
              b: R = Ph; NEt$_3$, 55–100 °C, 85%

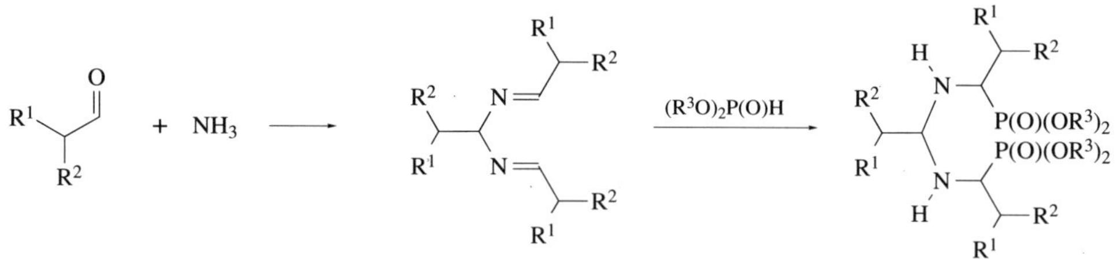

R^1 = Me, R^2 = Et
R^1R^2 = (CH$_2$)$_5$

R^3 = Me, Et, Pr, Pr^i (**79**)

Scheme 15

(d) Addition to Schiff's bases. The addition of nucleophilic phosphorus species to Schiff's bases is an important routine method for the synthesis of N—C—P compounds. Suitable trivalent phosphorus species include phosphorus acid (P(OH)$_3$) and its mono-, di- and trialkyl(aryl) esters, and tris(trimethylsilyl) ester and monoalkyl phosphonites. The imine may be preformed or formed *in situ*; it may be derived from an aldehyde or ketone component with ammonia or a primary or secondary amine; and it may be acyclic or cyclic. The most attractive imine substrates include those where a nitrogen substituent may be readily removed affording free amino groups, and those which have an optically active centre capable of introducing asymmetry into the product. The easy reductive cleavage of the benzyl group and 1-substituted benzyl groups is responsible for the wide use of Schiff's bases based on benzylamine. Thus diethyl phosphite reacts with a range of aldimines (**80**) (Scheme 16) to furnish, following acid hydrolysis, free α-aminophosphonic acids (**81**) in 46–66% yield ⟨77S239, 78PJC321⟩. Another method of nitrogen deprotection is heating with refluxing formic acid ⟨78PJC959⟩. Aminophosphonic acids with groups sensitive to hydrogenolysis/hydrolysis are readily obtained by addition of bis(*p*-methylbenzyl) hydrogen phosphite to aldimines. Selective removal of the ester group is observed upon solvolysis with formic acid. Thus, with judicious choice of protecting groups, aminophosphonic acids may be obtained in two steps ⟨78CL1103⟩. Free α-aminobenzyl-phosphonates and -phosphinates may also be obtained by an efficient route involving addition of trimethylsilyloxy phosphorus(III) derivatives, generated *in situ*, to *N*-benzylidineallylamines at room temperature, followed by catalytic deallylation ⟨90SL415⟩. Trityl protection has also been used in the synthesis of 1-aminoalkylphosphinic acids ⟨94S23⟩ by addition of bis(trimethylsilyl)phosphonites to an *N*-trityl alkanamine. The ease of removal of the trityl protecting group by dilute acid makes this an attractive route.

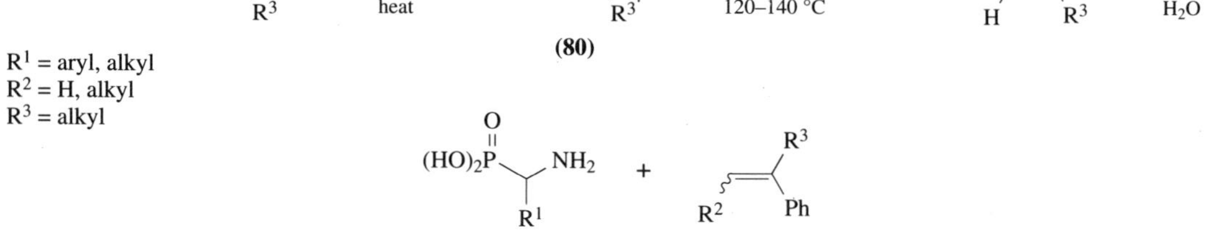

R^1 = aryl, alkyl
R^2 = H, alkyl
R^3 = alkyl

(**80**)

(**81**)

Scheme 16

With optically pure imines, asymmetric induction is observed. The addition of tris(trimethylsilyl) phosphite to (+)-(R)-(82) followed by methanolysis yields (83) (Scheme 17). Chemical yields are good and the diastereoselectivity is found to vary with the reaction conditions (see Table 3). The free amino acids (84) are obtained upon hydrogenolysis ⟨81PJC643⟩. Other examples include the use of galactosylamine or β-L-fructosylamine as a chiral auxiliary ⟨92S90⟩. Alternatively, optically pure aminoalkyl phosphonic acids may be obtained by resolving a diastereomeric mixture, employing, for example, dibenzoyl-L-tartaric acid anhydride ⟨83CJC2425, 92SC107⟩, or by enzymic hydrolysis. Thus penicillacylases have been successfully employed in the resolution of (R)- and (S)-amidophosphonic and phosphinic acids and their esters ⟨92ZOB1472⟩. Enantiomerically pure cyclic phosphites, stable upon storage, have been prepared and shown to react with a range of prochiral C=N groups, opening a general route to α-aminophosphonic acids ⟨85AG(E)1067⟩.

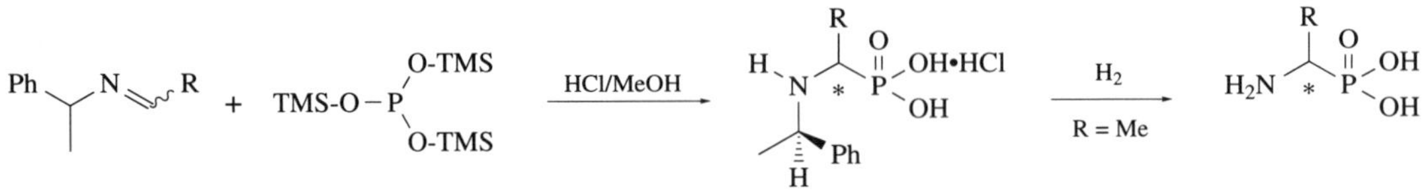

Scheme 17

Table 3 Addition of tris(trimethylsilyl) phosphite to the chiral imine (82).

R	Solvent	Temperature (°C)	Time (h)	Catalyst	Yield (%)	Diastereomeric mixture (83)
CH_3	Benzene	80	6		50	70:30
Pr^i	ether	20	10	$ZnCl_2$	70	66:34
Ph	ether	20	10	TsOH	80	85:15
Ph	ether	20	10	$ZnCl_2$	89	66:34
Ph	ether	20	10	ZnI_2	87	66:34
Ph	ether	20	10		88	90:10

Imines (85) derived from aryl aldehydes yield aminophosphonic acids (86) upon heating (100–115 °C) neat with phosphorous acid (yields 10–98%) (Table 4) (Equation (18)) ⟨78JOC992⟩. The same authors note that much lower yields of phosphonic acids are obtained from imines derived from aliphatic aldehydes or diaryl ketones; it is found that in these cases there is competition from reduction of imines to amines. The addition of dialkyl phosphites to aliphatic aldimines affords, after hydrolysis, the zwitterionic products (87) in 19–85% yield (Equation (19)) ⟨76ZOB1012⟩. Dialkyl phosphoriodites add two moles of aldimines, affording—as a consequence of an addition/cyclisation sequence—2,4-diazaphospholanes (88) in 82–95% yield (Equation (20)) ⟨85JGU404⟩. Tertiary phosphines react with iminium ions furnishing the corresponding phosphonium salts ⟨62CB2563, 63LA(665)91, 78JOM(149)167⟩.

Table 4 Addition of phosphorus acid to the imines of aryl aldehydes (85).

R^1	R^2	Yield (%)
$PhCH_2$	Ph	98
CH_3	Ph	61
Bu^t	Ph	40
$PhCH_2$	$p\text{-}ClC_6H_4$	87
$PhCH_2$	$o\text{-}HOC_6H_4$	10

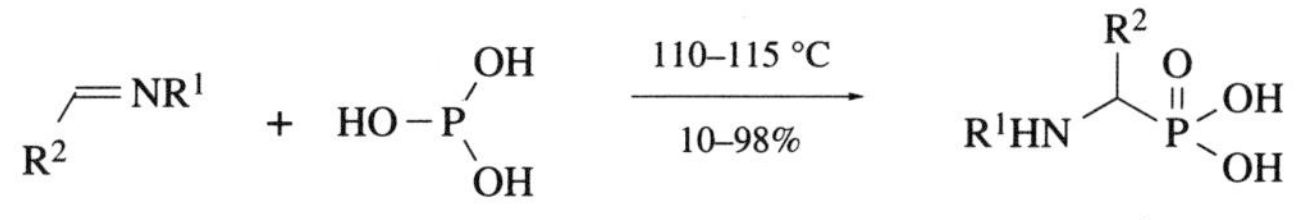

$$R^2\!\!-\!\!\!=\!\!NR^1 + HO-P(OH)(OH) \xrightarrow[10-98\%]{110-115\,°C} R^1HN-CH(R^2)-P(=O)(OH)(OH) \quad (18)$$

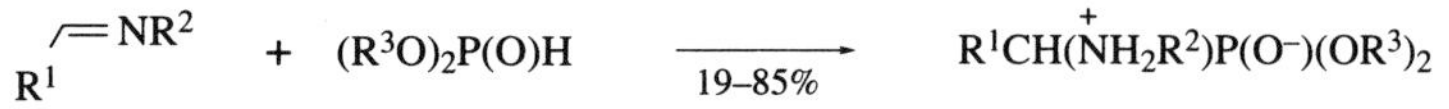

(85) → **(86)**

$$R^1\!\!-\!\!\!=\!\!NR^2 + (R^3O)_2P(O)H \xrightarrow{19-85\%} R^1CH(\overset{+}{N}H_2R^2)P(O^-)(OR^3)_2 \quad (19)$$

R^1 = Pr, hexyl
R^2 = Me, Pr, Bu
R^3 = Me, Et, Pr, Bu

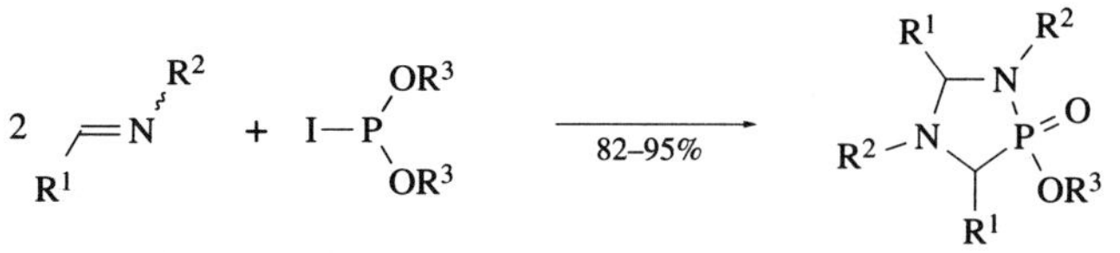

(87)

$$2\ R^1\!\!-\!\!\!=\!\!N-R^2 + I-P(OR^3)(OR^3) \xrightarrow{82-95\%} \text{cyclic product} \quad (20)$$

R^1 = Et, Pr^i
R^2 = Et, Bu^t
R^3 = Et, Bu

(88)

Ketimines undergo analogous reactions, for example, N-phenyl cyclohexylimine reacts with alkyl- and dialkylchlorophosphites and -chlorophosphines in the presence of alcohols to afford 1-aminoalkylphosphonates and -phosphinates ⟨88ZOB2456⟩. With the use of highly hindered carbonyl components, for example, 9-fluorenones, a longer reaction time is required and yields are variable (12–80%) ⟨80JPR213⟩.

Phosphonic and phosphinic acids and their esters add readily to imines, affording aminoalkyl derivatives in varying yield. These reactions were first discovered in the early 1950s ⟨52DOK(83)689, 52JA1528, 53DOK(92)959⟩ and their wide potential is illustrated by the following examples. Notably, deprotection of the nitrogen-substituted hydrophosphoryl product may not be carried out by catalytic hydrogenolysis since this results in poisoning of the catalyst. Protecting groups capable of being removed by acid hydrolysis include diphenylmethyl ⟨83PS(14)171⟩, Bu^t ⟨72SRI317⟩ and α,α-disubstituted benzyl groups ⟨77S239⟩. Ethyl phenylphosphonite **(89)** reacts with the imine **(90)**, furnishing **(91)** in the presence of an alkoxide catalyst (Equation (21)). Yields are low due to losses in workup. Similar reactions are observed for the thio ester analogues ⟨53DOK(92)773⟩. Similar reactivity is observed with phosphinic acids ⟨48CB477⟩, with product yields of 30–90%. Diaryl phosphites add directly to imines yielding α-aminophosphonates in excellent yield. The preparation of aminomethylphosphonic acid has been achieved by addition of diethyl phosphite to the imine **(92)**. Selective ester hydrolysis occurs in HCl, whilst treatment with HBr at 175 °C hydrolyses both amino and ester groups (Scheme 18) ⟨72SRI317⟩. In 1994, it was shown that sonochemical activation can assist in the formation of 1-aminophosphonic acids bearing pyrrole or thiophenic moieties ⟨94S51⟩. The Schiff's base **(94)** obtained *in situ* from the hexahydro-1,3,5-triazine **(93)** adds diethyl phosphite, furnishing the N-benzyl adduct **(95)**; hydrogenolysis then yields free amino phosphonates (Scheme 19) ⟨73TL4645, 80ZAAC(469)109⟩.

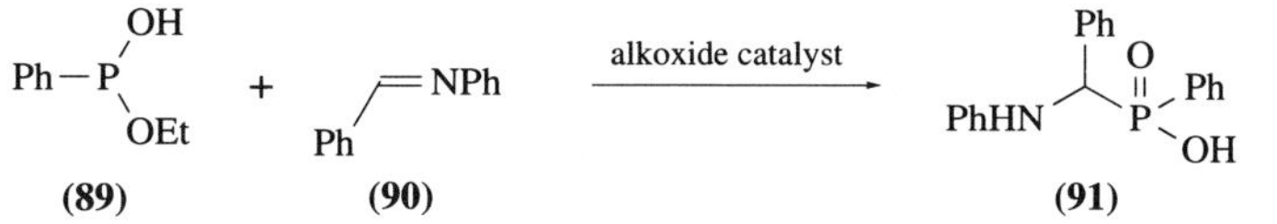

$$Ph-P(OH)(OEt) + Ph-CH\!\!=\!\!NPh \xrightarrow{\text{alkoxide catalyst}} PhHN-CH(Ph)-P(=O)(Ph)(OH) \quad (21)$$

(89) + **(90)** → **(91)**

$$Bu^t\!\!-\!\!N\!\!=\!\!CH_2 + (EtO)_2PHO \longrightarrow Bu^tHN\!-\!CH_2\!-\!P(=O)(OEt)_2 \xrightarrow[\text{ii, HBr, 175\,°C}]{\text{i, HCl}} H_2N\!-\!CH_2\!-\!P(=O)(OH)_2$$

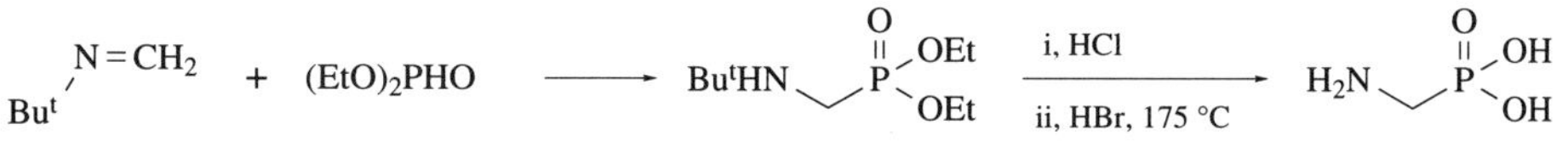

(92)

Scheme 18

Analogous reactions with phosphonites or dichlorophosphines furnish phosphinic acid derivatives ⟨81PS(14)139, 83PS(11)295⟩. Poor yields of product are observed if the reaction is carried out with

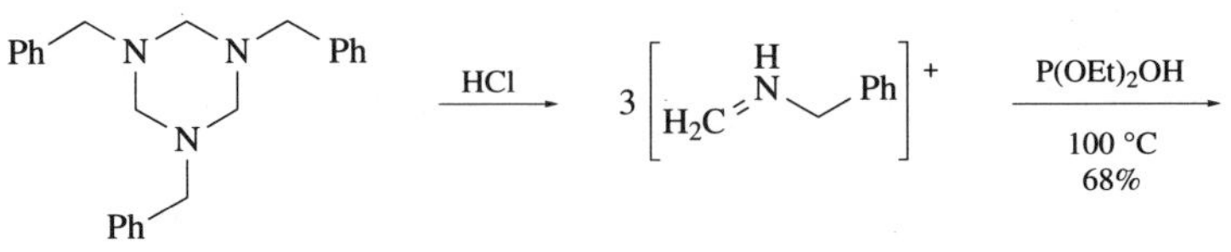

(93) (94) (95)

Scheme 19

phosphinic acids ⟨84JCS(P1)2845⟩. Cyclic imines also react with phosphorous acid ⟨78JOC992⟩; for example, the 3,4-dihydroisoquinoline (**96**) furnishes the 1-phosphonic acid (**97**) (Equation (22)), and the cyclic phosphinous acid (**98**) reacts with a Schiff's base (Equation (23)) yielding (**99**) ⟨77ZOB579⟩. Aryliminoxindoles react with trialkyl phosphites only in the presence of a protonating agent (H_2O or CH_3COOH), furnishing the corresponding phosphonate ⟨89JPR906⟩.

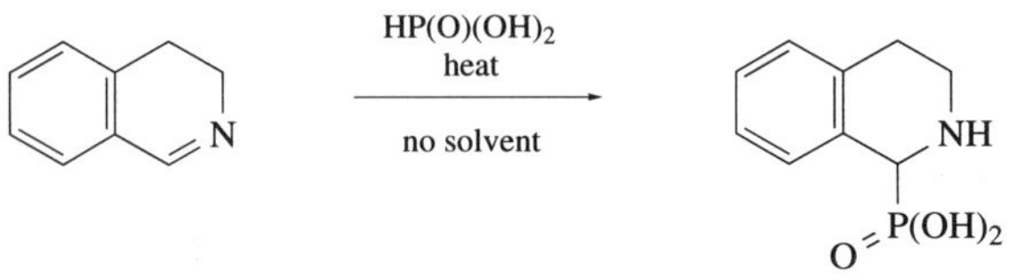

(96) (97)

(22)

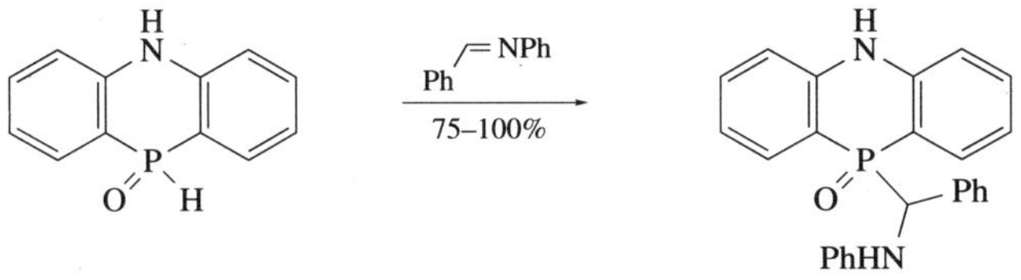

(98) (99)

(23)

(e) Addition to nitriles and oximes. The first example of the use of a nitrile for the preparation of α-aminoalkylphosphonic acids (**102**) involved addition of diethyl phosphite to the aldimine salt (**101**) (obtained by reduction of the parent nitrile (**100**) with tin(II) chloride) followed by hydrolysis (Scheme 20) ⟨77S625⟩. The ready availability of a diverse range of nitriles makes this an attractive route for investigation. Phosphinic acid reacts with oximes to give, upon subsequent modification, free aminophosphonic acids. Yields are rather low (10–30%) and products are purified by ion chromatography ⟨78IZV1951⟩.

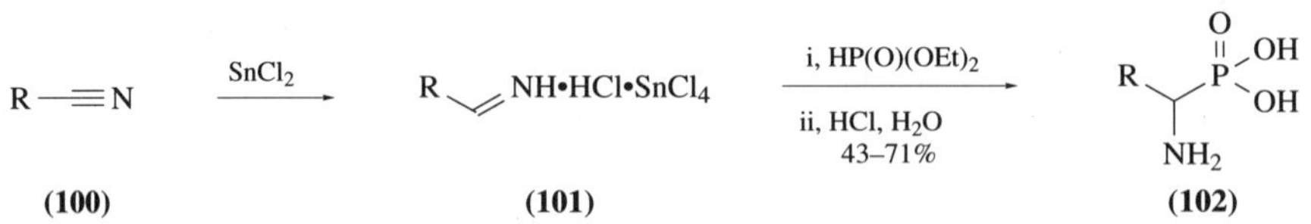

(100) (101) (102)

R = Me, Et, Pr, Ph, Bn, *m*-MeC_6H_4

Scheme 20

(ii) Tetracoordinate phosphorus functions by Arbuzov– and Michaelis–Becker-type reactions

Arbuzov– and Michaelis–Becker-type reactions are important for the synthesis of α-aminoalkyl phosphorus compounds. Historically this process represents the first synthetic route to aminomethylphosphonic acid. The classic reaction involves a trialkyl(aryl) phosphite with an α-halo amine. Valency expansion to form the stable P=O bond promotes the reaction. Less commonly employed tricoordinate phosphorus species include the chlorophosphines R_2PCl, $RPCl_2$ and PCl_3. Michaelis–Arbuzov reactions of trivalent phosphorus esters with *N*-bromomethylphthalimide (**103**) afford good yields of α-aminoalkyl phosphine oxides, phosphinates and phosphonates (**104**) (Scheme 21). Deprotection of the intermediate phthalimides may be achieved with hydrazine, aniline or sodium carbonate ⟨48BSF774, 63JOC2898, 87OS(65)119, 92TL77⟩. The reaction between trimethyl phosphite and the α-bromoglycine derivatives (**105**) furnishes, after hydrolysis of the resulting amides, the α-aminophosphonate (**106**) (Scheme 22) ⟨83MI 410-01⟩. α-Alkoxyglycine derivatives react successively with PCl_3 and $P(OEt)_3$ affording analogous products ⟨82AG(E)776⟩. A similar approach has been used in the synthesis of the alanine analogues (**107a**) and (**107b**) (Scheme 23) ⟨85S62⟩. Phosphonylation of N-chloromethylpyrrolidin-2-one (**108**) with triethyl phosphite gave (**109**) in 73% yield (Equation (24)) ⟨84JGU421⟩.

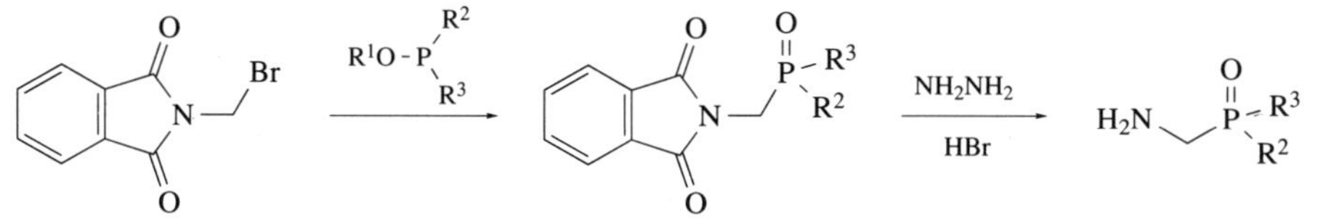

Scheme 21

R^1 = Et, Bu
R^2 = Ar, Me, OEt, OBu
R^3 = Ar, OEt, ONa

(**104**) R^2, R^3 = OH, Ar, Me

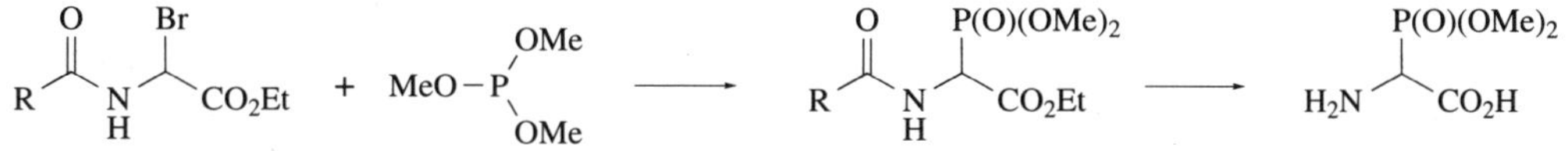

Scheme 22

R = Ph, p-ClC_6H_4, Me

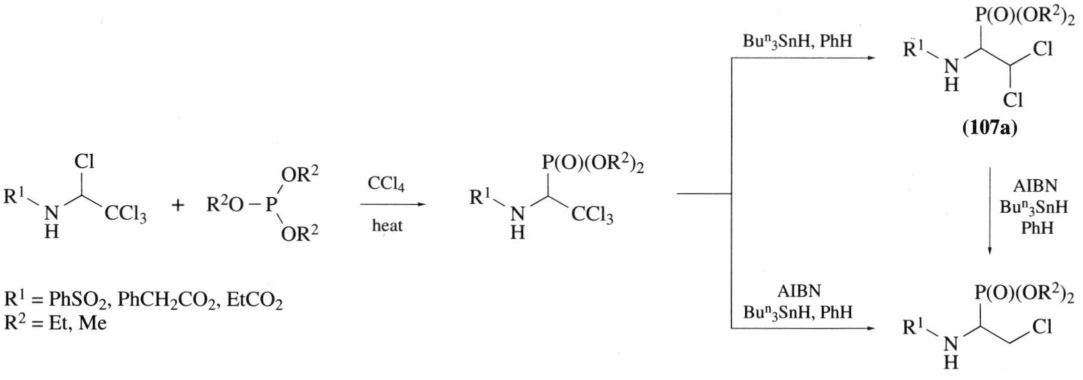

Scheme 23

R^1 = $PhSO_2$, $PhCH_2CO_2$, $EtCO_2$
R^2 = Et, Me

(24)

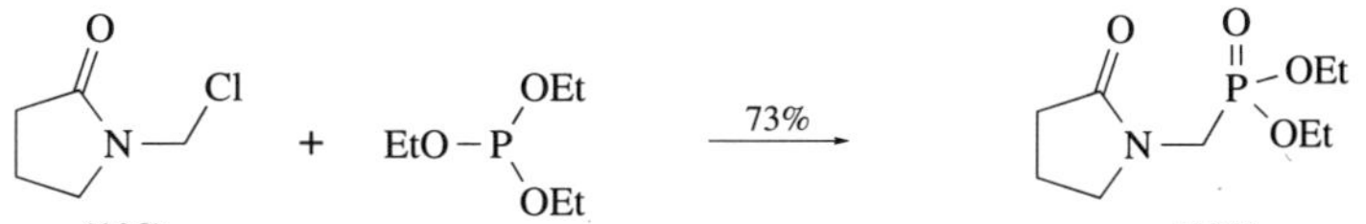

The α-halo amines required for the classic Arbuzov–Michaelis–Becker reaction may be difficult to obtain and may have poor reactivity, but nontraditional variants, in particular α-hydroxy amides and carbinolamines, have been successfully employed. Alkylaminomethyl phosphinic acids (**112**) can be prepared from the alkyldichlorophosphines (**110**) and an α-hydroxy amide (**111**) in AcOH. The same product results from the one-pot, three-component reaction shown in Scheme 24 ⟨88SC425⟩. Chlorophosphines also react with α-hydroxy amides (cyclic or acyclic) in AcOH and the products may be deacylated by treatment with HCl, yielding α-aminophosphonic acids in 70–100% yield ⟨89S547⟩. Tetraphenyl bisphosphines react with α-hydroxyamides, furnishing the expected phosphine oxides ⟨68HCA1608⟩. Elemental phosphorus, P_4, reacts with α-hydroxy amines yielding the range of products shown in Equation (25); with careful choice of reaction conditions and substrate ratios, tertiary phosphine oxides can represent up to 45% of this mixture ⟨65AG(E)527⟩.

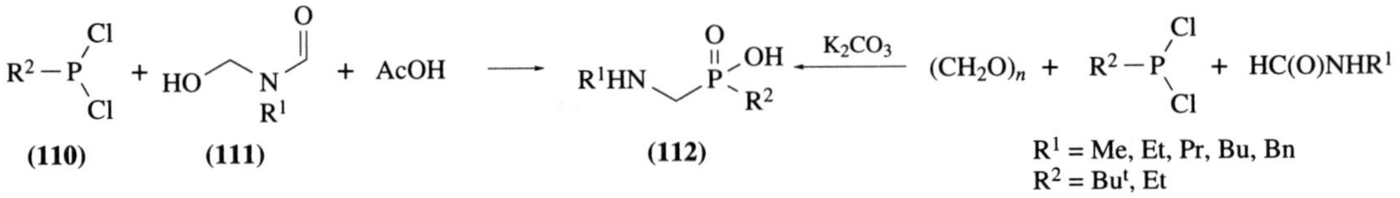

$$R^2-PCl_2 \text{ (110)} + HOCH_2N(R^1)CHO \text{ (111)} + AcOH \longrightarrow R^1HNCH_2P(O)(OH)R^2 \text{ (112)} \xleftarrow{K_2CO_3} (CH_2O)_n + R^2-PCl_2 + HC(O)NHR^1$$

R^1 = Me, Et, Pr, Bu, Bn
R^2 = Bu^t, Et

Scheme 24

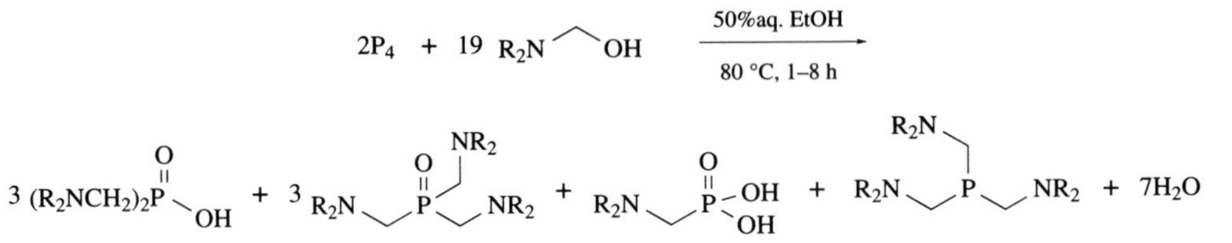

$$2P_4 + 19\ R_2NCH_2OH \xrightarrow[80\ °C,\ 1-8\ h]{50\%\ aq.\ EtOH}$$

$$3\ (R_2NCH_2)_2P(O)OH + 3\ R_2NCH_2P(O)(NR_2)CH_2NR_2 + R_2NCH_2P(O)(OH)CH_2NR_2 + R_2NCH_2P(NR_2)CH_2NR_2 + 7H_2O \quad (25)$$

It is widely reported that a diverse range of 1-aminoalkylphosphonates (both free and nitrogen-substituted) can be obtained upon heating a carbonyl component, an amino component and a trivalent phosphorus species. The amino component may be an amine, carbamate, urea or thiourea (or their substituted derivatives), or a phosphoramidothioate. The reaction may be catalysed by Lewis or Brønsted acids and hydrolysis of the primary products furnishes free α-aminophosphonic acids (**113**) (Scheme 25).

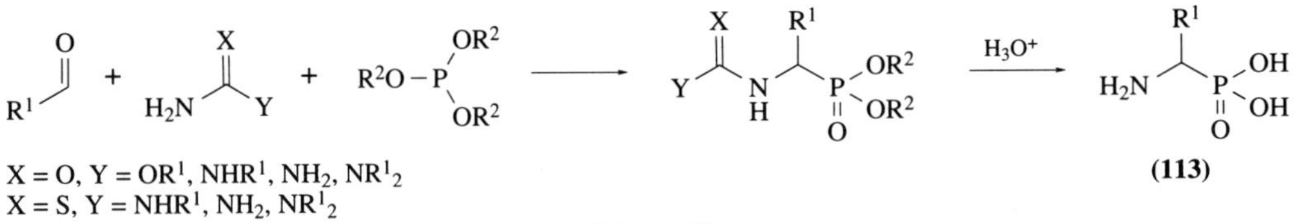

X = O, Y = OR^1, NHR^1, NH$_2$, NR$^1_2$
X = S, Y = NHR^1, NH$_2$, NR$^1_2$

Scheme 25

N-Benzylglycine reacts with formaldehyde and dichlorophosphines to afford α-aminophosphonites (52–100%) which can be *N*-deblocked by treatment with H_2–Pd/C ⟨81PS(11)139, 81PS(11)149⟩. The heterocyclic compound (**114**) arises in 90% yield from such a reaction between 2-chloroethyldichlorophosphine, *N*-benzylglycine and formaldehyde in acidic aqueous alcohol (Equation (26)).

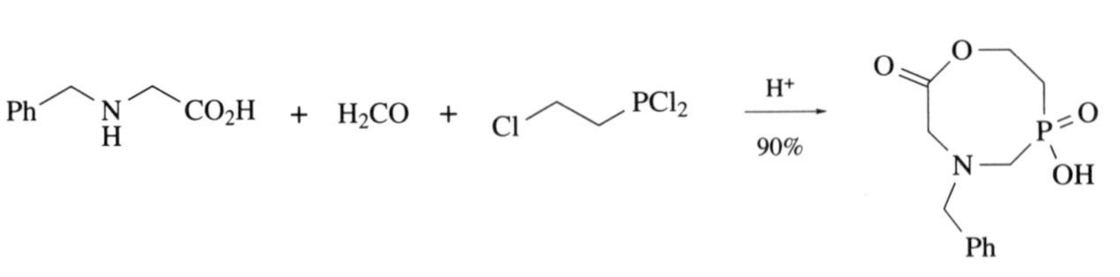

$$PhCH_2N(H)CH_2CO_2H + H_2CO + Cl(CH_2)_2PCl_2 \xrightarrow[90\%]{H^+} \text{(114)} \quad (26)$$

Birum ⟨74JOC209⟩ describes the first example of a condensation reaction between thiourea, an aldehyde and a trialkyl(aryl) phosphite; an important, unexpected observation was the more rapid conversion of triaryl phosphites than trialkyl phosphites to phosphonates, representing a reversal of the normal order of reactivity of phosphite esters. An analogous reaction between a nitrogen-substituted thiourea, triphenyl phosphite and the aldehyde (**115**) presents a suitable method for the large-scale preparation of phosphomethionine and its derivatives (**116**) ⟨82S188⟩. The condensation reaction takes place in AcOH and hydrolysis of the ester and nitrogen-substituent follows on thermal treatment with HCl (Scheme 26). It was further noted that the nature of the nitrogen-protecting group affects the yield of product. Phosphonic acid derivatives of cysteine and cystine ⟨81S643⟩, alanine, valine and phenylalanine ⟨78S469⟩ have also been prepared by this methodology. Aryl-glycines can be obtained in low yield by the reaction of aryl aldehydes, triphenyl phosphite and

N-methylthiourea in boiling toluene and acetic acid for 2 h ⟨78PJC1949⟩. Benzyl, methyl and ethyl carbamates can also be employed in this reaction. Thus phosphonic acid derivatives of a variety of α-amino acids are obtained from a two-step procedure involving, for example, triphenyl phosphite, benzyl carbamate (**117**) and aldehydes in the presence of an acid catalyst. Easy removal of the benzyloxycarbonyl group from (**118**) by treatment with HBr in AcOH followed by ammoniacal ether furnishes the diphenyl esters (**119**) (Scheme 27) ⟨79S985⟩. Overall yields are not high but the process compares favourably with other methods which require multistep synthesis or complex apparatus. The phosphonic acid analogue of serine may be simply obtained by this method; the racemic form has been resolved ⟨84S577⟩. The tricoordinate phosphorus species may, in addition to trialkyl and triaryl phosphites, be a mono-, di- or trichlorophosphine. Yuan and Wang ⟨90S256⟩ report a novel, one-pot, three-component procedure for the synthesis of the α-(benzyloxycarbonylamino)benzylphosphonic acids (**120**) and the monoesters (**121**), employing trichlorophosphine as the nucleophilic phosphorus species. The reaction was carried out in the presence of AcOH and SOCl$_2$ (Scheme 28); the role of the thionyl chloride in the reaction is not clear. An analogous reaction between *N*-phenylurea, triethyl phosphite and aldehydes in the presence of F$_3$B·OEt$_2$ furnishes (**122**), which upon hydrolysis yields analogues of valine and phenylglycine (Scheme 29) ⟨77S883⟩. A simple high-yielding procedure for the preparation of α-aminobenzylphosphonic and phosphinic acids (**126**) is described in Scheme 30. Reaction of a phosphoroamidothioate (**123**) with a substituted benzaldehyde and phosphorous or phosphonous esters in the presence of F$_3$B·OEt$_2$ furnished (**124**) in 34–85% yield ⟨86S821⟩. The nitrogen-protecting group may be selectively removed (HBr/AcOH), furnishing the hydrobromide (**125**), and the free acid (**126**) formed by treatment with hydrobromic acid followed by propylene oxide ⟨64JOC832⟩. A new route to *N*-arylaminomethylphosphonates (**128**), suitable even when complex functionality is present on the aromatic ring, has been described ⟨90TL1567⟩. The process which is simple and high yielding, involves condensation between a trialkyl phosphite and an *N*-methoxymethyl arylamine (**127**) catalysed by TiCl$_4$ (Equation (27)).

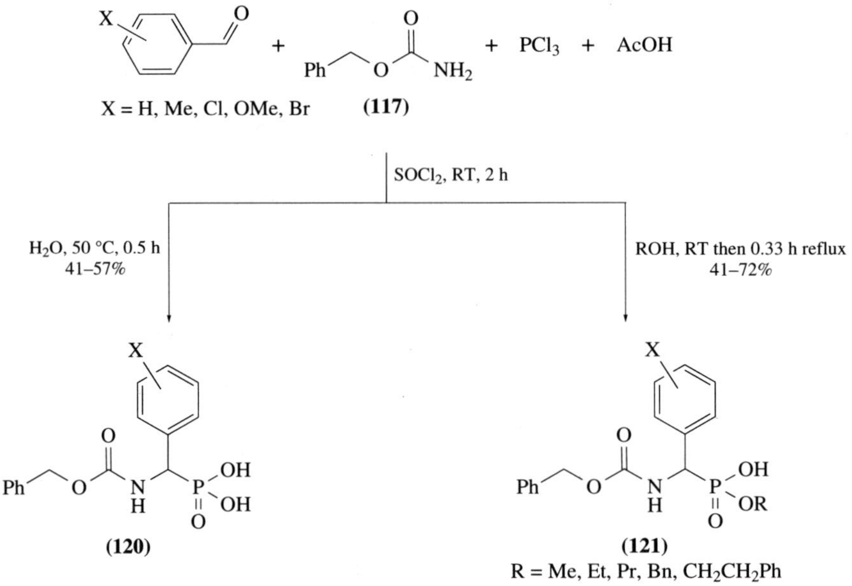

Scheme 28

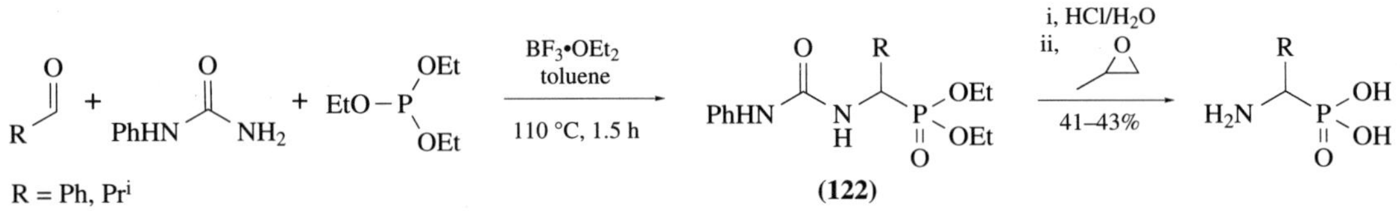

Scheme 29

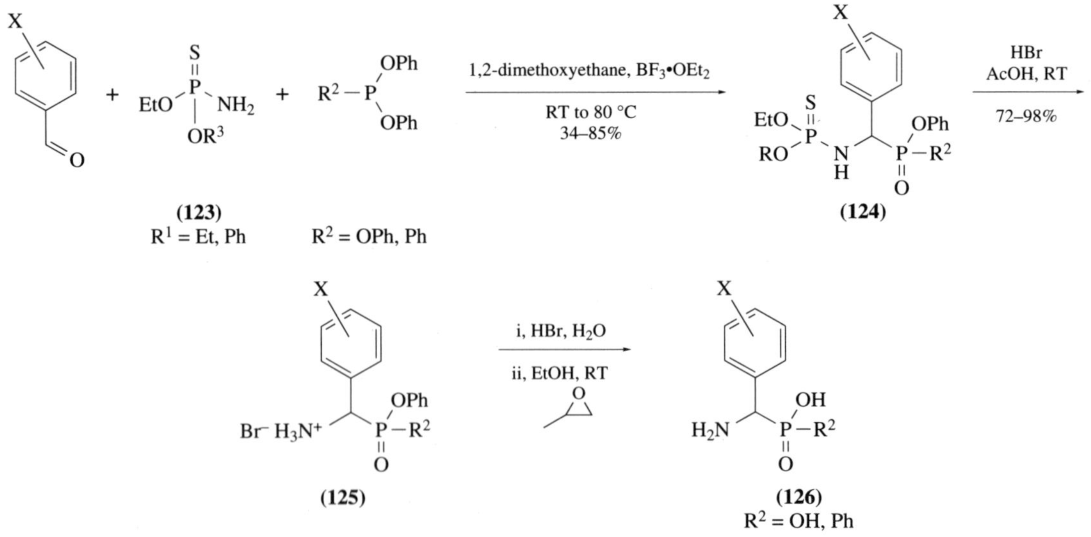

Scheme 30

The preparation of the 2-(diphenylphosphinoyl)pyrrolidines (**130**) involves, as a key step, an Arbuzov-type reaction. Thus, chlorodiphenylphosphine and *N*-substituted tetrahydro-1,3-oxazines furnish the intermediate (**129**) in quantitative yield; subsequent treatment with lithium diisopropylamide (LDA) at −70 °C yields the crystalline pyrrolidines (**130**) in 80–90% yield (Scheme 31) ⟨84TL4259⟩.

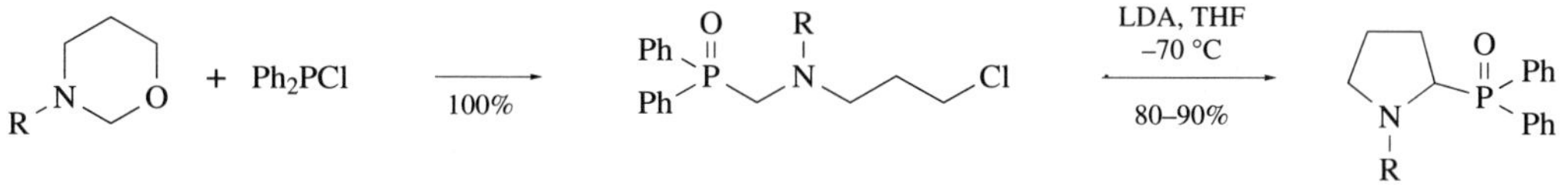

(iii) Tetracoordinate phosphorus functions by Kabachnik–Fields reactions

Prior to the 1950s, the preparation of α-aminomethylphosphonic acids and their derivatives was only possible by treatment of a hydroxymethyl amide with phosphorus trichloride in the presence of acetic acid, the free acids being obtained upon hydrolysis. This method required long reaction times (12 h to 7 days), was limited to the preparation of amino derivatives of methylphosphonic acid and was unsuitable for the preparation of dialkylaminomethylphosphonic acids. It was against this background that a new general method, a modified Mannich reaction, was reported for the synthesis of α-aminophosphonic and α-aminophosphinic acids. Two groups of researchers—Kabachnik and Medved ⟨52DOK(83)689⟩, and Fields ⟨52JA1528⟩—observed independently that the reaction of dialkyl phosphites with a carbonyl component and ammonia (Kabachnik and Medved) or an amine (Fields) led to the formation of α-aminophosphonates. The reactions are fast, high yielding and exothermic, and can be conducted in the presence or absence of solvent. Products are often isolated by distillation of their hydrochloride or picrate salts (Scheme 32). The primary products can be hydrolysed to furnish the mono ester or free acid, depending on the reaction conditions; overall yields tend to be below 45%. An improved experimental protocol suggests that yields can be increased by premixing the anhydrous ammonia and aldehydes in an alcoholic solvent before addition of the dialkyl phosphite ⟨53JA5278⟩. The range of suitable substrates is ever-expanding ⟨74RCR984, 87RCR859⟩, and the reaction is attractive in that it is a one-pot process and the starting materials are readily available. The amine component may be ammonia ⟨57IZV1357⟩, or a primary, secondary or aromatic amine; alternatively, a wide range of protected amino functions can be used. The range of suitable carbonyl components has expanded beyond formaldehyde to include higher aldehydes, ketones (cyclic, acyclic or aromatic), acetals, ketals and some amides. Two schools of thought exist concerning the mechanism of this reaction. The first suggests a nucleophilic attack of the amine component on the initially formed α-hydroxyalkylphosphonic acid ester ⟨68RCR7⟩; the second believes that an electrophilic intermediate is formed between the carbonyl and amine components, which is subsequently attacked by the phosphorus species (Schemes 33a and b) ⟨87RCR859⟩. Azacrown ethers may act as catalysts in this reaction and are especially useful when the amine is weakly basic (pH < 6); in such cases the yield of product may be increased by 20–40% ⟨92ZOB2708⟩. The catalytic activity has been attributed to an improved solubility of the amine in the presence of the crown ether and an increase in the reactivity of the phosphonate through hydrogen bonding. The parent reaction employing dialkyl(aryl) phosphites has been further expanded to include many other compounds with a P—H bond. The diesters of phosphonothioic acid react with ammonia and a carbonyl component, affording aminomethylthiophosphonic acid derivatives. The reaction is smoother than with the corresponding phosphonate and hydrolysis of the primary product furnishes 1-aminomethylphosphonic acids (Scheme 34) ⟨53DOK(92)959, 57IZV1357⟩.

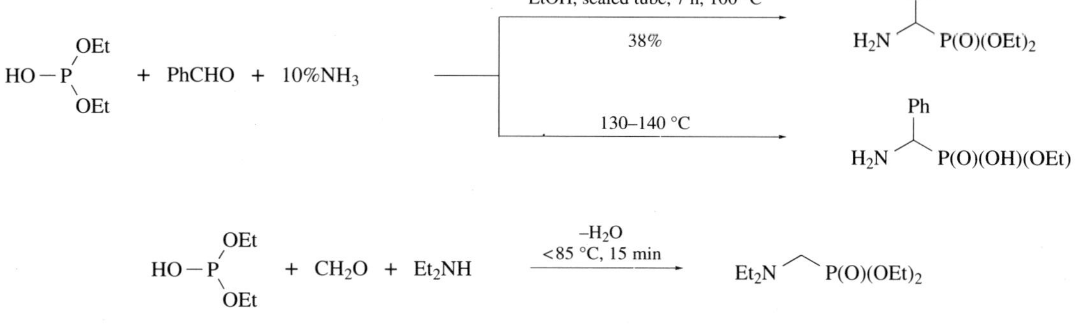

Scheme 32

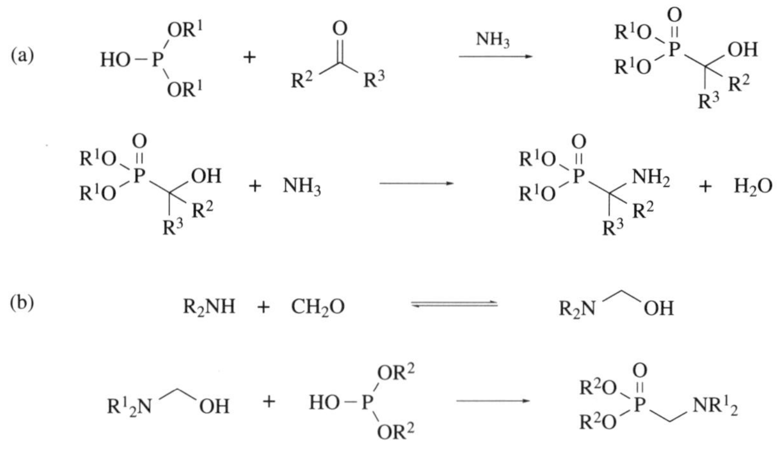

Scheme 33

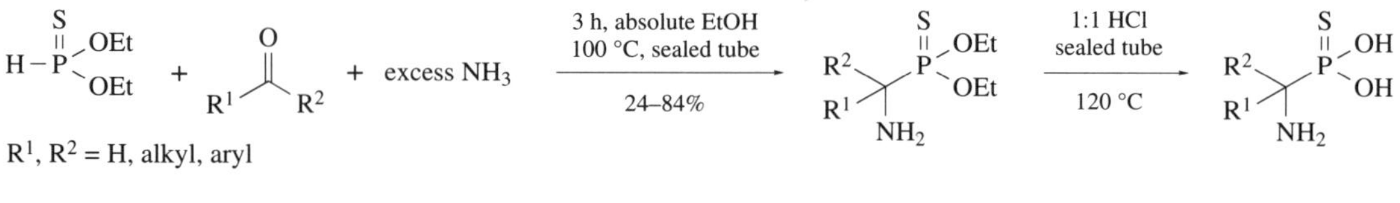

Scheme 34

The tervalent phosphorus acids hypophosphorus, phosphorous and phosphonous acids are suitable substrates in this reaction and their use leads directly to free α-aminophosphonous, α-aminophosphonic and α-aminophosphinic acids, respectively. This methodology was used in the first simple, direct route to free amino phosphonic acids ⟨66JOC1603⟩, by heating a mixture of amine, acid and formaldehyde (120–150 °C, 1–2 h) in a strongly acidic medium (pH 1–3). The general reaction is illustrated in Scheme 35; yields are in the range 50–100%. Monophosphinic acids (**131**) can be obtained by the reaction of primary amines, aldehydes or ketones and phosphinic acid ⟨48CB477⟩, while secondary amines, formaldehyde and phosphinic acid give corresponding products (**132**) ⟨67HCA1742⟩. In the presence of HCl the doubly substituted product (**133**) is obtained ⟨80PS(8)67⟩. Primary amines give the bisphosphinic acids (**134**) ⟨88JCR(S)34⟩.

Importantly, this procedure only succeeds for the preparation of aminomethylphosphonic acids, and primary amines and ammonia cannot be selectively converted into monophosphonic derivatives ⟨72SRI317, 73SRI75⟩. Thus cyclohexylamine reacts with one equivalent each of phosphorous acid and formaldehyde furnishing almost equal amounts of the mono- and diphosphonic acids (**135**) and (**136**) (Equation (28)). Other reports on the preparation of monophosphonic acids from primary amines ⟨73RZC929⟩ are considered to be incorrect ⟨78JOC996⟩.

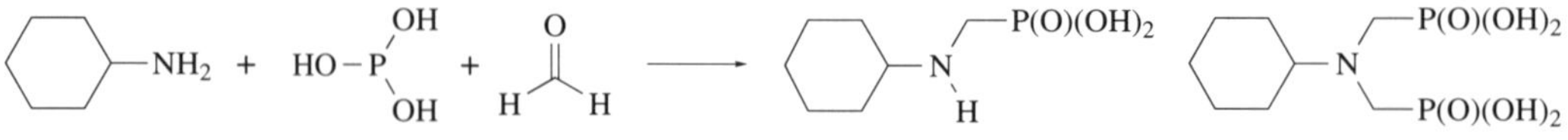

The preparation of 1-aminoalkylphosphonic acids by the three-component reaction involving phosphorus acid, an amide and a carbonyl compound in acetic anhydride has been described (10–75% yields) ⟨81TL3537⟩. The addition of acetyl chloride ⟨91S490⟩, ethanolic hydrogen chloride or toluene-*p*-sulfonic acid in acetic anhydride ⟨92S1124⟩ has been shown to improve the three-component condensation reaction leading to derivatives of aminophosphonic acids; for example, diethyl phosphoramidate (**137**) reacts with aldehydes and diphenyl phosphite at room temperature, furnishing (**138**) in 10–80% yields (Equation (29)) ⟨91S490⟩.

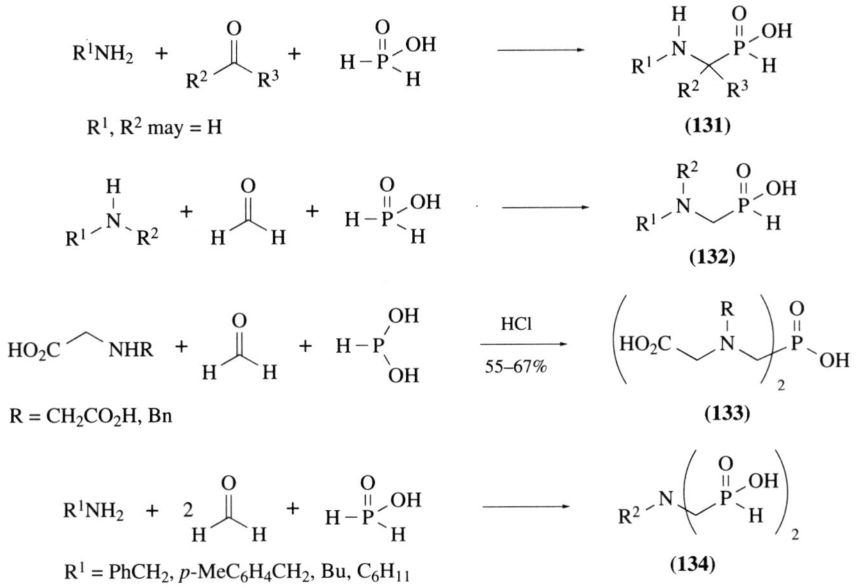

Nitrogen and Phosphorus

[Scheme 35 showing structures (131), (132), (133), (134)]

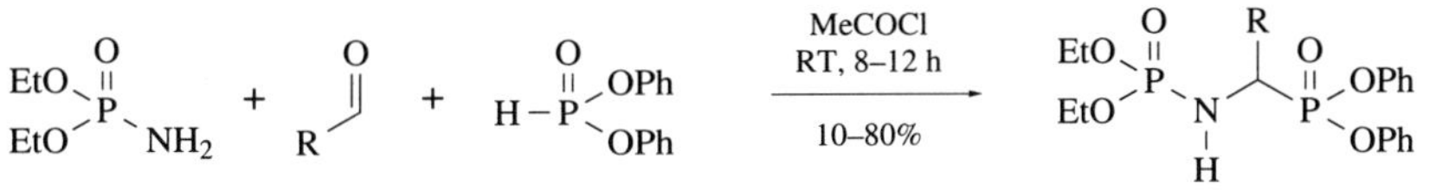

R¹, R² may = H → (131)

(132)

R = CH$_2$CO$_2$H, Bn ; HCl, 55–67% → (133)

R¹ = PhCH$_2$, p-MeC$_6$H$_4$CH$_2$, Bu, C$_6$H$_{11}$ → (134)

Scheme 35

$$\text{(137)} + \text{RCHO} + \text{H-P(O)(OPh)}_2 \xrightarrow[\text{10–80\%}]{\text{MeCOCl, RT, 8–12 h}} \text{(138)} \quad (29)$$

R = Pr^n, Ph, p-MeC$_6$H$_4$, p-MeOC$_6$H$_4$, p-Me$_2$NC$_6$H$_4$, p-ClC$_6$H$_4$, p-O$_2$NC$_6$H$_4$, m-O$_2$NC$_6$H$_4$, m-ClC$_6$H$_4$, o,p-Cl$_2$C$_6$H$_3$

Dibenzylphosphine oxide reacts with formaldehyde and aniline furnishing the anilinomethylphosphine oxides (**139**) in yields varying with the ratio of reactants employed (Scheme 36) ⟨76MI 410-02⟩. An intramolecular variant of this reaction gives the phosphine oxide (**140**) by a cyclocondensation reaction between (*o*-aminobenzyl)phenylphosphine oxide and benzaldehyde (Equation (30)) ⟨83AJC2095⟩.

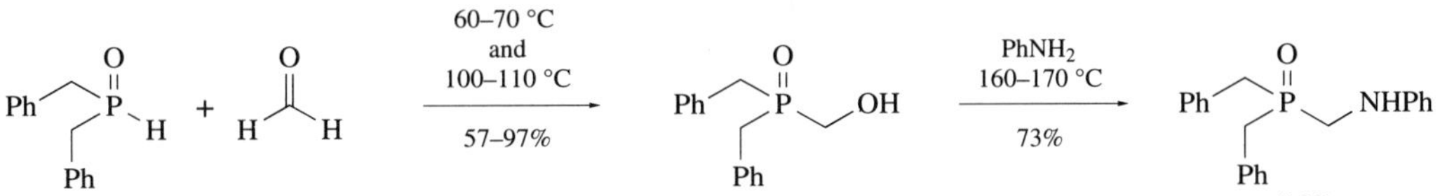

[Scheme 36: Ph$_2$P(O)H + HCHO → Ph$_2$P(O)CH$_2$OH (57–97%) at 60–70 °C and 100–110 °C; then PhNH$_2$, 160–170 °C, 73% → Ph$_2$P(O)CH$_2$NHPh (**139**)]

Scheme 36

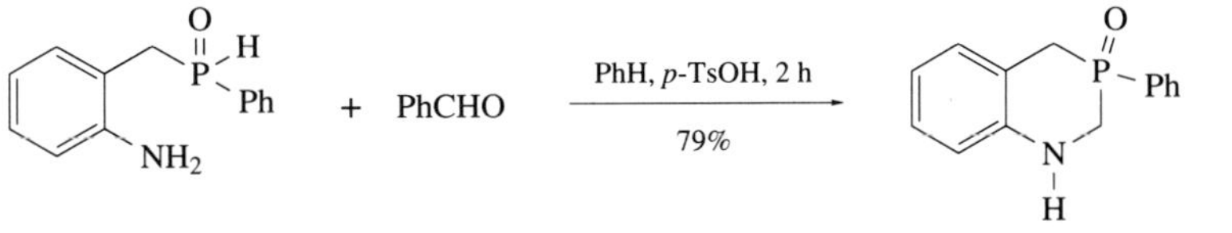

$$\text{(o-H}_2\text{N-C}_6\text{H}_4\text{-CH}_2\text{-P(O)(H)Ph)} + \text{PhCHO} \xrightarrow[\text{79\%}]{\text{PhH, }p\text{-TsOH, 2 h}} \text{(140)} \quad (30)$$

(iv) Tetracoordinate phosphorus functions by miscellaneous reactions

(a) Via α-oxo phosphorus compounds. α-Oxophosphonates and α-oxophosphinates (**141**) may be converted into the corresponding α-amino derivatives (**143**) by a two-step procedure involving oximation and subsequent reduction. This method, illustrated in Scheme 37, is sometimes known as the oxime route. This route benefits from the ready availability of the α-oxo compounds (treatment

of an acyl halide with a trialkylphosphite) and the facile reduction of the oxime group. A variety of reductants have been employed. The choice is dictated by the substituent on the α-carbon; thus aluminum amalgam works well with dialkyl amino(aryl)methylphosphonates ⟨68JOC3090, 90TL1759⟩, and diborane is superior for aminoalkylphosphonates and -phosphinates ⟨68JA4495, 93S955⟩. Catalytic reduction or zinc–formic acid ⟨81PJC713, 81S57⟩ have also been used. With the latter, part-formylation of the amino group may be observed but deformylation can be achieved with methanolic HCl. Conversion of α-oxophosphonates into their hydrazones (144) followed by reduction constitutes a parallel route to α-aminoalkyl phosphorus compounds. Again a number of reductants can be used including aluminum amalgam ⟨48JA1283⟩, zinc–formic or trifluoroacetic acid, or hydrogen used catalytically ⟨80S1028⟩. Hydrolysis affords the free α-aminophosphonic acid. Conversion of α-oxophosphonates into aminophosphonates can be achieved in a single step by treatment with ammonia or a primary amine and sodium tetrahydroborate (Equation (31)) ⟨79IZV1110, 79IZV2118⟩. With primary amines, mono- and diesters are unsuitable substrates.

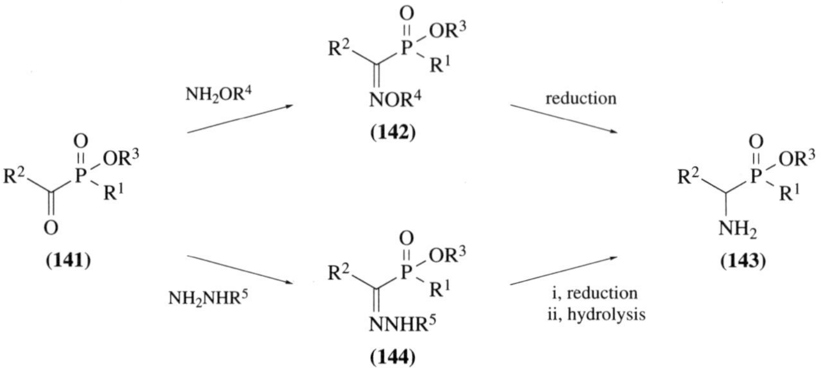

for example, R^1 = OEt, R^2 = p-BnOC$_6$H$_4$, R^3 = Et, R^4 = H gives 67% (142)

Scheme 37

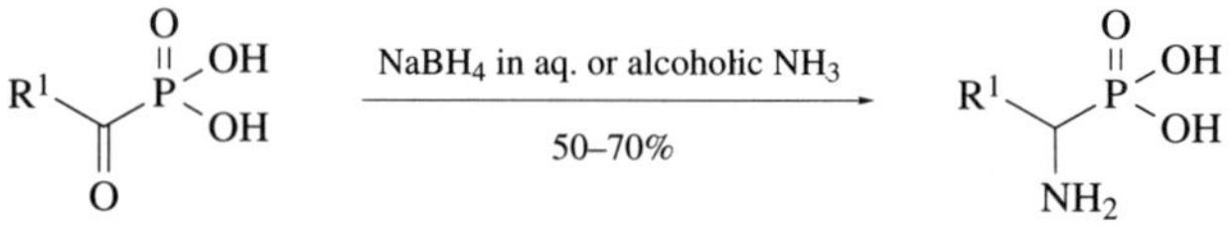

$$ \text{(31)} $$

R^1 = Me, Me$_2$CH, Me$_2$CHCH$_2$, PhCH$_2$, CH$_2$CH$_2$CO$_2$H, CH$_2$CH$_2$SMe

R^2 = H, BnOCO, F$_3$CO, HCO, [pyridoxyl group]

Diazophosphonates (145) eliminate nitrogen upon photochemical activation, and the carbenes thus produced react with aniline furnishing the corresponding *N*-phenyl α-aminophosphonates (146) (Equation (32)) ⟨69T5569⟩.

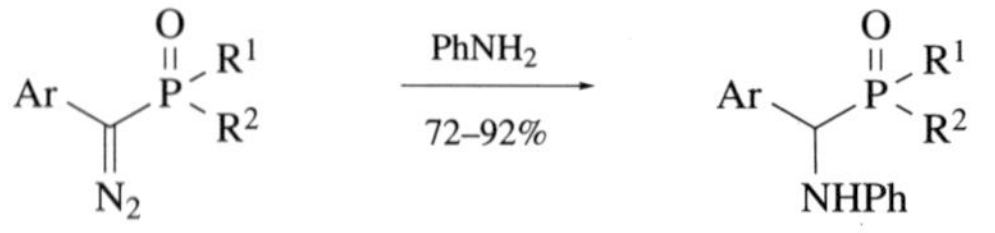

$$ \text{(32)} $$

(145) R^1 = OEt, Ph   (146)
        R^2 = OMe, Me, H, Cl, Br

(b) Addition of nucleophilic nitrogen species to P—C—X systems. The Hofmann reaction allows the preparation of amino phosphorus compounds via alkylation of ammonia and amines with haloalkyl organophosphorus compounds. The reaction is attractive in that the starting materials are widely available and a large number of amino and phosphorus components can be employed ⟨74RCR984⟩. The drawback is that a mixture of products often results, separation of which is not a trivial issue. Additionally, forcing reaction conditions may be necessary since these substrates are

not particularly susceptible to nitrogen-containing nucleophiles. A simple example is the preparation of 1-amino-1-phenylethylphosphonic acid (**148**) from the chloro analogue (**147**) with aqueous ammonia at 20 °C (Equation (33)) ⟨47JA2112⟩. Halomethylphosphonic and -phosphinic acids and their esters only undergo substitution by ammonia under severe reaction conditions, and furnish the desired products in moderate yield (Equation (34)) ⟨50IZV635, 51IZV95⟩. The reaction of aqueous ammonia with di(chloromethyl)phosphinic acid (**149**) in an autoclave has been reported to yield the *N*-methylamino derivative (**151**); the proposed route (Scheme 38) involves an intermediate phosphetidine oxide (**150**). With benzylamine, the chloro compound (**149**) furnishes the expected di(benzylamino)methyl derivative (**152**), catalytic reduction of which furnishes the free amino compound (**153**) ⟨79JOM(178)157⟩. With the use of high boiling point amines or in the presence of a Cu(II) catalyst, it is possible to obtain aminophosphonic acids in 25–90% yield; other metals are ineffective ⟨65ZC109⟩. In the presence of sodium hydroxide, chloromethylphosphonic acid reacts with a range of amines, for example, ethylenediamine, diethanolamine, glycine or iminodiacetic acid. Products are difficult to isolate in good yield ⟨49HCA1175, 65JA2567⟩. Chloromethylphosphinic acid reacts in a similar manner, furnishing α-aminophosphinic acid derivatives with ethylenediamine (Equation (35)). However, strongly basic media can give methylphosphonic acid (**154**) (Scheme 39) ⟨65JA4757⟩. An intramolecular variant of this reaction is seen in the formation of the 1,3-azaphosphane (**155**) (Scheme 40) ⟨84AJC205⟩. Intramolecular displacement is also proposed for the formation of the aziridinylphosphonic acid (**157**), obtained in moderate yield by heating diethyl α-bromovinylphosphonate (**156**) with liquid ammonia in a sealed tube (Scheme 41) ⟨72JOC3304⟩. The cyclic compound may be ring-opened with water or hydrogen sulfide to afford derivatives of serine or cysteine ⟨78PJC2271, 81PJC411, 81PJC713⟩. The addition of the lithiated (chloromethyl)phosphonic diesters (**158**) to aldimines followed by intramolecular displacment furnishes similar products (**159**) (Scheme 42) ⟨89H(28)1179⟩. Stereoselective synthesis of α-aminophosphonic acid derivatives (**161**) can be achieved by addition of carbanions of chiral bicyclic chloromethyl phosphonamides (**160**) to imines (Scheme 43) ⟨93SL35⟩. The Gabriel method is suitable for the preparation of certain unsubstituted aminoalkyl derivatives of phosphorus. Thus di(chloromethyl)phenyl-, tri(chloromethyl)- and diphenyltosyloxymethylphosphine oxides (**162**) react with potassium phthalimide, and the primary products (**163**) may be deprotected to furnish the aminomethylphosphine oxides (**164**) in reasonable yield (Scheme 44) ⟨74RCR984⟩.

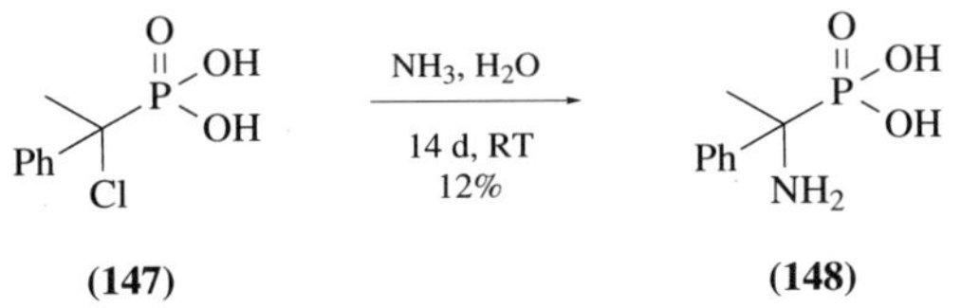

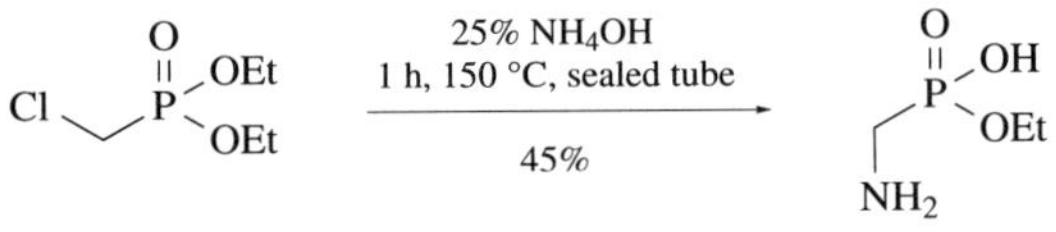

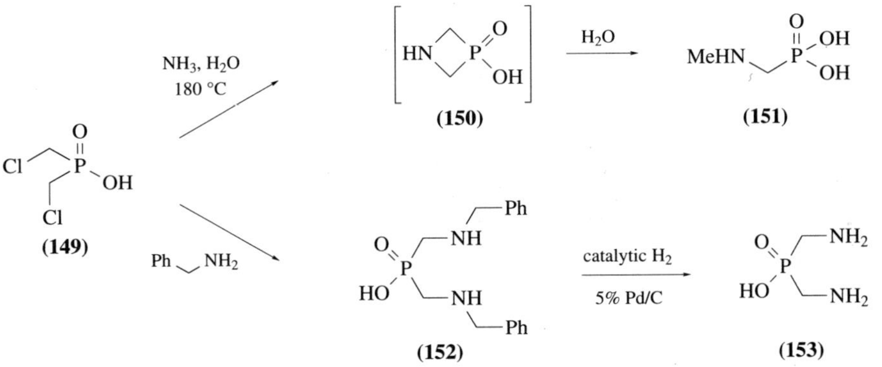

Scheme 38

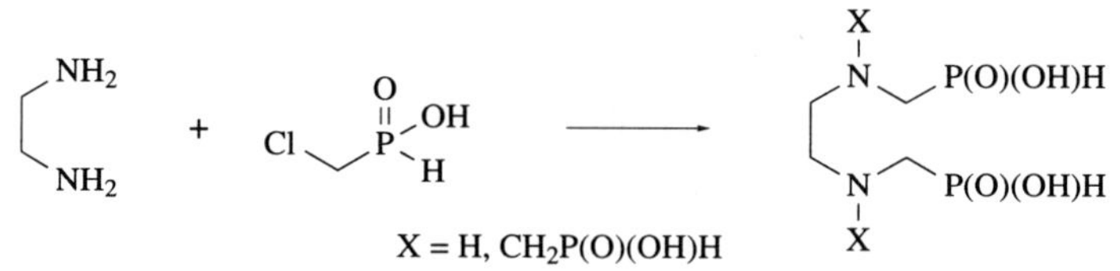 (35)

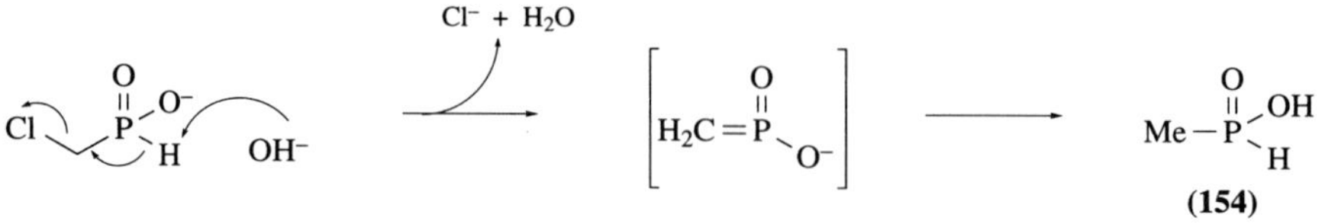

Scheme 39

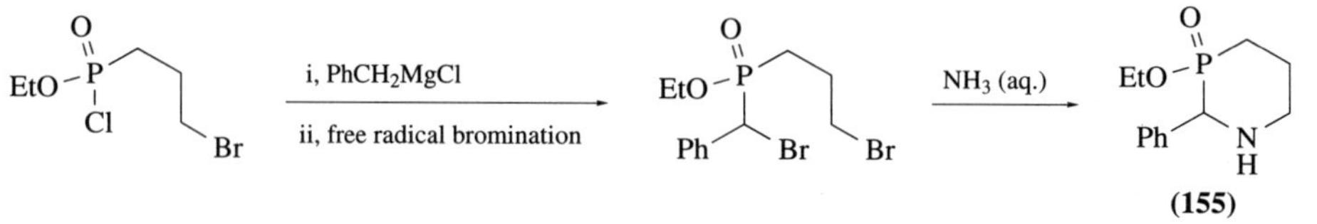

Scheme 40

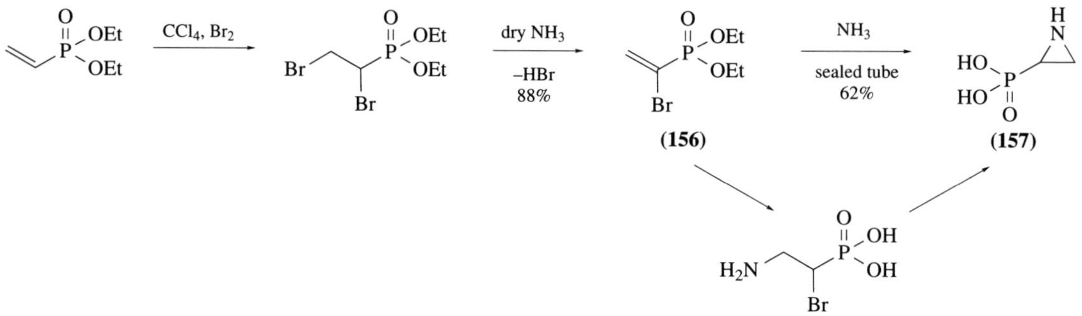

Scheme 41

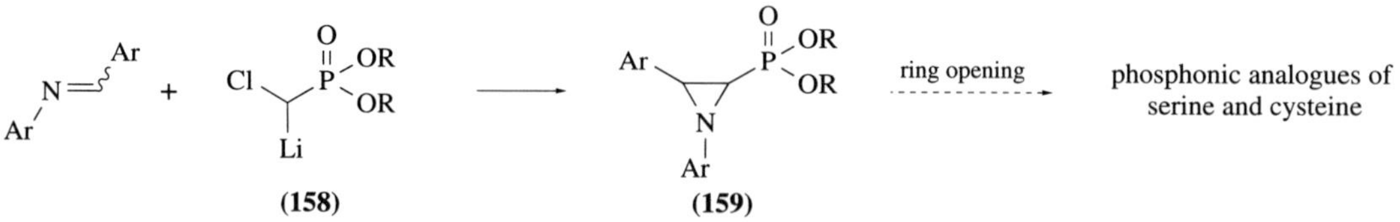

Scheme 42

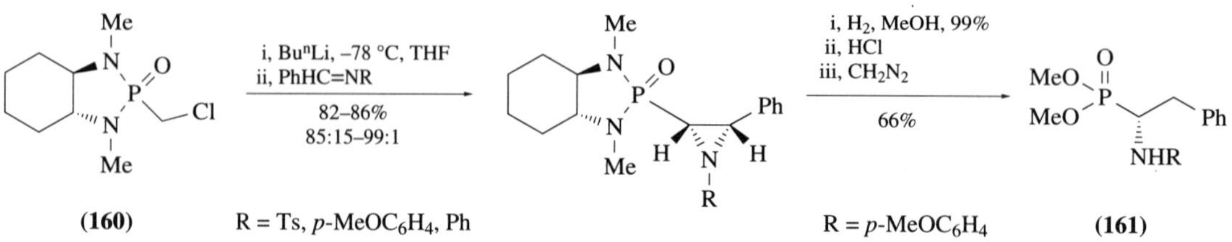

Scheme 43

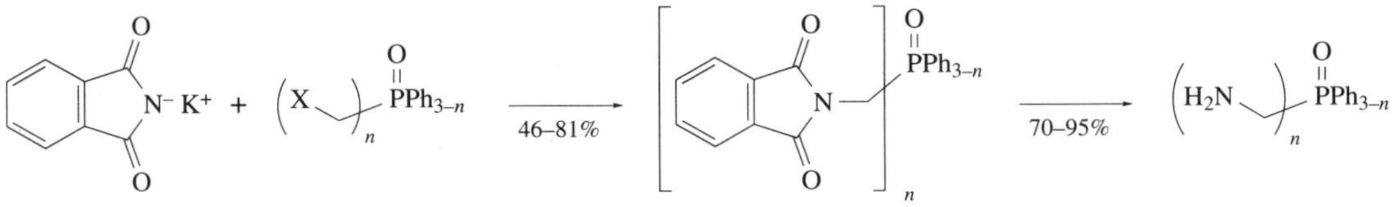

Scheme 44

The preparation of α-aminoalkyl phosphorus compounds via substitution of an α-hydroxy group by an α-amino group in the Mitsunobu reaction has been achieved. Thus the α-hydroxyphosphonates (**165**) react with phthalimide, triphenylphosphine and diethyl azodicarboxylate in the presence of triethylamine furnishing the nitrogen-protected compounds (**166**); hydrazinolysis followed by acid hydrolysis yields the free acid (Scheme 45) ⟨82S653⟩. A novel, enantioselective synthesis of α-aminophosphonic acids has been developed. The chiral phosphono alcohols (**167**), prepared by stereoselective opening of the precursor homochiral cyclic acetals by triethyl phosphite, have been converted via the α-hydroxy esters (**168**) into azides by a Mitsunobu reaction. Reduction gave the desired α-amino phosphonic acids (Scheme 46) ⟨92TA377⟩. Tetrakishydroxymethylphosphonium salts are readily transformed into the corresponding amino analogues upon treatment with ammonia or primary amines, as described previously ⟨72JOC2752, 81CJC27, 81PS(10)207⟩; for example, the sulfate salt (**169**) reacts with methyl carbamate furnishing the new salt (**170**) (Equation (36)).

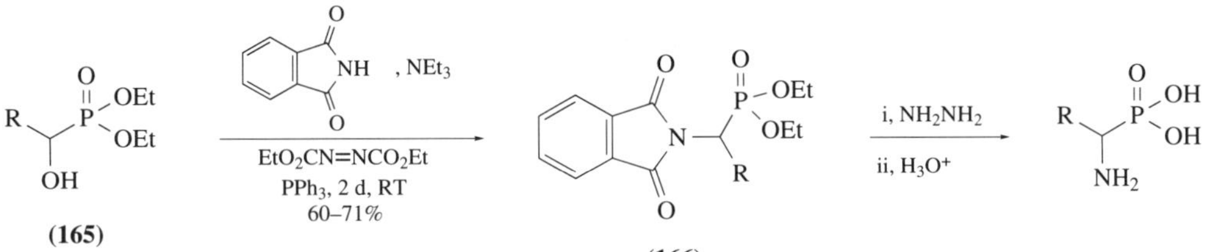

$R = H, Me, Pr, C_6H_{13}, C_7H_{15}$

Scheme 45

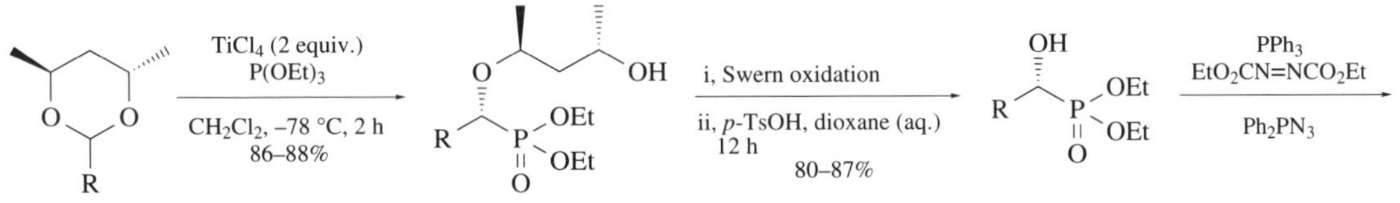

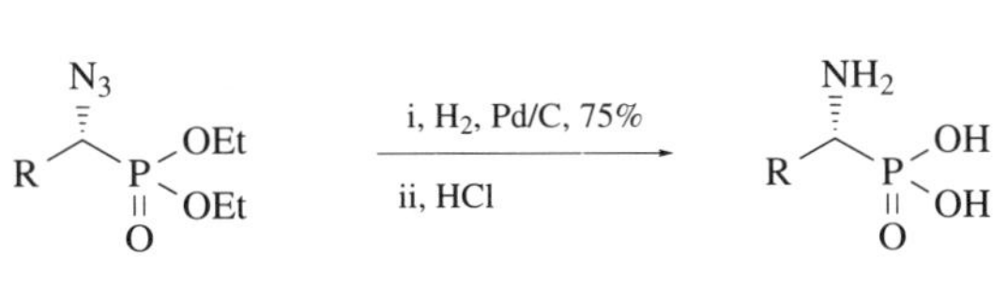

Scheme 46

$$[(HOCH_2)_4P^+]_2 \, SO_4^{2-} + NH_2CO_2Me \longrightarrow [(MeO_2CNHCH_2)_4P^+]_2 \, SO_4^{2-} \qquad (36)$$
(**169**)   (**170**)

(c) From phosphorus amides. The reaction of phosphoramidites ($P(NR^1R^2)_3$, $P(NR^1R^2)_2OR^3$, $P(NR^1R^2)(OR^3)OH$ and $P(NR^1R^2)(OR^3)_2$) with aldehydes or ketones gives α-aminoalkyl phosphorus derivatives ⟨74RCR984⟩. The products are generally obtained in poor to moderate yield and the availability of some of the starting materials may be low. These factors combine to lower the usefulness of this method for routine synthesis, but some processes are satisfactory. α-Aminophosphonic diamides (**172**) are readily obtained from the two-component reaction between aryl aldehydes

and hexamethylphosphorus triamide (**171**) (Equation (37)). The reaction occurs when the reagents are mixed neat or in the presence of a solvent; addition of an acid may assist the reaction ⟨66OS(46)31, 71ZOB2372, 79IZV2783, 91S225⟩. Significantly, aryl aldehydes substituted with an electron-withdrawing group react via a different path, forming the substituted epoxides (**173**) (Equation (38)). The *N*-phenylphosphoramidite (**174**) reacts analogously to the triamide, with addition to benzaldehyde furnishing the α-aminophosphonate (**175**) in good yield ⟨66JCS(B)789⟩. The same amidite reacts with *N*-phenyl benzaldimine yielding the phosphinimidic ester (**176**), hydrolysis of which affords (**175**). Treatment with CS_2 furnishes the thio derivative (**177**), from which (**175**) is obtained upon reaction with H_2O or AcOH (Scheme 47) ⟨69ZOB1235⟩.

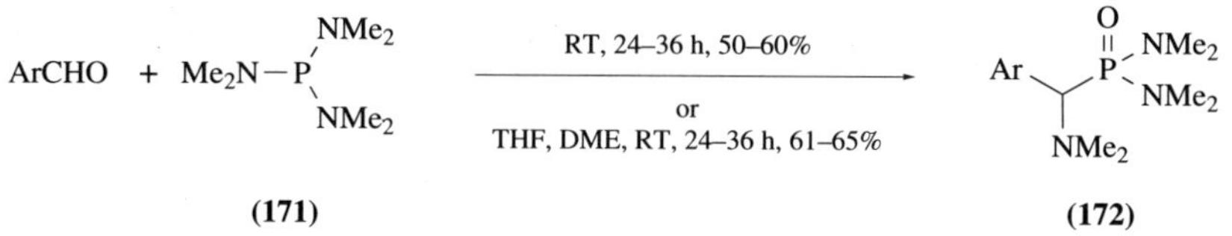

$$\text{ArCHO} + \text{Me}_2\text{N}-\overset{\text{NMe}_2}{\underset{\text{NMe}_2}{\text{P}}} \xrightarrow[\text{THF, DME, RT, 24–36 h, 61–65\%}]{\text{RT, 24–36 h, 50–60\%}} \text{Ar}\underset{\text{NMe}_2}{\overset{\overset{\text{O}}{\|}}{-}\text{P}}\underset{}{\overset{\text{NMe}_2}{\diagdown\text{NMe}_2}} \quad (37)$$

(**171**)   (**172**)

$$\text{XC}_6\text{H}_4\text{CHO} + \text{Me}_2\text{N}-\overset{\text{NMe}_2}{\underset{\text{NMe}_2}{\text{P}}} \longrightarrow \underset{(173)}{\text{XC}_6\text{H}_4 \triangle \text{C}_6\text{H}_4\text{X}} + \overset{\text{O}}{\underset{}{\|}}\text{P(NMe}_2)_3 \quad (38)$$

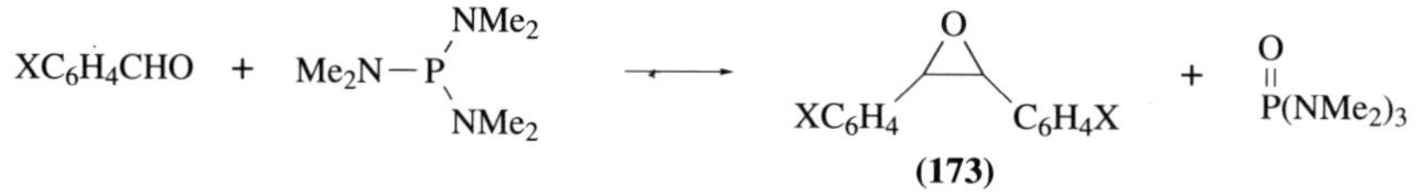

X = electron withdrawing group, for example, *o*-, *m*-, or *p*-NO_2, or *o*-Cl

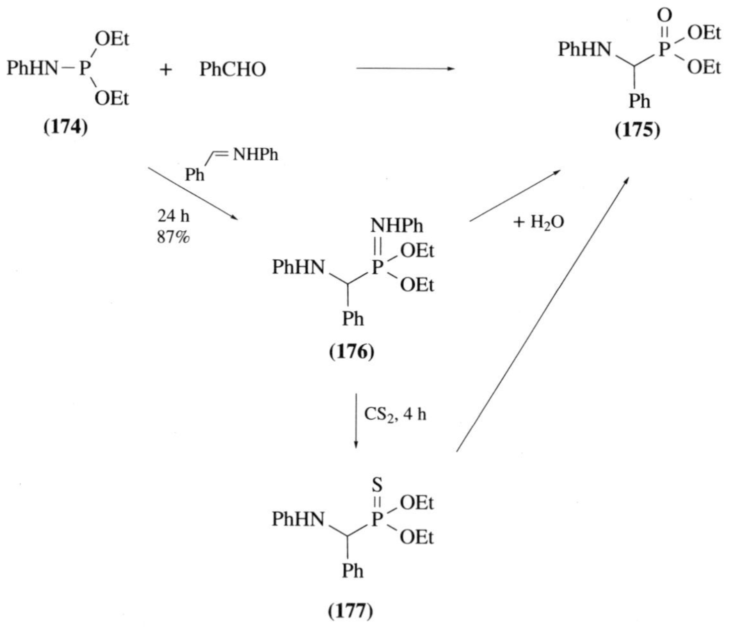

Scheme 47

(d) From rearrangement processes (Hofmann and Curtius rearrangements). The use of the Hofmann degradation for the synthesis of α-aminophosphorus compounds has not run a smooth course, and contradictory literature exists concerning the usefulness of this rearrangement as a synthetic tool. It has been reported that, depending on the structure of the phosphonoacetamides, the reaction follows one of two courses with formation of the α-amino- or α-bromophosphonates (**178**) (70–80%) or (**179**) (50–63%) (Scheme 48) ⟨73TL5201, 74ZC152⟩.

The Curtius rearrangement (Scheme 49) is a key step in the synthesis of α-aminophosphonic acids from the corresponding α-phosphono carboxylic acids; the phosphonic analogues of many naturally occurring α-amino acids have been obtained by this route ⟨64JOC832, 72JOC4397⟩. The reaction has broad applicability and although it is a multistep process it can be carried out in a single flask with moderate yields of product, the best experimental conditions involving the use of a 100% excess of hydrazine ⟨72JOC4397⟩.

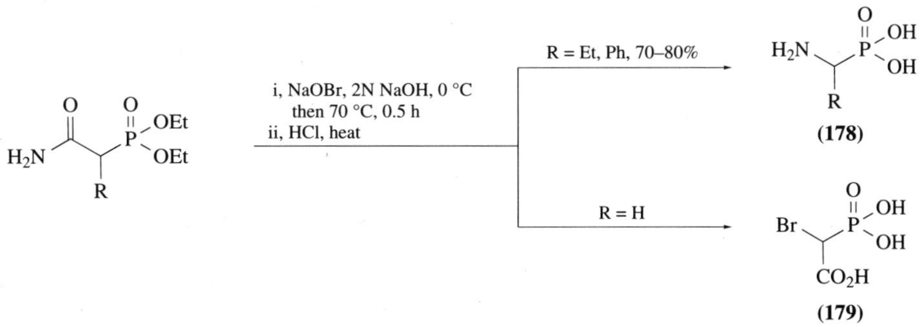

Scheme 48

when R = Me or Bn the major product is brominated

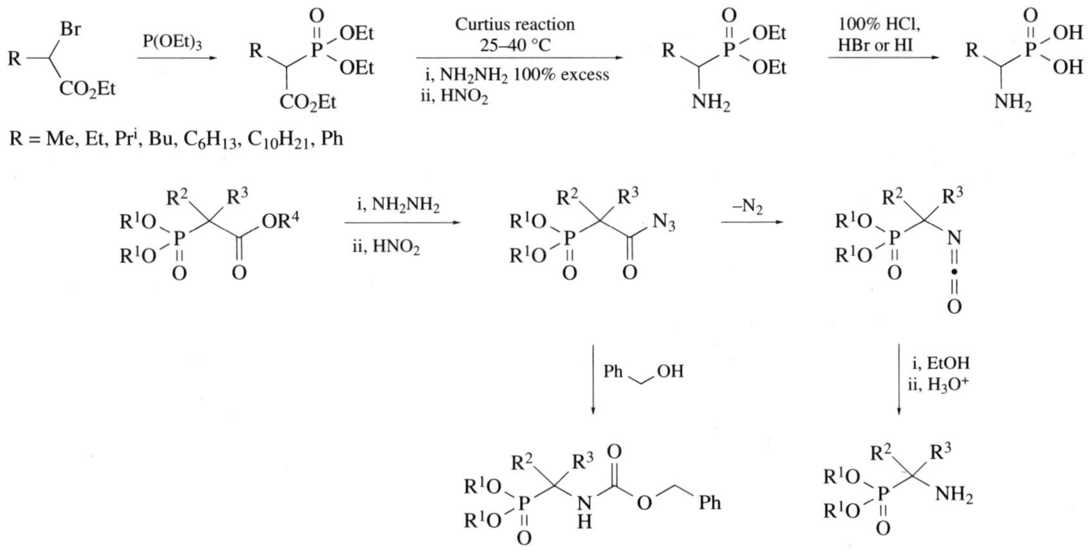

Scheme 49

(e) Modification of existing amino phosphorus compounds. Amino phosphorus compounds enter into many of the reactions characteristic of the functional groups which they contain. Thus cyclic phosphine oxides and phosphine sulfides are readily made by treatment of the corresponding tertiary phosphine with H_2O_2, air or sulfur (Scheme 50). Oxidation is most commonly carried out with H_2O_2 ⟨74JHC407, 80ZC152, 90SL261⟩, although slower air oxidation is also successful ⟨67JOC2383, 90SL261⟩. Various solvents can be employed and yields are generally excellent. Acetone/H_2O_2 mixtures are potentially explosive; where possible, other reaction media should be chosen. Treatment with elemental sulfur results in modest to good yields of phosphine sulfides ⟨67CB2685, 68CB3619, 68CB4032, 85ZAAC(523)7⟩. Ammonium polysulfide has also been used for this transformation ⟨81CJC27⟩; likewise, acyclic trisaminomethylphosphines participate in this type of chemistry ⟨90SL261⟩.

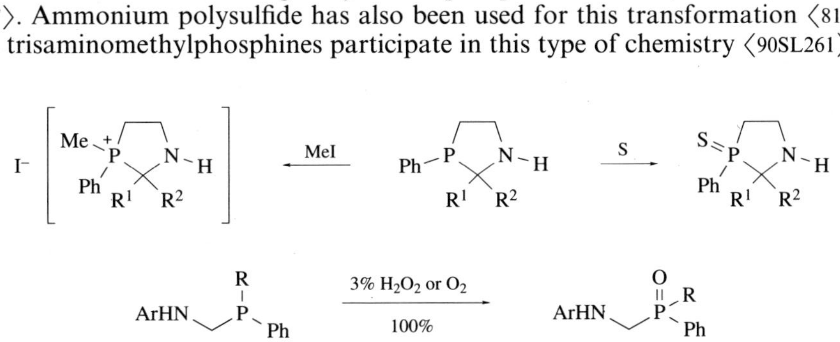

Scheme 50

α-Aminoalkylphosphines can be alkylated ⟨67CB2685, 83IZV1379⟩ or acylated ⟨78JOM(149)167⟩ on their phosphorus atom to furnish phosphonium salts. This is not a common reaction and a limited number of examples exist. Direct addition of trisubstituted phosphines to iminium ions is another route to 1-aminoalkyl phosphonium salts ⟨63LA(665)91, 71CB31, 72CB2233⟩.

Esterification, transesterification and deesterification reactions of α-aminoalkyl phosphorus acids and their derivatives take place under a variety of conditions, many of which have been included in the preceding account. This permits easy interchange of products. N-Acylated derivatives are the normal substrates for the esterification reaction. Monoester formation is commonly conducted with the hydroxy compound in pyridine or alcohol, with dicyclohexylcarbodiimide as the dehydrating reagent. Diesterification readily occurs with diazo alkanes, epoxy alkanes or *ortho*-formates. If a diester is required from a monoester, treatment of the ammonium salt with an alkylating agent furnishes the desired product. Transesterification takes place in alcoholic solution in the presence of an alkoxide catalyst. Deesterification to the free acid is carried out under severe reaction conditions, most commonly in concentrated acid solution at 100–140 °C for several hours. A hydrochloride salt results, from which the free acid may be obtained. The rate of reaction depends on the nature of the R group (decreasing rate with increasing size of R) and on the acid chosen (rate decreasing in the order HI > HBr > HCl). When the substrate is very resistant to hydrolysis, treatment with anhydrous HCl gas at elevated temperatures is usually successful. Those phosphonates which are unstable in acid (N-aryl and α-aryl) may be cleaved in alkaline media. *p*-Methylbenzyl esters can also be cleaved by formic acid, or deprotection may be conducted with trialkylsilyl halides ⟨92TL77⟩. Partial hydrolysis or nonhydrolytic methods may afford monoesters if required. Dichlorophosphine oxides can be obtained from the corresponding phosphonates in almost quantitative yields with PCl_5 ⟨76ZOB1246⟩.

4.10.1.1.4 Higher coordinate phosphorus functions

N—C—P compounds with a pentacoordinate or higher coordinate phosphorus atom comprise a rather small group of largely individual molecules.

A new range of phosphoranes can be obtained by the addition of mono- ⟨89MI 410-01⟩ or spirocyclic phosphoranes ⟨74MI 410-01, 76T2089, 93ZOB220⟩ with P—H bonds to activated C=N double bonds, for example, Schiff's bases. Compound (**181a**) was prepared in this way in racemic form by treatment of the spirophosphorane (**180a**) with N-methylbenzylimine (Equation (39)). The substituted and unsubstituted phosphoranes exist in equilibrium at elevated temperatures ⟨74MI 410-01⟩. 1-Aminoalkyl-substituted products result from addition of amines to the α-carbon atom of vinyl phosphorane (**180b**) ⟨82PS(13)85⟩ and from the reaction between (**182**) and α,ω-diols ⟨90JGU397⟩. The benzoxaphosphole (**182**) also reacts with unsubstituted tetraoxaspirononanes affording dimeric phosphonates; dioxadiazaspirononanes do not participate in this reaction (Equation (40)) ⟨93ZOB220⟩. Dichlorophenylphosphine and two equivalents of N-methyl(o-hydroxybenzylidene)amine participate in a head-to-tail cyclisation furnishing the tricyclic phosphorane (**184**) (Scheme 51); this is believed to be the first example of an oxidative cyclisation between an 8-P-3 centre and an imine function and is considered to involve (**183**) as an intermediate ⟨81PS(11)87, 82JA2497⟩. The benzo-1,3,2-oxaza-2-phospholene (**185**) exists in the dimeric form as the diazadiphosphetidine (**186**) both in solution and in the crystalline state; it is thermally unstable above its melting point (Equation (41)) ⟨86JGU632⟩. The 3-aminobenzophosphorane (**188**) arises from the reaction between $P(NEt_2)_3$ and the cyclic phosphonate (**187**) (Equation (42)) ⟨69IZV1757⟩.

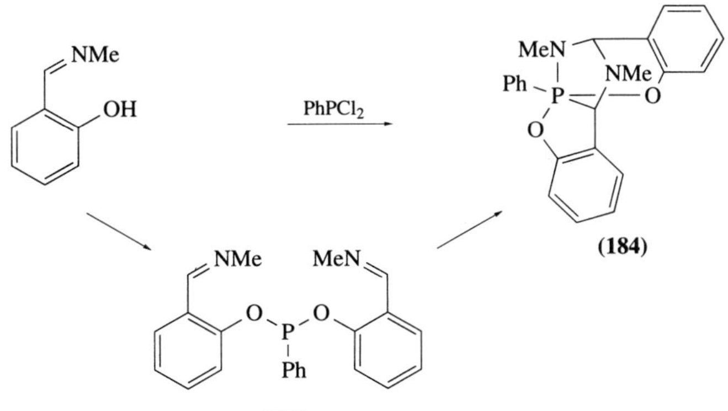

Nitrogen and Phosphorus 479

Scheme 51

(41)

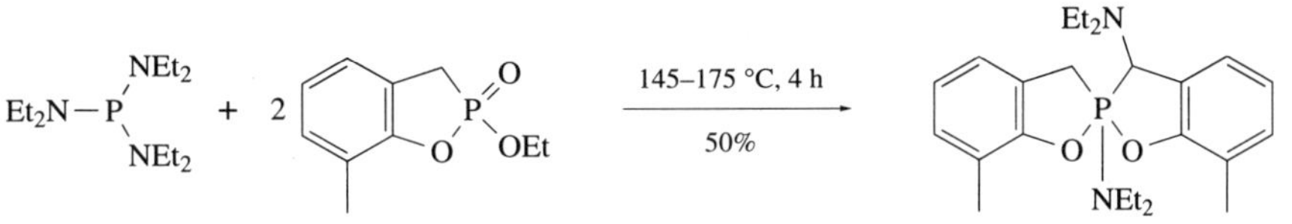

(42)

4.10.1.2 Other Nitrogen Functions: $R^1_2C(NY)PR^2_2$, $R^1_2C(NHX)PR^2_2$, $R^1_2C(NR^2X)PR^3_2$, $R^1_2C(NX_2)PR^2_2$, etc.

Many examples of functions containing a nitrogen and a phosphorus atom—where the nature of the nitrogen component is other than amino—have been illustrated in the preceding text. Such compounds are often used as protected N—C—P moieties; those included above will not be mentioned again in this section.

4.10.1.2.1 Dicoordinate phosphorus functions

Those compounds of general structure $R^1R^2C(NX)PY$, where phosphorus has coordination number 2, comprise a very small group and are without exception thermally unstable. Molecules in this group have a trivalent phosphorus atom and are largely derived from a dipolar cycloaddition reaction. The 3H-1,2,4-diazaphospholes (**189**) are considered to be the primary products from the [3 + 2]-cycloaddition reaction between phosphaalkynes and variously substituted diazomethanes and α-diazo carbonyl compounds (2,2-dimethylpropylidynylphosphine is the only phosphaalkyne described as stable and obtained on a preparative scale at room temperature). The initially formed products (**189**) undergo spontaneous aromatisation to furnish the 1,2,4-diazaphospholes (**190**) (Scheme 52); a number of examples are illustrated in Table 5. In those cases where the starting diazo alkane has an α-hydrogen, the rearrangement involves a 1,5-hydrogen shift from carbon to nitrogen (Table 5, entries 1–13). When there is no hydrogen atom available for migration and $R^2 = COR^1$ (Table 5, entries 14–18), a fast acyl shift is observed to lead to nitrogen-substituted diazaphosphole ⟨84CC1634, 86JOM(306)39, 87CB1645⟩. The spirocyclic 3H-1,2,4-diazaphosphole (**192**), a yellow oil, may be obtained at analytically pure grade, in quantitative yield, as the product of a [3 + 2]-cycloaddition reaction between the diazocyclohexane (**191**) and 2,2-dimethylpropylidynylphosphine. The reaction proceeds in pentane at $-40\,°C$ and ^{13}C and ^{31}P NMR spectral data of (**192**) are recorded. The compound is unstable with respect to both thermal and photochemical activation. The fused bicyclic product (**193**) arises as a consequence of a 1,5-sigmatropic migration from carbon to phosphorus at $20\,°C$ (Scheme 53). Irradiation of (**192**) at $\lambda \geq 280$ nm causes elimination of N_2 ⟨87AG(E)1257⟩. Cowley *et al.* have observed a similar 1,2-phenyl migration in the diazaphosphole (**194**), leading to the 4*H*-1,2,4-isomer (**195**) in good yield (Scheme 54). The existence of the dicoordinate phosphorus compound (**194**) is again supported by ^{31}P and ^{13}C NMR data ⟨88CC867⟩.

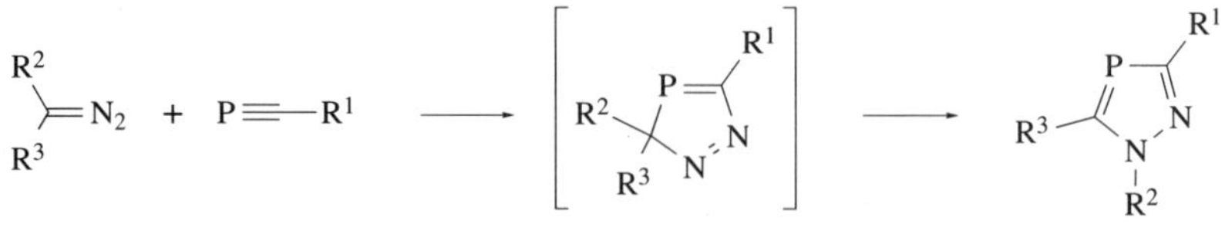

Scheme 52

Table 5 Reactions between phosphaalkynes $R^1C{\equiv}P$ and diazo compounds $R^2R^3CN_2$ to furnish 1,2,4-diazaphospholes (**190**).

No.	Phosphaalkyne R^1	Diazoalkane R^2	R^3	Reaction conditions Solvent	Temperature (0°C)	Time (h)	Yield (%)	Ref.
1	Bu^t	H	CO$_2$Me	CH$_2$Cl$_2$	RT	12	52	84CC1634
2	Bu^t	H	CO$_2$Me	ether	0	12	86	87CB1645
3	Pr^i	H	H	pentane	0	0.5	78	86JOM(306)39
4	CH$_2$Bu^t	H	H	pentane	0	0.5	93	86JOM(306)39
5		H	H	pentane	0	0.5	68	86JOM(306)39
6		H	H	ether	0	0.5	81	86JOM(306)39
7	Bu^t	H	H	pentane	0	0.5	93	87CB1645
8	Bu^t	H	Me	ether	0	0.5	92	87CB1645
9	Bu^t	H	Ph	ether	0	1	91	87CB1645
10	Bu^t	H	Bu^t	ether	0	0.5	93	87CB1645
11	Bu^t	H	CO$_2$Bu^t	ether	0		91	87CB1645
12	Bu^t	H	COPh	ether	0	14	92	87CB1645
13	Bu^t	H	P(O)Ph$_2$	ether	0	14	76	87CB1645
14	Bu^t	COMe	Me	benzene	RT	0.5	81	87CB1645
15	Bu^t	COMe	Ph	benzene	RT	6	86	87CB1645
16	Bu^t	COPh	Ph	benzene	RT	12	65	87CB1645
17	Bu^t	COMe	PhSO$_2$	benzene	RT	240	96	87CB1645
18	Bu^t	COMe	PhNHCO	benzene	RT	720	98	87CB1645

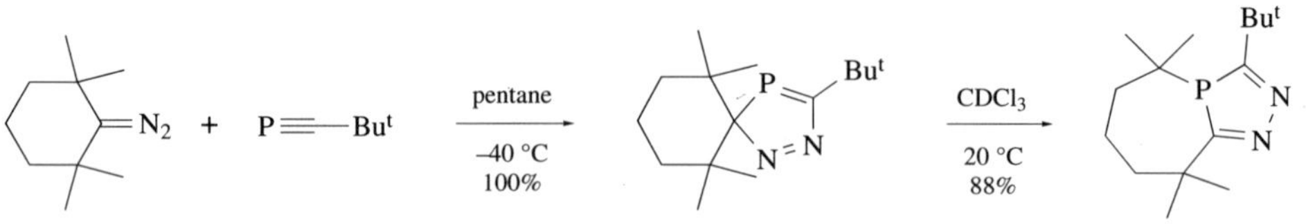

Scheme 53

4.10.1.2.2 Tricoordinate phosphorus functions

(i) Tricoordinate phosphorus functions by modification of existing α-amino phosphorus compounds

The amino group in aminoalkyl phosphorus derivatives enters into many of its characteristic reactions, for example, it is protonated by acids and can give excellent yields of quaternary ammonium salts ⟨67CB2685, 67TL1489, 68CB4032, 73JPR526, 76ZAAC(424)97⟩. The same salts (for example, (**197**)) are the first products of the reaction between iminium salts (**196**) and nucleophilic phosphines (Equation (43)) ⟨78JOM(149)167, 84ZC365⟩.

$$2\,[Me_2N=CH_2]^+\,Cl^- + H_2PPh \xrightarrow{83\%} Ph-P\begin{pmatrix}-NMe_2\cdot HCl\\ -NMe_2\cdot HCl\end{pmatrix} \quad (43)$$

(**196**)   (**197**)

In contrast to protonation, alkylation most commonly occurs on tertiary phosphorus in aminoalkyl phosphorus species. There are, however, some exceptions and the quaternary ammonium salts (**198**)–(**201**) result from treatment of the neutral compounds with methyl iodide ⟨74JHC407, 78JOM(149)167, 82IZV1611, 85ZAAC(523)7⟩. *N*-Acylated derivatives may be obtained by direct treatment of the tertiary amino derivatives with acetic anhydride, for example (**203**) results from acetylation of the triazaphosphaadamantane (**202**) (Equation (44)) ⟨77JHC337⟩.

(ii) Tricoordinate phosphorus functions by Mannich-type reactions

Attack of nucleophilic phosphorus species on compounds containing multiple C—N bonds has led to a number of nitrogen-substituted aminomethylphosphines. *N*-Arenesulfonyl imines (**204**) react with secondary phosphines in good yield (76–94%) to furnish the adducts (**205**) (Equation (45)) ⟨80ZC152⟩. Diphenylphosphine reacts with the stabilised *N*-fluorosulfonyl imine (**206**) giving (**207**) (83% yield) (Equation (46)) ⟨68M380⟩, and *N*-silylated products (**209**) result from attack of silylphosphines on the imines (**208**) (Equation (47)) ⟨90ZOB1718⟩; forcing conditions are required with the use of tertiary phosphines ⟨75HCA1316⟩. The attack of phenylphosphine on the nitrile (**210**) furnished a multitude of products including (**211**), the nature and amount of the products varying with the experimental conditions (Equation (48)) ⟨85JFC(30)269⟩.

R^1 = H, Me, Br; R^2 = Cy, Ph

Cy = cyclohexyl
c-C_6H_{11}

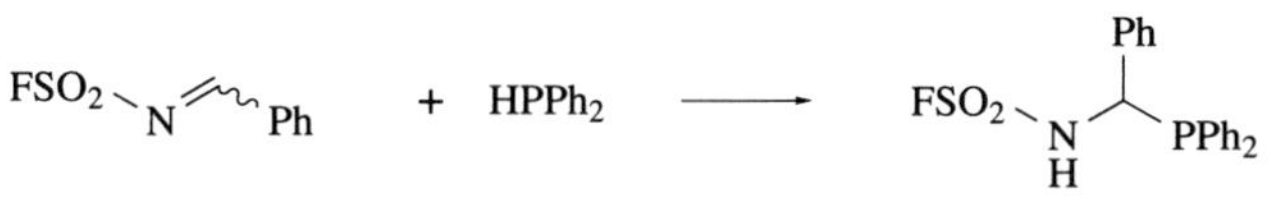

$$FSO_2-N=\text{CHPh} \quad + \quad HPPh_2 \quad \longrightarrow \quad FSO_2-N(H)-CH(Ph)-PPh_2 \tag{46}$$

(206)   (207)

$$PhHP-TMS \quad + \quad Ph-CH=N-R \xrightarrow[\text{b; R = Me, 38\%}]{\text{a; R = Ph, 43\%}} Ph(H)P-CH(Ph)-N(TMS)(R) \tag{47}$$

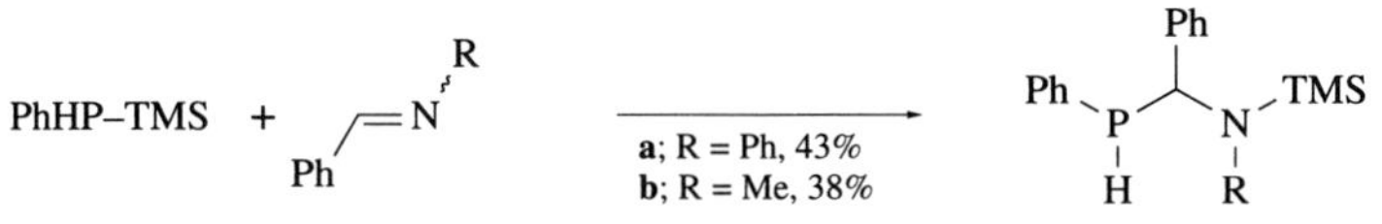

(208)   (209)

$$n\text{-}C_7F_{15}-\!\!\equiv\!\!N \quad + \quad PhPH_2 \quad \longrightarrow \quad n\text{-}C_7F_{13}\text{-}CHF\text{-}CH=N\text{-}CH(n\text{-}C_7F_{15})(PHPh) \tag{48}$$

(210)   (211)

The γ-lactams (**214a**) and (**214b**) are the products of nucleophilic attack on an sp^2 carbon followed by cyclisation. The 1-amino-1,3-azaphospholan-5-one (**214a**) is the product of reactions between the carboxyalkylphosphines (**212**) and the hydrazone (**213a**); likewise, the nitrogen-substituted hydrazone (**213b**) furnishes (**214b**) (Equation (49)) ⟨78PS(4)59⟩.

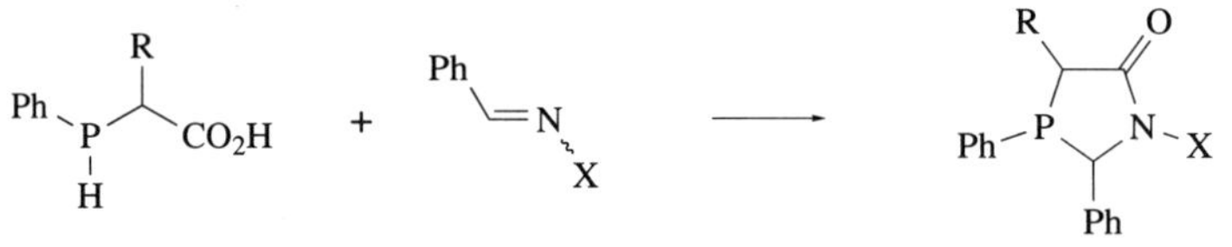

$$\text{(212)} \quad + \quad \text{(213)} \quad \longrightarrow \quad \text{(214)} \tag{49}$$

(212) R = H, Me   (213) a; X = NH$_2$; b; X = NHPh   (214)

The *N*-sulfonated, adamantane-type adducts (**215**) are obtained from tris(hydroxymethyl)phosphine, hexamethylenetetramine and $NH_2SO_2NH_2$ (Equation (50)) ⟨74JHC1085⟩. The N^1,N^2-dialkylhydrazines (**216**) react with bis(hydroxymethyl)phosphine furnishing the 1,2-diaza-4-phosphacyclopentanes (R = alkyl or aryl) (**217**) and 1,5-diaza-3,7-diphosphabicyclo[3.3.0]octanes (**218**) in good yield (Scheme 55). The monohydroxymethylphosphines (**219**) react similarly, affording the N^1,N^2-bisphosphinomethylene-N^1,N^2-dimethylhydrazines (**220**) in 85–97% yield (Equation (51)) ⟨81TL229, 81TL1105⟩.

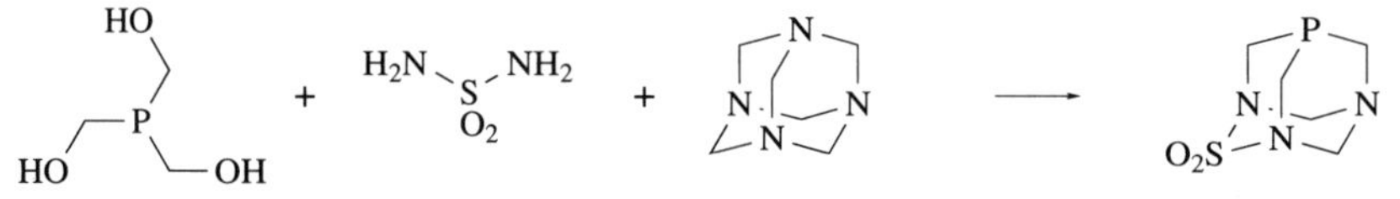

$$P(CH_2OH)_3 \quad + \quad H_2N\text{-}SO_2\text{-}NH_2 \quad + \quad (CH_2)_6N_4 \quad \longrightarrow \quad \text{(215)} \tag{50}$$

Scheme 55

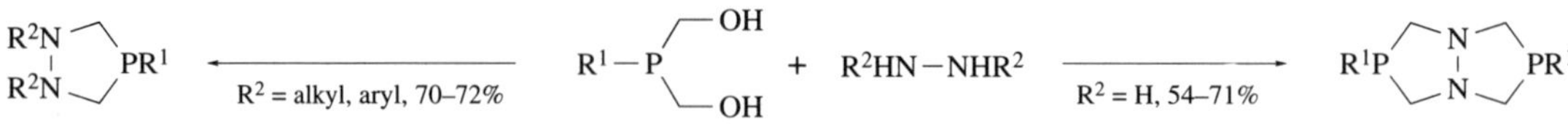

(217) R^2 = alkyl, aryl, 70–72%   (216) R^1 = alkyl, aryl   (218) R^2 = H, 54–71%

$$\text{Ph(Ar)P-CH}_2\text{OH} \quad + \quad \text{(216)}, R^2 = Me \quad \longrightarrow \quad \text{(220)} \tag{51}$$

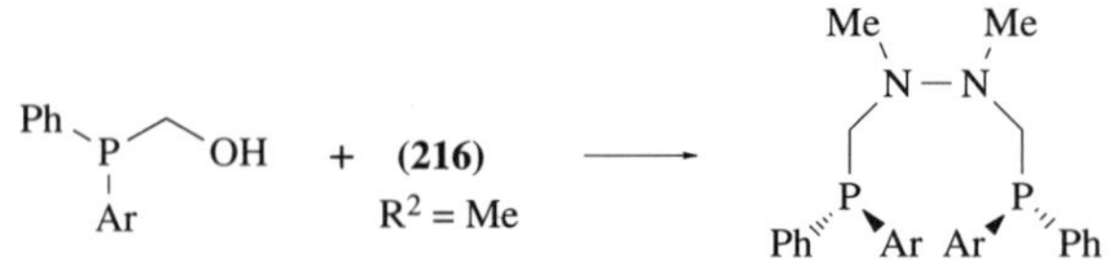

(219)   Ar = Ph, 97%; Ar = *o*-MeC$_6$H$_4$, 85%

(iii) Tricoordinate phosphorus functions by miscellaneous reactions

The (1-azidocycloprop-2-inylmethyl)phosphine (**222**), existing in equilibrium with the isomer (**223**) (35:65), is the product of the reaction between sodium azide and the cyclopropenylium ion (**221**) (Scheme 56) ⟨91S1099⟩. Phenyl azide reacted with the thermally stable mesityl(diphenylmethylene)phosphine (**224**) in CS_2 at 80 °C over 20 h, affording the [3 + 2]-cycloadduct (**225**) and (**226**) in 90% and 10% yield, respectively. In contrast, when the reaction was carried out in benzene or chloroform the cycloadduct was not observed and (**226**) was the only product (Scheme 57). Similarly, tosyl azide in benzene or CS_2 furnished only the cycloadduct (**227**). It is noteworthy that (**227**) is regioisomeric with (**225**) and that its formation is not solvent dependent ⟨84T991⟩.

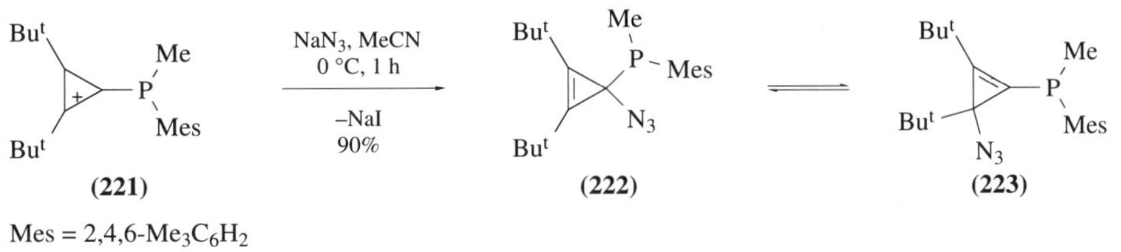

Scheme 56

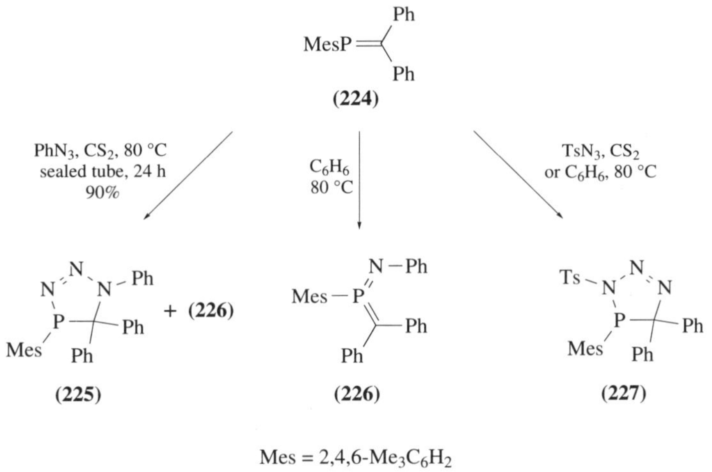

Scheme 57

Treatment of the chlorophosphine ester (**228**) with triethylamine yields the cycloadducts (**229**) and (**230**) in 45% and 15% yields, respectively (Equation (52)). Upon prolonged standing or repeated distillation, (**229**) dimerises to the 1,4-diaza-2,5-diphosphorinane (**230**) ⟨77ZOB1422, 78ZOR739⟩.

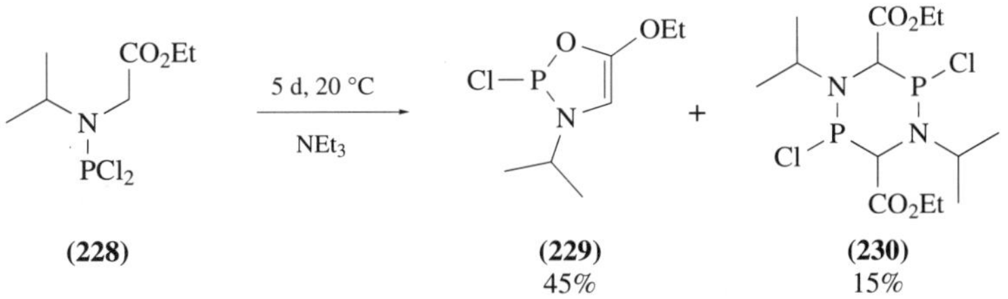

The *N*-phosphinomethylphthalimide (**232a**) results from attack by diphenylphosphine on the quaternary ammonium salt (**231a**) in a basic medium; succinimido derivatives undergo parallel chemistry. The adduct (**232b**), obtained similarly, undergoes reductive cleavage with $LiAlH_4$ to furnish the alcohol (**233**) (Scheme 58) ⟨74JPR851⟩. Analogous substitution reactions are observed with *N*-chloromethylsuccinimides and -phthalimides and the nucleophilic phosphorus species may be a sodium phosphide or a secondary phosphine in the presence of triethylamine. Tristrimethylsilylphosphine reacted with *N*-(chloroacetamide) (**234**) to furnish the phosphine (**235**) in moderate yield (Equation (53)) ⟨91ZOB1016⟩.

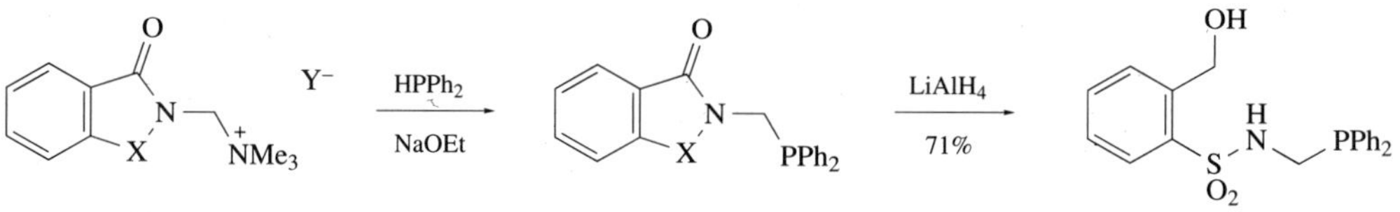

(231) a; X = CO, 86% b; X = SO_2, 73 % **(232)** X = SO_2 **(233)**

Scheme 58

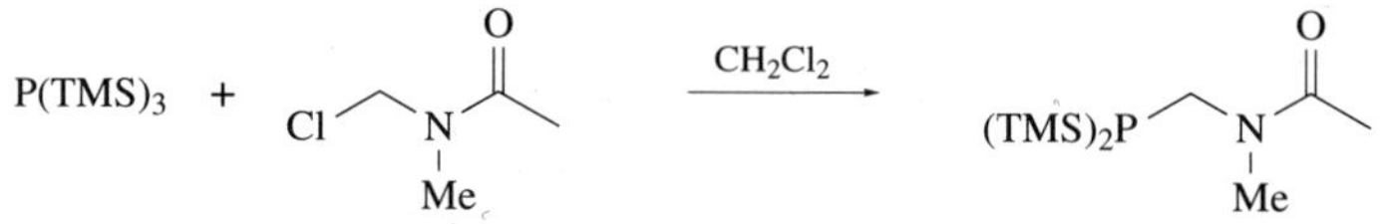

(234) **(235)** (53)

The 4,5-dihydro-3*H*-diazaphospholes (**238**) are products of [3 + 2]-cycloaddition between the phospha alkenes (**236**) and the diazo alkane (**237**) (Equation (54)) ⟨81AG(E)131⟩. A similar reaction was observed with the *P*-silyl phospha alkene (**239**) to give the stable adduct (**240**) in 76% yield (Equation (55)) ⟨86S31⟩.

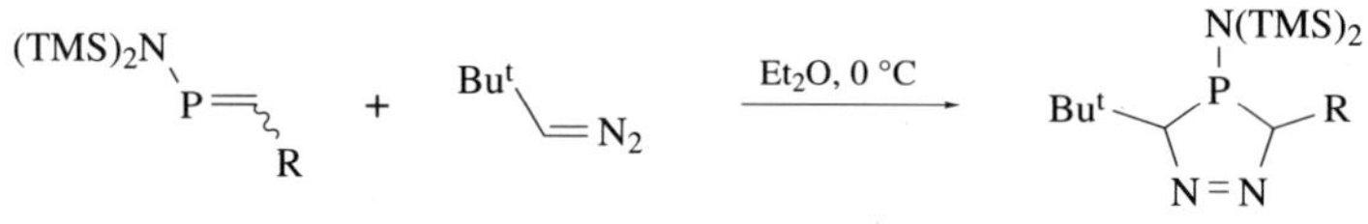

(236) **(237)** **(238)** (54)

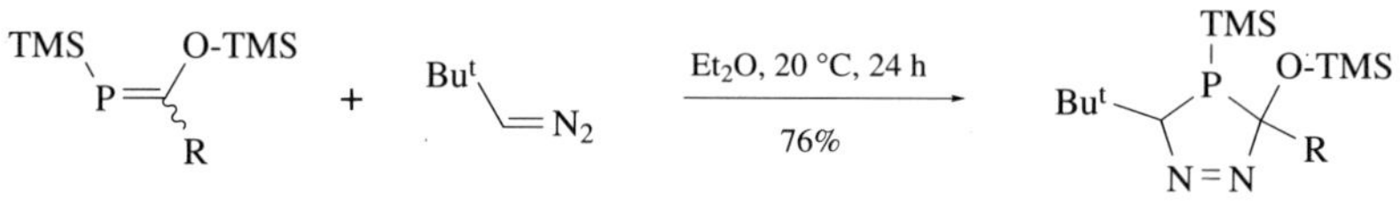

(239) **(240)** (55)

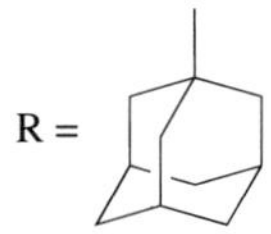

A most unusual N—C—P compound is described by Märkl *et al.* The 1,3λ^3-azaphosphinine (**241**) reacts very rapidly with diazomethane to afford the pentacyclic compound (**242**); a mechanism has been proposed for this rather complex reaction (Equation (56)) ⟨90TL6999⟩.

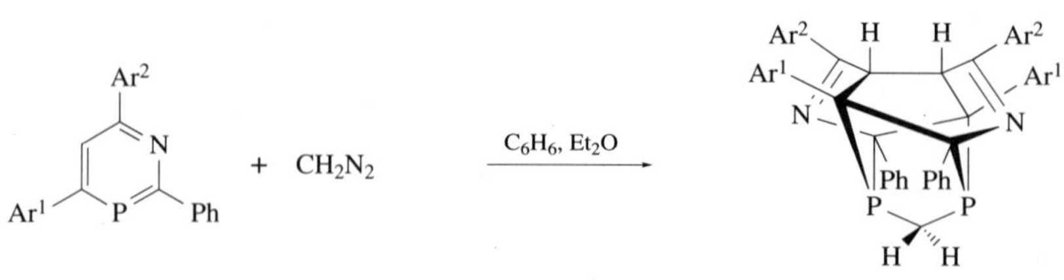

(241) Ar^1, Ar^2 = Ph, *p*-MeC_6H_4 **(242)** (56)

A novel, planar, asymmetric, four-membered azacarbaphosphaboretane (**244**) has been reported ⟨92JA9691⟩. An equimolar reaction between the diborylphosphine (**243**) and Bu^tLi furnishes (**244**) in good yield (Equation (57)).

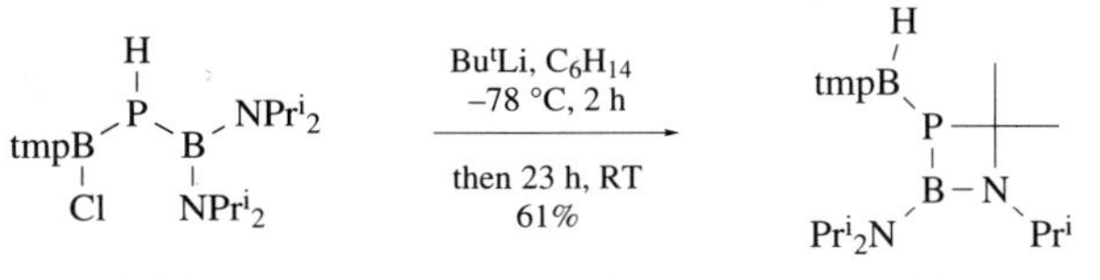

$$\text{(243)} \quad \text{tmp = 2,2,6,6-tetramethylpiperidino} \quad \text{(244)} \tag{57}$$

(243) with BuʻLi, C₆H₁₄, −78 °C, 2 h, then 23 h, RT, 61% → (244)

The novel, germanium-containing heterocycle (245) results from a [3 + 2]-reaction between the three-component mixture (Equation (58)); a second product (246) results from a [2 + 2]-reaction ⟨91JOM(415)327⟩.

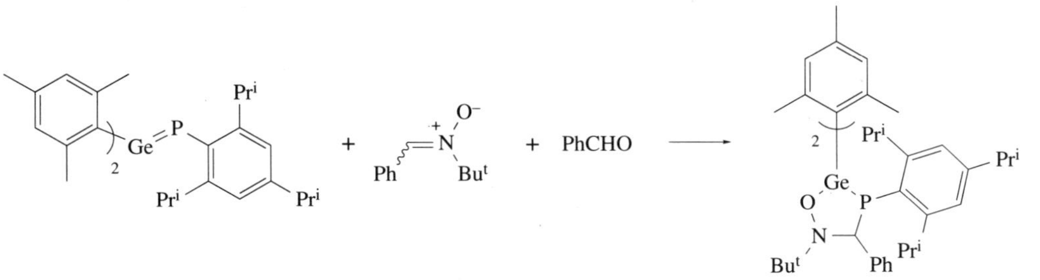

$$\tag{58}$$

(245)

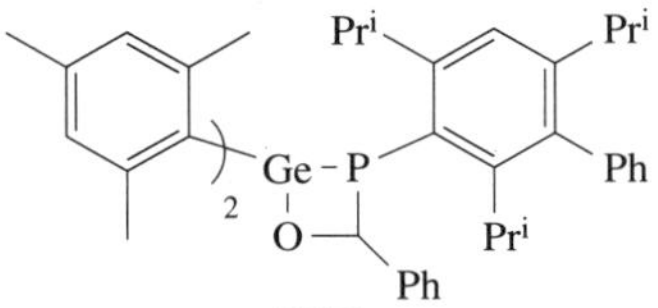

(246)

4.10.1.2.3 Tetracoordinate phosphorus functions

(i) Tetracoordinate phosphorus functions by reaction of phosphorus nucleophiles at sp² centres

The addition of trialkyl phosphites to oximes gives *N*-alkoxy-α-aminophosphonates in poor to moderate yield. For example, furan-2-carboxaldehyde oxime reacts with triethyl phosphite to afford (247) in 39% yield (Equation (59)) ⟨82JGU392⟩; under alkaline conditions with two mole equivalents of dialkyl phosphite, *N*-phosphorylated adducts (248) are formed (Equation (60)) ⟨80ZOB751⟩. *N*-Alkoxy-α-aminophosphonic acids may also be obtained by hydrolysis of the esters which result from the addition of diaryl phosphites to oxoiminium salts ⟨93LA955⟩. The addition of lithium or potassium salts of dialkyl phosphites to the chiral nitrones (249) has been employed in the asymmetric synthesis of analogues of serine, valine and alanine, but the highest diastereoselectivity was observed with lithium salts (Scheme 59) ⟨85HCA1730⟩. A further example is illustrated by the novel preparation of the β-phosphorylated, five-membered ring-nitroxide (250) (Scheme 60) ⟨91TL2125⟩. A push–pull type mechanism is suggested for the addition of trialkyl phosphites and alkyl halides to a nitrone. Thus trimethyl phosphite and methyl iodide react in benzene with Δ¹-pyrroline *N*-oxides, furnishing *N*-methoxy phosphonates in good yield (> 70%) ⟨88TL663⟩.

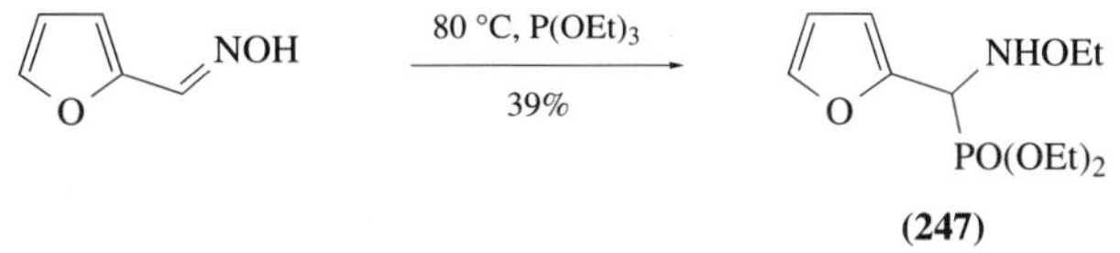

$$\tag{59}$$

(247)

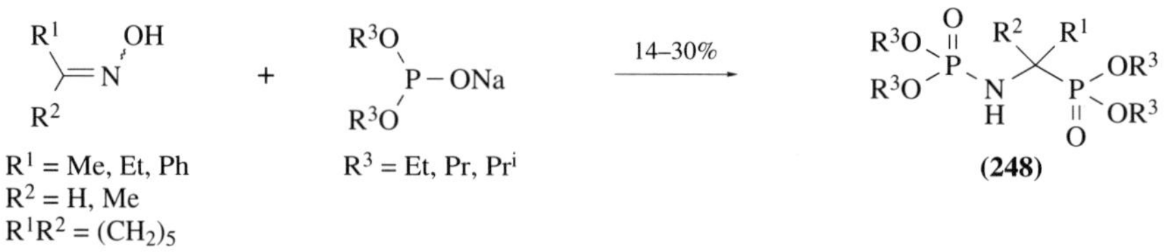

Scheme 59

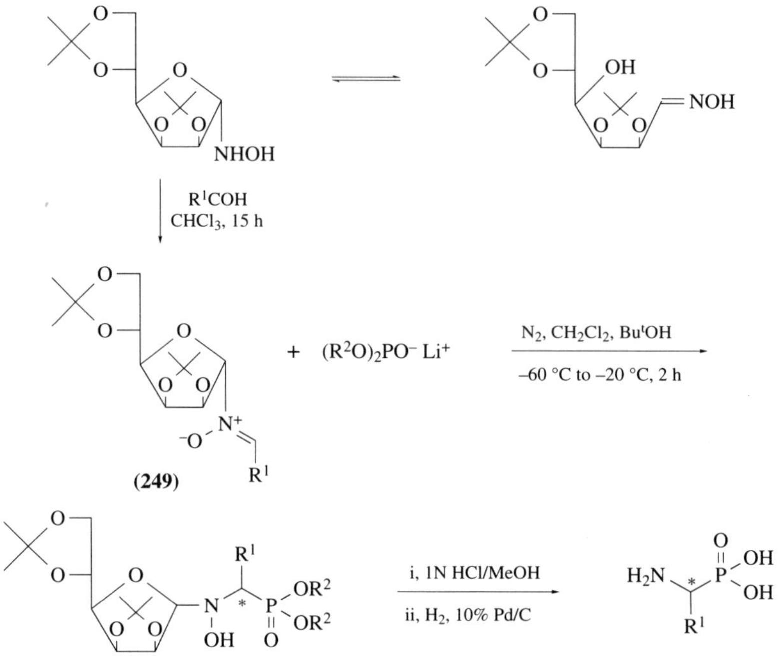

(equation 60)

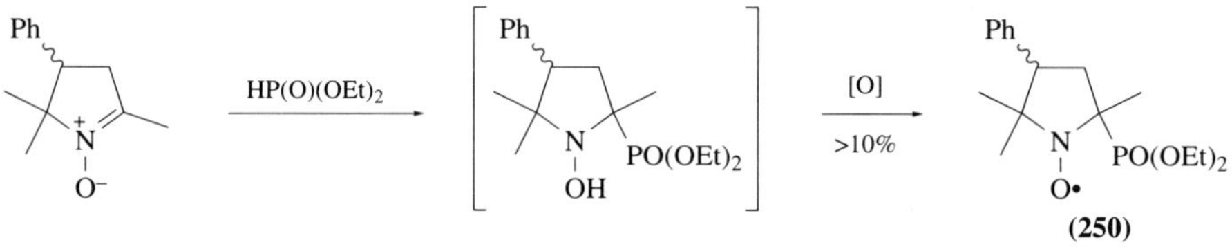

Scheme 60

The 1,4-diaza-2,5-phosphorinanes (**252a**)–(**252c**) are obtained in 20–35% yield by addition of chlorophosphonites to imines (Scheme 61) ⟨82ZOB930, 83IZV432⟩.

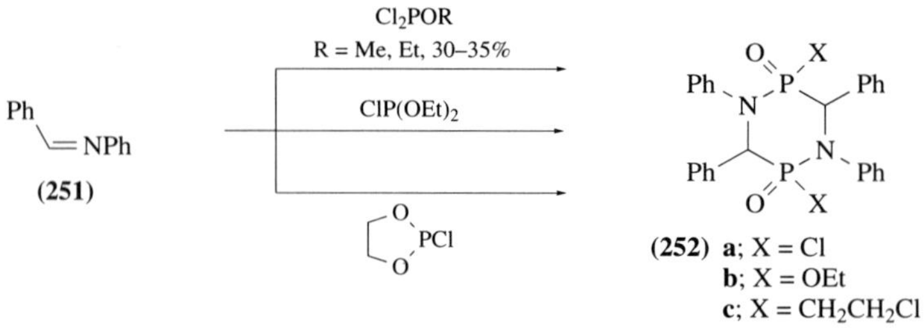

Scheme 61

Dialkyl phosphites add to *N*-acyl- and *N*-fluorosulfonyl-imines (**253**) furnishing, in high yield, nitrogen-substituted α-aminophosphonic acid derivatives (**254**). If desired, selective nitrogen deprotection can be conducted; alternatively, nitrogen-deprotection and ester hydrolysis may be carried out simultaneously with hydrochloric acid, yielding the free amino acid (**255**); formic acid effects selective nitrogen deprotection (Scheme 62) ⟨68M380, 69ZOB2192, 82PS(13)319⟩.

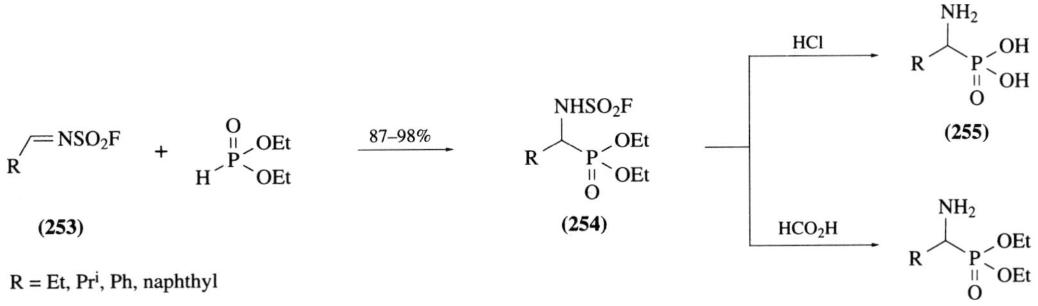

Scheme 62

The addition of tertiary phosphines to iminium ions affords moderate yields of N—C—P compounds with tetracoordinate, tetravalent phosphorus atoms ⟨63LA91, 71CB31, 72CB2233⟩. Thus triphenylphosphine reacts with the methyleneiminium salt (**256**) furnishing the adduct (**257**) (Equation (61)). These primary products are thermally and hydrolytically unstable but can be isolated in basic media at low temperatures.

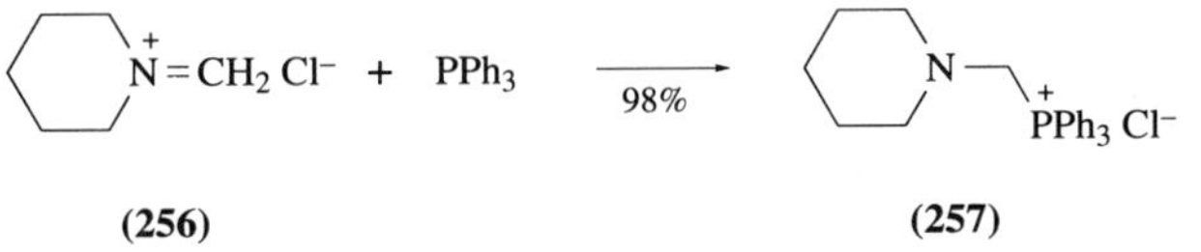

(ii) Tetracoordinate phosphorus functions from Arbuzov– and Michaelis–Becker-type reactions

The reaction of ethyl diphenylphosphinite with *N*-chloromethyl amide (**258**) prepared *in situ* is representative of an Arbuzov–Michaelis–Becker-type reaction leading to a functionalised amino phosphorus compound (Scheme 63) ⟨92SC2381⟩. A similar, three-component reaction between thiourea, an aldehyde and triphenyl phosphite gave the guanidinoalkyl phosphonic acids (**259**) (Scheme 64) ⟨77S571⟩. The overall yields are low but the process is fast and cheap, the starting materials being readily available.

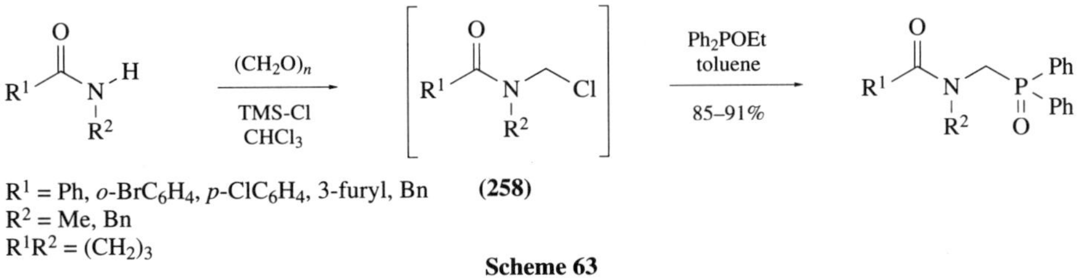

R^1 = Ph, *o*-BrC$_6$H$_4$, *p*-ClC$_6$H$_4$, 3-furyl, Bn   (**258**)
R^2 = Me, Bn
R^1R^2 = (CH$_2$)$_3$

Scheme 63

2-Oxoazetidin-4-yl phosphonates and phosphinates result from the rapid reaction of phosphites and phosphonites with 4-acetoxyazetidin-2-ones; for example, trimethyl phosphite reacted with (**260**) forming (**261**) in excellent yield, and (**261**) was converted in four steps into a phosphonic analogue of serine (Scheme 65) ⟨80CC730, 82T2513, 91SC1847⟩. Other nontraditional variants include the use of derivatives with trialkylammonium instead of halogen leaving groups ⟨73TL633⟩.

(iii) Tetracoordinate phosphorus functions from Kabachnik–Fields reactions

The three-component mixtures of amine carbonyl and nucleophilic phosphorus compounds used for the synthesis of 1-aminoalkyl organophosphorus compounds have usually contained amino derivatives other than simple amines. For example, ureas, thioureas, amides, carbamates and

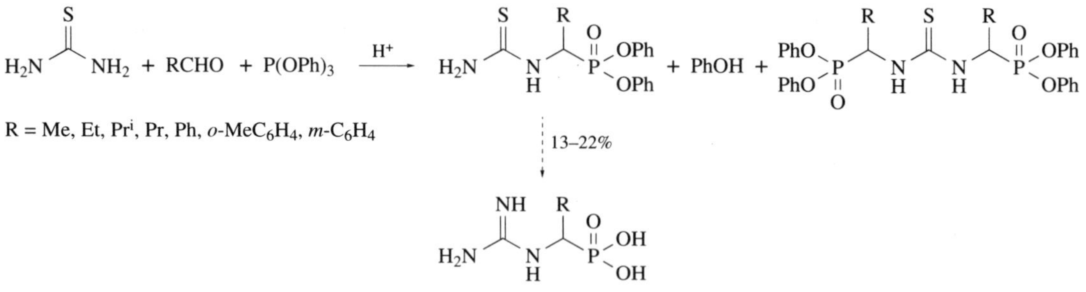

R = Me, Et, Pr^i, Pr, Ph, o-MeC$_6$H$_4$, m-C$_6$H$_4$

Scheme 64

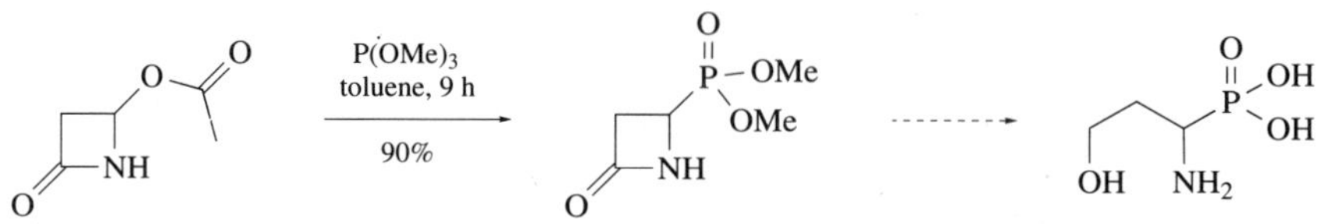

Scheme 65

phosphorimidates have been employed as the amino components ⟨91S490, 92S1124⟩. When phenyldichlorophosphine reacts with *N*-phenylurea and an aldehyde or ketone, the nature of the product depends on the conditions employed. For example, 1,3,4-diazaphospholidin-2-one 4-oxides have been formed in this manner (Equation (62)) ⟨93MI 410-01⟩. Phosphorus trichloride reacts with aldimines in the presence of *ortho*-formates to furnish *N*-formylaminoalkyl phosphonic acids; it is believed that the reaction between PCl$_3$ and the *ortho*-esters facilitates dialkyl phosphite formation. The yields obtained by this route are very favourable ⟨91SC1951, 92S263⟩.

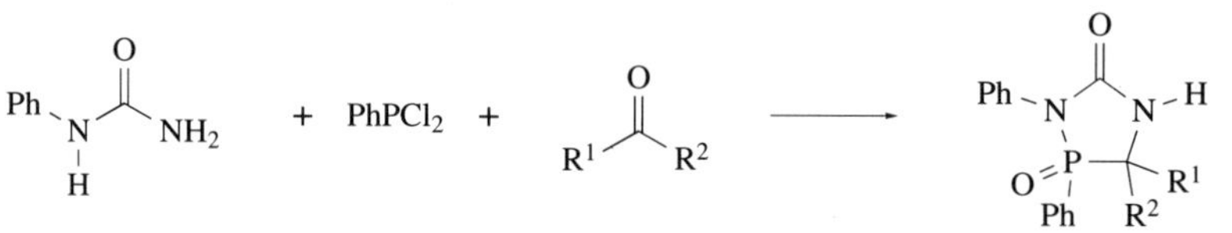

(iv) Tetracoordinate phosphorus functions by cycloaddition reactions

The 1,3-dipolar cycloaddition of diazo alkanes and nitrile imines to vinyl phosphonates and related phosphoryl compounds leads to a variety of phosphorus-containing heterocyclic compounds ⟨68ZOB1248, 70ZOB2618, 79ZOB493, 85CB3227⟩. Some representative examples of pyrazolines **(262)** prepared by this method (Equation (63)) are shown in Table 6. Chiral vinyl phosphorus compounds furnish adducts which are mixtures of *erythro* and *threo* isomers.

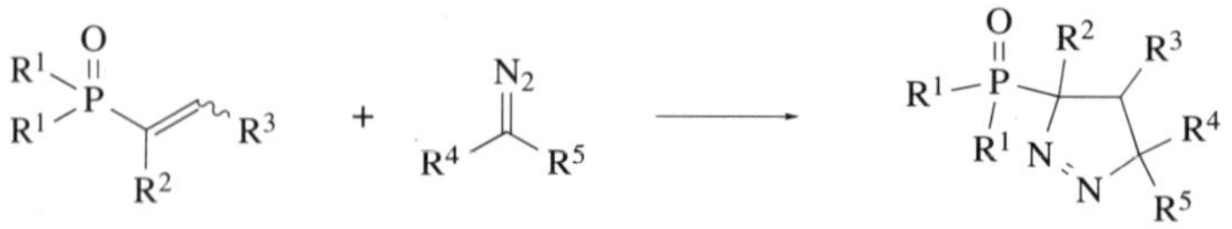

1-Cycloalkenyl phosphonium salts can also function as dipolarophiles ⟨85JOC1278⟩, furnishing the alkali-labile bicyclic products **(263)** with diazo alkanes (Equation (64)). With increasing ring size the reactivity of the cycloalkenyl phosphonium salts decreases and the effect of substitution on the diazoalkane is dramatic (Table 7). 1-Cyclobutenyldiphenylphosphine oxide reacts similarly with diazomethane to furnish the expected cycloadduct in 72% yield (10 h, 0 °C).

Table 6 Substituted pyrazolines (**262**) prepared by the reactions of diazo alkanes with phosphorus-containing dipolarophiles.

R^1	R^2	R^3	R^4	R^5	Yield (%)	Ref.
OEt	Ph	CN	H	H		70BSF1130
OMe	H	Me	Me	Me	81	72ZOB1227
OMe	H	Me	Ph	H		72ZOB1227
Ph	a	a	H	H	72	85JOC1278
Ph	H	=CP(O)Ph$_2$	H	H	41	90CB423

^{a} $R^2R^3 = (CH_2)_2$.

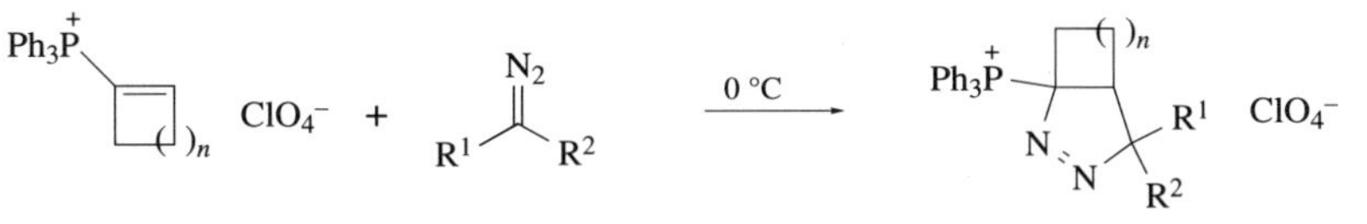

$$Ph_3\overset{+}{P}\text{-cycloalkenyl} \; ClO_4^- + R^1R^2C=N_2 \xrightarrow{0\,°C} \text{(263)} \quad ClO_4^- \quad (64)$$

Table 7 Reactivity of 1-cycloalkenyl triphenyl phosphonium salts with diazo alkanes furnishing the cycloadducts (**263**) (Equation (64)).

R^1	R^2	n	Time (h)	Temperature (°C)	Yield (%)
H	H	1	6	0	80
Ph	Ph	1	6	0	0
H	H	2	6	0	80
H	H	3	6	0	0

The phosphorus moiety of a cycloadduct may arise from the dipole as well as the dipolarophile. Thus α-diazoalkyl phosphonates and phosphine oxides add a range of substituted and unsubstituted dipolarophiles furnishing bicyclic products ⟨65AG1138, 69CB2216, 71JOC1379, 72CJC1078⟩. For example, (**264**) reacts with the vinyl phosphonates (**265**) to afford the cycloadducts (**266**), which are susceptible to thermal and photochemical decomposition with N$_2$ elimination (Scheme 66) ⟨82ZOB2444⟩. Phosphorus-substituted azomethine ylides obtained from *N*-benzylidene-α-(diphenylphosphinoyl)glycine esters undergo cycloaddition with cyclic and acyclic dipolarophiles at 70 °C, furnishing diastereomeric adducts arising from *endo* cycloaddition ⟨90JOC4063, 90JOC4069⟩. A further example is the cycloaddition of the *N*-glycosyl *C*-dialkoxyphosphoryl nitrones (**267**) to ethylene (Scheme 67); modest diastereoselection was observed ⟨82HCA1953, 85HCA1730⟩. This route has been used in the synthesis of an optically active asparagine analogue.

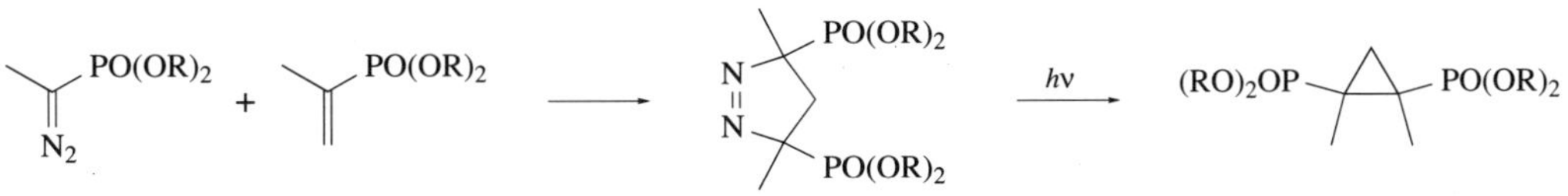

R = Me, Et, Pr^i

Scheme 66

Diels–Alder reactions furnishing new amino phosphorus compounds have also been reported. Seleno- and thiooxophospholes (**268**) function as the diene components, reacting with the heterodienophiles (**269**) to afford the bridged adducts (**270**) (Equation (65)). In the presence of water the formation of phosphine oxides competes ⟨85PS(25)201⟩. Diethyl (*N*-acyliminomethyl)phosphonates participate in Diels–Alder reactions as either dienes or dienophiles, depending on the nature of the second component. Some unusual 1-aminoalkylphosphonic acids have been prepared in this way ⟨90S1153⟩.

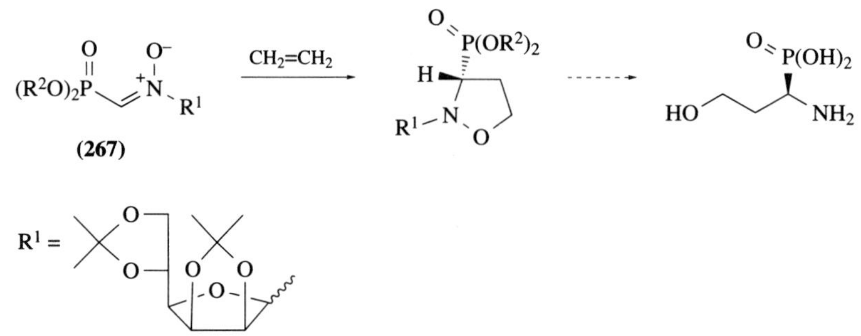

Scheme 67

(65)

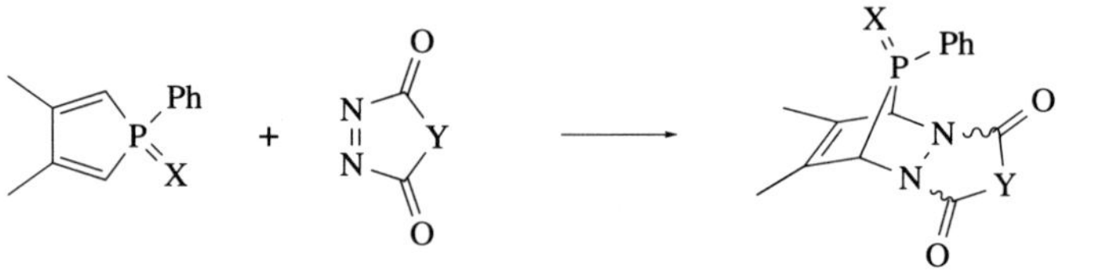

(268) X = Se, S
(269) Y = NMe, NPh, O
(270)

(v) Tetracoordinate phosphorus functions by miscellaneous reactions

(a) Reaction with electrophilic nitrogen compounds. The addition of sufficiently electrophilic nitrogen compounds to phosphorus ylides is a minor route to the synthesis of N—C—P compounds. Thus arenediazonium salts and the dehydrodithizone (271) add to phosphorus ylides to produce, for example, compound (272) formed in 72% yield (Equation (66)) the structure of which was established by ^{31}P NMR spectroscopy ⟨61TL809, 71CC490⟩. Ethyl nitrate and phenyl azide add to the anions of stabilised phosphine oxides furnishing stable products ⟨61TL9, 62CB2563, 92RCR1220⟩. The primary products from nucleophilic or slightly electrophilic nitrogen species are unstable.

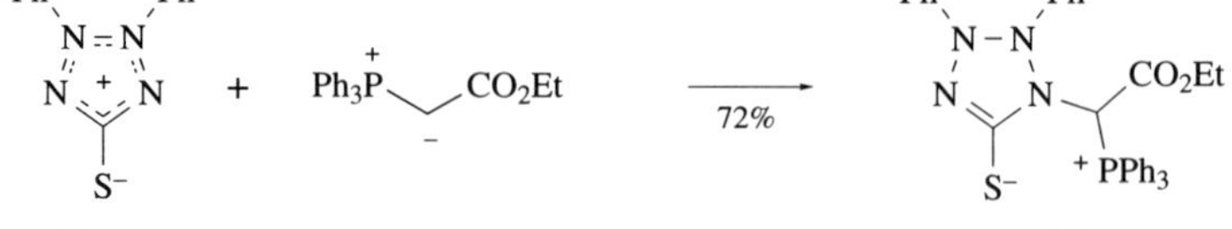

(66)

The addition of acetyl nitrate to the 2-alkoxyvinyl phosphonic esters (273) represents a convenient method for the synthesis of the 1-nitroalkylphosphonates (274). The reaction, which proceeds in adequate yields (45–70%) (Scheme 68), is catalysed by sulphuric acid and is nonstereospecific ⟨76ZOB1250⟩. The corresponding products (275) are obtained by analogous reactions with 2-alkoxyalkenylphosphinates and -phosphine oxides (Equation (67)) ⟨76ZOB1495⟩. In the latter case yields are often poor (28–61%) due to competing protonation of the relatively basic phosphoryl group. Substituted vinylphosphonates, for example styrylphosphonate, may be nitrated by dinitrogen tetroxide ⟨81ZOB757⟩. Preparation of amino phosphorus compounds by direct introduction of the nitrogen functionality into molecules with a stabilised carbanion is effected by treatment of phosphonoacetates with, for example, *O*-mesitylsulfonylhydroxyamines in the presence of a strong base (Equation (68)) (important technical information regarding the handling of this hydroxylamine is included in reference ⟨77JOC376⟩; see also ⟨82TL3835⟩.

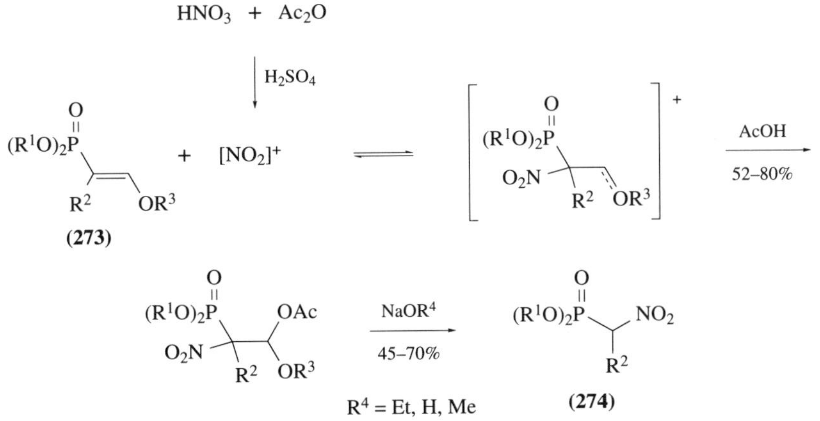

$HNO_3 + Ac_2O$

Scheme 68

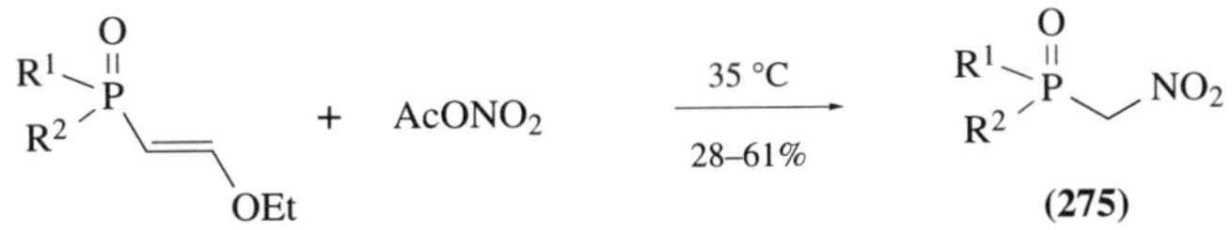

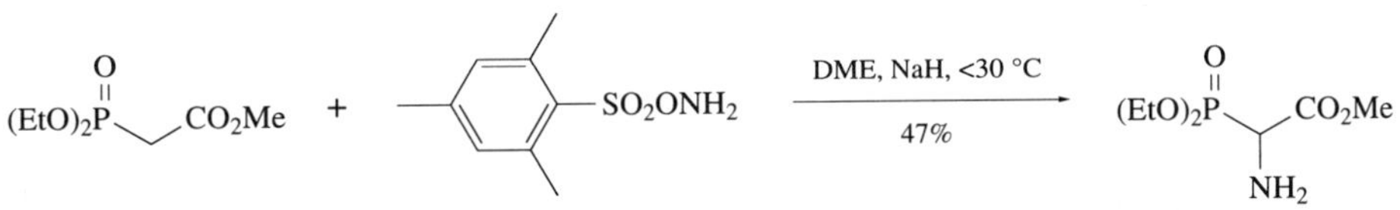

Trifluoromethanesulfonyl azide reacts similarly to give α-azido derivatives which may be hydrogenated catalytically to form the free amino compounds ⟨80SC429⟩.

(b) Via α-oxo phosphorus compounds. The preparation of 1-nitroalkylphosphonates has also been realised from α-oxophosphonates. Thus treatment of the esters **(276)** with hydroxylamine furnishes the 1-*N*-hydroxyiminoalkylphosphonates which can be oxidized with mcpba to the desired nitro compounds **(277)** (Scheme 69) ⟨84S661⟩. Ionic hydrogenation of (1-(benzyloximino)alkyl)phosphonates affords the corresponding amino derivatives in 54–79% yield ⟨93H(36)1925⟩.

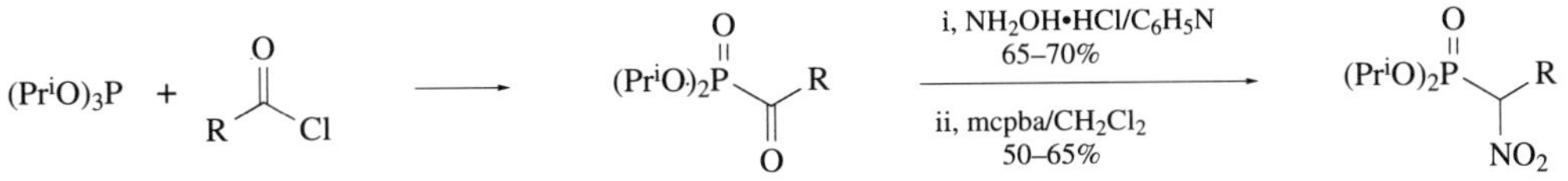

Scheme 69

(c) Mitsunobu reaction. The Mitsunobu reaction was employed in the early 1990s in a simple and efficient synthesis of diethyl 1-azidoalkylphosphonates **(278)** (Equation (69)) ⟨92S367⟩. The substrates are limited to primary and secondary diethyl 1-hydroxyalkylphosphonates and preformation of the betaine adduct from triphenyl phosphine and diethyl azodicarboxylate (dead) is essential for high-yield, high-purity products.

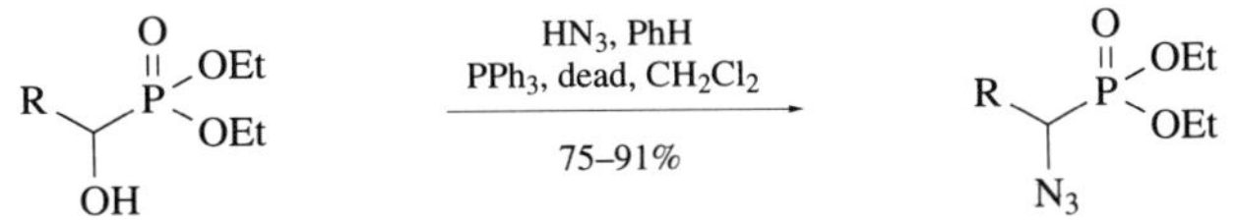

$R = H, Et, Pr, Pr^i, Ph, Bn$

(d) Addition of nucleophilic nitrogen species to P—C—X systems. A general protocol for the synthesis of optically active α-aminoalkylphosphonic acids involves, as a key step, treatment of the chloromethylphosphonamide (**279**) with sodium azide (Scheme 70) ⟨90TL6465⟩. 1-Azidoalkylphosphonates may also be obtained by the direct azidation of diethyl 1-chloromethylphosphonate ⟨67MIP1087066⟩. The *p*-nitro tosylate (**282**) reacts analogously with pyridine, which expels the tosylate anion (Equation (70)) ⟨72ZC60, 72ZC334⟩.

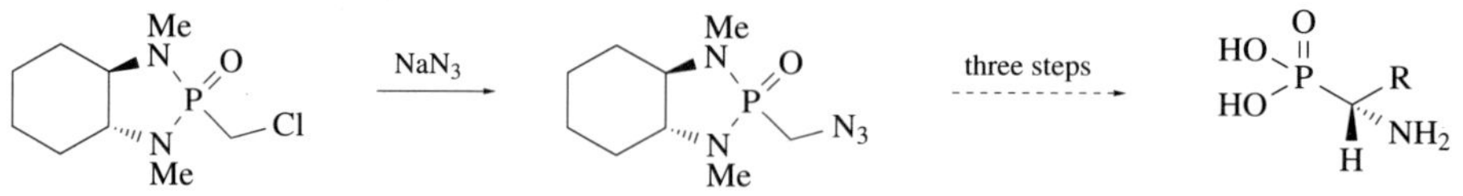

(279) **(280)** **(281)**

Scheme 70

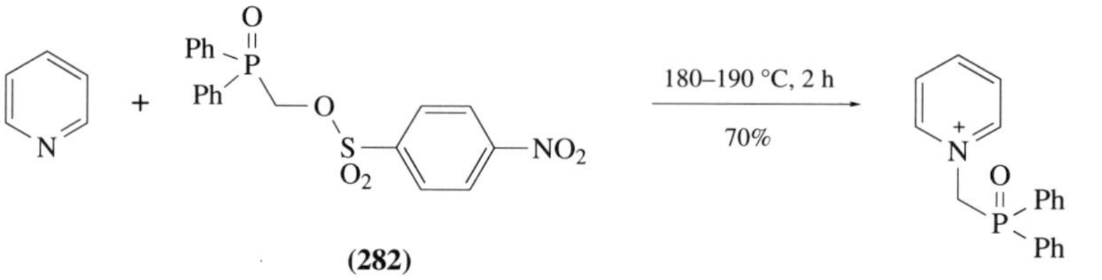

(70)

(282)

(e) Addition of nucleophilic phosphorus species to N—C—X systems. 1-Aminoalkylphosphonium salts result from the addition of trialkyl(aryl) or mixed phosphines to compounds containing N—C—X functionalities (X = halogen, OH, $N^+R_3X^-$). Several examples are included in a review article by Petersen dealing with the synthesis of cyclic ureas ⟨73S243⟩; for example, triphenylphosphine attacks the *N*-hydroxymethyl or -alkoxymethyl amides (**283**), furnishing the phosphonium salts (**284**) in good yield (Equation (71)). The same range of substrates reacts with dialkyl phosphites to afford phosphonic acid esters ⟨52JA1528⟩. Nitrilotriacetic acid (**285**) in the presence of a suitable dehydrating agent (PCl_3, Ac_2O or P_2O_5) reacts with phosphonic acid, furnishing the tris derivatives (**286**) in excellent yield (Equation (72)) ⟨72CZ691⟩. The addition of triphenylphosphine in C_2Cl_6, or Ph_3PCl_2 to trimethylsilylmethyl isocyanide in THF at room temperature, furnishes the corresponding phosphonium salts in 73% yield ⟨85AG(E)979⟩.

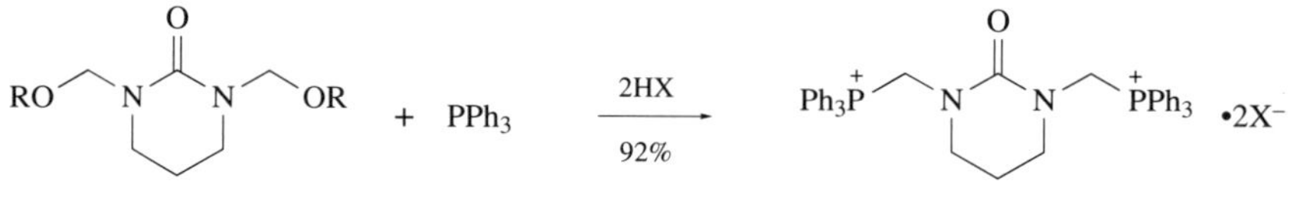

(71)

(283) R = H, Me **(284)**

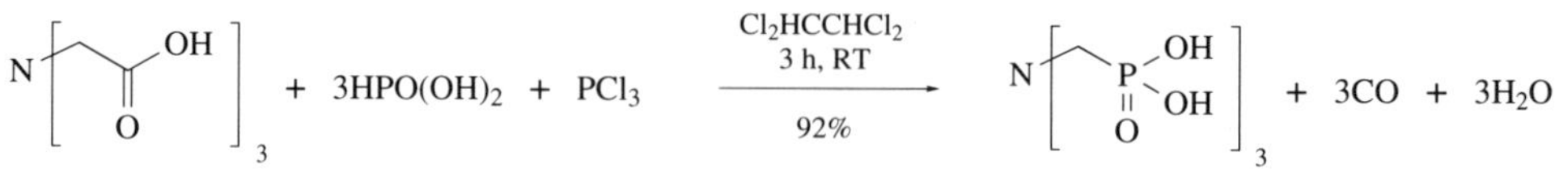

(72)

(285) **(286)**

(f) From rearrangement processes (Curtius and Lossen rearrangements). When the substrate (**287**) was subjected to a Curtius rearrangement in an alcoholic solution containing a phosphonoacetic acid, the carbamate derivative (**288**) was obtained (Equation (73)) ⟨79PJC541⟩. The phosphine oxide (**289**) has been subjected to the Lossen, hydroxamic acid rearrangement affording the isocyanate (**290**). In the presence of aqueous ethanol, the symmetrical urea (**291**) and the urethane (**292**) were obtained in 20% and 29% yield, respectively (Scheme 71).

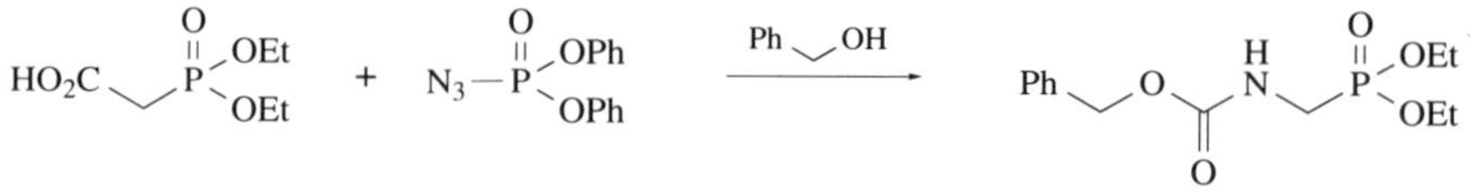

(73)

(287) **(288)**

(g) *Modification of existing amino phosphorus compounds.* A survey of the general literature indicates that the nitrogen functionality in 1-aminoalkyl phosphorus compounds retains much of the chemical reactivity expected of this functional group ⟨B-68MI 410-01, 74RCR984, 87RCR859⟩. Thus quaternization of aminoalkylphosphonates may be effected with dimethyl sulfate or alkyl halides, or acylation with acyl chlorides in pyridine. α-Aminoalkyl-phosphine oxides, -phosphinates and -phosphonic acids may also be *N*-acylated. Oxidation of tertiary amines with H_2O_2 gives *N*-oxides and oxidation of primary amines with NaOCl gives azo compounds. Oxidation of amino phosphorus compounds to the corresponding nitro derivatives may be realised, in fair yield (~50%), with $KMnO_4$ ⟨76MI 410-03, 76ZOB1246⟩. A review is available presenting a range of methods for the synthesis of α-nitroalkyl organophosphorus compounds ⟨92RCR1220⟩.

Additionally, amino phosphorus compounds may be modified at the N—C—P carbon.

4.10.1.2.4 Higher coordinate phosphorus functions

The number of compounds in this category is limited, the majority being prepared by [4 + 1]- or [3 + 2]-cycloaddition reactions.

Preparation of Δ^4-1,4,2,λ^5-oxazaphospholines involving a [4 + 1]-cycloaddition between phosphites and the substituted *N*-acylimine derivatives (**293**) was first described in 1971 ⟨71CB1826, 76LA36⟩. The new ring system (**294**) was obtained in good yield (60–90%) but was thermally, photochemically and hydrolytically unstable (Equation (74)) ⟨73CB3421, 74S816⟩. The crystal structures of two members of this class have been reported ⟨75JA38⟩. The thia analogue (**296**), also unstable at room temperature, is obtainable by deselenation of the heterocycle (**295**) by $P(OMe)_3$ (Equation (75)) ⟨78S44⟩. The 1,4,2-diazaphospholine (**298**) can be prepared by a [4 + 1]-cycloaddition between the corresponding trialkyl phosphites or dialkyl phosphonites and the appropriate benzamidines (**297**) (Equation (76)) ⟨78S526⟩. Similar treatment of the azine (**299**) does not yield the predicted [4 + 1]-adduct (**301**); instead, a new 1,2λ^5-azaphosphoridine system (**300**) results (Equation (77)) ⟨73AG(E)502⟩. The cyclic phosphorane (**305**) arises in 70–75% yield due to efficient trapping of the nonconjugated, 1,3-dipolar intermediate (**304**) across the C=O double bond of formaldehyde. The product (**305**) is extremely hydrolytically labile, opening to furnish (**306**). The betaine (**304**) arises by a two-step displacement of an alcohol molecule from the sulfonamide substrate (**302**) (Scheme 72) ⟨76JOC28⟩. Interestingly, di- and tricyclic products (**310**) and (**311**) result from the reaction between the phosphorocyanatidites (**307**) and *N*-alkylideneacetamide (**308**). Both products arise via the intermediate (**309**); (**310**) results from the addition of a second molecule of (**308**) to (**309**), and (**311**) is the product of a dimerisation reaction (Scheme 73) ⟨88JGU876, 88JGU1148⟩.

$R^2, R^{3,} R^4$ = OMe, OPr^i, OPh, Ph
X = Y = F or X = F, Y = Cl

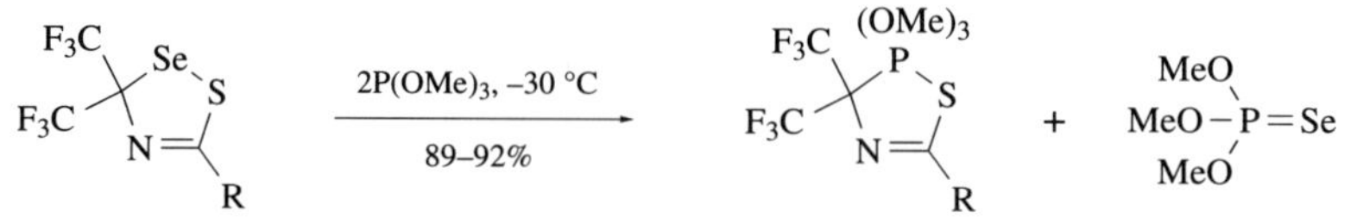

(75)

(295) R = Ph, *p*-MeC_6H_4, *p*-ClC_6H_4 **(296)**

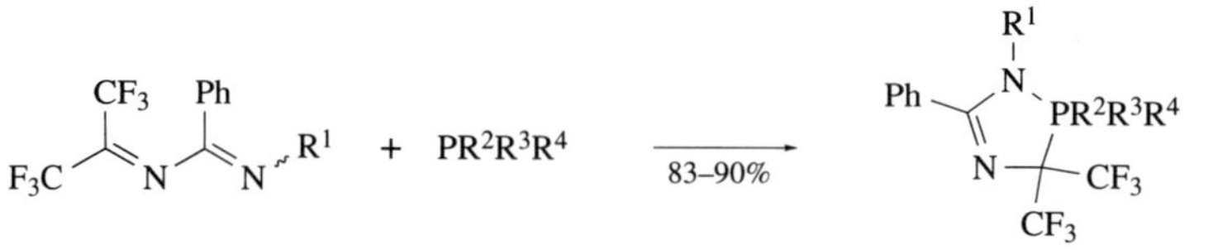

(76)

(297) **(298)**

R^1 = 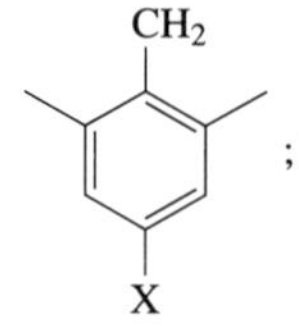 ; X = H or Me

$R^2 = R^3 = R^4$ = OMe, OEt
$R^2 = R^3$ = OMe; R^4 = Ph

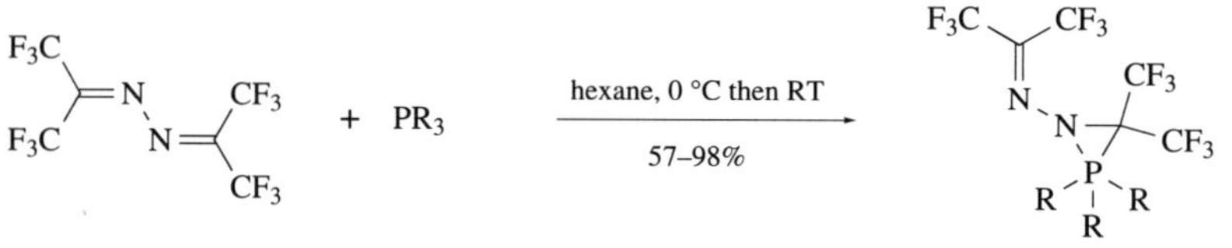

(77)

(299) R = OMe, OEt, NMe_2 **(300)**

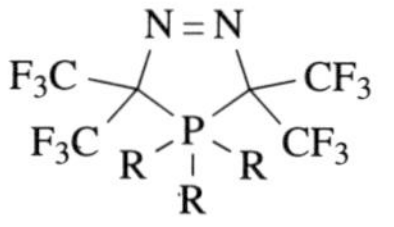

(301)

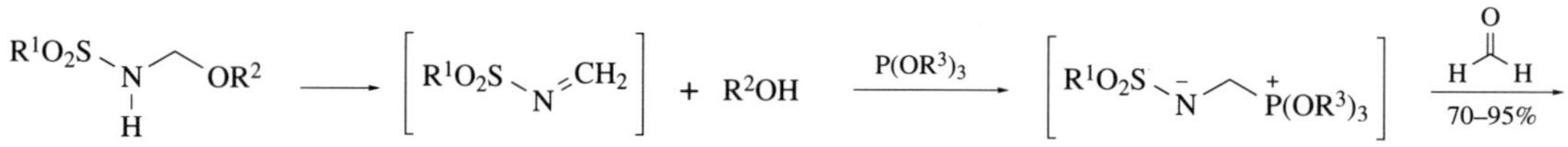

(302) **(303)** **(304)**

R^1 = Me, Ts
R^2 = H, alkyl
R^3 = Me, Et

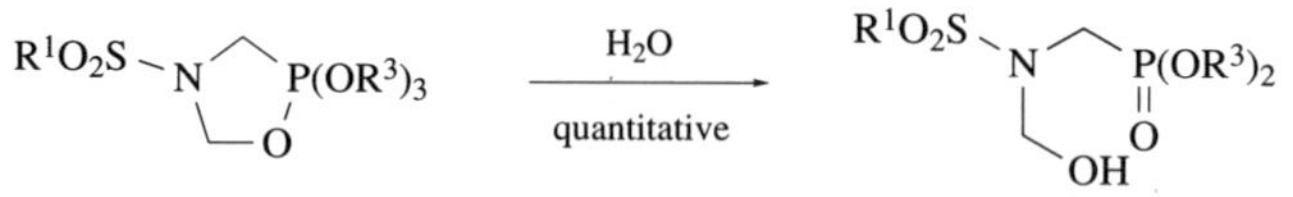

(305) **(306)**

Scheme 72

Hexacoordinate amino alkylphosphorus compounds are rare. An example is the molecule (**313**), obtained by cyclising (**312**) with secondary amines (Equation (78)) ((**313**) is one of three tautomeric forms). Crystal structure determination has been carried out on one of these adducts ⟨87DOK(297)1132, 87IZV1680, 88IZV936⟩.

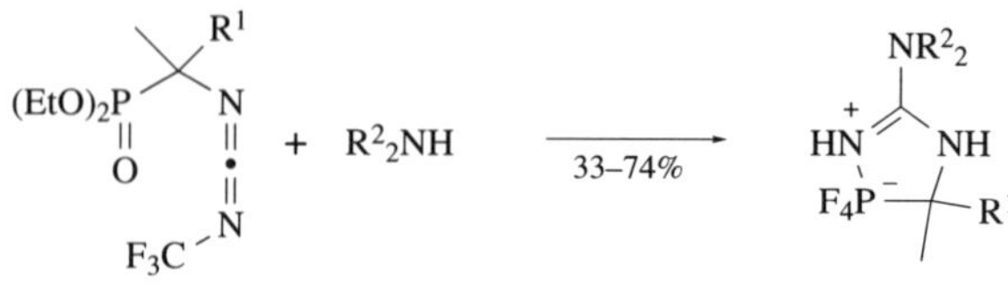

(78)

(312) R^1 = Me, Et, Bu^s **(313)**
R^2 = Et, Pr, Pr^i

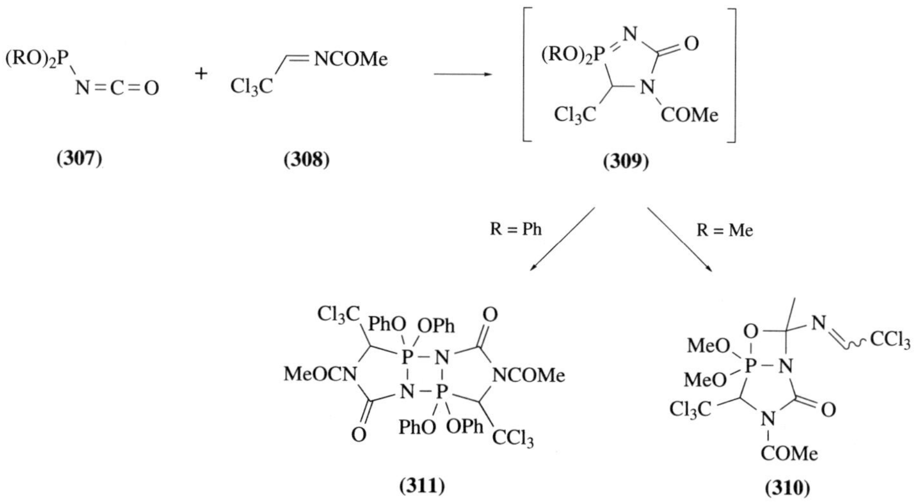

Scheme 73

4.10.2 FUNCTIONS CONTAINING NITROGEN AND ARSENIC, ANTIMONY OR BISMUTH

There appears to be no record of compounds containing the structural unit N—CR^{1}R^2—Bi and only very scanty reference to analogous compounds containing nitrogen and antimony. As can be seen from the following section, virtually all progress in the field of 1,3-azarsa compounds can be attributed to the efforts of Tzschach and co-workers over a number of years.

4.10.2.1 Amino Functions: R$^1_2$C(NH$_2$)ZR$^2_2$, R$^1_2$C(NHR^2)ZO(OR^3)$_2$, R$^1_2$C(NR$^2_2$)ZX, etc. (Z = As, Sb, Bi)

4.10.2.1.1 Dicoordinate Z functions

There appear to be no examples in this category.

4.10.2.1.2 Tricoordinate Z functions

(i) Tricoordinate Z functions by cyclocondensation reactions

This category comprises the largest single group of systems containing nitrogen and arsenic in a 1,3-relationship. A convenient general method for the synthesis of 1,3-azarsolidines (**315**) (*n* = 1) and the analogous azarsinanes (*n* = 2) (Equation (79)) involves the reaction of secondary aminoalkylarsines R^{1}AsHCH$_2$(CH$_2$)$_n$NHR^2 with an equimolar quantity of a suitable carbonyl component in a modification of the Mannich reaction. Typically, the carbonyl unit is an aldehyde or ketone (Table 8) ⟨70JOM(21)131, 74JOM(81)187, 76ZC64⟩. α-Keto acids (Table 9) and α-keto esters, β-keto esters and β-keto amides (Table 10) have also been employed ⟨73EGP99803, 73JPR65⟩. The secondary aminoalkylarsine (**317**) is readily prepared (Scheme 74). Treatment of an arsine with an equimolar amount of sodium in liquid ammonia affords the sodium arsenide, which reacts with a suitable aminoalkyl halide to afford the primary arsine (**316**), which can be further alkylated with a simple alkyl halide furnishing the desired secondary aminoalkylarsine (**317**) (39–73% yield) ⟨76ZC64⟩. With judicious choice of reagents a wide range of substituents R^1, R^2, R^3 and R^4 may be incorporated into the final product (**315**); the yields are generally very good (Table 8). The rate of the reaction is largely dependent on the nature of the carbonyl component. With aldehydes the reaction proceeds at room temperature and may be exothermic (freshly distilled aldehyde may be used in an equimolar amount or may be used in excess as the solvent). In contrast, several hours' heating may be required to promote the reaction with ketones. An equimolar quantity of the ketone is usually employed and the reaction may be carried out neat or in a solvent, often benzene (especially for the azarsinanes (**315**) where $n = 2$). When α-keto acids are employed ⟨73JPR65⟩, the resulting 1,3-azarsolidines (**315**) have $R^3 =$ COOH (Table 9) and IR analysis shows them to exist in zwitterionic form (**318**). The reactions are carried out in dioxane or ether and may require slight thermal activation. When the carbonyl component carries a δ- or γ-acid group, a second condensation may take place under the particular reaction conditions (xylene, 4 h, H_2O, acid catalyst) to afford the bicyclic lactams (**319**) in good yield. The reaction with α-keto esters proceeds in xylene or benzene in the presence of glacial acetic acid; heating is necessary and yields are not high (Table 10). Complications may also arise with the use of β-keto esters; in acid media (toluene, glacial acetic acid, 4 h, heat) the simple cyclocondensation reaction takes place to afford (**320**) in 56% yield, yet in neutral or basic media the initially formed imine isomerises to furnish the α,β-unsaturated β-amino ester (**321**) (Scheme 75) ⟨73JPR65⟩. When $R^2 = $ H the cyclic product (**315**) may be treated with phenyl isocyanate to yield the *N*-carbamoyl derivative ((**315**) where $R^2 = $ CONHPh) ⟨73JOM(60)95, 74JOM(81)187⟩. Additionally, unsubstituted 1-aminoalkyl arsines may be alkylated on arsenic (Scheme 76) ⟨76ZC64⟩.

$$R^1HAs\underset{}{\overset{(\ )_n}{\diagup}}NHR^2 \quad + \quad \underset{R^4}{\overset{O}{\underset{}{\diagup\!\!\!\diagdown}}}R^3 \quad \xrightarrow{30-90\%} \quad R^1As\underset{R^3\ \ R^4}{\overset{(\ )_n}{\diagup\!\!\!\diagdown}}NR^2 \quad + \quad H_2O \qquad (79)$$

$$n = 1 \text{ or } 2 \qquad\qquad (\mathbf{315})$$

Table 8 Representative examples of 1,3-azarsolidines and -azarsinanes (**315**) prepared from α,ω-aminoalkylarsines and aldehydes and ketones (Equation (79)).

No.	R^1	R^2	R^3	R^4	Yield (%)	Ref.
$n = 1$						
1	Ph	H	H	H	78	70JOM(21)131
2	Ph	H	Me	H	84	70JOM(21)131
3	Ph	H	Et	H	77	70JOM(21)131
4	Ph	H	Ph	H	88	70JOM(21)131
5	H	H	Me	H	54	76ZC64
6	H	Et	Pr^n	H	33	76ZC64
7	Ph	H	Me	Me	73	70JOM(21)131
8	Ph	H	Me	Et	74	70JOM(21)131
9	Ph	H	Et	Et	82	70JOM(21)131
10	Ph	H	a	a	80	70JOM(21)131
11	Ph	H	b	b	72	70JOM(21)131
$n = 2$						
12	Ph	H	H	H	54	74JOM(81)187
13	Ph	H	Me	H	70	74JOM(81)187
14	Bu	H	Me	H	77	74JOM(81)187
15	Ph	H	Et	H	78	74JOM(81)187
16	Ph	H	Ph	H	70	74JOM(81)187
17	Ph	H	Et	Me	91	74JOM(81)187
18	Ph	H	Bu^i	Me	86	74JOM(81)187
19	Ph	H	a	a	93	74JOM(81)187
20	Bu	H	a	a	95	74JOM(81)187
21	Bu	H	b	b	97	74JOM(81)187

^{a} $R^3R^4 = (CH_2)_4$. ^{b} $R^3R^4 = (CH_2)_5$.

The construction of a wide range of 1,3-benzazarsolines has been accomplished using the same cyclocondensation chemistry (Scheme 77) ⟨73JOM(60)95⟩. Secondary *o*-aminophenylarsines (**323**) have been prepared by reduction of *o*-nitrophenylarsonic acid (**322**), metallation of the *o*-aminophenylarsine formed and subsequent alkylation with an alkyl halide. Treatment with a carbonyl component (equimolar amount or excess) affords the fused products (**324**) in good yield (Table 11). The time required for complete reaction is dependent on the nature of the carbonyl component: aliphatic aldehydes or α-keto esters require 0.5–1 h; benzaldehyde requires 1 h; acetoacetic esters require 5 h; and aliphatic ketones and δ-keto esters require 8–12 h. When the chosen alkyl halide carries a terminal amino group (e.g., halogen–$CH_2CH_2NHR^4$), intermediates such as

Table 9 Representative examples of 1,3-azarsolidines (**315**; $n = 1$) prepared from α,ω-aminoarsines and α-keto acids.

No.	R^1	R^2	R^3	R^4	Yield (%)	Ref.
1	Ph	H	COOH	Me	90	73EGP99803
2	Ph	H	COO$^-$	Me	63	73JPR65
3	Ph	Et	COO$^-$	Me	40	73JPR65
4	Ph	H	COO$^-$	Ph	84	73JPR65
5	Cy	H	COO$^-$	Me	53	73JPR65
6	Cy	Et	COO$^-$	Me	70	73JPR65
7	Bu^n	H	COO$^-$	Me	40	73JPR65

Table 10 Representative examples of 1,3-azarsolidines (**315**; $n = 1$) prepared from α,ω-aminoarsines and β-keto esters, α-keto amides and β-keto esters (Equation (79)).

No.	R^1	R^2	R^3	R^4	Yield (%)	Ref.
1	Ph	H	CH$_2$CO$_2$Et	Me	56	73EGP99803
2	Ph	H	CH$_2$CONHPh	Me	86	73EGP99803
3	Ph	H	CO$_2$Et	Me	40	73JPR65
4	Ph	Et	CO$_2$Et	Me	39	73JPR65
5	Ph	H	CO$_2$Et	Ph		73JPR65
6	Ph	H	CH$_2$CO$_2$Et	Me	56	73JPR65
7	Ph	Et	(CH$_2$)$_2$CO$_2$Et	Me	27	73JPR65
8	Me	H	CH$_2$CO$_2$Me	CO$_2$Me	55	85ZC369
9	^{n}Bu	Et	CH$_2$CO$_2$Me	CO$_2$Me	56	85ZC369
10	H	H	CH$_2$CO$_2$Me	CO$_2$Me	30	85ZC369
11	Ph	H	CH$_2$CO$_2$Me	CO$_2$Me	30	85ZC369
12	Ph	Et	CH$_2$CO$_2$Me	CO$_2$Et	35	85ZC369

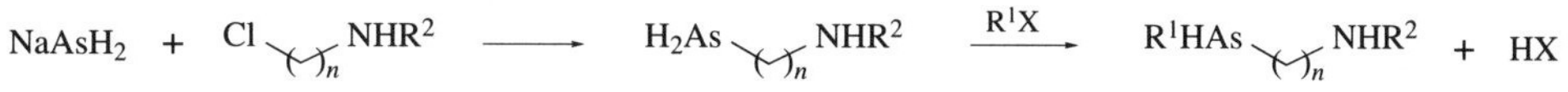

$$NaAsH_2 + Cl\frown_n NHR^2 \longrightarrow H_2As\frown_n NHR^2 \xrightarrow{R^1X} R^1HAs\frown_n NHR^2 + HX$$

(316) (317)

Scheme 74

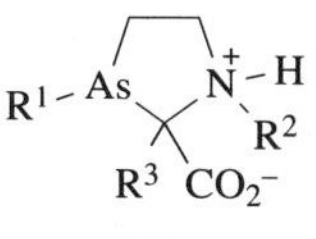

(318)

(**325**) result and the cyclocondensation reaction is observed to involve the pendant amino group, giving the *o*-aminophenylazarsolidine (**326**) (Equation (80)).

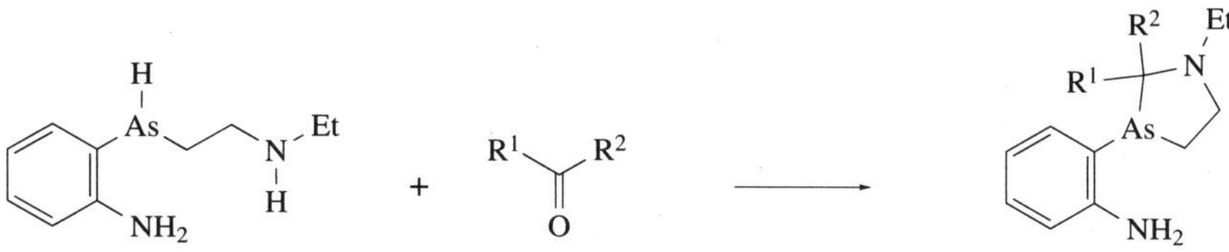

(**325**) + R^{1}COR^2 ⟶ (**326**) (80)

$R^1 = H$, $R^2 = Ph$, 68%
$R^1 = R^2 = Me$, 61%

1,3-Benzazarsolines which are unsubstituted at the arsenic may be alkylated on the arsenic via a

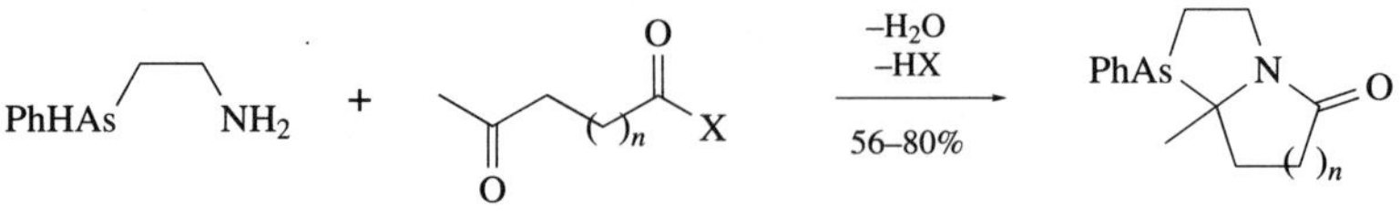

X = OH, O-alkyl (319)

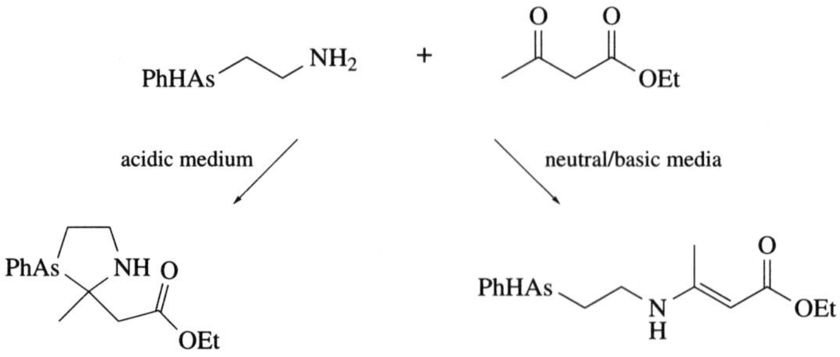

(320) (321)

Scheme 75

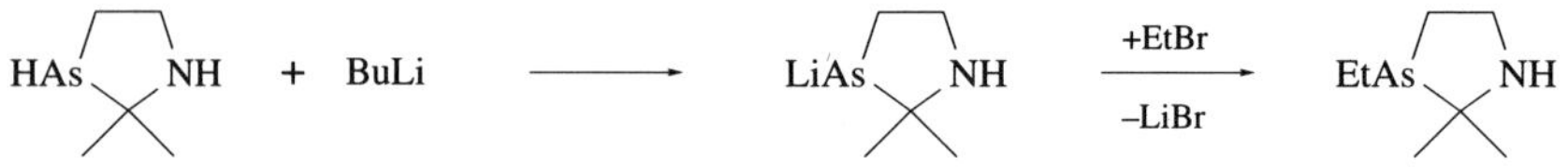

Scheme 76

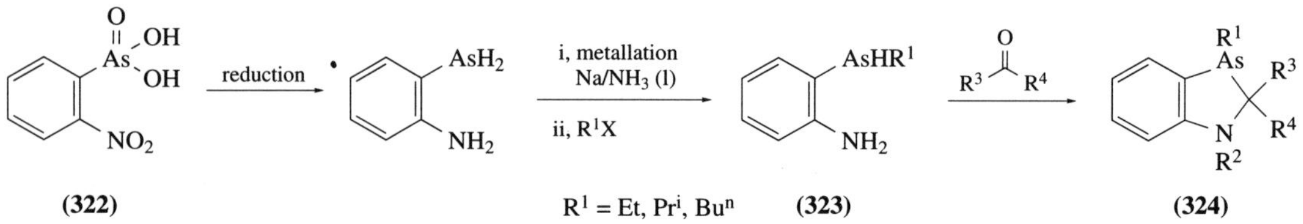

(322) R^1 = Et, Pr^i, Bu^n (323) (324)

Scheme 77

two-stage procedure: metallation (BuLi) followed by treatment with an alkyl halide. For example, ethyl bromide gave the 3-substituted product (**327**), whilst functionalisation at the 1-position was achieved with phenyl isocyanate affording the carbamoyl derivative (**328**) (Scheme 78) ⟨73JOM(60)95⟩.

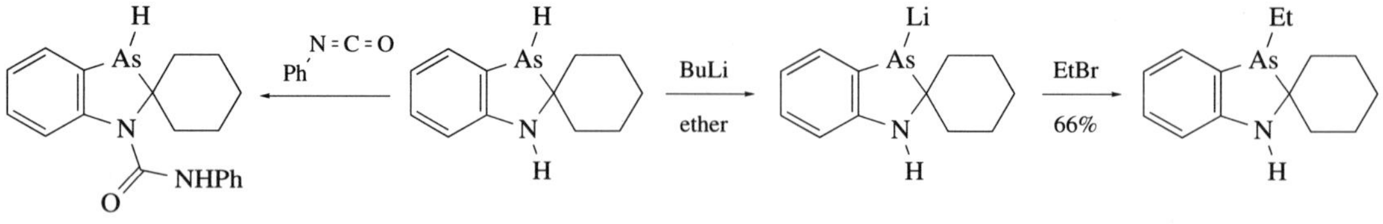

(328) (327)

Scheme 78

(ii) Tricoordinate Z functions by miscellaneous reactions

A second useful route to 1,3-azarsolidines (**331**) and their benzofused analogues (**332**) has been described. Substituted 2-aminoalkylarsines react with activated alkynes (**329**) furnishing the adducts (**330**), which readily cyclise to give 1,3-azarsolidines (**331**) under thermal or acid-catalysed conditions (Scheme 79) ⟨85ZC369⟩. 1,3-Benzazarsolines (**332**) are similarly obtained (Equation (81)). A variety of substituents R^1, R^2, and R^3 have been incorporated and yields are good (Table 12).

Table 11 Representative examples of 1,3-benzazarsolines (**324**) prepared from reaction of α-aminoalkylarsines (**323**) and carbonyl compounds (Scheme 77).

No.	R^1	R^2	R^3	R^4	Yield (%)	Ref.
1	Et	H	H	Et	70	73JOM(60)95
2	Et	H	H	Ph	68	73JOM(60)95
3	Et	H	Et	Et	74	73JOM(60)95
4[a]	Et	H	c	c	66	73JOM(60)95
5	Pr^i	H	H	Et	78	73JOM(60)95
6	Bu^n	H	H	Me	76	73JOM(60)95
7	H	H	Me	Me	72	73JOM(60)95
8	H	H	Et	Et	72	73JOM(60)95
9	H	H	c	c	87	73JOM(60)95
10	H	H	Me	CH_2CO_2Me	61	73JOM(60)95
11	Et	H	Me	$CH_2CH_2CO_2Et$	67	73JOM(60)95
12	Bu^n	H	Ph	CO_2Et	56	73JOM(60)95
13	Et	H	Me	CO_2Et	75	73JOM(60)95
14	Et	H	CH_2CO_2Me	CO_2Me	63	85ZC369
15	Me	Me	CH_2CO_2Me	CO_2Me	34	85ZC369
16[b]	Et	CONHPh	Et	H	73	73JOM(60)95
17[b]	Pr^i	CONHPh	Et	H	61	73JOM(60)95

[a] Compound (**323**) was prepared from its precursor with i, LiBu; ii, EtBr. [b] Compound (**324**) was prepared from (**323**) and PhNCO. [c] $R^3R^4 = (CH_2)_5$.

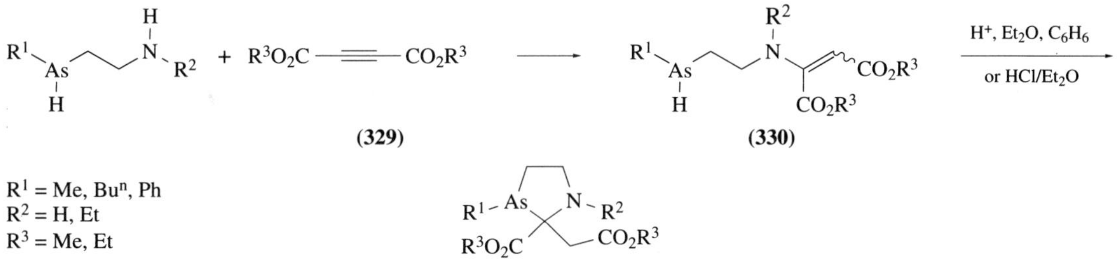

R^1 = Me, Bu^n, Ph
R^2 = H, Et
R^3 = Me, Et

Scheme 79

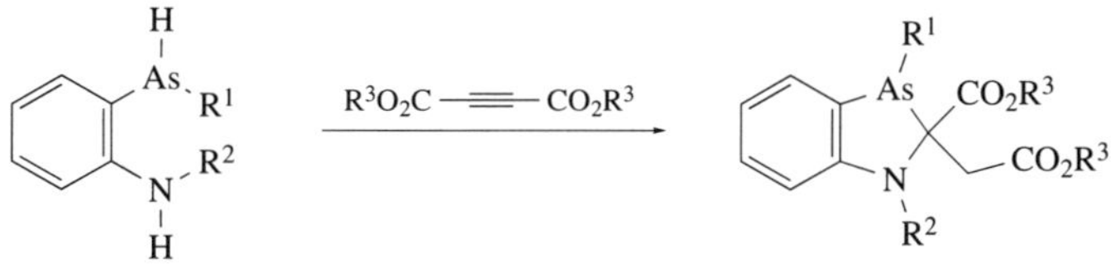

(81)

A most unusual 1,3-aminoalkylarsine compound—the arsine-capped, cage molecule (**335**)—was obtained by treatment of the trisaldimino salt (**334**) with arsine (Scheme 80). In this system the imine bonds are activated towards nucleophilic attack by their coordination to cobalt(III) ⟨89CC1648⟩.

(iii) Tricoordinate Z functions by Mannich-type reactions

The number of publications devoted to the preparation of acyclic 1-aminoalkylarsines is more limited. Such compounds are prepared by three main routes: direct Mannich ⟨78JOM(149)167, 81ZC403⟩ and indirect Mannich reactions ⟨78JOM(149)167⟩, and by treatment of an alkali arsene with an ammonium salt ⟨74JPR851, 78JOM(149)167⟩. Treatment of diphenylarsine with sodium affords the sodium arsene, which behaves as a nucleophilic arsenic species displacing the chlorine from

Table 12 Representative examples of 1,3-azarsolidines (**331**) and 1,3-benzazarsolines (**332**).

No.	R^1	R^2	R^3	Yield (%)	Ref.
(**331**)	Me	H	Me	55	85ZC369
	Bu^n	Et	Me	56	85ZC369
	H	H	Me	30	85ZC369
	Ph	H	Me	30	85ZC369
	Ph	Et	Et	35	85ZC369
(**332**)	Et	H	Me	63	85ZC369
	Me	Me	Me	34	85ZC369

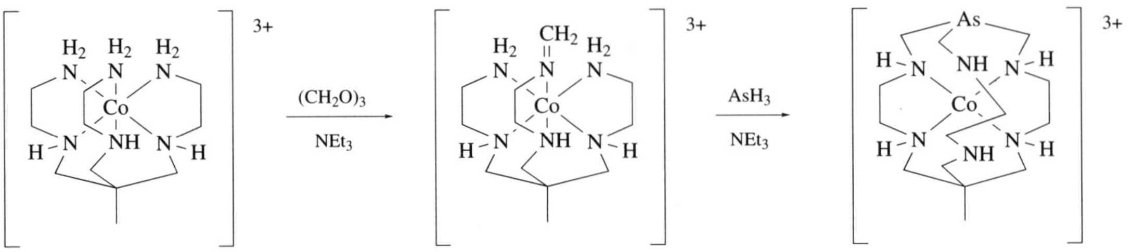

Scheme 80

N-chloromethylphthalimide (**336a**) (Equation (82)). An analogous preparation of (**337**) was realised when sodium diphenylarsenide generated *in situ* displaced the quaternary ammonium group in (**336b**); yields were poor to fair ⟨74JPR851⟩. The succinimido derivatives (**338**) may be prepared by the same chemistry.

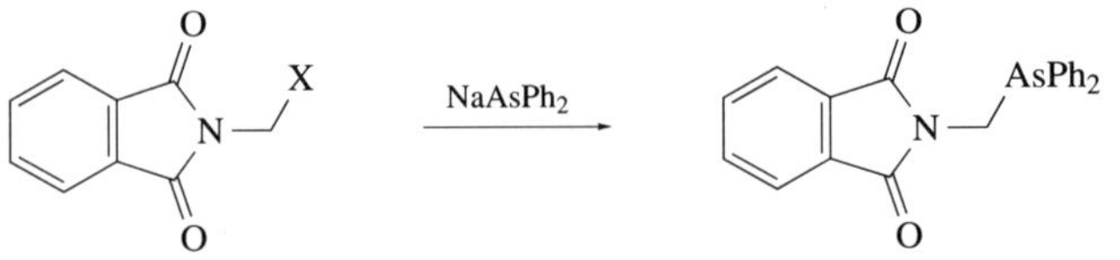

(**336**) a; X = Cl
b; X = NMe_3^+ I^-

The preparation of α-dialkylaminoalkylarsines has been realised by direct Mannich reactions. Equimolar amounts of a secondary amine, aqueous formaldehyde and a secondary arsine react in one pot to form the products (**339**) (Equation (83)). A variety of substituents may be incorporated and yields vary from moderate to good (Table 13). Higher aldehydes have also been used ⟨78JOM(149)167⟩. A Mannich-type reaction also gave the N,N-bis(diphenylarsinomethyl) amino acid ester (**340**) (Scheme 81) ⟨81ZC403⟩. The bisalkylated products are obtained in 62–74% yield. Interestingly, when the free amino acid is employed in this reaction only the monoalkylated product (**341**) is formed.

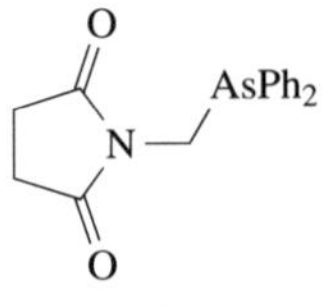

$$R^2{}_2NH + CH_2O + HAsR^1{}_2 \underset{-H_2O}{\overset{+H_2O}{\rightleftharpoons}} R^2{}_2N\diagdown AsR^1{}_2 \quad (83)$$

(**339**)

Table 13 Representative examples of acyclic aminomethylarsines (**339**) prepared by direct (Equation (83)) and indirect (Scheme 82) Mannich reactions.

No.	R^1	R^2	Yield (%)	Ref.
1[a]	Me	Ph	71	74JOM(81)187
2[a]	Et	Ph	73	74JOM(81)187
3[a]	pip[c]	Ph	86	74JOM(81)187
4[a]	morph[d]	Ph	58	74JOM(81)187
5[a]	Me	Cy	66	74JOM(81)187
6[a]	Et	Cy	78	74JOM(81)187
7[a]	pip	Cy	70	74JOM(81)187
8[a]	morph	Cy	74	74JOM(81)187
9[b]	Ph	Ph	65	74JOM(81)187
10[b]	Cy[e]	Cy	83	74JOM(81)187
11[b]	Bu^n	Bu^n	58	74JOM(81)187

[a] Equimolar amounts of a secondary amine and aqueous formaldehyde cooled in ice, and treated with a secondary arsine at RT for 3 h.   [b] Equimolar bis(dimethylamino)methane and secondary arsine, catalytic TsOH, 100–160 °C, 0.5–1 h.   [c] pip = $(CH_2)_5$.   [d] morph = $(CH_2CH_2)_2O$.   [e] Cy = cyclohexyl.

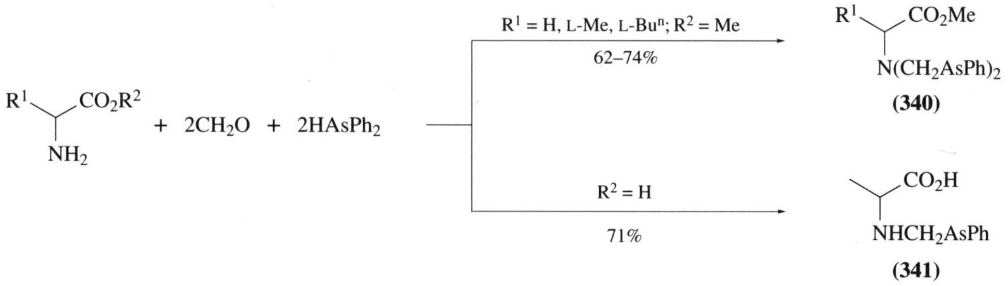

Scheme 81

Heating (100–160 °C, 10–60 min) equimolar quantities of a secondary arsine and bis(dimethylamino)methane in the presence of TsOH affords, by an indirect Mannich reaction, the α-dimethylaminomethylarsines (**342**) ⟨78JOM(149)167⟩. Secondary arsines, silylarsines and alkali arsenes react with *N*-hydroxymethylamines and quaternary ammonium aminals in acetonitrile at 0 °C to furnish the same aminomethylarsines (**342**) (Scheme 82). With sodium diphenylarsene the nature of the halide counterion has an effect on the yield of final product, the yield increasing in the order Cl < Br < I. The nature of the R groups on the secondary arsine HAsR$_2$ also influences the product yield (R = cyclohexyl, X = Cl, 93%; R = Ph, X = Cl, 77%; R = Ph, X = I, 92%).

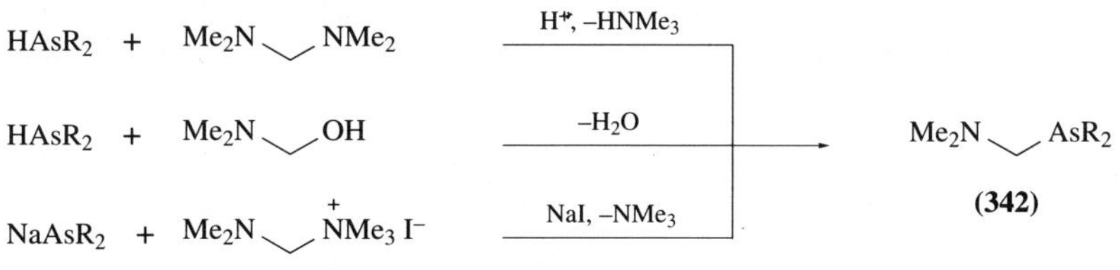

Scheme 82

4.10.2.1.3 Tetracoordinate Z functions

Only two examples of α-aminoalkyl arsenic compounds have been recorded where arsenic is present in the tetracoordinate state. Both result from reactions of spirocyclic-1,3-azarsinane. The derivative (**344**) is obtained simply by treating (**343**) with an equimolar amount of methyl iodide in ether (Equation (84)). Treatment of (**345**) with sulfur in refluxing benzene affords the tetracoordinate, pentavalent organoarsenic derivative (**346**) in moderate yield (Equation (85)) ⟨74JOM(81)187⟩.

4.10.2.1.4 Higher coordinate Z functions

No examples of higher coordinate Z functions have been identified.

4.10.2.2 Other Nitrogen Functions

4.10.2.2.1 Dicoordinate Z functions

No examples of dicoordinate Z functions have been located.

4.10.2.2.2 Tricoordinate Z functions

(i) Tricoordinate Z functions by cycloaddition reactions

A first example of cycloaddition across the As=C bond is the 1,3-dipolar reaction between the *C,N*-diarylnitrones (**348**) and the 2,5-diphenyl-1,2,3-diazarsole (**347**). The reactions take place in dichloromethane at room temperature over a 10-week period and the new arsenic-containing heterocycles (**349**) are obtained in modest yield (Equation (86)) ⟨79DOK(246)1130⟩. Similar chemistry is observed when the isomeric 1,2,4-diazarsole is allowed to react with *C,N*-diphenylnitrones. The [3.3.0]-bicyclic product (**351**) is obtained in 71% yield (Equation (87)). The dipolarophile (**350**) was prepared from the salt (**352**) with acetylhydrazine (**353**) (Equation (88)) ⟨86TL2957⟩. The preparation of the [3.1.0]-bicyclic arsines (**356**) has also been accomplished by a dipolar cycloaddition reaction ⟨90TL7607⟩. The 1*H*-1,2,4-diazarsoles (**354**) react with the sulfur ylide (**355**) to afford the products (**356**) in up to 75% isolated yield (DMF, 60 °C, 48 h) (Equation (89)). When R^1 = H (**354c**) the ylide may deprotonate the diazarsole and the resulting sulfonium salt may subsequently be the subject of nucleophilic attack by the so-formed heterocyclic anion; in such cases the product is the *N*-alkylated derivative (**357**) (Equation (90)). The ylides (**355b**) and (**355c**) are not sufficiently stable under the reaction conditions employed to form the bicyclic arsines (**356**).

(ii) Tricoordinate Z functions by miscellaneous reactions

The α-(N-arenesulfonamido)benzylarsines (**358**) can be formed by reaction of a secondary arsine with an N-arenesulfonylbenzaldimine under thermal or acid- (TsOH) or base- (NaOMe) catalysed conditions (Equation (91)). The yield of product is fair ⟨80ZC152⟩. The diphenylarsine (**359**) is prepared by an indirect Mannich-type reaction (Equation (92)). The appropriate quaternary ammonium iodide, diphenylarsine and sodium in a one-pot, exothermic reaction afford (**360**) in 28% yield.

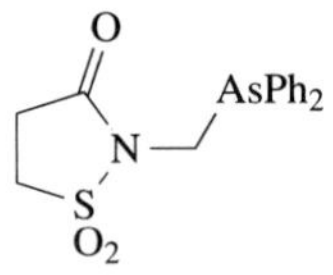

(360)

The solvolysis of SbCl$_3$ in nitromethane affords the complex (**361**) where antimony is present in the +5 oxidation state. The formation of this complex is rationalised on the basis of autoionization of the nitromethane ⟨67JIC995⟩.

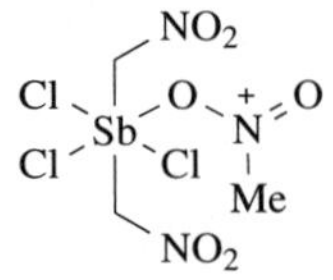

(361)

4.11
Functions Incorporating a Nitrogen and a Silicon, Germanium, Boron or Metal

JOHN STEELE
Pfizer Central Research, Sandwich, UK

and

MAX J. GOUGH
Technical Typesetters UK, Ashford, UK

4.11.1	FUNCTIONS CONTAINING A NITROGEN AND A METALLOID: $R^1_2C(NR^2_2)MR^3_n$ (M = Si, Ge, B), etc.	506
	4.11.1.1 Nitrogen and Silicon Functions	506
	4.11.1.1.1 α-Aminosilanes	506
	4.11.1.1.2 N-Acyl-α-aminosilanes	513
	4.11.1.1.3 C-Silylaziridines	514
	4.11.1.1.4 Trimethylsilylmethyl azide and related azides	516
	4.11.1.1.5 Trimethylsilylmethyl isocyanide, isocyanate and isothiocyanate	516
	4.11.1.1.6 α-Iminosilanes and related compounds	517
	4.11.1.1.7 N-(Silylmethyl) heterocycles	519
	4.11.1.2 Nitrogen and Germanium Functions	519
	4.11.1.3 Nitrogen and Boron Functions	522
	4.11.1.3.1 α-Aminoboranes and borohydrides	522
	4.11.1.3.2 α-Aminoboronic acids	525
4.11.2	FUNCTIONS CONTAINING A NITROGEN AND A METAL: $R^1_2C(NR^2_2)ML_n$ (M = Li, K, Mg, Sn, Zn), etc.	526
	4.11.2.1 α-Metallated Amine Functions	526
	4.11.2.1.1 Lithium, potassium and magnesium	527
	4.11.2.1.2 Tin and zinc	533
	4.11.2.2 α-Metallated Imine Functions	535
	4.11.2.2.1 Lithium	535
	4.11.2.2.2 Tin	536
	4.11.2.3 α-Metallated Isocyanides and Isothiocyanates	537
	4.11.2.3.1 Isocyanides	537
	4.11.2.3.2 Isothiocyanates	538
	4.11.2.4 Metallation of N-Methyl Heterocycles	539
	4.11.2.5 α-Metallated Nitroalkanes	540
4.11.3	ACKNOWLEDGEMENT	541

4.11.1 FUNCTIONS CONTAINING A NITROGEN AND A METALLOID: $R^1_2C(NR^2_2)MR^3_n$ (M = Si, Ge, B), etc.

4.11.1.1 Nitrogen and Silicon Functions

4.11.1.1.1 α-Aminosilanes

The chemistry of α-aminosilanes has not been reviewed since their first significant appearance in the 1950s, despite a growing role for these reactive entities in synthesis, most notably as ylide precursors. A range of general synthetic methods and several less general but nevertheless important routes are available. These are listed with details below. (Where common routes exist, *N*-acyl derivatives are also addressed in this section. Syntheses that apply only to *N*-acyl aminomethylsilanes are covered later, in Section 4.11.1.1.2.)

(i) Reaction of an amine with a halomethylsilane or a silyloxirane

The alkylation of ammonia or an amine by halomethylsilanes in an inert solvent is the most general and straightforward synthetic approach to α-aminosilanes, despite the potential for over-alkylation of the amine component. Speier and co-workers first prepared trimethylsilylmethylamine and several close analogues by the reaction of $ClCH_2$-TMS with excess ammonia ⟨51JA3867⟩. Anderson's useful review of the chemistry of $ClCH_2$-TMS also contains a small section on the synthesis of simple α-aminosilanes ⟨85S717⟩. A selection of relevant examples is included in Table 1. In general, no additional base is used and the amine component is used in excess to suppress multiple alkylation by-products. Chloro-, bromo- and iodomethylsilanes have all been used to good effect, although increasing substitution adjacent to the departing halide often necessitates a more reactive halogen (Table 1, entries 5 and 7). As part of their aminoalkylsilane stability studies, Duff and Brook ⟨77CJC2589⟩ have described detailed preparations for a large number of relatively simple derivatives (including Table 1, entry 8). Their studies indicate that aminomethylsilanes are markedly unstable with respect to the *N*-silyl structural isomer when exposed to catalytic quantities of a strong base such as BuLi (Equation (1)). This represents one expression of the Brook rearrangement.

Table 1 Monoalkylation of ammonia and amines with silylmethyl halides.

Entry	Amine	Halide	Yield (%)	Ref.
1	NH_3	TMS-CH$_2$-Cl	50	51JA3867
2	NH_3	$Ar_2(Me)Si$-CH$_2$-Cl	49–67	80BCJ789
3	$BnNH_2$	TMS-CH$_2$-Cl	72	89OS133
4	pyrrolidine-CO$_2$Me (N–H)	allyl-SiMe$_2$-CH$_2$-Br	86	89TL3041
5	morpholine (N–H)	TMS-CHPh-Br	90	85BCJ1991
6	tetrahydroisoquinoline-NH	TMS-CH$_2$-I	83	92JOC6711
7	NH_3	$PhMe_2Si$-CHMe$_2$-I (or CMe$_2$-I)	94	91JOM(415)39
8	$BnNH_2$	TMS-CPh$_2$-Br	68	77CJC2589

The silylmethylation process can usually be accomplished without the need for extensive protection of other functional groups. This is highlighted by the formation of the amino diol (**1**) (Equation (2)) ⟨91S996⟩. Diamines can also be selectively monofunctionalized; thus, ethylene diamine can be converted into a range of silylmethyl derivatives (**2**) in good yield, provided a large excess of amine is used ⟨84JOM(268)31⟩. Even under these conditions, small amounts of N,N'-disubstitution were observed.

Tertiary amines present a special case of this synthetic method in that the products are salts. Yields for the conversion are still reasonably good and, since no over-reaction is possible, equimolar quantities of reagents can be used. Literature reports suggest that the triflate reagent (TMS-CH$_2$OTf) may be superior to its halide counterparts because of the increased stability of the triflate salt products, and this is particularly so when the quaternary products are prone to Hofmann-type elimination or to migratory rearrangements. The quaternary ammonium triflate (**3**), a precursor to a nitrogen ylide, is produced in quantitative yield by the use of TMS-CH$_2$OTf (Equation (3)) ⟨90JA1999⟩. The reaction of diethylamine with ClSi(Me$_2$)CH$_2$Cl affords the unusual aminosilane (**4**) which has chemical properties suited to applications in photoresist construction but, unlike the foregoing examples, forcing conditions are required to drive this reaction to completion (Equation (4)) ⟨91EUP488681⟩.

The reaction of aromatic amines with chloromethylsilanes is less well documented although the vinyl silane (**5**) reacts with aniline via an alkylation–mercuration sequence to afford the azasiloline (**6**) after reductive demercuration (Equation (5)). The allyl analogue shows a marked preference for cyclization also to a five-membered product (Equation (6)) ⟨87JOM(326)159⟩.

$$\text{Me}_2\text{Si(allyl)(CH}_2\text{Cl)} \xrightarrow[\text{ii, NaBH}_4\text{, NaOH}]{\text{i, PhNH}_2\text{, Hg(OAc)}_2\text{, THF}} \text{1,1-dimethyl-3-methyl-2-phenyl-1-sila-2-azacyclopentane} \quad 29\% \tag{6}$$

Silyloxiranes are sufficiently electrophilic to react with unhindered amines. The *cis*-oxirane (**7**) is regioselectively ring opened by piperidine to generate an α-aminosilane (Equation (7)) ⟨85TL139⟩. Interestingly, the corresponding *trans*-isomer is completely unreactive towards amines, which obviates the need to use isomerically pure starting materials in the reaction. On exposure to KH, the carbinol products readily afford enamines via a Peterson elimination.

$$\underset{(\mathbf{7})}{\text{C}_5\text{H}_{11}\text{-oxirane-TMS}} \xrightarrow[63\%]{\text{piperidine, 85 °C}} \text{C}_5\text{H}_{11}\text{CH(OH)CH(TMS)(N-piperidyl)} \tag{7}$$

(ii) Further functionalization of existing α-aminosilanes

Primary, secondary and tertiary α-aminosilanes behave as typical amines and are readily alkylated by suitably reactive halides. In general, the reactions are accomplished in dipolar, aprotic solvents (MeCN, DMF) without additional base, thus ensuring that only monoalkylation occurs. Products are usually isolated as HX salts although the free bases are not markedly unstable. Table 2 contains a survey of α-aminosilanes prepared this way. Tertiary α-aminosilanes naturally generate quaternary alkylated products (Table 2, entries 6 and 7) and these have found widespread application as 1,3-dipole precursors in many cycloaddition processes.

Table 2 Mono-*N*-alkylation of α-aminosilanes.

Entry	Aminosilane	Halide	Yield (%)	Ref.
1	TMS$\sim$NH$_2$	Cl$\sim$CO$_2$Et	<30	74HCA1042
2	TMS$\sim$NH$_2$	Cl$\sim$CONH$_2$	78	74HCA1042
3	TMS$\sim$NHBn	3,5-(MeO)$_2$C$_6$H$_3$COCH$_2$Br	64	76CR(C)473
4	TMS$\sim$NHBn	allyl-Br	85	88S988
5	TMS-CH$_2$-N(H)-CH$_2$-CH=CH-Ph	BnBr	74	88S988
6	TMS$\sim$NMe$_2$	(R)$_n$-C$_6$H$_4$-CH$_2$Br		85CC1684
7	TMS$\sim$N(Me)CH(Me)Ph	MeI	90	91S117

Sato and co-workers have extensively investigated the intra- and intermolecular reactions of aminosilanes with benzynes, particularly as a means of creating new silicon-containing heterocycles. Lithiation of the *m*-chlorophenylsilane (**8**) presumably generates a benzyne intermediate which cyclizes to afford the azasiloline (**9**) (Scheme 1) ⟨76JOM(118)1⟩. Under identical conditions, the homologous benzylsilane also cyclizes to give a six-membered azasiline (Equation (8)) ⟨78JOM(153)193⟩. Both reactions are independent of the nitrogen substituent and proceed in synthetically useful yields. Intermolecular reactions between benzyne and tertiary α-aminosilanes proceed via *N*-arylation and subsequent Stevens rearrangement resulting in the transfer of an *N*-alkyl substituent to the adjacent carbon atom (Equation (9)) ⟨74JOM(82)21, 75CC640⟩. *N*-Alkyl groups bulkier than Me also migrate but competing Hofmann elimination is also observed in these cases.

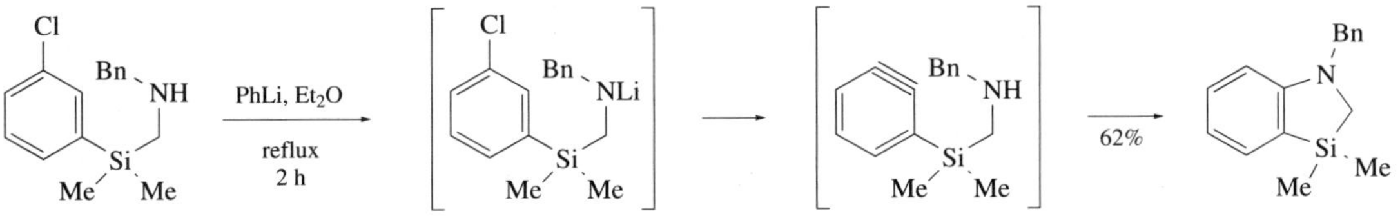

Scheme 1

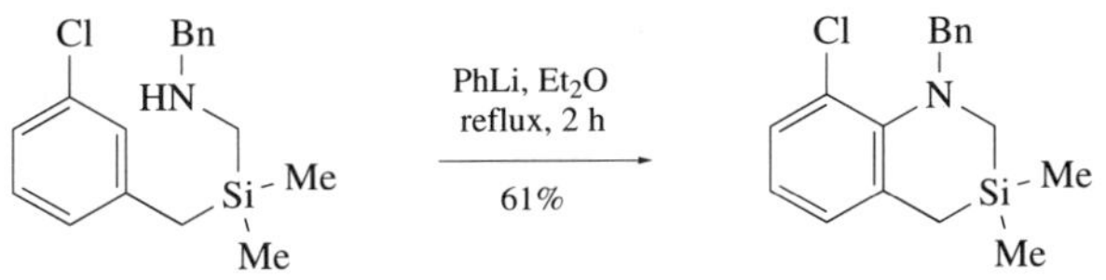

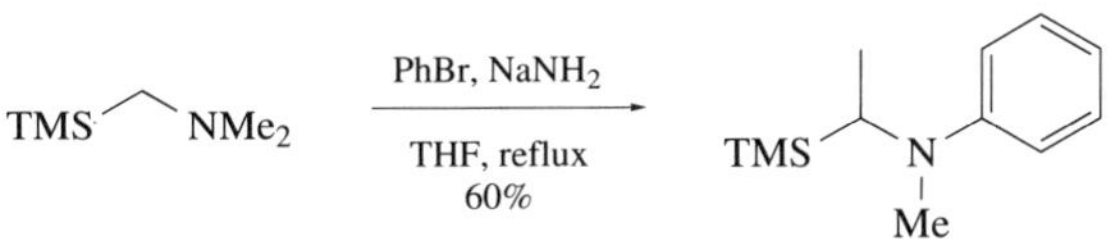

Further functionality can also be introduced by the reactions of aminosilanes with aldehydes following a number of traditional methods. High yields of *N*-alkoxymethyl derivatives such as (**10**) can be obtained by a Mannich-type reaction with formaldehyde in the appropriate alcohol (Equation (10)) ⟨91JCS(P1)1091⟩ and a detailed experimental procedure for the preparation of the methoxymethylaminosilane (**11**) has been reported by Padwa and Dent (Equation (11)) ⟨89OS(67)133⟩. Most primary amines afford triazines on reaction with aqueous formaldehyde and Achiwa and co-workers have described the formation of tris(TMS-methyl)triazine under these conditions (Equation (12)) ⟨85CPB4596⟩. Padwa *et al.* have also used a formal Strecker reaction to synthesize the cyanomethyl compound (**12**) using formaldehyde and KCN (Equation (13)). The authors claim this route to be suitable for other aldehydes, although no further examples are given ⟨85T3529⟩.

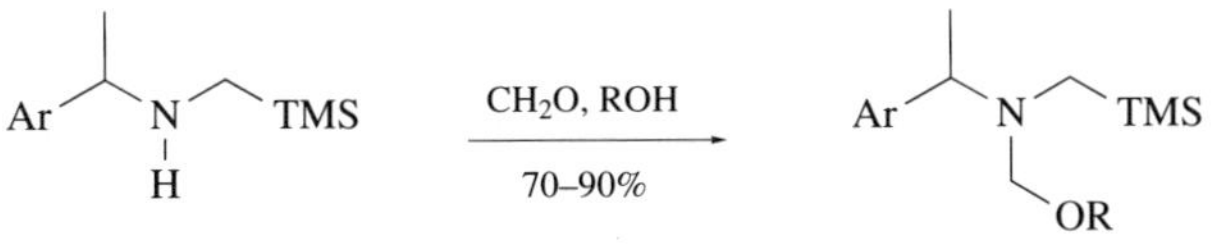

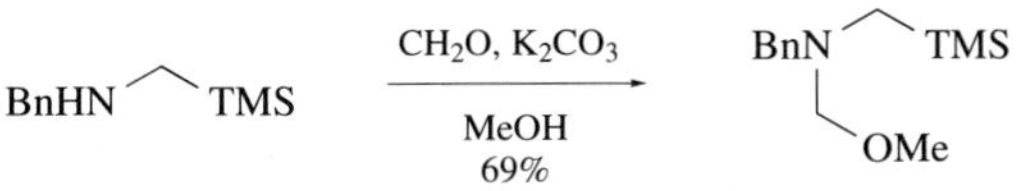

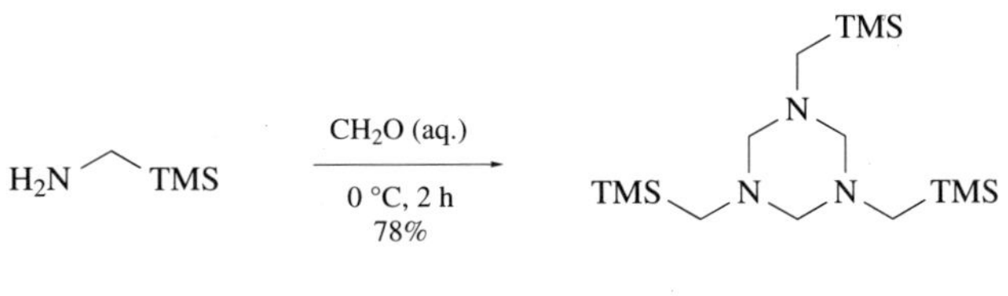

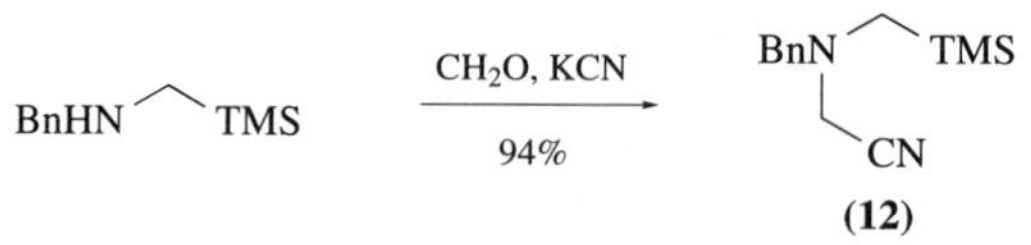

Michael reactions of α-aminosilanes with suitable acceptors are very scarce, possibly due to the propensity of the silanes to generate 1,3-dipoles under the reaction conditions. The only clear report of a Michael addition involves reaction with dimethyl acetylenedicarboxylate (DMAD) (Equation (14)) under conditions which the authors hoped would induce dipolar cycloadditions ⟨85JOC4006⟩.

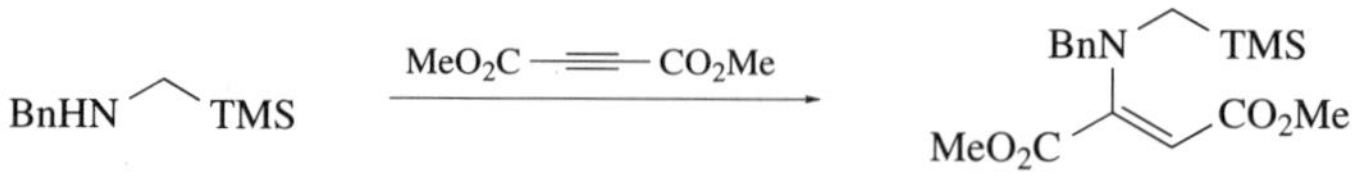

There are no general methods to achieve further substitution directly at the carbon atom bearing both silyl and amino functions. Mariano and co-workers have achieved the formal Michael addition of aminosilanes to cyclohexenones using a photo-induced single electron transfer (SET) process (Equation (15)). Evidence for radical cation intermediates on the reaction coordinate comes, in part, from the isolation of protodesilylated by-products ⟨87JA4421⟩.

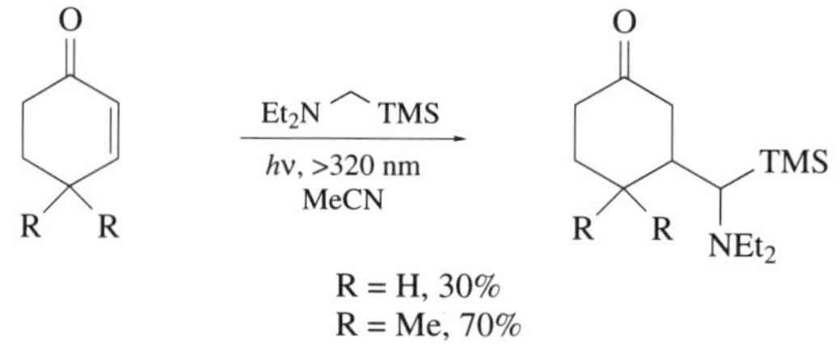

(iii) Hydrosilylation of enamines

Snyder has shown that trichlorosilane reacts regiospecifically with enamines, including relatively hindered examples, to afford hydrosilylated products such as (13) (Scheme 2). This very general transformation probably proceeds via initial *N*-silylation followed by hydride transfer and subsequent migration of silicon ⟨86JOM(301)137⟩. The products are readily desilylated by exposure to amine hydrochlorides in MeCN, constituting a useful means of formally hydrogenating enamines ⟨87JOM(320)163⟩. A related outcome has been achieved by rhodium-induced hydrosilylation of the enamide (14) in Et₃SiH, generating the silane (15) in 92% yield (Equation (16)) ⟨91JOM(408)297⟩.

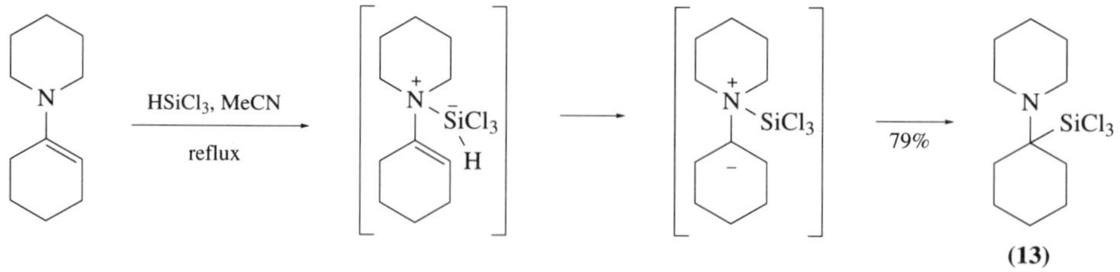

Scheme 2

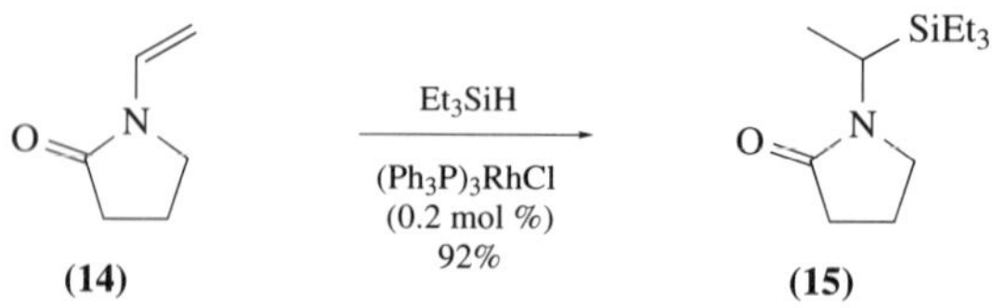

(iv) Reaction of halomethylamines with metallosilanes

The scope of this synthetic approach is understandably limited by the generally poor availability of α-chloroalkylamines and related compounds. Nevertheless, where precursors are available, the yields are good. To take an example from the only well-documented paper, tris(chloromethyl)amine (**16**) is exhaustively silylated to give the amine (**17**) in 74% yield (Equation (17)) ⟨73HCA1117⟩.

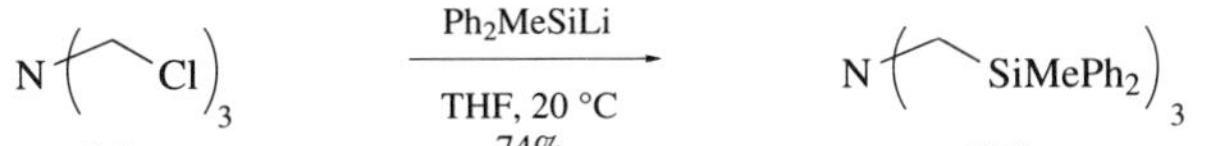

$$N(CH_2Cl)_3 \xrightarrow[\text{THF, 20 °C}]{\text{Ph}_2\text{MeSiLi}} N(CH_2SiMePh_2)_3 \quad (17)$$
(**16**) 74% (**17**)

(v) Silylation of aminomethyllithium reagents

The electrophilic silylation of organometallics is an enormously well-documented area, but the limited availability of α-amino metal derivatives has restricted this approach. In general, only organometallics where the formal carbanion or the metal itself is stabilized by further interactions are available for silylation (see section 4.11.2 for the preparation of metallated amines).

Lithiation of *N*-allylpyrrolidine and transmetallation with magnesium bromide gives the α-silylamine (**18**) after quenching with TMS-Cl. The transmetallation is essential to control the regiochemistry of the product; quenching the lithiated species affords only the γ-silylamine (**19**) (Scheme 3) ⟨83JOM(259)283⟩. Seebach and Enders showed that lithiation of *N*-nitrosoamines is a straightforward process, facilitating reaction with a range of electrophiles including TMS-Cl (Equation (18)) ⟨74JMC1225⟩. Meyers and Jagdmann have demonstrated that the lithiation and silylation of amidines is also an efficient process. Thus, the *N*-TMS-methyl amidine (**20**), a precursor to a range of enamidines, was isolated in 89% yield (Equation (19)) ⟨82JA877⟩.

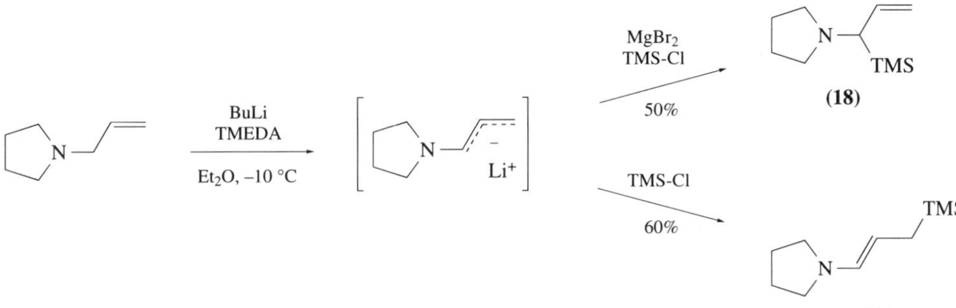

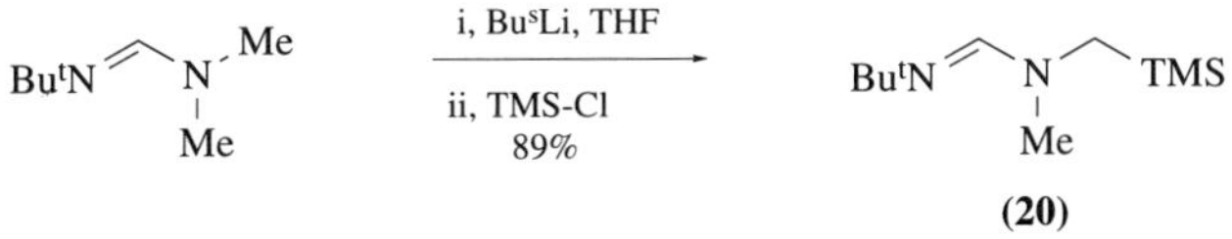

Scheme 3

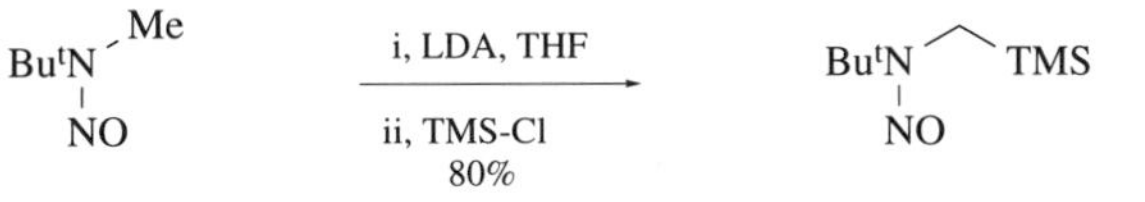

The use of directed metallation groups (DMGs) also facilitates the preparation of silanes via aminomethyllithium intermediates. BOC-piperidine is readily metallated by Bu^{s}Li/TMEDA to afford, after quenching, the 2-silylpiperidine (**21**) (Equation (20)) ⟨89TL1197⟩. In an extension of this methodology, Snieckus has described the utility of the bis(silyl)methylamine (**22**), prepared from *N*,*N*-dimethylbenzamide (Equation (21)) ⟨89TL5841⟩.

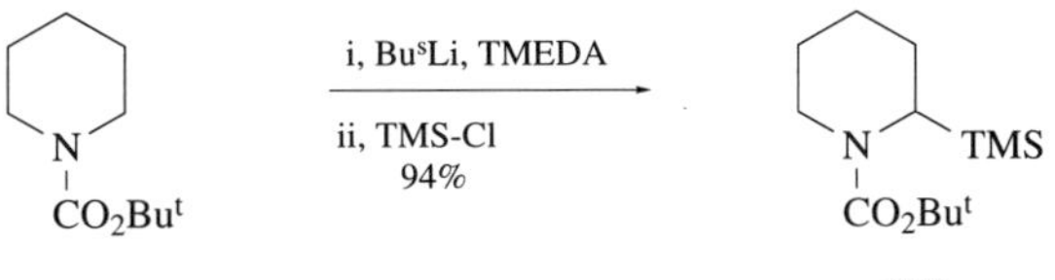

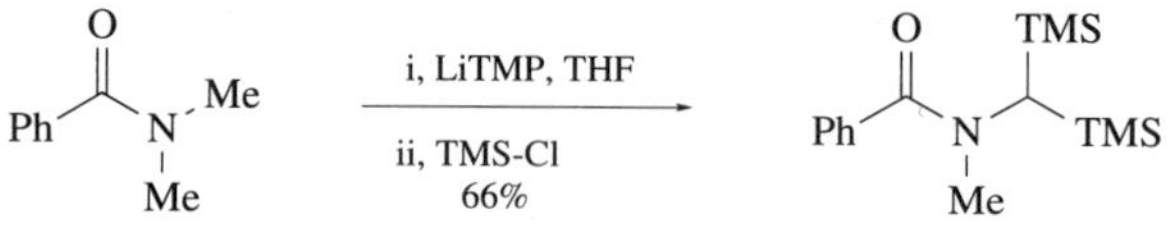

$$\text{Ph}\underset{\underset{\text{Me}}{|}}{\text{C(O)N}}\text{-Me} \xrightarrow[\text{ii, TMS-Cl}]{\text{i, LiTMP, THF}} \underset{(22)}{\text{Ph}\underset{\underset{\text{Me}}{|}}{\text{C(O)N}}\text{-CH(TMS)}_2} \quad 66\% \tag{21}$$

Finally in this section, are three methods that exploit α-amino anions but are less generally applicable. The silylamine (23) is the only isolated product after hydrolytic workup of the reaction between *p*-chlorobenzonitrile, magnesium and TMS-Cl (Scheme 4). The *N*-silylamine is an isolable intermediate ⟨79LA842⟩. In a useful extension of previous work by Padwa ⟨87JOC2427⟩, Sato and co-workers have described the quaternization of tertiary aminoacetonitriles and deprotonation of the resulting *N*-silylammonium salts (Scheme 5). The resulting ylides rearrange by Si 1,2-migration to afford the nitriles (24) ⟨90S36⟩. Further elaboration to the substituted silylmethylamines (25) has also been demonstrated. Ylides derived from (2-TMS-ethyl)ammonium salts such as (26) will also undergo Si migration, this time by a formal [2,3]-sigmatropic process, liberating the α-aminosilane (27) with concomitant extrusion of ethene (Scheme 6) ⟨76JOM(113)115⟩.

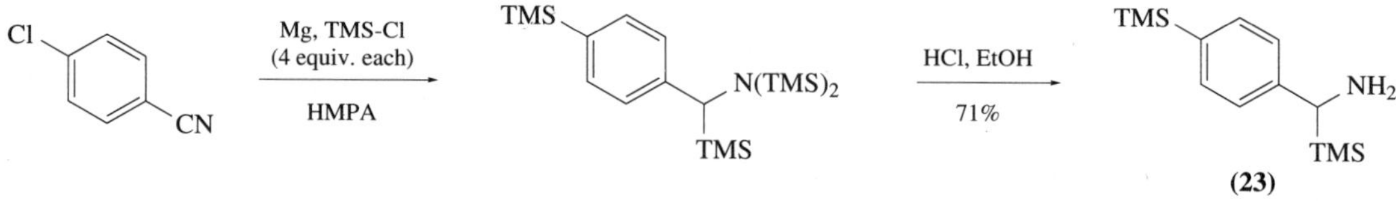

Scheme 4

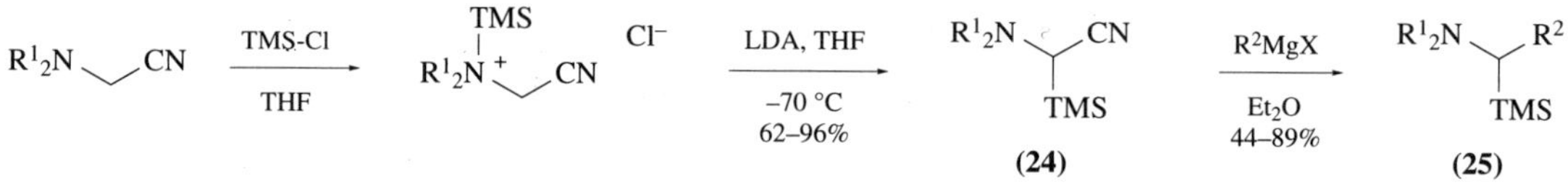

Scheme 5

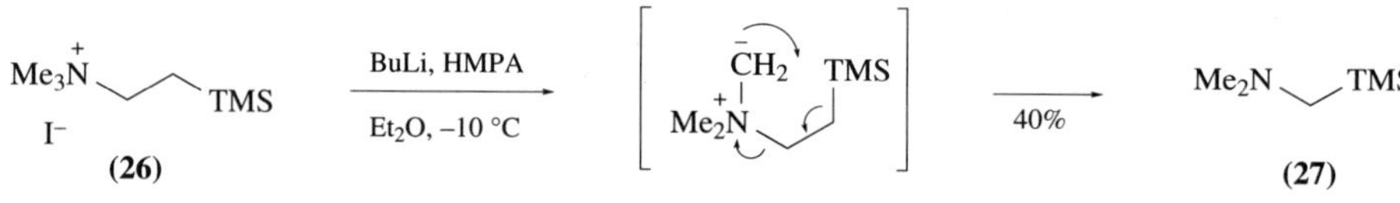

Scheme 6

(vi) Reduction of silylmethyl azides

Although in principle any of the methods available for azide reduction may be applied to silylmethyl azides, only two reports of such reductions have appeared. Both describe the conversion of trimethylsilylmethyl azide into trimethylsilylmethylamine (29) via an intermediate iminophosphorane (28) (Scheme 7). Both groups emphasize the mildness of this protocol and the utility of (28) for direct preparation of imines (see Section 4.11.1.1.6) ⟨86BCJ2537, 88SC1975⟩.

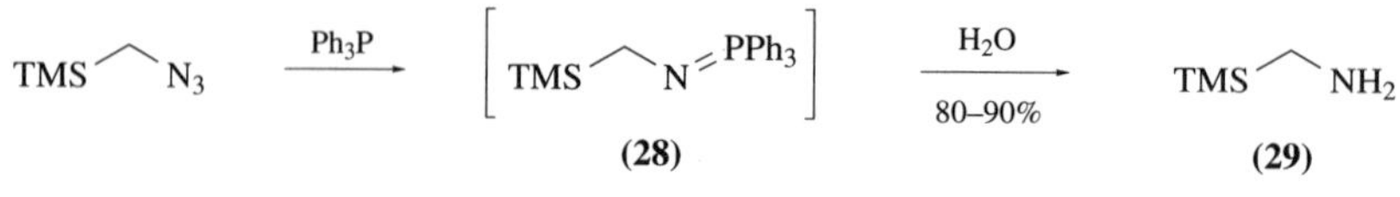

Scheme 7

4.11.1.1.2 N-Acyl-α-aminosilanes

Apart from the methods addressed in the previous section, several additional approaches to the title compounds are available, of which acylation of aminosilanes is the most straightforward. Table 3 contains a cross-section of examples covering a broad range of circumstances. Most conditions using acyl halides specify pyridine or Et_3N in a polar solvent although neutral (entry 6) and acidic (entry 7) conditions are also well represented. Entry 6 is particularly noteworthy as Tacke and co-workers proceeded to resolve the product of this reaction (30) by enantioselective enzymatic hydrolysis using pencillin G acylase (Scheme 8). The enantiomeric excess of the (R)-(30) so produced was 92% ⟨91JOM(415)39⟩.

Table 3 N-Acylation of α-aminosilanes.

Entry	Aminosilane	Acylating agent	Yield (%)	Ref.
1	TMS–CH(–)–NH₂	ArCOCl		89TL5837
2	TMS–CH(–)–NHMe	PhCOCl	90	84JOC3314
3	TMS–CH(–)–NHBn	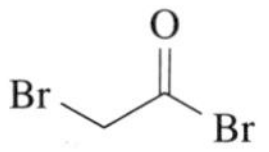 BrCH₂COBr	66	92JOC6037
4	TMS–CH(–)–NHMe	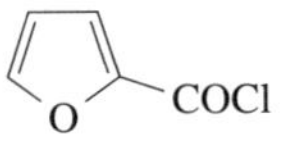 2-furyl-COCl	92	92JOC5419
5	TMS–CH(–)–NHMe	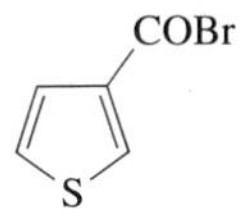 3-thienyl-COBr	96	92JOC5419
6	PhMe₂Si–C(Me)(H)–NH₂	PhCH₂CO₂H and Steglich's reagent	87	91JOM(415)39
7	TMS–CH(–)–NH₂	H₂NCONH₂/HCl		68JA1080

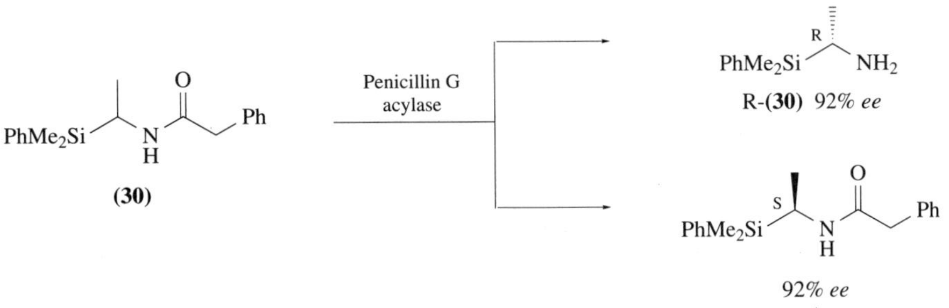

Scheme 8

Primary and secondary amides are readily alkylated by silylmethyl halides, usually via an alkali metal salt. Sommer and Rockett first demonstrated the conversion of phthalimide into the silane (31) (Equation (22)) ⟨51JA5130⟩, although commercially available potassium phthalimide permits lower reaction temperatures for examples such as (32) (Equation (23)) ⟨85BCJ1991⟩. N-Silyl and bis(silyl)amides such as N,O-bis(trimethylsilyl)acetamide (bsa) will also function as the nucleophilic component although the only well-documented report describes preparative details for the unusual (chloro)dimethylsilanes (33) (Equation (24)). Under the essentially neutral reaction conditions, the chlorosilane function is remarkably stable ⟨78JOM(153)369⟩. N-Silyl amides will also cleave silyloxiranes by regioselective attack adjacent to silicon ⟨84JOM(264)87⟩. Birkofer and Kopp have demonstrated that silyl amides react with the oxiranes (34) to afford only the carbinols (35). No reaction occurs with 1,1-disubstituted oxiranes (Equation (25)).

Palomo et al. have shown that acylation of the silylmethyl imine (**36**) proceeds without appreciable protodesilylation at low temperature to give the enamide (**37**) in 50% yield, although no further examples are quoted (Equation (26)) ⟨91CC524⟩. Finally, the reactive isocyanate chlorosulfonyl isocyanate (csi) and activated vinylsilanes generated 4-silylazetidinones (Equation (27)). For example the allene (**38**) was converted into the β-lactam (**39**) after hydrolysis of the intermediate N-sulfonyl species ⟨85TL5001⟩.

4.11.1.1.3 C-Silylaziridines

Syntheses of silylaziridines by intramolecular nucleophilic displacement of halide are uncommon. Lukevics et al. have prepared the N-ethoxycarbonylaziridine (**40**) by this method, using either phase-transfer catalysis or ultrasound irradiation to accelerate the reaction rate (Equation (28)) ⟨84JOM(268)C29⟩. In a related method, vinyl silanes have been shown to react with in situ-generated bromine azide to give a regioisomeric mixture of azides (**41**) (Scheme 9). Reduction of the crude mixture with LAH gave the silylaziridines (**42**) in modest yield ⟨78JOM(156)C25⟩.

Silylaziridines are more commonly assembled by the reaction of suitable alkenes with either nitrenes or azides. The *N*-ethoxycarbonylaziridine (**40**) can also be accessed by generation of *N*-ethoxycarbonylnitrene from (**43**) under phase-transfer catalysis and addition to vinyltrimethylsilane (Equation (29)). Trace amounts of the *N*-deacylated aziridine are also produced ⟨86JOM(316)249⟩. Atkinson and Kelly have used the nitrene formed by decomposition of the acetoxy quinazolone (**44**) to generate the isomeric aziridines (**45**) and (**46**) (Scheme 10) ⟨89TL2703⟩. The well-established dipolar cycloaddition of azides and alkenes is highly effective for silylaziridine synthesis. Vinylsilanes will react with alkyl and aryl azides, generating intermediate triazolines (**47**) which readily extrude nitrogen to leave the silylaziridines (**48**) (Scheme 11) ⟨68CB743, 91JCS(P1)2789⟩. Tsuge *et al.* have used TMS-methyl azide (see Section 4.11.1.1.4) as a nonexplosive cycloaddition reagent to generate triazolines such as (**49**) which readily give *N*-silylmethyl aziridines (**50**) on warming (Scheme 12). The syntheses of several related aziridines are also described ⟨84H(22)1955⟩.

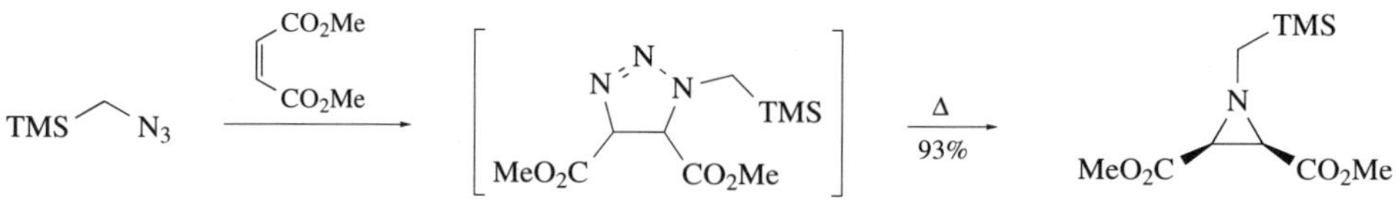

Scheme 12

4.11.1.1.4 Trimethylsilylmethyl azide and related azides

Trimethylsilylmethyl azide (TMS-MA, b.p. 58–61 °C/80 mmHg) is a relatively stable alternative synthon for reactions which formally require the shock-sensitive methyl azide. Two groups almost simultaneously reported the synthesis (Equation (30)) and reactivity of TMS-MA although the preparation described by Tsuge et al. ⟨83CL1131⟩ requires hexamethylphosphoramide (HMPA) as a solvent. Nishiyama and Tanaka used DMF as a solvent and obtained an equally good yield ⟨83CC1322⟩. TMS-MA is reportedly stable in refluxing toluene and is stable at refrigerator temperatures for at least six months. The chemistry of TMS-MA is discussed in part in Sections 4.11.1.1.3 and 4.11.1.1.6.

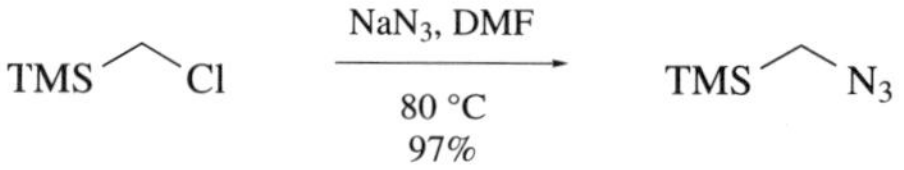

A general synthesis of β-hydroxy-α-silyl azides from silyl oxiranes has been described by Tomoda et al. (Scheme 13). Regioselective ring opening of the silyl oxiranes is achieved either by TMS azide or sodium azide under Lewis acid catalysis, generating the silyl azides (51) ⟨86CL1193⟩. This work has been extended by Chakraborty and Reddy who have demonstrated that the normal preference for C-3 ring opening of 2,3-epoxyalcohols by nucleophiles is reversed when a silyl group is attached to C-2 ⟨90TL1335, 91TL679⟩. Taking a specific example, the oxirane (52) affords only the silyl azide (53) on reaction with buffered sodium azide (Equation (31)).

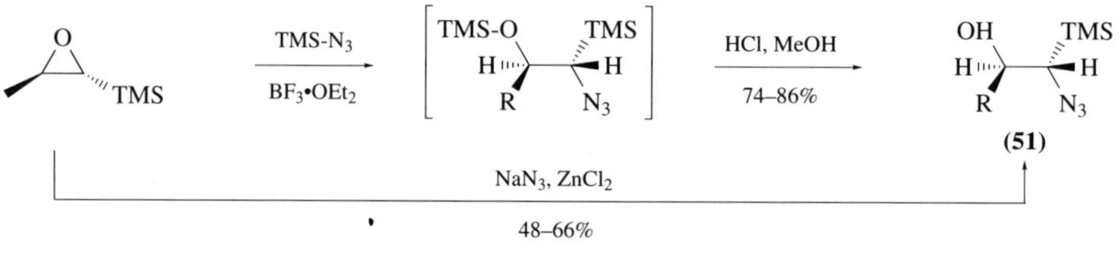

Scheme 13

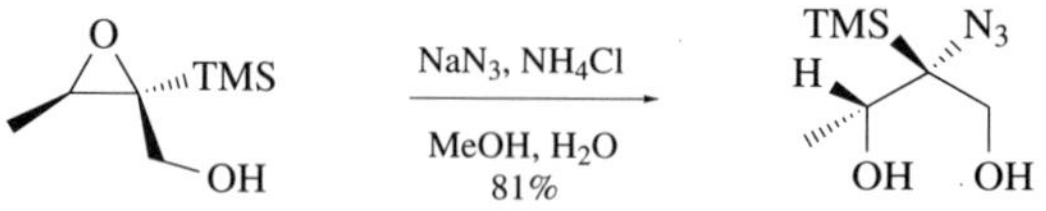

4.11.1.1.5 Trimethylsilylmethyl isocyanide, isocyanate and isothiocyanate

The title compounds have received considerable attention as small molecule synthons for a range of imine derivatives and heterocycles. Several groups have reported the synthesis of trimethylsilylmethyl isocyanide by lithiation of methyl isocyanide, since the first report ⟨70JOM(25)385⟩. The most detailed and useful practical account is that of Smith and Livinghouse (Equation (32)) who have modified the reaction conditions to achieve reproducibly acceptable yields ⟨84SC639⟩. Subsequently, an alternative preparation has emerged which avoids the use of the toxic, volatile and odorous MeNC (Scheme 14). Silylmethylation of formamide via its sodium salt gives the formamido silane (**54**) which is dehydrated under Ugi conditions to liberate trimethylsilylmethyl isocyanide ⟨86SC865⟩.

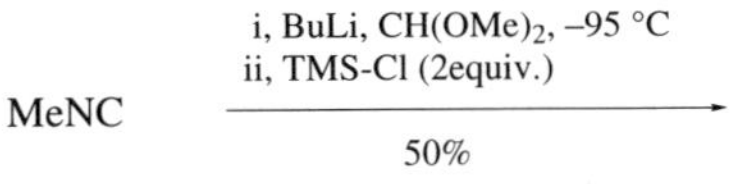

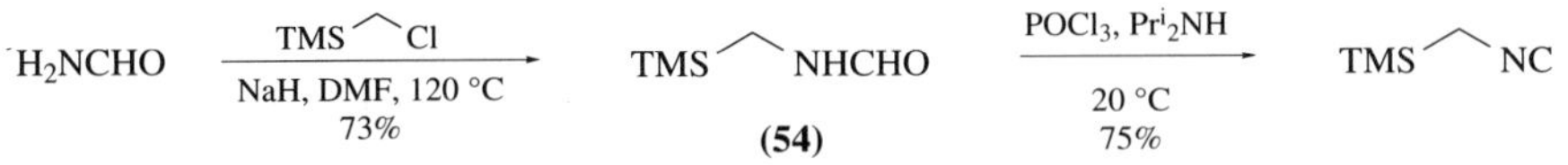

Scheme 14

Ohshiro and co-workers have prepared trimethylsilylmethyl isothiocyanate from the foregoing isocyanide by heating with elemental sulfur in an inert solvent (Equation (33)) ⟨82BCJ1163⟩. In a similar manner, the same group has prepared the analogous bis- and tris(trimethylsilyl)methyl isocyanates (**55**) and (**56**). An improved synthesis which uses CS_2 as the sulfur source is reported to give reproducibly better yields and is readily adapted to afford trimethylsilylmethyl isocyanate by use of CO_2 (Scheme 15) ⟨84JOC2688, 85H(23)2489⟩.

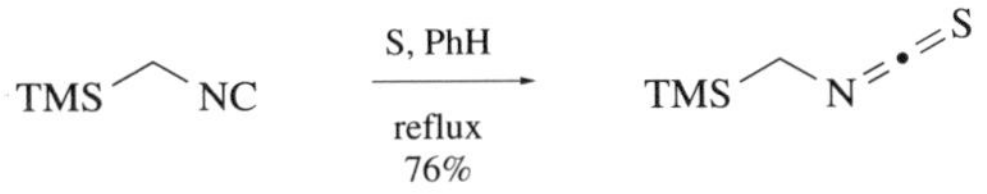

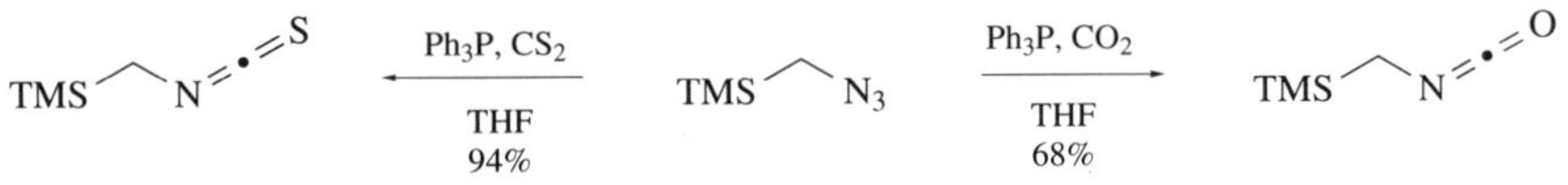

Scheme 15

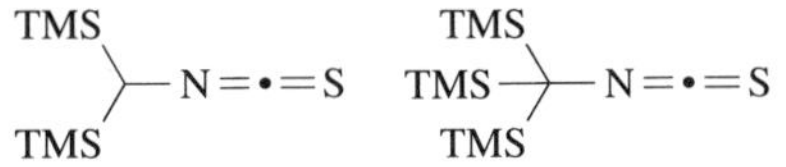

4.11.1.1.6 α-Iminosilanes and related compounds

Vedejs and Martinez first disclosed the potential of N-(trimethylsilylmethyl)iminium salts such as (**57**) for the generation of azomethine ylides and subsequently a range of nitrogen heterocycles (Scheme 16) ⟨79JA6452, 80JA7993⟩. Since then, this desilylation method has become the most frequently used means of producing azomethine ylides, prompting considerable interest in efficient routes to the iminosilane precursors. Vedejs and West have also prepared thioimidates such as (**58**) from simple amides and demonstrated the high reactivity of the ylides generated by their desilylation (Scheme 17) ⟨83JOC4773⟩. Other groups have subsequently prepared iminium salts by silylmethylation of imines. Synthetically useful examples include the α,β-unsaturated systems (**59**) and (**60**) used by Padwa and co-workers ⟨83TL4303⟩ and a range of substituted aryl derivatives (**61**) ⟨86JMC2241⟩.

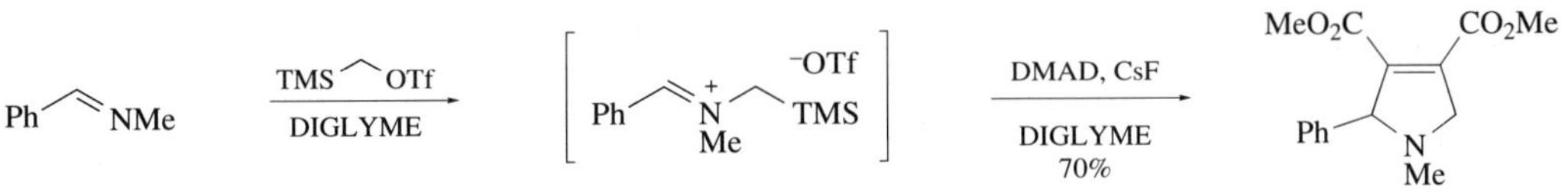

DMAD = dimethyl acetylenedicarboxylate

Scheme 16

Silylmethyl imines are also readily prepared from α-aminosilanes and carbonyl compounds, particularly trimethylsilylmethylamine, generated by hydrolysis of the iminophosphorane as discussed in Section 4.11.1.1.1(vi). *In situ* reaction of the carbonyl component with the iminophosphorane is successful but removal of the co-produced Ph$_3$PO is frequently difficult. Unreactive ketones including benzophenone require catalysis by Et$_3$Al to form imines (Scheme 18) ⟨86BCJ2537, 88SC1975⟩. Mariano and co-workers have synthesized the aza-1,3-diene precursors (**63**) by a similar method from the amines (**62**) (Equation (34)) ⟨85TL47⟩.

As discussed briefly in Section 4.11.1.1.1(v), the silylation of metallated imines represents an alternative approach to more complex iminosilanes. The amines (**62**), used earlier, are readily generated by metallation of a propargyl imine and quenching with TMS-Cl (Equation (35)) ⟨88T7013⟩. This approach is most successful when the intermediate organometallic derivative is further stabilized.

Livinghouse and Smith have prepared a range of amidines and thioimidates as ylide precursors by copper-catalysed additions to trimethylsilylmethyl isocyanide (Scheme 19) ⟨83CC210, 85T3559⟩. Although related isocyanide insertion reactions were previously known, the relatively mild conditions used here ensure that no protodesilylation occurs. The sequential addition of Grignard reagents and alkylating or acylating agents to trimethylsilylmethyl isothiocyanate has been used by Tsuge *et al.* to assemble the thioimidates (**64**) (Equation (36)). Although practical details are sketchy, the method represents a very flexible synthesis of these ylide precursors ⟨85H(23)2489⟩.

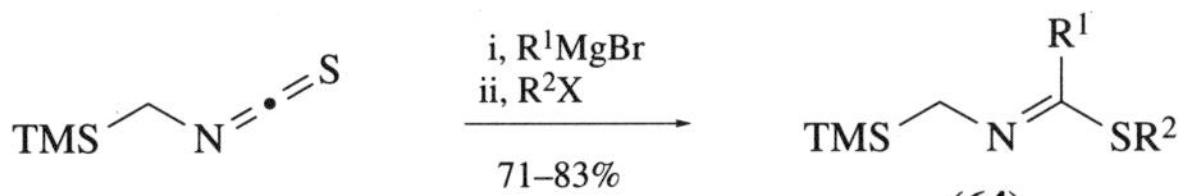

$$\text{TMS}\diagdown\text{N}{=}{\cdot}{=}\text{S} \xrightarrow[71-83\%]{\text{i, R}^1\text{MgBr} \atop \text{ii, R}^2\text{X}} \text{TMS}\diagdown\text{N}\diagdown\overset{\text{R}^1}{\underset{\text{(64)}}{\text{C}}}{-}\text{SR}^2 \qquad (36)$$

Finally, Whitham and co-workers have described the isolation of a silyloxazoline (**65**) following the acid catalysed reaction of a silyloxirane with acetonitrile (Equation (37)). Other products isolated from the reaction support a presumed nitrilium ion intermediate ⟨81JCS(P1)1934⟩.

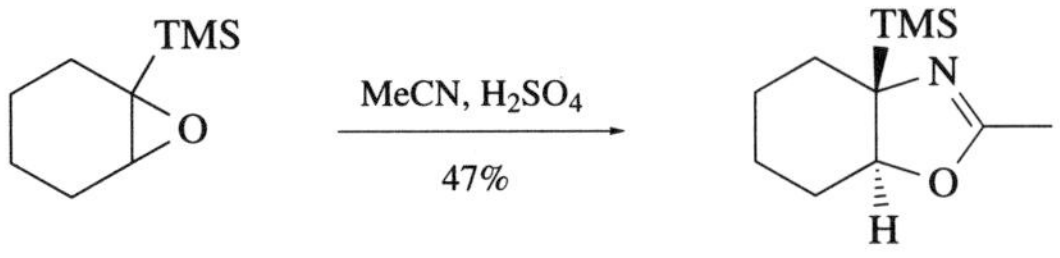

$$\text{(silyloxirane with TMS)} \xrightarrow[47\%]{\text{MeCN, H}_2\text{SO}_4} \text{(65)} \qquad (37)$$

4.11.1.1.7 N-(Silylmethyl) heterocycles

Vedejs' methodology for the generation of azomethine ylides from aminosilanes is readily extended to allow access to ylides formed from α-silyl heterocycles. The requisite *N*-(silylmethyl) heterocycles are in principle available by metallation of *N*-methyl derivatives and silylation, but the overriding preference for π-excessive nitrogen heterocycles to metallate on a ring carbon make this approach of little synthetic value. Alternatively, silylmethylation of a wide range of heterocycles under basic conditions is an efficient and popular process. Table 4 contains a synthetic summary of more-familiar substrates. The five-membered azoles (entries 2–4) are easily alkylated under relatively mild conditions (K₂CO₃, DMSO) whereas indole derivatives such as gramine (entry 6) require stoichiometric anion generation (BuLi-HMPA or NaH). Alper and Wolin have described two unusual intramolecular variants (entries 9 and 10) which afford novel silicon-containing heterocycles ⟨75JOC437, 75JOM(99)385⟩.

Two reports, which appeared almost simultaneously, have described the silylmethylation of pyridine and substituted pyridines by TMS-CH₂OTf, affording the pyridinium salts (**66**) (Scheme 20). In both cases the authors have desilylated the triflate salts and intercepted the intermediate ylides with DMAD and other dipolarophiles ⟨84CL279, 84H(22)701⟩.

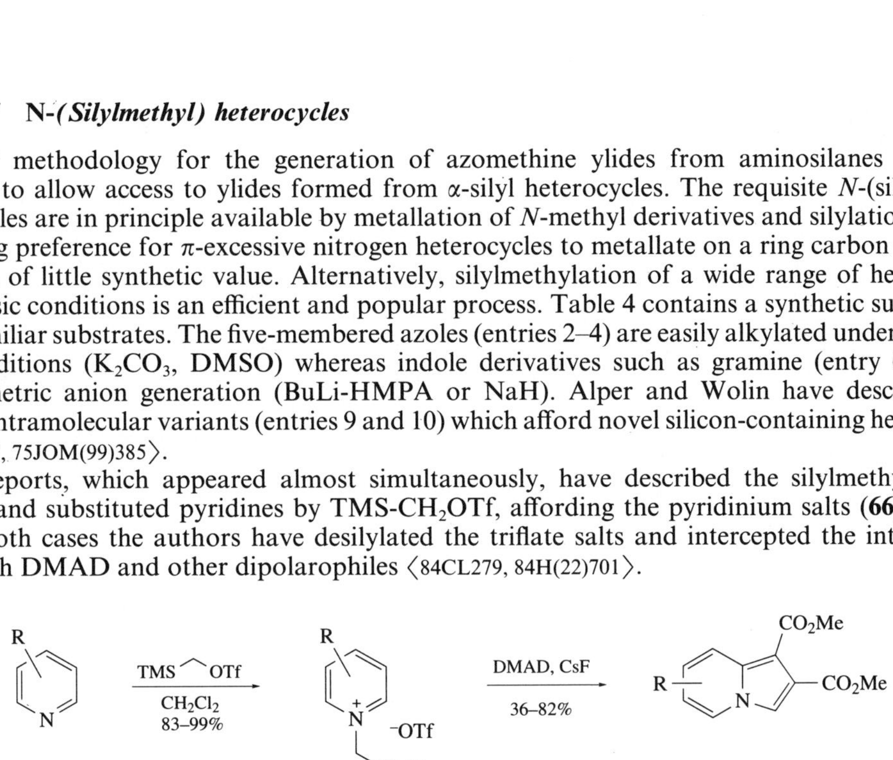

Scheme 20

4.11.1.2 Nitrogen and Germanium Functions

Synthetic routes to α-aminogermanes closely parallel those already described for the corresponding silanes. This parallel stretches across almost all of the established synthetic organogermanium chemistry and is discussed in depth in a useful review of that subject ⟨82COMC-I(2)365⟩.

Alkylation of amines by germylmethyl halides is by far the most frequently encountered preparative route. An early example is the synthesis of the germylmethylamine (**67**) (Scheme 21). Interestingly, the chlorogermane (**68**) which cannot be obtained by reaction with a single equivalent of lithium amide is accessible by treatment of (**67**) with TMS-Cl ⟨77JOM(132)77⟩. Terunuma *et al.* have prepared a series of germylmethylamines (Scheme 22) and demonstrated their base-catalysed fragmentation to trialkylgermanes, unlike silylmethylamines which undergo a Brook rearrangement

Table 4 *N*-silylmethylation of heterocycles.

Entry	Heterocycle	Halide	Product	Yield (%)	Ref.
1	pyrrole (NH)	TMS-CH2-Cl	N-CH2TMS pyrrole		67USP3346588
2	imidazole (NH)	TMS-CH2-Cl	N-CH2TMS imidazole		72USP3692798
3	pyrazole (NH)	TMS-CH2-Cl	N-CH2TMS pyrazole	51	86JOC3897
4	1,2,4-triazole (NH)	TMS-CH2-Cl	N-CH2TMS triazole	77	86JOC3897
5	tetrazole (NH)	TMS-CH2-Cl	2-CH2TMS and 1-CH2TMS tetrazoles	15 and 25	86JOC3897
6	3-(NMe2-CH2)indole (NH)	TMS-CH2-Cl	N-CH2TMS 3-(NMe2-CH2)indole	76	89JOC644
7	2-pyridone (NH)	TMS-CH2-Cl	N-CH2TMS 2-pyridone	70	87TL5419
8	pyrazole (NH)	allyl(Me)2Si-CH2-Cl	N-CH2Si(Me)2allyl pyrazole	63	89JOC317
9	2-mercaptoimidazole + BrCH2SiMe2		cyclized imidazo-thia-silacycle	85	75JOC437
10	2-aminobenzimidazole N-SiMe2CH2Br		cyclized benzimidazo-silacycle	48	75JOM(99)385

(see Section 4.11.1.1.1) under the same conditions ⟨91CL97⟩. The reaction of germylmethyl halides with potassium cyanate and ammonia affords α-germyl ureas which have been used to prepare germyldiazomethanes (Scheme 23). The transiently produced isocyanates are not isolated ⟨85TL5547, 86CPB3273⟩.

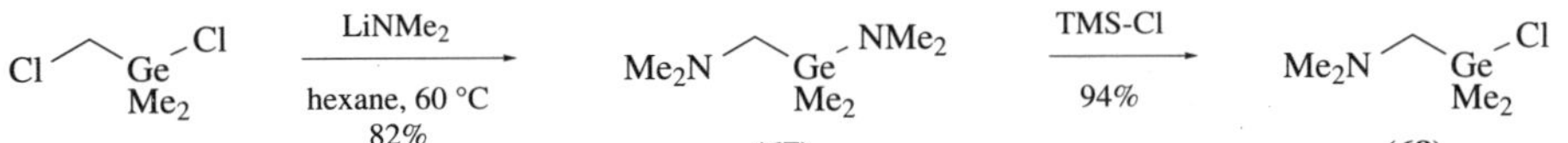

Scheme 21

Scheme 22

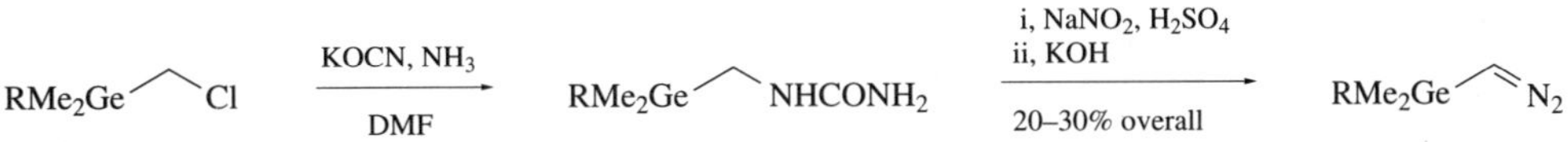

Scheme 23

Silylated amides will also react readily with chloromethylgermanes, typically under neutral conditions in a nonpolar solvent (Scheme 24). This chemistry and a subsequent radical-mediated, intramolecular germylation have been used to prepare the azagermine (**69**) ⟨88JOM(346)1⟩. A benzo-fused version of (**69**) is accessible by similar chemistry ⟨88JOM(339)259⟩.

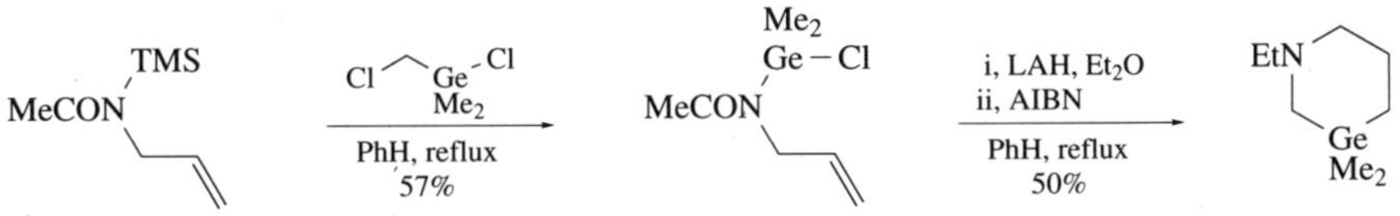

AIBN = 2,2'-azobisisobutyronitrile

Scheme 24

Germanium hydrides will add to aldimines in a non-regioselective manner to generate a mixture of germylmethylamines and aminogermanes (Equation (38)). The selectivity is reagent and substrate sensitive. The authors ascribe the product distribution to the operation of two competing (radical and ionic) mechanisms although the ionic mechanism, favouring formation of a C—Ge bond, generally predominates ⟨79JOM(168)43⟩. A similar addition to nitrones and related compounds has also been demonstrated ⟨72JOM(34)C18⟩.

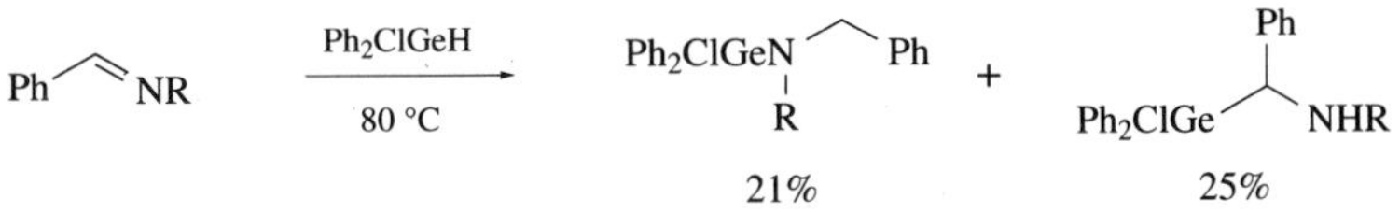

Lastly, and perhaps most usefully, Sato and co-workers have established a flexible lithiation–germylation of aminoacetonitriles which uses readily available reagents and gives access to a host of α-aminogermanes (**70**) (Scheme 25). The intermediate germyl nitriles can be isolated and purified.

Both alkyl and aryl substituents are tolerated although the yields of the Grignard addition vary dramatically with substitution. The Grignard reaction also generates variable amounts of the by-products (71) and (72) ⟨91S169⟩.

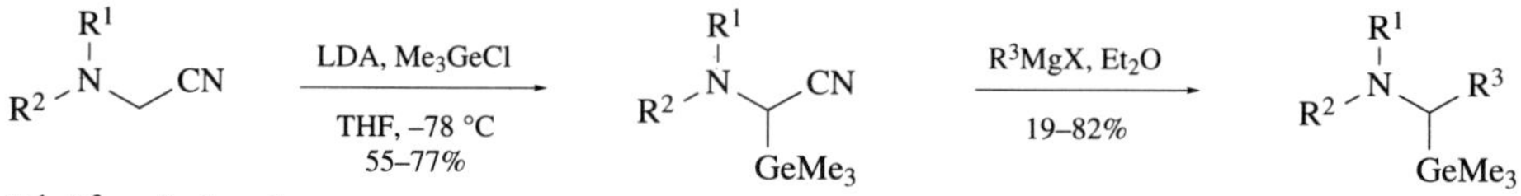

Scheme 25

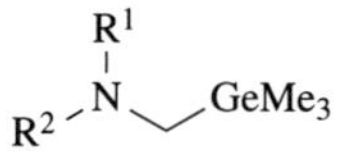

4.11.1.3 Nitrogen and Boron Functions

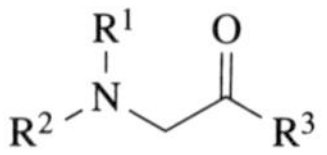

The chemistry of aminomethyl boron compounds has traditionally been regarded as a somewhat 'specialized' field, primarily due the reactivity of α-aminoboranes which are frequently air or moisture sensitive and readily undergo α-transfer rearrangements. However, a growing awareness of the value in medicinal chemistry of α-aminoboronic acids as serine protease inhibitors has prompted a steadily increasing output of synthetic work in this area since the first disclosure of significant biological relevance in 1977. This section is organized by the oxidation level of the boron substituent into two subsections between which there is, inevitably, a little overlap.

4.11.1.3.1 α-Aminoboranes and borohydrides

α-Aminoboranes have generally been synthesized by reactions between a nitrogen 'ylide' and a trialkyl- or triarylborane. The initial zwitterionic adducts are prone to several subsequent α-transfer reactions, frequently in tandem with loss of the amine substituent. Musker and Stevens first described the reaction of an unstabilized N-ylide with boranes (Scheme 26) in a process presumed to afford a borohydride intermediate which rapidly rearranged to a new borane. The overall process amounted to methylene insertion by the ylide ⟨67TL995⟩. Bickelhaupt et al. subsequently isolated the relatively stable adduct (73) (Scheme 27) and demonstrated its conversion into the boronic acid as part of a chemical proof of the structure of (73) ⟨68RTC188⟩.

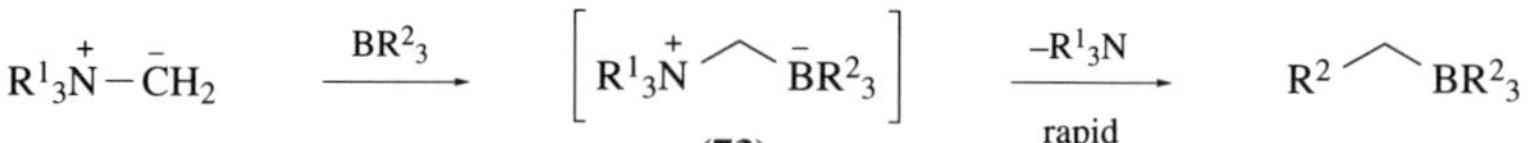

Scheme 26

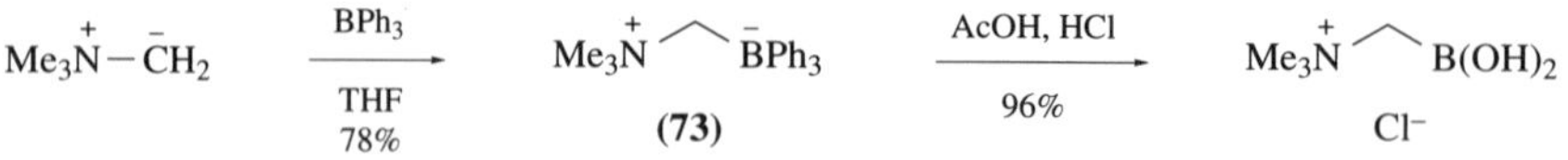

Scheme 27

Trialkylboranes will also react with isocyanides to generate dipolar adducts which are prone to thermal dimerization processes. Phenyl isocyanide and tributylborane react together to generate a zwitterion which quickly dimerizes to the isolated diboradihydropyrazine (74) (Scheme 28) ⟨65LA(687)1, 72LA(755)67⟩. Thermolysis of the pyrazine (74) prompts a second B–C migration, giving the diborapiperazine (75). Evidence for the intermediacy of an isonitrilium ion is supplied by conversion of the equivalent trimethylborane adduct into the iminoborane (76) (Scheme 29). Once again, a second B–C alkyl migration can be triggered, in this case by exposure to protic acids or alcohols ⟨69LA(722)21⟩.

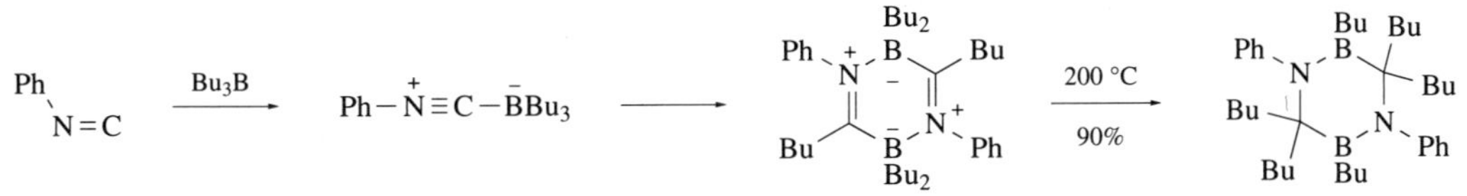

Scheme 28

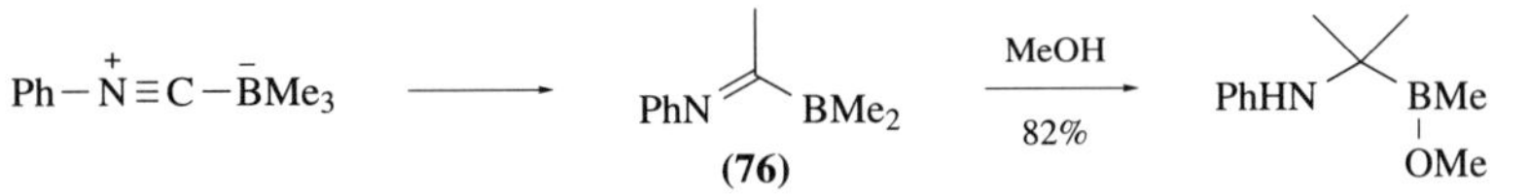

Scheme 29

Despite the handling difficulties experienced with many borane derivatives, the ease with which alkyl migration from boron to adjacent carbon can occur has led to some useful synthetic methodology. Pelter *et al.* have extensively characterized the utility of cyanoborate salts as precursors for α-migration chemistry. The salts (**77**) are readily prepared from trialkylboranes and KCN in organic solvents and react with any of a range of acylating agents (usually AcCl, PhCOCl or trifluoroacetic anhydride (TFAA)) to generate the oxazaborolines (**78**) in which two alkyl groups have transferred from boron to carbon (Scheme 30). Oxidative cleavage of the heterocycles (**78**) completes an overall synthesis of symmetrical ketones in excellent yield ⟨75JCS(P1)129⟩. A feature of the alkyl migration is the general insensitivity to steric congestion in the transition state, such that secondary alkyl groups also migrate readily. Use of TFAA as the acylating reagent in DIGLYME can induce the third and final B-alkyl substituent to migrate, giving the presumed intermediate boronate derivative (**79**) (Scheme 31). Oxidative cleavage liberates tertiary carbinols, again in good yield, usually associated with small amounts of ketone ⟨75JCS(P1)138⟩.

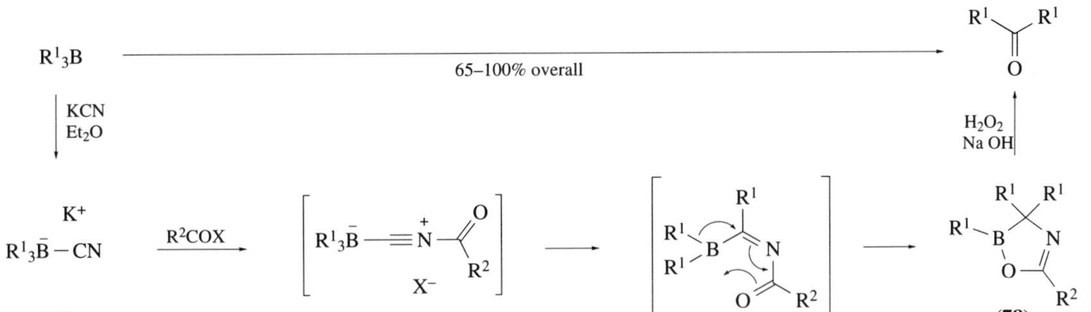

Scheme 30

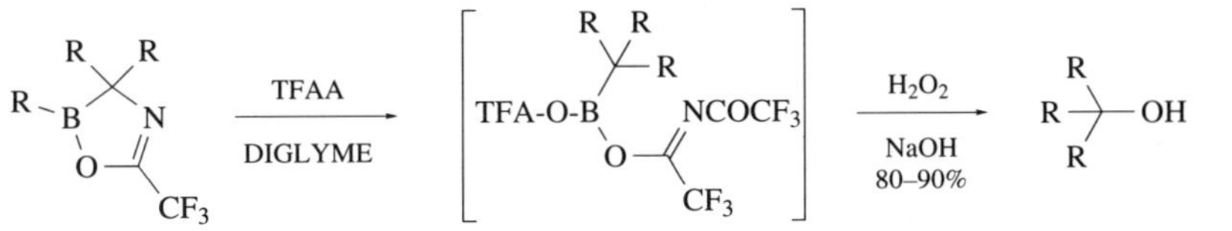

Scheme 31

TFAA = trifluoroacetic anhydride

The chemistry of low molecular weight α-aminoboranes is made more challenging by their extreme volatility but nevertheless, some synthetic details are available. Chloromethyldimethylborane (available by chlorination of BMe$_3$) undergoes straightforward S$_N$2 displacement with nucleophiles including lithium azide which generates the gaseous azidomethyl species (Scheme 32). Hydrogenation of the azide over platinum gave aminomethyldimethylborane which is essentially monomeric; NMR studies suggest the presence of a significant intramolecular B–N interaction ⟨65JA488⟩. Miller et al. have synthesized the isomeric α-aminoborane (81) as its trimethylamine complex by reaction of the cationic borane (80) with *t*-butyllithium (Equation (39)) and characterized it as a nonvolatile liquid which behaves as a nucleophile and a strong base ⟨64IC1196, 69IC275⟩. The Me_3N complex can be exchanged for other ligands including phosphines and other boranes ⟨81IC1328⟩. Products obtained by basic treatment of the cation (80) are extremely dependent on the nature of the base; even relatively subtle changes dramatically alter the product distribution. Thus, (80) reacts with butyllithium to give the oligomeric borohydride anion (82), which cyclizes to the BCN heterocycle (83) on exposure to aqueous base ⟨77JOM(137)131⟩. Reaction of (80) with NaH directly generates the parent heterocycle (84) albeit in low yield ⟨64IC1196⟩. The properties and chemistry of this heterocycle have been extensively surveyed by Miller et al. ⟨84JOM(269)123, 88JOM(349)11⟩.

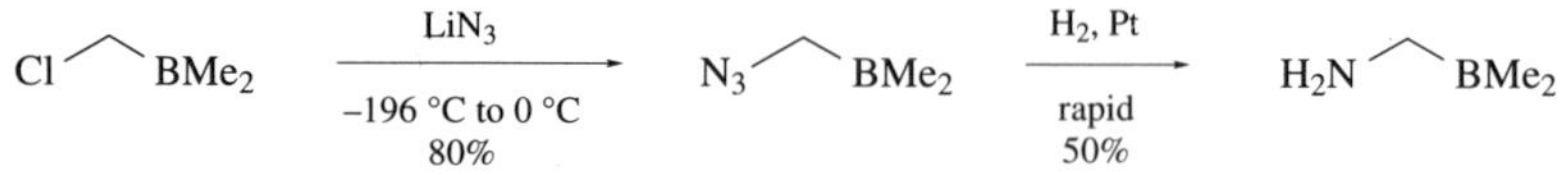

Scheme 32

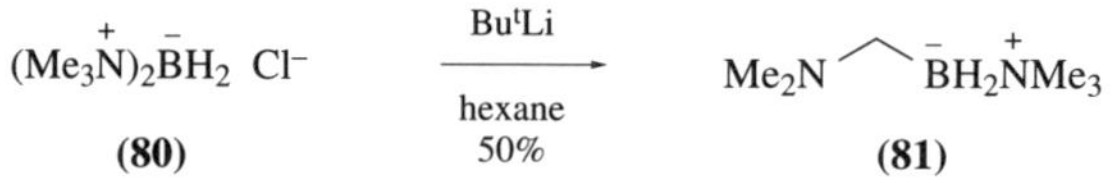

$$(Me_3N)_2\overset{+}{B}H_2 \; Cl^- \xrightarrow[\text{hexane} \; 50\%]{Bu^tLi} Me_2N\frown \overset{-}{B}H_2\overset{+}{N}Me_3 \quad (39)$$

(80)   (81)

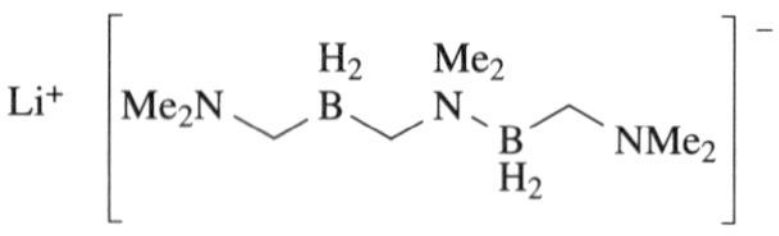

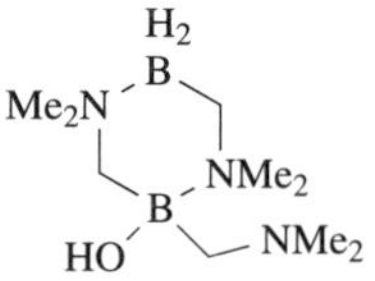

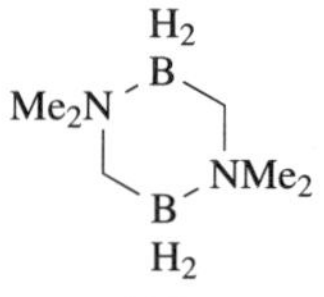

(82)   (83)   (84)

The hydroboration of unsubstituted enamines generally gives β-boron addition but Singaram et al. have demonstrated that β,β-substituted enamines can give good yields of α-boron addition products (Scheme 33) ⟨91JOC5691⟩. In the absence of an oxidative workup, the intermediate aminoboranes react further with borane to give decomposition products containing B–N bonds. Wipke and co-workers have used organic reaction prediction software (IGOR2) to compare computational and experimental observations of the outcome of enamine hydroboration and have achieved a good agreement ⟨93JA440⟩. Boronates can also be obtained by this methodology, using an aqueous workup protocol.

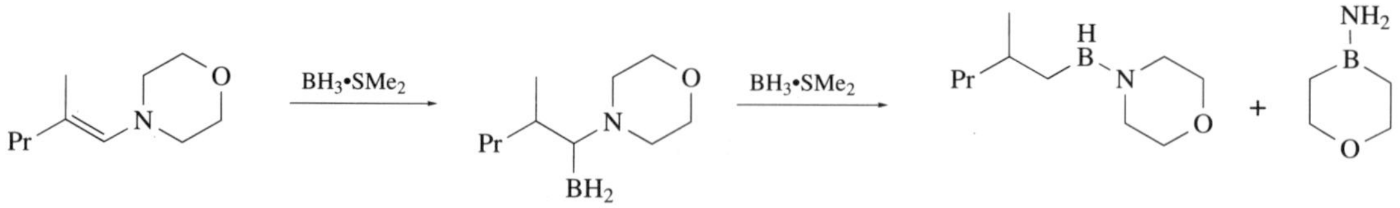

Scheme 33

In principle, α-aminoboranes can be obtained by reduction of the nitro equivalents. However, α-nitroboranes are exceedingly rare entities and their occurrences in the literature are associated with limited characterization. The nitroboranes (85) and (86) have been prepared by reaction of Et_2BCl with Na or K salts of the appropriate nitroalkane ⟨74MI 411-01⟩.

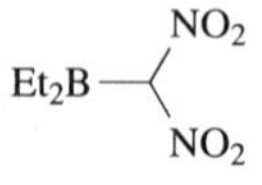

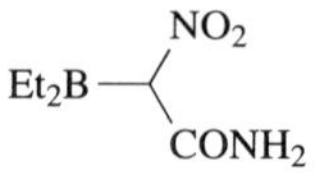

(85)   (86)

4.11.1.3.2 α-Aminoboronic acids

Aspects of aminomethylboronic acid chemistry including methods for their synthesis have appeared as subsections in three more-general reviews of boronic acids and boronates by Matteson ⟨89CRV1535, 89T1859, 91PAC339⟩.

Although several examples of aminomethylboronate syntheses were described up to 1977, the appearance of a landmark publication in that year documenting a boronate amino acid isostere as a useful inhibitor of the serine protease chymotrypsin ⟨77JA6435⟩ triggered a surge in synthetic and medicinal interest which is ongoing. Matteson and Cheng reported the first apparently general synthetic method (Scheme 34) relying on generation of an α-iodoboronate (**87**) and subsequent displacement by a secondary amine ⟨66JOM(6)100⟩. The free boronic acid products were generally isolated as catechol esters for characterization although, for most purposes, the readily removed pinacol ester protection has become the protective group of choice. Although the quaternization of tertiary amines by the iodide (**87**) was satisfactory, this preparative method failed when ammonia or primary amines were used ⟨78JA1325, 79JOM(170)259⟩. This failure can be overcome, at least in the case of primary aromatic amines by use of *N*-silyl-*N*-lithioamines (Scheme 35). Metallation and silylation of hindered primary amines such as (**88**) followed by remetallation gives a nucleophilic species which reacts cleanly with pinacol iodomethaneboronate to give a stable aminomethyl derivative (**89**) which can readily be distilled and directly acylated without the need to desilylate ⟨86JOC1610⟩.

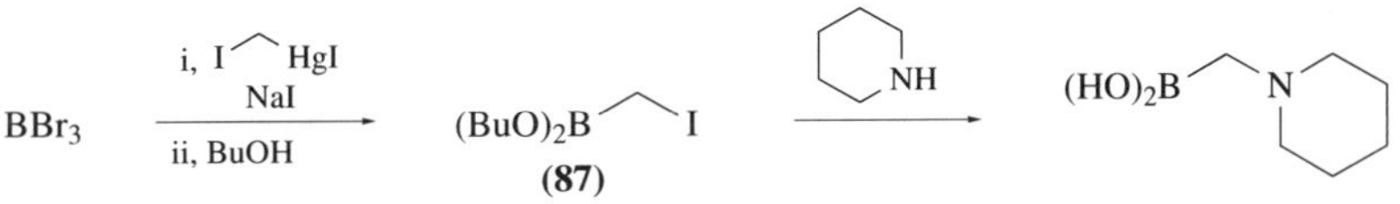

Scheme 34

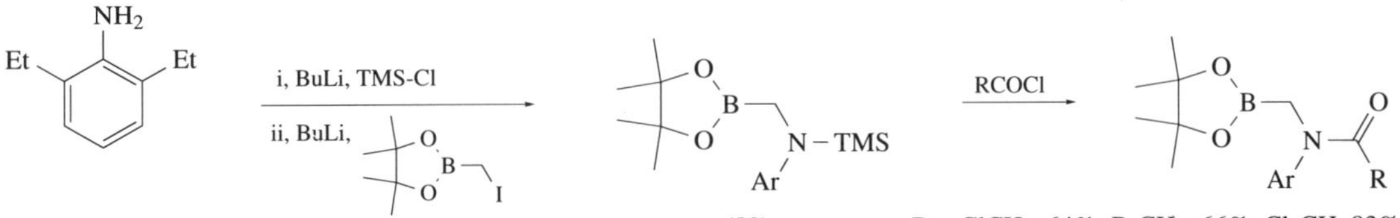

Scheme 35

In 1977, Lindquist and Nguyen described the synthesis of the boronate isostere of hippuric acid (**90**) and its inhibitory activity against chymotrypsin ⟨77JA6435⟩. The synthesis (Equation (40)) utilized the Na salt of benzamide which was apparently successfully alkylated by dibutyl iodomethylboronate to give (**90**) after aqueous hydrolysis. Although simple boronic acids had been established as competitive inhibitors of chymotrypsin and subtilisin, this report established the value of aminoboronic acids as isosteres of traditional amino acids. A subsequent reinvestigation of the benzamide reaction by Matteson *et al.* has strongly suggested that the structure (**90**) does not represent the isolated product and that alkylation occurs on oxygen to give the isomeric boronic acid (**91**) which probably exists as an intramolecular N–B chelate ⟨81JA5241⟩. It is all the more remarkable that the subsequent synthesis of authentic (**90**) produced a weaker chymotrypsin inhibitor than the O–C–B isomer.

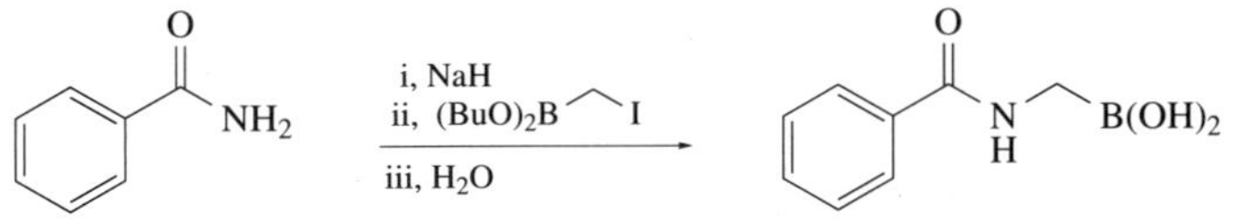

(40)

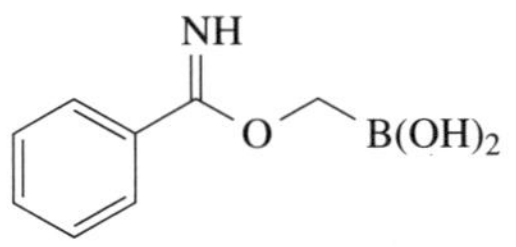

Almost all of the subsequent exploitation of aminomethylboronates has relied upon essentially the same strategy for assembly of the N–C–B functionality, specifically, the homologation, amination sequence pioneered by Matteson and co-workers. In summary, the method relies upon the B–C α-transfer reaction discussed earlier, followed by nucleophilic introduction of nitrogen (Scheme 36). Reaction of a boronate ester and dichloromethyllithium, typically using $ZnCl_2$ catalysis, gives an intermediate ate complex (92) which efficiently rearranges to the homologated α-chloroboronate (93) ⟨83OM1529⟩. The choice of boronate ester is large; in general, the achiral, cyclic boronates derived from ethylene glycol or pinacol have been used as precursors to achiral chloroboronates and the esters of (+)- or (−)-pinanediol give up to 99% *ee* in the equivalent diastereoselective chloroboronate syntheses ⟨83OM1536, 84OM1284⟩. The earlier difficulties in achieving reaction of α-haloboronates with primary amines are overcome by use of lithium hexamethyldisilazide (LHMDS) ⟨81JA5241, 84OM614⟩ or, rarely, sodium azide which produces an azidomethylboronate ⟨87TL4499⟩. Acylation on workup leads to a relatively stable α-amidoboronate. A modification to this approach, based on chemistry developed by Rathke *et al.*, generates the dichloroalkyl ate complex (92) by Grignard or alkyllithium addition to a dichloromethylboronate ester (Equation (41)) ⟨76JOM(122)145, 92TL4209⟩. Table 5 contains a summary of α-aminoboronic acids and boronates prepared by the foregoing procedure from chloromethylboronates. Entry 4, a proline isostere, was synthesized from a bromopropyl boronate which cyclizes *in situ* following LHMDS displacement. All of the entries in Table 5 have been used as protease inhibitors or as key components of protease inhibitors.

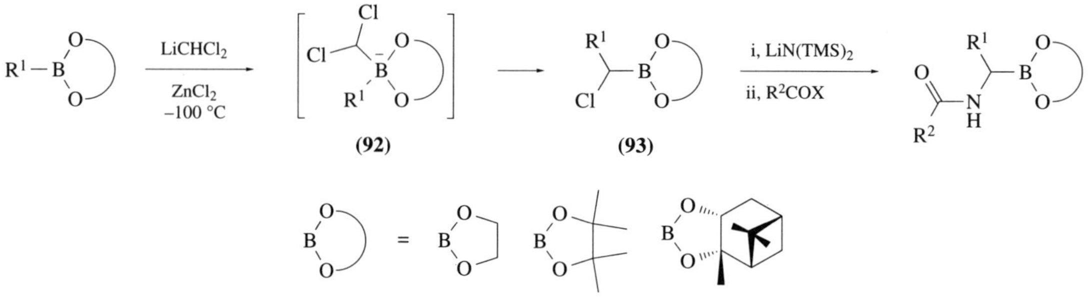

Scheme 36

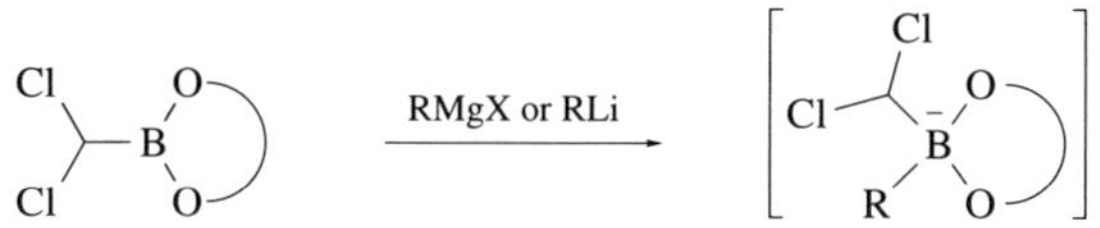

(41)

Since the initial disclosure of aminomethylboronates as chymotrypsin inhibitors, many disclosures of increasingly more complex structures designed to inhibit key serine proteases (including elastase, thrombin and cathepsin G) have appeared. Table 6 contains a qualitative and by no means comprehensive survey of exemplary aminomethylboronate inhibitors; the less-hindered boronate esters are labile at physiological pH and hydrolyse to give the active boronic acid.

4.11.2 FUNCTIONS CONTAINING A NITROGEN AND A METAL: $R^1_2C(NR^2_2)ML_n$ (M = Li, K, Mg, Sn, Zn), etc.

This section reviews the synthesis of α-metallated nitrogen compounds, focusing on those in which the nitrogen atom is a component of an amine, imine or equivalent species. The synthesis of metallated nitroalkanes (nitronates) is beyond the scope of this article to review in depth, but general methods are briefly surveyed, with emphasis on other comprehensive reviews.

4.11.2.1 α-Metallated Amine Functions

A thorough and well-organized review of nitrogen-stabilized carbanions by Gawley and Rein appeared in 1991 ⟨91COS(3)65⟩. This review underscores the need for additional carbanion-stabilizing

Table 5 α-Aminoboronic acids prepared from α-chloroboranes and LHMDS.

Entry	Boronic acid		Yield (%)[a]	Ref.
1	AcHN-CH(CH2Ph)-B(OH)2	(RS) (S) (R)	75 68 63	84OM614 81JA5241 84OM614
2	AcHN-C*HR(CH2Ph)-B(OH)2, * = ^{13}C		81	92MI 411-01
3	AcHN-CHR-CH2CH2-R-B(OH)2	R = Cl R = Br R = OBn	68 74 b	84OM1284
4	pyrrolidine-2-B(OH)2 (NH)		70[c]	90JBC(265)3738
5	R¹HN-CR(R)-B(OH)2 (R¹ = Ac, Bz, cbzAla)	R = Me R = Ph R = Pr^i R = Bu^i		85JMC1917
6	H2N-CHR-CH2CH2-R-B(OH)2	R = OMe R = Me		92TL4209
7	H2N-CHR-CH2CH2CH2-C(NH2)=NH - B(OH)2			92TL4209
8	H2N-CHR-CH2-(1,3-dioxolan-2-yl) - B(OH)2			92TL4209

[a] Overall yield for LHMDS displacement and acetylation. [b] Not obtained analytically pure. [c] Yield for LHMDS displacement and cyclisation.

groups to ensure the stability and utility of many α-lithiated amines, exemplifying this with amidines in particular. This review will aim to give an update on currently available methods for synthesis of unstabilized α-amino anions and their masked, stabilized equivalents. Some further specific examples, and discussions on the nature of the C—M bond will be found in two general reviews of carbanion chemistry ⟨80T2531, 87MI 411-01⟩.

The literature contains a vast array of α-metallated amines in which the metal is either tin or, particularly, lithium. Potassium, zinc and magnesium have all found occasional utility. The remaining metals which occur with regularity in most aspects of organic chemistry are very poorly represented in this area, despite the apparently obvious potential for transmetallation from accessible organolithium species.

4.11.2.1.1 Lithium, potassium and magnesium

(i) Unstabilized carbanion equivalents

Alkali metallation of unstabilized tertiary amines has been achieved using strong bases and generally proceeds in a nonspecific manner, producing the most thermodynamically stable anion. One of the most efficient examples, by Ahlbrecht, is the metallation of *N*-methylpiperidine by a

Table 6 α-Aminoboronate protease inhibitors.

Entry	Boronate	Enzymes inhibited	Ref.
1	H$_2$N–CH(CH$_3$)–B(OH)$_2$	Alanine racemase	89B3541
2	MeOSuc-Ala-Ala-Pro (pinacol boronate on pyrrolidine)	Bacterial IgA proteases	90JBC(265)3738
3	cbz-D-Phe-Pro-NH–CH(CH$_2$CH$_2$CH$_2$S–C(NH$_2$)=NH)–B(pinacol)	Thrombin	92TL4209
4	Ac-D-Phe-Pro-NH–CH(CH$_2$CH$_2$CH$_2$NH–C(NH$_2$)=NH)–B(OH)$_2$	Thrombin (kallikrein, plasmin)	92MI 411-02 88EUP293881
5	MeOSuc-Ala-Ala-Pro-NH–CH(CH$_2$Ph)–B(OH)$_2$	Elastase, cathepsin G chymotrypsin	84JBC(259)15106 85USP4499082
6	cbz-Ala-Ala-NH–CH(iPr)–B(OH)$_2$	Elastase	90USP4963655
7	Iva-Phe-Nva-NH–CH(CH$_2$-cyclohexyl)–B(pinacol)	Renin	89EUP315574
8	R-CH(NHBoc)-C(O)-N(pyrrolidine)-C(O)-NH-CH(CH$_2$CH$_2$CH$_2$NH-C(NH$_2$)=NH)-B(pinanediol) R = TMS, 2-naphthyl	Thrombin	92EUP471651

cocktail of Bu^{s}Li and KOBu^t, shown in Scheme 37 ⟨84TL1353⟩. The reaction is successful for most N-methyl tertiary amines. The frequently used activating agent TMEDA is itself lithiated by BuLi at or above room temperature in 1–3 hours, giving (**93**) ⟨87CB2081⟩, and (**94**) is more readily generated, presumably due to efficient intramolecular chelation of the metal ⟨90RTC305⟩. Peterson established one of the earliest site-specific protocols (Scheme 38) which relies on prior synthesis of the α-aminostannane and transmetallation by BuLi ⟨71JA4027⟩. Peterson's stability studies suggest that aminoalkyllithium reagents are more stable than the alkyllithium equivalent.

In 1991, Tsunoda *et al.* described an alternative, regiospecific method for the generation of aminomethyl carbanions (Scheme 39) which requires a thioaminal precursor (**95**). Reaction of (**95**) with lithium di-*t*-butylbiphenyl radical anion (LiDBB) in THF stoichiometrically liberates an α-lithio amine which is readily intercepted by aldehydes and alkyl halides ⟨91TL1975⟩. The method partly supersedes Peterson's earlier stannane transmetallation since the stannanes were synthesized from the readily available thioaminal analogues of (**95**).

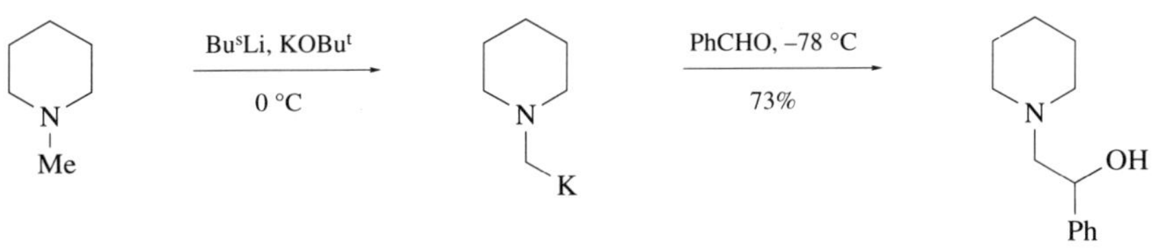

Scheme 37

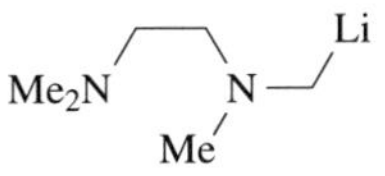

(93) 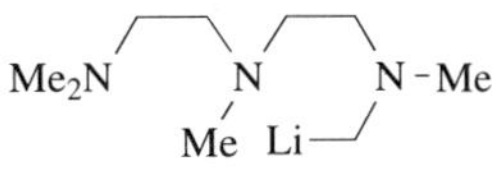   (94)

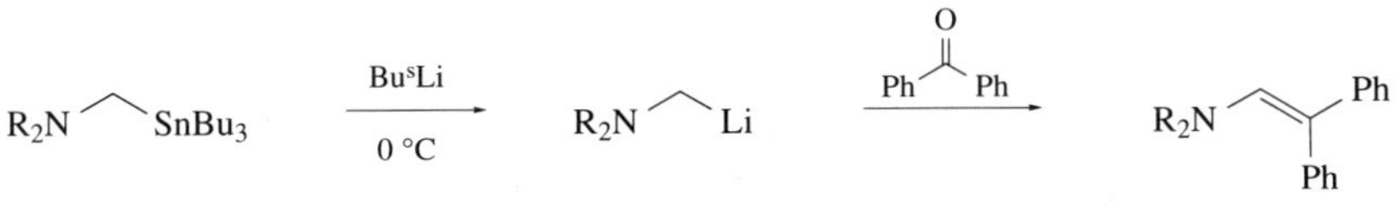

Scheme 38

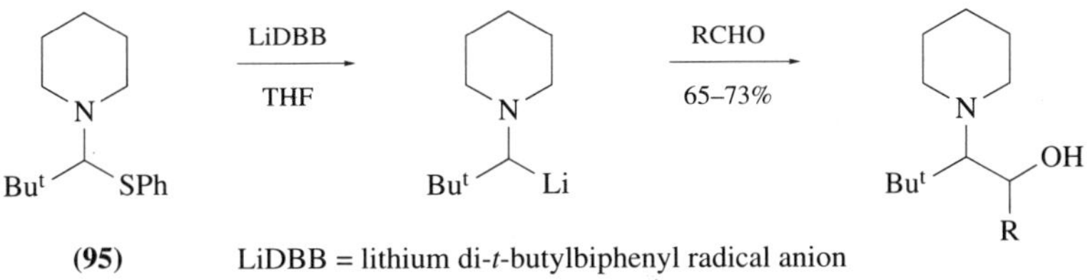

(95)   LiDBB = lithium di-*t*-butylbiphenyl radical anion

Scheme 39

(ii) Dipole-stabilized carbanion equivalents

A very wide range of carbanion stabilizing groups has been appended α to amino groups to stabilize the metallation process. The groups that achieve this most effectively by delocalizing electron density into enolate equivalents (carbonyl, cyano, etc.) are not easily removed to reveal the amino function and are not discussed in detail here. However, a range of alternative dipolar stabilizing groups is available and most are more readily removed after metallation, giving access to the generic sequence shown in Scheme 40, where MG denotes a metallation directing group. A 1984 review by Beak *et al.* comprehensively surveys essentially all of the known examples of α-metallo amine synthetic equivalents to that date ⟨84CRV471⟩ but as a guide to the available options, Table 7 contains examples of most directing groups that have proved to be of synthetic value. Further examples of each type can be found in Beak's review, but some of the entries require additional comments. The metallation of nitrosamines (entry 4) generally gives better results if LDA is substituted with KDA (Pr$^i_2$NH-KOBu^t-BuLi) which gives more reactive potassium derivatives, such as (96), although in some cases KOBu^t alone was suitable for addition reactions to non-enolizable ketones ⟨78CB2630⟩. Saavedra has extended the nitrosamine methodology to allow the generation of masked α-lithio primary amines (Scheme 41). This is achieved by lithiation of the methoxyalkyl nitrosamine (97), followed by trapping with an electrophile to give a new nitrosamine which can be further metallated *in situ* in some cases. Deprotection reveals a primary amine hydrochloride ⟨83JOC2388⟩.

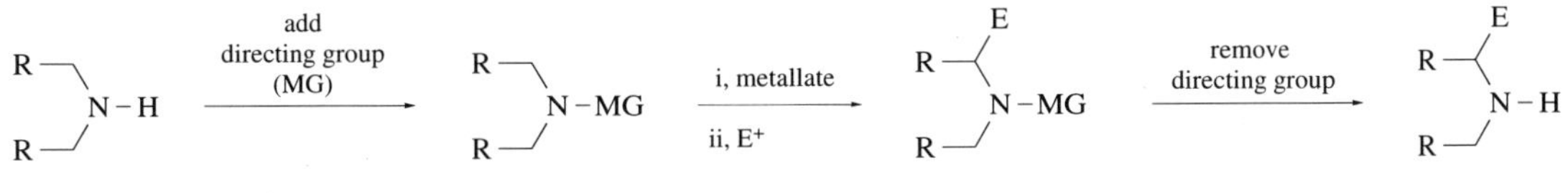

Scheme 40

Table 7 Reactions of dipole-stabilized, α-metallated amine equivalents.

Entry	Stabilizing group	Example	Yield (%)	Ref.
1	Amide	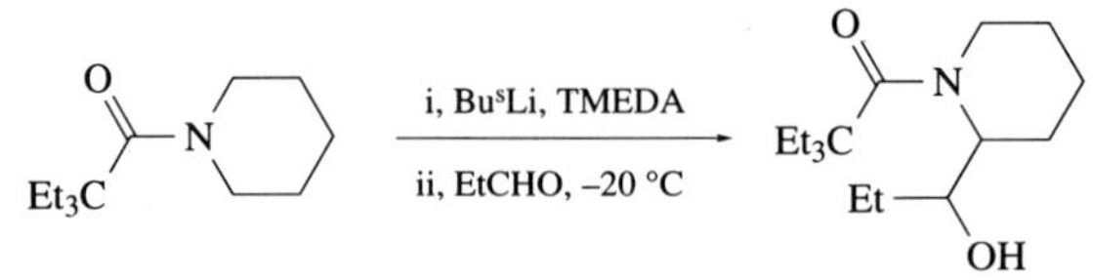 Et₃C-C(O)-N(piperidine), i, Bu^{s}Li, TMEDA; ii, EtCHO, −20 °C → 2-substituted product with CH(OH)Et	65	84JA1010
2	Carbamate	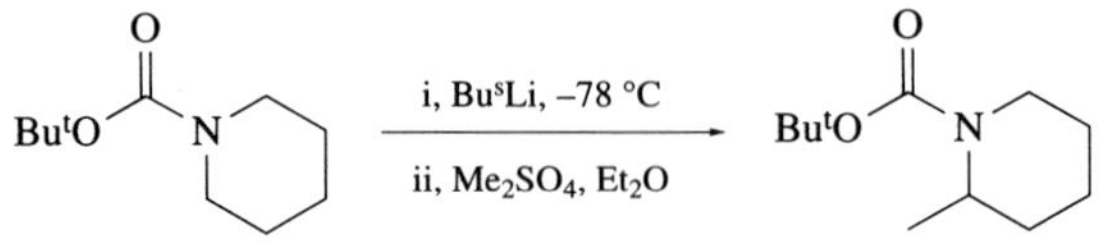 Bu^{t}O-C(O)-N(piperidine), i, Bu^{s}Li, −78 °C; ii, Me$_2$SO$_4$, Et$_2$O → 2-methyl product	53	89TL1197
3	Formamidine	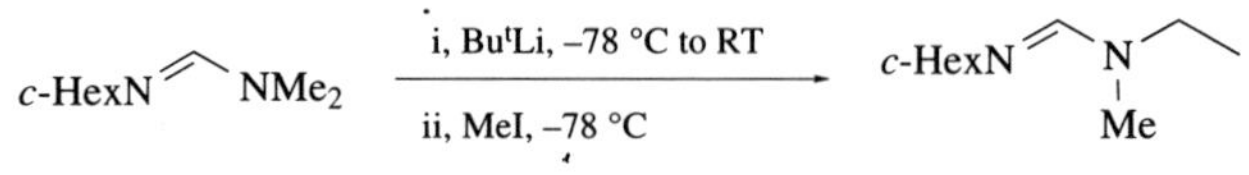 c-HexN=CH-NMe$_2$, i, Bu^{t}Li, −78 °C to RT; ii, MeI, −78 °C → c-HexN=CH-N(Et)(Me)	85	80JA7125, 85JOC1019
4	Nitrosamine	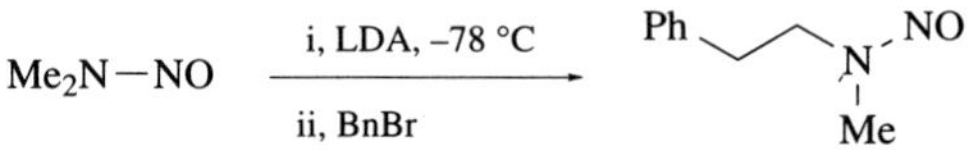 Me$_2$N−NO, i, LDA, −78 °C; ii, BnBr → Ph(CH$_2$)-N(Me)(NO)	95	72AG(E)301, 75AG(E)15
5	Dithiocarbamate	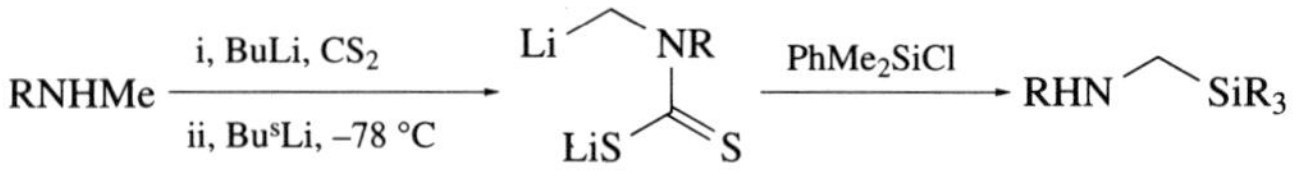 RNHMe, i, BuLi, CS$_2$; ii, Bu^{s}Li, −78 °C → Li-CH$_2$-NR-C(=S)-SLi, PhMe$_2$SiCl → RHN-CH$_2$-SiR$_3$	87–96	91S637, 88S775
6	Thioamide	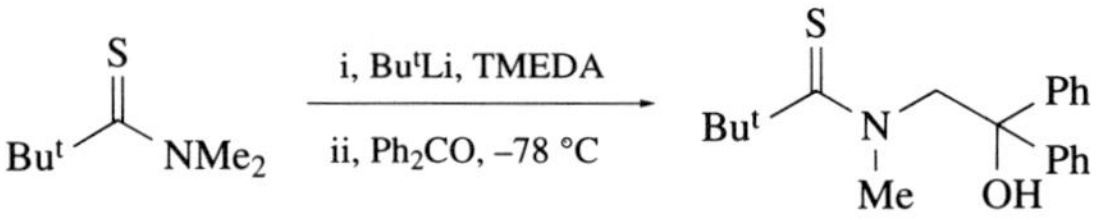 Bu^t-C(=S)-NMe$_2$, i, Bu^{t}Li, TMEDA; ii, Ph$_2$CO, −78 °C → Bu^t-C(=S)-N(Me)-CH$_2$-C(Ph)$_2$OH	78	80HCA102
7	Allylamine	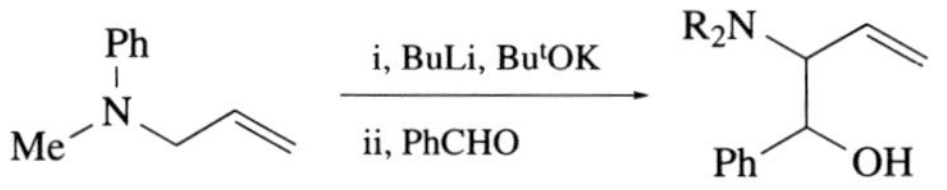 Me-N(Ph)-CH$_2$CH=CH$_2$, i, BuLi, Bu^{t}OK; ii, PhCHO → R$_2$N-CH(CH=CH$_2$)-CH(Ph)OH		74S672, 83S61

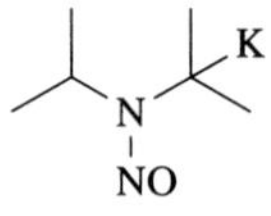

(Pr^i)$_2$N(K)-NO
(96)

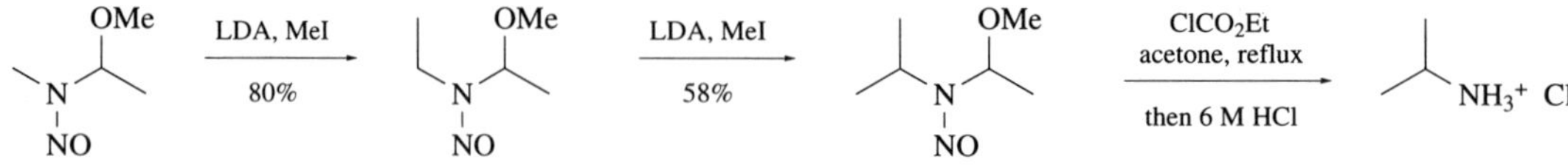

Scheme 41:
(97) MeN(NO)CH(OMe)Me →[LDA, MeI, 80%] EtN(NO)CH(OMe)Me →[LDA, MeI, 58%] Pr^{i}N(NO)CH(OMe)Me →[ClCO$_2$Et, acetone, reflux; then 6 M HCl] Pr^i-NH$_3^+$ Cl$^−$

Although the metallation of amides and carbamates (Table 7, entries 1 and 2) by directed deprotonation is a versatile reaction, its usefulness is limited for metallation of masked primary amine equivalents. Pearson *et al.* have demonstrated that this gap can be filled by the easy transmetallation of aminomethylstannanes (Scheme 42) to give the lithiated carbamate (**98**) ⟨89JOC5651⟩. After addition of the requisite electrophile, a transfer hydrogenation removes both of the protecting groups to liberate the functionalized primary amine. (See Section 4.11.2.1.2 for a discussion of synthetic routes to aminomethylstannanes.)

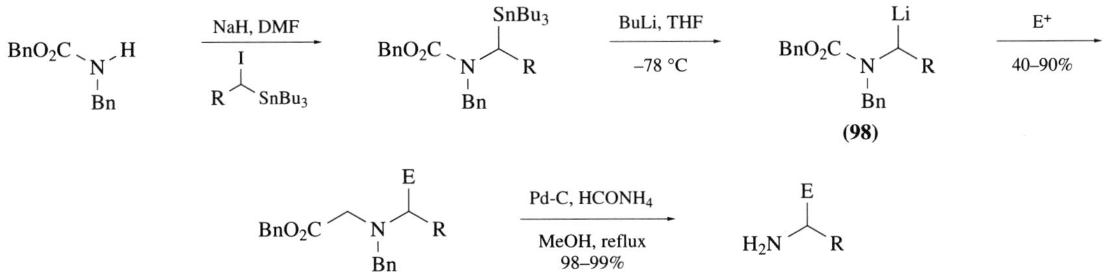

Scheme 42

Transmetallation from lithium to magnesium has infrequently been used to generate α-amino Grignard reagents in situations where the change of metal enhances yield or selectivity. For example, Seebach et al. have reported a chelation-controlled addition of the Grignard reagent (**99**) to acetophenone giving a 96:4 mixture of isomeric tetrahydroisoquinolines in favour of (**100**) (Equation (42)) ⟨85JOM(285)1⟩. In general, α-amino magnesium reagents have few advantages over their lithium counterparts but, if required, transmetallation with MgX_2 solvates is available. (See also Scheme 3 in Section 4.11.1.) The inaccessibility of halomethylamines which might seem to be the obvious Grignard precursors has discouraged extensive characterization of α-aminomagnesium derivatives.

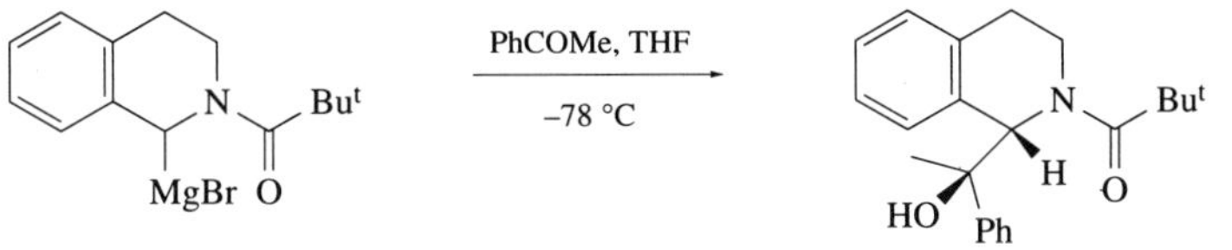

(iii) Chiral, non-racemic α-aminolithium reagents

A number of α-lithiated amine equivalents have been established in which the stabilizing group is chiral resulting in a configurationally stable carbanion, retaining its formal sp^3 hybridization. High *ee* values are readily achieved in cyclic systems, although when low *ee* values are obtained, they are probably the result of poor selectivity in removal of one diastereotopic proton from a pair. Consequently, methods which rely on transmetallation to lithium from more stable organometallics (usually stannanes) can achieve excellent selectivity. This section summarizes current methods.

Gawley et al. have made an elegant comparison of two relatively similar chiral auxiliaries which give vastly different selectivities ⟨89JOC3002⟩. Lithiation and methylation of the oxazoline (**101**) (Equation (43)) proceeds with poor selectivity but, in contrast, the oxazolidinone (**102**) (Equation (44)) is methylated with almost complete diastereocontrol. The lithiated oxazolidine (**103**) and urea (**104**) have been prepared by Pearson et al. via transmetallation of the corresponding stannanes ⟨91JA8546⟩. Both carbanions epimerize slowly even at −78 °C. In a similar vein, Chong and Park have synthesized the *t*-BOC protected amino stannane (**105**) and demonstrated that transmetallation with BuLi at −95 °C gives a carbanion that does not significantly epimerize provided the very low temperature is maintained (Scheme 43) ⟨92JOC2220⟩.

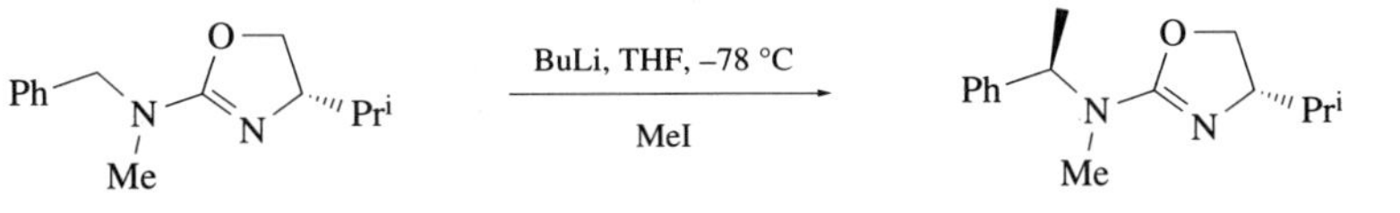

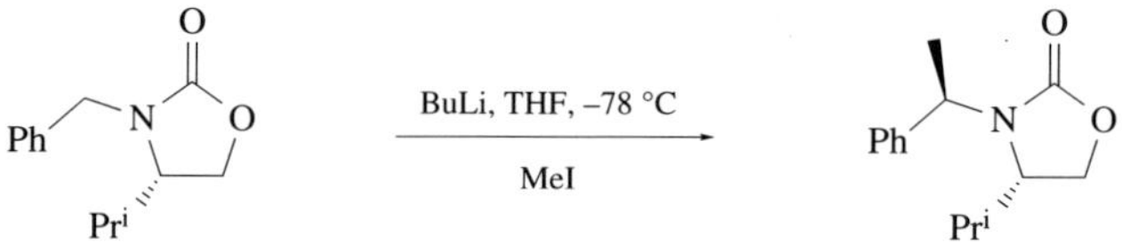

Scheme 43

The seminal work of Meyers et al. in developing chiral formamidines as dipolar, carbanion-stabilizing groups represents perhaps the most effective means of generating nonracemic α-amino lithium reagents ⟨84T1361⟩. Many representative examples are included in the aforementioned review by Gawley with a fuller discussion than is possible here. High induction is best achieved when the metallated amine forms part of a five-, six- or seven-membered ring, and benzofused systems such as tetrahydroisoquinolines give near perfect diastereoselection. Equations (45) and (46) show two examples of the value of this method as applied to natural product synthesis. In the first of these, metallation of the octalin (**106**) and benzylation of the resultant anion, followed by removal of the formamidine auxiliary gives a key intermediate in the synthesis of dextrorphan ⟨86JOC872⟩. Similar selectivities can be achieved when using less reactive alkyl halides as the electrophile (Equation (46)). Lithiation of the β-carboline (**107**) and reaction with the bromopropyl ortho ester gives a tetracyclic precursor to deplancheine, essentially as a single enantiomer ⟨86JOC3108⟩.

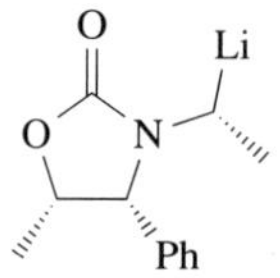

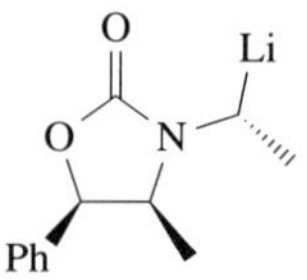

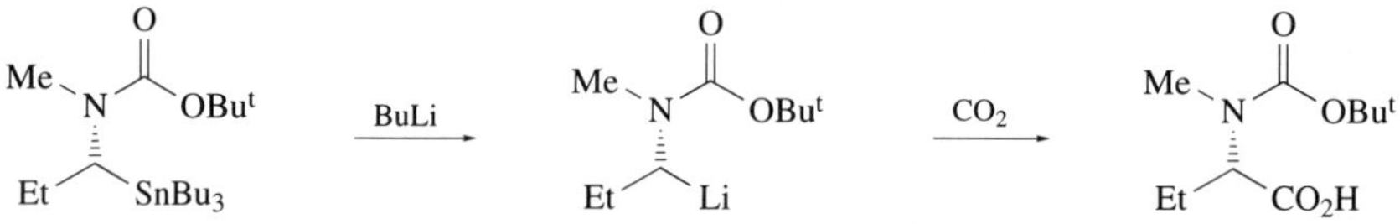

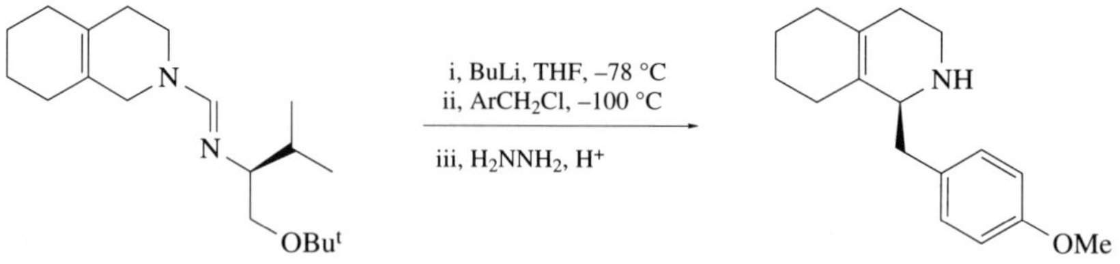

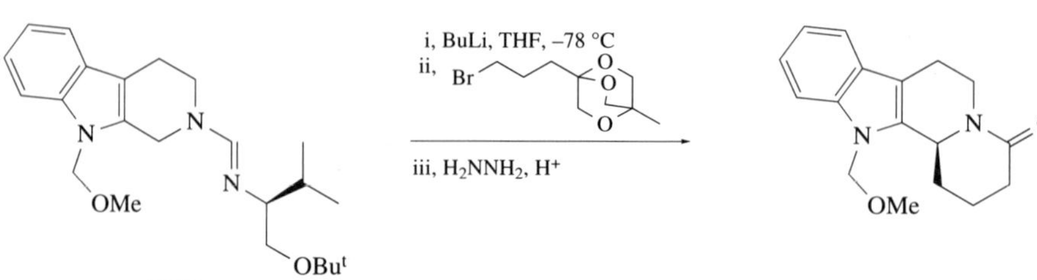

4.11.2.1.2 Tin and zinc

Peterson devised the first general route to tertiary amino derivatives which relies on reactions of tributylstannyllithium with aminomethyl phenyl sulfides. The yields for this process are relatively poor (Equation (47)) ⟨71JA4027⟩. Subsequently, Abel *et al.* have refined this method to offer improved yields ⟨75JOM(97)159⟩. The interchangeability of Sn and Li by transmetallation in either direction is a straightforward process and several α-aminostannanes have been prepared from lithium derivatives discussed in the previous section. For example, the lithiated TMEDA species (**93**) described earlier can be intercepted by Bu_3SnCl to generate the stannane (**108**) in reasonable yield (Scheme 44) ⟨87CB2081⟩.

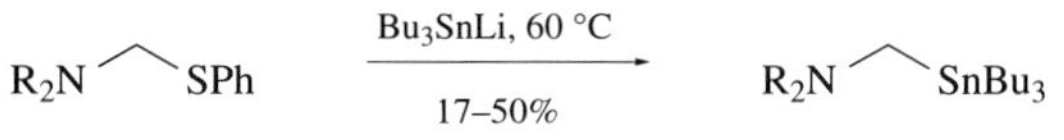

$$R_2N\text{—}SPh \xrightarrow[17-50\%]{Bu_3SnLi,\ 60\ °C} R_2N\text{—}SnBu_3 \qquad (47)$$

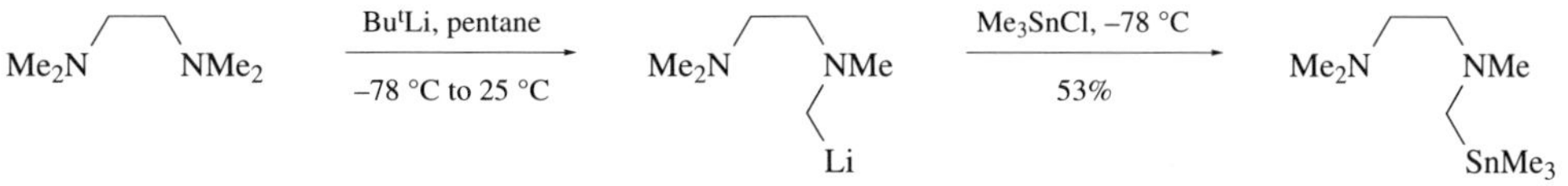

$$Me_2N\text{—}NMe_2 \xrightarrow[-78\ °C\ \text{to}\ 25\ °C]{Bu^tLi,\ pentane} \mathbf{(93)} \xrightarrow[53\%]{Me_3SnCl,\ -78\ °C} \mathbf{(108)}$$

Scheme 44

The limited availability of suitable aminomethyl phenyl sulfides restricts the value of the chemistry in Equation (47). To resolve this, Quintard *et al.* have used readily available aminals such as (**109**) which react with $Bu_3SnMgCl$ (prepared *in situ* from Bu_3SnH and Pr^iMgCl) to give excellent yields of the stannanes (Scheme 45). Yields for the two-step process are in the range 45–90% overall ⟨84S495⟩. The same group has shown that iminium ions (logical equivalents of aminals such as (**109**)) also react with $Bu_3SnMgCl$ to give stannanes; a typical example is shown in Scheme 46. The iminium ions are prepared by prior condensation of suitable aldehyde or ketones with aliphatic secondary amines, usually under acid catalysis ⟨86USP4617409, 88JOM(339)267⟩. Yields are generally comparable to the aminal process, although the versatility is undoubtedly greater.

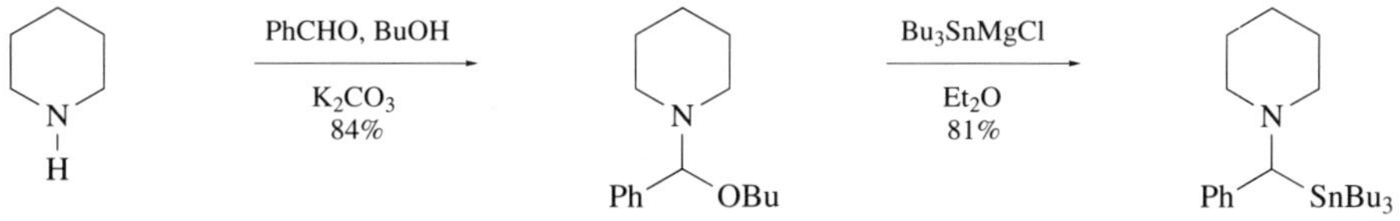

Scheme 45

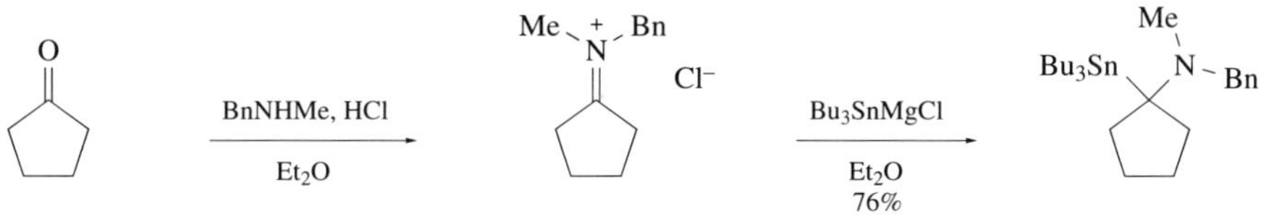

Scheme 46

Stannyl carbinols, prepared by the addition of Bu_3SnLi to aldehydes, can be converted into phthalimidomethylstannanes under Mitsunobu conditions as shown in Scheme 47. Hydrazinolysis and BOC protection gives an acylaminostannane which can be further alkylated at nitrogen under standard conditions ⟨92JOC2220⟩. Pearson *et al.* have also used this Mitsunobu protocol to generate the unprotected α-aminostannanes (**110**) and (**111**) as precursors to imino derivatives (see Section 4.11.2.2) ⟨92JOC6354⟩. Homochiral aminomethyltin species are accessible in a similar manner; enantioselective reduction of an acylstannane with (*S*)-BINAL-H gives a carbinol which is readily converted into the imide (**112**) with inversion of the carbinol stereochemistry (Equation (48)). The overall enantioselectivity of the process is dependent on the reduction step ⟨92JOC2220⟩.

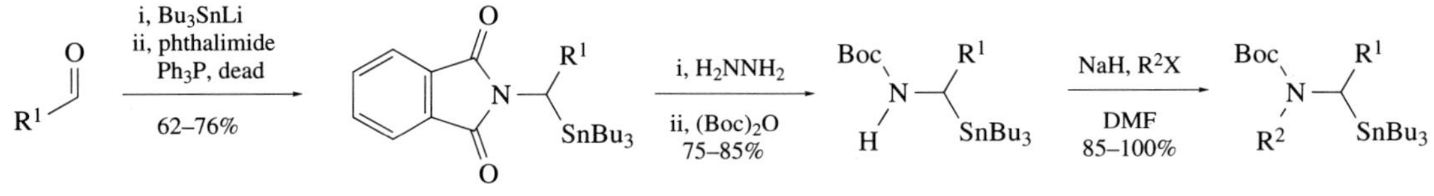

Scheme 47

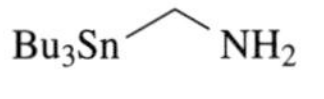

(110) Bu₃Sn–CH₂–NH₂

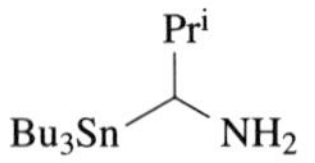

(111) Bu₃Sn–CH(Pr^i)–NH₂

$$Et-C(=O)-SnBu_3 \xrightarrow[\text{ii, (cbz)}_2\text{NH, Ph}_3\text{P, dead}]{\text{i, (S)-BINAL-H}} (cbz)_2N-CH(Et)-SnBu_3 \quad (112) \sim 90\% \; ee \tag{48}$$

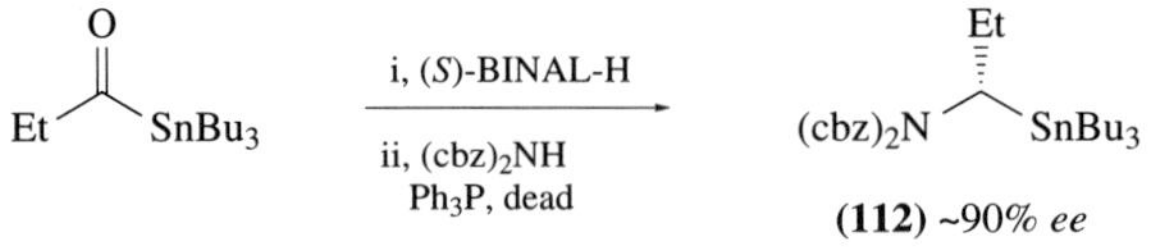

The synthesis of α-aminostannanes parallels to some extent the methods available for preparation of the analogous silanes (Section 4.11.1.1). Thus, Abel *et al.* have described a complementary synthetic approach to the foregoing methods which uses R_3SnCH_2I (*cf.* TMS-CH₂I) as the tin source ⟨75JOM(97)159⟩. Reaction with primary amines by simple mixing in ether gives a mixture of mono- and dialkylation, but higher yields and cleaner reactions ensue either when Et₃N is added, or when a metal salt of a secondary amine is used (see also ⟨76JOM(113)C13⟩). The last method is particularly well suited to α-stannylaziridine synthesis (see Scheme 48). Heteroaromatic systems are also readily alkylated by these methods; the pyrrole (113) is obtained in 80% yield. Alkylation of amide and carbamate NH functions is not yet clearly defined as a viable method. Chong and Park have attempted to alkylate secondary amides with (114) using NaH in DMF and obtained only the vinylstannane (115) ⟨93JOC523⟩.

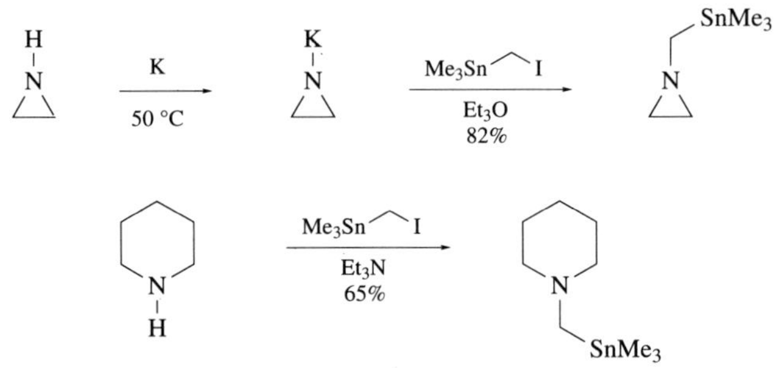

Scheme 48

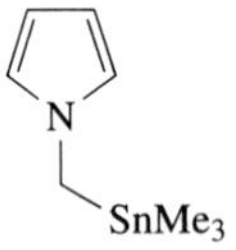

(113) N-CH₂SnMe₃ pyrrole

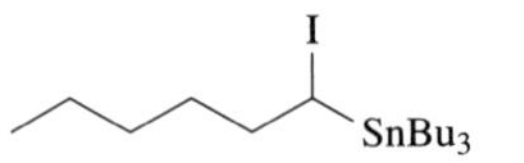

(114)

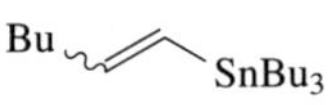

(115)

Few reports give details of α-aminozinc derivatives despite evidence in those papers that such species are relatively stable and can in some cases be isolated. Wittig and Schwarzenbach prepared bis(trimethylaminomethyl)zinc chloride (116) by the reaction of $ZnCl_2$, diazomethane and Me₃N ⟨61LA(650)1⟩. In the mid 1970s, a low yield (7%) of a species believed to be (117) was obtained by the reaction of the Simmons–Smith reagent (Zn-CH₂I₂) with Me₃N ⟨76IC1988⟩.

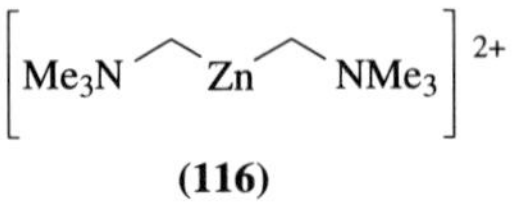

(116) [Me₃N–CH₂–Zn–CH₂–NMe₃]²⁺ 2Cl⁻

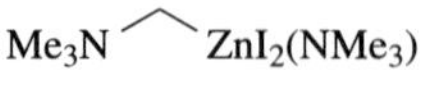

(117) Me₃N–CH₂–ZnI₂(NMe₃)

4.11.2.2 α-Metallated Imine Functions

4.11.2.2.1 Lithium

This section addresses lithiation by deprotonation; lithiation by metal exchange with organostannanes was briefly discussed in Section 4.11.2.1.2.

As mentioned in Section 4.11.2.1, the lithiation of otherwise unactivated amines can be achieved by first generating an aldimine or ketimine which is able to delocalize the new carbanion as an aza-allyl anion. The process is most successful with methylamine-derived imines which ensure that addition of the electrophile invariably occurs regioselectively at the site of deprotonation. Hydrolysis then releases the newly substituted amine as shown in generic form in Scheme 49. Much of the pioneering work in this field relates to benzophenone imines which offer the multiple advantages of increased stability, improved anion stabilizing capacity and complete regiocontrol of the alkylation site ⟨70AG(E)163, 73BSF2989, 77CB2659⟩. Two examples from this early work, shown in Scheme 50, illustrate the relative simplicity of the methodology and also demonstrate that both BuLi and lithium diisopropylamide (LDA) can be used as the base, with or without HMPA. Further examples of α-lithiated imines are tabulated in Beak's 1984 review ⟨84CRV471⟩.

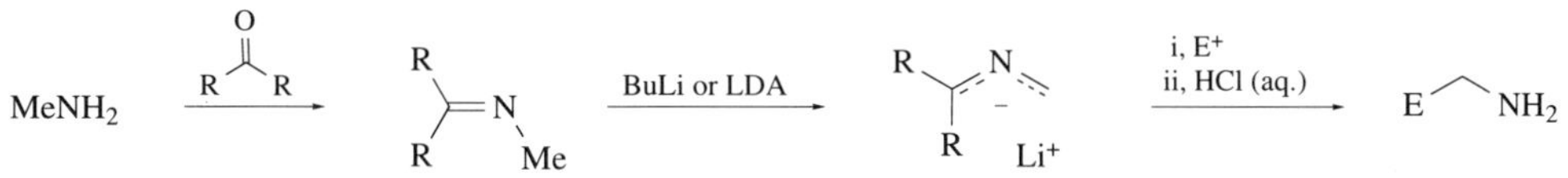

Scheme 49

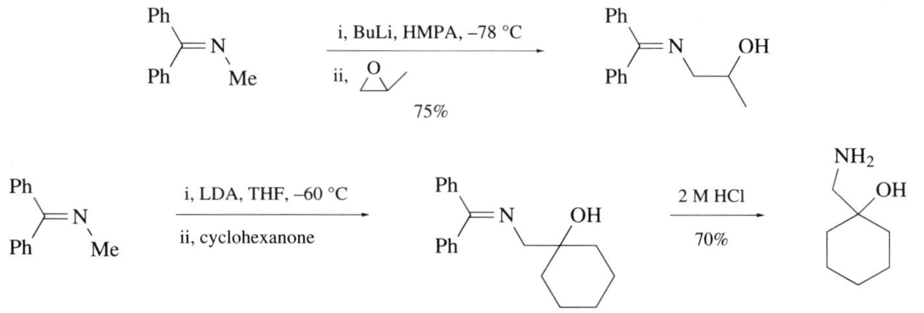

Scheme 50

Additional stabilization of the incipient aza-allyl carbanion increases the lifetime of the metallated intermediate. Lithiated imines of glycine have been used as glycine carbanion synthons, especially in the synthesis of unnatural α-amino acids. In two more recent examples from the late 1980s shown in Scheme 51, the metallo-imine (**118**) is coupled under palladium catalysis with allenes and allyl phosphates to give functionalized amino acid precursors ⟨88S983, 89TL3963⟩.

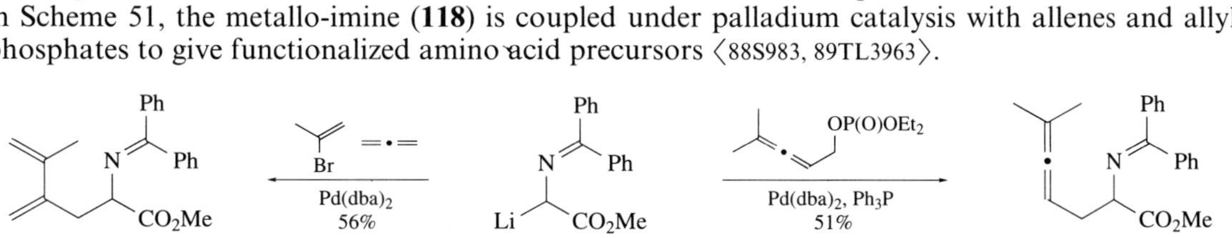

Scheme 51

The search for glycine synthons has culminated in Schöllkopf's elegant diastereocontrolled method which uses the bis-lactim ether (**119**) derived from (*S*)-valine (as the chiral control element) and glycine ⟨79AG(E)863, 83PAC1799, 83T2085⟩. Deprotonation by BuLi or LDA occurs exclusively at the glycine-derived methylene (C-5) and high selectivity for alkylation on the face opposite to the very bulky C-2 substituent is typically observed. The methodology is not restricted to glycine synthons; α,α-disubstituted amino acids derived from alanine can be obtained by alkylation and hydrolysis of the bis-lactim ether (**120**). The intermediate lithiated species behaves as a typical alkyllithium reagent in its reactivity towards electrophiles—Scheme 52 contains a diagrammatic summary of the potential of (**119**), highlighting in particular some of the less-familiar C—C bond forming reactions ⟨83LA1133, 86S737, 87AG(E)480, 88AG(E)433, 88AG(E)1194, 90JCS(P1)2251⟩. A cautionary practical note is that Schöllkopf's methodology inevitably leads to an intimate mixture of product and valine methyl ester which must be separated after hydrolysis.

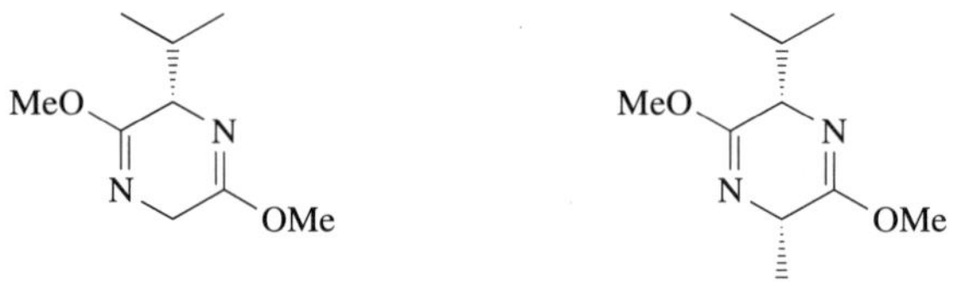

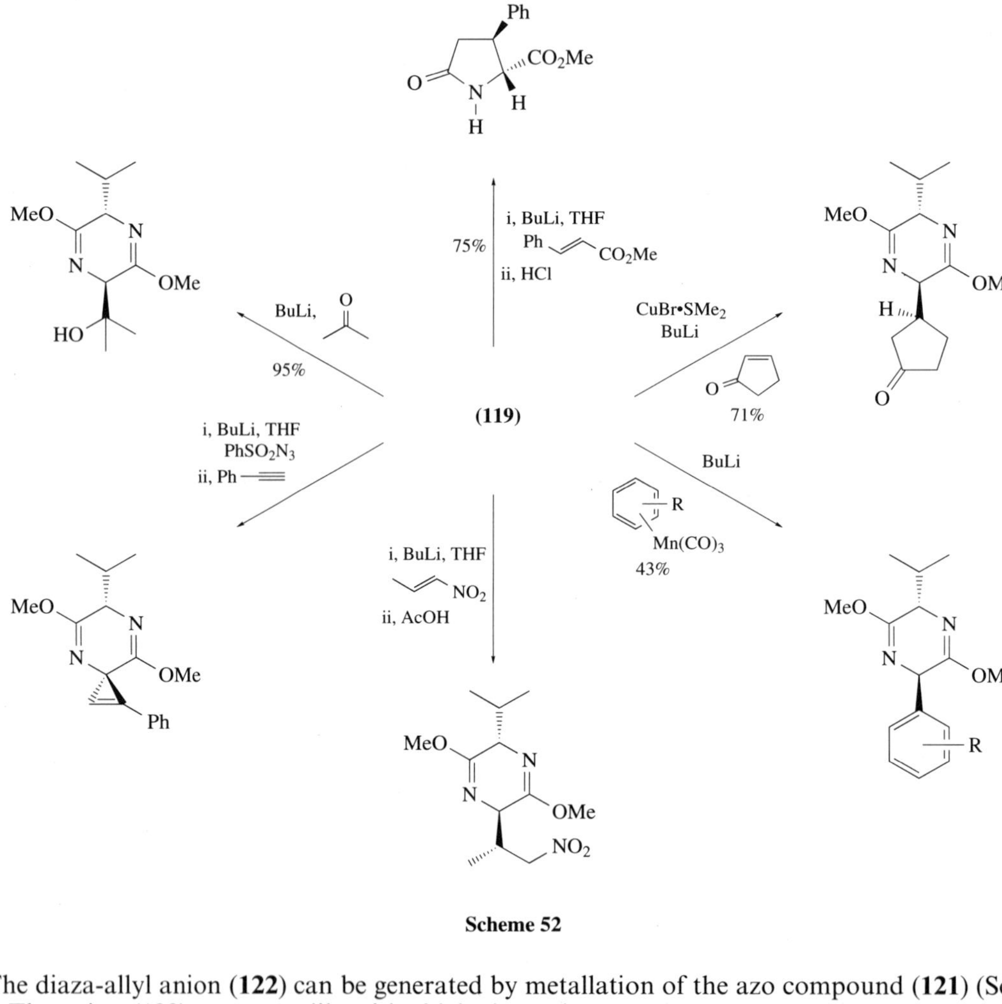

(**119**)

(**120**)

Scheme 52

The diaza-allyl anion (**122**) can be generated by metallation of the azo compound (**121**) (Scheme 53). The anion (**122**) reacts readily with aldehydes or ketones ⟨77CB3034⟩. Baldwin and co-workers have extensively investigated the metallation of *t*-butylhydrazones, as shown in Scheme 54. Lithiation of a *t*-butylhydrazone followed by reaction once again with an aldehyde or ketone produces an azo intermediate which is easily isomerized back to a hydrazone on exposure to BuLi. Hydrolysis of the hydrazone completes a useful and general synthesis of α-hydroxy ketones ⟨83CC1040, 86T4223⟩.

4.11.2.2.2 Tin

Pearson *et al.* have developed a useful, mild route to α-stannyl imines as precursors to lithio imines which undergo [π4s + π2s] cycloadditions to anionophilic alkenes, generating pyrrolidines.

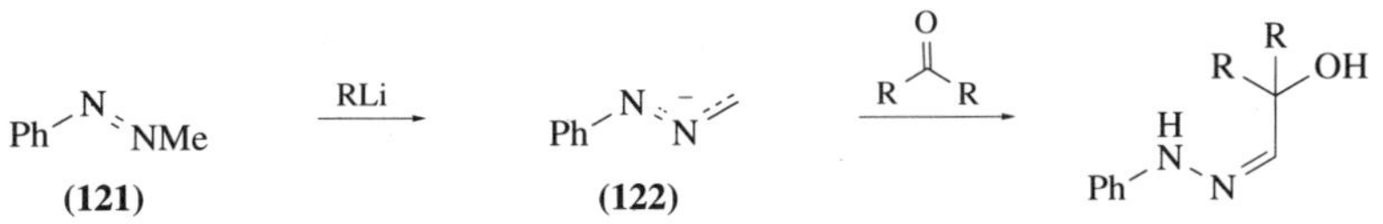

Scheme 53

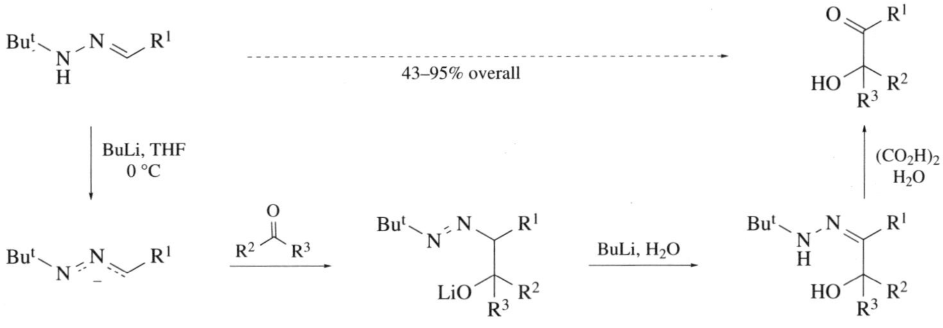

Scheme 54

Stannylmethyl iodides or mesylates readily undergo displacement by NaN$_3$ to give an azidomethylstannane (**123**). (**123**) undergoes a Staudinger reaction with Ph$_3$P and an aldehyde giving good yields of α-stannyl imines as shown in Scheme 55 ⟨88TL761, 92JA1329, 92JOC6354⟩. The ready transmetallation from tin to lithium is illustrated by the generation and stereoselective intramolecular cycloaddition reaction of an aza-allyl anion by this methodology (Scheme 56) ⟨88TL761⟩. The same group has also prepared α-stannyl imines by more traditional condensations of α-aminostannanes (Section 4.11.2.1.2) with aldehydes and ketones in good yield ⟨92JOC6354⟩.

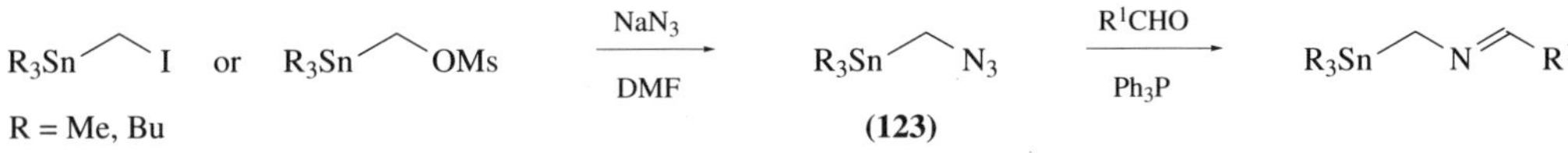

Scheme 55

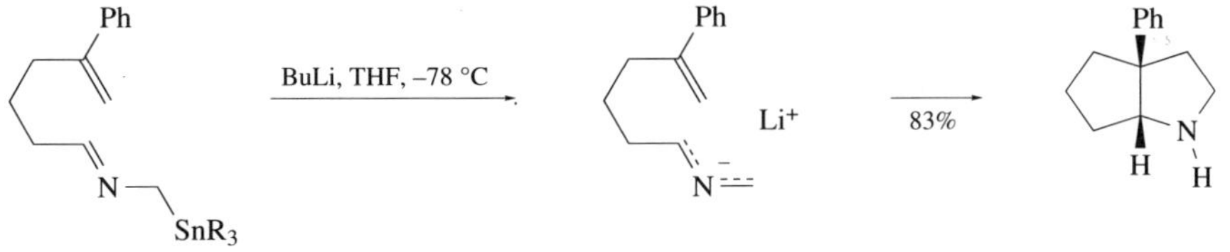

Scheme 56

4.11.2.3 α-Metallated Isocyanides and Isothiocyanates

4.11.2.3.1 Isocyanides

The synthesis and chemistry of α-lithiated isocyanides are the subject of two comprehensive reviews ⟨77AG(E)339, 84CRV471⟩ and well-described preparative details are available for the generation and silylation of LiCH$_2$NC ⟨84SC639⟩. Lithiated isocyanides are readily generated by alkyllithiums and are unusual stabilized carbanions in that the terminal isocyanide carbon is an electrophilic centre; immediate products of alkylation often react further, typically by a cyclization to generate heterocycles. This is summarized in Scheme 57 which highlights the potential of these reagents for the synthesis of oxazolines, thiazolines, imidazolines, pyrrole derivatives and related compounds.

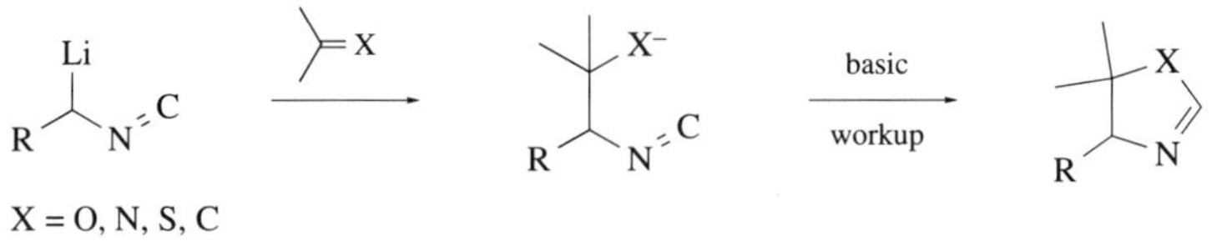

Scheme 57

$X = O, N, S, C$

Many explicit examples are cited in the quoted reviews and will not be duplicated here, but several points may usefully be made. Cyclization of lithiated isocyanides with carbonyl equivalents generates a new heterocyclic carbanion which is generally protonated on workup, but further, tandem reactions of these new anions are possible, as in Kozikowski's synthesis of the the disubstituted oxazoline (**124**) shown in Equation (49) ⟨80JOC2548⟩. Schöllkopf and co-workers have also detailed the chemistry of the alkylidene bis-isonitriles $CN(CH_2)_nNC$ ($n = 2$, 3 and 4) and their mono- and dianions ⟨80LA28⟩. Monolithiation of the isonitrile (**125**) results in interception by the unmetallated isocyanide to give a pyrroline after silylation to force the reaction to completion (Equation (50)).

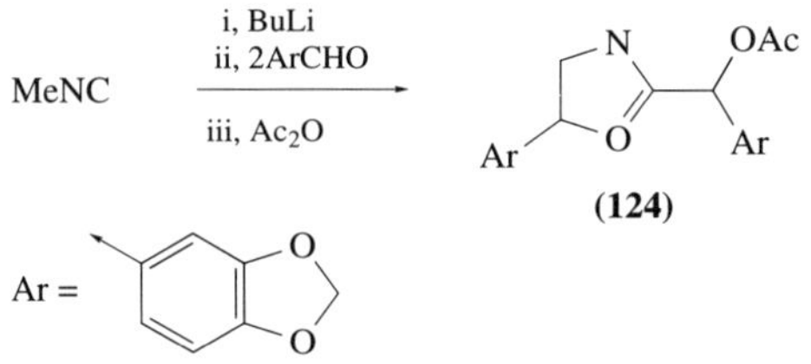

$$\text{MeNC} \xrightarrow[\text{iii, Ac}_2\text{O}]{\substack{\text{i, BuLi} \\ \text{ii, 2ArCHO}}} \text{(124)} \tag{49}$$

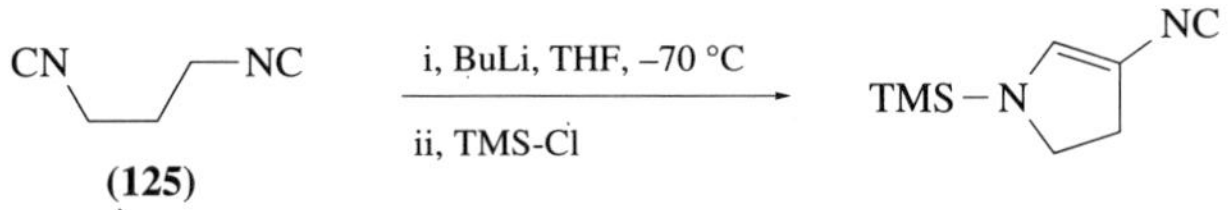

Ar =

$$\text{(125)} \xrightarrow[\text{ii, TMS-Cl}]{\text{i, BuLi, THF, }-70\,°\text{C}} \tag{50}$$

Van Leusen *et al.* have detailed the synthesis of tosylmethyl isocyanide (TosMIC) by lithiation of MeNC and reaction with TsF (Equation (51)) ⟨72TL2367⟩. TosMIC is an exceptionally versatile isonitrile, commercially available, already with a large literature describing the synthesis of substituted α-hydroxy aldehydes, nitriles, pyrroles, imidazoles and thiazoles. Details of the chemistry of TosMIC can be found in several useful leading references and overviews ⟨77TL4229, 79CPB2857, 80S325⟩.

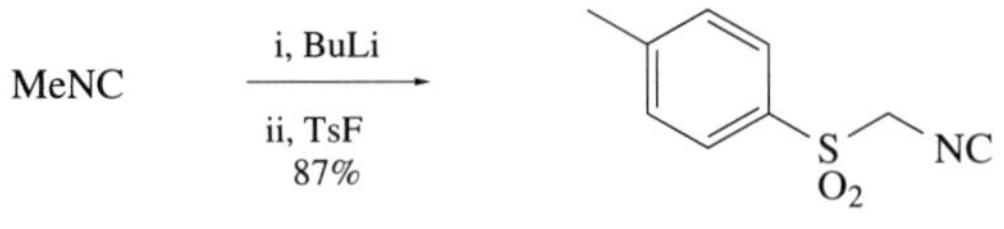

$$\text{MeNC} \xrightarrow[\substack{\text{ii, TsF} \\ 87\%}]{\text{i, BuLi}} \tag{51}$$

4.11.2.3.2 Isothiocyanates

Lithiated isothiocyanates without additional carbanion-stabilizing groups are only rarely invoked as intermediates in the literature. Metallation of MeNCS is reported to give the thiazolinethione (**126**) (Scheme 58) ⟨81AG(E)126⟩. A much more practical method for the generation of an equivalent carbanion has been realized by desilylation of TMS-CH$_2$NCS using tetraalkylammonium fluorides ⟨81AG(E)126, 82BCJ1163⟩. (See also Section 4.11.1.1.5.) The resultant salt (**127**) has been intercepted with a range of electrophiles, including carbonyl compounds which afford oxazolinethiones (Equation (52)). Similar isothiocyanate anions in which the formal charge is further stabilized by additional functionality have been prepared, and give similar results ⟨76CB3047, 76CB3062, 76TL609⟩.

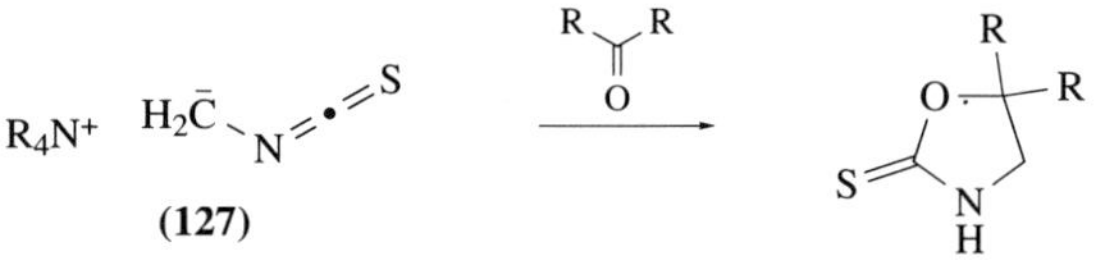

Scheme 58

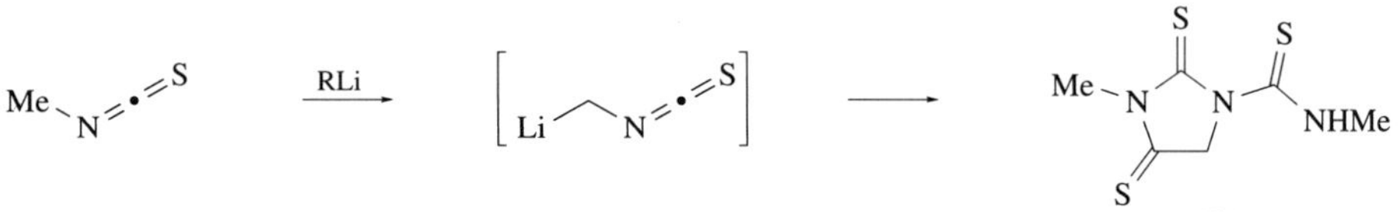

4.11.2.4 Metallation of *N*-Methyl Heterocycles

Nitrogen heterocycles, especially π-excessive heteroaromatic systems, which bear an *N*-methyl substituent can frequently be lithiated on that sp^3 carbon, either by deprotonation or transmetallation from Si and Sn species to give a dipole-stabilized carbanion. An important consideration and a major restriction is that the reaction is only practical if competing α-metallation of the aromatic ring is suppressed or impossible. This aspect is discussed in more detail in Gshwend and Rodriguez' monumental review of heteroatom-facilitated lithiations ⟨79OR(26)1⟩. This restriction is overcome when additional stabilizing functional groups are appended to the *N*-methyl group. Table 8 summarizes the conditions for lithiation of those heterocycles that afford synthetically useful anions. Only one example of the lithiation of *N*-methylindazoles appears in the literature (entry 2) and the extent of reaction with electrophiles is not fully defined. The metallation of tetrazoles reported by Moody *et al.* (entries 4 and 5) has been extended to higher *N*-alkyl derivatives. Lithiation and silylation of the *N*-butyl tetrazole (**128**) proceeds in useful yield (Equation (53)).

Table 8 Metallation of *N*-alkylheterocycles.

Entry	Substrate	Conditions	Metallated species	Yield (%)[a]	Ref.
1	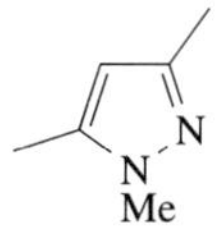	BuLi, THF, −78 °C	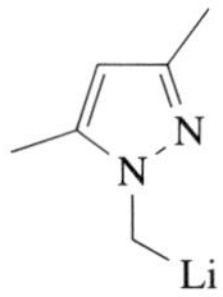	22–29	72JOC215 83T2023
2	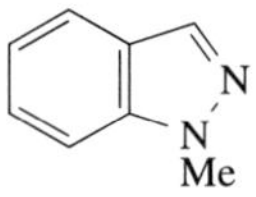	BuLi, −15 °C	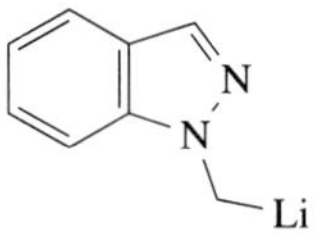		70MI 411-01
3		MeLi, THF, 0 °C	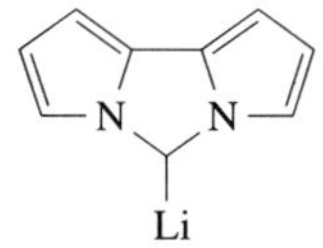		80HCA1190
4	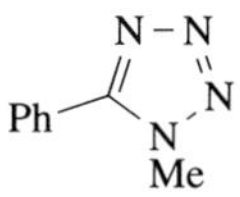	BuLi, THF, −78 °C	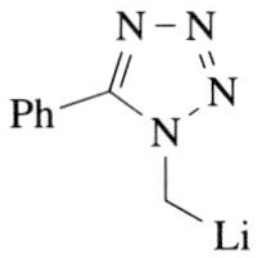	84–96	90SL413 91JCS(P1)323
5	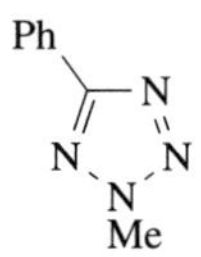	BuLi, THF, −78 °C	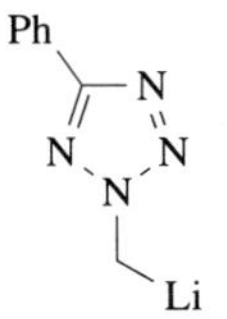	33–85	90SL413 91JCS(P1)323

[a] Yields represent maxima and minima for all electrophiles reported.

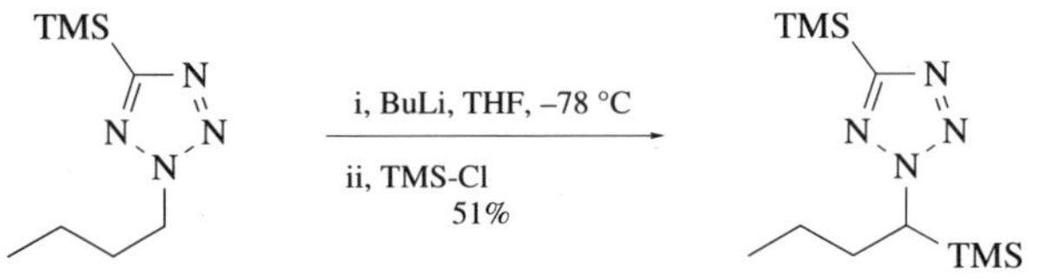

Katritzky *et al.* and Joule *et al.* have independently reported the lithiation of 1-methyl-2-pyridone with BuLi, generating (129) which is unstable and rapidly dimerizes at low temperatures ⟨82JCS(P1)143, 85CC1021⟩. Katritzky has subsequently reported the generation of a more useful equivalent of (129) by fluoride-mediated desilylation of the silylmethyl pyridone (130) ⟨87TL5419⟩ (for the synthesis of (130) see Section 4.11.1.1.7). The silane (130) reacts smoothly with aldehydes, alkyl halides and acylating agents.

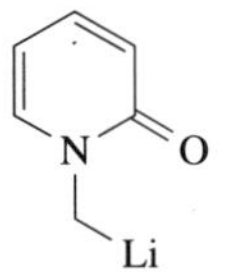

4.11.2.5 α-Metallated Nitroalkanes

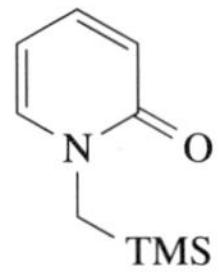

The synthesis and chemistry of α-anions derived from nitroalkanes is a vast, mature area which cannot be accommodated in this review. However, the field has been well surveyed by a number of excellent review articles and the aim of this section will be to point the reader to appropriate texts and reviews. Metallation of nitroalkanes by bases with hard counterions invariably generates nitronate (*aci*-nitro) intermediates (131) in which the metal is closely associated with the nitro group oxygen atoms ⟨88MI 411-01, 89AG(E)277⟩. Softer metals (SnR_3, for example) give more ambiguous outcomes in which both C-(132) and O-stannylated nitro compounds can be characterized. No C-silylated nitro compounds have been structurally confirmed. The relatively low pK_a (≈ 7–8) of nitroalkanes means that most reactions which require anion formation can be achieved under thermodynamic conditions using tertiary organic, or metal alkoxide bases.

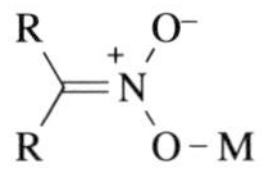

Two named reactions embrace the majority of nitroalkane chemistry, and both require α-nitro anions as intermediates. These are the Henry reaction and the Nef reaction. The reaction of nitronate metal salts with alkyl halides typically gives O-alkylation unless a soft counterion is used—this is often a silver ion ⟨79COC(2)305⟩. This restriction has been partly solved by double deprotonation of nitroalkanes using BuLi at low temperature to give a dianion which alkylates exclusively at the α-carbon atom as shown in Scheme 59 ⟨76AG(E)505⟩. The Henry reaction covers the condensation of aldehydes and ketones (and usually Michael acceptors) with nitroalkanes (Equation (54)) and has been reviewed ⟨59OR(10)179, 70MI 411-02⟩. In general, alkali metal hydroxides, alkoxides or carbonates are used to minimize Cannizzaro and aldol side reactions of the carbonyl component. In the late 1980s, an increasing number of Henry-type reactions, especially with nitromethane, appeared which use strong kinetic bases. For example, Equation (55) shows a diastereoselective version in which an arenetricarbonyl chromium complex induces face selectivity ⟨87JOM(330)357⟩.

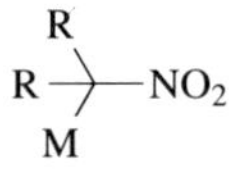

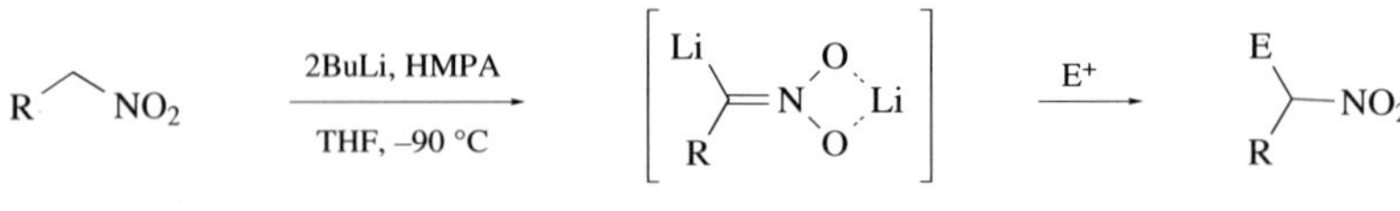

HMPA = hexamethylphosphoramide

Scheme 59

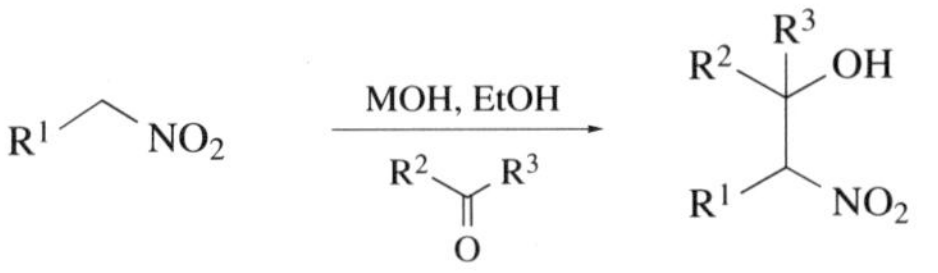

$$R^1\text{-}CH_2\text{-}NO_2 \quad \xrightarrow[R^2\text{-}CO\text{-}R^3]{\text{MOH, EtOH}} \quad R^2R^3C(OH)\text{-}CHR^1(NO_2) \tag{54}$$

$$MeNO_2 \quad \xrightarrow[\substack{\text{iii, } H_2O_2, CH_2Cl_2 \\ 50\%}]{\substack{\text{i, BuLi, } -40\,^\circ C \\ \text{ii, } o\text{-MeC}_6H_4\text{-}Cr(CO)_3\text{CHO}}} \quad O_2N\text{-}CH_2\text{-}CH(OH)\text{-}(o\text{-tolyl-Cr(CO)}_3) \tag{55}$$

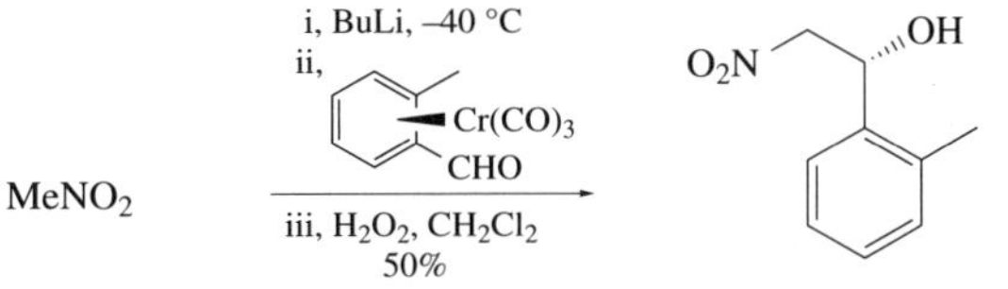

The Nef reaction covers all hydrolyses of nitronates to carbonyl compounds as shown in Scheme 60. The reaction is essentially independent of the base used to generate the nitronate intermediate, whereas much resource has been spent identifying reagents and conditions for the hydrolytic step. A review in 1990 comprehensively surveyed the known examples of, and conditions for the Nef reaction ⟨90OR(38)655⟩.

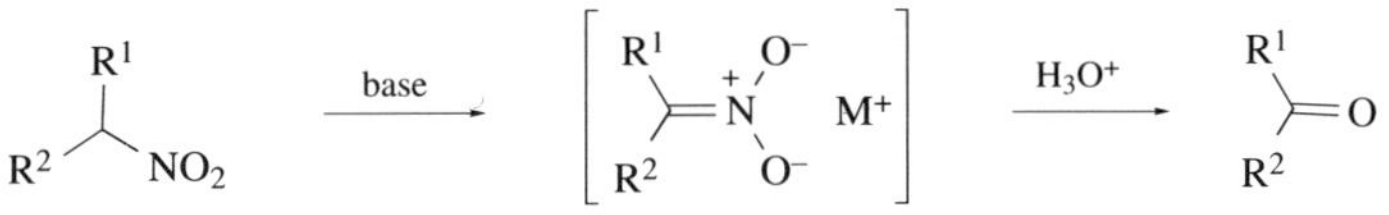

$$R^1R^2CH\text{-}NO_2 \quad \xrightarrow{\text{base}} \quad \left[R^1R^2C=N^+(O^-)_2 \; M^+ \right] \quad \xrightarrow{H_3O^+} \quad R^1R^2C=O$$

Most frequently used M = Na, K, R$_4$N, Li

Scheme 60

Finally, very few reports of heavy-metal derivatives of nitroalkanes have appeared in the literature. The nitrostannanes (**133**) and (**134**) have been provisionally characterized as minor components in the stannylation of nitroethane by R$_3$SnNEt$_2$ or (R$_3$Sn)$_2$O. Neither compound could be isolated pure ⟨87JOM(320)171⟩. Benzoylnitromethane is reported to react with alkynyltin compounds to give the nitrostannane (**135**) although only an English abstract of this report was available ⟨80ZOB1427⟩.

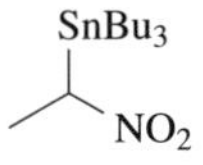

(**133**) Me$_2$C(SnBu$_3$)NO$_2$ (**134**) Me$_2$C(SnPh$_3$)NO$_2$ (**135**) PhC(O)CH(SnR$_3$)NO$_2$

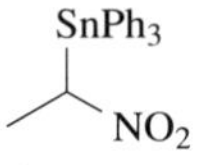

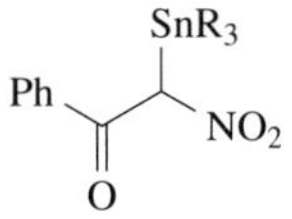

4.11.3 ACKNOWLEDGEMENT

The authors are indebted to Mrs. Sandra Wood of the Research Information Department, Pfizer Central Research, for her invaluable assistance in the preparation of this chapter.

4.12
Functions Containing One Phosphorus and Either Another Phosphorus or As, Sb, Bi, Si, Ge, B or a Metal

R. ALAN AITKEN
University of St Andrews, UK

4.12.1	FUNCTIONS CONTAINING TWO PHOSPHORUS ATOMS	545
4.12.1.1	*Symmetrical Dicoordinate Phosphorus Functions*	545
4.12.1.2	*Asymmetrical Systems Containing One Dicoordinate Phosphorus*	545
4.12.1.3	*Symmetrical Tricoordinate Phosphorus Functions*	545
4.12.1.3.1	From 1,1-dihalo alkanes	545
4.12.1.3.2	By nucleophilic substitution on 1-haloalkyl- and 1-aminoalkyl-phosphines	546
4.12.1.3.3	From 1,1-dimetallo alkanes	546
4.12.1.3.4	By electrophilic substitution on 1-metalloalkylphosphines	546
4.12.1.3.5	From (1-trialkylsilylalkyl)phosphines	547
4.12.1.3.6	From diphosphines	547
4.12.1.3.7	From diphosphenes	547
4.12.1.3.8	Other methods involving cleavage of a P—P bond	547
4.12.1.3.9	From diphosphinocarbenes	548
4.12.1.3.10	From 1,3-diphosphaalkenes	548
4.12.1.3.11	From 1,3-diphosphaallenes	548
4.12.1.3.12	From carbonyl compounds	548
4.12.1.3.13	From 1,1-diphosphino alkenes	548
4.12.1.3.14	From phosphaalkenes	549
4.12.1.3.15	From phosphaalkynes	549
4.12.1.3.16	Miscellaneous methods	549
4.12.1.3.17	By reduction of tetracoordinate systems	550
4.12.1.3.18	Interconversions	550
4.12.1.4	*Asymmetrical Systems Containing at Least One Tricoordinate Phosphorus*	551
4.12.1.4.1	From 1-aminoalkylphosphines	551
4.12.1.4.2	From 1-metalloalkylphosphorus compounds	551
4.12.1.4.3	By electrophilic attack on phosphorus ylides	551
4.12.1.4.4	From vinylphosphorus compounds	552
4.12.1.4.5	Miscellaneous methods	552
4.12.1.4.6	By oxidation of symmetrical tricoordinate systems	552
4.12.1.4.7	By reduction of symmetrical tetracoordinate systems	553
4.12.1.4.8	Interconversions	553
4.12.1.5	*Symmetrical Tetracoordinate Systems*	553
4.12.1.5.1	From reactions of 1,1-dihalo alkanes with phosphorus nucleophiles	553
4.12.1.5.2	From reactions of 1-haloalkylphosphorus compounds with phosphorus anions	554
4.12.1.5.3	From reactions of 1-haloalkylphosphorus compounds with phosphines	554
4.12.1.5.4	From Arbuzov reactions of 1,1-dihalo alkanes	555
4.12.1.5.5	From Arbuzov reactions of 1-haloalkylphosphorus compounds	555
4.12.1.5.6	From 1,1-dimetallo alkanes	556
4.12.1.5.7	From electrophilic substitution on 1-metalloalkylphosphorus compounds	557

4.12.1.5.8	From diphosphorus-substituted carbenes	557
4.12.1.5.9	By electrophilic attack on phosphorus ylides and from 1,3-diphosphaalkenes generated in other ways	557
4.12.1.5.10	From phosphaalkenes	558
4.12.1.5.11	From diphosphorus-substituted ketenes and related compounds	559
4.12.1.5.12	From 1,3-diphosphaallenes	559
4.12.1.5.13	From carbonyl compounds	559
4.12.1.5.14	From 1,1-diphosphorus-substituted alkenes	560
4.12.1.5.15	From vinylphosphorus compounds	561
4.12.1.5.16	From alkynes	561
4.12.1.5.17	Miscellaneous methods	561
4.12.1.5.18	By oxidation of tricoordinate species	562
4.12.1.5.19	Interconversions	563
4.12.1.6	Penta- and Hexacoordinate Systems	564

4.12.2 FUNCTIONS CONTAINING ONE PHOSPHORUS AND ONE ARSENIC, ANTIMONY OR BISMUTH GROUP — 564

4.12.2.1	Phosphorus and Arsenic Functions	564
4.12.2.2	Phosphorus and Antimony Functions	565
4.12.2.3	Phosphorus and Bismuth Functions	565

4.12.3 FUNCTIONS CONTAINING PHOSPHORUS AND A METALLOID — 565

4.12.3.1	Dicoordinate Phosphorus Derivatives	565
4.12.3.1.1	Dicoordinate phosphorus and silicon functions	565
4.12.3.1.2	Dicoordinate phosphorus and germanium functions	565
4.12.3.1.3	Dicoordinate phosphorus and boron functions	565
4.12.3.2	Tricoordinate Phosphorus Derivatives	565
4.12.3.2.1	Tricoordinate phosphorus and silicon functions	565
4.12.3.2.2	Tricoordinate phosphorus and germanium functions	568
4.12.3.2.3	Tricoordinate phosphorus and boron functions	569
4.12.3.3	Tetracoordinate Phosphorus Compounds	569
4.12.3.3.1	Tetracoordinate phosphorus and silicon functions	569
4.12.3.3.2	Tetracoordinate phosphorus and germanium functions	574
4.12.3.3.3	Tetracoordinate phosphorus and boron functions	574
4.12.3.4	Higher Coordinate Phosphorus Derivatives	575
4.12.3.4.1	Higher coordinate phosphorus and silicon functions	575
4.12.3.4.2	Higher coordinate phosphorus and germanium functions	575
4.12.3.4.3	Higher coordinate phosphorus and boron functions	575

4.12.4 FUNCTIONS CONTAINING PHOSPHORUS AND A METAL — 575

4.12.4.1	Group 1 and 2 Derivatives	575
4.12.4.1.1	Compounds containing phosphorus and lithium	575
4.12.4.1.2	Compounds containing phosphorus and sodium	576
4.12.4.1.3	Compounds containing phosphorus and potassium	577
4.12.4.1.4	Compounds containing phosphorus and beryllium	577
4.12.4.1.5	Compounds containing phosphorus and magnesium	577
4.12.4.1.6	Compounds containing phosphorus and heavier group 1 and 2 metals	577
4.12.4.2	Compounds Containing Phosphorus and a Lanthanide	577
4.12.4.3	Transition Metal Derivatives	578
4.12.4.3.1	Compounds containing phosphorus and scandium or yttrium	578
4.12.4.3.2	Compounds containing phosphorus and titanium, zirconium or hafnium	578
4.12.4.3.3	Compounds containing phosphorus and vanadium, niobium or tantalum	579
4.12.4.3.4	Compounds containing phosphorus and chromium, molybdenum or tungsten	579
4.12.4.3.5	Compounds containing phosphorus and manganese or rhenium	581
4.12.4.3.6	Compounds containing phosphorus and iron, ruthenium or osmium	581
4.12.4.3.7	Compounds containing phosphorus and cobalt, rhodium or iridium	582
4.12.4.3.8	Compounds containing phosphorus and nickel, palladium or platinum	583
4.12.4.3.9	Compounds containing phosphorus and copper, silver or gold	585
4.12.4.3.10	Compounds containing phosphorus and zinc, cadmium or mercury	586
4.12.4.4	Group 13 and 14 Derivatives	587
4.12.4.4.1	Compounds containing phosphorus and aluminum	587
4.12.4.4.2	Compounds containing phosphorus and gallium, indium or thallium	587
4.12.4.4.3	Compounds containing phosphorus and tin or lead	588
4.12.4.5	Actinide Derivatives	589

4.12.1 FUNCTIONS CONTAINING TWO PHOSPHORUS ATOMS

4.12.1.1 Symmetrical Dicoordinate Phosphorus Functions

A single example of this rare functional group is provided by the sodium salt (**1**) which is obtained, together with many other products, by the reaction of elemental phosphorus with sodium in DIGLYME ⟨87ZAAC(544)87⟩.

(**1**)

4.12.1.2 Asymmetrical Systems Containing One Dicoordinate Phosphorus

An example of this functional group, which is also extremely rare, is the bis(phosphonium) salt (**3**) which is formed by protonation of (**2**) followed by treatment with $AlCl_3$ ⟨93PS(76)13⟩.

(**2**)   (**3**)

4.12.1.3 Symmetrical Tricoordinate Phosphorus Functions

4.12.1.3.1 From 1,1-dihalo alkanes

This approach was first described in 1959 when it was reported that Ph_2PNa, formed by treatment of Ph_2PCl with sodium, reacted with CH_2Cl_2 to give the bis(phosphine) $Ph_2PCH_2PPh_2$ in 78% yield ⟨59CB3175⟩. The reaction can be achieved using sodium, potassium or lithium derivatives of secondary or primary phosphines with varying degrees of success. The range of dihalo alkanes used is limited. Reported examples include the reactions of CH_2Cl_2 with MePHNa ⟨63USP3086053, 63USP3086056⟩, EtPHK ⟨63ZAAC(324)259⟩, Ph_2PLi ⟨70ZAAC(376)37⟩, TMS_2PLi ⟨77ZAAC(431)76⟩, $(o\text{-}MeC_6H_4)_2PLi$ ⟨81JOM(217)51⟩, PhPHK ⟨89PS(42)97⟩ and (2-pyridyl)PhPNa ⟨90OM1222⟩, $MeCHCl_2$ with Ph_2PLi ⟨70ZAAC(376)37, 86OM2220⟩, $MeCHBr_2$ with Ph_2PNa ⟨76JOM(117)159⟩ and Me_2CCl_2 with Ph_2PNa ⟨62JCS1490⟩. The parent $PH_2CH_2PH_2$ can be formed in 60% yield as its bis-$(CO)_5W$ complex by treatment of $(CO)_5W(TMS)_2PLi$ with CH_2BrCl followed by hydrolysis ⟨87JOM(328)349⟩. Similarly, treatment of $Cp(CO)_2Mn-PPh_2Na$ with CH_2Cl_2 gives the corresponding dicoordinated bis(phosphine) ⟨76ZN(B)790⟩. Treatment of lithio(dialkyl)phosphines with CH_2Cl_2 led to lithium–halogen exchange giving R_2PPR_2 ⟨59CB3175⟩ but $Me_2PCH_2PMe_2$ can be produced in 58% yield by the reaction of CH_2Cl_2 with Me_2PCH_2Li ⟨77ZN(B)762⟩.

Insertion of a CH_2 unit into cyclopolyphosphines can be achieved by treatment with potassium metal followed by CH_2Cl_2. Treatment of cyclo-$(PhP)_5$ in this way affords (**4**) in 73% yield ⟨71AG(E)940⟩ and when this compound is further treated with 2 equivalents of potassium and CH_2Cl_2 it produces (**5**) in 18% yield ⟨73ZN(B)224⟩. In a similar way, treatment of the 1,2-diphosphetes (**6**) with lithium then CH_2Cl_2 gives (**7**) ⟨93PS(77)254⟩. The reactions of 1,2-dimetallodiphosphines with 1,1-dihalo alkanes have been used to form diphosphiranes. Thus PhP(K)P(K)Ph reacts with Me_2CCl_2 at −30 °C to give (**8**) ⟨77ZN(B)1490⟩ and similar reactions of $Bu^tP(K)P(K)Bu^t$ with CH_2Cl_2, Me_2CCl_2 ⟨78ZN(B)1208⟩ and $MeCHCl_2$ ⟨82ZAAC(486)39⟩ give the corresponding diphosphiranes. In a related process, TMS-P(Ph)P(Ph)-TMS reacts with CH_2I_2 to give (**9**) ⟨77ZN(B)1490⟩. Treatment of cyclo-

(MeP)₅ with potassium followed by ClCH=CHCl gives a mixture of two isomers of the bicyclic compound (10) ⟨78ZN(B)691⟩.

Early attempts to prepare R₂PCH₂PR₂ by the direct reaction of R₂PH with dihalomethanes followed by base were unsuccessful ⟨63CB2186⟩ but this has more recently been achieved in good yield, using CH₂Cl₂ and KOH in aqueous DMF or DMSO, for Ph₂PH ⟨86S198⟩, PhPH₂ ⟨87CB1707⟩ and (2-pyridyl)PH₂ ⟨90CB989⟩.

The valuable synthetic intermediate Cl₂PCH₂PCl₂ may be prepared by a Friedel–Crafts-type reaction of CH₂Cl₂ with aluminum followed by PCl₃. The original method required subsequent treatment with KCl and POCl₃ to free the product ⟨76URP539892, 77ZOB775⟩, but an improved procedure giving the product in 38% yield has appeared ⟨89IS120⟩. Minor products include Cl₂P-CH₂P(Cl)CH₂PCl₂ which can be isolated in 7% yield and Cl₂PCH₂P(Cl)CH₂Cl ⟨82AG(E)376⟩. Cl₂Al-CH₂Al(Cl)CH₂Cl and MeOPCl₂ give a 50% yield of Cl₂PCH₂PCl₂ ⟨81ZOB484⟩.

4.12.1.3.2 By nucleophilic substitution on 1-haloalkyl- and 1-aminoalkyl-phosphines

Me₂PCH₂Cl and Me₂PLi readily give Me₂PCH₂PMe₂ in 46% yield ⟨82CB823⟩. By adjusting the conditions, the reaction of Ph₂PCH₂Cl with PH₃ under phase-transfer conditions can be used to obtain either Ph₂PCH₂PH₂ in 84% yield or Ph₂PCH₂PHCH₂PPh₂ in 71% yield ⟨90CB995⟩. The related substitution of diethylaminomethylphosphines has also been described. Thus, Ph₂PCH₂NEt₂ and Ph₂PH at 180 °C give Ph₂PCH₂PPh₂ (31%) while Et₂NCH₂P(Ph)CH₂NEt₂ similarly gives Ph₂PCH₂P(Ph)CH₂PPh₂ (65%) ⟨65HCA1034, 66USP3253033⟩.

4.12.1.3.3 From 1,1-dimetallo alkanes

Treatment of 2,6-lutidine with 2 equivalents of PhLi followed by 2 equivalents of Ph₂PCl gives the bis(phosphine) (11) ⟨71JCS(A)3495⟩. CH₂Li₂ and Me₂PCl give Me₂PCH₂PMe₂ in 78% yield ⟨77ZN(B)762, 78GEP2658127⟩.

4.12.1.3.4 By electrophilic substitution on 1-metalloalkylphosphines

Metallation of R¹₂PMe followed by treatment with R²₂PCl gives the bis(phosphines) R¹₂PCH₂PR²₂ in moderate to good yield. Examples include the reaction of Ph₂PCH₂Li, formed by deprotonation of Ph₂PMe with BuⁿLi/TMEDA, with Ph₂PCl or Ph₂POPh to give Ph₂PCH₂PPh₂ (51%) ⟨67JOM(8)199⟩, and with PhP(Me)Cl or PhP(Prⁱ)Cl ⟨77IC1770⟩. Treatment of PhP(Me)CH₂Li with PhP(Me)Cl gives the symmetrical bis(phosphine) in 20% yield ⟨78JCR(S)368⟩. With aliphatic phosphines deprotonation requires BuᵗLi but phosphinylation proceeds similarly, as exemplified by the reaction of Me₂PCH₂Li with Me₂PCl (65%) ⟨78GEP2658127⟩, Buᵗ₂PCl and Ph₂PCl ⟨77ZN(B)762⟩, and of Buᵗ₂PCH₂Li with Buᵗ₂PCl (80%) ⟨92GEP4134772⟩. In the last example, the full range of alternative alkali metals and halogens was examined. Me₂PCH₂Li and MePCl₂ give the tris(phosphine) Me₂P-CH₂P(Me)CH₂PMe₂ ⟨82ZN(B)284⟩ while with P(OPh)₃ the tetrakis(phosphine) (Me₂PCH₂)₃P is formed ⟨77ZN(B)762⟩. The sodium enolates of α-phosphino esters such as $Pr^i_2PCH_2CO_2Et$ generally react with electrophiles such as Pr^i_2PCl and $(EtO)_2PCl$ at carbon to give $Pr^i_2PCH(CO_2Et)PR_2$, although there is competing reaction at oxygen in some cases ⟨76ZOB571⟩. The 2-zirconiophospholane formed by hydrozirconation of 1-phenyl-2-phospholine with Cp_2ZrHCl reacts with a variety of electrophiles, R_2PCl and R_2POTf (Tf = trifyl), to afford the *cis*-2-phosphinophospholanes (12) ⟨93AG(E)1735⟩.

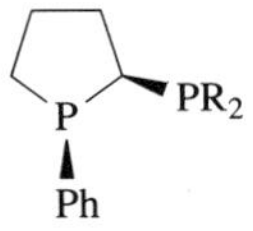

(12)

4.12.1.3.5 From (1-trialkylsilylalkyl)phosphines

The reaction of $R^1_2PCH_2$-TMS with R^2_2PCl proceeds with elimination of TMS-Cl to give the bis(phosphine) products $R^1_2PCH_2PR^2_2$ in good yield ⟨79CB648⟩. With PCl_3 as the chlorophosphine component, a variety of compounds $R_2PCH_2PCl_2$ useful for further transformations have been obtained ⟨83ZOB699⟩. An excess of Ph_2PCH_2-TMS with $Cl_2PCH_2CH_2PCl_2$ gives the hexakis(phosphine) $(Ph_2PCH_2)_2PCH_2CH_2P(CH_2PPh_2)_2$ ⟨88JOM(338)C31⟩.

4.12.1.3.6 From diphosphines

A variety of methods involving insertion of a carbon atom into the P—P bond of a diphosphine have been reported. The parent diphosphine, PH_2PH_2, and CH_2N_2 are reported to give $PH_2CH_2PH_2$ in 28% yield ⟨78GEP2705994⟩. Tetrakis(trifluoromethyl)diphosphine (3 equivalents) with CH_2I_2, $MeCHI_2$ or Me_2CI_2 in a sealed tube at 130 °C gives the corresponding products $(F_3C)_2PCR^1R^2P(CF_3)_2$ in good to excellent yield ⟨88IC4038⟩. In an unusual reaction, photolysis of a gaseous mixture of propyne and PF_2PF_2 gives a 5% yield of $MeC(PF_2)_2CH(PF_2)_2$ ⟨88JFC(40)41⟩.

4.12.1.3.7 From diphosphenes

The Yoshifuji diphosphene ArP=PAr (Ar = 2,4,6-$Bu^t_3C_6H_2$) reacts either thermally with CH_2N_2 at RT ⟨86TL1145⟩ or photochemically with Ph_2CN_2 or 9-diazofluorene ⟨87T1793⟩ to give the diphosphiranes (13).

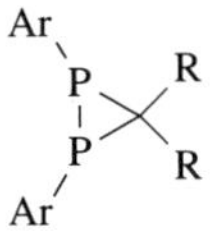

(13)
R = H, Ph; Ar = 2,4,6-$Bu^t_3C_6H_2$

4.12.1.3.8 Other methods involving cleavage of a P—P bond

The tricyclic zirconium compound (14) with HCl in Et_2O gives the 1,3-diphosphetane (15) ⟨B-90MI 412-01⟩. Treatment of (14) with $HgCl_2$ in Et_2O similarly affords (16) while Br_2 gives the corresponding dibromide ⟨B-90MI 412-01⟩.

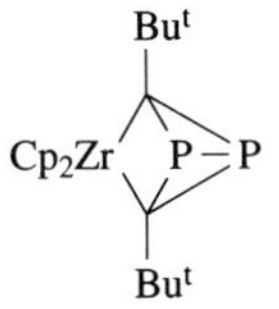

(14)

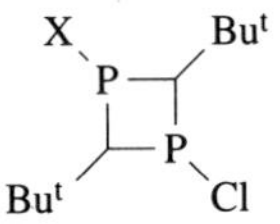

(15) X = H
(16) X = Cl

4.12.1.3.9 From diphosphinocarbenes

Intramolecular CH insertion of the carbene produced by heating $(Pr^i_2N)_2PC(=N_2)P(NPr^i_2)_2$ at 110 °C leads to the 1,2-azaphosphetane (**17**) ⟨90PS(47)327⟩.

4.12.1.3.10 From 1,3-diphosphaalkenes

The Diels–Alder cycloaddition of 1,4-diphenylbutadiene to the C=P double bond of 2-t-butyl-1-ethyl-1,3-diphosphaindene affords the diphosphatetrahydrofluorene (**18**) ⟨90PS(49/50)355⟩. Protonation of the mixture of cyclic anions $(Bu^tC)_3P_2^-$ and $(Bu^tC)_2P_3^-$ results in [4 + 2] cycloaddition to give the compound (**19**), which reacts with EtOH by addition to the C=P double bond to afford (**20**) ⟨90CC1307, 93PS(77)274⟩. Treatment of $Bu^tP(Cl)CH_2P(Cl)Bu^t$ with Et_3N results in elimination of HCl and head-to-head dimerization of the resulting 1,3-diphosphaalkene to produce (**21**) ⟨87ZOB2637⟩.

4.12.1.3.11 From 1,3-diphosphaallenes

Two consecutive 1,2-H migrations occur spontaneously in the carbodiphosphorane $(Pr^i_2N)_2PH=C=PH(NPr^i_2)_2$ at RT to afford $(Pr^i_2N)_2PCH_2P(NPr^i_2)_2$ ⟨93AG(E)1167⟩.

4.12.1.3.12 From carbonyl compounds

Certain aromatic aldehydes such as *p*-dimethylaminobenzaldehyde undergo acid-catalysed condensation with 2 equivalents of Ph_2PH to produce the acetal-like compounds $ArCH(PPh_2)_2$, although the reaction is not general ⟨79ZC416, 81T753⟩.

4.12.1.3.13 From 1,1-diphosphino alkenes

Addition to the double bond of various 1,1-diphosphino alkenes, notably $(Ph_2P)_2C=CH_2$, has been used to obtain 1,1-diphosphino alkanes. This compound with dimethylsulfoxonium methylide gives 1,1-bis(diphenylphosphino)cyclopropane ⟨84HCA2175⟩. Addition of Ph_2PH catalysed by $KOBu^t$ occurs readily to give $(Ph_2P)_2CHCH_2PPh_2$ in up to 94% yield ⟨86CC1041, 88CB1241, 88JCS(D)503⟩. The reaction is also successful for chiral phospholanes ⟨91JOM(413)55⟩. Twofold addition occurs for primary phosphines such as $PhPH_2$ which gives $(Ph_2P)_2CH-CH_2P(Ph)CH_2CH(PPh_2)_2$ in 85% yield ⟨88POL129⟩. The reaction has been extended to $(Ph_2P)_2C=CHMe$ ⟨88POL239, 92JCS(D)2353⟩, and is also possible using amine nucleophiles where the rate of addition is greatly enhanced by coordination to palladium, as exemplified by $Pd(OAc)_2$-catalysed addition of Me_2NNH_2 to give $Me_2NNHCH_2CH(PPh_2)_2$ ⟨86CC882⟩.

4.12.1.3.14 From phosphaalkenes

Photochemical isomerization of tetrakis(trifluoromethyl)-1,4-diphosphabenzene produces the benzvalene-type tricyclic compound (**22**) ⟨80JOC4683⟩. Various phosphaalkenes undergo spontaneous head-to-tail dimerization to afford 1,3-diphosphetanes such as (**23**) ⟨81ZAAC(479)41⟩ and (**24**) ⟨93AG(E)756⟩. The interaction of the C=P double bond of diazaphospholes with aliphatic diazo compounds, $R^1R^2CN_2$, to give products such as (**25**) has been examined in detail for Me_2CN_2 ⟨82IZV1196, 82IZV2718, 84IZV2023, 84ZOB1511⟩ and also Pr^iCHN_2 and Bu^tCHN_2 ⟨85ZOB1471⟩. The reaction of 3,5-diphenylphosphabenzene with diazo compounds follows an unexpected route to give the pentacyclic products (**26**), termed diphosphachiropteradienes ⟨87AG(E)236⟩.

4.12.1.3.15 From phosphaalkynes

The rich chemistry of phosphaalkynes, especially $Bu^tC{\equiv}P$, has led to many complex polycyclic phosphines containing one or more saturated carbons joined to two tricoordinate phosphorus atoms. Many of these are described in detail in a monograph ⟨B-90MI 412-01, B-90MI 412-02⟩ and only representative examples are given here. Depending on the conditions, treatment of $Bu^tC{\equiv}P$ with $AlCl_3$ may give either of the isomeric tetramers (**27**) or (**28**) ⟨92AG(E)1055⟩ while at 180 °C some of the compound (**29**) results from loss of isobutene ⟨93PS(76)1⟩. Reductive coupling of $Bu^tC{\equiv}P$ with potassium followed by air oxidation affords the pentamer (**30**) in up to 30% yield ⟨93PS(77)5⟩. Treatment with 1,3-dienes affords (**31**) ⟨87AG(E)1011⟩, while the corresponding addition to 4-alkyl-6-methylpyran-2-ones gives (**32**) ⟨B-90MI 412-02⟩. Treatment of $Bu^tC{\equiv}P$ with $TiCp_2(CO)_2$ affords (**33**) in 63% yield ⟨88AG(E)837⟩. Treatment with 2-aryl-4,6-diphenyl-3-phosphapyridines gives (**34**) ⟨87TL1093, 88AG(E)389⟩ while 9-fluorenylidenediarylgermanium gives (**35**) ⟨93CC569⟩. The mixture of cyclic anions, $(Bu^tC)_3P_2^-$ and $(Bu^tC)_2P_3^-$, undergoes a complex reaction with $PtCl_2(COD)$ to give (**36**), a hexamer of $Bu^tC{\equiv}P$ ⟨94AG(E)2202⟩.

4.12.1.3.16 Miscellaneous methods

Photolysis of F_2PCH_2I in the presence of mercury gives $F_2PCH_2PF_2$ ⟨76MI 412-01⟩. $(F_3C)_2PH$ and Me_3N react by a complex pathway to afford $(F_3C)_2PCH_2P(CF_3)P(CF_3)_2$, which upon treatment with

HCl gives $(F_3C)_2PCH_2P(CF_3)Cl$ ⟨81IC3734⟩. In a remarkable procedure, the direct reaction of phosphorus, CCl_4, PhOH and NaOPh is reported to give $(PhO)_2PCH_2P(OPh)_2$ in 10% yield ⟨83EGP201449⟩. Condensation of two molecules of $R^1_3P^+$—P=PR^2_3 with CH_2Cl_2 in the presence of $AlCl_3$ gives the diphosphirane bis(phosphonium salt) (**37**) ⟨86AG(E)253, 86ZN(B)444⟩. Addition of Ph_2PCl to carbon suboxide gives the diphosphinomalonyl chloride (**38**) ⟨90G53⟩.

$$R^1_3\overset{+}{P}-\overset{\triangle}{P-P}-\overset{+}{P}R^2_3 \quad 2AlCl_4^-$$
(**37**)

$$\begin{array}{c} Ph_2P \\ \\ Ph_2P \end{array} \!\!\times\!\! \begin{array}{c} COCl \\ \\ COCl \end{array}$$
(**38**)

4.12.1.3.17 By reduction of tetracoordinate systems

The method of choice for access to many compounds of this class is reduction of more readily available symmetrical tetracoordinate systems. Thus, $H_2PCH_2PH_2$ can be obtained by $LiAlH_4$ reduction of either $(EtO)_2P(O)CH_2P(O)(OEt)_2$ ⟨66HCA842, 68BRP1130487⟩ or $Cl_2P(O)CH_2P(O)Cl_2$ ⟨69USP3445522⟩. The latter method has also been extended to carbon substituted analogues, although yields are often low ⟨66JOC3391⟩. One of the best methods for formation of $H_2PCH_2PH_2$ is reduction of $(Pr^iO)_2P(O)CH_2P(O)(OPr^i)_2$ with Ph_2SiH_2 which proceeds in 30–35% yield ⟨84ZC261⟩. Reductive desulfurization of $Cl_2P(S)CH_2P(S)Cl_2$ with Ph_2PCl gives $Cl_2PCH_2PCl_2$ in 65% yield ⟨77CZ259⟩ and reduction of $Ph_2P(S)CH_2PR_2$ with Si_2Cl_6 or sodium has been used to prepare $Ph_2PCH_2PR_2$ ⟨77IC1770, 80IC3195⟩. Treatment of the bis(phosphonium) salts $R_2P^+(CH_2OH)CH_2P^+(CH_2OH)R_2$ with NaOH gives $R_2PCH_2PR_2$ in excellent yield ⟨65ZOB1602⟩.

4.12.1.3.18 Interconversions

A large proportion of the compounds in this class have been obtained not by *de novo* formation of the P—C—P function, but by conversion of a few key precursors using the standard reactions of phosphorus chemistry ⟨B-72MI 412-01⟩. Some of the more important of these are included here. Reduction of $Cl_2PCH_2PCl_2$ with $LiAlH_4$ gives $H_2PCH_2PH_2$ in 40% yield ⟨84CB3400⟩, while conversion into $F_2PCH_2PF_2$ is achieved with SbF_3 ⟨77CZ259⟩, and $Br_2PCH_2PBr_2$ can be formed with either $MgBr_2\cdot Et_2O$ ⟨77ZOB775⟩ or HBr ⟨82CB1947⟩. Conversion of $Cl_2PCH_2PCl_2$ into $R_2PCH_2PR_2$ is achieved by treatment with Grignard reagents ⟨77ZOB775, 80ZOB1881, 88CB2121, 91POL1713⟩ or MeLi ⟨77CZ259⟩ or Bu^tLi ⟨83ZN(B)1027⟩ and partial reaction to afford $RP(Cl)CH_2P(Cl)R$ ⟨77ZOB775, 80ZOB949⟩ can be followed by further reactions of the remaining chlorines ⟨80ZOB1881, 84CB3400, 85ZOB331, 88ZN(B)31⟩. Replacement of two or all four chlorines in $Cl_2PCH_2PCl_2$ by RO can be achieved by treatment with alcohols ⟨77CZ259, 77ZOB775, 79ZOB712⟩, alkoxides ⟨86CB2832⟩ or phenols ⟨83ZN(B)1027⟩, while replacement by R_2N occurs with secondary amines ⟨77CZ259, 77ZOB775, 79ZOB712, 86CB2832⟩ or R_2N-TMS ⟨83PS(18)279, 88ZAAC(561)157⟩. These newly introduced groups can then in turn be displaced by Grignard reagents and the yield of $Me_2PCH_2PMe_2$ from treatment of $(PhO)_2PCH_2P(OPh)_2$ with MeMgI, for example, is better than that obtained directly from Cl_2P-CH_2PCl_2 ⟨83ZN(B)1027⟩. Comproportionation reactions are occasionally of value and treatment of $Cl_2PCH_2PCl_2$ with $(Me_2N)_2PCH_2P(NMe_2)_2$, for example, allows convenient access to $(Me_2N)PClCH_2PCl(NMe_2)$ ⟨91ZAAC(594)66⟩. The corresponding reactions of $Cl_2PCH_2P(Cl)CH_2PCl_2$ with various nucleophiles have also been examined ⟨85JOM(296)411⟩.

Further transformations of $H_2PCH_2PH_2$ have also been studied in detail. A Mannich reaction gives $(Et_2NCH_2)_2PCH_2P(CH_2NEt_2)_2$ ⟨66HCA842, 70USP3496231⟩, while radical telomerization with terminal alkenes gives $RCH_2CH_2PHCH_2PHCH_2CH_2R$ ⟨70USP3518312⟩. Similar telomerization of $H_2PCH_2PH_2$, generated *in situ* from $(TMS)_2PCH_2P(TMS)_2$ and MeOH, with diethylvinylphosphine affords the hexakis(phosphine) $(Et_2PCH_2CH_2)_2PCH_2P(CH_2CH_2PEt_2)_2$ ⟨85JA7423⟩.

The tetraphenyl compound $Ph_2PCH_2PPh_2$ has also been much used as a starting material, for example with sodium metal to give $PhPHCH_2PHPh$ and with an excess of PCl_3 to give $Cl_2PCH_2PCl_2$ ⟨70ZAAC(376)37⟩, although other workers were unable to reproduce the latter result, instead obtaining $Ph_2PCH_2PCl_2$ ⟨76JCS(D)1113⟩. Radical telomerization of $PhPHCH_2PHPh$ with diethyl- and diphenyl-vinylphosphine produces $R_2PCH_2CH_2P(Ph)CH_2P(Ph)CH_2CH_2PR_2$ ⟨88JA5585, 89IC1872⟩. Attempts to alkylate $R_2PCH_2PR_2$ on the central carbon were initially frustrated by preferential alkylation at phosphorus to give the phosphino ylides $R^1R^2_2P=CHPR^2_2$ ⟨70JPR456, 79ZN(B)1178⟩ but deprotonation followed by slow addition of MeI at low temperature allowed conversion of $Me_2PCH_2PMe_2$ into $Me_2PCH(Me)PMe_2$ in 61% yield ⟨84ZN(B)1518⟩.

4.12.1.4 Asymmetrical Systems Containing at Least One Tricoordinate Phosphorus

4.12.1.4.1 From 1-aminoalkylphosphines

Treatment of $(Et_2NCH_2)_3P$ with diethyl phosphite results in elimination of Et_2NH to afford $(Et_2NCH_2)_2PCH_2P(O)(OEt)_2$ ⟨60BRP842593⟩.

4.12.1.4.2 From 1-metalloalkylphosphorus compounds

The ready availability of $Ph_2P(O)CH_2Li$, following the discovery that the reaction of Ph_3PO with MeLi proceeded by displacement of a phenyl group followed by deprotonation of the newly introduced methyl group by the PhLi produced ⟨63JA642⟩, has resulted in the reaction of this and related compounds with R_2PCl being widely used. The reaction with Ph_2PCl proceeds in good yield to afford $Ph_2P(O)CH_2PPh_2$ ⟨64JA1100, 75IC656⟩ and a wide range of electrophiles R^1R^2PCl react similarly ⟨69USP3426021, 80IC3195⟩. The reaction of Ph_3PO with other alkyllithiums, RCH_2Li, leads to $Ph_2P(O)CH(R)Li$ and thus to products $Ph_2P(O)CH(R)PR^1_2$ ⟨75IC656⟩. The corresponding reaction of Ph_3PS with MeLi to give $Ph_2P(S)CH_2Li$, followed by treatment with R_2PCl has been used to gain access to a wide variety of compounds $Ph_2P(S)CH_2PR_2$ ⟨69USP3426021, 74SRI221, 76IS195, 77IC1770, 80IC3195⟩ and extension to $Ph_2P(Se)CH_2PR_2$ has been described ⟨69USP3426021⟩. $Ph_2P(S)CH_2Li$ and $PhPCl_2$ give $Ph_2P(S)CH_2P(Ph)CH_2P(S)Ph_2$ (10%) while, with PCl_3, the symmetrical $(Ph_2P(S)CH_2)_3P$ is formed in 15% yield ⟨80ZN(B)832⟩. More conventional methods of metallation have also been used, and treatment of $(Et_2N)_2P(O)Me$ with Bu^nLi followed by $(Pr^iO)_2PCl$, for example, gives $(Et_2N)_2P(O)CH_2P(OPr^i)_2$ ⟨80ZOB989⟩. The sodium enolates of phosphonoacetates such as $(EtO)_2P(O)CH_2CO_2Et$ undergo mainly *C*-alkylation with $(EtO)_2PCl$ or Pr^i_2PCl to give $(EtO)_2P(O)CH(CO_2Et)PR_2$ although some *O*-alkylation is also observed ⟨78ZOB757⟩. The reaction of $F_2P(S)CH_2Li$ with chlorophosphines to give $F_2P(S)CH_2PR_2$ has also been described ⟨88ZAAC(566)90⟩. Derivatives of nonalkali metals have been less widely used, but $(EtO)_2P(O)CH_2SnR_3$ reacted with Ph_2PCl to give $(EtO)_2P(O)CH_2PPh_2$ and with $PhPCl_2$ to give $(EtO)_2P(O)CH_2P(Cl)Ph$ ⟨79JOM(182)465⟩.

4.12.1.4.3 By electrophilic attack on phosphorus ylides

Treatment of $Ph_3P=CH_2$ with Ph_2PBr was reported to give the expected product $Ph_3P^+CH_2PPh_2$ Br^- ⟨61JA2055⟩ but it later emerged that this process is generally accompanied by transylidation in which the initial ylide abstracts a proton from the product to give $Ph_3P=CHPR_2$ ⟨66LA(699)40, 70JPR135⟩. Where neither substituent on the ylide carbon is hydrogen, this problem does not arise, and $Ph_3P=CR^1_2$ and R^2_2PCl do give the expected products $Ph_3P^+C(R^1_2)PR^2_2$ Cl^- ⟨66LA(699)40, 85CB3105⟩. Even where transylidation does occur, the phosphinophosphonium salt is readily regenerated with HCl and this method has been used to obtain $Ph_3P^+CH(R^1)PR^2_2$ Cl^- ⟨67LA(707)120⟩ and $(Me_2N)_3P^+CH_2PR_2$ Cl^- ⟨70JPR135⟩. In the case of stabilized ylides $Ph_3P=CHCO_2Et$ ⟨80TL2931⟩ and $Ph_3P=CHCOR$ ⟨92ZOB263⟩, reaction with Ph_2PCl is accompanied by transylidation and the salts $Ph_3P^+CH(COR)PPh_2$ Cl^- resulting from subsequent protonation are subject to 'diadic tautomerism' involving forms such as $Ph_3P=C(COR)PHPh_2^+$ and $Ph_3P^+C(COR)=PPh_2H$. Reaction of $Ph_3P=CH_2$ with PCl_3 involves transylidation but subsequent treatment with HCl affords $(Ph_3P^+CH_2)_3P$ $(Cl^-)_3$ ⟨69JPR857⟩. Some cases have been discovered in which transylidation does not appear to occur. In particular, reaction of $(Me_2N)_3P=CH_2$ with $(Me_2N)_2PCl$ affords $(Me_2N)_3P^+CH_2P(NMe_2)_2$ Cl^- directly ⟨91CC302⟩. Treatment of phosphino ylides with other electrophiles such as MeI does not proceed on carbon to give $R^1_3P^+CHMePR^2_2$ but rather on phosphorus to give the delocalized salts $[R^1_3PCHPR^2_2Me]^+$ I^- ⟨68CB3545, 68LA(713)12⟩. Treatment of lithio diphosphino alkanes such as Ph$_2$PCH(Li)PPh$_2$ with PhCH$_2$Cl results in *P*-alkylation and subsequent treatment with HCl gives (PhCH$_2$)Ph$_2$P$^+$CH$_2$PPh$_2$ Cl$^-$ ⟨70JPR456⟩.

The reaction of (Me$_2$N)$_2$P(F)=CH$_2$ with PF$_3$ occurs by threefold addition to the C=P bond to give the tris(phosphoranomethyl)phosphine ((Me$_2$N)$_2$P(F)$_2$CH$_2$)$_3$P ⟨91ZAAC(596)139⟩.

4.12.1.4.4 From vinylphosphorus compounds

Diethyl vinylphosphonate adds to tetraethyl pyrophosphite to give (**39**) ⟨74ZOB276⟩, while a similar reaction with tetra(*n*-butoxy)diphosphine produces (**40**) ⟨76ZOB568⟩.

4.12.1.4.5 Miscellaneous methods

Treatment of Ph$_2$PCl with 'moist CH$_2$N$_2$' is reported to give Ph$_2$P(O)CH$_2$PPh$_2$ ⟨60MI 412-01⟩ while Ph$_2$PH with formic acid and concentrated HCl gives the same product in 76% yield ⟨91S125⟩. The latter method is applicable to other diarylphosphines. Treatment of (Me$_2$N)$_3$P$^+$—C≡P with hindered phenols results in twofold addition across the triple bond to give (Me$_2$N)$_3$P$^+$CH$_2$P(OAr)$_2$ ⟨91CC302⟩.

4.12.1.4.6 By oxidation of symmetrical tricoordinate systems

The conversion of bis(phosphines) such as Ph$_2$PCH$_2$PPh$_2$ into Ph$_2$P(X)CH$_2$PPh$_2$ for X = O, S and Se employing standard methods is straightforward ⟨80IC1982⟩ and comproportionation has also been successful for the preparation of Ph$_2$P(Se)CH$_2$PPh$_2$ from Ph$_2$PCH$_2$PPh$_2$ and Ph$_2$P(Se)CH$_2$P(Se)Ph$_2$ ⟨80JCS(D)871⟩. Ph$_2$PCH$_2$PPh$_2$ and MeBr give the bis(phosphonium salt) but with MeCl the salt Ph$_2$MeP$^+$CH$_2$PPh$_2$ Cl$^-$ is produced ⟨76ZN(B)721⟩ and MeOTf or Ph$_3$C$^+$ PF$_6^-$ also give corresponding monophosphonium salts ⟨86IC712⟩. A wide range of more unusual derivatives has also been prepared. Treatment with α-bromo ketones followed by Et$_3$N affords the phosphinomethyl ylides (**41**) ⟨75CL1259, 76JOM(122)113⟩. Treatment of Ph$_2$PCH$_2$PPh$_2$ with sulfonamides and Et$_3$N in CCl$_4$ affords the phosphinimines (**42**) ⟨71CB2250⟩ and such compounds can also be formed with azides, for example TMS-N$_3$ gives (**43**) ⟨89IC413⟩. The similar compound (**44**) formed with Me$_3$GeN$_3$ reacts with pentafluorobenzonitrile to give (**45**) ⟨90IC808⟩. Treatment with NaBH$_4$ and I$_2$ gives (**46**) while with BH$_3$·Me$_2$S and I$_2$, the salt (**47**) results ⟨86ICA(115)L29⟩. The bis(phosphonites) (RO)$_2$PCH$_2$P(OR)$_2$ undergo Arbuzov-type reactions with alkyl halides to afford (**48**) ⟨77ZOB2689⟩, with acid chlorides to give (**49**) ⟨79ZOB470⟩ and with Pr$^i_2$PI to give (**50**) ⟨79ZOB1446⟩. A similar process occurs on treatment of (**51**) with alkyl iodides to give (**52**) which loses RNH$_2$ on heating to produce (**53**) ⟨87ZOB949⟩.

(**42**) X = SO$_2$R
(**43**) X = TMS
(**44**) X = GeMe$_3$
(**45**) X = *p*-C$_6$F$_4$CN

(**48**) X = R^1
(**49**) X = COR^1
(**50**) X = PPr$^i_2$

4.12.1.4.7 By reduction of symmetrical tetracoordinate systems

Selective monodesulfurization of $Cl_2P(S)CH_2P(S)Cl_2$ with Ph_2PCl to afford $Cl_2P(S)CH_2PCl_2$ has been reported ⟨79PS(5)337⟩.

4.12.1.4.8 Interconversions

Sulfur transfer to the more basic phosphine site occurs with $Ph_2P(S)CH_2PMe_2$ at 160 °C to afford the isomer $Ph_2PCH_2P(S)Me_2$ ⟨75CC634⟩. Treatment of $Cl_2P(O)CH_2PCl_2$ with ethanol and pyridine results in replacement of all four chlorines to give $(EtO)_2P(O)CH_2P(OEt)_2$ in 63% yield ⟨87JA5544⟩, and stepwise replacement of chlorine by NMe_2 occurs on treatment with Me_2N-TMS ⟨88ZAAC(561)157⟩. Treatment of $Cl_2P(S)CH_2PCl_2$ with methanol or dimethylamine similarly results in replacement of some or all of the chlorines by OMe and NMe_2, respectively ⟨79PS(5)337⟩ and the latter process can again be achieved with Me_2N-TMS ⟨88ZAAC(561)157⟩. Conversion of $Cl_2P(S)CH_2PCl_2$ into $F_2P(S)CH_2PF_2$ is achieved with SbF_3 and further treatment with BCl_3 produces $F_2P(S)CH_2PCl_2$ ⟨87ZAAC(555)109⟩. Distillation of $F_2P(S)CH_2PF_2$ results in disproportionation to give $F_2P(S)CH_2PF_4$ and the tetrameric compound (**54**) ⟨88ZAAC(566)90⟩.

$$\begin{array}{c}\text{structure (54)}\end{array}$$

(**54**)

4.12.1.5 Symmetrical Tetracoordinate Systems

4.12.1.5.1 From reactions of 1,1-dihalo alkanes with phosphorus nucleophiles

Treatment of the sodium salts of dialkyl phosphites, $(RO)_2P(O)Na$, with CH_2I_2 followed by acid affords the diacid diesters (**55**) ⟨64JCS513⟩. Conditions which do not result in loss of the ester groups were later developed and the light-promoted reaction of sodium or ammonium dialkyl phosphites with CH_2Br_2 or CH_2BrCl in liquid ammonia, heptane, or a mixture of the two gives bis(phosphonates), $(RO)_2P(O)CH_2P(O)(OR)_2$, in 55–87% yield ⟨82JPR537, 84EGP206557⟩. $(EtO)_2P(O)Na$ and CH_2Cl_2 give $(EtO)_2P(O)CH_2P(O)(OEt)_2$ in 51% yield ⟨90SC1865⟩. Dialkyl thionophosphites react similarly and either $(EtO)_2P(S)K$ ⟨59FRP1185452⟩ or $(EtO)_2P(S)Na$ ⟨82JPR537⟩ and CH_2Br_2 afford $(EtO)_2P(S)CH_2P(S)(OEt)_2$ in up to 56% yield. Base is not always necessary and a range of dialkyl phosphites, $(RO)_2P(O)H$, reacted with CH_2Br_2 to give the bis(phosphonates) in 55–63% yield ⟨82EGP154700⟩. The anion of phenylphosphinic acid, $PhP(H)(O)OH$, can also be used and treatment of this acid with Bu^nLi followed by CH_2Cl_2 gives (**56**) in 32% yield ⟨79SC261⟩. $Ph_2P(O)H$ and KOH with CH_2Cl_2, CH_2Br_2, or CH_2I_2 in aqueous DMF or DMSO give $Ph_2P(O)CH_2P(O)Ph_2$ in 28–35% yield ⟨86S198⟩. The reactions sometimes take an unusual course, and $(EtO)_2P(O)Na$ reacts with benzyl chloride to give (**57**) ⟨69USP3463835⟩, and with $EtSO_2CBr_2SO_2Et$ or $EtSO_2C(Me)BrSO_2Et$ to give $(EtO)_2P(O)CH_2P(O)(OEt)_2$ or (**58**), respectively ⟨57ZOB2356⟩. A series of curious reactions of vinyl chlorides with phosphorus nucleophiles has been reported. Thus, $PhC(Cl)=CCl_2$ reacts with $(EtO)_2P(O)Na$ to give (**59**) ⟨85ZN(B)1152, 86GEP3444678⟩, while $(EtO)_2P(O)H$ reacts with $Cl_2C=CCl_2$ to give (**60**) and with either $CH_2=CCl_2$ or $ClCH=CCl_2$ to give (**61**) ⟨82GEP3111152⟩. In a similar way, $Cl_2C=CCl_2$ and the sodium salt of $PhPH(O)OPr^i$ gives (**62**) ⟨92ZN(B)725⟩, and treatment of $Ph_2P(O)H$ with $ClCH=CHCl$ under basic conditions gives (**63**) ⟨90AKZ537⟩.

The other major type of reaction in this class is formation of bis(phosphonium) salts from dihalo alkanes and tertiary phosphines. Treatment of CH_2Br_2 with Ph_3P at 150 °C gives the salt $Ph_3P^+CH_2P^+Ph_3$ $(Br^-)_2$ in 20% yield ⟨61JA3539⟩ and the same product can be formed with an excess of CH_2ClBr ⟨63USP3098878⟩. Ph_2PMe and CH_2Br_2 at 180 °C likewise give the salt in 71% yield ⟨77CB3501⟩. It was noted at an early stage that the corresponding reaction of dialkylphosphines with CH_2Br_2 or CH_2I_2 is not successful ⟨63CB2186⟩, but $Me_2PCH_2P(Me)CH_2PMe_2$ and CH_2Br_2 do give the cyclic salt (**64**) ⟨82ZN(B)284⟩. The reaction of hydroxymethylphosphines, R_2PCH_2OH, with CH_2Br_2 or CH_2I_2 is useful since the resulting bis(phosphonium) salts can subsequently be treated with NaOH to afford $R_2PCH_2PR_2$ ⟨65ZOB1602⟩. The product from the reaction of Me_2PCl with CH_2Br_2 at 180 °C, presumably a bis(chlorophosphonium) salt, is cleaved by Bu^nOH to afford $Me_2P(O)CH_2P(O)Me_2$ ⟨67ZOB2055⟩. Although attention has generally been confined to dihalomethanes, $PhCHCl_2$ and Ph_3P also give the bis(phosphonium) salt in 12% yield ⟨66ZC314⟩.

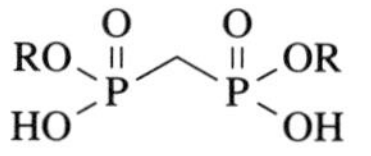

(55) 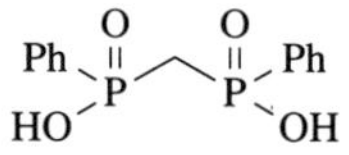 (56) 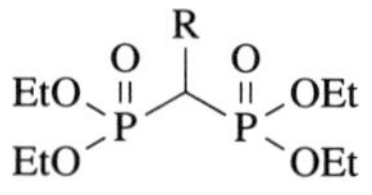 (57) R = Ph (58) R = Me (59) R = Ph (60) R = P(O)(OEt)$_2$ (61) R = H

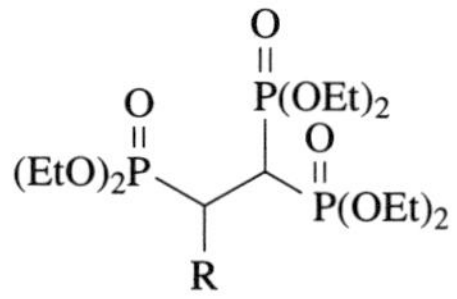

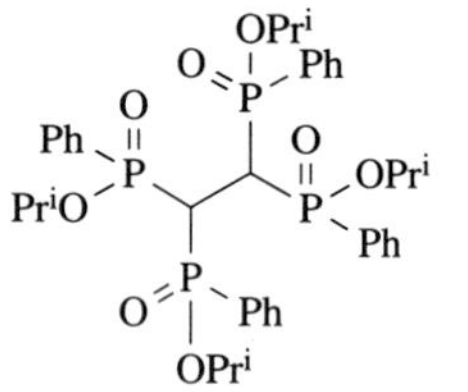

(62) 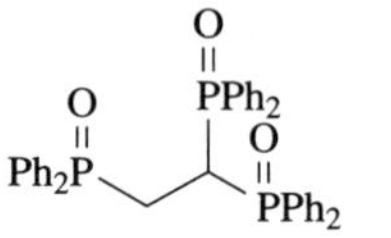 (63) 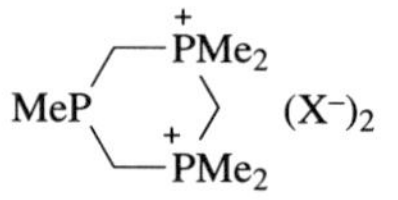 (64)

4.12.1.5.2 From reactions of 1-haloalkylphosphorus compounds with phosphorus anions

The reaction of $(EtO)_2P(O)CH_2Cl$ with $(EtO)_2P(O)Na$ proceeds readily to give the product $(EtO)_2P(O)CH_2P(O)(OEt)_2$ in 47% yield ⟨50M202⟩ and this route also gives $(RO)_2P(O)CH_2P(O)(OR)_2$ for $R = Bu^n$ ⟨60ZOB1602, 61JA1722⟩ and R = cyclohexyl ⟨69ZOB845⟩. $Bu^n_2P(O)MgBr$, which may be conveniently generated from $(EtO)_2P(O)H$ and an excess of Bu^nMgBr, and $Bu^n_2P(O)CH_2Cl$ afford $Bu^n_2P(O)CH_2P(O)Bu^n_2$ ⟨61JCS2423⟩ and similar reactions of $R_2P(O)MgBr$ with $(R^1O)_2P(O)CH_2Cl$ give a range of compounds, $R_2P(O)CH_2P(O)(OR^1)_2$ ⟨65JCED303⟩. Treatment of $Et(EtO)P(O)CH_2Cl$ with $(EtO)_2P(O)Na$ gives (**65**) ⟨65ZC419⟩ and the anion derived from (**66**) reacts with $(EtO)_2P(O)CH_2OTf$ to give (**67**) in 74% yield ⟨90TL5381⟩.

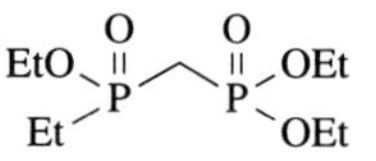

(65) 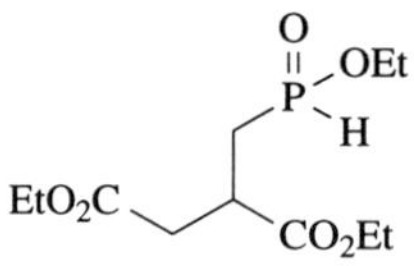 (66) 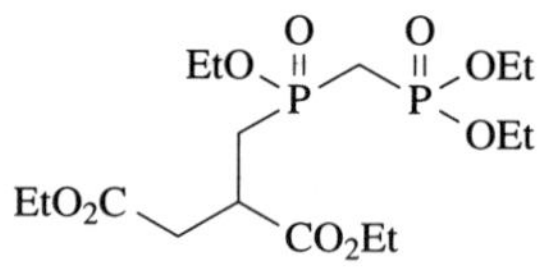 (67)

4.12.1.5.3 From reactions of 1-haloalkylphosphorus compounds with phosphines

The salt $Ph_3P^+CH_2Br\ Br^-$ obtained from CH_2Br_2 and Ph_3P reacted with further Ph_3P under forcing conditions to give the bis(phosphonium) salt $Ph_3P^+CH_2P^+Ph_3\ (Br^-)_2$, albeit in yields of only 30–40% ⟨64JOC2427, 65T5⟩, and the corresponding reactions with $PhMe_2P$ or Me_3P have also been reported ⟨84CB3374⟩.

Haloalkyl phosphonates and phosphine oxides undergo similar reactions and phosphonium salt formation from Ph_3P and both $(PhO)_2P(O)CH_2Cl$ ⟨68TL5731⟩ and $Ph_2P(O)CH_2OTs$ ⟨72ZC103⟩ has been described. Hydroxymethylphosphines have also been used and $R^1_2PCH_2OH$ reacts with $R^2_2P(O)CH_2Cl$ to afford $R^1_2P^+(CH_2OH)CH_2P(O)R^2_2Cl^-$ ⟨69ZOB1247⟩. Multiple reactions with poly(haloalkyl)phosphorus compounds are possible as illustrated by reaction of $P(CH_2OH)_3$ with

PO(CH$_2$Cl)$_3$ to give (**68**) and with (ClCH$_2$)$_4$P$^+$ Cl$^-$ to give (**69**) ⟨60USP2937207⟩. While MeP(O)(CH$_2$Cl)$_2$ reacts with Ph$_3$P to give the bis(phosphonium) salt (**70**) in 67% yield ⟨70ZOB285⟩, selective formation of the monophosphonium salts (**71**) is possible by treatment of PO(CH$_2$X)$_3$ with one equivalent of Ph$_3$P ⟨68JCED585, 71USP3607944⟩ and (**72**) has similarly been prepared from PhO-P(O)(CH$_2$Cl)$_2$ in 44% yield ⟨83JOC4775⟩. The cyclic bis(phosphonium) salt (**64**) mentioned earlier can alternatively be formed by treatment of MeP(CH$_2$Cl)$_2$ with Me$_2$PCH$_2$PMe$_2$ ⟨82ZN(B)284⟩. In some cases a protonated OH group can be displaced as illustrated by the reaction of (**73**) with Ph$_3$P in HBF$_4$ or HClO$_4$ to afford (**74**) ⟨83ZOB594⟩.

4.12.1.5.4 From Arbuzov reactions of 1,1-dihalo alkanes

This method provides one of the most direct and high yielding routes to methylene bis(phosphonates). In an early report, (EtO)$_3$P reacted with CH$_2$I$_2$ to afford (EtO)$_2$P(O)CH$_2$P(O)(OEt)$_2$ ⟨47JCS1465⟩ and a reliable large-scale procedure for the preparation of (Pr^{i}O)$_2$P(O)CH$_2$P(O)(OPr^i)$_2$ in over 90% yield from (Pr^{i}O)$_3$P and CH$_2$Br$_2$ at 185 °C was later described ⟨66BRP1026366, 66USP3251907⟩. This product is useful since it provides the bis(phosphonic acid), (HO)$_2$P(O)CH$_2$P(O)(OH)$_2$, in almost quantitative yield when heated with concentrated HCl. Preparation of the ^{14}C labelled bis(phosphonate) from (Pr^{i}O)$_3$P and ^{14}CH$_2$Br$_2$ has been reported ⟨91MI 412-01⟩. (CN)$_2$CCl$_2$ and (Ph)(CN)CCl$_2$ reacted with (EtO)$_3$P to give (**75**) and (**76**), respectively ⟨79ZOB2217⟩. Diisopropylphosphonites can also be used; thus, PhP(OPr^i)$_2$ and CH$_2$Br$_2$ ⟨65RZC1129⟩ or CH$_2$I$_2$ ⟨68USP3403176⟩ gave (**77**), and (Pr^{i}O)$_2$PCH$_2$P(OPr^i)$_2$ and CH$_2$Br$_2$ or CH$_2$Cl$_2$ gave (**78**) ⟨77ZOB2636⟩. In all these cases, treatment with HCl affords the corresponding phosphinic acids.

4.12.1.5.5 From Arbuzov reactions of 1-haloalkylphosphorus compounds

In its most general form, this very widely used reaction involves treatment of R$^1_2$P(O)CH$_2$X with R$^2_2$POR to give R$^1_2$P(O)CH$_2$P(O)R$^2_2$ with elimination of RX. In this context R^1 and R^2 can be alkyl, aryl, alkoxy or aryloxy in any combination ⟨67IZV591, 92PS(73)67⟩ and X can be Cl, Br or I. The only restriction is that R must be a small, alkyl group most commonly Et or Pr^i. The earliest reported reaction was between (EtO)$_2$P(O)CH$_2$Cl and (EtO)$_3$P in a sealed ampoule at 190 °C to give (EtO)$_2$P(O)CH$_2$P(O)(OEt)$_2$ ⟨58ZOB728⟩. Other reactions involving (EtO)$_3$P include those with a variety of compounds (RO)$_2$P(O)CH$_2$Cl ⟨66GEP1211200⟩ and with (**79**) to afford (**80**) ⟨68ZOB2071⟩. (PhO)$_2$P(O)CH$_2$I and (PhO)$_2$POEt gave the tetraphenyl diphosphonate, (PhO)$_2$P(O)CH$_2$P(O)(OPh)$_2$, in 47% yield ⟨82ZC307⟩. When this same phosphite reacted with (EtO)$_2$P(O)CH$_2$I, the initial product

(81) could be cleaved with TMS-Br to give the phosphonic acid (82) ⟨86EGP238612⟩. Similarly, the initial product (83) from (PhO)(EtO)P(O)CH₂I and (EtO)₃P gave the diphosphonic acid monoester (84) upon treatment with TMS-Br ⟨86ZAAC(536)187⟩.

(79)

(80)

(81) R¹ = Ph, R² = Et
(82) R¹ = Ph, R² = H
(83) R¹ = R² = Et
(84) R¹ = R² = H

Dialkyl phosphonites can also be used as the phosphorus(III) components and examples include the reactions of PhP(OEt)₂ with Ph₂P(O)CH₂Cl ⟨67USP3332986⟩, Me₂P(O)CH₂Cl and PhMeP(O)CH₂Cl ⟨65IC198⟩, EtP(OEt)₂ with (EtO)EtP(O)CH₂I ⟨65ZC419⟩, and MeP(OEt)₂ with Me₂P(O)CH₂Cl ⟨76GEP2523145⟩. In all these cases the components were heated at between 140 °C and 180 °C and yields were generally around 70%. The heterocyclic example (86) was prepared in 41% yield from (85) and diethyl 2-thienylphosphonite at 180 °C ⟨76ZOB266⟩.

(85)

(86)

The reaction is also successful with alkyl phosphinites, although yields are often lower (30–60%). Examples include treatment of Ph₂POEt with Ph₂P(O)CH₂Cl at 240 °C to give Ph₂P(O)CH₂P(O)Ph₂ in 60% yield ⟨62IZV2103⟩, and of a wide variety of Ar¹₂POEt with Ar²₂P(O)CH₂Cl to give the products in 30–40% yield ⟨66IZV1954⟩. Symmetrical tetraalkyl compounds R₂P(O)CH₂P(O)R₂ have been prepared in 50–60% yield from R₂P(O)CH₂Cl and R₂P(OEt) for R = cyclohexyl ⟨67IZV949⟩ and benzyl ⟨69ZOB839⟩. The mixed alkyl/aryl product Ph₂P(O)CH₂P(O)Buᵗ₂ is formed in 34% yield from Ph₂P(O)CH₂Cl and Buᵗ₂POEt ⟨78JCS(P1)947⟩. The reaction tolerates the presence of free OH groups and the phosphonic acid Ph₂P(O)CH₂P(O)(OH)₂ can be prepared directly from (HO)₂P(O)CH₂Cl and either Ph₂POEt or Ph₂POPrⁱ ⟨75JPR840⟩. The analogous reactions of R¹₂P(O)CH₂Cl with R²₂PO-TMS involving loss of TMS-Cl have been reported in a few cases ⟨87ZOB54⟩.

The basic reaction has been extended in a number of ways. Use of a bis- or tris(haloalkyl)phosphorus(V) component allows multiple reactions with the phosphorus(III) compound to form extended structures with several phosphorus atoms ⟨68IZV1417, 68IZV2062⟩. Thus, ROP(O)(CH₂Cl)₂ with phosphites, phosphonites and phosphinites gives products such as (87) ⟨67DOK(177)340, 68AG(E)384, 69HCA827, 70USP3534125⟩. In a similar way, RP(O)(CH₂Cl)₂ affords a range of structures (88) ⟨68AG(E)385, 69HCA845⟩ and PO(CH₂Cl)₃ gives (89) ⟨68AG(E)385, 69HCA858⟩. A bis(phosphonite) such as (EtO)₂P(CH₂)₅P(OEt)₂ affords (90) by a twofold reaction with (EtO)₂P(O)CH₂Cl ⟨70HCA1940, 74USP3845169⟩. Unexpectedly, a compound such as (EtO)₂P(O)CH₂Cl can actually fulfil the role of both components of the reaction and is reported to form an oligomeric structure of the type (EtO)₂P(O)CH₂[(EtO)P(O)CH₂]ₙCl with n ≈ 9 when heated at 170 °C ⟨73USP3734954⟩. In all these studies the corresponding polyphosphonic and -phosphinic acids derived from acid hydrolysis of the ester groups of the initial products have been the subject of great commercial interest.

(87) X = OR
(88) X = R
(89) X = CH₂P(O)(OEt)₂

(90)

4.12.1.5.6 From 1,1-dimetallo alkanes

The reaction of dilithiated malononitrile with (EtO)₂P(O)Cl to afford (75) has been described ⟨90ZAAC(586)25⟩. Lithiated 2-picoline is reported to react with Ph₂P(O)Cl to give the bis(phosphine oxide) corresponding to (11) in 49% yield ⟨73ZC55⟩.

4.12.1.5.7 From electrophilic substitution on 1-metalloalkylphosphorus compounds

The first report of this approach appeared in 1963 when the reaction of $Ph_2P(O)CH_2Li$, readily prepared from Bu^nLi and $Ph_2P(O)Me$, with $Ph_2P(O)Cl$ to afford $Ph_2P(O)CH_2P(O)Ph_2$ in 25% yield was described ⟨63JOC123⟩. The tetra-n-hexyl derivative was prepared in an analogous way in 35% yield. Similarly, the lithiated phosphine sulfide $Ph_2P(S)CH_2Li$, and $Me_2P(S)Br$ gave the asymmetrical bis(phosphine) sulfide in 12% yield ⟨72IC2340⟩. Most of the work in this area has, however, involved the use of α-metallated phosphonate esters. Treatment of $(EtO)_2P(O)CH_2Ph$ with Pr^iMgCl followed by $Ph_2P(O)Cl$, for example, affords $(EtO)_2P(O)CH(Ph)P(O)Ph_2$ ⟨64JOC2036⟩. Attempts to use lithiated phosphonates for this process brought to light a serious problem: that of self-condensation. Thus $(MeO)_2P(O)CH_2Li$, for example, reacts instantaneously at 0 °C to give $(MeO)_2P(O)CH_2P(O)(Me)OMe$. The self-condensation is much slower for more bulky ester groups and can be avoided at low temperature ⟨84JOM(264)9⟩. A more effective solution was found in the use of LDA which allows lithiation in the presence of the electrophile; this method afforded a wide variety of compounds $(EtO)_2P(O)CH(R^1)P(O)R^2_2$ in 70–85% yield ⟨85TL4435, 86JOM(304)283⟩. Detailed studies on the competition between the self-condensation of $(R^1O)_2P(O)CH(R^2)Li$ and its reaction with $R^3_2P(O)Cl$ have appeared and the reaction is now well understood ⟨86JOM(312)283, 87TL405⟩. Recent applications include the formation of isoprenoid (phosphinylmethyl)phosphonates ⟨90T6645⟩, the preparation of ^{13}C-labelled $(Pr^iO)_2P(O)^*CH_2P(O)(OPr^i)_2$ in 95% yield from $(Pr^iO)_2P(O)^*CH_3$, LDA and $(Pr^iO)_2P(O)Cl$ ⟨91MI 412-01⟩, and the reactions of the silver salts, $(RO)_2P(O)C(CN)_2^- Ag^+$ with $(EtO)_2P(O)Cl$ to give (75) ⟨90ZAAC(586)25⟩. The deprotonation of phosphine–borane adducts such as $Me_3P \cdot BH_3$ with Bu^sLi or Bu^tLi gives $Me_2P(BH_3)CH_2Li$ which can then react with $Me_2PCl \cdot BH_3$ to give (91) ⟨80JOM(200)287⟩. The application of this reaction to chiral phosphine–borane adducts is exemplified by treatment of (92) with Bu^sLi followed by (93) to afford (94) ⟨93PS(77)199⟩.

4.12.1.5.8 From diphosphorus-substituted carbenes

A single example of this type is provided by the photolysis of the diazo compound $(MeO)_2P(O)C(=N_2)P(O)(OMe)_2$ in benzene which initially gives an equilibrium mixture of the norcaradiene (95) and the cycloheptatriene (96) although these undergo subsequent photochemical rearrangements (Equation (1)) ⟨76CB2039⟩.

4.12.1.5.9 By electrophilic attack on phosphorus ylides and from 1,3-diphosphaalkenes generated in other ways

Ylides such as $Ph_3P=CMe_2$ react with $Ph_2P(O)Cl$ to give the stable salt (97), but ylides with a hydrogen atom on the ylide carbon undergo transylidation on treatment with $Ph_2P(O)Cl$ or $Ph_2P(S)Cl$ to give 1,3-diphospha alkenes (98) ⟨67LA(707)112⟩. When these products, obtained either in this way or with Ph_2PCl followed by oxidation or sulfurization, are treated with HX, the salts (99) are formed. A range of examples has been described both for Y = O ⟨88ZOB1987, 88ZOB1998, 88ZOB2012⟩ and for Y = S ⟨84ZOB36, 85ZOB1234, 86ZOB1220⟩. Treatment of (98) with MeI likewise affords (100) ⟨67LA(707)112⟩. $Ph_3P=CH_2$ and $POCl_3$ or $PSCl_3$ undergo transylidation but subsequent treatment with HCl gives (101) ⟨69JPR857⟩. When substituted benzoylphosphonates, $ArCO-P(O)(OR)_2$, are treated with an excess of trialkyl phosphite, deoxygenation is followed by ylide formation to give (**102**). When heated these undergo an internal Arbuzov reaction to give (**103**), while treatment with acids HX gives (**104**) ⟨86CC871, 92JCS(P1)479⟩.

(**97**) (**98**) (**99**) $R^1 = H$ (**100**) $R^1 = Me$ (**101**) Y = O, S

(**102**) (**103**) $R^2 = R^1$ (**104**) $R^2 = H$

Treatment of $Ph_2PCH_2PPh_2$ with an excess of MeBr affords the bis(phosphonium) salt (**105**) but with MeCl the reaction stops at the monophosphonium salt. Treatment of this with $NaNH_2$ gives the ylide (**106**) which then reacts with MeI to give (**107**) ⟨76ZN(B)721⟩. Treatment of the ylides (**108**) with an alkyl halide, R^3X, followed by HCl likewise affords (**109**) ⟨68LA(713)12⟩.

Photolysis of the phosphinyldiazo compound (**110**) results in loss of N_2 followed by 1,2-phenyl migration to generate (**111**) which then undergoes Diels–Alder dimerization to give (**112**) ⟨78CB705⟩. In a related process, photolysis of the diphosphinyldiazo compound (**113**) in the presence of a proton source such as water, MeOH or piperidine gives (**115**) by addition of HY to (**114**) ⟨81LA1044⟩. Hydrolysis of (**2**) gives (**116**) while treatment with sulfur and Et_2NH gives (**117**) ⟨93PS(76)13⟩.

(**105**) X = Br
(**107**) X = I

(**106**) (**108**) (**109**) (**110**) X = C
(**113**) X = PPh

(**111**) X = C
(**114**) X = PPh

(**112**) (**115**) Y = OH, OMe, NR_2 (**2**)

(**116**) (**117**)

4.12.1.5.10 From phosphaalkenes

The phosphaalkene (**118**) formed by treatment of $(TMS)_2N$—P=N-TMS with CH_2N_2 undergoes head-to-tail dimerization to form (**119**) ⟨80CB1549⟩. 1,2,3-Diazaphospholes and diazo compounds, $R^3_2CN_2$, such as diazomethane or 9-diazofluorene give phosphaalkenes (**120**) which form cyclic trimers (**121**) ⟨81IZV1113⟩.

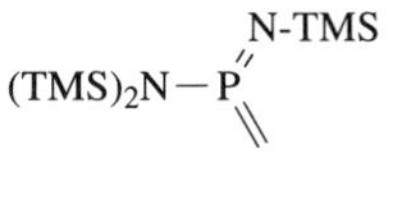

(118) (TMS)₂N—P(=N-TMS)

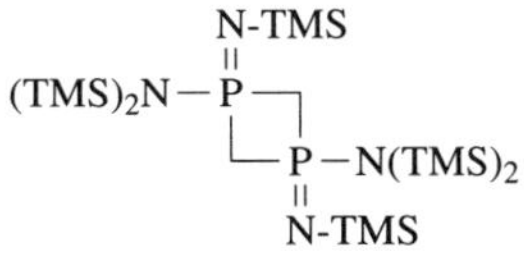

(119) Bis(TMS-amino)diphosphetane structure

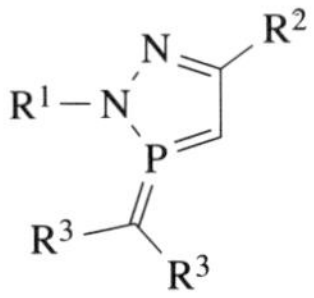

(120) Pyrazole-phosphorus structure with R¹, R², R³

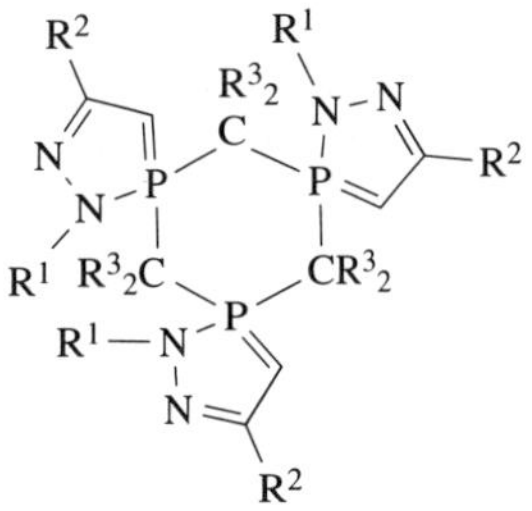

(121) Bicyclic structure with R¹, R², R³₂

4.12.1.5.11 *From diphosphorus-substituted ketenes and related compounds*

The bis(thiophosphinyl)ketene (**122**) undergoes dimerization over the course of 3 to 4 days to afford (**123**) ⟨90ZOB1539⟩. The bis(phosphoryl)ketenimine (**124**) similarly undergoes [2 + 2]-cycloaddition with Ph₃P=C=C=O to give (**125**) ⟨91TL4279⟩.

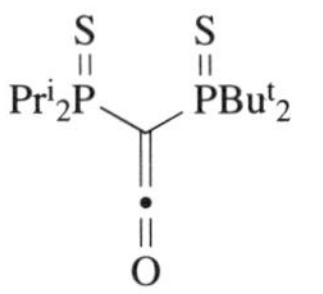

(122) $Pr^i_2P(=S)$—C(=C=O)—$PBu^t_2(=S)$

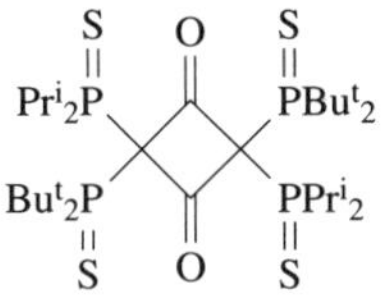

(123) Cyclobutanedione dimer with $Pr^i_2P(=S)$, $PBu^t_2(=S)$, $Bu^t_2P(=S)$, $PPr^i_2(=S)$, O, O

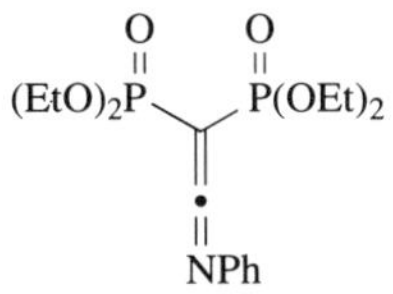

(124) $(EtO)_2P(=O)$—C(=C=NPh)—$P(OEt)_2(=O)$

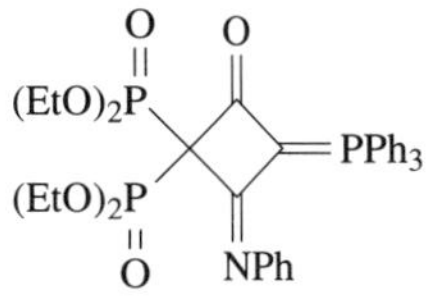

(125) Cyclobutane with $(EtO)_2P(=O)$ groups, =PPh₃, =NPh, O

4.12.1.5.13 *From 1,3-diphosphaallenes*

The diphosphaallene ArP=C=PAr (Ar = 2,4,6-tri-*t*-butylphenyl) undergoes hydrolysis at high temperature to afford ArHP(O)CH₂P(O)HAr ⟨88PS(37)241⟩. When this same compound is treated with sulfur and 1,5-diazabicyclo[5.4.0]undec-5-ene (dbu), the main product is the thiadiphosphetane (**126**) ⟨90CL643⟩. Careful examination of the product mixture also revealed the isomer (**127**), the dithiadiphospholane (**128**) ⟨91CL2213⟩ and the oxadiphosphetane (**129**) ⟨93PS(74)373⟩. Hydrolysis of (Me₂N)₃P=C=P(NMe₂)₃ proceeds with loss of Me₂NH to give (**130**; R = Me) ⟨83CB2275⟩. A range of compounds, (R₂N)₂P(Cl)=C=P(Cl)(NR₂)₂, are similarly hydrolysed to (**130**) while, with MeI present, (**131**) is formed ⟨88ZOB1665⟩.

(126) X = S
(129) X = O

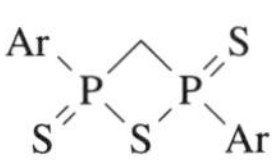

(127) Isomeric thiadiphosphetane

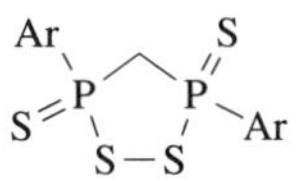

(128) Dithiadiphospholane

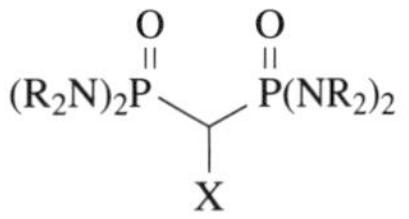

(130) X = H
(131) X = Me
$(R_2N)_2P(=O)$—CHX—$P(NR_2)_2(=O)$

4.12.1.5.13 *From carbonyl compounds*

Treatment of 3,5-di-*t*-butyl-4-hydroxybenzaldehyde with dialkyl phosphites, (R¹O)₂P(O)H, in the presence of Et₂NH gives the benzylidenediphosphonates (**132**) ⟨75JPR890⟩. If a primary amine is used, the initial product (**133**) can be treated with MeI to give (**134**) which then reacts with (EtO)₃P to give (**132**; R¹ = Et). Alternatively, (**133**), MeI and (EtO)₃P give (**132**; R¹ = Et) directly ⟨76JPR403⟩. The aldehyde, (EtO)₃P and diethyl malonate give (**132**; R¹ = Et) in 53% yield ⟨89ZOB1686⟩. Access to (**132**) can also be gained by treatment of the corresponding imines, ArCH=NPh ⟨85ZOB1865⟩ or

ArCH=NCH$_2$CH$_2$N=CHAr ⟨86ZOB2000⟩, with dialkyl phosphites. The bis(thionophosphonate) corresponding to (**132**) can also be prepared in 52% yield by the reaction of ArCHO with (EtO)$_2$P(S)H, Na and Et$_2$NH ⟨86ZOB2000⟩.

(**132**) X = P(O)(OR^1)$_2$
(**133**) X = NHR^2
(**134**) X = I

4.12.1.5.14 From 1,1-diphosphorus-substituted alkenes

Conjugate addition of Ph$_2$P(O)H to the double bond of (**135**) at 150 °C gives (**136**) ⟨70IZV1326⟩. Addition of various nucleophiles to the diphosphonic acid (**137**) affords (**138**). Examples include secondary amines ⟨83IZV2802⟩ and aliphatic and aromatic thiols ⟨84DOK(277)371, 85BBA(818)96, 87IZV860⟩. With primary amines and hydrazine, (**138**) is still formed, but is accompanied by the tetraphosphonic acids (**139**) and (**140**), respectively ⟨84IZV1122⟩. Addition to the diphosphonate (**141**) proceeds similarly to give (**142**) and nucleophiles used include secondary amines, thiols and dialkyl phosphites ⟨88JOM(346)341⟩, and carbanions such as CN$^-$ and those derived from diethyl malonate, PhCH=NCH$_2$CO$_2$Et, acetophenone and nitromethane ⟨91S661, 94PS(88)1⟩. Addition of alcohols to the bis(phosphonium) salt (**143**) to give (**144**) has also been described ⟨85CB3105⟩.

(**135**) (**136**) (**137**) (**138**) (**139**)

(**140**) (**141**) (**142**)

(**143**) (**144**)

4.12.1.5.15 From vinylphosphorus compounds

Addition of Ph$_2$P(O)H to the double bond of the α,β-unsaturated phosphine oxides (**145**) to afford (**146**) has been reported for E = CO$_2$Me and COPh ⟨83ZOB541, 84URP1067006⟩ and for E = P(O)Ph$_2$ ⟨67IZV591⟩. The corresponding conversion of α,β-unsaturated phosphonates (**147**) into (**148**) may be accomplished either with sodium dialkyl phosphites for E = COMe ⟨59CB1695⟩, CO$_2$R ⟨65ZOB354⟩ and CN ⟨68ZOB292⟩, or with (RO)$_3$P for E = COMe ⟨93PS(75)23⟩, COR ⟨79LA492⟩ and CO$_2$Et ⟨91ZOB2698⟩. The unsymmetrical phosphinate/phosphonate Et(MeO)P(O)CH(CH$_2$CO$_2$Me)P(O)(OMe)$_2$ has been obtained in 76% yield by a similar method ⟨66ZOB296⟩. The reaction of ClCH=CHSO$_2$F with MeP(OEt)$_2$ takes a most unusual course and the mechanism for formation of the product (**150**), presumably involving (**149**) as an intermediate, is not clear ⟨64ZOB2897⟩.

4.12.1.5.16 From alkynes

Many of the starting materials (**147**) mentioned in the previous section were obtained by addition of a phosphorus nucleophile to an alkyne or its equivalent. Direct twofold addition of dialkyl or trialkyl phosphites to HC≡C—E to give (**148**) has been reported for E = CO$_2$Me ⟨63USP3093672⟩ and E = CN ⟨71USP3622654⟩. The analogous twofold addition of (EtO)$_2$P(O)H ⟨63ZOB1045⟩, (RO)$_2$P(O)Na and (RO)$_2$P(S)Na ⟨64ZOB3938⟩ to the triple bond of MeC≡C—P(O)(OEt)$_2$ has also been described.

4.12.1.5.17 Miscellaneous methods

Either CCl$_4$ or CHCl$_3$ and 3 equivalents of (EtO)$_3$P give the bis(phosphonate) (EtO)$_2$P(O)CH$_2$P(O)(OEt)$_2$ in yields of 65% and 42%, respectively ⟨79ZOB1470⟩. The corresponding tetrabutyl ester (Bu^{n}O)$_2$P(O)CH$_2$P(O)(OBu^n)$_2$ is formed together with other products by treatment of white phosphorus with CCl$_4$ followed by NaOBu^n/Bu^{n}OH ⟨84EGP214610⟩. Treatment of Cl$_3$C—CH=C(CN)$_2$ with an excess of diethyl phosphite affords (**151**) in 80% yield ⟨87ZOB2138⟩. The α-oxophosphonate MeCOP(O)(OEt)$_2$ and Ph$_2$PH initially give (**152**) and, when this is heated, it rearranges to (**153**) ⟨78ZOB1001⟩. Conversion of the tricarboxylic acid (**154**) into the corresponding triphosphonic acid (**155**) can be achieved directly by treatment with H$_3$PO$_3$ in acetic anhydride ⟨73GEP2228928⟩. In a remarkable process, (**156**) is produced in 40% yield by treatment of PhCOCl with H$_3$PO$_3$ ⟨79ZAAC(457)203⟩. Even more remarkable are the products (**157**) which are formed from simple acetamides MeC(O)NHR by treatment with H$_3$PO$_3$/PCl$_3$ under certain conditions ⟨92ZN(B)1213⟩. Reactions of active methylene compounds such as diethyl malonate, malononitrile and benzyl cyanide, with dialkyl phosphonites, MeP(OR)$_2$, proceed with loss of the hydrocarbon RH to give (**158**) ⟨87ZOB2294⟩.

The bis(phosphonium) salt, $Ph_3P^+CH_2P^+Ph_3$ $(FeCl_4^-)_2$ was isolated in 11% yield by treatment of Ph_3P with $FeCl_3$ in chloroform followed by air oxidation ⟨89POL1293⟩. The related salt, $Me_3P^+C Me_2P^+Me_3$ $(I^-)_2$ can be obtained from (159) with an excess of MeI ⟨68IC709⟩.

(159)

Hydrolysis and other simple reactions of a variety of 1,3-diphosphetes provide unusual compounds of this class. Thus, hydrolysis of (160) affords (161) ⟨91ZAAC(601)65⟩, while (162) and (163) are hydrolysed to give (164) and (165), respectively ⟨86TL1903⟩. Treatment of (160) with electrophiles such as MeI or CS_2 gives the salts (166) and (167), but with two extra methyl groups present as in (168) the reaction with CS_2 follows a different course to give (169) ⟨93PS(77)258⟩.

(160) R = H
(168) R = Me

(161)

(162) $R^1 = R^2$ = Ph
(163) R^1 = OMe, $R^2 = NPr^i_2$

(164)
(165)

(166) X = Me I^-
(167) X = CS_2^-

(169)

4.12.1.5.18 By oxidation of tricoordinate species

Very many compounds in this class have been prepared from the symmetrical tricoordinate species of Section 4.12.1.3 and the asymmetrical species of Section 4.12.1.4 using the standard methods of oxidation, sulfurization and quaternization. Representative examples are given here for a few of the most important compounds. Treatment of $H_2PCH_2PH_2$ with nitric acid results in oxidation to $(HO)_2P(O)CH_2P(O)(OH)_2$ and a range of substituted examples $H_2PCH(R)PH_2$ react similarly ⟨69USP3445522⟩. The solid bis(borane) adduct of $H_2PCH_2PH_2$ is formed by treatment with $BH_3 \cdot Me_2S$ ⟨91CB275⟩. Hydrolysis of $Cl_2PCH_2PCl_2$ gives the diphosphinic acid $(HO)PH(O)CH_2P(O)H(OH)$ which can be oxidized stepwise to the diphosphonic acid $(HO)_2P(O)CH_2P(O)(OH)_2$ ⟨86IC1290⟩. Conversion of $Cl_2PCH_2PCl_2$ into $Cl_2P(O)CH_2P(O)Cl_2$ is possible with DMSO or P_2O_5/Cl_2 and $Cl_2P(S)CH_2P(S)Cl_2$ is formed by treatment with $S_8/AlCl_3$ ⟨70ZAAC(376)37, 77ZOB775⟩. By combining these methods, the asymmetrical $Cl_2P(O)CH_2P(S)Cl_2$ can be obtained ⟨87ZAAC(555)109⟩. More unusual reactions of $Cl_2PCH_2PCl_2$ include those with aldehydes to give (170), with ketones in acetic acid to give (171) ⟨83ZOB1673⟩ and with $CH_2(OEt)_2$ and $CH(OEt)_3$ to give (172) and (173), respectively ⟨90ZOB1420⟩. Arbuzov reactions of a variety of compounds $(EtO)_2PCH(R)P(O)(OEt)_2$ have been reported ⟨74ZOB276, 87JA5544⟩. $(R^1O)_2PCH_2P(OR^1)_2$ with 2 equivalents of R^2COCl or R^2_2PI gives (174) ⟨79ZOB470⟩ and (175) ⟨79ZOB1446⟩, respectively. The simple reactions of $Me_2PCH_2PMe_2$ including formation of bis(phosphonium) salts, the dioxide and disulfide have been described ⟨79ZN(B)31⟩. The diastereomers of $PhMeP(Se)CH_2P(Se)MePh$ formed from the bis(phosphine) and selenium have been partly separated by crystallization ⟨78JCR(S)368⟩.

(170)

(171)

(172) R = H
(173) R = OEt

(174) X = $C(O)R^2$
(175) X = PR^2_2

The conversion of $Ph_2PCH_2PPh_2$ into symmetrical tetracoordinate derivatives has been widely investigated. All possible compounds of $Ph_2P(X)CH_2P(Y)Ph_2$ have been prepared for X, Y = O, S, Se and Me^+ I^- ⟨80IC1982⟩. Symmetrical and asymmetrical bis(phosphinimines) $Ph_2P(=NR^1)CH_2P(=NR^2)Ph_2$ are formed by treatment with aryl azides ⟨66TL3187, 72IZV2612⟩. The bis(trimethylsilylphosphinimine) is obtained with $TMS-N_3$ and with $Me_2Si(N_3)_2$ the cyclic derivative (**176**) is produced ⟨74ZAAC(406)131⟩. $Ph_2PCH_2PPh_2$ and DMAD afford the 1,3-diphosphole (**177**) ⟨70JCS(C)504⟩. While treatment of $Ph_2PCH_2PPh_2$ with $PSCl_3$ gives the bis(phosphine sulfide), $POCl_3$ gives the bis(phosphonium) salt (**178**) ⟨72CB3261⟩. Lewis acid adducts formed include $Ph_2P(BH_3)CH_2P(BH_3)Ph_2$ from treatment with $NaBH_4/I_2$ ⟨69IC2671⟩, $Ph_2P(AlMe_3)CH_2P(AlMe_3)Ph_2$ ⟨88MI 412-01⟩, and $Ph_2P(AuCl)CH_2P(AuCl)Ph_2$ from treatment with $AuCl_3$ ⟨77CB1748⟩.

4.12.1.5.19 *Interconversions*

Conversion of $Cl_2P(O)CH_2P(O)Cl_2$ into $F_2P(O)CH_2P(O)F_2$ is readily achieved with AsF_3 ⟨73ZN(B)98⟩ and this reagent also converts $Cl_2P(S)CH_2P(O)Cl_2$ into a mixture of $F_2P(S)CH_2P(O)F_2$ and $Cl_2P(S)CH_2P(O)F_2$ ⟨87ZAAC(555)109⟩. Treatment of $Cl_2P(S)CH_2P(S)Cl_2$ with SbF_3 gives $F_2P(S)CH_2P(S)F_2$ ⟨87ZAAC(555)109⟩. The tetrachloride $Cl_2P(O)CH_2P(O)Cl_2$ can itself be formed from treatment of either $(Pr^iO)_2P(O)CH_2P(O)(OPr^i)_2$ ⟨69USP3445522, 81CB1082⟩, $(HO)_2P(O)CH_2P(O)(OH)_2$ ⟨69IC1775⟩, or a mixture of $(HO)_2P(O)CH_2P(O)(OH)_2$ and $(EtO)_2P(O)CH_2P(O)(OEt)_2$ ⟨65HCA133⟩ with PCl_5. This compound is a key intermediate since stepwise replacement of the chlorines by alkyl groups with Grignard reagents ⟨61JA1722, 65HCA133⟩ or by dialkylamino groups with R_2NH ⟨69IC1775⟩ or R_2N-TMS ⟨88ZAAC(561)157⟩ is then possible. Treatment with primary amines affords the azadiphosphetanes (**179**) ⟨76JCS(D)1113⟩. The diphosphinates $(R^3O)PH(O)CH_2P(O)H(OR^3)$ and carbonyl compounds initially give the hydroxyalkyl compounds (**180**) but these rearrange when heated with base to afford the alkoxy isomers (**181**) ⟨87ZOB525⟩. $(HO)_2P(O)CH_2P(O)(OH)_2$ and H_3PO_4 in a urea melt give the cyclic anhydride (**182**) ⟨89ZAAC(576)272⟩.

The other major class of interconversions involves functionalization on the central carbon atom. The first example of this transformation appeared in 1953 when treatment of $(EtO)_2P(O)CH_2P(O)(OEt)_2$ with potassium followed by Bu^nBr was reported to give the alkylated product $(EtO)_2P(O)CH(Bu^n)P(O)(OEt)_2$ in 20% yield ⟨53JA1500⟩. Mono- and disubstitution of diphosphonates has since been reported with a wide variety of electrophiles including alkyl halides ⟨66BRP1026366, 68JOM(13)199, 81AP(314)218⟩, benzyl bromides ⟨89USP4818774⟩, cinnamyl acetates ⟨92USP5103036⟩, benzoyl chloride ⟨88JAP63185993⟩, Me_3SnCH_2I ⟨89ZC409⟩, and CF_3CO_2D and CF_3CO_2T which afford the isotopically labelled products ⟨92TL3927⟩. The anions have most commonly been generated with NaH or sodium metal. Conjugate addition of the anions of methylenediphosphonates to acrylonitrile, acrylates and diethyl vinylphosphonate ⟨70ZOB499⟩ and to DMAD ⟨80JOC2698⟩ has also been reported. The conversion of $(EtO)_2P(O)CCl_2P(O)(OEt)_2$ into $(EtO)_2P(O)CMe_2P(O)(OEt)_2$ by twofold lithium–halogen exchange and alkylation with Bu^nLi and Me_2SO_4 has been described ⟨73JOM(59)237⟩. Threefold alkylation of the triphosphonate (**89**) with n-$C_8H_{17}Br$ gives a material useful as a calcium sequestering agent ⟨72USP3632634⟩. Treatment of the phosphonate/phosphinate (**183**) with NaH followed by Bu^nLi gives the dianion, which is preferentially alkylated on the more reactive site to give (**184**), whereas use of one equivalent of base results in alkylation on the central carbon ⟨89TL411⟩. The anion derived from $Ph_2P(O)CH_2P(O)Ph_2$

has similarly been alkylated with benzyl chloride ⟨70JPR456⟩ and alkyl bromides ⟨70IZV1326⟩ and by conjugate addition to acrylates, vinylphosphonates and vinylphosphine oxides ⟨71IZV2747⟩. Reactions with α,ω-dibromides generally result in simple twofold substitution, but with 1,5-dibromopentane the cyclic product (**185**) is formed ⟨70IZV1326⟩. The bis(borane) adduct $Me_2P(BH_3)CH_2P(BH_3)Me_2$ can be alkylated on the central carbon by successive treatment with Bu^nLi and MeI or $PhCH_2Cl$ ⟨79AG(E)782⟩.

(183)   (184)   (185)

4.12.1.6 Penta- and Hexacoordinate Systems

Access to one of the simplest compounds of this type, $F_4PCH_2PF_4$, was first gained by treating the disilacyclobutane (**186**) with PF_5 ⟨75CC468⟩. Other routes to this compound include the reactions of F_4PCH_2-TMS with PF_5, $F_2P(O)CH_2P(O)F_2$ with SF_4 ⟨81CB1082⟩ and $F_2PCH_2P(S)F_2$ with SbF_3 ⟨87ZAAC(555)109⟩. In this last reaction, $F_4PCH_2P(S)F_2$ is also formed and this compound can be prepared, together with (**54**) as mentioned earlier, by thermal disproportionation of $F_2PCH_2P(S)F_2$ ⟨88ZAAC(566)90⟩. $F_4PCH_2PF_4$ and KF give $(K^+)_2^-F_5PCH_2PF_5^-$, which is hydrolysed to give $K^+F_2P(O)CH_2PF_5^-$ ⟨87ZAAC(555)109⟩, while treatment with Me_2N-TMS affords $(Me_2N)_2FP^+CH_2PF_5^-$ ⟨79IC60⟩. Numerous methods have been developed for the preparation of $Ph_2F_2PCH_2PF_2Ph_2$ including treatment of $Ph_2PCH_2PPh_2$ with SF_4 (59%) ⟨88SRI727⟩, SF_4/NaF ⟨69ZN(B)1081⟩, F_2 in $CFCl_3$ ⟨77AG(E)718⟩ and F_3N^+—O^- (75%) ⟨90IC573⟩. Additional routes are available from the bis(phosphinimine) Ph_2P(N-TMS)CH_2P(N-TMS)Ph_2 and HF ⟨75CB919⟩ and from nickel(II) halide complexes $NiX_2(Ph_2PCH_2PPh_2)_2$ and COF_2 ⟨91JFC(52)1⟩.

(186)

The diphosphonites, $(R^1O)_2PCH_2P(OR^1)_2$, and aldehydes, R^2CHO, afford the diphosphoranes (**187**) ⟨89ZOB101⟩. Treatment of (**188**) with tetrachloro-*o*-benzoquinone results in a complex sequence of reactions to afford the remarkable structure (**189**) ⟨92CB1325, 93PS(75)233⟩.

(187)   (188)   (189)

4.12.2 FUNCTIONS CONTAINING ONE PHOSPHORUS AND ONE ARSENIC, ANTIMONY OR BISMUTH GROUP

4.12.2.1 Phosphorus and Arsenic Functions

Although many of the processes described in the previous sections are potentially also applicable to synthesis of phosphorus and arsenic functions, these have only been reported in a few cases. Thus, Ph_2PCH_2-TMS and Ph_2AsCl react with loss of TMS-Cl to give $Ph_2PCH_2AsPh_2$ ⟨79CB648⟩, and treatment of $PhAsCl_2$ with Ph_2PCH_2Li/TMEDA produces $Ph_2PCH_2As(Ph)CH_2PPh_2$ in 67%

yield ⟨85JA5272⟩. Treatment of (**190**) with lithium metal followed by CH_2Cl_2 gives (**191**) ⟨93PS(77)254⟩. 4-Aryl-2,6-diphenyl-3-arsapyridines and $Bu^tC\equiv P$ react with loss of benzonitrile to give (**192**) ⟨88AG(E)709⟩. Treatment of $Ph_2PCH_2AsPh_2$ with MeI gives $Ph_2MeP^+CH_2AsPh_2$ I^- and subsequent treatment of this with $NaNH_2$ followed by MeI affords (**193**). In contrast, $Ph_2PCH_2AsPh_2$ and $MeOSO_2F$ give $Ph_2MeP^+CH_2As^+MePh_2$ $(I^-)_2$ directly ⟨87CB1281⟩. The reaction of $Ph_3P{=}CH_2$ with Ph_2AsCl proceeds with double transylidation to give $Ph_3P{=}C(AsPh_2)_2$ and $Ph_3P^+CH_2AsPh_2$ Cl^- ⟨84ZN(B)1456⟩. Treatment of $(HO)_2P(O)CH_2Cl$ with sodium arsenite in alkaline solution affords $(HO)_2P(O)CH_2As(O)(OH)_2$ ⟨77MI 412-01, 78BJ(169)239⟩.

4.12.2.2 Phosphorus and Antimony Functions

Only isolated examples of these functions have been described. Treatment of $Ph_3P{=}CH_2$ with Me_2SbBr followed by MeBr affords $Ph_3P^+CH_2SbMe_3^+$ $(Br^-)_2$ ⟨61JA2055⟩. Ph_2SbCH_2Li and Ph_2PCl give $Ph_2SbCH_2PPh_2$ in 20% yield ⟨83CB473⟩ and $Ph_2P(O)CH_2Li$ and Ph_2SbBr similarly afford $Ph_2P(O)CH_2SbPh_2$ in 17% yield ⟨80TL2803⟩.

4.12.2.3 Phosphorus and Bismuth Functions

No compounds of this type appear to have been described.

4.12.3 FUNCTIONS CONTAINING PHOSPHORUS AND A METALLOID

4.12.3.1 Dicoordinate Phosphorus Derivatives

4.12.3.1.1 Dicoordinate phosphorus and silicon functions

A single example of this type is provided by the reaction of (**194**) with diazoalkanes, $RCHN_2$, such as diazomethane, diazoethane and *t*-butyl diazoacetate, which results in apparent carbene insertion into the P—Si bond to give (**195**) ⟨85ZN(B)1077⟩.

4.12.3.1.2 Dicoordinate phosphorus and germanium functions

No compounds of this type appear to have been described.

4.12.3.1.3 Dicoordinate phosphorus and boron functions

No compounds of this type appear to have been described.

4.12.3.2 Tricoordinate Phosphorus Derivatives

4.12.3.2.1 Tricoordinate phosphorus and silicon functions

(i) From 1-metalloalkylsilanes

The reaction of α-metallated silanes, particularly $TMS-CH_2MgCl$, with phosphorus electrophiles, R_2PCl, to give $TMS-CH_2PR_2$ has been widely used, and typical electrophiles include Ph_2PCl

⟨70USP3511865⟩, (Me₂N)₂PCl ⟨86IS110⟩, R₂NP(R)Cl ⟨81ZAAC(475)18⟩, TMS₂NP(Ar)Cl ⟨83OM921⟩, (TMS₂N)₂PCl ⟨83IC575⟩ and (EtO)₂PCl ⟨76ZOB575⟩. In some of these examples the metalloalkylsilanes have been generated by lithiation of such precursors as TMS-CH₂CO₂Me ⟨76ZOB575⟩ or TMS-CH₂Ph ⟨83IC575, 86IS110⟩, and so C-substituted products are formed. Treatment of CH₂=CHSiMe₂CH₂MgCl with Ph₂PCl gives the (vinyldimethylsilylmethyl)phosphine in 50% yield ⟨85MI 412-01, 88SRI163⟩. The bis(Grignard) reagent ClMgCH₂Si(Me₂)CH₂MgCl and Ph₂PCl give Ph₂PCH₂Si(Me₂)CH₂PPh₂ in 70% yield ⟨85MI 412-01, 88SRI163⟩. The allyllithium species derived from (196) and R¹R²PCl produced (197) ⟨89IC3219⟩.

TMS⎯⎯⎯⎯⎯TMS
         |
         X

(196) X = H
(197) X = PR¹R²

Treatment of TMS-CH₂MgCl or its lithium analogue with compounds of the general type RPX₂ gives TMS-CH₂P(R)CH₂-TMS and this process has been reported for PhPBr₂ and Buⁿ PBr₂ ⟨60USP2964550⟩, ArPCl₂ ⟨83OM921⟩, MeP(OPh)₂ and Buᵗ PCl₂ ⟨82ZN(B)284⟩, Me₂NPCl₂ ⟨86ZAAC(535)47⟩, and TMS₂NPCl₂ ⟨81IC1679, 82IC3568⟩. In some cases the reaction can be stopped at the stage of TMS-CH₂P(R)Cl if desired ⟨81IC1679⟩ and monosubstitution with the anion derived from (196) and TMS₂NPCl₂ has also been achieved ⟨88OM572⟩.

Treatment of PCl₃ with one equivalent of TMS-CH₂MgCl gives TMS-CH₂PCl₂ ⟨61JOC2604⟩, but the reaction proceeds to completion to give P(CH₂-TMS)₃ when an excess of TMS-CH₂MgCl reacts with PBr₃ ⟨58JA1336, 60USP2964550⟩. The latter product can also be formed from TMS-CH₂Li and P(OPh)₃ ⟨70USP3511865⟩.

(ii) From 1-haloalkylsilanes

The reaction of TMS-CH₂Cl with metal phosphides to give TMS-CH₂PR₂ occurs readily and has been reported for Me₂PLi ⟨69IC862⟩, Ph₂PNa, and various mixed alkylarylphosphides, PhP(R)Na ⟨81ZAAC(475)18, 89CZ349⟩. Other silanes have been used, including EtOSiMe₂CH₂Br ⟨88SRI695⟩ and PhOSiMe₂CH₂Cl ⟨88JOM(356)285⟩ with Ph₂PLi, and Buⁿ₂SiMeCH₂Cl and (EtO)₂SiMeCH₂Cl with Ph₂PNa ⟨70BRP1179242⟩. Treatment of ClSiMe₂CH₂Cl with one equivalent of Me₂PLi gives ClSiMe₂CH₂PMe₂, but with an excess of the reagent, Me₂PSiMe₂CH₂PMe₂ is formed. Treatment of ClSiMe₂CH₂Cl with Me₂NH followed by Me₂PLi gives Me₂NSiMe₂CH₂PMe₂ ⟨73JOM(61)133, 77ZAAC(433)157⟩. Treatment of the bis(chloromethylsilane) ClCH₂SiMe₂NHSiMe₂CH₂Cl with an excess of R₂PLi affords R₂PCH₂SiMe₂N(Li)SiMe₂CH₂PR₂ ⟨85IC642⟩.

(iii) From 1-metalloalkylphosphines

Simple lithioalkylphosphines such as Ph₂PCH₂Li ⟨83JCS(P1)861⟩ and Me₂PCH₂Li ⟨83ZN(B)1399⟩ react readily with R₃SiCl to give the expected products. Interestingly, the first compound reacts only once with R₂SiCl₂ or RSiCl₃ to give mainly Ph₂PCH₂SiR₂Cl or Ph₂PCH₂SiRCl₂ ⟨79JA7410, 81IC3200, 83JCS(P1)861⟩, whereas Me₂PCH₂Li undergoes multiple reactions to give Me₂Si(CH₂PMe₂)₂, MeSi(CH₂PMe₂)₃, and with SiCl₄, Si(CH₂PMe₂)₄ ⟨83ZN(B)1399⟩. Lithiation followed by treatment with TMS-Cl has also been reported for (Me₂N)₂PCH₂Ph ⟨86ZAAC(535)47⟩ and Ph₂PCH₂Ar ⟨76ZAAC(422)237⟩. Attempts to apply the same approach to the α-phosphino esters R¹₂PCH₂CO₂R² are frustrated by the preferential O-alkylation of the enolates with TMS-Cl ⟨76ZOB1218⟩. Lithiation of the chiral borane adduct (92) derived from PhP(Buᵗ)Me, with Buⁿ Li/TMEDA, then treatment with MeSiCl₃ and removal of the borane with morpholine affords the enantiomerically pure tripodal ligand, MeSi(CH₂P(Buᵗ)Ph)₃ ⟨91HCA983⟩.

(iv) From 2-silyl phospha alkenes

The reactivity of the double bond in compounds of the general structure (198) has been examined in detail. Diels–Alder reactions, both with 2,3-dimethylbutadiene to give (199) ⟨83TL3591, 84CB2693, 85CB814, 87PS(30)523⟩, and with cyclopentadiene to give the corresponding bicyclic adducts ⟨85CB3419,

85CB4068⟩ have been reported. Carbene addition to compounds (**198**) with R = Cl or alkyl gives the 2-silylphosphiranes (**200**) ⟨87TL2693, 89TL3951⟩. The other main compound used in this context is TMS$_2$N—P=CH-TMS. This reacts with Bu^{i}I to give (**201**), with MeOH to give (**202**), and by [2 + 2]-cycloaddition with Me$_2$Si=CHCH$_2$Bu^t to give (**203**). Treatment of (**201**) with MeLi affords (**204**) ⟨83PS(18)43, 85IC1993⟩. Treatment of the same phospha alkene with TMS-CHN$_2$ followed by photochemical extrusion of N$_2$ gives the disilylphosphirane (**205**) ⟨83CC1171⟩.

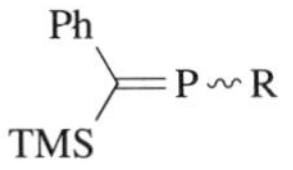

(**198**)

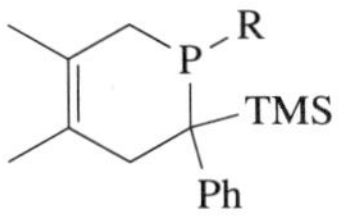

(**199**)

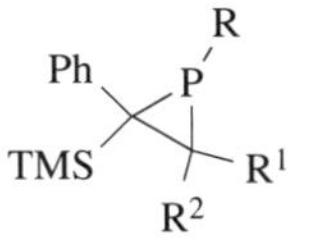

(**200**)

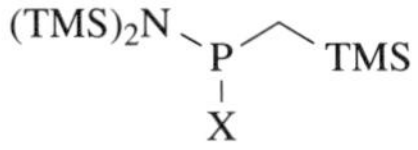

(**201**) X = I
(**202**) X = OMe
(**204**) X = Me

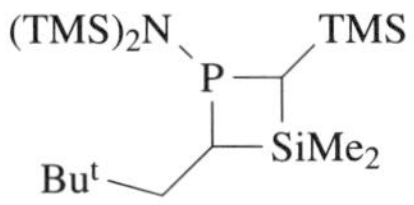

(**203**)

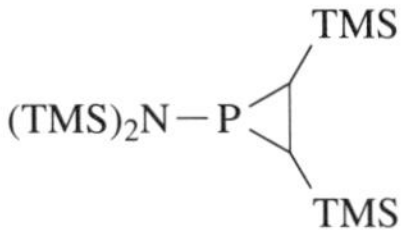

(**205**)

(v) From 1-silyl-1-phosphino alkenes

Hydrostannylation of (**206**) with Me$_3$SnH affords (**207**) in 35% yield and this can be converted into the corresponding phosphine oxide with KMnO$_4$ ⟨89JOM(368)167⟩.

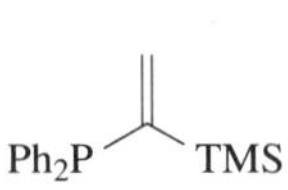

(**206**)

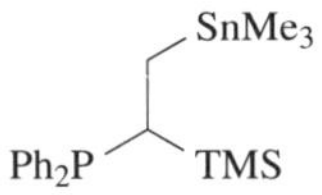

(**207**)

(vi) From phosphino(silyl)carbenes

Intramolecular CH insertion of the carbenes (**208**) ⟨88JA6463⟩ and (**209**) ⟨90AG(E)1429⟩ formed by pyrolysis of the corresponding diazo compounds gives the 1,2-azaphospholanes (**210**) and (**211**), respectively. Addition of (**208**) to the double bond of methyl acrylate or dimethyl fumarate affords the cyclopropanes (**212**) ⟨89AG(E)621⟩.

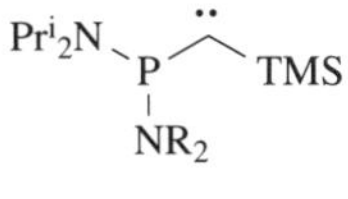

(**208**) R = Pr^i
(**209**) R = Me

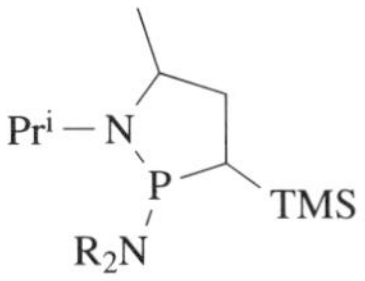

(**210**)
(**211**)

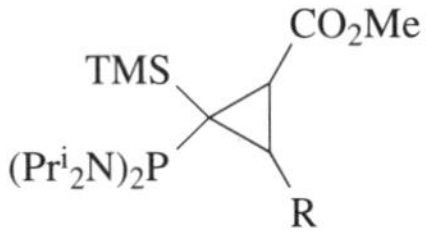

(**212**)
R = H, CO$_2$Me

(vii) Miscellaneous methods

Treatment of the tetramethyldisilacyclobutane (**186**) with PCl$_3$ or PBr$_3$ results in cleavage to give (**213**) and the tetraphenyl derivative behaves similarly ⟨83ZAAC(500)132⟩. Treatment of (**214**) with PCl$_3$ results in loss of TMS-Cl to afford (**215**) ⟨86TL5611⟩. The phospha alkenes (**216**) undergo dimerization by a novel mechanism when heated to 180–190 °C to produce (**217**) ⟨93S285⟩. Heating (**218**) (Ar = 2,4,6-triisopropylphenyl) at 110 °C results in rearrangement and loss of N$_2$ to give the silaphosphirane (**219**) ⟨93PS(76)57⟩. Treatment of TMS$_2$CHPCl$_2$ with the carbenoid TMS$_2$C(Cl)Li at

−100 °C affords the trisilylphosphirane (**220**) and this reacts with RLi to produce the *P*-alkyl analogue ⟨89CB453⟩. A series of compounds TMS-CH=P(R)=CH-TMS undergo thermal isomerization to the more stable phosphiranes (**221**) ⟨92CB771⟩. The remarkable dimeric structures (**222**) and (**223**) can be formed from $(TMS)_2C=P(Cl)=C(TMS)_2$ by treatment with magnesium and by heating, respectively ⟨86CB535⟩.

(viii) Interconversions

Stepwise displacement of the chlorines in compounds such as $TMS-CH_2PCl_2$ with ArLi ⟨82ZAAC(488)75⟩, Me_2NH and AgCN ⟨87JSP(121)304⟩ has been described, and a $P-NMe_2$ group can be converted back into P—Cl with HCl gas ⟨86IS110, 87JSP(121)304⟩. Replacement of the Si—Cl in $Ph_2PCH_2SiMe_2Cl$ by CpLi has also been achieved ⟨79JA7410, 81IC3200⟩, and hydrolysis of the related compound $Ph_2PCH_2SiMe_2(OEt)$ affords the disiloxane, $Ph_2PCH_2Si(Me_2)OSi(Me_2)CH_2PPh_2$ ⟨88MIP4121⟩.

4.12.3.2.2 Tricoordinate phosphorus and germanium functions

The reaction of $(EtO)_2PCH_2CO_2Et$ with $NaN(TMS)_2$ followed by Me_3GeBr results in *C*-alkylation to give $(EtO)_2PCH(CO_2Et)GeMe_3$ ⟨76ZOB1218⟩. Treatment of $ClMe_2GeCH_2Cl$ with Me_2PLi results first in displacement of the Ge—Cl rather than the C—Cl to give $Me_2PGe(Me_2)CH_2Cl$, but this product isomerizes when heated to give $ClMe_2GeCH_2PMe_2$. Use of an excess of Me_2PLi gives $Me_2PGe(Me_2)CH_2PMe_2$ directly and $Me_2AsGe(Me_2)CH_2PMe_2$ has also been prepared ⟨77JOM(132)77⟩. $Bu^tC≡P$ and $GeCl_4$ react in a ratio of 2:1 to afford the compound (**224**) ⟨87PS(30)349⟩.

4.12.3.2.3 Tricoordinate phosphorus and boron functions

A single example of this type is provided by the reaction of PhP=C(Bu^t)O-TMS with an excess of dicyclohexylborane which affords the distillable (α-phosphinoalkyl)borane (**225**) in 22% yield ⟨90PS(53)1⟩.

$$\text{PhHP} - \underset{\underset{\textbf{(225)}}{\text{B}(C_6H_{11})_2}}{\overset{Bu^t}{\text{C}}}$$

4.12.3.3 Tetracoordinate Phosphorus Compounds

4.12.3.3.1 Tetracoordinate phosphorus and silicon functions

(i) From 1-metalloalkylsilanes

The reaction of TMS-OSi(Me$_2$)CH$_2$MgBr with diphenylphosphoryl chloride, (PhO)$_2$P(O)Cl, affords TMS-OSi(Me$_2$)CH$_2$P(O)(OPh)$_2$ ⟨59BRP815231, 59USP2889349⟩. PhP(O)Cl$_2$ and Et$_3$SiCH$_2$MgCl likewise give Et$_3$SiCH$_2$P(O)(Ph)CH$_2$SiEt$_3$ ⟨74ZAAC(404)204⟩. The threefold reaction of TMS-CH$_2$MgBr with POCl$_3$ gives PO(CH$_2$-TMS)$_3$, and the corresponding reaction of Me$_2$ArSiCH$_2$MgBr proceeds similarly ⟨59JCS3751⟩.

(ii) From reactions of 1-haloalkylsilanes with phosphorus nucleophiles

The reaction of TMS-CH$_2$Cl or TMS-CH$_2$Br with tertiary phosphines to give the corresponding phosphonium salts, TMS-CH$_2$PR$_3^+$ X$^-$, proceeds in good yield and has been widely used. Examples of the phosphines used include Me$_3$P ⟨65JA390, 78IS135⟩, Ph$_3$P ⟨59CI(L)849, 61JA1610, 61JA1613, 83ZOB1831⟩, Me$_2$PEt ⟨68CB595⟩, PhPMe$_2$ ⟨74ZN(B)485⟩, PhP(CH$_2$Bu^t)$_2$ and Ph$_2$P(CH$_2$Bu^t) ⟨79JOC1057⟩. Treatment of TMS-CH$_2$Cl with PCl$_3$ and AlCl$_3$ gives TMS-CH$_2$PCl$_3^+$ AlCl$_4^-$ ⟨66JOC4288⟩. Other haloalkylsilanes have also been used including H$_3$SiCH$_2$Cl and Me$_2$Si(H)CH$_2$Cl with Me$_3$P and Et$_3$P ⟨69AG(E)372, 70CB3007⟩ and (EtO)$_3$SiCH$_2$I with Ph$_3$P and Bu$^n_3$P ⟨83ZOB1831⟩. Phosphines already bearing silylmethyl groups can also be used as illustrated by the reaction of Me$_2$PCH$_2$-TMS and MeP(CH$_2$-TMS)$_2$ with TMS-CH$_2$Cl to give Me$_2$P$^+$(CH$_2$-TMS)$_2$ Cl$^-$ and MeP$^+$(CH$_2$-TMS)$_3$ Cl$^-$, respectively ⟨69CB83⟩. The tris(silylalkyl)phosphine P(CH$_2$-TMS)$_3$ and TMS-CH$_2$I likewise afford (TMS-CH$_2$)$_4$P$^+$I$^-$ ⟨58JA1336⟩. Stepwise reactions of Me$_2$Si(CH$_2$Cl)$_2$ with PMe$_3$ first give ClCH$_2$Si(Me$_2$)CH$_2$PMe$_3^+$ Cl$^-$ and then Me$_3$P$^+$CH$_2$Si(Me$_2$)CH$_2$PMe$_3^+$ (Cl$^-$)$_2$ under more forcing conditions ⟨70CB97⟩. Treatment of ClCH$_2$Si(Me$_2$)OSi(Me$_2$)CH$_2$Cl with Ph$_2$PNHBu^t to give the bis(phosphonium) salt has also been described ⟨63JOC272⟩.

Reactions of haloalkylsilanes with the sodium salts of dialkyl phosphites, (RO)$_2$P(O)Na, have been reported in a few cases. Thus, TMS-CH$_2$Br, TMS-Si(Me$_2$)CH$_2$Cl ⟨59JCS3751, 62JCS592⟩, TMS-OSi(Me$_2$)CH$_2$Br and BrCH$_2$Si(Me$_2$)OSi(Me$_2$)CH$_2$Br ⟨59BRP815231, 59USP2889349⟩ have all been converted into the corresponding silylmethylphosphonates. With ClSi(Me$_2$)CH$_2$Cl both chlorine atoms are displaced to afford (RO)$_2$P(O)Si(Me$_2$)CH$_2$P(O)(OR)$_2$ ⟨60USP2920094⟩.

(iii) From Arbuzov reactions of 1-haloalkylsilanes

This approach, typified by the reaction of TMS-CH$_2$Cl with (EtO)$_3$P in a sealed vessel at 150–200 °C to give TMS-CH$_2$P(O)(OEt)$_2$, has been much used since it was first reported in 1956 ⟨56USP2768193⟩. Most studies have involved (EtO)$_3$P, and the range of haloalkylsilanes used with this phosphite includes TMS-CH$_2$Cl, (EtO)$_3$SiCH$_2$Cl ⟨58ZOB728⟩, TMS-OSi(Me$_2$)CH$_2$Cl ⟨56USP2768193⟩, EtSi(Me$_2$)CH$_2$Cl, EtOSi(Me$_2$)CH$_2$Cl, ClCH$_2$Si(Me$_2$)CH$_2$Si(Me$_2$)CH$_2$Cl ⟨62DOK(143)840⟩, Et$_3$SiCH$_2$Cl ⟨63DOK(148)875⟩, PhSi(Me$_2$)CH$_2$Cl ⟨65IZV286⟩ and (EtO)$_3$SiCH$_2$I, which gives an excellent yield of 95% ⟨75ZOB2010⟩. While most work has been confined to (halomethyl)silanes, some reactions of, for example, TMS-CH(Ph)Br have been reported ⟨72ZOB112⟩. Other phosphites have been successfully used ⟨62USP3019248⟩ including (MeO)$_3$P, which reacts with TMS-CH$_2$Cl to afford TMS-CH$_2$P(O)(OMe)$_2$ in 65% yield ⟨62JCS592⟩, and also with BrCH$_2$Si(Me$_2$)OSi(Me$_2$)CH$_2$Br and brominated octamethylcyclotetrasiloxane ⟨87JPS(A)1967⟩, and (Et$_3$SiO)$_3$P, which reacts with (EtO)$_3$SiCH$_2$Cl to afford (EtO)$_3$SiCH$_2$P(O)(OSiEt$_3$)$_2$ ⟨71MI 412-01⟩.

Reactions of a variety of (halomethyl)silanes with arylphosphonites, ArP(OEt)$_2$ to give the products R$_3$SiCH$_2$P(O)(OEt)Ar have been reported for Ar = Ph ⟨62DOK(147)117, 62MI 412-01⟩, *p*-chlorophenyl ⟨70ZOB609⟩ and 2-thienyl ⟨71ZOB2186⟩. Only one report of the use of a phosphinite has appeared: Ph$_2$P(OEt) and either TMS-CH$_2$Cl or PhSi(Me$_2$)CH$_2$Cl give the corresponding products, R$_3$SiCH$_2$P(O)Ph$_2$ ⟨65IZV286⟩.

(iv) From 1-metalloalkylphosphorus compounds

α-Lithiated phosphine oxides, phosphonates and phosphinates react with trialkylsilyl halides readily in good yield. For example, Ph$_2$P(O)CH$_2$Li or Ph$_2$P(S)CH$_2$Li, formed either from Ph$_3$PX (X = O or S) and MeLi ⟨64JA1100, 69USP3426021⟩ or by deprotonation of Ph$_2$P(X)Me with Bu^{n}Li ⟨68CJC2115⟩ reacted with TMS-Cl to give the products Ph$_2$P(X)CH$_2$-TMS. Other trialkylsilyl and triarylsilyl halides have been used ⟨69USP3426021, 74ZAAC(404)204⟩ and the reaction has also been extended to the substituted phosphine oxides Ph$_2$P(O)CH(R)Li ⟨64JA1100, 78ZAAC(447)53, 84JOC263⟩ and Ar$_2$P(O)CH$_2$Li and to the sulfides Ar$_2$P(S)CH(R)Li ⟨69USP3426021⟩.

The corresponding reaction with phosphonates occurs in high yield as illustrated by treatment of (MeO)$_2$P(O)Me with Bu^{n}Li or LDA followed by Bu^{t}Me$_2$SiCl to give (MeO)$_2$P(O)CH$_2$SiMe$_2$Bu^t in 90% yield ⟨87TL5121⟩ and of (MeO)$_2$P(O)CH$_2$Ph with Bu^{n}Li followed by TMS-Cl to afford (MeO)$_2$P(O)CH(Ph)TMS in 92% yield ⟨87BCJ1831⟩. The procedure has been extended to (Et$_2$N)$_2$P(O)Me, which gives (Et$_2$N)$_2$P(O)CH$_2$-TMS in 43% yield ⟨87BCJ1831⟩, and to a wide variety of phosphonates (R^{1}O)$_2$P(O)CH$_2$R^2 with LDA and TMS-Cl ⟨86S934⟩. Treatment of diethyl allylphosphonate with either LiN(TMS)$_2$ ⟨90ZOB695⟩ or LDA ⟨91MI 412-02⟩ and TMS-Cl first gives (**226**) which then reacts to give (**227**). In contrast, diethyl pent-2-enylphosphonate reacts directly with either LDA or Bu^{n}Li and TMS-Cl to afford (**228**) in 77% yield ⟨91PS(55)41⟩. In a few cases, metals other than lithium have been used. Thus, for example, treatment of (EtO)$_2$P(O)CH$_2$Li with CuI gives (EtO)$_2$P(O)CH$_2$Cu which reacts with ClSiMe$_2$CH$_2$Cl to afford (EtO)$_2$P(O)CH$_2$SiMe$_2$CH$_2$Cl ⟨92JOM(423)339⟩. A range of copper reagents, (R^{1}O)$_2$P(O)CH(R^2)Cu, prepared similarly, also reacted with TMS-Cl ⟨82S725⟩. The reaction of (EtO)$_2$P(O)CH$_2$SnEt$_3$ with PhLi and TMS-Cl to give (EtO)$_2$P(O)CH$_2$-TMS has also been described ⟨79JOM(182)465⟩. With TMS-OTf as the electrophile, silylation can be achieved with as weak a base as Et$_3$N, and the phosphonate (PhO)$_2$P(O)Me or the phosphinate PhP(O)(OMe)Me in this way afford (PhO)$_2$P(O)CH$_2$-TMS and PhP(O)(OMe)CH$_2$-TMS in yields of 51% and 56%, respectively ⟨87EGP251134, 87EGP251136⟩.

TMS ⟶ P(OEt)$_2$ (=O)

(**226**)

R ⟶ P(OEt)$_2$ (=O), TMS

(**227**) R = TMS
(**228**) R = Et

Lithiation of the phosphine–borane adduct Ph$_2$P(BH$_3$)Me with Bu^{s}Li followed by silylation with TMS-Cl gives Ph$_2$P(BH$_3$)CH$_2$-TMS in almost quantitative yield ⟨90JA5244⟩. Similar lithiation of the chiral phosphine–borane adducts (**92**) and treatment with R$^2_2$SiCl$_2$ affords (**229**) ⟨93PS(77)199⟩. ClSi(Me$_2$)CH$_2$Si(Me$_2$)Cl and CH$_2$=P(Me)(CH$_2$Li)$_2$ afford (**230**) as a minor product, and this is converted into (**231**) with HCl. The same silicon compound and Me$_3$P=C(TMS)$_2$ gives the remarkable structure (**232**) ⟨78CB2696⟩. Treatment of (**233**) with Bu^{n}Li and RMe$_2$SiCl gives (**234**) ⟨87OM959⟩. Partial silylation of the polyphosphazene (**235**) to give (**236**) has been reported ⟨86MM2089⟩.

(v) From reactions of phosphorus ylides with silyl halides and from 2-silyl phosphaalkenes generated in other ways

The reaction of TMS-Br with Ph$_3$P=CH$_2$ gives the corresponding phosphonium salt, Ph$_3$P$^+$CH$_2$-TMS Br$^-$ ⟨59CI(L)849, 61JA1610, 61JA1613⟩, and the corresponding reactions of TMS-Cl with Et$_3$P=CHMe ⟨70CB3448⟩, Ph$_3$P=CHCONHCONH$_2$ ⟨77ZOB1715⟩, and a variety of ylides, R$_3$P=CH$_2$ and (R$_2$N)$_3$P=CH$_2$ ⟨87CB789⟩ proceed similarly. In the last report Bu$^t_2$SiCl$_2$ and

Bu^{t}Si(Me)Cl$_2$ were also used to give monophosphonium salts with one Si—Cl remaining unreacted. In other cases, however, the reaction proceeds with transylidation, for example, Ph$_3$P=CHR^1 reacts with TMS-Cl in a 2:1 ratio to give Ph$_3$P=C(R^1)-TMS and Ph$_3$P$^+$CH$_2$R^1 Cl$^-$ ⟨65IC1458, 82AG(E)545⟩. Subsequent treatment of the silyl ylide with an acid or alkyl halide, R^{2}X, then affords the salts Ph$_3$P$^+$CH(R^1)-TMS X$^-$ or Ph$_3$P$^+$C(R^1)(R^2)TMS X$^-$, respectively ⟨65JA4156, 67CB1032, 82AG(E)545, 85TL2769⟩. Studies have revealed that, while treatment of Ph$_3$P=CHR with TMS-Cl or TMS-Br results in transylidation, this can be prevented with TMS-I at −78 °C and the salts Ph$_3$P$^+$CH(R)-TMS I$^-$ can be isolated in over 90% yield ⟨92S787⟩. Transylidation has also been prevented by the use of an anion which makes the initially formed salt so insoluble that it precipitates and cannot react further. Thus, for example, Me$_3$P=CH$_2$ reacts with F$_3$SiMo(CO)$_3$Cp to give Me$_3$P$^+$CH$_2$SiF$_3$ [Mo(CO)$_3$Cp]$^-$ directly, and silyl chromium complexes such as MeF$_2$SiCr(CO)$_3$Cp have been used similarly ⟨74JOM(77)C15⟩. Treatment of hexamethylsilacyclopropane with Me$_3$P=CH$_2$ results in initial ring opening to give (237) but this undergoes intramolecular proton transfer to give (238), which can then be converted into (239) with HCl gas ⟨78JOM(159)137⟩. The intermediate phosphonium carboxylate (240) is formed from Ph$_3$P=C(R)-TMS and CO$_2$, but it is unstable and undergoes a 1,3-silyl migration to afford Ph$_3$P=C(R)CO$_2$TMS ⟨92CB2081⟩. Treatment of Me$_3$P=C(TMS)$_2$ with an excess of MeCl proceeds with elimination of TMS-Cl to afford Me$_3$P$^+$C(Me)$_2$-TMS Cl$^-$ ⟨70CB3448⟩.

In the reaction of Me$_3$P=CH$_2$ with silyl halides, not only the ylide carbon but also one of the methyl groups can be functionalized and this gives access to a variety of disilyl ylides and phosphonium salts ⟨70AG(E)737, 72CB3173⟩. Thus, TMS-Cl or TMS-Si(Me$_2$)Cl in a ratio of 3:2 afford (241) and (242), respectively. These ylides then react with MeCl ⟨70CB3448⟩ or MeI ⟨72CB3173⟩ to give the phosphonium salts (243). In a similar way the bis(silyl chloride) ClSi(Me$_2$)Si(Me$_2$)Cl reacts with 3 equivalents of Me$_3$P=CH$_2$ to give (244) and with Me$_3$P=CH-TMS and 2 equivalents of Bu^{n}Li to give (245) ⟨70AG(E)737, 72CB3173⟩.

A 1,3-silyl migration from nitrogen to carbon occurs in ylides bearing a trialkylsilylamino group on phosphorus. The transformation was first observed when Me$_3$P$^+$N(TMS)$_2$ I$^-$ was treated with Bu^{n}Li to give TMS-CH$_2$P(Me$_2$)=N-TMS ⟨77CC308⟩. The corresponding reaction of (246) leads to ring expansion to give (247) ⟨79IC347⟩. In a related process, attempted formation of phosphonium salts from (248) and ethyl bromoacetate instead gave the rearranged salts (249) ⟨82OM623⟩. The bis(trimethylsilyl)aminophosphines (**250**) and CCl$_4$ initially give the *P*-chloro ylides (**251**) which similarly rearrange to (**252**) ⟨83OM921, 84IC2063⟩.

The *P*-chloro silyl ylides, TMS-CH=P(Cl)(OR)$_2$ undergo [2 + 2]-cycloaddition with benzaldehyde or trifluoroacetophenone to give the cyclic phosphonium salts (**253**) which are hydrolysed to afford (**254**) ⟨87ZOB2640⟩. The *P*-fluoro ylides, TMS-CH=P(F)R$^1_2$ react similarly with CO$_2$, PhNCO and a variety of aldehydes and ketones to give the cyclic phosphoranes (**255**) and, where R^3 contains a β-hydrogen atom, heating these results in loss of HF to give (**256**) ⟨87ZOB831⟩. Phosphonium salts such as Bu$^t_2$P$^+$=CH-TMS AlCl$_4^-$ undergo Diels–Alder cycloaddition with 2,3-dimethylbutadiene to afford (**257**) ⟨91AG(E)709⟩. The corresponding reaction of Bu$^t_2$P$^+$=C(Ph)-TMS AlCl$_4^-$ gives not only (**258**) but also the ene reaction product (**259**), and the situation is further complicated by the fact that the starting salt exists partly in the cyclic form (**260**) which isomerizes thermally to (**261**) ⟨93AG(E)1359⟩. Addition of dialkylamines and trifluoroethanol to the P=C double bond of (**262**) affords (**263**) and (**264**), respectively ⟨89IC899⟩.

(vi) From 1-silyl-1-phosphorus-substituted alkenes

A single example of this approach is provided by the conjugate addition of alkyllithiums and Grignard reagents to the double bond of (**265**) to afford (**266**) upon aqueous workup, or (**267**) by treatment with MeI, in overall yields of 70–99% ⟨89SC1891⟩.

(vii) From insertion into the P—Si bond

In an unusual reaction, Ph$_2$P-TMS and hexafluoroacetone afford (268) as the main product ⟨68JCS(A)1105⟩.

(viii) Miscellaneous methods

Passing oxygen through a mixture of PCl$_3$ and TMS-Me affords Cl$_2$P(O)CH$_2$-TMS ⟨58USP2835651⟩. Application of the same method to TMS-Cl and Me$_2$SiCl$_2$ produces Cl$_2$P(O)CH$_2$SiMe$_2$Cl (27%) and Cl$_2$P(O)CH$_2$SiMeCl$_2$, respectively ⟨78URP351438, 84ZOB59⟩, and further treatment of Cl$_2$P(O)CH$_2$-TMS under the same conditions gives Cl$_2$P(O)CH$_2$Si(Me$_2$)CH$_2$P(O)Cl$_2$ ⟨77URP403314⟩.

Base-catalysed thermal rearrangement of (269) leads to ring expansion to form (270) ⟨74AG(E)540⟩. Treatment of (271) with TMS-N$_3$ results in cyclization to afford (272) ⟨87TL6121⟩. The compound (273), related to (221), can be prepared by treatment of R$_2$N—P=CH-TMS with two equivalents of the carbenoid TMS-CH(Cl)Li ⟨92CB771⟩.

(ix) By oxidation of tricoordinate phosphorus functions

Convenient access to compounds of this type can often be gained from the corresponding phosphinoalkylsilanes by conversion into the oxides, sulfides and selenides using standard methods. Thus, for example, the conversion of Ph$_2$PCH$_2$-TMS into Ph$_2$P(O)CH$_2$-TMS and into Ph$_3$P(S)CH$_2$-TMS has been described ⟨65IZV286⟩ and treatment with TMS-N$_3$ affords Ph$_2$P(=N-TMS)CH$_2$-TMS ⟨89CZ349⟩. Sulfurization of the corresponding phosphines has likewise been used to obtain Ph$_2$P(S)CH$_2$SiMe$_2$(OPh) ⟨88JOM(356)285⟩ and (EtO)$_2$P(S)CH$_2$-TMS ⟨87BCJ1831⟩. The conversion of Ph$_2$PCH$_2$Si(Me$_2$)CH$_2$PPh$_2$ into both the bis(phosphine oxide) and the bis(phosphine selenide) has also been reported ⟨85MI 412-01⟩.

(x) Interconversions

Hydrolysis of TMS-CH$_2$PCl$_3^+$ AlCl$_4^-$ gives the versatile intermediate TMS-CH$_2$P(O)Cl$_2$ ⟨66JOC4288⟩ and the chlorines can then be replaced by a wide variety of other atoms or groups including OH, OR, NCO, NCS, NH$_2$, F, NHCO$_2$R, NHC(O)NHPh and NHC(S)NHPh ⟨72ZOB293⟩. Alkylation of TMS-CH$_2$P(O)(OEt)$_2$ is also possible, for example Bu^{n}Li and MeI give TMS-CH(Me)P(O)(OEt)$_2$ ⟨72JOC939⟩.

4.12.3.3.2 Tetracoordinate phosphorus and germanium functions

Many of the methods for synthesis of the corresponding phosphorus and silicon functions are also applicable here, but only a few examples have been described. The reaction of Me$_3$GeCH$_2$Cl with Me$_3$P gives the phosphonium salt Me$_3$GeCH$_2$PMe$_3^+$ Cl$^-$ ⟨68CB3545⟩ while with (EtO)$_3$P an Arbuzov reaction affords Me$_3$GeCH$_2$P(O)(OEt)$_2$ in 48% yield, and this is readily hydrolysed to the corresponding phosphonic acid ⟨63IZV1563⟩.

The reaction of Ph$_3$P=CH$_2$ with Ph$_3$GeBr or Ph$_3$GeCl gives the expected phosphonium salts Ph$_3$P$^+$CH$_2$GePh$_3$ X$^-$ ⟨61JA1610, 61JA1613, 82BCJ3025⟩, but with Me$_3$GeCl double transylidation takes place to give Ph$_3$P=C(GeMe$_3$)$_2$ ⟨65IC1458, 67CB1032, 77CB677⟩. Treatment of Me$_3$P=CH$_2$ with three equivalents of both an alkyllithium and Me$_3$GeCl affords Me$_3$GeCH$_2$P(Me$_2$)=C(GeMe$_3$)$_2$ ⟨77CB677⟩. GeCl$_4$ with an excess of (Me$_2$N)$_3$P=CH$_2$ gives the tetra(phosphonium salt) Ge[CH$_2$P$^+$(NMe$_2$)$_3$]$_4$ (Cl$^-$)$_4$ ⟨84BCJ43⟩.

Addition of Me$_3$Ge—O—P(OEt)$_2$ to diethyl vinylphosphonate gives (**274**) ⟨75ZOB1486⟩, while either Et$_3$GePEt$_2$ or Et$_3$GePPh$_2$ with hexafluoroacetone affords (**275**) ⟨70CR(C)351, 70JOM(24)633⟩.

(**274**)

(**275**) R = Et, Ph

4.12.3.3.3 Tetracoordinate phosphorus and boron functions

Treatment of Ph$_3$P=CHR with diborane gives the stable, solid, zwitterionic products Ph$_3$P$^+$CH(R)BH$_3^-$ ⟨58JA3480, 61JA367⟩ and these can be formed more conveniently from either BH$_3$·THF ⟨81AG(E)1038⟩ or BH$_3$·Me$_2$S ⟨88CB1509⟩. Aliphatic ylides, R$_3$P=CH$_2$, and BH$_3$·THF similarly give R$_3$P$^+$CH$_2$BH$_3^-$ for R = Me, Et, Pr^i, Bu^n, and Bu^t ⟨80CB1480⟩. Although the compounds Ph$_3$P$^+$C(R^1)(R^2)BH$_3^-$ are generally stable at RT, they undergo rearrangement at 100–130 °C to give the phosphine/borane adducts Ph$_3$P·H$_2$BCH(R^1)(R^2) ⟨81AG(E)1038, 88CB1509⟩, although some examples such as Ph$_3$P$^+$CH(CN)BH$_3^-$ show no tendency to rearrange ⟨91CB199⟩. Reactions of ylides with BF$_3$ to give products of the type R$_3$P$^+$CH$_2$BF$_3^-$ have also been reported for Ph$_3$P=CH$_2$ ⟨59CI(L)849⟩, Me$_3$P=CH$_2$ and Et$_3$P=CH$_2$ ⟨76ZAAC(421)1⟩, and the *P*-halo ylides R$^1_2$P(F)=CHR^2 and R$^1_2$P(Cl)=CHR^2 behave similarly ⟨80TL3984⟩. Halogen exchange of Ph$_3$P$^+$CH$_2$BF$_3^-$ to give Ph$_3$P$^+$CH$_2$BX$_3^-$ by treatment with BCl$_3$, BBr$_3$ or BI$_3$ has been reported ⟨70ZN(B)314⟩. Ph$_3$P=CHR^1 and R$^2_2$BCl undergo transylidation to afford Ph$_3$P=C(R^1)BR$^2_2$ ⟨84AG(E)381⟩, but with R^{2}BCl$_2$ transylidation only occurs if a second equivalent of ylide is added. Otherwise, adducts of the general type (**276**) are stable and do not rearrange, but can be reduced with LiAlH$_4$ to give (**277**) ⟨86AG(E)559⟩.

(**276**)

(**277**)

Treatment of the cyclic compound (**278**), obtained from [Me$_3$PBH$_2$PMe$_3$]$^+$ Br$^-$ and Bu^{n}Li, with Me$_2$BBr affords (**279**) ⟨80JOM(200)287⟩.

(278)   (279)

4.12.3.4 Higher Coordinate Phosphorus Derivatives

4.12.3.4.1 Higher coordinate phosphorus and silicon functions

The only compounds of this type to be reported are the cyclic phosphoranes (**255**) mentioned earlier in this chapter ⟨87ZOB831⟩.

4.12.3.4.2 Higher coordinate phosphorus and germanium functions

No compounds of this type appear to have been described.

4.12.3.4.3 Higher coordinate phosphorus and boron functions

No compounds of this type appear to have been described.

4.12.4 FUNCTIONS CONTAINING PHOSPHORUS AND A METAL

4.12.4.1 Group 1 and 2 Derivatives

4.12.4.1.1 Compounds containing phosphorus and lithium

The preparation of lithiated phosphines has been reviewed ⟨81ZC341⟩. Deprotonation of trialkylphosphines is not an easy process and the strongest bases are required. Conversion of compounds of the general type R^1R^2PMe into $R^1R^2PCH_2Li$ has been achieved using either Bu^nLi or Bu^tLi ⟨65JOC1939⟩ or Bu^nLi/TMEDA ⟨67JOM(8)199⟩ for such examples as Ph_2PMe, $n\text{-}C_{12}H_{25}PMe_2$, $PhPMe_2$ and $(n\text{-}C_6H_{13})_2PMe$, although with the last two phosphines Bu^tLi gives some dilithiation ⟨66JOC2373⟩. On the other hand, treatment of Bu^n_3P with even Bu^tLi gives essentially no reaction ⟨65JOC1939⟩. Lithiation of Me_3P requires Bu^tLi but the process is efficient affording Me_2PCH_2Li in 93% yield ⟨77ZN(B)762, 78GEP2658127⟩. Treatment of the enantiomerically pure chiral phosphine $PhP(Pr^n)Me$ with Bu^tLi gave $PhP(Pr^n)CH_2Li$ without epimerization at the phosphorus centre, indicating no significant π-overlap between the formal carbanion and the phosphorus ⟨66TL3315⟩. In a few cases α-lithiophosphines have been generated by addition of alkyllithiums to vinylphosphines. Thus, for example, $Ph_2PCH=CH_2$ reacts with Bu^nLi to give $Ph_2PCH(Li)CH_2Bu^n$ and Bu^tLi reacts similarly ⟨66JOC950⟩. Treatment of the same phosphine with $PhP(Pr^i)Li$, derived from Ph_2PPr^i and lithium, affords $Ph_2PCH(Li)CH_2PPh(Pr^i)$ ⟨76IS192⟩. Attempts to extend this approach to dialkylvinylphosphines such as $Bu^n_2PCH=CH_2$ are frustrated by oligomerization of the phosphine on addition of Bu^nLi ⟨66JOC950⟩. The reaction of a series of (o-halobenzyl)phosphines with Bu^nLi leads to lithiation at the benzylic position if the halogen is chlorine but to lithium/halogen exchange where it is bromine ⟨76ZAAC(422)237, 78ZAAC(447)53, 82ZAAC(494)55⟩. In the case of the ylide $(Me_2N)_2P(F)=CH_2$, treatment with Bu^nLi results in lithium/halogen exchange and rearrangement to give $(Me_2N)_2PCH_2Li$ ⟨88ZAAC(567)23⟩.

Lithiation of the methyl group of phosphine-borane adducts such as $Ph_2PMe\cdot BH_3$ and (**92**) is readily achieved using Bu^sLi ⟨85JA5301, 90JA5244, 93PS(77)199⟩, while $Me_3P\cdot BH_3$ and Bu^tLi give $Me_2P(BH_3)CH_2Li$ ⟨80JOM(200)287⟩.

Formation of α-lithiophosphine oxides, for example by treatment of $Ph_2P(O)CH_2Ph$ with PhLi to afford $Ph_2P(O)CH(Li)Ph$, was reported at an early date ⟨59CB2499, 61CB1987⟩. Other studies have generally employed Bu^nLi, and this lithiates less activated positions to give products such as $Ph_2P(O)CH_2Li$ and $(n\text{-}C_6H_{13})_2P(O)CH_2Li$ ⟨63JOC123⟩. Treatment of $Ph_2P(O)CH_2Ar$ with Bu^nLi

results in exclusive α-lithiation even where there is the possibility of competing lithium/halogen exchange as for Ar = o-BrC$_6$H$_4$ ⟨78ZAAC(447)53⟩. Conversion of the related phosphine imines, Ph$_2$P(=NPh)CH$_2$R, into Ph$_2$P(=NPh)CH(R)Li by using PhLi has been reported for R = H and Pr^n ⟨75JOC1173⟩.

An important route to compounds of this type is by the reaction of Ph$_3$PX with RCH$_2$Li which proceeds by displacement of one phenyl group by the alkyl group, followed by deprotonation in the α-position by the PhLi produced, to give Ph$_2$P(X)CH(R)Li directly. This has been widely used for X = O and S ⟨63JA642, 64JOM(2)1, 76IS195, 80IC3195⟩ and for X = NPh ⟨75JOC1173⟩.

The phosphonates (R^{1}O)$_2$P(O)CH$_2$R^2 react readily with Bu^{n}Li ⟨84JOM(264)9⟩ or lithium diisopropylamide (LDA) ⟨86JOM(304)283⟩ even at low temperatures to afford (R^{1}O)$_2$P(O)CH(R^2)Li, and the related phosphonic diamides, (R$^1_2$N)$_2$P(O)CH$_2$R^2 are similarly lithiated by Bu^{n}Li ⟨68JA6816⟩. The reaction of trialkyl phosphates (R^{1}O)$_3$PO with two equivalents of R^{2}CH$_2$Li proceeds by displacement of one alkoxy group followed by α-lithiation to afford (R^{1}O)$_2$P(O)CH(R^2)Li ⟨87TL405⟩, and the cyclic compounds (**280**) react under the same conditions to give (**281**) ⟨93PS(75)143⟩. The ylides R$_2$MeP=CH$_2$ and Bu^{n}Li give products which can formally be represented as (**282**) although NMR evidence shows that they are more accurately described by the symmetrical structure (**283**) ⟨68CB3556⟩.

(**280**) X = Cl, OEt
(**281**) X = CH(R^2)Li

(**282**)

(**283**)

4.12.4.1.2 Compounds containing phosphorus and sodium

Metallation of compounds such as Ph$_2$P(O)Me and (EtO)$_2$P(O)CH$_2$Ph with NaNH$_2$ was reported at an early stage ⟨58CB61⟩, and conversion of Ph$_2$P(O)CH$_2$Ph into Ph$_2$P(O)CH(Ph)Na can be achieved with NaOEt ⟨59CB2499⟩, NaNH$_2$ ⟨61CB1987⟩, or sodium metal ⟨59JA5519⟩. The same sodium compound can be obtained from Ph$_3$PO and sodium in toluene ⟨60CB1371⟩, and is also implicated in the reaction of Ph$_2$P(O)Na with PhCHO to give stilbene ⟨61CB1323⟩. Treatment of Et$_3$PO with PhNa gives Et$_2$P(O)CH(Me)Na ⟨60CB1339⟩. Deprotonation of (MeO)$_2$P(S)CH(CN)$_2$ by NaOEt to afford (MeO)$_2$P(S)C(CN)$_2$Na has been described ⟨73EGP94992⟩. With more stabilizing groups present, the extent, if any, of covalent character and location of the sodium on carbon as opposed to the oxygen of an enolate form become more questionable. Thus, for example, the sodium derivatives from the reactions of (EtO)$_2$PCH$_2$CO$_2$Et or Pr$^i_2$PCH$_2$CO$_2$Et with NaN(TMS)$_2$ can be alkylated on carbon or on oxygen depending on the electrophile used ⟨76ZOB1218⟩ and for both (EtO)$_2$P(O)CH$_2$CO$_2$Et ⟨27CB291⟩ and (EtO)$_2$P(O)CH$_2$P(O)(OEt)$_2$ ⟨63JA2394⟩ it is clear that the sodium derivatives have the cyclic chelate structures (**284**) with the metal bound to two oxygens. Chelation by a suitably placed phosphine can, however, sometimes stabilize covalent C—Na bonds as in the compound (**286**), formed by treatment of (**285**) with two equivalents of NaNH$_2$, whose structure has been confirmed by x-ray methods ⟨80CB902⟩, and the related fluorene derivative (**287**) ⟨82ZN(B)950⟩.

(**284**) X = C, P(OEt)

(**285**)

(**286**)

(**287**)

4.12.4.1.3 Compounds containing phosphorus and potassium

Treatment of phosphine oxides, $Ph_2P(O)CH_2R$ ⟨59CB2499, 61CB1987, 62CB581⟩, phosphinates, $Ph(EtO)P(O)CH_2Ar$ ⟨62CB581⟩, and phosphonates, $(EtO)_2P(O)CH_2Ar$ ⟨62CB581⟩, with $KOBu^t$ gives potassium derivatives which may be thought of as containing a 1-potassioalkylphosphorus function, although as discussed for sodium in the previous section, their covalent nature is questionable. Treatment of Et_3PO with potassium metal gives some of the expected product, $Et_2P(O)CH(Me)K$ ⟨60CB1339⟩. The potassium derivatives of β-oxophosphonates and methylenediphosphonates have the cyclic chelate structure analogous to (284) ⟨63JA2394⟩. The potassium derivative corresponding to (287), which is likely to have high covalent character, has been prepared with KH ⟨82ZN(B)950⟩.

4.12.4.1.4 Compounds containing phosphorus and beryllium

$Ph_3P=CH_2$ and $BeCl_2$ give the dimeric salt (288) ⟨83BCJ1772⟩. Treatment of (289) with $BeCl_2$ gives (290) ⟨85OM1233⟩, and the carbon analogue (291) is formed from (278) and $BeCl_2$ ⟨80JOM(200)287⟩.

(288) M = Be
(292) M = Mg

(289)

(290) M = Be, X = O
(291) M = Be, X = CH_2
(294) M = Mg, X = CH_2

(293)

4.12.4.1.5 Compounds containing phosphorus and magnesium

Treatment of phosphonates such as $(EtO)_2P(O)CH_2Ph$ with Pr^iMgCl results in metallation to give $(EtO)_2P(O)CH(Ph)MgCl$ ⟨64JOC2036⟩, while treatment of $(EtO)_2P(O)CH_2I$ with the same reagent leads to $(EtO)_2P(O)CH_2MgCl$ by metal–halogen exchange ⟨86JOM(316)13⟩. Conjugate addition of Grignard reagents to $Ph_2P(O)CH=CH_2$ first gives $Ph_2P(O)CH(Mg)CH_2R$ but these react further with the vinylphosphine oxide to give polymers ⟨63JPS(A)3627⟩. $Ph_3P=CH_2$ and $MgCl_2$ give the dimeric phosphonium salt (292) ⟨83BCJ1772⟩, while treatment of $Me_3P=CH_2$ with Et_2Mg results in loss of ethane to give the polymeric compound (293) ⟨77ZAAC(434)145⟩. Treatment of (278) with $MgCl_2$ affords (294) ⟨80JOM(200)287⟩, while the nitrogen analogue (295) can be obtained from $Me_3P=N-P(=CH_2)Me_2$ and Et_2Mg ⟨77CB3528⟩.

(295)

4.12.4.1.6 Compounds containing phosphorus and heavier group 1 and 2 metals

The use of $Bu^t_2PCH_2Rb$ and $Bu^t_2PCH_2Cs$ has been described ⟨92GEP4134772⟩, and the barium analogue of (287) has been prepared with barium metal ⟨82ZN(B)950⟩.

4.12.4.2 Compounds Containing Phosphorus and a Lanthanide

The reaction of $Ph_2P(O)CH_2Li$ with $CeCl_3$ to afford $Ph_2P(O)CH_2CeCl_2$ has been reported ⟨86CB2150⟩. Treatment of Me_2PCH_2Li with $Cp_2LuCl \cdot THF$ gives the three-membered ring compound (296) ⟨85JOM(297)C30⟩. The majority of the known compounds of this type have, however, been obtained by the reactions of ylides with lanthanide species. Thus, $Cp_3Lu \cdot THF$ and $Ph_3P=CH_2$ give $Ph_3P^+CH_2Lu(Cp)_3^-$ ⟨84JOM(269)21⟩, and the corresponding reactions with $Cp_2LuX \cdot THF$ to give (**297**) have been reported for X = Cl, Bu^t and CH_2TMS ⟨83JOM(255)305⟩. Treatment of (**297**; X = Cl) with MeLi gives (**297**; X = Me) and this spontaneously eliminates methane to afford (**298**) ⟨84JOM(269)21⟩, a product which can also be obtained by treatment of (**297**; X = Cl) with NaH, of $Cp_2LuCl \cdot THF$ with $Ph_3P=CH(Li)$ ⟨84JOM(269)21⟩, or of either Cp_2LuH or Cp_2LuMe with $Ph_3P=CH_2$ ⟨83CC276⟩.

(296)   (297)   (298)

The trimethyl ylide $Me_3P=CH_2$ reacts with MCl_3 to give the tri(phosphonium) salts $(Me_3P^+CH_2)_3M$ $(Cl^-)_3$ for M = La, Pr, Nd, Sm, Gd, Ho, Er and Lu, and when these are treated with Bu^nLi the tricyclic compounds (**299**) are formed ⟨76CZ336⟩. The hexa-*t*-butyl analogue of (**299**) can be prepared directly from $Bu^t_2P(=CH_2)CH_2Li$ and $LuCl_3$ and the same phosphorus compound with Cp_2LuCl gives (**300**) ⟨82JOM(235)287⟩. The reaction of $R_2P(=CH_2)CH_2Li$ with $(Cp^*)_2MCl \cdot LiCl$ to give the similar products (**301**) has been reported for M = Lu (R = Me) ⟨84AG(E)522⟩, and for M = Nd and Sm (R = Me, Bu^t, Ph) ⟨90POL875⟩. When $Me_2P(=CH_2)CH_2Li$ (2 equivalents) reacts with $LuCl_3$ in the presence of Cp^*Na, the bicyclic product (**302**) is formed ⟨84AG(E)522⟩.

(299)   (300)   (301)   (302)

4.12.4.3 Transition Metal Derivatives

The interaction of phosphorus ylides with transition metal derivatives leads to a rich chemistry and has been extensively studied with well over 200 publications between 1975 and 1995. Several reviews exist ⟨75ACR62, 83AG(E)907⟩. For reasons of space, only representative examples are given in the following sections and the interested reader is referred to a comprehensive review ⟨B-93MI 412-01⟩.

4.12.4.3.1 Compounds containing phosphorus and scandium or yttrium

A single example of this type is provided by the cyclic compound (**303**) formed upon treatment of $Ph_2P(=CH_2)CH_2Li$ with Cp_2ScCl ⟨76IC2567⟩.

No compounds containing phosphorus and yttrium appear to have been described.

(**303**) M = Sc, R = Ph
(**304**) M = Ti, R = Me
(**305**) M = Ti, R = Ph
(**313**) M = V, R = Ph

4.12.4.3.2 Compounds containing phosphorus and titanium, zirconium or hafnium

The reaction of $Ph_2P(O)CH_2Li$ with $(Pr^iO)_3TiCl$ gives $Ph_2P(O)CH_2Ti(OPr^i)_3$ and with $Ti(OPr^i)_4$ the salt $Ph_2P(O)CH_2Ti^-(OPr^i)_4Li^+$ is formed ⟨86CB2150⟩. Cp_2TiCl_2 and $Me_3P=CH_2$ give the bis(phosphonium) salt $Cp_2Ti(CH_2PMe_3^+)_2 (Cl^-)_2$ while, with the lithiated form $Me_2P(=CH_2)CH_2Li$, the cyclic compound (304) is produced ⟨77ZN(B)858⟩. The diphenyl analogue, $Ph_2P(=CH_2)CH_2Li$, reacts similarly with Cp_2TiCl to give (305) and, with $Et_2P(=CHMe)CH_2Li$, (306) can be obtained ⟨76IC2567⟩. Treatment of Cp_2TiCl with Bu^nLi and $Me_3P=N-P(=CH_2)Me_2$ gives the six-membered ring compound (307) ⟨77ZN(B)858⟩.

(306)   (307)

Hydrozirconation of 1-phenyl-2-phospholine with Cp_2ZrHCl gives the 2-zirconiophospholane (308) ⟨93AG(E)1735⟩. Cp_2ZrCl_2 and one equivalent of Me_2PCH_2Li gives (309; X = Cl), while with two equivalents (309; X = CH_2PMe_2) is formed and Cp_2ZrCl_2 with MeLi followed by Me_2PCH_2Li gives (309; X = Me) ⟨84JOM(273)195⟩. Treatment of $Me_3P=CH_2$ with Cp_2ZrCl_2 gives the bis(phosphonium) salt $Cp_2Zr(CH_2PMe_3^+)_2 (Cl^-)_2$ ⟨77ZN(B)858⟩ and with the cyclohexylmethyl compound, $Cp_2Zr(H)CH_2C_6H_{11}$, the cyclic product (310) is formed ⟨80IC3207⟩. A similar process occurs upon treatment of Cp_2ZrCl_2 with $(Et_2N)_2P(Me)=CH_2$ to give (311) in 78% yield ⟨85ZN(B)352⟩. $Zr(CH_2Bu^t)_4$ and three equivalents of $Me_3P=CH_2$ give a product with the remarkable structure (312) ⟨83OM154⟩.

(308)   (309)   (310) R = Me, X = H   (312)
                    (311) R = NEt_2, X = Cl

No compounds containing phosphorus and hafnium appear to have been described.

4.12.4.3.3 Compounds containing phosphorus and vanadium, niobium or tantalum

A single example of a compound containing phosphorus and vanadium is provided by (313) which is formed from $Ph_2P(=CH_2)CH_2Li$ and Cp_2VCl ⟨76IC2567⟩.

No compounds containing phosphorus and niobium appear to have been described.

Treatment of Ph_2PCH_2Li with Me_3TaCl_2 gives $Me_3Ta(CH_2PPh_2)_2$ ⟨80JOM(187)331⟩ and similarly Cp^*TaCl_4 and either $Ph_2P(O)CH_2Li$ or $Ph_2P(S)CH_2Li$ give the monosubstitution products $Cp^*Cl_3TaCH_2P(X)Ph_2$ ⟨89OM1604⟩. When Cp^*TaCl_4 is treated with $Ph_2P(=CH_2)CH_2Li$, the four-membered ring product (314) is formed but, for less bulky phosphorus analogues such as $Me_2P(=CH_2)CH_2Li$, the dimeric eight-membered ring form (315) is preferred ⟨89OM1604⟩.

(314)   (315)

4.12.4.3.4 Compounds containing phosphorus and chromium, molybdenum or tungsten

Treatment of $Cp(NO)_2CrCH_2I$ with Ph_3P results in simple substitution to give $Cp(NO)_2CrCH_2PPh_3^+ I^-$ ⟨90OM2683⟩, while two equivalents of $Ph_2P(O)CH_2Li$ with $CrCl_3$ give $Ph_2P(O)CH_2Cr(Cl)CH_2P(O)Ph_2$ ⟨86CB2150⟩. The majority of compounds of this type have, however, been prepared either by treatment of phosphorus ylides with metal compounds or of metal carbene complexes with phosphines. Thus, ylides such as $Ph_3P=CH_2$, $Ph_3P=CHPh$, and $Ph_3P=CMe_2$ react with $Cr(CO)_6$ ⟨72JOM(42)C35⟩, $Cr(CO)_5 \cdot THF$ ⟨78JOM(148)C25⟩, $Cr(CO)_5 \cdot RCN$ ⟨79CB2709⟩, and $Et_4N^+Cr(CO)_5Br^-$ ⟨79JOM(182)77⟩ to give the zwitterionic products (316), and with $(C_6H_6)Cr(CO)_2 \cdot THF$, the adducts (317) are formed ⟨80JOM(193)47⟩. The compound (318), stabilized by chelation of the phosphine group, has also been prepared ⟨82CB1379⟩. Photolysis of the stabilized ylide (319) results in loss of $Me_3P=CH_2$ and CO to afford (320) ⟨83ZN(B)711⟩. The tricyclic compound (321) can be prepared either from Li_3CrPh_6 and $Me_4P^+ Cl^-$ or from $CrCl_3 \cdot 3THF$ and $Me_2P(=CH_2)CH_2Li$ ⟨73AG(E)854, 78JOM(160)35⟩.

Alternative access to compounds of the type (316) can be gained by treatment of the sulfoxonium salts (322) with phosphines or phosphites ⟨77JOM(131)49⟩ and kinetic evidence has shown the intermediacy of the carbene complex $(CO)_5Cr=CH_2$ in these processes ⟨84CB3340⟩. If a phosphine bearing a second nucleophilic group is used, one molecule of CO is displaced to afford cyclic structures such as (323) where X can be PR^2_2 ⟨84CB1103, 85CB541, 85CB3570⟩, SPh ⟨85CB541⟩, SMe, or $SbPh_2$ ⟨85ZN(B)968⟩.

As described above for chromium, compounds of the type $Ph_3P^+CH(Ar)Mo(CO)_5^-$ can be formed either from $Ph_3P=CHAr$ and $Mo(CO)_6$ ⟨72JOM(42)C35⟩, or $(CO)_5Mo=CHAr$ and Ph_3P ⟨90JOM(385)351⟩. The dimolybdenum compound $Cp(CO)_2Mo\equiv Mo(CO)_2Cp$ reacts with $Ph_3P=CHR$ to afford (324) ⟨88ZN(B)1293⟩, while the dioxomolybdenum(VI) compound $Ar_2Mo(=O)_2$ reacts with $Bu^n_3P=CH_2$ to give (325) ⟨86CC1208⟩. When $MoCl_3$ is treated with $Me_2P(=CH_2)CH_2Li$ a compound of overall formula $Mo_2(CH_2P(Me_2)CH_2)_4$ is formed ⟨75JOM(84)C13⟩. $Cp(CO)_2MoCH(Me)OMe$ and Ph_3P react by way of a metal carbene intermediate to give (326) ⟨91JCS(D)1117⟩. Treatment of $Cp(CO)_2Mo=P=C(TMS)_2$ with two equivalents of dimethyl sulfoxonium methylide affords (327) ⟨90CB739⟩.

Compounds of the type $Ph_3P^+CH(Ar)W(CO)_5^-$ can be formed either from $Ph_3P=CHAr$ and $W(CO)_6$ ⟨72JOM(42)C35⟩, or $(CO)_5W=CHAr$ and Ph_3P ⟨77JA6097, 79JA7282⟩. Related compounds have also been obtained by addition of Me_3P to $(CO)_5W=CAr_2$ ⟨75JOM(86)C10, 77CB799⟩, and of

(MeO)$_3$P to (CO)$_5$W=CHAr ⟨86JOM(306)203⟩. Addition of Me$_2$S or Et$_2$S to (CO)$_5$W=CHPh gives R$_2$S$^+$CH(Ph)W(CO)$_5^-$ from which the sulfide is immediately displaced on addition of Ph$_3$P ⟨87CB583⟩. The reaction of (CO)$_5$W=CHPh with PhC≡CH followed by Ph$_3$P takes an unusual course to produce (**328**) in 65% yield ⟨91AG(E)998⟩. The compound (**329**) related to (**318**) has been reported ⟨82CB1379⟩, and (**330**) is formed from Cp(CO)$_2$W=P=C(TMS)$_2$ in an analogous way to (**327**) ⟨90CB739⟩. Treatment of [(Me$_3$P)$_4$W(Cl)=CH$_2$]$^+$ OTf$^-$ with CO under pressure results in intramolecular transfer of Me$_3$P to give (**331**) ⟨82IC3913, 84OM476⟩. The novel isomerization (**332**) to (**333**) occurs at 90 °C ⟨86OM2555⟩.

4.12.4.3.5 Compounds containing phosphorus and manganese or rhenium

Treatment of two equivalents of Ph$_2$P(O)CH$_2$Li with MnCl$_2$ affords Ph$_2$P(O)CH$_2$MnCH$_2$P(O)Ph$_2$ ⟨86CB2150⟩. Compounds of the type Cp(CO)$_2$Mn$^-$CH$_2$PR$_3^+$ have been formed both from Cp(CO)$_2$Mn·THF and R$_3$P=CH$_2$ ⟨80JOM(193)47⟩, and by photolysis of the Cp(CO)$_2$Mn analogue of (**319**) ⟨83CB690⟩. High pressure carbonylation of the four-membered ring compounds (**334**) results in ring opening and rearrangement to afford (CO)$_5$MnCH(R)P(O)Ph$_2$ ⟨81CB413, 83CB1070⟩. The *cis* isomer of Ph$_3$PMn(CO)$_4$CH$_2$I undergoes spontaneous rearrangement in solution at RT to give Ph$_3$P$^+$CH$_2$Mn$^-$(CO)$_4$I ⟨90JOM(397)313⟩. Treatment of (**278**) with MnCl$_2$ gives the bicyclic manganese analogue of (**294**) ⟨83OM257⟩. Treatment of (Na$^+$)$_2$ [(CO)$_4$MnPPh$_2$]$^{2-}$ with CH$_2$Cl$_2$ or MeCHCl$_2$ results in formation of the metallaphosphiranes (**335**) ⟨83CB1209⟩. Protonation of (**336**) with HBF$_4$ results in ring closure to produce (**337**) ⟨89JOM(362)117⟩.

The rhenium carbene complex Cp(Ph$_3$P)(NO)Re$^+$=CH$_2$ PF$_6^-$ reacts with Ph$_3$P or Bu$^n_3$P to give (**338**) ⟨79JA5440⟩, and these compounds may be methylated on the central carbon atom by treatment with Bu^{n}Li followed by MeOTf ⟨86CC1154⟩. Addition of Me$_3$P to the carbene complex Cp(CO)$_2$Re=CHCH$_2$CH$_2$Bu^t gives (**339**) ⟨90JA3713⟩.

4.12.4.3.6 Compounds containing phosphorus and iron, ruthenium or osmium

Treatment of the ylides Ph$_3$P=CHR with CpFe(CO)$_2$I gives the expected products, Cp(CO)$_2$FeCH(R)PPh$_3^+$ I$^-$ ⟨77ZN(B)1268⟩, while with [CpFe(CO)$_2$]$_2$ and two equivalents of Bu^{n}Li, the cyclic compounds (**340**) are formed ⟨89JOM(371)329⟩. Treatment of Ph$_3$P with the iron carbene complexes (CO)$_4$Fe=CHR, generated by a variety of methods, provides ready access to (CO)$_4$Fe$^-$CH(R)PPh$_3^+$ ⟨79JA6433, 83OM1846, 90IC1674⟩ and the corresponding reaction of Cp*(CO)$_2$Fe=CH$_2^+$ BF$_4^-$ proceeds similarly ⟨90OM1036⟩. Perhaps the simplest route to

$(CO)_4Fe^-CH_2PPh_3^+$ is the reaction of $Fe(CO)_5$ with CH_2Cl_2, Ph_3P and aqueous NaOH under phase-transfer conditions ⟨85JOM(280)C31, 89JOM(359)205⟩. Treatment of $Cp(CO)_2FeCH_2Cl$ with a variety of tertiary phosphines gives the simple substitution products $Cp(CO)_2FeCH_2PR_3^+BF_4^-$ ⟨83JCS(D)1495⟩, while with Et_2PH the initially formed salt can be deprotonated with NaOMe to afford $Cp(CO)_2FeCH_2PEt_2$. This compound is rather unstable and, upon photolysis, it loses CO to give (341) ⟨91OM369⟩. Reduction of $FeCl_2$ in the presence of Me_3P results in the formation of $(Me_3P)_4Fe$ which exists in solution almost entirely in the isomeric form (342) ⟨77CB2200⟩.

(340) (341) (342)

Treatment of trans-$(Me_3P)_4RuCl_2$ with $Me_3P=CH_2$ results in displacement of one chlorine to form $(Me_3P)_4Ru(Cl)CH_2PMe_3^+$ Cl^-, while with two equivalents of $Me_2P(=CH_2)CH_2Li$ (343) is produced ⟨80ZN(B)584⟩. Complexation of $PhC\equiv CPh$ or $F_2C=CF_2$ to the ruthenium carbene complex $Cl(NO)(Ph_3P)_2Ru=CH_2$ induces migration of one phosphine to the carbon centre to produce (344) ⟨91JCS(D)609⟩.

(343) (344)

No compounds containing phosphorus and osmium appear to have been described.

4.12.4.3.7 Compounds containing phosphorus and cobalt, rhodium or iridium

Treatment of $CoCl_2$ with Me_3P, CH_2Cl_2 and Mg gives $Me_3P^+CH_2CoCH_2PMe_3^+$ $(Cl^-)_2$. This is also one of the products from the reaction of $(Me_3P)_4Co$ with CH_2Cl_2, and can be reduced by sodium in the presence of an excess of Me_3P to afford the metallaphosphirane (345) in 60% yield ⟨76AG(E)42⟩. $(Ph_3P)_3CoCl$ and Me_2PCH_2Li afford the analogue (346) ⟨77ZN(B)762⟩. Photolysis of the $NO(CO)_2Co$ analogue of (319) to produce $NO(CO)_2Co^-CH_2PMe_3^+$ has been described ⟨83ZN(B)711⟩. Treatment of $(Me_3P)_3Co(Me)_2Br$ with two equivalents of $Me_3P=CH_2$ gives (347) ⟨74CB3692, 74CB3706⟩, and the reaction of (278) with $CoCl_2$ affords the bicyclic cobalt analogue of (294) ⟨83OM257⟩.

(345) R = Me
(346) R = Ph

(347)

Treatment of $(Me_3P)_4RhCl$ with CH_2Cl_2 produces $(Me_3P)_3Rh(Cl)_2CH_2PMe_3^+$ Cl^- ⟨87CC1543⟩. (cod-RhCl)$_2$ and a variety of ylides, $R_2P(Me)=CH_2$ give (348) in which the cod ligand can then be replaced by CO or Me_3P, and the same rhodium compound with $Me_3P=N-P(=CH_2)Me_2$ gives the cod-Rh analogue of (307) ⟨77IC3187, 78JOM(160)41⟩. The reaction of compounds (349), formed by oxidative addition of CH_2I_2 to rhodium(I) species, with phosphines such as Ph_3P or Pr^i_3P results in substitution to give (350) ⟨81JOM(219)C29, 85CB3032, 85JOM(281)317⟩, but where L = Me_3P, base catalyses a rearrangement to give (351) ⟨85CB261⟩. A similar process is observed for the cationic complexes (352) which rearrange to (353) and for (354) whose rearrangement to (355) is catalysed by Et$_3$N ⟨82JOM(236)C65, 88OM1106⟩. Treatment of [(C$_2$H$_4$)$_2$RhCl]$_2$ with Pr$^i_3$P=CH$_2$ followed by CpLi gives (**356**) which undergoes oxidative addition of MeI to give (**357**) ⟨91JOM(417)149⟩.

(**348**)

(**349**) X = I
(**350**) X = Ph$_3$P$^+$

(**351**)

(**352**)

(**353**)

(**354**)

(**355**)

(**356**)

(**357**)

Treatment of [(cod)IrCl]$_2$ with Me$_2$P(=CH$_2$)CH$_2$Li gives the iridium analogue of (**348**) and with Me$_3$P=N—P(=CH$_2$)Me$_2$ the cod-Ir analogue of (**307**). The latter phosphorus reagent and (Ph$_3$P)$_2$Ir(CO)Cl afford (**358**) ⟨79ZN(B)1218⟩. An unusual rearrangement is observed for I(CO)(Ph$_3$P)$_2$Ir=CH$_2$ leading to (**359**) ⟨84JOM(273)C17⟩.

(**358**)

(**359**)

4.12.4.3.8 Compounds containing phosphorus and nickel, palladium or platinum

Treatment of Cp(CO)NiI with Ph$_2$PCH$_2$Cl followed by Na/Hg affords the cyclic structure (**360**) ⟨82ZN(B)1146⟩. In solution the complex (**361**) equilibrates with the nickelacyclopropane isomer (**362**) ⟨84OM1438⟩. Ni-cod$_2$ reacts with Ph$_2$PCH$_2$PPh$_2$ and CH$_2$Br$_2$ directly to afford (**363**) ⟨92OM1392⟩. The majority of compounds of this type have, however, been obtained from reactions of ylides or their metallated derivatives with nickel species. Thus, a variety of ylides, R$_3$P=CH$_2$, react with Ni(CO)$_4$ to give R$_3$P$^+$CH$_2$Ni(CO)$_3^-$ ⟨72IJ293⟩, while with Ni(H$_2$C=CH$_2$)$_3$, the analogues R$_3$P$^+$CH$_2$Ni(H$_2$C=CH$_2$)$_2^-$ are formed ⟨85CB298⟩. In a similar way, Me$_3$P=CH$_2$ reacts with Me$_2$Ni(PMe$_3$)$_2$ to give Me$_3$P$^+$CH$_2$NiMe$_2$(PMe$_3$)$^-$ and with (Me$_3$P)$_2$Ni(Me)Cl to give (Me)(PMe$_3$)Ni$^-$(CH$_2$PMe$_3^+$)$_2$ Cl$^-$ ⟨74CB93⟩. When the same ylide is treated with (Me$_3$P)$_2$NiCl$_2$, however, the two polycyclic products (**364**) and (**365**) are formed ⟨73AG(E)853, 74CB3684, 74CB3706⟩. (Me$_3$P)$_2$NiCl$_2$ and Ph$_2$PCH=P(Ph)$_2$CH$_2^-$Na$^+$ afford (**366**) ⟨81AG(E)586⟩, while Ni-cod$_2$ and Pr$^i_3$P=CH$_2$ and Ph$_3$P=CHC(O)Me directly gives (**367**) ⟨92OM2701⟩. The (**368**) series of compounds have been prepared, from NiCl$_2$ and suitable phosphorus compounds, for X = CH ⟨77CB3517⟩, N ⟨77CB3536⟩, BH$_2$ ⟨80JOM(200)287⟩, and BMe$_2$ ⟨83OM257⟩.

Treatment of Pd(PPh$_3$)$_4$ with (EtO)$_2$P(O)CH$_2$I results in oxidative addition to give (Ph$_3$P)$_2$Pd(I)CH$_2$P(O)(OEt)$_2$ ⟨86JOM(309)225⟩. Thermal rearrangement of (Ph$_3$P)$_2$Pd(Cl)CH=CHCO$_2$R affords (**369**) ⟨85JOM(288)119⟩. Treatment of either NaPdCl$_4$ or PdCl$_2$·2PhCN with Ph$_2$PCH$_2$PPh$_2^+$CH$_2$C(O)R Br$^-$ under basic conditions affords (**370**) ⟨90JCS(D)2509⟩. The reaction of (Me$_3$P)$_2$PdCl$_2$ with Me$_3$P=CH$_2$ affords the palladium analogue of (**365**) ⟨78CB797⟩, while treatment of (Me$_2$N)$_3$P=CH$_2$ with PdCl$_2$ results in simple substitution to give (Me$_2$N)$_3$P$^+$CH$_2$PdCH$_2$P$^+$(NMe$_2$)$_3$ (Cl$^-$)$_2$ ⟨84BCJ43⟩. Compound (**371**) analogous to (**329**) has been described ⟨82CB1379⟩ and the bicyclic compounds (**372**) have been formed from PdCl$_2$ and the appropriate phosphorus compounds for X = CH ⟨77CB3517⟩ and BH$_2$ ⟨80JOM(200)287⟩. Stabilized phosphorus ylides form adducts of various types with PdCl$_2$ and related palladium compounds, and the structures reported include (**373**), which undergoes cyclopalladation to afford (**374**) when heated ⟨89JOM(364)407⟩, (**375**) ⟨76JOM(111)361⟩, (**376**) ⟨89JOM(375)265⟩, (**377**) ⟨85JOM(290)125⟩, and (**378**) ⟨85IJC(A)398⟩.

(Ph$_3$P)$_4$Pt and CH$_2$ClI give (Ph$_3$P)$_2$Pt(Cl)CH$_2$PPh$_3$$^+$ I$^-$ ⟨79JOM(182)C20, 86JOM(315)255⟩, and (Et$_3$P)$_3$Pt reacts similarly with CH$_2$I$_2$ to give (Et$_3$P)$_2$Pt(I)CH$_2$PEt$_3$$^+$ I$^-$ ⟨81CC698, 82JOM(228)C71⟩.

Treatment of $(Ph_3P)_4Pt$ with $(EtO)_2P(O)CH_2I$ gives $(Ph_3P)_2Pt(I)CH_2P(O)(OEt)_2$ ⟨86JOM(309)225⟩. With $(Ph_3P)_2Pt(H_2C=CH_2)$, CH_2ClI undergoes oxidative addition to give $(Ph_3P)_2Pt(I)CH_2Cl$ ⟨82JOM(232)C78, 86JOM(315)255⟩, and either this or $(Ph_3P)_2Pt(Cl)CH_2Cl$ ⟨81ZN(B)1663⟩ can react with Ph_3P to afford $(Ph_3P)_2Pt(Cl)CH_2PPh_3^+ X^-$. Treatment of cod-$Pt(CH_2I)_2$ with four equivalents of Ph_3P affords (379) while with five equivalents of Me_3P, (380) is formed ⟨88OM2082⟩. Treatment of (381) with Ph_3P or Ph_2PH gives (382), while with $Ph_2PCH_2PPh_2$, (383) is formed ⟨90JCS(D)1553⟩. $Me_3P=CH_2$ and cod-$PtMe_2$ give (384) and $Bu^t_2MeP=CH_2$ behaves similarly ⟨79JOM(182)251⟩, while treatment of $(Me_2N)_3P=CH_2$ with $PtCl_2$ results in simple substitution to give $(Me_2N)_3P^+CH_2Pt$ $CH_2P^+(NMe_2)_3$ $(Cl^-)_2$ ⟨84BCJ43⟩. This last ylide and PtI_4 afford $[(Me_2N)_2P^+CH_2]_4Pt$ $(I^-)_4$ ⟨84BCJ43⟩. Compound (385) analogous to (329) has been described ⟨82CB1379⟩ and the bicyclic compounds (386) have been formed from $PtCl_2$ and appropriate phosphorus compounds for $X = CH$ ⟨77CB3517⟩, N ⟨77CB3536⟩ and BH_2 ⟨80JOM(200)287⟩. Formal insertion of CH_2 into a Pt—P bond is involved in the unusual reaction of (387) with an excess of CH_2N_2 to afford (388) ⟨82CC614⟩.

4.12.4.3.9 Compounds containing phosphorus and copper, silver or gold

Treatment of Ph_2PCH_2Li with CuCl gives Ph_2PCH_2Cu and in a similar way $Ph_2P(O)CH_2Cu$, $Ph_2P(O)CH(Pr^n)Cu$ and $Ph_2P(=NPh)CH_2Cu$ have been generated from the corresponding lithium compounds ⟨77CB3930⟩. Treatment of phosphonates such as $(EtO)_2P(O)CH_2R$ and thionophosphonates such as $(EtO)_2P(S)CH_2R$ with Bu^nLi followed by CuI yields the synthetically useful copper derivatives $(EtO)_2P(X)CH(R)Cu$ ⟨76S766, 82S725, 92JOM(423)339⟩. Conjugate addition of cuprates such as Bu^n_2CuLi to diethyl vinylphosphonate affords $(EtO)_2P(O)CH(CH_2Bu^n)Cu^-Bu^n Li^+$ ⟨80PS(9)121⟩. Reactions with ylides can give a range of different structures depending on the ylide, the copper compound and the conditions. Thus, $Me_3P=CH_2$ reacts with $Me_3P \cdot CuCl$ to give (389), with TMS-CH_2Cu to give (390) and with CuCl to give the cyclic structure (391) ⟨73AG(E)415, 74CB3697, 74JOM(74)C23⟩. Products analogous to (391) have also been obtained from CuCl and $Ph_3P=CHR$ ⟨75JOM(96)133⟩, $Bu^n_3P=CH_2$ ⟨77MI 412-02⟩, and $(Me_2N)_3P=CH_2$ ⟨84BCJ43⟩, while with a 1:1 ratio of $Ph_3P=CHR$ to CuCl, products of the formula $[Ph_3P^+CH(R)Cu^-Cl]_4$ with a cubane type structure are obtained ⟨75JOM(97)479⟩.

Silver compounds analogous to (389) have been obtained by treatment of $Ph_3P=CHR$ with AgCl ⟨75JOM(96)133⟩ or $AgClO_4$ ⟨88JCS(D)341⟩ and from $(Me_2N)_3P=CH_2$ with AgCl ⟨84BCJ43⟩. On the other hand, $Me_3P=CH_2$ and $Me_3P \cdot AgCl$ give the cyclic structure analogous to (391) ⟨73AG(E)415, 74CB3697, 78IS140⟩. As for copper, a 1:1 ratio of $Ph_3P=CH_2$ to AgCl gives a product of the formula $[Ph_3P^+CH_2Ag^-Cl]_4$ with a cubane type structure ⟨75JOM(97)479⟩. Compounds of the type $R_3P^+CH_2Ag^-C_6F_5$ can be obtained either by treatment of $R_3P=CH_2$ with C_6F_5Ag or of R_3P^+Me $CF_3CO_2^-$ with C_6F_5Li and $Ag^+ CF_3CO_2^-$ ⟨88JCS(D)341⟩. Stabilized ylides $Ph_3P=CHC(O)R$ react with AgClO$_4$ or AgNO$_3$ to give (**392**) ⟨87JOM(331)409⟩, and when these products are treated with Ph$_3$P$^+$CH$_2$COCH$_2$PPh$_3^+$ (ClO$_4^-$)$_2$, exchange takes place to give (**393**) ⟨88OM997⟩.

(**392**)

(**393**)

Treatment of tetrahydrothiophene gold(I) chloride with Ph$_2$P(S)CH$_2$Li gives (**394**) ⟨87OM1992⟩, and this reacts with Br$_2$ to afford (**395**) ⟨88IC836⟩. Reactions of two equivalents of ylides with Me$_3$P·AuCl to give gold compounds analogous to (**389**) have been reported for Me$_3$P=CH$_2$ ⟨73AG(E)416, 75CB1321⟩, Et$_3$P=CH$_2$ ⟨78ZN(B)1325⟩, (Me$_2$N)$_3$P=CH$_2$ ⟨80CL311⟩, and Ph$_3$P=CHR ⟨75CB1321, 79BCJ2560⟩. Use of an excess of Me$_3$P=CH$_2$ gives the cyclic product analogous to (**391**) ⟨73AG(E)416⟩, while with a 1 : 1 ratio the reaction can be stopped at the stage of R$_3$P$^+$CH$_2$Au·PMe$_3$ Cl$^-$ for R = Me or Ph ⟨75CB1321⟩. Me$_3$P=CH$_2$ and Me$_3$P·AuMe give Me$_3$P$^+$CH$_2$Au$^-$Me while with TMS-CH$_2$Au·PMe$_3$ the analogue of (**390**) is formed ⟨75CB1321⟩. Ph$_3$P=CH$_2$ and AuCN gives Ph$_3$P$^+$CH$_2$Au$^-$CN ⟨84JCS(D)2859⟩, and various routes to pentahalophenyl compounds Ph$_3$P$^+$CH(R)Au$^-$C$_6$X$_5$ have been developed for X = F ⟨83ICA63⟩ and Cl ⟨88SRI69⟩. Treatment of gold(III) compounds such as Me$_3$P·AuMe$_3$ with Me$_3$P=CH$_2$ gives Me$_3$P$^+$CH$_2$Au$^-$Me$_3$ while the same reaction with Me$_2$AuBr$_2$ gives (**396**) which reductively eliminates ethane when heated to give Me$_3$P$^+$CH$_2$Au$^-$CH$_2$PMe$_3^+$ Br$^-$ ⟨75ICA(13)79⟩. The gold compounds analogous to (**392**) have been reported ⟨87JOM(333)129⟩ including examples with R = NMe$_2$ ⟨91JCS(D)2579⟩ and exchange to give the analogue of (**393**) also occurs as for silver ⟨88OM997⟩. The cyclic compounds (**397**) have been prepared for X = CH ⟨77CB3517, 77IJ149⟩, N ⟨77CB3536⟩ and BH$_2$ ⟨80JOM(200)287⟩. The last product results from treating (**278**) with (Me$_2$AuCl)$_2$ but, if (CO)AuCl is used instead, the 12-membered ring compound (**398**) is formed ⟨81CB441⟩.

(**394**)

(**395**)

(**396**)

(**397**)

(**398**)

4.12.4.3.10 Compounds containing phosphorus and zinc, cadmium or mercury

The reaction of ClCH$_2$ZnCH$_2$Cl with two equivalents of Ph$_3$P to give Ph$_3$P$^+$CH$_2$ZnCH$_2$PPh$_3^+$ (Cl$^-$)$_2$ was reported at an early stage ⟨61LA(650)1⟩. Most compounds in this class have, however, as for the other transition metals, been formed by reactions of ylides. Treatment of ylides R$^1_3$P=CHR^2 with ZnCl$_2$ and CdCl$_2$ to give R$^1_3$P$^+$CH(R^2)MCH(R^2)PR$^1_3{}^+$ (Cl$^-$)$_2$ has been reported for Ph$_3$P=CH$_2$, Ph$_3$P=CHMe ⟨80BCJ3176⟩, and (Me$_2$N)$_3$P=CH$_2$ ⟨84BCJ43⟩. With the more hindered ylide, Ph$_3$P=CHPr^i, the reaction stops at the stage of Ph$_3$P$^+$CH(Pr^i)MCl Cl$^-$ ⟨80BCJ3176⟩. Treatment of Me$_3$P=CH$_2$ with Et$_2$Zn or Et$_2$Cd gives the polymers (**399**) ⟨77ZAAC(434)145⟩. Me$_3$P=C=PMe$_3$ and Et$_2$Zn or Et$_2$Cd afford the cyclic compounds (**400**; X = CH) ⟨76AG(E)542, 77CB3517⟩, and the corresponding reactions of Me$_3$P=N—P(=CH$_2$)Me$_2$ give the nitrogen analogues (**400**; X = N)

⟨76AG(E)541, 77CB3536⟩. Treatment of (**278**) with $ZnCl_2$ or $CdCl_2$ gives (**400**; X = BH_2) ⟨80JOM(200)287⟩, while the sulfur analogue (**401**) similarly affords (**402**) ⟨85OM1233⟩.

(**399**) M = Zn, Cd

(**400**) M = Zn, Cd, Hg

(**401**)

(**402**) M = Zn, Cd

The reaction of $Ph_2P(S)CH_2Li$ with $HgCl_2$ proceeds in a straightforward manner to give $Ph_2P(S)CH_2HgCH_2P(S)Ph_2$ ⟨88OM2415⟩, but $Ph_2P(O)CH_2Li$ and $HgBr_2$ give a polymeric complex of the expected product containing LiBr ⟨91OM3392⟩. Treatment of $(MeO)_2P(S)CH(CN)_2$ with $Hg(NO_3)_2$ gives (**403**) ⟨73EGP94992⟩, while conjugate addition of Bu^tHgCl to diethyl vinylphosphonate affords $(EtO)_2P(O)CH(HgCl)CH_2Bu^t$ ⟨86JOC5498⟩. The reaction of ylides, $R^1_3P=CHR^2$ with $HgCl_2$ or $HgBr_2$ to give products of the type $R^1_3P^+CH(R^2)HgCH(R^2)PR^1_3{}^+ (X^-)_2$ is well known and the ylides used include $Ph_3P=CH_2$ ⟨59CI(L)849, 61JA1610, 61JA1613⟩, $Ph_3P=CHMe$ and $Ph_3P=CHPr^i$ ⟨80BCJ3176⟩, $Me_3P=CH_2$ ⟨74CB102, 78IS140⟩, and $(Me_2N)_3P=CH_2$ ⟨84BCJ43⟩. By treating $Me_3P=CH_2$ with MeHgCl at low temperature, $Me_3P^+CH_2HgMe\ Cl^-$ can be obtained, but this disproportionates to give Me_2Hg and $Me_3P^+CH_2HgCH_2PMe_3{}^+ (Cl^-)_2$ when warmed to RT ⟨74CB102⟩. A range of stabilized ylides $Ph_3P=CHC(O)R$ form adducts with $HgCl_2$ which most likely have the salt structure (**404**). Examples reported include those with R = Ph ⟨64IZV772⟩, H and OEt ⟨66ZOR942⟩, as well as other groups ⟨65JOM(4)202⟩, and in some cases the adducts dissociate to the starting components in solution ⟨68ZOR1685⟩. Bis(ylides) may form the cyclic structures (**405**) with $HgCl_2$ ⟨85IJC(A)398⟩. Treatment of the resonance stabilized ylide $Ph_3P=CH-CH=CH-PPh_3{}^+ Cl^-$ with an excess of $HgBr_2$ gives (**406**) ⟨66DOK(171)111⟩. The reaction of (**278**) with $HgCl_2$ proceeds in the same way as with $ZnCl_2$ and $CdCl_2$ to give (**400**; X = BH_2) ⟨80JOM(200)287⟩.

(**403**)

(**404**)

(**405**)

(**406**)

4.12.4.4 Group 13 and 14 Derivatives

4.12.4.4.1 Compounds containing phosphorus and aluminum

The reaction of Me_2PCH_2Li with $AlCl_3$ is rather complex and, depending on the reacting ratios, can give products of formula $Me_2PCH_2AlCl_2$, $(Me_2PCH_2)_2AlCl$, $(Me_2PCH_2)_3Al$, or $Li^+ (Me_2PCH_2)_4Al^-$, although the first two exist as dimers while the last has a polymeric structure ⟨85OM231⟩. Some ylides react with aluminium species in a simple way to give, for example, $Me_3P^+CH_2AlMe_3{}^-$ from $Me_3P=CH_2$ and Me_3Al ⟨68CB595⟩ and $R^1_2P(Cl)^+CH(R^2)AlCl_3{}^-$ or $R^1_2P(F)^+CH(R^2)AlCl_3{}^-$ from the appropriate P-halo ylides and $AlCl_3$ ⟨80TL3984⟩. $Ph_3P=CH_2$ and $AlCl_3$ on the other hand, gives (**407**) which can continue to react with the ylide to form a chlorine-bridged dimer of $(Ph_3P^+CH_2)_3Al$ ⟨83BCJ1772⟩, and ylides $Ph_3P=CHR$ react with Et_2AlCl to give the stable adducts (**408**) ⟨93PAC617⟩. Treatment of (**278**) with Me_2AlCl gives the cyclic compound (**409**) ⟨80JOM(200)287⟩.

4.12.4.4.2 Compounds containing phosphorus and gallium, indium or thallium

The reaction of $Me_3P=CH_2$ with R_3M to give the stable products, $Me_3P^+CH_2MR_3{}^-$ occurs very readily for Me_3Ga, Et_3Ga, Me_3In and Me_3Tl ⟨75JOM(99)353⟩. The same ylide reacts with the halides

Me$_2$MCl or Me$_2$MBr for M = Ga, In and Tl to afford the cyclic products (**410**) ⟨74CB3674⟩. The cyclic compounds (**411**) have been obtained by treating Me$_4$Ga, Me$_3$Ga or LiGaMe$_4$ with appropriate phosphorus compounds for X = CH ⟨76AG(E)542, 77CB3517⟩, N ⟨76AG(E)541, 77CB3528⟩ and BH$_2$ ⟨80JOM(200)287⟩, and the thallium analogue has also been obtained for X = N ⟨77CB3528⟩. Treatment of Br$_2$InCH$_2$Br with Ph$_3$P gives Ph$_3$P$^+$CH$_2$InBr$_3^-$ ⟨91OM2159⟩.

4.12.4.4.3 Compounds containing phosphorus and tin or lead

A good number of compounds containing phosphorus and tin are known and the methods for their preparation largely parallel those used for analogues containing phosphorus and silicon. Me$_3$SnCH$_2$I and tertiary phosphines give the rather unstable salts Me$_3$SnCH$_2$PR$_3^+$ I$^-$ ⟨76JOM(107)73⟩, while R$_3$SnCH$_2$I and phosphites such as (EtO)$_3$P or phosphonites such as PhP(OEt)$_2$ undergo an Arbuzov process to give R$_3$SnCH$_2$P(O)(OEt)$_2$ and R$_3$SnCH$_2$P(O)(OEt)Ph respectively ⟨79JOM(182)465, 80EGP142887⟩. Reduction of products such as Et$_3$SnCH$_2$P(O)(OEt)$_2$ obtained in this way, with LiAlH$_4$, provides the phosphine Et$_3$SnCH$_2$PH$_2$ in 55% yield ⟨79JOM(182)465⟩. Me$_2$PCH$_2$Li and SnCl$_2$ give a good yield of Sn(CH$_2$PMe$_2$)$_4$ by disproportionation of the initially formed Me$_2$P-CH$_2$SnCH$_2$PMe$_2$ with formation of tin metal ⟨86OM1664⟩. Treatment of the appropriate methyl tin chlorides, Me$_n$SnCl$_{4-n}$, with Me$_2$PCH$_2$Li gave the series of compounds Me$_3$SnCH$_2$PMe$_2$, Me$_2$Sn(CH$_2$PMe$_2$)$_2$, MeSn(CH$_2$PMe$_2$)$_3$, and Sn(CH$_2$PMe$_2$)$_4$ ⟨83ZN(B)1399⟩. The α-lithiated phosphine oxides such as Ph$_2$P(O)CH$_2$Li react readily with Ph$_3$SnCl to give Ph$_2$P(O)CH$_2$SnPh$_3$ ⟨63JA642, 64JA1100, 69USP3426021⟩, and Ph$_2$P(S)CH$_2$Li and Ph$_2$P(S)CH(Me)Li similarly afford Ph$_2$P(S)CH(R)SnPh$_3$ ⟨64JOM(2)1⟩. Use of Me$_2$SnCl$_2$ as the electrophile with lithiated phosphine oxides gives products of the type Ar$_2$P(O)CH$_2$Sn(Me$_2$)CH$_2$P(O)Ar$_2$ ⟨69USP3426021⟩. α-Lithiated phosphonates ⟨91S31, 93PS(75)99⟩ or their copper derivatives ⟨82S725⟩ react with electrophiles such as Me$_3$SnCl, Bu$^n_3$SnCl and Ph$_3$SnCl to give products of the type (R^{1}O)$_2$P(O)CH$_2$SnR$^2_3$.

Ylides and Ph$_3$SnCl or Ph$_3$SnBr generally give the simple substitution products, R$^1_3$P$^+$CH(R^2)SnPh$_3$ X$^-$, as exemplified by Ph$_3$P=CH$_2$ ⟨59CI(L)849, 61JA1610, 61JA1613⟩, Ph$_3$P=CHMe and Ph$_3$P=CHPr^i ⟨82BCJ3025⟩, (Me$_2$N)$_3$P=CH$_2$ ⟨84BCJ43⟩, and Ph$_3$P=CHC(O)Ar ⟨73JOM(51)167⟩. The last two ylides react with Me$_3$SnCl similarly, but other ylides such as Ph$_3$P=CH$_2$ react with double transylidation to give Ph$_3$P=C(SnMe$_3$)$_2$ ⟨67CB1032⟩. Transylidation is also observed with Ph$_3$P=CHR and Bu$^n_3$SnCl ⟨93PAC617⟩. Ph$_3$P=CH$_2$ and Me$_2$SnCl$_2$ react in the expected way to afford Ph$_3$P$^+$CH$_2$Sn(Me$_2$)CH$_2$PPh$_3^+$ (Cl$^-$)$_2$ ⟨61JA1610, 61JA1613⟩, while (Me$_2$N)$_3$P=CH$_2$ reacts only three times with SnCl$_4$ to give [(Me$_2$N)$_3$P$^+$CH$_2$]$_3$SnCl (Cl$^-$)$_3$ ⟨84BCJ43⟩. The adducts formed between bis(ylides) and Me$_2$SnCl$_2$, Me$_3$SnCl and Ph$_3$SnCl apparently have structures of the type (**412**) with hexavalent tin, rather than the ionic alternatives ⟨86JOM(303)351⟩. Treatment of Ph$_3$P=CHSn(OPr^i)Bu$^t_2$ with BF$_3$·Et$_2$O gives products including Ph$_3$P$^+$CH$_2$Sn(F)Bu$^t_2$ BF$_4^-$ ⟨91OM938⟩.

Treatment of Cl$_2$P(O)CH$_2$Cl with SnCl$_2$ at 165–170 °C gives Cl$_2$P(O)CH$_2$SnCl$_3$ in 60% yield ⟨77ZOB1664⟩. Addition of (EtO)$_2$P—OSnEt$_3$ to the double bond of diethyl vinylphosphonate gives (**413**) ⟨75ZOB1486⟩. Hydrostannylation of (MeO)$_2$P(S)CH=CH$_2$ with Pr^{n}SnH gives both (**414**) and the regioisomeric adduct (**415**) ⟨78ZOB1422, 79ZOB1772⟩. Photolysis of (Pr$^i_2$N)$_2$PC(=N$_2$)SnR$_3$ with methyl acrylate results in carbene addition to afford (**416**) for R = Me, Ph and cyclohexyl ⟨93PS(76)49⟩.

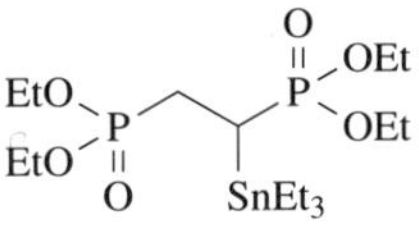

(**413**) 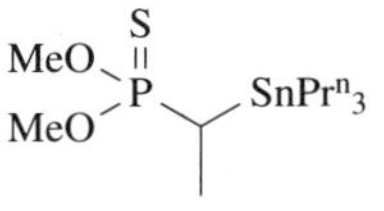 (**414**) 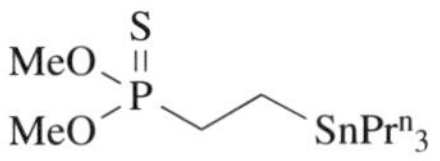 (**415**) 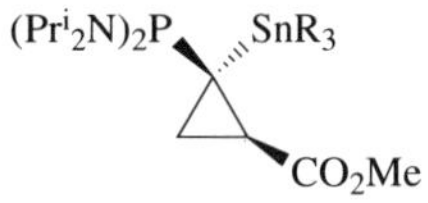 (**416**)

Reactions of the ylides R^1_3P=CHR^2 with Ph_3PbCl to afford $R^1_3P^+CH(R^2)PbPh_3$ Cl^- have been reported for Ph_3P=CH_2, Ph_3P=$CHMe$, Ph_3P=$CHPr^i$ ⟨82BCJ3025⟩ and $(Me_2N)_3P$=CH_2 ⟨84BCJ43⟩.

4.12.4.5 Actinide Derivatives

Studies in this area have been confined to compounds of thorium and uranium. $(Cp^*)_2ThCl_2$ and Me_2PCH_2Li initially give (**417**) and, when this is heated at 60 °C, Me_3P is lost to afford (**418**). The latter product may also be obtained from $(Cp^*)_2Th(Cl)CH_2Bu^t$ and Me_2PCH_2Li or by treatment of (**419**) with Me_3P at 60 °C ⟨86JA425⟩. Treatment of Cp_3UCl with one equivalent of Ph_2P(=CH_2)CH_2Li gives the dimeric structure (**420**) ⟨78JA5562, 81IC2466⟩, while with three equivalents of the same reagent, (**421**) is formed ⟨81IC2466, 84JA5920⟩. The corresponding reactions of $PhMeP$(=CH_2)CH_2Li give the compounds analogous to (**420**) and (**421**) ⟨81IC2466, 90IS177⟩. $(Cp^*)_2ThCl_2$ or $(Cp^*)_2UCl_2$ and the same phosphorus compounds give (**422**) ⟨89OM1192⟩. The remarkable structure of (**423**), formed by chance, has been confirmed by x-ray methods ⟨88OM1465⟩.

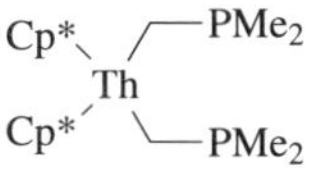

(**417**) 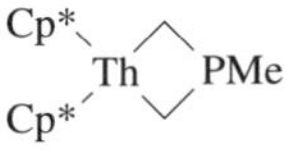 (**418**) 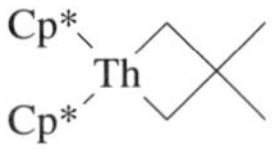 (**419**) 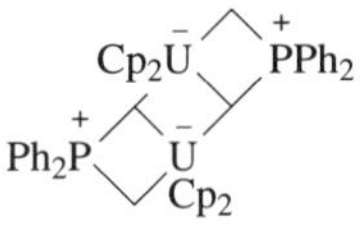 (**420**) 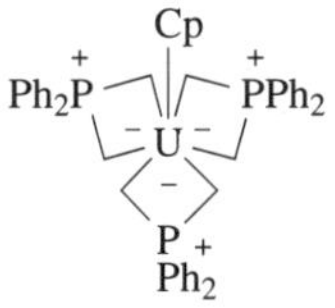 (**421**)

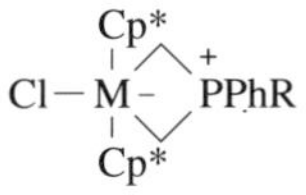

(**422**) M = Th, U

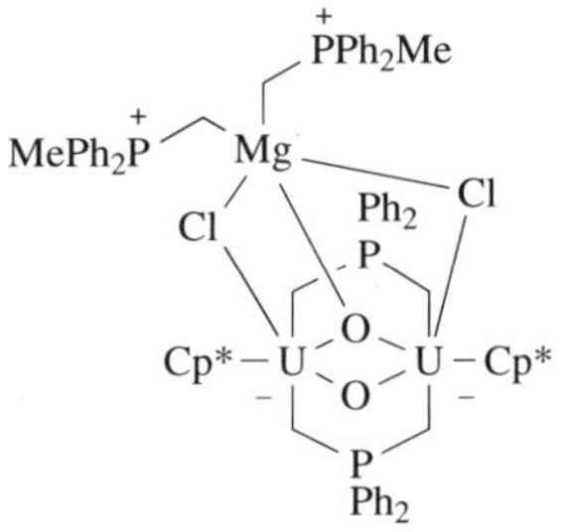

(**423**)

4.13
Functions Containing at Least One As, Sb or Bi with or without a Metalloid (Si or Ge) or a Metal

W. M. HORSPOOL
University of Dundee, UK

4.13.1	FUNCTIONS CONTAINING TWO ARSENIC, ANTIMONY OR BISMUTH GROUPS	591
	4.13.1.1 Functions with Two Similar Elements: $R^1_2AsCR^2_2AsR^3_2$, etc.	592
	4.13.1.1.1 Arsenic functions	592
	4.13.1.1.2 Antimony functions	595
	4.13.1.1.3 Bismuth functions	595
	4.13.1.2 Functions Containing Dissimilar Elements: $R^1_2AsCR^2_2SbR^3_2$, etc.	596
	4.13.1.2.1 Arsenic and antimony functions	596
4.13.2	FUNCTIONS CONTAINING ARSENIC, ANTIMONY OR BISMUTH AND A METALLOID: $R^1_2AsCR^2_2SiR^3_3$, etc.	597
	4.13.2.1 Arsenic Derivatives	597
	4.13.2.1.1 Arsenic and silicon functions	597
	4.13.2.2 Antimony Derivatives	598
	4.13.2.2.1 Antimony and silicon functions	598
	4.13.2.2.2 Bismuth derivatives	599
	4.13.2.2.3 Bismuth and germanium functions	599
4.13.3	FUNCTIONS CONTAINING ARSENIC, ANTIMONY OR BISMUTH AND A METAL	599
	4.13.3.1 Arsenic Derivatives	599
	4.13.3.1.1 Arsenic and group 1 metals	599
	4.13.3.1.2 Arsenic and group 14 metals	600
	4.13.3.2 Antimony Derivatives	600
	4.13.3.2.1 Antimony and group 1 metals	600
	4.13.3.3 Bismuth Derivatives	600
	4.13.3.3.1 Bismuth and group 1 metals	600
	4.13.3.3.2 Bismuth and group 14 metals	600

4.13.1 FUNCTIONS CONTAINING TWO ARSENIC, ANTIMONY OR BISMUTH GROUPS

Two major texts ⟨B-71MI 413-01, 82COMC-I⟩ give some coverage of this area and readers are directed to these for general reference.

4.13.1.1 Functions with Two Similar Elements: $R^1_2AsCR^2_2AsR^3_2$, etc.

4.13.1.1.1 Arsenic functions

(i) Bisarsinomethane and related compounds

The reaction of arsenic(III) oxide (As_2O_3) with acetyl chloride and aluminum chloride at 170 °C followed by treatment of the reaction mixture with thionyl chloride afforded the bis(dichloroarsino)methane derivative (**1**; R = H) (b.p. 138–140 °C) in good yield ⟨49CB152⟩. The bisarsine (**1**; R = H) was readily hydrolysed to afford $(AsO)_2CH_2$, which probably exists as a dimer (m.p. 265 °C) ⟨70ZAAC(377)120⟩. Oxidation by hydrogen peroxide of dimeric $(AsO)_2CH_2$ gave the bisarsenic acid derivative (**2**) (m.p. 168 °C) ⟨49CB152⟩. It is reported that a variety of derivatives analogous to (**2**) can be prepared by the reaction of As_2O_3 or As_2O_6 with aluminum trichloride and acid chlorides. This affords polymeric materials that can be readily oxidised to yield the final products. A specific example is the reaction of As_2O_3 with EtCOCl to yield $(AsO)_2CHMe$ (m.p. 76 °C). Alternatively, this compound can be prepared by reaction of propanoic acid with $KAsO_2$ ⟨69ZAAC(370)31⟩. Both $MeAsO(OH)_2$ and $PhAsO(OH)_2$ can be converted into the chloroarsines (**3**; R = Me) and (**3**; R = Ph), respectively, on treatment with acetyl chloride and aluminum chloride ⟨49CB152⟩.

As mentioned above, the chloroarsine derivatives (**1**) and (**3**) are readily hydrolysed and yield dimeric species ⟨70ZAAC(377)120⟩. The methylene bis(phenylarsenic acid) derivative (**4**) was synthesized by the reaction of phenylarsenic oxide with dibromomethane.

Alternative methods for the synthesis of bisarsinomethanes using appropriate lithium or sodium derivatives have been reported. Thus the reaction of the lithium derivative (**5**) with diphenylchloroarsine afforded a path to the bis(diphenylarsino)methane (**6**; Ar = Ph, R = H) ⟨85CB2353⟩. A derivative of this (**6**; Ar = *p*-MeOC$_6$H$_4$, R = H) was synthesized by the reaction of one derivative (**5**; Ph = *p*-methoxyphenyl) with dichloromethane ⟨90ZOB2291⟩. Alternatively, reaction of sodium diphenylarsenic with 1,1-dichloroethane in THF gave a high yield (80–85%) of the crystalline derivative (**6**; Ar = Ph, R = Me) (m.p. 102 °C) ⟨69ZAAC(370)31⟩.

All of the previously mentioned compounds can also be used as synthetic precursors to other derivatives of bisarsinomethanes. Thus the dimeric bisarsenic oxide $(AsO)_2CH_2$ can be methylated on arsenic with MeI to afford, after aqueous workup, the acid derivative (**7**) ⟨70ZAAC(377)120⟩. Bis(diphenylarsino)methane (**6**; Ar = Ph, R = H) can also undergo loss of two phenyl groups by reduction with sodium; oxidation with hydrogen peroxide then yields methylene bisphenylarsenic acid (**8**) ⟨70ZAAC(377)120⟩. Dephenylation of (**6**; Ar = Ph, R = H) can also be brought about with arsenic trichloride when the tetrachloro derivative (**1**; R = H) is obtained. Reductive dephenylation of (**6**; Ar = Ph, R = Me) has also been reported as a path to the bisoxide (AsO)$_2$CHMe ⟨69ZAAC(370)31⟩. Furthermore, treatment with HBr and bromine gives the bis(dibromoarsino)methane (b.p. 150–152 °C/0.01 mmHg) ⟨69ZAAC(370)31⟩.

$$\begin{array}{c} \text{O} \\ \| \\ \text{As(OH)Me} \\ \diagdown \\ \text{As(OH)Me} \\ \| \\ \text{O} \\ (\mathbf{7}) \end{array} \qquad \begin{array}{c} \text{O} \\ \| \\ \text{As(OH)Ph} \\ \diagdown \\ \text{As(OH)Ph} \\ \| \\ \text{O} \\ (\mathbf{8}) \end{array}$$

Reduction by LAH can be used to convert bis(dichloroarsino)methane (**1**; R = H) into bisarsinomethane, obtained as a crystalline solid (m.p. 91–96 °C). Other arsines such as (**9**; R = H or Me) were obtained by reductive removal of phenyl groups by sodium from bis(diphenylarsino)methane (**6**; Ar = Ph, R = H) ⟨70ZAAC(377)120⟩ and the methyl derivative (**6**; Ar = Ph, R = Me) ⟨69ZAAC(370)31⟩, respectively.

$$\text{R} \diagup\!\!\!\diagdown \begin{array}{c} \text{AsPhH} \\ \text{AsPhH} \end{array}$$

(**9**)

Bis(dichloroarsino)methane (**1**; R = H) is a particularly useful substrate for the preparation of a variety of derivatives. Thus, the synthesis of tetraalkyl derivatives, e.g., tetramethyl and tetraethyl, can be effected by their reactions with the corresponding Grignard reagent ⟨87CB1281⟩. The use of secondary amines as nucleophiles opens high yield routes to the tetraamino derivatives (**10**) ⟨75JOM393⟩; similarly structure (**3**; R = Me) affords the corresponding amine derivatives (**11**) ⟨75ZAAC202⟩. Reactions of this type are usually carried out in ether at −30 °C. Primary amines also react with compound (**1**; R = H) although the outcome is somewhat different providing the 2,4,6,8-tetraaza-1,3,5,7-tetraarsaadamantanes (**12**) are obtained as oils ⟨75JOM393⟩.

$$\text{R}_2\text{As}\diagdown\!\!\diagup\text{AsR}_2$$

(**10**)

R	b.p. (°C)
Me$_2$N	105/0.2 torr
Et$_2$N	120/0.2 torr
Pr$^n_2$N	120/0.3 torr
⌬N (pyrrolidine)	130/0.2 torr
⌬N (piperidine)	140/0.9 torr
Bu$^n_2$N	130/0.6 torr

The tetraamino derivative (**10**; R = NMe$_2$) can undergo As—N bond fission on reaction with appropriate nucleophiles. Thus, the reaction with 2-(methylamino)ethanol afforded the cyclic arsenic derivative (**13**) (b.p. 155 °C/0.01 mmHg) in 45% yield, while ethane-1,2-dithiol gave (**14**) (175 °C/0.001 mmHg) in 40% yield. These cyclic derivatives can also be obtained directly from bis(dichloroarsino)methane (**1**; R = H), e.g., the reaction with 1,2-bis(*N*-methylamino)ethane afforded (**15**) (b.p. 155 °C/0.01 mmHg) in 55% yield ⟨91ZAAC(605)151⟩. The chemistry of amino arsine derivatives has been reviewed ⟨82S173⟩.

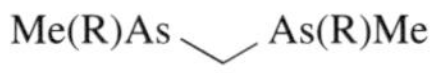

Me(R)As⌒As(R)Me

(11)

R	b.p. (°C)
Me$_2$N	76/0.4 torr
Et$_2$N	86/0.8 torr
Pr$_2$N	98/0.6 torr
 pyrrolidinyl-N	105/0.5 torr
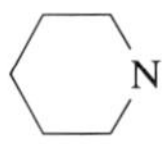 piperidinyl-N	125/0.6 torr

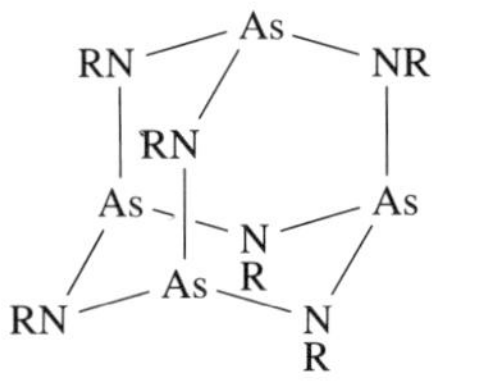

(12) R = Me, Et, Pr^n, Pr^i, Bu^n, Bu^i

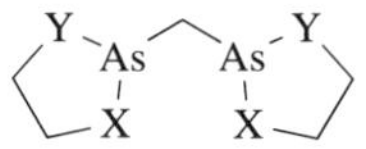

(13) Y = O, X = MeN
(14) Y = X = S
(15) Y = X = MeN

(ii) Cyclic derivatives

Alkyl and aryl arsenic dichlorides react readily with malonic esters. This process affords the 1,3-dialkyl- and 1,3-diaryl-1,3-diarsacyclobutanes (16) in yields ranging from 10% to 75% ⟨76AG(E)56, 78JCR(S)252⟩. Another approach to the synthesis of 1,3-diarsacyclobutanes (17) has been reported ⟨80ZAAC(470)144, 80ZAAC(470)157⟩ and involves the [2 + 2]photodimerisation of the arsoranes (18). The crystalline 1,3-diarsacyclobutanes (17; R = Me), (m.p. 94 °C) and (17; R = Et) (m.p. 123 °C) were obtained in yields of 86 and 95%, respectively. The thermal dimerisation of the cumulene (19), generated by the treatment of (20) with Bu^{t}Li in THF at −78 °C, also afforded a route to a 1,3-diarsacyclobutane, *viz.* derivative (21) (m.p. 224 °C, 47% yield) ⟨90TL6331⟩. Other ring compounds can be formed by the reaction of the chloroarsine (22) with K$_2$SN$_2$. Thus, (23) was formed from (22) and (24) from a double reaction with the tetrachloro derivative (1; R = Me) ⟨87ZN(B)118⟩. Reductive elimination of chlorine from bis(methylchloroarsino)methane by sodium amalgam provided a convenient route to 1,2,4,5-tetramethyl-1,2,4,5-tetraarsacyclohexane (25) ⟨75ZAAC202⟩.

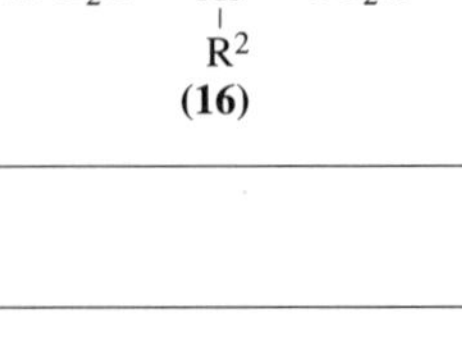

(16)

R^1	R^2	m.p. (°C)	Yield (%)
Me	Ph	168	55
Et	Ph	145	40
Me	Me	176	75
Et	Me	102	60
Bu^t	Me	208	10

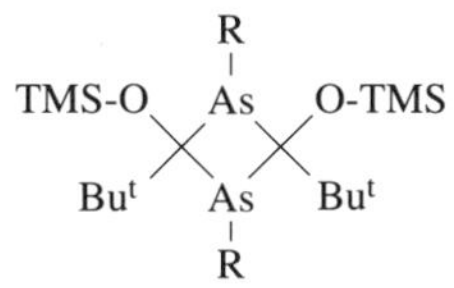

4.13.1.1.2 Antimony functions

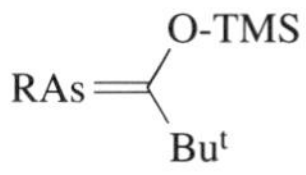

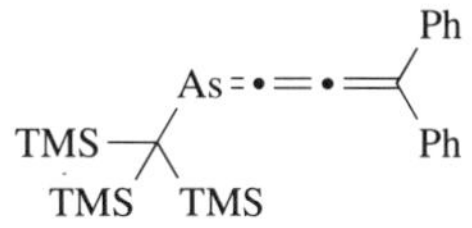

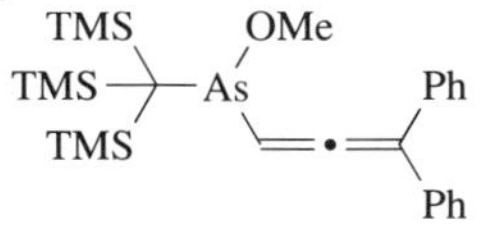

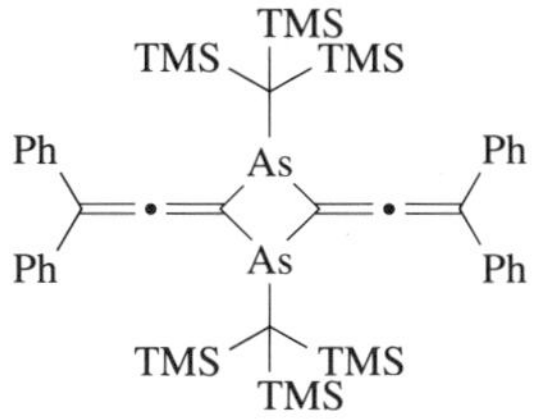

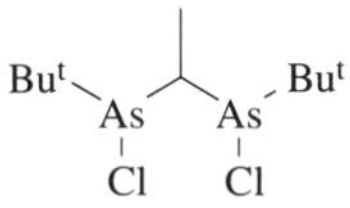

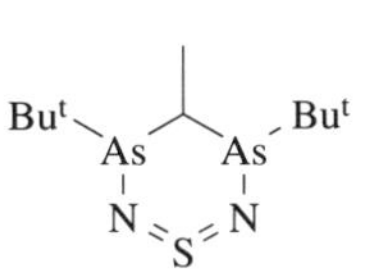

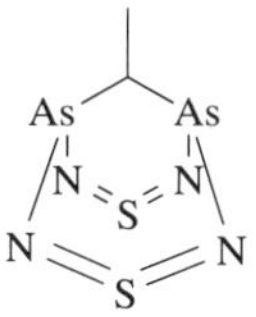

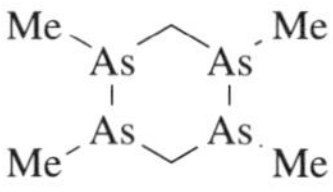

(i) Distibines

The common method for the synthesis of groups flanked by two antimony atoms is illustrated by the reaction of phenylmethylantimonysodium (**26**) with dichloromethane whereby the bisstibinomethane (**27**) (b.p. 130–168 °C/0.01 mmHg) is obtained (Scheme 1) ⟨72JOM(43)333⟩. An alternative approach to a variety of derivatives is the use of dephenylation of derivative (**28**) ⟨71MI 413-01⟩ with HCl in chloroform to afford bis(dichlorostibino)methane (**29**). The presence of chlorine in (**29**) permits readily the synthesis of derivatives such as (**30**) by the reaction with methylmagnesium chloride ⟨71MI 413-01⟩.

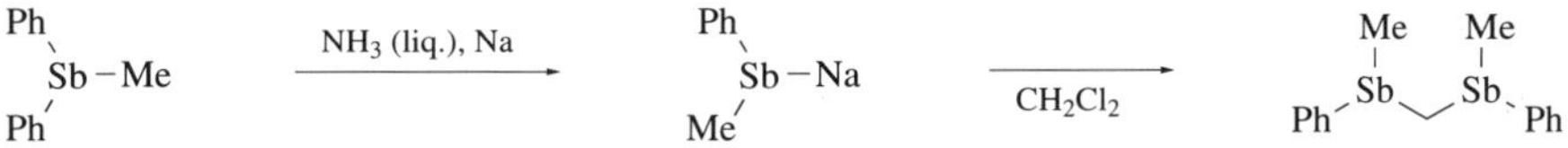

Scheme 1

Alternatively, the chlorines can be displaced with dithioamides and dithioesters (dithiocarbamates and xanthogenates) to afford the derivatives (**31**) and (**32**). Another displacement reaction of this type, using dithiophosphate derivatives, gave the compounds (**33**) (Table 1) ⟨92ZAAC(607)164⟩. Complex salts such as (**34**) can also be obtained from (**29**) by the reactions with the corresponding quaternary bromides ⟨92ZAAC(607)157⟩.

Further reactions can be carried out on derivatives such as (**28**). For example, it was converted into the monothio compound (**35**) by treatment with sodium sulfide in methanol at room temperature under nitrogen ⟨75JOM57⟩.

4.13.1.1.3 Bismuth functions

The preparation, reactions and physical properties of organobismuth compounds have been reviewed ⟨82CRV15⟩.

Table 1 Distibines from reaction of chloro distibines with dithio-amides and dithioesters.

Structure	R	Yield (%)	m.p. (°C)
$CH_2[Sb(S_2CNR_2)_2]_2$	Me	89	184
(31)	Et	87	197
$CH_2[Sb(S_2COR)_2]_2$	Et	96	142
(32)	Pr^i	92	135
$CH_2[Sb(S_2P(OR)_2)_2]_2$	Me	72	78
(33)	Pr^i	71	57

$$(R_4E)_2(BrCl_2Sb)_2CH_2 \equiv \left[\begin{array}{c} X \diagdown \diagup X \\ X - Sb \diagup \diagdown Sb - X \\ X \diagdown X \diagup X \\ X \end{array} \right]^{2-} (R_4E^+)_2$$

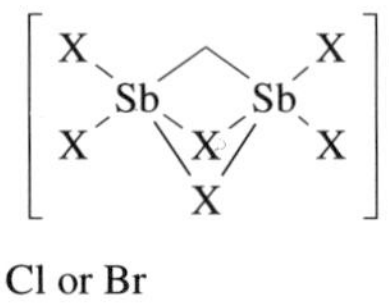

(34)

X = Cl or Br

R_4E	m.p. of salt (°C)
Et_4N	226
Ph_4P	203
Ph_4As	190
Ph_4Sb	155

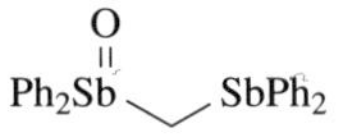

$$Ph_2Sb\diagdown\diagup SbPh_2$$
$$\overset{O}{\underset{\|}{}}$$

(35)

(i) Dibismuthines

The dibismuthinomethane derivative (**36**) (m.p. 91 °C, 53% yield) can be prepared analogously to the corresponding stibine derivative by the reaction of diphenylbismuthsodium with dichloromethane ⟨80AG(E)723, 85CB1039⟩.

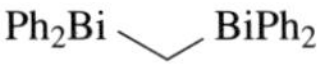

$$Ph_2Bi\diagdown\diagup BiPh_2$$

(36)

4.13.1.2 Functions Containing Dissimilar Elements: $R^1{}_2AsCR^2{}_2SbR^3{}_2$, etc.

4.13.1.2.1 *Arsenic and antimony functions*

The formation of functions with dissimilar elements is readily brought about by the reactions of an appropriate alkyllithium with a suitable chloro derivative. This is illustrated for the reaction of the stibinomethyllithium (**37**) and diphenylchloroarsine to give the derivative (**38**) in a yield of 23% ⟨83CB473, 85CB2353⟩.

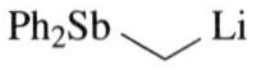

$$Ph_2Sb\diagdown\diagup Li \qquad\qquad Ph_2As\diagdown\diagup SbPh_2$$

(37)    (38)

4.13.2 FUNCTIONS CONTAINING ARSENIC, ANTIMONY OR BISMUTH AND A METALLOID: $R^1_2AsCR^2_2SiR^3_3$, etc.

4.13.2.1 Arsenic Derivatives

4.13.2.1.1 *Arsenic and silicon functions*

(i) Arsines

In general, derivatives of this type are prepared by reaction of a chloroarsine with a trimethylsilylmethyllithium or a Grignard reagent. Typical of this is the reaction of trimethylsilylmethylmagnesium chloride with arsenic trichloride in THF to yield tris(trimethylsilylmethyl)arsine (**39**) (m.p. 67–68.5 °C, 80% yield) ⟨58JA1336⟩. Another report details the reaction of the same Grignard reagent with arsenic trichloride in a 1:1 molar ratio when bis(trimethylsilylmethyl)chloroarsine (**40a**) (this arsine can be reduced by LAH to yield $(TMS\text{-}CH_2)_2AsH$) was formed ⟨91PS(57)1⟩. Arsenic trichloride also reacted with bis(trimethylsilyl)methyllithium in ether to afford the chloroarsine (**40b**) (m.p. 70–72 °C) in a reasonable yield of 61% ⟨80JCS(D)2428⟩. Other derivatives can be obtained in the same manner. Thus the reaction of bis(diethylamino)chloroarsine with trimethylsilylmethylmagnesium chloride at −78 °C followed by aqueous workup afforded dichloro(trimethylsilylmethyl)arsine (**41**) in 81% yield ⟨91MI 413-01, 91POL319⟩. The chlorine atoms in (**41**) can be substituted with hydrogen by reduction with LAH at low temperature (−78 °C). The arsine, $TMS\text{-}CH_2AsH_2$, was obtained in 47% yield as a colourless liquid ⟨91MI 413-01, 91POL319⟩.

                (**39**)   (**40**)a; R = TMS-CH$_2$ (**41**)
                                        b; R = (TMS)$_2$CH

As well as reacting with lithium or Grignard reagents, arsenic trichloride also reacts with the electron-rich alkene (**42**) at 20 °C in THF or ether, the final product, obtained quantitatively, was compound (**43**) ⟨89TL349⟩.

              (**42**)   (**43**)

(ii) Arsoranes

Interestingly the arsorane (**44**) is thermally unstable and at 170 °C rearrangement to the trivalent derivative (**45**) (69% yield) occurred with loss of bromotrimethylsilane probably via the salt (**46**) ⟨83CB348⟩. The reaction of tris(trimethylsilylmethyl)arsine with bromine affords the arsorane (**44**) (m.p. 118–120 °C) ⟨58JA1336⟩.

    (TMS-CH$_2$)$_3$AsBr$_2$
        (**44**)   (**45**)   (**46**)

(iii) Quaternary salts

Quaternary salts incorporating silicon and arsenic are generally prepared by the reaction of the corresponding trialkylarsine with, for example, chloromethyltrimethylsilane ⟨68IC168⟩. A specific example of this is the reaction of trimethylarsine and chloromethyltrimethylsilane in a sealed tube at 130 °C that gives a high yield of the corresponding salt (**47**) ⟨65IC1458⟩. Triethylarsine reacted in an analogous fashion with chloromethyltrimethylsilane to give the arsonium salt (**48**) in a yield of 79% ⟨77CB1312⟩. Salts of this type (**49**) can also be obtained from the methylenearsorane (**50**) with trimethylsilyl chloride at low temperature ($-70\,°C$ in ether under nitrogen) ⟨84TL4425⟩.

$Me_3\overset{+}{As}$⌒TMS Cl⁻ $Et_3\overset{+}{As}$⌒TMS Cl⁻ $Ph_3\overset{+}{As}$⌒TMS Cl⁻ $Ph_3As=$

(**47**) (**48**) (**49**) (**50**)

(iv) Ylides and cumulenes

All salts of the type mentioned above in 4.13.2.1.1(iii) react readily with BuLi to afford ylides (**51**) in high yield. Arsonium ylides have been reviewed ⟨82AOC(20)115⟩ and reports have dealt with various spectroscopic studies ⟨75CB2649, 76CB473⟩. Such ylides, for example, (**51**; R = Et), also react readily with alkyl halides such as methyl bromide to afford, for example, the salt (**52**) which again can be converted into the corresponding ylide (**53**). An unstable arsorane (**54**) can be obtained by dehydrohalogenation of the dichloroarsine (**43**) by 1,4-diazabicyclo[2.2.2]octane (dabco) ⟨89TL349⟩.

(**51**) R = Me, Et or Ph   (**52**)   (**53**)   (**54**)

(v) Photoreactions

The chloroarsine (**40b**) is photochemically labile and irradiation brings about As—Cl bond fission to afford the persistent arsenic-centred radical (**55**) with a half-life of 1 month at $20\,°C$ ⟨76CC623, 80JCS(D)2428⟩. Irradiation of the acid (**2**) has also been studied; a carbon-centred radical, flanked by two arsenics, is produced ⟨81HCA329⟩.

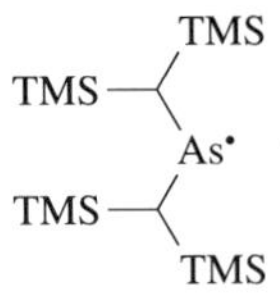

(**55**)

4.13.2.2 Antimony Derivatives

4.13.2.2.1 Antimony and silicon functions

(i) Stibines

A variety of stibines can be prepared by the reaction of substituted alkyllithiums or Grignard reagents with antimony trichloride in varying molar ratios. In the late 1950s, Seyferth ⟨58JA1336⟩ reported that trimethylsilylmethylmagnesium chloride (**56**) reacted with antimony trichloride in THF to afford tris(trimethylsilylmethyl)stibine (m.p. 64–$65\,°C$) in 74% yield.

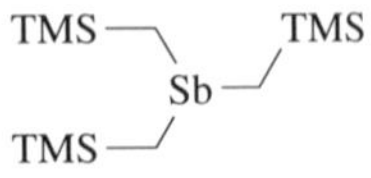

(**56**)   (TMS-CH$_2$)$_3$SbBr$_2$ (**57**)   (TMS-CH$_2$)$_3$SbMe$_2$ (**58**)   (TMS-CH$_2$)$_5$Sb (**59**)

Me$_3$SbBr$_2$ (**60**)   Me$_3$Sb(CH$_2$-TMS)$_2$ (**61**)   Me$_4$Sb⌒TMS (**62**)   (TMS-CH$_2$)$_3$$\overset{+}{Sb}$Me (**63**)   (TMS-CH$_2$)$_4$SbMe (**64**)

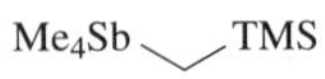

(ii) Stiboranes

A series of stiboranes can be obtained by the reaction of suitable lithium reagents with the appropriate antimony derivative ⟨78CB2702⟩. The reaction of the stibine (**56**) with bromine yields the stiborane (**57**) (m.p. 158–160 °C) ⟨58JA1336⟩. This can be used as substrate for the synthesis of other stiboranes. Thus, the reaction of (**57**) with methyllithium gave the stiborane (**58**) (m.p. 51–53 °C, 78% yield), while the reaction with trimethylsilylmethyllithium gave the pentakis derivative (**59**) (m.p. 93 °C, 89% yield) ⟨78CB2702⟩. The stiborane (**60**) also underwent reaction with trimethylsilylmethyllithium to yield the trimethyl derivative (**61**) (b.p. 65 °C/0.1 torr, 83% yield) while the tetramethylstiborane (**62**) (b.p. 68 °C/0.1 torr, 37% yield) was prepared from tetramethyliodostiborane with the same lithium reagent ⟨78CB2702⟩. The salt (**63**) was transformed into (**64**) (m.p. 33 °C, 68% yield) by treatment with trimethylsilylmethyllithium ⟨78CB2702⟩.

4.13.2.2.2 Bismuth derivatives

(i) Bismuth and silicon functions

The reaction of bismuth trichloride with trimethylsilylmethylmagnesium chloride in THF gave the bismuthine (**65**) (35% yield), which underwent extensive decomposition on attempted purification ⟨58JA1336⟩. Other derivatives can be synthesized in a similar manner by the reaction of bismuth trichloride, as a suspension in ether, with suitable alkyllithium reagents. Dimethyl-(trimethylsilylmethyl)bismuthine (**66**) is formed by reaction of dimethylbismuthinosodium with chloromethyltrimethylsilane ⟨88ZN(B)739⟩.

$$\text{TMS} \diagdown \atop \text{TMS} \diagup \!\!\!\! \text{Bi} \diagup\!\! \text{TMS} \qquad\qquad \text{Me}_2\text{Bi} \diagdown\!\!\diagup \text{TMS}$$

(**65**)   (**66**)

4.13.2.2.3 Bismuth and germanium functions

The germanium derivative (**67**) has been synthesized from dimethylbismuthinosodium and chloromethyltrimethylgermane ⟨88ZN(B)739⟩.

$$\text{Me}_2\text{Bi} \diagdown\!\!\diagup \text{GeMe}_3$$

(**67**)

4.13.3 FUNCTIONS CONTAINING ARSENIC, ANTIMONY OR BISMUTH AND A METAL

4.13.3.1 Arsenic Derivatives

4.13.3.1.1 Arsenic and group 1 metals

Transmetallation reactions of the arsenic derivatives (**68**) have provided a route for the synthesis of the lithium derivative (**5**). This involves the reaction of (**68**; R = Ph) with phenyllithium in ether at −40 °C ⟨82CB1810⟩ (36% yield) or the reaction of the tributyl derivative (**68**; R = Bu) with BuLi in THF at −78 °C ⟨85CB2353⟩ (92% yield). Alternatively, bis(diphenylarsino)methane reacted with *n*-butyllithium at −40 °C in THF to afford a 72% yield of diphenylarsinomethyllithium (**5**) ⟨78TL4391⟩. The same lithium derivative can be prepared by the reaction of iodomethyldiphenylarsine with BuLi or PhLi at −78 °C when (**5**) is obtained quantitatively ⟨85CB2353⟩. An alternative approach to derivatives of this general type involves the reaction of lithium diisopropylamide at $-40\,°C$ in THF with the oxide (**69**). The lithium derivative (**70**) was obtained in 95% yield ⟨82CB645⟩.

Ph$_2$As⌒SnR$_3$ Ph$_2$As(=O)⌒Me   Ph$_2$As(=O)⌒Li

(**68**) R = Ph or Bu 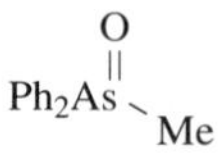   (**69**) 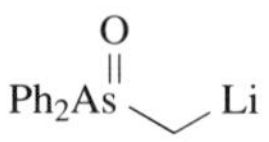   (**70**)

4.13.3.1.2 Arsenic and group 14 metals

Derivatives containing both arsenic and tin can be obtained by the reaction of tributylstannylmethyllithium (**71**) with diphenylchloroarsine or the reaction of diphenylarsinomethyllithium with tributyltin chloride to yield (**68**; R = Bu) ⟨85CB2353⟩. A similar route has been used to synthesize the lead derivative (**72**) ⟨85CB2353⟩.

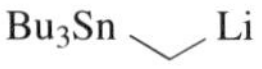

Bu$_3$Sn⌒Li   Ph$_2$As⌒PbPh$_3$

(**71**)    (**72**)

4.13.3.2 Antimony Derivatives

4.13.3.2.1 Antimony and group 1 metals

The reaction of phenyllithium in THF at $-70\,°C$ with bis(diphenylstibino)methane afforded a quantitative yield of diphenylstibinomethyllithium (**37**) ⟨78TL4391, 83CB473, 85CB2353⟩.

4.13.3.3 Bismuth Derivatives

4.13.3.3.1 Bismuth and group 1 metals

As with the arsenic and antimony systems, transmetallation was also used to synthesize the lithium derivative (**73**) (72% yield) by reaction of the dibismuth compound (**36**) with phenyllithium in ether at $-78\,°C$ ⟨80AG(E)723, 85CB1039⟩.

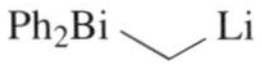

Ph$_2$Bi⌒Li

(**73**)

4.13.3.3.2 Bismuth and group 14 metals

The reaction of dimethylbismuthinosodium with chloromethyltrimethyltin provided a route to the derivative (**74**) ⟨88ZN(B)739⟩.

Me$_2$Bi⌒SnMe$_3$

(**74**)

4.14
Functions Containing at Least One Metalloid (Si, Ge or B) Together with Another Metalloid or Metal

CHRISTOPHER G. BARBER
Pfizer Central Research, Sandwich, UK

4.14.1 FUNCTIONS CONTAINING TWO METALLOIDS	602
4.14.1.1 *Functions Bearing Two Silicons: $R^1_2C(SiR^2_3)_2$, etc.*	602
4.14.1.1.1 Formation of the Si—C—Si linkage	602
4.14.1.1.2 Changing the groups attached to the central methylene	620
4.14.1.1.3 Changing the ligands on silicon	621
4.14.1.2 *Functions Bearing Two Germaniums: $R^1_2C(GeR^2_3)_2$*	625
4.14.1.2.1 Formation of the Ge—C—Ge linkage	625
4.14.1.2.2 Changing the groups attached to the central methylene	628
4.14.1.2.3 Changing the ligands on germanium	629
4.14.1.3 *Functions Bearing Two Borons: $R^1_2C(BR^2_2)_2$, etc.*	629
4.14.1.3.1 Formation of the B—C—B linkage	629
4.14.1.3.2 Changing the groups on the central methylene	634
4.14.1.3.3 Changing the ligands on boron	634
4.14.1.4 *Functions Bearing a Silicon and a Germanium Group: $R^1_2CSiR^2_3GeR^3_3$, etc.*	636
4.14.1.4.1 Formation of the Si—C—Ge linkage	636
4.14.1.4.2 Changing the groups attached to the central methylene	640
4.14.1.4.3 Changing the groups attached to the metalloids	640
4.14.1.5 *Functions Bearing a Silicon and a Boron Group: $R^1_2CSiR^2_3BR^3_2$, etc.*	642
4.14.1.5.1 Formation of the Si—C—B linkage	642
4.14.1.5.2 Changing the groups attached to the central methylene	645
4.14.1.5.3 Changing the groups attached to the metalloids	645
4.14.1.6 *Functions Bearing a Germanium and a Boron Group: $R^1_2CBR^2_2GeR^3_3$, etc.*	646
4.14.2 FUNCTIONS CONTAINING A METALLOID AND A METAL	646
4.14.2.1 *Silicon and a Metal: $R^1_2CSiR^2_3M$ etc.*	646
4.14.2.1.1 Silicon and a Group 1 or Group 2 metal: $R^1_2CSiR^3_3Li$, etc.	646
4.14.2.1.2 Silicon and a transition metal: $R^1_2CSiR^2_3CuX$, etc.	655
4.14.2.1.3 Silicon and a group 13 or Group 14 metal: $R^1_2CSiR^2_3SnR^3_3$, etc.	659
4.14.2.1.4 Silicon and other elements	662
4.14.2.2 *Germanium and a metal: $R^1_2CGeR^2_3M$*	662
4.14.2.2.1 α-Lithiogermyl alkanes	662
4.14.2.2.2 Other α-metallogermyl alkanes	663
4.14.2.3 *Boron and a Metal: $R^1_2CBR^2_2M$*	664
4.14.2.3.1 α-Lithioboryl alkanes	664
4.14.2.3.2 Other α-metalloboryl alkanes	665

4.14.1 FUNCTIONS CONTAINING TWO METALLOIDS

4.14.1.1 Functions Bearing Two Silicons: $R^1{}_2C(SiR^2{}_3)_2$, etc.

Methods for the preparation of a group consisting of an sp^3-hybridised carbon connected to two silyl groups and either protons or carbon fragments are described. This account has been divided into three main parts reflecting the routes that may be used. Firstly, the Si—C—Si linkage can be formed through generation of either one or both C—Si bonds. Once this basic system has been established the functionality of either the central methylene or the silicons may be manipulated. These latter two methods will be considered in much less detail than routes leading to Carbon–Silicon bond formation.

4.14.1.1.1 Formation of the Si—C—Si linkage

Routes to 1,1-bis(silyl)alkanes have been established for some time but the area has not been well served with reviews, with the exception of one which was published in 1965 ⟨65MI 414-01⟩.

The methods that have been used to generate either one or both of the C—Si bonds can be broken down into a number of categories. One of the most important routes is the quenching of a carbanion with an electrophilic silylating reagent. The reverse procedure—that of quenching a silyl anion with a carbon electrophile—has not been commonly used. Silenes may be used as intermediates in the preparation of 1,1-bis(silyl)alkanes, particularly cyclic systems. Hydrosilylations and rearrangements have also been successfully used along with routes proceeding through the intermediacy of radicals.

(i) Quenching a carbanion with a silyl electrophile

The most common route for the preparation of 1,1-bissilylated alkanes is through the generation of a carbanion α to silicon and the subsequent quenching of this by an appropriately substituted halosilane. A number of organometallic derivatives have been used and these will be considered in turn. This route can be used to add one silyl group to an α-metallosilyl alkane, or to add both silyl groups to the central carbon. The preparation of α-metallosilanes are discussed in detail later in this chapter (see 4.14.2).

(a) Organolithium reagents. The most common organometallic derivatives that have been reacted with silyl electrophiles are organolithium reagents. This is a reflection of their relative ease of preparation and availability of suitable precursors. The most frequently used route is deprotonation—a process facilitated by the presence of one silicon bound to the central methylene. Other routes such as halogen–metal exchange have also been used to generate products ranging from simple, unsubstituted acyclic systems to substituted derivatives and those containing Si—C—Si linkages within a ring.

Organolithium reagents will readily react with silyl electrophiles to generate a C—Si bond. Many bis(silyl)methanes have been generated through the intermediacy of α-lithiosilylmethanes. These may be formed through direct metallation ⟨90ZAAC(583)195⟩. In the simplest case, treatment of tetramethylsilane with *n*-butyllithium-pentamethyldiethylenetriamine followed by TMS-Cl generated bis(trimethylsilyl)methane ⟨82CC1323⟩. Similarly, 2-methoxy-2,4,4-trimethyl-2,4-disilapentane was prepared from methoxytrimethylsilane by sequential treatment with *t*-butyllithium then TMS-Cl ⟨89JOM(359)285⟩. There have been several reported cases of α-metallation and subsequent silylation to give a 1,1-bis(silyl) alkane occurring unexpectedly, when, for example a silyl protecting group is metallated (see 4.14.2.1.1(i)(a)) ⟨88JOM(341)293⟩.

The intermediate α-metallosilylmethane need not be generated by direct deprotonation; halogen–metal exchange has also been used. Thus, treatment of chloromethyltrimethylsilane with lithium dispersion gave the anion which, upon quenching with TMS-Cl, yielded bis(trimethylsilyl)methane in 75% yield ⟨80JA1584⟩. This lithio derivative has also been quenched with tetrachlorosilane resulting in the substitution of two, three or all four chlorides upon the use of increasingly vigorous conditions ⟨54JA1619, 84OM354⟩. Similarly, reaction with *t*-butyltrimethoxysilane gave *t*-butyldimethoxy(trimethylsilylmethyl)silane ⟨92CB607⟩. Likewise, bis(lithiomethyl)dimethylsilane, prepared from bis(chloromethyl)dimethylsilane with lithium shot, after treatment with TMS-Cl, gave the corresponding 2,4,6-trisilaheptane ⟨55JA907⟩. Many other simple derivatives have been similarly prepared ⟨90ZAAC(583)195⟩.

Both silyl groups can be directly added to the central methylene. For example, dilithiomethane has been prepared and reacted with TMS-Cl to generate bis(trimethylsilyl)methane in 88% yield ⟨71CB1347⟩. A process that has also been performed following *in situ* halogen–metal exchange when dihalomethane was treated with TMS-Cl and lithium to give bis(trimethylsilyl)methane ⟨65JOM(4)98, 88SC85⟩.

This process of metallation followed by a silyl quench may also be performed when the central carbon is substituted. For example, treatment of (*E*)-1,3-bis(trimethylsilyl)propene with *s*-butyllithium in pentane with TMEDA at −78 °C generated *exo,exo*-[1,3-bis(trimethylsilyl)allyl]lithium (**1**) ⟨83OM21, 90JA1382⟩ which upon quenching with TMS-Cl gave (*E*)-1,3,3-tris(trimethylsilyl)-1-propene (**2**) exclusively (Equation (1)) ⟨90OM1314⟩. Similarly (*E*)-1,4-di(trimethylsilyl)-2-butene was dimetallated with butyllithium in the presence of TMEDA and the lithio derivative then treated with TMS-Cl to give (*Z*)-1,1,4,4-tetra(trimethylsilyl)-2-butene in 80% yield ⟨91OM3167⟩. In a further example, trimethylsilylcyclopentadienyllithium has been quenched with TMS-Cl to give 1,1-bis(trimethylsilyl)cyclopentadiene ⟨71JOM(30)C57⟩ and with dichlorodimethylsilane to give, after methanolysis, 1-[dimethyl(methoxy)silyl]-1-trimethylsilylcyclopentadiene ⟨81TL7⟩. Other related systems have also been prepared ⟨91CB2391, 92JOM(429)C14⟩. Addition of the halosilane under certain conditions has been shown to be reversible (Scheme 1). Presumably the stability of the intermediate cyclopentadienyl anion allows silyl exchange ⟨85OM2206⟩. The migration of silyl groups on a cyclopentadienyl system has also been observed ⟨91CB1575⟩. α-Silylation can be accomplished when one of the silicons is within a ring. For example (**4**) was prepared from 1-*t*-butyl-1-sila-2,4-cyclohexadiene (**3**) by treatment with BuLi then TMS-Cl (Equation (2)) ⟨84CB1885⟩. In another example, treatment of 2,3-dihydro-1,1-dimethyl-1-silaphenalene (**5**) with butyllithium-TMEDA followed by TMS-Cl gave (**6**) (Equation (3)) ⟨82CC1023⟩. However when 1,8-bis(trimethylsilylmethyl) naphthalene (**7**) was treated with two equivalents of butyllithium in the presence of TMEDA although the isolable dilithio compound (**8**) was formed, subsequent treatment with dichlorodimethylsilane gave the anticipated siline (**9**) in only 5% yield (Equation (4)). The major product was that formed from decomposition of the dianion with loss of methane. A subsequent silyl quench then gave (**10**) ⟨84JCS(D)311⟩.

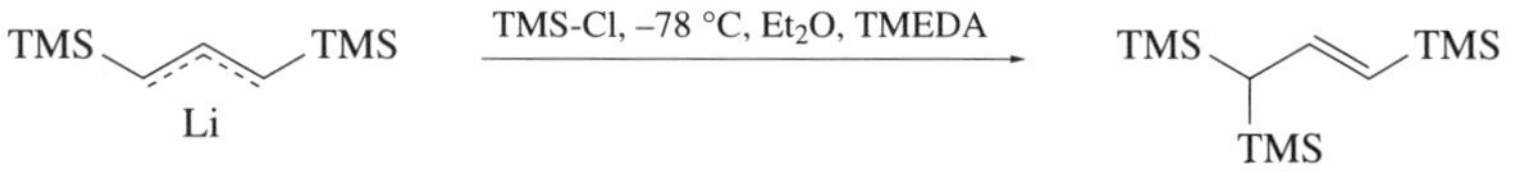

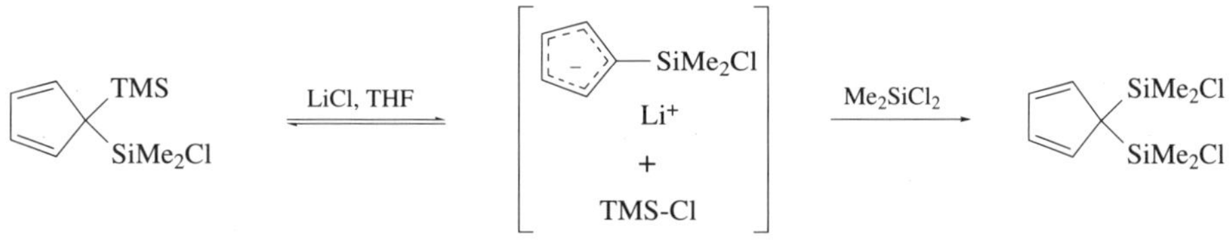

Scheme 1

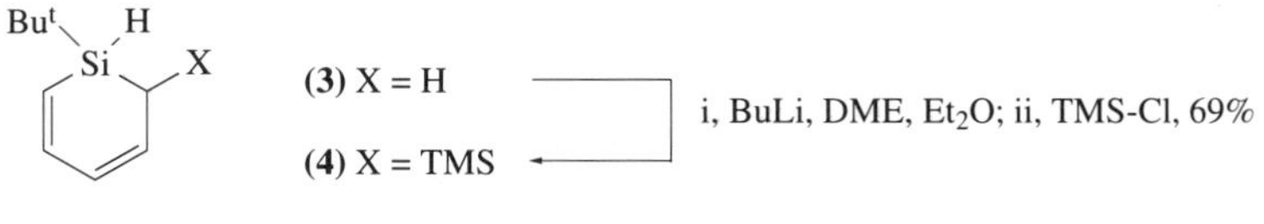

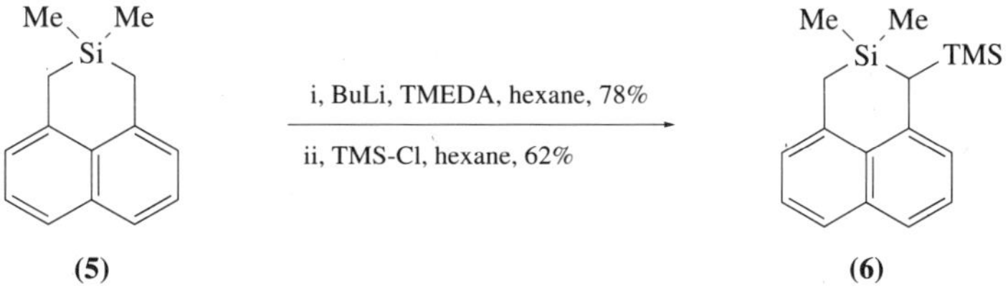

(7) X = H
(8) X = Li
BuLi, TMEDA, hexane, 83%

(9) R = TMS, 5%
(10) R = H, 75%

(4)

Both silyl groups can be added to the central methylene to generate a 1,1-bis(silyl) alkyl moiety. This can be performed through two cycles of metallation followed by a silyl quench. The monosilane intermediate need not be isolated. In fact attempts to isolate some monosilylated products have been hindered by significant bissilylation as a result of the increased acidity of the product over the starting material. This results in competitive deprotonation of the monosilylated product by unquenched metallated starting material. For example treatment of cyclopropa[b]naphthalene (**11**; $n = 0$) with *n*-butyllithium and subsequent quenching with TMS-Cl gave, not the anticipated monosilylated derivative (**11**; $n = 1$), but 1,1-bis(trimethylsilyl)-1*H*-cyclopropa[b]naphthalene (**11**; $n = 2$) in 40% yield and recovered starting material in 49% yield ⟨84JA6108⟩. The bis-silyl derivative (**12**) has similarly been prepared by two cycles of metallation followed by silylation without isolation of intermediates in 33% yield ⟨87CB471⟩. By repeating this process three times (**11**; $n = 2$) was prepared in 66% and (**12**) in 75% yield ⟨86JA5949⟩. The disilylation of benzylic positions through a stepwise process was also demonstrated for (**13**) (Equation (5)) ⟨89JOC4372⟩. Many other bis(trimethylsilyl)methylbenzene derivatives have similarly been prepared ⟨82CC14, 84JCS(D)321, 84JCS(D)1801, 88JCS(D)2403⟩ as has 9,9-bis(trimethylsilyl)fluorene ⟨66JOC2036⟩.

(11)

(12)

(13)

i, Bu^{s}Li, THF, TMEDA
ii, TMS-Cl
iii, i, then ii
91%

(5)

Methylpyridines have been both mono- and bissilylated in good yields using this same protocol ⟨90JCS(D)1161, 91S1221⟩. For example, 2-trimethylsilylmethylpyridine was prepared in 82% yield when the anion generated from 2-methylpyridine and *n*-butyllithium-TMEDA was added to an excess of TMS-Cl. When the process was repeated and the anion quenched by the addition of TMS-Cl, then 2-bis(trimethylsilyl)methylpyridine was obtained as the major product ⟨83CC1419, 90JCS(D)1161⟩. The acidity of the monosilylated product is greater than that of the starting material. Thus, if TMS-Cl is added slowly, then continuous re-equilibration results in the formation of the thermodynamically favoured anion and thence the generation of the disilylated product. It should be noted that although the acidity of the bissilylated product is greater than that of either the starting material or the monosilylated derivative, no tris(trimethylsilyl)methylpyridine was formed.

In some cases it has been found that treatment of the substrate with two equivalents of base and TMS-Cl can give yields of the bissilylated material superior to that obtained through a two-step process. An example is the preparation of α,α-bis(trimethylsilyl)-*t*-butylacetaldimine (**14**) (Scheme 2). Other more highly substituted bissilylated imines were also prepared using this single-step process ⟨93JOC2517⟩. Likewise α,α-bis(silyl) amides may be prepared by treatment of an amide with two equivalents of base and silyl electrophile although the amide enolate preferentially silylates on oxygen rather than on carbon resulting in low yields ⟨88JOM(354)155⟩.

Treatment of monolithioacetonitrile with TMS-Cl generates not only the anticipated trimethylsilylacetonitrile, but also significant quantities of bis- and tris(trimethylsilyl)acetonitrile ⟨71JA1714⟩.

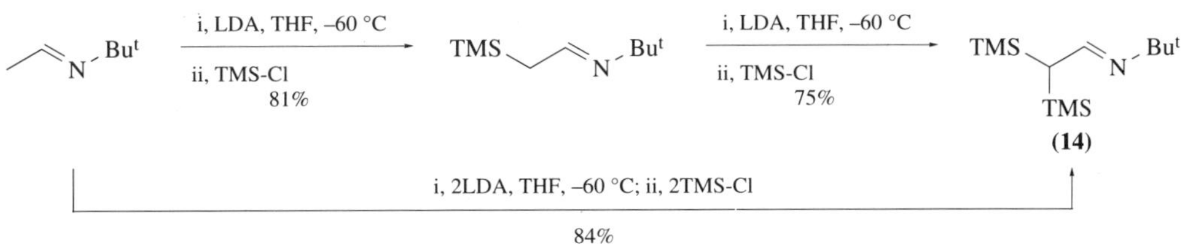

Scheme 2

In a similar fashion, α,α-bis(silyl)sulfonates are one of the products formed by the treatment of simple alkyl methanesulfonates with sodium hexamethyldisilazide and silylating reagents ⟨84ZOB1842⟩.

Reactions to generate polysilylated products need not only proceed through a stepwise sequence of metallation followed by a silyl quench. Polylithiated precursors have been generated which can be treated with excess silylating reagents. For example, treatment of 2-butyne with *n*-butyllithium and TMEDA (2:1) or 1-butyne with *n*-butyllithium in hexane at reflux for 5 h both generated trilithiobutyne (MeC_3Li_3) which upon quenching with TMS-Cl gave equal amounts of two isomeric trissilylated products, one of which was 1,3,3-tris(trimethylsilyl)-1-butyne ⟨76JA8412⟩.

Halogen–metal exchange has also been used to generate substituted 1,1-bis(silyl)alkanes in a manner similar to the unsubstituted cases described earlier. For example, treatment of several *gem*-dichlorocyclopropanes with TMS-Cl in the presence of lithium gave the corresponding 1,1-bis(trimethylsilyl) derivatives ⟨85JOM(280)313⟩.

Formation of a C—Si bond via an organolithium intermediate has also been used to generate many cyclic systems containing a Si—C—Si linkage within the ring. For example treatment of (15) with lithium dispersion generated the diorganolithium intermediate (16; X = Ph), which was treated with dichlorodimethylsilane to generate (17; X = Ph) (Scheme 3). It should be noted that lithium thiophenoxide generated during the formation of the intermediate (16) can compete for reaction with the dichlorosilane ⟨88JOM(338)159⟩. The analogous disilacyclobutane (17; X = H) has been prepared from bis(chloromethyl)dimethylsilane by treatment with lithium dispersion and subsequent quenching of the dianion with dichlorodimethylsilane ⟨70JOM(21)103⟩. Another cyclisation to prepare a 1,3-disilacyclobutane was demonstrated by treatment of (18) with lithium metal to generate (19) (Equation (6)) ⟨70ZAAC(372)21⟩. Five-membered ring systems—both saturated and unsaturated—have been prepared using similar cyclisation procedures. One example is shown in Scheme 4 where the spirocycle (21) was prepared in two steps from the hydrosilation product (20) ⟨70ZAAC(372)21⟩.

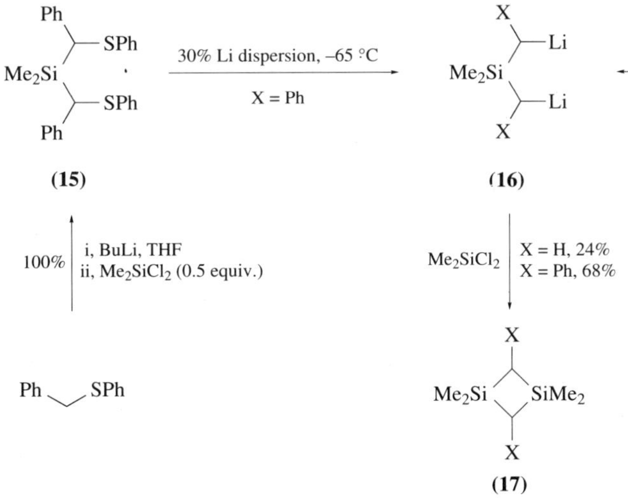

Scheme 3

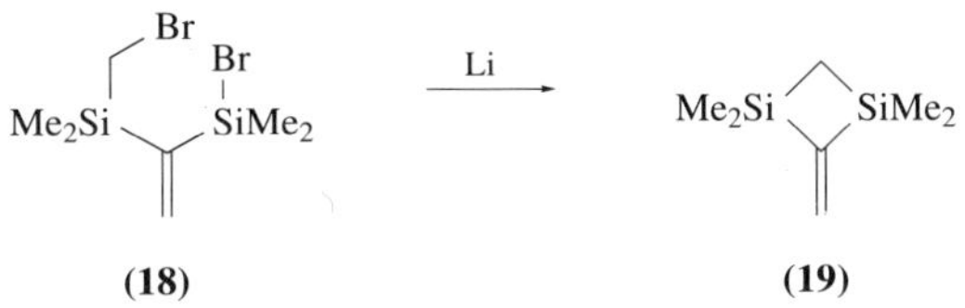

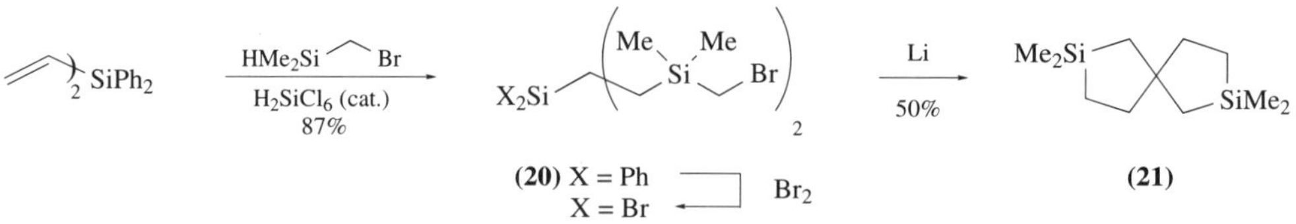

Scheme 4

The means of generation of an α-metallosilyl alkane need not be limited to either metallation or halogen–metal exchange; it has also been performed by nucleophilic addition to a vinyl silane. Treatment of a 1:1 mixture of vinyltrimethylsilane and TMS-Cl with *t*-butyllithium gave a mixture of products dependent upon the solvent system used (Scheme 5). The reaction is believed to proceed via the α-lithiosilane (**22**) which can either be quenched with TMS-Cl to give (**23**) or eliminate LiH to generate (**24**). Metallation of (**23**) followed by reaction with TMS-Cl would yield (**25**). The relative rates of these processes is solvent dependent with the bis(silyl) alkane (**23**) being favoured in THF ⟨77JA8447⟩. Other bis(silyl) alkanes have been prepared in a similar fashion ⟨92OM1137⟩.

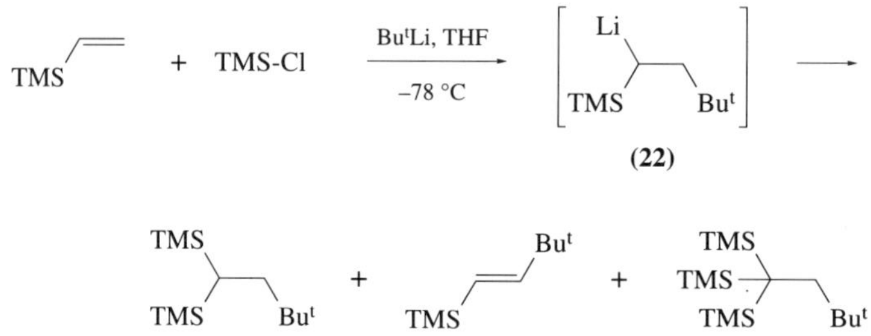

Solvent	Yield (**23**) (%)	Yield (**24**) (%)	Yield (**25**) (%)
hexane	40	30	10
ether	30	60	5
THF	76	0	0

Scheme 5

Another route to α-metallosilyl alkanes was demonstrated by treatment of 1,1,3,3-tetramethyl-1,3-disilacyclobutane with methyllithium ⟨90OM2677⟩. Ring cleavage gave 1-lithio-2,2,4,4-tetramethyl-2,4-disilapentane (**26**) which was successfully quenched with TMS-Cl to give the trisilahexane (**27**) (Equation (7)).

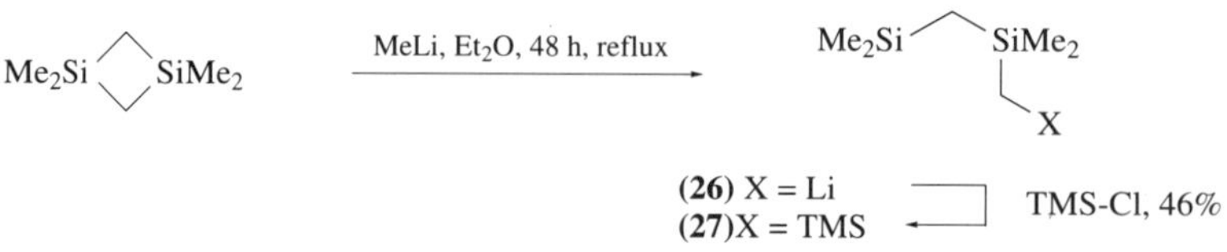

Four-membered rings containing a Si—C—Si linkage are not limited to 1,3-disilacyclobutanes; the fourth member of the ring has been a nitrogen. The cyclisation of a carbanion onto a silicenium intermediate resulted in the generation of a 1,3-disila-2-azacyclobutane ⟨76AG(E)619⟩.

(b) Grignard reagents. 1,1-Bis(silyl) alkanes have been prepared by quenching the appropriate α-magnesiosilyl alkane with a chlorosilane. Some acyclic systems that have been generated via such Grignard reagents are listed in Table 1. As with the organolithio derivatives described above, this route is restricted to systems containing functional groups with low reactivity towards organometallic reagents ⟨93JOM(451)C1⟩. A surprising exception to this is a report that the formation of the Grignard reagent derived from trichloro(chloromethyl)silane was stable for several days in ether at reflux, only decomposing to polymeric material after the addition of methanol ⟨91OM1336⟩, a

result contrary to that reported elsewhere ⟨93TL2111⟩. Entry 20 in Table 1 shows that triethoxysilylmethylmagnesium chloride is sufficiently stable at low temperatures to be synthetically useful. However, upon standing at room temperature, polycondensation resulted in the formation of (28) and (29) in 64% and 9% yields, respectively. Attempted use of trimethoxysilylmethylmagnesium chloride gave rapid polycondensation ⟨93TL2111⟩.

As with the organolithium reagents described above, both silyl groups may be added to the

Table 1 Preparation of acyclic bis(silyl)methanes from Grignard reagents.

Entry	Grignard Reagent	Electrophile	Product	Yield (%)
1[a]	TMS⌒MgCl	TMS-Cl	TMS⌒TMS	67
2		Me$_2$SiCl$_2$	TMS⌒SiMe$_2$Cl	55[b], 33[c], 47[d]
3[e]		SiCl$_4$	TMS⌒SiCl$_3$	58
4[c]		MeSi(OMe)$_3$	TMS⌒SiMe(OMe)$_2$	86
5[c]		MeSi(OEt)$_3$	TMS⌒SiMe(OEt)$_2$	37
6[f]	TMS⌒MgBr	Me$_2$SiClH	TMS⌒SiMe$_2$H	20
7[f]		MeCl$_2$SiH	(TMS⌒)$_2$SiMeH	62
8[f]		Cl$_3$SiH	(TMS⌒)$_3$SiH	65
9[g]	HMe$_2$Si⌒MgCl	TMS-Cl	HMe$_2$Si⌒TMS	39
10[g]		Ph$_3$SiCl	HMe$_2$Si⌒SiPh$_3$	80
11[g]		Ph$_2$MeSiCl	HMe$_2$Si⌒SiMePh$_2$	44
12[c]	H(Et)$_2$Si⌒MgCl	MeSi(OMe)$_3$	H(Et)$_2$Si⌒SiMe(OMe)$_2$	77
13[h]		PhMe$_2$SiCl	H(Et)$_2$Si⌒SiMe$_2$Ph	35
14[g]	H(Et)PhSi⌒MgBr	PhMe$_2$SiCl	H(Et)PhSi⌒SiMe$_2$Ph	75
15[b]	PhMe$_2$Si⌒MgCl	Me$_2$SiCl$_2$	PhMe$_2$Si⌒SiMe$_2$Cl	85
16[i]		Me$_2$SiClH	PhMe$_2$Si⌒SiMe$_2$H	71
17[j]	≡―SiMe$_2$⌒MgCl	TMS-Cl	≡―SiMe$_2$⌒TMS	85
18[i]	Me$_2$(EtO)Si⌒MgCl	Me$_2$SiCl$_2$	(Me$_2$(EtO)Si⌒)$_2$SiMe$_2$	41
19[k]	TMS-O(Me)$_2$Si⌒MgCl	Me$_2$ClSi⌒Cl	TMS-O(Me)$_2$Si⌒SiMe$_2$⌒Cl	73
20[l]	(EtO)$_3$Si⌒MgCl	TMS-Cl	(EtO)$_3$Si⌒TMS	65
21[m]	(Pr^{i}O)$_3$Si⌒MgCl	TMS-Cl	(Pr^{i}O)$_3$Si⌒TMS	82
22[n]	Me$_2$(NEt$_2$)Si⌒MgBr	Me$_2$SiCl$_2$	Me$_2$(NEt$_2$)Si⌒SiMe$_2$Cl	—
23[o]	N(CH$_2$CH$_2$O)$_3$Si⌒MgCl	TMS-Cl	N(CH$_2$CH$_2$O)$_3$Si⌒TMS	65
24[o]	N(CH$_2$CH$_2$O)$_3$Si⌒MgCl	MePh$_2$SiCl	N(CH$_2$CH$_2$O)$_3$Si⌒SiMePh$_2$	78
25[o]	N(CH$_2$CH$_2$O)$_3$Si⌒MgCl	Me$_2$HSiCl	N(CH$_2$CH$_2$O)$_3$Si⌒SiMe$_2$H	31
26[o]	N(CH$_2$CH$_2$O)$_3$Si⌒MgCl	H(Me)PhSiCl	N(CH$_2$CH$_2$O)$_3$Si⌒SiMe(Ph)H	54

[a] 91JOM(421)175. [b] 67JOM(10)111. [c] 54JOC250. [d] 88JA2611. [e] 54JA1619. [f] 90ZAAC(583)195. [g] 73JOM(57)261. [h] 75JCS(D)1434. [i] 89OM1585. [j] 86JPS(A)1839. [k] 58JOC1392. [l] 93TL2111. [m] 93JOM(451)Cl. [n] 76ZAAC(419)157. [o] 91JOM(418)C21.

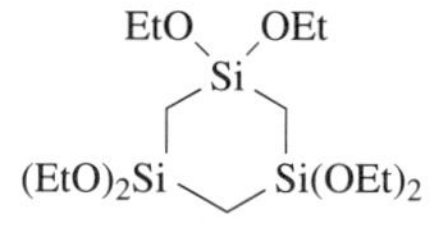

                    EtO   OEt   (EtO)₃Si   OEt
                       \ /   \ /
                        Si   Si
                       / \   / \
              (EtO)₂Si   Si(OEt)₂   (EtO)₂Si   Si(OEt)₂
                       (28) 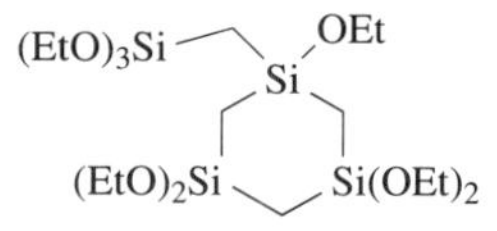   (29)

central methylene via Grignard reagents generated in the presence of TMS-Cl. For example 1,3,5-tri[bis(trimethylsilyl)methyl]benzene was prepared in 45% yield in a single step from 1,3,5-tri(dichloromethyl)benzene by treatment with a mixture of TMS-Cl and magnesium ⟨87CL2293⟩. Other systems have been prepared in a similar manner ⟨74JOM(71)377⟩. Di-Grignard reagents may be used to prepare 1,1-bis(silyl) alkanes. For example, the isolable di-Grignard reagent $CH_2(MgBr)_2$, prepared from dibromomethane and magnesium amalgam, was treated with excess TMS-Cl to give bis(trimethylsilyl)methane in 55% yield ⟨85JOM(288)13⟩. This magnesium reagent was not very reactive and several days were needed for complete reaction. Similarly $PhC(MgCl)_3$ may be treated with two equivalents of TMS-Cl and the remaining organomagnesium compound protonolysed to generate bis(trimethylsilyl)phenylmethane ⟨80JOM(201)197⟩.

Grignard reagents have been the most commonly used organometallic reagents for the generation of cyclic systems, of which 1,3-disilacyclobutanes have been the most frequently prepared. 1,1,3,3-Tetramethyl-1,3-disilacyclobutane (31) was first prepared by the reaction of magnesium with 1-chloro-4-fluoro-2,2,4-trimethyl-2,4-disilapentane (30) ⟨58JOC1392⟩. However, a multistep procedure was needed to generate the required starting material. The preparation could be simplified by using chloro(chloromethyl)dimethylsilane (32) ⟨64JOM(2)277⟩. The reaction is believed to proceed through a similar intermediate (33) which gave (31) in 20% overall yield (Scheme 6). The poor yields observed were due to substantial amounts of polymeric material which have since been reduced by changing the solvent and the order of addition of reactants—although yields still remain dependent upon the range of substituents incorporated ⟨64JOC1601⟩. A further improvement was observed with bromomethyl(chloro)dimethylsilane. This general method subsequently opened up the preparation of a large range of substituted 1,3-disilacyclobutanes which have been described elsewhere ⟨72MI 414-01, 80JOM(188)151⟩. Unsymmetrically substituted systems have been generated through the condensation of mixed halo(halomethyl)silanes, but the yields are poor being limited by the expected statistical distribution of possible products. By modification of the reagents and conditions used above to prepare 1,3-disilacyclobutanes, significant amounts of larger ring systems can be generated (Equation (8)). Replacing chloride by ethoxide sterically disfavoured ring closure to the four-membered ring enabling the addition of another equivalent of Grignard reagent before ring closure, thus generating the six-membered ring system (34). The remainder of the reactant was accounted for as being polymeric ⟨64JOC1601, 93TL2111⟩. Other ring systems have also been prepared ⟨69ZOB2281⟩. For example, the di-Grignard reagent of (35) was treated with tetrachlorosilane to give 1,1-dichloro-3,3-dimethyl-1,3-disilacyclohexane (36) in 62% yield (Equation (9)) ⟨64JOC1601⟩.

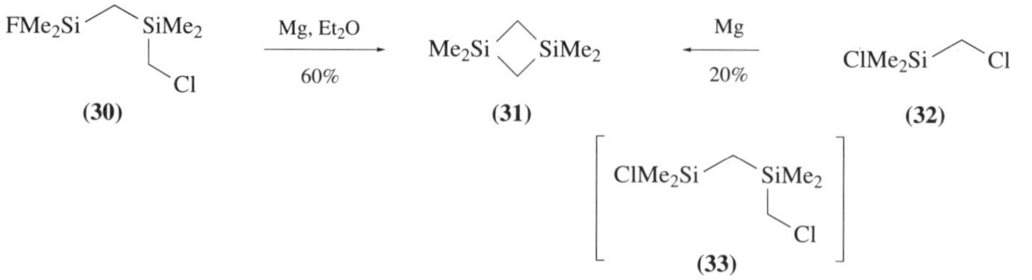

Scheme 6

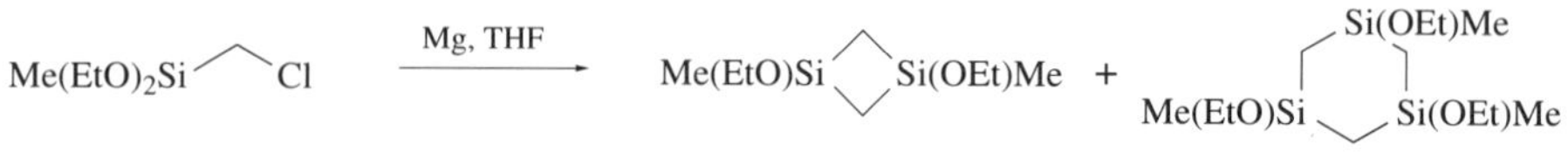

$$Me(EtO)_2Si\frown Cl \xrightarrow{Mg,\ THF} Me(EtO)Si\langle\rangle Si(OEt)Me\ +\ \text{[six-membered ring (34)]} \quad (8)$$

12%   (34) 40%

(c) Other organometallic intermediates. Organosodium derivatives have also been used to generate 1,1-bis(silyl) alkanes ⟨47JA2247, 56JA2274⟩. For example treatment of sodium in refluxing toluene with a mixture of chloromethyltriphenylsilane and chlorotriphenylsilane yielded bis(triphenylsilyl)methane in 52% yield along with some hexaphenyldisilane ⟨68CJC2119⟩. The latter was presumably generated by competing metallation of chlorotriphenylsilane. Polymeric material has been generated by the treatment of chloro(chloromethyl)dimethylsilane with sodium through the presumed intermediacy of an α-silyl organosodium intermediate ⟨49USP2483972, 52USP2607791⟩.

Copper may also be used to generate 1,1-bis(silyl) alkanes, such as the cyclisation of di(chloromethyl)disiloxane (**37**) to (**38**) by thermolysis over a silicon–copper amalgam (Equation (10)) ⟨72ZOB1361⟩. In a similar fashion, polymeric systems have been prepared by passing (chloromethyl)silanes over silicon and copper in the presence of HCl ⟨93OM2360⟩. Ring closure through the intermediacy of an organotitanium species has been reported ⟨81IZV1603⟩ and other cyclisations have been promoted by aluminum tribromide, an example of which is shown in Equation (11). The mechanism for this process has been investigated using polydeuterated materials showing that cleavage occurred at the bonds indicated (by arrows) as a result of polarisation of the Si—C bond by the Lewis acid ⟨83ZAAC(497)134, 88JOM(341)109⟩.

(ii) Quenching a silyl anion with a carbon electrophile

The preparation of 1,1-bis(silyl) alkanes using a silyl anion has been much less used than those involving carbanions described above. In one example dimethylphenylsilyllithium reacted with chloromethyl(dimethyl)phenylsilane to give the expected 1,1-bis(silyl) methane in 37% yield. This anion also reacted with dichloromethane to give the same product in slightly greater yield ⟨81ZAAC(473)59⟩. In another example, dimethylphenylsilyllithium reacted with the chlorosilane (**39**) to give two products (Equation (12)), showing that attack by the silyl anion occurred at both the silicon and carbon ⟨89OM1237⟩. The formation of disilanes could be minimised by reducing the reaction temperature and using the corresponding bromide as the nucleofugal group. Thus treatment of a suspension of lithium dispersion in ether–THF at −50 °C with TMS-Cl followed by dibromomethane gave bis(trimethylsilyl)methane in 55% yield, and only traces of hexamethyldisilane ⟨88SC85⟩.

(iii) From silenes or disilenes

Silenes are compounds containing a Si=C double bond while disilenes contain a Si=Si double bond. Both are generally highly reactive intermediates that are not isolated, although silenes may be studied by matrix isolation techniques. Disilenes have not been frequently used to generate 1,1-bis(silyl) alkanes unlike silenes which have been used extensively, particularly to prepare 1,3-disilacyclobutanes.

(a) Silenes. Pyrolysis of a silacyclobutane will generate an alkene and a silene through a cyclo-elimination process. Rapid dimerisation of the silene in a head-to-tail fashion can then form a 1,3-disilacyclobutane ⟨93JA3322⟩. For example, thermolysis of 1,1-dimethyl-1-silacyclobutane generated 1,3-disilacyclobutane (**31**) in high yield (Equation (13)). This method has been applied to a wide variety of substrates but clearly is dependent for its utility upon the availability of the silacyclobutane precursors ⟨72MI 414-02, 87BCJ2263⟩. Unsymmetrical 1,3-disilacyclobutanes have been prepared by the copyrolysis of a mixture of two different silacyclobutanes (Equation (14)), but this route will always generate a mixture of products and yields cannot better the statistical ratio of possible products ⟨80JOM(188)151⟩. It should be noted that one system that cannot be generated under these conditions is the unsubstituted (parent) system *viz.* 1,3-disilacyclobutane as only polymeric material is produced ⟨75JA7371⟩. However, it may be prepared by matrix isolation techniques: 1-silaethene (**41**; X = H) was isolated in an argon matrix at 10 K following the vacuum flash pyrolysis of the adduct (**40**; X = H) through a thermal cyclo-elimination process (Scheme 7). When the matrix was warmed, 1,3-disilacyclobutane (**42**; X = H) was formed ⟨81AG(E)597⟩. These conditions were also used to generate the tetrachloro derivative (**42**; X = Cl) ⟨72JOM(42)C21⟩. Cyclo-elimination of silenes from a number of other precursors has also been used to prepare 1,3-silacyclobutanes and these have been reviewed elsewhere ⟨79CRV529⟩.

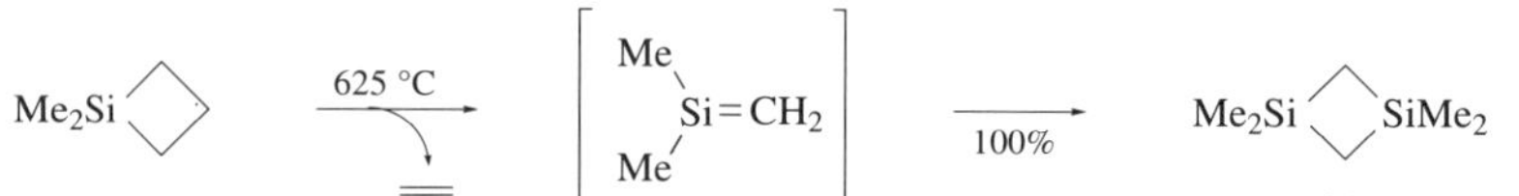

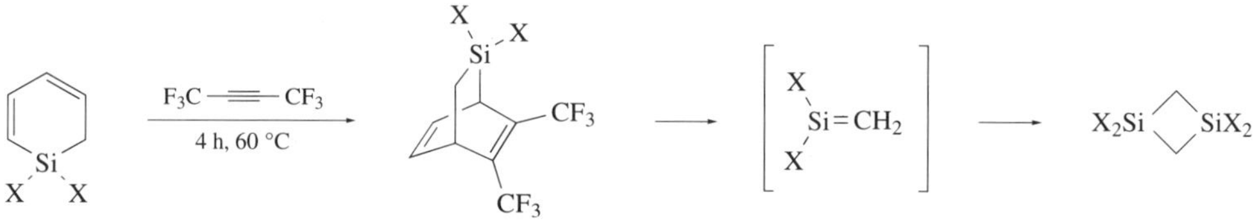

Scheme 7

Larger ring systems may also be generated through the intermediacy of silenes. For example, trimerisation of 1,1-dimethylsilaethene gave 1,1,3,3,5,5-hexamethyl-1,3-5-trisilacyclohexane ⟨79CRV529⟩. This product was also prepared from 1,1,3,3-tetramethyl-1,3-disilacyclobutane and zirconium tetrachloride; the reaction was again believed to involve 1,1-dimethylsilaethene ⟨85IZV1917⟩.

Polymeric carbosilanes have also been produced through the intermediacy of silenes generated from silacyclobutanes. This has proved possible under a variety of transition metal catalysts including platinum, palladium, iridium, ruthenium and gold ⟨65JOC2618, 66JCS(C)1137, 72MI8011⟩.

Pyrolysis of a silacyclobutane in the presence of methanol suppresses dimerisation to a 1,3-disilacyclobutane giving instead two acyclic products (Scheme 8). Trimethylmethoxysilane (**43**) is generated by addition of methanol to the silene intermediate and the bis(silyl) methane (**44**) by addition of (**43**) to a second equivalent of silene. Increased yields of (**44**) can be obtained by the pyrolysis of 1,1-dimethylsilacyclobutane in the presence of trimethylmethoxysilane ⟨81CC806⟩. Later experiments have shown that trimethylmethoxysilane can also trap more highly substituted silenes ⟨83JA6725, 85OM581⟩. For example, thermolysis of (**45**) generated 1-methyl-1-phenyl-2-neopentylsilaethene (**46**) which in the presence of trimethymethoxysilane formed (**47**) (Scheme 9). With either (*E*)- or (*Z*)-(**45**) it was shown that the intermediate silene is configurationally stable generating only one diastereoisomer of (**47**) when reacting with trimethylmethoxysilane through a stereospecific *syn*-addition ⟨83JA6725⟩. 1,1-Dimethylsilaethene has been trapped with several halosilanes, for example, with trichlorosilane, it yielded 3-methyl-1,1,3-trichloro-1,3-disilabutane in 22% yield ⟨74JA7105⟩. When the trapping agent used is the cyclic siloxane (**48**), then insertion into the ring generates (**49**) in reasonable yield (Equation (15)) ⟨75JA7371⟩. The reverse process has also been observed; cyclocarbosiloxanes upon heating can extrude silenes to produce systems of reduced ring size ⟨79CRV529⟩.

Scheme 8

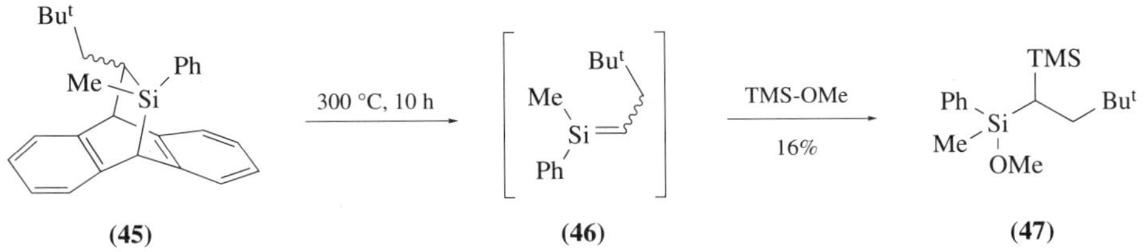

Scheme 9

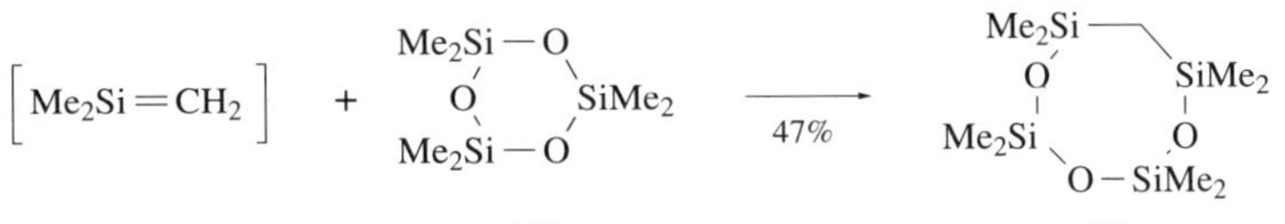

Silenes have also been prepared following addition of *t*-butyllithium to an α-chlorovinylsilane after elimination of lithium chloride. The first example of this as a route to 1,3-disilacyclobutanes appeared in 1977 ⟨77JA2013⟩. It was anticipated that treatment of chloro(dimethyl)vinylsilane (**50**; X = Me) with *t*-butyllithium would yield *t*-butyldimethyl(vinyl)silane following displacement of the chloride. However, 1,1,3,3-tetramethyl-2,4-dineopentyl-1,3-disilacyclobutane (**53**; X = Me) was generated in 46% yield (Scheme 10). Similarly, *t*-butyllithium reacted with chloro(methyl)phenyl(vinyl)silane (**50**; X = Ph) to yield a mixture of separable isomers (**53**; X = Ph) ⟨77JA2013, 82JOM(232)33⟩. When the reaction was repeated in the presence of trimethylmethoxysilane or methoxytriphenylsilane, cyclisation was inhibited and the two diastereomeric products (**54**) and (**55**) were formed instead ⟨85OM581, 86JA3122⟩. Treatment of a hexane solution of (**50**; X = Me) with *t*-butyllithium in the presence of 2,3-dimethyl-1,3-butadiene generated the anticipated [2 + 4] cycloadduct (**56**). However if the solvent was changed to tetrahydrofuran (THF) then none of this adduct was generated, but the 1,1-bis(silyl) alkane (**57**) was formed in good yield. These, and other results, led to the proposal that the silene intermediate (**51**) is only generated in hydrocarbon solvents. In THF, the reactive intermediate was believed to be the α-lithiosilane (**52**) formed by addition of *t*-butyllithium to the vinylic bond of the silane (**50**) without elimination of lithium chloride (Scheme 8) ⟨77JA8447, 80JA4970, 92ZN(B)805⟩. This preparation of 1,1-bis(silyl)alkanes through the intermediacy of silenes has since been a much studied reaction ⟨88ZAAC(558)55, 90JOM(393)33, 93CB575⟩.

A similar process of addition of *t*-butyllithium to a chlorovinylsilane and the subsequent generation of a 1,3-disilacyclobutane has been demonstrated when the silicon was bound to an iron complex. It was not unambiguously established whether the reaction proceeded through an intermediate silene or whether the intermediate was an α-lithiosilylalkane ⟨89JOM(363)7⟩.

Silene (**59**) was generated by treatment of the chlorovinylsilane (**58**) with *t*-butyllithium and could be subsequently converted into Diels–Alder products with dienes, but favoured an intramolecular reaction yielding (**60**) following migration of the trimethylsilyl group from nitrogen to carbon (Scheme 11) ⟨92ZN(B)1377⟩. Diels–Alder reactions of silenes have also been used to prepare 1,1-bis(silyl) alkanes ⟨92ZN(B)805⟩. A radical process was proposed for a related process that transformed the α-silyl silene (**61**) into the Diels–Alder adduct (**62**) and an isomer (**63**). The silene was

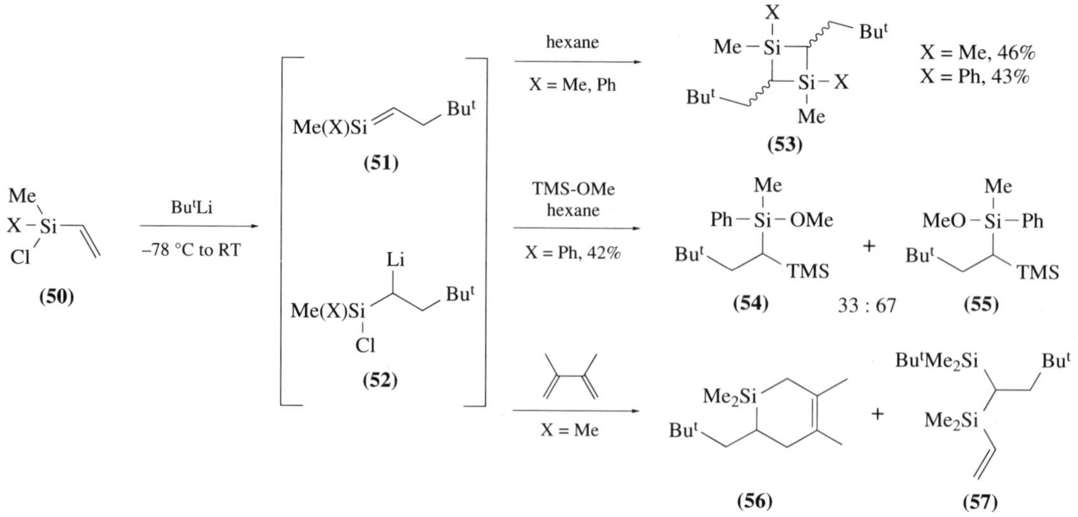

Scheme 10

Solvent	Yield (56) (%)	Yield (57) (%)
hexane	45	0
THF	0	71

generated by thermolysis of bis(trimethylsilyl)diazomethane (Scheme 12). This α-silyl silene may also be generated photochemically and subsequently trapped with D_2O to give the disiloxane (64) ⟨80JA1584⟩.

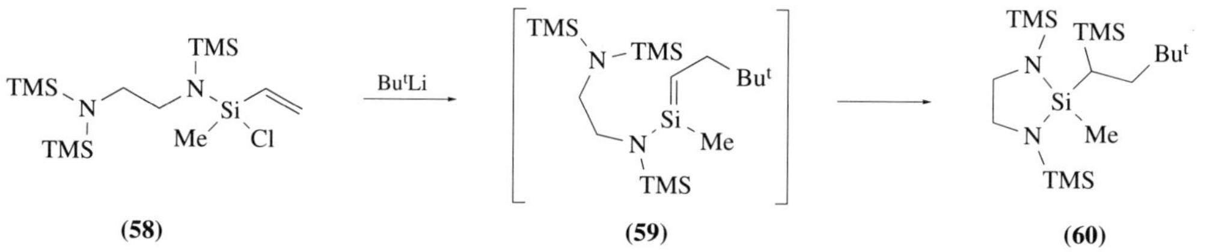

Scheme 11

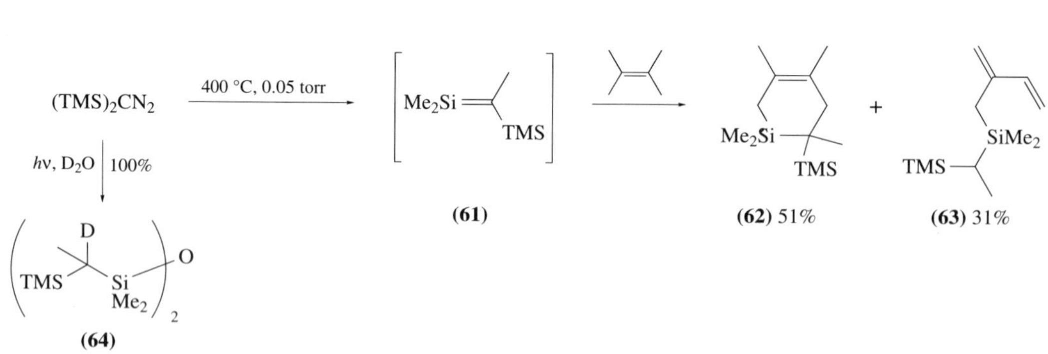

Scheme 12

Other silenes have also been prepared through photolysis, such as (66) which was generated by irradiation of the 1,2-disilacyclohexadiene (65). This subsequently isomerised to the 1,3-disilacyclobutane (67) which itself could be reversibly transformed into (68) (Scheme 13) ⟨74JA5623, 76JA7424⟩. In another example of photolysis generating a silene, extrusion of nitrogen from (69) generated (70) which was subsequently trapped with t-butanol to give (71) (Scheme 14) ⟨87OM1857⟩. Other photocatalysed reactions to produce 1,1-bis(silyl) alkanes that proceed through silenes have been reported ⟨91OM2695⟩.

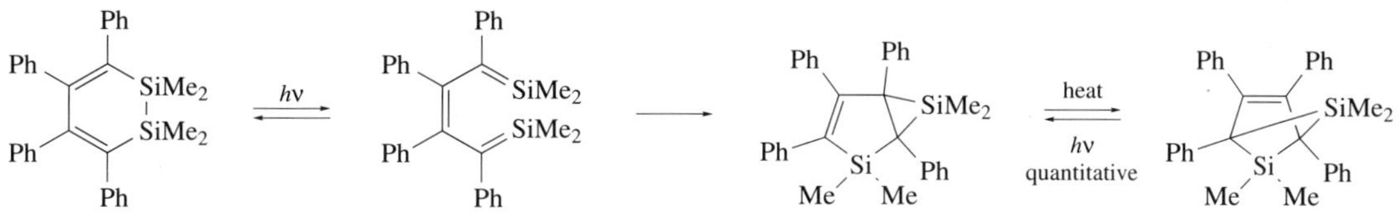

(65) **(66)** **(67)** **(68)**

Scheme 13

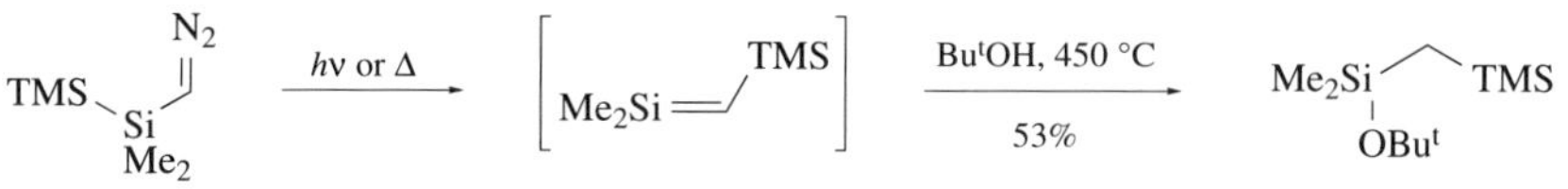

(69) **(70)** **(71)**

Scheme 14

Thermal or photocatalysed 1,3-silyl migrations of alkenyl disilanes may also lead to the generation of silenes, which again, in the absence of any trapping agents will form 1,3-disilacyclobutanes ⟨79JOM(168)23⟩. For example, photolysis of 1-phenyl-2-vinyltetramethyldisilane (**72**) generated the silene (**73**), which readily dimerised to afford an almost equal mixture of *cis*- and *trans*-1,3-disilacyclobutanes (Scheme 15) ⟨78JOM(149)37⟩. The cumulene (**74**), a stable silene equivalent, yielded the [2 + 2] cyclo-adduct (**75**) upon reaction with hexafluoro-2-butyne and a [2 + 4] cyclo-adduct (**76**) following reaction with 2,3-dimethyl-1,3-butadiene (Scheme 16) ⟨88CL1441⟩. Other stable silenes have been prepared and subsequently treated with Grignard reagents to give 1,1-bis(silyl) alkanes (Equation (16)) ⟨91OM3292⟩.

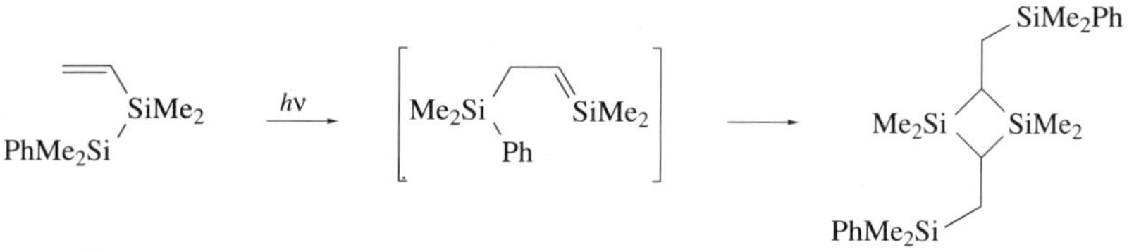

(72) **(73)** *cis*, 21% *trans*, 22%

Scheme 15

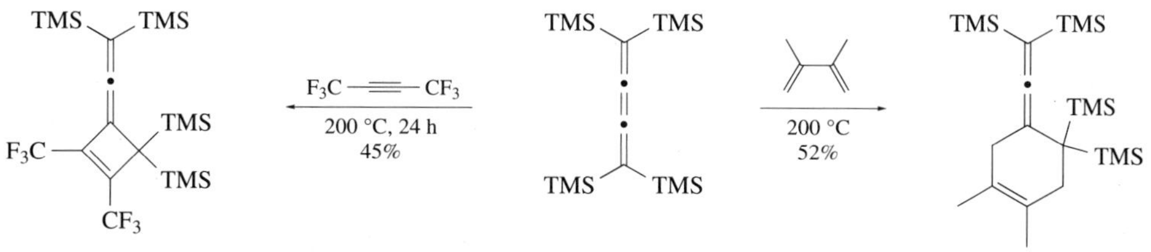

(75) **(74)** **(76)**

Scheme 16

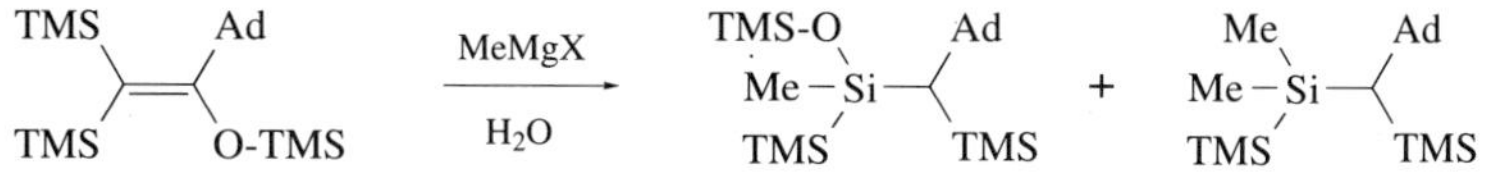 (16)

Ad = adamantyl

pentanes	10	:	1
THF	1	:	20

Many other silenes have been implicated as intermediates during the preparation of 1,1-bis(silyl) alkanes and these have been described in various reviews ⟨79CRV529, 80TL429, 81JA6788⟩.

(b) Disilenes. Disilenes have been used as intermediates in the preparation of 1,1-bis(silyl) alkanes ⟨93OM289⟩. Like silenes, these can be formed by extrusion under thermolytic conditions ⟨72JA5837⟩. For example, heating **(77)** generated disilene **(78)** which gave a number of products, the major ones being the 1,3-disilacyclobutanes **(79)** and **(80)**. An alternative route to this silylene intermediate was through thermolysis of 2-chloroheptamethyltrisilane **(81)**. Again, the same two 1,3-disilacyclobutanes were generated in similar yields (Scheme 17) ⟨78JA6236⟩. If the pyrolysis was performed in the presence of 2-butyne then the 1,3-disilacyclopentene **(82)** was generated, while in the presence of trimethylsilane, the product was 1,1-bis(dimethylsilyl)methane **(83)** as a result of reaction via **(78)** ⟨76JA7746, 78JA6236⟩.

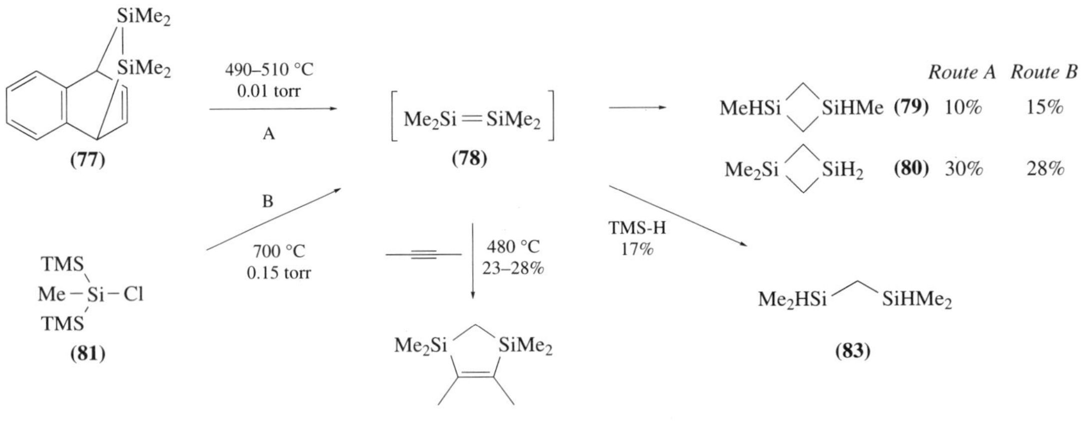

Scheme 17

(iv) Rearrangements

There have been many reported examples of the generation of 1,1-bis(silyl) alkanes through rearrangements (those that proceed through the intermediacy of silenes were discussed above). The most common precursors for such rearrangements are disilanes. Thus treatment of chloromethylpentamethyldisilane **(84**; X = Me) with a catalytic quantity of anhydrous aluminum chloride resulted in a vigorous, exothermic reaction to give 2-chloro-2,4,4-trimethyl-2,4-disilapentane **(85**; X = Me) in 82% yield (Scheme 18). This 1,2-silyl shift will also occur when the migrating group is chlorodimethylsilyl to give **(85**; X = Cl) ⟨58JOC292⟩. A 1,2-silyl shift was also observed when **(84**; X = Me) was treated with an alkoxide; however, the use of an iodide, amine or a thiol resulted only in the displacement of chloride ⟨66JOM(5)226⟩. Treatment of **(84**; X = Me) with potassium acetate generated the analogous product **(87)** although here the reaction was demonstrated to occur in two steps via the disilane **(86)** (Scheme 18). Silver oxide has also been shown to induce a 1,2-silyl shift of a disilane ⟨66JOM(5)226⟩.

A very similar rearrangement has been described through what is formally a 1,1-dyotropic rearrangement ⟨74JOM(78)C35, 77AOC(16)33⟩. Photolysis of **(88)** resulted in a 1,2-silyl shift with concomitant 1,2-migration of the iron complex to give the 1,1-bis(silyl) methylene complex **(89)** (Equation (17)). In another photo-induced migration, the styryl disilanes **(90)** were transformed through the presumed intermediacy of silacyclopropanes **(91)** to substituted 1,1-bis(silyl) alkanes **(92)** (Scheme 19) ⟨76JA7424⟩. An alternative reaction course was observed following irradiation of 2-phenylethenylpentamethyldisilane **(93)** when 1,1-dimethyl-2,3-benzo-5-trimethylsilyl-1-silacyclopentene **(94)** was generated (Scheme 20) ⟨78JOM(149)37, 93OM2536⟩.

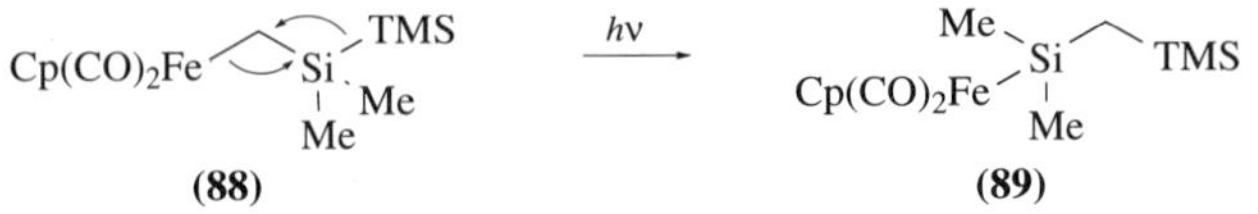

$$\text{Cp(CO)}_2\text{Fe}\diagdown\text{Si(Me)}_2\text{-TMS} \xrightarrow{h\nu} \text{Cp(CO)}_2\text{Fe-Si(Me)}_2\text{-CH}_2\text{-TMS} \quad (17)$$

(88) → **(89)**

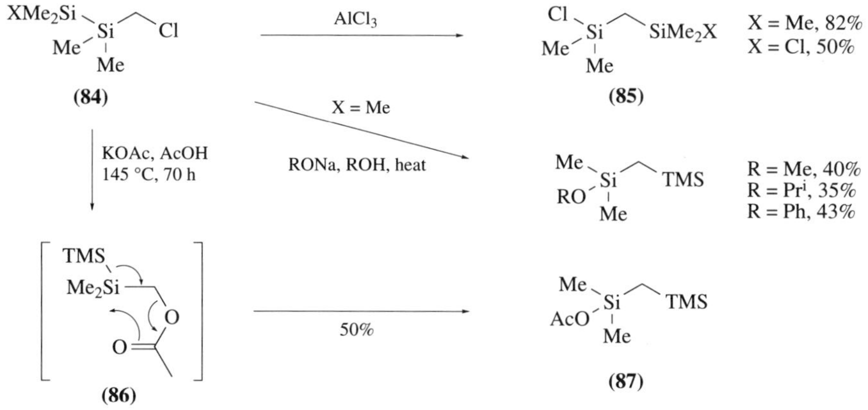

Scheme 18

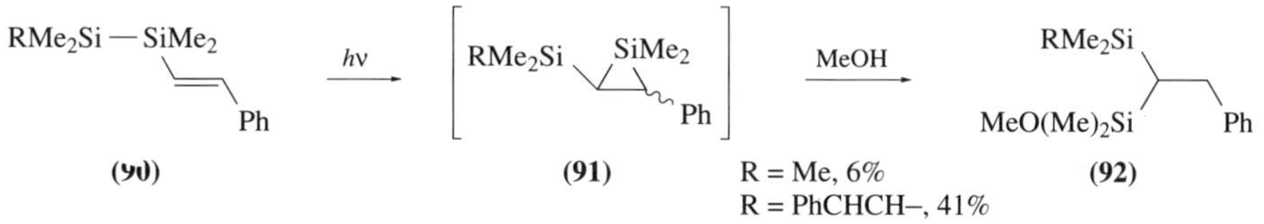

Scheme 19

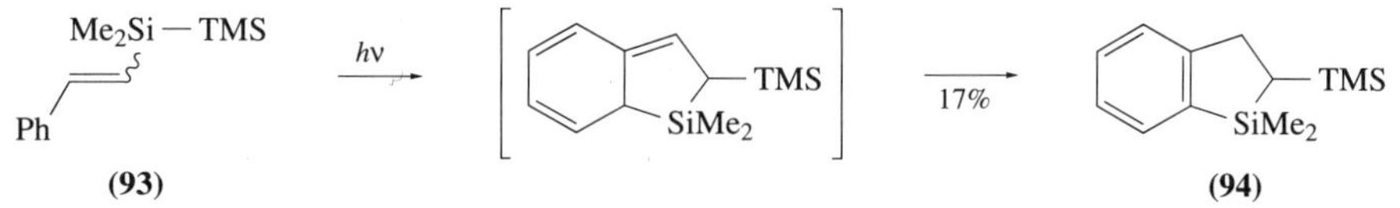

Scheme 20

In a further example of rearrangements from a disilane, a five-coordinate silicon intermediate (**95**) was proposed during the preparation of (**97**) through a 1,2-shift of a trimethylsilyl group from the pentavalent silicon to an adjacent carbon (Scheme 21). Evidence supporting this route included the detection by NMR of the intermediate anion (**96**) ⟨83JOM(250)109⟩. This process has also been demonstrated when a silyl anion was used to generate what was believed to be a similar pentavalent silicon intermediate ⟨83OM351⟩.

Migration of silicon from oxygen to carbon may also occur. For example, thermolysis of the

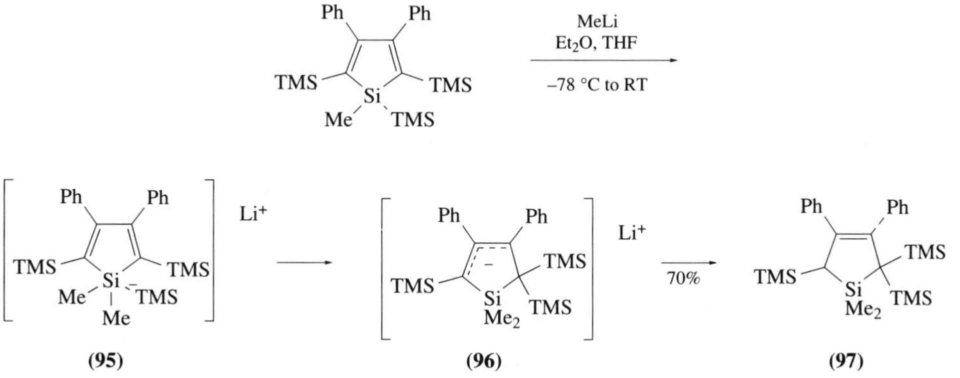

Scheme 21

cyclopropyl silyl ether (**98**) resulted in the formation of the ring-opened 1,1-bis(silyl) derivative (**99**) (Scheme 22) ⟨79ZOB2785⟩.

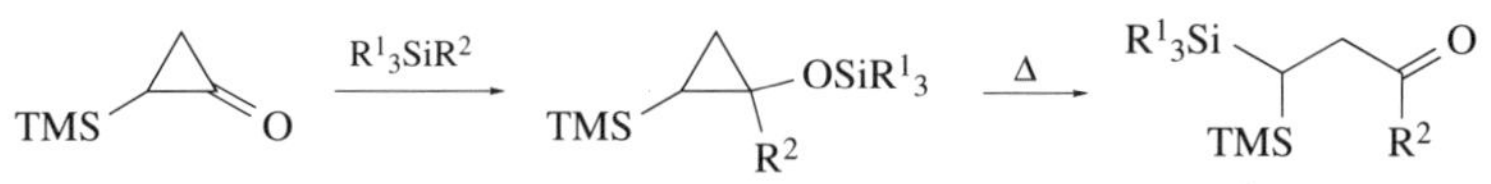

R^1 = Me, Et, Bu; R^2 = OMe, Et$_3$N, Me$_2$N

Scheme 22

1,3-Silyl-allyl rearrangements have also resulted in the generation of 1,1-bis(silyl) alkanes. Thus treatment of 1,1-dimethyl-2,5-diphenyl-1-silacyclohex-3,5-diene (**100**) with butyllithium followed by TMS-Cl gave exclusively (**101**) which upon heating to 170 °C for 38 h rearranged to give (**103**) (Scheme 23) ⟨81JA6788, 86BCJ1509⟩.

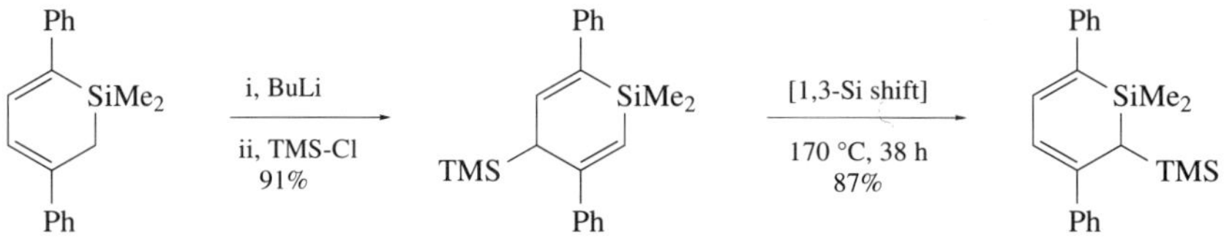

Scheme 23

Another example of a rearrangement generating bis(silyl) alkanes was shown when a ferrocene-substituted 1,2-disilylethane was transformed into the cyclic 1,1-bis(silyl) alkane (**104**) under thermal conditions. This rearrangement was believed to occur via the betaine (**103**) (Scheme 24) ⟨69CC207⟩. Other rearrangements involving transition metals are known. For example, the transfer to butadiene of the disilyl ligand on molybdenum complex (**105**) resulted in the formation of the 1,3-disilacyclopentene (**106**) (Equation (18)). The analogous iron complex, and various substituted dienes have also been used ⟨87OM1861⟩. Another 1,3-disilacyclopentene, (**107**) was generated following a tungsten-catalysed rearrangement of a silacyclopropene ⟨91JOM(407)157⟩.

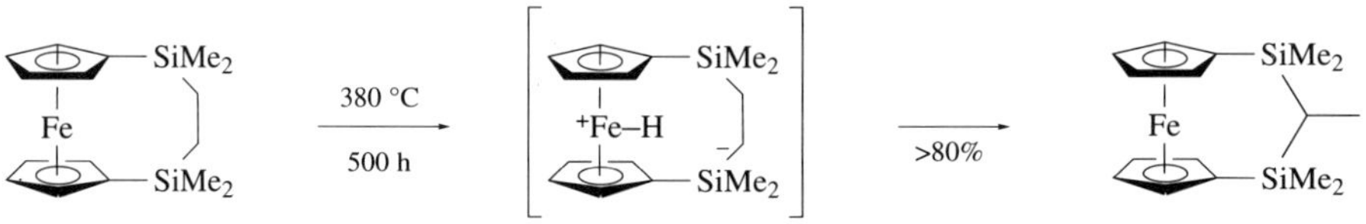

Scheme 24

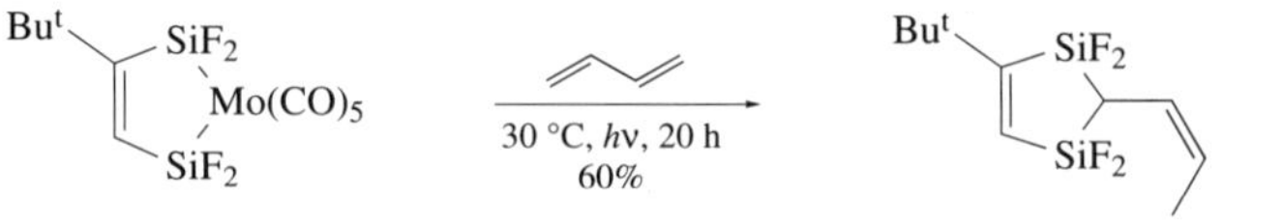

(18)

Rearrangements to generate 1,1-bis(silyl) alkanes that proceed through the intermediacy of radicals have been described ⟨80JOM(201)197⟩. For example, thermolysis of hexamethyldisilane at 450 °C for 1 h gave 2,2,4-trimethyl-2,4-disilapentane in quantitative yield ⟨68CC930⟩. Later experiments used xylene as a solvent and demonstrated that a range of products could be formed depending upon the reaction conditions. A radical nonchain sequence has been proposed for the process

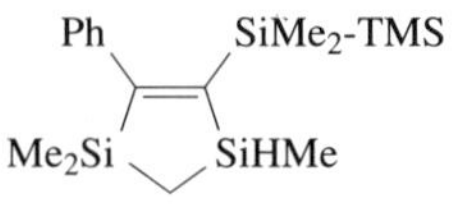

⟨73CC323⟩. In a similar fashion, a large number of polycyclic carbosilanes have been prepared by thermolysis of tetramethylsilanes ⟨84ZAAC(512)103⟩. Other pyrolyses to give materials containing a Si—C—Si linkage have also been reported ⟨61JA3345, 62ZAAC(315)157, 67AG(E)677⟩.

A radical sequence was proposed for the photolytic transformation of 1,2-bis(1-naphthyl)tetramethyldisilane into (**108**) (Equation (19)) ⟨91OM2695⟩. Equation (20) shows a second example where 1,2-diallyl-1,1,2,2-tetramethyldisilane (**109**) was transformed into 1,1,3,3-tetramethyl-1,3-disilacyclopentene (**110**). A radical mechanism has been described to account for this and several related transformations ⟨80JA7979⟩.

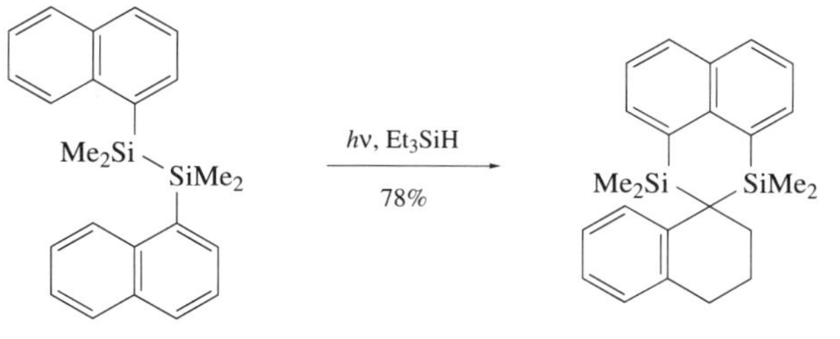

(19)

(108)

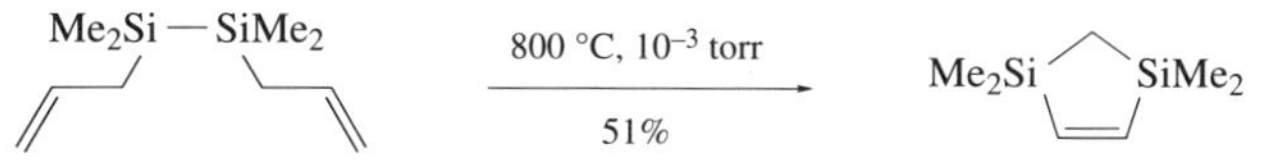

(20)

(109) **(110)**

(*v*) *Hydrosilylation*

Hydrosilylation of a vinylsilane under rhodium catalysis has been reported, however this gave the corresponding 1,1-bis(silyl) ethane as the minor product favouring instead the production of a 1,2-bis(silyl) alkane ⟨87IZV1424⟩. The platinum-catalysed dihydrosilylation of the alkyne (**111**) was used to form a mixture of 1,1-bis(silyl) alkanes (Equation (21)) ⟨87TL3719⟩. Hydrosilylation of 1,3-disilacyclobutanes have also been shown to generate new 1,1-bis(silyl) alkanes. For example, treatment of 1,1,3,3-tetramethyldisilacyclobutane (**31**) with trichlorosilane in the presence of a catalytic amount of hexachloroplatinic(IV) acid generated a new 1,1-bis(silyl) methyl species (**112**) (Scheme 25). The formation of this product could conceivably occur through either chlorine or hydrogen exchange. When trichlorosilane is used the dominant reaction pathway is via chlorine exchange. Some hydrogen exchange can take place if a chlorine is replaced by an electron-releasing group such as a methyl. Products arising from both chlorine exchange (**113**) and hydrogen exchange (**114**) are then observed although the former still dominates ⟨75JCS(D)1832⟩. Platinum on carbon has also been used as the catalyst ⟨87BCJ2263⟩.

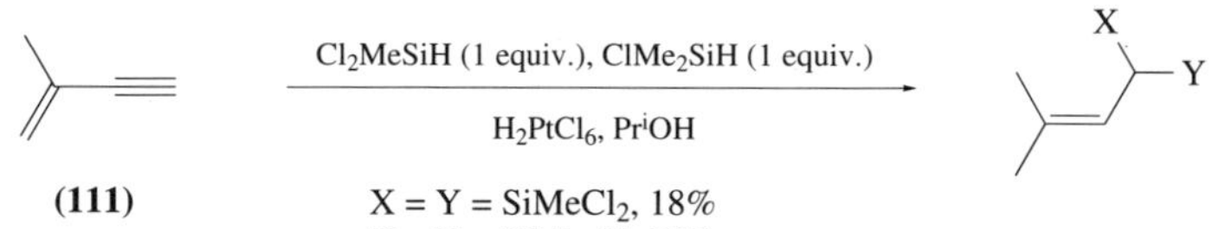

(21)

(111)

X = Y = $SiMeCl_2$, 18%
X = Y = $SiMe_2Cl$, 29%
X = $SiMeCl_2$, Y = $SiMe_2Cl$, 53%

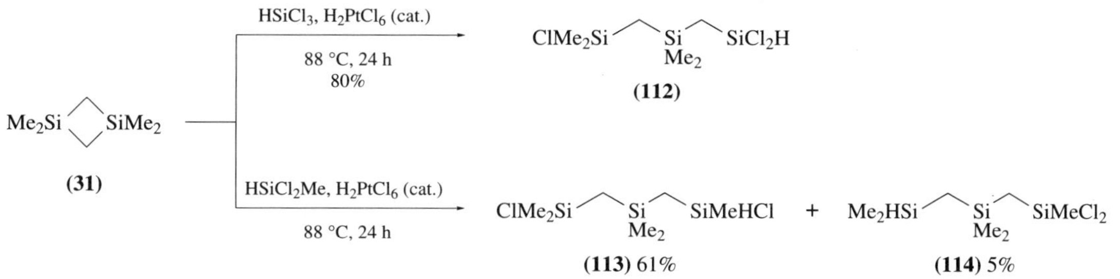

(112) **(31)** **(113)** 61% **(114)** 5%

Scheme 25

The dehydrogenative double silylation of alkenes under platinum catalysis has also been shown to generate 1,1-bis(silyl) alkanes in good yields. For example, treatment of styrene with 1,2-bis(dimethylsilyl)benzene in the presence of a platinum catalyst gave (**115**) in high yield (Equation (22)). Mechanisms have been proposed for the reaction ⟨92JOM(428)1⟩.

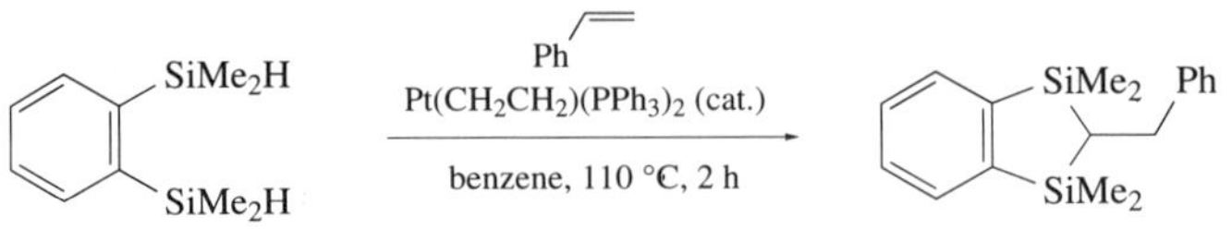

$$\text{(22)}$$

(vi) Other routes

Many routes to the generation of 1,1-bis(silyl) alkanes involve the replacement of functionality on the central methylene by either protons or carbon fragments. For example, metal–halogen exchange has been used to generate α-metallated 1,1-bis(silyl) alkanes which may then be further functionalised. Thus treatment of diiododi(trimethylsilyl)methane with zinc in deuteriomethanol gave the corresponding dideuterio derivative ⟨88POL2023⟩. These dimetallated intermediates have also been isolated, such as dilithiodi(trimethylsilyl)methane which was prepared by treatment of dichlorobis(trimethylsilyl)methane with lithium metal vapour. This was also converted to the dideuterio derivative upon quenching with D_2O ⟨84CC1664⟩. The dilithiated intermediate was also prepared with lithium 4,4′-di-*t*-butylbiphenyl, whereupon it was quenched with a variety of carbon electrophiles ⟨88TL5237, 89TL6195⟩.

The corresponding di-Grignard reagents have also been prepared, such as $(TMS)_2C(MgBr)_2$ which was generated by treatment of dibromodi(trimethylsilyl)methane with magnesium ⟨89TL6195⟩. Unlike the dilithio intermediate described above, which is not stable in THF at room temperature, the di-Grignard reagent can be isolated, and only reacts slowly at room temperature even with strong electrophiles.

1,1-Bis(silyl) alkanes may be generated by the addition of organolithium reagents to 1,1-bis(silyl) alkenes (**116**) in a fashion similar to that described during the preparation of α-lithio silyl alkanes (see 4.14.1.1.1(i)(a)) (Equation (23)) ⟨74AC(E)83, 77LA830⟩.

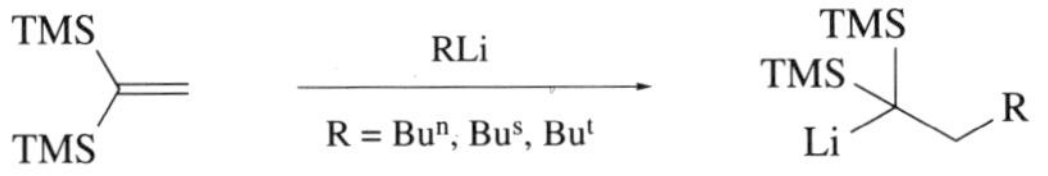

$$\text{(23)}$$

Radical reactions have also generated 1,1-bis(silyl)-1-lithio alkanes. For example, treatment of the 1,1-bis(silyl) ethylene (**117**) with lithium metal gave the dianion (**118**) which was subsequently quenched with a range of electrophiles (Scheme 26) ⟨88TL6939⟩. Similarly, treatment of tetrakis(trimethylsilyl)ethene with lithium gave the diorganolithium derivative (**119**) which has been protonated ⟨89JA3748⟩. Slow, intramolecular rearrangement of (**119**) was found to give a number of 1,1-bis(silyl) alkanes ⟨93CL267⟩. In another example, reductive lithiation of 1,1-bis(trimethylsilyl)-1-phenylthioalkanes with lithium naphthalide gave the corresponding bis(silyl) anion (**120**) which could be further functionalised (Equation (24)) ⟨84JOC168⟩.

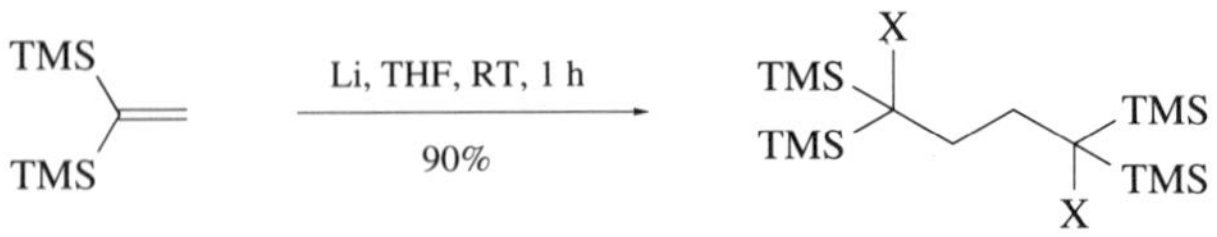

E = H, 96%, Me, 83%, MOM, 73%

MOM = methoxymethyl

Scheme 26

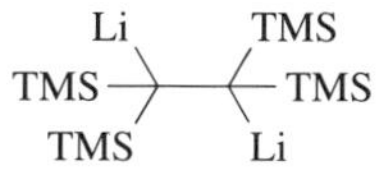

(119)

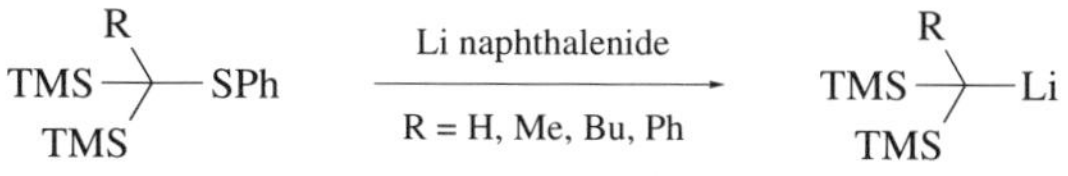

$$\text{TMS} \underset{\text{TMS}}{\overset{\text{Li}}{\diagup}} \underset{\text{Li}}{\overset{\text{TMS}}{\diagdown}} \text{TMS}$$

$$\underset{\text{TMS}}{\overset{R}{\diagup}}\text{SPh} \xrightarrow[R = H, Me, Bu, Ph]{\text{Li naphthalenide}} \underset{\text{TMS}}{\overset{R}{\diagup}}\text{Li} \quad (24)$$

(120)

Other routes which do not proceed through the intermediacy of α-lithio-1,1-bis(silyl) alkanes have also been demonstrated. For example, hydride reduction of dichlorobis(trimethylsilyl)methane gave a mixture of the monochloro derivative ((TMS)$_2$CHCl) and bis(trimethylsilyl)methane ⟨72JOC2662⟩. The monochloro derivative has also been reduced with LAH to bis(trimethylsilyl)methane ⟨80JA1584⟩.

1,1-Disilylated epoxides have also been transformed into 1,1-bis(silyl) alkanes. For example treatment of (121) with hexylmagnesium chloride in ether gave the β-hydroxysilane (122; R = hexyl), while the analogue (122; R = Ph) was prepared from (123) by hydride reduction (Scheme 27) ⟨89JOC5613⟩. Another epoxide was implicated in the transformation of 1,1,4,4-tetrakis(trimethylsilyl)butatriene (124) into the α,α-disilyl ketone (125). This triene was also transformed into bis(silyl) alkanes through both hydrogenation and hydrosilylation (Scheme 28) ⟨87TL1811⟩. A Diels–Alder reaction was used during the preparation of the bis-silylated norbornene (126) from 1,1-bis(trichlorosilyl)ethene and cyclopentadiene ⟨89JA4127⟩. Bis(silyl) ketenes may be precursors for the generation of 1,1-bis(silyl) alkanes, such as the formation of (128) from (127) (Equation (25)) ⟨70ZOB939⟩.

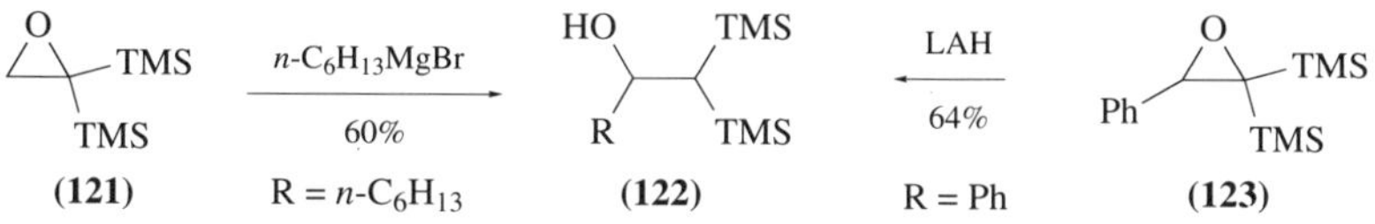

Scheme 27

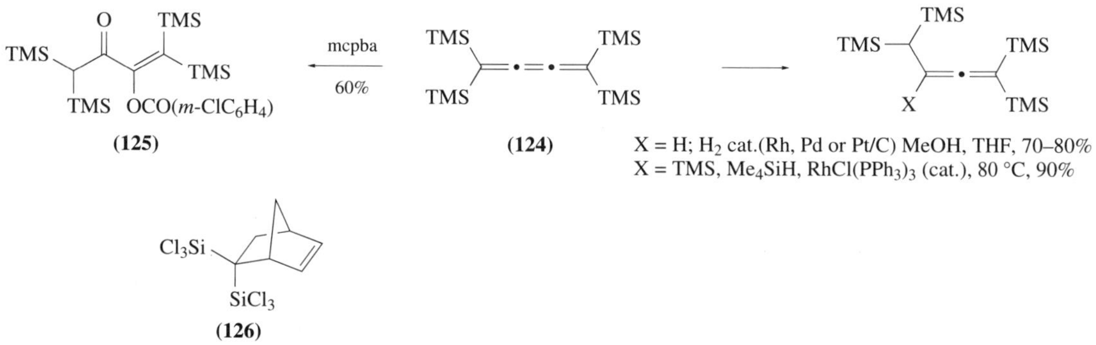

Scheme 28

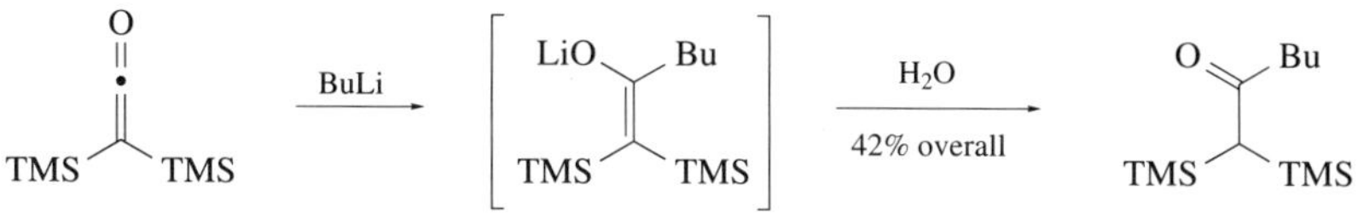

$$\underset{(127)}{\overset{O}{\underset{\text{TMS}\ \ \text{TMS}}{\|}}} \xrightarrow{\text{BuLi}} \left[\underset{\text{TMS}\ \ \text{TMS}}{\overset{\text{LiO}\diagdown\ \diagup\text{Bu}}{}}\right] \xrightarrow[42\% \text{ overall}]{H_2O} \underset{(128)}{\overset{O\diagdown\ \diagup\text{Bu}}{\underset{\text{TMS}\ \ \text{TMS}}{}}} \quad (25)$$

1,1,1-Tris(trimethylsilyl) alkanes have been transformed into 1,1-bis(silyl) alkanes by the replacement of one of the silyl groups by either a proton or an alkyl group. The cleavage of one of the C—Si bonds has been accomplished using base. Thus, treatment of tris(trimethylsilyl)methane with sodium methoxide in hexamethylphosphoramide (HMPA) gave, after protonation, bis(trimethylsilyl)methane in 83% yield ⟨73TL4193⟩. This cleavage was also accomplished during the migration of a silyl group from carbon to oxygen. Deprotonation of 3,3,3-tris(trimethylsilyl)propanol with sodium hydride gave the alkoxide (**129**) which underwent a 1,4-silyl shift to give the silyl ether (**130**) after protonation (Equation (26)). The 1,4-silyl shift is facilitated by stabilisation of the carbanion by the two remaining silyl groups. The corresponding 1,6-migration of the trimethylsilyl group has also been demonstrated ⟨81JCS(P1)969⟩. A similar rearrangement was observed following the addition tris(trimethylsilyl)methyllithium to styrene oxide via (**131**) (Scheme 29). The adduct (**131**) could react with another molecule of styrene oxide to give (**132**) following protonation, or by an intramolecular displacement to give the bis(silyl) cyclopropane (**133**) ⟨81JCS(P1)969⟩.

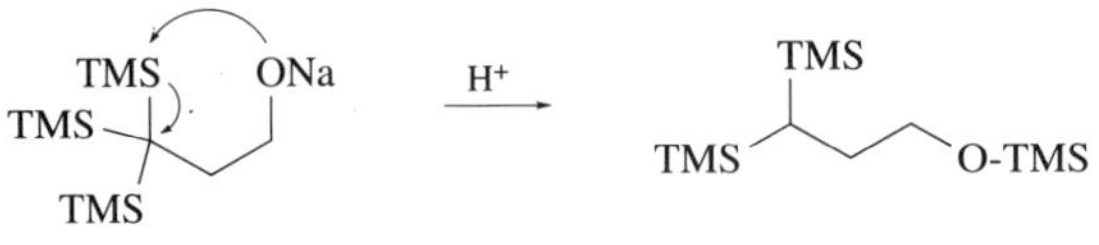

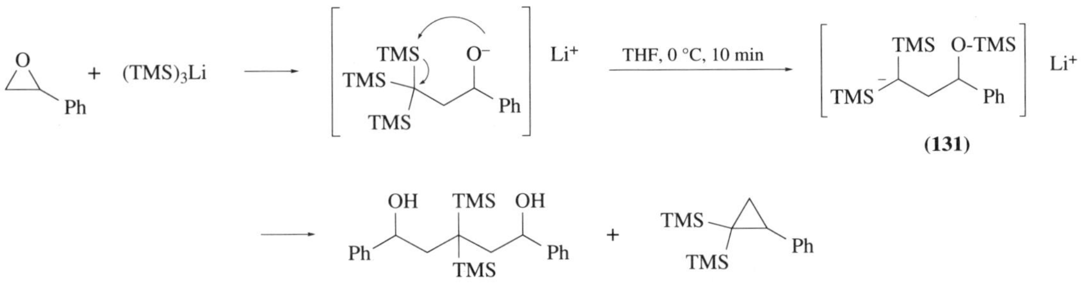

Scheme 29

4.14.1.1.2 Changing the groups attached to the central methylene

New 1,1-bis(silyl) alkanes may be generated by manipulation of the central methylene group of an existing 1,1-bis(silyl)methane. The simplest route to such a change is through metallation followed by an electrophilic quench ⟨74AG(E)83, 87CB1695⟩. For example, treatment of 2-methoxy-2,4,4-trimethyl-2,4-disilapentane with *t*-butyllithium gave the monodeuteriated product (**135**) in 98% yield upon quenching with D_2O (Equation (27)) ⟨89JOC1784⟩. Similarly, bis(trimethylsilyl)methane has been metallated with *t*-butyllithium ⟨77CB852, 89JOC5613⟩. The most effective base for deprotonation of the central methylene group is *n*-butyllithium–potassium *t*-butoxide ('superbase'). Bis(trimethylsilyl)methane was metallated at $-70\,°C$ more completely and more rapidly than with *n*-butyllithium-TMEDA. Polymeric material may also be metallated but only one in every four methylenes could be deprotonated, even with up to five equivalents of 'superbase' ⟨91OM551⟩.

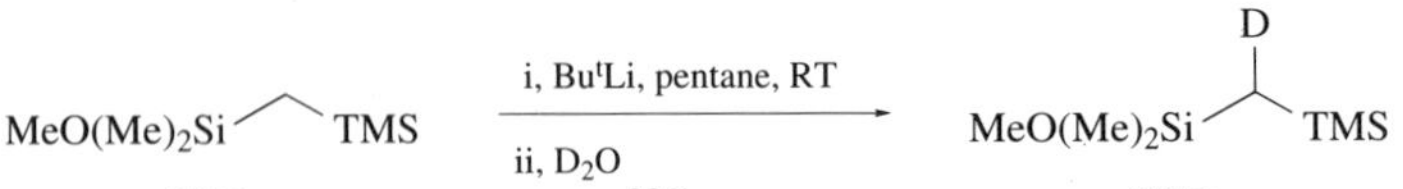

Cyclic systems can be metallated, although care has to be taken in choosing the base as 1,3-disilacyclobutanes are more susceptible to ring opening if challenged with a nucleophilic base (see 4.14.1.1.1(i)(a)). However with the correct choice of base and conditions, efficient metallation can be performed. *t*-Butyllithium-TMEDA has been used to metallate 1,1,3,3-tetramethyl-1,3-disilacyclobutane in high yield. An electrophilic quench can then furnish a more highly substituted system. Only one of the ring methylenes will be deprotonated, and this is also true of 1,1,3,3,5,5-hexamethyl-1,3,5-trisilacyclohexane which was also metallated under similar conditions. 'Superbase' has also been used ⟨90OM2677, 91OM551⟩.

4.14.1.1.3 Changing the ligands on silicon

The preparation of 1,1-bis(silyl) alkanes need not be restricted to the formation of one (or both) of the C—Si bonds, or to the modification of substituents on the central methylene of an existing system. Silicon-based substitution reactions on 1,1-bis(silyl) alkanes have also been used to generate new systems. Clearly, many reactions are known which will change the ligands on silicon and only those which have been demonstrated on 1,1-bis(silyl)alkanes will be considered here.

It was shown earlier that a chlorine bound to silicon can be displaced by a nucleophilic carbon (see 4.14.1.1.1(i)). This is the most common process by which an existing 1,1-bis(silyl) alkane may be further functionalised. For example, bis[dimethyl(9-anthryl)silyl]methane was prepared by the reaction of 9-anthryllithium with bis(chlorodimethylsilyl)methane ⟨89JCR(S)146⟩. Similarly, the 1,4-disilacyclopentane (136) was prepared by the treatment of 2,3-dilithio-1,3-butadiene with bis(chlorodimethylsilyl)methane for subsequent use in a Diels–Alder reaction ⟨93JA6625⟩. Grignard reagents may be used in a similar fashion ⟨70ZAAC(372)59, 89JA4127⟩. This methodology has also been applied to cyclic systems. 1,3-Dichloro-1,3-dimethyl-1,3-disilacyclobutane (137) has been transformed into a variety of substituted derivatives through treatment with organometallics (Equation (28)) ⟨80JOM(188)151⟩. Other halides may also be displaced, for example, 1-methoxyvinyllithium has been shown to displace a fluoride bound to silicon ⟨92CB607⟩.

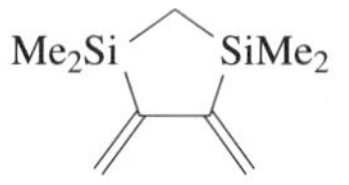

(136)

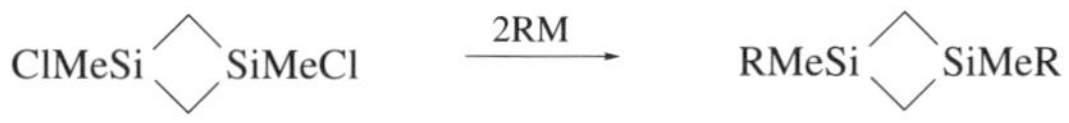

$$\text{ClMeSi} \diamondsuit \text{SiMeCl} \xrightarrow{2RM} \text{RMeSi} \diamondsuit \text{SiMeR} \quad (28)$$

(137)

M	R	Yield (%)
MgCl	Ph	50
MgI	Me	85
MgCl	vinyl	67
Li	Me	79
Na	η-cyclopentadienyl	64

Other nucleophiles that may displace halides include acetate, hydroxide ⟨75ZOB2672, 81ZAAC(478)94⟩, alkoxides ⟨88JA2611, 92(JOM(429)C14⟩ and hydrides ⟨85OM581, 89OM1585⟩. For example, reduction of 1,1,3,3-tetrachloro-1,3-disilacyclobutane with LAH generated 1,3-disilacyclobutane ⟨76ZAAC(426)28, 77JA3273⟩. Direct reaction of an amine can result in the displacement of a halide bound to silicon ⟨76ZAAC(419)157, 80JOM(188)151⟩. Several 1,1-bis(silyl) alkanes containing a Si—Si linkage have also been prepared by displacement of a halide, such as in the preparation of trisilacyclopentane (138) (Equation (29)) ⟨81ZAAC(473)59⟩.

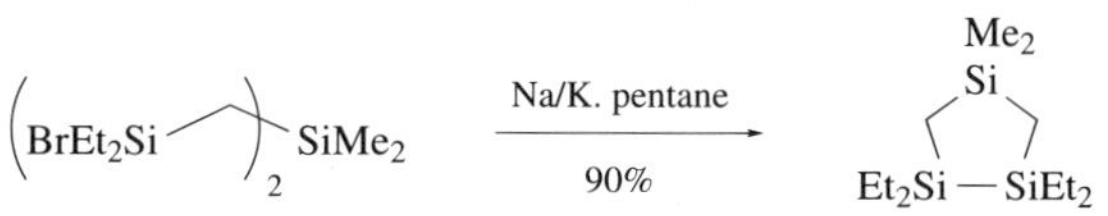

$$(\text{BrEt}_2\text{Si}\frown\text{SiMe}_2)_2 \xrightarrow[90\%]{\text{Na/K. pentane}} \begin{array}{c}\text{Me}_2\\ \text{Si}\\ \diamondsuit\\ \text{Et}_2\text{Si}-\text{SiEt}_2\end{array} \quad (29)$$

(138)

Silicon—halogen bonds have been formed, for example when zinc fluoride induced exchange of chloride by fluoride ⟨72ZAAC(391)219⟩. Amines have been replaced by chloride by the action of HCl ⟨76ZAAC(419)157⟩, or of PhPCl$_2$ ⟨80JOM(188)151⟩. Silicon-bound chlorides can be generated by the displacement of ethoxy groups following reaction with phosphorus trichloride or benzoyl chloride ⟨64JOC1601⟩. Hydride has been replaced by a halide by the action of chlorine ⟨72JA5837, 85OM581⟩ or bromine ⟨77ZAAC(433)61⟩. Treatment of a phenylsilyl group with HBr or bromine can result in the formation of a Si—Br bond ⟨70ZAAC(372)21, 81ZAAC(478)94, 89OM1585⟩. Similarly, methyl groups have been replaced by a halide with chlorine, iodine, ICl or HBr in the presence of catalytic aluminum tribromide ⟨88ZAAC(556)23⟩. Fluoride has replaced trimethylsilyloxy ⟨58JOC1392⟩ or methoxy ⟨92CB607⟩ by the action of BF$_3$·OEt$_2$. Other reactions leading to changes of the substituents on silicon of 1,1-bis(silyl) alkanes have been reviewed elsewhere ⟨72MI 414-01, 80JOM(188)151⟩.

Metals may be introduced as ligands on silicon. Several transition-metal substituted 1,3,5-trisilacyclohexanes have been prepared by displacement of a bromide by the metal anion (Equation (30)) ⟨80ZAAC(464)107⟩. A mercury-containing cyclic system (**139**) was prepared by displacement of a chloride (Equation (31)) ⟨70IC2372⟩. Hydrides have also been displaced by transition metals (Scheme 30) ⟨78IC2324, 89OM1585⟩. Analogous metallacycles containing palladium and iridium have also been prepared, although the latter was insufficiently stable to enable complete characterisation ⟨78JA6362⟩.

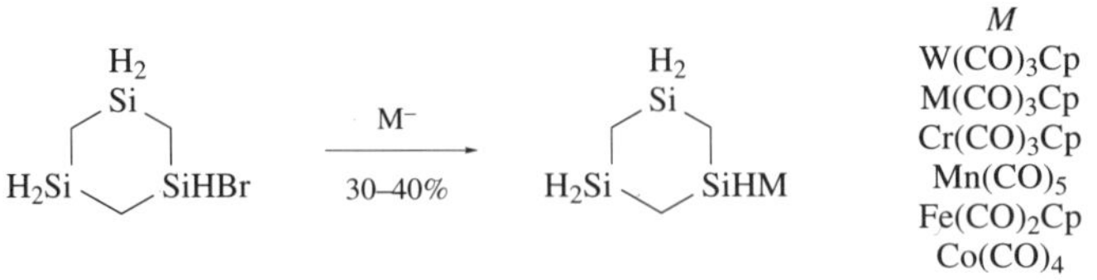

H_2Si SiHBr
M⁻
30–40%
SiHM
M
$W(CO)_3Cp$
$M(CO)_3Cp$
$Cr(CO)_3Cp$
$Mn(CO)_5$
$Fe(CO)_2Cp$
$Co(CO)_4$

(30)

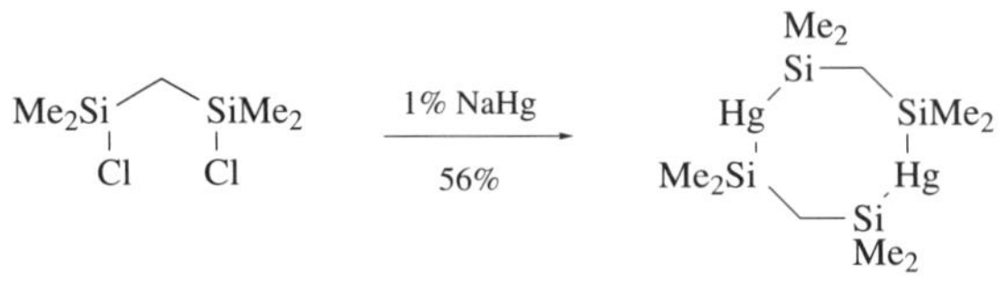

1% NaHg
56%

(31)

(139)

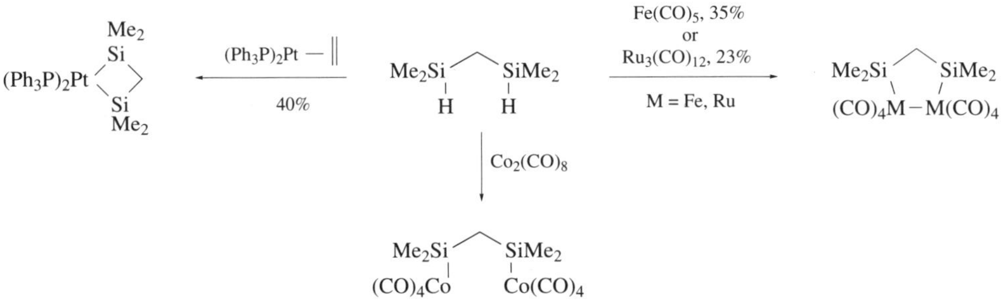

$(Ph_3P)_2Pt$
40%
$Fe(CO)_5$, 35%
or
$Ru_3(CO)_{12}$, 23%
M = Fe, Ru
$Co_2(CO)_8$

Scheme 30

Modification of the ligands on the silyl groups of cyclic carbosilanes can result in ring cleavage or ring expansions as a route to the generation of new 1,1-bis(silyl) alkanes. The earliest report of ring cleavages to generate acyclic systems was by Knoth in 1958 ⟨58JOC1392⟩ who observed that bromine would slowly react with 1,1,3,3-tetramethyl-1,3-disilacyclobutane to give what was later found to be the acyclic dibromo derivative (**140**) (Equation (32)) ⟨62CB3030⟩. Later it was observed that HCl could open 1,3-disilacyclobutanes to give two products the ratios of which depended upon the substituents of the 1,3-disilacyclobutane (Equation (33)) ⟨69DOK(189)334⟩. LAH will also open 1,3-disilylcyclobutanes (Equation (34)) ⟨87BCJ2263⟩. Many other reagents both nucleophilic and electrophilic will ring-open disilacyclobutanes and these have been reviewed elsewhere ⟨72MI 414-01, 75JCS(D)1434⟩. A study in the early 1980s of ring opening by Lewis acids has been performed. Some of these reactions are summarised in Equation (35) ⟨83ZAAC(500)132⟩.

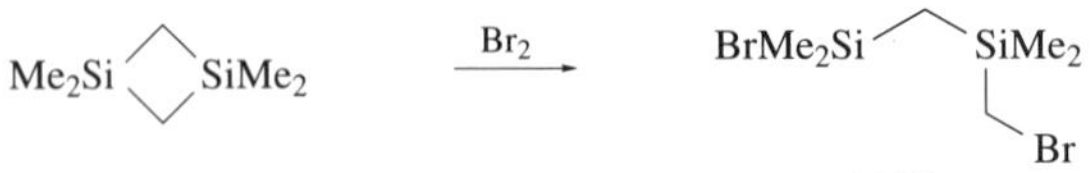

Br_2

(32)

(31) **(140)**

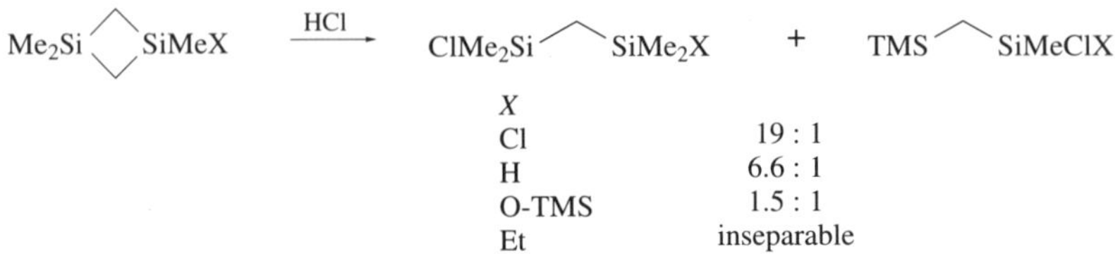

HCl
+
X
Cl 19 : 1
H 6.6 : 1
O-TMS 1.5 : 1
Et inseparable

(33)

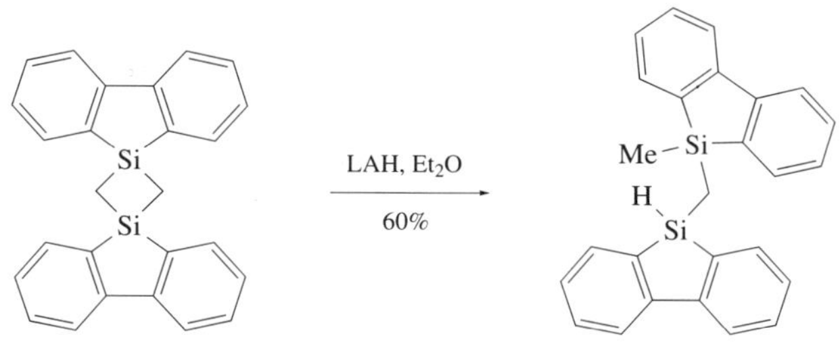

$$\text{(structure)} \xrightarrow{\text{LAH, Et}_2\text{O}}_{60\%} \text{(structure)} \quad (34)$$

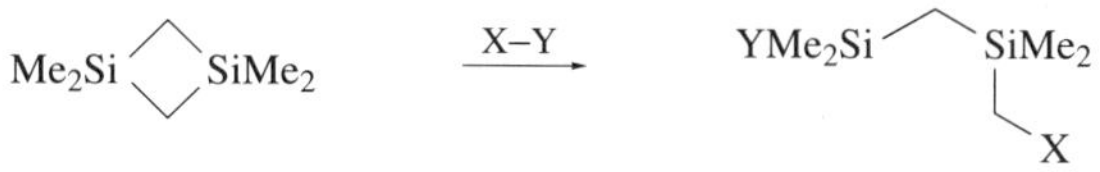

$$\text{Me}_2\text{Si}\langle\rangle\text{SiMe}_2 \xrightarrow{X-Y} \text{YMe}_2\text{Si}\frown\underset{X}{\text{SiMe}_2} \quad (35)$$

X–Y = BCl₂–Cl, BCl₂–Cl
BBr₂–Br, Br–Br
BMe₂–Br, AsCl₂–Cl
SnCl₃–Cl, AsMeCl–Cl
SnCl–Cl, AsMe₂–Cl
SnMe₂Cl–Cl, SbCl₂–Cl
SnMe₃–Cl

Methanol will also open 1,1,3,3-tetramethyldisilacyclobutane, generating 2-methoxy-2,4,4-trimethyl-2,4-disilapentane (**141**) in excellent yield or the siloxane (**142**) upon prolonged heating ⟨83ZAAC(500)132⟩. Silanols can also successfully cleave 1,3-disilacyclobutanes. For example 1,1,3,3-tetramethyl-1,3-disilacyclobutane will react with trimethylsilanol to generate the adduct (**143**) in 91% yield (Scheme 31). This reaction has also been extended to other systems. When triphenylsilanol was used, the disiloxane (**144**) was generated in high yield ⟨75JCS(D)1832⟩.

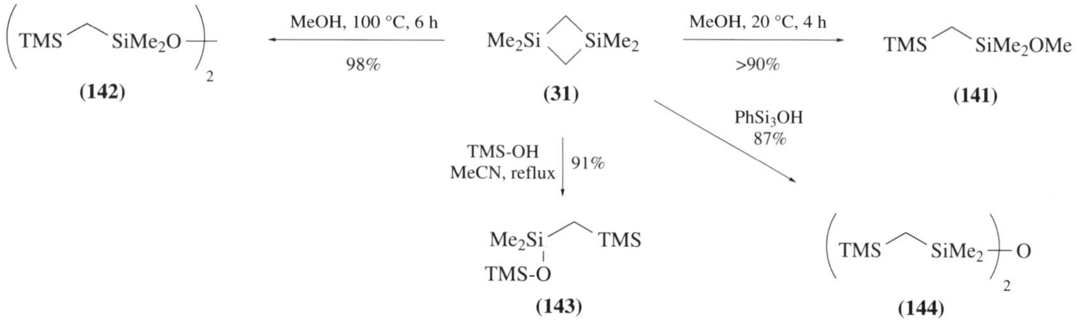

Scheme 31

Although nucleophilic organometallic reagents can open the 1,3-disilacyclobutane ring (see 4.14.1.1.1(i)(a)), metallation of a ring carbon can be achieved under some conditions if the base is sufficiently hindered (see 4.14.1.1.2). For example, *t*-butyllithium in the presence of TMEDA will metallate the central methylene without ring opening. If the metallated disilacyclobutane (**145**) is quenched with an aldehyde, then ring opening is observed through a modified Peterson alkeneation process yielding (**146**) (Scheme 32) ⟨90OM2677⟩. In a similar manner, 2-lithio-1,1,3,3,5,5-hexamethyl-1,3,5-trisilacyclohexane will react with benzaldehyde to give ring-opened products ⟨92OM3464⟩. If the electrophile used is benzoyl chloride or benzonitrile, then ring expansions are observed. With benzoyl chloride, the intermediate (**147**) could not be isolated, but immediately underwent a Brook rearrangement to give (**148**). When the electrophile was benzonitrile, (**150**) was obtained following a methanolic quench ⟨90OM2677⟩. It is believed that the intermediates (**147**) and (**149**) could not be observed due to the large driving force to continue reaction in order to relieve the ring strain inherent in the disilacyclobutanes. The corresponding trisilacyclohexane does not suffer such ring strain and analogous intermediates could be isolated ⟨92OM3464⟩.

Several other examples of ring expansions of 1,3-disilacyclobutanes have been recorded. For example, an organomercury carbenoid inserted into 1,1,3,3-tetramethyl-1,3-disilacyclobutane gave the disilacyclopentane (**151**), which after reduction with lithium gave (**152**) (Equation (36)). Thermal decomposition of (**151**) gave (**153**) ⟨71JA3709⟩.

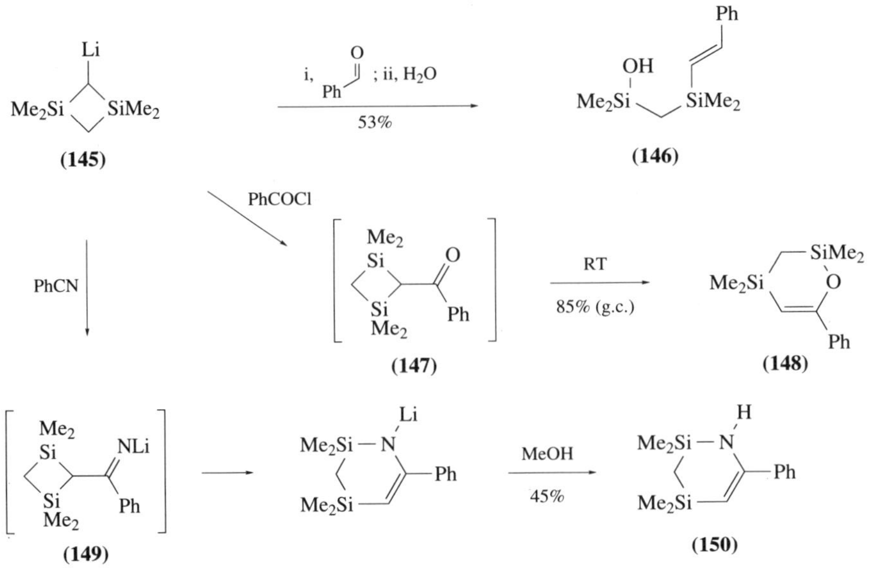

Scheme 32

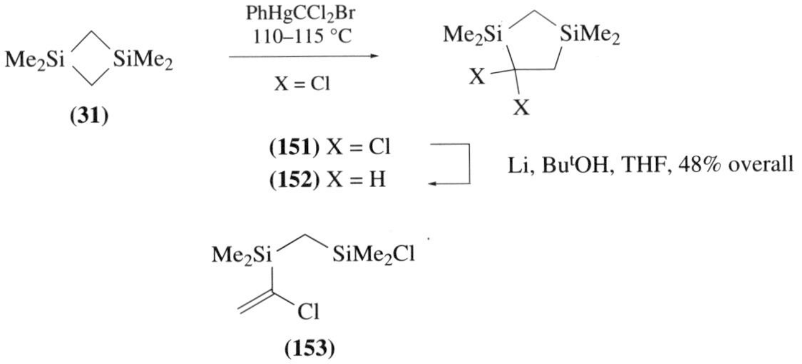

Heteroatoms have also been inserted into 1,3-disilacyclobutanes. Both sulfur dioxide and sulfur trioxide have been inserted into 1,1,3,3-tetramethyl-1,3-disilacyclobutane (Equation (37)) ⟨72MI 414-01⟩. Insertion into the Si—Si bond of a disilirane can also generate new 1,1-bis(silyl) alkanes such as during the photooxidation of (**154**) to give 3,3,5,5-tetramesityl-1,2,3,5-dioxadisilolane (**155**) in 86% isolated yield (Equation (38)) ⟨93OM1514⟩. Other ring-insertion reactions have also been reviewed here.

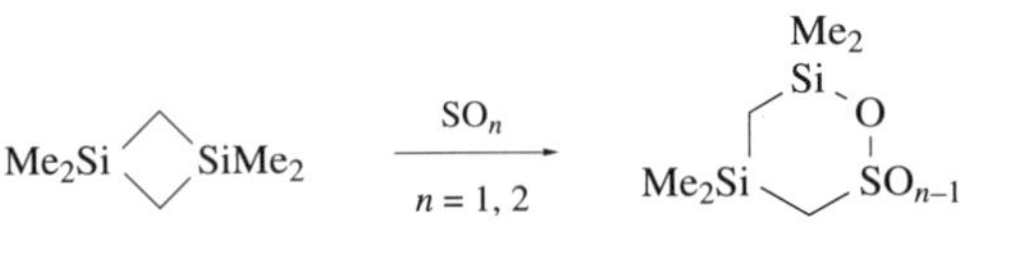

Ring contractions have been demonstrated, such as the preparation of the cyclic carbosilane (**157**) from the dimercuracycle (**156**) by photolysis (Equation (39)) ⟨70IC2372⟩. Rearrangement of polysilanes has also generated 1,1-bis(silyl) alkyl moieties (Equation (40)) ⟨88JOM(341)125⟩. Trans-silylation was used to prepare a new 1,1-bis(silyl) alkane during the formation of (**159**) from trimethylbis(trimethylsilyl)methylenephosphorane (**158**) (Equation (41)) ⟨76AG(E)367⟩.

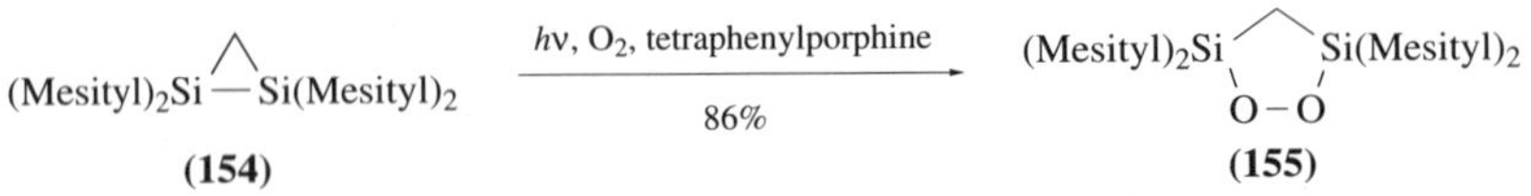

$$\text{(156)} \xrightarrow{\text{pentane, } h\nu, \text{ 5 d}} \text{(157)} \quad (39)$$

$$\left(\underset{\text{Me}_2}{\text{Si}}\right)_n \xrightarrow[40-63\%]{380\,°\text{C, catalyst}} {}^{n}/_{n'}\left(\underset{\text{Me}_2}{\text{Si}}\right)_{n'} \quad (40)$$

catalyst = B(OMe)$_3$, Me$_2$BN(TMS)$_2$, B(NEt)$_3$ ...

$$\text{(158)} \xrightarrow[58\%]{\text{Me}_2\text{Si–SiMe}_2,\ \text{Cl Cl}} \text{(159)} \quad (41)$$

4.14.1.2 Functions Bearing Two Germaniums: R$^1_2$C(GeR$^2_3$)$_2$

Many of the methods used for the generation of an sp^3-hybridised carbon connected to two germanium atoms and either protons or carbon fragments are similar to those used to generate the analogous bis(silyl) systems (see 4.14.1.1). 1,1-Bisgermyl alkanes may be prepared by the formation of either one or both of the Ge—C bonds, or by the modification of groups attached to either the germaniums or to the central carbon.

4.14.1.2.1 Formation of the Ge—C—Ge linkage

There are a number of ways by which the Ge—C bonds may be formed. A carbon nucleophile can displace a leaving group (usually halogen) attached to the germanium. Conversely, a germanium anion can displace a carbon-bound leaving group. This bond may also be formed through the intermediacy of a germene, a digermene or as a result of a rearrangement.

(i) Quenching a carbanion with a germyl electrophile

The most common method for the formation of a Ge—C bond is through the quenching of a carbanion with a germyl chloride. In this fashion, either both germyl groups may be added to the central methylene carbon or one germyl group may be added to an α-metallogermyl alkane (see 4.14.2.2 for preparations). Deprotonation of an activated methylene and subsequent germylation has frequently been performed. For example, cyclopentadiene was metallated with *n*-butyllithium and the anion treated with chlorotrimethylgermane. Subsequent deprotonation of the monosubstituted product and quenching with chlorotrimethylgermane generated 1,1-bis(trimethylgermyl)cyclopentadiene in 67% yield ⟨71JOM(30)C57⟩.

Contrary to the chemistry exhibited by silicon, a carbonyl enolate anion will react with a germyl chloride to give the *C*-germylated product preferentially. Thus treatment of dimethylacetamide (dma) with lithium diisopropylamide (LDA) followed by chlorotrimethylgermane gave the mono *C*-germylated product. Addition of further base and another equivalent of chlorotrimethylgermane gave *N,N*-dimethyl-bis(trimethylgermyl)acetamide. The same digermylated product was also obtained directly by treating dma with two equivalents of LDA followed by two equivalents of chlorotrimethylgermane ⟨88JOM(354)155⟩. This process has also been used to prepare α,α-digermylated esters ⟨88OM739, 90OM1325⟩, ketones ⟨87OM2568⟩ and nitriles ⟨86OM1197⟩. In the last case, significant quantities of the bisgermyl nitrile (**162**) was formed from the α-lithionitrile (**161**) as a result of rapid transmetallation with the starting material (**160**) which was found to occur even at

−78 °C (Equation (42)). The distribution of products obtained demonstrates that the more highly germylated anion is favoured ⟨87SC1273⟩.

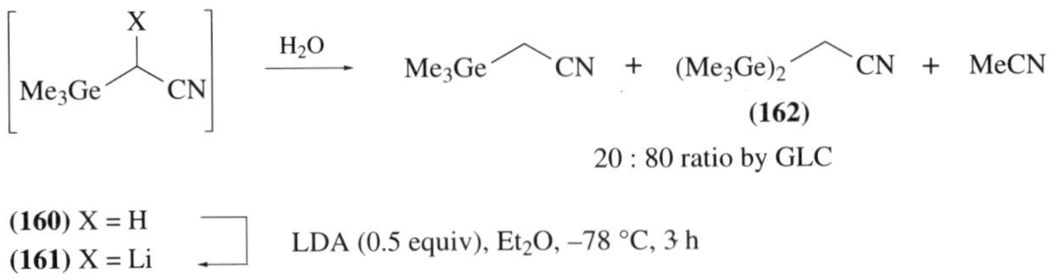

$$[Me_3Ge-CH(X)-CN] \xrightarrow{H_2O} Me_3Ge\frown CN + (Me_3Ge)_2\frown CN + MeCN \quad (42)$$
$$\qquad\qquad\qquad\qquad\qquad\qquad \textbf{(162)}$$
$$\qquad\qquad\qquad\qquad\qquad\qquad 20:80 \text{ ratio by GLC}$$

(160) X = H
(161) X = Li
LDA (0.5 equiv), Et$_2$O, −78 °C, 3 h

Halogen–metal exchange has also been used to generate the intermediate α-lithiogermyl alkane. Chloromethyl(trimethyl)germane reacted with lithium dispersion to form the organolithium intermediate which was subsequently quenched with chlorotrimethylgermane to form bis(trimethylgermyl)methane in 69% yield ⟨80JA1584⟩.

Cyclic systems may also be prepared using the chemistry described for the corresponding silyl systems in 4.14.1.1. For example, 1,3-digermylcyclobutanes have been prepared following the chemistry outlined for the silyl analogues (4.14.1.1.1(i)(a), Scheme 3) ⟨88JOM(338)159⟩. Similarly, the germyl analogue of (**9**; R = H; Equation (4)) has also been prepared ⟨87JCS(D)1647⟩.

The metallated intermediate which reacts with a germyl chloride need not be an organolithium reagent; direct transmetallation from an organostannane has been demonstrated ⟨91ZOB921⟩. Grignard reagents may also be used such as during the preparation of 1,1,3,3-tetramethyl-1,3-digermacyclobutane (Scheme 33) ⟨69ZOB2601⟩. The *in situ* formed di-Grignard reagent CH$_2$(MgBr)$_2$ will slowly react (16 h, room temperature) with two equivalents of chlorotrimethylgermane to give the acyclic 1,1-bisgermyl alkane (**163**) in 61% yield. A more reactive form of this reagent, obtained by isolation as a complex with MgBr$_2$, reacted to give the same bisgermyl alkane in a few minutes in high yield. Alternatively, reaction with dichlorodimethylgermane gave a mixture of germacycles (**164**), the ratio of which was dependent upon the composition of the complex (Scheme 34) ⟨85JOM(288)13⟩.

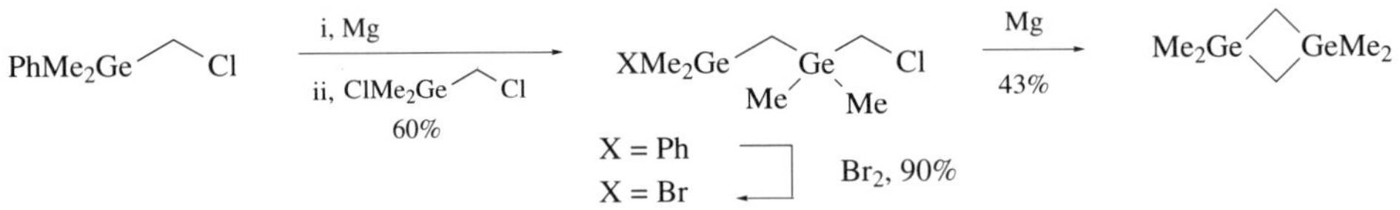

Scheme 33

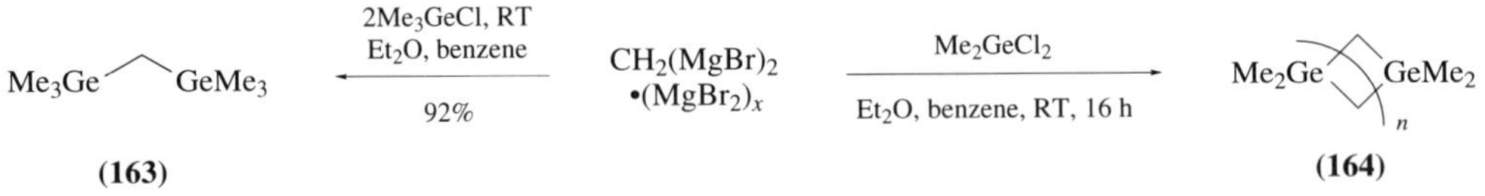

Scheme 34

n	Yield (%)
1	5–35
2	3–36
3	0–2

(ii) Quenching a germyl anion with a carbon electrophile

As with the silicon chemistry (see 4.14.1.1.1(ii)) very little work has been performed using this methodology to construct the Ge—C bonds when forming 1,1-bisgermyl alkanes. The requisite trialkylgermyl anions can be prepared by treatment of either a trialkylchloromethylgermane or a hexaalkyldigermane with an alkali metal (Li, Na, K). The trimethylgermyl anion reacted with dichloromethane to give bis(trimethylgermyl)methane in 65% yield ⟨67TL1443, 71JOM(29)409⟩.

(iii) From germenes or digermenes

Germenes are compounds containing a Ge═C double bond while digermenes contain a Ge═Ge double bond; these have both been used to generate 1,1-bisgermyl alkanes.

The most frequently used route to silenes was through the thermolysis of silacyclobutanes; however, pyrolysis of the corresponding germacyclobutanes tends to give products derived from alkenic and carbenoid intermediates ⟨79CRV529, 92OM3307⟩. Germenes may however be prepared under pyrolytic conditions. For example heating the Diels–Alder adduct (**165**) at 450°C gave the 1,3-digermacyclobutane (**167**) (Scheme 35). This was formed through the germene (**166**) through a head-to-tail dimerisation ⟨73JA3078⟩. Other 1,1-bisgermyl alkanes have also been formed via germenes, for example during the pyrolysis of the thiagermetane (**168**) (Scheme 36) ⟨88JOM(345)39⟩. Similarly, several 1,1-bisgermyl alkanes have been prepared via the intermediacy of diethylgermafulvene ⟨80IZV2668⟩. The 1,3-digermacyclobutane (**172**) was generated through head-to-tail dimerisation of the germene (**171**) following treatment of (**170**) with *t*-butyllithium (Equation (43)) ⟨80TL1405⟩. Germene (**174**) was shown to be an intermediate during the preparation of the bisgermyl alkane (**175**) following photolysis of bis(trimethylgermyl)diazomethane (**173**) in the presence of deuteriomethanol (Equation (44)) ⟨80JA1584⟩.

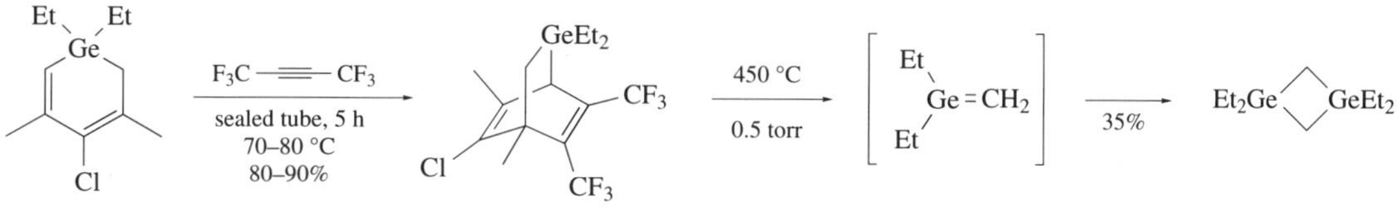

(**165**) (**166**) (**167**)

Scheme 35

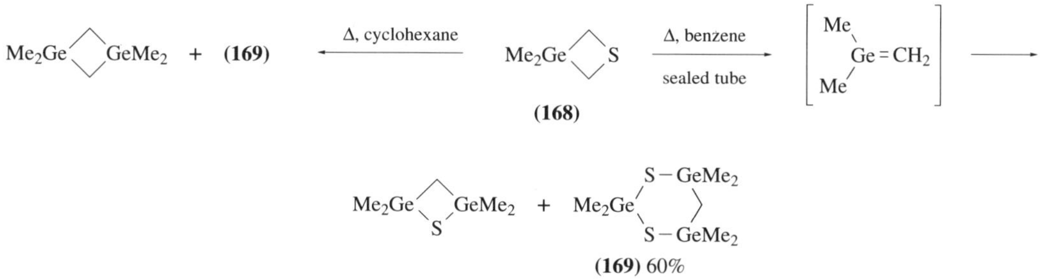

Scheme 36

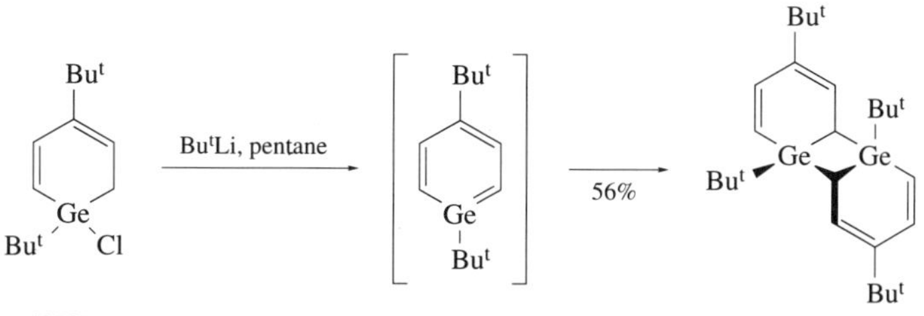

(**170**) (**171**) (**172**) (43)

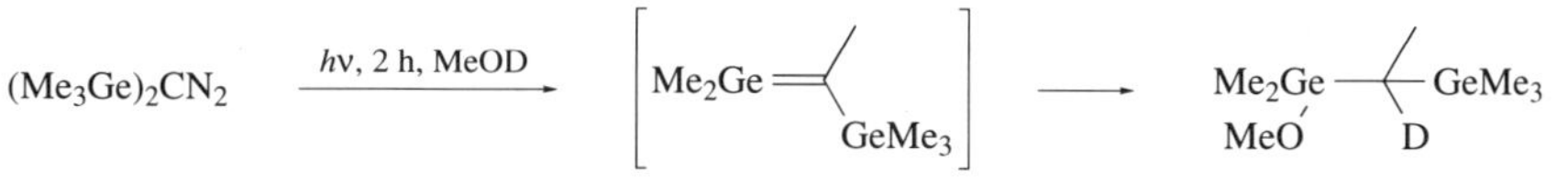

(**173**) (**174**) (**175**) (44)

1,1-Bisgermyl alkanes have also been generated from digermenes. Thus treatment of tetrakis (2,6-diethylphenyl)digermene (**176**) with diazomethane gave the [2 + 3] cyclo-adduct (**177**) which eliminated nitrogen to yield the digermacyclopropane (**178**) (Scheme 37) ⟨88TL3383⟩. Other derivatives have also been prepared ⟨90OM2061⟩.

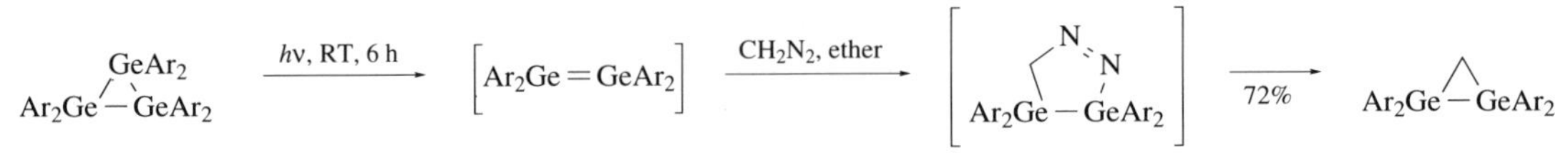

$$Ar_2Ge\underset{GeAr_2}{\overset{GeAr_2}{\diagup\!\!\diagdown}}GeAr_2 \xrightarrow{h\nu,\,RT,\,6\,h} [Ar_2Ge=GeAr_2] \xrightarrow{CH_2N_2,\,ether} \left[\underset{Ar_2Ge-GeAr_2}{\overset{N\diagdown N}{\diagup}}\right] \xrightarrow{72\%} Ar_2Ge\overset{\triangle}{-}GeAr_2$$

(**176**)   (**177**)   (**178**)

Ar = 2,6-diethylphenyl

Scheme 37

(iv) Other routes

The fact that germanium can exist in more than one oxidation state has enabled oxidative addition to be used as a route to 1,1-bisgermyl alkanes. For example, treatment of diiodomethane with germanium diiodide gave bis(triiodogermyl)methane ⟨87OM659, 89OM1585⟩.

1,1-Bisgermyl alkanes have also been generated through rearrangements, for example treatment of (**179**) with boron trichloride resulted in a mixture of (**180**) and (**181**) following a 1,2-migration of a metalloid from germanium to the adjacent carbon (Equation (45)) ⟨86TL4015⟩.

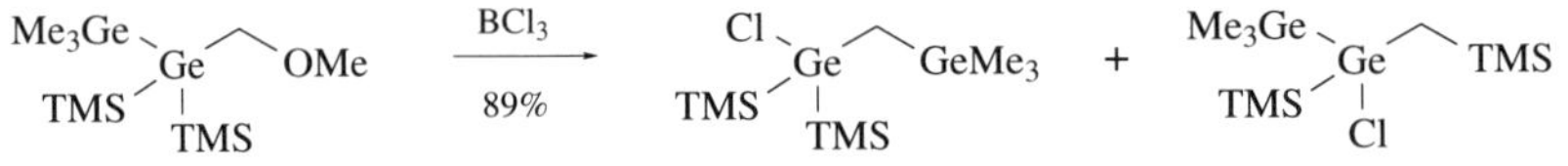

$$\underset{TMS}{\overset{Me_3Ge}{\diagdown}}\underset{|}{\overset{}{Ge}}\underset{TMS}{\overset{}{\diagup}}OMe \xrightarrow[89\%]{BCl_3} \underset{TMS}{\overset{Cl}{\diagdown}}\underset{|}{\overset{}{Ge}}\underset{TMS}{\overset{}{\diagup}}GeMe_3 + \underset{TMS}{\overset{Me_3Ge}{\diagdown}}\underset{|}{\overset{}{Ge}}\underset{Cl}{\overset{}{\diagup}}TMS \qquad (45)$$

(**179**)   (**180**)   (**181**)

Carbon suboxide has also been used as a precursor to 1,1-bisgermyl alkanes during the preparation of *N*,*N*-dimethylbis(trimethylgermyl)acetamide following reaction with pentamethylgermylamine ⟨88OM210⟩. An analogous reaction with methoxy(trimethyl)germane has also been reported ⟨88G577⟩.

The Lewis acid-catalysed cyclisations of various carbosilanes during the preparation of bis(silyl) alkanes were described earlier (4.14.1.1.1(i)(c)). This process has also been applied to carbogermanes using aluminum tribromide ⟨72ZOB1521⟩.

Removal of additional functionality from the central methylene has also been demonstrated as a route to 1,1-bisgermyl alkanes. For example metal–halogen exchange has been used to remove a chloride and generate 1-lithio-1,1-bis(trimethylgermyl)methane by treatment with *s*-butyllithium ⟨80JA1584⟩.

A digermyl-substituted ketene was demonstrated as a precursor to a 1,1-bisgermyl alkane following treatment of (**182**) with *n*-butyllithium to give the lithium enolate (**183**) which upon protonation gave the α,α-digermyl ketone (**184**) ⟨70ZOB939⟩ while direct reaction of (**182**) with ethanol gave (**185**) (Scheme 38) ⟨76IZV1887⟩.

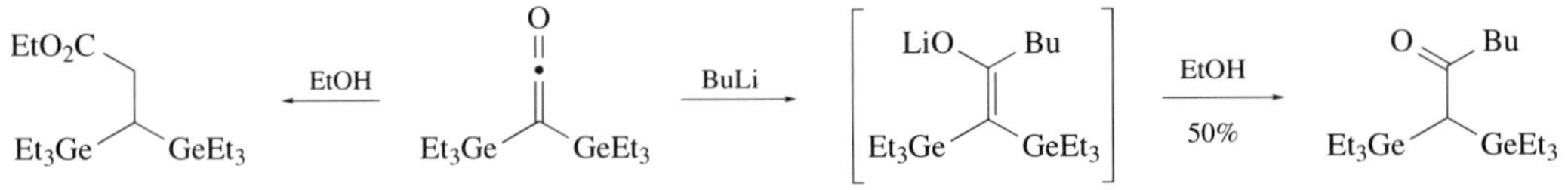

$$\underset{Et_3Ge}{\overset{EtO_2C}{\diagdown}}\overset{}{\diagup}\underset{GeEt_3}{\overset{}{\diagdown}} \xleftarrow{EtOH} \underset{Et_3Ge}{\overset{O}{\diagdown}}\underset{}{\overset{\|}{C}}\underset{GeEt_3}{\overset{}{\diagup}} \xrightarrow{BuLi} \left[\underset{Et_3Ge}{\overset{LiO}{\diagdown}}\underset{}{\overset{}{C=C}}\underset{GeEt_3}{\overset{Bu}{\diagup}}\right] \xrightarrow[50\%]{EtOH} \underset{Et_3Ge}{\overset{O}{\diagdown}}\underset{}{\overset{\|}{C}}\underset{GeEt_3}{\overset{Bu}{\diagup}}$$

(**185**)   (**182**)   (**183**)   (**184**)

Scheme 38

4.14.1.2.2 *Changing the groups attached to the central methylene*

The only method demonstrated for the modification of the central methylene group was metallation of bis(trimethylgermyl)methane with *t*-butyllithium and subsequent reaction with an electrophile ⟨80JA1584⟩.

4.14.1.2.3 Changing the ligands on germanium

Novel 1,1-bisgermyl alkanes have been prepared by manipulation of the functionality on the germanium atoms. Much work has been described changing the ligands on germanium but only the studies on 1,1-bisgermyl alkanes will be described here.

Displacement of a halogen from germanium by organometallic reagents may be used to modify existing 1,1-bisgermyl alkanes such as the preparation of bis(trimethylgermyl)methane from bis(triiodogermyl)methane following reaction with methylmagnesium iodide. Subsequent treatment with concentrated sulfuric acid then ammonium chloride generated the dichloro derivative (**186**) (Scheme 39). Further reaction with sodium sulfide gave a mixture of the digermathiacyclobutane (**187**) and a larger ring analogue ⟨85JOM(282)315⟩. The dichloro derivative (**186**) can be reduced with LAH to 2,4-dimethyl-2,4-digermapentane. An analogous reduction with LAH was also demonstrated on 1,1,3,3,5,5-hexachloro-1,3,5-trigermacyclohexane ⟨88ZN(B)727, 89ZN(B)285⟩. The exchange of alkoxide ligands on a bis(trialkoxygermyl)methane has been demonstrated ⟨85ZOB2396⟩.

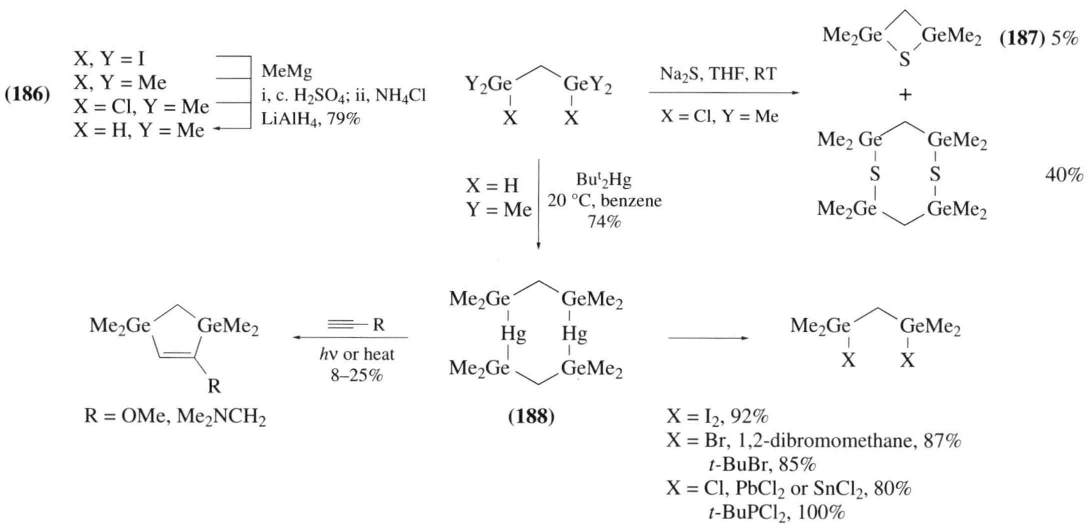

Scheme 39

Many reactions of 2,4-dimethyl-2,4-digermapentane and the mercuracycle (**188**) have been reported. A few of these transformations are summarised in Scheme 39 ⟨85JOM(282)315, 87OM659, 89OM1585⟩.

Several ring expansions of digermacyclopropanes by insertion into the Ge—Ge bond have been reported; some of these are shown in Scheme 40 ⟨88OM1882, 90OM2061⟩. Insertion of palladium has also been demonstrated to occur, to give (**189**), the proposed intermediate for several palladium-catalysed reactions ⟨89OM2286⟩. Analogous reactions of complexes containing ruthenium ⟨90ICA159, 91JOM(406)123⟩ and cobalt ⟨90ICA141, 91NJC657⟩ have also been reported. The metallacycle (**191**) was prepared by photolysis of (**190**) with $Ru_3(CO)_{12}$. Subsequent reactions with a number of electrophiles were described, some of which are shown in Scheme 41 ⟨90JOM(395)27⟩.

4.14.1.3 Functions Bearing Two Borons: $R^1_2C(BR^2_2)_2$, etc.

This section describes the preparation of systems containing two boryl groups and either protons or carbon fragments bound to an sp^3-hybridised carbon. These can be prepared by generating either one or both C—B bonds, or by the manipulation of systems already containing two borons bound to a carbon. The large volume of work covering carboranes will not be described here as this area has been adequately reviewed elsewhere ⟨92CRV175⟩.

4.14.1.3.1 Formation of the B—C—B linkage

The predominant method for the preparation of *gem*-diboryl alkanes is through hydroboration of a vinylborane, or bis(hydroboration) of an alkyne. Other methods analogous to those described

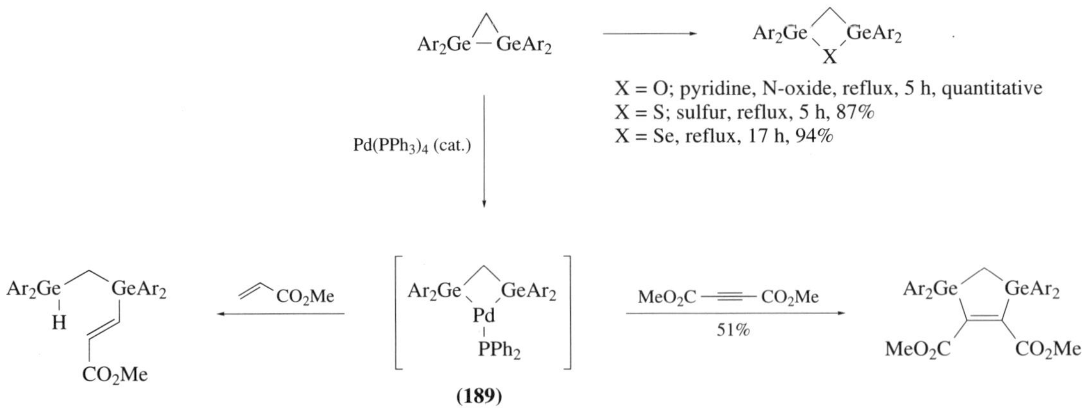

Ar = 2,6-diethylphenyl

Scheme 40

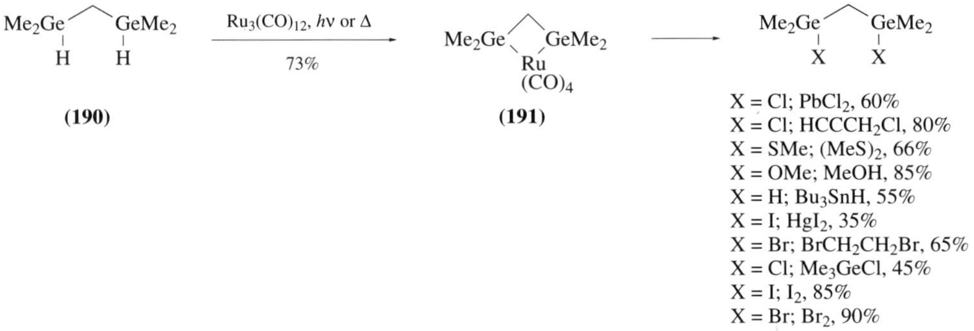

Scheme 41

earlier during the preparation of the bis(silyl) and bisgermyl analogues have also been used such as displacement of a leaving group on boron by a carbon nucleophile. Modification to the groups attached to either the boron or to the central methylene to generate new and otherwise inaccessible systems have also been demonstrated.

The stability of the *gem*-diboryl alkane moiety is dependent upon the substituents on boron. If these are alkyl, then unless there is a large amount of steric bulk, decomposition to a borinic acid and a trialkylborane may be facile (Equation (46)) ⟨82JCR(S)132⟩. Replacement of the alkyl groups with alkoxy groups makes the boron less acidic and reduces the rate of hydrolysis.

$$R_2B\frown BR_2 \xrightarrow{H_2O} R_2BOH + R_2B-Me \qquad (46)$$

(i) Hydroboration of vinylboranes or alkynes

(a) Terminal alkynes. Monohydroboration of a terminal alkyne will generate a *trans*-vinylborane which may react with further hydroborating reagent to add a second boron to the alkyne. Most hydroborating reagents will react more readily with alkynes than with alkenes and can thus be used to generate vinylboranes selectively. This is not the case with 9-borabicyclo[3.3.1]nonane (9-BBN-H) where selective monohydroboration has proved difficult ⟨82JA531, 89JA4873, 93JA6065⟩. The second hydroboration could occur in two ways to give a product with either the borons on the same carbon or on adjacent carbons. Early workers inferred that the former was favoured, resulting in the formation of a 1,1-bisboryl derivative. For example, when 1-hexyne was treated with dicyclohexylborane and the adduct oxidised with alkaline hydrogen peroxide, the major product was 1-hexanol (90% yield). It was reasoned that this resulted from the decomposition of the terminal bisboryl derivative (**192**) upon workup (Scheme 42). Nucleophilic attack at one of the borons

generates the boron 'ate' complex (**193**) which collapses cleaving the B—C bond to give a carbanion which is stabilised by the vacant *p*-orbital of the remaining boron. Only traces of 1,2-hexanediol was obtained as a result of addition of a boron to each end of the alkyne ⟨64JA3039⟩. Diborane has also been used ⟨61JA3834⟩ as has deuteriodiborane which demonstrated that dihydroboration of alkynes proceeded predominantly (>80%) by addition of both boron atoms to the same carbon ⟨67JA291⟩.

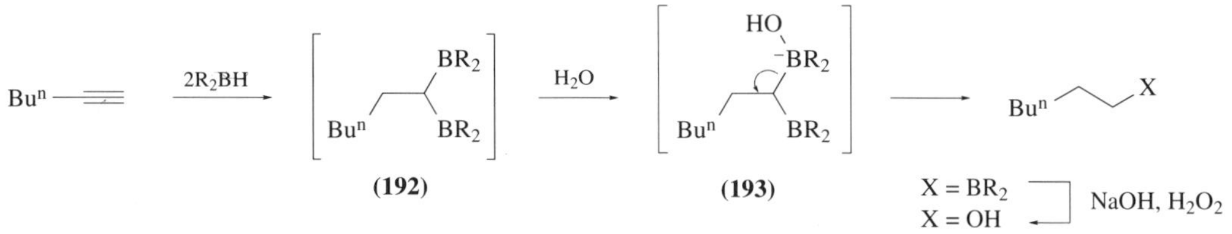

Scheme 42

Many other examples of the dihydroboration of terminal alkynes to produce 1,1-diboryl alkanes are known. When the hydroborating reagent is 9-BBN-H, the products can often be isolated ⟨93JA6065⟩. 3,6-Dimethylborepane has also been used as the dihydroborating reagent ⟨73S37⟩. Functionalised alkynes may also be dihydroborated—such as trimethylsilylacetylene, which with 9-BBN-H yields the product with both borons attached to the terminal carbon ⟨89JA4873⟩. This result is believed to be sterically dictated as the silyl group will electronically direct the boron to the α-carbon (see 4.14.1.5). Propargyl chloride reacts with diethylborane in a similar manner to yield 1,1-bis(diethylboryl)-3-chloropropane ⟨62AG(E)508⟩ and 9-BBN-H reacts likewise with the corresponding bromide ⟨69JA4306⟩. Alkynes may also be dihydroborated with dichloroborane (Equation (47)) ⟨76JA1798, 90ZN(B)15⟩. Dichloroborane has also been proposed as the reactive intermediate in the formation of 1,1-bis(dichloroboryl)hexane (**194**) following the combination of triethylsilane, boron trichloride and 1-hexyne (Scheme 43) ⟨90JOC2274⟩.

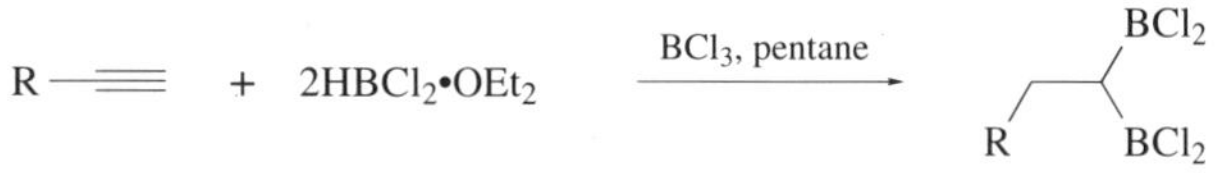

$$R = H, 31\%$$
$$R = Me, 38\%$$
$$R = Bu^t, 57\%$$

(47)

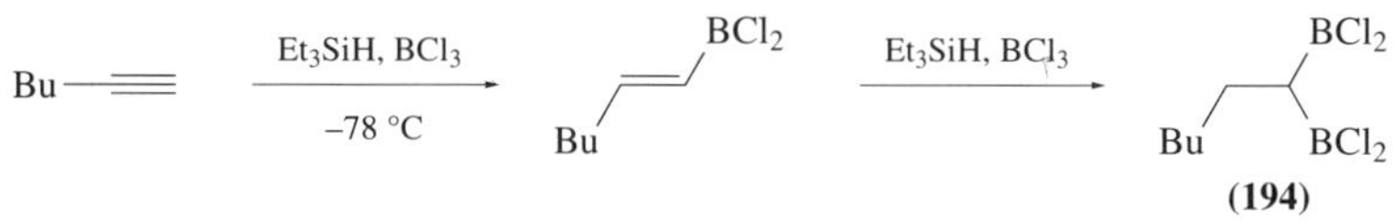

Scheme 43

(b) Internal alkynes. Internal alkynes may also be dihydroborated ⟨61JA3834, 64JA3039, 76JA1798⟩. Addition of the first equivalent can give a mixture of regioisomers, the ratio of which is dependent upon steric and electronic factors; however, the second equivalent will preferentially place the second boron on the same carbon as the first ⟨79JA96⟩. Contrary to the dihydroboration of terminal alkynes with 9-BBN-H where hydroboration will readily occur twice, reaction with internal alkynes can be stopped at the vinylborane stage ⟨82JA531⟩.

(c) Vinylboranes. Clearly, 1,1-bisboryl alkanes need not be prepared directly from an alkyne; hydroboration of the intermediate vinylborane is also possible. This two-step process has also enabled the isolation of many 1,1-bisboryl alkanes by the hydroboration of a vinylboronate ester. One of the earliest examples of hydroboration of a vinylboronate ester was demonstrated in 1963 when the vinylboronate (**195**) was treated with diborane and then butanol to generate tetrabutylethylidenediboronate and a small quantity of the 1,2-isomer. These were readily hydrolysed with water to give the separable diboronic acids (Scheme 44) ⟨63JA2684, 64JOC2742⟩. Dibutyl β-styrylboronate has been similarly transformed into tetrabutyl 2-phenylethylidenebisboronate. The corresponding 1,2-diboronate isomer could not be detected; however, hydroboration of dibutyl

α-styrylboronate gave the 1,1-bisboronate as the minor product; the major product obtained was as a result of addition of boron to the terminal carbon ⟨67TL723, 69T1557⟩. Difluoroborane has been used to hydroborate difluorovinylborane to give 1,1-bis(difluoroboryl)ethane which subsequently polymerised ⟨68JOM(12)269⟩. In a further example, *cis*-3,4-bis(diethylboryl)-3-hexene was hydroborated with diethylborane to give 3,3,4-tris(diethylboryl)hexane ⟨66TL2675⟩.

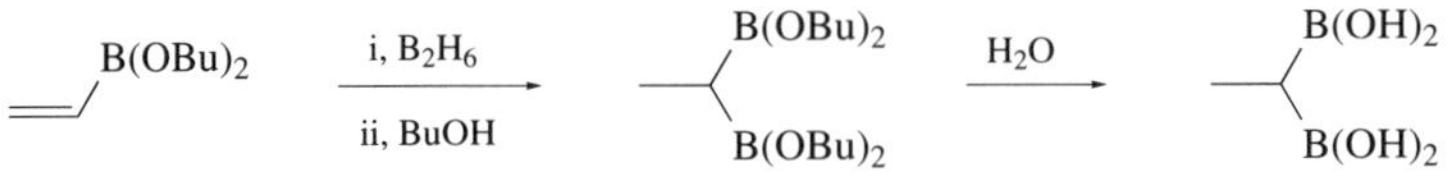

$$\text{(195)} \quad \overset{B(OBu)_2}{=\!\!\!\!=\!\!\!\!/} \quad \xrightarrow[\text{ii, BuOH}]{\text{i, B}_2\text{H}_6} \quad \overset{B(OBu)_2}{\underset{B(OBu)_2}{\diagdown\!\!\!/}} \quad \xrightarrow{\text{H}_2\text{O}} \quad \overset{B(OH)_2}{\underset{B(OH)_2}{\diagdown\!\!\!/}}$$

Scheme 44

(ii) Quenching a carbanion with a boryl electrophile

One of the earliest preparations of 1,1-bisboryl alkanes by the displacement of a leaving group on boron with a carbanion was reported in 1966. Methylene bis(mercuric iodide) was treated with boron tribromide to yield tetrabromomethylenediboronate ⟨66JOM(6)100⟩. The addition of dichloromethane to a mixture of lithium dispersion and trimethylborane in THF generated tetramethylmethylenediboronate in modest yield. The yield was later increased to 28% when the reaction temperature was reduced to −30 °C and further improved to 35% through the use of chlorodimethoxyborane ⟨69JOM(20)19, 75S147⟩. α,α-Dichlorotoluene has been used in place of dichloromethane ⟨69JOM(20)19⟩. An alternative procedure which proved less successful was preformation of dilithiomethane and the subsequent addition of boron trichloride, boron tribromide or trimethoxyborane. Those reactions performed with solvents would not proceed far enough to produce workable quantities of product whilst the absence of solvent caused either no reaction or a violent one leading to polymeric materials. More success was achieved with the use of bis(dimethylamino)boron chloride whereupon reaction with dilithiomethane gave yields of 1,3,5-(dimethylamino)-1,3,5-triborocyclohexane in yields ranging from 20 to 70% ⟨71CB1347⟩. Tebb's reagent has been used to generate the highly reactive diborocyclopentene (**197**) from (**196**) (Equation (48)) ⟨83JA2582⟩.

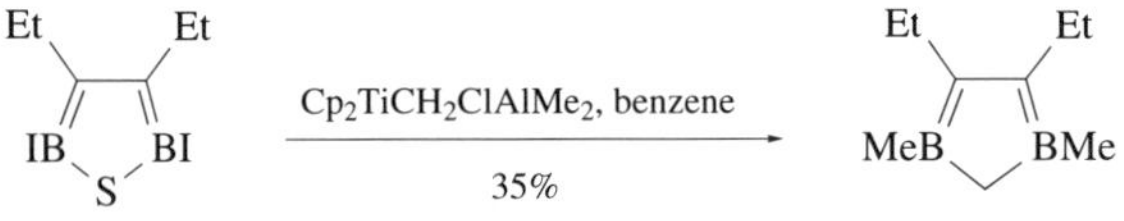

$$\underset{(196)}{\overset{\text{Et} \quad \text{Et}}{\underset{\text{S}}{\text{IB}\diagdown\!\!\!/\text{BI}}}} \quad \xrightarrow[35\%]{\text{Cp}_2\text{TiCH}_2\text{ClAlMe}_2,\text{ benzene}} \quad \underset{(197)}{\overset{\text{Et} \quad \text{Et}}{\text{MeB}\diagdown\!\!\!/\text{BMe}}} \quad (48)$$

The displacement of a fluoride on boron with an α-lithiated borane has also been used to prepare 1,1-bisboryl alkanes, such as the formation of (**198**) (Equation (49)) ⟨82JCR(S)132⟩.

$$(\text{Mesityl})_2\text{B}\diagup\!\!\!\diagdown\text{Li} \;+\; (\text{Mesityl})_2\text{BF} \quad \xrightarrow{65\%} \quad (\text{Mesityl})_2\text{B}\diagup\!\!\!\diagdown\text{B}(\text{Mesityl})_2 \quad (49)$$

$$(\textbf{198})$$

(iii) Other routes

In addition to hydroboration, vinylboranes have been boroborated to yield 1,1,2-trisboryl alkanes. Treatment of (dichloro)vinylborane with tetrachlorodiborane gave 1,1,2-tris(dichloroboryl)ethane in yields up to 90%. More forcing conditions were needed for the corresponding fluoro analogue ⟨68JOM(12)269⟩.

Thermal rearrangements have also been used to prepare *gem*-diboryl alkanes. For example, thermolysis of (**199**) gave the isomeric 1,3-diboretane (**200**) in 62% yield (Equation (50)) ⟨93AG(E)295⟩. In another rearrangement, pyrolysis of *cis*-3,4-di(diethylboryl)-3-hexene (**201**) gave 1,3-diboracyclopent-4-ene (**202**) (Equation (51)) ⟨68AG(E)286⟩. Other examples of this thermal ring closure have since been described ⟨85AG(E)943⟩. A thermal rearrangement was also employed during the preparation of (**204**) from the 1,2,5,6-tetraborocine (**203**) (Equation (52)) ⟨89ZN(B)96⟩. Dihydroboration of alkyl-substituted silyl alkynes by 9-BBN-H is strongly governed by steric effects (see 4.14.1.5.1(iii)). Dihydroboration of 1-(triethylsilyl)propyne gave the 1,2-adduct (**205**) which upon heating smoothly isomerised to give the 3,3-diborylpropylsilane (**206**) (Equation (53)) ⟨89JA4873⟩.

$$\text{(199)} \xrightarrow{\text{1 h, benzene, 65 °C}} \text{(200)} \qquad (50)$$

$$\text{(201)} \xrightarrow[54\%]{160\,°\text{C, 5 h}} \text{(202)} \qquad (51)$$

$$\text{(203)} \xrightarrow[\substack{\text{toluene, reflux}\\98\%}]{h\nu,\ 90\%\ \text{or}} \text{(204)} \qquad (52)$$

$$\text{(205)} \xrightarrow[84\%]{1\ \text{h, 165 °C}} \text{(206)} \qquad (53)$$

⌬B = 9-BBN

1,3-Diboretanes (**207**) have been prepared from the corresponding dilithium compounds by reactions with methyl iodide (Equation (54)) ⟨89AG(E)784⟩.

$$\left[\,\text{TMS}_2\text{C}=\text{B(Ar)}-\equiv\text{B}^--\text{Ar}\,\right]\ 2\text{Li}^+ \xrightarrow{2\text{MeI}} \text{(207)} \qquad (54)$$

Ar = 2,4,6-Me$_3$C$_6$H$_2$, 89%
Ar = 2,3,5,6-Me$_4$C$_6$H, quantitative

Earlier it was noted that 1,1-bisboryl alkanes may cleave a B—C bond when treated with a nucleophile to generate an α-metalloalkylborane (Scheme 42). This procedure has been also used to remove a boron from a triboryl alkane, generating a 1,1-diboryl-1-metalloalkane which can then be functionalised. For example tris(dimethoxyboryl)methane upon treatment with methyllithium was transformed into the bis(dimethoxyboryl)methide anion and quenched with alkyl halides in yields up to 42% ⟨69JOM(20)19, 70ACR186⟩. Lithium methoxide may also be used ⟨70JOM(24)263⟩. Tris-(ethylenedioxyboryl)methane will react similarly when treated with either butyllithium, methyllithium ⟨76JOM(110)25⟩, Grignard reagents or sodium methoxide ⟨68JA2194, 75S147⟩. Similarly, tetra-(dimethoxyboryl)methane initially gave the tris(dimethoxyboryl)methide anion which could be quenched and the process repeated ⟨75S147⟩. When only two boryl groups remain attached, no further cleavage occurs. Presumably two or more borons are required to provide sufficient stabilisation of the anion formed upon loss of a boryl group.

Direct hydrogenation of 1,1-diboryl-alkenes has also been performed to generate bisboryl alkanes in high yields (Equation (55)) ⟨82OM20⟩. Reduction of the double bond of (**208**) has been performed using a Na–K alloy to generate the stable dianion (**209**) which can be further functionalised (Scheme 45) ⟨85AG(E)788⟩.

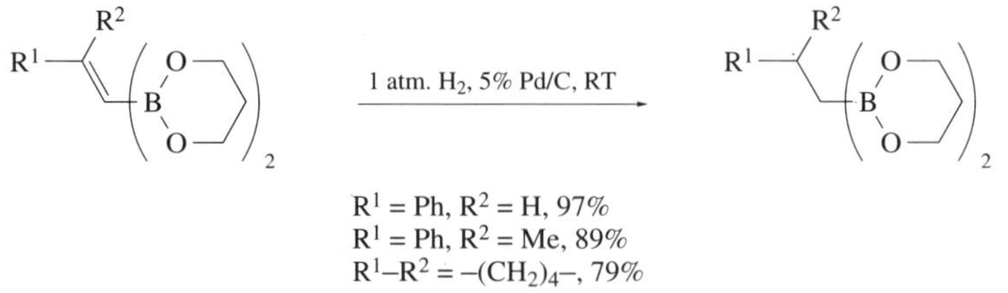

$$R^1 = Ph, R^2 = H, 97\%$$
$$R^1 = Ph, R^2 = Me, 89\%$$
$$R^1-R^2 = -(CH_2)_4-, 79\%$$

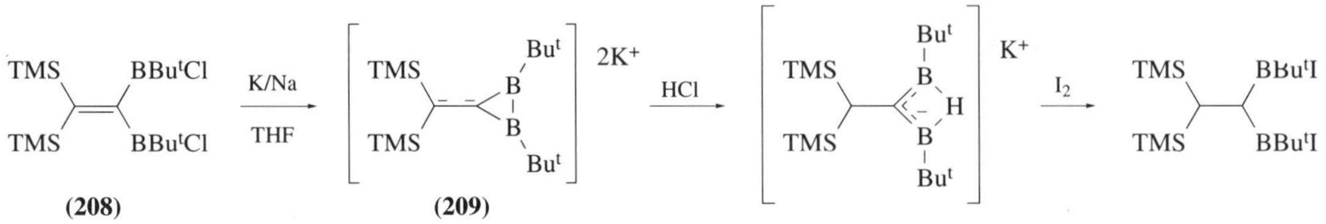

Scheme 45

4.14.1.3.2 Changing the groups on the central methylene

As with the corresponding bis(silyl) and bisgermyl alkanes described earlier, the simplest method for adding functionality to the central methylene is by direct deprotonation followed by an electrophilic quench.

Bis(trimethylenedioxyboryl)methane (**210**) can be deprotonated with lithium tetramethylpiperidide (LITMP) and the resultant anion alkylated. Metallation of the central methylene is not restricted to unsubstituted systems and so the process may be repeated with another equivalent of base and a second alkylating agent (Scheme 46) ⟨77JA3196, 82OM20⟩. Closer scrutiny of the optimum conditions revealed that TMEDA was beneficial and that LITMP was the most effective base. Only primary alkyl halides or tosylates were found to add efficiently. Metallation of (dimesitylboryl)methane could be achieved using potassium hydride, but the resulting anion was so hindered that no electrophile other than a deuteron could be added ⟨82JCR(S)132⟩.

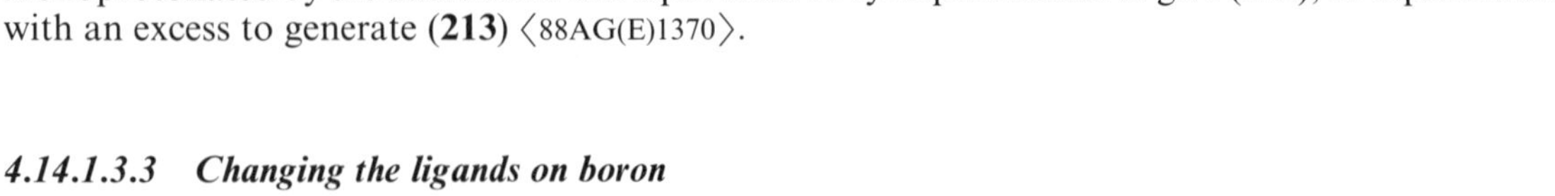

Scheme 46

The diboryldilithiomethane (**211**) has also been prepared (Scheme 47). This dianion could be monoprotonated by the addition of one equivalent of cyclopentadiene to give (**212**), or diprotonated with an excess to generate (**213**) ⟨88AG(E)1370⟩.

4.14.1.3.3 Changing the ligands on boron

Another route to the generation of novel 1,1-bisboryl alkanes is through the modification of ligands on boron. Several such reactions have been demonstrated on bisboryl alkanes. Hydride ligands on the boron can be replaced by alkoxides ⟨67TL723⟩. Alkoxides in turn have been replaced with hydroxide following treatment with water ⟨63JA2684⟩. Alkoxide ligands can themselves be exchanged. For example, tetrabutyl 2-phenylethylidene-1,1-bisboronate was heated with ethylene glycol to generate the diethylene boronate ester (**214**) ⟨69T1571⟩. Halides bound to boron can be replaced with alkyne ⟨93OM2423⟩, alkoxide ⟨76JA1798⟩ or with hydroxide ligands ⟨66JOM(6)100, 90JOC2274⟩. Chloride has also been substituted by nitrogen ligands such as during the preparation of (**215**) (Scheme 48) ⟨92CB2213⟩. Carbon nucleophiles can also displace halides, such as the reaction of (**216**) with lithium phenylacetylide (Equation (56)) ⟨93CB2003⟩. In another example, the diborole (**218**) was prepared from (**217**) and 2-butyne. Subsequently, trimethylaluminum caused displacement of the remaining iodides to give (**219**) (Scheme 49) ⟨90ZN(B)15⟩. Halides can be exchanged; for example 1,1-bis(dichloroboryl)ethane was transformed into the corresponding tetraiodo derivative by boron triodide ⟨90ZN(B)15⟩. In another example of ligand exchange, 1,3,5-trichloro-1,3,5-triboracyclohexane was prepared from the corresponding 1,3,5-tri(dimethylamino) derivative (**220**) and BCl$_3$ (Equation (57)) ⟨71CB1347⟩.

4.14.1.4 Functions Bearing a Silicon and a Germanium Group: $R^1_2CSiR^2_3GeR^3_3$, etc.

Systems containing both a silyl and a germyl group attached to the same sp^3-hybridised carbon along with either hydrogen or carbon fragments are well known. The preparation of such systems follows closely those described earlier in this chapter for the formation of 1,1-bis(silyl) or 1,1-bisgermyl alkanes (see 4.14.1.1 and 4.14.1.2). Once the Si—C—Ge backbone has been formed, further systems may be prepared by the modification of groups attached to the central carbon or to the metalloids.

4.14.1.4.1 Formation of the Si—C—Ge linkage

The most common route for the preparation of α-germylsilyl alkanes is through the addition of a germyl electrophile to an α-metallosilyl alkane. Similarly, products may be formed by the reaction of an α-metallogermyl alkane with a silyl electrophile. Less common methods that have been described include those proceeding through rearrangements, via germenes, or as a result of hydrogermylation.

(i) Quenching a carbanion with a germyl electrophile

Methods for the generation of the requisite α-metallosilyl alkanes are described in greater detail elsewhere (see 4.14.1.1.1(i) and 4.14.2.1). Once generated, reaction with a germyl electrophile will result in the formation of the desired Si—C—Ge backbone. Systems that have been prepared in this manner are listed in Table 2. Entries 11–15 demonstrate that the reaction between the enolate of an α-silyl carbonyl compound and a germyl electrophile results in C-germylation. Reaction of a germyl enolate with a silyl electrophile gives the O-silylated product ⟨76MI8501⟩, a result which reflects the chemistry described earlier for the bisgermyl and bis(silyl) alkanes.

An α-silyl organomercury intermediate was used to generate a 1-germyl-1-silyl alkane system when (**221**) was reacted with trichlorogermane to give (**222**) which upon thermolysis eliminated mercury to give trichlorogermyl(trimethylsilyl)methane (Scheme 50) ⟨70IC1060⟩.

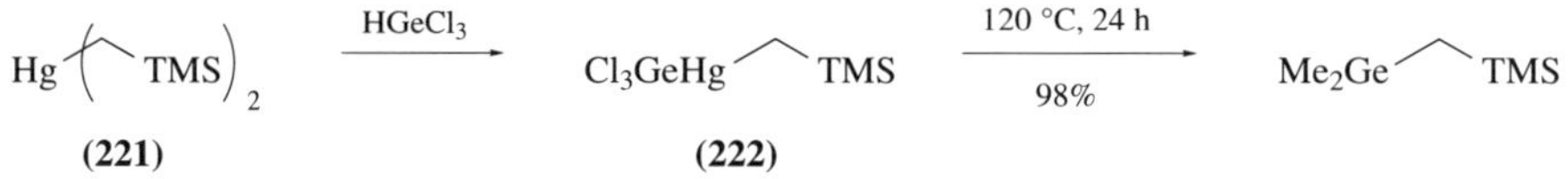

Scheme 50

Many cyclic systems have been prepared using similar chemistry. Thus treatment of the diorganolithium reagent (**223**) with dimethylgermanium dichloride gave the 1-germa-3-silacyclobutane (**224**) (Equation (58)) ⟨88JOM(338)159⟩. Di-Grignard reagents may also be used (Equation (59)) ⟨70JOM(21)103⟩.

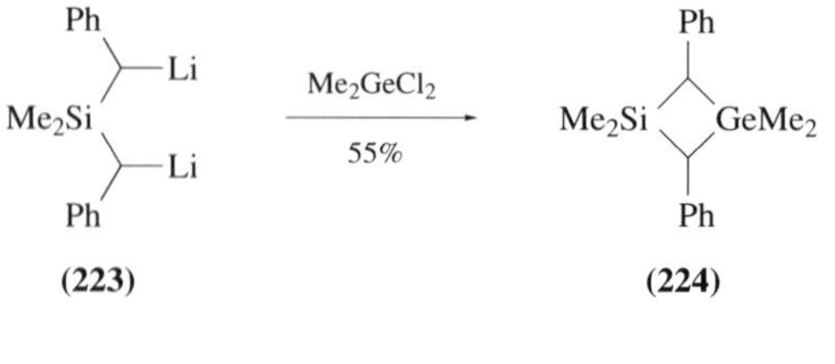

Table 2 Preparation of α-germylsilylalkanes from α-metallosilyl alkanes.

Entry	α-Metallosilylalkanes	Electrophile	Product	Yield (%)
1[a]	TMS–CH₂–Li	Me₃GeCl	TMS–CH₂–GeMe₃	72
2[b]		PhMe₂GeCl	TMS–CH₂–GePhMe₂	64
3[b]		MePhGeBr₂	(TMS–CH₂)₂GeMePh	
4[b]		GeCl₄	(TMS–CH₂)₄Ge	33
5[c]	TMS–CH₂–MgCl	GeCl₄	(TMS–CH₂)₄Ge	80
6[d]	Me₂HSi–CH₂–MgBr	Me₃GeBr	Me₂HSi–CH₂–GeMe₃	30
7[e]		Me₃GeCl	Me₂HSi–CH₂–GeMe₃	20
8[f]		TPPGeCl₂	(Me₂HSi–CH₂)₂GeTPP	90
9[g]	PhMe₂Si–CH₂–MgCl	Me₂Ge(CH₂Cl)Cl	PhMe₂Si–CH₂–GeMe₂(CH₂Cl)	67
10[h]	TMS–CH(CN)–Li	Me₃GeBr	TMS–CH(CN)–GeMe₃	86
11[i]	TMS–CH(CO₂Me)–Na	Me₃GeBr	TMS–CH(CO₂Me)–GeMe₃	
12[j]	TMS–CH(CO₂Et)–Li	Me₃GeCl	TMS–CH(CO₂Et)–GeMe₃	79
13[k]	TMS–CH(CO₂Bu^t)–Li	Me₃GeCl	TMS–CH(CO₂Bu^t)–GeMe₃	83
14[l]	TMS–CH(CONMe₂)–Li	Me₃GeCl	TMS–CH(CONMe₂)–GeMe₃	95
15[l]	TMS–CH(CO₂N(CH₂)₅)–Li	Me₃GeCl	TMS–CH(CO₂N(CH₂)₅)–GeMe₃	84
16[m]	cyclopentadienyl(Li)(TMS)	Me₂GeCl₂	cyclopentadienyl(GeMe₂Cl)(TMS)	64
17[n]		Me₃GeCl	cyclopentadienyl(GeMe₃)(TMS)	86
18[o]	Li–CHPh–SiMe₂–CHPh–Li	Me₃GeCl	Me₃Ge–CHPh–SiMe₂–CHPh–GeMe₃	70

[a] 80JA1584. [b] 84BCJ796. [c] 75JOM(96)213. [d] 90ZAAC(583)195. [e] 93JA6025. [f] 77JOM(137)157. [g] 69ZOB2601. [h] 86OM1197. [i] 76MI 414-01. [j] 88OM739. [k] 88JOM(348)25. [l] 88JOM(354)155. [m] 85OM2206. [n] 71JOM(30)C57. [o] 88JOM(338)159. TPP = tetraphenylporphyryl.

(ii) Quenching a carbanion with a silyl electrophile

The preparation of α-germyl silyl alkanes through the intermediacy of α-metallogermyl alkanes has not been a frequently used method. Those examples that have been reported are shown in Table 3. The preparation of α-metallogermyl alkanes are detailed elsewhere (see 4.14.1.2.1(i) and 4.14.2.2).

Table 3 Preparation of α-germysilyl alkanes from α-metallogermyl alkanes.

Entry	α-Metallosilyl alkanes	Electrophile	Product	Yield (%)
1[a]	Me$_3$Ge−CH$_2$−MgCl	MeSi(OMe)$_3$	Me$_3$Ge−CH$_2$−SiMe(OMe)$_2$	76
2	Me$_3$Ge−CH(Li)−CO$_2$Bu^t	TMS-Cl	Me$_3$Ge−CH(TMS)−CO$_2$Bu^t	79[b], 91[c]
3[d]	Me$_3$Ge−CH(Li)−C(O)N(piperidine)	TMS-Cl	Me$_3$Ge−CH(TMS)−C(O)N(piperidine)	50

[a] 54JOC250. [b] 88JOM(348)25. [c] 91JOC347. [d] 88JOM(354)155.

Again, cyclic systems have also been prepared through the intermediacy of α-metallogermyl alkanes. For example, the Grignard reagent of (**225**) was formed and subsequently cyclised to give 1,1,3,3-tetramethyl-1-germa-3-silacyclobutane (Equation (60)) ⟨69DOK(188)120, 69ZOB2281⟩.

$$Me_2BrSi-CH_2-GeMe_2-CH_2-Cl \xrightarrow{Mg} Me_2Si\square GeMe_2 \qquad (60)$$

(**225**)

(iii) Quenching a metalloid anion with a carbon electrophile

As with the preparation of 1,1-bis(silyl) and 1,1-bisgermyl alkanes described earlier, a metalloid anion can displace electrofugal groups bound to the central carbon to form the Si—C—Ge backbone. Although both silicon and germanium anions can be formed, only the latter have been demonstrated as intermediates in the preparation of α-germylsilyl alkanes. Those examples are listed in Table 4. The yield for Entry 3 is low due to competing reaction at silicon giving the disubstituted product shown and other products formed as a result of the route used to prepare the germyllithium reagent.

Table 4 Preparation of α-germylsilyl alkanes from germyl anions.

Entry	Electrophile	Germyl anion	Product	Yield (%)
1[a]	H$_3$Si−CH$_2$−Cl	NaGeH$_3$	H$_3$Si−CH$_2$−GeH$_3$	35
2[b]	TMS−CH$_2$−Cl	NaGeH$_3$	TMS−CH$_2$−GeH$_3$	35
3[c]	Me$_2$ClSi−CH$_2$−Cl	LiGeMe$_3$	Me$_2$(Me$_3$Ge)Si−CH$_2$−GeMe$_3$	29
4[d]	TMS−CH$_2$−Cl	LiGe(TMS)Et$_2$	TMS−CH$_2$−Ge(TMS)Et$_2$	
5[e]	TMS−CH$_2$−Cl	LiGe(vinyl)$_3$	TMS−CH$_2$−Ge(vinyl)$_3$	89

[a] 72IC408. [b] 70MI414-01. [c] 89OM1237. [d] 85ZOB2396. [e] 80IZV1451.

(iv) Other routes

α-Germylsilyl alkanes have been prepared by a number of other routes. For example, passing a stream of di(chloromethyl)dimethylsilane over Ge—Cu resulted in the formation of a number of materials containing a Si—C—Ge linkage (Equation (61)) ⟨72ZOB1521, 73ZOB625⟩. In another case, trichloro(chloromethyl)silane was shown to react with germanium powder at 350 °C to form (trichlorosilylmethyl)germanium trichloride and bis(trichlorosilylmethyl)germanium dichloride ⟨90ZN(B)961⟩. The former was also generated from (iodomethyl)trimethylsilane and $Et_3NHGeCl_3$ ⟨70ZOB2601⟩.

$$Me_2Si(CH_2Cl)_2 \xrightarrow{\text{Ge–Cu, 370–390 °C}} (Me_2Si\text{–}CH_2\text{–}GeMe_2)_n + \text{cyclic} + TMS\text{–}CH_2\text{–}GeCl_3 \quad (61)$$

45%   6%   20%

α-Germylsilyl alkanes may also be generated by rearrangements. For example, treatment of the silyl germanes **(226)** with boron trichloride in pentane resulted in a 1,2-silyl shift with loss of methoxide to form the 1-germyl-1-silyl alkanes **(227)** (Equation (62)) ⟨86TL4015⟩. A rearrangement resulting in the cleavage of a Si—Ge bond and the formation of an α-germylsilyl alkane has also been observed in transition metal chemistry. Equations (63) and (64) show the photolytic rearrangement of both tungsten and iron complexes containing a (germylsilyl)methyl or a (silylgermyl)methyl ligand ⟨93OM3979⟩.

$$(TMS)_2Ge(R^1)(R^2)(OMe) \xrightarrow{BCl_3,\ pentane} TMS\text{-}ClGe(R^1)(R^2)(TMS) \quad (62)$$

(226) → **(227)**

R^1	R^2	Yield (%)
TMS	H	99
TMS	Bu^t	71
TMS	Ph	99
Ph	H	56
Et	H	75
Vinyl	H	71

$$M\text{–}CH_2\text{–}SiMe_2\text{–}GeMe_3 \xrightarrow{h\nu} M\text{–}SiMe_2\text{–}CH_2\text{–}GeMe_3 \quad (63)$$

M	Yield (%)
CpFe(CO)$_2$	59
CpW(CO)$_3$	65

$$M\text{–}CH_2\text{–}GeMe_2\text{–}TMS \xrightarrow{h\nu} M\text{–}GeMe_2\text{–}CH_2\text{–}TMS \quad (64)$$

M	Yield (%)
CpFe(CO)$_2$	80
CpW(CO)$_3$	50

Hydrogermylation has also been shown to produce α-germylsilyl alkanes ⟨70IZV2280⟩. Under palladium catalysis, addition of triphenylgermane to the silylallene **(228)** was shown to produce two regioisomers, one of which was 3-dimethylphenylsilyl-3-triphenylgermyl-1-propene **(229)** (Equation (65)) ⟨88BCJ2693⟩.

$$PhMe_2Si\text{–}CH\text{=}C\text{=}CH_2 \xrightarrow{Ph_3GeH,\ Pd(PPh_3)_4} \underset{\textbf{(229)}\ 38\%}{Ph_3Ge,\ PhMe_2Si\ \text{allyl}} + \underset{57\%}{PhMe_2Si\text{–}CH\text{=}CH\text{–}CH_2GePh_3} \quad (65)$$

(228)

The use of coordinatively unsaturated metalloid intermediates such as silenes or disilenes to produce 1-germyl-1-silyl alkanes has not been widely explored. In one example, the digermene intermediate (**230**), generated under thermal conditions, reacted with a silacyclopentadiene to give (**231**) in 67% yield (Scheme 51) ⟨82CL1855⟩.

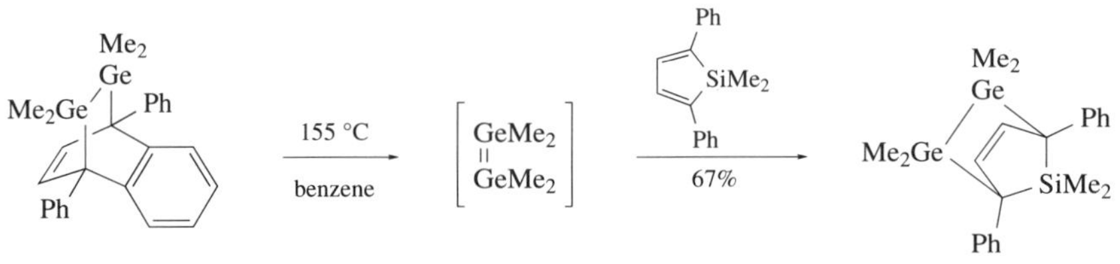

Scheme 51

A ketene also gave a 1-silyl-1-germyl derivative following treatment with either water or an amine (Scheme 52) ⟨76ZOB594⟩.

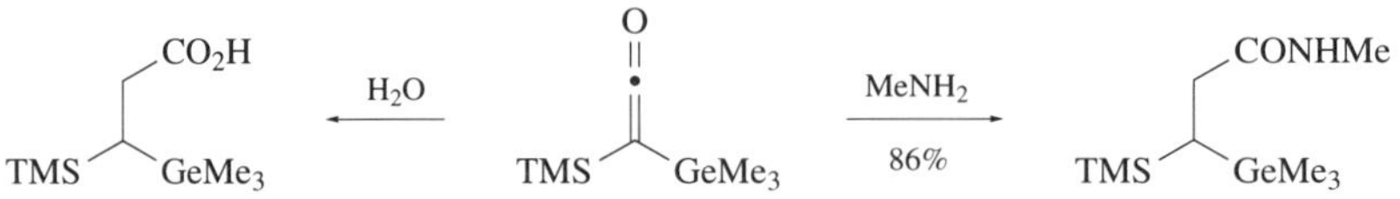

Scheme 52

4.14.1.4.2 Changing the groups attached to the central methylene

Novel α-germylsilyl alkanes can be produced by the modification of the groups attached to the central carbon. Although the chemistry required to do this is identical to that widely used with the analogous bis(silyl) and bisgermyl alkanes described earlier (4.14.1.1.2 and 4.14.1.2.2), little work has been performed in the area.

Metallation of the central carbon has been demonstrated by the deprotonation of trimethylgermyl(trimethylsilyl)methane with *t*-butyllithium. The same intermediate may also be prepared from the corresponding α-chloro derivative through metal–halogen exchange with lithium metal ⟨80JA1584⟩.

4.14.1.4.3 Changing the groups attached to the metalloids

Groups on both the silicon and the germanium may be modified to generate novel α-germyl silyl alkanes, and again the chemistry to do this reflects that used more extensively with the analogous bis(silyl) and bisgermyl alkanes (see 4.14.1.1.3 and 4.14.1.2.3). For example, one of the most frequently used transformations was the replacement of a halogen by a carbon nucleophile—a reaction which has also been applied to 1-silyl-1-germyl alkanes. Thus the chlorides bound to germanium in (**232**) have been displaced following treatment with methyl magnesium chloride to generate the octamethyl derivative ⟨72ZOB1521⟩. Sulfur has also displaced halogens bound to both metalloids ⟨90SRI1373⟩.

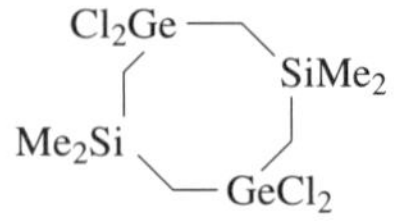

(**232**)

The formation of halogen—silicon or halogen—germanium bonds has also been well described for these systems. For example the cleavage of Si—O bonds by $BF_3 \cdot OEt_2$ generated fluorides bound to silicon ⟨76ZOB837⟩ while the use of thionyl chloride resulted in substitution by chloride ⟨73ZOB625⟩. In a further example, the displacement of a phenyl group on silicon by bromide was accomplished using bromine ⟨69DOK(188)120, 69ZOB2601⟩. The replacement of a germanium-bound phenyl group by a halogen has also been demonstrated. Iodine monochloride replaced the phenyl group of

(**233**) with chloride which was subsequently replaced by a hydride following treatment with LAH (Equation (66)) ⟨84BCJ796⟩. A methyl group bound to either silicon or germanium can be replaced by a chloride by treatment with concentrated sulfuric acid then ammonium chloride ⟨90JOM(387)65⟩. Reaction with halogens can result in the cleavage of the Si—C—Ge linkage such as in the reaction of tetrakis(trimethylsilylmethyl)germane with bromine which gave tris(trimethylsilylmethyl)bromogermane in 79% yield ⟨84BJC796⟩. Subsequent reduction with LAH gave tris(trimethylsilylmethyl)germane which was treated with diethylmercury and then lithium foil to generate the germyllithium (**234**) (Equation (67)) ⟨75JOM(96)213⟩.

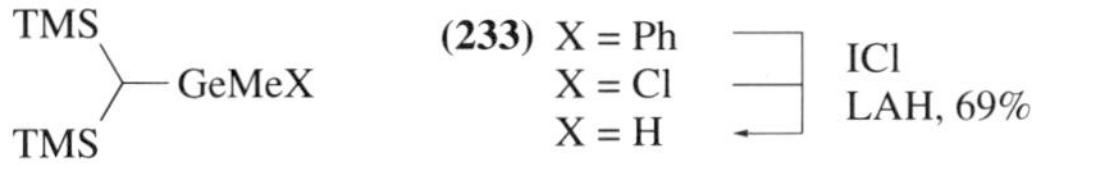
TMS
GeMeX
TMS
(233) X = Ph
X = Cl
X = H
ICl
LAH, 69%

(66)

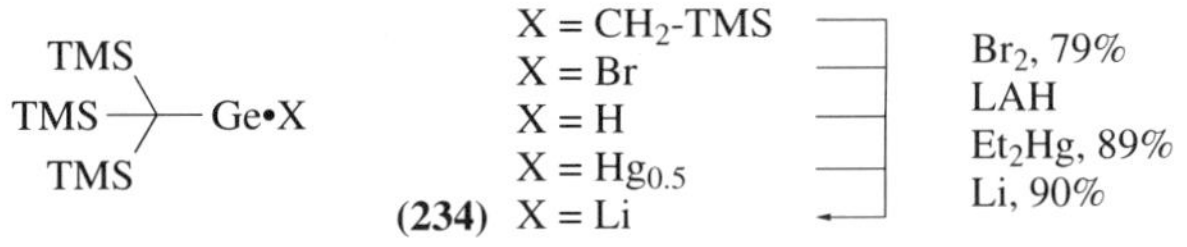
TMS
TMS
Ge•X
TMS
X = CH2-TMS
X = Br
X = H
X = Hg0.5
(234) X = Li
Br2, 79%
LAH
Et2Hg, 89%
Li, 90%

(67)

Alkoxide ligands on the metalloids have been displaced by both oxygen and nitrogen ligands ⟨80ZOB1764⟩. Some selectivity between the reaction of ligands on silicon over those on germanium has been demonstrated. Reaction of germyl(silyl)methane (**235**) with HCl in the presence of aluminum trichloride gave the monochloro derivative (**236**) in 65% yield (Equation (68)). This reactivity difference was not anticipated as methylgermane reacts more readily with halides than its silyl analogue. Increasing the quantity of HCl resulted in the replacement of two of the hydrides on silicon with chlorides in 36% yield ⟨72IC408⟩.

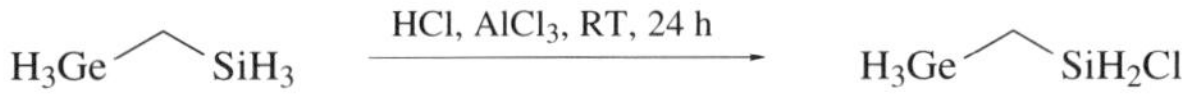
H3Ge
SiH3
HCl, AlCl3, RT, 24 h
H3Ge
SiH2Cl

(68)

(**235**) (**236**)

The silyl(germyl)methane moiety has been used as a ligand for transition metals. For example, the iron complex (**237**) was generated by photochemical addition of iron pentacarbonyl to dimethylgermyl(dimethylsilyl)methane. This has been shown to exhibit many reactions resulting in substitution on both silicon and germanium, some of which are shown in Scheme 53 ⟨90JOM(387)65⟩. Other bidentate silyl(germyl)methyl ligands have been bound to cobalt ⟨75ZOB1905⟩ and to thallium ⟨76IZV1342⟩.

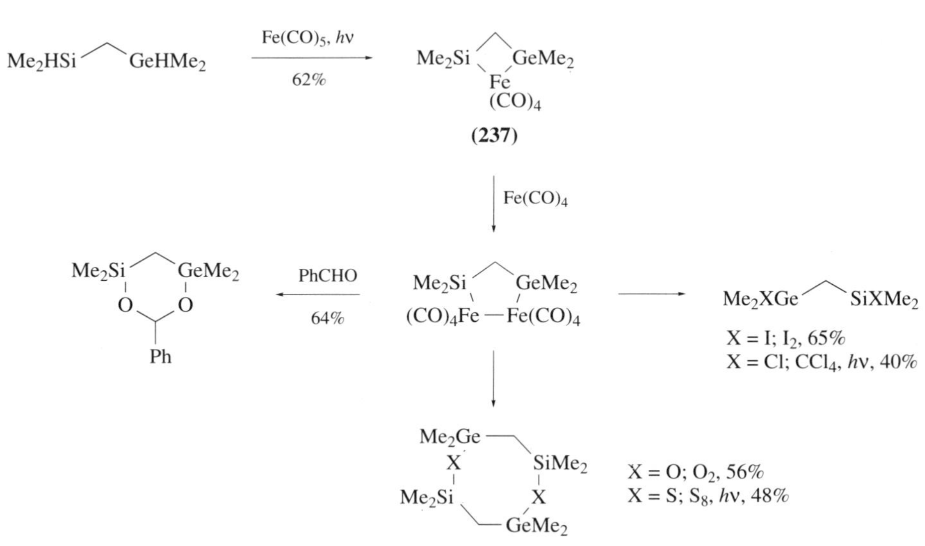
Me2HSi
GeHMe2
Fe(CO)5, hν
62%
Me2Si
GeMe2
Fe
(CO)4
(237)
Fe(CO)4
Me2Si
GeMe2
(CO)4Fe—Fe(CO)4
PhCHO
64%
Me2Si
GeMe2
O
O
Ph
Me2XGe
SiXMe2
X = I; I2, 65%
X = Cl; CCl4, hν, 40%
Me2Ge
X
SiMe2
Me2Si
X
GeMe2
X = O; O2, 56%
X = S; S8, hν, 48%

Scheme 53

4.14.1.5 Functions Bearing a Silicon and a Boron Group: $R^1{}_2CSiR^2{}_3BR^3{}_2$, etc.

The preparation of systems containing both a silyl and a boryl group attached to the same sp^3-hybridised carbon along with either hydrogen or carbon fragments are described here. Most of the methods used are an extension of the chemistry described for the formation of 1,1-bis(silyl) and 1,1-bisboryl alkanes (see 4.14.1.1 and 4.14.1.3). α-Borylsilyl alkanes may be generated by the formation of either a B—C bond or a Si—C bond, or by the modification of the groups on either the central methylene or the two metalloids.

4.14.1.5.1 Formation of the Si—C—B linkage

The most common route for the preparation of α-borylsilyl alkanes is through the addition of a boryl electrophile to an α-metallosilyl alkane. The same products may also be generated by the reaction of an α-metalloboryl alkane with a silyl electrophile. Other methods have also been described, such as hydroboration or boryl-boration of vinyl silanes.

(i) Quenching a carbanion with a boryl electrophile

General methods for the generation of the requisite α-metallosilyl alkanes are described in greater detail elsewhere (see 4.14.1.1.1(i) and 4.14.2.1). Once generated, reaction with a boryl electrophile will result in the formation of the desired Si—C—Ge framework. For example, reaction of trimethylsilylmethylmagnesium chloride with trimethylborate yielded dimethoxy(trimethylsilylmethyl)borane in 65–75% yield ⟨80CC39, 83OM230⟩. This Grignard reagent and a number of others have displaced a fluoride bound to boron to generate other α-borylsilyl alkanes ⟨59JA1844, 80IC1021, 86CB3253⟩. The corresponding organolithium reagents may also be used ⟨82CC970⟩. Quaternary borate salts have been prepared by reaction of trimethylsilylmethyllithium with trialkylboranes ⟨80IZV2429, 84JOC1096⟩.

α-Borylsilyl alkanes have been prepared through rearrangements, for example reaction of (**238**) with chloro(trimethylsilyl)methyllithium generated (**240**) through a 1,2-migration from the initially formed boron 'ate' complex (**239**) (Scheme 54). Some stereoselectivity was observed, for example (**240**; R = Ph) was formed in a 73:27 ratio of diastereomers ⟨83OM236⟩.

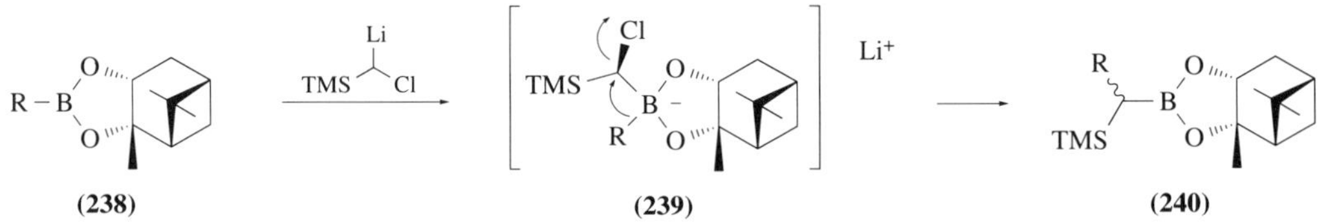

(**238**)   (**239**)   (**240**)

R = CH₂CH, 45%; Me(H)CCH, 64%; Ph, 88%; Me₂HCCH₂, 60%

Scheme 54

(ii) Quenching a carbanion with a silyl electrophile

Carbanions α to boron are readily synthesised and upon quenching with silyl electrophiles will yield 1-boryl-1-silyl alkanes. The general preparation of α-metalloboryl alkanes is discussed in more detail elsewhere (see 4.14.2.3).

Dimesitylborylmethyllithium has been prepared and quenched with TMS-Cl to yield dimesitylboryl(trimethylsilyl)methane ⟨83TL637⟩. In another example, the 2-benzoborole (**241**) upon treatment with lithium 2,2,6,6-tetramethylpiperidide gave the isolable dianion (**242**) which reacted with TMS-Cl to generate the α-borylsilyl alkane (**243**) in good yield (Equation (69)) ⟨93CB1397⟩. In a similar fashion, 1-(diisopropylamino)-3-borolene has been dimetallated with *t*-butyllithium and subsequently treated with TMS-Cl to give 1-(diisopropylamino)-2,3-di(trimethylsilyl)-4-borolene ⟨83ZN(B)1388⟩.

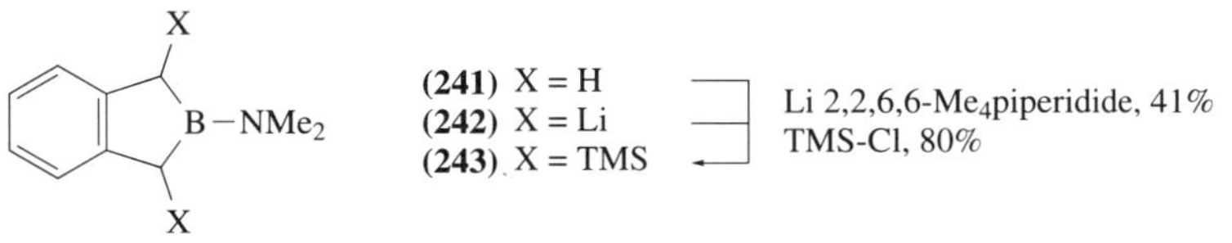

$$\begin{array}{c} \text{(241) X = H} \\ \text{(242) X = Li} \\ \text{(243) X = TMS} \end{array} \quad \begin{array}{c} \text{Li 2,2,6,6-Me}_4\text{piperidide, 41\%} \\ \text{TMS-Cl, 80\%} \end{array} \quad (69)$$

(iii) Hydroboration

Another route for the preparation of α-boryl silyl alkanes that has been demonstrated is hydroboration of a vinyl silane. For example, hydroboration of vinyltrimethylsilane with borane is a high yielding process generating a mixture of regioisomers favouring formation of the α-boryl silyl alkane. Substitution on the β-carbon greatly increases the selectivity observed; a single methyl group increased the selectivity to 95:5 (Equation (70)). The intermediate α-borylsilyl alkanes (**244**) were not isolated although NMR evidence has been reported ⟨88T4033⟩. Increasing the steric bulk on the hydroborating reagent was found to shift the preference of boron to addition to the β-carbon (Equation (71)) ⟨78JOM(156)C12, 80JOC3571, 84JOC2565⟩. Quenching the hydroboration mixture with methanol enabled silaborolane (**245**) to be isolated from the reaction mixture ⟨84JOC2565⟩. Other hydroborating reagents have also been used ⟨59JA1844⟩. Asymmetric hydroboration of vinyl silanes has been shown using monoisopinocamphenylborane, but the diastereomeric excesses observed were not great ⟨88T4033⟩. Homochiral α-boryl silyl alkanes have also been prepared through the resolution of diastereomeric pairs. For example, hydroboration of 1-(trimethylsilyl)-1,3-butadiene with borane dimethylsulfide provided *B*-methoxy-2-(trimethylsilyl)borolane (**246**) in 45% yield. The products having a new stereocentre were subsequently resolved by crystallisation using (1*S*,2*S*)-(+)-*N*-methylpseudoephedrine (Scheme 55) ⟨89JA1892⟩.

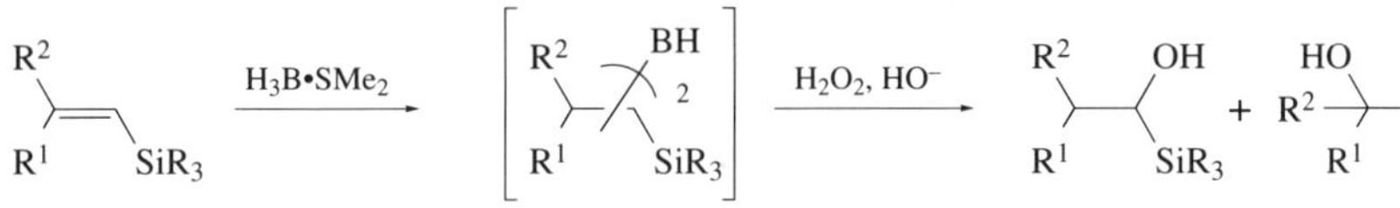

SiR_3	R^1	R^2	Yield (%)			
TMS	H	H	98	57	:	43
TMS	H	Me	77	95	:	5
TMS	Me	Me	79	>98	:	2
SiMe₂Ph	Me	Me	80	>98	:	2

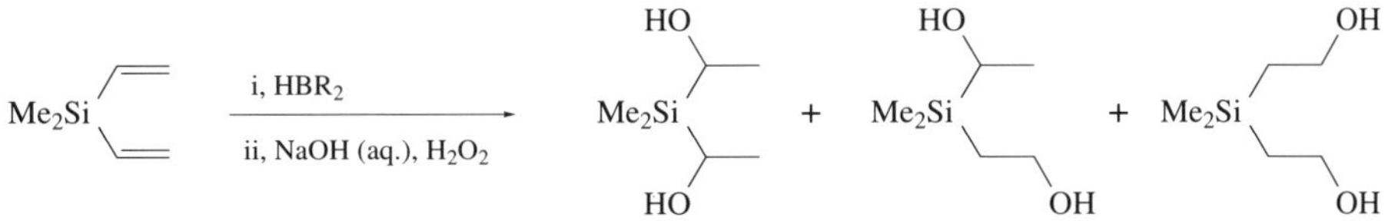

HBR₂	Yield (%)					
H₃B·SMe₂	100	38	:	56	:	6
H₂B(thexyl)	82	17	:	68	:	15
HB(cyclohexyl)₂	100	0	:	5	:	95
9-BBN-H	100	0	:	0	:	100

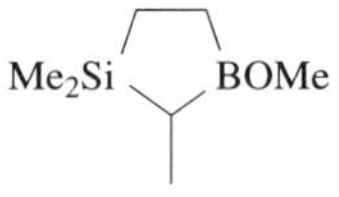

(**245**)

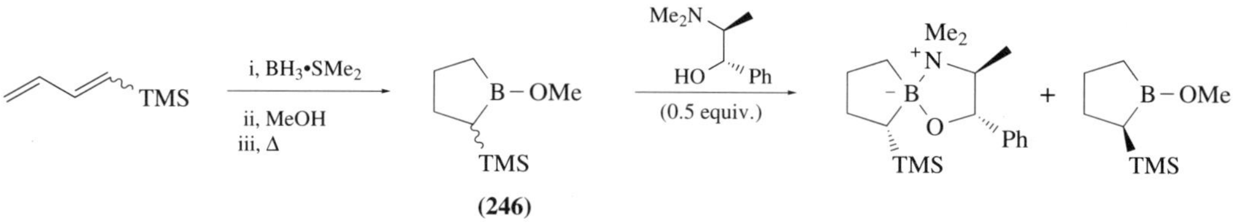

Vinyl silanes need not be the only precursors for preparation of 1-boryl-1-silyl alkanes through hydroboration. The double hydroboration of silyl acetylenes has also been demonstrated. 1-Trimethylsilyl-1-propyne may be monohydroborated with 9-BBN-H to place the boron α to the silicon. The steric bulk on the α-carbon then forces the second boron to add the β-carbon to produce the 1,2-diborylated product (**247**) (Equation (72)) ⟨89JA4873⟩.

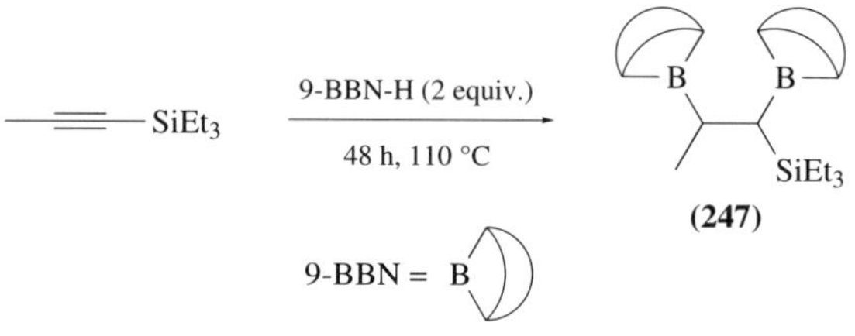

(iv) Other routes

In addition to the hydroboration of vinyl silanes, borylboration has also been accomplished to yield 1,2-diboryl-1-silyl alkanes. Tetrachlorodiborane will readily react with trimethyl(vinyl)silane to give 1,2-bis(dichloroboryl)-1-(trimethylsilyl)ethane ⟨68JOM(12)269⟩.

1,3-Disilacyclobutanes have also been used as precursors for the preparation of α-borylsilyl alkanes following ring opening with boron halides. The products obtained were dependent upon the reaction temperature (Scheme 56).

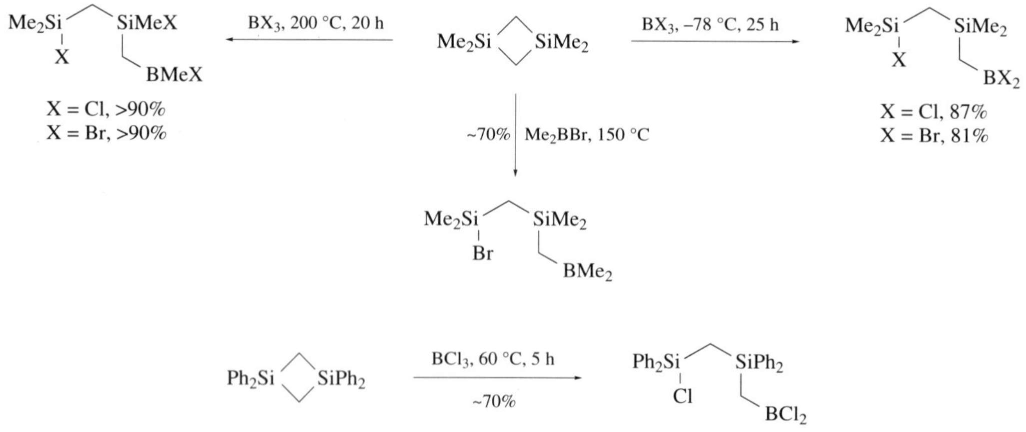

Scheme 56

The nucleophilic addition to 1-(dimesitylboryl)-1-trimethylsilyl-1-ethene (**248**) has been demonstrated as a powerful method for the preparation of α-borylsilyl alkanes. Cuprates and organolithium reagents would readily add, but Grignard reagents did not (Equation (73)). Substituents on the β-carbon of (**248**) are tolerated in the absence of any allylic protons in the substituents which are removed by organolithium reagents. The intermediate anion generated has also been quenched with methyl iodide ⟨87JA931⟩.

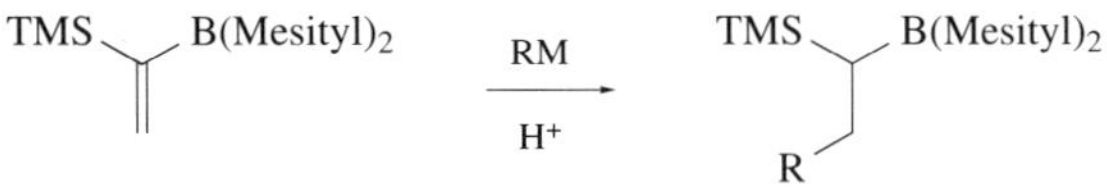

$$\text{TMS}\diagdown\text{C}=\text{CH}_2\diagup\text{B(Mesityl)}_2 \quad\xrightarrow[\text{H}^+]{\text{RM}}\quad \text{TMS}\diagdown\text{CH(R)}\diagup\text{B(Mesityl)}_2 \qquad (73)$$

(248)

RM	Yield(%)
BuLi	96
Bu$_2$Cu(CN)Li$_2$	66
PhLi	95
Bu^{t}Li	86

4.14.1.5.2 Changing the groups attached to the central methylene

The only method demonstrated for the manipulation of functionality on the central methylene of α-borylsilyl alkanes is through deprotonation followed by an electrophilic quench. This deprotonation has been accomplished with mesityl lithium ⟨83TL637⟩. LITMP has also been used and the resulting anions, such as (249) and (250) were quenched with several primary alkyl halides ⟨80CC39, 83OM230⟩. Secondary alkyl halides did not react in good yields, and 1-substituted-1-boryl-1-silyl alkanes could not be deprotonated under these conditions ⟨83OM230⟩.

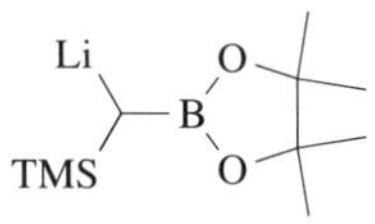

(249) 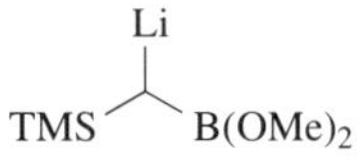   (250)

4.14.1.5.3 Changing the groups attached to the metalloids

The little work that has been performed to generate new α-borylsilyl alkanes through the modification of groups attached to either the silicon or the boron resembles that used with 1,1-bis(silyl) or 1,1-bisgermyl alkanes described earlier (see 4.14.1.1.3 and 4.14.1.3.3).

Ester exchange provides a simple route to the manipulation of the functionality on boronic acids, and was used to transform (251) to the (trimethylsilyl)methane boronate of pinacol via trimethylsilylmethane boronic acid ⟨83OM230⟩. Other groups on boron may also be modified. Methanolysis and hydrolysis of bis-[(1-trimethylsilyl)isobutyl]borane (252) gave the corresponding methoxy or hydroxy derivatives (253) and (254), respectively ⟨88T4033⟩.

TMS—CH$_2$—B(OMe)$_2$

(251)

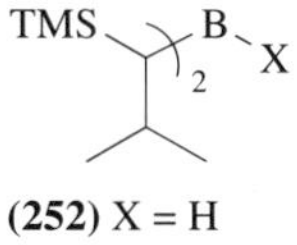

(252) X = H
(253) X = OMe
(254) X = OH

A C—B bond may be formed by the displacement of an alkoxide ligand, such as the reaction of an allyl Grignard reagent with (255) (Equation (74)) ⟨89JA1892⟩. *t*-Butyllithium was used to displace the methoxy ligand of (245) ⟨84JOC2565⟩.

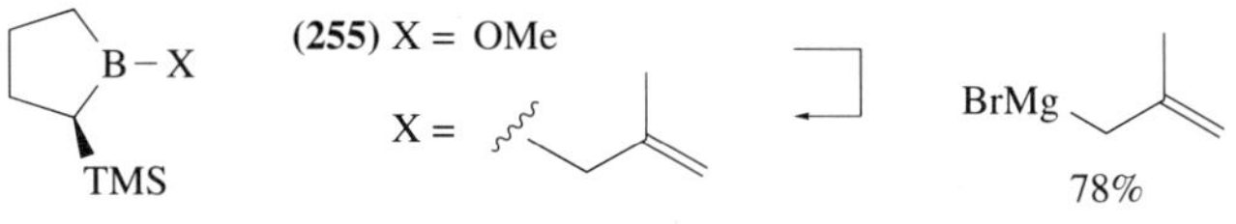

(255) X = OMe

$$\qquad\qquad\qquad\qquad\qquad\qquad\qquad\qquad\qquad\qquad\qquad\qquad\qquad\qquad 78\% \qquad (74)$$

No chemistry has been reported whereby the groups on the silicon of an α-borylsilyl alkane have been manipulated.

4.14.1.6 Functions Bearing a Germanium and a Boron Group: $R^1_2CBR^2_2GeR^3_3$, etc.

Very little work has been reported of the generation of systems containing both a boron and a germanium on the same sp^3-hybridised carbon with no other groups except carbon or hydrogens. One example was the hydroboration of dimethyl(divinyl)germane with borane dimethylsulfide which resulted in a mixture of regioisomers as evidenced by the alcohols produced upon oxidative workup (Equation (75)) ⟨84JOC2565⟩.

$$Me_2Ge(CH=CH_2)_2 \xrightarrow[\text{ii, NaOH (aq.), H}_2O_2]{\text{i, H}_3B\cdot SMe_2, \text{THF}} Me_2Ge(CH(OH)CH_3)_2 + Me_2Ge(CH(OH)CH_3)(CH_2CH_2OH) + Me_2Ge(CH_2CH_2OH)_2 \quad (75)$$

g.c. ratio   38 : 56 : 100

4.14.2 FUNCTIONS CONTAINING A METALLOID AND A METAL

4.14.2.1 Silicon and a Metal: $R^1_2CSiR^2_3M$ etc.

Methods for the preparation of a group consisting of an sp^3-hybridised carbon connected to a silyl group, a metal and either protons or carbon fragments are well known. Several papers or reviews have been published which include coverage of this area ⟨70CJC561, 72MI 414-01, 74JOC3264, 75JOM(93)71, 80T2531, 85S717, B-88MI 414-01⟩. An α-silyl carbanion is the precursor for the Peterson alkenation reaction and reviews of this subject have discussed in detail the preparation of these precursors ⟨84S384, 90SRI1373⟩. Preparations of α-metallosilyl alkanes as intermediates for the formation 1,1-bis(silyl) alkanes were discussed earlier in this chapter (see 4.14.1.1.1).

Of the several routes to the preparation of 1-metallo-1-silyl alkanes, the most common is deprotonation. This has been most generally performed when the resultant carbanion is stabilised, by for example aryl, allyl or carbonyl groups. Halogen–metal exchange has also been frequently used, as has the addition of an organolithium or organomagnesium reagent to a vinylsilane. Transmetallation of the resultant organometallic can substitute the accessible organometallic reagent with a number of other metals, especially transition-metal derivatives.

4.14.2.1.1 Silicon and a Group 1 or Group 2 metal: $R^1_2CSiR^3_3Li$, etc.

Several of the Group 1 and 2 metals have been used to prepare α-metallosilyl alkanes. Of these, the most common are lithium and magnesium; in fact, trimethylsilylmethyllithium and trimethylsilylmethylmagnesium chloride are now commercially available as solutions in pentane and ether, respectively.

(i) α-Lithiosilyl alkanes

There are several main methods for the preparation of α-lithiosilyl alkanes, the most common of which is deprotonation of an alkyl silane. Addition of an organolithium reagent to a vinylsilane has also been frequently used as has halogen–lithium exchange.

(a) By deprotonation. The simplest route to 1-lithio-1-silyl alkanes is by deprotonation. This has most frequently been applied to systems where the resultant carbanion may be stabilised by the remaining groups. This is particularly the case for benzylic systems, where deprotonation is readily performed. *n*-Butyllithium has usually been the base of choice in conjunction with a Lewis base such as TMEDA ⟨68JOC780, 84JCS(D)321, 84JCS(D)1801, 85JCS(D)337, 86JPS(A)1839, 87CB1695, 89JCS(D)105, 91CB543⟩, pentamethyldiethylenetriamine ⟨82CC14, 85JCS(D)337⟩ or HMPA ⟨70TL1137⟩. Other bases have been used, including methyllithium ⟨74JOC3264⟩, *s*-butyllithium ⟨81CC179, 89JOC4372⟩ and even LDA ⟨88CC807⟩. The additional stabilisation of the nascent anion by silicon enables selective benzylic deprotonation α to the silyl group in the presence of other nonsilylated benzylic sites. In examples where two identical benzylic sites are present, selective monolithiation or dilithiation may be controlled by the choice of additive. Pentamethyldiethylenetriamine favours monolithiation

whereas TMEDA enables dilithiation to occur in the presence of two equivalents of base (Equation (76)) ⟨85JCS(D)337⟩. This process has also been demonstrated with analogous systems ⟨80CC1284, 82CC14, 84JCS(D)321, 84JCS(D)1801, 85NJC249, 86JCS(D)603⟩. In many cases, the alkyllithium derivatives are stable solids, some of which have been studied by x-ray crystallography.

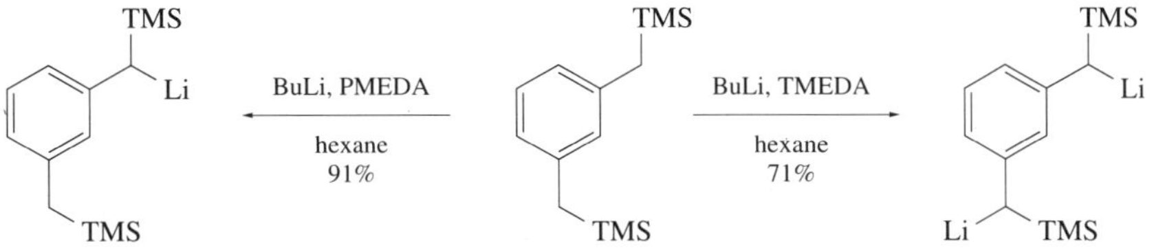

(76)

Deprotonation is not limited to trimethylbenzylsilanes; more highly substituted derivatives may also be deprotonated ⟨51JA5878, 70JA7567⟩ including cyclic derivatives such as (**256**) ⟨84JOM(269)C40⟩. 7-Lithio-7-trimethylsilylcyclopropabenzene was formed by deprotonation with *n*-butyllithium ⟨87CB471⟩. The corresponding cyclopropanaphthylene derivative has also been prepared ⟨84JA6108, 86JA5949⟩. Certain 9-silylfluorenes may also be deprotonated with either *n*-butyllithium or *t*-butyllithium to give the lithio derivatives (**257**) which were isolable ⟨79JOM(172)11, 92JOM(440)233⟩. Other aryl-stabilised systems that may be similarly prepared are shown in Figure 1 ⟨82CC1023, 84JCS(D)311, 84TL4645, 88JCS(D)2403, 92JCS(D)775⟩. Treatment of (**258**) with 1 equivalent of *n*-butyllithium in the presence of pentamethyldiethylenetriamine resulted in monolithiation, whereas the use of two equivalents of base and TMEDA gave the dianion (**259**) which eliminated methane to give the stable dianion (**260**) (Equation (77)) ⟨84JCS(D)311⟩.

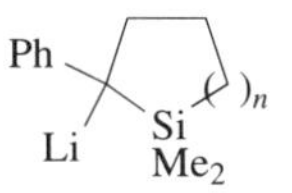

(**256**)
n = 1; MeLi, THF, 38%
n = 2; BuLi

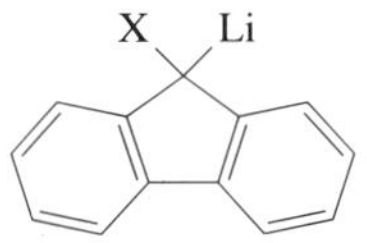

(**257**)
X = But$_2$BrSi
X = Pri$_2$ClSi

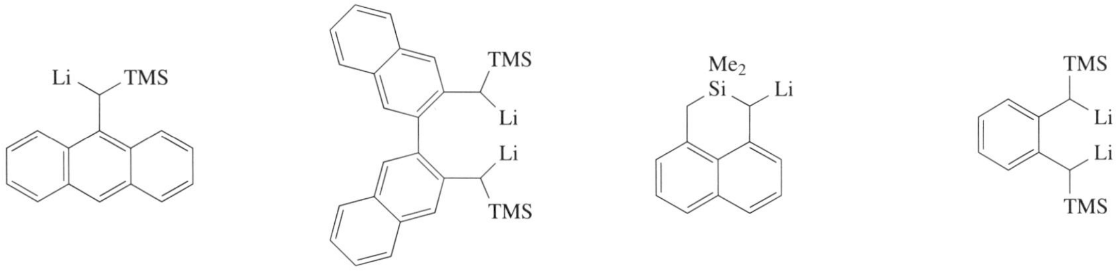

Figure 1

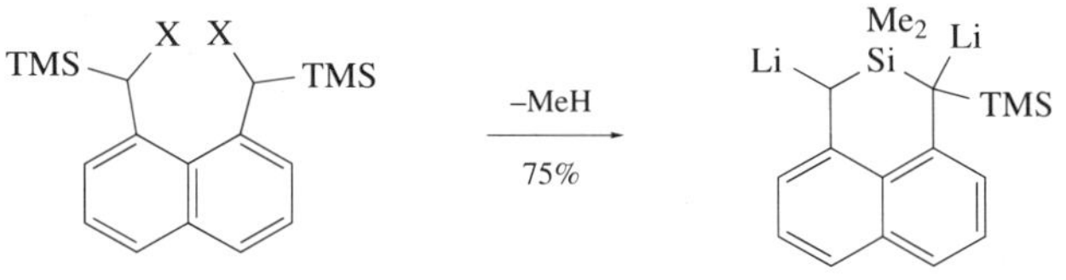

(**258**) X = H
(**259**) X = Li   2 BuLi, TMEDA, hexane, 83%

(**260**)

(77)

2- and 4-Substituted trimethylsilylmethylpyridines have been metallated with *n*-butyllithium and TMEDA to give the corresponding lithio derivatives which were characterised by x-ray crystallography ⟨84CC1708, 86JOM(316)C4, 86JCS(D)603⟩. This lithiation may also be performed with LDA ⟨91S1221⟩. Metallation of the corresponding 3-derivative has also been accomplished, but in much poorer yield ⟨90JCS(D)1161⟩.

Some work has been directed towards asymmetric deprotonation of benzyl silanes using a chiral ligand on silicon derived from (S)-(+)-2-(methoxymethyl)pyrrolidine. Deprotonation with s-butyllithium resulted in the formation of only one diastereomer (**261**) ⟨92JOC6107⟩. The use of a chiral Lewis base to direct deprotonation has also been investigated; however the asymmetric induction observed was not great. The presence of (−)-sparteine during the deprotonation of 2-trimethylsilylmethylpyridine with n-butyllithium gave after methylation at best only 20% enantiomeric excess ⟨90JCS(D)1161⟩.

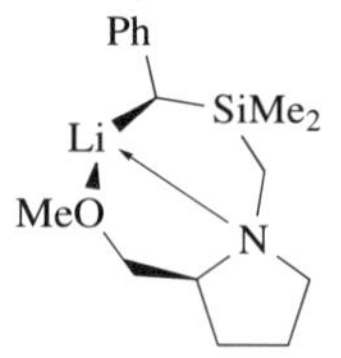

(**261**)

Allylsilanes may also be readily deprotonated with organolithium reagents to give silyl-substituted allyllithiums. These have been the focus of much study to understand both their structure and their chemical reactivity—especially the factors governing addition of an electrophile to either the α- or the γ-carbon ⟨78JOM(144)155, 84JOC1096, 90JA2582, 93JA1551⟩. Tables 5 and 6 list those anions derived from allylsilanes and substituted allylsilanes, respectively.

Table 5 Silyl-substituted allylithium reagents.

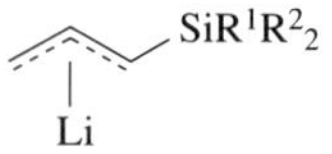

R^1	R^2	Metallating reagent	Ref.
Me	Me	BuLi, THF	77CC772
Me	Me	BuLi, TMEDA, Et$_2$O	73JOM(57)C5
Me	Me	BuLi, PMEDA, pentane	90JA2582
Me	Me	Bu^{s}Li, TMEDA, THF	80TL11, 88TL4281
CMe$_2$Pr^i	Me	Bu^{s}Li, TMEDA, THF	90JA1382
Me	Ph	Bu^{s}Li, THF	88TL4281
Ph	Ph	Bu^{s}Li, THF	88TL4281
Ph	Ph	BuLi, TMEDA, Et$_2$O	88TL4281
allyl	Ph	MeLi, TMEDA	73JOM(57)C5
CH$_2$N(CH$_2$CH$_2$OMe)$_2$	Me	BuLi, Et$_2$O	77JOM(127)281, 93JA1551

PMEDA = pentamethyldiethylenetriamine.

Deprotonation of propargylic systems may be accomplished in a similar fashion, for example the dianion (**262**) was prepared following reaction of 1-chloro-3-(trimethylsilyl)-1-propyne with n-butyllithium by a process of halogen–metal exchange followed by deprotonation ⟨78JOM(156)299⟩.

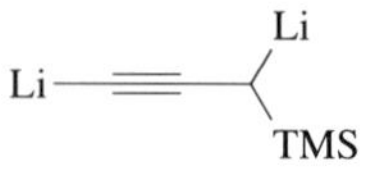

(**262**)

A third class of 1-lithio-1-silyl alkanes that are readily obtained by deprotonation are α-silyl enolates following reaction of of the corresponding α-silyl carbonyl compound by treatment with a base, usually LDA. The metallation of α-silyl carbonyl compounds has been frequently applied to esters or lactones during the preparation of α-β-unsaturated esters through a Peterson alkeneation process ⟨74JA1620, 74TL1403, 81CC877, 88JOC2274, 91JOC347⟩. In a similar manner, α-silyl amide derivatives may also be metallated ⟨78S746⟩. α-Silyl carboxylic acids have been deprotonated; for example, trimethylsilylacetic acid was dimetallated with LDA ⟨75CC536⟩. LDA was used to effect the deprotonation of trimethylsilylacetonitrile ⟨71JA1714⟩. α-Silyl imines and related species are also readily deprotonated (Table 7).

Table 6 Preparation of substituted silylallyllithium reagents.

Entry	Product	Reagents	Ref.
1	TMS–CH=CH–CH(Li)–TMS	Bu^{n}Li, TMEDA, Et$_2$O-THF	81JCS(P1)2415
		Bu^s, THF	90JA1382
		Bu^{s}Li, TMEDA, pentane	90JA1382
		Bu^{s}Li, PMEDA, Et$_2$O	90JA1382
2	TMS–CH=CH–CH(Li)–SiPh$_3$	Bu^{n}Li, TMEDA, pentane	73JOM(57)C5
3	(TMS)$_2$C–CH=CH–TMS with Li	Bu^{n}Li, TMEDA, pentane	90OM1314
4	cyclopropyl-CH(Li)–TMS (with =CH$_2$)	Bu^{n}Li, THF	85TL301
5	X=Y=H	LDA, HMPA, THF	80HCA555
		LiN(cyclohexyl)$_2$, HMPA, DME	80HCA555
		Bu^{n}Li, THF	83OM21
	X=H, Y=TMS	LiN(cyclohexyl)$_2$, HMPA, DME	80HCA555
	X=Y=TMS	LiN(cyclohexyl)$_2$, HMPA, DME	80HCA555
6	TMS–CHR–C(Li)=C(R)–CH(Li)–TMS, R=H, Me	2Bu^{n}Li, TMEDA	91OM3167

PMEDA = pentamethyldiethylenetriamine.

Table 7 α-lithio-α-silylimines.

Entry	SiR$^1_3$	R^2	R^3	Metallating reagent	Ref.
1	TMS	Bu^t	Me	LDA	76TL7
				Bu^{s}Li	78JA2916
2	SiEt$_3$	Bu^t	Me	Bu^{s}Li	85TL2391
3	TMS	NMe$_2$	H	LDA	78CB1362
4	TMS	NMe$_2$	Me	LDA	78CB1362
				LiNEt$_2$	76TL7

The removal of an acidic proton has also been extended to other systems, such as the lithiation of 1,2-diaza-3-sila-5-cyclopentenes with *n*-butyllithium to give (**263**) ⟨81ZAAC(482)65, 83ZN(B)953⟩. The 2-lithio(silyl)methyl-1,3-oxazine (**264**) has also been successfully generated from the protio compound with *n*-butyllithium ⟨76TL4041⟩.

Deprotonation is not only possible when the proton is rendered acidic by an electron-withdrawing group as above; direct deprotonation of an unactivated alkyl silane is also possible, although generally harsher conditions are required. Several examples were described earlier during discussion on the preparation of 1,1-bis(silyl)alkanes (see 4.14.1.1.1).

In 1967 it was shown that tetramethylsilane may be partially deprotonated by *n*-butyllithium in the presence of **TMEDA**, a process requiring several days ⟨67JOM(9)373⟩. In the early 1980s it was found that changing the Lewis base to pentamethyldiethylenetriamine enabled almost quantitative

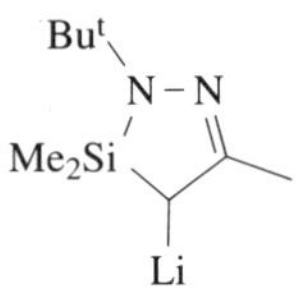

(263) 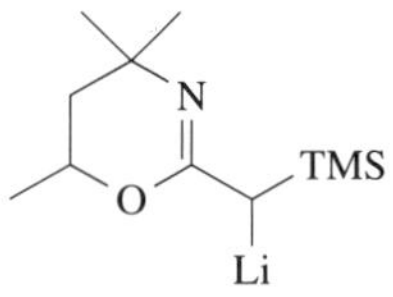 **(264)**

metallation to be achieved in just 12 h ⟨82CC1323⟩. Other simple alkyl silanes have also been metallated including *n*-butyl(trimethyl)silane by *n*-butyllithium—TMEDA and *s*-butyl(trimethyl)silane by *s*-butyllithium—TMEDA. 'Superbase' (*n*-butyllithium and potassium *t*-butoxide) has also been successfully used as the metallating agent ⟨86T1845⟩. Competing ligand exchange was observed during the attempted metallation of 1,1-dimethylsilacyclopentane **(265)** with *n*-butyllithium (Equation (78)) ⟨84JOM(273)C57⟩.

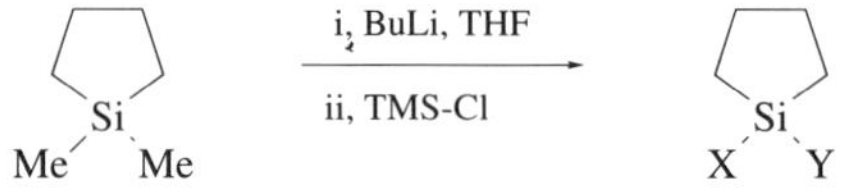

$$\text{(265)} \quad \xrightarrow[\text{ii, TMS-Cl}]{\text{i, BuLi, THF}} \quad \text{(78)}$$

X	Y	Yield(%)
Me	Me	4.3
Bu	Me	14.1
Bu	Bu	16.3
Me	CH$_2$-TMS	32.2
Bu	CH$_2$-TMS	33.1

The rate enhancement observed following addition of a Lewis base may be mimicked by the presence of other groups. For example, metallation of **(266)** by *n*-butyllithium unexpectedly resulted in metallation α to the silicon as a result of chelation by the phosphinoyl group (Equation (79)). This chelation could be destroyed by the use of TMEDA resulting in the generation of the thermodynamically favoured anion **(267)** ⟨77JCS(P1)550⟩. Activation of the metallating agent and subsequent stabilisation of the anion has been demonstrated by a β-amino group. This enabled the anions **(268)** to be generated in high yield with *n*-butyllithium in ether ⟨91JCS(P1)2276⟩. A similar effect has also been observed with the γ-nitrogen of **(269)**. In this case, metallation is also facilitated by the alkoxy group on silicon. Inductive electron withdrawal by the oxygen facilitates deprotonation as a result of the increased acidity of the α-proton ⟨87CPB1663⟩. This effect also enabled the deprotonation of a methyl group of methoxy(trimethyl)silane to be readily accomplished by *t*-butyllithium without the need for a Lewis base such as TMEDA to be used ⟨89JOM(359)285⟩. If *s*-butyllithium is used then the product obtained is methoxy(trimethylsilyl)methyllithium ⟨82OM553⟩. Siloxanes may also be metallated, for example hexamethyldisiloxane was deprotonated with *t*-butyllithium over a period of several days ⟨70JOC1308, 93OM338⟩.

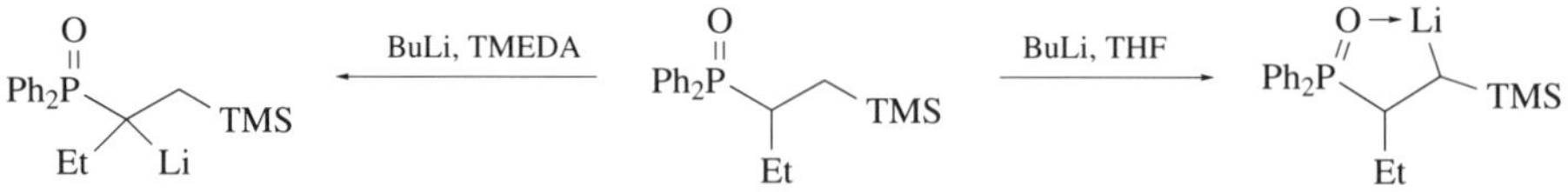

(267) ⇌ **(266)** → product (79)

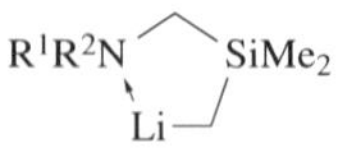

(268)   R^{1}R^2 = –(CH$_2$)–$_5$, BuBu

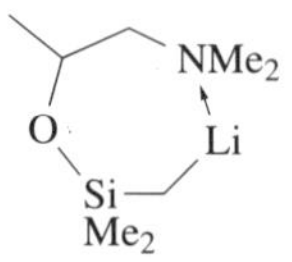

(269)

The increased susceptibility of alkoxy(alkyl)silanes to undergo deprotonation has in some cases led workers to unexpectedly observe metallation of silyl ether protecting groups. For example, attempted C(1) metallation of 3,4,6-tri-*O*-(TBDMS)-D-glucal with *t*-butyllithium led to a mixture of vinylic and α-silyl anions ⟨91JOC1944⟩. A similar occurrence resulted in the generation of (**270**) following almost exclusive metallation of the TBDMS group of an enol ether containing a silyl ether protecting group ⟨92TL543⟩. The trimethylsilyl group has also been unexpectedly metallated when used as a blocking group on an aromatic ring. Metallation of (**271**) gave not the anticipated 2-lithio derivative, but resulted in metallation of the trimethylsilyl group to give (**272**) (Scheme 57) ⟨87TL1851⟩, an effect also observed with the amino-substituted aryl(dimethyl)silane (**273**) where metallation of the methyl groups bound to silicon competed with 2-lithiation of the aromatic ring ⟨90TL2925⟩.

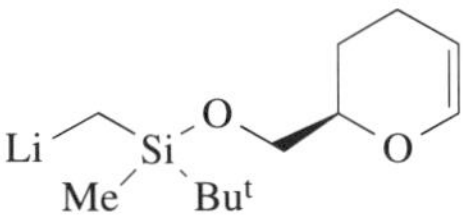

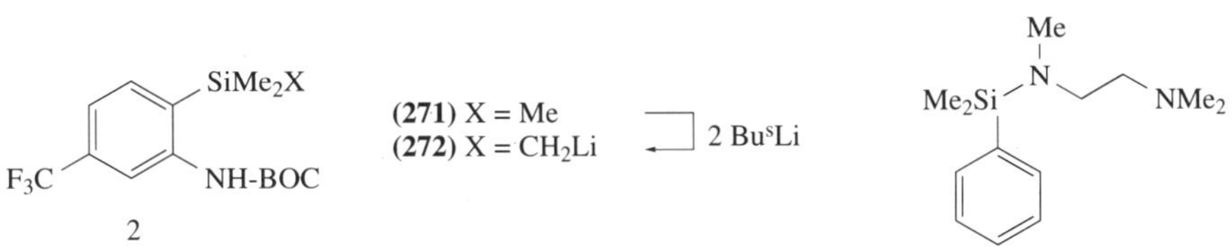

Scheme 57

(b) By addition of an organolithium to a vinylsilane. α-Lithiosilyl alkanes may be regioselectively prepared by the addition of organolithium reagents to vinylsilanes generating an α-lithiosilyl alkane. This was described earlier (4.14.1.1.1(iii)(a)) during the preparations of silenes, as intermediates to 1,1-bis(silyl) alkanes.

Several organolithium reagents have been shown to add to trimethyl(vinyl)silane and triphenyl-(vinyl)silane. The yields obtained generally reflected the reactivity of the organolithium reagent and the steric hindrance of the vinylsilane ⟨54JOC1278, 70CJC561⟩. This was also true when several chloro(vinyl)silanes were used where increased steric bulk of substituents on either the silicon or the α-carbon of the vinyl group resulted in reduced yields ⟨74JOC3264, 88TL6697, 92JOM(426)1, 93OM4135⟩. Even greater steric effects were observed on the β-carbon of several chloro(vinyl)silanes when any substitution resulted in the displacement of the chloro group rather than addition of *t*-butyllithium to the vinyl silane ⟨93OM4123⟩.

Attempts to achieve asymmetric induction during this addition using an epheridine-derived group on silicon have been reported (Scheme 58) ⟨84TL1913⟩.

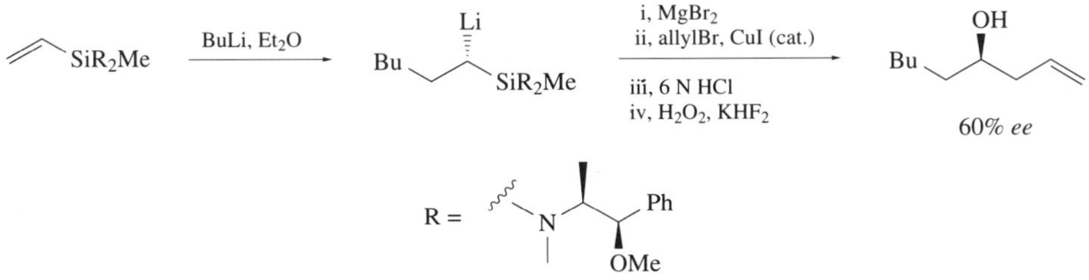

Scheme 58

Further reaction of the nascent α-lithio alkylsilane has been reported such as competition with the remaining organolithium reagent for addition to further vinylsilane. Alternatively, if the resultant 1-lithio-1-silyl alkane contains a silicon—halogen bond, then elimination of the lithium halide could give the corresponding silene. A similar 1,2-elimination can occur if there is a hydrogen on silicon. The nature of any further reaction is governed not only by the substitution pattern, but also by the reaction conditions (4.14.1.1.1(iii)(a)) ⟨87JOM(336)59, 88ZAAC(558)87⟩.

The addition of organolithium reagents to silylated α,β-unsaturated amidate anions derived from (**274**) also resulted in the generation of α-lithiosilyl alkanes (**275**) (Equation (80)). The placing of an anion α to the carbonyl group suppresses the usual 1,2-addition to the carbonyl group enabling conjugate addition to occur. Although β-substitution on the vinyl group is tolerated, β,β-disubstituted α,β-unsaturated amides failed to react ⟨93JOC7474⟩.

$$\underset{(274)}{\underset{R^2{}_3Si}{\overset{O}{\underset{H}{\overset{\|}{\underset{}{C}}}}}\diagdown N\diagup Ph} \xrightarrow{R^1Li\ (3\ equiv.),\ THF,\ -78\ °C,\ 4\ h} \underset{(275)}{\underset{R^2{}_3Si}{\overset{R^1}{\underset{Li}{C}}}\diagdown \underset{Li}{\overset{O}{C}}\diagdown N\diagup Ph} \qquad (80)$$

R¹Li = PhLi, BuLi, BuᵗLi, 2-Li-1,3-dithiane
R² = Me, Ph

(c) By halogen–lithium exchange. Halogen–lithium exchange has frequently been used as a route to α-lithiosilyl alkanes from the corresponding halo derivatives. This transformation has been accomplished in good yield using lithium dispersion ⟨70ZAAC(377)37, 80JA1584, 86IS95⟩. Lithium shot has also been used, for example during the preparation of bis(lithiomethyl)dimethylsilane from bis(chloromethyl)dimethylsilane ⟨55JA907⟩. Alternatively, an organolithium reagent may be used to effect halogen–lithium exchange. *t*-Butyllithium ⟨90JOC5406, 91JOC638⟩, *n*-butyllithium ⟨70JA7567, 70CJC561⟩ and phenyllithium ⟨88CB1393⟩ have all been used. Halogen–lithium exchange was also used to prepare several 1-trimethylsilylcyclopropyllithium reagents (**276**) and (**277**). These were found to be configurationally unstable in THF at −95 °C ⟨82TL1279⟩.

(**276**)

(**277**) R = Ph, PhCH₂OCH₂–

(d) By transmetallation. The transmetallation of organostannanes into organolithium reagents is a well-used technique and has been applied to the preparation of α-lithiosilyl alkanes. Tributyl(trimethylsilylmethyl)tin upon reaction with *n*-butyllithium in THF will readily transmetallate to generate trimethylsilylmethyllithium ⟨80TL3451⟩. In another example, the chiral organolithium reagent (**278**) was generated from the corresponding tributylstannane; however subsequent reaction with ketones or aldehydes only led to low *des* ⟨88T3781⟩.

(**278**)

Transmetallation of α-selenosilyl alkanes was demonstrated when treatment of (**279**) with *n*-butyllithium gave the corresponding organolithium (**280**) (Equation (81)) ⟨76AG(E)161⟩. In a similar fashion, tellurium–lithium exchange has been used to generate trimethylsilylmethyllithium from butyl trimethylsilyl telluride ⟨87AG(E)1187⟩.

$$\underset{(279)}{R^1{}_3Si\overset{R^2}{\underset{}{\diagdown}}\diagup SeMe} \xrightarrow[R^1\ =\ Me,\ Et;\ R^2\ =\ Me,\ cyclohexyl]{BuLi} \underset{(280)}{R^1{}_3Si\overset{R^2}{\underset{}{\diagdown}}\diagup Li} \qquad (81)$$

(e) By sulfur–lithium exchange. There have been several reports of the preparation of α-lithiosilyl alkanes by reductive lithiation of 1-phenylthio-1-silyl alkanes with either lithium 1-(dimethylamino)naphthalenide or lithium 4,4′-di-*t*-butylbiphenylide (Equation (82)) ⟨81TL2923, 84JOC168⟩ (see also 4.14.1.1.1(i)(a), Scheme 3). Some 1-silylcyclopropyllithium reagents have also been generated by such a process ⟨88TL25, 93JA10754⟩.

$$\text{TMS-CHR-SPh} \xrightarrow{\text{Li, naphthalenide}} \text{TMS-CHR-Li} \quad (82)$$

R = H, 86%; Me, 90%; Pr, 82%; Pr^i, 87%; Bu, 75%; Bu^s, 79%; pentyl, 72%; Ph, 84%

(f) By other routes. Other routes to the preparation of α-lithiosilyl alkanes have been reported. For example, ring opening of 1,1,3,3-tetramethyl-1,3-disilacyclobutane with methyllithium gave (**281**). This was described earlier during preparations of 1,1-bis(silyl) alkanes (4.14.1.1.1(i)) ⟨90OM2677⟩. Cleavage of a Si—C bond to generate an organolithium reagent may also be effected by lithium *t*-butoxide. Thus 1,1-bis(silyl) alkanes may be transformed into α-lithio derivatives ⟨84TL2705⟩.

$$\text{TMS} \diagdown \text{SiMe}_2 \diagup \text{Li}$$

(**281**)

A 1,3-silicon rearrangement was shown to generate an α-lithiosilyl alkane when (**282**) rearranged to give (**283**) (Equation (83)) ⟨91TL2049⟩. Other rearrangements have also been reported ⟨87CB1695⟩.

(**282**) $\xrightarrow{\text{THF, RT}}$ (**283**) (83)

(ii) α-Sodio- or α-potassiosilyl alkanes

Compared to the vast quantity of published literature on the preparations of α-lithiosilyl alkanes, there has been relatively little reported work describing the use of other Group 1 metals. The work which has been done reflects the methods used to generate the corresponding organolithium derivatives.

For example, direct deprotonation with sodium hydride is possible and several benzylsilanes were metallated during the preparation of a series of indoles ⟨88CC807⟩. Trimethylsilylmethylsodium has been prepared by the cleavage of one C—Si bond of bis(trimethylsilyl)methane with sodium methoxide in HMPA (Equation (84)) ⟨73TL4193, 84TL2705⟩. Use of potassium *t*-butoxide gave the potassio analogue ⟨84TL2705⟩. Trimethylsilylmethylpotassium has also been prepared by transmetallation from bis(trimethylsilylmethyl)mercury with potassium sand ⟨87S645⟩.

$$\text{TMS-CH}_2\text{-TMS} \xrightarrow{\text{NaOMe, HMPA}} \text{TMS-CH}_2\text{-Na} \quad (84)$$

(iii) α-Magnesiosilyl alkanes

Many α-silylalkyl magnesium reagents are known. Trimethylsilylmethylmagnesium chloride has been the most frequently used and it is now a commercially available material. Methods for the preparation of α-silyl Grignard reagents reflect broadly those used to prepare the corresponding organolithium reagents, except that deprotonation of alkyl silanes by magnesium reagents cannot be performed. The simplest route is through halogen–magnesium exchange, a standard route for the preparation of Grignard reagents. Other well-tried routes include the addition of an organomagnesium reagent to a vinylsilane or via transmetallation. Several examples of the preparation of 1-magnesio-1-silyl alkanes were described earlier during the preparation of 1,1-bis(silyl) alkanes (see 4.14.1.1.1(i)(b)).

(a) By halogen–magnesium exchange. The simplest derivative—trimethylsilylmethylmagnesium chloride—is readily obtained from trimethylsilylmethyl chloride and magnesium ⟨74JOC3264⟩. This

methodology has been applied to generate other (trialkylsilyl)methyl Grignard reagents ⟨84TL4249, 92CL2047⟩. For example, treatment of 1,2-bis(bromomethyl)tetramethyldisilane with magnesium in ether gave the di-Grignard reagent (**284**) in 60% yield ⟨77IJ265⟩. Vinyl groups on silicon are tolerated ⟨89TL403, 89JOM(361)123⟩ as are aryl groups ⟨87MI 414-01⟩, although the stability of triarylsilylmethyl Grignard reagents has been questioned ⟨70CJC561⟩.

Alkoxy substituents on silicon are also tolerated during the formation of the corresponding silylmethyl Grignard reagents despite their lability to organometallic reagents ⟨67JOM(10)111, 84TL4245, 92MI 414-01, 92MIP173178, 93TL2111⟩. This is not true of halo substituents which will readily undergo further reaction. For example, trichloro(chloromethyl)silane rapidly polymerised in the presence of magnesium ⟨91OM1336⟩. This ensuing reaction has been put to good use, for example in the preparation of 1,3-disilacyclobutanes from chloromethyldimethylsilyl halides and magnesium (see 4.14.1.1.1(ii)(b)) ⟨75JCS(D)1837⟩. Siloxy groups are stable to the generation of α-silyl Grignard reagents as was demonstrated by the preparation of the corresponding Grignard reagent from chloromethylpentamethyldisiloxane ⟨58JOC1392⟩. Silyl-substituted allyl Grignard reagents have also been prepared by halogen–magnesium exchange, for example triphenylsilylallylmagnesium bromide ⟨75JOM(93)71⟩.

Magnesium anthracene (**285**) is an alternative source of magnesium for the preparation of Grignard reagents, especially benzylic Grignard reagents which may be difficult to prepare by classical methods due to side reactions. This reagent was used to generate the α-silylbenzylmagnesium halides (**286**) and (**287**) (Scheme 59) ⟨88JOC3134⟩. Alternative silylated derivatives of this reagent (**288**) have been prepared by silylation of (**285**) or directly from the silylanthracenes with magnesium in THF ⟨87OM2110, 88CC652⟩.

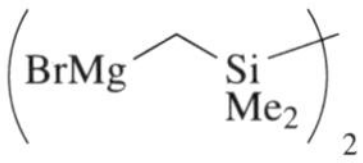

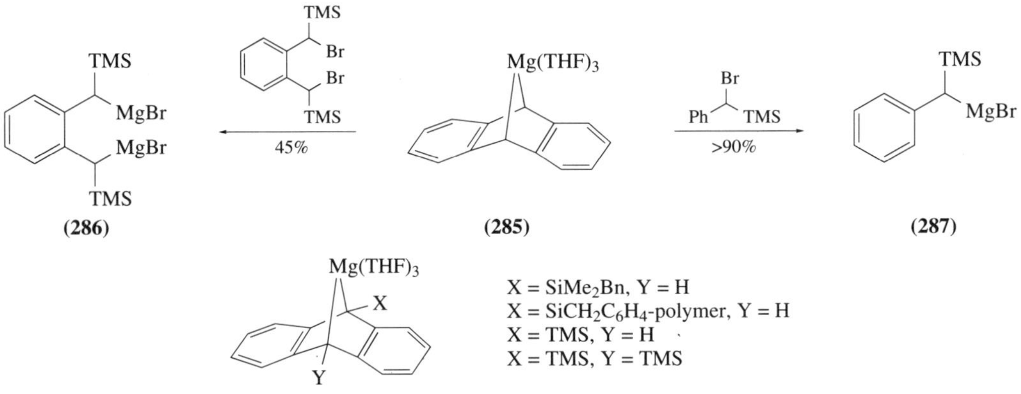

Scheme 59

(b) By addition of a Grignard reagent to a vinylsilane. Grignard reagents will add to vinyl silanes to generate the corresponding α-magnesiosilyl alkanes (Equation (85)). As with the preparation of α-lithiosilyl alkanes above, the reaction is sensitive to the reagents used. Good yields for addition to the vinyl group were only achieved with either chloro or alkoxide groups on silicon—presumably as a result of reducing the electron density of the double bond. Displacement of the halide or alkoxy group on silicon only became significant when either a primary Grignard reagent was used or more than one halide or alkoxy group was present ⟨70JA7424⟩. Amino groups on silicon have also been shown to facilitate the Grignard reagent addition ⟨84TL1905⟩. The addition of Grignard reagents to trimethyl(vinyl)silane has been reported when forcing conditions were used ⟨84CB383⟩.

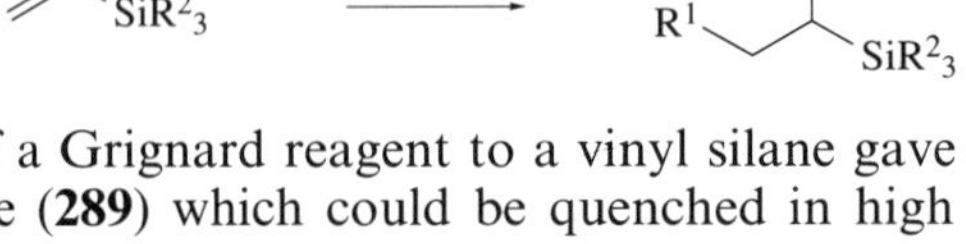

Intramolecular addition of a Grignard reagent to a vinyl silane gave a highly diastereoselective synthesis of the cyclopentane (**289**) which could be quenched in high yield with an electrophile (Equation (86)) ⟨85TL2101⟩.

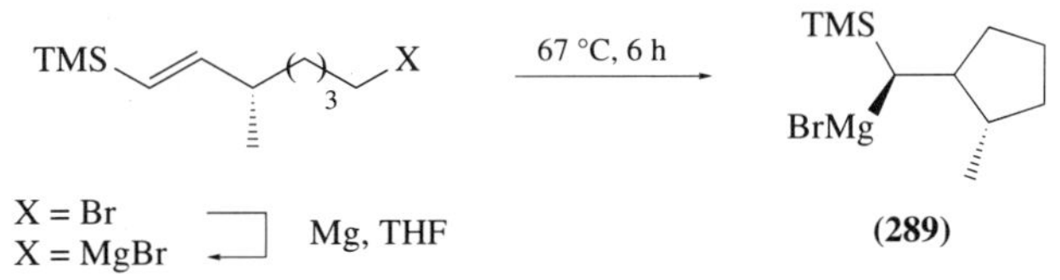

$$\text{TMS} \diagup\kern-0.5em\diagdown \underset{3}{\diagup\kern-0.5em\diagdown} X \quad \xrightarrow{67\,°C,\,6\,h} \quad \text{TMS–CH(BrMg)–cyclopentyl} \quad (289) \tag{86}$$

X = Br
X = MgBr ⎤ Mg, THF

The addition of an organomagnesium reagent to the silylated α,β-amidate anion of (290) resulted in the generation of an α-magnesio-α-silylamide (Scheme 60). The normal 1,2-addition of the Grignard reagent to the carbonyl group was suppressed by the α-anion generated during the reaction. As with the addition of organolithium reagents described above, the reaction was sensitive to substitution on the β-carbon of the vinylic group ⟨93JOC7474⟩.

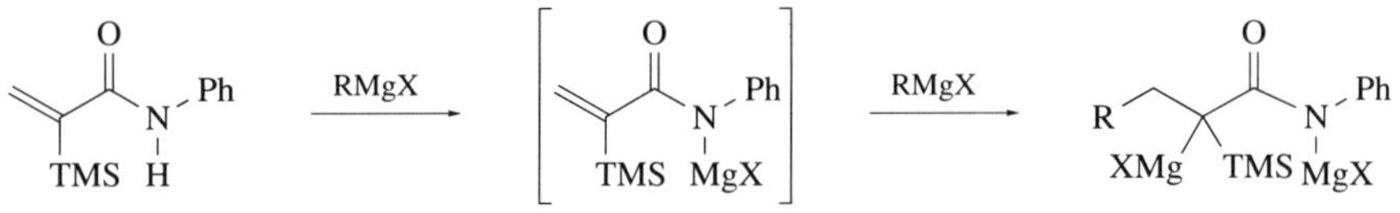

RMgX = MeMgBr, PhMgBr, cyclopentylMgBr, vinylMgBr, allylMgCl, n-BuMgCl, t-BuMgCl

Scheme 60

(c) By transmetallation. Transmetallation has been used to generate α-silyl alkyl Grignard reagents. Transmetallation from the corresponding organolithium reagent with magnesium bromide has been reported ⟨82CC970, 84TL1913⟩ although different reactivity of the resultant organometallic solution when generated in this manner has been reported ⟨78TL2383⟩.

The dialkyl Grignard reagent bis(trimethylsilylmethyl)magnesium may be generated from trimethylsilylmethylmagnesium chloride by precipitation from an ethereal solution with 1,4-dioxane ⟨79MI15262⟩.

(d) By rearrangement. An α-silylmethyl Grignard reagent has been generated as a result of a 1,2-migration ⟨90PAC1933⟩ of a group from magnesium to carbon (Scheme 61) ⟨88JA646⟩.

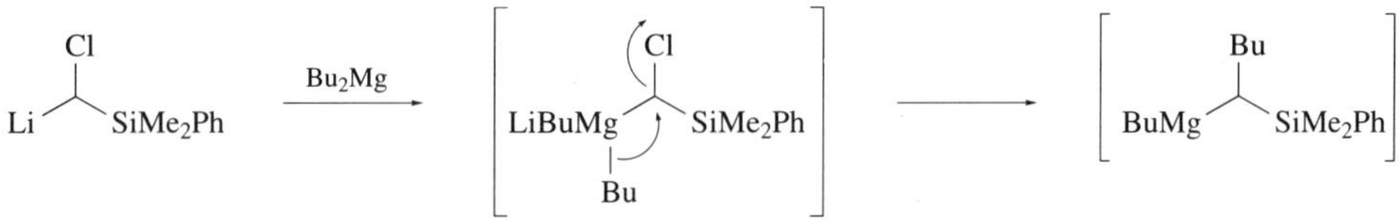

Scheme 61

4.14.2.1.2 Silicon and a transition metal: $R^1{}_2CSiR^2{}_3CuX$, etc.

There have been many reports of systems containing both a silicon and a transition metal (including zinc) bound to the same carbon together with either protons or carbon groups. This work has been largely focussed on the use of the trimethylsilylmethyl ligand.

Some attention has, however, been directed towards the moderation of reactivity of the α-silylmethyl anion when bound to transition metals. The most common example is the generation of a cuprate for subsequent addition to electrophiles ⟨81AJC181, 81TL2985, 84TL4249, 86T1389, 89TL5693, 91JOC638, 91JOM(414)295, 92CL2047, 93S202⟩. Other transition-metal derivatives have also been prepared for reaction with electrophiles including those of cerium ⟨87TL6261, 88TL4281⟩, titanium ⟨88TL4281⟩ and zinc ⟨80TL11⟩.

Many simple transition-metal alkyl complexes are unstable, but the use of ligands with an α-silyl group can result in stable complexes. These are stable to β-elimination, often a facile process for alkyl ligands with a β-hydrogen. Also, the steric bulk of these ligands may block vacant coordination sites—another potential route to decomposition. Further stabilisation may be gained through secondary interactions between the ligand and the metal centre, such as the inductive electron donation by these ligands resulting in the stabilisation of high oxidation states. Another advantage that has been cited for the use of these ligands is the increased solubility of the resultant complexes in organic solvents ⟨73JOM(57)269⟩. The subsequent use of these ligands in transition-metal complexes has become widespread ⟨70CC1369, 78PAC703⟩.

(i) Reaction of a transition-metal complex with an α-silylmethyl anion

The generation of transition-metal complexes with silylmethyl ligands is generally performed through a transmetallation process, typically the displacement of a transition-metal nucleofuge by an α-lithiosilyl alkane or the corresponding Grignard reagent. The ligand used almost to the exclusion of others is trimethylsilylmethyl. Table 8 lists those complexes containing a silylmethyl ligand while Table 9 lists more substituted systems which have also been generated by reaction of an electrophilic complex with an α-metallosilyl alkane.

Table 8 Transition-metal complexes with silylmethyl ligands (L_nM—CH_2SiR_3).

M	Ref.
Ti	70CC1369, 70JOM(25)C36, 71SAP7004922, 73JCS(D)445, 74IZV2861, 81JCS(D)1593, 89JOM(375)59
V	82ZN(B)957
Cr	70CC1369
Mn	70CC1369, 76JCS(D)2204
Fe	74JOM(78)C35, 93OM3979
Cu	77JCS(D)999, 88OM1208, 90OM1720
Zn	91JOM(421)175, 93JA7215, 93OM3624
Y	78CC140, 93OM298
Zr	70JOM(25)C36, 71SAP7004922, 73JCS(D)445, 78CC1081, 81JCS(D)1593, 91OM2191, 92CSR271, 93JA8493, 93OM184, 93OM338, 93OM1936
Nb	71SAP7004922,
Mo	70CC1369, 70JOM(25)C36, 71CC1079, 91OM2857, 93BCJ1849
Ru	92CSR271, 93JA5527
Rh	71SAP7004922
Ce	87TL6261
Tb	78CC140
Er	78CC140, 78PAC703
Yb	78CC140, 78PAC703
Hf	70JOM(25)C36, 73JCS(D)445, 81JCS(D)1593, 93JA8493
Ta	71CC1477
W	70JOM(25)C36, 76AC(E)609, 91OM2857, 92CSR271, 93OM3955, 93OM3979, 93OM4572
Re	82POL31
Os	86JA7964
Ir	92CSR271
Pt	72CC613, 88POL1953, 89JOM(361)123, 89POL57
Au	75CB1321
Hg	61JOC2604, 75JCS(D)1832, 76DOK(231)1138, 77JCR(S)116, 79JCS(D)767, 84JOM(269)249, 84ZOB2025
Th	86JA40, 86JA425, 88OM1828
U	77JCS(D)812

Table 9 Transition-metal complexes with substituted silylmethyl ligands (L_nM—$CH(R)SiR_3$).

M	R	Ref.
Ni	2-$Me_2NC_6H_4$	85NJC249
Zr	4-MeC_6H_4, 2-$Ph_2PC_6H_4$, 2-pyridyl, 9-anthracyl	86JCS(D)603
Cu	2-pyridyl	87JCS(D)3085

Cyclic systems containing an α-silylmethyl group have also been prepared. These have often been formed by transmetallation of dimetallic reagents such as the di-Grignard reagent derived from bis(chloromethyl)dimethylsilane to generate 1,3-dimetallocyclobutanes. Systems that have been generated in this way are listed in Table 10.

The generation of cyclic systems containing an α-silylmethyl ligand may also be through a two-step process. For example, lithiation of the bis(trimethylsilyl)methyl ligand on the platinum complex **(291)** gave metallocycle **(292)** (Equation (87)). Steric hindrance prevented deprotonation at the

Table 10 Cyclic systems prepared by transmetallation of dimetallic ligands.

Entry	Nucleophile	Product	ML$_n$	Ref.
1	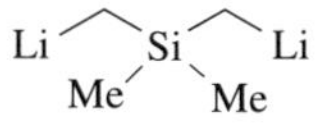 Li–CH$_2$–SiMe$_2$–CH$_2$–Li	Me$_2$Si(∆)ML$_n$	Cp$_2$Ti	77JOM(142)C49, 77JAP7973759
2	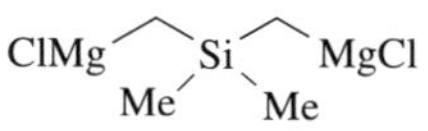 ClMg–CH$_2$–SiMe$_2$–CH$_2$–MgCl	 Me$_2$Si(∆)ML$_n$	Cp$_2$Ti, Cp$_2$Zr, Cp$_2$Nb, Cp$_2$Mo	84OM825
3	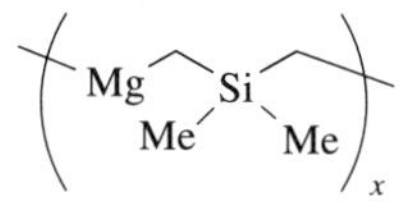 (–Mg–CH$_2$–SiMe$_2$–CH$_2$–)$_x$		(Cp$_2$Me$_2$Si)Ti, (Cp$_2$Me$_2$Si)Zr, Cp$_2$Zr	89JOM(364)105, 87OM2007
4	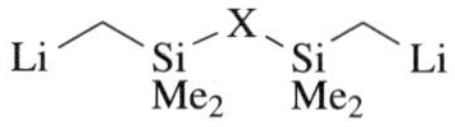 Li–CH$_2$–SiMe$_2$–X–SiMe$_2$–CH$_2$–Li	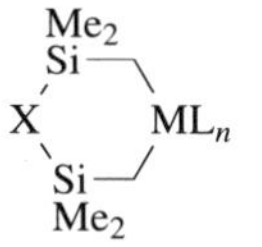 X(SiMe$_2$CH$_2$)$_2$ML$_n$	Cp$_2$Ti	77JOM(142)C49
5	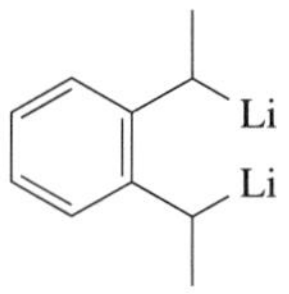 1,2-bis(α-lithioethyl)benzene	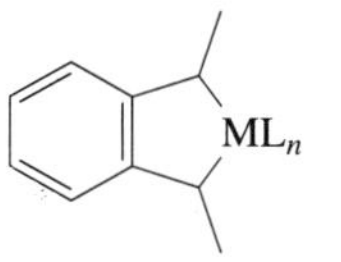 1,3-dimethyl-2-indanyl-ML$_n$	Cp$_2$Ti, Cp$_2$Zr, Cp$_2$Hf	80CC1284, 80CC1284, 82CC1023, 80CC1284, 82CC1023

thermodynamic site between the two silyl groups ⟨86JOM(299)C35⟩. A similar process of metallation of a ligand followed by intramolecular cyclisation onto the transition-metal centre resulted in the generation of (**293**) (Equation (88)) ⟨74CC29⟩. Addition of α-silylmethyl Grignard reagent to (**294**) gave the acyclic derivative (**295**) which subsequently cyclised through insertion of the transition metal into the silicon—hydride bond to form η^2-silene complex (**296**) (Scheme 62) ⟨93JA5527⟩.

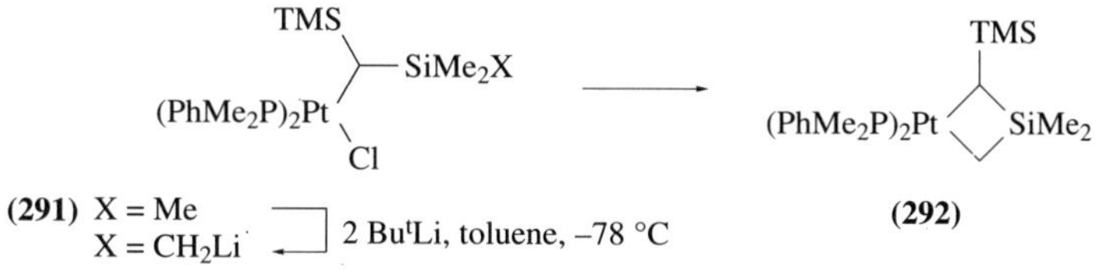

$$\text{(PhMe}_2\text{P)}_2\text{Pt(Cl)(C(TMS)(SiMe}_2\text{X})) \longrightarrow (\text{PhMe}_2\text{P})_2\text{Pt}\underset{\text{SiMe}_2}{\overset{\text{TMS}}{\square}} \quad (87)$$

(**291**) X = Me
X = CH$_2$Li ⟵ 2 Bu^{t}Li, toluene, −78 °C

(**292**)

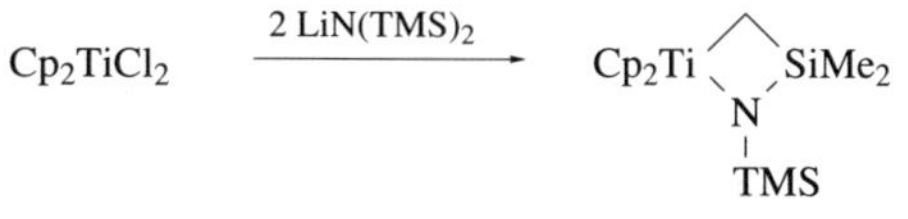

$$\text{Cp}_2\text{TiCl}_2 \xrightarrow{2\text{ LiN(TMS)}_2} \text{Cp}_2\text{Ti}\underset{\underset{\text{TMS}}{\text{N}}}{\overset{}{\square}}\text{SiMe}_2 \quad (88)$$

(**293**)

$$\text{Pr}^i_3\text{P–Ru(Cp*)(Cl)} \xrightarrow{\text{ClMg–CH}_2\text{–SiHX}_2,\ X = \text{Me, Ph}} [\text{Pr}^i_3\text{P–Ru(Cp*)(CH}_2\text{SiHX}_2)] \longrightarrow \text{Pr}^i_3\text{P–Ru(Cp*)}\underset{\text{H}}{\overset{}{\square}}\text{SiX}_2$$

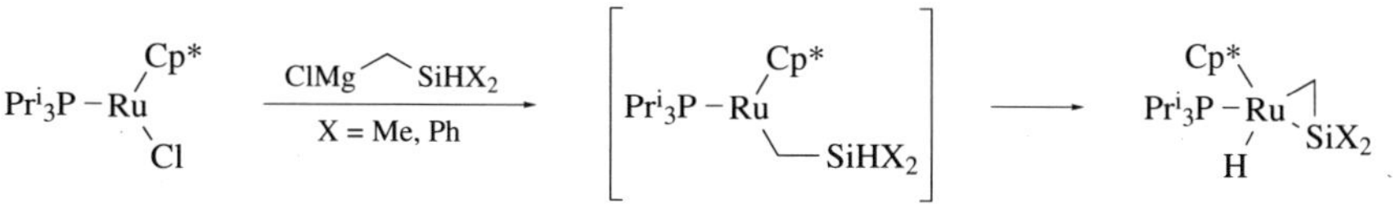

(**294**) (**295**) (**296**)

Scheme 62

(ii) Reaction of an α-metallomethyl transition metal with a silyl electrophile

The generation of a transition-metal complex containing an α-silylmethyl ligand may also be performed by the reaction of an anion α to the transition metal, for example the generation of 1-sila-3-titanacyclobutane (**297**) (Scheme 63) ⟨84TL5191⟩.

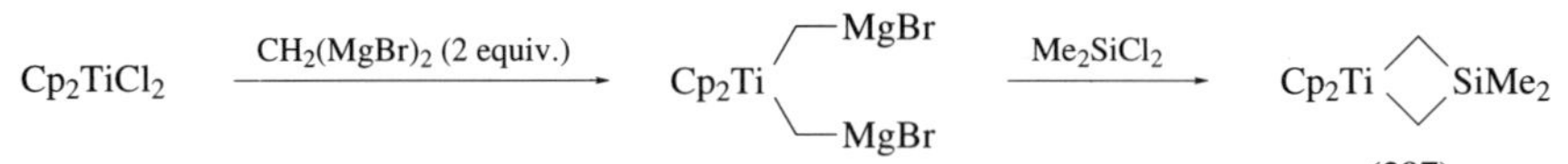

Scheme 63

(iii) Other routes

Other methods have been used to generate transition-metal complexes containing an α-silylmethyl ligand. One of these is the ring opening of disilacyclobutanes. For example, the ring opening of 1,1,3,3-tetramethyldisilacyclobutane with titanium tetrachloride gave chloro(dimethyl)silylmethyltitanium trichloride ⟨74IZV2861⟩. Analogous mercury ⟨75IZV1600, 75JCS(D)1837⟩, zirconium and hafnium ⟨85ZOB1629⟩ complexes have also been prepared.

Rearrangements can result in the formation of systems containing a silyl group α to a transition-metal complex. For example, rearrangement of the zirconocene (**298**) gave the 1-phenyl-1-trimethylsilyl derivative (**299**) (Equation (89)) ⟨85OM1308⟩. 1,2-Migration reactions have been used to generate new systems containing a silylmethyl ligand on a number of transition metals. Reaction of 1-silyl-1-chloro-1-lithiomethane with a transition-metal complex resulted in the formation of (**300**). Subsequent rearrangement is through, what is believed to be, a 1,2-migration of a ligand on the metal centre to the adjacent carbon, with concomitant loss of chloride generated (**301**) (Scheme 64). This process has been applied to a large number of transition metals ⟨88JA646, 89JA3089⟩.

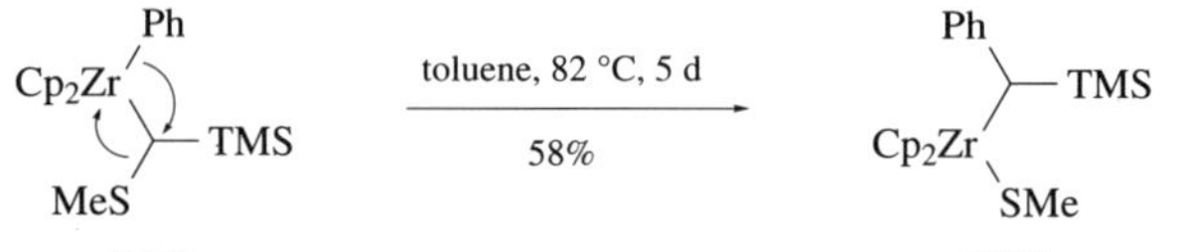

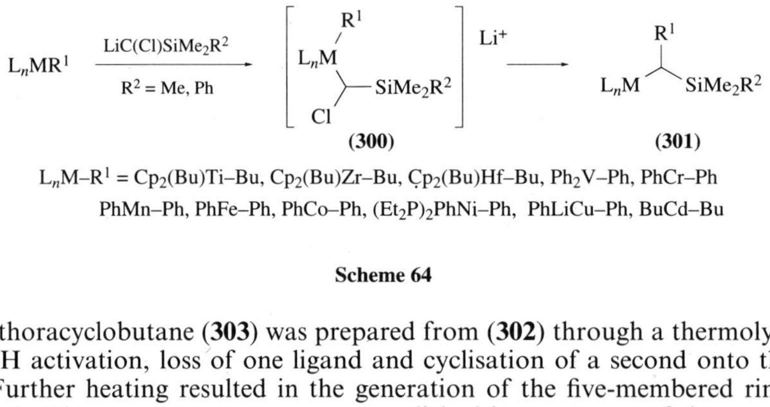

L_nM–R^1 = Cp$_2$(Bu)Ti–Bu, Cp$_2$(Bu)Zr–Bu, Cp$_2$(Bu)Hf–Bu, Ph$_2$V–Ph, PhCr–Ph
PhMn–Ph, PhFe–Ph, PhCo–Ph, (Et$_2$P)$_2$PhNi–Ph, PhLiCu–Ph, BuCd–Bu

Scheme 64

The 1-sila-3-thoracyclobutane (**303**) was prepared from (**302**) through a thermolytic process as a result of γ-C—H activation, loss of one ligand and cyclisation of a second onto the metal centre (Scheme 65). Further heating resulted in the generation of the five-membered ring isomer (**304**) ⟨82JA7357, 86JA40⟩. The reverse process was accomplished by treatment of thoracyclobutane (**305**) with an excess of tetramethylsilane to give the acyclic derivative (**306**) (Equation (90)) ⟨86JA425⟩. Further studies have revealed that the cyclisation is entropically driven ⟨86JA6805⟩.

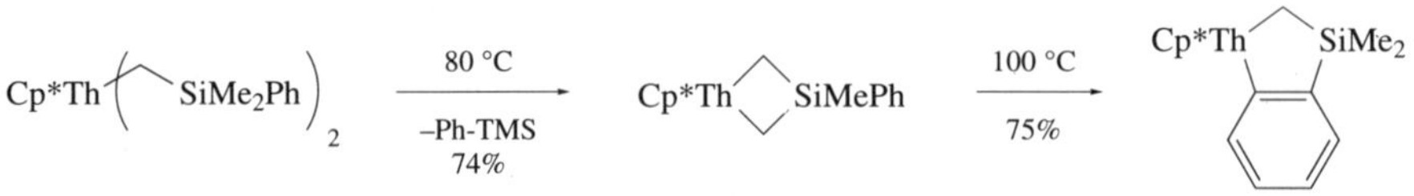

Scheme 65

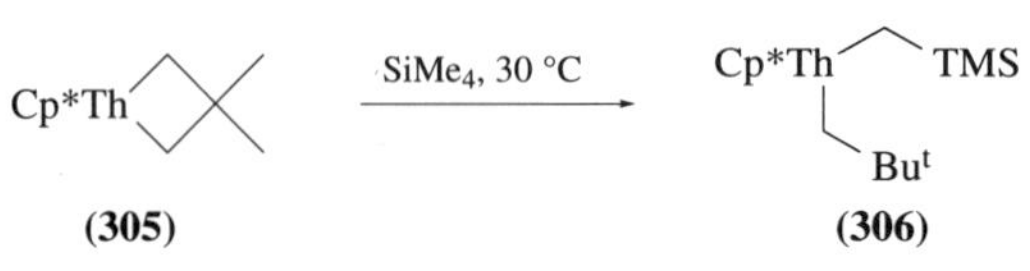

(iv) Changing the groups attached to the central methylene or to the metal

Ring expansions have been used to generate new α-silylmethyl ligands on transition metal complexes, for example, the insertion of carbon monoxide ⟨87OM2007⟩ or isonitriles ⟨89OM2461⟩ into (**307**) (Equation (91)). This insertion chemistry has also been applied to the corresponding hafnium complex ⟨93OM3890⟩.

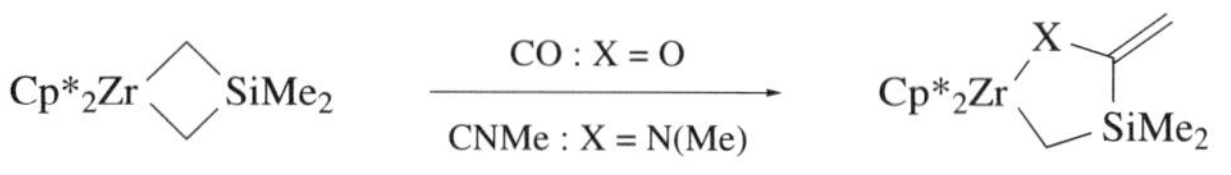

$$\text{Cp*}_2\text{Zr} \underset{(\mathbf{307})}{\overset{}{\triangle}} \text{SiMe}_2 \quad \xrightarrow[\text{CNMe : X = N(Me)}]{\text{CO : X = O}} \quad \text{Cp*}_2\text{Zr} \diagup\!\!\!\diagdown \text{SiMe}_2 \tag{91}$$

4.14.2.1.3 Silicon and a group 13 or Group 14 metal: $R^1{}_2CSiR^2{}_3SnR^3{}_3$, etc.

Preparations of functional groups containing both a silicon and a Group 13 or 14 metal bound to the same carbon along with either protons or carbon atoms have been reported. The most common methods for their preparation is through transmetallation from the easily accessible α-lithio or α-magnesiosilyl alkane derivatives, although other routes such as rearrangement have been reported.

(i) Group 13 metals

After boron (section 4.14.1.5), the most common Group 13 element found bound to the same carbon as silicon is aluminum. These compounds have been prepared by transmetallation, particularly from either an organolithium ⟨91JOM(415)181⟩ or organomagnesium precursor ⟨85CB4248⟩. For example, treatment of aluminum trichloride with an α-silylmethyl Grignard reagent generated the corresponding organoalane, the structure of which depended upon the stoichiometry of the reagents used (Scheme 66) ⟨76ZC65⟩. Tetracoordinated organoaluminum 'ate' complexes containing an α-silylmethyl ligand can similarly be generated ⟨84JOC1096⟩. This process has been applied to other Group 13 metals. For example, thallium trichloride will react with either trimethylsilylmethyllithium ⟨74JOM(70)C21, 78BCJ1397⟩ or with the corresponding Grignard reagent ⟨80JOM(193)21⟩ to give bis(trimethylsilylmethyl)thallium chloride. Thallium tribromide will generate the corresponding dialkylthallium bromide ⟨74JOM(70)C21⟩. Tris(trimethylsilylmethyl)gallium was prepared by reaction of trimethylsilylmethylmagnesium chloride with gallium trichloride in 92% yield ⟨80IC1021⟩. Subsequent reaction with either 0.5 or 2 equivalents of gallium trichloride generated bis(trimethylsilylmethyl)gallium chloride or trimethylsilylmethylgallium dichloride, respectively ⟨88OM1516⟩.

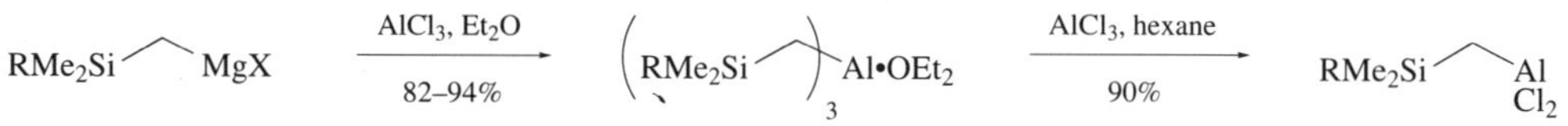

$$\text{RMe}_2\text{Si}\frown\text{MgX} \quad \xrightarrow[\text{82–94\%}]{\text{AlCl}_3,\ \text{Et}_2\text{O}} \quad \left(\text{RMe}_2\text{Si}\frown\right)_3\!\text{Al}\cdot\text{OEt}_2 \quad \xrightarrow[\text{90\%}]{\text{AlCl}_3,\ \text{hexane}} \quad \text{RMe}_2\text{Si}\frown\text{AlCl}_2$$

R = Me, Bu, octyl, TMS-O

Scheme 66

Rearrangements can result in the generation of α-silylmethyl alanes. Scheme 67 shows such an example where a 1,2-migration of a group from aluminum to the adjacent carbon with concomitant displacement of the chloride generated (**308**) ⟨88JA646⟩.

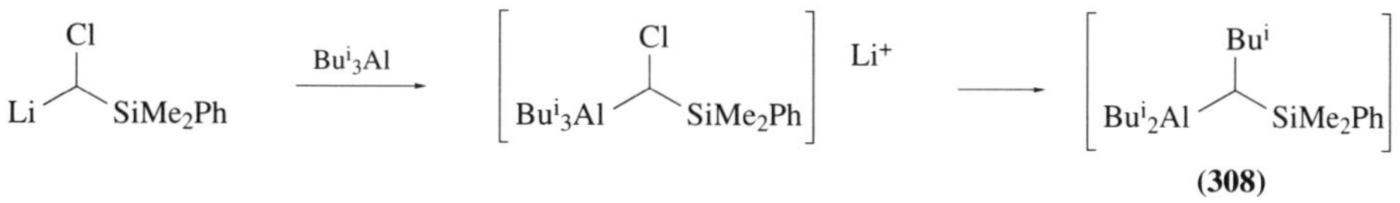

$$\underset{\text{Li}}{\overset{\text{Cl}}{\diagup}}\!\!\!\diagdown\text{SiMe}_2\text{Ph} \quad \xrightarrow{\text{Bu}^i{}_3\text{Al}} \quad \left[\underset{\text{Bu}^i{}_3\text{Al}}{\overset{\text{Cl}}{\diagup}}\!\!\!\diagdown\text{SiMe}_2\text{Ph}\right]\text{Li}^+ \quad \longrightarrow \quad \left[\underset{\text{Bu}^i{}_2\text{Al}}{\overset{\text{Bu}^i}{\diagup}}\!\!\!\diagdown\text{SiMe}_2\text{Ph}\right]$$

(**308**)

Scheme 67

(ii) Group 14 metals

Much work has been focussed on the preparation and chemistry of Group 14 elements bound to the same carbon as a silyl group. The preparations of 1,1-bis(silyl) alkanes and α-germylsilyl alkanes were discussed earlier (4.14.1.1 and 4.14.1.4). Only one preparation of a lead derivative containing an α-silyl alkyl ligand has been reported. Tetrakis(trimethylsilylmethyl)lead was generated by the reaction of trimethylsilylmethylmagnesium chloride with lead dichloride in 37% yield ⟨61JOC2604⟩. In contrast, there has been a vast amount of work directed towards the preparation of α-silylmethylstannanes. Many routes to the preparation of these derivatives have been described, the most common of which is transmetallation of α-lithio or magnesiosilyl alkanes.

(a) Transmetallation of α-metallosilyl alkanes. Many routes to the preparation of α-lithio and α-magnesiosilyl alkanes were discussed earlier (4.14.2.1.1) and these derivatives have been transmetallated to give α-stannylsilyl alkanes. Thus the reaction of trimethylsilylmethylmagnesium chloride with tin tetrachloride gave the tetraalkyl derivative (**309**) in 76% yield ⟨72JOM(46)51⟩. A number of analogues have been prepared in a similar fashion ⟨73ZOB801, 77JOM(137)157, 88JAP0267291, 90ZAAC(583)195, 92JOM(423)339, 93OM2788⟩. Organolithium reagents will react similarly. For example, the addition of trimethylsilylmethyllithium to a solution of triphenyltin chloride gave (trimethylsilylmethyl)triphenyltin in 88% yield ⟨81SC673⟩. Several 1-lithio-1-aryl-1-silyl alkanes have been treated with tin halides to generate the corresponding stannanes ⟨82JOM(233)C28, 86OM1551, 90BCJ3036, 91NJC307, 92LA725⟩. α-Silyl ester enolates are *C*-stannylated following treatment with tributyltin chloride ⟨93JOM(448)69⟩. Dimetallic intermediates may be used in the preparation of stannacycles such as (**310**) (Equation (92)) ⟨84JCS(D)331, 92JCS(D)775⟩.

$$\left(TMS\frown\right)_4 Sn$$

(**309**)

(92)

(**310**)

(b) Quenching a stannyl anion with a carbon electrophile. Another well-used route to α-silyl stannyl alkanes is through the reaction of a stannyl anion with a halomethylsilane. For example, tributylstannyllithium will react with (chloromethyl)trimethylsilane to generate (trimethylsilyl)tributyltin in high yield ⟨80TL3451, 81S557⟩. Many other derivatives have been formed in this manner ⟨91TL3285⟩. The corresponding trimethylstannylsodium derivative will also react ⟨85OM821⟩.

(c) Transmetallation of α-metallostannyl alkanes. The quenching of an α-metallostannyl alkane with a silyl electrophile has been used to generate α-silyl stannyl alkanes. For example, trimethylsilyl chloride will react with tributylstannylmethyllithium to generate tributyl(trimethylsilylmethyl)stannane in good yield ⟨90OM1355⟩, and with (**311**) to generate (**312**) (Equation (93)) ⟨78JOM(148)247⟩.

(93)

(**311**)   (**312**)

(d) Other routes. Oxidative addition of (halomethyl)silyl alkanes to both tin ⟨72ZOB631, 75ZOB2448⟩ and tin dichloride ⟨74ZOB806, 76ZOB1043, 89MI 414-01, 91MINJC301⟩ have been demonstrated to generate α-silylstannyl alkanes. For example, reaction of (**313**) with tin gave the bis(silylmethyl) stannane (Equation (94)) ⟨77ZOB1751⟩.

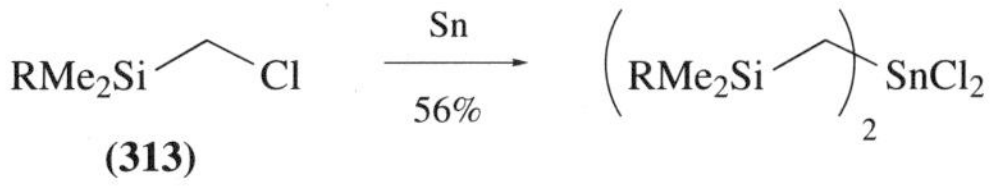

$$RMe_2Si-CH_2-Cl \xrightarrow[56\%]{Sn} (RMe_2Si-CH_2)_2SnCl_2 \quad (94)$$

(313)

R = cyclopropyl

Several methods have been used to transform 1-sila-1-stanna-1-alkenes into their saturated derivatives. Direct hydrogenation has proved successful (Equation (95)) ⟨92AG(E)232⟩. Cyclopropanation may also be performed and this has also shown itself to be highly diastereoselective (Equation (96)) ⟨92JOC798⟩. Another reaction that has enabled the chirality of the dimetallic centre to be controlled is the Claisen rearrangement. Thus **(315)** was generated when allylic alcohol **(314)** was heated with the dimethyl acetal of DMA (Equation (97)) ⟨90TL5829⟩. Hydrostannylation of these systems was demonstrated when 1-trimethylsilyl-1-trimethylstannylethene was transformed into **(316)** in good yield (Equation (98)). The high selectivity for addition of the stannyl group to the β-carbon was also observed when the reaction was repeated with trimethyl(vinyl)silane resulting in only small amounts of 1-trimethylsilyl-1-trimethylstannylethane ⟨85OM1044⟩. Silyl ketenes have been transformed into α-silyl-α-stannyl esters following reaction with alkoxytributylstannanes (Equation (99)) ⟨70ZOB2607, 91SL911, 92JCS(P1)2813⟩.

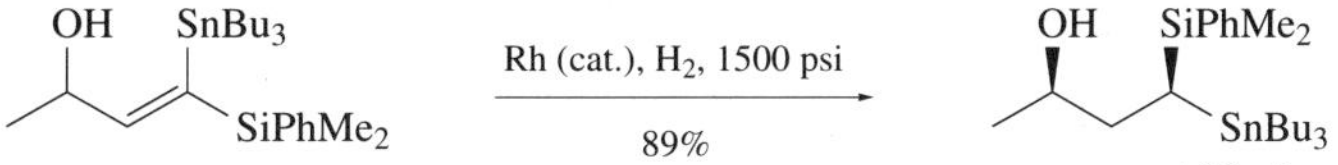

Equation (95): Rh (cat.), H_2, 1500 psi, 89%, syn : anti 60 : 1

Equation (96): Sm($HgCl_2$), CH_2I_2, THF, R = cyclohexyl, 80%; Pr, 67%; Me, 65%

Equation (97): **(314)** → MeC(OMe)$_2$NMe$_2$, 110 °C, 74% → **(315)**

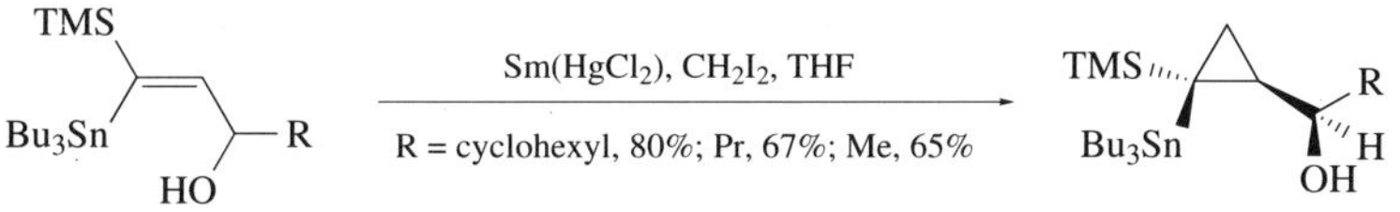

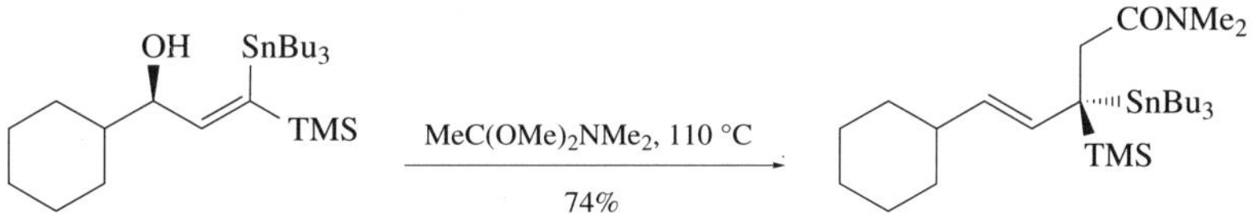

Equation (98): Me$_3$SnH, hν, AIBN, 82% → **(316)**

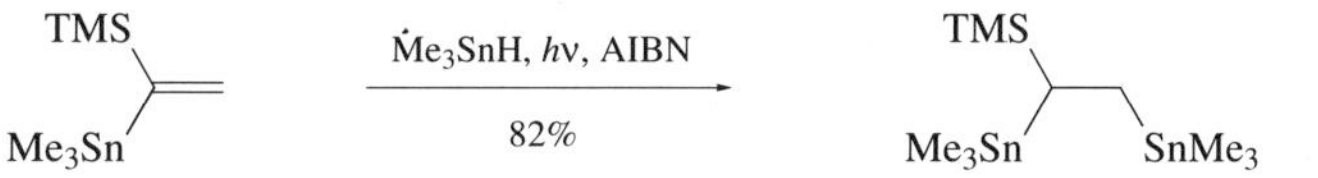

Equation (99): ROSnBu$_3$; R = Me, Et, Pr^i, cyclohexyl, CH$_2$Bu^t, CHEt$_2$

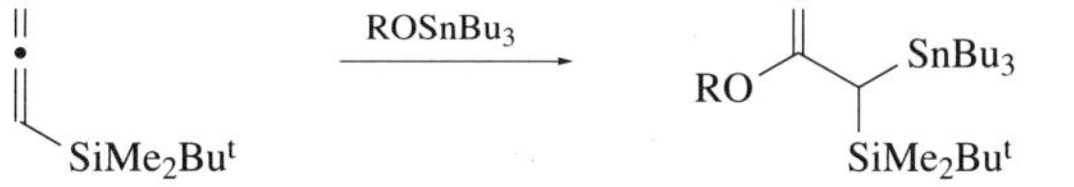

A number of other reactions have been found to generate α-silylstannyl alkanes ⟨86JOM(310)151, 87JOM(321)C1⟩. For example **(318)** was obtained along with several other products during the radical cyclisation of **(317)** (Equation (100)) ⟨92TL7895⟩. In another example, tin tetrachloride has been shown to ring-open 1,3-disilacyclobutanes (Equation (101)) ⟨75JCS(D)1832⟩.

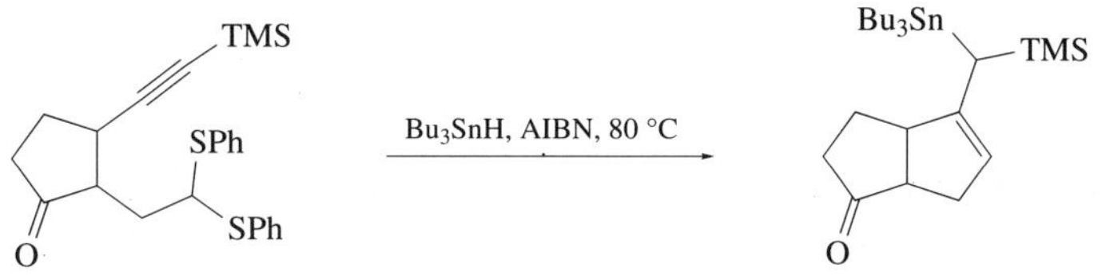

Equation (100): **(317)** → Bu$_3$SnH, AIBN, 80 °C → **(318)**

AIBN = 2,2'-azobisisobutyronitrile

$$\text{Me}_2\text{Si}\underset{}{\overset{}{\triangleleft}}\text{SiMe}_2 \quad \xrightarrow{\text{SnCl}_4,\ \text{RT}} \quad \text{Me}_2\underset{\text{Cl}}{\text{Si}}\diagdown\underset{\text{Me}_2}{\text{Si}}\diagdown\text{SnCl}_3 \qquad (101)$$

Functionalisation of an existing system was demonstrated when tributyl(trimethylsilylmethyl)stannane was deprotonated with either potassium diisopropylamide or LDA with HMPA to generate the α-silyl-α-stannyl anion which was subsequently functionalised with electrophiles ⟨86OM1906⟩.

4.14.2.1.4 Silicon and other elements

Several elements from Groups 15 and 16 have been bound to the same carbon as a silicon together with either hydrogen or carbon. Of these, selenium has been the most explored although reports of systems containing tellurium, arsenic and antimony have also been published. Compounds containing silicon and arsenic and antimony on the same sp^3-hybridised carbon will be found in chapter 4.13, whilst those containing silicon and selenium and tellurium are covered in chapter 4.08.

4.14.2.2 Germanium and a metal: $R^1{}_2CGeR^2{}_3M$

There have been many reports of the generation of α-metallogermyl alkanes where the central carbon carries only either hydrogens or carbon fragments. These have generally stemmed from the preparation of α-lithio derivatives, which in turn may be generated under similar conditions to those used to prepare the corresponding silyl derivatives (see 4.14.2.1). Transmetallation can then provide access to a variety of α-metallo derivatives.

4.14.2.2.1 α-Lithiogermyl alkanes

Deprotonation α to the germyl group provides the most efficient route to these systems. As with the corresponding silyl derivatives, this is greatly facilitated by a second group which can stabilise the anion. Thus α-germyl enolates have been readily generated by treatment of either an α-germyl ester ⟨91JOC347⟩ or ketone ⟨87OM2568⟩ with LDA. α-Germyl nitriles may similarly be deprotonated with LDA, although α-lithiotrimethylgermylacetonitrile was found to be unstable even at −78 °C in ether (Equation (42)) ⟨87SC1273⟩.

Stabilisation of an anion α to the germyl group by an aryl group also enables easy deprotonation; for example, the preparation of (**319**) with *t*-butyllithium ⟨87JA4411, 91MI 414-01⟩. Similarly, deprotonation of (**320**) generated the dianion which subsequently underwent further reaction to form germacycle (**321**) (Equation (102)) ⟨87JCS(D)1647⟩. The analogous bis(silylmethyl)naphthalene dianion was shown to react similarly (see 4.14.2.1.1(i)(a)).

(**319**)
R = mesityl, (TMS)$_2$CH, Bu^t

(**320**) X = H
X = Li   2 BuLi, TMEDA

(**321**)   (102)

Many systems that are not readily generated by deprotonation have been generated by halogen–

lithium exchange. Thus trimethylgermylmethyllithium was prepared from (chloromethyl)trimethylgermane and lithium dispersion or *s*-butyllithium ⟨80JA1584⟩. Halogen–metal exchange with *n*-butyllithium enabled the organolithium reagents corresponding to bromo(triphenyl)germane ⟨70CJC561⟩ and the chiral germane (**322**) to be prepared ⟨70JA7567⟩.

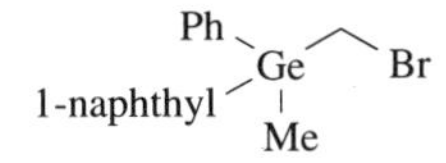

$$\text{1-naphthyl} - \underset{\underset{Me}{|}}{Ge}(Ph) - CH_2Br$$

(**322**)

Addition of an organolithium reagent to a vinyl germane has also been shown to generate α-lithiogermyl alkanes, a process extensively explored with vinylsilanes (see 4.14.2.1.1(i)(b)). Thus the reaction of (**323**) with *t*-butyllithium generated the α-lithiogermane (**324**) which was quenched with electrophiles (Equation (103)) ⟨92OM3176⟩. Triphenylgermylmethyllithium was prepared in 87% yield by transmetallation from triphenyl(triphenylgermylmethyl)lead using phenyllithium ⟨78TL4391⟩.

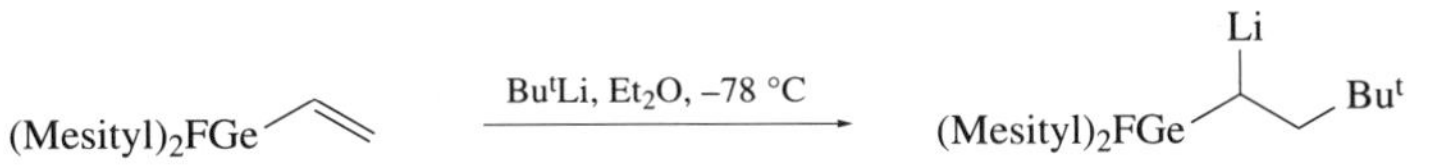

$$(Mesityl)_2FGe-CH=CH_2 \xrightarrow{Bu^tLi, Et_2O, -78\,°C} (Mesityl)_2FGe-CH(Li)-CH_2Bu^t \quad (103)$$

(**323**)   (**324**)

4.14.2.2.2 Other α-metallogermyl alkanes

Transmetallation of an α-lithio derivative can provide a rapid entry into other metal derivatives. For example, the displacement of a halide on a titanocene complex by trimethylgermylmethyllithium has been performed ⟨81JCS(D)1593⟩. A number of mercury and tin derivatives have been prepared similarly ⟨84JOM(269)249⟩.

The quenching an anion α to a metal with a germyl electrophile may also be performed such as during the preparation of the germylmethyltitanocenes (**326**) and (**327**) from the titanocene di-Grignard reagent (**325**) (Scheme 68) ⟨84TL5191⟩. Germyl electrophiles have been reacted with a number of other α-metallo Grignard reagents including derivatives containing tin ⟨84JOM(269)249, 85OM821⟩, mercury ⟨84JOM(269)249⟩ and lead ⟨78TL4391⟩. Metallomethyllithium reagents will react similarly ⟨78JOM(148)247⟩. For example, deprotonation of *t*-butyl trimethylgermylacetate with LDA gave the lithium enolate which upon quenching with trialkyltin chlorides gave (**328**). This can be metallated again enabling further functionalisation via the enolate anion ⟨88JOM(348)25⟩.

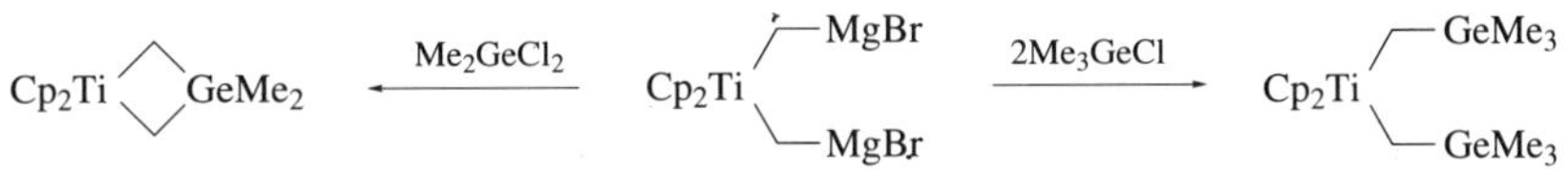

$$Cp_2Ti(CH_2GeMe_2) \xleftarrow{Me_2GeCl_2} Cp_2Ti(CH_2MgBr)_2 \xrightarrow{2Me_3GeCl} Cp_2Ti(CH_2GeMe_3)_2$$

(**326**)   (**325**)   (**327**)

Scheme 68

$$Me_3Ge-CH(SnR_3)-CO_2Bu^t$$

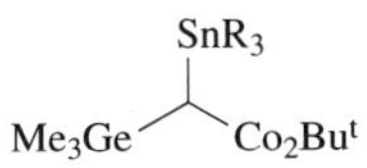

(**328**)

R = Me, 48%; Et, 51%; Bu, 98%

A further route to α-metallogermyl alkanes was demonstrated by the displacement of the halide of (halomethyl)trimethylgermanes with anionic iron and tungsten complexes resulting in the generation of the derivatives (**329**) ⟨93OM3979⟩.

$$L_nM-CH_2-GeMe_2\text{-TMS}$$

(**329**)

L_n = Cp(CO)$_2$Fe, Cp(CO)$_3$W

4.14.2.3 Boron and a Metal: $R^1_2CBR^2_2M$

This section describes preparations of compounds containing a carbon bound to a metal, a boron and either hydrogen or carbon atoms. The most readily available derivatives are the α-lithioorganoboranes which have most often been prepared by deprotonation, by halogen–metal exchange or by transmetallation. Other metallo derivatives are also known and their methods of preparation tend to reflect those used to form the α-lithio derivatives. Some examples of the generation of α-metalloalkylboranes were discussed during the preparation of bisboryl alkanes (see 4.14.1.3.1(ii)).

4.14.2.3.1 α-Lithioboryl alkanes

The deprotonation of alkylboranes must compete against the addition of the base to the boron to generate a tetracoordinate boron 'ate' complex. Formation of the 'ate' complex may be sterically disfavoured by the use of bulky groups on boron, or electronically disfavoured by increasing the electron density on boron by using alkoxy or amino substituents.

Alkyl-9-BBN derivatives may be deprotonated when hindered bases such as LITMP are used; however, the reaction is slow and the yields not high. For example, B-methyl-9-BBN was only 65% monodeuteriated in following deprotonation over 48 h with LITMP and deuteriation with D_2O ⟨72JA6854⟩. Alkyl dimesityl boranes may also be α-metallated when treated with hindered bases such as lithium dicyclohexylamide or mesityl lithium ⟨83TL623⟩. Metallation of other alkyl derivatives has been accomplished ⟨93T2988⟩ as has metallation of an allyl derivative ⟨83TL631⟩.

The use of alkoxy groups on boron to favour α-metallation over addition to boron was demonstrated when propanediol benzylboronate was deprotonated with LITMP to generate (**330**). Propanediol methaneboronate was, however, not deprotonated under these conditions ⟨82OM20⟩.

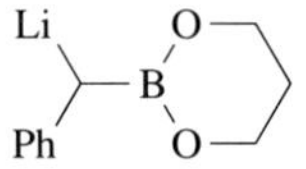

(**330**)

Routes other than deprotonation have also been used successfully to generate α-boryl alkyl-lithiums, such as lithium–boron exchange from *gem*-diboryl alkanes. The reaction of (**331**) with *n*-butyllithium gave α-lithioborane (**332**) (Scheme 69) ⟨65TL3429, 66TL2535⟩.

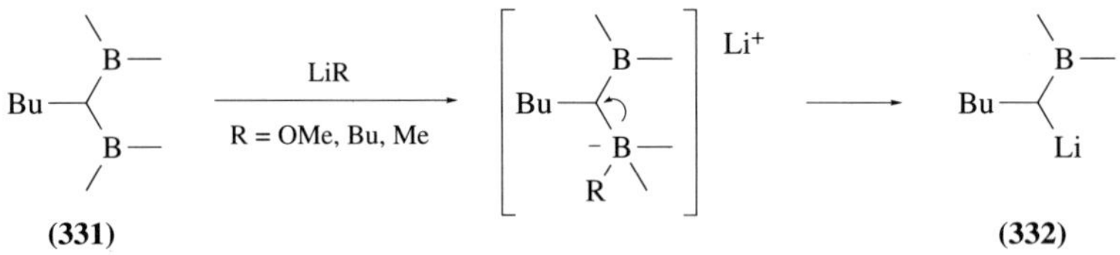

Scheme 69

Lithium–halogen exchange was demonstrated when pinacol (1-iodoethyl)boronate was treated with *t*-butyllithium at $-100\,°C$ to generate (**333**). This was also prepared by transmetallation from the corresponding α-boryl organostannane following treatment with methyllithium ⟨85OM1690, 89JA4399⟩.

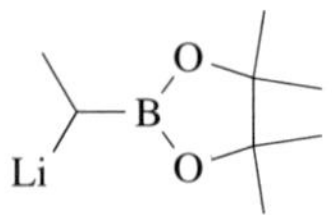

(**333**)

4.14.2.3.2 Other α-metalloboryl alkanes

Few other organometallic derivatives have received as much attention as the organolithium derivatives although they may be prepared under similar conditions ⟨76JOM(114)1⟩. For example, the treatment of *gem*-diboryl alkanes with organolithium reagents to generate α-lithioboryl alkanes (Scheme 69) has been duplicated using sodium methoxide ⟨66TL2535⟩. Similarly, metal–halogen exchange may be used to generate other organometallic derivatives. Thus treatment of pinacol (1-chloroethyl)boronate with trimethylstannyllithium generated pinacol [1-(trimethylstannyl)ethyl]boronate in good yield ⟨85OM1690⟩, a process repeated with other α-haloalkylboronic esters ⟨89JA4399⟩. Organozinc derivatives have been prepared by the reaction of α-haloboronic esters with zinc dust. Subsequent reaction with copper cyanide generated the corresponding cuprates ⟨90JA7431⟩.

Transmetallation from organolithium derivatives may also be performed as during the preparation of mercury, lead and tin derivatives of dimesitylmethylborane ⟨93T2979⟩.

Rearrangements have been implicated during the preparation of a number of cyclic organotin compounds containing an α-boryl tin moiety. These were formed by the reaction of alkynyltin compounds with organoboranes ⟨83CB3182, 84OM1, 92CB1341, 93CB1361⟩. For example the 1-stanna-3-boracyclopentene (**335**) was generated from triisopropylborane and the alkynylstannane (**334**) (Equation (104)) ⟨85CC1199⟩. Similar reactions have also been performed with lead derivatives ⟨91ZN(B)1207⟩.

$$ClMe_2Sn\text{—}\!\!\equiv\!\!\text{—} + BPr^i_3 \xrightarrow[\text{ii, LiNEt}_2]{\text{i, hexane, reflux}}_{\text{quantitative}} \underset{(\mathbf{335})}{\text{Me}_2Sn\diagdown\!\!\diagup BPr^i \text{ with } Pr^i} \quad (104)$$

(**334**)   (**335**)

4.15
Functions Containing Two Atoms of the Same Metallic Element

WILLIAM J. KERR and PETER L. PAUSON
University of Strathclyde, Glasgow, UK

4.15.1	FUNCTIONS CONTAINING TWO GROUP 1 METALS	667
4.15.2	FUNCTIONS CONTAINING TWO GROUP 2 METALS	669
4.15.3	FUNCTIONS CONTAINING TWO TRANSITION METALS	669
4.15.3.1	*Introduction*	669
4.15.3.2	*Functions Containing Two Sc, Y or La Atoms*	670
4.15.3.3	*Functions Containing Two Ti, Zr or Hf Atoms*	670
4.15.3.4	*Functions Containing Two V, Nb or Ta Atoms*	671
4.15.3.5	*Functions Containing Two Cr, Mo or W Atoms*	672
4.15.3.6	*Functions Containing Two Mn, (Tc) or Re Atoms*	676
4.15.3.7	*Functions Containing Two Fe, Ru or Os Atoms*	678
4.15.3.8	*Functions Containing Two Co, Rh or Ir Atoms*	685
4.15.3.9	*Functions Containing Two Ni, Pd or Pt Atoms*	691
4.15.3.10	*Multidecker Sandwich Compounds of Transition Metals*	694
4.15.3.11	*Functions Containing Two Cu, Ag or Au Atoms*	695
4.15.3.12	*Functions Containing Two Zn, Cd or Hg Atoms*	696
4.15.4	FUNCTIONS CONTAINING TWO Al, Ga, In OR Tl ATOMS	698
4.15.5	FUNCTIONS CONTAINING TWO Sn OR Pb ATOMS	699
4.15.6	FUNCTIONS CONTAINING TWO LANTHANIDE OR ACTINIDE ATOMS	702

4.15.1 FUNCTIONS CONTAINING TWO GROUP 1 METALS

It should be noted that compounds in which carbon is formally pentavalent are included in this chapter.

The considerable interest in 1,1-dilithio alkanes and their structures is hardly matched by experimental data. Direct reactions of *gem*-dihalo alkanes with lithium are limited by the tendency of the intermediate monolithio compounds $R^1R^2C(Li)X$ to eliminate lithium halide leaving carbene species ⟨85BSF825⟩. Moreover, all *gem*-dilithio compounds possessing β-hydrogens are unstable losing lithium hydride to leave vinyllithium compounds (see, for example (Equation (1)) ⟨83AG(E)733⟩).

$$\underset{\underset{Li}{|}}{\overset{\overset{R}{|}}{\text{C}}}\text{—Li} \longrightarrow \overset{R}{\underset{Li}{\diagdown}}\!\!=\!\! \tag{1}$$

The simplest methylenedilithium, CH_2Li_2, can be generated in reportedly excellent yields by any of three methods:

(i) Whereas early work using dibromomethane plus free metal ⟨53JOC1739⟩ led to a low yield (6%) admixed with polylithio species, use of diiodomethane and the 4,4′-di-*t*-butyldiphenyl adduct of lithium is now regarded as the simplest route (Equation (2)). Subsequent reaction with Me_3SnCl gives $CH_2(SnMe_3)_2$ in 93% overall yield ⟨87RTC514⟩.

$$CH_2I_2 \ + \ Li(4\text{-}Bu^tC_6H_4\text{-}C_6H_4\text{-}4\text{-}Bu^t) \ \longrightarrow \ CH_2Li_2 \ + \ LiI \qquad (2)$$

(ii) Thermal decomposition of methyllithium gives the compound as a pale brown solid by loss of methane ⟨55ZAAC(282)345⟩. High yields (95–96%) require the use of halide-free methyllithium at an optimum temperature of 223–226 °C; the product is reported to be a mixture of monomer and tetramer ⟨82JA2637⟩ and this route has been employed to generate both $CH_2^6Li_2$ and, at a slightly higher temperature, CD_2Li_2 for ^{13}C NMR and structural studies ⟨84IC3717⟩.

(iii) The third and in principle much more general method is the transmetallation of the corresponding bis(mercury halide). In the case of methylene, the latter was generated directly by photolysis of diiodomethane with mercury. For the transmetallation step *t*-butyllithium was found to be better than lithium metal (Scheme 1) ⟨83AG(E)733⟩.

$$CH_2I_2 \ + \ Hg \ \xrightarrow{h\nu} \ CH_2(HgI)_2 \ \xrightarrow[\text{or Li}]{Bu^tLi} \ CH_2Li_2$$

Scheme 1

Maercker *et al.* also treated ethylidenebis(mercury chloride) (generated from the corresponding diboronic ester) with lithium metal in ether (Equation (3)) ⟨83AG(E)733⟩ and reported quantitative formation of 1,1-dideuterioethane on deuteriolysis, despite the facile loss of LiH from CH_3CHLi_2.

$$MeCH(HgCl)_2 \ + \ 2Li \ \xrightarrow{Et_2O, RT} \ MeCHLi_2 \ + \ 2Hg \qquad (3)$$

Structural information about methylenedilithium has been obtained by x-ray scattering and by neutron diffraction of CD_2Li_2. The rather complex arrangement found fits neither a planar nor a tetrahedral model and shows the importance of metal/H—C interactions ⟨90JA2425⟩.

More complex *gem*-dilithio compounds have received relatively little attention. Even at −110 °C the halogen/metal exchange with *t*-butyllithium (in THF) has given only very modest yields of Ph_2CLi_2 (31%) from Ph_2CCl_2, and $PhCHLi_2$ (7.6%) from $PhCHCl_2$ ⟨90JA9415⟩.

Like CH_2Li_2, other 1,1-dilithio compounds lacking β-hydrogens are accessible by the pyrolysis route, for example, Bu^tCHLi_2 (88%) from neopentyllithium at 180 °C ⟨88POL2023⟩.

Direct lithiation of hydrocarbons is possible for allylic and propargylic dilithio species. Thus, propene itself is metallated by butyllithium in the presence of TMEDA (Equation (4)) ⟨75CC877⟩. Additionally, the analogous, stepwise formation of $PhC{\equiv}CCHLi_2$ and $PhC{\equiv}CCLi_2CH_3$ from 1-phenylpropyne and 1-phenyl-1-butyne respectively and BuLi in Et_2O at 25 °C has been followed kinetically by the appearance of the characteristic absorption at λ_{max} 382 nm of the dilithio compounds ⟨78JOM(157)1⟩.

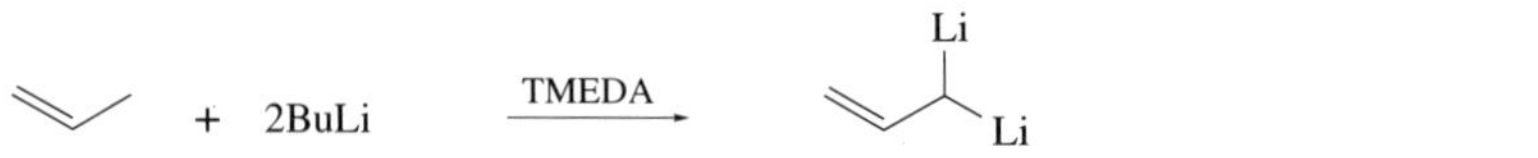

$$\diagdown\!\!\!\diagup \ + \ 2BuLi \ \xrightarrow{TMEDA} \ CH_2{=}CH{-}CHLi_2 \qquad (4)$$

Other geminal dilithio compounds arise when two formally separated lithium atoms form bridging structures as illustrated by 2,2′-dilithiobiphenyl whose bis-TMEDA complex was predicted theoretically and confirmed by x-ray crystallography to have the structure (**1**) ⟨82CC1184⟩.

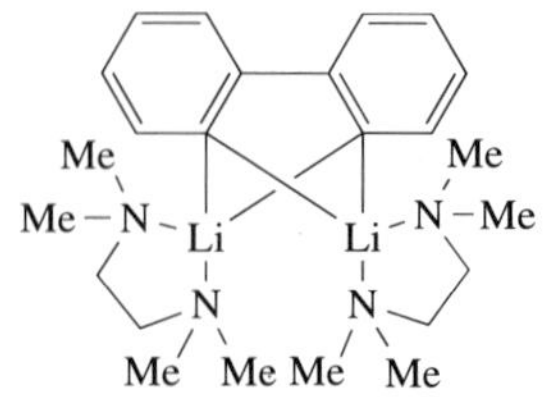

(**1**)

Formation of a small amount (~6%) of 1,1-disodio-1-phenylethane, PhCNa$_2$CH$_3$, as a transient product of metallation of ethylbenzene by *n*-pentylsodium and TMEDA, has been demonstrated by dicarboxylation ⟨78JOC2170⟩.

4.15.2 FUNCTIONS CONTAINING TWO GROUP 2 METALS

The ability of methylene halides to form di-Grignard reagents was first demonstrated in 1926 ⟨26CR(183)665⟩. The initially very poor yields were much improved when methylene bromide (or iodide) was treated with magnesium amalgam (rather than just magnesium) in 1:1 ether–benzene ⟨70T1281⟩. Use of a rather dilute (1%) amalgam minimizes formation of MeMgBr as a by-product ⟨85JOM(288)13⟩ and diisopropyl ether as solvent further improves the yield (to 80–90%) of a product which precipitates from this solvent and can be extracted into 1:1 Et$_2$O/C$_6$H$_6$ to give a solution of nearly pure CH$_2$(MgBr)$_2$ with a concentration of ~0.03 M ⟨86JOM(308)1⟩.

The reaction of diiodomethane with a mixture of the lithium adduct of 4,4'-di-*t*-butylbiphenyl and magnesium bromide leads to CH$_2$(MgX)$_2$ as the major product mixed with substantial amounts of the higher di-Grignard reagents XMg(CH$_2$)$_{2-4}$MgX, formed by reduction ⟨88RTC393⟩. Also, a methylenating reagent obtained from diiodomethane, magnesium powder and chlorodiethylaluminum has been tentatively formulated as shown in Equation (5) ⟨89IZV2562⟩.

$$\text{Et}_2\text{AlCl} + \text{CH}_2\text{I}_2 + \text{Mg} \longrightarrow \underset{\text{I}-\text{Mg}\diagdown\diagup\text{Mg}-\text{Cl}}{\overset{\text{Et}\qquad\qquad\text{Et}}{\text{Et}-\text{Al}-\text{Cl}\quad\text{I}-\text{Al}-\text{Et}}} \qquad (5)$$

Equivalent calcium species have been much less widely studied and, indeed, diiodomethane and calcium in THF afford CH$_2$(CaI)$_2$ in only 17% yield ⟨76BCJ1177⟩.

The structures of dimethylmagnesium and related compounds show that adjacent metal atoms are bridged by methyl or other alkyl groups (i.e., by formally five-coordinate carbon atoms). For the formation of these functions see Chapter 2.11.2.

4.15.3 FUNCTIONS CONTAINING TWO TRANSITION METALS

4.15.3.1 Introduction

Transition metal complexes in which one or more tetracoordinate carbons are linked to two metal atoms are now known in great number. An excellent if now dated review, *The Methylene Bridge*, dealing with the main class of such compounds has been published by Herrmann, himself a main contributor to the field ⟨82AOC(20)159⟩.

Few of the methods used for the synthesis of these functions possess wide generality. One such is the use of diazomethane and higher diazoalkanes. These can act as carbene sources and either displace CO, and similarly weakly held ligands, or add to metal–metal multiple bonds. In the former case monomeric carbene complexes may be intermediates. Alternative carbene sources have not been widely used.

Transmetallation, usually involving lithio or Grignard precursors, a widely used route to simple metal alkyls (see Chapter 2.11) has not been systematically applied, but examples of its use for miscellaneous 1,1-dimetal complexes will be found in the following pages. They include formation of simple methylene complexes of the titanium group ⟨86JOM(308)1⟩ and mercury ⟨85JOM(288)13⟩ and also the linking of a second metal atom to a π-bonded aromatic ligand as in the reaction of ferrocenyllithium with haloiron complexes (Equation (6)) ⟨73IZV2796⟩.

$$\text{FcLi} + \text{BrFeL}_n \longrightarrow \text{Fc-FeL}_n + \text{LiBr} \qquad (6)$$

A large and very varied group of carbon-bridged complexes result when reactive complexes are treated with aromatic, alkenyl and, especially, alkynyl hydrocarbons. Many of these reactions are accompanied by other changes of the hydrocarbons. It is difficult to classify and systematize this group of reactions and their variety is such that only a selection can be included in this chapter. One of the few common patterns is provided by alkynes reacting with certain metal carbonyls according to Equation (7). This applies to L$_n$M = Cp(CO)$_2$Mo, Cp(CO)$_2$W, (CO)$_3$Co and CpNi. In particular, extensive study of the cobalt and molybdenum series has been made. Further transformations of such complexes (e.g., via the stabilized α-cations) are outside the scope of this chapter unless they create new bridging carbon atoms.

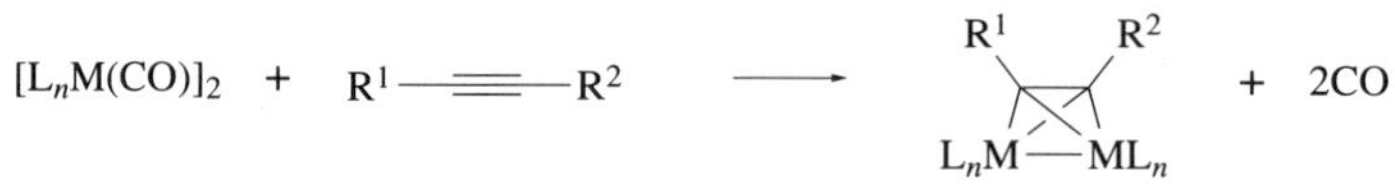

$$[L_nM(CO)]_2 + R^1\!\!\equiv\!\!R^2 \longrightarrow \underset{L_nM-ML_n}{\overset{R^1\quad R^2}{\bowtie}} + 2CO \qquad (7)$$

The treatment of transition metal compounds is subdivided according to the individual groups with the exception of the multidecker sandwich compounds which are more conveniently collected as a class in Section 4.15.3.10.

4.15.3.2 Functions Containing Two Sc, Y or La Atoms

No methylene-bridged compounds of these elements have been found. On the other hand, at least one methyl-bridged yttrium compound $Cp^*_2Y(\mu\text{-}CH_3)YCp^*_2$ has been reported ⟨83JA6491⟩.

4.15.3.3 Functions Containing Two Ti, Zr or Hf Atoms

Methylene-bridged dititanium complexes are formed when trimethylenetitanocene derivatives are warmed above 0 °C (Equation (8)). The reaction involves the alkene cleavage step of a metathesis reaction and cyclodimerization of the Cp_2TiCH_2 fragment ⟨81JA5922⟩. The product (**2**) precipitates from the toluene solution of the precursor as moderately air-stable purple/red platelets.

$$2\;\underset{Cp}{\overset{Cp}{\bowtie}}Ti\underset{Pr^i}{\bowtie}\;\xrightarrow{85\%}\;2\;\underset{Pr^i}{=}\!\!\!\!< + \;\underset{Cp}{\overset{Cp}{\bowtie}}Ti\underset{}{\bowtie}Ti\underset{Cp}{\overset{Cp}{\bowtie}} \qquad (8)$$
$$\mathbf{(2)}$$

An alternative route to the same product involves treatment of Cp_2TiCl_2 with $CH_2(MgBr)_2$ in ether/benzene (1 : 1) to give solutions containing '$Cp_2Ti(X)CH_2MgX \cdot MgX_2$'. Evaporation followed by addition of toluene leaves a precipitate of $MgCl_2$ and gives a solution believed to contain the complex (**3**), which decomposes slowly at room temperature, but instantly and quantitatively on addition of THF or dioxane, to give complex (**2**) ⟨87TL6493⟩.

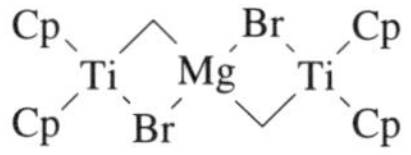

$$\underset{Cp}{\overset{Cp}{\bowtie}}Ti\underset{Br}{\overset{}{\bowtie}}Mg\underset{}{\overset{Br}{\bowtie}}Ti\underset{Cp}{\overset{Cp}{\bowtie}}$$
$$\mathbf{(3)}$$

Stepwise treatment of the metallocene dichlorides $Cp_2M^1Cl_2$ (M^1 = Ti, Zr, Hf) with methylenebis(magnesium bromide) followed by a second $Cp_2M^2Cl_2$ allows not only the dititanium compound (**2**) but also the orange/red dizirconium, the pale yellow dihafnium and the mixed metal (Ti—Zr, Ti—Hf and Zr—Hf) analogues to be formed (Scheme 2) ⟨86JOM(308)1⟩.

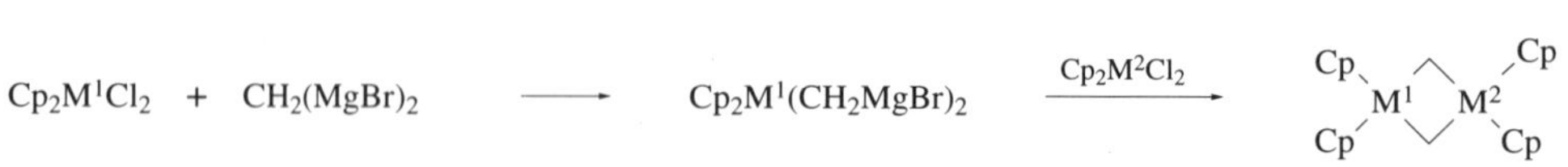

$$Cp_2M^1Cl_2 + CH_2(MgBr)_2 \longrightarrow Cp_2M^1(CH_2MgBr)_2 \xrightarrow{Cp_2M^2Cl_2} \underset{Cp}{\overset{Cp}{\bowtie}}M^1\underset{}{\bowtie}M^2\underset{Cp}{\overset{Cp}{\bowtie}}$$

Scheme 2

Many complexes are known in which formally sp^2 hybridized carbon atoms are σ- or π-bonded to two metal atoms. An early example was tris(cyclooctatetraene)dititanium shown by crystallography to have the structure (**4**) with the two asterisked carbons bonded to both metal atoms

⟨66AG(E)899⟩. It is formed either from Ti(OBu)$_4$, C$_8$H$_8$ and Et$_3$Al in a 1:2:10 ratio or from Ti(C$_8$H$_8$)$_2$ and Et$_2$AlH ⟨66AG(E)898⟩.

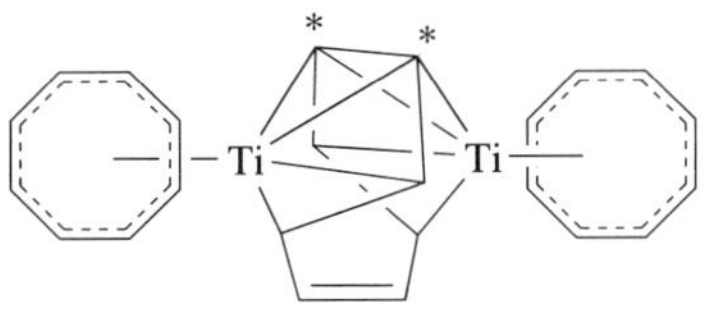

(4)

Reduction of Cp$_2$TiCl$_2$ by potassium naphthalenide yields, after addition of THF, the complex of structure (5) with one cyclopentadienyl carbon as a bridging atom ⟨76JA8072⟩, whereas the analogous reduction of Cp$_2$ZrCl$_2$ introduces a naphthyl group with one β-carbon taking a bridging position (6) ⟨79JA6933⟩.

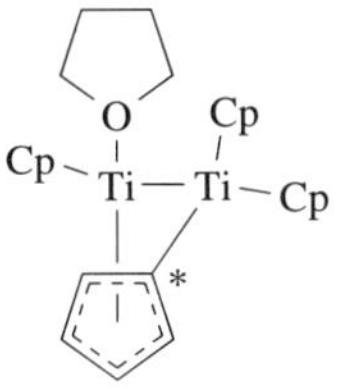

(5) 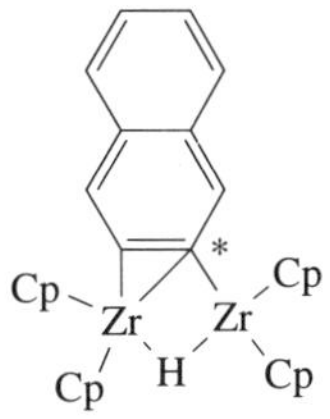 (6)

The complex cation (7) with a planar tetracoordinate carbon (C^2) is obtained according to Equation (9). Its fluxional behaviour demonstrated by NMR spectroscopy suggests that it is in equilibrium at or above room temperature with the symmetrical structure resulting from cleavage of the Zr^1—C^2 bond, thus allowing C^1 and C^2 to alternate as metal-bridging atoms ⟨93AG(E)1623⟩.

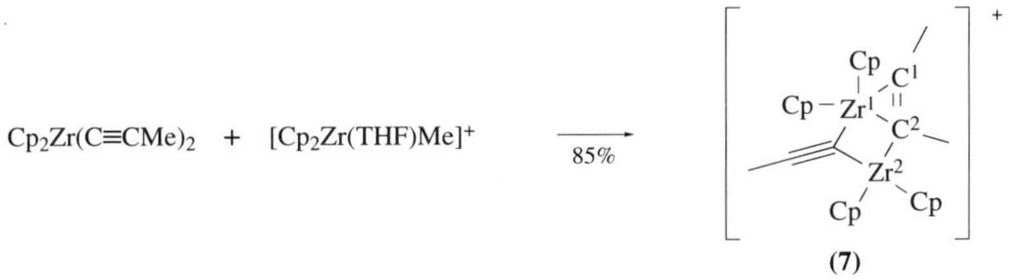

$$Cp_2Zr(C \equiv CMe)_2 + [Cp_2Zr(THF)Me]^+ \xrightarrow{85\%} (7) \qquad (9)$$

4.15.3.4 Functions Containing Two V, Nb or Ta Atoms

The dialkylcyclopentadienylvanadium complex CpV(PMe$_3$)(CH$_2$CMe$_2$Ph)$_2$ loses trimethylphosphine in cyclohexane solution at 20 °C to give a dimeric complex whose structure (8) shows that the two vanadium atoms are bridged by two formally pentacoordinate carbon atoms of CH$_2$ groups and also by carbon atoms of two benzene rings ⟨93OM2268⟩. A closely related complex, Cp$_2$V$_2$(C$_4$H$_8$)$_2$, with the four terminal CH$_2$ groups of two tetramethylene chains bridging the metal atoms, results when ethene displaces the arene ligand from (η^5-cyclopentadienyl)(η^6-naphthalene)vanadium ⟨86AG(E)925⟩.

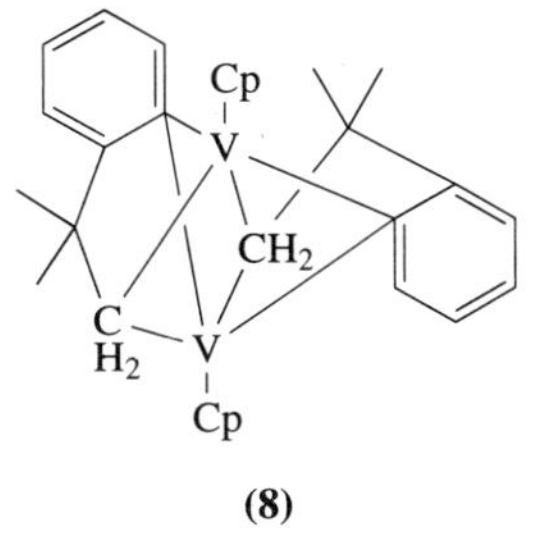

(8)

The bis(cyclopentadienyl)trihydrides, Cp_2MH_3, of both niobium and tantalum lose hydrogen in refluxing benzene to give 'dimeric niobocene' and the isomorphous 'dimeric tantalocene' ⟨71JA3793⟩. The hydride structure (**9**) with bridging cyclopentadienyl groups has been confirmed by x-ray crystallography on the yellow niobium compound ⟨73IC294⟩.

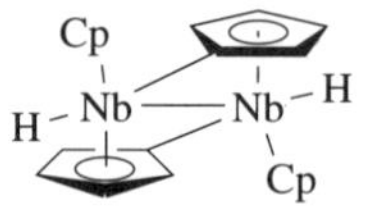

(**9**)

Loss of carbon monoxide at 80 °C converts the tolan-niobium complex $Cp_2Nb(CO)_2(PhC\equiv CPh)$ into the dinuclear complex (**10**) with four tetracoordinate carbon atoms bridging the two niobium atoms ⟨68CC1365, 69IZV100⟩. This type of alkyne bridging will be found to be common to many of the transition metals discussed in succeeding sections.

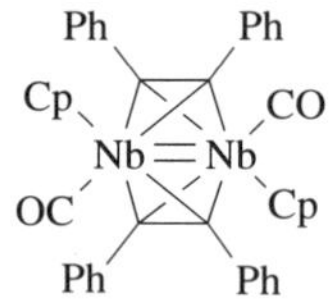

(**10**)

Treatment of $VCl_2 \cdot 2THF$ with CpNa and $K_2C_8H_8$ (2:2:1) yields the cyclooctatetraene-bridged complex (**11**; no X, M = V) in which the two central carbons (*) are bridging ⟨83JA2905⟩. The analogous complexes (**11**; M = V) with X = CH_2, $SiMe_2$, or $GeMe_2$ have been obtained similarly from $Li_2(C_5H_4XC_5H_4)$ ⟨90ZN(B)221⟩.

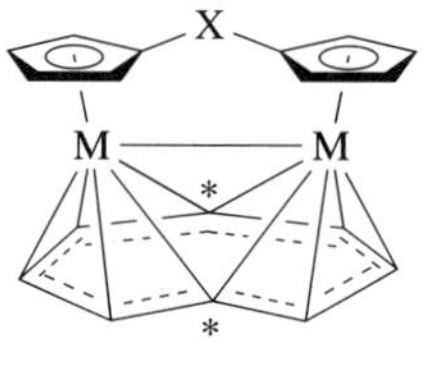

(**11**)

4.15.3.5 Functions Containing Two Cr, Mo or W Atoms

The reaction of $Cr_2(OAc)_4$ with $(TMS-CH_2)_2Mg$ in the presence of trimethylphosphine yielded ⟨78JCS(D)446⟩ a complex shown by x-ray crystallography ⟨78JCS(D)1314⟩ to have the structure (**12**) with formally five-coordinate bridging carbons.

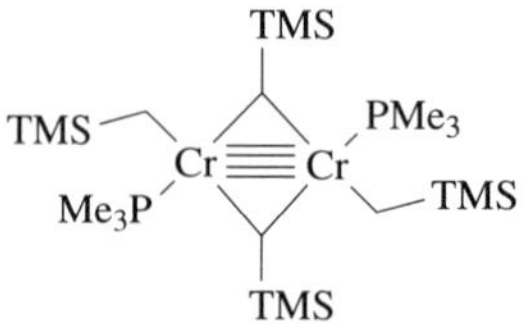

(**12**)

Reactions of molybdenum and tungsten halides and oxyhalides (e.g., Mo_2Cl_{10}, $MoOCl_3(THF)_3$ and $WOCl_4$) with methyllithium generate a series of products which are useful for *in situ* application in carbonyl methylenation. Although initially formulated as monomeric carbenes (e.g., $Cl_2W(O)=CH_2$) ⟨86AG(E)909⟩ they are now, on the basis of NMR spectra, regarded as dimeric with bridging methylene groups (e.g., structures (13) and (14)) ⟨92CB143, 93CB79, 93CB89⟩.

$Cl_3Mo\cdots MoCl_3$ $(THF)_2Cl(O)Mo\cdots Mo(O)Cl(THF)_2$

(13)   (14)

The dimeric allyls of chromium and molybdenum, obtainable respectively by thermal or photochemical decomposition of $(C_3H_5)_2Cr$ or directly from allylmagnesium bromide and chromium chloride ⟨69BCJ545⟩ and from allylmagnesium chloride and molybdenum(V) chloride ⟨71JA5441⟩, have both been shown crystallographically to have structures (15; M = Cr or Mo) in which the asterisked atoms bridge both metal atoms. Somewhat similar bridging situations are found in tris(cyclooctatetraene)dimolybdenum (16; M = Mo) and -ditungsten (16; M = W), both obtained from the corresponding tetrachlorides with $K_2C_8H_8$ ⟨78IC2093⟩, and in the chromium complexes (11; M = Cr, no X or X = CH_2, TMS, or $GeMe_2$), formed from $CrCl_2$ (as the THF or TMEDA adduct), CpNa and $K_2C_8H_8$ ⟨83AG(E)330, 88JOM(342)45, 90ZN(B)221⟩. Of these, the unsubstituted complex (11; M = Cr, no X) is not formed directly but only on pyrolysis of an intermediate (17) in which the cyclooctatetraene has undergone ring opening ⟨88JOM(342)45⟩.

(15)   (16)   (17)

A single, tetracoordinate carbon bridges the two molybdenum atoms in the complex (18), formed from allene according to Equation (10) ⟨78JA802⟩. The pentamethylcyclopentadienyl analogue of this allene complex has been formed by adding diazomethane to the corresponding vinylidene-bridged species ⟨85CC170⟩.

(10)

(18)

The activation of aromatic rings by a $Cr(CO)_3$ group towards nucleophilic substitution has been exploited by attaching a second chromium atom to one of the ring carbons as shown in Equation (11) ⟨89OM1199⟩.

(11)

As in the titanium and vanadium groups, attempts to form simple metallocenes, Cp_2M, from molybdenum and tungsten (unlike chromium) fail to give stable products of this structure, yielding instead binuclear compounds in which cyclopentadienyl rings are simultaneously η^5-bonded to one and η^1-bonded to a second metal atom (cf. structures (5) and (9)). Thus photolysis of Cp_2MoH_2 or oxidation of $(Cp_2Mo(H)Li)_4$ by N_2O yields the complex (19; M = Mo, $X^1 = X^2 = H$); further photolysis results in the loss of hydrogen to give (19; M = Mo, no X^1, X^2) which can, in turn, add I_2, MeBr, etc. yielding analogues with different X^1 and X^2 groups (e.g., $X^1 = X^2 = I$ or $X^1 = Me$, $X^2 = Br$) ⟨80JCS(D)29⟩. Photolysis of Cp_2WH_2 or thermal decomposition (70 °C) of $Cp_2W(CH_3)H$ similarly yields the tungsten analogue (19; M = W, $X^1 = X^2 = H$) ⟨80JCS(D)29⟩. All such processes are believed to involve the metallocene, Cp_2M, as a transient intermediate. The phosphine-bridged complex (**20**) on photolysis or heating reversibly and quantitatively loses CO to yield the product (**21**) ⟨93JA3786⟩.

(**19**)   (**20**)   (**21**)

Alkyne-bridged molybdenum complexes of the type (**23**; M = Mo) are known with a wide variety of substituents on both the alkyne bridge and the cyclopentadienyl rings. They range from the simplest (**23**; M = Mo, $R^1 = R^2 = H$) ⟨77ICA85, 78JA5764, 78JOM(161)23⟩ to compounds in which the alkyne is a 17-ethynylsteroid ⟨90M2993⟩ or an ethynylmetal complex such as $CpFe(CO)_2C\equiv CH$ ⟨93OM2925⟩. They can be obtained from the alkyne and hexacarbonylbis(η^5-cyclopentadienyl)dimolybdenum, $Cp_2Mo_2(CO)_6$, but it is generally found preferable to decarbonylate the latter in a separate step to give the tetracarbonyl (**22**). Addition of the alkyne to the Mo–Mo triple bond then proceeds under relatively mild conditions. Several tungsten analogues (**23**; M = W) have been obtained in similar fashion using the tetracarbonyl species, $Cp_2W_2(CO)_4$ ⟨77ICA85, 78JOM(157)41⟩. The diester (**23**; M = W, $R^1 = R^2 = CO_2Me$) has also been obtained by treating the alkyne with $CpW(CO)_3H$ in refluxing THF ⟨77JOM(124)29⟩ and by irradiation of an isolated intermediate (**24**) which itself contains one bridging carbon atom (*) ⟨82JCS(D)1783⟩. All crystal structure determinations on the complexes (**23**; M = Mo or W) have found the somewhat asymmetric arrangement shown, characterized by one semi-bridging CO ligand.

(**22**)   (**23**)   (**24**)

Modification of the alkyne substituents, R^1 and R^2, in these complexes (**23**; M = Mo) has been achieved by making use of the stability of the α-carbocations (**23**; M = Mo, $R^1 = CH_2^+$ or $CHMe^+$ etc.). These are most commonly generated by protonation of alcohols or ethers (e.g., $R^1 = CH_2OH$ or CH_2OMe) (see, for example ⟨94OM2244⟩) but are equally accessible by protonation of, or other electrophilic addition to, vinyl substituents (e.g., **23**; M = Mo, $R^1 = CH=CH_2$, $R^2 = H \rightarrow R^1 = {}^+CHMe$) ⟨92OM721⟩. Additionally they are available by pyrolysis (100 °C) of tetraalkylammonium salts (e.g., **23**; $R^1 = CH_2NMe_3^+BF_4^-$, $R^2 = H$) ⟨93JOM(455)C9⟩. The simplest of these salts (**23**; M = Mo, $R^1 = {}^+CH_2$, $R^2 = H$) has been obtained from the corresponding propyne complex ($R^1 = CH_3$) or the isomeric allene complex (**18**) by hydride abstraction with Ph_3CBF_4 ⟨86ZOR2457⟩. Furthermore, they can not only be generated *in situ* as intermediates in nucleophilic substitution processes, but isolated in crystalline form and characterized by x-ray crystallography ⟨91JOM(418)C24, 92OM721⟩. Even the dication (**23**; M = Mo, $R^1 = R^2 = {}^+CH_2$) has been isolated as its tetrafluoroborate salt ⟨85JOM(297)C25⟩.

Protonation of complexes of the type (**23**) leads to cations (**25**) retaining only one bridging carbon or, after capture of a suitable nucleophile X^- (e.g., $CF_3CO_2^-$ when protonated with trifluoroacetic acid), the corresponding neutral complex (**26**) ⟨80JOM(202)C49, 83OM1172⟩.

(**25**)   (**26**)

The closely related phosphine-bridged complexes (**27**) are the products of alkyne addition to $Cp_2Mo_2(CO)_4(\mu\text{-}H)(\mu\text{-}PMe_2)$ ⟨88NJC559⟩ (cf. also ⟨86CC542⟩ and ⟨89CC688⟩).

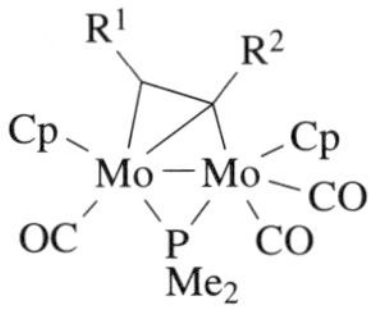

(**27**)

The cyclopentadienylmetal series notwithstanding, bridged alkyne–molybdenum complexes include the (substituted) phenylalkyne–ethylenediamine complexes (**28**), obtained by adding both ligands to $Mo_2(OAc)_4$ ⟨91IC156⟩, and the sulfide-bridged dithiocarbamate complexes (**29**), obtained from $(\eta^2\text{-}R^1C\equiv CR^2)_2Mo(S_2CNR^3{}_2)_2$ by heating with phosphines ⟨83JA2599⟩.

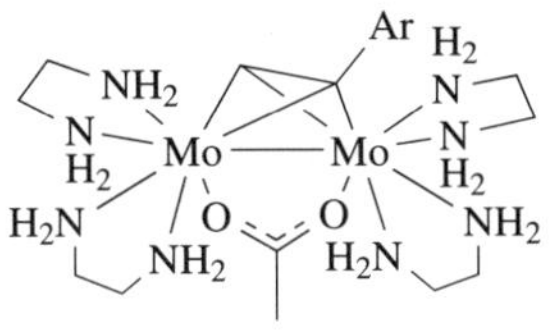

(**28**) 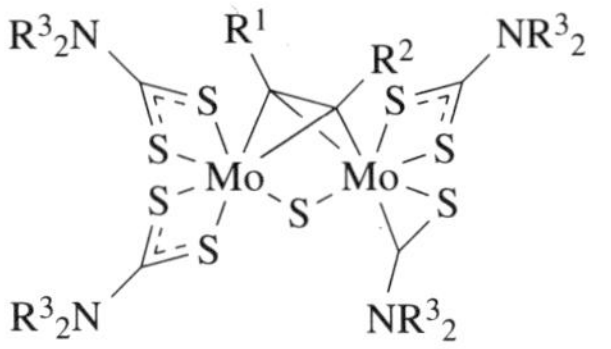 (**29**)

Various complexes with tungsten–tungsten multiple bonds (e.g., $W_2Cl_2(NMe_2)_4$, $W_2(OBu^t)_6$, $W_2(OPr^i)_6(py)_2$ (py = pyridine), $W_2(CH_2R)_2(OPr^i)_2$, etc.) have been shown to add alkynes to form bridged complexes, sometimes with simultaneous addition of other donors. The products are generally of the type (**30**), for example $X^1 = Y = NMe_2$, $X^2 = Cl$, $L = py$ ⟨86OM2171⟩; $X^1 = X^2 = Cl$, $Y = NMe_2$, $L = py$ ⟨85OM1312⟩; $X^1 = X^2 = Y = OPr^i$, $L = py$ ⟨84JA6794⟩; $X^1 = X^2 = Cl$, $L = THF$ ⟨89CC418⟩ or (**31**), for example $X = OCH_2Bu^t$, $L = py$ ⟨84JA6794, 86OM2457⟩.

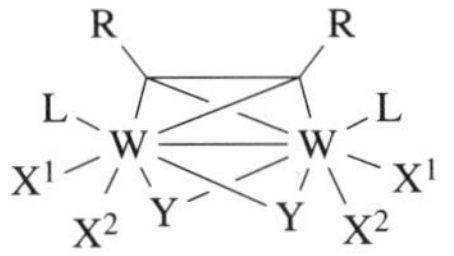

(**30**) 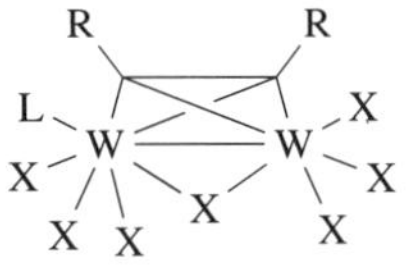 (**31**)

A more remarkable route to such compounds involves the formation of the bridging alkyne from a carbyne ligand as in the reactions shown in Equations (12) ⟨89OM1626⟩ and (13) ⟨91OM535⟩ (cf. also ⟨85JCS(D)905⟩).

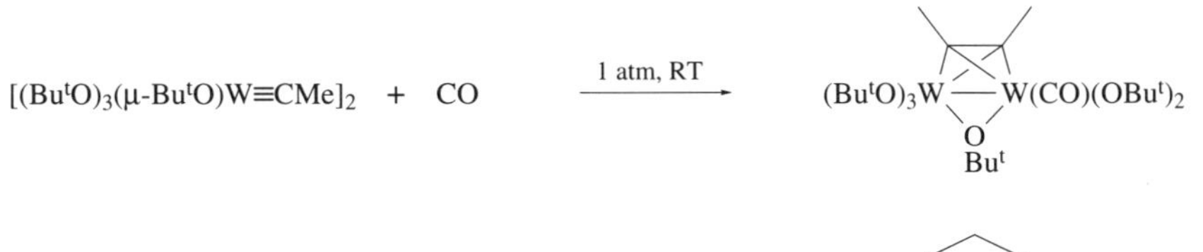

$[(Bu^tO)_3(\mu\text{-}Bu^tO)W\equiv CMe]_2$ + CO $\xrightarrow{\text{1 atm, RT}}$ $(Bu^tO)_3W\text{—}W(CO)(OBu^t)_2$ (12)

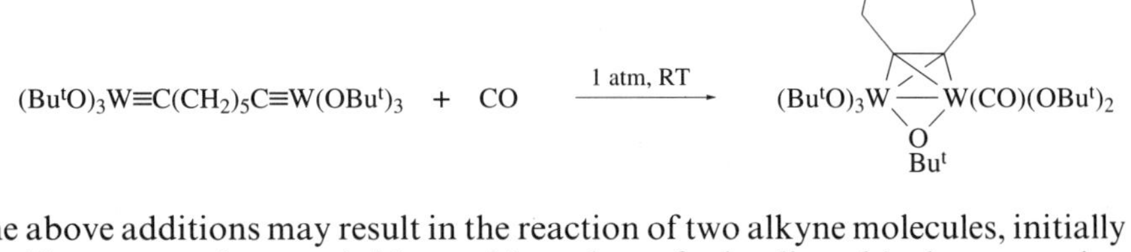

$(Bu^tO)_3W\equiv C(CH_2)_5C\equiv W(OBu^t)_3$ + CO $\xrightarrow{\text{1 atm, RT}}$ $(Bu^tO)_3W\text{—}W(CO)(OBu^t)_2$ (13)

Some of the above additions may result in the reaction of two alkyne molecules, initially terminally (η^2) bonded to the two metal atoms (with or without loss of other ligands); these may than rearrange when warmed to either doubly bridged complexes (**32**), for example $R^1 = Me$, $R^2 = Pr^i$ ⟨89OM80⟩ or $R^1 = Ph$, $R^2 = Bu^t$ ⟨83OM1167⟩, or to complexes (**33**) with a four-carbon ligand (tungstacyclopentadienes) formed by linking the two alkyne units, sometimes with addition of a third alkyne (**33**; L = R^{1}C≡CR^1) ⟨84JA6806, 86OM2384, 92AG(E)66⟩ (cf. also ⟨89OM67⟩ and ⟨90OM886⟩).

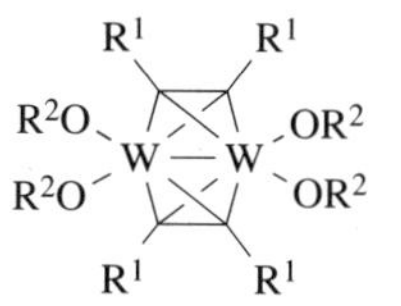

(**32**) 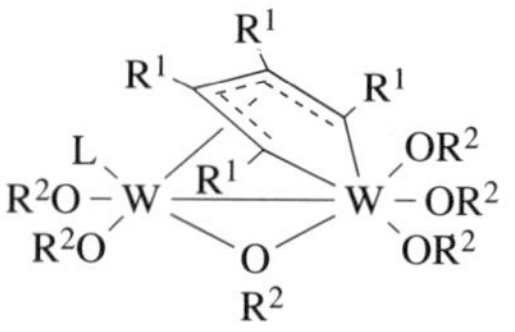   (**33**)

Closely related molybdenacyclopentadienes have been made by similar alkyne additions ⟨86OM602, 88POL919⟩. Moreover, the chromacyclopentadienes (**34**), obtained from CpCr(CO)$_2$≡Cr(CO)$_2$Cp and PhC≡CH ⟨78JOM(150)C1⟩, and (**35**), obtained from Cp*CrCl$_2$, MeC≡CMe and Mg ⟨92ICA741⟩, seem to be the simplest known alkyne-derived bridged chromium complexes. When this reaction is carried out at higher temperature the product has a six-carbon bridge, which means it is the corresponding chromacycloheptatriene complex. The more symmetrical 'flyover' six-carbon bridge of the molybdenum complex (**36**) is formed by sodium or magnesium amalgam reduction of the salt [C$_9$H$_7$Mo(η^2-MeC≡CMe)$_2$(NCMe)]BF$_4$ ⟨86JCS(D)657⟩ but reduction of the cyclopentadienyl analogue [CpMo(η^2-MeC≡CMe)$_2$(NCMe)]BF$_4$ by CpFe(CO)$_2$Na gave the eight-carbon-bridged (molybdenacyclononatriene) complex (**37**) ⟨84JCS(D)2455⟩ (cf. ⟨88JCS(D)1843⟩ and ⟨89OM412⟩).

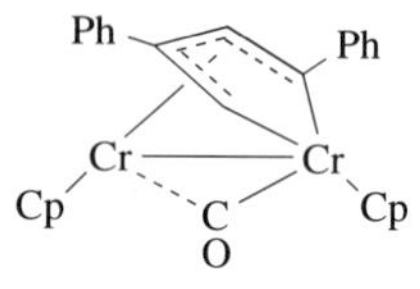

(**34**) 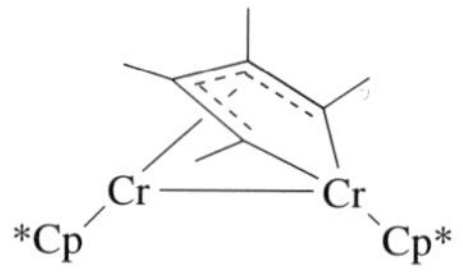   (**35**)

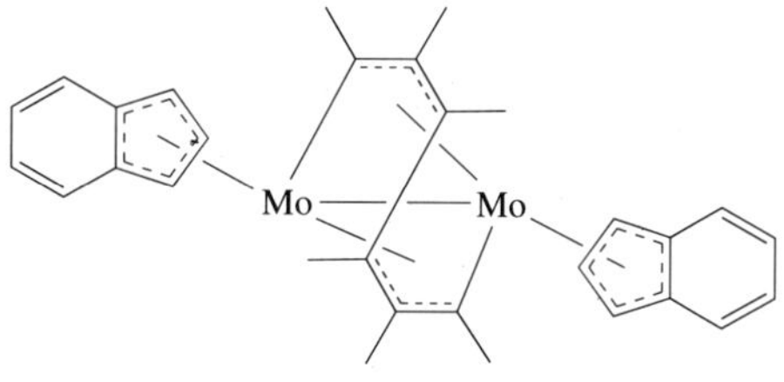

(**36**) 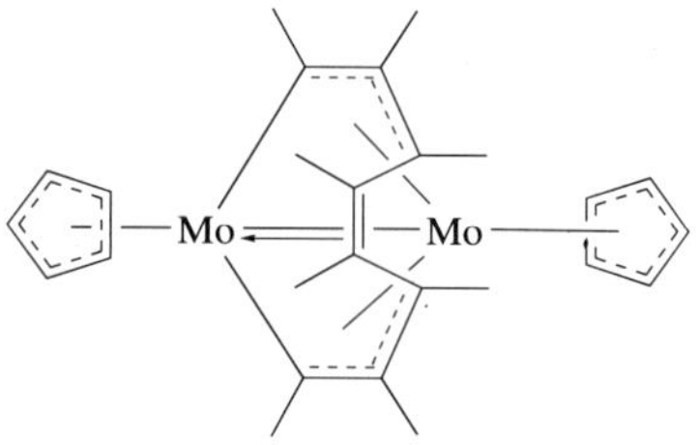   (**37**)

In contrast to the reaction shown in Equation (12), irradiation of the cyclopentadienyltungsten carbyne complex CpW(CO)$_2$≡CC$_6$H$_4$-p-Me under CO yields the three-carbon-bridged ketone (**38**) ⟨88OM289⟩; this may be protonated by HBF$_4$·OEt$_2$ to form the salt (**39**) ⟨88IC3248⟩.

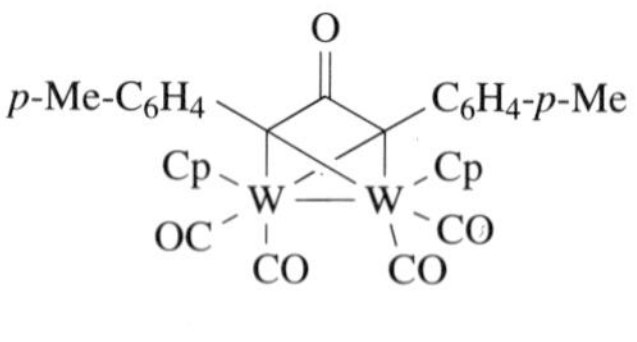

(**38**) 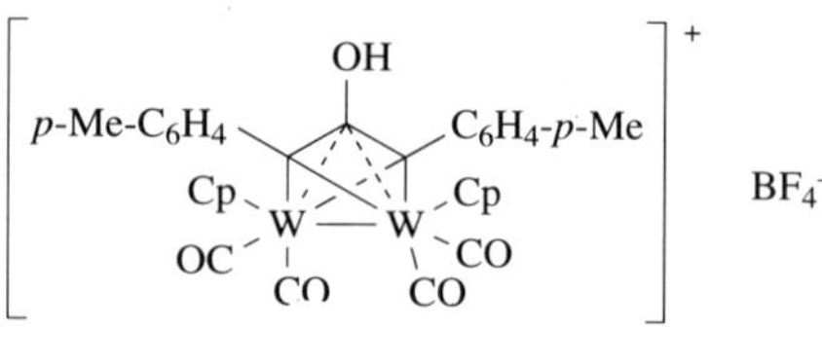   (**39**)

4.15.3.6 Functions Containing Two Mn, (Tc) or Re Atoms

The reaction shown in Scheme 3 provided the first example of the diazoalkane route to methylene-bridged complexes ⟨75JOM(97)245⟩. The intermediacy of the monomeric carbene, which is not observed directly, is supported by the formation of the mixed (C$_5$H$_5$/C$_5$H$_4$Me) complex when (C$_5$H$_4$Me)Mn(CO)$_2$(THF) is added after allowing CpMn(CO)$_2$(THF) to react with an equimolar quantity of diazomethane ⟨82AOC(20)159⟩. X-ray crystallography has confirmed the structures of both $(C_5H_4Me)_2Mn_2(CO)_4(\mu\text{-}CH_2)$ ⟨82IC645⟩ and $Cp_2Mn_2(CO)_4(\mu\text{-}CH_2)$ ⟨81AG(E)887⟩.

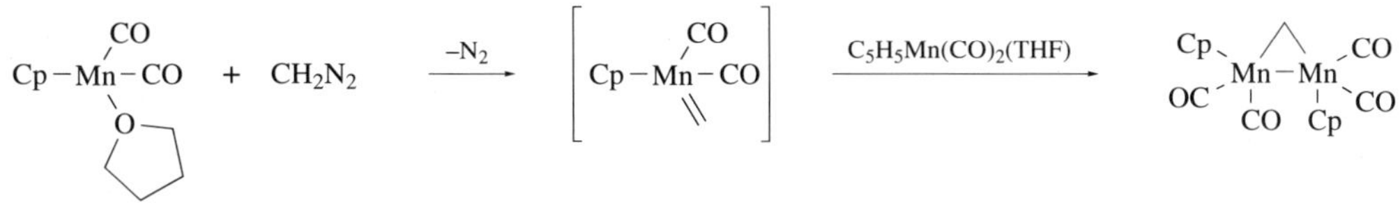

Scheme 3

The dimeric product from MnI_2 and $o\text{-}LiCH_2C_6H_4NMe_2$ was shown to have the curiously unsymmetrical structure (**40**) with bridging, formally pentacoordinate carbons ⟨77JOM(139)C34⟩ while the product from $MnCl_2$ and $PhCMe_2CH_2MgCl$ has similar bridges but the more symmetrical structure (**41**) ⟨76JCS(D)2204⟩.

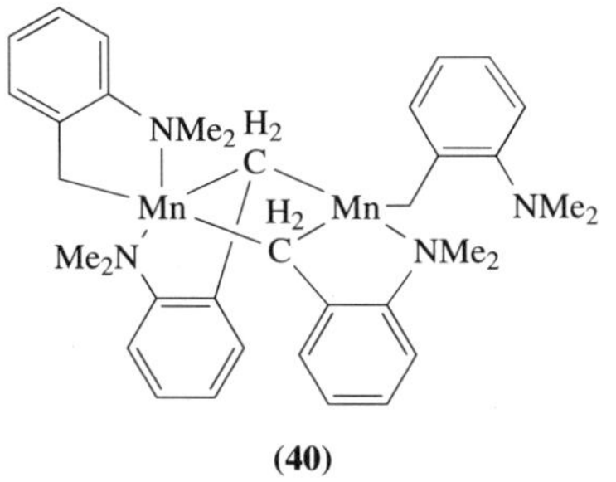

(**40**) 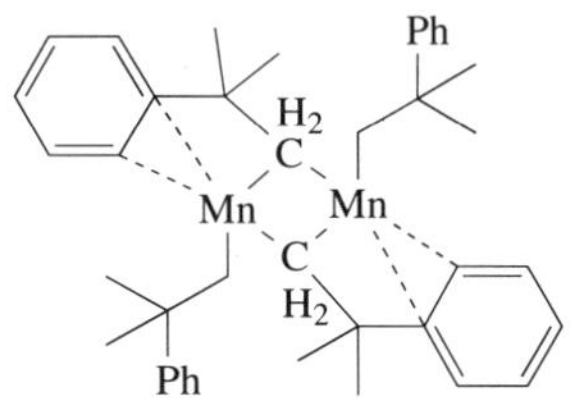   (**41**)

An example of a complex with a second manganese atom σ-bonded to the ring of a π-cyclopentadienylmanganese derivative has been obtained according to Equation (14) ⟨93JOM(458)181⟩.

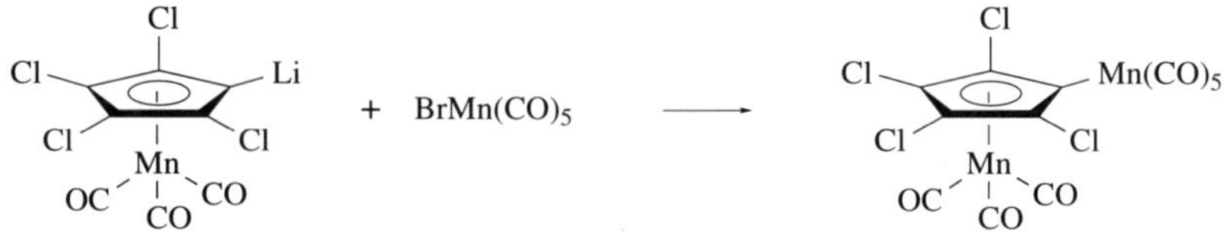

(14)

Decacarbonyldirhenium adds terminal alkenes in high yield on UV photolysis to give the dinuclear complexes (**42**) with one bridging carbon (*) ⟨82JA4955⟩. The process is probably reversible since treatment with, for example, 1-dimethylaminopropyne leads to the loss of the alkene and formation of complex (**43**; M = Re, R = Me) which is, in turn, converted on warming with a second mole of alkyne into the rhenacyclopentadiene complex (**44**) ⟨91OM1278⟩. The manganese analogue (**43**; M = Mn, R = Et) was formed by photochemical reaction of $Mn_2(CO)_{10}$ with $MeC\equiv CNEt_2$. Furthermore, both the manganese and rhenium complexes (**43**) were found to isomerize on warming giving successively products with the bridged structures (**45**), (**46**) and then, with loss of CO, (**47**) ⟨92OM1473, 92OM1480⟩.

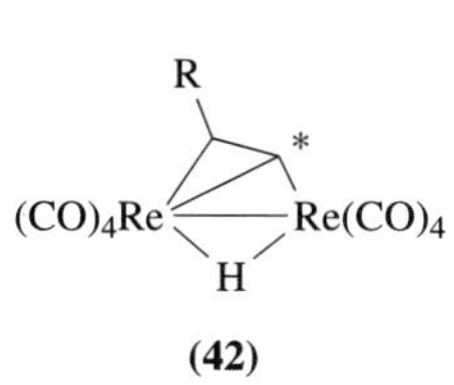

(**42**) 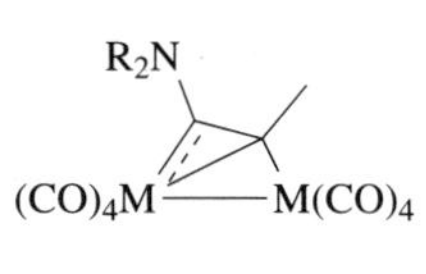   (**43**) 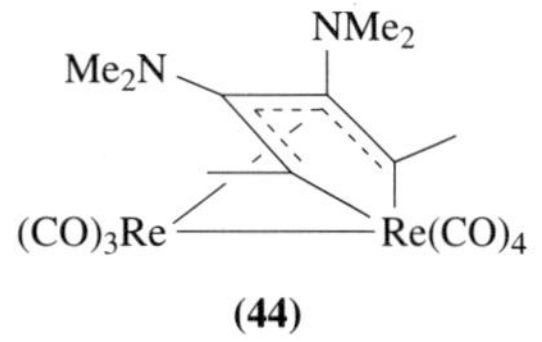   (**44**)

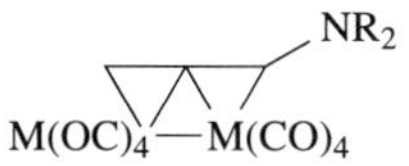

(**45**) 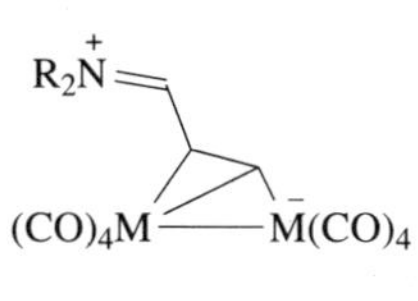   (**46**) 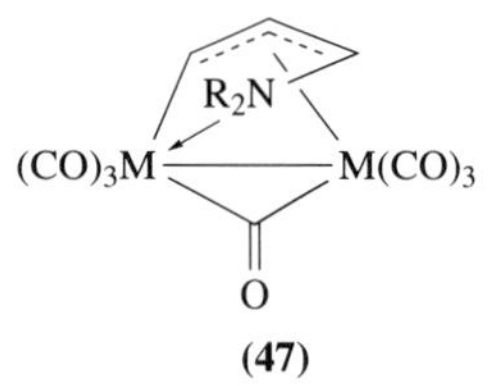   (**47**)

The rhenacyclopentadiene complex (**48**; L = (CO)$_2$) can be prepared from tolan and Re$_2$(CO)$_8$(MeCN)$_2$ ⟨94JOM(464)191⟩ and had previously been obtained, together with (**48**; L = PhC≡CPh) and the 'flyover' complex (**49**) by photolysis of H$_3$Re$_3$(CO)$_{12}$ with the alkyne ⟨84JOM(273)333⟩.

A rather complex rearrangement with CO insertion must accompany the formation, from Mn$_2$(CO)$_9$(MeCN) and MeC≡CNMe$_2$, of the amide complex (**50**) which can add a second mole of the alkyne to give the bridged product (**51**) ⟨91OM2541⟩.

Only some terminal alkynes (e.g., R = H, Ph) add to the dihydride (**52**) to yield the bridged products of the type (**53**) (Equation (15)) whereas other alkynes add with loss of H$_2$ to form bridging vinylcarbene complexes ⟨92OM370⟩.

The cyclopentadienylidene-bridged imino-rhenium complex (**54**) is formed from (Bu^{t}N)$_3$ReCl and cyclopentadienylsodium and is a rare example of a transition metal complex with only σ-bonded cyclopentadienyl groups ⟨91CC181⟩. Both terminal carbons of the η^5-pentadienyl ligand are bridging in the manganese complex (**55**) formed from MnCl$_2$ and a 1:1 mixture of cyclopentadienylsodium and (2,4-dimethylpentadienyl)potassium ⟨92OM3617⟩.

4.15.3.7 Functions Containing Two Fe, Ru or Os Atoms

Various relevant ruthenium complexes which fit this category are included in a review ⟨90JOM(400)255⟩.

An example of the simplest possible type of complex in this category, with a linking methylene group unsupported by metal–metal bonding or additional bridging groups, is the compound [(η^5-C$_5$H$_5$)Ru(CO)$_2$]$_2$CH$_2$ obtained from dichloromethane and the salt Na[(η^5-C$_5$H$_5$)Ru(CO)$_2$] ⟨83JA1679⟩. Complexes with a metal–metal bond and one carbon bridge are known for all three metals of this group. Thus, the methylenediiron complex (**56**; M = Fe, R = H) is obtained from $[Fe_2(CO)_8]^{2-}$ with diiodomethane ($Me_2CO/0\,°C$) as air-stable golden yellow prisms in 60% yield. It is believed to have this structure in solution (no bridging CO peaks in the IR spectrum), whereas its crystal structure and solid-state IR spectrum show it to adopt the triply bridged form (**57**), analogous to $Fe_2(CO)_9$, in the solid-state ⟨80JA1752⟩. This method is general; the diiodides $RCHI_2$ with, for example, R = Me, Pr^i, CH=CH_2 or CO_2Et giving the corresponding complexes (**56**) and Me_2CI_2 giving the CMe_2-bridged analogues ⟨82OM1350⟩. The osmium compound (**56**; M = Os, R = H) is an air-stable white crystalline solid prepared from $Na_2Os_2(CO)_8$ and $CH_2(OTs)_2$ ⟨82JA7325⟩. It is also formed by the reversible decarbonylation of $[(OC)_4Os]_2(\mu\text{-}CH_2CO)$ at 80 °C ⟨88JA7868⟩ and at 120–130 °C from $[Me(OC)_4Os]_2$ or $[(OC)_4Os]_2(\mu\text{-}CH_2CH_2CH_2)$ but is unstable at these higher temperatures ⟨82JA7325⟩. Its homologue $[(OC)_4Os]_2(\mu\text{-}CHCH_3)$ was prepared from $Na_2Os_2(CO)_8$ and $CH_3CH(OTf)_2$ ⟨88JA7868⟩.

(**56**) (**57**)

Diazomethane addition (Equation (16)) has been used to obtain a methylene-bridged iron nitrosyl complex ⟨81JOM(204)C21⟩ while another carbene source, the ketene $(CF_3)_2C=C=O$, displaces one CO from $Fe_2(CO)_9$ to yield $[(OC)_4Fe]_2[\mu\text{-}C(CF_3)_2]$ ⟨94JOM(476)101⟩.

$$Cp-Fe\underset{N\atop O}{\overset{N\atop O}{=}}Fe-Cp \;+\; CH_2N_2 \longrightarrow \underset{Cp}{ON}Fe\text{—}Fe\underset{Cp}{NO} \qquad (16)$$

Complexes with more than one bridge have been obtained by a variety of routes. Thus, the iron complexes (**58**; M = Fe, X = CO, R = H or Me) are formed on treating the bridged carbonyl, $Cp_2Fe_2(CO)_2(\mu\text{-}CO)_2$ with the corresponding Wittig reagents, $Ph_3P=CHR$, with loss of CO ⟨81AG(E)1049⟩. Alternatively, the alkoxymethylene complexes $Cp_2Fe_2(CO)_2(\mu\text{-}CO)[\mu\text{-}CR(OMe)]$ react with Ph_3CBF_4 to give stable cations (**59**; M = Fe) which are reduced with $NaBH_4$ or $[HFe(CO)_4]^-$ to (**58**; M = Fe, X = CO) ⟨82OM911⟩. A third route to (**58**; M = Fe, X = CO, R = H) is the reaction of $CpFe(CO)_2CH_2OAc$ with $K[CpFe(CO)_2]$ ⟨82JA1134⟩. The best route to the ruthenium analogue (**58**; M = Ru, X = CO, R = H) is the direct reduction of the tetracarbonyl $Cp_2Ru_2(CO)_2(\mu\text{-}CO)_2$ with $LiHBEt_3$ ⟨88POL759⟩ whereas reduction with, for example, H_2SiEt_2 can proceed further (at 150 °C) to yield the bismethylene complex (**58**; M = Ru, R = H, X = CH_2) ⟨92CC1031⟩. The Wittig route also succeeds provided that the rather unreactive tetracarbonyl is first converted into the tolan adduct (**60**; M = Ru, $R^1 = R^2 = Ph$) which reacts with loss of the alkyne. The Wittig reagents $Ph_3P=CHR$ (R = H, Me, Et, Ph) (but not $Ph_3P=CMe_2$) then succeed in varying yields to give the corresponding complexes (**58**; M = Ru, X = CO) ⟨84JCS(D)2293⟩. Reduction of dodecacarbonyltriosmium with $K[BH(OPr^i)_3]$ followed by acidification (H_3PO_4 or CF_3CO_2H) has yielded a trinuclear complex, $Os_3(CO)_{10}(\mu\text{-}CO)(\mu\text{-}CH_2)$ believed to have a structure related to $Fe_3(CO)_{12}$ ⟨81JA1278⟩.

(**58**) (**59**) (**60**) (**61**)

The alkyne complexes (**60**) form readily from both the iron and ruthenium tetracarbonyls, $Cp_2M_2(CO)_4$, on irradiation with a wide range of alkynes ⟨82JCS(D)1297⟩. Protonation by $HBF_4\cdot OEt_2$ leads to the salts (**61**) with retention of the carbon bridge which, in turn, can add hydride to yield complexes of type (**58**) ⟨83JCS(D)1417, 94OM2527⟩. Another variant leading to a ruthenium complex of this type (R = Me) is the reduction ($NaBH_4$) of the ^+CMe-bridged complex (**59**; M = Ru, R = Me) obtained from $Cp_2Ru_2(CO)_4$ with MeLi followed by HBF_4 ⟨83JCS(D)2661⟩. The complex (**61**; M = Ru, R^1 = Me, R^2 = H) whose $NaBH_4$ reduction leads to the CMe_2-bridged $Cp_2Ru_2(CO)_2(\mu\text{-}CMe_2)$

complex was itself obtained by protonating another type of carbon-bridged molecule: the η^1, η^3-allyl complex (**62**) ⟨83JCS(D)2099⟩.

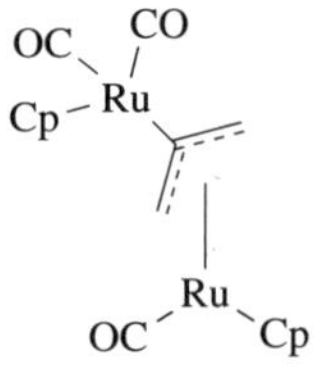

(**62**)

The cations (**59**; M = Fe, R = H or Me) add *t*-butyl isocyanide to yield the bridged cations (**63**; X = NBu^t) and add carbon monoxide at 0 °C and 1 atm. to give the cation (**63**; R = H, X = O) which, in turn, adds H$_2$O or NH$_3$ yielding (**58**; M = Fe, X = CO, R = CO$_2$H or CONH$_2$) or is reduced by [HFe(CO)$_4$]$^-$ to (**58**; M = Fe, X = CO, R = CHO) ⟨88JA6070⟩. The salts (**61**; M = Fe, R^1 = R^2 = H or R^1 = Me, R^2 = H) add various carbanions or silyl enol ethers (e.g., $^-$CH(CO$_2$Et)$_2$) to yield (**58**; M = Fe, X = CO, R = CH$_2$CH(CO$_2$Et)$_2$) ⟨88OM934⟩.

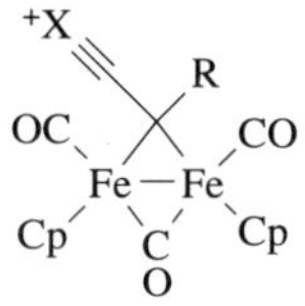

(**63**)

[Cp*RuCl]$_4$ and Mg(CH$_2$-TMS)$_2$ yield the methylene and chloride bridged compound (**64**) ⟨94OM2309⟩ which, in turn, reacts with two moles of ethyne to give the ruthenabenzene complex (**65**) with two bridging carbons (*) ⟨93CC284⟩.

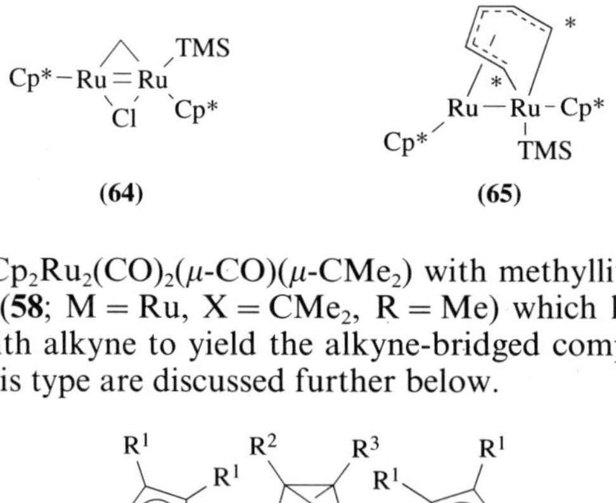

(**64**)   (**65**)

Successive treatment of Cp$_2$Ru$_2$(CO)$_2$(μ-CO)(μ-CMe$_2$) with methyllithium, fluoroboric acid and hydride gives the complex (**58**; M = Ru, X = CMe$_2$, R = Me) which loses both bridging carbene groups when photolysed with alkyne to yield the alkyne-bridged complex (**66**; R^1 = R^2 = R^3 = H) ⟨81CC862⟩. Complexes of this type are discussed further below.

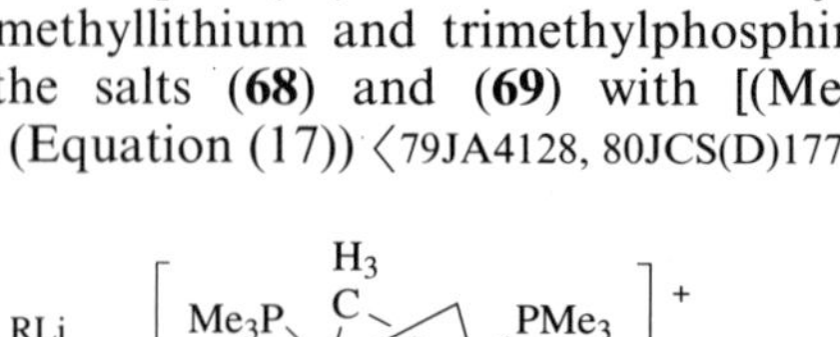

(**66**)

The trismethylenediruthenium complex (**67**) forms in moderate yield (30%) when [Ru$_3$O(OAc)$_6$(OH$_2$)$_3$][OAc] is treated with methyllithium and trimethylphosphine. Its protonation leads successively and reversibly to the salts (**68**) and (**69**) with [(Me$_3$P)$_4$Ru(μ-CH$_2$)$_2$Ru(μ-CH$_2$)$_2$Ru(PMe$_3$)$_4$][BF$_4$]$_2$ as a by-product (Equation (17)) ⟨79JA4128, 80JCS(D)1771⟩.

(**67**)   (**68**)   (**69**)   (17)

Photochemical decarbonylation of complexes (**60**; M = Ru) provides a general route to the alkyne-bridged compounds (**66**; e.g., R^1 = H, R^2 = R^3 = Ph). These, in turn, undergo reactions which lead to numerous other types of bridged complexes. Thus, protonation yields cations (**70**), diazoethane adds to form compound (**71**), while two moles of diazomethane add to form (**72**), etc. ⟨90JCS(D)761, 92CC310⟩.

(**70**)   (**71**)   (**72**)

Heating [Cp*RuCl]$_4$ with MeCH=CMeCHO and K_2CO_3 leads to the alkyne complex (**66**; $R^1 = R^2 = R^3$ = Me) together with the (intermediate) (η^5-Cp*)Ru(η^5-CH_2CHCMeCHO) complex ⟨94OM2423⟩.

An alkyne-bridged ruthenium hydride (**73**) is formed from Cp*$_2$Ru$_2$(μ-H)$_4$ and tolan ⟨92OM989⟩.

(**73**)

An extended family of carbon-bridged complexes can be obtained from alkynes and the three iron carbonyls, Fe(CO)$_5$, Fe$_2$(CO)$_9$, and Fe$_3$(CO)$_{12}$. Early work has been well reviewed ⟨B-68MI 415-01⟩. In general, complexation involves coupling of two or more alkyne units with or without one or more carbonyl groups and leads to mixtures of products. Only in special cases have iron carbonyl complexes incorporating unmodified alkynes been isolated from such reactions. Examples are with the bulky di-*t*-butylalkyne which reacts with Fe$_2$(CO)$_9$ at room temperature to give the mono-bridged complex (**74**) ⟨76JA1774⟩ and with Fe$_3$(CO)$_{12}$ in refluxing methylcyclohexane forming the corresponding bis-bridged compound (**75**) ⟨71CC608⟩. An analogous product was obtained from 3,3,6,6-tetramethyl-1-thiacyclohept-4-yne ⟨73ZN(B)508⟩, while amino-alkynes (ynamines) bridge in a different mode as illustrated for PhC≡CNEt$_2$ by structure (**76**) ⟨92JOM(441)81⟩. This however will smoothly add a second ynamine, RC≡CNEt$_2$, when heated to yield a ferracyclopentadiene ('ferrole') complex (**77**) ⟨92JOM(441)81⟩, a member of a group of complexes commonly obtained from other alkynes. Thus, the unsubstituted ($R^1 = R^2$ = H), disubstituted (R^1 = Ph, R^2 = H) and tetrasubstituted ($R^1 = R^2$ = Ph) ferracyclopentadienes (**78**) are formed from ethyne, phenyl- and diphenylethyne respectively ⟨59JINC250, 62CB1155⟩ and are accompanied by *inter alia* complexes of the type (**79**; X = O or CH_2) and sometimes (e.g., R = Ph) the violet trinuclear complex (**80**) which isomerizes thermally to the 'black' (or dark green) compound (**81**). The latter (in which the metal-bridging carbons are marked:*) can be formed from the ferracyclopentadienes (**77**) and (**78**) with Fe$_2$(CO)$_9$ ⟨92JOM(441)81⟩.

Tolan is also the precursor of the benzoferracyclopentadiene (**82**; R = Ph) ⟨59JINC250⟩ whose parent (**82**; R = H) results both from photolysis of 2-bromostyrene with Fe(CO)$_5$ ⟨71CC1241⟩ and, together with its isomer (**83**), from tricarbonylbenzocyclobutadieneiron and Fe$_3$(CO)$_{12}$ at 120 °C ⟨74JA7108⟩. The complex (**82**; R = Ph) also results, together with (**78**; $R^1 = R^2$ = Ph), from the reaction of Fe$_3$(CO)$_{12}$ with diphenylcyclopropenone ⟨67JCS(C)1862⟩.

Reactions of 1,4-dibromocyclooctatetraene with Fe$_2$(CO)$_9$, Ru$_3$(CO)$_{12}$ and Me$_2$Os(CO)$_4$ give the product (**82**) (25%) and its ruthenium (18%) and osmium (21%) analogues respectively ⟨76JCS(D)377⟩.

An alternative route, giving the unsubstituted ferrole (**78**; $R^1 = R^2$ = H) in up to 17% yield, is the reaction of Fe$_3$(CO)$_{12}$ with thiophene ⟨60JA4749, 76JOM(108)213⟩ while relatively efficient syntheses of tetrasubstituted ferroles (**78**; $R^1 = R^2$ = alkyl or aryl), apparently as the sole organometallic products, have been accomplished by treating the corresponding alkynes (including macrocyclic alkynes) with (benzylideneacetone)tricarbonyliron ⟨73JOM(60)C57⟩. A high yield (60%) of the complex (**79**; R = Ph) is reported from the room temperature reaction of tolan with Fe$_2$(CO)$_9$ whereas Fe$_3$(CO)$_{12}$ yields the complex (**80**) under these conditions ⟨75JOM(99)281⟩.

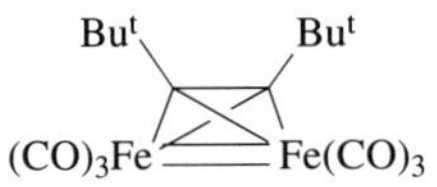

Since the only convenient starting carbonyls of ruthenium and osmium are the trinuclear $M_3(CO)_{12}$ or the corresponding hydrides (e.g., $H_2Os_3(CO)_{11}$) most of the initially formed products are themselves trinuclear. These include species with unmodified alkyne ligands. Thus, various alkynes have been shown to react with $Os_3(CO)_{12}$ ⟨75G939, 75ICA155⟩, or $H_2Os_3(CO)_{10}$ ⟨75JA7172, 75JCS(D)1614, 89JOM(365)163⟩, or (preferably) $Os_3(CO)_{10}(C_6H_8)$ (where C_6H_8 = 1,3-cyclohexadiene), or $Os_3(CO)_{10}(MeCN)_2$ ⟨88JCS(D)1421⟩ to yield complexes (**84**) ⟨77JCS(D)1328⟩. Some of these (e.g., R = Et) have the symmetrical structures shown ⟨89JOM(365)163⟩, but others (e.g., R = Ph) have unsymmetrical bridging ⟨77IC636⟩. If the hydrido complex is used as precursor, the 'vinyl' complexes (**85**) are by-products, intermediates or, sometimes, (e.g., R = Ph) the only products ⟨75JCS(D)1614⟩. Various benzyne and substituted benzyne complexes, bridged like type (**84**), have been obtained, for example by pyrolysis of $Os_3(CO)_{11}(Me_2AsC_6H_4R)$ ⟨82JCS(D)1155, 89AG(E)1296⟩. A second alkyne bridge may be introduced into the complexes (**84**) by treatment with an *N*-oxide (Me_3NO) in acetonitrile in the presence of the alkyne, RC≡CR, giving the bis(alkyne)octacarbonyl complexes (**86**) under mild conditions ⟨87JOM(319)C51⟩. On the other hand, simply heating the monoalkyne complexes (**84**) with alkynes leads via the (isolable) trinuclear osmacyclopentadiene complexes (**87**) to the binuclear analogues of the ferracyclopentadienes (**78**) (see, for example ⟨75G939⟩).

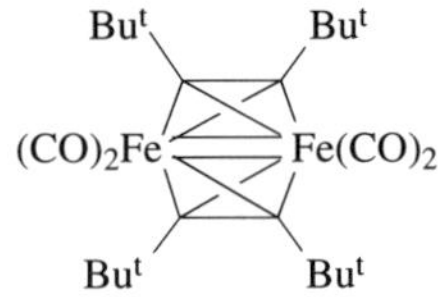

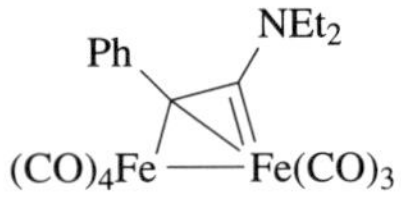

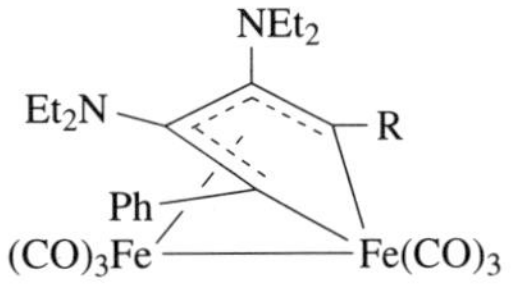

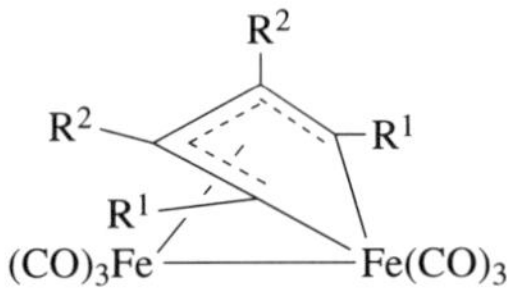

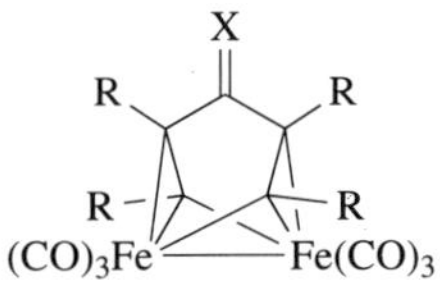

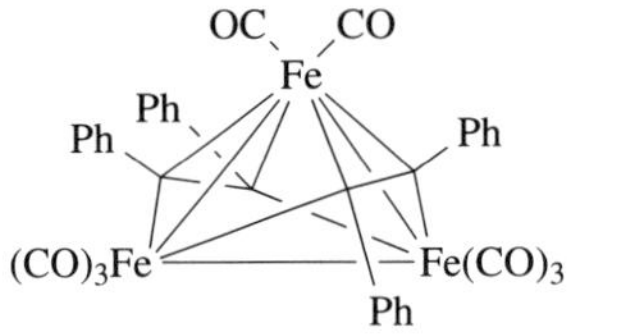

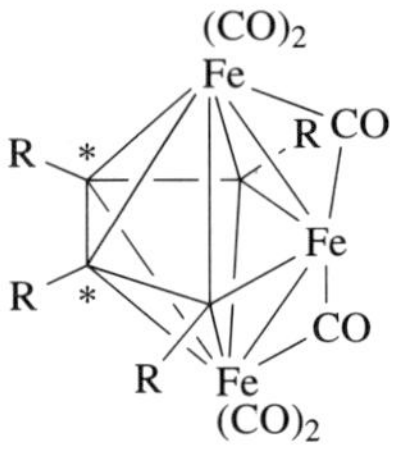

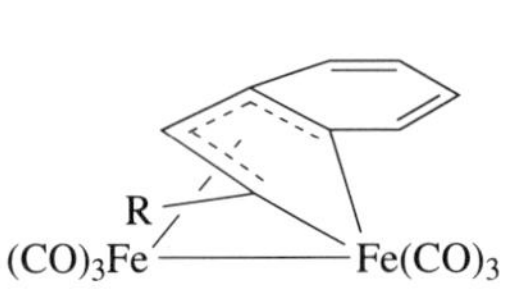

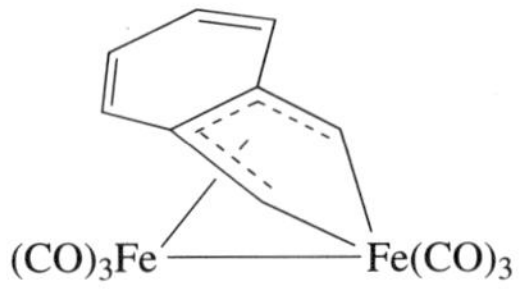

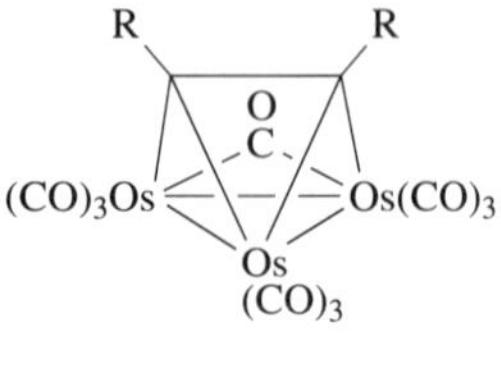

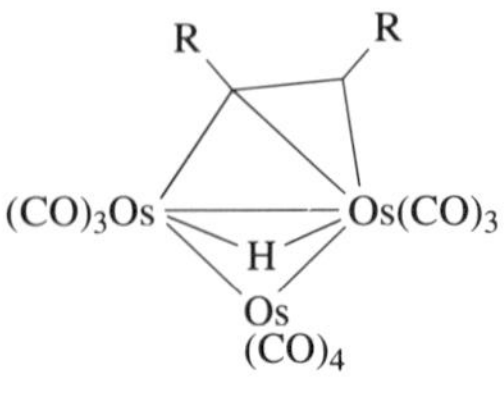

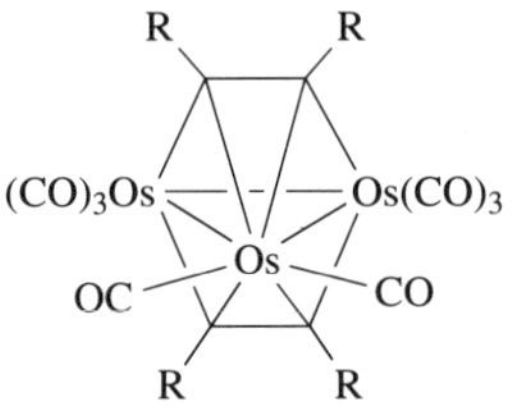

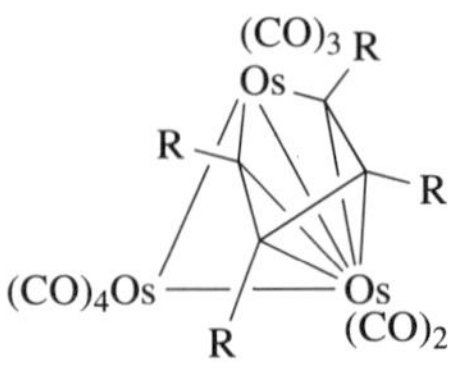

Like $Fe_3(CO)_{12}$, $Ru_3(CO)_{12}$ reacts with thiophene or 2-methylthiophene to give ruthenacyclopentadiene complexes. With thiophene itself the ruthenium analogue of (**78**; $R^1 = R^2 = H$) is the major product isolated (20%) whereas methylthiophene yields chiefly a tetranuclear derivative $Ru_4(CO)_{11}(\mu_3\text{-}S)(C_4H_3Me)$ and also gives two trinuclear products with an intact thiophene ring ⟨92JCS(D)2423⟩.

The ruthenium analogue of the complex (**78**; $R^1 = R^2 = Ph$) has been obtained by heating $Ru_3(CO)_{12}$ with tolan to 200 °C ⟨68JOM(11)644⟩. Thermal decarbonylation of the complexes (**85**) is accompanied by hydrogen migration or loss leading to other bridged systems including type (**88**; M = Os) (see, for example ⟨75JCS(D)1614⟩). Ruthenium analogues (**88**; M = Ru) are obtained directly from $Ru_3(CO)_{12}$ and terminal alkynes, RC≡CH, with high efficiency when R is a bulky substituent ⟨72JOM(39)169, 80JOM(187)81⟩.

$$\underset{(\mathbf{88})}{(CO)_3M\text{—}\overset{R}{\diagup\!\!\diagdown}\text{—}M(CO)_3 \atop M(CO)_3}$$

Trinuclear hydrazino-bridged ruthenium hydridocarbonyls have been shown to add phenyl- and diphenylethyne to give bridged Ru_3-complexes, for example (**89**) ⟨93CB2017⟩.

$$\underset{(\mathbf{89})}{\text{(complex 89)}}$$

When the tolan complex (**84**; R = Ph) was treated with phenyllithium followed by methyl triflate, addition to the bridging carbonyl group resulted in formation of the complex (**90**; $R^1 = Ph$, $R^2 = OMe$) which, on heating in hexane, lost the elements of benzaldehyde with conversion into the methylene-bridged compound (**90**; $R^1 = R^2 = H$). Acid treatment of the same intermediate (**90**; $R^1 = Ph$, $R^2 = OMe$) led to the cationic bridged allyl complex (**91**) by an insertion process, and hence on hydride addition (from $LiBHEt_3$) to a mixture of (**90**; $R^1 = Ph$, $R^2 = H$) and (**92**) ⟨93JOM(461)207⟩. The latter (**92**) had previously been obtained ⟨88JOM(353)103⟩ from the reaction of $(\mu\text{-}H)Os_3(CO)_{10}(\mu_3\text{-}CPh)$ with tolan, which also gave complex (**93**) and three other related products. Similar alkyne insertions into triply-bridged iron complexes, notably into $Fe_3(CO)_9(\mu_3\text{-}CF)_2$, yield the green complexes (**94**) ⟨88CB1413, 90IC4396⟩, which on further alkyne insertion give ferracyclopentadiene complexes of the type (**81**) ⟨92OM2916⟩.

The propargyl-iron complex, $CpFe(CO)_2CH_2C≡CPh$, rearranges during its room temperature reaction with $Fe_2(CO)_9$ giving the product (**95**), which has one bridging carbon ⟨92OM154⟩.

Reactions of $Fe_3(CO)_{12}$ with allenyl thiol derivatives, $CH_2=C=CH(SR)$, give a series of complexes of type (**96**) with one carbon (*) linked to both metal atoms ⟨92OM3736⟩. Furthermore, $Ru_3(CO)_{12}$ reacts with $Me_3P=CHCOPh$ to yield the complexes (**97**; both L = PMe_3 and L = CO), which give related bridged complexes by alkyne insertion ⟨93CB373⟩.

Several bridged iron and ruthenium complexes have been obtained from the reactions of their carbonyls and certain cyclic polyenes, notably cyclooctatetraene. Whereas, under mild conditions this yields η^4-complexes $(C_8H_8)M(CO)_3$ and $(C_8H_8)[M(CO)_3]_2$, these on heating or photolysis decarbonylate to give complexes of the type (**98**) with two bridging carbons ⟨69JA6598, 72CC814⟩. Moreover, with $Ru(CO)_4(TMS)_2$, cyclooctatetraene undergoes ring closure to give a pentalene complex (**99**) also with two bridging carbons ⟨78JCS(D)403⟩. A minor product from $Ru_3(CO)_{12}$ has a related trinuclear structure ⟨79JCS(D)1801⟩.

The cation $[CpRu(NCMe)_3]^+$ and dipotassium cyclooctatetraenide, $K_2C_8H_8$, yield the complex (**100**) in which, again, two of the carbons (*) of the twisted C_8-ligand are bridging ⟨86OM2413⟩ and which undergoes reversible ring opening to the dication (**101**) on two-electron oxidation ⟨90JA7113⟩.

As was the case with the preceeding groups, various iron-group compounds have been prepared having one metal atom σ- and the other π-bonded to the same carbon. Indeed, some of the above alkyne derived compounds (e.g., type (**78**)) can be regarded as fitting this category. The most obvious route to such substances is exemplified by Equation (18); the reaction of ferrocenyllithium with a halo-iron compound ⟨73IZV2796⟩.

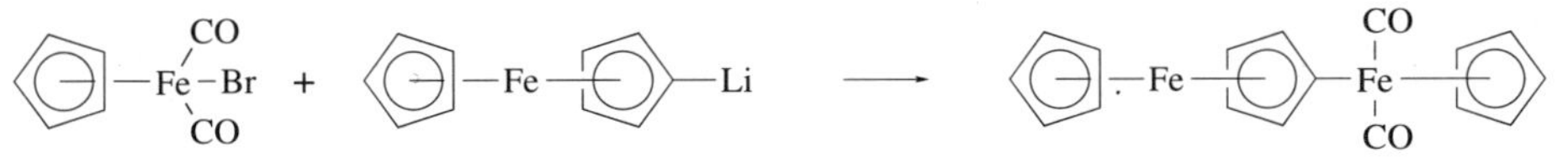

$$\text{Cp-Fe(CO)}_2\text{Br} + \text{Cp-Fe-C}_5\text{H}_4\text{-Li} \longrightarrow \text{Cp-Fe-C}_5\text{H}_4\text{-Fe(CO)}_2\text{Cp} \qquad (18)$$

(102)

The same product (**102**) was obtained in similar yield (42–51%) from the iodo-iron compound CpFe(CO)$_2$I and diferrocenylmercury, Fc$_2$Hg (Fc = ferrocenyl) ⟨72IZV1600⟩, and a dichloro-substituted compound was made using 1′,2-dichloroferrocenyllithium ⟨79JOM(181)425⟩. A less efficient route to compound (**102**) is by photolytic or thermal (99 °C) decarbonylation of FcCOFe(CO)$_2$Cp giving 36% or 23% yields respectively ⟨76IC2671⟩.

The trinuclear ruthenium complexes (**103**; L = CO: 49%) and (**103**; L = PPh$_3$: 29%) are the principal products formed on refluxing dodecacarbonyltriruthenium with cyclopentadienylidene-triphenylphosphorane ⟨93AG(E)1048⟩.

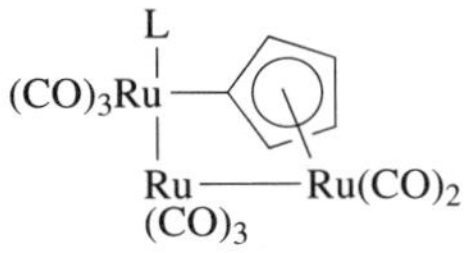

(103)

X-ray diffraction has revealed the dimeric structure (**104**), with two bridging tetracoordinate aryl carbons ⟨93JOM(445)133⟩, of dimesityliron, synthesized (61%) from mesitylmercuric chloride and FeCl$_2$ ⟨76ZC116⟩.

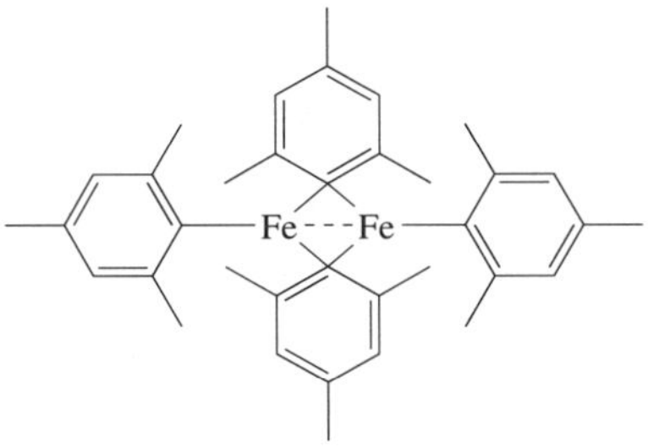

(104)

4.15.3.8 Functions Containing Two Co, Rh or Ir Atoms

Diiodomethane adds to the coordinatively unsaturated dimeric dicarbonyl(μ-pyrazolyl)iridium giving a product which rearranges on warming to the methylene-bridged product (**105**) (Scheme 4) ⟨86CC285, 89CC498⟩. An exactly equivalent sequence converts [Ir(CO)(PPh$_3$)]$_2$[μ-1,8-C$_{10}$H$_6$(NH)$_2$] by treatment with CH$_2$I$_2$ in dichloromethane followed by refluxing in toluene into the complex (**106**) ⟨90JCS(D)2587⟩.

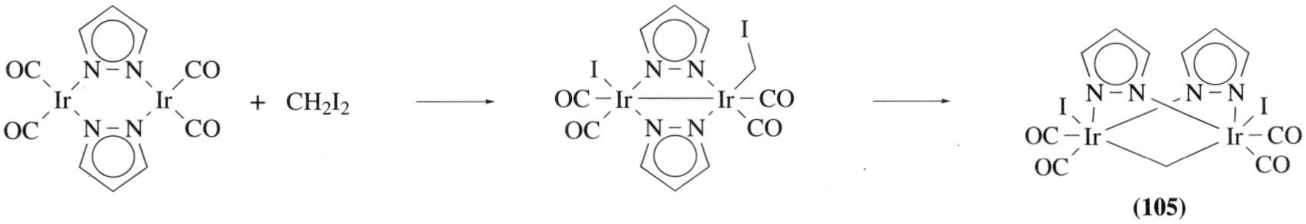

(105)

Scheme 4

The air and light sensitive, red, hexafluoroisopropylidene complex (**107**) is formed from octacarbonyldicobalt on treatment with (CF$_3$)$_2$CN$_2$ ⟨69JCS(A)1872⟩. Much more extensive application of this diazoalkane route has led to numerous methylene and substituted methylene-bridged

(106)

cyclopentadienylcobalt and rhodium complexes. Both the unsubstituted and the pentamethylcyclopentadienyl series have been studied with a range of diazoalkanes, $R^1R^2CN_2$ (R^1, R^2 = H, alkyl, aryl, CO_2Me, etc.). The simplest procedure involves refluxing the diazoalkanes with the monomeric carbonyls $CpM(CO)_2$ (M = Co or Rh) in benzene to give the corresponding products (**108**) ⟨78CB1077, 78ZN(B)911⟩. When low temperature photolysis is used, the same products (**108**) may be accompanied by intermediates ⟨79JOM(165)C17⟩. Frequently, especially for the pentamethyl derivatives, $Cp^*M(CO)_2$, it has been found advantageous to thermally decarbonylate the organometallic starting materials to give the much more reactive $Cp^*M_2(\mu\text{-}CO)_2$, thus allowing the carbene addition (Equation (19)) to proceed under much milder conditions (e.g., Co ⟨82JOM(226)C59, 83JOM(258)81⟩, Rh ⟨80JOM(201)C31, 82CB878, 82JCS(D)1309⟩). Depending on the nature of the substituents R^1, R^2, the products exist preferentially in the mono- (**109a**) or tribridged form (**109b**); in the case of the cyclohexylidene complexes [R^1, $R^2 = (CH_2)_5$] the cobalt compound adopts the tribridged structure (**109b**), but the rhodium compound adopts the monobridged form (**109a**) ⟨82JOM(236)C18⟩. These complexes may be further decarbonylated by heating (e.g., in THF) to give (**110**), which can, in turn, react smoothly with a (different) diazo alkane, $R^3R^4CN_2$, to give the tribridged compounds (**111**; M = Co or Rh) with all steps occurring in ⩾90% yields ⟨82CB878, 82JOM(226)C59⟩. The related $Cp_2Rh_2(\mu\text{-}CPh_2)_2$ has been formed by heating [Rh(CO)$_2$Cl]$_2$ with diphenylketene and then treating the resultant [Rh(CO)(CPh$_2$)Cl]$_n$ with cyclopentadienylsodium ⟨72CC993, 77BCJ2250⟩.

(107) **(108)**

$$Cp^*-M{=}M-Cp^* + \underset{R^1}{\overset{N_2}{\|}}R^2 \longrightarrow \text{(109a) or (109b)} \quad (19)$$

(110) **(111)**

A successful alternative procedure involved the decarbonylation of $CpRh(CO)_2$ to [CpRh(CO)]$_2(\mu\text{-}CO)$ followed by refluxing in benzene with the diazoalkane precursors R^1CH_2N-(NO)CONH$_2$ (R^1 = H or Me) to give compounds (**108**; M = Rh, $R^1 = R^2 =$ H (86%) or $R^1 =$ H, $R^2 =$ Me (57%)) ⟨77JOM(140)73⟩.

The cobalt complexes (**108**; M = Co), including the unsubstituted compound ($R^1 = R^2 =$ H), may be obtained by reactions of the salt [CpCo(μ-CO)]$_2^-$Na$^+$ with diiodoalkanes $R^1R^2CI_2$. Heating the product (**108**; M = Co, $R^1 = R^2 =$ H) with $CpRh(CO)_2$ at 61 °C leads to partial methylene transfer to give (**108**; M = Rh, $R^1 = R^2 =$ H) together with the corresponding mixed CoRh complex.

Moreover, in the same work it was found that the isopropylidene complex (**108**; M = Co, $R^1 = R^2$ = Me) isomerizes thermally to the propylidene complex (R^1 = H, R^2 = Et) ⟨81JA2489, 82OM219⟩.

The complex (**110**; M = Co, $R^1 = R^2$ = H) was formed by an unknown mechanism according to Equation (20), the lithium enolate of acetaldehyde having been formed adventitiously from the solvent, tetrahydrofuran, and the butyllithium used in generating lithium pentamethyl-cyclopentadienide ⟨80JA5101⟩.

$$CoCl_2 + C_5Me_5Li + Li[CH_2CHO] \longrightarrow (C_5Me_5Co)_2(\mu\text{-}CO)(\mu\text{-}CH_2) \quad (20)$$

$$(\textbf{110}; M = Co, R^1 = R^2 = H)$$

When dichloro(pentamethylcyclopentadienyl)rhodium dimer, $Cp^*_2Rh_2(\mu\text{-}Cl)_2Cl_2$, reacts with hexamethyldialuminum, Al_2Me_6, an initial adduct is formed which in the presence of Lewis acids or on oxidation with acetone yields the bismethylene-bridged $cis\text{-}Cp^*_2Rh_2Me_2(\mu\text{-}CH_2)_2$. The latter slowly isomerizes to the *trans* form (**112**), a red crystalline substance obtained in ⩾85% yield overall. The pure *cis* form is more readily obtained, albeit in poor yield, by treating the chloride precursor with methyllithium ⟨83JCS(D)1441⟩.

(**112**)

Alkynes react with octacarbonyldicobalt in much more predictable fashion than with carbonyls of iron. The first product, usually obtainable under ambient conditions and in near quantitative yield, is invariably an alkyne-bridged complex of the type (**113**). These have been prepared in large numbers as they have proved synthetically useful as noted briefly below. Examples include not only compounds in which the substituents R^1 and R^2 are alkyl, alkenyl or aryl groups but also many heteroatoms including transition metals. Moreover, easy replacement of one or two carbonyl groups by phosphines or other donor ligands further extends the range of these bridged compounds ⟨B-73MI 415-01, 74AOC323⟩.

(**113**)

Apart from $Co_2(CO)_8$ itself, the tetranuclear carbonyl, $Co_4(CO)_{12}$ ⟨69AJC1143⟩, the mercury compound, $Hg[Co(CO)_4]_2$ ⟨57CB1259, 61CB2829, 67G1327⟩, and the germanium-bridged $Co_2(CO)_6(\mu\text{-}CO)(\mu\text{-}GePh_2)$ ⟨74JCS(D)1856⟩ have yielded complexes (**113**) on treatment with alkynes. The reactions of phosphine-substituted carbonyls with alkynes may involve partial phosphine as well as CO substitution (e.g., Equation (21)) ⟨74JCS(D)607⟩.

$$Co_2(CO)_6(PBu_3)_2 + Ph\!\!=\!\!=\!\!Ph \longrightarrow$$

$$(CO)_3Co(\mu\text{-}PhC\!\!\equiv\!\!CPh)Co(CO)_2(PBu_3) + (PBu_3)(CO)_2Co(\mu\text{-}PhC\!\!\equiv\!\!CPh)Co(CO)_2(PBu_3) \quad (21)$$

In addition to their formation from the free alkynes some complexes of (unstable) alkynes have been obtained by generating these ligands *in situ*. The earliest example was the hexafluorocyclohexenyne complex (**114**) formed from $Co_2(CO)_8$ and perfluoro-1,3-cyclohexadiene ⟨64PCS401, 65IC1270, 68JCS(A)1293⟩. Perfluorocyclooctatetraene reacts similarly with $Co_2(CO)_6L_2$ or, better, $Na[Co(CO)_3L]$ (L = CO or PR_3) ⟨90OM2745⟩.

A further route to the complexes (**113**) is provided by the exchange of one alkyne by another (Equation (22)). The following order of ease of displacement has been established for typical alkyne substituents (those to the left in the sequence displacing any further to the right): $(CF_3)_2$ > $(CO_2Me)_2$ > Ph_2 > (Me, Ph) > Me_2 > (H, Ph) ~ (H, Me) > H_2 > $(CH_2NEt_2)_2$ ⟨67JOM(8)149⟩. Such exchange is however in competition with alkyne insertion which ultimately leads to arenes (i.e., alkyne trimerization products). The isolable intermediates in this process have a six-carbon chain

(114)

linking the two cobalt atoms, with the terminal carbons attached to both metal atoms, a structure first elucidated crystallographically for the complex (115) obtained from $Co_2(CO)_6(\mu\text{-}HC\equiv CH)$ and $HC\equiv CBu^t$ ⟨64PCS187⟩ cf. for example ⟨67AJC77, 93JOM(444)203⟩. While carbon monoxide insertion into the alkyne-bridged complexes (113) yields lactone complexes (e.g., (116)) in which the original cobalt–carbon bonds have been severed, subsequent insertion of two alkyne molecules gives complexes with one tetracoordinate carbon (*) shared by the cobalt atoms. The structure (117) illustrates the product from the complex (116) and two moles of propyne ⟨76JOM(120)C13, 80JCS(D)435⟩ cf. ⟨81JOM(206)119⟩.

$$Co_2(CO)_6(\mu\text{-}R^1C\equiv CR^2) + R^3\!\!=\!\!\!=\!\!\!=\!\!R^4 \longrightarrow Co_2(CO)_6(\mu\text{-}R^3C\equiv CR^4) + R^1\!\!=\!\!\!=\!\!\!=\!\!R^2 \quad (22)$$
(113)

(115)   **(116)**   **(117)**

Apart from these insertion reactions and reactions with alkenes to produce (metal-free) cyclopentenones, the complexes (113) are utilized in a wide range of reactions in which the cobalt–alkyne fragment remains intact, but the substituents R^1 and R^2 are modified. These processes depend on two key features: the cobalt–alkyne bonding itself and the stabilization of α-carbocations. The bonding has the effect of lowering the C—C bond order, hence changing the linear alkyne to a bent structure with the substituents pointing away from the metal atoms and making the π-electrons unavailable. This makes the $Co_2(CO)_6$ group useful as a general alkyne protecting group which allows, for example, electrophilic substitution (Friedel–Crafts acylation) of aryl substituents (e.g., $R^1 = R^2 = Ph$) ⟨70JA5520⟩ or hydroboration of attached alkene groups ⟨71TL3475⟩ without interference from the alkyne fragment.

The stabilization of α-carbocations, seen by the isolation of salts after protonation of the propargyl alcohol complex, (e.g., (113; $R^1 = H$, $R^2 = CH_2OH$) + $HBF_4 \rightarrow H_2O$ + (113; $R^1 = H$, $R^2 = CH_2^+)BF_4^-$), allows facile substitution of OH, OR, etc. by a wide range of nucleophiles or addition (e.g., of acyl cations) to α-vinyl derivatives (e.g., (113; $R^1 = H$, $R^2 = CH=CH_2$)). These changes of reactivity together with the U-shape of the carbon chain facilitate reactions which link the substituents R^1, R^2, thus forming $Co_2(CO)_6$ complexes of both small-ring (cyclohexyne) and large-ring alkynes ⟨87ACR207⟩. Since none of these reactions involve formation of new carbon–metal bridging bonds only the general features are mentioned here to indicate how they lead to a much wider range of compounds still retaining the type (113) structure.

Relatively little is known concerning rhodium and iridium analogues of the complexes (113). Octakis(phosphorus trifluoride)dirhodium, $Rh_2(PF_3)_8$ has been treated with a variety of alkynes yielding the complexes (118; R = F) ⟨76IC90⟩; and this structure was confirmed by x-ray analysis for the complex (118; R = Ph) obtained by replacing two PF_3 groups ⟨76IC97⟩. The iridium complex (119) is formed in fair yield from $[Ir(CO)_3PPh_3]_2$ and tolan ⟨84JOM(267)199⟩. In part, the scarcity of this type of rhodium and iridium complex is due to the tendency of both metals to bond with the alkyne parallel to the metal–metal bond, that is as 1,2-dirhodia- or 1,2-diiridiacyclobutenes. Sometimes the two types are interconvertible. Thus, the cyclopentadienyl dirhodiacyclobutene complex $Cp_2(CO)_2Rh_2(CF_3C_2CF_3)$ (from $CpRh(CO)_2$ and C_4F_6) loses CO on treatment with Me_3NO yielding the complex (120) ⟨82JOM(224)377⟩ but this reverts to the precursor or analogues on treatment with CO or other donor ligands and yields the methylene-bridged compound (**121**) on treatment with diazomethane ⟨85OM355⟩. Thermally, dicarbonylcyclopentadienylcobalt has yielded analogues of complex (**120**) only with disilyl alkynes ⟨72JOM(39)365⟩ since most other alkynes react rapidly with CpCo(CO)$_2$ to give cobaltacyclopentadienes or further to give cyclotrimerization products. However, low temperature photolysis of CpCo(CO)$_2$ with tolan has given Cp$_2$Co$_2$(μ-CO)(μ-PhC≡CPh) in modest yield (20%) together with the trinuclear, alkyne-bridged, major product, Cp$_3$Co$_3$(μ-CO)(μ-PhC≡CPh) (**122**; M = Co) (57%) ⟨77JOM(127)93⟩. Related rhodium compounds have been identified ⟨73IC2396⟩ as minor by-products in the cyclotrimerization of tolan or other diarylalkynes by CpRh(CO)$_2$.

Reactions of the metal–metal double bonded cobalt and rhodium complexes (**123**) with alkynes proceed according to Equation (23), reversibly when M = Co and R^1 = R^2 = H. The products (**124**) can expand the bridge by inserting SO$_2$ between the C=O and CR^1 groups ⟨83JOM(256)147⟩. Closely related cobalt and rhodium complexes (**125**) have been obtained from the same precursors (**123**) and 3,3-dimethylcyclopropene ⟨85JCS(D)2483⟩.

Formation of binuclear metallacyclopentadiene complexes, similar to the iron compounds (**78**), from alkynes and cobalt, rhodium, or iridium complexes has been noted with a variety of systems. A rare example with Co$_2$(CO)$_8$ occurs in its reaction with cyclooctyne which yields 66% of the orange complex (**126**) along with only 22% of the expected complex (**113**; R^1, R^2 = (CH$_2$)$_6$) ⟨78IC1995⟩. The related complex (**127**) was formed in 10% yield (along with Co$_2$(CO)$_6$(PhC≡CH) and Fe$_3$(CO)$_{12}$) from Na[Co(CO)$_4$], PhC≡CH, Fe$_2$(CO)$_9$ and CF$_3$CO$_2$H ⟨95JOM(489)C65⟩ and the tetracobalt complex (**128**) is one of the products from Hg[Co(CO)$_4$]$_2$ and PhC≡CH ⟨61CB2829, 89JOM(361)231⟩. Rhodium analogues include the blue complex (**129**), one of several products from the reaction of [Rh(CO)$_2$Cl]$_2$ with hex-3-yne ⟨69JA7292, 69JOM(19)169⟩. In the above-mentioned reaction with Rh$_2$(PF$_3$)$_8$ only a few alkynes, notably RC≡CCO$_2$Me (where R = H or CO$_2$Me), when used in excess, go beyond the products of type (**118**) to give the rhodiacyclopentadiene complexes (**130**) ⟨76IC107⟩. Whereas but-2-yne reacts with [(cod)IrCl]$_2$ and isopropyllithium to yield a mononuclear product and the dihydrido diiridium complex (**131**; M = Ir, R = Me, L = H$_2$, diene = cod) ⟨93ZN(B)1558⟩, the parent compound (R = H) was obtained by the apparently more general route of treating [(cod)IrCl]$_2$ with butadienylmagnesium, C$_4$H$_6$Mg(THF)$_2$. The simultaneously formed trinuclear complex (**132**; M = Ir, diene = cod) is cleaved by donors (e.g., PMe$_3$) to give the complex (**131**; M = Ir, R = H, L = PMe$_3$, diene = cod). Exactly analogous reactions occur when the magnesium derivative of 2,3-dimethylbutadiene is used and also when the (cod)Ir complex is replaced by the (norbornadiene)rhodium (but not the (cod)Rh) analogue ⟨93JOM(459)325⟩.

The cyclopentadienyl–cobaltacyclopentadiene complex (**133**; M = Co, R = H) is formed together with the cyclobutadiene complex CpCo(C$_4$H$_4$) when CpCo(CO)$_2$ is treated with photopyrone ⟨68JA1060⟩ while its tetramethyl derivative (**133**; M = Co, R = Me) is obtained as dark green crystals when CpCo(CO)I$_2$ is reduced with sodium amalgam in the presence of but-2-yne ⟨77JOM(127)93⟩. Aryl substituted rhodium analogues (e.g., (**133**; M = Rh, R = Ph)) have been identified, along with the above-mentioned trinuclear species, among the products from (substituted) diphenylalkynes and CpRh(CO)$_2$, e.g. ⟨73IC2396⟩. A rhodium compound (**134**) reminiscent of the iron complexes (**79**) is one of the products formed from the same precursor with hex-3-yne ⟨76AJC2189⟩.

Loss of one molecule of the triene ligand from two of the cobalt complex CpCo(η^4-CH$_2$=CH-CH=CHCH=CH$_2$) at 60 °C yields the bridged complex (**135**) ⟨83JA4846⟩. The complex (**136**) is formed from Cp*Co(η^4-cycloheptatriene) and Cp*Co(C$_2$H$_4$)$_2$ ⟨90AG(E)686⟩, and the last complex reacts with thiophene with ring-opening to yield the bridged compound (**137**) ⟨92OM2698⟩.

Two other types of complex are dimesitylcobalt, obtained from CoCl$_2$ with mesityllithium and shown by x-ray crystallography to have the dimeric structure (**138**) ⟨89OM2001⟩, and the compounds of the type (**139**), obtained from CpCo(η^4-C$_5$H$_6$) with (C$_5$R$_5$)Co(C$_2$H$_4$)$_2$. The initially formed ethene complexes (**139**; L = C$_2$H$_4$) undergo ready replacement of ethene not only by donor ligands (CO, PMe$_3$) but also by halides (using CoBr$_2$ or C$_2$Cl$_4$). Furthermore, whereas the structure of the phosphine complex (L = PMe$_3$, R = H) suggests a major contribution from the resonance form (a), that of the bromo compound (L = Br, R = Me) corresponds with a much greater contribution of form (b) in which C^1 is linked to both metal atoms ⟨87AG(E)127, 93CC1459⟩.

4.15.3.9 Functions Containing Two Ni, Pd or Pt Atoms

Diazomethane reacts with the 'A-frame' platinum complex Pt$_2$(μ-Ph$_2$PCH$_2$PPh$_2$)$_2$Cl$_2$ inserting a methylene group into the Pt—Pt bond ⟨79IC2808⟩ giving the complex (**140**; M = Pt, X = Cl), and cationic platinum–hydride complexes of the same type react according to Equation (24) ⟨82CC614⟩.

L = CO or PMe$_2$Ph

Palladium analogues (**140**; M = Pd, X = Br or I) and the corresponding ethylidene-bridged complex have been prepared from Pd$_2$(Ph$_2$PCH$_2$PPh$_2$)$_3$ with methylene dihalides and ethylidene dihalides (CH$_3$CHI$_2$) ⟨81JA3764⟩.

Perfluoropropene rearranges on reaction with (cod)$_2$Pt to give the carbene-bridged product (**141**) ⟨75CC451⟩. Perfluorophenyl groups act as one-carbon bridges in a series of anionic platinum and palladium complexes, exemplified by the salt (**142**) obtained either from [Bu$_4$N]$_2$[Pt(C$_6$F$_5$)$_4$] and *cis*-[Pt(C$_6$F$_5$)$_2$(THF)$_2$] or [Bu$_4$N]$_2$[Pt(C$_6$F$_5$)$_3$Cl] and AgClO$_4$ (1 : 1) ⟨88OM2279, 89JCS(D)169, 93JA4145⟩.

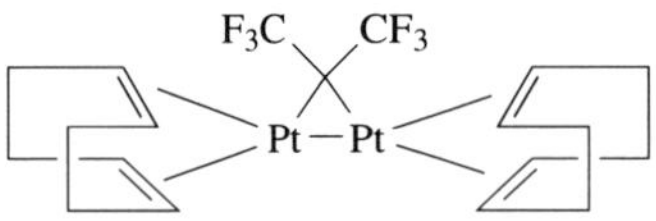

(**141**)

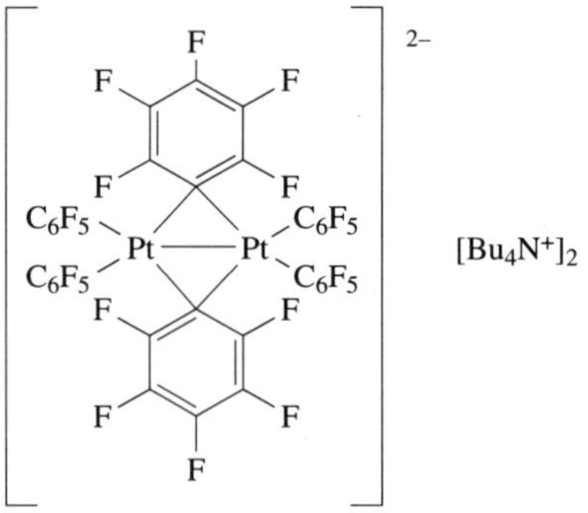

(**142**)

Diphenylethyne forms a bridge between the two nickel atoms in complex (**143**) obtained by treating (cod)$_2$Ni with this alkyne ⟨76JA8289, 78JA2090⟩. The long metal–metal distances (0.2617 nm) observed in this dark red, diamagnetic compound and its platinum analogues suggest that they are best regarded as 16-electron compounds (metal–metal double bonds would be required for 18-electron structures). Several related di- and trinuclear platinum complexes (e.g., Pt$_2$(μ-PhC≡CPh)(PMe$_3$)$_4$ and Pt[(μ-PhC≡CPh)Pt(PMe$_3$)$_2$]$_2$) have been obtained by treating Pt$_2$(μ-cod)(PR$_3$)$_4$ or mixtures of (cod)$_2$Pt and PR$_3$ with PhC≡CPh. A related transformation is shown in Equation (25) ⟨80JCS(D)2182⟩ cf. ⟨80CC1281⟩.

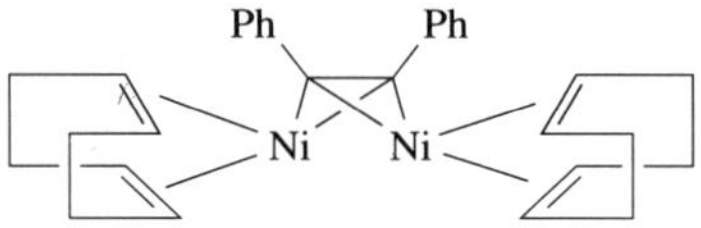

(**143**)

$$\text{Pt(PhC≡CPh)}_2 + \text{(C}_2\text{H}_4\text{)Pt(PR}_3\text{)}_2 \longrightarrow \text{(R}_3\text{P)}_2\text{Pt(}\mu\text{-PhC≡CPh)Pt(PhC≡CPh)} \quad (25)$$

Both nickelocene, Cp$_2$Ni, and the dinuclear carbonyl [CpNi(μ-CO)]$_2$ react with ethyne and a wide range of other alkynes to yield the green, metal–metal bonded complexes of the type (**144**) ⟨59JA4757, 60JA502, 61JCS577, 74JOM(64)271⟩, including complexes of macrocyclic dialkynes ⟨73JOM(47)145⟩ and complexes with perfluoroalkyl substituents ⟨74JOM(80)C39⟩ and with phosphine-oxide substituents ⟨77IC172⟩. Mechanistic studies on the formation of complexes from the bridged dicarbonyl have been reported ⟨72IC2279, 74ICA5⟩. Exchange of one alkyne by another provides an alternative synthetic method for which the following order of displacement has been established (any alkyne displacing one further to the right in the sequence) ⟨74JOM(64)271⟩: C$_2$(CF$_3$)$_2$ ≫ C$_2$(CO$_2$Me)$_2$ > C$_2$Ph$_2$ > PhC$_2$H > C$_2$H$_2$ ⩾ PhC$_2$Me > MeC$_2$H > C$_2$Me$_2$. A palladium analogue (**145**; R = Ph) was obtained from Pd(OAc)$_2$ and tolan in an alcoholic medium. The reaction proceeds via an alkoxytetraphenylcyclobutenyl complex [(C$_4$Ph$_4$OR^1)Pd(μ-Cl)]$_2$ and if this is isolated and treated with a different alkyne 'mixed' products result; for example with hex-3-yne the complex (**145**; R = Et) is obtained ⟨77JA4707⟩.

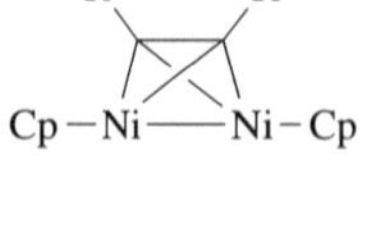

(**144**) 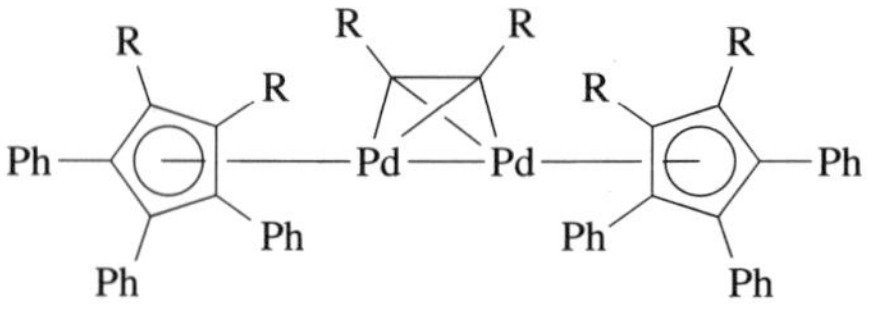   (**145**)

Cationic trinuclear palladium and platinum complexes have been obtained according to Equation (26) (M = Pd or Pt) ⟨92OM2224⟩.

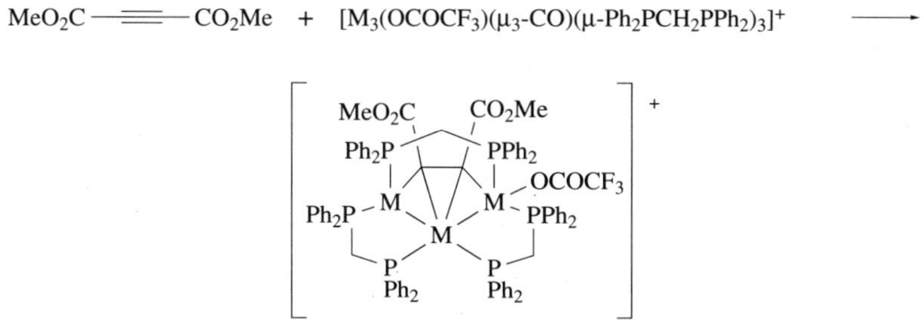

$$\text{MeO}_2\text{C}=\!=\!=\text{CO}_2\text{Me} + [\text{M}_3(\text{OCOCF}_3)(\mu_3\text{-CO})(\mu\text{-Ph}_2\text{PCH}_2\text{PPh}_2)_3]^+ \longrightarrow$$

$$\begin{bmatrix} \text{structure} \end{bmatrix}^+ \quad (26)$$

When a mixture of pentaphenylaluminacyclopentadiene and its dilithium salt was treated with nickel(II) bromide a deep violet complex was formed which was shown to have the remarkable triphenylpropenetriyl-bridged structure (146) ⟨77AG(E)183⟩.

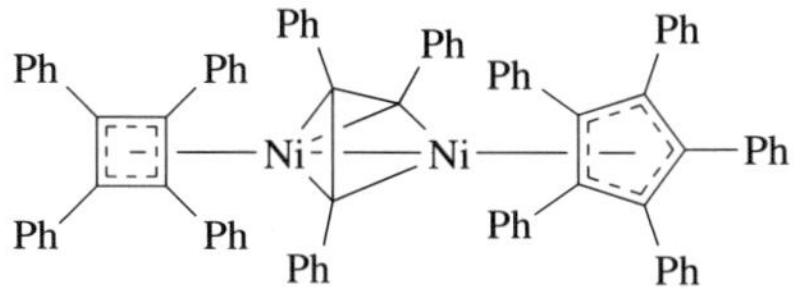

(146)

When triarylcyclopropenium bromides, $[\text{C}_3\text{R}^1{}_2\text{R}^2]\text{Br}$, react with the palladium(0) or platinum(0) complexes $\text{M}_2(\text{PhCH}\!=\!\text{CHCOCH}\!=\!\text{CHPh})_3$, trinuclear complexes $\text{M}_3(\text{C}_3\text{R}^1{}_2\text{R}^2)_2\text{Br}_2$ are formed; after treatment with thallium acetylacetonate, the resultant palladium compounds were shown to have the structure (147) by x-ray crystallography of the dianisyl phenyl compound ($\text{R}^1 = 4\text{-MeOC}_6\text{H}_4$, $\text{R}^2 = \text{Ph}$) ⟨78JCS(D)1825, 78JCS(D)1830⟩.

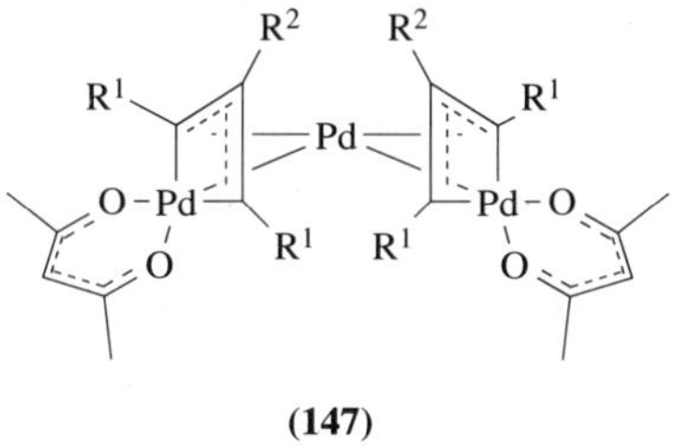

(147)

A series of allyl-bridged palladium complexes is known in which (in contrast to the preceding propenetriyl compounds) the three-carbon bridge lies parallel to the Pd—Pd bond with only the central carbon atom linked to both metals. In early work cyclopentadienylpalladium compounds $\text{CpPd}(\text{PR}_3)\text{X}$ ($\text{R} = \text{Pr}^i$, Ph, or C_6H_{11}, X = Cl, Br, or I) were reduced by a variety of reagents, but preferably LiAlH_4, to the complexes (148). Replacement of the η^3-cyclopentadienyl by the unsubstituted allyl unit, with allylmagnesium bromide, yielded the analogous compounds (149) or (depending on R) the bisallyl bridged (150) ⟨75CC615, 77JOM(129)429⟩ and reduction (Na-K) of the acetates $\text{CpPd}(\text{PR}_3)\text{OAc}$ (e.g., $\text{R} = \text{Pr}^i$) gave the bis(cyclopentadienyl)-bridged complexes (151) ⟨79CC814⟩. Other variants, including allyl/cyclopentadienyl-bridged compounds and platinum analogues, have been described ⟨77CB1763, 79JOM(179)421, 79JOM(179)439, 80CB1072⟩. Furthermore, $\text{CpPd}(\eta^3\text{-allyl})$ reacts with $(\text{C}_6\text{H}_{11})_2\text{PH}$ to give the allyl-bridged complex (149; $\text{R}_3\text{P} = (\text{C}_6\text{H}_{11})_2\text{PH}$, $\text{X} = (\text{C}_6\text{H}_{11})_2\text{P}$) ⟨95JOM(488)39⟩.

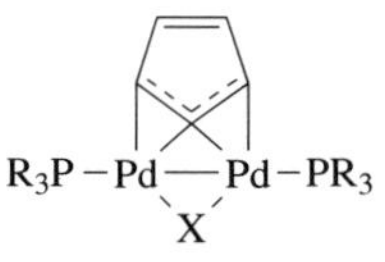

(148) 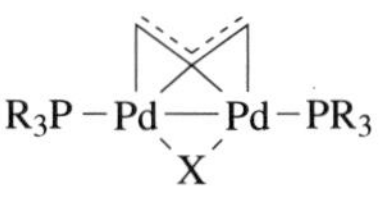   **(149)** 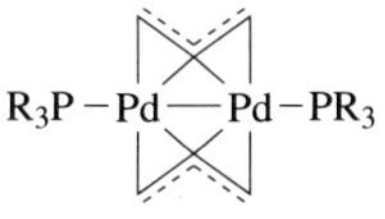   **(150)** 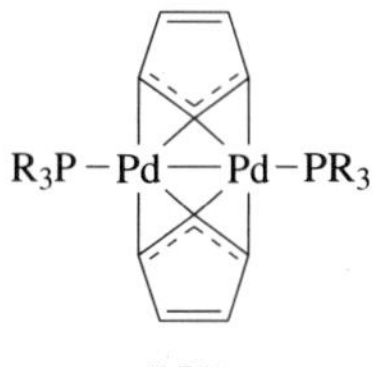   **(151)**

Another variant of this type of bonding is found in bis(pentadienyl)dinickel (**152**), the very air-sensitive, golden-yellow product of the reaction of nickel(II) chloride with 1,4-pentadiene and triethylaluminum at room temperature ⟨69AG(E)677, 69AG(E)678⟩.

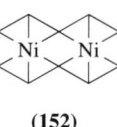

(**152**)

4.15.3.10 Multidecker Sandwich Compounds of Transition Metals

Although strictly the alkyne-derived iron complexes (**81**) should probably be regarded as the earliest examples, directed study of multidecker sandwich compounds has its origin in Werner and Salzer's synthesis of tris(cyclopentadienyl)dinickel salts (**153**) and alkyl derivatives thereof (these compounds were originally, correctly, described as 'double deckers' but have now generally become known as 'triple deckers') ⟨72AG(E)930, 72SRI239⟩. It was prepared by treating nickelocene [or derivatives, $(RC_5H_4)_2Ni$] with fluoroboric acid in propanoic anhydride and can also be formed in equally high yield by treating nickelocene in the same medium with its oxidized form, $[Cp_2Ni]BF_4$ ⟨77AG(E)1, 77JOM(141)339⟩. An electron-deficient cation, probably $[(C_5H_5)Ni]^+$, is regarded as the key intermediate which adds to the neutral metallocene. This approach works with iron and ruthenium provided the more electron-rich permethylated metallocenes are used as the electron donors. With $[CpFe(C_6H_6)]PF_6$ as the cationic precursor and $C_{10}Me_{10}Fe$ this yields the blue-green salt (**154**; M = Fe, R = H) and from $[(C_5R_5)Ru(MeCN)_3]PF_6$ and $C_{10}Me_{10}Ru$, the yellow-orange salts (**154**; M = Ru, R = H or Me) were prepared ⟨87JOM(336)187⟩.

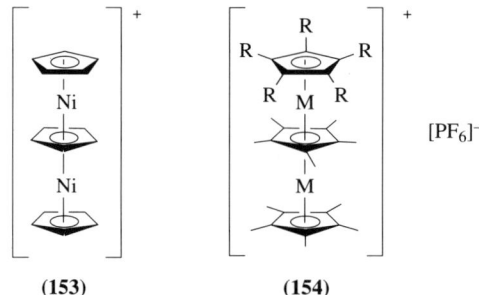

(**153**)   (**154**)

Heating the bis(η^3-allyl) complex $CpV(C_3H_5)_2$ to 100 °C with cyclohexadiene results in the formation of the paramagnetic benzene-bridged compound (**155**) whose central ring can be displaced by other arenes (toluene, mesitylene) ⟨83JA5479⟩.

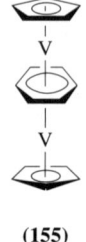

(**155**)

At high metal to ligand ratio, the reaction of chromium vapour with mesitylene yields tris(mesitylene)dichromium (**156**) alongside $(C_6H_3Me_3)_2Cr$ ⟨86JA2096, 87OM1583⟩. The metal vapour route was also used to generate the cobalt complex (**157**), albeit in very low yield, from pentamethylcyclopentadiene ⟨91AG(E)1124⟩.

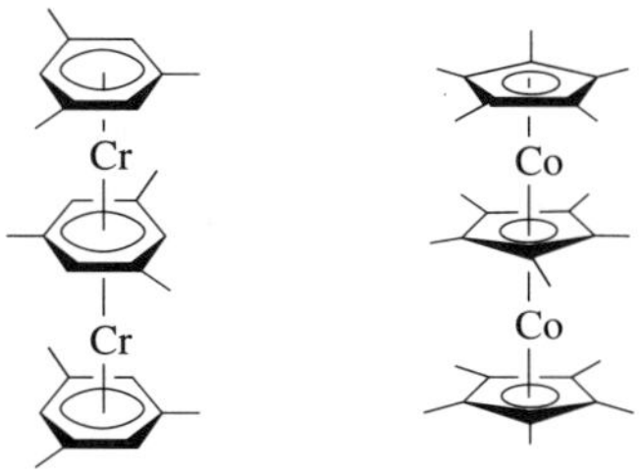

(**156**)   (**157**)

Numerous triple-decker borole complexes have been reported by Herberich and co-workers. They are prepared from metal carbonyls, or similarly reactive precursors, either by heating with 1-phenyl-1-boracyclohepta-2,6-diene which contracts to the 1-phenyl-2-ethylborolyl ligand or from 1-methyl- or 1-phenyl-2,5- or 2,3-dihydroboroles which suffer dehydrogenation ⟨89JOM(372)53⟩. These processes yield complexes (**158**; $ML_n = Mn(CO)_3$, $R^1 = Ph$, $R^2 = Et$) and (**158**; $ML_n = CpFe$, $R^1 = Ph$, $R^2 = Et$) with $Mn_2(CO)_{10}$ and $Cp_2Fe_2(CO)_4$ respectively ⟨83JOM(246)141⟩, (**158**; $ML_n = Mn(CO)_3$, $R^1 = Ph$, $R^2 = H$) with $Mn_2(CO)_{10}$ ⟨86JOM(308)153⟩, (**158**; $ML_n = V(CO)_4$, $R^1 = Ph$, $R^2 = H$) with $V(CO)_6$ ⟨89AG(E)319⟩, (**158**; $ML_n = CoC_4H_4BR^3$, $R^1 = Me$ or Ph, $R^2 = H$) with $Co_2(CO)_8$ when the intermediate $Co_2(C_4H_4BR^3)_2(\mu\text{-}CO)_2(CO)_2$ is heated to 160–180 °C ⟨87JOM(319)9⟩, and the analogous rhodium complex (**158**; $ML_n = RhC_4H_4BR^3$, $R^1 = Me$ or Ph, $R^2 = H$) with $[(C_2H_4)_2RhCl]_2$ ⟨86JOM(312)13⟩.

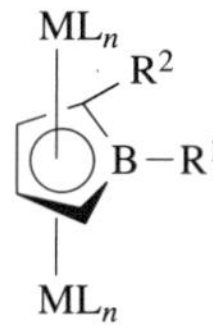

(**158**)

The boratabenzene complex $(C_5Me_5)Ru(C_5H_5BMe)$ adds a second $(C_5Me_5)Ru$ group to form the salt $[(C_5Me_5)Ru(\mu\text{-}C_5H_5BMe)Ru(C_5Me_5)][OTf]$ when treated with the ruthenium electrophile produced by the action of AgOTf on $(C_5Me_5RuCl)_4$ ⟨93JOM(459)1⟩.

4.15.3.11 Functions Containing Two Cu, Ag or Au Atoms

Addition of a dihalomethane (X = Cl, Br, I) to the cyclic precursor (**159**) (Equation (27)) yields the methylene-bridged gold complexes (**160**) ⟨82AG(E)73⟩.

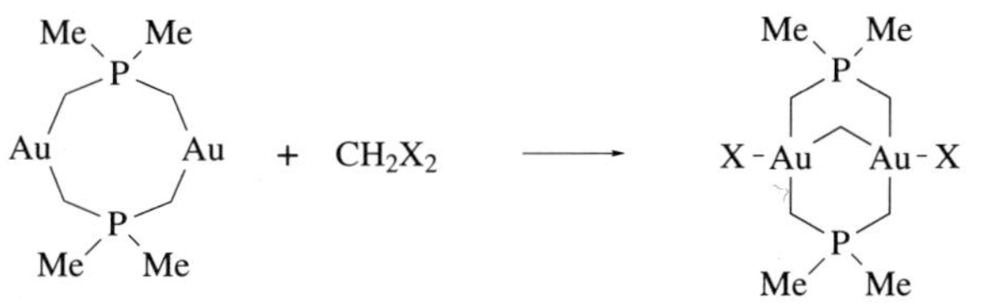

(**159**)   (**160**)   (27)

The 'simple' alkyl, alkenyl and aryl derivatives of copper and silver are oligomeric in the solid state and those which are sufficiently stable and crystalline to have been examined crystallographically contain bridging (formally pentacovalent) carbon atoms. Most are prepared from the metal(I) halides and organolithium or -magnesium reagents. Thus, CuI reacts with $LiCH_2$-TMS to yield the tetranuclear alkyl (**161**) ⟨77JCS(D)999⟩. Copper(I) bromide was used in reactions with 1-(dimethylaminomethyl)-2-lithiobenzene derivatives to yield tetrameric and *N*-coordinated aryls, for example (**162**) ⟨75JOM(84)117, 75JOM(84)129⟩ while the analogous (2-dimethylaminomethyl)ferrocenylcopper ⟨77JOM(137)217⟩ and -silver tetramers ⟨78JOM(153)115⟩, obtained from CuI and AgI respectively, lack metal–nitrogen coordination.

(161)   **(162)**

Mesitylsilver obtained from silver chloride and mesitylmagnesium bromide is likewise tetrameric with the structure (**163**) ⟨83CC1087⟩ whereas the copper compound, prepared from the same Grignard reagent with CuCl ⟨81JOC192⟩, is a pentamer with a similar structure based on a pentagon of copper atoms ⟨83CC1156⟩. The arrangement of the five copper atoms in the anion $[Cu_5(\mu\text{-Ph})_6]^-$, obtained from CuBr with PhLi, is described as a 'squashed' trigonal bipyramid ⟨82JA2072⟩ and an octahedral arrangement of copper atoms is found in the (2-dimethylaminophenyl)-bridged $Cu_6(\mu\text{-Br})_2(C_6H_4NMe_2)_4$ ⟨75JOM(102)551⟩. The octanuclear $(2\text{-MeOC}_6H_4Cu)_8$, obtained from 2-lithioanisole and copper(I) bromide, adopts a square antiprismatic structure with the same aryl-bridging between adjacent copper atoms and Cu—O bonding to a different anisyl grouping ⟨79JOM(174)121⟩.

(163)

Silver fluoride addition to hexafluoro-2-butyne is the route to the tetranuclear alkenyl complex (**164**) ⟨93JOM(449)203⟩. A series of aryl-, vinyl- and ferrocenyl-bridged salts $[R^1(AuPR^2_3)_2]BF_4$ has been obtained from the mononuclear gold compounds, $R^1AuPR^2_3$, by treatment with HBF_4 and has been shown to have structures exemplified by the phenyl compound (**165**) ⟨74JOM(65)131⟩.

(164)   **(165)**

4.15.3.12 Functions Containing Two Zn, Cd or Hg Atoms

The reaction of metallic zinc with diiodomethane yields IZn—CH_2—ZnI ⟨73NKK381, 83TL2043⟩. The mercury analogue, IHg—CH_2—HgI, was first prepared by Sakurai in 1881 by the mercury(I) iodide catalysed photochemical reaction of mercury metal and CH_2I_2 ⟨1881JCS485⟩. The corresponding bromide has been made from the bis-Grignard reagent $CH_2(MgBr)_2$ by metal exchange with $HgBr_2$ ⟨85JOM(288)13⟩ while an apparently more general route to $CH_2(HgX)_2$ (X = Cl, Br, I, CN, OAc, etc.) consists of treating the diboronic acid ester $CH_2[B(OMe)_2]_2$ with the appropriate mercury salt HgX_2 in aqueous sodium hydroxide or with RHgOAc (e.g., R = Me) to obtain

$CH_2(HgR)_2$ ⟨73JOM(54)35, 80JOM(191)7, 83JOM(243)245, 86JOM(301)1⟩. The structure has been confirmed for $CH_2(HgCl)_2$ ⟨81ZN(B)188⟩. The homologue $CH_3CH(HgCl)_2$ has been obtained similarly from $CH_3CH[B(OMe)_2]_2$ either directly or via polymeric $X-[CHMe-Hg]_n-X$ ⟨64JOC2742⟩ or by trans-metallation from $CH_3CH(AlCl_2)_2$ and $HgCl_2$ ⟨88JA3231⟩. The higher homologues, $RCH_2CH[B(OMe)_2]_2$, required to obtain for example $EtCH(HgCl)_2$, $(n\text{-}C_3H_7)CH(HgCl)_2$, $PhCH_2CH(HgCl)_2$ and $Bu^tCH_2CH(HgCl)_2$ have been generated *in situ* from the terminal alkynes $RC{\equiv}CH$ by bis-hydroboration followed by methanolysis and treated with $HgCl_2$ without isolation ⟨73JOM(61)27⟩.

Addition of mercuric acetate to 1-propenylmercury acetate, $MeCH{=}CHHgOAc$, in water or acetic acid provides a route to $MeCH(OX)CH(HgOAc)_2$ (where $X = H$ or Ac depending on the solvent) ⟨78JOM(159)237⟩.

The susceptibility of enolizable molecules, including carboxylic acids, RCH_2CO_2H, to undergo bis-mercuration of the methylene group has been widely used as a route to carbon-bridged bis-chloromercury compounds. But whereas the structure $(MeCO)_2C(HgCl)_2$, for the product from pentan-2,4-dione and $HgCl_2$ ⟨70JOM(22)5, 77JOM(136)7⟩ cf. ⟨76SA(A)1459⟩ is firmly established by x-ray crystallography ⟨75JOM(99)31⟩, the alternative structure (**166**), deduced for the products from dibenzoylmethane ⟨74JOM(67)C61⟩, suggests that the formulation of some other related compounds must be viewed with caution. The reaction with mercuric acetate or halides has been used with ethyl acetoacetate ⟨67IJC210, 77JOM(136)7⟩, malonic acid and malononitrile (using $Hg(OAc)_2$ or $HgCl_2$ but not $HgBr_2$) and so on ⟨77JOM(136)7⟩. Propanal and butanal react similarly with $HgCl_2$ to give $RC(HgCl)_2CHO$ (R = Me or Et) and x-ray crystallography has confirmed the structure of the latter product ⟨84CCA689⟩.

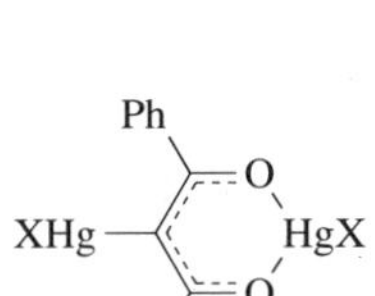

(**166**)

Mercuration of simple carboxylic acids, which requires elevated temperatures, has been carried out by heating their mercury(II) salts with the corresponding acid anhydride, followed by treatment of the (oligomeric) intermediate with dilute aqueous HCl (Scheme 5) ⟨91JOM(405)59, 93OM4708⟩ and the structure of the product from propanoic acid (R = Me) was confirmed by x-ray crystallography ⟨92CSC1116⟩.

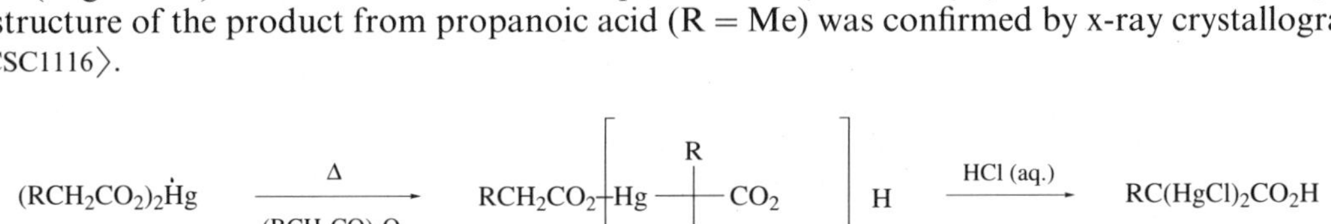

Scheme 5

Mercuration of the salt $[(CF_3)_2CHCO_2]_2Hg$ by $Hg(OAc)_2$ is followed by decarboxylation yielding the trinuclear $[(CF_3)_2C(HgOAc)]_2Hg$ ⟨71IZV620⟩ together with some $(CF_3)_2C(HgOAc)_2$. The ketene, $(CF_3)_2C{=}C{=}O$, on addition to mercury(II) trifluoroacetate yields an intermediate ['$(CF_3)_2C(CO_2^-)Hg^+$'] which loses CO_2 on pyridine treatment to give polymeric $[-C(CF_3)_2-Hg-]_n$, isolated as a pyridine complex ⟨79IZV1831⟩.

The complete mercuration of cyclopentadiene by mercuric acetate or chloride, to give the hexa-substituted derivatives (**167**; X = OAc or Cl), may also be thought of as involving mercuration of an active methylene group ⟨64JINC1531⟩.

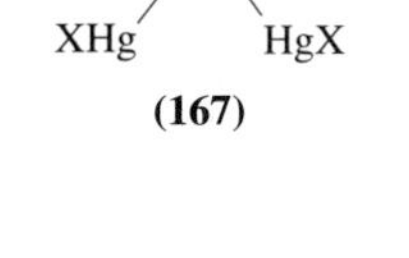

(**167**)

4.15.4 FUNCTIONS CONTAINING TWO Al, Ga, In OR Tl ATOMS

Metallic aluminum reacts with dibromomethane according to Equation (28) (X = Br) and the corresponding reaction with dichloromethane is catalysed by CH_2Br_2 or $AlBr_3$ ⟨66TL2315, 73JOM(50)47, 74JOM(72)C4, 88JOC2829⟩.

$$Al + CH_2X_2 \longrightarrow X_2AlCH_2AlX_2 + (CH_2)_n \qquad (28)$$

Alkyllithiums convert the products to the rather unstable alkyls, $R_2AlCH_2AlR_2$ (e.g., R = Me, Et), which may be stabilized by coordination to electron donors such as **(168)** ⟨82JOM(225)71⟩ cf. ⟨90POL277, 91CB1511, 91JOM(415)181⟩. Tetrachloro(μ-methylene)diindium, as the chelate **(169)**, is obtained from a mixture (1:1) of InCl and $InCl_3$ with dichloromethane and TMEDA ⟨86OM525⟩. Ethylene reacts with $AlCl_3$ when heated with a reducing metal (Al or K), initially forming $Cl_2Al\text{-}CH_2CH_2AlCl_2$, but on prolonged reaction this isomerizes to the more stable $CH_3CH(AlCl_2)_2$ as the final product ⟨85ZN(B)182, 88JA3231⟩.

(168) **(169)**

The higher homologues, $RCH_2CH(AlCl_2)_2$ and $R^1CH_2CH(AlR^2_2)_2$ (R^2 = Me, Et), have been obtained by the addition of the appropriate aluminum hydride to 1-alkynes, $RC{\equiv}CH$ ⟨60LA(629)222, 75CA(83)10238, 80JOM(201)97, 88JOC2829⟩. Remote double bonds remain unaffected; for example hept-6-en-1-yne yields $CH_2{=}CH(CH_2)_3CH(AlR_3)_2$ (R = Et or Bu^i) ⟨77LA1633⟩. The method fails with symmetrical internal alkynes but 1-phenylpropyne has been shown to react according to Equation (29) ⟨88IZV1448⟩. The compound $CH_3CH(AlEt_2)_2$ is also slowly formed by photolysis of trimethylaluminum in the presence of mercury ⟨70JOM(25)C6⟩.

$$Ph{-}{\equiv}{-} + HAlBu^i_2 \longrightarrow \underset{\text{Ph}}{\overset{Bu^i_2Al}{\diagdown}}\!\!\!\bigg\backslash\text{AlBu}^i_2 \qquad (29)$$

As is well known, 'trimethylaluminum' and other simple trialkyl- and triarylaluminums exist as dimers linked by formally pentacovalent bridges (see also Chapter 2.11.6). Organometallic compounds with such electron-deficient bonds have been reviewed ⟨77AOC(15)235⟩. In addition to the classic x-ray structure determination on Al_2Me_6 **(170)** ⟨53JCP(21)986⟩, analogous structures have been confirmed *inter alia* for hexakis(cyclopropyl)dialuminum ⟨71JA1035⟩ and hexakis(*o*-tolyl)dialuminum **(171)** ⟨82OM1307⟩ while the x-ray structure of $[Me_2Al(\mu\text{-}C_6H_5)]_2$ confirms the preference for aryl- over alkyl-bridging ⟨72JCS(D)2649⟩. Alkyl-bridging involves rather weak bonding as shown *inter alia* by NMR evidence and by the rapid exchange reactions between two different alkyls. Even Pr_3Al and higher *n*-alkyls contain significant amounts of monomer and most branched alkyl-aluminums are largely or wholly (e.g., Bu^t_3Al) monomeric, for example ⟨65JPC3418, 87JOM(333)155⟩.

(170) **(171)**

The most general synthetic method is the treatment of aluminum chloride or intermediate alkylaluminum halides with alkyl- or aryllithiums ⟨73LA708⟩. The reaction of organomercurials R_2Hg with metallic aluminum not only provides a route to strictly halide-free trialkylaluminums, but remains one of the best methods for triaryl derivatives ⟨82OM1307⟩. Sesquihalides $R_3Al_2X_3$ can be reduced to R_3Al by sodium or other reducing metals and the salts $Na[AlR_2F_2]$ decompose when heated (250–300 °C for R = Me, but 150–210 °C for R = Et or Pr.) giving the alkyls, R_3Al, in high yields ⟨57LA(608)1⟩. Branched alkyls isomerize on heating ($Pr^i_3Al \to Pr^n_3Al$; $Bu^t_3Al \to Bu^i_3Al$) ⟨73LA715⟩. This process depends on the facile β-elimination followed by the regioselective addition (Equation (30)). This step (Equation (30)) together with the reaction shown in Equation (31) is responsible for the success of the commercially important synthesis of the trialkyls (e.g., Et_3Al) from the metal, hydrogen and alkene, which as Equation (31) shows, requires the initial presence of the trialkylaluminum ⟨57LA(608)1, 60LA(629)1⟩. The easy exchange of alkyl groups, as in Equation (32), also depends on the reversibility of the reaction shown in Equation (30) ⟨60LA(629)14, 60LA(629)53⟩.

$$(RCH_2CH_2)_2AlH + \underset{R}{\diagup\!=} \quad \rightleftharpoons \quad (RCH_2CH_2)_3Al \qquad (30)$$

$$2(RCH_2CH_2)_3Al + Al + 1.5\,H_2 \longrightarrow 3(RCH_2CH_2)_2AlH \qquad (31)$$

$$Bu^i_3Al + 3\,\underset{H\;\;H}{\overset{H\;\;H}{>\!=\!<}} \longrightarrow Et_3Al + \diagdown\!= \qquad (32)$$

Whereas most donors (L) react with the dimeric alkyls R_6Al_2 to cleave both bridges yielding R_3AlL, a few reactions are known in which one bridge is retained. Thus, Me_6Al_2 reduces aldehydes and ketones, R^1COR^2, to complexes (172), also formed from the corresponding alcohols, R^1R^2CHOH ⟨70AJC715⟩. Me_6Al_2 and $(TMS)_2NX$ (X = H or Cl) yield the related n-bridged complex (173) ⟨72JOM(36)267, 72JOM(36)277⟩.

(172)

(173)

4.15.5 FUNCTIONS CONTAINING TWO Sn OR Pb ATOMS

The long-known methylenebis(trimethylstannane), $CH_2(SnMe_3)_2$, was first made from trimethyltinsodium with dichloromethane ⟨30JA695⟩ and this or analogous routes remain the best for generating a wide range of related tin and lead compounds. Thus, trivinyltinsodium with CH_2Cl_2 in liquid ammonia gives a 70% yield of $CH_2[Sn(CH=CH_2)_3]_2$ ⟨83OM106⟩, triphenylleadlithium, Ph_3PbLi, with CH_2Cl_2 or CH_2Br_2 yields $CH_2(PbPh_3)_2$ (72%) ⟨67JOM(9)117, 70JOM(23)471⟩, and the magnesium chloride derivatives $Me_3PbMgCl$ and $(TMS\text{-}CH_2)_3PbMgCl$ yield $CH_2(PbMe_3)_2$ (80%) ⟨70JOM(22)141⟩ and $CH_2[Pb(CH_2\text{-}TMS)_3]_2$ (90%) ⟨70JOM(23)465⟩ respectively.

Me_3SnNa and $CHCl_3$ in liquid ammonia also yield $CH_2(SnMe_3)_2$ according to Equation (33) ⟨61JA1514⟩.

$$4Me_3SnNa + CHCl_3 \xrightarrow{NH_3} CH_2(SnMe_3)_2 + Me_6Sn_2 + 3NaCl + NaNH_2 \qquad (33)$$

Similar reactions of other dihaloalkanes, for example 1,1-dichloroethane or 2,2-dichloropropane with Me_3SnLi yield the expected compounds $R^1R^2C(SnMe_3)_2$ (e.g., $R^1, R^2 = H, Me$) ⟨79JOM(172)293⟩ but appear to have been little used. The polarity of the methylene and metal components can be reversed as in the high-yielding reaction of dilithiomethane, CH_2Li_2, with Me_3SnCl, at low temperature, to form $CH_2(SnMe_3)_2$ ⟨87RTC514⟩, and of $(NC)_2CLi_2$ to yield $(NC)_2C(SnMe_3)_2$ ⟨90ZAAC(586)25⟩.

Another approach involves the triethylantimony (or triethylamine) catalysed addition of tin(II) bromide to diiodomethane (Scheme 6). The intermediate halide compound need not be isolated if it is to be converted into the methylenebis(trialkyltin) (e.g., R = Me or Bu) ⟨76JOM(117)329, 91T3281⟩.

On the other hand, only a 14% overall yield of MeCH(SnMe$_3$)$_2$ was obtained when this route was applied to CH$_3$CHI$_2$ ⟨92SL891⟩.

$$SnBr_2 + CH_2I_2 \xrightarrow[\text{cat.}]{120-140\ °C} CH_2(SnBr_2I)_2 \xrightarrow{RMgX} CH_2(SnR_3)_2$$

Scheme 6

Cleavage of carbon–tin bonds with replacement of alkyl or aryl groups by halogens (but retention of the methylene bridge) can be carried out in stepwise and apparently easily controlled transformations. Thus, CH$_2$(SnMe$_3$)$_2$ reacts with Me$_2$SnCl$_2$ to give CH$_2$(SnMe$_2$Cl)$_2$ and the latter gives ClMe$_2$SnCH$_2$SnMeCl$_2$ when treated with 0.5 mol SnCl$_4$ but gives CH$_2$(SnMeCl$_2$)$_2$ with excess of the same reagent ⟨83OM106⟩. Bromine cleaves CH$_2$(SnMe$_3$)$_2$ to CH$_2$(SnMe$_2$Br)$_2$. The easier Ph—Sn cleavage of CH$_2$(SnPh$_3$)$_2$ to give CH$_2$(SnPh$_2$X)$_2$ is effected with HCl/MeOH, Br$_2$ or I$_2$, but an excess of Br$_2$ or use of HCl/Et$_2$O leads to CH$_2$(SnPhX$_2$)$_2$ and acetic acid yields CH$_2$[Sn(OAc)$_3$]$_2$ ⟨84JCR(S)152, 84JOM(273)303⟩. Furthermore, HgCl$_2$ has also been used to cleave Ph—Sn bonds, for example to convert CH$_2$(SnPh$_2$R)$_2$ into CH$_2$(SnRCl$_2$)$_2$ ⟨93OM2788⟩. Methylenebis(trichlorostannane), CH$_2$(SnCl$_3$)$_2$, is efficiently formed (87%) on treatment of the vinyltin compound, CH$_2$[Sn(CH=CH$_2$)$_3$]$_2$, with tin(IV) chloride ⟨83OM106⟩.

The mixed alkyl-halogeno tin compounds add various oxygen and nitrogen donor ligands giving a variety of both five- and six-coordinated tin complexes, such as the dimethyl sulfoxide adducts **(174)**–**(176)** ⟨82OM404⟩ and the pyridazine complexes **(177)** and **(178)** ⟨87OM834⟩.

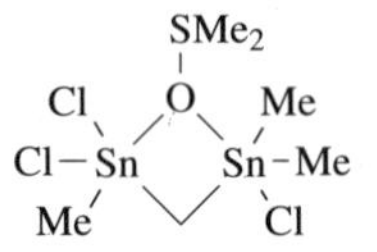

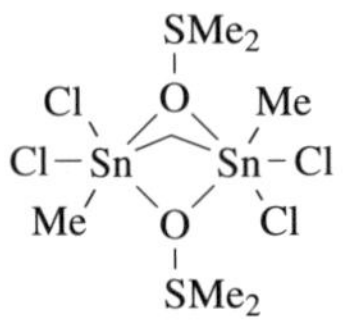

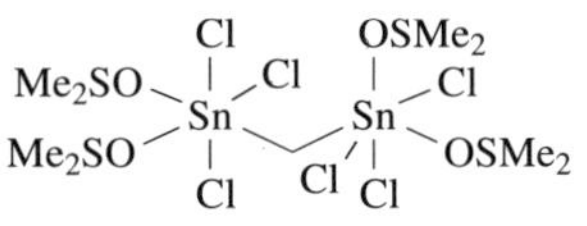

The chief importance of these haloalkyl compounds is undoubtedly as precursors of other methylene(ditin) compounds which result from halogen replacement. Thus, CH$_2$(SnMe$_2$Br)$_2$ is reduced to the tin hydride CH$_2$(SnMe$_2$H)$_2$ by Bu$_3$SnH ⟨91OM936⟩. The chlorotin complexes CH$_2$(SnMe$_2$Cl)$_2$, CH$_3$CH(SnMe$_2$Cl)$_2$ and (CH$_3$)$_2$C(SnMe$_2$Cl)$_2$ have been converted by BrMg(CH$_2$)$_4$MgBr into the corresponding distannacycloheptanes **(179**; R^1, R^2 = H, Me) ⟨92JOM(429)311⟩ and the bromo- or chlorodiphenyltin derivatives, CH$_2$(SnPh$_2$X), have been reduced electrochemically ⟨84JCR(S)152⟩ or by sodium/liquid ammonia ⟨82JOM(236)69⟩ to the tetrastannacyclohexane **(180)** and converted by Na$_2$S into the eight-membered ring compound **(181)** ⟨84JOM(273)303⟩. Moreover, the tin–sodium compounds, CH$_2$(SnR$_2$Na)$_2$ (R = Me or Ph), derivable from these halides, have been treated with the dichlorides Cl(CH$_2$)$_n$Cl (n = 1–4) in liquid ammonia to give the corresponding rings **(182**; R = Me or Ph) ⟨85BSB299⟩. Alkaline hydrolysis of CH$_2$(SnPhBr$_2$)$_2$ yields CH$_2$[SnPh(O)]$_2$ which reacts with MeN(CH$_2$CH$_2$SH)$_2$ to give the methylene-bridged, pentacoordinated tin complex **(183)** ⟨84JOM(277)335⟩.

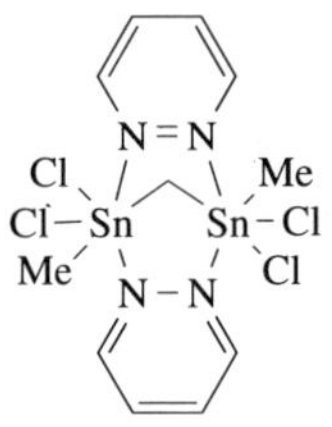

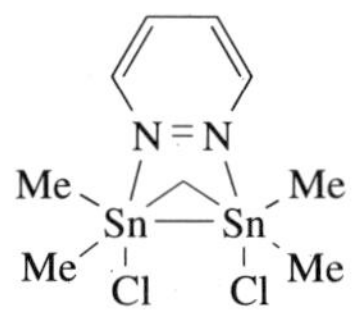

Routes to the tritin compound **(184)** include the reaction of Me$_2$SnCl$_2$ with the di-Grignard reagent CH$_2$(MgBr)$_2$. This yields **(184)** as the major product (33%) along with the cyclic tetramer (CH$_2$SnMe$_2$)$_4$ (9%) and the linear products (Me$_3$SnCH$_2$)$_2$SnMe$_2$ (15%) and (Me$_3$SnCH$_2$SnMe$_2$)$_2$CH$_2$ (8%) ⟨85JOM(288)13⟩. ISnMe$_2$CH$_2$I and magnesium metal also give compound **(184)** ⟨74SRI515⟩ but the corresponding dichloride or dibromide XSnMe$_2$CH$_2$X (X = Cl or Br) fails to give this product ⟨73ZOB801⟩. The Grignard reagent Me$_3$SnCH$_2$MgBr reacts with tin(IV) chloride to yield the tetramethylene-bridged Sn(CH$_2$SnMe$_3$)$_4$ ⟨84JOM(269)249⟩.

The other general route to substituted methylene-bridged tin compounds utilizes hydrostannation of alkynes or vinylstannanes. Alkynes must react in two steps, as in Equations (34) and (35); that is vinylstannanes are necessarily intermediates and the direction of the second step, leading to the ditin compounds (185) and (186), is strongly dependent on the nature of R^1 and probably also R^2. The reactions are radical processes, commonly catalysed by 2,2′-azobisisobutyronitrile (AIBN), and examples which follow suggest that both electronic and steric effects are important in determining the regiochemical direction. Thus, phenylethyne with triphenylstannane gives exclusively (185; $R^1 = R^2 = Ph$) ⟨69JOM(16)83⟩ and addition of Me_3SnH to $Me_3SnCH=CHCO_2Me$ likewise gives a single product (185; $R^1 = CO_2Me$, $R^2 = Me$) ⟨69JOM(16)91⟩. Similar selectivity is reported for the addition of the same stannane to the propenylstannane $CH_3CH=CHSnMe_3$ yielding (185; $R^1 = R^2 = Me$) ⟨79JOM(172)293⟩ although the vinylstannane $CH_2=CHSnMe_3$ gives a preponderance of the 1,2-ditin compounds (186; $R^1 = H$, $R^2 = Me$ or Et) with both Me_3SnH and Et_3SnH ⟨76JOM(111)C33⟩. Addition of the former to the vinylidenedistannane, $CH_2=C(SnMe_3)_2$, gives a 3:2 mixture of the CHR-bridged $Me_3SnCH_2CH(SnMe_3)_2$ and the more hindered $CH_3C(SnMe_3)_3$ ⟨85OM1044⟩.

$$R^1-\!\!\!\equiv\!\!\! + R^2{}_3SnH \longrightarrow \begin{array}{c} SnR^2{}_3 \\ \diagup \\ R^1 \end{array} \qquad (34)$$

$$\begin{array}{c} SnR^2{}_3 \\ \diagup \\ R^1 \end{array} + R^2{}_3SnH \longrightarrow R^1\!\!\diagdown\!\!\begin{array}{c} SnR^2{}_3 \\ SnR^2{}_3 \end{array} + \begin{array}{c} R^1 \\ \diagdown \\ SnR^2{}_3 \\ SnR^2{}_3 \end{array} \qquad (35)$$

$$\qquad\qquad\qquad\qquad\qquad\qquad (185) \qquad\qquad (186)$$

1,3-Enynes with Me_3SnH give fairly complex mixtures of products, some of which form with double bond migration. Those from $CH_2=CMeC\equiv CH$ include 15% of compound (185; $R^1 = CH_2=CMe$, $R^2 = Me$) but $Me_2C=CHC\equiv CH$ yields $Pr^iCH=CH-CH(SnMe_3)_2$ ⟨83OM1099⟩.

The photochemical addition of Me_3SnH to propargyl acetate, $HC\equiv C-CH_2OAc$, leads cleanly (70%) to the single adduct (185; $R^1 = CH_2OAc$, $R^2 = Me$) a product which has been transformed into numerous related compounds by manipulation of the CH_2OAc group and by halodealkylation ⟨83OM909, 84OM1687⟩ without disrupting the carbon-bridged ditin system.

Like the tin hydrides, the amino, Me_3SnNMe_2, and alkoxytin compounds, Me_3SnOMe, can add to vinyltin systems. Depending on the substituents such additions to the boranes $R^1{}_2BCR^1=CR^2SnMe_3$ have yielded $(Me_3Sn)_2CR^2-CR^1{}_2BR^1X$ (X = NMe_2 or OMe) ⟨91CB503⟩. The same and related tin compounds (Bu_3SnOMe, Me_3SnPPh_2) add to carbon suboxide, C_3O_2, to yield $(XCO)_2C(SnR_3)_2$ (X = Me_2N, Ph_2P, MeO; R = Me, Bu) ⟨86G471, 88G577, 88OM210, 90G53⟩.

Cyclopentadiene reacts with two equivalents of Me_3SnNEt_2 to yield the 5,5-ditin derivative (187) (90%) ⟨71JOM(30)C57⟩.

Reactions leading to bridged ditin compounds which involve replacement of other elements by tin include reactions of allylselenium compounds (Equation (36) and Scheme 7) ⟨88JOC455⟩, the replacement of zinc and copper (e.g., Scheme 8) ⟨86TL4427⟩ and the displacement of titanium as shown by the reaction in Equation (37) ⟨84TL5191⟩.

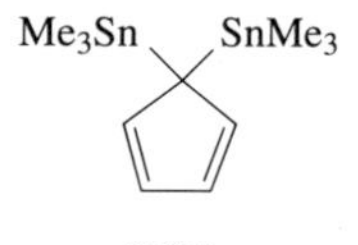

(187)

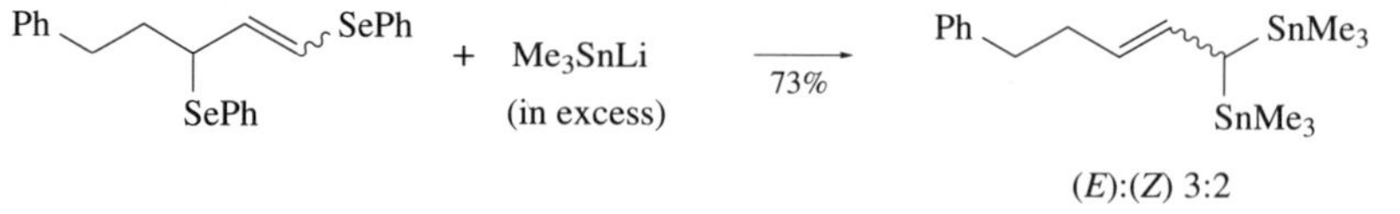

$$\text{Ph}\diagdown\diagdown\diagup^{\text{SePh}}\diagup\diagdown^{\text{SePh}} + \text{Me}_3\text{SnLi} \xrightarrow[73\%]{} \text{Ph}\diagdown\diagdown\diagup^{\text{SnMe}_3}\diagup\diagdown^{\text{SnMe}_3} \quad (E):(Z)\ 3:2 \quad (36)$$

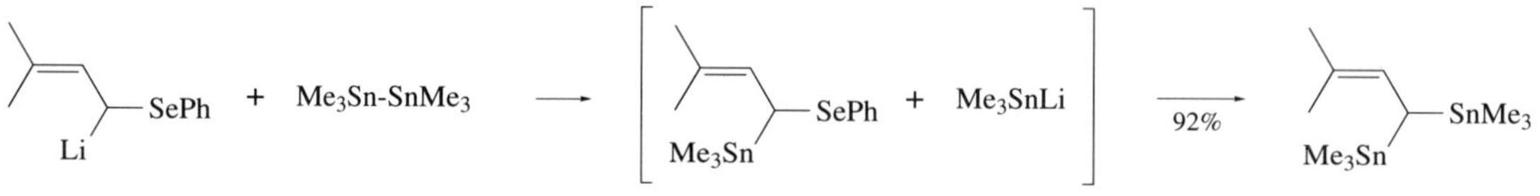

Scheme 7

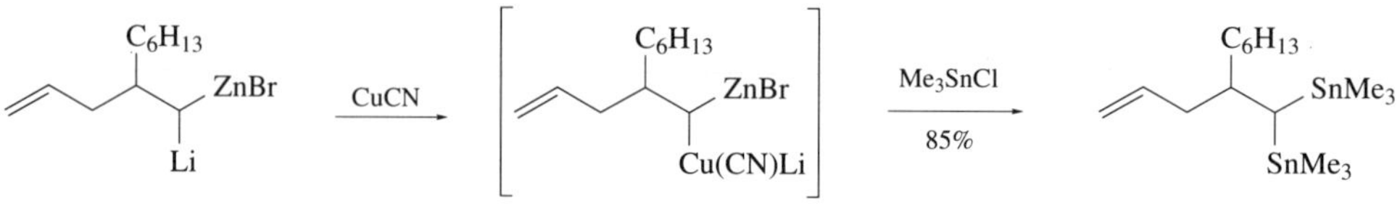

Scheme 8

$$\text{Cp}_2\text{Ti}\diagup\!\!\diagdown\text{SnMe}_2 + \text{Me}_3\text{SnCl} \longrightarrow \text{Cp}_2\text{TiCl}_2 + (\text{Me}_3\text{SnCH}_2)_2\text{SnMe}_2 \quad (37)$$

The reaction shown in Equation (38), which produces a bis(tributyltin) compound bridged by a tetrahedral carbon from a vinylidene-ditin precursor, involves an example of the Eschenmoser modification of the Claisen rearrangement ⟨90TL5829⟩.

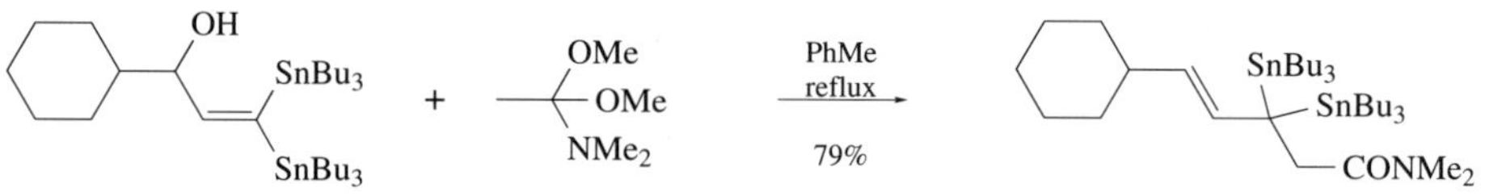

(38)

4.15.6 FUNCTIONS CONTAINING TWO LANTHANIDE OR ACTINIDE ATOMS

Dimeric bis(cyclopentadienyl)metal–methyl complexes (**188**; M = Y, Dy, Ho, Er, Tm or Yb), bridged (like Al_2Me_6) by formally pentacoordinate carbon atoms, are formed when the mixed-metal compounds $Cp_2M(\mu\text{-Me})_2AlMe_2$ are treated with pyridine ⟨79JCS(D)54⟩. The corresponding pentamethylcyclopentadienyl complexes of yttrium and lutetium are reported to have the asymmetric structure (**189**) with a single methyl bridge in the solid state ⟨83JA6491⟩.

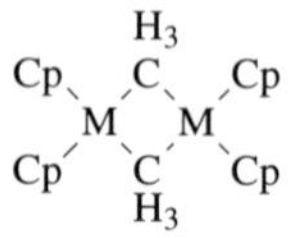

(188) 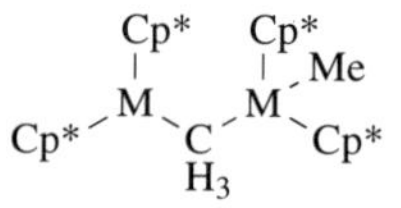   (189)

When heated to 170 °C butyltris(cyclopentadienyl)thorium, Cp_3ThBu, decomposes to $[Cp_2Th(\mu\text{-}C_5H_4)]_2$ (**190**) ⟨74JA7586⟩.

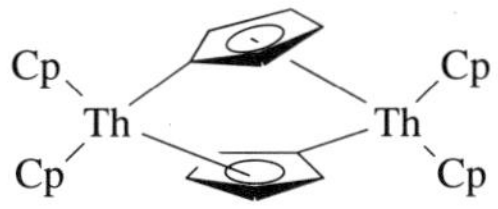

(**190**)

Condensation of cyclooctatetraene at −196 °C with the vapours of Er, Nd or La, followed by extraction with THF yields the complexes $(C_8H_8)_3M_2(THF)_2$; the structure (**191**) was established for the neodymium compound ⟨78IC625⟩.

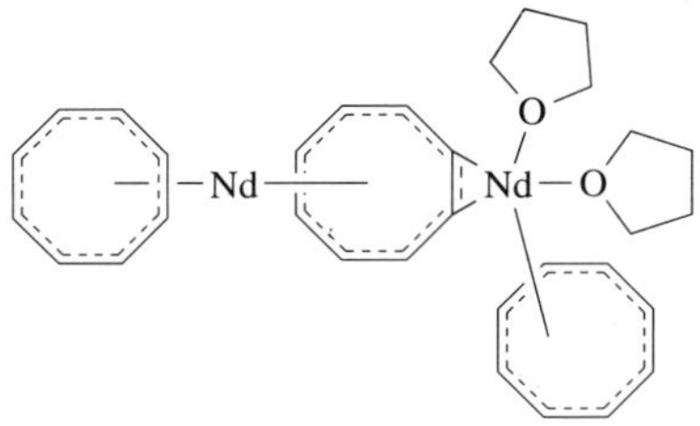

(**191**)

The Ce, La and Sm compounds $(C_5Me_5)_2M\text{—}C\equiv CR$ (R = Me or Bu^t) dimerize at room temperature to molecules with structures (**192**) having two bridging carbons ⟨93OM2609, 93OM2618⟩.

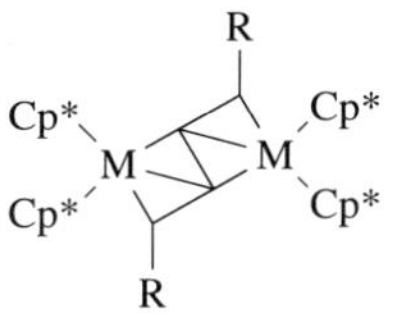

(**192**)

4.16
Functions Containing Two Atoms of Different Metallic Elements

WILLIAM J. KERR and PETER L. PAUSON
University of Strathclyde, Glasgow, UK

4.16.1	FUNCTIONS CONTAINING AT LEAST ONE GROUP 1 METAL	705
4.16.2	FUNCTIONS CONTAINING AT LEAST ONE GROUP 2 METAL (AND NO GROUP 1 METALS)	709
4.16.3	FUNCTIONS CONTAINING AT LEAST ONE TRANSITION METAL (AND NO GROUP 1 OR 2 METALS)	709
	4.16.3.1 Two Different Transition Metals	709
	4.16.3.1.1 Two 'genuine' transition metals (Ti–Pt)	709
	4.16.3.1.2 A 'genuine' transition metal linked to a 'late' transition (i.e. group 11 (Cu, Ag, Au) or 12 (Zn, Cd, Hg)) metal	722
	4.16.3.1.3 Two 'late' transition metals	724
	4.16.3.2 A Transition Metal and a Group 13 or 14 Metal	724
	4.16.3.3 A Transition Metal and Other Metals	726
4.16.4	FUNCTIONS CONTAINING AT LEAST ONE GROUP 13 METAL (AND NO GROUP 1, 2 OR TRANSITION METAL)	726

4.16.1 FUNCTIONS CONTAINING AT LEAST ONE GROUP 1 METAL

(*Note*: Compounds in which carbon is formally pentacovalent are included here as they do not fit into any other chapter. The following abbreviation is used in this chapter: Fc = ferrocenyl $[(C_5H_5)Fe(C_5H_4)]$.)

No examples involving two different group 1 metals or one group 1 and one group 2 metal have been found.

Compounds containing a group 1 and a transition metal arise by metallation of π-complexes of the transition metal, most commonly with butyllithium as metallating agent. They are rarely isolated in pure form, but are of great value as intermediates. Typical examples are the mono- and 1,1'-dilithio derivatives of ferrocene, (**1**) and (**2**), and (lithiobenzene)tricarbonylchromium (**3**).

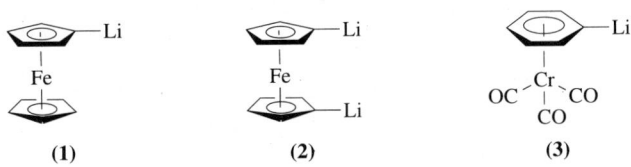

(**1**)   (**2**)   (**3**)

In the case of ferrocene many variations in reaction conditions have been examined with a view to controlling the extent of lithiation. *n*-Butyllithium alone (in a variety of solvents) always gives mixtures of mono- and dilithio derivatives (Equation (1)). Replacement of *n*- by *t*-butyllithium in THF gives predominantly (mono-)lithioferrocene (**1**) ⟨90TL3121⟩, whereas the reaction of *n*-butyllithium in the presence of TMEDA yields, even with limiting amounts of metallating agent, chiefly the 1,1′-dilithio compound (**2**) ⟨73JOM(51)1⟩. Smooth dilithiation has also been achieved with *t*-butyllithium in pentane ⟨84JA2337⟩. TMEDA and $(Me_2NCH_2CH_2)_2NMe$ adducts of 1,1′-dilithioferrocene have been isolated and characterised by x-ray crystallography ⟨78JA6382, 85OM2196⟩. The reaction of ferrocene with pentylsodium, preferably with addition of TMEDA, also leads to predominant dimetallation ⟨57DOK(112)439, 58DOK(120)1263, 78JOM(161)C17⟩.

$$Cp_2Fe + BuLi \longrightarrow C_4H_{10} + CpFe(C_5H_4Li) + Fe(C_5H_4Li)_2 \quad (1)$$

The reaction of bromo- (or iodo)ferrocene with butyllithium (Equation (2)) provides a good alternative route to the monosubstituted compound (**1**) ⟨69TL4011⟩. Transmetallation of (chloromercuri)ferrocene (Equation 3) ⟨62IC227⟩ or diferrocenylmercury ⟨62IC414⟩ is only slightly less efficient. Using a combination of these two methods, Roling and Rausch ⟨74JOC1420⟩ have prepared 1,2-dilithioferrocene by treating bis(2-iodoferrocenyl)mercury with butyllithium.

$$FcBr + BuLi \longrightarrow BuBr + FcLi \quad (2)$$

$$FcHgCl + BuLi \longrightarrow FcLi \quad (3)$$

In contrast to bromo- and iodoferrocenes (Equation (2)), chloroferrocene is metallated α to chlorine according to Equation (4) ⟨69TL4011; cf. also 65JOC1600⟩ and 1,1′-dichloroferrocene is similarly monolithiated ⟨79JOM(181)425⟩ or, in the presence of TMEDA, dilithiated ⟨79JOM(181)425⟩. Like chlorine, other polar substituents, notably those which can chelate the metal ($CHRNMe_2$, CHROR′ etc.), direct lithiation largely or wholly to the 2-position, as, for example, in Equation (5) ⟨65JA1241, 73JOM(51)1⟩ and Equation (6) ⟨68JCS(C)656, 70IZV2133⟩.

$$FcCl + BuLi \xrightarrow[85\%]{THF} C_4H_{10} + \text{[Fe(C}_5\text{H}_4\text{)(C}_5\text{H}_3\text{ClLi)]} \quad (4)$$

$$FcCH_2NMe_2 + BuLi \longrightarrow \text{[2-lithio-(dimethylaminomethyl)ferrocene, Li---NMe}_2\text{]} \quad (5)$$

$$\text{Fc-(2-pyridyl)} + BuLi \longrightarrow \text{[2-lithio-(2-pyridyl)ferrocene, Li----N]} \quad (6)$$

Analogous lithiation with *n*-butyllithium has been employed to convert ruthenocene to mono- ⟨72IZV1823⟩ and 1,1′-dilithio derivatives ⟨60JA76, 93JOM(463)163⟩, tricarbonylcyclopentadienylmanganese and -rhenium to the corresponding lithio derivatives, $(C_5H_4Li)M(CO)_3$ where M = Mn or Re ⟨68IZV686, 88OM2566⟩ and $(C_4Ph_4)Co(C_5H_4HgCl)$ albeit inefficiently to $(C_4Ph_4)Co(C_5H_4Li)$ ⟨70JOC3888⟩; tricarbonylcyclobutadieneiron reacted in more complex fashion (probably via attack at CO), but $(OC)_3Fe(C_4H_3HgCl)$ with BuLi gave smooth conversion to $(OC)_3Fe(C_4H_3Li)$ ⟨68PAC235, 77CRV691⟩. Methyllithium was used for the 2-lithiation of $(OC)_3Mn(C_5H_4SO_2NMe_2)$ ⟨70CC555⟩ and the chelation effect is seen to be equally strong in the lithiation of the dimethylaminomethyl- (Scheme 1) ⟨79JOM(168)C33⟩ and hydroxymethyl-substituted manganese compounds (Scheme 2) ⟨81JOM(209)233⟩. In both of these cases the isotope effect is sufficient to lead a second metallation step predominantly into the 5-position of the 2-deuterio compounds and, to a remarkable extent, even into the 3-position (i.e. the β-position) when both 2- and 5-positions are deuteriated (Scheme 1). Steric hindrance must be responsible for the predominant 3-lithiation of the diphenylphosphido-cyclopentadienyl compound (Equation (7)) ⟨81JOM(209)233⟩.

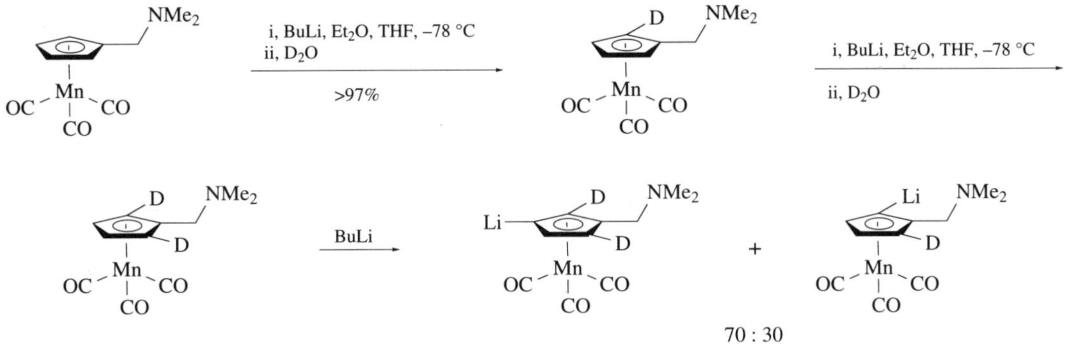

Scheme 1

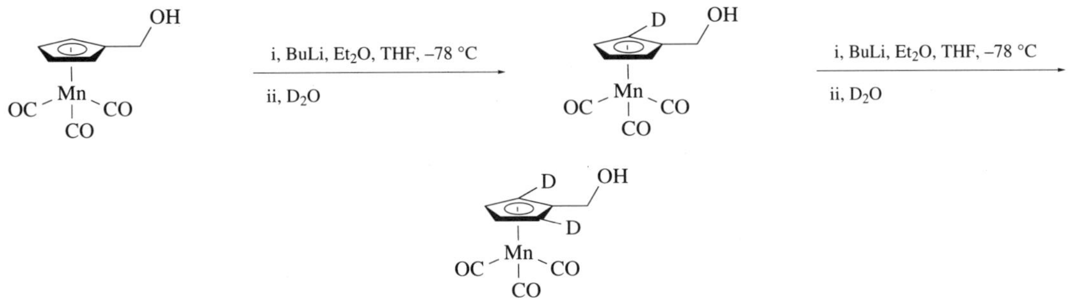

Scheme 2

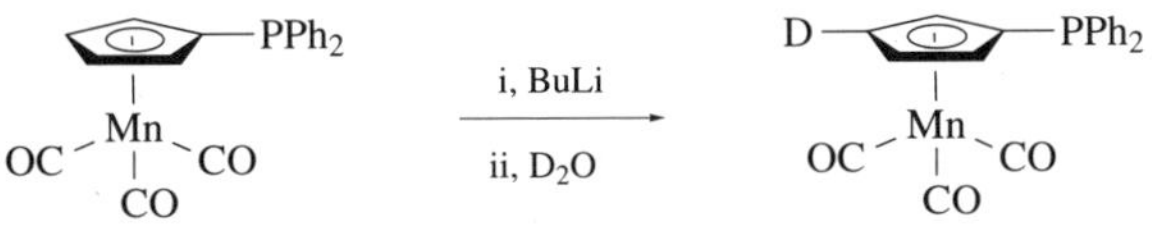

Halogen–lithium exchange using butyllithium allowed the smooth synthesis of $(OC)_3Mn(C_5Cl_4Li)$ from $(OC)_3Mn(C_5Cl_4Br)$ ⟨88CB799, 93JOM(458)181⟩. After treatment of the product with, for example dimethyl disulfide to yield $(OC)_3Mn(C_5Cl_4SMe)$ further halide replacement occurs preferentially at the 3-position (Equation (8)) and repetition of this sequence allows complete halide replacement to yield $(OC)_3Mn[C_5(SMe)_5]$ ⟨88AG(E)939⟩. Butyllithium also replaced one chlorine in (pentachlorocyclopentadienyl)(1,5-cyclooctadiene)rhodium ⟨89JOM(368)67⟩.

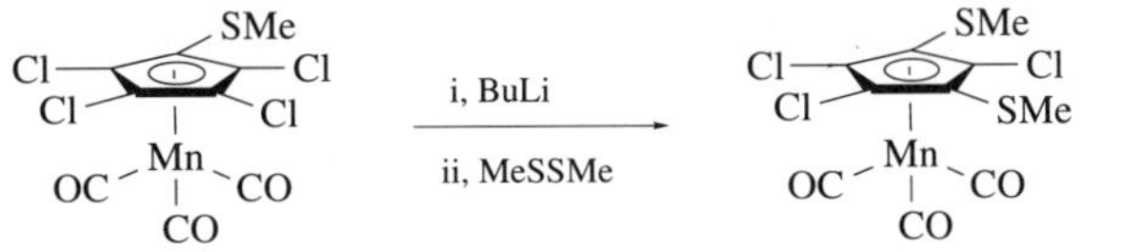

In the arenechromium series, benzenetricarbonylchromium is most efficiently lithiated by butyllithium in THF at $-78\,°C$ in the presence of TMEDA, yielding $(OC)_3Cr(C_6H_5Li)$ ⟨78JOM(153)59, 79JA768⟩. Tricarbonylthiophenechromium, like the free arene, is metallated in the α-position (Equation (9)) ⟨83JOM(244)C21⟩ and $(C_6H_6)Cr(C_6F_5H)$ is selectively lithiated in the fluorinated ring to give $(C_6H_6)Cr(C_6F_5Li)$ ⟨77JOM(141)85⟩. Polar substituents, X in $(OC)_3Cr(C_6H_5X)$, direct lithiation specifically into the *ortho*-position, the anisole (X = OMe) and chlorobenzene (X = Cl) complexes reacting smoothly with BuLi at $-35\,°C$ and the fluorobenzene complex (X = F) at $-78\,°C$. The efficiency of some of these reactions is illustrated by the example of Equation (10) ⟨79JA768⟩.

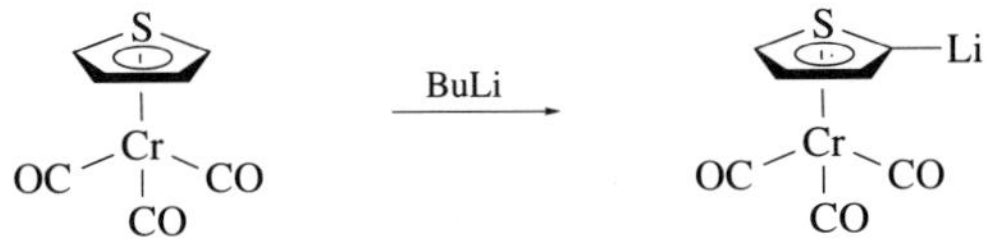

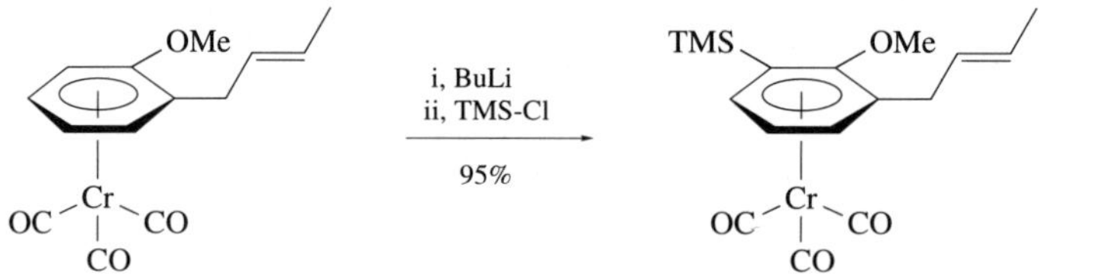

The addition of two equivalents of 2-lithio(dimethylaminomethyl)benzene to silver bromide yields the dilithio–disilver complex (**4**) ⟨73JOM(55)419⟩.

A complex containing chromium atoms linked to lithium by bridging methyl groups is formed on treating chromium(III) chloride with methyllithium in ether followed by dioxane (Equation (11)) ⟨65JOM(4)114⟩.

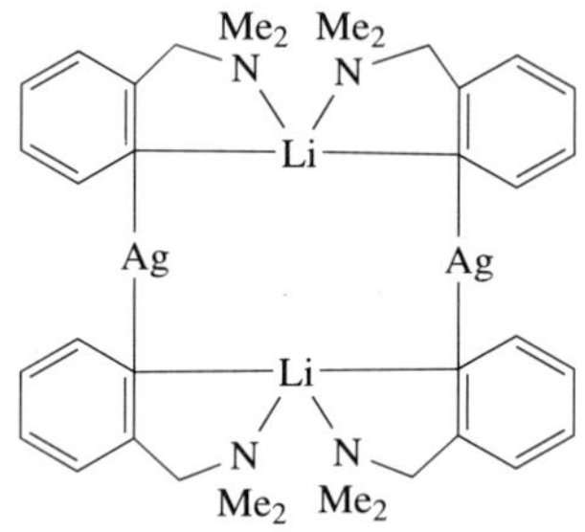

Examples of group 1 metals linked to group 13 or 14 metals via aliphatic (e.g. CH_2) groups are known for lithium: (trimethylstannylmethyl)lithium is obtained from the corresponding ditin compound (Equation (12)) ⟨93AG(E)105⟩ or from Me_3SnCH_2I ⟨82CB1810, 84ZN(B)798⟩. Analogous methods have been employed to prepare other trialkyl- or triaryltin ⟨82CB1810, 86TL4339⟩ or -lead compounds, for example Ph_3PbCH_2Li from either $(Ph_3Pb)_2CH_2$ or Ph_3PbCH_2I in quantitative yield ⟨85CB370⟩.

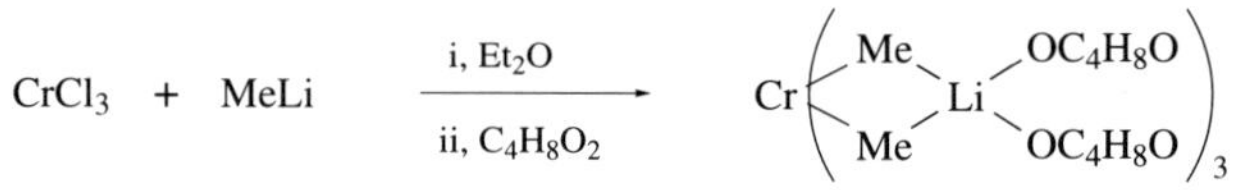

Methylenedialuminum tetrabromide reacts with butyllithium according to Equation (13) ⟨74JOM(72)C4⟩. That the structure of this product may be much more complex is suggested by closely related examples (Scheme 3) ⟨91CB1511⟩.

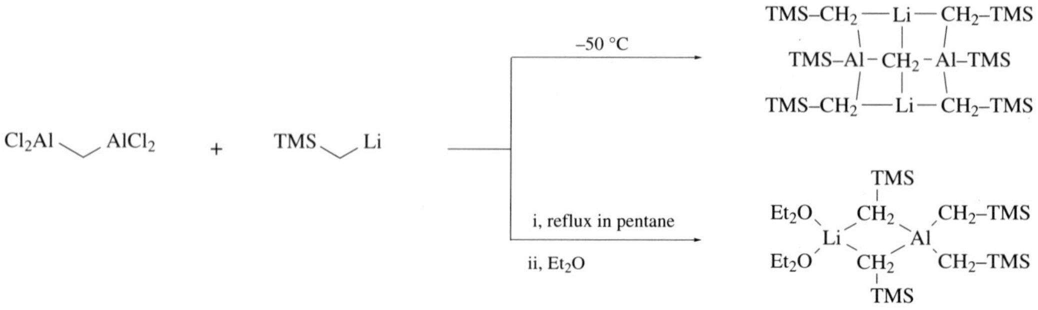

Scheme 3

4.16.2 FUNCTIONS CONTAINING AT LEAST ONE GROUP 2 METAL (AND NO GROUP 1 METALS)

The typical aromatic Grignard reagent, ferrocenylmagnesium bromide is generated from bromoferrocene in the usual manner and used without isolation ⟨61JOC1034, 74JOM(78)405⟩; the 1,1′-di Grignard reagent has been obtained similarly ⟨61JOC1034⟩ or from disodioferrocene with magnesium bromide ⟨62IC227⟩.

Magnesium–zinc compounds are regarded as the intermediates in the addition of allylzinc bromide to vinylmagnesium, vinyllithium or vinylaluminum compounds (Scheme 4) and subsequent reactions with electrophiles ⟨71CR(C)1669, 86TL1039, 90JOC5446, 93S530⟩.

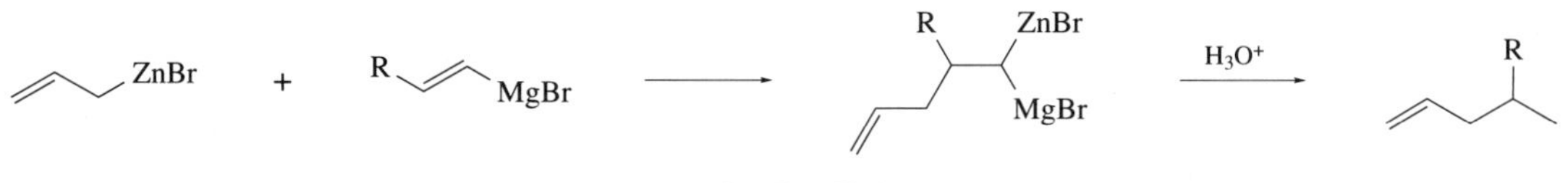

Scheme 4

The trimethylstannylmethyl halides Me_3SnCH_2X react with magnesium under normal conditions for forming Grignard reagents (Equation (14)) ⟨59JA1844, 81JCS(D)1593⟩.

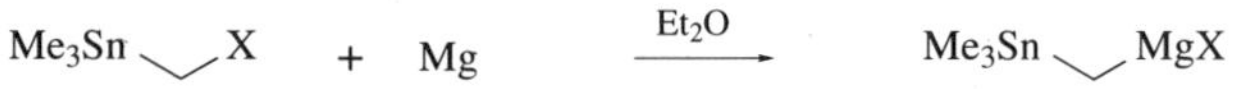

$$Me_3Sn{\frown}X + Mg \xrightarrow{Et_2O} Me_3Sn{\frown}MgX \quad (14)$$

Dimethylmagnesium and -beryllium add to trimethylaluminum. In the former case, the structure (5) of the principal product has been confirmed by x-ray crystallography ⟨69JA2538⟩. The ethyl analogue, Et_8Al_2Mg is also known ⟨57LA(605)93⟩.

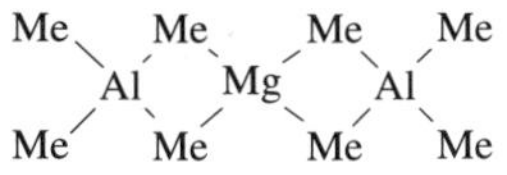

(5)

4.16.3 FUNCTIONS CONTAINING AT LEAST ONE TRANSITION METAL (AND NO GROUP 1 OR 2 METALS)

4.16.3.1 Two Different Transition Metals

4.16.3.1.1 Two 'genuine' transition metals (Ti–Pt)

(i) Formation by the diazoalkane method and other carbene routes

Several 'carbene' routes to obtain 1-carbon-bridged metal systems, both with the same (see chapter 4.15) and different metals, have been explored and reviewed ⟨82AOC(20)159⟩. Of these, the uses of diazoalkanes as carbene sources have proved most generally applicable; they may involve replacement of labile ligands (e.g. CO) and/or addition to multiple metal–metal bonds as shown in, for example Equation (15), where M^1 = Co, Rh, Ir; M^2 = Rh, Ir; R^1, R^2 = H or Me; $R^3 = R^4$ = H, Me or R^3 = H; $R^4 = CO_2Et$ or $R^3R^4C = (CH_2)_5C$ or cyclopentadienylidene, and where some of the diazoalkanes are most conveniently generated *in situ* from $R^3R^4C{=}N \cdot NH_2 + MnO_2$ ⟨87JOM(321)257⟩. Analogous methylene addition occurs with coordinatively unsaturated nickel–molybdenum and –tungsten complexes (Equation (16)) ⟨90OM1345, 92OM2128⟩.

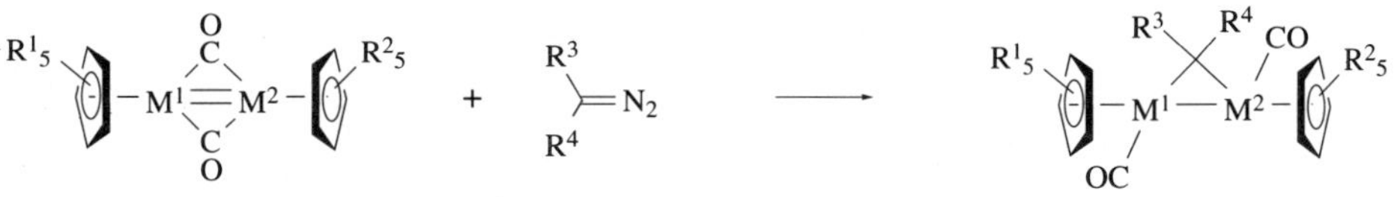

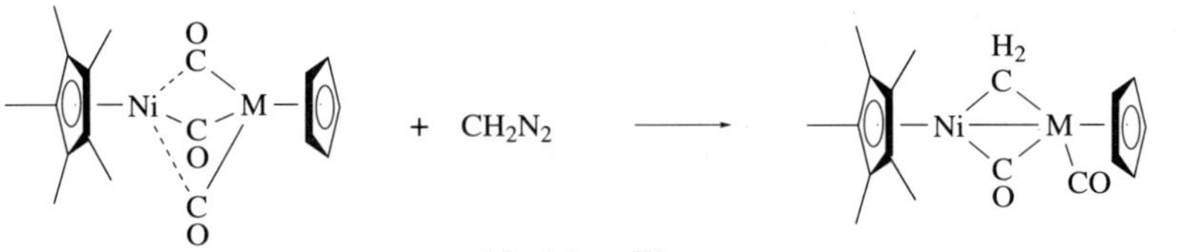

Carbene generation by ring opening of 3,3-dimethylcyclopropene appears to be involved in the reaction shown in Equation (17) and those of analogous Rh–Ir and Co–Ir complexes ⟨85JCS(D)2483⟩.

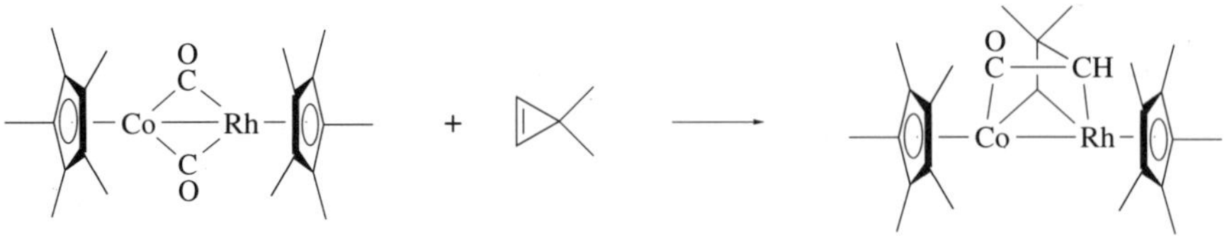

Mononuclear metal carbene complexes may add coordinatively unsaturated fragments as in the examples of Equations (18) and (19) ⟨92ICA377, 93JA2743, 93OM65, 94OM2668⟩. In a closely similar reaction, $Cp_2Ta(=CH_2)CH_3$ reacts with $CpPd(\eta^3-C_3H_5)$ with loss of propene to give $Cp_2Ta(\mu-CH_2)_2PdCp$ ⟨93OM4269, 94OM1899⟩.

The complex (**6**) ⟨82OM1607⟩, but not its synthesis, has been described; it was probably prepared by the reaction of $CpFe(CO)_2CH_2OAc$ with $Mn(CO)_5^-$ by analogy with the synthesis of $\{[CpFe(CO)]_2(\mu-CO)(\mu-CH_2)\}$; if this is an S_N1 reaction, the intermediate is the cationic carbene complex $[CpFe(CO)_2=CH_2]^+$.

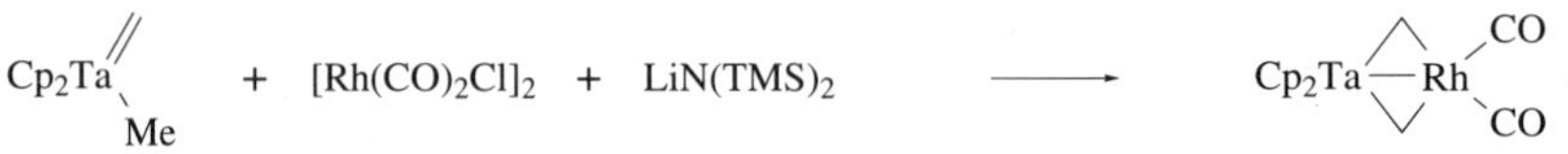

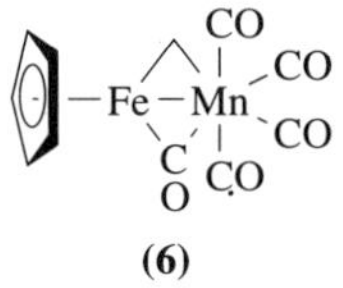

(**6**)

(ii) Formation from metal acetylides, metal alkyne complexes and free alkynes (or allenes)

The cationic molybdenum alkyne complexes, $[CpMo(CO)(L)(RC_2R)]^+$ add pentacarbonylrhenate (Equation (20)) ⟨86AG(E)734⟩ and, less efficiently, the related anions $Mn(CO)_5^-$ and $CpW(CO)_3^-$ ⟨89CB1901, 93AG(E)923⟩.

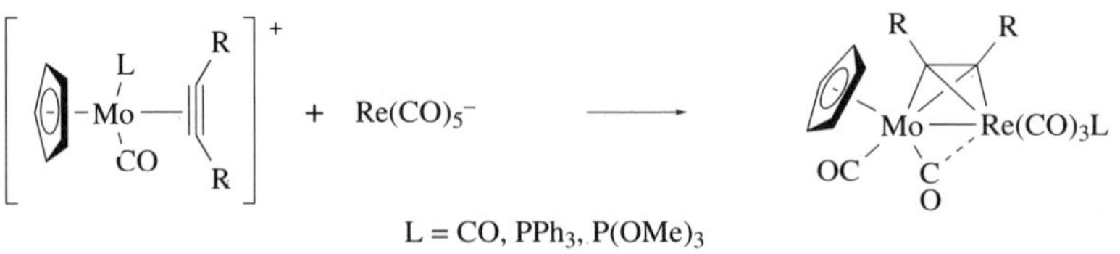

L = CO, PPh_3, $P(OMe)_3$

The reaction of tetracarbonylethyneosmium with carbonyl (η^5-cyclopentadienyl) phosphinerhodium proceeds according to Equation (21) when the phosphine is PMe_3 yielding a zwitterionic product, but according to Equation (22) involving CO–alkyne coupling when the PMe_2Ph or $PMePh_2$ complexes are used ⟨94OM1078⟩.

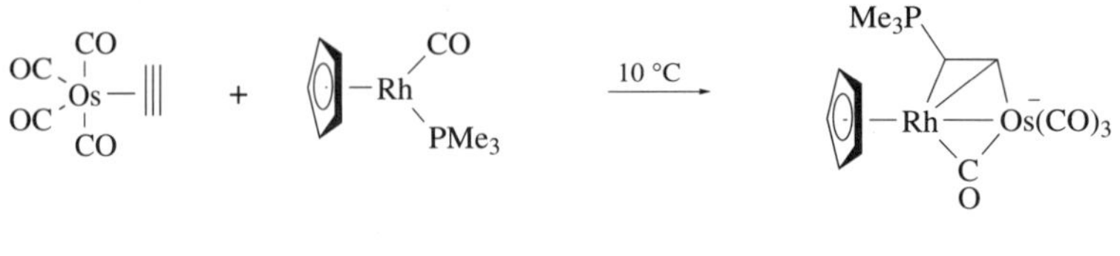

$$\text{(22)}$$

Depending on reaction conditions, the propargylruthenium complex, $Cp(CO)_2RuCH_2CCPh$, reacts with enneacarbonyldiiron to yield several trinuclear complexes in addition to the dinuclear Fe–Ru compound of Equation (23) ⟨94OM1999⟩. Metal migration accompanies the reactions of propargylic tungsten (Equation (24)) and manganese complexes (Equation (25)) with $Fe_2(CO)_9$ and $Co_2(CO)_8$, respectively ⟨93OM108⟩.

$$\text{(23)}$$

$$\text{(24)}$$

$$\text{(25)}$$

The alkyne-bridged hexacarbonyldicobalt complexes can add iron or ruthenium carbonyls to give bridged trinuclear mixed-metal complexes (e.g. Equation (26)) ⟨90CB661⟩. Rhodium and iridium/iron analogues are among the products formed from rhodium and iridium acetylides ⟨91JOM(407)391⟩. An anionic vinyliron complex reacts with cobalt carbonyl according to Equation (27) ⟨86JOM(315)C22⟩.

The products obtained from a phosphido–Ni/Fe complex with alkynes were found to have added the PPh_2 group to the alkyne moiety (e.g. Equation (28)) ⟨72JOM(35)367, 73CSC347⟩. Similar alkyne insertions have been observed with mixed Ru/Co complexes ⟨90OM2234⟩.

$$\text{(26)}$$

$$[Fe_2(CO)_6(\mu\text{-}CO)(\mu\text{-}CH\text{=}CHPh)]^- + Co_2(CO)_8 \longrightarrow \quad \text{(27)}$$

$$\text{(28)}$$

The mixed cyclopentadienylmolybdenum or -tungsten–cobalt carbonyls react with alkynic esters as shown in Equation (29) ⟨92JOM(427)335⟩, while the 'carbyne-bridged' Mo–Fe complexes add various alkynes to give *inter alia* the three-carbon bridged products (Equation (30)) ⟨88JCS(D)2431⟩. Analogous additions are known for tungsten–rhodium and –cobalt complexes (Equation (31)) ⟨81CC867, 81CC1255, 84JCS(D)2553⟩ while the trinuclear complex $CpWFe_2(CO)_8(\mu_3\text{-}CAr)$ (Ar = 4-MeC_6H_4) reacts with alkynes to give the same type of product (**7**; [M] = $Fe(CO)_3$) with loss of one iron atom ⟨81CC867⟩. Protonation of this type of complex (**7**; $R^1 = R^2$ = Me; [M] = $Co(C_5Me_5)$) is accompanied by rearrangement (Scheme 5), the same final product being obtained if protonation precedes alkyne addition ⟨81CC1255, 84JCS(D)1581⟩. The protonation step leading to the M—CH-Ar—M′ bridge has also been reported for a tungsten–rhodium complex, a tungsten–chromium complex (with [M] = Cr(CO)(NO)Cp) and a tungsten–platinum complex. In the last case hydride, instead of reversing the protonation, displaces CO to give a hydrogen-bridged complex (Scheme 6) ⟨81CC1255, 82OM1597, 84JCS(D)1563⟩. The alkyne addition step of Scheme (5) finds a parallel in the addition of butyne to the methylene tungsten–nickel complex ⟨90OM1345⟩, whereas a different structural type results with phenylethyne (Scheme 7) ⟨91OM3003⟩.

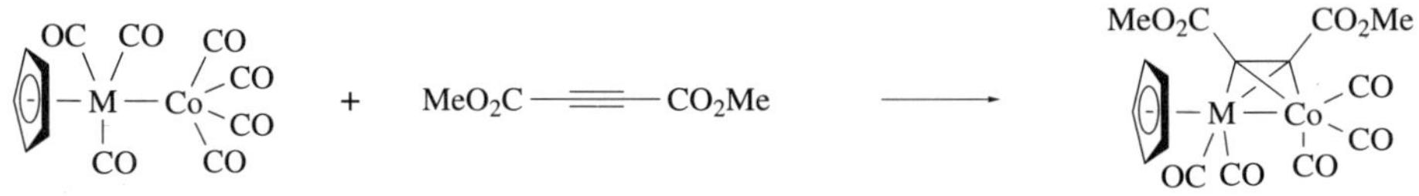

(29)

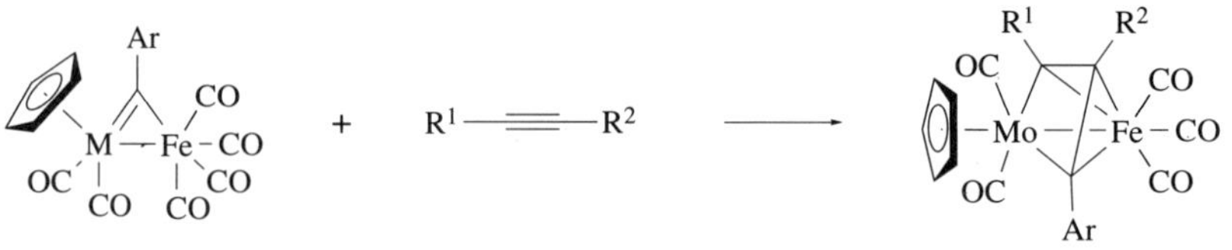

(30)

Ar = 4-MeC_6H_4

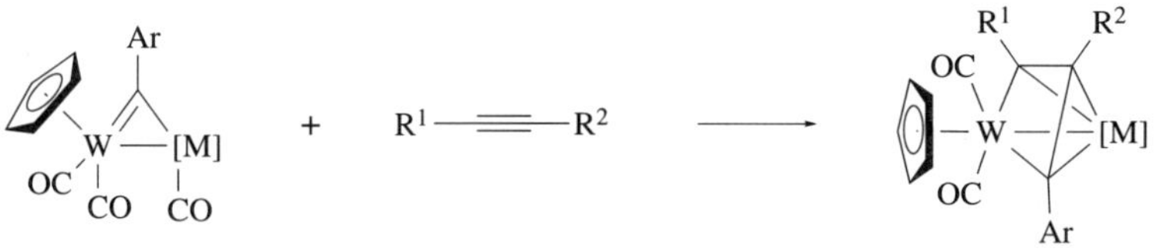

(31)

(**7**)

Ar = 4-MeC_6H_4; [M] = $Rh(C_9H_7)$ or $Co(C_5Me_5)$

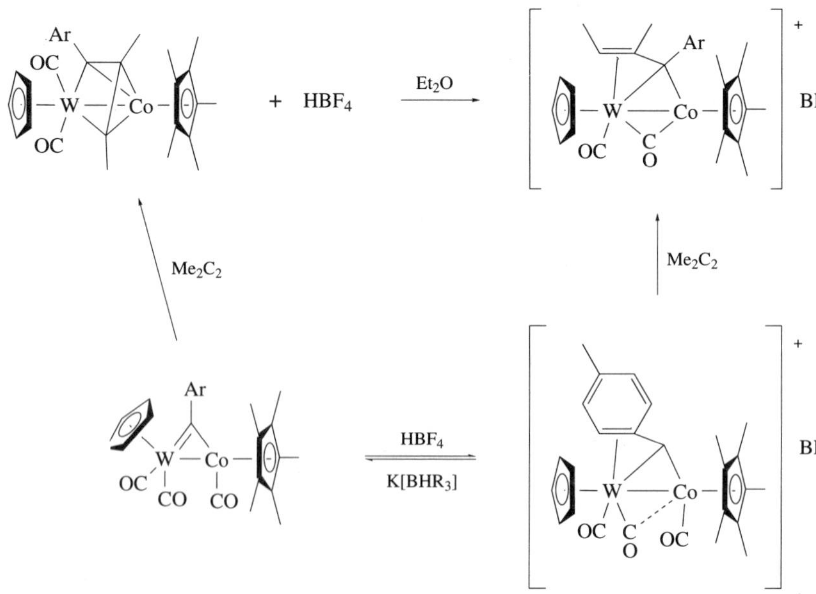

Scheme 5

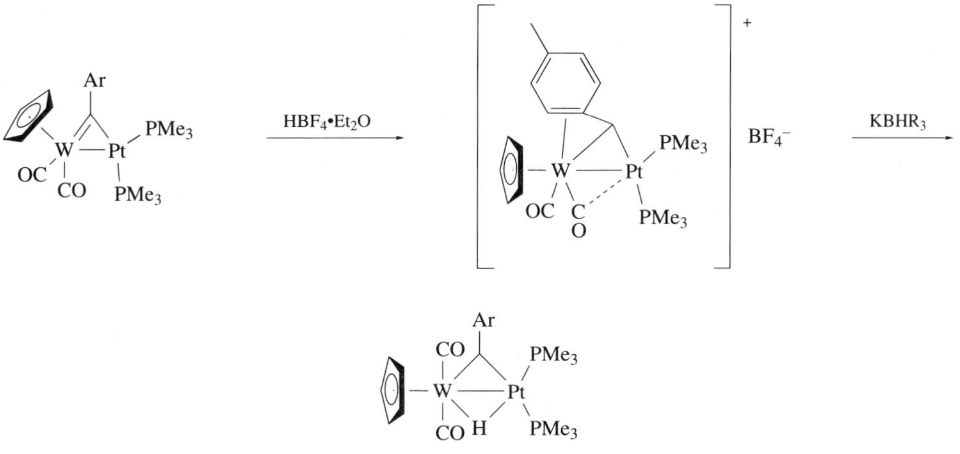

Scheme 6

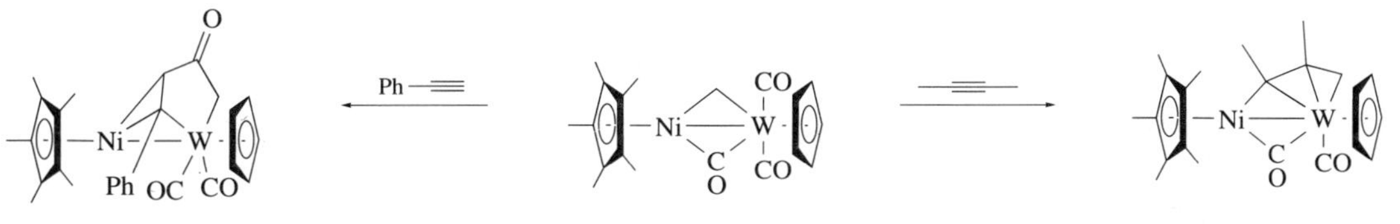

Scheme 7

Bridged nickel–cobalt alkyne complexes are formed from pentacarbonylcyclopentadienylcobalt-nickel or nickelocene plus octacarbonyldicobalt with alkynes (Scheme 8) or by metal exchange processes from the corresponding symmetrical alkyne complexes (Scheme 9). Alkyne-bridged cyclopentadienylnickel–cyclopentadienylmolybdenum complexes are obtained similarly, for instance from $(R^1C_5H_4)Ni(CO)_4Mo(C_5H_4R^2)$ ⟨80IC693, 82OM225, 88OM2450; cf. also 90IC1295⟩.

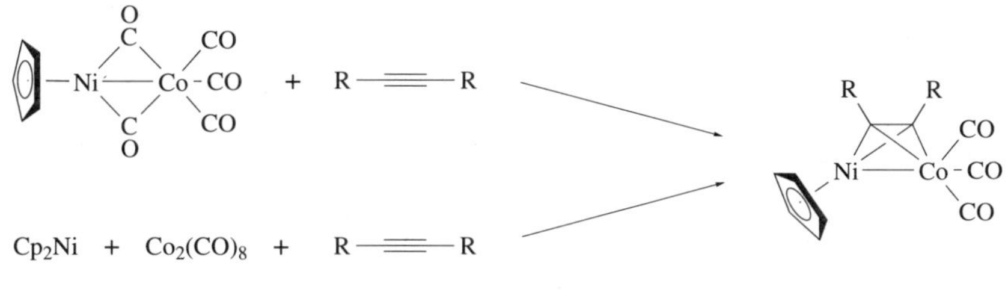

Scheme 8

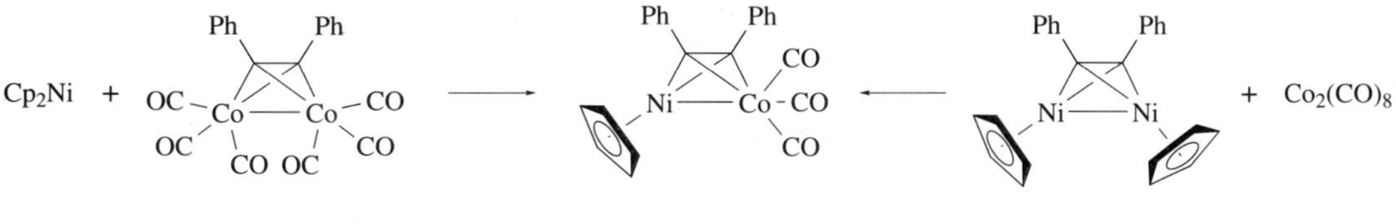

Scheme 9

Allenes can react analogously as in the example shown in Scheme 10 ⟨89OM2077⟩.

The symmetrical dinickel–alkyne and mixed nickel–cobalt complexes can add an iron carbonyl fragment as in Equation (32) ⟨82JOM(240)299; cf. also 80JOM(197)335, 85OM1123, 91OM1907⟩ and Equation (33) ⟨82CC371⟩ and a ruthenium tricarbonyl group can be added similarly ⟨81JOM(221)93⟩. Iron–nickel complexes of related structure are formed from alkynes on reaction with $CpNi(PPh_3)Br$ and the anion $HFe(CO)_4^-$ (giving intermediate acryloyl complexes $[Fe(R^1CH=CR^2CO)(CO)_3]^+$) (e.g. Equation (34)) ⟨90JOM(395)305⟩.

The phenylethyne addition shown in Equation (35) is also accompanied by substantial rearrangement including cleavage of the Mo—P bond ⟨93JOM(447)C1⟩.

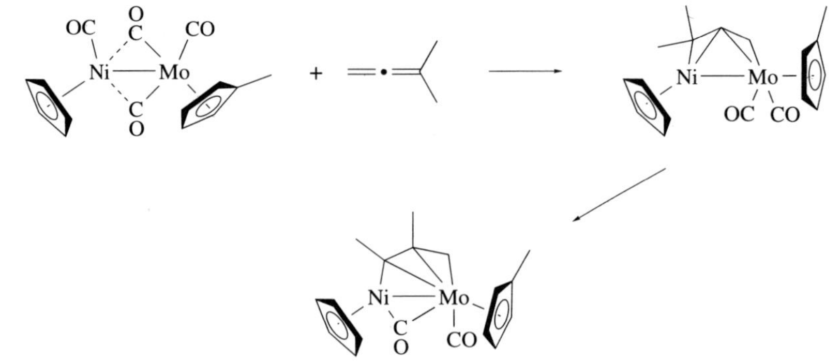

Scheme 10

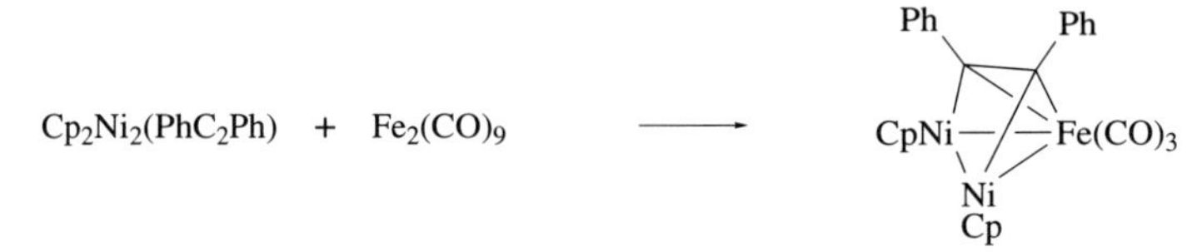

$$Cp_2Ni_2(PhC_2Ph) + Fe_2(CO)_9 \longrightarrow [\text{CpNi—Fe(CO)}_3 \text{ cluster}] \quad (32)$$

$$[\text{CpNi—Co(CO)}_3 \text{ with Ph, Ph}] + Fe_2(CO)_9 \longrightarrow [\text{CpNi—Fe—Co cluster}] \quad (33)$$

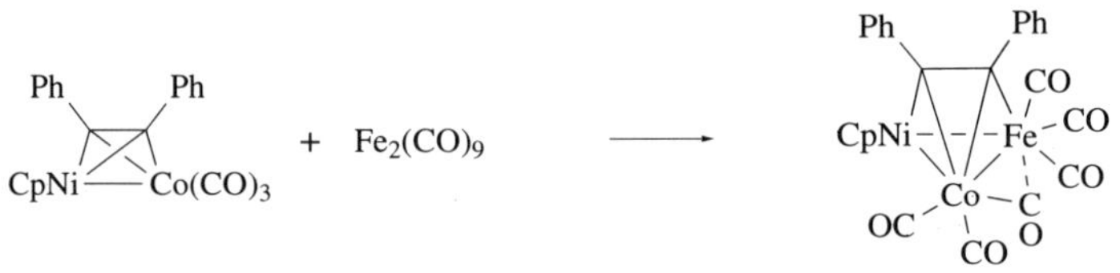

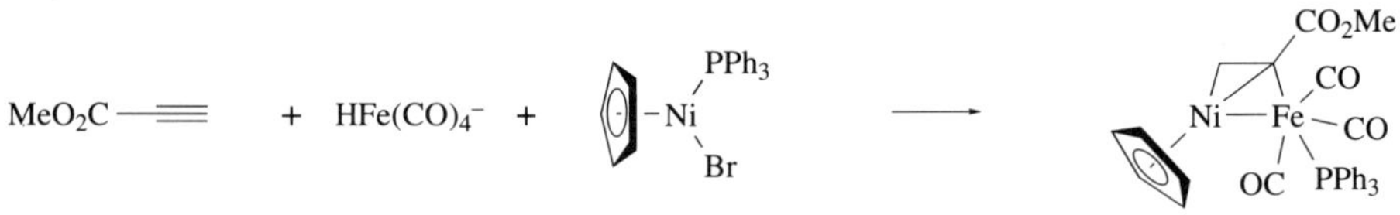

$$MeO_2C\!\!\equiv\!\!\equiv + HFe(CO)_4^- + [\text{CpNi(PPh}_3\text{)Br}] \longrightarrow [\text{Ni—Fe product}] \quad (34)$$

$$[\text{Fe–Mo complex with P(OAr), vinyl}] + Ph\!\!\equiv\!\!\equiv \longrightarrow [\text{rearranged Fe–Mo product}] \quad (35)$$

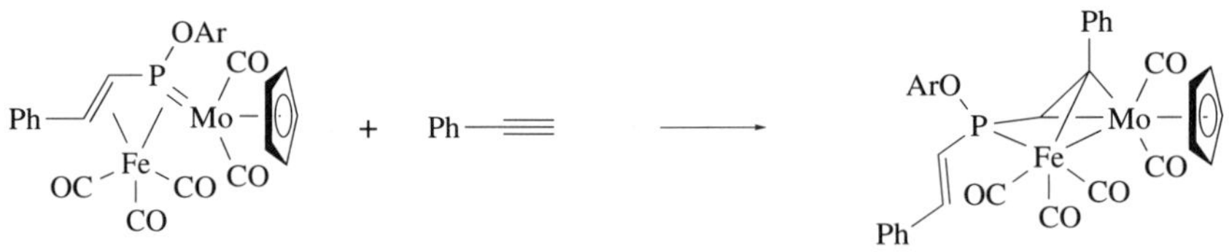

Ar = 2,4,6-Bu$^t_3$C$_6$H$_2$

Several alkyne-derived cobaltacyclopentadiene complexes have been shown to add tricarbonyliron groups on treatment with nonacarbonyldiiron (Equation (36)) ⟨83OM726⟩. The transformation of a cobalta- into a ferracyclopentadiene in this process finds analogy in the facile equilibration of the related diiron complexes (Equation (37), [M^1] = [M^2] = Fe(CO)$_3$). Tungstacyclopentadiene–cobalt complexes, for example (**8**) form with disruption of tetranuclear Co—W clusters on addition of, for example phenylethyne ⟨91IC4710⟩ and both molybda- and tungstacyclopentadiene–cobalt complexes of this type are also generated when the preformed bis(hexafluorobutyne) complex of the metal reacts with octacarbonyldicobalt (Equation (38)) ⟨80CC749⟩.

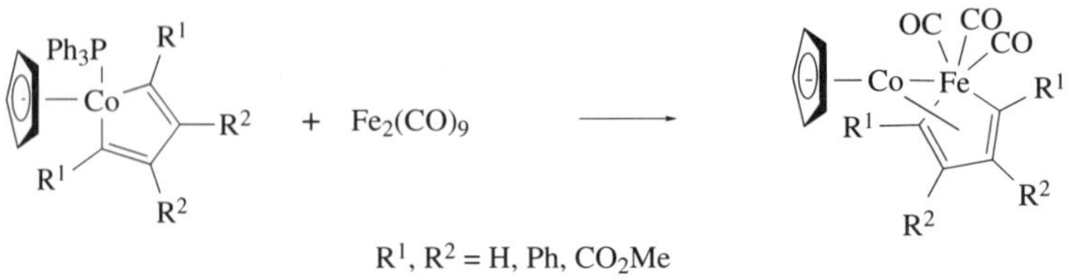

$$[\text{Cp(Ph}_3\text{P)Co cyclopentadiene with }R^1, R^2] + Fe_2(CO)_9 \longrightarrow [\text{Cp-Co-Fe product}] \quad (36)$$

R^1, R^2 = H, Ph, CO$_2$Me

$$\text{(37)}$$

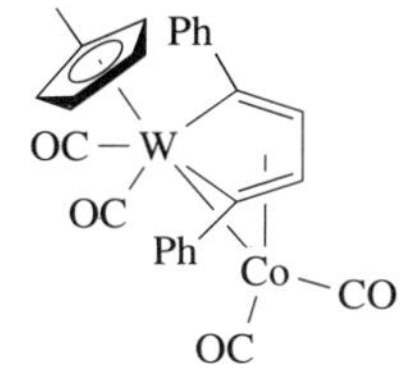

(8)

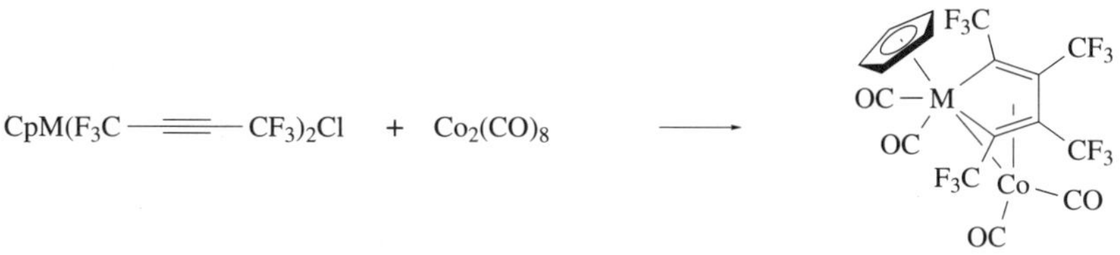

$$\text{CpM(F}_3\text{C}\equiv\text{CF}_3)_2\text{Cl} + \text{Co}_2\text{(CO)}_8 \longrightarrow \quad (38)$$

M = Mo or W

A more complex transformation involving both transfer and ring-opening of a tetramethylcyclobutadiene ligand is involved in the formation, albeit in low yield, of a ferracyclopentadiene–nickel complex from the 2-butyne derived dimeric $(C_4Me_4)NiCl_2$ (Equation (39)) ⟨67CJC2011, 70JA502⟩.

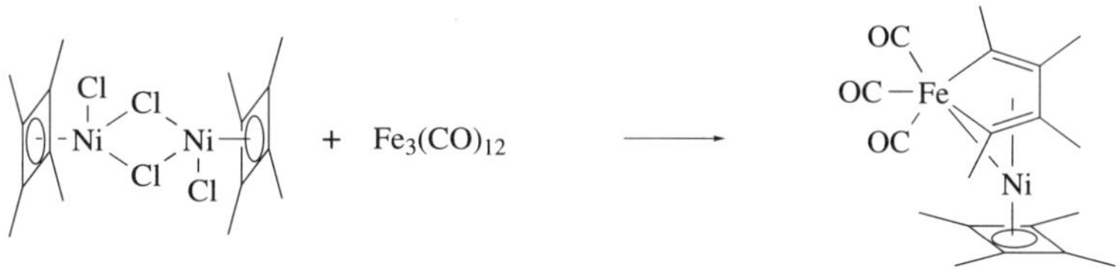

$$+ \text{Fe}_3(\text{CO})_{12} \longrightarrow \quad (39)$$

(iii) Formation by miscellaneous addition and transition metal exchange reactions

In addition to examples cited in the preceding section which involve alkyne or alkyne-derived ligands, various similar processes involve addition of organometallic fragments to other organometallic species. Exchange reactions whereby one metal in a homobinuclear complex is replaced by a fragment containing a different metal are included here as they may well involve initial addition.

Simple additions occur when, for example a tricarbonylchromium fragment is added to a metal-bound phenyl group. Thus, the dicarbonylcyclopentadienyl(substituted phenyl)iron complexes (R = Ph or 4-$C_6H_4Fe(CO)_2Cp$) react with hexacarbonylchromium according to Equation (40) ⟨92OM864⟩. This reaction has been applied to several closely related complexes ⟨89OM1118, 90CJC41⟩ while other examples of the same product class have been made by nucleophilic substitution of haloarenechromium complexes as described below (see (vi) below).

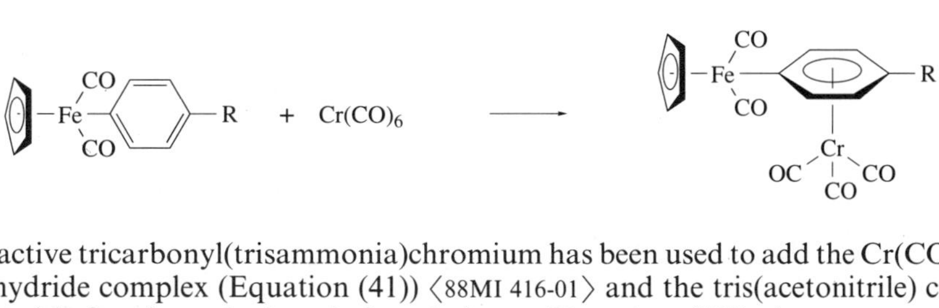

$$+ \text{Cr(CO)}_6 \longrightarrow \quad (40)$$

The more reactive tricarbonyl(trisammonia)chromium has been used to add the $Cr(CO)_3$ fragment to a tungsten hydride complex (Equation (41)) ⟨88MI 416-01⟩ and the tris(acetonitrile) complex was similarly used with fluorine-substituted titanaindenes (Equation (42)) ⟨93JOM(463)C6⟩ and with bis(diene)phenyliridium ⟨94JOM(471)249⟩.

Molybdenum and tungsten carbyne complexes insert into the Pd—C bond on adding to cyclopalladated N,N-dimethylbenzylamine (Equation (43)) ⟨91CC1274⟩. The same carbyne tungsten complexes add the $Cp_2Ti=CH_2$ fragment derived by loss of Me_2AlCl from the 'Tebbe reagent' (Equation (44)) ⟨85JCS(D)2009⟩. Addition of a (cod)Pt fragment to a ketene–tungsten complex proceeds according to Equation (45) ⟨82OM1597⟩.

The cationic rhenium carbyne complex [CpRe(CO)$_2$CPh]BBr$_4$ on reaction with [Fe$_2$(CO)$_8$]$^{2-}$ gives *inter alia* complex (**9**) ⟨93OM1213⟩ and carbene complexes CpRe(CO)$_2$=CHR add zirconocene chlorohydride to yield the bridged products, for example (**10**) ⟨90JOM(387)C31⟩.

Simple additions of organometallic anions to a cationic ruthenium enimine complex yield the mixed Ru–Co and Ru–Fe complexes shown in Scheme 11 ⟨93OM315; cf. 92ICA689⟩. Related diruthenium compounds yield Ru–Fe analogues by metal exchange with Fe$_2$(CO)$_9$ ⟨93OM3187⟩.

The anion [CpRe(CO)$_2$CH$_3$]$^-$ adds to the vinylplatinum complex shown in Equation (46), presumably by initial substitution of triflate followed by loss of propene ⟨92ICA557⟩.

The rhodium bis(triflate) complex shown adds ethylenebis(triphenylphosphine)platinum with

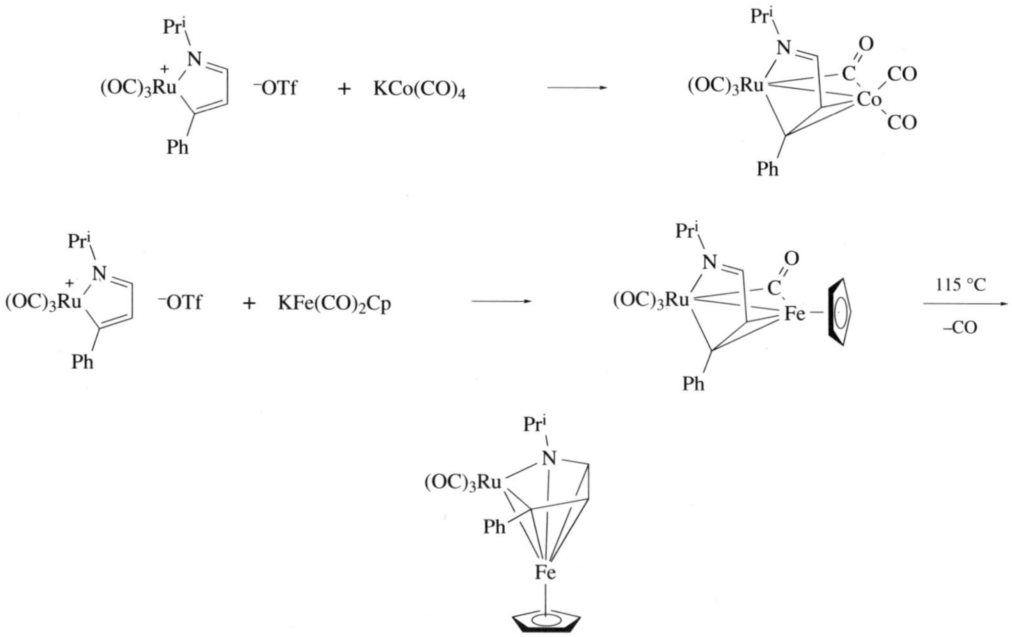

Scheme 11

hydrogen migration (Equation (47)) ⟨93OM996⟩. Bis(cyclopentadienyl)hydridoniobium reacts with dicarbonylcyclopentadienylmethyliron, apparently with loss of methane (Equation (48)) ⟨80JOM(201)269⟩.

Addition reactions are also involved in the formation of most 'triple decker sandwich' complexes (this term has gained common usage although 'double decker' would be more appropriate).

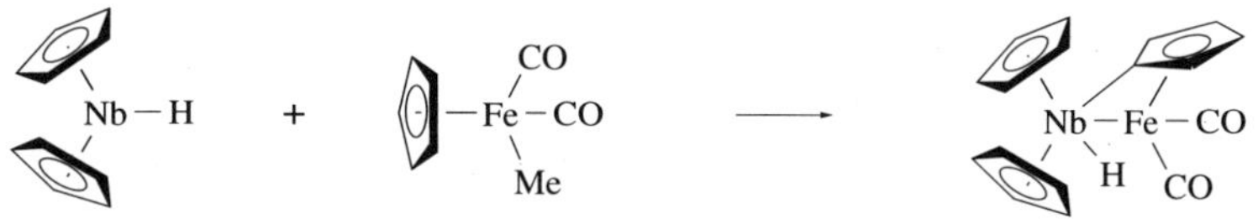

Thus, penta- and decamethylmetallocenes of the iron group add the cyclopentadienylmetal cation derived from, for example [(C$_5$R$_5$)Ru(MeCN)$_3$]PF$_6$ or [CpFe(C$_6$H$_6$)]PF$_6$ to give the dimetal tris(cyclopentadienyl) cations (Equation (49)) ⟨87DOK(293)1137, 87JOM(336)187⟩. Similarly cyclopentadienyl-(1-phenylborole)cobalt adds dicationic ruthenium or iridium fragments (Scheme 12) ⟨89AG(E)737⟩. The corresponding iron hydride complex adds a tetracarbonylvanadium fragment with loss of the hydrido-hydrogen (Equation (50)) ⟨89AG(E)319⟩, while its anion reacts with subgroup 16 carbonyls ⟨86AG(E)165⟩ or the halocarbonyls of niobium, tantalum ⟨89AG(E)319⟩ and rhenium ⟨89AG(E)737⟩ according to Scheme 13 ⟨89JOM(372)53⟩.

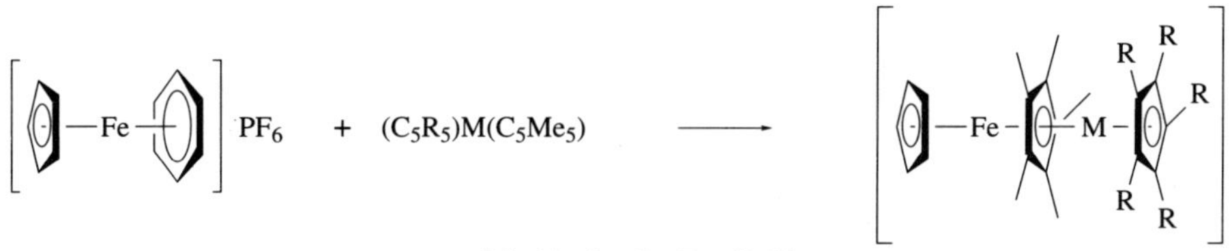

$$[Cp-Fe-C_6H_6]PF_6 + (C_5R_5)M(C_5Me_5) \longrightarrow [Cp-Fe-M-C_5R_5(C_5Me_5)]PF_6 \quad (49)$$

M = Fe, Ru, Os; R = H, Me

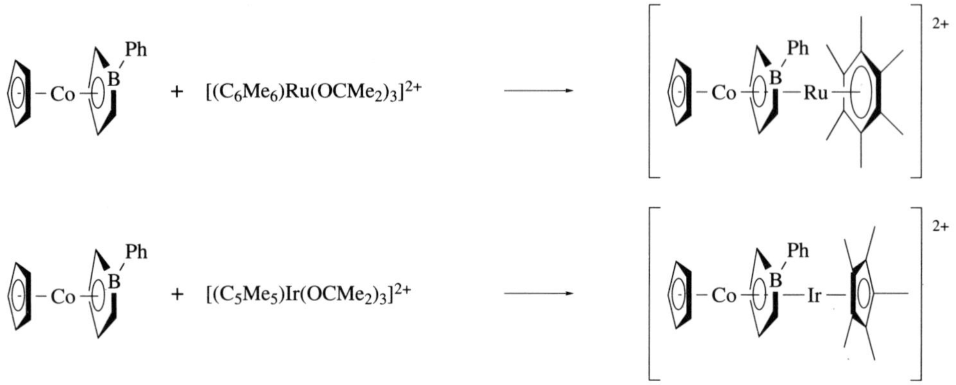

Scheme 12

$$Cp-Fe(H)(B-Ph) + V(CO)_6 \longrightarrow Cp-Fe-B(Ph)-V(CO)_4 \quad (50)$$

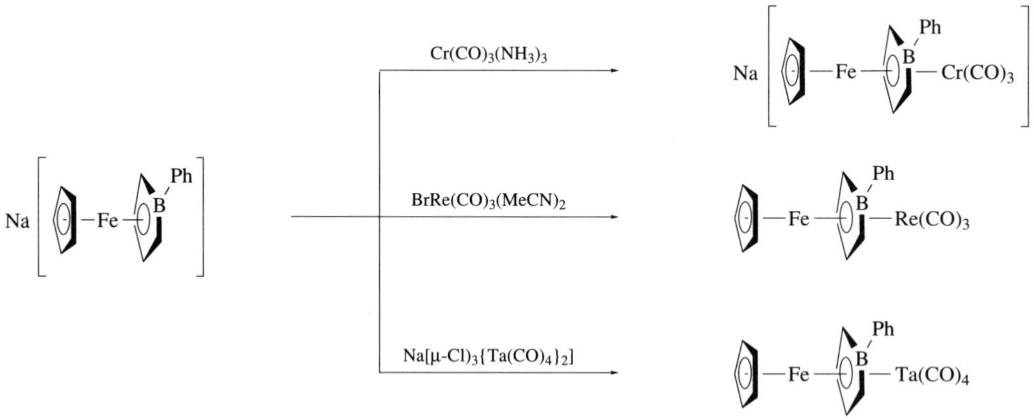

Scheme 13

Exchange of platinum for one of the cobalt moieties occurs in the example shown in Equation (51) ⟨87JOM(334)117⟩. Exchange of iron (as $Fe(CO)_5$) for one cobalt in hexacarbonyl(propargylalcohol)dicobalt complexes is accompanied by dehydration and follows the general pattern of Equation (52). It has not only been demonstrated for complexes of very simple alcohols (e.g. $R^1 = Me$, $R^2 = R^3 = H$) but also for complexes of 17-ethynyl-17-hydroxysteroids ⟨83POL77, 91ZN(B)1169, 93OM4545⟩.

$$\text{[Cp-Co(CO)-Co(CO)-Cp]} + \text{Ph}_3\text{P-Pt(PPh}_3\text{)(}\eta^2\text{-alkyne)} \xrightarrow[68\%]{\text{THF, reflux}} \text{Ph}_3\text{P-Pt(PPh}_3\text{)(}\mu\text{-alkyne)Co(CO)Cp} \quad (51)$$

$$\text{[(HO)CR}^1\text{R}^2\text{R}^3\text{-Co}_2(\text{CO})_6] + \text{Fe(CO)}_5 \longrightarrow \text{[R}^1\text{R}^2\text{R}^3\text{C-CoFe(CO)}_6] \quad (52)$$

(iv) Formation by replacement of lithium by a transition metal

Lithiated ferrocenes, tricarbonyl(lithiocyclopentadienyl)manganese and similar compounds can react with halo-transition metal complexes replacing lithium as, for example according to Equation (53) where X is a halide, usually Cl or Br, and $ML_n = Ti(NEt_2)_3$ ⟨73JOM(56)269⟩, $Mn(CO)_5$ ⟨87JOM(334)347⟩, $Ru(CO)_2(C_5R_5)$ or $Ru(CO)(Cl)(C_6Me_6)$ ⟨92JOM(436)333⟩, UCp_3 ⟨75IC78⟩, $W(NO)_2Cp$ ⟨86ZN(B)1431⟩, $ZrCl(C_5Me_5)_2$ ⟨88ZN(B)1461⟩ etc. and similarly with the dihalides Cp_2MCl_2 (M = Ti, Zr, Hf) to yield Cp_2MFc_2 ⟨77JOM(141)313, 88ZN(B)1461⟩ or with $(Et_2N)_2TiBr_2$ to yield $(Et_2N)_2TiFc_2$ ⟨73JOM(56)269⟩ and with $WOCl_4$ to yield $Fc_3W(O)Cl$ ⟨89JOM(371)205⟩. Similarly 1′,2-dichlorolithioferrocene has yielded Ti, Mn and Ir complexes (Equation (54)) on treatment with Cp_2TiCl_2, $BrMn(CO)_5$ and $Ir(CO)(PPh_3)_2Cl$ respectively ⟨79JOM(181)425⟩. 1,1′-Dilithioferrocene and -ruthenocene react with $(Et_2N)_3TiBr$ to yield $M[C_5H_4Ti(NEt_2)_3]_2$ (M = Fe, Ru) ⟨76ZAAC(423)112⟩. 2-Chlorolithioferrocene and tricarbonyl(lithiocyclopentadienyl)manganese have both been treated with chlorocyclopentadienyl(triphenylphosphine)nickel to give the mixed metal complexes (**11**) and (**12**) in 46 and 66% yields respectively ⟨73DOK(209)869⟩.

$$\text{FcLi} + \text{XML}_n \longrightarrow \text{FcML}_n + \text{LiX} \quad (53)$$

$$\text{[1′,2-dichloro-1-lithioferrocene]} + [M]X \longrightarrow \text{[1′,2-dichloro-1-M-ferrocene]} \quad (54)$$

$[M] = [Cp_2Ti]_{0.5}$, $Mn(CO)_5$, $Ir(CO)(PPh_3)_2$

(**11**) (**12**)

Tricarbonyl(lithiocyclopentadienyl)manganese reacts with titanocene dichloride according to Equation (55) ⟨80JCS(D)2315⟩, while the corresponding cuprate, whose solutions are stable at room temperature, has been used to react with bromodicarbonylcyclopentadienyliron (Scheme 14) ⟨90MI 416-01⟩. More surprisingly, the same product results by the reaction shown in Equation (56) ⟨74IZV2645⟩.

$$Cp_2TiCl_2 + \text{Li-C}_5H_4\text{-Mn(CO)}_3 \xrightarrow{\text{Et}_2\text{O-THF, }-78\,°\text{C}} Cp_2Ti[\text{C}_5H_4\text{-Mn(CO)}_3]_2 \quad (55)$$

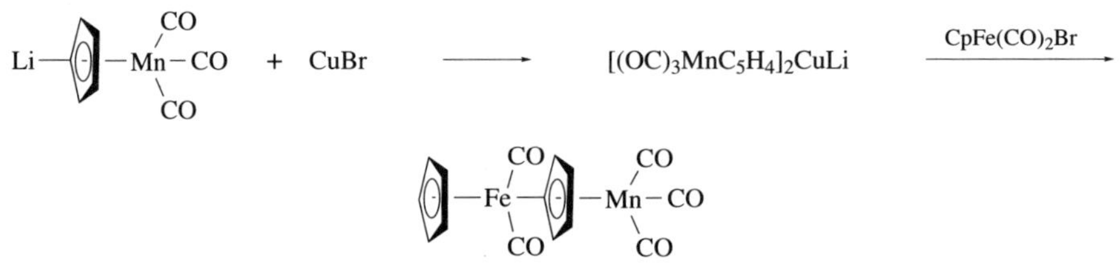

Scheme 14

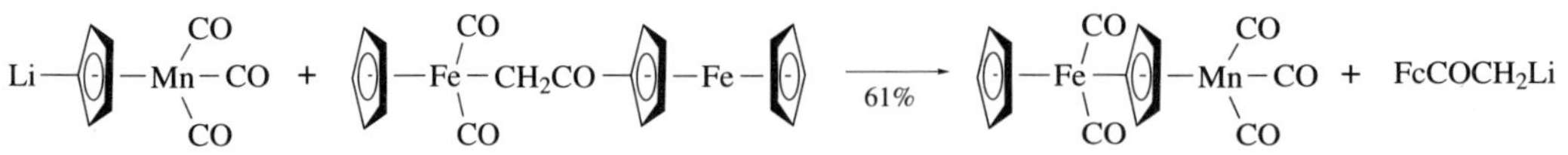

[Reaction scheme showing Li-CpMn(CO)₃ + CuBr → [(OC)₃MnC₅H₄]₂CuLi, then with CpFe(CO)₂Br → Fe(Cp)(CO)₂–CpMn(CO)₃]

[Reaction: Li-CpMn(CO)₃ + FcCH₂CO-Fc ferrocenyl compound $\xrightarrow{61\%}$ Fc–Mn(CO)₃ product + FcCOCH₂Li] (56)

1,1′-Dithioferrocene reacts with (substituted) bis(cyclopentadienyl)dichloro complexes of the subgroup 14 metals to give the corresponding metallaferrocenophanes (Equation (57)) ⟨90IC1817⟩, of which the zirconium compound (R = Bu^t) may be hydrolysed to the ferrocenylhydroxy–zirconium complex (Equation (58)) ⟨89JOM(362)C27, 92JOM(427)231⟩.

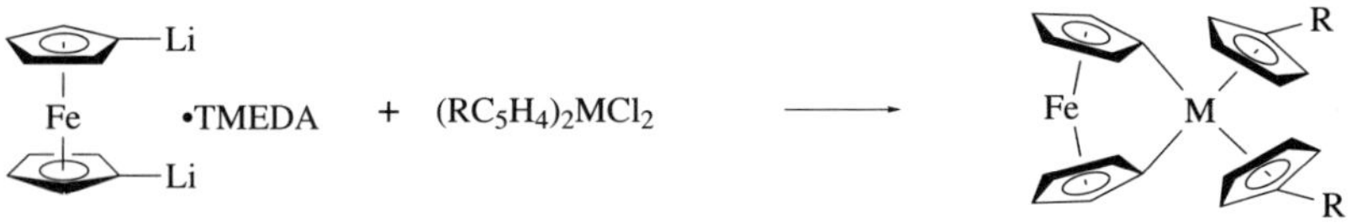

[Equation (57): 1,1′-dilithioferrocene·TMEDA + (RC₅H₄)₂MCl₂ → metallaferrocenophane; M = Ti, Zr, Hf] (57)

[Equation (58): Zirconium ferrocenophane (Bu^t) + H₂O → ferrocenyl-Zr(OH)(Cp-Bu^t)₂] (58)

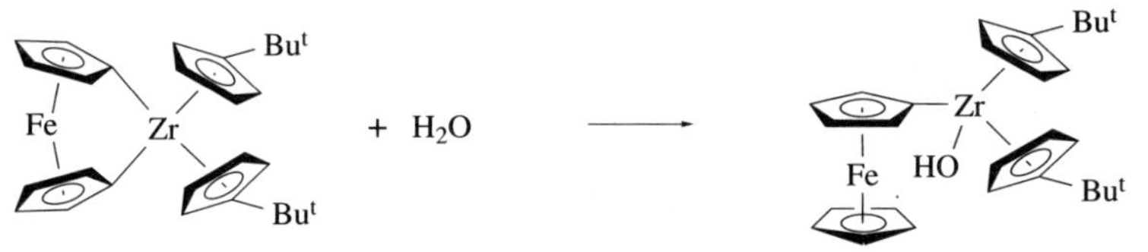

Lithiated tricarbonylthiophenechromium reacts with bromopentacarbonylmanganese; the expected and initially formed π-chromium compound slowly rearranges to the π-manganese isomer (Scheme 15) ⟨93AG(E)710⟩. Only low yields of the expected metallation products result when benzene(pentafluorolithiobenzene)chromium reacts with IRe(CO)₅ or IFe(CO)₂Cp ⟨77JOM(141)85⟩.

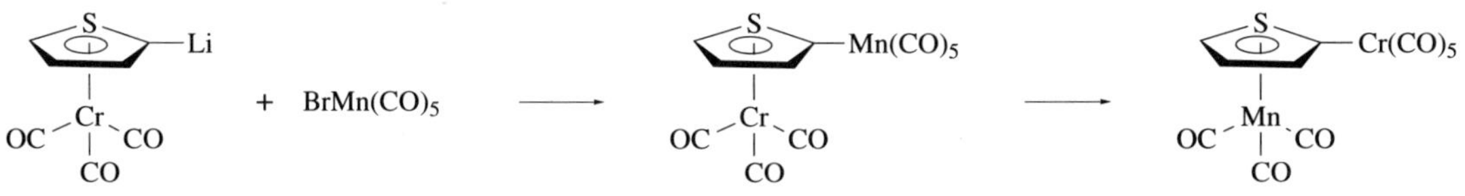

[Scheme 15: Li-thiophene-Cr(CO)₃ + BrMn(CO)₅ → thiophene(Mn(CO)₅)-Cr(CO)₃ → thiophene(Cr(CO)₅)-Mn(CO)₃]

Scheme 15

(v) Formation by metallation or cyclometallation by a transition metal

Unsubstituted ferrocene has been metallated using a neopentylruthenium complex (Equation (59)) ⟨90ZAAC(581)41⟩. More commonly, metallations follow initial coordination to a donor group (cyclometallations). Thus, acylferrocenes, like other aromatic ketones, react with methylmanganese or -rhenium pentacarbonyls giving the cyclometallated products according to Equation (60), whereas only the rhenium compound reacts similarly with (dimethylaminomethyl)ferrocene to give the analogous product (**13**) ⟨75JOM(91)C57, 77IC3193⟩. The same amine or its derivatives (**14**) undergo efficient cyclopalladation by sodium tetrachloropalladate in the presence of sodium acetate (Equation (61), R² = Me) ⟨75JOM(102)511⟩. If the latter salt is replaced by the salt of an asymmetric acid, preferably *N*-acetyl-L-valine, asymmetric induction occurs and the palladated product (**15**; R¹ = H or Me) is formed with high *ee* ⟨79JOM(182)537⟩. The palladated ferrocenylethylamine (**15**; R¹ = Me) has also been prepared from the corresponding chloromercuri compound, (ClHgC$_5$H$_3$CHMeNMe$_2$)Fe(C$_5$H$_5$) with Pd$_2$(dba)$_3$ ⟨79IZV1528⟩ and in homochiral form by starting with the optically resolved precursor (14) ⟨77DOK(236)371⟩. Analogous cycloplatinated complexes have been obtained using *cis*-(Me$_2$SO)$_2$PtCl$_2$ as metallating agent ⟨94OM500, 94OM511⟩. The thioamide Fe(C$_5$H$_4$CSNMe$_2$)$_2$ has been chloropalladated in both rings to yield a product, isolated after chloride replacement, for example as compound (16) ⟨93JOM(463)169⟩.

Bis(cyclopentadienyl)molybdenum, -tungsten and -rhenium hydrides are metallated by pentacarbonylmethylmanganese according to Equations (62) and (63) ⟨79IC3453⟩.

(vi) Formation by other methods

Direct replacement of activated halide by a transition metal occurs in the reaction of tricarbonyl(chlorobenzene)chromium with palladium(0) (Scheme 16) ⟨91OM4005⟩. Chloro- and fluorobenzene and many substituted halobenzenetricarbonylchromium complexes also react with some anionic organometallics, notably $CpFe(CO)_2^-$ (Equation (64)) whereas $Mn(CO)_5^-$ and $CpMo(CO)_3^-$ fail to react ⟨88OM1715⟩. Use of the dianions $[Fe(CO)_4]^{2-}$, $[Cr(CO)_5]^{2-}$ or $[W(CO)_5]^{2-}$ gives the anionic products $[(OC)_3Cr(\eta^6\text{-}C_6H_5Fe(CO)_4)]^-$ and $[(OC)_3Cr(\eta^6\text{-}C_6H_5M(CO)_5)]^-$ where M = Cr or W ⟨89OM1199⟩.

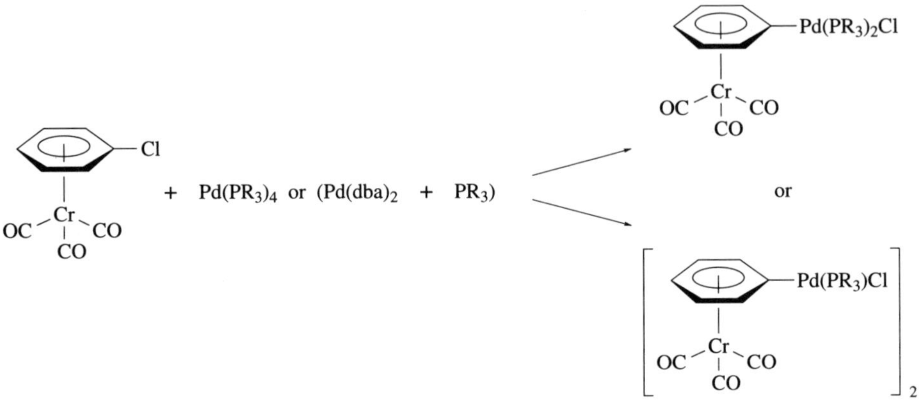

Scheme 16

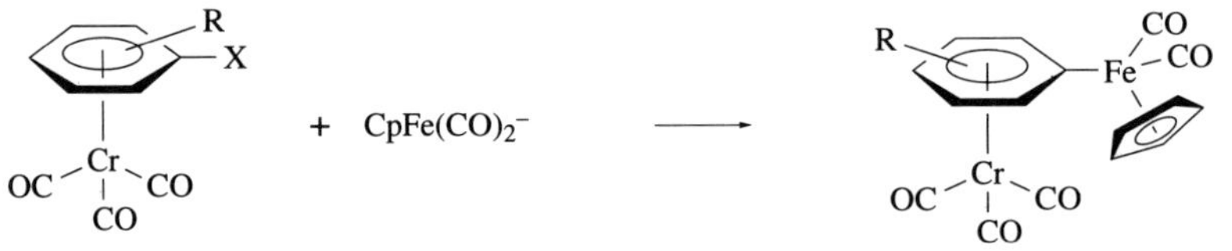

(64)

Since acylmetal complexes are generally readily decarbonylated by photolysis or heating, the reaction of anionic metal complexes with such acid halides as ferrocenoyl chloride (Scheme 17) ⟨73IZV2796⟩ provides a potentially good general route. In some cases, for example $[ML_n]^- = [CpMo(CO)_3]^-$, decarbonylation occurs spontaneously during workup so that $CpMo(CO)_3(C_5H_4)FeCp$ is obtained directly ⟨76IC2671⟩, whereas the tungsten analogue has to be heated to 130 °C ⟨73IZV2796⟩.

Replacement of tin by platinum has been used to generate ferrocenylplatinum complexes (e.g. Equation (65)); the 1,1′-bisplatinum analogue is generated similarly from $Fe(C_5H_4SnMe_3)_2$ ⟨79JOM(170)95⟩.

$$FcCOCl + [ML_n]^- \longrightarrow FcCOML_n \xrightarrow{\text{heat or } h\nu} FcML_n$$

Scheme 17

(65)

4.16.3.1.2 *A 'genuine' transition metal linked to a 'late' transition (i.e. group 11 (Cu, Ag, Au) or 12 (Zn, Cd, Hg)) metal*

Arguably the simplest route to such compounds is the addition of the group 11 or 12 metal halide to a transition metal carbene complex, as in Scheme 18 ⟨83JA5939⟩, where $HgCl_2$ adds as $HgCl^+$ Cl^- while AuI (as Et_4NAuI_2) adds intact.

Cp_2ZrHCl adds to vinylzinc compounds $CHR^1{=}CR^2ZnX$ to yield mixed Zn–Zr complexes

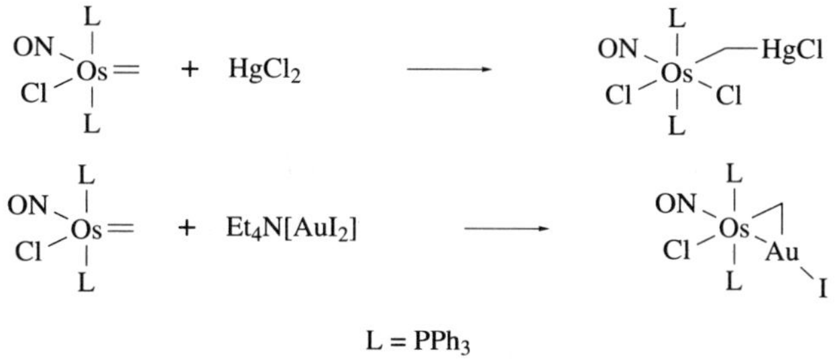

[Scheme 18]

$R^1CH_2CR^2(ZnX)ZrCp_2Cl$ which react with aldehydes to yield alkenes ⟨94OM94⟩. Several complexes containing two different transition metals linked to the same aryl carbon atom have been formed from the σ-aryl of one of the metals and a halide of the other. Thus, a tantalum/zinc compound is obtained by the reaction shown in Equation (66) ⟨93OM2227⟩. Formation of metallocene -Cd and -Hg complexes has been reviewed ⟨92JOM(441)241⟩.

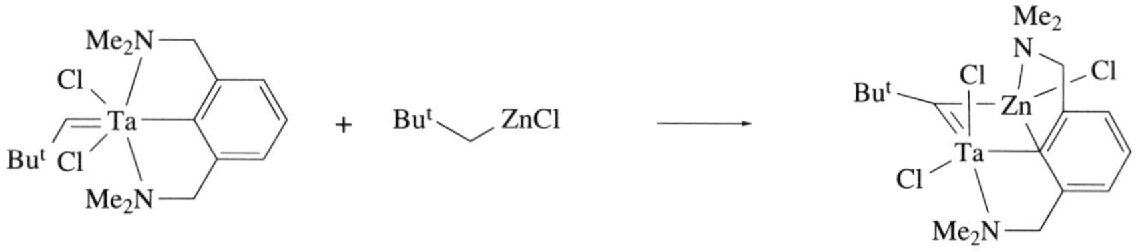

[Equation (66)]

(i) Formation by replacement of lithium by a late transition metal

Reactions of lithiated sandwich or 'half-sandwich' ('piano-stool') complexes with late transition metal halides provide the most general and most widely used route. While lithioferrocene derivatives furnish many examples, the method is equally applicable to tricarbonyl(lithiocyclopentadienyl)manganese, tricarbonyl(lithiobenzene)chromium, tricarbonyl(lithiocyclobutadiene)iron and many similar systems. Thus, both tricarbonyl(lithiocyclopentadienyl)manganese ⟨69IZV2030, 69IZV2032⟩ and its tetrachloro derivative ⟨93JOM(458)181⟩ have been converted with $ClAuPPh_3$ into $(OC)_3Mn(C_5R_4AuPPh_3)$ where R = H or Cl respectively.

Examples with lithioferrocene (Equation (67)) include reaction with $ClAuPPh_3$ ⟨69IZV2030⟩ and $(BrCuPPh_3)_4$ (giving $(FcCu)_n$) ⟨80JOM(185)C6, 87ZOB2606⟩. The 1,1'-dicopper and -gold complexes have been obtained similarly ⟨72IZV2594, 82JOM(224)C53⟩. 2-Lithio(dimethylaminomethyl)ferrocene has been converted into the 2-cupro and -argento derivatives using CuI (or its $FcCH_2NMe_2$ adduct) and AgI respectively ⟨76DOK(226)1092, 78JOM(153)115⟩.

$$FcLi + XML_n \longrightarrow FcML_n \qquad (67)$$

The indirect route via the boronic acid shown in Scheme 19 has been used in the tricarbonylcyclopentadienylmanganese series ⟨71DOK(198)590, 72DOK(202)362⟩ and also to convert ruthenocenyllithium into diruthenocenylmercury ⟨72IZV1823⟩.

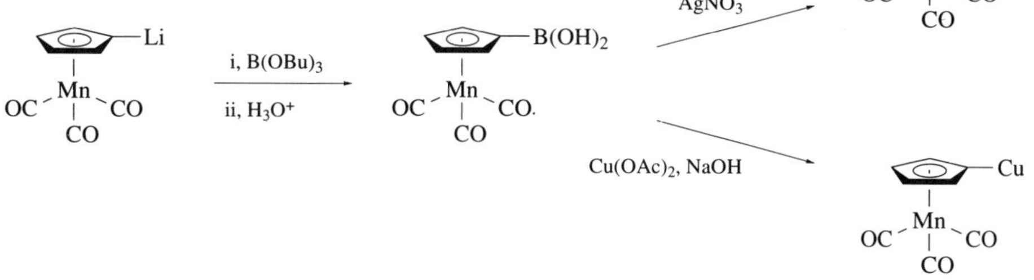

[Scheme 19]

(ii) Formation by mercuration

Direct mercuration with mercury(II) acetate usually followed by exchange of acetoxy for chloro is normally the method of choice for obtaining all such 'aromatic' organometal–mercury compounds as FcHgCl, 1,1'-Fe(C_5H_4HgCl)$_2$, ⟨54DOK(97)459, 57JOC900⟩, $(OC)_3Mn(C_5H_4HgCl)$ ⟨62IZV1683⟩, CpCo-(C_4H_3HgOAc) ⟨68JA1057⟩, $(OC)_3Cr(C_6H_5HgCl)$ ⟨72ZOB2450⟩ and $(OC)_3Fe(C_4H_3HgCl)$ ⟨65JA3254⟩. Iodoferrocene yields a mixture of 3-, 1'- and 2-HgCl derivatives ⟨74JOC1420⟩. The corresponding 'bis(arene)'mercury compounds are formed by reduction (with e.g. $Na_2S_2O_3$), for example Fc_2Hg (from FcHgCl) ⟨57JOC900, 63JOC3337⟩, and are sometimes the only form isolated, for example $[CpRu(C_5H_4)]_2Hg$ from ruthenocene by successive treatment with $Hg(OAc)_2$, $CaCl_2$, and $Na_2S_2O_3$ ⟨72IZV1823⟩.

Polymercuration is not uncommon: thus, tricarbonylcyclobutadieneiron ⟨70CC161⟩ and pentamethylruthenocene ⟨92OM3169⟩ have given the fully substituted acetoxymercuri derivatives, $(OC)_3Fe[C_4(HgOAc)_4]$ and $(C_5Me_5)Ru[C_5(HgOAc)_5]$ respectively. Replacement of other metals (e.g. Cu, Equation (68)) ⟨62IZV1683⟩ by mercury appears of little practical interest.

$$\text{CpCu-Mn(CO)}_3 + HgCl_2 \longrightarrow \text{ClHgCp-Mn(CO)}_3 \tag{68}$$

4.16.3.1.3 Two 'late' transition metals

The gold–zinc compound (17) is the product from diphenylzinc with $AuCl_3$ or Au(CO)Cl ⟨77JOM(127)391⟩ and the gold–copper compound (18) results from the reaction of Li[(2-Me_2N-$CH_2C_6H_4)_2Au$] with CuI ⟨80JOM(186)427⟩.

(17)   (18)

4.16.3.2 A Transition Metal and a Group 13 or 14 Metal

Reaction of tin-containing Grignard or organolithium reagents with various halides has been used both for metals of the titanium group (Scheme 20) ⟨81JCS(D)1593⟩ and for mercury (Equations (69) and (70)) ⟨84JOM(269)249⟩ and copper (Equation (71)) ⟨82CB1810⟩. In very similar fashion, treatment of the pent-4-enylzinc–magnesium compounds, for example CH_2=$CHCH_2$CHRCH(ZnBr)MgBr (cf. Scheme 4) or their Zn–Li analogues with chlorotrimethylstannane or with copper(I) cyanide gives the corresponding Zn–Sn and Zn–Cu compounds ⟨86TL1039, 88TL6697⟩. Trialkyl- or -aryltinlithium can be used with sufficiently activated haloaryl complexes, for example the cationic cyclopentadienyliron complexes in Equation (72) ⟨86JOM(307)231⟩.

$$Me_3Sn\text{–}CH_2CH_2\text{–}MgX + Bu^iHgX \longrightarrow Me_3Sn\text{–}CH_2CH_2\text{–}HgBu^i + MgX_2 \tag{69}$$

$$2\,Me_3Sn\text{–}CH_2CH_2\text{–}MgX + HgX_2 \longrightarrow (Me_3SnCH_2)_2Hg + 2MgX_2 \tag{70}$$

$$Ph_3Sn\text{–}CH_2CH_2\text{–}Li + CuCl \longrightarrow Ph_3Sn\text{–}CH_2CH_2\text{–}Cu + LiCl \tag{71}$$

$$Cp_2MCl_2 + Me_3Sn\diagdown MgX \longrightarrow Cp_2M\diagup_{X}^{CH_2SnMe_3}$$

$$\Big\downarrow HX$$

$$Cp_2MCl_2 + 2\ Me_3Sn\diagdown MgX \longrightarrow Cp_2M(CH_2SnMe_3)_2$$

M = Ti, Zr, Hf; X = Cl, Br

Scheme 20

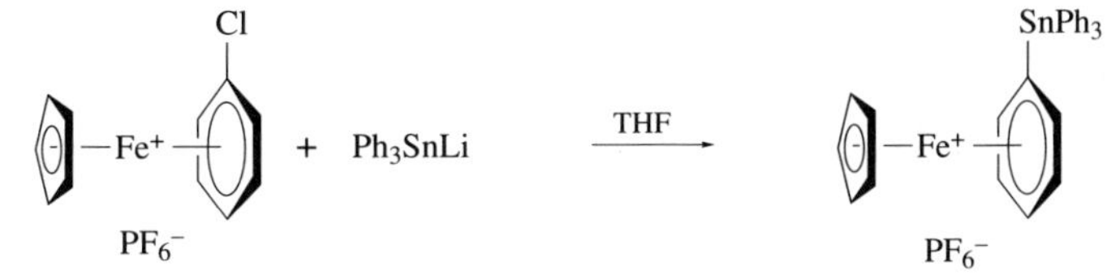

$$[CpFe(C_6H_5Cl)]^+PF_6^- + Ph_3SnLi \xrightarrow{THF} [CpFe(C_6H_5SnPh_3)]^+PF_6^- \qquad (72)$$

Aluminum alkyls form adducts with zirconium alkyls (e.g. Equation (73)) ⟨71JOM(26)357⟩. A carbene complex of the type $Cp_2Ti=CH_2$ may be intermediate in the formation of the 'Tebbe reagent' (**19**) from $Cp_2TiMe_2 + ClAlMe_2$ or $Cp_2TiCl_2 + AlMe_3$ ⟨78JA3611⟩ and of $Cp_2Ti(\mu\text{-}CH_2)(\mu\text{-}CH_3)AlMe_2$ from $Cp_2TiMe_2 + AlMe_3$ ⟨75CC621, 79JCS(D)45⟩. A closely related titanium(III)–aluminum complex bridged by two methyl groups (**20**) is formed from Cp_2TiCl with $Li[AlMe_4]$ and its scandium and yttrium analogues are formed similarly ⟨79JCS(D)45⟩. A unique group of zirconium complexes in which a square planar carbon atom is bridging zirconium to aluminum or gallium is formed either from zirconocene–alkyne complexes (e.g. Equation (74)) or from Cp_2ZrMe_2 or Cp_2ZrHCl with alkynylaluminum or -gallium compounds (e.g. Equation (75)) ⟨90JA9620, 91OM3791, 92CB1953, 92JA8531, 92OM3517, 93OM4979⟩. Hafnium analogues can also be made ⟨92JOM(427)C21⟩.

$$Zr(CH_2Ph)_4 + Al(CH_2Ph)_3 \rightleftharpoons (PhCH_2)_4Zr-CH_2\diagup_{Al(CH_2Ph)_2}^{Ph} \qquad (73)$$

(**19**) Cp₂Ti(μ-Cl)(μ-Me)AlMe₂ (**20**) Cp₂Ti(μ-Me)₂AlMe₂

$$Cp_2Zr(\eta^2\text{-cyclohexyne})(PMe_3) + GaMe_3 \longrightarrow Cp_2Zr(\mu\text{-cyclohexenyl})GaMe_2 + Me_3GaPMe_3 \qquad (74)$$

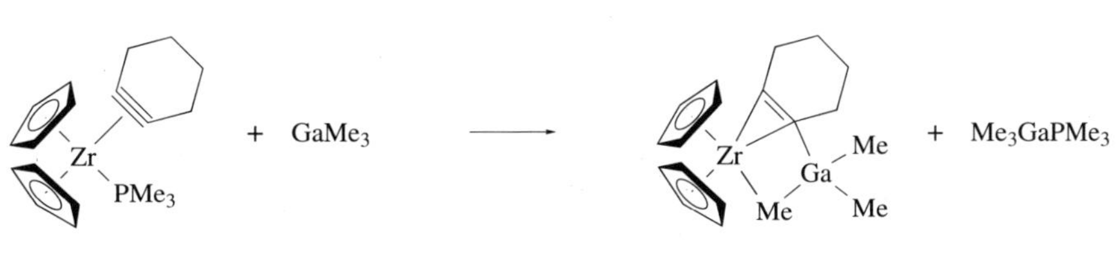

$$Cp_2ZrMe_2 + [MeC\equiv CAlMe_2]_2 \longrightarrow \text{products} \qquad (75)$$

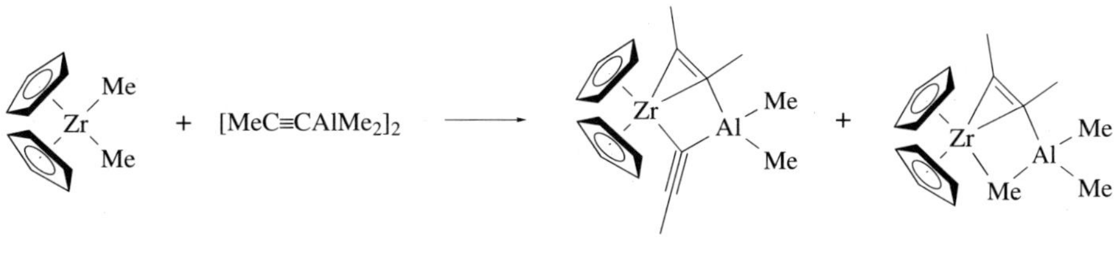

Replacement of lithium on a π-bonded ligand (cf. Equation (53)) is again a major route. Thus lithio- and 1,1′-dilithioferrocene react with $ClSnMe_3$ to give $FcSnMe_3$ and $Fe(C_5H_4SnMe_3)_2$ respectively ⟨79JOM(170)95⟩. Similarly, tricarbonyl(lithiocyclopentadienyl)manganese and -rhenium react with chlorotriphenyltin or -lead (Equation (76)) to yield the expected bimetallic products

⟨73IZV2815⟩. Mercury is also replaced by tin(II), for example from the Mn–Hg complex (Equation (77)) and in analogous fashion from diferrocenylmercury to yield Fc_2SnCl_2 ⟨73DOK(209)1113⟩.

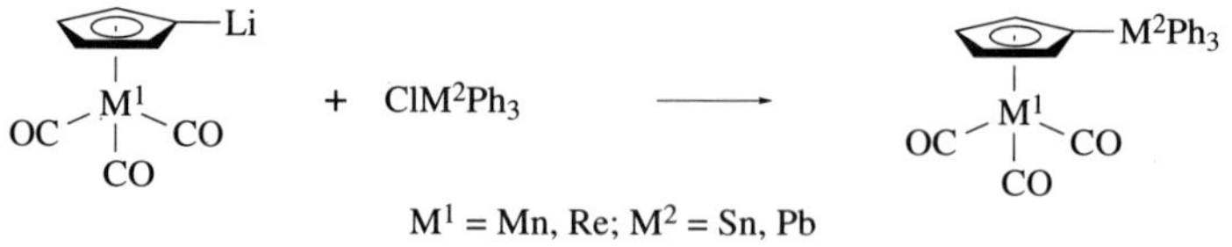

$$M^1 = Mn, Re; M^2 = Sn, Pb \quad (76)$$

$$(77)$$

Triphenylstannane, $SnHPh_3$, reacts with $CpCo(CO)_2$, probably via a Co–Sn complex followed by migration of the $SnPh_3$ group, to yield $(Ph_3SnC_5H_4)Co(CO)_2$ and, as principal product (70%), [1,3-$(Ph_3Sn)_2C_5H_3]Co(CO)_2$ ⟨94JOM(468)235⟩.

Aryltin compounds, for example $PhSnMe_3$, Ph_2SnMe_2, 1,4-$C_6H_4(SnMe_3)_2$, add the tricarbonylmetal fragment to give the corresponding η^6-compounds, for example $(OC)_3M(\eta^6\text{-}C_6H_5SnMe_3)$ when heated with the hexacarbonyls of chromium or molybdenum ⟨71IC522⟩.

Reaction of the cobaltacyclopentadiene complex (21) with aluminum hydride-triethylamine has yielded an unusual cobalt–aluminum compound (Equation (78)) ⟨91OM2726⟩.

The methyl-bridged aluminum–yttrium complex (22) has been prepared from $AlMe_3$ and $[(C_5H_4SiMe_3)Y(OBu^t)(\mu\text{-}OBu^t)]_2$ ⟨93JOM(462)141⟩.

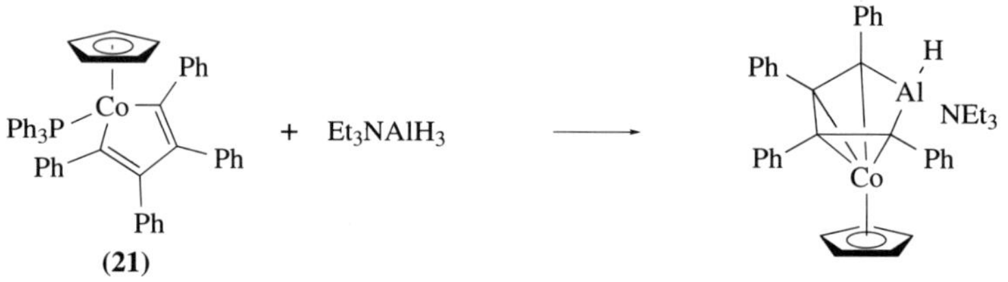

$$(78)$$

(21)

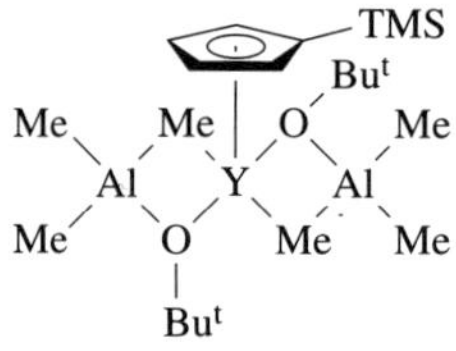

(22)

4.16.3.3 A Transition Metal and Other Metals

Lithio- and 1,1'-dilithioferrocene have been treated with chlorotricyclopentadienyluranium to yield Cp_3UFc and $Fe(C_5H_4UCp_3)_2$ respectively ⟨75IC78⟩.

4.16.4 FUNCTIONS CONTAINING AT LEAST ONE GROUP 13 METAL (AND NO GROUP 1, 2 OR TRANSITION METAL)

Trimethylstannylmethyllithium reacts with indium chlorides according to Equation (79) ⟨84ZN(B)798⟩.

A series of methyl-bridged aluminum–lanthanide compounds has been obtained from lithium tetramethylaluminate and the (dimeric) chlorobis(cyclopentadienyl)metal complexes (Equation (80) where M = Sc, Y, Gd, Dy, Ho, Er, Tm or Yb) ⟨79JCS(D)45⟩. The ethyl analogues are obtained similarly using $Li[AlEt_4]$ and the yttrium–indium compound $Cp_2Y(\mu\text{-}Me)_2InMe_2$ has also been made.

$$Me_3Sn\diagup Li\ +\ R_nInCl_{3-n}\ \longrightarrow\ R_nIn(CH_2SnMe_3)_{3-n} \quad (79)$$

$$R = Me, Bu^t;\ n = 0\text{--}2$$

$$Cp_2M\text{--}Cl\ +\ Li[AlMe_4]\ \longrightarrow\ Cp_2M(\mu\text{-}Me)_2AlMe_2 \quad (80)$$

M = Sc, Y, Gd, Dy, Ho, Er, Tm, Yb

4.17
Functions Incorporating Two Halogens or a Halogen and a Chalcogen

PETER D. KENNEWELL
Hoechst Roussel Ltd, Swindon, UK

ROBERT WESTWOOD
Roussel-Uclaf, Romainville, France

and

NICHOLAS J. WESTWOOD
University of Oxford, Oxford, UK

4.17.1 DIHALO FUNCTIONS: $R_2C=C(Hal)_2$	730
4.17.1.1 General Methods	730
4.17.1.2 Difluoro Alkenes	730
4.17.1.2.1 Elimination reactions	730
4.17.1.2.2 Wittig reactions	732
4.17.1.2.3 Ylide–carbene reactions	735
4.17.1.2.4 Miscellaneous routes	735
4.17.1.3 1,1-Dichloro Alkenes	737
4.17.1.3.1 Wittig and related reactions	737
4.17.1.3.2 Carbene reactions	740
4.17.1.3.3 Carbenoid reactions	741
4.17.1.3.4 Elimination reactions	742
4.17.1.3.5 Rearrangements	742
4.17.1.3.6 Miscellaneous methods	743
4.17.1.4 Dibromo Alkenes	745
4.17.1.4.1 Wittig and related reactions	745
4.17.1.4.2 Carbenoid reactions	746
4.17.1.4.3 Ring-opening reactions of cyclopropanes	747
4.17.1.4.4 Miscellaneous reactions	748
4.17.1.5 Diiodo Alkenes	749
4.17.1.6 Mixed Halo Derivatives	750
4.17.1.6.1 Fluoro halo alkenes	750
4.17.1.6.2 Chloro halo alkenes	756
4.17.1.6.3 Diazo alkanes with carbenes	759
4.17.1.6.4 Bromo iodo alkenes	759
4.17.2 FUNCTIONS INCORPORATING A HALOGEN AND A CHALCOGEN: $R_2C=C(Hal)(Chalc)$	760
4.17.2.1 Halogen and Oxygen Derivatives	760
4.17.2.1.1 α-Halo enols and ethers	760

4.17.2.1.2	α-Halo acyloxy and other O-substituted derivatives	768
4.17.2.2	Halogen and Sulfur Derivatives	769
4.17.2.2.1	Dicoordinate sulfur derivatives	769
4.17.2.2.2	Tricoordinate sulfur derivatives	776
4.17.2.2.3	Tetracoordinate sulfur derivatives	778
4.17.2.3	Halogen and Selenium or Tellurium Derivatives	784
4.17.2.3.1	Wittig reaction	784
4.17.2.3.2	Addition–elimination reactions	785
4.17.2.3.3	Anion reactions	785
4.17.2.3.4	Ring opening of selenophenes	785
4.17.2.3.5	Addition to substituted alkynes	786
4.17.3	ACKNOWLEDGEMENT	788

4.17.1 DIHALO FUNCTIONS: $R_2C=C(Hal)_2$

4.17.1.1 General Methods

As might be expected from such similar structures, the most important synthetic methods are common to all 1,1-dihalo alkenes. Thus elimination reactions, Wittig and related condensations and ylidene–carbenoid reactions will be found represented in all sections of this chapter along with a selection of more esoteric transformations peculiar to specific groups. However, the different properties of the various carbon–halogen bonds mean that the relative importance of the synthetic methods varies.

Also, it should be noted that there is a complication in the synthesis of the mixed dihalo compounds in that those bearing different R groups will show stereoisomerism and this has also influenced the design of synthetic strategies.

To avoid unnecessary duplication, these general methods will not be discussed further here.

4.17.1.2 Difluoro Alkenes

As indicated above, the methods of preparing this system can be conveniently grouped into those involving eliminations, Wittig condensations, ylidene–carbene reactions and a miscellaneous collection of reactions.

4.17.1.2.1 Elimination reactions

Whilst a considerable range of halogens and hydrogen halides have been eliminated from suitable precursors to produce 1,1-difluoro alkenes, the general applicability of the method is limited by the availability of these precursors. Table 1 lists examples of the base-catalysed elimination of hydrogen halides and, as expected, the acidity of the eliminated hydrogen atom dictates the strength of the base required to achieve the transformation.

Alternatively, halogens, including fluorine, have been eliminated from appropriate substrates. However, the elimination of fluorine itself is very unfavourable due to the low fluorine–fluorine bond strength. Nevertheless, steel gauze at 600 °C converted decafluoroethylbenzene (**1**) into a number of products from which octafluorostyrene (**2**) was isolated in 15% yield by gas chromatography ⟨62CI(L)1472⟩. Perhaps more practical is the use of high surface area carbon at 400 °C which converts the cyclobutane (**4**) or perfluoro-2,3-dimethylbutane (**5**) into perfluoro-2,3-dimethylbuta-1,3-diene (**3**) in 59% yield (based on material converted) and 20% yield respectively ⟨93JFC(65)67⟩.

Much more synthetically useful is the zinc-mediated loss of chlorine from the pentahaloethylbenzenes (**6**) and (**7**) to give (**8**) and (**9**) in yields of 48% and 70% respectively ⟨49JA3439⟩. Similarly, the polyhalo alkene (**10**) gave (**11**) in 40% yield ⟨81IZV1920⟩ whilst a mixture of zinc and zinc chloride removes ClF from (**12**) to give 58% of (**13**) ⟨75ZOR961⟩. Alternatively, ClF was eliminated from (**14**) to give (**15**) in 58% yield by treatment with butyllithium at −78 °C ⟨90JOC4448⟩. The tertiary iodide (**16**) loses IF on treatment with zinc yielding 71% of (**17**) ⟨87JFC(37)223⟩ whilst BrCl is readily lost from a number of alkenes, for example (**18**) to give the dienes (**19**) ⟨54JA5423⟩.

Dihalo Functions

Table 1 Elimination of hydrogen halides to produce difluoro alkenes.

Starting material	HX	Base	Product	Yield (%)	Ref.
PhCH$_2$CH=CHCH$_2$OCH$_2$CF$_3$	HF	LDA, −70 °C TMS-Cl	PhCH$_2$CH=CHCH$_2$OC(TMS)=CF$_2$	88	85TL2861
(CF$_3$)$_2$CHCONMe$_2$	HF	Et$_3$N·BF$_3$	CF$_2$=C(CF$_3$)CONMe$_2$	85.5	75JFC(6)227
CF$_3$CF$_2$COCH(CF$_3$)$_2$	HF	Et$_3$N·BF$_3$	CF$_2$=C(CF$_3$)COCF$_2$CF$_3$		80CA71025t
(CF$_3$)$_2$CHC$_3$F$_7$	HF	Et$_3$N·BF$_3$	CF$_2$=C(CF$_3$)C$_3$F$_7$		86IZV91
(CF$_3$)$_2$CHC(SEt)=C(CF$_3$)$_2$	HF	Et$_3$N·BzCl	CF$_2$=C(CF$_3$)C(SEt)=C(CF$_3$)$_2$	62.5	74IZV2545
(CF$_3$)$_2$CHCF$_2$OMe	HF	NaOH	CF$_2$=C(CF$_3$)CF$_2$OMe	13.2	85JFC(29)471
PhCClFCHF$_2$	HCl	NaOH	CF$_2$=CFPh	12.6	53JA968
CF$_2$BrCH$_2$CO$_2$Et	HBr	Et$_3$N	CF$_2$=CHCO$_2$Et	74	87JOC290
CF$_2$ClCH$_2$COCHMe$_2$	HCl	Et$_3$N	CF$_2$=CHCOCHMe$_2$	55	90JOC3562
CF$_2$ClCH$_2$CO$_2$CHMe$_2$	HCl	Et$_3$N	CF$_2$=CHCO$_2$CHMe$_2$	49	90JOC3562
CF$_3$CHFCOF	HF	NaF	F$_2$C=CFCOF	90	73JFC(3)63
RCOCHFCF$_2$Cl	HCl	Et$_3$N	RCOCF=CF$_2$	85	85JFC(29)445
p-ClC$_6$H$_4$C(CF$_2$Cl)HCO$_2$Et	HCl	dbu	p-ClC$_6$H$_4$C(CO$_2$Et)=CF$_2$		80TL2555
BrCF$_2$CH$_2$CBrF$_2$	HBr	KOH	CF$_2$BrCH=CF$_2$	72	55JA2783
CF$_3$CH(CH$_3$)CF$_2$I	HI	KOH	CF$_3$C(CH$_3$)=CF$_2$	83	53JCS3565
(p-NO$_2$C$_6$H$_4$)$_2$CHCF$_3$	HF	KOBu^t	(p-NO$_2$C$_6$H$_4$)$_2$C=CF$_2$		85CJC576

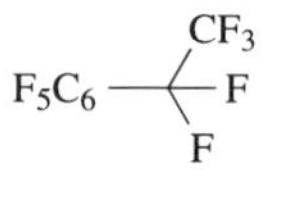

(1) 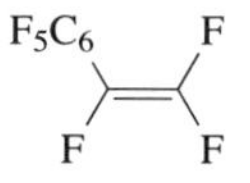 (2) 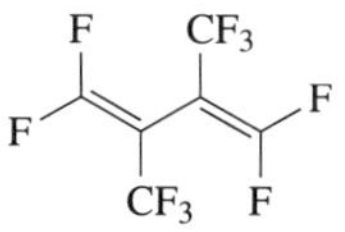 (3) 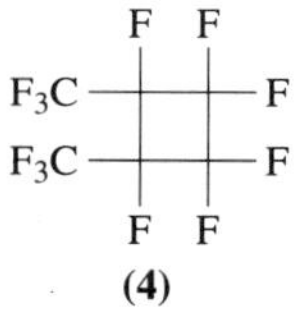 (4) 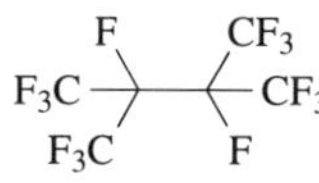 (5)

Perhaps the most unusual elimination reaction is that of the ethoxythietane (**20**) with phenylmagnesium bromide to give (**21**) ⟨76BCJ2491⟩.

Finally, the thermal decarboxylation of sodium perfluoropropionate has been used as a source of tetrafluoroethylene ⟨55JA4168⟩.

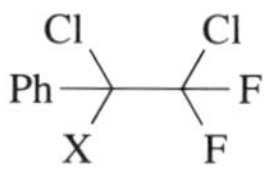

(6) X = F
(7) X = Cl

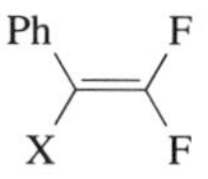

(8) X = F
(9) X = Cl

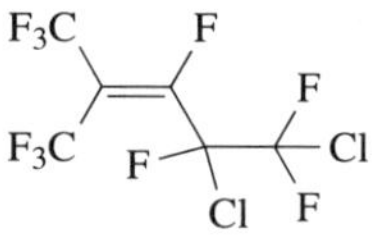

(10) 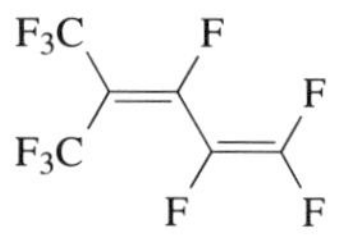 (11) 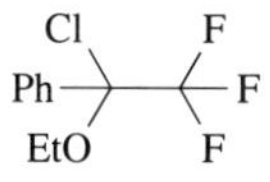 (12)

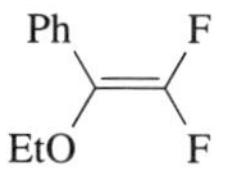

(13) 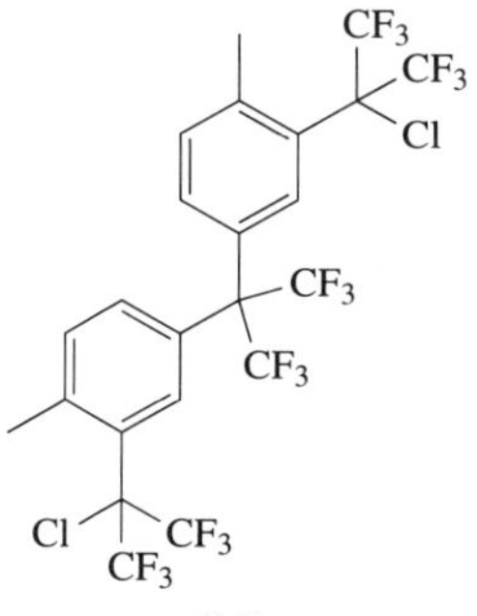 (14) 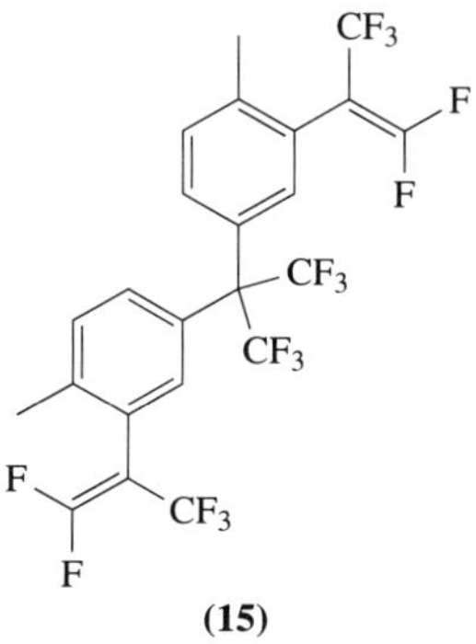 (15) 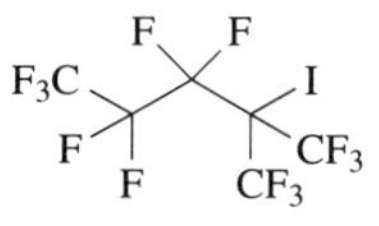 (16) 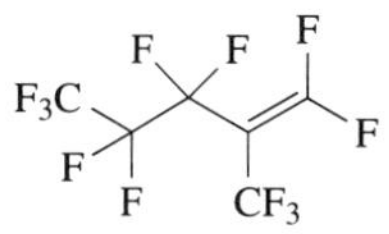 (17)

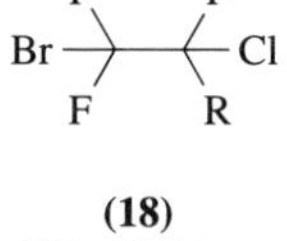

(18)
R = CH=CHMe
R = CH$_2$CH=CH$_2$
R = CH=CH(CH$_2$)$_5$Me

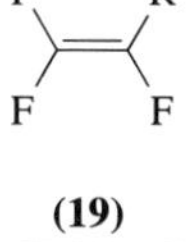

(19)
R = CH=CHMe, 60%
R = CH$_2$CH=CH$_2$, 65%
R = CH=CH(CH$_2$)$_5$Me, 72%

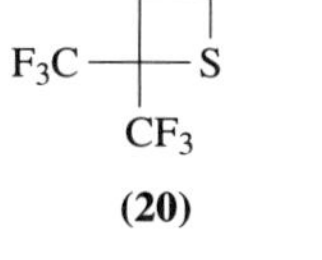

(20) 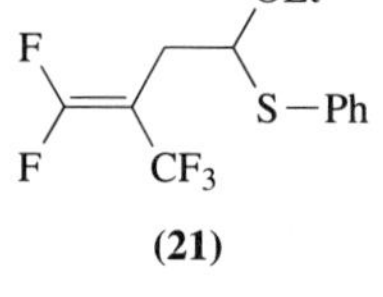 (21)

4.17.1.2.2 Wittig reactions

Fuqua, Duncan and Silverstein first reported that a mixture of an aldehyde, triphenylphosphine and sodium chlorodifluoroacetate in either DIGLYME or MONOGLYME gave 1,1-difluoro alkenes in moderate to good yields, (Equation (1)) ⟨64TL1461⟩. Subsequently, the same authors modified the reaction conditions and showed that the optimal reagent proportions were 1 m of aldehyde, 1.1 m of triphenylphosphine and 1.5 m of sodium chlorodifluoroacetate ⟨65JOC1027⟩.

$$ClF_2CO_2Na + Ph_3P + RCHO \longrightarrow \underset{R}{\overset{F}{\diagup}}=\underset{F}{\overset{F}{\diagdown}} \qquad (1)$$

R = n-hexyl, 16%
R = Ph, 74%
R = p-MeOC$_6$H$_4$, 60%

The extension of this reaction to ketones has been more problematic. Initially triphenylphosphine in DIGLYME gave no product with cyclohexanone but (**22**) and (**23**) could be produced in 46% and 34% yields respectively with tributylphosphine and N-methylpyrrolidone ⟨65TL521⟩.

(**22**)   (**23**)

No trace of the expected 1,1-difluoro alkene was produced from phenyl 2-pyridyl ketone, trifluoroacetophenone, benzoquinone, hexachloroacetone or camphor ⟨65JOC2543⟩. However, Burton and Herkes used a 2:2:1 mole ratio of triphenylphosphine, sodium chlorodifluoroacetate and ketone in dry DIGLYME and obtained good to excellent yields of substituted trifluoroacetophenones (Table 2) ⟨65TL1883⟩. The reaction was subsequently extended to other ketones and its mechanism examined ⟨67JOC1311⟩. Control experiments showed that little if any free difluorocarbene is formed under the conditions of the reaction; in contrast to what is found for dichlorocarbenes and dibromocarbenes (see below) ⟨68TL71⟩. It is suggested that the betaine (**24**) is the intermediate and that this subsequently decarboxylates to give the difluoromethylene ylide (**25**).

Table 2 Preparation of 2-arylperfluoropropenes, $R^1C(R^f)$=CF$_2$.

R^1	R^f	Yield (%)	Ref.
C$_6$H$_5$	CF$_3$	75.9	65TL1883, 66CI(L)2054
p-MeC$_6$H$_4$	CF$_3$	81.9	65TL1883
p-MeOC$_6$H$_4$	CF$_3$	90.7	65TL1883
p-FC$_6$H$_4$	CF$_3$	66.6	65TL1883
p-ClC$_6$H$_4$	CF$_3$	64	67JOC1311
p-Me$_2$NC$_6$H$_4$	CF$_3$	16	67JOC1311
n-C$_4$H$_9$	CF$_3$	59	67JOC1311
C$_6$H$_5$CH$_2$	CF$_3$	61	67JOC1311
C$_6$H$_{11}$	CF$_3$	65	67JOC1311
C$_6$H$_5$	C$_2$F$_5$	48	67JOC1311
C$_6$H$_5$	C$_3$F$_7$	42	67JOC1311

(**24**)   (**25**)

It was further conjectured that the difficulties experienced in extending the reaction to more substituted ketones is due to fluoride ion catalysed isomerisation (Scheme 1) and that the use of lithium chlorodifluoroacetate in DMF should be preferred ⟨68JOC1854⟩. However, even these conditions give only isomerised or addition products when a carbanion-stabilising group such as Br, I, CN or NO$_2$ is present on the aromatic ring (Equation (2)) ⟨72JFC123⟩.

It was obvious that another technique was needed and this was provided by the use of dibromodifluoromethane with triphenylphosphine in DIGLYME at 70 °C (Table 3) ⟨72JFC123⟩.

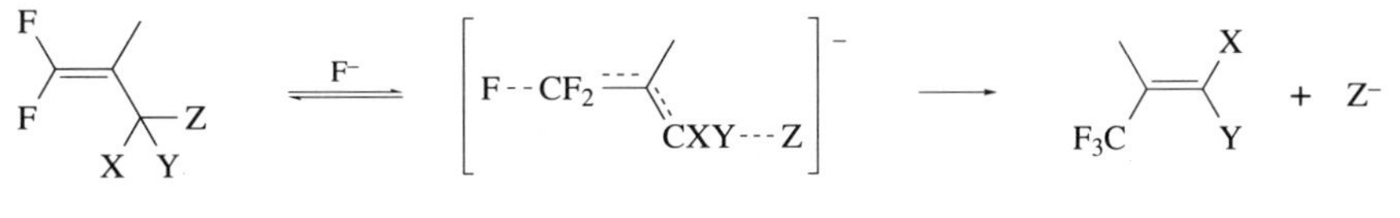

Scheme 1

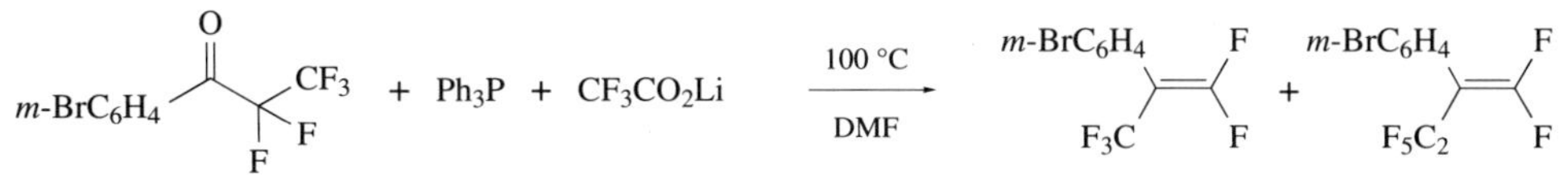

$$m\text{-BrC}_6\text{H}_4\text{COCF}_2\text{F} + \text{Ph}_3\text{P} + \text{CF}_3\text{CO}_2\text{Li} \xrightarrow[\text{DMF}]{100\,°\text{C}} m\text{-BrC}_6\text{H}_4(\text{F})\text{C}=\text{C}(\text{F}_3\text{C})\text{F} + m\text{-BrC}_6\text{H}_4(\text{F})\text{C}=\text{C}(\text{F}_5\text{C}_2)\text{F} \quad (2)$$

Table 3 Comparison of the use of dibromodichloromethane and chlorodifluoroacetate for the preparation of 1,1-difluoro alkenes.

$$\text{RCOR}^f + \text{CF}_2\text{ClCO}_2\text{M} \text{ or } \text{CF}_2\text{Br}_2 \xrightarrow{\text{Ph}_3\text{P}} \text{F}_2\text{C}=\text{C}(\text{R})\text{R}^f + \text{CF}_3\text{CF}=\text{C}(\text{R})\text{R}^f + \text{CF}_3\text{CH}(\text{R})\text{R}^f$$

		Terminal alkenes		Internal alkenes		HF additions	
R	R^f	CF_2Br_2	CF_2ClCO_2M	CF_2Br_2	CF_2ClCO_2M	CF_2Br_2	CF_2ClCO_2M
Ph	CF_3	85	68				
p-ClC$_6$H$_4$	CF_3	86	32				37
p-BrC$_6$H$_4$	CF_3	83	2			<1	65
Ph	C_2F_5	82	42	<1	2		
m-BrC$_6$H$_4$	C_2F_5	87	42	<1	54	<1	

These conditions were also used with biphenyl analogues of trifluoroacetophenone, for example to give (**26**) ⟨77JOC1780⟩, and a range of aldehydes and ketones (Equation (3)) ⟨83JFC(23)339⟩. Triphenylphosphine was also replaced by tris(dimethylamino)phosphine (Equation (4)) ⟨83JFC(23)339⟩. Thus, the use of tris(dimethylamino)phosphine in TRIGLYME with dibromodifluoromethane gives a solution which retains its alkenation ability for over 300 hours and which is able to react cleanly with aldehydes and nonactivated ketones (Table 4) ⟨73SC197⟩.

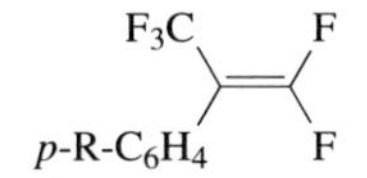

(**26**) R = 4-ClC$_6$H$_4$, 45%

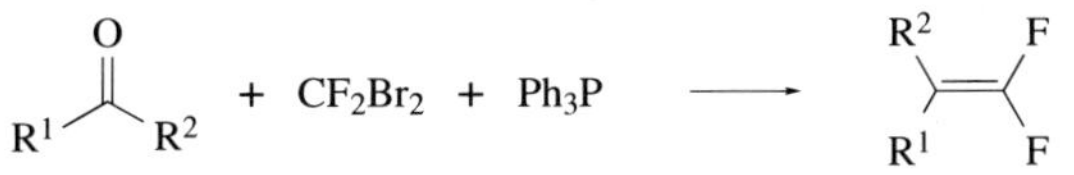

$$R^1\text{COR}^2 + \text{CF}_2\text{Br}_2 + \text{Ph}_3\text{P} \longrightarrow R^1(R^2)\text{C}=\text{CF}_2 \quad (3)$$

R^1 = Ph, R^2 = H, 65%
R^1 = C$_6$F$_5$, R^2 = H, 20%
R^1 = Ph, R^2 = C$_2$F$_5$, 82%
R^1 = n-C$_6$H$_{13}$, R^2 = H, 72%

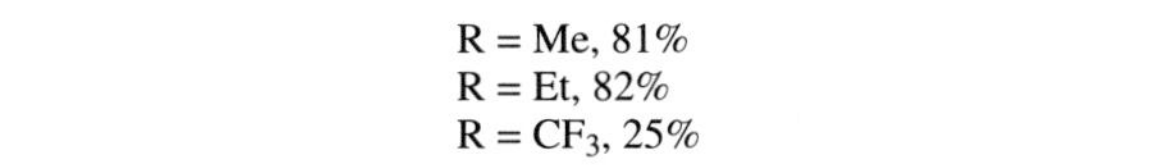

$$\text{PhCOR} + \text{CF}_2\text{Br}_2 + (\text{Me}_2\text{N})_3\text{P} \longrightarrow \text{Ph}(R)\text{C}=\text{CF}_2 \quad (4)$$

R = Me, 81%
R = Et, 82%
R = CF$_3$, 25%

Table 4 Formation of 1,1-difluoro alkenes by reactions with dibromodifluoromethane and tris-(dimethylamino)phosphine.

Aldehyde or ketone	Product	Yield (%)	Ref.
PhC(O)Et	Ph-C(=CF$_2$)-Et	82	73SC197
cyclohexanone	cyclohexylidene=CF$_2$	71	73SC197
Et-C(O)-Et	Et$_2$C=CF$_2$	70	73SC197
TMS-O-C(Et)(Me)-CHO	TMS-O-C(Et)(Me)-CH=CF$_2$	65	83JOC4661
n-C$_{11}$H$_{23}$-CH(OAc)-CHO	n-C$_{11}$H$_{23}$-CH(OAc)-CH=CF$_2$	62	85CPB5137
Ph-CH$_2$CH$_2$-C(O)-CH$_2$OAc	Ph-CH$_2$CH$_2$-C(=CF$_2$)-CH$_2$OAc	87	85CPB5137
TrOCH$_2$-cyclohexyl-O-CHO	TrOCH$_2$-cyclohexyl-O-CH=CF$_2$		84TL4329

Subsequently, Burton and co-workers have shown that 1 equiv. of phosphorus compound can be replaced by metals such as zinc, zinc–copper couples, cadmium or mercury to give improved yields from aldehydes, for example (**27**) ⟨79CL983⟩ or from ketones containing other functional groups which might react with triphenylphosphine, for example (**28**) ⟨81JFC(18)293⟩ or (**29**) ⟨81TL1421⟩.

n-C$_6$H$_{13}$CHO   PhCOCF$_2$Cl   (dimethyl-substituted butyrolactone-dione)

(**27**)   (**28**)   (**29**)

Alternative approaches have used the anions derived from diethyl difluoromethylphosphonate (**30**) (Wittig–Horner technique) ⟨82TL2323⟩, or its silylated derivative (**31**) ⟨82TL2327⟩, or from (difluoromethyl)diphenylphosphine oxide (**32**) ⟨90TL5571⟩. All of these are capable of giving good to excellent yields.

(EtO)$_2$P(O)-CF$_2$R   Ph$_2$P(O)-CF$_2$H

(**30**) R = H   (**32**)
(**31**) R = TMS

4.17.1.2.3 Ylide–carbene reactions

The limitations of the Wittig reactions described above lie in the loss of costly phosphorus reagents converted to oxides and the tendency of enolisable carbonyl compounds to undergo condensation reactions in the presence of strong bases. To overcome these problems Wheaton and Burton used nonstabilised alkylidene- and (arylalkylidene)triphenylphosphoranes with chlorodifluoromethane, (Equation (5)) to give good yields of 1,1-difluoroalkenes (Table 5) ⟨76TL895, 83JOC917⟩. The poor yields from the pentafluorophenyl ylide (33) is believed to be due to extensive charge delocalisation into the pentafluorophenyl group. This technique has also been used to prepare the silyl alkene (34) and the stannyl derivative (35) ⟨79JOM(182)455, 81JOM(205)301⟩.

$$2\ Ph_3\overset{+}{P}-\overset{R^1}{\underset{R^2}{\overset{|}{C}^-}} + HCF_2Cl \longrightarrow \left[Ph_3\overset{+}{P}-\overset{R^1}{\underset{R^2}{\overset{|}{C}}}\right] Cl^- + Ph_3P + \overset{R^1}{\underset{R^2}{}}\!>=<\!\!\!\!\!\!\!\!\!\!\!\!\!\!\!\!\!\!\!\overset{F}{\underset{F}{}} \tag{5}$$

Table 5 Preparation of 1,1-difluoro alkenes, $R^1R^2C=CF_2$ (Equation (5)).

R^1	R^2	Isolated yield (%)
H	n-C_3H_7	58
H	Ph	60
H	C_6F_5	trace
Me	Me	70
—$(CH_2)_2$—		80
Ph	Ph	57
H	$CH=CH_2$	trace

$PH_3\overset{+}{P}-\overset{-}{C}HC_6H_5$

(33)

TMS–C(F)=CF₂ (34)

Me₃Sn–C(F)=CF₂ (35)

4.17.1.2.4 Miscellaneous routes

3-Chloro-3,3-difluoro-2-trifluoromethylprop-1-ene (36) and closely related compounds readily react with lithium aluminum hydride to give (37) in good to excellent yields presumably by an S_N2' reaction ⟨67JOC2749⟩. Further examples of this versatile reaction are shown in Table 6. Chlorodifluoromethyl ketones (38) readily form enol silyl ethers or enol phosphates (39), which can undergo further transformations ⟨83TL507, 83TL5657⟩, whilst trifluoromethyltriphenylsilane (40) reacts with n-butyllithium to give (39; $R^1 = Bu^n$, $R^2 = SiPh_3$) in 95% yield ⟨92TL1221⟩. Full details of the reaction of (40) with a wide range of Grignard reagents to give (37; $R^1 = OSiPh_3$, R = Me, Et, allyl, TMS-CH_2, p-MeOC_6H_4, Me_2CH) in yields of 88–100% have been given ⟨93JCS(P1)795⟩.

(36) $CH_2=C(CF_3)-CF_2Cl$

(37) $R^1R^2C=CF_2$
$R^1 = Me, R^2 = CF_3, 84\%$
$R^1 = Et, R^2 = CF_3, 43\%$
$R^1 = Me, R^2 = CClF_2, 75\%$

Nakai showed that 2,2,2-trifluoroethyl tosylate (41) is readily deprotonated by lithium diisopropylamide (LDA) at $-78\,°C$ and that the anion reacts quantitatively with aldehydes and ketones to give the carbinols (42) ⟨78TL4809⟩. However, the great potential of this reaction was only realised when Ichikawa and co-workers treated the anion with trialkyl boranes to give initially (43) but then

Table 6 S_N2' reactions on fluoro alkenes.

Fluoro alkene	Nucleophile	Product	Yield (%)	Ref.
$CH_2=C(CF_3)CF_2Cl$	$LiAlH_4$	$CH_3C(CF_3)=CF_2$	84	67JOC2749
$CH_2=CHCH=C(CF_3)(CF_2X)$	BuLi	$CF_2=C(CF_3)CH=CHCH_2Bu$	45	70JOC2096
$(MeO)_2C=C(CF_3)CO_2Me$[a]	$BF_3 \cdot OEt_2$[a]	$CF_2=C(CO_2Me)_2$	82	76IZV895
$H_2C=C(CF_2Cl)_2$	PhMgBr	$PhCH_2(CClF_2)=CF_2$	50	68JOC2173
$CF_3CH=CH_2$	$PhMe_2SiSiPhMe_2$	$CF_2=CHCH_2SiPhMe_2$	84	83TL4113, 88T4135
$ClCH_2HCClCF_3$	Ph_2CCN	$Ph_2C(CN)CH_2CCl=CF_2$	77	84JFC(24)387
$CH_2=C(CF_3)CO_2Na$	$LiAlH_4$	$CF_2=C(CH_3)CO_2H$	54	86TL3173
$CH_2=C(CF_3)CO_2Menth$	Ph_2CuLi	$CF_2=C(CH_2Ph)CO_2Menth$	64	88S614
$CH_2=C(CF_3)CO_2Me$	Bu_2CuLi	$CF_2=C(CH_2Bu)CO_2Me$	56	89JOC5630
$(CF_3)_2C=CHCO_2Et$	NaSPh	$CF_2=C(CF_3)CH(SPh)CO_2Et$	67	92JOC5530
$CH_2=C(CF_3)CO_2H$	$CH_2CHMgBr$	$CF_2=C(CO_2H)CH_2CH=CH_2$	30	93JFC(62)201

[a] Lewis acid catalysed fragmentation.

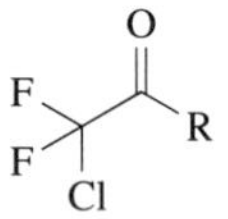

(38) R = alkyl, aryl

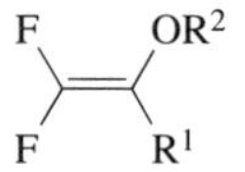

(39) R^1 = Ph, R^2 = OP(O)(OPh)$_2$, 75%
R^1 = n-C$_6$H$_{13}$, R^2 = TMS, 63%

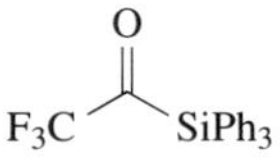

(40)

exploited the great versatility of borane compounds to give in excellent yields the alkenes (**44**) ⟨89TL1641⟩ and (**45**) ⟨89TL6379⟩, α,β-unsaturated ketones (**46**) ⟨91CL961, 92TL337⟩, styrenes (**47**) ⟨92TL3779⟩, dienes (**48**) ⟨92SL739⟩ and alkynes (**49**) ⟨93JFC(63)281⟩.

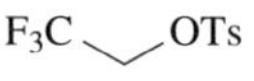

(41) 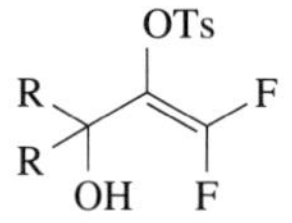 (42) 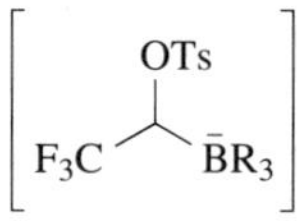 (43) 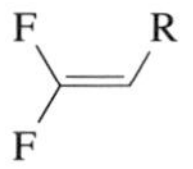 (44) R = (CH$_2$)$_4$Ph, (CH$_2$)$_2$Ph (45) R^2 = R^1, I

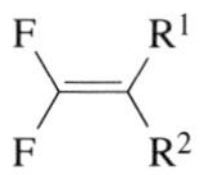

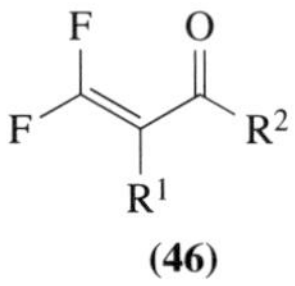

(46) 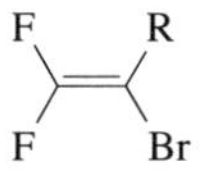 (47) 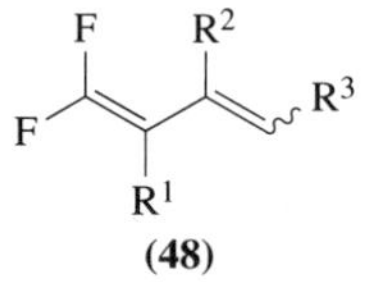 (48) 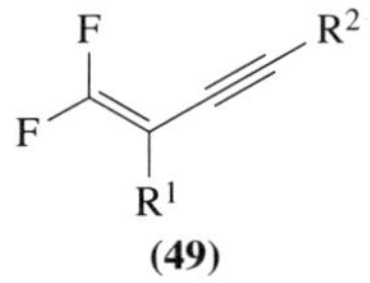 (49)

Palladium-catalysed elimination of (**50**) has been reported to give 49% of (**51**) along with other products ⟨90CPB1104⟩ whilst a Reformatsky reaction of the crotonate (**52**) with ketones gives good to excellent yields of (**53**) as diastereomeric mixtures ⟨92SL977⟩.

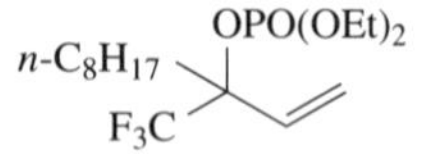

(50) 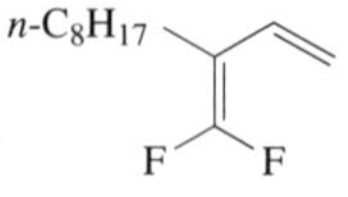 (51) 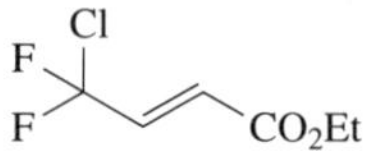 (52) 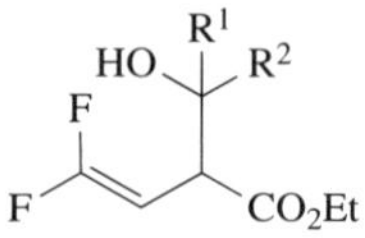 (53)

The vinylstannane (**54**) (see Section 4.18.5.3) reacts with 1-chloromethyl-4-fluorodiazoniabicyclo[2.2.2]octane bis(tetrafluoroborate) to give a 1,1-difluoro alkene, for example (**55**) in 45% yield ⟨93TL3057⟩.

Allenes, for example (**56**), react with methanol to give a low yield of, for example (**57**) ⟨78JCS(P1)422⟩, whilst (**58**) reacts with silver fluoride to give the stable complex (**59**) which readily undergoes further conversions ⟨70TL5215⟩. The fluoroketene (**60**) thermolyses over sodium fluoride

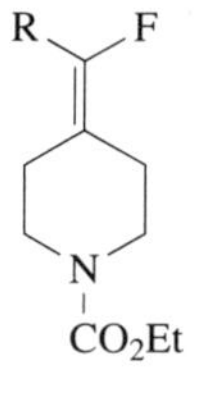

(54) R = SnBu$_3$
(55) R = F

at 300 °C to give 34% of pentafluoromethylacryloyl fluoride (61) ⟨66JA5582⟩ whilst the ketene silyl acetal (62) condenses with dibromodifluoromethane to give (63) in 84% yield ⟨91CL1319, 92BCJ1513⟩.

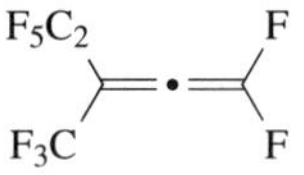

(56)

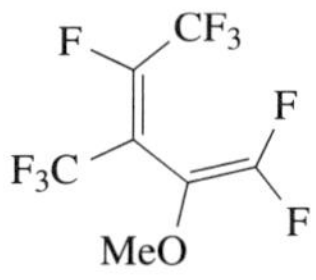

(57)

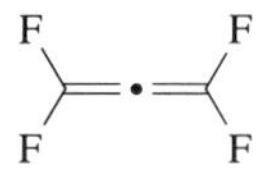

(58)

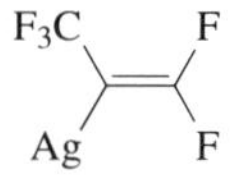

(59)

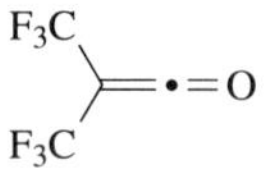

(60)

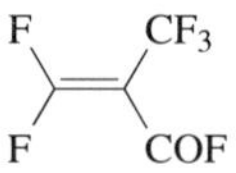

(61)

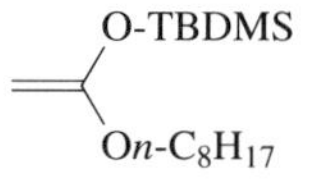

(62)

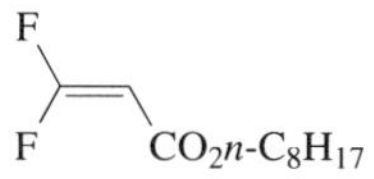

(63)

Difluorocarbene, generated from (bromodifluoromethyl)triphenylphosphonium bromide, gave only a very low (8%) yield of (64) on reaction with 1,2-diphenylcyclopropene ⟨92TL4537⟩. A rather more successful reaction which may also involve carbenes is the lithium chloride catalysed decomposition of methyl chlorodifluoroacetate in the presence of α-chloro-α,α-difluoroacetophenone to give 2,2-difluoro-1-chlorostyrene (65) in 50% yield ⟨78JOC2643⟩.

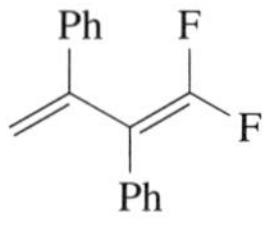

(64)

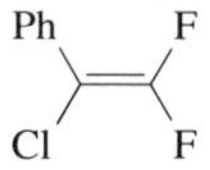

(65)

Finally, a miscellaneous group of thermal decompositions and rearrangements has been reported as leading to this system. Thus, the potassium enolate (66) at 100 °C gives 61% of (67) ⟨81JFC(17)441⟩ whilst copyrolysis of pentafluorobenzotrifluoride with chlorodifluoromethane gave 80% of 1-chloro-perfluorostyrene (68) ⟨85JFC(28)99⟩. Pyrolysis of (69) over platinum at 670 °C gave 59% of the diene (70) ⟨78CC475⟩, while the homologue (71) underwent quantitative conversion into (72) ⟨80JCS(P1)487⟩. UV irridation of perfluorocyclohexane gave perfluoro(methylenecyclopentane) (73) as one of a number of products ⟨71JCS(C)925⟩. Perfluorocyclobutane gave perfluoroisobutene (74) in 70% yield ⟨53JA2698⟩.

4.17.1.3 1,1-Dichloro Alkenes

4.17.1.3.1 *Wittig and related reactions*

In 1960 Speziale *et al.* showed that triphenylphosphine can trap dichlorocarbene, generated by the action of potassium *t*-butoxide on chloroform, to give dichloromethylenetriphenylphosphine (75) which then condensed with benzophenone to give (76) in 46% yield ⟨60JA1260⟩. Subsequently, it was shown that the preparative usefulness of the procedure was greatly improved by rapid removal of the formed *t*-butyl alcohol by azeotropic distillation with *n*-pentane (Table 7) ⟨62JA854, 65OR(14)270⟩. In 1988, the ylide (75) generated by this technique, has been shown to react with fluorinated alkenyl phosphates (77) to give the polyfluorinated (*E*)-1,3-dienes (78) ⟨88JFC(41)435⟩.

Rabinowitz and Marcus found that triphenylphosphine reacts with carbon tetrachloride either at

Table 7 Wittig preparation of 1,1-dichloro alkenes, $R^1R^2C\!\!=\!\!CCl_2$ from Ph_3PCCl_2 **(75)**.

R^1	R^2	Yield (%)	Ref.
Ph	Ph	59	62JA854
—(CH$_2$)$_6$—		33	62JA854
PhCH=CH	H	77	62JA854
p-NO$_2$C$_6$H$_4$	H	83	62JA854
p-NMe$_2$C$_6$H$_4$	H	81	65OS(45)33
2,6-Cl$_2$C$_6$H$_3$	H	46	65OS(45)33

Another product of the preparation is ⟨70CA21539j⟩.

room temperature during 48 hours or at 60 °C during 2–3 hours to give a solution containing **(75)** and **(79)** ⟨62JA1312⟩. This solution reacted with benzaldehyde to give 1,1-dichlorostyrene and benzal chloride and with benzophenone to give **(76)** along with an equivalent of unchanged benzophenone. Despite the requirement for a two molar excess of phosphine, the ease of the method has lead to its widespread use (Table 8). However, there are other drawbacks and Isaacs and Kirkpatrick reported that a number of ketones give mixtures of enyl chlorides **(80)** and 1,1-dichloro alkenes (Table 9) ⟨72CC443⟩. Subsequently, it was shown that the use of acetonitrile as solvent largely suppressed the formation of enyl chloride and that replacement of carbon tetrachloride by bromotrichloromethane gave 100% of the 1,1-dichloroalkene **(81)** ⟨88TL3003⟩. In an extension of this methodology, Suda and Fukushima found that esters, for example **(82)**, and ketones give 2-alkoxy 1,1-dichloro alkenes **(83)** in varying yields ⟨81TL759⟩.

An alternative method of improving this reaction has been the replacement of triphenylphosphine by tris(dimethylamino)phosphine (Table 10) ⟨64LA(679)51, 71TL1035⟩ or by concomitant replacement of tetrachloromethane by bromotrichloromethane ⟨77TL1239⟩. Appel has explained the difference in activity induced by these phosphines as a result of the greater basicity of the tris(dimethylamino)phosphine which gives not the ylide **(75)** but the salt **(84)** which then reacts further (Scheme 2) ⟨75AG(E)801⟩.

(79) Ph₃PCl₂ **(80)** **(81)** **(82)** **(83)** 82%

Table 8 Preparation of 1,1-dichloro alkenes by treatment of aldehydes or ketones with **(75)** generated from triphenylphosphine and tetrachloromethane, $R^1R^2C=CCl_2$.

R^1	R^2	Yield (%)	Ref.
Ph	CN	67	73JOC479
p-MeC$_6$H$_4$	CN	68	73JOC479
p-ClC$_6$H$_4$	CN	20	73JOC479
p-MeOC$_6$H$_4$	CN	60	73JOC479
Ph	CH(OEt)$_2$	71.5	70CR(C)1467
CH$_3$	CO$_2$Et	39	70CR(C)1467
Ph	CO$_2$Et	67	73MI 417-01

Table 9 Reaction of triphenylphosphine in tetrachloromethane with various ketones.

Ketone	Yield (% total product per mole Ph₃P)	Product ratio	
		enyl chloride	1,1-dichloro alkene
cyclohexanone	65	93	7
4-t-butylcyclohexanone	70	95	5
cyclopentanone	64	5	95
cycloheptanone	80	50	50
acetophenone	55	18	82
norcamphor	80	5	95
cholestan-3-one	80	0	100

Table 10 Use of tris(dimethylamino)phosphine and either tetrachloromethane or bromotrichloromethane with aldehydes, RCHO, to give 1,1-dichloro alkenes (Scheme 2).

R	Yield (%)	Ref.
Pr^i	55	71TL1035
Bu^t	50	71TL1035
Pr^{i}CHOHCCl$_3$[a]	70	71TL1035
n-C$_5$H$_{11}$	94	77TL1239
Ph	86	77TL1239
(bicyclic lactone)	85	77TL1239
EtO$_2$C-cyclopropyl(Me,Me)	80	80S554

[a] This alcohol may be used in place of the aldehyde Pr^{i}CHO.

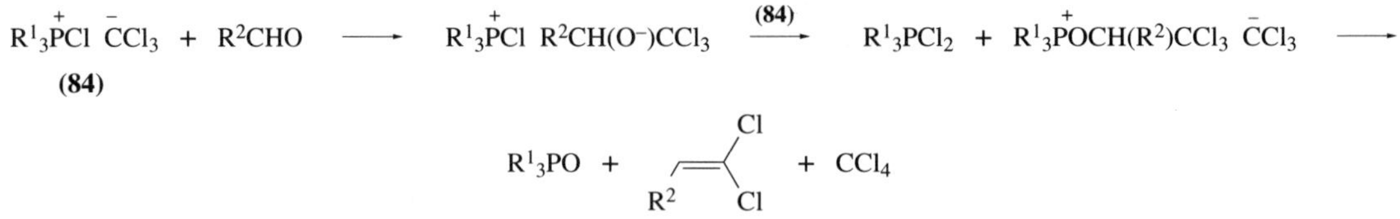

$$R^1_3\overset{+}{P}Cl\ \overset{-}{C}Cl_3 + R^2CHO \longrightarrow R^1_3\overset{+}{P}Cl\ R^2CH(O^-)CCl_3 \overset{(84)}{\longrightarrow} R^1_3PCl_2 + R^1_3\overset{+}{P}OCH(R^2)CCl_3\ \overset{-}{C}Cl_3 \longrightarrow$$

(84)

$$R^1_3PO + \underset{R^2}{\overset{}{\diagup}}=\underset{Cl}{\overset{Cl}{\diagdown}} + CCl_4$$

Scheme 2

Clement and Soulen however, showed that the simplest procedure is to replace tetrachloromethane by bromotrichloromethane and to run the reaction in dry benzene at 0 °C when the reaction proceeds rapidly and in high yield (Table 11) ⟨76JOC556⟩. Overall, it is apparent that Wittig reactions producing 1,1-dichloro alkenes proceed more readily than those giving the analogous fluoro compounds.

Table 11 Wittig reaction with bromotrichloromethane and R^1R^2CO to form 1,1-dichloro alkenes, $R^1R^2C{=}CCl_2$.

R^1	R^2	Yield	Ref.
CH_3	CN	90 (61)[a]	76JOC556
Et	CN	82(45)	76JOC556
Pr^i	CN	87(40)	76JOC556
Bu^t	CN	91(48)	76JOC556
p-ClC_6H_4	CN	43(20)	76JOC556
p-$MeOC_6H_4$	CN	85(60)	76JOC556
$MeO_2C(CH_2)_2$	H	41	81JOC4290

[a] Yield using CCl_4.

The Wittig–Horner route with diethyl lithiodichloromethylphosphonate (**85**) has been exploited by several groups (Table 12) ⟨73JOM(59)237, 75S458, 75S535, 75TL609⟩. From Table 12 it can be seen that high yields are generally obtained even in the case of hindered substrates. Routes to (**85**) involve the reaction of butyllithium on either diethyl trichloromethylphosphonate or on diethyl dichloromethylphosphonate.

$$(EtO)_2P(O)CLiCCl_2$$

(**85**)

4.17.1.3.2 *Carbene reactions*

Dichlorocarbene, generated by the action of potassium *t*-butoxide on chloroform, reacts with diaryldiazomethanes, for example (**86**) to give (**87**) in good yields ⟨62AG153, 64CB3503⟩. Similar condensations also take place with Wittig compounds (**88**) and carbenes generated by decomposition of sodium trichloroacetate ⟨64MI 417-01, 64TL7, 66T2615⟩.

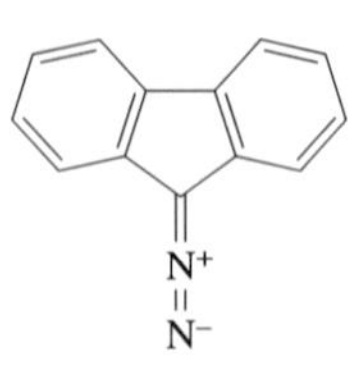

(**86**)

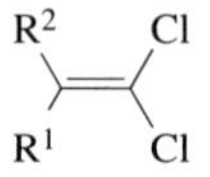

(**87**)

R^1, R^2 = (biphenyl), 80%
$R^1 = R^2 = p$-CNC_6H_4, 23%
$R^1 = R^2 = p$-BrC_6H_4, 80%

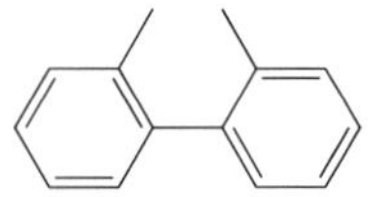

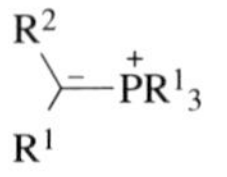

(**88**)
R^1, R^2 = fluorenyl
$R^1 = R^2 = CO_2Et$

Table 12 1,1-Dichloro alkenes, $R^1R^2C=CCl_2$, by Wittig–Horner reactions from the phosphonate (**85**).

R^1	R^2	Yield (%)	Ref.
Me	Me	47	73JOM(59)237
Ph	Ph	69	73JOM(59)237
C_9H_{19}	H	92	75S458
Et_2CH	H	82	75S458
c-C_6H_{11}	H	84	75S458
$PhCH=CH$	H	92	75S458
2-$MeOC_6H_4$	H	94	75S458
$EtSCH=CH$	H	87	75S458
p-$MeOC_6H_4$	H	97	75S458
p-FC_6H_4	H	80	75S535
(cycloheptyl with methyl)		78	75S535
(cycloheptyl with gem-dimethyl)		84	75S535
(Pr^i-substituted cycloheptyl with methyl)		70	75S535
(2,2'-dimethylbiphenyl)		70	75S535
(4-methylcyclohexylidene with ethyl)		90	75S535

4.17.1.3.3 Carbenoid reactions

In this section, the term 'carbenoid' is used in the sense defined by Kobrich who used the term '...without mechanistic implications to denote α-haloorganolithium compounds, and compounds in general that have a metal atom and a leaving group on the same carbon atom...' ⟨72AG(E)473⟩. Thus, 1-chloro-2,2-diphenylethene (**89**) can be deprotonated by lithiodichloromethane at −85 °C and the resultant ion (**90**) quenched with tetrachloromethane to give (**76**) in 91% yield ⟨67AG(E)41⟩. Lithiodichloromethane also reacts with ketones to give the adducts (**91**) which dehydrate to (**92**) (from methyl ketones) on treatment with either chlorotrimethylsilane alone ⟨76CB2175⟩ or, subsequently, with lithium diisopropylamide ⟨89JCS(P1)691⟩. Lithiotrichloromethane reacts at −100 °C with lithiated *n*-heptylphenyl sulfone in THF : ether (50 : 50) to give a 58% yield of (**93**)

⟨89JOM(379)201⟩ whilst (trimethylsilyl)dichloromethyllithium reacts with aldehydes at −100 °C followed by hydrolysis below −80 °C to give (**94**). Boron trifluoride etherate or sulfuric acid in dichloromethane then readily gives (**95**) ⟨83TL4727⟩.

The sulfoxide-stabilised anion (**96**) reacts with primary alkyl halides to give good to excellent yields of the 1,1-dichloro alkenes (**97**) ⟨83TL527⟩.

(**89**) R = H
(**90**) R = Li

(**91**)

(**92**)
R = Me$_2$=CHCH$_2$CH$_2$–, 56%
R = Me$_2$C=CHCH$_2$–, 35%
R = c-C$_3$H$_5$–, 48%
R = Ph, 51%

(**93**)

(**94**)

(**95**)
R = Ph, 77%
R = n-C$_5$H$_{11}$, 97%
R = PhCH$_2$CH$_2$, 73%
R = p-ClC$_6$H$_4$, 85%

PhS(O)C(Cl)$_2$Li
(**96**)

(**97**)
R = n-C$_8$H$_{17}$, 67%
R = n-C$_{10}$H$_{21}$, 80%
R = n-C$_{12}$H$_{25}$, 83%
R = 2-Cl,3-MeOC$_6$H$_3$, 76%

4.17.1.3.4 Elimination reactions

There have been many reports of the elimination of a chlorine atom from the trichloromethyl group (Table 13). Also, for example, the trichloro compound (**98**) readily eliminates hydrogen chloride to give 49% of (**99**) ⟨67JOC1941, 93MM916⟩. A closely related synthesis involves the reaction of thiophene 1,1-dioxide with chloroform in sodium hydroxide–benzyltrimethylammonium chloride to give 3-dichloromethylene-2,3-dihydrothiophene 1,1-dioxide (**100**) ⟨89AJC1307⟩. Further, the enol sulfonate (**101**) gave 85% of (**102**) in bromotrichloromethane at 100 °C ⟨70JA3203⟩. Vernier and coworkers have shown that thiofluorenone S-oxide (**103**) reacts in refluxing dimethoxyethane with sodium trichloroacetate to give 80% of dichloromethylenefluorene (**104**) ⟨74JOC501⟩. At first sight this appears to be a carbene reaction but the authors produce good evidence that the high electrophilicity of the sulfine group results in initial attack by the trichloromethyl anion on the sulfur atom followed by a Ramberg–Backlund rearrangement involving chloride ion loss to give (**105**), which then loses sulfur monoxide.

Enamines such as (**106**) react with tetrachloromethane to give good yields of α-dichloromethylene ketones (**107**) ⟨66CC567, 70JOC1211⟩ whilst phenyl acetate reacts with phosphorus trichloride at 120 °C to give 70% of 1,2,2-trichlorovinylphenylether ⟨56M323⟩.

4.17.1.3.5 Rearrangements

The bicyclic trihalocyclopropanes (**108**) ring open in quinoline at 100 °C to give the cyclic dienes (**109**) ⟨82TL3795⟩ whilst thermolysis of cyclopropanes (**110**) gives the dienes (**111**) in high yields ⟨85JCR(S)182, 90JCS(P1)1881⟩. The reactions can also, as expected, be catalysed by silver ion. Thus, the cyclopropane (**112**) with silver perchlorate in methanol gave (**113**) in 80% yield ⟨88JCR(S)292⟩. A similar silver ion catalysed ring opening of the cyclopropa[c]chromene (**114**) gave 93% of (**115**) ⟨90JCS(P1)139⟩. The dichlorocyclobutene (**116**), ring opened in refluxing butanol to give, via the ketene, good yields of the β,γ-ketone (**117**) ⟨83JOC3382⟩. The reaction can be performed photolytically and the ketene trapped by alkenes to give, for example (**118**) with styrene ⟨86TL6389⟩. Compound

Table 13 Preparation of 1,1-dichloro alkenes from trichloromethyl derivatives.

Starting material	Conditions	Yield (%)	Ref.
Ph$_2$CHCCl$_3$	KOH-EtOH	80	26JA3144
PhCCl$_2$CCl$_3$	Zn-Et$_2$O	81	50JA3952
CCl$_3$CHClCHCl$_2$	AlCl$_3$-CCl$_4$	84	63BSF2147
[dihydrofuran with CCl$_3$]	Cr^{2+}	80	86JCS(P1)733
CCl$_3$CH$_2$CHClOEt[a]	H$_2$O	41	59BSF1828
(ClC$_6$H$_4$)$_2$CHCCl$_3$	KO$_2$, DMF	95	82TL5003
Cl$_3$CCONMe$_2$[b]	(EtO)$_3$P	53	60JA903
(p-ClC$_6$H$_4$)$_2$CHCCl$_3$	KOH	70	55OSC(3)270
Cl$_3$CC(CO$_2$Et)=CHCO$_2$Et[c]	(EtO)$_3$P, toluene, Δ	40	89AJC301
[steroid with CCl$_3$ and OH]	pyridine	83	60T149
[4-(4-BrC$_6$H$_4$)-C$_6$H$_5$]$_2$CHCCl$_3$	KOH, EtOH	90	62JOC1426
[tetrachlorocyclopentadiene with CCl$_3$ and Cl]	P(OEt)$_3$	94	72JOC683
(4-NO$_2$C$_6$H$_4$)$_2$CHCCl$_3$	KMnO$_4$, acetone	88	48CB422
2,4-Me$_2$,5-Cl-C$_6$H$_2$CHCCl$_3$	piperidine	93	62JA995
[p-tolyl-CH(CCl$_3$)CO$_2$H]	KOH, H$_2$O, EtOH	93	56JOC638
CCl$_3$(Me)C=CH$_2$[d]	2N NaOH	78	58JA5988

[a] Product is Cl$_2$C=CHCHO. [b] Product is Cl$_2$C=C(Cl)NMe$_2$. [c] Product is Cl$_2$C=C(CO$_2$Et)CH(CO$_2$Et)PO(OEt)$_2$.
[d] Product is Cl$_2$C=C(Me)CH$_2$OH.

(**116**) has also been ring opened to (**117**) in 50% yield by treatment with oxovanadium(V) species ⟨94JCS(P1)3⟩.

4.17.1.3.6 Miscellaneous methods

Chlorination of ketene dithioacetals, for example (**119**), in acetic acid gives low to moderate yields of the corresponding dichloro derivative (**120**) ⟨66CB2900⟩ whilst Oshima and co-workers have shown that the ketene silyl acetates (**121**; R = trialkylsilyl) react with tetrachloromethane or bromotrichloromethane in the presence of triethylborane to give good yields of the 3,3-dichloroacrylate (**122**) (Table 14) ⟨91CL1319, 92BCJ1513⟩.

Chloropropiolic acid adds hydrogen chloride in essentially quantitative yield to give 3,3-dichloro-

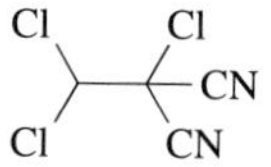

Two Halogens or a Halogen and a Chalcogen

(98) 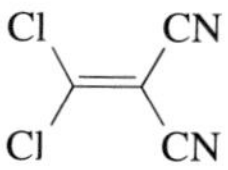 (99) 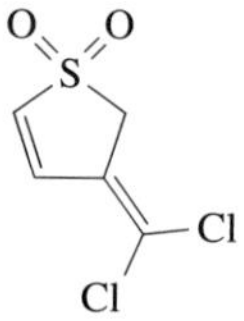 (100) 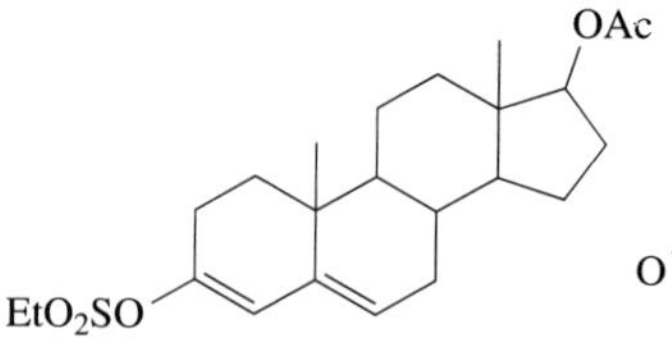 (101) 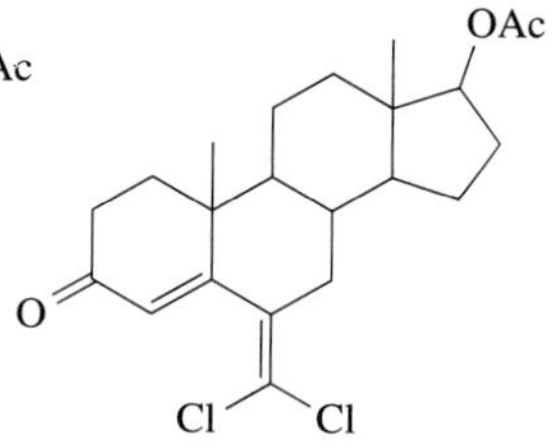 (102)

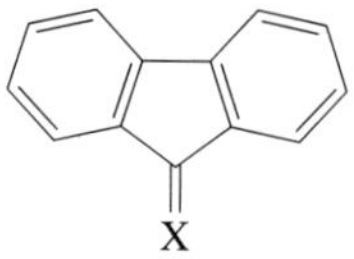

(103) X = S = O
(104) X = CCl$_2$

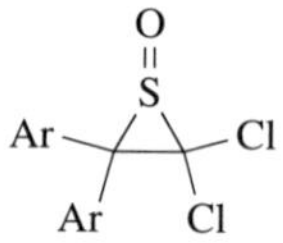

(105)

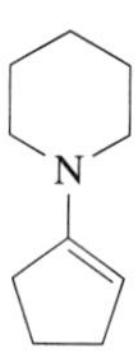

(106) 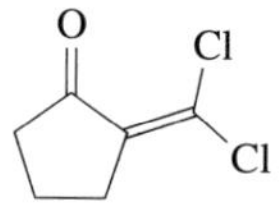 (107)

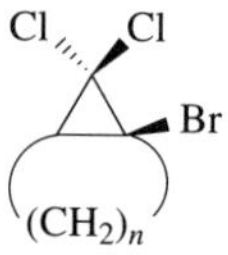

(108) n = 4, 5, 6

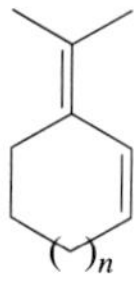

(109) n = 1, 90%
n = 2, 70%
n = 3, 65%

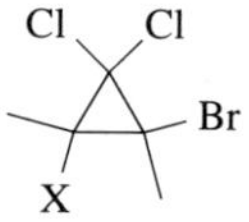

(110) X = Br, Me

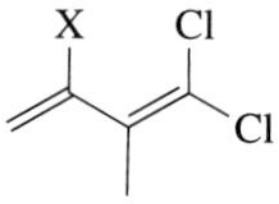

(111) X = Br, Me

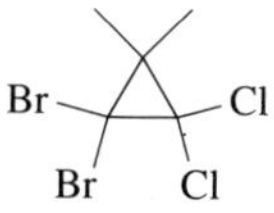

(112) 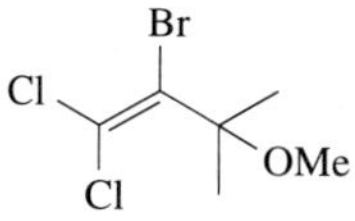 (113) 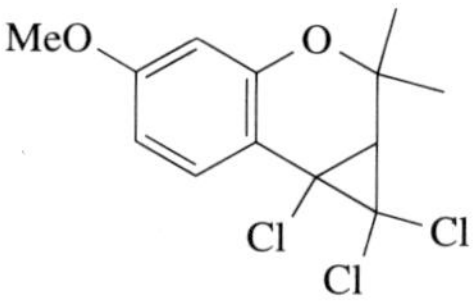 (114) 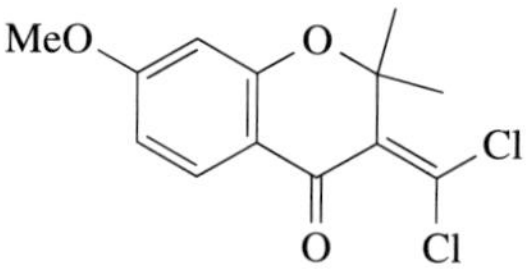 (115)

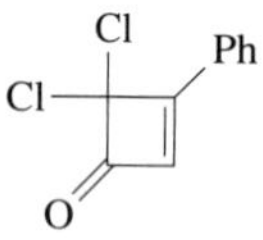

(116) 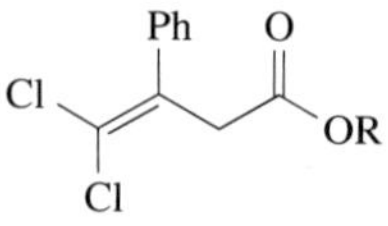 (117) R = Bu^n, 97% 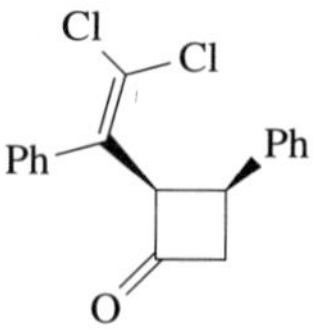 (118) 75%

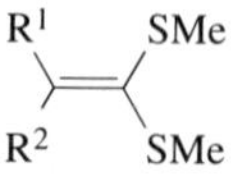

(119) 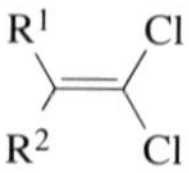 (120)
R^1 = R^2 = CN, 15%
R^1 = CN, R^2 = CO$_2$Me, 33%
R^1 = R^2 = CO$_2$Me, 63%

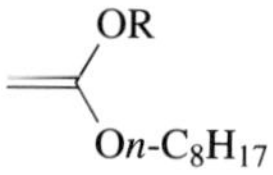

(121) 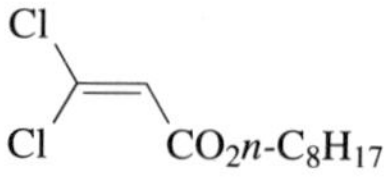 (122)

Table 14 Reactions of the ketene silyl acetals (121) with polyhalogenomethanes to form n-octyl 3,3-dichloroacrylate (122).

R	Polyhalogenomethanes	Yield of (122) (%)
SiMe$_2$Bu^t	CCl$_4$	79
SiMe$_2$Bu^t	CBrCl$_3$	93
TMS	CCl$_4$	71
TMS	CBrCl$_3$	78

propenolic acid ⟨30CB1868⟩. p-Cresol condenses with 3,3,3-trichloro-2-methylpropene to give a low yield of (123) ⟨56JA6101⟩ whilst 1,1-diaryl-2,2-dichloroethenes, for example (124), were prepared in moderate yields from mixtures of anisole, phenetole and dichloroacetyl chloride ⟨88JOC5554⟩.

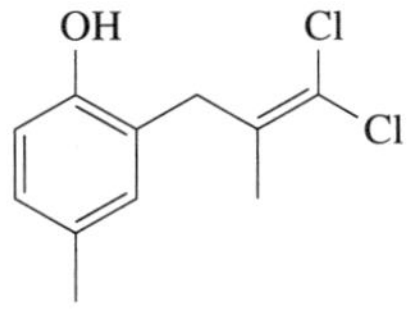

(123)

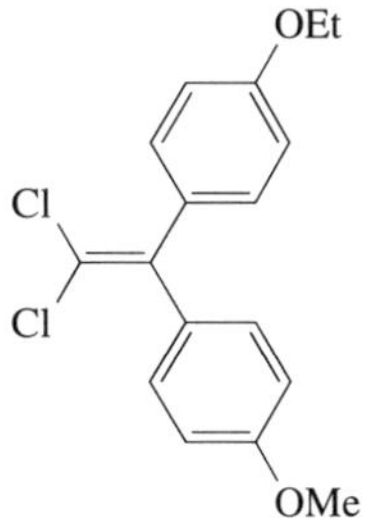

(124)

4.17.1.4 Dibromo Alkenes

4.17.1.4.1 Wittig and related reactions

Ramirez et al. showed that 2 m of triphenylphosphine reacted with tetrabromomethane in dry dichloromethane to give the dibromophosphorane (125) which then rapidly reacted with benzaldehyde to give 84% of 1,1-dibromostyrene (126) ⟨62JA1745⟩. Subsequently, Corey and Fuchs showed that addition of zinc dust removed the requirement for excess triphenylphosphine and produced the derivatives (127) in yields of 80–90% ⟨72TL3769⟩.

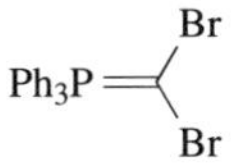

(125) 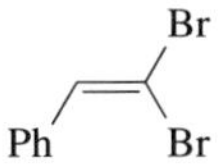 (126) 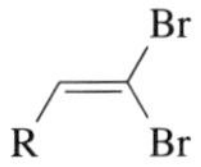 (127) R = Ph, n-C$_7$H$_{15}$,

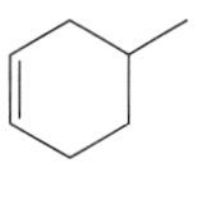

Table 15 lists some representative aldehydes and ketones which reacted with triphenylphosphine and carbon tetrabromide. It is clear that a wide range of conditions have been employed depending on the reactivity of the substrate, with steric effects being particularly important. Thus 1-decalone is recovered under conditions in which 2-decalone reacts in high yield ⟨75TL1373⟩.

The steric problem can be overcome by the use of the lithium salt of diethyl dibromomethylphosphonate (128) which reacts with a wide range of aldehydes and ketones to give 1,1-dibromo alkenes (129) (Table 16) ⟨76S197⟩.

It is also possible to use the potassium t-butoxide bromoform system analogous to that described above in Section 4.17.1.3.1 ⟨62JA854⟩.

Combret et al. have reported good yields of the dibromo alkenes (129) when tris(dimethylamino)phosphine is used in place of triphenylphosphine ⟨71TL1035⟩.

Table 15 Part 1: Preparation of 1,1-dibromo alkenes, $R^1R^2C=CBr_2$, from PH_3P/CBr_4 and an aldehyde or ketone.

R^1	R^2	Yield (%)	Conditions	Ref.
$CCl_2=CH$	CO_2Me	75	CH_2Cl_2 RT	72CR(C)1357
(cyclopropane with H, CO_2Bu^t, Me, Me)	H	'satisfactory'		74JCS(P1)2470
(cyclohexylidene)		86	toluene, 65°C, 15 h	84CB1877
Me	Me	48	benzene, Δ	86OM1991
Me	Et	22	benzene, Δ	86OM1991
Me	$EtOCH_2$	64	benzene, Δ	86OM1991
Me	Et_2NCH_2	41	benzene, Δ	86OM1991
Me	$MeOCH_2CH_2$	41	benzene, Δ	86OM1991
Me	$MeO(Me)CH$	23	benzene, Δ	86OM1991
$MeOCH_2$	$MeOCH_2$	75	benzene, Δ	86OM1991
$MeOCH_2$	$MeOCH_2CH_2$	24	benzene, Δ	86OM1991
CF_3	Ph	80	benzene, Δ	86OM1991
—$(CH_2)_4$—		55	benzene, Δ	86OM1991
—$(CH_2)_5$—		50	benzene, Δ	86OM1991
$4\text{-}MeC_6H_4$	H	57	CH_2Cl_2, 5°C	80LA2061
$4\text{-}NO_2C_6H_4$	H	80	CH_2Cl_2, 5°C	80LA2061
2-Thienyl	H	32	CH_2Cl_2, 5°C	80LA2061
$n\text{-}C_9H_{19}$	H	85	CH_2Cl_2, 5°C	80LA2061
TMS-C≡C	C≡C-TMS	66	benzene, Δ	91AG(E)698
PhC≡C	C≡C-Ph		2 equiv. CBr_4 5 equiv. Ph_3P CH_2Cl_2	94CC205
C_2F_5	Ph	67	CH_2Cl_2, −48°C	93JA5430
CF_3	C_6F_5	76	CH_2Cl_2, −48°C	93JA5430
CF_2Cl	Ph	67	CH_2Cl_2, −48°C	93JA5430
CF_3	CO_2Et	26	CH_2Cl_2, −48°C	93JA5430
Me	$CH_2CH_2CH=CH_2$	39	benzene, Δ	82JA2223
$Cl_2C=CH$	H	76	CH_2Cl_2, 0°C	70CR(C)1467
$Br_2C=CH$	H	72	CH_2Cl_2, 0°C	70CR(C)1467
$Cl_2C=CCl$	H	62	CH_2Cl_2, 0°C	70CR(C)1467

Table 15 Part 2.

$R^1R^2C=O$	Yield (%)	Conditions	Ref.
4-*t*-butylcyclohexane	70	benzene, Δ	75TL1373
(decalone structure)	86	benzene, Δ	75TL1373
trans-2-decalone	80	benzene, Δ	75TL1373
3,3-dimethylcyclopentanone	78	benzene, Δ	75TL1373
cycloheptanone	28	benzene, Δ	75TL1373
2-octanone	35	benzene, Δ	75TL1373

4.17.1.4.2 Carbenoid reactions

Metallation of bromoform with lithium diisopropylamide at −105°C in THF:ether (50:50) produces a solution of lithium tribromomethylide which reacts with the lithiated sulfone (**130**) to give good yields of (**131**) ⟨89JOM(379)201⟩. Dibromomethyllithium also reacts with ketones in dry

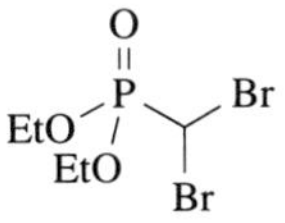

(128) — (EtO)₂P(O)CHBr₂

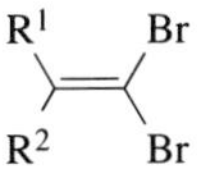

(129) R¹R²C=CBr₂
R¹ = Pr^i, R² = H, 54%
R¹ = n-C₅H₁₁, R² = H, 54%
R¹ = Ph, R² = H, 70%
R¹R²C = fluorenylidene, 50%

Table 16 Reactions of lithium diethyl dibromomethyl phosphonate with aldehydes and ketones to give the 1,1-dibromo alkenes (**129**).

R^1	R^2	Yield (%)
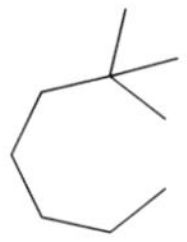 (cyclohexyl, t-Bu substituted)		45
 (2,6-dimethylcyclohexyl)		53
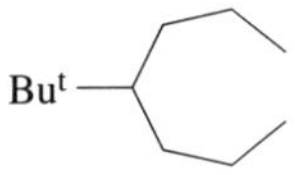 4-t-Bu-cyclohexyl		70
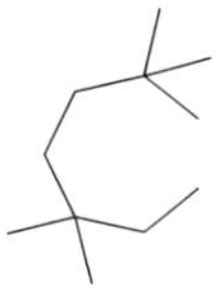 (3,3,5,5-tetramethylcyclohexyl)		61
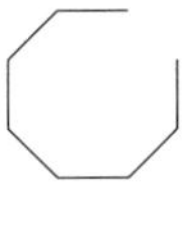 cyclooctyl		67
PhCH=CH	Me	80
Ph	Ph	50
Ph	H	40
p-Me₂NC₆H₄	H	61

ether to give initially the alkoxides (**132**) which are then trapped by chlorotrimethylsilane to produce (**133**); elimination of trimethylsilanol by LDA at −100 °C then yields (**134**) ⟨76CB2175⟩. Barluenga and co-workers have produced further examples of the use of this reaction to give, for example, the dibromo derivatives (**135**) ⟨89JCS(P1)691⟩.

4.17.1.4.3 Ring-opening reactions of cyclopropanes

Baird and Hussain reported that the tetrabromocyclopropane (**136**) ring opens at 60–65 °C during 6 h to give a 67% yield of (**137**) ⟨88JCR(S)292⟩.

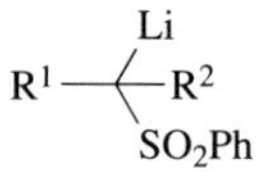

748

(130) R¹–C(Li)(R²)(SO₂Ph)

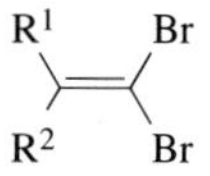

(131) R¹R²C=CBr₂
R¹ = Buⁿ, R² = H, 66%
R¹ = n-C₆H₁₃, R² = H, 67%
R¹ = Me₂C=CH, R² = H, 37%
R¹ = Ph, R² = Me, 60%
R¹ = n-C₆H₁₃, R² = Me, 67%

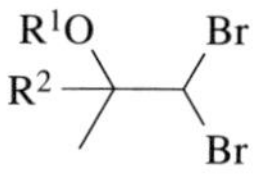

(132) R¹ = Li
(133) R¹ = TMS
R¹O–C(R²)(Br)–CH(Br)

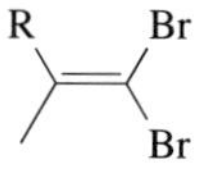

(134) RMeC=CHBr
R = Me₂C=CHCH₂CH₂, 58%
R = Me₂C=CHCH₂, 65%
R = c-C₃H₅, 35%
R = Ph, 59%

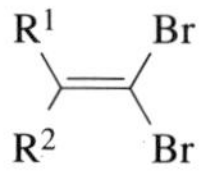

(135) R¹R²C=CBr₂
R¹ = R² = Et, 60%
R¹, R² = (CH₂)₅, 70%

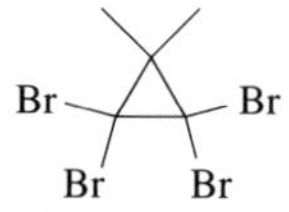

(136) tetrabromocyclopropane

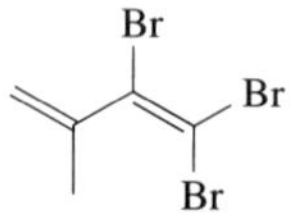

(137) 2-methyl-3,4,4-tribromo-1,3-butadiene (CH₂=C(Me)–C(Br)=CBr₂)

4.17.1.4.4 Miscellaneous reactions

Potassium hydroxide in methanol dehydrobrominated (138) to give (139) in 57% yield ⟨91JCS(P1)2575⟩ whilst potassium *t*-butoxide gave a 30% yield of 1,1-dibromo-2-fluorostyrene (140) from (141) ⟨93JA5430⟩. Similar dehydrobrominations of the steroids (142) and (143) gave (144) and (145), respectively, in high yields ⟨60T149, 65RTC904⟩. The dibromothiolane 1,1-dioxide (146) condenses with bromoform under basic conditions to give 53% of (147) ⟨89AJC1307⟩. Bromination with loss of trimethylsilyl bromide from 1-bromo-1-trimethylsilyl-2-ethylhex-1-ene (148) gives 92% of (149) ⟨86MI 417-01⟩.

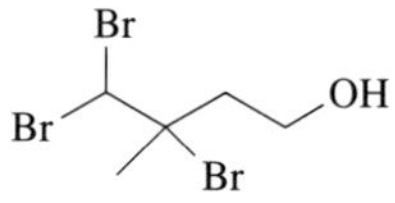

(138) BrCH(Br)–C(Me)(Br)–CH₂CH₂OH

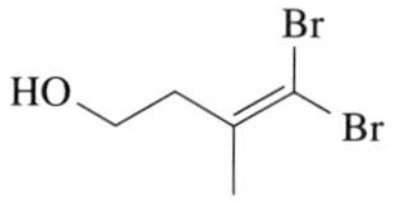

(139) (Z/E)-HOCH₂CH₂–C(Me)=CBr₂

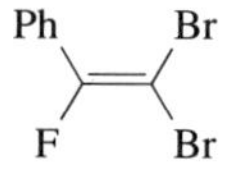

(140) PhCF=CBr₂

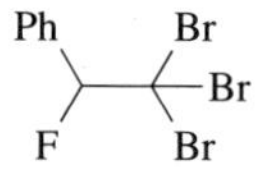

(141) PhCHF–CBr₃

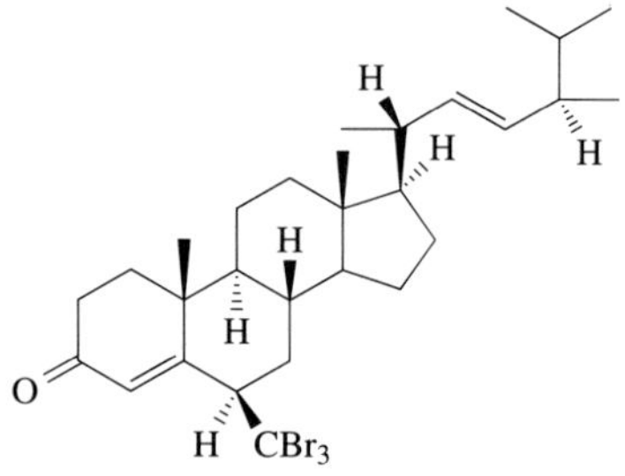

(142) steroid with CHBr₂... CBr₃ substituent

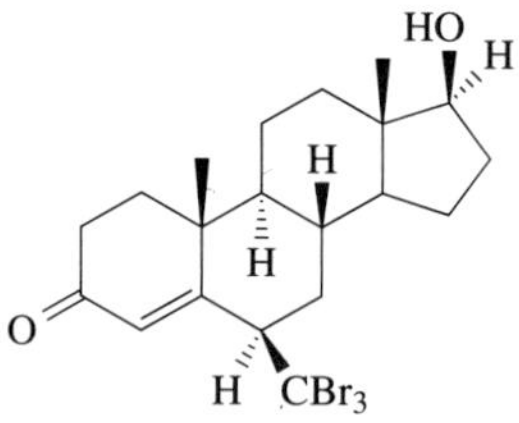

(143) steroid with CBr₃ group

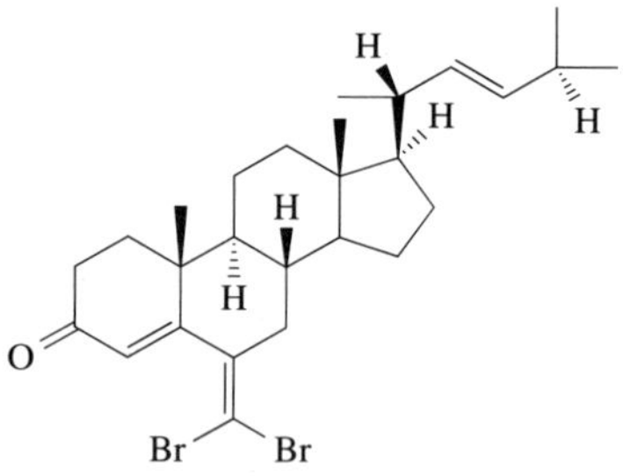

(144) steroid with =CBr₂ exocyclic

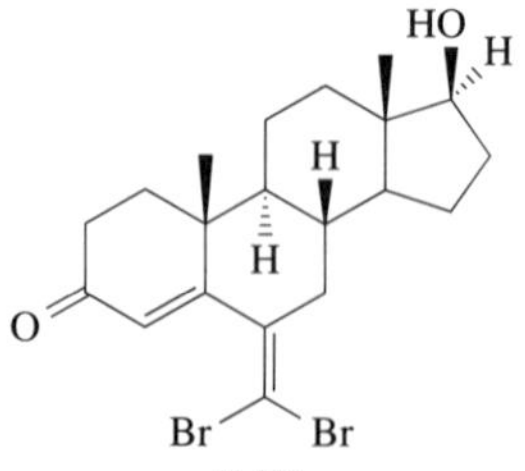

(145) steroid with =CBr₂ exocyclic, 17-OH

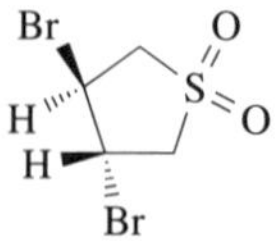

(146) 3,4-dibromothiolane 1,1-dioxide

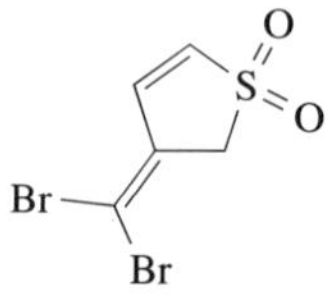

(147) 3-(dibromomethylene)thiolane 1,1-dioxide

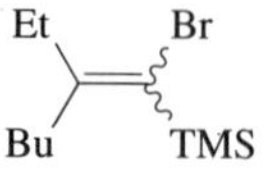

(148) (Et)(Bu)C=C(Br)(TMS)

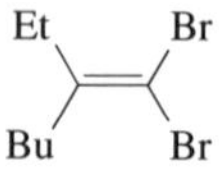

(149) (Et)(Bu)C=CBr₂

Dibromocarbene, formed from the decomposition of phenyl tribromomethylmercury, reacts with alkenes, for example (150), to give mixtures of cyclopropanes and the alkenes (151) ⟨78TL4253, 80JA6615, 84JA3584⟩. Longifolene (152) reacted with mercuric acetate followed by sodium chloride to give the dimercury compound (**153**) in 30% yield, which then gave the dibromo derivative (**154**) in 92% yield on treatment with pyridinium perbromide ⟨79IJC(B)500⟩. A similar reaction with camphene gave (**155**).

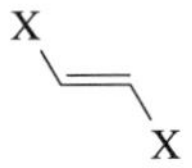

(**150**) X = Br, Cl

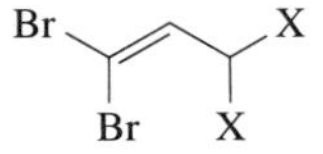

(**151**) X = Br, Cl

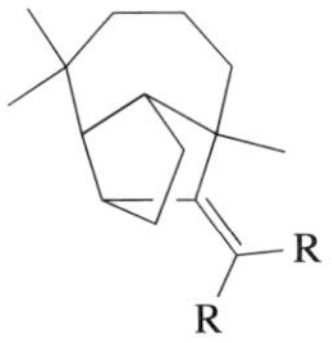

(**152**) R = H
(**153**) R = HgCl
(**154**) R = Br

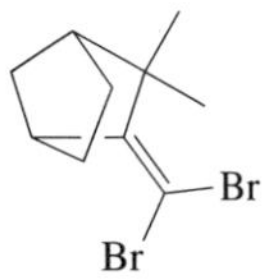

(**155**)

4.17.1.5 Diiodo Alkenes

In 1993, Duhamel and co-workers published a general route to 1,1-diiodo alkenes ⟨93S1071⟩. They showed that diethyl diiodomethylphosphonate (**156**), which is produced *in situ* from either diethyl iodomethylphosphonate (**157**) or diethyl methylphosphonate (**158**), readily condenses with a wide range of ketones and aldehydes to give good yields of 1,1-diiodo alkenes (Table 17). Previous studies showed that aldehydes, but not ketones, when treated with tetraiodomethane and triphenylphosphine at 0 °C then with zinc powder, gave high yields of (**159**) ⟨85CC296, 86JCR(M)2843⟩.

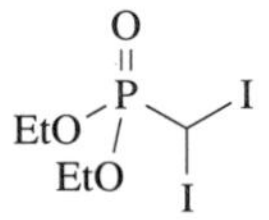

(**156**)

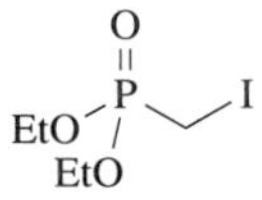

(**157**)

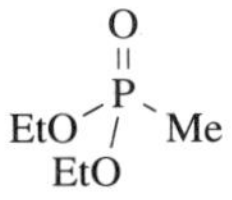

(**158**)

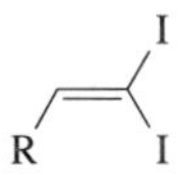

(**159**)
R = Bu^n, 75%
R = Ph, 87%
R = *p*-ClC$_6$H$_4$, 80%
R = 3,4-MeO$_2$C$_6$H$_3$, 60%

Table 17 Preparation of 1,1-diiodo alkenes, R^{1}R^{2}C=CI$_2$, from aldehydes or ketones and diethyl diiodomethyl phosphonate (**156**).

R^1	Aldehyde or ketone R^2	Starting material	Base	Yield (%)
C$_5$H$_{11}$	H	(**157**)	(TMS)$_2$NLi	71
C$_6$H$_{13}$	H	(**157**)	(TMS)$_2$NK	60
(EtO)$_2$CHCH$_2$C(Me)=CH	H	(**157**)	(TMS)$_2$NK	60
Ph	H	(**157**)	(TMS)$_2$NK	60
Ph	H	(**158**)	(TMS)$_2$NH	53
2-Furyl	H	(**157**)	(TMS)$_2$NK	78
—(CH$_2$)$_5$—		(**157**)	(TMS)$_2$NK	74
—(CH$_2$)$_5$—		(**158**)	(TMS)$_2$NH	63
Ph	Ph	(**157**)	(TMS)$_2$NLi	51
Ph	Me	(**157**)	(TMS)$_2$NLi	42

1-Iodoalkynes (**160**) react with bis(pyridine)iodine-(1) tetrafluoroborate in the presence of a variety of nucleophiles such as halide ions, alcohols, thiols and acids to give the 1,1-diiodo alkenes (**161**) in good to very good yields ⟨88JA5567⟩. In the absence of an added nucleophile, head-to-tail dimers (**162**) are produced ⟨93AG(E)893⟩. Treatment of phenylpropiolic acid with a mixture of iodine and iodine pentoxide in methanol gave a number of products from which the enol ether (163) was isolated in 59% yield ⟨84JOC515⟩.

(160) R≡I

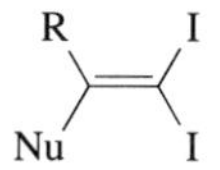

(161) R₁R₂C=C(Nu)(I), with Nu on R-side and I on both carbons
Nu = Cl, Br, I, NCS, Pr^{i}OH, PhSH, anisole

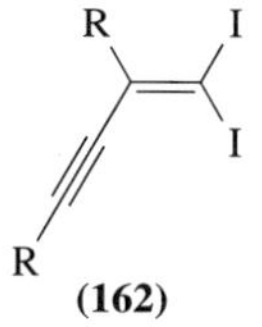

(162)

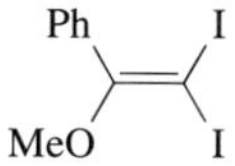

(163) Ph, MeO / I, I

Direct nucleophilic displacement reactions have also been described. Thus 1,1-dibromo alkenes (164) react with potassium iodide–cupric iodide in HMPA at 120 °C to give (165a) and (165b) in 74% and 84% yields respectively ⟨88S236⟩, and the aluminum–zirconium compound (166) gives 92% of (167) with iodine ⟨81JA1276⟩. Further, bis(trimethylenedioxyboryl)methylenecyclohexane (168) reacts with iodine in the presence of sodium hydroxide to give 71% of (165a) ⟨78JOM(152)1⟩. The organomercurials derived from longifolene and camphene described in Section 4.17.1.4.4 react with iodine monochloride in pyridine to give excellent yields of the diiodomethylene compounds ⟨79IJC(B)500⟩.

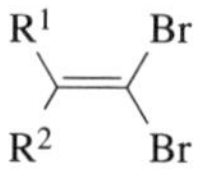

(164) R¹R²C=CBr₂
(165) R¹R²C=CI₂
  a; R¹,R² = (CH₂)₅
  b; R¹ = c-C₆H₁₁, R² = H
(166) Pr^n(AlMe₂)C=C(ZrCp₂Cl) 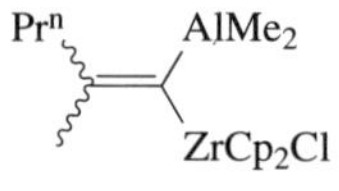
(167) Pr^n(I)C=CI
(168) cyclohexylidene=C(B(OCH₂CH₂CH₂O))₂

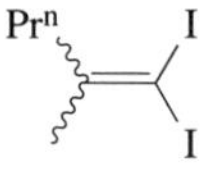

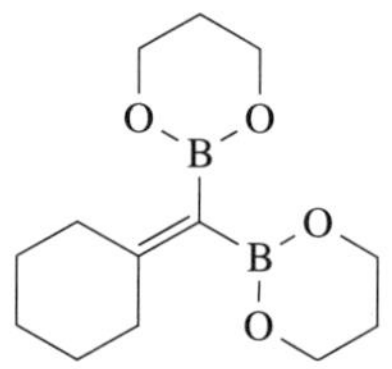

4.17.1.6 Mixed Halo Derivatives

As might be expected, many of the general reactions used in the previous sections have also been employed to prepare mixed halo derivatives, but the unsymmetrical derivatives R¹R²C=CXY may form mixtures of geometrical isomers.

4.17.1.6.1 Fluoro halo alkenes

(i) Elimination reactions

Both zinc-catalysed dechlorination and base-catalysed dehydrohalogenations of appropriately substituted compounds have been used to prepare these systems (Equations (6)–(8), Scheme 3). Thus, 1,1-diaryl-1,2,2-trichloro-2-fluoroethanes (169) give 1,1-diaryl-2-chloro-2-fluoroethenes (170) in good to high yields whilst 1-aryl-1,2,2-trichloro-2-fluoroethane (171) gives good yields of mixtures of (*E*)- and (*Z*)-1-aryl-2-chloro-2-fluoroethenes (172) ⟨50JA3952, 67BCJ1275⟩.

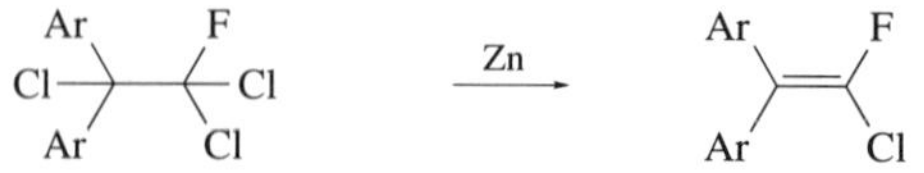

$$Ar_2C(Cl)-CFCl_2 \xrightarrow{Zn} Ar_2C=CFCl \qquad (6)$$

(169) → (170)

Ar = Ph, 78%
Ar = *p*-Tol, 55%
Ar = *p*-ClC₆H₄, 56%
Ar = *p*-FC₆H₄, 47%

Dihalo Functions

Scheme 3

Equation (7): Compound (171) with Zn gives (172) (two isomers)

Ar		ratio		
Ar = Ph, 68%		1.65	:	1
Ar = p-Tol, 66%		1.5	:	1
Ar = p-ClC$_6$H$_4$, 62%		1.46	:	1
Ar = p-FC$_6$H$_4$, 76%		1.60	:	1
Ar = p-MeOC$_6$H$_4$, 57%		1.46	:	1

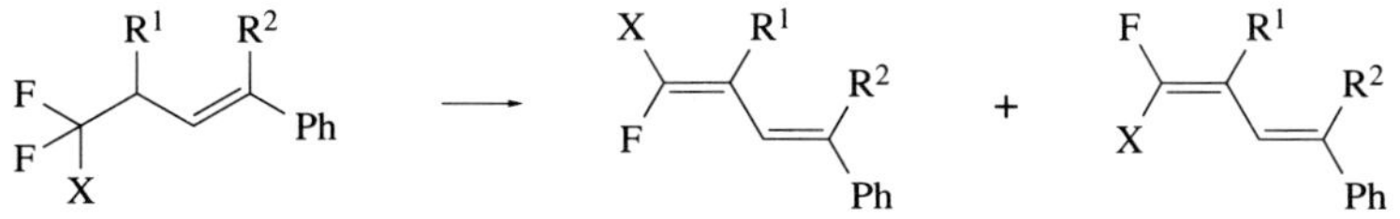

Equation (8): Compound (176) gives (E,E)-(177) + (Z,E)-(177)

		(E,E)		(Z,E)
X = Cl, R^1 = R^2 = H, 87%		56	:	44
X = Br, R^1 = R^2 = H, 69%		99	:	1
X = Cl, R^1 = Me, R^2 = H, 77%		49	:	51
X = Br, R^1 = H, R^2 = Me, 88%		60	:	40

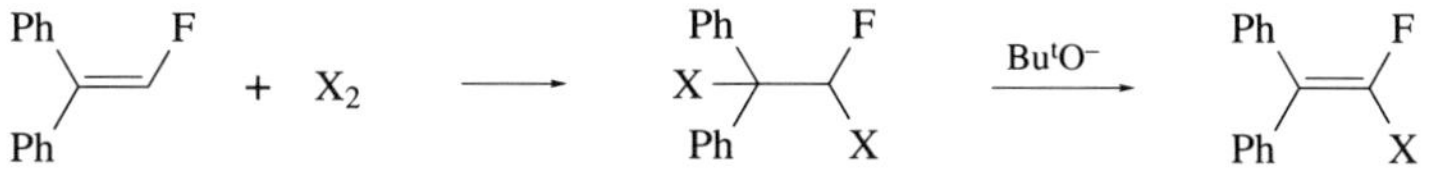

(173) + X$_2$ → (174) →[Bu^{t}O$^-$] (175)

X = Cl, 75%
X = Br, 47%

Halogenation of 1,1-diphenyl-2-fluoroethene (**173**) with either chlorine or bromine gives the adducts (**174**) which form the alkenes (**175**) on treatment with potassium *t*-butoxide in *t*-butyl alcohol ⟨87JFC(37)313⟩. The same base in ether at 0 °C converted (**176**) to a mixture of the (*E,E*) and (*Z,E*) dienes (**177**) ⟨88JFC(41)425⟩ whereas KOH in hot methanol gave a 85% yield of (**170**; Ar = *p*-ClC$_6$H$_4$) from (**178**) ⟨57JA4174⟩, but in mineral oil KOH converted (**179**) into the perfluorovinyl iodide (**180**) in 76% yield ⟨56JA59⟩.

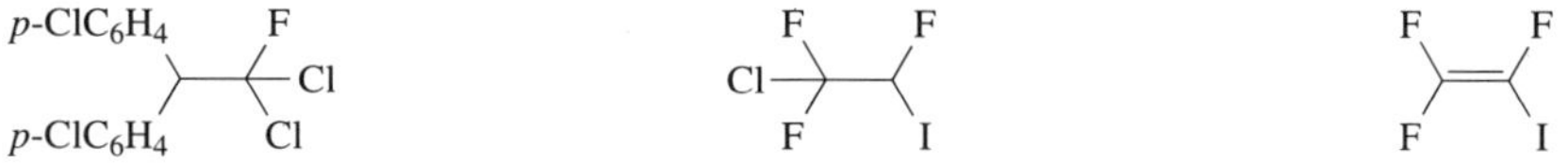

Compounds (178), (179), (180) structures shown.

Treatment of vinylsilanes (**181**) (Scheme 4) with bromine in tetrachloromethane or dichloromethane gave both the polyhalo addition products (**182**) and (**183**) by *anti*, *syn* addition respectively, the ratio of the two processes being dependent on the experimental conditions ⟨90JOM(394)37⟩. Subsequent bromodesilylation by tetra-*n*-butylammonium fluoride in THF at −50 °C gave the corresponding (*E*)- and (*Z*)-fluoro bromo alkenes (**184**) in a ratio that reflected the stereoselectivity of the addition reaction. Bromination of (**181**; R = C$_7$H$_{15}$) in dichloromethane was found to be less stereoselective than in tetrachloromethane suggesting lower stability for the bromonium ion in dichloromethane. The stereoselectivity of the bromination of (**181**; R = Ph) was found to be very dependent on solvent and temperature with either (*E*) or (*Z*) isomers being predominant depending on the experimental conditions (Table 18).

(ii) Wittig and related reactions

The reactions of mixed halo phosphorus ylides with aldehydes and ketones has been applied to the synthesis of fluoro chloro alkenes. Three methods for the synthesis of the fluoro chloro ylide have been reported:
(a) the reaction of dichlorofluoromethane with potassium *t*-butoxide in the presence of triphenylphosphine ⟨62JA854, 68TL71, 70JOC2125⟩;

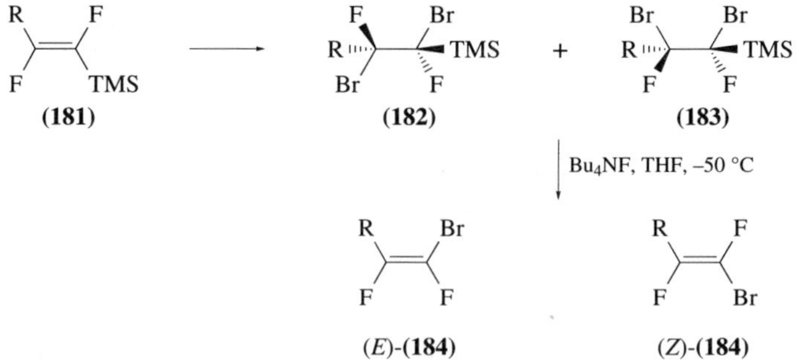

Scheme 4

R = n-C$_7$H$_{15}$, Ph

Table 18 Bromination of the fluorovinylsilanes (**181**) (Scheme 4).

R	Experimental conditions	Ratio (**182**):(**183**)	Yield of addition (%)	Ratio (E):(Z) of (**184**)
C$_7$H$_{15}$	CCl$_4$, 0 °C	90:10	100	90:10
	CH$_2$Cl$_2$, −20 °C	63:37	35(+ other products)	≈50:50
Ph	CCl$_4$, 55 °C	75:25	100	77:23
	CH$_2$Cl$_2$, −80 °C	18:82	71	13:87

(b) sodium hydride treatment of methyl dichlorofluoroacetate ⟨67TL1123⟩ in the presence of triphenylphosphine ⟨68BCJ756⟩;
(c) decarboxylation of sodium dichlorofluoroacetate in the presence of triphenylphosphine ⟨68TL71, 70JOC2125⟩.

In general for aldehydes or perfluoro ketones, the decarboxylation method (c) gives the higher yields. However, the ylide from all three methods reacts with these substrates, but not, in general with ketones, to give moderate yields of fluoro chloro alkenes (Table 19). A mole ratio of 1.5:1.5:1.0 of triphenylphosphine:sodium dichlorofluoroacetate:carbonyl compound in TRIGLYME was found to be optimal in method (c) where it appears that the generation of the ylide occurs via the phospho betaine Ph$_3$P$^+$CFClCO$_2^-$ rather than from a carbene species ⟨70JOC2125⟩.

Table 19 Wittig preparations of PhRC=CFCl.

Carbonyl compound	Method of ylide generation[a]	Yield of alkene (%)	Ratio (**185**):(**186**)
PhCHO	(a)	39	56:42
	(b)	40	43:57
	(c)	49	56:44
PhCOCF$_3$	(a)	31 → 55[b]	51:49
	(b)	40	48:52
	(c)	56	53:47
PhCOCF$_2$CF$_3$	(c)	42	59:41
PhCOCF$_2$CF$_2$CF$_3$	(c)	41	60:40
⬠=O	(a)	27	
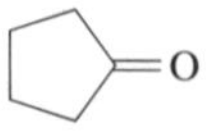	(c)	9	
PhCOMe	(b)	8 → 36[c]	[d]

[a] See text. [b] Increased ratio of ylide:carbonyl compound from 1.5:1 to 3:1 gave an increase in yield ⟨62JA854⟩. [c] Rapid cooling after ylide formation resulted in an increase in yield ⟨68BCJ756⟩. [d] Estimated isomer ratio 1:1.

The isomer ratio in the product has been determined by ^{19}F NMR spectroscopy but neither the method of generation of the ylide, nor the solvent used in method (c) have a great effect on this ratio. Also, the presence of lithium or iodide ions had no dramatic effect on the isomer ratio as has been reported for other Wittig reactions ⟨63T149, 64TL2669⟩. As expected the ratio of isomers (**185**):(**186**) (Table 19) for the perfluoro ketones increases with increase in size of R. The isomer ratio from the reaction with benzaldehyde by method (c) has been accounted for by invoking the existence of hydrogen bonding between the aldehydic proton and the ylide fluorine in the betaine-forming step ⟨70JOC2125⟩. In contrast, the reaction of benzaldehyde with the ylide prepared by method (b) gave the opposite isomer (**186**) as the most abundant species.

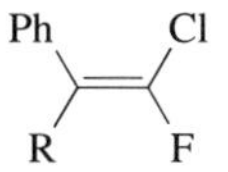

Ph   Cl   Ph   F

R   F   R   Cl

(**185**) 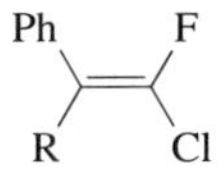   (**186**)

Burton and van Hamme criticised these preparative methods either because of the low yields resulting from method (a) or the hazardous, expensive or difficult routes needed to prepare the precursors for methods (b) or (c) ⟨77JFC(10)131⟩. As an improvement, they treated trichlorofluoromethane with triphenylphosphine, zinc dust and aldehydes or ketones in DMF at 60 °C to give good yields of the alkenes (**187**) (Equation (9), Table 20) ⟨77JFC(10)131⟩. However, this route also gave very low yields with nonactivated ketones and a two-stage procedure involving initial preparation of the salt (**188**) followed by ylide generation was introduced (Scheme 5, Table 21) ⟨79JFC(13)407⟩. Note that under these conditions a reasonable yield of product from acetophenone could be achieved.

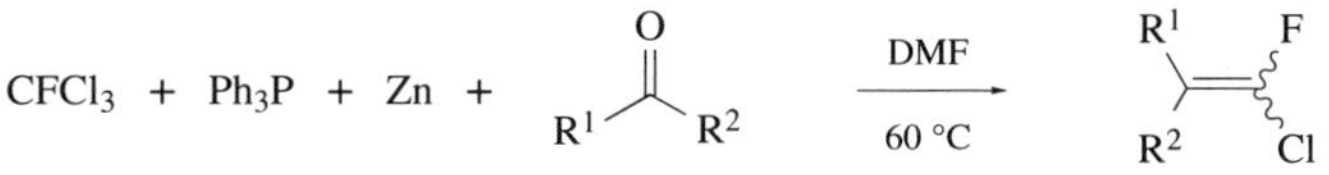

$$CFCl_3 + Ph_3P + Zn + R^1COR^2 \xrightarrow[60\,°C]{DMF} R^1R^2C=CFCl \quad (187) \qquad (9)$$

Table 20 Preparation of $R^1R^2C=CFCl$ (**187**) from $CFCl_3$ and R^1R^2CO (Equation (9)).

R^1	R^2	Yield (%)	Reaction time (h)	$R^1R^2C=CClF$	$R^1R^2C=CFCl$
Ph	CF_3	71	24	53	47
Ph	H	64	115	53	47
n-C_6H_{13}	H	49	89	41	59
Ph	H	15	96	43	57
—$(CH_2)_4$—		4	208		

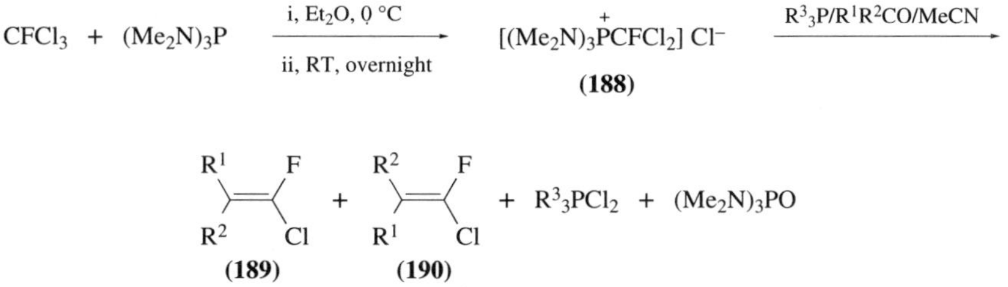

$$CFCl_3 + (Me_2N)_3P \xrightarrow[\text{ii, RT, overnight}]{\text{i, Et}_2O, 0\,°C} [(Me_2N)_3PCFCl_2]^+ Cl^- \xrightarrow{R^3_3P/R^1R^2CO/MeCN}$$

(**188**)

$$\underset{(189)}{R^1R^2C=CFCl} + \underset{(190)}{R^2R^1C=CFCl} + R^3_3PCl_2 + (Me_2N)_3PO$$

Scheme 5

In a further modification of this procedure ⟨79JOM(169)123⟩ an initial reaction of tris(dimethylamino)phosphine with trichlorofluoromethane gave the phosphonium salt (**188**), treatment of which with zinc dust or, more reliably, a zinc–copper couple in the presence of a carbonyl compound at 60 °C produced fluoro chloro alkenes in improved yields (Equation (10), Table 22). In the absence of a carbonyl compound, the solution resulting from the dechlorination of (**188**) was found to retain comparable alkenating capacity for 30 days.

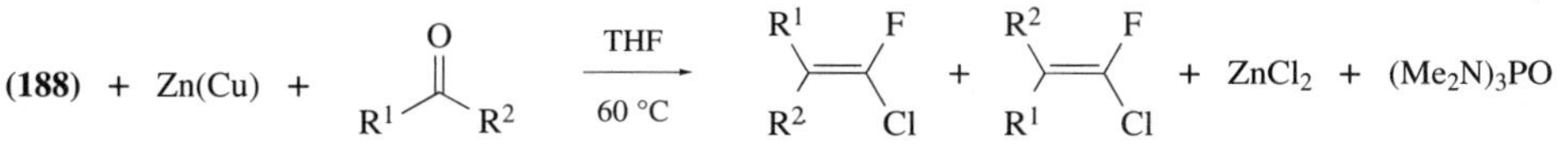

$$(\mathbf{188}) + Zn(Cu) + R^1COR^2 \xrightarrow[60\,°C]{THF} \underset{(189)}{R^1R^2C=CFCl} + \underset{(190)}{R^2R^1C=CFCl} + ZnCl_2 + (Me_2N)_3PO \qquad (10)$$

In a parallel approach by Oda *et al.*, chlorofluorocarbene, generated *in situ* from dichlorofluoromethane and potassium *t*-butoxide, reacted with triphenylphosphonium fluorenylide (**191**) to

Table 21 Use of the two-stage procedure to prepare $R^1R^2C{=}CFCl$ (**189**) and (**190**) from $CFCl_3$ and R^1R^2CO (Scheme 5).

R^1	R^2	R^3	Reaction temp (°C)	Reaction duration (h)	Yield (%)	(**189**) : (**190**)
Ph	CF_3	Ph	60	1.5	83	42:58
Ph	H	Ph	100	3	60	56:44
Ph	CH_3	Ph	60	11	3[a]	50:50
Me_2N	CH_3	Ph	60	3	56	52:48

[a] A possible explanation put forward by the authors for the decrease in yield with Ph_3P is that production of the required ylide is a reversible process in this case (i.e. recapture of chlorine from Ph_3PCl_2 by the ylide to form (**188**) is faster than the ylide's reaction with acetophenone).

Table 22 Reactions of the phosphonium salt (**188**) and R^1R^2CO with zinc to give $R^1R^2C{=}CFCl$ (Equation (10)).

R^1	R^2	Reaction time (h)	Yield (%)	(E):(Z)
Ph	CF_3	1	100	44:56
Ph	H	1	100	47:53
C_6H_{13}	H	0.5	100	59:41
Ph	Me	4.5	70	52:48
CF_3	CF_3	1.5	75	
—$(CH_2)_4$—		25	18	
Ph	Ph	28	0	

yield the 1-chloro-1-fluoro alkene (**192**) (43%). Applications of Wittig routes to bromo fluoro alkenes have also been reported ⟨71JFC(1)381⟩. Bromotrifluoromethane and 2 equiv. of triphenylphosphine gave the ylide (**193**), which is significantly more stable in solution than the corresponding difluoro and fluoro chloro ylides enabling a solution to be prepared and stored in the absence of a reactive carbonyl compound. In contrast to the difluoro ylide, (**193**) also reacts with nonactivated ketones (Table 23).

(**191**)   (**192**)   (**193**)

Table 23 Preparation of 1-bromo 1-fluoro alkenes $R^1R^2C{=}CBrF$ from $Ph_3P{=}CFBr$ (**193**) and R^1R^2CO.

R^1	R^2	Yield (%)	$R^1R^2C{=}CFBr : R^1R^2C{=}CBrF$
Ph	Me	40	41:59
Ph	CF_3	97	46:54
Ph	CF_3CF_2	77	60:40
CF_3	CF_3	38	
$PhCH_2$	CF_3	71	43:57

(iii) From 1-halo 1-metallo alkenes

trans 1-Chlorostyrene (**194**) can be deprotonated by *n*-butyllithium at $-100\,°C$ to give (**195**) which is then fluorinated by perchloryl fluoride ($FClO_3$) to give 45% of (*E*)-1-chloro-1-fluorostyrene (**196**)

⟨69CB1944⟩. Likewise, (Z)-2-phenyltetrafluoropropene (**197**) and the bromo analogue (**198**) were converted into the lithio derivative (**199**) by lithium tetramethylpiperidide or n-butyllithium, respectively, which was brominated or iodinated to give in high yields and with complete retention of stereochemistry, the (E) isomers (**200**) ⟨75TL773⟩. The (E) isomer of (**197**) was converted similarly into the (Z) isomers of (**200**).

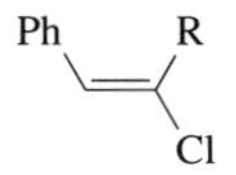

(**194**) R = H
(**195**) R = Li
(**196**) R = F

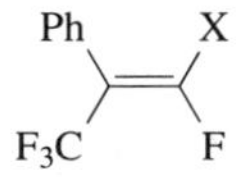

(**197**) X = H
(**198**) X = Br
(**199**) X = Li
(**200**) X = Br, 95%(Z), 96%(E)
X = I, 66%(Z), 80%(E)

The synthesis of organotin reagents, for example (**201**) (Scheme 6) as a mixture of isomers by treatment of (**202**), or its (Z) isomer, with tributyltin hydride and AIBN in toluene at reflux was carried out in 75% yield. Subsequent treatment of the mixture of isomers with iodine in ether at reflux gave the fluoro iodo alkenes (E)- and (Z)-(**203**) in an overall yield of 86% after separation and an (E):(Z) ratio of 3:2 ⟨92JA360, 93TL7197⟩.

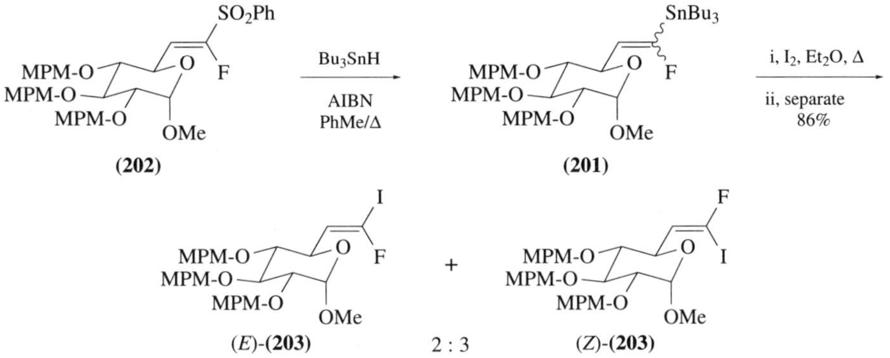

AIBN = 2,2-azobisisobutyronitrile
MPM = 4-MeOC$_6$H$_4$CH$_2$

Scheme 6

(iv) Allylic rearrangements

A series of publications by Miller *et al.* noted that compounds such as (**204**) (Table 24) rearrange when treated with halide ions in formamide by an SN2' mechanism to form 1-chloro 1-fluoro alkenes of undefined stereochemistry (Table 24) ⟨57JA4164, 59JA2078, 60JA3091⟩. The observed rates of reaction were in the order F$^-$ > Cl$^-$ > I$^-$·. Analogous, 1-chloro 1-fluoro alkenes were inert under these conditions but they did react under more forcing ones ⟨60JA3091⟩

(v) Reaction of diazo alkanes with carbenes

Reimlinger has shown that diphenyldiazomethane (**205**) and diazofluorene (**206**) may be converted, in moderate yields, into mixed 1,1-dihalo alkenes (Table 25) by treatment with the required carbene, produced from the haloform and potassium *t*-butoxide ⟨62AG153, 64CB339⟩.

(vi) Miscellaneous

During work on the synthesis and chemistry of highly fluorinated bicyclo[2.2.0]hexenones, the curious reaction shown in Equation (11) was discovered ⟨89JOC5502⟩. Treatment of the ketones

Table 24 Allylic rearrangements of polyfluoro chloro alkenes.

Starting material	Experimental conditions	Product	Yield (%)
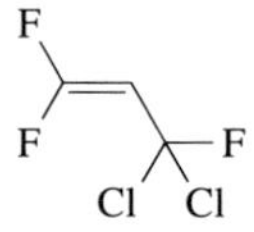 (204)	KF, formamide 17 h, 50 °C	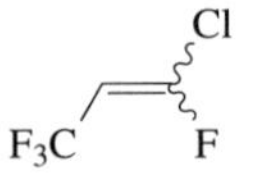	62
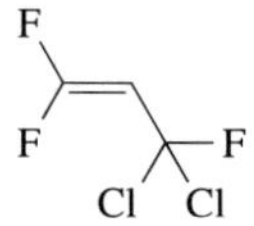 (204)	LiCl, acetone 24 h, 27 °C	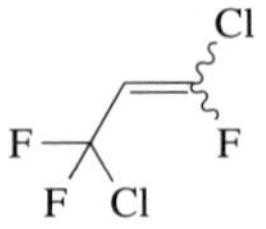	
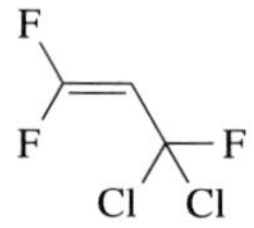 (204)	NaI, acetone 45 h, 27 °C	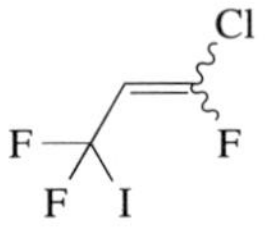	63
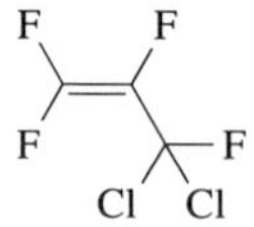	KF, formamide 14 h, 32 °C	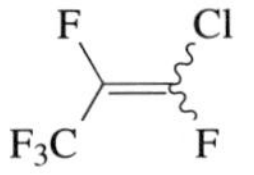	71

(207) with potassium *t*-butoxide gave the corresponding esters (208) and (209) as single isomers. Alternatively, treatment of the hydrate of (207; X = Br) with 2 equiv. of LDA followed by acid yielded the acid (210). The reaction of (211) with thiobenzophenone at 80 °C produced (212) probably via an unstable carbene addition product (213) ⟨73JOM(49)117⟩.

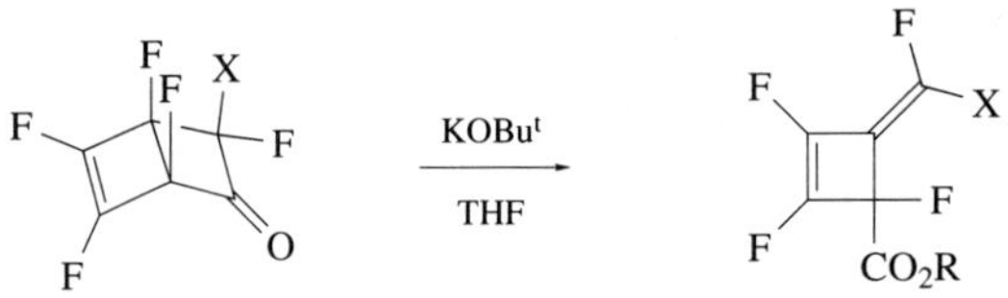

(11)

(207) X = Br, Cl

(208) X = Br, R = Bu^t
(209) X = Cl, R = Bu^t
(210) X = Br, R = H

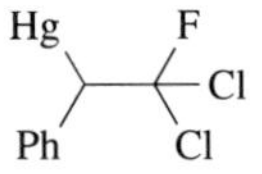

(211) 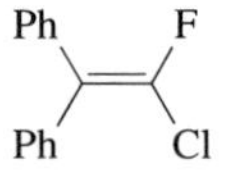   (212) 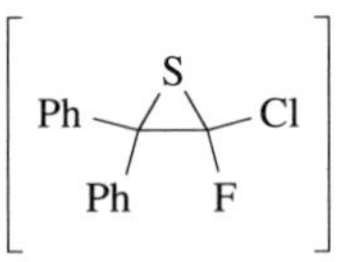   (213)

4.17.1.6.2 Chloro halo alkenes

(i) Elimination reactions

Treatment of (214) with LDA gave, by a β-elimination process, the chloro bromo alkene (215) in 80% yield ⟨89JCS(P1)691⟩ whilst (216) with triethylamine gave a mixture of *cis* and *trans* 1-bromo 1-chloro alkenes ⟨67CB3893⟩. A stereoselective synthesis of this system has also been reported by Zweifel *et al.* ⟨82S127⟩. Regioselective addition of a disubstituted borane derivative to 1-chloro alkynes gave the adducts (217) conversion of which into the borinic esters (218) with trimethylamine *N*-oxide was followed by a bromine addition–sodium methoxide elimination procedure to give the (*Z*)-1-chloro 1-bromo alkenes (219) with high stereochemical purity. The corresponding (*E*) alkenes

Table 25 Condensation of diazo alkenes with carbenes.

Starting material	Carbene precursor	Carbene	Product	Yield (%)
Ph₂C=N⁺≡N (205)	CHCl₂F	CFCl	Ph(Ph)C=CFCl	63
Ph₂C=N⁺≡N (205)	CHBr₂F	CFBr	Ph(Ph)C=CFBr	38
Ph₂C=N⁺≡N (205)	CHBr₂Cl	CClBr	Ph(Ph)C=CClBr	63
Fluorenylidene-N⁺≡N (206)	CHCl₂F	CFCl	Fluorenylidene=CFCl	67
Fluorenylidene-N⁺≡N (206)	CHBr₂F	CFBr	Fluorenylidene=CFBr	82
Fluorenylidene-N⁺≡N (206)	CHBr₂Cl	CClBr	Fluorenylidene=CClBr	48

(**220**) (Table 26) were prepared by either *trans*-bromination of (**221**) followed by *anti*-debromosilylation or by *trans*-chlorination of (**222**) followed by *anti*-dechlorosilylation. The stereoselectivity of this process was found to depend on the size of the group R. The bromination–debromosilylation of (**223**) to give chloro bromo alkenes in high yields has also been reported ⟨86MI 417-01⟩.

Dehydrochlorination of (**224**) by potassium hydroxide in mineral oil gave 60% of (**225**) ⟨58JOC1661⟩.

(ii) Wittig and related reactions

A patent has described the use of the mixed chloro bromo Wittig reagent (**226**), prepared from triphenylphosphine and dibromochloromethane in the presence of potassium *t*-butoxide in *t*-butyl alcohol, with (**227**) to give (**228**) in 70% yield as a mixture of (*E*) and (*Z*) isomers ⟨81EUP0030887⟩.

(iii) Quenching of organometallic agents

Slow addition of butyllithium to 1,1-dibromo alkenes resulted in stereoselective formation of the thermodynamically favoured organolithium reagent (**229**) (Table 27) by a rapid geometrical

758   Two Halogens or a Halogen and a Chalcogen

(214) TMS-O, cyclohexyl with CH(Cl)Br
(215) cyclohexylidene=C(Cl)Br
(216) PhCH(Cl)-CH(Br)Br (Ph-CHCl-CHBr₂ pattern: Br,Br on one C; Ph,Cl on other)
(217) R,Cl vinyl–B(thexyl)(iBu)
(218) R¹,Cl vinyl–B(thexyl)(OR²)
(219) R,Br / Cl vinyl
(220) R,Cl / Br vinyl
(221) R,TMS / Cl vinyl
(222) R,Br / TMS vinyl
(223) Et,TMS / Bu,Cl vinyl
(224) ClCF₂–CHFCl
(225) F,I / F,Cl vinyl

Table 26 Stereoselective addition–elimination route to 1-bromo 1-chloro alkenes.

Starting material	R	Reagent	Product	Yield (%)	Isomeric purity (%)
(221)	$n\text{-}C_4H_9$	i, Br_2; ii, NaOMe/MeOH	(220)	80	99
(221)	$n\text{-}C_6H_{13}$	i, Br_2; ii, NaOMe/MeOH	(220)	69	99
(221)	$c\text{-}C_6H_{11}$	i, Br_2; ii, NaOMe/MeOH	(220)	96	91
(221)	$t\text{-}C_4H_9$	i, Br_2; ii, NaOMe/MeOH	(220)	82	57
(222)	$n\text{-}C_4H_9$	i, Cl_2; ii, NaOMe/MeOH	(220)	83	98
(222)	$c\text{-}C_6H_{11}$	i, Cl_2; ii, NaOMe/MeOH	(220)	84	95
(222)	$t\text{-}C_4H_9$	i, Cl_2; ii, NaOMe/MeOH	(220)		73

(226) $Ph_3P=C(Cl)Br$
(227) X = O ; (228) X = CBrCl (cyclopropane carboxylate with CN, OPh aryl, X=vinyl substituent)

isomerisation involving bromine–lithium exchange with the starting alkene ⟨87TL5145, 91CB1379, 92JOC5805⟩. Quenching of (229) with 1,1-trichlorodifluoroethane gave (230) stereoselectively in high yield (Table 27) ⟨93JOC4897⟩.

(229) R^1,R^2 / Li,Br vinyl
(230) R^1,R^2 / Br,Cl vinyl

Stereo- and regioselective addition of $LiAlH_4$ in THF to 1-chloro 1-alkynes gave the lithium (*E*)-1-chloro-1-alkenyl alanates (231) which could be converted by addition of acetone into the triisopropoxyalumino derivatives (232) further reaction of which with bromine gave the bromides (233) or with iodine monochloride gave the iodides (234) in high yields ⟨79JA5101⟩.

The reaction of the α-chlorovinyltrimethylsilane (235) with anhydrous tetrabutylammonium fluoride in the presence of bromotrichloromethane gave (236) in 76% yield (based on recovered starting material) ⟨78JOM(162)1⟩ whilst addition of bromine to the lithium reagent (237), prepared from (238) by treatment with *n*-butyllithium, gave (239) in 94% yield ⟨66CB670, 69BSF2712⟩.

Cyclohexanone reacted with bromochloromethyllithium to give the alcohol (240) which was trapped by chlorotrimethylsilane to give (241). Treatment of this with LDA gave 80% of (242) ⟨89JCS(P1)691⟩.

Table 27 Preparation of (**230**) from (**229**) and CCl_3CHF_2.

R^1	R^2	Yield (%)	Ratio $(E):(Z)$
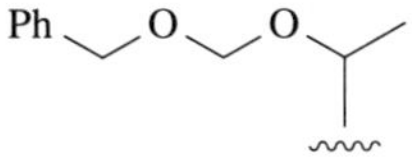 Ph–O–O–⟨⟩ (wavy)	H	41	> 40 : 1
Ph	Et	88	5.2 : 1
Ph	MeOCH$_2$	94	1 : 10
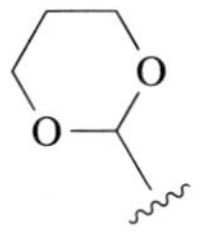 1,3-dioxane-2-yl	Me	94	10 : 1

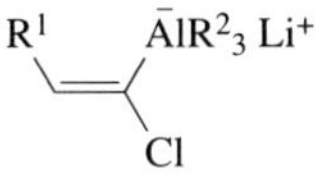

(**231**) $R^2 = H$
(**232**) $R^2 = OPr^i$
$R^1 = Bu^n$, n-Hex, c-Hex, Bu^t

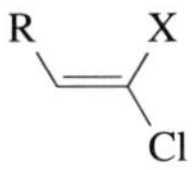

(**233**) X = Br
(**234**) X = I

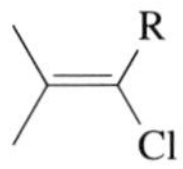

(**235**) R = TMS
(**236**) R = Br

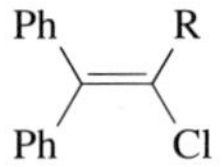

(**237**) R = Li
(**238**) R = H
(**239**) R = Br

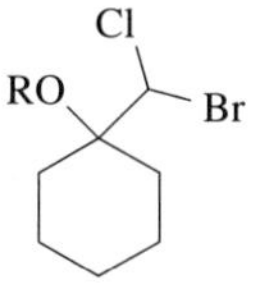

(**240**) R = H
(**241**) R = TMS

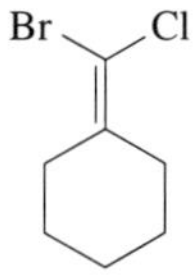

(**242**)

4.17.1.6.3 Diazo alkanes with carbenes

The methodology of Reimlinger (see Section 4.17.1.6.1(v)) has also been utilised for the synthesis of bromo chloro alkenes from diazo alkanes and *in situ* generated bromochlorocarbene (**243**) ⟨87JA6687⟩.

(**243**)

4.17.1.6.4 Bromo iodo alkenes

(i) Elimination reactions

Dehydrochlorination of (**244**) with potassium hydroxide gave a 78% yield of (**245**) ⟨58JOC1661⟩ whilst (**246**) added iodine and then underwent an iododesilylation reaction to give 80% of (**247**) ⟨86MI 417-01⟩.

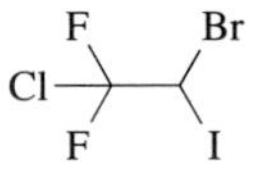

760   Two Halogens or a Halogen and a Chalcogen

(244)

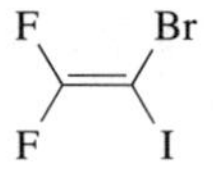

(245)

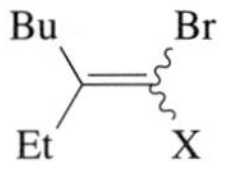

(246) X = TMS
(247) X = I

(ii) Quenching of organometallic reagents

The (*E*)-zinc enolate (**248**), prepared stereoselectively by the reaction of (**249**) with zinc-TMEDA in DMF, was quenched with iodine to give a 46% yield of (**250**) ⟨93JA5430⟩.

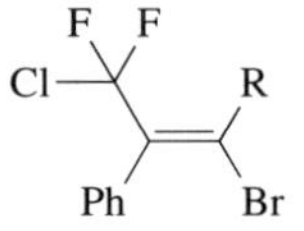

(**248**) R = ZnX
(**249**) R = Br
(**250**) R = I

(iii) Addition reactions to 1-iodo alkynes

The reaction of the alkynylmethanol (**251**) with *N*-bromosuccinimide or *N*-iodosuccinimide (NIS) with catalytic amounts of silver salts gave (**252**) and (**253**) respectively whilst reaction of (**252**) with NIS and catalytic amounts of Koser's reagent [hydroxy(tosyloxy)iodobenzene] in acetonitrile gave the (*Z*)-bromo iodo alkene (**254**) via the iodonium ion (**255**) with 95% selectivity in 72% yield. The reaction of (**253**) with NBS gave only low yields of (**256**) as a mixture of compounds but the use of bromine and Koser's reagent gave a mixture of (**256**) and (**257**) in the ratio of 6:1 in low yield and (**256**) could not be isolated in pure form ⟨92TL2285⟩.

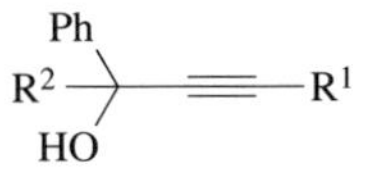

(**251**) R^1 = H, R^2 = Me
(**252**) R^1 = Br, R^2 = Me
(**253**) R^1 = I, R^2 = Me
(**257**) R^1 = Br, R^2 = H

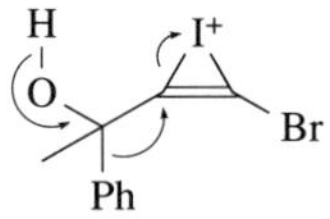

(**255**)

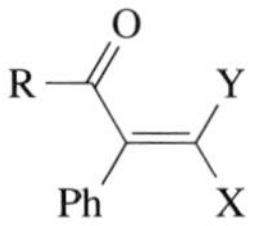

(**254**) Y = I, X = Br, R = Me
(**256**) Y = Br, X = I, R = Me
(**258**) Y = I, X = Br, R = H

The treatment of (**257**) with iodine and Koser's reagent in acetonitrile gave exclusively (**258**) in high yield ⟨92TL7705⟩ whilst treatment of the 1-chloro alkyne (**259**; R = *n*-C$_6$H$_{13}$) with bis(pyridine)iodine(1) tetrafluoroborate and a range of nucleophiles in the presence of 2 equiv. of tetrafluoroboric acid gives (**260**)–(**262**) in good yields as single stereoisomers ⟨90TL2751⟩. Only with (**259**; R = Ph) are a mixture of isomers (**263**) and (**264**) produced.

(**259**)

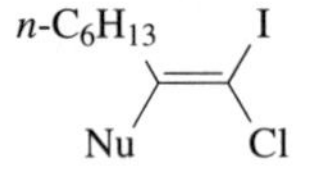

(**260**) Nu = OAc, 75%
(**261**) Nu = OCHO, 90%
(**262**) Nu = Cl, 60%

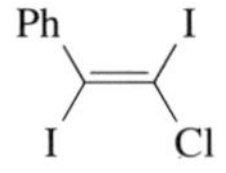

(**263**)   4.5 : 1 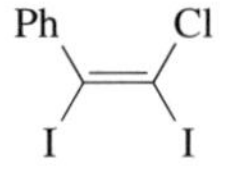   (**264**)
             42% overall

4.17.2 FUNCTIONS INCORPORATING A HALOGEN AND A CHALCOGEN: R$_2$C=C(Hal)(Chalc)

4.17.2.1 Halogen and Oxygen Derivatives

4.17.2.1.1 α-Halo enols and ethers

The synthesis and chemistry of these compounds has been reviewed by Bakker and Scheeren ⟨93HOU(E15/2)1599⟩. Four major methods of synthesis have been described:

(a) displacement of halogens in alkyl and alkylidene halides by alkoxides;
(b) addition to substituted alkynes;
(c) reaction of acetoxy groups with phosphorus pentachloride; and
(d) elimination reactions.

(i) Displacement of halogen atoms in alkyl and alkylidene halides by alkoxides and phenols

Tanimoto et al. studied the reactions of the polychlorinated ethanes (**265**)–(**266**) with an excess of sodium phenoxide in DMSO at 70 °C, which give the alkenes (**267**)–(**268**) ⟨76BCJ1931⟩. The authors present good evidence that the reaction proceeds via initial dehydrochlorination to (**269**) and (**270**) and indeed (**269**) and (**270**) under the same reaction conditions give (**267**) and (**268**) in 78% and 28% yields respectively. Since the reactions of alkylidene halides with nucleophiles are much slower than those of alkyl halides, it is possible that the reactions of (**265**) and (**269**) involve the intermediary of the dichloro alkyne which then undergoes *trans*-addition of phenol to give the *trans*-1,2-dichloro-1-phenoxyethylene (**267**) as the sole product ⟨55JA3886⟩. Further evidence for this route is provided by the observation of 'small explosions' during the reaction and the known propensity of this alkyne to explode.

The reaction of phenoxide with polyhalogenated ethylenes is in fact an old one. Slimmer reported in 1903 that potassium phenolate reacts with tribromoethylene to give a dibromo compound to which he initially gave the structure 1,1-dibromo-2-phenoxyethene (**271**) ⟨03CB289⟩. Much later, Jacobs and Whitcher showed that the product was actually the 1,2-dibromo-1-phenoxyethene (**272**) ⟨42JA2635⟩.

Sodium ethoxide reacts with trichloroethylene (**273**) in hot ethanol to give 1,2-dichloro-1-ethoxyethene (**274**) in 70% yield ⟨20JCS691⟩. In the light of the structural problems described above it is of interest to note that the product of the reaction of sodium phenolate with trichloroethylene was proven to be (**267**) ⟨52M1⟩. This means that the dichlorovinyl ethers described in a patent and whose structure was not defined, most probably are 1,2-dichloro-1-alkoxyethenes ⟨49BRP617820⟩. This reaction was repeated by Normant and extended to that of sodium phenolate where it was necessary to use DMF as the solvent ⟨63BSF1876⟩. Normant also reported that care had to be taken to prevent the release of the explosive and inflammable dichloroalkyne. The reaction of substituted phenols on (**273**) has also been conducted under phase-transfer conditions ⟨88PJC483⟩. The reaction was stereoselective with only (*E*) isomers being produced; a fact which was interpreted as showing that the reaction proceeded via dichloroalkyne (Table 28).

Tetrafluoroethylene reacts with sodium methoxide at 50–60 °C and 40 psi to give 1,2,3-trifluoro-1-methoxyethene (**275**) in 69% yield and with sodium phenoxide to give the phenyl ether (**276**) in 21% yield ⟨66USP3277068, 67ZOR1006, 68JOC816⟩. Table 29 lists further examples of this type of displacement reaction.

Notably, the reaction of alkoxides with polyfluoro alkenes often leads not to displacement but rather to addition across the double bond. Thus (**277**), (**278**), (**279**) and (**280**) all give saturated ethers, for example (**281**), as the main products ⟨56JA1685⟩.

Table 28 Phase transfer catalysed reactions of substituted phenols with trichloroethylene to give (E)-1,2-dichloro-1-aryloxyethenes.

ArOH	Catalyst	Yield (%)
$(4\text{-}HOC_6H_4)_2CMe_2$	$PhCH_2^+NEt_3^-Cl$	62[a]
$(4\text{-}HOC_6H_4)_2CMe_2$	$PhCH_2^+NPr^n_3{}^-Cl$	60[a]
$3,4\text{-}Me_2C_6H_3OH$	$PhCH_2^+NPr^n_3{}^-Cl$	51
$4\text{-}EtC_6H_4OH$	$PhCH_2^+NPr^n_3{}^-Cl$	50
$3\text{-}MeC_6H_4OH$	$PhCH_2^+NPr^n_3{}^-Cl$	58
$4\text{-}NO_2C_6H_4OH$	$PhCH_2^+NPr^n_3{}^-Cl$	50

[a] Product is Cl₂C=CH–O–C₆H₄CMe₂C₆H₄–O–CH=CCl₂

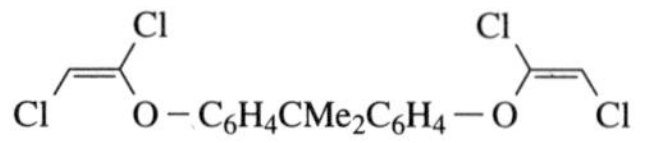

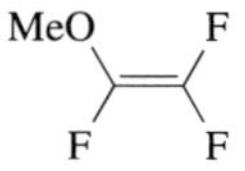

(275) MeO, F / F, F
(276) PhO, F / F, F

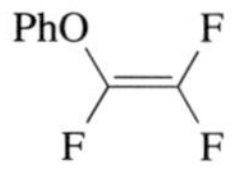

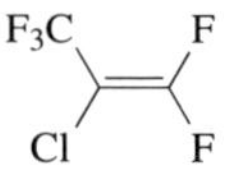

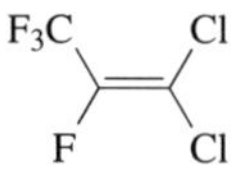

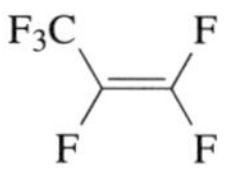

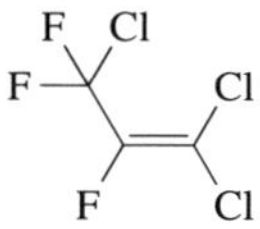

(277) F₃C, F / Cl, F
(278) F₃C, Cl / F, Cl
(279) F₃C, F / F, F
(280) F, Cl / F–C–Cl / F, Cl
(281) F₃C, F–OEt / Cl, F

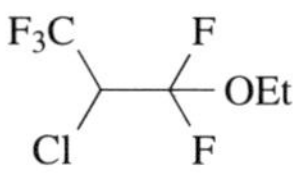

Instead of halide ions, a nitro group, for example in (**282**), may be displaced, as nitrite giving, for example the ethers (**283**) ⟨91ZOB56⟩.

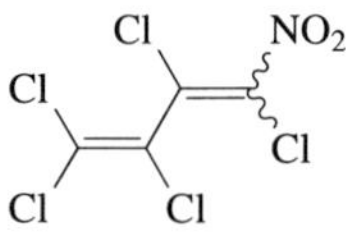

(282) Cl, NO₂ / Cl, Cl / Cl, Cl
(283) Cl, OR / Cl, Cl / Cl, Cl
R = Me, Et, Pr^i, 45–56%

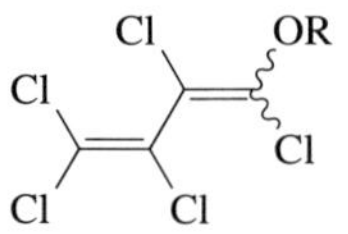

Clearly, displacement reactions should be much enhanced if the double bond also carries an electron-withdrawing group which would assist an initial Michael addition. This indeed is found. However, stopping the reaction after only one halogen has been displaced is difficult. Thus β,β-dichloroacrylonitrile (**284**) gives only the cyanoketene acetals (**285**) ⟨70JOC828⟩. With care, however, it is possible to isolate by GLC the monosubstitution product (**287**) from 1,2,2-trichloroacrylonitrile (**286**) and methoxide ⟨71JOC3386⟩.

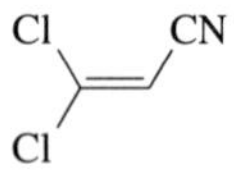

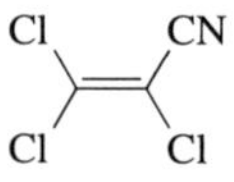

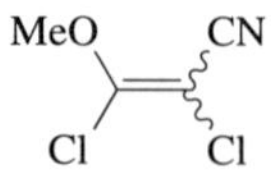

(284) Cl, CN / Cl
(285) RO, CN / RO
(286) Cl, CN / Cl, Cl
(287) MeO, CN / Cl, Cl

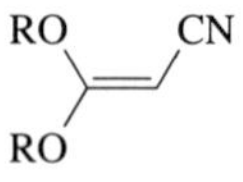

Some insights into the mechanism of this reaction were provided by Burton and Krutzsch who showed that the halogenated styrenes (**288**) react with methoxide to give monomethyl ethers in up to 77% yield by displacement of chlorine with 90–96% stereospecificity ⟨71JOC2351⟩. The results were rationalised in terms of an initial, irreversible, *trans*-addition of the nucleophile and electron pair across the double bond to give the short-lived carbanionic species. This then undergoes a rapid *cis* elimination of the chloride ion. Further examples of this reaction are given in Table 30. A slightly different approach is that of Rossman and Muller who treated perfluoroisobutene (**289**) with trimethyltin methoxide to give 1,3,3,3-tetrafluoro-2-trifluoromethyl-1-methoxy-1-propene (**290**) in 62% yield ⟨93JFC(60)61⟩.

Table 29 Displacement reactions of unactivated polyhalo alkenes.

Haloalkene	RONa (K)	Product	Yield (%)	Ref.
$(CF_3)_2CF-CF=CF-CF_3$	EtONa	(Z)- and (E)-(EtO)CF=C(CF_3)CF(OEt)CF(OEt)CF_3	32	76BCJ502
$(CF_3)_2C=CFC_2F_5$	EtONa, NaOH	(Z)- and (E)-(EtO)CF=C(CF_3)CF(OEt)C_2F_5	52	76BCJ502
$Cl_2C=CCl_2$	$MeOCH_2CH_2ONa$	$MeOCH_2CHOCCl=CCl_2$	12	56M319
$ClFC=CF_2$	MeONa	1:1(Z):(E), Cl(F)C=CF(OMe)	75	67ZOR1006
$ClFC=CF_2$	EtONa	1:1(Z):(E), Cl(F)C=CF(OEt)	97	67ZOR1006
$Cl_2C=CHCl$	[MeS-substituted bicyclic alcohol]	[MeS-substituted bicyclic vinyl ether with =CCl-CCl]	78	94JA2153
$Cl_2C=CHCl$	EtONa	EtO(Cl)C=CHCl	34–54	
$Cl_2C=CHCl$	Pr^{i}ONa	Pr^{i}O(Cl)C=CHCl	34–54	92S495
$Cl_2C=CHCl$	BuONa	BuO(Cl)C=CHCl	34–54	
BrCF=CF(Br)	MeONa (1 equiv.)	(E)- MeO(F)C=C(F)Br	30	93JOC3421
		(Z)- MeO(F)C=C(Br)F	18	
		+ others		
BrCF=CF(Br)	$EtOCH_2CH_2ONa$ (1 equiv.)	3:2 (E):(Z), (EtOCH_2CHO)FC=CBr(F)	95	93JOC3421
$Br_2C=CF(Br)$	MeONa (1 equiv.)	$Br_2C=C(F)OMe$	68	93JOC3421
PhCF=CF(Cl) 72:28 (Z):(E)	BuONa	PhCF=CF(OBu) (E) 15% (Z) 35%		75T891
		PhCF=CCl(OBu) (E) 15% (Z) 35%		
$CF_3CCl=CCl_2$	MeOK	$CF_3ClC=Cl(OMe)$		52JA4104
$CF_3CCl=CCl_2$	EtOK	$CF_3ClC=Cl(OEt)$		52JA4104
$CF_3CCl=CCl_2$	PrOK	$CF_3CCl=CCl(OPr)$		52JA4104
$CF_3CCl=CCl_2$	BuOK	$CF_3CCl=CCl(OBu)$		52JA4104
$CF_3CCl=CFCl$	EtOK	$CF_3CCl=CF(OEt)$	34	56JA1685
BuSCF=CFCl 1:1 (Z):(E)	BuONa	BuSCF=CF(OBu), 87:13 (Z):(E) BuSCF=CCl(OBu), 85:15 (Z):(E)		75T897
$Cl_2C=CHCl$	[cyclopentenyl]–ONa	[cyclopentenyl-O–C(Cl)=CHCl]		87JA5029
$CF_3CF_2CF=CCl_2$	PhONa	$CF_3CF_2CF=C(Cl)OPh$ + others	44	89JFC69
$Cl_2CCClCCl=CCl_2$	NaOEt	$Cl_2C=CClCCl=C(OEt)Cl$		80MI 417-01

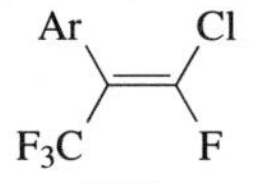

(288) Ar⟩C=C⟨Cl / F₃C⟩ F
Ar = Ph, p-ClC₆H₄, p-MeOC₆H₄

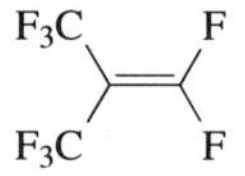

(289) F₃C⟩C=C⟨F / F₃C⟩ F

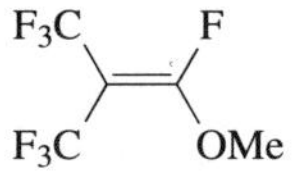

(290) F₃C⟩C=C⟨F / F₃C⟩ OMe

Table 30 Displacement reactions of activated halo alkenes.

$$\underset{R^3}{\overset{R^2}{>}}C=C\underset{Y}{\overset{X}{<}} + R^1OM \longrightarrow \underset{R^3}{\overset{R^2}{>}}C=C\underset{Y}{\overset{OR^1}{<}}$$

Substrate	R¹OM	Yield (%)	Ref.
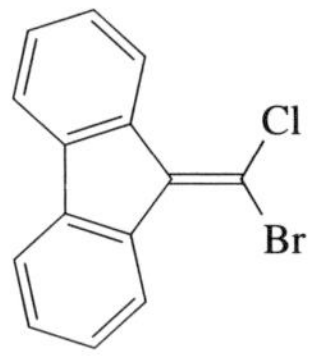 9-(chlorobromomethylene)fluorene	p-MeC₆H₄ONa	Y = Cl (29)	87JA6687
(p-NO₂-C₆H₄)₂C=CClBr	p-MeC₆H₄ONa	Y = Br (10) + 24 acetal	87JA6687
3-(difluoromethylene)-γ-butyrolactone	PhC(=CH₂)OTMS	36 (Z)	81TL1421
PhSO(Me)C=CF₂ [a]	MeONa	11 (E) / 57 (E,Z)	85CAR177
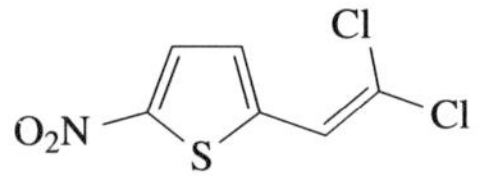 5-nitro-2-(2,2-dichlorovinyl)thiophene	MeONa	42 (Z)	88MI 417-01
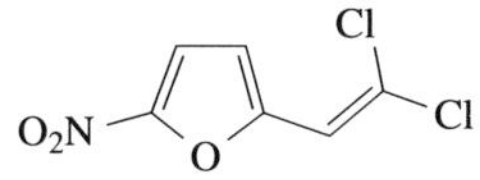 5-nitro-2-(2,2-dichlorovinyl)furan	MeONa	28 (Z) [b]	88MI 417-01
p-Cl-C₆H₄S(O)CH=CCl₂	EtOLi	75 (E)	91JOC6987

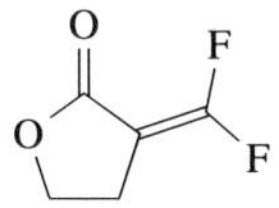

[a] Generated *in situ* by addition of a mole of NaOMe to CF₃CH(SOPh)CH₃.   [b] The (E) isomer was also isolated but no yield was given.

(ii) Addition to substituted alkynes

Haloalkynes have not been widely used as precursors of this system partly because of their instability but also because the usual synthetic route involves the action of alkoxides on polyhalo alkenes which, as seen above, are favoured direct precursors of 1-halo enol ethers ⟨69MI 417-01⟩. Nevertheless there have been reports of their use, particularly in perfluoro compounds. 1,2-Dichloroalkyne, usually prepared *in situ* from trichloroethylene, adds alkoxides to give (**274**) ⟨88PJC483⟩. 1-Chlorotrifluoropropyne (**291**) reacts with 'silver trifluoromethoxide', a reagent which is made *in situ* from silver fluoride, carbonyl difluoride and hydrogen fluoride in adiponitrile, to give, in 45% yield, (**292**) as a single, (E) isomer ⟨78JOC43⟩. The reaction was extended to the perfluoropropoxy analogue (**293**). Hexachlorophenylalkyne (**294**) reacts with sodium methoxide in refluxing methanol to give a mixture from which 24% of (**295**) can be isolated ⟨86JOC1413⟩.

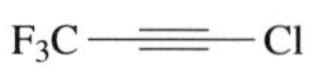

(291) F₃C–C≡C–Cl

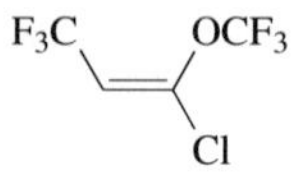

(292) F₃C⟩C=C⟨OCF₃ / Cl

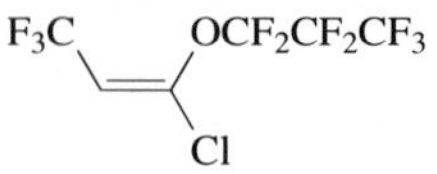

(293) F₃C⟩C=C⟨OCF₂CF₂CF₃ / Cl

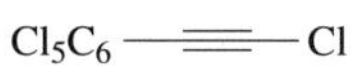

(294) Cl₅C₆–C≡C–Cl

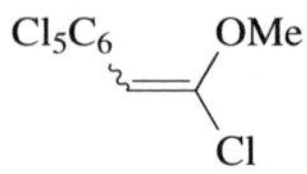

(295) Cl₅C₆⟩C=C⟨OMe / Cl

Russian workers have examined the reactivity of alkylthiochloroalkynes (**296**) ⟨88IZV514⟩. As predicted from ^{13}C NMR shifts, reactions with alkoxides give 1-alkoxy 1-halo 2-alkylthio alkenes (**297**) and 1,2-dialkoxy 1-alkylthio alkenes (**298**). It is presumed that (**298**) arise from the alkyne (**299**) resulting from loss of HCl from (**297**). Table 31 lists the yields and isomer distribution of compounds (**297**).

Table 31 Addition of alkoxides to alkylthiochloroalkynes.

R^1	R^2	Yield (%)	Isomer
C$_3$H$_7$	Ph	40	1:1(E):(Z)
C$_3$H$_7$	p-BrC$_6$H$_4$	30	(E)
C$_3$H$_7$	p-ClC$_6$H$_4$	53	(E)
C$_3$H$_7$	p-F-C$_6$H$_4$	41	(E)
C$_4$H$_9$	Ph	50	1:1(E):(Z)
C$_4$H$_9$	p-ClC$_6$H$_4$	44	(E)
C$_4$H$_9$	m-MeC$_6$H$_4$	33	(E)
Ph	Ph	21	(E)

1-Alkoxyalkynes (**300**) are more readily prepared and have been widely used as precursors to 1-halo enol ethers ⟨60MI 417-01, 69MI 417-02⟩. Bakker and Scheeren reviewed the addition of a wide variety of halo compounds, for example COCl$_2$, RCOCl, HCl, (Hal)$_2$, SCl$_2$, PCl$_5$, GeCl$_4$, SnCl$_4$, BCl$_3$, HgCl$_2$ and PCl$_3$ (Equation (12)), to (**300**) ⟨93HOU(E15/2)1599⟩. In general the products appear to be mixtures of (E) and (Z) isomers but this may be due to thermal isomerisation during isolation and purification by distillation.

$$R^1 = C_4H_9; R^2 = Me, 50\%$$
$$R^1 = H; R^2 = Bu, 68\%$$
$$R^1 = Me, R^2 = Me, 82\%$$
$$R^1 = Pr^i, R^2 = Me, 50\%$$

Thus, for example, PCl$_3$ reacts with alkoxyalkynes to give reasonable yields of the adducts (**301**) as apparently single isomers but the reaction mixtures were heated at 60 °C for several hours ⟨77ZOR1675⟩.

Phenoxyalkyne reacts with bromine and iodine to give 1-(bromo)iodo-2-bromo(iodo)-1-phenoxyethene (**302**), (neither yields nor stereochemistry defined) ⟨40JA1849⟩. Filippova also prepared (**302**; X = Br) in 63% yield and queried the structure from the earlier paper ⟨71IZV162⟩. Simultaneous addition of phenoxyalkyne and iodine in potassium iodide to cold potassium hydroxide gave 28% of phenoxytriiodoethene ⟨42JA2635⟩. A better yield (60%) of this compound arises from the reaction of phenoxyethynylmagnesium bromide and iodine.

Table 32 lists a selection of the less common reagents which have been added across the triple bond.

1-Iodo 1-alkoxy alkenes have also been prepared, without isolation, as intermediates in synthesis. Thus 1-ethoxyalkyne reacted at −78 °C with B-iodo-9-borabicyclo[3.3.1]nonane to give the adduct (**303**), which then reacted further with aldehydes in overall excellent yields ⟨89TL5153⟩. Likewise, 1-ethoxy-1-octyne reacts with iodotrimethylsilane to give (**304**) ⟨89TL1833⟩.

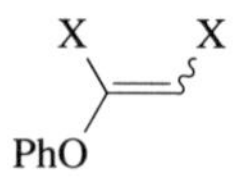

(302) X = Br, I

Table 32 Addition to alkoxyalkynes.

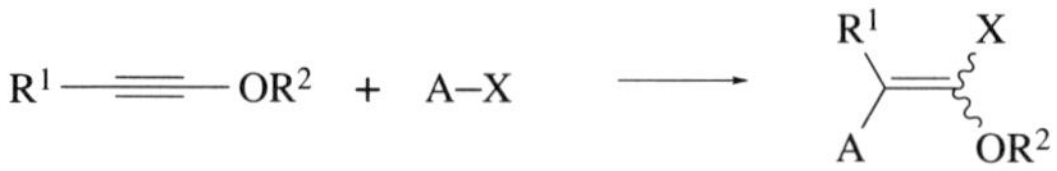

R^1	R^2	A	X	Yield (%)	Ref.
H	Ph	GeCl$_3$	Cl	70 ((E) isomer)	83IZV2153
(Me$_3$C)$_2$P(S)	Et	Ac	Br	70	91ZOB1014
Me	Et	(a)	Br	80	89ZOB955
Pr$^i_2$P	Et	TMS	Br	80–90(unstable)	90ZOB1539
Me$_3$Sn	Me	CF$_3$	I	95	89ZOB2145

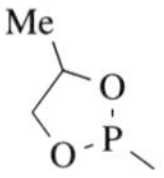

(a)

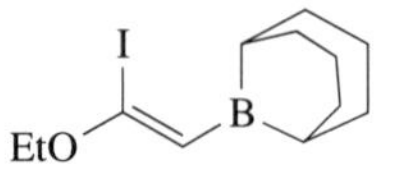

(303) 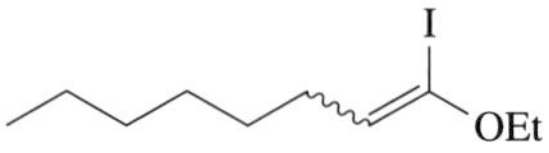 **(304)**

(iii) Reaction of acetoxy groups with phosphorus pentachloride

In 1886 Michael showed that phenyl acetate reacted with phosphorus pentachloride to give 1,2,2-trichloro-1-phenoxyethene **(305)** ⟨1886CB845⟩. In 1989 it was shown that vinyl acetate reacts with PCl$_5$ in dry benzene to give in 66% yield 2-chloro-2-(1-chloroethenyloxy)ethylphosphonodichloridate **(306)** ⟨89ZOB997⟩. The acetylated sugars **(307)** and **(308)** reacted with PCl$_5$ in refluxing tetrachloromethane to give mixtures from which **(309)** and **(310)** were isolated in yields of 43% and 62% respectively ⟨92CPB3261⟩.

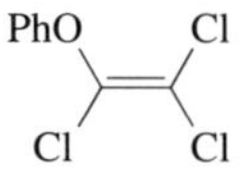

(305) 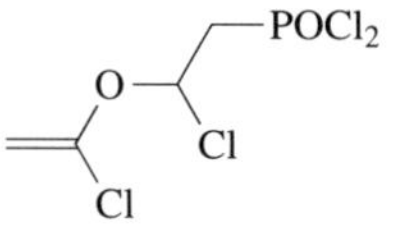 **(306)**

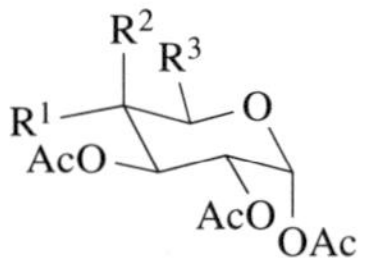

(307) R^1 = OAc, R^2 = H, R^3 = CH$_2$OAc
(308) R^1 = H, R^2 = OAc, R^3 = CH$_2$OAc
(309) R^1 = OAc, R^2 = H, R^3 = CH$_2$OC(Cl)=CCl$_2$
(310) R^1 = H, R^2 = OAc, R^3 = CH$_2$OC(Cl)=CCl$_2$

(iv) Elimination reactions

Vatele has synthesised the interesting, strong Michael acceptor **(311)** from 1,1,1-trifluoroacetone **(312)** via the sequence shown in Scheme 7 ⟨83TL1239⟩. Subsequently, the same approach was used to prepare the methoxyethyl derivative **(313)** ⟨86T4443⟩. The yield of the final displacement–elimination reaction was 71% and the product was a mixture of stereoisomers. Fluorinated vinyl ethers, for example **(314)**, have been prepared in low yield by the thermal decarboxylation of the appropriate acid **(315)** ⟨69JOC1841⟩. A rather more successful pyrolysis is that of perfluoro-9-azido-2,5-dimethyl-8-n-propoxy-3,6-dioxanonanoyl fluoride **(316)**, which was hydrolysed to give its sodium salt, which was then pyrolysed at 220–235°C for 5 h to give 48% of perfluoro-9-azido-5-methyl-8-n-propoxy-3,6-dioxanon-1-ene (**317**) ⟨86JOC326⟩.

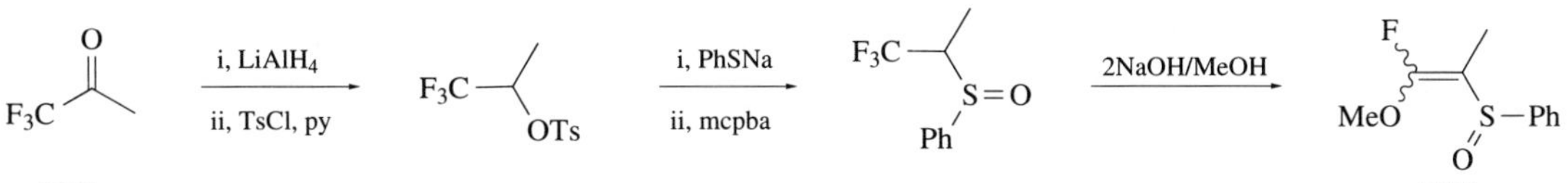

(312) **(311)**

Scheme 7

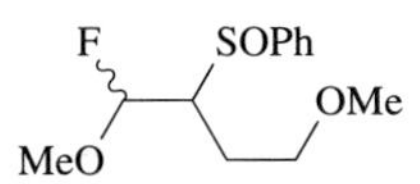

(313)

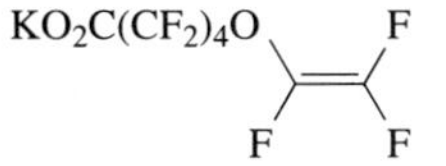

(314)

$[O_2C(CF_2)_4OCF(CF_3)CO_2]^{2-}\ H^+K^+$

(315)

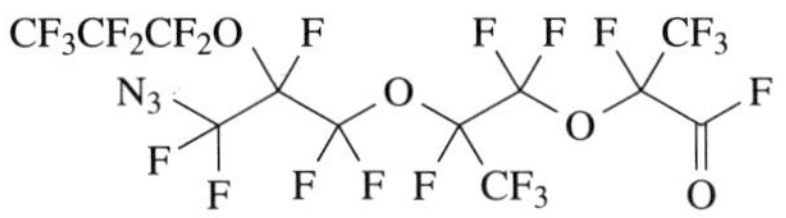

(316)

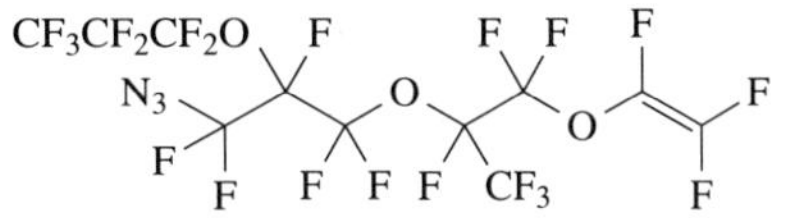

(317)

Somewhat more conventional elimination procedures were described by O'Connor and Smithers ⟨68JOC1991, 85S556⟩. 1,2-Dichloro-1,2-dimethoxyethane, prepared by chlorination of 1,2-dimethoxyethene was dehydrochlorinated by powdered potassium hydroxide to give a 50% yield of a 52 : 48, (*E*) : (*Z*) mixture of (**318**). Likewise, the 1-alkoxy-2,2,2-tribromo-1-chloroethanes (**319**) readily lose HBr on treatment with sodium ethoxide in ether at 0–20°C to give the 1-alkoxy-2,2-dibromo-1-chloroethenes (**320**) in yields of 53–62%. This reaction was subsequently used to produce the di(2,2-dibromo-1-chlorovinyl) ether (**321**) in 87% yield ⟨89JOC1479⟩.

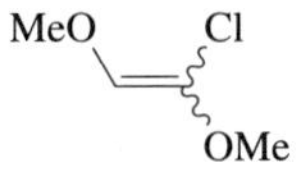

(318)

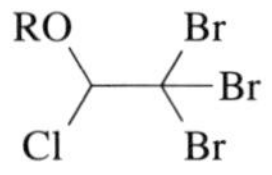

(319)

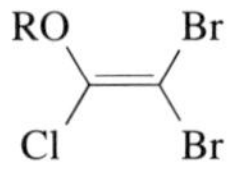

(320) R = Me, 62%
R = Et, 60%
R = *n*-C_8H_{17}, 54%
R = $HOCH_2CH_2$, 53%

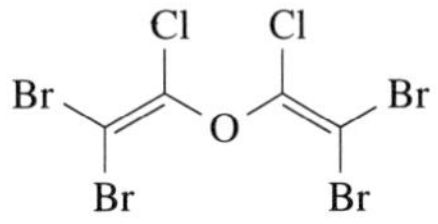

(321)

The elimination of HCl from *meso*- and (±)-1,2-dichloro-1,2-diisopropoxyethane (**322**) was studied by Percias and Serratosa ⟨77TL4433⟩. Treatment with potassium *t*-butoxide in pentane gave 65% of a mixture of (*E*)- and (*Z*)-1-chloro-1,2-diisopropoxyethene (**323**). Alternatively, (**322**) could be dechlorinated by magnesium metal to give (*E*)- and (*Z*)-1,2-diisopropoxyethene in 80% yield. Bromination–dehydrobromination then gave pure (*E*)-1-bromo-1,2-diisopropoxyethene (**324**) in 80% yield from (**322**). Other reported eliminations are of HCl from the ethers (**325**) and (**326**) ⟨57USP2803665, 57USP2803666⟩, of IF from (**327**) ⟨80BP2029827⟩ and of HF from (**328**) ⟨85JFC(29)471⟩.

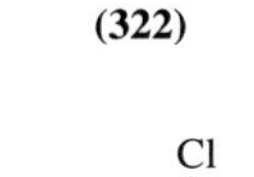

(322)

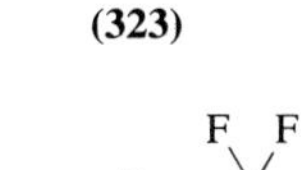

(323)

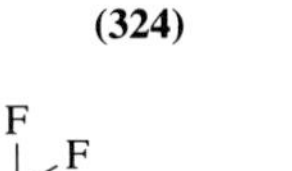

(324)

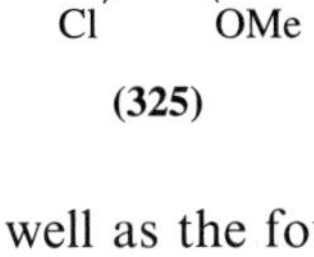

(325)

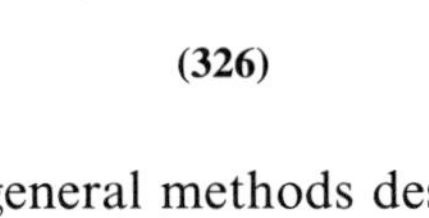

(326)

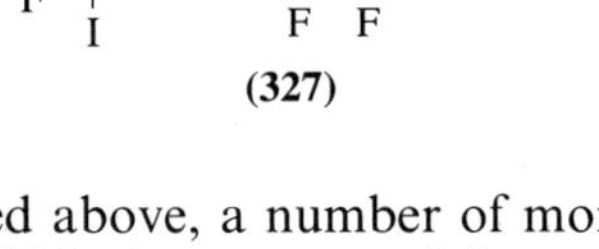

(327)

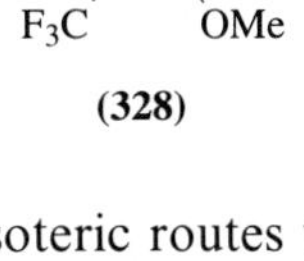

(328)

As well as the four general methods described above, a number of more esoteric routes to this system have been described. Thus Dehmlow and Neuhaus reported that primary aliphatic alcohols

ROH (R = Et, CH$_2$CHMe$_2$, CH$_2$Bu^t) react with dichlorocarbene under phase-transfer conditions to give the cyclopropane derivatives (**329**) as mixtures of stereoisomers ⟨87ZN(B)796⟩. Hydrolytic thermolysis of the compounds (**329**) at 120 °C gives the esters (**330**) in ca. 20% yield. Eapen and co-workers have reported that the perfluoro iodo ether (**331**) reacts with aluminum chloride at 100–105 °C during 60 h to give a mixture which contains 19% each of the enol ethers (**332**) and (**333**) ⟨86JFC(31)405⟩.

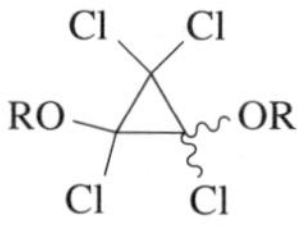

(**329**) 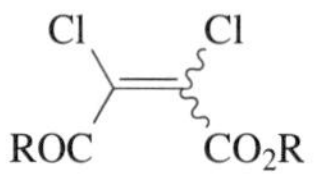   (**330**) 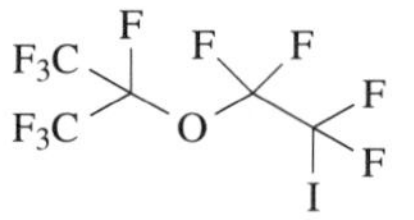   (**331**) 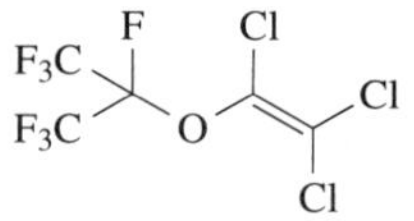   (**332**) 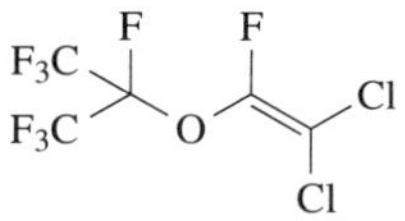   (**333**)

4.17.2.1.2 α-Halo acyloxy and other O-substituted derivatives

One of the potentially most versatile routes to α-halo acyloxy alkenes is one of the most recent. Tsuckazaki and Snieckus reported that O-(1-lithiovinyl) carbamates (**334**), which are generated from the O-vinyl carbamate by S-butyllithium at −78 °C, are readily brominated by 1,2-dibromo-tetrafluoroethane to give (**335**) ⟨93TL411⟩. This would appear to offer scope of extension to other systems and other halogens provided that suitable halogenating agents could be identified.

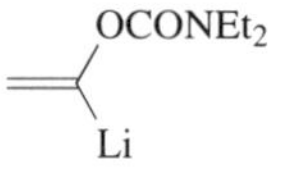

(**334**) 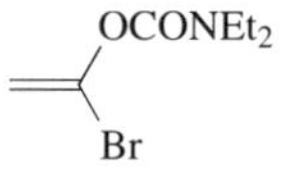   (**335**)

In contrast, an early synthesis failed to generate the vinyl acetate structure. Thus, Brady and Roe treated α-chloropropionyl chloride (**336**) with 0.5 M triethylamine in dry hexane at 0–5 °C to give, in 64% yield, a compound described as the β-keto acid chloride (**337**) ⟨68TL1977⟩. However, it was quickly shown by Dreiding et al. that the product was actually a mixture of stereoisomers of 1,2-dichloropropenyl α-chloropropionate (**338**) ⟨68HCA1466⟩. They, and Lavanish, extended the reaction to trichloroacetyl chloride where the product, in 40% yield, was trichlorovinyl dichloroacetate (**339**) ⟨68TL6003⟩. Brady then extended the reaction further to give the results shown in Table 33 ⟨70JOC1515⟩. It is assumed that this reaction proceeds via an enolate anion which is O-acylated by an excess of acid chloride. A similar, but potentially more versatile approach is to generate a zinc enolate from either trichloro- or tribromoacetyl chloride and then trap this with a different carboxylic acid chloride (Scheme 8, Table 34) ⟨92ZOR522⟩.

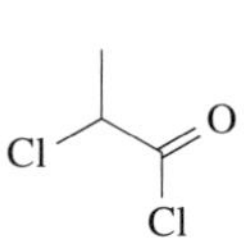

(**336**) 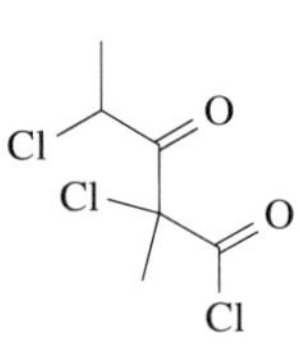   (**337**) 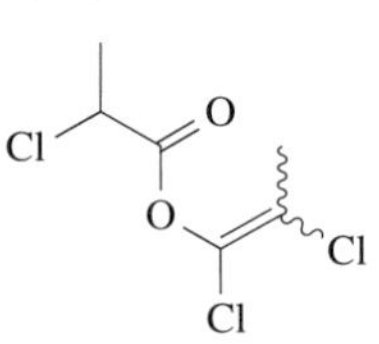   (**338**) 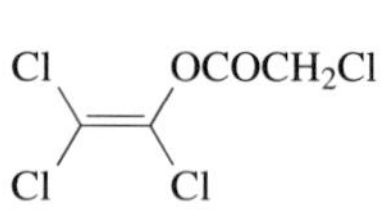   (**339**)

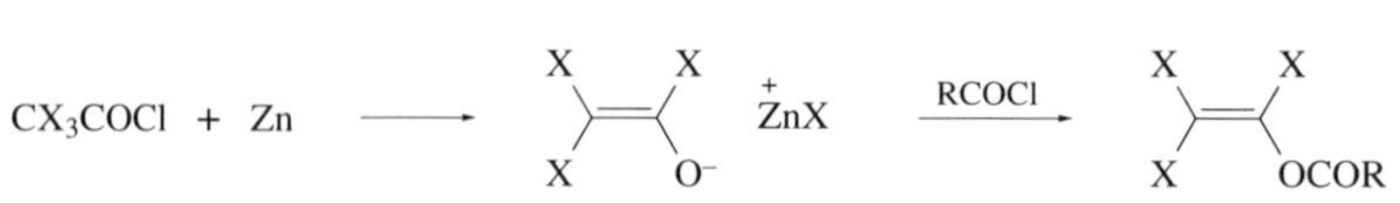

CX$_3$COCl + Zn ⟶ [X$_2$C=C(X)O$^-$ $\overset{+}{Z}$nX] $\xrightarrow{\text{RCOCl}}$ X$_2$C=C(X)OCOR

Scheme 8

Diacetoxyalkyne (**340**) reacts in a biphasic medium with a caesium fluoride in the presence of tetrabutylammonium dihydrogen trifluoride as a phase transfer catalyst to give 67% of the fluorinated enol acetate (**341**) ⟨89MI 417-01⟩. The product was a 85:15 mixture of (Z) and (E) isomers and the reaction also worked with the analogous propionate although the yield was then only 41% of an 83:17, (Z):(E) mixture.

Tri(2-chloroethyl) phosphate (**342**) reacts with chloroacetyl chloride to give 37% of (**343**) ⟨75ZOB2374⟩. A more conventional addition–elimination route has also been described. Diethyl

Table 33 Self-*O*-acylation of α-halo acid halides.

Acid halide	Product	Yield (%)
α-bromopropionyl chloride	$CH_3CH(Br)COOC(Cl)=C(Br)CH_3$	45
α-chlorobutyryl chloride	$C_2H_5CH(Cl)COOC(Cl)=C(Cl)CH_3$	41
α-chlorobutyryl bromide	$C_2H_5\text{-}CH(Cl)COOC(Br)=C(Cl)C_2H_5$	31
α-chloro-α-phenylacetyl chloride	$Ph\,CH(Cl)COOC(Cl)=C(Cl)Ph$	54
α-chloropropionyl bromide	$CH_3CH(Cl)COOC(Br)=C(Cl)CH_3$	43
dibromoacetyl chloride	$CH\,Br_2COOC(Cl)=C\,Br_2$	11

Table 34 Reaction of zinc enolates with acid chlorides (Scheme 8).

CX_3COCl	$RCOCl$	Product	Yield (%)
CCl_3COCl	$MeCOCl$	$CCl_2=C(Cl)OCOMe$	82
CCl_3COCl	$EtCOCl$	$CCl_2=C(Cl)OCOEt$	74
CCl_3COCl	Bu^tCOCl	$CCl_2=C(Cl)OCOBu^t$	41
CCl_3COCl	$PhCOCl$	$CCl_2=C(Cl)OCOPh$	68
CBr_3COCl	$MeCOCl$	$CBr_2=C(Cl)OCOMe$	64
CBr_3COCl	$EtCOCl$	$CBr_2=C(Cl)OCOEt$	57
CBr_3COCl	Bu^tCOCl	$CBr_2=C(Cl)OCOBu^t$	38

$$MeCO_2-\!\!\!\equiv\!\!\!-OCOMe$$
(340)

(341) — $MeCO_2$, F, $OCOMe$ substituted alkene

vinyl phosphate **(344)** can be brominated to give the dibromo compound **(345)** which is then dehydrobrominated to give **(346)** by a variety of bases, of which the best in terms of yield and selectivity of product, appears to be lithium bis(trimethylsilyl)amide in toluene ⟨94JA789⟩. The reaction was also extended to **(347)** and **(348)**.

(342) $(Cl\text{\textasciitilde}O)_3P$

(343) $Cl\text{\textasciitilde}O\text{-}P(O)(Cl)\text{-}O\text{-}C(=CH_2)$

(344) $R = (EtO)_2P(O)$; $RO\text{-}CH=CH_2$

(345) $RO\text{-}CHBr\text{-}CH_2Br$

(346) $R = (EtO)_2P(O)$, 91%
(347) $R = PhCO$, 86%
(348) $R = Tos$, 94%

4.17.2.2 Halogen and Sulfur Derivatives

4.17.2.2.1 *Dicoordinate sulfur derivatives*

Many of the reactions described in Section 4.17.2.1.1 for the preparation of α-halo enols have, of course, also been applied to the synthesis of this system. In addition, reactions peculiar to sulfur have also been exploited. The most important and versatile methods are displacement reactions on polyhalogeno alkenes, elimination reactions, Wittig condensations, halogenation of organometallics and additions to substituted alkynes.

(i) Displacement reactions on polyhalogeno alkenes

The reaction of sodium arenethiolates with 1,1,2-trichloroethene has been described by a number of authors ⟨37JCS767, 57CA(51)5721f, 57CA(51)12849g, 57G1061, 58JA1916, 63BSF1876, 90G569⟩. The reaction needs the presence of both sodium thiolate and sodium ethoxide and the product (**349**) formed in 80% yield, is a single isomer in which the chlorine atoms are *trans*. The simplest interpretation of these results is that the reaction proceeds by initial elimination of HCl to give dichloroalkyne which then undergoes *trans* addition of thiophenol. The reaction appears to be quite general and goes with thiophenol, *o*- and *p*-thiocresol and aliphatic thiols such as methane, ethane and propane thiol and the *n*- and *t*-butyl derivatives ⟨53RTC813, 62BP896373⟩.

$$\underset{\text{Cl}}{\text{ArS}}\diagup\hspace{-0.5em}=\hspace{-0.5em}\diagdown\underset{}{\text{Cl}}$$
(**349**)

When the trichloroethene is replaced by tetrachloroethene the elimination route via alkyne is no longer available and more forcing conditions are needed. Truce and Kassinger initially reported that the product of the reaction of sodium toluene-*p*-thiolate with tetrachloroethene was *cis*-1,2-dichloro-1,2-di-(*p*-tolylthio)ethene (**350**) ⟨58JA6450⟩. Subsequently, Truce *et al.* revised the stereochemistry of the product to *trans* (**351**) ⟨65T2899⟩. An unexpectedly high dipole moment of the *trans* isomer was one cause of the initial error.

(**350**)   (**351**)

Prilezhaeva *et al.* have shown how modification of the reaction conditions can direct the displacement reaction with tetrachloroethene to give the replacement of 2, 3 or 4 chlorines ⟨90G235⟩. Thus, use of an excess of dry sodium thiolate in DMF replaces all 4 chlorine atoms; a 3 molar excess of thiolate in DMF-EtOH (1:2) gives (**352**) whilst 2 molar excess in DMF-EtOH (1:5) gives the (*E*)-isomer (**353**) (Table 35).

(**352**)   (**353**)

Table 35 Polydisplacement reactions by RSNa on tetrachloroethene.

RS(RS)C=CClSR		RSClC=CClSR	
R	Yield (%)	R	Yield
Et	67	Et	78
Pr^n	62	Pr^n	79
Pr^i	50	Bu^n	75
Hex^n	60	Bu^t	61
Oct^n	69	Oct^n	83
$HOCH_2CH_2$	51	$HOCH_2CH_2$	67
		Ph	55

The replacement of halogens from a wide range of other haloalkenes has also been reported and a representative selection of these are shown in Table 36. More complex systems include the polychlorobutene (**354**) which reacts with sodium phenylmethanethiolate to give a mixture of products including 24% of (**355**) whilst hexachlorobutadiene (**356**) reacts with 1,3-propanedithiol to give (**357**) in 26% yield, (**358**) (3%) and (**359**) (18%) ⟨92PS(72)225, 92SUL251⟩. The reaction can also be conducted under free radical conditions. Thus, an excess of 1,2,2-tribromo-1-fluoroethene when treated under reflux with thiophenol, with UV irradiation, reacts to give (**360**) in 75% yield ⟨90SUL109⟩.

Table 36 Nucleophilic thiol displacement reactions of halo akenes.

Halo alkene	Thiolate	Conditions	Product	Yield (%)	Ref.
BrC(F)=C(F)Br	p-TolSNa	MeOH/Δ	p-TolSC(F)=CHF 23 : 77, (Z) : (E)	37	93JOC3421
BrC(F)=C(F)Br	p-TolSNa	DMSO, RT	p-TolSC(F)=C(F)Br 16 : 84, (E) : (Z)	30	93JOC3421
			p-TolSC(F)=C(F)STol-p	35	93JOC3421
$(CF_3)_2C=CBr_2$	CF_3SCu	−78 °C	$(CF_3)_2C=CBr(SCF_3)$	51	93JFC(63)253
$Br_2C=CHF$	PhSNa	RT, EtOH	PhSCF=CHBr[a] 64 : 36, (E) : (Z)	70–80	91ZOR2096
BrFC=CHBr 72 : 28, (Z) : (E)	PhSNa	RT, EtOH	PhSCF=CHBr 80 : 20, (Z) : (E)		91ZOR2096
$PhCONHC(\overset{+}{P}Ph_3)=CCl_2$	p-ClC$_6$H$_4$SNa	MeOH, Et$_3$N	$PhCONHC(\overset{+}{P}h_3P)=C(Cl)SC_6H_4$-pCl	85	91ZOB874
$CCl_2=CCl—C(NO_2)=CCl_2$	PhSNa	EtOH, 0°	$CCl_2=CClC(NO_2)=C(Cl)SBu$	89	93ZOR56
PhCF=CFCl	BuSNa	THF, 60°	(E,Z)-PhCF=CFSBu (1) + (E,Z)-PhCF=CClSBu (2.5)		75T891
BuSCF=CFCl	BuSNa	THF, 0 °C	(E)-BuSCF=CFSBu	15	75T897
			BuSCF=CClSBu	30	75T897
			$(BuS)_2C=CClSBu$	20	75T897

[a] The structure of this product indicates that the reaction is actually an addition–elimination one rather than a direct displacement.

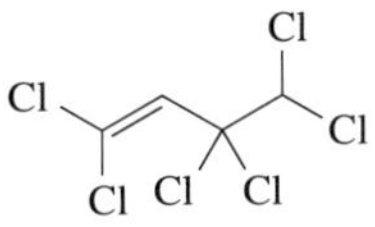

(354) 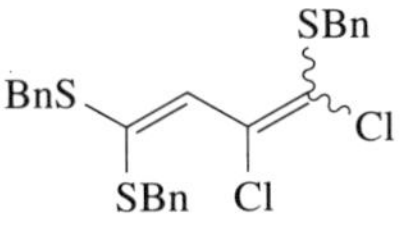 (355) 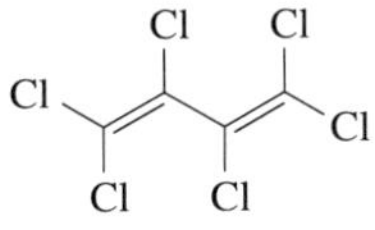 (356) 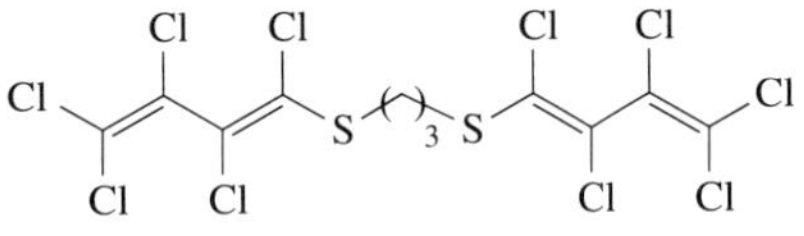 (357)

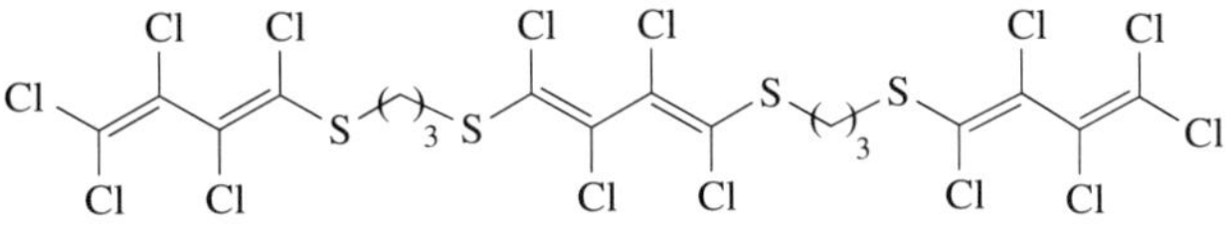

(358)

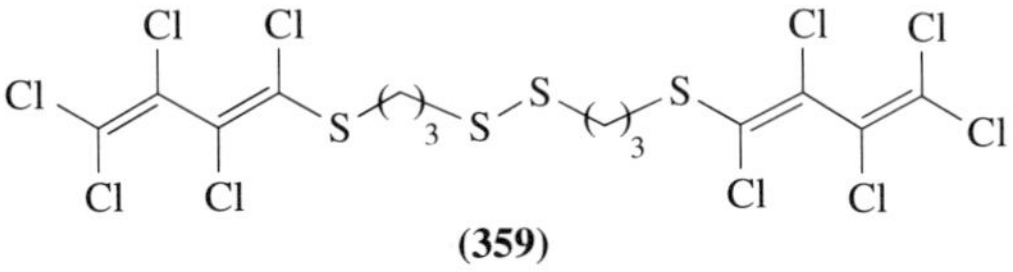

(359) 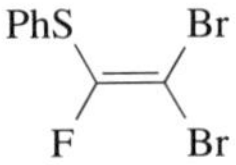 (360)

The presence of an electron-withdrawing group on the 2-position of the haloethene will, of course, enhance the halogen displacement by initial Michael addition followed by elimination. Therefore, triodonitroethene reacts rapidly with trifluoromethylthiocopper to give a mixture of mono- (**361**) and disubstituted (**362**) products ⟨90ZOR740⟩. However, it is not easy to stop the reaction at the monosubstitution stage and, for instance, the dichloroacrylonitriles (**363**) give only the disubstituted products (**364**) ⟨71JOC3386⟩.

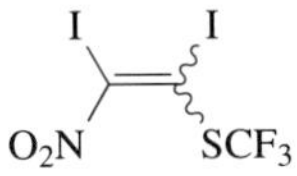

(361) 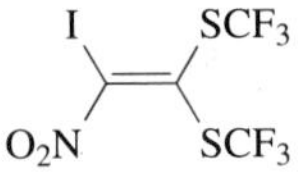 (362) 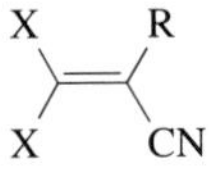 (363) X = Cl
(364) X = SR[1]

Finally, the halo alkene can be produced *in situ*. For example, 1,1,2,2-tetrachloroethane reacts with sodium thiophenoxide in DMF at 70 °C over 5 h to give 52% of (**365**) ⟨76BCJ1931⟩.

(365)

(ii) Addition to substituted alkynes

It was noted above that the reaction of sodium arenethiolates with trichloroethene most probably goes via dichloroalkyne. This reaction also proceeds in good yield with preformed dichloroalkyne to give (**366**) ⟨57G1061⟩. The alternative isomer (**367**) is produced by the addition of an arenesulfenyl chloride to chloroalkyne whilst the analogous reactions of fluoroalkyne with thiols give the fluoro compounds (**368**) ⟨63CA(59)11313c⟩.

(**366**) Ar = Ph, 90%
Ar = o-Tol, 79%
Ar = p-Tol, 77%

(**367**)

(**368**) 59%

Alkynyl sulfides (**369**) have also been used as starting materials. For example, the derivatives (**369**; R = S-p-Tol, S-p-NO$_6$C$_6$H$_4$, R^1 = p-Tol, p-NO$_2$C$_6$H$_4$) react with bromine in chloroform to give the dibromides (**370**) presumably as the (E) isomers but the yield was not stated, whilst, in a later report the sulfides (**369**) reacted with bis(pyridine)iodine(1) tetrafluoroborate (**371**) and nucleophiles, including halide ions, to give single geometrical isomers (**372**) in high yields ⟨12LA(394)325, 22CB1014, 90TL7375⟩. The propargylic alcohol (**369**; R = HOCH$_2$, R^1 = Ph) is reduced and iodinated by LiAlH$_4$-NaOMe-I$_2$ to give (**373**) as a single isomer in 85% yield ⟨90TL6137⟩. Organocuprates add to alkynes and the intermediate copper group can be displaced by iodine to give (**374**) (Scheme 9) ⟨91JA5735⟩. Russian workers have added phosphorus pentachloride to the sulfides (**375**) to give the adducts (**376**) as mixtures of isomers in good yield. Similarly, addition of phosphorus trichloride followed by treatment with sulfur dioxide gave the derivatives (**377**) in high yields, again as mixtures of (E) and (Z) forms ⟨90ZOB227, 91ZOB1084⟩.

(**369**)

(**370**)

(**371**)

(**372**)
R = Ph, Nu = Cl, 84%
R = Bu^n, Nu = Cl, 76%
R = Ph, Nu = Br, 82%
R = Bu^n, Nu = I, 67%
R = Ph, Nu = I, 64%

(**373**)

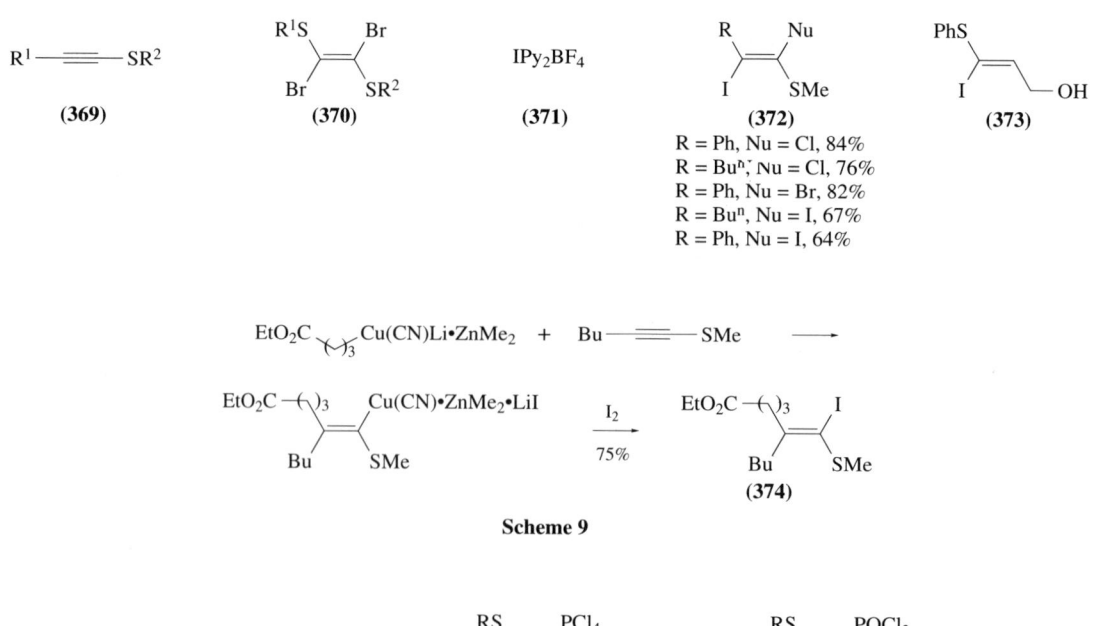

Scheme 9

(**375**)

(**376**)

(**377**)

(iii) Elimination reactions

A number of elimination reactions leading to 1-halo 1-thioalkyl alkenes have been described. Thus, trimethylsilylbromide has been eliminated from (**378**), HBr from (**379**) and halogens from (**380**) reductively by metals ⟨82TL1945, 87S1034, 87T4309⟩. In addition, chloramine-T reacts with the sulfur mustards (**381**) to give the thioenol ethers (**382**) presumably by a chlorination–dehydrochlorination process ⟨93TL7645⟩.

(**378**)   (**379**)   (**380**)   (**381**) X = Cl, Br   (**382**)

In a similar addition–elimination reaction, (**383**) reacts with silver fluoride, iodine and triethylamine to give (**384**) as a mixture of isomers ⟨89USP4877899⟩. α,α-Dichloro sulfoxides, for example (**385**), readily undergo eliminative deoxygenation to give α-chlorovinyl sulfides, for example (**386**), on treatment with zinc and titanium(IV) chloride ⟨83TL531⟩, whilst Miller and Hassig have shown that the α-halo sulfoxides (**387**) react with trimethylsilyl triflate in ether in the presence of triethylamine to give 1-halovinyl sulfides (**388**) in high yields (Table 37) ⟨84SC1285⟩.

(**383**)   (**384**)   (**385**) X = H, OH

(**386**) R¹ = H, R² = n-C₇H₁₅, 57%
R¹ = H, R² = p-MeOC₆H₄, 78%
R¹, R² = (CH₂)₅, 85%

(**387**)

(**388**) R¹, R² = (CH₂)₅

Table 37 Dehydration of α-halo sulfoxides to 1-halovinyl sulfides with TMS-OTf in Et₂O.

Sulfoxide (387)		Conditions	Product (388)				Yield (%)
R¹	R²		X	R³	R⁴		
Ph	Me	1 h, 25 °C	Cl	H	H		86
Ph	Prⁱ	0.5 h, 0 °C	Cl	Me	Me		92
Ph	CH₂CH=CH₂	1 h, 25 °C	Cl	H	CH=CH₂	(Z):(E), 2.0	72
Ph	Prⁿ	0.5 h, 0 °C	Cl	H	Et	(Z):(E), 1.0	91
Ph	CH₂Ph	0.5 h, 0 °C	Cl	H	Ph	(Z):(E), 2.5	
Ph	CHMePh	1 h, 25 °C	Cl	Ph	Me	(Z):(E), 1.2	92
Buᵗ	Me	0.25 h, 0 °C	Cl	H	H		78
Ph	Me	0.5 h, 0 °C	Br	H	H		77
Ph	Prⁿ	0.25 h, 0 °C	Br	H	Et	(Z):(E), 1.2	89
Ph	Prⁱ	0.25 h, 0 °C	Br	Me	Me		89
Ph	CH₂Ph	0.5 h, 0 °C	Br	H	Ph	(Z) only	86

(iv) Wittig reactions

The Wittig–Horner reaction has been applied to the synthesis of 1-halo 1-phenylthio alkenes (Scheme 10, Table 38) ⟨76S107, 85S676⟩. The reaction can be conducted in 'one-pot' and the yields are generally good, but the stereochemistry of the products was not defined. A triphenyl phosphonium ylide was also generated *in situ* by the reaction of chlorofluoromethyl phenyl sulfide (**389**) with *n*-butyllithium in the presence of triphenylphosphine at −78 °C and allowed to react with aldehydes such as 3-phenylpropanal to give (**390**) in 63% yield but with low (Z) selectivity

($E:Z$, 2:3) ⟨92BCJ210⟩. Changing the phosphine to tributylphosphine and the base to methyllithium–lithium bromide gave predominantly (42%) the bromo compound (**391**). Kunugi *et al.* employed the same technique to produce a range of aryl alkenes (**392**) in good yields, again as mixtures of stereoisomers ⟨93MI 417-01⟩.

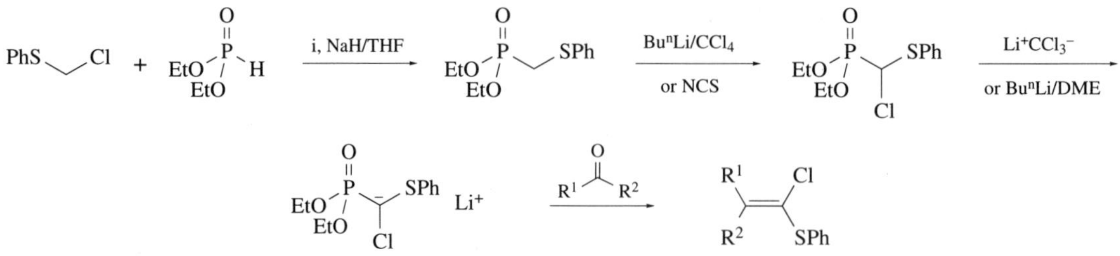

Scheme 10

Table 38 Wittig–Horner preparation of 1-chloro 1-phenylthio alkenes, $R^1R^2C=CClPh$ (Scheme 10).

R^1	R^2	Yield (%)
p-FC$_6$H$_4$	H	64
p-ClC$_6$H$_4$	H	65
p-MeOC$_6$H$_4$	H	68
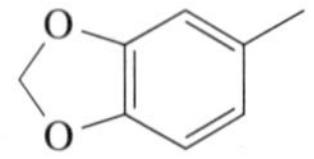 (methylenedioxy-methylphenyl)	H	53
PhCH=CH	H	46
—(CH$_2$)$_5$—		60
(CH$_2$)$_2$CHBu^t(CH$_2$)$_2$		51
Ph	H	64

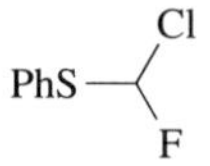

(**389**)  (**390**)  (**391**) 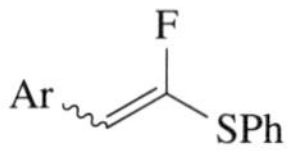 (**392**)

Ar = 4-C$_6$H$_4$C$_6$H$_4$
Ar = Ph
Ar = 4-MeC$_6$H$_4$, 80%
Ar = 4-MeOC$_6$H$_4$
Ar = 4-ClC$_6$H$_4$
Ar = 4-NCC$_6$H$_4$

(v) Halogenation of organometallics

The preparation of 1-alkylthio and 1-arylthio 1-metallo alkenes (**393**) is described in Section 4.20.4.2 and it might be expected that halogenation of these organometallics could be an efficient route to 1-halo 1-thio alkenes. In fact, apparently only one publication reports this. Thus, Takeda *et al.* prepared the organostannane (**394**) and showed that this was chlorinated by cupric chloride in high yield with complete retention of configuration (Table 39) ⟨91TL6563⟩. *N*-Bromosuccinimide or cupric bromide gave the corresponding bromide, again in good yield but with a loss of stereospecificity.

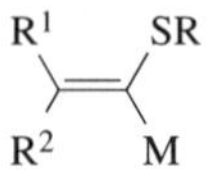

(**393**) 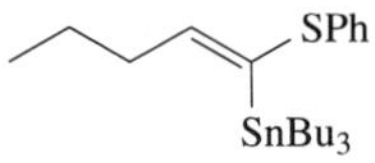 (**394**)

Table 39 Reactions of 1-phenylthiovinylstannanes with halogenating reagents.

$$\text{R}\diagup\!\!=\!\!\diagdown\begin{smallmatrix}\text{SPh}\\\text{SnBu}_3\end{smallmatrix} + [\text{X}^+]$$

R	Halogenating agent	Product	Yield (%)	
Pr (Z)	NBS	Pr–C(SPh)=CH–Br	84	(E):(Z) 14:86
Pr (Z)	CuBr$_2$	Pr–C(SPh)=CH–Br	82	(E):(Z) 41:59
Pr (Z)	CuCl$_2$	Pr–C(SPh)=CH–Cl	90	(Z) only
Pr (E)	CuCl$_2$	Pr–C(SPh)=CH–Cl	83	(E) only
Pr^i (Z)	CuCl$_2$	Pr^i–C(SPh)=CH–Cl	97	(Z) only

(vi) Miscellaneous

A US patent describes how methylthioacetone (**395**) was treated first at 85 °C with an excess of sulfuryl chloride and then with triethyl phosphite to give (**396**) ⟨60USP2954316⟩. Somewhat related is the reaction of the 2-arylthio-2,2-dichloroethanols (**397**) with zinc in the presence of carboxylic acid chlorides to give the enol esters (**398**) ⟨91ZOR1796⟩. In general, the products are mixtures of (E) and (Z) forms but the product having Ar = p-Tol and R = p-BrC$_6$H$_4$, is a single isomer, probably with the (E) configuration.

(**395**) MeS–CH$_2$–C(=O)–CH$_3$
(**396**) MeS–C(Cl)=C(CH$_3$)–O–P(=O)(OEt)$_2$
(**397**) ArS–CCl$_2$–CH$_2$OH
(**398**) ArS–C(Cl)=CH–OCOR

Ar = Ph, R = Me, 72%
Ar = Ph, R = Et, 77%
Ar = Ph, R = Bu^t, 55%
Ar = p-Tol, R = p-BrC$_6$H$_4$, 60%

The reactions of sulfides, for example (**399**) and (**400**), with dichlorocarbene under phase-transfer conditions to give initially ylides (**401**), which then undergo 2,3-sigmatropic shifts to yield (**402**) and dehydrochlorination to (**403**) and (**404**), have been reported by Russian workers ⟨88ZOR443, 92ZOR1780⟩. Likewise, dichlorocarbene inserts into the thioacetals (**405**) to give 24–25% of the thioenol ethers (**406**) ⟨90CA230883w⟩.

The trifluoromethyl dithioacetals (**407**) react with the electron transfer reagents lithium naphthalenide or lithium biphenylide to give the product (**408**) in yields of 30–50% whilst (**407**) with thiolates, RSM, gave (**409**) in excellent yields ⟨92MI 417-01⟩. In all cases, the products were mixtures of (E) and (Z) isomers.

5-Halo-4-isothiazolin-3-ones (e.g. (**410**)) react with nucleophiles (**411**) to give the ring-opened structures (**412**) ⟨91USP5023275⟩.

Diethyl tetrathiomalonate (**413**) reacts with N,N-dimethylchloromethaniminium chloride (**414**) to give 48% of ethyl (E,Z)-3-chloro-3-ethylthiodithioacrylate (**415**) ⟨88AP903⟩.

Dichlorocarbene reacts with allyl benzyl sulfide (**416**) to give a mixture of (**417**) and (**418**), whilst the cyclopropene (**419**) gives the diene (**420**), of unknown stereochemistry, in 43% yield when treated with benzenesulfenyl chloride ⟨90MI 417-01, 93JCS(P1)1945⟩.

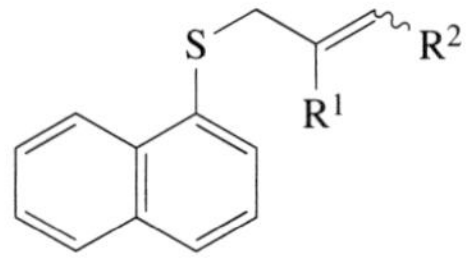

(399) 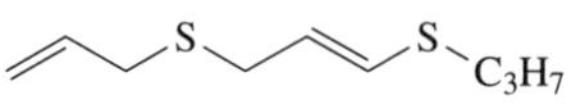 (400) 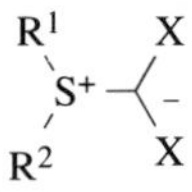 (401) 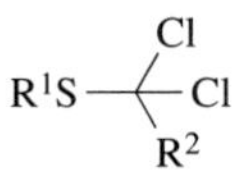 (402)

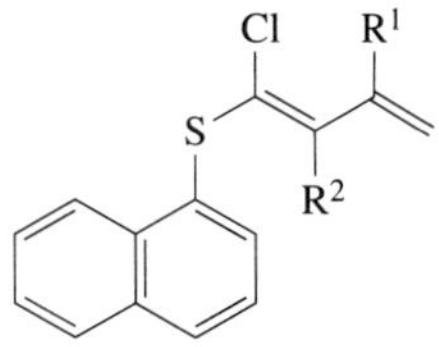

(403) 50% 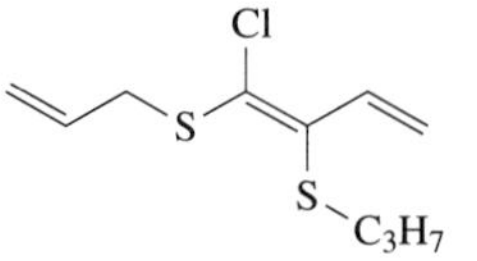 (404) 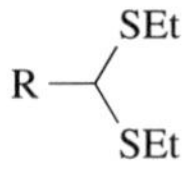 (405) 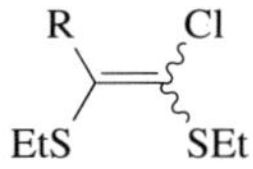 (406)

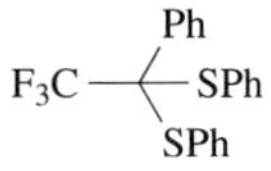

(407) 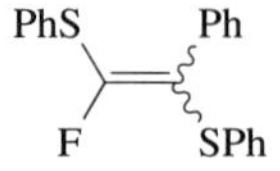 (408) 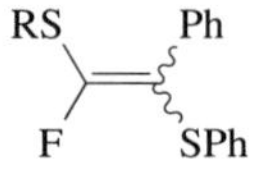 (409)
R = Ph, 98%
R = CH$_2$=CHCH$_2$, 92%
R = Pr^n, 93%

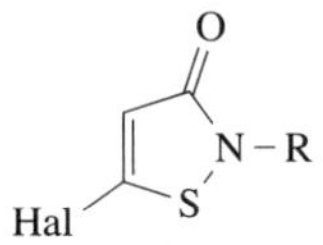

(410) 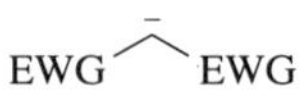 (411) 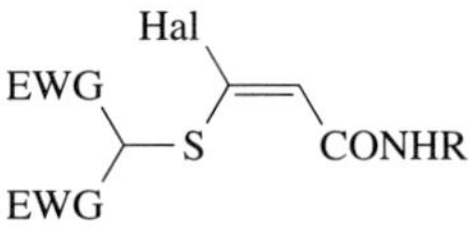 (412)

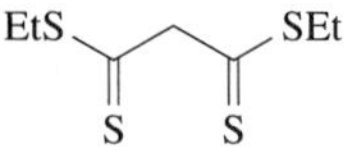

(413) 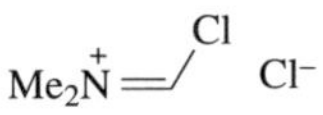 (414) 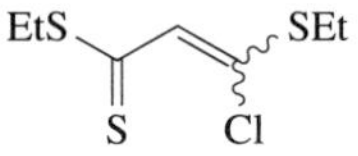 (415)

(416) 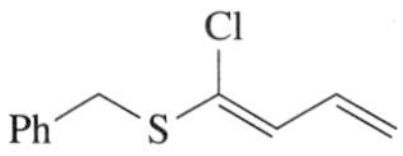 (417) 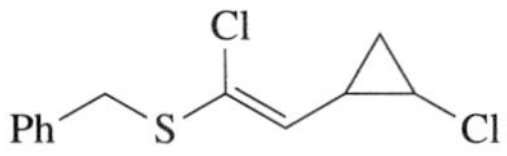 (418) 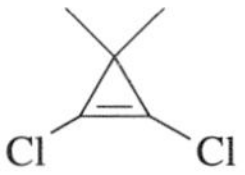 (419) 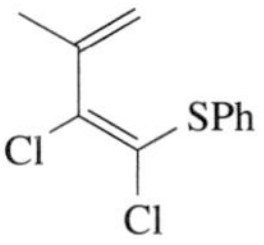 (420)

4.17.2.2.2 Tricoordinate sulfur derivatives

1-Halo 1-sulfinyl alkenes (e.g. (**421**)–(**424**)) are readily synthesised by the oxidation of 1-halo 1-alkylthio alkenes (see Section 4.17.2.2.1) with hydrogen peroxide in acetic acid or 3-chloroperbenzoic acid ⟨53RTC813, 81CB684, 85CA(102)78589t, 92PS(72)225⟩. Despite the ease of this oxidation, and the existence of the synthetic sequences shown below, this particular functional group has not been widely reported. One reason probably is that the sulfinyl group introduces the complicating feature of a chiral centre into the molecule and, if all that is required is an electron-withdrawing group, then this can be more conveniently achieved by further oxidation to a sulfone.

Of course, in some situations the chirality of the sulfinyl group has been exploited. Thus, (+)-(S)-1-bromovinyl p-tolyl sulfoxide (**425**) was synthesised in 81% overall yield by bromination–dehydrobromination of (+)-(R)-vinyl p-tolyl sulfoxide (Scheme 11) which was itself prepared by the Andersen procedure ⟨62TL93, 90PAC1987, 92TL5121⟩.

The same halogenation–dehydrohalogenation procedure has been used to produce the analogues shown in Table 40.

The other major synthetic route to these compounds is based on anion addition to aldehydes. Thus, Yamakawa generated the anion of chloromethyl phenyl sulfoxide (**426**) with LDA at $-60\,°C$

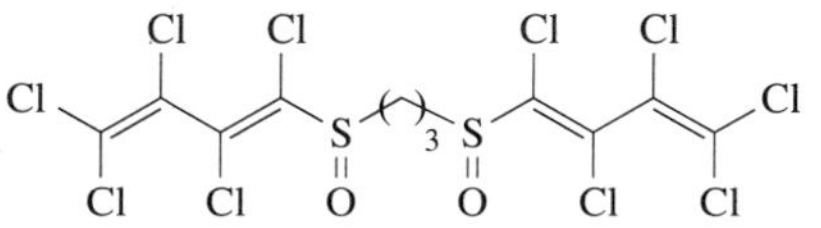

Scheme 11

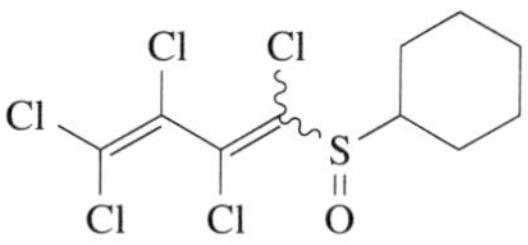

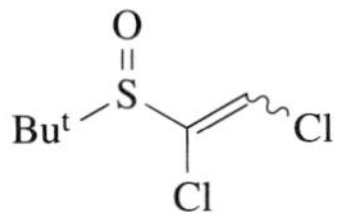

Table 40 Halogenation–dehydrohalgenation of sulfinyl alkenes.

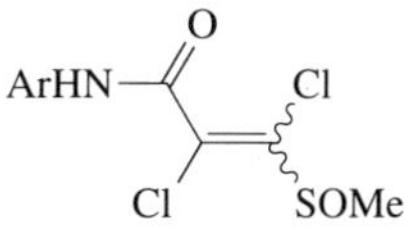

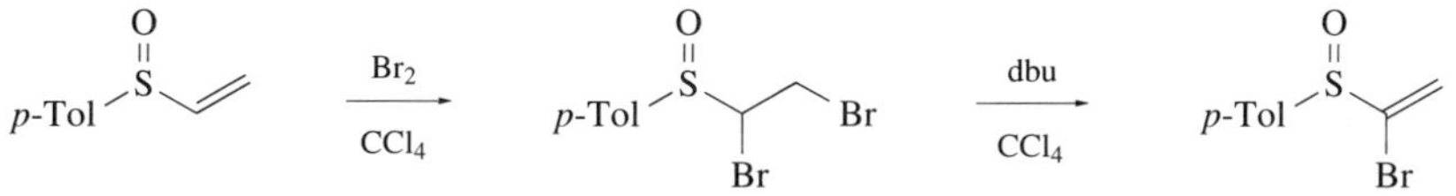

Sulfinyl alkene or its dihalide	Product	Base	Yield (%)	Ref.
CH$_2$=CHSOMe	2-Br-CH$_2$=C(SOMe)			75CC107
Carbapenem-SO-CH=CH-NHAc (RCOCMe$_2$, CO$_2$H)	Carbapenem-SO-C(X)=CH-NHAc (X = Cl)	Et$_3$N	10	83CA(98)16507e
	X = Br	Et$_3$N	20	82CA(97)127400p
	X = I	Et$_3$N	29	82CA(97)215888j
CF$_3$SOCHClCHCl$_2$	CF$_3$SOC(Cl)=CHCl	Et$_3$N	76	73ZOR69
(Cl$_2$CHCHCl)$_2$SO	(CHCl=CCl)$_2$SO	Et$_3$N		71USP3592896
(BrCH$_2$CHBr)$_2$SO	(CH$_2$=CBr)$_2$SO	PhNEt$_2$	45	90ZOR2056
(CH$_2$XCHX)$_2$SO	(CH$_2$=CX)$_2$SO (X = Br, Cl)	Et$_3$N		68RTC49

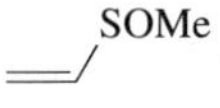

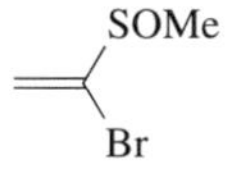

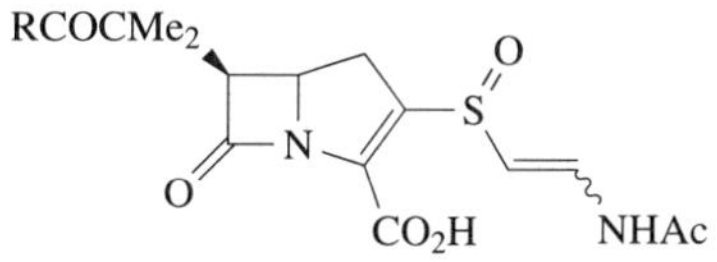

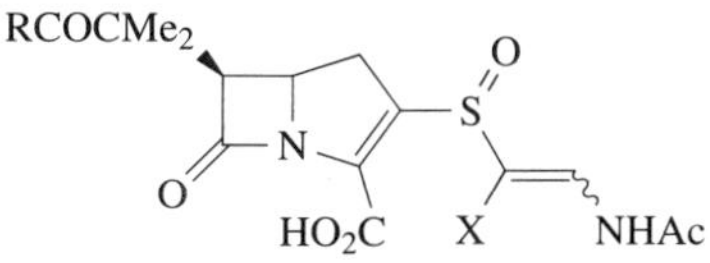

in THF and condensed it with a range of aldehydes to give the alcohols (**427**). Mesylation of (**427**) followed by dbu induced elimination gave, in good yields, the 1-chloro 1-phenylsulfinyl alkenes (**428**) (Scheme 12, Table 41) ⟨93BCJ1866⟩. In all cases, the final product was a mixture of (*E*)- and (*Z*)-isomers.

A rather more unexpected elimination was reported by Uno *et al*. ⟨92BCJ210⟩. The α-fluoro

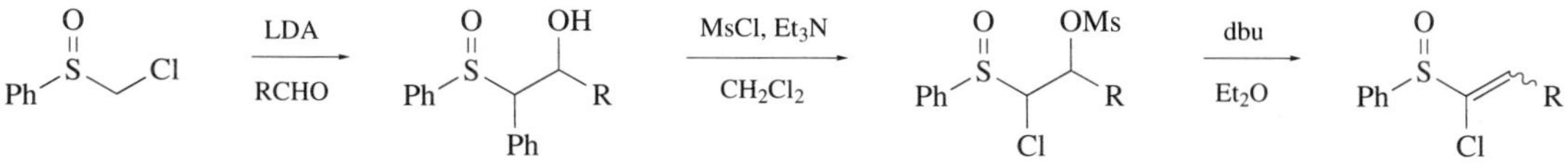

LDA = lithium diisopropylamide

Scheme 12

Table 41 Preparation of (**428**) from RCHO and PhS(O)CH$_2$Cl (**426**).

RCHO	Yield of the adol (**427**)[a]	Yield of (**428**) (%)
MeOC$_6$H$_4$(CH$_2$)$_2$CHO	99	91
MeOC$_6$H$_4$CH$_2$CH(Me)CHO	88	77
PhCHO	92	86
3-MeO-C$_6$H$_4$CHO	92	89
1-naphthyl-CHO	86	73
PhCH=CHCHO	85	71

[a] Mixture of diastereomers.

sulfoxide (**429**) reacts with lithium diphenylcuprate to give 1-fluoro-2-(4-biphenylyl)vinyl phenyl sulfoxide (**430**) in 55% yield as a 7:5, (E:Z) mixture of isomers.

(**429**)   (**430**)

Coutrot, however, has reported that the sulfoxide (**431**) generates an anion on treatment with Bu^{n}Li in THF but that this does not react with tetrachloromethane, thus ruling out a Wittig–Horner approach analogous to that described in Section 4.17.2.2.1 for 1-halo 1-phenylthio alkenes ⟨76S107⟩. Further, Reutrakul has described the synthesis of 1-fluoro-2-hydroxyalkyl phenyl sulfoxides (**432**) but did not attempt to dehydrate them. This would appear to be a possible approach to the synthesis of the 1-fluoro 1-phenylsulfinyl alkenes (**433**) ⟨83TL725⟩.

Finally, the β-lactam (**434**) reacts with xenon difluoride under phase transfer conditions to give (**435**) in 25% yield ⟨85CA(103)6148w⟩.

(**431**)   (**432**)   (**433**)   (**434**) R = H
                                      (**435**) R = F

4.17.2.2.3 Tetracoordinate sulfur derivatives

As was noted in Section 4.17.2.2.2, a favoured route for the preparation of 1-halovinyl sulfones has been the oxidation of the corresponding sulfides (Table 42), prepared as described in Section 4.17.2.2.1. The favoured oxidants have been 3-chloroperbenzoic acid and hydrogen peroxide in acetic acid. However, the routes leading to the construction of the alkene group appear to have been investigated more than those for the sulfoxides. The main procedures are elimination reactions, Wittig-type condensations, additions to substituted alkynes and substitution reactions.

Table 42 Oxidation of 1-halo 1-alkyl or 1-arylthio alkenes to sulfones.

Starting material	*Product*	*Oxidant*	*Yield (%)*	*Ref.*
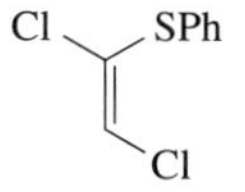 Cl SPh Cl	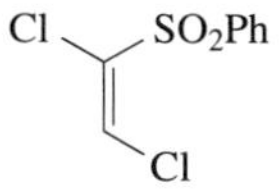 Cl SO2Ph Cl	H_2O_2	75 overall from trichloroethylene	90G569
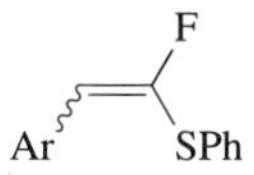 F Ar SPh	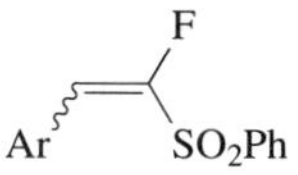 F Ar SO2Ph	mcpba	81	93MI 417-01
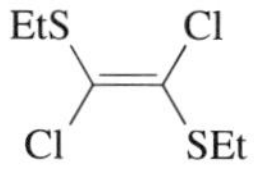 EtS Cl Cl SEt	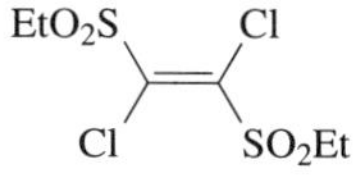 EtO2S Cl Cl SO2Et	H_2O_2, HOAc	78	90G235
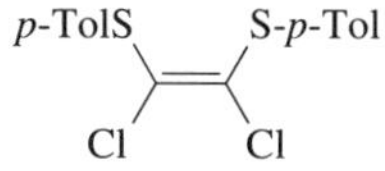 p-TolS S-p-Tol Cl Cl	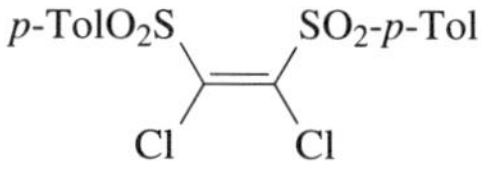 p-TolO2S SO2-p-Tol Cl Cl	H_2O_2, HOAc	77	65T2899

(i) Elimination reactions

Philips and Oka prepared 1-bromovinyl methyl sulfone (**436**) in 80% yield by the bromination–dehydrobromination of methylvinyl sulfone (**437**) ⟨72JA1012⟩. Similar reaction sequences (Schemes 13–15) have been used by Vessiere *et al.* to prepare (**438**), (**439**) and (**440**) ⟨77CJC3190⟩. Compound (**438**) was also prepared directly by bromination–dehydrobromination of phenylvinyl sulfone ⟨87JOC4943⟩. Yields throughout all the sequences were very good. Compound (**439**) was prepared as a mixture of stereoisomers ($Z:E$, 8 : 2) whilst (**440**) was produced solely as the (Z)-isomer. A detailed study of the stereochemistry of the bromination–dehydrobromination sequence was reported by Philips *et al.* ⟨74TL4157⟩. In particular, pure (E)- and (Z)-isomers of the sulfones (**441**) give mixtures of *erythro* and *threo* dibromides on treatment with bromine but stereospecific *anti*-bromination results from treatment with pyridinium hydrobromide perbromide. Triethylamine-induced elimination proceeds exclusively *anti* from the *threo* isomers, to give the (Z)-isomers, but the *erythro* isomers give mixtures. However, a catalytic amount of molecular bromine induces (E)→(Z)-isomerisation. The (E)-isomer is best obtained with about 90% specificity by dehydrobromination of the *erythro* form with tetrabutylammonium acetate.

Takeuchi *et al.* prepared the fluoro compound (**442**) from the corresponding C—H compound

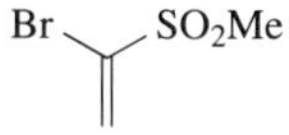
Br
SO2Me

(436)

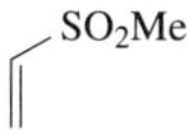
SO2Me

(437)

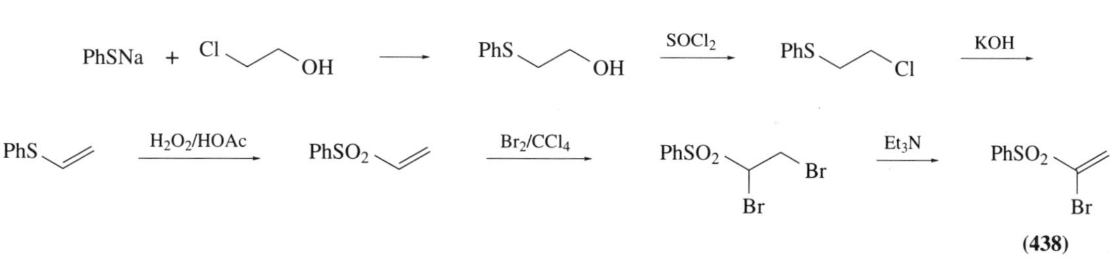
PhSNa + Cl OH
PhS OH
SOCl2
PhS Cl
KOH
PhS
H2O2/HOAc
PhSO2
Br2/CCl4
PhSO2 Br Br
Et3N
PhSO2 Br
(438)

Scheme 13

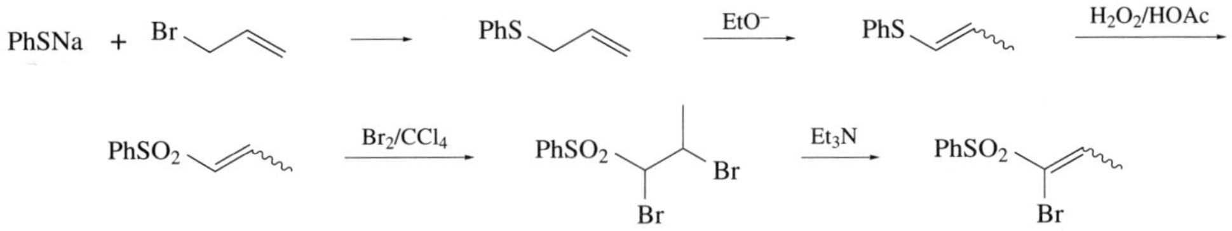

Scheme 14

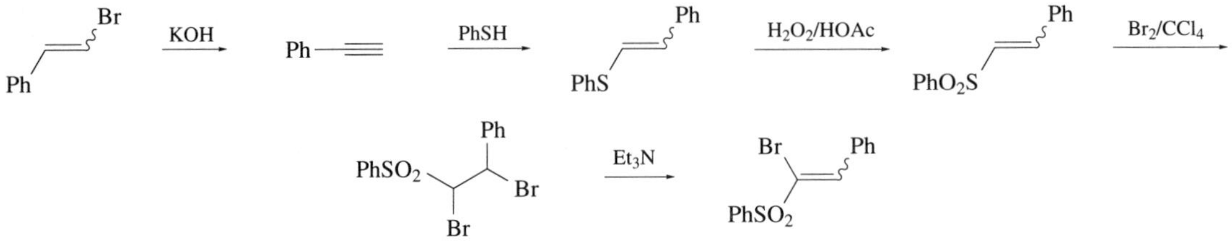

Scheme 15

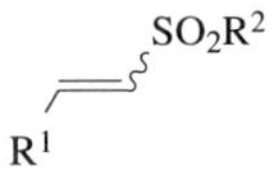

(**441**) R^1, R^2 = Me, Ph

(**443**) with freshly prepared perchloryl fluoride, FClO$_3$, in greater than 80% yield ⟨89JOC5453⟩. This compound readily lost nitrous acid on treatment with silica gel to give, in 65% yield, (**444**) solely the (*E*)-isomer.

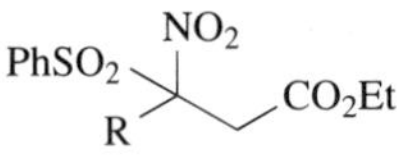

(**442**) R = F
(**443**) R = H

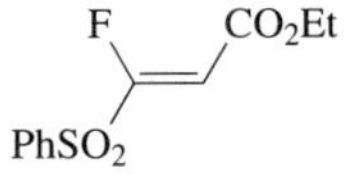

(**444**)

(ii) Wittig-type reactions

McCarthy *et al.* prepared fluoromethyl phenyl sulfone (**445**) by the route shown in Scheme 16 and showed that its lithium salt reacts with aromatic aldehydes to give alcohols of general structure (**446**) ⟨85CC678, 85JA735⟩. Treatment of (**446**) with methanesulfonyl chloride gives the (*E*)-1-fluoro 2-aryl sulfones (**447**) (Table 43). Whilst the reaction works well with the compounds shown in Table 43, it was found that acetophenone gives only the allyl sulfone (**448**). Another group has utilised the reaction of (**445**) with (**449**) to give (**450**) but with no yield quoted ⟨91JMC2525⟩.

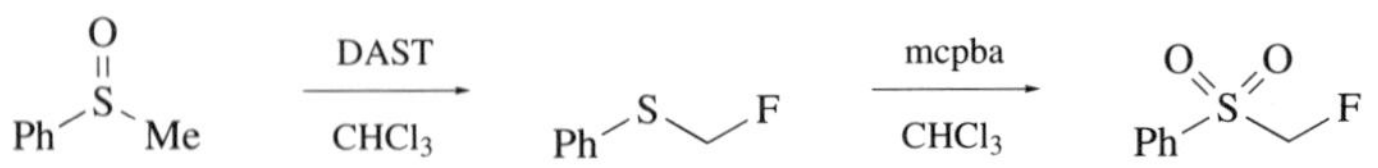

Scheme 16

To overcome the problems caused by isomerisations such as were found with acetophenone, McCarthy prepared the carbanion of diethyl 1-fluoro-1-(phenylsulfonyl)methylphosphonate (**451**) *in situ* by treatment of fluoromethyl phenyl sulfone, diethyl chlorophosphate and 2 equiv. of either lithium diisopropylamide or lithium hexamethyldisilazide at −78 °C ⟨90TL5449⟩. The carbonyl

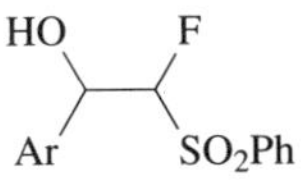

(446) 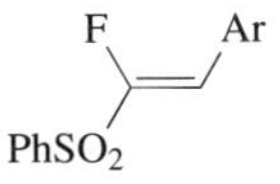 (447) 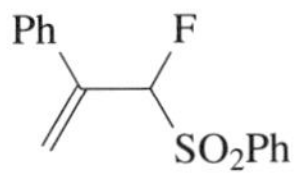 (448) 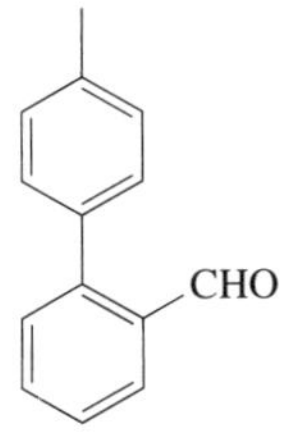 (449) 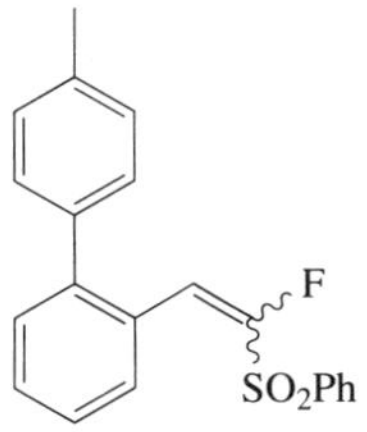 (450)

Table 43 Preparation of (**447**) from PhSO$_2$CHFLi and ArCHO.

Ar	Yield (%)
Ph	67
4-ClC$_6$H$_4$	80
3,4-(MeO)$_2$C$_6$H$_3$	71
C$_6$F$_5$	78

compound was added at this temperature and then the solution allowed to warm to room temperature to give good yields of the alkene (Table 44).

A variation on this approach has been reported by Koizuni *et al.* who generated (**451**) by fluorination of (**452**) by perchloryl fluoride ⟨87CPB3959⟩. Compound (**451**) was then deprotonated at −70 °C in THF by sodium hydride and the anion treated with carbonyl compounds over the temperature range of −78 °C to 0 °C (Table 45). The reaction has also been extended to the chloro analogue (**453**; R = Li) which is either generated *in situ* by the reaction of the dianion (**454**) with diethyl chlorophosphate ⟨90SC273⟩ or by deprotonation of (**453**; R = H) ⟨85S676⟩. As before, (**453**; R = Li,Na) has been condensed with a variety of aldehydes and ketones to give 1-chloroalkenyl phenyl sulfones (Table 46).

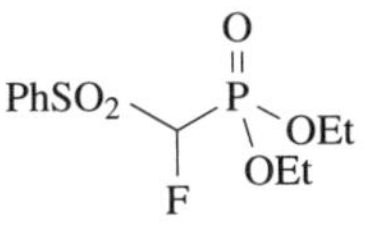

(451) 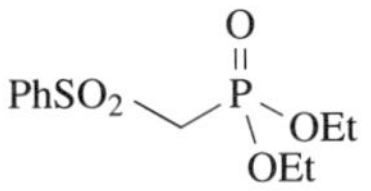 (452) 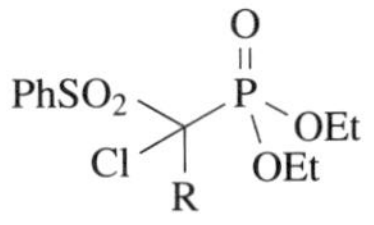 (453) 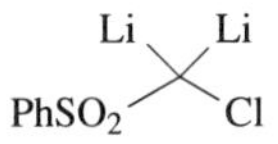 (454)

(iii) Addition to substituted alkynes

Curran and Kim have found that benzenesulfonylalkyne (**455**) reacts with alkyl halides including *n*-butyl iodide, isopropyl iodide and *t*-butyl iodide in benzene in the presence of hexabutylditin following initiation by sun lamp irradiation to give the 1-iodo 1-sulfonyl alkenes (**456**) ⟨91T6171⟩.

(455) (456)

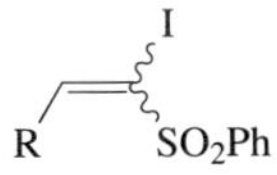

R = Bu^n, 18%; (*E*):(*Z*) 1:14
R = Pr^i, 61%; (*E*):(*Z*) 1:17
R = Bu^t, 83%; (*E*):(*Z*) 1:190

(iv) Substitution reactions

Sodium toluene-*p*-sulfinates react nucleophilically on sulfur with activated alkenyl halides, for example dichloroacrylonitrile, to give solely the disubstituted derivative, presumably because the second halogen atom is further activated by a sulfone group and is thus more rapidly displaced than

Table 44 Emmons–Homer reaction of (**451**) with carbonyl compounds.

R^1R^2CO	$(E):(Z)$	Yield (%)	Ref.
2,5-(MeO)₂C₆H₃COMe	1:1	87	90TL5449
Me₂CO		84	90TL5449
MeCOPh	1.3:1	95	90TL5449
MeCO*p*-ClC₆H₄	4.4:1	85	90TL5449
2,2-dimethyl-1,3-dioxa-spiro[5.5]cyclohexan-9-one		75	90TL5449
CH₃(CH₂)₁₀CHO	2.7:1	71	90TL5449
4-ethoxy uridine derivative	8:1		91JA7349
polyprenyl aldehyde		98	92JA360

Table 45 Emmons–Horner reaction of (**451**) with carbonyl compounds to give $R^1R^2C{=}CFSO_2Ph$.

Carbonyl compound	Yield (%)	$(E):(Z)$
HCHO	70	
PhCHO	85	(E)
p-MeOC₆H₄CHO	81	49
CH₃COCO₂Et	86	0.7
PhCOMe	74	24
MeCOMe	71	
Cyclohexanone	92	
Pr^{n}CHO	71	1
p-NO₂C₆H₄CHO	95	14

Table 46 Preparation of $R^1R^2C=CClSO_2Ph$ from R^1R^2CO and $PhSO_2C(Cl)MP(O)(OEt)_2$ (**453**; R = Li or Na).

R^1	R^2	Yield (%)	$(E):(Z)$	Ref.
Ph	H	82	93:7	90SC273
p-ClC$_6$H$_4$	H	85	94:6	90SC273
p-MeOC$_6$H$_4$	H	83	95:5	90SC273
2-Furyl	H	83	92:8	90SC273
CH$_3$(CH$_2$)$_5$	H	79	62:38	90SC273
CH$_3$	H	80	59:41	90SC273
CH$_3$	CH$_3$	76		90SC273
p-NO$_2$C$_6$H$_4$	H	80	(Z) only	85S676
m-NO$_2$C$_6$H$_4$	H	90	(Z) only	85S676
p-NMe$_2$C$_6$H$_4$	H	85	(Z) only	85S676
p-ClC$_6$H$_4$	H	84	(Z) only	85S676
Ph	H	77	(Z) only	85S676
PhCH=CH	H	90	(Z) only	85S676

the first. However, it has been found that the substituted acrylonitriles (**457**) react with sodium toluene-p-sulfinate in DMF at room temperature to give the monosulfones (**458**) ⟨84JOC1125⟩.

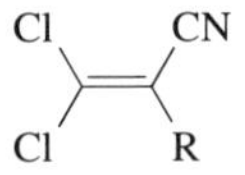

(**457**) R = Me, Ph, Bu^t

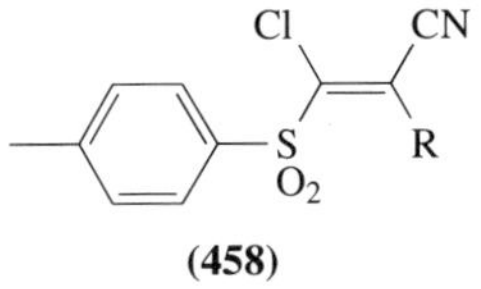

(**458**)
R = Me, 64%; $(Z):(E)$ >90:10
R = Ph, 67%; (Z)
R = Bu^t, 55%; $(Z):(E)$ 5:2

(v) Miscellaneous routes

Russian workers have discovered that 2,2-dihalovinyl sulfones, for example (**459**), isomerise to the 1,2-dihalo isomers (**460**) on treatment with potassium fluoride in the presence of 18-crown-6-ether ⟨83ZOR1344, 84ZOR972, 85ZOR965⟩. The benzenesulfonyl product was solely the (Z)-isomer (**461**) ⟨86ZOR1727⟩.

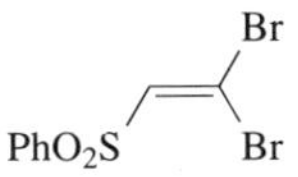

(**459**) 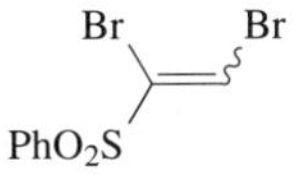 (**460**) 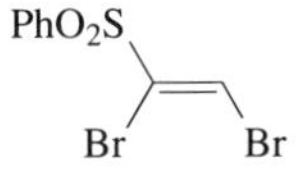 (**461**)

Hewkin and Jackson have shown that the oxiranes (**462**) rearrange in refluxing THF in the presence of magnesium bromide etherate to give a mixture of the α-bromoacylsilane (**463**) and the pure (Z)-α-bromovinyl sulfone (**464**) (Table 47) ⟨91JCS(P1)3103⟩.

Table 47 Ring opening of the oxiranes (**462**) to give α-bromoacylsilanes (**463**) and (Z)-α-bromovinyl sulfones.

R	Yield (**463**) (%)	Yield (**464**) (%)
Me	48	30
Et	40	40
Pr	27	48
Bu	41	50
Pr^i	7	64
Ph	0	48

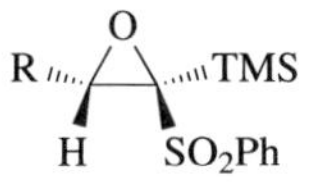

(462), 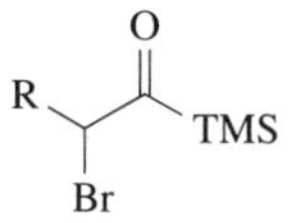 **(463)**, 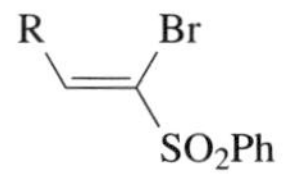 **(464)**

Finally, an American patent reports that 2-methanesulfonyl-1-(2-chlorophenyl)ethanone **(465)** is phosphorylated on the carbonyl oxygen by diethyl chlorothiophosphate to give a 75:25 mixture of (*E*) and (*Z*) **(466)** ⟨86USP4621078⟩.

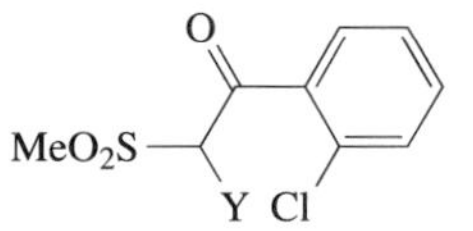

(465), 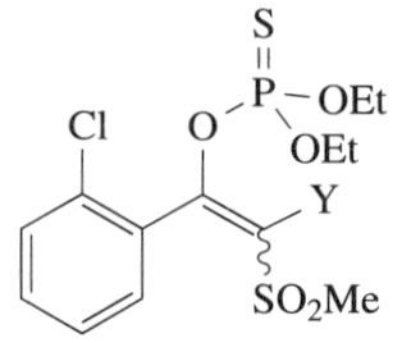 **(466)**

4.17.2.3 Halogen and Selenium or Tellurium Derivatives

Whilst there does not appear to be a particularly extensive literature on these compounds, synthetic routes have been described which are sufficiently general that further examples could be prepared. Five general synthetic methods have been described; Wittig alkenation eliminations, anion reactions, ring opening of selenophenes and additions to triple bonds. Of these, the Wittig alkenation appears to be the most general.

4.17.2.3.1 *Wittig reaction*

Coutrot and co-workers ⟨87S169⟩ have employed the sequence shown in Scheme 17. Diethyl phenylselenomethylphosphonate **(467)**, prepared as shown from diethyl iodomethylphosphonate, can be deprotonated by butyllithium and then chlorinated with tetrachloromethane. Under the conditions of the reaction the chloro derivative **(468)** is deprotonated by trichloromethyllithium to give a solution of the lithio species **(469)** which readily reacts with aldehydes and ketones thus producing the 1-chlorovinyl phenyl selenides **(470)** (Table 48). The reaction with aldehydes gives only a single, undefined isomer, whilst butanone gives a 1:1, (*E*):(*Z*) mixture. The transformation can also be performed by isolation of **(468)** and subsequent deprotonation by *n*-butyllithium at −78 °C but the yields were essentially identical to the 'one-pot' technique.

On the evidence of this paper, this method would seem to be capable of extension to a wider range of derivatives.

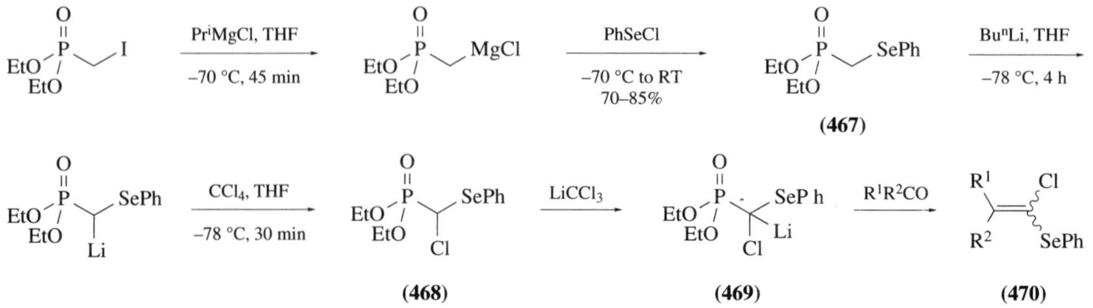

Scheme 17

Table 48 Preparations of (**470**) from (**469**) and R^1R^2CO (Scheme 17).

R^1R^2CO	Yield (%)
p-CH₃C₆H₄CHO	58
p-MeOC₆H₄CHO	60
piperonal (3,4-methylenedioxybenzaldehyde)	67
m-ClC₆H₄CHO	56
p-ClC₆H₄CHO	55
Pr^iCHO	57
n-C₆H₁₃CHO	51
EtCOMe	50
cyclopentanone	62

4.17.2.3.2 Addition–elimination reactions

The use of addition–elimination reactions to produce sulfur-substituted alkenes was discussed in Section 4.17.2.2.1. Analogous reactions have been reported for the selenium compounds but with rather less success. In general, treatment of polyhalogenated ethyl phenyl selenide with metals such as magnesium gives only complex mixtures of products ⟨87T4309⟩. However, 1,2-dichloro-1,2,2-trifluoroethyl phenyl selenide (**471**) reacts with magnesium in THF to give 41% of (*E*)-1,2-difluoro-2-chlorovinyl phenyl selenide (**472**) and a 'small amount' of 1,2,2-trifluorovinyl phenyl selenide (**473**).

(**471**)   (**472**)   (**473**)

4.17.2.3.3 Anion reactions

The fluoro alkenes (**474**) readily produce the alkenyllithiums (**475**) on treatment with alkyllithium at low temperature ⟨87T4309⟩. These lithio derivatives react with benzeneselenenyl bromide at −100 °C to give (**476**).

(**474**) R = F, Cl   (**475**)   (**476**) R = F, 17%; R = Cl, 9%

4.17.2.3.4 Ring opening of selenophenes

Frejd has reported that 2,3,5-tribromoselenophene (**477a**) reacts with ethyllithium at −110 °C to give (**478**) as the main, but impure, product ⟨76CS133⟩ whilst 2,5-dichloro-3-iodoselenophene (**477b**) gave the unstable (**479**) in 50–70% yield ⟨76ACS(B)439⟩.

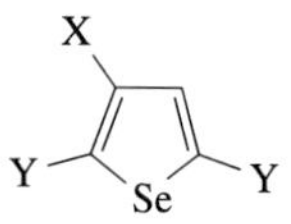

(477) a; X = Y = Br
     b; X = I, Y = Cl

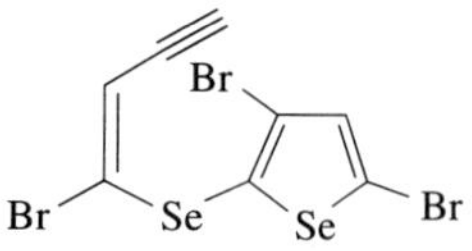

(478)

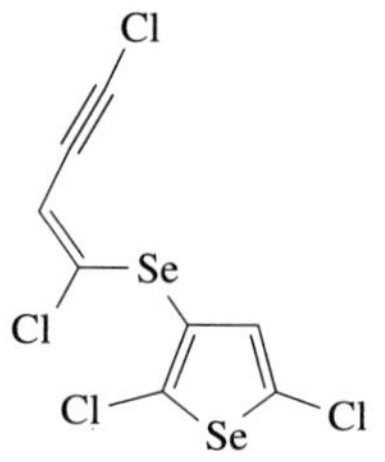

(479)

4.17.2.3.5 Addition to substituted alkynes

(i) Addition of halogens to selenyl-substituted alkynes

Bromine adds to 1-(phenylseleno)pent-1-yne (**480**) at 12 °C to −12 °C to give a 81% yield of the dibromo compound (**481**) ⟨89SA(A)1011⟩ and to 1-methylselenobut-3-en-1-yne at −60 °C to give 24% of (**482**) ⟨74ZOR1986⟩. The analogous tellurium compound gives solely 87% of (**483**). Likewise, hydrogen chloride adds across the triple band of the 1-alkylselenobut-3-en-1-ynes (**484**) to give the 1-chlorobutadienyl selenides (**485**) ⟨77ZOR254⟩. The same reaction with the tellurium analogue resulted in the loss of the telluride group.

The similar reaction with 1-methylselenohexa-1,3-diyne (**486**) gave a good yield of the product (**487**) arising from the addition of two moles of hydrogen chloride ⟨85ZOR244⟩.

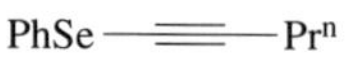

(480) 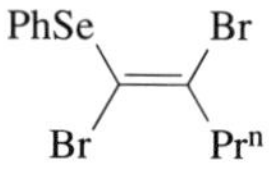   (481) 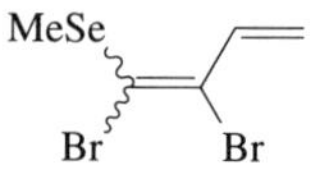   (482) 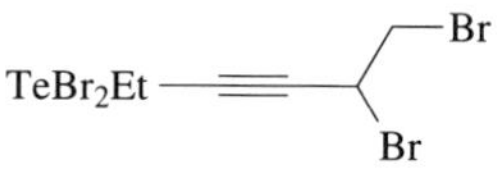   (483)

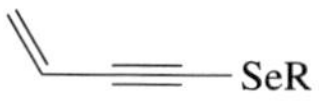

(484) 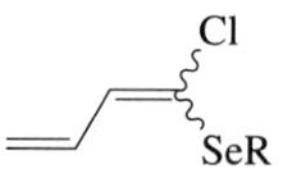   (485)   (486)   (487)
         R = Me, 93%
         R = Et, Pr^n, 85%
         R = Ph, 87%

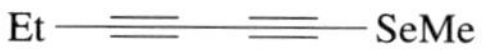

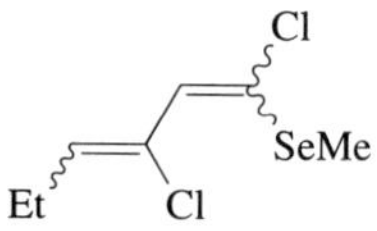

(ii) Addition of seleno compounds to halo-substituted alkynes

The Russian group of Martynov and co-workers has published a number of papers describing the addition of a variety of selenyl derivatives to haloalkynes. Thus treatment of *cis*-1,2-dichloroethene with benzeneselenol with phase-transfer catalysis lead mainly to *cis*-β-chlorovinyl phenyl selenide (**488**) with smaller amounts of the α-chlorovinyl selenide (**489**) and the *trans*-β-chlorovinyl selenide (**490**) ⟨87ZOR60⟩. The reaction is presumed to proceed via the *in situ* generation of chloroalkyne. Trichloroethylene and benzeneselenol under the same conditions gave 51% of (Z)-α,β-dichlorovinyl phenyl selenide (**491**) ⟨85ZOR2467⟩. More detailed experiments were reported subsequently and extended to tetrachloroethene when the product was a mixture of (**492**) and (E,Z)- (**491**) ⟨88ZOR509⟩. The reaction of dimethyl diselenide with tetrachloroethene under phase-transfer conditions in the presence of 50% sodium hydroxide solution gave 50% of (**493**) (assumed to be *trans*) but mixtures of products if the concentration of hydroxide was reduced ⟨89ZOR1773⟩.

These reactions were all assumed to progress via chloroalkyne intermediates. When chloro- or dichloroalkyne (stabilised as their diethyl etherates) react with benzeneselenol or diphenyl or dimethyl diselenide in ether, compounds of the general structure (**494**) are produced in good yields (Table 49) ⟨89ZOR1470⟩.

It was subsequently shown that the reaction mixture also contains small amounts of the mono-

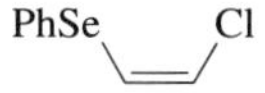

                    PhSe   Cl   PhSe   PhSe
                      __//   __   __
                                                   Cl   \\Cl
                     (488) 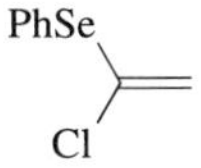   (489) 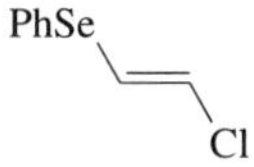   (490)

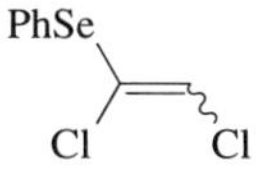

PhSe   PhSe   Cl   MeSe   Cl   RSe   Cl
   __   __//   __//   __//
  //   \\   // \\   // \\   // \\
 Cl   Cl   Cl   Cl   Cl   SeMe   X   Y
 (491) 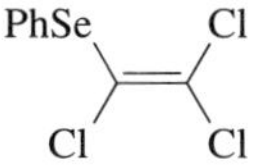   (492) 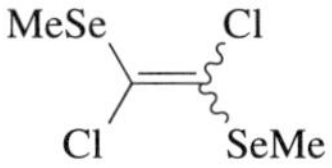   (493) 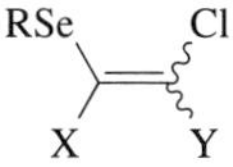   (494)

Table 49 Preparation of (494) from chloroalkynes.

R	X	Y	Yield (%)	(E):(Z)	Conditions
Ph	Cl	H	84	20:80	Et₂O, 0°, 2 h
Ph	H	SePh	60	predominantly trans	Et₂O, 20°, 6 h
Ph	Cl	SePh	40	predominantly trans	Et₂O, Δ, 14 h
Me	H	SeMe	87	95:5	20°, 17 h
Me	Cl	SeMe	67	77:23	Δ, 2 h

selenyl compounds (495) ⟨90ZOR978⟩. The authors invoked a series of free radical reactions involving hydrogen extraction from the ether solvent to explain this result.

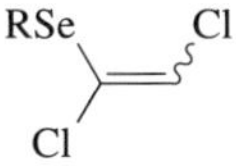

RSe   Cl
   __//
   // \\
  Cl

(495)

Selenols react exothermically with alkylthiochloroalkynes in diethyl ether to give good yields of 1-alkylthio-2-seleno-2-chloroethenes (496) in approximately 1:1, (E):(Z) ratios ⟨90IZV1693⟩. Diselenides give good yields of the (E)-isomer of (497) ⟨92ZOB2730⟩. Finally, benzeneselenenyl bromide and chloride react with dichloroalkyne to give the adducts (498) in the presence of Lewis acids ⟨91IZV2106⟩. The adducts (498) could be oxidised to the selenoxides by hydrogen peroxide.

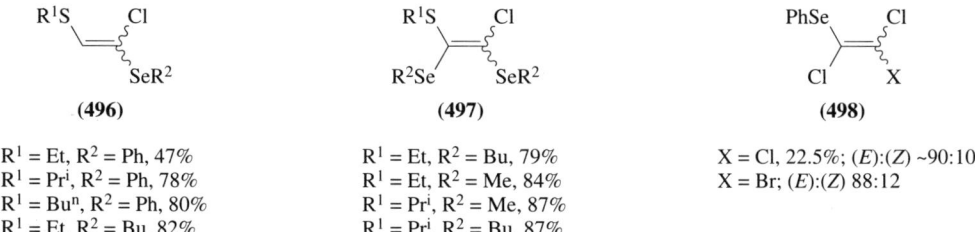

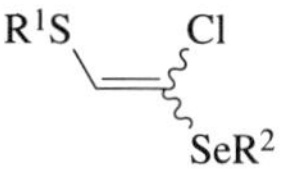

(496)

R¹ = Et, R² = Ph, 47%
R¹ = Prⁱ, R² = Ph, 78%
R¹ = Buⁿ, R² = Ph, 80%
R¹ = Et, R² = Bu, 82%

(497)

R¹ = Et, R² = Bu, 79%
R¹ = Et, R² = Me, 84%
R¹ = Prⁱ, R² = Me, 87%
R¹ = Prⁱ, R² = Bu, 87%

(498)

X = Cl, 22.5%; (E):(Z) ~90:10
X = Br; (E):(Z) 88:12

(iii) Addition of both the halogen and selenyl group to alkynes

Anker *et al.* treated *N*-phenylselenophthalimide with terminal alkynes in the presence of Et₃N·3HF to give mixtures of products including (499) ⟨90TL2127⟩.

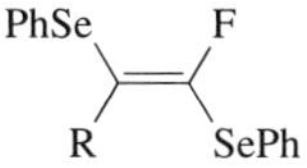

PhSe   F
    __//
    // \\
   R   SePh

(499)
R = Buⁿ, 28%
R = *n*-Hex, 48%

4.17.3 ACKNOWLEDGEMENT

The superb information collecting skills of Dr. A. Barnes and Mrs. F. Shepherd and the word processing talents of Mrs. S. Hughes are gratefully recognised.

4.18
Functions Incorporating a Halogen or Another Group other than a Halogen or a Chalcogen

DAVID I. SMITH
Sanofi Research Division, Alnwick, UK

4.18.1 HALOGEN AND NITROGEN DERIVATIVES	790
4.18.1.1 gem-Amino Halo Alkenes	790
4.18.1.1.1 By addition of halide	790
4.18.1.1.2 By amination	792
4.18.1.1.3 By electrophilic attack	793
4.18.1.1.4 Other methods of alkene formation	794
4.18.1.2 gem-Halo Nitro Alkenes	795
4.18.1.2.1 By halogenation	795
4.18.1.2.2 By nitration	797
4.18.1.2.3 By condensation between aromatic aldehydes and halonitromethanes	798
4.18.1.3 gem-Halo Nitroso Alkenes	798
4.18.1.4 gem-Azido Halo Alkenes	799
4.18.1.5 Diazonium and Diazo Derivatives	799
4.18.1.6 Iminophosphorane, Sulfimide and Metallonitrene Complexes	800
4.18.2 DERIVATIVES OF PHOSPHORUS AND OTHER GROUP 15 ELEMENTS	801
4.18.2.1 α-Haloalkenylphosphorus Derivatives	801
4.18.2.1.1 From carbonyl compounds	801
4.18.2.1.2 From fluorinated alkenes	802
4.18.2.1.3 By halogenation	804
4.18.2.2 α-Haloalkenyl Derivatives of Arsenic, Antimony and Bismuth	805
4.18.3 DERIVATIVES OF SILICON AND OTHER GROUP 14 ELEMENTS	805
4.18.3.1 α-Haloalkenylsilicon Derivatives	805
4.18.3.1.1 From carbonyl compounds	805
4.18.3.1.2 By silylation of α-haloalkenyllithium and -magnesium species	806
4.18.3.1.3 By halogenation	807
4.18.3.1.4 By halogenation of α-silylalkenyl metal derivatives	808
4.18.3.1.5 By addition of alkyl halides to alkynylsilanes	810
4.18.3.2 α-Haloalkenylgermanium Derivatives	811
4.18.3.3 α-Haloalkenyl Derivatives of Tin and Lead	812
4.18.4 DERIVATIVES OF BORON AND OTHER GROUP 13 ELEMENTS	813
4.18.4.1 α-Haloalkenylboron Derivatives	813
4.18.4.1.1 By ligand exchange and by halogenation reactions	813
4.18.4.1.2 By hydroboration of halo alkynes	814
4.18.4.2 α-Haloalkenyl Derivatives of Aluminum, Gallium, Indium and Thallium	815
4.18.5 DERIVATIVES OF LITHIUM AND OTHER GROUP 1 AND GROUP 2 METALS	816

4.18.5.1 α-Haloalkenyllithium Derivatives	816
4.18.5.1.1 α-Fluoroalkenyllithium compounds	817
4.18.5.1.2 α-Chloroalkenyllithium compounds	817
4.18.5.1.3 α-Bromoalkenyllithium compounds	818
4.18.5.1.4 α-Iodoalkenyllithium compounds	819
4.18.5.2 α-Haloalkenyl Derivatives of Sodium and Potassium	819
4.18.5.3 α-Haloalkenylmagnesium Derivatives	820
4.18.6 DERIVATIVES OF THE TRANSITION METALS	820
4.18.6.1 By Transmetallation Reactions	820
4.18.6.2 From Alkenyl Halides	821
4.18.6.3 From Alkynes	822

4.18.1 HALOGEN AND NITROGEN DERIVATIVES

4.18.1.1 *gem*-Amino Halo Alkenes

The synthesis of α-halo enamines has been reviewed by Ghosez and Marchand-Brynaert ⟨B-76MI 418-01⟩, and later updated ⟨88OSC(6)282⟩. These reviews concentrated on *N*-alkyl enamines and did not cover those compounds in which the amine electron pair is involved in further functionality, such as the halo enamides or halo heterocycles. This review aims to summarise and update the earlier works and also to include compounds bearing further functionalisation on nitrogen, though lack of space precludes any mention of compounds in which the halogen atom is directly attached to a heterocyclic ring.

4.18.1.1.1 By addition of halide

The reactions described in this section are those in which the halogen atom is added, in the form of halide anion, to an activated precursor in which the nitrogen atom is already present. The most common method of synthesis starts with a tertiary amide.

The first report of the synthesis of an *N*,*N*-dialkyl α-halo enamine was by von Braun and Heymans who treated a tertiary amide with phosphorus pentachloride ⟨29CB409⟩. This general method of synthesis has since been further developed, using phosgene as a chlorinating agent and triethylamine as base (Equation (1)) ⟨88OSC(6)282⟩. A wide variety of α-halo enamines have been prepared by this method.

$$\underset{\text{O}}{\overset{\text{NMe}_2}{\diagup\hspace{-0.5em}\diagdown}} \quad \xrightarrow[\substack{\text{ii, Et}_3\text{N, CH}_2\text{Cl}_2 \\ \text{distil} \\ 69-77\%}]{\text{i, COCl}_2, \text{CH}_2\text{Cl}_2} \quad \underset{\text{Cl}}{\overset{\text{NMe}_2}{\diagup\hspace{-0.5em}\diagdown}} \tag{1}$$

The products are reactive, readily hydrolysed materials that self-condense on standing, especially where the nitrogen atom is more basic or the β-position is not fully substituted (Scheme 1). This self-condensation is accelerated by relatively polar solvents, such as dichloromethane and chloroform, and the less stable compounds are therefore best prepared in dilute solution in strictly nonpolar solvents such as carbon tetrachloride or alkanes ⟨80TL223⟩. A consequence of their mode of reactivity is that α-halo enamines bearing a basic nitrogen atom are usually found in a mixture of (*E*) and (*Z*) forms, equilibrating via the keteniminium halide. This property is also used in the preparation of the fluoro, bromo and iodo derivatives from the readily available chlorides (Scheme 2) ⟨77NJC369, 79CC1180⟩. These halides behave in a fashion similar to the chloro compounds, differing only in their degree of reactivity and stability. When a fluoro derivative was prepared by an alternative method, addition of potassium hydrogen difluoride to *N*,*N*-diethyl-1-propynamine, the crude product was found in a 9:1 (*E*):(*Z*) ratio, consistent with a *cis* addition proceeding by fluoride attack on the less-hindered face of the intermediate keteniminium ion. After the mixture was kept for one week in chloroform, the ratio was found to be 10:1 in favour of the (*Z*) isomer, indicative of a slow isomerisation under conditions in which the chloro compound would be expected to undergo rapid self-condensation (Scheme 3). By contrast, the *N*,*N*-diisopropylamino iodo derivative was found to exist exclusively as the keteniminium salt (**1**), though less sterically compressed compounds, such as the corresponding *N,N*-dimethylamino iodo derivative (**2**) were found in the normal, covalently bound, α-halo enamine form ⟨77NJC369⟩.

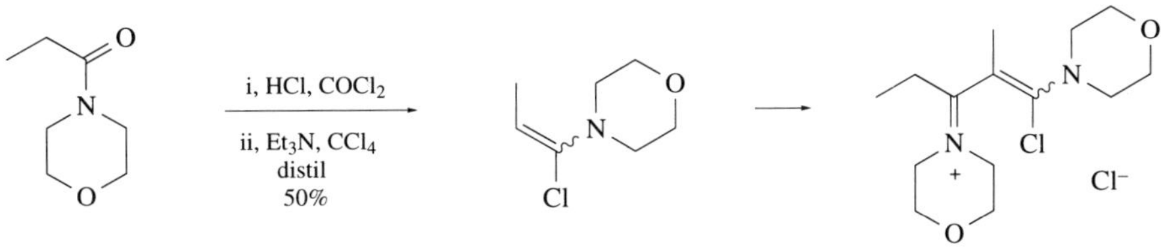

Scheme 1

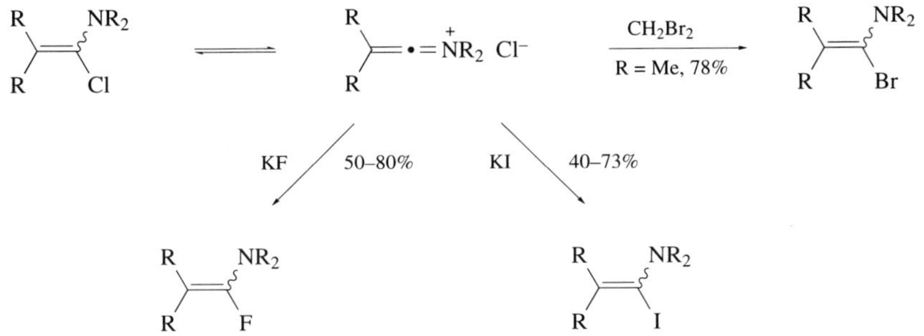

Scheme 2

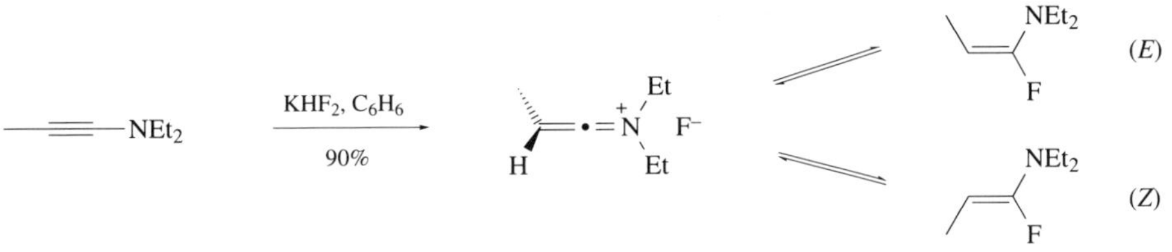

Scheme 3

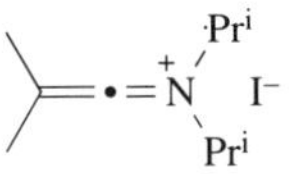

(**1**) 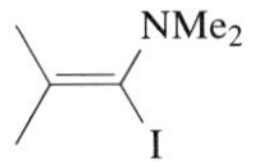   (**2**)

A second route to α-halo enamines, briefly referred to above, is by addition of halide ion to a keteniminium salt derived by reaction of an ynamine with an electrophile. Ghosez *et al.* added alkane- and arenesulfenyl chlorides to *N,N*-diethyl ynamines, giving α-chloro enamines in high yields (80–100%, Equation (2)), although this method showed no advantage over the chlorination of the corresponding tertiary amides ⟨81H(15)1179⟩.

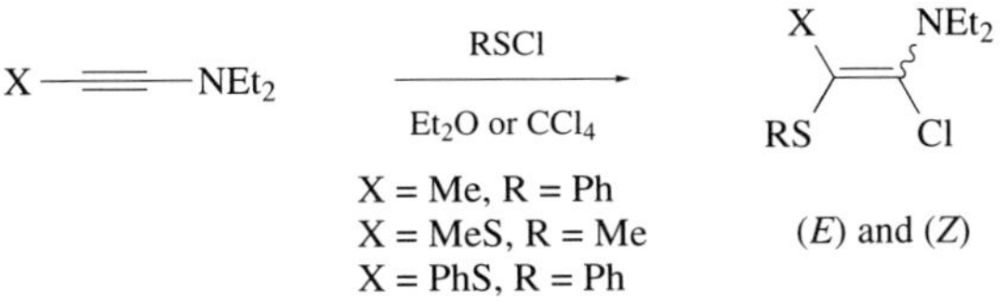

X = Me, R = Ph
X = MeS, R = Me   (*E*) and (*Z*)
X = PhS, R = Ph

(2)

When it is desirable for the β-position to carry further functionality, for example, a halogen atom, the appropriate α-functionalised amide may be employed ⟨B-76MI 418-01⟩. Alternatively, trichlorovinyl amines may be prepared by the action of tributylphosphine on trichloroacetamides (Equation (3)) ⟨60JA903, 73OSC(5)387⟩. In some cases, the action of chlorinating agents on tertiary acetamides may result in further functionalisation due to electrophilic attack on the resultant enamine. Thus, Okamoto and Kundu prepared *N*-(1,2-dichlorovinyl)carbazole (**3**) (Equation (4)) by heating *N*-acetylcarbazole with phosphorus pentachloride in benzene. The product was obtained in moderate yield and was found to be a single isomer by proton NMR spectroscopy, the increased configurational stability being due to the greatly reduced basicity of the carbazole nitrogen atom ⟨70JOC4250⟩.

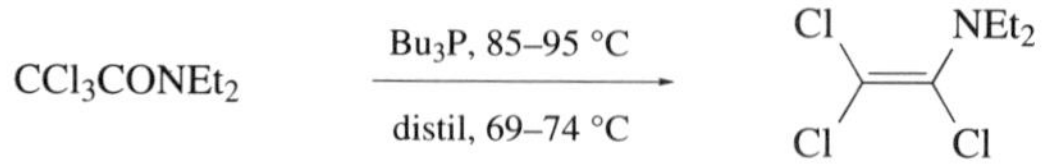

$$CCl_3CONEt_2 \xrightarrow[\text{distil, 69–74 °C}]{\text{Bu}_3\text{P, 85–95 °C}} \begin{array}{c} Cl \quad NEt_2 \\ \diagup\!\!=\!\!\diagdown \\ Cl \quad\;\; Cl \end{array} \qquad (3)$$

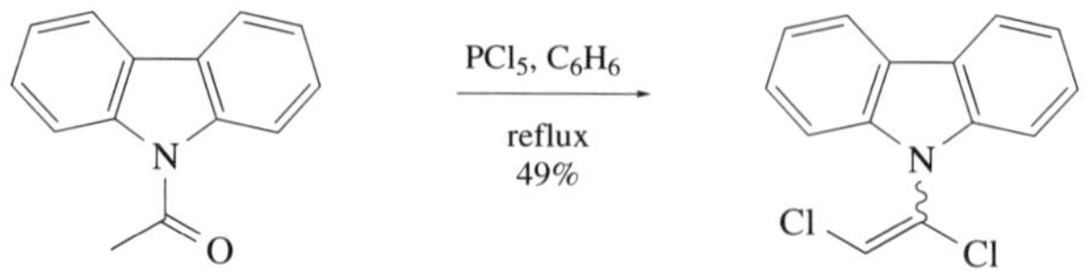

$$\text{(carbazole-N-acetyl)} \xrightarrow[\text{reflux} \atop 49\%]{\text{PCl}_5, \text{C}_6\text{H}_6} \text{(carbazole-N-CCl=CHCl)} \quad (E)\text{ or }(Z) \qquad (4)$$
(3)

When the β-position carries strongly electron-withdrawing substituents, any tautomeric equilibrium favours the enamine over the corresponding imine. In this way, primary or secondary α-halo enamines become sufficiently stable to isolate.

Primary α-halo enamines are usually prepared by addition of hydrogen halides across acetonitrile derivatives bearing further electron-withdrawing groups (Scheme 4). For example, addition of hydrogen chloride, bromide or iodide to the potassium salt of tricyanomethane proceeds in very high yields (93–98%, ⟨63CB3230⟩), as does the corresponding addition of hydrogen chloride to ethyl dicyanoacetate ⟨63CB1035⟩ and to 2,2-dicyano-*N*-phenylacetamide ⟨70TL1937⟩. Secondary α-chloro enamines have been prepared by interception of a ketenimine intermediate obtained by thermolysis of 2,2-dicyanovinyl azides (Scheme 5). This unusual reaction path is observed when the double bond has insufficient electron density to allow aziridine formation, enabling a Curtius route to be followed instead ⟨67AG(E)959⟩.

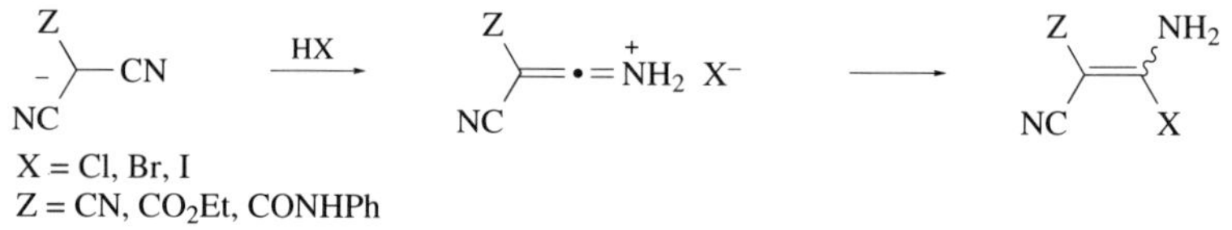

$$\begin{array}{c} Z \\ \diagdown \\ \text{–}\!\!-\!\!CN \\ \diagup \\ NC \end{array} \xrightarrow{HX} \begin{array}{c} Z \\ \diagdown \\ \!\!=\!\!\overset{+}{NH_2} \; X^- \\ \diagup \\ NC \end{array} \longrightarrow \begin{array}{c} Z \quad NH_2 \\ \diagup\!\!=\!\!\diagdown \\ NC \quad\;\; X \end{array}$$

X = Cl, Br, I
Z = CN, CO₂Et, CONHPh

Scheme 4

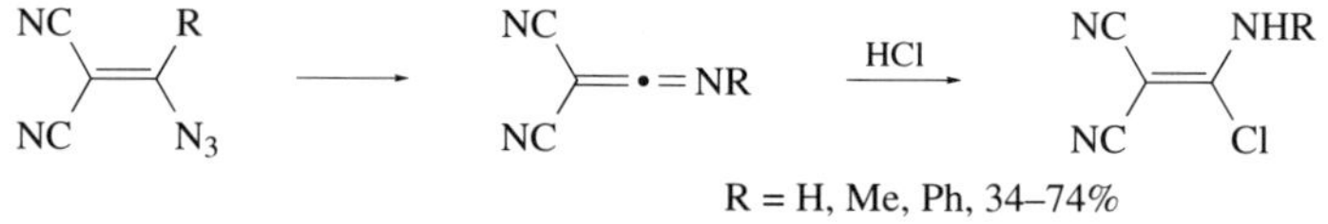

$$\begin{array}{c} NC \quad R \\ \diagup\!\!=\!\!\diagdown \\ NC \quad N_3 \end{array} \longrightarrow \begin{array}{c} NC \\ \diagdown \\ \!\!=\!\!NR \\ \diagup \\ NC \end{array} \xrightarrow{HCl} \begin{array}{c} NC \quad NHR \\ \diagup\!\!=\!\!\diagdown \\ NC \quad\;\; Cl \end{array}$$

R = H, Me, Ph, 34–74%

Scheme 5

4.18.1.1.2 By amination

A complementary approach to the products described above is by addition of nitrogen nucleophiles to unsaturated halides.

Dimethyl- and diethylamine add to dichloroalkyne, which was prepared *in situ* immediately prior to addition of the amine (Scheme 6). The products were obtained in high yield as air and moisture sensitive oils and were found to have well-defined proton NMR spectra, consistent with the presence of only one geometric isomer ⟨87S76⟩, although it may be that such α-halo enamines exist as a mixture of isomers, rapidly interconverting on the NMR timescale. Heterocyclic nitrogen nucleophiles have also been added to dichloroalkyne, in this case giving stable products (owing to the reduced basicity of the nitrogen atom) as single isomers, usually formulated as the (*E*) isomer by the assumption of *trans*-addition across the multiple bond. Pielichowski and Bogdal employed phase transfer conditions to generate dichloroalkyne in the presence of carbazole, obtaining the adduct (3) in 80% yield specifically as the (*E*) isomer (benzyltriethylammonium chloride, cyclohexane, 70 °C; note that this gives a better yield than the chlorination method described in the previous section) ⟨88LA595⟩. The phase-transfer catalyst was subsequently replaced by dimethyl sulfoxide, enabling the addition of relatively base-sensitive imidazoles at lower temperatures (ether, 0 °C) but in only 60% yield ⟨89BAP123⟩. Monohaloalkynes may also be employed, but only where the remaining substituent stabilises the developing negative charge more efficiently than the halogen atom. Tikhomirov *et al.* found that aziridine added to methyl bromopropiolate in methanol, affording methyl (*E*)-β-aziridino-β-bromoacrylate in 78% yield (Equation (5)). This compound was described as being thermodynamically stable ⟨84CHE1231⟩.

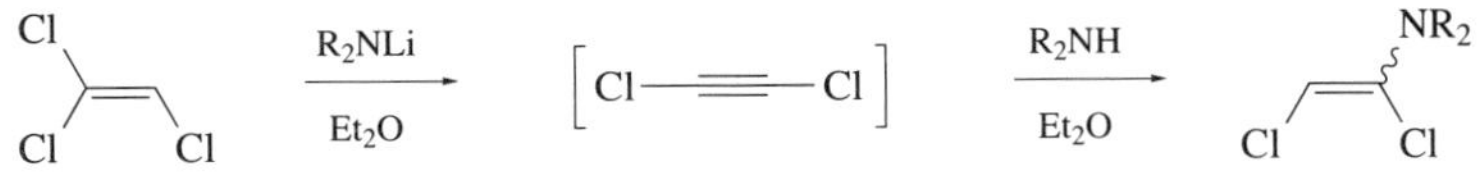

Scheme 6

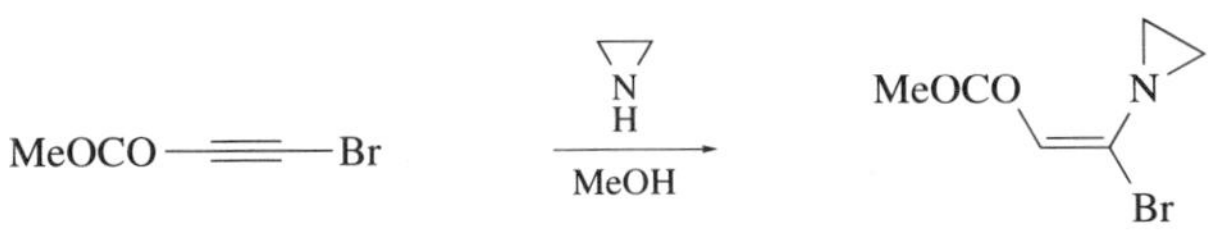

Monosubstitution of *gem*-dihalo alkenes has also been reported. Yakubovich and co-workers allowed lithium dialkylamides to react with an excess of polyfluoro alkenes, giving the α-fluoro enamines (**4**) in only moderate yield (20–58%). Lithium acetanilide gave the corresponding enamides (**5**) ⟨70JOU750⟩. Yagupol'skii and co-workers comment that free secondary amines tend to *add* to fluorinated alkenes but induce substitution with the corresponding chlorinated species. They isolated polyhalogenated α-chloro enamines (**6**) in reasonable yield (45–66%) ⟨77JOU1918⟩.

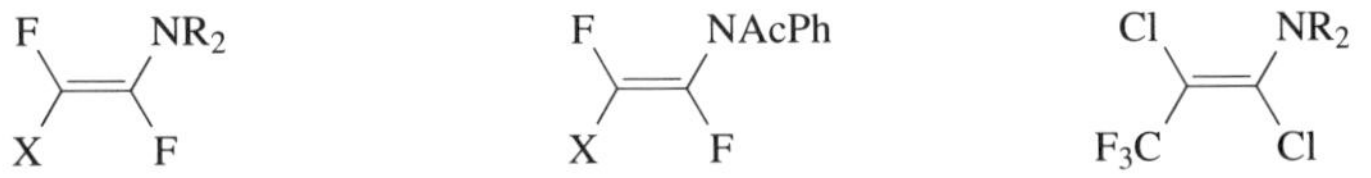

R_2N = dialkylamine, cycloalkylamine
X = F, Cl, CF_3

Schroth *et al.* isolated *gem*-chloro pyrrolidino alkenes after low temperature addition of pyrrolidine to a β,β-dichloroalkenyl aryl ketone (Equation (6)). These products are configurationally unstable, the proton NMR spectrum having a broad signal at 35 °C (δ 5.80), becoming a sharp single peak at −60 °C (δ 5.66). This instability was attributed to a lowering of the barrier to rotation about the double bond by donor–acceptor interactions, rather than to any tendency to form a ketenimine ⟨82S199⟩.

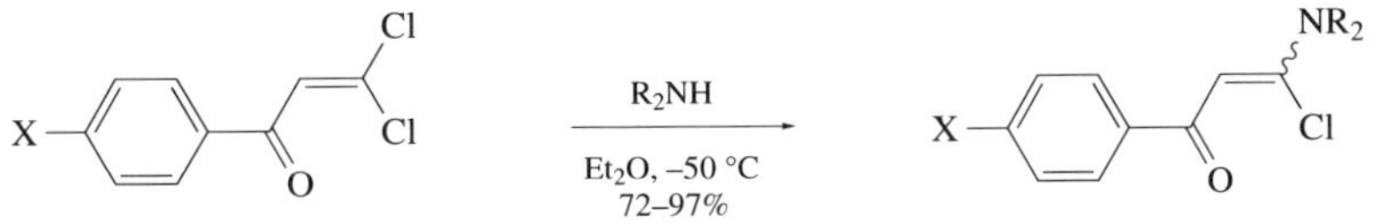

R_2N = pyrrolidino
X = H, Me, Br, NO_2

4.18.1.1.3 By electrophilic attack

This section covers those reactions in which the multiple bond participates as a nucleophile, the heteroatoms being introduced by formal electrophilic attack.

Electrophilic halogenation is not well represented here as enamines direct such substitution into their 2-position. α-Halo enamines are only formed under unusual conditions, such as during halogenation of *vic*-diamino alkenes ⟨67CB1087⟩. In this case, bromine and chlorine were allowed to react with a series of 1-diethylamino-2-phthalimidostyrenes. Normal 2-substitution furnished the *gem*-halo phthalimido alkenes in good yield (68–95%, Equation (7)). An unusual example of 1-bromination has been observed with perfluoroamines. Perfluoro-*N*-bromodimethylamine was mixed with 1,1,3-tris(hexafluorodimethylamino)propadiene, affording the corresponding 3-bromo allene in high yield (Equation (8)) ⟨73JCS(P1)1066⟩.

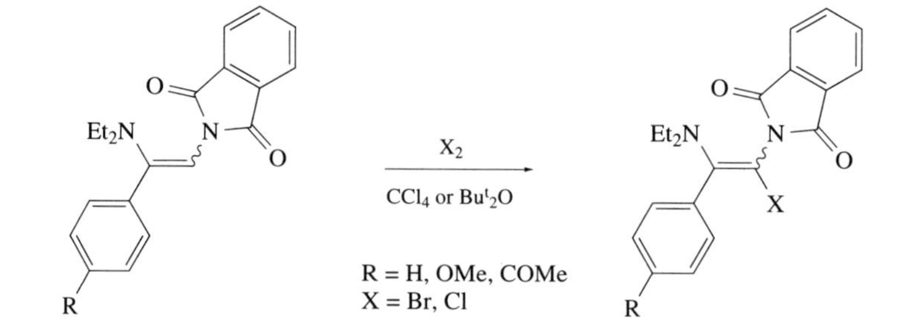

Electrophilic amination has been reported by Bal'on and Moskaleva. These workers added alkyl *N*,*N*-dichlorocarbamates to phenylalkyne. Intermediate *N*-chloro adducts were isolated, which isomerised to, the *gem*-chloro carbamates on trituration (Scheme 7) ⟨78JOU135⟩.

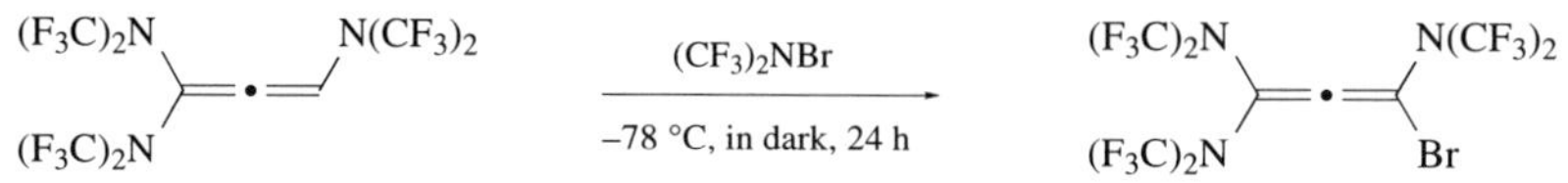

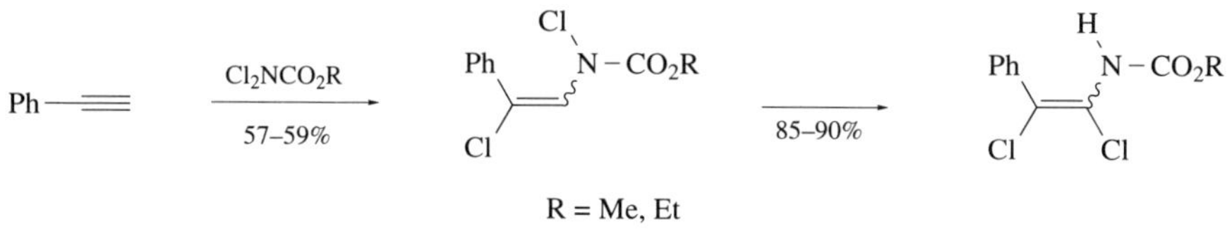

Scheme 7

4.18.1.1.4 Other methods of alkene formation

This section includes those preparative methods in which the alkene is formed with the heteroatom functionality already present in one reactant. These methods include condensation and elimination sequences.

Phosgeniminium salts are very reactive electrophiles that readily condense with mildly activated methyl and methylene groups. A series of reactions affording highly functionalised products has been reported by Gorissen and Viehe. For example, fluoroacetyl chloride was allowed to react with dimethylphosgeniminium chloride, producing α-chloro enamine isomers in good yield (68%) (Equation (9)) ⟨78BSB391⟩.

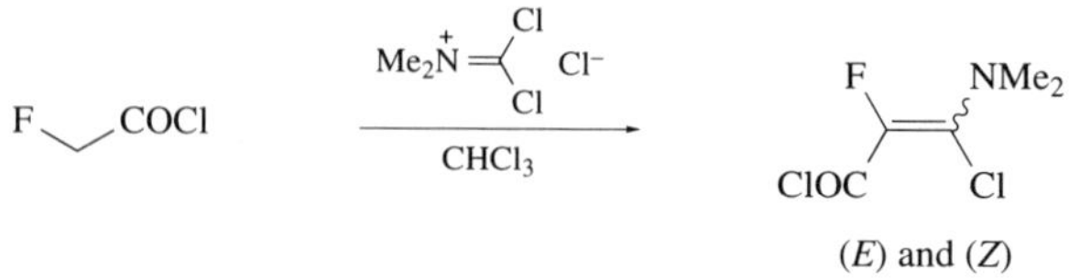

Drach and co-workers have prepared *N*-trichlorovinylbenzimidoyl chloride and -benzylideneamine by 1,2- and 1,4-elimination of hydrogen chloride (Equations (10) and (11)) ⟨75JOU119, 76JOU2252, 80JOU1762⟩. More recently, a wide range of perfluorinated *N*-trifluorovinyl secondary amines have been prepared by elimination of carbon dioxide from potassium perfluoro 2- or 3-aminopropanoates (Equation (12)). These reactions work well with both cyclic and acyclic secondary amines (yield 68–98%, determined by GLC) ⟨88CL1887, 89CL905⟩.

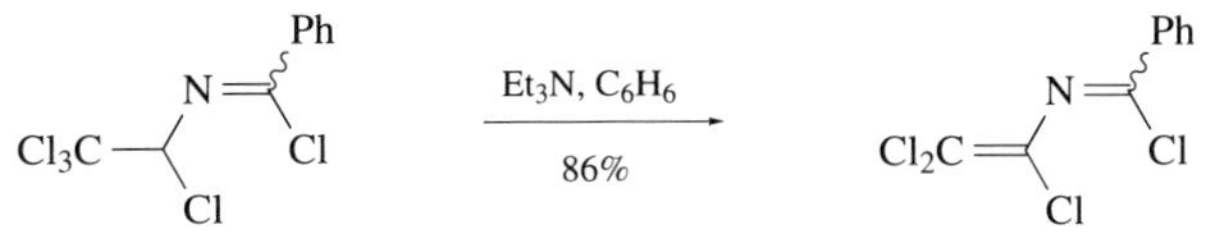

$$Cl_3C-\underset{Cl}{\underset{|}{C}}(=N-Ph)-Cl \xrightarrow[86\%]{Et_3N, C_6H_6} Cl_2C=C(Cl)(N=S-Ph) \quad (10)$$

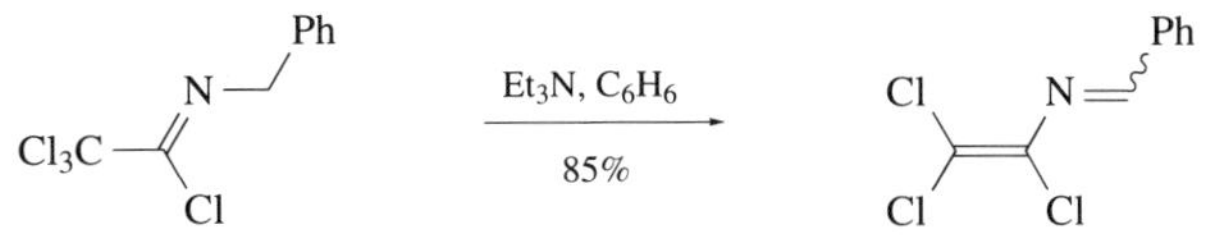

$$Cl_3C-C(Cl)=N-CH_2Ph \xrightarrow[85\%]{Et_3N, C_6H_6} (Cl)(Cl)C=C(Cl)(N=S-Ph) \quad (11)$$

$$\begin{matrix} R_2N-CH(CH_3)-CO_2H \\ R_2N-CH_2CH_2-CO_2H \end{matrix} \xrightarrow[\text{iii, heat}]{\text{i, perfluorination; ii, }K_2CO_3} (F)(F)C=C(F)(NRf_2) + CO_2 + KF \quad (12)$$

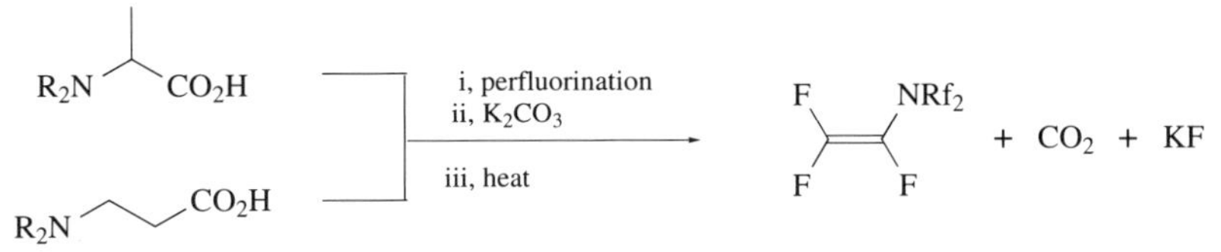

4.18.1.2 *gem*-Halo Nitro Alkenes

The preparation and synthetic utility of 1-heterosubstituted nitro alkenes has recently been reviewed by Barrett, covering ether, thioether, halogen and nitro substitutents ⟨91CSR95⟩. The present review aims to extend and update that work, while restating the more important aspects of the chemistry of these compounds.

4.18.1.2.1 *By halogenation*

The reactions described in this section are those in which halogen is added as an electrophilic species. Bromine or chlorine is usually required for unactivated alkenes while the *N*-halosuccinimides are adequate for more nucleophilic substrates such as the nitro enamines. Iodine can only be added to the more activated systems while fluorine has been introduced by direct fluorination of nitronate anions.

gem-Fluoro nitro alkenes are very rare in the literature. 1,1,2,2-Tetranitroethane has been fluorinated and the product subjected to flash vacuum pyrolysis, eliminating dinitrogen tetraoxide to give (*E*)-1,2-difluoro-1,2-dinitroethene (**7**) in 92% yield ⟨91JOC537⟩. A more general reaction is the fluorination of the dianion of 1,1,3,3-tetranitropropane ⟨70JOC4236⟩ followed by alcohol-induced elimination leading to a series of alkyl *β*-fluoro-*β*-nitroacrylate esters in 40–45% yield (Scheme 8) ⟨89BAU635⟩. Surprisingly, these compounds represent the only examples of the general class of *β*-halo-*β*-nitro-*α*,*β*-unsaturated carbonyl compounds described in the literature.

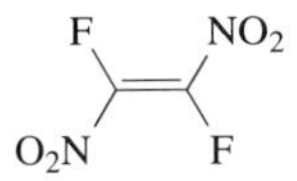

$$(F)(O_2N)C=C(NO_2)(F)$$
(**7**)

$$O_2N-CH(N^+O_2^-)-CH_2-CH(N^+O_2^-)-NO_2 \xrightarrow[H_2O]{F_2} (O_2N)(O_2N)C(F)-C(F)(NO_2)(NO_2) \xrightarrow[-HNO_2,\,-NOF]{ROH} (ROCO)(H)C=C(NO_2)(F)$$

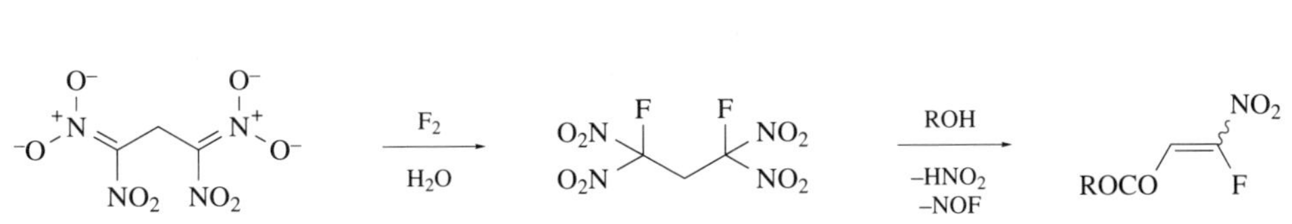

Scheme 8

An important method of preparation of α-bromo or α-chloro nitro alkenes is from nitro alkenes by dihalogenation and subsequent elimination (Scheme 9). The stereochemistry of the products is usually irrelevant to subsequent transformations and is not normally elucidated. A valuable exception is provided by the work of Miller *et al.* who showed that the proton NMR spectra of the (*Z*) isomer of 2-bromo- or 2-chloro-2-nitrostyrene exhibited a characteristic downfield shift for both the alkene and the phenyl protons compared with those of the corresponding (*E*) isomer. It was further shown that substantial (*Z*) to (*E*) isomerisation occurred upon irradiation with sunlight, giving a separable photostationary mixture composed of 60–80% (*E*) for the chloro derivative but only 10–20% (*E*) for the bromo compound (Scheme 10). Treatment with iodine or triethylamine brought about quantitative reversion to the (*Z*) form ⟨76JOC2112⟩. Clearly there is a strong preference for the (*Z*) isomer in these systems, especially with the bromo derivative.

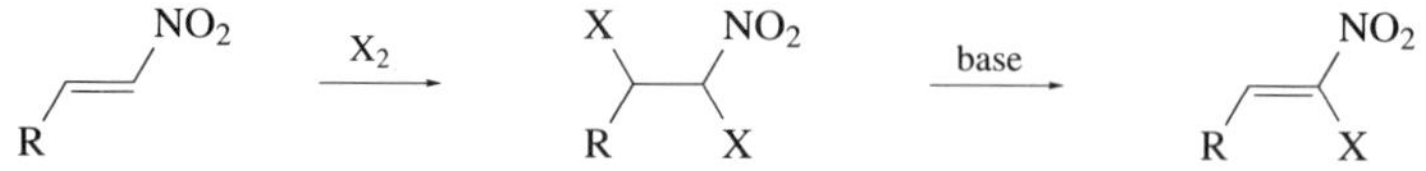

Scheme 9

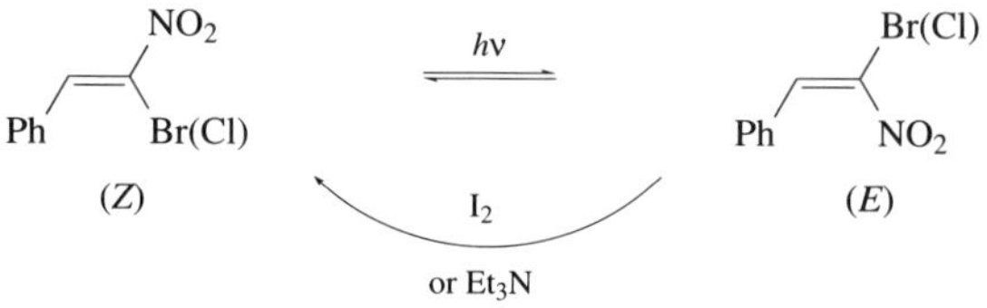

Scheme 10

One interesting reaction is the addition of an *N*-chloroguanidine across a nitro alkyne, giving a *gem*-chloro nitro alkene in one step, although in poor yield (Equation (13)) ⟨75C512⟩.

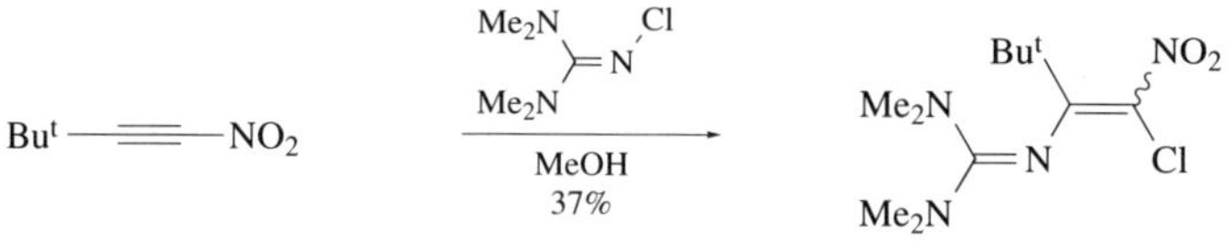

$$Bu^t\!-\!\!\equiv\!\!-NO_2 \xrightarrow[\text{37%}]{\text{MeOH}} \qquad (13)$$

When the alkene is activated by an electron-donating group the halogenation proceeds directly with no need for an extra elimination step. A comprehensive study has been carried out by Tokumitsu and Hayashi using primary, secondary, tertiary and aromatic 2-nitro enamines and *N*-halosuccinimides in benzene–chloroform (Equation (14)). The 2-nitro enamines bearing a 1-phenyl group gave generally good yields (50–90%) while those with a 1-methyl substituent gave lower yields (30–60%) or, in the case of a tertiary (piperidino) derivative, unstable products. Chlorine, bromine or iodine could be introduced with equal efficiency. It was considered that the products with primary and secondary amines exist as the intramolecularly hydrogen bonded (*E*) isomer, on the basis of a characteristic peak at 3150 cm^{-1} in their infrared spectra ⟨85JOC1547⟩. However, it is likely that, in common with other 2-nitro enamines, they are configurationally unstable and tend to exist in the (*Z*) form in polar media ⟨81T1453⟩. Such a configurational instability has recently been demonstrated for the 'parent' enamine 2-anilino-1-bromo-1-nitroethene ⟨90SC3339⟩.

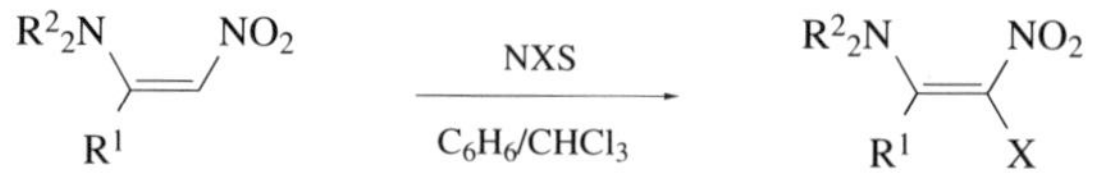

$$\qquad (14)$$

R^2 = H, Ph, alkyl, cycloalkyl
R^1 = Me, Ph
X = Cl, Br, I

An alternative route to 2-chloro-2-nitro enamines has been demonstrated by Böhme and Weisel who chlorinated nitroacetone and condensed the product with aniline in the presence of titanium tetrachloride (Scheme 11). The same products were also obtained, in better yield, by a more normal (NCS) chlorination of the corresponding nitro enamine ⟨77AP(310)30⟩.

Iodine has been added to the apparently unactivated 2-position of 1-ferrocenyl-2-nitroethene (Scheme 12). However, this transformation was carried out in sodium methoxide and the authors speculate that the methoxide initially added to the double bond, producing a nitronate anion which was then iodinated, and the alkene system was then re-established by elimination of methanol.

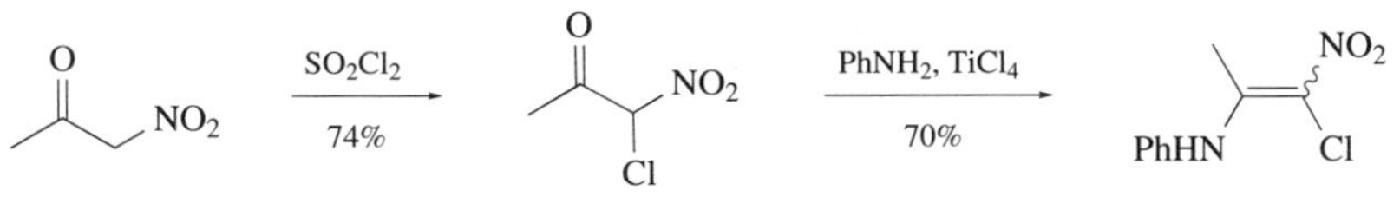

Alternatively, the double bond was chloromercuriated and the mercury replaced by subsequent reaction with potassium iodide–iodine in aqueous dioxan. Other methods of halogenation destroyed the ferrocenyl ring system. The authors propose, on the basis of the ultraviolet spectrum, that the nitro group is *trans* to the ferrocenyl ring, that is the product (**8**) has the (*Z*) configuration ⟨69BCJ3270⟩.

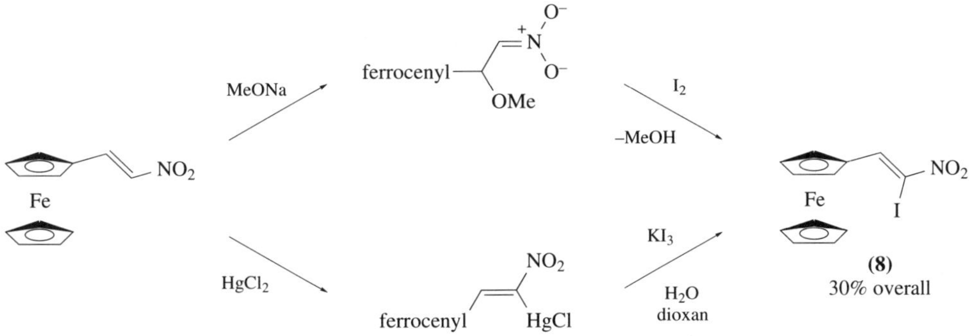

Scheme 12

4.18.1.2.2 By nitration

Alkenes have been nitrated by the action of nitric acid or of the oxides of nitrogen. These reactions are not as general as those described in the previous section unless the precursors are heavily substituted or polyhalogenated. This route therefore provides material suitable for β-halogen displacements, yielding compounds which may be hard to synthesise by other means.

1*H*-Perchloro-1,3-butadiene has been nitrated by concentrated nitric acid, affording compound (**9**) in 49% yield. The product was taken on to 2-amino derivatives (**10**) by displacement of chloride ion with an excess of the appropriate primary, secondary or aromatic amine, but it is interesting to note that the action of alkoxides led only to displacement of the *nitro* group ⟨91JOU48⟩.

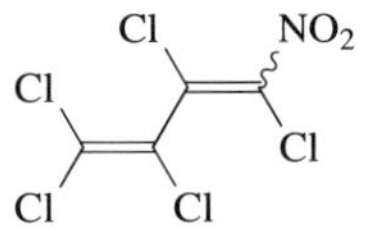

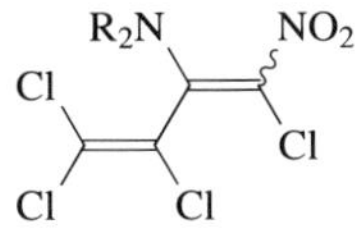

Nitrous acid has been added to diiodoalkyne in an excess of ethereal iodine to yield nitro-triiodoethene in variable yield (40–70%, Equation (15)) ⟨08CB2190⟩. Nitrogen dioxide has been added to dichloroalkyne in ether solution, yielding 1,2-dichloronitroethene in unspecified yield and geometry (Scheme 13). The alkenic proton was presumably derived from the solvent and both ethyl nitrate and ethyl nitrite were by-products ⟨43CB88⟩. This compound was subsequently taken on to an aryl vinyl sulfone by β-halogen displacement (Scheme 13) ⟨77UKZ1000⟩.

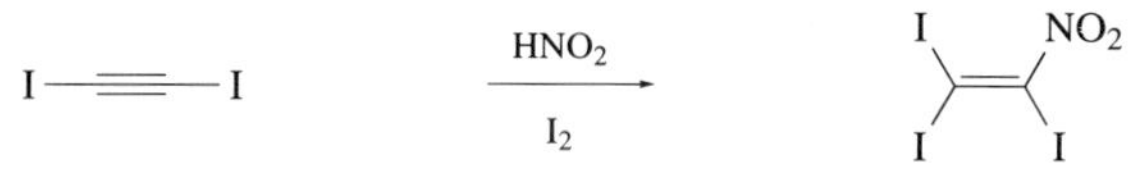

Scheme 13

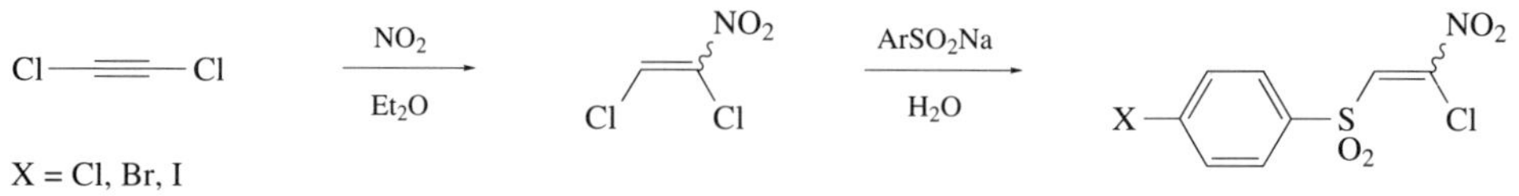

4.18.1.2.3 By condensation between aromatic aldehydes and halonitromethanes

Extension of the nitro-aldol reaction to halonitromethanes has not been commonly employed. However, some notable examples exist and this reaction may prove to be the most efficient entry into the general class of 2-chloro-2-nitrostyrene derivatives, complementing the halogenation approach which is possibly more efficient for bromonitrostyrenes.

Dauzonne *et al.* found that treatment of a variety of aromatic aldehydes with bromonitromethane, a large excess of dimethylammonium chloride and a catalytic amount of potassium fluoride in xylene or dibutyl ether gave good yields of the corresponding 2-chloro-2-nitrostyrenes (Equation (16)) ⟨90S66⟩. Apparently, initial conversion into chloronitromethane took place. If dimethylammonium bromide was used, chloronitromethane could not be formed and the reaction became very slow, only 10% conversion under conditions which otherwise gave 61% of the 2-chloro-2-nitrostyrene ⟨87S1020⟩. Similarly, Fishwick *et al.* found that bromonitromethane failed to react with benzaldehyde and catalytic amine in ethanolic sodium carbonate, even after four days at reflux ⟨86JCS(P1)1171⟩. 2-Bromo-2-nitrostyrenes have been made in good yields by the slow addition of bromonitromethane to a neat suspension of aldehyde and tri-*n*-butylarsine (Equation (17)). The authors do not speculate on the mechanism of this transformation but do point out that 2 equivalents of bromonitromethane are consumed, one of which is converted into nitromethane ⟨93SC1⟩. All the above compounds were found to have exclusively the (Z) configuration about the double bond, as determined by proton NMR or x-ray methods.

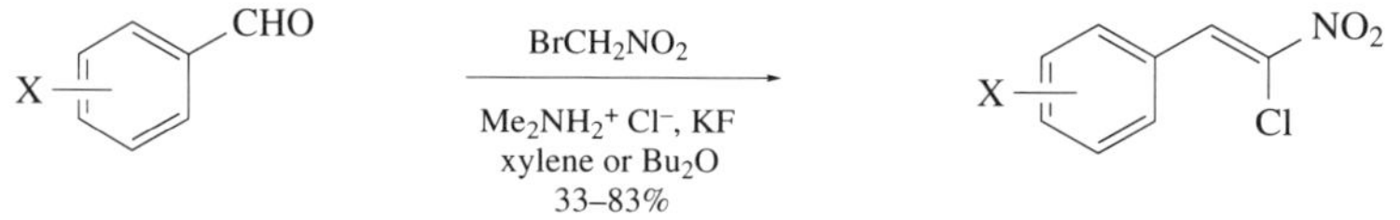

4.18.1.3 *gem*-Halo Nitroso Alkenes

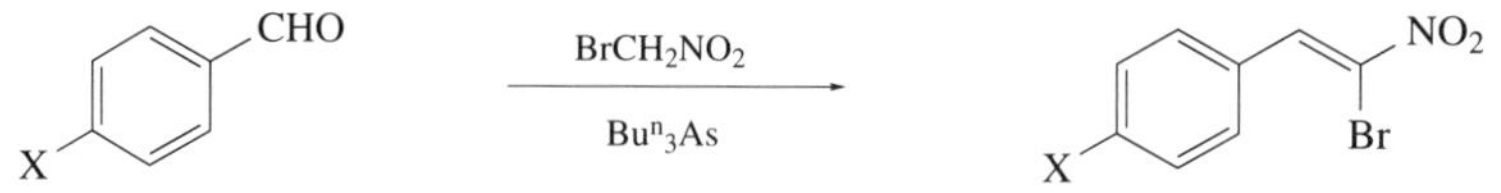

There is much interest in nitroso alkenes as reactive intermediates but they are not normally isolated ⟨83CSR53⟩. A notable exception is provided by the work of Griffin and Haszeldine who allowed iodotrifluoroethene to react with nitric oxide in the presence of elemental mercury and under ultraviolet irradiation (Equation (18)). They obtained nitrosotrifluoroethene in an optimised yield of only 9% as a blue gas (boiling at $-23\,^\circ$C but stable up to $100\,^\circ$C) showing no evidence of dimerisation in the liquid phase ⟨60JCS1398⟩. Francotte *et al.* prepared a series of *gem*-chloro nitroso alkenes by elimination of HCl from α-chloro oximes (Equation (19)). The yield was not determined but the products were stable enough to afford spectral data ⟨78AG(E)936, 81HCA1208⟩.

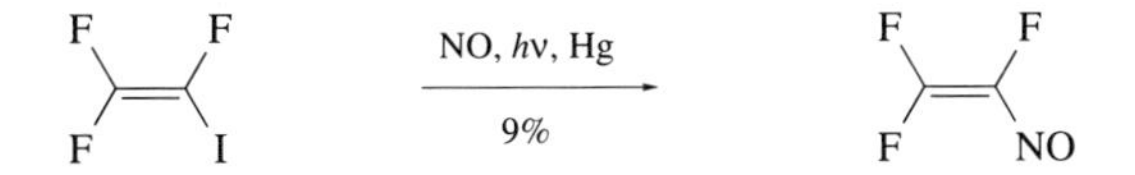

$$\text{(18)}$$

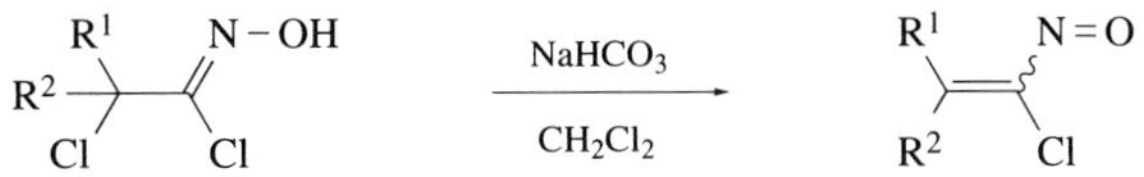

$$\text{(19)}$$

$R^1 = R^2 = Cl$
$R^1 = R^2 = H, Cl$
$R^1, R^2 = H, Ph$

4.18.1.4 *gem*-Azido Halo Alkenes

A number of workers have prepared 1-azidopentafluoropropene by addition of azide anion to hexafluoropropene (Equation (20)). This was obtained as an unstable yellow–green liquid and was shown to have the (*E*) geometry by fluorine NMR spectroscopy ⟨60MI 418-01, 65JA3716, 66JCS(C)2304⟩.

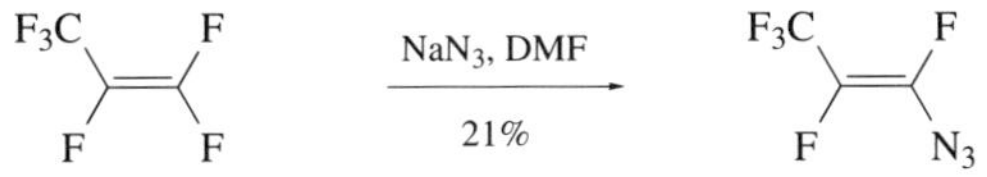

$$\text{(20)}$$

4.18.1.5 Diazonium and Diazo Derivatives

Arylazo derivatives of alkenes are more stable than the nitroso and azido species described above and are commonly isolated as coloured solids. The alkylazo equivalents are much less stable and are not isolated but are used as reactive intermediates ⟨79JCS(P1)249⟩. Chattaway *et al.* found that the aryl hydrazones of chloral spontaneously lose HCl and the resulting azo compounds crystallise from acetic acid solution as orange solids. Treatment with bromine or chlorine in acetic acid gave the corresponding α-haloalkenyl azo compounds in unspecified yields (Scheme 14) ⟨27JCS2850, 28JCS2756, 31JCS1088⟩. In a similar fashion Knunyants and co-workers found that phenylhydrazine reacts with perfluoro-1-alkenes to give the (*E*) isomer of *gem*-fluoro phenylazo alkenes (Equation (21)) ⟨91BAU1705⟩. Hanack and co-workers prepared crystalline 2-chloro-2-tosylazostyrene derivatives in four steps from arylacetic acids (Scheme 15). Treatment of the *p*-chlorophenyl derivative with Lewis acids at −40 °C gave a solution containing the α-chloroalkenyl diazonium ion (**11**) ⟨85CB1008, 89CB331⟩. Finally, Colquhoun *et al.* have reported the preparation and x-ray crystal structure of an α-chloroalkenyl tungsten diazo complex (**12**) ⟨80CC879, 88JCS(D)2781⟩.

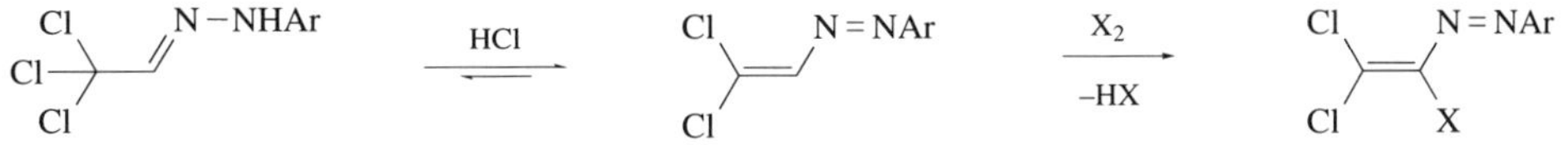

$X = Cl, Br$

Scheme 14

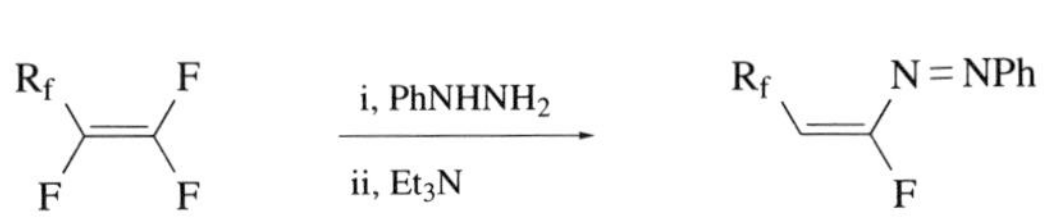

$R_f = C_4F_9, 91\%$

$$\text{(21)}$$

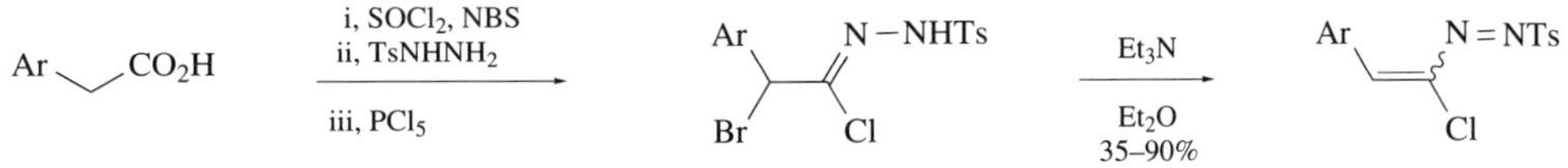

Scheme 15

4.18.1.6 Iminophosphorane, Sulfimide and Metallonitrene Complexes

A variety of phosphine imides have been prepared by the net addition of a phosphorus species across a cyanide triple bond. Foucaud and co-workers added phosphites across the triple bond of α-bromomalononitrile derivatives to produce the corresponding *gem*-bromo iminophosphoranes in high yield (Equation (22)) ⟨69TL2441, 72T4039, 72T5149⟩. Phosphorus halides also undergo addition across nitriles (Scheme 16) ⟨78JGU711, 79JGU1947⟩. The sulfinylamine (**13**) ⟨79JOU2207⟩ and dichlorosulfimides have been prepared by the action of sulfur halides on a variety of nitrile-derived starting materials (Scheme 17) and the products transformed into other sulfimides, such as compound (**14**) ⟨76JOU782⟩. In a similar fashion Roesky *et al.* have added molybdenum and tungsten chlorides across 1,1-dicyanobis(trifluoromethyl)ethene (Equation (23)) ⟨86CB3150⟩.

4.18.2 DERIVATIVES OF PHOSPHORUS AND OTHER GROUP 15 ELEMENTS

4.18.2.1 α-Haloalkenylphosphorus Derivatives

4.18.2.1.1 From carbonyl compounds

While there are many reported syntheses of α-haloalkenylphosphorus species, surprisingly few proceed by direct condensation with carbonyl compounds. The majority of those that do are by the Wadsworth–Emmons reaction, utilising halomethanediphosphonate esters (Scheme 18).

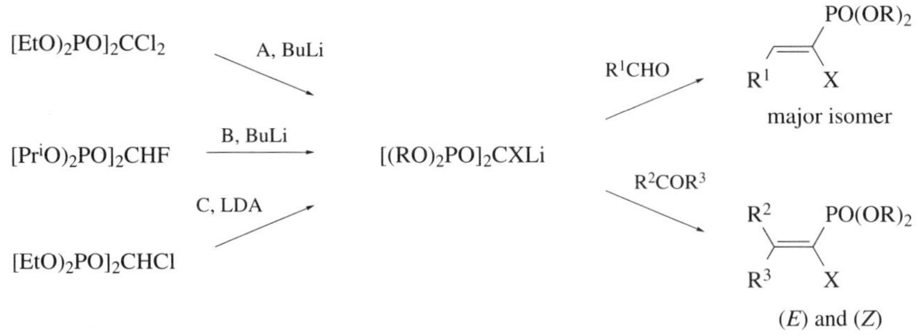

Scheme 18

The first such reaction was described by Seyferth and Marmor, who found that the anion derived by the action of *n*-butyllithium on tetraethyl dichloromethanediphosphonate gave high yields of *gem*-chloro phosphono alkenes on treatment with pivaldehyde or acetone, but failed to react with benzophenone (Route A of Scheme 18) ⟨73JOM(59)237⟩.

This initial study was not followed up until Blackburn and Parratt reported the synthesis of *gem*-fluoro phosphono alkenes, again using aldehydes and ketones, though in this case both *t*-butyl methyl ketone and camphor failed to react (Route B of Scheme 18) ⟨82CC1270, 86JCS(P1)1417⟩. Almost simultaneously, Savignac and co-workers carried out similar reactions with the anion previously employed by Seyferth (Route C of Scheme 18) ⟨86JOM(304)283⟩. These products are formed in generally high yields (56–95%, except for the examples noted). More recently, Yuan *et al.* reported the action of the thallium salt of bromomethanediphosphonate esters on α,β-unsaturated ketones or aldehydes. Efficient Wadsworth–Emmons reaction took place with acrolein or methacrolein (90–93%, affording exclusively the (Z) isomer) (Equation (24)) ⟨91S854⟩.

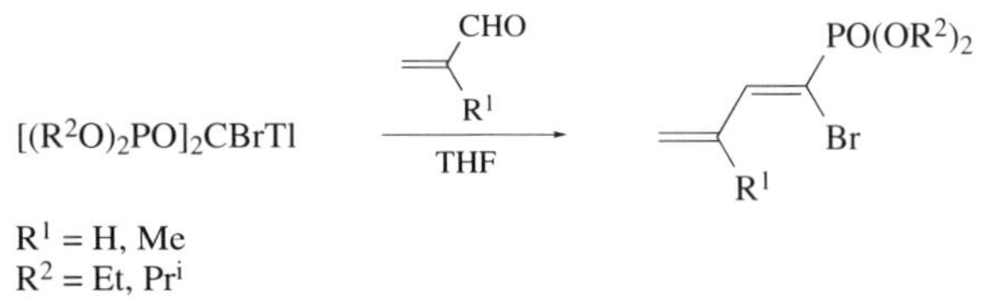

$$R^1 = H, Me \qquad R^2 = Et, Pr^i \tag{24}$$

There are only two reports of similar condensations using the Wittig reaction. Burton and Cox found that a bisphosphonium ylide underwent condensation with perfluoro acid fluorides to give (Z)-perfluorovinylphosphonium species in 70–82% yield (Scheme 19). This reaction is very specific, only ylide acylation being observed if the nature of either reactant is significantly altered, although functionality in the fluoroalkyl chain is well tolerated ⟨83JA650⟩. The same bisphosphonium ylide gave the normal, (E)-selective, Wittig reaction with alkyl aldehydes, but gave (Z)-selective alkenation with aryl aldehydes (Scheme 19). The authors speculate that the (Z) geometry is encouraged by aryl to phosphorus charge transfer in the oxaphosphetane intermediate ⟨85JA2811⟩.

A trimethylsilyl group may be substituted for a phosphonate unit, *gem*-halo phosphono alkenes

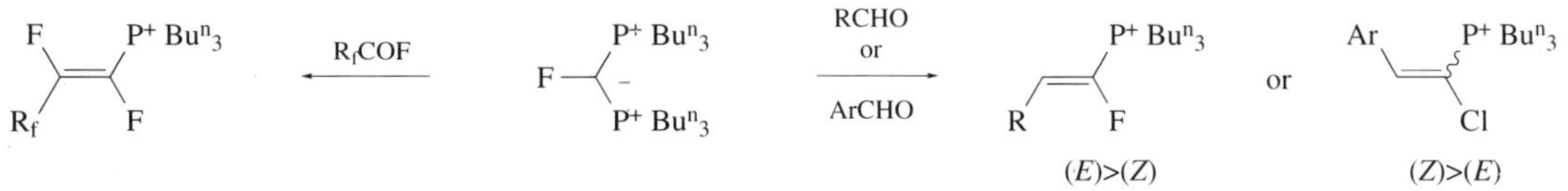

Scheme 19

being formed by Peterson elimination (Equation (25)). The products were formed in very high yield (90–100%), but with no stereochemical control ⟨86JCS(P1)1425, 88JOM(338)295⟩.

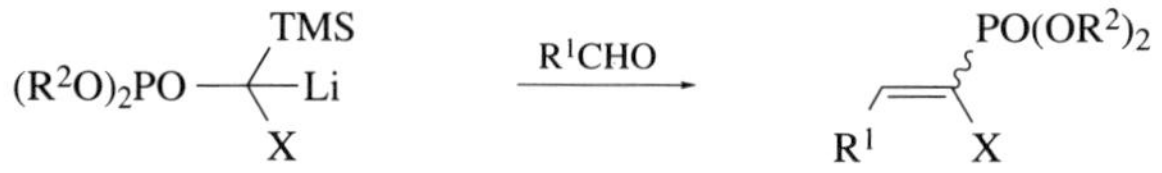

$$(R^2O)_2PO-\underset{X}{\overset{TMS}{C}}-Li \xrightarrow{R^1CHO} \underset{R^1}{\overset{PO(OR^2)_2}{C}}=\underset{X}{C} \qquad (25)$$

$R^1 = Bu^t, Ph, OEt$
$R^2 = Et, Pr^i$
$X = F, Cl$

The product of ylide acylation may exist preferentially or exclusively in the enol form, for example compound (15), and this form has been fixed by alkylation (16; R = OEt) and by acylation (16; R = OCOMe) ⟨69AG(E)216, 86JCS(P1)1425, 93JOC1531⟩. Similarly the interesting *gem*-chloro phosphono ketene (17) has been prepared by base-induced elimination from the corresponding acid chloride ⟨88CL211⟩.

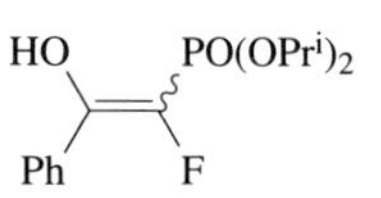

(15) HO, PO(OPr^i)_2 / Ph, F

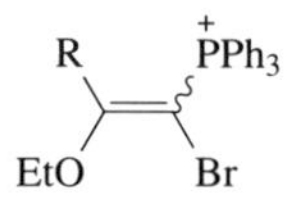

(16) R, $\overset{+}{P}Ph_3$ / EtO, Br

(17) O=•=C(PO(OEt)_2)(Cl)

R = MeCO_2 or EtO

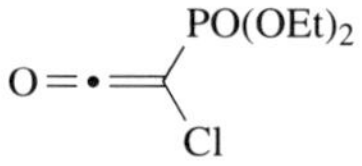

Although not derived directly from carbonyl compounds, it is worth noting that *gem*-halo phosphorus allenes have been prepared from propargyl alcohols (Scheme 20) ⟨67JGU2024, 71CR(C)1107, 75BSF2259, 87JGU844⟩.

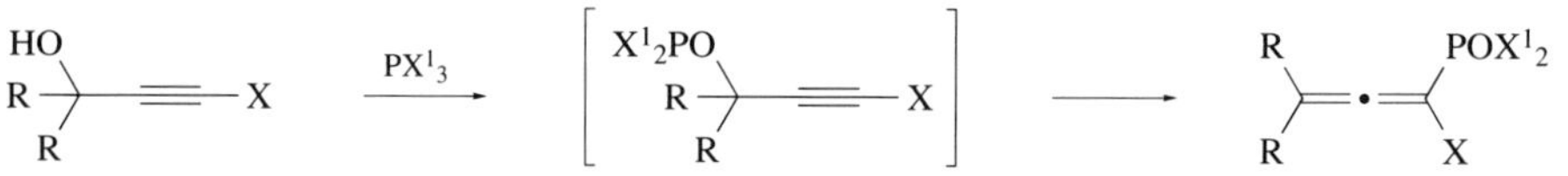

$$\underset{R}{\overset{HO}{\underset{R}{>}}}-C\equiv C-X \xrightarrow{PX^1_3} \left[\underset{R}{\overset{X^1_2PO}{\underset{R}{>}}}-C\equiv C-X \right] \longrightarrow \underset{R}{\overset{R}{>}}=C=\underset{X}{\overset{POX^1_2}{C}}$$

R = H, Ph, alkyl
X^1 = Cl, OEt, Ph, H, OH
X = Cl, Br, I

Scheme 20

4.18.2.1.2 *From fluorinated alkenes*

A number of α-trifluorovinylphosphorus derivatives have been prepared by the action of trifluorovinyllithium or -magnesium on phosphorus(III) halides (Equation (26)) ⟨60BAU1851, 69JA1929, 69JA1934, 78ZN(B)1422, 88BAU1686⟩. The success of such reactions is strongly dependent on experimental conditions and variable results have been obtained by different groups of workers (compare, e.g., ⟨60BAU1851⟩ and ⟨69JA1929⟩).

$$F_2C=CFM \xrightarrow{R_{3-n}PX_n} (F_2C=CF)_nPR_{3-n} \qquad (26)$$

R = Cl, Br, Ph, OEt, OBu^n, NMe_2
M = Li, Mg
X = Cl, Br

A second route to α-fluorovinylphosphorus compounds is by the action of a phosphorus(III) species on a polyfluoro alkene (see the references quoted in ⟨90JFC(49)75⟩). An Arbusov reaction on iodotrifluoroethene gave 67% of the phosphonate (18), as the (*E*) isomer, in spite of the fact that a phosphonate group activates the double bond to further nucleophilic attack ⟨75JCS(P1)702, 81PS(10)127⟩. Tetrafluoroethene appears to be too reactive to form a monophosphonate, contrary to earlier reports (⟨88BAU1686⟩ and references quoted therein). Phosphorus(III) species have been added to perfluoro alkenes. Both dimethylphosphine and tetramethyldiphosphine react with hexa-fluoropropene in the dark (Route A of Scheme 21) ⟨75JCS(P1)702⟩. Burton *et al.* employed tertiary phosphines and long chain fluoro alkenes (Route B of Scheme 21) ⟨79JA3689, 80JFC(15)543⟩. Von

Allworden and Röschenthaler describe examples of all the Routes of Scheme 21, including mixed methyl, phenyl *P*-fluorophosphoranes and the vinyl phosphonate product of an Arbusov reaction ⟨88CZ89⟩. Michael addition of phosphites to perfluoromethacrylate esters gave rare examples of *gem*-halo phosphono α,β-unsaturated carbonyl compounds, but in this case an Arbusov reaction was not observed, the major products being the corresponding *P*-fluorophosphoranes obtained as single isomers of undefined geometry (e.g., Equation (27)) ⟨76BAU853, 76BAU873⟩. More recently, palladium-catalysed displacement of iodine by diethyl phosphite has furnished *gem*-fluoro phosphono alkenes with retention of stereochemistry but in only moderate yield (Equation (28)) ⟨93TL7197⟩.

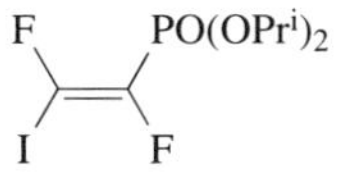

(18)

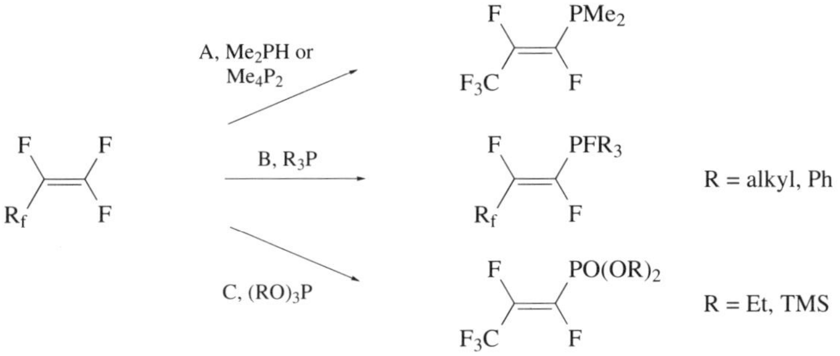

Scheme 21

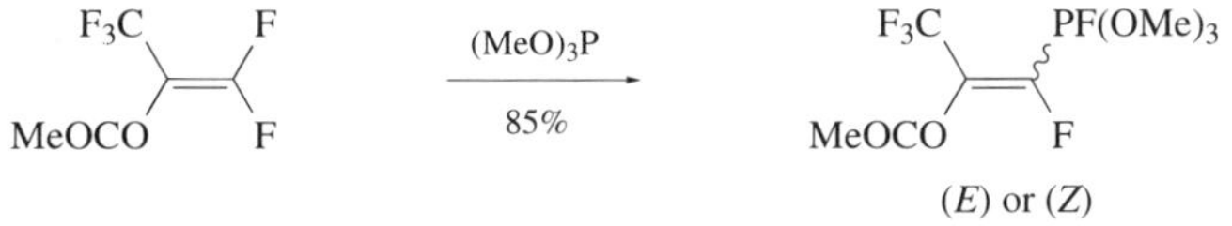

(27)

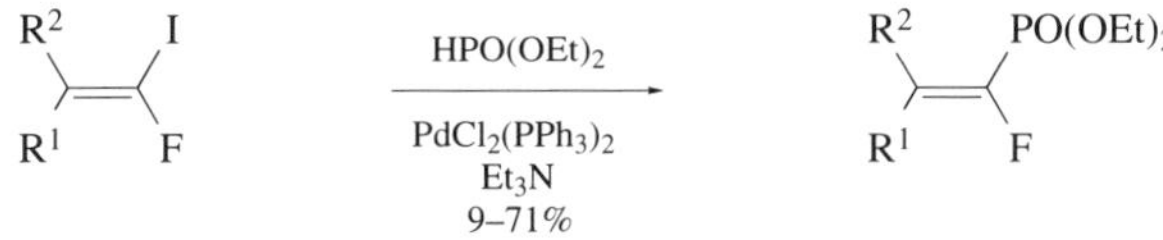

(28)

R^1, R^2 = H, Ph, alkyl, cycloalkyl

The synthesis of alkene diphosphonates has generated much interest (see ⟨90JA3152⟩). For example, an (*E*,*Z*)-diphosphonate mixture has been prepared by double reaction of phosphites with polyfluoro alkenes (Equation (29)) ⟨81PS(10)127, 88BAU1686⟩. Products of this type are not wholly new, however, Maier having reported the synthesis of (*Z*)-halo alkene diphosphonic acids (**19**) by aqueous hydrogen halide hydrolysis of alkyne diphosphonates ⟨73MI 418-01⟩.

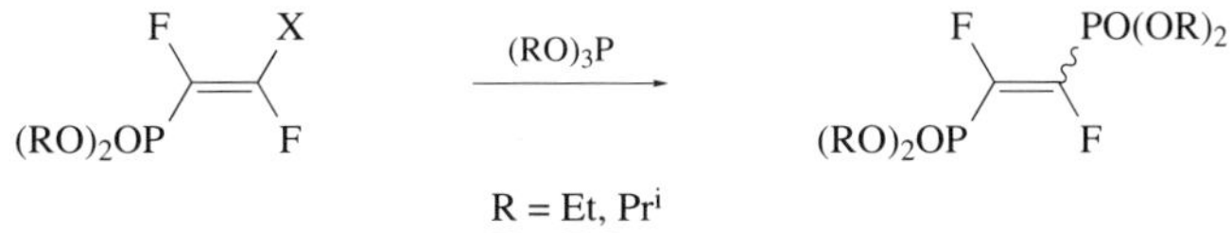

(29)

R = Et, Pr^i
X = F, I

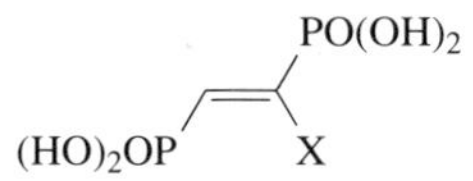

X = Cl, Br, I
(19)

4.18.2.1.3 By halogenation

A number of α-haloalkenylphosphorus oxides have been prepared by direct halogenation of vinylphosphorus(V) compounds. Hägele and Dolhaine reported the halogenation of diethyl vinylphosphonate with chlorine, bromine and their diatomic interhalogen compounds, including those with iodine (Scheme 22). The resultant α,β-dihaloethanes were produced in an anti-Markownikov fashion, with the more electrophilic halogen geminal to the phosphorus atom. Elimination afforded the *gem*-halo(phosphono)ethenes in 30–82% yield ⟨77PS(3)47⟩. The same sequence has been applied to homochiral (*S*)-(−)-methylphenylvinylphosphine oxide, giving the (*R*)-(−)-α-bromo and chloro compounds (20) (83–90% yield) without loss of stereochemical integrity ⟨89T337⟩. Boyce *et al.* have studied the stereochemistry of the elimination step, noting that the products (21), and unhalogenated analogues, always have a β group bearing a *trans* relationship to the phosphorus atom. They comment that the slow *syn*, or more rapid *anti*, mode of elimination is determined largely by the ground state conformational properties of the substrate ⟨74JCS(P1)1644⟩.

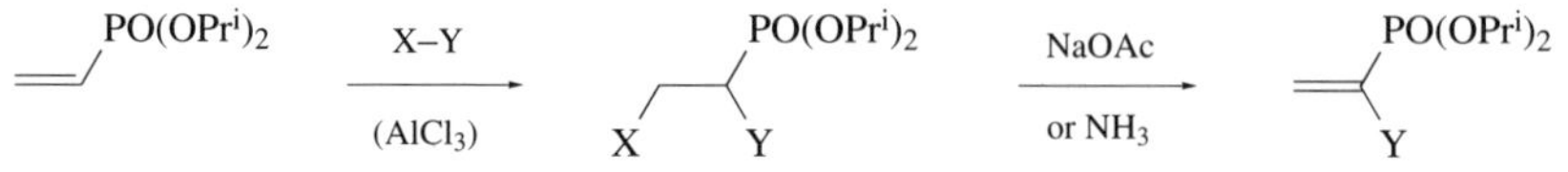

Scheme 22

X = Cl, Br
Y = Cl, Br, I

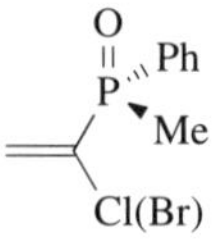

(20)

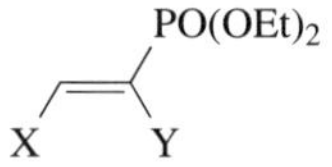

(21) X, Y = F, Cl

By contrast, when the β-position bears an electron-releasing substituent, halogenation proceeds readily. Buzykin *et al.* report the halogenation of a 2-ethoxyvinylphosphonamide in aqueous medium affording *gem*-halo phosphonamido aldehydes, predominantly in the (*Z*)-enol form, and describe subsequent halogen exchange to give a *gem*-iodo phosphonamido enol (Scheme 23) ⟨92JGU1222⟩. Similarly, 2-(alkylthio)vinylphosphonates have been brominated to afford β-bromo-β-phosphonovinyl thioethers (22) ⟨83T1189, 87JCS(P1)1275⟩.

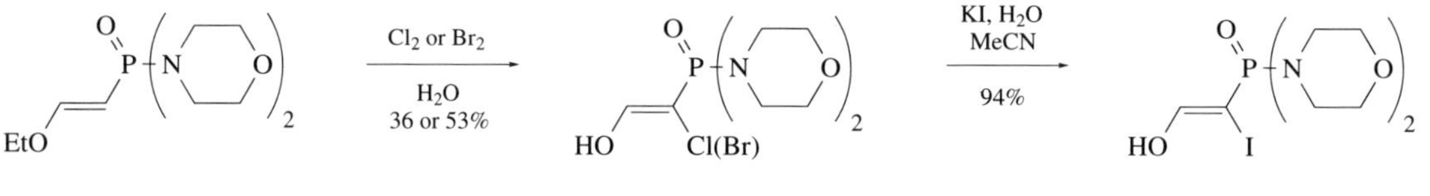

Scheme 23

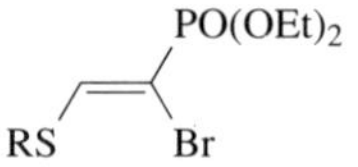

R = Et, *p*-tolyl
(22)

Electrophilic attack by phosphorus has been observed with electron-rich alkenes and alkynes. Thus Seredkina *et al.* described the addition of phosphorus pentachloride to (alkylthio)alkyne chlorides affording predominantly (*Z*)-thioether derivatives of (α-chlorovinyl)tetrachlorophosphoranes (Equation (30)) ⟨91JGU983⟩.

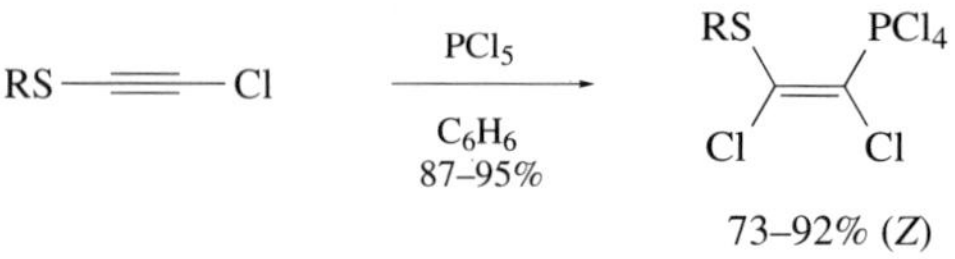

$$RS\text{—}\!\!\equiv\!\!\text{—}Cl \xrightarrow[\substack{C_6H_6 \\ 87\text{–}95\%}]{PCl_5} \underset{\text{73–92\% (Z)}}{\overset{RS\quad PCl_4}{\underset{Cl\quad Cl}{\diagup\!\!=\!\!\diagdown}}} \quad (30)$$

4.18.2.2 α-Haloalkenyl Derivatives of Arsenic, Antimony and Bismuth

In spite (or perhaps because) of the use of Lewisite (23), (β-chlorovinylarsine dichloride) as a chemical warfare agent, there has been very little reported work on the α-haloalkenyl derivatives of arsenic, and less on those of antimony and bismuth.

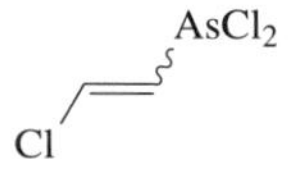

(23)

Goldwhite et al. provided a relatively high yielding route to a *gem*-arsino fluoro alkene by the prolonged reaction of dimethylarsine with 2,3-dichlorotetrafluoropropene followed by isomerisation with caesium fluoride (Scheme 24) ⟨68JOM(12)133⟩. Dichloro(trifluorovinyl)arsine (24) was prepared in high yield (92%) by treatment of the *As*-trifluorovinyl heterocycle (25) with condensed hydrogen chloride gas. Finally, in a similar fashion to the phosphorus derivative previously described (see Section 4.18.2.1.2), the action of trifluorovinylmagnesium iodide on arsenic, antimony and bismuth trichloride afforded the corresponding tris(trifluorovinyl) species (26) in moderate yields ⟨60BAU1851, 61MI 418-01⟩.

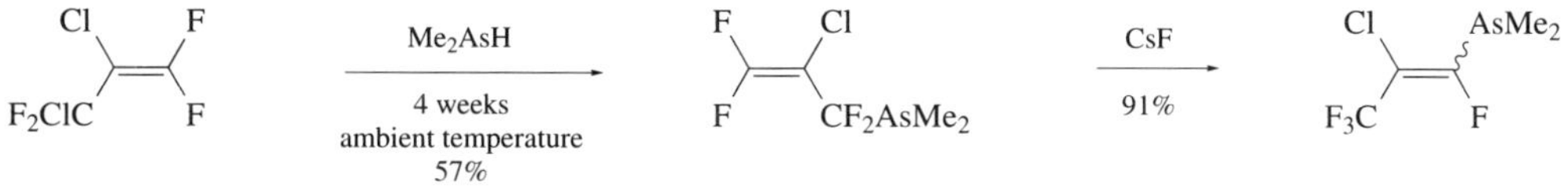

Scheme 24

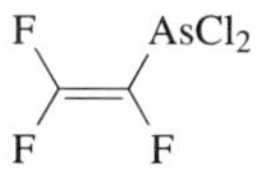

(24) 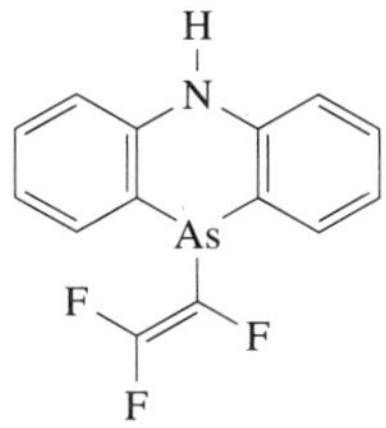   (25)   (F$_2$C=CF)$_3$M   M = As, Sb, Bi   (26)

4.18.3 DERIVATIVES OF SILICON AND OTHER GROUP 14 ELEMENTS

Although elemental silicon and germanium are often classed as metalloids and tin and lead are metals, the synthesis of their covalent derivatives resemble one another so closely that they are conveniently dealt with in one section. Further examples of the preparation of α-haloalkenylsilicon derivatives will be found under the part dealing with germanium (see Section 4.18.3.2), as these elements are invariably considered together, while tin derivatives (see Section 4.18.3.3) are more likely to be synthesised independently of any reference to silicon.

4.18.3.1 α-Haloalkenylsilicon Derivatives

4.18.3.1.1 From carbonyl compounds

Perhaps surprisingly, the Peterson equivalent of the Wadsworth–Emmons reaction, commonly encountered in *gem*-fluoro phosphono alkene chemistry, is hardly represented in the synthesis of the silyl alkenes. Seyferth et al. published a study of the reaction of bis(trimethylsilyl)bromomethyllithium with aldehydes and ketones. Aldehydes afforded *gem*-bromo silyl alkenes in good yields (52–78%), the (Z) isomer being preferred for more bulky alkyl groups (Equation (31)).

By contrast, ketones gave either deprotonation or oxirane formation, no alkenes being observed ⟨77JOM(142)39⟩.

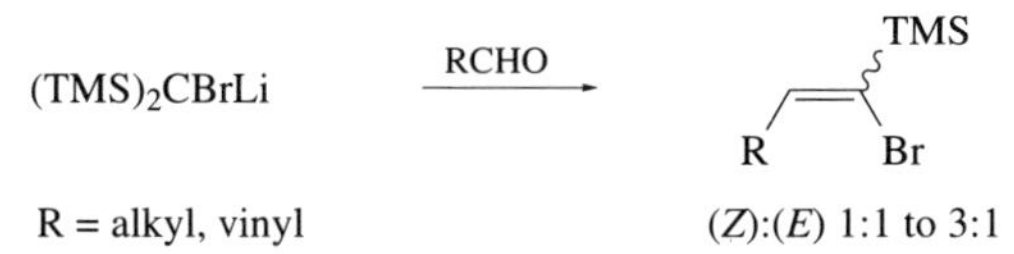

$$(TMS)_2CBrLi \xrightarrow{RCHO} \underset{R = alkyl, vinyl}{\underset{R \quad\quad Br}{\text{TMS}}} \quad (Z):(E) \text{ 1:1 to 3:1} \tag{31}$$

4.18.3.1.2 By silylation of α-haloalkenyllithium and -magnesium species

This is the most general method for the preparation of *gem*-fluoro silyl alkenes, although it has also been applied to the synthesis of α-chloro and α-bromo alkenes. Iodo derivatives have not been prepared by this method. Early examples of the use of organomagnesium compounds in the preparation of trifluorovinylsilanes have been reviewed by Seyferth, who also reported subsequent nucleophilic displacement of fluoride by triphenylsilyllithium (Scheme 25) ⟨62IC78, 62MI 418-01, 68JOC472⟩. More recent examples emphasise the use of lithium derivatives, a comprehensive series of silanes being reported by Hiyama *et al.*, who carried out metal–halogen exchange at very low temperatures in the presence of the appropriate chlorosilanes (Scheme 26) ⟨84CL1765⟩. Further β-functionalisation by displacement of fluoride has also been reported (Scheme 26) ⟨84JOM(264)155, 89JOM(367)1⟩.

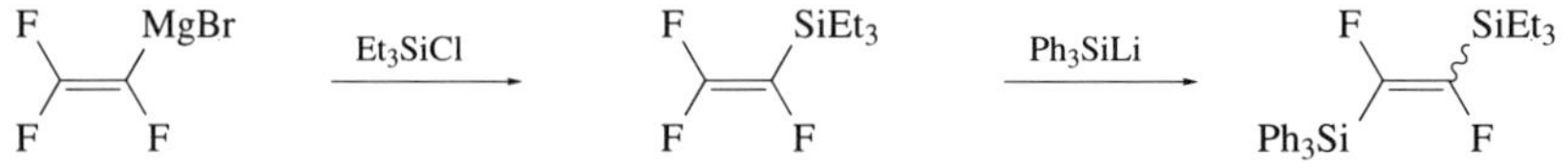

Scheme 25

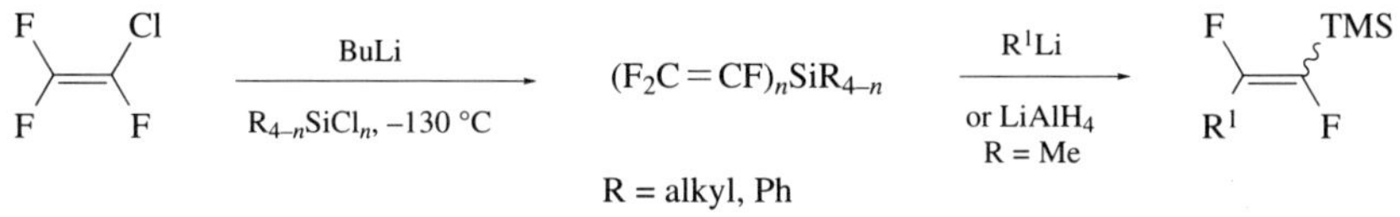

Scheme 26

gem-Chloro and bromo silyl alkenes have also been prepared by metallation and subsequent silylation. Sequential lithiation and silylation of (*E,E*)-1,2,3,4-tetrachlorobutadiene gave the (*E,E*)-disilyltetrachlorobutadiene (**27**) ⟨73CB1601⟩. Shimizu *et al.* treated a cold (less than −75 °C) mixture of lithium diisopropylamide (LDA) and TMS chloride with a variety of alkenyl halides, affording mixed *gem*-bromo or chloro silyl alkene isomers (**28**) in 23–84% yields. *gem*-Halo silyl enol ethers (**29**) and ketene acetals (**30**) were also produced by this versatile but nonstereoselective method, although the allyl chloride isomers (**31**) were best prepared by an 'inverse' addition of LDA to a mixture of the allyl chloride and TMS chloride ⟨87BSJ777⟩.

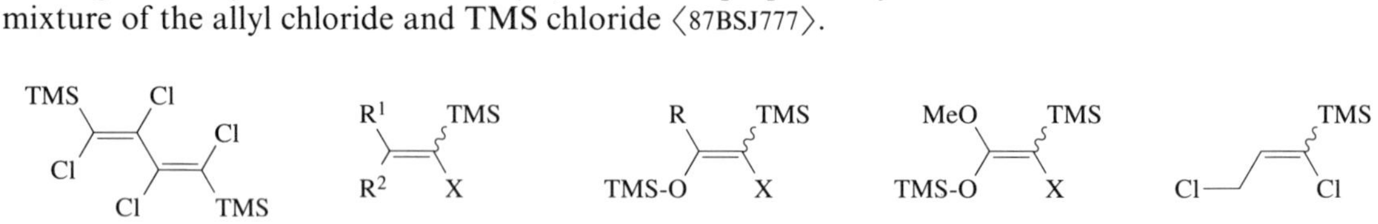

(27)   (28)   (29)   (30)   (31)

R^1, R^2 = H, Me, Ph
X = Cl, Br

Finally, direct silylation of a β-chloroketene acetal with trimethylsilyl triflate has been reported (Equation (32)) ⟨90LA745⟩. The thermal or radical induced reaction of chlorosilanes with chloro alkenes should also be mentioned here (e.g., Equation (33)) ⟨62MI 418-01⟩.

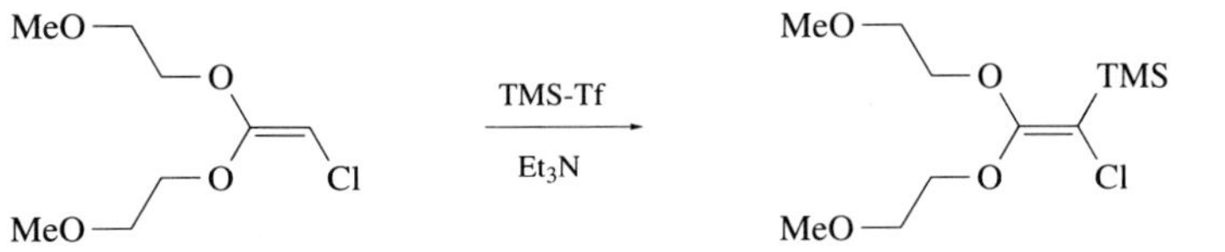

4.18.3.1.3 By halogenation

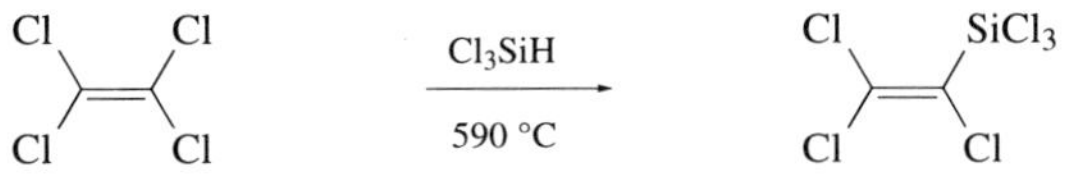

This section deals with those reactions in which a halogen atom is introduced by electrophilic attack on an unsaturated silane. This process is most appropriate for the introduction of bromine and chlorine, although some examples of iodination are also presented. The single example of fluorination is that shown by compound (**32**), prepared in 50% yield by addition of nitryl fluoride to bis(trimethylsilyl)alkyne ⟨86S132⟩.

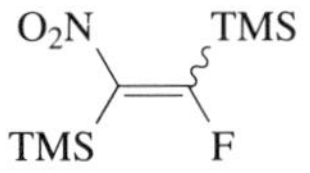

Much of the early work, mostly covering chlorination of vinyl silanes, has been reviewed by Seyferth ⟨62MI 418-01⟩, while another early example, the synthesis of 1-bromo-1-trimethylsilylethene, has since been reported as part of an organic synthesis preparation (Scheme 27) ⟨63CJC2977, 88OSC(6)1033⟩.

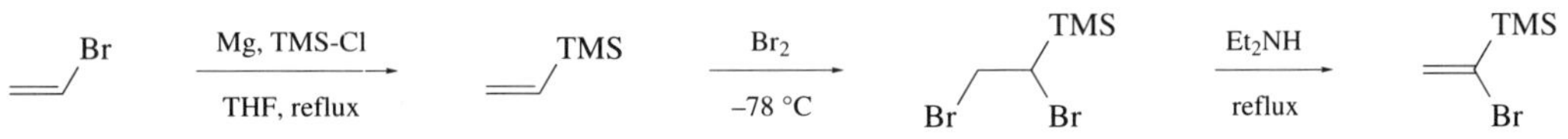

Scheme 27

Bromination of trimethylsilylketene, followed by base elimination, has been reported to afford the bromoketene (**33**) ⟨76TL1553, 77DIS(B)1214⟩. The preparatively useful (Z)-1,3-dibromo-1-trimethylsilylpropene (**34**) has been produced by the action of two equivalents of NBS on allyltrimethylsilane ⟨93SL189⟩.

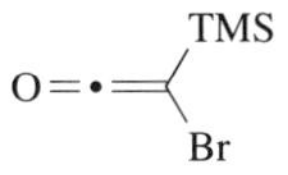

Bromination of alkynylsilanes has been shown to afford a range of (E)-1,2-dibromo silyl alkenes (**35**) in generally good yields (42–82%) ⟨89JOM(372)183⟩. In certain circumstances, the action of brominating agents on silyl alkynes may afford *gem*-bromosilylpropadienes. Both Ivanov and Stadnichuk, and Bogoradovskii *et al.*, derived propadienes by 1,4-addition to a silyl enyne and by loss of a γ-germyl group, respectively (Equations (34) and (35)) ⟨90JGU967, 91JGU1295⟩.

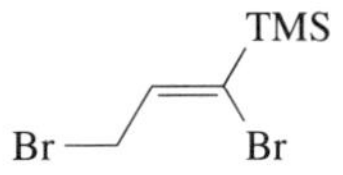

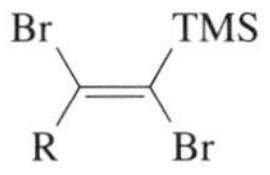

R = H, alkyl, Ph, TMS

(**35**)

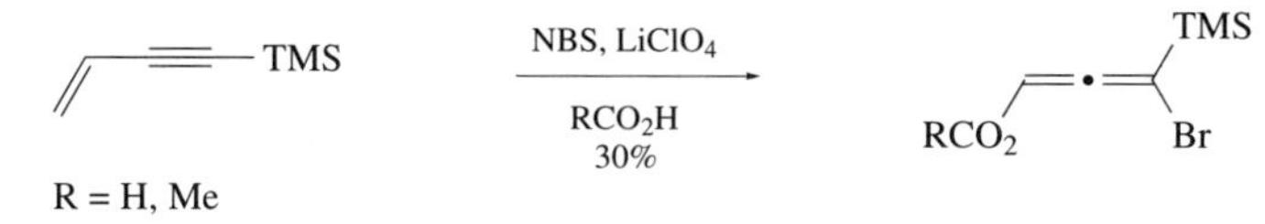

Iodine has been incorporated in a variety of ways. *gem*-Iodo silyl alkenes are produced by mercury(II)-catalysed addition of iodine to phenyl(trimethylsilyl)acetylene, the counterion providing the β-substituent (Equation (36), the authors suggest that the initial attack is by iodine, not by mercury) ⟨87JCS(P1)1017⟩. Finally, iodine derivatives have been made by the action of iodosylbenzene and boron trifluoride etherate on alkynylsilanes (Equations (37) and (38)) ⟨88CC1076, 89TL6701⟩.

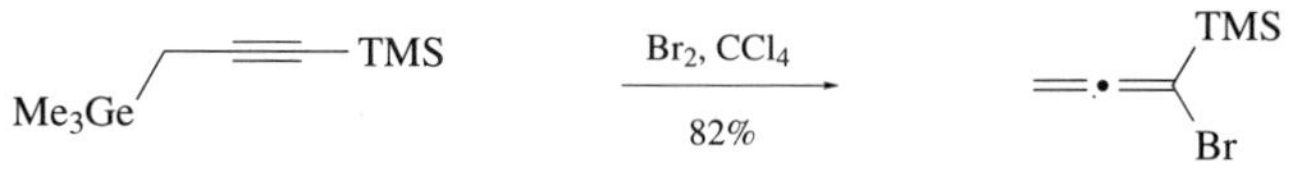

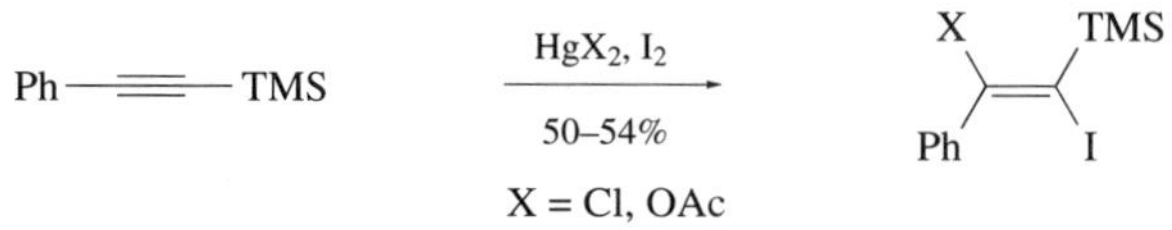

4.18.3.1.4 By halogenation of α-silylalkenyl metal derivatives

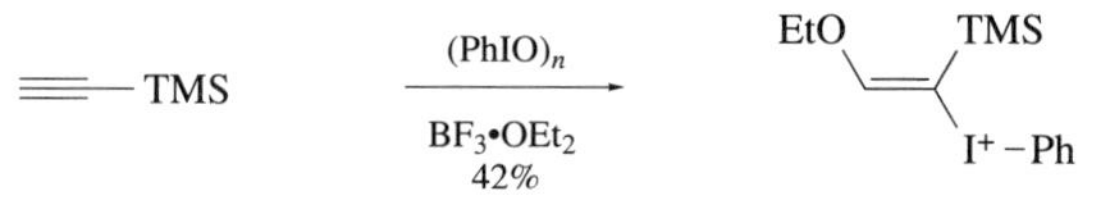

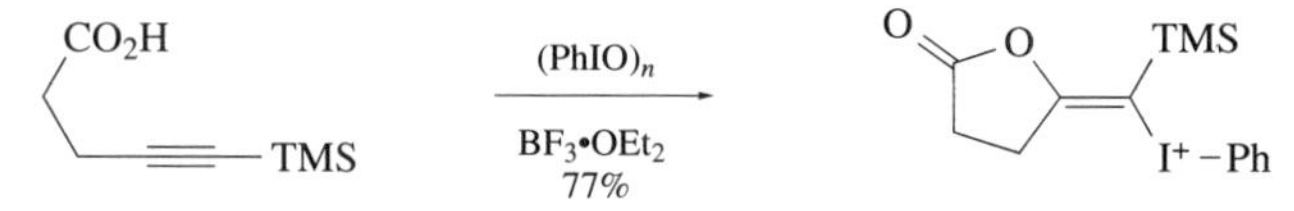

This section covers those reactions in which a halogen atom replaces a metal atom on an alkenylsilane. The metal atom is usually introduced by addition to an alkynylsilane (see also Chapter 4.23), while chlorine, bromine or iodine have been employed as the halogen. These reactions represent the first general method for the preparation of *gem*-iodo silyl alkenes.

Hydroalumination of alkynylsilanes leads to trisubstituted alkenes, subsequent halogenation furnishing (E)-*gem*-halo silyl alkenes in good yield (Scheme 28, NBS or bromocyanogen being used in place of bromine if it is necessary to avoid competitive bromination of alkenes) ⟨78JOC2739⟩. Such alumination has been reviewed ⟨84OR(32)375⟩, but it is worth repeating that the resultant (E)-α-chloro- or bromoalkenylsilanes may be isomerised to their (Z) isomers by catalytic bromine under UV irradiation (Scheme 28, X = Cl, Br). The (E)-α-iodoalkenylsilanes were similarly isomerised to their (Z) isomers by treatment with *t*-butyllithium (5 mol%) (Scheme 28, X = I) making use of the well-described configurational instability of *gem*-metallo silyl alkenes ⟨81JOC1292, 86JA3402⟩. It is worth noting that (E) to (Z) isomerisation of α-iodoalkenylsilanes has also been carried out with triethylborane in hexane solution (⟨89TL3155⟩, see Section 4.18.3.1.5 for more details of this method). In certain circumstances, (Z) isomers may be formed directly by *trans*-hydroalumination (e.g., Equation (39)), possibly by isomerisation after initial *cis*-hydroalumination ⟨93CC1309, 93CC1817⟩.

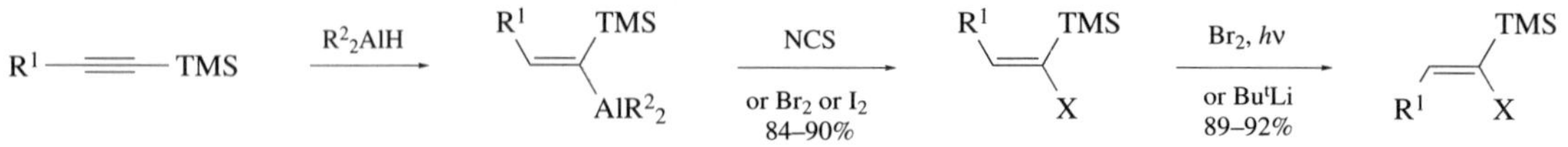

Scheme 28

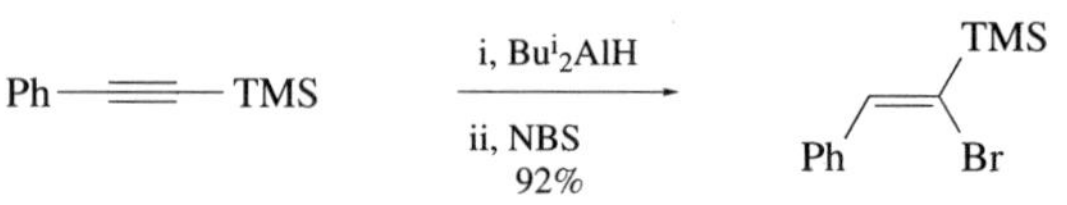

Carbometallation–halogenation of ethynylsilanes provides another route to trisubstituted alkenes. Organocopper reagents have been added to triphenyl- and trimethylsilylethyne, followed by quenching with NCS, NBS, bromocyanogen or iodine, giving (*Z*)-*gem*-halo silyl alkenes (**36**) in high yield (80–95%) ⟨77TL1823, 90TL4937⟩.

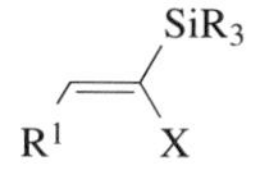

X = Cl, Br, I
(36)

Tetrasubstituted alkenes may be prepared by carbometallation of alkynylsilanes. Halogenation of a titanocene-catalysed alkylalumination product has been achieved (Equation (40), *N*-iodosuccinimide proving superior to iodine in this case) ⟨84JOC725⟩. Titanium- or, preferably, zirconium-catalysed *trans*-carboalumination of bis(trimethylsilyl)butadiyne, followed by halogenation, has been reported to yield a (*Z*)-α-haloalkenylsilane, the stereochemical outcome presumably being due to the instability of *gem*-metallo silyl alkenes noted above (Equation (41)) ⟨90BCJ1947⟩. Similar halogenation of zirconacycles has been accomplished, ⟨89JA3336⟩ (see also ⟨88JA5383⟩ and references cited therein, for further examples of halogenation of the products derived from zirconocene-promoted carbometallation).

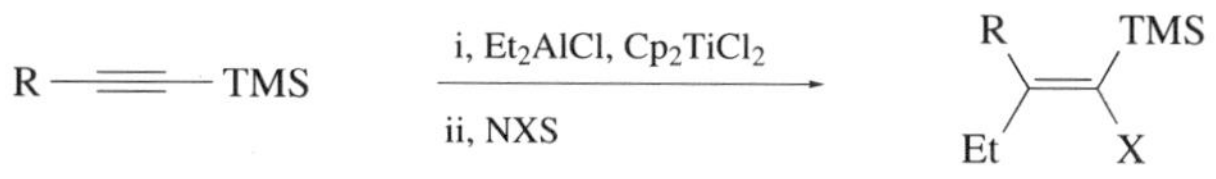

R—≡—TMS
i, Et_2AlCl, Cp_2TiCl_2
ii, NXS
R
TMS
Et
X

(40)

R = Bu^n, *c*-C_6H_{11}
X = Cl, Br, I

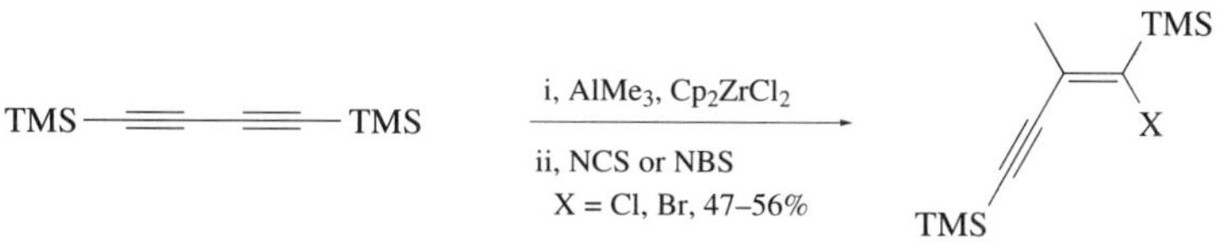

TMS
TMS
i, $AlMe_3$, Cp_2ZrCl_2
ii, NCS or NBS
X = Cl, Br, 47–56%
TMS
X
TMS

(41)

Allylmetallation-iodination has been reported by Negishi and Miller, who added allylzinc bromide to 1-trimethylsilyl-1-octyne followed by iodine. The (*Z*)-1,4-diene (**37**) was produced in 83% yield with a significant amount of (*E*) isomer (15%) ⟨83JA6761⟩. Allylzinc bromide has also been added to a lithiated propynol ether (Scheme 29). The resultant dimetal derivative was treated with a wide range of electrophiles prior to being quenched by the rapid addition of iodine crystals. *gem*-Iodo silyl alkenes were obtained in good yields (50–78%), except that some vinylzinc intermediates failed to react with elemental iodine, yielding hydrolysis products instead ⟨92JOC6903⟩.

Finally, Bogoradovskii *et al.* reported an interesting series of reactions in which a halogen atom

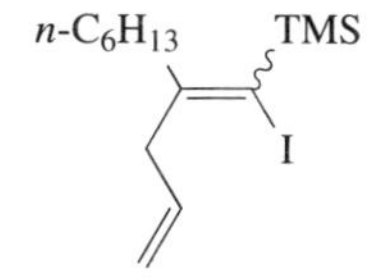

n-C_6H_{13}
TMS
I

(37)

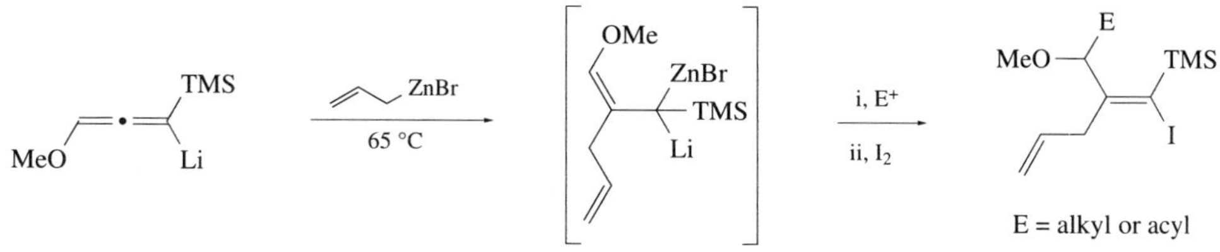

TMS
MeO
Li
ZnBr
65 °C
OMe
ZnBr
TMS
Li
i, E^+
ii, I_2
E
MeO
TMS
I
E = alkyl or acyl

Scheme 29

replaced a tin atom in propadienes. Chlorine and bromine gave mixtures of propadienes and propynes that were not separated, but iodine gave exclusive destannylation in very high yield (compounds (**38**), 89.7% and 86.3% respectively) ⟨91JGU1306⟩.

$$X = TMS, I$$
(**38**)

4.18.3.1.5 By addition of alkyl halides to alkynylsilanes

This final section covers those reactions in which the halogen atom is derived from an alkyl halide. Addition is promoted by transition metal catalysis or by radical-initiated chain reaction. Although alkyl bromides have been used, this sequence is almost exclusive to the transfer of alkyl iodides. The bulk of reported reactions employ trimethylsilylalkyne, with some reports providing a few examples of internal cyclisation onto a substituted alkynylsilane. Notable is the surprising but consistent reversal of selectivity between alkyl transfer (*cis*-selective) and perfluoroalkyl transfer (*trans*-selective), regardless of the method of catalysis.

Allyl bromide has been added to trimethylsilylalkyne with the aid of a palladium(II) catalyst, providing exclusively the (*Z*) isomer in high yield (Equation (42)) ⟨82SC1027, 88TL5811⟩. Palladium(0) catalysis has also afforded (*E,Z*)-*gem*-iodo silyl alkene isomers bearing perfluoroalkyl groups (Equation (43)) ⟨86CL1895⟩. Perfluoroalkyl groups have similarly been transferred with the aid of a variety of metal carbonyl compounds (iron, cobalt and ruthenium). Iron pentacarbonyl was the most effective catalyst, allowing the addition of perfluoroalkyl iodides to trimethylsilylalkyne in 92–93% yield, to give 3:1 (*E*):(*Z*) mixtures of isomers (**39**). Perfluoroalkyl bromides were also transferred, in more modest yield (26%) but with a much higher (*E*)-selectivity (>20:1) ⟨84TL303⟩.

$$\equiv\!-\text{TMS} \xrightarrow[\substack{\text{PdBr}_2(\text{PhCN})_2 \\ 0\,°\text{C},\,20\,\text{h} \\ 94\%}]{\diagup\!\!\diagdown\text{Br}} \quad \text{(TMS/Br alkene)} \tag{42}$$

$$\equiv\!-\text{TMS} \xrightarrow[\substack{\text{Pd(PPh}_3)_4 \\ 60\text{–}67\,°\text{C},\,24\,\text{h}}]{\text{CF}_3(\text{CF}_2)_n\text{I}} \quad F_3C\!-\!(CF_2)_n\,\,\text{TMS alkene with I}$$

(*E*):(*Z*) 70:30
n = 3, 68%
n = 5, 76%
(43)

$$F_3C\!-\!(CF_2)_n\,\,\text{TMS, I}$$

n = 2 and 7
(**39**)

Di-*t*-butyl peroxide has been used to initiate radical addition of perfluorinated α,ω-diiodo alkanes to trimethylsilylalkyne at elevated temperature (120 °C). Only the longer chain diiodides (hexadiyl or greater) gave high yields of mixed (*E*) and (*Z*) isomers (**40**) (up to 92%) ⟨82JOC2251⟩. A series of reactions has been reported in which a solution of triethylborane in hexane was used as the radical initiator, allowing the (*Z*)-specific addition of a wide range of alkyl iodides to trimethylsilylalkyne, affording compounds (**41**; R = alkyl or cycloalkyl) in 66–88% yields. The (*Z*)-specificity is noteworthy and is due to post-addition isomerisation to the more thermodynamically stable isomer. When (*E*)-α-iodoalkenylsilanes were treated with a solution of triethylborane in hexane, the corresponding (*Z*) isomers were formed quantitatively, representing an alternative to *t*-butyllithium for this isomerisation (⟨81JOC1292, 89TL3155⟩, see also ⟨92TL7031⟩). It should be noted that oxygen may be necessary for triethylboron to behave as a radical initiator ⟨94TA961⟩.

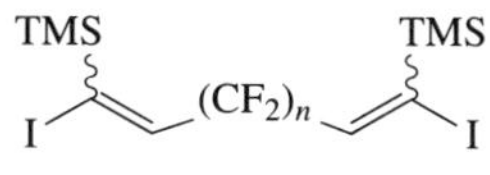
TMS
TMS
$(CF_2)_n$
I
I

$n = 6, 8, 10, 12$

(40)

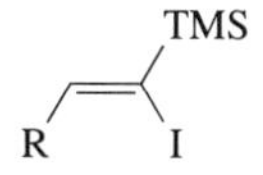
TMS
R
I

(41)

Benzoyl peroxide-initiated radical cyclisation has led to exocyclic *gem*-iodo silyl alkenes (Equation (44)) (⟨93T4229⟩ and references cited therein). Curran and Chang have reported the use of hexa-alkylditin derivatives to initiate very similar cyclisations ⟨89JOC3140⟩. Rate and product distribution studies have shown that a terminal alkenyl radical inverts much more rapidly than subsequent quenching by iodine atoms. A consequence of this is that the same product ratio is observed when a radical is generated from either a primary or a tertiary centre in a complementary pair of molecules (Scheme 30) ⟨89JA6265⟩.

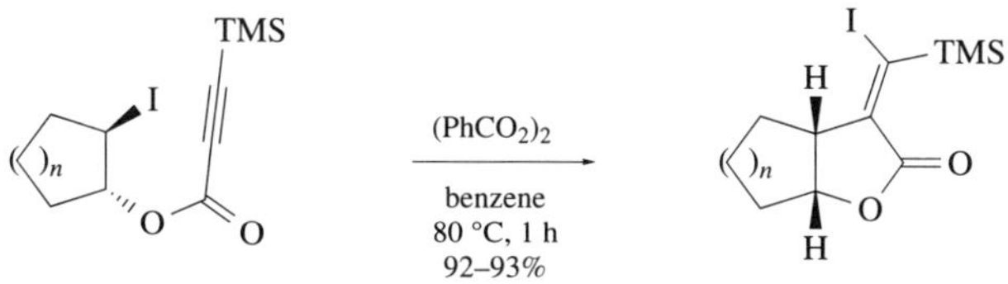
TMS
I
$(\;)_n$
O
O
$(PhCO_2)_2$
benzene
80 °C, 1 h
92–93%
I
TMS
H
$(\;)_n$
O
O
H

$n = 1, 2$ (44)

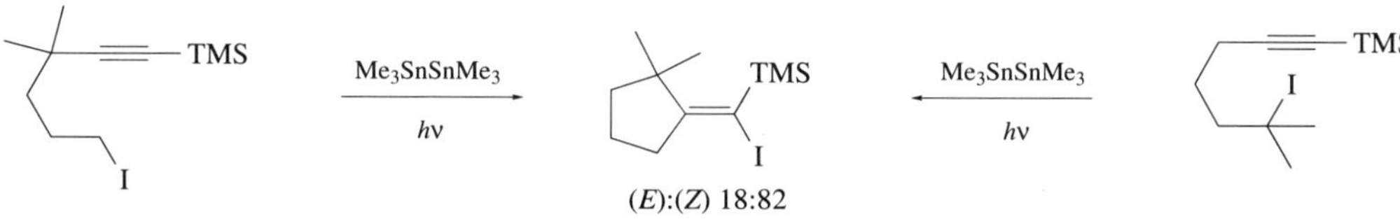
TMS
I
$Me_3SnSnMe_3$
hν
TMS
I
$Me_3SnSnMe_3$
hν
TMS
I

(*E*):(*Z*) 18:82

Scheme 30

4.18.3.2 α-Haloalkenylgermanium Derivatives

The preparation of *gem*-germyl halo alkenes closely follows that of the corresponding silicon compounds, but far fewer examples are known. Trifluorovinylmagnesium halides react with alkyl- and arylgermanium halides, giving the corresponding trifluorovinylgermanium derivatives **(42)**, while triphenylgermyllithium displaces fluoride from a trifluorovinylsilane, affording (*E*) products **(43)** in 55% yield ⟨60AG77, 62IC78, 68CJC2165⟩.

$(CF_2{=}CF)_nGeR_{4-n}$

$n = 1, 2$

(42)

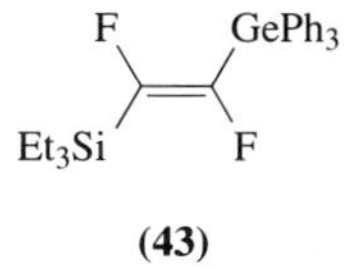
F
$GePh_3$
Et_3Si
F

(43)

Bromination of vinylgermanes, followed by base-induced elimination, has been used to produce *gem*-bromo germyl alkenes, for example compound **(44)** ⟨76JOM(122)31, 84TL3221⟩. 1-Bromo-1-(trimethylgermyl)propadiene **(45)** is formed by bromodesilylation of a (1-germylpropargyl)silane ⟨91JGU1295⟩.

$$\underset{(44)}{\overset{\text{GePh}_3}{\underset{\text{Br}}{\diagup\!\!\!\!\diagdown}}} \qquad \underset{(45)}{\overset{\text{GeMe}_3}{\underset{\text{Br}}{=\!\!\bullet\!\!=}}}$$

Germyl alkynes have been shown to undergo hydroalumination and hydroboration, but subsequent halogenation has not yet been reported ⟨84TL3221⟩. Voronkhov *et al.* have published a very short note concerning the photolytic addition of perfluoroalkyl iodides to trialkylgermylacetylenes. *gem*-Iodo trialkylgermyl alkenes, such as compound (46), were reported to be formed in good yield with the same (*E*)-preference as observed with the alkynylsilanes ⟨79JGU625⟩.

$$\underset{\substack{(E):(Z)\ 2:1 \\ (46)}}{\overset{\text{F}_3\text{C}\quad\text{GeMe}_3}{\underset{\text{I}}{\diagup\!\!\!\!=\!\!\!\!\diagdown}}}$$

4.18.3.3 α-Haloalkenyl Derivatives of Tin and Lead

The synthesis of *gem*-halo stannyl alkenes follows the preparation of the corresponding silicon and germanium derivatives, but is limited by their greater tendency to undergo halodestannylation. Unlike germanium, however, the preparation of tin compounds is rarely referred to in the same publication as the preparation of silicon compounds, emphasising the increasing differences between them.

α-Fluorovinyltin compounds were first prepared by stannylation of trifluorovinylmagnesium halides. The optimum method was found to be Barbier-type addition of a mixture of the tin halide and bromotrifluoroethene to magnesium turnings suspended in THF at −15 °C. α-Fluorovinyllithium species have also been employed, either technique affording a range of compounds (47) ⟨60AG77, 60JA6232, 61JOC2934, 61MI 418-01, 62JA4266⟩ (see ⟨82JFC(20)699⟩ for a longer chain perfluoroalkenyl derivative). An approach reported in 1991 involves the reaction of an α-fluoroalkenyl phenyl sulfone with 2 equivalents of tributyltin hydride and a radical initiator (Equation (45)). α-Fluoroalkenyltin derivatives were prepared with retention of configuration about the double bond ⟨91JA7439⟩. A silyl derivative (48) has been prepared by a similar route and employed as an α-fluorovinyl anion equivalent ⟨94TL1027⟩.

$$(CF_2\!=\!CF)_n SnR_{4-n}$$

R = alkyl, vinyl
n = 1–4
(47)

$$\underset{\text{AIBN} = 2,2'\text{-azobisisobutyronitrile}}{\text{Ph}\diagdown\!\!\!\!\diagup\!\!\overset{\text{SO}_2\text{Ph}}{\underset{\text{F}}{=}}} \xrightarrow[\text{AIBN, C}_6\text{H}_6]{2\text{Bu}_3\text{SnH}} \underset{\substack{(E)\ 82\% \\ (Z)\ 76\%}}{\text{Ph}\diagdown\!\!\!\!\diagup\!\!\overset{\text{SnBu}_3}{\underset{\text{F}}{=}}} \qquad (45)$$

$$\underset{(48)}{\overset{\text{SnBu}_3}{\underset{\text{TMS}\quad\text{F}}{\diagup\!\!\!=\!\!\!\diagdown}}}$$

Palladium-catalysed *cis*-addition of tributyltin hydride to 1-chlorooctyne has yielded an (*E*)-*gem*-chloro stannyl alkene in 73% yield (Equation (46)). By contrast, alkynyl bromides yield primarily

(*E*)-alkenylstannanes by concomitant reduction of the assumed intermediate *gem*-bromo stannyl alkene ⟨90JOC1857⟩.

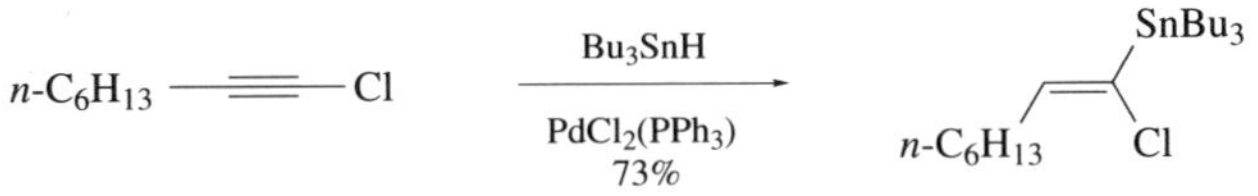

$$n\text{-}C_6H_{13}\text{—}\equiv\text{—Cl} \xrightarrow[73\%]{\text{Bu}_3\text{SnH}, \text{PdCl}_2(\text{PPh}_3)} \underset{n\text{-}C_6H_{13}}{\overset{\text{SnBu}_3}{\diagdown}}\!\!=\!\!\underset{\text{Cl}}{\diagup} \qquad (46)$$

Bromine and iodine have been used to replace a stannyl group in unsaturated bisstannyl species. Monohalogenation of *gem*-distannyl alkenes has been achieved with iodine or NBS at low temperatures, while NCS chlorinates only on heating (Scheme 31). Iodine always gives some *gem*-diiodo alkene. The bromo and iodo compounds isomerised to the less crowded isomer upon UV irradiation or thermolysis, in the latter case even in the presence of magnesium, when no Grignard reagent was observed ⟨83JOM(256)37, 86OM1991⟩. In a fashion similar to the corresponding silicon and germanium species, γ-halodestannylation of a 1,3-bis(trimethylstannyl)propyne leads to *gem*-halo(trimethylstannyl)propadienes (**49**) ⟨91JGU1295⟩. Free-radical addition of alkyl bromides has been briefly described by Voronkov *et al.* who reported the UV-promoted *cis*-addition of bromotrichloromethane to ethynyltin derivatives, affording *gem*-bromo stannanes (**50**) in 78–80% yield ⟨73JGU679⟩. In a similar fashion, triethylborane has been used to catalyse the free radical *cis*-addition of alkyl iodides to ethynyltriphenyltin, affording moderate yields of (*Z*)-adducts (**51**), in an analogous fashion to the corresponding silicon derivatives ⟨89TL3155⟩.

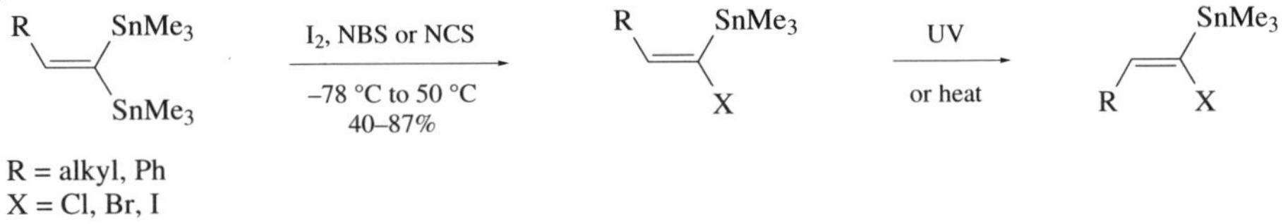

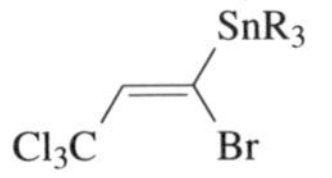

Scheme 31

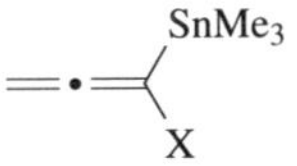

(**49**) X = Cl, Br, I

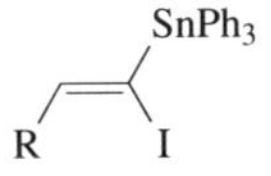

(**50**) R = Et, Pr^n, Bu^n

(**51**) R = Pr^i, Bu^t

Finally, tetrakis(trifluorovinyl)lead (**52**) is the only reported lead derivative falling within the scope of this survey and was prepared by the action of trifluorovinylmagnesium iodide upon lead(IV) salts ⟨61MI 418-01⟩.

$$(CF_2\!=\!CF)_4Pb$$

(**52**)

4.18.4 DERIVATIVES OF BORON AND OTHER GROUP 13 ELEMENTS

4.18.4.1 α-Haloalkenylboron Derivatives

4.18.4.1.1 By ligand exchange and by halogenation reactions

The first reported synthesis of an α-haloalkenylboron derivative was by Stafford and Stone ⟨60JA6238⟩ using a method followed later by Stampf and Odom ⟨76JOM(108)1⟩, both treating the same perfluorovinyltin reagent with boron halides (Equation (47)). Matteson and Beedle produced a pinanediol 1-chlorovinylboronate ester by treatment of 1,1-dichloroethane with LDA and trimethylborate, followed by *trans*-esterification and subsequent elimination (Scheme 32, 2ROH = pinanediol) ⟨90HAC135⟩. Surprisingly, this is the first example of an otherwise unsubstituted 1-haloalkenylboronic ester to appear in the literature.

$$(CF_2=CF)_2SnMe_2 \xrightarrow[\text{or } Me_nBBr_{3-n}]{BCl_3} (CF_2=CF)_nBX_{3-n} \qquad (47)$$

X = Cl, n = 1 or 2, 93% or 85%
X = Me, n = 1 or 2, 98% or 95%

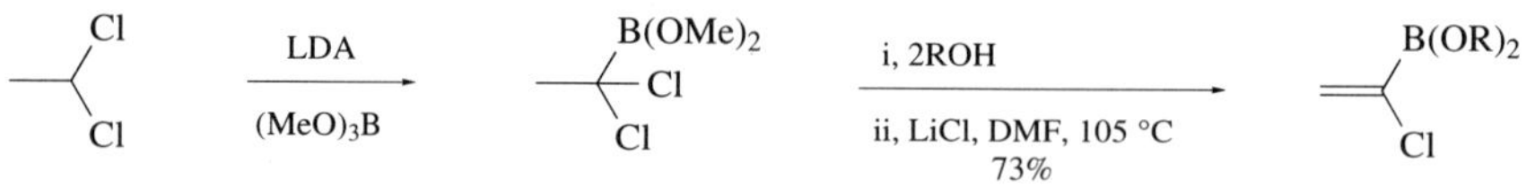

Scheme 32

Preparation of 1-haloalkenyl derivatives by halogenation of vinylboronates and subsequent elimination is limited by the relatively good electrofugal-leaving ability of tetracoordinate boron anions compared to protons. However, bromodeboronation has afforded a *gem*-bromo boroalkene, starting from a *gem*-diboro alkene, which was produced in turn by condensation of a triborylmethane anion with acetone (with a chloroborane to induce elimination) (Scheme 33) ⟨74JOM(69)53⟩. Electrophilic halogenation may take place if the substrate is stabilised, for example iodination of one of the boradiazine derivatives described by Gronowitz and Maltesson (53) ⟨75ACS(B)461⟩. Bromine has been added to an activated alkyne containing a boron atom stabilised by B–N bonds, affording a (probably (*E*)) adduct (54) in 57% yield ⟨89JGU2040⟩. Finally, radical addition to an alkyneboronate has yielded the *gem*-bromo boron alkene adducts (55) in high yield but of unknown geometry (carbon tetrachloride failed to react) ⟨63JOC369⟩.

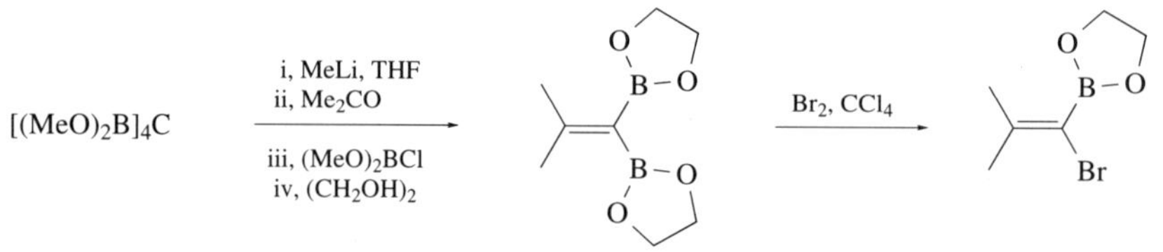

Scheme 33

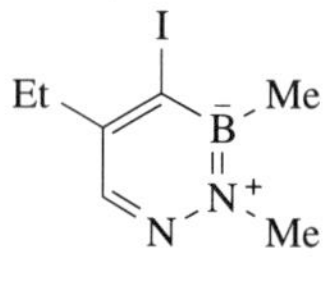

(53)

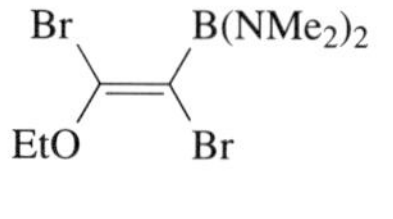

(54)

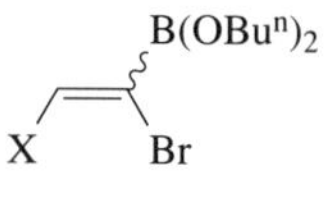

(55)

X = Br or CCl$_3$

4.18.4.1.2 By hydroboration of halo alkynes

The hydroboration of 1-bromo and 1-iodo alkynes was first reported by Zweifel and Arzoumanian using dicyclohexylborane ⟨67JA5086⟩. The products were obtained as (*Z*) isomers by exclusive *cis*-hydroboration and were described as being relatively stable in tetrahydrofuran solution (although it is worth noting that they are usually not isolated). Double hydroboration, occasionally observed with terminal alkynes, has not been reported with 1-halo alkynes (Scheme 34). The reaction has been extensively investigated and has been shown to be applicable to a wide variety of 1-halo alkynes, substituted with *t*-butyl, cycloalkyl or (more usually) *n*-alkyl chains, but with the notable exception of phenyl. The halogen atom may be chlorine, bromine or iodine, reactivity increasing in that order. 1-Halo-3-enynes undergo *cis*-hydroboration of the halo alkyne portion in preference to that of the alkene (Equation (48)) ⟨77JA5192, 93SC2937⟩.

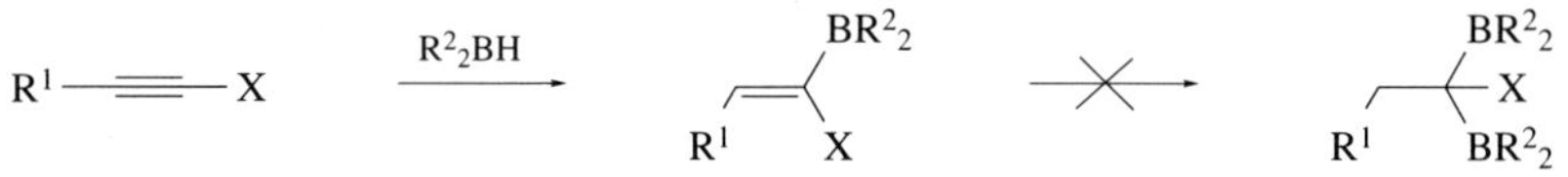

$$R^1 \equiv\!\!\!\equiv X \xrightarrow{R^2_2BH} \underset{R^1 \quad X}{\overset{BR^2_2}{\diagup\!\!\!\diagdown}} \xrightarrow{\times} \underset{R^1 \quad BR^2_2}{\overset{BR^2_2}{\diagup\!\!\!\diagdown X}}$$

R^1, R^2 = alkyl, cycloalkyl
X = Cl, Br, I

Scheme 34

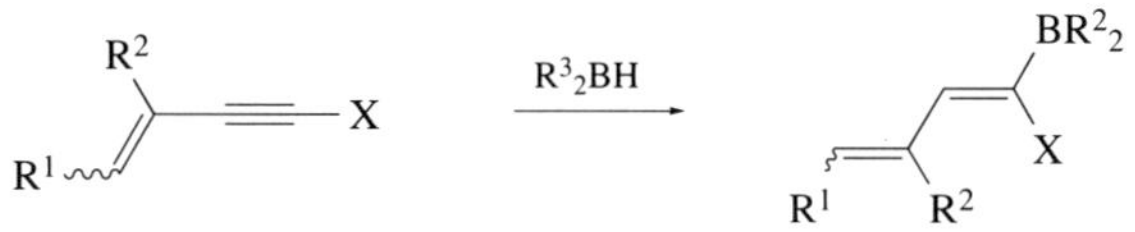

$$\underset{R^1}{\overset{R^2}{\diagup\!\!\!\diagdown}}\!\!\!\equiv\!\!\!\equiv X \xrightarrow{R^3_2BH} \text{diene product with } BR^2_2, X \tag{48}$$

R^1, R^2, R^3 = alkyl, cycloalkyl
X = Br, I

Thexylborane has been used in an attempt to achieve selectivity in subsequent alkyl group migrations; however, it is prone to undergo double alkenylation, leading to bis-1-haloalkenylboron derivatives (**56**) as studied by Negishi *et al.* ⟨91T343⟩. Chlorothexylborane has been developed by Brown as an alternative, stable reagent, activated by *in situ* reduction ⟨82S193, 82S195⟩. Dithexylborane has also been used ⟨88BSJ3764⟩. A number of reactions have been carried out in which mono- or dibromoborane dimethylsulfide complexes add to halo alkynes, producing compounds such as (**57**; Y = Br), which may be taken on to boronic or borinic acids or esters ⟨82JOC3808, 84OM1392⟩. Another route to derivatives such as (**57**) is by the reaction of halo alkynes with isopinocamphenylborane (IpcBH$_2$) or diisopinocampheylborane (Ipc$_2$BH), producing for example (**57**; Y = Ipc), then displacing the isopinocampheyl units with acetaldehyde (acting as a hydride acceptor). Boronic ethyl esters (**57**; Y = OEt), and the corresponding borinic esters, are thus formed directly, without need for hydroxylic reagents and without generating acidic by-products ⟨91TA277, 93SC2851⟩.

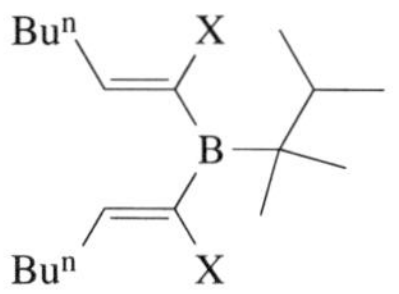

(**56**) X = Cl, Br, I

(**57**) R = alkyl, cycloalkyl
X = Br, I
Y = Br, OEt, Ipc

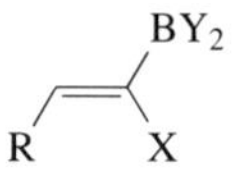

Finally, a series of reactions carried out by Knochel and co-workers emphasises the stability of these agents. After an ω-chloro-α-iodoalkenylboronate ester had been prepared by the dibromoborane method outlined above (80% yield after column chromatography), the chlorine atom was replaced by iodine (81% yield), then the product was selectively converted into the zinc and then copper organometallic species prior to cyanation with tosyl cyanide (60% overall yield, Equation (49)) ⟨92TL3717⟩.

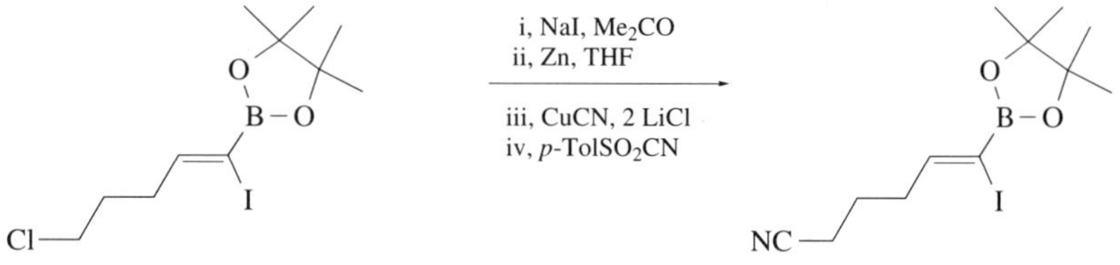

$$\text{Cl-(CH}_2\text{)}_n\text{-C(I)=CH-B(pin)} \xrightarrow[\substack{\text{iii, CuCN, 2 LiCl} \\ \text{iv, }p\text{-TolSO}_2\text{CN}}]{\substack{\text{i, NaI, Me}_2\text{CO} \\ \text{ii, Zn, THF}}} \text{NC-(CH}_2\text{)}_n\text{-C(I)=CH-B(pin)} \tag{49}$$

4.18.4.2 α-Haloalkenyl Derivatives of Aluminum, Gallium, Indium and Thallium

α-Halovinylaluminum compounds are not commonly encountered. A compound formulated as tris(trifluorovinyl)aluminum trimethylamine complex (**58**) was prepared by the reaction of the corresponding alane complex with bis(trifluorovinyl)mercury. The compound is moderately stable in the absence of air. It was described as being highly associated in solution ⟨61JA2202⟩.

$$(CF_2=CF)_3Al\cdot NMe_3$$
(58)

Aluminum hydride reduction of 1-halo alkynes has yielded *trans*-halo alkenes after workup; these transformations probably involve *gem*-halo aluminum alkenes. Detailed studies were first undertaken by Zweifel *et al.* who treated a series of chloro alkynes with lithium aluminum hydride to produce compounds formulated as lithium alanates (**59**) by *trans*-addition ⟨79JA5101⟩. The products were described as being moderately stable at 0 °C and were characterised by their subsequent reactions. In one case reduction of the chlorine atom occurred (giving *t*-butylacetylene in 23% yield). A later paper described the addition of trialkylaluminum hydrides to chloro alkynes, affording the corresponding α-chloroalkenylaluminum anions (**60**), characterised by proton NMR spectroscopy ⟨89JOC998⟩.

Finally, the author is unaware of any reports of α-haloalkenylgallium, -indium or -thallium derivatives.

$$\text{(59)} \quad R\text{-C(AlH}_3\bar{)}=\text{CCl} \quad Li^+ \qquad \text{(60)} \quad R\text{-C(AlR}^1{}_3\bar{)}=\text{CCl} \quad Na^+$$

4.18.5 DERIVATIVES OF LITHIUM AND OTHER GROUP 1 AND GROUP 2 METALS

Derivatives of lithium and other group 1 and group 2 metals, and α-haloalkenyl metals in general, constitute examples of what have been described as carbenoids, specifically alkylidene carbenoids ⟨64JA4042⟩. Such species may exhibit carbene-like behaviour but with different selectivity to that found for free carbenes (for lead references to reviews see ⟨88JOC3089, 93AG(E)1023, 93JOC546⟩). Although the term 'carbenoid' has been applied to the compounds themselves (usually to the α-chloro, bromo or iodo alkenes, rarely to the fluoro derivatives), it may be that this expression should denote the electrophilic aspects of their reactivity ⟨72AG(E)473, 88JOC3089⟩. These compounds readily behave as ordinary organometallic nucleophiles at sufficiently low temperatures (generally below $-80\,°C$ for the lithium species, while other metals may tolerate much higher temperatures, for example tin 'carbenoids' may be distilled *in vacuo*). Examples of their preparation and nucleophilic behaviour are to be found in other areas of this chapter (especially for silicon, see Section 4.18.3; also phosphorus, see Section 4.18.2 and the transition metals, see Section 4.18.6). A review of fluorinated vinyl organometallic reagents has been published ⟨94T2993⟩.

The alkylidene carbenoids of lithium are the most thoroughly studied of the group 1 and 2 species. Carbon-13 NMR spectra have been obtained by Seebach and co-workers, illustrating chemical shift differences, *deshielded* relative to the unlithiated vinyl halide, and also large C—Li coupling constants relative to unhalogenated vinyllithium species ⟨83HCA308⟩. Theoretical calculations have been performed by Schleyer and co-workers ⟨84JA6467⟩. In 1993, an x-ray crystal structure determination was carried out on a THF.TMEDA complex of 1-chloro-2,2-bis(4-chlorophenyl)vinyllithium (**61**), held at $-115\,°C$ ⟨93AG(E)1032⟩. These methods agree that the C—Li bonds have more *s*-character while the bonds to the halogen atoms carry more *p*-character than predicted from a superficial study of their formulae. The crystal structure mentioned above has a shorter than expected C—Li bond, bent towards the axis of the alkene C—C bond, but a longer C—Cl bond adopting a position more nearly at right angles to the line of the alkene bond. Similarly, the two aryl groups were found to be bent towards the lithium atom and away from the chlorine atom, respectively. It is interesting to note that the most stable carbenoids generally carry the metal atom *cis* to the most bulky group, especially where that group is an aryl ring (see, e.g., Equation (50)).

$$\underset{F}{\overset{Ph}{>}}\!\!=\!\!\underset{Cl}{\overset{Cl}{<}} \quad \xrightarrow[\text{Et}_2\text{O}]{\text{Bu}^n\text{Li}} \quad \underset{F}{\overset{Ph}{>}}\!\!=\!\!\underset{Cl}{\overset{Li}{<}} \qquad (50)$$

4.18.5.1 α-Haloalkenyllithium Derivatives

The presence of a halogen atom on an alkene makes a geminal proton more acidic and also stabilises the resultant *gem*-halo lithio alkene. These compounds are therefore generally accessible

(61)

through metallation as well as lithium–halogen exchange, in either case using strong bases such as the butyllithiums or, for metallation, lithium amides. The choice of solvent is important, THF accelerating either reaction but having unpredictable effects on the stability of the product.

Routes of thermal decomposition are often hard to predict as apparently similar structures tend to react by different pathways, making direct comparison difficult. Potential leaving groups in the β-position may undergo elimination to yield a halo alkyne, although subsequent lithium–halogen exchange generally leads to a lithio alkyne. These reactions may be suppressed by low temperatures and β-chloro, fluoro and alkoxy groups are commonly found in carbenoids ⟨72AG(E)473, 93JA5430⟩.

It is worth noting that quenching a carbenoid with an added electrophile may be slower than lithium–halogen exchange in the resultant vinyl halide (particularly for the bromide or iodides, or if the added group is itself anion-stabilising). The presence of excess butyllithium, or in unfavourable cases the carbenoid itself, may thus allow further exchange, loosing the desired halogenated product and giving the appearance of a *gem*-dilithio alkene ⟨72AG(E)473⟩.

4.18.5.1.1 α-Fluoroalkenyllithium compounds

Early work was carried out by Tarrant and co-workers who employed metallation (trifluoroethene) and lithium–halogen exchange (trifluorovinyl bromide) with alkyllithium reagents in ether ⟨68JOC286⟩. Normant, Sauvêtre and co-workers also used lithium–halogen exchange, on the more readily available trifluorovinyl chloride, and found that the stability of the product was enhanced by the absence of THF. As lithium–chlorine exchange may be prohibitively slow in purely ethereal solution, they found an acceptable compromise in keeping the amount of THF to low levels ⟨90JOM(400)19⟩.

Delavarenne and Viehe reported the use of lithium diethylamide to metallate 1,1-dichloro-2-fluoroethene (ether, −78 °C) ⟨70CB1198⟩. Hiyama *et al.* carried out lithium–chlorine exchange on trifluorovinyl chloride at very low temperatures in the presence of chlorosilanes ⟨84CL1765⟩ (see Section 4.18.3.1.2 for further details). Finally, a paper by Yagupol'skii *et al.* showed that the rates of metallation and of lithium–chlorine exchange were comparable when the isomers of 1,2-difluorovinyl chloride were treated with butyllithium (Equation (51)) ⟨87JOU246⟩.

$$\text{F-CH=CCl-F} \xrightarrow{\text{BuLi}} \text{F-CH=CLi-F} + \text{F-CCl=CLi-F} \quad (51)$$

67 : 33

4.18.5.1.2 α-Chloroalkenyllithium compounds

Köbrich and Trapp first reported the metallation of 1-chloro-2,2-diarylethenes (to give products such as (**61**)), using *n*-butyllithium at low temperatures ⟨63ZN(B)1125⟩. The area has since been reviewed ⟨67AG(E)41, 72AG(E)473, 90JOM(400)19⟩.

Ficini and Depezay treated (*E*)- and (*Z*)-2-ethoxyvinyl chlorides with *n*-butyllithium at −100 °C, affording the corresponding lithiated enol ethers (**62**) and (**63**) stereospecifically ⟨68TL937⟩. Shimizu *et al.* added chloro (and bromo) alkenes and enol ethers to a mixture of LDA and TMS-Cl ⟨87BSJ777⟩ (see Section 4.18.3.1.2 for further details).

(62)   (63)

An example of lithium–chlorine exchange is presented in Equation (50), the (*E*)-chlorine atom being selectively exchanged in a 94:6 ratio ⟨90JOM(400)19⟩.

4.18.5.1.3 α-Bromoalkenyllithium compounds

gem-Bromo lithio alkenes have not been as intensively studied as the corresponding fluoro and chloro species. Attempts to metallate a monobromo alkene with butyllithium may lead to competition from lithium–bromine exchange, affording non-halogenated products. Although (*E*)-2-ethoxyvinyl bromide gives the corresponding *gem*-bromo lithio enol ether (**64**), the (*Z*) isomer, unlike (*Z*)-ethoxyvinyl chloride (above), affords only (*Z*)-2-ethoxyvinyllithium (**65**) ⟨78JOC1595⟩. Metallation is nevertheless accelerated by the presence of the bromine atom and there is a fine balance of reactivity. Thus the (*E*)-3-bromo-3-lithioacrylate salt (**66**) is formed by metallation while the chiral bromo alkene (**67**) affords the products of lithium–bromine exchange in preference to metallation, especially in THF (Equation (52)) ⟨85JOC2195, 93JOC546⟩. Lithium amides may provide alternative bases in difficult cases.

(64)   (65)   (66)

(67)
58 : 17 (Et$_2$O)
90 : 5 (THF)

(52)

Lithium–bromine exchange on a *gem*-dibromo alkene is a very fast reaction, even at low temperatures, and usually affords isomers when the alkene substituents are not symmetrical. Treatment of 2,2-dibromovinyl ethyl ether with *n*-butyllithium (ether, −78 °C) gave mixed *gem*-bromo lithio enol ethers without the stereospecificity associated with metallation (Equation (53)) ⟨83JOC2095⟩.

(53)

The question of kinetic vs. thermodynamic control of *gem*-dibromo alkene exchange has occasionally been raised. Mahler and Braun showed that a high level of chelated (*E*) isomer (**68**) could be obtained if the lithium–bromine exchange was carried out with a deficiency of *n*-butyllithium (0.95 equivalents in ether, −105 °C, 30 min). In the presence of excess *n*-butyllithium a high proportion (2:1) of the non-chelated (*Z*) isomer was formed and this ratio did not change after prolonged reaction. The presence of the (*Z*) isomer was considered to be due to exchange of the less hindered bromine atom under conditions favouring kinetic control. Excess *gem*-dibromo alkene thus allows lithium exchange resulting in thermodynamic control of the product mixture. It is worth noting that the shorter ether (**69**) did not allow such complete control, emphasising the need for polyether complexation ⟨91CB1379⟩. An attempt to differentiate kinetic from thermodynamic control has also been reported by Harada *et al.*, using the same principle of excess *n*-butyllithium (kinetic) vs. excess *gem*-dibromo alkene (thermodynamic), in this case using 1-alkyl-2,2-dibromostyrene derivatives. 1-Ethyl- and 1-methoxymethylstyrenes gave lithium–bromine exchange *cis* to the phenyl group under thermodynamic control (Equation (54)). The two substrates behaved similarly, again illustrating the marginal effect that a single oxygen atom has upon lithium–bromine exchange ⟨92JOC5805⟩.

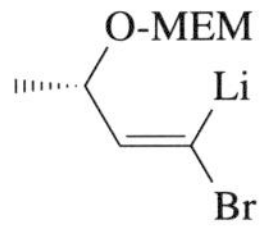

(68) 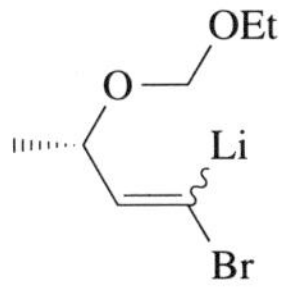 (69)

MEM = methoxyethoxymethyl

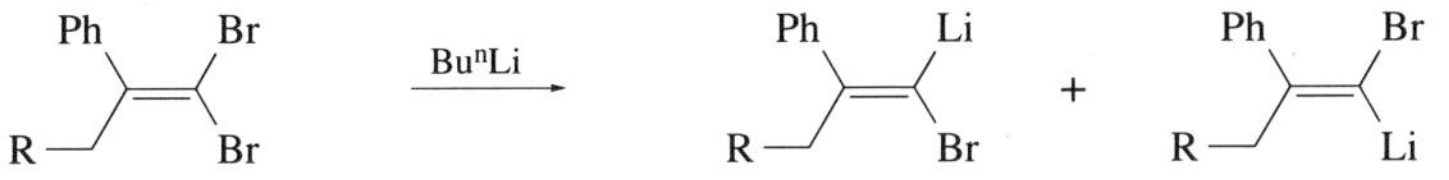

$$\text{R}=\text{Me, OMe} \quad \xrightarrow{\text{Bu}^n\text{Li}} \quad \text{R}=\text{Me} \quad 40 : 60 \quad 2.0 \text{ equiv, Bu}^n\text{Li} \quad (54)$$
$$\qquad\qquad\qquad\qquad\qquad\qquad\qquad 97 : 3 \quad 0.5 \text{ equiv. Bu}^n\text{Li}$$

4.18.5.1.4 α-Iodoalkenyllithium compounds

There is only one report of a stable solution of a *gem*-iodo lithio alkene. Treatment of the geminal diiodides of acetophenone enol ethers with an excess of *s*-butyllithium produced (Z)-*gem*-iodo lithio alkene solutions which were stable at up to −20 °C and which reacted smoothly with an excess of electrophile (Equation (55)). Treatment of the intermediate carbenoid with an excess of methyl-lithium, followed by an electrophile, produced a *gem*-disubstituted product, again in high yield ⟨88JA5567⟩. The latter reaction was shown not to involve an example of a *gem*-dilithio alkene ⟨93AG(E)1023⟩.

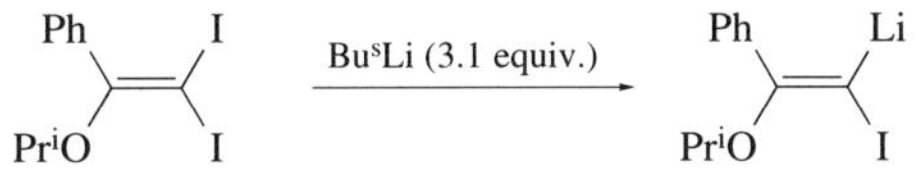

$$\xrightarrow{\text{Bu}^s\text{Li (3.1 equiv.)}} \qquad (55)$$

4.18.5.2 α-Haloalkenyl Derivatives of Sodium and Potassium

The greater reactivity of the other alkali metals has long been recognised and there have been very few attempts to use them in this series. Two reports are worth noting, however. Delavarenne patented the use of sodium amide in the preparation of the trifluorovinyl anion (70). He reported the use of a 10% mixture of hexamethylphosphoramide (HMPA) in THF to produce a stable solution at −10 °C. Addition to ketones took place in 30–82% yield ⟨73USP3751492⟩.

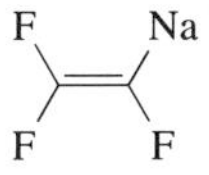

(70)

Transient formation of *gem*-fluoro potassio alkenes possibly accounts for the desilylation of *gem*-fluoro silyl alkenes with potassium fluoride in wet DMSO (Scheme 35). The reaction takes place with complete retention of configuration and constitutes the most efficient route to monosubstituted *trans*-difluoro alkenes ⟨84JOM(264)155, 85JA4085⟩.

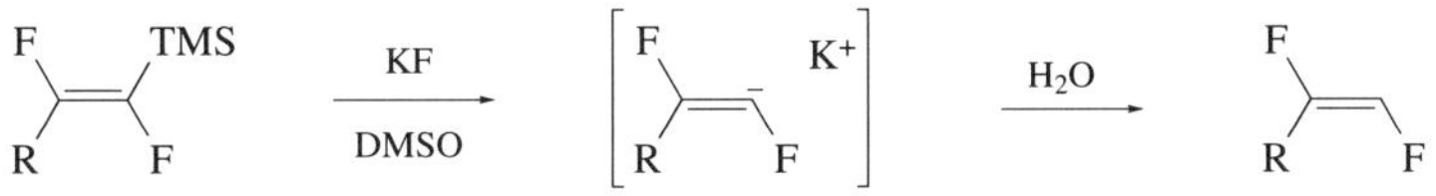

R = alkyl, alkenyl, aryl

Scheme 35

4.18.5.3 α-Haloalkenylmagnesium Derivatives

Trifluorovinylmagnesium iodide was first prepared by Park *et al.*, using magnesium metal in ether, along with 1-chloro- and 1-bromo-2,2-difluorovinylmagnesium iodides (Equation (56)) ⟨58JOC1661⟩. More recently, a longer chain perfluorovinyl bromide has been converted into the Grignard reagent (**71**) in THF ⟨76CC174⟩. The Grignard approach has been extensively employed ⟨62MI 418-01⟩ but has now been almost entirely superseded by organolithium agents.

$$\underset{F}{\overset{F}{>}}=\underset{X}{\overset{I}{<}} \quad \xrightarrow[\text{Et}_2\text{O}]{\text{Mg}} \quad \underset{F}{\overset{F}{>}}=\underset{X}{\overset{\text{MgI}}{<}} \tag{56}$$

X = F, Cl, Br

$$\underset{C_6F_{13}}{\overset{F}{>}}=\underset{F}{\overset{\text{MgBr}}{<}}$$

(**71**)

4.18.6 DERIVATIVES OF THE TRANSITION METALS

There are many reports of α-haloalkenyl derivatives of the transition elements but most concentrate only on ligand modification. Manganese and the iron group (groups 8 to 10) have been extensively studied, but chemistry of direct relevance to organic synthesis is largely limited to copper and the Group 12 elements, especially zinc. There are no reports of α-haloalkenyl derivatives of the lanthanide and actinide elements. A review of fluorinated vinyl organometallic reagents has been published ⟨94T2993⟩.

This chapter aims only to briefly summarise the available routes, concentrating on those of greatest relevance to organic synthesis.

4.18.6.1 By Transmetallation Reactions

Metal exchange between metal salts and lithium carbenoids is the most general route to α-haloalkenyl transition metals (Equation (57)). Köbrich and co-workers studied the generation and fate of a variety of α-chloroalkenyl species, some of which were surprisingly stable. Conversely, the iron(III) chloride σ-complex (**72**) was unstable and was found to dimerise to a tetrachloro diene with retention of configuration ⟨72AG(E)473⟩. Normant, Sauvêtre and co-workers concentrated on the synthesis of zinc derivatives ⟨90JOM(400)19⟩, while mercury derivatives have been made by transmetallation of both lithium and magnesium carbenoids ⟨76JGU2367, 76JOM(104)145, 82BAU2498, 82JFC(20)699⟩. Finally, an α-bromoalkenylcopper(I) ate complex (**73**) has been reported as an intermediate in the synthesis of radialenes ⟨86CC1794⟩.

$$\underset{R^2}{\overset{R^1}{>}}=\underset{X}{\overset{\text{Li}}{<}} \quad \xrightarrow[\text{THF or Et}_2\text{O}]{\text{MX}_n} \quad \underset{R^2}{\overset{R^1}{>}}=\underset{X}{\overset{\text{MX}_{n-1}}{<}} \tag{57}$$

M = Ti, Cr, Fe, Cu, Zn, Ag, Hg
X = F, Cl, Br
R^1, R^2 = H, F, Cl, Ph, alkyl, R_f

(**72**) Cl-CH=C(FeCl₂)(Cl)

(**73**) bis-bromoalkenyl CuLi complex

Transmetallation is not restricted to lithium and magnesium carbenoids. Burton and co-workers formed α-haloalkenylcopper species by treatment of vinylzinc and -cadmium compounds (see Section 4.18.6.2 and Scheme 36) with copper(I) bromide in DMF. α-Fluoro- and α-chloro-alkenylcopper species reacted normally but the α-bromo compounds dimerised to 1,2,3-butatrienes even at low temperatures ⟨86JA4229, 93JA5430⟩.

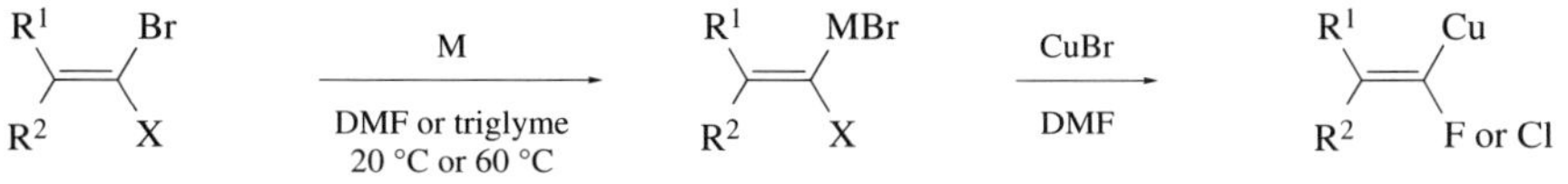

$$R^1\text{, }R^2\text{=Ph, F, }R_f$$

$M = Zn, Cd$
$X = F, Cl, Br$
$R^1, R^2 = Ph, F, R_f$

Scheme 36

4.18.6.2 From Alkenyl Halides

Transmetallation reactions using the lithium carbenoids has the disadvantage of requiring low temperatures, with consequent difficulty when the reactions are carried out on a large scale. Burton and co-workers developed a route to overcome this problem by direct treatment of a variety of fluorinated alkenyl bromides with metallic zinc and cadmium in polar aprotic solvents (Scheme 36). The reactions proceed readily, often generating a mild exotherm, but the reactants may be heated if necessary. Mercuric chloride has been used to catalyse the reaction. Zinc inserts into the less hindered bond of a geminal dibromide, with little tendency to adopt a *cis* relationship with a phenyl ring, contrary to the reactivity of lithium (compare Section 4.18.5) ⟨93JA5430⟩.

Harada, Oku and co-workers have studied the zinc–bromine exchange of alkenyl bromides with lithium tributylzincate (Equation (58)). The exchange took place at low temperatures, under kinetic control and with little (*E,Z*) selectivity. Chlorine atoms were not exchanged ⟨92JOC5805, 93JOC4897⟩. Bis(trimethylsilyl)mercury has been shown to insert into the C—Br bond of trifluorovinyl bromide, producing the silylmercury species (**74**) ⟨71TL1879⟩.

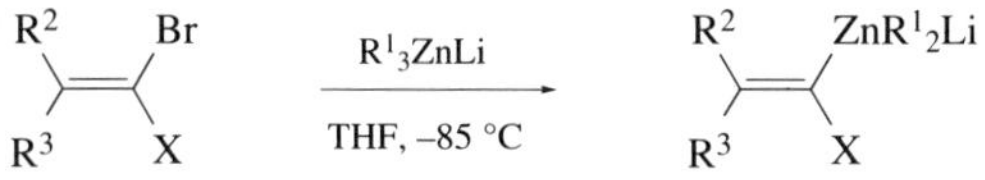

$R^1 = Bu^n, Bu^s, Bu^t$
$X = Cl, Br$
$R^2, R^3 = Ph, alkyl$ (58)

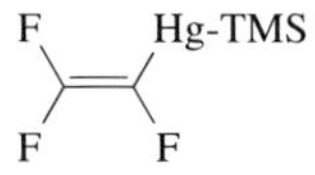

(**74**)

Low-valent metal anions react with polyfluoro alkenes to produce α-fluoroalkenyl metal species, either by net addition–elimination or by allylic substitution followed by rapid fluoride ion migration (Equation (59)) ⟨68AG(E)747, 70ACR417, 89CC1159⟩. King and Saran have described a similar net addition–elimination to 1,1-dichloro-2,2-dicyanoethene. *gem*-Chloro metallo alkenes (**75**) were produced in 43–95% yield ⟨73JA1811⟩. Sodium tetracarbonylcobaltate reacts with difluorofumaric dichloride, giving the *trans*-dicobalt species (**76**) after extrusion of carbon monoxide ⟨90AG(E)315⟩.

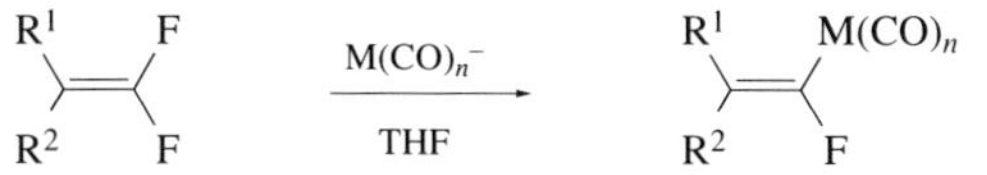

$M = Mn, FeCp, Co, Re$
$R^1, R^2 = F, CF_2=CF, R_f$ (59)

4.18.6.3 From Alkynes

α-Haloalkenyl metal compounds have been prepared from metallo alkynes or from halo alkynes. An example of the first mode of addition is that of HCl to an ethynylplatinum complex (Equation (60)). The addition is thought to proceed through a stabilised vinyl cation and is readily reversible, leading eventually to a dichloroplatinum complex and alkyne ⟨76CC200, 76JA6046⟩. An example of the second mode of addition is provided by the recently described reaction of two moles of dichloroacetylene with Group 9 carbonyl complexes. Tetrachloro metalloles, such as compound (77), were produced in 11–68% yield ⟨91CB2449⟩.

4.19
Functions Bearing Two Chalcogens

GARY N. SHELDRAKE
The Queen's University of Belfast, UK

4.19.1	INTRODUCTION	824
4.19.2	FUNCTIONS CONTAINING TWO OXYGENS $R^1_2C{=}C(OR^2)_2$, etc.	824
	4.19.2.1 Ketene Acetals—$R^1_2C{=}C(OR^2)_2$	824
	4.19.2.1.1 *From carboxylic acids and esters*	824
	4.19.2.1.2 *From ortho-esters and analogues*	825
	4.19.2.1.3 *From α-haloacetals and analogues*	826
	4.19.2.1.4 *From cycloaddition reactions of α,β-unsaturated esters*	827
	4.19.2.1.5 *By addition to 1-alkoxy alkynes*	827
	4.19.2.1.6 *By Horner–Wittig chemistry*	828
	4.19.2.1.7 *From ketenes*	829
	4.19.2.1.8 *From gem-dihalogeno alkenes and analogues*	829
	4.19.2.1.9 *By miscellaneous transformations*	830
	4.19.2.2 Other Related Ketene Derivatives—$R^1_2C{=}COR^2OX$, etc.	831
	4.19.2.2.1 *Ketene silyl acetals—$R^1_2C{=}COR^2OSiR^3_3$, etc.*	831
	4.19.2.2.2 *Boryloxy derivatives—$R^1_2C{=}COR^2OBR^3_2$, etc.*	832
	4.19.2.2.3 *Miscellaneous derivatives*	833
4.19.3	FUNCTIONS CONTAINING OXYGEN AND SULFUR—$R^1_2C{=}COR^2SR^3$, etc.	833
	4.19.3.1 Dicoordinated Sulfur Derivatives—$R^1_2C{=}COR^2SR^3$	833
	4.19.3.1.1 *From monothiocarboxylic acids and esters*	833
	4.19.3.1.2 *By alkeneation methods*	834
	4.19.3.1.3 *By miscellaneous methods*	836
	4.19.3.2 Tricoordinated Sulfur Derivatives—$R^1_2C{=}COR^2S(O)R^3$	837
	4.19.3.3 Tetracoordinated Sulfur Derivatives—$R^1_2C{=}COR^2SO_2R^3$	838
	4.19.3.3.1 *By sulfoxidation*	838
	4.19.3.3.2 *From α-sulfonyl ethers*	838
	4.19.3.3.3 *From 1-sulfonyl alkynes*	840
	4.19.3.3.4 *By miscellaneous methods*	840
4.19.4	FUNCTIONS CONTAINING OXYGEN AND EITHER SELENIUM OR TELLURIUM—$R^1_2C{=}COR^2SeR^3$, etc.	841
4.19.5	FUNCTIONS CONTAINING TWO SULFURS—$R^1_2C{=}CSR^2S(O)_2R^3$, etc.	842
	4.19.5.1 Two Dicoordinated Sulfurs—$R^1_2C{=}C(SR^2)_2$, etc.	842
	4.19.5.1.1 *From dithiocarboxylic acids and derivatives*	842
	4.19.5.1.2 *By double bond formation via elimination*	844
	4.19.5.1.3 *By alkenation methods*	847
	4.19.5.1.4 *By rearrangement of α,β-unsaturated dithioacetals*	849
	4.19.5.1.5 *From 1,3-dithioles and analogues*	850
	4.19.5.1.6 *From gem-dihalogeno alkenes and analogues*	851
	4.19.5.1.7 *By miscellaneous methods*	852
	4.19.5.2 Higher Coordinated Sulfur Derivatives by Oxidation—$R^1_2C{=}CS(O)_mR^2S(O)_nR^3$	853
	4.19.5.3 One Dicoordinated and One Higher Coordinated Sulfur—$R^1_2C{=}CSR^2S(O)R^3$, etc.	856
	4.19.5.3.1 *From methyl methylthiomethyl sulfoxide and analogues*	856
	4.19.5.3.2 *Other methods for compounds containing one tricoordinated sulfur*	858
	4.19.5.3.3 *Other methods for compounds containing one tetracoordinated sulfur*	861

4.19.5.4 Two Tricoordinated Sulfurs—$R^1_2C=C[S(O)R^2]_2$, etc.		862
4.19.5.5 One Tricoordinated and One Tetracoordinated Sulfur—$R^1_2C=CS(O)R^2S(O)_2R^3$, etc.		864
4.19.5.6 Two Tetracoordinated Sulfurs—$R^1_2C=C[S(O)_2R^2]_2$, etc.		864
4.19.5.6.1 From disulfonylmethanes		864
4.19.5.6.2 By miscellaneous methods		866
4.19.6 FUNCTIONS CONTAINING SULFUR AND EITHER SELENIUM OR TELLURIUM—$R^1_2C=CSR^2SeR^3$, etc.		868
4.19.6.1 Dicoordinated Sulfur Derivatives		868
4.19.6.1.1 Selenium derivatives		868
4.19.6.1.2 Tellurium derivatives		872
4.19.6.2 Tri- and Tetracoordinated Sulfur Derivatives		873
4.19.7 FUNCTIONS CONTAINING SELENIUM AND/OR TELLURIUM—$R^1_2C=C(SeR^2)_2$, etc.		874
4.19.7.1 Diselenium Derivatives		874
4.19.7.1.1 Tetraselenafulvalenes and analogues		874
4.19.7.1.2 Dimerisation of lithium ethyneselenolates		874
4.19.7.1.3 Miscellaneous methods		875
4.19.7.2 Other Derivatives		876

4.19.1 INTRODUCTION

This review covers compounds in which there is a true carbon–carbon double bond with one of the carbons bearing two chalcogen atoms. For the most part, compounds in which one or both of the chalcogens form part of an aromatic heterocyclic ring have not been included. The preparation of these compounds is more logically accessed through reference works such as *Comprehensive Heterocyclic Chemistry*. Also, compounds which technically achieve the defined substructure as a minor part of a tautomeric mixture, or by coordination to a metal ion, are considered to lie outside the scope of this chapter. The literature has been reviewed up to entries appearing in *Chemical Abstracts* in July 1993.

4.19.2 FUNCTIONS CONTAINING TWO OXYGENS $R^1_2C=C(OR^2)_2$, etc.

4.19.2.1 Ketene Acetals—$R^1_2C=C(OR^2)_2$

Compounds in this section are defined as those in which both oxygen atoms of the acetal are further bonded to carbon. Early work on this important class of compounds was comprehensively reviewed by McElvain in 1949 ⟨49CRV453⟩. Reviews include those by Swetkin in 1970 ⟨70MI 419-01⟩, Brassard in 1980 ⟨B-80MI 419-01⟩ and in 1993 Scheeren in *Houben-Weyl* ⟨93HOU(E15/2)1674⟩.

4.19.2.1.1 From carboxylic acids and esters

Although simple in concept, this method for the preparation of ketene acetals has the problem of carbon vs. oxygen regioselectivity. Much work has been carried out towards understanding the factors affecting the regioselectivity of alkylation reactions of ambident nucleophiles ⟨B-72MI 419-01⟩ and the reported preparations of ketene acetals from esters follow the normal empirical rules for regioselectivity. The preparations are most successful when they are intramolecular and selectivity for *O*-alkylation is achieved primarily by geometrical constraint. In this situation, yields and selectivities can be very high, as shown by the preparations of compounds of Structures (**1**) and (**2**) (Scheme 1) ⟨56JA4944, 85JOC1117⟩. Formation of a five-membered ring is usually favoured but Bentley *et al.* found that the chloro β-lactam (**3**) gave a higher proportion of the oxazepinone (**4**) than the oxazolidine (**5**) ⟨79TL391⟩. Triacylmethanes have a high charge on oxygen in the deprotonated form and are usually much more easily alkylated on oxygen than carbon. Examples of ketene acetals derived from these species are the furanone (**6**) ⟨72JCS(P1)1225⟩ and the malonate derivative (**7**), formed from the triester enolate (**8**) which was produced *in situ* ⟨88S981⟩. *O*-Alkylation of the acyl phosphoranes (**9**) was used by Bestmann *et al.* to produce the phosphonium salts (**10**)

⟨69AG(E)216⟩. When one of the α-substituents on the phosphorane was hydrogen (**9**; R = H), then deprotonation of the intermediate phosphonium salt (**10**; R = H) gave the phosphorane (**11**) (Scheme 1) ⟨77AG(E)349⟩.

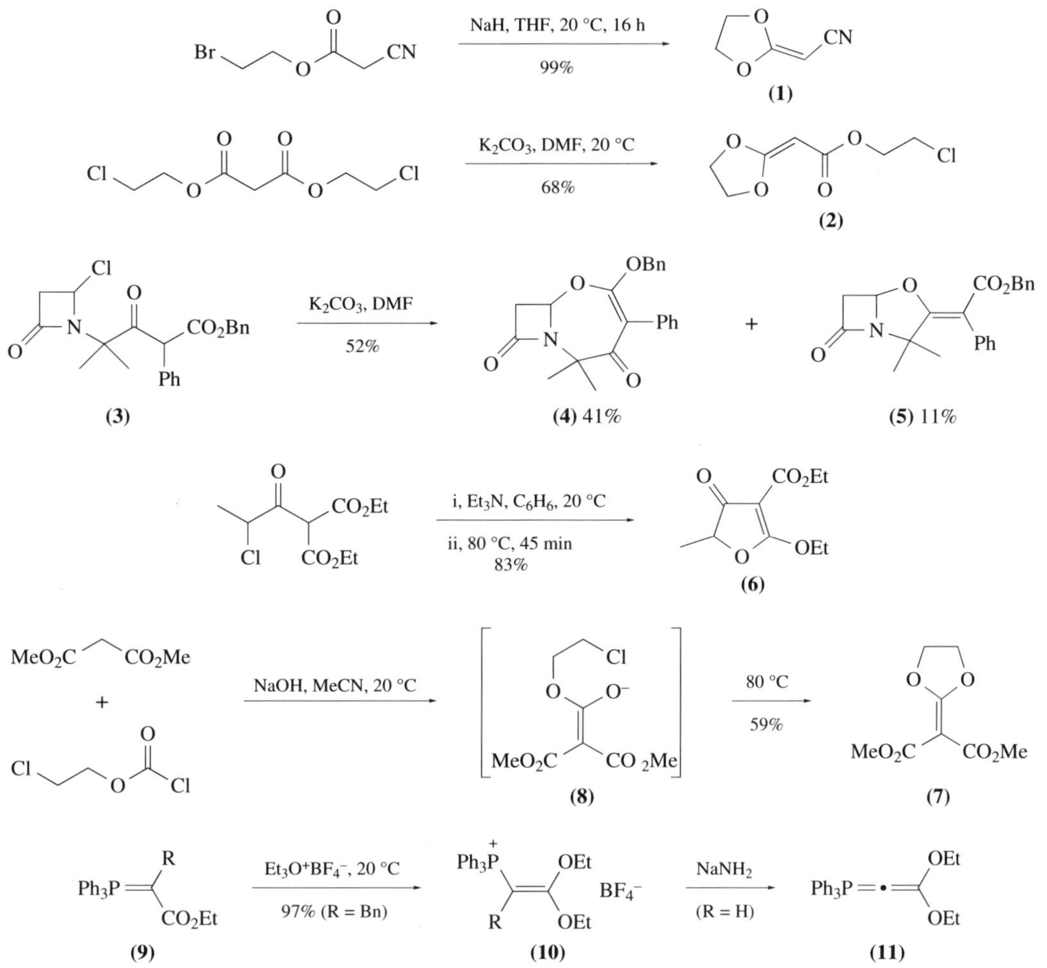

Scheme 1

4.19.2.1.2 *From* ortho-*esters and analogues*

The utility of this approach depends on the accessibility of the *ortho*-esters. A typical reaction is the elimination of methanol from trimethyl *ortho*-(phenylacetate) (**12**) using aluminum methoxide at high temperature to give the acetal (**13**) (Equation (1)) ⟨50JA1661⟩. Similar conditions were used to prepare the heterocyclic ketene acetals (**14**) ⟨55JA5601⟩ and (**15**) ⟨79TL2925⟩. Heat alone was sufficient to prepare the tetrasubstituted alkene (**16**) by elimination of ethanol from the corresponding *ortho*-ester ⟨83JOC298⟩. A related reaction is the rearrangement of the substituted β-lactam (**17**) to the ring-opened amide (**18**) (Equation (2)) ⟨88TL2327⟩.

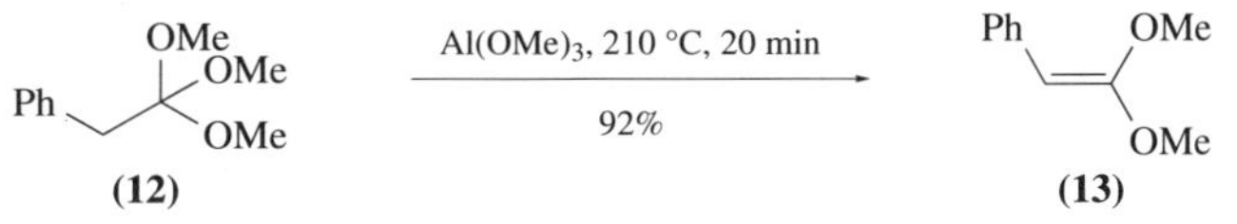

Certain trihalomethanes give ketene acetals by substitution–elimination when treated with alkoxides or phenoxides. Good examples are the preparations of the diarylketene acetal (**19**) (Equation (3)) ⟨56JOC801, 65ZOB1639⟩ and the nitro compounds (**20**) (Equation (4)) ⟨81ZOR1550⟩. In a similar reaction, McBee and Bolt prepared the mixed acetal (**21**) from the α,α-difluoro ether (**22**) (Equation (5)) ⟨47IEC412⟩.

4.19.2.1.3 From α-haloacetals and analogues

Historically, this method has probably been most widely used for preparing acetals of the parent, unsubstituted ketene and it continues to be reported regularly. A fairly strong, non-nucleophilic base is commonly used to effect the elimination. McElvain and Kundiger used potassium *t*-butoxide to convert the α-bromoacetal (**23**) into ketene diethylacetal (**24**) (Equation (6)) ⟨43OS(23)45⟩ and these conditions, with minor variations, have been satisfactory in most cases. The α-halogen may also be iodine ⟨89SC21⟩ and tosylate has also been used as the leaving group ⟨82GEP3240287⟩. Although further alkyl substitution on the α-carbon could lead to mixtures of products, Aben and Scheeren found that the α,α-dimethyl compound (**25**) gave exclusively the ketene acetal (**26**), which they propose is the result of a kinetically controlled elimination (Equation (7)) ⟨90RTC399⟩. Bailey and Zhou showed that the normal *t*-butoxide elimination reactions could be greatly accelerated by 2 mol% Aliquat 336, a phase transfer catalyst. The elimination of HBr from the α-bromo acetal (**27**) to give the cyclic ketene acetal (**28**) was essentially complete within 2 h at 0 °C (Equation (8)) ⟨91TL1539⟩. In the early 1990s, Diez-Barra and co-workers reported the use of ultrasound, in combination with a phase transfer catalyst, to promote the elimination of HBr from an α-bromo acetal with potassium hydroxide ⟨92SL893, 93SC1935⟩.

A variation on this strategy is the use of α-halo *ortho*-esters as precursors of ketene acetals. Belanger and Brassard induced ring-opening of the cyclic *ortho*-ester (**29**) by elimination of HCl followed by a pericyclic reaction to give the 5,5-dimethoxy-2,4-pentadienal (**30**) ⟨72CC863⟩. In a rather different reaction, the α-bromo *ortho*-ester (**31**) was treated with sodium sand in benzene at reflux for 10 h to give the acetal (**32**) in 68% yield ⟨46JA1922⟩.

4.19.2.1.4 From cycloaddition reactions of α,β-unsaturated esters

Heterodienes which react in this way usually have an extra electron-withdrawing group conjugated to the alkenic double bond. A good example is the *exo*-methylene compound (**33**), prepared *in situ* from Meldrum's acid (**34**) and formaldehyde, which was treated with the substituted styrene (**35**) to give the thermally labile adduct (**36**) (Scheme 2) ⟨88H(27)1929⟩. These cycloaddition reactions are often intramolecular. The diene–dienophile system (**37**), prepared *in situ* from Meldrum's acid (**34**) and the aldehyde (**38**), cyclised at room temperature to give the adduct (**39**) ⟨88CC189⟩. A similar example was reported by Snider *et al*. ⟨79JA6023⟩. Alkynes ⟨80H(14)15⟩ and ynamines ⟨70TL885⟩ have also been used as the dienophiles. Hall and Rasoul *et al*. proposed a stepwise mechanism for the reaction of the α-cyanoacrylate (**40**) with the enol ether (**41**), proceeding through a (nonisolated) zwitterionic intermediate, to give the thermally unstable dihydropyran (**42**) (Scheme 2) ⟨82JOC2080, 82TL603⟩.

4.19.2.1.5 By addition to 1-alkoxy alkynes

Broekma showed that hindered phenols, such as 2,6-dimethylphenol (**43**), add once to 1-alkoxy alkynes (e.g., Structure (**44**)) in the presence of mercurous acetate to give ketene acetals (e.g., Structure (**45**)) (Equation (11)) ⟨75RTC209⟩. Less hindered phenols added twice under these conditions to give *ortho*-esters. Under similar conditions, Wassermann and Wharton reported the addition of a carboxylic acid (**46**) to 1-methoxyethyne (**47**) to give the methoxyvinyl ester (**48**) (Equation (12)) ⟨58T321⟩. This methodology was also utilised by Livantsova *et al*. to prepare silyl- and germanyl-substituted ethenes ⟨84ZOB1925⟩. In the early 1990s, Kita *et al*. effected a similar transformation with an aliphatic alcohol (**49**) using a ruthenium catalyst to give the acetal (**50**) (Equation (13)) ⟨93SL273⟩.

4.19.2.1.6 By Horner–Wittig chemistry

The Wittig reaction and related chemistry, which features so prominently in the preparation of sulfur and selenium analogues (*vide infra*), has not been as widely utilised to prepare ketene acetals. One example is the reaction of the (diethoxymethyl)phosphine oxide (**51**) with benzaldehyde to give the acetal (**52**) (Equation (14)). The same paper gives 17 other examples of the reactions of similar

1,1-dialkoxyalkylphosphine oxides with a range of aldehydes and ketones, in moderate to good yields, although the reaction failed with cyclopentanone ⟨83TL1303⟩.

LDA = lithium diisopropylamide

4.19.2.1.7 From ketenes

Feiler *et al.* showed that diphenylketene (**53**) reacted with diphenylacetic anhydride (**54**) in the presence of triethylamine to give the diacyloxy alkene (**55**) at room temperature, although 1 month was required for an 80% yield (Equation (15)) ⟨74CC405⟩. Also using diphenylketene (**53**), Woerner *et al.* reported the preparation of a 2:1 adduct (**56**) with the azirine (**57**) (Equation (16)) ⟨71CB2786⟩. A very similar reaction was reported by Hassner *et al.* ⟨72TL1353⟩ and Aue and Thomas extended the scope to include 2:1 adducts of ketenes with azetines ⟨75JOC2552⟩. Diphenylketene (**53**) has also been shown to react with oxiranes, as in the preparation of the acetal (**58**) (Equation (17)). The authors reported that the reaction gave ketene acetals only with alkyl-substituted oxiranes; vinyl and aryl substituents promoted the formation of lactones ⟨88JOC5974⟩.

4.19.2.1.8 From *gem*-dihalogeno alkenes and analogues

Direct displacement of vinyl halogen by alkoxide ions has been used for a few isolated examples of ketene acetal preparation. Conditions are often vigorous, as in the synthesis of the bis(methoxyethyl) acetal (**59**) (Equation (18)) ⟨67OS(47)78⟩, but in some cases the reaction can be very high-yielding under moderate conditions, as in the conversion of the perchloropentadienoic acid nitrile (**60**) into the triphenoxy derivative (**61**) (Equation (19)) ⟨82CB1733⟩. Other examples of similar vinyl halogen displacements are detailed in the review by Scheeren ⟨93HOU(E15/2)1674⟩. A related reaction is the preparation of the dicyanoketene acetal (**62**) from tetracyanoethylene (**63**) (Equation (20)) ⟨58JA2788, 59OS(39)13⟩.

4.19.2.1.9 *By miscellaneous transformations*

The [3 + 2]-cycloaddition reaction of the vinyl cyclopropane (**64**) with the methacrylate ester (**65**) gave the ketene acetal (**66**) (Equation (21)) ⟨89JA7285⟩. Isomerisations of α,β-unsaturated acetals have been carried out with strong bases such as potassium amide. Scheeren *et al.* reported the preparation of a series of ketene acetals (**67**) in this way (Equation (22)) ⟨77JOC3128⟩. A photochemical isomerisation, catalysed by iron carbonyl compounds, has also been reported ⟨74GEP2331675⟩. The potassium amide method has also been extended to the preparation of 1,1,4,4-tetraalkoxy-1,3-butadienes from 1,1,4,4-tetraalkoxy-2-butynes ⟨74TL1019⟩.

The dimerisation of *ortho*-formates has been used to prepare tetraalkoxyethenes. Scheeren *et al.* reported that treatment of the *ortho*-formates (**68**) with sodium hydride gave the tetraalkoxyethenes (**69**) in moderate yield (Equation (23)) ⟨73RTC11⟩. Bellus *et al.* published an improved procedure using similar substrates in tetraglyme solution ⟨78HCA1784⟩. Moss *et al.* prepared a 1:1 (*E*):(*Z*) mixture of 1,2-dimethoxy-1,2-diphenoxyethene (**70**) by dimerisation of the carbene generated from the diazo compound (**71**) (Equation (24)) ⟨87JA3811⟩.

4.19.2.2 Other Related Ketene Derivatives—$R^1_2C{=}COR^2OX$, etc.

4.19.2.2.1 *Ketene silyl acetals—$R^1_2C{=}COR^2OSiR^3_3$, etc.*

The silylation of ester enolates to give ketene silyl acetals was first demonstrated by Ireland and Mueller in 1972 ⟨72JA5897⟩ and this class of compounds has continued to be of interest to the organic chemist because of the synthetically important Mukaiyama and Claisen rearrangement reactions. A comprehensive review by Pawlenko of ketene silyl acetals appeared in the *Houben-Weyl* series ⟨93HOU(E15/2)1742⟩. Various aspects of ketene silyl acetal chemistry have been reviewed frequently, often as part of wider-ranging reviews on silyl enol ethers in general ⟨83S1, B-88MI 419-01, 91COS(2)604⟩.

The original Ireland method, involving deprotonation of an ester (**72**) with a strong base and trapping of the ester enolate (**73**) with a silylating agent, is the most common procedure although many variations on the conditions have been applied in order to control the regioselectivity and stereochemistry of the reaction (Equation (25)). In a rigorous study, Ireland *et al.* investigated the effects of solvent, the nature of the base, ester : base ratios and oxygen substituents elsewhere in the molecule on the stereochemistry of the ketene silyl acetal. The main conclusions were that the "kinetic" (*E*) ketene silyl acetal is formed preferentially when the counterion to the base is not chelated by the solvent and that the "thermodynamic" (*Z*) acetal is formed in the presence of strong chelaters, such as hexamethylphosphoramide (HMPA), as a result of a kinetic resolution of the initial mixture of ester enolate ions rather than through an equilibrium process ⟨91JOC650⟩. In most cases, careful choice and control of conditions enables high selectivity of either the (*E*) or (*Z*) ketene silyl acetal as required, as demonstrated by the complementary reactions to give the two isomers (**74**) and (**75**) (Scheme 3) ⟨79TL4029⟩. Control of *O*- vs. *C*-silylation can be effected by the nature of the silylating agent, the bulkiness of the esterifying alcohol and by the counterion. Bulky silylating agents (e.g., *t*-butyldimethylsilyl chloride) tend to favour *O*-silylation, as does substitution at the position α to the carboxyl carbon, but bulky ester groups (e.g., *t*-butyl) tend to promote *C*-silylation ⟨88JOC633⟩.

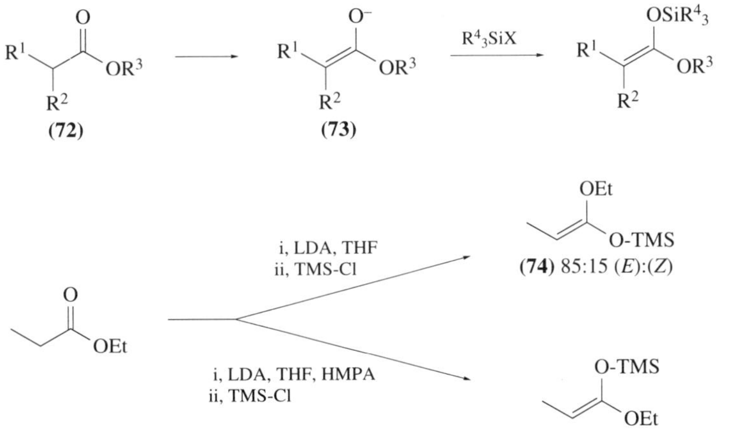

$$\begin{array}{c}\text{(72)}\end{array} \longrightarrow \begin{array}{c}\text{(73)}\end{array} \xrightarrow{R^4_3SiX} \begin{array}{c}\end{array} \quad (25)$$

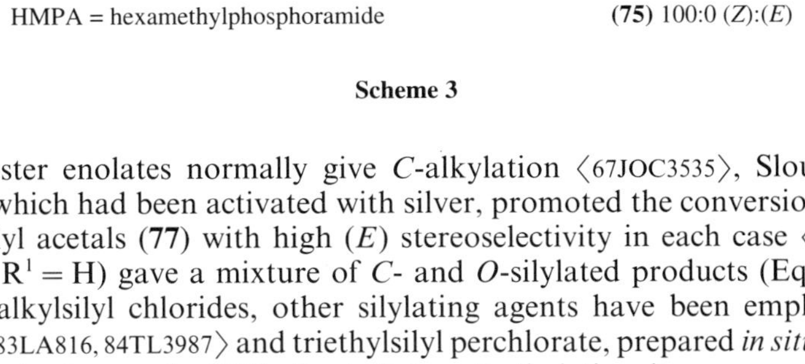

Scheme 3

HMPA = hexamethylphosphoramide

(**74**) 85:15 (*E*):(*Z*)
(**75**) 100:0 (*Z*):(*E*)

Although zinc ester enolates normally give *C*-alkylation ⟨67JOC3535⟩, Slougui and Rousseau reported that zinc which had been activated with silver, promoted the conversion of α-bromo esters (**76**) into ketene silyl acetals (**77**) with high (*E*) stereoselectivity in each case ⟨87SC1⟩. Only ethyl bromoacetate (**76**; $R^1 = H$) gave a mixture of *C*- and *O*-silylated products (Equation (26)). Apart from the usual trialkylsilyl chlorides, other silylating agents have been employed including trialkylsilyl triflates ⟨83LA816, 84TL3987⟩ and triethylsilyl perchlorate, prepared *in situ* from triethylsilane and trityl perchlorate ⟨84TL699⟩. With the triflates, *C*-alkylation becomes a competing reaction unless the ester is activated by an α-electron withdrawing group (e.g., cyano (**78**; $R^1 = CN$) ⟨83LA816⟩ or trifluoromethyl (**78**; $R^1 = CF_3$)) ⟨84TL3987, 84TL3991⟩ and *C,O*-disilylation has been observed with lactones (e.g., butyrolactone (**79**)) (Scheme 4) ⟨77S867⟩. There have been several reports of the use of silanes with rhodium catalysts to generate silylating agents *in situ* ⟨74CPB2767, 87SC1, 90JOC2972⟩.

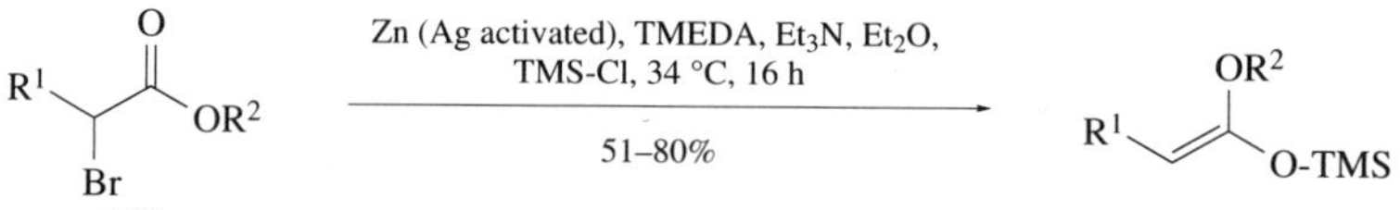

Scheme 4

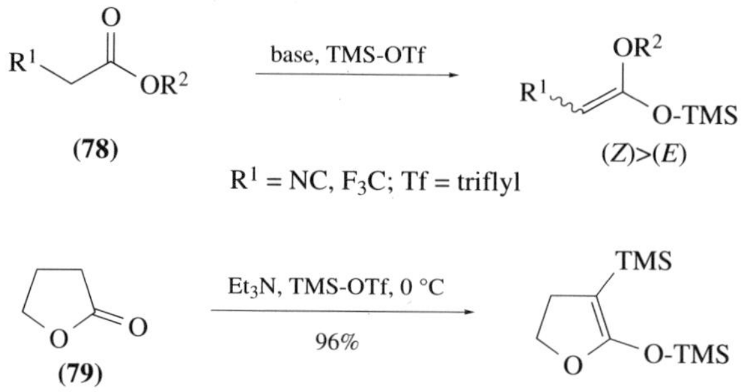

The conversion of carboxylic acids into ketene disilyl acetals has been reported by several groups, exemplified by the diphenylketene disilyl acetal (**80**) ⟨77JOC38⟩ and tris(trimethylsilyloxy)ethene (**81**) ⟨78TL2749⟩. Trialkylsilyl esters (**82**) also give ketene disilyl acetals, by the reaction of either the corresponding lithium enolate with a trialkylsilyl chloride ⟨72JOM(46)73, 84JOM(263)C21⟩ or the silyl ester itself with a trialkylsilane (Scheme 5) ⟨71JOM(30)C64, 72JOM(46)73⟩.

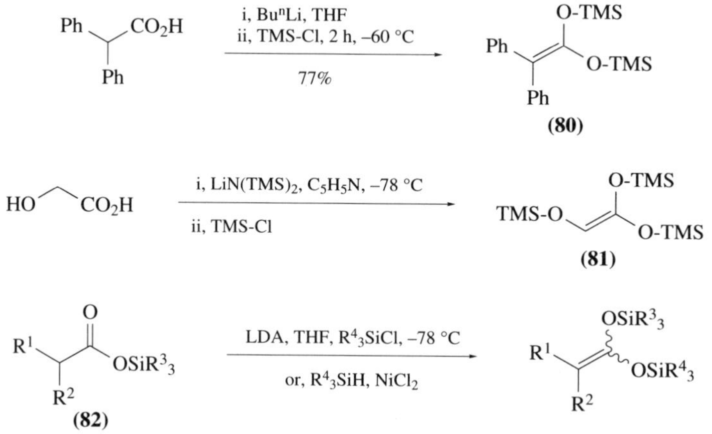

Scheme 5

4.19.2.2.2 Boryloxy derivatives—$R^1{}_2C{=}COR^2OBR^3{}_2$, etc.

This area has been reviewed by Pelter ⟨93HOU(E15/2)1737⟩.

1-Alkoxy-1-(dialkylboryloxy) alkenes are usually prepared by treatment of an α-bromo ester enolate with a trialkylborane. The ester enolate coordinates with the boron and one of the alkyl groups is transferred to the α-carbon, replacing the bromine atom. Brown *et al.* have published extensively in this area, a typical example being the reaction of tris(cyclopentyl)borane (**83**) with ethyl bromoacetate (**84**) (Equation (27)) ⟨68JA818⟩. In most cases, the boryloxy compound is not isolated and is converted into an ester with overall α-alkylation. The analogous bis(dialkylboryloxy)ethenes may be prepared from carboxylic acids by treatment with a dialkylboryl triflate, as in the preparation of the compound (**85**) by Evans *et al.* (Equation (28)) ⟨81JA3099⟩. Related preparations from α-sulfuranylidene esters ⟨67JA6804⟩ and from α-diazo esters ⟨69JA6195⟩ have also been reported.

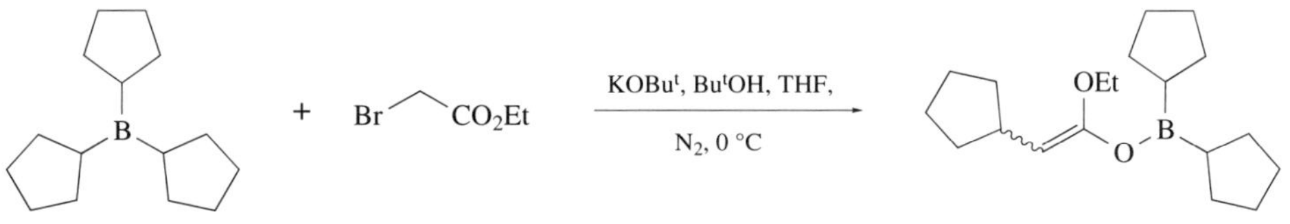

Oxygen and Sulfur

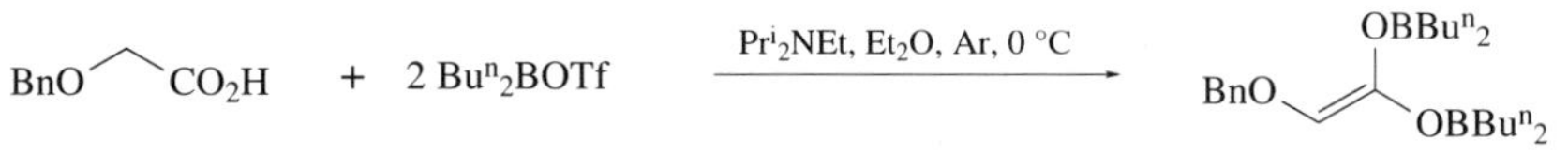

[Equation (27): compound (83) tricyclopentylborane + (84) BrCH₂CO₂Et → with KOBuᵗ, BuᵗOH, THF, N₂, 0 °C → enol borinate product]

[Equation (28): BnOCH₂CO₂H + 2 Bunₙ₂BOTf → with Pri₂NEt, Et₂O, Ar, 0 °C → (85)]

4.19.2.2.3 Miscellaneous derivatives

There have not been many examples of ketene acetals in which either of the oxygen atoms is further bonded to atoms other than carbon, silicon or boron. Two examples, which were both prepared directly from bis(trifluoromethyl)ketene (**86**), are the cyclic sulfate (**87**) ⟨71KGS1645⟩ and the dioxazole (**88**) ⟨70IZV2140⟩. The hydroxamic vinyl ester (**89**) was formed in very good yield when the ketene acetal (**90**) was treated with the hydroxamic acid (**91**) (Scheme 6) ⟨62BCJ71⟩.

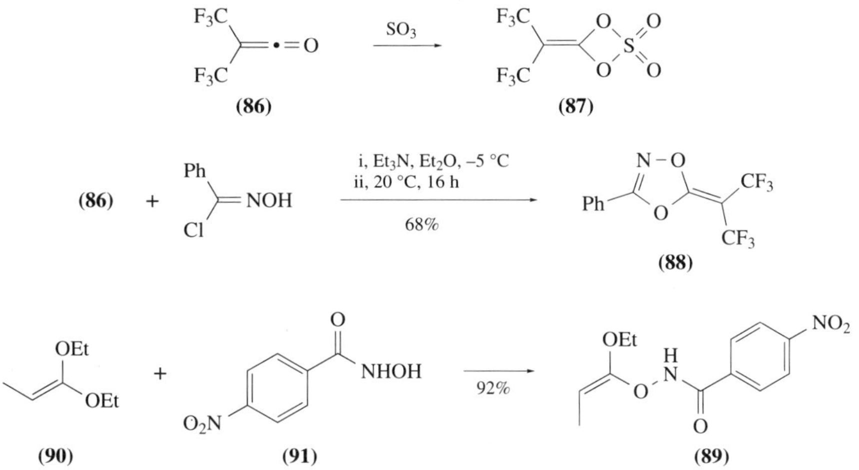

Scheme 6

4.19.3 FUNCTIONS CONTAINING OXYGEN AND SULFUR—$R^1_2C\!=\!COR^2SR^3$, etc.

4.19.3.1 Dicoordinated Sulfur Derivatives—$R^1_2C\!=\!COR^2SR^3$

The chemistry of this functional group was reviewed by Schaumann in 1985 ⟨85HOU(E11/1)255⟩.

4.19.3.1.1 From monothiocarboxylic acids and esters

As for ketene acetals (Section 4.19.2.1.1), the reaction of a thiol ester enolate with an electrophile has been used frequently to prepare this functional group. In this case, however, there is also the possibility of using the enethiolate of a thiono ester. Enolisation and alkylation of 2-oxopenams (e.g., Structure (**92**)) has been used extensively in β-lactam chemistry, exemplified by the preparation of compound of Structure (**93**) by Ghosez and co-workers, in which diazomethane was used as the alkylating agent (Equation (29)) ⟨83T2493⟩. In a more generally applicable example, Schafer and Gewald reported the reaction of activated methylene compounds, such as ethyl cyanoacetate (**94**), with carbonyl sulfide followed by methylation, first on sulfur with methyl iodide and then on oxygen with diazomethane, to give the ketene monothioacetal (**95**) (Equation (30)) ⟨75JPR337⟩. Ketene *O*-silyl monothioacetals may be prepared from the corresponding thiol ester by methods similar to those for the dioxa analogues (see Section 4.19.2.2) ⟨78CC478⟩. Similar factors, such as the nature of the silylating agent and the chelating power of the solvent, determine the (*E*) or (*Z*) geometry of the product, as exemplified by the preparation (Scheme 7) of the monothioacetals (**96**) ⟨78CC478⟩ and (**97**) ⟨88JOC4015⟩. The *O*-silylation of thiol ester enolates has also been carried out using trimethylsilyl triflate ⟨77S247⟩.

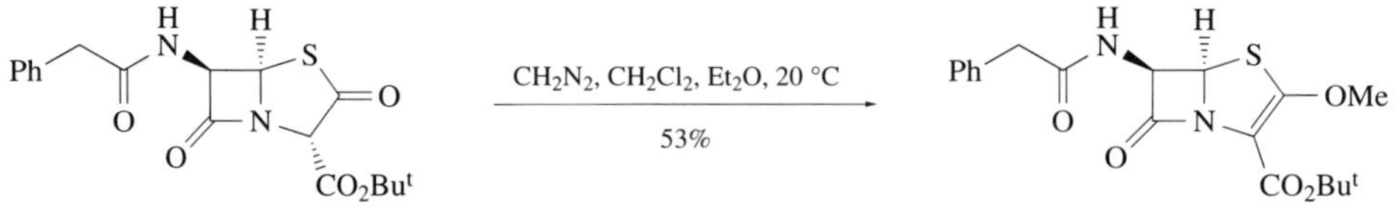

(92) **(93)** (29)

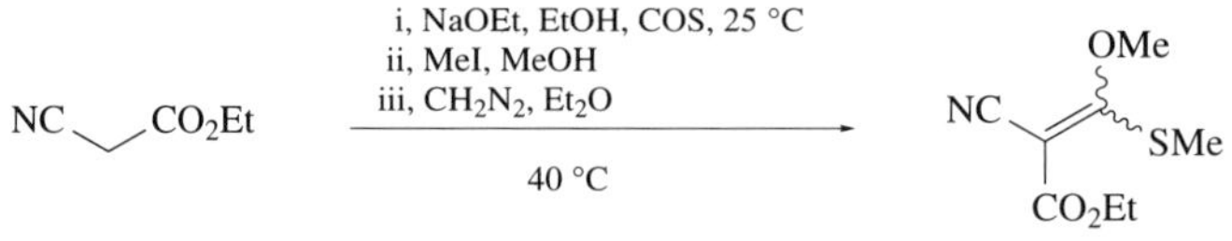

(94) **(95)** (30)

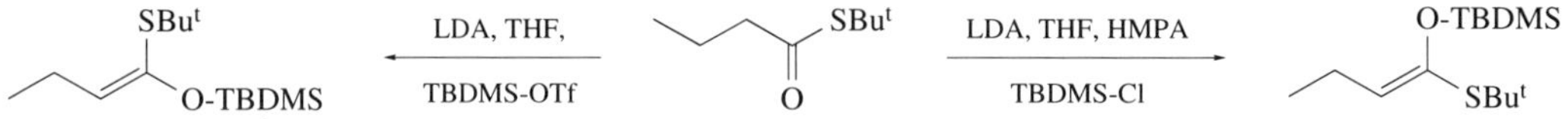

(**96**) >95:5 (*Z*):(*E*) (**97**) >95:5 (*E*):(*Z*)

TBDMS = *t*-butyldimethylsilyl

Scheme 7

The *S*-alkylation of thiono esters is usually effected by enethiolization, using a strong base at low temperature, followed by addition of an alkyl halide. A good example is the preparation of the thionopropionate derivative (**98**), which was obtained as a mixture of (*E*) and (*Z*) isomers (Equation (31)) ⟨68RTC929⟩. In contrast, the dithionosuccinate (**99**) was converted exclusively into the (*Z*,*Z*) isomer of the conjugated diene (**100**) (Equation (32)) ⟨88LA933⟩. The greater nucleophilicity of an enethiolate, compared with an ester enolate, means that competing *C*-alkylation is not usually a problem (*cf.* Section 4.19.2.1.1); even α-thioacyl phosphoranes react at sulfur rather than carbon ⟨77AG(E)349⟩. *S*-Acylation of thiono ester enethiolates has also been reported. The enethiolate (**101**) of the penam intermediate (**102**) was prepared *in situ* by treatment with lithium hexamethyldisilazide and phenyl chlorothioformate (**103**). Workup with pivaloyl chloride produced the thiol ester (**104**) ⟨83TL3373⟩. In a variation of the functionalization of thiono esters, dimethyloxosulfonium methylide (**105**) and *O*,*O'*-diethyl trithiopyrocarbonate (**106**) gave the zwitterionic enethiolate (**107**). Treatment of this betaine with carbon disulfide gave the 1,2-dithiole 3-thione (**108**) by *C*-thioacylation followed by intramolecular thiophilic attack and elimination of DMSO (Scheme 8) ⟨76BCJ3128⟩. Petrov and co-workers reported a similar reaction ⟨80ZOR13⟩.

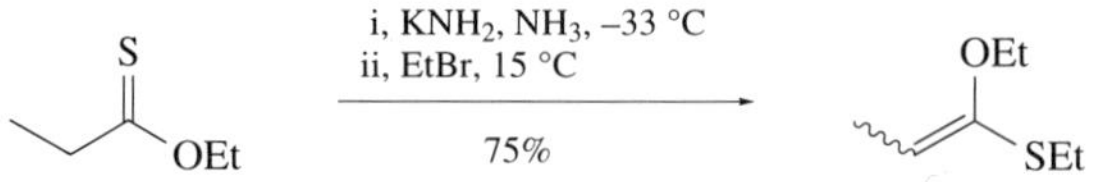

(98) (31)

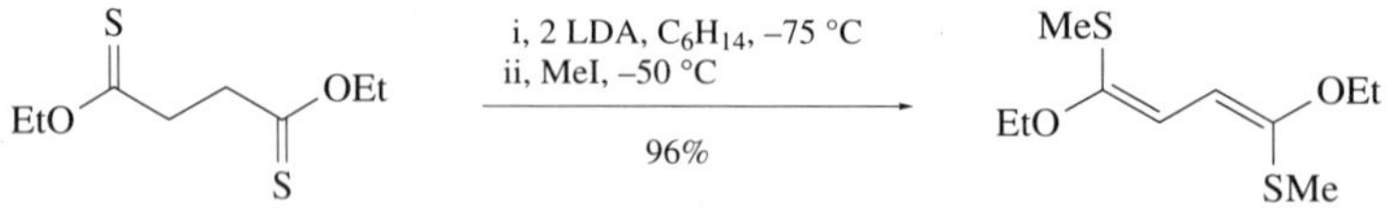

(99) **(100)** (32)

4.19.3.1.2 By alkeneation methods

Hackett and Livinghouse used a Peterson alkeneation to convert aldehydes (e.g., Structure (**109**)) and ketones (e.g., Structure (**110**)) into the corresponding ketene monothioacetals (**111**) and (**112**) (Scheme 9). Deprotonation was found to be a competing reaction for some ketones and (*E*)/(*Z*) product mixtures were formed in all unsymmetrical examples ⟨84TL3539, 86JOC879⟩. Mikolajczyk *et*

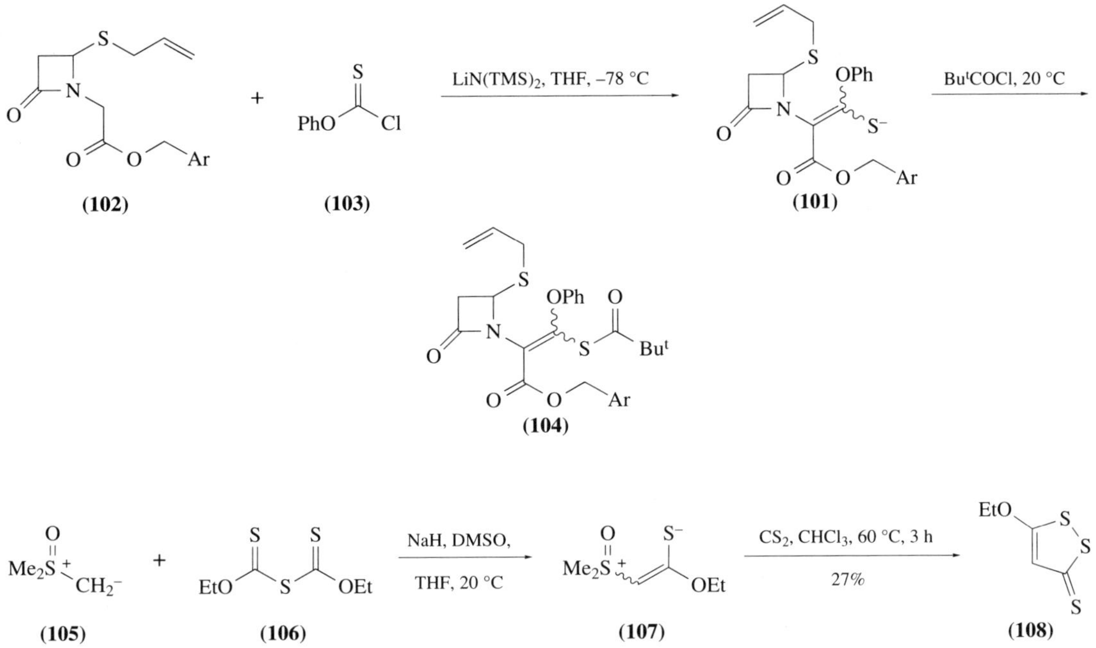

Scheme 8

al. employed Horner–Wittig chemistry to prepare the series of ketene *O*,*S*-acetals (**113**) (Equation (33)) ⟨78T3081⟩. Hirai and Ishiba reported the use of iminium salts (e.g., Structure (**114**)) as electrophiles in the condensation with 1,3-dicarbonyl compounds (e.g., Structure (**115**)) to give 1,3-oxathiafulvenes such as Structure (**116**) (Equation (34)) ⟨71CPB2194, 71TL1137⟩. A similar method was used by Gotthardt and Oppermann to prepare the fulvene (**117**) (Equation (35)) ⟨85CC1154⟩. The reaction was reported to be unsuccessful when other active methylene compounds were used, such as cyanoacetamide or ethyl cyanoacetate, which gave ketene *S*,*N*-acetals ⟨71TL1137⟩.

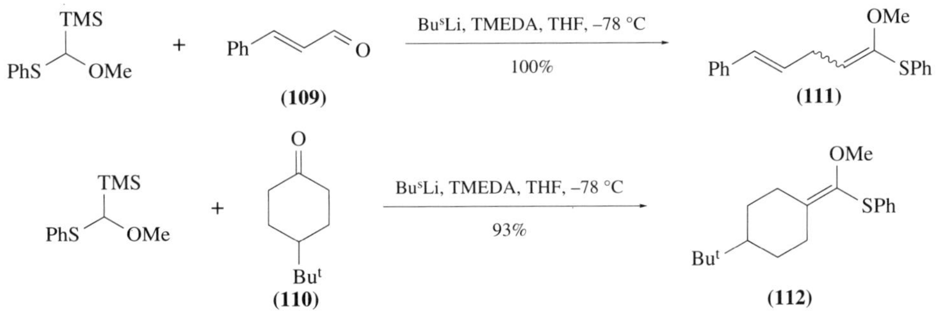

Scheme 9

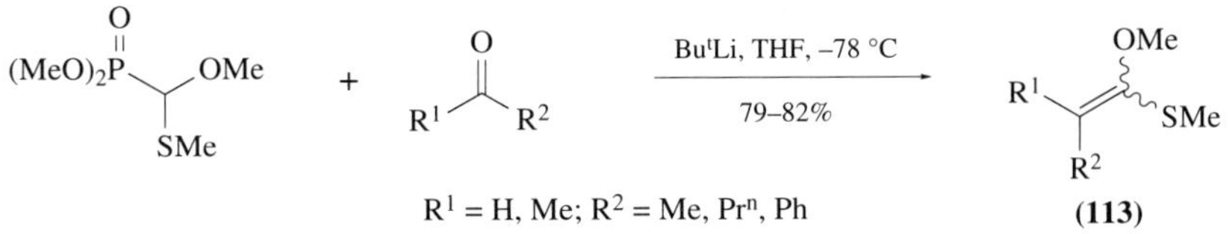

R^1 = H, Me; R^2 = Me, Pr^n, Ph (33)

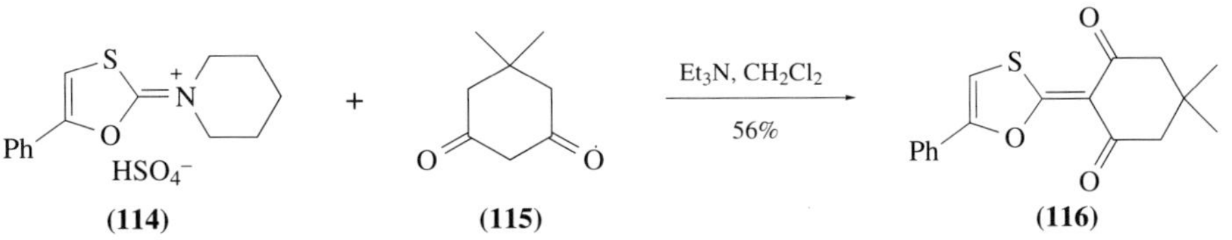

(34)

4.19.3.1.3 By miscellaneous methods

Brandsma and co-workers have demonstrated that α-anions of enol ethers (e.g., Structures (**118**) and (**119**)) can be sulfenylated efficiently with dialkyl disulfides to give the corresponding ketene monothioacetals (**120**) and (**121**) (Scheme 10) ⟨68RTC1179, 87JOM(332)99⟩. Martin *et al.* described the curious reaction of 2 mol equiv dimethylketene (**122**) with carbon disulfide, with triphenylphosphine catalysis, to give the heterocyclic compound (**123**) (Equation (36)) ⟨71JOC2205⟩. Also, diphenylketene (**53**) was treated with tetramethylallene-1,2-episulfide (**124**) to give the monothioacetal (**125**) (Equation (37)) ⟨84TL4011⟩.

Hackett and Livinghouse found that the α,β-unsaturated *O*,*S*-acetal (**126**) and the allylic iodide (**127**) gave the ketene *O*,*S*-acetal (**128**) (Equation (38)) ⟨86JOC879⟩. Treatment of the α-chloro thiol ester (**129**) with triethyl phosphite gave the enol phosphate (**130**) in a variant of the Perkow reaction (Equation (39)) ⟨68JOC4470⟩. Ketene *O*,*S*-acetals have been prepared from ketene dithioacetals in acetal exchange reactions. In most of these cases, the double bond is conjugated to a carbonyl or nitrile group and the reaction takes place by Michael addition of the alkoxide and elimination of a thiolate. When an acyclic dithioacetal is converted into a cyclic *O*,*S*-acetal, the reaction is favoured entropically, as in the preparation of the malononitrile derivative (**131**). In this example, the reverse reaction was also prevented by removal of the liberated methanethiol by distillation (Equation (40))

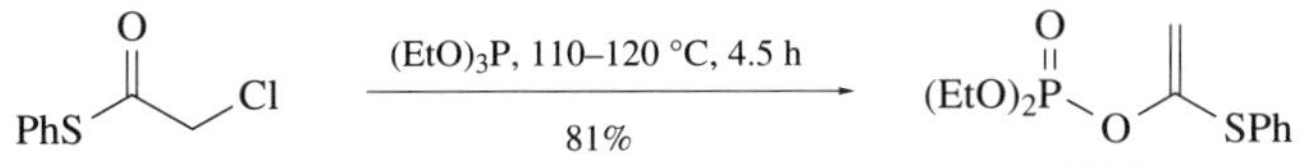

$$\underset{(129)}{\text{PhS}\overset{\text{O}}{\underset{\|}{\text{C}}}\text{Cl}} \xrightarrow[81\%]{(\text{EtO})_3\text{P}, 110-120\,°\text{C}, 4.5\,\text{h}} \underset{(130)}{(\text{EtO})_2\overset{\text{O}}{\underset{\|}{\text{P}}}-\text{O}-\text{C}(=\text{CH}_2)\text{SPh}} \quad (39)$$

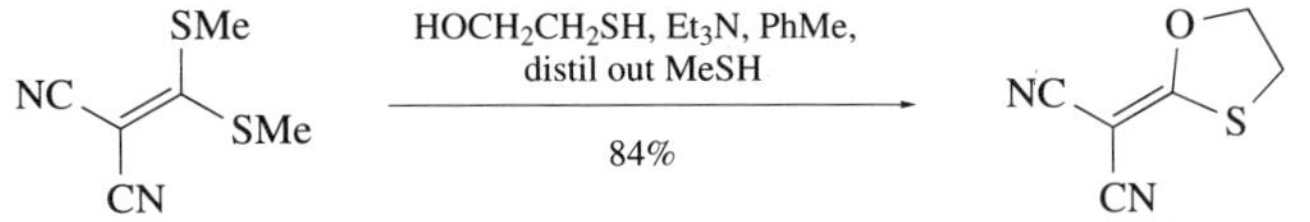

$$\underset{(\text{NC})(\text{CN})\text{C}=\text{C}(\text{SMe})_2}{(129...)} \xrightarrow[84\%]{\text{HOCH}_2\text{CH}_2\text{SH}, \text{Et}_3\text{N}, \text{PhMe}, \text{distil out MeSH}} \underset{(131)}{(\text{NC})(\text{CN})\text{C}=\text{C}(\text{OCH}_2\text{CH}_2\text{S})} \quad (40)$$

⟨84MI 419-01⟩. The reaction is also favoured when it is intramolecular ⟨84T381⟩, and the alkylthio group may also be activated towards substitution by *S*-alkylation to form a sulfonium ion ⟨82S206⟩.

4.19.3.2 Tricoordinated Sulfur Derivatives—$R^1{}_2C\!\!=\!\!COR^2S(O)R^3$

This group of compounds has not been studied as widely as the di- and tetracoordinated sulfur analogues. Indeed, a *Chemical Abstracts* substructure search discovered only four papers with structures within this category. Two of these involved oxidation of the corresponding divalent sulfur species. Thus, mcpba converted the 1,3-oxathiolan-5-ones (132) into the sulfoxides (133) (Equation (41)) ⟨85CPB2256⟩ and peracetic acid oxidised the 1-thiovinyl phosphates (134) to the corresponding 1-sulfinylvinyl phosphates (135) (Equation (42)). Although the yields were only moderate, there was no evidence of overoxidation to sulfonyl products when only one molar equivalent of oxidant was used ⟨68JOC4470⟩. Recently, Kerbage *et al.* reported the reaction of the α-lithiated methoxyallene (136) with the optically active sulfinate ester (137) to give the allenyl sulfoxide (138) with inversion of stereochemistry at sulfur (Scheme 11) ⟨92SL493⟩. Optimum yields and enantiomeric purity were obtained when the reaction was carried out at −115 °C. Ando and co-workers reported that when di-*t*-butylthioketene (139) was treated with 2 mol equiv. dimethyl diazomalonate (140), in the presence of a catalytic amount of rhodium diacetate, the stable sulfonium ylide (141) was produced (Equation (43)) ⟨89TL1249⟩.

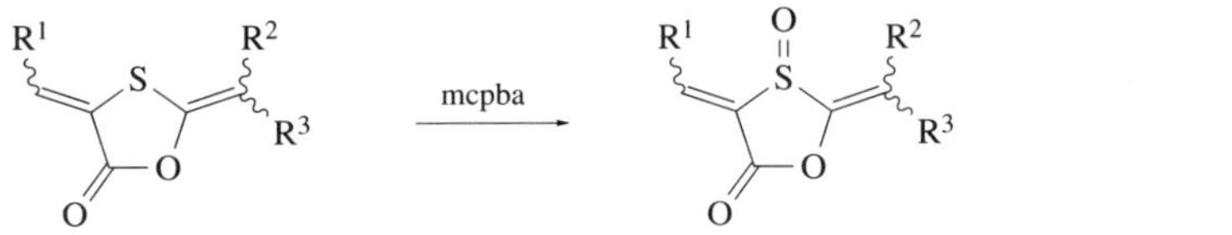

$$(132) \xrightarrow{\text{mcpba}} (133) \quad (41)$$

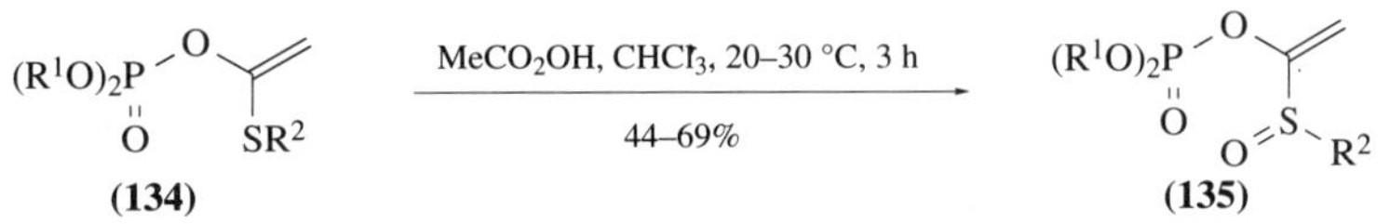

$$(134) \xrightarrow[44-69\%]{\text{MeCO}_3\text{H}, \text{CHCl}_3, 20-30\,°\text{C}, 3\,\text{h}} (135) \quad (42)$$

$R^1 = \text{Me, Et}; R^2 = \text{Ph, 4-ClC}_6\text{H}_4, \text{Bn, Me}$

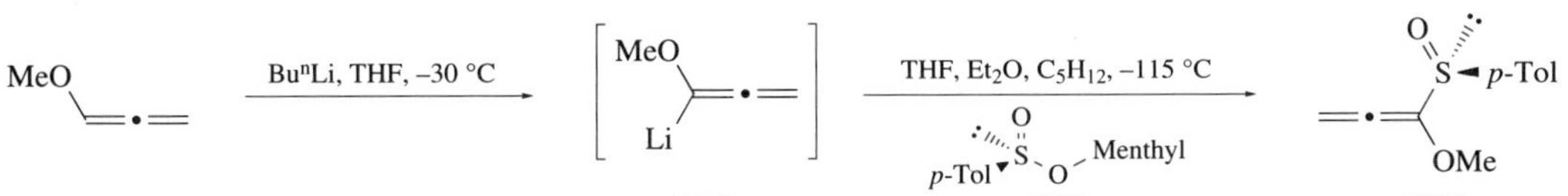

$$\text{MeO-CH=C=CH}_2 \xrightarrow{\text{Bu}^n\text{Li}, \text{THF}, -30\,°\text{C}} [(136)] \xrightarrow[90\%]{\text{THF}, \text{Et}_2\text{O}, \text{C}_5\text{H}_{12}, -115\,°\text{C}; (137)} (138)$$

Scheme 11

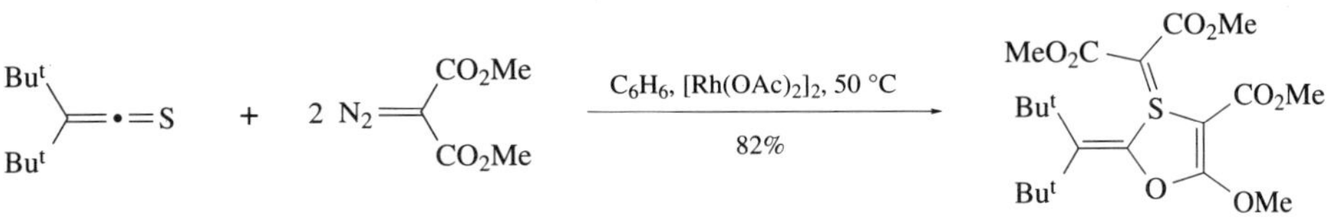

$$\underset{(139)}{\text{Bu}^t{}_2\text{C}=\text{C}=\text{S}} + 2\,\underset{(140)}{\text{N}_2\text{=C}(\text{CO}_2\text{Me})_2} \xrightarrow[82\%]{\text{C}_6\text{H}_6, [\text{Rh}(\text{OAc})_2]_2, 50\,°\text{C}} (141) \quad (43)$$

4.19.3.3 Tetracoordinated Sulfur Derivatives—$R^1{}_2C{=}COR^2SO_2R^3$

4.19.3.3.1 By sulfoxidation

Oxidation of the corresponding divalent sulfur species to a sulfone has not been widely studied. Oxidation of the 1-thiovinyl phosphates (**142**), with 2 mol equiv. peracetic or monoperoxyphthalic acid, gave the 1-sulfonylvinyl phosphates (**143**) in generally better yields than those of the analogous monoxidation reactions (see Section 4.19.3.2) (Equation (44)) ⟨68JOC4470⟩.

R^1 = Me, Et; R^2 = Ph, 4-ClC$_6$H$_4$, 2,4,6-Cl$_3$C$_6$H$_2$, Bn, Me; X = H, Cl

4.19.3.3.2 From α-sulfonyl ethers

A considerable amount of the published material in this area comes from a series of papers on α-sulfonyl ethers by Schank and co-workers. Deprotonation of the α-acyl α-sulfonyl ether (**144**), and trapping of the enolate with acylating agents, gave mixtures of the (*E*) and (*Z*) 1-methoxyvinyl sulfones (**145**) (Equation (45)) ⟨73CB1107⟩. The authors found that *C*-alkylation was generally observed when similar enolates were treated with alkyl halides, with the exception of chloromethyl methyl ether. Analogously, the potassium enolates (**146**), which were isolated quantitatively from the reactions of the α-sulfonyl ethers (**147**) with methyl formate and potassium *t*-butoxide, were treated with dimethyl sulfate, benzoyl chloride and aniline hydrochloride to give the 1-methoxyvinyl sulfones (Structures (**148**)–(**150**)) respectively (Scheme 12) ⟨76S406⟩. Condensation of similar α-sulfonyl ethers (**151**) with aldehydes not bearing an α-hydrogen gave the trisubstituted alkenes (**152**), with elimination of a carboxylate anion (Equation (46)). The geometry of the double bond in the products was dependent on the nature of the aldehyde and examples are given of exclusive (*Z*)- or (*E*)-selectivity as well as mixtures ⟨75LA1484, 76S408⟩. This condensation was less successful with α-alkanesulfonyl ethers; α-alkoxycinnamaldehydes or 1-acylvinyl ethers became significant by-products, resulting from competitive deprotonation from the other α-carbon ⟨76S404⟩.

R = MeOCH$_2$, Ac, Bz, 4-ClC$_6$H$_4$CO; solvent = DMF or MeCN or THF

When the α-sulfonyl ethers carried a halogen on the β-carbon (e.g., Structure (**153**)), then HX elimination gave the disubstituted alkenes (**154**) (Equation (47)) ⟨77CB3235⟩. Silylation of the α-sulfonyl ether (**155**), followed by deprotonation with *n*-butyllithium and treatment of the anion with aldehydes and ketones, gave the tri- and tetrasubstituted alkenes (**156**) by Peterson alkeneation reactions (Equation (48)). Where geometrical isomerism was possible, the *E* product predominated ⟨77LA1676⟩. Meichle and Otto reported a similar reaction of the disilylation compounds (**157**) with acetaldehyde to give the alkenes (**158**) either directly or by dehydration of the aldol product with thionyl chloride and pyridine (Equation (49)) ⟨89AP(322)263⟩. Elimination of HX from the α-halo α-sulfonyl ethers (**159**) generated the carbenes (**160**), which were trapped with diazomethane to give the 1-methoxyvinyl sulfones (**161**) (Scheme 13). When diazoethane was used as the trapping agent, the homologous 1-methoxyvinyl sulfone product was consumed by excess diazoethane, ultimately leading to a pyrazole ⟨89T6667⟩.

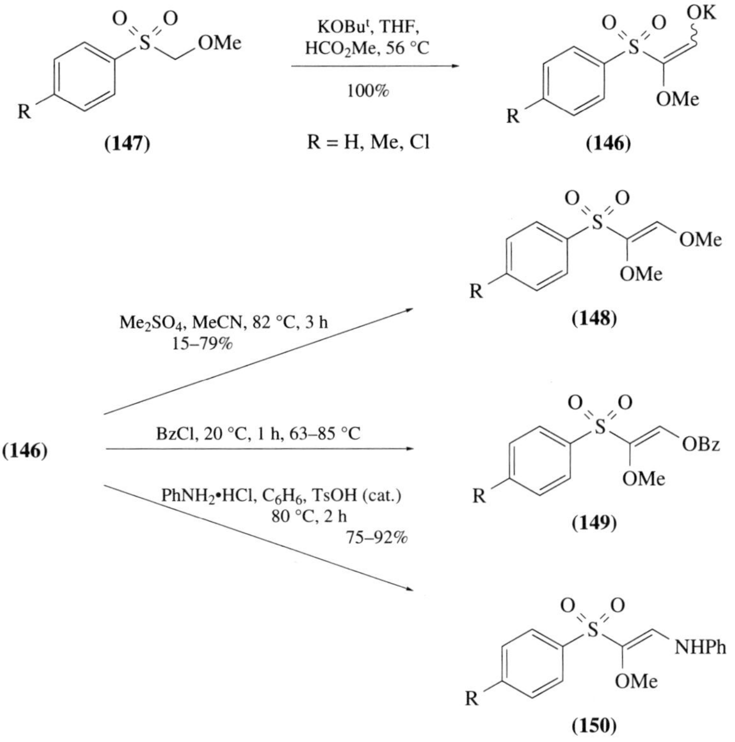

Scheme 12

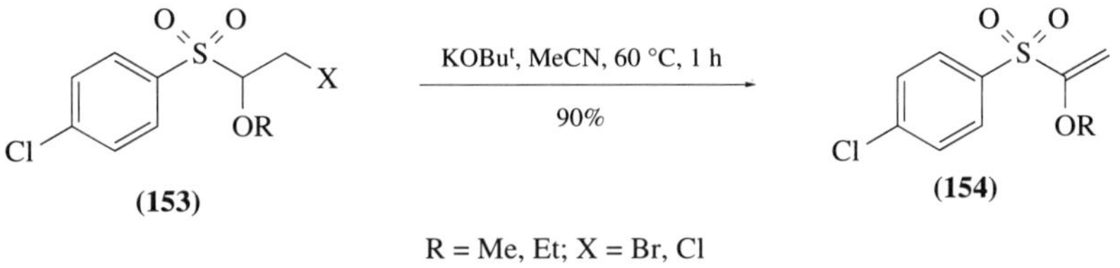

R = H, Me, Cl

(147) → (146): KOBu^t, THF, HCO$_2$Me, 56 °C, 100%

(146) → (148): Me$_2$SO$_4$, MeCN, 82 °C, 3 h, 15–79%

(146) → (149): BzCl, 20 °C, 1 h, 63–85 °C

(146) → (150): PhNH$_2$·HCl, C$_6$H$_6$, TsOH (cat.), 80 °C, 2 h, 75–92%

(153) → (154): KOBu^t, MeCN, 60 °C, 1 h, 90%   (47)

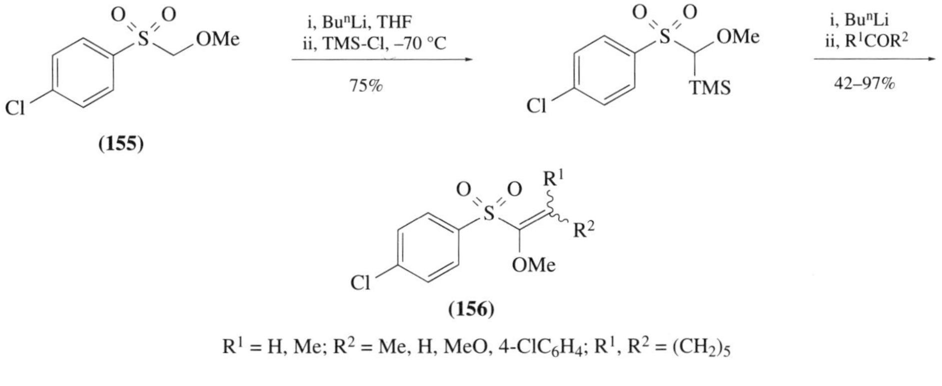

R = Me, Et; X = Br, Cl

(155) → intermediate: i, Bu^{n}Li, THF; ii, TMS-Cl, −70 °C, 75%

intermediate → (156): i, Bu^{n}Li; ii, R^{1}COR^2, 42–97%   (48)

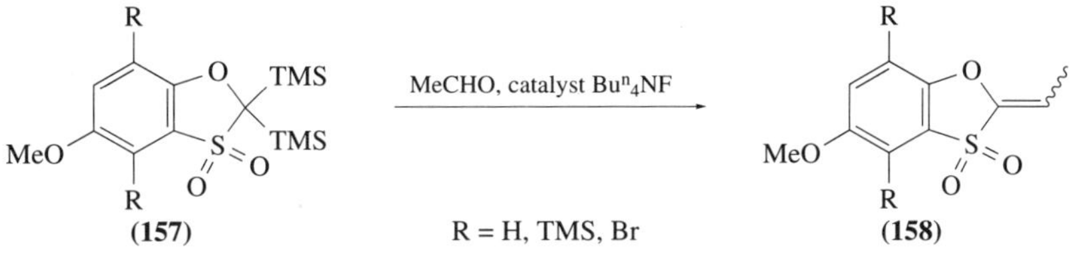

R^1 = H, Me; R^2 = Me, H, MeO, 4-ClC$_6$H$_4$; R^1, R^2 = (CH$_2$)$_5$

(157) → (158): MeCHO, catalyst Bu$^n_4$NF   (49)

R = H, TMS, Br

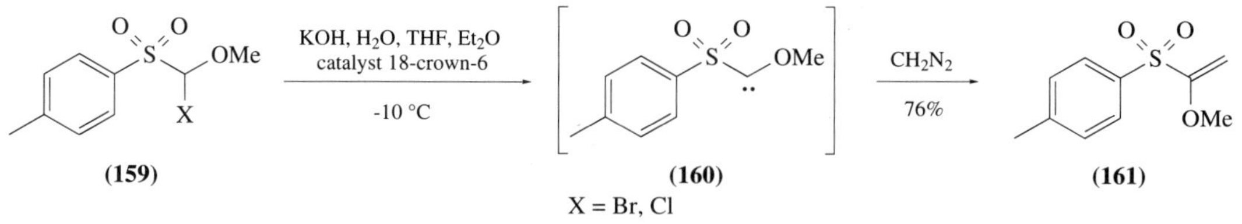

Scheme 13

An alternative method for the alkeneation of α-sulfonyl ethers, a one-pot Horner–Wittig reaction, was reported in 1990 by Lee and Oh ⟨90SC273⟩. The sulfone (**162**) was dilithiated with lithium diisopropylamide (LDA) and then treated sequentially with diethyl phosphorochloridate and carbonyl compounds to give the alkenes (**163**) (Scheme 14). Mixtures of isomers were obtained in each case for products derived from aldehyde substrates (only symmetrical ketones were used). $(E):(Z)$ ratios were approximately 1 : 1 for aliphatic and 9 : 1 for aromatic aldehydes.

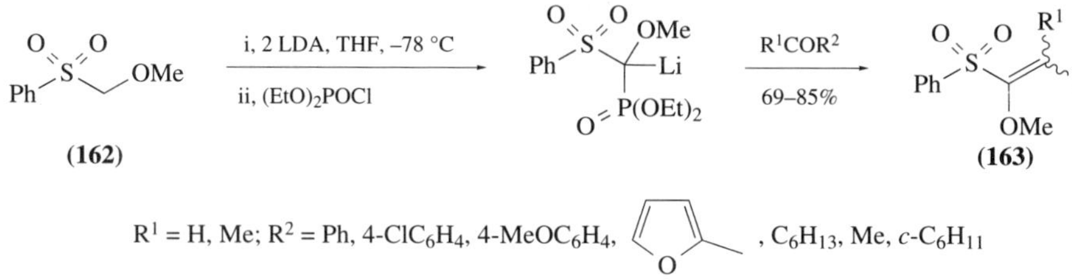

Scheme 14

4.19.3.3.3 From 1-sulfonyl alkynes

Addition of 3-phenylglycidol (**164**) to the 1-(benzenesulfonyl) alkyne (**165**) gave approximately equal amounts of the 1- and 2-alkoxyvinyl sulfones (**166**) and (**167**) (Equation (50)). However, selectivity for the 2-alkoxy regioisomer (**167**) increased when electron-donating groups were introduced into the 2-phenyl group ⟨90JCS(P1)2775⟩. Acheson and Ansell studied the reactions of an ethyne (**168**) which had strongly electron-withdrawing groups at both ends. They deduced from spectroscopic data that methanol or ethanol attack the carbon α to the sulfonyl group to give the 1-alkoxyvinyl sulfones (**169**), but they were unable to assign unambiguously the geometry of the double bond (Equation (51)) ⟨87JCS(P1)1275⟩. In a related reaction, de Lucchi and co-workers described the addition of methanol to bis(arenesulfonyl) alkynes to give trisubstituted alkenes, although the paper contains few details ⟨91TL2177⟩.

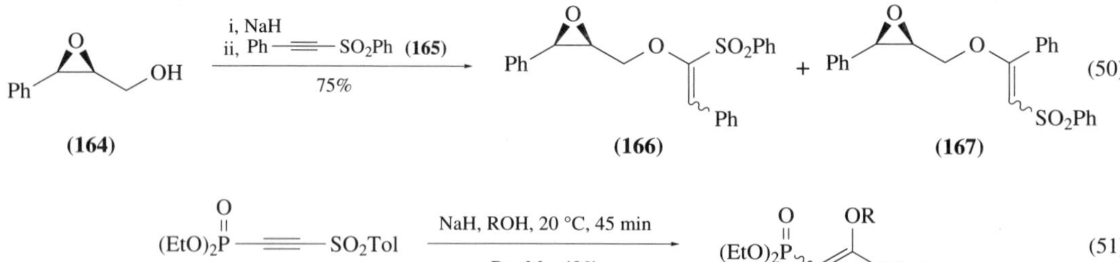

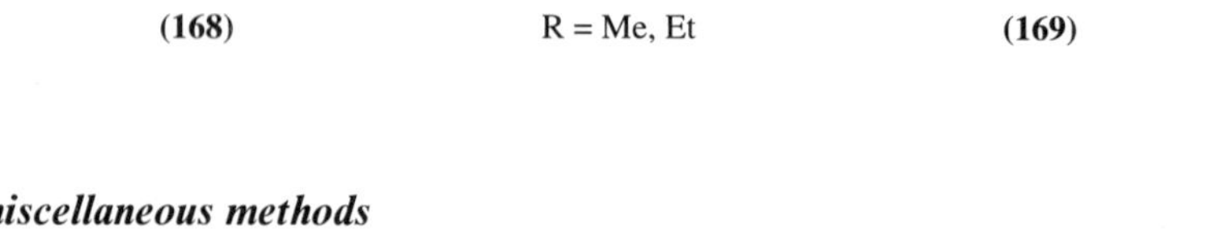

4.19.3.3.4 By miscellaneous methods

Dehydrochlorination of the 1-sulfonylalkyl carbamates (**170**) gave the 1-sulfonylvinyl carbamates (**171**) (Equation (52)) ⟨68JCS(C)3011⟩. Ring-opening of the tetrahydrofuranyl derivative (**172**) to give a mixture of the (E) and (Z) alkenes (173) provided another example of the use of elimination to prepare this functional group (Equation (53)) ⟨84CC1028⟩.

$$\text{(170)} \xrightarrow{\text{Et}_3\text{N, Et}_2\text{O, 20 °C, 1 h}}_{84-93\%} \text{(171)} \qquad (52)$$

R = Bu^n, Pr^i, Et

$$\text{(172)} \xrightarrow{\text{Bu}^n\text{Li or LDA}} \text{(173)} \qquad (53)$$

Hoppe and co-workers reported that deprotonation of the 1-sulfonylallyl carbamate (174), followed by addition of 2-methylpropanal in the presence of chlorotriisopropoxytitanium, gave exclusively the (E)-anti-adduct (175), resulting from attack at C-3 (Equation (54)) ⟨89TL2915⟩. In contrast, tetraisopropoxytitanium catalysed attack exclusively at C-1. Sengupta and Snieckus have studied the α-metallation of the enol carbamate (176) and the reactions of the metallated intermediate with electrophiles (see also Section 4.19.4) ⟨90JOC5680⟩. They found that the 1-sulfonylvinyl carbamate (177) was obtained when the 1-lithio derivative of Structure (176) was treated with tosyl fluoride (Equation (55)).

$$\text{(174)} \xrightarrow[85\%]{\text{i, Bu}^n\text{Li, THF, -78 °C} \atop \text{ii, Pr}^i\text{CHO, ClTi(OPr}^i)_3} (\pm)\text{-(175)} \qquad (54)$$

$$\text{(176)} \xrightarrow[75\%]{\text{i, Bu}^s\text{Li, TMEDA, THF, -78 °C} \atop \text{ii, TolSO}_2\text{F}} \text{(177)} \qquad (55)$$

4.19.4 FUNCTIONS CONTAINING OXYGEN AND EITHER SELENIUM OR TELLURIUM—$R^1_2C{=}COR^2SeR^3$, etc.

True ketene monoseleno- or monotelluroacetals, rather than substituted heterocyclic compounds, are rare. Radchenko and his group reported the addition of acetic acid to the heterosubstituted alkynes (178) and (179) to give the monoseleno- and monotelluroacetals (180) and (181) (Scheme 15) ⟨72ZOR1329, 74ZOR2269⟩. Later work from the same group showed that acetic acid added to the bis(methylseleno)alkyne (182) to give the (Z) adduct (183) (Equation (56)) but that the ditelluro analogue of (182) was simply protonated by acetic acid ⟨80ZOR720⟩. The potentially more general strategy of eneselenolisation and Se-alkylation of O-alkyl selenoesters was reported by Barton et al. ⟨77JCS(P1)1723⟩. Deprotonation of the selenoester (184) with potassium hexamethyldisilazide, followed by work-up with methyl iodide, gave the ketene monoselenoacetal (185) in 54% yield after distillation (Equation (57)). In a different approach, Sengupta and Snieckus treated the α-lithiated enol carbamate (186) with a range of electrophiles, including diphenyl diselenide, which gave the selenoacetal (187) (Scheme 16) ⟨90JOC5680⟩. This methodology was recently extended to the synthesis of the 2,2-difluoro analogue of (187) ⟨93TL415⟩.

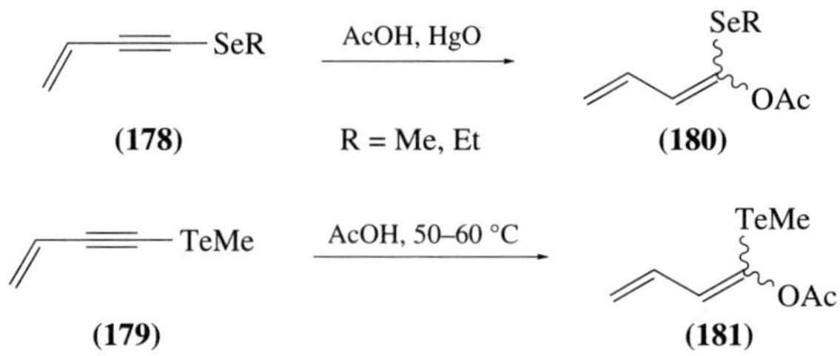

Scheme 15

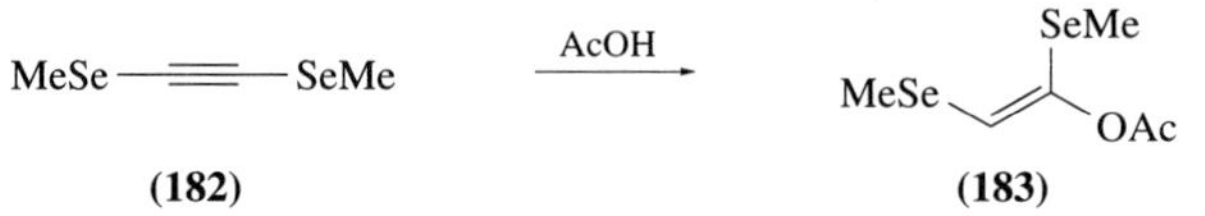

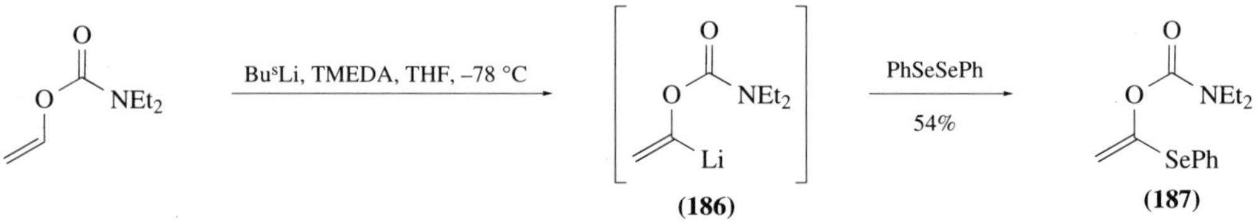

Scheme 16

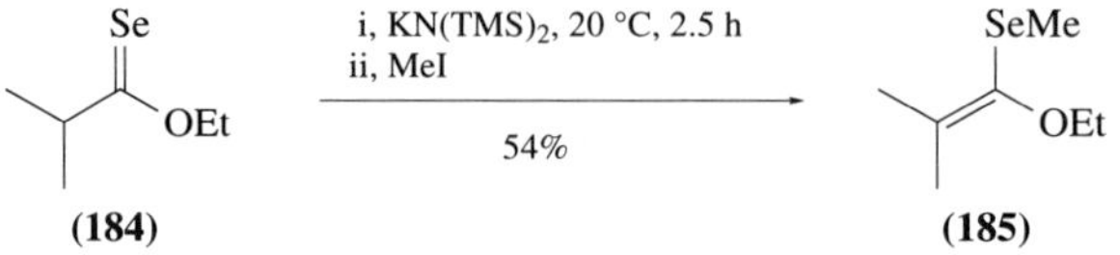

4.19.5 FUNCTIONS CONTAINING TWO SULFURS—$R^1{}_2C\!=\!CSR^2S(O)_2R^3$, etc.

The preparation and properties of ketene dithioacetals and their higher-coordinated analogues were reviewed in 1980 by Kolb ⟨B-80MI 419-01⟩ and by Schaumann ⟨85HOU(E11/1)260⟩.

4.19.5.1 Two Dicoordinated Sulfurs—$R^1{}_2C\!=\!C(SR^2)_2$, etc.

4.19.5.1.1 From dithiocarboxylic acids and derivatives

Conceptually, one of the simplest ways to generate a ketene dithioacetal is to deprotonate α to a dithiocarboxylic acid or ester and to trap the anion by attack on sulfur with an appropriate electrophile. This subsection also includes examples of dithiocarboxylates produced *in situ* from nucleophilic addition to carbon disulfide. This extra dimension to the method makes it one of the most widely used and versatile for preparing ketene dithioacetals.

Good examples of the classical deprotonation–alkylation reaction are the preparations of the tetra(methylthio) compound (**188**) ⟨89JCS(P1)1793⟩ and the cyclic dithioacetal (**189**) ⟨78JOC3065⟩ (see Scheme 17). Milder bases may be used when the α-hydrogen in the reactant is more acidic ⟨75CPB2390, 83TL2563⟩, and potassium fluoride on alumina was used without a solvent to prepare the (Z)/(E) isomeric mixture (**190**) (Equation (58)) ⟨85CC870⟩. The strong nucleophilicity of the thiolate anion means that S-alkylation is preferred when there is choice of sites for attack, for example, in the preparations of Structure (**188**) and the dithiomalonate derivative (**191**) (Equation (59)) ⟨76T2507⟩. While alkyl halides and dialkyl sulfates are most commonly used to trap the thiolate, examples of other electrophiles include acid halides ⟨82TL1075⟩, trimethylsilyl chloride ⟨79S455⟩, alkyl tosylates ⟨88ACS629⟩, phosgene ⟨79LA1715⟩, *p*-benzoquinone as a Michael acceptor ⟨70JOC283⟩, salicylaldehyde ⟨75JCS(P1)1277⟩ and α-bromo ketones ⟨82TL1075⟩. Intramolecular additions to allenes

⟨69RTC597⟩ and alkynes ⟨73RTC1067⟩ have also been reported, giving, for example, the cyclic products **(192)** and **(193)** respectively (Scheme 18).

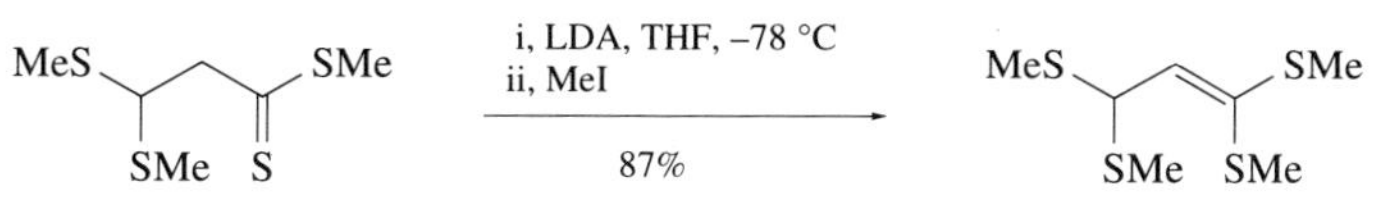

Scheme 17

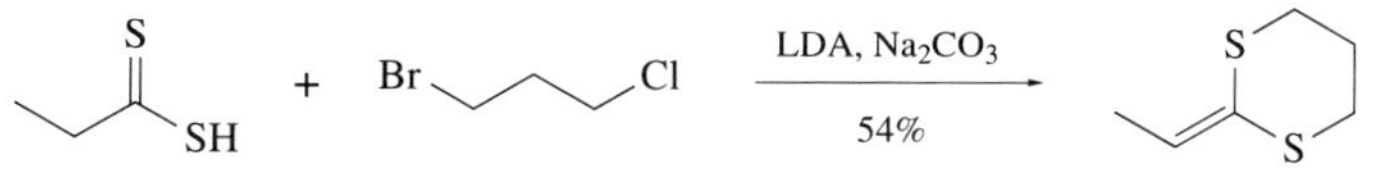

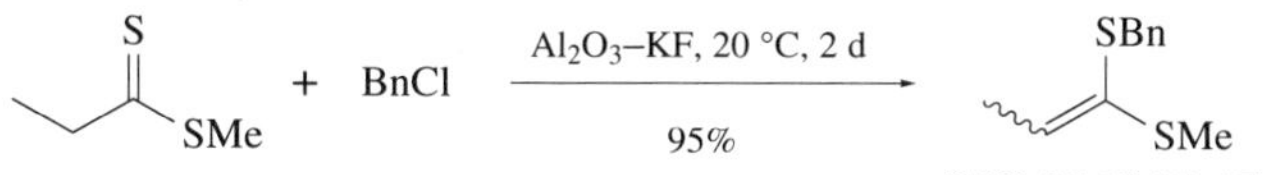

Scheme 18

Production of dithiocarboxylates *in situ* usually entails the deprotonation of an active methylene group followed by its reaction with carbon disulfide. Typical examples are the conversion of methyl cyanoacetate **(194)** into the ketene dithioacetal **(195)** ⟨62CB2861⟩ and the preparation of the product **(196)** from Meldrum's acid **(197)** (Scheme 19) ⟨86S967⟩. The choice of base depends to a large extent on the acidity of the methylene hydrogens. A diverse range of structures has been prepared under a variety of conditions (e.g., ⟨85S958, 86OS(64)189, 89H(29)1877, 89JHC1771, 90S991, 91S481⟩). Nucleophilic bases can occasionally cause problems, as in the unexpected formation of the *S*-butyl derivative **(198)** resulting from thiophilic attack by *n*-butyllithium on the dithiocarboxylate intermediate **(199)** (Scheme 20) ⟨87TL3805⟩. Dithiocarboxylation of Grignard reagents, followed by alkylation, has also been used in the preparation of ketene dithioacetals ⟨88CB1165⟩. Synthesis of the styrene derivative **(200)** started with the reaction of a phosphorane **(201)** with carbon disulfide (Equation (60)) ⟨79CB28⟩.

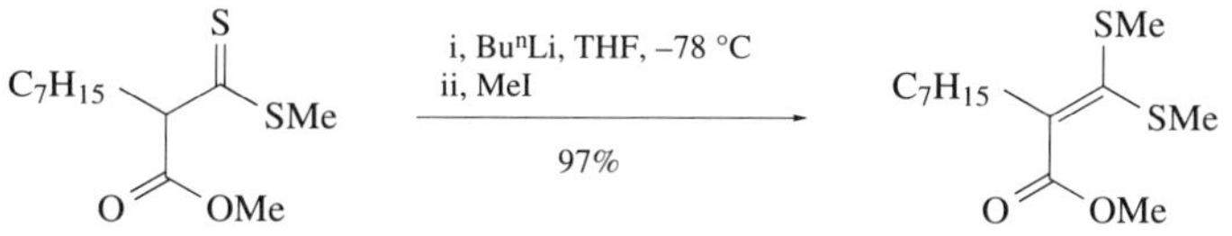

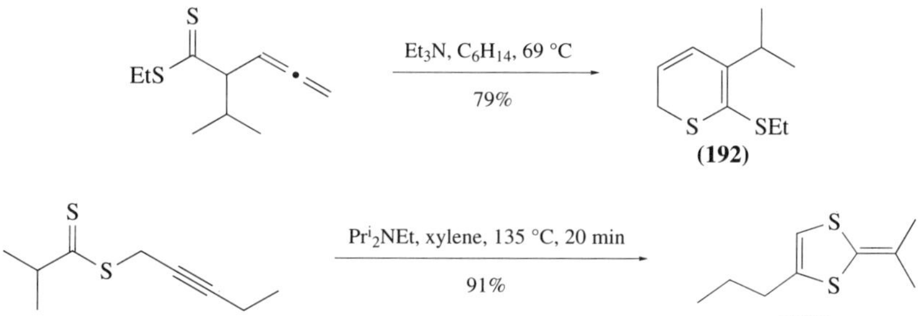

Other conversions of dithiocarboxylic acid derivatives into ketene dithioacetals are not as common. Wesdorp *et al.* reported the thiophilic attack of diethylamine on the α-bromo dithiocarboxylate **(202)** to generate the sulfenamide **(203)** (Equation (61)) ⟨74RTC184⟩. [4 + 2]-Cycloadditions of dienophiles to α,β-unsaturated dithioesters can generate ketene dithioacetals. Whitham and co-workers published a typical example in which the dithioester **(204)** was produced *in situ* and trapped with maleic anhydride **(205)** to give the adduct **(206)** (Equation (62)) ⟨84JCS(P1)859⟩. A similar reaction was reported by Kalish *et al.* ⟨71TL2241⟩. Analogous cycloadditions, using dimethyl acetylenedicarboxylate (DMAD) as the dienophile, have been described by Tominaga and co-workers ⟨73CPB2770, 75CPB2749, 91JHC1245⟩.

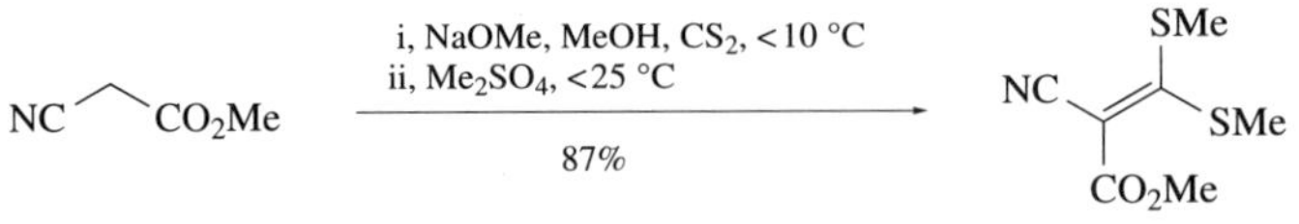

(194) **(195)**

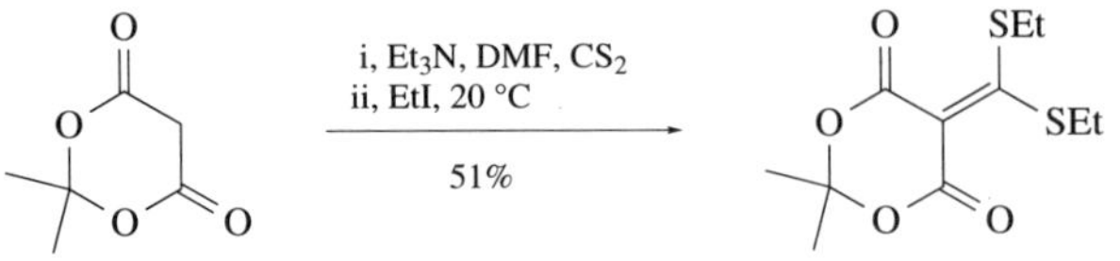

(197) **(196)**

Scheme 19

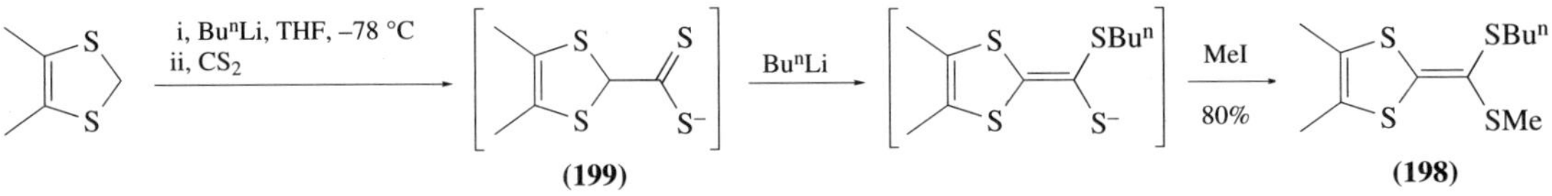

(199) **(198)**

Scheme 20

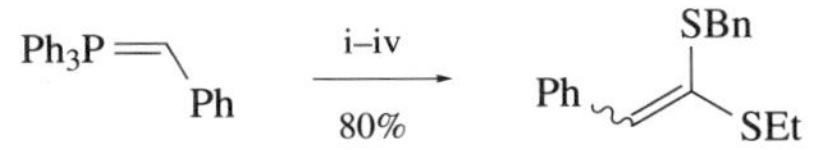

(201) **(200)** (60)

i, CS_2, 0–5 °C; ii, BnCl, reflux; iii, EtI, 70 °C; iv, KOH, H_2O

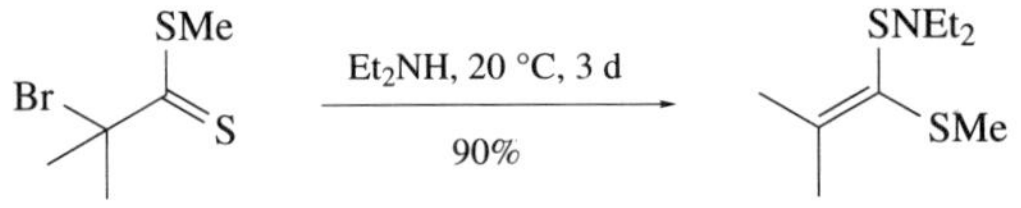

(202) **(203)** (61)

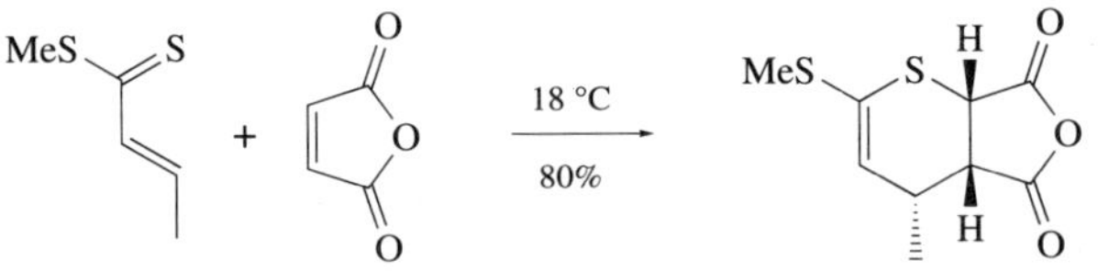

(204) **(205)** **(206)** (62)

4.19.5.1.2 By double bond formation via elimination

In general terms, this section covers four types of elimination reaction: those in which the leaving group is on the dithioacetal carbon (Type A) or the α-carbon (Type B), reductive eliminations (Type C) and oxidative 'eliminations' where hydrogen is removed from both carbons (Type D) (Scheme 21).

The elimination of a thiol from a trithio *ortho*-ester, typified by the elimination of ethanethiol to give the ketene dithioacetal (**207**) (Equation (63)) ⟨59RTC354⟩, has become a less common method since the 1960s. In 1985 Nsunda and Hevesi reported the Lewis acid-catalysed elimination of alkanethiols at low temperature using tertiary amines, to give the simple ketene dithioacetals (**208**) (Equation (64)) ⟨85CC1000⟩. Tin or titanium tetrachlorides were effective catalysts but aluminum chloride gave no reaction. Phosphorus iodides with triethylamine have also been used to effect this transformation ⟨82TL3407⟩. Substitution on the dithioacetal carbon, to provide the leaving group,

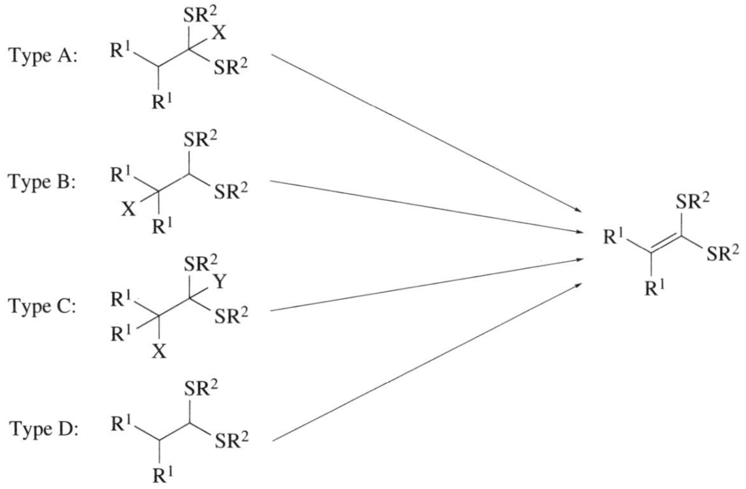

Scheme 21

Type A, B, C, D structures leading to ketene dithioacetal.

followed by elimination has been used by several groups. The bis(dithianyl)methane (**209**) was deprotonated and the anion was trapped with 2,2′-dipyridyl disulfide (**210**) to give the intermediate (**211**). 2-Pyridinethione was eliminated on treatment with aqueous ammonia to give the ketene dithioacetal (**212**) (Scheme 22) ⟨89JCS(P1)1793⟩.

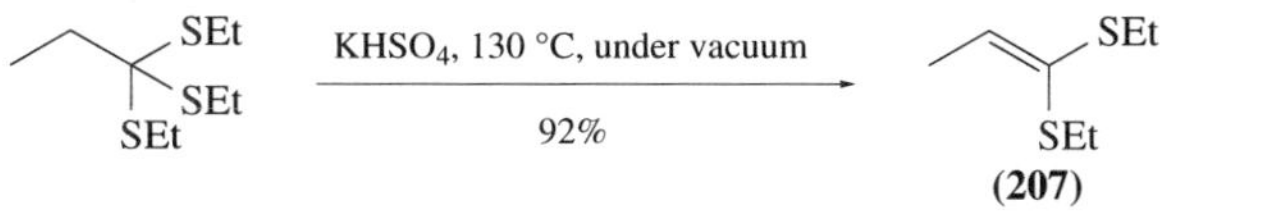

$$\text{CH}_3\text{CH}_2\text{C}(\text{SEt})_3 \xrightarrow[\text{92\%}]{\text{KHSO}_4,\ 130\ °\text{C, under vacuum}} \text{CH}_3\text{CH}=\text{C}(\text{SEt})_2 \quad (207) \tag{63}$$

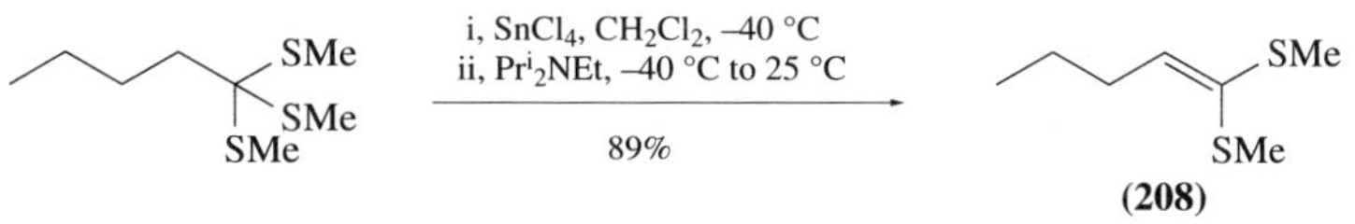

$$\text{Bu}\text{CH}_2\text{C}(\text{SMe})_3 \xrightarrow[\text{89\%}]{\substack{\text{i, SnCl}_4,\ \text{CH}_2\text{Cl}_2,\ -40\ °\text{C} \\ \text{ii, Pr}^i_2\text{NEt},\ -40\ °\text{C to 25 °C}}} \text{BuCH}=\text{C}(\text{SMe})_2 \quad (208) \tag{64}$$

(**209**) + (**210**) → [Bu^{n}Li, THF, −78 °C] → (**211**) → [i, 20 °C; ii, NH$_4$OH, H$_2$O; 75%] → (**212**)

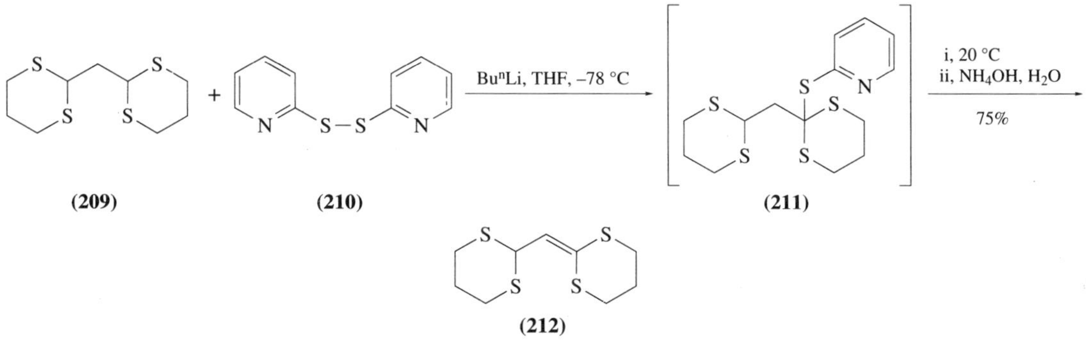

Scheme 22

The corollary to this strategy is to start with the leaving group in place and then introduce the thiol groups by substitution. The reaction developed by Corey and Beames, originally as a method for protecting esters and lactones against nucleophilic attack, has been used preparatively for converting carboxylic acids and esters into ketene dithioacetals ⟨73JA5829⟩. In principle, the method is attractive, given the ready availability of carboxylic acids as starting materials, but the reaction has not found wide favour for this purpose since the mid 1970s when the work was first published. Yields can be excellent, as shown by the preparations of the compounds (**213**) and (**214**). When the substrate is a lactone (e.g., Structure (**215**)) then the eliminated alcohol forms part of the product and can add to the double bond of the dithioacetal under acidic conditions (Scheme 23). A modification to the standard conditions, using carboxylic acids as substrates, was published by Cohen *et al.* ⟨79JOC4744⟩. Other variations on this theme include the use of Lewis acid catalysis to effect elimination ⟨88CB1165⟩ and thiolation–elimination sequences starting from acid chlorides ⟨59RTC354, 87G227⟩.

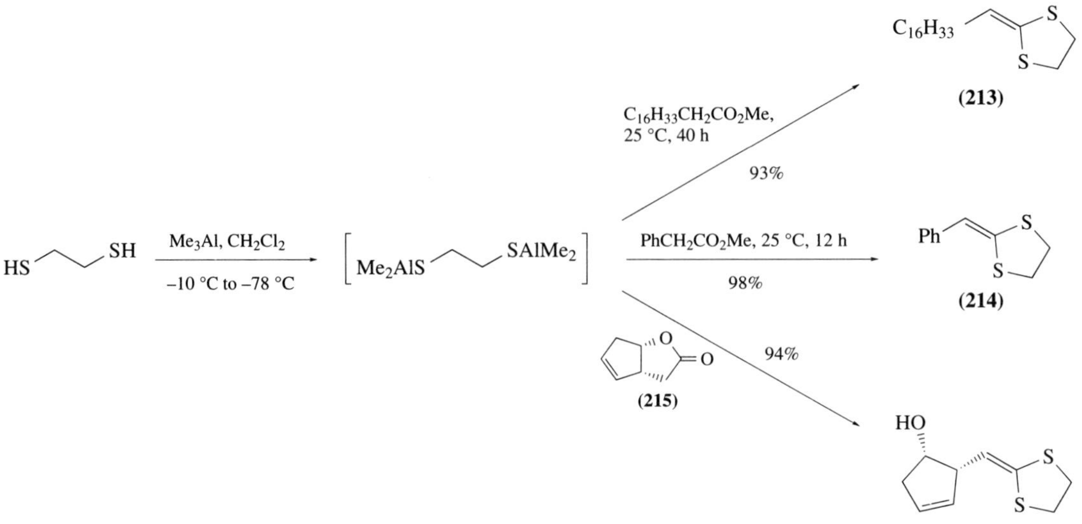

Scheme 23

Type B eliminations (see Scheme 21) often require more elaborate synthesis of the substrate to set up the correct arrangement of substituents. In one example, the elimination of butanethiol from the α-thio dithioacetal (**216**), acetylene was used to trap the thiolate as it was formed and drive the equilibrium in favour of the product (**217**) (Equation (65)) ⟨61JOC1987⟩. Taschner and Kraus found that treatment of the furan derivative (**218**) with *n*-butyllithium at −40 °C gave the aldehyde (**219**), after ring-opening and rearrangement, although the reaction was suppressed at −78 °C (Equation (66)) ⟨78JOC4235⟩. A vinylogous example of this type of elimination is the Grob fragmentation of the tosylate (**220**) to give the ring-opened product (**221**) (Equation (67)) ⟨71TL871⟩.

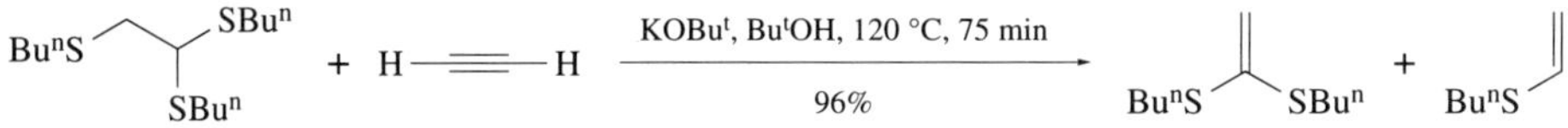

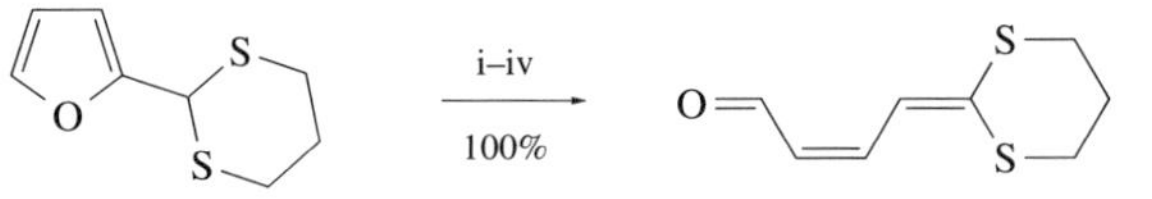

i, Bu^{n}Li, THF, −40 °C; ii, −20 °C, 2 h; iii, TMS-Cl, −78 °C; iv, H$_2$O

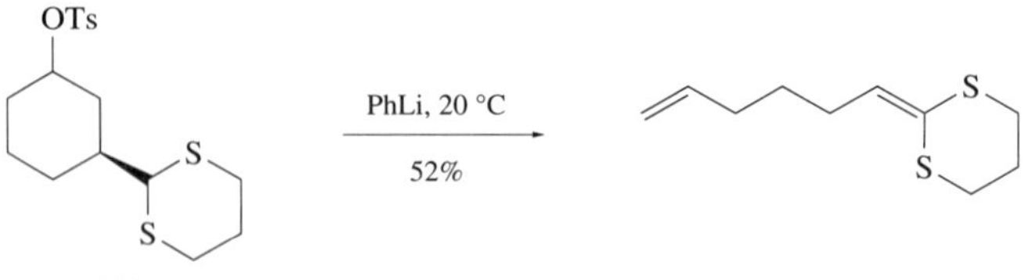

There have been a few isolated examples of Type C (reductive) eliminations (see Scheme 21). Moses and Chambers reported that thermal elimination of diethyl disulfide from the polythio compound (**222**) gave the tetra(methylthio)tetrathiafulvalene (**223**) (Equation (68)) ⟨74JA945⟩. An analogous reaction, under acidic conditions, was published by Brown and co-workers ⟨84TL995⟩. Demethoxycarbonylation and elimination of lithium pivalate from the diester (**224**) gave the ketene dithioacetal (**225**) in quite good yield (Equation (69)) ⟨84TL5729⟩.

Some extended conjugation systems have been accessed by a final oxidative step, or Type D 'elimination' (see Scheme 21). One example is the preparation of the Weitz-type donor (**226**) ⟨89TL2091⟩. A similar example was reported by Ueno *et al.* (Equation (70)) ⟨78CC74⟩. Using a different approach, Yoshida *et al.* converted the dithioacetal (**227**) into the ketene dithioacetal (**228**) via the sulfimide (**229**) (Scheme 24) ⟨76S552⟩.

4.19.5.1.3 By alkenation methods

Reactions in this category involve condensation of a carbonyl compound, or its equivalent, with a dithioacetal anion, usually stabilized by an activating group.

Reaction conditions, and the range of substrates available for Wittig and Horner–Emmons-type reagents, are fairly typical of this type of reaction in general. The two α-thio groups increase the acidity of the methine hydrogen to make the reagents more like their resonance-stabilised counterparts in the standard Wittig reactions. Thus, α,α-dithiophosphoranes do not generally react with ketones ⟨81S53⟩ and the more reactive phosphonate esters must be used ⟨76TL2731⟩. This method for the synthesis of ketene dithioacetals has been used extensively and the examples in Scheme 25 illustrate the typical range of reagents and conditions which have been employed ⟨76TL2631, 78T3081, 79JOC930, 80CJC2780, 80H(14)271, 91S26⟩.

Some variations of the normal reaction sequence have been reported. Bulpin *et al.* used the phosphonodithioformate (**230**), which was readily prepared from diisopropyl phosphite and carbon disulfide ⟨89TL3415⟩. Thiophilic attack by methyllithium produced the anion (**231**), which was then treated with benzaldehyde to give the ketene dithioacetal (**232**) (Scheme 26). While phosphoranes are unreactive towards ketones, Cava and co-workers showed that the Wittig reagent (**233**) would react with a dithiolium salt to give the tetrathiofulvalene (**234**) by elimination of triphenylphosphine (Equation (71)) ⟨78JOC369⟩. A similar procedure was reported by Fabre *et al.* ⟨87BSF823⟩. DiNinno *et al.* used the reverse of the normal strategy to prepare the penem precursor (**235**), although the reaction was fairly slow (Equation (72)) ⟨82TL3535⟩.

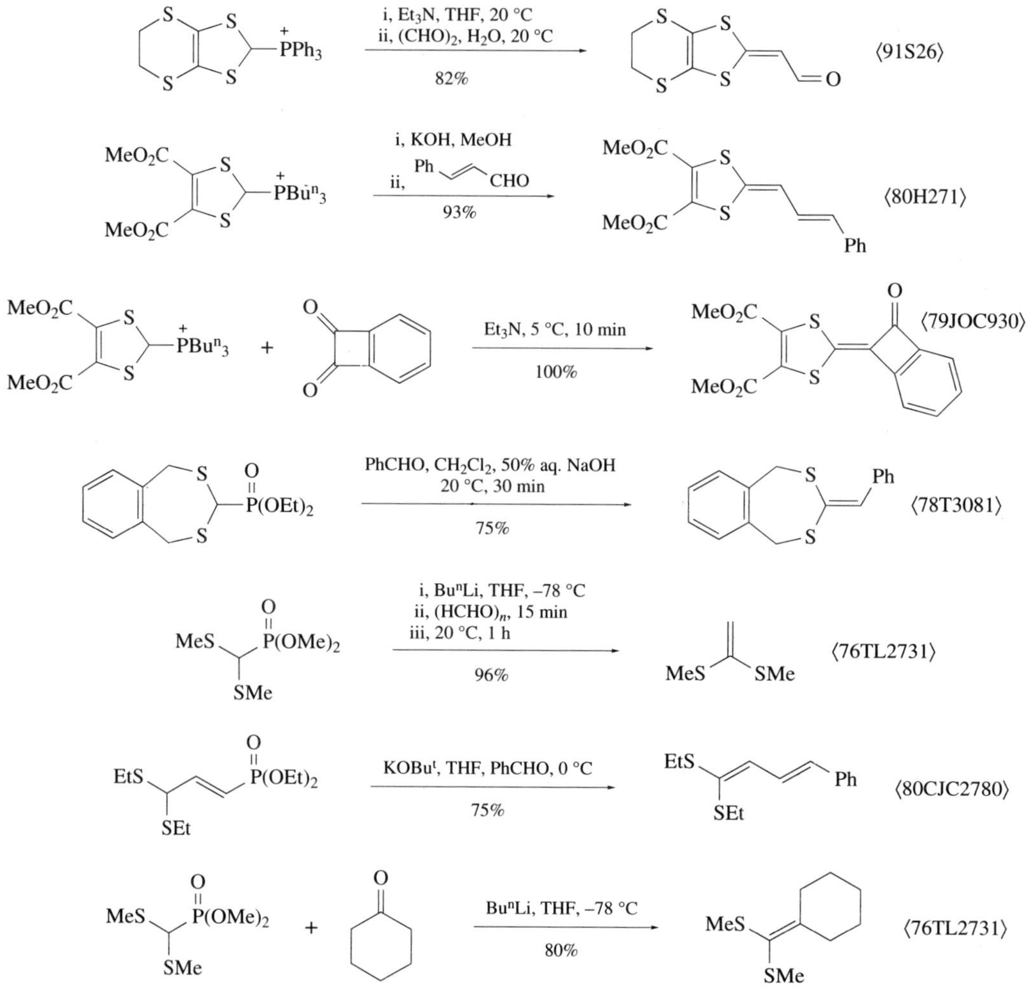

Scheme 25

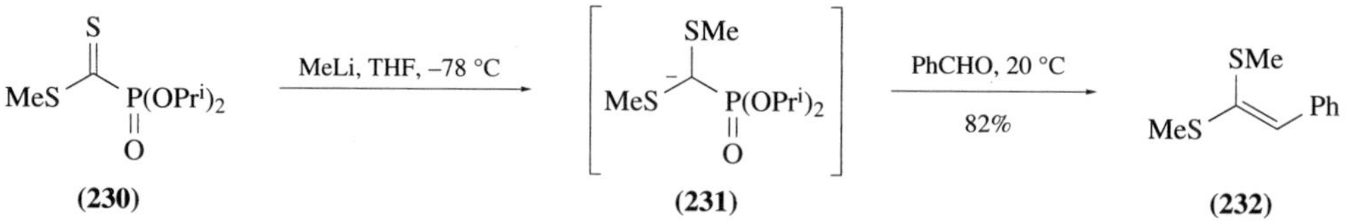

Scheme 26

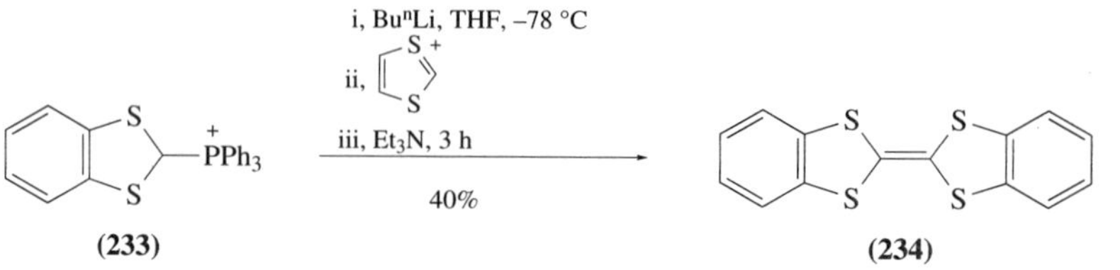

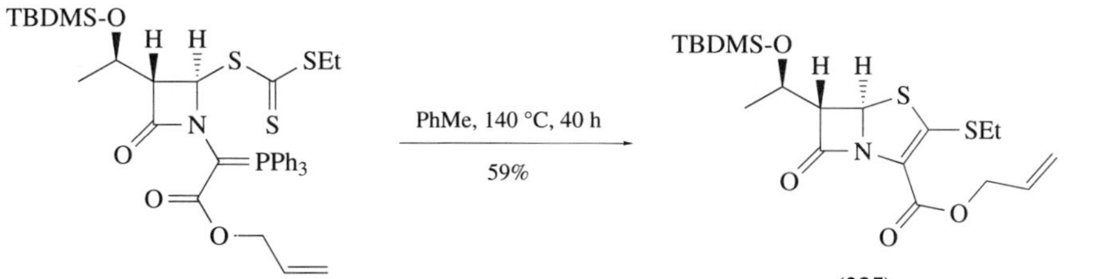

(72)

(235)

The Peterson alkenation reaction, exemplified in Scheme 27, is often the method of choice for ketone substrates. The method has been applied to α,β-unsaturated aldehydes ⟨73CB2277⟩ but was unsuccessful for conjugated enones, with Michael addition occurring almost exclusively ⟨75S461⟩. Other functional groups, such as epoxides, can also be vulnerable to attack ⟨90JOC5555⟩. Yields are generally very high and there has been little variation on the standard conditions ⟨73JCS(P1)2272, 75S461, 82S579⟩.

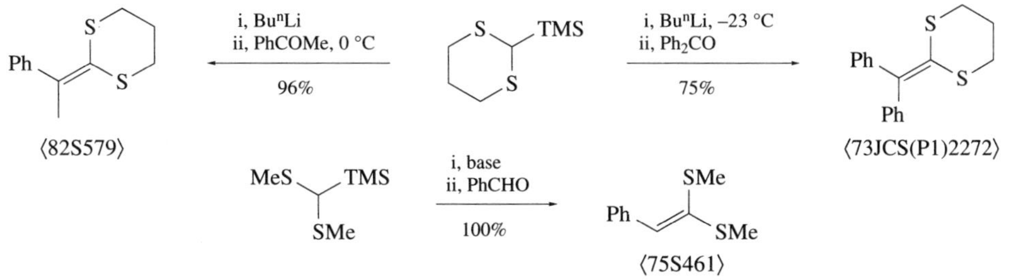

Scheme 27

Apart from phosphorus and silicon, boron has also been used in an activating group for this type of alkenation, for example, the boronate ester **(236)** which was used to prepare the dithioacetals **(237)** (Equation (73)) ⟨79JOC1352⟩. No extra stabilisation of the anion of the formaldehyde dithioacetal **(238)** was required when the electrophile was benzonitrile, which gave the enamine product **(239)** (Equation (74)) ⟨88JCS(P1)269⟩.

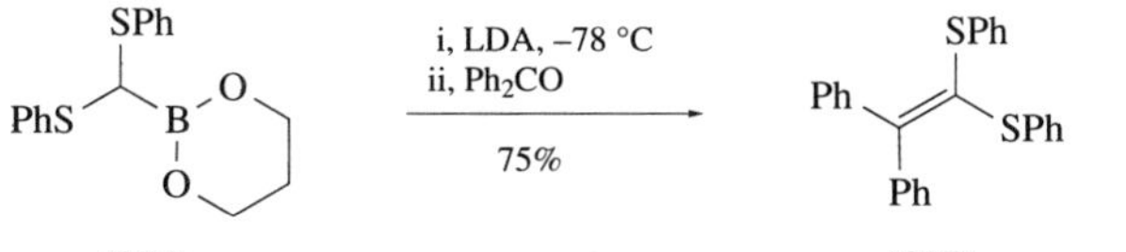

(73)

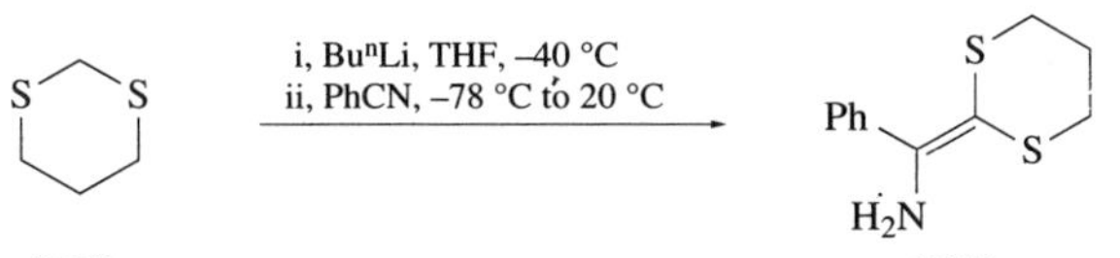

(74)

4.19.5.1.4 By rearrangement of α,β-unsaturated dithioacetals

This reaction is most common when combined with addition to a carbonyl compound in an analogy of the enamine reaction. A good example, demonstrating the regio- and diastereoselective control possible with this method, is the reaction of the crotonaldehyde dithioacetal **(240)** with benzaldehyde to give, predominantly, the *anti*-isomer **(241)** (Equation (75)) ⟨87JOC855⟩. Two other examples involve Michael additions to conjugated enones ⟨82JA7174, 84LA450⟩.

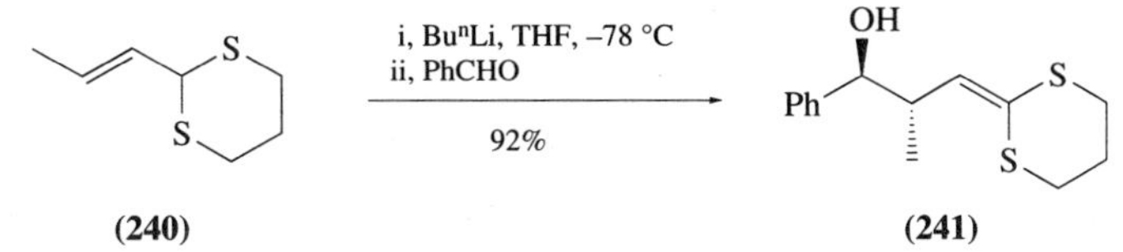

(75)

4.19.5.1.5 From 1,3-dithioles and analogues

This chemistry may be divided into four main themes: self- or cross-coupling reactions of 1,3-dithiole-2-thiones, with loss of sulfur; 1,3-dipolar cycloadditions to 1,2-dithiole-3-thiones; nucleophilic additions to 1,3-dithiolium salts; dimerisation of 1,3-dithiolium salts via carbene intermediates.

The first of these four categories has been widely used for the preparation of tetrathiafulvalenes in the organic conductors area. The reaction normally involves heating a 1,3-dithiole-2-thione with a trialkyl phosphite or triphenylphosphine and examples of products are the bisthieno compound (**242**) ⟨83JOC4713⟩ and the tetracyano derivative (**243**) (Scheme 28) ⟨75JOC2577⟩. Some groups have claimed superior yields by conversion of the thiono group to an oxo group prior to coupling ⟨74CC751, 87S837⟩. The reaction has also been carried out under photolytic ⟨85CC1803⟩ and radical conditions ⟨76JA7440, 78JA3868⟩. Attempts at cross-coupling reactions have not met with great success unless the two thiones have very similar reactivity, otherwise two symmetrical products are formed at different rates ⟨78ANY355⟩. In some cases, an entirely different species has been used as the other substrate, such as a ketone ⟨88CC822⟩ or aldehyde ⟨67TL3201⟩.

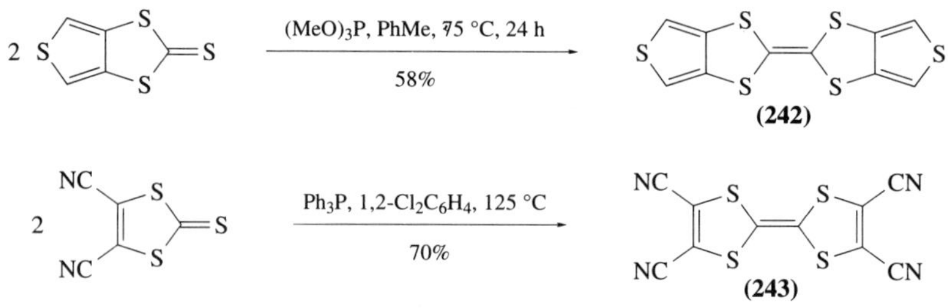

Scheme 28

1,2-Dithiole-3-thiones (e.g., Structure (**244**) ⟨72JCS(P1)41⟩) and isothiazoline-5-thiones (e.g., Structure (**245**) ⟨80JCS(P1)2693⟩) react as 1,3-dipoles with a variety of 1,3-dipolarophiles (Scheme 29). This chemistry has been reviewed extensively ⟨84CHEC(6)798, 84CHEC(6)848⟩ and will not be covered further here.

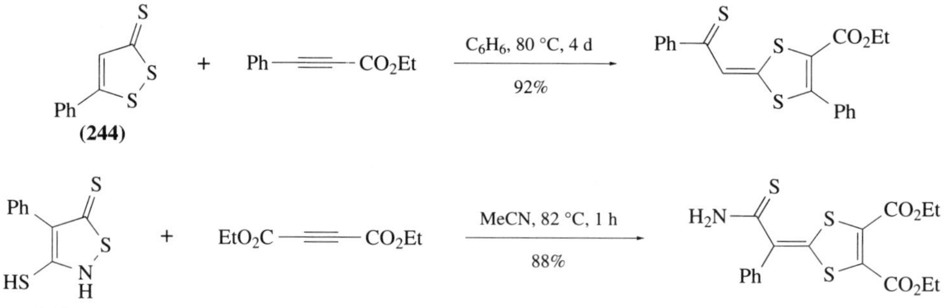

Scheme 29

Preparation of ketene dithioacetals by nucleophilic attack of a carbanion on a dithiolium salt is usually combined with the elimination of methanethiol. A typical range of active methylene compounds have been condensed with dithiolium salts in this way, for example, to give the malononitrile derivative (**246**) (Equation (76)) ⟨64CB1298⟩ (see also ⟨75JPR137⟩ and ⟨77CL287⟩. There have also been vinylogous analogues ⟨86CL1623⟩, other enamine reactions ⟨80BCJ1661, 81CC565⟩ and one example of a dithiazolium salt as the electrophile ⟨85CC696⟩.

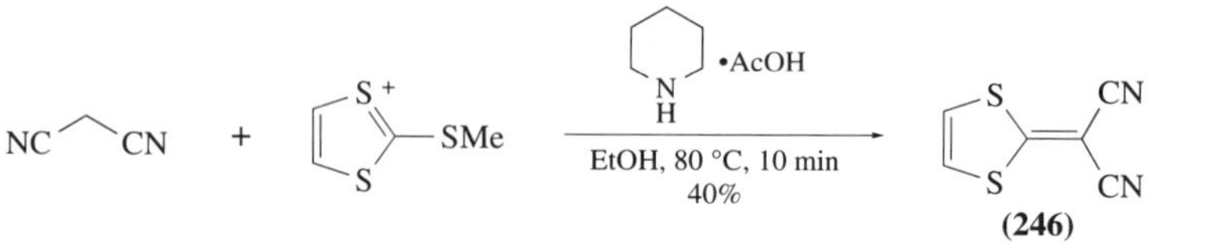

(76)

Dithiolium salts which are unsubstituted at the 2-position give resonance-stabilised carbenes on treatment with base. These carbenes can then dimerise to give symmetrical tetrathiafulvalenes. One example is the very high-yielding preparation of the parent compound (**247**) (Equation (77)) ⟨76S489⟩ and an intramolecular version of this reaction was reported by Staab *et al.* ⟨80AG(E)66⟩. Meli and Bianchini prepared the dithiolium carbene (**248**), from a nickel–sulfur complex (**249**) and the substituted alkyne (**250**). The carbene (**248**) dimerised to the tetrathiafulvalene (**251**) (Scheme 30) ⟨83CC1309⟩.

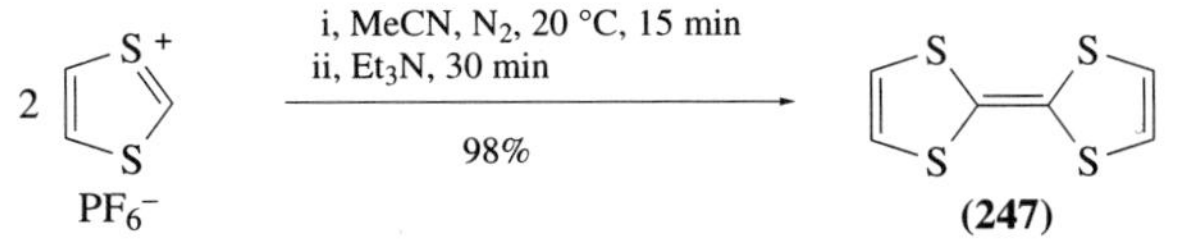

(247) (77)

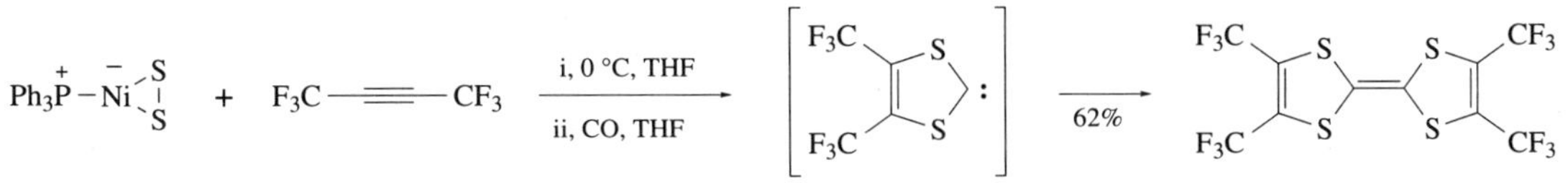

(249) **(250)** **(248)** **(251)**

Scheme 30

4.19.5.1.6 *From* gem-*dihalogeno alkenes and analogues*

Vinylic chloride displacement by sulfur nucleophiles has been used in a few specific cases. The cumulenes (**252**) were prepared by the treatment of perchloro-1,3-butadiene (**253**) with the arenethiolate (**254**) (Equation (78)) ⟨77CB1484⟩ and 3,3-dichloroacrolein (**255**) gave the cyclic dithioacetal (**256**) in excellent yield when treated with propane-1,3-dithiol (Equation (79)) ⟨89JCS(P1)1793⟩. In a related reaction, the dichloro alkene (**257**) was thiolated with tetraethylammonium hydrosulfide and then treated with alkyl halides to give dithioacetals (e.g., Structure (**258**) (Equation (80)) ⟨80S907⟩). Treatment of dichloroacetylene (**259**) with ethanethiol, under basic conditions, initially gave a 1,2-dithio alkyne which reacted with more thiol to give the trithio alkene (**260**) (Equation (81)) ⟨90TL2169⟩.

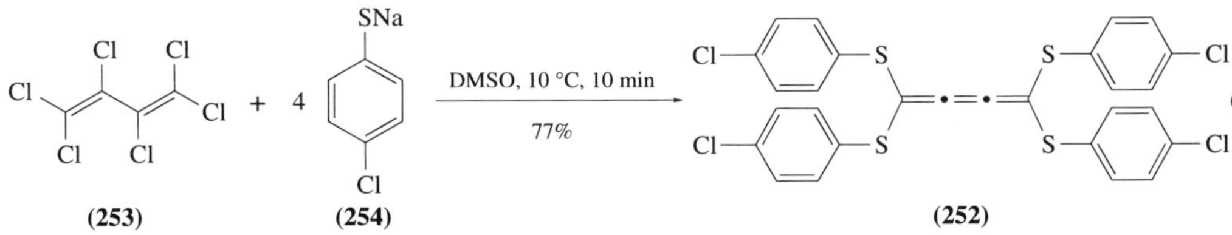

(78)

(253) **(254)** **(252)**

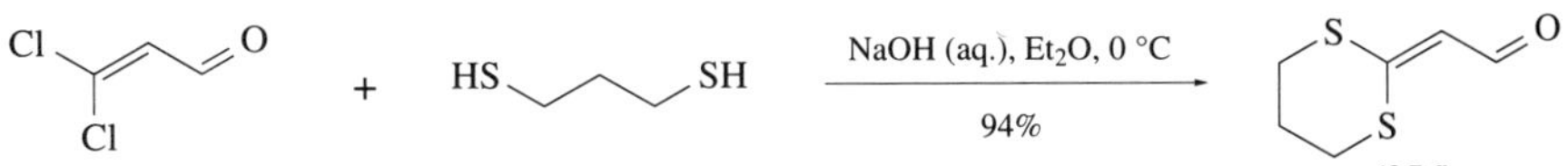

(255) **(256)** (79)

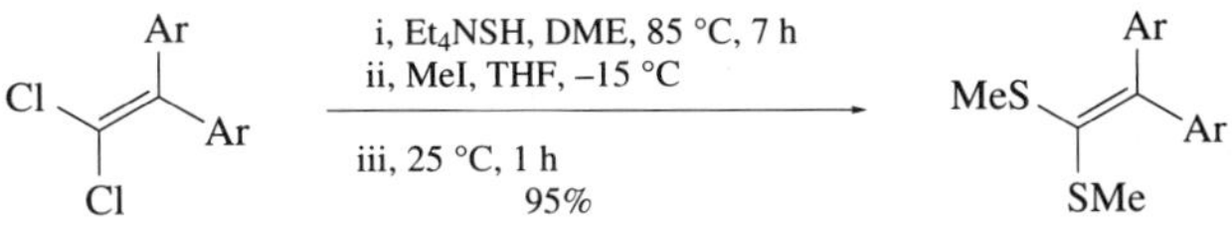

(80)

(257) **(258)**

Ar = 4-ClC$_6$H$_4$

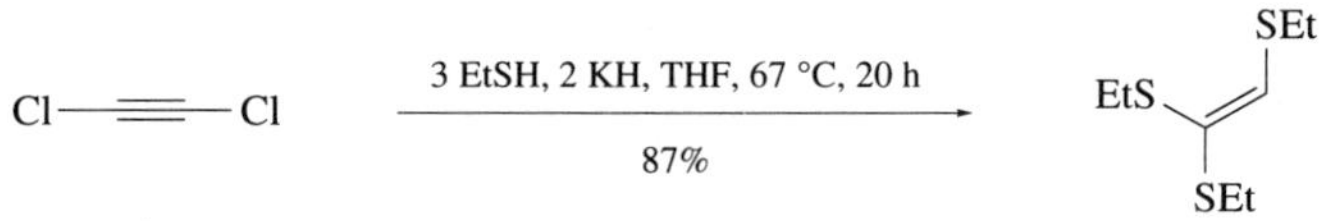

$$\text{Cl}\text{—}\!\!\equiv\!\!\text{—}\text{Cl} \quad \xrightarrow[87\%]{\text{3 EtSH, 2 KH, THF, 67 °C, 20 h}} \quad \underset{\text{SEt}}{\overset{\text{SEt}}{\text{EtS}\diagup\!\!=\!\!\diagdown}} \qquad (81)$$

(259)   (260)

4.19.5.1.7 By miscellaneous methods

A myriad of isolated preparations of ketene dithioacetals has been published over the years, an indication of the seemingly endless versatility and variety of organosulfur chemistry. Many of these reactions have little more than curiosity value but some, which may have a more general synthetic application, have been included here.

There has been considerable interest in 3,5-bismethylene-1,2,4-trithioles and the area was reviewed comprehensively in 1984 ⟨84CHEC(6)890⟩. Almost all recorded preparations have involved the oxidative dimerisation of enethiolizable dithiocarboxylic acids, as in the formation of (261) (Equation (82)) ⟨77JCS(P1)1273⟩. The oxidant is usually iodine ⟨65LA(684)37, 72JOC3226, 76JCS(P1)1706⟩ but bromine ⟨70BCJ2938⟩, ammonium persulfate ⟨62CB2861, 71CJC1477⟩ and sulfuric acid ⟨73BSF581⟩ have also been used. The intermediate oxidation products have also been converted separately into the trithioles thermally ⟨66T3001, 74AC(R)305⟩.

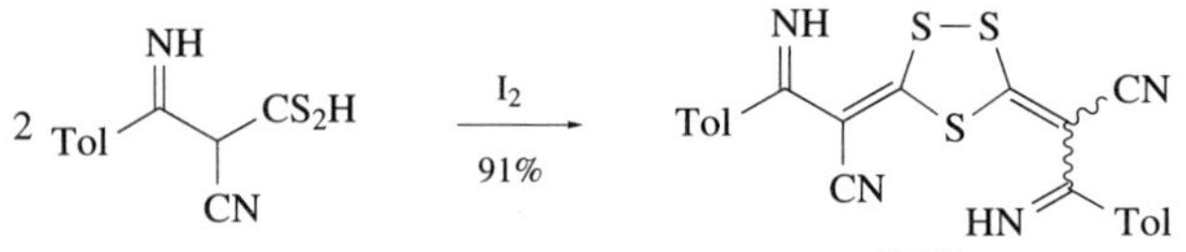

$$(82)$$

(261)

An analogue of the electrophilic reactions of 2-alkylthio-1,3-dithiolium salts (see Section 4.19.5.1.5) is the reaction of the trithiocyclopropenium salt (262) with the cyclopentadienyl anion (263) to give the dithioacetal (264) (Equation (83)). The authors propose that a vinyl carbene is involved ⟨89HCA1506⟩.

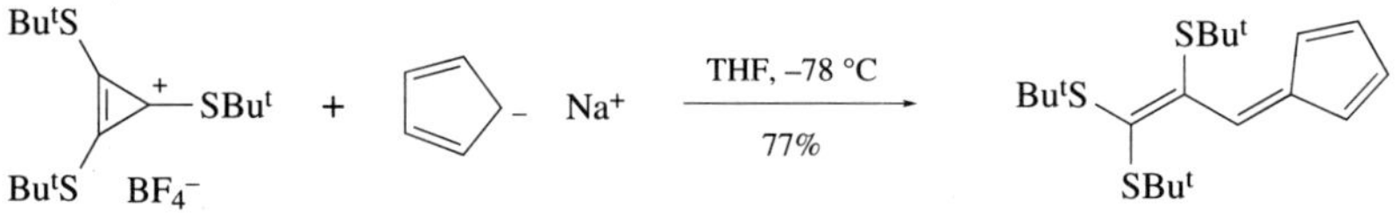

$$(83)$$

(262)   (263)   (264)

[2 + 2]-Cycloadditions of thiocarbonyl compounds have also been applied to the preparation of ketene dithioacetals. Schaumann showed that the thioketene (265), generated *in situ* from the β-thiolactam (266), underwent a [2 + 2]-cycloaddition with thiobenzophenone to give the dithietane (267) (Scheme 31) ⟨76CB906⟩.

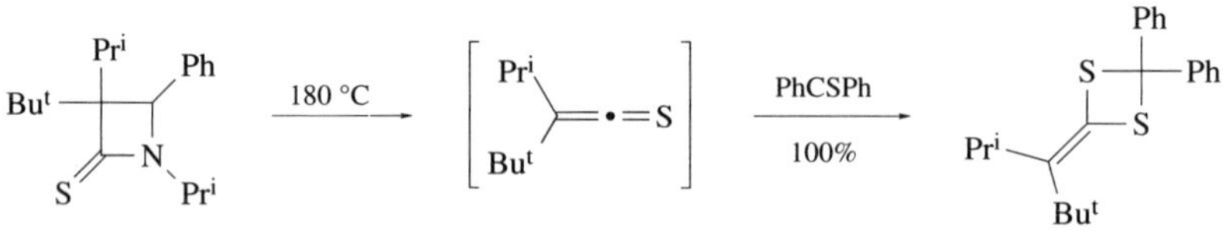

(266)   (265)   (267)

Scheme 31

1,2,3-Thiadiazoles (268) dimerise, with loss of nitrogen, to give ketene dithioacetals (269) under basic conditions (Equation (84)) ⟨68CJC2251, 73JHC11⟩ or, if the thiadiazole is 4,5-disubstituted, by photolysis ⟨58LA(614)4⟩. Treatment of the *p*-tosylhydrazone (270) with 2 mol equiv. sodium hydride gave the carbene (271) which underwent a [2,3]-sigmatropic rearrangement to give the dithioester (272). Work up with methyl iodide gave 1,1-bis(methylthio)-1,3-butadiene (273) (Scheme 32) ⟨78CL1243⟩.

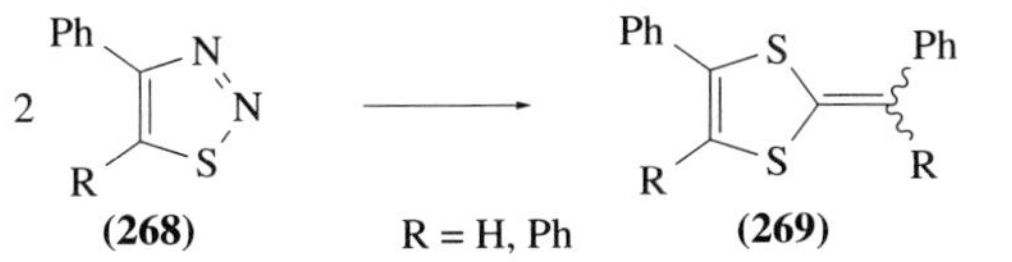

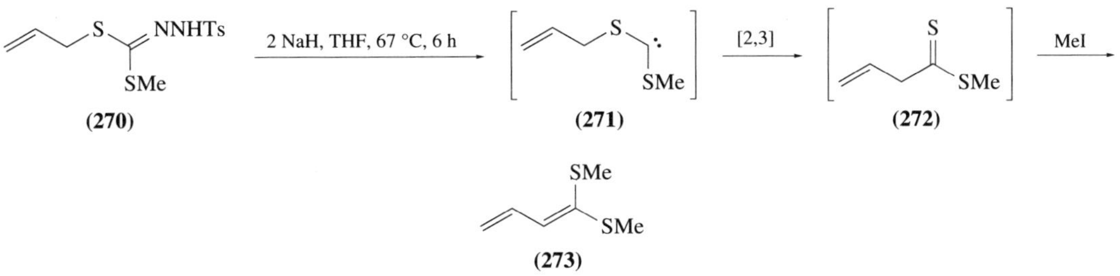

Scheme 32

4.19.5.2 Higher Coordinated Sulfur Derivatives by Oxidation—$R^1_2C=CS(O)_mR^2S(O)_nR^3$

Oxidative methods for preparing ketene dithioacetals bearing higher coordinated sulfurs have been brought into one section for two main reasons. First, reagents and methods are often similar for a variety of substrates and a range of oxidation levels may be reported in a single reference. Coverage of these aspects in one section thus avoids unnecessary duplication. Second, the selectivity of higher oxidations is not always predictable and it may be of help to the reader to compare and contrast cases which would otherwise be subdivided among different sections.

Most frequently, peracids or sodium periodate are used to effect the sulfoxidations and mcpba is often chosen for monoxidations because it is mild, selective and usually quick (e.g. ⟨77CC522⟩). There are many examples of the monoxidation of ketene dithioacetals and a representative selection of the products is shown in Table 1.

Table 1 Preparation of ketene dithioacetal S-monoxides $R^1R^2C:CSR^3S(O)R^4$ by sulfoxidation.

R^1	R^2	R^3	R^4	Oxidant	Yield (%)	Comments	Ref.
CO_2Me	CO_2Me	-CH_2-		mcpba	47		90CPB3242
CO_2Me	CO_2Me	-CH_2CH_2-		mcpba	63		90CPB3242
H	H	Me	Me	mcpba	85		81JOC196
H	NH_2	4-ClC_6H_4	4-ClC_6H_4	mcpba	80	(Z) isomer	80JOC2597
H	H	-$(CH_2)_3$-		$NaIO_4$	82		78JOC3922
H	H	Et	Et	$NaIO_4$	84		78JOC3922
H	Pr^n	Me	Me	mcpba		Mainly (E) isomer	78JCR(S)68
Me	$CH_2:CH$	Ph	Ph	mcpba		(E)/(Z) mixture	78SC87
H	H	Ph	Ph	mcpba	>90		77CC522
H	H	Ph	Me	mcpba	>90	Single regioisomer	77CC522
H	C_5H_{11}	Me	Me	$NaIO_4$	ca. 75	(E):(Z) 5:1	77LA830

There is, however, conflicting evidence in the literature about the order of selectivity of the oxidations and about which oxidation levels are cleanly accessible from this method. The groups of de Lucchi and Potts have analysed the oxidation products obtained by using different stoichiometries of mcpba on ketene dithioacetal substrates. Potts *et al.* found that oxidation of the dithioacetal (**274**) with 1 mol equiv. mcpba cleanly gave a single sulfoxide product which was assigned the (Z) configuration (**275**) ⟨82JOC3027⟩. With 2 mol equiv. mcpba, the other sulfur atom was attacked to give the disulfoxide (**276**). However, with 3 or 4 mol equiv. oxidant, mixtures were obtained containing mainly the disulfoxide (**276**) and the disulfonyl epoxide (**277**) whereas five mol. equiv. of mcpba gave Structure (**277**) as the only product. In contrast, and using a rather different substrate (**278**), de Lucchi *et al.* found that all sequential levels of sulfur oxidation Structures (**279**)–(**282**) were obtained by using the exact stoichiometric amounts of mcpba and that no epoxidation of the double bond was observed ⟨91SL565⟩. The selectivity of the second oxidation for the remaining divalent sulfur to give disulfoxide products seems to be usual when organic peracids or sodium periodate are used ⟨79BCJ466, 81CB684, 84T2951, 86SC233, 90CPB3242⟩ although the oxidation of the dithietane (**283**) to the monosulfone (**284**) provides a notable exception with mcpba as the oxidant (Equation (85)) ⟨85CB2852⟩. The use of potassium permanganate under phase transfer conditions to convert ketene dithioacetal monosulfoxides into the corresponding monosulfones has been demonstrated to be effective for a wide range of substrates by Schaumann et al. ⟨88CB1165⟩. Potassium permanganate was also used, this time in aqueous acetone, to oxidise a series of monosulfoxides (e.g., Structure (**285**)) to the corresponding monosulfones (e.g., Structure (**286**)) (Equation (86)) ⟨81JAP(K)56002956⟩.

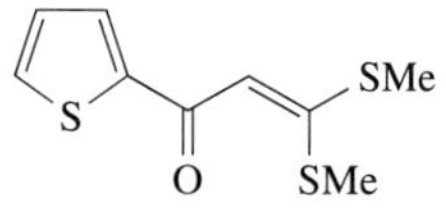

(**274**)

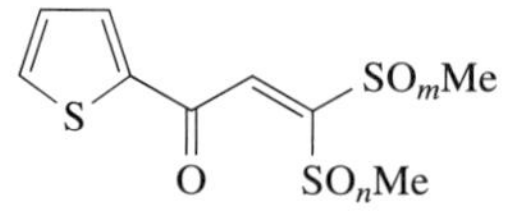

(**275**) $m = 0, n = 1$
(**276**) $m = n = 1$

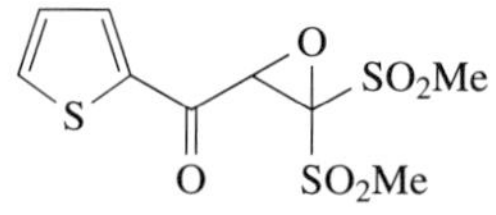

(**277**)

(**278**)

(**279**) $m = 0, n = 1$
(**280**) $m = n = 1$
(**281**) $m = 1, n = 2$
(**282**) $m = n = 2$

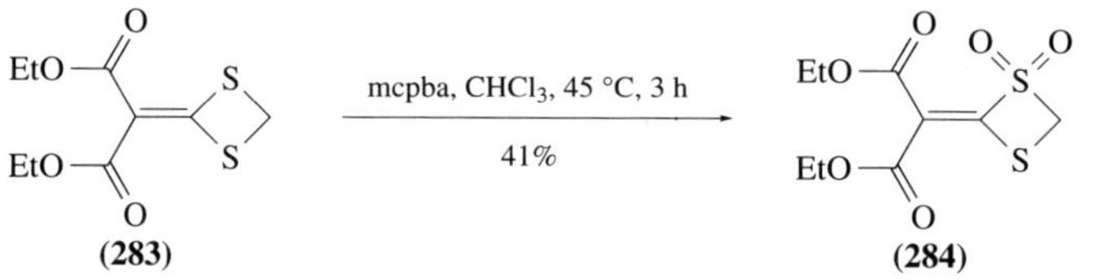

(**283**) → mcpba, CHCl$_3$, 45 °C, 3 h, 41% → (**284**) (85)

(**285**) → KMnO$_4$, Me$_2$CO, H$_2$O, 0 °C to 20 °C, 19 h, 97% → (**286**) (86)

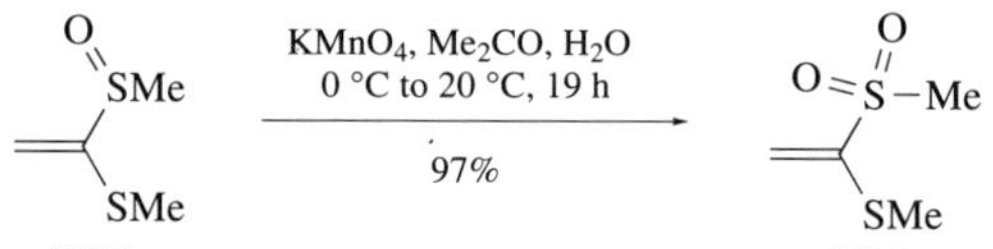

Starting from a monosulfone monosulfide, partial oxidation to the corresponding trioxide using peracetic acid is also known, for example, the preparation of the series of compounds of Structure (**287**) (Equation (87)) ⟨82JOC5404, 83CL767⟩.

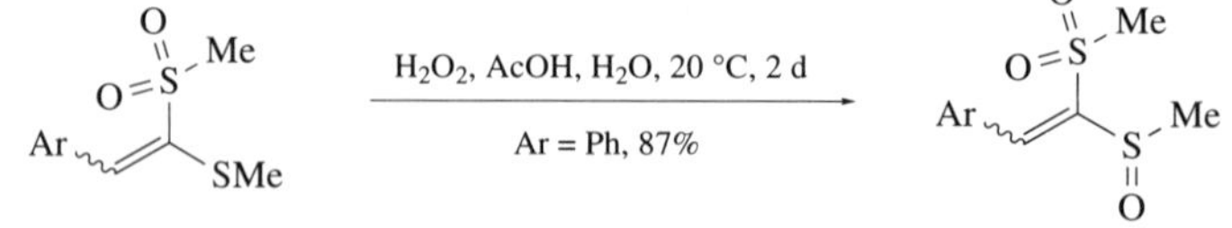

H$_2$O$_2$, AcOH, H$_2$O, 20 °C, 2 d
Ar = Ph, 87% (87)

Ar = Ph, 4-ClC$_6$H$_4$, 4-MeOC$_6$H$_4$, 3,5- or 2,4-(MeO)$_2$C$_6$H$_3$,

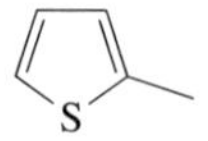

Peracids, usually mcpba ⟨92T1485⟩ or trifluoroperacetic acid ⟨78IZV2654, 82IZV2327⟩, are the reagents of choice for full oxidation to 1,1-disulfonyl alkenes. The method has been applied to a wide range of substrates and can give very high yields, especially when the substituents on sulfur are aromatic, e.g., Structure (**288**) (Equation (88)) ⟨92T1485⟩. Incomplete oxidation is sometimes observed, for example, with the benzodithioles (**289**) ⟨87G227⟩. In this case, the partial oxidation products (**290**) were separated from the products and fully oxidised to the disulfones (**291**) using fresh reagents (Equation (89)). Competing epoxidation of the double bond is sometimes observed ⟨82JOC3027, 84T2951⟩.

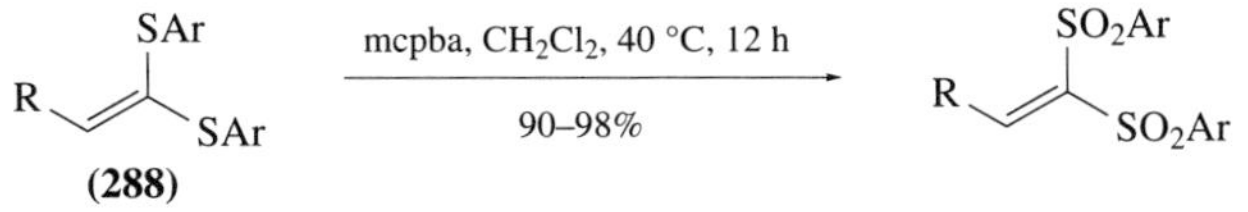

$$R = Me, Ph, H_2C=CH; \ Ar, Ar = Ph, Ph; \text{(2,2'-dimethylbiphenyl)}; \text{(2,2'-dimethyl-1,1'-binaphthyl)}$$

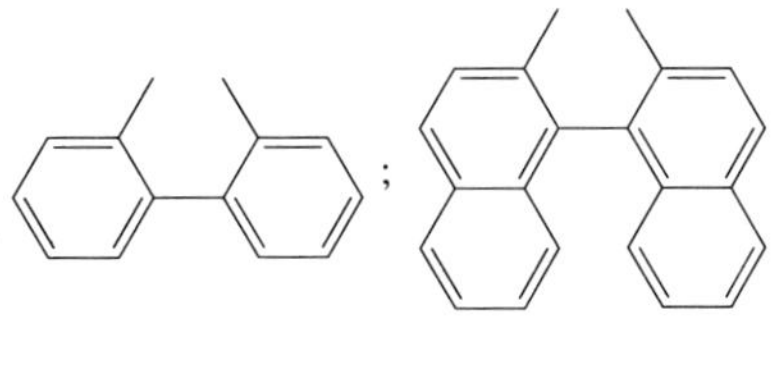

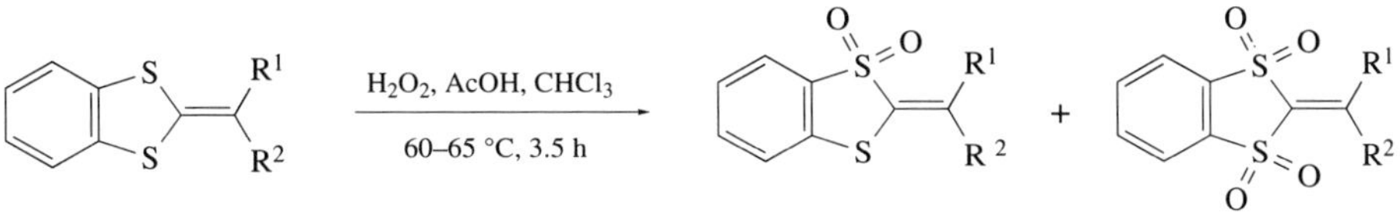

$$R^1, R^2 = Ph, Ph; \ Ph, Et; \ (CH_2)_{11}$$

A combination of sulfoxidation and elimination has been used to convert α-substituted dithioacetals into ketene dithioacetals at various oxidation levels. The sulfoxide (**292**) was prepared from the chloral dithioacetal (**293**) by oxidation and then elimination of HCl on basic alumina (Scheme 33) ⟨90S271⟩. Schlessinger and co-workers prepared the sulfoxides (**294**) by oxidation of the corresponding α-acetoxy dithioacetals (**295**) and then elimination of acetic acid with potassium hydroxide in benzene (Scheme 34) ⟨73TL4711⟩. The quinonoid compound (**296**) was prepared by the formal dehydration of the *S,S'*-dioxide (**297**) (Equation (90)) ⟨78JOC82⟩. At the disulfone oxidation level, the BOC-protected α-hydroxy dithioacetal (**298**) gave the ketene dithioacetal tetroxide (**299**) on treatment with mcpba (Equation (91)) ⟨81USP4348529A⟩. A similar example, in sugar chemistry, was reported by Armarego *et al.* ⟨82AJC785⟩.

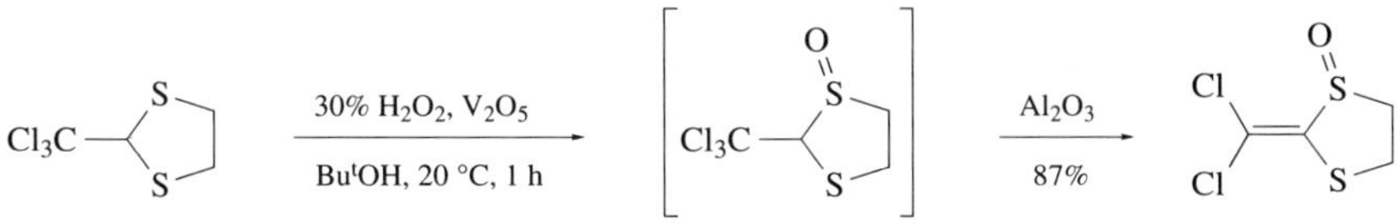

Scheme 33

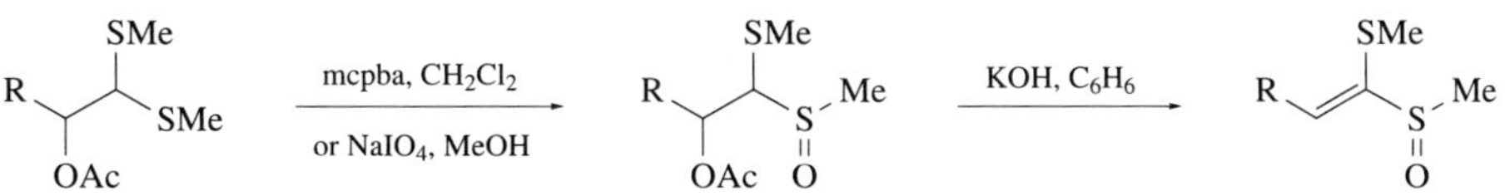

Scheme 34

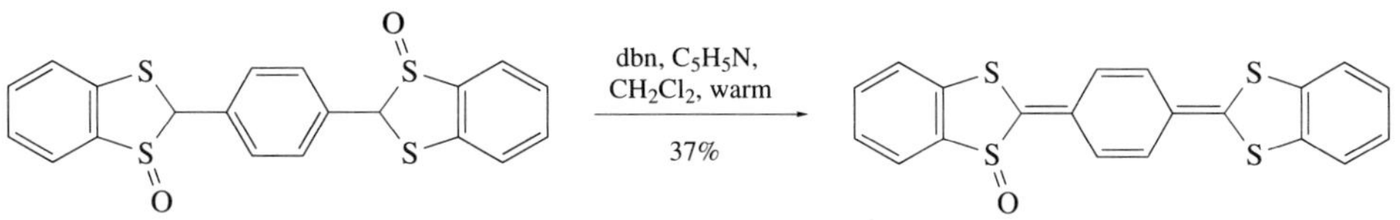

(dbn = 1,5-diazabicyclo [4.3.0.] non-5-ene)

$$\text{(91)}$$

(298) → (299)

Boc = *t*-butoxycarbonyl

4.19.5.3 One Dicoordinated and One Higher Coordinated Sulfur—$R^1_2C{=}CSR^2S(O)R^3$, etc.

The methods used to prepare ketene dithioacetal monosulfoxides often have very similar counterparts for the monosulfones. Oxidation methods for both functional groups are discussed in the preceding section.

4.19.5.3.1 *From methyl methylthiomethyl sulfoxide and analogues*

A large proportion of the preparative methods in this section concern the condensation of methyl methylthiomethyl sulfoxide (300) with carbonyl compounds and analogues to give ketene dithioacetal monoxides. These reactions, and the use of the products in homologation strategies, were the subject of a prodigious series of more than 50 patents during the 1970s and early 1980s by Ogura and co-workers. Much of their work has also been published in the academic literature ⟨72TL1383, 74CL659, 74JA1960, 75S385, 78TL375, 79BCJ2013, 80S736⟩ and many other groups in academia and industry have utilised this chemistry.

The earliest examples of the use of methyl methylthiomethyl sulfoxide (300) were in Knoevenagel-type condensation reactions with nonenolisable (often aromatic or heteroaromatic) aldehydes. Reaction conditions usually involved heating the aldehyde and the sulfoxide together in a polar solvent in the presence of a nonnucleophilic base. In the original work by Ogura and co-workers on simple benzaldehydes (301), benzyltrimethylammonium hydroxide (Triton-B) was employed as the base in methanol–THF solution. In later work, sodium or potassium hydroxides were alternatively used as the base. The ketene dithioacetal monoxides (302) were obtained in good to excellent yields and were assigned (*E*) stereochemistry on the basis of spectroscopic evidence (Equation (92)). A roughly equal mixture of (*E*) and (*Z*) isomers was obtained by irradiation of the benzaldehyde product (302; Ar = Ph) in methanol with a low pressure mercury vapour lamp ⟨72TL1383, 79BCJ2013⟩. Using the single enantiomer (303), Scolastico and co-workers found that treatment with benzaldehyde gave a product (304) which retained optical purity at sulfur (Equation (93)) ⟨81JCS(P1)1278⟩.

$$\text{MeS-CH}_2\text{-SOMe} + \text{ArCHO} \xrightarrow[\text{47-93\%}]{\text{triton-B, solvent, 60 °C–80 °C}} \text{Ar-CH=C(SMe)(SOMe)} \quad (92)$$

(300)   (301)   (302)

$$\text{(+)-(S)-(303)} + \text{PhCHO} \xrightarrow[66\%]{\text{triton-B, THF, 20 °C, 24 h}} \text{(+)-(S)-(304)} \quad (93)$$

While similar reaction conditions for this transformation have been used by many groups (e.g., ⟨75S385, 76CS90, 78CPB685, 80AJC1073, 80S736, 82JHC1493, 84T3677, 85CC618, 87JCS(P1)2017, 90JCS(P1)2967, 92T7527⟩), variations have been reported. Rees and co-workers found that *n*-butyllithium in THF at low temperature was required to prepare the azidothiophene (305) (Equation (94)) ⟨86JCS(P1)501⟩. Schuda and Price used solid sodium hydroxide and no solvent ⟨87JOC1972⟩ and Kunungi *et al.* demonstrated the use of electrogenerated anions as bases ⟨86MI 419-01⟩.

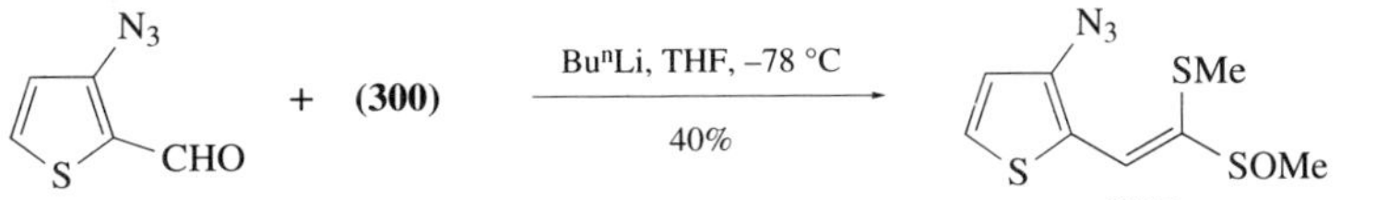

$$\underset{\text{CHO}}{\underset{S}{\bigcirc}}\hspace{-0.5em}^{N_3} + (300) \xrightarrow[40\%]{\text{Bu}^n\text{Li, THF, }-78\,°\text{C}} \underset{S}{\underset{}{\bigcirc}}\hspace{-0.5em}^{N_3}\hspace{-0.5em}\diagdown\hspace{-0.2em}=\hspace{-0.2em}\diagup_{\text{SOMe}}^{\text{SMe}} \tag{94}$$

(305)

At about the same time as the work from Ogura's group first appeared, Schlessinger and co-workers also published a series of papers on the use of alkyl alkylthiomethyl sulfoxides (306) as carbonyl anion equivalents. The Schlessinger team showed that deprotonation of the reagents (306) with n-butyllithium followed by treatment with an aldehyde at low temperature and acetyl chloride workup gave the 2-acetoxy dithioacetal monoxides (307) which were readily converted into the ketene dithioacetal monoxides (308) by treatment with solid potassium hydroxide in benzene (Scheme 35) ⟨73TL4707⟩. In contrast to Ogura's strategy, this method was not restricted to aldehydes lacking an α-hydrogen. Ogura et al. found that a similar method had to be adopted when they extended the scope of the original reaction to include aryl ketone substrates ⟨79S880⟩.

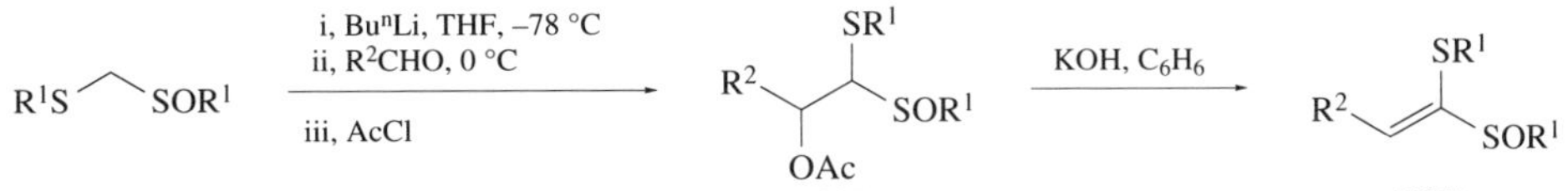

$$R^1S\diagdown SOR^1 \xrightarrow[\text{iii, AcCl}]{\substack{\text{i, Bu}^n\text{Li, THF, }-78\,°\text{C}\\\text{ii, R}^2\text{CHO, 0 °C}}} R^2\underset{\text{OAc}}{\overset{SR^1}{\diagdown}}SOR^1 \xrightarrow{\text{KOH, C}_6\text{H}_6} R^2\diagdown=\diagup\underset{SOR^1}{\overset{SR^1}{}}$$

(306)   (307)   (308)

Scheme 35

The principle was also extended to the conversion of nitriles into 2-aminoketene dithioacetal monosulfoxides (e.g., Structure (309)) (Equation (95)) which could be utilised as precursors of α-amino acids. Reaction conditions were more vigorous than for aldehyde analogues and a mixture of geometrical isomers was formed, tentatively assigned to be predominantly Z ⟨74JA1960, 78TL375⟩. Esters (310) were treated with the anion of methyl methylthiomethyl sulfoxide (300) to give the methanesulfinyl(methylthio)methyl ketones (311) which were converted into the ketene dithioacetal monoxides (312) by sodium borohydride reduction, acetylation and elimination (Scheme 36) ⟨74CL659⟩.

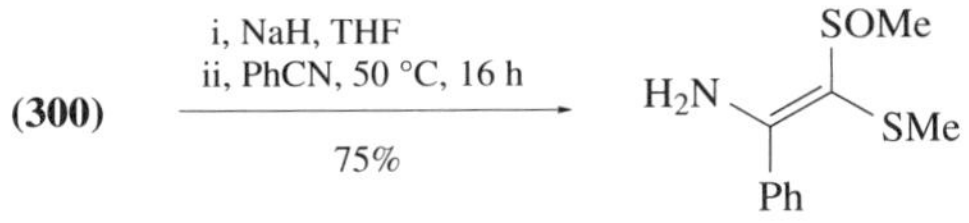

$$(300) \xrightarrow[75\%]{\substack{\text{i, NaH, THF}\\\text{ii, PhCN, 50 °C, 16 h}}} \underset{\text{Ph}}{\underset{|}{H_2N}}\diagdown=\diagup\underset{\text{SMe}}{\overset{\text{SOMe}}{}} \tag{95}$$

(309) 9:1 (Z):(E)

$$(300) \xrightarrow[50-83\%]{\substack{\text{i, NaH}\\\text{ii, RCO}_2\text{Et, (310)}}} \underset{\underset{\text{SMe}}{|}}{R\overset{O}{\underset{}{\|}}{-}}\hspace{-0.5em}\text{SOMe} \xrightarrow[\text{iii, base}]{\substack{\text{i, NaBH}_4\\\text{ii, Ac}_2\text{O, C}_5\text{H}_5\text{N}}} \underset{\text{SMe}}{\overset{R}{}}\diagdown=\diagup\text{SOMe}$$

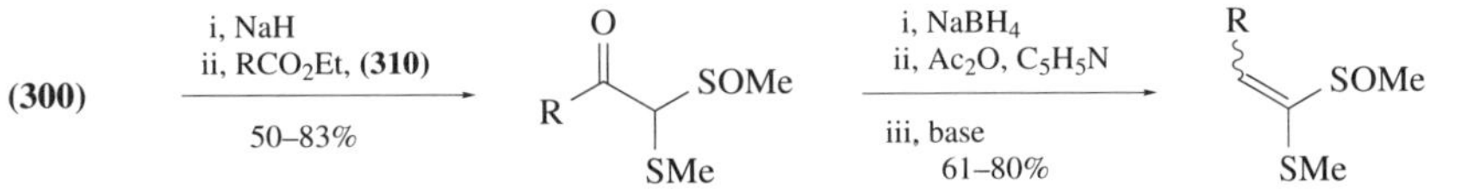

             (311)   61–80%   (312)

R = Ph, 4-ClC$_6$H$_4$, Ph(CH$_2$)$_2$, [methylenedioxytolyl]

Scheme 36

Ogura and co-workers have also investigated the condensation reactions of the analogous methyl methylthiomethyl sulfone (313; R = Me). Heating the sulfone (313; R = Me) in propan-2-ol with benzaldehyde in the presence of anhydrous potassium carbonate gave the ketene dithioacetal S,S-dioxide (314; R = Me) in quite good yield (Scheme 37). As with the sulfoxide analogues, aromatic aldehydes gave exclusively the E isomers in the condensation reaction ⟨82JOC5404⟩. In extending the sulfone chemistry to parallel the sulfoxide work, Ogura and co-workers had more success with methylthiomethyl p-tolyl sulfone (313; R = p-Tol) as the reagent. This compound reacted similarly with aromatic aldehydes to give the p-tosyl compounds (314; R = p-Tol) ⟨83CL767⟩ and was also used to convert primary alkyl halides into ketene dithioacetal S,S-dioxides (315) by chlorination and elimination ⟨86JOC700⟩. The anion of the sulfone (313; R = Me) gave Michael addition products with most α,β-unsaturated ketones and esters but with cinnamaldehyde the condensation product (316) was formed as an $(E)/(Z)$ mixture ⟨86JOC508⟩. Isomerisation with iodine readily converted the mixture into the (*E*,*E*) product (**317**) (Scheme 37). The condensation reaction was also shown to be applicable to other conjugated enals ⟨90TL4621⟩.

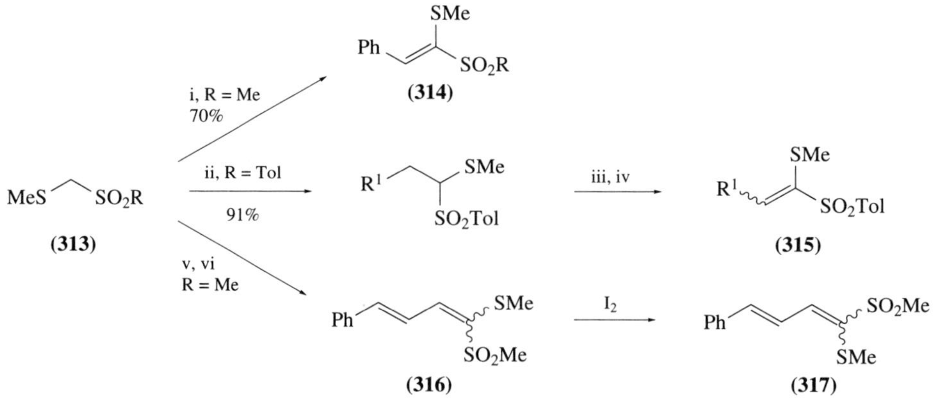

i, PhCHO, K_2CO_3, Pr^iOH, 82 °C; ii, R^1CH_2Br, TOMAC, PhMe, NaOH, H_2O, 20 °C; iii, SO_2Cl_2, $CHCl_3$; iv, PhMe, 111 °C; v, Bu^nLi, THF, −78 °C; vi, PhCH=CHCHO, 20 °C.

Scheme 37

The synthesis of ketene dithioacetal *S*,*S*-dioxides from aliphatic aldehydes and methylthiomethyl *p*-tolyl sulfone (**313**; R = Tol) usually requires a separate activation and elimination step from the initial aldol product ⟨88TL3125⟩. However, Jackson and co-workers reported a one-pot condensation and elimination to give the aliphatic aldehyde derivatives (**318**) (Equation (96)) ⟨88TL4889, 91JCS(P1)3091⟩. In an interesting corollary to this work, Alcaraz *et al.* reported that the condensation of the monosulfone monosulfoxide (**319**) with aldehydes, in the presence of piperidine, gave the vinyl sulfones (**320**) in which elimination of toluene-*p*-sulfinic acid had occurred rather than dehydration ⟨91TL1385⟩. After protection of the alcohol, the vinyl sulfone (**320**; R^2 = Me, R^2 = H) was converted into the ketene dithioacetal *S*,*S*-dioxides (**321**; R^1 = Me) by lithiation α to the sulfonyl group and treatment with *S*-methyl methanethiosulfonate (Scheme 38).

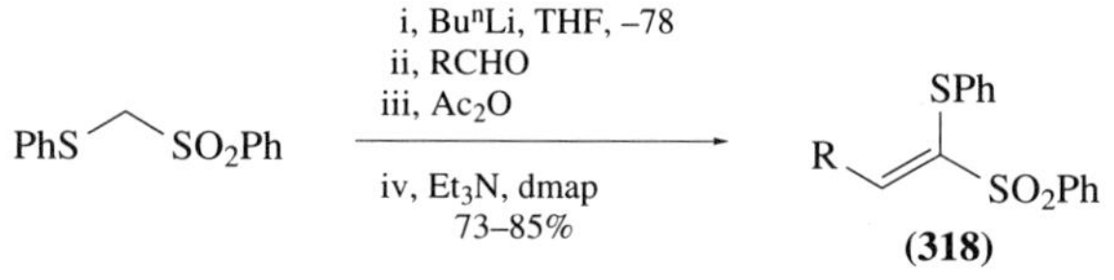

(96)

R = Me, Pr^i, Bu^n

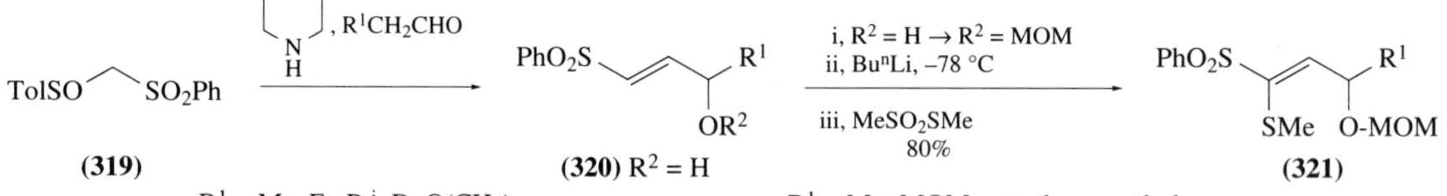

R^1 = Me, Et, Pr^i, $BnO(CH_2)_2$ R^1 = Me; MOM = methoxymethyl

Scheme 38

4.19.5.3.2 Other methods for compounds containing one tricoordinated sulfur

The elimination of HX from a precursor to generate the double bond in ketene dithioacetal monosulfoxides is, of course, part of the condensation reactions covered in the preceding section but there are also other specific examples. Dehydration of the carbohydrate derivatives of Structures (**322**) and (**323**) gave a single (*E*) sulfoxide (**324**) which retained 100% optical purity at the chiral sulfur (Scheme 39) ⟨93JCR(S)60⟩; see also ⟨88CJC2975⟩. The addition of benzenesulfenyl chloride to the vinyl sulfoxide (**325**), followed by elimination of HCl, gave the alkene (**326**) (Scheme 40) ⟨85T2527⟩. In a similar vein, addition of toluene-*p*-sulfenyl chloride to the optically pure vinyl sulfoxide (**327**), followed by elimination of HCl with triethylamine, gave the optically active sulfoxide (**328**) (Scheme 41) ⟨86SC233⟩.

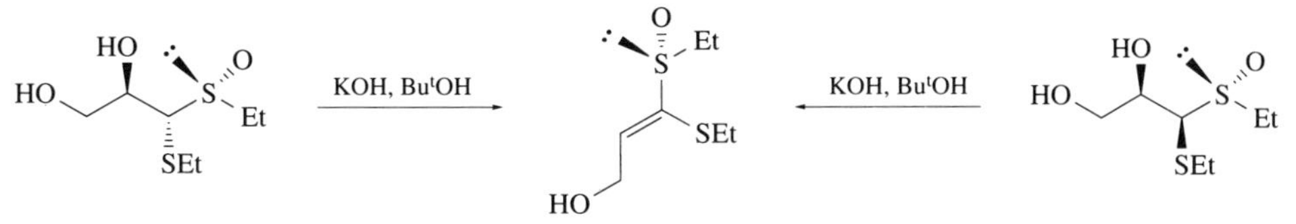

(322) **(324)** **(323)**

KOH, Bu^tOH

Scheme 39

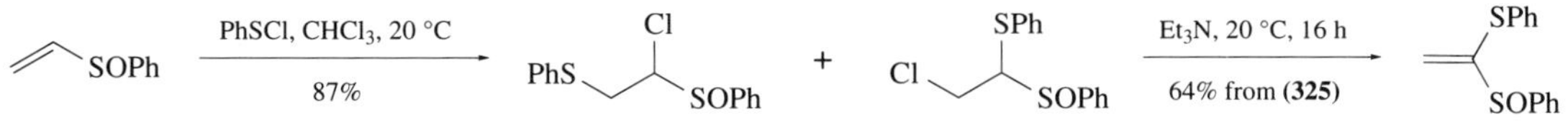

(325) 24 : 63 **(326)**

$PhSCl, CHCl_3$, 20 °C, 87%; Et_3N, 20 °C, 16 h, 64% from (**325**)

Scheme 40

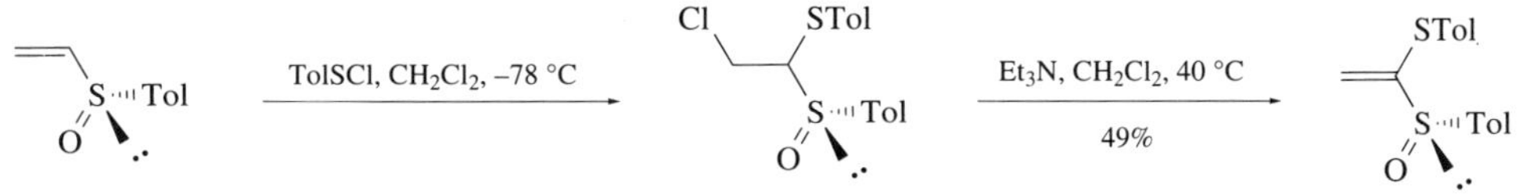

(+)-(*R*)-(**327**) + regioisomer **(328)**

$TolSCl, CH_2Cl_2$, −78 °C; Et_3N, CH_2Cl_2, 40 °C, 49%

Scheme 41

Zwanenburg and co-workers have reported that ketene dithioacetal *S*-oxides are accessible from α-alkylthio sulfines. Treatment of the dithioester *S*-oxide (**329**) with thallium ethoxide followed by iodomethane gave the ketene dithioacetal monoxide (**330**) (Equation (97)). Furthermore, Michael addition of carbanions to this product (**330**), followed by elimination of benzenesulfinic acid, gave access to a wider range of substituted ketene dithioacetal *S*-oxides ⟨76RTC202⟩. In an extension of this study, the sulfine (**331**) was treated with a range of nucleophiles which attacked at the sulfine sulfur atom leading to elimination of toluene-*p*-sulfinate to give the sulfoxide products (**332**) (Scheme 42) ⟨80RTC45⟩. In the early 1990s, the same group reported that the α-(allylthio) sulfine (**333**) initially tautomerised to the sulfenic acid (**334**) which then underwent intramolecular addition to the double bond to give the *syn*-sulfoxide (**335**) (Scheme 43) ⟨92TL6383⟩.

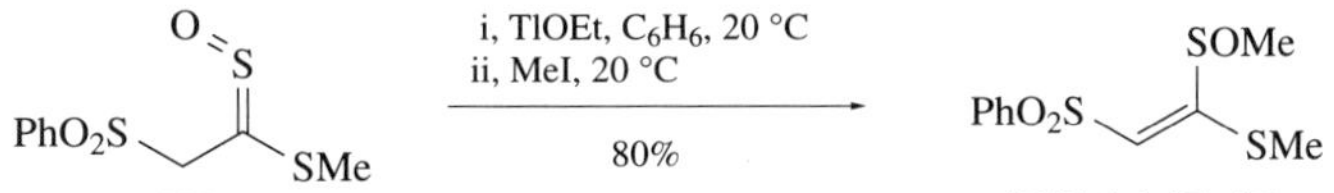

(329) **(330)** 1:1 (*E*):(*Z*) (97)

i, TlOEt, C_6H_6, 20 °C; ii, MeI, 20 °C; 80%

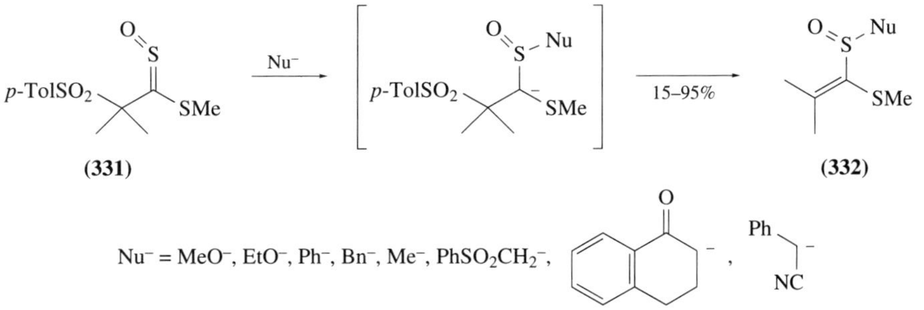

(331) **(332)**

Nu^-; 15–95%

Nu^- = MeO^-, EtO^-, Ph^-, Bn^-, Me^-, $PhSO_2CH_2^-$,

Scheme 42

Cutting and Parsons reported that the allenes (**336**) could be prepared by treatment of the alkyne (**337**) with 2 mol equiv. *n*-butyllithium followed by 2 mol equiv. benzenesulfenyl chloride (Equation

(98)) ⟨83CC1209⟩. In a rare example of the use of Wittig-type reactions in this section, Arai et al. reported the reaction of the phosphonate ester (**338**) with the optically pure menthyl sulfinate (**339**); the intermediate anion (**340**) was trapped with paraformaldehyde to give the optically active sulfoxide (**328**) (Scheme 44) ⟨86SC233⟩.

Ketene dithioacetal *S*-imides are the nitrogen analogues of the *S*-oxides and have been the subject of a handful of papers. The first recorded preparation of this species was by Claus and Setzer who treated the cyclic ketene dithioacetal (**341**) with substituted anilines (**342**) and *t*-butyl hypochlorite to give the *S*-imides (**343**) (Equation (99)) ⟨85M413⟩. The *N*-tosyl analogue (**343**; Ar = Ts) was prepared using chloramine-T. Gallagher and co-workers found that the *S*-imides (**344**) were formed, although in poor yields, when the dithioacetals (**345**) were heated with tosyl azide (Equation (100)) ⟨88TL6475, 92T7551⟩.

A further subcategory of tricoordinated sulfur derivatives are 1-thiovinyl sulfonium salts (**346**) (Equation (101)). Preparations of such compounds are quite rare and, as might be expected, involve the reactions of ketene dithioacetals, or their precursors, with powerful alkylating agents (e.g., ⟨72BCJ198, 82LA1022, 82S206⟩).

4.19.5.3.3 Other methods for compounds containing one tetracoordinated sulfur

Addition–elimination reactions of vinyl sulfones have been used to good effect for a number of compounds in this area. Addition of benzenesulfenyl chloride to the vinyl sulfone (**347**), followed by elimination of HCl, gave the alkene (**348**) (Scheme 45) ⟨85T2527⟩. There have been other examples of the use of arene– and alkane–sulfenyl chlorides to effect this type of transformation and the (*E*) product predominates when isomerism is possible (e.g., ⟨78ZOR478, 87JCS(P2)1253, 88MI 419-02⟩). More commonly in this type of addition–elimination reaction, the leaving group is already present in the starting material. Examples of ketene dithioacetal *S*,*S*-dioxides formed in this way include the trifluoromethyl-substituted alkene (**349**) (Equation (102)) ⟨68ZOB1503⟩, the trisulfonyl alkenes (**350**) (Equation (103)) ⟨84IZV194⟩ and the trisubstituted alkene (**351**) (Equation (104)) ⟨92ZOR1711⟩. In a study of the mechanism of this type of reaction, for arenethiolate nucleophiles, it was suggested that intramolecular migration of the arylthio group is involved ⟨93JPO59⟩.

A related strategy is the addition of thiols to alkynic sulfones. Pasquato *et al.* found that thiophenol added to bis(benzenesulfonyl)ethyne to give a mixture of (*E*) and (*Z*) alkene products (**352**) (Equation (105)) ⟨91TL2177⟩. Selling reported that a mixture of products was formed by the addition of alkanethiols to aryl sulfonyl alkynes but that the course of the reaction could be influenced by the properties of the aryl substituent and the thiol ⟨75T2387⟩. Conversion exclusively to the ketene dithioacetal derivatives (**353**) occurred when the aryl ring was electron-deficient and the nucleophile was *n*-butanethiol (Equation (106)).

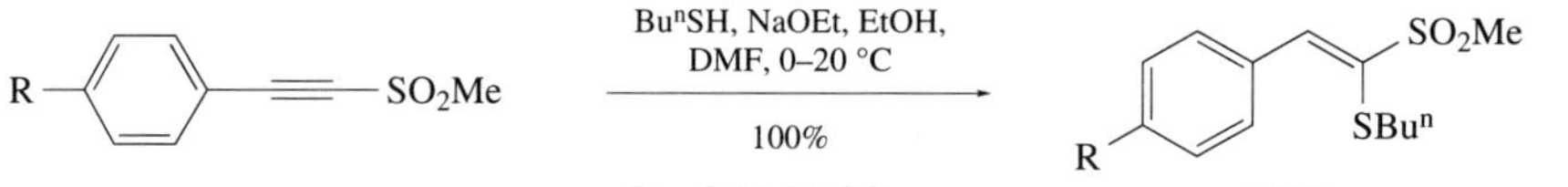

R = O_2N, $MeSO_2$ **(353)** (106)

A small number of miscellaneous preparations of this functional group have been reported. Senning and co-workers studied the reactions of the electron-deficient thiocarbonyl compound (**354**). Treatment of the compound (**354**) with carbenes produced the thiiranes (**355**) which were desulfurised quantitatively with triethyl phosphite to give the corresponding ketene dithioacetal *S,S*-dioxides (**356**) (Scheme 46) ⟨73TL2389⟩. The authors were unable to determine the geometry of the double bond in Structure (**356a**) but did establish that only one isomer was formed. When a range of similar trithiocarbonate *S,S*-dioxides (**357**) was treated with Grignard reagents, the authors reported that the tetrasubstituted alkenes (**358**) were formed as (*E*)/(*Z*) mixtures (Equation (107)). Two possible mechanisms were proposed, both of which involved thiophilic attack of the Grignard reagent at the thiocarbonyl group and reaction of an intermediate with excess starting material (**357**) ⟨74CB2345⟩. In a similar vein, Tamagaki and Ichihara found that the benzo-1,3-dithiole-2-thione (**359**) gave the thiosulfonate ester (**360**) when treated with the diazo compounds (**361**) in the presence of copper(II) (Equation (108)) ⟨76H(4)963⟩. The authors found that the isomeric benzo-1,2-dithiole-3-thiones behaved in a similar manner.

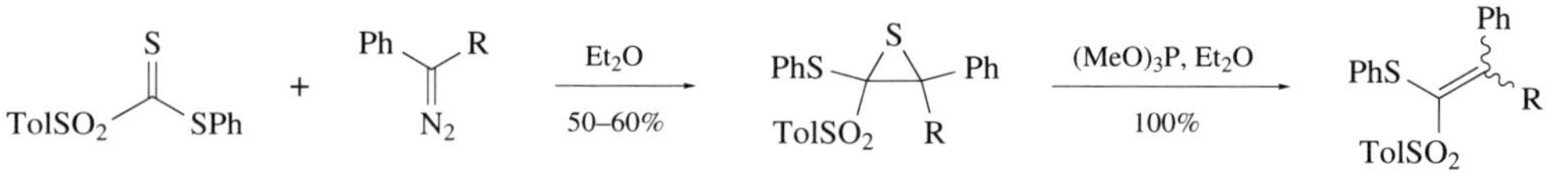

(354) R = H, Ph **(355)** **(356) a**; R = H **(356) b**; R = Ph

Scheme 46

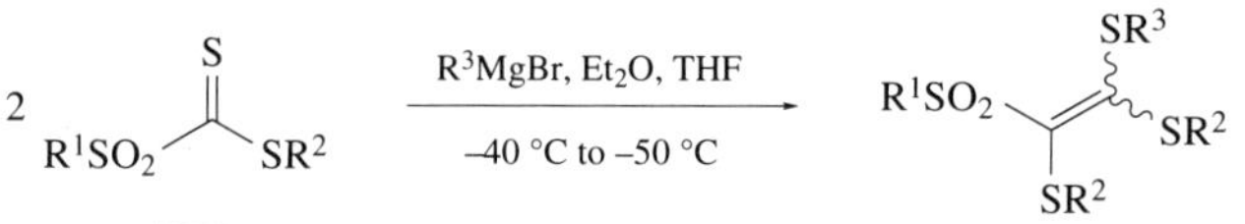

(357) **(358)** (107)

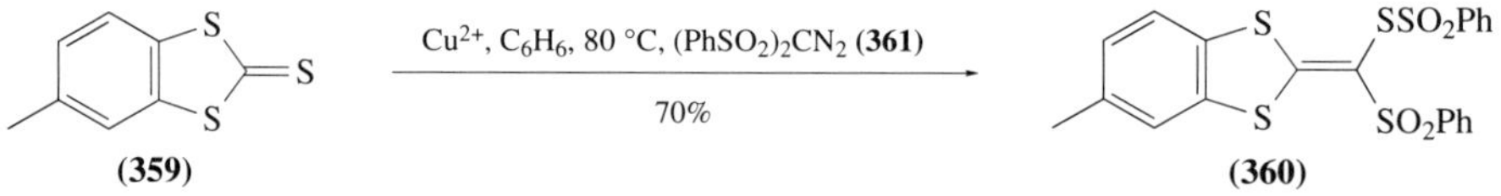

(359) **(360)** (108)

Finally in this section, in 1993 Jackson and co-workers reported that the chiral vinyl sulfoximide (**362**) was lithiated and then treated with diphenyl disulfide to give the sulfenylated product (**363**) without affecting the optical purity of the sulfoximide group (Equation (109)) ⟨93JCS(P1)577⟩. A similar lithiation–sulfenylation reaction, starting from vinyl sulfones, was described earlier (Equation (96)).

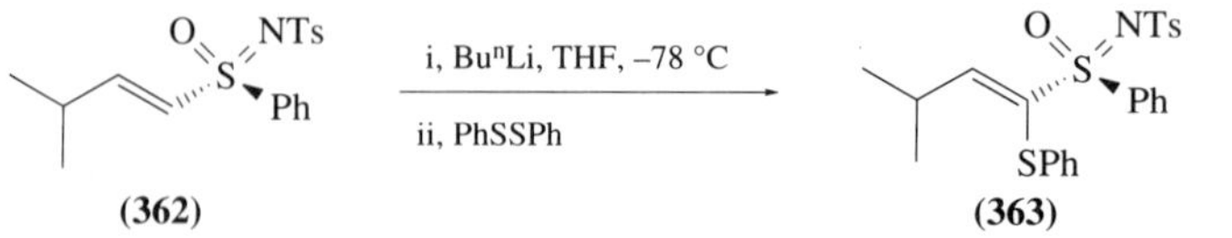

(362) **(363)** (109)

4.19.5.4 Two Tricoordinated Sulfurs—$R^1_2C{=}C[S(O)R^2]_2$, etc.

This functional group does not have an extensive literature and is usually formed by the oxidation of lower coordinated sulfur species, which is covered in Section 4.19.5.2.

There is only one paper recording examples of the Knoevenagel-type condensation reaction which has been so successfully applied to the preparation of ketene dithioacetal monoxides (Section 4.19.5.3.1). Solladié *et al.* reported that deprotonation of (+)-(*S*,*S*)-bis(toluene-*p*-sulfinyl)methane (**364**) with *n*-butyllithium and then treatment with α,β-unsaturated aldehydes gave the optically active 1,1-bis(toluene-*p*-sulfinyl)butadienes (**365**) in good yields (Equation (110)). Nonconjugated aldehydes gave the corresponding aldol products ⟨91TL3695⟩. There have been other examples of preparations starting from disulfinylmethanes. Beer *et al.* reported the preparation of tetra-substituted alkenes (**366**) by treatment of bis(benzenesulfinyl)methane (**367**) with sodium hydride and carbon disulfide and then dialkylation of the dithiocarboxylate intermediate (Equation (111)) ⟨80ACS(A)577⟩. The Mannich reaction of the benzo-1,3-dithiolane *S*,*S*′-dioxide (**368**), followed by quaternisation and elimination of trimethylamine, gave the *exo*-methylene compound (**369**) (Scheme 47). The same paper also contains details of an alternative method which was used to prepare the dithiane analogue (**370**) ⟨92SL730⟩.

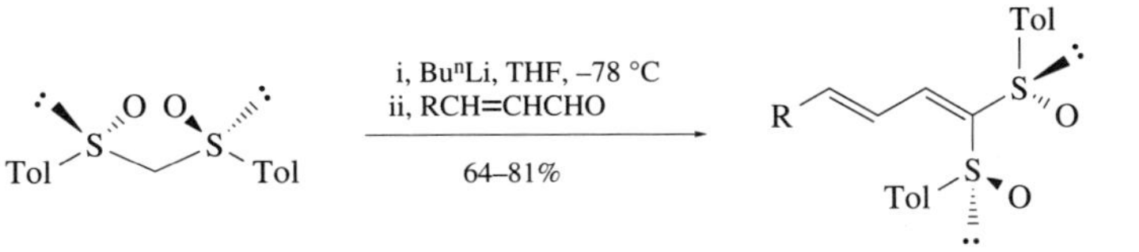

(+)-(*S*,*S*)-(**364**) R = Me, Et, Ph, 2-$MeOC_6H_4$ (**365**) (110)

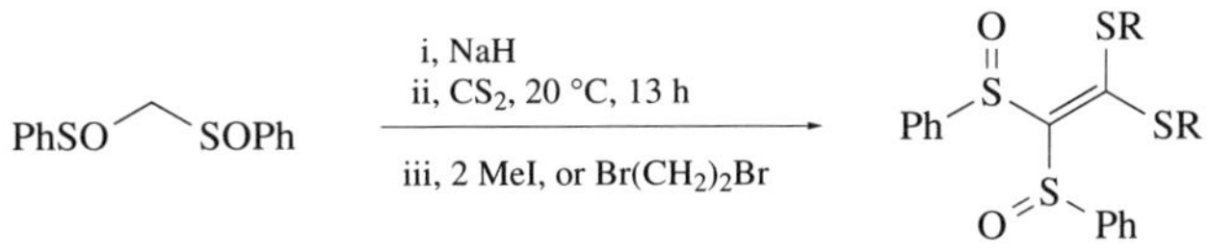

(**367**) R, R = Me, Me; $(CH_2)_2$ (**366**) (111)

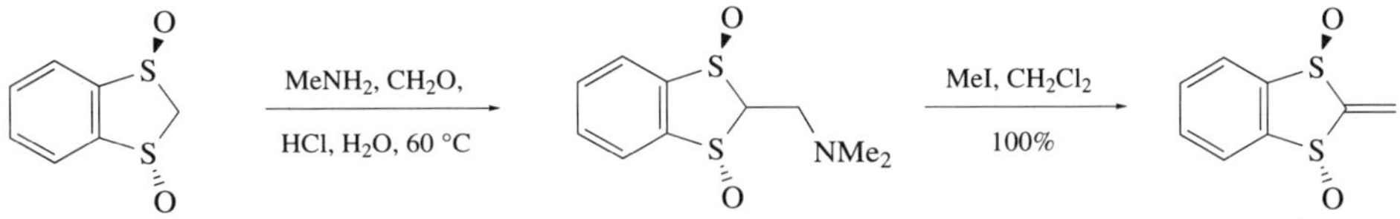

(**368**) (**369**)

Scheme 47

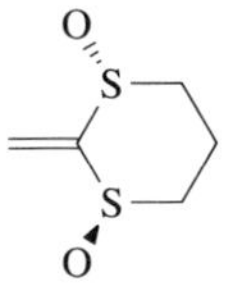

(**370**)

Uchida *et al.* found that treatment of phenylethynylmagnesium bromide with 1 mol equiv. thionyl chloride at −78 °C gave the curious products of Structure (**371**) (Equation (112)) ⟨92MI 419-01⟩. A few examples have appeared with two sulfonium groups. Two examples are the reaction of the pyrylium salt (**372**) with DMSO and acetic anhydride to give the disulfonium salt (**373**) ⟨72T3545⟩ and the treatment of the ketene dithioacetal (**374**) with methyl fluorosulfonate ('magic methyl') to give the dimethylated derivative (**375**) (Scheme 48) ⟨73JOC2747⟩.

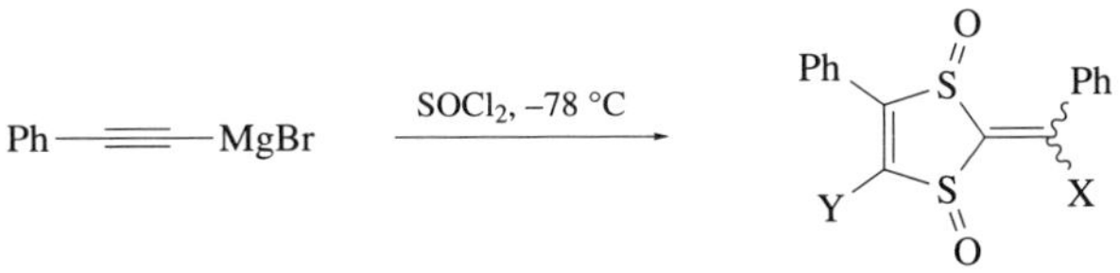

(**371**) X = Br, Y = Cl,
X = Cl, Y = Br (112)

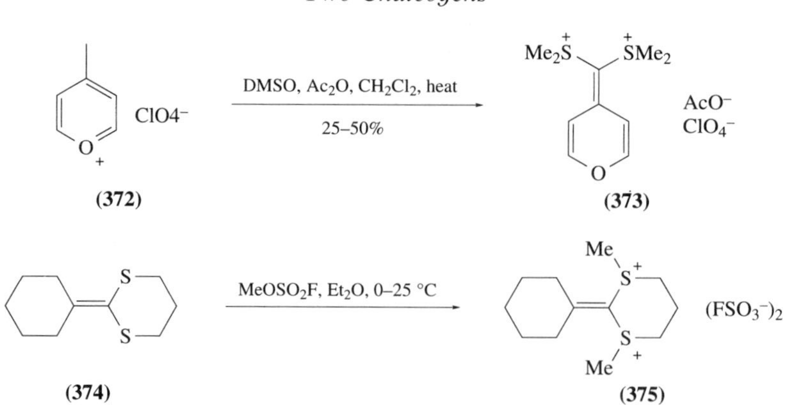

Scheme 48

4.19.5.5 One Tricoordinated and One Tetracoordinated Sulfur—$R^1_2C=CS(O)R^2S(O)_2R^3$, etc.

This functional group is rarer even than that with two tricoordinated sulfurs (see Section 4.19.5.4). Once again, most of the published work concerns oxidation reactions which are covered in Section 4.19.5.2. Only one other example was found. (+)-(S)-(1,1-Dimethylethanesulfonyl)toluene-p-sulfinylmethane (**376**) was converted into the optically pure disubstituted alkene (**377**) by a Mannich reaction followed by quaternisation and elimination of trimethylamine (Equation (113)) ⟨91TA93⟩.

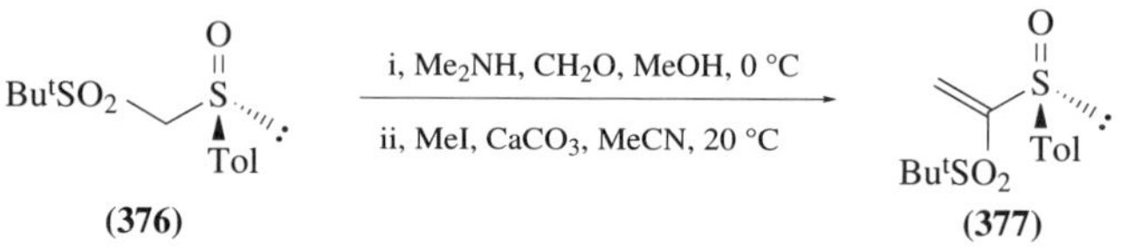

4.19.5.6 Two Tetracoordinated Sulfurs—$R^1_2C=C[S(O)_2R^2]_2$, etc.

The preparation of this functional group is dominated by two methods:
(i) the oxidation of ketene dithioacetals containing di-and/or tricoordinated sulfurs, which is covered in Section 4.19.5.2 and
(ii) the condensation of disulfonylmethanes with aldehydes or other electrophiles.

The oxidative method is the method of choice for tetrasubstituted alkene products. Other miscellaneous preparations have been proven to be of synthetic value but have not been exploited to the same extent.

4.19.5.6.1 From disulfonylmethanes

Knoevenagel-type, base-catalysed condensation of disulfonylmethanes with aldehydes has been widely used to prepare this functional group, as exemplified by the products (**378**) ⟨90UKZ1310⟩ and (**379**) (Scheme 49) ⟨67USP3335188⟩; see also ⟨75ZOR415⟩ and ⟨86MI 419-02⟩. A word of caution is appropriate concerning the condensation under Knoevenagel-type conditions. Friedman and Graber found that condensation of bis(ethanesulfonyl)methane (**380**) with benzaldehyde in the presence of catalytic quantities of piperidine did not give the expected condensation product (**381**), as had previously been reported, but instead gave the rearranged product (**382**) (Scheme 50) ⟨72JOC1902⟩. In some instances, particularly when bis(perfluoroalkanesulfonyl)methanes are the nucleophiles, no base is required and the reaction is driven to completion by removal of water either azeotropically ⟨76USP3932526, 76USP3933914, 76USP3984357, 77USP4018810⟩ or with acetic anhydride ⟨92MI 419-02⟩.

Another nucleophile which has been used in analogous fashion is the disulfonate ester (**383**), which was treated with a range of aldehydes to give the products (**384**) (Equation (114)) ⟨69MI 419-01⟩. Similarly, Maletina *et al.* found that the disulfonyl fluoride (**385**) gave the product (**386**) when it was treated with *p*-(dimethylamino)benzaldehyde (Equation (115)) ⟨79ZOR2416⟩. There are very few reports of ketones being used as the electrophiles in this type of condensation and yields are low. One typical example of a ketone electrophile is the pyrone derivative (**387**) (Equation (116)) ⟨73AP(306)389⟩.

Of the other electrophiles employed in condensation reactions with disulfonylmethanes, the most common is an *ortho*-ester, often triethyl *ortho*-formate (**388**). The reaction is usually carried out with Lewis acid catalysis and can be very high yielding. The product is a vinyl ether, for example, Structure (**389**) (Equation (117)) ⟨89ZOR1102⟩. These vinyl ether products are themselves reactive towards nucleophiles and have been frequently used as starting points for further synthetic elaboration (e.g., ⟨78ZOR1947⟩). A related reaction is that of the 2-ethoxyvinyl sulfones (**390**) with the disulfones (**391**) to give the 1,1-bis(sulfonyl) alkenes (**392**) (Scheme 51) ⟨79ZOR1477⟩. Note that the double bond is in tautomeric equilibrium in Structure (**392**) but lies predominantly towards conjugation with the *gem*-disulfonyl groups. This series of products could thus have been approached equally from the reversed electrophile–nucleophile pair (**393**) and (**394**) (Scheme 51). The reaction of bis(benzenesulfonyl)methane (**395**) with the silylated formamide (**396**) gave the enamine (**397**) (Equation (118)) ⟨83ZOR2417⟩, but this class of compound has been more frequently approached by substitution of vinyl ethers such as Structure (**394**) with amines (for example, see ⟨77ZOR508, 78UKZ183, 84GEP3408757A1, 86EUP174832A2⟩. 2,2-Diamino-1,1-disulfonylethenes have been prepared by treatment of the tetrasulfones (**398**) with secondary amines to give, among other products, the disubstituted compounds (**399**) (Equation (119)) ⟨84IZV611⟩. A carbodiimide (**400**) has also been used as an electrophile with bis(fluorosulfonyl)methane (**401**), which gave the disulfonyl fluoride (**402**) (Equation (120)) ⟨88CZ107⟩. There are no reported examples of substituted ureas being used as electrophiles in this type of reaction.

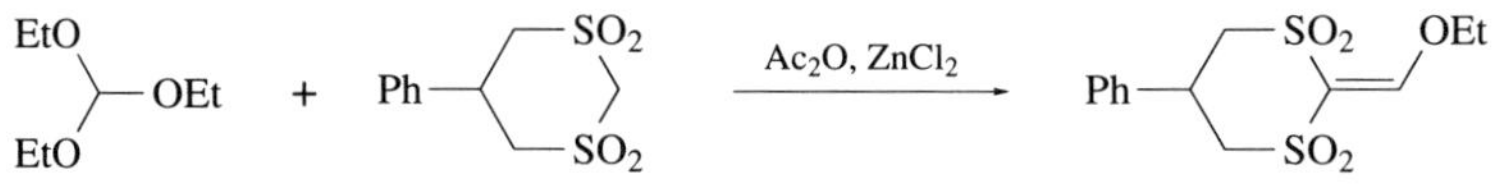

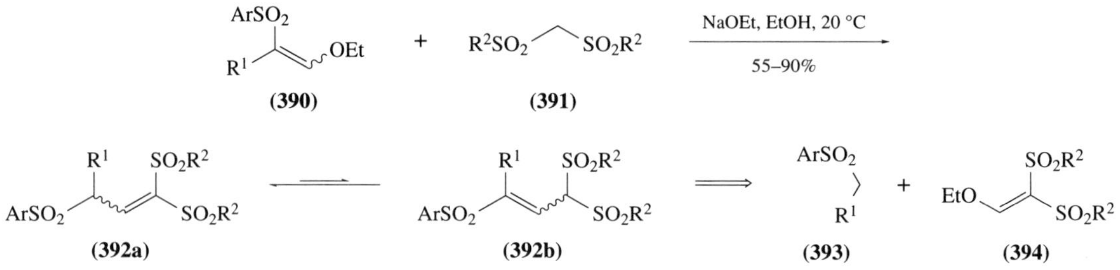

Ar = Ph, 4-MeC$_6$H$_4$, 4-ClC$_6$H$_4$; R^1 = NC, PhCO; R^2 = Et, Ph, 4-MeC$_6$H$_4$, 4-ClC$_6$H$_4$

Scheme 51

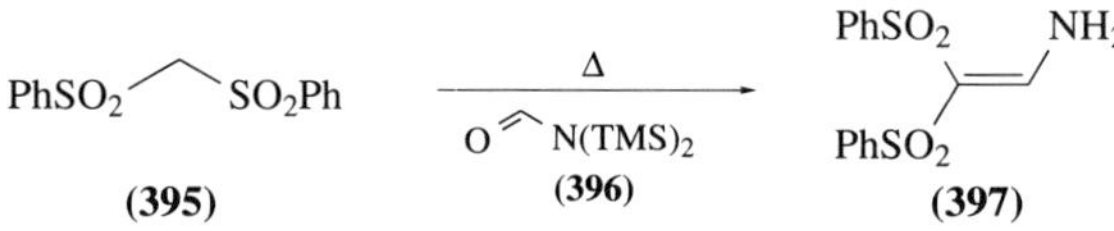

R^1 = Et, Bu^n, Ph; R^2 = Bu^n, Ph, 4-EtC$_6$H$_4$, 4-NO$_2$C$_6$H$_4$, 2-naphthyl; R^3 = H, Et; R^2,R^3 = (morpholine)

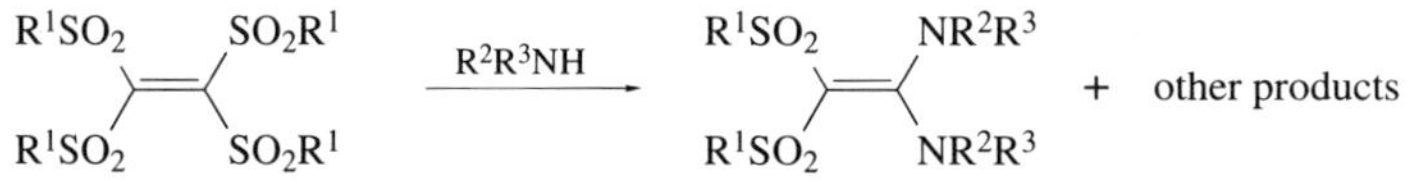

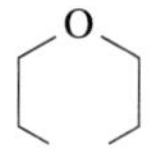

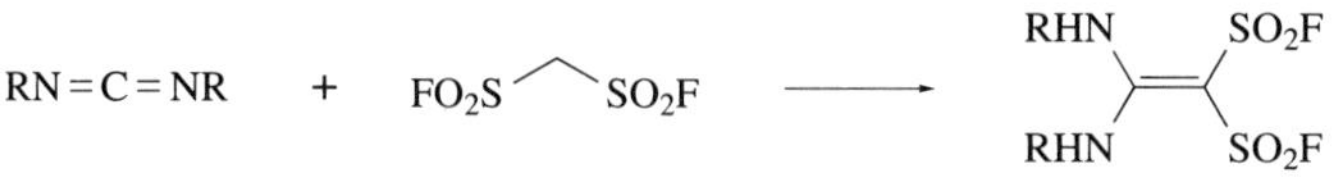

R = c-C$_6$H$_{11}$, Pr^i

4.19.5.6.2 By miscellaneous methods

Phenyliodinium ylides (**403**) have been found to give 1,1-disulfonyl alkenes in a number of different reactions. While the main products from the thermolysis of phenyliodinium ylides in the presence of thiobenzophenones were substituted benzo[b]thiophenes or *gem*-bis(arenesulfonyl) thiiranes, in one case the disulfone (**404**) was isolated (Equation (121)) ⟨89JCS(P1)379⟩. Hatjiarapoglou *et al.* reported that the reaction of a different phenyliodinium ylide (**403**; R = Ph) with carbon disulfide initially gave the α-dithiolactone (**405**) which then dimerised to the tetrasulfide (**406**). Partial desulfuration of Structure (**406**) with copper powder in DIGLYME gave the dithietane (**407**)

(Scheme 52) ⟨85JA7178⟩. A third example is the curious reaction of the same ylide (**403**; R = Ph) with diethylamine in the dark at room temperature to give the 'push–pull' diene (**408**) (Equation (122)) ⟨91PS(60)131⟩.

2,2-Disulfonylketenes have been produced in two different reactions. In the first example, 2,2-bis(trifluoromethylthio)acetic acid (**409**) was oxidised and dehydrated in chromic acid to give 2,2-bis(trifluoromethanesulfonyl)ketene (**410**) in 5% yield after liquid nitrogen trapping and fractional distillation (Equation (123)) ⟨92CB571⟩. In the second example, treatment of 2,2-difluoroethene-1-sulfonyl fluoride (**411**) with sulfur trioxide initially gave the adduct (**412**), which then rearranged to the acid fluoride (**413**). This was then dehydrofluorinated with potassium fluoride to give 2,2-bis(fluorosulfonyl)ketene (**414**) in 91% overall yield from Structure (**411**) (Scheme 53) ⟨80IZV892⟩. In the same paper, the authors report that the conversion of the ketene (**415**) into the disulfonyl fluoride (**416**) was also accomplished with sulfur trioxide (Equation (124)).

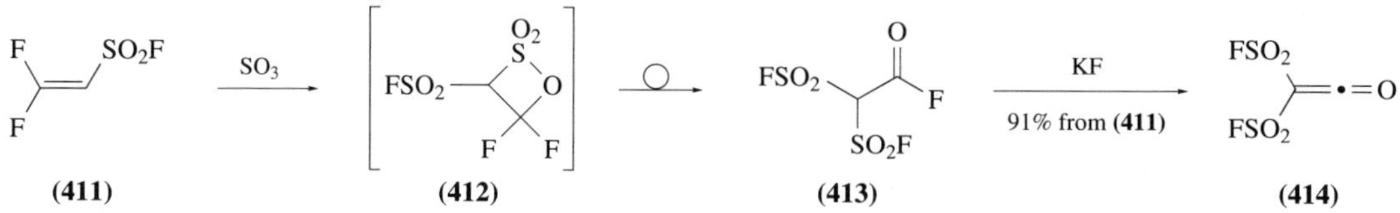

Scheme 53

There are very few references to ketene dithioacetals bearing hexavalent sulfur atoms which are not disulfones. All compounds so far reported which lie within this classification are S-polyfluoro derivatives and have been prepared in the laboratories of Seppelt in Germany and Gard in the USA. The chemistry is summarized in Scheme 54 ⟨88AG(E)1534, 88IC4329, 89CB463, 89IC3766, 92CB557⟩.

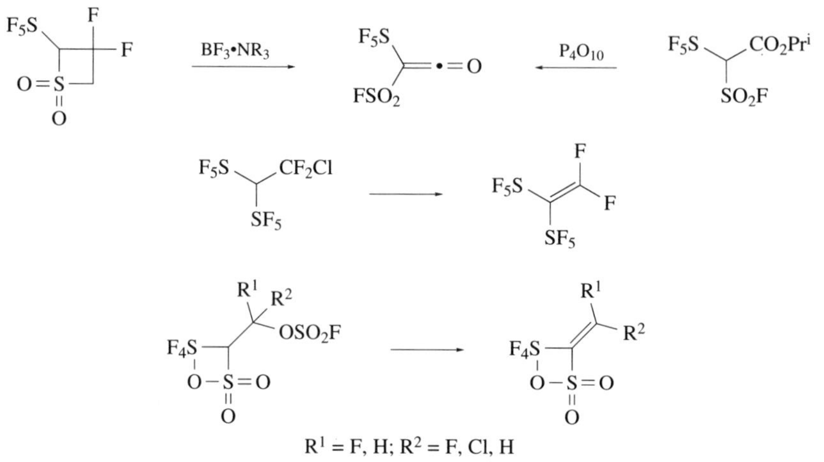

Scheme 54

4.19.6 FUNCTIONS CONTAINING SULFUR AND EITHER SELENIUM OR TELLURIUM—$R^1_2C=CSR^2SeR^3$, etc.

4.19.6.1 Dicoordinated Sulfur Derivatives

4.19.6.1.1 *Selenium derivatives*

A large proportion of the compounds with this substructure that have appeared since the early 1980s have been in the heterafulvalene area. Specific methods have been developed for the preparation of mixed thiaselenafulvalenes, which have found little synthetic utility outside this area and so these will be covered as a separate subsection.

(i) Tetraheterafulvalenes

Symmetrical dithiadiselenafulvalenes (**417**) are readily prepared by the coupling of the appropriate 1,3-thiaselenole-2-selenone (**418**) or -thione (**419**) in the presence of a trialkyl phosphite (Scheme 55). This method has been used by a number of groups to generate a wide range of substituted dithiadiselenafulvalenes, often in quite good yields, as exemplified in Table 2. The stereochemistry around the central double bond is normally an approximately equal mixture of (*E*) and (*Z*) isomers and has been examined in detail by NMR spectroscopy by Engler and Patel for the parent, unsubstituted compound (**417**; $R^1 = H$) ⟨75CC671⟩. Electron-withdrawing groups (e.g., $R^1 = CO_2Me$) facilitate the reaction, and the coupling also proceeds much more easily when the selenones (**418**) are used, although these compounds are harder to prepare directly. However, a general method for the conversion of the more accessible thiones (**419**) into the corresponding selenones has been published by Engler and Patel (Scheme 56) ⟨76TL423⟩. Using the same coupling reaction, monothiatriselenafulvalenes (**420**) have been prepared from 1,3-diselenole-2-thiones (**421**) by scrambling of the sulfur and selenium atoms (Equation (125)). When the substituent is electron-withdrawing (e.g., $R^1 = CF_3$ or CO_2Me) the rearrangement is complete and only the monothiatriselena product is observed, but when R^1 is electron-rich, such as methyl, then only partial scrambling is seen and the tetraselena compound (**422**) predominates (Equation (126)) ⟨77CC835⟩.

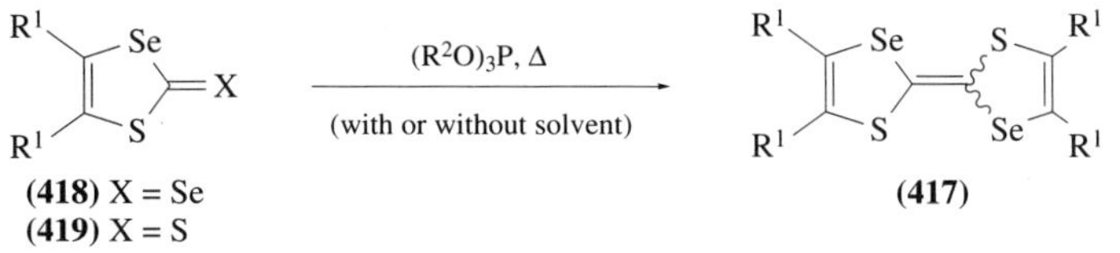

Scheme 55

Table 2 Preparation of symmetrical dithiadiselenafulvalenes (**417**) from 1,3-thiaselenole-2-selenones (**418**) and -2-thiones (**419**).

(**418**), (**419**) R^1 R^1	X	$(R^2O)_3P$ R^2	Yield (%) of (**417**)	Ref.
$(CH_2)_3$	Se	Et	50	77CC505
H H	Se	Me	ca. 60	77JA5909
H H	S	Me	1	77JA5909
$(CH_2)_4$	S	Et	11	76JA3916
CH=CHCH=CH	Se	Et	39	75CC 867
CH=CHCH=CH	S	Et	4	75CC867

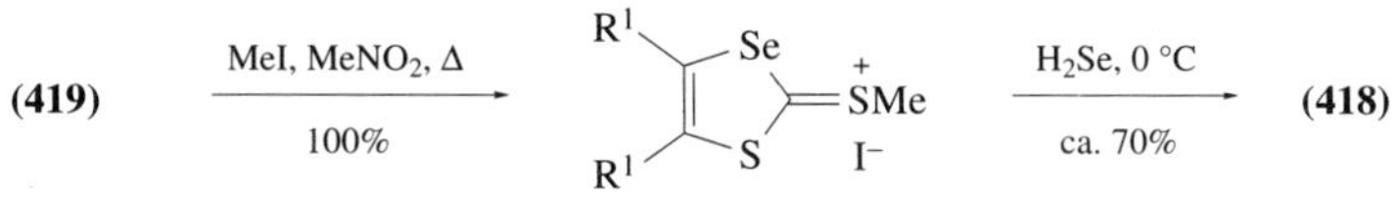

Scheme 56

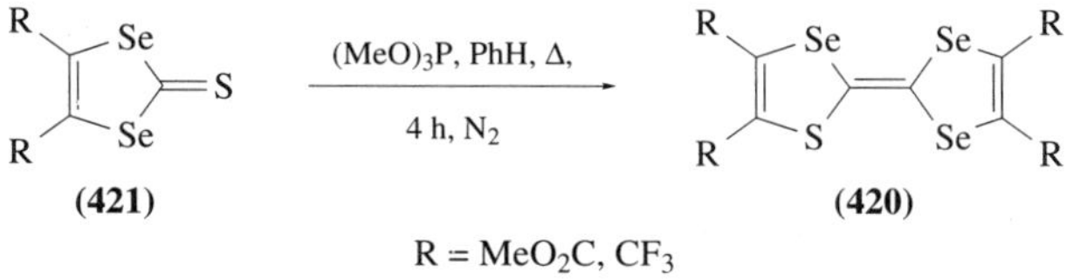

$$(421) \xrightarrow[\text{4 h, N}_2]{(MeO)_3P, PhH, \Delta,} (420) \quad R = MeO_2C, CF_3 \tag{125}$$

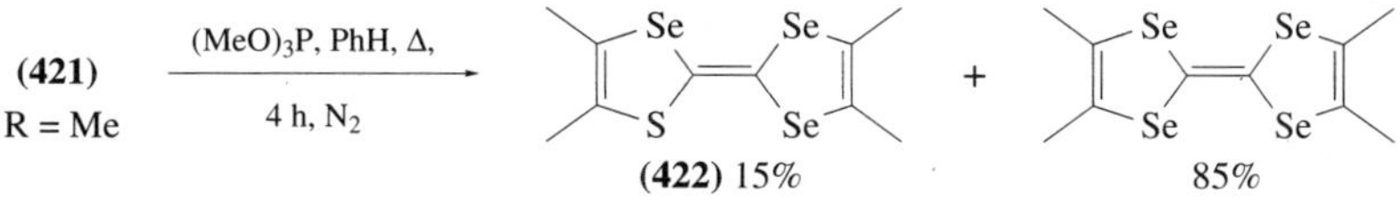

$$(421) \xrightarrow[\text{4 h, N}_2]{(MeO)_3P, PhH, \Delta,} (422)\ 15\% \ + \ 85\% \tag{126}$$
R = Me

Cross-coupling, using two different substrates in the same reaction, has been used as a way of obtaining unsymmetrical tetraheterafulvalenes. The reaction is successful only if the two components are of similar reactivity, otherwise only a mixture of the two symmetrical products is produced ⟨78ANY355⟩. The reaction is also only of preparative value if the unsymmetrical product is sufficiently different in polarity (or some other physical property such as solubility) from the two symmetrical by-products to allow easy separation ⟨91SM2535⟩. Okamoto and Wojciechowski reported another method for the preparation of dithiadiselenafulvalenes of this type in a patent which described the high pressure reaction of DMAD (**423**) with carbon selenide sulfide (or a mixture of carbon diselenide and disulfide) to give the tetrasubstituted fulvalene (**424**) in 89% yield (Equation (127)) ⟨84USP4465845A⟩. This method has also been used to prepare tetraselena analogues (see Section 4.19.7.1).

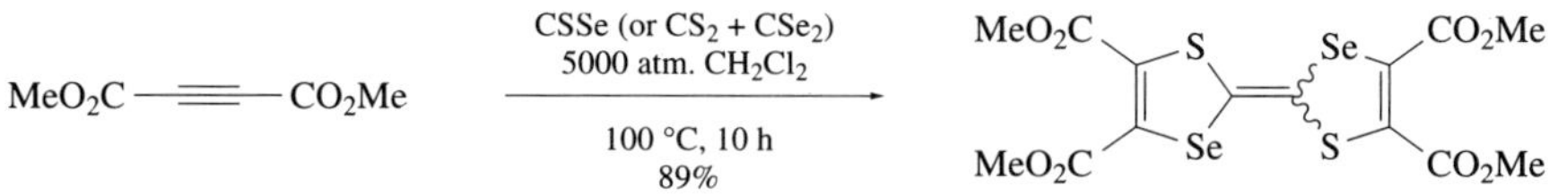

870   *Two Chalcogens*

$$\text{MeO}_2\text{C}-\!\!\!\equiv\!\!\!-\text{CO}_2\text{Me} \quad \xrightarrow[\substack{100\,°C,\,10\,h \\ 89\%}]{\substack{\text{CSSe (or CS}_2 + \text{CSe}_2) \\ 5000\,\text{atm. CH}_2\text{Cl}_2}} \quad \text{(424)} \tag{127}$$

(423)

Following up an observation that the charge transfer complexes such as **(425)** were unstable towards recrystallisation, Weiss and Compper found that the heterafulvalenes **(426)** could be prepared by elimination of ethanethiol from **(425)** by heating in acetonitrile followed by deprotonation (Scheme 57). The methodology was extended to the imidazole analogues **(427)** (Scheme 64) ⟨70TL481⟩.

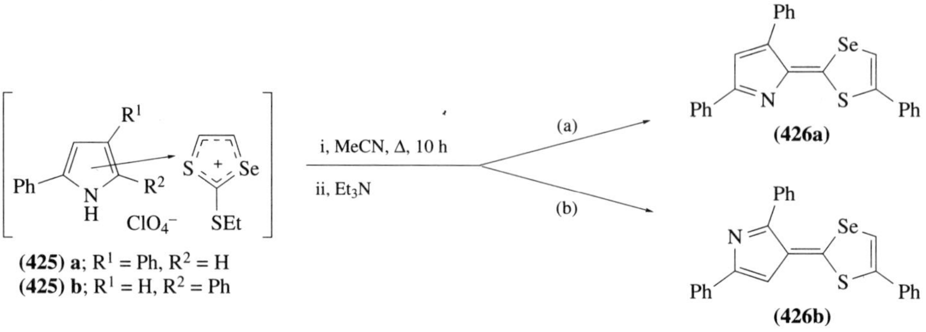

Scheme 57

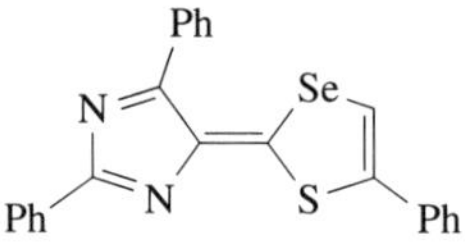

(425) a; R^1 = Ph, R^2 = H
(425) b; R^1 = H, R^2 = Ph

(426a)
(426b)

(427a) 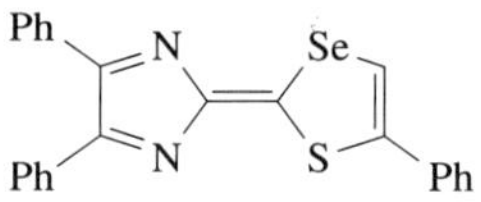   (427b)

(ii) Thiaselenafulvenes

The 1,3-thiaselenole-2-selenones **(418)** and -thiones **(419)** have also been used as starting materials for thiaselenafulvenes. Shafiee *et al.* reported the reaction of a range of thiones (e.g., Structure **(428)**) with ethyl diazoacetate to give the fulvenes **(429)** in moderate yields (Equation (128)) ⟨80JHC117⟩. The stereochemistry of the products was assigned by NMR spectroscopy and was found to be exclusively (*E*) in each case. Wittig and Horner–Wittig reactions have also been used to prepare thiaselenafulvenes. Bryce *et al.* treated anthraquinone with 2 mol equiv. phosphonate esters **(430)** in a sequential manner to give the dithiodiseleno compound **(431)** and also the trithiomonoseleno analogue **(432)** (Scheme 58) ⟨90CC470, 90PAC473⟩. Similar examples starting from triphenylphosphonium salts ⟨80H(14)271⟩ and 1,3-thiaselenolium salts ⟨83S840, 84EGP211356A1⟩ have also been reported.

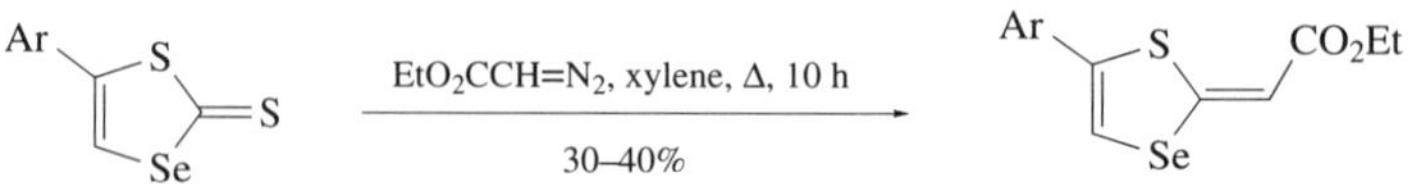

$$\text{(428)} \quad \xrightarrow[\substack{30–40\%}]{\text{EtO}_2\text{CCH}=\text{N}_2,\,\text{xylene},\,\Delta,\,10\,h} \quad \text{(429)} \tag{128}$$

Ar = Ph, 4-BrC$_6$H$_4$, 4-ClC$_6$H$_4$, 4-MeC$_6$H$_4$, 4-MeOC$_6$H$_4$, 2-naphthyl

(iii) Other methods

Reports of acyclic ketene thioselenoacetals are relatively sparse. One fairly general synthesis is the alkylation of thionoseleno esters **(433)**, first reported in 1992 by Lemarie *et al.* ⟨92TL6131⟩.

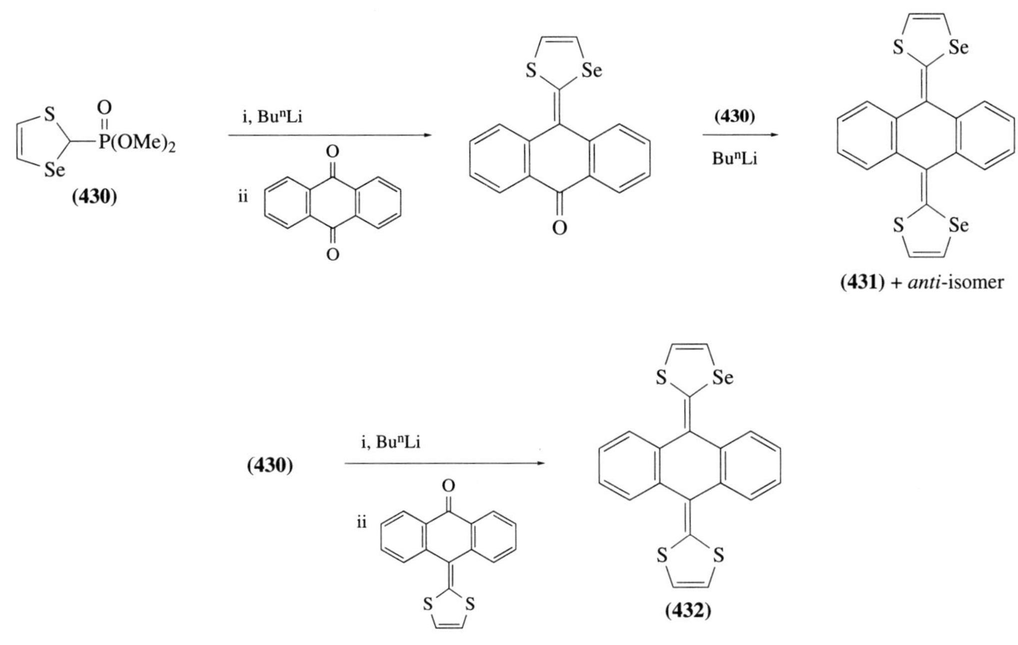

Scheme 58

Deprotonation of the position α to the thiocarbonyl group in Structure (**433**) with LDA at low temperature, followed by quenching of the enethiolate with primary alkyl halides, gave the thioselenoacetals (**434**) in moderate to excellent yields. In all cases the (*E*) isomer predominated (Scheme 59). In another, potentially general method, deprotonation of aryl vinyl selenides (**435**) with LDA at low temperature and treatment of the resulting anions (**436**) with dimethyl disulfide gave the thioselenoacetals (**437**) in good yield (Scheme 60) ⟨81JOC2775⟩. Similar compounds have been prepared by the alternative strategy of lithiation of an enethiol ether (**438**) and quenching of this anion with benzeneselenenyl bromide to give the thioselenoacetal (**439**) in 84% yield (Equation (129)) ⟨77CC522⟩.

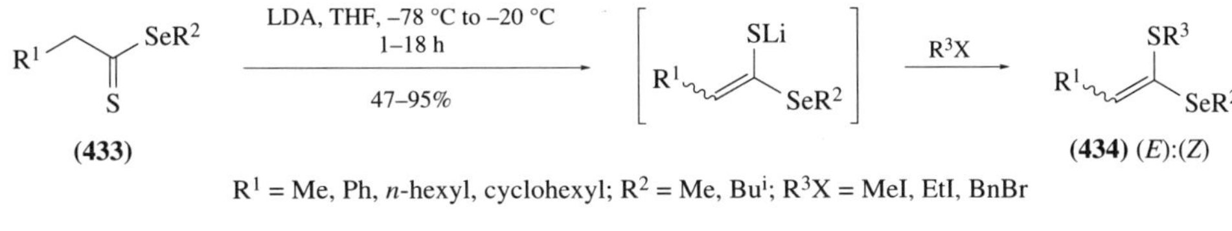

Scheme 59

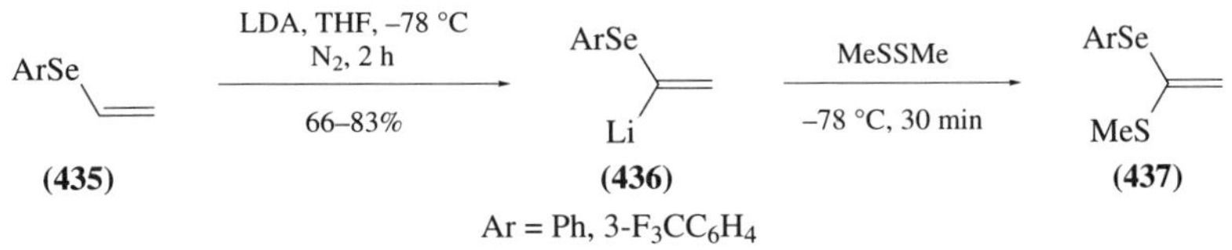

Scheme 60

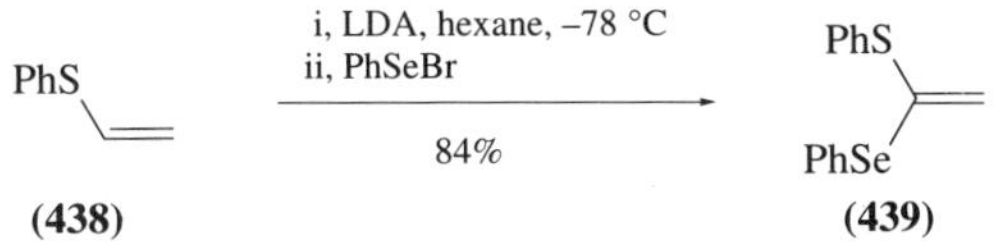

(129)

A number of more specific methods have also been reported. The reactions of the alkylthio alkynes (**440**) with diselenides (**441**) in diethyl ether at reflux gave the addition compounds (**442**)

(Equation (130)) ⟨91MI 419-01⟩. Deprotonation of cyanoacetamide (**443**) with sodium hydride and then treatment with *S,S'*-dimethyl dithioselenocarbonate (**444**) gave the eneselenol (**445**), which was isolated in 77% yield. This compound was, in turn, converted into the heterocycles (**446**) and (**447**) in 66% and 41% yields, respectively (Scheme 61) ⟨86BCJ2909⟩. The reaction of the enyne (**448**) with ethanethiol ⟨74ZOR2456⟩ in the presence of di-*t*-butyl peroxide gave a 2:3 mixture of the 1,1-disubstituted diene (**449**) and the trisubstituted compound (**450**) (Equation (131)). Fragmentation of the selenole (**451**) with ethyllithium and quenching of the intermediate with ethyl bromide gave the alkynic ketene thioselenoacetal (**452**) as an (*E*)/(*Z*) mixture in 84% yield (Equation (132)) ⟨73ACS2242⟩. Finally, in this collection of more esoteric transformations, photolysis of bis(trifluoromethyl) diselenothionocarbonate (**453**) in hexane gave a complex mixture containing all possible tetrasubstituted alkenes including the four Structures (**454a–d**) which meet the substructure requirements for this section (Equation (133)) ⟨84T4963⟩.

4.19.6.1.2 *Tellurium derivatives*

Compounds containing dicoordinated sulfur and tellurium attached to an sp^2 carbon fall into three categories: thienyl tellurols and their alkylated analogues (**455**) ⟨82JOC3946, 84SC1119, 85S497,

90JFC265⟩, a single example of an alkylthiotellurophene (**456**) ⟨77JCS(P2)775⟩ and alkyltelluro derivatives of tetrathiafulvalenes. Arguably, only the last of these three categories falls within the definition of this review.

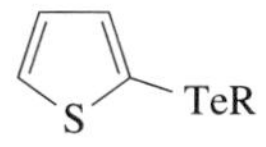

(**455**) 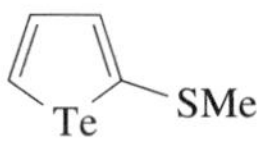 (**456**)

The hydrogens in tetrathiafulvalene (**457**) are quite acidic and all four positions can be lithiated using LDA or, less efficiently, *n*-butyllithium. Treatment of the tetralithiated intermediate (**458**) with tellurium metal followed by an alkyl halide gives the tetra(alkyltelluro)tetrathiafulvalenes (**459**) (Scheme 62). In an exhaustive study Okada *et al.* looked at the alkyl series methyl to octadecyl and found that yields were in the range 33–77% but without a consistent correlation with chain length ⟨86CL1861⟩. With only 1 mol equiv. LDA under the same conditions, the monoalkyltelluro derivative (**460**) was formed (Scheme 62) ⟨90CC816⟩. In an attempt to fuse further heterocyclic rings onto the tetrathiafulvalene structure, Becker *et al.* found that the tetratelluro intermediate (**461**) gave the monotelluride (**462**) in 20% yield when treated with *cis*-1,2-dichloroethene (Scheme 63) ⟨88TL6177⟩ and that the ditelluride (**463**) was obtained in 36% yield when (**461**) was treated with 1,2-dichlorobenzene (Scheme 63) ⟨92CC1048⟩.

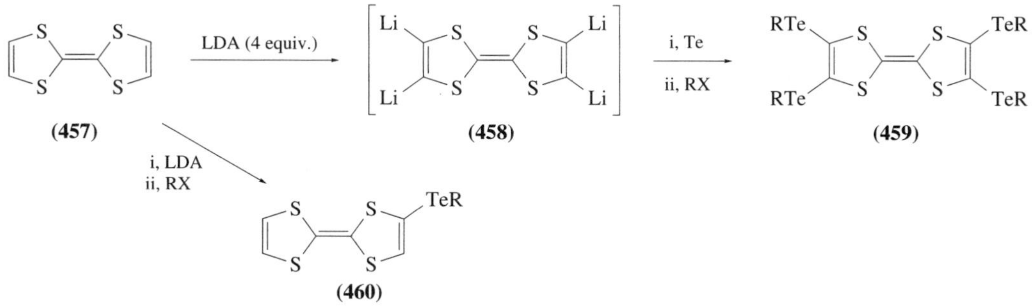

Scheme 62

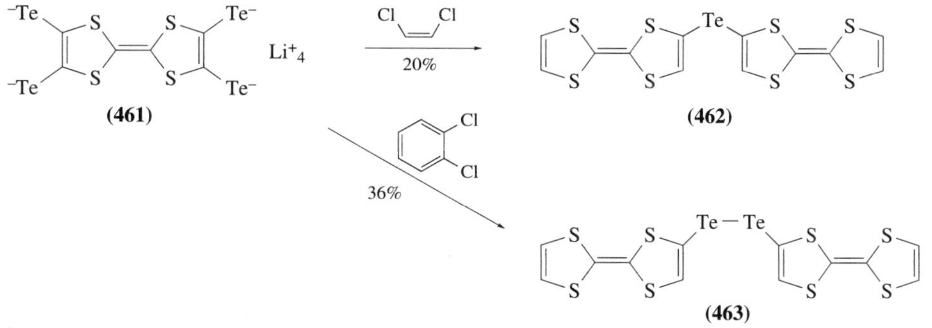

Scheme 63

4.19.6.2 Tri- and Tetracoordinated Sulfur Derivatives

Mixed ketene thioselenoacetals in which either heteroatom is in an oxidation state higher than 2 are rare indeed. A *Chemical Abstracts* substructure search (July 1993) found only 18 examples from seven publications and found no structures for tellurium analogues. Indeed, only Harirchian and Magnus have reported the results of treating a ketene thioselenoacetal with an oxidising agent ⟨77CC522⟩. 1-Phenylthio-1-phenylselenoethene (**439**) was treated with mcpba in dichloromethane at −78 °C to give exclusively the selenoxide (**464**) in 90% yield (Equation (134)). Treatment of the 2-(benzeneseleninyl)vinyl sulfones (**465**) with secondary amines led to the rearranged products (**466**)

(Equation (135)) ⟨85CJC2313⟩. The yield was improved by the addition of benzeneselenenyl chloride to the reaction mixture. Danilenko and co-workers have investigated the thermal rearrangement of salts such as Structure (**467**). Heating Structure (**467**) at 190–230 °C under an argon atmosphere resulted in an internal transfer of an alkyl group from selenium to the enolate oxygen to give the products (**468**) (Equation (136)) ⟨79MI 419-01, 80ZOR654⟩.

$$
\begin{array}{c}
\text{PhS} \\
\diagdown \\
\text{PhSe}
\end{array}
\xrightarrow[>90\%]{\text{mcpba, CH}_2\text{Cl}_2,\, -78\,°\text{C},\, 5-10\,\text{min}}
\begin{array}{c}
\text{PhS} \\
\diagdown \\
\text{PhSe(=O)}
\end{array}
\quad (134)
$$

(**439**) → (**464**)

$$
\begin{array}{c}
\text{R}^1 \quad\text{SO}_2\text{Ar} \\
\diagdown\diagup \\
\text{PhSe(=O)}
\end{array}
\xrightarrow[50-86\%]{\substack{1-24\,\text{h, THF, 20 °C} \\ \text{R}^2\text{R}^3\text{NH + PhSeCl}}}
\begin{array}{c}
\text{R}^1 \quad\text{SO}_2\text{Ar} \\
\diagdown\diagup \\
\text{R}^2\text{R}^3\text{N}\quad\text{SePh}
\end{array}
\quad (135)
$$

(**465**) → (**466**)

R¹ = H, Ph, n-octyl; Ar = 4-MeC₆H₄; R²R³NH = morpholine, pyrrolidine

$$
\begin{array}{c}
^-\text{O}\quad\,\overset{+}{\text{SMe}_2} \\
\diagdown\diagup \\
\text{Ph}\quad\overset{+}{\text{SeR}^1\text{R}^2}
\end{array}
\text{BF}_4^-
\xrightarrow{190-230\,°\text{C, Ar}}
\begin{array}{c}
\text{R}^1\text{O}\quad\overset{+}{\text{SMe}_2} \\
\diagdown\diagup \\
\text{Ph}\quad\text{SeR}^2
\end{array}
\text{BF}_4^-
\quad (136)
$$

(**467**) → (**468**)

R¹ = Ph, Me; R² = Ph, Me

4.19.7 FUNCTIONS CONTAINING SELENIUM AND/OR TELLURIUM—$R^1{}_2C{=}C(SeR^2)_2$, etc.

4.19.7.1 Diselenium Derivatives

4.19.7.1.1 Tetraselenafulvalenes and analogues

Several related strategies have been used to synthesise the family of tetraselenafulvalenes and there is much overlap between these methods and those used for the analogous tetrathiafulvalenes (see Section 4.19.5.1).

Trialkyl phosphites have been used to abstract selenium and sulfur from 1,3-diselenole-2-selenones and -thiones (**469**) respectively in poor to moderate yields to give the fulvalenes (**470**) (Scheme 64) ⟨76CC148, 77JA5909, 80CC866, 80CC867⟩. In one case, selenium–sulfur positional exchange took place and the product was the triselenamonothiafulvalene (**471**) ⟨77CC835⟩, a structure which belongs to both this section and Section 4.19.6.1. In one, poor-yielding example, photolysis was used to remove sulfur from the thiones (**472**) ⟨79ZC192⟩. Using a rather different strategy, Okamato and Wojciechowski prepared (methoxycarbonyl)tetraselenafulvalenes in excellent yields by the high pressure reaction of carbon diselenide with the corresponding ethynecarboxylate esters (**473**) (Scheme 64) ⟨81CC669⟩.

4.19.7.1.2 Dimerisation of lithium ethyneselenolates

In the related diselenafulvene area, there have been a number of reported syntheses of 2-benzylidene- and 2-vinylidene-1,3-diselenoles (**474**) using substituted ethyneselenolate salts (**475**). These salts are produced by the base-induced fragmentation of 1,2,3-selenadiazoles (**476**) (Scheme 65). In some examples the diselenoles were produced by deliberate dimerisation ⟨72TL445, 73JOC338, 74JOC3906, 79JHC1405⟩ while in other examples the diselenoles were major by-products of the Michael reaction of the selenolate salts (**475**) with ethynecarboxylate esters ⟨77S765, 81JOU578⟩.

4.19.7.1.3 Miscellaneous methods

A number of miscellaneous methods for the generation of ketene diselenoacetals have been published, varying considerably both in yield and synthetic utility.

Potentially one of the most versatile methods is the elimination of methaneselenol, induced by a tertiary amine, from the triseleno *ortho*-esters (**477**) after initial treatment with tin tetrachloride. The ketene diselenoacetals (**478**) were produced in moderate to good yields (Equation (137)). The authors also report that the tin reagent can be replaced by titanium tetrachloride but that aluminum chloride is ineffective ⟨85CC1000⟩. The nickel-catalysed substitution of *gem*-dibromo alkenes (**479**) with selenolates is also quite general in its application, depending on the availability of the dibromo substrates. The reaction proceeds in moderate to good yields to give the diselenoacetals (**480**) (Equation (138)) ⟨86JOC875⟩. A Horner–Wittig reaction of the 2-phosphoryl benzo-1,3-diselenole (**481**) with benzophenone gave the diselenoacetal (**482**) in 77% yield (Equation (139)). This reaction was extended to two other ketones ⟨88HCA1242⟩. Vinyl carbenes (**483**) were generated by *gem*-elimination of triflic acid from the corresponding vinyl triflates (**484**) and trapped *in situ* with diphenyl diselenide to give the diselenoacetals (**485**) and (**486**) in yields of 47% and 61% respectively (Scheme 66) ⟨84JOC1653⟩. This paper also reports the preparation of the analogous ditelluroacetals (see Section 4.19.7.2). Deprotonation of malononitrile with potassium ethoxide in ethanol followed by treatment with carbon diselenide gave the 1,1-diselenolate (**487**) which then reacted with alkyl halides to give straight-chain and cyclic diselenoacetals (**488**) (Scheme 67) ⟨70ACS3213⟩. Lastly, a potential starting material (**489**) for a series of tetraselenafulvalenes is a ketene diselenoacetal in its own right. Electrochemical reduction of carbon diselenide using a platinum electrode in DMF solution, followed by methylation of the intermediate diselenolate (**490**), gave the selenium heterocycle (**489**) in 55% yield (Scheme 68) ⟨76CC148⟩.

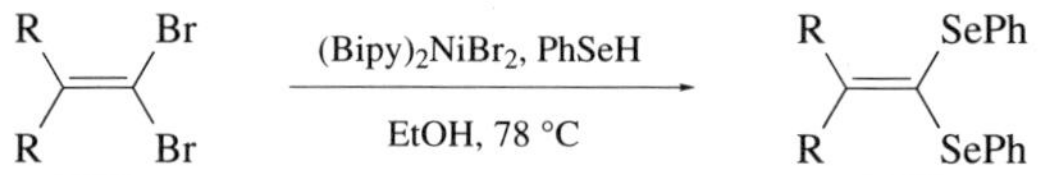

$$\text{(479)} \xrightarrow[\text{EtOH, 78 °C}]{\text{(Bipy)}_2\text{NiBr}_2,\ \text{PhSeH}} \text{(480)} \qquad (138)$$

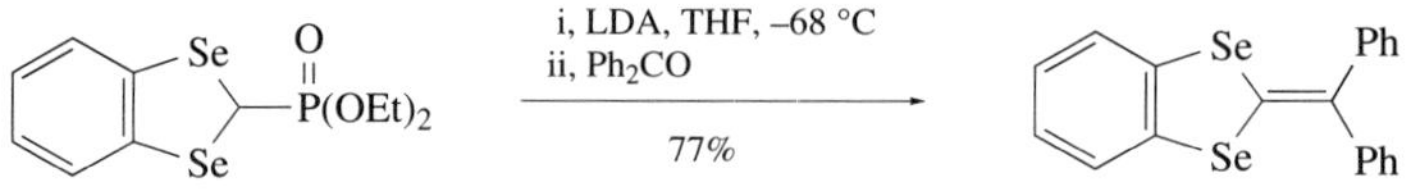

Bipy = 2,2'-bipyridyl

$$\text{(481)} \xrightarrow[77\%]{\substack{\text{i, LDA, THF, }-68\text{ °C} \\ \text{ii, Ph}_2\text{CO}}} \text{(482)} \qquad (139)$$

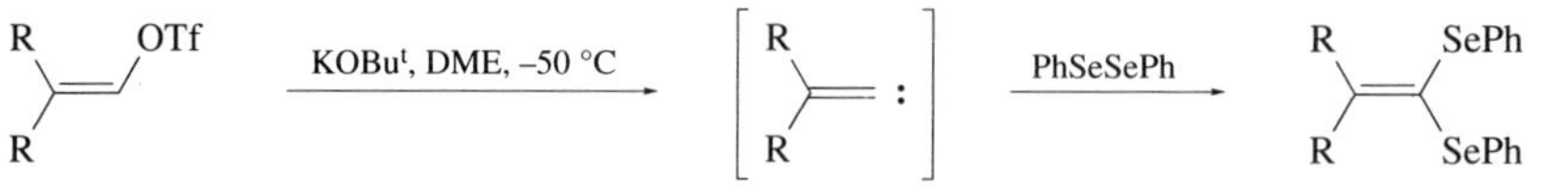

$$\text{(484)} \xrightarrow{\text{KOBu}^t,\ \text{DME},\ -50\text{ °C}} \text{(483)} \xrightarrow{\text{PhSeSePh}} \text{(485) R = Me} \quad \text{(486) R, R = (CH}_2)_5$$

Scheme 66

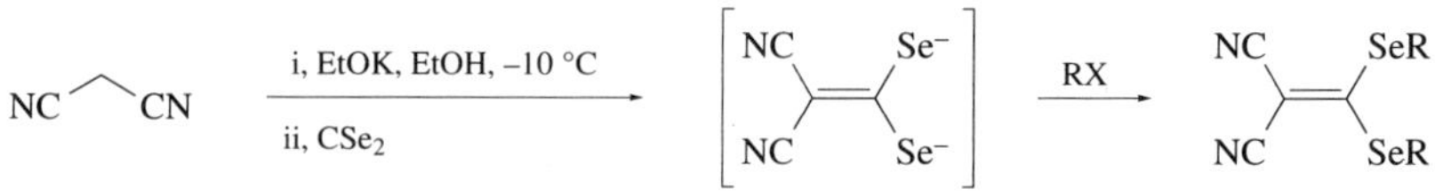

$$\text{NC-CN} \xrightarrow[\text{ii, CSe}_2]{\text{i, EtOK, EtOH, }-10\text{ °C}} \text{(487)} \xrightarrow{\text{RX}} \text{(488)}$$

Scheme 67

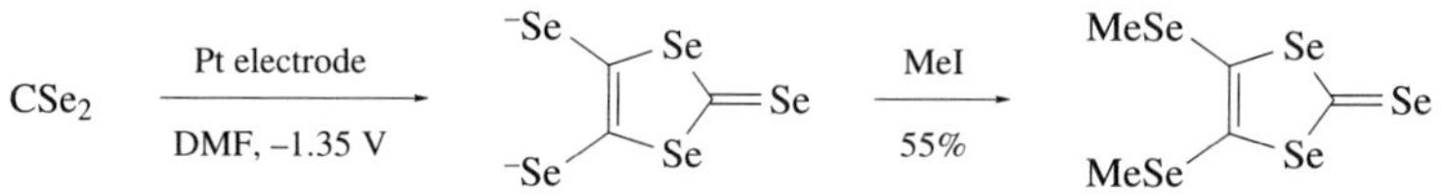

$$\text{CSe}_2 \xrightarrow[\text{DMF, }-1.35\text{ V}]{\text{Pt electrode}} \text{(490)} \xrightarrow[55\%]{\text{MeI}} \text{(489)}$$

Scheme 68

4.19.7.2 Other Derivatives

There are no reported syntheses of monoseleno monotelluro derivatives and only a few recorded preparations of ketene ditelluroacetals, most of which have arisen from work in the area of tetra-tellurafulvalenes. The methods used to prepare tetrathia- and tetraselena-fulvalenes (see Sections 4.19.5.1 and 4.19.7.1) have not been successfully extended to the tellurium analogues and new approaches have been developed. The most successful strategy has been the stepwise dilithiation of a 1,2-dibromide followed by treatment with elemental tellurium to give a 1,2-ditelluride (**491**). *In situ* condensation of this lithio derivative with tetrachloroethene gives the tetratellurafulvalenes (**492**) in moderate to high yields (Scheme 69). The parent tetratellurafulvalene (**493**) has been prepared in an analogous fashion from acetylene via the distannane (**494**) (Equation (140)) ⟨87JA4115⟩ and in an improved method by the same authors ⟨88SMB425⟩, who used tetra-bromoethene in the final condensation. A total of five other tetratellurafulvalenes have been reported in 1994, all prepared by similar methods but with varying yields ⟨82CC336, 82JA1154, 84JA8303, 87SM647⟩.

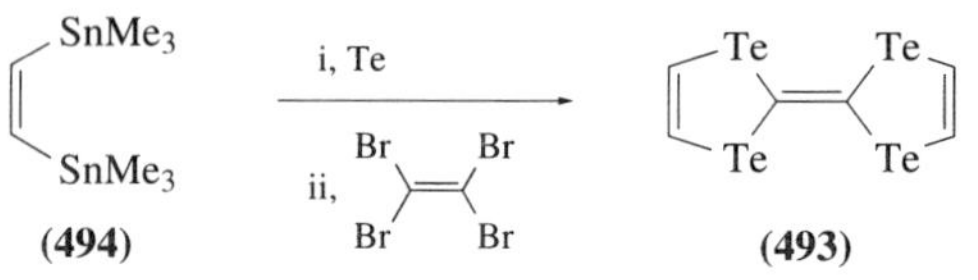

$$\text{(494)} \xrightarrow[\text{ii, Br}_2\text{C=CBr}_2]{\text{i, Te}} \text{(493)} \qquad (140)$$

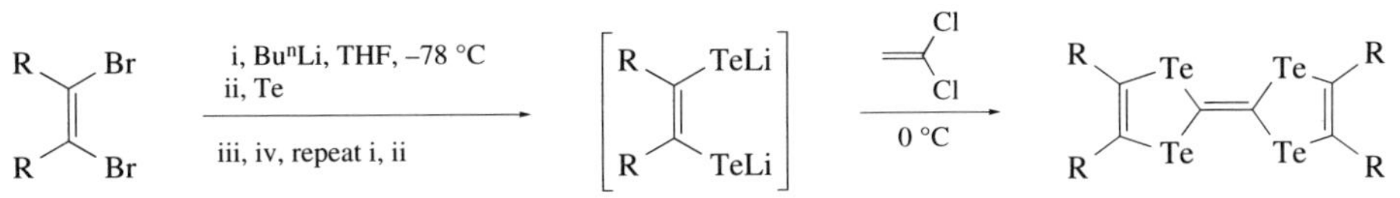

(491)

(492)

Scheme 69

The final condensation of the 1,2-ditelluride intermediates (**491**) with tetrachloroethene could give rise to two isomeric products: the desired fulvalenes (**492**) or a tetratelluranaphthalene, such as Structure (**495**), which also has a structure appropriate for this section. This latter compound has been identified as a by-product in the preparation of Structure (**493**) ⟨87JA4115⟩, and the hexamethylene analogue (**496**) has been synthesised intentionally using high dilution techniques (Equation (141)) ⟨86CL311⟩.

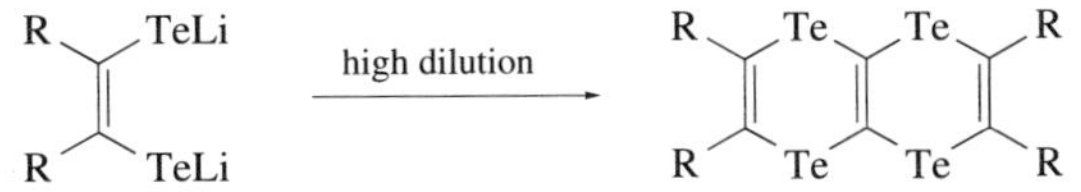

(141)

(491)

(**495**) R = H
(**496**) R, R = $(CH_2)_3$

The earliest report of the synthesis of a ketene ditelluroacetal (**497**), by Petrov *et al.* in 1979 ⟨79ZOR2596⟩, was re-examined first by Bender *et al.* ⟨81TL1495⟩ and then by Cava and co-workers ⟨81TL4199⟩. Both groups identified the material isolated by Petrov as the *trans*-dibenzylidene-1,3-ditelluretane (**498**) (the *cis*-isomer (**499**) was also formed) and Cava's group also isolated the 1,2,4-tritellurole (**500**) (Scheme 70). An approach to ketene ditelluroacetals of a more general structure was published by Stang *et al.* ⟨84JOC1653⟩. Vinyl carbenes (**483**) were generated by *gem*-elimination of triflic acid from the corresponding enol triflates (**484**) and trapped *in situ* with diphenyl ditelluride (**501**) to give the ditelluroacetals (**502**) and (**503**) in yields of 26% and 30% respectively (Scheme 71). This paper also reports the preparation of the analogous ketene diselenoacetals (see Section 4.19.7.1).

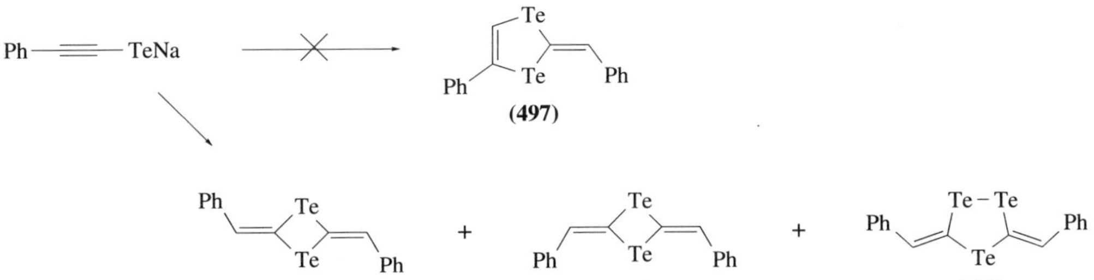

(497)

(498) (499) (500)

Scheme 70

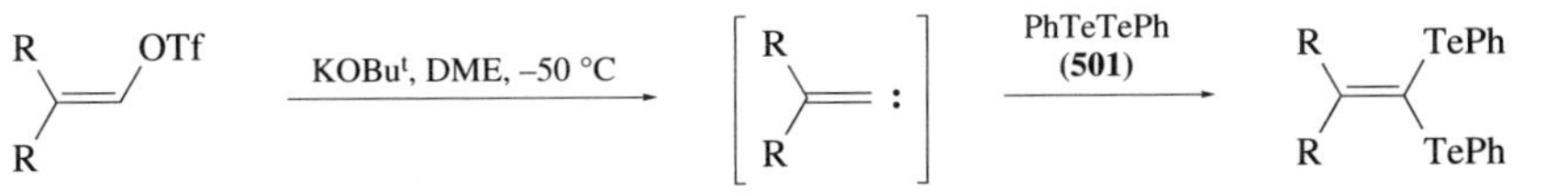

(484) (483)

(**502**) R = Me
(**503**) R, R = $(CH_2)_5$

Scheme 71

4.20
Functions Containing a Chalcogen and Any Group Other Than a Halogen or a Chalcogen

PETER D. KENNEWELL
Hoechst Roussel Ltd, Swindon, UK

ROBERT WESTWOOD
Roussel-Uclaf, Romainville, France

and

NICHOLAS J. WESTWOOD
University of Oxford, UK

4.20.1 FUNCTIONS CONTAINING A CHALCOGEN AND A NITROGEN FUNCTION	880
4.20.1.1 Functions Bearing Oxygen and Nitrogen	880
4.20.1.1.1 Ketene hemiaminal derivatives	880
4.20.1.1.2 Other nitrogen derivatives	896
4.20.1.2 Functions bearing Sulfur and Nitrogen	898
4.20.1.2.1 Dicoordinate sulfur derivatives	898
4.20.1.2.2 Tricoordinate sulfur derivatives	907
4.20.1.2.3 Tetra – and higher – coordinate sulfur derivatives	908
4.20.1.3 Functions Bearing Selenium or Tellurium Together with Nitrogen	909
4.20.2 FUNCTIONS CONTAINING A CHALCOGEN AND A PHOSPHORUS, ARSENIC, ANTIMONY OR BISMUTH	911
4.20.2.1 Functions Bearing Oxygen	911
4.20.2.1.1 Elimination reactions	911
4.20.2.1.2 Displacement reactions	912
4.20.2.1.3 Condensation of alcohols with α-diazo alkylphosphonates	913
4.20.2.1.4 Addition to ketenes	914
4.20.2.1.5 Condensation reactions	916
4.20.2.1.6 Isomeration of β,γ-unsaturated α-oxyphosphorus compounds	917
4.20.2.1.7 Enolisation of α-carbonylphosphorus compounds	920
4.20.2.1.8 Rearrangements	923
4.20.2.2 Functions Bearing Sulfur	923
4.20.2.2.1 Elimination reactions	923
4.20.2.2.2 Condensation reactions	925
4.20.2.2.3 Addition to substituted alkynes	926
4.20.2.2.4 From phosphorus-substituted sulfines	926
4.20.2.2.5 From ketenes	926
4.20.2.2.6 From 1-lithio-1-phosphorylated alkenes	927

4.20.2.2.7	By substitution of a 1-chlorovinylphosphonate	927
4.20.2.2.8	From alkyl dithiocarboxylates	927
4.20.2.3	Functions Bearing Selenium	928

4.20.3 FUNCTIONS CONTAINING A CHALCOGEN AND A METALLOID — 929

4.20.3.1	Functions Bearing Oxygen	929
4.20.3.1.1	Oxygen and silicon or germanium	929
4.20.3.1.2	Oxygen and boron	934
4.20.3.2	Functions Bearing Sulfur, Selenium or Tellurium	936
4.20.3.2.1	Sulfur and silicon or germanium	936
4.20.3.2.2	Selenium or tellurium together with silicon or germanium	946

4.20.4 FUNCTIONS CONTAINING A CHALCOGEN AND A METAL — 948

4.20.4.1	Oxygen Functions	948
4.20.4.1.1	Lithium	948
4.20.4.1.2	Tin	951
4.20.4.1.3	Copper	953
4.20.4.1.4	Zinc	954
4.20.4.1.5	Cerium	954
4.20.4.1.6	Transition metals	955
4.20.4.2	Sulfur Functions	955
4.20.4.2.1	Lithium compounds	956
4.20.4.2.2	Tin compounds	961
4.20.4.2.3	Copper compounds	963
4.20.4.2.4	Zinc compounds	964
4.20.4.3	Selenium and Tellurium Functions	964
4.20.4.3.1	Selenium	964
4.20.4.3.2	Tellurium	965

4.20.5 ACKNOWLEDGEMENT — 965

4.20.1 FUNCTIONS CONTAINING A CHALCOGEN AND A NITROGEN FUNCTION

4.20.1.1 Functions Bearing Oxygen and Nitrogen

4.20.1.1.1 *Ketene hemiaminal derivatives*

Compounds of general structure (**1**) can exist as a tautomeric equilibrium of imidic ester (**1a**) and 1-alkoxyenamine (**1b**) forms (Equation (1)). Hantzsch and Osswald originally assigned structure (**1a**; $X = Y = CN$, $R^1 = Et$, $R^2 = H$) to the product of the addition of ethanol to tricyanomethane ⟨1899CB641⟩. Much later, this was corrected to the enamino form (**1b**) on the basis of its infrared spectrum ⟨58JA2788⟩. It is to be expected that electron-withdrawing substituents on C-2 would favour the enamine tautomer and this has been confirmed by a variety of spectroscopic studies of, inter alia, (**2**) ⟨45JA1017⟩, (**3**) and (**4**) ⟨49JA40, 76LA1762, 85JPR297⟩.

The actual position of the tautomeric equilibrium depends on structure, solvent and temperature, and an NMR study of (**4**) shows that the enamine structure represents 92% of the mixture in acetone, 88% in carbon disulfide, 86% in benzene and 75% in chloroform ⟨85JPR297⟩.

The seven main methods of synthesising this system are:
(i) from amides,
(ii) from nitriles,
(iii) by addition to substituted alkynes,
(iv) by addition to the carbonyl group of ketenes,
(v) from ketene derivatives,
(vi) from 1-haloenamines, and
(vii) from aryl cyanates.

In addition, a number of rather singular and nongeneral methods have been described and these will be listed under 'miscellaneous'.

(i) From amides

Primary, secondary and tertiary amides, lactams and imides are readily alkylated on the carbonyl oxygen by trialkyloxonium tetrafluoroborates, dialkyl sulfates and alkyl halides to give alkoxymethylene iminium salts (**5**) ⟨56CB2060, 61AG493, 63CB1350⟩, and see the comprehensive reviews by Kantlehner ⟨B-79MI 420-01, 91COS(6)485⟩. When Y is an electron-withdrawing group, for example CN ⟨71CB3475⟩, CONMe$_2$ ⟨64CB3081, 85S1062⟩ and NO$_2$ ⟨85EUP230127⟩, the C-2 proton is readily removed by bases to give the desired ketene *O,N*-acetal (**6**). The yields, as shown in Table 1, are generally high.

Table 1 The synthesis of ketene hemiaminals by alkylation of amides.

X	R^1	R^2	R^3	Alkylating agent	Base	Yield (%)	Ref.
CN	Me	Me	Et	Et$_3$O·BF$_4$		51	71CB3475
NO$_2$	Me	Me	Me	Me$_3$O·BF$_4$	NaHCO$_3$	35.2	85EUP230127
CONR$_2$	Me	Me	Et	Et$_3$O·BF$_4$	(TMS)$_2$NNa	85	85S1062
CONR$_2$	Et	Et	Et	Et$_3$O·BF$_4$	(TMS)$_2$NNa	91	85S1062
CONR$_2$	Pr^i	Pr^i	Et	Et$_3$O·BF$_4$	(TMS)$_2$NNa	86	85S1062
CONR$_2$	Ph	Ph	Et	Et$_3$O·BF$_4$	(TMS)$_2$NNa	77	85S1062

In other cases, the salts are converted into the dialkoxydialkylaminoalkanes (**7**) by treatment with sodium alkoxides in methylene chloride at 0 °C ⟨71CB3475⟩. Subsequent heating of (**7**) over sodium or, better, calcium metal followed by distillation gave (**6**) ⟨64CB3081, 68TL497⟩. Interestingly, if Y^4 in (**5**) is methyl, the alkoxy group in the product (**6**) is derived from the alkoxide ⟨64CB3081⟩. Examples of these reactions are given in Table 2.

The effect of an α-electron-withdrawing group is clearly shown by the contrast between the lactam (**8**), which is resistant to the loss of ethanol, and (**9**), which is readily produced by the action of sodium ethoxide on (**10**) ⟨64CB3081, 69MI 420-01⟩.

Tertiary amides can be acylated by acetyl or benzoyl chlorides in dichloromethane at room temperature in the presence of silver trifluoromethane sulfonate to give, in 'essentially quantitative' yields, the acyloxyiminium triflates (**11**) ⟨80CC790⟩. Treatment of (**11**; R^4 = Ph) with triethylamine or pyridine gives the products (**12**; R^3 = Me or Et, R^1 = R^2 = H, Ph), which are unstable and could only be obtained in solution. Much more stable are the cyclic analogues (**13**) produced by the exothermic reaction of monoalkyl amides with dimethylmalonyl dichloride in toluene in the presence of triethylamine (Equation (2), Table 3) ⟨66JOC2966⟩.

Table 2 The synthesis of ketene hemiaminals via dialkylamino alkanes.

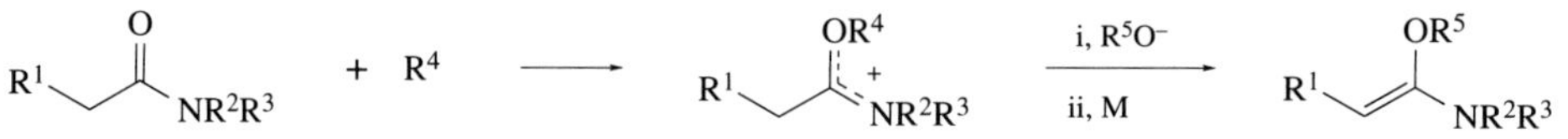

R^1	R^2	R^3	R^4	Alkylating agent	R^5	M	Yield (%)	Ref.
H	Me	Me	Me	Me$_2$SO$_4$	Et	Ca	74	64CB3081
H	Me	Me	Me	Me$_2$SO$_4$	Me	Ca	58	64CB3081
H	Et	Et	Me	Me$_2$SO$_4$	Et	Ca	67	64CB3081
H	Et	Et	Me	Me$_2$SO$_4$	Me	Ca	48	64CB3081
Me	Me	Me	Me	Me$_2$SO$_4$	Et	Ca	59	64CB3081
Me	Me	Me	Me	Me$_2$SO$_4$	Me	Ca	57	64CB3081
Et	Me	Me	Me	Me$_2$SO$_4$	Et	Ca	69	64CB3081
H	Me	Me	Me	Me$_2$SO$_4$	Me	Na	70	64HCA2425
H	Me	Me	Me	Me$_2$SO$_4$	Me	Na	?	65GEP1212069
H	Me	Me	Et	Et$_3$O·BF$_4$	Et	Na		61LA(641)1

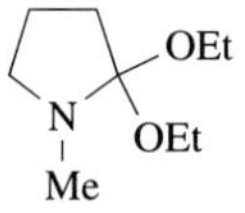

(8) 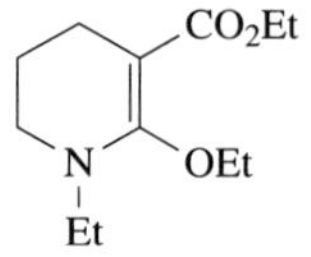 (9) 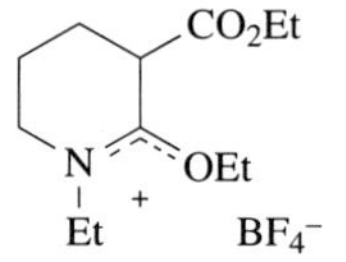 (10)

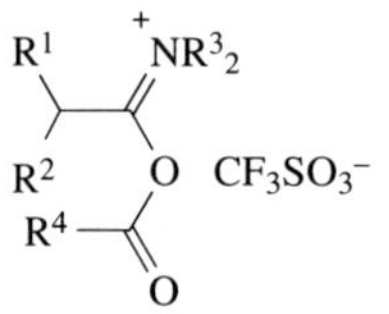

(11) 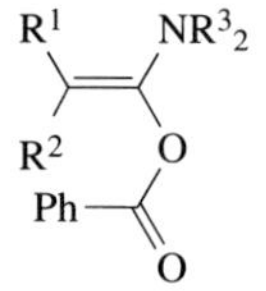 (12)

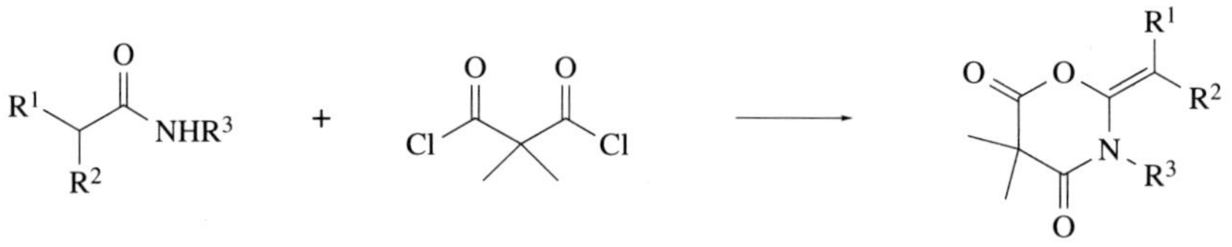

(13) (2)

Table 3 Dihydro-2-methylene-4H-1,3-oxazine-4,6(5H)diones.

R^1	R^2	R^3	Yield (%)	Ref.
H	H	Me	72	66JOC2966
H	H	Ph	21	66JOC2966
Me	H	Me	80	66JOC2966
Me	Me	Me	69	66JOC2966
Me	Me	Bu^n	67	66JOC2966
Me	Me	Ph	50	66JOC2966
Ph	H	Me	75	66JOC2966
Ph	Ph	Me	98	66JOC2966

Related is the reaction of N,N-dimethylacetamide with ethyl chloroformate to give an adduct which then decomposes to (**6**; Y^1 = H, Y^4 = Et, Y^2 = Y^3 = Me) when treated with triethylamine ⟨83MI 420-01⟩.

(ii) From nitriles

The acid-catalysed addition of alcohols to the nitrile group to give imidates was first reported by Ritter in the last century ⟨1895CB473⟩. With the appropriate electron-withdrawing 2-substituent, the initial product can tautomerise to a ketene-hemiaminal (Scheme 1), although two early examples (**14a**) from Haller ⟨1887BSF23⟩ and Hantzsch ⟨1899CB641⟩ were originally assigned the incorrect imino structures (**14b**). These structures were later corrected by, *inter alia*, Elnaydi *et al.* ⟨89S775⟩ and Middleton and Engelhardt ⟨58JA2788⟩.

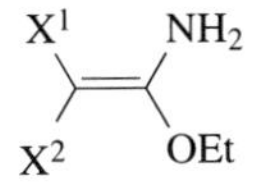

(**14a**) $X^1 = Ph, X^2 = H$
$X^1 = X^2 = CN$

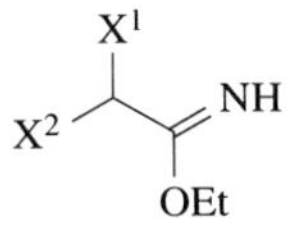

(**14b**) $X^1 = Ph, X^2 = H$
$X^1 = X^2 = CN$

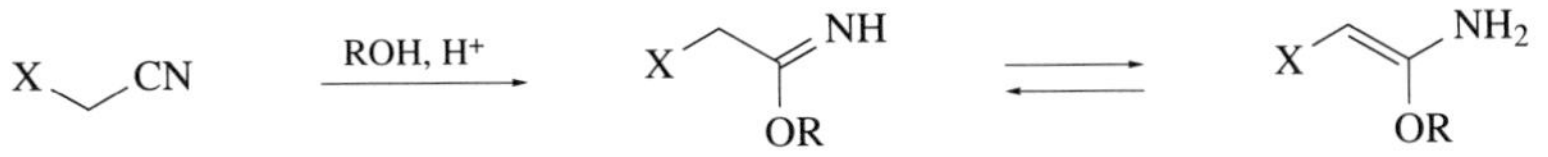

Scheme 1

In general, an ethereal solution of the nitrile is treated at 0 °C with a slight excess of ethanol saturated with dry hydrogen chloride gas and left for 24 h (Scheme 1). Treatment with aqueous sodium bicarbonate or triethylamine liberates the free base (**15**) (X = PhCO, 76% ⟨89S775⟩; X = CO_2Et ⟨86LA533⟩, 90% as the HCl salt ⟨83HCA809⟩, 94% ⟨45JA1017⟩; X = CN, 97% ⟨49JA40⟩). A variation on the technique was developed by Niedlein and Kikelj who used boron trifluoride etherate as the catalyst (Table 4) ⟨88CB1817⟩.

Table 4 Acid-catalysed addition of alcohols to nitriles.

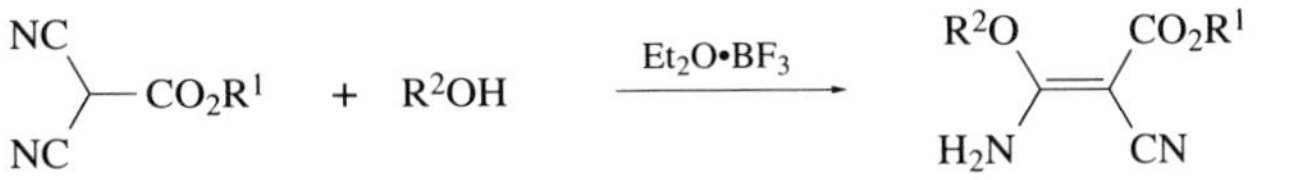

R^1	R^2	Conditions	Yield (%)
Me	Me	$CHCl_3$, RT, 48 h	83
Me	Et	$CHCl_3$, RT, 48 h	82
Et	Me	$CHCl_3$, RT, 48 h	89
Me	Ph	$CHCl_3$, RT, 48 h	50
CH_2Ph	Me	$CHCl_3$, RT, 48 h	80
Me	CH_2CH_2Ph	$CHCl_3$, RT, 48 h	63

The reaction has also been performed under neutral conditions. Friedman and Kim showed that ammoniopropanedinitrile toluene-*p*-sulfonate (**16**) reacts with aromatic aldehydes in methanol buffered with ammonium acetate to give, in generally excellent yields, the (*E,E*)-4-amino-1-aryl-3-cyano-4-methoxy-2-azabutadienes (**17**) (Equation (3) and Table 5) ⟨89S698, 91JOC657⟩. The configuration (**17**) was determined in the crystalline state by x-ray crystallography ⟨91MI 420-01⟩.

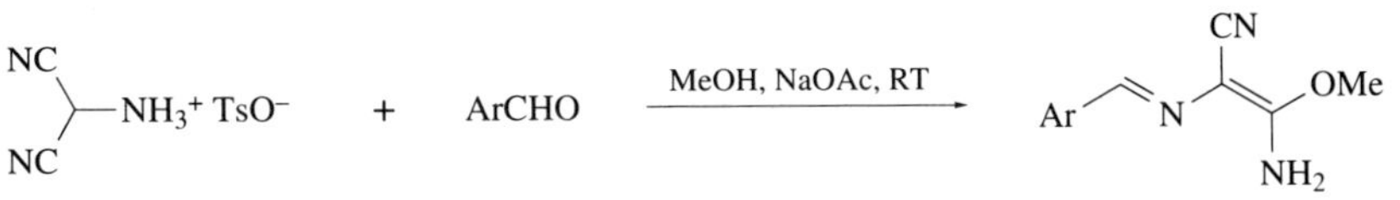

The base-catalysed Ritter reaction has been reported not to work with nitriles carrying a strongly acidic α-proton, which are exactly the requirements needed to shift the tautomeric equilibrium to the ketene hemiaminal form ⟨61JOC412⟩. However, Gilman and Cope showed that ethyl cyanoacetate condenses with isobutylene oxide under basic conditions to give the lactone (**18**) ⟨45JA1012⟩.

Table 5 Reaction of ammoniopropanedinitrile with aromatic aldehydes to give the azabutadienes (**17**) (Equation (3)).

Ar (in ArCHO)	Reaction time (h)	Yield (%)
Ph	24	90
2-BrC$_6$H$_4$	3	92
2-ClC$_6$H$_4$	6	99
2-NO$_2$C$_6$H$_4$	72	80
4-NO$_2$C$_6$H$_4$	13	99
2-naphthyl	24	69
4-NO$_2$C$_6$H$_4$	24	23

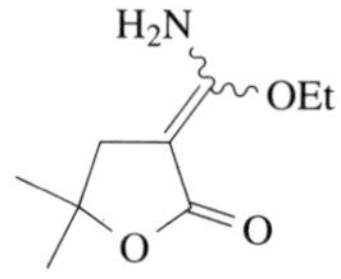

(**18**)

(iii) By addition to substituted alkynes

(a) Addition of amines to alkynic ethers. Secondary amines react with the ethoxyalkynes (**19**; X = H or P(OEt)$_2$) in the absence of water to give ketene hemiaminals (Equation (4)) ⟨54MI 420-01, 56RTC1377, 91ZOB1600⟩. Further, the allylamines (**20**) reacted with (**19**; R^1 = H) in the absence of solvent to give (**21**), but these adducts were very unstable to water and it proved impossible to obtain the hemiaminal uncontaminated with the amide (**22**) ⟨70CR(C)1890⟩.

$$X-\!\!\equiv\!\!-OEt \;+\; R^1R^2NH \;\longrightarrow\; \underset{NR^1R^2}{\overset{X\quad OEt}{\diagup\!\!\diagdown}} \quad (4)$$

(**19**)

(**20**) (**21**) (**22**)

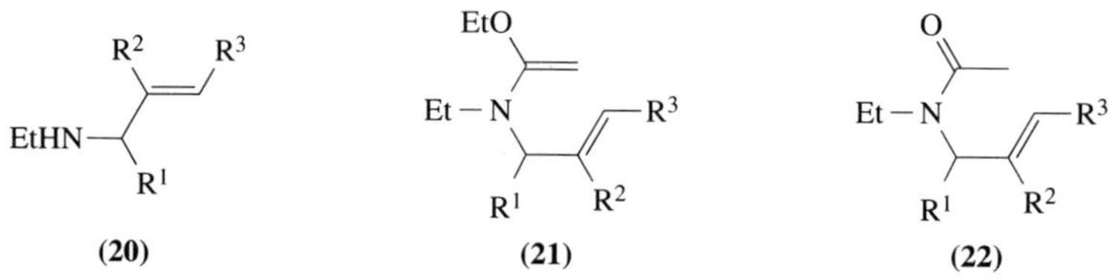

Dorokhov and Mikhailov have shown that α-pyridylaminodialkylboranes (**23**) react with (**19**; X = H) in hexane at 20 °C to give 70% of the chelate (**24**) ⟨72IZV1895⟩.

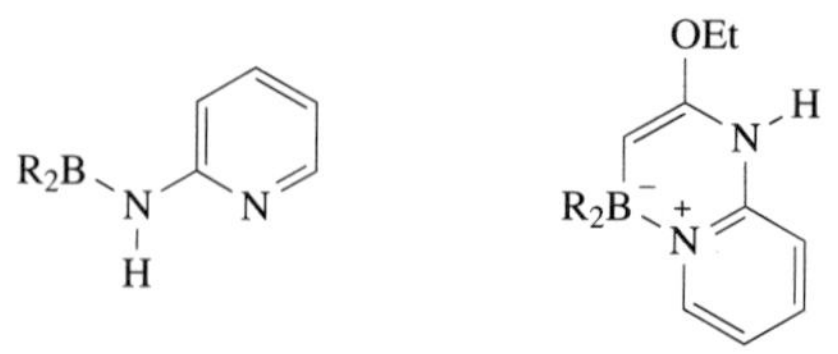

(**23**) (**24**)

Strong bases, such as sodium bis(trimethylsilyl)amine, react with diphenoxyethyne (**25**) to give the adduct (**26**) ⟨79TL3329⟩.

In appropriate circumstances, the alkynic ether can be generated *in situ* and then treated with the required amine (Equation (5)) ⟨77CA135599v⟩.

$$\text{PhO}\!\!-\!\!\!\equiv\!\!\!-\!\!\text{OPh} \qquad \underset{\text{N(TMS)}_2}{\overset{\text{PhO}\quad\text{OPh}}{\diagdown\!\!=\!\!\diagup}}$$

(25)   (26)

$$\underset{\text{Cl}}{\overset{\text{O}}{(\text{EtO})_2\text{P}}}\!\!\diagdown\!\!=\!\!\diagup\text{OEt} \xrightarrow[65\%]{\text{i, NaOEt} \atop \text{ii, Et}_2\text{NH}} \underset{\text{NEt}_2}{\overset{\text{O}}{(\text{EtO})_2\text{P}}}\!\!\diagdown\!\!=\!\!\diagup\text{OEt} \quad (5)$$

(b) Addition of alcohols to ynamines. Viehe *et al.* showed that ethanol reacts with the ynamine (27) in ether at 20 °C under acid catalysis to give the hemiaminal (28) (Equation (6)) ⟨64AG571⟩.

$$\text{Ph}\!\!-\!\!\!\equiv\!\!\!-\!\!\text{NMe}_2 + \text{EtOH} \xrightarrow{\text{Et}_2\text{O, H}^+} \underset{\text{OEt}}{\overset{\text{Ph}\quad\text{NMe}_2}{\diagdown\!\!=\!\!\diagup}} \quad (6)$$

(27)   (28)

The reactions of ynamines also carrying an electron-withdrawing group, for example (29; X = H, Me or MeO), with a variety of nucleophiles have been reported in a series of papers by Neuenschwander and co-workers ⟨69HCA2641, 73HCA944, 73HCA1318, 73HCA1331⟩. The rate of the reaction with phenol increases in the order H < Me < MeO. Potassium *t*-butoxide was required to initiate addition of methanol to (29; X = H) whilst the reaction proceeded readily under neutral conditions with the other analogues (Table 6).

Table 6 The addition of alcohols to ynamines.

$$\underset{\text{(29)}}{\text{Me}_2\text{N}\!\!-\!\!\!\equiv\!\!\!-\!\!\overset{\text{O}}{\underset{\text{X}}{\diagdown}}} + \text{ROH} \longrightarrow \underset{\text{RO}}{\overset{\text{Me}_2\text{N}\quad\overset{\text{O}}{\diagdown}\!\!\text{X}}{\diagdown\!\!=\!\!\diagup}}$$

X	R	Conditions	Yield (%)	Ref.
H	Me	MeOH, RT, 24 h	92	69HCA2641
H	Et	EtOH, RT, 48 h	86	69HCA2641
Me	Me	KOBu^t, Δ, 0.5 h	81	73HCA944
MeO	Me	MeOH, Δ, 2 h	83	73HCA944
H	Et	EtOH, Δ, 1 h	75	73HCA944
Me	Et	EtOH, Δ, 1 h	85	73HCA944
MeO	Et	EtOH, Δ, 2 h	82[a]	73HCA944
H	Ph	CCl$_4$, Δ, 1 h	89	73HCA1331
Me	Ph	CCl$_4$, RT, 4 h	88	73HCA1331
MeO	Ph	CCl$_4$, RT, 4 h	85	73HCA1331

[a] The product is the ethyl ester (R = OEt).

Organic acids also add to (29), but the adducts are very unstable and rapidly rearrange to acrylamides and crotonamides ⟨73HCA1318⟩.

Tolchinskii *et al.* have investigated the addition of alcohols to ynamines substituted with alkenyl and alkynyl groups ⟨80ZOB984, 91ZOB1192⟩. These are much less reactive, and a mixture of HCl and boron trifluoride etherate is required to catalyse the addition of primary alcohols such as methanol and propanol. Phenols and fluoro-substituted alcohols are, however, much more reactive and the reaction proceeds readily in ether (Table 7) ⟨80ZOB984⟩. Aliphatic alcohols gave a mixture of *cis* and *trans* products whilst phenol gave a single stereoisomer with the phenoxy group *cis* to the C-2 hydrogen. Phenol also readily added to the dialkynic ynamines (30; R = Me, Et) at room temperature in benzene to give (31; R = Me, Et) ⟨91ZOB1192⟩.

An interesting variation of this addition to ynamines was provided by Takamatsu and Sekiya who treated the *N,N*-disubstituted trichloroethylamine (32) with 5 molar equivalents of potassium *t*-butoxide in THF over 1–6 h at 30–40 °C to give (33), presumably via the chloroynamine (34) ⟨80CPB3098⟩.

Table 7 The addition of alcohols to alkenyl alkynes.

R^1	R^2	R^3	Conditions	Yield (%)
H	Et	Et	HCl, BF$_3$OEt$_2$, EtOH, Et$_2$O, 60–70 °C, 2 h	52
H	Et	Pr^n	HCl, BF$_3$OEt$_2$, EtOH, Et$_2$O, 60–70 °C, 2 h	56
H	Et	F$_2$CHCF$_2$CH$_2$	Et$_2$O, RT	75
H	Et	Ph	Et$_2$O, RT	69
H	Et	4-ClC$_6$H$_4$	Et$_2$O, RT	72
H	Pr^n	4-ClC$_6$H$_4$	Et$_2$O, RT	75
Me	Et	Pr^n	HCl, BF$_3$OEt$_2$, EtOH, Et$_2$O, 60–70 °C, 21 h	57
Me	Et	F$_2$CHCF$_2$CH$_2$	Et$_2$O, RT	71
Me	Et	4-ClC$_6$H$_4$	Et$_2$O, RT	74
Me	Pr^n	Ph	Et$_2$O, RT	92

(iv) Addition to the carbonyl group of ketenes

A few examples of the addition of substituted amines across the carbonyl group of ketenes have been reported. The simplest is perhaps the reaction of the silylamine (**35**) with ketene at −20 °C (Equation (7)) ⟨69JOM(17)241⟩. The product (**36**) can be isolated with care but is unstable and undergoes spontaneous O–C silyl group transfer to give a *C*-silylated acetamide. In contrast, the silylated amide (**37**) reacted exothermically with ketene and the product (**38**) was isolated by distillation (Table 8).

$$H_2C=\bullet=O \;+\; TMS\text{-}NMe_2 \xrightarrow{89\%} \text{(36)} \quad (7)$$

(**35**)   (**36**)

The diamines (**39**) also reacted with bistrifluoromethylketene (**40**) to give the ketene hemiaminals (**41**) ⟨69DOK(186)835⟩. This ketene (**40**) also reacts with two molecules of imines (e.g., with (**42**)) to give the oxadiazine (**43**). The reaction of C=N bonds with ketenes was first reported by Staudinger, who investigated the reaction of dimethylketene with a number of heterocycles. Unfortunately, he incorrectly identified the structure of the products, and Taylor and co-workers showed that quinoline gives (**44**) in 74% yield and acridine (**45**) in 78% yield ⟨65CC574, 65JCS5877, 67JCS(C)1569⟩. Further

Table 8 The reaction of silylated amides with ketene.[a]

X	R^1	R^2	Yield (%)
Me	Me	Me	89
Me	Et	Me	81
Me	Bu	Me	90
Et	Me	Me	80
Me	Me	Pr^n	77
OMe	Me	Me	80

[a] Reaction conditions: ketene bubbled through the neat amide.

examples of the reaction of imines were given by Martin et al. (Table 9) ⟨65TL3589, 71JOC2211⟩; the condensation was extended to ketene imines (Equation (8)) ⟨71JOC2205⟩.

(v) From ketene derivatives

(a) Ketene acetals. The reaction of ketene diethyl acetal (**46**) with *N*-ethylaniline to give 1-ethoxy-1-(*N*-ethylanilino)ethylene (**47**), when heated neat on a steam bath for 5 h, was originally described by McElvain and co-workers ⟨40JA1281⟩.

Subsequently, substituted ketene acetals have been treated with a wide range of amines (Table 10)

Table 9 The reaction of imines with ketenes.

R^1	R^2	R^3	R^4	Yield (%)
Me	Me	Ph	Me	80
Me	Me	Ph	Bu^t	35
Me	Me	Pr^i	Pr^n	80
Me	Me	Et	Pr^n	59
Me	Me	Pr^i	Et	71
Me	Me	(CH$_2$)$_5$^{a}	Pr^n	86
Me	Me	H	Bu^t	75
Et	Bu^n	Ph	Me	90

^{a} cyclohexylidene=NPr gave the corresponding product with R^3, H = (CH$_2$)$_5$ at C-4.

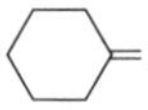

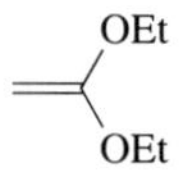

(46) ketene diethyl acetal (OEt, OEt)
(47) 1-ethoxy-1-(N-ethyl-N-phenylamino)ethylene (NEtPh, OEt)

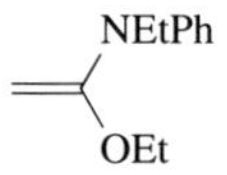

⟨49CR453⟩. Since the reaction involves initial nucleophilic addition of the amine to C-1, the rate and ease of the displacement is strongly dependent on the nature of the substituents on C-2 (Scheme 2).

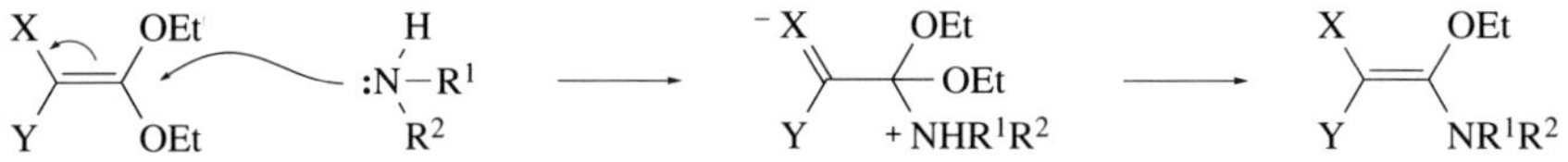

Scheme 2

Similarly, the alkoxy group in the product can also be displaced by amines and, therefore, the best yields are obtained by careful control of the reaction temperature and the relative proportions of the reagents. Table 11 shows the wide variety of examples of this reaction.

(b) Ketene aminals. McElvain and Tate showed that 1,1-bis(diethylamino)ethylene (**48**) reacts with refluxing alcohol during 4 h to give 1-ethoxy-1-diethylaminoethylene (**49**) in 62% yield ⟨45JA202⟩. However, the limitations of this method should be obvious in the light of the preceding section.

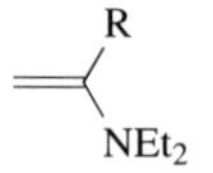

(**48**) R = NEt$_2$
(**49**) R = OEt

(c) From ketene dithioacetals. Ethanolamine condenses with the dithioacetal (**50**) at 130–150 °C during 3 h to give the hemiaminal (**51**) ⟨69YZ203⟩.

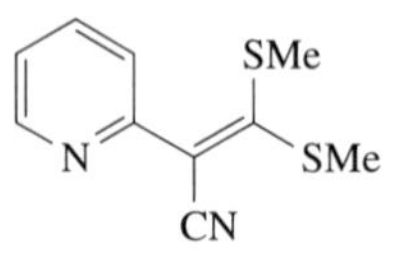

(**50**) (**51**)

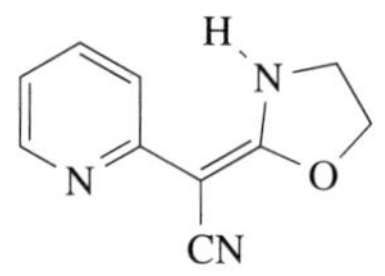

(d) By exchange of amines in ketene hemiaminals. The 1,2,3-triazole ring in (**52**) can be replaced

Table 10 The reaction of amines with ketene acetals.

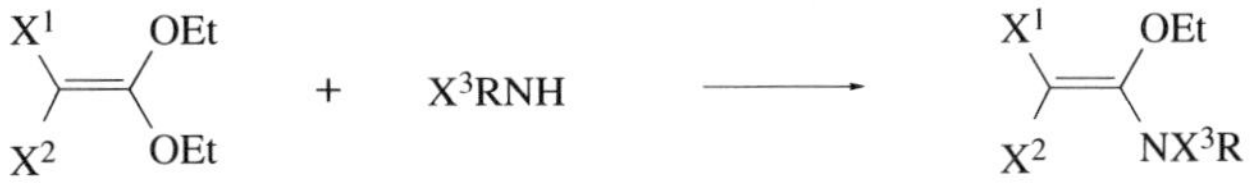

X^1	X^2	X^3	R	Conditions	Yield (%)	Ref.
H	H	Et	Ph	Δ, 5 h	47	40JA1281
H	H	Bu	Bu	130 °C, 13 h	5	45JA202
H	$COCF_3$	H	H	MeCN, RT	~100	90S195
H	$COCF_3$	Me	Me	MeCN, RT	93	90S195
H	$COCF_3$	Ph	H	MeCN, RT	~100	90S195
H	$COCF_3$	4-MeC_6H_4	H	MeCN, RT	~100	90S195
H	$COCF_3$	4-$MeOC_6H_4$	H	MeCN, RT	~100	90S195
H	$COCF_3$	4-ClC_6H_4	H	MeCN, RT	~100	90S195
H	$COCF_3$	1-naphthyl	H	MeCN, RT	~100	90S195
H	$COCF_3$	2-naphthyl	H	MeCN, RT	~100	90S195
H	$COCF_3$	Me	Me	MeCN, RT	84	86S1013
CN	CN	H	H	NH_4OH, 100 °C	67	58JA2788
CN	CN	H	Me	$MeNH_2$, H_2O, 100 °C	96	58JA2788
H	Ant[a]	H	H	Et_2O, Δ	45	80AP65
H	Ant[a]	Bn	H	Et_2O, Δ	60	80AP65
H	Ant[a]	Ph	H	Et_2O, Δ	57	80AP65
H	Ant[a]	2-HOC_6H_4	H	Et_2O, Δ	62	80AP65

[a] Ant =

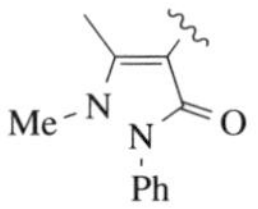

by other amines, for example by *N*-methylaniline and diethylamine in yields of 61% and 39%, respectively ⟨83AG(E)553⟩.

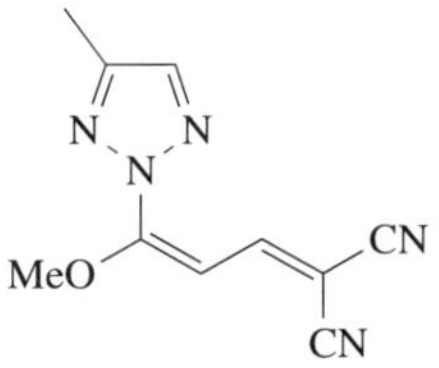

(52)

(vi) From 1-halo-1-dialkylaminoalkenes

Phenols and sodium alkoxides readily displace the chlorine atom of 1-chloro-1-dialkylaminoalkenes **(53)** to give ketene hemiaminals. Table 12 gives a number of examples of this reaction. Carboxylate ions from sodium or, better, silver carboxylates or generated *in situ* from the acid and triethylamine also react with **(53)** to give 1-acyloxyenamines (Table 13). The fluorine atom in **(54)** has been displaced by lithium methoxide in ether at 0 °C ⟨77NJC371⟩.

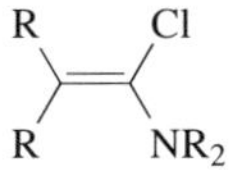

(53) 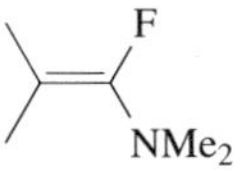   **(54)**

(vii) From aryl cyanates

Reich *et al.* have shown that under carefully controlled conditions aryl cyanates react with active methylene groups to give ketene hemiaminals (Equation (9) and Table 14) ⟨66CB2302, 75BSF2089⟩.

Table 11 Further reactions of ketene acetals (reaction scheme in Table 10).

Ketene acetal	Amine	Conditions	Yield (%)	Ref.
NC, NC-C=C(OCH$_2$CH$_2$O) (1,3-dioxolane)	NH$_3$	NH$_4$OH, 100 °C	60	58JA2788
NC, NC-C=C(OCH$_2$CH$_2$CH$_2$O) (1,3-dioxane)	NH$_3$	NH$_4$OH, 100 °C	66	58JA2788
(EtO)$_2$P(O)-C(CN)=C(OCH$_2$CH$_2$O)	pyrrolidine	neat, RT	49	92HCA124
(EtO)$_2$P(O)-C(CN)=C(OCH$_2$CH$_2$O)	allylamine	THF, RT	49	92HCA124
Bu^t-CH=C(OAr)$_2$ Where Ar is:				
Ph	1,2,4-triazole	180–190 °C 0.5% dodecylbenzenesulfonic acid	93	93LA207
p-Tol	1,2,4-triazole	180–190 °C 0.5% dodecylbenzenesulfonic acid	90	93LA207
2-ClC$_6$H$_4$	1,2,4-triazole	180–190 °C 0.5% dodecylbenzenesulfonic acid	91	93LA207
4-ClC$_6$H$_4$	1,2,4-triazole	180–190 °C 0.5% dodecylbenzenesulfonic acid	92	93LA207
2,4-Cl$_2$C$_6$H$_3$	1,2,4-triazole	180–190 °C 0.5% dodecylbenzenesulfonic acid	95	93LA207
4-PhC$_6$H$_4$	imidazole	180–190 °C 0.5% dodecylbenzenesulfonic acid	98	93LA207
2-ClC$_6$H$_4$	imidazole	180–190 °C 0.5% dodecylbenzenesulfonic acid	26	93LA207
2,4-Cl$_2$C$_6$H$_4$	imidazole	180–190 °C 0.5% dodecylbenzenesulfonic acid	13	93LA207
4-PhC$_6$H$_4$	benzimidazole	180–190 °C 0.5% dodecylbenzenesulfonic acid	93	93LA 207
4-ClC$_6$H$_4$	benzimidazole	180–190 °C 0.5% dodecylbenzenesulfonic acid	85	93LA207
(EtO)(EtO)C=CH-C(O)-C(O)-CH=C(OEt)(OEt)	NH$_3$	NH$_4$OH	80	60CB1059
(EtO)(EtO)C=CH-C(O)-C(O)-CH=C(OEt)(OEt)	n-propylamine	Et$_2$O	82	60CB1059
(EtO)(EtO)C=CH-C(O)-C(O)-CH=C(OEt)(OEt)	benzylamine	dioxan	72	60CB1059

Chalcogen and Nitrogen

Table 11 (continued).

Ketene acetal	Amine	Conditions	Yield (%)	Ref.
EtO, EtO, O, O, OEt, OEt (bis-ketene acetal)	aniline	Et$_2$O	82	60CB1059
EtO, EtO, O, O, OEt, OEt (bis-ketene acetal)	piperidine	Et$_2$O	65	60CB1059

Table 12 The reaction of 1-halo-1-dialkylaminoalkenes with alkoxides.

$$\underset{R^1}{\overset{X^1}{\diagdown}}C=C\underset{Cl}{\overset{NX^2X^3}{\diagup}} + R^2ONa \longrightarrow \underset{R^1}{\overset{X^1}{\diagdown}}C=C\underset{OR^2}{\overset{NX^2X^3}{\diagup}}$$

X^1	R^1	X^2, X^3	R^2	Conditions	Yield (%)	Ref.
Me	Me	(CH$_2$)$_5$	Et	THF, 20 °C, 1–3 h	80	69AG(E)454
Me	Me	(CH$_2$)$_5$	Bn	?	?	70CR(C)1890
CO$_2$Et	H	Et, Et	Ph	Δ, 90 °C	15	71G269
CO$_2$Et	H	Et, Et	p-Tol	Δ, 90 °C	16	71G269
COCl	Me	(CH$_2$)$_5$	Me	Et$_2$O, 25 °C, 10 h	74[a]	69T3447
COCl	Me	Et, Et	Me	Et$_2$O, 25 °C, 10 h	63[a]	69T3447
Me	Me	Me, Me	Et	Et$_2$O, 20 °C, 3 h	86	B-76MI 420-01
Me	Me	Me, Me	Me	Et$_2$O, 20 °C, 3 h	82	B-76MI 420-01
Ph	Et	Me, Me	Me	Et$_2$O, 20 °C, 3 h	83	B-76MI 420-01
H	H	H, CONHAr	Me	Et$_2$O, 12 h	61[b]	71ZOB482

[a] Product is methyl ester, $X^1 = CO_2Me$ in the product. [b] Cl replaced by F in the substrate.

Table 13 Reactions of 1-halo 1-dialkylaminoalkenes with carboxylate anions.

$$\underset{R^2}{\overset{R^1}{\diagdown}}C=C\underset{NR^3R^4}{\overset{Cl}{\diagup}} + R^5CO_2^- M^+ \longrightarrow \underset{R^2}{\overset{R^1}{\diagdown}}C=C\underset{NR^3R^4}{\overset{O_2CR^5}{\diagup}}$$

R^1	R^2	R^3, R^4	R^5	M	Conditions	Yield (%)	Ref.
Me	Me	Me, Me	Bn	Na	PhH, 80 °C	70	B-76MI 420-01
Me	Me	Me, Me	Ph	Ag	CCl$_4$, 20 °C, 0.5 h	80	B-76MI 420-01
Me	Me	Me, Me	Ph	H	NEt$_3$, CCl$_4$, 1 h	84	B-76MI 420-01
Me	Me	(CH$_2$)$_5$	Bn	Ag	CCl$_4$, 20 °C, 1 h	77	B-76MI 420-01
Ph	Et	Me	Me	Ag	CCl$_4$, 20 °C, 1 h	85	B-76MI 420-01

$$ArOCN + \underset{Y}{\overset{X}{\diagdown}}CH_2 \xrightarrow{Et_3N, Et_2O, 0 \text{ to } -5 \,°C} \underset{Y}{\overset{X}{\diagdown}}C=C\underset{OAr}{\overset{NH_2}{\diagup}} \quad (9)$$

(viii) Miscellaneous preparations of ketene hemiaminals

Finally, examples of these compounds have been produced in ways that are interesting, though they lack the versatility of the previous methods.

(a) Ring-opening reactions. The azetidines **(55)** ring opened to **(56)** in 31% yield when kept at room temperature in chloroform–methanol ⟨83JOC481⟩, whilst the tetra-substituted cyclobutadienes

Table 14 Reactions of aryl cyanates with active methylene groups.

$$X \diagdown Y + ArOCN \longrightarrow \begin{matrix} X & OAr \\ & \diagup \\ Y & NH_2 \end{matrix}$$

Ar	X	Y	Yield (%)
Ph	CN	CN	85
p-ClC$_6$H$_4$	CN	CN	87
p-Tol	CN	CN	82
p-MeOC$_6$H$_4$	CN	CN	99
p-NO$_2$C$_6$H$_4$	CN	CN	7
p-Tol	CN	CO$_2$Et	74
p-MeOC$_6$H$_4$	CN	CO$_2$Et	75

(57) readily added methanol to give the butadienes (58) as mixtures of geometric isomers ⟨71AG(E)67⟩. The latter products also arise quantitatively from the addition of methoxide to the tetrafluoroborate salts (59; $X = BF_4^-$).

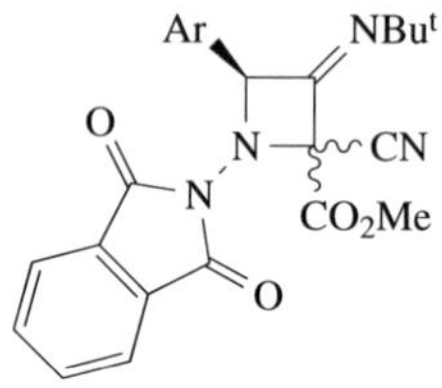

(55) 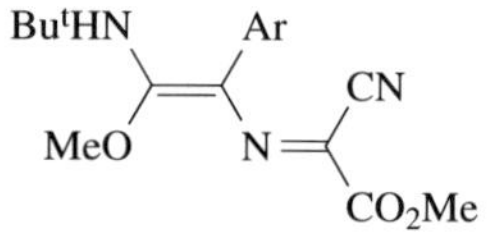 (56) 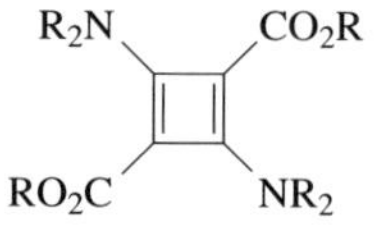 (57)

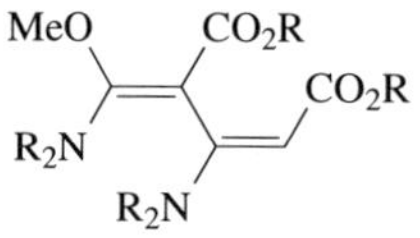

(58) 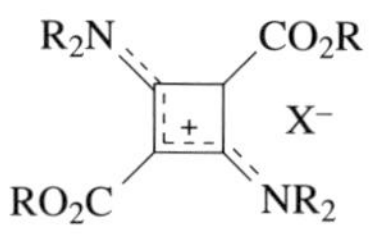 (59)

The isoxazoles (60) photolyse in methanol to give a mixture from which the ketene hemiaminals (61) can be isolated ⟨73CL111⟩. The reaction is cleaner in ethanol, i-propanol and t-butanol, although in the last case the products hydrolysed to the amides (62) on attempted purification ⟨76JCS(P1)783⟩.

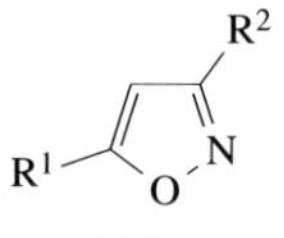

(60) 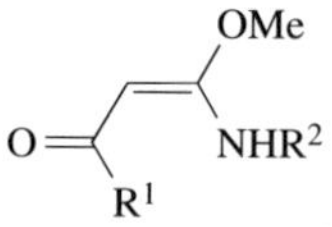 (61) 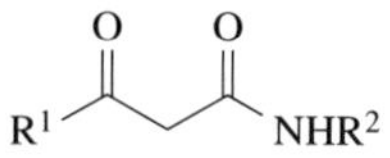 (62)

Jones and Phipps reported that the isoxazole salts (63; R = H, Me) ring open in boiling methanol–trimethylamine to give the hemiaminals (64; R = H, Me) in 56% and 67% yields, respectively ⟨76JCS(P1)1241⟩.

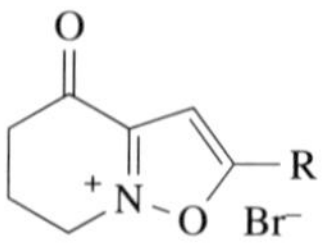

(63) 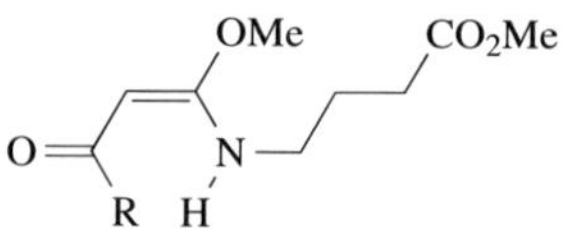 (64)

The enamine structure (65) was assigned to the product of the reaction of N-acetyl-L-tyrosinamide with Woodward's reagent K (N-ethyl-5-phenylisoxazolium-3'-sulfonate) (66) on the basis of spectroscopic evidence ⟨86JA5543⟩.

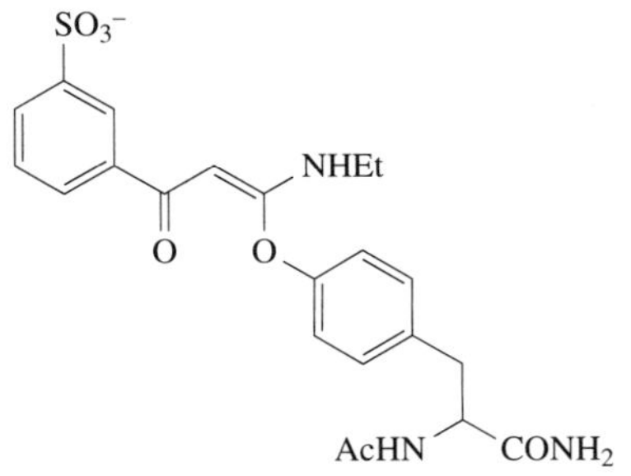

The 1,3-diethyl-1,4(3,4)dihydro-4-oxopyrimidinium tetrafluoroborates (**67**; R = H, Ph), ring open with concentrated aqueous ammonia at room temperature to give (**68**; R = H, Ph) in 42% and 17% yields, respectively ⟨77RTC68⟩.

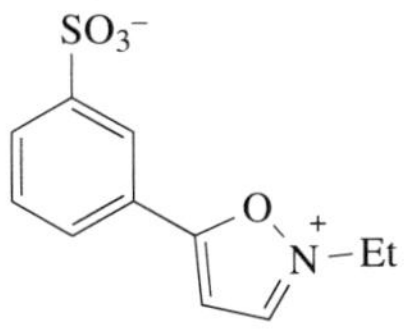

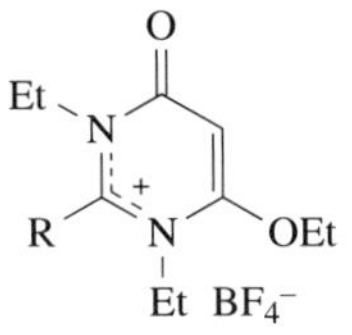

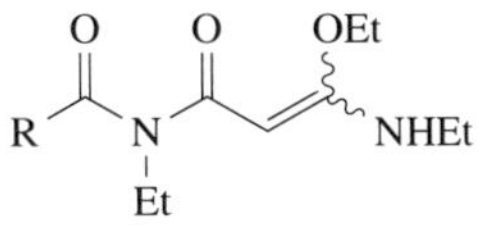

Anderson and Brown showed that 6-amino-1,3-dimethyluracil (**69**; R = H) ring opens on reaction with dimethylethyne dicarboxylate in DMSO at room temperature to give the isolable but unstable enamine (**70**) in 70% yield. They suggested Scheme 3 as the mechanism for this unexpected reaction ⟨77JOC4159⟩.

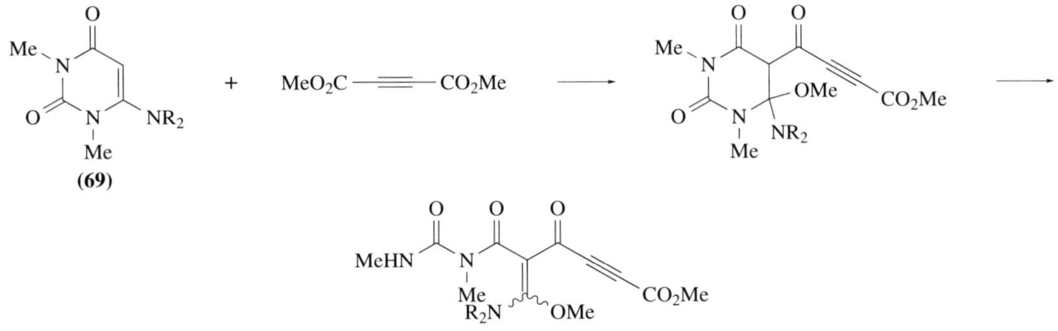

Scheme 3

Subsequently, Kawahara et al. showed that with the uracils (**69**; R = alkyl, e.g., R_2 = $(CH_2)_4$ and R = Me) the products (**70**) are now stable and readily isolated in yields of 81% and 43%, respectively ⟨80H(14)619, 82CPB63⟩.

(b) Other displacement reactions on ketene derivatives. 7,7,8,8-Tetracyanoquinodimethane (**71**) reacts with ethanolamine at room temperature to give (**72**) in a yield of 55% ⟨62JA3387⟩. Another nitrile displacement was seen in 1-[bis-(2-chloroethyl)amino]-1,2,2-tricyanoethene (**73**), which reacted with ethanol in the presence of triethylamine to give 67% of (**74**) ⟨67CB3460⟩.

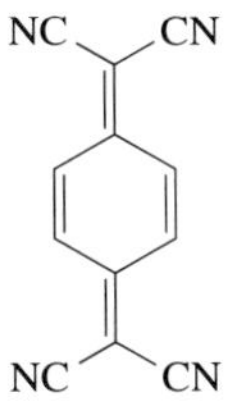

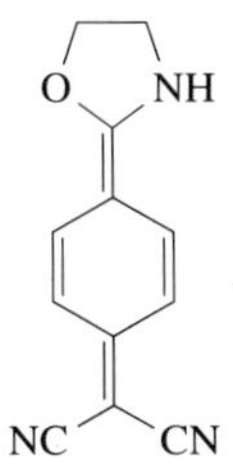

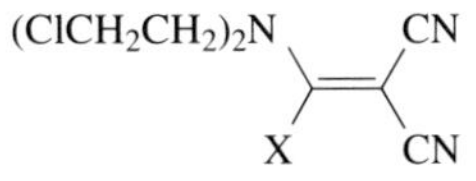

(**73**) X = CN
(**74**) X = OEt

(c) Other displacement reactions. The fluorine atom in the 1-fluoroenol ethers (**75**) can be displaced by amines to give (**76**) ⟨72IZV1347, 74MI 420-01⟩.

(**75**) R = alkyl, aryl

(**76**)

Activated methylene groups, for example that in (**77**), react with orthocarbamate esters (**78**) with elimination of two moles of alcohol to give, for example, 72% of the ketene hemiaminal (**79**; R = Pr^i) ⟨79LA2096⟩. Similarly, (**77**) also reacts with the imine (**80**) to give (**79**; R^1 = Et, R^2 = R^3 = H) in 51% yield ⟨67CB2604⟩.

(**77**)   (**78**)   (**79**)   (**80**)

(d) Addition to iminium ions. Honzl showed that *t*-butyl isocyanide reacts with hydrogen chloride in dry ether to give 2,3-di(*t*-butylimino)propiononitrile dihydrochloride (**81**), which reacts with lithium ethoxide in THF to give (**82**) in 86% yield ⟨70TL2357, 71CCC3314⟩.

(**81**)   (**82**)

(e) Addition to ketenimine. 2-Cyanoesters, for example (**83**; R = Ph$_2$C(CN)), are oxidised by phenyliodoso diacetate in methanol–chloroform to give (**84**) via the ketenimine (**85**) ⟨77TL3349⟩. A similar oxidation was reported for (**83**; R = mesityl) with oxygen, copper and an amine as the oxidant ⟨72JOC1960⟩.

(**83**)   (**84**)   (**85**)

Vinyl azides (**86**), prepared by displacement of chlorine from the corresponding vinyl chlorides, decompose in the presence of alcohols to give, for example, (**87**; X^1 = X^2 = CN, R^1 = H, R^2 = Et) in 61% yield via the ketenimine (**88**) ⟨67AG(E)959, 70CB1982, 82JOC1397⟩.

(**86**)   (**87**)   (**88**)

A similar material can also be produced in 40% yield by condensation of phenylhydroxylamine and benzoyl(phenyl)ethanol in alcohol (Scheme 4) ⟨71JOC1685, 78JCS(P1)1113⟩.

More directly, aniline reacts with perfluoroacrylic ester (**89**) in the presence of pyridine to give the ketenimine (**88**; X^1 = CF$_3$, X^2 = CO$_2$R, R^1 = Ph), which readily adds ethanol to give (**90**; X^1 = X^2 = Et), although this spontaneously tautomerises to the iminoester form ⟨75IZV1274⟩.

(f) Rearrangements. The reaction of alcohols with the dimethyl amide of perfluoromethylacrylic acid (**91**) gives initially the dimethylamides of the 2-trifluoromethyl-3-alkoxy-3-fluoroacrylic acids (**92**). These readily rearrange to give 92% of (**93**) at 120 °C ⟨74IZV358⟩.

Flash vacuum pyrolysis of the Meldrum's acid derivatives (**94**) and the pyrroline diones (**95**) give initially ketenes, which, following a 1,3-shift of the methylthio group and trapping with ethanol, yield in 78% the ketene hemiaminals (**96**) (Scheme 5) ⟨92CC487⟩.

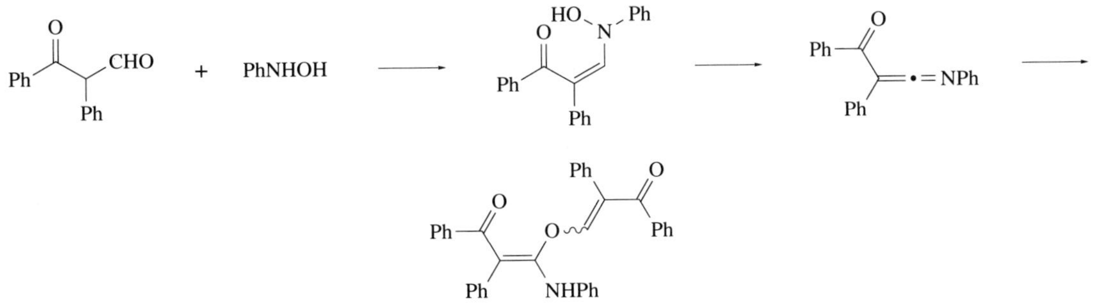

Scheme 4

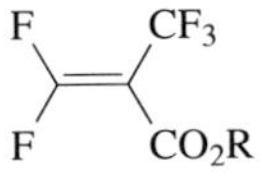

(89), 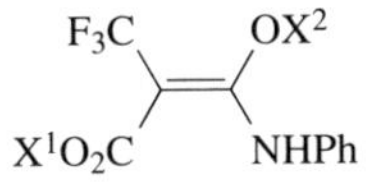 **(90)**, 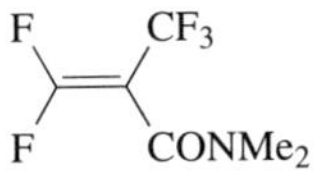 **(91)**, 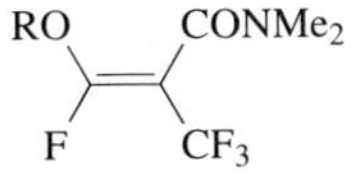 **(92)**, 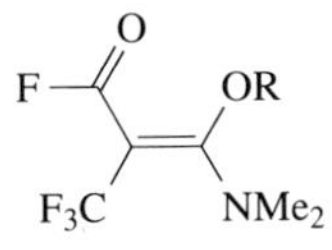 **(93)**, 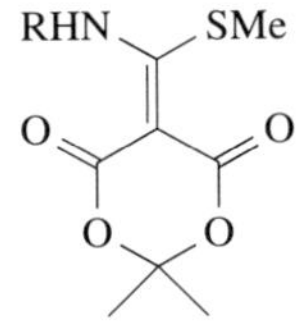 **(94)**, 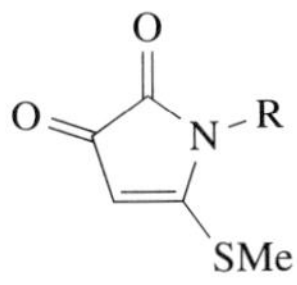 **(95)**, 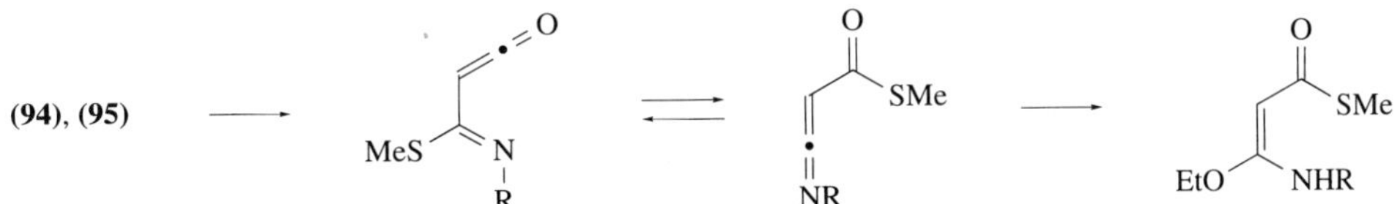 **(96)**

Scheme 5

(g) From thioesters. Ammonia and primary and secondary amines react with dithionomalonic esters (**97**) to give the (*E*)-3-amino-3-alkoxythioacrylic-*O*-alkyl esters (**98**) in 66–91% yields ⟨88AP(321)863⟩.

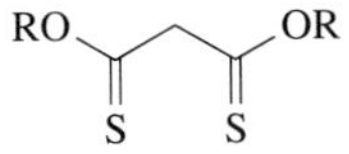

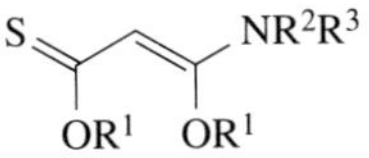

(**98**) R^1 = Me, Et; $R^2 = R^3$ = H
R^1 = Et; R^2 = H; R^3 = Me, *c*-C_6H_{12}, Ph
R^1 = Et; $R^2 = R^3$ = *c*-C_6H_{12}, *c*-C_5H_{10}

Similarly the cyanoacetic thionoester (**99**) condenses with benzylamine to give 3-benzylamino-3-ethoxyacrylonitrile (**100**) in 74% yield, although in this case the product contains some 33% of the iminoester (**101**) (Equation (10)) ⟨88AP(321)879⟩.

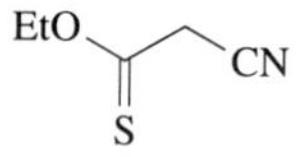

(99)

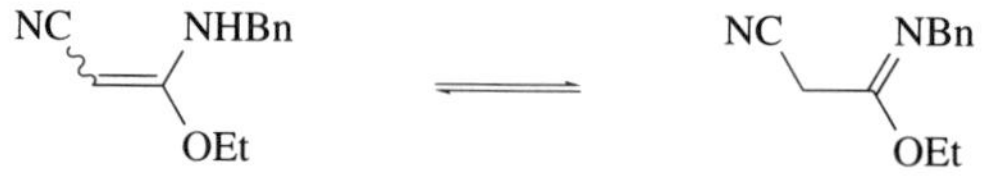

(100) **(101)** (10)

A rather different reaction of the dithionomalonic ester (**97**) is initial salt formation with *t*-butoxy-bis(dimethylamino)methane (**102**) to give 77% of (**103**) which decomposes in ethanol to give (**98**; $R^1 = Et$, $R^2 = R^3 = Me$) ⟨88AP(321)873⟩.

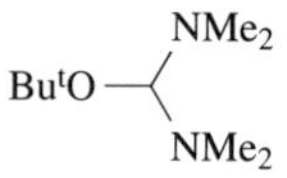

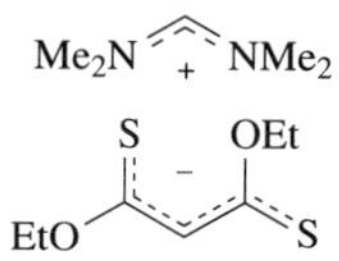

(102) **(103)**

4.20.1.1.2 *Other nitrogen derivatives*

(i) N-*Nitroso derivatives*

Kupper and Michejda reported a preparation of enol ethers of *N*-nitrosoamides (**104**) ⟨80JOC2919⟩. Thus *N*-nitroso(methyl)vinylamine (**105**) produced *N*-nitroso(1-methoxy-2-chloroethyl)methylamine (**106**) with *t*-butyl hypochlorite in methanol. This was dehydrohalogenated with potassium *t*-butoxide in ether in the presence of a catalytic quantity of 18-crown-6 ether (Scheme 6).

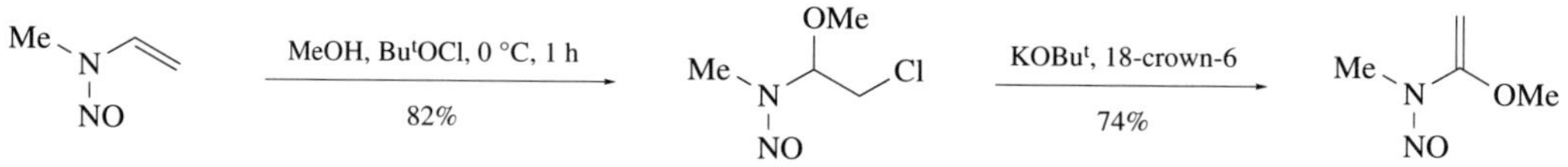

(105) **(106)** **(104)**

Scheme 6

(ii) Hydroxylamines

Neidlein *et al.* showed that *O*-benzylhydroxylamine reacts with 2-(methoxycarbonylcyanomethylene)-1,3-dioxolane (**107**) to give methyl 3-benzyloxyamino-2-cyano-3-(2-hydroxyethoxy) acrylate (**108**) ⟨89JHC1335⟩.

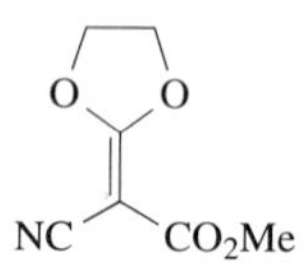

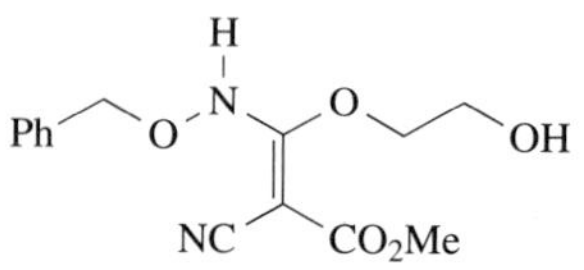

(107) **(108)**

(iii) Hydrazines

The same authors also showed that 1,1-dimethylhydrazine reacts with (**107**) to give (**109**) ⟨89JHC1335⟩.

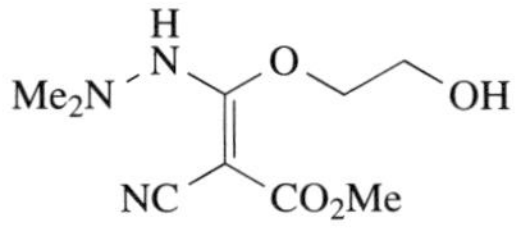

(**109**)

In a different approach, Monge *et al.* *O*-acetylated the hydrazines (**110**) and (**111**) to give (**112**) and (**113**) in 30% and 21% yields, respectively ⟨85JHC1445⟩.

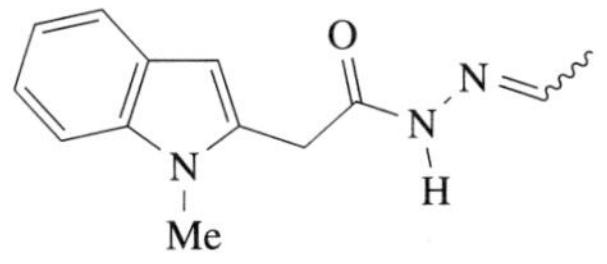

(**110**) 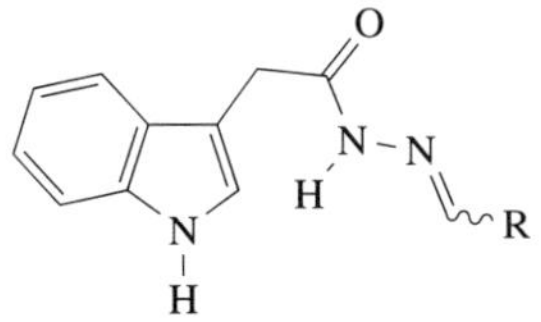   (**111**) 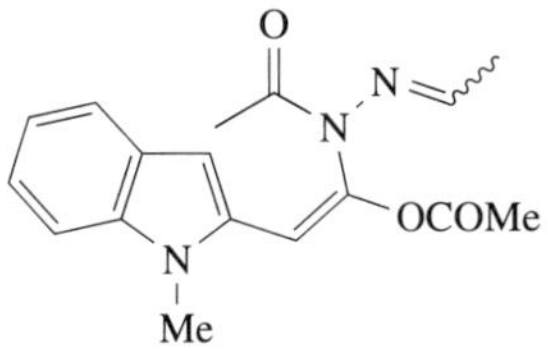   (**112**)

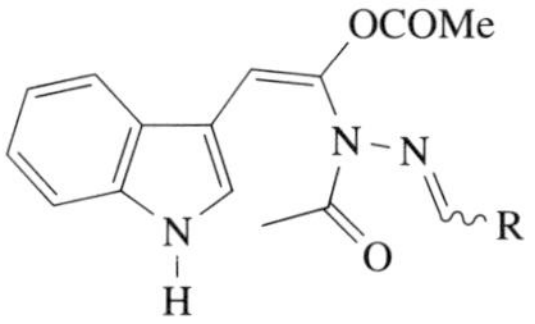

(**113**)

(iv) Azides

Ghosez *et al.* reported the transient existence of the azide (**114**), prepared as shown in Scheme 7 ⟨86RTC456⟩. Although (**114**) could be detected by IR spectroscopy, warming in pertane rapidly converted it into the azirine (**115**).

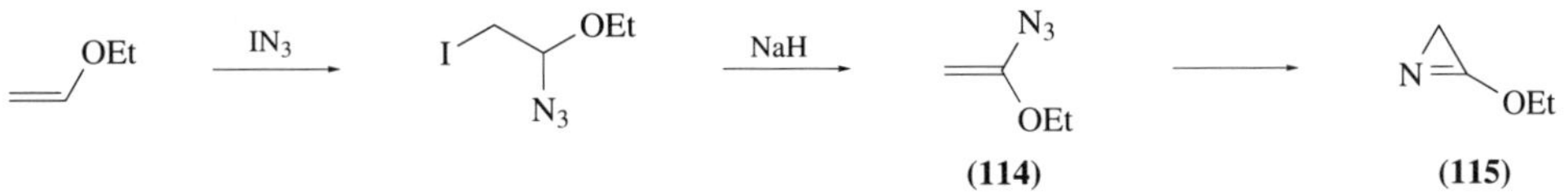

Scheme 7

(v) Nitro derivatives

Barrett and co-workers showed that benzyloxynitromethane (**116**) readily reacts with aldehydes in the present of potassium *t*-butoxide in *t*-butanol–THF ⟨89TL2349⟩. The presumed alcohol intermediate (**117**) was not isolated but dehydrated directly with acetyl chloride followed by triethylamine at 10 °C. The resultant benzyloxynitroalkenes (**118**), which were formed in 55% yield, are electrophilic and were used in penicillin synthesis.

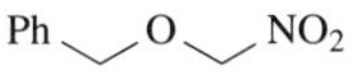

(**116**) 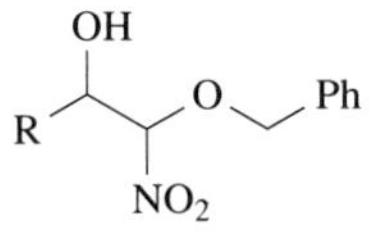   (**117**) 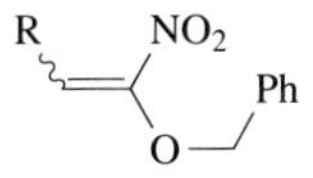   (**118**)

The elimination reaction (117)→(118) was subsequently improved, and yields increased to 81%, by the use of methanesulfonyl chloride, triethylamine and 1,5-diazabicyclo[5.4.0]undec-5-ene (dbu) in dichloromethane ⟨89JOC992⟩. In this case, the sole product was the (Z)-isomer. Full details of the use of this reagent to synthesise penicillanic acid S,S-dioxide and 6-aminopenicillanic acid are given in ⟨90JOC5110⟩.

Russel and Dedolph have investigated the reaction of 1,1-dinitro-2,2-diphenylethylene (**119a**) with a number of nucleophiles, including a variety of alkoxides and phenoxides ⟨85JOC3878⟩. Compound (**119a**) can undergo a number of reactions and only phenoxide gives a substantial yield (49%) of the nitro displacement product (**119b**), along with 31% of benzophenone.

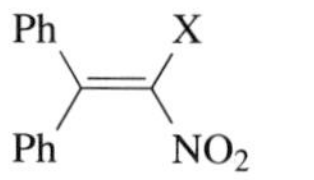

(**119**) a; X = NO$_2$
b; X = OPh

4.20.1.2 Functions bearing Sulfur and Nitrogen

4.20.1.2.1 *Dicoordinate sulfur derivatives*

(i) 1-Amino-1-thioalkenes

These compounds, of general structure (**120**), are colloquially known as ketene-*S,N*-acetals and more systematically named as derivatives of alkenamines. Therefore, the compound (**120**; $R^1 = H$, $R^2 = R^3 = Me$) is *N,N*-dimethyl-1-methylthioethenamine.

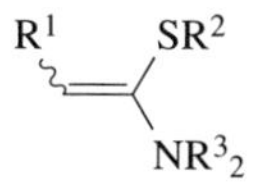

(**120**)

As for the ketene hemials described in Section 4.20.1.1.1, compounds unsubstituted or monosubstituted on nitrogen are potentially tautomeric with thioimidates (**121**). Walter and Krohn examined the tautomerism of the *N*-acyl compounds (**121**; $R^3 = COR^3$) and showed that the tautomeric equilibrium depends on the α substituents ⟨73LA443⟩. Thus the compound (**121**; $R^1 = R^2 = H$) exists solely in the thioimidate form, monosubstituted derivatives ($R^1 = H$, $R^2 = $ alkyl) as a mixture of tautomers and dialkyl derivatives ($R^1 = R^2 = $ alkyl), except where one of the groups is Bu^t exclusively in the ketene *S,N*-acetal form. Otherwise the tautomerism appears to be similar to that of ketene hemiaminals with electron-withdrawing groups on C-2 stabilising the ketene tautomer.

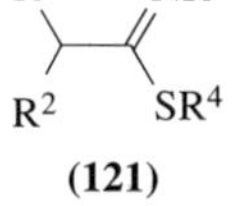

(**121**)

The synthesis of this system was reviewed by Borrmann and the three main routes described remain the most important ⟨68HOU(7/4)434⟩. These are:
(a) displacement of an alkylthio group from 1,1-bis(alkylthio)alkenes (**122**);
(b) *S*-alkylation, either of a thioamide followed by deprotonation, or of an initially formed thioamide anion; and
(c) treatment of a carbanion with a bis(methylthio)methaneimine (**123**).

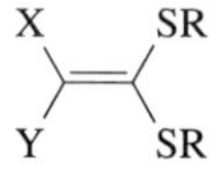

(**122**) 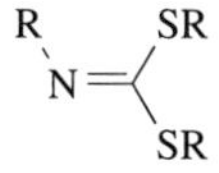   (**123**)

(a) Reaction of amines with bis(1,1-alkylthio)alkenes. Bis(1,1-alkylthio)alkenes (**122**), whose synthesis is discussed in Section 4.19.4.1, readily react with a wide variety of amines (Equation (11)).

$$\underset{Y}{\overset{X}{>}}=\underset{SR^1}{\overset{SR^1}{<}} + R^2R^3NH \longrightarrow \underset{Y}{\overset{X}{>}}=\underset{NR^2R^3}{\overset{SR^1}{<}} \quad (11)$$

Since the reaction involves the initial addition to the C-1 position of the acetal (**122**), its rate and ease will depend on both the nucleophilicity of the amine and the electron-withdrawing abilities of X and Y. Thus the dicyano derivative (**122**; X = Y = CN, R = Me) reacted with aziridine in ether at room temperature within 2–3 h to give the mono displacement product in 85% yield, ⟨76CC592⟩, whilst the cyano ester (**122**; X = CO₂Me, Y = CN, R = Me) failed to react with *N*-methylaniline even in boiling ethylene glycol ⟨69T4649⟩. Table 15 lists representative examples of this reaction.

The nucleophilic displacement of alkylthio groups can also take place intramolecularly. The thiol (**124**) readily reacts with aziridine and the resultant amine cyclises rapidly in refluxing methanol (Scheme 8) ⟨62CB2861⟩. Similarly, Chinese workers have shown that substituted (2-amino)thiophenols react with the thioacetals (**122**; R = Me) to give the benzthiazole derivatives (**125**) in yields of 75–90% ⟨90CB541⟩. When X and Y are both electron-withdrawing groups, the reaction proceeds in boiling ethanol; in contrast, when only one such group is present, the anion of 2-aminothiophenol has to be generated first with sodium in boiling dioxan.

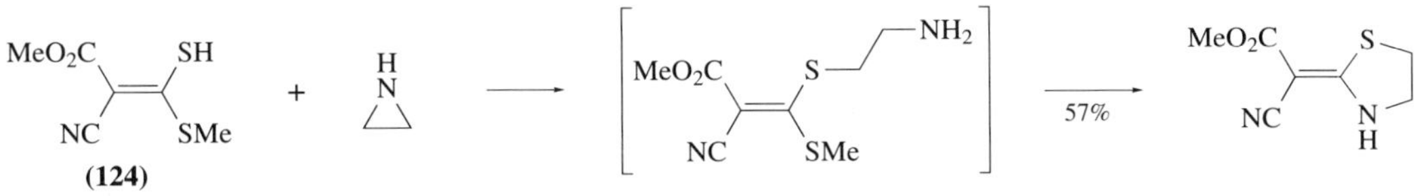

Scheme 8

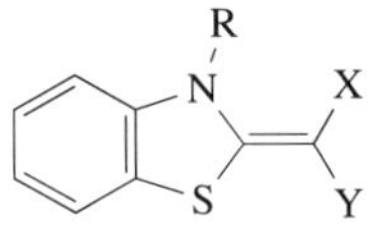

(**125**)

The ketene *S,S*-dithioacetal group can also be part of a heterocyclic ring (Equation (12)) ⟨79JPR699⟩.

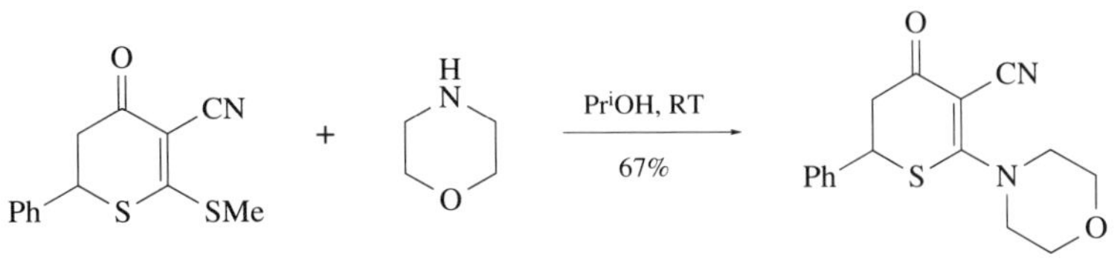

$$(12)$$

The second alkylthio group can also be displaced by, for example, carbon and nitrogen nucleophiles and thus the initial condensation products of (**122**) with sulfoximines and sulfodiimines (Table 16) have been used as intermediates on a route to novel heterocycles.

A related reaction is described by Kashima *et al.* who displaced either an imidazole or an *N*-methylimidazole group from the enamines (**126a,b**) with thiophenol to give (**126c**) in 31% and 39% yields, respectively ⟨84JHC133⟩.

(b) From thioamides. Tertiary thioamides readily react with alkyl halides to give *S*-alkyl salts. If the thioamide carries an α-proton, this can be removed by base to give ketene *S,N*-acetals (Equation (13) and Table 17) ⟨52JCS4067, 68JA3781⟩. Russian workers have alkylated thioamides using a radical reaction ⟨81CA96556w⟩.

$$R^1\underset{NR^2_2}{\overset{S}{\diagup\!\!\!\diagdown}} + R^3X \longrightarrow R^1\underset{NR^2_2}{\overset{SR^3}{\diagup\!\!\!\diagdown}}\ X^- \quad (13)$$

Table 15 Displacement of methylthio groups from 1,1-alkylthioalkenes.

$$\underset{(122)}{\underset{Y}{X}\!\!>\!\!=\!\!<\!\!\underset{SMe}{SMe}} + ZRNH \longrightarrow \underset{Y}{X}\!\!>\!\!=\!\!<\!\!\underset{NZR}{SMe}$$

X	Y	Z	R	Conditions	Yield (%)	Ref.
CN	CN	H	H	Δ, EtOH	93	62CB2871
CN	CN	H	Ph	Δ, EtOH	97	62CB2871
CN	CO$_2$Et	H	CH$_2$CO$_2$Et	Δ, EtOH, NaOEt	70	75CB174
CN	CO$_2$Et	CH$_2$CH$_2$		Et$_2$O, 5–20 °C, 2 h	87	76CC592
CN	CONH$_2$	CH$_2$CH$_2$		Et$_2$O, 5–20 °C, 2 h	89	76CC592
MeCO	CO$_2$Et	CH$_2$CH$_2$		Et$_2$O, 5–20 °C, 12 h	70	76CC592
H	PhCO	H	CH$_2$CH(OEt)$_2$	EtOH, Δ, 25–40 h		76CC593
H	p-EtOC$_6$H$_4$CO	H	CH$_2$CH(OEt)$_2$	EtOH, Δ, 25–40 h		76CC593
CN	CONH$_2$	H	H	EtOH, Δ	92	67CB2577
CN	PhSO$_2$	H	Ph	BuOH, Δ	72	77ZC289
CN	CN	\multicolumn{2}{c}{tetrahydropyranyl (O)}	Δ, EtOH	93	82CL1933
CN	CN	\multicolumn{2}{c}{cyclohexyl}	Δ, EtOH	86	82CL1933
H	NO$_2$	H	Bn	Δ, EtOH	82	86AP161
H	NO$_2$	H	PhCH$_2$CH$_2$	Δ, EtOH	84	86AP161
CN	CONH$_2$	H	Ph			85CC279
H	COCF$_3$	H	H	MeCN, RT, 18 h	69	90S195
H	COCF$_3$	H	Ph	MeCN, RT, 18 h	100	90S195
H	COCF$_3$	Me	H	MeCN, RT, 18 h	65	90S195
H	COCF$_3$	Et	Et	MeCN, RT, 18 h	96	90S195
H	COCF$_3$	Bn	H	MeCN, RT, 18 h	55	90S195
CN	COPh	CH$_2$CH$_2$		Et$_2$O, RT	86	80T1791
CN	2-COthienyl	CH$_2$CH$_2$		Et$_2$O, RT	94	80T1791
H	PhCO	Ph	H	160 °C, 15 h	40	80S748
CO$_2$Me	CN	\multicolumn{2}{c}{1-indolinyl}	MeOH, Δ	77	80S748
\multicolumn{2}{c}{isopropylidene dioxydiacetyl}	H	COPh	DMSO, KOH, 90 °C, 7–8 h	80	91SC1213
		H	Bu^t	EtOH, RT, 24 h	60	91JOC970
CN	CO$_2$Me	H	COPh	NaH, PhH, MeCONMe$_2$, RT	76	88JHC959
CN	CO$_2$Me	H	COCH$_2$Cl	NaH, PhH, MeCONMe$_2$, RT	40	88JHC959
CN	2-benzimidazolyl	H	Ph	EtOH, Δ	82	78ZC345
CN	CN	H	TMS-CH$_2$	MeOH, Δ, 10 min	96	90JOC5308
H	COPh	\multicolumn{2}{c}{CH$_2$CH$_2$SH}	EtOH, Δ	88[a]	90S162
H	NO$_2$	\multicolumn{2}{c}{cyclohexyl}	EtOH, Δ, 2 h	39	93MI 420-01
2-pyridyl	CN	Bn	H	THF, RT, 18 h	93	89T1801
2-pyridyl	CN	Ph	H	Δ, 3 h	90	69YZ203

[a] Product is 2-(phenacylidene)thiazolidine

Table 16 Displacement of methylthio groups from 1,1-methylthioalkenes with sulfoximides.

X	Y	Z	Conditions	Yield (%)	Ref.
CN	PhCO	O	100 °C, 1 h, no solvent	70	83S926
CN	p-BrC$_6$H$_4$CO	O	100 °C, 1 h, no solvent	59	83S926
CN	p-MeC$_6$H$_4$SO$_2$	O	100 °C, 1 h, no solvent	44	83S926
2-(MeSO$_2$)C$_6$H$_4$CO		O	Δ, PhH, 1 h	94	87JHC519
2-MeC$_6$H$_4$CO		NH	MeCN, RT	71	88CB805
CN	CONH$_2$	NH	Δ, no solvent	35	86CB1745

(126) a; R = (N-methylimidazolyl)
  b; R = (N-methylimidazolium, Me)
  c; R = SPh

The α,β-unsaturated thioamides, for example (127), can be acylated on nitrogen whereupon the product readily undergoes Diels–Alder reactions with alkenes to give dihydrothiopyrans (Scheme 9) ⟨89TL4449, 91TL405⟩. Alternatively, the anion can be generated first and then alkylated by the required alkyl halide (Table 18).

A variation on this is due to Gompper and Schmidt, who removed a proton from conjugated phenols to give in high yield the corresponding naphthoquinone or quinone derivatives (128)–(130), e.g. (129; X = SO$_2$Ph, R^1 = R^2 = R^3 = Me, 90%) (Scheme 10) ⟨62ZN851, 65CB1385, 67AG147, 68AG38⟩.

A versatile route to thioamides and hence to ketene S,N-acetals is provided by thiocarbamoylation of the carbanions from C-H acids (Scheme 11) ⟨74S433⟩, or by condensation with isothiocyanates (Scheme 12 and Table 19) ⟨81T1453⟩.

(c) Carbanion reactions. The bis(methylthio)methanimine (123; R = Me), its ammonium salts (131) or the open-chain or cyclic iminium salts (132), (133) react readily with stabilised carbanions to give, after elimination of methanethiol, the required S,N-acetals (Equations (14)–(16) and Table 20) ⟨64AG781, 64TL2737, 65CB1293, 68CB1131, 89AP(322)593, 90AP(323)619, 90JCS(P1)1217, 93T2101⟩.

Table 17 Preparation of ketene S,N-acetals from tertiary thioamides.

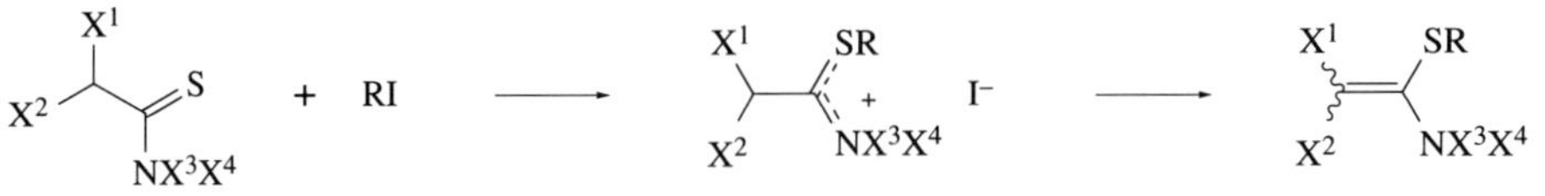

X^1	X^2	X^3	X^4	R	Conditions	Yield (%)	Ref.
Ph	H	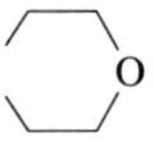 CH$_2$CH$_2$OCH$_2$CH$_2$		Me	i, MeI, acetone; ii, KHCO$_3$	78	50JCS3350
H	H	Me	Me	Me	i, MeI, Et$_2$O; ii, Bu^{t}OK		64TL1971
H	H	Me	Ph	Me	i, MeI, Et$_2$O; ii, Bu^{t}OK		64TL1971
Me	Me	Me	Me	Me	i, MeI, Et$_2$O; ii, Bu^{t}OK		64TL1971
H	Ph	CH$_2$CH$_2$OCH$_2$CH$_2$		Me	i, MeI, Et$_2$O; ii, Bu^{t}OK		64TL1971
Ph	Ph	Et	Et	Ph	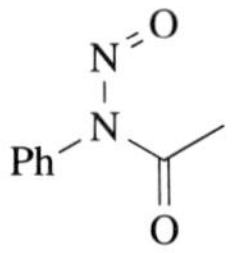 Ph-N(N=O)-C(=O)-Me		81CA96556w
H	H	Me	Me	Et	EtI, NaSEt	75	67BCJ2641
Ph	H	Me	Me	Me	i, MeI, acetone; ii, Bu^{t}OH	60	78ACS(B)141
CO$_2$Me	CO$_2$Me	Me	Me	Me	i, Me$_2$SO$_4$; ii, K$_2$CO$_3$	60	69T4649
Me	H	Me	Ph	Me	i, MeI; ii, KOBu^t		87JCS(P1)1501

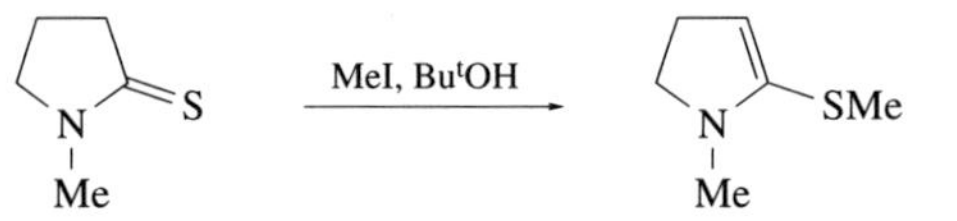

N-methyl pyrrolidinethione $\xrightarrow{\text{MeI, Bu}^t\text{OH}}$ 2-SMe-N-methyl-2,3-dihydropyrrole, 81, 68OS97

N-methyl dihydroquinoline-2-thione → 2-SMe-N-methyl quinoline derivative, 58, 82CPB3959

Scheme 9: EtHN-C(=S)-CH=CH-Ph **(127)** $\xrightarrow{\text{AcX}}$ [Ac-N(Et)-C(=S)-CH=CH-Ph] $\xrightarrow[\text{81\%, ratio A:B 7:1}]{X=Y=CO_2Me}$ **A** and **B**

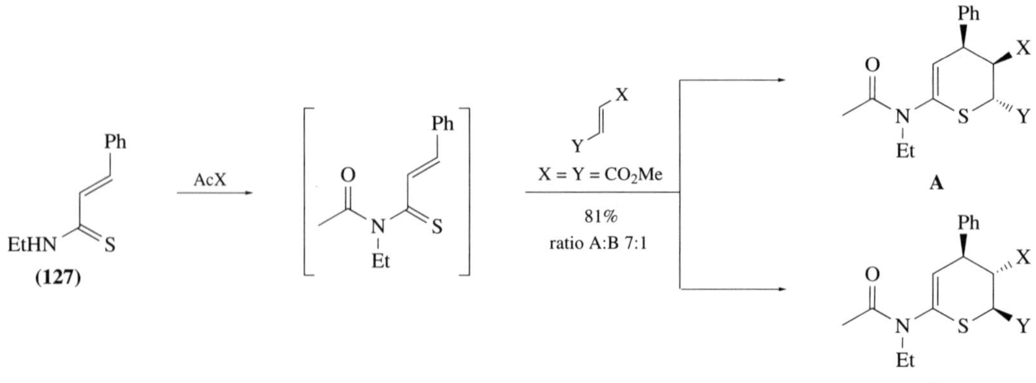

Scheme 9

Table 18 Preparation of ketene S,N-acetals from tertiary thioamides by initial deprotonation then alkylation.

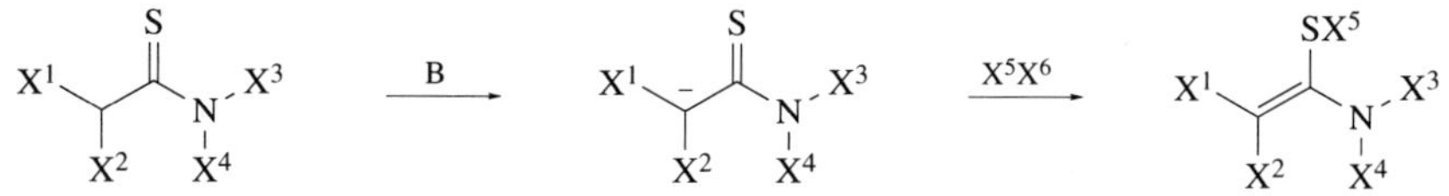

X^1	X^2	X^3	X^4	X^5	B	Yield (%)	Ref.
ArCO	H	CH$_2$CH$_2$CH$_2$		Me	TlOEt		74T1283
ArCO	H	CH$_2$CH$_2$CH$_2$		allyl	TlOEt		74T1283
ArCO	H	CH$_2$CH$_2$CH$_2$		crotyl	TlOEt		74T1283
CO$_2$Me	H	Et	Et	Me	NaH, THF	75	68CJC2255
CO$_2$Et	CN	H	Ph	CH$_2$CO$_2$R	NaOEt	75	66ZC417
PhCO	H	H	CH$_2$CO$_2$Et	Me	K$_2$CO$_3$	75	84S250
CO$_2$Et	CO$_2$Et	H	Ph	Me	EtNPr$^i_2$	55	75LA19
CO$_2$Et	CO$_2$Et	Me	Me	Me	K$_2$CO$_3$	60	69T4649
MeS	H	Me	Me	Me	NaH, DMF	74	82LA585
Me$_2$N	H	Me	Me	Me	LiBu, THF, $-70\,°$C	69	82LA585

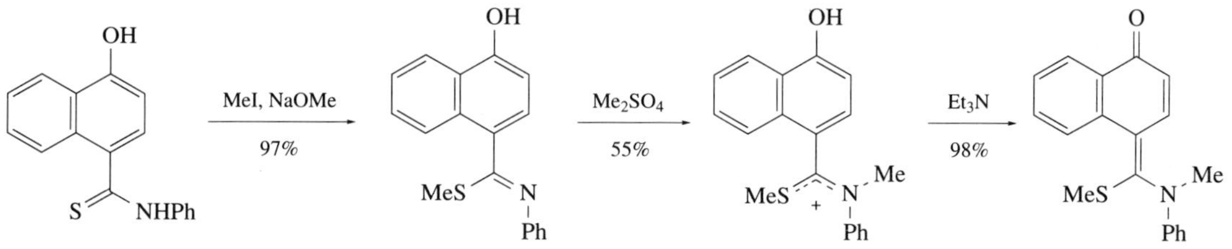

Scheme 10

(128)

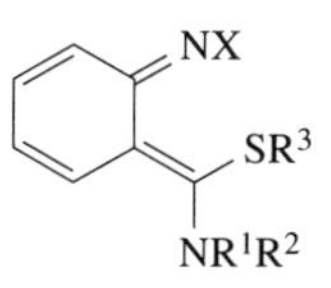

(129)
R^1 = R^2 = R^3 = Me, X = SO$_2$Ph, 90%

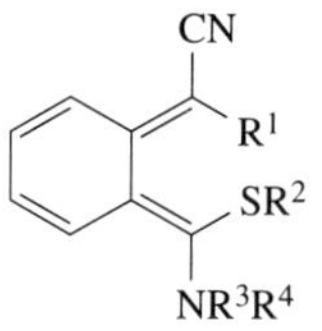

(130)

Scheme 11

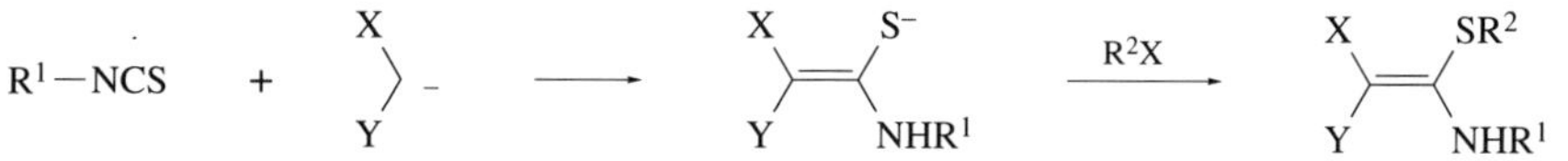

Scheme 12

Table 19 Thiocarbamoylation of active methylene groups.

$$R^1NCS + X^1\text{---}Y \xrightarrow{R^2X^2} \underset{Y\quad NHR^1}{\overset{X^1\quad SR^2}{\diagup\!\!\!\diagdown}}$$

R^1	X^1	Y	R^2	Conditions	Yield (%)	Ref.
Ph	2-ClC$_6$H$_4$CO	H	Me	i, NaH, DMF; ii, MeI	69	79T551
Ph	CN	CN	CH$_2$CN			62AG251
Ph	CN	benzimidazol-1-yl (NR)	Me	i, NaH, DMF; ii, MeI	82	78ZC345
Ph	NO$_2$	H	Me	i, NaH, DMF; ii, MeI	74	67CB591
Ph	CO$_2$Me	CO$_2$Me	Me	NaH, DMA	79	69T4649

$$\underset{(123)\ R=Me}{\overset{MeS\quad SMe}{\underset{NMe}{\diagup\!\!\!\diagdown}}} + NC\text{---}CN \xrightarrow{85\%} \underset{NC\quad NHMe}{\overset{NC\quad SMe}{\diagup\!\!\!\diagdown}} \qquad (14)$$

$$\underset{(131)}{\overset{MeS\quad SMe}{\underset{Me\text{---}N^+\text{---}Me}{\diagup\!\!\!\diagdown}}} + 2\ \text{(cyclopentadienide)} \xrightarrow[72\%]{THF,\ RT} \text{(fulvene-MeS, NMe}_2\text{)} \qquad (15)$$

$$\underset{(132)}{\text{(2-chloro-3-R-dihydrothiazine)}} \xrightarrow[Et_3N]{X^2\text{---}CH(X^1)\text{---}CO\text{---}X^1\cdots X^2} \text{product} \qquad (16)$$

R = Me, X^1 = O, X^2 = OEt, 4%

$$\underset{(133)}{\overset{R^1R^2N^+\text{---}SMe}{\underset{Cl}{\diagup\!\!\!\diagdown}}\ Cl^-}$$

Table 20 Reaction of active methylene groups with bis(methylthio)methanimines.

Methanimine	CH$_2$XY		Conditions	Yield (%)	Ref.
	X	Y			
123	CN	CN	warm EtOH	95	64AG781, 68CB1131
123	H	NO$_2$	excess CH$_3$NO$_2$, zeolite, Δ	50	90JCS(P1)1217, 93T2101
131	CO$_2$Et	CO$_2$Et	CH$_2$Cl$_2$, Et$_3$N	4	89AP(322)593, 90AP(323)619
123	CN	CO$_2$Et	120 °C	42	68CB1131
123	CN	CN	70 °C, pyridine	27	65CB1293
133; $R^1 = R^2 = Ph$	CN	CN	PhH, Et$_3$N, 40 °C	24	65CB1293

In addition to these, five other methods have been described.

(a) From 1-chloroenamines. 1-Chloroenamines readily react with metal salts of thiols and thiophenols to give ketene S,N-acetals (Equation (17) and Table 21) ⟨69AG(E)454, 69T3447⟩.

$$\underset{R^2}{\overset{R^1}{>}}=\underset{NR^3R^4}{\overset{Cl}{<}} + R^5S^- \longrightarrow \underset{R^2}{\overset{R^1}{>}}=\underset{NR^3R^4}{\overset{SR^5}{<}} \qquad (17)$$

Table 21 Reaction of thiolates with 1-chloro-1-dialkylaminoalkenes.

$$\underset{R^2}{\overset{R^1}{>}}=\underset{NR^3R^4}{\overset{Cl}{<}} + R^5S^-\,M^+ \longrightarrow \underset{R^2}{\overset{R^1}{>}}=\underset{NR^3R^4}{\overset{SR^5}{<}}$$

R^1	R^2	R^3, R^4	R^5	Conditions	Yield (%)	Ref.
Me	Me	Et, Et	Et	THF, 20 °C	90	69AG(E)454
Me	COCl	(CH$_2$)$_5$	Et	THF, 20 °C	68[a]	69T3447
Me	Me	(CH$_2$)$_5$	Et	THF, 20 °C	90	B-76MI 420-02
Me	Me	Me	Me	Et$_2$O, 20 °C	87	B-76MI 420-02

[a] The product has $R^2 = $ EtSCO.

(b) From nitriles. The thio equivalent of the Ritter reaction (4.20.1.1.1) involves the addition of ethanethiol to ethyl cyanoacetate in the presence of hydrogen chloride to give, after neutralisation, ethyl 3-amino-3-(ethylthio)-2-propenoate in 37% yield (Equation (18)) ⟨90JMC1510⟩. The reaction has also been catalysed by boron trifluoride etherate ⟨88CB1817⟩.

$$NC\!\!-\!\!CO_2Et + EtSH \xrightarrow[36\%]{HCl,\,Et_2O,\,RT} \underset{H_2N}{\overset{EtS}{>}}=\overset{CO_2Et}{<} \qquad (18)$$

(c) By addition to alkynes. Radchenko has shown that diethylamine will add to symmetrical alkylthioalkynes (**134**) to give 43% of a 1:1 mixture of *cis* and *trans* products (**135**) and (**136**), and to unsymmetrical alkynes (**137**) to give a 65% yield of a mixture of all four adducts (**138**)–(**141**) ⟨77ZOR504⟩.

RS—≡—SR   (134)

$\underset{Et_2N}{\overset{RS}{>}}=\overset{SR}{<}$ (135)

$\underset{Et_2N}{\overset{RS}{>}}=\overset{}{<}{SR}$ (136)

R^{1}S—≡—SR^2 (137)

$\underset{NEt_2}{\overset{R^1S}{>}}=\overset{SR^2}{<}$ (138)

$\underset{SR^2}{\overset{R^1S}{>}}=\overset{NEt_2}{<}$ (139)

$\underset{Et_2N}{\overset{R^1S}{>}}=\overset{SR^2}{<}$ (140)

$\underset{Et_2N}{\overset{R^1S}{>}}=\overset{}{<}{SR^2}$ (141)

1-Diethylaminoprop-1-yne (**142**) reacts with *N*-benzoyl-*S*-phenylthiobenzimidate (**143**) in boiling dichloromethane to give (**144**; X = COPh) in 52% yield along with the amide (**145**) ⟨85CC845⟩. The compound (**144**; X = SO$_2$Ph) is readily produced from the reaction of (**142**) with *N*-benzenesulfonyl-*S*-phenylthiobenzimidate (**146**). Finally (**142**) also reacts with 3-methylthio-1,2-benzisothiazole 1,1-dioxide (**147**) to give (**148**) in 'good yields'.

(d) By addition of imines to thioketenes. The polyfluorinated phenyl imine (**149**) undergoes a 1:1 cycloaddition with bis(trifluoromethyl)thioketene (**150**) to give the 1,3-thiazetidine (**151**) in 79% yield ⟨78JOC2500⟩. With the nonfluorinated imine (**152**), the product in 78% yield is the 2:1 adduct (**153**).

(e) By ring-opening reactions. Grignard reagents react with the 1,2-dithioles (**154**) to give directly the corresponding ketene *S,N*-acetal (Equation (19)) ⟨73LA247⟩.

(ii) 1-Nitro-1-thioalkenes

Nitro(phenylthio)methane (**155**) reacts with aldehydes in the presence of 5% methanolic potassium hydroxide at −10 °C for 1 h then at 0 °C for 7 h to give, after neutralisation with acetic acid, the nitro alcohols (**156**) ⟨90OS(68)8⟩. These were dehydrated by addition of methanesulfonyl chloride and triethylamine in dichloromethane at −78 to 0 °C to give the nitroalkenes (**157**) ⟨78CC362⟩.

Barrett extended the reaction using potassium hydroxide in methanol as the base ⟨84CC670, 86JOC1012⟩ for the initial condensation reaction and then applied the concept to the synthesis of bicyclic β-lactams ⟨85JOC2603, 87JOC4693⟩ and S-phenyl thioesters ⟨86JOC1012, 89JOC4723⟩. Under the conditions described here, the sole products from the dehydration reactions were the (Z)-alkenes (**157**).

The reaction of 1,1-dinitro-2,2-diphenylethene (**119a**) with sulfur nucleophiles gives the 1-thioalkyl (or thioaryl)-1-nitro-2,2-diphenylethenes (**158**) and proceeds more readily and successfully than the reaction with the oxygen nucleophiles ⟨85JOC3878⟩. The yields of the product (**158**) were 85% for X = Ph, 87% (X = p-ClC$_6$H$_4$), 47% (X = p-NO$_2$C$_6$H$_4$), 87% (X = p-Tol) and 32% (X = Bu^t). In all cases, small quantities of benzophenone were also produced.

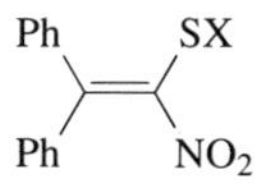

(**158**)
X = Ph, p-ClC$_6$H$_4$, p-NO$_2$C$_6$H$_4$, p-MeC$_6$H$_4$, Bu^t

Additionally, nitrophenylthioalkenes have also resulted from two ring-opening reactions. Thus 2-nitrothiophene (**159**) reacts with secondary amines in ethanol at 0 °C for 8–10 days to give the disulfides (**160**) in 50–80% yields (Equation (20)) ⟨74JCS(P1)2357⟩. 1-Methyl-2,5-diphenyl-6-nitro-1,4-dithiinium tetrafluoroborate (**161**) ring opens both in aqueous phosphate buffer at pH 6.8 and 37 °C and with triethylamine in acetonitrile at 25 °C to give the nitroalkene (**162**) (Equation (21)) ⟨80JOC933⟩.

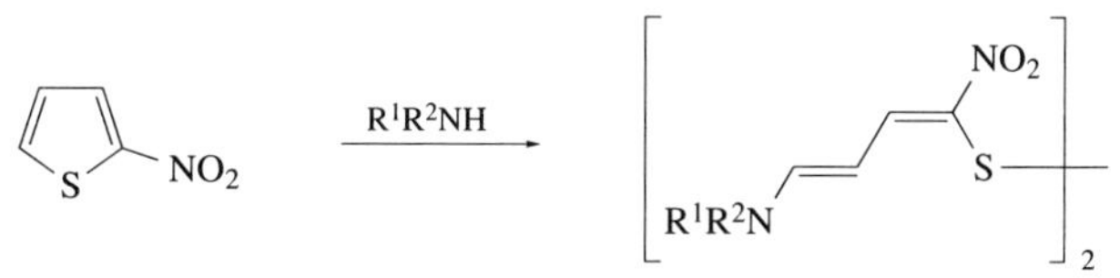

(20)

(**159**) (**160**)

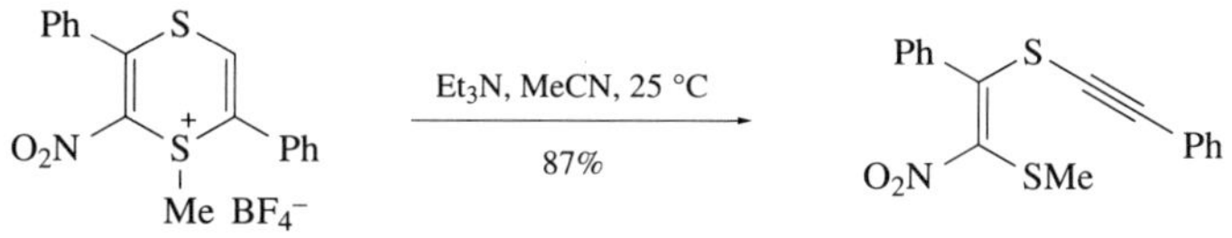

(21)

(**161**) (**162**)

Nitroenamines react with benzenesulfonyl chloride to give 2-nitro-2-phenylthioenamines (Equation (22)) ⟨77CA121332u⟩.

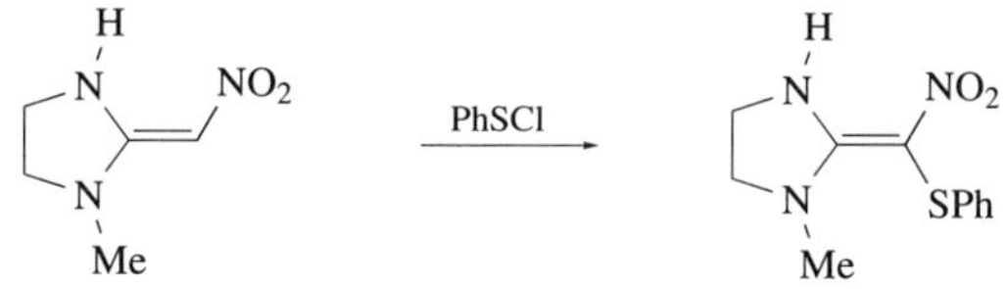

(22)

4.20.1.2.2 Tricoordinate sulfur derivatives

Only two examples of this system have been found. Compound (**163**) is used as an intermediate in the preparation of the histamine H$_2$ antagonist ranitidine (**164**) ⟨83CA(99)22296x⟩. No synthetic details were given, but it is expected that this would be more reactive than the analogous sulfide which has also been used in this preparation.

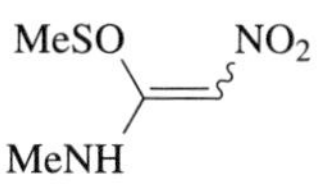

(**163**) 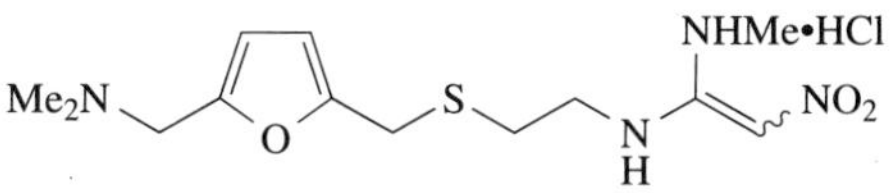 (**164**) 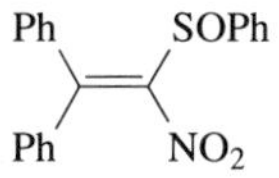 (**165**)

1-Benzenesulfinyl-1-nitro-2,2-diphenylethene (**165**) was prepared by m-oxidation with mcpba of the corresponding sulfide, prepared as described in the previous section ⟨85JOC3878⟩.

4.20.1.2.3 Tetra – and higher – coordinate sulfur derivatives

The rich and varied chemistry of arenesulfonylmethyl isocyanides, for example (**166**; X = *p*-Tol, TsMIC), has been explored by the groups of van Leusen and Schöllkopf ⟨80MI 420-01⟩. Amongst many reactions described by these authors are condensations with aldehydes and ketones under aprotic conditions to give *N*-(1-arenesulfonylalkenyl)formamides (**167**) (Table 22) ⟨72AG(E)311, 72LA(766)130, 79RTC258⟩.

Table 22 Reaction of arenesulfonylmethyl isocyanides with ketones.

R^1	R^2	Ar	Base	Yield (%)	Ref.
Ph	H	Ph	KOBu^t, THF, 7–10 °C	72	72AG(E)311, 72LA(766)130
Ph	H	*p*-Tol	KOBu^t, DME, −30 °C	87	79RTC258
Ph	Me	*p*-Tol	KOBu^t, THF, 7–10 °C	42	72AG(E)311, 72LA(766)130
α-naphthyl	Me	*p*-Tol	KOBu^t, THF, 7–10 °C	50	72AG(E)311, 72LA(766)130
Pr^i	H	*p*-Tol	KOBu^t, THF, 7–10 °C	10	72AG(E)311, 72LA(766)130
PhCH=CH	H	*p*-Tol	KOBu^t, THF, 7–10 °C	73	72AG(E)311, 72LA(766)130
Bu^t	H	*p*-Tol	KOBu^t, THF, 7–10 °C	63	72AG(E)311, 72LA(766)130
Bu^t	H	*p*-Tol	KOBu^t, DME, −30 °C	65	79RTC258
Me	Me	*p*-Tol	KOBu^t, DME, −30 °C	83	79RTC258
Me	H	*p*-Tol	KOBu^t, THF, 7–10 °C	83	72AG(E)311, 72LA(766)130
(CH$_2$)$_5$	(CH$_2$)$_5$	*p*-Tol	KOBu^t, THF, 7–10 °C	61	72AG(E)311, 72LA(766)130
(CH$_2$)$_5$	(CH$_2$)$_5$	*p*-Tol	KOBu^t, DME, −30 °C	80	79RTC258
Ph	Ph	*p*-Tol	KOBu^t, DME, −30 °C	50	79RTC258
p-NO$_2$C$_6$H$_4$	H	*p*-Tol	KOBu^t, DME, −30 °C	77	79RTC258

This reaction has been applied to 17-oxosteroids (**168**) to functionalise the 17 position ⟨84TL2581⟩ and to the pyrrolidone-4-aldehyde (**169**) ⟨93CPB217⟩.

The action of POCl$_3$, Et$_3$N rapidly converts the initial formamide (**167**) into the isonitrile (**170**) (Table 23) ⟨79RTC258⟩.

Subsequently, van Leusen and Wildeman showed that the silyl lithio TsMIC (**171**) reacts with aldehydes and ketones to give the isocyanates (**170**) directly (Table 24) ⟨82RTC202⟩.

The nitro sulfide (**158**; R = Ph) has been oxidised by 30% hydrogen peroxide at 25 °C for 12 h to give the sulfone in 65% yield ⟨85JOC3878⟩.

Table 23 Reaction of TsMIC with ketones and subsequent dehydration.

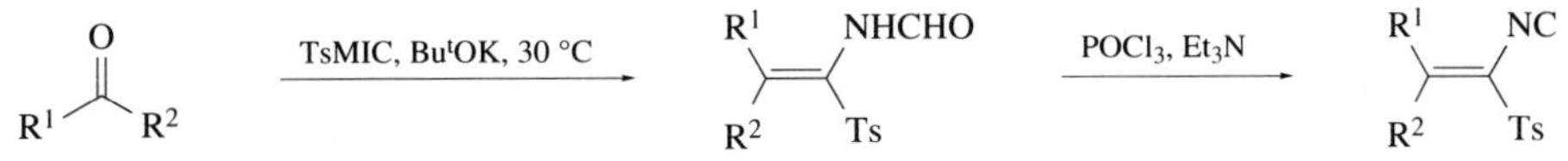

R^1	R^2	Yield (second stage) (%)
Ph	H	54
p-$NO_2C_6H_4$	H	55
Bu^t	H	77
Me	Me	68
Ph	Ph	68

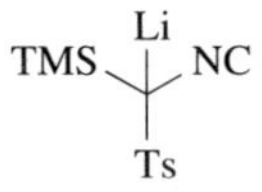

(171)

Table 24 Reactions of lithio trimethylsilyl TsMIC with aldehydes and ketones.

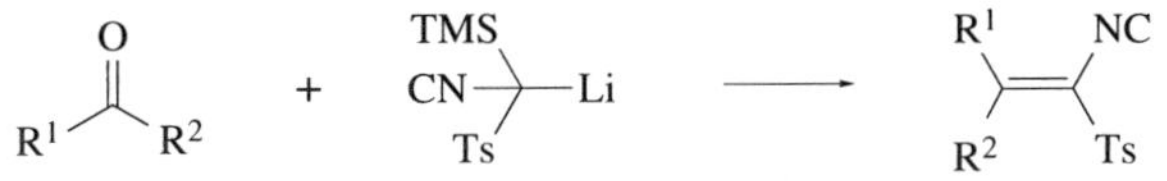

R^1	R^2	Yield (%)
Ph	H	82
p-$Me_2NC_6H_4$	H	76
p-$MeOC_6H_4$	H	93
p-ClC_6H_4	H	85
p-$NO_2C_6H_4$	H	55
2-furyl	H	88
5-NO_2-2-furyl	H	10
2-thienyl	H	87
(E)-PhCH=CH	H	83
(E,Z)-Me_2C=CH(CH$_2$)$_2$CMe=CH	H	90
(CH$_2$)$_3$		64

4.20.1.3 Functions Bearing Selenium or Tellurium Together with Nitrogen

A limited number of examples of these systems has been reported. Five general procedures have been used for their synthesis.

(i) From selenoamides by alkylation (Scheme 13)

Treatment of the selenoamide (**172**) with 2.2 equivalents of phenyllithium at $-78\,°C$ followed by treatment with 3 equivalents of methyl iodide at $-78\,°C$ to $0\,°C$ gave 59% of the 1-(methylseleno)enamine (**173**) along with 29% of the 1-phenylenamine (**174**) formed by addition of phenyllithium to (**172**) ⟨90CL2053⟩.

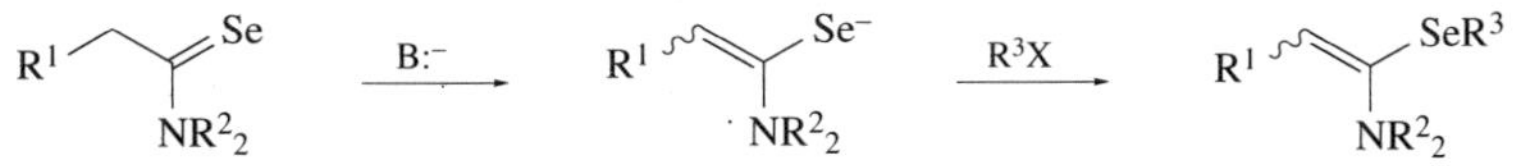

Scheme 13

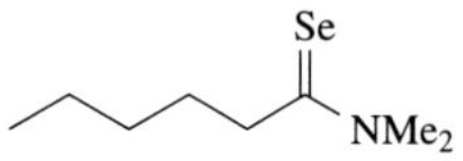

(172)

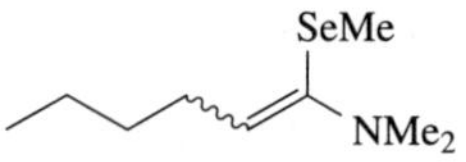

(173)

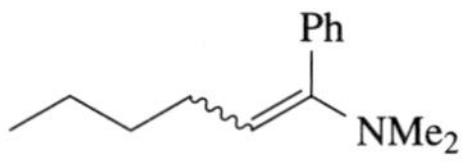

(174)

(ii) By addition of amines to selenyl-substituted alkynes

Radchenko and co-workers have shown that diethylamine adds to the symmetrically substituted selenoalkynes (**175**) to give a mixture of *cis* and *trans* adducts (**176**), (**177**), with the *cis* compound predominating (73:27) ⟨77ZOR504⟩. With unsymmetrical alkynes (**178**), all four possible isomers (**179**)–(**182**) are produced. Similar results were obtained with the tellurium analogue (**183**).

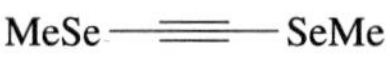

(175)

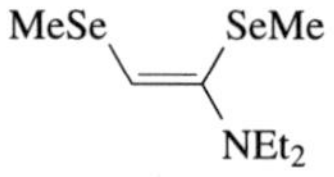

(176)

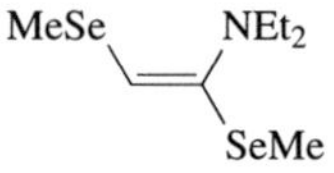

(177)

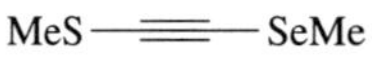

(178)

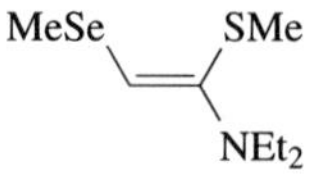

(179)

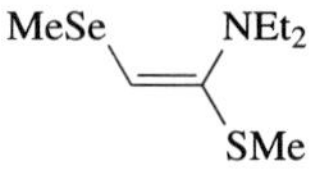

(180)

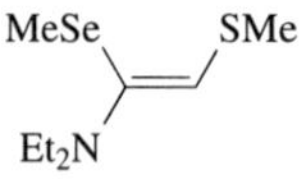

(181)

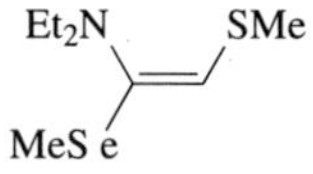

(182)

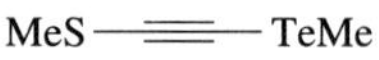

(183)

(iii) Condensation with aldehydes

Nitro(phenylseleno)methane (**184**) is readily prepared from nitromethane and benzeneselenenyl bromide. This condenses with aldehydes to give the alcohols (**185**), which, following acetylation to form (**186**), are dehydrated to the 1-seleno 1-nitro alkenes (**187**) (Scheme 14 and Table 25) ⟨83S920⟩.

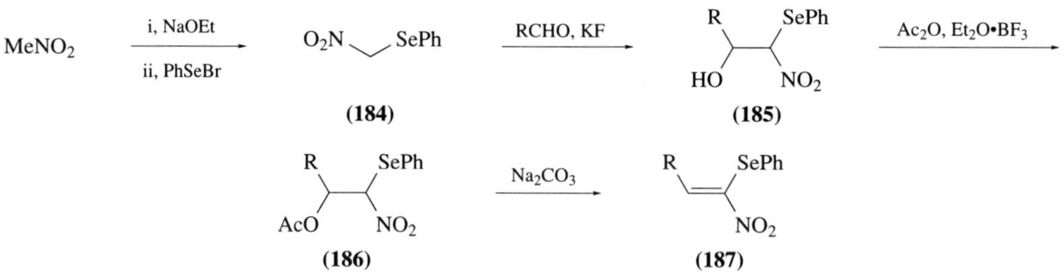

Scheme 14

Table 25 Preparation of the 1-phenylseleno-1-nitro-alkenes RCH=C(NO_2)SePh (**187**) (Scheme 14).

R	*Yield* (%)
n-C_9H_{19}	83
n-C_7H_{15}	80
n-C_4H_9CH(Et)	88
Et	88
Ph	79
2-$NO_2C_6H_4$	67

(iv) Displacement of a nitro group from 1,1-dinitroethenes

1,1-Dinitro-2,2-diphenylethene (**119a**) reacts with potassium phenyl selenide in DMSO to give (**188**) in 48% yield along with 31% of benzophenone (Scheme 15) ⟨85JOC3878⟩. It is possible that the benzophenone arises from hydrolysis of (**188**) during work up.

Scheme 15

(v) Condensation

1-Diethylaminoprop-1-yne (**189**) condenses with 3-phenylseleno-1,2-benzisothiazole 1,1-dioxide (**190**) to give (**191**) in 'good yield' ⟨85CC845⟩.

4.20.2 FUNCTIONS CONTAINING A CHALCOGEN AND A PHOSPHORUS, ARSENIC, ANTIMONY OR BISMUTH

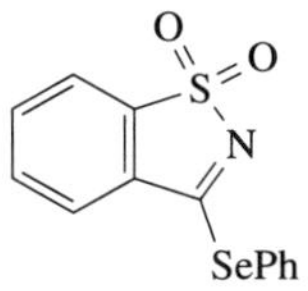

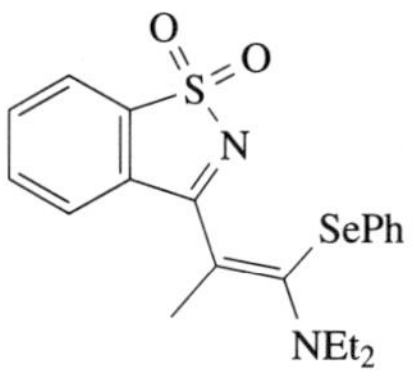

4.20.2.1 Functions Bearing Oxygen

Nine distinct reactions have been used to produce these systems:
(i)   elimination reactions,
(ii)   displacement reactions,
(iii)   reactions with diazophosphonates,
(iv)   addition to alkynes,
(v)   addition to ketenes,
(vi)   condensation reactions,
(vii)   isomerisations,
(viii) enolisations, and
(ix)   rearrangements.

4.20.2.1.1 Elimination reactions

Acetic anhydride and chlorodiethyl phosphine (**192**) condense at 100 °C during 2 h to give 77% of the acetoxy derivative (**193**) possibly via the route shown in Scheme 16 ⟨68ZOB1523⟩.

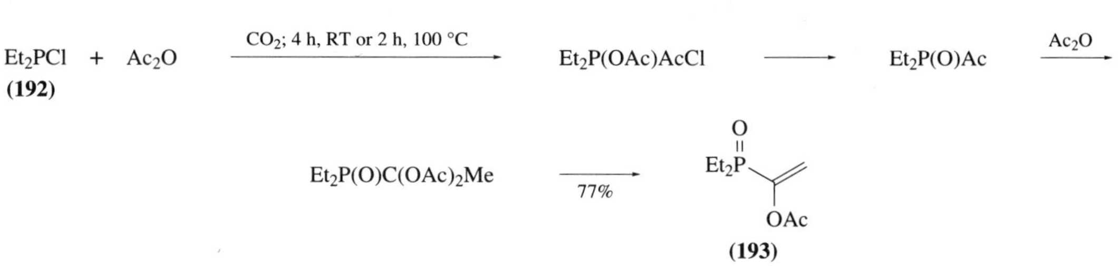

Scheme 16

Golborn prepared dialkyl (1,1-dialkoxyethyl)phosphonates (**194**) by the condensation of trialkyl orthoacetates with trialkyl phosphites and phosphorus trichloride (Scheme 17). Heating (**194**) with sodium methoxide or sodium carbonate at 160–240 °C gave the diethyl 1-alkoxyvinylphosphonates (**195**) ⟨73CA(78)45000b, 73S547⟩.

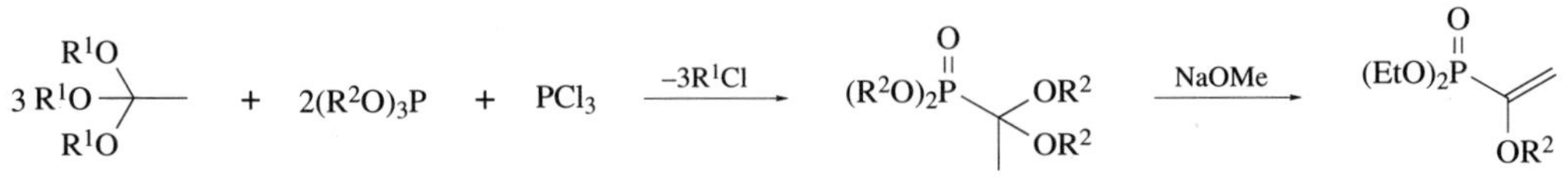

(**194**)

(**195**)
R^2 = Me, 66%
R^2 = Et, 84%

Scheme 17

It has also been reported that (**195**; R^2 = Me) can be produced by condensation of diethyl phosphonate with trimethyl orthoacetate in the presence of sodium ethoxide in a sealed tube at 165 °C over 5 h ⟨67ZOB1623⟩. 1,2-Dibromo-1-ethoxyethane (**196**) reacted with triphenylphosphine to give a hygroscopic salt which was treated *in situ* with triethylamine to give (**197**) in 54% yield ⟨74S862⟩.

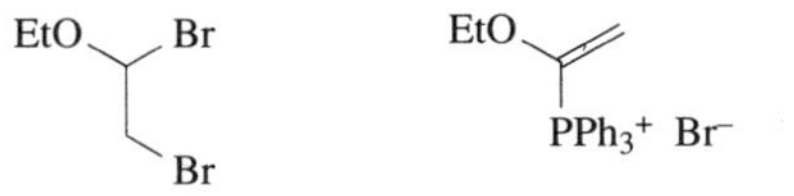

(**196**) (**197**)

Similarly, the methoxyl analogue of (**196**) reacted with trimethyl phosphite to give (**198**), which was readily dehydrobrominated by potassium hydroxide to give dimethyl 1-methoxyvinylphosphonate (**199**) in 54% yield ⟨75MI 420-01⟩. The elimination of hydrogen chloride from the insecticide trichlorofon (**200**) by treatment with diazomethane (!) and from acetylated trichlorofon (**210**) by dbu gave (**202**) and (**203**), respectively, in 71% yield ⟨81MI 420-01, 82MI 420-01⟩.

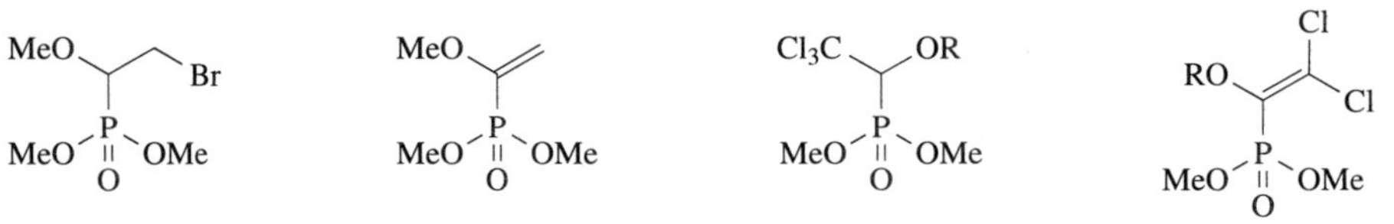

(**198**) (**199**) (**200**) R = H (**201**) R = Ac (**202**) R = Me (**203**) R = Ac

Finally, Zn/Cu in ethanol reductively eliminates acetate and bromide ions from the phosphorylated sugar (**204**) to give (**205**) in 76% yield ⟨91HCA451⟩.

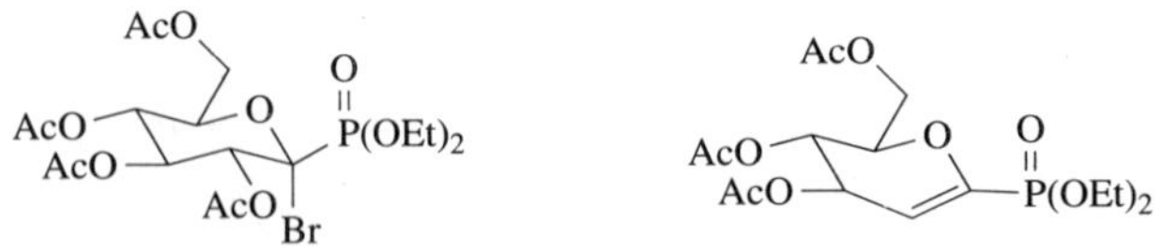

(**204**) (**205**)

4.20.2.1.2 *Displacement reactions*

Direct displacement of halides by phosphorus compounds to form 1-alkoxy-1-haloalkenes has been reported, but the yields are low. Thus triethyl phosphite gives (**206**) from 1-ethoxy-1,2-

dichloroethene in 30% yield ⟨76ZOB1652⟩ and (**207**) from 1,2-dibromo-1,2-diethoxyethane in 10% yield ⟨77CA135599v⟩.

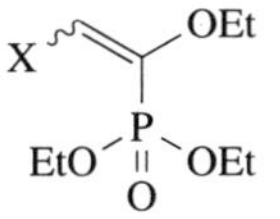

(**206**) X = Cl
(**207**) X = OEt·

Pudovick has shown that triethyl phosphite reacts with (**208**) to give (**209**) in 20% yield, presumably by a 1,2-shift ⟨76ZOB2385⟩.

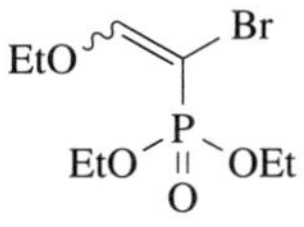

(**208**)   (**209**)

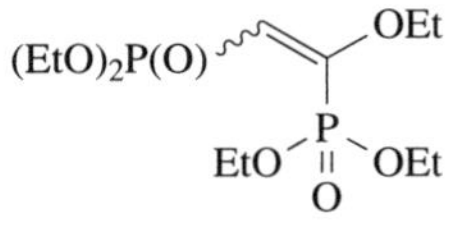

4.20.2.1.3 *Condensation of alcohols with α-diazo alkylphosphonates*

Ganem and co-workers have described the sequence shown in Scheme 18 whereby alcohols react with α-diazo alkylphosphonates in the presence of rhodium acetate to give (**210**). These products are then converted into the enol ethers (**211**) either (when X = CO$_2$Et) by condensation with Eschenmoser's salt followed by methylation and base-catalysed loss of trimethylamine or (when X = PO(OR)$_2$), in much better yields, by generation of the α-anion with LDA followed by treatment with formaldehyde ⟨92JOC178⟩.

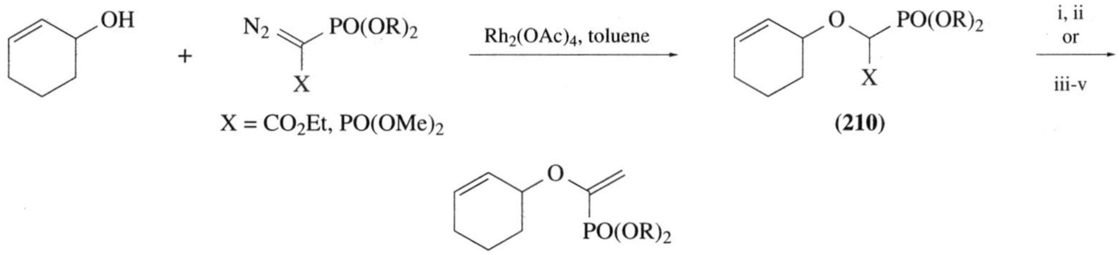

X = CO$_2$Et, PO(OMe)$_2$ (**210**)

(**211**)

i, LDA; ii, CH$_2$O; X = PO(OR)$_2$, 77%; iii, H$_2$C=NMe$_2$$^+$ I$^-$, 46%; iv, MeI; v, HO$^-$; X = CO$_2$Et, 32%

Scheme 18

(i) *Addition of phosphorus compounds to alkoxyalkynes*

A very wide variety of phosphorus compounds add to the 1-position of alkoxyalkynes. Thus, triethyl phosphite undergoes the Arbusov reaction with 1-ethoxyethyne to give (**212**) in 70% yield when heated in ethanol for 3 h ⟨74ZOB2067⟩. The compound (**212**) is also produced, albeit in only 12% yield, from 1-ethoxyethyne and diethyl phosphite in the presence of sodium ethoxide at 110 °C ⟨84CB2622⟩. The rate of addition of trialkyl phosphites was shown to increase in the order Pr^i > Pr > Et > Me ⟨76CA192821t⟩. Dialkoxyphosphines (**213**) give (**214**) at 20 °C over 3 days ⟨73S547⟩. Tetrabutoxydiphosphines are more reactive and cooling is required to hold the temperature below 15 °C so that a 54% yield of the bis(dibutoxyphosphino)ethene (**215**) is produced ⟨76ZOB565⟩.

A later publication confirmed the *cis* nature of the product (**215**) by ^{31}P NMR spectroscopy ⟨79ZOB1910⟩.

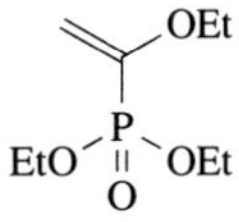

(**212**)

(**213**)

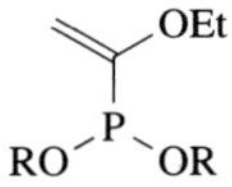

(**214**)
R = Bu^n, 50%
R = Bu^i, 41%

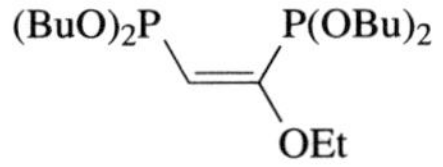

(**215**)

A different result arises from the addition of alkoxy(trimethylsiloxy)phosphines (**216**) to 1-ethoxyethyne in ether, which gives a 55% yield of a mixture of the (*E*)- and (*Z*)-isomers of the 2:1 adducts (**217**) ⟨87ZOB1406⟩.

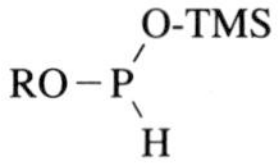

(**216**)

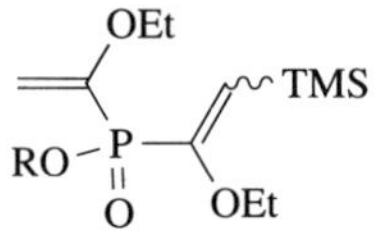

(**217**)

The reaction of tributylphosphine with 1-ethoxyethyne is more complicated, as the initial product has been variously assigned the structure (**218**) and (**219**) but, in any event, the addition of *inter alia* methyl iodide then gives the salt (**220**) in 78% yield ⟨80CA6622a, 80CA186471c, 83CA(98)4614z, 86CA(104)19636z⟩.

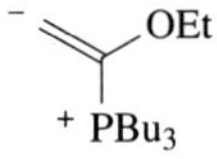

(**218**)

(**219**)

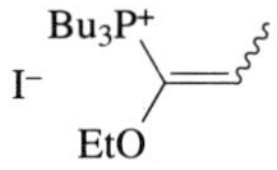

(**220**)

Dialkylhalophosphines (**221**) react with alkoxy alkynes in ether below 20 °C giving the 2:1 adducts (**222**) in 'quantitative' yield ⟨84ZOB416⟩. The rate of reaction is markedly dependent on the halogen atom with the halophosphine (**221**; X = I) reacting at −35 °C and (**221**; X = Cl) requiring 30–40 h to react to completion at 20 °C.

R$_2$PX

(**221**)

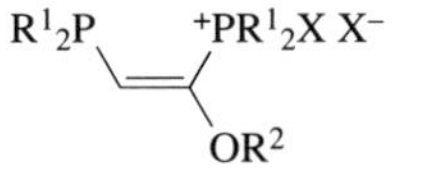

(**222**)
R^1 = Pr^i, Et; R^2 = Bu, Et, Me
X = Cl, Br, I

Tribromophosphine reacts photolytically with 1-ethoxyethyne to give (*E*)-(**223**) in 60% yield ⟨84ZOB457⟩.

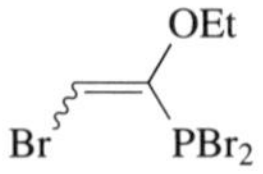

(**223**)

4.20.2.1.4 Addition to ketenes

Dimethylketene reacts at −70 °C with a wide variety of trivalent phosphorus compounds to give the 2:1 adducts (**224**) (Table 26) ⟨68JA5924, 72JA3058⟩. A different type of 2:1 adduct (**225**) arises from the reaction of triethyl phosphite with diphenylketene in ether at room temperature ⟨70JOC3583⟩.

Compound (**224**; X = Y = Z = OMe) readily reacts with water, bromine, methyl iodide, carbon dioxide and carbon disulfide to give the ring-opened structures (**226**) as essentially the only products.

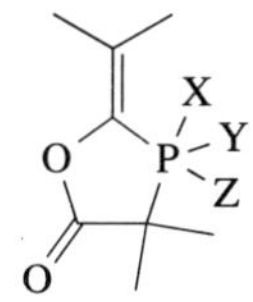

(**224**)

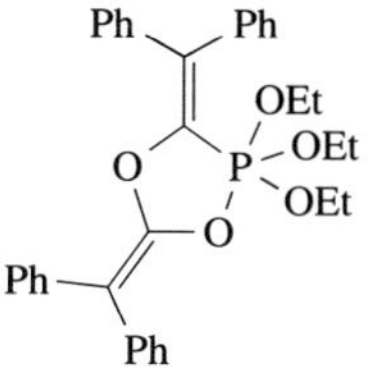

(**225**)

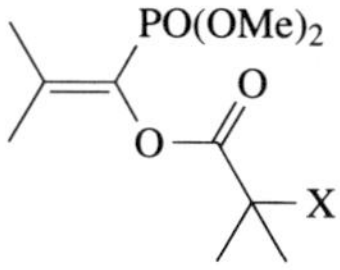

(**226**) X = H, Br, Me, CO_2Me, CS_2Me

Table 26 Adducts of dimethylketene and trivalent phosphorus compounds.

X	Y	Z	Yield (%)
MeO	MeO	MeO	97
Pr^nO	Pr^nO	Pr^nO	87
MeO	OCH_2CH_2O		85
Me_2N	MeO	MeO	100
MeO	$OCH_2CH_2N(Me)$		~80
Me_2N	$OCH_2CH_2N(Me)$		96
Me_2N	OCH_2CH_2O		100
Ph	MeO	MeO	100
Ph	OCH_2CH_2O		100

Ketene itself reacts with *t*-butyl diethyl phosphite to give 48% of (**227**; R = Me) ⟨88CA211147t⟩ and with mixed anhydrides of dialkyl hydrogen phosphites and carboxylic acids, for example (**228**), to give exothermically 1:1 adducts (**227**; R = Me, $CH=CH_2$) (Table 27) ⟨68CA(68)49691u, 68ZOB139, 70ZOB27⟩.

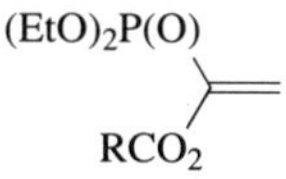

(**227**)

(**228**) R = Me, $HC=CH_2$

Similar enol acetates (**229**) result from the action of the mercuric halide complexes of trialkyl phosphites (**230**) with 2 mol ketene in benzene at 10–15 °C ⟨78ZOB267⟩.

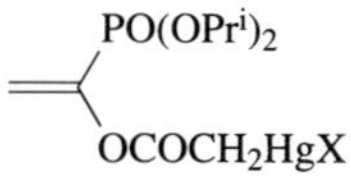

(**229**)
X = Cl, 89%
X = Br, 70%
X = I, 77%

(**230**) $(Pr^iO)_3PHgX_2$

Metallated ketenes, for example (**231**; R^2_3M = TMS, R^1 = H) react exothermically with dialkylphosphine oxides in the presence of triethylamine to give the adducts (**232**) (Table 28) ⟨75ZOB556⟩. The bis-silylated enol (**233**) results in 67% yield from the reaction of the bis-silylated ketene (**231**; R^2_3M = R^1 = TMS) with dimethyl hydrogen phosphite (**234**).

Table 27 Adducts of ketenes with the mixed anhydrides of dialkyl hydrogen phosphites and carboxylic acids.

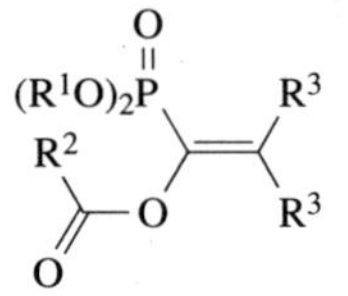

R^1	R^2	R^3	Conditions	Yield (%)
Et	Me	H	RT	76
Pr	Me	H	RT	76
Bu	Me	H	RT	52
Bu^i	Me	H	RT	19
Et	Et	H	RT	54
Et	Bu	H	RT	51
Et	Bu^i	H	RT	65
Et	Ph	H	RT	83
Et	Me	Ph	110 °C, 1 h	52
Et	Et	Ph	110 °C, 1 h	23
Et	Bu	Ph	110 °C, 1 h	56
Et	Bu^i	Ph	110 °C, 1 h	41
Et	Ph	Ph	110 °C, 1 h	26
Pr	Me	Ph	110 °C, 1 h	63
Bu	Ph	Ph	110 °C, 1 h	58

Table 28 Adducts of metallated ketenes with dialkyl phosphine oxides.

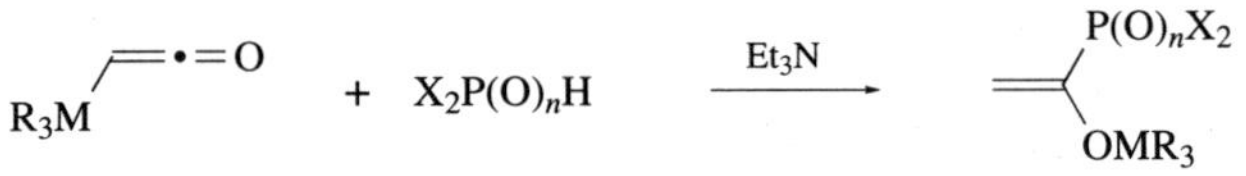

M	X	R	n	Yield (%)
Si	Ph	Me	0	75
Si	Ph	Me	1	88
Si	OMe	Me	1	74
Si	OEt	Me	1	63
Si	OMe	Et	1	62
Si	OEt	Et	1	62
Ge	OMe	Et	1	68

(231) $R^2{}_3MCR^1=\bullet=O$

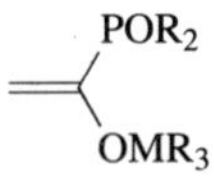

(232)

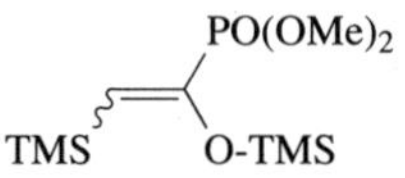

(233)

(234) $(MeO)_2P(O)H$

4.20.2.1.5 Condensation reactions

Zbiral and Binder showed that diethyl (2-trimethylsilylethoxy)methylphosphonate (235) can be deprotonated and silylated to form (236), which readily undergoes Peterson alkenation reactions ⟨68JOC780⟩ to give (237) as a mixture of (*E*)- and (*Z*)-isomers (Scheme 19 and Table 29) ⟨86TL5829⟩. (Methoxymethylene)bisphosphonic esters (238), prepared from sodium dibutyl phosphite and dibutyl (chloromethoxymethyl)phosphonate, condensed with benzaldehyde in the presence of sodium butoxide to give (239) (Scheme 20) ⟨80ZOB1225⟩.

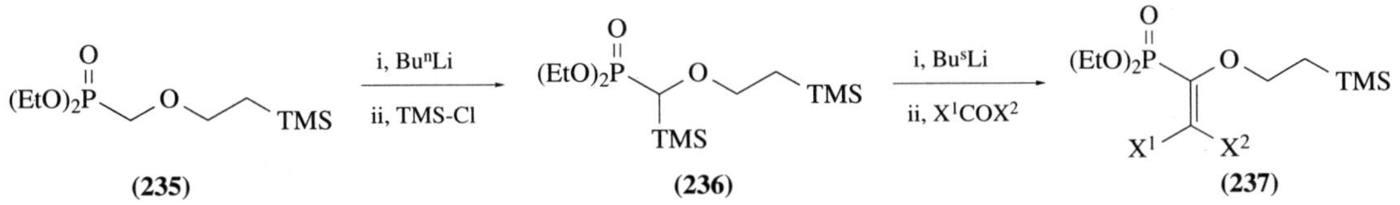

Scheme 19

Table 29 Peterson alkenation with diethyl(2-trimethylsilylethoxy)methyl phosphonate and the aldehydes and ketones, R^1COR^2 (Scheme 19).

X^1	X^2	Yield (%)
(CH$_2$)$_5$		88
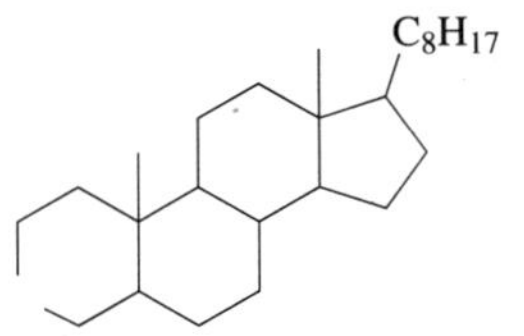 [steroid with C$_8$H$_{17}$]		85
Pr^i	H	61
4-MeOC$_6$H$_4$	H	80
3-MeOC$_6$H$_4$	H	77
Ph	Me	40

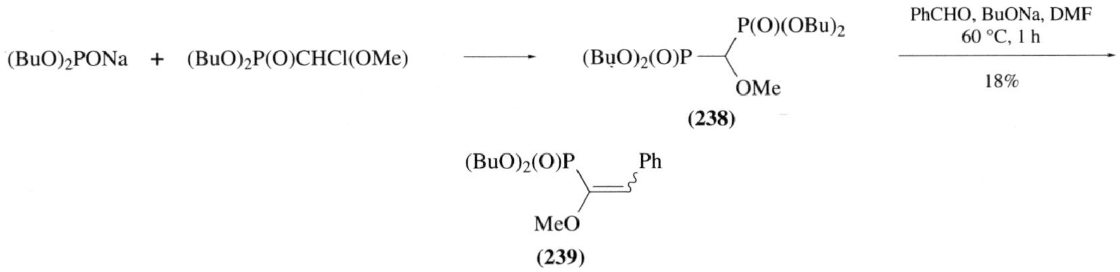

Scheme 20

4.20.2.1.6 Isomeration of β,γ-unsaturated α-oxyphosphorus compounds

Evans and co-workers deprotonated the α-silyloxyalkylphosphonamides (**240**) with *n*-butyllithium at −78 °C, alkylated the intermediate anion with methyl iodide and worked up the resultant solution with sodium methoxide to give the ester (**242**) in 75% overall yield (Scheme 21) ⟨79JA371⟩. From the structure of the product, it was evident that the initial β,γ-unsaturated system in (**240**) had been isomerised into the α,β system in (**241**), which could be isolated but was not purified. Subsequently, Krief *et al.* used a mixture of the β,γ- and α,β-unsaturated phosphonates (**243**) and (**244**) (Scheme 22) as a source of the α-anion, as treatment with LDA followed by aldehydes gives exclusively the α-substituted derivative ⟨84TL2883⟩.

The synthetic utility of anions derived from α-alkoxy alkyl phosphorus compounds has been extensively studied by the Dundee group of Miller and, subsequently, Murray, who conducted detailed examinations of the behaviour of different phosphorus substituents and established that the preferred system is that in the α-methoxyallylphosphine oxide (**245**) ⟨B-81MI 420-02⟩. Thus, treatment of (**245**) with LDA in THF at −70 °C generates an anion that can react with electrophiles either at the α- or γ-positions ⟨81TL3789⟩. The actual distribution of products, which are typically formed in yields of 70–97%, depends on both the electrophile and the nature of R^1 and R^2, but protonation of the anion derived from (**245**; $R^1 = R^2 = H$) gives (**246**; R = Me) as only the (*E*)-isomer whilst benzaldehyde and acetaldehyde react with (**245**; $R^1 = R^2 = H$) to give (**247**) in 70%

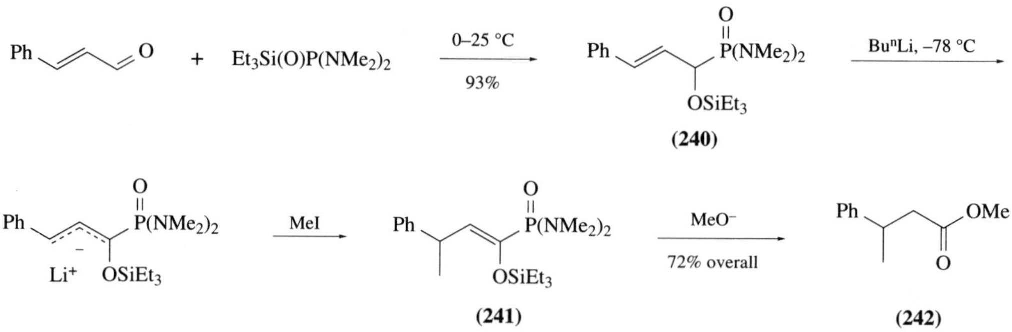

Scheme 21

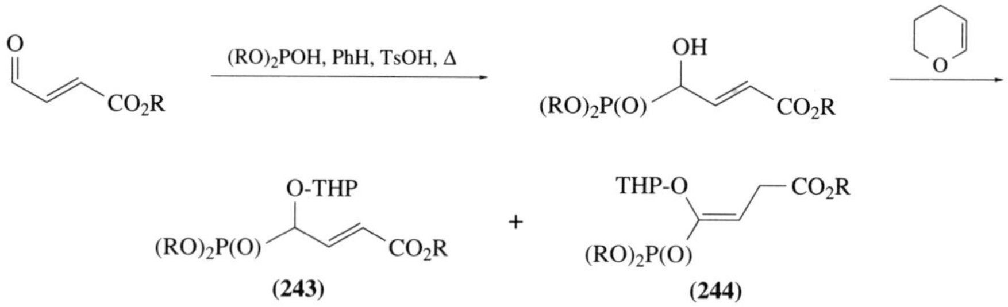

Scheme 22

yield. Similarly, the alkylation of the anion with TMS chloride or the sulfur electrophiles $MeSSO_2Me$, $PhSSO_2Ph$ and PhSSPh gives solely the γ-substituted products (**248**) (Table 30) ⟨86TL4635⟩. The use of titanium ate complexes of the ambident ion generally leads to α-substitution, but acyl halides give high yields of the (*E*)-γ-substituted ketones (**249**) (Table 31) ⟨88JCS(P1)1039⟩. Finally, carbon dioxide or alkyl chloroformates react to give the acids and esters (**250**) ⟨90JCS(P1)2811⟩. To produce good yields of the esters, it was necessary to use 2.2 equivalents of LDA, presumably because of the acidity of the γ-proton in the product which results in protonation of the starting material.

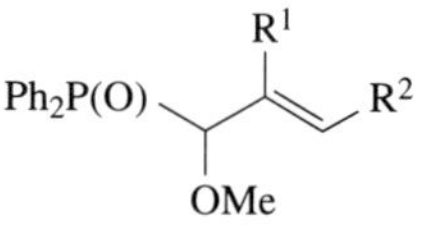

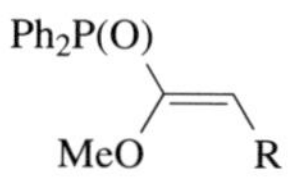

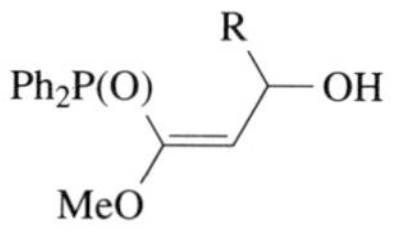

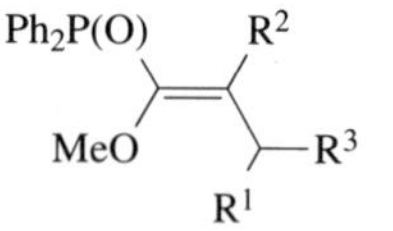

(**248**) R^1 = TMS, SPh, SMe

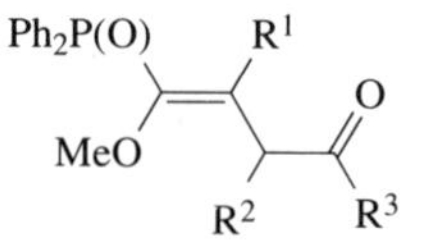

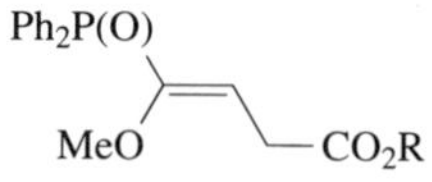

(**250**) R = H, 74%
R = Me, 83%
R = Pr^i, Et, allyl, 67%

Similar isomerisations have been reported for the 1-alkoxyallyltriphenylphosphoranes (**251**) (Scheme 23) ⟨90SL115⟩ and the 1-acyloxyallylphosphonate (**252**) (Equation (23)) ⟨84CB2622⟩ whilst diethyl TMS phosphite (**253**) condenses with, for example, cinnamaldehyde to give (**254**), which is then deprotonated and alkylated to give exclusively, in 43% yield, the unstable (*E*)-silyl enol ether (**255**) (Scheme 24) ⟨79TL2047⟩.

Table 30 Electrophilic γ-substitution of (1-methoxyallyl)diphenylphosphine oxides.

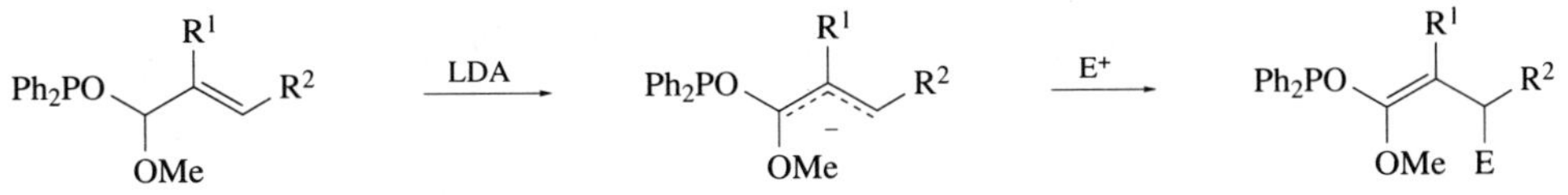

R^1	R^2	E	*Yield* (%)
H	H	TMS	84
Me	H	TMS	82
H	Me	TMS	88
H	H	SMe	79
H	Me	SMe	84
H	Me	SPh	82
Me	H	SMe	72
H	Ph	SMe	69
Me	H	$P(O)(OEt)_2$	14
H	Me	$P(O)Ph_2$	61

LDA = lithium diisopropylamide.

Table 31 γ-Acylation of (1-methoxyallyl)diphenylphosphine oxides.

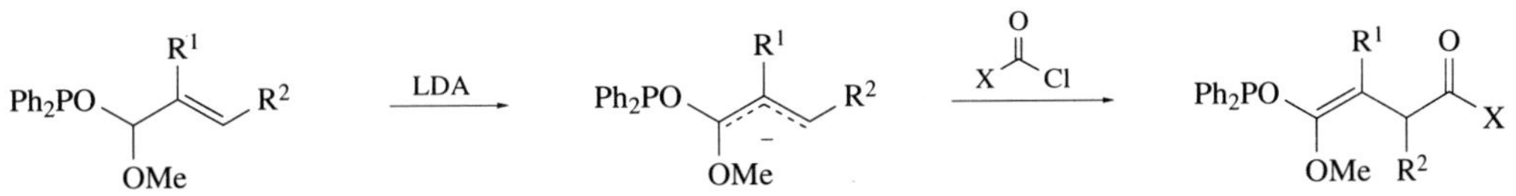

R^1	R^2	X	*Yield* (%)
H	H	p-$MeOC_6H_4$	75
H	H	Me_2CH	72
H	H	$Me(CH_2)_3$	74
H	Me	Ph	68
H	Me	Me_2CH	71
H	Me	$MeCH_2$	78

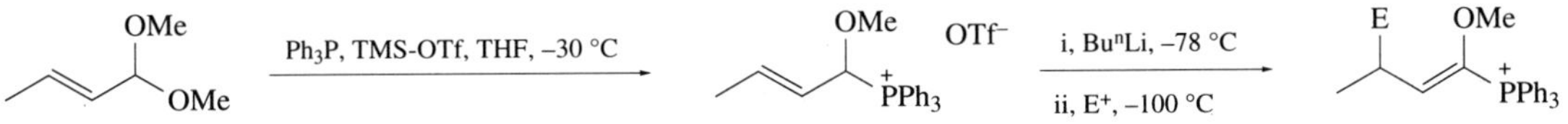

(251)

E = Bz, 76%

Scheme 23

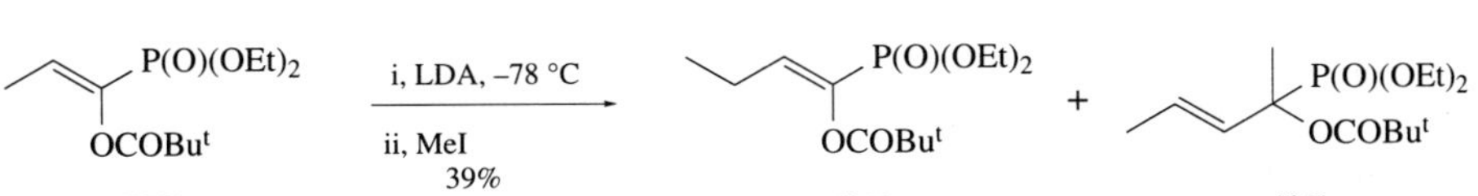

(23)

(252) 60% 40%

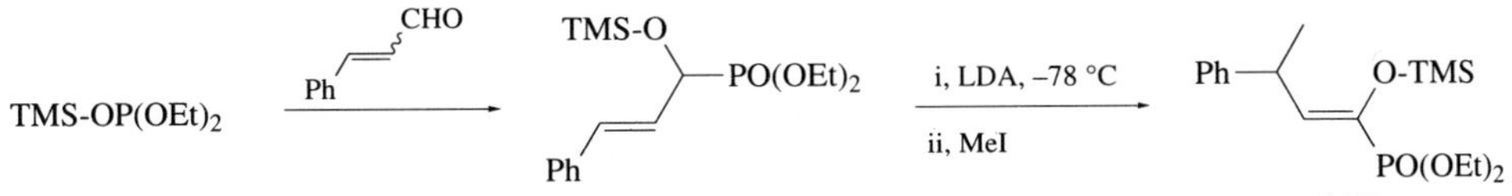

(253) **(254)** **(255)**

Scheme 24

4.20.2.1.7 Enolisation of α-carbonylphosphorus compounds

The standard Michaelis–Arbuzov reaction results from the alkylation of trialkyl phosphites to give dialkyl alkylphosphonates (Equation (24)). With α-halo ketones, the reaction proceeds differently to give a product which was shown by Perkow to be an enol phosphate (**256**) (Equation (25)), whilst with α-halo acid halides both reactions take place and the product is an α-phosphorylated enol phosphate (**257**) (Equation (26) and Table 32) ⟨61CRV607⟩.

$$(R^1O)_3P + R^2X \longrightarrow (R^1O)_2P(O)R^2 \quad (24)$$

$$(RO)_3P + \text{[α-halo ketone]} \longrightarrow \text{(256)} \quad (25)$$

$$2\,(RO)_3P + \text{[α-halo acyl halide]} \xrightarrow[\text{ii, 100 °C, 2 h}]{\text{i, < 40 °C}} \text{(257)} \quad (26)$$

Table 32 The Perkow reaction of α-halo acyl halides.

X^1	X^2	X^3	X^4	Yield (%)	Ref.
Me	H	H	Cl	86	57CA(51)10366g, 60CA(54)18356h
Et	H	H	Cl	86	57ZOB2161, 60CA(54)18356h
Et	H	Me	Br	79	57ZOB2161, 60CA(54)18356h
Et	H	Cl	Cl	86	59CA(53)10040f
Me	Cl	Cl	Cl	78	57ZOB2161, 60CA(54)18356h, 90JOC5982
Et	Cl	Cl	Cl	79	59CA(53)10040f, 90JOC5982
Me	Me	Me	Br	92	57ZOB2161
Et	Me	Me	Br	90	57ZOB2161
Pr^i	H	H	Cl	quantitative	60CA(54)18356h
Me	Me	Me	Cl	quantitative	60CA(54)18356h
CH$_2$CH$_2$Cl	H	H	Cl	quantitative	60CA(54)18356h
Me	H	Cl	Cl	quantitative	60CA(54)18356h
Et	Me	Me	Cl	quantitative	60CA(54)18356h
CH$_2$CH$_2$Cl	Cl	Cl	Cl	quantitative	60CA(54)18356h
CH$_2$CH$_2$Cl	H	Cl	Cl	quantitative	60CA(54)18356h

Baldwin and Swallow showed that triethyl phosphite reacts with diphenylacetyl chloride in refluxing diethyl ether to give the stable, isolable enol, diethyl 1-hydroxy-2,2-diphenylethenylphosphonate (**258**) ⟨70JOC3583⟩.

(**258**)

The reaction in Equation (26) presumably goes via an acylphosphonate (**259**), and preformed derivatives (**259**; X = H) react with acylating agents under acidic conditions at 130–140 °C to give, for example, the enol acetate (**260**) (Table 33) ⟨73CA(78)85856s⟩. Zinc and phosgene convert the

2-bromo-2-methylpropanoylphosphonate (**261**) into the unstable chloroformate (**262**) (Scheme 25) ⟨89FP2610926, 90JOC5982⟩.

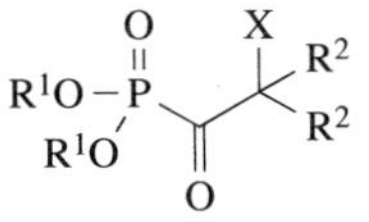

(**259**) 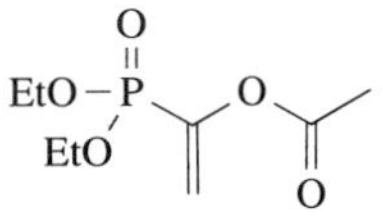 (**260**)

Table 33 *O*-Acylation of acyl phosphonates.

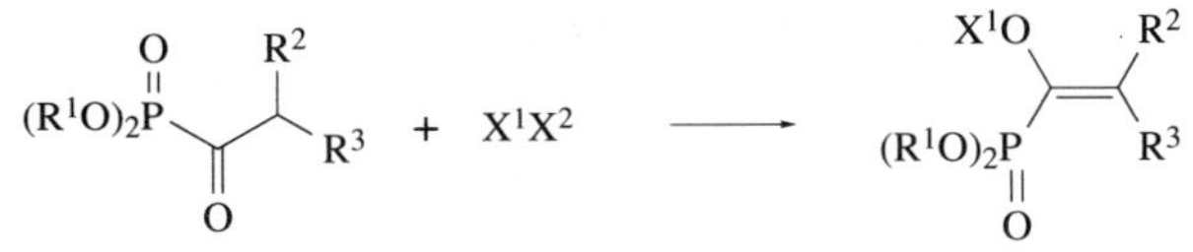

R^1	R^2	R^3	X^1	X^2	Yield (%)	Ref.
Et	H	H	MeCO	Cl	36	73CA85856s, 76CA122034u, 84ZOB1324
Et	H	H	(EtO)$_2$P	Cl	42	84ZOB1324
Et	H	H	TMS	Cl	43	84ZOB1324
Me	H	H	TMS	Cl	50	84ZOB1324
Et	H	H	EtCO	OCOEt	83	76CA122034u
Et	H	Me	MeCO	OCOMe	77	76CA122034u
Et	H	Me	EtCO	OCOEt	72	76CA122034u

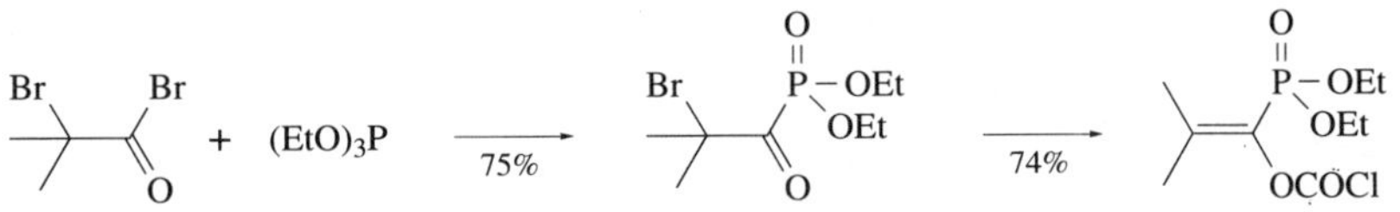

Scheme 25

The Perkow reaction also proceeds with other phosphorus compounds, for example triphenylphosphine (Equation (27)) ⟨74BSF2263⟩, methyl diphenylphosphinate (Equation (28)) ⟨83CB3141, 83ZN(B)726⟩ and ethyl tetraethyldiamidophosphite (Equation (29)) ⟨90ZOB1940⟩.

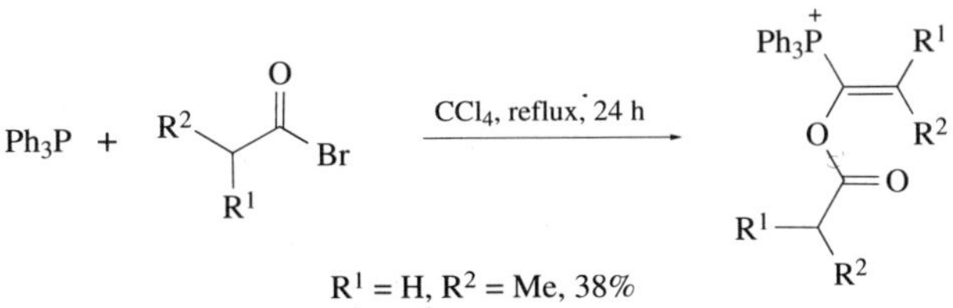

R^1 = H, R^2 = Me, 38%
R^1 = R^2 = Me, 23%

(27)

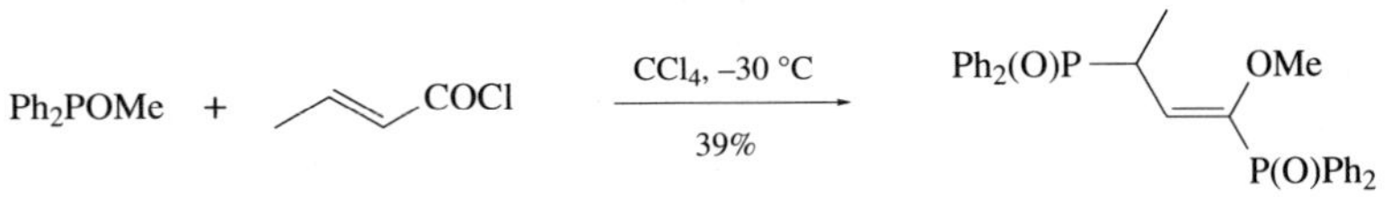

(28)

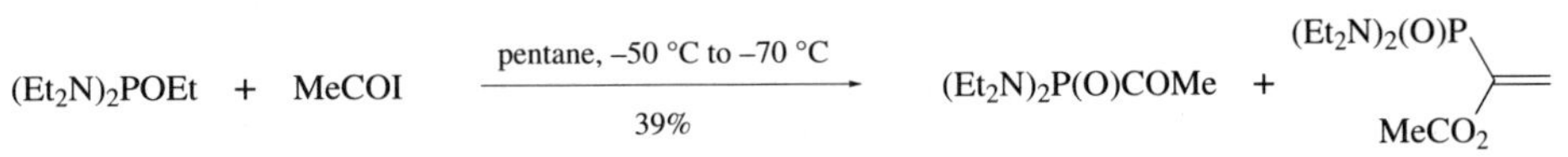

(29)

Tyryshkin and co-workers have shown that dialkyl phosphites react with acetic anhydride in refluxing acetonitrile to give mixtures of the acetyl phosphonates (**263**) and 1-acetoxyvinylphosphonates (**264**) and that the use of metal catalysts, particularly iron(II), iron(III) and cobalt(II), greatly increases the yield of (**264**) (Table 34) ⟨92HAC127⟩.

(RO)$_2$P(O)COMe

(**263**)

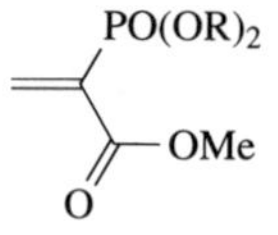

CH$_2$=C(PO(OR)$_2$)–C(OMe)=O — structure (**264**): 2-(dialkoxyphosphoryl)-1-methoxyprop-2-en-1-... enol acetate

(**264**)

Table 34 Reactions of dialkyl phosphites with acetic anhydride in the presence of metal catalysts to give acetyl phosphonates (RO)$_2$P(O)COCH$_3$ (**263**) and their enol acetates (**264**).

R	Catalyst	Time (h)	Yield (%) 263	Yield (%) 264
Me	CoCl$_2$	4	10	75
	FeCl$_3$	2	10	88
	FeCl$_3$·6H$_2$O	2	5	60
Et	CoCl$_2$	4	10	80
	FeCl$_2$	2	10	85
	FeCl$_3$·6H$_2$O	1.8	5	60
Pr	CoCl$_2$	4	15	80
	FeCl$_3$	2	10	90
	FeCl$_3$·6H$_2$O	2	10	60
Pr^i	CoCl$_2$	4	10	85
	FeCl$_2$	2	10	90
	FeCl$_3$·6H$_2$O	2	10	80
Bu	CoCl$_2$	3	15	75
	FeCl$_2$	2	10	85
	FeCl$_3$·6H$_2$O	1.5	5	70

With unsaturated acid chlorides, the reaction can take a different course. Thus, *trans*-but-2-enoyl chloride reacts with trimethyl phosphite to give, *inter alia*, (**265**) in about 27% yield (Scheme 26) ⟨81JCS(P1)1363⟩. *p*-Chlorocinnamoyl chloride, however, reacts with trimethyl phosphite in the absence of solvent to give the 'trimer' (**266**) in 37% yield ⟨80HCA402⟩.

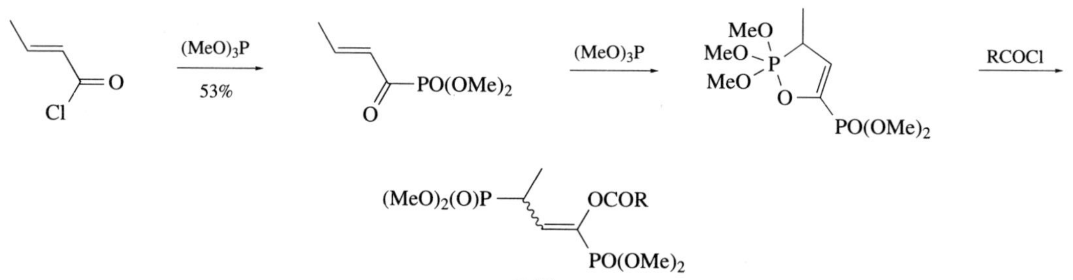

Scheme 26

(**265**) R = *trans*-MeCH=CH

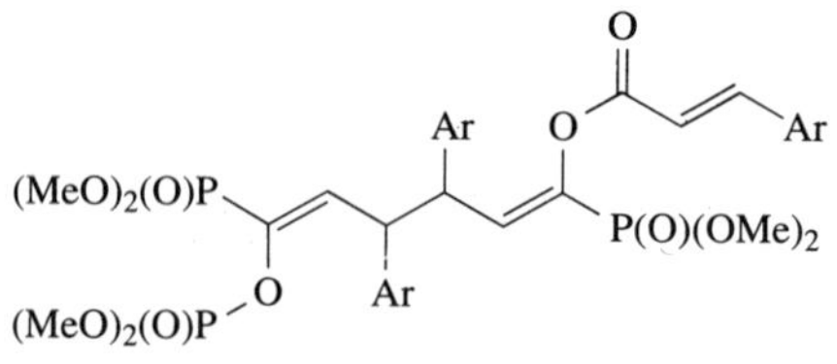

(**266**) Ar = *p*-ClC$_6$H$_4$

4.20.2.1.8 Rearrangements

Sturtz and Corbel have shown that the phosphorodiimidate (267) undergoes a Wittig-type carbanion rearrangement with *n*-butyllithium to give an enolate anion which can be trapped by acetyl chloride, with care, at −70 °C to give (268) ⟨73CR(C)395⟩.

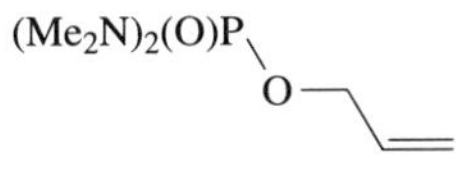

4.20.2.2 Functions Bearing Sulfur

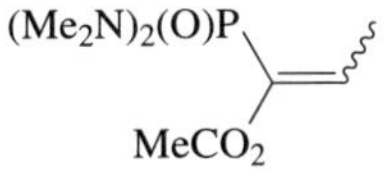

A range of reactions similar to those described in the previous section have been reported for the production of these compounds.

4.20.2.2.1 Elimination reactions

The elimination reactions include loss of hydrogen halides, sulfenic and selenenic acids, and water by the Pummerer rearrangement. Representative base-catalysed losses of hydrogen chloride are listed in Table 35. Warren has particularly studied the elimination of hydrogen chloride from the derivative (269) (Scheme 27) with silica or zinc bromide ⟨82TL4167, 87JCS(P1)967⟩.

Table 35 Base-catalysed elimination of HCl from 2-chloro-1-phosphinyl 1-thioalkanes.

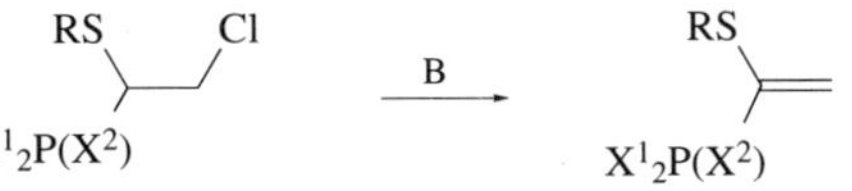

X^1	X^2	R	B	Yield (%)	Ref.
MeO	O	Ph	Et$_3$N, Et$_2$O, RT, 12 h	?	83PS345
Pr^{i}O	O	Ph	Et$_3$N, Et$_2$O, RT, 12 h	?	83PS345
Cl	O	Ph	Et$_3$N, Et$_2$O, RT, 12 h	?	83PS345
Pr^{i}O	S	Ph	Et$_3$N, Et$_2$O, RT, 12 h	?	83PS345
Cl	S	Ph	Et$_3$N, Et$_2$O, RT, 12 h	?	83PS345
EtO	O	Et	Et$_3$N, Et$_2$O, 4 h, 30 °C	94	67ZOB454
EtO	O	Me	Et$_3$N, Et$_2$O, 4 h, 30 °C	86	83S332
Pr^{i}O	O	Me	Et$_3$N, Et$_2$O, 4 h, 30 °C	?	71CA(75)88759m
EtO	O	Me	dbn^a	72	81TL191

^{a} dbn = 1,5-diazobicyclo[4.3.0]non-5-ene.

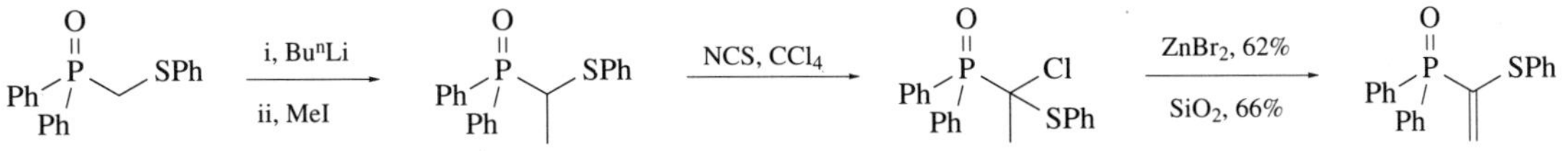

Scheme 27

Structure (**271**) reacts with dbu (Scheme 28) to give a mixture of the phosphirane oxide (**272**) and the rearranged alkene (**273**) ⟨86BCJ3293⟩.

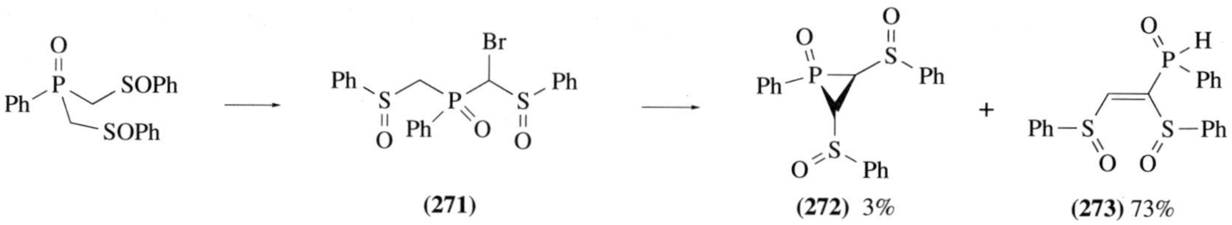

Scheme 28

Compound (**270**) was also prepared by the Pummerer elimination of the sulfoxide (**274**) with acetic anhydride and methanesulfonic acid, although the reaction required 8 days to go to completion ⟨87JCS(P1)967⟩. Other Pummerer reactions are listed in Table 36 ⟨87JCS(P1)1095, 87MI 420-01⟩.

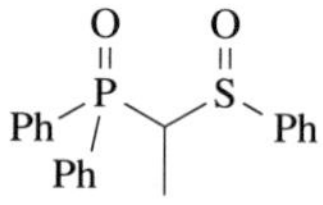

(**274**)

Table 36 Pummerer rearrangements.

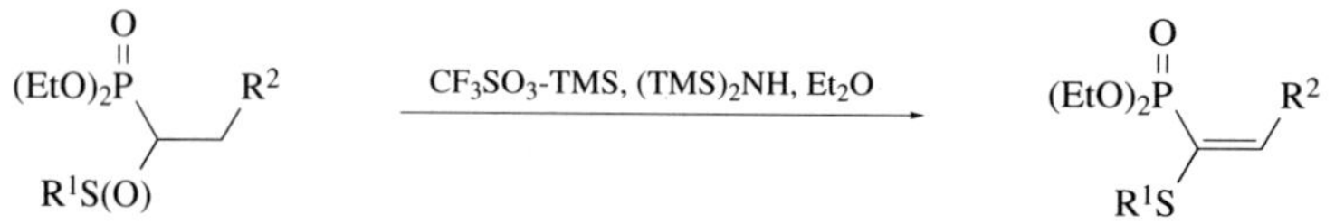

R^1	R^2	Yield (%)
Me	H	75
Me	Me	75
Me	CH=CH$_2$	85
Me	Ph	85
Ph	H	75
Ph	Me	75
4-MeC$_6$H$_4$	H	70
4-ClC$_6$H$_4$	H	70

Mikolajczyk has described efficient elimination reactions of selenoxides (Schemes 29 and 30) with the chiral sulfoxide (**275**) resulting from the chiral precursor (**276**) ⟨81TL3097, 92TA1515⟩.

Methanesulfenic acid can also be eliminated from, for example, the sulfoxide (**277**) in refluxing benzene (Scheme 31) ⟨82S394⟩. Only the (Z)-isomer of the diene (**278**) was produced.

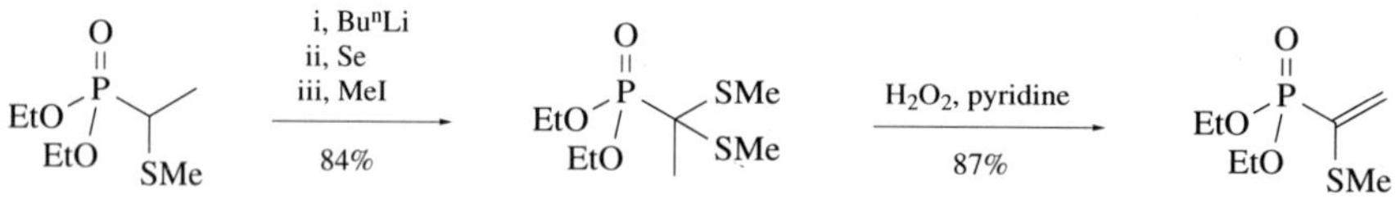

Scheme 29

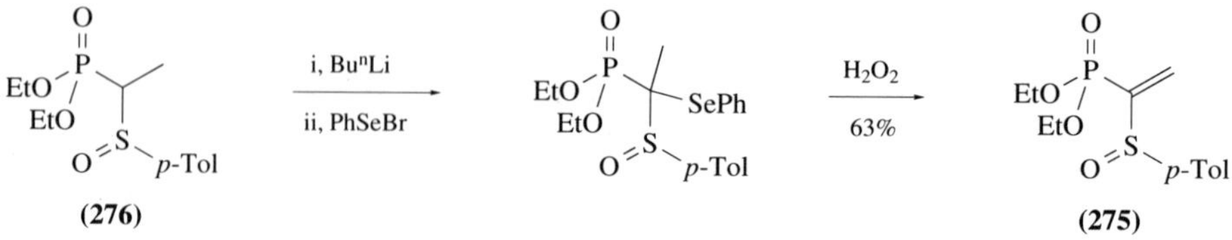

Scheme 30

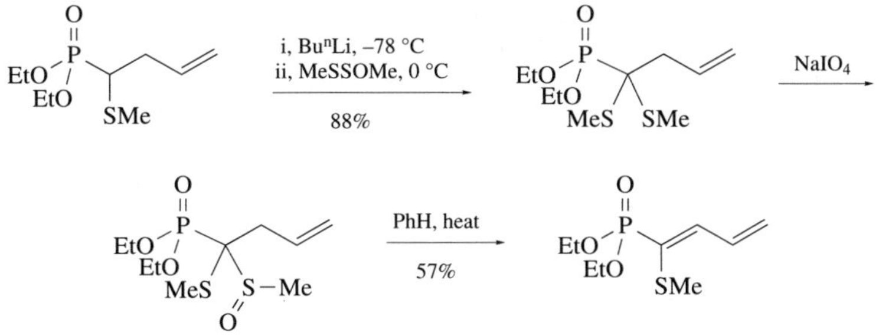

(277) (278)

Scheme 31

4.20.2.2.2 Condensation reactions

Hewson and MacPherson have described the use of the vinylphosphonium salts (**279**) in the synthesis of, for example, cyclopentanoid natural products ⟨85JCS(P1)2625⟩. The salt (**279**; R = Ph) was prepared in 89% yield by condensation of the phosphonium salt (**280**) with two equivalents of Eschenmoser's salt (**281**) in refluxing acetonitrile during 15 h ⟨78TL3267⟩.

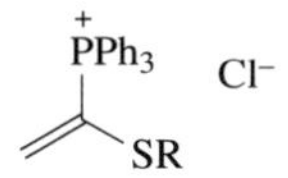

(**279**) R = Ph, Me

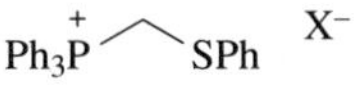

(**280**)

$H_2C=NMe_2^+ Cl^-$

(**281**)

Subsequently, the reaction was extended to give the *S*-methyl analogue (**279**; R = Me) in 84% yield ⟨83JCS(P1)2979⟩. Diethyl benzenesulfonylmethylphosphonate (or the methanesulfonyl derivative) condenses with paraformaldehyde, benzaldehyde and cinnamaldehyde under basic conditions to give the products (**282**) ⟨83TL767, 84BCJ2127, 84CB1424⟩.

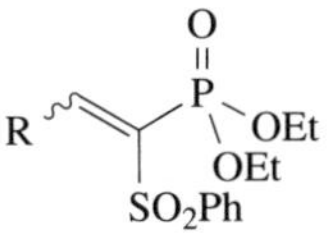

(**282**)

R = H, 92%
R = Ph (Z-isomer), 31%
R = *trans*-PhCH=CH(*E*-isomer), 93%

The condensation can also be performed under Peterson conditions with the silyl derivatives (**283**) ⟨84CB2622, 89S101⟩. Deprotonation is achieved with Bu^nLi in THF, hexane at −70 °C and the resultant anion condenses with aldehydes, for example paraformaldehyde, acetaldehyde, benzaldehyde and cinnamaldehyde, to give the products (**284**). The transformation can also be performed in one pot from diethyl methylthiomethylphosphonate (**285**) by successive treatment with butyllithium, chlorotrimethylsilane and acetaldehyde to give (**284**; R = Me) in 52% overall yield.

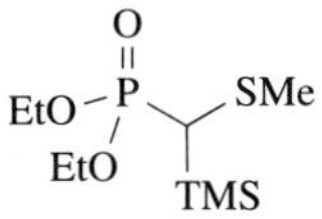

(**283**)

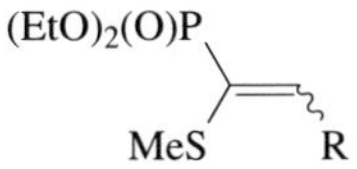

(**284**)

R = H, 60%
R = Me (*E*)- and (*Z*)-isomers, 85%
R = Ph (*E*)- and (*Z*)-isomers, 83%
R = PhCH=CH (*E*)- and (*Z*)-isomers, 81%

$(EtO)_2P(O)CH_2SMe$

(**285**)

4.20.2.2.3 *Addition to substituted alkynes*

Russian workers reported that bis(alkylthio)ethynes (**286**) react exothermically in benzene with phosphorus pentachloride to give the salts (**287**) in 83% yield ⟨82ZOB2375⟩. Where R^1 and R^2 differ the products are formed in roughly equal amounts. Treatment of the salts (**287**) with sulfur dioxide gives the phosphonyl derivatives (**288**) in 94% yield and with tetrabutylammonium iodide the phosphinyl derivatives (**289**) in 85% yield.

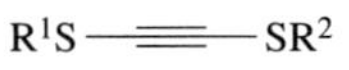

(286)

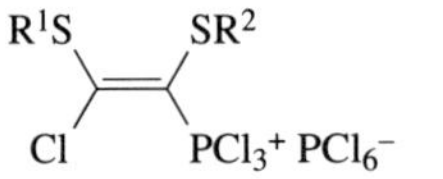

(287)

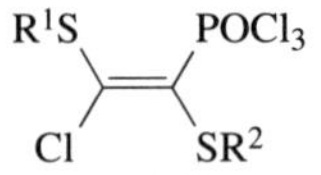

(288)

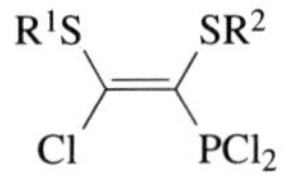

(289)

Alternatively, benzenesulfenyl chloride (**290**) reacts in dichloromethane at −10 °C with the phosphorus-substituted ethyne (**291**) to give, in 49% yield, a mixture of (**292**) and (**293**), formed by *trans*-addition ⟨84ZOB1758⟩.

PhSCl

(290)

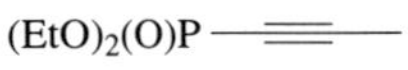

(291)

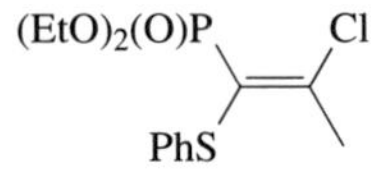

(292)

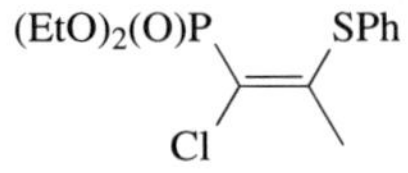

(293)

4.20.2.2.4 *From phosphorus-substituted sulfines*

The modified Peterson reaction shown in Scheme 32 gives the phosphoryl sulfines (**295**) ⟨84JOC263⟩. If the R group contains an α-hydrogen this can be removed by sodium hydride or thallous ethoxide and the resultant anion (**296**) alkylated by alkyl halides to give a mixture of the sulfoxides (**297**) and sulfides (**298**) (Scheme 33).

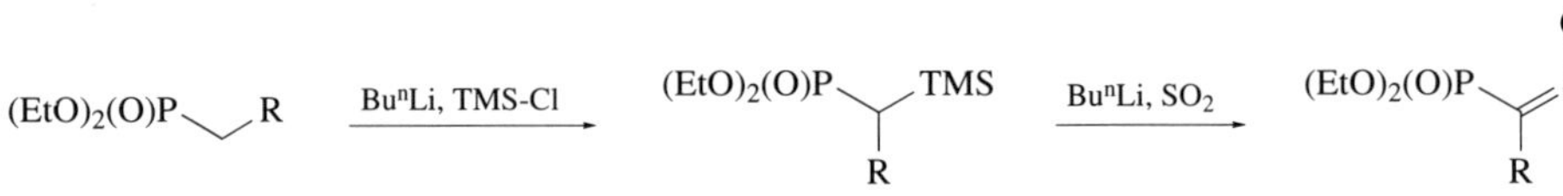

Scheme 32

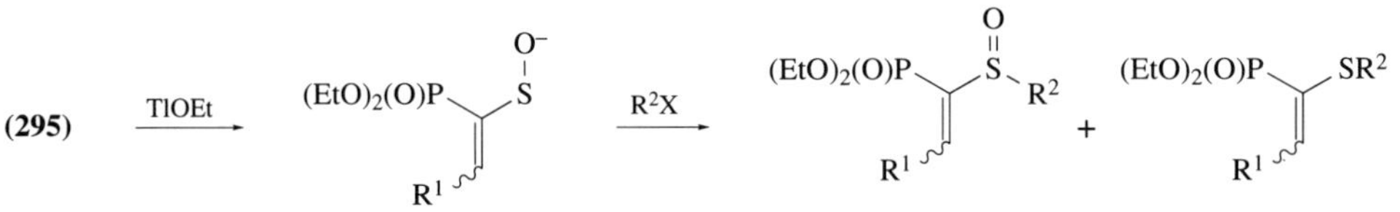

	(297)	(298)
R^1 = Ph, R^2 = Me	61%	18%
R^1 = Ph, R^2 = Et	37%	45%

Scheme 33

4.20.2.2.5 *From ketenes*

Heating triethyl phosphite at 100 °C for 5 h with the dimer of bis(trifluoromethyl)thioketene gave the alkene (**299**) (90%) ⟨69CA(70)57950j⟩.

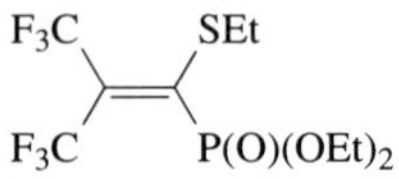

(299)

4.20.2.2.6 From 1-lithio-1-phosphorylated alkenes

(E)-Dimethyl 2-phenylvinylphosphonate (**300**) in dry THF at $-100\,°C$ was treated with LDA for 45 min and then with dimethyl disulfide. After 1 h at $-100\,°C$, the mixture was allowed to warm to $0\,°C$ and then worked up to give 55% of the product (**301**) ⟨86CB472⟩.

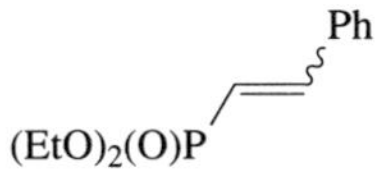

(EtO)$_2$(O)P–CH=CH–Ph
(**300**)

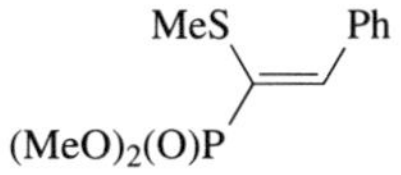

(MeO)$_2$(O)P–C(SMe)=CH–Ph
(**301**)

The anion of diethyl methylphosphonate adds to aryl cyanides to generate the corresponding anions (**302**), which then react at $-78\,°C$ with a disulfide or sulfenyl chloride to generate the substituted enamines (**303**) (Table 37) ⟨91SC279⟩.

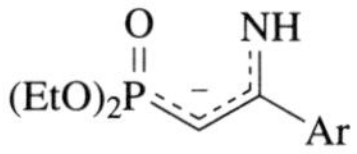

(EtO)$_2$P(O)–CH=C(NH)–Ar
(**302**)

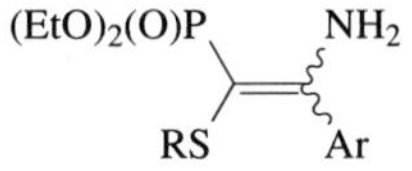

(EtO)$_2$(O)P\\C(NH$_2$)=C(SR)/Ar
(**303**)

Table 37 Condensation of diethylmethyl phosphonate with aryl cyanides.

Ar	R	Yield (%)	
Ph	Ph	78	(from PhSSPh)
Ph	Ph	87	(from PhSCl)
p-ClC$_6$H$_4$	Ph	82	
p-MeC$_6$H$_4$	Ph	82	
p-ClC$_6$H$_4$	Me	66	(from MeSO$_2$SMe)

4.20.2.2.7 By substitution of a 1-chlorovinylphosphonate

Sodium methanethiolate reacts at room temperature overnight with S,S-diethyl 1-chlorovinylphosphonodithioate (**304**) to give S,S-diethyl 1-ethylthiovinylphosphonodithioate (**305**) in 81% yield ⟨70BRP1183130⟩.

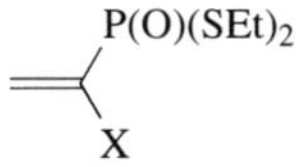

CH$_2$=C(X)–P(O)(SEt)$_2$

(**304**) X = Cl
(**305**) X = SEt

4.20.2.2.8 From alkyl dithiocarboxylates

Methyl dithioisobutyrate (**306**; R^1 = Me) condenses with diethylphosphinous chloride in the presence of triethylamine to give (**307**; R^1 = Et, R^2 = Me) in 87% yield. Heating (**307**) under argon causes a rearrangement to give (**308**; R^1 = Et, R^2 = Me) ⟨85ZOB838⟩. Other derivatives (**308**) have been prepared likewise.

Me$_2$CHS$_2$R^1

(**306**)

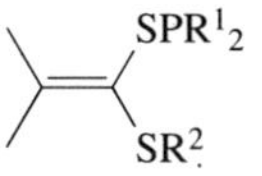

Me$_2$C=C(SPR$^1_2$)(SR^2)

(**307**)

Me$_2$C=C(P(S)R$^1_2$)(SR^2)

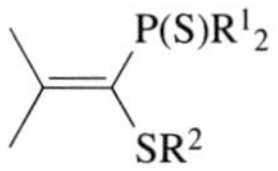

(**308**)
R^1 = Et, R^2 = Me, 43%
R^1 = Et, R^2 = Et, 28%
R^1 = Et$_2$N, R^2 = Me, 26%

4.20.2.3 Functions Bearing Selenium

Magdesieva and Danilenko showed that the benzoylmethylphosphonium salts (**309**) react with selenoxides in refluxing, dry dichloromethane in the presence of acetic anhydride to give the salts (**310**) (Scheme 34) ⟨79ZOB332⟩.

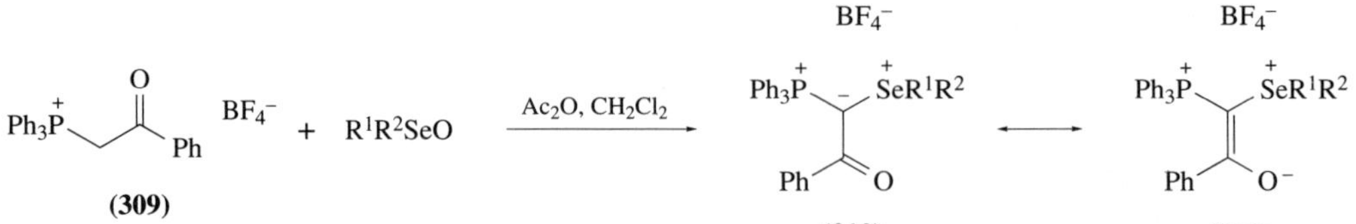

$R^1 = R^2 = Ph, 85\%$
$R^1 = R^2 = Me, 83\%$
$R^1 = Ph, R^2 = Me, 86\%$
$R^1 = R^2 = (CH_2)_5, 82\%$

Scheme 34

Initially, the products (**310**) were drawn as the carbonyl structure but later this was changed to the enolate form (**311**), and it was found that protonation by trifluoroacetic acid gave the bis salts (**312**) ⟨79ZOB2198⟩.

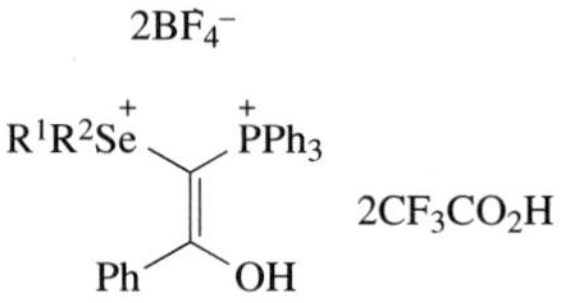

In addition, alkylation of the anions (**310**) by triethyloxonium tetrafluoroborate takes place on the oxygen atom to give the products (**313**) ⟨79ZOB1857⟩. Heating the salts (**310**) in an argon atmosphere for brief periods at 200–230 °C results in migration of an alkyl or aryl group from selenium to oxygen to give the enol ethers (**314**) (Equation (30)) ⟨80ZOB573⟩.

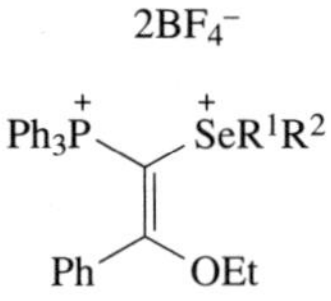

$R^1 = R^2 = Me, 87\%$
$R^1 = R^2 = Ph, 82\%$

(30)

$R^1 = R^2 = Ph, 52\%$
$R^1 = R^2 = Me, 87\%$
$R^1 = Ph, R^2 = Me, 91\%$

Diethyl methylphosphonate was deprotonated with *n*-butyllithium in dry THF at −78 °C and treated with benzonitrile or *i*-propyl cyanide to give, at −5 °C, a solution of the anion (**315**) ⟨91SC279⟩. The solution was cooled again to −78 °C and benzeneselenyl bromide added to give the enamine (**316**) (Scheme 35; *cf.* Table 37 and 4.20.2.2.6).

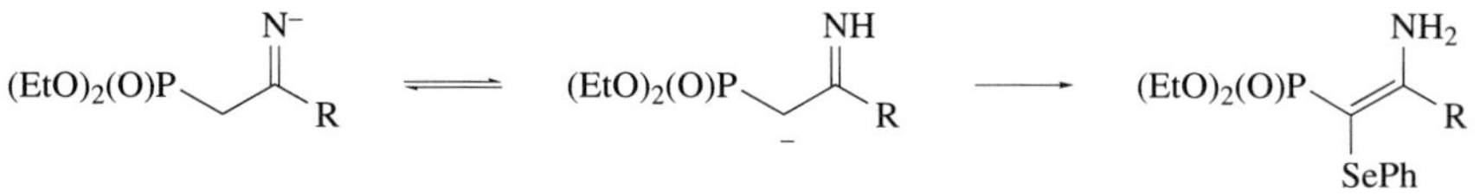

Scheme 35

4.20.3 FUNCTIONS CONTAINING A CHALCOGEN AND A METALLOID

4.20.3.1 Functions Bearing Oxygen

4.20.3.1.1 *Oxygen and silicon or germanium*

Compounds of general structure (**317**) are enol ethers of acylsilanes (**318**), a group of molecules whose synthesis and chemistry have been the subject of two reviews ⟨89S647, 90CSR147⟩.

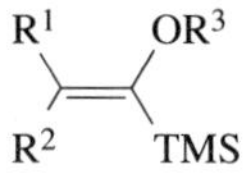

(i) Preparation from vinyl ethers

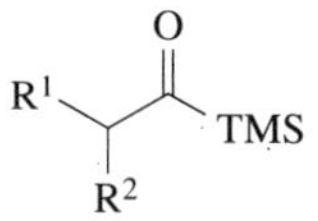

Soderquist *et al.* have studied in considerable detail the deprotonation and metallation of vinyl ethers (**319**) (Scheme 36 and Table 38) ⟨80JA1577, 82OM830, 88TL1899, 90OS(68)25⟩. The best method appears to be to treat the vinyl ether with an excess of *t*-butyllithium dropwise at −70 °C in dry THF and to allow the solution to warm to 0 °C over 3 h. The solution is then cooled again to −70 °C, chlorotrimethylsilane added dropwise and the mixture warmed to room temperature. Final work up involves a quick water wash and distillation to give the desired silyl enol ether, for example (**317**; $R^2 = R^3 = H$, $R^3 = R^1 = Me$) in 84% yield; with chlorotrimethylgermane the germanium analogue was produced.

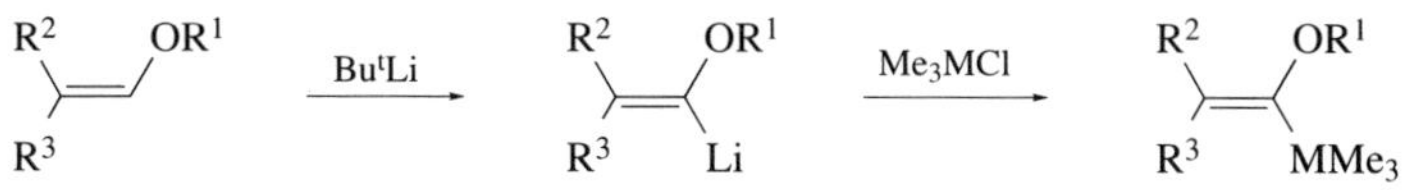

Scheme 36

The reaction is stereospecific, with the metalloid group in (**320**) replacing the hydrogen atom in (**319**) with retention of configuration.

The reaction has also been used with other chlorosilanes to give (**321**) in 65% yield ⟨92CB607⟩ and (**322**) in 50% yield ⟨84H(22)987⟩. Compound (**322**) reacts with alkenes and alkynes in the presence of platinum catalysts to give the more substituted silanes (**323**) and (**324**) in 73% and 95% yields, respectively ⟨84H(22)987⟩.

Table 38 Deprotonation and metallation of vinyl ethers.

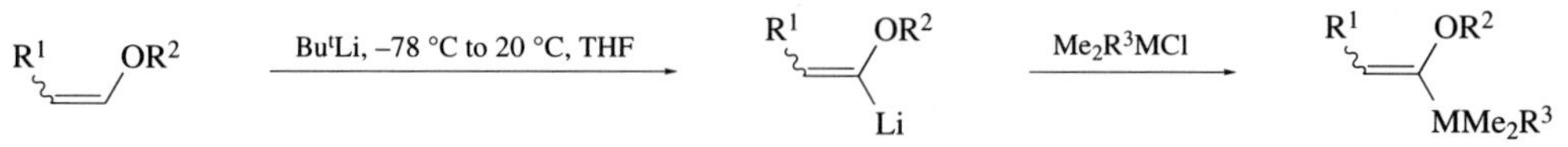

R^1	R^2	M	R^3	Yield (%)	Ref.
H	Me	Si	Me	~90	90OS(68)25
H	Me	Si	Me	~100 (with the enol ether in excess of BuLi)	82OM830
trans-vinyl	Me	Si	Me	59	80JA1577
trans-vinyl	Me	Ge	Me	53	80JA1577
trans-vinyl	Me	Si	Me	92	80JA1577
trans-vinyl	Me	Ge	Me	56	80JA1577
cis-Ph	Me	Si	Me	80	80JA1577
cis-Ph	Me	Ge	Me	60	80JA1577
H	Et	Si	H	44	76JOM(107)229
H	Et	Si	Bu^t	75	88T4113
trans-vinyl	Me	Si	H	77	80JOC541
trans-vinyl	Me	Ge	H	66	80JOC541
H	Me	Si	Ph_3Si	49	79JA83
H	Et	Si	Me	67	85JOC5410

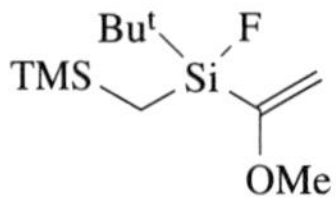

(321)

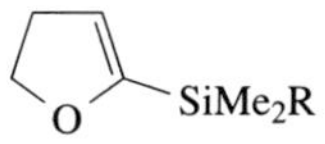

(322) R = H
(323) R = CH_2CH_2Ph
(324) R = $-C(=CH_2)Ph$

(ii) From acylsilanes

A number of methods based on the enolisation of acylsilanes has been described. Thus, silylated nucleophiles (**325**) react 1,4 with α,β-unsaturated acylsilanes (**326**) (Equation (31)) to give high yields of the enol ethers (**327**) with greater than 90% (*E*) stereochemistry (Table 39) ⟨87TL4093⟩. Intramolecular cyclisations have also been reported (Equation (32) and Table 40) ⟨92JOC7010⟩.

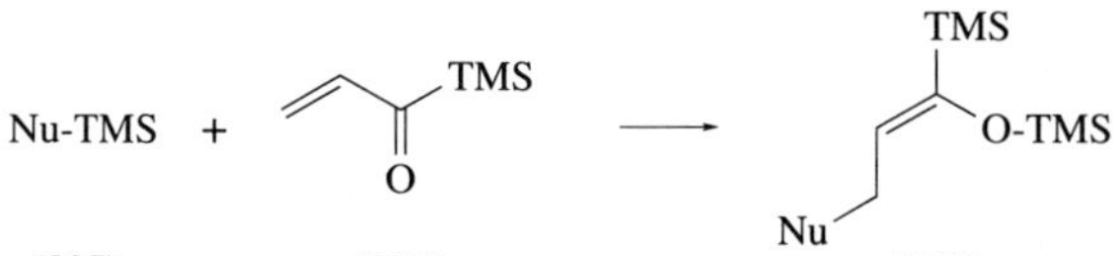

$$\text{Nu-TMS} + \text{(326)} \longrightarrow \text{(327)} \qquad (31)$$

(325) (326) (327)

Table 39 Addition of silylated nucleophiles to α,β-unsaturated acylsilanes.

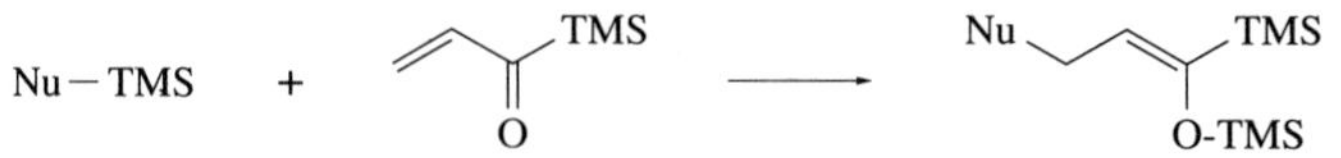

Nu	Yield (%)	(E):(Z) ratio
Et_2N	98	95:5
TMSNH	20	?
1-Imidazolyl	90	95:5
PhS	98	94:6
MeS	81	95:5
TMSS	95	91:9
Br	95	95:5

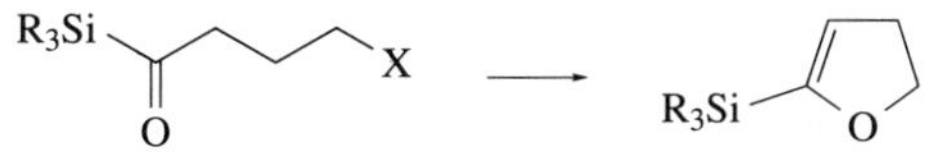

$$R_3Si-C(=O)-CH_2CH_2CH_2-X \longrightarrow R_3Si\text{-furanyl} \quad (32)$$

Table 40 Cyclization of acylsilanes.

$$R_3Si-C(=O)-(CH_2)_n-X \longrightarrow R_3Si\text{-oxacycle}-(CH_2)_{n-1}$$

R_3	n	X	Yield (%)	Conditions
Ph_2Me	3	Cl	97	N-methylpyrrolidine, 100 °C, 48 h
Me_3	3	Cl	71	N-methylpyrrolidine, 90 °C, 20 h
Bu^tMe_2	3	Cl	85	N-methylpyrrolidine, 90 °C, 14 h
Ph_2Me	2	Cl	72	N-methylpyrrolidine, 66 °C, 30 h
Ph_2Me	4	Br	trace	N-methylpyrrolidine, 100 °C, 40 h

The β-oxoacylsilanes (**328**) are readily prepared by acylation of acylsilanes and exist in the keto–enol tautomeric form shown (Equation (33)) ⟨88T4113⟩.

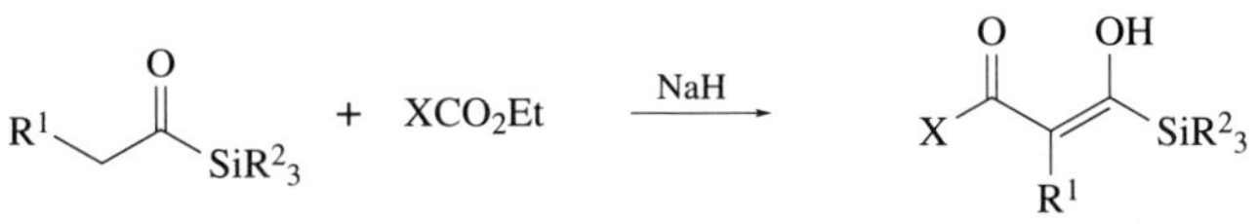

$$R^1-CH_2-C(=O)-SiR^2{}_3 + XCO_2Et \xrightarrow{NaH} X-C(=O)-C(R^1)=C(OH)-SiR^2{}_3 \quad (33)$$

(**328**)

$R^1 = X = H$, $R^2 = Bu^tMe_2$, 35–48%
$R^1 = H$, $X = OEt$, $R^2 = Bu^tMe_2$, 40%
$R^1 = X = H$, $R^2 = Bu^tPh_2$, 68–81%
$R^1 = Me$, $X = H$, $R^2 = Bu^tMe_2$, 4–20%

The cyclopropylacylsilanes (**329**) ring expand to (**330**) on treatment with Lewis acids, such as stannic chloride, titanium tetrachloride or borontrifluoride etherate (Equation (34)) ⟨86CL181⟩.

$$\text{cyclopropyl-C(=O)-TMS} \longrightarrow \text{dihydrofuran-TMS} \quad (34)$$

(**329**) (**330**)

$R^1 = R^2 = Me$	$SnCl_4$, –10 °C	74%
$R^1 = R^2 = Me$	$BF_3\cdot Et_2O$, 0 °C	71%
$R^1 = H$, $R^2 = Ph$	$TiCl_4$, –70 °C	56%
$R^1 = H$, $R^2 = Ph$	HCl, 25 °C	70%

(iii) Insertion of carbon monoxide

A most interesting reaction involving the insertion of carbon monoxide and a rearrangement has been reported by Murai and co-workers ⟨84JA2440⟩. The silylmethyllithium derivatives (**331**) in ether at room temperature were exposed to 1 atmosphere of carbon monoxide and the reaction mixtures quenched with chlorotrimethylsilane to give the enol ethers (**332**), possibly via the pathway shown in Scheme 37 (Table 41) ⟨84JA2440⟩. Interestingly, substituted silyllithiums give either selectively or predominantly the (E)-isomers (**332**). An alternative, and possibly more versatile process using carbon monoxide condenses the gas with alkenes and hydrosilanes in the presence of the iridium complexes $[IrCl(CO)_3]_n$ and $Ir_4(CO)_{12}$, (Equation (35) and Table 42) ⟨92JA9710⟩. In this case, the products are mixtures of (E)- and (Z)-isomers.

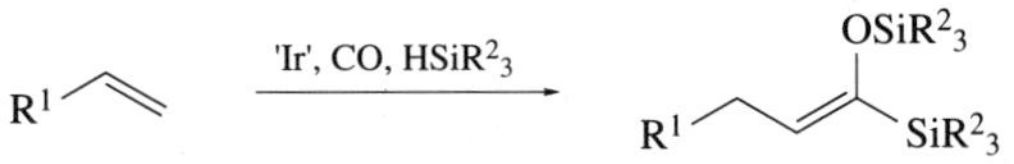

$$R^1-CH=CH_2 \xrightarrow{\text{'Ir', CO, } HSiR^2{}_3} R^1-CH_2-CH=C(OSiR^2{}_3)(SiR^2{}_3) \quad (35)$$

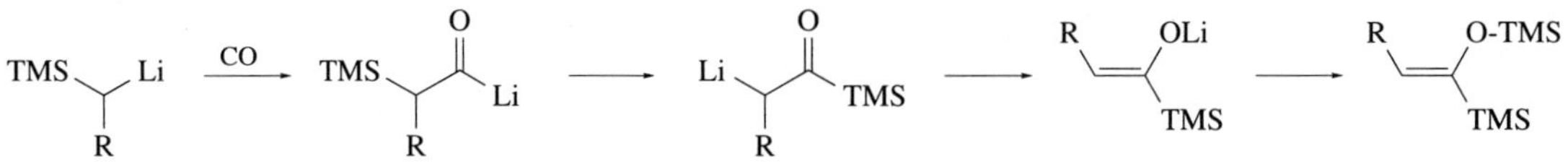

Scheme 37

Table 41 Carbonylation of silylalkyllithium derivatives.

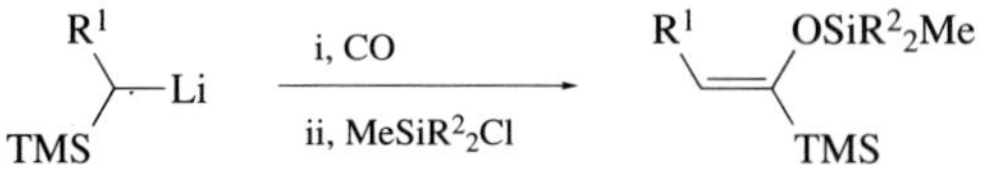

R^1	R^2	Yield (%)
H	Me	86
H	Et	80
Me	Me	88
Pr^n	Me	63
Bu^tCH_2	Me	96
Ph	Me	92

Table 42 Carbonylation: silylation of alkenes.

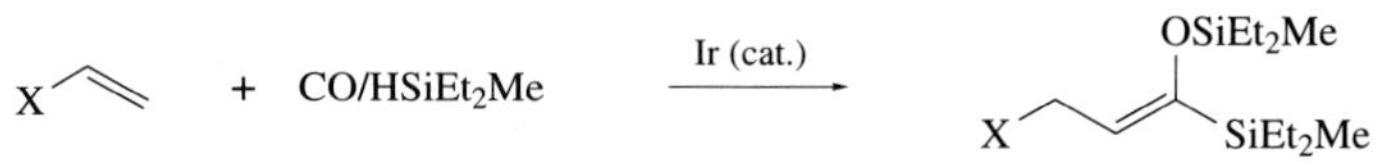

X	Yield (%)	$(E):(Z)$ ratio
Ph	50	72:36
BuO	67	57:43
Me_3Si	73	73:27
$BuOCH_2$	75	68:32
$TMS-CH_2$	53	79:21
$TMS-OCH_2$	67	67:33
$(EtO)_2CH$	58	73:27
$NCCH_2$	45	73:27
CH_2CH_2	56	65:35

(iv) Oxidation of lithium 1,1-bis(trimethylsilyl)alkoxides

The lithium 1,1-bis(trimethylsilyl)alkoxides (**333**) were oxidised by benzophenone in refluxing *n*-hexane during 14 h to give the related enol ethers (Equation (36)) ⟨77JA4181⟩.

$R = Bu^n, 64\%$
$R = C_6H_{13}, 67\%$
$R = Bn, 74\%$

(36)

(v) From carboxylic acid derivatives

Kuwajima *et al.* showed that thiophenyl esters (**334**) are converted into trimethylsilyl enol ethers (**335**) by treatment with LDA followed by quenching with chlorotrimethylsilane. The compounds

(335) were then reduced by sodium in benzene in the presence of more chlorotrimethylsilane to give (336) in yields in excess of 80% (Scheme 38 and Table 43) ⟨78CC478⟩.

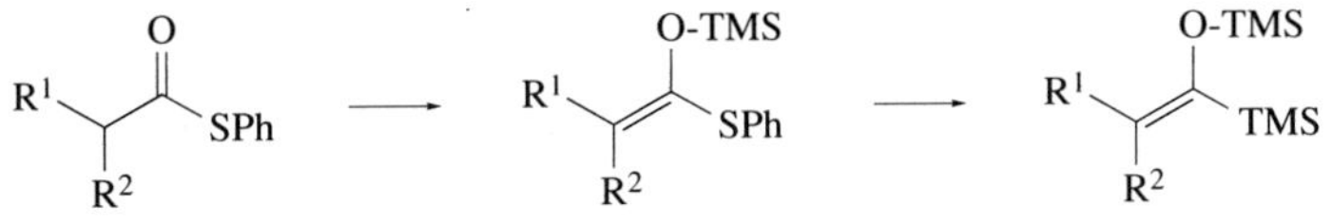

Scheme 38

Table 43 Preparation of the enol ethers (336) from thiophenyl esters (Scheme 38).

R^1	R^2	Yield (%)
Bu^n	H	95
$n\text{-}C_6H_{13}$	H	96
$n\text{-}C_8H_{17}$	H	96
$CH_2{=}CH(CH_2)_7$	H	95
$PhCH_2$	H	92
$(CH_2)_5$		92

Acylimidazoles (337) react with magnesium and chlorotrimethylsilane in hexamethylphosphamide to give directly the enol ethers (338) in yields of 30–35% ⟨74JOM(80)C25⟩.

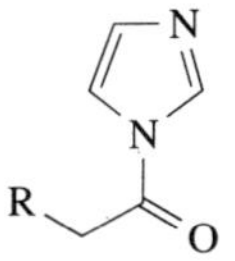

(337)

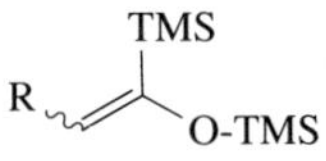

(338)
R = H, Et, 30–35%

Cinnamoyl, methylacryloyl and crotonoyl chlorides react under similar conditions to give mixtures from which the (E) and (Z) products (339) can be isolated ⟨72JOM(43)157⟩.

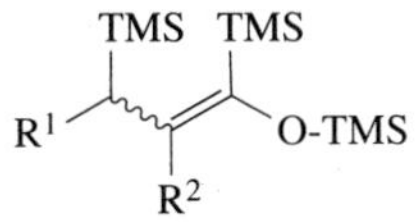

(339)
R^1 = Ph, R^2 = H, 55%
R^1 = Me, R^2 = H, 40%
R^1 = H, R^2 = Me, 75%

(vi) Addition to alkynes

Diethyl silane and germane add to 1-phenoxyprop-1-yne to give mixtures of the isomers (340) and (341), although no experimental details were given (Equation (37)) ⟨81IZV654⟩.

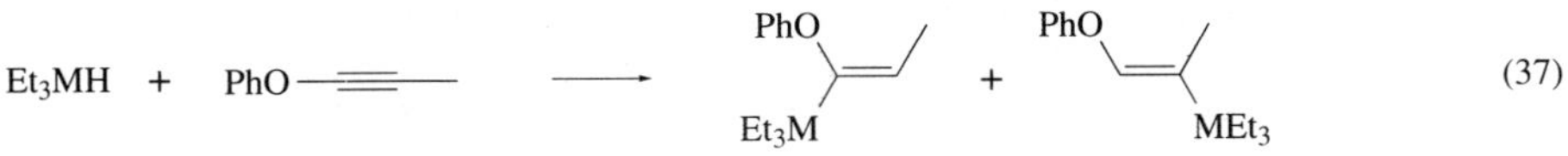

Et_3MH + PhO—≡— → (340) + (341)   (37)

M = Si, Ge   (340)   (341)

(vii) Rearrangement of tungsten complexes

An interesting reaction and rearrangement of tungsten complexes has been reported (Equation (38)) ⟨91OM2121⟩.

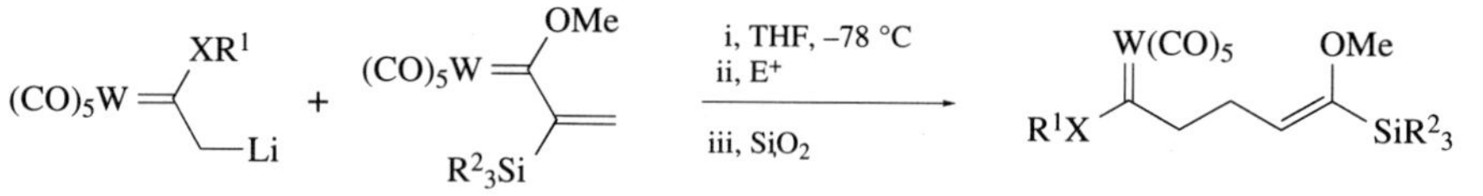

$$R^1X = OMe, R^2 = Me, 53\%$$
$$R^1X = NMe_2, R^2 = Me, 89\%$$
$$R^1X = OMe, R^2 = Ph, 23\%$$
$$R^1X = NMe_2, R^2 = Ph, 56\%$$

(38)

(viii) By a Perkow reaction

Trimethyl phosphite reacted with 1-(t-butyldimethylsilyl)-2-chloro-1-ethanone (**342**) to give the enol phosphate (**343**), although no experimental details were given ⟨89JOC2798⟩.

(**342**) 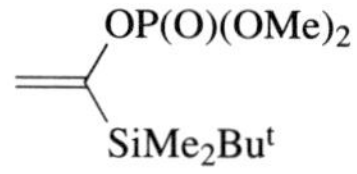   (**343**)

4.20.3.1.2 Oxygen and boron

Only two reports of the preparation of stable, neutral systems with this functionality have been reported. Matteson and Beedle treated 1-methoxyvinyllithium (see Section 4.20.4.1.1) with triisopropyl borate at −78 °C, allowed the mixture to warm to room temperature and carefully neutralised it with acetic acid ⟨90HAC135⟩. Distillation gave the air-sensitive diisopropyl (1-methoxyvinyl)boronate (**344**) in 49% yield, and further reaction with pinanediol produced (S)-pinanediol (1-methoxyvinyl)boronate (**345**) (Scheme 39).

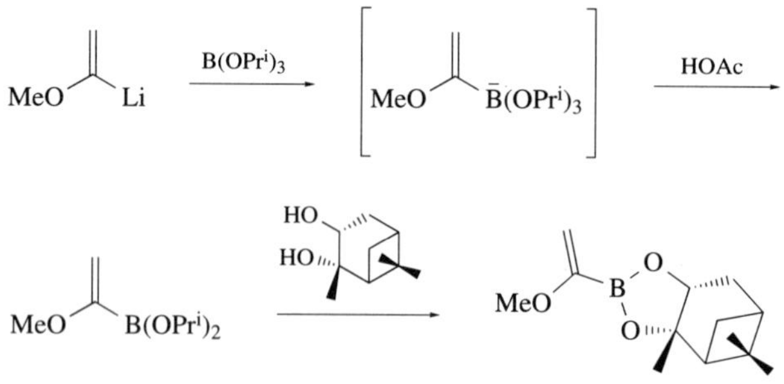

Scheme 39

Birkinshaw and Kocienski prepared the carbamate (**346**) in 47% yield from the lithium compound (**347**) and triethylborane followed by protonation with acetic acid at 25 °C. The borane (**346**) rather unexpectedly turned out to be stable and purifiable by chromatography ⟨91TL6961⟩. This stability was unexpected since Levy and Schwartz had previously shown that the reaction of 1-methoxyvinyllithium with trialkylboranes (e.g., tri-n-hexylborane) at −78 °C gives, after oxidation of the products with hydrogen peroxide, ketones (Scheme 40) ⟨76TL2201⟩.

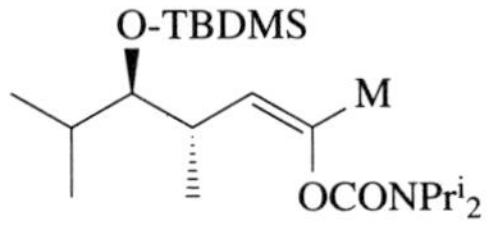

(346) M = BEt$_2$
(347) M = Li

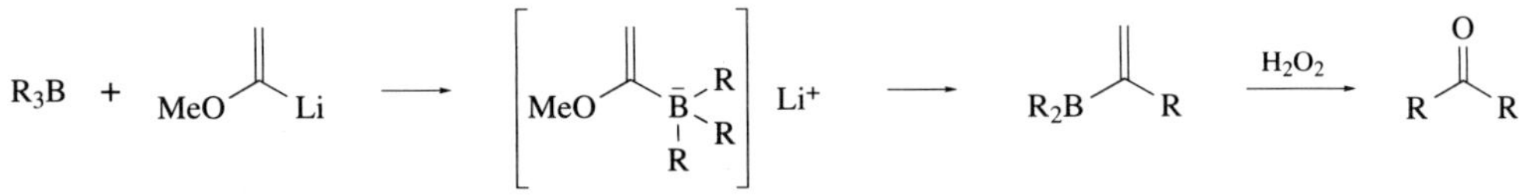

Scheme 40

If acid is added to the medium before the peroxide, the product is the tertiary carbinol (349). Apart from the structures of the reaction products, no evidence for the existence of the 'ate' complexes (348) was given. However, direct evidence was provided by the low temperature ^{11}B NMR spectrum of (348; R = Bu^i), which showed a sharp absorption at $\delta + 16.3$ relative to boron trifluoride etherate, a value compatible with other lithium-boron 'ate' complexes of defined structures ⟨78JOM(156)123⟩. This type of reaction was further examined by Suzuki, who treated 1,2-dimethoxy-1-lithioethene (350) with trialkylboranes to give aldehydes, 1-methoxy ketones and ketones via the complex (351) ⟨80CL591, 81CL1059, 83CL933, 83SC1149⟩.

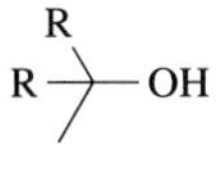

(349) 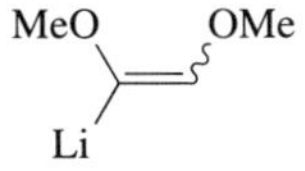   (350) 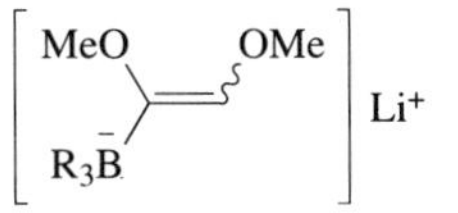   (351)

Therefore, the existence of these complexes and some useful chemistry was established, but it appeared that they were of very limited stability. However, Soderquist and Rivera were able to prepare very pure 1-methoxyvinyllithium by lithium/tin exchange from the pure tin compound (352) ⟨89TL3919⟩. This reacted with triisobutylborane to give an 'ate' complex (353) that was stable in solution at room temperature for at least 2 days. Addition of chlorotrimethylsilane initiated the 1,2 shift to give good (>70%) yields of the vinylboranes (354).

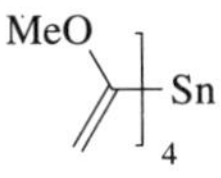

(352) 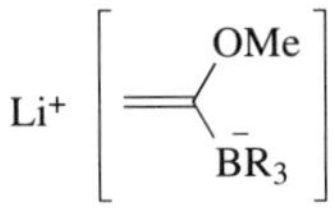   (353) 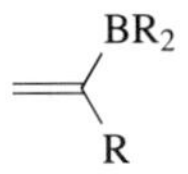   (354)
R = Pr^i, Bu^n, Bu^i, Bu^s

Following this lead, Kocienski treated 5-lithio-2,3-dihydrofuran with a number of trialkylboranes ⟨91TL6961⟩. The immediate 'ate' complexes (355) were stable in refluxing THF for 24 h. Treatment with electrophiles, for example aqueous acetic acid, induced the 1,2-alkyl shift to give (356; R = Et) in 69% overall yield. As noted earlier, these authors also prepared (357) but observed that treatment with acetic acid at 25 °C caused loss of one ethyl group, rather than the 1,2 shift, to give the product (346). This finding does not, in 1995, appear to have been followed up, but it is expected that many analogues of (346) could be prepared and would be isolable.

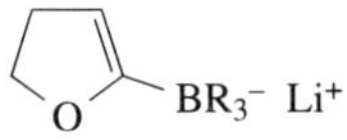

(355)   (356)   (357)

R = Et, n-C$_6$H$_{13}$, c-C$_6$H$_{12}$, Bu^i

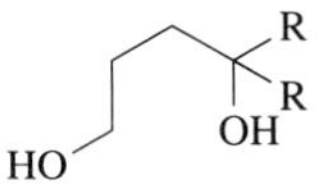

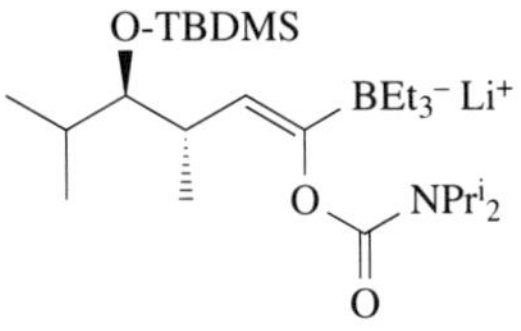

Another stable system is that in the oxaborole betaines (**359**)–(**361**) formed as shown in Equation (39) ⟨87CB213⟩. An x-ray crystallographic study of (**361**) confirms the close B—O bond distance and it is suggested that the stability of the oxaborole ring is responsible for the loss of one of the alkyl groups intermolecularly rather than by the intramolecular 1,2 shift seen with other systems.

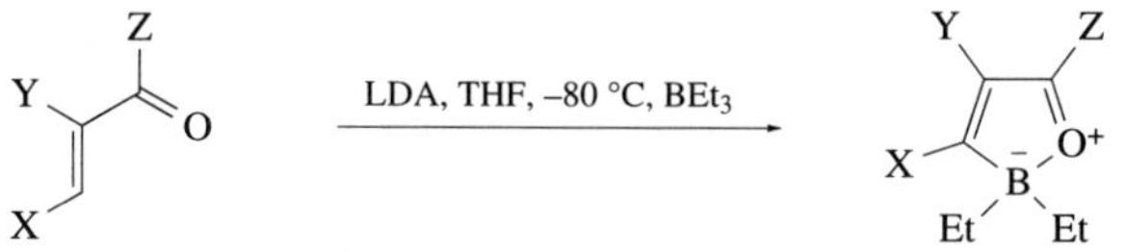

$$\text{(358)} \xrightarrow{\text{LDA, THF, } -80\ °C,\ BEt_3} \text{(359)-(361)} \qquad (39)$$

(**358**)

(**359**) X = OMe, Y = H, Z = OMe, 48%
(**360**) X = SEt, Y = H, Z = OMe, 75%
(**361**) X = OMe, Y = OMe, Z = NHCH(Me)Ph, 63%

4.20.3.2 Functions Bearing Sulfur, Selenium or Tellurium

4.20.3.2.1 *Sulfur and silicon or germanium*

A survey of the literature reveals more references to these systems than were found for the analogous oxygen compounds described in Section 4.20.3.1.1: a fact which presumably reflects the greater synthetic utility of the 1-thio 1-silyl alkenes. Therefore, the system has been used for the synthesis of ketones (Scheme 41) ⟨83TL95⟩, for the stereoselective conjugate addition of alkyllithiums to sulfones (Equation (40)) ⟨93JOC1596⟩, which is exploited in the total synthesis of maytansinoids ⟨82JA4997⟩, and in cyclisations to give cyclopentanones (Equation (41)) ⟨82OM1243⟩.

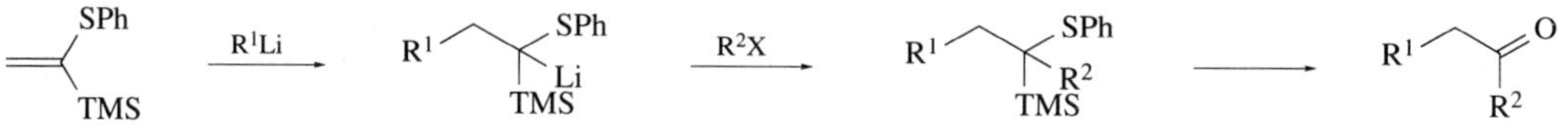

Scheme 41

$$\text{(40)}$$

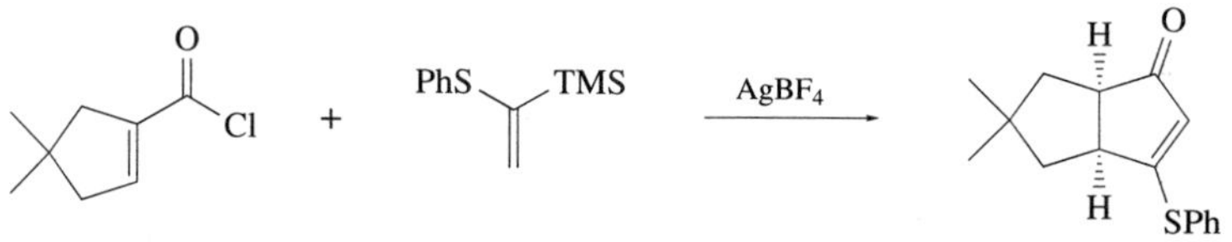

$$\text{(41)}$$

Many of the reactions described herein can be performed with alkyl or aryl sulfides, sulfoxides or sulfones; in other cases, the higher oxidation states can be better produced by oxidation of the initially formed sulfides by mcpba to give sulfoxides ⟨77CC522, 83TL4387, 88BCJ3957, 89CC1872⟩ or by sodium metaperiodate to give sulfones ⟨78TL3383, 90CPB902⟩.

(i) Condensation by a Peterson reaction

The alkylthio or arylthio bis(trimethylsilyl)methyllithium derivatives (**362**) react with aldehydes and ketones to give mixtures of the (*E*)- and (*Z*)-1-silyl 1-alkylthio or 1-arylthio alkenes (**363**) (Equation (42), Table 44) ⟨85AG(E)696⟩.

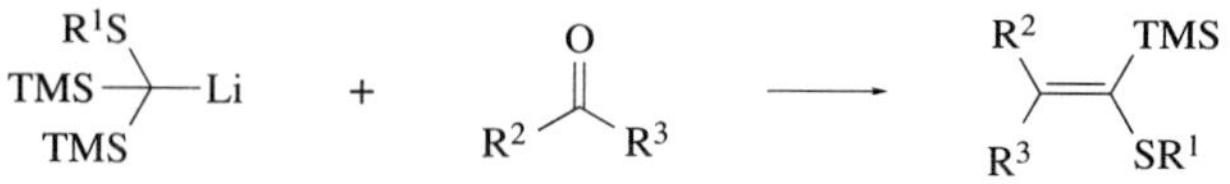

$$\text{(362)} + R^2COR^3 \longrightarrow \text{(363)} \qquad (42)$$

Table 44 The Peterson reaction (Equation (42)).

R^1	R^2	R^3	Temperature (°C)[a]	(Z):(E) ratio	Yield (%)	Ref.
Ph	H	(MeO, OMe, OBn pyranose)	−45	2:1[b] 3:1	40[b] 72	81CL457 84JA3252
Ph	Ph	H	−78		88[b]	80CL331
Ph	H	RO(CH$_2$)$_3$	−78	1:1	54	84JA8209
Ph	H	(dihydropyran)	−78	57:43	72	84JOC3517 81TL239
Ph	H	(2,6-dimethyl-OEt tetrahydropyran)	−78		60[b]	81TL4287
Ph	H	C$_8$H$_{17}$	0	2:1	?	93T2011
Ph	H	PhCH$_2$CH$_2$	0	3:2	?	93T2011
Ph	H	c-C$_6$H$_{11}$	0	3:2	?	93T2011
Ph	H	(spiroketal)	−78	10:1	> 56	87T4759
Ph	H	(dioxolane)	−78	?	80	86T2863
Ph	H	(RO dihydropyran OPr^i)	−78	?	> 61	86T2863
Ph	H	H	−78	?	84	77CB852
Ph	H	n-C$_5$H$_{11}$	−78	?	70	77CB852
Ph	H	CHEt$_2$	−78	?	80	77CB852
Ph	H	(cyclohexenyl)	−78	?	85	77CB852
Ph	H	Ph	−78	?	76	77CB852
Ph	H	(methylenedioxyphenyl)	−78	?	74	77CB852
Me	H	H	−78	?	79	77CB852
Me	H	Me	−78	?	53	77CB852
Me	H	n-C$_6$H$_{13}$	−78	?	71	77CB852
Ph	H	CF$_3$	−78	28:72	68	85JFC357 86BSF937
CH$_2$-TMS	H	Ph	?	?	84	86CL2089
Ph	H	(Ph(CH$_2$)$_2$CH(O-MEM))	?	4:1	?	79TL3465

[a] Reaction carried out in THF at various temperatures. [b] Yield of the sulfone produced by oxidation with mcpba.

The reaction has also been performed with the bislithio compound (**364**), which reacts with benzaldehyde to give the silyl sulfone (**365**) as a by-product in only 4% yield ⟨85AG(E)696⟩. However, this arose from quenching with D_2O; quenching with chlorotrimethylsilane may, therefore, give (**365**) in good yield.

(**364**) 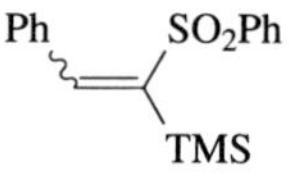   (**365**)

(ii) Anion reactions

The α-protons of arylthioalkenes (aryl vinyl sulfides) (**366**) can be readily removed by *t*-butyllithium at −78 °C in THF, and the resultant anions silylated by chlorotrimethylsilane to give (**367**) (Table 45) ⟨83JCS(P1)1131⟩ (see also Section 4.20.4.2). The technique also works with higher oxidation states, for example for the sulfone (**368**) (Table 46) ⟨89JOC1757⟩ and the sulfoximide (**369**) ⟨93JCS(P1)343⟩.

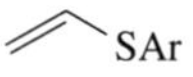

(**366**) 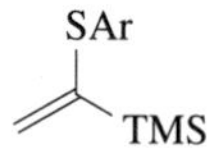   (**367**) 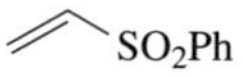   (**368**) 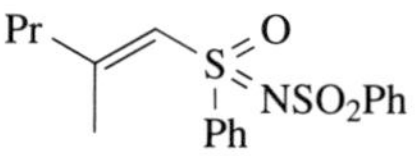   (**369**)

Table 45 Silylation of 1-lithio-1-alkylthioalkenes.

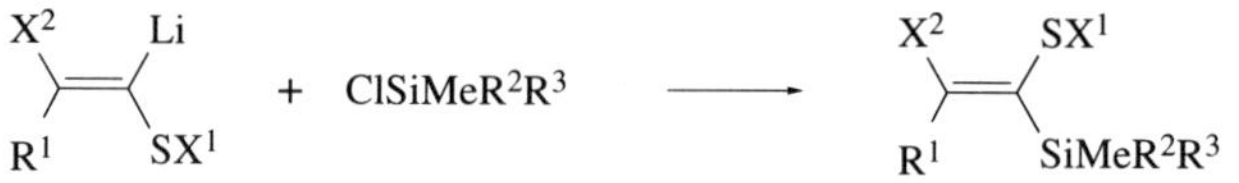

X^1	X^2	R^1	R^2	R^3	Conditions (base, °C)	Yield (%)	Ref.
Ph	H	H	Me	Me	LDA, −78	97	77CC522, 80JOC1046
Ph	H	H	Me	Me	Bu^nLi, TMEDA, −90	82	81TL587, 83JCS(P1)1131
Ph	H	H	Ph	Ph	LDA, −78	77	88BCJ3957
Ph	H	H	Bu^t	Me	LDA, −78	61	88BCJ3957
$CONMe_2$	Me	Me	Me	Me	Bu^tLi, −78	76	80AG(E)303
Bu^t	CHMeOMe	H	Me	Me	Bu^sLi, −78	97	83TL4825
Ph	H	Me_2CHCH_2	Me	Me	Bu^nLi, −30	84	84BCJ1863
Bu^t	Me_2COH	H	Me	Me	Bu^nLi, −78	55	85BSF1250

Alternatively, 3-phenylthio-3-(trimethylsilyl)-1-propene (**370**) can be deprotonated by Bu^sLi in HMPA–THF at −78 °C to give, after equilibration, the allylic anion (**371**), which readily reacts with a variety of carbonyl compounds and epoxides to give predominately the products of γ-alkylation, for example (**372**), in yields of 50–78% (Scheme 42) ⟨81JOC5182, 83JA619, 83JOC34, 83JOC383, 91JOC717⟩.

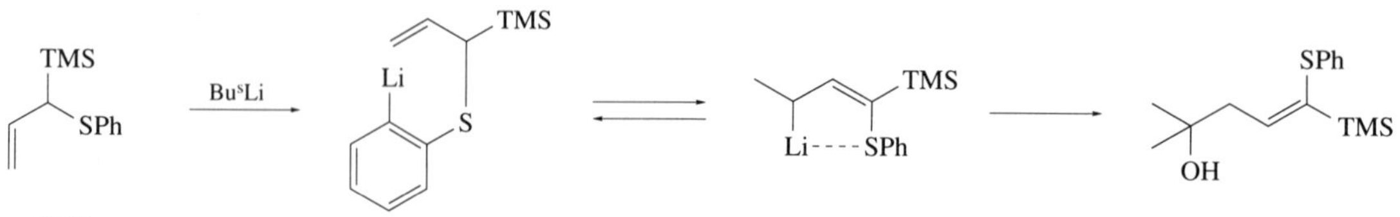

(**370**)   (**371**)   (**372**)

Scheme 42

Table 46 Silylation of 1-lithio-1-arenesulfonyl alkenes.

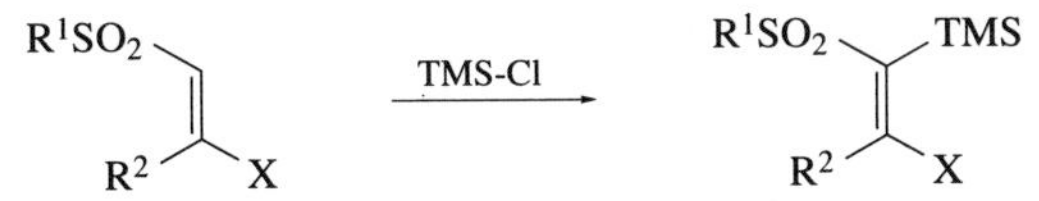

R^1	R^2	X	Conditions	Yield (%)	Ref.
p-Tol	H	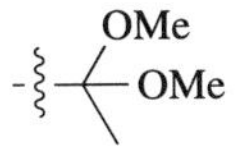 -C(OMe)₂Me	MeLi, LiBr, −20 °C	85	87TL6709, 88JOC4708
Ph	H	OMe	Bu^{n}Li, THF	84	87TL989
Ph	SMe	SMe	Bu^{n}Li, THF, −25 °C	53	89JOC1757
Ph	H	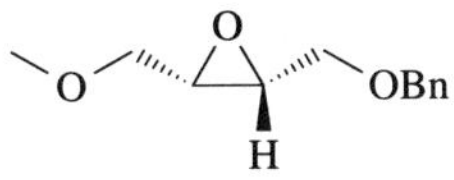 (epoxide-CH₂OBn)	LDA, THF, −78 °C	53	91JOC3556
Ph	H	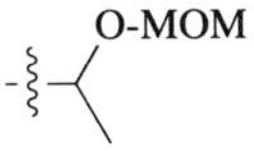 -CMe₂(O-MOM)	Bu^{n}Li, THF, −78 °C	76	91TL1385, 91TL5159, 93JOC1596
Ph	H	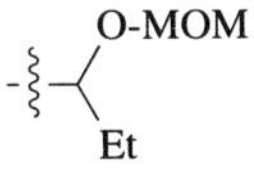 -C(Et)(Me)(O-MOM)	Bu^{n}Li, THF, −78 °C	84	93JOC1596
Ph	H	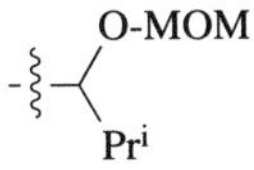 -C(Pr^i)(Me)(O-MOM)	Bu^{n}Li, THF, −78 °C	62	93JOC1596

This reaction has been used, for example, to construct regio- and stereoselectively side chains on pregnenolone (**373**), i.e., (**374**) ⟨83JA619⟩.

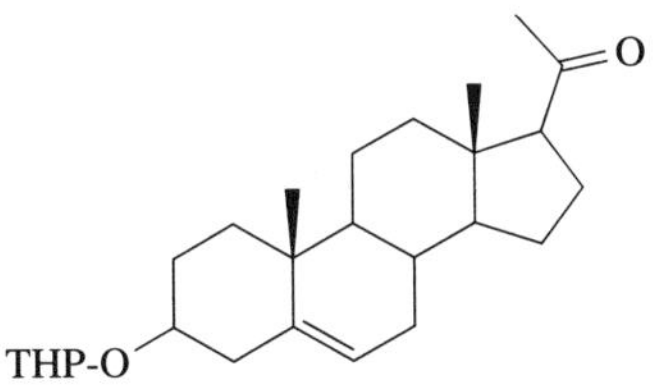

(**373**) 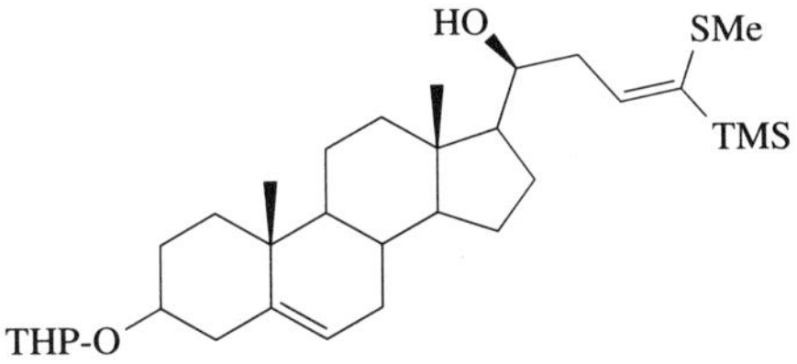   (**374**)

Simpkins has shown that the acetal (**375**) eliminates methanol and is lithiated to give (**376**) in a single step on treatment with two equivalents of *n*-butyllithium at −78 °C ⟨87TL989⟩. The anion (**376**) can react with electrophiles including chlorotrimethylsilane to give (**377**) as a single isomer in 84% yield.

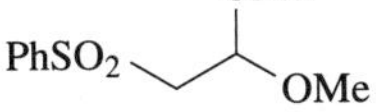

(**375**) 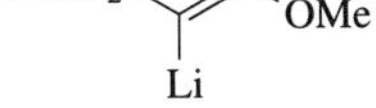   (**376**)    (**377**)

Another reaction is the anionic, 1,4 O→C silyl migration of (**378**) induced by *t*-butyllithium at −78 °C to give (**379**) as a pure (*Z*)-isomer in 70% yield (Scheme 43) ⟨90TL6137⟩.

(iii) Addition to alkynes

Russian workers have investigated the addition of thiols and thiol derivatives to variously substituted alkynes. In general, mixtures of products are formed from which the required 1,1-disubstituted alkenes can be separated. Thus 2-aminobenzenethiol reacts with the silylethyne (**380**)

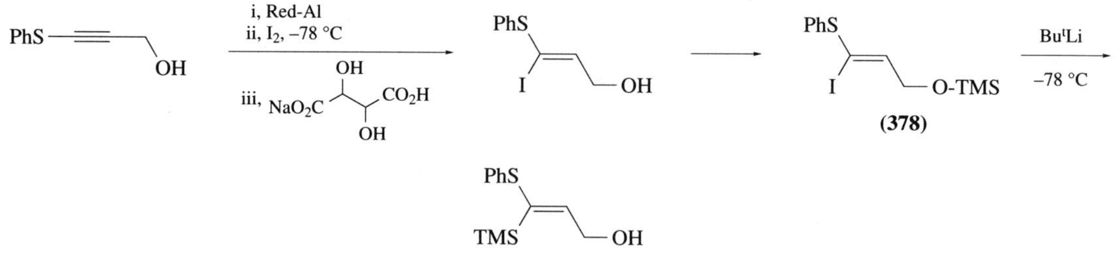

(378) **(379)**

Scheme 43

in ether to give 80% of (**381**) ⟨91ZOB1893⟩. Representative reactions of sulfenyl halides are given in Table 47 and those of thiols in Table 48.

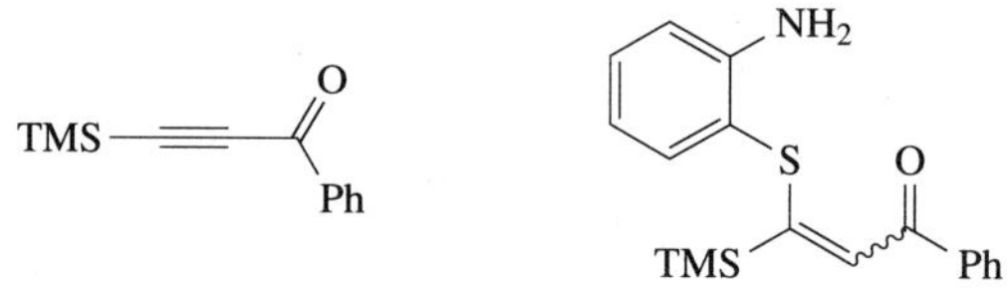

(380) **(381)**

Ogawa, Sonoda and co-workers have studied the palladium-catalysed reactions of alkynes with disulfides (Equation (43)) and thiols (Equation (44)) ⟨91JA9796, 92JA5902⟩. Thiophenol also adds spontaneously to certain chromium-substituted alkynes (Equation (45)) ⟨92CB2051⟩.

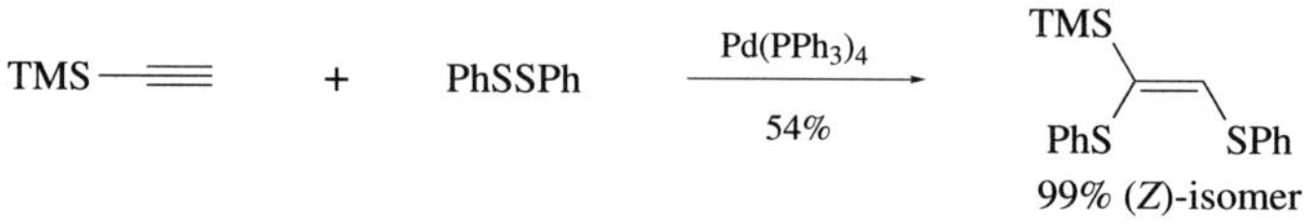

(43)

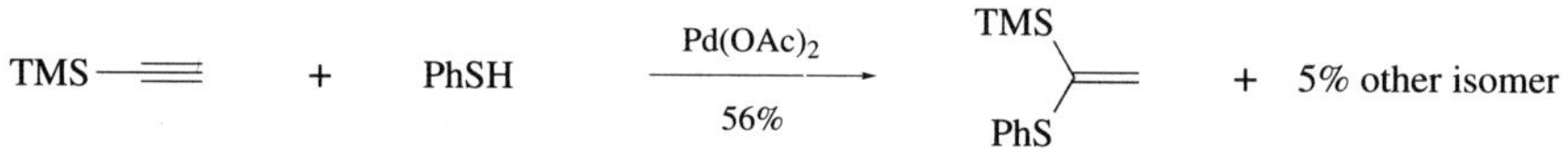

(44)

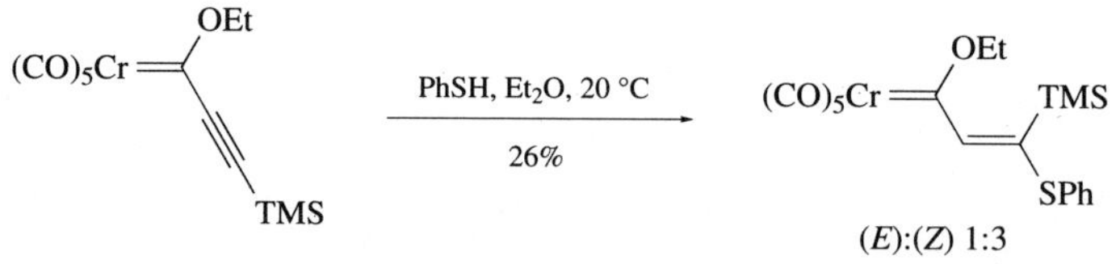

(45)

1-Ethynylsilanes react with organocopper reagents to give relatively stable vinylcopper(I) compounds (**382**), which react with, *inter alia*, *S*-methyl methanethiosulfonate to give the pure (*Z*)-isomers (**383**) ⟨84JOM(276)317⟩.

(382) **(383)**

R = Pr^i, Bu^t, > 90%

Hydrosilation of phenylthioalkynes catalysed by platinum(IV) has been described by Isobe and co-workers (Scheme 44) ⟨88TL4773, 90SL701, 91T3727⟩. In these examples the initial arylthio product was oxidised to the sulfone (**384**) before isolation. In contrast, triethylgermane reacted with phenylthioethyne in the presence of the same catalyst to give a mixture from which 4% of the 1-phenylthio-1-triethylgermylethene (**385**) was isolated ⟨77IZV596⟩.

Table 47 Addition of sulfenyl halides to alkynes (Part 1).

$$Me_3M-\!\!\!\equiv\!\!\!-R^1 + R^2SX \longrightarrow \underset{R^2S}{\overset{Me_3M}{}}\!\underset{X}{\overset{R^1}{}}$$

M	R^1	R^2	X	Condition	Yield (%)	Ref.
Si	CH=CH$_2$	Me	Cl	CH$_2$Cl$_2$, −10 °C	41	71ZOB719
Si	CH=CH$_2$	Me	Cl	CH$_2$Cl$_2$, −10 °C	52	72ZOB870
Si	CH=CH$_2$	Et	Cl	CH$_2$Cl$_2$, −10 °C	53	72ZOB870
Si	CH=CH$_2$	Ph	Cl	CH$_2$Cl$_2$, −10 °C	35	72ZOB870
Si	CH=CH$_2$	Me	Br	CH$_2$Cl$_2$, −10 °C	37	72ZOB870
Si	CH=CH$_2$	Et	Br	CH$_2$Cl$_2$, −10 °C	36	72ZOB870
Ge	CH=CH$_2$	Me	Cl	CH$_2$Cl$_2$, −10 °C	43	72ZOB870
Ge	CH=CH$_2$	Et	Cl	CH$_2$Cl$_2$, −10 °C	45	72ZOB870
Ge	CH=CH$_2$	Me	Br	CH$_2$Cl$_2$, −10 °C	43	72ZOB870
Si	CH=CH$_2$	Ph	Cl	CH$_2$Cl$_2$, −10 °C	54	79ZOB805
Si	CH=CH$_2$	p-Tol	Cl	CH$_2$Cl$_2$, −10 °C	52	79ZOB805
Si	CH=CH$_2$	p-NO$_2$C$_6$H$_4$	Cl	CH$_2$Cl$_2$, −10 °C	32	79ZOB805
Ge	CH=CH$_2$	Ph	Cl	CH$_2$Cl$_2$, −10 °C	69	80ZOB722
Ge	CH=CH$_2$	p-Tol	Cl	HOAc	50	80ZOB722
Ge	CH=CH$_2$	p-NO$_2$C$_6$H$_4$	Cl	HOAc	62	80ZOB722
Si	CH=CHCN	Me	Cl	CH$_2$Cl$_2$, −10 °C	35	79ZOB812
Si	CH=CHCOMe	Me	Cl	CH$_2$Cl$_2$, −10 °C	42	79ZOB812
Si	CH=CHCO$_2$Et	Me	Cl	CH$_2$Cl$_2$, −10 °C	40	79ZOB812
Si	CH=CMeCO$_2$Et	Me	Cl	CH$_2$Cl$_2$, −10 °C	37	79ZOB812
Si	CH=CHNO$_2$	Me	Cl[a]	CH$_2$Cl$_2$, −20 °C	?	80ZOB1748
Si	CH=CHNO$_2$	Et	Cl[a]	CH$_2$Cl$_2$, −20 °C	?	80ZOB1748
Si	CH=CHNO$_2$	p-Tol	Cl[a]	CH$_2$Cl$_2$, −20 °C	?	80ZOB1748

[a] Products are exclusively the result of *trans*-addition.

Table 47 Addition of sulfenyl halides to alkynes (Part 2).

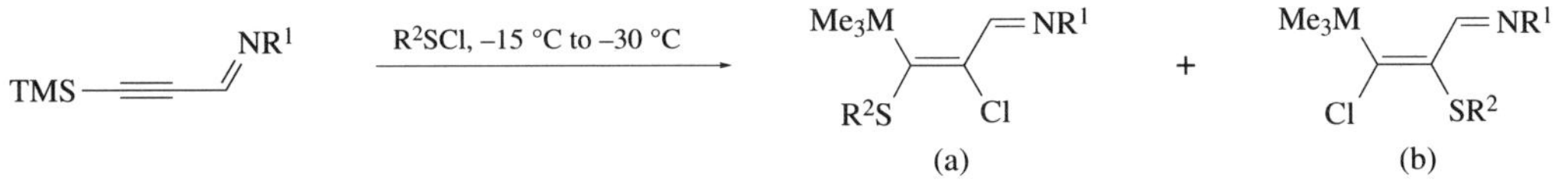

$$TMS-\!\!\!\equiv\!\!\!-\!\!/\!\!=\!\!NR^1 \xrightarrow{R^2SCl, -15\,°C \text{ to } -30\,°C} \underset{R^2S\quad Cl}{\overset{Me_3M\quad =\!\!NR^1}{\diagdown\!\!\!\diagup}} + \underset{Cl\quad SR^2}{\overset{Me_3M\quad =\!\!NR^1}{\diagdown\!\!\!\diagup}}$$

(a)   (b)

R^1	R^2	Yield (%)	(a):(b)	Ref.
Pr^i	Me	42	70:30	84ZOB309
Bu^t	Me	61	85:15	84ZOB309
Bu^t	Ph	61	65:35	84ZOB309
Bu^t	p-NO$_2$C$_6$H$_4$	16	100:0	84ZOB309
2,6-Me$_2$C$_6$H$_3$	Me	64	100:0	84ZOB309

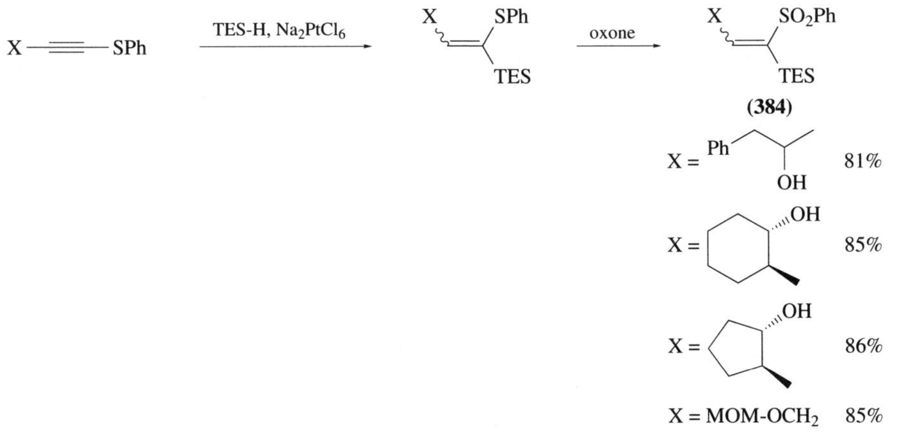

$$X-\!\!\!\equiv\!\!\!-SPh \xrightarrow{TES-H, Na_2PtCl_6} \underset{TES}{\overset{X\quad SPh}{\diagdown\!\!\!\diagup}} \xrightarrow{oxone} \underset{TES}{\overset{X\quad SO_2Ph}{\diagdown\!\!\!\diagup}}$$

(384)

X = PhCH(OH)CH−   81%

X = *trans*-2-methylcyclohexan-1-ol   85%

X = *trans*-2-methylcyclopentan-1-ol   86%

X = MOM-OCH$_2$ 85%

Scheme 44

Table 48 Addition of thiols to silylated alkynes.

$$TMS\!\!=\!\!=\!\!X \;+\; RSH \longrightarrow \begin{array}{c} TMS \quad X \\ \diagdown\!\!=\!\!\diagup \\ RS \end{array}$$

X	R	Conditions	Yield (%)	(Z):(E) ratio	Ref.
CH=NBu^t	Ph	PhH, RT	93	80:20	84ZOB116
CH=NPr^i	Ph	PhH, RT	96	80:20	84ZOB116
CH=NBn	Ph	PhH, RT	68	70:30	84ZOB116
CH=NPh	Ph	PhH, RT	39	20:80	84ZOB116
CH=N-o-Tol	Ph	PhH, RT	58	65:35	84ZOB116
CH=N-p-Tol	Ph	PhH, RT	43	25:75	84ZOB116
CH=N-p-ClC$_6$H$_4$	Ph	PhH, RT	41	50:50	84ZOB116
CH=N(2,6-Me$_2$C$_6$H$_3$)	Ph	PhH, RT	26	100:0	84ZOB116
CH=N(2,4,6-Me$_3$C$_6$H$_2$)	Ph	PhH, RT	54	75:25	84ZOB116
COCl	Bu^a	hexane, −25 °C	44		83ZOB1292
COPh	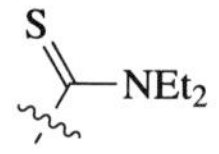 C(=S)NEt$_2$	CS$_2$, MeCN, −20 °C	48		90ZOB526
COPh	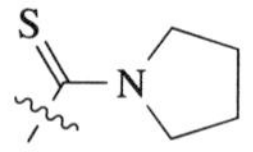 C(=S)N(pyrrolidine)	CS$_2$, MeCN, −20 °C	55		90ZOB526
CO$_2$Bu^n	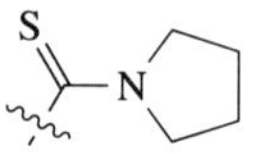 C(=S)N(pyrrolidine)	CS$_2$, MeCN, −20 °C	65		90ZOB526
COSBu^n	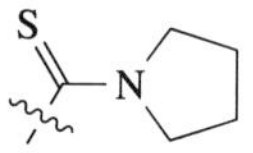 C(=S)N(pyrrolidine)	CS$_2$, MeCN, −20 °C	85		90ZOB526
COPh	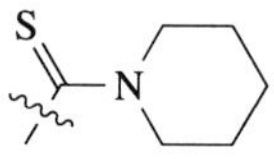 C(=S)N(piperidine)	CS$_2$, MeCN, −20 °C	48		90ZOB526

^{a} BuSNa was used

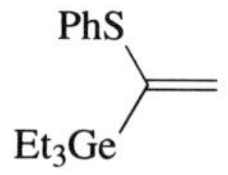

$$\begin{array}{c} PhS \\ \diagdown\!\!=\!\!\diagup \\ Et_3Ge \end{array}$$
(385)

(iv) Elimination reactions

Mandai *et al.* have described an alkylation–elimination sequence which is capable of producing (Z)-1-phenylthio-1-trialkylsilyl alkenes (**386**) with a high degree of stereoselectivity (Scheme 45 and Table 49) ⟨90T4553⟩. The authors present evidence that the dominance of the (Z)-isomer is the result of postreaction, radical, (E) to (Z) isomerisation in the presence of oxygen.

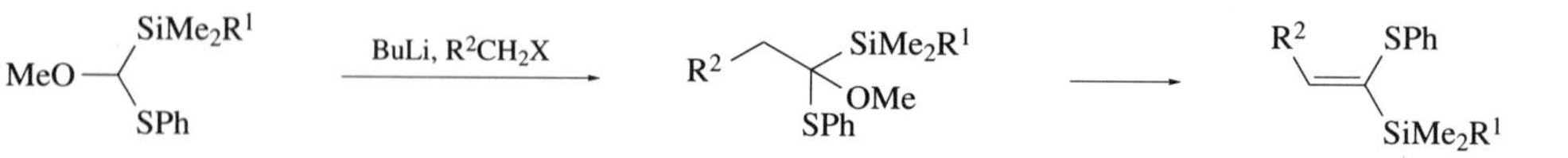

$$\underset{SPh}{\overset{SiMe_2R^1}{MeO\diagdown\!\!\diagup}} \xrightarrow{BuLi,\, R^2CH_2X} \underset{SPh}{\overset{SiMe_2R^1}{R^2\diagdown\!\!\diagup OMe}} \longrightarrow \underset{SiMe_2R^1}{\overset{R^2 \quad SPh}{\diagdown\!\!=\!\!\diagup}}$$
(386)

Scheme 45

Arenesulfonyl chlorides readily add to trimethylvinylsilane in dichloromethane at −78 °C to give an adduct which loses hydrogen chloride on treatment with either dbu or 1,5-diazabicyclo-[4.3.0]non-5-ene (dbn) (Scheme 46) ⟨80JOC1046, 85JOC1621⟩.

Table 49 Elimination reactions (Scheme 45).

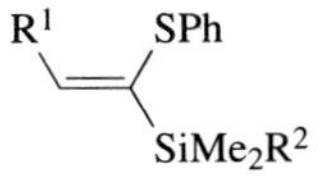

$$R^1\text{–CH=C(SPh)(SiMe}_2R^2\text{)}$$
(386)

R^1	R^2	Solvent	Temperature	Time (h)	Et_3B (equiv.)	Yield (%)	(Z):(E)
H	But	CH_2Cl_2	reflux	0.5	none	85	
H	Ph	CH_2Cl_2	reflux	10	none	82	
H	Me	toluene	reflux	0.5	none	90	
C_5H_{11}	But	toluene	reflux	6	none	74	96:4
C_5H_{11}	But	benzene	RT	24	1.2	62	98:2
C_5H_{11}	But	THF	reflux	3	0.5	72	97:3
C_5H_{11}	Me	toluene	reflux	6	none	84	94:6
C_5H_{11}	Me	benzene	RT	24	1.2	38	97:3
C_5H_{11}	Me	THF	reflux	5	0.5	77	95:5
C_5H_{11}	Me	THF	RT	22	0.5	80	95:5
C_5H_{11}	Ph	xylene	RT	24	1.2	49	96:4
C_5H_{11}	Ph	THF	reflux	3	0.5	75	94:6
Ph	Me	THF	reflux	3	0.5	72	92:8
$Bu^tPh_2SiO(CH_2)_3$	Me	THF	reflux	3	0.5	89	94:6
$Bu^tMe_2SiO(CH_2)_{10}$	Me	THF	reflux	3	0.5	78	95:5
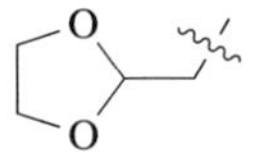 1,3-dioxolan-2-yl	Me	THF	reflux	3	0.5	95	95:5
$Bu^tMe_2SiO(CH_2)_{10}$	But	THF	reflux	3	0.5	89	97:3
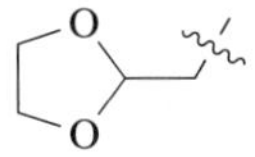 1,3-dioxolan-2-yl	But	THF	reflux	3	0.5	84	92:8
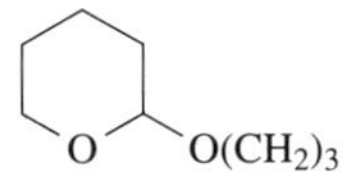 tetrahydropyran-2-yl-O(CH$_2$)$_3$	Me	THF	reflux	3	0.5	84	93:7

ArSCl + CH$_2$=CH–TMS $\xrightarrow{CH_2Cl_2,\ -78\ °C}$ ArS–CH$_2$–CH(TMS)Cl $\xrightarrow{\text{dbu or dbn}}$ ArS–CH=CH–TMS

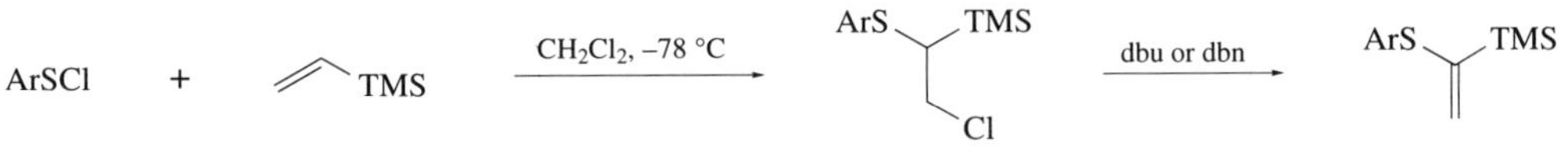

Ar = 2,4-(NO$_2$)$_2$C$_6$H$_3$, 81%; Ph, 92%; 4-ClC$_6$H$_4$, 80%

dbu = 1,5-diazabicyclo[5.4.0]undec-5-ene; dbn = 1,5-diazabicyclo[4.3.0]non-5-ene

Scheme 46

Cyclohexene oxide reacts with trimethylsilyl(phenylthio)methyllithium to give *trans*-2-trimethylsilyl(phenylthio)methylcyclohexanol (**387**), which ring opens to (**388**) on treatment with *N*-chlorosuccinimide and triethylamine ⟨87JOC1256⟩. A similar chlorination–elimination sequence was reported for the compounds (**389**), which with NCS and dbu gave the dienes (**390**) as single, but undefined, stereoisomers ⟨87JCS(P1)589⟩.

Remarkably, the simplest and, as claimed by the authors, most stereoselective procedure is one of the most recent ⟨90SC267⟩. 1-Phenylthio-1-silyl alkanes and NCS in tetrachloromethane at room temperature give good yields of pure (*Z*)-isomers (Table 50). Thermolysis of the highly substituted ethane (**391**) at 70 °C, or in chlorobenzene at 125 °C, gives the two ethenes (**392**) and (**393**) in yields of 68% and 64%, respectively ⟨77CB2880⟩.

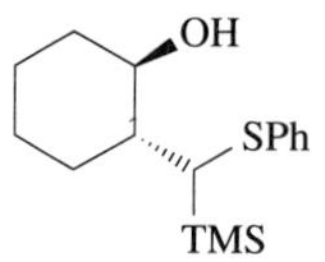

(387)

(388) 80% (E):(Z) 80:20

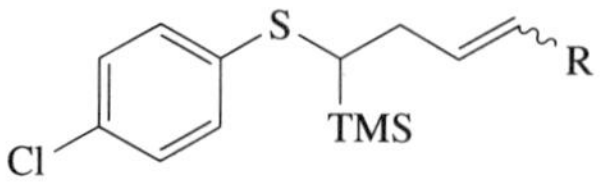

(389)

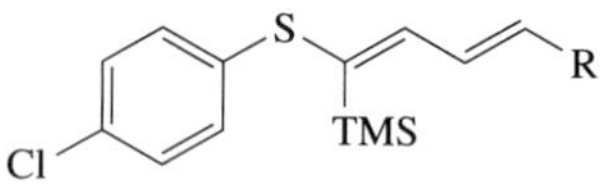

(390)
R = Et, 55%
R = C_5H_{11}, 52%

Table 50 Halogenation–elimination of 1-phenylthio-1-silyl-alkanes.

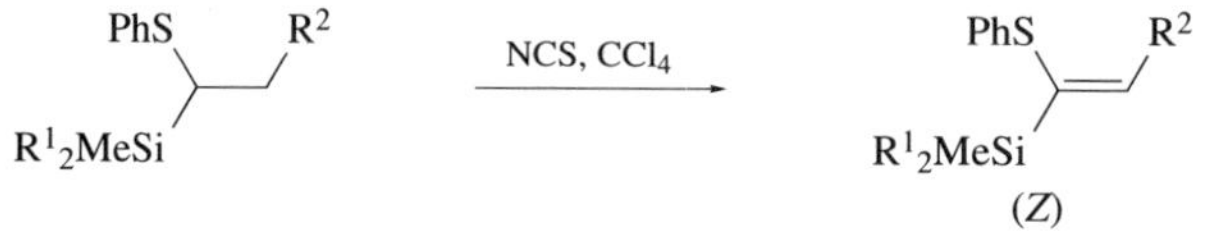

R^1	R^2	Reaction time (h)	Yield (%)
Me	H	5	94
Me	Me	6	85
Me	Ph	8	91
Me	CH_2CH_2	5	65
Me	CH=CHPh	12	69
Ph	H	3	88
Ph	$CH=CH_2$	8	84

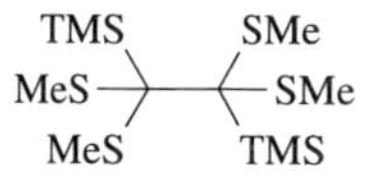

(391) 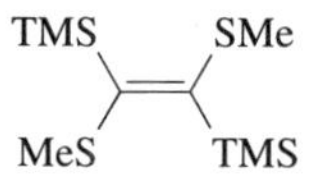   **(392)** 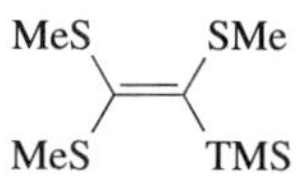   **(393)**

(v) From epoxides

The triphenylsilyloxirane (**394**) reacts at −78 °C in THF with s-butyllithium to give the anion (**395**), which condenses with butylthiodimethylaluminum to give 52% of (**396**) ⟨91TL2783⟩.

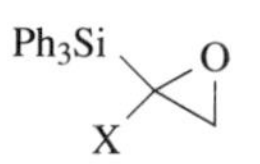

(**394**) X = H
(**395**) X = Li

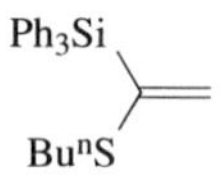

(**396**)

(vi) From acylsilanes

The silyl vinyl ketone (**397**) reacts with two equivalents of trimethylphenylthiosilane (**398**) at 30 °C in the presence of boron trifluoride etherate to give the (Z)-isomers (**399**) ⟨92SL499, 92SL883⟩.

Methyl trimethylsilyl ketone was treated at −30 °C in dry ether with hydrogen chloride and hydrogen sulfide, the solution washed under CO_2 with 5% sodium hydrogen carbonate and the product chromatographed to give both the disulfide (**400**) and the hemidithioacetal (**401**) ⟨89JCS(P1)2083⟩.

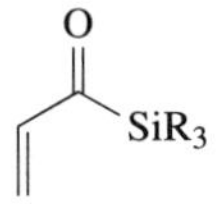

(397) (398) (399) R = Me, 74%; R = Ph, 70%

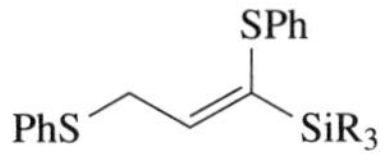

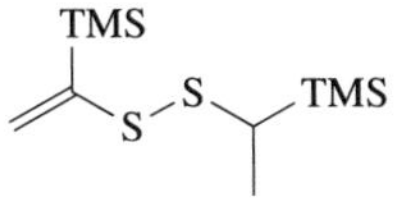

(400) 47% 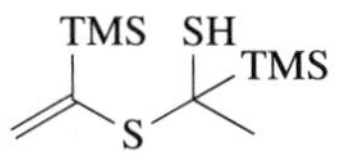   (401) 16%

(vii) From sulfoxides

Miller and Hassig have shown that the alkyl phenyl sulfoxides (**402**) react with 3 mol LDA in THF at −10 °C and chlorotrimethylsilane to give the 1-trimethylsilylvinyl sulfides (**403**) (Equation (46) and Table 51) ⟨84TL5351⟩.

$$\text{(402)} \xrightarrow{\text{LDA, TMS-Cl}} \text{(403)} \tag{46}$$

Table 51 Condensation of sulfoxides with chlorotrimethylsilane.

R^1	R^2	R^3	$(E):(Z)$ ratio	Yield (%)
Ph	Et	H		75
Ph	n-C$_4$H$_9$	Et	2:1	80
Ph	CH$_2$CH=CHMe	CH=CH$_2$	3:2	70
Ph	CH$_2$CH$_2$Ph	Ph	3:1	75
Bu^t	Et	H		75

(viii) By replacement of phenylthio groups

The cyclohexylidene derivative (**404**) reacts with lithium 1-dimethylaminonaphthalenide at −45 °C to give the lithium compound (**405**) which, on quenching with chlorotrimethylsilane, yields (**406**) in 95% yield ⟨84JA3245⟩. Analogously, Ager has shown that bis(phenylthio)bis(trimethylsilyl)methane (**407**) reacts with lithium naphthalenide to give the lithium compound (**408**), which condenses with aldehydes and nonenolisable ketones to give the products (**409**) ⟨84JOC168⟩.

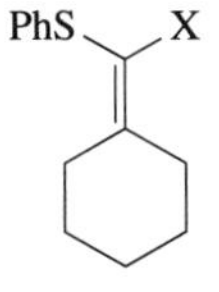

(**404**) X = SPh
(**405**) X = Li
(**406**) X = TMS

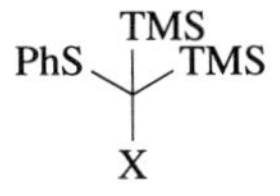

(**407**) X = SPh
(**408**) X = Li

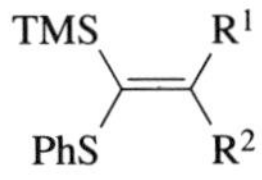

(**409**) R^1 = Ph, R^2 = H, 78%
$R^1 = R^2$ = Ph, 49%
$R^1 = R^2$ = H, 80%

(ix) From ketene derivatives

Tanaka and co-workers have examined the reaction of phenyl (1-trimethylsilyl)propadienyl sulfide (**410**) with organometallic compounds ⟨90BCJ51⟩. Lithium butyl(diisobutyl)aluminium hydride gives initially a complex (**411**), which can be trapped by ketones to give the products (**412**) along with by-products and the starting material (**410**). Likewise, the same authors showed that methyl-(methyldiphenylsilyl)magnesium or lithium (methyldiphenylsilyl)cuprate initially add the silyl group to give a single isomer (**413**), which can be either protonated by water or treated with benzaldehyde to give (**414**), 67% as a single isomer, or (**415**), 51% as a 1:1 mixture of isomers, respectively.

(**412**) $R^1 = H, R^2 = Ph$, 62%
$R^1 = H, R^2 = Pr^n$, 44%
$R^1 = H, R^2 = Bu^t$, 57%
$R^1 = Me, R^2 = Ph$, 37%
$R^1 = R^2 = (CH_2)_5$, 51%

(x) Displacement of bromine from (1-bromovinyl)trimethylsilane

Trialkylstannyl phenyl sulfides react with (1-bromovinyl)trimethylsilane (**416**) in refluxing benzene in the presence of 3 mol % tetra(triphenylphosphino)palladium to give an 80% yield of a 76:24 mixture of (**417**) and (**418**) ⟨89TL2699⟩.

(**416**) X = Br
(**417**) X = SPh

(xi) From sulfinates

(−)-Menthyl *p*-toluenesulfinate reacts with 1-trimethylsilylvinylmagnesium bromide to give a 78% yield of the enantiomer (**419**) ⟨90TL673⟩.

4.20.3.2.2 Selenium or tellurium together with silicon or germanium

The small number of synthetic studies devoted to these compounds can be grouped into two main classes: reactions of selenium- and tellurium-stabilised anions and addition to substituted alkynes.

(i) Reactions of selenium- and tellurium-stabilised anions

Aryl vinyl selenides (**420**) and aryl vinyl tellurides (**421**) can be deprotonated by treatment with lithium amides, for example LDA or lithium dicyclohexylamide, in THF–HMPA at −78 °C and the resultant anions condensed with chlorotrimethylsilane to give (**422**) and (**423**) ⟨81JOC2775, 83CB1001⟩.

(**420**) X = Se
(**421**) X = Te

(**422**) X = Se, Ar = Ph, 77%; X = Se, Ar = *m*-CF$_3$C$_6$H$_4$, 99%
(**423**) X = Te, Ar = Ph, 80%

(1-Bromovinyl)trimethylsilane (**424**) can be converted into the Grignard reagent (**425**), which condenses with benzeneselenenyl bromide to give (**422**; Ar = Ph, X = Se) in 54% yield ⟨91JCS(P1)1555, 92JOC5610⟩. Likewise, (**424**) can be converted into its lithium derivative, by Bu^{n}Li at −78 °C, and gives the cyanocuprate (**426**) on reaction with cuprous cyanide. Further treatment with selenocyanogen [Se(CN)$_2$] gives (**427**) in 64% yield ⟨88JOC3632⟩.

(**424**) X = Br
(**425**) X = MgBr
(**426**) X = CuCN
(**427**) X = SeCN

The deprotonation of β,γ-unsaturated 1-phenylseleno-1-trimethylsilyl compounds, for example (**428**), proceeds rapidly with lithium diethylamide at −78 °C in THF and the resultant allylic anion is regioselectively alkylated at the γ-position to give the products (**429**) ⟨82JOC1618⟩. These can be isolated but are more usually oxidised and eliminated to give the α,β-unsaturated acylsilanes (**430**) in yields of 60–75% overall.

(**428**)   (**429**)   (**430**)

(ii) Addition to alkynes

(a) Silylalkynes. Back and co-workers have reported that *Se*-phenyl *p*-tolueneselenosulfonate (**431**) undergoes a radical-induced thermal addition to alkynes in benzene at 80 °C during 96 h to give, in high yields, the alkene (**432**) in a highly regio- and stereoselective manner involving anti-Markovnikov *trans*-addition (Equation (47)) ⟨83JOC3077⟩.

$$\text{PhSeSO}_2\text{Tol} + \text{TMS}{-}{\equiv}{-} \xrightarrow[94\%]{\text{AIBN, 80 °C, 96 h}} \begin{array}{c}\text{TMS} \quad \text{SO}_2\text{Ph} \\ \diagdown / \\ \text{PhSe} \end{array} \quad (47)$$

(**431**)   (**432**)

Subsequently, Ogawa, Sonoda and co-workers showed that the photoinduced addition of diphenyl diselenide to trimethylsilylethyne in the absence of solvent at 40 °C for 24 h gives (**433**) in good yield (73%), but the product is a mixture of (*E*)- and (*Z*)-isomers ⟨91JOC5721⟩. However, if the reaction is conducted in the presence of tetrakis(triphenylphosphine)palladium, the sole product in 66% yield is the (*Z*)-isomer of (**433**) ⟨91JA9796⟩. The same group has shown that *n*-butyl *t*-butyl telluride reacts with alkynes by a free radical mechanism (AIBN initiation) to give, for example, (**434**) in 63% yield as a pure (Z)-isomer ⟨92JA7591⟩.

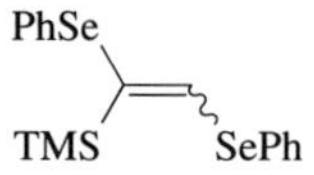

(**433**) 73%, (E):(Z) 28:72

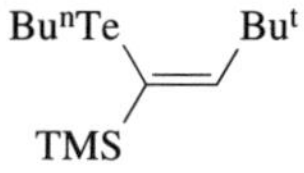

(**434**)

Russian workers have shown that areneselenenyl chlorides react with the triple bonds of (but-2-en-1-ynyl)trimethyl-silane and -germane (**435**) to give the adducts (**436**) ⟨79ZOB829, 80ZOB1053⟩.

(**435**) M = Si, Ge

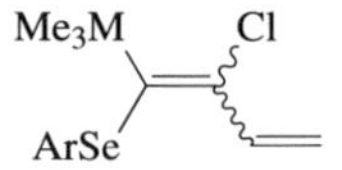

(**436**)
M = Si, Ar = Ph, 64%
M = Si, Ar = p-Tol, 60%
M = Ge, Ar = Ph, 65%

The propargylic alcohol (**437**) was reduced by sodium bis(2-methoxyethoxy)aluminum hydride and iodinated to give (**438**) in 92% yield ⟨90TL6137⟩. This was silylated to afford (**439**), which was treated with *t*-butyllithium to give an anion which underwent a stereoselective, 1,4-silyl shift to give (**440**).

(**437**)

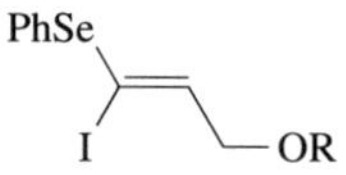

(**438**) R = H
(**439**) R = TMS

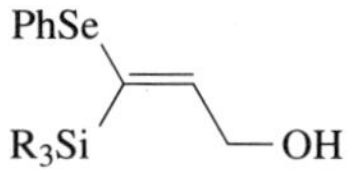

(**440**)
R = Me, 82%
R_3 = Bu^tMe_2, 76%
R = Pr^i, 50%

4.20.4 FUNCTIONS CONTAINING A CHALCOGEN AND A METAL

A wide range of compounds with metal atoms bonded to C-1 of vinyl ethers have been prepared and found to have the stabilities and reactivities expected from other organometallics. Thus, lithium, tin and copper compounds are the most important, but, as will be seen below, compounds with other metals have been prepared and their reactivities investigated.

4.20.4.1 Oxygen Functions

Compounds of general structure (**441**) are acylanion (**442**) equivalents and have been used in this manner in a number of syntheses including those of 4-hydroxyisopyrazoles ⟨76JOC2874⟩, lathrane diterpenes ⟨88JOC3647⟩, steroid side chains ⟨76JOC2312⟩ and leinamycin ⟨92TL5701⟩.

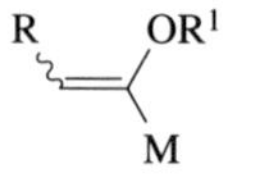

(**441**)

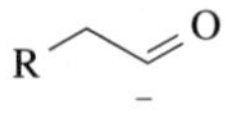

(**442**)

4.20.4.1.1 *Lithium*

The synthesis and use of 1-oxyalkenyllithiums were amongst topics reviewed by Gschwend and Rodriguez ⟨76OR(26)1⟩ and Krief ⟨80T2531⟩.

Baldwin *et al.* were the first to report the preparation of (**441**; R = H, R^1 = Me, M = Li) by treatment of a solution of methyl vinyl ether (**441**; R = M = H, R^1 = Me) in dry tetrahydrofuran at

−65 °C dropwise with *t*-butyllithium in THF ⟨74JA7125⟩. A yellow precipitate of a 2:1 complex of *t*-butyllithium and THF forms and then redissolves if the temperature is allowed to rise to −5 °C to 0 °C, when the solution becomes colourless. Conversion into (**441**; R = H, R^1 = Me, M = Li) is essentially quantitative and the anion is generally used after recooling the solution to −65 °C. Subsequently, a number of authors have used essentially the same conditions ⟨76TL2201, 78JOM(156)123, 80CJC130⟩, but the most detailed examination of optimal reaction conditions was made by Soderquist and Hsu ⟨82OM830⟩. They found that the original *t*-butyllithium–THF system is indeed the best, but if extremely pure (**441**; R = H, R^1 = Me, M = Li) is needed, this is best produced by first transmetallation of the lithium alkyl to give the tin compound (**352**) by treatment with tin tetrachloride. This is a stable crystalline material that can be purified by recrystallisation from ethanol. The pure material is dissolved in dry pentane and treated from 0 °C to room temperature with *n*-butyllithium in pentane. The organolithium derivative (**441**; R = H, R^1 = Me, M = Li) settles out, the supernatant liquor can be removed and the solid washed to give pure, if extremely pyrophoric, 1-methoxyvinyllithium.

Table 52 lists other enol ethers that have been metallated by analogous reactions.

The structure of the lithium compounds is generally inferred from the structure of the products of their reactions with electrophiles. However, more direct structural information has been provided by observations of their ^{13}C NMR spectra ⟨80JOC4959, 82JOC3094, 84AG(E)366⟩. In the case of enol ethers more highly substituted than ethenyl, there are potential problems of regio- and stereoselectivity. Oakes and Sebastian could find no evidence for the production of allylic anions from dihydrofurans and dihydropyrans, reflecting an anticipated relative instability of an allylic anion with a terminal oxygen atom ⟨80JOC4959⟩. Further, Knorr and von Roman investigated the isomerisation of 1-ethoxy-1-propenyllithium (**443**) by ^{1}H NMR spectroscopy (Equation (48)) ⟨84AG(E)366⟩. The ratio of a:b remains constant over several days at 25 °C and, therefore, it is concluded that these anions are stereochemically stable.

$$\text{(443a)} \quad \rightleftharpoons \quad \text{(443b)} \tag{48}$$

Warren and co-workers have shown that the geometry of the initial vinyl ether plays a major role in determining whether allyl or vinyl anions are formed ⟨79JCS(P1)3099⟩. Thus the (*E*)-isomer (**444**) tends to give the vinyl anion whilst the (*Z*)-isomer produces a chelated allylic anion (**445**). Similar isomerisations have been discussed by Schlosser ⟨74AG(E)701⟩.

(**444**)   (**445**)

A different isomerisation occurs with the acrylic acid derivative (**446**), which gives initially the β-lithiated compound (**447**) on treatment with LDA at −90 °C ⟨79TL4273⟩. However, at −90 °C during 5 h, an equilibrium with the α-anion (**448**) is established (Equation (49)).

$$\text{(446) X = H} \quad \rightleftharpoons \quad \text{(448)} \tag{49}$$
$$\text{(447) X = Li}$$
$$\text{57\%} \qquad\qquad \text{43\%}$$

A more potent, basic reagent is formed by mixing equimolar amounts of *t*-butyllithium, potassium *t*-butoxide and TMEDA in hexane at −30 °C ⟨87JOM(332)99⟩. This can deprotonate ethyl vinyl ether, 2,3-dihydrofuran and 2,3-dihydropyran within 10 min at −20 to −50 °C to give the potassium derivatives in essentially quantitative yields.

It is also possible to generate the lithium compounds, for example (**449**), by treatment of the corresponding bromo compound (**450**) with *n*-butyllithium at −78 °C ⟨80CL591⟩.

An entirely different synthesis is possible from the fluorinated ethers (**451**). Thus, treatment with LDA in THF at −78 °C gives directly the fluorinated 1-lithiovinyl ethers (**452**) ⟨77CL1379, 85TL2861, 90TL3931, 92SL483⟩.

Table 52 Production of 1-lithio-1-alkoxyalkanes by deprotonation of enol ethers.

Enol ether	Conditions	Ref.
CH$_2$=CH–OEt	Bu^{t}Li, THF, –78 °C to 0 °C	84TL2021, 85JCS(P1)1201, 87LA311, 91T6539
CH$_2$=CH–OEt	Bu^{t}Li, THF, 0 °C Bu^{t}Li, TMEDA, pentane, –30 °C	81JA5259 72LA(763)208
CH$_2$=CH–O–C(O)NR$_2$	Bu^{s}Li, THF, TMEDA, –78 °C	90JOC5680
(E)-CH$_2$=CH–CH=CH–OMe	Bu^{t}Li, THF, –78 °C to –20 °C	88T3139
(Z)-CH$_2$=CH–CH=CH–OMe	Bu^{t}Li, THF, –78 °C to –20 °C	80JA1577
3,4-dihydro-2H-pyran	Bu^{n}Li, THF, TMEDA, RT	72HCA594, 80JOC4959
3-chloro-3,4-dihydro-2H-pyran	Bu^{n}Li, hexane, THF	77CR(C)281
2,3-dihydrofuran	Bu^{n}Li, hexane, THF	77CR(C)281
2,3,4,5-tetrahydrooxepine	Bu^{t}Li, THF^a, –65 °C to –5 °C Bu^{t}Li, pentane, THF, –78 °C	77TL4187 82JOC3094
Ph–CH(Me)–O–CH=C(Me)–CO$_2$Et	LDA, THF, –100 °C	92SL429
Ph–CH(Me)–O–CH=CH–CO$_2$Et	LDA, THF, –100 °C	92TL8035
(Z)-CH$_3$CH=CH–OEt	Bu^{t}Li, THF, –78 °C	84AG(E)366
(Z)-CH$_3$CH=CH–OPh	Bu^{s}Li, THF, –78 °C	74S888
(MeO)(Cl)C=C(Cl)(OMe)	Bu^{n}Li, THF, –100 °C	68JOC1991
tri-O-TBDMS glycal	Bu^{t}Li, THF, –78 °C to 0 °C	86TL6201
3,3-dimethyl-1,5-dioxaspiro[5.5] enol ether	Bu^{t}Li, THF	87JA1269
MeO–CH=CH–CO$_2$Me	LDA, THF, –80 °C	87CB213

Table 52 (continued).

Enol ether	Conditions	Ref.
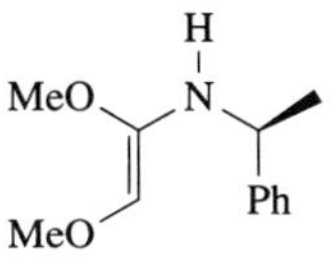 MeO, N(H)(CHMePh), C=C(OMe) (ketene N,O,O-acetal with α-methylbenzylamine)	LDA, THF, −80 °C	87CB213
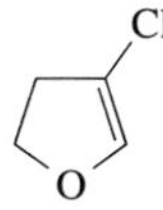 3-chloro-2,5-dihydrofuran	'butyllithium', THF, −78 °C	73HCA2166

a The authors stress the need to keep the amount of THF to a minimum.

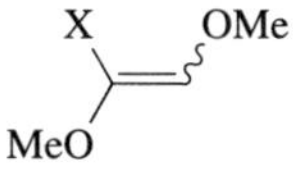

(449) X = Li
(450) X = Br

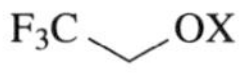

(451) X = CH$_2$CH=CHPh, MEM, CONEt$_2$, Ph

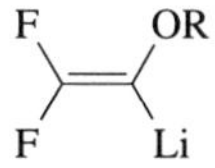

(452)

4.20.4.1.2 Tin

Four main routes to 1-alkoxyvinylstannanes have been described with metal exchange with the analogous lithium compounds being the most common. Thus, the sequence alkene to alkenyllithium to alkenylstannane shown in Scheme 47 has been established (Table 53) ⟨80JA1577, 82OM830, 83TL2361, 85JOC3255⟩.

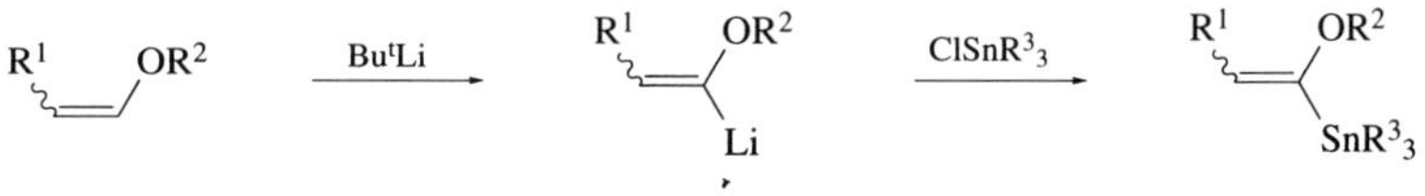

Scheme 47

In contrast to the lithium compounds, the stannanes are stable, isolable and purifiable materials ((453) is even available commercially). Consequently, they have found important synthetic roles, for example for coupling with vinyl halides in the presence of palladium(0) ⟨89JOC5828⟩.

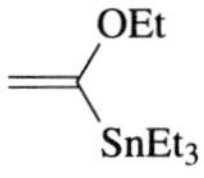

(453)

The other routes to these compounds are less versatile and have been less widely used. The acylstannane (454) readily reacts conjugatively with nucleophiles and the resulting anions can be trapped on oxygen. Thus, phenylthiotrimethylsilane reacts at room temperature to give (455) in virtually quantitative yield ⟨91TL1899⟩.

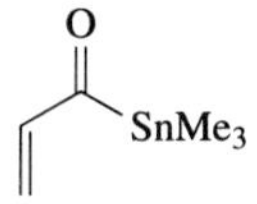

(454)

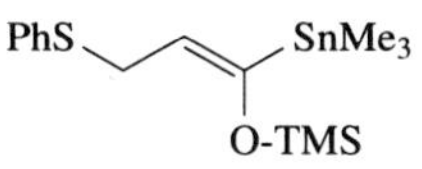

(455) (E):(Z) > 95:5

Table 53 Preparation of 1-trialkylstannyl-1-alkoxyalkanes from vinyl ethers by successive treatment with Bu^{t}Li and a tin halide (Scheme 47).[a]

Vinyl ether	Tin halide	Product	Yield (%)	Ref.
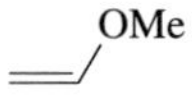 =\OMe	Me$_3$SnCl	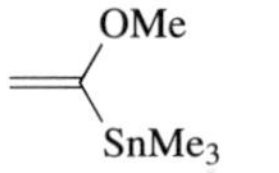 =C(OMe)(SnMe$_3$)	85	82OM830
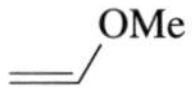 =\OMe	Bu$_3$SnCl	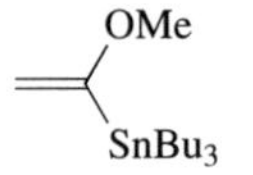 =C(OMe)(SnBu$_3$)	92	82OM830
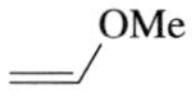 =\OMe	SnCl$_4$	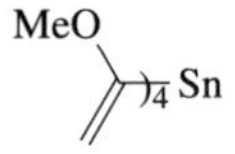 (MeO-C(=CH$_2$))$_4$Sn	72	82OM830
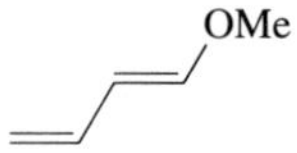 =\-\OMe	Me$_3$SnCl	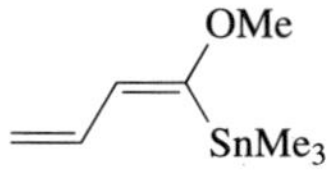 =\-C(OMe)(SnMe$_3$)	74	80JA1577
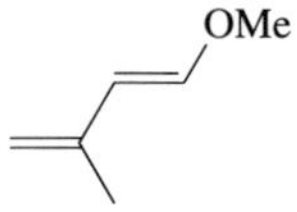 =\-\OMe (methyl)	Me$_3$SnCl	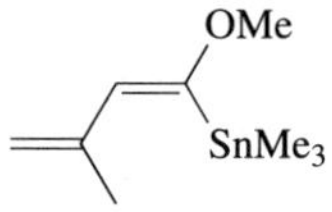 =\-C(OMe)(SnMe$_3$) (methyl)	60	80JA1577
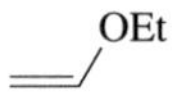 =\OEt	Bu$_3$SnCl	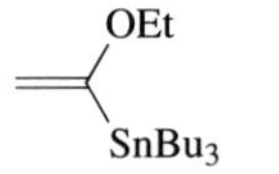 =C(OEt)(SnBu$_3$)	97	85JOC3255
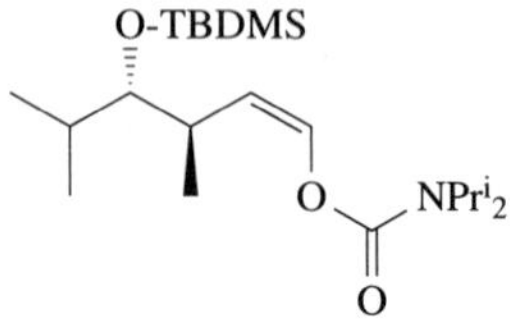 O-TBDMS carbamate vinyl ether	Me$_3$SnCl	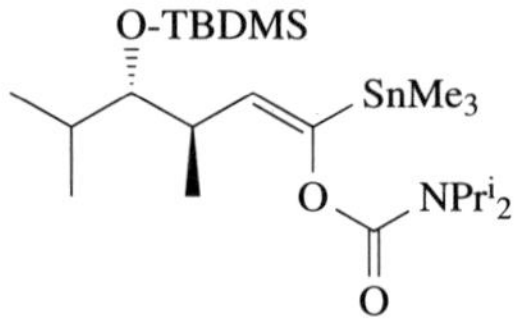 O-TBDMS carbamate stannyl vinyl ether	95	89SL52
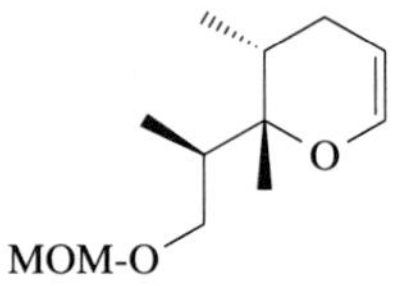 MOM-O dihydropyran	Bu$_3$SnCl	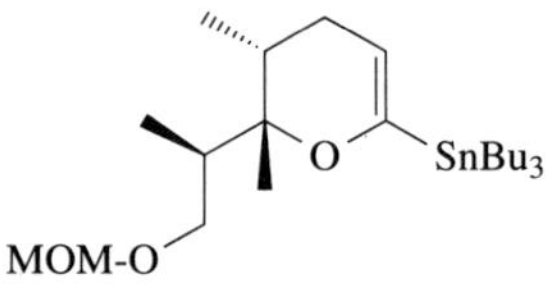 MOM-O dihydropyran-SnBu$_3$	'quantitative'	87JA7553

[a] Generally in THF at −78 °C.

McGarvey and Bajwa have described an ingenious, reverse Diels–Alder route to these compounds (Scheme 48) ⟨84JOC4091⟩. Bicyclo[2,2,1]hept-5-en-2-one (**456**) was condensed with tri-*n*-butylstannyllithium and the resultant alcohol alkylated to give the ethers (**457**). These were thermolysed *in vacuo* at 400 °C over quartz to give the stannyl vinyl ethers (**458**) in the yields shown in Table 54.

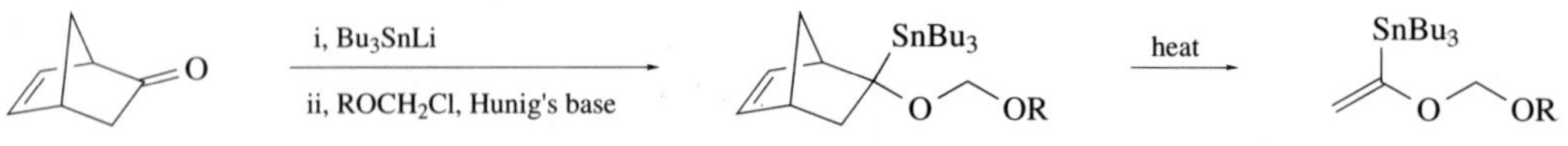

Scheme 48

Finally, Beau and co-workers displaced the benzenesulfonyl group in (**459**) with tri-*n*-butyltin hydride to give the corresponding stannyl derivative (**460**) in 77% yield ⟨86TL6201⟩.

Table 54 Reverse Diels–Alder route to 1-alkoxy-1-tributylstannylalkanes.

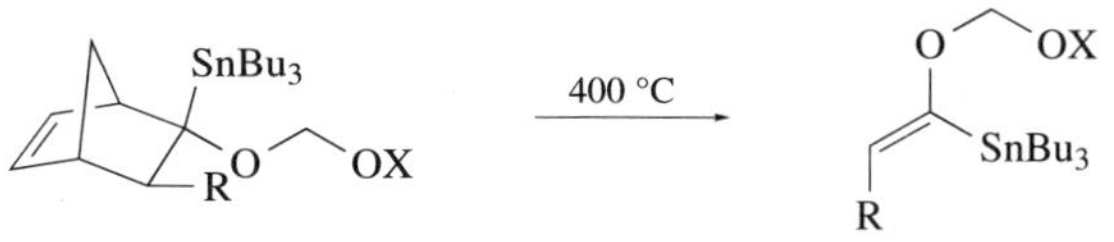

R	X	Yield (%)
H	Me	98
H	Bu^n	95
H	(CH$_2$)$_2$OMe	95
Me	Me	85
CH$_2$OBn	Me	98

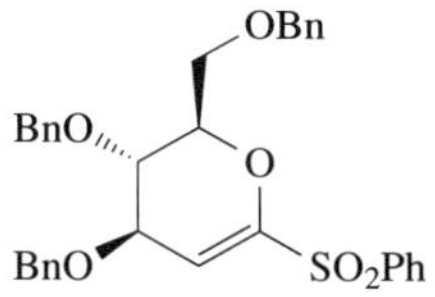

(459) 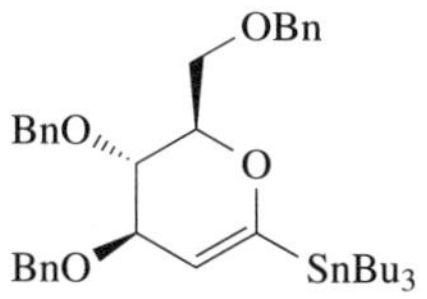   (460)

4.20.4.1.3 Copper

Unstable copper(I) compounds have been described by a number of authors and used, *inter alia*, for alkylation or Michael coupling. These are usually formed and used at low temperatures and, therefore, yields of formation are not measured. However, from the yields of the ultimate products, it would appear that most conversions into the copper compounds proceed in high yield.

Chavdarian and Heathcock treated 1-methoxyvinyllithium with purified cuprous iodide and dimethyl sulfide in dry THF at −40 °C to give a deep red solution, believed to contain 1-methoxyvinylcopper(I), which gradually turns yellow as lithium di(1-methyoxyvinyl)cuprate (**461**) is formed ⟨75JA3822⟩. Back *et al.* have also described this technique ⟨87JOC4258⟩.

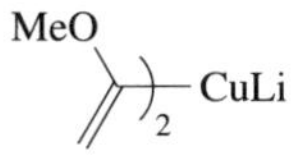

(461)

The ethyl analogue of (**461**) was prepared by Boeckman, Baldwin and co-workers by the low temperature (−78 °C) reaction of 2 mol 1-ethoxyvinyllithium with purified copper iodide ⟨75CC519⟩ whilst Kocienski used the same process to prepare (**462**) ⟨83TL3905⟩.

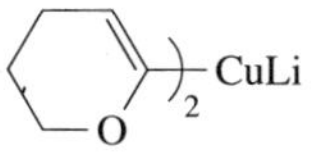

(462)

Lower-order cyanocuprates react with stannyl or lithio alkenes (Equations (50)–(52)) to give the corresponding lithiocuprates, for example (**463**) ⟨87JCS(P1)2189, 89SL52⟩.

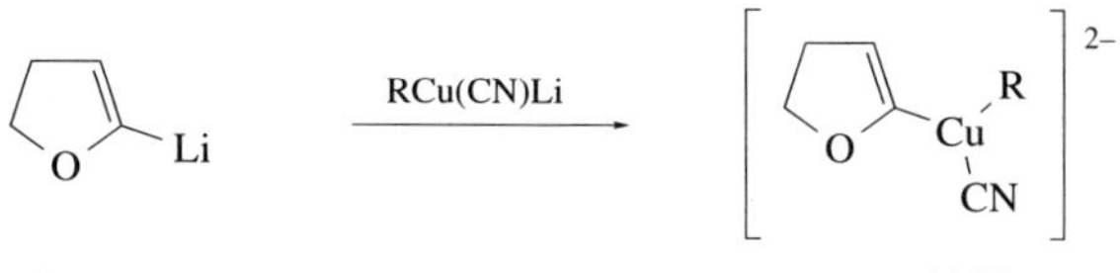

(463)   (50)

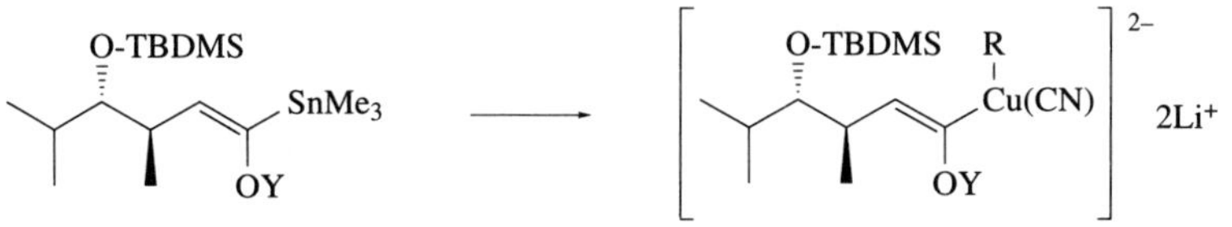

Tetrakis(1-methoxyvinyl)tin was treated successively with butyllithium and cuprous cyanide at −78 °C in THF to give a solution of the organocuprate (**464**) ⟨92JOC5844⟩.

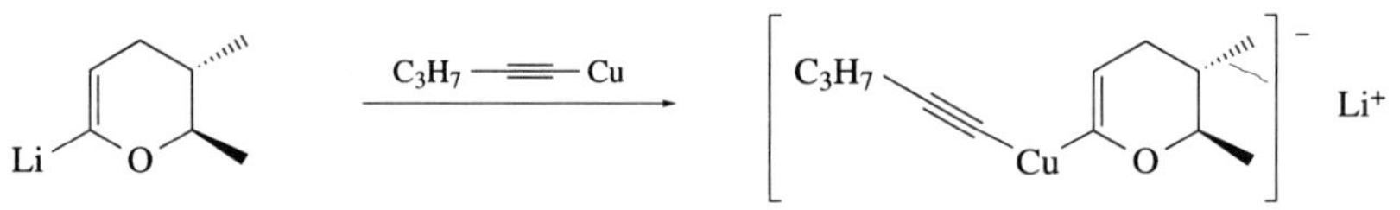

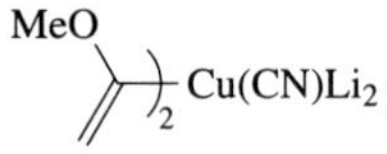

4.20.4.1.4 Zinc

Organozinc compounds have also been prepared as intermediates which react with vinyl iodides in the presence of palladium(0) to give high yields of coupled products. Thus organolithium derivatives, including 1-ethoxyvinyllithium, react with 1 equivalent of zinc chloride in THF at −78 °C to 0 °C to give, for example, the zinc compound (**465**) (Scheme 49) ⟨83JOC1560⟩.

Similarly prepared were the methoxy analogues (**466**) and (**467**) ⟨89CL1959, 92BCJ2303⟩.

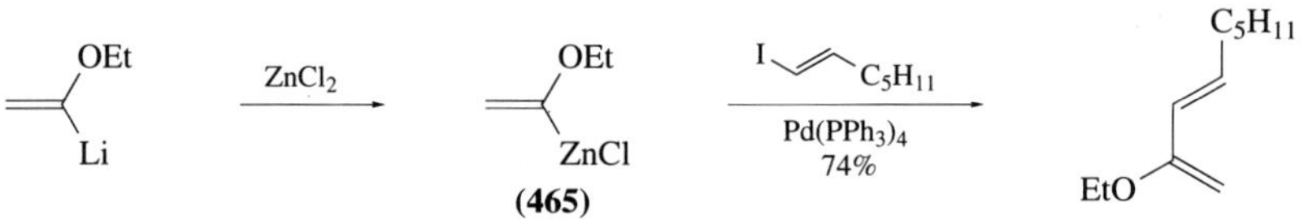

Scheme 49

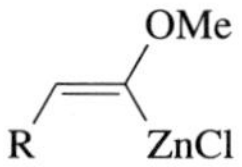

(**466**) R = H
(**467**) R = n-C$_4$H$_9$

4.20.4.1.5 Cerium

The organocerium compounds (**468**) and (**469**) were prepared at −78 °C by treatment with ceric chloride of the corresponding organolithium derivatives and were then used at low temperature without isolation ⟨84TL4233, 88T3139⟩.

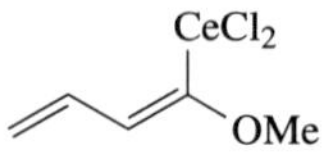

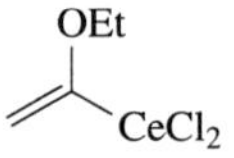

4.20.4.1.6 Transition metals

A considerable number of complexes of transition metals are known, for example of chromium, nickel, manganese, platinum, palladium and ruthenium.

Casey et al. showed that the anion of (methoxy)methylcarbenepentacarbonylchromium(0) (**470**), generated by the action of n-butyllithium, exists mainly as the vinylchromium anion (**471**) ⟨74JA1230⟩.

(**470**) M = Co
(**472**) M = W

(**471**)

This was confirmed in studies of the reactions of (**471**) and the related tungsten derivative (**472**) was also prepared ⟨85JA503, 87JOC3263⟩.

Stable nickel, palladium and platinum complexes (**473**) were prepared by treatment of the appropriate chlorometal complex with 1-methoxyvinyllithium in THF–benzene at $-78\,°C$ to $0\,°C$ in yields of 65%, 50% and 35%, respectively ⟨75CC899, 80JOM(201)477⟩. Alternatively the nickel complex (**473**; M = Ni) can be prepared by the sequence shown in Scheme 50 ⟨78JOM(159)417⟩.

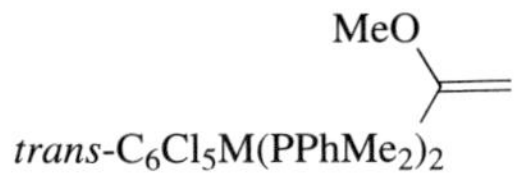

(**473**) M = Pd, Ni, Pt

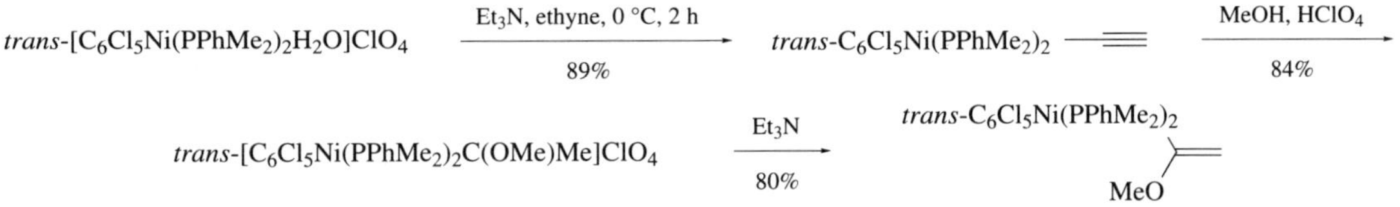

Scheme 50

The manganese complex (**474**) is produced in poor yields (<15%) by either the reaction of $Mn(CO)_9NCMe$ and ethoxyethyne or the photochemical reaction of $Mn_2(CO)_{10}$ with the alkyne ⟨91JA9406, 93OM3431⟩.

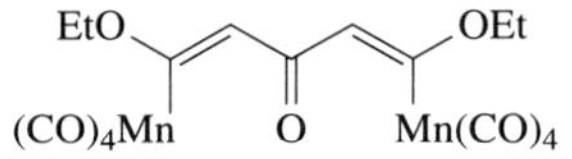

(**474**)

Finally, Russian workers have produced the lead compound (**475**) by the addition of triethylplumbyllithium or triethylplumbylsodium to ethoxyethyne ⟨81ZOB1857⟩.

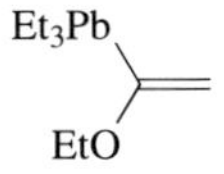

(**475**)

4.20.4.2 Sulfur Functions

As with the oxygen systems just described, a substantial number of organolithium, -tin and -copper compounds of general structure (**476**) has been reported. An additional feature of the sulfinyl group is its potential chirality, and this has been used to introduce diastereoselectivity into certain reactions.

4.20.4.2.1 Lithium compounds

(i) Sulfide derivatives

The most widely used route to these compounds is deprotonation of the corresponding alkenyl sulfide (**477**) by a variety of strong bases at low temperature (Table 55). It should be noted that the existence of the lithio compound is largely inferred from the structures of the products formed by subsequent reactions with electrophiles. To date, no one appears to have isolated 1-methylthiovinyllithium.

The activating effect of sulfur is greater than that of oxygen, as is shown by the compounds (**478**), which are regioselectively lithiated to give (**479**) ⟨76JA2008⟩. In addition, although the ethers (**478**) are 35:65 mixtures of *cis–trans* isomers, the products of low-temperature quenching of the lithio derivatives (**479**) by H_2O or D_2O are 95% *trans*. It is believed that this isomerisation results from intramolecular complexation of the lithium atom by the *cis*-ethoxy group (**480**).

(**478**) X = H, R = Ph, n-C_5H_{11}
(**479**) X = Li, R = Ph, n-C_5H_{11}

There are a number of questions posed by these reactions: what are the yields of the lithiated products; how stable are they; what is the regioselectivity of the deprotonation reaction and what is the stereochemical stability of the anions? It is difficult to assess the yields of reactions where the product is not isolated or quantified, but, judging from the yields of subsequent reactions with a variety of electrophiles, they must be very high. Moreover, Yamomoto and co-workers reported that (**481**) was stable at −78 °C in THF–HMPA for at least 6 h ⟨73JA2694⟩.

Uda *et al.* showed that (**482**) is deprotonated by LDA at −80 °C to give quantitatively the vinyl carbanion, with no indication of the formation of the allylic anion ⟨79CL785⟩. Conflicting evidence is found regarding the stereochemical stability of the anion. The derivatives (**478**) give 95% of the *trans*-isomers after lithiation and protonation, whilst (**483**) shows no evidence of isomerisation ⟨84AG(E)723⟩, probably because of stabilisation by the adjacent oxygen atom. The *cis*-butadiene derivatives (**484**) isomerize to the extent shown over 30 min at −30 to −20 °C but the *trans*-isomers are stable and, therefore, the isomerisation ratio does not reflect thermodynamic control ⟨78RTC69⟩.

In addition to deprotonation, three other synthetic schemes have been reported. Thus, aryl- and alkyllithium derivatives add to thioketenes (Equation (53)) ⟨74CB3562⟩. The ketene thioacetals (**485**) can be reduced by lithium naphthalenide to give the 1-alkylthiovinyllithium compounds (Equation (54)), which are then treated with a variety of electrophiles ⟨79JOC3601⟩. In Equations (53) and (54), the yields are of products formed by reaction with electrophiles.

Table 55 Alpha-lithiation of vinyl sulfides.

Starting material	Conditions	Ref.
PhS-C(Ph)=CH-Ph	Bu^{n}Li, THF, –30 °C	66JOC4097
PhS-CH=CH$_2$	LDA, THF-HMPA, –60 °C	76CC990
PhS-CH=CH$_2$	KOBu^t, BuLi, petrol ether, 25 °C	76HCA13
	Bu^{t}Li, TMEDA-THF, –90 °C	78CC821
	Bu^{n}Li, TMEDA-THF, –90 °C	83JCS(P1)1131
EtS-CH=CH$_2$	Bu^{s}Li, THF-HMPA, –78 °C	73JA2694
MeS-CH=CH-n-C$_8$H$_{17}$	Bu^{s}Li, THF-HMPA, –78 °C	73JA2694
MeS-CH=CH-Ph	Bu^{s}Li, THF-HMPA, –78 °C	73JA2694
PhS-CH=CH-OEt	Bu^{t}Li, THF, –70 °C	76JA2008
EtS-CH=CH-SEt	Bu^{t}Li, THF, –80 °C	77TL3583
MeS-CH=CH-CH=CH-SR	Bu^{n}Li, THF-TMEDA, –40 °C	77CC801
PhS-CH=CH-CO$_2$Me	LDA, THF, hexane, –80 °C	79CL785
Me$_2$N-C(O)-S-C(Me)=CH$_2$	Bu^{t}Li, THF, hexane, –70 °C	80AG(E)303
MeS-CH=CH-CH(Me)-O-MEM	Bu^{n}Li, hexane, –78 °C	84AG(E)723
TTF (tetrathiafulvalene)	LDA, Et$_2$O, –90 °C	79JOC1476
RS-CH=CH-CH=CH$_2$	KOBu^t, Bu^{n}Li, THF, –70 °C	78RTC69
2H-thiopyran	Bu^{n}Li, THF-TMEDA, –40 °C	78RTC69

An elimination–deprotonation route to the compounds (**486**) from the 2-methoxyalkyl sulfides (**487**) with 2 mol *n*-butyllithium in the presence of TMEDA at −30 to 0 °C has been reported by Takeda and co-workers (Equation (55)) ⟨84BCJ1863⟩. This reaction usually appears to give pure (*E*)-lithioalkenyl sulfides.

(ii) Sulfinyl derivatives

The sulfinyl group is a stronger electron-withdrawing substituent than the sulfide group; a fact neatly illustrated by direct competition in the substrate (**488**) which, after deprotonation by *t*-butyllithium in THF at −120 °C and treatment with a variety of electrophiles, for example D$^+$, MeX, MeSX, gave, as judged by NMR spectroscopy, at most 15% of (**490**), with the majority being (**489**) ⟨79TL4277⟩. However, (**489**) was not configurationally stable and was converted into (**491**), whereas (**490**) was configurationally stable.

Similar isomerisations have been noted by a number of authors and ascribed to the lowering of the alkenyl anion isomerisation barrier by the sulfinyl group. Thus deprotonation of the mixtures (*E/Z*) (**492**) by LDA at −78 °C in THF gives only the (*E*)-2-alkenyl sulfoxides (**493**) with there being no evidence for the formation of an allylic anion (Equation (56)) ⟨78TL3995⟩.

Likewise, no allylic anions were seen when the sulfoxides (**494**) were deprotonated, in contrast with the analogous sulfides (**495**), which produced only (**496**) after deprotonation and alkylation ⟨78CL517, 80CL1209⟩.

In all cases, predominant *cis→trans* isomerisation took place rapidly even at $-100\,°C$. With the sulfoxides (**495**), it is possible that chelation as in (**497**) enhances the stability of the (*E*)-form.

The sulfoxide (**498**) was produced in optically pure form, and the desired lithium anion generated by LDA in THF at $-100\,°C$ and condensed with the substituted benzaldehyde (**499**) to give 75% of the alcohol (**500**) as a single diastereomer ⟨84JA6097⟩. In this sequence, the initial sulfoxide (**498**) was a 55:45 mixture of (*E*:*Z*)-isomers, but the alcohol (**500**) was a pure (*E*)-isomer. Chelation to the oxygen of the acetal probably assists this isomerisation, since the close analogue (**501**) only changed from 50:50 (*E*:*Z*) to 66:33 (*E*:*Z*) isomers after a deprotonation–protonation sequence ⟨84C233⟩.

Similarly a mixture of (*E*,*Z*) (**502**) isomerised to predominantly (*E*) (**503**) on treatment with LDA in THF at $-78\,°C$, with no racemisation ⟨86TL2191⟩.

Treatment of (**504**) with methyllithium followed by reprotonation gave no double bond isomerisation and no racemisation (Equation (57)). However, the (Z)-isomer (**505**) was entirely converted into the (E)-isomer with some racemisation under the same conditions (Equation (58)) ⟨81PAC2307⟩.

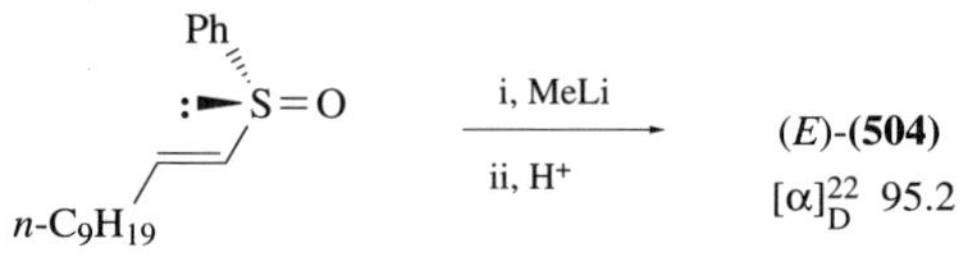

$$\text{(E)-(504)} \quad [\alpha]_D^{22}\ 95.7 \xrightarrow[\text{ii, H}^+]{\text{i, MeLi}} \text{(E)-(504)} \quad [\alpha]_D^{22}\ 95.2 \tag{57}$$

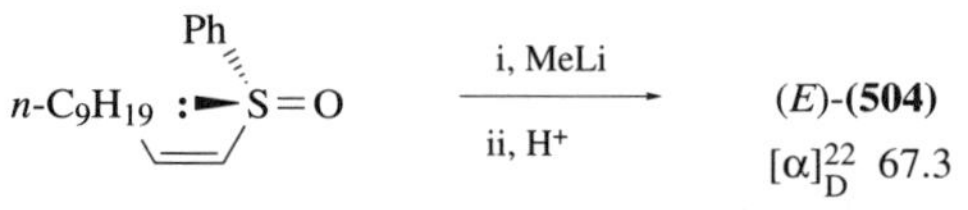

$$\text{(Z)-(505)} \quad [\alpha]_D^{22}\ -199.9 \xrightarrow[\text{ii, H}^+]{\text{i, MeLi}} \text{(E)-(504)} \quad [\alpha]_D^{22}\ 67.3 \tag{58}$$

Most recently, the deprotonation of (Z)-(**506**) with LDA in THF at $-78\,°C$ followed by reprotonation gave (E)-(**506**) with no racemisation. In reactions of the anion with aldehydes, moderate diastereoselectivity was noted ⟨93JCS(P1)67⟩.

(Z) (**506**)   (E) (**506**)

(iii) Sulfones

Simpkins has pointed out that the parent, unsubstituted vinyl sulfone (**507**; R = X = H) does not form a usable vinylic anion because of rapid polymerisation initiated by Michael addition to the carbon–carbon double bond ⟨90T6951⟩.

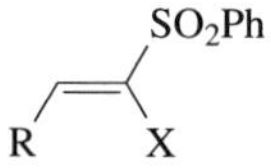

(**507**)
R = H, Ph, Me
X = H, Li, MgBr

This notwithstanding, the substituted sulfones (**507**) readily produced the α-lithiated alkenes in high yield on treatment with methyllithium at $-95\,°C$ in THF ⟨79JOC3279⟩. There was no exchange of the *ortho* protons on the phenyl ring, or of the β-vinyl or the γ-allyl proton of (**507**; R = Me), and the original (E)-stereochemistry was retained in the anion. Amongst the reactions undergone by the anion is condensation with magnesium bromide to give the Grignard reagent. Subsequently (**507**; R = Ph, X = Li) was also generated by the reaction of *n*-butyllithium at $-78\,°C$ in THF and was stereospecifically alkylated with methyl iodide at the same temperature to give the (E)-product (**508**) ⟨87BCJ1523⟩.

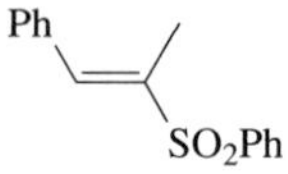

(**508**)

However, the stereochemical result is not because of the stability of the anion since the (E)- and (Z)-isomers of (**507**; R = Ph, X = H) give the same (E)-isomer (**508**) ⟨86JOM(302)1⟩. Other vinyl sulfones of general structure (**509**) give equilibrium mixtures of (E)- and (Z)-isomers of the lithio derivatives, with the equilibrium being rapidly achieved at $-60\,°C$ at a position dependent upon the nature of R and X. With the correct substituents on the β-position, the stereochemistry of the anion can be controlled. Thus, the ketal (**510**) gives stereo- and regiospecifically the anion (**511**) on treatment with methyllithium–lithium bromide in THF at $-20\,°C$ ⟨87TL6709, 88JOC4708⟩. Similar stabilisation is provided by the methoxymethyl (MOM) group in the derivative (**512**) ⟨91TL1385⟩.

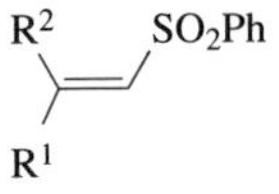

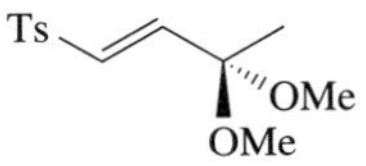

The 1-sulfonylvinyllithium derivatives of the sulfones (**513**), generated by the action of *n*-butyllithium in THF at $-70\,°C$, react intramolecularly to give the dihydrofurans (**514**) (Scheme 51) ⟨85TL6301, 87TL4123, 90JCS(P1)2775⟩. A similar intramolecular condensation has been used to produce furans by Equation (59) ⟨87TL4123⟩. It should be noted that the anion α to the sulfone group is produced despite the presence of the enolisable ketone.

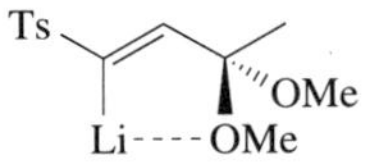

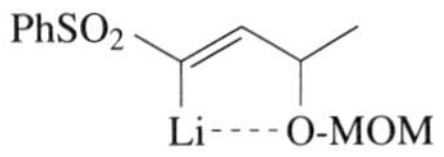

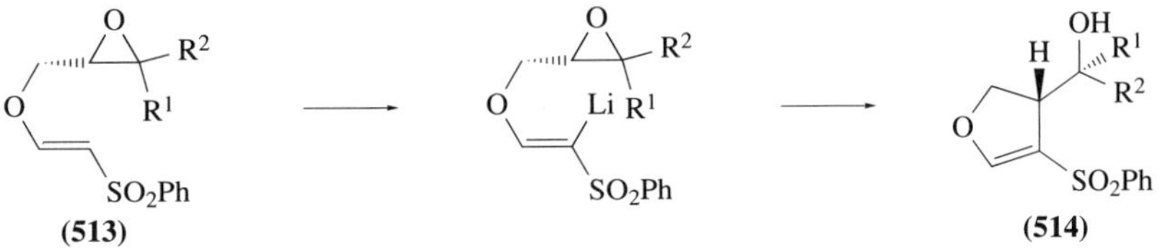

Scheme 51

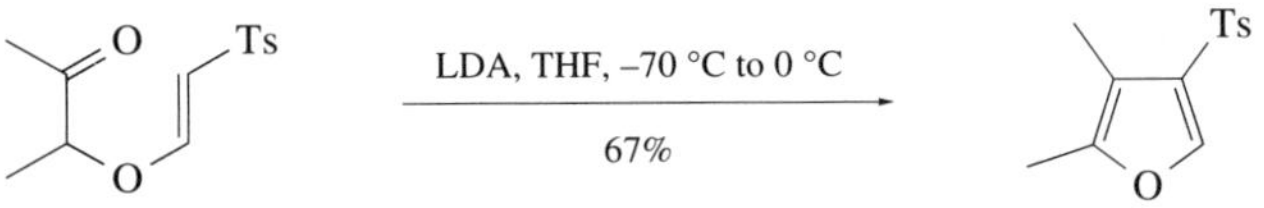

Despite deprotonation by strong bases of alkenyl sulfoxides being the most common method of producing the lithium compounds, two alternative routes have been reported. Simpkins showed that elimination and deprotonation of the acetal (**515**) occurred on treatment with 2 equivalents of *n*-butyllithium at $-78\,°C$ (Equation (60)) ⟨87TL989⟩ and the β-keto sulfoxide (**516**) was also deprotonated by 2 equivalents of LDA to give the dianion (**517**) (Equation (61)), which was subsequently allylated solely on the α-carbon atom ⟨85TL169⟩.

4.20.4.2.2 Tin compounds

In contrast to the lithium compounds, tin derivatives are stable, isolable compounds. Perhaps surprisingly, the most obvious route to these compounds, namely the reaction of 1-lithiovinyl phenyl sulfides with tin chlorides has only been reported twice ⟨71JOM(26)215, 77CC522⟩. Thus, the corresponding lithio compounds reacted with tributylchlorostannane to give (**518**) in 87% yield and

(**519**) in ~50% yield. Subsequently (**518**) was oxidised to the sulfoxide (**520**) ⟨77CC522⟩ or sulfone (**521**) by mcpba at −78 °C or by 30% hydrogen peroxide–glacial acetic acid, respectively ⟨93T6483⟩.

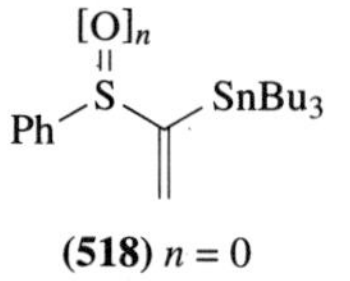
[O]n
Ph
S
SnBu3

(**518**) $n = 0$
(**520**) $n = 1$
(**521**) $n = 2$

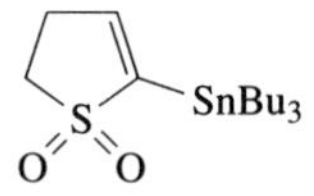
S
SnBu3
O
O

(**519**)

The Peterson reaction (Equation (62) and Table 56) has also been used ⟨77CB852, 86OM1906⟩. Whilst the yields are generally good, there is no stereochemical control.

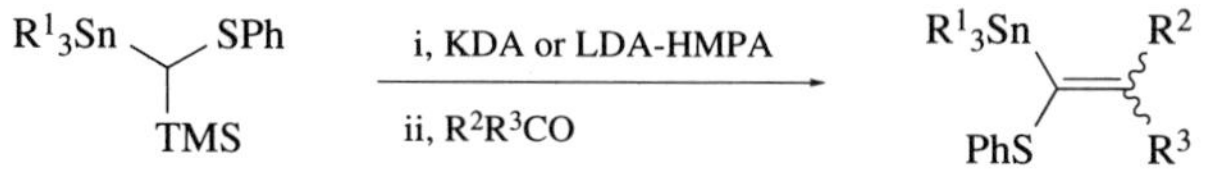
R1 3Sn
SPh
TMS
i, KDA or LDA-HMPA
ii, R2R3CO
R1 3Sn
R2
PhS
R3

(62)

Table 56 Peterson reaction of 1-alkylstannyl-1-thioalkylalkanes with aldehydes and ketones (Equation (61)).

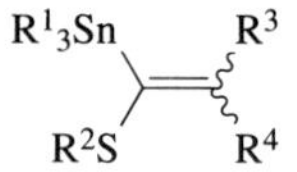
R1 3Sn
R3
R2S
R4

R^1	R^2	R^3	R^4	*Yield* (%)	*Ref.*
Me	Me	H	H	33	77CB852
Me	Me	H	Ph	60	77CB852
Me	Ph	H	H	71	77CB852
Me	Ph	H	n-C_3H_7	81	77CB852
Me	Ph	H	n-C_5H_{11}	74	77CB852
Me	Ph	H	Ph	82	77CB852
Me	Ph	Ph	Ph	60	77CB852
Bu	Ph	H	Ph	72	86OM1906
Bu	Ph	Ph	Ph	46	86OM1906

Addition of tin compounds to sulfur-substituted alkynes gives better stereo- and regioselectivity. Thus tributylstannylcuprate reacts with the phenylthioethyne (**522**) to give, after protonation, solely the (*E*)-tributylstannylalkene (**523**) in 65% yield (Equation (63)) ⟨90TL2541⟩. The intermediate cuprate formed from phenylthioethyne can react with haloalkynes in the presence of palladium(0) catalysts to give, for example, solely the (*E*)-isomer (**524**) in 40% yield ⟨91TL6085⟩.

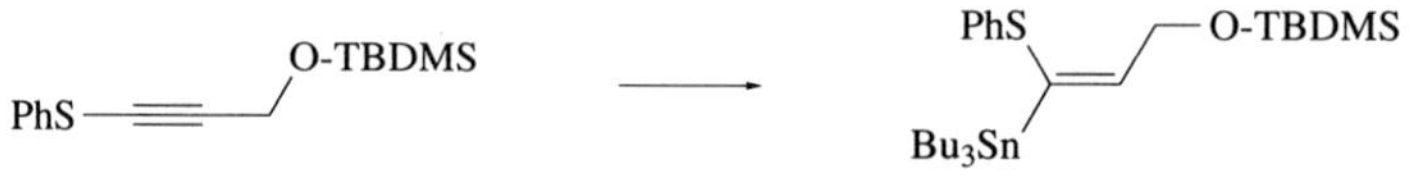
O-TBDMS
PhS
PhS
O-TBDMS
Bu3Sn

(**522**) (**523**) (63)

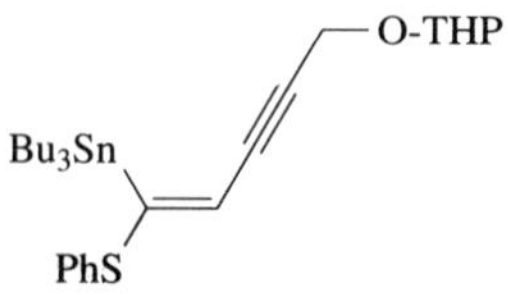
O-THP
Bu3Sn
PhS

(**524**)

Tributyltin hydride adds to phenylthioalkynes in the presence of palladium(0) with good regio- and stereoselectivity (Table 57) ⟨91TL5047, 92SL886⟩.

Table 57 Addition of tributyltin hydride to phenylthioalkynes.

$$\text{PhS}\diagdown\text{X} \atop \text{Bu}_3\text{Sn}\diagup$$

X	Yield (%)	Ref.
CH$_2$O-TBDMS	90	91TL5047
CH$_2$OH	84	91TL5047
Me	79	91TL5047
CH(OH)CH$_2$O-TBDMS	80	91TL5047
CH$_2$OCH$_2$CON(Me)OMe	90	91TL5047
H	87	91TL5047
COCH$_2$O-TBDMS	88	91TL5047
Ph	75	91TL5047
Bu	82	91TL5047
TMS	85	91TL5047
(CH(O-TBDMS)CH(iPr)...)	95	92SL886

4.20.4.2.3 Copper compounds

These are known as reactive intermediates which are particularly valuable for the palladium-catalysed, coupling reactions just described. Organocuprates readily add to alkynic sulfides and sulfones at −70 °C (Equation (64) and Table 58).

$$R\!\!=\!\!=\!\!S(O)_nR \;+\; \text{'R}_2\text{Cu}^-\text{'} \;\longrightarrow\; \left[\begin{array}{cc} R & S(O)_nR \\ R & Cu \end{array}\right] \tag{64}$$

Table 58 Addition of organocuprates to alkynic sulfides and sulfones.

$$\left[\begin{array}{c} R^1 \quad [O]_n \\ \diagdown \;\;\|\; \\ R^2 \diagup \diagdown S{-}R^3 \\ \text{'Cu'} \end{array}\right]$$

R^1	R^2	R^3	n	Copper species	Yield[a] (%)	(E):(Z) ratio	Ref.
H	Et	Me	2	R^{2}MgX : CuBr	80–90	99 : 1	75RTC14
Me	Et	Me	2	R^{2}MgX : CuBr	80–90	94 : 6	75RTC14
Me	Ph	Me	2	R^{2}MgX : CuBr	80–90	99 : 1	75RTC14
Me	Et	Ph	2	R^{2}MgX : CuBr		89 : 11	75RTC14
Ph	Me	Me	0	R$^2{}_2$CuMgX		100E	75TL2923
Ph	Et	Me	0	R$^2{}_2$CuMgX		100E	75TL2923
H	Ph	Ph	2	R$_2$Cu	>90	stereospecific E	86JOM(302)1
H	n-C$_6$H$_{13}$	Ph	2		>90	stereospecific E	86JOM(302)1
Ph	n-C$_6$H$_{13}$	Ph	2		>90	stereospecific E	86JOM(302)1
Bu^{t}CH$_2$	Ph	Ph	2		>90	stereospecific E	86JOM(302)1
H	n-C$_7$H$_{15}$	Et	0	R^{2}MgX : CuBr	95	stereospecific E	77BSF693
Bu	Et	Me	0	EtCu(CN)ZnEt	>70	stereospecific E	91JA5735

[a] Yields overall of products generated from a variety of electrophiles and the copper complex.

In addition, Posner and co-workers showed that dimethylcopperlithium did not add conjugatively to the vinyl sulfone (525) but instead generated an organometallic compound, presumed to be (526) (Equation (65)) ⟨81PAC2307⟩.

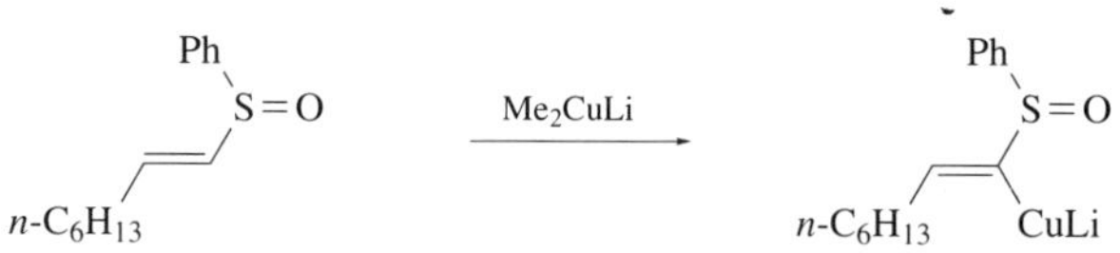

$$\text{(525)} \xrightarrow{\text{Me}_2\text{CuLi}} \text{(526)} \quad (65)$$

4.20.4.2.4 Zinc compounds

Kocienski and co-workers showed that the organostannane (527) could be exchanged to the corresponding lithium compound (528) by treatment with *n*-butyllithium at −78 °C. This then reacted with zinc bromide in THF at −78 °C to 0 °C to give the zinc compound (529) ⟨92SL886⟩.

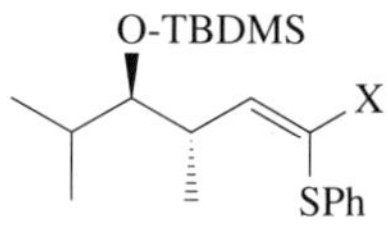

(527) X = SnBu₃
(528) X = Li
(529) X = ZnBr

4.20.4.3 Selenium and Tellurium Functions

4.20.4.3.1 Selenium

Ketene diselenoacetals (530) react quantitatively with *n*-butyllithium at −78 °C in THF to give 1-lithio 1-selenoalkenes (531) ⟨77CB867, 82TL3411⟩. These are unstable and are treated with electrophiles at the same temperature.

(530)

(531)
R¹ = Ph, R² = H, R³ = Ph
R¹ = *n*-C₁₀H₂₁, R² = H, R³ = Ph

With (530; R¹ = Ph, R² = H, R³ = Ph) the reaction is stereospecific, only the phenylseleno group *cis* to the phenyl group being replaced. For (530; R¹ = *n*-C₁₀H₂₁, R² = H, R³ = Ph), the products are described as a 40:60 mixture of (Z:E)-stereoisomers, but the stereochemistry of the intermediate (531) was not defined.

Phenyl vinyl selenide (532) is more acidic than its sulfur analogue and this reactivity can be further increased by the presence of a *m*-trifluoromethyl group on the phenyl ring (533) ⟨80JOC5227⟩. The best base for the deprotonation is 1 equivalent of *n*-butyllithium in the presence of a catalytic amount of diisopropylamine in THF at −78 °C. Under these conditions, deprotonation is complete in a few minutes. Butyllithium alone in THF largely displaces the phenyl group, whilst in ether it adds to the vinyl group ⟨78AG(E)526, 78JOC4252⟩. Alternatively, a mixture of potassium diisopropylamide and lithium *t*-butoxide (KDA) rapidly deprotonates (533) but with 2-methyl substituted compounds gives the allylic ion ⟨78JOC3794⟩.

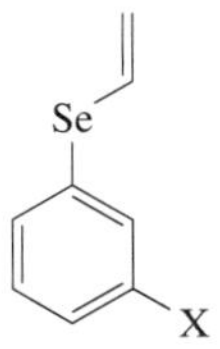

(532) X = H
(533) X = CF$_3$

Reich and co-workers recommend the use of LDA in the presence of HMPA, which gives complete deprotonation in 5 min at $-78\,°C$ ⟨81JOC2775⟩. However, this mixture also deprotonates allylic positions. With lithium tetramethylpiperidide in THF at $-50\,°C$, the 1-butenyl and 1-butenyl-3-methyl selenides (534) and (535) were deprotonated solely in the vinyl position. Alkylation with methyl iodide gave the derivatives (536) and it is interesting that pure (*E*)- and (*Z*)-(535) give the same, 7:3, (*E*:*Z*)-mixture of products, implying that the anion undergoes isomerisation.

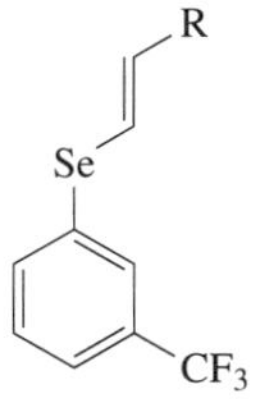

(534) R = Et
(535) R = Pr^i

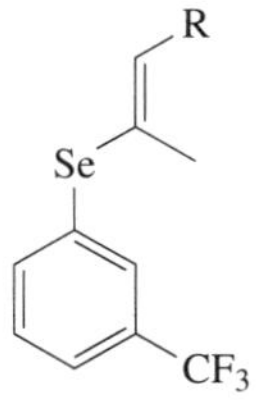

(536)

Phenyl vinyl selenoxide could be deprotonated by the LDA method, but the product was a mixture of selenide and decomposition products. It was concluded that the half life of 1-lithiovinyl phenyl selenoxide is under 30 min at $-78\,°C$ in THF.

4.20.4.3.2 *Tellurium*

Phenyl vinyl telluride (537) was deprotonated by lithium 2,2,6,6-tetramethylpiperidide or lithium dicyclohexylamide to give the lithium compound (538), which readily condensed with electrophiles such as benzaldehyde or chlorotrimethylsilane ⟨83CB1001⟩.

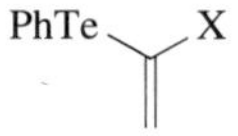

(537) X = H
(538) X = Li

4.20.5 ACKNOWLEDGEMENT

It is a pleasure to acknowledge the high professional literature-searching abilities of Dr Alan Barnes and Mrs Fran Shepherd and the typing skills of Mrs Susan Hughes. Without them, this task would have been impossible.

4.21
Functions Containing at Least One Nitrogen and No Halogen or Chalcogen

GRAHAM L. PATRICK
University of Paisley, UK

4.21.1	FUNCTIONS CONTAINING TWO NITROGENS: $R^3_2C= CNR^1_2NR^2_2$, etc.	969
4.21.1.1	Ketene Aminals, $R^3_2C= CNR^1_2NR^2_2$	969
4.21.1.1.1	From ketene dithioacetals	969
4.21.1.1.2	From vinylidene dihalides	970
4.21.1.1.3	From amides via chloro enamines	971
4.21.1.1.4	From ketene acetals	971
4.21.1.1.5	From imino esters	972
4.21.1.1.6	From amides and other carboxylic acid derivatives	972
4.21.1.1.7	From thioformamidinium salts, thioureas, imidazoles, and isothioureas	973
4.21.1.1.8	From alkoxytris(dimethylamino)methanes	973
4.21.1.1.9	From urea acetals	974
4.21.1.1.10	From tetramethylchloroformamidinium chloride	974
4.21.1.1.11	From aminals, ortho-esters, and amide acetals	975
4.21.1.1.12	From miscellaneous 'head group' synthons	975
4.21.1.1.13	From cyanamides	975
4.21.1.1.14	From halo alkynes	976
4.21.1.1.15	From ynamines	976
4.21.1.1.16	From tricyanomethanide, dicyanoacetates, dicyanothioacetates, and dicyanoacetamides	977
4.21.1.1.17	From ring opening of pyrimidines	978
4.21.1.1.18	Miscellaneous syntheses from ring systems	979
4.21.1.1.19	Miscellaneous methods	979
4.21.1.1.20	By modification of other ketene aminals	979
4.21.1.2	Derivatives of Ketene Aminals	980
4.21.1.2.1	α-Amido enamines, $R^4_2C= C(NR^1_2)(NR^2COR^3)$	980
4.21.1.2.2	α-Alkoxycarbonylamino enamines, $R^3_2C= C(NR^1_2)(NHCO_2R^2)$	981
4.21.1.2.3	α-Ureido enamines, $R^3_2C= C(NR^1_2)(NHCONR^2_2)$	981
4.21.1.2.4	N-Acylated α-ureido enamines	982
4.21.1.2.5	gem-Diureido alkenes, $R^2_2C= C(NHCONR^1_2)_2$	982
4.21.1.3	Derivatives Bearing at Least One NRX Function	983
4.21.1.3.1	α-Hydrazino enamines, $R^4_2C= C(NR^1_2)(NR^2NR^3_2)$	983
4.21.1.3.2	gem-Dihydrazino alkenes, $R^3_2C= C(NR^1NR^2_2)_2$	985
4.21.1.3.3	Derivatives of α-hydrazino enamines.	986
4.21.1.4	Derivatives Bearing One NY or NZ Function and One NR_2 Function	987
4.21.1.4.1	α-Aminoalkylideneamino enamines, $R^4_2C= C(NR^1_2)(N= CR^2NR^3_2)$, $R^2_2C= C(NR^1_2)(N= CXNR^3_2)$	987
4.21.1.4.2	α-Diaminomethyleneamino enamines, $R^3_2C= C(NR^1_2)(N= C(NR^2_2)_2)$	988
4.21.1.4.3	α-Bis(methylthio)methyleneamino enamines, $R^2_2C= C(NR^1_2)(N= C(SMe)_2)$	988
4.21.1.4.4	α-Azo enamines, $R^3_2C= C(NR^1_2)(N= NR^2)$	989
4.21.1.4.5	α-Azido enamines, $R^2_2C= C(NR^1_2)N_3$	989
4.21.1.4.6	α-Isocyano enamines, $R^2_2C= C(NR^1_2)NC$	990
4.21.1.4.7	α-Phosphimino enamines, $R^3_2C= C(NR^1_2)(N= PR^2_3)$	990
4.21.1.4.8	α-(N,N',N''-Triphenylphosphorimidic triamido) enamines, $R_2C= C(NHPh)(N= P(NHPh)_3)$	991

	4.21.1.5 Derivatives Bearing One NY Function and One NRX Function	991
	4.21.1.5.1 gem-*Acylhydrazino azido alkenes*, $R^2{}_2C=C(NHNHCOR^1)N_3$	991
	4.21.1.6 Derivatives Bearing Two NY Functions	992
	4.21.1.6.1 gem-*Alkylideneamino isocyanato alkenes*, $R^2{}_2C=C(NCO)(N=CR^1{}_2)$	992
	4.21.1.6.2 The gem-*diisothioureido alkene* (**187**)	992
	4.21.1.6.3 gem-*Bisazo alkenes*, $R^2{}_2C=C(N=NR^1)_2$	992
	4.21.1.6.4 gem-*Azido phosphazido alkenes*, $R_2C=C(N=N—N=PPh_3)N_3$	993
	4.21.1.6.5 gem-*Diazido alkenes*, $R_2C=C(N_3)_2$	993
	4.21.1.6.6 gem-*Dinitro alkenes*, $R_2C=C(NO_2)_2$	993
	4.21.1.6.7 gem-*Arylazo nitro alkenes*, $R_2C=C(NO_2)_2(N=NAr)$	995
	4.21.1.6.8 gem-*Dimethoxazonyl alkenes*, $R_2C=C(NO=NOMe)_2$	996
	4.21.1.6.9 gem-*Imino phosphimino alkenes*, $R^3{}_2C=C(N=PR^1{}_3)(N=CR^2{}_2)$	996
	4.21.1.6.10 gem-*Diphosphazido alkenes* and gem-*diphosphimino alkenes*, $R^2{}_2C=C(N=N—N=PR^1{}_3)_2$, $R^2{}_2C=C(N=PR^1{}_3)_2$	996
	4.21.1.6.11 gem-*Disulfoximido alkenes*, $R^2{}_2C=C(N=SOR^1{}_2)_2$	997
4.21.2	FUNCTIONS CONTAINING ONE NITROGEN AND ONE PHOSPHORUS, $R^3{}_2C=CNR^1{}_2PO(OR^2)_2$, $R^3{}_2C=CNR^1{}_2PR^2{}_2$, etc.	997
	4.21.2.1 Derivatives Bearing a Phosphino Function	997
	4.21.2.1.1 α-*Phosphino enamines*, $R^2{}_2C=C(PPh_2)NHR^1$	997
	4.21.2.1.2 gem-*Amido phosphino alkenes*, $R^3{}_2C=C(NHCOR^1)PR^2{}_2$	998
	4.21.2.2 Derivatives Bearing a Phosphonium Group	998
	4.21.2.2.1 α-*Trialkylphosphonio enamines*, $R^3{}_2C=C(NR^1{}_2)P^+R^2{}_3$	998
	4.21.2.2.2 gem-*Amido phosphonio alkenes*, $R^2{}_2C=C(NHCOR^1)P^+Ph_3$	998
	4.21.2.2.3 gem-*Alkoxycarbonylamino phosphonio alkenes*, $R_2C=C(NHCO_2R)P^+Ph_3$	999
	4.21.2.2.4 gem-*Phosphonio ureido alkenes*, $R_2C=C(NHCONR^1{}_2)P^+Ph_3$	999
	4.21.2.2.5 gem-*Arylchloromethyleneamino phosphonio alkenes*, $R^2{}_2C=C(N=C(Cl)Ar)P^+R^1{}_3$	1000
	4.21.2.3 Derivatives Bearing a Phosphinoyl or Thiophosphinoyl Group	1000
	4.21.2.3.1 α-*Phosphinoyl enamines*, $R^3{}_2C=C(NR^1{}_2)POR^2{}_2$	1000
	4.21.2.3.2 gem-*Amido diphenylphosphinoyl alkenes*, $R^2{}_2C=C(NHCOR^1)POPh_2$	1000
	4.21.2.3.3 The gem-*methoxycarbonylamino diphenylphosphinoyl alkene* (**262**)	1000
	4.21.2.3.4 gem-*Diphenylphosphinoyl ureido alkenes*, $R^2{}_2C=C(NHCONR^1{}_2)POPh_2$	1001
	4.21.2.3.5 α-*Thiophosphinoyl enamines*, $R^3{}_2C=C(NR^1{}_2)P(=S)R^2{}_2$	1001
	4.21.2.3.6 Derivatives bearing a diphenylphosphinoyl group and a $N=C(X)Ar$ group, $Cl_2C=C(N=C(X)Ar)POPh_2$	1001
	4.21.2.4 Derivatives Bearing a Phosphonoyl Group, POXR	1001
	4.21.2.4.1 gem-*Amido phosphonoyl alkenes*, $R^3{}_2C=C(NHCOR^1)POXR^2$	1001
	4.21.2.4.2 The gem-*phosphonoyl ureido alkene* (**275**)	1002
	4.21.2.5 Derivatives Containing a Dialkoxy Phosphoryl Group	1002
	4.21.2.5.1 α-*Phosphoryl enamines*, $R^3{}_2C=C(NR^1{}_2)PO(OR^2)_2$	1002
	4.21.2.5.2 gem-*Formamido phosphoryl alkenes*, $R^3{}_2C=C(NHCHO)PO(OR^1)_2$	1004
	4.21.2.5.3 gem-*Amido phosphoryl alkenes* $R^3{}_2C=C(NHCOR^1)PO(OR^2)_2$	1005
	4.21.2.5.4 gem-*Alkoxycarbonylamino phosphoryl alkenes*, $R^3{}_2C=C(NHCO_2R^1)PO(OR^2)_2$	1006
	4.21.2.5.5 The gem-*phosphoryl ureido alkene* (**310**)	1007
	4.21.2.5.6 gem-*Diphenylphosphinoylamino phosphoryl alkenes*, $R^2{}_2C=C(NHPOPh_2)PO(OR^1)_2$	1008
	4.21.2.5.7 gem-*Phosphorylamino phosphoryl alkenes*, $R^4{}_2C=C[NR^2PO(OR^1)_2][PO(OR^3)_2]$	1008
	4.21.2.5.8 gem-*Alkylideneamino phosphoryl alkenes*, $R^3{}_2C=C(N=CR^1{}_2)PO(OR^2)_2$	1009
	4.21.2.5.9 The gem-*isocyanato phosphoryl alkene* (**333**)	1009
	4.21.2.6 Derivatives Bearing a $PS(OR)_2$, $PO(OR^1)(NR^2{}_2)$ or POX_2 Group	1010
	4.21.2.6.1 The α-*thiophosphoryl enamine* (**340**)	1010
	4.21.2.6.2 gem-*Amido phosphoramidoyl alkenes*, $Cl_2C=C(NHCOPh)PO(OMe)(NR_2)$	1010
	4.21.2.6.3 gem-*Phosphoramidoyl ureido alkenes*, $R^4{}_2C=C(NHCONR^1R^2)PO(OR^3)NR^1R^2$	1011
	4.21.2.6.4 α-*Phosphorodiamidoyl enamines*, $R^3{}_2C=C(NR^1{}_2)PO(NR^2{}_2)_2$	1011
	4.21.2.6.5 gem-*Phosphorodiamidoyl ureido alkenes*, $R_2C=C(NHCONHAr)PO(NHAr)_2$	1011
	4.21.2.6.6 gem-*Aminobenzylideneamino phosphorodiamidoyl alkenes*, $R^2{}_2C=C(N=C(NR^1{}_2)Ph)PO(NR^1{}_2)_2$	1012
	4.21.2.6.7 gem-*Difluorophosphoryl ureido alkenes* (**351**)	1012
	4.21.2.6.8 gem-*Arylchloromethyleneamino dichlorophosphoryl alkenes*, $R_2C=C(N=C(Cl)Ar)POCl_2$	1012
	4.21.2.6.9 gem-*Dihalophosphoryl isocyanato alkenes*, $R_2C=C(NCO)POX^1X^2$	1012
4.21.3	FUNCTIONS CONTAINING ONE NITROGEN AND ONE METALLOID, $R^3{}_2C=CNR^1{}_2SiR^2{}_3$, etc.	1013
	4.21.3.1 Nitrogen and Silicon Derivatives, $R^3{}_2C=CNR^1{}_2SiR^2{}_3$	1013
	4.21.3.1.1 α-*Silyl enamines*	1013
	4.21.3.1.2 gem-*Isocyano silyl alkenes*	1014
	4.21.3.1.3 The gem-*isothiocyanato silyl alkene* (**370**)	1014
	4.21.3.1.4 gem-*Aminoborylamino silyl alkenes*, $PrCH=C(TMS)N(R)B(X)NMe_2$	1015
	4.21.3.2 Nitrogen and Boron Derivatives, $R^3{}_2C=CNR^1{}_2BR^2{}_2$	1015
4.21.4	FUNCTIONS CONTAINING ONE NITROGEN AND ONE METAL, $R^2{}_2C=CNR^1{}_2M$, etc.	1015
	4.21.4.1 Group 1 and 2 Metals, $R^2{}_2C=CNR^1{}_2Li$, etc.	1015
	4.21.4.1.1 gem-*Isocyano lithio alkenes*	1015

4.21.4.1.2	α-Lithio allenamines	1015
4.21.4.1.3	α-Lithio enamines	1016
4.21.4.1.4	gem-*Iminomethylamino lithio alkenes*	1016
4.21.4.1.5	α-Sodio enamines	1017
4.21.4.1.6	α-Potassio allenamines	1017
4.21.4.1.7	α-Magnesio enamines	1017
4.21.4.2	Transition Metals, $R^2{}_2C=C(NR^1{}_2)PdX$, etc.	1017
4.21.4.2.1	Palladium and platinum	1017
4.21.4.2.2	Copper	1018
4.21.4.2.3	Molybdenum and tungsten	1018
4.21.4.2.4	Manganese	1019
4.21.4.2.5	Iron	1019
4.21.4.2.6	Rhenium	1020
4.21.4.3	Other Metals	1020
4.21.4.3.1	Tin	1020

4.21.1 FUNCTIONS CONTAINING TWO NITROGENS: $R^3{}_2C=CNR^1{}_2NR^2{}_2$, etc.

4.21.1.1 Ketene Aminals, $R^3{}_2C=CNR^1{}_2NR^2{}_2$

Ketene aminals ((2) in Scheme 1) are tautomeric with amidines but are less stable. Therefore, ketene aminals reported in the literature either contain stabilizing electron-withdrawing groups at the β position or contain tertiary amino groups which prevent the formation of the amidine tautomer. Ketene aminals of the former type often have low energy barriers to rotation about the formal double bond such that geometrical isomerism is not observed unless other factors such as intramolecular hydrogen bonding are present.

A large variety of methods have been reported for the synthesis of ketene aminals.

4.21.1.1.1 From ketene dithioacetals

The reaction of ketene dithioacetals (1) with two or more equivalents of amine is a popular method used for the synthesis of ketene aminals (2) (Scheme 1) ⟨68MI 421-01, 70ACS3102, 73BSF581, 80T1791, 84MI 421-01, 85JHC937, 88MI 421-01, 88T1667, 90JPR1035⟩. The reaction rate depends on the strength of the electron-withdrawing group(s) at the β position as well as the strength of the attacking amine. Thus, the ketene dithioacetal (3) reacted with dimethylamine at room temperature to give a quantitative yield of ketene aminal, whereas the dithioacetal (4) required heating at 150 °C in a sealed tube to give a yield of 68% ⟨70ACS3102⟩. The ketene dithioacetal (5) reacted readily with aliphatic amines such as n-butylamine but reacted only sluggishly with anilines. Ketene dithioacetals bearing two ester groups, for example (6), did not react with amines. The use of phenol or of acetic acid as solvent has proved advantageous in specific cases ⟨68HOU(7)341, 79T551⟩.

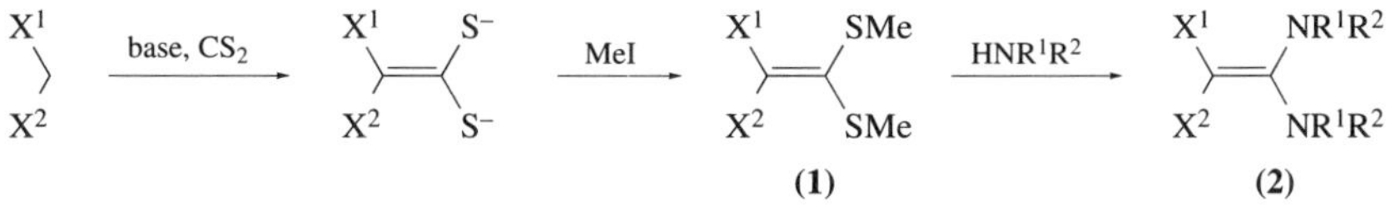

$$\begin{array}{c} X^1 \\ \diagup \\ X^2 \end{array} \xrightarrow{\text{base, CS}_2} \begin{array}{c} X^1 \\ \diagup \\ X^2 \end{array}\!\!=\!\!\begin{array}{c} S^- \\ \diagdown \\ S^- \end{array} \xrightarrow{\text{MeI}} \begin{array}{c} X^1 \\ \diagup \\ X^2 \end{array}\!\!=\!\!\begin{array}{c} \text{SMe} \\ \diagdown \\ \text{SMe} \end{array} \xrightarrow{\text{HNR}^1\text{R}^2} \begin{array}{c} X^1 \\ \diagup \\ X^2 \end{array}\!\!=\!\!\begin{array}{c} \text{NR}^1\text{R}^2 \\ \diagdown \\ \text{NR}^1\text{R}^2 \end{array}$$
$$\hspace{5.5cm}(1)\hspace{4cm}(2)$$

Scheme 1

$$\begin{array}{c} X^1 \\ \diagup \\ X^2 \end{array}\!\!=\!\!\begin{array}{c} \text{SMe} \\ \diagdown \\ \text{SMe} \end{array}$$

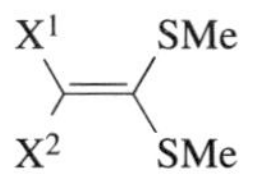

(3) $X^1 = X^2 = \text{MeCO}$
(4) $X^1 = \text{Ph}, X^2 = \text{MeCO}$
(5) $X^1 = \text{CO}_2\text{Me}, X^2 = \text{CN}$
(6) $X^1 = X^2 = \text{CO}_2\text{Me}$

The reaction of ketene dithioacetals with amines involves nucleophilic substitution of the alkylthio groups. It has been observed that the first alkylthio group is substituted considerably more rapidly than the second. Therefore, it is possible to isolate the intermediate ketene *N,S*-acetal (7) and then substitute the second alkylthio group with a different amine ⟨80EUP0010396, 82AP(970)680, 82EUP0049618, 83JCS(P1)1741, 90S195, 91EUP0413343, 91EUP0425030⟩.

$$\begin{array}{c} X^1 \\ X^2 \end{array}\!\!=\!\!\begin{array}{c} NR^1R^2 \\ SMe \end{array}$$

(7)

Reaction of ketene dithioacetals with diamines gives ketene aminals containing a diaza heterocyclic ring (8) ⟨77JPR545, 84MI 421-01, 91EUP0413343, 90JPR1035⟩. The reaction proceeds more smoothly than with monoamines such that even the weakly reactive ketene dithioacetal (6) reacts ⟨70ACS3102⟩.

$$\begin{array}{c} X^1 \\ X^2 \end{array}\!\!=\!\!\begin{array}{c} R^1 \\ N \\ N \\ R^2 \end{array}\!\!(CH_2)_n$$

(8) $n = 2$ or 3

Drawbacks with the foregoing synthetic methods include the release of methanethiol during the reaction. Furthermore, pressure is often required with volatile amines in order to displace the second alkylthio group ⟨62CB2871⟩.

4.21.1.1.2 From vinylidene dihalides

The reaction of amines with vinylidene dihalides is another popular method of synthesizing ketene aminals ⟨66CRV161, 70MI 421-01, 79ZOR1099, 82CB1733, 82ZOR1835⟩. Vinylidene dichlorides (9) are mainly used, but there are instances of ketene aminals being synthesized from vinylidene dibromides ⟨88ZOR2374⟩, vinylidene diiodides ⟨92JOC235⟩, and vinylidene difluorides ⟨60JA5116, 71DOK(201)1359, 81CCC1389⟩. Normally four to six equivalents of amine are employed since two equivalents are required to neutralize the two equivalents of hydrogen chloride released during the reaction. Alternatively, a tertiary amine such as triethylamine can be added for this purpose. The reaction is normally carried out in ether but a variety of other solvents have been used successfully.

$$\begin{array}{c} X^1 \\ X^2 \end{array}\!\!=\!\!\begin{array}{c} Cl \\ Cl \end{array}$$

(9)

The reaction rate is increased by electron-withdrawing groups at the β position. Vinylidene dihalides are more reactive than ketene dithioacetals and so the β substituents permitted are more variable. For example, vinylidene dihalides with two β ester substituents react with aniline, whereas this reaction does not take place with the corresponding ketene dithioacetal ⟨66CB2900⟩. Substitution of the vinylic halogens is even possible with weakly electron-withdrawing β substituents such as a nitrofuran ring or a sulfonyl group. However, in such cases, the nature of the product is affected by the nature of the amine. Ketene aminals are obtained from reactions with secondary amines, whereas either an amidine or a ketene aminal may be obtained with primary amines. The ketene aminal is more likely with stronger amines ⟨88ZOR2374⟩.

Under carefully controlled conditions (i.e., cooling to $-50\,°C$ in an aprotic solvent) the chloro enamine intermediate (10) was obtained from the reaction of a dichlorovinyl ketone with two equivalents of a cyclic amine. This chloro enamine reacted further with a different amine to produce a mixed ketene aminal ⟨82S199⟩.

The reactions of vinylidene dihalides with diamines gave ketene aminals containing a diaza heterocyclic ring (8) ⟨82ZOR1835, 92JOC235⟩.

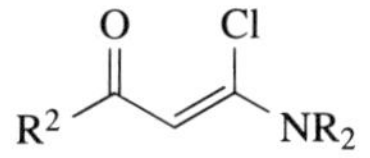

4.21.1.1.3 From amides via chloro enamines

The conversion of tertiary amides into chloro enamines by treatment with phosgene is a useful method of obtaining chloro enamines without electron-withdrawing substituents at the β position. The chloro enamines can then be treated with lithium amides (or free amines) to give ketene aminals (Scheme 2) ⟨73ZOR39, 76MI 421-01, 79ZOR1381, 82S199⟩.

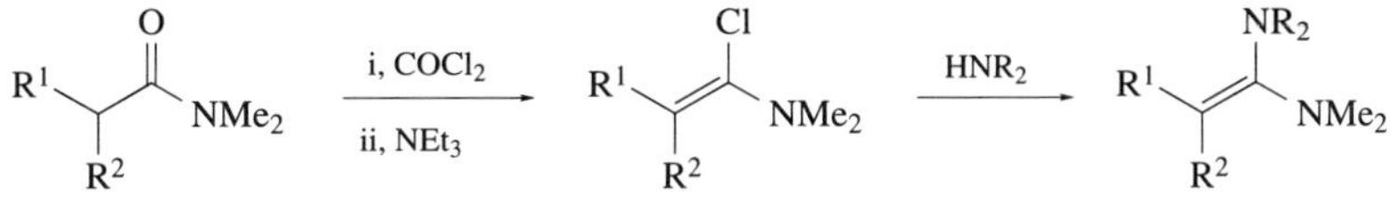

Scheme 2

The rate of the substitution reaction depends on the nature of the β substituents as described above. However, the reaction is still possible in the absence of electron-withdrawing groups. For example, the enamine (11; $R^1 = R^2 = Me$) reacted with secondary amines to give the ketene aminals (12; $R^1 = R^2 = Me$). Reaction also took place with primary amines but the substitution products tautomerized to amidines.

Substitution is reported to be faster using the solvents dimethylformamide or acetonitrile rather than diethyl ether or benzene. The rate also depends upon the base strength of the amino portion of the halo enamine. For example, the rate of substitution increases dramatically from N-methylanilino to morpholino to dimethylamino. This corresponds to the order of stability of the corresponding keteniminium ions (13) which have been proposed as intermediates in the reaction ⟨76MI 421-01⟩.

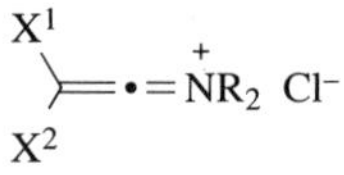

Chloro enamines bearing an unsubstituted amino group, such as (14), react by a different mechanism involving loss of chloride to give an intermediate tricyanomethanide salt (15). The salt is converted into the ketene aminal by strong heating ⟨63CB3230⟩.

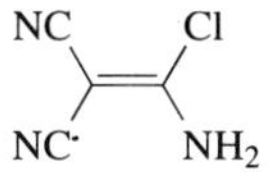

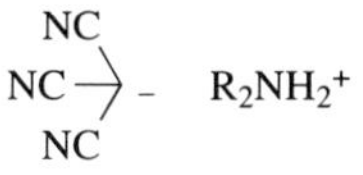

4.21.1.1.4 From ketene acetals

There have been fewer reports on the use of ketene acetals (16) as precursors of ketene aminals than for ketene dithioacetals or vinylidene dihalides ⟨58JA2788, 59USP2883368, 65JOC71, 90S195⟩.

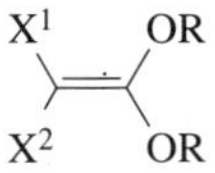

Like ketene dithioacetals and vinylidene dihalides, the ketene acetals can be treated with an excess of ammonia or a primary amine to give ketene aminals having two identical amino groups. However, the reaction of ketene acetals with secondary amines is more difficult and proceeds in low yield or not at all. For example, the reaction of the ketene acetal (**17**) with diethylamine gave a complex mixture, whereas the same reaction carried out with the corresponding dithioacetal gave the ketene aminal in 92% yield ⟨90S195⟩. With diamines, ketene aminals containing a diaza heterocyclic ring (**8**) are formed ⟨90S195⟩.

The first substitution by the amino group is faster than the second and it is possible to synthesize ketene N,O-acetals (**18**) in good yield. Aliphatic amines are more reactive than aromatic amines, but quantitative yields are possible for both ⟨58JA2788, 90S195⟩. The reaction of ketene N,O-acetals with a different amine has given mixed ketene aminals ⟨58JA2788, 59USP2883368, 66CB2302, 68CB1232, 77LA1895, 80LA372, 88FES103⟩. However, exchange of amino substituents is often a competing reaction (Equation (1)) ⟨64CB3081, 90S195⟩.

4.21.1.1.5 From imino esters

Imino esters and their salts are tautomers of ketene N,O-acetals and likewise can be converted into ketene aminals ⟨63CB3306, 64CB3081, 73LA573, 87S357⟩. For example, the cyano ketene aminal (**20**) was obtained from the imino ester (**19**) in 80% yield ⟨88FES103⟩.

4.21.1.1.6 From amides and other carboxylic acid derivatives

Tertiary amides react under mild conditions with 0.5–0.6 equivalents of $Ti(NMe_2)_4$ in an ether solvent to give ketene aminals in good yield (Equation (2)) ⟨66JA850⟩. The method is particularly useful in synthesizing ketene aminals which do not have electron-withdrawing groups at the β carbon. The reaction was also attempted with secondary amides. However, the products rearranged to the more stable amidine tautomers ⟨71JOC1613⟩.

Esters, anhydrides and carboxylic acids can similarly be converted into ketene aminals using a larger quantity of the titanium reagent ⟨66JOC2874, 68JOC1506⟩.

4.21.1.1.7 From thioformamidinium salts, thioureas, imidazoles, and isothioureas

Carbanions of active methylene compounds react with N,N,N^1,N^1-tetramethylmethylthioformamidinium iodide (22) to give ketene aminals (23) ⟨70ACS3102⟩. The salt (22) was prepared from the thiourea (21) (Scheme 3). This method has proved useful for obtaining ketene aminals (23) where X^1 and X^2 are ester groups. These compounds are unobtainable from ketene dithioacetals ⟨78USP4075001⟩.

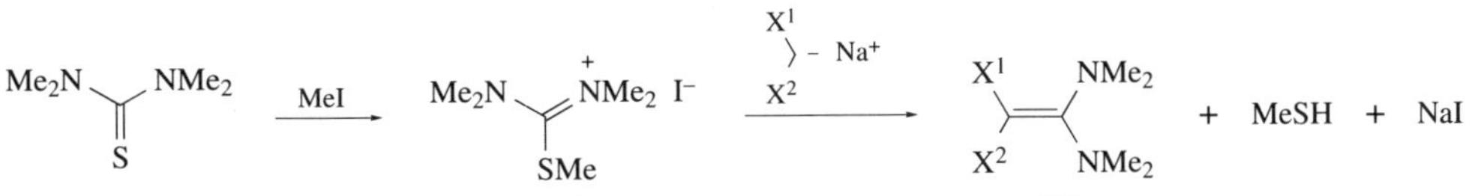

Scheme 3

Ketene aminals with two different amino groups have been synthesized from unsymmetrical thioureas ⟨70CB133, 77IJC(B)297, 88EUP0302389⟩. Diaza heterocyclic ketene aminals (8) have been synthesized from the heterocyclic structures (24), (25), (26), and (27) ⟨75JST(24)373, 78USP4075001, 86CB2208⟩.

4.21.1.1.8 From alkoxytris(dimethylamino)methanes

The alkoxymethanetriamines (28) react with compounds containing acidic methylene or methyl groups to give ketene aminals of general structure (23) ⟨83LA290⟩. The method works best with weakly acidic methylene groups, bearing substituents such as nitrile, ester, and tertiary amide groups. Early work used the isopropoxymethanetriamine (28a) synthesized *in situ* from the chloroformamidinium chloride (29) by treatment with sodium isopropoxide and dimethylamine. This reagent worked well with simple acetonitriles, malononitrile, and ethyl cyanoacetate. With less-acidic methylene or methyl groups the yields were low to moderate. For example, the reaction of isopropyl acetate with (28a) gave only 25% yield of product ⟨79LA2096⟩. The strongly acidic disulfone (30), on the other hand, gave the guanidinium salt (31) and not the ketene aminal ⟨79LA2096⟩.

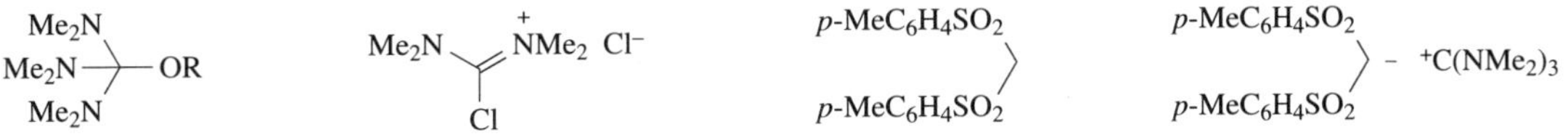

(28a) R = OPr^i
(28b) R = OEt

It is proposed that the reagents (28) dissociate to the ion pairs (32). The resulting alkoxide ion then reacts with the acidic methylene or methyl compound to generate a second ion pair (33). With a strong acid such as the disulfone (30), the ion pair (33) is favoured over the final ketene aminal. With weaker acidic groups, the anions are more nucleophilic and combine with the cation to form the intermediate (34). Subsequent loss of an amino group generates the ketene aminal.

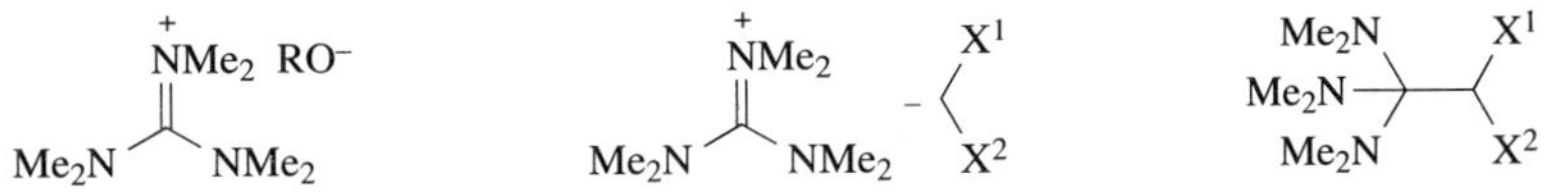

Work in the 1980s has used the ethoxy compound (**28b**). The ethoxy compound reacts with a large range of weakly acidic compounds such as ketones, nitriles, thioamides, lactams, thiolactams, and lactones to give ketene aminals ⟨83LA290⟩.

4.21.1.1.9 From urea acetals

The chloroformamidinium chloride (**29**) was treated with two equivalents of sodium isopropoxide in THF to give tetramethylurea diisopropyl acetal (**35**) which was then treated with one equivalent of nitromethane to give the nitroketene aminal (**36**) in 82% yield ⟨79LA2096⟩. Compound (**35**) is less reactive than the triamino derivatives (**28**) and reacts with less-acidic methylene groups to give ketene N,O-acetals rather than ketene aminals. The cyclic urea acetals (**37**) and (**38**) have also been reported to give ketene aminals ⟨61LA(641)1⟩.

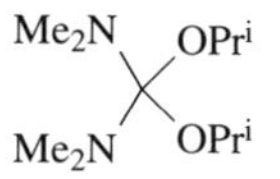

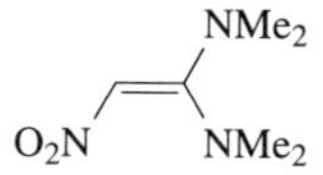

4.21.1.1.10 From tetramethylchloroformamidinium chloride

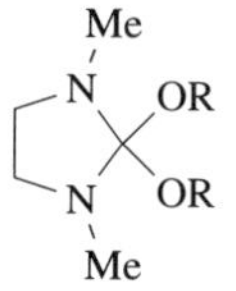

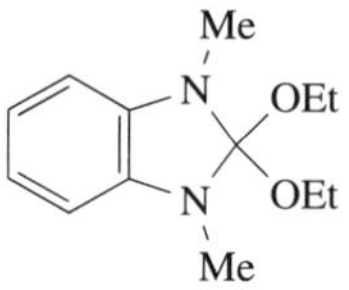

In the methods above, tetramethylchloroformamidinium chloride (**29**) was first converted into an active reagent. However, it is possible to synthesize ketene aminals directly from tetramethylchloroformamidinium chloride. 6,6-Bis(dimethylamino)fulvenes (**39**) have been made from the reaction of cyclopentadienes with one equivalent of tetramethylchloroformamidinium chloride in the presence of two equivalents of triethylamine ⟨70CB147⟩. The same method has also been used to prepare a series of 1,1-bis(dimethylamino)-2-aryl-2-cyanoethylenes. Yields are best when electron-withdrawing groups are present on the aromatic ring ⟨70CB973⟩. The reaction has also been carried out by generating the chloroformamidinium chlorides *in situ* from ureas and phosgene ⟨61CB2278⟩.

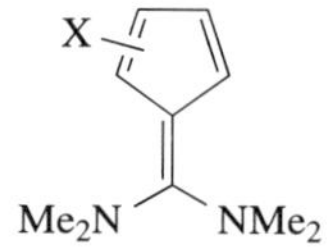

Several chloroformamidinium chlorides apart from the tetramethyl derivative (**29**) have been synthesized but rarely used. The tetraethyl analogue derived from the urea (**40b**) is one such example. The only other two derivatives which have been used are the piperidine analogue (**41**) and the N,N,N^1-trimethyl-N^1-phenyl analogue (**42**), which reacted with cyclopentadiene ⟨61CB2278, 70CB133⟩.

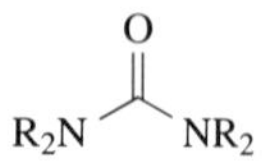

(40a) R = Me
(40b) R = Et

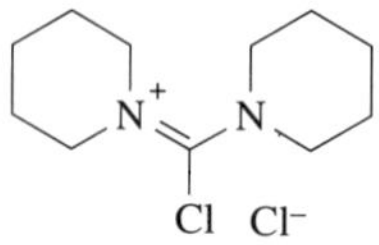

(41)

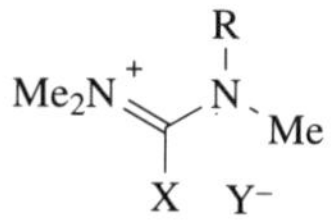

(42) X = Cl, R = Ph, Y = Cl
(43) X = OEt, R = Me, Y = BF$_4$

4.21.1.1.11 From aminals, ortho-esters, and amide acetals

The treatment of aminals (**44**) with heat, acid catalysts, or strong bases results in the elimination of acid (HX) and the generation of carbenes which couple together to give tetraaminoethylenes (**45**) and (**46**). The reaction of *ortho*-esters (HC(OR)$_3$) or amide acetals R$^1_2$NCH(OR^2)$_2$ with amines or diamines gives tetraaminoethylenes via the above aminals (**44**) which are formed *in situ* ⟨68AG(E)766⟩.

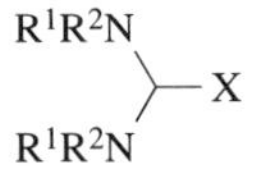

(**44**) X = NMe$_2$, OR, CN, CCl$_3$, I

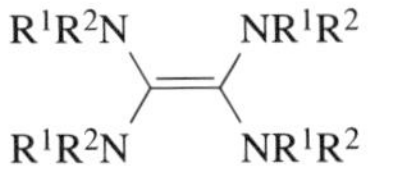

(**45**)

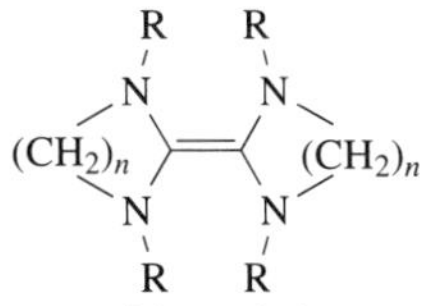

(**46**) n = 2, 3

4.21.1.1.12 From miscellaneous 'head group' synthons

Reagents such as tetramethylchloroformamidinium chloride (**29**) can be considered as synthons for the ketene aminal 'head group'. Other synthons which have been used less frequently include N,N,N^1,N^1-tetramethylethoxyformamidinium fluoroborate (**43**), 1,3-dimethyl-2-ethoxy-4,5-dihydroimidazolium tetrafluoroborate (**47**), tetrakis(dimethylamino)methane (**48**), bis(dialkylamino) malononitriles (**49**), and the bis(formamidinium) ethers (**50**) and (**51**) ⟨67JOC3293, 68T2767, 84SC1073, 89EUP0297872, 90LA965⟩.

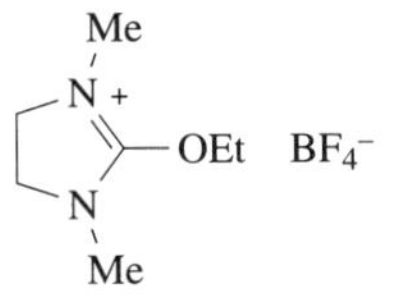

(**47**) 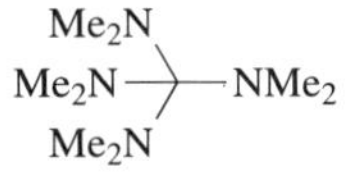 (**48**) 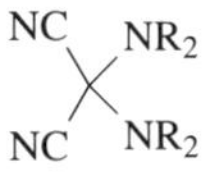 (**49**) 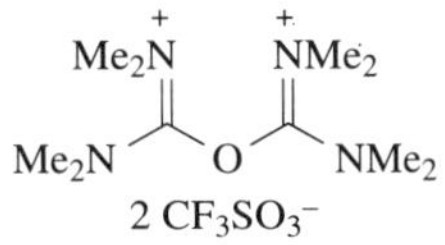 (**50**) 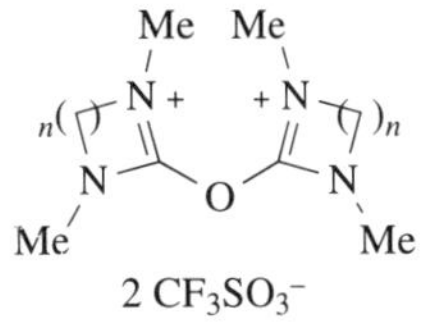 (**51**) n = 2 or 3

4.21.1.1.13 From cyanamides

Cyanamide and monosubstituted cyanamides (**52**) react with a fourfold excess of an activated methylene compound such as a β-diketone, β-keto ester, β-diester, β-nitro ketone, or β-nitro ester in the presence of 1–5 mol% nickel acetylacetonate to give ketene aminals (**53**) ⟨89IZV1806⟩. The nickel catalyst is essential to the reaction and is thought to coordinate with, and hence enhance the electrophilicity of, the CN groups. The method has been useful in the synthesis of β-nitro ketene aminals but is of particular interest in the synthesis of ketene aminals with β-keto and acyl substituents. However, the reaction of acetylacetone with cyanamide itself gave only 13% of the ketene aminal (**54**) owing to further reactions such as *N*-acylation to form compound (**55**). Ketene aminals such as (**54**) are better synthesized via the *N*-benzoyl derivatives as shown in Scheme 4. The *N*-benzoyl group can be removed by treatment with one equivalent of sodium methoxide.

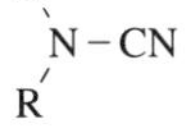

(**52**) 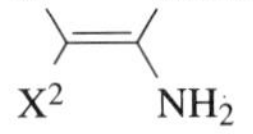 (**53**) 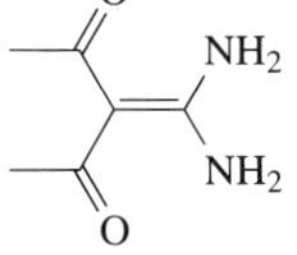 (**54**) 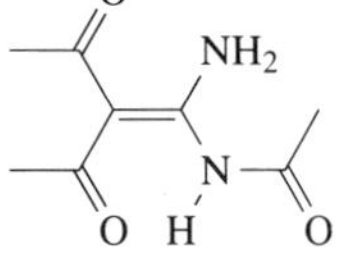 (**55**)

4.21.1.1.14 From halo alkynes

Chloro alkynes are generally unreactive to amines unless they contain an activating, electron-withdrawing substituent such as a keto or cyano group, or a chlorine atom. When this condition is fulfilled, chloro alkynes react with secondary amines to give ketene aminals of general structure (**56**) ⟨B-69MI 421-01, 71JCS(C)196, 86ZOR2256⟩. Since only one electron-withdrawing group can be present in the final product, this method is most suitable for the synthesis of ketene aminals containing tertiary amino functions (i.e., those derived from secondary amines). Reactions will take place with ammonia and with primary amines, but the products obtained usually undergo elimination to give nitriles and tautomerization to give amidines, respectively ⟨B-69MI 421-01⟩.

Chloro alkynes have been synthesized *in situ*. For example, lithium phenylacetylide reacted with N-chlorodiethylamine to give the corresponding chloro alkyne, which reacted with diethylamine to give the ketene aminal (**57**) in 47% yield ⟨60LA(638)33⟩. Dichloroethyne has also been generated *in situ* by the reaction of trichloroethene with a lithium amide, then treated with an amine to give a ketene aminal ⟨B-69MI 421-01⟩.

Chloro alkynes are normally used for the synthesis of ketene aminals. Fluoro alkynes tend to be unstable and are potentially explosive. Iodo alkynes react with secondary amines, but the products have not been identified. However, bromo alkynes have been used successfully ⟨B-69MI 421-01, 86ZOR2256⟩.

4.21.1.1.15 From ynamines

The reaction mechanism for the conversion of halo alkynes into ketene aminals involves an ynamine intermediate (**58**). The intermediate ynamines have been isolated in certain cases and are available by a variety of other routes. Although ynamines have nucleophilic properties they can also behave as electrophiles in the presence of an acid catalyst. The acid catalyst protonates the β carbon of the alkyne to give an eniminium ion (**59**) which then directs the subsequent attack of the amine at the α carbon atom to give ketene aminals ⟨B-69MI 421-01, B-69MI 421-02, 71JCS(C)196, 73HCA944, 81ZOR1169⟩. Thus, treatment of the cyano ynamines (**60**) with an amine in the presence of boron trifluoride–diethyl ether complex gave the cyano ketene aminals (**61**) ⟨70JCS(C)476⟩.

Ynamines with an ester or vinylic substituent react successfully with secondary amines to give ketene aminals ⟨79ZOR1824⟩ as do ynediamines ⟨67AG(E)767, B-69MI 421-01⟩. However, ynamines substituted with a keto or aldehyde group give a side product (**62**) resulting from attack at the β carbon. This side reaction is suppressed by the use of polar solvents ⟨73HCA944⟩. The reaction has been carried out mainly with secondary acyclic or cyclic amines but has been successful in a few cases with ammonia and with primary amines. In general, strongly basic amines appear to react more slowly than weakly basic amines. It has been suggested that strongly basic amines block acid catalysis and hence hinder the formation of the electrophilic ketene iminium intermediate.

(62)

Not all amines and ynamines react to give ketene aminals. For example, ynamines which lack an electron-withdrawing substituent react with primary amines to give the tautomeric amidines ⟨67AG(E)767⟩.

4.21.1.1.16 From tricyanomethanide, dicyanoacetates, dicyanothioacetates, and dicyanoacetamides

The tricyanomethanide route is a useful alternative to the ketene dithioacetal route for the synthesis of ketene aminals where one of the amino groups is NH_2. The use of ammonia in the latter route often requires pressure and that, together with the attendant problem of methanethiol release during the reaction, can make the tricyanomethanide route more convenient.

Amines react directly with the potassium salt of tricyanomethanide to give aryl- or alkylammonium tricyanomethanide salts (63) which are then heated to give the dicyano ketene aminals (64) ⟨83JCS(P1)1741⟩. The aryl- or alkylammonium salts can also be synthesized from silver tricyanomethanide and an alkylamine hydrobromide in methanol ⟨92ZAAC(611)68⟩. The conversion of the salts into ketene aminals works better for arylammonium salts than for alkylammonium salts, such that it is often necessary to heat the latter in the related amine as solvent. Alternatively, a solvent such as xylene or 1,2-dimethoxyethane can be used ⟨76USP4075001, 80EUP0010396, 83JCS(P1)1741⟩. In many preparations the intermediate salt was isolated before further reaction. However, this is not necessary and in aqueous acidic solution the formation of the salt and subsequent conversion into the ketene aminal can be done in one pot ⟨80EUP0010396, 83JCS(P1)1741⟩.

(63)   (64) X = NR^1R^2
        (65) X = NH_2

A variation which has proved useful for aliphatic amines has been to mix potassium tricyanomethanide with the methanesulfonate salt of an aliphatic amine and to heat the dry mixture ⟨78USP4075001⟩. Another variation which has been used with aliphatic amines is the conversion of potassium tricyanomethanide into the keteneimine (66) by treatment with concentrated sulfuric acid. This keteneimine is then treated *in situ* with an alkylamine to give the relevant ketene aminal in good yield ⟨78USP4075001⟩. This method appears to be preferable for the more strongly basic alkylamines. However, it was not possible to synthesize the diamino structure (65) by this method since the intermediate salt (67) polymerized when heated.

(66)   (67)

The potassium salts of dicyanoacetates (68), dicyanothioacetates (69), *N*,*N*-dimethyldicyanoacetamides (70), and *N*-(alkyl)cyanoacetamides (71) have been prepared and treated with amines in the presence of one equivalent of hydrogen chloride to give the corresponding ketene aminals (**72**) ⟨80EUP0010396, 83JCS(P1)1741, 91HCA579⟩. The keto-substituted methanide salts (**73**) have been prepared but were not converted into the corresponding ketene aminals ⟨92ZAAC(611)68⟩.

(**68**) X = OEt
(**69**) X = SEt
(**70**) X = NMe$_2$
(**71**) X = NHR

(**72**)

(**73**)

It is proposed that the previously mentioned ammonium methanide salts are in equilibrium with a mixture of free amine and free acid (tricyanomethane, dicyanoacetate, etc.) ⟨83JCS(P1)1741⟩. The free amine attacks a cyano group on the free acid and the rate of the reaction depends on the ease with which the free amine and acid can be formed. Thus, the rate of the reaction is inversely proportional to the basicity of the amine. Tricyanomethane is a strong acid as are the dicyanoacetates. Therefore, the equilibrium is very much biased towards the salt rather than the free base and acid. The degree of dissociation is inherently low and is further reduced for salts of the more basic aliphatic amines such that the rate of reaction becomes unacceptably slow and strong heating is required. Similarly, the reaction is inhibited if there are strongly electron-withdrawing groups such as a keto group on the methanide ion. Generation of the keteneimine (**66**), as described above, has proved a useful method of avoiding this problem.

4.21.1.1.17 From ring opening of pyrimidines

Pyrimidines are normally resistant to ring opening but electron-withdrawing groups (X) at the five-position decrease the stability of the pyrimidine ring such that pyrimidines of general structure (**74**) are prone to nucleophilic attack. Ring opening occurs to give ketene aminals of general structure (**75**) ⟨76JCS(P1)1004⟩. Electron-withdrawing groups (X) such as formyl, acetyl, nitro, and cyano aid ring cleavage, whereas substituents such as chloro, bromo, and methyl do not. Pyrimidine rings containing the latter class of substituents resist ring cleavage even under forcing conditions. A good leaving group such as Cl or SCH$_2$CO$_2$H at position 6 is important. Pyrimidine rings with a nitro, aldehyde, cyano, or acetyl group at position 5 (**74**) are prepared from the corresponding 4,6-dichloro-5-substituted pyrimidine (**76**) by treatment with amines.

(**74**)

(**75**)

(**76**)

The fragmentation reaction can be done under acidic or basic conditions, but basic conditions are preferable. Under acidic conditions, the scope of the reaction is limited to pyrimidine derivatives which have a bulky dialkylamino group at position four and a nitro group at position five, owing to a competing hydrolysis reaction which leads to pyrimidones (**77**).

By incorporating a suitable amine substituent at position four, it is possible to create an intra-

(**77**)

molecular cyclization which produces a positively charged quaternary nitrogen. This can then facilitate the degradation of the pyrimidine ring so that the reactions can occur under neutral conditions to give a variety of mono- and polycyclic heterocycles (Scheme 5) ⟨74JCS(P1)1611⟩.

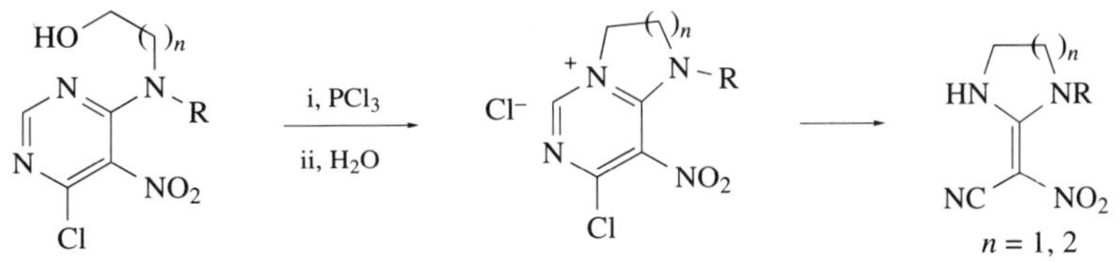

Scheme 5

4.21.1.1.18 Miscellaneous syntheses from ring systems

Ketene aminals have also been obtained by the ring opening of benzodiazepines ⟨91JHC485⟩, trichlorocyclopropenes ⟨76TL2405⟩, 5-indolones ⟨92AP(325)551⟩, 1,2-dithiolimines ⟨86CB162⟩, 2-amino-2H-pyrido[1,2-a]pyrimidines ⟨88JCS(P1)975⟩, isoxazoles ⟨82FES387, 86MI 421-01⟩, isoxazolium salts ⟨69JOC3451, 79CPB2787, 86JA5543⟩, 1,2-benzodithiole-3-thione ⟨79BCJ3640⟩, and diaziridines ⟨74JOC3198⟩, but these methods appear to have no general synthetic utility.

4.21.1.1.19 Miscellaneous methods

A variety of other procedures have been used for the synthesis of specific ketene aminals. These include syntheses from ethoxy alkynes ⟨64JOC2932⟩, nitriles ⟨69JOC3451, 91PHA741⟩, *gem*-dicyano alkenes ⟨62JA3387, 70JPC2722, 71JHC241, 77HCA1781⟩, *gem*-alkylthio nitromethyl alkenes ⟨73YZ612⟩, keteninimines ⟨69AP(302)81, 71JOC3442, 86IZV2344, 92JOC3331⟩, β-aminodichloro alkanes ⟨68JCS(C)1726⟩, enamines ⟨68JCS(C)796⟩, tetraamines ⟨68JCS(C)2721⟩, oxazoles ⟨76BSF2053⟩, aminopyridines ⟨88JCS(P1)975⟩, vinylidene azides ⟨93MM916⟩, amidines and imidazolines ⟨61CB3109, 63CB2100, 78ACS141, 82IJC(B)1, 82T1673, 88GEP3639877A1⟩, azines ⟨71CB792⟩, trihalomethanes ⟨91CB2897, 91ZOR382⟩, orthoesters ⟨62CB2095⟩, thioesters ⟨67CB1661⟩, vinylidene diazoles ⟨84JHC133⟩, carbodiimides ⟨65CB1391, 66M695, 86JOC1997⟩, and dichlorocarbenes ⟨66TL2311⟩.

4.21.1.1.20 By modification of other ketene aminals

(i) Introduction of β substituents

Ketene aminals have nucleophilic properties and can be elaborated to introduce groups at the β carbon by electrophilic substitution. Both nitrogen groups have to be tertiary and there must necessarily be a hydrogen at the β position. The other β substituent is not crucial to the success of the reaction. Keto groups can be introduced by reactions with acid chlorides, acid anhydrides or substituted ketenes ⟨64JOC2513, 66USP3227714, 80LA372⟩, amido groups with substituted isocyanates ⟨64JOC2932, 66CB3892, 71LA(748)59⟩, and sulfonylureas with arenesulfonyl isocyanates ⟨79IJC(B)478⟩. Sulfonyl groups can be introduced with alkanesulfonyl chlorides ⟨65JOC71, 65LA(684)103, 67JOC990⟩. Methyl ester groups can be introduced by treating ketene aminals with phosgene to give acid chlorides, which are then converted into the corresponding methyl esters by treatment with methanol ⟨70JCS(C)881⟩.

(ii) Deacylation

Treatment of β,β-diacylketene aminals with methanol in the presence of 5 mol% Co(OAc)$_2$·4H$_2$O results in monodeacylation to give β-acylketene aminals which are difficult to synthesize directly ⟨90IZV401⟩.

4.21.1.2 Derivatives of Ketene Aminals

4.21.1.2.1 α-Amido enamines, $R^4{}_2C\!=\!C(NR^1{}_2)(NR^2COR^3)$

Although *N*-acylation of ketene aminals would appear to be the obvious method of obtaining α-amido enamines, the reaction has rarely been used ⟨77ZOR954, 88EUP0302389⟩ and is limited in scope owing to the nucleophilic character of the β carbon in ketene aminals. It is therefore more usual to obtain these structures by procedures similar to those used for the synthesis of ketene aminals.

(i) From ynamines

The lactam (**79**) was added to the ynamine (**78**) to give the ketene aminal (**80**) ⟨B-69MI 421-01, B-69MI 421-02⟩.

(ii) From N-acylketenimines

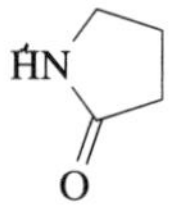

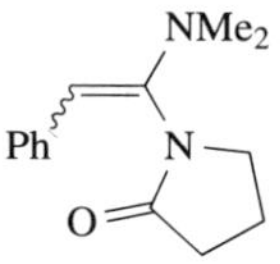

The *N*-acylketenimine (**81**; R = Ph) reacted with *p*-chloroaniline to give the acylated ketene aminal (**82**; R = Ph, Ar = 4-ClC$_6$H$_4$) in 69% yield (Scheme 6) ⟨81CB1976⟩.

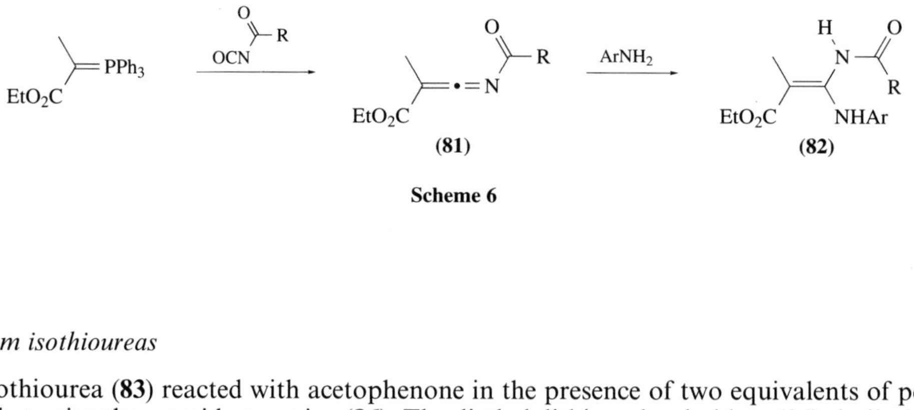

Scheme 6

(iii) From isothioureas

The isothiourea (**83**) reacted with acetophenone in the presence of two equivalents of potassium *t*-butoxide to give the α-amido enamine (**86**). The diethyl dithiocarbonimidate (**84**) similarly reacted with 2-acetylthiophene to give (**85**), which was treated with diethylamine to give the α-amido enamine (**87**) in 42% overall yield ⟨83JOC623⟩.

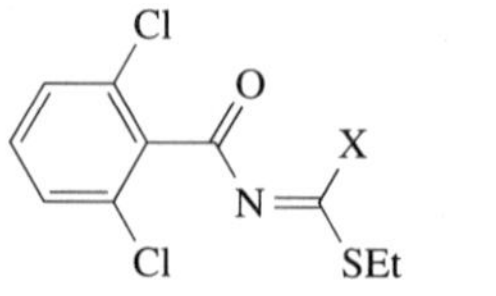

(**83**) X = NEt$_2$
(**84**) X = SEt
(**85**) X = 2-thienyl-C(O)CH$_2$

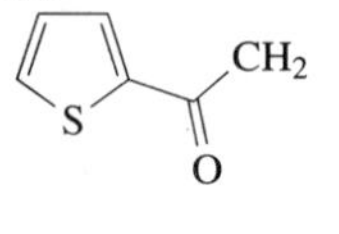

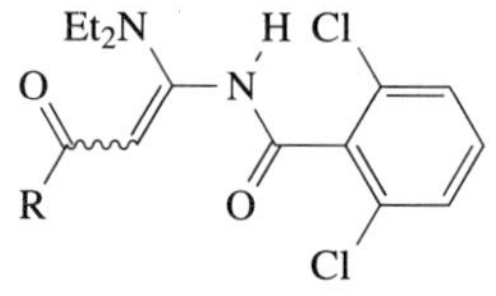

(**86**) X = Ph
(**87**) X = 2-C$_4$H$_3$S

(iv) From N-acylated ketene N,S-acetals

The α-amido enamines (**89**) have been prepared in 70–79% yield from the *N*-acylated ketene *N,S*-acetals (**88**) by treatment with primary amines ⟨82JPR915⟩.

4.21.1.2.2 α-Alkoxycarbonylamino enamines, $R^3_2C=C(NR^1_2)(NHCO_2R^2)$

The *N*-acylketenimine (**81**; R = OEt) reacted with *p*-chloroaniline to give the acylated ketene aminal (**82**; R = OEt, Ar = 4-ClC$_6$H$_4$) in good yield (Scheme 6) ⟨81CB1976⟩.

4.21.1.2.3 α-Ureido enamines, $R^3_2C=C(NR^1_2)(NHCONR^2_2)$

(i) From ketene aminals

Reaction of the ketene aminal (**90**) with one equivalent of methyl or phenyl isocyanate gave the ureido enamines (**91**) in 63–67% yield ⟨87LA1101⟩. It was not possible to synthesize ureido enamines with larger alkyl groups (R) since these structures cyclized during the purification procedure.

(ii) From the 1,3-oxazin-6-one (92)

The 1,3-oxazin-6-one (**92**) reacted with butylamine to form the ring-opened ureido enamine (**93**) in 76% yield. The reaction was not general and did not take place with diethylamine or with aniline ⟨90S959⟩.

*(iii) From the imino ester (**94**)*

The imino ester hydrochloride salt (**94**) reacted with urea to give the ureido enamine (**95**) in 78% yield ⟨63CB3306⟩.

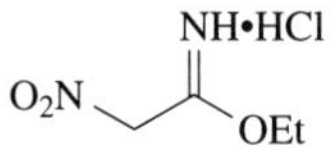

(94)

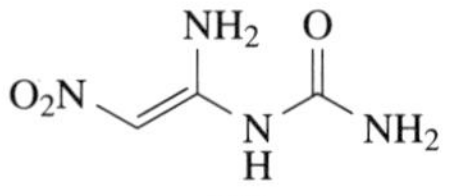

(95)

4.21.1.2.4 N-*Acylated α-ureido enamines*

(i) From N-*acylketenimines*

The *N*-acylketenimine (**96**) reacted with carbodiimides to give the oxadiazines (**97**), which, on treatment with acid, ring-opened to give the *N*-acylated ureido enamines (**98**) (Scheme 7) ⟨81CB1976⟩.

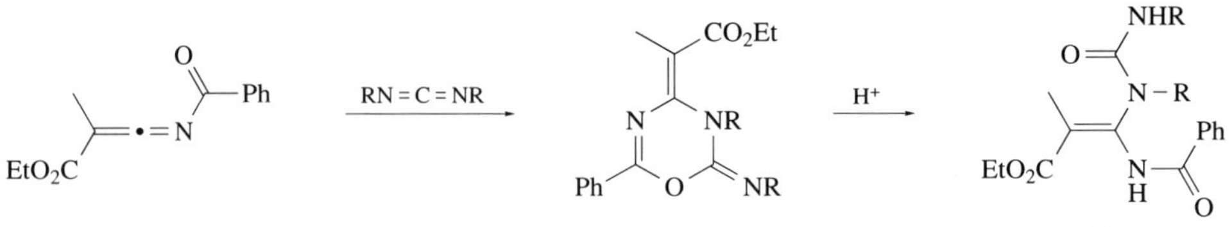

(96) **(97)** **(98)**

Scheme 7

(ii) From a Curtius rearrangement

The acyl azide (**99**) was heated with anilines and underwent a Curtius rearrangement to give *N*-acylated ureido enamines (**100**) in 65–85% yield ⟨66JIC650⟩.

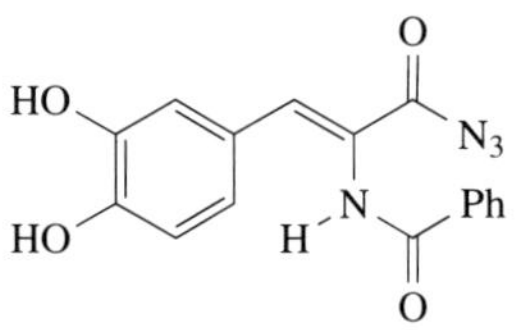

(99)

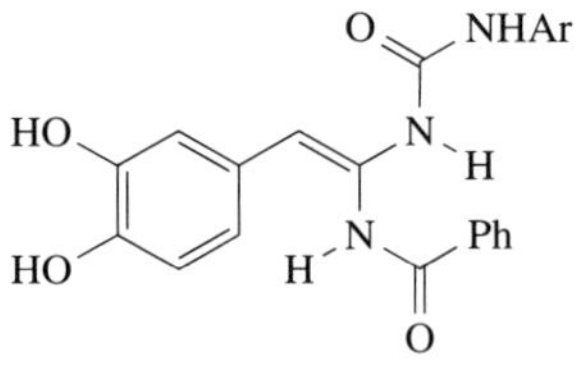

(100)

4.21.1.2.5 gem-*Diureido alkenes, $R^2_2C{=}C(NHCONR^1_2)_2$*

The reactions of the ketene aminal (**90**) with two equivalents of methyl or phenyl isocyanate gave *gem*-diureido alkenes (**101**) ⟨87LA1101⟩.

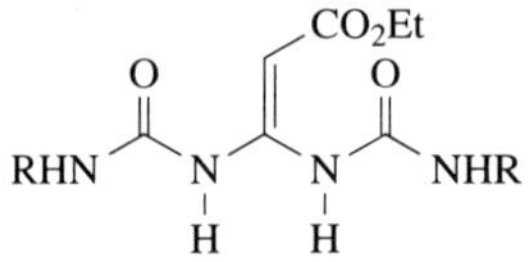

(101)

4.21.1.3 Derivatives Bearing at Least One NRX Function

4.21.1.3.1 α-Hydrazino enamines, $R^4{}_2C\!=\!C(NR^1{}_2)(NR^2NR^3{}_2)$

α-Hydrazino enamines are less common than ketene aminals owing to the ease with which a hydrazino group can react with a β substituent such as a keto group or a cyano group to form a ring-closed heterocycle. There are several papers proposing α-hydrazino enamines as intermediates in the synthesis of heterocycles such as pyrazoles ⟨78JPR585⟩; a small number report the isolation of α-hydrazino enamines by a variety of methods similar to those used in the synthesis of ketene aminals. No one synthetic method has gained favour.

(i) From ketene N,S-acetals

Ketene *N,S*-acetals containing one or more electron-withdrawing groups have been treated with hydrazine or phenylhydrazine to give α-hydrazino enamines of general structure (**102**) ⟨77JPR149, 77TL3619, 78JPR585, 92JPR190⟩.

(ii) From ketene dithioacetals

The nitroketene dithioacetal (**103**) was treated with one equivalent of a hydrazine followed by an excess of an amine to give α-hydrazino enamines ⟨85JHC937, 88EUP0302389⟩. The same dithioacetal has also been treated with aminoalkylhydrazines to give the diaza heterocyclic structures (**104**) ⟨75BEP821282⟩.

(iii) From ketene acetals, N,O-acetals, and imino esters

The phenylhydrazino enamine structure (**105**) was synthesized by the reaction of the ketene acetal (**106**) with ammonia, to give an intermediate ketene *N,O*-acetal, followed by treatment with phenylhydrazine ⟨81AP65⟩.

(**105**) $X^1 = NH_2$, $X^2 = NHNHPh$
(**106**) $X^1 = X^2 = OEt$

(iv) By addition to nitrile groups

The addition of an aryl hydrazine to a nitrile group has been reported to give the α-hydrazino enamine (**107**) (Equation (3)) ⟨86AP(319)1098⟩.

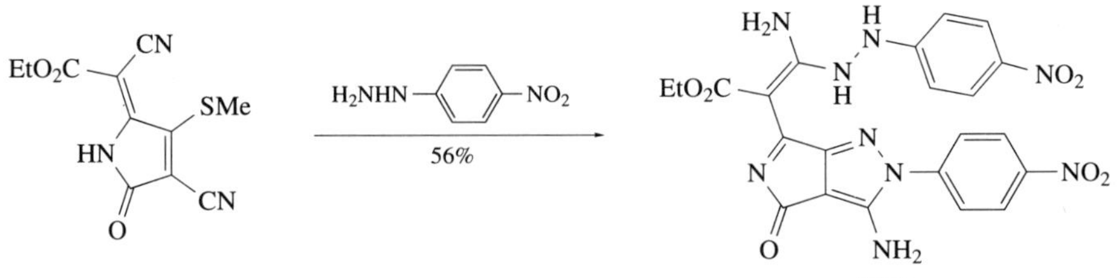

$$\text{(3)}$$

(v) From α-trichloromethyl enamines

The trichloromethyl enamines (**108**) reacted with hydrazines with loss of the CCl_3 group to give the α-hydrazino enamines (**109**) in 73–83% yield ⟨86M201, 75JOC2720⟩.

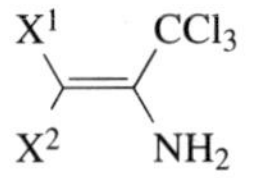

(**108a**) X^1 = PhCO, X^2 = CN
(**108b**) X^1 = X^2 = CO_2R

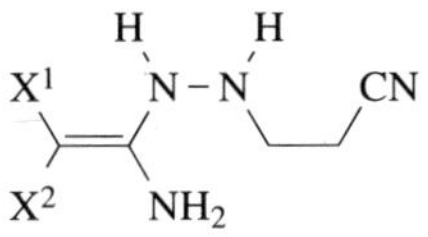

(**109**)

(vi) From ynamines

α-Hydrazino enamines were reported as addition products from the reaction of ynamines with hydrazine, but were not isolated ⟨71BSF283⟩.

(vii) From thioamides

The reaction of the thioamide (**110**) with hydrazine yielded the hydrazino enamine (**113**) in good yield. By varying the experimental conditions, it was also possible to synthesize structures (**114**) and (**115**) in good yield. The thioamides (**111**) and (**112**) cyclized spontaneously on treatment with hydrazine to give heterocycles ⟨77G555⟩.

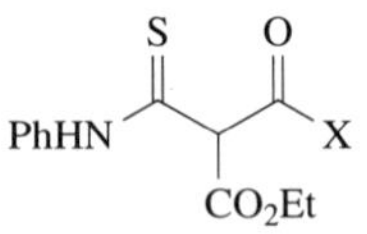

(**110**) X = OEt
(**111**) X = Me, Ph
(**112**) X = N=NPh

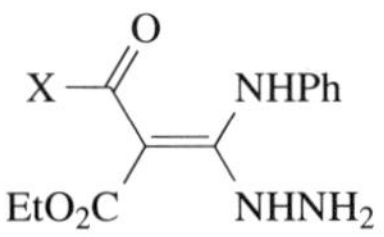

(**113**) X = OEt
(**114**) X = $NHNH_2$
(**115**) X = OH

(viii) From thiosemicarbazides

The thiosemicarbazide (**116**) was prepared as shown in Scheme 8 and treated with methyl iodide to give the isothiosemicarbazide (**117**). Treatment with nitromethane gave the hydrazino enamine (**118**) ⟨88EUP0302389⟩.

(ix) From the tricyanoquinodimethane (119)

The tricyanoquinodimethane (119) was treated with one equivalent of hydrazine to give the α-hydrazino enamine (120) in 71% yield ⟨62JA3387⟩.

4.21.1.3.2 gem-Dihydrazino alkenes, $R^3{}_2C=C(NR^1NR^2{}_2)_2$

(i) From the azobenzene (123)

The gem-dihydrazino alkene (126; Ar = 3,5-$(CF_3)_2C_6H_3$) was obtained in poor yield from the reaction of the azobenzene (123) with 2-methyl-1-propenyl triflate (124) and potassium t-butoxide (Scheme 9). The reaction involves the generation of isopropylidenecarbene (125) ⟨84JA6015⟩.

(ii) From the tricyanoquinodimethane (119)

Treatment of the tricyanoquinodimethane (119) with an excess of hydrazine resulted in the substitution of one cyano group and the amino group to give the gem-dihydrazino alkene (121) in 55% yield. The tetracyanoquinodimethane (122) was reduced to an aromatic product under the same reaction conditions and did not give (121) ⟨62JA3387⟩.

(iii) From bis(dialkylamino)-1-methoxymethanes, $CH(NR_2)_2OMe$

Treatment of bis(dialkylamino)-1-methoxymethanes with trimethylhydrazine gave the tetrahydrazinoethene (127) ⟨68AG(E)766⟩.

4.21.1.3.3 Derivatives of α-hydrazino enamines

α-Hydrazino β-nitro enamines have been treated with a range of reagents to give various derivatives. The *N*-acylated hydrazine (**128**) was obtained from the corresponding α-hydrazino enamine by treatment with acetic anhydride ⟨77JPR149, 85JHC937⟩, while treatment of the α-hydrazino enamine (**129**) with methyl chloroformate or methyl isocyanate gave the methoxycarbonylhydrazine (**130**) and the methylaminocarbonylhydrazine (**131**), respectively ⟨88EUP0302389⟩. *N*-Acylated hydrazino enamines have also been obtained by treating imino esters with arylcarbonylhydrazines (Equation (4)) ⟨91JPR333, 92JHC1631⟩. The imino ester can be employed either as the free base or as the hydrochloride salt in the presence of triethylamine.

The hydrazino enamine (**132**) reacted with carbon disulfide followed by methyl iodide to give the *N*-bis(methylthio)methylene derivative (**133**) in 58% yield ⟨77JPR149⟩.

The hydrazone (**136**) was synthesized from the 5-pyrazolone (**134**) and the ynamine (**135**) as shown in Equation (5). The reaction goes well if X is hydrogen or an electron-withdrawing substituent. Yields drop, however, if X is an electron-donating group ⟨79JPR43⟩.

4.21.1.4 Derivatives Bearing One NY or NZ Function and One NR₂ Function

4.21.1.4.1 α-Aminoalkylideneamino enamines, $R^4_2C=C(NR^1_2)(N=CR^2NR^3_2)$, $R^2_2C=C(NR^1_2)(N=CXNR^3_2)$

(i) From the ketene aminal (137)

The primary amino group of the dicyanoketene aminal (**137**) was found to react with trichloromethyldimethylamine to give the aminochloromethyleneamino structure (**138**) (Equation (6)) ⟨77ZOR954⟩.

(ii) From the iminium perchlorate (139)

The iminium perchlorate (**139**) was treated with two equivalents of a series of anilines to give the α-aminomethyleneamino enamines (**140**) (Equation (7)). Yields were high except for *p*-nitroaniline (36%) ⟨79CB484⟩.

The same iminium perchlorate (**139**) was hydrolyzed with potassium hydroxide to give the perchlorate salt (**141**). Dimethylamine is released during the hydrolysis and reacts with unhydrolyzed (**139**) with substitution of chlorine. Partial hydrolysis of (**139**) with aqueous ethanol gave the aldehyde (**142**), which on treatment with two equivalents of an aliphatic secondary amine or one equivalent of an aromatic amine in the presence of triethylamine gave the substitution products (**143**) in 62–91% yield ⟨81CB2001⟩. The chloro alkene (**144**) similarly reacted with arylamines to give the α-aminomethyleneamino enamines (**145**) in 62–71% yield ⟨74ZOR36⟩.

(iii) From acetals and enamines

Reaction of the enamine (**146**) with *N,N*-dimethylacetamide diethyl acetal (**147**) gave the α-aminoethylideneamino enamine (**148**) in low yield ⟨87KGS1477⟩. Reaction with the cyclic acetals (**149**) proceeded in much better yield (70–80%) to give the corresponding products (**150**).

The mechanism involves the formation of the nucleophilic intermediate (**151**) which reacts with the enamine (**146**) with substitution of the primary amino group. Subsequent reactions with ammonia lead to the observed product. The reaction will only proceed if the enamine contains an NH₂ group. With NMe₂ present, ketene aminals are obtained.

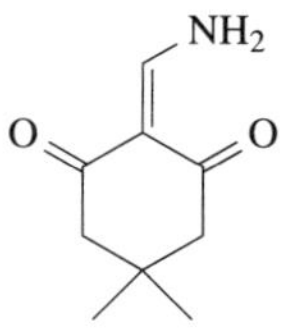

(146) 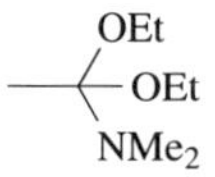 (147) 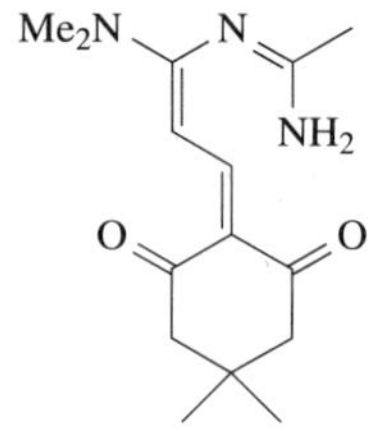 (148) 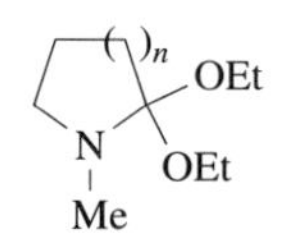 (149) n = 1,2 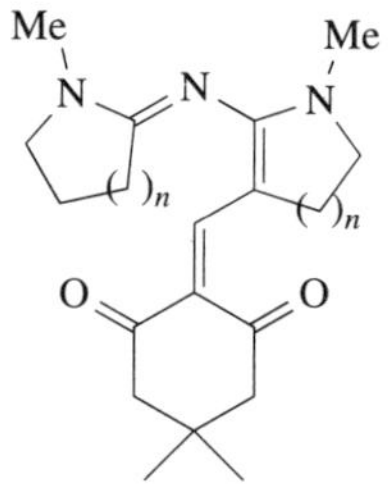 (150)

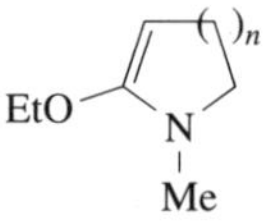

(151)

(iv) From ketene N,S-acetals

Benzamidine (153) condensed with the ketene *N,S*-acetals (152) in NaOEt/EtOH with heating to give pyrimidines, but at room temperature to give the α-aminobenzylideneamino enamine intermediates (154) in good yield (62–70%) (Equation (8)) ⟨84TL1291⟩.

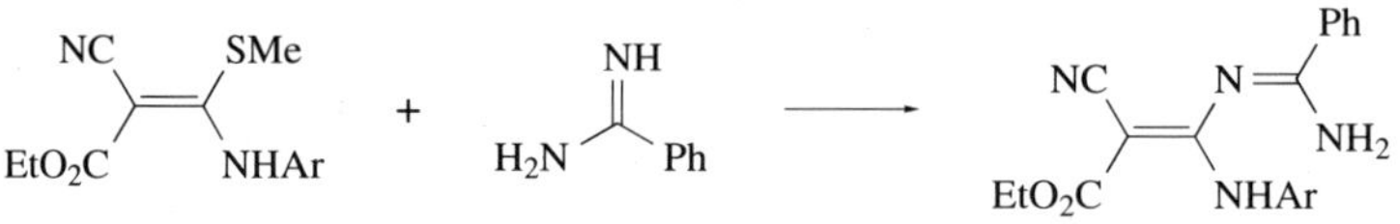

$$(152) + (153) \longrightarrow (154) \tag{8}$$

(v) Miscellaneous

α-Aminoalkylideneamino enamines have also been reported as intermediates in the syntheses of pyrimidines from *N*-acyl amides ⟨71JOC1613⟩ and of pyridines from amidinium salts ⟨80AG208⟩.

4.21.1.4.2 α-Diaminomethyleneamino enamines, $R^3_2C{=}C(NR^1_2)(N{=}C(NR^2_2)_2)$

The chloro alkenes (155) were treated with four equivalents of aniline to give, with substitution of both chlorine atoms, the α-diaminomethyleneamino enamines (156) in high yield (Equation (9)) ⟨74ZOR36⟩.

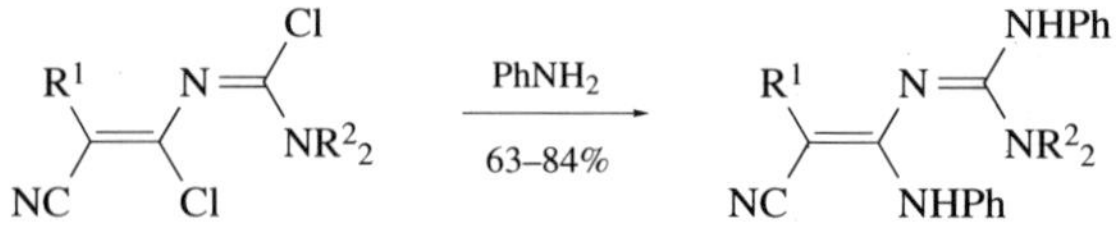

$$(155) \xrightarrow[63-84\%]{PhNH_2} (156) \tag{9}$$

$R^1 = (NC)_2C{=}CH$ or CN

4.21.1.4.3 α-Bis(methylthio)methyleneamino enamines, $R^2_2C{=}C(NR^1_2)(N{=}C(SMe)_2)$

N,N-Dimethylthiourea was converted into the iminium salt (157) as shown in Scheme 10 ⟨82JCS(P1)1059⟩. The salt acts as an electrophile and with malononitrile gave the bis(methylthio)methyleneamino enamine (158) in 37% yield.

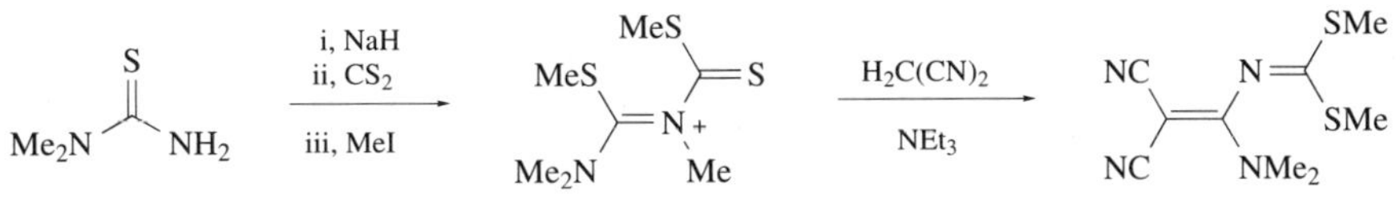

(157) (158)

Scheme 10

4.21.1.4.4 α-Azo enamines, $R^3_2C{=}C(NR^1_2)(N{=}NR^2)$

The α-azo enamines (**160**) were synthesized from the pyrazole ring systems (**159**) by treatment with base (Scheme 11) ⟨86TL4281⟩. The azo enamines (**160**) are unstable in acid and, if the reaction is carried out with heating in acetic acid, the products contain an aryl ring at the β position (**161**) resulting from partial decomposition of the initial azo enamine. The reaction can also be carried out at room temperature in the presence of the diazene (**162**) to give products with two different aryl rings.

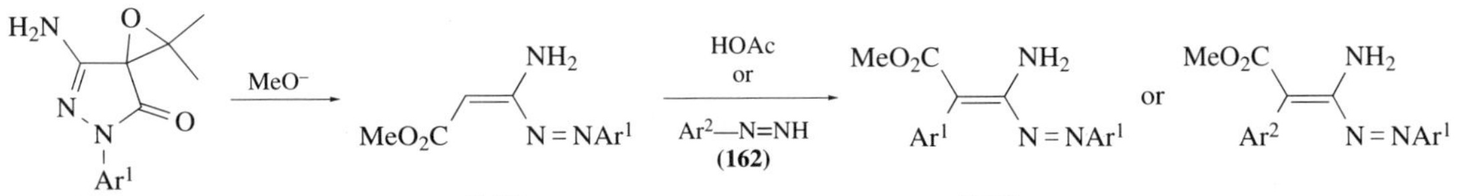

(159) (160) (161)

Scheme 11

Ring opening of the dimeric bipyrazole (**163**) gave the azo substituted enamine (**164**) (Equation (10)) ⟨82JPR309⟩.

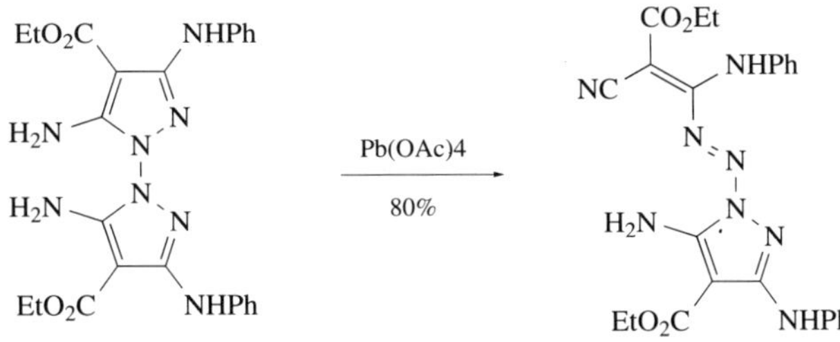

(10)

(163) (164)

4.21.1.4.5 α-Azido enamines, $R^2_2C{=}C(NR^1_2)N_3$

(i) From gem-*diazido alkenes*

α-Azido enamines (**165**) have been made in good yield from the corresponding *gem*-diazido alkenes (**166**) by monosubstitution with a primary aliphatic, primary aromatic, or cyclic amine ⟨91CB595⟩. Reactions with ammonia and diethylamine also took place but the products were not isolated.

(165) (166)

(ii) From N-alkyl-5-phenylisoxazolium fluoroborates (167)

Treatment of the *N*-alkyl-5-phenylisoxazolium fluoroborates (**167**) with sodium azide gave two α-azido enamines ⟨80JOC4302⟩.

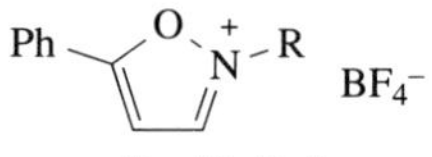

R = Et, Bu^t
(**167**)

(iii) From α-chloro enamines or ketene N,S-acetals

Treatment of α-chloro enamines or ketene N,S-acetals with sodium azide was reported to give α-azido enamines but, apart from the unstable azido enamine (**168**) ⟨76MI 421-01⟩, these spontaneously cyclized and were not isolated ⟨80CC940⟩.

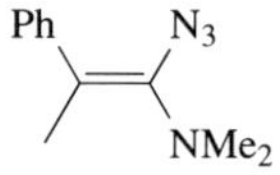

(**168**)

4.21.1.4.6 α-Isocyano enamines, $R^2{}_2C{=}C(NR^1{}_2)NC$

The α-isocyano enamine (**169b**) was synthesized from the chloro enamine (**169a**) by treatment with silver cyanide ⟨76MI 421-01⟩.

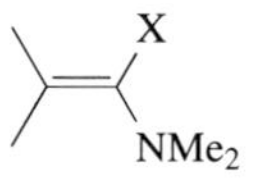

(**169a**) X = Cl
(**169b**) X = NC

4.21.1.4.7 α-Phosphimino enamines, $R^3{}_2C{=}C(NR^1{}_2)(N{=}PR^2{}_3)$

(i) From ketene aminals

The ketene aminals (**170**) were treated with one equivalent of dibromotriphenylphosphorane (**171**) to give the phosphimino enamines (**172**) (Equation (11)) ⟨78ZOB803⟩.

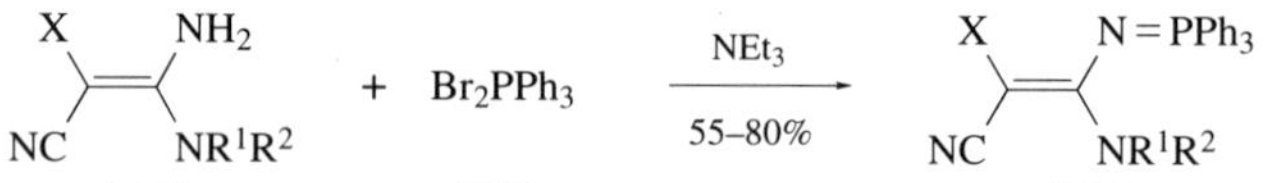

(**170**)   (**171**)   (**172**)   (11)
X = CN, CO$_2$Et

(ii) From an α-azido enamine

An α-phosphimino enamine was synthesized from the α-azido enamine (**173**), by treatment with one equivalent of triphenylphosphine. The reaction goes through the intermediate (**174**), which can be isolated. Heating results in loss of nitrogen and formation of the product (**175**) (Scheme 12) ⟨90CB115⟩.

(iii) From a bisphosphimino alkene

The phosphimino enamine (**176**) was obtained from the hydrolysis of the bisphosphimino alkene (**177**) ⟨85BSB475⟩.

4.21.1.4.8 α-(N,N′,N″-Triphenylphosphorimidic triamido) enamines, $R_2C{=}C(NHPh)(N{=}P(NHPh)_3)$

The phosphorimidic tribromide (**178**) was treated with aniline to give the α-(triphenylphosphorimidic triamido) enamine (**179**) (Scheme 13) ⟨78ZOB778⟩.

4.21.1.5 Derivatives Bearing One NY Function and One NRX Function

4.21.1.5.1 gem-Acylhydrazino azido alkenes, $R^2{}_2C{=}C(NHNHCOR^1)N_3$

gem-Hydrazino azido alkenes (**180**) were synthesized from the corresponding *gem*-diazido alkene by treatment with hydrazines. The acylated derivatives (**181**) were not isolated since they spontaneously cyclized ⟨89CB519, 91JHC1863⟩.

4.21.1.6 Derivatives Bearing Two NY Functions

4.21.1.6.1 gem-*Alkylideneamino isocyanato alkenes, $R^2_2C{=}C(NCO)(N{=}CR^1_2)$*

The *gem*-diisocyanato alkane (**182**) reacted with phosphorus ylides to give *gem*-alkylideneamino isocyanato alkenes, e.g., (**184**), instead of the expected, isomeric α-isocyanatoalkylketene imines, e.g., (**183**), since the latter are unstable and isomerize with migration of the NCO group to give the observed products (Scheme 14) ⟨78ZOR2624⟩.

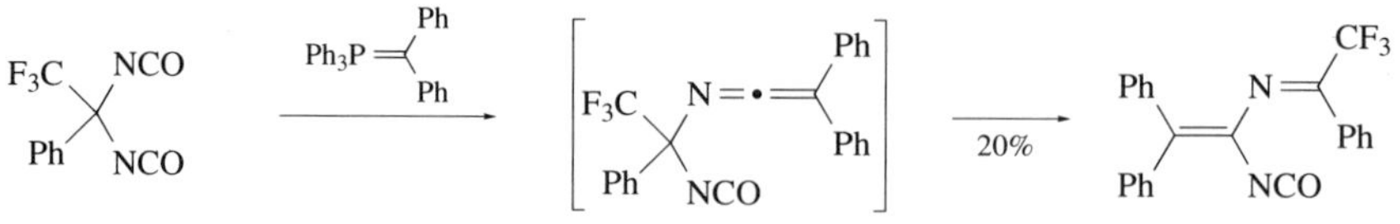

(182) (183) (184)

Scheme 14

4.21.1.6.2 The gem-*diisothioureido alkene (187)*

The only *gem*-diisothioureido alkene which has been synthesized (**187**) was obtained by the reaction of the vinylidene difluoride (**185**) with 3-methylbenzthiazoloneimine (**186**) (Equation (12)) ⟨76UKZ204⟩.

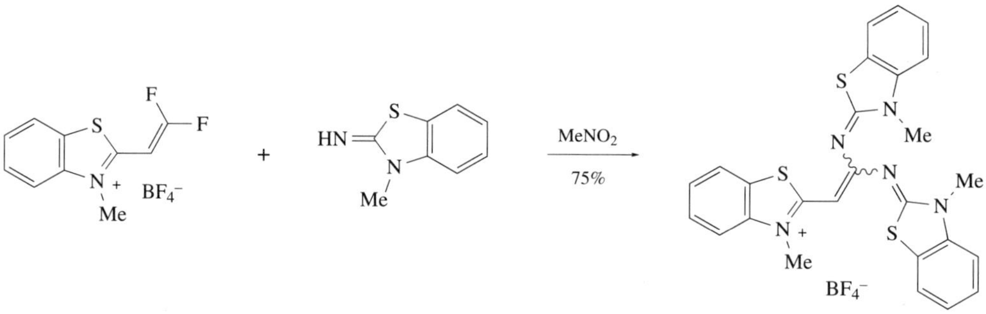

(12)

(185) (186) (187)

4.21.1.6.3 gem-*Bisazo alkenes, $R^2_2C{=}C(N{=}NR^1)_2$*

There are few reports of *gem*-bisazo alkenes and yields have been poor ⟨79ZOB1087, 80CB1226⟩. Treatment of γ-keto carboxylic acids with four equivalents of arenediazonium hydroxides gave the betaine structures (**188**) in 9–25% yield ⟨79CB2369⟩. Treatment of 3-formyl-1,5-diphenylformazan (**189**) with acetic anhydride gave the dimer (**190**), while treatment with phenylhydrazine, benzenediazonium hydroxide, then tetrakis(4-methylphenyl)hydrazine gave the betaine (**191**) in low yield ⟨80CB1226⟩.

gem-Bisazo alkenes (**192**) were obtained in low yield (4–9%) from the reaction of diazonium salts with acetone ⟨79ZOB1087⟩.

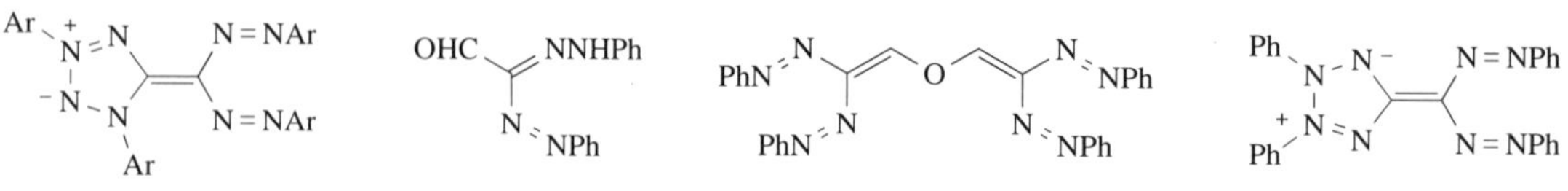

(188) (189) (190) (191)

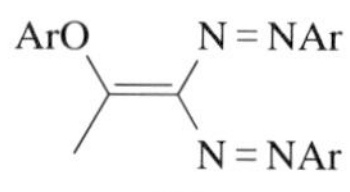

(192)

4.21.1.6.4 gem-Azido phosphazido alkenes, $R_2C=C(N=N-N=PPh_3)N_3$

The *gem*-azido phosphazido alkene (**193**) was reported as an unstable intermediate in the reaction of 3,3-diazido-2-cyanoacrylate with triphenylphosphine ⟨85BSB475⟩.

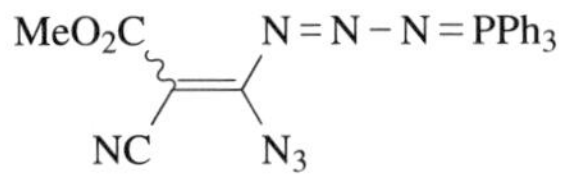

$$\text{MeO}_2\text{C}\diagup\text{C}=\text{C}\diagdown\text{N}=\text{N}-\text{N}=\text{PPh}_3$$
$$\text{NC} \qquad \text{N}_3$$

(**193**)

4.21.1.6.5 gem-Diazido alkenes, $R_2C=C(N_3)_2$

gem-Diazido alkenes (**194**) were synthesized from vinylidene dichlorides by treatment with 5–10 equivalents of sodium azide in a polar solvent. The products are explosive and the reaction should only be done on a small scale. Good electron-withdrawing substituents such as the cyano group at the β position are important. With two ester substituents, no reaction was observed ⟨90JOC1768⟩.

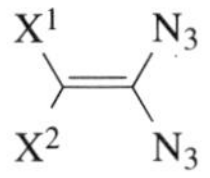

$$X^1\diagup\text{C}=\text{C}\diagdown N_3$$
$$X^2 \qquad N_3$$

(**194**)

4.21.1.6.6 gem-Dinitro alkenes, $R_2C=C(NO_2)_2$

(i) Nitrations of alkenes and alkanes with nitric acid

2,2-Diaryl-1,1-dinitro alkenes were synthesized in poor to moderate yield from the reaction of *gem*-diaryl alkenes with fuming nitric acid and acetic acid ⟨92CJC1022⟩.

Treatment of the ylides (**195**) with acetic anhydride and nitric acid resulted in deacylation as well as nitration to give the dinitro alkenes (**196**) in 30–60% yield ⟨92H(34)1005⟩.

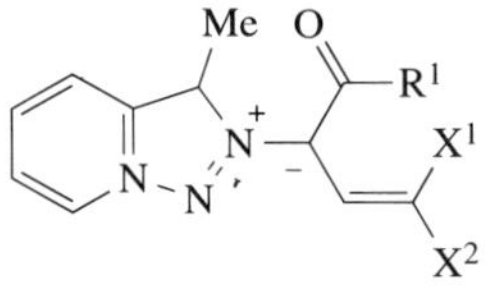

(**195**) $X^1 = H$, $X^2 = COR^2$
(**196**) $X^1 = X^2 = NO_2$

With the correct proportions of nitric acid and water, 1,1-diarylethanes were converted into dinitro alkenes in preference to a competing oxidation which gives benzophenones ⟨76MI 421-02⟩.

Tetraiodoethene was nitrated with HNO_3 to give 2,2-diiodo-1,1-dinitroethene. The reaction was found to be sensitive to temperature, the particle size of C_2I_4 and the strength of the nitric acid ⟨92JOC235⟩.

(ii) Nitrations with dinitrogen tetroxide

Nitrations with dinitrogen tetroxide have been carried out on *gem*-diaryl alkenes ⟨68ZOR2116, 71ZOR1670⟩, the arylnitroethenes (**197**) ⟨87BSF325⟩, the diarylethanols (**198**) ⟨68ZOR2116⟩, and the naphthylethanols (**199**) ⟨75ZOR452⟩ to give the corresponding dinitro alkenes. The alkene (**200**) and the alkanol (**201**) undergo decarboxylation and nitration to give dinitro alkenes in moderate yield ⟨71ZOR1670⟩. The reaction is limited to the production of dinitro alkenes containing at least one aromatic ring at the β position and is inhibited if electron-withdrawing substituents are present on the aromatic ring. For example, the diaryl alkenes (202) and (203) only give mononitrated products ⟨71ZOR1670⟩. The use of chlorinated solvents is important. In diethyl ether further reaction, including nitration of any aryl group present, occurs.

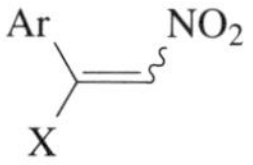

(197) X = H, Ar
(198) X = H, NO₂
(199) 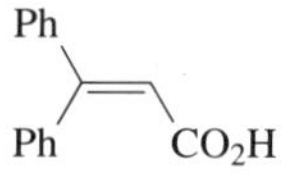
(200) 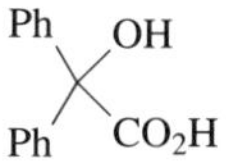
(201) 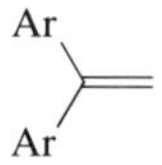
(202) Ar = *p*-chlorophenyl
(203) Ar = *p*-biphenyl

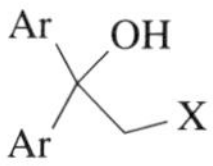

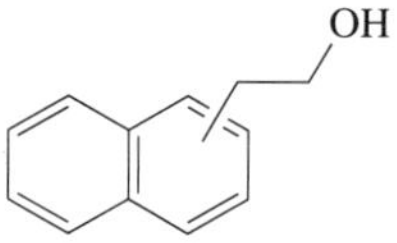

(iii) Thermal extrusion of N_2O_4 from nitro alkanes

Heating hexanitroethane in benzene leads to decomposition and loss of dinitrogen tetroxide to give the highly reactive tetranitroethene, which can be isolated in a cold trap set at 10 °C or trapped *in situ* with agents such as anthracene ⟨85JOC2736⟩. 2,2-Dibromo-1,1-dinitroethene has also been generated by pyrolysis of 1,2,2-tribromo-1,2,2-trinitroethane and trapped *in situ*. However, competing reactions take place which reduce yields ⟨92TL2949⟩.

(iv) Thermal or acid/base-catalyzed dehydrations

Similar highly reactive dinitro alkenes have been generated but not isolated by dehydration of the dinitro alkanols (204) and (205) ⟨57JOC1665, 85ZOR1111⟩.

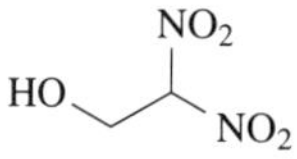

(204)
(205) 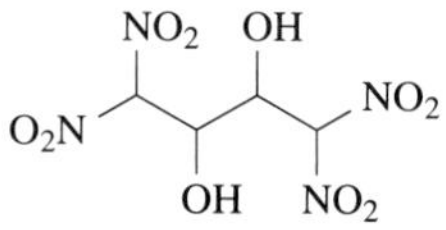

(v) Base-catalyzed extrusions

A variety of methods and reagents have been used to generate the highly reactive parent compound 1,1-dinitroethene. This compound is too reactive to be isolated but it can be generated by dechlorination of (206) with KI ⟨66JOC369⟩, debromoacetoxylation of (207) with sodium phthalimide ⟨63T49⟩, loss of HNO₂ from (208) with a base ⟨57JA4708⟩, and loss of HOAc from (209) in the presence of KI ⟨58JOC813⟩.

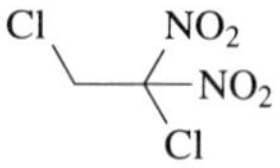

(206)
(207) 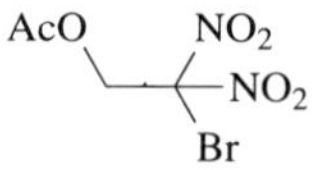
(208) 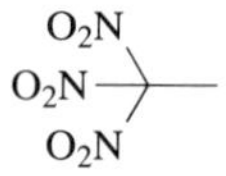
(209) 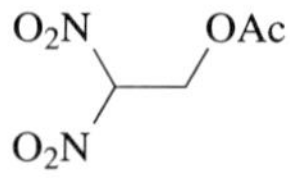

(vi) From dinitromethane

Dinitromethane was deprotonated with a suitable base and treated with suitable electrophiles to give dinitro alkenes. For example, reactions with the formimidates (210) (generated *in situ* from amines and triethyl *ortho*-formate) gave the 2-arylamino-1,1-dinitroalkenes (211). Yields were low (10–20%) but were doubled when two equivalents of the triethyl *ortho*-formate were used and pure dinitromethane was used instead of the salt. The synthesis is limited to primary arylamines without electron-withdrawing groups ⟨79JOC633⟩.

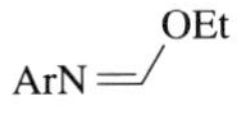

(210) 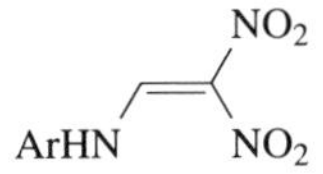 (211)

The carbanion of dinitromethane reacted with the amido nitrile (212) in the presence of Ni(OAc)$_2$ to give the dinitro substituted ketene aminal (213) in 52% yield. The reaction of dinitromethane with dimethylformamide dimethyl acetal generated the thermally labile dinitro enamine (214), which was not isolated but was treated further with KOH ⟨91IZV1849⟩.

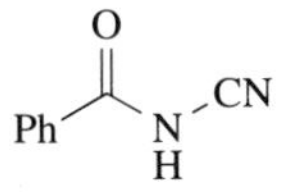

(212) 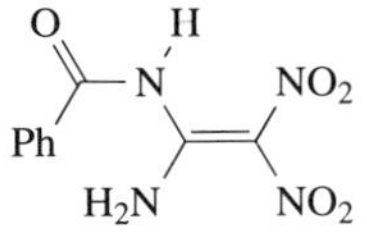 (213) 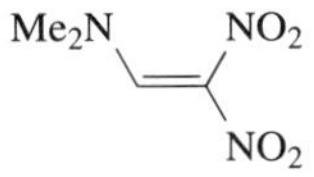 (214)

(vii) From diazo compounds

This method has a high risk of explosion and should be treated with caution. The diazo compounds (215) and (216) were treated with Cl(NO$_2$)$_3$ to give the corresponding dinitroalkenes (217) and (218) in high yield ⟨72ZOR2457⟩. However, the reaction is not general and other diazo compounds either failed to give dinitro alkenes or gave them in low yields.

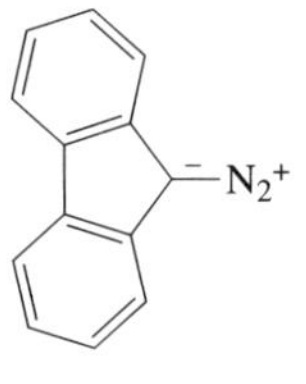

(215) 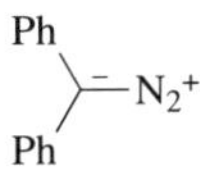 (216) 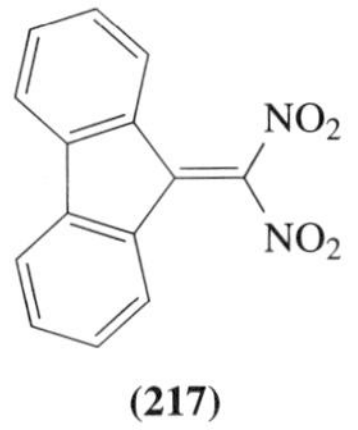 (217) 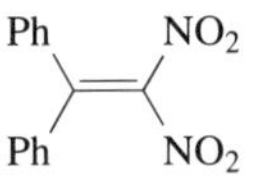 (218)

(viii) Miscellaneous

Dinitro alkenes have also been reported from the nitration of a nitro alkene with cupric nitrate and acetic anhydride ⟨70JMC1248⟩, the reaction of dibromodinitromethane with tetramethylethene ⟨89AX(C)1751⟩, the reaction of hydrazines with tetranitromethane ⟨22CB644, 79JOC633⟩, and the reaction of pyridinium salts with potassium trinitromethanide ⟨75IZV2342⟩.

4.21.1.6.7 gem-Arylazo nitro alkenes, $R_2C=C(NO_2)_2(N=NAr)$

Alkenes containing *gem*-nitro and arylazo groups (219) were synthesized in reasonable yields from the nitroformaldehyde arylhydrazone (220) by treatment with ethyl *ortho*-formate ⟨79ZOR2356⟩. It has also been reported that N,N' disubstituted 2,2-diamino-1-nitroethenes undergo azo coupling to form the *gem*-arylazo nitro ketene aminals (221) in yields of 60–94% ⟨80JPR87⟩.

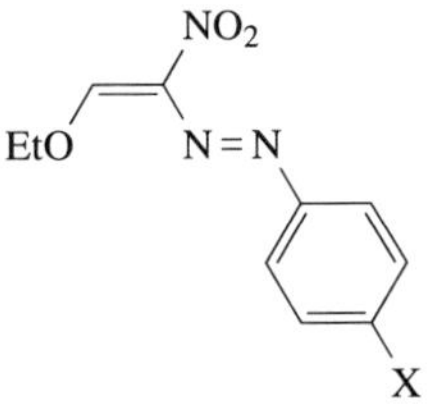

(219) 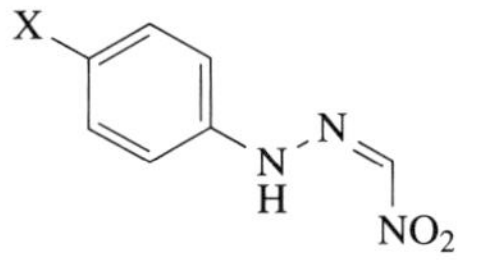 (220) 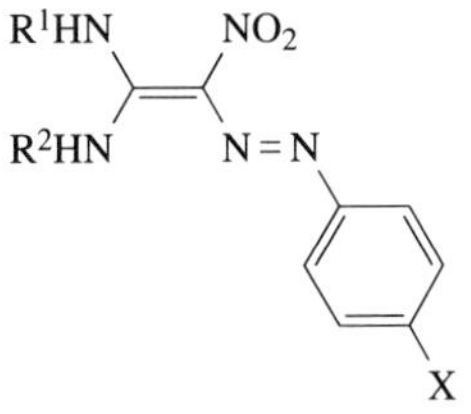 (221)

4.21.1.6.8 gem-Dimethoxazonyl alkenes, $R_2C{=}C(NO{=}NOMe)_2$

An alkene containing two novel methoxazonyl groups (**223**) was prepared as shown in Scheme 15 by treating bismethoxazonylmethane (**222**) with benzaldehyde. The resulting intermediate was acetylated then treated with triethylamine to give the final product. No yield was quoted ⟨69TL2689⟩.

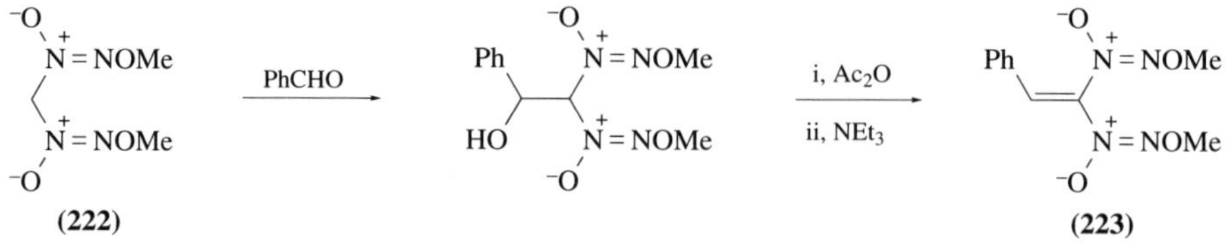

Scheme 15

4.21.1.6.9 gem-Imino phosphimino alkenes, $R^3{}_2C{=}C(N{=}PR^1{}_3)(N{=}CR^2{}_2)$

The reaction of the α-bromo esters (**224**) with triisopropyl phosphite followed by treatment with triethylamine yielded the 1,3-oxazine derivatives (**225**), which on treatment with methanol yielded the *gem*-imino phosphimino alkenes (**226**) in 30–40% yield (Scheme 16) ⟨72T5055⟩.

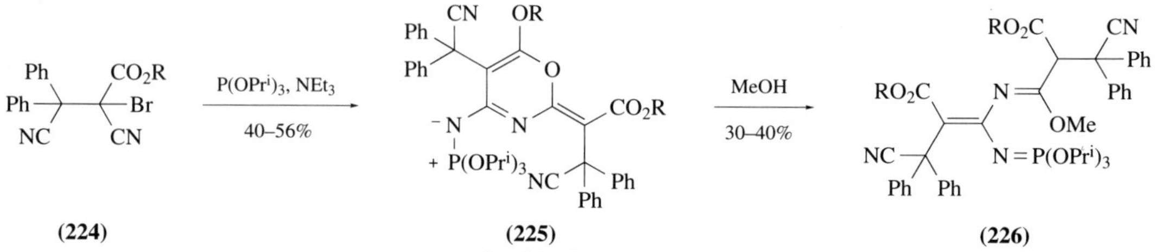

Scheme 16

4.21.1.6.10 gem-Diphosphazido alkenes and gem-diphosphimino alkenes, $R^2{}_2C{=}C(N{=}N{-}N{=}PR^1{}_3)_2$, $R^2{}_2C{=}C(N{=}PR^1{}_3)_2$

The bisphosphimino alkenes (**228**) and (**229**) were synthesized in high yields from methyl 3,3-diazido-2-cyanoacrylate (Scheme 17). The bisphosphazido intermediate (**227**) was also isolated ⟨85BSB475⟩.

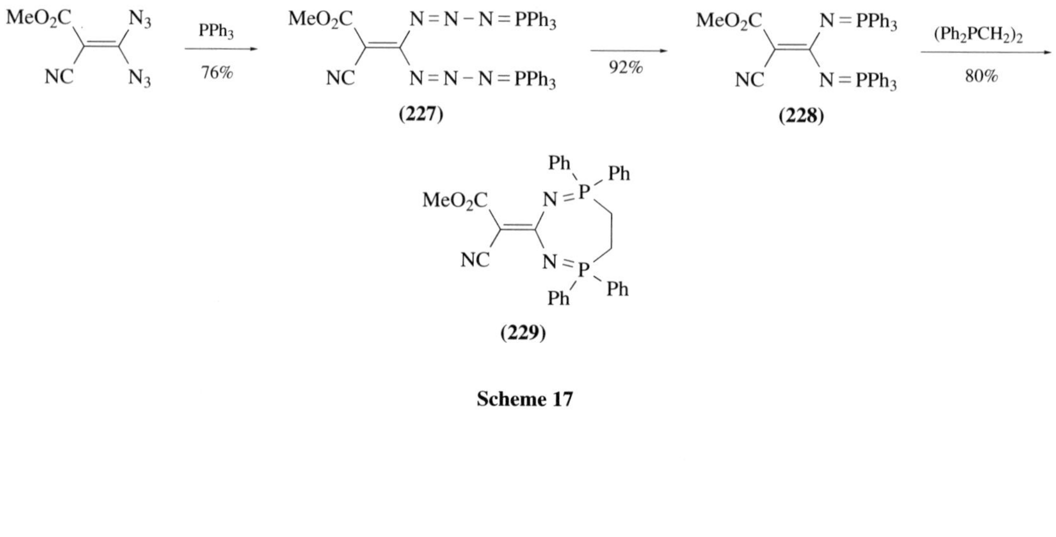

Scheme 17

4.21.1.6.11 gem-Disulfoximido alkenes, $R^2{}_2C=C(N=SOR^1{}_2)_2$

gem-Disulfoximido alkenes (**232**) have been synthesized by treating the β-acyl-β-cyanoketene dithioacetals (**230**) with S,S-dimethylsulfoximide (**231**) (Equation (13)) ⟨83S926⟩. The substitution is reported to proceed more easily than the corresponding reaction with amines.

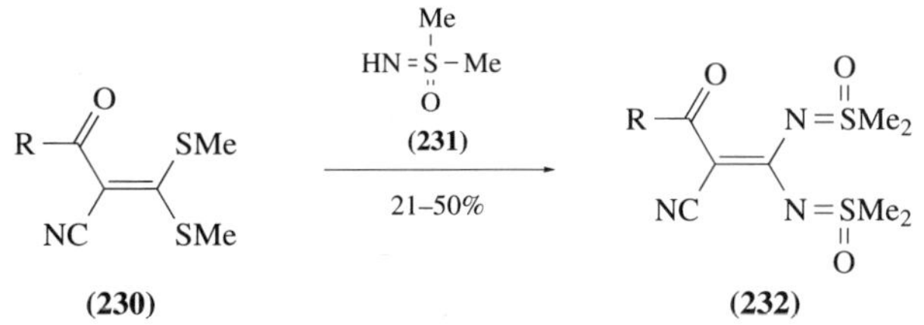

4.21.2 FUNCTIONS CONTAINING ONE NITROGEN AND ONE PHOSPHORUS, $R^3{}_2C=CNR^1{}_2PO(OR^2)_2$, $R^3{}_2C=CNR^1{}_2PR^2{}_2$, etc.

4.21.2.1 Derivatives Bearing a Phosphino Function

4.21.2.1.1 α-Phosphino enamines, $R^2{}_2C=C(PPh_2)NHR^1$

(i) From dihalo alkenes

The dichloro alkene (**233**) reacted with trimethylsilyldiphenylphosphine to give the *gem*-chloro phosphino alkene (**234**). Treatment with aniline gave the α-phosphino enamine (**235**) (Scheme 18) ⟨79ZAAC(459)131⟩.

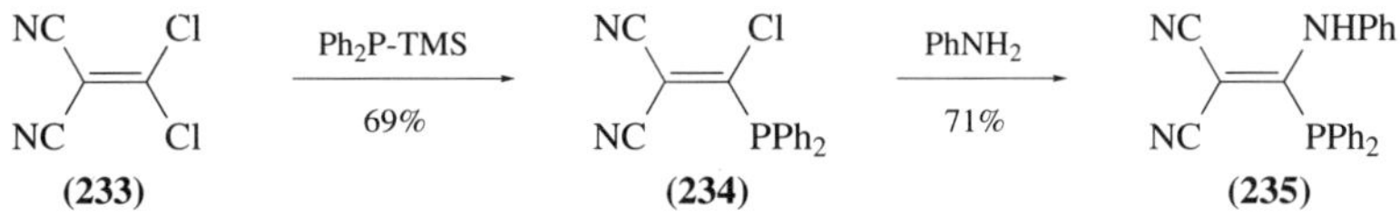

Scheme 18

(ii) From halo enamines

The chloro enamine (**236**) reacted with sodium diphenylphosphide and sodium phenylphosphide to give the α-phosphino enamines (**237**) and (**239**) respectively (Scheme 19) ⟨79ZAAC(459)131⟩.

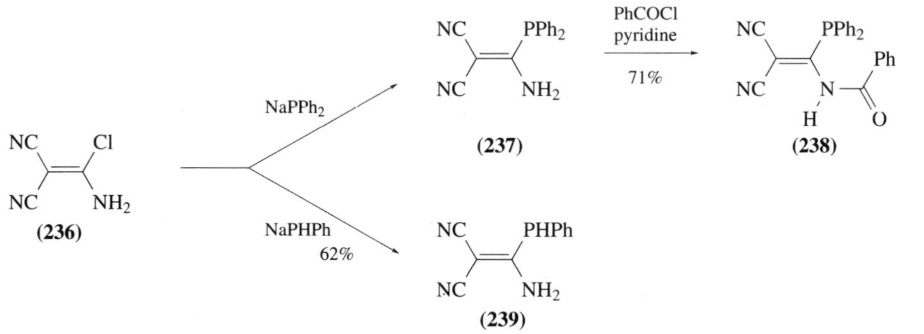

Scheme 19

(ii) From ketimines

The silyl or germyl phosphines (**241**) reacted with the C=N group of the ketimines (**240**) to give the addition products (**242**) in good yield. The metal–nitrogen bond is stable owing to the delocalization of the nitrogen lone pair over both the phenyl ring and the double bond adjacent to it. Hydrolysis with base breaks the metal–nitrogen bond to give predominantly the *C*-phosphorylated imine (**243**) as the major tautomer (Scheme 20) ⟨75HCA1316⟩.

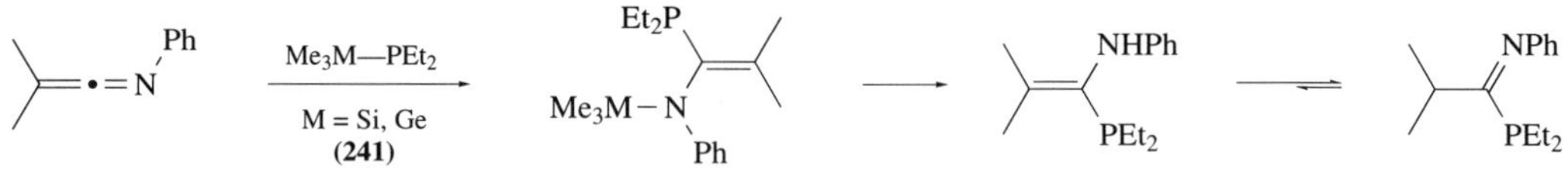

Scheme 20

4.21.2.1.2 gem-*Amido phosphino alkenes*, $R^3{}_2C=C(NHCOR^1)PR^2{}_2$

The α-phosphino enamine (**237**) was treated with benzoyl chloride and pyridine to give the *gem*-amido phosphino alkene (**238**) ⟨79ZAAC(459)131⟩.

4.21.2.2 Derivatives Bearing a Phosphonium Group

4.21.2.2.1 α-*Trialkylphosphonio enamines*, $R^3{}_2C=C(NR^1{}_2)P^+R^2{}_3$

The halo enamine (**244**) was treated with triphenyl- and tributylphosphine to give the corresponding α-phosphonio enamines (**245**) in moderate yield (31–56%) ⟨82AG559⟩.

4.21.2.2.2 gem-*Amido phosphonio alkenes*, $R^2{}_2C=C(NHCOR^1)P^+Ph_3$

(i) From gem-amido carboxy alkenes

The *gem*-amido carboxy alkenes (**246**) were treated with chlorine to give the dichloro intermediates (**247**) which were treated with triphenylphosphine. Substitution followed by spontaneous decarboxylation and loss of hydrogen chloride gave the title compounds (**248**) in good yield, 45–92% (Scheme 21). In general, the compounds synthesized were unsubstituted or had one substituent at the β position. The dimethyl derivative (**249**) was obtained in low yield (30%) ⟨92ZOB1084, 92ZOB1423⟩.

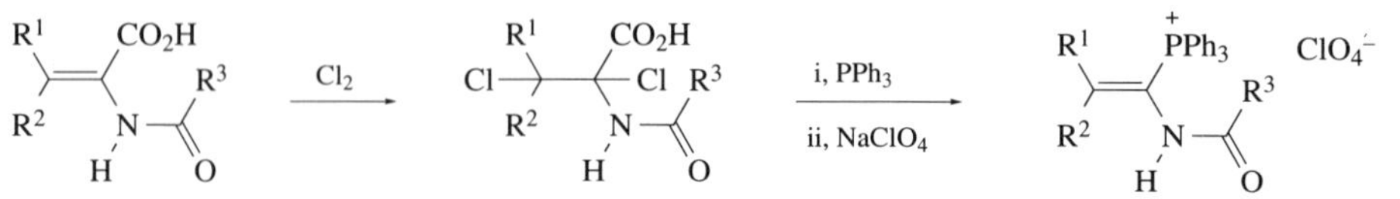

(**248**) R^1 = H, R^2 = alkyl, aryl
(**249**) R^1 = R^2 = Me

Scheme 21

(ii) From N-trichloroethyl amides and N-tetrachloroethyl amides

The *N*-chloroethyl amides (**250**) were treated with triphenylphosphine to give the triphenylphosphonium chlorides (**251**) in 76–93% yields ⟨83ZOB2015, 84ZOB288⟩.

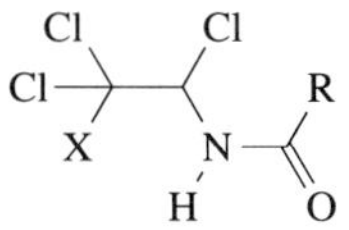

(**250a**) X = H
(**250b**) X = Cl

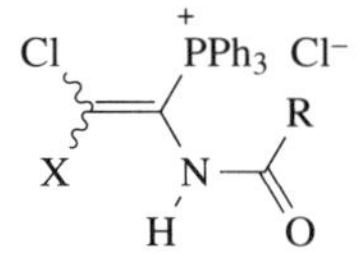

(**251a**) X = H
(**251b**) X = Cl

(iii) By modification of gem-amido phosphonio alkenes

gem-Amido phosphonio alkenes are electrophilic in character and undergo nucleophilic substitution at the β position to give triphenylphosphonium salts such as (**252**) and (**253**) in high yield ⟨83ZOB2015, 84ZOB288, 86ZOB321⟩. *N*-Methylation is also possible to give the products (**254**) ⟨80ZOB2248⟩.

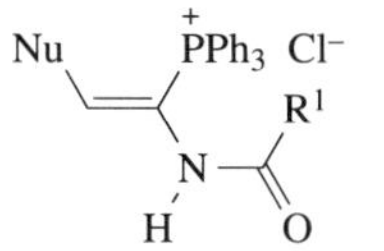

(**252**)
Nu = R^{2}S, R^{2}R^{3}N, MeSe

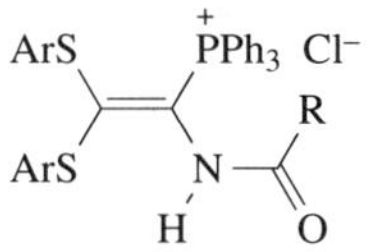

(**253**)

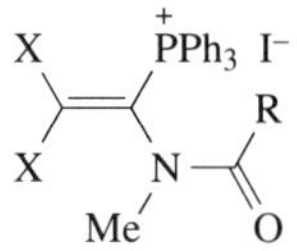

(**254**)
X = Cl, ArS

4.21.2.2.3 gem-Alkoxycarbonylamino phosphonio alkenes, $R_2C{=}C(NHCO_2R)P^+Ph_3$

The *gem*-alkoxycarbonylamino phosphonio alkenes (**255a**) were synthesized in 88–92% yield from the reaction of the tetrachloroethyl urethanes (**256a**) with triphenylphosphine ⟨75ZOB12⟩.

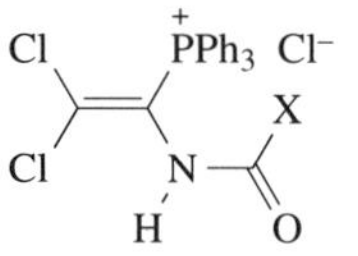

(**255a**) X = OR
(**255b**) X = NR^{1}R^2

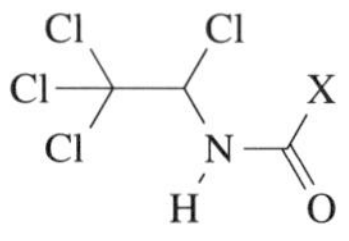

(**256a**) X = OR
(**256b**) X = NR^{1}R^2

4.21.2.2.4 gem-Phosphonio ureido alkenes, $R^2{}_2C{=}C(NHCONR^1{}_2)P^+Ph_3$

Treatment of the *N*-tetrachloroethyl ureas (**256b**) with triphenylphosphine gave the *gem*-phosphonio ureido alkenes (**255b**) ⟨84ZOB2186⟩.

4.21.2.2.5 gem-Arylchloromethyleneamino phosphonio alkenes, $R^2{}_2C{=}C(N{=}C(Cl)Ar)P^+R^1{}_3$

The *gem*-arylchloromethyleneamino phosphonio alkenes (**257**) were made in high yields (70–95%) by treating the appropriate amides (**258**) with $PCl_5/POCl_3$ or with PCl_5 ⟨86ZOB321⟩.

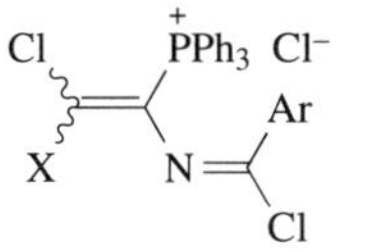

4.21.2.3 Derivatives Bearing a Phosphinoyl or Thiophosphinoyl Group

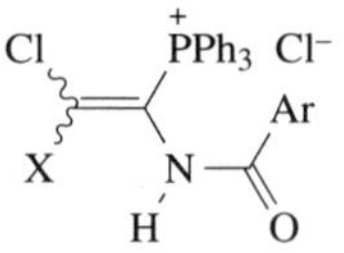

4.21.2.3.1 α-Phosphinoyl enamines, $R^3{}_2C{=}C(NR^1{}_2)POR^2{}_2$

Malononitrile was treated with an excess of the dialkyl phosphinous acids (**259a**) in the presence of an alkoxide to give the addition products (**260a**) in moderate yield (51–74%) ⟨72ZOB1727⟩.

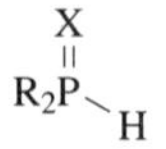

(**259a**) X = O
(**259b**) X = S

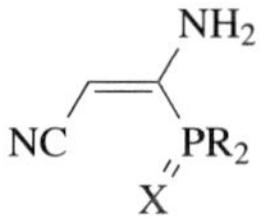

(**260a**) X = O
(**260b**) X = S

4.21.2.3.2 gem-Amido diphenylphosphinoyl alkenes, $R^2{}_2C{=}C(NHCOR^1)POPh_2$

The *N*-tetrachloroethyl amides (**250b**) were treated with ethyl diphenylphosphinite followed by triethylamine to give the *gem*-amido diphenylphosphinoyl alkenes (**261**) in 74–79% yield ⟨75ZOB12, 78ZOB1994⟩.

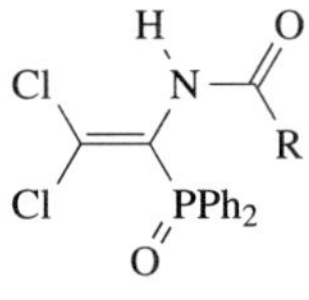

(**261**)

4.21.2.3.3 The gem-methoxycarbonylamino diphenylphosphinoyl alkene (262)

The carbamate (**256a**; X = OMe) was treated with ethyl diphenylphosphinite then triethylamine to give the *gem*-methoxycarbonylamino diphenylphosphinoyl alkene (**262**) ⟨75ZOB12, 78ZOB1994⟩.

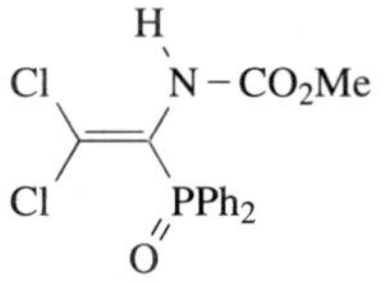

(**262**)

4.21.2.3.4 gem-Diphenylphosphinoyl ureido alkenes, $R^2{}_2C{=}C(NHCONR^1{}_2)POPh_2$

Treatment of the *gem*-isocyanato phosphinoyl alkene (**263**) with amines gave the *gem*-diphenylphosphinoyl ureido alkenes (**264**) in 76–80% yield ⟨84ZOB2186⟩.

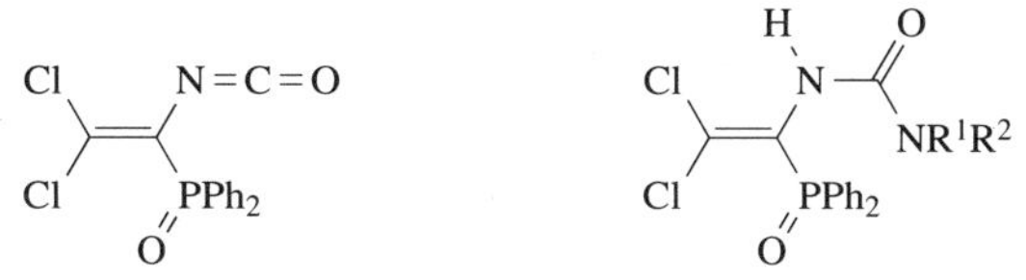

(263)   (264)

4.21.2.3.5 α-Thiophosphinoyl enamines, $R^3{}_2C{=}C(NR^1{}_2)P({=}S)R^2{}_2$

Malononitrile was treated with an excess of the thiophosphinous acids (**259b**) in the presence of an alkoxide to give the addition products (**260b**) in moderate yield (51–74%) ⟨72ZOB1727⟩.

4.21.2.3.6 Derivatives bearing a diphenylphosphinoyl group and a N=C(X)Ar group, $Cl_2C{=}C(N{=}C(X)Ar)POPh_2$

The *gem*-amido phosphinoyl alkenes (**265**) were converted into the corresponding benzimidoyl chlorides (**266**) by treatment with phosphorus pentachloride in 80–84% yield (Scheme 22). Further treatment with 4-chlorothiophenol resulted in substitution of the chlorine atom to give the thiobenzimidate esters (**267**) in 65–82% yield ⟨78ZOB1994⟩.

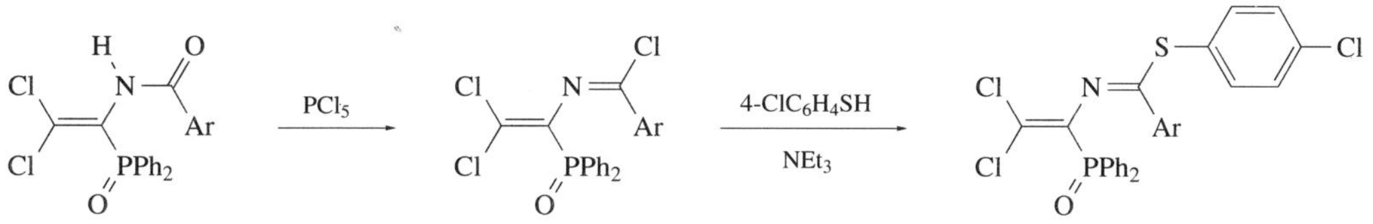

(265) → PCl₅ → (266) → 4-ClC₆H₄SH, NEt₃ → (267)

Scheme 22

4.21.2.4 Derivatives Bearing a Phosphonoyl Group, POXR

4.21.2.4.1 gem-Amido phosphonoyl alkenes, $R^3{}_2C{=}C(NHCOR^1)POXR^2$

(i) From the acylaminophosphinic ester (268)

The acylaminophosphinic ester (**268**) was brominated, then treated with the vinyl cuprate reagent (**269**) to generate the amido phosphonoyl alkene (**270**) in 43% yield (Scheme 23) ⟨89S97⟩.

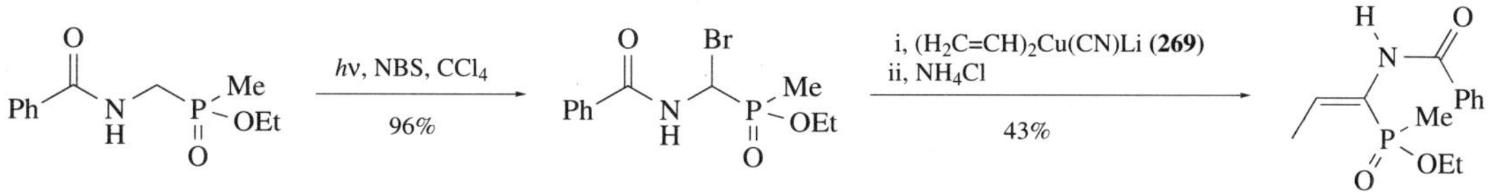

(268) → hv, NBS, CCl₄, 96% → i, (H₂C=CH)₂Cu(CN)Li (**269**); ii, NH₄Cl, 43% → (270)

Scheme 23

(ii) From N-tetrachloroethyl amides

The N-tetrachloroethyl amides (**250b**) were treated with diethyl phenylphosphonite PhP(OEt)$_2$ then triethylamine to give the *gem*-amido phosphonoyl alkenes (**271**) in 48–64% yield. Treatment of the products with pyridine then phosphorus pentachloride gave the azlactones (**272**). The azlactones could be ring-opened by treatment with water or sodium methoxide to give the *gem*-amido phosphonoyl alkenes (**273**) (Scheme 24) ⟨77ZOB1994⟩.

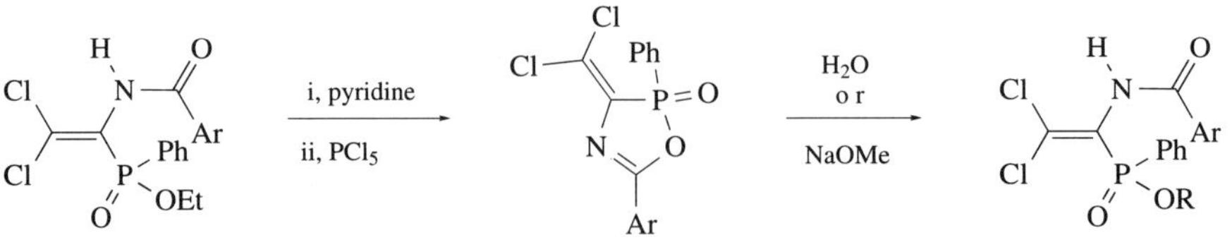

Scheme 24

4.21.2.4.2 The gem-phosphonoyl ureido alkene (275)

The isocyanate (**274**) was treated with aniline then water to give the *gem*-phosphonoyl ureido alkene (**275**) in 68% yield ⟨84ZOB2186⟩.

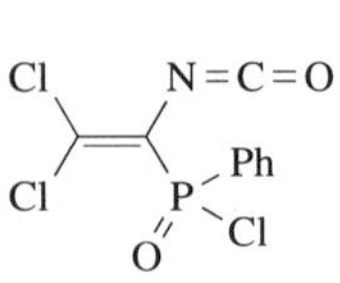

4.21.2.5 Derivatives Containing a Dialkoxy Phosphoryl Group

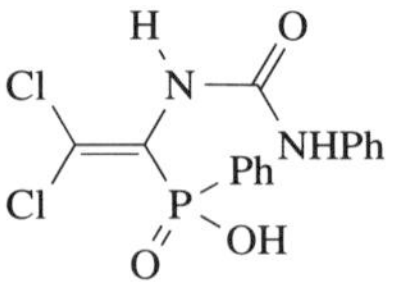

4.21.2.5.1 α-Phosphoryl enamines, $R^3{}_2C{=}C(NR^1{}_2)PO(OR^2)_2$

(i) From tetraethyl dimethylaminomethylenediphosphonate (276)

The amino diphosphonate reagent (**276**) was treated with sodium hydride then with an aliphatic, aromatic or heteroaromatic aldehyde to give the α-phosphoryl enamines (**277**) ⟨81T1227, 82SC415⟩. Yields are best when the aldehyde has no α hydrogens since self-aldol condensation is a competing reaction. Thus, aromatic and heteroaromatic aldehydes were converted in good yields (61–75%), whereas aliphatic aldehydes were converted in yields of typically 32–34%. Products having the (*E*) geometry were formed exclusively from aromatic aldehydes, whereas a mixture of isomers was obtained from aliphatic aldehydes. Better yields (46–48%) were obtained from aliphatic aldehydes if the enamine phosphonate was first converted into the Grignard reagents (**278**). The amino diphosphonate reagents (**279**) have also been prepared and allowed to react with aromatic aldehydes ⟨81T1227⟩.

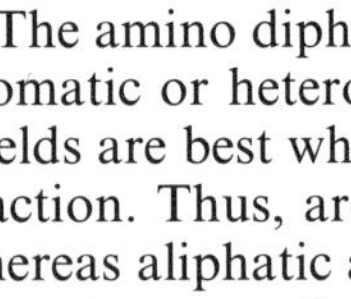

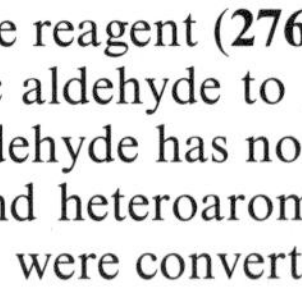

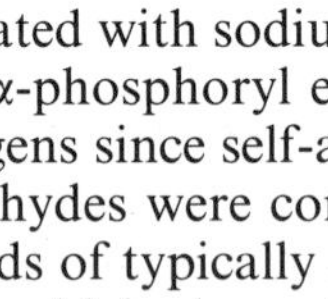

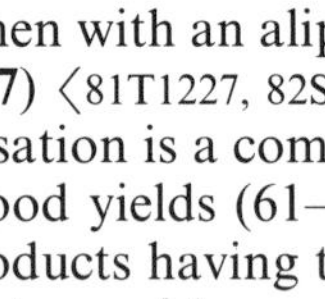

(**279a**) NR$_2$ = morpholino
(**279b**) NR$_2$ = NHPh

X = Br, OEt

(ii) From ynamines

α-Phosphoryl enamines were synthesized in one step by the addition of dialkyl phosphites to ynamines. Products with a β alkyl substituent were obtained as a mixture of isomers in 25–46% yield ⟨71CB2021⟩.

(iii) From ketene N,O-*acetals,* N,S-*acetals, and ketene aminals*

β-Alkyl-α-phosphoryl enamines are obtained in better yield from the reaction of dialkyl phosphites with ketene *N,X*-acetals where X = OMe, SEt or NMe_2. The reaction proceeds under mild and simple conditions and with good yields (47–81%). Only the (*E*) isomer is obtained ⟨75BCJ2103⟩.

(iv) From tertiary amides

The iminium ester (**280**) obtained from the reaction of *N,N*-dialkylacetamides and dimethyl sulfate was treated with sodium diethyl phosphite to give α-phosphoryl enamines (Scheme 25) in 29–38% yield. The reaction most likely goes through the same intermediate involved in the reaction of ketene *N,X*-acetals with dialkyl phosphites but proceeds in poorer yield.

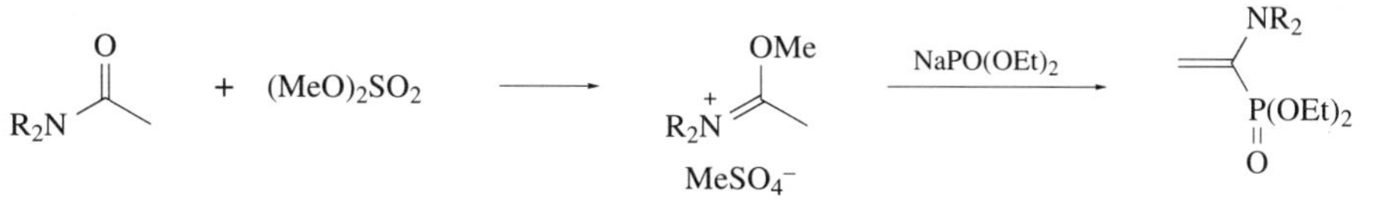

(**280**)

$NR_2 = NMe_2, NEt_2$

Scheme 25

Tertiary amides were also converted into α-phosphoryl enamines by treatment with phosgene to give iminium chlorides, which were treated with two equivalents of a trialkyl phosphite. Yields were low or unquoted ⟨76JPR116, 85USP4501695⟩.

(v) From 1-aminoalkyl phosphonates

Treatment of the 1-aminoalkyl phosphonates (**281**) with *n*-butyllithium and dimethyl sulfide gave the thermally labile adduct (**282**), which decomposed with the loss of methanethiol to give the α-phosphoryl enamines (**283**) (Scheme 26) ⟨84CB1⟩. A mixture of (*E*) and (*Z*) isomers was obtained. Yields were high (69–84%) where the β substituent was alkyl and moderate for alkenyl and phenyl substituents (58–69%).

The reaction proceeds best for compounds with only one β substituent. Further substitution results in a drop in yield. For example, the α-phosphoryl enamine (**284**) was obtained in only 36% yield. This α-phosphoryl enamine was obtained in higher yield from the corresponding α-halo enamine (see below).

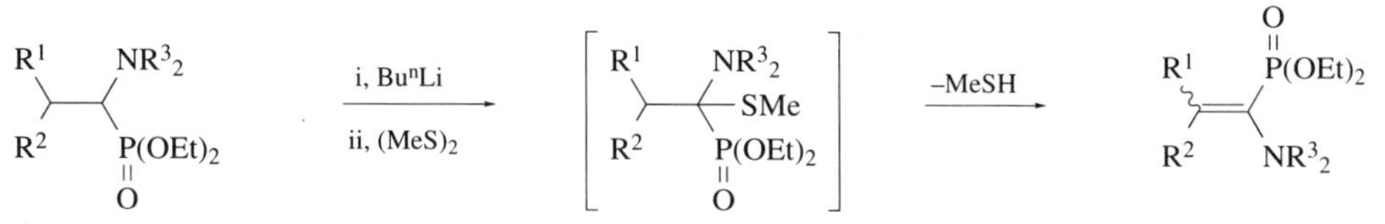

(**281**) $NR^3_2 = NMe_2$
NR^3_2 = pyrrolidino

(**282**)

(**283**) $R^2 = H$
(**284**) $R^1 = R^2 = Me$

Scheme 26

(vi) From α-halo enamines

Three α-halo enamines (**285**) were treated with triethyl phosphite to give α-phosphoryl enamines in good yield (65–80%) ⟨77S336, 79ZOB2217⟩.

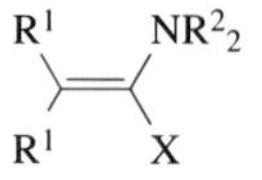

(**285**)

(vii) By modification of α-phosphoryl enamines

Simple α-phosphoryl enamines such as (**286a,b**) were treated with *t*-butyllithium to give the allylic carbanions (**287**) which were treated with a range of electrophiles such as ketones, aldehydes, epoxides, and alkyl halides to give more complex α-phosphoryl enamines (**288**). The best yields were obtained with alkyl halides and in particular with primary bromides, primary iodides, and secondary iodides ⟨84CB1, 84T733, 85USP4501695⟩. The alkylation of (**286b**) proceeds specifically at the methyl group *anti* to the phosphoryl substituent. The mechanism is interesting in that it is the *cis* methyl group which is deprotonated preferentially. Bond rotation round the C1—C2 bond then occurs to place the carbanion *anti* to the phosphoryl group before alkylation takes place. It is thus possible to alkylate each of the methyl groups in (**286b**) sequentially. A third alkylation is also possible and three sequential alkylations were carried out as a one-pot preparation in 80% yield (Scheme 27).

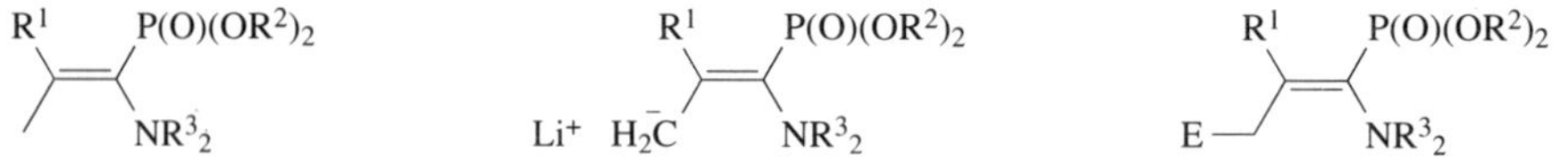

(**286a**) R = H
(**286b**) R = Me

(**287**)

(**288**)

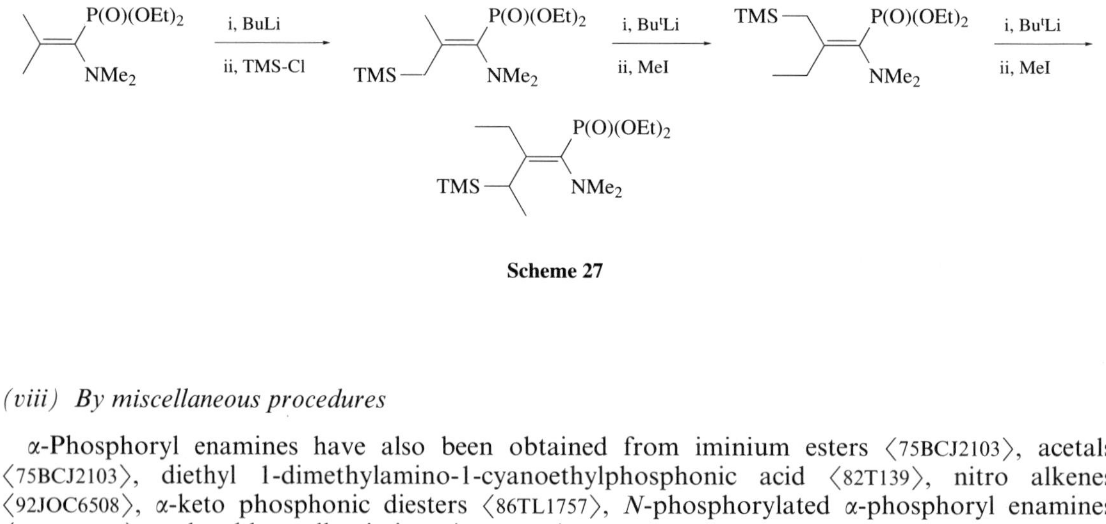

Scheme 27

(viii) By miscellaneous procedures

α-Phosphoryl enamines have also been obtained from iminium esters ⟨75BCJ2103⟩, acetals ⟨75BCJ2103⟩, diethyl 1-dimethylamino-1-cyanoethylphosphonic acid ⟨82T139⟩, nitro alkenes ⟨92JOC6508⟩, α-keto phosphonic diesters ⟨86TL1757⟩, *N*-phosphorylated α-phosphoryl enamines ⟨79ZOB2217⟩, and α-chloro alkanimines ⟨86ZOB805⟩.

4.21.2.5.2 gem-Formamido phosphoryl alkenes, $R^2_2C{=}C(NHCHO)PO(OR^1)_2$

*(i) From tetraethyl formamidomethylenediphosphonate (**289**)*

A range of alkyl, aromatic, and heteroaromatic aldehydes were treated with the reagent (**289**) to give the corresponding *gem*-formamido phosphoryl alkenes (**290**) in variable yields (Equation (14)) ⟨85LA555⟩.

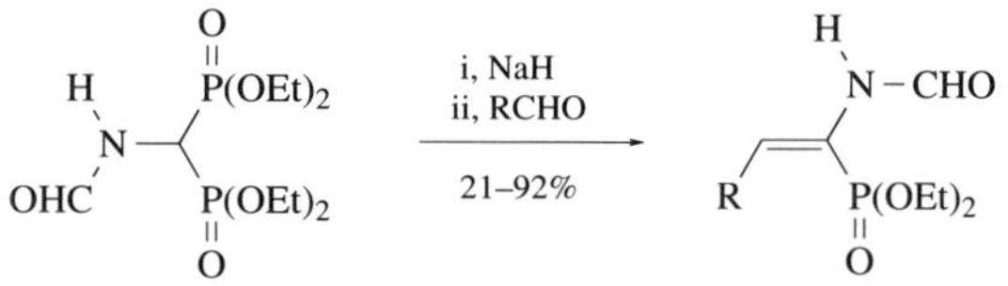

(289) (290)

(14)

(ii) From oxazolines

The oxazolines (**291**) synthesized as shown in Scheme 28 underwent ring opening in the presence of base. Treatment with acid then generated the *gem*-formamido phosphoryl alkenes (**292**) in high yield (78–92%) regardless of whether the β substituent was an alkyl or aryl group ⟨74LA44⟩.

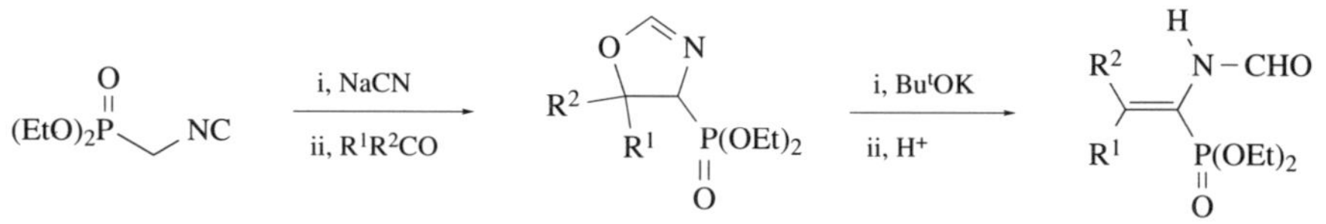

(291) (292)

Scheme 28

4.21.2.5.3 gem-*Amido phosphoryl alkenes* $R^3_2C{=}C(NHCOR^1)PO(OR^2)_2$

(i) From azlactones

A single reference describes the conversion of the azlactones (**293**) into *gem*-amido phosphoryl alkenes. The azlactones were treated with water then chlorine. The ring-opened products (**294**) were then treated with triethyl phosphite to give the final products (**295**) in high yield (60–90%) (Scheme 29) ⟨92ZOB707⟩.

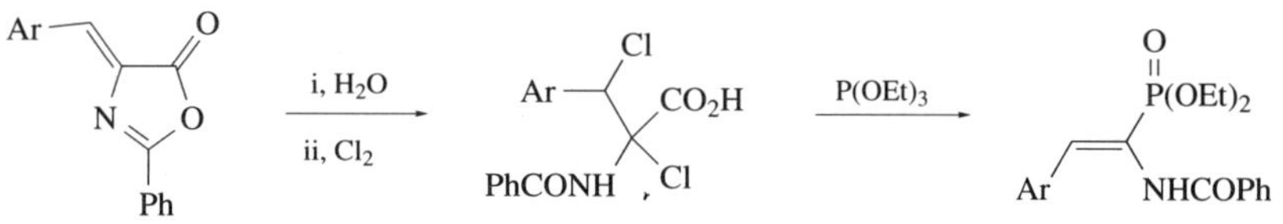

(293) (294) (295)

Scheme 29

(ii) From amides and α-oxo phosphonates

The reaction of primary amides with diethyl 1-oxoethylphosphonate (**296**) gave the *gem*-amido phosphoryl alkenes (**297**) in low yield (12–20%). Diethyl phosphite was formed as an undesirable side product ⟨81S324⟩.

(296) (297)

(iii) From N-tetrachloroethyl amides

The Arbuzov reaction of the *N*-tetrachloroethyl amides (**250b**) with triethyl phosphite yielded the phosphonates (**298**) in 75–86% yield. These were treated with triethylamine to give the *gem*-amido phosphoryl alkenes (**299a**) in 70–91% yield ⟨74ZOR1712⟩.

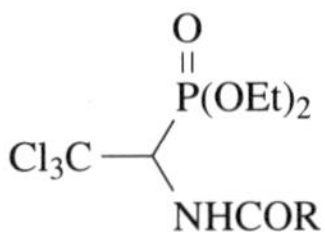

(**298**)

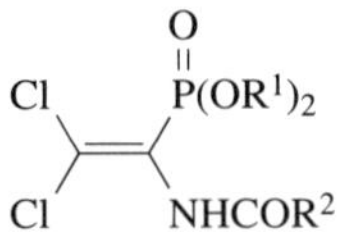

(**299a**) R^1 = Et
(**299b**) R^1 = Me

(iv) From N-trichloroethylidene amides

The *N*-trichloroethylidene amides (**300**) were treated with trimethyl phosphite to give the saturated products (**298**) which on treatment with triethylamine gave the *gem*-amido phosphoryl alkenes (**299b**) ⟨73IZV1112, 73ZOR211⟩.

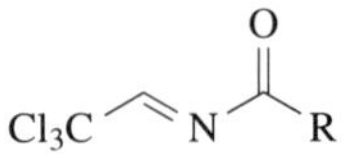

(**300**)

(v) Modification of the title compounds

The *gem*-amido phosphoryl alkenes (**299a**; R^2 = Ar) formed the azlactones (**301**) in high yields (75–96%) on treatment with two equivalents of PCl$_5$. The azlactones were ring-opened under a variety of reaction conditions to give, for example, 1-benzamido-2,2-dichlorovinylphosphonic acid (**302a**), the corresponding monoalkyl esters (**302b**), the diphenyl ester (**302c**), and the dialkyl esters (**302d**) in good yield (71–85%) ⟨82ZOB1122⟩.

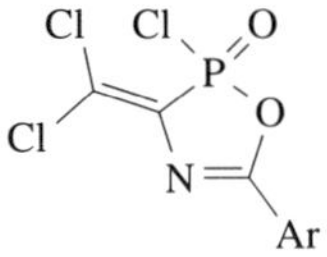

(**301**)

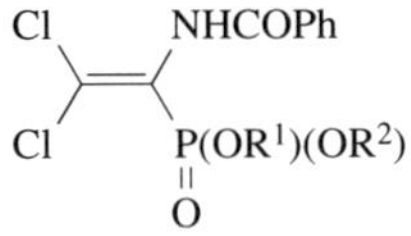

(**302a**) $R^1 = R^2$ = H
(**302b**) R^1 = H, R^2 = Me, Et, Ph
(**302c**) $R^1 = R^2$ = Ph
(**302d**) $R^1 = R^2$ = Me, Et

4.21.2.5.4 gem-Alkoxycarbonylamino phosphoryl alkenes, $R^3{}_2C{=}C(NHCO_2R^1)PO(OR^2)_2$

(i) By alcoholic hydrolysis of isocyanato or chlorophosphoryl groups

Isocyanato, chlorophosphoryl, and dichlorophosphoryl groups are reactive towards alcohols in the presence of triethylamine. Thus, *gem*-alkoxycarbonylamino phosphoryl alkenes of general structure (**303**) were synthesized from a range of saturated or unsaturated starting materials containing such groups, (304), (305), (306), (307), (308). Saturated structures such as (304), (306), (308) dehydrohalogenate under the reaction conditions ⟨83ZOB548, 89ZOB571⟩.

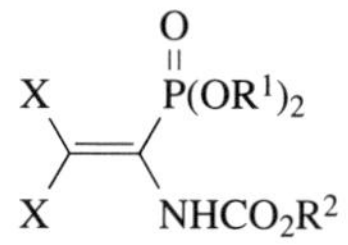

(303) X = H or Cl

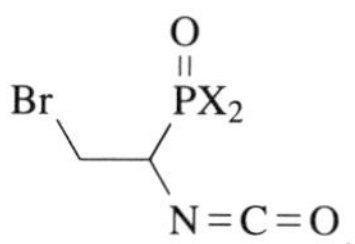

(304) X = Cl, OEt

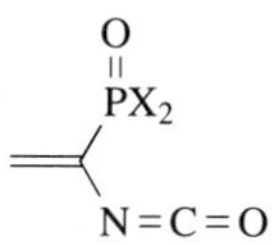

(305) X = Cl, OEt

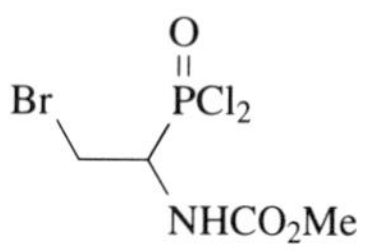

(306)

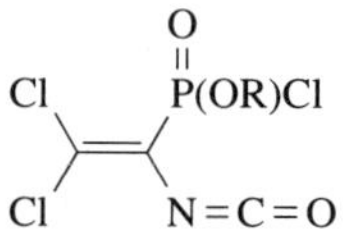

(307)

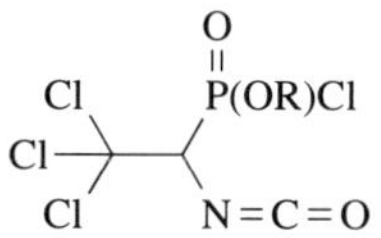

(308)

It is not normally possible to attack the isocyanato group in preference to the dichlorophosphoryl group. The only example where this has been achieved was in the reaction of the isocyanate (305; X = Cl) with one equivalent of *t*-butanol, followed by treatment with ethanol in the presence of triethylamine to give the product (309) in 89% yield.

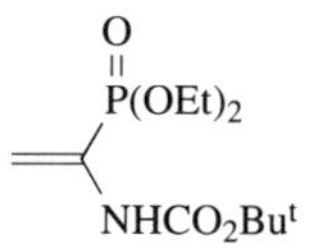

(309)

4.21.2.5.5 The gem-phosphoryl ureido alkene (310)

(i) From the isocyanate (305; X = OEt)

The isocyanate (305; X = OEt) was treated with an excess of aniline to give the title compound (310) ⟨77ZOB75⟩.

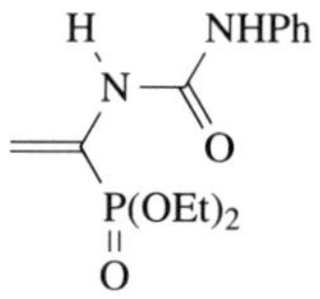

(310)

(ii) From N-tetrachloroethyl ureas

The *N*-tetrachloroethyl ureas (311) were treated with triethyl phosphite to give the Arbuzov products (312) in 64–97% yield, which on treatment with triethylamine gave the *gem*-phosphoryl ureido alkenes (313) in 73–89% yield ⟨84ZOB2186⟩.

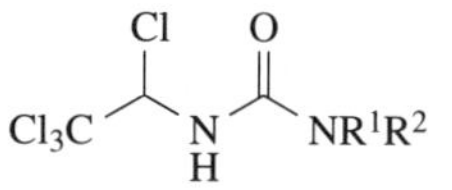

(311)

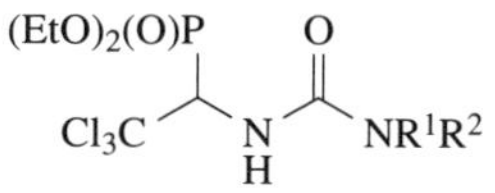

(312)

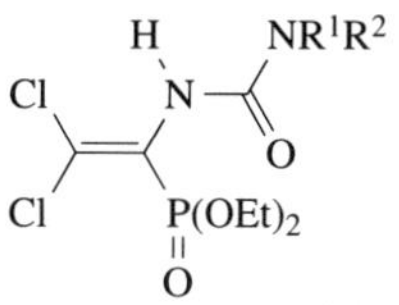

(313)

4.21.2.5.6 gem-*Diphenylphosphinoylamino phosphoryl alkenes*, $R^2{}_2C{=}C(NHPOPh_2)PO(OR^1)_2$

The only synthesis of compounds of this structure involves the reaction of the hydroxyiminoethylphosphonate (**314**) with chlorodiphenylphosphine. The intermediate (**315**) undergoes a free-radical rearrangement to give the final product (**316**) in 44% yield (Scheme 30) ⟨88PJC165⟩.

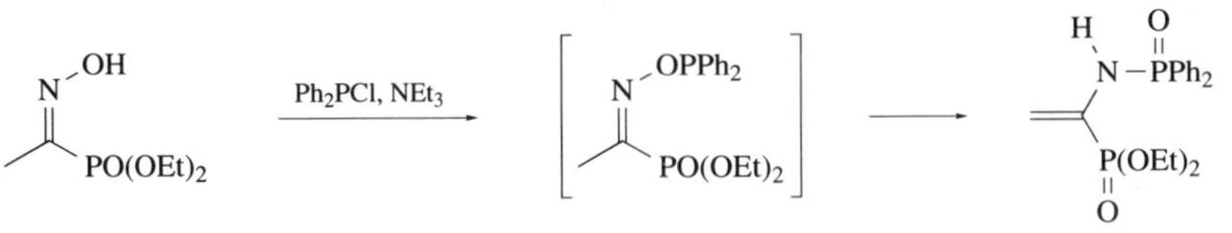

Scheme 30

4.21.2.5.7 gem-*Phosphorylamino phosphoryl alkenes*, $R^4{}_2C{=}C[NR^2PO(OR^1)_2][PO(OR^3)_2]$

(i) From benzoylhydrazones

The cinnamaldehyde benzoylhydrazone (**317**) reacted with diethyl phosphite and sodium diethyl phosphite to give the *gem*-phosphorylamino phosphoryl alkene (**318**) in 35% yield ⟨91PS(55)97⟩.

(ii) From halotricyanomethanes

The reactions of chlorotricyanomethane or bromotricyanomethane with triethyl phosphite generated the electrophilic *N*-phosphorylated ketenimine (**319**), which reacted rapidly with a second molecule of triethyl phosphite to give the *gem*-phosphorylamino phosphoryl alkene (**320**) in 64–68% yield ⟨79ZOB2217⟩.

(iii) From 2,2,2-trihaloacetimidoyl chlorides

The 2,2,2-trichloroacetimidoyl chloride (**321**; R = Me) was treated with 2.5 equivalents of trialkyl phosphites to give the *gem*-phosphorylamino phosphoryl alkenes (**322**) in good yield (71–84%) ⟨86ZOB2681⟩. Yields were significantly reduced if either the benzyl derivative (**321**; R = PhCH$_2$) or triisopropyl phosphite was used. The products (**322**) obtained from this reaction were treated with an excess of a secondary aliphatic amine to substitute both vinylic chlorine atoms and give the *gem*-phosphorylamino phosphoryl alkenes (**323**) in good yield (67–80%).

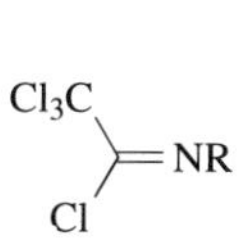

(**321**) 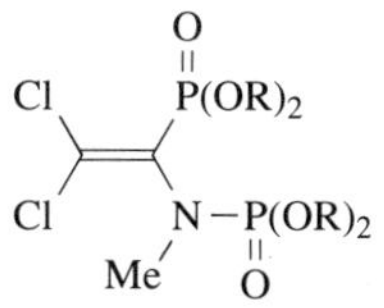   (**322**) 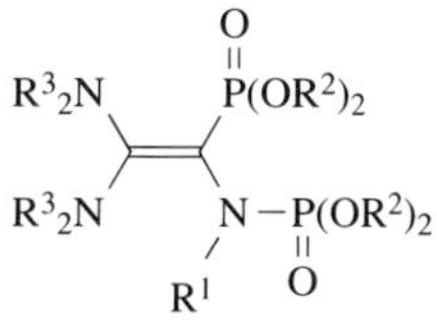   (**323**)

4.21.2.5.8 gem-Alkylideneamino phosphoryl alkenes, $R^3{}_2C{=}C(N{=}CR^1{}_2)PO(OR^2)_2$

(i) From imidoyl chlorides

The imidoyl chlorides (**324**) were treated with a range of trialkyl phosphites to give the title compounds (**325**) in low to medium yield (13–42%). The reaction mechanism involves substitution of the chlorine by the trialkyl phosphite group, an Arbuzov reaction, then isomerization to the final product ⟨89ZOB2492, 90ZOB229⟩. Again, the 2,2,2-trichloroacetimidoyl chloride (**326**) was dehydrochlorinated with base to give the azadiene (**327**), which with triethyl phosphite gave the product (**328**) in 34% yield ⟨86ZOB2681⟩.

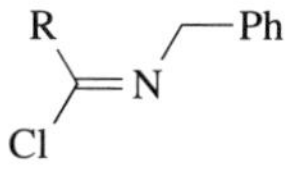

(**324**)   (**325**) 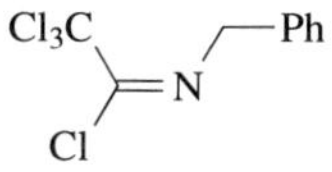   (**326**) 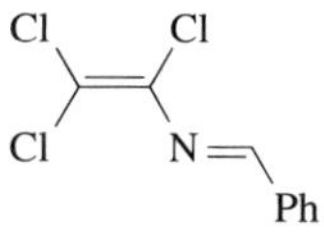   (**327**) 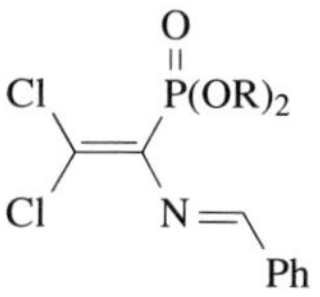   (**328**)

R = H$_2$C=C(Me), (Me)$_2$CCl

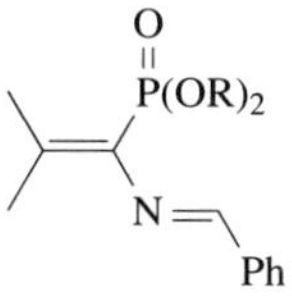

(ii) From N-*acylamides*

The reaction of *N*-acetyltrifluoroacetamide (**329**) with two equivalents of a chlorophosphite in the presence of triethylamine gave the *gem*-imino phosphoryl alkenes (**330**) in moderate yield (42–50%) ⟨90ZOB1187⟩.

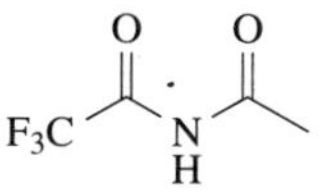

(**329**) 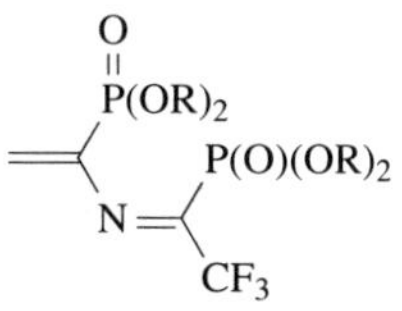   (**330**)

4.21.2.5.9 The gem-isocyanato phosphoryl alkene (333)

(i) From 1,2-dibromoethyl isocyanate

1,2-Dibromoethyl isocyanate (**331**) reacted with triethyl phosphite specifically at the α position. The subsequent Arbuzov reaction gave the saturated isocyanate (**332**) in 53% yield, which on treatment with triethylamine gave the *gem*-isocyanato phosphoryl alkene (**333**) in 69% yield ⟨89ZOB571⟩.

(ii) From 1-chlorovinyl isocyanate

The reaction of 1-chlorovinyl isocyanate (**334**) with triethyl phosphite gave the *gem*-isocyanato phosphoryl alkene (**333**) in low yield (20–25%). Yields were low owing to the ready polymerization of both starting material and product ⟨77ZOB75⟩.

(iii) From tetrachloroethyl isocyanate

Treatment of the tetrachloroethyl isocyanate (**335**) with triethyl or trimethyl phosphite gave a mixture of the saturated and unsaturated substitution products (**336**) and (**337**) ⟨83ZOB548⟩. The crude mixture was heated or treated with base to encourage elimination of hydrogen chloride and give higher yields of the unsaturated product (**337**). Nevertheless, yields were low.

4.21.2.6 Derivatives Bearing a PS(OR)$_2$, PO(OR^1)(NR$^2{}_2$) or POX$_2$ Group

4.21.2.6.1 *The α-thiophosphoryl enamine (340)*

The α-thiophosphoryl enamine (**340**) was synthesized in 35% yield by the addition of the dialkyl thiophosphonate (**339**) to the ynamine (**338**) (Equation (15)) ⟨71CB2021⟩.

4.21.2.6.2 *gem-Amido phosphoramidoyl alkenes, Cl$_2$C=C(NHCOPh)PO(OMe)(NR$_2$)*

There has been one report of the synthesis of *gem*-amido phosphoramidoyl alkenes, which is of the compounds (**342**) from the azlactones (**341**) (Scheme 31) ⟨82ZOB1122⟩

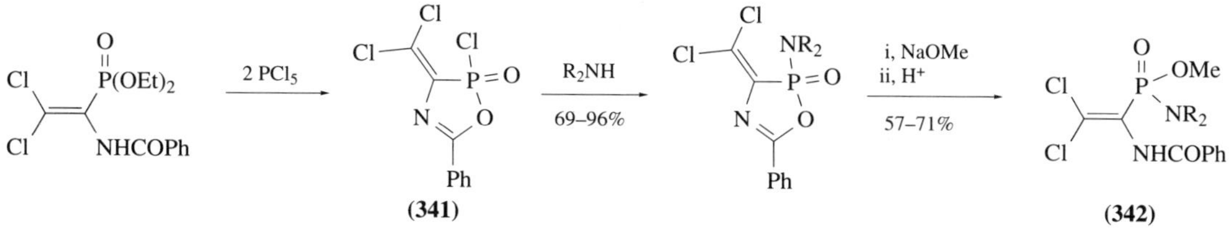

Scheme 31

4.21.2.6.3 gem-Phosphoramidoyl ureido alkenes, $R^4_2C=C(NHCONR^1R^2)PO(OR^3)NR^1R^2$

The *gem*-phosphoramidoyl ureido alkene (**343**; NR_2 = morpholino) was synthesized from the isocyanate (**305**; X = Cl) in 52% yield by treatment with morpholine and methanol ⟨80ZOB343⟩. The isocyanate (**344**) was treated with four equivalents of a primary amine then with water to give the *gem*-phosphoramidoyl ureido alkenes (**345**) in 67–75% yield ⟨84ZOB2186⟩.

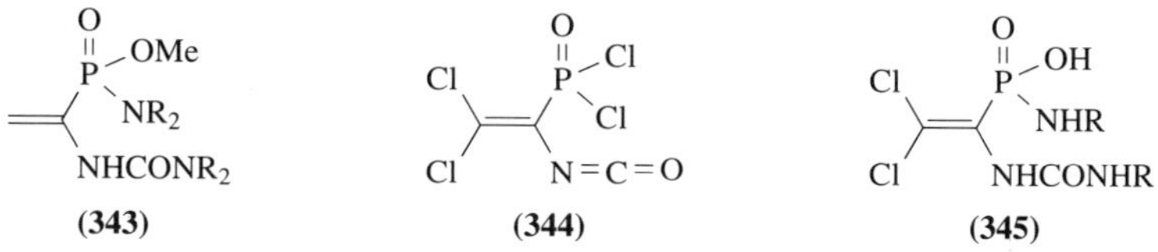

4.21.2.6.4 α-Phosphorodiamidoyl enamines, $R^3_2C=C(NR^1_2)PO(NR^2)_2$

The α-phosphorodiamidoyl enamines (**347**) were synthesized from imino chlorides and the alkoxy phosphorus diamine (**346**) (Scheme 32). The products were treated with base and then with electrophiles as shown in Scheme 32 ⟨85USP4501695⟩.

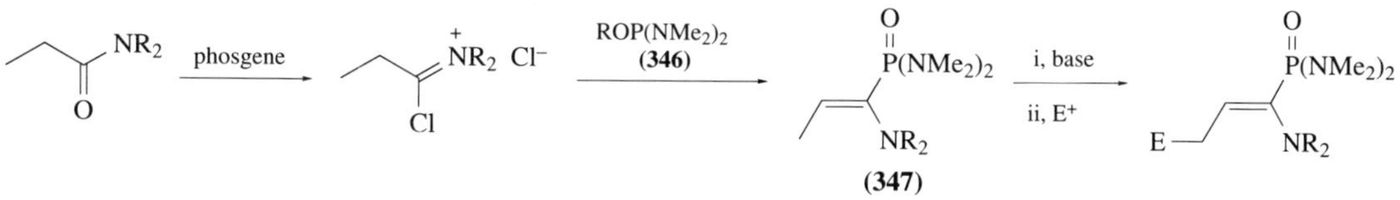

Scheme 32

4.21.2.6.5 gem-Phosphorodiamidoyl ureido alkenes, $R_2C=C(NHCONHAr)PO(NHAr)_2$

The isocyanato dichlorophosphoryl reagents (**305**; X = Cl) and (**304**; X = Cl) were used to synthesize the title compounds (**348**) in 91–96% yield by treatment with excess aniline or with three equivalents of an aniline in the presence of triethylamine ⟨80ZOB343, 89ZOB571⟩.

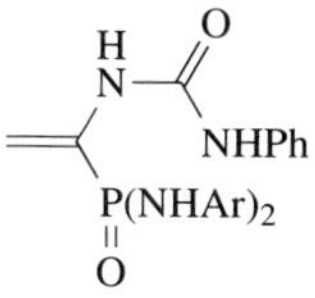

4.21.2.6.6 gem-*Aminobenzylideneamino phosphorodiamidoyl alkenes,* $R^2{}_2C=C(N=C(NR^1{}_2)Ph)PO(NR^1{}_2)_2$

Treatment of the dichlorophosphoryl reagent (**349**) with six equivalents of R^1R^2NH gave the *gem*-aminobenzylideneamino phosphorodiamidoyl alkenes (**350**) in 65–74% yield ⟨82ZOB1122⟩.

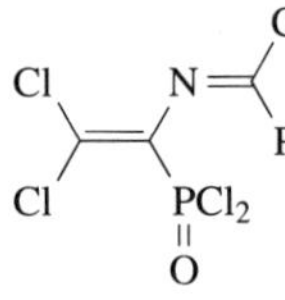

4.21.2.6.7 gem-*Difluorophosphoryl ureido alkenes (351)*

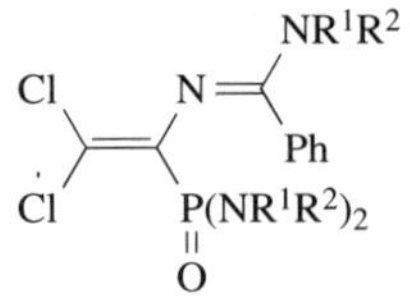

The *gem*-difluorophosphoryl isocyanato reagent (**305**; X = F) was treated with an excess of aniline to give the compound (**351**) in 81% yield ⟨89ZOB571⟩.

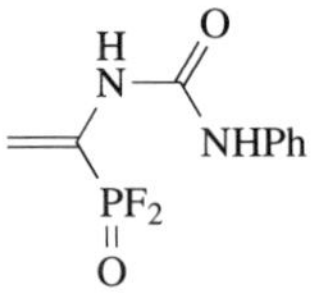

4.21.2.6.8 gem-*Arylchloromethyleneamino dichlorophosphoryl alkenes,* $R_2C=C(N=C(Cl)Ar)POCl_2$

The *gem*-amido phosphoryl alkenes (**352**) were treated with PCl_5 to generate the azlactones (**353**), which on heating with one equivalent of PCl_5 gave the title compounds (**354**) in high yield (77–89%) (Scheme 33) ⟨82ZOB1122⟩.

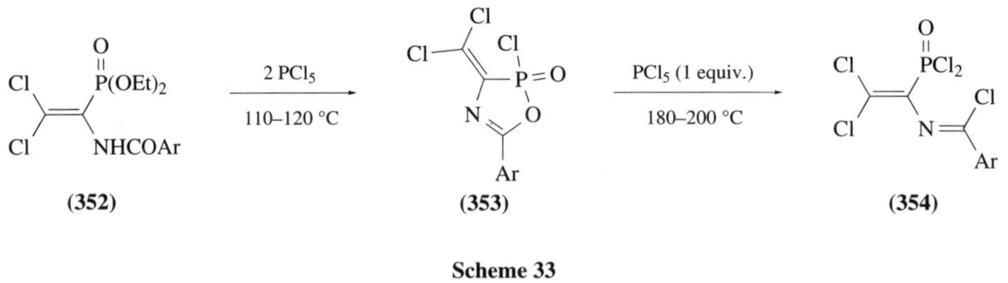

Scheme 33

4.21.2.6.9 gem-*Dihalophosphoryl isocyanato alkenes,* $R_2C=C(NCO)POX^1X^2$

(i) From haloalkyl isocyanates

(a) From dihaloethyl isocyanates. *gem*-Dihalophosphoryl isocyanato alkenes (**357a,b**) were synthesized from dihaloalkyl isocyanates (**355**). The isocyanate was treated first with a phosphite reagent $P(OEt)X^1X^2$ to give the phosphorylated isocyanate (**356**), which was then dehydrohalogenated with triethylamine as shown in Scheme 34 in good yield (48–71%) ⟨89ZOB571⟩. The *gem*-difluoro-

phosphoryl isocyanato alkene (**357d**) was obtained from the *gem*-dichlorophosphoryl isocyanato alkene (**357c**) by treatment with antimony trifluoride.

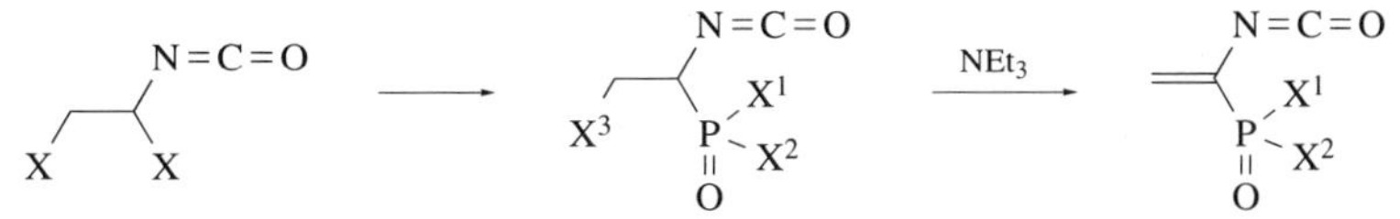

(**355**) X = Cl, Br

(**356**)

(**357a**) X^1 = Cl, X^2 = OEt
(**357b**) X^1 = F, X^2 = OEt
(**357c**) X^1 = X^2 = Cl
(**357d**) X^1 = X^2 = F

Scheme 34

(b) From the tetrachloroethyl isocyanate (335). Treatment of the tetrachloroethyl isocyanate (**335**) with phosphites ($PX^1X^2(OR)$) gave a mixture of the saturated and unsaturated substitution products (**358**) and (**359**) ⟨83ZOB548⟩. The crude mixture was heated or treated with base to encourage elimination of hydrogen chloride and give higher yields of the unsaturated product (**359**). Nevertheless, yields were low.

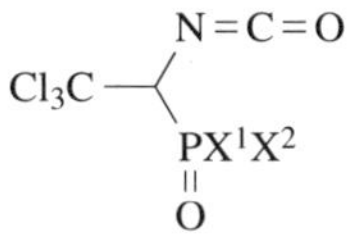

(**358**)

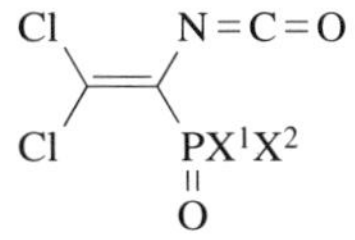

(**359a**) X^1 = X^2 = F
(**359b**) X^1 = Cl, X^2 = OEt

The dichlorophosphoryl product (**359**; $X^1 = X^2 = $ Cl) was obtained by treating the chlorophosphonoyl compound (**359b**) with phosphorus pentachloride ⟨83ZOB548⟩.

(ii) By modification of the gem*-dichlorophosphoryl isocyanato alkene (357c)*

The *gem*-dichlorophosphoryl isocyanato alkene (**357c**) was treated with bromine to give the addition product (**360**), which was dehydrobrominated by treatment with triethylamine to give the vinylidene bromide (**361**) ⟨89ZOB571⟩.

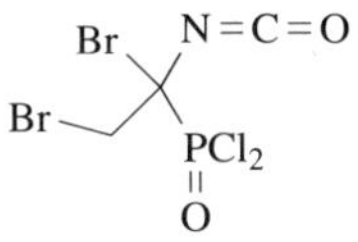

(**360**)

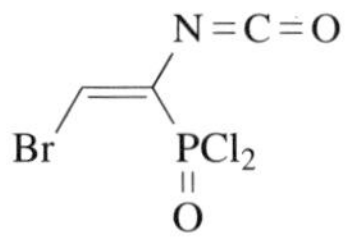

(**361**)

4.21.3 FUNCTIONS CONTAINING ONE NITROGEN AND ONE METALLOID, $R^3{}_2C{=}CNR^1{}_2SiR^2{}_3$, etc.

4.21.3.1 Nitrogen and Silicon Derivatives, $R^3{}_2C{=}CNR^1{}_2SiR^2{}_3$

4.21.3.1.1 α-Silyl enamines

(i) From cyanohydrins

The most promising synthetic approach has been the reductive silylation of silylated cyanohydrins with lithium and chlorotrimethylsilane, using hexamethylphosphoramide (HMPA) as solvent (Scheme 35) ⟨90JOM(332)13⟩. This method gives the *N,N*-bissilylated silyl enamines (**362**) in good yield (68–82%). The use of HMPA as solvent is important. The reaction is thought to proceed by a radical anion mechanism which is favoured by polar solvents. The use of THF as solvent results in a mixture of products and reduced yields. The β substituents have little influence on the reaction. Removal of the N-silyl groups can be achieved in quantitative yield by treatment with dry hydrogen chloride in ether or with chlorotrimethylsilane in methanol.

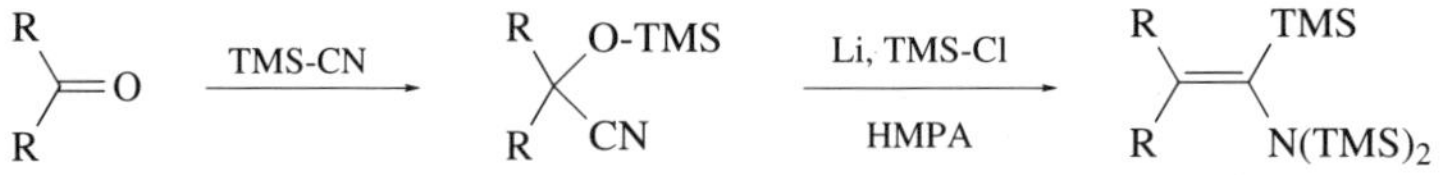

Scheme 35

(ii) From propargylic alcohols

The base-catalyzed addition of the propargyl alcohols (**363**) to trichloroacetonitrile gave the secondary and tertiary propargylic trichloroacetimidates (**364**), which rearranged on heating to give the dienes (**365**). The yield was good (63%) for the more substituted diene (**365**; R = Me) but poor (29%) for the less substituted diene (**365**; R = H) (Scheme 36) ⟨81JA2807⟩.

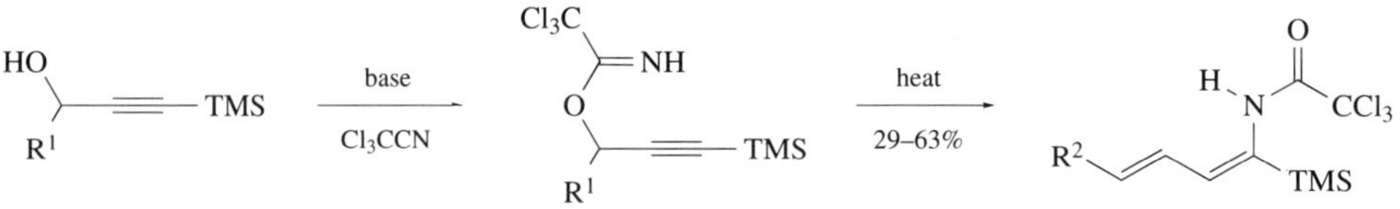

Scheme 36

(iii) Miscellaneous syntheses

Several miscellaneous examples of these compounds have been reported but the yields are either not given or are too low to be of synthetic value. These reports include the silylation of an epoxynitrile ⟨90MI 421-01⟩, the treatment of an alkynylcarbene chromium complex with an amine ⟨92CB2051⟩, silylation of methacrylic amides ⟨72JOM(43)139⟩ and the treatment of aromatic nitriles with chlorotrimethylsilane and magnesium ⟨84OM1660⟩. A silylated enamine was also proposed as an intermediate in a rearrangement reaction ⟨78CJC2286⟩.

4.21.3.1.2 gem-Isocyano silyl alkenes

A single paper reports the synthesis of three compounds of this type (**367**) by lithiation of the vinylidene isocyanide (**366**) followed by treatment with chlorotrimethylsilane. Yields were moderate to good (53–78%) ⟨84CRV471⟩.

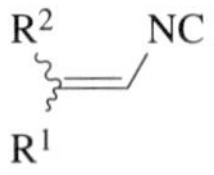

4.21.3.1.3 The gem-isothiocyanato silyl alkene (370)

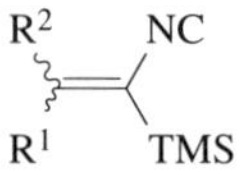

A single paper refers to the synthesis in low yield of the *gem*-isothiocyanato silyl alkene (**370**). The isocyanide (**368**) was treated with sulfur to give the isothiocyanate (**369**) in 76% yield, which was then treated with benzaldehyde in the presence of a catalytic amount of *n*-Bu₄NF to give the product (**370**) in 26% yield ⟨82BCJ1163⟩. Tetrabutylammonium fluoride provides a fluoride ion which catalyzes the generation of the isothiocyanate carbanion (**371**), which is the intermediate which reacts with benzaldehyde.

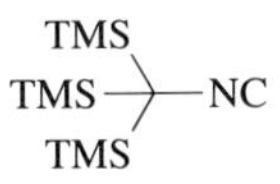

(**368**) 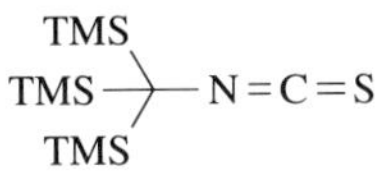   (**369**) 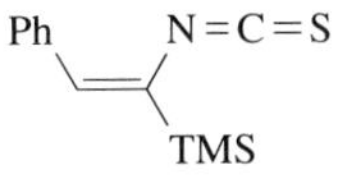   (**370**) 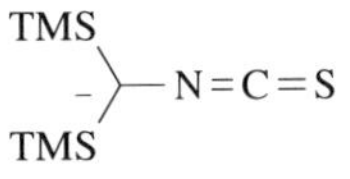   (**371**)

4.21.3.1.4 gem-Aminoborylamino silyl alkenes, PrCH=C(TMS)N(R)B(X)NMe₂

Three compounds having the general structure (**372**) were formed in low yield from the carbiminosilane (**373**) and the amino halogeno boranes (**374**) at 200 °C ⟨82JOM(231)191⟩.

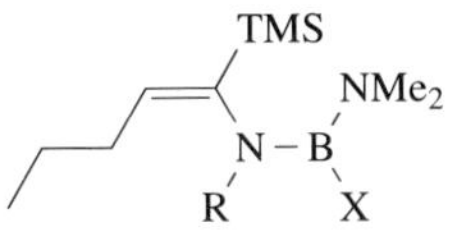

(**372**) X = Br, NMe₂
R = Pr^i, Bu^n

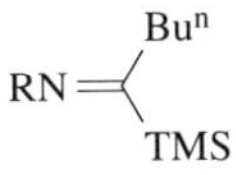

(**373**)

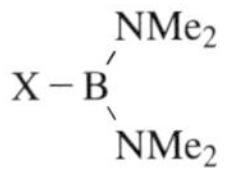

(**374**)

4.21.3.2 Nitrogen and Boron Derivatives, R³₂C=CNR¹₂BR²₂

No satisfactory general synthesis of these structures has been devised. The only reported method is the addition of trialkylboranes to ynamines to give compounds of general structure (**375**), but yields are poor (16–22%) ⟨74M684⟩.

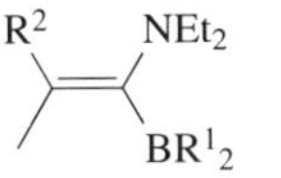

(**375**) R² = Et, allyl

4.21.4 FUNCTIONS CONTAINING ONE NITROGEN AND ONE METAL, R²₂C=CNR¹₂M, etc.

4.21.4.1 Group 1 and 2 Metals, R²₂C=CNR¹₂Li, etc.

4.21.4.1.1 gem-Isocyano lithio alkenes

gem-Isocyano lithio alkenes (**376**) can be obtained by treating the corresponding vinylidene isocyanides with butyllithium. They have not been isolated but have been treated with electrophiles ⟨84CRV471⟩.

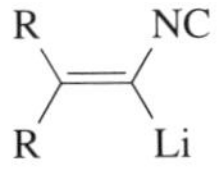

(**376**)

4.21.4.1.2 α-Lithio allenamines

The lithio allenamine (**377**) has been reported but with no details of its synthesis ⟨89OM1371⟩.

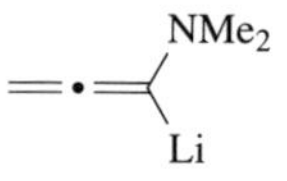

(377)

4.21.4.1.3 α-Lithio enamines

(i) From α-chloro enamines

Treatment of the α-chloro enamines (378) with lithium in THF generated the α-lithio enamines (379; M = Li), which were deuteriated to test the success of the reaction. The rate of lithium/chlorine exchange was greater for an *N*-methylanilino, than for a morpholino than for a dimethylamino group ⟨76AG417⟩.

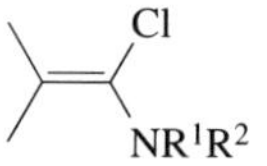

(378) 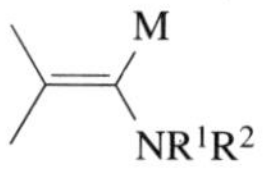   (379)

(ii) From β-amino α,β-unsaturated amides and esters

The amide (380; X = H) was treated with *t*-butyllithium to give the α-lithio enamine (380; X = Li), which was stable at room temperature ⟨76AG(E)171⟩ and was treated with alkyl halides, esters, aldehydes, ketones, and imines ⟨76AG(E)171, 77S869⟩.

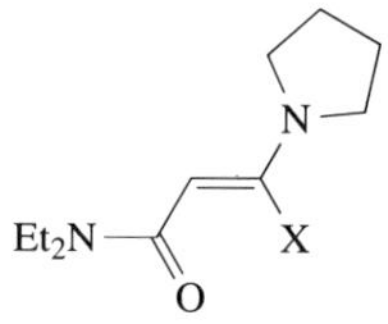

(380)

The esters (381; X = H) were similarly treated with lithium diisopropylamide to give the corresponding α-lithio enamines (381; X = Li), which reacted with aldehydes, esters and alkenes ⟨78AG(E)204, 82S748⟩.

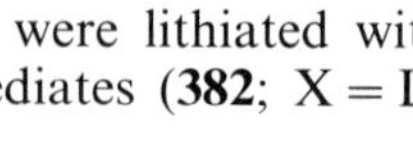

(381) X^1 = OMe, SMe

4.21.4.1.4 gem-Iminomethylamino lithio alkenes

The *N*-vinyl amidines (382; X = H) were lithiated with *t*-butyllithium or with lithium diisopropylamide to give the lithio intermediates (382; X = Li), which were then treated with halogenoalkanes and aldehydes ⟨84CRV471⟩.

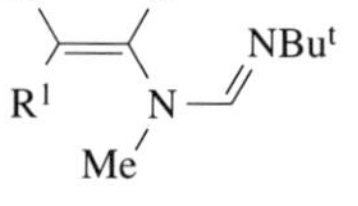

(382)

4.21.4.1.5 α-Sodio enamines

Treatment of the enamine (**383**; X = H) with sodium dimethylamide gave the α-sodio enamine (**383**; X = Na), which underwent further reactions ⟨86ZOR582⟩.

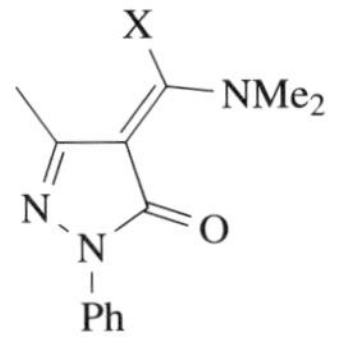

(**383**)

The α-chloro enamines (**378**) were treated with sodium in THF to give the α-sodio enamines (**379**; M = Na). The rate of sodium/chlorine exchange was related to the amino substituent in the same way as the lithium/chlorine exchange mentioned above ⟨76AG417⟩. However, sodium/chlorine exchange did not require the basic conditions required for lithium/chlorine exchange.

4.21.4.1.6 α-Potassio allenamines

The metallation of ynamines with potassium is reported to give tautomeric mixtures of the potassiomethyl alkynes (**384**) and the α-potassio allenamines (**385**) ⟨87MI 421-01⟩.

$K^+ \ ^-H_2C{\equiv}\!\!\equiv\!\!{-}NR_2$

(**384**)

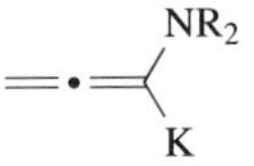

(**385**)

4.21.4.1.7 α-Magnesio enamines

The α-chloro enamines (**378**) were treated with magnesium in THF to give the Grignard reagents (**379**; M = MgCl). The rates of formation of the reagents were related to the amino substituents in the same way as the aforesaid exchanges with sodium and lithium. Like lithium, basic conditions were necessary for the success of the reaction. The Grignard reagents were treated with acid anhydrides, carbon dioxide, and aldehydes ⟨76AG417⟩.

4.21.4.2 Transition Metals, $R^2{}_2C{=}C(NR^1{}_2)PdX$, etc.

4.21.4.2.1 Palladium and platinum

Active methylene groups have been added to various isocyanides coordinated to palladium(II) or platinum(II) (**386**) and (**387**) to give the complexes (**388**) and (**389**), respectively. Yields were greater than 90% ⟨75JOM(97)61, 82ZN(B)1044⟩.

The reactions of the carbon-bonded metal imine complexes (**390**) with the alkyne (**391**) were

$(PPh_3)_2Pt(Cl){\equiv}NR\ BF_4^-$

(**386**)

$[M(CNMe)_4]^{2+}$

(**387**) M = Pd, Pt

(**388**) X = CN, CO₂Me

(**389**) M = Pd, Pt
R = CN, CO₂Et

reported to give the α-palladio enamines (**392**) as unstable products ⟨78JA1164⟩. No yields were reported.

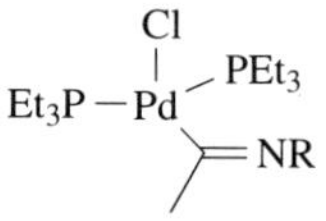

(**390**)    (**391**) 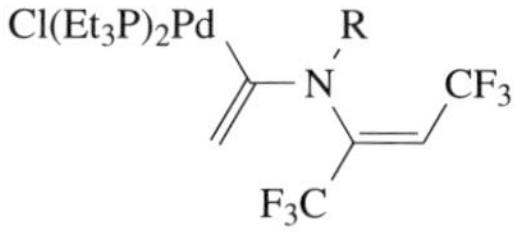   (**392**)

Migratory insertion of alkyl or aryl isocyanides into the Pd—C sigma bond of *trans*-[PdCl(CH$_2$COMe)(PPh$_3$)$_2$] yielded the 1-amino-3-oxo-1-butenyl complexes *trans*[PdCl{C(NHR)=CHCOMe}(PPh$_3$)$_2$] where R = *p*MeOC$_6$H$_4$ or Me. No yields were reported ⟨84JOM(269)C15⟩.

4.21.4.2.2 Copper

Treatment of the ynamine (**393**) with (TMS)$_2$CuLi·LiCN generated the copper complex (**394**). The complex was not isolated but was transformed by protonation or alkylation to give enamines ⟨92PAC439⟩.

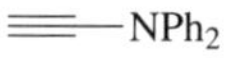

(**393**) 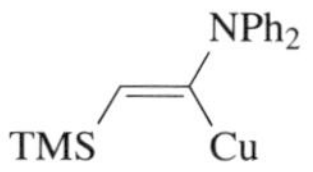   (**394**)

4.21.4.2.3 Molybdenum and tungsten

The molybdenum- and tungsten-substituted enamines (**395**; X = NR$_2$) were proposed as intermediates in the synthesis of the dicyanoketeneiminium derivatives (**396**) from the reaction of the α-chloro alkenes (**395**; X = Cl) and secondary amines ⟨75IC1018⟩.

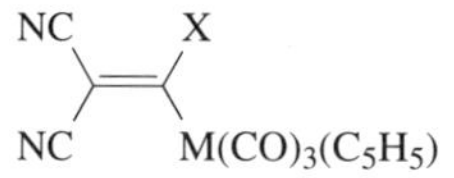

(**395**)
M = W, Mo

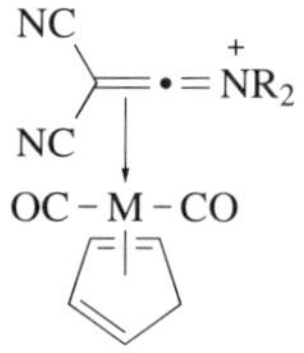

(**396**)

The reaction of pentacarbonylbenzylidenetungsten (**397**) with triphenylketenimine (**398**) gave an addition product (**399**) which rearranged to give the imine adduct of pentacarbonyl(diphenylvinylidene)tungsten (**400**) in 80% yield (Scheme 37) ⟨91OM389⟩.

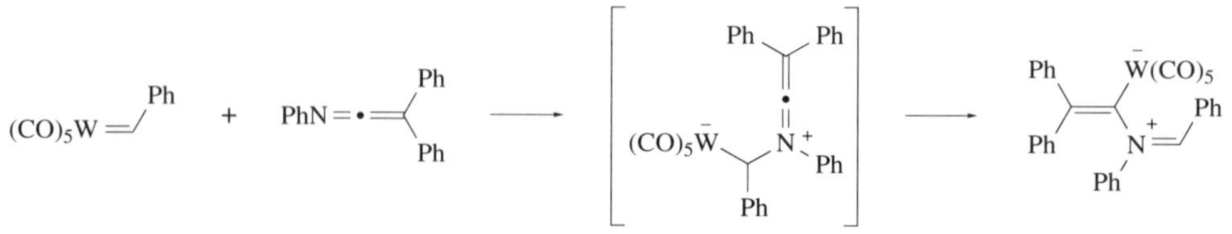

(**397**)   (**398**)   (**399**)   (**400**)

Scheme 37

The reaction of the tungsten complex (**401**) and the ynamine (**402**) gave an α-metallo enamine intermediate (**403**), which was isolated (Equation (16)) ⟨76JOM(121)211⟩. No yields were reported.

$$(MeO)_3\overset{|}{\underset{|}{PW}}-H \quad + \quad =\!\!=\!\!=\!-NMe_2 \quad \xrightarrow{PhH} \quad \overset{W(Cp)(CO)_2P(OMe)_3}{\underset{NMe_2}{\diagdown\!\!=\!\!\diagup}} \qquad (16)$$
$$(CO)_2$$
(**401**)   (**402**)   (**403**)

4.21.4.2.4 Manganese

The reactions of the (1-chloro-2,2-dicyanovinyl)manganese derivatives (**404**) with diethylamine with UV irradiation gave a dicyanoketeniminium derivative, but with other, less hindered amines such as dimethylamine, pyrrolidine, piperidine, and morpholine, the dialkylcarbamoyl derivatives (**405**) were obtained ⟨79IC69⟩.

    NC   Cl   NC   NR$_2$
         C=C   C=C
    NC   Mn(CO)$_3$(ligand)$_2$ NC   Mn(CO)$_2$(CONR$_2$)(ligand)$_2$
            (**404**)   (**405**)

4.21.4.2.5 Iron

(i) From α-chloro enamines

Treatment of the α-chloro enamine (**406**; X = Cl) with the iron complex NaFe(CO)$_2$Cp gave the α-metallo enamine (**406**; X = Fe(CO)$_2$Cp) as the major product of an air-sensitive mixture. No yield was reported. ⟨75JA2702⟩.

$$\underset{NMe_2}{\overset{X}{\diagdown\!\!=\!\!\diagup}}$$
(**406**)

(ii) From the iron vinylidene complex (**407**)

The iron vinylidene complex (**407**) reacted readily with an imine to give the iron complex (**408**) in 80% yield ⟨88OM2553⟩.

               +   Me   Ph
         =•=Fe(Cp)(CO)P(OMe)$_3$ $\overset{+}{N}$=   CF$_3$SO$_3^-$
                                                 C=C
            CF$_3$SO$_3^-$ Fe(Cp)(CO)P(OMe)$_3$
              (**407**)   (**408**)

(iii) Miscellaneous syntheses

α-Iron-substituted enamines have been proposed as reaction intermediates on a couple of occasions but have not been isolated or identified ⟨82JOM(224)153, 86TL3811⟩.

4.21.4.2.6 Rhenium

Treatment of the α-chloro enamine (**406**; X = Cl) with the rhenium reagent NaRe(CO)$_5$ gave the mono hapto vinyl derivative (**406**; X = Re(CO)$_5$) in low yield (10%). Treatment of the piperidine analogue (**409**; X = Cl) with the same rhenium reagent gave only the CC dihapto cyclic derivative (**410**) in 65% yield. When heated in hexane, the derivative (**410**) was converted into the monohapto vinyl derivative (**409**; X = Re(CO)$_5$) in 42% yield ⟨75JA2702⟩.

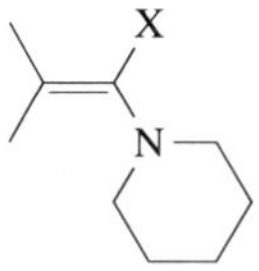

(**409**) 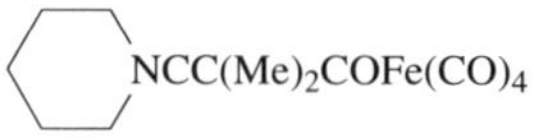   (**410**)

4.21.4.3 Other Metals

4.21.4.3.1 Tin

The diamine (**411**) was treated with one equivalent each of *n*-butyllithium and *t*-butyllithium to deprotonate a methylene group (**412**). Loss of a dimethylamide anion generated the vinylic anion, which on treatment with trimethyltin chloride gave the α-trimethylstannyl enamine (**413**) in 42% yield ⟨87CB2081⟩. Both bases were required. Treatment with one equivalent of *t*-butyllithium deprotonated the methyl substituent, leading to the saturated product (**414**) in 53% yield.

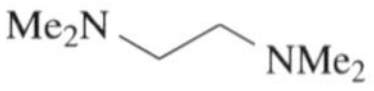

(**411**) 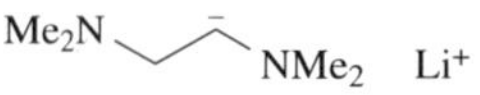   (**412**) 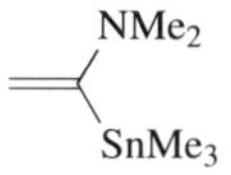   (**413**) 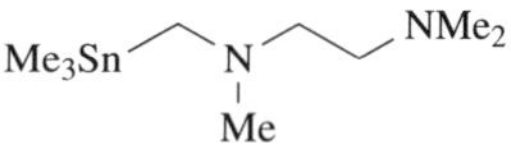   (**414**)

4.22
Functions Containing at Least One Phosphorus, Arsenic, Antimony or Bismuth and No Halogen, Chalcogen or Nitrogen

JOHN M. BERGE
SmithKline Beecham Pharmaceuticals, Epsom, UK

4.22.1	DIPHOSPHORUS FUNCTIONS — $R^1_2C=C(PR^2_2)_2$ etc.	1022
	4.22.1.1 Acyclic Bisphosphino Alkenes	1022
	4.22.1.2 Cyclic Bisphosphino Alkenes	1025
	4.22.1.2.1 Three-membered rings	1025
	4.22.1.2.2 Four-membered rings	1026
	4.22.1.2.3 Five-membered rings	1026
	4.22.1.3 Transition Metal Complexes of Bisphosphino Alkenes	1028
4.22.2	FUNCTIONS CONTAINING ONE PHOSPHORUS AND EITHER ARSENIC, ANTIMONY OR BISMUTH, $R^1_2C=C(PR^2_2)AsR^3_2$, etc.	1030
4.22.3	FUNCTIONS CONTAINING ONE PHOSPHORUS AND A METALLOID, $R^1_2C=C(PR^2_2)SiR^3_3$, etc.	1031
	4.22.3.1 Silicon and Germanium Derivatives	1031
	4.22.3.2 Boron Derivatives	1031
4.22.4	FUNCTIONS CONTAINING PHOSPHORUS AND A METAL, $R^1_2C=C(PR^2_2)M$, etc.	1031
	4.22.4.1 Group 1 and 2 Metals, $R^1_2C=C(PR^2_2)M$, etc.	1031
	4.22.4.2 Transition Metals, $R^1_2C=C(PR^2_2)PdX_2$, etc.	1032
	4.22.4.2.1 σ-Bonded compounds	1032
	4.22.4.2.2 Phosphino metallocenes	1032
	4.22.4.2.3 Phosphametallocenes	1034
	4.22.4.2.4 Phosphorus-containing cyclopentadienylmanganese (cymantrene) derivatives	1036
	4.22.4.3 Tin Derivatives, $R^1_2C=C(PR^2_2)SnR^3_3$, etc.	1038
4.22.5	FUNCTIONS CONTAINING TWO ARSENIC, ANTIMONY OR BISMUTH FUNCTIONS, $R^1_2C=C(AsR^2_2)_2$, etc.	1038
4.22.6	FUNCTIONS CONTAINING TWO DISSIMILAR COMBINATIONS OF ARSENIC, ANTIMONY OR BISMUTH, $R^1_2C=C(AsR^2_2)SbR^3_4$, etc.	1038
4.22.7	FUNCTIONS CONTAINING ARSENIC, ANTIMONY OR BISMUTH WITH A METALLOID, $R^1_2C=C(AsR^2_2)SiR^3_3$ etc.	1038
4.22.8	FUNCTIONS CONTAINING ARSENIC, ANTIMONY OR BISMUTH AND A METAL, $R^1_2C=C(AsR^2_2)M$ etc.	1039
	4.22.8.1 Arsa-, Stiba- and Bisma-ferrocene Derivatives	1039
	4.22.8.2 Arsenic and Antimony Analogues of Cymantrene	1042

4.22.1 DIPHOSPHORUS FUNCTIONS—$R^1_2C{=}C(PR^2_2)_2$ etc.

4.22.1.1 Acyclic Bisphosphino Alkenes

Compounds of this type are named as derivatives of phosphines in *Chemical Abstracts*, for example, ethenylidenebisdimethylphosphine. However, for the purpose of this review, the structures will be treated as derivatives of alkenes; thus the parent system is named as 1,1-bisdimethylphosphino ethene. In a fashion analogous to the more widely studied bisphosphino alkanes, bisphosphino alkenes act as ligands for transition-metal complexes. The inherent geometry of the sp^2-hybridised carbon bearing the phosphine groups imparts a larger 'bite' angle to the alkene derivative compared to that of the saturated analogue. Exactly what this means in terms of chemical reactivity and stability of the resulting complexes is unclear. Up to 1995, the literature lacks a comprehensive review of the synthesis of alkenes of this type. A number of reports from the primary literature cover the two most direct methods of preparation, namely, formation of the C—P bonds and formation of the C=C bond. However, these methods are not well exemplified and, for the former, the yields are generally very low as would be expected from nucleophilic displacement from an sp^2 carbon. An example (Scheme 1) of this is the method reported by Colquhoun and McFarlane ⟨82JCS(D)1915, 92JCS(D)2353⟩ which involves the reaction of the lithium salt of diphenylphosphine with a geminal dichloro alkene in tetrahydrofuran. Using this method, bisdiphenylphosphino propene (**1**) was prepared in 12% yield.

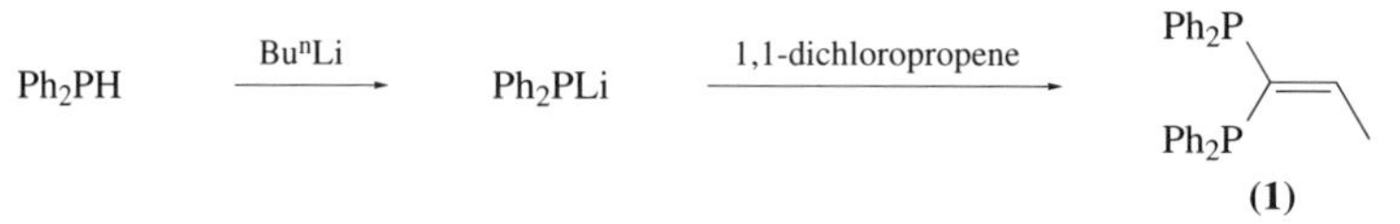

Scheme 1

An alternative and higher-yielding approach (Scheme 2) was described by Goli and Grim ⟨91TL3631⟩, the key step of which involved the formation of the C=C of the alkene by the condensation of the anion of bisdiphenylthiophosphinoyl methane (**3**) with an aldehyde. The synthesis is then completed by reduction of the pentavalent phosphorus derivatives (**4a–c**), to the trivalent compounds (**5a–c**). Surprisingly, if either or both of the phosphine sulfide groups were replaced by the phosphine oxides, the condensation with the aldehyde failed. The nature of the base, or at least the cation, also appears to be critical to the outcome of the reaction. Replacement

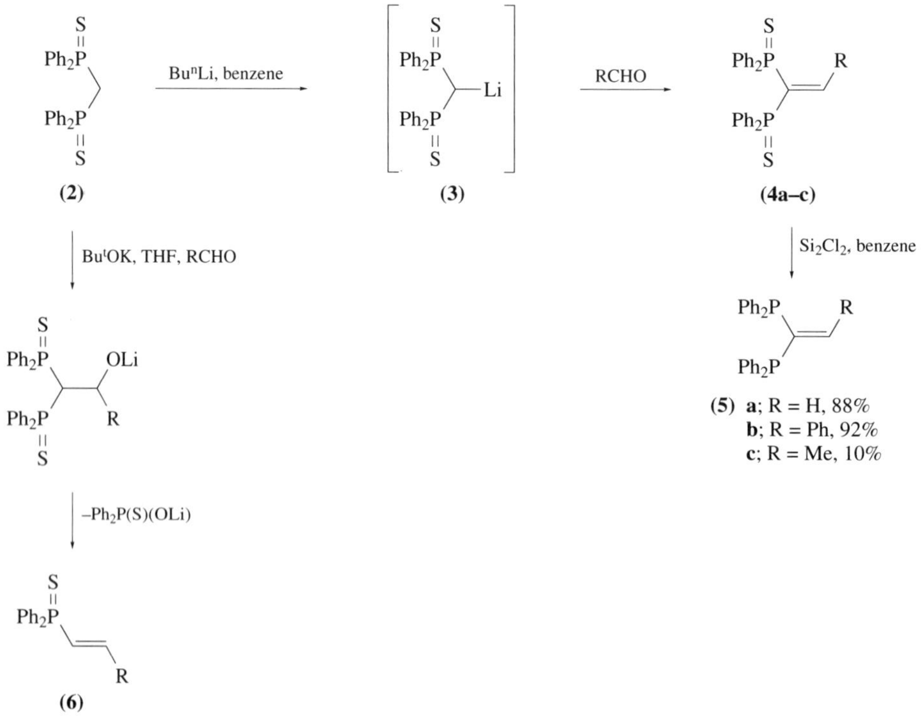

(**5**) a; R = H, 88%
b; R = Ph, 92%
c; R = Me, 10%

Scheme 2

of *n*-butyllithium with potassium *t*-butoxide in THF or a Group 1 metal in benzene ⟨66AG(E)847, 66ZC28⟩ resulted in exclusive formation of the (*E*) vinylphosphine sulfide (**6**).

The phosphine substituents in 1,1-bisdiphenylphosphino ethene can be sequentially alkylated (Scheme 3) with iodomethane or methyl trifluoromethanesulfonate to form the corresponding mono- (**7**) and diphosphonium salts (**8**) ⟨84AG(E)247, 85CB3105⟩. The former compound was found to be unstable in alcohol and underwent a head-to-tail [3 + 3] cycloaddition to form compound (**9**).

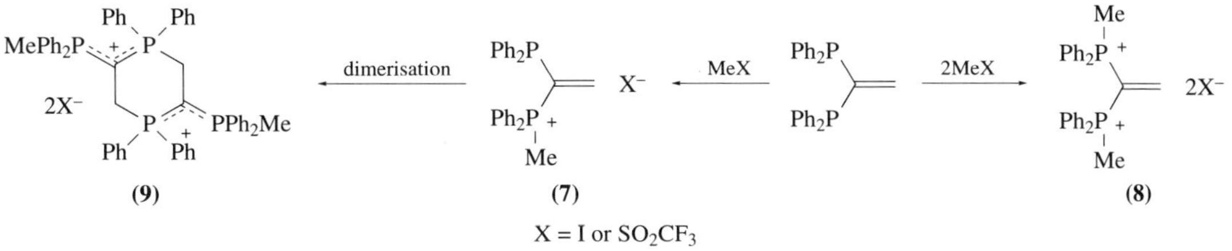

Scheme 3

Knunyants *et al.* reported the preparation of asymmetrical compounds bearing dissimilar alkyl groups on the phosphorus atoms (Scheme 4) ⟨72CA(77)126770j, 72ZOR346⟩. Reaction of bistrifluoromethyl ketene (**10**) with a dialkylphosphine gave a mixture of 35% α-ketophosphine (**11**) and 32% 1,1-bis(diphenylphosphino)-2,2-bistrifluoromethyl ethene (**12**). The former material, on treatment with di-*i*-isopropylphosphine and hydrogen chloride in diethyl ether yielded the asymmetrically substituted compound (**13**).

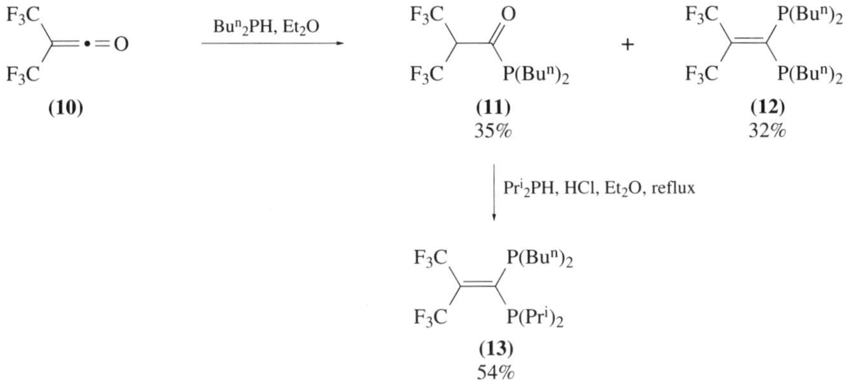

Scheme 4

A potential alternative procedure for the preparation of asymmetrical compounds, as yet untried, is illustrated by the synthesis of trisdiphenylphosphino ethene (**15**) (Scheme 5). This was readily prepared, in moderate yield, by potassium *t*-butoxide-catalysed Michael addition of diphenylphosphine to 1,2-bisdiphenylphosphino ethyne (**14**) as reported by Schmidbaur *et al.* ⟨88CB1241⟩. A minor by-product was also isolated, and the structure initially reported as 1,2,3,4-tetrakis(diphenylphosphino)-1,3-butadiene, (**16**). In a later paper ⟨89CB1857⟩ the same authors report the structure of this by-product to be 1,1,4,4-tetrakisdiphenylphosphino-1,3-butadiene (**17**). This structure was confirmed by x-ray analysis on the tetrasulfide (**18**). The bismolybdenumtetracarbonyl complex (**19**) was formed when (**17**) was treated with molybdenum hexacarbonyl in THF under reflux (Scheme 5). The formation of complexes between 1,1-bisphosphino alkenes and transition metal atoms is a general reaction, and examples are known for platinum ⟨85CC1635⟩, palladium ⟨86CC882⟩, tungsten ⟨85CC614⟩, iron ⟨88CC673⟩ and chromium ⟨82CC286⟩ (Section 4.22.1.3).

Several derivatives of (**15**) have been prepared (Scheme 6). Thus, borane in THF yielded the bisboranato compound (**20**) and iodomethane caused monoquaternisation, the structure of the resultant phosphonium salt (**21**) being confirmed by x-ray crystallography.

Schmidbaur *et al.* have also reported the synthesis of tetrakisdiphenylphosphino allene (**23**) by the reaction of 3-lithio-1-diphenylphosphinopropyne (**22**) with chlorodiphenylphosphine (Scheme 7) ⟨86AG(E)348⟩. The structure of (**23**) was confirmed by IR spectroscopy; as would be expected the allene functionality showed an absorption at 1890 cm^{-1}. The isomeric tetrakisdiphenylphosphino propyne (**24**) was discounted as a product since the ^{31}P-NMR spectrum exhibited only a single

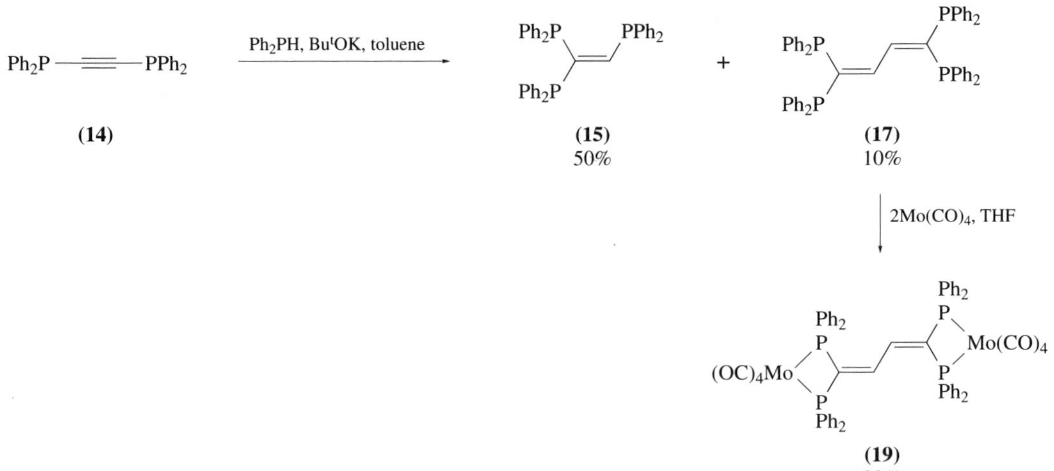

(14) (15) 50% (17) 10% (19) 92%

Scheme 5

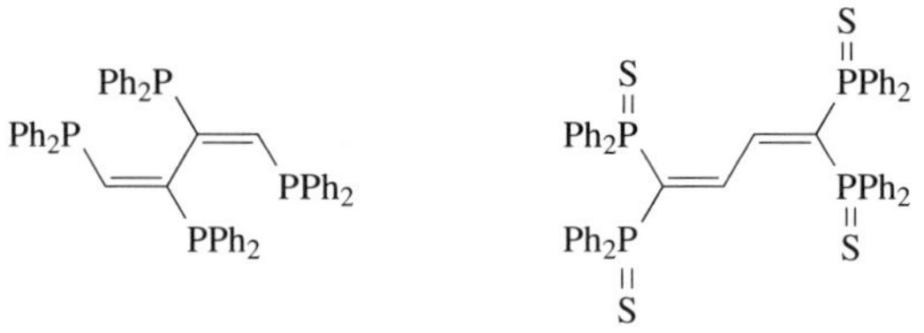

(16) (18)

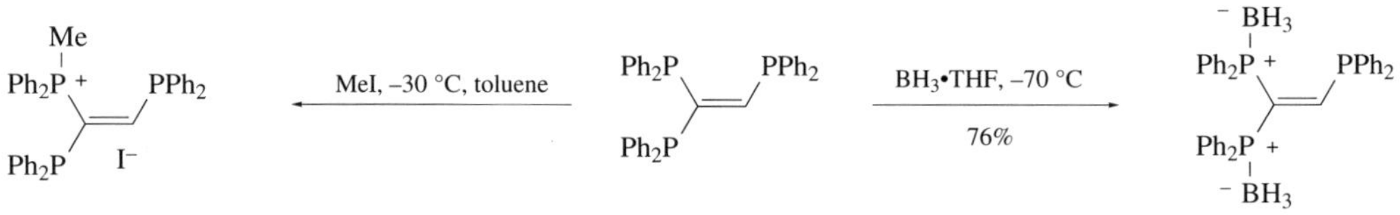

(21) (15) (20)

Scheme 6

phosphorus signal. In a similar fashion to the bisdiphenylphosphino ethene (**5a**) and tris(diphenylphosphino) ethene (**15**), the allene (**23**) was converted into the monophosphonium salt (**25**) by iodomethane (Scheme 8). Treatment of (**23**) with excess iodomethane led to rearrangement of the allene functionality and elimination of dimethyldiphenylphosphonium iodide to form the propyne (**26**).

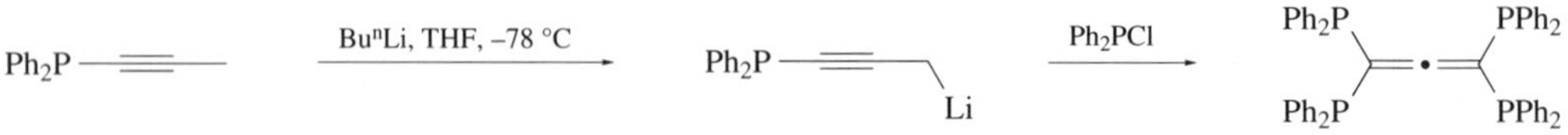

(22) (23)

Scheme 7

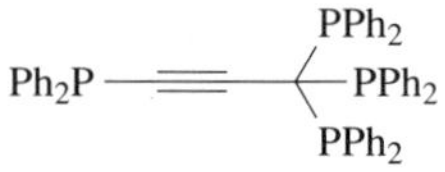

(24)

4.22.1.2 Cyclic Bisphosphino Alkenes

Only a handful of examples of cyclic bisphosphino alkenes with at least one *endo*-cyclic phosphorus have appeared in the literature, and these are restricted to three-, four- and five-membered ring derivatives. It is not clear whether this is due to a genuine lack of appropriate synthetic methodology or whether it simply reflects an inherent instability in this class of compound.

4.22.1.2.1 Three-membered rings

The first members of this class described were the methylenediphosphiranes (Equation (1)) (diphosphamethylenecyclopropanes) (**28a–c**), which were reported by Baudler *et al.* ⟨84CB1542⟩. The synthesis of these derivatives involved the cyclo-condensation of 1,2-dipotassium-1,2-di-*t*-butyldiphosphide (**27**) with 1,1-dichloro alkenes. According to Baudler *et al.*, the dimethyl (**28b**) and di(chlorophenyl) methylenediphosphiranes (**28c**) can be isolated in a pure state and are surprisingly stable entities. On the other hand, the terminally unsubstituted methylenediphosphirane is thermally unstable and was only characterised as a mixture contaminated with di-*t*-butyldiphosphide. X-ray crystallography studies on (**28a**) indicated that the structure had approximate C_2-symmetry and one of the smallest experimentally measured P—P—C bond angles of 52.4°.

The preparation of a methylenediphosphirane (**30**) by addition of isopropylidene carbene to a *trans*-diaryldiphosphene (**29**) (Equation (2)) has also been reported ⟨91TL3687⟩. However, the compound underwent partial decomposition under the reaction conditions to yield 70% of (**30**) and 30% of the phosphacumulene (**31**). The latter is probably produced by a concerted rupture of the P—P and P—C bonds, by a cheletropic reaction, involving loss of ArP.

4.22.1.2.2 Four-membered rings

The symmetrical 1,3-diphosphetane (**34**) was inadvertently prepared by the head-to-tail cyclo-dimerisation of the phosphallene (**33**) (Scheme 9) ⟨86CB2466⟩. The latter compound was prepared by alkaline hydrolysis of the phosphino silyl enolate (**32**). In contrast to the general behaviour of three- and five-membered ring phosphorus compounds, (**34**) was a stable solid (m.p. 248 °C) and could be purified by recrystallisation from chloroform. The structure was proved conclusively by x-ray crystallography. The four-membered ring was virtually planar, and the phosphorus substituents adopted a *trans* conformation relative to the plane of the ring. In the ^{31}P-NMR spectrum both phosphorus atoms were equivalent.

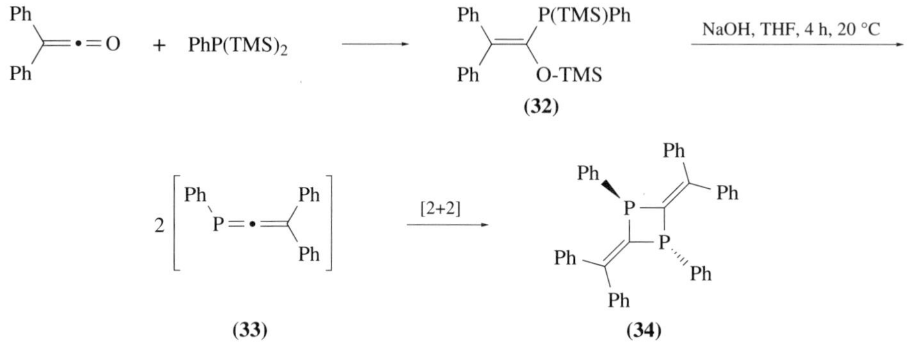

Scheme 9

4.22.1.2.3 Five-membered rings

(i) Phospholes

The first synthesis of a phosphole was reported in 1959 ⟨88CRV429⟩. However, the synthesis of a phosphole with a phosphorus substituent in the 2-position was not described until 1992 by Deschamps and Mathey ⟨92OM1411⟩. The synthesis involved (Scheme 10), as the key step, the selective phosphinylation of the anion of the cyclic phosphine sulfide (**35**) in the 2-position by chlorodiphenylphosphine. Under the reaction conditions the sulfur atom migrates from the *endo*- to the *exo*-cyclic phosphorus atom. This internal reduction is not surprising, since it is well known that phosphole sulfides are readily reduced by tertiary phosphines ⟨73TL3255, 74T3127, 76T2395⟩. Subsequent treatment of the *exo*-cyclic sulfide (**36**) with lithium naphthalide resulted in reduction of the P=S bond and concomitant cleavage of the *endo*-cyclic P—Ph bond. The anion (**37**) was characterised by ^{31}P NMR spectroscopy, the P—P coupling of 122 Hz indicating the enhanced double bond nature of the ring P—C bonds ⟨84TL3687⟩.

In situ treatment of the anion (**37**) with iron(II) chloride yielded a mixture of the asymmetrical and symmetrical phosphaferrocene derivatives (**38**) and (**39**) in a ratio of 85:15, respectively, in 40% yield (Scheme 10). The ^{31}P NMR spectra of the compounds (**38**) and (**39**) showed a typical upfield shift for both the *endo*- and *exo*-cyclic phosphorus resonances.

Both isomers coordinated with transition metal carbonyls, for example (**38**) with molybdenum hexacarbonyl gave (**40**) and (**39**) with norbornadiene–molybdenum tetracarbonyl gave (**41**). The x-ray crystal structure of (**41**) was shown to possess a C_2-centre of symmetry, and thus establishing unambiguously the stereochemistry of (**38**) and (**39**).

(ii) Diphospholes

The first example of this class of compound, 1-*t*-butyl-2-ethylidene-2,3-dihydro-4-methyl-3-phenyl-1*H*-1,3-diphosphole (**44a**) (Scheme 11) was reported as a by-product of the radical mediated cycloaddition reaction of phenylphosphine with *t*-butyldipropynylphosphorine ⟨75TL3171⟩. Compound (**44a**) could not be obtained in a pure state, and thus the only evidence for its formation was

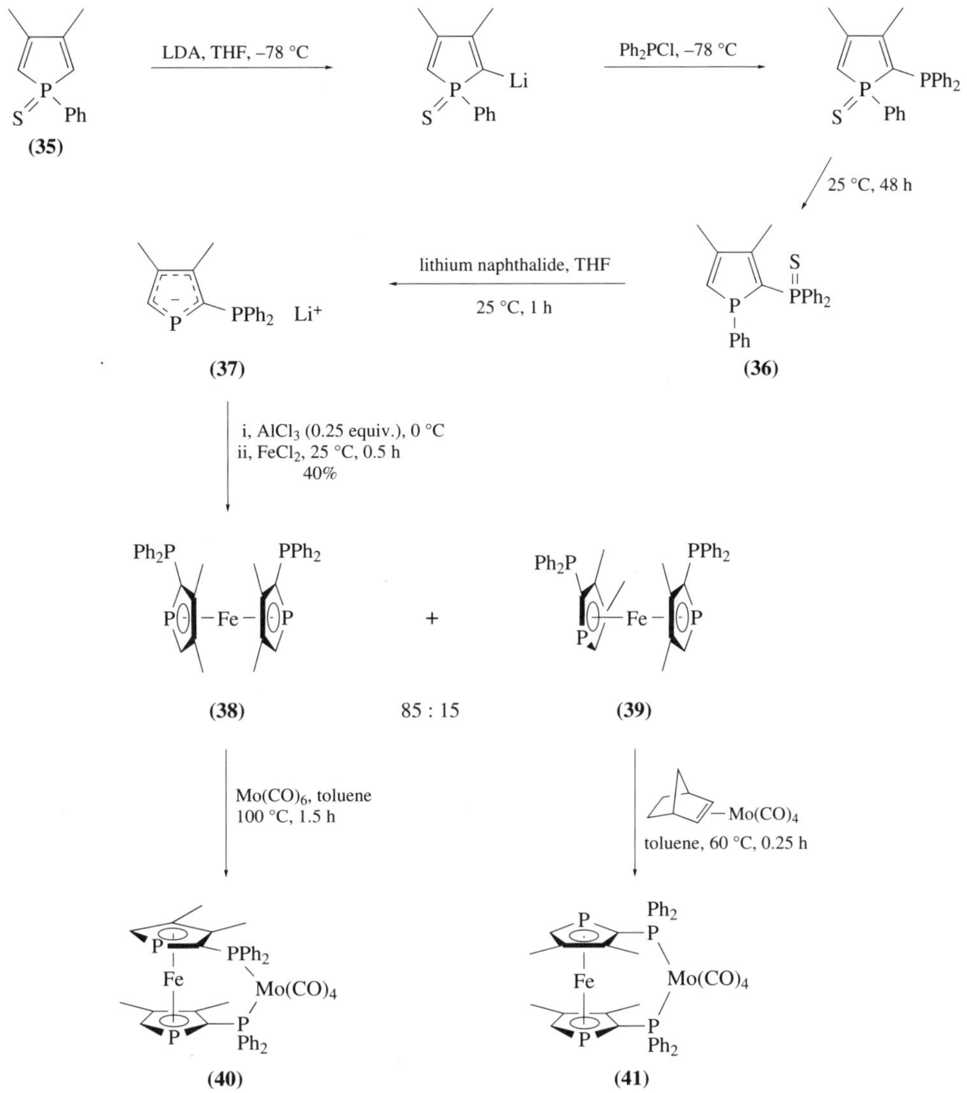

Scheme 10

a ^{1}H-NMR spectrum. The reaction was subsequently reinvestigated ⟨85CB2365⟩ and further examples prepared (**44a–d**), but once again the 5-membered ring diphospholes could not be separated from the isomeric six-membered diphosphorines (**43a–d**). The ratio of (**44**) to (**43**) remained constant at 66 : 34 in all examples. Furthermore, assignment of the stereochemistry of the exocyclic double bond in (**44a–d**) was not determined, although by implication all compounds were single geometric isomers.

Under the same conditions the reaction of phenylphosphine with *t*-butyldiethynylphosphine (**45**) gave only a mixture of the isomeric diphosphorines (**46**) and none of the corresponding diphosphole (**47**) (Equation (3)).

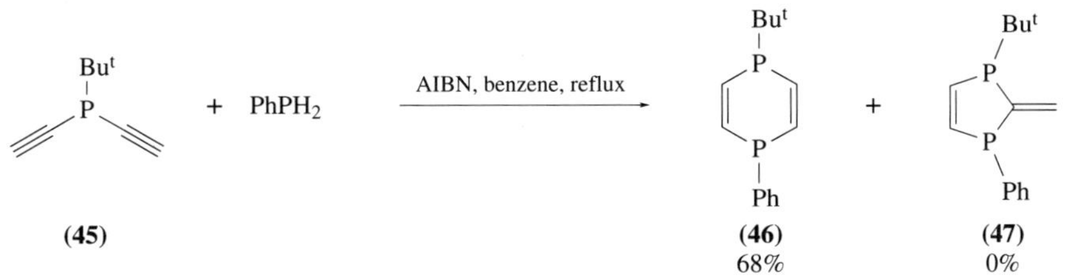

AIBN = 2,2′-azobisisobutyronitrile

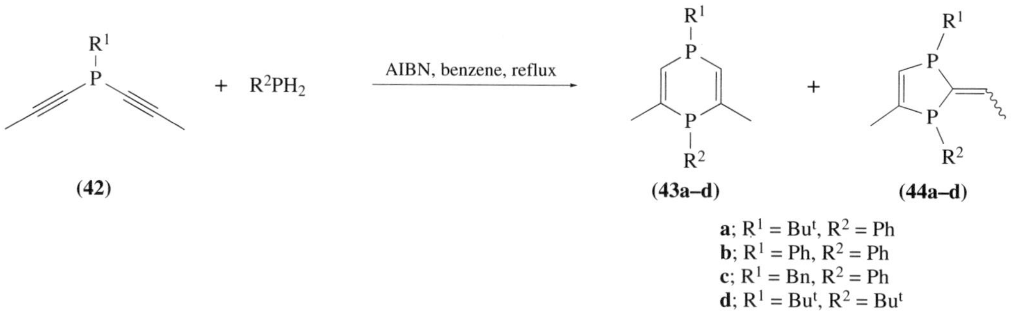

Scheme 11

a; R¹ = Buᵗ, R² = Ph
b; R¹ = Ph, R² = Ph
c; R¹ = Bn, R² = Ph
d; R¹ = Buᵗ, R² = Buᵗ

AIBN = 2,2'-azobisisobutyronitrile

4.22.1.3 Transition Metal Complexes of Bisphosphino Alkenes

The complexing ability of 1,1-bisphosphino alkenes with transition metal compounds was referred to earlier (4.22.1.1). Several examples of this type of structure have already been described in previous sections, where they were used to characterise bisphosphino alkene structures.

Additionally, there are a number of reports devoted exclusively to the synthesis of transition metal–bisphosphine complexes. Particularly noteworthy is the contribution of Shaw and co-workers who have studied the complexes of tungsten, chromium, molybdenum ⟨82CC286, 82CC287, 85CC614⟩, platinum ⟨85CC1635⟩ and palladium ⟨86CC882⟩. Complexation was generally achieved by treatment of a suitable transition metal compound, usually containing carbonyl or diene ligands, with a bisphosphino alkene in a solvent. For example, the dimethylplatinum(II) complex (**49**) was prepared from cyclo octa-1,5-dienyldimethylplatinum (**48**) and 1,1-bisdiphenylphosphino ethene in toluene at reflux (Equation (4)). The tungsten, molybdenum and chromium analogues were also prepared in a similar fashion. Alternatively, the ionic salts of platinum can be used, platinum(II) chloride and bisdiphenylphosphino ethene in dichloromethane at room temperature yielded the corresponding dichloroplatinum complex (**50**). Analogously, palladium acetate reacts to form the complex (**51**) in 85% yield.

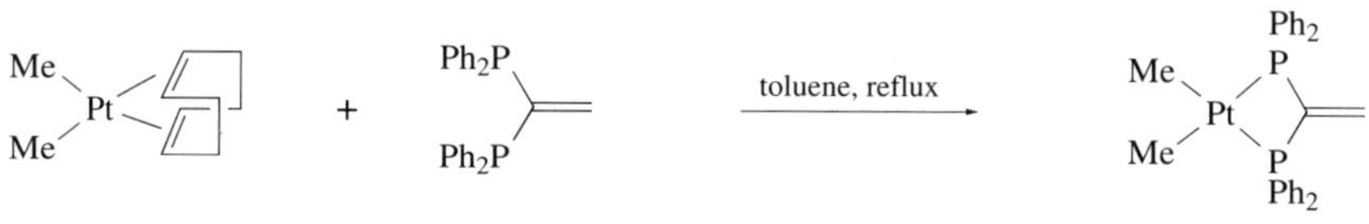

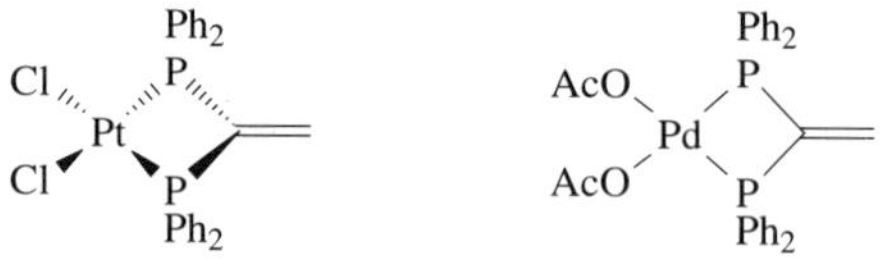

The effect of complexation is to enhance the reactivity of the alkene towards nucleophilic attack. This is clearly demonstrated when uncomplexed bisdiphenylphosphino ethene was treated with a Michael donor and no reaction occurred at the double bond. However, the complexed alkene reacted instantaneously with amines and carbon nucleophiles at room temperature. The complex did not need to be preformed, a catalytic amount of palladium acetate added to a mixture of uncomplexed alkene and nucleophile resulting in rapid reaction ⟨85CC614⟩. Most of the additions documented involved the formation of compounds where the phosphorus atoms are eventually bonded to a sp^3 carbon centre, and thus fall outside the scope of this chapter. However, one novel bisphosphino alkene was prepared by this procedure (Scheme 12) ⟨85CC614⟩. Reaction of the tungsten complex (**52**) with the lithium salts of alkynes, in the presence of TMEDA, afforded the Michael addition products (**53a–b**). The butyne (**53a**), when treated with potassium *t*-butoxide in propan-2-ol isomerized to the corresponding butadiene (**54**). The stereochemistry of the double bond was not reported, but Cooper *et al.* do allude to a x-ray crystal structure for the compound.

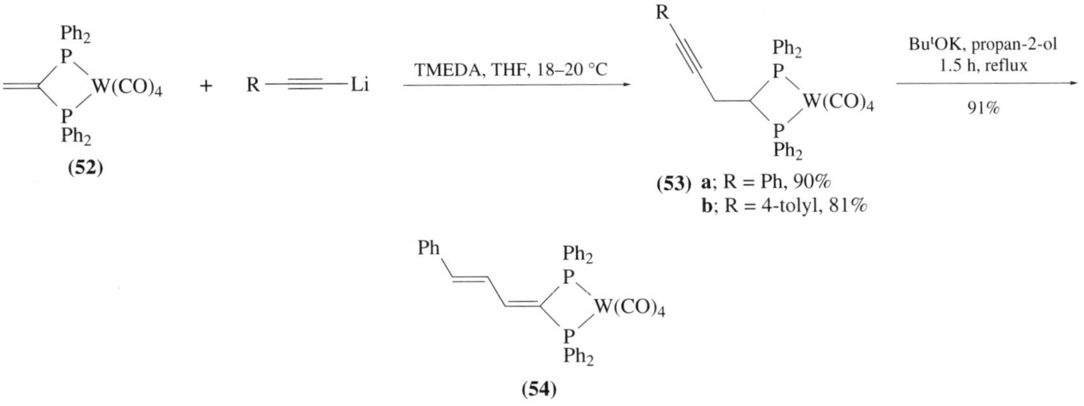

Scheme 12

Analogous reactions also occurred when the bisphosphino alkene was complexed with the corresponding molybdenum and chromium tetracarbonyls. In conclusion, this sequence of reactions represents a new method for the functionalisation of the double bond of bisphosphino alkenes and could provide routes to compounds not available by existing methodologies.

So far the reactions of the transition metal–alkene complex have concentrated on the chemistry of the alkene, but some examples are known where the reaction takes place at the metal centre. Treatment of the platinum(II) complex (**49**) with excess iodomethane at room temperature resulted in oxidation to the platinum(IV) complex (**55**) (Equation (5)). In contrast, uncomplexed bisdiphenylphosphino ethene undergoes quaternisation of the phosphorus atoms to give compound (**8**) (see Scheme 3).

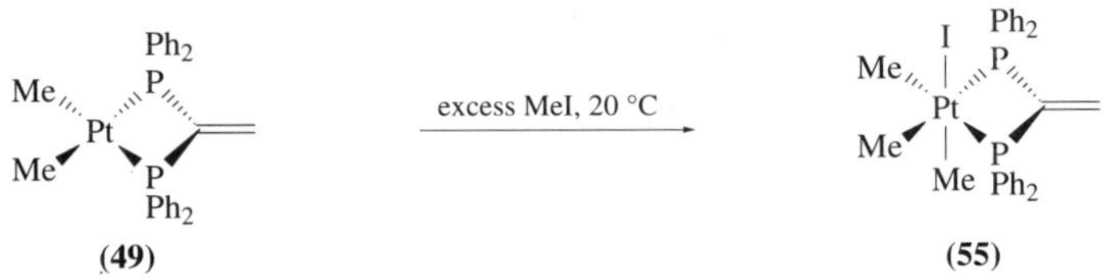

Metalla-heterocycles can also be formed from the reaction of bisphosphino alkenes with organometallic reagents. Weber and Wewers ⟨85CB3560⟩ reported the formation of a 5-membered ring containing two phosphorus atoms and a chromium atom from the reaction of pentacarbonyl-[(dimethyloxosulfono)methanide]chromium (**56**) with bisdiphenylphosphino ethene in toluene (Equation (6)). In common with the complexes already discussed, this product (**57**) showed enhanced reactivity towards Michael addition.

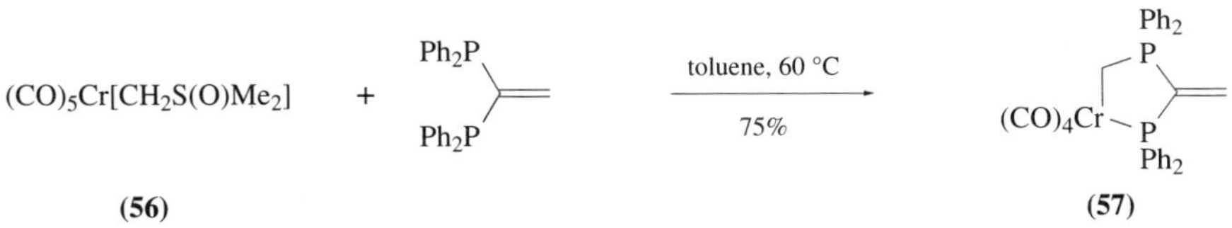

The ruthenium–bisphosphino alkene complex ⟨91JCS(D)351⟩ (**58**) was prepared from cyclopentadienylbisdiphenylphosphino ruthenium chloride and bisdiphenylphosphino ethene by heating in benzene (Scheme 13). The structure was confirmed by x-ray crystallography, the salient feature being a ring P—C—P bond angle of 95.3° compared with 119° in the free ligand. The cause of the decrease in bond angle was attributed to chelation of the phosphorus atoms in the ligand to the ruthenium atom. Such chelation is indicative of increased strain within the ring system. The increased ring strain is reflected in the chemical reactivity; unlike all the other complexes discussed so far (**58**) underwent ring cleavage when treated with bischlorodicarbonyl rhodium in toluene to give a heterobimetallic complex (**59**) in 80% yield as a yellow crystalline solid. A similar sequence of reactions has been reported to yield a heterobimetallic complex involving iron and ruthenium ⟨86JOM(309)C56, 87CC826⟩.

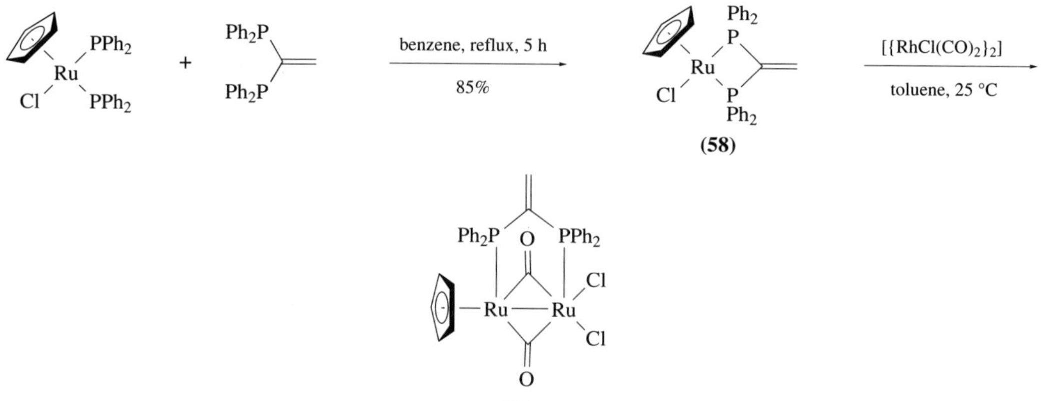

Scheme 13

Finally, Al-Jibori, McDonald and Shaw ⟨82CC286⟩ and ⟨82CC287⟩ have described a bisdiphenylphosphino ethene–chromium complex with an oxygen substituent on the β-carbon atom of the alkene. Thus the reaction of the chromium complex (**60**) with *n*-butyllithium yielded the anion (**61**) and quenching with benzoyl chloride gave the benzoyl intermediate (**62**) (Scheme 14). With sodium ethoxide (**62**) formed the corresponding enolate (**63**), and subsequent trapping with benzoyl chloride afforded the benzoyl enol ester (**64**). Unfortunately, no further details were given about the chemistry of this interesting structure.

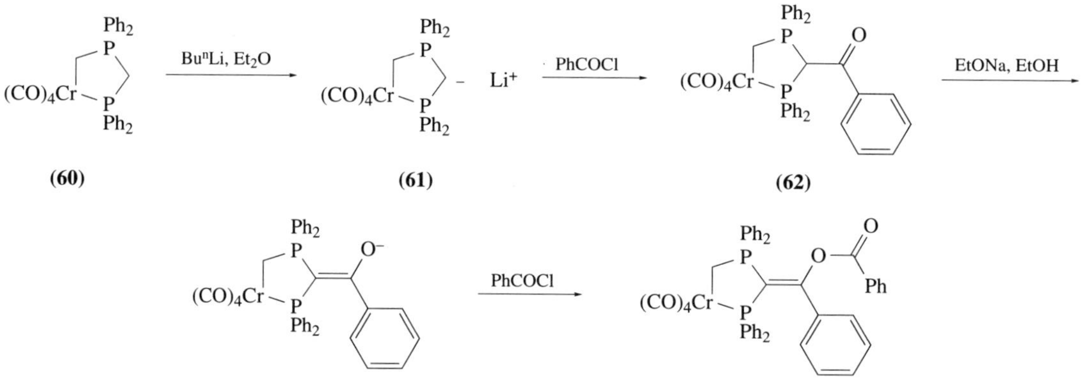

Scheme 14

4.22.2 FUNCTIONS CONTAINING ONE PHOSPHORUS AND EITHER ARSENIC, ANTIMONY OR BISMUTH, $R^1_2C=C(PR^2_2)AsR^3_2$, etc.

Few examples of this type of compound are known, the only members of the class are the arsenic analogues of the diphospholes (**44**), namely the 1-substituted 2-ethylidene-2,3-dihydro-4-methyl-3-phenyl-1*H*-1,3-phospharsoles (**67a–c**). These were prepared by a reaction analogous to that used in the formation of the diphospholes (**44**), (Section 4.22.1.2.3.ii) involving radical cycloaddition of phenylarsine (**65**) to *t*-butyldipropynylphosphine (**66**) at 140 °C (Equation (7)) ⟨75TL3171, 85CB2365⟩. The crude reaction product was then chromatographed and vacuum distilled to yield a mixture that consisted of the five-ring (**67a–c**) and six-ring (**68a–c**) compounds in a 66:34 ratio, in an overall yield of 54% for (**67a**) and (**68a**). As reported for the preceding diphospholes (**44**), the geometry of the exocyclic double bond was not determined.

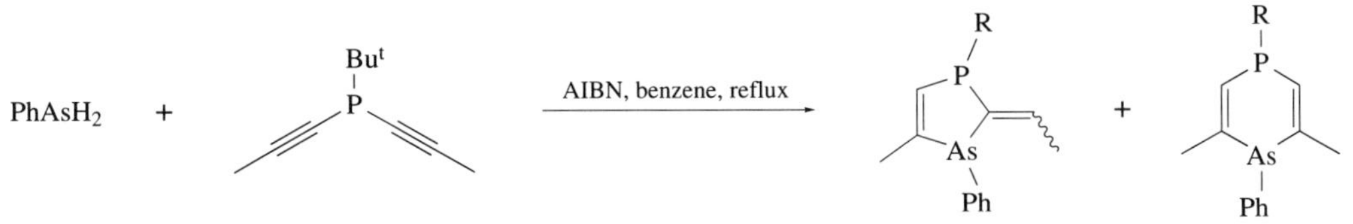

$$PhAsH_2 + \text{(66)} \xrightarrow{\text{AIBN, benzene, reflux}} \text{(67)} + \text{(68a–c)} \quad (7)$$

(**65**)   (**66**)   (**67**) a; R = Bu^t 2 : 1   (**68a–c**)
b; R = Ph
c; R = Bn

4.22.3 FUNCTIONS CONTAINING ONE PHOSPHORUS AND A METALLOID, $R^1_2C{=}C(PR^2_2)SiR^3_3$, etc.

4.22.3.1 Silicon and Germanium Derivatives

In 1995, there are no examples of this type of structure reported in the literature. This is probably due to a lack of appropriate synthetic methods rather than to a lack of stability, as the corresponding antimony and bismuth compounds are known, (Section 4.22.7). The most obvious route to prepare this type of compound would involve the deprotonation of a suitable phosphino alkene and quenching with a trisubstituted silicon or germanium electrophile. In the case of phosphorus aryl or alkyl phospholes, the most widely studied phosphino alkenes, treatment with strong bases results exclusively in cleavage of the exocyclic P—C bond to furnish the phosphorus anion. For a full discussion of this chemistry, see Section 4.22.4.1.

4.22.3.2 Boron Derivatives

Like the silicon and germanium derivatives, phosphino alkenes α-substituted with a dialkylboryl group are unknown, and even the antimony and bismuth analogues have yet to be prepared.

4.22.4 FUNCTIONS CONTAINING PHOSPHORUS AND A METAL, $R^1_2C{=}C(PR^2_2)M$, etc.

4.22.4.1 Group 1 and 2 Metals, $R^1_2C{=}C(PR^2_2)M$, etc.

Compounds containing an alkali or alkali earth metal attached to the α-carbon of a phosphino alkene are not known. In the case of P—aryl or P—alkyl phospholes (**69**), ⟨70JHC1, 85JHC513, 88CR(88)429⟩ treatment with phenyllithium ⟨88CRV429⟩ gave the anion (**70**) (Scheme 15). The driving force for the cleavage of the exocyclic P—phenyl bond is the increased electronic delocalisation achieved in going from the weakly aromatic phosphole (**69**) to the strongly aromatic phospholyl anion (**70**). Treatment of the anions with an electrophile formed new P—electrophile bonds (**71**). For synthetic purposes, the lithium is often exchanged for a metal such as magnesium or aluminum giving a less electrophilic derivative ⟨88CR(88)429⟩.

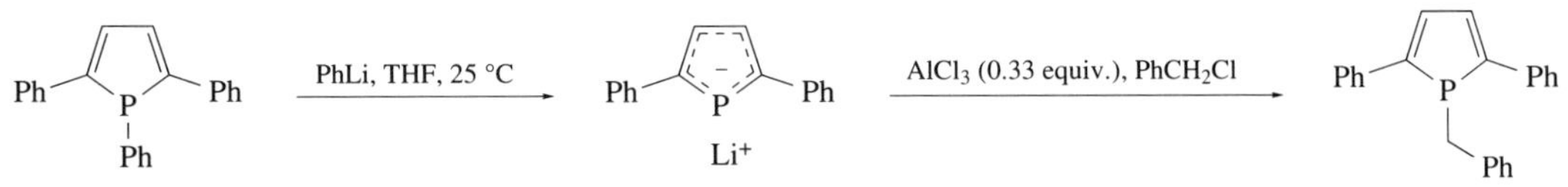

(**69**) $\xrightarrow{\text{PhLi, THF, 25 °C}}$ (**70**) $\xrightarrow{\text{AlCl}_3 \text{ (0.33 equiv.), PhCH}_2\text{Cl}}$ (**71**)

Scheme 15

1032   *Phosphorus, Arsenic, Antimony or Bismuth and No Halogen, Chalcogen or Nitrogen*

4.22.4.2 Transition Metals, $R^1_2C=C(PR^2_2)PdX_2$, etc.

4.22.4.2.1 σ-Bonded compounds

In contrast to derivatives of the Group 1 and 2 metals, phosphino alkenes α-substituted with a transition metal are well known. Nearly all the examples are derivatives of phosphinoferrocenes or phosphaferrocenes, (see Section 4.22.4.2.2) but one system has been reported where the metal was σ-bonded to a sp^2-carbon already bearing a phosphine group ⟨88CC673⟩. Thus thermal rearrangement of the iron complex (**72**) resulted in ligand exchange (Scheme 16); one of the carbon monoxide groups being displaced by the diphenylphosphine substituent of the alkene, to form (**73**). Photolysis of (**73**) resulted in the loss of another carbon monoxide ligand and concomitant coordination of the iron atom to the π-system of the carbon—carbon double bond to give (**74**). Further photolysis of (**74**) in the presence of iron pentacarbonyl yielded (**76**). The same result was achieved in two steps, although in lower overall yield, by photolysis of the original complex (**72**) with iron pentacarbonyl to give the intermediate (**75**) which underwent thermal rearrangement to form (**76**).

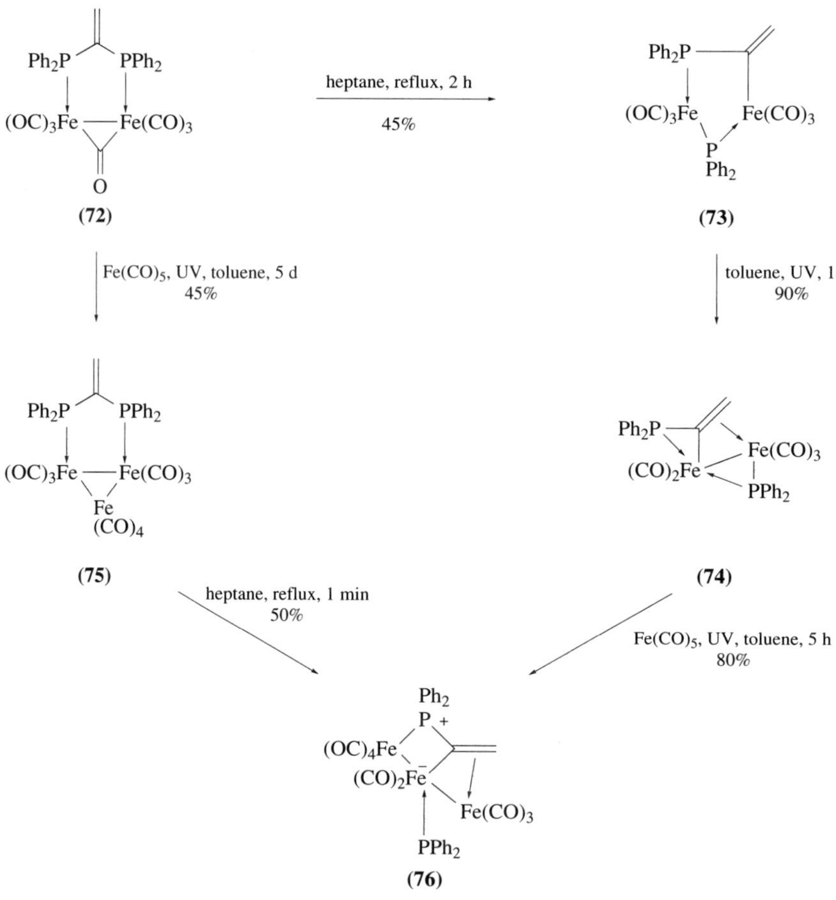

Scheme 16

X-ray crystallography studies on (**76**) showed the ring P—C bond (17.4 nm) to be shorter than anticipated for a P—C single bond, indicating substantial double bond character. Thus these compounds could also be considered as phosphallene derivatives.

4.22.4.2.2 Phosphino metallocenes

The most widely reported phosphino metallocenes are, not surprisingly the phosphinoferrocenes, the best known compounds being Kumada's ligands ⟨74TL4405⟩. These were originally prepared by deprotonation of the chiral *N,N*-dimethyl-1-ferrocenyl-2-propanamine (**77**) with *n*-butyllithium and quenching the resultant anion with chlorodiphenylphosphine or chlorodimethylphosphine to yield phosphinoferrocene derivatives (**78a–b**) (Equation (8)). Phosphine substituents were introduced sequentially into both of the cyclopentadienyl rings by treating (**77**) stepwise with *n*-butyllithium, *n*-butyllithium-TMEDA followed by quenching with two equivalents of chlorodiphenylphosphine to yield (**79**) (Equation (9)).

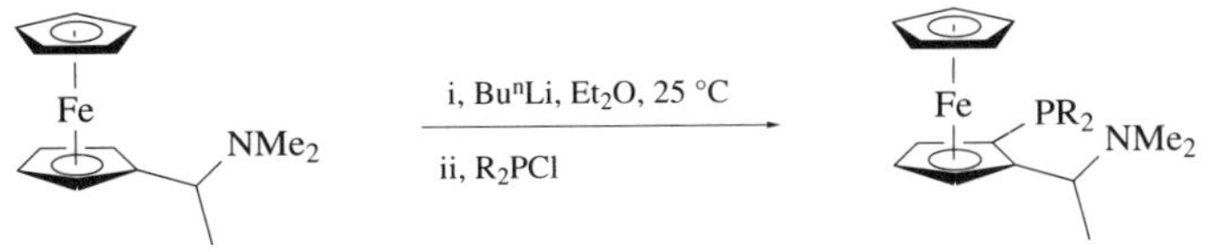

Kumada and co-workers also described the synthesis of the analogous ethylferrocene compound (**82**) (Scheme 17) ⟨74TL4405⟩. Conversion of (**78**) into the phosphine oxide (**80**) followed by quaternisation and elimination of trimethylamine afforded the vinylferrocene (**81**), which was reduced sequentially with rhodium–hydrogen and lithium aluminum hydride to give (**82**).

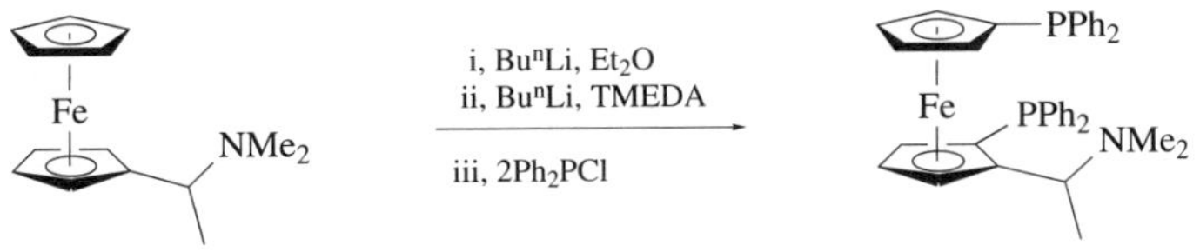

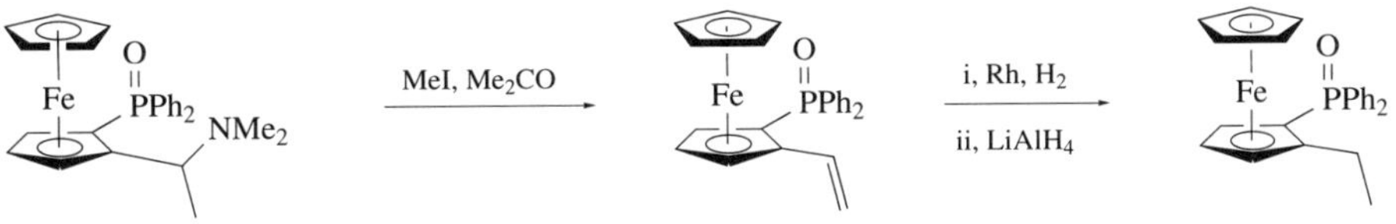

Scheme 17

Kumada and co-workers have used these phosphinoferrocenes as asymmetric catalysts in Grignard cross-couplings ⟨76JA3718⟩ or nickel and palladium-catalysed cross-couplings ⟨82JA180, 86JOC3772, 88JA8153, 90TA151⟩. The ligand has also been linked to a polymer support for the catalytic hydrosilylation of alkenes ⟨87JOM(333)269⟩.

Several other structurally similar catalysts, where the phosphorus is substituted with functionalised aryl rings, have been reported ⟨80BCJ1138⟩, and are prepared in an analogous way to the original compounds prepared by Kumada and co-workers.

An alternative chiral synthetic approach (Equation (10)) to the Kumada structures has been reported ⟨80JOM(202)C58⟩ using the chiral palladium compound (**83**) to prepare (**85**) via the palladium(II) phosphine intermediate (**84**).

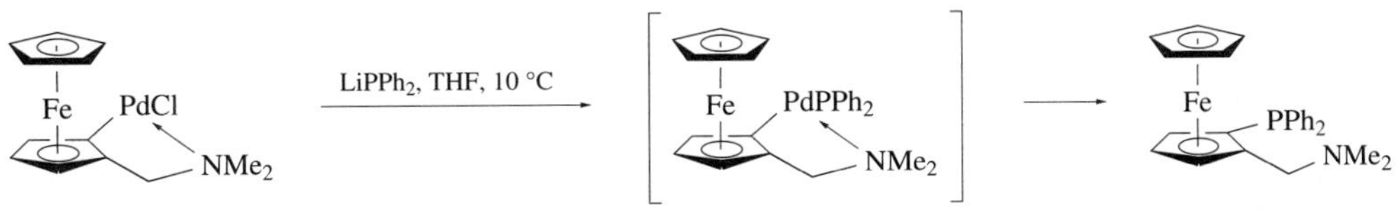

Introduction of further substituents into phosphinoferrocenes is possible (Equation (9)), but this resulted in substitution in the second cyclopentadienyl ring. However, sulfur substituents were introduced (Equation (11)), in low yield, into the more substituted cyclopentadienyl ring by the method of Deus, Hubener and Herrmann ⟨90JOM(384)155⟩. Treatment of (*R*,*S*)-1-diphenylphosphino-2-(1-dimethylaminoethyl)ferrocene (**86**) with *n*-butyllithium in diethyl ether followed by quenching with dimethyl disulfide gave (**87**) in 29% yield.

1034   *Phosphorus, Arsenic, Antimony or Bismuth and No Halogen, Chalcogen or Nitrogen*

$$\underset{(86)}{\text{Fe-CH(Me)NMe}_2\text{-PPh}_2} \xrightarrow[\text{29\%}]{\text{i, Bu}^n\text{Li, Et}_2\text{O, reflux, 2 h}\atop \text{ii, MeSSMe}} \underset{(87)}{\text{Fe(SMe)-CH(Me)NMe}_2\text{-PPh}_2} \quad (11)$$

Other modifications of Kumada's initial structures have been reported in which the dimethylamino group is replaced by a methylthio group ⟨92OM613⟩. In addition, more heavily functionalised derivatives with various amine or phosphorus substituents are also known ⟨90JOC1649, 91CC1390⟩, but the synthetic methodology is identical to that used by Kumada.

One area where new methodology developed was in the synthesis of the ferrocenophanes. Thus treatment of the ferrocenes (**88a–c**) with two equivalents of *n*-butyllithium–TMEDA resulted in formation of the remarkably stable dilithioferrocene–TMEDA complexes (**89a–c**) which are isolable (Scheme 18). Quenching of the dianions (**89a–c**) with phenyl or *t*-butyldichlorophosphine yielded the ferrocenophanes (**90a–e**). When the ferrocenophane (**90c**) reacted with phenyl lithium, it underwent cleavage of one of the P—cyclopentadienyl bonds to form (**91**), a regioisomer of the original Kumada ligand ⟨83CJC2354, 86OM1320⟩.

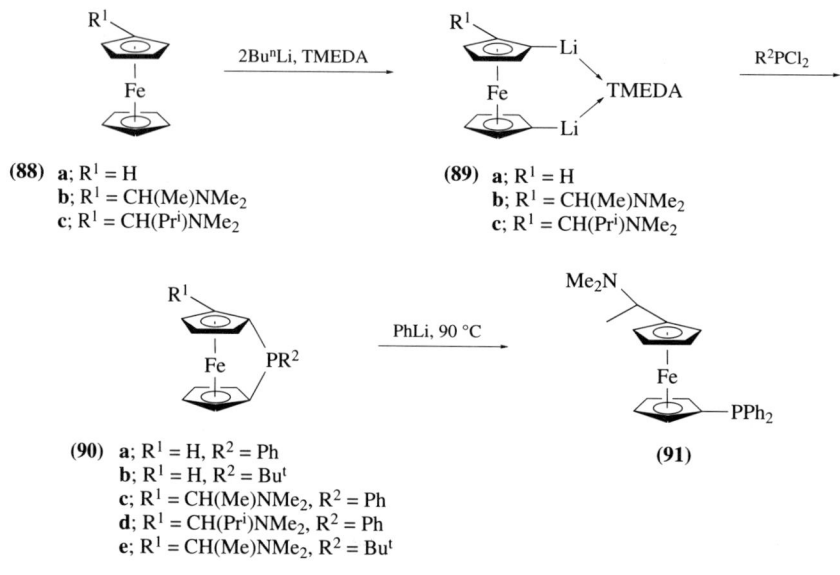

Scheme 18

4.22.4.2.3 *Phosphametallocenes*

Several publications have described the preparation of phosphaferrocenes ⟨77JA3537, 77JOM(139)77, 79JCS(D)1552⟩. Typically, the iron cyclopentadienyl dicarbonyl complex (**92**) loses both carbon monoxide ligands on prolonged heating to form the phosphaferrocene derivative (**93**) (Equation (12)). The resultant tetraphenylphosphaferrocene was isolated as a dark red solid and the absence of metal carbonyl absorptions in the IR spectrum confirmed the proposed structure. An alternative synthesis of phosphaferrocenes (**95a–b**) ⟨77JA3537⟩ involved the reaction of the phospholes (**94a–b**) with dicyclopentadienyltetracarbonyldiiron in boiling xylene (Scheme 19). An earlier attempt at the preparation of a phosphaferrocene by the reaction of tetraphenylphospholyl anion with dicarbonylcyclopentadienyl iron iodide failed to give the required π-sandwich structure ⟨71BSB651⟩.

Phosphaferrocenes are fully aromatic and react with electrophiles (Scheme 19); thus acetylation of (**95b**) with acetyl chloride and aluminum(III) chloride resulted in exclusive substitution in the phosphole ring to yield the acetylphosphaferrocene (**96**).

Diphosphaferrocenes are also known and were prepared (Scheme 20) in moderate yields from 1,2,3-triphenylphosphole (**69**) ⟨80JA994⟩. This involved initial cleavage of the exocyclic P—Ph bond with lithium metal to afford the anion (**70**) as described previously ⟨88CRV429⟩. Treatment of (**70**) with anhydrous iron(II) chloride gave the tetraphenyldiphosphaferrocene (**97**) in 53% yield. The yield of the corresponding dimethyl compound was very low using this procedure and subsequent experimentation established a more efficient procedure (Scheme 21). Initial cleavage to the P—phenyl bond was best achieved with sodium naphthalide. However, the resultant sodium derivative is very basic and a powerful reducing agent, and probably reduces the iron(II) chloride to iron. To circumvent these problems, the metallic cation was changed to magnesium to give (**98**) by addition of magnesium bromide to the reaction mixture. Complexation of (**98**) with iron(II) chloride gave the 2,2′,5,5′-tetramethyl compound (**99**) in 60% yield.

The chemistry of diphosphaferrocenes reflects the aromatic nature of these structures, and closely resembles the behaviour of ferrocenes. Compound (**99**) underwent facile electrophilic substitution when treated with acetyl or benzoyl chloride and aluminum(III) chloride catalyst to give the acyl compounds (**100a–b**) (Equation (13)). The acylated derivative (**100a**) can be treated with an excess of Grignard reagent to prepare the corresponding tertiary alcohols. In the presence of an excess of acetyl chloride, disubstitution can occur to give the symmetrical diacetyltetraphenyl-diphosphaferrocene (101) in 64% yield (Equation (14)).

In addition to the numerous phosphole-containing ferrocene derivatives, there is one example of an iron π-sandwich complex containing a phosphacyclohexadienyl ligand ⟨86OM877⟩. The synthesis of this structure involved the initial displacement of a cyclopentadienyl ligand from ferrocene by 2,4,6-triphenylphosphabenzene (102) in the presence of aluminum(III) chloride and aluminum powder (Scheme 22). Hydrolysis of the resultant cationic intermediate (103) afforded the phosphine oxide (104). This was reduced, with trichlorosilane in toluene, to the phosphadienyl complex (105), which when heated gave a mixture of (106) and (107). Prolonged heating under reflux eventually gave the phosphacyclohexadienyl structure (107). This sequence of events was deduced from the ^{31}P-NMR spectra of the reduction mixture over the reaction time course. The starting material exhibited a ^{31}P resonance at −27 ppm, which after reaction with trichlorosilane for 5 mins. at room temperature was replaced by a signal at −84 ppm consistent with (105). Heating the reaction mixture for 45 mins. led to the appearance of two new signals, at −18 ppm and −173 ppm, arising from (106) and (107), respectively. Finally, after 3 hours, only the resonance at −173 ppm remained, and no further change took place.

4.22.4.2.4 Phosphorus-containing cyclopentadienylmanganese (cymantrene) derivatives

The previous section dealt with π-sandwich compounds where the ligands coordinated to the metal were either phosphino cyclopentadienyl or phospholyl rings. However, structures are known in which the metal is coordinated to a single ring of either type, and the remaining coordination sites on the metal are occupied by carbonyl groups. Thus treatment of (dimethylaminomethyl)cyclopentadienyltricarbonylmanganese (dimethylaminomethylcymantrene) (108) with n-butyllithium followed by diphenylphosphinoyl chloride gave the 1,2-disubstituted derivative (109) (Equation (15)) ⟨79JOM(168)C33⟩. Furthermore, the parent ring system, cyclopentadienyltricarbonyl manganese (110), was deprotonated with n-butyllithium and quenched with diphenylphosphinoyl chloride to give the diphenylphosphinocymantrene (111) (Scheme 23). Subsequent deprotonation–quenching treatments of (111) with a variety of electrophiles yielded compounds (112a–c) (Scheme 23). Curiously proton abstraction occurred at the 3- rather than the 2-position of the cyclopentadienyl ring ⟨80JOM(209)233⟩ in contrast with the directing effect of the dimethylaminomethyl substituent.

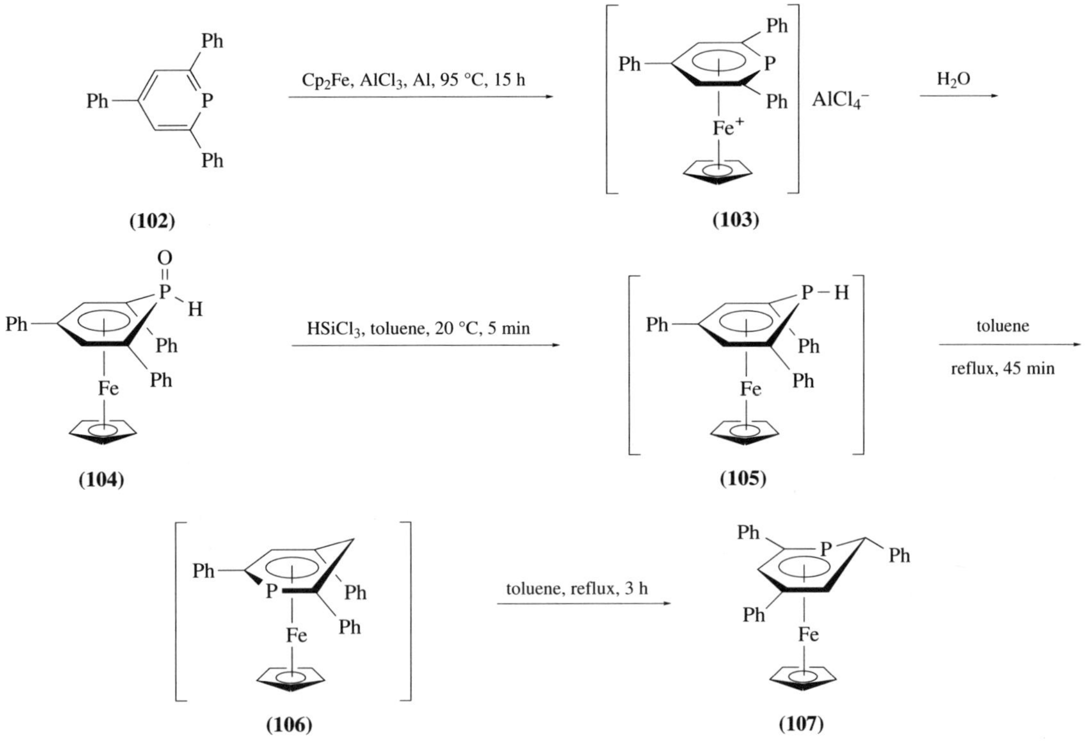

Scheme 22

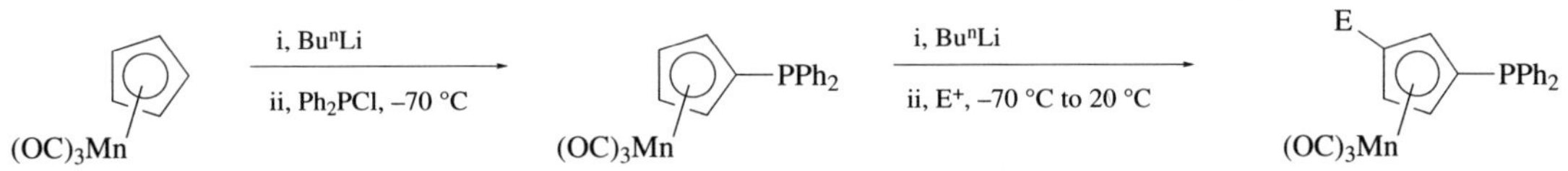

(110) (111) (112) **a**; E = D
b E = CHO
c; E = PPh_2

Scheme 23

The corresponding phosphole analogues of cymantrenes, phosphacymantrenes are known ⟨79JCS(D)1552⟩. The manganese or rhenium tricarbonyl complexes of 2,3,4,5-tetraphenylphosphole (**114a–b**) can be prepared by heating the σ-bonded metal–phosphorus pentacarbonyl compounds (**113a–b**) (Equation (16)). A more direct preparation of the manganese derivative involves the reaction of 1-phenylphosphole (**94b**) with manganese pentacarbonyl at 150 °C to yield (**115**) (Equation (17)) ⟨76TL4155, 78JA5748⟩. The phosphacymantrenes have some aromatic character, but unlike the phosphaferrocenes they do not undergo Friedel–Crafts alkylation or electrophilic formylation. Introduction of substituents via an anionic intermediate of phosphacymantrenes is not known; attempted deprotonation of (**115**) with *n*-butyllithium resulted in decomplexation of the manganese and formation of 1-butyl-3,4-dimethylphosphole (**116**) (Equation (18)).

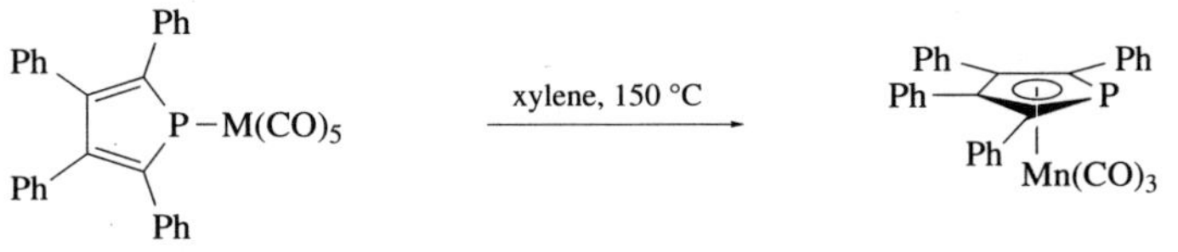

(16)

(**113**) **a**; M = Mn
b; M = Re

(**114**) **a**; M = Mn
b; M = Re

$$\text{(94b)} \xrightarrow{[Mn(CO)_5]_2,\ 150\ ^\circ C} \text{(115)} \quad (17)$$

$$\text{(115)} \xrightarrow{Bu^nLi} \text{(116)} \quad (18)$$

4.22.4.3 Tin Derivatives, $R^1{}_2C{=}C(PR^2{}_2)SnR^3{}_3$, etc.

The solitary example of this class of compound reported so far is 1-*t*-butyl-2-ethylidene-2,3-dihydro-4-methyl-3,3-di-*n*-butyl-1*H*-1,3-phosphastannole (**117**) ⟨76TL2599⟩. The preparation (Equation (19)) is analogous to the synthesis of the diphospholes (**44**), (Section 4.22.1.2.3.ii) and the phospharsoles (**67**), (Section 4.22.2) namely by radical cycloaddition of di-*n*-butyltin hydride to *t*-butyldipropynylphosphine (**66**). Purification by Kugelrohr distillation gave a 3:1 mixture of the 1,4-dihydro-1-phospha-4-stannabenzene (**117**) and the phosphastannole (**118**). The geometry of the exocyclic double was not established, and the only evidence for the phosphastannole structure is the ^{1}H-NMR spectrum of the above mixture.

$$\text{(66)} + Bu^n{}_2SnH_2 \xrightarrow{AIBN,\ benzene,\ reflux} \text{(117)} + \text{(118)} \quad (19)$$

4.22.5 FUNCTIONS CONTAINING TWO ARSENIC, ANTIMONY OR BISMUTH FUNCTIONS, $R^1{}_2C{=}C(AsR^2)_2$, etc.

Up to 1995, somewhat surprisingly, these analogues of the bisphosphino alkenes, discussed in Section 4.22.1, have not been reported. It is difficult to see why the bisarsenio alkenes, for example, should be significantly less stable than the corresponding phosphines.

4.22.6 FUNCTIONS CONTAINING TWO DISSIMILAR COMBINATIONS OF ARSENIC, ANTIMONY OR BISMUTH, $R^1{}_2C{=}C(AsR^2{}_2)SbR^3{}_4$, etc.

In view of the comments made in the previous section, the fact that compounds of this type have yet to appear in the literature causes less surprise.

4.22.7 FUNCTIONS CONTAINING ARSENIC, ANTIMONY OR BISMUTH WITH A METALLOID, $R^1{}_2C{=}C(AsR^2{}_2)SiR^3{}_3$ etc.

In the early 1990s a series of papers ⟨92OM1491, 92JA372⟩ reported the first synthesis of 2,5-bis(trimethylsilyl)-3,4-dimethylstibole (**122a**) and the corresponding bismole (**122b**). These were subsequently used in the synthesis of the corresponding stiba and bismaferrocenes, (Section 4.22.8.1).

The preparation of (**122a**) and (**122b**) involved the initial reaction of two equivalents of 1-trimethylsilylpropyne (**119**) with zirconocene dichloride (ZrCp$_2$Cl$_2$) in the presence of magnesium amalgam, according to the method of Fagan and Nugent ⟨88JA2310⟩, to yield the zirconocycle (**120**)

(Scheme 24). On treatment with iodine, (**120**) was converted stereospecifically into the diiodide (**121**), which in turn could be transmetallated with *n*-butyllithium, to form the dilithiodiene. Quenching with phenylantimony dichloride (PhSbCl$_2$) or phenylbismuth diiodide (PhBiI$_2$) yielded the 2,5-bis(trimethylsilyl)stibole (**122a**) or corresponding bismole (**122b**), respectively.

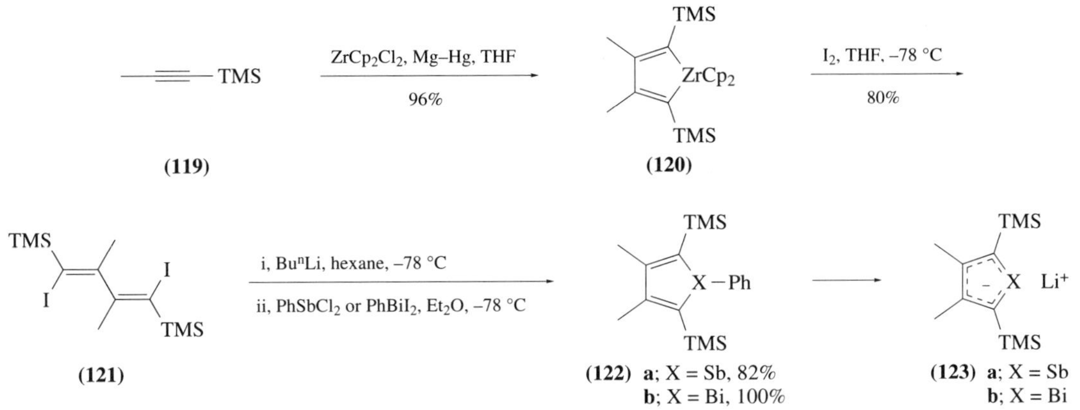

Scheme 24

The detailed chemistry of these derivatives is still obscure, but the first indications are they will be similar to the phospholes. Thus treatment of (**122a**) or (**122b**) with lithium metal led to cleavage of the heteroatom—Ph bond to form the stibolyl (**123a**) or bismolyl anions (**123b**). This is analogous to the reaction of the phosphole (**69**) with phenyllithium (see Scheme 15).

4.22.8 FUNCTIONS CONTAINING ARSENIC, ANTIMONY OR BISMUTH AND A METAL, R$^1_2$C=C(AsR$^2_2$)M etc.

4.22.8.1 Arsa-, Stiba- and Bisma-ferrocene Derivatives

The heteroferrocene series incorporating the Group 15 elements provides a complete progression down an entire column of the Periodic Table. This has allowed a comparison of the π-bonding characteristics of these molecules as a function of increasing atomic number.

The synthetic strategy for the preparation of these derivatives is similar to that described for the phosphaferrocenes (Section 4.22.4.2.3). Typically, reaction of the anion derived from an appropriately substituted arsole (**124**), stibole (**122a**) or bismole (**122b**) with iron(II) chloride led to formation of the π-sandwich compound. The exact experimental conditions vary according to the nature of the heteroatom. Thus the 1,1'-diarsaferrocene was prepared by initial treatment of 2,5-dimethyl-1-phenylarsole (**124**) with potassium, then treatment of the resultant potassium arsolyl (**125**) with iron(II) chloride to form the 1,1'-diarsaferrocene (**126**) in 40% yield (Scheme 25) ⟨79ICAL67⟩. Monoarsaferrocenes were also prepared from the arsolyl anion (**125**) by treatment with cyclopentadienyldicarbonyliron iodide (CpFe(CO)$_2$I) to form initially the σ-bonded complex (**127**), which when heated rearranged to 2,5-dimethylarsaferrocene (**128**) in 30% yield. The dimeric σ-bonded complex (**129**) was formed as a minor by-product in the reaction.

The diarsa- and monoarsaferrocenes were characterised by spectroscopy after purification by chromatography. X-ray crystallography on (**126**) confirmed the structure ⟨80AX(B)1344⟩.

The parent system, 1,1'-diarsaferrocene (**133**), is also available by a similar procedure ⟨87AG(E)229⟩. Cyclisation of a mixture of the geometrical isomers of 1,4-dichlorobuta-1,3-diene (**130**) with dilithiophenylarsine (PhAsLi$_2$) (Scheme 26) gave 1-phenylarsole (**131**) which was then sequentially converted into the arsolyl anion (**132**) and complexed with iron(II) chloride to form (**133**) in 50% yield as a crystalline, air stable solid. This method was an improvement over the original preparation involving the explosive 1,4-diiodobutadiyne (**134**) ⟨85OM1478⟩.

1,1'-Diarsaferrocene (**133**) underwent rapid, acid-catalysed deuterium exchange in deuterated trifluoroacetic acid to yield the 2,2',5,5'-tetradeuterio-1,1'-arsaferrocene (**135**) (Equation (20)). Ashe et al. ⟨87AG(E)229⟩ also reported that Friedel–Crafts acetylation of (**133**) gave 2-acetyl-1,1'-diarsaferrocene (**136**) (Equation (21)).

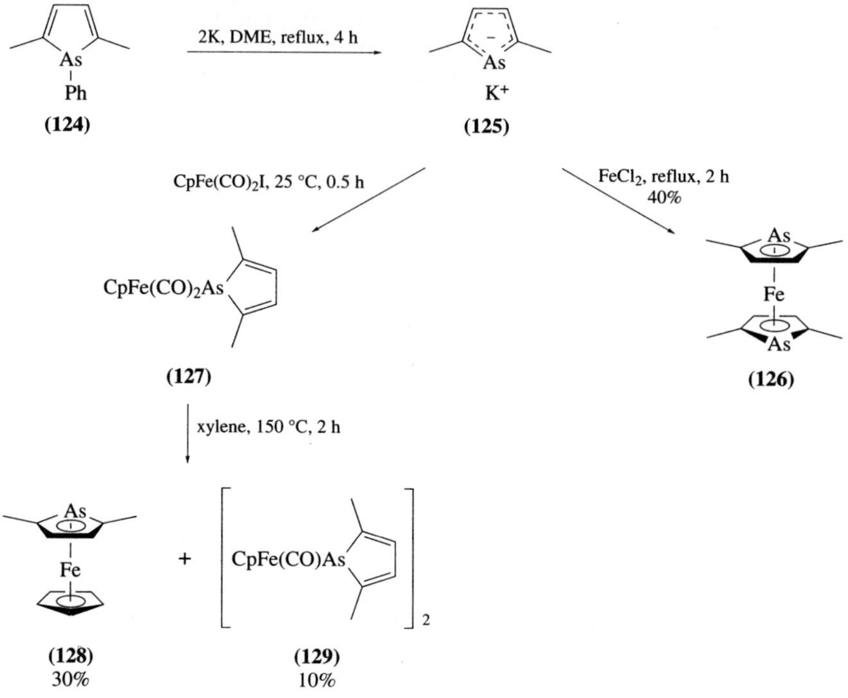

Scheme 25

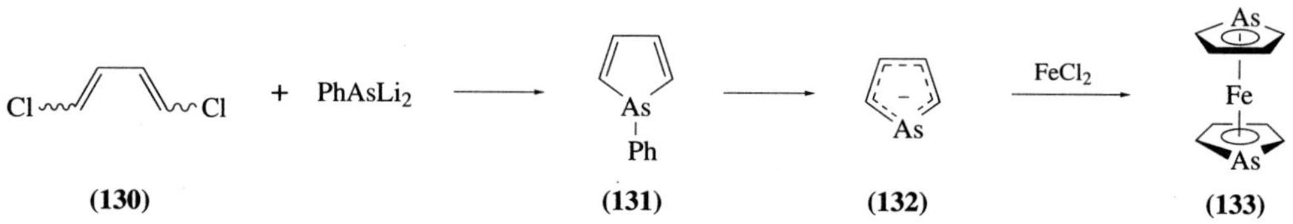

Scheme 26

(134)

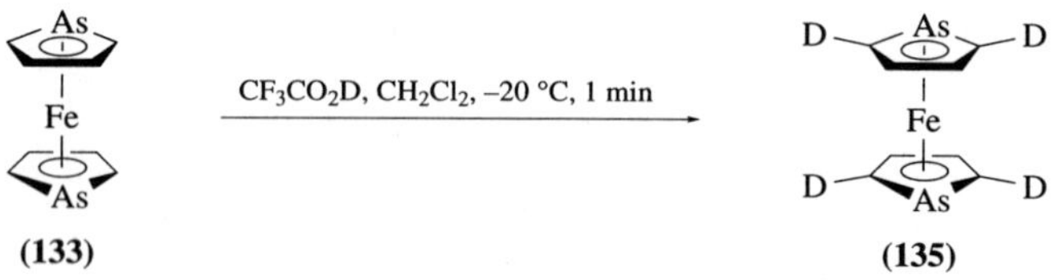

$$(133) \xrightarrow{CF_3CO_2D, CH_2Cl_2, -20\ ^\circ C,\ 1\ min} (135) \quad (20)$$

$$(133) \xrightarrow{MeCOCl, AlCl_3, CH_2Cl_2, 25\ ^\circ C} (136) \quad (21)$$

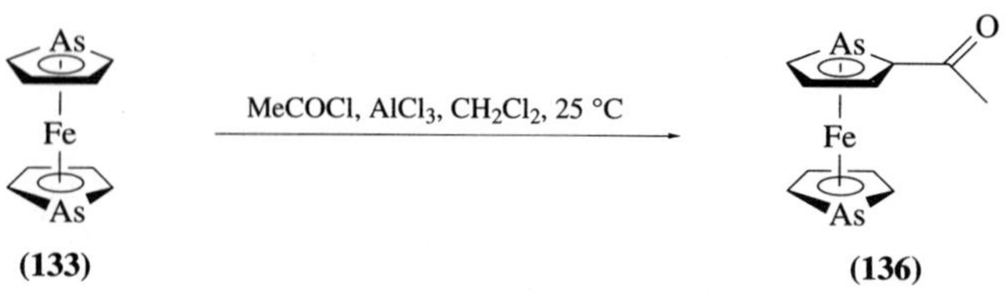

The distibaferrocenes and dibismaferrocenes are prepared in an analogous way to the diarsaferrocenes (Scheme 27) ⟨92OM1491⟩. Treatment of the stibolyl (**123a**) or bismolyl (**123b**) anions with iron(II) chloride in the presence of aluminum(III) chloride (to remove phenyllithium) gave the corresponding π-sandwich compounds (**137a**) and (**137b**), respectively.

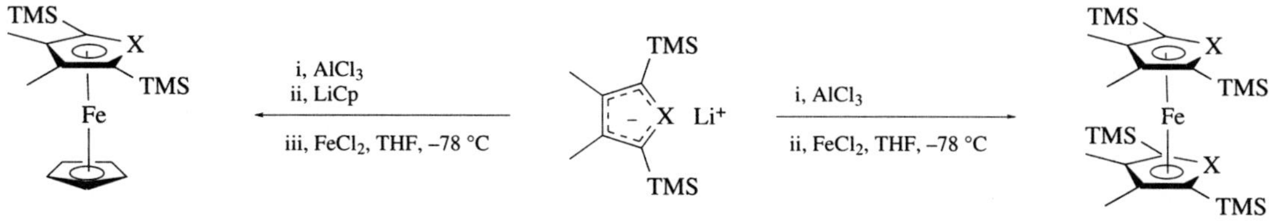

(138) **a**; X = Sb, 30%
b; X = Bi, 25%

(123) **a**; X = Sb
b; X = Bi

(137) **a**; X = Sb, 20%
b; X = Bi, 17%

Scheme 27

The monostiba- (**138a**) and monobismaferrocenes (**138b**) were also available from the intermediates (**123a**) and (**123b**) by treatment with lithium cyclopentadienide (CpLi) followed by complexation ⟨92OM1491⟩.

Less highly substituted distibaferrocenes were prepared from the 1,1-di-*n*-butyl-2,5-dimethylstannole (**139**), which reacted with phenylantimony dichloride ($PhSbCl_2$) to form 1-phenyl-2,5-dimethylstibole (**140**) in 63% yield (Scheme 28). The standard conditions for the cleavage of the Sb—Ph bond and complexation of the ion (**141**) afforded 2,2′,5,5′-tetramethyl-1,1′-distibaferrocene (**142**) ⟨91OM2068⟩. X-ray crystallography showed that the two antimony-containing rings were completely eclipsed, and the structure closely resembled that of the arsenic analogue (**126**). The 2,2′,5,5′-tetramethyldibismaferrocene (**143**) was prepared in 37% yield using an identical synthetic strategy. X-ray crystallography on this compound indicated the η^5-aromatic nature of the bismolyl rings, but unlike the arsenic analogue there was evidence for a direct bismuth—bismuth bond between the two rings ⟨92OM2743⟩. The distibaferrocenes are intermediate in this respect; the interatomic distance between the two antimony atoms may indicate that some bonding is possible ⟨91OM2068⟩.

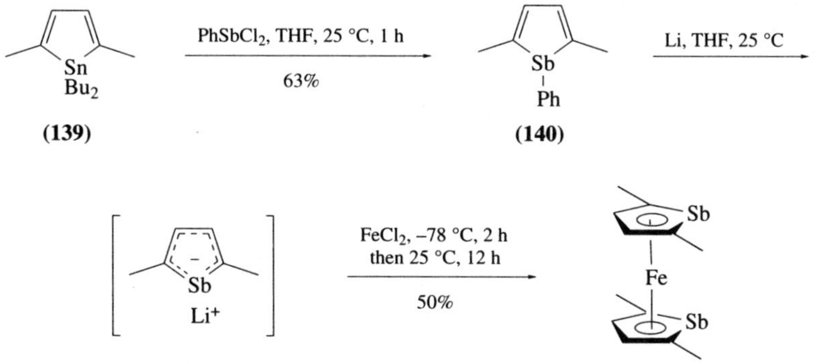

(139) **(140)** **(141)** **(142)**

Scheme 28

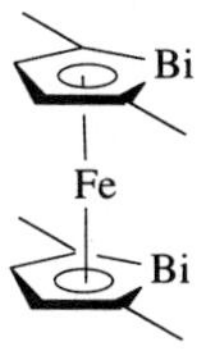

(143)

In addition to the arsaferrocenes, one example of a chromium π-sandwich compound involving a six-membered arsenic-containing ring is known ⟨86AG(E)571⟩. Bis(η^6-arsabenzene)chromium (**145**) is prepared by the co-condensation of the vapours of chromium and arsabenzene (**144**) under vacuum at $-196\,°C$ (Equation (22)) ⟨82TCC125⟩. Recrystallisation of the crude product from toluene yielded 15% of the air-sensitive compound (**145**).

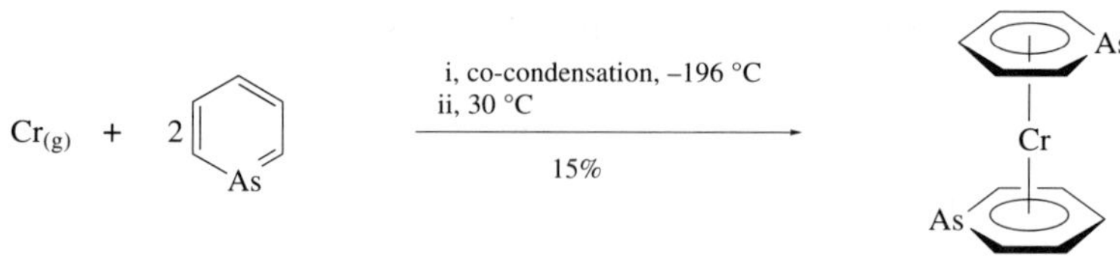

$$Cr_{(g)} + 2\,(\mathbf{144}) \xrightarrow[15\%]{\text{i, co-condensation, }-196\,°C;\ \text{ii, }30\,°C} (\mathbf{145}) \qquad (22)$$

4.22.8.2 Arsenic and Antimony Analogues of Cymantrene

The arsacymantrenes, analogues of the phosphacymantrenes, (Section 4.22.4.2.4) have been prepared by several research groups. In the earliest report in 1973 the tricarbonyl manganese and rhenium complexes of arsole were described, and the compounds were shown to have a π-arsole structure similar to the phosphole derivative (**112**), but no synthetic details were given ⟨73CC258⟩. The preparations of the tetraphenylarsole complexes with manganese and rhenium were described by Abel, Clark and Towers ⟨79JCS(D)1552⟩, who showed that the σ-bonded manganese pentacarbonyl complex (**146a–b**) underwent thermally induced loss of two carbonyl ligands to form the cymantrenes (**147a–b**) (Equation (23)) ⟨79JCS(D)814⟩. Alternatively, the 2,5-dimethylarsacymantrene (**148**) was prepared by the reaction of 2,5-dimethyl-1-phenylarsole (**122**) with decacarbonyldimanganese (Scheme 29) ⟨78ICAL294⟩. Unlike the phosphacymantrenes, the arsacymantrenes underwent Friedel–Crafts acetylation to form the 3-acetyl derivative (**149**) in 40% yield.

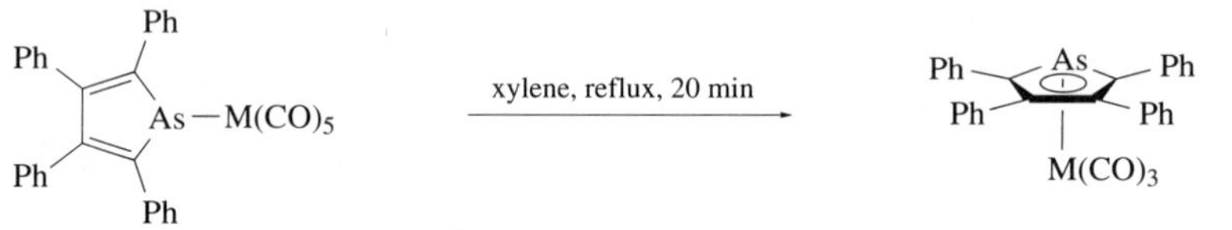

$$(\mathbf{146})\ \mathbf{a};\ M = Mn \qquad \xrightarrow{\text{xylene, reflux, 20 min}} \qquad (\mathbf{147})\ \mathbf{a};\ M = Mn,\ 58\% \qquad (23)$$
$$\qquad\qquad \mathbf{b};\ M = Re \qquad\qquad\qquad\qquad\qquad\qquad\qquad \mathbf{b};\ M = Re,\ 61\%$$

$$(\mathbf{122}) \xrightarrow[50\%]{Mn_2(CO)_{10},\ 150\,°C} (\mathbf{148}) \xrightarrow[40\%]{MeCOCl,\ AlCl_3,\ CH_2Cl_2,\ 5\ h} (\mathbf{149})$$

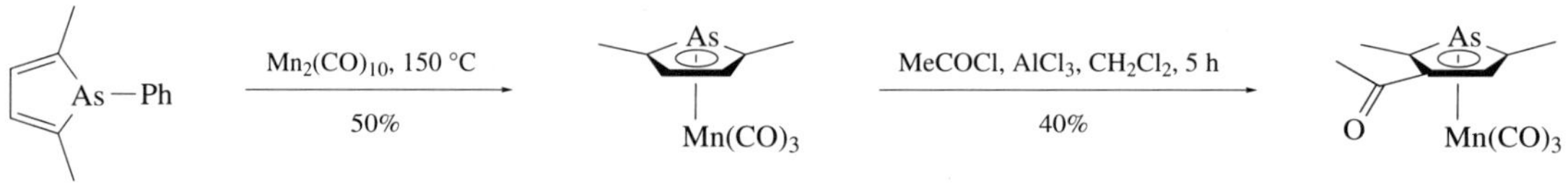

Scheme 29

The synthesis of the only known stibacymantrene (**150**) was reported by Ashe and Diephouse ⟨80JOM(202)C95⟩. Thus treatment of the dimethylstibolyl anion (**141**) with bromopentacarbonylmanganese (MnBr(CO)$_5$) gave the tricarbonyl complex (**150**) in 24% yield (Equation (24)). The compound was stable to chromatography and was fully characterised by spectroscopic techniques. As yet, the chemistry of this derivative has not been explored.

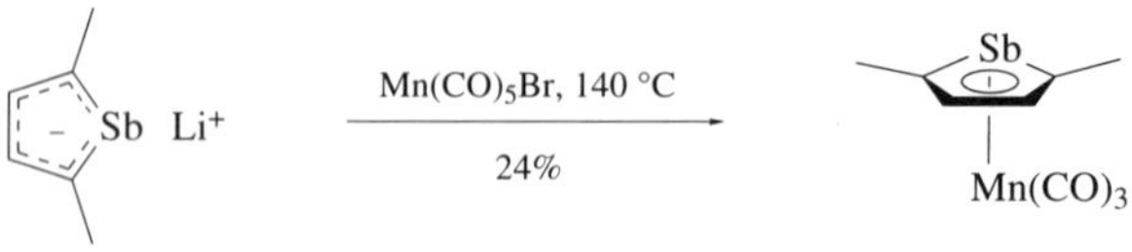

$$(\mathbf{141}) \xrightarrow[24\%]{Mn(CO)_5Br,\ 140\,°C} (\mathbf{150}) \qquad (24)$$

4.23
Functions Containing at Least One Metalloid (Si, Ge or B) and No Halogen, Chalcogen or Group 15 Element; also Functions Containing Two Metals

RICHARD A. B. WEBSTER
Pfizer Central Research, Sandwich, UK

4.23.1	FUNCTIONS CONTAINING TWO METALLOIDS—$R^1_2C\!=\!C(SiR^2_3)_2$, etc	1044
4.23.1.1	*Functions Bearing Two Silicons—$R^1_2C\!=\!C(SiR^2_3)_2$, etc.*	1044
4.23.1.1.1	Functions with no ring silicons	1044
4.23.1.1.2	Functions with one ring silicon	1052
4.23.1.1.3	Functions with two ring silicons	1053
4.23.1.2	*Functions Bearing Two Borons—$R^1_2C\!=\!C(BR^2_3)_2$, etc.*	1054
4.23.1.3	*Functions Bearing Two Germaniums—$R^1_2C\!=\!C(GeR^2_3)_2$, etc.*	1055
4.23.1.4	*Other Functions—$R^1_2C\!=\!CSiR^2_3BR^3_3$, etc.*	1056
4.23.1.4.1	Functions containing one silicon and one boron	1056
4.23.1.4.2	Functions containing one silicon and one germanium	1058
4.23.1.4.3	Functions containing one boron and one germanium	1058
4.23.2	FUNCTIONS CONTAINING A METALLOID AND A METAL	1059
4.23.2.1	*Silicon Functions—$R^1_2C\!=\!CSiR^2_3M$*	1059
4.23.2.1.1	Functions with one silicon and one group 1 metal	1059
4.23.2.1.2	Functions with one silicon and one group 2 metal	1060
4.23.2.1.3	Functions with one silicon and one transition group metal	1061
4.23.2.1.4	Functions with one silicon and one group 13 or group 14 metal	1062
4.23.2.2	*Boron Functions—$R^1_2C\!=\!CBR^2_3M$*	1063
4.23.2.3	*Germanium Functions—$R^1_2C\!=\!CGeR^2_3M$*	1063
4.23.3	FUNCTIONS CONTAINING A GROUP 1 METAL—$R_2C\!=\!CLiM$	1064
4.23.4	FUNCTIONS CONTAINING A GROUP 2 METAL (AND NO GROUP 1 METAL)—$R_2C\!=\!CMgXM$, etc.	1065
4.23.5	FUNCTIONS CONTAINING A TRANSITION METAL (AND NO GROUP 1 OR GROUP 2 METAL)—$R_2C\!=\!CTiM$, etc.	1066
4.23.6	FUNCTIONS CONTAINING A GROUP 13 OR GROUP 14 METAL (AND NO GROUP 1, 2 OR TRANSITION METAL)—$R_2C\!=\!CAlM$, etc.	1068
4.23.7	OTHER METAL DERIVATIVES	1070

4.23.1 FUNCTIONS CONTAINING TWO METALLOIDS—$R^1_2C=C(SiR^2_3)_2$, etc

4.23.1.1 Functions Bearing Two Silicons—$R^1_2C=C(SiR^2_3)_2$, etc.

4.23.1.1.1 Functions with no ring silicons

The regioselective hydrosilylation of alkynylsilanes as a route to 1,2-disilyl alkenes has been widely used in synthesis ⟨B-81MI 423-01⟩. A number of metal catalysts have been used in the process, with hexachloroplatinic(IV) acid (Speier's catalyst) being favoured. The corresponding regioselective hydrosilylation of 1-silyl alkynes to form 1,1-disilyl alkenes has also been reported. Careful selection of both the transition metal catalyst and the silicon substituent on the reactants is needed to ensure optimum yields of 1,1-disilyl compounds while limiting the formation of the 1,2-disilyl analogues. The use of Speier's catalyst leads to (E)-1,2-disilyl alkenes in high yield, the 1,1-disilyl alkenes being present as a minor component if at all ⟨82CL1663, 88JOM(346)297⟩. Chloro tris(triphenylphosphine)rhodium(I) resulted in a mixture of 1,1-disilyl and 1,2-disilyl products, while tetrakis(triphenylphosphine)platinum(0) afforded the 1,1-disilyl product (**1**) almost exclusively, in high yield (Equation (1)). It was noted that successive replacement of chlorine for the methyl groups in the alkynylsilane led to a decrease in the proportion of the 1,1-disilyl product (**1**) (57:43 for dimethyl and 20:80 for trimethyl (1,1:1,2-disilyl products) ⟨82CL1663⟩.

$$Cl_2MeSi-\!\!\!\equiv\!\!\!- \xrightarrow[120\,°C,\,10\,h]{Me_2ClSi,\,(Ph_3P)_4Pt}_{95\%} \begin{array}{c} SiMeCl_2 \\ =\!\!\!\!< \\ SiMe_2Cl \end{array} \quad (1)$$

(**1**)
95% regioselectivity

The hydrosilylation of 1,2-disilylethynes has been reported on a number of occasions; in general, Speier's catalyst was the catalyst of choice. When performed at 90 °C the reaction can be slow, but good yields of the 1,1,2-trisilylethenes are obtained. The addition is predominantly *syn*. At 130 °C many different products were observed, not only isomers of the expected adducts but also products arising from elimination and further addition reactions ⟨90JOM(391)19, 90JOM(396)299⟩. A series of reactions was performed in the presence and absence of air. Different products and product ratios were observed (Table 1).

Table 1 Hydrosilylation of bis(trimethylsilyl)ethyne (**2**) with hexyldimethyl silane.

Compound ($R = C_6H_{11}$)	Yield determined by gas chromatography (%)			
	90 °C, 24 h, N_2	130 °C, 7 h, N_2	130 °C, 7 h, air, open system	130 °C, 8 h, air, sealed system
TMS—≡—TMS (**2**)	16	13	16	11
TMS, TMS / RMe$_2$Si (**3**)	50	39	10	20
TMS / RMe$_2$Si, TMS (**4**)	0	1	7	4
RMe$_2$Si—≡—SiMe$_2$R (**5**)	0	7	17	7
TMS—≡—SiMe$_2$R (**6**)	7	18	33	19

A range of chloro- and methyl-substituted 1,2-disilylethynes reacted with differently substituted chloro methyl hydrosilanes at 120–130 °C in the presence of Speier's catalyst. Yields ranged from 3% to 87%. Poor yields (3–15%) were obtained in the reaction of chlorinated hydrosilanes with 1,2-di(TMS)ethyne. In contrast, a 65% yield was obtained with trimethylsilane. In general, yields are poorer if the silane contains more chlorine than the silyl groups of the coreacting silylethyne ⟨86ZOB2743⟩.

The reaction of 1,4-bis(trimethylsilyl)-1,3-butadiene (**7**) with trialkylsilanes or arylalkylsilanes in the presence of a transition metal catalyst initially gave a 1,2-*cis* adduct (**8**). After longer reaction times this underwent a 1,4-addition to afford the 1,3-disilyl-1,4-bis(TMS)-1,2-butadienes (**9**) (Scheme 1). Speier's catalyst, RhCl(PPh$_3$)$_3$ and Pt(PPh$_3$)$_4$ all gave >80% yield of the allene (Table 2). For the addition of Et$_3$SiH, Speier's catalyst at 80 °C for 30 min was preferable, but with TMS-H at 100 °C for 2 h this gave a 1:1 mixture of products. For this addition, RhCl(PPh$_3$)$_3$ at 100 °C for 1 h was most efficient (90% yield) ⟨85CL1405⟩. The palladium-catalysed double silylation of (**7**) with methylated chlorosilanes has been investigated (Scheme 1). A series of PdCl$_2$(PR$_3$)$_2$ catalysts were studied (Table 3). Their reactivity and selectivity changed depending upon the phosphine ligands, but in general the butatriene (**10**) was the major product. The highest combined yields of (**10**) and the 1-buten-3-yne (**11**) were obtained with ethyl and benzyl ligands. Different chlorinated silanes, Me$_n$Cl$_{3-n}$SiSiMe$_m$Cl$_{m-3}$, were studied and it was shown that the more highly chlorinated disilanes effected 1,4-disilylation preferentially, while with less chlorinated disilanes the butenyne (**11**), arising from the 1,2-addition, was the major product ⟨87TL1807⟩.

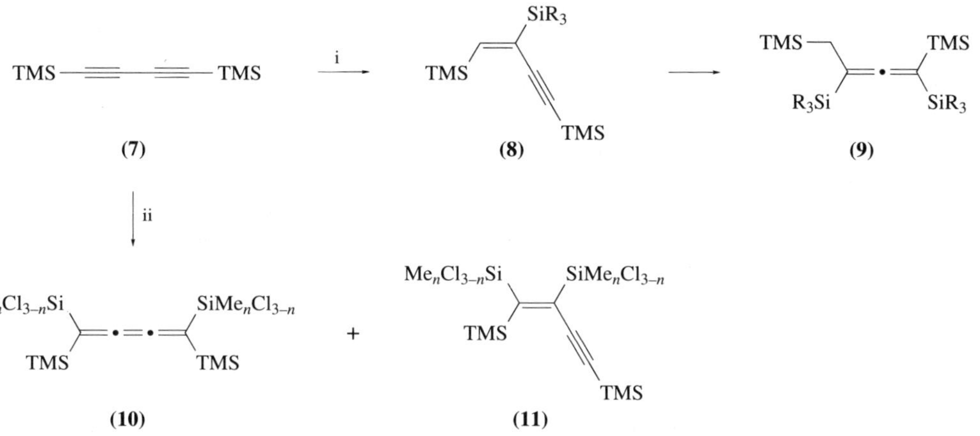

i, R$_3$SiH (see Table 2); ii, (Me$_n$Cl$_{3-n}$Si)$_2$ (see Table 3)

Scheme 1

Table 2 Hydrosilylation of 1,4-bis(trimethylsilyl)buta-1,3-diyne (**7**) with silanes.

Silane	Catalyst	Conditions	Yield of (**8**) (%)	Yield of (**9**) (%)
TMSH	H$_2$PtCl$_6$	100 °C, 2 h	40	46
TMSH	RhCl(PPh$_3$)$_3$	100 °C, 1 h	0	90
TMSH	Pt(PPh$_3$)$_4$	90 °C, 12 h	1	94
Et$_3$SiH	H$_2$PtCl$_6$	80 °C, 0.5 h	0	100
Et$_3$SiH	RhCl(PPh$_3$)$_3$	90 °C, 5 h	0	83
PhMe$_2$SiH	RhCl(PPh$_3$)$_3$	100 °C, 2 h	0	86

Regio- and stereoselective addition of silicon–metal reagents to alkynes has attracted some attention. Silylalumination, silylcupration and silyltitanation of alkynes lead exclusively to the 2-metallo 1-silyl 1-alkenes, except in certain special cases described below. In contrast, the reaction of silyl-manganese reagents gave the rather unexpected 1,2-disilylated products. The Si–Mn reagent was generated from TMS-Li, MeMgI and MnCl$_2$; although the role of the MeMgI is unclear, without it only monosilylated products were obtained. The use of this reagent with 1-silyl alkynes and 1,2-disilylethynes led to high yields of 1,1,2-trisilyl alkenes and 1,1,2,2-tetrasilylethenes, respectively (Equation (2)), where HMPA represents hexamethylphosphoramide ⟨88T4277⟩.

Table 3 Double silylation of 1,4-bis(trimethylsilyl)buta-1,3-diyne (7) with disilanes.

Disilane	Catalyst	Conditions	Yield of (10) (%)	Yield of (11) (%)
$MeCl_2SiSiCl_2Me$	$PdCl_2(PEt_3)_2$	100 °C, 5 h	49	22
$MeCl_2SiSiCl_2Me$	$PdCl_2[P(CH_2Ph)_3]_2$	120 °C, 1 h	45	36
$MeCl_2SiSiClMe_2$	$PdCl_2(PEt_3)_2$	120 °C, 3 h	48	14
$MeCl_2SiSiClMe_2$	$PdCl_2[P(CH_2Ph)_3]_2$	120 °C, 5 h	12	11
$Me_2ClSiSiClMe_2$	$PdCl_2(PMe_3)_2$	120 °C, 16 h	3	36
$Me_2ClSiSiClMe_2$	$PdCl_2(PEt_3)_2$	120 °C, 4 h	5	72
$Me_2ClSiSiClMe_2$	$PdCl_2[P(CH_2Ph)_3]_2$	120 °C, 11 h	4	74
$Me_2ClSiSiClMe_2$	$PdCl_2(PPh_3)_2$	140 °C, 29 h	3	36

$$R-\!\!\!\equiv\!\!\!-TMS \xrightarrow[\text{THF/HMPA, 0 °C, 3 h}]{\text{TMS-Li/MeMgI/MnCl}_2} \begin{array}{c} R \quad\quad TMS \\ \diagup\!\!=\!\!\diagdown \\ TMS \quad\quad TMS \end{array} \quad (2)$$

R	Yield(%)
H	72
TMS	76
$n\text{-}C_6H_{13}$	80
$PhCH_2OCH_2CH_2$	59
$THP\text{-}OCH_2CH_2$	83
$HOCH_2CH_2$	58

Monosilylation of a 1-silyl 3-en-1-yne (12) using TMS-Cl/Li or TMS-Cl/Mg afforded the 1,1-disilyl vinylallenes (13) (Equation (3)). The presence and position of the chlorine are important if the reaction is to occur ⟨81TL3179⟩. The reaction of the silyl cuprate $(PhMe_2Si)_2CuLi$ with the 3-bromo-1-(dimethylphenylsilyl)but-1-yne gave the 1,1-disilylallene (14) in 45% yield (Equation (4)) ⟨87JCS(P1)2269⟩.

$$\text{(12)} \xrightarrow[\text{THF/Et}_2\text{O, } -20\ °C \atop 65\%]{\text{TMS-Cl/Li}} \text{(13)} \quad (3)$$

$$\text{(PhMe}_2\text{Si)} \xrightarrow[0\ °C,\ 3\ h \atop 45\%]{(PhMe_2Si)_2CuLi} \text{(14)} \quad (4)$$

A novel rearrangement of the alkynyldisilanes (15) to the 1,1-disilylallenes (16), effected by $FeCl_3$ or $TiCl_4$ in the presence of TMS-Cl, has been reported (Equation (5)) ⟨90TL5607⟩. While $FeCl_3$ worked catalytically, $TiCl_4$ needed to be used in stoichiometric amounts. The rearrangement is thought to involve Lewis acid-aided fission of the propargylic C—O bond followed by transfer of chlorine to the α-silicon. The Si—Si bond breaks, the TMS group migrates and reorganisation of the π bond gives the allene (16).

$$\text{(15)} \xrightarrow[\text{TMS-Cl}]{FeCl_3 \text{ or } TiCl_4} \text{(16)} \quad (5)$$

R	Yield(%)
$R^1 = R^2 = Me$	82
$R^1, R^2 = (CH_2)_5$	73
$R^1 = Ph, R^2 = H$	52
$R^1 = Bu^t, R^2 = H$	71
$R^1 = n\text{-}C_5H_{11}, R^2 = H$	56

The reaction of BuLi/TMEDA and TMS-Cl with the 1-silyl-3-phenylpropynes (**17**) gave a mixture of the 1,3-bis(TMS)-1-silyl-3-phenylpropadienes (**18**) (14–46% yield) and 3-TMS-1-silyl-3-phenyl-propynes (**19**) (12–24% yield) (Scheme 2). Photolysis of compounds (**19**) resulted in their conversion into the propadiene isomer (**19a**) in high yields (>70%). The corresponding monosilyl derivatives (**17**) were found to be photochemically stable under the conditions used ⟨78JOM(161)299⟩. BuLi/TMS-Cl reacted with a number of different substrates resulting in the formation of 1,1-disilylallenes. The reaction of the 1-TMS alkyne (**20**) gave a 65% yield of the 1,1-bis(TMS)allene (**21**) (Equation (6)) ⟨83T3073⟩. With 1-TMS-2-alkynes the yield of 1,1-bis(TMS)allene (**22**) was dependent upon the group in the 3-position (Equation (7)). If this was a small alkyl or phenyl group, allenes (**22**) were minor products. When this group was Pr^i or Bu^t, (**22**) was the major product in the former case (36% yield) and the sole product in the latter case (52% yield) ⟨82TL4083⟩. Polysilylation of 2-butynoic acid with TMS-Cl and six equivalents of BuLi/TMEDA at $-78\,°C$ gave a 70% yield of the allene (**23**) (Equation (8)). This product is thought to be formed by stepwise silylation and not via rearrangement of 2-butynoic acid to 3-butynoic or allenic acids ⟨79TL87⟩. 1,3,3-Tris(trimethyl-silyl)propyne is claimed to give an intermediate lithioallene (**24**) when treated with BuLi in THF ⟨86AG1020⟩.

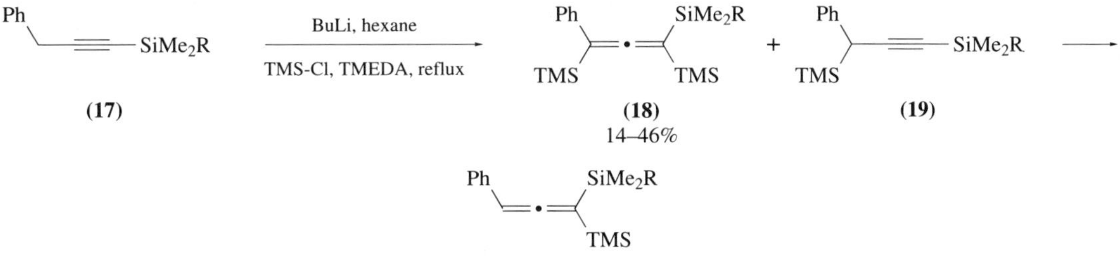

R	Yield(%)
TMS	21
PhMe$_2$Si	23
Et	14
Pr^i	46
TMS-CH$_2$	18

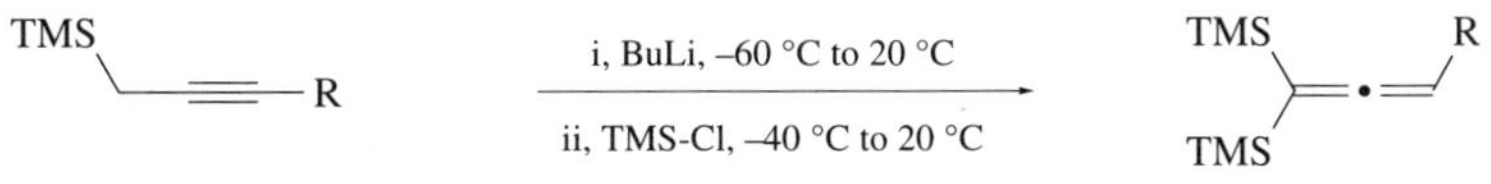

Scheme 2

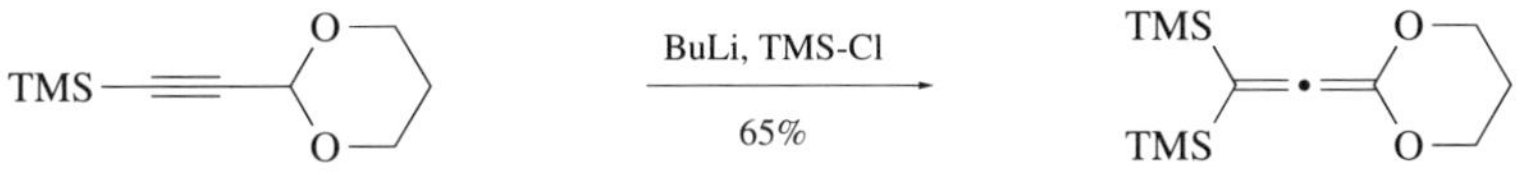

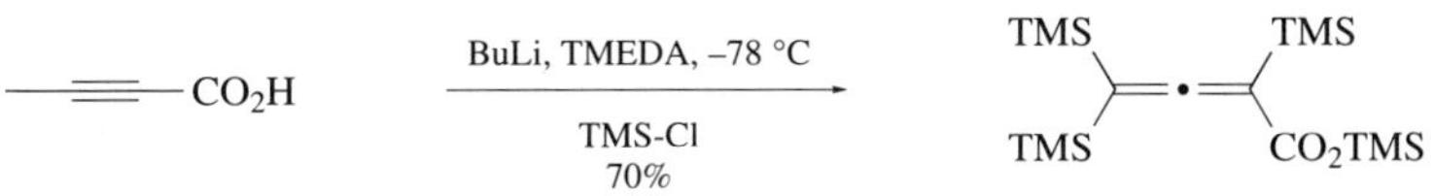

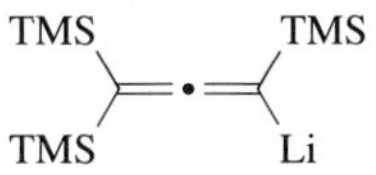

The reaction of chlorodiboranes with 1,2-bis(TMS)ethyne gave 1,1-diboro-2,2-bis(TMS)ethenes (**25**), one of the silyl groups having migrated from one end of the π-bond to the other (Equation (9)) ⟨84ZN(B)19042⟩. This reaction has been extended to other 1-silyl alkynes (see Section 4.23.1.2).

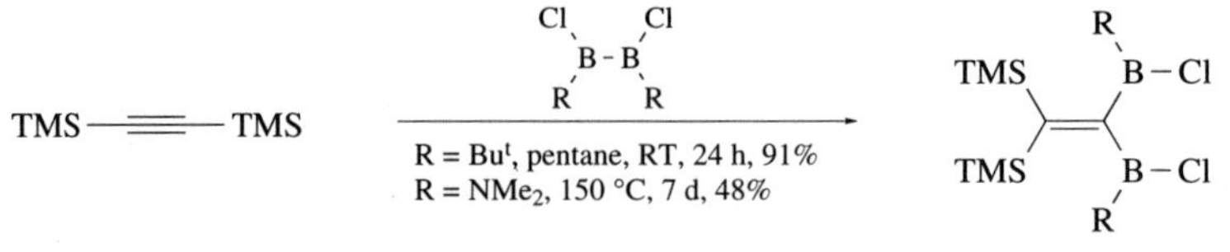

$$TMS-\!\!\equiv\!\!-TMS \xrightarrow[\substack{R = Bu^t, \text{ pentane, RT, 24 h, 91\%} \\ R = NMe_2, 150\ °C, 7\ d, 48\%}]{\substack{Cl\ \ \ \ Cl \\ B-B \\ R\ \ \ R}} \begin{array}{c} R \\ TMS \diagdown \diagup B-Cl \\ \diagup\!\!=\!\!\diagdown \\ TMS\ \ \ \ B-Cl \\ R \end{array} \quad (9)$$

(**25**)

The photolysis of alkynyl di- and polysilanes with low- and high-pressure mercury lamps has been reported. In the absence of a trapping agent such as methanol, silacyclopropenes (silirenes) were isolated as the major products along with starting material. The stability of the silirene is dependent upon the nature of the ring substituents. The 1,1-dimesityl derivative (**26**) was isolated as a crystalline solid. This material is unaffected by moisture and exposure to air ⟨80JOM(194)147⟩. When the photolysis reaction was carried out in the presence of methanol, the recovery of the silirene was greatly reduced and one of the major products was the 1,1-disilyl alkene (**27**) (Scheme 3). The products (**27**) arise from the methanol opening of the silirene or addition of methanol to a silapropadiene, a possible but not isolated intermediate. Yields are generally low with ethyne derivatives ⟨79JOM(179)377⟩. With phenylethynes yields of up to 90% of 1,1-disilyl-2-phenylethenes (**27**; R^2 = Ph) are obtained as 1:1 mixtures of *cis* and *trans* isomers ⟨77JA3879⟩. The reaction favours the *trans* isomer in the early stages of the irradiation, but the ratio eventually becomes 1:1. When each isomer was irradiated separately, a 1:1 mixture was again obtained.

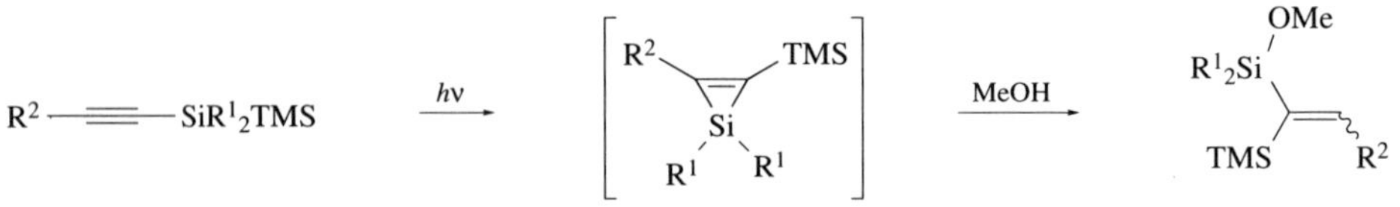

$$R^2-\!\!\equiv\!\!-SiR^1_2TMS \xrightarrow{h\nu} \left[\begin{array}{c} R^2 \diagdown\ \ \ \diagup TMS \\ \diagup\!\!=\!\!\diagdown \\ Si \\ R^1\ \ R^1 \end{array} \right] \xrightarrow{MeOH} \begin{array}{c} OMe \\ R^1_2Si\diagdown\ \diagup \\ \diagup\!\!=\!\!\diagdown \\ TMS\ \ \ R^2 \end{array}$$

R^1 = Me, Ph, 2,4,6-Me$_3$C$_6$H$_2$ (**26**)   (**27**)
R^2 = H, Ph, TMS R^1 = 2,4,6-Me$_3$C$_6$H$_2$
    R^2 = TMS

Scheme 3

The displacement of halogen by a silyl group in 1-halo 1-silyl alkenes, via a metal–halogen exchange reaction, has been reported (Equation (10)). The nature of the silyl group attached to the alkene does not appear to affect the outcome of the reaction. 1-Bromo-1-diphenylmethylsilyl-1-propene has been used in one study ⟨86OM2274⟩. The bromine was removed with Bu^{s}Li and the second silyl group added with its chloride with high stereospecificity. Bu^{t}Li has also been used for the displacement of bromides ⟨85JA3935, 87JA913⟩ and iodides ⟨84JA6105⟩. The 1,1-disilyl alkenes are obtained in reasonable yields (>40%) with retention of configuration. The displacement of fluoride by silyl groups has been performed by irradiating fluoroalkenes in the presence of disilylmercury. In the reaction of 1,2,3,3,3-pentafluoro-1-TMS-propene and bis(trimethylsilyl)mercury, the tri-silylpropene formed was identified by its ^{19}F and ^{1}H NMR and mass spectra. The use of a large excess of the mercury compound gave the highest yields (44%). With equimolar quantities the yield dropped to 12% ⟨80JCR(M)235⟩.

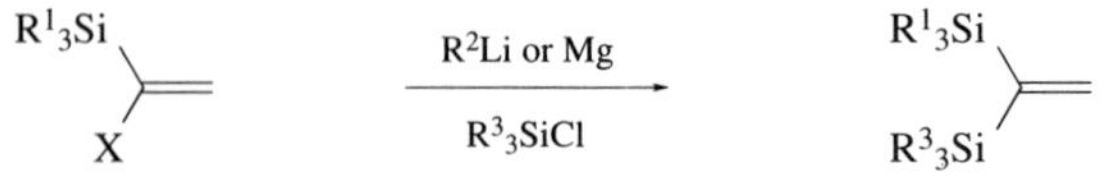

$$\begin{array}{c} R^1_3Si\diagdown \\ \diagup\!\!=\!\! \\ X \end{array} \xrightarrow[R^3_3SiCl]{R^2Li\ or\ Mg} \begin{array}{c} R^1_3Si\diagdown \\ \diagup\!\!=\!\! \\ R^3_3Si \end{array} \quad (10)$$

X = Cl, Br

Tris(trimethylsilyl)methyl lithium (**28**) ⟨70JOM(24)529, 84JOC168⟩ has been treated with a range of carbon electrophiles (Scheme 4). The reaction of (**28**) with methoxymethyl chloride and ethoxymethoxymethyl chloride gave the corresponding tris(trimethylsilyl)ethyl ethers. These ethers on treatment with BCl$_3$ and ZnBr$_2$, respectively, gave 1,1-bis(trimethylsilyl)ethene ⟨81JCS(P1)969⟩. Following reaction of (**28**) with benzoyl chloride, the bis(trimethylsilyl) silyl enol ether (**29**) could not be isolated although it is claimed to be present in solution. In one reaction, performed at reflux, a crystalline product was isolated. This was shown to be a 4,4-bis(trimethylsilyl)-1-phenylbutadiene (**30**) which arises from the reaction of two equiv. of benzoyl chloride, one equiv. of (**28**) and one equiv. of the enolate of acetaldehyde (which is formed when strong bases are present in THF) ⟨81JCS(P1)969⟩. The reaction of (**28**) with nonenolisable carbonyl compounds gave the 1,1-bis(trimethylsilyl)methylene derivatives (**31**) in yields ranging from 20% to 73% ⟨84JOC168, 85CB2493⟩. With enolisable ketones no condensation products were isolated, due to (**28**) acting as a base ⟨84JOC168⟩. Phenylmethanal can be converted into 1,1-bis(trimethylsilyl)-2-phenylethene by reaction at 450 °C with pentamethyldisilanyl(trimethylsilyl)diazomethane, giving a 58% yield. The reaction is thought to proceed via a carbene and silene intermediate ⟨87OM1857⟩. Treatment of (**28**) with 1-ethyl thioformate gave a mixture of bis(trimethylsilyl)vinyl ethyl ether (**32**) and thioaldehyde (**33**) (Scheme 4) ⟨87JA279⟩. The thermolysis (80 °C) of (**33**) gave exclusively the thioether (**34**) via a Brook-type rearrangement. The corresponding aldehyde (**35**) gave a quantitative yield of the silyl enol ether (**36**). Photolysis of (**33**) gave (**34**) (66%) and 1,1,2-tris(trimethylsilyl)ethene (**37**). The latter compound arises by a 1,2-TMS shift and concomitant loss of sulfur ⟨87JA279⟩. Under similar conditions, (**35**) gave only tris(trimethylsilyl)methane. Treatment of (**33**) with MeLi or Bu^{t}Li gave (**37**) in 79% and 34% yields, respectively.

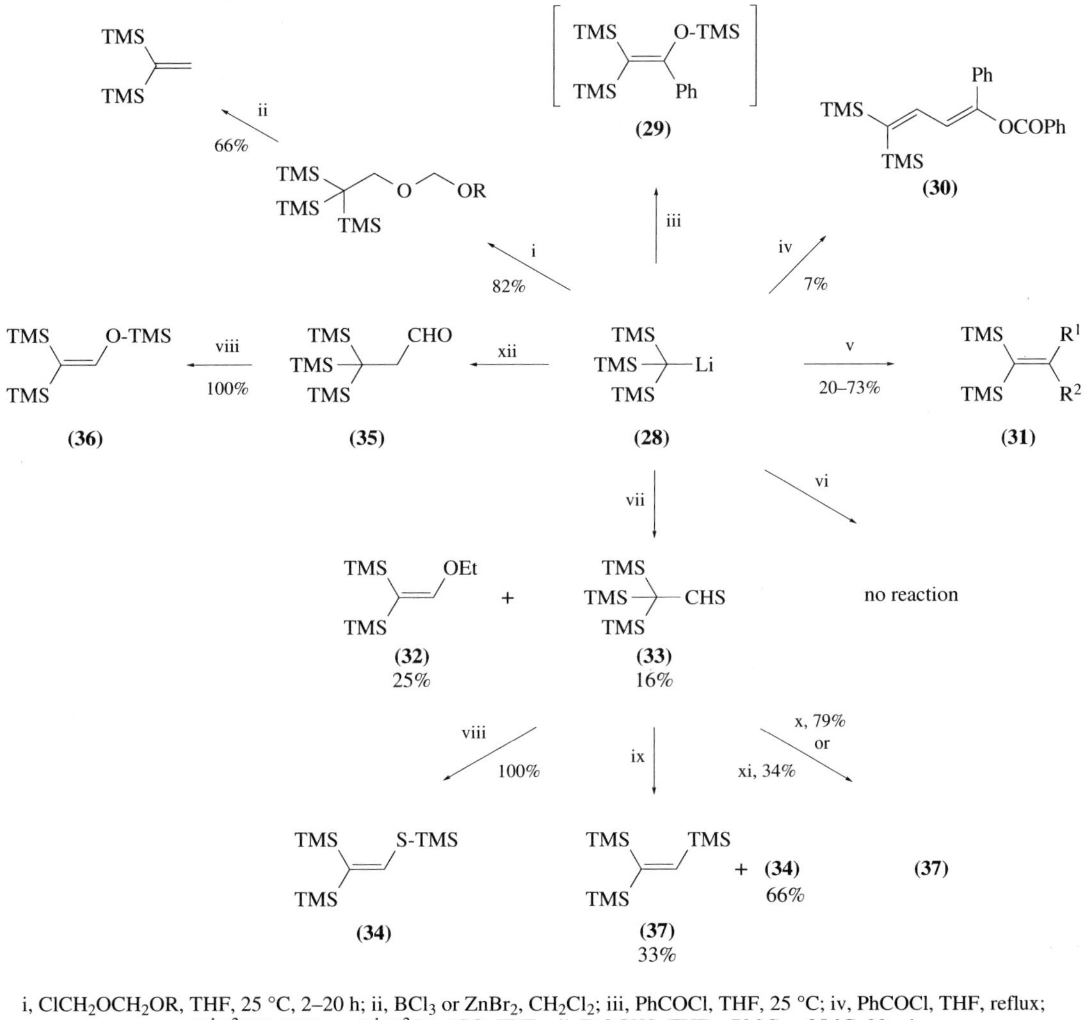

i, ClCH$_2$OCH$_2$OR, THF, 25 °C, 2–20 h; ii, BCl$_3$ or ZnBr$_2$, CH$_2$Cl$_2$; iii, PhCOCl, THF, 25 °C; iv, PhCOCl, THF, reflux; v, R^{1}R^{2}CO, THF; vi, R^1(R^{2}CH$_2$)CO, THF; vii, EtOCHS, THF, −78 °C to 25 °C, 90 min; viii, toluene, 80 °C; ix, $h\nu$, benzene, 5 °C, 17 h; x, MeLi, THF; xi, Bu^{t}Li, THF; xii, EtOCHO, THF, −78 °C to 25 °C, 90 min

Scheme 4

The reaction of TMS-CCl$_2$SiMe$_2$Cl with one equiv. of BuLi, initially at low temperature, gave the *cis*- and *trans*- 1,3-disilacyclobutanes (**38**) and (**39**) as major products. These compounds reacted with a further equiv. of BuLi to give a mixture of tetrasilylethenes ((**40**), (**41**) and (**42**)) and the 1,3-

disilabicyclo[1.1.0]butane (**43**) (Scheme 5) ⟨84JOM(271)107⟩. The bicyclobutane (**43**) has been shown to be an intermediate in the formation of (**40**), (**41**) and (**42**), as these compounds were formed when (**43**) was treated with BuLi. The tetrasilylethenes (**40**) and (**41**) are inseparable. The mechanism for the formation of (**40**) and (**41**) has not been described, but it has been postulated that (**42**) may be formed by exchange of silyl substituents as described by Sakurai et al. ⟨82JA4288⟩. The yields and ratio of (**40**), (**41**) and (**42**) have been shown to be solvent dependent. A similar reaction occurs between TMS-CCl$_2$SiMeCl$_2$ and BuLi in THF, with the tetrasilylethenes (**44a**) and (**45a**) being isolated, by sublimation, in 74% yield (based on recovery of starting material). Reduction of (**44a**) and (**45a**) with LiAlH$_4$ gave (**44b**) and (**45b**). No yields were quoted for these reactions ⟨84JOM(271)107⟩. Tetra(trimethylsilyl)ethene has been prepared from the mixture of (**38**) and (**39**) and from (**43**), by treatment with MeLi, in 66% and 70% yields, respectively (based on recovery of starting material). With MeMgCl the mixture of (**38**) and (**39**) was converted into the same product in 45% yield (based on recovery of starting material).

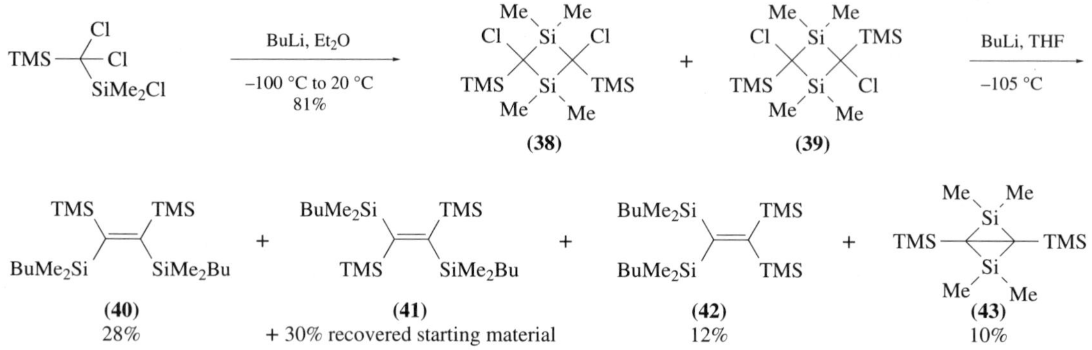

Scheme 5

(TMS)$_2$CCl$_2$ reacted with halosilanes in the presence of lithium to give low yields (10–20%) of the expected trisilylmethanes (**47**) but substantial yields of the 1,1-disilylpentenes (**46**) (Table 4 and Equation (11)). These compounds are thought to arise by reductive cleavage of the THF ring with concomitant silyl group migration to the oxygen ⟨85JOM(291)277⟩.

Table 4 Formation of 1,1-disilylpentenes (**46**) from bis(trimethylsilyl)dichloromethane and halosilanes.

R^1	R^2	X	Yield (%)
Me	N(TMS)$_3$	F	56
Me	N(TMS)$_3$	Cl	74
CMe$_3$	CMe$_3$	F	40
2,4,6-Me$_3$C$_6$H$_2$	2,4,6-Me$_3$C$_6$H$_2$	F	52

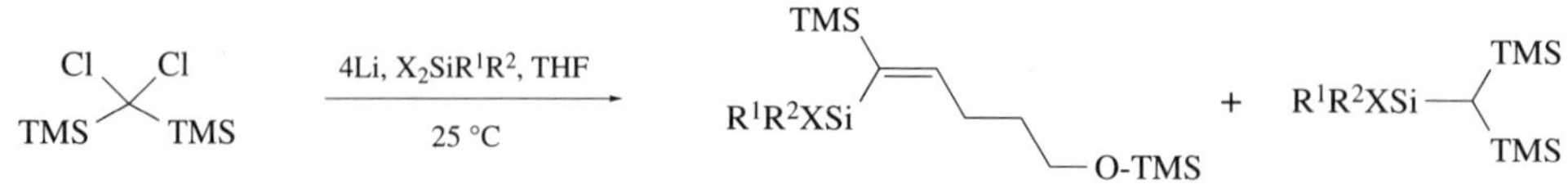

The cleavage reactions of the 3,4-disilyl-1,2-disilacyclobut-3-enes (**48**) (whose preparation is described below) and 4-silyl-1,3-disilapent-4-ene (**49**) have been described ⟨88CL965⟩. The disilabut-3-enes (**48**) may be cleaved with bromine at 0 °C giving unstable 1,2-(bromosilyl) derivatives (**50**), which were identified following treatment with alkyl lithiums or LiAlH$_4$ (Scheme 6) ⟨88CL965⟩. X-ray analysis of 1,1,2-tri(TBDMS)-1-(TMS)ethene (**51**; $R^2 = Bu^t$, $R^1 = Me$) showed that it is highly twisted, with the C-1 carbon planar but the C-2 carbon showing a significant extent of pyramidalisation. Tetra(dimethylsilyl)ethene (**51**; $R^1 = R^2 = H$) shows no twisting about the double bond. Cleavage of (**49**) with dry hydrogen chloride in benzene followed by addition of MeLi gave a 49% yield of the trisilyl alkene (**52**) (Equation (12)) ⟨85OM2040⟩. Extrapolation of this reaction to other disilapent-2-enes has not been investigated but should be possible.

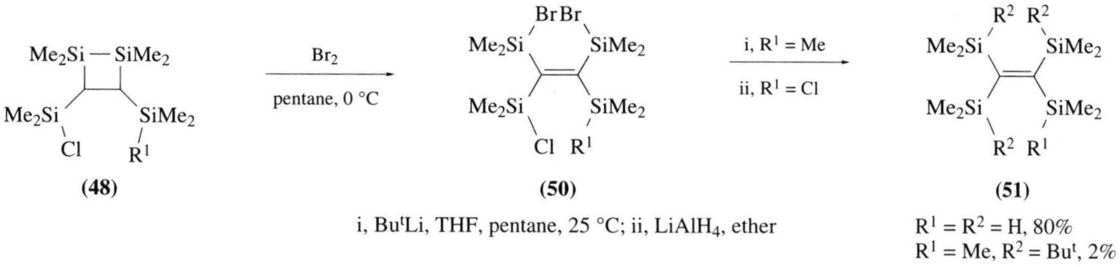

Scheme 6

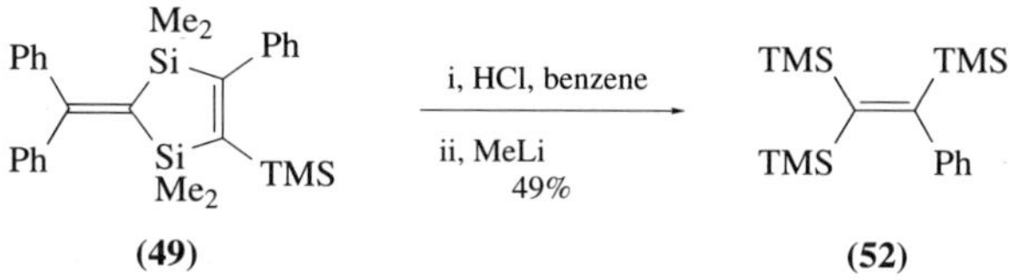

The literature relating to the synthesis and reactions of silirenes such as (**53**) has been reviewed up until 1981 ⟨82COMC-I(1)221⟩. The ring opening of (**53**) with alcohols and alkynes was shown to give 1,1,2-trisilylethenes. Further examples of these reactions have now been reported and the scope of the ring-opening reaction has been extended to include alkenes and carboxylic acids. Other silirenes have also been employed. A strongly exothermic reaction was observed between (**53**) and acetic acid giving TMS-CH=C(TMS)(SiMe$_2$OAc) in 79% yield. Alkenes required heating to 70–75 °C for 18 h in order for reaction to occur (Scheme 7). Yields of 13–52% were obtained of the products (**54**) and (**55**), often as a mixture of (*E*) and (*Z*) isomers. It is proposed that the reaction occurs by a radical mechanism in which the 1,5-diradical (**56**) initially formed can freely rotate about the C—C bond, giving both (**54**) and (**55**) after hydrogen atom transfer ⟨84OM1897⟩. The behaviour of other silyl silirenes (e.g., (**57**)) resembles that of (**53**), reacting exothermically with water and alcohols. Asymmetrically substituted silirenes could give two products ((**58**) and (**59**)) from the ring opening. Some reactions are regioselective and it would appear that steric factors play an important role in determining the outcome (Equation (13)) ⟨84JOM(272)123⟩.

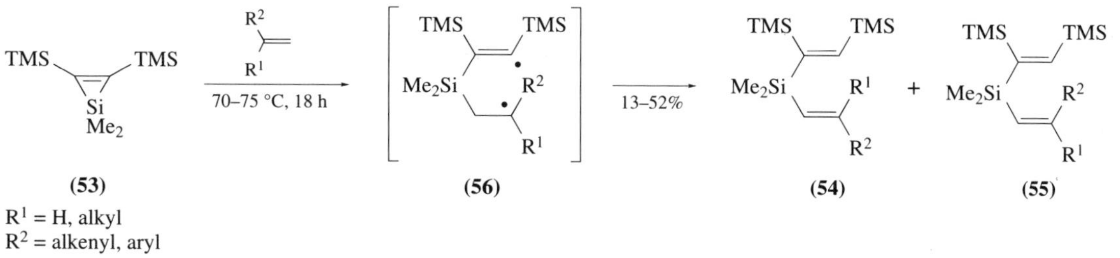

Scheme 7

	(58)	(59)
R = Me	56%	0%
R = Et	59%	0%
R = Me$_2$CH	16%	37%

Modifications of the substituents attached to the silyl groups in tri- or tetrasilylethenes have been performed. Bromine and chlorine have been replaced in good yield (>80%) with hydrogen using LiAlH$_4$ ⟨84JOM(271)107, 88CL965⟩. Diisobutylaluminum hydride (dibal-H) has also been used ⟨87ZN(B)142⟩. Displacement of bromine by carbon using phenyl lithium ⟨80AG632⟩ or methyl lithium ⟨80TL3077⟩ gave moderate yields. Transhalogenation of chlorine for fluorine with CuF$_2$ occurred in 54% yield ⟨83JOM(254)13⟩.

4.23.1.1.2 Functions with one ring silicon

A number of these systems have been reviewed ⟨82COMC-I(1)221⟩ and only a few representative examples of later work are included here. Silylsilirenes (e.g., (57)) have been prepared by the reaction of siliranes with alkynes ⟨84JOM(272)123⟩ and the photolysis of alkyldisilanes ⟨81AOC51⟩. The synthesis of this ring system continues to attract interest; compounds bearing a vinylic hydrogen have been isolated in quantitative yield ⟨93TL6541⟩. These compounds are extremely sensitive towards moisture.

When an excess of 1,1-dimethyl-2,3-bis(trimethylsilyl)silirane (60) was used in the reaction with bis(trimethylsilyl)ethyne the Me$_2$Si formed by thermal decomposition of (60) inserted into the silirene (53), giving 1,2-disilyl-1,2-disilacyclobut-3-ene (61) (Equation (14)) ⟨84JOM(272)123⟩. Alternatively, the reaction of tetramethyldisilene, formed *in situ* from a disilabicyclo-octadiene system, with acetylenes may be employed (Equation (15)). The reaction was performed at elevated temperatures in a sealed tube, and the disilacyclobutane products were isolated by distillation. These compounds are extremely sensitive to air ⟨78JOM(162)C43, 88CL965⟩.

The silirene (53) has been found to undergo a two-atom insertion reaction with alkenes, alkynes, aldehydes, ketones and imines to give disilacyclopentenes (62) and (63) (Scheme 8) ⟨84OM1897⟩. Conjugated terminal alkynes reacted at room temperature to afford (62) in 30–34% yields; unconjugated alkynes failed to react. 1,3-Dienes required heating to 70 °C to achieve yields up to 45% (only terminal 1,3-dienes appear to react). With both alkynes and alkenes, acyclic products can also be formed (Scheme 7). In general, conjugated aldehydes and ketones undergo an exothermic reaction at room temperature. The high reactivity of these substrates is thought to be due to their ability to stabilise the radical at the carbonyl carbon of the 1,5-diradical intermediate (64). In contrast, nonconjugated derivatives required UV irradiation, or prolonged heating at 70 °C, in order to react. Benzaldehyde N-methylimine undergoes insertion at room temperature giving a 77% yield of (63; X = NMe).

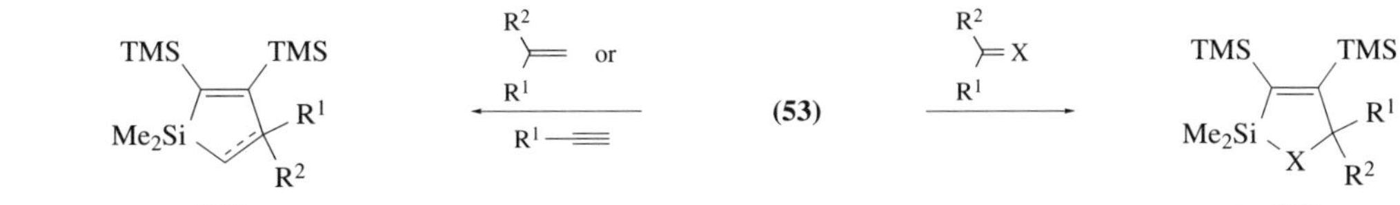

Scheme 8

(62) R¹ = Ph, alkenyl, alkynyl
R² = H, Me

X = O, R¹ = Me, Pr^i, Bu^t, CF$_3$, MeCO, alkenyl, alkynyl, Ph, heteroaryl;
R² = H, Me, Et, CF$_3$; R¹, R² = (CH$_2$)$_4$, (CH$_2$)$_5$
X = NMe, R¹ = Ph, R² = H

4.23.1.1.3 Functions with two ring silicons

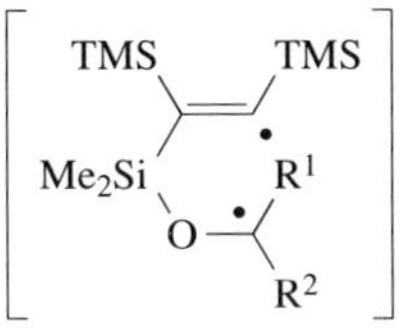

The reaction of the silylsilirene (**65**; R = TMS) and 1-trimethylsilyl-2-phenylethyne at 135 °C in the presence of NiCl$_2$(PEt$_3$)$_2$ catalyst gave a 51% yield of the 1,4-disilacyclohexa-2,5-diene (**66**; R = TMS), whose structure was confirmed by x-ray analysis (Equation (16)) ⟨85OM2040⟩. The use of this catalyst would appear not to be generally applicable. No cyclohexadiene product was isolated if one or both of the TMS groups attached to silicon in (**65**) were changed for methyl or phenyl groups. The existence of an intermediate nickelasilacyclobutene compound is proposed, which in the case of the 1,1-bis(trimethylsilyl) analogue cannot undergo reaction to the usual silole due to the bulky nature of the TMS group. In contrast, in the absence of an alkyne the 1,4-disilacyclohexa-2,5-diene (**66**; R = Me) was formed in 47% yield by thermolysis of (**65**; R = Me) at 250 °C for 18 h ⟨77JOM(142)C45⟩. Similar dimerisation of silirenes has been reported using PdCl$_2$(PEt$_3$)$_2$ as catalyst ⟨82OM1473⟩. Study of the reaction of (**65**; R = TMS) under these conditions was not included. Dimerisation of the 1-methyl-1-phenylsilirene could lead to the formation of *cis* and *trans* isomers (**67**) and (**68**). X-ray studies revealed that only the *trans* isomer (**68**) was formed (30% yield), in which the phenyl groups on the silicon are both equatorial, with the ring adopting a chair conformation. The 1,3-disilacyclobutane (**69**) has been prepared in a 78% yield by a Ni(0)-catalysed dimerisation of (phenylethynyl)pentamethyldisilane at high temperatures (Equation (17)). The reaction is thought to proceed via a 1,3-silyl shift followed by head-to-tail dimerisation ⟨85OM2040⟩. No stereoisomer of (**69**) was detected.

(65) R = Me, TMS, Ph

(16)

$$Ph\text{—}\equiv\text{—}SiMe_2TMS \xrightarrow[78\%]{NiCl_2(PEt_3)_2, 180\,°C, 20\,h} \textbf{(69)} \quad (17)$$

Intramolecular hydrosilylation of silyl(alkylsilyl) alkynes using Speier's catalyst under high dilution conditions is reported to give 1,3-disilacyclopentanes (**70**; $n = 2$) and 1,3-disilacyclohexanes (**70**; $n = 3$) (Equation (18)) ⟨89JOM(378)1⟩. Slow addition (6 h) of the alkynes to a refluxing solution of the catalyst was required to give optimum yields. A similar reaction occurs with the disilane (**71**) in the presence of Pd(OAc)$_2$, giving the sterically crowded alkene (**72**) (Equation (19)) ⟨91JOC1948⟩.

$$(18)$$

n	R	Yield(%)
2	H	58
2	H	52
3	Ph	42
3	Ph	60

$$(71) \xrightarrow[88\%]{i} (72) \quad (19)$$

i, Pd(OAc)$_2$ (0.01 equiv.), 1-adamantyl-NC (0.15 equiv.), mesitylene, 180 °C, 8 h

4.23.1.2 Functions Bearing Two Borons—$R^1{}_2C\!\!=\!\!C(BR^2{}_3)_2$, etc.

Despite the broad spectrum of organoboron literature, surprisingly little work has been carried out on the preparation of 1,1-diboryl alkenes. The hydroboration of 1-boryl alkynes has not been reported. The diboronation of alkynes has been described. Reactions of diboranes with silyl alkynes led to high yields of the 1,1-diboryl alkenes (**73**) (Equation (20)); when $R^3 = H$, however, 1,2-diboryl alkenes were formed (see Section 4.23.1.4.1). The 1-silyl group migrates to C-2 during the course of the reaction ⟨84ZN(B)19042⟩. Dialkylaminodiboranes required much more forcing conditions (150 °C, 7 d) than alkyldiboranes. When stannyl alkynes rather than silyl alkynes reacted with the diboranes, the products were the 2-borylborienes (**74**). In these cases a stannyl group was lost during the cyclisation ⟨84AG(E)313⟩. Crude yields are virtually quantitative and the products could be distilled. The addition of tris(dialkoxyboryl)methyl lithium derivatives to alkyl or aryl aldehydes and ketones gave 1,1-diboryl alkenes (**75**) (Equation (21)) ⟨75JOM(93)21, 78JOC950⟩. In general, high yields were obtained although the product from cinnamaldehyde could not be purified. The stereochemical outcome of the reaction has not been explored as achiral borylmethanes have been used. An example of a 2-borylborole (**76**) has been published ⟨86JOM(307)157⟩.

$$(20)$$

$R^1 = Me, CH_2Bu^t, Ph$
$R^2 = alkyl, aryl$

$R^3 = Bu^t$, 25 °C, 24 h, 75–93%
$R^3 = NMe_2$, 150 °C, 7 d, 48–90%

(74)

R = Bu^t, SnMe$_3$

$$\text{Li}\begin{array}{c}\text{B(OR}^1)_2\\\text{B(OR}^1)_2\\\text{B(OR}^1)_2\end{array} \xrightarrow[\text{THF, 25 °C, 18 h}]{\text{R}^2\text{COR}^3} \begin{array}{c}(\text{R}^1\text{O})_2\text{B}\quad\text{R}^3\\ \diagup\!\!\!\!=\!\!\!\!\diagdown \\(\text{R}^1\text{O})_2\text{B}\quad\text{R}^2\end{array} \quad (75) \tag{21}$$

$(R^1O)_2B =$ [cyclic boronate] B or (alkylO)$_2$B

$R^2 = H$; $R^3 = H$, alkyl, Ph, 66–93%
$R^2 = Me$; $R^3 = Me$, Ph, 70–90%
$R^2R^3 = (CH_2)_5$, 90%

(76)

4.23.1.3 Functions Bearing Two Germaniums—R$^1{}_2$C=C(GeR$^2{}_3$)$_2$, etc.

Few examples of the preparation of 1,1-digermyl alkenes have been reported. The addition of germanium tetrachloride to 1-germylalkynyl ethers was reported to give the 1,1-digermyl 2-chloro alkenes (**77**) in good yields as distillable oils (Equation (22)) ⟨80ZOB692⟩. The reaction of two equiv. of lithium trialkylgermanium with propynoyl chloride derivatives gave a mixture of 1,1-digermyl alkenes (**78**) and 1,1-digermylallenes (**79**) formed by addition of a second molecule of alkynoyl chloride to the intermediate allenyl alkoxide (Scheme 9) ⟨80ZOB2044⟩. It has been shown that 1,1-digermyl alkanes, containing a β-oxygen function, can be converted into 1,1-digermyl alkenes under the influence of a base such as lithium diisopropylamide (LDA) (Equation (23)) ⟨88OM739⟩. 2,2,2-Tris(trimethylgermyl)ethanol was converted into 1,1-bis(trimethylgermyl)ethene with BF$_3$·Et$_2$O (Equation (24)) ⟨88OM739⟩. The flash vacuum pyrolysis and photolytic conversion of bis(trimethylgermyl)diazomethane has been reported ⟨80PAC615⟩. In the former case two products were isolated, one (**80**) arising from rearrangement then dimerisation of the germene intermediate (**82**) and the second, the germazene (**81**), believed to result from the reaction of the starting material with (**82**). The photochemical reaction gave an 89% yield of (**81**) (Scheme 10) ⟨80PAC615⟩.

$$\text{Me}_3\text{Ge}-\!\!\!\equiv\!\!\!-\text{OR} \xrightarrow{\text{GeCl}_4, 0\,°\text{C}} \begin{array}{c}\text{Me}_3\text{Ge}\quad\text{Cl}\\\diagup\!\!\!\!=\!\!\!\!\diagdown\\\text{Cl}_3\text{Ge}\quad\text{OR}\end{array} \tag{22}$$

R = Et, 82%
R = Bu, 53%
(77)

$$\begin{array}{c}\text{Me}_3\text{Ge}\\\text{Me}_3\text{Ge}\!\!-\!\!\text{C}\!-\!\text{CO}_2\text{Et}\\\text{Me}_3\text{Ge}\end{array} \xrightarrow[\substack{0\,°\text{C, 1 h}\\76\%}]{\text{LDA, Et}_2\text{O}} \begin{array}{c}\text{Me}_3\text{Ge}\quad\text{OGeMe}_3\\\diagup\!\!\!\!=\!\!\!\!\diagdown\\\text{Me}_3\text{Ge}\quad\text{OEt}\end{array} \tag{23}$$

$$\begin{array}{c}\text{Me}_3\text{Ge}\\\text{Me}_3\text{Ge}\!\!-\!\!\text{C}\!-\!\text{OH}\\\text{Me}_3\text{Ge}\end{array} \xrightarrow[\substack{0\,°\text{C, 1 h}\\83\%}]{\text{BF}_3\cdot\text{Et}_2\text{O, CH}_2\text{Cl}_2} \begin{array}{c}\text{Me}_3\text{Ge}\\\diagdown\!\!\!\!=\\\text{Me}_3\text{Ge}\end{array} \tag{24}$$

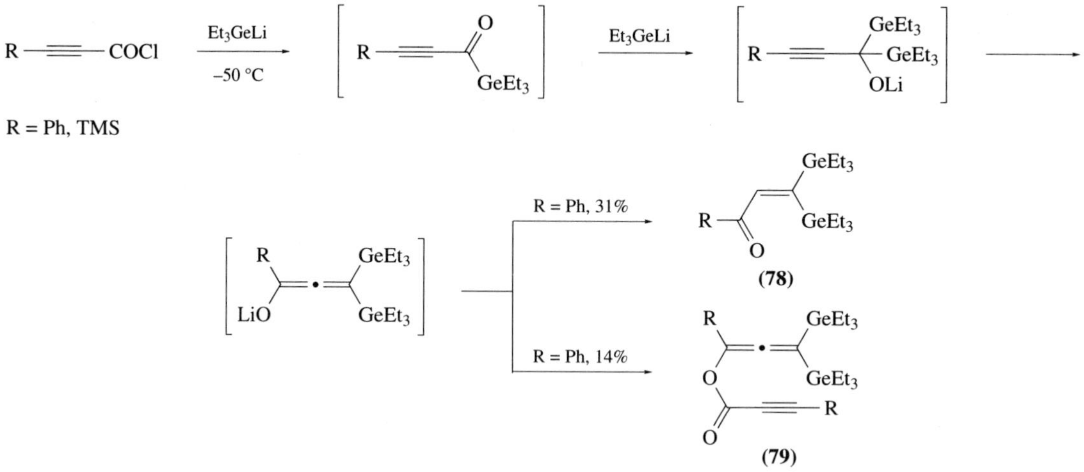

Scheme 9

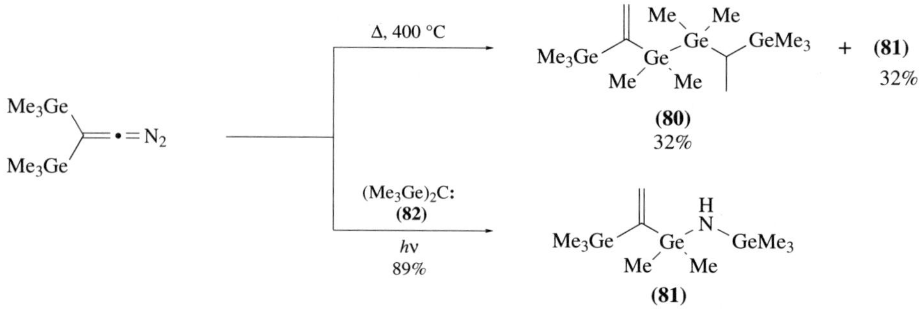

Scheme 10

4.23.1.4 Other Functions—$R^1{}_2C$=$CSiR^2{}_3BR^3{}_3$, etc.

This section covers those functions containing boryl silyl, germyl silyl and boryl germyl alkenes.

4.23.1.4.1 Functions containing one silicon and one boron

Hydroboration of 1-silyl alkynes has been reported to proceed in a stereoselective and regioselective manner. The boron was almost always added to the carbon bearing the silyl group (Scheme 11) ⟨77JA3184, 86BCJ659, 86JOC1330, 87ZOB1741⟩. The boryl groups of 1-boryl 1-silyl alkenes have been modified. Oxidation with trimethylamine N-oxide can lead to ring expansion (Equation (25)) ⟨86JOC1330⟩. The boronate (**83**) was formed if 2,3-dihydroxy-2,3-dimethylbutane was present during the oxidation reaction (Equation (26)) ⟨87LA977⟩.

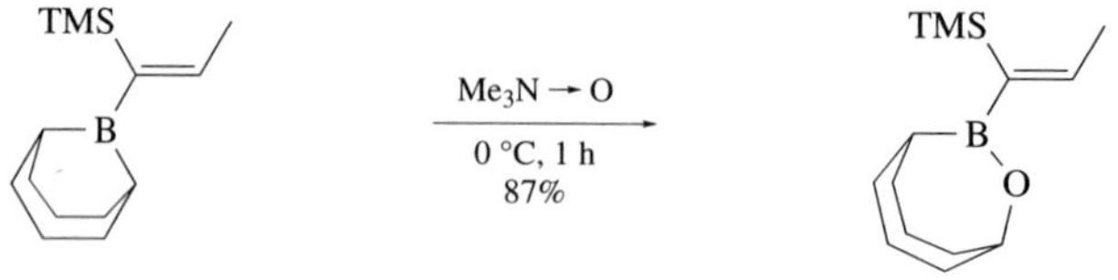

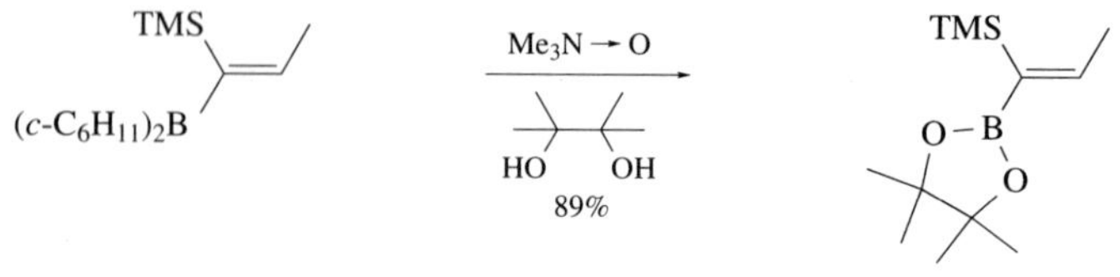

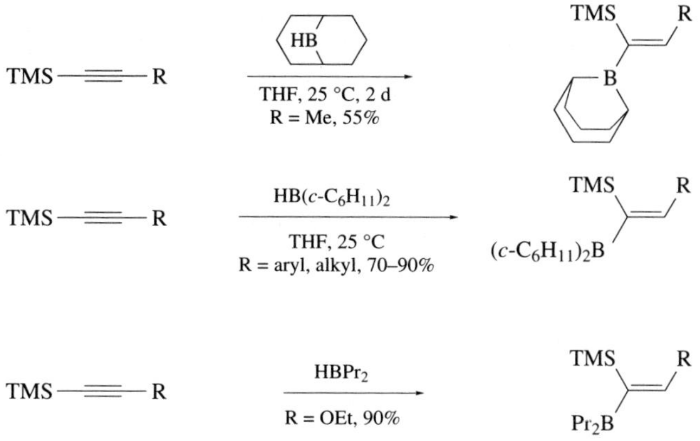

Scheme 11

Addition of one equiv. of trialkoxyboranes to 1-silyl alkynes having a γ-hydrogen, in the presence of Bu^tLi, is reported to give the 1-dialkoxyboryl 1-silyl allenes (**84**) in high yields. These compounds are too unstable to be isolated but can be used *in situ* (Equation (27)) ⟨86JOC886⟩. The reaction of trihaloboranes with 1-silyl 2-alkoxy alkynes afforded the 1-dihaloboryl-1-silylketenes (**85**) (Equation (28)) ⟨84ZOB1817⟩. Similarly, dialkylaminohaloboranes gave the dialkylaminoboryl analogues (**85**; $X = NR_2$). The NEt_2 analogue may also be obtained, in 70% yield, from the 1-triethylsilyl 1-dibromoboryl analogue by treatment with diethylamine ⟨85ZOB2801⟩.

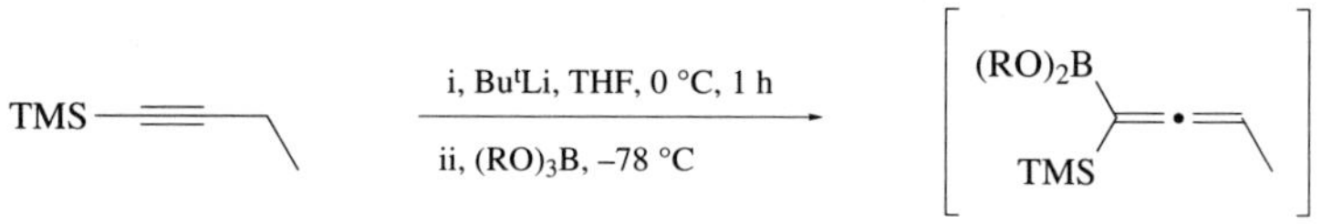

(27)

(84)
R = alkyl

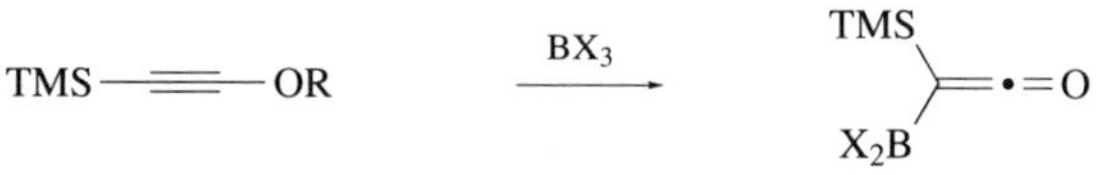

(28)

(85)

X = Cl, 67%
X = Br, 67%
X = NMe_3, 43%

Preparation of the 1,1-diboryl alkenes (**73**) from alkynes and dialkyldihalodiboranes is described above (Section 4.23.1.2). With 1-trimethylsilylethyne, 1,2-addition occurred giving the trimethylsilyl-1,2-diborylethene (**86**) (Equation (29)) ⟨84ZN(B)19042⟩. A *cis* configuration is postulated for the product.

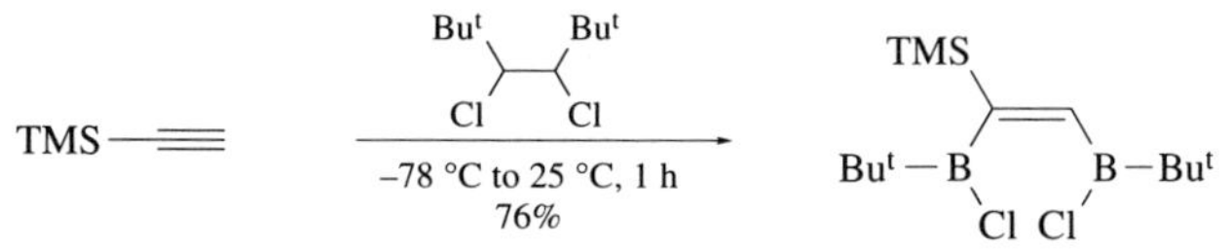

(29)

(86)

Dimesitylfluoroboranes reacted with 1-lithio 1-trimethylsilyl alkenes to give 1-dimesitylboryl 1-trimethylsilyl alkenes (Equation (30)) ⟨87JA931; cf. 81JOM(209)1⟩.

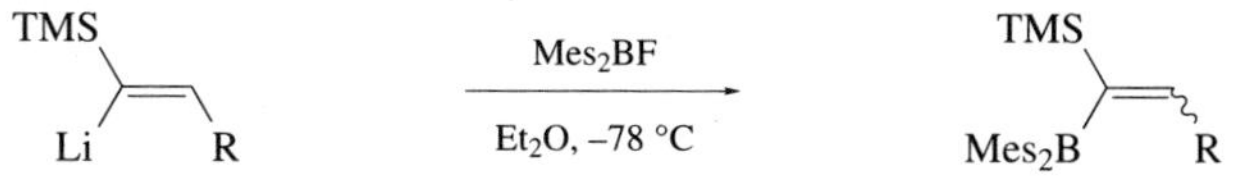

(30)

Mes = 2,4,6-$Me_3C_6H_2$

R = H
R = CH_2=CH (*Z*)
R = Bu (*E*)
R = Ph (*Z*)

There is a very limited body of literature relating to borocycles containing an α-silyl group. Triarylsilylboranes have been irradiated at −196 °C for 5 h in the presence of 1,2-bis(trimethylsilyl)ethyne to give borirenes of the type (**87**) ⟨84AG(E)454⟩.

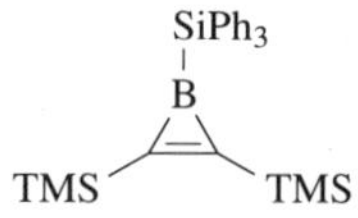

(**87**)

4.23.1.4.2 Functions containing one silicon and one germanium

1-Silyl alkynes have been treated with BuLi to generate the intermediate allenic anion which was trapped by the addition of trialkylgermanium chloride giving 1-trialkylgermyl-1-silylallenes in high yields (e.g., Equation (31)) ⟨83T3073⟩. Addition of germanium tetrachloride to 1-silyl-2-alkoxyethynes occurred in a similar fashion to the addition to 1-germyl alkynes described above (Section 4.23.1.3). The 1-trichlorogermyl 1-silyl alkenes ((**88**; M = Ge) were isolated in high yields, in most reported cases, as distillable oils. Similarly, silicon tetrachloride added to 1-germyl-2-alkoxyethynes to give the 1-trichlorosilyl-1-germyl alkenes (**88**; X = Si) in high yields (Equation (32)) ⟨80ZOB692⟩. The insertion of germanium dibromide dioxan complex into the carbon–bromine bond of 1-bromo-1-trimethylsilylethene has been employed to prepare 1-tribromogermyl-1-trimethylsilylethene (Equation (33)) ⟨86ZOB1535⟩. The displacement of lithium by germanium has been reported to be an efficient process. For example, 1-lithio 1-diphenylmethylsilyl alkenes, generated from the bromo analogues with Bu^{s}Li at −78 °C, reacted with trimethylgermanium chloride to give, for example, 1-trimethylgermyl 1-diphenylmethylsilyl alkene (**89**) (Equation (34)) ⟨86OM2274⟩.

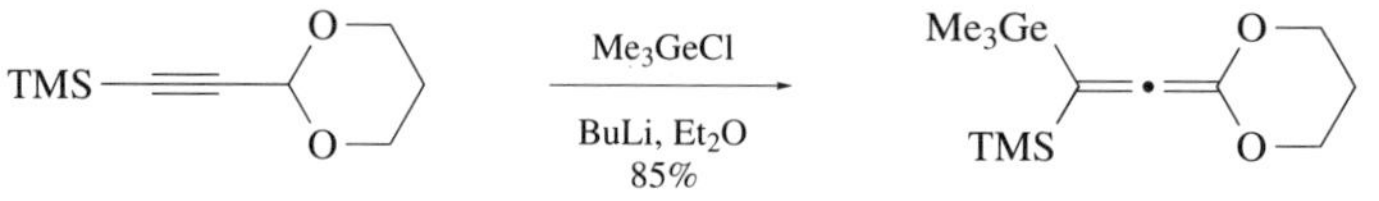

(31)

R = TMS, M = Ge, 85%
R = Me$_3$Ge, M = Si, 78%

(32)

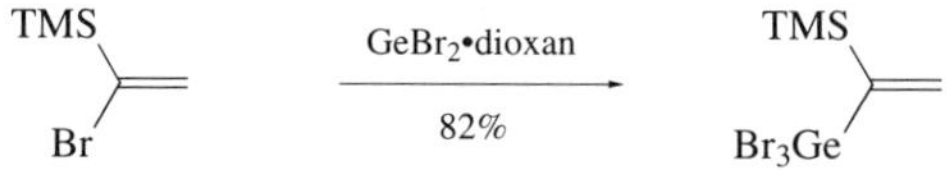

(33)

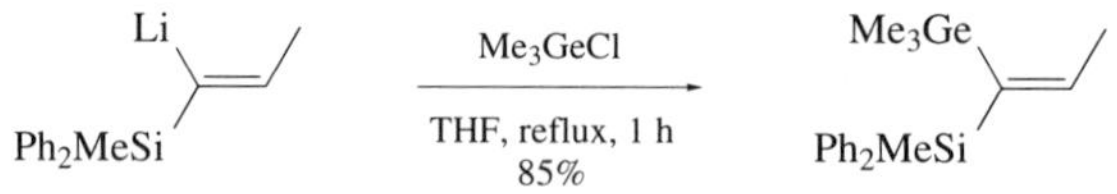

(**89**)

(34)

1-Germyl 1-silyl alkanes possessing a β-oxygen functionality have been converted into 1-germyl 1-silyl alkenes in high yield using LDA ⟨88OM739⟩, in a fashion analogous to the formation of bisgermyl alkenes described above (Equation (23)).

4.23.1.4.3 Functions containing one boron and one germanium

Only a single paper relating to the preparation of 1-boryl 1-germyl alkenes has been found. In principle, those methods described for the synthesis of 1-silyl 1-germyl alkenes should also be applicable for the boryl analogues. Hydroboration of 1-trimethylgermyl-2-alkoxyethynes has been reported to give 1-boryl-1-germyl-2-alkoxyethynes (Equation (35)) ⟨87ZOB1741⟩.

$$\text{Me}_3\text{Ge}-\!\!\!\equiv\!\!\!-\text{O}\diagdown \quad \xrightarrow[66\%]{\text{Pr}_2\text{BH}} \quad \begin{array}{c} \text{Pr}_2\text{B} \\ \diagdown\!\!=\!\!\diagup \\ \text{Me}_3\text{Ge} \quad \text{O}\diagdown \end{array} \tag{35}$$

4.23.2 FUNCTIONS CONTAINING A METALLOID AND A METAL

4.23.2.1 Silicon Functions—$R^1_2C=CSiR^2_3M$

4.23.2.1.1 Functions with one silicon and one group 1 metal

The silylcupration of alkynes has been described by Fleming et al. ⟨81JCS(P1)2527⟩. When a 1-lithio alkyne reacted with a bis silylated copper species (**90**) at low temperatures, a 1-lithio 1-silyl alkene intermediate (**91**) was obtained in solution (Equation (36)). The addition is not thought to be regioselective as on quenching with water a 10:1 ratio of 1-silyl alkene to 2-silyl alkene was observed. 1,3-Bis(trimethylsilyl)propyne was metallated by Bu^tLi or Bu^nLi/TMEDA (1:1) at $-78\,°C$ in essentially quantitative yield (Equation (37)) ⟨81JA5568⟩. The resulting allene (**92**; $R^1 = TMS$) is chiral but was found to be too unstable to be resolved ⟨84TL5711⟩. Addition of aldehydes to (**92**; $R^1 = TMS$, $R^2 = H$) led to a mixture of (Z) and (E) enynes (3:1 in the case of hexanal). The use of the 3-TBDMS propyne analogue (**92**; $R^1 = TBDMS$, TBDMS represents t-butyldimethylsilyl) further increased the diastereoselectivity in favour of (Z); the change of counter ion from Li to Mg increased the selectivity to 50:1 (Z):(E) ⟨81JA5568⟩. 1,3,3-Tris(trimethylsilyl)propyne reacted similarly ⟨86AG1020⟩.

$$\text{Bu}-\!\!\!\equiv\!\!\!-\text{Li} \quad \xrightarrow[80\%]{(\text{PhMe}_2\text{Si})_2\text{CuLi (90)}} \quad \left[\begin{array}{cc} (\text{Cu}) & \text{SiMe}_2\text{Ph} \\ \diagdown\!\!=\!\!\diagup \\ \text{Bu} & \text{Li} \end{array}\right] + \left[\begin{array}{cc} \text{PhMe}_2\text{Si} & (\text{Cu}) \\ \diagdown\!\!=\!\!\diagup \\ \text{Bu} & \text{Li} \end{array}\right] \tag{36}$$

(**91**)   10:1

$$\begin{array}{c} R^2 \\ \diagdown \\ R^1 \end{array}\!\!=\!\!=\!\!-\text{TMS} \quad \xrightarrow{Bu^tLi \text{ or } Bu^nLi/TMEDA} \quad \left[\begin{array}{cc} R^2 & \text{Li} \\ \diagdown & \diagup \\ \quad =\!\!\cdot\!\!= \\ \diagup & \diagdown \\ R^1 & \text{TMS} \end{array}\right] \tag{37}$$

$R^1 = TMS, PhMe_2Si$
$R^2 = H, TMS$

(**92**)

Metal–halogen exchange has been used in the preparation of 1-lithio 1-silyl alkenes. Direct metallation of 1-halo 1-silyl alkenes with lithium metal proved to be unsatisfactory, while the reaction with alkyl lithium reagents proceeded in high yield ⟨78JOC2739⟩. The bromo or iodo alkene was generally used as the chloro derivatives may have promoted α-hydrogen metallation when the reaction was performed in THF. Lithiation of the (E) isomer (**93**) was quantitative with Bu^nLi at $-78\,°C$ in THF. However, no reaction was observed in Et_2O (Scheme 12). Problems have been encountered with the analogous (Z) isomer (**94**) under these conditions, in that substantial alkylation of the lithio product with Bu^nBr, produced in the reaction, was observed (Scheme 12). The problem was overcome to some extent by performing the reaction at $-100\,°C$, although undesired alkylation was still a potential problem in subsequent reactions, with both (**93**) and (**94**) ⟨79JOC4623⟩. The use of one or two equiv. of Bu^tLi for the exchange also caused problems. The configurational instability of 1-lithio 1-trimethylsilyl alkenes formed using Bu^tLi has been investigated by Negishi and Takahashi ⟨86JA3402⟩. In contrast, with one equiv. of Bu^sLi in THF at $-78\,°C$ the exchange reaction was a highly efficient process, without the problems encountered with the other butyl lithium reagents ⟨79JOC4623⟩. The stereointegrity of the 1-lithio 1-silyl alkenes was gradually destroyed as the temperature was increased. The presence of a bulky silyl group led to greater stereointegrity ⟨86OM2274⟩. In the late 1980s a 1-lithio 1-silyl alkene containing a chiral naphthylphenylmethylsilyl group was prepared. The reactions of this reagent with prochiral carbonyl species led to the formation of diastereomeric products ⟨88TL1355⟩. Reductive lithiation of the 1-phenylthio 1-trimethylsilyl alkenes (**93**, with PhSe in place of Br) with lithium 1-(dimethylamino)naphthalenide provided an alternative route to 1-lithio 1-silyl alkenes ⟨84JA3245⟩. Transmetallation of the 1-silyl 1-stannyl alkenes (**95**) with methyl lithium is described by Mitchell and Reimann (Equation (38)) ⟨85JOM(281)163⟩. When bulky substituents are present at the C-2 position of the alkene, a single species is formed which is stable for 24 h at −78 °C. With simple alkyl substituents in this position, a mixture of (*E*) and (*Z*) isomers was formed, the ratio of which could only be determined after quenching.

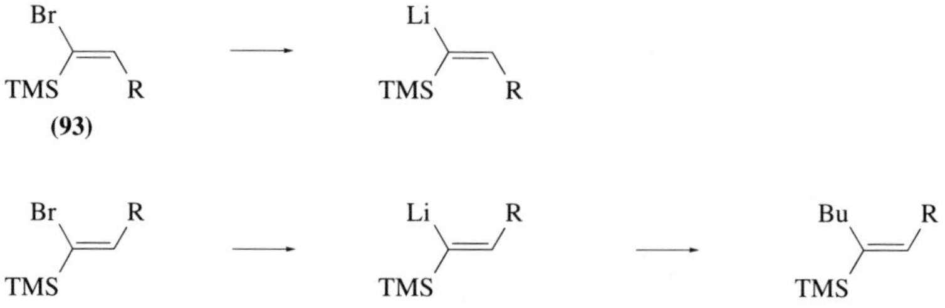

Scheme 12

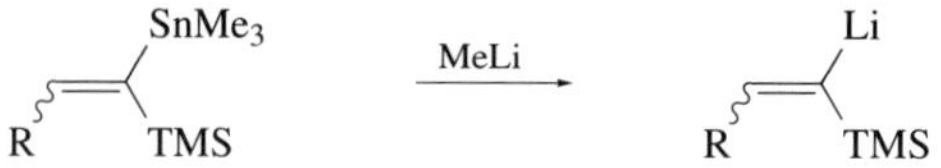

(38)

4.23.2.1.2 Functions with one silicon and one group 2 metal

Nickel-catalysed addition of MeMgBr to 1-silyl alkynes in the presence of trimethylaluminum has been reported to give 1-magnesio 1-silyl alkenes (**96**) ⟨78JA4624⟩. Hydromagnesiation of 1-silyl alkynes may be effected with Bu^{i}MgBr using Cp$_2$TiCl$_2$ as catalyst. This reaction proceeded in a regioselective and stereoselective manner (Equation (39)) ⟨84CC1130⟩. A method of stereoselective, intramolecular, carbamagnesiation of the ω-bromo 1-silyl alkynes (**97**) has been described ⟨84TL1999⟩. The resulting exocyclic 1-magnesio 1-silyl alkenes (**98**) had their stereochemistry defined (Equation (40)). It is proposed that the initially formed Grignard intermediate adds to the silyl alkyne in a suprafacial manner. The cyclopentyl and cyclohexyl products are formed in high yield but only a trace of the cyclobutyl derivative could be isolated. Transmetallation of 1-lithio 1-silyl alkenes with MgBr$_2$ at 0 °C gave the expected 1-magnesio 1-silyl alkenes (**99**) whose stereointegrity was maintained (Equation (41)). However, when magnesium reacted with the 1-bromo derivative (**100**) a mixture of the (*E*) and (*Z*) isomers of (**99**) was obtained ⟨86OM2274⟩. In a similar manner, treatment of 1,3-bis(trimethylsilyl)-1-lithioallene with MgBr$_2$ gave the 1-magnesio derivative (**101**) ⟨81JA5568⟩. The metal–halogen exchange between magnesium and 1-bromo-1-trimethylsilylethene is the subject of a standard preparation ⟨88OSC(6)1033⟩.

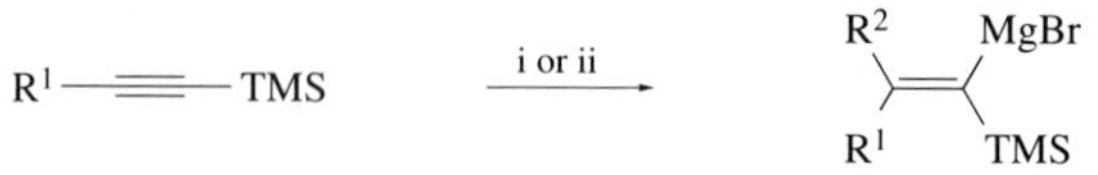

i, MeMgBr, Ni(acac)$_2$, AlMe$_3$, 25 °C, 24 h   R^2 = Me
ii, Bu^{i}MgBr, CpTiCl$_2$, Et$_2$O, 25 °C, 6 h   R^2 = H

(39)

(40)

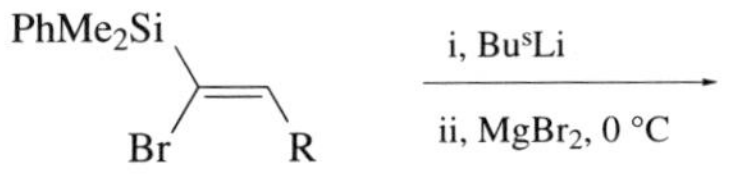

$$\text{(100)} \xrightarrow[\text{ii, MgBr}_2, 0\,°\text{C}]{\text{i, Bu}^s\text{Li}} \text{(99)} \quad (41)$$

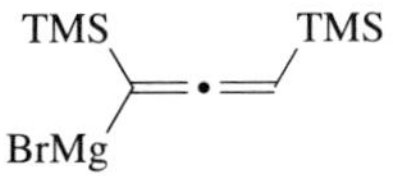

(100): PhMe₂Si, Br, R on alkene
(99): PhMe₂Si, BrMg, R on alkene

(101): TMS, TMS, BrMg allene

4.23.2.1.3 Functions with one silicon and one transition group metal

A number of 1-metallo 1-silyl alkenes incorporating transition metals are described in the literature. In the majority of cases they are prepared by transmetallation of the corresponding lithium or magnesium derivative. Lithiation of 1-trimethylsilyl-1-butyne followed by addition of Ti(OPr^i)$_4$ generated the 1-titano-1-silylallene (102) *in situ*, which was analysed by IR spectroscopy (Equation (42)) ⟨84BCJ2768⟩. Under similar conditions, 1-trimethylsilyl-1-propyne only gave a 3-titano 1-trimethylsilyl propyne. Carbatitanation of 1-trimethylsilyl-2-phenylethyne has been performed using TiCp$_2$Cl$_2$ and AlCl$_2$Me at −20 °C, giving the 1-trimethylsilyl 1-titano alkene (103) whose structure was confirmed by x-ray analysis ⟨85JA7219⟩. Zirconation of alkynes, via transmetallation, with ZrCp$_2$Cl$_2$ has been performed in a similar fashion to titanation. The use of Bu^{n}Li ⟨86TL2829⟩, Bu^{t}Li ⟨87TL917⟩, HgCl$_2$/Mg ⟨85JA2568⟩ and MeLi ⟨89JA2870⟩ to prepare the initial lithio species has been described.

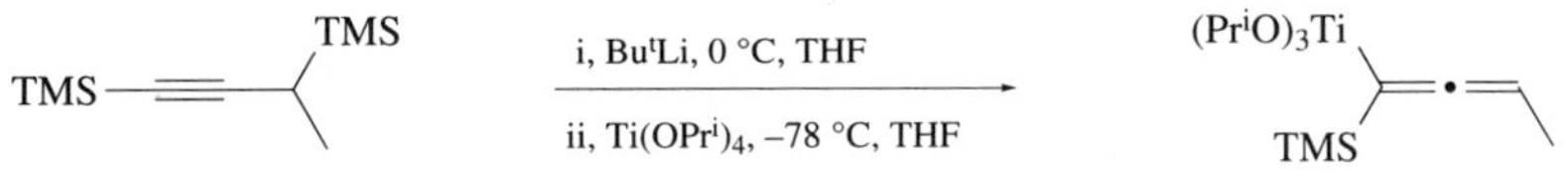

$$\text{TMS}-\equiv-\text{C(TMS)} \xrightarrow[\text{ii, Ti(OPr}^i)_4, -78\,°\text{C, THF}]{\text{i, Bu}^t\text{Li, 0\,°C, THF}} \text{(102)} \quad (42)$$

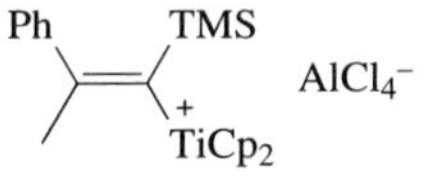

(103): Ph, TMS, TiCp$_2^+$, AlCl$_4^-$

The reaction of trimethylsilylethyne with RCu·MgBr$_2$ in Et$_2$O at 0 °C gave 1-trimethylsilylethene, via the 1-trimethylsilylethene copper compound (105; R^1 = H) which cannot be isolated. This reaction has only been found to work for the ethyne. The cuprates (105) could also be prepared by addition of a Grignard reagent to the intermediates (104) (Scheme 13) ⟨79JOM(177)145⟩. The addition of CuBr·SMe$_2$ to terminal alkynes, in the presence of magnesium, has been reported to generate (105) ⟨84JOC1574⟩. *In situ* transmetallation of 1-lithio 1-silyl alkenes or 1-boryl 1-silyl alkenes with CuI in THF at 0 °C gave (105). CuBr and CuCl were found to be less efficient in this process ⟨86BCJ659⟩. The *in situ* formation of 1-silyl 1-zinco alkenes (106) by reaction of RZnBr with 1-silyl alkynes at 100 °C for 30 h in a sealed tube has been reported ⟨87TL2889⟩. Analysis of the product following deuteriolysis indicated the presence of two geometric isomers. The reaction of 1-silyl alkynes having a γ-hydrogen with Bu^{t}Li and ZnCl$_2$ at −30 °C generated 1-silyl-1-zincoallenes (107) *in situ* ⟨84JOC4565⟩.

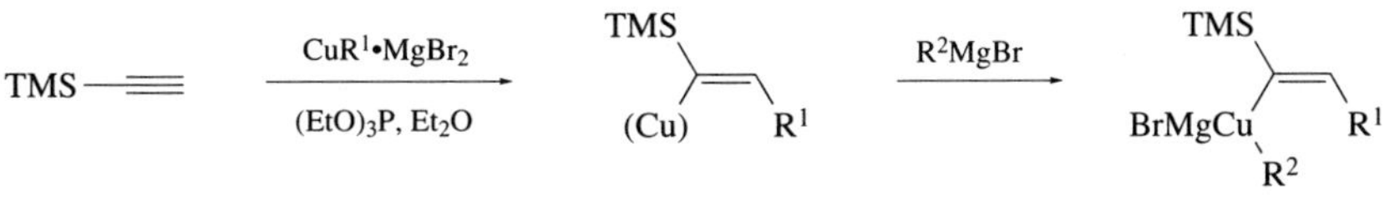

$$\text{TMS}-\equiv \xrightarrow[\text{(EtO)}_3\text{P, Et}_2\text{O}]{\text{CuR}^1\cdot\text{MgBr}_2} \text{(104)} \xrightarrow{\text{R}^2\text{MgBr}} \text{(105)}$$

(104): TMS, (Cu), R^1 — not isolated
(105): TMS, BrMgCu, R^1, R^2 — not isolated

Scheme 13

4.23.2.1.4 Functions with one silicon and one group 13 or group 14 metal

The preparation of 1-alumino 1-silyl alkenes has been reviewed by Zweifel and Miller ⟨84OR(32)375⟩. Hydralumination of 1-trimethylsilyl alkynes was effected using dialkylaluminum hydrides, for example dibal-H ⟨83JOC1560⟩. Carbalumination was performed with trialkylaluminum reagents, for example AlMe₃ ⟨84JA6105⟩. 1-Silyl 1-stannyl alkenes can in principle be prepared by hydrometallation of 1-silyl or 1-stannyl alkynes. Hydrosilylation was not found to be an efficient process while hydrostannylation of 1-trimethylsilylpent-1-yne gave a mixture of regioisomers (**108**; R = Pr and **109**; R = Pr). The regioselectivity increased to 9:1, in favour of (**109**; R = Ph) with 1-trimethylsilyl-2-phenylethyne ⟨81JOM(210)C17⟩. The reaction of BPr$^i_3$ with 1-trimethylsilyl-2-trimethylstannylethyne resulted in the migration of the stannyl group and addition of the borane at the C-2 position to give (**110**) (Equation (43)) ⟨86JOM(303)73⟩. Transmetallation of lithio ⟨86OM2274⟩ or magnesio silyl alkenes ⟨85OM1044⟩ with trialkylstannyl halides, or lithio stannyl alkenes with trialkylsilyl halides have also been used to prepare 1-silyl 1-stannyl alkenes (Equation (44)) ⟨85JOM(281)163⟩. The first two processes proceeded in good yield with high stereoselectivity, while the last gave mixtures of isomers. The titanium-catalysed hydromagnesiation of 1-trimethylsilyl-1-propyn-3-ol (**111**; R^1 = R^2 = H) followed by trapping of the intermediate with a trialkylstannyl chloride was highly regioselective, with the (Z) isomer of the 1-trimethylsilyl 1-trialkylstannylpropene (**112**) being formed (Equation (45)) ⟨90TL3105⟩. The reaction of (**103**) with R₃SnCl led to 1-trimethylsilyl 1-stannyl alkenes in good yield (>50%) with retention of configuration ⟨79JOM(177)145⟩. BuLi/Et₃N has also been used to effect the addition of R₃SnCl to terminal alkynes. If a γ-hydrogen is present, the product is an allene of the type (**113**) rather than an alkene ⟨83T3073⟩.

Reactions similar to those described above for the preparation of stannyl alkenes may serve for preparation of the plumbyl alkenes. Such reactions have generally not been reported. However, transmetallation of 1-lithio 1-plumbyl alkenes (**114**) with trialkylsilyl halides gave the 1-silyl 1-plumbyl alkenes (**115**) in good yields (Equation (46)) ⟨87JOM(322)151⟩. The products, if oils, may be purified by distillation.

$$\underset{(\mathbf{114})}{\underset{\text{Li}\quad\text{R}}{\overset{\text{Me}_3\text{Pb}\quad\text{R}}{\diagup\!\!\!\diagdown}}} \xrightarrow[\text{THF, }-78\,^\circ\text{C}]{\text{TMS-Cl}} \underset{(\mathbf{115})}{\underset{\text{TMS}\quad\text{R}}{\overset{\text{Me}_3\text{Pb}\quad\text{R}}{\diagup\!\!\!\diagdown}}} \qquad (46)$$

R = H, 61%
R = MeOCH$_2$, 56%
R = Ph, 82%

4.23.2.2 Boron Functions—R$^1_2$C=CBR$^2_3$M

Hydroboration of 1-stannyl alkynes is reported to give 1-boryl 1-stannyl alkenes (e.g., (**116**)) in moderate yields (Equation (47)) ⟨87ZOB1741⟩. This reaction is similar to that described in Scheme 11 for the silyl analogues. The presence of a trimethylstannyl group as the terminal substituent of an alkyne activates the alkyne to attack by trialkylboranes. The reaction of three equiv. of triethylboron with a trimethylstannylethynylboron (**117**) or bis(trimethylstannylethynyl)boron (**118**) gave the 1-boryl 1-stannyl alkenes (**119**) and (**120**) (Scheme 14) ⟨86JOM(307)157⟩. No mechanism is proposed for the reaction, but in essence the stannyl group has migrated from one end of the triple bond to the other, and the triethylboron has added at the vacant terminal position. Under the reaction conditions the stannyl and diethylboryl groups of the products adopt a *cis* configuration. Similar reactions were observed with trimethylboron. Interestingly, the bisalkyne (**118**) reacted only once with triethylboron even when an excess of reactant was used. The *cis* to *trans* isomerisation of (**119**) was achieved in benzene at 60 °C over 24 h. The *cis* to *trans* isomerisation of (**120**) was effected similarly, but the resulting product cyclised to give the borole (**121**). On further heating (60 °C, 7 d), (**121**) underwent isomerisation to (**122**) (Scheme 14) ⟨86JOM(307)157⟩.

$$\text{Me}_3\text{Sn}-\!\!\!\equiv\!\!\!-\text{OR} \xrightarrow[\text{R = C}_4\text{H}_9,\,48\%]{\text{HB}\langle\,\rangle} \underset{(\mathbf{116})}{\text{(Me}_3\text{Sn, OR, B, cyclohexyl)}} \qquad (47)$$

The reaction between 1,2-bis(trimethylstannyl)ethyne and 1,2-di-*t*-butyl-1,2-dichlorodiborane gave the borirene (**74**) described in Section 4.23.1.2 ⟨84AG(E)313⟩.

4.23.2.3 Germanium Functions—R$^1_2$C=CGeR$^2_3$M

The synthesis of this class of compound has been poorly documented. In theory the majority of the methods described in Section 4.23.2.1 for the preparation of the silyl analogues should be applicable. Hydrostannylation of 1-germyl alkynes by the addition of trialkyltin hydrides gave 1-germyl 1-stannyl alkenes in high yields (Equation (48)) ⟨78ZOB2147⟩. The reaction of 1-trimethylgermyl-2-phenylethyne with dibal-H produced high yields of either the *cis* (**123**) or *trans* (**124**) products, depending upon the solvent used ⟨71JOC3520⟩. Transmetallation of 1,1-distannyl alkenes with methyl lithium followed by quenching with trimethylgermanium chloride gave 1-germyl 1-stannyl alkenes ((**125**; M = Sn) in moderate yields (Equation (49)). The products were purified by distillation, except for the 2,2-diphenyl derivative (**125**; R^2 = R^3 = Ph, M = Sn) which is a solid ⟨87JOM(322)151⟩. The product of the reaction of a 1,1-distannyl-2-phenylethene with Me$_3$GeCl exists predominantly (74%) in the (*E*) form (**125**; R^2 = Ph, R^3 = H, M = Sn). In a similar fashion, 1,1-diplumbyl alkenes may be transmetallated with methyl lithium and quenched with trimethylgermanium chloride giving the 1-germyl 1-plumbyl alkenes (**125**; M = Ph) (Equation (49)) ⟨87JOM(322)151⟩. The majority of these products are stable enough to be purified by distillation or crystallisation.

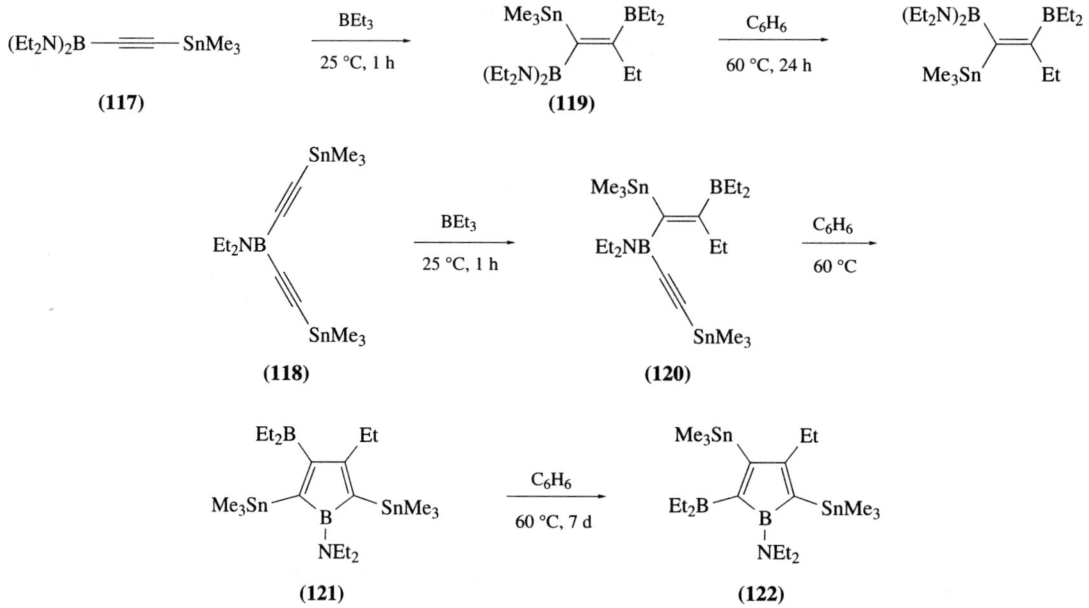

Scheme 14

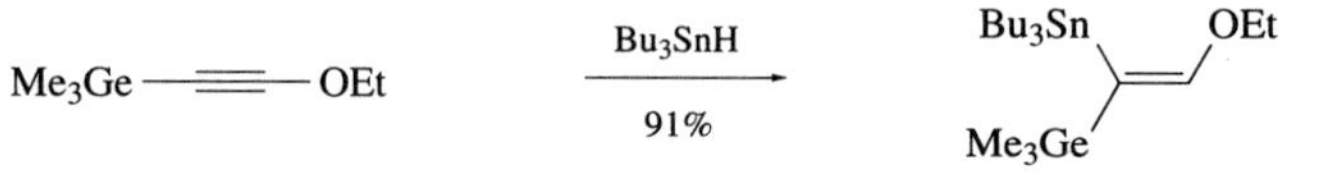 (48)

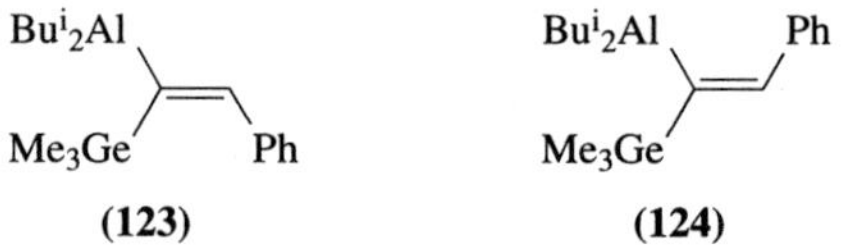

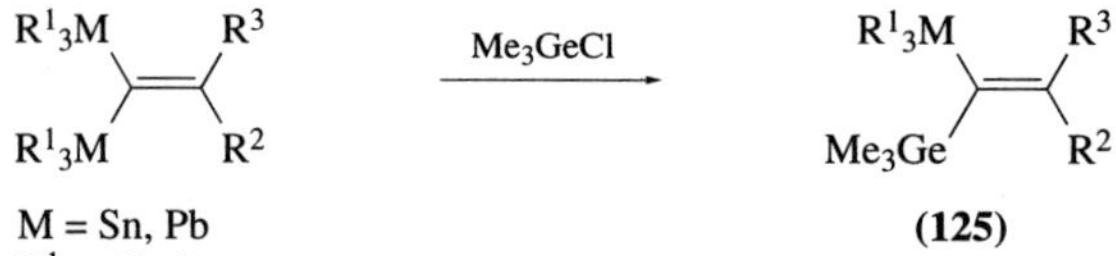 (49)

M = Sn, Pb
R^1 = alkyl
R^2 = Me, CH_2OMe, Ph
R^3 = H, Me, CH_2OMe, Ph

4.23.3 FUNCTIONS CONTAINING A GROUP 1 METAL—$R_2C{=}CLiM$

The synthesis of 1,1-dilithioallenes (**127**) has been accomplished by addition of two equiv. of alkyl lithiums to a terminal alkyne bearing both a γ-hydroxy and γ-TMS group (**126**). If only one equiv. of alkyl lithium was used, a mixture of non- and dilithiated allenes ((**128**) and (**127**), respectively) was obtained (Equation (50)) ⟨87TL1299⟩. Metal–halogen exchange of 1,1-diiodo alkenes has been reported ⟨88JA5567⟩. Removal of the first iodine was achieved using Bu^sLi at $-70\,^\circ C$; addition of MeLi then gave the 1,1-dilithio alkenes (e.g., Equation (51)). These compounds are stable in solution at $-70\,^\circ C$. Maercker and Dujardin have shown that the replacement of Hg by Li is an effective way to prepare dilithio alkenes ⟨85AG(E)571⟩. 1,1-Dilithioethene was predicted to be unstable; however, 1,1-dilithio-2-methyl-1-propene (**130**; $R^1 = R^2 =$ Me) could be prepared from the bis(chloromercurio) derivative (**129**; X = Cl) with lithium powder ⟨85AG(E)571⟩. In an analogous manner, dilithiomethylenecyclohexane (**130**; $R^1R^2 = —(CH_2)_5—$) was obtained from the bromomercurio derivative (**129**; X = Br) and Bu^tLi (Equation (52)) ⟨84AG(E)224⟩. The failure of similar reactions with distannyl alkenes may be the result of the use of monoalkyl rather than dialkyl substituted 1,1-distannyl alkenes ⟨83JOM(252)47⟩ and insufficient MeLi ⟨87JOM(322)141⟩.

$$\underset{\underset{\text{(126)}}{\text{R = alkyl}}}{\overset{\text{HO}}{\underset{\text{TMS}}{\text{R}}}\!\!\!\!\!\!\!\!\!\searrow\!\!\!\equiv} \xrightarrow{\text{BuLi, THF}} \underset{(127)}{\overset{\text{TMS-O}}{\underset{\text{R}}{\searrow}}\!\!=\!\!\bullet\!\!=\!\!\overset{\text{Li}}{\underset{\text{Li}}{\swarrow}}} + \underset{(128)}{\overset{\text{TMS-O}}{\underset{\text{R}}{\searrow}}\!\!=\!\!\bullet\!\!=} \quad (50)$$

$$\underset{\text{I}}{\overset{\text{Pr}^i\text{O}}{\searrow}}\!\!=\!\!\underset{\text{I}}{\swarrow} \xrightarrow[\text{ii, MeLi}]{\text{i, Bu}^s\text{Li}} \underset{\text{Li}}{\overset{\text{Pr}^i\text{O}}{\searrow}}\!\!=\!\!\underset{\text{Li}}{\swarrow} \quad (51)$$

$$\underset{\underset{(129)}{R^2}}{\overset{R^1}{\searrow}}\!\!\underset{\text{HgX}_2}{\overset{\text{HgX}_2}{\swarrow}} \xrightarrow{\text{Li or Bu}^t\text{Li}} \underset{\underset{(130)}{R^2}}{\overset{R^1}{\searrow}}\!\!\underset{\text{Li}}{\overset{\text{Li}}{\swarrow}} \quad (52)$$

$$R^1 = R^2 = \text{Me}$$
$$R^1R^2 = -(CH_2)_5-$$
$$X = Cl, Br$$

Transmetallation of 1,1-bis(trimethylstannyl) alkenes with MeLi led to the monolithiated derivatives $R^1R^2C=CLiSnMe_3$ (131) and $RCH=CLiSnMe_3$ (132) which gradually decomposed at 25 °C ⟨87JOM(322)141⟩. It is important to have the same alkyl group on both the lithium and tin reactants otherwise the alkyl groups scramble, resulting in mixtures of products. The derivatives (131) are more readily prepared and are more stable than (132). Both classes of compounds can exist in either the (E) or (Z) form. For (132), large R groups such as Ph or Bu^t gave the (E) isomer exclusively. The orientations of the alkenes (131) have been discussed; large or coordinating groups such as methoxyethyl appear to be *trans* to the stannyl group in the major isomer ⟨87JOM(322)141⟩. The 1-plumbyl vinyl lithiums (133) have been prepared from either the 1,1-diplumbyl alkenes or 1-plumbyl 1-stannyl alkenes by treatment with methyl lithium in THF at −78 °C (Equation (53)). In the latter case only the stannyl group was replaced by lithium. These compounds are more stable than the analogous 1-stannyl vinyl lithiums ⟨87JOM(322)151⟩. A similar (E)/(Z) isomer pattern to that seen with the 1-stannyl vinyl lithiums described above was observed. With monoalkylsubstituted 1-lithio 1-plumbyl alkenes (133; $R^2 = H$), the presence of a coordinating substituent such as dimethylaminomethyl was shown to lead to the (E) isomer, as determined by ^{1}H NMR spectroscopy.

$$\underset{R^1}{\overset{R^2}{\searrow}}\!\!=\!\!\underset{M}{\overset{\text{PbMe}_3}{\swarrow}} \xrightarrow[\text{THF, −78 °C}]{\text{MeLi}} \underset{R^1}{\overset{R^2}{\searrow}}\!\!=\!\!\underset{\text{Li}}{\overset{\text{PbMe}_3}{\swarrow}} \quad (53)$$

(133)

R^1 or R^2 = Bu^n, Bu^t, c-C$_6$H$_{11}$, Ph, Me$_2$NCH$_2$, MeO(C$_6$H$_{10}$), EtOCH$_2$
M = PbR$_3$, SnR$_3$

4.23.4 FUNCTIONS CONTAINING A GROUP 2 METAL (AND NO GROUP 1 METAL)—$R_2C=CMgXM$, etc.

The reaction of Grignard reagents with propargyl alcohol resulted in the formation of the 1,1-bis(bromomagnesio) alkenes (134) (Equation (54)) ⟨79SC53⟩. Similarly, addition of the Grignard reagents to conjugated enynes gave a mixture containing 1,1-dimagnesioallenes (135) (Equation (55)) ⟨77JOM(131)321, 77JOM(140)237⟩. Addition of transition metals to the reaction mixture may direct the attack towards the alkene rather than the alkyne π-bond ⟨77ZOR457⟩.

$$\text{HO}\!\!\searrow\!\!\equiv \xrightarrow{\text{RMgBr}} \underset{R}{\overset{\text{BrMgO}}{\searrow}}\!\!=\!\!\underset{\text{MgBr}}{\overset{\text{MgBr}}{\swarrow}} \quad (54)$$

(134)
R = Me, allyl

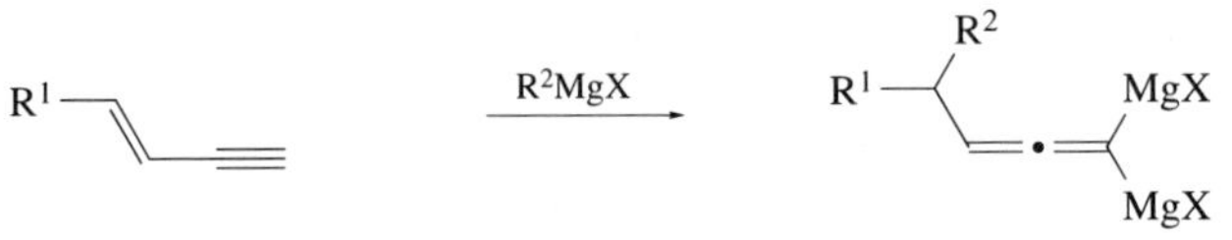

1-Lithio 1-stannyl alkenes react with $MgBr_2$ at $-78\,°C$ to afford the 1-magnesio 1-stannyl alkenes (**136**). These are stable enough in solution at $25\,°C$ to allow ^1H NMR analysis, which showed that (**136**) existed with the magnesio and alkyl groups *cis* to each other. These compounds could not be prepared from the analogous distannyl alkenes even with prolonged heating ⟨83JOM(252)47⟩. The derivatives (**136**) were also prepared from 1-iodo 1-stannyl alkenes and magnesium in THF at $70\,°C$ for 20 h ⟨83JOM(256)37⟩. For the reaction between the bis(bromomagnesium) derivative of prop-2-yn-1-ol (**137**; R = MgBr) and allyl zinc derivatives a cyclic intermediate 1-magnesio 1-zinco alkene (**138**) was proposed, formed via the bis(magnesio)derivative (**139**). For the methyl ether (**137**; R = Me), an acyclic intermediate (**140**) was proposed (Scheme 15) ⟨77JOM(142)9⟩.

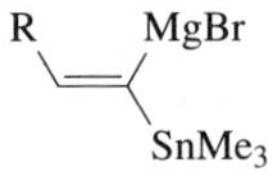

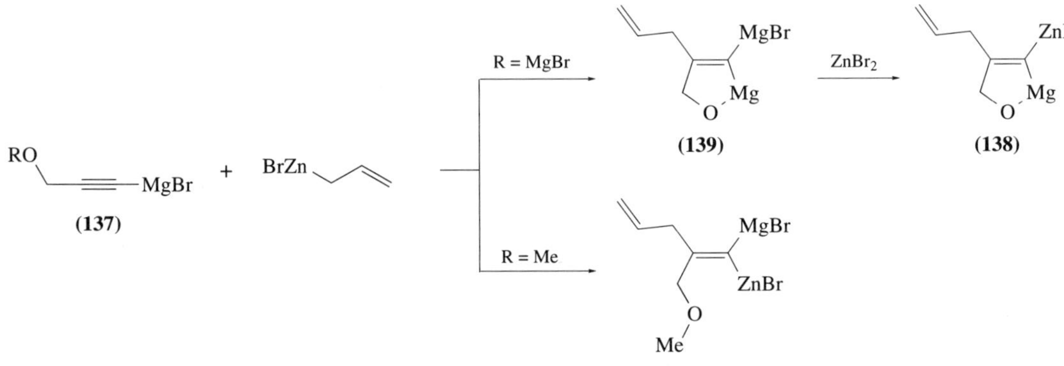

Scheme 15

4.23.5 FUNCTIONS CONTAINING A TRANSITION METAL (AND NO GROUP 1 OR GROUP 2 METAL)—$R_2C{=}CTiM$, etc.

$Cl(Me)ZrCp_2$ reacted with 1-pentynyldimethylalane (**141**) at $25\,°C$ to produce a zirconio alumino alkene (**142**; M = Zr) in virtually quantitative yield as a single stereoisomer which was not isolated (Equation (56)) ⟨81JA1276, 85JA6639⟩. The alkene (**142**; M = Zr) did not undergo isomerisation at room temperature (>48 h). The rigidity of (**142**; M = Zr) may prove a useful property, as it would appear possible to differentiate between the two metal groups. The reaction of (**141**) with a mixture of $TiCl_2Cp_2$ and trimethylaluminum, or the preformed $Cl(Me)TiCp_2$, gave the analogous titanium aluminum species (**142**; M = Ti) in excellent yield (90–100%) (Equation (56)). The reaction was ⩾99% regioselective and the presence of the two metal groups on the terminal position of the alkene was confirmed by deuteration ⟨81JA1276⟩. Unlike the zirconium analogue, the titanium species underwent slow isomerisation at $25\,°C$ to give a mixture of (*E*) and (*Z*) isomers. A similar type of reaction occurs between $ZrCl_2Cp_2$ and trimethylsilyl or trimethylstannyl alkynes (**143**) containing an ω-alkene group. In these cases a zirconabicyclic system (**144**) was isolated (Equation (57)) ⟨87TL917⟩.

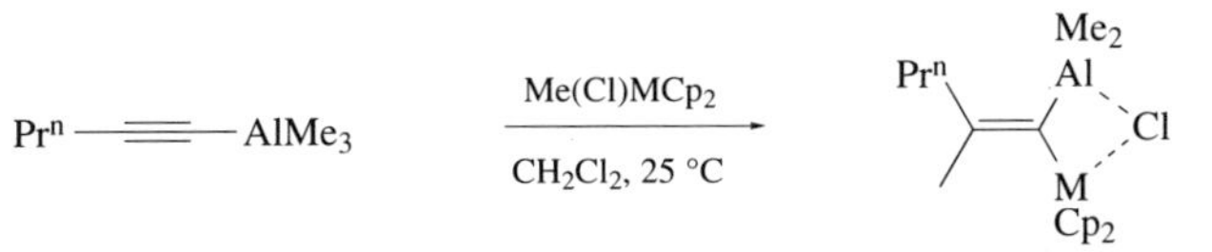

$$Pr^n \equiv\!\!\!= \!\!\!\equiv AlMe_3 \quad \xrightarrow{Me(Cl)MCp_2}_{CH_2Cl_2,\ 25\ °C} \quad \begin{array}{c} Pr^n \\ \diagup\!\!\!\diagdown \\ (142) \\ M = Zr,\ Ti \end{array} \quad (56)$$

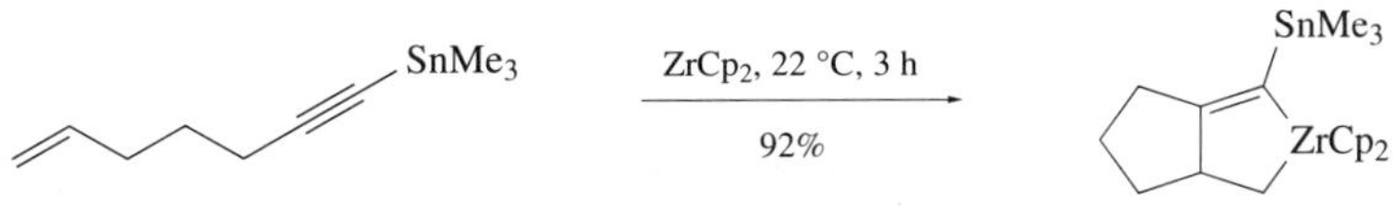

(141)

$$\text{(143)} \quad \xrightarrow[92\%]{ZrCp_2,\ 22\ °C,\ 3\ h} \quad \text{(144)} \quad (57)$$

ZrCp$_2$ = ZrCp$_2$Cl$_2$, BuLi, THF, −78 °C

Addition of 2.5 equiv. of an allyl zinc bromide to the methyl ether of prop-2-yn-1-ol in THF is reported to give the 1,1-dizinco alkene intermediate **(145)**, which could not be isolated (Equation (58)). Similarly, an intermediate formed from the addition of allylzinc bromide to the zinc derivative of prop-2-yn-1-ol **(146)** was reported to be the metallocycle **(147)**, in which two zinc atoms are attached to the same end of an alkene (Equation (59)) ⟨77JOM(142)9⟩.

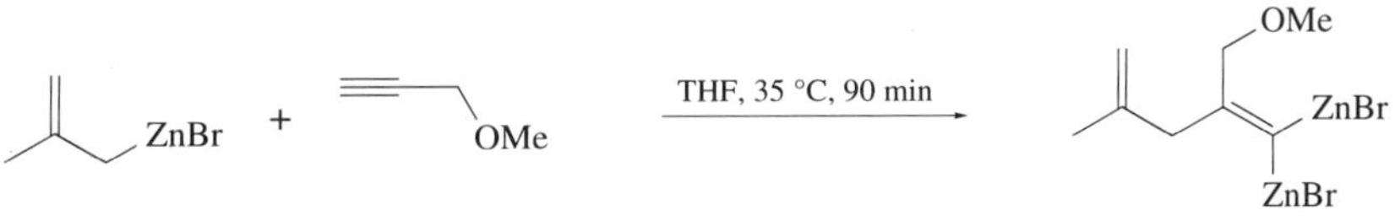

$$\text{alkene-ZnBr} + \equiv\!\!\!-\!\!\!\text{OMe} \quad \xrightarrow{THF,\ 35\ °C,\ 90\ min} \quad \text{(145) not isolated} \quad (58)$$

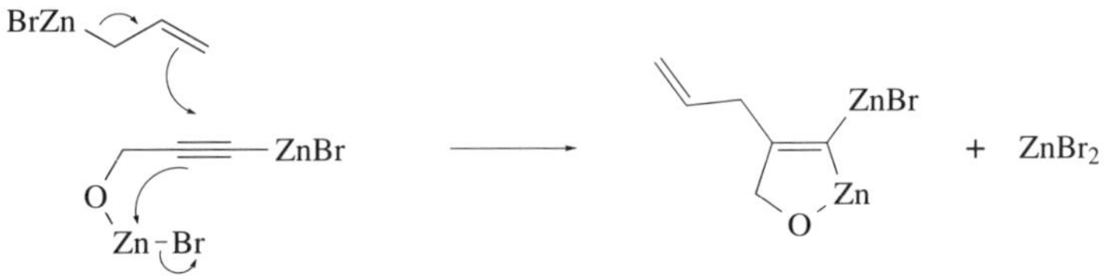

$$\text{(146)} \quad \longrightarrow \quad \text{(147)} + ZnBr_2 \quad (59)$$

Dimercuration of alkynes has been accomplished with mercuric acetate in CHCl$_3$ at 25 °C. With phenylethyne the 2,2-bis(acetoxymercurio)alkene **(148)** was obtained in quantitative yield as a stable solid ⟨83JCS(P1)1087⟩. A similar mercuration of alkenes has been effected in moderate yield using mercury(II) salts. 1,1-Diarylethenes have been treated with mercuric nitrate and sodium chloride to give the mono(chloromercurio) derivatives **(149)**, which in a few cases reacted further to afford the di(chloromercurio) derivatives **(150)** (16–54% yield; Scheme 16). The alkenes **(150)** are relatively stable solids and may be crystallised ⟨78JOM(162)271⟩. A number of alkene 1,1-diboronic esters **(151)**, whose preparation is described in Section 4.23.1.2, have been treated with mercuric chloride and sodium acetate in methanol. 1,1-Di(chloromercurio) alkenes **(152)** were precipitated immediately in high yields (70–97%; Equation (60)) ⟨78JOM(152)1⟩.

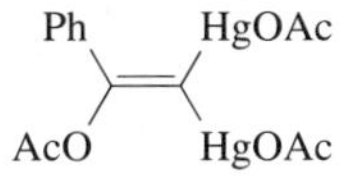

(148)

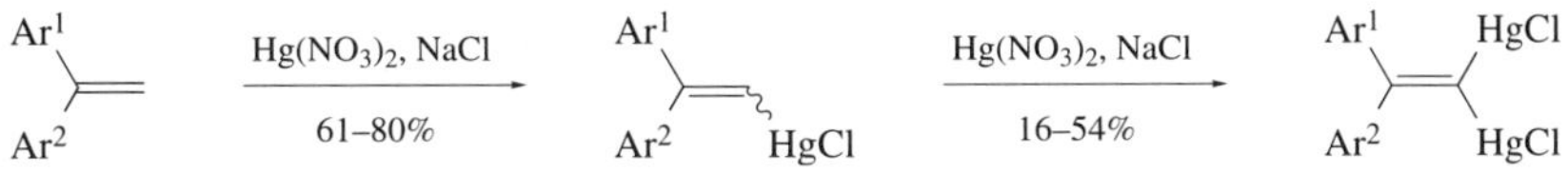

$$Ar^1\!\!\diagup\!\!Ar^2 \quad \xrightarrow[61-80\%]{Hg(NO_3)_2,\ NaCl} \quad \text{(149)} \quad \xrightarrow[16-54\%]{Hg(NO_3)_2,\ NaCl} \quad \text{(150)}$$

Scheme 16

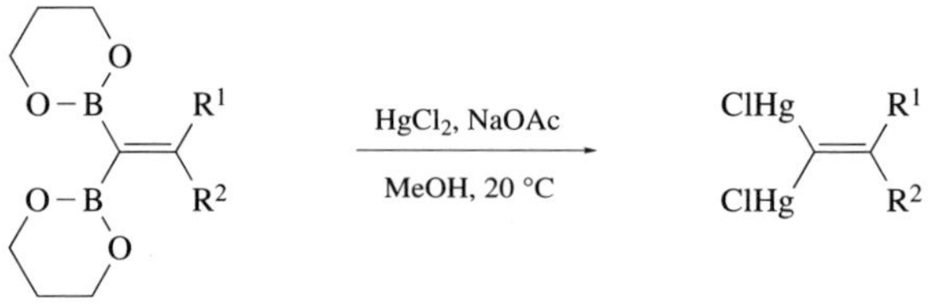

$$R^1 = Ph, R^2 = H, Me, 85\%$$
$$R^1 = Me, R^2 = CO_2Et, 80\%$$
$$R^1, R^2 = -(CH_2)_5-, 97\%$$

(151) → (152)   (60)

4.23.6 FUNCTIONS CONTAINING A GROUP 13 OR GROUP 14 METAL (AND NO GROUP 1, 2 OR TRANSITION METAL)—$R_2C=CAlM$, etc.

The general principles of hydralumination have been discussed by Eisch ⟨82COMC-I(2)641⟩. Hydralumination of heterosubstituted alkynes has been reviewed by Zweifel and Miller ⟨84OR(32)375⟩. Hydralumination of terminal alanyl alkynes with dibal-H gave 1,1-dialumino alkenes (153) ⟨75LA565, 88ZOB1567⟩. The addition of transition metals such as titanium, zirconium and nickel is known to promote hydralumination reactions and may be of use in this transformation. Carbalumination of phenylethyne gave a bridged bisalumino species (154) after the diphenyl-(phenylethynyl)aluminum (155) initially formed had been heated above 145 °C ⟨74JA1941⟩. Similarly, terminal alkynes reacted with thallium acetate at 20 °C to afford the bridged diorgano thallium compounds (156) in good yields (73–94%; Scheme 17). These solids decomposed if kept in air for a few weeks ⟨81JCS(P1)991⟩. An alkynylthallium diacetate (157) was proposed as an intermediate (cf. (155)).

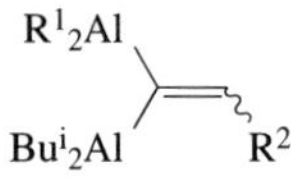

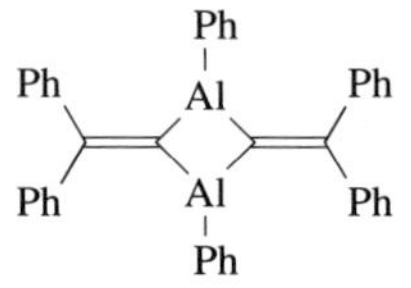

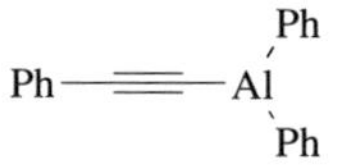

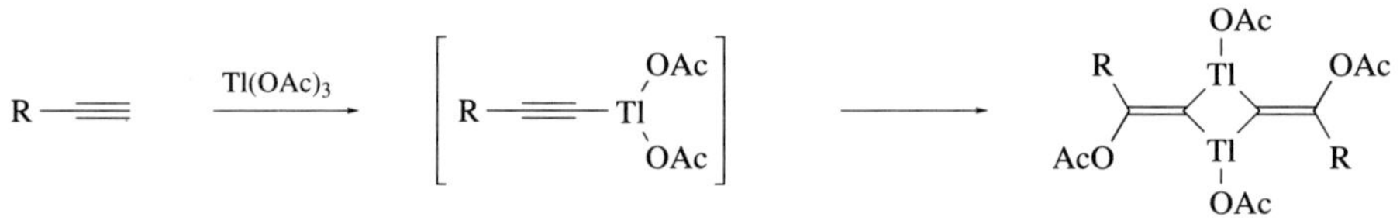

Scheme 17

Photolytic hydrostannylation of 1-stannyl alkynes with trimethyltin hydride in the presence of 2,2′-azobisisobutyronitrile (AIBN) afforded 1,1-distannyl alkenes (158) as the major products in high yield after distillation (Equation (61)). The reaction with 1-trimethylstannylethene gave the 1,2 adduct (159; R = H) as the sole product. When an oxygen-containing C-2 substituent, for example, methoxymethyl, was present, substantial amounts of (159) were formed ⟨83JOM(252)47⟩. Treatment of 1-(trimethylstannyl)ethyne with triethylboron led to the 1-boro-2-(trimethylstannyl) alkene (160), which with excess alkyne underwent further addition and rearrangement to give the 1,1-bis(trimethylstannyl)allene (161) in 85% yield (Scheme 18) ⟨81JOM(205)1⟩. In certain cases a 2-stannyl stannacyclopentadiene compound (162) was formed (Equation (62)) ⟨88JOM(338)195⟩.

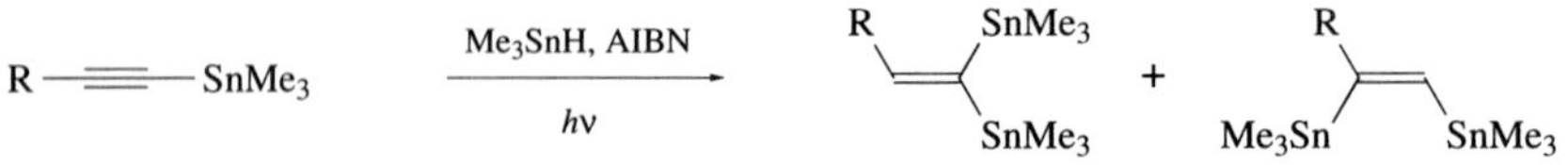

	Bu^n	Bu^t	Ph	$MeOCH_2$	$EtOCH_2$	$PhOCH_2$	H
R				Yield (%)			
158	94	98	95	63	54	70	0
159	6	2	5	37	46	30	100

The hydrostannylation of 1-stannyl alkynes led to the formation of 2-alkyl or 2-aryl 1,1-distannyl alkenes. In order to form 2,2-dialkyl or 2,2-diaryl 1,1-distannyl alkenes, geminal dihaloalkenes have been used. The reaction of 1,1-dibromo alkenes (**163**) with a trialkyl or triaryl lithium gave 1,1-distannyl alkenes (**164**) in 36–78% yields (Scheme 19). The use of the analogous 1,1-dichloro compounds led only to the formation of an alkyne ⟨86OM1991⟩. The majority of these compounds were purified by distillation. The carbostannylation of the trichloro alkynyltin compound (**165**) has been reported (Scheme 20) ⟨93CL1881⟩. A low yield (19%) of the product (**166**), which may be crystallised, was obtained. The x-ray analysis of (**166**) showed that one of the tin atoms coordinates to the carbonyl oxygen. This is thought to stabilise the molecule as the alkene bond was not isomerised in the presence of $SnCl_4$. The introduction of functionalised tin groups could be achieved by treating (**164**) with two equiv. of dibromo- or dichlorodimethylstannane in the absence of a solvent, to give (**167**) in yields of 53–95%. Treatment of the chlorodimethylstannyl derivative (**168**) with one equiv. of $SnCl_4$ gave the unstable compound (**169**), which with one further equivalent of $SnCl_4$ was converted into the 1,1-bis(trichlorostannyl) alkene (**170**) which was also unstable (Scheme 21) ⟨86OM1991⟩. Tristannyl alkanes are usually relatively stable. However, the phenoxy derivative (**171**) at 140 °C gave 1,1-bis(trimethylstannyl)ethene, which was identified by ^1H NMR spectroscopy (Equation (63)) ⟨83JOM(252)47⟩.

The preparation of 1-plumbyl 1-stannyl alkenes (**172**) has been carried out by transmetallation of the corresponding 1,1-distannyl alkenes with MeLi, followed by addition of trimethyllead chloride (Scheme 22) ⟨87JOM(322)151⟩. The products may be distilled without apparent decomposition. The choice of solvent can be important. THF was used for compounds with chelating substituents while DME was required for nonchelating species. If DME was not used a substantial amount of the 1,1-diplumbyl alkene (**173**) was formed as well as the desired product. When the starting alkenes are asymmetrical, the $(E):(Z)$ ratio varies widely. The 1,1-diplumbyl alkenes may be prepared directly

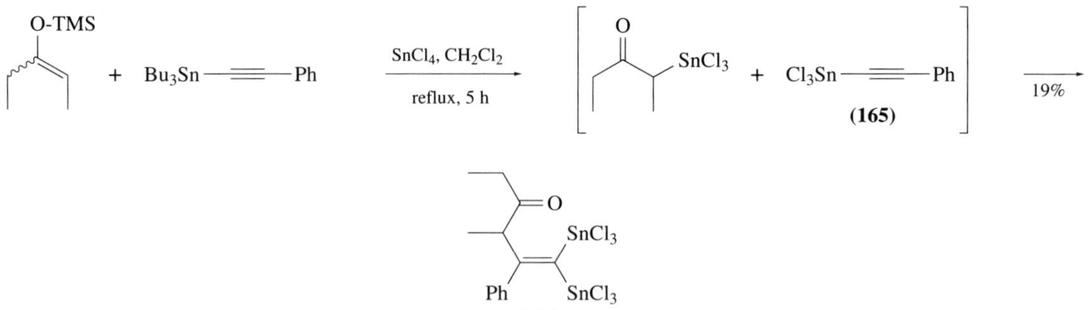

(166)

Scheme 20

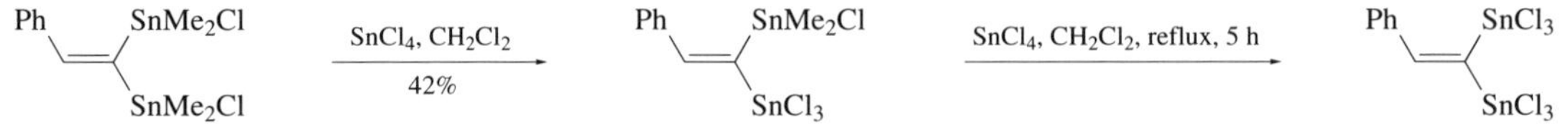

(168) **(169)** **(170)**

Scheme 21

in good yields (37–73%) via a stepwise double transmetallation of the corresponding 1,1-distannyl alkenes (Scheme 22). These compounds are high boiling point, light-sensitive liquids which are best stored below room temperature ⟨87JOM(322)151⟩.

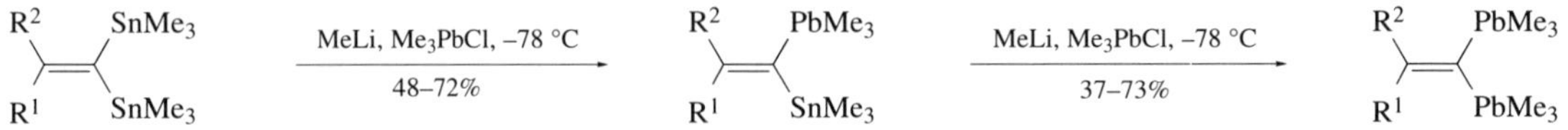

R^1 = H, alkyl, Ph, $MeOCH_2$
R^2 = H, Me, Ph

(172) **(173)**

Scheme 22

4.23.7 OTHER METAL DERIVATIVES

No examples not already covered in other sections of this chapter were found.

4.24
Tri- and Dicoordinated Ions, Radicals and Carbenes Bearing Two Heteroatoms ($RC^+X^1X^2$, $RC^-X^1X^2$, $RC \cdot X^1X^2$, $:CX^1X^2$)

WILLIAM M. HORSPOOL
University of Dundee, UK

4.24.1	INTRODUCTION	1072
4.24.2	CATIONS	1072
4.24.2.1	*Cationic Centres Flanked by a Nitrogen and an Oxygen*	1072
4.24.2.2	*Cationic Centres Flanked by Two Oxygens*	1072
4.24.2.3	*Cationic Centres Flanked by an Oxygen and a Sulfur*	1073
4.24.2.4	*Cationic Centres Flanked by Two Sulfur Atoms*	1073
4.24.2.4.1	*Cyclisation of dithiols*	1073
4.24.2.4.2	*Group abstraction methods*	1074
4.24.2.4.3	*Photochemical methods*	1075
4.24.2.5	*Cationic Centres Flanked by Two Selenium Atoms*	1075
4.24.3	ANIONS	1075
4.24.3.1	*Anion Centres Flanked by Oxygen and Phosphorus*	1075
4.24.3.2	*Anion Centres Flanked by Two Sulfur Atoms*	1075
4.24.3.3	*Anion Centres Flanked by Two Phosphorus Atoms*	1076
4.24.3.4	*Anion Centres Flanked by Two Halogens or a Halogen and an Oxygen*	1076
4.24.3.5	*Miscellaneous Anions*	1077
4.24.4	RADICALS	1077
4.24.4.1	*Radical Centres Flanked by Nitrogen and Oxygen*	1077
4.24.4.2	*Radical Centres Flanked by Nitrogen and Sulfur*	1077
4.24.4.3	*Radical Centres Flanked by Two Oxygens*	1078
4.24.4.3.1	*Intermolecular hydrogen abstraction*	1078
4.24.4.3.2	*Intramolecular hydrogen abstraction*	1079
4.24.4.4	*Radical Centres Flanked by Two Halogens*	1080
4.24.4.5	*Radical Centres Flanked by Two Sulfurs*	1080
4.24.5	CARBENES	1080
4.24.5.1	*Carbenes Flanked by Two Nitrogens*	1080
4.24.5.2	*Carbenes Flanked by Two Oxygens*	1080
4.24.5.3	*Carbenes Flanked by Two Sulfurs or One Sulfur and One Oxygen*	1081
4.24.5.3.1	*One sulfur and one oxygen*	1081
4.24.5.3.2	*Two sulfur atoms*	1081
4.24.5.4	*Miscellaneous Carbenes*	1082
4.24.5.4.1	*Stabilisation by phosphorus and silicon*	1082

4.24.5.4.2 Stabilisation by two boron atoms — 1082
4.24.5.5 Stabilisation by Halogens — 1082
   4.24.5.5.1 Miscellaneous methods for the generation of dihalocarbenes — 1082

4.24.1 INTRODUCTION

While the approach of *Comprehensive Organic Functional Group Transformations* is to emphasise isolable and persistent species, when selecting material for this chapter it became clear that several important processes would be excluded if the choice was restricted. Therefore, although not giving comprehensive coverage, this chapter aims to illustrate the wide variety of carbon species having two heteroatomic ligands. Many of the systems included have considerable value in synthesis. The emphasis is placed on examples since the 1970s and important reviews are cited.

4.24.2 CATIONS

The synthesis of cations in general and the synthesis of heteroatom flanked carbocations have been the subjects of a variety of review articles and treatises ⟨B-65MI 424-01, B-68MI 424-01, B-74MI 424-01, B-92MI 424-01⟩. A variety of cations are described in the literature where the cationic site is flanked by two heteroatoms. These are described below by periodic arrangement.

4.24.2.1 Cationic Centres Flanked by a Nitrogen and an Oxygen

A study has reported the formation of the cations (**1**) by reaction of the keto azides (**2**) in superacid medium at $-78\,°C$ ⟨83CB3926⟩. The presence of the cation was established using NMR spectroscopy.

4.24.2.2 Cationic Centres Flanked by Two Oxygens

The fission of C—C bonds in compounds can be brought about by electron transfer to a suitable reagent. An example of the formation of a cation flanked by two oxygen atoms using this method arises with the acetals (**3**). Here, the oxidising agent is the sulfate radical anion formed by irradiation of $K_2S_2O_8$ in a borate buffer. Electron transfer is rapid and affords the radical cation of (**3**) which undergoes facile C—C heterolysis to afford the cation (**4**) (where $R^1 = Et$ and $R^2 = H$) and a benzyl radical ⟨89JA4967⟩.

(**3**) $R^1 = Et, R^2 = H$
$R^1\text{-}R^1 = (CH_2)_3, R^2 = Ph$

(**4**) $R^1 = Et, R^2 = H$
$R^1\text{-}R^1 = (CH_2)_2, R^2 = Ph$

Another mode of C—C bond fission as a route to cations is that described by Wan and Muralidharan ⟨86CJC1949, 88JA4336⟩. Here, irradiation of suitably substituted acetals (**5**) resulted in loss of the *p*-nitrobenzyl anion and the formation of the cation (**4**) (where $R^1 = (CH_2)_3$ and $R^2 = Ph$).

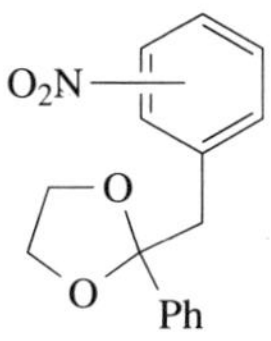

(**5**)

4.24.2.3 Cationic Centres Flanked by an Oxygen and a Sulfur

The reaction of ethereal solutions of 2-hydroxythiophenol with acid chlorides in the presence of tetrafluoroboric acid provides an efficient route to the synthesis of the ions (**6**), which can be isolated as the tetrafluoroborate salts. The yields are high; typical yields are given in Table 1 ⟨76JCS(P1)323⟩.

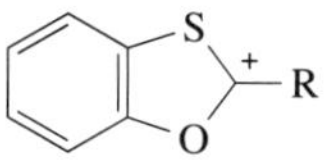

(**6**)

Table 1 Yield of cations (**6**) from 2-hydroxy-thiophenol and acid chlorides.

R	Yield (%)
Ph	91
p-MeC$_6$H$_4$	93
p-MeOC$_6$H$_4$	94
o-ClC$_6$H$_4$	90
p-ClC$_6$H$_4$	95
m-NO$_2$C$_6$H$_4$	96
p-EtO$_2$CC$_6$H$_4$	92
2-naphthyl	92
Pr^n	84
Bu^t	84
n-C$_{15}$H$_{31}$	100

4.24.2.4 Cationic Centres Flanked by Two Sulfur Atoms

4.24.2.4.1 *Cyclisation of dithiols*

The dithio cations of structure (**7**) are generally quite stable and can be isolated as the corresponding salts. Typical of the synthetic approach to these compounds is the conversion of the dithiol alkanes $HS(CH_2)_{n+1}SH$ in the presence of acid chlorides ⟨85CB1798⟩. The reactions are carried out in the presence of boron trifluoride etherate under argon, whereupon the cations (**7**) are obtained and isolated as the tetrafluoroborate salts. Typical yields are shown in Table 2.

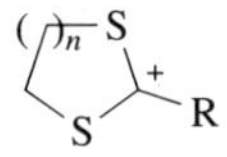

(**7**)

Table 2 Yield of cations (**7**) from dithiol alkanes $(HS(CH_2)_{n+1}SH)$.

R	Yield (%)
Ph	72
p-MeOC$_6$H$_4$	56
p-MeC$_6$H$_4$	31
p-ClC$_6$H$_4$	67
p-NO$_2$C$_6$H$_4$	85
Bu^t	75
1-adamantyl	64
2-naphthyl	64
cyclobutyl	60

Other routes have been used, such as the reaction of dithiol alkanes $HS(CH_2)_{n+1}SH$ with ethanoic anhydride to yield the thioesters $CH_3COS(CH_2)_{n+1}SH$, which can be subsequently cyclised to the cations (**8**) (where $n = 1$ or 2, R = Me) in high yield by reaction with perchloric acid ⟨82TL2665⟩, or the cations (**8**), where $n = 2$, R = Et (80% yield) or $n = 2$, R = Ph (88% yield), with acid chlorides under similar conditions ⟨86BCJ453⟩. Benzo-substituted derivatives (**9**) can also be prepared by a similar reaction ⟨59HCA1733⟩.

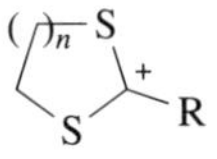

(**8**)
$n = 1$ or 2, R = Me
$n = 2$, R = Et or Ph
$n = 1$, R = Ph

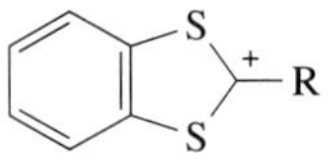

(**9**)

Alternatively, the ion of the type represented by the salt (**10**), referred to as 1,3-dithienium fluoroborate, can be prepared by reaction of the 1,3-dithiane with trityl fluoroborate in dry dichloromethane under reflux for 30 min. Workup and trituration of the crude product with cold ether afforded the product (**10**) in 92% yield ⟨72JA8932⟩.

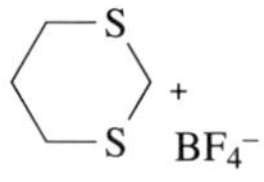

(**10**)

4.24.2.4.2 *Group abstraction methods*

The abstraction of a group other than a hydride can also be used as a synthetic approach to cations of this type. Thus trityl tetrachloroantimonate has been used to convert the *ortho*-esters (**11**) into the corresponding salts (**12**), again in high yield ⟨84JA3785⟩.

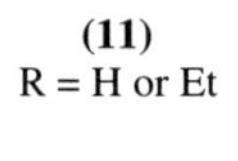

(**11**)
R = H or Et

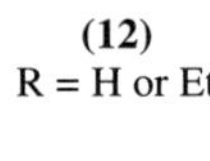

(**12**)
R = H or Et

4.24.2.4.3 Photochemical methods

Photochemical methods for the generation of cations of this type have also been described. Irradiation of the *ortho*-ester (13) (where R = MeO), for example, brings about O—C bond fission and formation of the cation (8), where $n = 1$ and R = Ph. Interestingly, this bond fission occurs even though the nucleofuge does not absorb light at the wavelength used ⟨91BCJ2751⟩. More commonly the nucleofuge is light-absorbing, as with the toluenesulfonyl derivative (13), where R = TolSO$_2$ ⟨90BCJ3056⟩. Generation of cations such as (8) (where $n = 1$ and R = Ph) in water provides an efficient route to the ring-opened thioesters in yields as high as 90%.

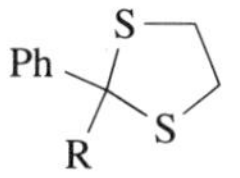

(13)
R = OMe, TolSO$_2$

4.24.2.5 Cationic Centres Flanked by Two Selenium Atoms

Group abstraction reactions, such as that described in Section 4.24.2.4.2, have also been used for the efficient formation of the cation (14). Here, the reaction was carried out under argon in dichloromethane at $-20\,°$C using Ph$_3$C$^+$SbCl$_4^-$ as the abstracting reagent. The substrate was the *ortho*-ester (15). The reaction was efficient, with a yield of 75% ⟨84JA3785⟩.

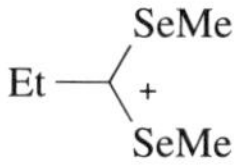

(14) 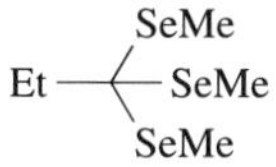   (15)

4.24.3 ANIONS

Several reviews have dealt with the formation, stability and reactivity of carbon acids ⟨84MI 424-01, 85MI 424-01, 87AR(C)197, 87MI 424-01, B-92MI 424-01⟩.

4.24.3.1 Anion Centres Flanked by Oxygen and Phosphorus

The anions (16) flanked by oxygen and phosphorus can be generated in high yield by reaction of their conjugate acids (17) with lithium diisopropylamide (LDA) in THF at $-70\,°$C ⟨86TL4635⟩.

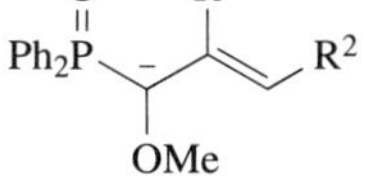

(16) 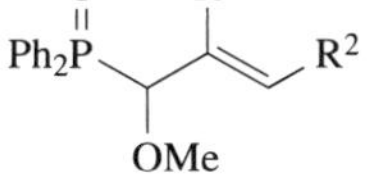   (17)

4.24.3.2 Anion Centres Flanked by Two Sulfur Atoms

The formation of anionic centres flanked by two sulfur atoms is a commonly used sequence in many synthetic strategies. Typically, a dithioacetal such as (18) in THF is treated with *n*-butyl-lithium at low temperature. This process affords the corresponding anion (19) in virtually quantitative yield ⟨75JOC231⟩. Several reviews on the formation of these anions and their synthetic applications have appeared ⟨69AG(E)639, 69S17, 77S357⟩. Typical reactions are illustrated for the dithioacetal (20) where the anion (21), generated by treatment with butyllithium in THF at $-30\,°$C, undergoes an intramolecular displacement reaction to afford the product (22) ⟨91JOC6038⟩; they are also given for the complex **(23)** where the anion **(24)** is formed on reaction with lithium di(trimethylsilyl)amide ⟨86JOM(315)59⟩. It is clear that a variety of substituents can be present on the central atom of the dithioacetal without impeding the reaction to any extent, and that the anion **(25)** is formed readily under the conditions described for the formulation of **(19)** ⟨85CC742⟩. Furthermore, changes in the oxidation level of the sulfur do not affect the efficacy of the reaction, and formation of **(26)** can also be brought about ⟨86CC1191⟩.

4.24.3.3 Anion Centres Flanked by Two Phosphorus Atoms

One example of this type of anion has been reported in the early 1990s. Thus the salt **(27)** can be obtained by the reaction of the neutral compound **(28)** with butyllithium in thiophene-free benzene under a nitrogen atmosphere. Alternatively, **(27)** can be obtained by reaction with potassium *t*-butoxide in THF at 50 °C ⟨91TL3631⟩.

4.24.3.4 Anion Centres Flanked by Two Halogens or a Halogen and an Oxygen

The formation of stabilised anions flanked by two fluorine atoms has been known about since the mid-1960s. These species are produced by addition of a nucleophile such as a cyanide ion to a perfluoro alkene such as tetrafluoroethene, yielding **(29)** ⟨65AFC(4)50, 67FCR359, 69APO(7)1⟩. The same approach has been used to provide examples of a variety of anions stabilised by fluorine and chlorine, or fluorine and an alkoxy group. The anions **(30)** are obtained by addition of the ions N_3^-, RS^-, $(RO)_2PO^-$, CF_3O^-, $CF_3CH_2O^-$ or Cl^- to the alkenes **(31)** ⟨84JA5544⟩.

4.24.3.5 Miscellaneous Anions

Anions flanked by boron and sulfur or silicon have become of interest. Examples of these have been prepared by the reaction of mesityllithium or lithium dicyclohexylamide on the neutral hydrocarbon (**32**) ⟨83TL637, 86TL5033⟩.

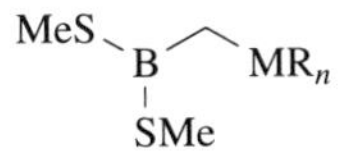

$$\text{MeS-B(SMe)-CH}_2\text{-MR}_n$$

(**32**)

M = Si, R_n = Me$_3$
M = Sn, R_n = Me$_3$
M = S, R_n = Ph

4.24.4 RADICALS

There are many review articles ⟨B-92MI 424-01⟩, treatises ⟨B-73MI 424-01⟩ and monographs ⟨B-74MI 424-02, B-79MI 424-01⟩ devoted to the synthesis and reactivity of free radicals.

4.24.4.1 Radical Centres Flanked by Nitrogen and Oxygen

One of the most common methods for the formation of such systems is found in photochemical processes. A typical illustration involves photochemical hydrogen abstraction within a phthalimide system. This provides a route to biradicals, one component of which is stabilised by a nitrogen and an oxygen atom. An example is that involving the cyclisation of the phthalimide (**33**) ⟨72TL4517⟩. This process is brought about by irradiation of the phthalimide in solution and involves the excited triplet state. The resultant biradical (**34**) cyclises in good yield to the product (**35**). Since the discovery of this reaction, many examples of its synthetic utility have been published and some of these have been reviewed by Kanaoka ⟨78ACR407, 82S1078⟩.

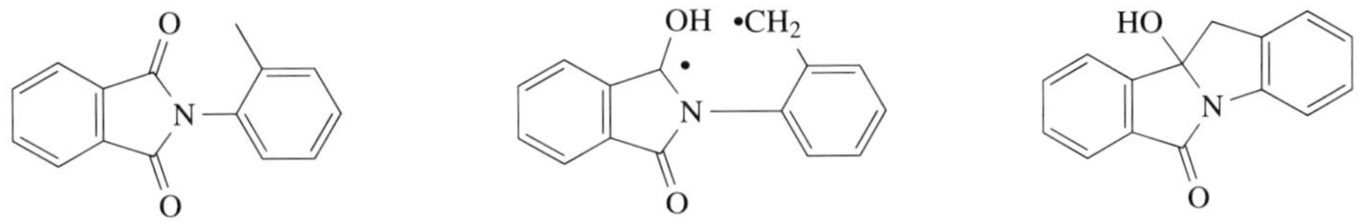

(**33**)   (**34**)   (**35**)

4.24.4.2 Radical Centres Flanked by Nitrogen and Sulfur

Radicals flanked by nitrogen and sulfur are involved in several transformations. A common method for their formation is by the addition of another radical to the sulfur atom of a thioamide. Thus, the addition of *t*-butylthiyl radicals, formed by the irradiation of 2-methylpropane-2-thiol using visible light, to the thioamide (**36**) results in the formation of the new radical (**37**) ⟨91TL1035⟩. The radical so produced has a transient existence and fragments to yield (**38**) and (**39**). A similar approach has been described by Barton *et al.* ⟨83CC939⟩ in the highly efficient, radical-chain decarboxylation of carboxylic acids. Scheme 1 shows that addition of Bu^{t}S• radicals to the ester derivative brings about decarboxylation and the formation of derivative (**40**) in high yield. A similar system has been used ⟨83TL4979⟩ as an efficient method for the formation of bromo- or chloroalkanes from carboxylic acids. This reaction path involves the trichloromethyl radical as the chain carrier, as illustrated in Scheme 2. The yields are again high, affording the chloroalkane in 70% yield and the bromoalkane in 95% yield.

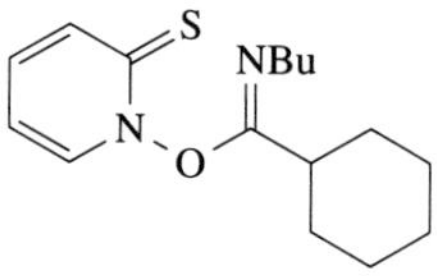

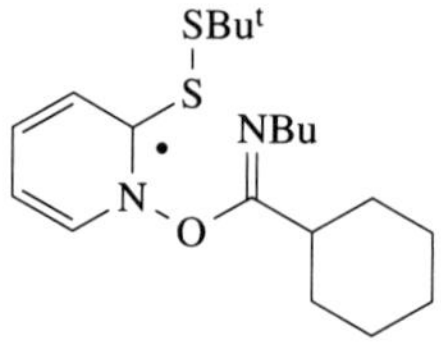

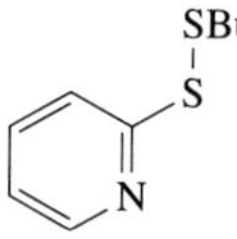

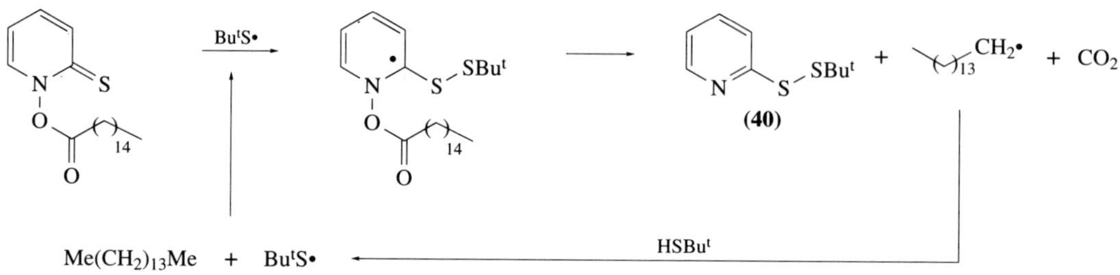

Scheme 1

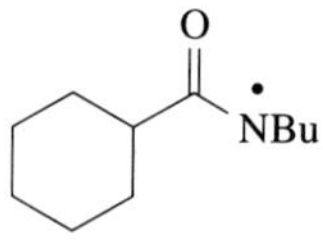

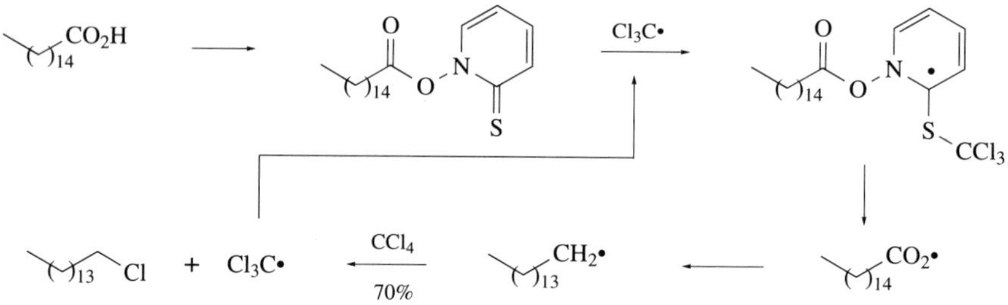

Scheme 2

4.24.4.3 Radical Centres Flanked by Two Oxygens

4.24.4.3.1 Intermolecular hydrogen abstraction

Interest in radicals of this type, represented by (**41**), is fairly widespread. The formation of these radicals is easily brought about in low concentrations—usually in an ESR spectrometer—by the reaction of *t*-butoxy radicals (formed by the irradiation of di-*t*-butyl peroxide) in Freon 13 with the appropriate acetal ⟨90JA4284⟩. Many reports of such systems describe evidence for the existence of the radicals (**42**) ⟨83JCS(P2)1071⟩, (**43**) ⟨83AG(E)500⟩ and (**44**) ⟨80JCS(P2)883, 81JCS(P2)143, 86MI 424-01⟩. Radicals of this type (e.g., (**41**), where R = H) can be trapped by methanal to yield hydroxymethyl derivatives (e.g., (**45**)) ⟨87JOC3243⟩.

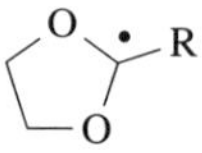

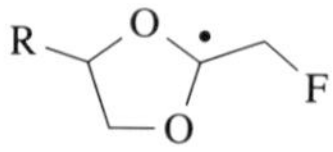

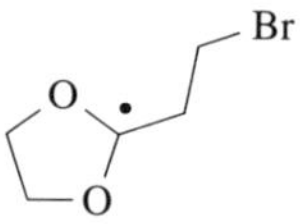

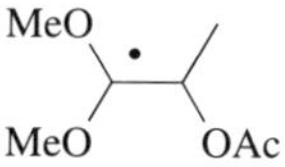

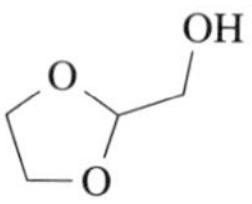

4.24.4.3.2 Intramolecular hydrogen abstraction

Radicals stabilised by two oxygens, as part of a biradical system, can also be produced photochemically ⟨B-91MI 424-01, B-92MI 424-02⟩ by irradiation of suitable derivatives. A review of such hydrogen abstraction reactions has been published ⟨91MI 424-02⟩. The majority of the reactions are Norrish type II processes involving 1,5-hydrogen transfer, that is, abstraction of a γ-hydrogen, yielding a 1,4-biradical. Related reactions are those involving 1,6- or higher hydrogen transfers by abstraction of a δ or ε hydrogen. Within the biradicals produced from these processes, cyclisations are the usual outcome. γ-Hydrogen abstraction leads to a biradical that can undergo elimination to bring about the formation of a carbonyl group, as in the example with a sugar molecule (Scheme 3) ⟨78NJC79⟩. Selective δ-hydrogen abstraction is observed on irradiation of the acetal (**46**). No γ-hydrogen abstraction from the methyl groups is observed. This reaction leads to the heterocycle (**47**). The reaction has been developed as a simple approach to the synthesis of a pheromone (Scheme 4) ⟨81T1875⟩. Cyclisation is also observed in the carbohydrate derivatives (**48**) and (**49**) ⟨81CAR(C)1, 81T2515⟩, and in the acetophenone derivative (**50**) ⟨90TL1819⟩. Yields from reactions of this sort can often be high.

Hydrogen abstraction from acetals by an enone (**51**) has also been reported. Here the hydrogen abstraction occurs from the ε site. Cyclisation affords the product (**52**) ⟨72JCS(P1)1103⟩.

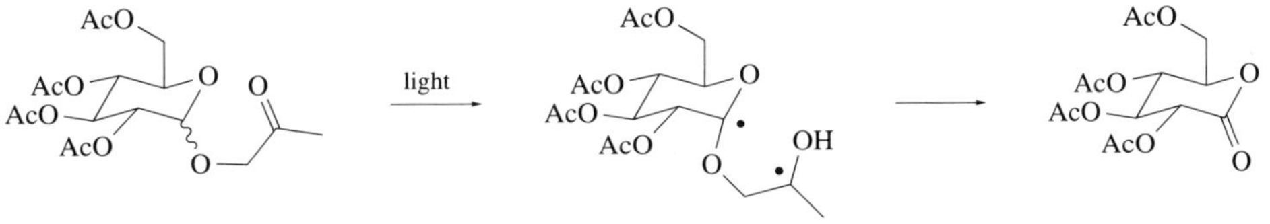

Scheme 3

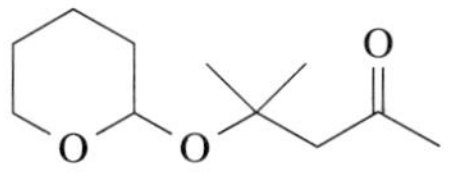

(**46**) 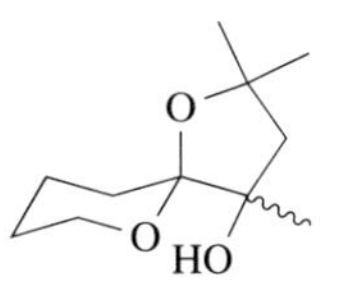   (**47**)

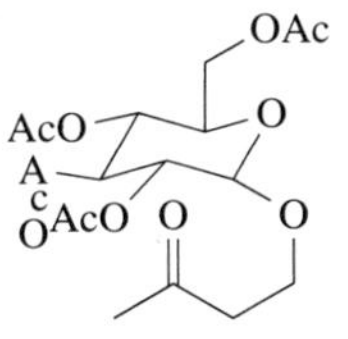

(**48**) 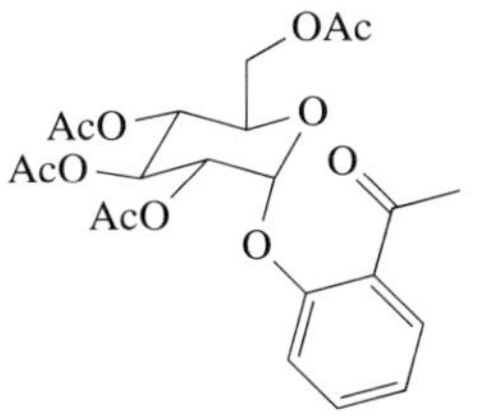   (**49**) 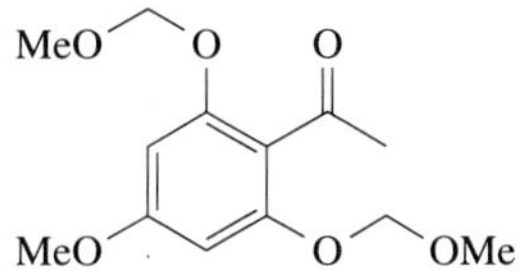   (**50**)

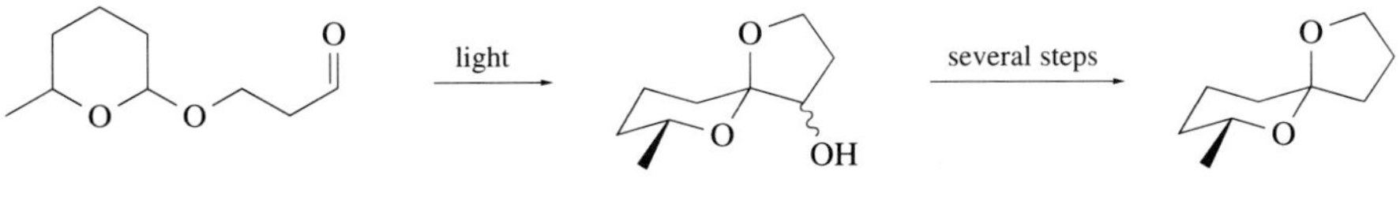

Scheme 4

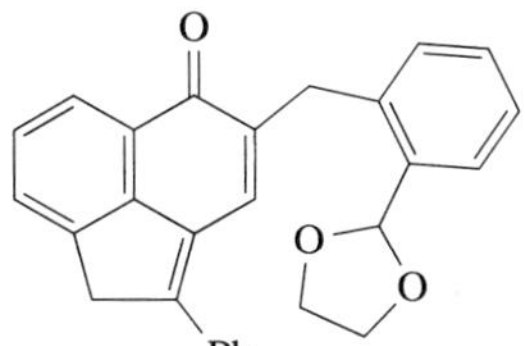

(**51**) 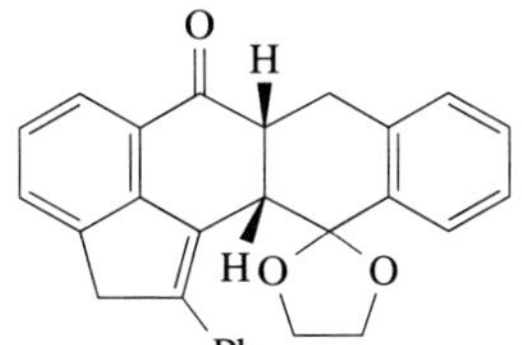   (**52**)

4.24.4.4 Radical Centres Flanked by Two Halogens

The formation of a radical centre flanked by two halogens is a common process. The resultant radical is usually short-lived and is part of a radical chain. The formation of these species can be brought about by the addition of almost any other radical to a halogenated alkene; a detailed summary was made by Sosnovsky ⟨B-64MI 424-01⟩.

Dihalo functionalised radicals can also be formed in a variety of other fashions, for example by the Norrish type I fission of suitably substituted acetone derivatives. Thus, the irradiation of decafluoropentan-3-one provides a route to the pentafluoroethyl radical ⟨88MI 424-01⟩.

4.24.4.5 Radical Centres Flanked by Two Sulfurs

Radicals are readily formed from a variety of substrates by hydrogen abstraction with other radicals. Thus, the dithioacetals (**53**) and $RCH(SEt)_2$ (where R = H, Me or EtS) yielded the corresponding radical centres flanked by two sulfur atoms ⟨83ZOB416⟩.

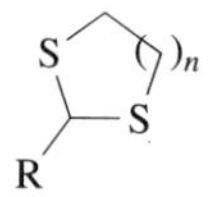

(**53**)
R = H or Me
n = 1, 2 or 3

4.24.5 CARBENES

There are many methods by which carbenes flanked by two heteroatoms can be prepared ⟨B-69MI 424-01, B-71MI 424-01, B-73MI 424-02, B-92MI 424-01⟩. Much of the reactivity of such systems can be modified and controlled by the presence of phase-transfer reagents ⟨B-84MI 424-02⟩. Readers are directed to several that affect the reactivity of dichlorocarbene ⟨85CB2137, 87CC31⟩, dibromocarbene ⟨84JCR(S)396, 87TL5489⟩, difluorocarbene ⟨87CC469⟩ and bromochlorocarbene ⟨89A187⟩. Interestingly, it has been argued that free dibromo- and dichlorocarbenes are involved in phase-transfer environments ⟨93CC1241⟩. A selection of some novel, interesting species and new methods for synthesis are described in the following sections.

4.24.5.1 Carbenes Flanked by Two Nitrogens

A successful route to the formation of carbenes flanked by two nitrogens involves flash vacuum thermolysis at 700 °C. An example of this approach is illustrated in Equation (1) ⟨88JOC1806⟩.

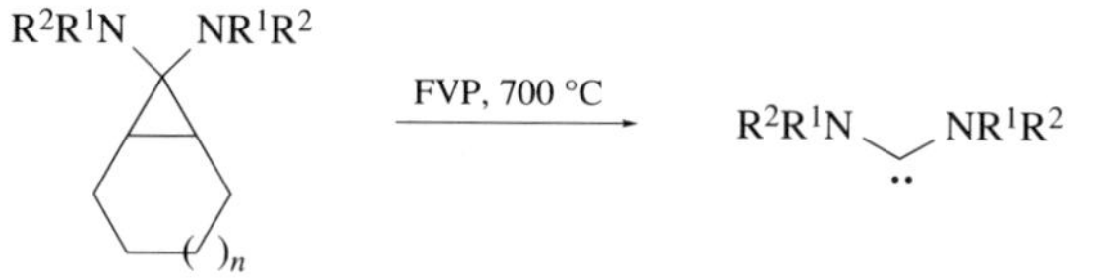

$R^2R^1N \quad NR^1R^2$ $\xrightarrow{\text{FVP, 700 °C}}$ $R^2R^1N \quad NR^1R^2$ (1)

$R^1–R^2$	n
$(CH_2)_2O(CH_2)_2$	1
$(CH_2)_2O(CH_2)_2$	6
$(CH_2)_4$	1
$(CH_2)_5$	1

4.24.5.2 Carbenes Flanked by Two Oxygens

An example of a carbene flanked by two oxygens has been reported following the thermolysis of hexamethoxycyclopropane. When this was carried out at 200 °C, dimethoxycarbene was extruded and tetramethoxyethene was formed as a by-product ⟨85TL1931⟩. Higher yields of dimethoxycarbene were obtained from thermolysis of the norbornadiene derivative (**54**). Another example of such a carbene (**55**) is reported to be produced on reaction of the *ortho*-esters (**56**) with trifluoro- or trichloro-acetic acid ⟨88KGS852⟩.

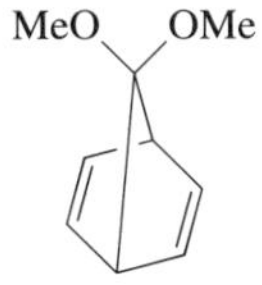

(**54**) 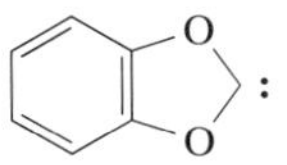   (**55**) 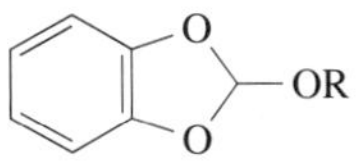   (**56**)
R = Me or Et

4.24.5.3 Carbenes Flanked by Two Sulfurs or One Sulfur and One Oxygen

4.24.5.3.1 One sulfur and one oxygen

α-Elimination is a common path to carbenes and has been used as a route to the carbenes (**57**) using Bu^{n}Li in THF at −78 °C as the base to eliminate HCl or HBr from the substrate (**58**) ⟨89T6667⟩.

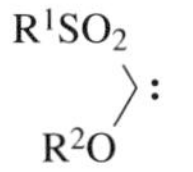

R^1SO_2
R^2O
(**57**)

R^1SO_2
R^2O —X
(**58**)

R^1 = *p*-MeC$_6$H$_4$
R^2 = Me, X = Cl or Br
R^2 = Et, X = Br
R^2 = Me, X = F

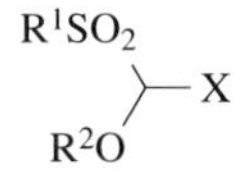

R^1 = *p*-MeC$_6$H$_4$
R^2 = Me, X = Cl or Br
R^2 = Et, X = Br
R^2 = Me, X = F

4.24.5.3.2 Two sulfur atoms

A variety of thermal methods are available for the formation of carbenes flanked by two sulfurs and a review has indicated the use of such species ⟨90S431⟩. One path involves the thermolysis of iodonium salts such as (**59**), where R = Ph ⟨86JOC3453, 88JCS(P1)2839⟩, or (**59**) where R = C$_4$H$_9$ ⟨91CC470⟩. Thermal methods have also been reported for the generation of the carbenes (**60**) and (**61**) from the cyclic compounds (**62**) and (**63**), respectively ⟨83JCS(P2)1687⟩. Photolysis has also been employed, as in the irradiation of the thermally stable iodonium salt (**59**), where R = CF$_3$, C$_4$F$_9$ or R—R = (CF$_2$)$_3$. The use of a low-pressure mercury arc lamp is required in this instance to yield the corresponding carbenes ⟨90CC1459⟩.

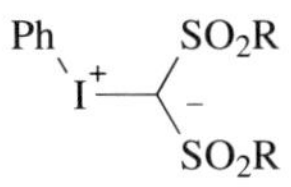

(**59**)   (**60**)   (**61**)   (**62**)   (**63**)

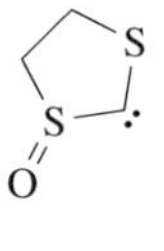

R = Ph
R = C$_4$H$_9$
R = CF$_3$, C$_4$F$_9$ or R–R = (CF$_2$)$_3$

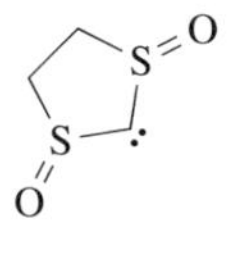

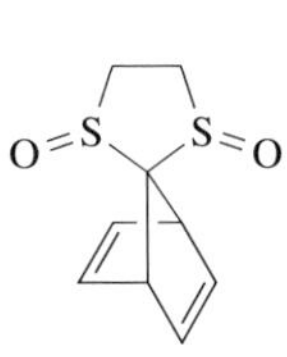

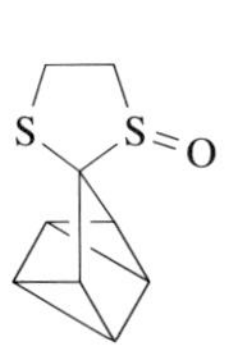

4.24.5.4 Miscellaneous Carbenes

4.24.5.4.1 Stabilisation by phosphorus and silicon

Photochemically-induced loss of nitrogen from diazo compounds is a common route to the synthesis of carbenes. This path has been used with great success in the synthetic approaches to carbene systems where the carbene atom is flanked by a silicon and a phosphorus. Typically, the irradiation of diazo compounds such as (64) results in the production of the corresponding carbene. However, there has been some doubt as to the carbene character of the product and some suggestion that a triply bonded P—Si system had been produced ⟨89AG(E)621⟩. However, the carbene nature of the intermediate produced on photochemical loss of nitrogen was demonstrated by trapping with *t*-butyl isocyanide, whereupon the adduct (65) was obtained ⟨90AG(E)1429⟩. In the absence of such a trapping agent, (2 + 2) dimerisation results. Thermal routes to carbenes of this type have also been described and the stable carbenes obtained can be characterised by NMR spectroscopy. A commentary on such carbenes has been reported ⟨B-95MI 424-01⟩.

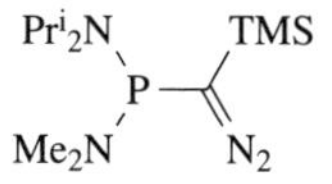

(64) 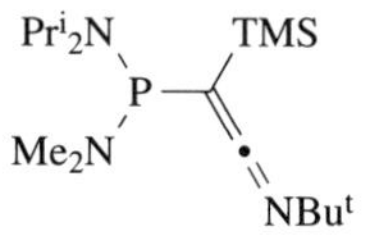   (65)

4.24.5.4.2 Stabilisation by two boron atoms

In the mid-1980s it was discovered that heating the boracyclopropane (66) in the presence of di(trimethylsilyl)ethyne induced thermal conversion into the carbene (67) prior to reaction with the alkyne ⟨84AG(E)826⟩. Others have used this carbene in other trapping reactions ⟨87AG(E)546⟩.

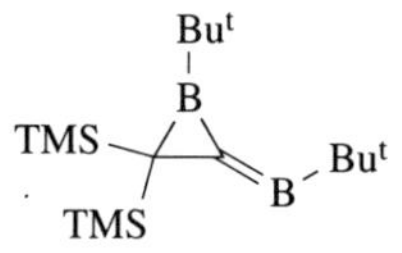

(66) 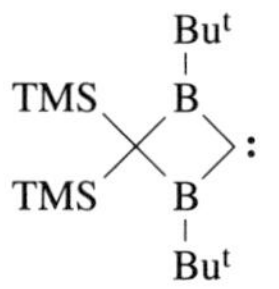   (67)

4.24.5.5 Stabilisation by Halogens

Carbene centres flanked by two halogens are probably the most widely studied. Indeed, dichlorocarbene was the first divalent carbon species to be identified following the reaction of base with chloroform. Since those early days, a plethora of methods to synthesize these intermediates and to control their reactivity have become available to the synthetic chemist. Reviews of the general methods of synthesis are available ⟨89UK1122⟩.

Numerous other methods have been developed over the years. Some of these are outlined in the following section.

4.24.5.5.1 Miscellaneous methods for the generation of dihalocarbenes

(i) Reduced titanium

Reduced titanium, generated by the reaction of titanium tetrachloride with LAH in THF at 0 °C, is an efficient reagent for the formation of chlorofluorocarbene from trichlorofluoromethane. Yields of cyclopropanes from the trapping of the carbene by alkenes can be high ⟨88TL6749, 90JOC589⟩.

(ii) Carbenoids

The Simmons–Smith process for the generation of metallated carbenes (carbenoids) has proved to be an extremely versatile *modus operandi* ⟨58JA5323⟩. More modern variants have also proved useful, such as the route described for the generation of difluorocarbene. This involves the reaction of dibromodifluoromethane with zinc and iodine under an atmosphere of nitrogen. The process is carried out in THF at room temperature and yields of cyclopropanes from alkenes can be as high as 96% ⟨90JOC5420⟩.

(iii) Nucleophilic attack

Difluorocarbene can be produced from trifluoromethylphenylmercury by reaction with a suitable nucleophile, for example, iodide. The efficiency of the reaction is not affected by the presence of phase-transfer catalysts ⟨84JFC(24)503⟩.

(iv) Reactions with hydroxide and alkoxides

Dichlorocarbene is formed reportedly with good efficiency from the reaction of carbon tetrachloride in a mixture of potassium hydroxide, *t*-butyl alcohol and dimethyl sulfone ⟨84JCS(P1)1561⟩. Difluorocarbene can be produced by the reaction of a variety of alkoxides in alcoholic media from CHF_2SO_2F. Decomposition to the carbene is followed by trapping to yield trifluoromethyl alkyl ethers ⟨84MI 424-03⟩.

(v) Irradiative methods

Photochemical methods for the formation of dichlorocarbene have involved the irradiation at ambient temperatures of strained cyclic compounds such as the cyclopropanes (**68**), using a 450 W mercury arc lamp with a pyrex filter ⟨86TL5907⟩. These compounds can be used as the precursors for both dichloro- and dibromocarbene. The yields of cyclopropanes by this path are reported to be high. Dibromocarbene can also be extruded on photolysis of the tricyclic compound (**69**) ⟨89JA8491⟩. Photolysis of ketenes is also effective as a route to dichlorocarbene, as demonstrated by the reaction of dichloroketene ⟨87DOK(296)403⟩.

Multiphoton irradiation at 9.47 μm has been reported as a method for the formation of chlorofluorocarbene from dichlorofluoromethane ⟨88CPL439⟩.

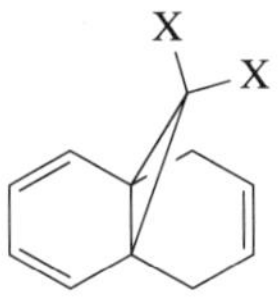

(**68**)
X = Br or Cl

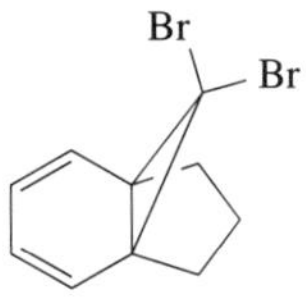

(**69**)

(vi) Electrochemical methods

Electrochemical methods for the production of a variety of carbenes including the dihalo species have been reviewed ⟨89UK1105⟩. The method has also been described for the generation of dinitrocarbene by electrochemical reduction of the ylide (**70**) ⟨89IZV356⟩.

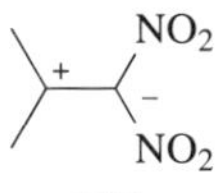

(**70**)

References

EXPLANATION OF THE REFERENCE SYSTEM

Throughout this work, references are designated by a number–lettering coding of which the first two numbers denote tens and units of the year of publication, the next one to three letters denote the journal, and the final numbers denote the page. This code appears in the text each time a reference is quoted; the advantages of this system are outlined in the Introduction. The system has been used previously in "Comprehensive Heterocyclic Chemistry," eds A. R. Katritzky and C. W. Rees, Pergamon, Oxford, 1984 and is based on that used in the following two monographs: (a) A. R. Katritzky and J. M. Lagowski, "Chemistry of the Heterocyclic N-Oxides," Academic Press, New York, 1971; (b) J. Elguero, C. Marzin, A. R. Katritzky and P. Linda, "The Tautomerism of Heterocycles," in "Advances in Heterocyclic Chemistry," Supplement 1, Academic Press, New York, 1976.

The following additional notes apply:

1. A list of journal codes in alphabetical order, together with the journals to which they refer, is given immediately following these notes. Journal names are abbreviated throughout using the CASSI (Chemical Abstracts Service Source Index) system.

2. Each volume contains all the references cited *in that volume*; no separate lists are given for individual chapters.

3. The list of references is arranged in order of (a) year, (b) journal in alphabetical order of journal code, (c) part letter or number if relevant, (d) volume number if relevant, (e) page number.

4. In the reference list the code is followed by (a) the complete literature citation in the conventional manner and (b) the number(s) of the page(s) on which the reference appears, whether in the text or in tables, schemes, etc.

5. For nontwentieth-century references the year is given in full in the code.

6. For journals which are published in separate parts, the part letter or number is given (when necessary) in parentheses immediately after the journal code letters.

7. Journal volume numbers are *not* included in the code numbers unless more than one volume was published in the year in question, in which case the volume number is included in parentheses immediately after the journal code letters.

8. Patents are assigned appropriate three-letter codes.

9. Frequently cited books are assigned codes.

10. Less common journals and books are given the code "MI" for miscellaneous with the whole code for books prefixed by the letter "B-".

11. Where journals have changed names, the same code is used throughout, e.g. CB refers to both *Chem. Ber.* and to *Ber. Dtsch. Chem. Ges.*

Journal Codes

AAC	Antimicrob. Agents Chemother.
ABC	Agric. Biol. Chem.
AC	Appl. Catal.
AC(P)	Ann. Chim. (Paris)
AC(R)	Ann. Chim. (Rome)
ACH	Acta Chim. Acad. Sci. Hung.

ACR	Acc. Chem. Res.
ACS	Acta Chem. Scand.
ACS(A)	Acta Chem. Scand., Ser. A
ACS(B)	Acta Chem. Scand., Ser. B
AF	Arzneim.-Forsch.
AFC	Adv. Fluorine Chem.
AG	Angew. Chem.
AG(E)	Angew. Chem., Int. Ed. Engl.
AHC	Adv. Heterocycl. Chem.
AHCS	Adv. Heterocycl. Chem. Supplement
AI	Anal. Instrum.
AJC	Aust. J. Chem.
AK	Ark. Kemi
AKZ	Arm. Khim. Zh.
AM	Adv. Mater. (Weinheim, Ger.)
AMLS	Adv. Mol. Spectrosc.
AMS	Adv. Mass. Spectrom.
ANC	Anal. Chem.
ANL	Acad. Naz. Lncei
ANY	Ann. N. Y. Acad. Sci.
AOC	Adv. Organomet. Chem.
AP	Arch. Pharm. (Weinheim, Ger.)
APO	Adv. Phys. Org. Chem.
AQ	An. Quim.
AR	Annu. Rep. Prog. Chem.
AR(A)	Annu. Rep. Prog. Chem., Sect. A
AR(B)	Annu. Rep. Prog. Chem., Sect. B
ARP	Annu. Rev. Phys. Chem.
ASI	Acta Chim. Sin. Engl. Ed.
ASIN	Acta Chim. Sin.
AX	Acta Crystallogr.
AX(A)	Acta Crystallogr., Part A
AX(B)	Acta Crystallogr., Part B
B	Biochemistry
BAP	Bull. Acad. Pol. Sci., Ser. Sci. Chim.
BAU	Bull. Acad. Sci. USSR, Div. Chim. Sci.
BBA	Biochim. Biophys. Acta
BBR	Biochim. Biophys. Res. Commun.
BCJ	Bull. Chem. Soc. Jpn.
BEP	Belg. Pat.
BJ	Biochem. J.
BJP	Br. J. Pharmacol.
BMC	Bioorg. Med. Chem. Lett.
BP	Biochem. Biopharmacol.
BPJ	Br. Polym. J.
BRP	Br. Pat.
BSB	Bull. Soc. Chim. Belg.
BSF	Bull. Soc. Chim. Fr.
BSF(2)	Bull. Soc. Chim. Fr., Part 2
C	Chimia
CA	Chem. Abstr.
CAN	Cancer
CAR	Carbohydr. Res.
CAT	Chim. Acta Turc.
CB	Chem. Ber.

CBR	Chem. Br.
CC	J. Chem. Soc., Chem. Commun.
CCA	Croat. Chem. Acta
CCC	Collect. Czech. Chem. Commun.
CCR	Coord. Chem. Rev.
CE	Chem. Express
CEN	Chem. Eng. News
CHE	Chem. Heterocycl. Compd. (Engl. Transl.)
CHEC	Comp. Heterocycl. Chem.
CI(L)	Chem. Ind. (London)
CI(M)	Chem. Ind. (Milan)
CJC	Can. J. Chem.
CJS	Can. J. Spectrosc.
CL	Chem. Lett.
CLY	Chem. Listy
CM	Chem. Mater.
CMC	Comp. Med. Chem.
COC	Comp. Org. Chem.
COMC-I	Comp. Organomet. Chem., 1st edn.
COS	Comp. Org. Synth.
CP	Can. Pat.
CPB	Chem. Pharm. Bull.
CPH	Chem. Phys.
CPL	Chem. Phys. Lett.
CR	C. R. Hebd. Seances Acad. Sci.
CR(A)	C. R. Hebd. Seances Acad. Sci., Ser. A
CR(B)	C. R. Hebd. Seances Acad. Sci., Ser. B
CR(C)	C. R. Hebd. Seances Acad. Sci., Ser. C
CRAC	Crit. Rev. Anal. Chem.
CRV	Chem. Rev.
CS	Chem. Scr.
CSC	Cryst. Struct. Commun.
CSR	Chem. Soc. Rev.
CT	Chem. Tech.
CZ	Chem.-Ztg.
CZP	Czech. Pat.
DIS	Diss. Abstr.
DIS(B)	Diss. Abstr. Int. B.
DOK	Dokl. Akad. Nauk SSSR
DP	Dyes Pigm.
E	Experientia
EC	Educ. Chem.
EF	Energy Fuels
EGP	Ger. (East) Pat.
EJM	Eur. J. Med. Chem.
EUP	Eur. Pat.
FCF	Forschr. Chem. Forsch.
FCR	Fluorine Chem. Rev.
FES	Farmaco Ed. Sci.
FOR	Forschr. Chem. Org. Naturst.
FRP	Fr. Pat.
G	Gazz. Chim. Ital.
GAK	Gummi Asbest Kunstst.
GEP	Ger. Pat.
GEP(O)	Ger. Pat. Offen.

GSM	Gen. Synth. Methods
H	Heterocycles
HAC	Heteroatom Chem.
HC	Chem. Heterocycl. Compd.
HCA	Helv. Chim. Acta
HOU	Methoden Org. Chem. (Houben-Weyl)
HP	Hydrocarbon Process
IC	Inorg. Chem.
ICA	Inorg. Chim. Acta
IEC	Ind. Eng. Chem. Res.
IJ	Isr. J. Chem.
IJC	Indian J. Chem.
IJC(A)	Indian J. Chem., Sect. A
IJC(B)	Indian J. Chem., Sect. B
IJM	Int. J. Mass Spectrom. Ion Phys.
IJQ	Int. J. Quantum Chem.
IJS	Int. J. Sulfur Chem.
IJS(A)	Int. J. Sulfur Chem., Part A
IJS(B)	Int. J. Sulfur Chem., Part B
IS	Inorg. Synth
IZV	Izv. Akad. Nauk SSSR Ser. Khim.
JA	J. Am. Chem. Soc.
JAN	J. Antibiot.
JAP	Jpn. Pat.
JAP(K)	Jpn. Kokai
JBC	J. Biol. Chem.
JC	J. Chromatogr.
JCC	J. Coord. Chem.
JCE	J. Chem. Ed.
JCED	J. Chem. Eng. Data
JCI	J. Chem. Inf. Comput. Sci.
JCP	J. Chem. Phys.
JCPB	J. Chim. Phys. Physico-Chim. Biol.
JCR(M)	J. Chem. Res. (M)
JCR(S)	J. Chem. Res. (S)
JCS	J. Chem. Soc.
JCS(A)	J. Chem. Soc. (A)
JCS(B)	J. Chem. Soc. (B)
JCS(C)	J. Chem. Soc. (C)
JCS(D)	J. Chem. Soc., Dalton Trans.
JCS(F1)	J. Chem. Soc., Faraday Trans. 1
JCS(F2)	J. Chem. Soc., Faraday Trans. 2
JCS(P1)	J. Chem. Soc., Perkin Trans. 1
JCS(P2)	J. Chem. Soc., Perkin Trans. 2
JCS(S2)	J. Chem. Soc. (Suppl. 2)
JEC	J. Electroanal. Chem. Interfacial Electrochem.
JEM	J. Energy Mater.
JES	J. Electron. Spectrosc.
JFA	J. Sci. Food. Agri.
JFC	J. Fluorine Chem.
JGU	J. Gen. Chem. USSR (Engl. Transl.)
JHC	J. Heterocycl. Chem.
JIC	J. Indian Chem. Soc.
JINC	J. Inorg. Nucl. Chem.
JLC	J. Liq. Chromatogr.

JMAS	J. Mat. Sci.
JMC	J. Med. Chem.
JMOC	J. Mol. Catal.
JMR	J. Magn. Reson.
JMS	J. Mol. Sci.
JOC	J. Org. Chem.
JOM	J. Organomet. Chem.
JOU	J. Org. Chem. USSR (Engl. Transl.)
JPC	J. Phys. Chem.
JPJ	J. Pharm. Soc. Jpn.
JPO	J. Phys. Org. Chem.
JPP	J. Pharm. Pharmacol.
JPR	J. Prakt. Chem.
JPS	J. Pharm. Sci.
JPS(A)	J. Polym. Sci., Polym. Chem., Part A
JPU	J. Phys. Chem. USSR (Engl. Transl.)
JSC	J. Serbochem. Soc.
JSP	J. Mol. Spectrosc.
JST	J. Mol. Struct.
K	Kristallografiya
KFZ	Khim. Farm. Zh.
KGS	Khim. Geterotsikl. Soedin.
KO	Kirk-Othmer Encyc.
KPS	Khim. Prir. Soedin.
L	Langmuir
LA	Liebigs Ann. Chem.
LC	Liq. Cryst.
LS	Life Sci.
M	Monatsh. Chem.
MAC	Macromol. Chem.
MC	Mendeleev Chem. J. (Engl. Transl.)
MCLC	Mol. Cryst. Liq. Cryst.
MI	Miscellaneous [book/journal]
MIP	Miscellaneous Pat.
MM	Macromolecules
MP	Mol. Phys.
MRC	Magn. Reson. Chem.
N	Naturwissenschaften
NAT	Nat.
NEP	Neth. Pat.
NJC	Nouv. J. Chim.
NKK	Nippon Kagaku Kaishi (J. Chem. Soc. Jpn.)
NKZ	Nippon Kagaku Zasshi
NZJ	N. Z. J. Sci. Technol.
OCS	Organomet. Synth.
OM	Organometallics
OMR	Org. Magn. Reson.
OMS	Org. Mass Spectrom.
OPP	Org. Prep. Proced. Int.
OR	Org. React.
OS	Org. Synth.
OSC	Org. Synth., Coll. Vol.
P	Phytochemistry
PA	Polym. Age
PAC	Pure Appl. Chem.

PAS	Pol. Acad. Sci.
PB	Polym. Bull.
PC	Personal Communication
PCS	Proc. Chem. Soc.
PHA	Pharmazi
PHC	Prog. Heterocycl. Chem.
PIA	Proc. Indian Acad. Sci.
PIA(A)	Proc. Indian Acad. Sci., Sect. A
PJC	Pol. J. Chem.
PJS	Pak. J. Sci. Ind. Res.
PMH	Phys. Methods Heterocycl. Chem.
PNA	Proc. Natl. Acad. Sci. USA
POL	Polyhedron
PP	Polym. Prepr.
PRS	Proceed. Roy. Soc.
PS	Phosphorus Sulfur
QR	Q. Rev., Chem. Soc.
QRS	Quart. Rep. Sulfur. Chem.
QSAR	Quant. Struct. Act. Relat. Pharmacol. Chem. Biol.
RC	Rubber Chem. Technol.
RCM	Rapid Commun. Mass Spectrom.
RCP	Rec. Chem. Prog.
RCR	Russ. Chem. Rev. (Engl. Transl.)
RHA	Rev. Heteroatom Chem.
RJ	Rubber J.
RP	Rev. Polarogr.
RRC	Rev. Roum. Chim.
RS	Ric. Sci.
RTC	Recl. Trav. Chim. Pays-Bas
RZC	Rocz. Chem.
S	Synthesis
SA	Spectrochim. Acta
SA(A)	Spectrochim. Acta, Part A
SAP	S. Afr. Pat.
SC	Synth. Commun.
SCI	Science
SL	Synlett
SM	Synth. Met.
SR	Sulfur Reports
SRI	Synth. React. Inorg. Metal-Org. Chem.
SS	Sch. Sci. Rev.
SST	Org. Compd. Sulphur, Selenium, Tellurium [R. Soc. Chem. series]
SUL	Sulfur Letters
SZP	Swiss Pat.
T	Tetrahedron
T(S)	Tetrahedron, Suppl.
TA	Tetrahedron Asymmetry
TAL	Talanta
TCA	Theor. Chim. Acta
TCC	Top. Curr. Chem.
TCM	Tetrahedron, Comp. Method
TFS	Trans. Faraday Soc.
TH	Thesis
TL	Tetrahedron Lett.
TS	Top. Stereochem.

UK	Usp. Khim.	
UKZ	Ukr. Khim. Zh. (Russ. Ed.)	
UP	Unpublished Results	
URP	USSR Pat.	
USP	US Pat.	
WCH	Wiadom. Chem.	
YGK	Yuki Gosei Kagaku Kyokaishi	
YZ	Yakugaku Zasshi (J. Pharm. Soc. Jpn.)	
ZAAC	Z. Anorg. Allg. Chem.	
ZAK	Zh. Anal. Khim.	
ZC	Z. Chem.	
ZN	Z. Naturforsch.	
ZN(A)	Z. Naturforsch., Teil A	
ZN(B)	Z. Naturforsch., Teil B	
ZOB	Zh. Obshch. Khim.	
ZOR	Zh. Org. Khim.	
ZPC	Hoppe–Seyler's Z. Physiol. Chem.	
ZPK	Zh. Prikl. Khim.	

VOLUME 4 REFERENCES

1837LA(21)130	M. A. Laurent; *Justus Liebigs Ann. Chem.*, 1837, **21**, 130.	437
1855MI 402-01	M. A. Riche; *Ann. Chim. Phys.*, iii, 1855, **43**, 283.	63
1875LA(175)88	V. Meyer; *Justus Liebigs Ann. Chem.*, 1875, **175**, 88.	439
1876LA(181)1	E. ter Mer; *Justus Liebigs Ann. Chem.*, 1876, **181**, 1.	432
1881JCS485	J. Sakurai; *J. Chem. Soc.*, 1881, **39**, 485.	696
1885CB883	E. Baumann; *Ber. Dtsch. Chem. Ges.*, 1885, **18**, 883.	244
1886CB845	A. Michael; *Ber. Dtsch. Chem. Ges.*, 1886, **19**, 845.	766
1887BSF23	A. Haller; *Bull. Chim. Soc. Fr.*, 1887, **2**, 23.	883
1894CB673	E. Fischer; *Ber. Dtsch. Chem. Ges.*, 1894, **27**, 673.	244
1895BSF157	L. Henry; *Bull. Soc. Chim. Fr.*, 1895, **13**, 157.	405
1895CB473	A. Pinner; *Ber. Dtsch. Chem. Ges.*, 1895, **28**, 473.	883
1899CB641	A. Hantzsch and G. Osswald; *Ber. Dtsch. Chem. Ges.*, 1899, **32**, 641.	880, 883
00LA353	E. Thiele and E. Winter; *Justus Liebigs Ann. Chem.*, 1900, 353.	728
02CB984	O. Dimroth; *Ber. Dtsch. Chem. Ges.*, 1902, **35**, 984.	304
03CB289	M. Slimmer; *Ber. Dtsch. Chem. Ges.*, 1903, **36**, 289.	761
03CB434	J. Meisenheimer; *Ber. Dtsch. Chem. Ges.*, 1903, **36**, 434.	430, 431, 432
05CB1057	M. Bazlen; *Ber. Dtsch. Chem. Ges.*, 1905, **38**, 1057.	234
05LA(343)207	A. Einhorn; *Justus Liebigs Ann. Chem.*, 1905, **343**, 207.	296
06CB2543	J. Meisenheimer; *Ber. Dtsch. Chem. Ges.*, 1906, **39**, 2543.	431, 432
06JPR494	G. Ponzio; *J. Prakt. Chem.*, 1906, **73**, 494.	429
07BSF394	M. Sommelet; *Bull. Soc. Chim. Fr.*, 1907, **1**, 394.	
07JCS816	W. H. Perkin, Jr. and J. L. Simonsen; *J. Chem. Soc.*, 1907, **91**, 816.	24, 29
08CB24	A. Einhorn and A. Hamburger; *Ber. Dtsch. Chem. Ges.*, 1908, **41**, 24.	297
08CB2190	H. Biltz and E. Kedesdy; *Ber. Dtsch. Chem. Ges.*, 1908, **33**, 2190.	797
08LA(361)113	A. Einhorn; *Justus Liebigs Ann. Chem.*, 1908, **361**, 113.	296

09CB1476	E. Fischer and K. Delbrück; *Ber. Dtsch. Chem. Ges.*, 1909, **42**, 1476.	223
09CB3966	H. Staudinger; *Ber. Dtsch. Chem. Ges.*, 1909, **42**, 3966.	18
09CB4634	M. Bazlen; *Ber. Dtsch. Chem. Ges.*, 1909, **42**, 4634.	234
10RTC296	A. P. N. Franchimont; *Recl. Trav. Chim. Pays-Bas*, 1910, **29**, 296.	442
12BSB323	J. de Lattre; *Bull. Soc. Chim. Belg.*, 1912, **26**, 323.	218
12LA(394)325	E. Fromm, H. Benzinger and F. Schäfer; *Justus Liebigs Ann Chem.*, 1912, **394**, 325.	772
14JA1899	L. Clarke and E. K. Bolton; *J. Am. Chem. Soc.*, 1914, **36**, 1899.	28
19CR(169)1143	V. Grignard, G. Rivat and E. Urbain; *C. R. Hebd. Seances Acad. Sci.*, 1919, **169**, 1143.	59
20JCS691	H. Crompton and P. L. Vanderstichele; *J. Chem. Soc.*, 1920, 691.	761
21JA341	A. Lowy and E. H. Balz; *J. Am. Chem. Soc.*, 1921, **43**, 341.	297
21JA651	H. E. French and R. Adams; *J. Am. Chem. Soc.*, 1921, **43**, 651.	53
21JA660	L. H. Ulich and R. Adams; *J. Am. Chem. Soc.*, 1921, **43**, 660.	53
21JCS1470	C. M. McLeod and G. M. Robinson; *J. Chem. Soc.*, 1921, **119**, 1470.	300, 315
22CB644	S. Goldschmidt and K. Renn; *Ber. Dtsch. Chem. Ges.*, 1922, **55**, 644.	995
22CB1014	E. Fromm and E. Siebert; *Ber. Dtsch. Chem. Ges.*, 1922, **55**, 1014.	772
22JCS2202	W. Davies and W. H. Perkin, Jr.; *J. Chem. Soc.*, 1922, **121**, 2202.	11
23JCS576	H. Compton and K. M. Carter; *J. Chem. Soc.*, 1923, **123**, 576.	39
23LA(431)270	H. Lindemann; *Justus Liebigs Ann. Chem.*, 1923, **431**, 270.	25
24CB795	B. Helferich and J. Hausen; *Ber. Dtsch. Chem. Ges.*, 1924, **57**, 795.	178
24LA(435)219	H. Lindemann and H. Forth; *Justus Liebigs Ann. Chem.*, 1924, **435**, 219.	18
25MI 403-01	R. Hunt and R. R. Renshaw; *J. Pharmacol. Exp. Ther.*, 1925, **25**, 315.	135
26CR(183)665	G. Emschwiller; *C. R. Hebd. Seances Acad. Sci.*, 1926, **183**, 665.	669
26JA520	E. H. Shaw, Jr. and E. E. Reid; *J. Am. Chem. Soc.*, 1926, **48**, 520.	287
26JA1093	R. C. Fuson; *J. Am. Chem. Soc.*, 1926, **48**, 1093.	20
26JA3144	E. E. Harris and G. B. Frankforter; *J. Am. Chem. Soc.*, 1926, **48**, 3144.	743
27CB291	A. E. Arbuzov and A. A. Dunin; *Ber. Dtsch. Chem. Ges.*, 1927, **60**, 291.	576
27JCS538	A. M. McMath and J. Read; *J. Chem. Soc.*, 1927, 538.	39
27JCS2850	F. D. Chattaway and R. Bennett; *J. Chem. Soc.*, 1927, 2850.	799
27LA(453)113	J. von Braun, F. Jostes and W. Münch; *Justus Liebigs Ann. Chem.*, 1927, **453**, 113.	37
28JCS2125	H. J. Backer and H. W. Mook; *J. Chem. Soc.*, 1928, 2125.	38
28JCS2756	F. D. Chattaway and F. G. Daldy; *J. Chem. Soc.*, 1928, 2756.	799
29CB409	J. von Braun and A. Heymans; *Ber. Dtsch. Chem. Ges.*, 1929, **62**, 409.	790
30CB1868	F. Straus, L. Kollek and W. Heyn; *Ber. Dtsch. Chem. Ges.*, 1930, **63**, 1868.	745
30JA651	L. C. Swallen and C. E. Boord; *J. Am. Chem. Soc.*, 1930, **52**, 651.	51
30JA695	C. A. Kraus and A. M. Neal; *J. Am. Chem. Soc.*, 1930, **52**, 695.	699
30JA2995	A. Hoffman; *J. Am. Chem. Soc.*, 1930, **52**, 2995.	113, 118, 121
30JA4573	H. Gilman and J. B. Dickey; *J. Am. Chem. Soc.*, 1930, **52**, 4573.	319
31CB2696	M. Gehrke and W. Kohler; *Ber. Dtsch. Chem. Ges.*, 1931, **64**, 2696.	223
31JA1594	H. W. Doughty and G. J. Derge; *J. Am. Chem. Soc.*, 1931, **53**, 1594.	14
31JCS1088	F. D. Chattaway and T. E. W. Browne; *J. Chem. Soc.*, 1931, 1088.	799
31JCS1913	J. R. Alexander and H. McCombie; *J. Chem. Soc.*, 1931, 1913.	77
31RTC753	T. De Crauw; *Recl. Trav. Pays-Bas*, 1931, **50**, 753.	18
32JA4172	T. D. Stewart and W. E. Bradley; *J. Am. Chem. Soc.*, 1932, **54**, 4172.	300
32LA(493)191	F. Straus and H. Heinze; *Justus Liebigs Ann. Chem.*, 1932, **493**, 191.	48
32LA(498)101	F. Straus and H.-J. Weber; *Justus Liebigs Ann. Chem.*, 1932, **498**, 101.	48, 49, 50
32OSC(1)357	W. W. Hartman and E. E. Dreger; *Org. Synth. Coll. Vol.*, 1932, **1**, 357.	23
32OSC(1)358	R. Adams and C. S. Marvel; *Org. Synth. Coll. Vol.*, 1932, **1**, 358.	29
32OSC(1)369	C. S. Marvel and P. K. Porter; *Org. Synth., Coll. Vol.*, 1932, **1**, 369.	47
33JA3741	M. T. Bogert and R. O. Roblin, Jr.; *J. Am. Chem. Soc.*, 1933, **55**, 3741.	181
33JA4588	G. Dougherty and W. H. Taylor; *J. Am. Chem. Soc.*, 1933, **55**, 4588.	317
33JCS496	D. E. Armstrong and D. H. Richardson; *J. Chem. Soc.*, 1933, 496.	18
34BSF990	E. Hug; *Bull. Soc. Chim. Fr.*, 1934, 990.	427
34JA712	M. S. Kharasch and C. W. Hannum; *J. Am. Chem. Soc.*, 1934, **56**, 712.	38, 39
34JA2730	G. B. Bachman and A. J. Hill; *J. Am. Chem. Soc.*, 1934, **56**, 2730.	18

35JA1088	G. B. Bachman; *J. Am. Chem. Soc.*, 1935, **57**, 1088.	23, 38
35JA2360	W. M. Lauer and C. M. Langkammerer; *J. Am. Chem. Soc.*, 1935, **57**, 2360.	86
35JA2364	K. Summerbell and L. N. Bauer; *J. Am. Chem. Soc.*, 1935, **57**, 2364.	47, 48, 49, 51
35JA2463	M. S. Kharasch, J. G. McNab and M. C. McNab; *J. Am. Chem. Soc.*, 1935, **57**, 2463.	24
35MI 402-01	A. G. Kostova; *Tr. Voronezh. Gos. Univ.*, 1935, **88**, 92 (*Chem. Abstr.*, 1938, **32**, 6618).	85
35USP2000252	W. Reppe and K. Baur; *US Pat.* 2 000 252 (1935) (*Chem. Abstr.*, 1935, **29**, 4029).	186, 187
36CB1610	H. Bohme; *Ber. Dtsch. Chem. Ges.*, 1936, **69**, 1610.	69
36CB2106	C. Mannich and H. Davidsen; *Ber. Dtsch. Chem. Ges.*, 1936, **69**, 2106.	404, 405
36JA80	D. B. Killian, G. F. Hennion and J. A. Nieuwland; *J. Am. Chem. Soc.*, 1936, **58**, 80.	195
36JA889	A. L. Henne and M. W. Renoll; *J. Am. Chem. Soc.*, 1936, **58**, 889.	3, 30, 33
36JA892	D. B. Killian, G. F. Hennion and J. A. Nieuwland; *J. Am. Chem. Soc.*, 1936, **58**, 892.	195
36JA1806	C. A. Young, R. R. Vogt and J. A. Nieuwland; *J. Am. Chem. Soc.*, 1936, **58**, 1806.	24
36JGU283	A. E. Arbusov and N. P. Kushkova; *J. Gen. Chem. USSR* (*Engl. Transl.*), 1936, **6**, 283 (*Chem. Abstr.*, 1936, **30**, 48 139).	133
37JA1400	A. L. Henne; *J. Am. Chem. Soc.*, 1937, **59**, 1400.	30
37JA1407	J. H. Simons and L. P. Block; *J. Am. Chem. Soc.*, 1937, **59**, 1407.	2
37JCS767	N. W. Cusa and H. McCombie; *J. Chem. Soc.*, 1937, 767.	770
37JGU2495	S. N. Ushakov and A. M. Itenberg; *J. Gen. Chem. USSR* (*Engl. Transl.*), 1937, **7**, 2495 (*Chem. Abstr.*, 1938, **32**, 20 838).	137
37USP2091373	J. Nelles, *US Pat.* 2 091 373 (1937) (*Chem. Abstr.*, 1937, **31**, 7444).	196
38CB1803	E. J. Salmi and I. Mitteil; *Ber. Dtsch. Chem. Ges.*, 1938, **71**, 1803.	177, 178
38JA864	A. L. Henne and H. M. Leicester; *J. Am. Chem. Soc.*, 1938, **60**, 864.	4
39JA938	A. L. Henne, M. W. Renoll and H. M. Leicester; *J. Am. Chem. Soc.*, 1939, **61**, 938.	18
39JA1460	R. O. Norris, R. R. Vogt and G. F. Hennion; *J. Am. Chem. Soc.*, 1939, **61**, 1460.	16
39JA2142	M. S. Kharasch and H. C. Brown; *J. Am. Chem. Soc.*, 1939, **61**, 2142.	11
39JOC234	J. J. Spurlock and H. R. Henze; *J. Org. Chem.*, 1939, **4**, 234.	48
40CRV297	M. M. Sprung; *Chem. Rev.*, 1940, **26**, 297.	294, 295
40CRV351	F. R. Mayo and C. Walling; *Chem. Rev.*, 1940, **27**, 351.	23
40JA1171	D. S. Young, N. Fukuhara and L. A. Bigelow; *J. Am. Chem. Soc.*, 1940, **62**, 1171.	3
40JA1281	H. M. Barnes, D. Kundiger and S. M. McElvain; *J. Am. Chem. Soc.*, 1940, **62**, 1281.	887, 889
40JA1849	T. L. Jacobs, R. Cramer and F. T. Weiss; *J. Am. Chem. Soc.*, 1940, **62**, 1849.	765
40JA2596	W. M. Ziegler and R. Connor; *J. Am. Chem. Soc.*, 1940, **62**, 2596.	83, 84
40JOC244	H. W. Post; *J. Org. Chem.*, 1940, **5**, 244.	191
B-40MI 402-01	C. M. Suter; "The Organic Chemistry of Sulfur," Wiley, New York, 1940; Reprinted: Inter-Science Foundation, Santa Monica, 1969, p. 678.	81
41CB145	G. Willfang; *Ber. Dtsch. Chem. Ges.*, 1941, **74**, 145.	181
41JA1537	M. S. Newman and C. D. McLeary; *J. Am. Chem. Soc.*, 1941, **63**, 1537.	211
41JA1571	T. B. Johnson and I. B. Douglass; *J. Am. Chem. Soc.*, 1941, **63**, 1571.	85
41JGU41	G. I. Braz and A. Ya. Yakubovich; *J. Gen. Chem. USSR* (*Engl. Transl.*), 1941, **11**, 41 (*Chem. Abstr.*, 1941, **35**, 54 591).	135
41JOC888	R. L. Shriner and A. H. Land; *J. Org. Chem.*, 1941, **6**, 888.	237
42CB13	B. Sjoberg; *Ber. Dtsch. Chem. Ges.*, 1942, **75B**, 13.	229
42JA2635	T. L. Jacobs and W. J. Whitcher; *J. Am. Chem. Soc.*, 1942, **64**, 2635.	761, 765
42OS(22)65	H. R. Snyder, R. G. Handrick and L. A. Brooks; *Org. Synth.*, 1942, **22**, 65.	407
43CB88	E. Ott and W. Bossaller; *Ber. Dtsch. Chem. Ges.*, 1943, **76**, 88.	797
43JA389	J. H. Simons and E. O. Ramler; *J. Am. Chem. Soc.*, 1943, **65**, 389.	18
43JA1477	N. K. Richtmyer, C. J. Carr and C. S. Hudson; *J. Am. Chem. Soc.*, 1943, **65**, 1477.	223
43OS(23)45	S. M. McElvain and D. Kundiger; *Org. Synth.*, 1943, **23**, 45.	826
43OSC(2)181	A. C. Cope, J. R. Clark and R. Connor; *Org. Synth., Coll. Vol.*, 1943, **2**, 181.	14
43OSC(2)312	M. W. Farlow; *Org. Synth., Coll. Vol.*, 1943, **2**, 312.	14
44JA222	W. I. Weaver, J. K. Simons and W. E. Baldwin; *J. Am. Chem. Soc.*, 1944, **66**, 222.	423
44JIC112	P. S. Varma, K. S. V. Raman and P. M. Nilkantiah; *J. Indian Chem. Soc.*, 1944, **21**, 112.	11
44OR(2)49	A. L. Henne; *Org. React.*, 1944, **2**, 49.	2, 3, 4
45JA202	S. M. McElvain and B. E. Tate; *J. Am. Chem. Soc.*, 1945, **67**, 202.	888, 889
45JA655	L. A. Walter, L. H. Goodson and R. J. Fosbinder; *J. Am. Chem. Soc.*, 1945, **67**, 655.	69
45JA657	L. A. Walter, L. H. Goodson and R. J. Fosbinder; *J. Am. Chem. Soc.*, 1945, **67**, 657.	69
45JA1012	S. A. Glickman and A. C. Cope; *J. Am. Chem. Soc.*, 1945, **67**, 1012.	883
45JA1017	G. A. Glickman and A. C. Cope; *J. Am. Chem. Soc.*, 1945, **67**, 1017.	880, 883
46JA496	A. L. Henne and T. P. Waalkes; *J. Am. Chem. Soc.*, 1946, **68**, 496.	30
46JA734	M. Senkus; *J. Am. Chem. Soc.*, 1946, **68**, 734.	212
46JA1922	R. E. Kent, C. L. Stevens and S. M. McElvain; *J. Am. Chem. Soc.*, 1946, **68**, 1922.	827
46JA2252	L. F. Fieser and W. von E. Doering; *J. Am. Chem. Soc.*, 1946, **68**, 2252.	429, 430, 431

46JOC55	P. H. Dirstine and F. W. Bergstrom; *J. Org. Chem.*, 1946, **11**, 55.	23
47G375	P. Galimberti; *Gazz. Chim. Ital.*, 1947, **77**, 375.	317
47IEC412	E. T. McBee and R. O. Bolt; *Ind. Eng. Chem. Res.*, 1947, **39**, 412.	826
47JA254	S. R. Buc; *J. Am. Chem. Soc.*, 1947, **69**, 254.	728
47JA279	A. L. Henne and R. P. Ruh; *J. Am. Chem. Soc.*, 1947, **69**, 279.	5, 32
47JA281	A. L. Henne and W. J. Zimmerschied; *J. Am. Chem. Soc.*, 1947, **69**, 281.	6
47JA500	W. E. Cass; *J. Am. Chem. Soc.*, 1947, **69**, 500.	207
47JA1976	F. C. Whitmore, L. H. Sommer and J. Gold; *J. Am. Chem. Soc.*, 1947, **69**, 1976.	137
47JA2112	G. M. Kosolapoff; *J. Am. Chem. Soc.*, 1947, **69**, 2112.	473
47JA2136	W. J. Burke; *J. Am. Chem. Soc.*, 1947, **69**, 2136.	416
47JA2233	N. C. Deno; *J. Am. Chem. Soc.*, 1947, **69**, 2233.	193
47JA2246	G. F. Woods and D. N. Kramer; *J. Am. Chem. Soc.*, 1947, **69**, 2246.	186
47JA2247	J. T. Goodwin Jr., W. E. Baldwin and R. R. McGregor; *J. Am. Chem. Soc.*, 1947, **69**, 2247.	609
47JA2358	D. T. Mowry, W. H. Yanko and E. L. Ringwald; *J. Am. Chem. Soc.*, 1947, **69**, 2358.	205, 213
47JCS1465	A. H. Ford-Moore and J. H. Williams; *J. Chem. Soc.*, 1947, 1465.	555
48AG267	A. M. Panquin; *Angew. Chem.*, 1948, **A60**, 267.	416
48BSF774	V. Chavane; *Bull. Soc. Chim. Fr.*, 1948, 774.	465
48CB422	W. Lorenz; *Chem. Ber.*, 1948, **81**, 422.	743
48CB477	H. Schmidt; *Chem. Ber.*, 1948, **81**, 477.	463, 470
48JA1244	L. R. Williams and A. Ravve; *J. Am. Chem. Soc.*, 1948, **70**, 1244.	88
48JA1283	G. M. Kosolapoff; *J. Am. Chem. Soc.*, 1948, **70**, 1283.	472
48JA1550	J. D. Park, D. K. Vail, K. R. Lea and J. R. Lacher; *J. Am. Chem. Soc.*, 1948, **70**, 1550.	32
48JA2115	R. E. Benson and T. L. Cairns; *J. Am. Chem. Soc.*, 1948, **70**, 2115.	325
48JA2435	W. A. Bonner and R. W. Drisko; *J. Am. Chem. Soc.*, 1948, **70**, 2435.	235
48JA2624	D. P. Langlois and H. Wolff; *J. Am. Chem. Soc.*, 1948, **70**, 2624.	211
48JA2805	W. J. Croxall, F. J. Glavis and H. T. Neher; *J. Am. Chem. Soc.*, 1948, **70**, 2805.	188, 208
48JA2813	E. H. Huntress and F. Sanchez-Nieva; *J. Am. Chem. Soc.*, 1948, **70**, 2813.	14
48JA2827	M. Sulzbacher, E. Bergmann and E. R. Pariser; *J. Am. Chem. Soc.*, 1948, **70**, 2827.	178
48JA3426	D. Davidson and S. A. Bernhard; *J. Am. Chem. Soc.*, 1948, **70**, 3426.	204
48JA3659	H. H. Richmond, G. S. Myers and G. F. Wright; *J. Am. Chem. Soc.*, 1948, **70**, 3659.	294
48JCS699	B. C. Saunders, G. J. Stacey, F. Wild and I. G. E. Wilding; *J. Chem. Soc.*, 1948, 699.	76
48JCS2174	A. F. Childs, L. J. Goldsworthy, G. F. Harding, F. E. King, A. W. Nineham, W. L. Norris, S. G. P. Plant, B. Selton and A. M. Tompsett; *J. Chem. Soc.*, 1948, 2174.	309
48JOC223	D. D. Coffman, G. H. Kalb and A. B. Ness; *J. Org. Chem.*, 1948, **13**, 223.	195
48RTC451	H. J. Backer, W. Stevens and N. Dost; *Recl. Trav. Chim. Pays-Bas*, 1948, **67**, 451 (*Chem. Abstr.*, 1949, **43**, 559).	83
49BRP617820	Ciba Ltd.; *Br. Pat.* 617 820 (1949) (*Chem. Abstr.*, 1949, **43**, 7045f).	761
49CB152	F. Pöpp; *Chem. Ber.*, 1949, **82**, 152.	592
49CRV453	S. M. McElvain; *Chem. Rev.*, 1949, **45**, 453.	824, 888
49HCA1175	G. Schwartzenbach, H. Ackermann and P. Ruckstuhl; *Helv. Chim. Acta*, 1949, **32**, 1175.	129, 473
49JA40	S. M. McElvain and J. P. Schroeder; *J. Am. Chem. Soc.*, 1949, **71**, 40.	880, 883
49JA258	H. W. Lucien and C. T. Mason; *J. Am. Chem. Soc.*, 1949, **71**, 258.	50
49JA490	D. D. Coffman, P. L. Barrick, R. D. Cramer and M. S. Raasch; *J. Am. Chem. Soc.*, 1949, **71**, 490.	5
49JA2271	R. A. Henry and W. M. Dehn; *J. Am. Chem. Soc.*, 1949, **71**, 2271.	405, 406
49JA2736	W. J. Croxall, J. O. Van Hook and R. Luckenbaugh; *J. Am. Chem. Soc.*, 1949, **71**, 2736.	195, 195
49JA2741	W. J. Croxall, J. O. Van Hook and R. Luckenbaugh; *J. Am. Chem. Soc.*, 1949, **71**, 2741.	195
49JA3439	S. G. Cohen, H. T. Wolosinski and P. J. Scheuer; *J. Am. Chem. Soc.*, 1949, **71**, 3439.	18, 730
49JA4007	C. D. Hurd and H. L. Wehrmeister; *J. Am. Chem. Soc.*, 1949, **71**, 4007.	47, 50, 51
49JCS1638	F. Chapman, P. G. Owston and D. Woodcock; *J. Chem. Soc.*, 1949, 1638.	442, 443
49JCS2342	A. H. Cook and S. F. Cox; *J. Chem. Soc.*, 1949, 2342.	405
49JOC754	C. F. H. Allen and J. A. Van Allan; *J. Org. Chem.*, 1949, **14**, 754.	47
49LA(563)54	H. Böhme, H. Fischer and R. Frank; *Justus Liebigs Ann. Chem.*, 1949, **563**, 54.	63, 69, 80, 220
B-49MI 407-01	A. H. Cook and I. M. Heilbron; in "The Chemistry of Penicillin," eds. H. T. Clarke, J. R. Johnson and R. Robinson, Princeton University Press, 1949, chap. 25, p. 921.	318
49MI 409-01	F. J. Brockman, D. C. Downing and G. F. Wright; *Can. J. Res., Sec. B*, 1949, **27**, 469.	434, 440
49USP2483972	J. T. Goodwin Jr.; *US Pat.* 2 483 972 (1949) (*Chem. Abstr.*, 1950, **44**, 2011e).	609
50CB87	H. Brintzinger, K. Pfannstiel, H. Koddebusch and K. E. Kling; *Chem. Ber.*, 1950, **83**, 87.	71
50DOK(75)219	M. I. Kabachnik and E. S. Shepleva; *Dokl. Akad. Nauk SSSR*, 1950, **75**, 219 (*Chem. Abstr.*, 1951, **45**, 6569i).	130
50IZV39	M. I. Kabachnik and E. S. Shepleva; *Izv. Akad. Nauk SSSR, Ser. Khim.*, 1950, 39 (*Chem. Abstr.*, 1950, **44**, 7257f).	130
50IZV635	M. I. Kabachnik and T. Ya. Medved; *Izv. Akad. Nauk SSSR Otd. Khim. Nauk*, 1950, 635 (*Chem. Abstr.*, 1951, **45**, 8444b).	473
50JA354	W. A. Bonner and A. Robinson; *J. Am. Chem. Soc.*, 1950, **72**, 354.	239
50JA514	W. S. Johnson, A. L. McCloskey and D. A. Dunnigan; *J. Am. Chem. Soc.*, 1950, **72**, 514.	211
50JA1661	S. M. McElvain and J. T. Venerable; *J. Am. Chem. Soc.*, 1950, **72**, 1661.	825
50JA2535	D. T. Mowry; *J. Am. Chem. Soc.*, 1950, **72**, 2535.	205

50JA2613	C. C. Price and J. A. Pappalardo; *J. Am. Chem. Soc.*, 1950, **72**, 2613.	196
50JA3843	W. E. Parnham and J. L. Bleasdale; *J. Am. Chem. Soc.*, 1950, **72**, 3843.	440
50JA3952	S. G. Cohen, H. T. Wolosinski and P. J. Scheuer; *J. Am. Chem. Soc.*, 1950, **72**, 3952.	743, 750
50JA4378	A. L. Henne and M. A. Smook; *J. Am. Chem. Soc.*, 1950, **72**, 4378.	44
50JA5409	W. R. Nes and A. Burger; *J. Am. Chem. Soc.*, 1950, **72**, 5409.	297
50JCS2689	R. N. Haszeldine and F. Smith; *J. Chem. Soc.*, 1950, 2689.	3
50JCS2920	K. W. Dunning and W. J. Dunning; *J. Chem. Soc.*, 1950, 2920.	435
50JCS2925	K. W. Dunning and W. J. Dunning; *J. Chem. Soc.*, 1950, 2925.	97
50JCS3350	M. A. T. Rogers; *J. Chem. Soc.*, 1950, 3350.	902
50JOC715	G. E. Hall and F. M. Ubertini; *J. Org. Chem.*, 1950, **15**, 715.	47, 48
50JOC795	I. B. Douglass and F. T. Martin; *J. Org. Chem.*, 1950, **15**, 795.	71
50JOC1285	I. Dvoretzky and G. H. Richter; *J. Org. Chem.*, 1950, **15**, 1285.	728
50JPP764	E. M. Bavin, R. J. W. Rees, J. M. Robson, M. Seiler, D. E. Seymour and D. Suddaby; *J. Pharm. Pharmacol.*, 1950, **2**, 764.	206
50M202	G. Schwarzenbach and J. Zurc; *Monatsh. Chem.*, 1950, **81**, 202.	554
51CB648	W. Mathes and W. Sauermilch; *Chem. Ber.*, 1951, **84**, 648.	237
51CB834	H. Stetter and W. Brockmann; *Chem. Ber.*, 1951, **84**, 834.	408
51CJC1079	R. U. Lemieux; *Can. J. Chem.*, 1951, **29**, 1079.	224
51G80	G. Rodighiero; *Gazz. Chim. Ital.*, 1951, **81**, 80.	427
51IZV95	M. I. Kabachnik and T. Ya. Medved; *Izv. Akad. Nauk SSSR Otd. Khim. Nauk*, 1951, 95 (*Chem. Abstr.*, 1952, **46**, 421a).	473
51JA212	H. C. Brown, R. S. Fletcher and R. B. Johannesen; *J. Am. Chem. Soc.*, 1951, **73**, 212.	163
51JA711	J. D. Park, W. R. Lycan and J. R. Lacher; *J. Am. Chem. Soc.*, 1951, **73**, 711.	31, 34
51JA822	F. Kipnis and J. Ornfelt; *J. Am. Chem. Soc.*, 1951, **73**, 822.	223
51JA824	J. L. Speier; *J. Am. Chem. Soc.*, 1951, **73**, 824.	140
51JA826	J. L. Speier; *J. Am. Chem. Soc.*, 1951, **73**, 826.	140
51JA1329	J. D. Park, M. L. Sharrah, W. H. Green and J. R. Lacher; *J. Am. Chem. Soc.*, 1951, **73**, 1329.	5
51JA1668	A. C. Cope and S. W. Fenton; *J. Am. Chem. Soc.*, 1951, **73**, 1668.	165
51JA1879	J. D. Roberts and S. Dev; *J. Am. Chem. Soc.*, 1951, **73**, 1879.	137
51JA2596	J. S. Pierce and C. D. Lunsford; *J. Am. Chem. Soc.*, 1952, **74**, 2596.	302
51JA2769	W. E. Bachmann, W. J. Horton, E. L. Jenner, N. W. MacNaughton and L. B. Scott; *J. Am. Chem. Soc.*, 1951, **73**, 2769.	435, 443
51JA2773	W. E. Bachmann and E. L. Jenner; *J. Am. Chem. Soc.*, 1951, **73**, 2773.	434, 435
51JA2777	W. E. Bachmann and N. C. Deno; *J. Am. Chem. Soc.*, 1951, **73**, 2777.	434, 435, 443
51JA3867	J. E. Noll, J. L. Speier and B. F. Daubert; *J. Am. Chem. Soc.*, 1951, **73**, 3867.	506
51JA4241	E. J. Barber, L. L. Burger and G. H. Cady; *J. Am. Chem. Soc.*, 1951, **73**, 4241.	3
51JA4393	L. F. Hatch and J. J. D'Amico; *J. Am. Chem. Soc.*, 1951, **73**, 4393.	14
51JA4476	R. L. Letsinger and C. W. Kammeyer; *J. Am. Chem. Soc.*, 1951, **73**, 4476.	28
51JA5005	C. J. Claus and J. L. Morgenthau, Jr.; *J. Am. Chem. Soc.*, 1951, **73**, 5005.	193
51JA5130	L. H. Sommer and J. Rockett; *J. Am. Chem. Soc.*, 1951, **73**, 5130.	513
51JA5184	F. G. Bordwell and G. D. Cooper; *J. Am. Chem. Soc.*, 1951, **73**, 5184.	80
51JA5187	F. G. Bordwell and G. D. Cooper; *J. Am. Chem. Soc.*, 1951, **73**, 5187.	64
51JA5252	H. Heymann and L. F. Feiser; *J. Am. Chem. Soc.*, 1951, **73**, 5252.	209
51JA5870	L. Field and D. H. Settlage; *J. Am. Chem. Soc.*, 1951, **73**, 5870.	53
51JA5878	H. Gilman and H. Hartzfeld; *J. Am. Chem. Soc.*, 1951, **73**, 5878.	647
51JCS1145	B. R. Brown, D. L. Hammick and B. H. Thewlis; *J. Chem. Soc.*, 1951, 1145.	13, 23
51JIC675	A. N. Kappanna and E. R. Talaty; *J. Indian Chem. Soc.*, 1951, **28**, 675.	28
51JOC1507	M. H. Gold and H. H. Levine; *J. Org. Chem.*, 1951, **16**, 1507.	440
51RTC269	J. B. Polya and T. M. Spotswood; *Recl. Trav. Chim. Pays-Bas*, 1951, **70**, 269.	424, 425
52DOK(83)689	M. I. Kabachnik and T. Ya. Medved; *Dokl. Akad. Nauk SSSR*, 1952, **83**, 689 (*Chem. Abstr.*, 1953, **47**, 724h).	463, 469
52DOK(83)865	A. N. Pudovik; *Dokl. Akad. Nauk SSSR*, 1952, **83**, 865 (*Chem. Abstr.*, 1953, **47**, 4300g).	455
52JA1068	F. Kipnis and J. Ornfelt; *J. Am. Chem. Soc.*, 1952, **74**, 1068.	216
52JA1528	E. K. Fields; *J. Am. Chem. Soc.*, 1952, **74**, 1528.	463, 469, 492
52JA3594	W. E. Truce, G. H. Birum and E. T. McBee; *J. Am. Chem. Soc.*, 1952, **74**, 3594.	64, 65, 80
52JA3868	J. H. Burckhalter, V. C. Stephens and L. A. R. Hall; *J. Am. Chem. Soc.*, 1952, **74**, 3868.	728
52JA3885	G. W. Stacy and R. J. Morath; *J. Am. Chem. Soc.*, 1952, **74**, 3885.	319
52JA3895	C. L. Agre and W. Hilling; *J. Am. Chem. Soc.*, 1952, **74**, 3895.	38
52JA3919	L. Field; *J. Am. Chem. Soc.*, 1952, **74**, 3919.	395
52JA4104	J. D. Park, E. Halpern and J. R. Lacher; *J. Am. Chem. Soc.*, 1952, **74**, 4104.	763
52JA6189	W. W. Kaeding and L. J. Andrews; *J. Am. Chem. Soc.*, 1952, **74**, 6189.	189
52JCS3437	A. M. Kinnear and E. A. Perren; *J. Chem. Soc.*, 1952, 3437.	130
52JCS4067	D. A. Peak and F. Stansfield; *J. Chem. Soc.*, 1952, 4067.	899
52JCS4259	R. N. Haszeldine; *J. Chem. Soc.*, 1952, 4259.	32, 33, 35, 37
52JGU1569	A. Ya. Yakubovich and S. P. Makarov; *J. Gen. Chem. USSR* (*Engl. Transl.*), 1952, **22**, 1569.	135, 136
52JGU1575	A. Ya. Yakubovich and V. A. Ginsburg; *J. Gen. Chem. USSR* (*Engl. Transl.*), 1952, **22**, 1575.	114
52JGU1827	A. Ya. Yakubovich, S. P. Makarov and G. I. Gavrilov; *J. Gen. Chem. USSR* (*Engl. Transl.*), 1952, **22**, 1827.	154, 155

52M1	E. Ziegler, W. Kaufmann, W. Klementschitz, N. Kreisel and E. Wiesenberger; *Monatsh., Chem.*, 1952, **83**, 1.	761
52USP2607791	J. T. Goodwin Jr.; *US Pat.* 2 607 791 (1952) (*Chem. Abstr.*, 1954, **48**, 13 732h).	609
52ZOB1783	A. Ya. Yakubovich and V. A. Ginsburg; *Zh. Obshch. Khim.*, 1952, **22**, 1783 (*Chem. Abstr.*, 1953, **47**, 9256e).	140
53CB1463	H.-W. Wanzlick and W. Löchel; *Chem. Ber.*, 1953, **86**, 1463.	404, 405, 408, 410
53DOK(92)773	A. N. Pudovik; *Dokl. Akad. Nauk SSSR*, 1953, **92**, 773 (*Chem. Abstr.*, 1953, **49**, 3050a).	463
53DOK(92)959	M. I. Kabachnik, T. Ya. Medved and T. A. Mastryukova; *Dokl. Akad. Nauk SSSR*, 1953, **92**, 959 (*Chem. Abstr.*, 1955, **49**, 839g).	463, 469
53IZV763	M. I. Kabachnik and E. S. Shepeleva; *Izv. Akad. Nauk SSSR, Ser. Khim.*, 1953, 763.	119, 125, 126
53JA968	M. Prober; *J. Am. Chem. Soc.*, 1953, **75**, 968.	731
53JA1148	J. C. Conly; *J. Am. Chem. Soc.*, 1953, **75**, 1148.	27
53JA1500	G. M. Kosolapoff; *J. Am. Chem. Soc.*, 1953, **75**, 1500.	563
53JA2698	T. J. Brice, J. D. Lazerte, L. J. Hals and W. H. Pearlson; *J. Am. Chem. Soc.*, 1953, **75**, 2698.	737
53JA3297	E. J. Corey; *J. Am. Chem. Soc.*, 1953, **75**, 3297.	28
53JA3469	E. A. Popenoe, D. G. Doherty and K. P. Link; *J. Am. Chem. Soc.*, 1953, **75**, 3469.	427
53JA4091	E. T. McBee, O. R. Pierce and H. W. Kilbourne; *J. Am. Chem. Soc.*, 1953, **75**, 4091.	33
53JA5278	M. E. Chamlers and G. M. Kosolapoff; *J. Am. Chem. Soc.*, 1953, **75**, 5278.	469
53JA5738	P. C. Crofts and G. M. Kosolapoff; *J. Am. Chem. Soc.*, 1953, **75**, 5738.	130
53JCP(21)986	P. H. Lewis and R. E. Rundle; *J. Chem. Phys.*, 1953, **21**, 986.	698
53JCS2083	B. Atkinson and A. B. Trenwith; *J. Chem. Soc.*, 1953, 2083.	5
53JCS3565	R. N. Haszeldine; *J. Chem. Soc.*, 1953, 3565.	731
53JOC1739	R. West and E. G. Rochow; *J. Org. Chem.*, 1953, **18**, 1739.	668
53LA(581)133	H. Bohme and H. J. Gran; *Justus Liebigs Ann. Chem.*, 1953, **581**, 133.	64
53RTC813	H. J. Backer, J. Strating and J. F. A. Hazenberg; *Recl. Trav. Chim. Pays-Bas*, 1953, **72**, 813.	770, 776
54CB129	H. Bredereck and E. Bäder; *Chem. Ber.*, 1954, **87**, 129.	234
54CB300	H. Brintzinger and H. Ellwanger; *Chem. Ber.*, 1950, **87**, 300.	71
54CB784	H. Bredereck, E. Bäder and G. Höschele; *Chem. Ber.*, 1954, **87**, 784.	235
54CB1690	H. Hellmann, I. Löschmann and F. Lingens; *Chem. Ber.*, 1954, **87**, 1690.	322
54DOK(97)459	A. N. Nesmeyanov, E. G. Perevalova, R. V. Golovnya and O. A. Nesmeyanova; *Dokl. Akad. Nauk SSSR*, 1954, **97**, 459.	724
54HCA90	B. Prijs, R. Gall, R. Hinderling and H. Erlenmeyer; *Helv. Chim. Acta*, 1954, **37**, 90.	21
54JA479	A. L. Henne and C. J. Fox; *J. Am. Chem. Soc.*, 1954, **76**, 479.	23, 33
54JA693	S. Akiyoshi and K. Okuno; *J. Am. Chem. Soc.*, 1954, **76**, 693.	212
54JA743	H. V. Anderson, E. R. Garrett, F. H. Lincoln, A. H. Nathan and J. A. Hogg; *J. Am. Chem. Soc.*, 1954, **76**, 743.	209
54JA1359	H. J. Dauben, Jr., B. Löken and H. J. Ringold; *J. Am. Chem. Soc.*, 1954, **76**, 1359.	191
54JA1619	L. H. Sommer, R. M. Murch and F. A. Mitch; *J. Am. Chem. Soc.*, 1954, **76**, 1619.	602, 607
54JA2502	H. Gilman and T. C. Wu; *J. Am. Chem. Soc.*, 1954, **76**, 2502.	354
54JA2943	N. S. Leeds, D. K. Fukushima and T. F. Gallagher; *J. Am. Chem. Soc.*, 1954, **76**, 2943.	209
54JA3489	H. E. Baumgarten and M. R. DeBrunner; *J. Am. Chem. Soc.*, 1954, **76**, 3489.	440
54JA3577	W. E. Truce and K. R. Buser; *J. Am. Chem. Soc.*, 1954, **76**, 3577.	391, 393
54JA3969	G. F. Grillot, H. R. Felton, B. R. Garrett, H. Greenberg, R. Green, R. Clementi and M. Moskowitz; *J. Am. Chem. Soc.*, 1954, **76**, 3969.	316
54JA4962	W. E. Parham and D. M. DeLaitsch; *J. Am. Chem. Soc.*, 1954, **76**, 4962.	223
54JA5161	W. B. Hughes and R. D. Kleene; *J. Am. Chem. Soc.*, 1954, **76**, 5161.	212
54JA5423	P. Tarrant and E. G. Gillman; *J. Am. Chem. Soc.*, 1954, **76**, 5423.	730
54JA6113	E. P. Oliveto, C. Gerold and E. B. Hershberg; *J. Am. Chem. Soc.*, 1954, **76**, 6113.	178
54JA6162	W. von E. Doering and A. K. Hoffmann; *J. Am. Chem. Soc.*, 1954, **76**, 6162.	16
54JCS520	E. B. Reid and J. R. Seigel; *J. Chem. Soc.*, 1954, 520.	162
54JCS3747	R. N. Haszeldine and B. R. Steele; *J. Chem. Soc.*, 1954, 3747.	31
54JOC250	D. Seyferth and E. G. Rochow; *J. Org. Chem.*, 1954, **20**, 250.	607, 638
54JOC1278	L. F. Cason and H. G. Brooks; *J. Org. Chem.*, 1954, **20**, 1278.	651
54MI 420-01	J. F. Arens and Th. R. Rix; *Proz. Koniskl. Akad. Wetenschap.*, 1954, **57B**, 275 (*Chem. Abstr.*, 1955, **49**, 8798i).	884
54OSC(4)807	J. C. Bill and D. S. Tarbell; *Org. Synth., Coll. Vol.*, 1954, **4**, 807.	20
55AJC512	A. J. Birch, J. Cymerman-Craig and M. Slaylor; *Aust. J. Chem.*, 1955, **8**, 512.	413
55BSF669	J. Lichtenberger, J.-P. Fleury and B. Barette; *Bull. Soc. Chim. Fr.*, 1955, 669.	423, 427
55CB41	E. Bader and H. D. Hermann; *Chem. Ber.*, 1955, **88**, 41.	331
55CJC923	C. C. Bombardieri and A. Taurins; *Can. J. Chem.*, 1955, **33**, 923.	442
55CRV301	L. Summers; *Chem. Rev.*, 1955, **55**, 301.	46, 47
55JA572	F. G. Bordwell and B. M. Pitt; *J. Am. Chem. Soc.*, 1955, **77**, 572.	64, 65, 66, 69
55JA745	W. B. Whalley, E. L. Anderson, F. Duban, J. W. Wilson and G. E. Ullyot; *J. Am. Chem. Soc.*, 1955, **77**, 745.	420, 421, 426
55JA907	D. Seyferth and E. G. Rochow; *J. Am. Chem. Soc.*, 1955, **77**, 907.	602, 652
55JA1098	A. T. Stewart, Jr. and C. R. Hauser; *J. Am. Chem. Soc.*, 1955, **77**, 1098.	300, 405
55JA2559	C. W. Sauer and R. J. Bruni; *J. Am. Chem. Soc.*, 1955, **77**, 2559.	424
55JA2783	P. Tarrant, A. M. Lovelace and M. R. Lilyquist; *J. Am. Chem. Soc.*, 1955, **77**, 2783.	731
55JA3278	K. G. Rutherford and C. L. Stevens; *J. Am. Chem. Soc.*, 1955, **77**, 3278.	12
55JA3465	H. L. Goering and L. L. Sims; *J. Am. Chem. Soc.*, 1955, **77**, 3465.	23, 38

55JA3869	G. W. Stacy, R. I. Day and R. J. Morath; *J. Am. Chem. Soc.*, 1955, **77**, 3869.	319
55JA3886	J. Hine, C. H. Thomas and S. J Ehrenson; *J. Am. Chem. Soc.*, 1955, **77**, 3886.	761
55JA4168	T. F. McGrath and R. Levine; *J. Am. Chem. Soc.*, 1955, **77**, 4168.	731
55JA5601	S. M. McElvain and G. R. McKay; *J. Am. Chem. Soc.*, 1955, **77**, 5601.	825
55JCS4280	M. M. Frazer and R. A. Raphael; *J. Chem. Soc.*, 1955, 4280.	427
55JOC1695	C. A. Mackenzie and J. H. Stocker; *J. Org. Chem.*, 1955, **20**, 1695.	177, 179
55OSC(3)270	O. Grummitt, A. Buck and R. Egan; *Org. Syn. Coll. Vol.*, 1955, **3**, 270.	743
55OSC(3)502	M. Renoll and M. S. Newman; *Org. Synth., Coll. Vol.*, 1955, **3**, 502.	178
55RTC1429	J. M. Vander Zanden and G. de. Vries; *Recl. Trav. Chim. Pays-Bas*, 1955, **74**, 1429.	206
55ZAAC(282)345	K. Ziegler, K. Nagel and M. Patheiger; *Z. Anorg. Allg. Chem.*, 1955, **282**, 345.	668
55ZN296	F. Weygand and H. J. Bestmann; *Z. Naturforsch.*, 1955, **106**, 296.	69
56CB81	H. Hellmann and G. Opitz; *Chem. Ber.*, 1956, **89**, 81.	295
56CB1134	H. Hellmann and K. Teichmann; *Chem. Ber.*, 1956, **89**, 1134.	442
56CB1254	A. Rieche and E. Schmitz; *Chem. Ber.*, 1956, **89**, 1254.	212
56CB2060	H. Meerwein, P. Borner, O. Fuchs, H. J. Sasse, H. Schrodt and J. Spille; *Chem. Ber.*, 1956, **89**, 2060.	881
56CB2873	H. Böhme and F. Eiden; *Chem. Ber.*, 1956, **89**, 2873.	405
56CRV219	R. G. Johnson and R. K. Ingham; *Chem. Rev.*, 1956, **56**, 219.	27, 35, 37
56JA59	J. D. Park, R. J. Seffl and J. R. Lacher; *J. Am. Chem. Soc.*, 1956, **78**, 59.	36, 751
56JA1685	J. D. Park, W. M. Sweeney, S. L. Hopwood, Jr. and J. R. Lacher; *J. Am. Chem. Soc.*, 1956, **78**, 1685.	761, 763
56JA2167	N. G. Gaylord and D. J. Kay; *J. Am. Chem. Soc.*, 1956, **78**, 2167.	306
56JA2264	C. L. Stevens, T. K. Mukherjee and V. J. Traynelis; *J. Am. Chem. Soc.*, 1956, **78**, 2264.	28, 38
56JA2274	M. Prober; *J. Am. Chem. Soc.*, 1956, **78**, 2274.	609
56JA4944	C. O. Parker; *J. Am. Chem. Soc.*, 1956, **78**, 4944.	824
56JA5430	P. S. Skell and A. Y. Garner; *J. Am. Chem. Soc.*, 1956, **78**, 5430.	24
56JA6101	D. G. Kundiger and H. Pledger, Jr.; *J. Am. Chem. Soc.*, 1956, **78**, 6101.	745
56JCS61	R. N. Haszeldine and J. E. Osborne; *J. Chem. Soc.*, 1956, 61.	5
56JOC638	R. L. Tse and M. S. Newman; *J. Org. Chem.*, 1956, **21**, 638.	743
56JOC801	R. Mechoulam, S. Cohen and A. Kaluszyner; *J. Org. Chem.*, 1956, **21**, 801.	826
56M319	W. Klementschitz and K. Gitschthaler; *Monatsh. Chem.*, 1956, **87**, 319.	763
56M323	W. Klementschitz and K. Gitschthaler; *Monatsch. Chem.*, 1956, **87**, 323.	742
56RTC1377	J. C. W. Postma and J. F. Arens; *Rec. Trav. Chim. Pays-Bas*, 1956, **75**, 1377.	884
56USP2768193	R. Gilbert; *US Pat.* 2 768 193 (1956) (*Chem. Abstr.*, 1957, **51**, 5816).	569
57CA(51)5721f	F. Montanari and A. Negrini; *Chem. Abstr.*, 1957, **51**, 5721f (*Boll. sci. fac. chim. ind. Bologna*, 1956, **14**, 68).	770
57CA(51)10366g	M. I. Kabachnik and P. A. Rossiiskaya; *Chem. Abstr.*, 1957, **51**, 10 366g (*Izv. Akad. Nauk. SSSR Otdel. Khim. Nauk*, 1957, 48).	920
57CA(51)12849g	F. Montanari and A. Negrini; *Chem Abstr.*, 1957, **51**, 12 849g (*Boll. sci. fac. chim. ind. Bologna*, 1957, **15**, 27).	770
57CB50	H. Hellmann and G. Haas; *Chem. Ber.*, 1957, **90**, 50.	322, 417, 424, 427
57CB53	H. Hellmann and G. Haas; *Chem. Ber.*, 1957, **90**, 53.	417
57CB444	H. Hellmann and G. Haas; *Chem. Ber.*, 1957, **90**, 444.	322
57CB1259	W. Hieber and R. Breu; *Chem. Ber.*, 1957, **90**, 1259.	687
57DOK(112)439	A. N. Nesmeyanov, E. G. Perevalova and Z. A. Beinoravichute; *Dokl. Akad. Nauk SSSR*, 1957, **112**, 439.	706
57G1061	F. Montanari and A. Negrini; *Gazz. Chim. Ital.*, 1957, **87**, 1061.	770, 772
57IZV1357	T. Ya. Medved and M. I. Kabachnik; *Izv. Akad. Nauk SSSR Otd. Khim. Nauk*, 1957, 1357 (*Chem. Abstr.*, 1958, **52**, 7361b).	469
57JA376	F. G. Bordwell, G. D. Cooper and H. Morita; *J. Am. Chem. Soc.*, 1957, **79**, 376.	63, 64
57JA456	E. M. Chamberlin, E. Tristram, T. Utne and J. M. Chemerda; *J. Am. Chem. Soc.*, 1957, **79**, 456.	213
57JA2825	W. H. Watanabe and L. E. Conlon; *J. Am. Chem. Soc.*, 1957, **79**, 2825.	306
57JA2833	W. H. Watanabe; *J. Am. Chem. Soc.*, 1957, **79**, 2833.	306
57JA4164	W. T. Miller, Jr. and A. H. Fainberg; *J. Am. Chem. Soc.*, 1957, **79**, 4164.	755
57JA4170	A. H. Fainberg and W. T. Miller, Jr.; *J. Am. Chem. Soc.*, 1957, **79**, 4170.	34
57JA4174	E. D. Bergmann, P. Moses, M. Neeman, S. Cohen, A. Kaluszyner and S. Reuter; *J. Am. Chem. Soc.*, 1957, **79**, 4174.	751
57JA4708	L. Zeldin and H. Shechter; *J. Am. Chem. Soc.*, 1957, **79**, 4708.	430, 431, 432, 994
57JA6270	H. L. Goering and L. L. Sims; *J. Am. Chem. Soc.*, 1957, **79**, 6270.	23
57JA6562	N. Kornblum, J. W. Powers, G. J. Anderson, W. J. Jones, H. O. Larson, O. Levand and W. M. Weaver; *J. Am. Chem. Soc.*, 1957, **79**, 6562.	161
57JCS158	J. W. Cornforth and R. H. Cornforth; *J. Chem. Soc.*, 1957, 158.	211
57JCS618	J. L. Hales, J. I. Jones and W. Kynaston; *J. Chem. Soc.*, 1957, 618.	58
57JOC730	W. E. Parham and R. R. Twelves; *J. Org. Chem.*, 1957, **22**, 589.	32
57JOC900	M. D. Rausch, M. Vogel and H. Rosenberg; *J. Org. Chem.*, 1957, **22**, 900.	724
57JOC984	T. S. Gardner, E. Wenis and J. Lee; *J. Org. Chem.*, 1957, **22**, 984.	424
57JOC1662	H. E. Ungnade and L. W. Kissinger; *J. Org. Chem.*, 1957, **22**, 1662.	440
57JOC1665	M. H. Gold, E. E. Hamel and K. Klager; *J. Org. Chem.*, 1957, **22**, 1665.	430, 431, 432, 994
57LA(602)135	L. Horner and E. Jürgens; *Justus Liebigs Ann. Chem.*, 1957, **602**, 135.	220
57LA(604)168	J. Goubeau and K. H. Rohwedder; *Justus Liebigs Ann. Chem.*, 1957, **604**, 168.	143

57LA(605)93	K. Ziegler and E. Holzkamp; *Justus Liebigs Ann. Chem.*, 1957, **605**, 93.	709
57LA(608)1	K. Ziegler and R. Köster; *Justus Liebigs Ann. Chem.*, 1957, **608**, 1.	699
57OR(9)332	C. V. Wilson; *Org. React.*, 1957, **9**, 332.	27
57USP2803665	C. B. Miller and C. Woolf (Allied Chemical & Dye Corp.); *US Pat.* 2 803 165 (1957) (*Chem. Abstr.*, 1958, **52**, 2047e).	767
57USP2803666	C. B. Miller and C. Woolf (Allied Chemical & Dye Corp.); *US Pat.* 2 803 666 (1957) (*Chem. Abstr.*, 1958, **52**, 2047g).	767
57ZOB2161	A. N. Pudovik and C. G. Birktimirova; *Zh. Obshch. Khim.*, 1957, **27**, 2161.	920
57ZOB2356	B. A. Arbuzov and N. P. Bogonostseva; *Zh. Obshch. Khim.*, 1957, **27**, 2356.	553
57ZOB2360	B. A. Arbuzov and N. P. Bogonostseva; *Zh. Obshch. Khim.*, 1957, **27**, 2360.	346
58CB61	L. Horner, H. Hoffmann and H. G. Wippel; *Chem. Ber.*, 1958, **91**, 61.	576
58CB410	E. Schmitz; *Chem. Ber.*, 1958, **91**, 410.	183
58CB598	H. Stetter, J. Schäfer and K. Dieminger; *Chem. Ber.*, 1958, **91**, 598.	408
58CB660	H. Böhme and D. Morf; *Chem. Ber.*, 1958, **91**, 660.	444
58CB1432	H. Zinner and B. Spangenberg; *Chem. Ber.*, 1958, **91**, 1432.	327
58CB2432	H. Hellmann and K. Teichmann; *Chem. Ber.*, 1958, **91**, 2432.	417, 423
58DOK(120)1263	A. N. Nesmeyanov, E. G. Perevalova, Z. A. Beinoravichute and I. L. Malygina; *Dokl. Akad. Nauk SSSR*, 1958, **120**, 1263.	706
58JA1083	D. H. Hirsh, R. I. Hoaglin and D. G. Kubler; *J. Am. Chem. Soc.*, 1958, **80**, 1083.	188
58JA1336	D. Seyferth; *J. Am. Chem. Soc.*, 1958, **80**, 1336.	566, 569, 597, 598, 599
58JA1916	W. E. Truce and R. Kassinger; *J. Am. Chem. Soc.*, 1958, **80**, 1916.	770
58JA1942	H. Snyder and C. W. Kruse; *J. Am. Chem. Soc.*, 1958, **80**, 1942.	20
58JA2680	H. Gilman and G. D. Lichtenwalter; *J. Am. Chem. Soc.*, 1958, **76**, 2680.	354
58JA2788	W. J. Middleton and V. A. Engelhardt; *J. Am. Chem. Soc.*, 1958, **80**, 2788. 829, 880, 883, 889, 890, 971, 972	
58JA3480	M. F. Hawthorne; *J. Am. Chem. Soc.*, 1958, **80**, 3480.	574
58JA3765	G. F. Holland and L. A. Cohen; *J. Am. Chem. Soc.*, 1958, **80**, 3765.	223
58JA4072	C. L. Stevens and A. J. Weinheimer; *J. Am. Chem. Soc.*, 1958, **80**, 4072.	183
58JA4607	B. W. Howk and J. C. Sauer; *J. Am. Chem. Soc.*, 1958, **80**, 4607.	192
58JA4921	S. R. Safir and R. J. Lopresti; *J. Am. Chem. Soc.*, 1958, **80**, 4921.	448
58JA5323	H. E. Simmons and R. D. Smith; *J. Am. Chem. Soc.*, 1958, **80**, 5323.	1083
58JA5837	F. F. Caserio and J. D. Roberts; *J. Am. Chem. Soc.*, 1958, **80**, 5837.	179
58JA5988	D. G. Kundiger and G. F. Morris; *J. Am. Chem. Soc.*, 1958, **80**, 5988.	743
58JA6233	R. J. Wineman, E.-P. T. Hsu and C. E. Anagnostopoulos; *J. Am. Chem. Soc.*, 1958, **80**, 6233.	12
58JA6450	W. E. Truce and R. Kassinger; *J. Am. Chem. Soc.*, 1958, **80**, 6450.	770
58JA6533	C. E. Inman, R. E. Oesterling and E. A. Tyczkowski; *J. Am. Chem. Soc.*, 1958, **80**, 6533.	3
58JA6613	C. Piantadosi, C. E. Anderson, E. A. Brecht and C. L. Yarbro; *J. Am. Chem. Soc.*, 1958, **80**, 6613.	178, 184
58JOC292	M. Kumada, J.-I. Nakajima, M. Ishikawa and Y. Yamamoto; *J. Org. Chem.*, 1958, **23**, 292.	614
58JOC813	M. B. Frankel; *J. Org. Chem.*, 1958, **23**, 813.	994
58JOC1368	R. L. Cobb; *J. Org. Chem.*, 1958, **23**, 1368.	28
58JOC1392	W. H. Knoth Jr. and R. V. Lindsey Jr.; *J. Org. Chem.*, 1958, **23**, 1392.	607, 608, 621, 622, 654
58JOC1661	J. D. Park, J. Abramo, M. Hein, D. N. Gray and J. R. Lacher; *J. Org. Chem.*, 1958, **23**, 1661.	757, 759, 820
58LA(614)4	W. Kirmse and L. Horner; *Justus Liebigs Ann. Chem.*, 1958, **614**, 4.	852
58LA(615)34	W. Schneider and B. Müller; *Justus Liebigs Ann. Chem.*, 1958, **615**, 34.	442
58LA(619)169	F. Asinger, M. Thiel and V. Tesar; *Justus Liebigs Ann. Chem.*, 1958, **619**, 169.	317
58QR341	P. Crofts; *Q. Rev. Chem. Soc.*, 1958, **12**, 341.	341
58T321	H. H. Wasserman and P. S. Wharton; *Tetrahedron*, 1958, **3**, 321.	827
58USP2835651	A. R. Gilbert and F. Precopio; *US Pat.* 2 835 651 (1958) (*Chem. Abstr.*, 1958, **52**, 15 125).	573
58ZOB728	V. A. Ginsburg and A. Ya. Yakubovich; *Zh. Obshch. Khim.*, 1958, **28**, 728.	555, 569
59AJC97	D. L. Ingles; *Aust. J. Chem.*, 1959, **12**, 97.	237
59BRP815231	W. D. Garden and J. M. C. Thompson; *Br. Pat.* 815 231 (1959) (*Chem. Abstr.*, 1960, **54**, 7134).	569
59BSF459	H. Normant and C. Crisan; *Bull. Soc. Chim. Fr.*, 1959, 459.	383
59BSF1828	M. Julia and J. Bullot; *Bull. Chim. Soc. Fr.*, 1959, 1828.	743
59CA(53)10040f	H. Wilms, O. Bayer and R. Wegler; *Chem. Abstr.*, 1959, **53**, 10 040f (*Ger. Pat.* 949 948 to Farbenfabriken Bayer Akt-Ges).	920
59CB1695	N. Kreutzkamp and H. Schindler; *Chem. Ber.*, 1959, **92**, 1695.	561
59CB1818	F. Klages and E. Mühlbauer; *Chem. Ber.*, 1959, **92**, 1818.	47
59CB2499	L. Horner, H. Hoffmann, H. G. Wippel and G. Klahre; *Chem. Ber.*, 1959, **92**, 2499.	575, 576, 577
59CB3175	K. Issleib and D.-W. Müller; *Chem. Ber.*, 1959, **92**, 3175.	545
59CCC64	M. Cerny, J. Vrkoc and J. Stanek; *Collect. Czech. Chem. Commun.*, 1959, **24**, 64.	223
59CCC2566	M. Cerny and J. Pacák; *Collect. Czech. Chem. Commun.*, 1959, **24**, 2566.	225
59CI(L)849	S. O. Grim and D. Seyferth; *Chem. Ind.* (*London*), 1959, 849.	569, 570, 574, 587, 588
59FRP1185452	*Fr. Pat.* 1 185 452 (1959) (*Chem. Abstr.*, 1961, **55**, 19 786).	553
59HCA1733	L. Soder and R. Wizinger; *Helv. Chim. Acta*, 1959, **42**, 1733.	1074
59JA992	K. Hofmann, S. F. Orochena, S. M. Sax and G. A. Jeffrey; *J. Am. Chem. Soc.*, 1959, **81**, 992.	25
59JA1844	D. Seyferth; *J. Am. Chem. Soc.*, 1959, **81**, 1844.	356, 642, 643, 709

59JA2078	J. H. Fried and W. T. Miller, Jr.; *J. Am. Chem. Soc.*, 1959, **81**, 2078.	755
59JA3148	G. A. Berchtold, B. E. Edwards, E. E. Campaigne and M. Carmack; *J. Am. Chem. Soc.*, 1959, **81**, 3148.	245
59JA3789	S. R. Breshears, S. S. Wang, S. G. Bechtolt and B. E. Christensen; *J. Am. Chem. Soc.*, 1959, **81**, 3789.	413
59JA4757	J. F. Tilney-Bassett and O. S. Mills; *J. Am. Chem. Soc.*, 1959, **81**, 4757.	692
59JA5008	W. T. Miller, Jr. and C. S. Y. Kim; *J. Am. Chem. Soc.*, 1959, **81**, 5008.	16
59JA5519	A. K. Hoffmann and A. G. Tesch; *J. Am. Chem. Soc.*, 1959, **81**, 5519.	576
59JA5937	H. L. Goering and D. W. Larson; *J. Am. Chem. Soc.*, 1959, **81**, 5937.	23
59JCS3751	A. E. Canavan and C. Eaborn; *J. Chem. Soc.*, 1959, 3751.	569
59JINC250	W. Hübel and E. H. Braye; *J. Inorg. Nucl. Chem.*, 1959, **10**, 250.	681
59JOC284	G. B. Payne and P. H. Williams; *J. Org. Chem.*, 1959, **24**, 284.	213
59JOC865	T. E. Stevens; *J. Org. Chem.*, 1959, **24**, 865.	428
59JOC1035	G. F. Grillott and R. E. Schaffrath; *J. Org. Chem.*, 1959, **24**, 1035.	316
59JOC1768	H. Moe and B. B. Corson; *J. Org. Chem.*, 1959, **24**, 1768.	212
59JPR150	H. Zinner, U. Zelck and G. Rembarz; *J. Prakt. Chem.*, 1959, **8**, 150.	728
59JPR285	K. Rühlmann; *J. Prakt. Chem.*, 1959, **8**, 285.	318
59LA(620)1	H. Böhme and H. P. Teltz; *Justus Liebigs Ann. Chem.*, 1959, **620**, 1.	215
59LA(622)83	F. Asinger, M. Thiel and H. G. Hauthal; *Justus Liebigs Ann. Chem.*, 1959, **622**, 83.	318
59LA(622)94	K. Rühlmann, M. Thiel and F. Asinger; *Justus Liebigs Ann. Chem.*, 1959, **622**, 94.	318
59MI 410-01	N. Kreutzkamp and G. Cordes; *Annalen*, 1959, **623**, 103.	460
59OR(10)179	E. D. Bergmann, D. Ginsberg and R. Pappo; *Org. React.*, 1959, **10**, 179.	540
59OS(39)13	C. L. Dickinson and L. R. Melby; *Org. Synth.*, 1959, **39**, 13.	829
59PCS229	W. M. Wagner; *Proc Chem. Soc.*, 1959, 229.	16
59RTC354	L. C. Rinzema, J. Stoffelsma and J. F. Arens; *Recl. Trav. Chim. Pays-Bas*, 1959, **78**, 354.	844, 846
59T269	H. J. E. Loewenthal; *Tetrahedron*, 1959, **6**, 269.	191
59USP2883368	W. J. Middleton, *US Pat.* 2 883 368 (1959)	971, 972
59USP2889349	W. D. Garden and J. M. C. Thompson; *US Pat.* 2 889 349 (1959) (*Chem. Abstr.*, 1960, **54**, 418).	569
60AG77	D. Seyferth, K. A. Brändle and G. Raab; *Angew. Chem.*, 1960, **72**, 77.	811, 812
60AP(293)203	K. Winterfeld and H. Schuller; *Arch. Pharm.* (*Weinheim, Ger.*), 1960, **293**, 203.	408
60BAU1851	R. N. Sterlin, R. D. Yatsenko, L. N. Pinkina and I. L. Knunyants; *Bull. Acad. Sci. USSR, Div. Chem. Sci.*, 1960, 1851.	802, 805
60BRP842593	H. Coates and P. A. T. Hoye; *Br. Pat.* 842 593 (1960) (*Chem. Abstr.*, 1961, **55**, 4363).	551
60CA(54)18356h	J. W. Baker and G. A. Saul; *Chem. Abstr.*, 1960, **54**, 18 356h (*US Pat.* 2 934 469 to Monsanto Chemical Co.).	920
60CB1059	H.-D. Stachel; *Chem. Ber.*, 1960, **93**, 1059.	890, 891
60CB1249	G. Hesse and M. Förderreuther; *Chem. Ber.*, 1960, **93**, 1249.	177, 178
60CB1339	F. Hein and H. Hecker; *Chem. Ber.*, 1960, **93**, 1339.	576, 577
60CB1371	L. Horner and P. Beck; *Chem. Ber.*, 1960, **93**, 1371.	576
60CB2340	F. Weygand, H. J. Bestmann and H. Fritzsche; *Chem. Ber.*, 1960, **93**, 2340.	69
60CCC2334	J. Denkstein and V. Kadeřábek; *Collect. Czech. Chem. Commun.*, 1960, **25**, 2334.	311
60CI(L)656	J. J. Cawley and F. H. Westheimer; *Chem. Ind. (London)*, 1960, 656.	189
60DOK(135)853	R. G. Kostyanovskii; *Dokl. Akad. Nauk SSSR*, 1960, **135**, 853.	295
60IS37	D. Seyferth and E. G. Rochow; *Inorg. Synth.*, 1960, **6**, 37.	140
60JA76	M. D. Rausch, E. O. Fischer and H. Grubert; *J. Am. Chem. Soc.*, 1960, **82**, 76.	706
60JA502	M. Dubeck; *J. Am. Chem. Soc.*, 1960, **82**, 502.	692
60JA543	W. R. Hasek, W. C. Smith and V. A. Engelardt; *J. A. Chem. Soc.*, 1960, **82**, 543.	8
60JA903	A. J. Speziale and R. C. Freeman; *J. Am. Chem. Soc.*, 1960, **82**, 903.	743, 791
60JA1260	A. J. Speziale, G. J. Marco and K. W. Ratts; *J. Am. Chem. Soc.*, 1960, **82**, 1260.	737
60JA1888	H. C. Clarke and C. J. Willis; *J. Am. Chem. Soc.*, 1960, **82**, 1888.	7
60JA2288	H. E. Simmons and D. W. Wiley; *J. Am. Chem. Soc.*, 1960, **82**, 2288.	164, 183
60JA2731	E. Cohen, B. Klarberg and J. R. Vaughan; *J. Am. Chem. Soc.*, 1960, **82**, 2731.	420
60JA3091	W. T. Miller, Jr., J. H. Fried and H. Goldwhite; *J. Am. Chem. Soc.*, 1960, **82**, 3091.	755
60JA4749	H. D. Kaesz, R. B. King, T. A. Manuel, L. D. Nichols and F. G. A. Stone; *J. Am. Chem. Soc.*, 1960, **82**, 4749.	681
60JA5116	D. C. England, L. R. Melby, M. A. Dietrich and R. V. Lindsey, Jr.; *J. Am. Chem. Soc.*, 1960, **82**, 5116.	970
60JA5298	R. D. Chambers, H. C. Clarke and C. J. Willis; *J. Am. Chem. Soc.*, 1960, **82**, 5298.	7
60JA6115	J. Hine and R. J. Rosscup; *J. Am. Chem. Soc.*, 1960, **82**, 6115.	18
60JA6232	H. D. Kaesz, S. L. Stafford and F. G. A. Stone; *J. Am. Chem. Soc.*, 1960, **82**, 6232.	812
60JA6238	S. L. Stafford and F. G. A. Stone; *J. Am. Chem. Soc.*, 1960, **82**, 6238.	813
60JCS1398	C. E. Griffin and R. N. Haszeldine; *J. Chem. Soc.*, 1960, 1398.	36, 798
60JCS3058	W. V. Farrar; *J. Chem. Soc*, 1960, 3058.	85, 86
60JCS4503	R. N. Haszeldine and J. C. Young; *J. Chem. Soc.*, 1960, 4503.	32
60JGU3986	L. M. Yagupol'skii and Zh. M. Ivanova; *J. Gen. Chem. USSR* (*Engl. Transl.*), 1960, **30**, 3986.	123
60JOC521	N. B. Lorette and W. L. Howard, *J. Org. Chem.*, 1960, **25**, 521.	178, 184
60JOC525	N. B. Lorette and W. L. Howard, *J. Org. Chem.*, 1960, **25**, 525.	184
60JOC899	G. Sosnovsky; *J. Org. Chem.*, 1960, **25**, 899.	206, 207
60JOC1699	D. L. Heywood and B. Phillips; *J. Org. Chem.*, 1960, **25**, 1699.	203
60JOC2016	C. W. Tullock and D. D. Coffman; *J. Org. Chem.*, 1960, **25**, 2016.	130

60JOC2108	M. E. Freed and A. R. Day; *J. Org. Chem.*, 1960, **25**, 2108.	414
60LA355	J. Thiele and E. Winter; *Justus Liebigs Ann. Chem.*, 1960, 355.	203
60LA(629)1	K. Ziegler, H.-G. Gellert, H. Lehmkuhl, W. Pfohl and K. Zosel; *Justus Liebigs Ann. Chem.*, 1960, **629**, 1.	699
60LA(629)14	K. Ziegler, H. Martin and F. Krupp; *Justus Liebigs Ann. Chem.*, 1960, **629**, 14.	699
60LA(629)53	K. Ziegler, W.-R. Kroll, W. Larbig and O.-W. Steudel; *Justus Liebigs Ann. Chem.*, 1960, **629**, 53.	699
60LA(629)222	G. Wilke and H. Müller; *Justus Liebigs Ann. Chem.*, 1960, **629**, 222.	698
60LA(638)33	V. Wolf and F. Kowitz; *Justus Liebigs Ann. Chem*, 1960, **638**, 33.	976
B-60MI 403-01	H. Hellman; in "Neuere Methoden der Praparativen Organishcen Chemie," ed. W. Foerst, Verlag Chemie, Weinheim/Bergstr., 1960, vol. II, p. 190.	96
60MI 412-01	K. Issleib and L. Baldrauf; *Pharm. Zentralhalle*, 1960, **99**, 329 (*Chem. Abstr.*, 1961, **55**, 900).	552
60MI 417-01	J. F. Arens; *Adv. Org. Chem.*, 1960, **2**, 117.	765
60MI 418-01	I. L. Knunyants and E. G. Bykhovskaya; *Proc. Acad. Sci. USSR*, 1960, **131**, 411.	799
60OSC(5)239	D. C. England and L. R. Melby; *Org. Synth., Coll. Vol.*, 1960, **5**, 239.	5
60PCS81	J. M. Birchall, G. W. Cross and R. N. Haszeldine; *Proc. Chem. Soc.*, 1960, 81.	8
60T149	S. Liisberg, W. O. Godtfredsen and S. Vangedal; *Tetrahedron*, 1960, **9**, 149.	743, 748
60USP2920094	F. Fekete; *US Pat.* 2 920 094 (1960) (*Chem. Abstr.*, 1960, **54**, 24 399).	569
60USP2937207	M. Reuter, L. Orthner, F. Jacob and E. Wolf; *US Pat.* 2 937 207 (1960) (*Chem. Abstr.*, 1960, **54**, 22 362).	555
60USP2954316	E. E. Gilbert, J. A. Otto, E. J. Rumanowski (Allied Chemical Corp.); *US Pat.* 2 954 316 (1960) (*Chem. Abstr.*, 1961, **55**, 2003c).	775
60USP2964550	D. Seyferth; *US Pat.* 2 964 550 (1960) (*Chem. Abstr.*, 1961, **55**, 6439).	566
60ZOB1602	K. A. Petrov, F. L. Maklyaev and N. K. Bliznyuk; *Zh. Obshch. Khim.*, 1960, **30**, 1602.	554
61AG493	H. Bredereck, F. Effenberger and G. Simchen; *Angew. Chem.*, 1961, **73**, 493.	881
61AP(294)404	K. Winterfeld and G. B. Singh; *Arch. Pharm.* (*Weinheim, Ger.*), 1961, **294**, 404.	416
61BSB77	M. Renson; *Bull. Soc. Chim. Belg.*, 1961, **70**, 77.	54
61CB929	B. Eistert and F. Geiss; *Chem. Ber.*, 1961, **94**, 929.	204
61CB1323	L. Horner, P. Beck and V. G. Toscano; *Chem. Ber.*, 1961, **94**, 1323.	576
61CB1373	G. Wittig and M. Schlosser; *Chem. Ber.*, 1961, **94**, 1373.	347
61CB1987	L. Horner, H. Hoffmann, G. Klahre, V. G. Toscano and H. Ertel; *Chem. Ber.*, 1961, **94**, 1987.	575, 576, 577
61CB2209	H. Zinner and H. Wigert; *Chem. Ber.*, 1961, **94**, 2209.	416
61CB2258	R. Kuhn and H. Trischmann; *Chem. Ber.*, 1961, **94**, 2258.	183
61CB2278	H. Bredereck and K. Bredereck; *Chem. Ber.*, 1961, **94**, 2278.	974
61CB2829	U. Krüerke and W. Hübel; *Chem. Ber.*, 1961, **94**, 2829.	687, 689
61CB3109	H. Bohme and F. Soldan; *Chem. Ber.*, 1961, **94**, 3109.	979
61CCC2206	M. Cerny, D. Zachystalová and J. Pacák; *Collect. Czech. Chem. Commun.*, 1961, **26**, 2206.	226
61CRV607	F. W. Lichtenthaler; *Chem. Rev.*, 1961, **61**, 607.	920
61DOK(139)877	R. G. Kostyanovskii; *Dokl. Akad. Nauk SSSR*, 1961, **139**, 877.	297
61JA367	M. F. Hawthorne; *J. Am. Chem. Soc.*, 1961, **83**, 367.	574
61JA382	J. L. Anderson, R. E. Putman and W. H. Sharkey; *J. Am. Chem. Soc.*, 1961, **83**, 382.	5
61JA747	D. C. Iffland, L. Salisbury and W. R. Schafer; *J. Am. Chem. Soc.*, 1961, **83**, 747.	314
61JA1514	H. D. Kaesz; *J. Am. Chem. Soc.*, 1961, **83**, 1514.	699
61JA1610	D. Seyferth and S. O. Grim; *J. Am. Chem. Soc.*, 1961, **83**, 1610.	569, 570, 574, 587, 588
61JA1613	D. Seyferth and S. O. Grim; *J. Am. Chem. Soc.*, 1961, **83**, 1613.	569, 570, 574, 587, 588
61JA1722	J. J. Richard, K. E. Burke, J. W. O'Laughlin and C. V. Banks; *J. Am. Chem. Soc.*, 1961, **83**, 1722.	554, 563
61JA1811	L. C. D. Groenweghe and J. H. Payne; *J. Am. Chem. Soc.*, 1961, **83**, 1811.	125, 128
61JA2055	D. Seyferth and K. A. Brändle; *J. Am. Chem. Soc.*, 1961, **83**, 2055.	551, 565
61JA2202	B. Bartocha and A. J. Bilbo; *J. Am. Chem. Soc.*, 1961, **83**, 2202.	816
61JA2205	D. C. England; *J. Am. Chem. Soc.*, 1961, **83**, 2205.	160
61JA2299	E. Uhing, K. Rattenbury and A. D. F. Toy; *J. Am. Chem. Soc.*, 1961, **83**, 2299.	115, 123, 133
61JA2383	M. Hauptschein and M. Braid; *J. Am. Chem. Soc.*, 1961, **83**, 2383.	39
61JA2574	R. K. Robins, E. F. Godefroi, E. C. Taylor, L. R. Lewis and A. Jackson; *J. Am. Chem. Soc.*, 1961, **83**, 2574.	306
61JA3345	A. L. Smith and H. A. Clark; *J. Am. Chem. Soc.*, 1961, **83**, 3345.	617
61JA3535	R. B. Kaplan and H. Shechter; *J. Am. Chem. Soc.*, 1961, **83**, 3535.	428
61JA3539	F. Ramirez, N. B. Desai, B. Hansen and N. McKelvie; *J. Am. Chem. Soc.*, 1961, **83**, 3539.	553
61JA3834	H. C. Brown and G. Zweifel; *J. Am. Chem. Soc.*, 1961, **83**, 3834.	631
61JA4105	W. T. Miller, M. B. Freedman, J. H. Fried and H. F. Koch; *J. Am. Chem. Soc.*, 1961, **83**, 4105.	36
61JA4381	K. Moedritzer; *J. Am. Chem. Soc.*, 1961, **83**, 4381.	125, 127, 128
61JA4670	S. Andreades and D. C. England; *J. Am. Chem. Soc.*, 1961, **83**, 4670.	43, 164
61JCS577	J. F. Tilney-Bassett; *J. Chem. Soc.*, 1961, 577.	692
61JCS2423	G. M. Kosolapoff and R. F. Struck; *J. Chem. Soc.*, 1961, 2423.	554
61JCS2813	S. Trippett; *J. Chem. Soc.*, 1961, 2813.	337
61JCS3452	E. D. Bergmann, S. Cohen, E. Hoffman and Z. Rand-Meir; *J. Chem. Soc.*, 1961, 3452.	33
61JCS3779	R. D. Chambers, W. K. R. Musgrave and J. Savory; *J. Chem. Soc.*, 1961, 3779.	34, 36
61JCS3825	H. Goldwhite, R. N. Haszeldine and R. N. Mukherjee; *J. Chem. Soc.*, 1961, 3825.	32
61JGU3189	K. A. Petrov and V. A. Parshina; *J. Gen. Chem. USSR* (*Engl. Transl.*), 1961, **31**, 3189.	121
61JOC412	F. C. Schaefer and G. A. Peters; *J. Org. Chem.*, 1961, **26**, 412.	883

61JOC949	D. H. Clemens and W. D. Emmons; *J. Org. Chem.*, 1961, **26**, 949.	421
61JOC1034	H. Shechter and J. F. Helling; *J. Org. Chem.*, 1961, **26**, 1034.	709
61JOC1649	H. Bauer; *J. Org. Chem.*, 1961, **26**, 1649.	425
61JOC1822	R. H. Hasek, E. U. Elam and J. C. Martin; *J. Org. Chem.*, 1961, **26**, 1822.	438
61JOC1987	H. J. Schneider, J. J. Bagnell and G. C. Murdoch; *J. Org. Chem.*, 1961, **26**, 1987.	846
61JOC2273	Z.-I. Horii, C. Iwata and Y. Tamura; *J. Org. Chem.*, 1961, **26**, 2273.	303
61JOC2478	K. C. Schreiber and V. P. Fernandez; *J. Org. Chem.*, 1961, **26**, 2478.	68
61JOC2604	D. Seyferth and W. Freyer; *J. Org. Chem.*, 1961, **26**, 2604.	566, 656, 660
61JOC2934	D. Seyferth, G. Raab and K. Brändle; *J. Org. Chem.*, 1961, **26**, 2934.	812
61JOC3591	C. P. Lo; *J. Org. Chem.*, 1961, **26**, 3591.	327
61JOC4029	E. F. Pratt and M. J. Kamlet; *J. Org. Chem.*, 1961, **26**, 4029.	297
61JOC4480	E. Schipper and E. Chinery; *J. Org. Chem.*, 1961, **26**, 4480.	420
61JOC4709	M. B. Frankel; *J. Org. Chem.*, 1961, **26**, 4709.	442, 443
61JPR72	H. Zinner and W. Schritt; *J. Prakt. Chem.*, 1961, **15**, 72.	416, 424
61LA(641)1	H. Meerwein, W. Florian, N. Schon and G. Stopp; *Justus Liebigs Ann. Chem.*, 1961, **641**, 1.	882, 974
61LA(650)1	G. Wittig and K. Schwarzenbach; *Justus Liebigs Ann. Chem.*, 1961, **650**, 1.	534, 586
61MI 418-01	R. N. Sterlin, S. S. Dubov, W. K. Li, L. P. Vakhomchik and I. L. Knunyants; *Zh. Vses. Khim. Obshch. im. D. I. Mendeleeva*, 1961, **6**, 110 (*Chem. Abstr.*, 1961, **55**, 15 336).	805, 812, 813
61TL9	L. Horner, H. Hoffmann, H. Ertel and G. Klarhe; *Tetrahedron Lett.*, 1961, **2**, 9.	490
61TL809	G. Märkl; *Tetrahedron Lett.*, 1961, **2**, 809.	490
61USP2991315	C. W. Plummer; *US Pat.* 2 991 315 (1961) (*Chem. Abstr.*, 1962, **56**, 2330).	429, 431
61ZOB3411	K. A. Petrov, V. A. Parshina and V. A. Gaidamak; *Zh. Obshch. Khim.*, 1961, **31**, 3411.	337
61ZOB3417	K. A. Petrov and V. A. Parshina; *Zh. Obshch. Khim.*, 1961, **31**, 3417.	336
62ACS2467	C. Rappe; *Acta Chem. Scand.*, 1962, **16**, 2467.	20
62AG153	H. Reimlinger; *Angew. Chem.*, 1962, **74**, 153.	740, 755
62AG251	R. Gompper and E. Kutter; *Angew. Chem.*, 1962, **74**, 251.	728
62AG872	H. V. Brachel and R. Merten; *Angew. Chem.*, 1962, **74**, 872.	310
62AG(E)508	P. Binger and R. Köster; *Angew. Chem., Int. Ed. Engl.*, 1962, **1**, 508.	631
62AG(E)592	H. V. Brachel and R. Merten; *Angew. Chem., Int. Ed. Engl.*, 1962, **1**, 592.	426
62BCJ71	T. Mukaiyama, H. Nohira and S. Asano; *Bull. Chem. Soc. Jpn.*, 1962, **35**, 71.	833
62BP896373	Chem Agro Corp.; *Br. Pat.* 896 373 (1962) (*Chem. Abstr.*, 1962, **57**, 13 615b).	770
62BSB379	M. Renson and L. Christiaens; *Bull. Soc. Chim. Belg.*, 1962, **71**, 379.	211
62CB581	L. Horner, H. Hoffmann, W. Klink, H. Ertel and V. G. Toscano; *Chem. Ber.*, 1962, **95**, 581.	577
62CB1155	E. Weiss, W. Hübel and R. Merényi; *Chem. Ber.*, 1962, **95**, 1155.	681
62CB1764	J. Jentzsch, J. Fabian and R. Mayer; *Chem. Ber.*, 1962, **95**, 1764.	245
62CB2095	H. Baganz and L. Domaschke; *Chem. Ber.*, 1962, **95**, 2095.	979
62CB2514	G. Wittig, W. Böll and K.-H. Krück; *Chem. Ber.*, 1962, **95**, 2514.	340
62CB2563	H. Hoffmann; *Chem. Ber.*, 1962, **95**, 2563.	462, 490
62CB2861	R. Gompper and W. Toepfl; *Chem. Ber.*, 1962, **95**, 2861.	843, 852, 899
62CB2871	R. Gompper and W. Topfl; *Chem. Ber.*, 1962, **95**, 2871.	900, 970
62CB3030	R. Müller, R. Köhne and H. Beyer; *Chem. Ber.*, 1962, **95**, 3030.	622
62CI(L)1472	B. R. Letchford, C. R. Patrick, M. Stacey and J. C. Tatlow; *Chem. Ind. (London)*, 1962, 1472.	730
62CI(L)1868	B. Wladislaw; *Chem. Ind. (London)*, 1962, 1868.	197
62DOK(143)211	M. I. Kabachnik and E. N. Tsvetkov; *Dokl. Akad. Nauk SSSR*, 1962, **143**, 211.	120
62DOK(143)840	E. F. Bugerenko, E. A. Chernyshev and A. D. Petrov; *Dokl. Akad. Nauk SSSR*, 1962, **143**, 840.	569
62DOK(147)117	E. A. Chernyshev, E. F. Bugerenko, N. A. Nikolaeva and A. D. Petrov; *Dokl. Akad. Nauk SSSR*, 1962, **147**, 117.	570
62IC78	D. Seyferth and T. Wada; *Inorg. Chem.*, 1962, **1**, 78.	806, 811
62IC227	D. Seyferth, H. P. Hofmann, R. Burton and J. F. Helling; *Inorg. Chem.*, 1962, **1**, 227.	706, 709
62IC414	M. D. Rausch; *Inorg. Chem.*, 1962, **1**, 414.	706
62IZV1683	A. N. Nesmeyanov, K. N. Anisimov and Z. P. Valueva; *Izv. Akad. Nauk SSSR, Ser. Khim.*, 1962, 1683.	724
62IZV2103	M. I. Kabachnik and T. Ya. Medved'; *Izv. Akad. Nauk SSSR, Ser. Khim.*, 1962, 2103.	556
62JA684	R. Breslow and E. Mohacsi; *J. Am. Chem. Soc.*, 1962, **84**, 684.	271
62JA813	W. E. Parham and L. D. Huestis; *J. Am. Chem. Soc.*, 1962, **84**, 813.	170, 209
62JA854	A. J. Speziale and K. W. Ratts; *J. Am. Chem. Soc.*, 1962, **84**, 854.	737, 738, 745, 751
62JA995	M. S. Newman, D. Pawellek and S. Ramachandran; *J. Am. Chem. Soc.*, 1962, **84**, 995.	743
62JA1312	R. Rabinowitz and R. Marcus; *J. Am. Chem. Soc.*, 1962, **84**, 1312.	738
62JA1745	F. Ramirez, N. B. Desai and N. McKelvie; *J. Am. Chem. Soc.*, 1962, **84**, 1745.	745
62JA3387	W. R. Hertler, H. D. Hartzler, D. S. Acker and R. E. Benson; *J. Am. Chem. Soc.*, 1962, **84**, 3387.	893, 979, 985
62JA4266	D. Seyferth, D. E. Welch and G. Raab; *J. Am. Chem. Soc.*, 1962, **84**, 4266.	812
62JA4715	M. E. D. Hillman; *J. Am. Chem. Soc.*, 1962, **84**, 4715.	364
62JA4789	R. G. Hiskey and W. P. Tucker; *J. Am. Chem. Soc.*, 1962, **84**, 4789.	223
62JCS592	A. E. Canavan and C. Eaborn; *J. Chem. Soc.*, 1962, 592.	569, 570
62JCS1490	W. Hewertson and H. R. Watson; *J. Chem. Soc.*, 1962, 1490.	545
62JCS1993	R. D. Chambers, W. K. R. Musgrave and J. Savory; *J. Chem. Soc.*, 1962, 1993.	146
62JCS3686	A. T. Kader and C. J. M. Stirling; *J. Chem. Soc.*, 1962, 3686.	218, 235
62JOC272	C. R. Dick; *J. Org. Chem.*, 1962, **27**, 272.	12

62JOC281	B. Wladislaw and A. M. J. Ayres; *J. Org. Chem.*, 1962, **27**, 281.	197
62JOC649	H. D. Finch; *J. Org. Chem.*, 1962, **27**, 649.	237
62JOC748	J. Sonnenberg and S. Winstein; *J. Org. Chem.*, 1962, **27**, 748.	24
62JOC757	M. P. Cava and K. Muth; *J. Org. Chem.*, 1962, **27**, 757.	21
62JOC1426	A. B. Galun, A. Kaluszyner and E. D. Bergmann; *J. Org. Chem.*, 1962, **27**, 1426.	743
62JOC1813	C. G. Krespan; *J. Org. Chem.*, 1962, **27**, 1813.	36
62JOC3598	H. Feuer, C. E. Colwell, G. Leston and A. T. Nielsen; *J. Org. Chem.*, 1962, **27**, 3598.	430, 431, 432
62JOC3664	D. H. Clemens, E. Y. Shropshire and W. D. Emmons; *J. Org. Chem.*, 1962, **27**, 3664.	414
62LA(658)128	F. Weygand, W. Steglich and H. Tanner; *Justus Liebigs Ann. Chem.*, 1962, **658**, 128.	320
62LA(659)49	H. Hellmann, J. Bader, H. Birkner and O. Schumacher; *Justus. Liebigs Ann. Chem.*, 1962, **659**, 49.	112
62MI 412-01	E. F. Bugerenko; *Sintez i Svoistva Monomerov, Akad. Nauk SSSR, Inst. Neftekhim. Sinteza, Sb. Rabot* 12-*oi* [*Ovenadstsatoi*] *Konk. po Vysokomolekul. Soedin.*, 1962, 145 (*Chem. Abstr.*, 1965, **62**, 5294).	570
62MI 418-01	D. Seyferth; *Progr. Inorg. Chem.*, 1962, **3**, 129.	806, 807, 820
62RTC925	W. M. Wagner, H. Kloosterziel and A. F. Bickel; *Recl. Trav. Chim. Pays-Bas*, 1962, **81**, 925.	16
62RTC966	J. Strating and A. M. Van Leusen; *Recl. Trav. Chim. Pays-Bas*, 1962, **81**, 966.	332
62T15	G. Sosnovsky; *Tetrahedron*, 1962, **18**, 15.	220, 226
62TL93	K. K. Andersen; *Tetrahedron Lett.*, 1962, **3**, 93.	776
62TL103	E. L. Eliel and L. A. Pilato; *Tetrahedron Lett.*, 1962, **3**, 103.	229
62USP3019248	F. Fekete; *US Pat.* 3 019 248 (1962) (*Chem. Abstr.*, 1962, **57**, 11 238).	569
62USP3024284	W. L. Howard and N. B. Lorette; *US Pat.* 3 024 284 (1962) (*Chem. Abstr.*, 1962, **57**, 8441).	186
62ZAAC(315)157	G. Fritz and J. Grobe; *Z. Anorg. Allg. Chem.*, 1962, **315**, 157.	617
62ZN851	R. Gompper and R. R. Schmidt; *Z. Naturforsch.*, 1962, **17**, 851.	901
63AG1050	F. Asinger, W. Schäfer, K. Halcour, A. Saus and H. Triem; *Angew. Chem.*, 1963, **75**, 1050.	328
63AHC(2)311	Z. Eckstein and T. Urbanski; *Adv. Heterocycl. Chem.*, 1963, **2**, 311.	302
63AK(20)225	C. Berglund and S. O. Lawesson; *Ark. Kemi*, 1963, **20**, 225 (*Chem. Abstr.*, 1963, **59**, 2624).	207
63BRP934090	Stauffer Chemical Co.; *Br. Pat.* 934 090 (1963) (*Chem. Abstr.*, 1964, **60**, 559c).	123
63BSF1868	J. Normant; *Bull. Soc. Chim. Fr.*, 1963, 1868.	18
63BSF1876	J. Normant; *Bull. Soc. Chim. Fr.*, 1963, 1876.	761, 770
63BSF2147	C. Raulet and M. Levas; *Bull. Chim. Soc. Fr.*, 1963, 2147.	743
63CA(59)11313c	V. Concialini, G. Modena and F. Taddei; *Chem. Abstr.*, 1963, **59**, 11 313c (*Boll. sci. fac. chim. ind. Bologna*, 1963, **21**, 207).	772
63CB77	K. Weissermel and M. Lederer; *Chem. Ber.*, 1963, **96**, 77.	187
63CB595	H. Böhme, K. Hartke and A. Müller; *Chem. Ber.* 1963, **96**, 595.	99
63CB600	H. Böhme and K. Hartke; *Chem. Ber.* 1963, **96**, 600.	96
63CB604	H. Böhme and K. Hartke; *Chem. Ber.*, 1963, **96**, 604.	305
63CB1035	E. Allenstein and P. Quis; *Chem. Ber.*, 1963, **96**, 1035.	792
63CB1350	H. Bredereck, F. Effenberger and G. Simchen; *Chem. Ber.*, 1963, **96**, 1350.	881
63CB1630	J. Goerdeler and H. Ruppert; *Chem. Ber.*, 1963, **96**, 1630.	412
63CB2100	C. Jutz and H. Amschler; *Chem. Ber.*, 1963, **96**, 2100.	979
63CB2186	K. Issleib, K. Krech and K. Gruber; *Chem. Ber.*, 1963, **96**, 2186.	546, 553
63CB2266	U. Schöllkopf, A. Lerch and J. Paust; *Chem. Ber.*, 1963, **96**, 2266.	49, 86
63CB3230	E. Allenstein; *Chem. Ber.*, 1963, **96**, 3230.	792, 971
63CB3306	W. Ried and A. Sinharay; *Chem. Ber.*, 1963, **96**, 3306.	972, 982
63CI(L)1397	J. Kocourek, J. Klenha and V. Jiracek; *Chem. Ind. (London)*, 1963, 1397.	239
63CJC2977	A. Ottolenghi, M. Fridkin and A. Zilkha; *Can. J. Chem.*, 1963, **41**, 2977.	807
63DOK(148)875	E. A. Chernyshev, E. F. Bugerenko and A. D. Petrov; *Dokl. Akad. Nauk SSSR*, 1963, **148**, 875.	569
63HCA1818	H. Hopff and G. Valkanas; *Helv. Chim. Acta*, 1963, **46**, 1818.	30
63IZV1563	V. F. Mironov and A. L. Kravchenko; *Izv. Akad. Nauk SSSR, Ser. Khim.*, 1963, 1563.	574
63JA642	D. Seyferth, D. E. Welch and J. K. Heeren; *J. Am. Chem. Soc.*, 1963, **85**, 642.	551, 576, 588
63JA2394	F. A. Cotton and R. A. Schunn; *J. Am. Chem. Soc.*, 1963, **85**, 2394.	576, 577
63JA2684	D. S. Matteson and J. G. Shido; *J. Am. Chem. Soc.*, 1963, **85**, 2684.	631, 634
63JA4004	R. N. McDonald and P. A. Schwab; *J. Am. Chem. Soc.*, 1963, **85**, 4004.	50
63JCED600	J. E. Fernandez, C. Powell and J. S. Fowler; *J. Chem. Eng. Data*, 1963, **8**, 600.	300
63JCS1324	M. Green; *J. Chem. Soc.*, 1963, 1324.	347
63JIC777	G. B. Singh and S. P. Agrawal; *J. Indian Chem. Soc.*, 1963, **40**, 777.	416
63JOC112	J. T. Maynard; *J. Org. Chem.*, 1963, **28**, 112.	30
63JOC123	J. J. Richard and C. V. Banks; *J. Org. Chem.*, 1963, **28**, 123.	557, 575
63JOC272	N. L. Smith and H. H. Sisler; *J. Org. Chem.*, 1963, **28**, 272.	569
63JOC369	D. S. Matteson and K. Peacock; *J. Org. Chem.*, 1963, **28**, 369.	814
63JOC494	D. A. Rausch, R. A. Davis and D. W. Osborne; *J. Org. Chem.*, 1963, **28**, 494.	31, 34, 44
63JOC1144	H. R. Snyder and J. G. Michels; *J. Org. Chem.*, 1963, **28**, 1144.	437
63JOC1420	H. Fukuda, F. J. Frank and W. E. Truce; *J. Org. Chem.*, 1963, **28**, 1420.	281
63JOC1825	F. W. Hoover, H. B. Stevenson and H. S. Rothrock; *J. Org. Chem.*, 1963, **28**, 1825.	102, 313
63JOC2082	F. W. Hoover and H. S. Rothrock; *J. Org. Chem.*, 1963, **28**, 2082.	314
63JOC2171	D. S. Matteson and R. W. H. Mah; *J. Org. Chem.*, 1963, **28**, 2171.	144
63JOC2494	B. Farah and S. Horensky; *J. Org. Chem.*, 1963, **28**, 2494.	32
63JOC2898	I. C. Popoff, L. K. Huber, B. P. Block, P. D. Morton and R. P. Riordan; *J. Org. Chem.*, 1963, **28**, 2898.	465
63JOC3337	M. D. Rausch; *J. Org. Chem.*, 1963, **28**, 3337.	724

63JPS(A)3627	H. R. Allcock and R. L. Kugel; *J. Polym. Sci., Polym. Chem.: Part A*, 1963, **1**, 3627.	577
63LA(664)130	H. Böhme, H. Böhn, E. Köhler and J. Roehr; *Justus Liebigs Ann. Chem.*, 1963, **664**, 130.	309, 310
63LA(665)91	O. von Günter, A. Griesinger and H. W. Schubert; *Justus Liebigs Ann. Chem.*, 1963, **665**, 91.	462, 478
63M290	M. Cerny, J. Stanek and J. Pacák; *Monatsh. Chem.*, 1963, **94**, 290.	223, 225
63MI 403-01	R. Schmutzler; *Adv. Chem. Ser.*, 1963, **37**, 150.	115
B-63MI 404-01	J. F. W. Keana; in "Steroid Reactions, an Outline for Organic Chemists," ed. C. Djerassi, Holden-Day, San Francisco, 1963, p. 1.	191
63MI 404-02	H. Vorbrueggen; *Steroids*, 1963, **1**, 45.	191
63MI 405-01	D. Horton and D. H. Hutson; *Adv. Carbohydr. Chem. Biochem.*, 1963, **18**, 123.	224, 227
63OR(13)55	W. E. Parham and E. E. Schweizer; *Org. React.*, 1963, **13**, 55.	7, 16, 24
63OSC(4)21	J. A. VanAllan; *Org. Synth., Coll. Vol.*, 1963, **4**, 21.	179
63OSC(4)101	S. R. Buc; *Org. Synth., Coll. Vol.*, 1963, **4**, 101.	47, 51
63OSC(4)184	B. Englund; *Org. Synth. Coll. Vol.*, 1963, **4**, 184.	32
63OSC(4)254	J. W. Wilt and J. L. Diebold; *Org. Synth. Coll. Vol.*, 1963, **4**, 254.	28
63OSC(4)427	R. B. Moffett; *Org. Synth., Coll. Vol.*, 1963, **4**, 427.	189
63OSC(4)489	R. T. Bertz; *Org. Synth., Coll. Vol.*, 1963, **4**, 489.	203
63OSC(4)558	C. C. Price and J. A. Pappalardo; *Org. Synth., Coll. Vol.*, 1963, **4**, 558.	196
63OSC(4)713	J. Nishimura; *Org. Synth., Coll. Vol.*, 1963, **4**, 713.	206
63OSC(4)748	O. Grummitt, E. P. Budewitz and C. C. Chudd; *Org. Synth., Coll. Vol.*, 1963, **4**, 748.	47, 50
63OSC(4)801	B. W. Howk and J. C. Sauer; *Org. Synth., Coll. Vol.*, 1963, **4**, 801.	192
63T49	L. J. Winters and W. E. McEwen; *Tetrahedron*, 1963, **19** (Suppl. 1), 49.	994
63T149	L. D. Bergelson and M. M. Shemyakin; *Tetrahedron*, 1963, **19**, 149.	752
63T(S)3	C. M. Wright and D. R. Levering; *Tetrahedron*, 1963, **19** (Suppl. 1), 3.	428
63T(S)23	W. E. Noland and R. Libers; *Tetrahedron*, 1963, **19** (Suppl. 1), 23.	439
63USP3086053	R. I. Wagner; *US Pat.* 3 086 053 (1963) (*Chem. Abstr.*, 1963, **59**, 10 124).	545
63USP3086056	R. I. Wagner; *US Pat.* 3 086 056 (1963) (*Chem. Abstr.*, 1963, **60**, 559).	545
63USP3093672	L. A. Miller; *US Pat.* 3 093 672 (1963) (*Chem. Abstr.*, 1963, **59**, 13 823).	561
63USP3098878	J. E. Harris and C. N. Matthews; *US Pat.* 3 098 878 (1963) (*Chem. Abstr.*, 1963, **59**, 14 024).	553
63ZAAC(324)259	K. Issleib and G. Döll; *Z. Anorg. Allg. Chem.*, 1963, **324**, 259.	545
63ZN(B)1125	G. Köbrich and H. Trapp; *Z. Naturforsch., Teil B*, 1963, **18**, 1125.	145, 817
63ZOB1045	A. M. Pudovik, N. G. Khusainova and I. M. Aladzheva; *Zh. Obshch. Khim.*, 1963, **33**, 1045.	561
63ZOB3353	A. N. Pudovik and M. A. Pudovik; *Zh. Obshch. Khim.*, 1963, **33**, 3353 (*Chem. Abstr.*, 1964, **60**, 9308a).	455
64AG382	R. Neidlein; *Angew. Chem., Int. Ed. Engl.*, 1964, **3**, 382.	439
64AG571	H. G. Viehe, R. Fuks and M. Reinstein; *Angew. Chem.*, 1964, **76**, 571.	885
64AG781	K. Hartke; *Angew. Chem.*, 1964, **76**, 781.	901, 904
64AP(297)461	G. Wagner and P. Nuhn; *Arch. Pharm.* (*Weinheim, Ger.*), 1964, **297**, 461.	239
64CB179	H. Böhme, L. Tils and B. Unterhalt; *Chem. Ber.*, 1964, **97**, 179.	69
64CB339	H. Reimlinger; *Chem. Ber.*, 1964, **97**, 339.	755
64CB483	R. Albrecht, G. Kresze and B. Mlaker; *Chem. Ber.*, 1964, **97**, 483.	304
64CB490	G. Kresze and R. Albrecht; *Chem. Ber.*, 1964, **97**, 490.	304, 311
64CB1298	R. Mayer and B. Gebhardt; *Chem. Ber.*, 1964, **97**, 1298.	850
64CB3081	H. Bredereck, F. Effenberger and H. P. Beyerlin; *Chem. Ber.*, 1964, **97**, 3081.	881, 882, 972
64CB3503	H. Reimlinger; *Chem. Ber.*, 1964, **97**, 3503.	740
64CI(L)1461	N. Petragnani and M. de Moura Campos; *Chem. Ind.* (*London*), 1964, 1461.	348
64CR(259)830	H. Normant and B. Castro; *C. R. Acad. Sci. Paris*, 1964, **259**, 830.	395
64G595	G. Pala and A. Mantegani; *Gazz. Chim. Ital.*, 1964, **94**, 595.	416
64GEP1183513	H. Böhme; *Ger. Pat.*, 1 183 513 (1964) (*Chem. Abstr.*, 1965, **62**, 14 505).	310
64HCA2425	A. E. Wick, D. Felix, K. Steen and A. Eschenmoser; *Helv. Chim. Acta*, 1964, **47**, 2425.	882
64IC410	R. Schmutzler; *Inorg. Chem.*, 1964, **3**, 410.	133
64IC1196	N. E. Miller and E. L. Muetterties; *Inorg. Chem.*, 1964, **3**, 1196.	524
64IZV772	N. A. Nesmeyanov, V. M. Novikov and O. A. Reutov; *Izv. Akad. Nauk SSSR, Ser. Khim.*, 1964, 772.	587
64JA1100	D. Seyferth, D. E. Welch and J. K. Heeren; *J. Am. Chem. Soc.*, 1964, **86**, 1100.	551, 570, 588
64JA1701	H. Shechter, S. S. Rawalay and M. Tubis; *J. Am. Chem. Soc.*, 1964, **86**, 1701.	420, 424, 449
64JA2183	J. J. Brown, R. H. Lenhard and S. Bernstein; *J. Am. Chem. Soc.*, 1964, **86**, 2183.	179
64JA3039	D. J. Pasto; *J. Am. Chem. Soc.*, 1964, **86**, 3039.	631
64JA4042	G. L. Closs and R. A. Moss; *J. Am. Chem. Soc.*, 1964, **86**, 4042.	816
64JA4085	L. A. Paquette; *J. Am. Chem. Soc.*, 1964, **86**, 4085.	64
64JA4089	L. A. Paquette; *J. Am. Chem. Soc.*, 1964, **86**, 4089.	64
64JA4383	L. A. Paquette; *J. Am. Chem. Soc.*, 1964, **86**, 4383.	80
64JA4645	F. G. Bordwell and W. T. Brannen, Jr.; *J. Am. Chem. Soc.*, 1964, **86**, 4645.	74, 75, 77
64JA4851	J. T. Suji, M. Morikawa and J. Kiji; *J. Am. Chem. Soc.*, 1964, **86**, 4851.	14
64JCS513	H. Gorcian and D. Grdenic; *J. Chem. Soc.*, 1964, 513.	553
64JCS766	D. M. Green, A. G. Long, P. J. May and A. F. Turner; *J. Chem. Soc.*, 1964, 766.	56, 57
64JCS1217	H. B. Henbest, J. A. W. Reid and C. J. M. Stirling; *J. Chem. Soc.*, 1964, 1217.	207
64JCS3832	P. J. E. Brownlee, M. E. Cox, B. O. Handford, J. C. Marsden and G. T. Young; *J. Chem. Soc.*, 1964, 3832.	218
64JCS4962	M. Asscher and D. Vofsi; *J. Chem. Soc.*, 1964, 4962.	82
64JINC1531	G. W. Watt and L. J. Baye; *J. Inorg. Nucl. Chem.*, 1964, **26**, 1531.	697
64JOC407	W. J. Burke, W. A. Nasutavicus and C. Weatherbee; *J. Org. Chem.*, 1964, **29**, 407.	301

64JOC724	G. E. McCasland, S. Furuta and A. Furst; *J. Org. Chem.*, 1964, **29**, 724.	229
64JOC832	J. R. Chambers and I. R. Isbell; *J. Org. Chem.*, 1964, **29**, 832.	467
64JOC1358	H.-D. Becker; *J. Org. Chem.*, 1964, **29**, 1358.	221
64JOC1371	Y. Hachihama, T. Shono and S. Ikeda; *J. Org. Chem.*, 1964, **29**, 1371.	211
64JOC1591	E. R. Bissell and D. B. Fields; *J. Org. Chem.*, 1964, **29**, 1591.	31
64JOC1601	W. A. Kriner; *J. Org. Chem.*, 1964, **29**, 1601.	608, 621
64JOC1956	D. P. Wyman and P. R. Kaufman; *J. Org. Chem.*, 1964, **29**, 1956.	13
64JOC1985	Y. Ogata, A. Kawasaki and N. Okumura; *J. Org. Chem.*, 1964, **29**, 1985.	437
64JOC2036	F. F. Bicke and S. Raines; *J. Org. Chem.*, 1964, **29**, 2036.	557
64JOC2427	J. S. Driscoll, D. W. Grisley, J. V. Pustinger, J. E. Harris and C. N. Matthews; *J. Org. Chem.*, 1964, **29**, 2427.	554
64JOC2513	R. H. Hasek, P. G. Gott and J. C. Martin; *J. Org. Chem.*, 1964, **29**, 2513.	979
64JOC2742	D. S. Matteson and J. G. Shdo; *J. Org. Chem.*, 1964, **29**, 2742.	631, 697
64JOC2932	D. H. Clemens, A. J. Bell and J. L. O'Brien; *J. Org. Chem.*, 1964, **29**, 2932.	979
64JOC2951	L. Skattebøl; *J. Org. Chem.*, 1964, **29**, 2951.	24
64JOC3070	H. Becker; *J. Org. Chem.*, 1964, **29**, 3070.	172
64JOC3114	P. E. Newallis and E. J. Rumanowvski; *J. Org. Chem.*, 1964, **29**, 3114.	297
64JOM(2)1	D. Seyferth and D. E. Welch; *J. Organomet. Chem.*, 1964, **2**, 1.	576, 588
64JOM(2)277	H. Gilman and W. H. Atwell; *J. Organomet. Chem.*, 1964, **2**, 277.	608
64LA(679)51	W. Ried and H. Appel; *Justus Liebigs Ann. Chem.*, 1964, **679**, 51.	738
B-64MI 401-01	W. Kirmse; "Carbene Chemistry," Academic Press, New York, 1964.	16, 24
B-64MI 402-01	G. A. Olah and W. S. Tolgyesi; in "Friedel-Crafts and Related Reactions," ed. G. A. Olah, Interscience, New York, 1964, vol. 3, p. 737.	41
B-64MI 410-01	M. Grayson and E. J. Griffith (eds); "Topics in Phosphorus Chemistry," Wiley, New York, 1964, vols 1–11.	451
64MI 417-01	M. Okano, Y. Ito and R. Oda; *Bull. Int. Chem. Res. (Kyoto, Japan)*, 1964, 217.	740
B-64MI 424-01	G. Sosnovsky; "Free Radical Reactions in Preparative Organic Chemistry," Collier-Macmillan, Toronto, 1964.	1080
64PCS187	O. S. Mills and G. Robinson; *Proc. Chem. Soc.*, 1964, 187.	688
64PCS401	N. A. Bailey, M. R. Churchill, R. Hunt, R. Mason and G. Wilkinson; *Proc. Chem. Soc.*, 1964, 401.	687
64T1567	G. J. Benoy; *Tetrahedron*, 1964, **20**, 1567.	11
64T2181	H. H. Freedman and G. A. Doorakian; *Tetrahedron*, 1964, **20**, 2181.	165
64TL7	R. Oda, Y. Ito and M. Okano; *Tetrahedron Lett.*, 1964, **5**, 7.	740
64TL543	A. M. van Leusen, R. J. Mulder and J. Strating; *Tetrahedron Lett.*, 1964, **5**, 543.	218, 235
64TL547	B. Zwanenburg, J. B. F. N. Engberts and J. Strating; *Tetrahedron Lett.*, 1964, **5**, 547.	234
64TL1461	S. A. Fuqua, W. G. Duncan and R. M. Silverstein; *Tetrahedron Lett.*, 1964, **5**, 1461.	8, 732
64TL1503	U. Schollkopf and H. Kuppers; *Tetrahedron Lett.*, 1964, **5**, 1503.	381
64TL1693	R. A. Corral and O. O. Orazi; *Tetrahedron Lett.*, 1964, **5**, 1693.	417
64TL1971	R. Gompper and W. Elser; *Tetrahedron Lett.*, 1964, **5**, 1971.	902
64TL2669	L. D. Bergelson, V. A. Vaver, L. I. Barsukov and M. M. Shemyakin; *Tetrahedron Lett.*, 1964, **5**, 2669.	752
64TL2737	K. Hartke; *Tetrahedron Lett.*, 1964, **5**, 2737.	901
64USP3127450	N. B. Lorette and W. L. Howard; *US Pat.* 3 127 450 (1964) (*Chem. Abstr.*, 1964, **60**, 15 737).	184
64USP3161607	A. Y. Garner; *US Pat.* 3 161 607 (1964) (*Chem. Abstr.*, 1965, **62**, 5405e).	114, 117
64ZC401	H. Gross and E. Hoeft; *Z. Chem.*, 1964, **4**, 401.	46
64ZOB2897	B. M. Gladshtein, E. I. Babkina, V. V. Fedotova and L. Z. Soborovskii; *Zh. Obshch. Khim.*, 1964, **34**, 2897.	561
64ZOB3938	A. M. Pudovik, N. G. Khusainova and A. B. Ageeva; *Zh. Obshch. Khim.*, 1964, **34**, 3938.	561
65AFC(4)50	R. D. Chambers and R. H. Mobbs; *Adv. Fluorine Chem.*, 1965, **4**, 50.	1076
65AG1134	E. J. Corey and D. Seebach; *Angew. Chem.*, 1965, 1134.	250
65AG1138	N. Kreutzkamp, E. Schmidt, T. Samoa and A. K. Herberg; *Angew. Chem.*, 1965, 1138.	489
65AG(E)527	L. Maier; *Angew. Chem., Int. Ed. Engl.*, 1965, **4**, 527.	452, 466
65CB487	F. Weygand and W. Steglich; *Chem. Ber.*, 1965, **98**, 487.	305
65CB1293	H. Eilingsfeld and L. Möbius; *Chem. Ber.*, 1965, **98**, 1293.	901, 904
65CB1385	R. Gompper and R. R. Schmidt; *Chem. Ber.*, 1965, **98**, 1385.	901
65CB1391	R. Gompper and R. Kunz; *Chem. Ber.*, 1965, **98**, 1391.	979
65CB4036	G. Zinner and W. Kliegel; *Chem. Ber.*, 1965, **98**, 4036.	405
65CC574	R. N. Pratt, S. A. Procter and G. A. Taylor; *J. Chem. Soc., Chem. Commun.*, 1965, 574.	886
65CR(260)2252	A. Sekera and P. Rumf; *C. R. Hebd. Seances Acad. Sci.*, 1965, **260**, 2252.	428
65CR(261)2232	M. Kerfanto and D. Jegou; *C. R. Hebd. Seances Acad. Sci.*, 1965, **261**, 2232.	411
65GEP1212069	H. Bredereck; *Ger. Pat.* 1 212 069 (1965) (*Chem. Abstr.*, 1966, **64**, 19 421c).	882
65HCA133	L. Maier; *Helv. Chim. Acta*, 1965, **48**, 133.	563
65HCA1034	L. Maier; *Helv. Chim. Acta*, 1965, **48**, 1034.	546
65IC198	J. P. King, B. P. Block and I. C. Popoff; *Inorg. Chem.*, 1965, **4**, 198.	556
65IC1270	R. Hunt and G. Wilkinson; *Inorg. Chem.*, 1965, **4**, 1270.	687
65IC1458	N. E. Miller; *Inorg. Chem.*, 1965, **4**, 1458.	571, 574, 597
65IZV286	E. F. Bugerenko, E. A. Chernyshev and A. D. Petrov; *Izv. Akad. Nauk SSSR, Ser. Khim.*, 1965, 286.	569, 570, 573
65IZV2046	Yu. V. Zeifman, N. P. Gambaryan and I. L. Knunyants; *Izv. Akad. Nauk SSSR, Ser. Khim.*, 1965, 2046.	304

65JA390	N. E. Miller; *J. Am. Chem. Soc.*, 1965, **87**, 390.	569
65JA488	R. Schaeffer and L. J. Todd; *J. Am. Chem. Soc.*, 1965, **87**, 488.	142, 524
65JA1241	D. W. Slocum, B. W. Rockett and C. R. Hauser; *J. Am. Chem. Soc.*, 1965, **87**, 1241.	706
65JA1345	E. J. Corey and M. Chaykovsky; *J. Am. Chem. Soc.*, 1965, **87**, 1345.	391, 393
65JA2567	S. K. Westerback, S. Rajan and A. E. Martell; *J. Am. Chem. Soc.*, 1965, **87**, 2567.	473
65JA3151	K. Griesbaum, W. Naegele and G. G. Wanless; *J. Am. Chem. Soc.*, 1965, **87**, 3151.	15, 24, 29
65JA3254	J. D. Fitzpatrick, L. Watts, G. F. Emerson and R. Pettit; *J. Am. Chem. Soc.*, 1965, **87**, 3254.	724
65JA3716	C. S. Cleaver and C. G. Krespan; *J. Am. Chem. Soc.*, 1965, **87**, 3716.	799
65JA4147	D. F. Hoeg, D. I. Lusk and A. L. Crumbliss; *J. Am. Chem. Soc.*, 1965, **87**, 4147.	13
65JA4156	D. Seyferth and G. Singh; *J. Am. Chem. Soc.*, 1965, **87**, 4156.	571
65JA4259	D. Seyferth, J. M. Burtlitch, R. J. Minasz, J. Y.-P. Mui, H. D. Simmons, Jr., A. J. H. Treiber and S. R. Dowd; *J. Am. Chem. Soc.*, 1965, **87**, 4259.	16
65JA4757	C. E. Griffin, E. H. Uhing and A. D. F. Toy; *J. Am. Chem. Soc.*, 1965, **87**, 4757.	473
65JA5498	A. Rauk, E. Buncel, R. Y. Moir and S. Wolfe; *J. Am. Chem. Soc.*, 1965, **87**, 5498.	391
65JCED303	T. H. Siddall, III and M. A. Davis; *J. Chem. Eng. Data*, 1965, **10**, 303.	554
65JCS1298	M. Kyaw and L. N. Owen; *J. Chem. Soc.*, 1965, 1298.	229
65JCS2601	J. R. Bull, E. R. H. Jones and G. D. Meakins; *J. Chem. Soc.*, 1965, 2601.	429, 430, 431, 439
65JCS3075	P. S. Steyn, W. J. Conradie, C. F. Garbers and M. J. de Vries; *J. Chem. Soc.*, 1965, 3075.	56
65JCS5630	R. Schmutzler; *J. Chem. Soc.*, 1965, 5630.	116
65JCS5877	S. A. Procter and G. A. Taylor; *J. Chem. Soc.*, 1965, 5877.	886
65JGU2053	K. A. Petrov, V. A. Parshina and A. F. Manuilov; *J. Gen. Chem. USSR (Engl. Transl.)*, 1965, **35**, 2053.	119
65JOC71	W. E. Truce and P. N. Son; *J. Org. Chem.*, 1965, **30**, 71.	971, 979
65JOC74	R. Grashey, R. Huisgen and K. K. Sun; *J. Org. Chem.*, 1965, **30**, 74.	436
65JOC855	E. L. Eliel, E. W. Della and M. Rogic; *J. Org. Chem.*, 1965, **30**, 855.	229
65JOC1027	S. A. Fuqua, W. G. Duncan and R. M. Silverstein; *J. Org. Chem.*, 1965, **30**, 1027.	732
65JOC1241	B. S. Farah and E. E. Gilbert; *J. Org. Chem.*, 1965, **30**, 1241.	18
65JOC1313	N. P. Neureiter; *J. Org. Chem.*, 1965, **30**, 1313.	83
65JOC1384	W. J. Middleton and W. H. Sharkey; *J. Org. Chem.*, 1965, **30**, 1384.	256
65JOC1398	W. J. Middleton and C. G. Krespan; *J. Org. Chem.*, 1965, **30**, 1398.	97, 444
65JOC1600	J. W. Huffman, L. H. Keith and R. L. Asbury; *J. Org. Chem.*, 1965, **30**, 1600.	706
65JOC1939	D. J. Peterson and H. R. Hays; *J. Org. Chem.*, 1965, **30**, 1939.	575
65JOC2190	J. F. Harris; *J. Org. Chem.*, 1965, **30**, 2190.	215
65JOC2195	S. F. Reed, Jr.; *J. Org. Chem.*, 1965, **30**, 2195.	16
65JOC2543	S. A. Fuqua, W. G. Duncan and R. M. Silverstein; *J. Org. Chem.*, 1965, **30**, 2543.	732
65JOC2618	D. R. Weyernberg and L. E. Nelson; *J. Org. Chem.*, 1965, **30**, 2618.	610
65JOC2769	P. M. Quan, T. K. B. Karns and L. D. Quin; *J. Org. Chem.*, 1965, **30**, 2769.	424
65JOC3834	P. E. Newallis and P. Lombardo; *J. Org. Chem.*, 1965, **30**, 3834.	164, 209
65JOM(4)98	R. L. Merker and M. J. Scott; *J. Organomet. Chem.*, 1965, **4**, 98.	603
65JOM(4)114	E. Kurras and J. Otto; *J. Organomet. Chem.*, 1965, **4**, 114.	708
65JOM(4)151	W. R. Cullen and G. E. Styan; *J. Organomet. Chem.*, 1965, **4**, 151.	344
65JOM(4)202	N. A. Nesmeyanov, V. M. Novikov and O. A. Reutov; *J. Organomet. Chem.*, 1965, **4**, 202.	587
65JPC3418	K. C. Ramey, J. F. O'Brien, I. Hasegawa and A. E. Borchert; *J. Phys. Chem.*, 1965, **69**, 3418.	698
65LA(684)37	R. Gompper, R. R. Schmidt and E. Kutter; *Justus Liebigs Ann. Chem.*, 1965, **684**, 37.	852
65LA(684)103	G. Opitz and H. Schempp; *Justus Leibigs Ann. Chem.*, 1965, **684**, 103.	979
65LA(686)102	H. Biener; *Justus Liebigs Ann. Chem.*, 1965, **686**, 102.	310
65LA(687)1	G. Hesse and H. Witte; *Justus Liebigs Ann. Chem.*, 1965, **687**, 1.	522
65LA(687)187	W. Ried and E. Torok; *Justus Liebigs Ann. Chem.*, 1965, **687**, 187.	307
B-65MI 403-01	P. A. S. Smith; "The Chemistry of Open-chain Organic Nitrogen Compounds," W. A. Benjamin, New York, 1965, vol. 1, p. 291.	96
65MI 414-01	G. Fritz, J. Grobe and D. Kummer; *Adv. Inorg. Chem., Radiochem.*, 1965, **7**, 349.	602
B-65MI 424-01	D. Bethell and V. Gold; "Carbonium Ions," Academic Press, New York, 1965.	1072
65OR(14)52	H. E. Zaugg and W. B. Martin; *Org. React.*, 1965, **14**, 52.	96, 98, 99, 308, 416, 417, 424, 425, 427
65OR(14)270	A. Maercker; *Org. React.*, 1965, **14**, 270.	737
65OS(45)33	A. J. Speziale, K. W. Ratts and D. E. Bissing; *Org. Synth.*, 1965, **45**, 33.	738
65RTC904	H. van Kamp and S. J. Halkes; *Recl. Trav. Chim. Pays-Bas*, 1965, **84**, 904.	748
65RZC1129	P. Mastalerz; *Rocz. Chem.*, 1965, **39**, 1129 (*Chem. Abstr.*, 1966, **64**, 6684).	555
65T5	D. W. Grisley, J. C. Alm and C. N. Matthews; *Tetrahedron*, 1965, **21**, 5.	554
65T871	G. Sosnovsky; *Tetrahedron*, 1965, **21**, 871.	207
65T2743	H. Goldwhite, M. S. Gibson and C. Harris; *Tetrahedron*, 1965, **21**, 2743.	281, 282
65T2899	W. E. Truce, M. G. Rossmann, F. M. Perry, R. M. Burnett and D. J. Abraham; *Tetrahedron*, 1965, **21**, 2899.	770, 779
65TL521	S. A. Fuqua, W. G. Duncan and R. M. Silverstein; *Tetrahedron Lett.*, 1965, **6**, 521.	732
65TL1883	D. J. Burton and F. E. Herkes; *Tetrahedron Lett.*, 1965, **6**, 1883.	732
65TL3429	G. Cainelli, G. Dal Bello and G. Zubiani; *Tetrahedron Lett.*, 1965, **6**, 3429.	664
65TL3589	J. C. Martin, V. A. Hoyle, Jr. and K. C. Brannock; *Tetrahedron Lett.*, 1965, **6**, 3589.	887
65ZAAC(336)234	K. Issleib and R. J. Bleck; *Z. Anorg. Allg. Chem.*, 1965, **336**, 234.	455
65ZC109	E. Uhlig and W. Achilles; *Z. Chem.*, 1965, **5**, 109.	473
65ZC419	H. G. Henning and G. Petzold; *Z. Chem.*, 1965, **5**, 419.	554, 556
65ZOB354	A. N. Pudovik and R. G. Kuzovleva; *Zh. Obshch. Khim.*, 1965, **35**, 354.	561
65ZOB1602	K. A. Petrov, V. A. Porchina and A. F. Manuilov; *Zh. Obshch. Khim.*, 1965, **35**, 1602.	550, 554

65ZOB1639	B. E. Gruz and L. M. Yagupol'skii; *Zh. Obshch. Khim.*, 1965, **35**, 1639 (*Chem. Abstr.*, 1965, **63**, 18 062).	826
66ACS253	C. Rappe and B. Albrecht; *Acta Chem. Scand.*, 1966, **20**, 253.	19
66AG(E)314	W. Verbeek and W. Sundermeyer; *Angew. Chem., Int. Ed. Engl.*, 1966, **5**, 314.	3
66AG(E)420	H. Baganz and H. J. May; *Angew. Chem., Int. Ed. Engl.*, 1966, **5**, 420.	198
66AG(E)594	G. Opitz, M. Kleeman, D. Bücher, G. Walz and K. Rieth; *Angew. Chem., Int. Ed. Engl.*, 1966, **5**, 594.	282
66AG(E)847	D. Gloyna and H. G. Henning; *Angew. Chem., Int. Ed. Engl.*, 1966, **5**, 847.	1023
66AG(E)898	H. Breil and G. Wilke; *Angew. Chem., Int. Ed. Engl.*, 1966, **5**, 898.	671
66AG(E)899	H. Dietrich and H. Dierks; *Angew. Chem., Int. Ed. Engl.*, 1966, **5**, 899.	671
66AP(299)906	N. Kreutzkamp and H. Y. Oei; *Arch. Pharm. (Weinheim. Ger.)*, 1966, **299**, 906.	321
66BCJ219	S. Nagase, K. Tanaka, H. Baba and T. Abe; *Bull. Chem. Soc. Jpn.*, 1966, **39**, 219.	3
66BRP1026366	C. H. Roy; *Br. Pat.* 1 026 366 (1966) (*Chem. Abstr.*, 1966, **65**, 5685).	555, 563
66CAR(2)461	E. Zissis, A. L. Clingman and N. K. Richtmyer; *Carbohydr. Res.*, 1966, **2**, 461.	235
66CB48	K. Schank; *Chem. Ber.*, 1966, **99**, 48.	234
66CB670	G. Köbrich and H. Trapp; *Chem. Ber.*, 1966, **99**, 670.	758
66CB724	K. H. Büchel, A. K. Bocz and F. Korte; *Chem. Ber.*, 1966, **99**, 724.	413
66CB868	H. R. Hensel; *Chem. Ber.*, 1966, **99**, 868.	406
66CB930	G. Domschke; *Chem. Ber.*, 1966, **99**, 930.	307
66CB1678	G. Zinner, W. Kliegel, W. Ritter and H. Böhlke; *Chem. Ber.*, 1966, **99**, 1678.	437
66CB1932	F. Weygand, W. Steglich, I. Lengyel and F. Fraunberger; *Chem. Ber.*, 1966, **99**, 1932.	309
66CB1944	F. Weygand, W. Steglich, I. Lengyel, F. Fraunberger, A. Maierhofer and W. Dettmeier; *Chem. Ber.*, 1966, **99**, 1944.	297, 327
66CB2127	H. Böhme and G. Berg; *Chem. Ber.*, 1966, **99**, 2127.	310
66CB2302	D. Martin, K.-H. Schwartz, S. Rackow, P. Reich and E. Gründemann; *Chem. Ber.*, 1966, **99**, 2302.	889, 972
66CB2686	G. Zinner and W. Kliegel; *Chem. Ber.*, 1966, **99**, 2686.	435
66CB2900	R. Gompper and R. Kunz; *Chem. Ber.*, 1966, **99**, 2900.	743, 970
66CB3892	F. Effenberger, R. Gleiter and G. Kiefer; *Chem. Ber.*, 1966, **99**, 3892.	979
66CB3903	F. Effenberger and R. Gleiter; *Chem. Ber.*, 1966, **99**, 3903.	312
66CC567	J. Wolinsky and D. Chan; *J. Chem. Soc., Chem. Commun.*, 1966, 567.	742
66CI(L)1555	D. L. Tuleen and T. B. Stevens; *Chem. Ind. (London)*, 1966, 1555.	66
66CI(L)2054	P. M. Barna; *Chem. Ind. (London)*, 1966, 2054.	732
66CJC2593	A. W. Frank and I. Gordon; *Can. J. Chem.*, 1966, **44**, 2593.	130
66CRV161	A. E. Pohland and W. R. Benson; *Chem. Rev.*, 1966, **66**, 161.	970
66DOK(171)111	N. A. Nesmeyanov and O. A. Reutov; *Dokl. Akad. Nauk SSSR*, 1966, **171**, 111.	587
66GEP1211200	S. J. Fitch and S. K. Liu; *Ger. Pat.* 1 211 200 (1966) (*Chem. Abstr.*, 1966, **64**, 15 925).	555
66HCA842	L. Maier; *Helv. Chim. Acta*, 1966, **49**, 842.	453, 550
66IZV1108	I. L. Knunyants, Yu. V. Zeifman and N. P. Gambaryan; *Izv. Akad. Nauk SSSR, Ser. Khim.*, 1966, 1108.	297, 304
66IZV1954	K. S. Yudina, T. Ya. Medved' and M. I. Kabachnik; *Izv. Akad. Nauk SSSR, Ser. Khim.*, 1966, 1954.	556
66JA850	H. Weingarten and W. A. White; *J. Am. Chem. Soc.*, 1966, **88**, 850.	972
66JA865	N. Kornblum and H. W. Frazier; *J. Am. Chem. Soc.*, 1966, **88**, 865.	161
66JA1140	E. J. Boros, K. J. Coskran, R. W. King and J. G. Verkode; *J. Am. Chem. Soc.*, 1966, **88**, 1140.	342
66JA2481	S. W. Tobey and R. West; *J. Am. Chem. Soc.*, 1966, **88**, 2481.	4, 22, 23, 30
66JA3572	H. Goldwhite and D. G. Rowsell; *J. Am. Chem. Soc.*, 1966, **88**, 3572.	123, 125
66JA3672	N. J. Turro and W. B. Hammond; *J. Am. Chem. Soc.*, 1966, **88**, 3672.	162
66JA5044	P. W. Jolly and R. Pettit; *J. Am. Chem. Soc.*, 1966, **88**, 5044.	148
66JA5582	D. C. England and C. G. Krespan; *J. Am. Chem. Soc.*, 1966, **88**, 5582.	737
66JA5656	E. J. Corey and T. Durst; *J. Am. Chem. Soc.*, 1966, **88**, 5656.	394
66JA5682	L. A. Carpino and R. H. Rynbrandt; *J. Am. Chem. Soc.*, 1966, **88**, 5682.	83
66JCE527	C. D. Hurd; *J. Chem. Educ.*, 1966, **43**, 527.	163
66JCED620	B. D. Vineyard; *J. Chem. Eng. Data*, 1966, **11**, 620.	315
66JCS(B)789	R. F. Hudson, R. J. G. Searle and F. H. Devitt; *J. Chem. Soc. (B)*, 1966, 789.	476
66JCS(C)533	D. J. Cooper and L. N. Owen; *J. Chem. Soc., C*, 1966, 533.	20
66JCS(C)862	J. A. Bell and I. Dunstan; *J. Chem. Soc. (C)*, 1966, 862.	440, 442
66JCS(C)867	J. A. Bell and I. Dunstan; *J. Chem. Soc. (C)*, 1966, 867.	440
66JCS(C)870	J. A. Bell and I. Dunstan; *J. Chem. Soc. (C)*, 1966, 870.	440
66JCS(C)1137	W. R. Bamford, J. C. Lovie and J. A. C. Watt; *J. Chem. Soc., (C)*, 1966, 1137.	610
66JCS(C)2075	R. Fields, H. Goldwhite, R. H. Haszeldine and J. Kirman; *J. Chem. Soc. (C)*, 1966, 2075.	111
66JCS(C)2304	R. E. Banks and G. J. Moore; *J. Chem. Soc. (C)*, 1966, 2304.	799
66JHC531	I. C. Nordin; *J. Heterocycl. Chem.*, 1966, **3**, 531.	306
66JIC650	S. P. Gupta, S. N. Rastogi and R. K. Arora; *J. Indian Chem. Soc.*, 1966, **43**, 650.	982
66JMC127	W. R. Hardie, J. Hidalgo, I. F. Halverstadt and R. E. Allen; *J. Med. Chem.*, 1966, **9**, 127.	184, 189
66JOC369	H. E. Ungnade and L. W. Kissinger; *J. Org. Chem.*, 1966, **31**, 369.	994
66JOC556	A. I. Meyers and J. M. Greene; *J. Org. Chem.*, 1966, **31**, 556.	321
66JOC853	K. G. Shipp and M. E. Hill; *J. Org. Chem.*, 1966, **31**, 853.	178, 179
66JOC916	H. Gershon, J. A. A. Renwick, W. K. Wynn and R. D'Ascoli; *J. Org. Chem.*, 1966, **31**, 916.	3
66JOC950	D. J. Peterson; *J. Org. Chem.*, 1966, **31**, 950.	575

66JOC1232	L. A. Paquette and H. Stucki; *J. Org. Chem.*, 1966, **31**, 1232.	307
66JOC1603	K. Moedritzer and R. R. Irani; *J. Org. Chem.*, 1966, **31**, 1603.	470
66JOC1857	R. C. Neuman, Jr. and M. L. Rahm; *J. Org. Chem.*, 1966, **31**, 1857.	29, 40
66JOC2036	A. E. Bey and D. R. Weyenberg; *J. Org. Chem.*, 1966, **31**, 2036.	604
66JOC2373	D. J. Peterson and J. H. Collins; *J. Org. Chem.*, 1966, **31**, 2373.	575
66JOC2773	D. J. Pasto and Sr. R. Snyder; *J. Org. Chem.*, 1966, **31**, 2773.	142
66JOC2874	*J. Org. Chem.*, 1966, **31**, 2874.	972
66JOC2966	J. C. Martin, K. C. Brannock and R. H. Meen; *J. Org. Chem.*, 1966, **31**, 2966.	881
66JOC3032	I. J. Borowitz, G. Gonis, R. Kelsey, R. Rapp and G. J. Williams; *J. Org. Chem.*, 1966, **31**, 3032.	174
66JOC3391	H. R. Hays and T. J. Logan; *J. Org. Chem.*, 1966, **31**, 3391.	550
66JOC3514	I. E. Pollak and G. F. Grillot; *J. Org. Chem.*, 1966, **31**, 3514.	333
66JOC4041	H. Weingarten and W. A. White; *J. Org. Chem.*, 1966, **31**, 4041.	405
66JOC4097	E. J. Corey and D. Seebach; *J. Org. Chem.*, 1966, **31**, 4097.	250, 391, 392, 957
66JOC4288	R. J. Hartle; *J. Org. Chem.*, 1966, **31**, 4288.	569, 574
66JOM(5)226	M. Kumada, M. Ishikawa and K. Tamao; *J. Organomet. Chem.*, 1966, **5**, 226.	614
66JOM(5)486	G. J. D. Peddle; *J. Organomet. Chem.*, 1966, **5**, 486.	386
66JOM(6)100	D. S. Matteson and T.-C. Cheng; *J. Organomet. Chem.*, 1966, **6**, 100.	525, 632, 635
66JOM(6)633	W. R. Cullen and G. E. Styan; *J. Organomet. Chem.*, 1966, **6**, 633.	137, 141, 153
66JPR144	G. Domschke; *J. Prakt. Chem.*, 1966, **32**, 144.	307
66LA(697)171	W. Seeliger and W. Diepers; *Justus Liebigs Ann. Chem.*, 1966, **697**, 171.	296
66LA(699)40	K. Issleib and R. Lindner; *Justus Liebigs Ann. Chem.*, 1966, **699**, 40.	551
66LA(700)1	H. Kropf and R. Lambeck; *Justus Liebigs Ann. Chem.*, 1966, **700**, 1.	314
66M695	A. Stephen; *Monatsh. Chem.*, 1966, **97**, 695.	728
66MI 403-01	J. Pielichowski; *Rozniki Chem.*, 1966, **40**, 1765 (*Chem. Abstr.*, 1967, **66**, 75 868).	97
B-66MI 405-01	L. C. Schroeter; "Sulfur Dioxide," Pergamon, Oxford, 1966, p. 105 *et seq.*	237
B-66MI 409-01	H. Hellmann and G. Opitz; "α-Amino-alkylierung," Verlag Chemie, Weinheim, 1966, p. 1.	416
66OS(46)21	R. Schmutzler; *Org. Synth.*, 1966, **46**, 21.	133
66OS(46)31	V. Mark; *Org. Synth.*, 1966, **46**, 31.	476
66T2615	Y. Ito, M. Okano and R. Oda; *Tetrahedron*, 1966, **22**, 2615.	740
66T3001	A. J. Kirby; *Tetrahedron*, 1966, **22**, 3001.	852
66TL2311	W. E. Parham and J. R. Potoski; *Tetrahedron Lett.*, 1966, **7**, 2311.	979
66TL2315	H. Lehmkuhl and R. Schäfer; *Tetrahedron Lett.*, 1966, **7**, 2315.	698
66TL2535	G. Zweifel and H. Arzoumanian; *Tetrahedron Lett.*, 1966, **7**, 2535.	664, 665
66TL2675	P. Binger; *Tetrahedron Lett.*, 1966, **7**, 2675.	632
66TL3187	A. M. Aguiar, H. J. Aguiar and T. G. Archibald; *Tetrahedron Lett.*, 1966, **7**, 3187.	563
66TL3315	L. Horner, W. P. Balzer and D. J. Peterson; *Tetrahedron Lett.*, 1966, **7**, 3315.	575
66TL5263	G. Opitz and D. Bücher; *Tetrahedron Lett.*, 1966, **7**, 5263.	282
66TL6145	J. J. Tufariello and W. J. Kissel; *Tetrahedron Lett.*, 1966, **7**, 6145.	206
66USP3227714	J. C. Martin, *US Pat.* 3 227 714 (1966).	979
66USP3251907	C. H. Roy; *US Pat.* 3 251 907 (1966) (*Chem. Abstr.*, 1966, **65**, 3908).	546
66USP3253033	L. Maier; *US Pat.* 3 253 033 (1966) (*Chem. Abstr.*, 1966, **65**, 5488).	546
66USP3277068	L. A. Wall and W. J. Pummer (US Dept. of the Navy); *US Pat.* 3 277 068 (1966) (*Chem. Abstr.*, 1967, **66**, 18 591r).	761
66ZC28	H. G. Henning and D. Gloyna; *Z. Chem.*, 1966, **6**, 28.	1023
66ZC314	H. G. Henning and K. Forner; *Z. Chem.*, 1966, **6**, 314.	554
66ZC417	G. Barnikow and H. Niclas; *Z. Chem.*, 1966, **6**, 417.	903
66ZOB296	V. K. Khairullin, T. I. Sobchuk and A. N. Pudovik; *Zh. Obshch. Khim.*, 1966, **36**, 296.	561
66ZOR942	N. A. Nesmeyanov, V. M. Novikov and O. A. Reutov; *Zh. Org. Khim.*, 1966, **2**, 942.	587
67AG147	R. Gompper, E. Kutter and M. Kast; *Angew. Chem.*, 1967, **79**, 147.	901
67AG(E)41	G. Köbrich; *Angew. Chem., Int. Ed. Engl.*, 1967, **6**, 41.	145, 741, 817
67AG(E)74	F. Gerhart, U. Schöllkopf and H. Schumacher; *Angew. Chem. Int. Ed. Engl.*, 1967, **6**, 74.	142
67AG(E)677	G. Fritz; *Angew. Chem., Int. Ed. Engl.*, 1967, **6**, 677.	728
67AG(E)767	H. G. Viehe; *Angew. Chem., Int. Ed. Engl.*, 1967, **6**, 767.	976, 977
67AG(E)907	F. Asinger and H. Offermanns; *Angew. Chem., Int. Ed. Engl.*, 1967, **6**, 907.	328
67AG(E)959	K. Friedrich; *Angew. Chem., Int. Ed. Engl.*, 1967, **6**, 959.	792, 894
67AJC77	R. S. Dickson and D. B. W. Yawney; *Aust. J. Chem.*, 1967, **20**, 77.	688
67AP(300)241	H. Böhme and H. H. Hotzel; *Arch. Pharm.* (*Weinheim, Ger.*), 1967, **300**, 241.	297
67AP(300)647	H. Böhme and H.-H. Otto; *Arch. Pharm.* (*Weinheim, Ger.*), 1967, **300**, 647.	321
67BCJ594	S. Furukawa, K. Naruchi, S. Matsui and M. Yuuki; *Bull. Chem. Soc. Jpn.*, 1967, **40**, 594.	23, 27
67BCJ1275	T. Ando, F. Namigata, M. Kataoka, K. Yachida and W. Funasaka; *Bull. Chim. Soc. Jpn.*, 1967, **40**, 1275.	750
67BCJ2641	T. Mukaiyama, S. Aizawa and T. Yamaguchi; *Bull. Chem. Soc. Jpn.*, 1967, **40**, 2641.	902
67BRP1087066	Commonwealth Scientific and Industrial Research Organisation; *Br. Pat.* 1 087 066 (1967) (*Chem. Abstr.*, 1968, **68**, 95 956g).	131
67BSF571	P. A. Laurent; *Bull. Soc. Chim. Fr.*, 1967, 571.	728
67BSF1533	B. Castro; *Bull. Soc. Chim. Fr.*, 1967, 1533.	383
67BSF4172	L. Maguet and M. Lerer; *Bull. Soc. Chim. Fr.*, 1967, 4172.	212
67BSF4289	E. Gryszkiewicz-Trochimowski; *Bull. Chim. Soc. Fr.*, 1967, 4289.	129, 130, 131
67CB591	R. Gompper and H. Schaefer; *Chem. Ber.*, 1967, **100**, 591.	728
67CB1032	H. Schmidbaur and W. Tronich; *Chem. Ber.*, 1967, **100**, 1032.	571, 574, 588
67CB1087	L. Paul, E. Schuster and G. Hilgetag; *Chem. Ber.*, 1967, **100**, 1087.	793

67CB1289	H. von Hirsch; *Chem. Ber.*, 1967, **100**, 1289.	405
67CB1661	G. Barnikow and G. Strickmann; *Chem. Ber.*, 1967, **100**, 1661.	979
67CB2120	R. Kuhn and E. Teller; *Chem. Ber.* 1967, **100**, 2120.	53, 56
67CB2131	H. Böhme and D. Eichler; *Chem. Ber.*, 1967, **100**, 2131.	412
67CB2515	G. Zinner and W. Kliegel; *Chem. Ber.*, 1967, **100**, 2515.	405
67CB2577	A. Dornow and K. Dehmer; *Chem. Ber.*, 1967, **100**, 2577.	900
67CB2604	E. Allenstein and R. Fuchs; *Chem. Ber.*, 1967, **100**, 2604.	894
67CB2685	K. Issleib and H. Oehme; *Chem. Ber.*, 1967, **100**, 2685.	453, 454, 477, 478, 481
67CB3460	W. Schulze and M. Willitzer; *Chem. Ber.*, 1967, **100**, 3460.	893
67CB3893	M. Schlosser and V. Ladenberger; *Chem. Ber.*, 1967, **100**, 3893.	756
67CC889	H. Sakurai, M. Kira and M. Kumada; *J. Chem. Soc., Chem. Commun.*, 1967, 889.	395
67CJC2011	R. Bruce, K. Moseley and P. M. Maitlis; *Can. J. Chem.*, 1967, **45**, 2011.	715
67DOK(177)340	V. E. Bel'skii, T. A. Zyablikova, A. R. Panteleeva and I. M. Shermergorn; *Dokl. Akad. Nauk SSSR*, 1967, **177**, 340.	556
67FCR359	J. A. Young; *Fluorine Chem. Rev.*, 1967, **1**, 359.	1076
67G1327	G. Peyronel, A. Ragni and E. F. Trogu; *Gazz. Chim. Ital.*, 1967, **97**, 1327.	687
67HCA1723	L. Maier; *Helv. Chim. Acta*, 1967, **50**, 1723.	452
67HCA1742	L. Maier; *Helv. Chim. Acta*, 1967, **50**, 1742.	470
67IC1751	S. Cradock, G. A. Gibbon and C. H. Van Dyke; *Inorg. Chem.*, 1967, **6**, 1751.	360
67IC1989	G. A. Gibbon, J. T. Wang and C. H. Van Dyke; *Inorg. Chem.*, 1967, **6**, 1989.	360
67IJC210	D. N. Sen and N. Thankarajan; *Indian J. Chem.*, 1967, **5**, 210.	697
67IS63	R. Schmutzler; *Inorg. Synth.*, 1967, **9**, 63.	133
67IZV591	M. I. Kabachnik, T. Ya. Medved', Yu. M. Polikarpov and K. S. Yudina; *Izv. Akad. Nauk SSSR, Ser. Khim.*, 1967, 591.	555, 561
67IZV949	M. I. Kabachnik, L. P. Zhuravleva, M. G. Suleimanova, Yu. M. Polikarpov and T. Ya. Medved'; *Izv. Akad. Nauk SSSR, Ser. Khim.*, 1967, 949.	556
67IZV1535	E. Kh. Mukhametzyanova, A. R. Panteléeva and I. M. Shermergorn; *Izv. Akad. Nauk SSSR, Ser. Khim.*, 1967, 1535.	127
67JA291	G. Zweifel and H. Arzoumanian; *J. Am. Chem. Soc.*, 1967, **89**, 291.	631
67JA3366	C. B. Reese, R. Saffhill and J. E. Sulston; *J. Am. Chem. Soc.*, 1967, **89**, 3366.	186
67JA4099	F. Wudl, D. A. Lightner and D. J. Cram; *J. Am. Chem. Soc.*, 1967, **89**, 4099.	78
67JA4456	N. J. Leonard and B. Zwanenburg; *J. Am. Chem. Soc.*, 1967, **89**, 4456.	305
67JA4483	L. A. Paquette and L. S. Wittenbrook; *J. Am. Chem. Soc.*, 1967, **89**, 4483.	69
67JA4487	L. A. Paquette, L. S. Wittenbrook and V. V. Kane; *J. Am. Chem. Soc.*, 1967, **89**, 4487.	82
67JA5086	G. Zweifel and H. Arzoumanian; *J. Am. Chem. Soc.*, 1967, **89**, 5086.	814
67JA5838	M. S. Simon, J. B. Rogers, W. Saenger and J. Z. Gougoutas; *J. Am. Chem. Soc.*, 1967, **89**, 5838.	71
67JA6573	R. N. McDonald and T. E. Tabor; *J. Am. Chem. Soc.*, 1967, **89**, 6573.	50
67JA6804	J. J. Tufariello, L. T. C. Lee and P. Wojtkowski; *J. Am. Chem. Soc.*, 1967, **89**, 6804.	832
67JCED282	L. H. Chance, D. G. Daigle and G. L. Drake, Jr.; *J. Chem. Eng. Data*, 1967, **12**, 282.	113, 118, 120, 122
67JCS(C)58	M. A. Rehman and G. R. Proctor; *J. Chem. Soc.(C)*, 1967, 58.	105
67JCS(C)562	J. A. Bell and I. Dunstan; *J. Chem. Soc. (C)*, 1967, 562.	442, 443
67JCS(C)735	M. J. Harrison, R. O. C. Norman and W. A. F. Gladstone; *J. Chem. Soc. (C)*, 1967, 735.	314
67JCS(C)1130	J. M. Cox and L. N. Owen; *J. Chem. Soc. (C)*, 1967, 1130.	226
67JCS(C)1411	G. M. Clarke and P. Sykes; *J. Chem. Soc. (C)*, 1967, 1411.	321
67JCS(C)1569	R. N. Pratt, G. A. Taylor and S. A. Proctor; *J. Chem. Soc (C)*, 1967, 1569.	886
67JCS(C)1862	C. W. Bird, E. M. Briggs and J. Hudec; *J. Chem. Soc. (C)*, 1967, 1862.	681
67JCS(C)2696	G. R. Proctor and M. A. Rehman; *J. Chem. Soc. (C)*, 1967, 2696.	300
67JGU1768	B. E. Ivanov, A. R. Panteleeva, R. R. Shagidullin and I. M. Shermergorn; *J. Gen. Chem. USSR (Engl. Transl.)*, 1967, **37**, 1768.	127
67JGU2024	V. M. Ignat'ev, B. I. Ionin and A. A. Petrov; *J. Gen. Chem. USSR (Engl. Transl.)*, 1967, **37**, 2024.	802
67JHC625	H. Ahrens and W. Korytnyk; *J. Heterocycl. Chem.*, 1967, **4**, 625.	164
67JIC995	R. C. Paul, R. Kaushal and S. S. Pahil; *J. Indian Chem. Soc.*, 1967, **44**, 995.	504
67JOC204	D. L. Tuleen and V. C. Marcum; *J. Org. Chem.*, 1967, **32**, 204.	63, 69
67JOC213	W. A. White and H. Weingarten; *J. Org. Chem.*, 1967, **32**, 213.	405
67JOC607	C. B. Anderson and D. T. Sepp; *J. Org. Chem.*, 1967, **32**, 607.	49, 51
67JOC990	W. E. Truce, D. J. Abraham and P. Son; *J. Org. Chem.*, 1967, **32**, 990.	979
67JOC1311	F. E. Herkes and D. J. Burton; *J. Org. Chem.*, 1967, **32**, 1311.	732
67JOC1717	D. J. Peterson; *J. Org. Chem.*, 1967, **32**, 1717.	365, 391, 392
67JOC1941	A. D. Josey, C. L. Dickinson, K. C. Dewhirst and B. C. McKusick; *J. Org. Chem.*, 1967, **32**, 1941.	742
67JOC2383	A. M. Aguiar, K. C. Hansen and J. T. Mague; *J. Org. Chem.*, 1967, **32**, 2383.	455, 477
67JOC2595	M. E. Hill and L. O. Ross; *J. Org. Chem.*, 1967, **32**, 2595.	170
67JOC2651	W. H. Mueller, P. E. Butler and K. Griesbaum; *J. Org. Chem.*, 1967, **32**, 2651.	15
67JOC2669	H. Hart and L. R. Lerner; *J. Org. Chem.*, 1967, **32**, 2669.	24
67JOC2749	M. H. Kaufman, J. D. Braun and J. G. Shdo; *J. Org. Chem*, 1967, **32**, 2749.	735
67JOC2891	I. E. Pollak and G. F. Grillot; *J. Org. Chem.*, 1967, **32**, 2891.	317
67JOC3101	I. E. Pollak and G. F. Grillot; *J. Org. Chem.*, 1967, **32**, 3101.	333
67JOC3293	H. Weingarten and N. K. Edelmann; *J. Org. Chem.*, 1967, **32**, 3293.	975
67JOC3535	R. J. Fessenden and J. S. Fessenden; *J. Org. Chem.*, 1967, **32**, 3535.	831
67JOC4034	R. C. Petry and J. P. Freeman; *J. Org. Chem.*, 1967, **32**, 4034.	104

67JOM(8)149	G. Cetini, O. Gambino, R. Rossetti and E. Sappa; *J. Organomet. Chem.*, 1967, **8**, 149.	687
67JOM(8)199	D. J. Peterson; *J. Organomet. Chem.*, 1967, **8**, 199.	546, 575
67JOM(8)361	P. R. Jones, C. J. Jarboe and R. Nadeau; *J. Organomet. Chem.*, 1967, **8**, 361.	53, 54
67JOM(9)117	H. Gorth and M. C. Henry; *J. Organomet. Chem.*, 1967, **9**, 117.	699
67JOM(9)373	D. J. Peterson; *J. Organomet. Chem.*, 1967, **9**, 373.	649
67JOM(10)111	M. Kumada, H. Tsunemi and S. Iwasaki; *J. Organomet. Chem.*, 1967, **10**, 111.	607, 654
67JSP(23)32	D. G. Rowsell; *J. Mol. Spectr.*, 1967, **23**, 32.	728
67LA85	F. Nerdel, J. Buddrus, G. Scherowsky, D. Klamann and M. Fligge; *Justus Liebigs Ann. Chem.*, 1967, 85.	181
67LA(702)24	G. Wittig and M. Jautelat; *Justus Liebigs Ann. Chem.*, **702**, 1967, 24.	385
67LA(703)1	H. Hoberg; *Justus Liebigs Ann. Chem.*, 1967, **703**, 1.	153
67LA(704)120	U. Schollkopf, H. Kueppers, H. J. Traenckner and W. Pitteroff; *Justus Liebigs Ann. Chem.*, 1967, **704**, 120.	381
67LA(706)107	F. A. Neugebauer and H. Trischmann; *Justus Liebigs Ann. Chem.*, 1967, **706**, 107.	436
67LA(707)112	K. Issleib and R. Lindner; *Justus Liebigs Ann. Chem.*, 1967, **707**, 112.	557
67LA(707)120	K. Issleib and R. Lindner; *Justus Liebigs Ann. Chem.*, 1967, **707**, 120.	551
B-67MI 404-01	E. Schmitz and I. Eichhorn; in "The Chemistry of the Ether Linkage," ed. S. Patai, Wiley, New York, 1967, p. 309.	163, 176
67MIP1087066	Commonwealth Scientific and Industrial Research Organisation and Imperial Chemical Industries of Australia and New Zealand, *Pat.* 1 087 066 (1967) (*Chem. Abstr.*, 1968, **68**, 9595b).	492
67OS(47)23	C. S. Davis and G. S. Lougheed; *Org. Synth.*, 1967, **47**, 23.	47
67OS(47)78	W. C. Kuryla and J. E. Hyre; *Org. Synth.*, 1967, **47**, 78.	829
67RTC417	W. J. M. van Tilborg, S. E. Schaafsma, H. Steinberg and Th. J. De Boer; *Recl. Trav. Chim. Pays-Bas*, 1967, **86**, 417.	406
67T2549	R. A. Moss and R. Gerstl; *Tetrahedron*, 1967, **23**, 2549.	32
67T2781	L. Birkofer, F. Müller and W. Kaiser; *Tetrahedron*, 1967, **23**, 2781.	313
67T2869	S. W. Breuer, T. Bernath and D. Ben-Ishai; *Tetrahedron*, 1967, **23**, 2869.	304
67TL723	D. J. Pasto, J. Chow and S. K. Arora; *Tetrahedron Lett.*, 1967, **8**, 723.	632, 634
67TL995	W. K. Musker and R. R. Stevens; *Tetrahedron Lett.*, 1967, **8**, 995.	522
67TL1123	T. Ando, H. Yamanaka, S. Terabe, A. Horike and W. Funasaka; *Tetrahedron Lett.*, 1967, **8**, 1123.	752
67TL1443	E. J. Bulten and J. G. Noltes; *Tetrahedron Lett.*, 1967, **8**, 1443.	626
67TL1489	K. Issleib and H. Oehme; *Tetrahedron Lett.*, 1967, **8**, 1489.	453, 481
67TL3057	R. J. Mulder, A. M. van Leusen and J. Strating; *Tetrahedron Lett.*, 1967, **8**, 3057.	218, 235
67TL3061	R. J. Mulder, A. M. van Leusen and J. Strating; *Tetrahedron Lett.*, 1967, **8**, 3061.	235
67TL3085	N. J. Turro and W. B. Hammond; *Tetrahedron Lett.*, 1967, **8**, 3085.	406
67TL3201	E. J. Corey and G. Markl; *Tetrahedron Lett.*, 1967, **8**, 3201.	850
67TL3769	O. Tsuge, M. Tashiro and Y. Nishihara; *Tetrahedron Lett.*, 1967, **8**, 3769.	307
67USP3306937	R. B. Clampitt, G. H. Birum and R. M. Anderson; *US Pat.* 3 306 937 (1967) (*Chem. Abstr.*, 1967, **66**, 105 051h).	122
67USP3314900	E. H. Uhing; *US Pat.* 3 314 900 (1967) (*Chem. Abstr.*, 1967, **67**, 12 086d).	117
67USP3316311	C. W. Plummer; *US Pat.* 3 316 311 (1967) (*Chem. Abstr.*, 1967, **67**, 53650).	429, 431
67USP3332986	B. P. Block, I. C. Popoff, J. P. King and L. K. Huber; *US Pat.* 3 332 986 (1967) (*Chem. Abstr.*, 1967, **67**, 64 534).	556
67USP3335188	M. L. Oftedahl (Monsanto Co.); *US Pat.* 3 335 188 (1967) (*Chem. Abstr.* 1968, **68**, 69 078).	864
67USP3346588	B. A. Ashby; *US Pat.* 3 346 588 (1967) (*Chem. Abstr.*, 1968, **68**, 78 131s).	520
67USP3360556	K. Moedritzer; *US Pat.* 3 360 556 (1967) (*Chem. Abstr.*, 1968, **68**, 49 755t).	123
67ZC421	G. Rembarz and B. Ernst; *Z. Chem.*, 1967, **7**, 421.	440
67ZOB454	S. Z. Ivin, V. K. Promonenkov and B. I. Tetel'baum; *Zh. Obshch. Khim.*, 1967, **37**, 454.	923
67ZOB1623	V. V. Moskva, A. I. Maikova and A. I. Razumov; *Zh. Obshch. Khim.*, 1967, **37**, 1623.	912
67ZOB2055	B. M. Gladstein and V. M. Zimin; *Zh. Obshch. Khim.*, 1967, **37**, 2055.	554
67ZOR1006	B. L. Dyatkin, Yu. S. Konstantinov, L. T. Lantseva, R. A. Bekker and I. L. Knunyants; *Zh. Org. Khim.*, 1967, **3**, 1006.	761, 763
68ACR299	H. G. Kuivila; *Acc. Chem. Res.*, 1968, **1**, 299.	13
68ACS3256	A. Senning, S. Kaae, C. Jacobsen and P. Kelly; *Acta Chem. Scand.*, 1968, **22**, 3256.	77
68AG38	R. Gompper and H.-D. Lehmann; *Angew. Chem.*, 1968, **80**, 38.	901
68AG(E)172	R. Graf; *Angew. Chem., Int. Ed. Engl.*, 1968, **7**, 172.	310
68AG(E)286	P. Binger; *Angew. Chem., Int. Ed. Engl.*, 1968, **7**, 286.	632
68AG(E)293	H. Schildknecht and G. Hatzmann; *Angew. Chem., Int. Ed. Engl.*, 1968, **7**, 293.	441
68AG(E)384	L. Maier; *Angew. Chem., Int. Ed. Engl.*, 1968, **7**, 384.	556
68AG(E)385	L. Maier; *Angew. Chem., Int. Ed. Engl.*, 1968, **7**, 385.	556
68AG(E)747	M. I. Bruce and F. G. A. Stone; *Angew. Chem., Int. Ed. Engl.*, 1968, **7**, 747.	821
68AG(E)766	N. Wiberg; *Angew. Chem., Int. Ed. Engl.*, 1968, **7**, 766.	975, 985
68AJC1197	M. J. Gallagher; *Aust. J. Chem.*, 1968, **21**, 1197.	119
68AP(301)867	D. Matthies; *Arch. Pharm. (Weinheim, Ger.)*, 1968, **301**, 867.	311
68BCJ756	H. Yamanaka, T. Ando and W. Funasaka; *Bull. Chem. Soc. Jpn.*, 1968, **41**, 756.	752
68BRP1130487	Monsanto Co.; *Br. Pat.* 1 130 487 (1968) (*Chem. Abstr.*, 1969, **70**, 37 915).	550
68BSB379	M. Renson and L. Christiaens; *Bull. Soc. Chim. Belg.*, 1968, **71**, 379.	205
68CA(68)49691u	A. N. Pudovik and T. Kh. Gazizov; *Chem. Abstr.*, 1968, **68**, 49 691u (*Dokl. Akad. Nauk SSSR*, 1967, **175**, 1073).	915

68CB595	H. Schmidbaur and W. Tronich; *Chem. Ber.*, 1968, **101**, 595.	569, 587
68CB743	E. Ettenhuber and K. Rühlmann; *Chem. Ber.*, 1968, **101**, 743.	515
68CB1131	C. Metzger and R. Wegler; *Chem. Ber.*, 1968, **101**, 1131.	901, 904
68CB1232	E. Allenstein and R. Fuchs; *Chem. Ber.*, 1968, **101**, 1232.	972
68CB3545	H. Schmidbaur and W. Tronich; *Chem. Ber.*, 1968, **101**, 3545.	551, 574
68CB3556	H. Schmidbaur and W. Tronich; *Chem. Ber.*, 1968, **101**, 3556.	576
68CB3604	M. Regitz and H.-G. Adolph; *Chem. Ber.*, 1968, **101**, 3604.	55, 57
68CB3612	K. Issleib, R. Kümmel, H. Oehme and I. Meißner; *Chem. Ber.*, 1968, **101**, 3612.	453
68CB3619	K. Issleib, H. Oehme, R. Kümmel and E. Leißring; *Chem. Ber.*, 1968, **101**, 3619.	453, 454, 477
68CB4032	K. Issleib, H. Oehme and E. Leißring; *Chem. Ber.*, 1968, **101**, 4032.	453, 454, 477, 481
68CC930	H. Sakurai, A. Hosomi and M. Kumada; *J. Chem. Soc., Chem. Commun.*, 1968, 930.	616
68CC1365	A. N. Nesmeyanov, A. I. Gusev, A. A. Pasynskii, K. N. Anisimov, N. E. Kolobova and Yu. T. Struchkov; *J. Chem. Soc., Chem. Commun.*, 1968, 1365.	672
68CJC385	R. N. Renaud and L. C. Leitch; *Can. J. Chem.*, 1968, **46**, 385.	311
68CJC2115	A. G. Brook and D. G. Anderson; *Can. J. Chem.*, 1968, **46**, 2115.	222, 370, 371, 570
68CJC2119	A. G. Brook, P. F. Jones and G. J. D. Peddle; *Can. J. Chem.*, 1968, **46**, 2119.	609
68CJC2165	M. Akhtar and H. C. Clark; *Can. J. Chem.*, 1968, **46**, 2165.	153, 811
68CJC2251	R. Raap; *Can. J. Chem.*, 1968, **46**, 2251.	852
68CJC2255	R. Raap; *Can. J. Chem.*, 1968, **46**, 2255.	903
68GEP1924135	B. G. Christensen and R. A. Firestone; *Ger. Pat.* 1 924 135 (1968) (*Chem. Abstr.*, 1970, **72**, 43 870).	346
68HCA1466	R. Giger, M. Rey and A. S. Dreiding; *Helv. Chim. Acta*, 1968, **51**, 1466.	768
68HCA1608	L. Maier; *Helv. Chim. Acta*, 1968, **51**, 1608.	466
68HOU(7)341	D. Borrmann; *Methoden Org. Chem.* (*Houben-Weyl*), 1968, **7**, 341.	969
68HOU(7/4)434	D. Borrmann; *Methoden Org. Chem.* (*Houben-Weyl*), 1968, **7/4**, 434.	898
68IC168	H. Schmidbaur and W. Tronich; *Inorg. Chem.*, 1968, **7**, 168.	597
68IC709	D. R. Mathiason and N. E. Miller; *Inorg. Chem.*, 1968, **7**, 709.	562
68IZV270	R. G. Kostyanovskii and A. K. Prokof'ev; *Izv. Akad. Nauk SSSR, Ser. Khim.*, 1968, 270.	155
68IZV686	A. N. Nesmeyanov, K. N. Anisimov, N. E. Kolobova and Yu. V. Makarov; *Izv. Akad. Nauk SSSR, Ser. Khim.*, 1968, 686.	706
68IZV1417	T. Ya. Medved', Yu. M. Polikarpov, S. A. Pisareva and M. I. Kabachnik; *Izv. Akad. Nauk SSSR, Ser. Khim.*, 1968, 1417.	556
68IZV1557	A. R. Panteleeva and I. M. Shermergorn; *Izv. Akad. Nauk SSSR, Ser. Khim.*, 1968, 1557.	127, 128
68IZV1625	N. V. Ivasyuk, Ekh. Mukhametzyanova and I. M. Shermergorn; *Izv. Akad. Nauk SSSR, Ser. Khim.*, 1968, 1625.	346
68IZV2062	T. Ya. Medved', Yu. M. Polikarpov, S. A. Pisareva, E. I. Matrosov and M. I. Kabachnik; *Izv. Akad. Nauk SSSR, Ser. Khim.*, 1968, 2062.	556
68JA435	F. G. Bordwell and J. M. Williams, Jr.; *J. Am. Chem. Soc.*, 1968, **90**, 435.	83, 84
68JA818	H. C. Brown, M. M. Rogić, M. W. Rathke and G. W. Kabalka; *J. Am. Chem. Soc.*, 1968, **90**, 818.	832
68JA1057	R. G. Amiet and R. Pettit; *J. Am. Chem. Soc.*, 1968, **90**, 1057.	724
68JA1060	M. Rosenblum and B. North; *J. Am. Chem. Soc.*, 1968, **90**, 1060.	690
68JA1080	D. Seyferth, A. W. Dow, H. Menzel and T. C. Flood; *J. Am. Chem. Soc.*, 1968, **90**, 1080.	513
68JA2194	R. B. Castle and D. S. Matteson; *J. Am. Chem. Soc.*, 1968, **90**, 2194.	633
68JA2915	H. C. Brown and R. L. Sharp; *J. Am. Chem. Soc.*, 1968, **90**, 2915.	142
68JA3781	G. E. Lienhard and T.-C. Wang; *J. Am. Chem. Soc.*, 1968, **90**, 3781.	899
68JA4495	K. D. Berlin, N. K. Roy, R. T. Claunch and D. Bude; *J. Am. Chem. Soc.*, 1968, **90**, 4495.	472
68JA4496	M. Hojo and Z. Yoshida; *J. Am. Chem. Soc.*, 1968, **90**, 4496.	74, 80
68JA5548	E. J. Corey and T. Durst; *J. Am. Chem. Soc.*, 1968, **90**, 5548.	394
68JA5924	W. G. Bentrude and W. D. Johnson; *J. Am. Chem. Soc.*, 1968, **90**, 5924.	914
68JA6259	D. J. Pasto, J. Hickman and T-C. Cheng; *J. Am. Chem. Soc.*, 1968, **90**, 6259.	144
68JA6816	E. J. Corey and G. T. Kwaitkowski; *J. Am. Chem. Soc.*, 1968, **90**, 6816.	576
68JA7083	K. Baum; *J. Am. Chem. Soc.*, 1968, **90**, 7083.	439
68JA7367	W. T. Miller, Jr. and R. J. Burnard; *J. Am. Chem. Soc.*, 1968, **90**, 7367.	152
68JCED585	D. J. Daigle, L. H. Chance and G. L. Drake; *J. Chem. Eng. Data*, 1968, **13**, 585.	555
68JCS(A)1105	E. W. Abel and I. H. Sabherwal; *J. Chem. Soc. (A)*, 1968, 1105.	573
68JCS(A)1293	N. A. Bailey and R. Mason; *J. Chem. Soc. (A)*, 1968, 1293.	687
68JCS(C)7	J. S. McConaghy and J. J. Bloomfield; *J. Chem. Soc. (C)*, 1968, 7.	168
68JCS(C)656	D. J. Booth and B. W. Rockett; *J. Chem. Soc. (C)*, 1968, 656.	706
68JCS(C)796	E. S. Alexander, R. N. Haszeldine, M. J. Newlands and A. E. Tipping; *J. Chem. Soc. (C)*, 1968, 796.	979
68JCS(C)1726	A. Halleux and H. G. Viehe; *J. Chem. Soc. (C)*, 1968, 1726.	979
68JCS(C)2721	P. Ferruti, A. Segre and A. Fere; *J. Chem. Soc (C)*, 1968, 2721.	301, 309, 979
68JCS(C)3011	I. T. Kay and N. Punja; *J. Chem. Soc. (C)*, 1968, 3011.	840
68JGU1276	B. S. Drach and A. D. Sinitsa; *J. Gen. Chem. USSR (Engl. Transl.)*, 1968, **38**, 1276.	106
68JGU1284	Zh. M. Ivanova, S. K. Mikhailik and G. I. Derkach; *J. Gen. Chem. USSR (Engl. Transl.)*, 1968, **38**, 1284.	130
68JGU2006	U. K. Khairullin, M. A. Vasyanina and A. N. Pudovik; *J. Gen. Chem. USSR (Engl. Transl.)*, 1968, **38**, 2006.	125, 127, 128
68JGU2563	N. N. Mel'nikov, L. V. Razvodovskaya and A. F. Grapov; *J. Gen. Chem. USSR (Engl. Transl.)*, 1968, **38**, 2563.	131
68JINC1715	H. G. Ang, G. Manoussakis and Y. O. El-Nigumi; *J. Inorg. Nucl. Chem.*, 1968, **30**, 1715.	116
68JOC25	F. Ramirez, K. Tasaka, N. B. Desai and C. P. Smith; *J. Org. Chem.*, 1968, **33**, 25.	27

Ref	Citation	Pages
68JOC286	F. G. Drakesmith, R. D. Richardson, O. J. Stewart and P. Tarrant; *J. Org. Chem.*, 1968, **33**, 286.	817
68JOC472	F. G. Drakesmith, O. J. Stewart and P. Tarrant; *J. Org. Chem.*, 1968, **33**, 472.	806
68JOC780	D. J. Peterson; *J. Org. Chem.*, 1968, **33**, 780.	646, 916
68JOC816	D. C. England and C. G. Krespan; *J. Org. Chem.*, 1968, **33**, 816.	761
68JOC1080	L. A. Paquette, L. S. Wittenbrook and K. Schreiber; *J. Org. Chem.*, 1968, **33**, 1080.	64, 69, 80
68JOC1506	H. Weingarten and M. G. Miles; *J. Org. Chem.*, 1968, **33**, 1506.	972
68JOC1854	D. J. Burton and F. E. Herkes; *J. Org. Chem.*, 1968, **33**, 1854.	732
68JOC1861	S. F. Reed, Jr.; *J. Org. Chem.*, 1968, **33**, 1861.	104
68JOC1991	B. R. O'Conner; *J. Org. Chem.*, 1968, **33**, 1991.	767, 950
68JOC2104	I. B. Douglass and R. V. Norton; *J. Org. Chem.*, 1968, **33**, 2104.	78
68JOC2133	G. E. Wilson, M. G. Huang and W. W. Schlomann; *J. Org. Chem.*, 1968, **33**, 2133.	229
68JOC2173	J. G. Shdo, M. H. Kaufman and D. W. Moore; *J. Org. Chem.*, 1968, **33**, 2173.	736
68JOC2330	G. N. Sausen and A. L. Logothetis; *J. Org. Chem.*, 1968, **33**, 2330.	104
68JOC2368	N. P. Marullo and J. A. Alford; *J. Org. Chem.*, 1968, **33**, 2368.	162
68JOC2692	L. G. Anello, A. K. Price and R. F. Sweeney; *J. Org. Chem.*, 1968, **33**, 2692.	160
68JOC2934	R. N. McDonald and T. E. Tabor; *J. Org. Chem.*, 1968, **33**, 2934.	50
68JOC3055	D. S. Matteson and T.-C. Cheng; *J. Org. Chem.*, 1968, **33**, 3055.	144, 373
68JOC3090	K. D. Berlin, N. K. Roy and E. T. Gaudy; *J. Org. Chem.*, 1968, **33**, 3090.	472
68JOC3335	M. Shamma, L. Novak and M. G. Kelly; *J. Org. Chem.*, 1968, **33**, 3335.	47
68JOC4470	L. F. Ward Jr., R. R. Whetstone, G. E. Pollard and D. D. Phillips; *J. Org. Chem.*, 1968, **33**, 4470.	836, 837, 838
68JOM(11)644	C. T. Sears, Jr. and F. G. A. Stone; *J. Organomet. Chem.*, 1968, **11**, 644.	683
68JOM(12)133	H. Goldwhite, D. G. Rowsell and C. Valdez; *J. Organomet. Chem.*, 1968, **12**, 133.	805
68JOM(12)269	T. D. Coyle and J. J. Ritter; *J. Organomet. Chem.*, 1968, **12**, 269.	632, 644
68JOM(13)199	O. T. Quimbly, J. D. Curry, D. A. Nicholson, J. B. Prentice and C. H. Roy; *J. Organomet. Chem.*, 1968, **13**, 199.	563
68JPS715	B. B. Thompson; *J. Pharm. Sci.*, 1968, **57**, 715.	299
68LA(713)12	K. Issleib and R. Lindner; *Justus Liebigs Ann. Chem.*, 1968, **713**, 12.	551, 558
68M380	H. Hoffmann and H. Förster; *Monatsh. Chem.*, 1968, **99**, 380.	481, 487
68M990	T. Kappe, E. Lender and E. Ziegler; *Monatsh. Chem.*, 1968, **99**, 990.	411
68MI 403-01	Y. Matsumura and R. Okawara; *Inorg. Nucl. Chem. Lett.*, 1968, **4**, 219.	136
68MI 407-01	M. H. Balba, M. S. Singer, M. Slade and J. E. Casida; *J. Agric. Food. Chem.*, 1968, **16**, 821 (*Chem. Abstr.*, 1968, **69**, 86 522).	310
B-68MI 407-02	A. O. S. Fitton and R. K. Smalley; "Practical Heterocyclic Chemistry," Academic Press, London, 1968, p. 31.	312
B-68MI 410-01	D. W. Hutchinson and B. J. Walker (Senior Reporters); "Organophosphorus Chemistry," Chemical Society, Specialist Periodicals, London, 1968–1995.	451, 493
B-68MI 415-01	W. Hübel; in "Organic Synthesis via Metal Carbonyls," ed. I. Wender and P. Pino, Interscience, New York, 1968, vol 1.	681
68MI 421-01	F. Clesse, M. Le Goff and H. Quiiou; *C. R. Acad. Sci. Paris Ser. C*, 1968, **266**, 1799.	969
B-68MI 424-01	G. A. Olah and P. Von R. Schleyer (ed.); "Carbonium Ions," Wiley, New York, 1968–1976, vols 1–5.	1072
68OS(48)97	R. Gompper and W. Elser; *Org. Synth.*, 1968, **48**, 97.	902
68PAC235	R. Pettit; *Pure Appl. Chem.*, 1968, **17**, 235.	706
68RCR7	M. I. Kabachnik, T. Ya. Medved, N. M. Dyatlova, O. G. Arkhipova and M. V. Rudomino; *Russ. Chem. Rev.* (*Engl. Transl.*), 1968, **38**, 7.	451, 469
68RTC49	E. Molenaar and J. Strating; *Recl. Trav. Chim. Pays-Bas.*, 1968, **87**, 49.	777
68RTC188	F. Bickelhaupt and J. W. F. K. Barnick; *Recl. Trav. Chim. Pays-Bas*, 1968, **87**, 188.	522
68RTC929	P. J. W. Schuijl and L. Brandsma; *Recl. Trav. Chim. Pays-Bas*, 1968, **87**, 929.	834
68RTC1179	S. Hoff, L. Brandsma and J. F. Arens; *Recl. Trav. Chim. Pays-Bas*, 1968, **87**, 1179.	836
68T2767	H. Weingarten; *Tetrahedron*, 1968, **24**, 2767.	975
68TL71	D. J. Burton and H. C. Krutzsch; *Tetrahedron Lett.*, 1968, **9**, 71.	732, 751, 752
68TL497	T. Oishi, M. Ochiai, M. Nagai and Y. Ban; *Tetrahedron Lett.*, 1968, **9**, 497.	881
68TL937	J. Ficini and J. C. Depezay; *Tetrahedron Lett.*, 1968, **9**, 937.	817
68TL1977	W. T. Brady and R. Roe, Jr.; *Tetrahedron Lett.*, 1968, **9**, 1977.	768
68TL5415	R. N. Loeppky and D. C. K. Chang; *Tetrahedron Lett.*, 1968, **9**, 5415.	74
68TL5731	G. H. Jones, E. K. Hamamura and J. G. Moffatt; *Tetrahedron Lett.*, 1968, **9**, 5731.	554
68TL6003	J. M. Lavanish; *Tetrahedron Lett.*, 1968, **9**, 6003.	768
68USP3403176	B. P. Block and G. H. Dahl; *US Pat.* 3 403 176 (1968) (*Chem. Abstr.*, 1968, **69**, 106 870).	555
68ZOB139	A. N. Pudovic and T. Kh. Gazuzov; *Zh. Obshch. Khim.*, 1968, **38**, 139.	915
68ZOB292	A. N. Pudovik, G. E. Yastrbova, V. I. Nikitina and Yu. Yu. Samitov; *Zh. Obshch. Khim.*, 1968, **38**, 292.	561
68ZOB1248	I. G. Kolokol'tseva, V. N. Chistoklletov, B. I. Ionin and A. A. Petrov; *Zh. Obshch. Khim.*, 1968, **38**, 1248 (*Chem. Abstr.*, 1968, **69**, 96 834y).	488
68ZOB1503	L. M. Yagupol'skii and A. M. Aleksandrov; *Zh. Obshch. Khim.*, 1968, **38**, 1503 (*Chem. Abstr.* 1969, **70**, 3 436).	861
68ZOB1523	V. S. Tsvunin, Yu. Afanas'ev, R. G. Ivanova, T. A. Zyablikova and Kh. G. Kamai; *Zh. Obshch. Khim.*, 1968, **38**, 1523.	911
68ZOB2071	V. K. Khairullin, M. A. Vasyanina and A. N. Pudovik; *Zh. Obshch. Khim.*, 1968, **38**, 2071.	555
68ZOR1685	N. A. Nesmeyanov, S. T. Berman, L. D. Ashkinadze, L. A. Kazitsyna and O. A. Reutov; *Zh. Org. Khim.*, 1968, **4**, 1685.	587

68ZOR2116	A. D. Nikolaeva, A. I. Sitkin and G. Kh. Kamai; *Zh. Org. Khim.*, 1968, **4**, 2116 (Engl. Transl. 2043).	993
69AG(E)216	H. J. Bestmann, R. Saalfrank and J. P. Snyder; *Angew. Chem., Int. Ed. Engl.*, 1969, **8**, 216.	802, 825
69AG(E)372	H. Schmidbaur and W. Malisch; *Angew. Chem., Int. Ed. Engl.*, 1969, **8**, 372.	569
69AG(E)450	D. Seebach and N. Peleties; *Angew. Chem., Int. Ed. Engl.*, 1969, **8**, 450.	289, 398, 400
69AG(E)454	L. Ghosez, B. Haveaux and H. G. Viehe; *Angew. Chem., Int. Ed. Engl.* 1969, **8**, 454.	891, 904, 905
69AG(E)639	D. Seebach; *Angew. Chem., Int. Ed. Engl.*, 1969, **8**, 639.	1075
69AG(E)677	R. Rienäcker and H. Yoshiura; *Angew. Chem., Int. Ed. Engl.*, 1969, **8**, 677.	694
69AG(E)678	C. Krüger; *Angew. Chem., Int. Ed. Engl.*, 1969, **8**, 678.	694
69AJC1143	R. S. Dickson and G. R. Tailby; *Aust. J. Chem.*, 1969, **22**, 1143.	687
69AP(302)81	H. Bohme and W. Pasche; *Arch. Pharm. (Weinheim, Ger.)*, 1969, **302**, 81.	979
69APO(7)1	Z. Rappoport; *Adv. Phys. Org. Chem.*, 1969, **7**, 1.	1076
69BCJ545	T. Aoki, A. Furusaki, Y. Tomie, K. Ono and K. Tanaka; *Bull. Chem. Soc. Jpn.*, 1969, **42**, 545.	673
69BCJ3270	H. Kono, M. Shiga, I. Motoyama and K. Hata; *Bull. Chem. Soc. Jpn*, 1969, **42**, 3270.	797
69BSF2712	G. Köbrich; *Bull. Soc. Chim. Fr.*, 1969, 2712.	758
69CA(70)57950j	B. L. Dyatkin, S. R. Sterlin, L. G. Zhuravkova and I. L. Knunyants; *Chem. Abstr.*, 1969, **70**, 57 950j (*Dokl. Akad. Nauk SSSR*, 1969, **183(3)**, 598).	926
69CB83	H. Schmidbaur and W. Malisch; *Chem. Ber.*, 1969, **102**, 83.	569
69CB1944	M. Schlosser and G. Heinz; *Chem. Ber.*, 1969, **102**, 1944.	755
69CB2216	M. Regitz and W. Anschutz; *Chem. Ber.*, 1969, **102**, 2216.	489
69CB2651	H. Böhme and P. Wagner; *Chem. Ber.*, 1969, **102**, 2651.	305
69CC207	M. Kumada, M. Ogura, H. Tsunemi and M. Ishikawa; *J. Chem. Soc. Chem. Commun.*, 1969, 207.	616
69CC369	H. Newman and R. B. Angier; *J. Chem. Soc., Chem. Commun.*, 1969, 369.	312
69CC1175	J. W. Scheeren, J. E. W. van Melick and R. J. F. Nivard; *J. Chem. Soc., Chem. Commun.*, 1969, 1175.	177, 190
69CJC2875	P. F. Vogt and D. F. Tavares; *Can. J. Chem.*, 1969, **47**, 2875.	236
69CJC3983	G. B. Gubelt and J. Warkentin; *Can. J. Chem.*, 1969, **47**, 3983.	314
69CPB32	M. Sekiya and H. Sakaï; *Chem. Pharm. Bull.*, 1969, **17**, 32.	405, 418
69CR(C)1718	Y. Le Floc'h, A. Brault and M. Kerfanto; *C. R. Hebd. Seances Acad. Sci., Ser. C*, 1969, **269**, 1718.	317, 322
69DOK(185)311	O. V. Kuz'min, N. S. Nametkin, T. I. Chernysheva and S. P. Shapov; *Dokl. Akad. Nauk SSSR*, 1969, **185**, 311.	140
69DOK(186)835	R. G. Kostyanovskii, Z. E. Samoilova and I. O. Chervin; *Dokl. Akad. Nauk SSSR*, 1969, **186**, 835 (*Chem. Abstr.*, 1969, **71**, 61 356h).	886
69DOK(188)120	V. F. Mironov, T. K. Gar and S. A. Mikhailyants; *Dokl. Akad. Nauk SSSR*, 1969, **188**, 120 (*Chem. Abstr.*, 1970, **72**, 3541z).	638, 640
69DOK(189)334	N. S. Nametkin, E. D. Babich, V. N. Karel'skii and V. M. Vdovin; *Dokl. Akad. Nauk SSSR*, 1969, **189**, 334 (*Chem. Abstr.*, 1970, **72**, 55 571d).	622
69HCA827	L. Maier; *Helv. Chim. Acta*, 1969, **52**, 827.	127, 556
69HCA845	L. Maier; *Helv. Chim. Acta*, 1969, **52**, 845.	556
69HCA858	L. Maier; *Helv. Chim. Acta*, 1969, **52**, 858.	113, 121, 556
69HCA2641	H.-J. Gais, K. Hafner and M. Neuenschwander; *Helv. Chim. Acta*, 1969, **52**, 2641.	885
69IC275	N. E. Miller and D. L. Reznicek; *Inorg. Chem.*, 1969, **8**, 275.	524
69IC862	N. E. Miller, D. L. Reznicek, R. J. Rowatt and K. R. Lundberg; *Inorg. Chem.*, 1969, **8**, 862.	566
69IC1336	K. L. Lundberg, R. J. Rowatt and N. E. Miller; *Inorg. Chem.*, 1969, **8**, 1336.	455
69IC1775	K. P. Lannert and M. D. Joesten; *Inorg. Chem.*, 1969, **8**, 1775.	563
69IC2671	K. C. Nainan and G. E. Ryschkewitsch; *Inorg. Chem.*, 1969, **8**, 2671.	563
69IZV100	A. N. Nesmeyanov, K. N. Anisimov, N. E. Kolobova and A. A. Pasynskii; *Izv. Akad. Nauk SSSR, Ser. Khim.*, 1969, 100.	672
69IZV181	G. K. Genkina and V. A. Gilyarov; *Izv. Akad. Nauk, SSSR, Ser. Khim.*, 1969, 181.	116
69IZV1757	B. E. Ivanov, A. B. Ageeva, S. V. Pasmanyuk, R. R. Shagidullin, S. G. Salikhov and E. I. Loginova; *Izv. Akad. Nauk SSSR Ser. Khim.*, 1969, 1757 (*Chem. Abstr.*, 1970, **72**, 3531w).	478
69IZV2030	A. N. Nesmeyanov, E. G. Perevalova, D. A. Lemenovskii, A. N. Kosina and K. I. Grandberg; *Izv. Akad. Nauk SSSR, Ser. Khim.*, 1969, 2030.	723
69IZV2032	A. N. Nesmeyanov, K. I. Grandberg, T. V. Baukova, A. N. Kosina and E. G. Perevalova; *Izv. Akad. Nauk SSSR., Ser. Khim.*, 1969, 2032.	723
69JA1034	T. Durst; *J. Am. Chem. Soc.*, 1969, **91**, 1034.	74, 75, 77, 82, 233
69JA1386	J. E. Anderson, E. S. Glazer, D. L. Griffith, R. Knorr and J. D. Roberts; *J. Am. Chem. Soc.*, 1969, **91**, 1386.	9
69JA1860	S. J. Benkovic, P. A. Benkovic and D. R. Comfort; *J. Am. Chem. Soc.*, 1969, **91**, 1860.	412
69JA1929	A. H. Cowley and M. W. Taylor; *J. Am. Chem. Soc.*, 1969, **91**, 1929.	114, 802
69JA1934	A. H. Cowley and M. W. Taylor; *J. Am. Chem. Soc.*, 1969, **91**, 1934.	802
69JA2538	J. L. Atwood and G. D. Stucky; *J. Am. Chem. Soc.*, 1969, **91**, 2538.	709
69JA2803	T. Shono and Y. Matsumura; *J. Am. Chem. Soc.*, 1969, **91**, 2803.	201
69JA3870	L. A. Paquette and R. W. Houser; *J. Am. Chem. Soc.*, 1969, **91**, 3870.	80
69JA4306	H. C. Brown and S. P. Rhodes; *J. Am. Chem. Soc.*, 1969, **91**, 4306.	631
69JA5027	D. Seyferth, R. M. Turkel, M. A. Eisert and L. J. Todd; *J. Am. Chem. Soc.*, 1969, **91**, 5027.	152
69JA6195	J. Hooz and D. M. Gunn; *J. Am. Chem. Soc.*, 1969, **91**, 6195.	832
69JA6598	F. A. Cotton, A. Davison, T. J. Marks and A. Musco; *J. Am. Chem. Soc.*, 1969, **91**, 6598.	683
69JA7292	L. R. Bateman, P. M. Maitlis and L. F. Dahl; *J. Am. Chem. Soc.*, 1969, **91**, 7292.	689
69JA7510	C. Y. Meyers, A. M. Malte and W. S. Matthews; *J. Am. Chem. Soc.*, 1969, **91**, 7510.	81

69JCS(A)1872	J. Cooke, W. R. Cullen, M. Green and F. G. A. Stone; *J. Chem. Soc. (A)*, 1969, 1872.	685
69JCS(C)652	A. Lawson and R. B. Tinkler; *J. Chem. Soc. (C)*, 1969, 652.	86
69JCS(C)1202	W. Bonthorne and J. W. Cornforth; *J. Chem. Soc. (C)*, 1969, 1202.	190
69JCS(C)2334	F. E. Hardy, P. R. H. Speakman and P. Robson; *J. Chem. Soc. (C)*, 1969, 2334.	74, 75
69JGU577	V. K. Khairullin, A. N. Pudovik and N. I. Kharitonova; *J. Gen. Chem. USSR (Engl. Transl.)*, 1969, **39**, 577.	124, 125
69JGU1490	E. N. Tsvetkov, R. A. Malevannaya and M. I. Kabachnik; *J. Gen. Chem. USSR (Engl. Transl.)*, 1969, **39**, 1490.	118, 132, 133
69JHC115	D. L. Tuleen and R. H. Bennett; *J. Heterocycl. Chem.*, 1969, **6**, 115.	66
69JHC187	N. Rabjohn and K. B. Sloan; *J. Heterocycl. Chem.*, 1969, **6**, 187.	437
69JHC429	M. von Strandtmann, M. P. Cohen and J. Shavel, Jr.; *J. Heterocycl. Chem.*, 1969, **6**, 429.	304
69JHC921	G. Garcia-Muñoz, R. Madroñero, M. Rico and M. C. Saldaña; *J. Heterocycl. Chem.*, 1969, **6**, 921.	314
69JOC31	D. L. Tuleen and T. B. Stephens; *J. Org. Chem.*, 1969, **34**, 31.	66, 67, 68, 69
69JOC1233	E. J. Corey and E. Block; *J. Org. Chem.*, 1969, **34**, 1233.	80, 83, 84
69JOC1799	L. Field and B. J. Sweetman; *J. Org. Chem.*, 1969, **34**, 1799.	216
69JOC1841	R. Sullivan; *J. Org. Chem.*, 1969, **34**, 1841.	766
69JOC2046	K. Baum; *J. Org. Chem.*, 1969, **34**, 2046.	439
69JOC2049	K. Baum; *J. Org. Chem.*, 1969, **34**, 2049.	439
69JOC2080	M. P. Mertes, H.-K. Lee and R. L. Schowen; *J. Org. Chem.*, 1969, **34**, 2080.	229
69JOC2618	D. L. Ransley; *J. Org. Chem.*, 1969, **34**, 2618.	18
69JOC2720	P. Aeberli and W. J. Houlihan; *J. Org. Chem.*, 1969, **34**, 2720.	415
69JOC2728	A. Padwa and D. Eastman; *J. Org. Chem.*, 1969, **34**, 2728.	154
69JOC3104	W. E. Truce, T. C. Klingler, J. E. Paar, H. Feuer and D. K. Wu; *J. Org. Chem.*, 1969, **34**, 3104.	332
69JOC3230	A. M. Cameron, P. R. West and J. Warkentin; *J. Org. Chem.*, 1969, **34**, 3230.	314
69JOC3451	D. J. Woodman and Z. L. Murphy; *J. Org. Chem.*, 1969, **34**, 3451.	979
69JOC3949	W. G. Lloyd and B. J. Luberoff; *J. Org. Chem.*, 1969, **34**, 3949.	193
69JOM(16)83	M. Delmas, J. C. Maire and R. Pinzelli; *J. Organomet. Chem.*, 1969, **16**, 83.	701
69JOM(16)91	A. J. Leusink and J. G. Noltes; *J. Organomet. Chem.*, 1969, **16**, 91.	701
69JOM(17)161	E. W. Abel and S. M. Illingworth; *J. Organomet. Chem.*, 1969, **17**, 161.	344
69JOM(17)241	I. F. Lutsenko, Yu. I. Baukov, A. S. Kostyuk, N. I. Savelyeva and V. K. Krysina; *J. Organomet. Chem.*, 1969, **17**, 241.	886
69JOM(19)169	S. McVey and P. M. Maitlis; *J. Organomet. Chem.*, 1969, **19**, 169.	689
69JOM(20)19	R. B. Castle and D. S. Matteson; *J. Organomet. Chem.*, 1969, **20**, 19.	632, 632, 633
69JPR857	K. Issleib and M. Lischewski; *J. Prakt. Chem.*, 1969, **311**, 857.	551, 557
69LA(722)21	H. Witte, P. Mischke and G. Hesse; *Justus Liebigs Ann. Chem.*, 1969, **722**, 21.	523
69LA(728)12	E. Vilsmaier; *Justus Liebigs Ann. Chem.*, 1969, **728**, 12.	47, 48
69LA(729)73	I. Butula; *Justus Liebigs Ann. Chem.*, 1969, **729**, 73.	425
69M469	O. Hromatka, M. Knollmüller and H. Deschler; *Monatsh. Chem.*, 1969, **100**, 469.	416
B-69MI 401-01	H. J. Emeleus; "The Chemistry of Fluorine and its Compounds," Academic Press, New York, 1969.	7
B-69MI 403-01	J. V. Paukstelis; in "Enamines," ed. A. G. Cook, Marcel Dekker, New York, 1969, p. 169.	96
69MI 417-01	S. V. Delavarenne and H. G. Viehe; in "The Chemistry of Acetylene," ed. H. G. Viehe, Marcel Dekker, New York, 1969, p. 651.	764
69MI 417-02	L. Brandoma, H. J. T. Bos and J. F. Arens; in "The Chemistry of Acetylene," ed. H. G. Viehe, Marcel Dekker, New York, 1969, p. 751.	765
69MI 419-01	L. N. Lutsenko and B. G. Boldirev; *Visn. L'viv. Politekh Inst.*, 1969, **36**, 73 (*Chem. Abstr.* 1970, **73**, 98 544).	864
69MI 420-01	B. M. Pyatin and R. G. Glushkov; *Pharm. Chem. J.*, 1969, **5**, 256.	881
B-69MI 421-01	S. Y. Delavarenne and H. G. Viehe; in "Chemistry of Acetylenes," ed. H. G. Viehe, Marcel Dekker, New York, 1969, p. 651.	976, 980
B-69MI 421-02	H. G. Viehe; in "Chemistry of Acetylenes," ed. H. G. Viehe, Marcel Dekker, New York, 1969, p. 861.	976, 980
B-69MI 424-01	T. L. Gilchrist and C. W. Rees; "Carbenes, Nitrenes and Arynes," Nelson, London, 1969.	1080
69OPP5	F. Minisci and A. Quilico; *Org. Prep. Proced. Int.*, 1969, **1**, 5.	439
69RTC597	P. J. W. Schuijl, H. J. T. Bos and L. Brandsma; *Recl. Trav. Chim. Pays-Bas*, 1969, **88**, 597.	843
69S17	D. Seebach; *Synthesis*, 1969, 17.	244, 1075
69T1557	D. J. Pasto, J. Chow and S. K. Arora; *Tetrahedron*, 1969, **25**, 1557.	632
69T1571	D. J. Pasto, S. K. Arora and J. Chow; *Tetrahedron*, 1969, **25**, 1571.	634
69T1679	J. Libman, M. Sprecher and Y. Mazur; *Tetrahedron*, 1969, **25**, 1679.	204
69T3447	R. Buyle and H. G. Viehe; *Tetrahedron*, 1969, **25**, 3447.	904, 905
69T4257	M. Kocor, A. Kurek and J. Dabrowski; *Tetrahedron*, 1969, **25**, 4257.	209
69T4649	Y. Shvo and I. Belsky; *Tetrahedron*, 1969, **25**, 4649.	899, 902, 903
69T5569	W. Jugelt and D. Schmidt; *Tetrahedron*, 1969, **25**, 5569.	472
69TFS1697	E. W. Abel, M. A. Cooper, R. J. Goodfellow and A. J. Rest; *Trans. Faraday Soc.*, 1969, **65**, 1697.	344
69TL173	R. M. Carlson and P. M. Helquist; *Tetrahedron Lett.*, 1969, **10**, 173.	250
69TL1957	C. Jefford and D. T. Hill; *Tetrahedron Lett.*, 1969, **10**, 1957.	35
69TL2441	R. Leblanc and A. Foucaud; *Tetrahedron Lett.*, 1969, **10**, 2441.	800
69TL2689	R. B. Woodward and C. Wintner; *Tetrahedron Lett.*, 1969, **10**, 2689.	996
69TL4011	F. L. Hedberg and H. Rosenberg; *Tetrahedron Lett.*, 1969, **10**, 4011.	706
69TL4647	N. N. Girotra and N. L. Wendler; *Tetrahedron Lett.*, 1969, **10**, 4647.	130

69TL5259	S. Iriuchijima and G. Tsuchihashi; *Tetrahedron Lett.*, 1969, **10**, 5259.	74, 75
69USP3426021	D. Seyferth; *US Pat.* 3 426 021 (1969) (*Chem. Abstr.*, 1969, **71**, 39 167).	551, 570, 588
69USP3445522	H. R. Hays and G. M. Kosolapoff; *US Pat.* 3 445 522 (1969) (*Chem. Abstr.*, 1969, **71**, 81 520).	550, 562, 563
69USP3463835	E. G. Budnick; *US Pat.* 3 463 835 (1969) (*Chem. Abstr.*, 1969, **71**, 91 645).	553
69YZ203	G. Kobayahi, S. Farakawa, Y. Matsuda and S. Matsumaga; *Yakugaku Zasshi* (*J. Pharm. Soc. Jpn.*), 1969, **89(2)**, 203 (*Chem. Abstr.*, 1969, **70**, 106 328y).	888, 900
69ZAAC(370)31	K. von Sommer and M. Becke-Goehring; *Z. Anorg. Allg. Chem.*, 1969, **370**, 31.	592, 593
69ZN(B)1081	T. A. Blazer, R. Schmutzler and I. K. Gregor; *Z. Naturforsch., Teil B*, 1969, **24**, 1081.	564
69ZOB839	Yu. I. Baranov, S. V. Gorelenko and S. P. Kochkol'da; *Zh. Obshch. Khim.*, 1969, **39**, 839.	556
69ZOB845	K. A. Petrov, M. A. Raksha and V. P. Korotkova; *Zh. Obshch. Khim.*, 1969, **39**, 845.	554
69ZOB1235	A. N. Pudovik, E. S. Batyeva and O. A. Raevskii; *Zh. Obshch. Khim.*, 1969, **39**, 1235 (*Chem. Abstr.*, 1969, **71**, 113 039d).	476
69ZOB1247	K. A. Petrov, V. A. Parshina and G. M. Petrova; *Zh. Obshch. Khim.*, 1969, **39**, 1247.	554
69ZOB2192	B. S. Drach, A. D. Sinitsa and A. V. Kirsanov; *Zh. Obshch. Khim.*, 1969, **39**, 2192 (*Chem. Abstr.*, 1970, **72**, 42 706b).	487
69ZOB2281	V. F. Mironov, S. A. Mikhailyants and T. K. Gar; *Zh. Obshch. Khim.*, 1969, **39**, 2281 (*Chem. Abstr.*, 1970, **72**, 43 815y).	608, 638
69ZOB2601	V. F. Mironov, S. A. Mikhaillyants and T. K. Gar; *Zh. Obshch. Khim.*, 1969, **39**, 2601 (*Chem. Abstr.*, 1970, **72**, 67 064d).	626, 637, 640
69ZOR2181	B. S. Drach, A. D. Sinitsa and A. V. Kirsanov; *Zh. Org. Khim.*, 1969, **5**, 2181.	304
70ACR186	D. S. Matteson; *Acc. Chem. Res.*, 1970, **3**, 186.	633
70ACR417	R. B. King; *Acc. Chem. Res.*, 1970, **3**, 417.	821
70ACS3102	E. Ericsson, J. Sandstrom and I. Wennerbeck; *Acta Chem. Scand.*, 1970, **24**, 3102.	969, 970, 973
70ACS3213	K. A. Jensen and L. Henriksen; *Acta Chem. Scand.*, 1970, **24**, 3213.	875
70AG(E)163	T. Kauffmann, E. Köppelmann and H. Berg; *Angew. Chem., Int. Ed. Engl.*, 1970, **9**, 163.	535
70AG(E)737	H. Schmidbaur and W. Vornberger; *Angew. Chem., Int. Ed. Engl.*, 1970, **9**, 737.	571
70AJC715	E. A. Jeffery and T. Mole; *Aust. J. Chem.*, 1970, **23**, 715.	699
70AJC989	A. Pross and S. Sternhell; *Aust. J. Chem.*, 1970, **23**, 989.	29
70BCJ1223	M. Oki and K. Kobayashi; *Bull. Chem. Soc. Jpn.*, 1970, **43**, 1223.	68
70BCJ2271	G. Tsuchihashi and S. Iriuchijima; *Bull. Chem. Soc. Jpn.*, 1970, **43**, 2271.	74
70BCJ2938	M. Yokoyama; *Bull. Chem. Soc. Jpn.*, 1970, **43**, 2938.	852
70BRP1179242	W. J. Owen and B. E. Cooper; *Br. Pat.* 1 179 242 (1970) (*Chem. Abstr.*, 1970, **72**, 100 877).	566
70BRP1183130	Chemagro; *Br. Pat.*, 1 183 130 (1970) (*Chem. Abstr.*, 1970, **72**, 121 703b).	927
70BSF332	M. Farines and J. Soulier; *Bull. Soc. Chim. Fr.*, 1970, 332.	213
70BSF1130	D. Danion and R. Carrie; *Bull. Soc. Chim. Fr.*, 1970, 1130.	489
70CB97	H. Schmidbaur and W. Malisch; *Chem. Ber.*, 1970, **103**, 97.	569
70CB104	H. Böhme and M. Hilp; *Chem. Ber.*, 1970, **103**, 104.	96
70CB133	K. Hartke and G. Salamon; *Chem. Ber.*, 1970, **103**, 133.	973, 974
70CB147	K. Hartke and G. Salamon; *Chem. Ber.*, 1970, **103**, 147.	974
70CB643	H. Stetter and E. Reske; *Chem. Ber.*, 1970, **103**, 643.	192
70CB973	H. Kessler, *Chem. Ber.*, 1970, **103**, 973.	974
70CB1198	S. Y. Delavarenne and H. G. Viehe; *Chem. Ber.*, 1970, **103**, 1198.	817
70CB1982	K. Friedrich and H. K. Thieme; *Chem. Ber.*, 1970, **103**, 1982.	894
70CB2271	N. Petragnani and G. Schill; *Chem. Ber.*, 1970, **103**, 2271.	290
70CB2314	W. Oettmeier; *Chem. Ber.*, 1970, **103**, 2314.	311
70CB2775	H. Böhme and G. Fuchs; *Chem. Ber.*, 1970, **103**, 2775.	327
70CB2984	L. Horner and H. Roder; *Chem. Ber.*, 1970, **103**, 2984.	343
70CB3007	H. Schmidbaur and W. Malisch; *Chem. Ber.*, 1970, **103**, 3007.	569
70CB3205	W. Flitsch; *Chem. Ber.*, 1970, **103**, 3205.	299
70CB3448	H. Schmidbaur and W. Malisch; *Chem. Ber.*, 1970, **103**, 3448.	570, 571
70CB3459	L. Capuano and W. Ebner; *Chem. Ber.*, 1970, **103**, 3459.	417
70CB3918	H. Böhme and W. Höver; *Chem. Ber.*, 1970, **103**, 3918.	310
70CC161	G. Amiet, K. Nicholas and R. Pettit; *J. Chem. Soc., Chem. Commun.*, 1970, 161.	724
70CC555	R. G. Sutherland and A. K. V. Unni; *J. Chem. Soc., Chem. Commun*, 1970, 555.	706
70CC1369	G. Yagupsky, W. Mowart, A. Shortland and G. Wilkinson; *J. Chem. Soc., Chem. Commun.*, 1970, 1369.	656
70CC1441	M. Cinquini, S. Colonna and F. Montanari; *J. Chem. Soc., Chem. Commun.*, 1970, 1441.	233
70CC1689	K. Ogura and G. Tsuchihashi; *J. Chem. Soc., Chem. Commun.*, 1970, 1689.	233
70CJC561	A. G. Brook, J. M. Duff and D. G. Anderson; *Can. J. Chem.*, 1970, **48**, 561.	138, 141, 646, 651, 652, 654, 663
70CPB1245	S. Ishiwata and Y. Shiokawa; *Chem. Pharm. Bull.*, 1970, **18**, 1245.	296
70CR(C)351	J. Satgé, C. Couret and J. Escudié; *C. R. Hebd. Seances Acad. Sci., Ser. C*, 1970, **270**, 351.	574
70CR(C)436	Y. LeFloc'h, A. Brault and M. Kerfanto; *C. R. Hebd. Seances Acad. Sci., Ser. C*, 1970, **270**, 436.	322
70CR(C)1467	C. Raulet and E. Levas; *C. R. Hebd. Seances Acad. Sci., Ser. C*, 1970, **270**, 1467.	739, 746
70CR(C)1890	J. Corbier, P. Cesson and P. Jelenc; *C. R. Hebd. Seances Acad. Sci. Ser. C*, 1970, **270**, 1890.	884, 891
70CZ215	G. Ege and G. Jooss; *Chem.-Ztg.*, 1970, **94**, 215.	432
70HCA251	W. Skorianetz and E. sz Kováts; *Helv. Chim. Acta*, 1970, **53**, 251.	437
70HCA1330	P. Baudet and C. Otten; *Helv. Chim. Acta*, 1970, **53**, 1330.	203
70HCA1598	H. Dahn, H. Gowal and H. P. Schlunke; *Helv. Chim. Acta*, 1970, **53**, 1598.	162
70HCA1940	L. Maier; *Helv. Chim. Acta*, 1970, **53**, 1940.	556

70HCA2069	L. Maier; *Helv. Chim. Acta.*, 1970, **53**, 2069.	120
70IC975	K. Niedenzu, K. E. Blick and C. D. Miller; *Inorg. Chem.*, 1970, **9**, 975.	107
70IC1060	C. R. Bettler, J. C. Sendra and G. Urry; *Inorg. Chem.*, 1970, **9**, 1060.	636
70IC2372	C. R. Bettler and G. Urry; *Inorg. Chem.*, 1970, **9**, 2372.	622, 624
70IZV1326	Yu. M. Polikarpov, K. Zh. Kulunbetova, T. Ya. Medved' and M. I. Kabachnik; *Izv. Akad. Nauk SSSR, Ser. Khim.*, 1970, 1326.	560, 564
70IZV2133	A. N. Nesmeyanov, V. A. Sazonova and V. E. Fedorov; *Izv. Akad. Nauk SSSR, Ser. Khim.*, 1970, 2133.	706
70IZV2140	D. P. Del'tsova, S. D. Koshtoyan, Y. V. Zeifman and V. Yu; *Izv. Akad. Nauk SSSR, Ser. Khim.*, 1970, 2140 (*Chem. Abstr.* 1971, **74**, 76 378)	833
70IZV2280	M. A. Kadina, A. V. Ignatenko and V. A. Ponomarenko; *Izv. Akad. Nauk SSSR Ser. Khim.*, 1970, 2280 (*Chem. Abstr.*, 1971, **75**, 129 873q).	639
70JA343	M. Ohno, T. F. Spande and B. Witkop; *J. Am. Chem. Soc.*, 1970, **92**, 343.	424
70JA502	E. F. Epstein and L. F. Dahl; *J. Am. Chem. Soc.*, 1970, **92**, 502.	715
70JA3203	N. Frydman and Y. Mazur; *J. Am. Chem. Soc.*, 1970, **92**, 3203.	742
70JA5118	S. Trofimenko; *J. Am. Chem. Soc.*, 1970, **92**, 5118.	437
70JA5464	T. H. Fife and E. Anderson; *J. Am. Chem. Soc.*, 1970, **92**, 5464.	218
70JA5469	W. R. Moore and W. R. Moser; *J. Am. Chem. Soc.*, 1970, **92**, 5469.	24
70JA5520	D. Seyferth and A. T. Wehman; *J. Am. Chem. Soc.*, 1970, **92**, 5520.	688
70JA5664	R. N. McDonald and R. N. Steppel; *J. Am. Chem. Soc.*, 1970, **92**, 5664.	50
70JA7359	F. Weigert, M. B. Winstead, J. I. Garrels and J. D. Roberts; *J. Am. Chem. Soc.*, 1970, **92**, 7359.	38
70JA7424	G. R. Buell, R. Corriu, C. Guerin and L. Spialter; *J. Am. Chem. Soc.*, 1970, **92**, 7424.	654
70JA7567	A. G. Brook, J. M. Duff and D. G. Anderson; *J. Am. Chem. Soc.*, 1970, **92**, 7567.	647, 652, 663
70JCS(C)178	G. Camaggi and F. Gozzo; *J. Chem. Soc.* (C), 1970, 178.	7
70JCS(C)476	T. Sasaki and A. Kojima; *J. Chem. Soc.* (C), 1970, 476.	976
70JCS(C)504	M. A. Shaw, J. C. Tebby, R. S. Ward and D. H. Williams; *J. Chem. Soc.* (C), 1970, 504.	563
70JCS(C)744	R. Fields, R. N. Haszeldine and N. F. Wood; *J. Chem. Soc.(C)*, 1970, 744.	111
70JCS(C)881	A. Halleux and H. G. Viehe; *J. Chem. Soc.* (C), 1970, 881.	979
70JGU255	E. N. Tsvetkov, G. Borisov, Kh. Sivriev, R. A. Malevannaya and M. I. Kabachnik; *J. Gen. Chem. USSR (Engl. Transl.)*, 1970, **40**, 255.	113, 120
70JHC1	A. N. Hughes and C. Srivanavit; *J. Heterocycl. Chem.*, 1970, **7**, 1.	1031
70JHC355	H. J. Beim and A. R. Day; *J. Heterocycl. Chem.*, 1970, **7**, 355.	408
70JHC439	R. E. Harmon, V. L. Rizzo and S. K. Gupta; *J. Heterocycl. Chem.*, 1970, **7**, 439.	442
70JHC1311	M. von Strandtmann, M. P. Cohen and J. Shavel, Jr.; *J. Heterocycl. Chem.*, 1970, **7**, 1311.	307
70JMC1248	P. J. Mulligan and S. LaBerge; *J. Med. Chem.*, 1970, **13**, 1248.	995
70JOC215	R. G. Hiskey and J. T. Sparrow; *J. Org. Chem.*, 1970, **35**, 215.	218
70JOC283	M. Yokoyama; *J. Org. Chem.*, 1970, **35**, 283.	842
70JOC468	D. J. France, J. J. Hand and M. Los; *J. Org. Chem.*, 1970, **35**, 468.	166
70JOC508	T. Koenig and J. Wieczorek; *J. Org. Chem.*, 1970, **35**, 508.	13
70JOC597	Y. Ogata and H. Tomioka; *J. Org. Chem.*, 1970, **35**, 597.	343
70JOC828	N. Hashimoto, Y. Kawano and K. Morita; *J. Org. Chem.*, 1970, **35**, 828.	762
70JOC1211	J. Wolinsky and R. A. Kasubick; *J. Org. Chem.*, 1970, **35**, 1211.	742
70JOC1297	D. Seyferth and K. V. Darragh; *J. Org. Chem.*, 1970, **35**, 1297.	32
70JOC1308	C. L. Frye, R. M. Salinger, F. W. G. Fearon, J. M. Klosowski and T. DeYoung; *J. Org. Chem.*, 1970, **35**, 1308.	650
70JOC1501	C. E. Diebert; *J. Org. Chem.*, 1970, **35**, 1501.	328
70JOC1515	W. T. Brady, F. H. Parry, III, R. Roe, Jr., E. F. Hoff and L. Smith; *J. Org. Chem.*, 1970, **35**, 1515.	768
70JOC2096	V. A. Pattison; *J. Org. Chem.*, 1970, **35**, 2096.	736
70JOC2125	D. J. Burton and H. C. Krutzsch; *J. Org. Chem.*, 1970, **35**, 2125.	751, 752, 753
70JOC2310	J. W. Rathke, J. W. Guyer and J. G. Verkade; *J. Org. Chem.*, 1970, **35**, 2310.	342
70JOC2551	H. Feuer and M. Auerbach; *J. Org. Chem.*, 1970, **35**, 2551.	332
70JOC3002	G. E. Wilson, Jr. and M. G. Huang; *J. Org. Chem.*, 1970, **35**, 3002.	64, 71, 80
70JOC3140	J. M. Kliegman and R. K. Barnes; *J. Org. Chem.*, 1970, **35**, 3140.	298
70JOC3375	W. W. Zajac and K. J. Byrne; *J. Org. Chem.*, 1970, **35**, 3375.	178
70JOC3583	J. E. Baldwin and J. C. Swallow; *J. Org. Chem.*, 1970, **35**, 3583.	914, 920
70JOC3888	M. D. Rausch and R. A. Genetti; *J. Org. Chem.*, 1970, **35**, 3888.	706
70JOC4201	L. O. Moore, J. P. Henry and J. W. Clark; *J. Org. Chem.*, 1970, **35**, 4201.	31
70JOC4236	F. G. Borgardt, P. Noble, Jr., W. L. Reed and A. K. Seeler; *J. Org. Chem.*, 1970, **35**, 4236.	795
70JOC4250	Y. Okamoto and S. K. Kundu; *J. Org. Chem.*, 1970, **35**, 4250.	792
70JOM(21)103	D. Seyferth and C. J. Attridge; *J. Organomet. Chem.*, 1970, **21**, 103.	605, 636
70JOM(21)131	A. Tzschach and D. Drohne; *J. Organomet. Chem.*, 1970, **21**, 131.	495, 496
70JOM(22)5	F. Bonati and G. Minghetti; *J. Organomet. Chem.*, 1970, **22**, 5.	697
70JOM(22)141	K. C. Williams; *J. Organomet. Chem.*, 1970, **22**, 141.	699
70JOM(23)99	D. Seyferth and S. P. Hopper; *J. Organomet. Chem.*, 1970, **23**, 99.	138
70JOM(23)465	K. C. Williams; *J. Organomet. Chem.*, 1970, **23**, 465.	699
70JOM(23)471	L. C. Willemsens and G. J. M. van der Kerk; *J. Organomet. Chem.*, 1970, **23**, 471.	699
70JOM(24)263	D. S. Matteson and J. R. Thomas; *J. Organomet. Chem.*, 1970, **24**, 263.	633
70JOM(24)529	M. A. Cook, C. Eaborn, A. E. Jukes and D. R. M. Walson; *J. Organomet. Chem.*, 1970, **24**, 529.	1048
70JOM(24)633	J. Satgé, C. Couret and J. Escudié; *J. Organomet. Chem.*, 1970, **24**, 633.	574
70JOM(25)C6	W. H. Glaze, T. L. Brewer and A. C. Ranade; *J. Organomet. Chem.*, 1970, **25**, C6.	698

70JOM(25)C36	M. R. Collier, M. F. Lappert and M. M. Truebik; *J. Organomet. Chem.*, 1970, **25**, C36.	656
70JOM(25)385	R. West and G. A. Gornowicz; *J. Organomet. Chem.*, 1970, **25**, 385.	516
70JOU750	A. P. Sergeev, T. I. Novozhilova and A. Ya. Yakubovich; *J. Org. Chem. USSR (Engl. Transl.)*, 1970, **6**, 750.	793
70JPC2722	W. J. Lautenberger and J. G. Miller; *J. Phys. Chem.*, 1970, **74**, 2722.	979
70JPR135	K. Issleib and M. Lischewski; *J. Prakt. Chem.*, 1970, **312**, 135.	551
70JPR456	K. Issleib and H. P. Abicht; *J. Prakt. Chem.*, 1970, **312**, 456.	551, 552, 564
70JPR683	D. Martin, A. Berger and R. Peschel; *J. Prakt. Chem.*, 1970, **312**, 683.	74
70M568	J. Schantl; *Monatsh. Chem.*, 1970, **101**, 568.	441
B-70MI 402-01	C. A. Buehler and D. E. Pearson; "Survey of Organic Syntheses," Wiley, New York, 1970, vol. 1, p. 310.	48
B-70MI 403-01	S. Dayagi and Y. Degani (p. 61); K. Harada (p. 255); J. P. Anselme (p. 299); in "The Chemistry of the Carbon–Nitrogen Double Bond," ed. S. Patai, Wiley-Interscience, New York, 1970.	728
B-70MI 404-01	C. A. Buehler and D. E. Pearson; "Survey of Organic Syntheses," Wiley, New York, 1970, p. 513.	176
B-70MI 404-02	R. H. DeWolfe; "Carboxylic Ortho Acid Derivatives," Academic Press, New York, 1970, p. 154.	177
B-70MI 404-03	R. H. DeWolfe; "Carboxylic Ortho Acid Derivatives," Academic Press, New York, 1970, p. 223.	192
B-70MI 409-01	B. C. Challis and J. A. Challis; in "The Chemistry of Amides," ed. J. Zabicky, Interscience, London, 1970, p. 731.	416
70MI 411-01	B. A. Tertov and P. P. Onishchenko; *Khim. Get. Soed.*, 1970, 1435. (English translation; *Chem. Heterocycl. Compd.*, 1970, 1339).	539
70MI 411-02	H. H. Baer and L. Urbas; in "The Chemistry of the Nitro and Nitroso Groups," ed. H. Feuer, Wiley Interscience, New York, 1970, p. 75.	540
70MI 414-01	G. A. Gibbon, E. W. Kifer and C. H. Van Dyke; *Inorg. Nucl. Chem. Lett.*, 1970, **6**, 617.	638
70MI 419-01	J. W. Swetkin; *Wiss. Z. Martin-Luther-Univ., Halle-Wittenberg, Math.-Naturwiss. Reihe*, 1970, **19**, 19 (*Chem. Abstr.* 1971, **75**, 109 413).	824
70MI 421-01	S. Tanimoto, S. Yusuda and M. Okawo; *J. Synth. Org. Chem. Jpn.*, 1970, **28**, 1041.	970
70OMR(2)81	J. E. Bissey, H. Goldwhite and D. G. Rowsell; *Org. Magn. Reson.*, 1970, **2**, 81.	116
70OPP235	M. L. Kee and I. B. Douglass; *Org. Prep. Proceed. Int.*, 1970, **2**, 235.	78
70QRS67	A. M. van Leusen and J. Strating; *Quart. Rep. Sulfur Chem.*, 1970, **5**, 67.	235
70S49	H. E. Zaugg; *Synthesis*, 1970, 49.	96, 98
70S588	S. Iriuchijima and G. Tsuchihashi; *Synthesis*, 1970, 588.	77
70T1281	F. Bertini, P. Grasselli, G. Zubiani and G. Cainelli; *Tetrahedron*, 1970, **26**, 1281.	669
70T1311	B. N. Blackett, J. M. Coxon, M. P. Hartshorn, A. J. Lewis, G. R. Little and G. J. Wright; *Tetrahedron*, 1970, **26**, 1311.	182
70T2555	J. M. Kliegman and R. K. Barnes; *Tetrahedron*, 1970, **26**, 2555.	298
70TL481	R. Weiss and R. Gompper; *Tetrahedron Lett.*, 1970, **11**, 481.	870
70TL885	J. Ficini and A. Krief; *Tetrahedron Lett.*, 1970, **11**, 885.	827
70TL935	B. Zwanenburg and J. ter Wiel; *Tetrahedron Lett.*, 1970, **11**, 935.	237
70TL1137	T. H. Chan, E. Chang and E. Vinokur; *Tetrahedron Lett.*, 1970, **11**, 1137.	646
70TL1729	H. H. Wasserman and M. S. Baird; *Tetrahedron Lett.*, 1970, **11**, 1729.	299, 406
70TL1937	H. Stamm and G. Führling; *Tetrahedron Lett.*, 1970, **11**, 1937.	792
70TL2357	J. Honzl and P. Krivinka; *Tetrahedron Lett.*, 1970, **11**, 2357.	894
70TL4095	P. Äyräs and K. Pihlaja; *Tetrahedron Lett.*, 1970, **11**, 4095.	213, 213
70TL4643	K.-C. Tin and T. Durst; *Tetrahedron Lett.*, 1970, **11**, 4643.	74, 75, 76
70TL5215	R. E. Banks, R. N. Haszeldine, D. R. Taylor and G. Webb; *Tetrahedron Lett.*, 1970, **11**, 5215.	736
70USP3479373	J. Speziale and K. W. Ratts (Monsanto Chem. Corp.); *US Pat.* 3 479 373 (1969) (*Chem. Abstr.*, 1970, **72**, 21 539).	728
70USP3496231	L. Maier; *US Pat.* 3 496 231 (1970) (*Chem. Abstr.*, 1970, **72**, 90 625).	550
70USP3511865	D. J. Peterson; *US Pat.* 3 511 865 (1970) (*Chem. Abstr.*, 1970, **73**, 35 502).	566
70USP3518312	L. Maier; *US Pat.* 3 518 312 (1970) (*Chem. Abstr.*, 1970, **73**, 88 015).	550
70USP3534125	K. O. Knollmüller; *US Pat.* 3 534 125 (1970) (*Chem. Abstr.*, 1971, **74**, 53 989).	556
70ZAAC(372)21	G. Fritz and P. Schober; *Z. Anorg. Allg. Chem.*, 1970, **372**, 21.	605, 621
70ZAAC(372)59	G. Fritz and P. Schober; *Z. Anorg. Allg. Chem.*, 1970, **372**, 59.	621
70ZAAC(376)37	K. Sommer; *Z. Anorg. Allg. Chem.*, 1970, **376**, 37.	545, 550, 562
70ZAAC(377)37	G. Fritz and P. Schober; *Z. Anorg. Allg. Chem.*, 1970, **377**, 37.	652
70ZAAC(377)120	K. von Sommer; *Z. Anorg. Allg. Chem.*, 1970, **377**, 120.	592, 593
70ZAAC(377)128	K. Sommer; *Z. Anorg. Allg. Chem.*, 1970, **377**, 128.	135
70ZN(B)314	W. Siebert; *Z. Naturforsch., Teil B*, 1970, **25**, 314.	574
70ZOB27	T. Kh. Gazizov, A. P. Pashinkin and A. N. Pudovik; *Zh. Obshch. Khim.*, 1970, **40**, 27.	915
70ZOB285	E. N. Tsvetkov, G. Borisov, Kh. Sivriev, R. A. Malevannaya and M. I. Kabachnik; *Zh. Obshch. Khim.*, 1970, **40**, 285.	555
70ZOB499	A. N. Pudovik, G. E. Yastrebova and O. A. Pudovik; *Zh. Obshch. Khim.*, 1970, **40**, 499.	563
70ZOB609	E. F. Bugerenko, A. S. Petukhova and E. A. Chernyshev; *Zh. Obshch. Khim.*, 1970, **40**, 609.	570
70ZOB939	S. V. Ponomarev and S. A. Lebedev; *Zh. Obshch. Khim. Lett.*, 1970, **40**, 939 (*Chem. Abstr.*, 1970, **73**, 35 478w).	619, 628
70ZOB2560	V. S. Tsivunin, L. N. Krutskii, Ernazarov and G. K. Kamai; *Zh. Obshch. Khim.*, 1970, **40**, 2560.	340

70ZOB2601	T. K. Gar, E. M. Berliner, A. V. Kisin and V. F. Mironov; *Zh. Obshch. Khim.*, 1970, **40**, 2601 (*Chem. Abstr.*, 1971, **75**, 5143g).	639
70ZOB2607	T. A. Manukina, G. P. Kolyukhina, G. S. Burlachenko, I. F. Lutsenko and I. Y. Baukov; *Zh. Obshch. Khim.*, 1970, **40**, 2607 (*Chem. Abstr.*, 1971, **75**, 36 272c).	661
70ZOB2618	I. G. Kolokol'tseva, V. N. Chistokletov and A. A. Petrov; *Zh. Obshch. Khim.*, 1970, **40**, 2618 (*Chem. Abstr.*, 1971, **75**, 88 689p).	488
71AG(E)67	R. Gompper and G. Seybold; *Angew. Chem., Int. Ed. Engl.*, 1971, **10**, 67.	892
71AG(E)267	P. Riviere and J. Satge; *Angew. Chem., Int. Ed. Engl.*, 1971, **10**, 267.	361
71AG(E)940	M. Baudler, J. Vesper, P. Junkes and H. Sandmann; *Angew. Chem., Int. Ed. Engl.*, 1971, **10**, 940.	545
71BCJ1726	G. Tsuchihashi and K. Ogura; *Bull. Chem. Soc. Jpn.*, 1971, **44**, 1726.	74, 77
71BCJ2864	S. Miyano and H. Hashimoto; *Bull. Chem. Soc. Jpn.*, 1971, **44**, 2864.	39, 40
71BCJ3445	T. Toda, S. Morimura, E. Mori, H. Horiuchi and K. Murayama; *Bull. Chem. Soc. Jpn.*, 1971, **44**, 3445.	420
71BSB651	E. H. Bray and K. K. Joshi; *Bull. Soc. Chim. Belg.*, 1971, **80**, 651.	1034
71BSF283	J. Chapelle, J. Elguero, R. Jacquier and G. Tarrago; *Bull. Soc. Chim. Fr.*, 1971, 283.	984
71BSF942	J. Abblard and A. Meynaud; *Bull. Soc. Chim. Fr.*, 1971, 942.	103
71CA(75)88759m	B. G. Christenson, T. R. Beattie and D. W. Graham; *Chem. Abstr.*, 1971, **75**, 88 759m (*Fr. Pat.* 2 034 480 to Merk and Co.).	923
71CB31	F. Knoll and A. Krumm; *Chem. Ber.*, 1971, **104**, 31.	478, 487
71CB792	H. Bredereck, G. Simchen and G. Kapaun; *Chem. Ber.*, 1971, **104**, 792.	979
71CB1104	G. Büttner and S. Hünig; *Chem. Ber.*, 1971, **104**, 1104.	313
71CB1347	P. Krohmer and J. Goubeau; *Chem. Ber.*, 1971, **104**, 1347.	603, 632, 635
71CB1826	K. Burger, J. Fehn and E. Moll; *Chem. Ber.*, 1971, **104**, 1826.	493
71CB1921	M. Schosser, G. Heinz and L. V. Chau; *Chem. Ber.*, 1971, **104**, 1921.	32
71CB2021	N. Schindler and W. Ploger; *Chem. Ber.*, 1971, **104**, 2021.	1003, 1010
71CB2250	R. Appel, R. Kleinstück and K.-D. Ziehn; *Chem. Ber.*, 1971, **104**, 2250.	552
71CB2786	F. P. Woerner, H. Reimlinger and R. Merenyi; *Chem. Ber.*, 1971, **104**, 2786.	829
71CB3354	H. Böhme, G. Auterhoff and W. Höver; *Chem. Ber.*, 1971, **104**, 3354.	406
71CB3475	H. Bredereck, W. Kantlehner and D. Schweizer; *Chem. Ber.*, 1971, **104**, 3475.	881
71CC490	P. Rajagopalan and P. Penev; *J. Chem. Soc., Chem. Commun.*, 1971, 490.	490
71CC608	K. Nicholas, L. S. Bray, R. E. Davis and R. Pettit; *J. Chem. Soc., Chem. Commun.*, 1971, 608.	681
71CC1079	F. Huq, W. Mowat, A. Shortland, A. C. Skapski and G. Wilkinson; *J. Chem. Soc., Chem. Commun.*, 1971, 1079.	656
71CC1241	R. Victor, R. Ben-Shoshan and S. Sarel; *J. Chem. Soc., Chem. Commun.*, 1971, 1241.	681
71CC1477	F. Huq, W. Mowat, A. C. Skapski and G. Wilkinson; *J. Chem. Soc., Chem. Commun.*, 1971, 1477.	656
71CCC1867	O. Paleta, A. Pošta and K. Tesařik; *Collect. Czech. Chem. Commun.*, 1971, **36**, 1867.	5
71CCC3314	J. Honzl, P. Krivinka and M. Ryska; *Collect. Czech. Chem. Commun*, 1971, **36**, 3314.	894
71CJC1477	P. Yates and T. R. Lynch; *Can. J. Chem.*, 1971, **49**, 1477.	852
71CPB2194	K. Hirai and T. Ishiba; *Chem. Pharm. Bull.*, 1971, **19**, 2194.	835
71CR(C)1107	J. Berlan, M. L. Capmau and W. Cholkiewicz; *C. R. Hebd. Seances Acad. Sci., Ser. C*, 1971, **273**, 1107.	802
71CR(C)1669	M. Gaudemar; *C. R. Hebd. Sceances Acad. Sci., Ser. C*, 1971, **273**, 1669.	709
71DOK(196)85	S. P. Kolesnikov, B. L. Perl'mutter and O. M. Nefedov; *Dokl. Akad. Nauk SSSR*, 1971, **196**, 85.	142
71DOK(198)590	A. N. Nesmeyanov, V. A. Sazonova and N. N. Sedova; *Dokl. Akad. Nauk SSSR*, 1971, **198**, 590.	723
71DOK(201)1359	Y. V. Zeifman, V. V. Tyuleneva and I. L. Knunyants; *Dokl. Akad. Nauk SSSR*, 1971, **201**, 1359 (Engl. Transl. 1051).	970
71FCR77	P. Tarrant, C. G. Allison, K. P. Barthold and E. C. Stump, Jr.; *Fluorine Chem. Rev.*, 1971, **5**, 77.	45
71G269	A. Ermili and G. Roma; *Gazz. Chim. Ital.*, 1971, **101**, 269.	891
71GEP2114367	D. J. Peterson; *Ger. Pat.* 2 114 367 (1971) (*Chem. Abstr.*, 1971, **76**, 25 425z).	397
71HCA1037	R. Kyburz, H. Schaltegger and M. Neuenschwander; *Helv. Chim. Acta*, 1971, **54**, 1037.	54, 56
71HCA1651	L. Maier; *Helv. Chim. Acta*, 1971, **54**, 1651.	112, 115, 121, 123, 127, 128
71IC522	T. P. Poeth, P. G. Harrison, T. V. Long, B. R. Willeford and J. J. Zuckermann; *Inorg. Chem.*, 1971, **10**, 522.	726
71IZV162	A. Kh. Filippova, A. I. Borisova and G. S. Lyashenko; *Izv. Akad. Nauk. SSSR, Ser. Khim.*, 1971, **1**, 162.	765
71IZV620	V. R. Polishchuk, L. S. German and I. L. Knunyants; *Izv. Akad. Nauk SSSR, Ser. Khim.*, 1971, 620.	697
71IZV1159	V. K. Khairullin, G. V. Dmitrieva and A. N. Pudovik; *Izv. Akad. Nauk SSSR, Ser. Khim.*, 1971, 1159.	124, 125, 128
71IZV2747	K. Zh. Kulumbetova, T. Ya. Medved' and M. I. Kabachnik; *Izv. Akad. Nauk SSSR, Ser. Khim.*, 1971, 2747.	564
71JA476	L. A. Carpino, L. V. McAdams, III, R. H. Rynbrandt and J. W. Spiewak; *J. Am. Chem. Soc.*, 1971, **93**, 476.	83, 85, 86
71JA746	G. Büchi and S. M. Weinreb; *J. Am. Chem. Soc.*, 1971, **93**, 746.	209
71JA781	G. A. Olah and G. D. Mateescu; *J. Am. Chem. Soc.*, 1971, **93**, 781.	42, 44

71JA1035	J. W. Moore, D. A. Sanders, P. A. Scherr, M. D. Glick and J. P. Oliver; *J. Am. Chem. Soc.*, 1971, **93**, 1035.	698
71JA1701	E. Anderson and T. H. Fife; *J. Am. Chem. Soc.*, 1971, **93**, 1701.	189
71JA1714	G. A. Gornowicz and R. West; *J. Am. Chem. Soc.*, 1971, **93**, 1714.	604, 648
71JA2481	J. B. Sieja; *J. Am. Chem. Soc.*, 1971, **93**, 2481.	59
71JA2796	H. C. Brown and Y. Yamamoto; *J. Am. Chem. Soc.*, 1971, **93**, 2796.	142, 143
71JA3709	D. Seyferth, R. Damrauer, S. B. Andrews and S. S. Washburne; *J. Am. Chem. Soc.*, 1971, **93**, 3709.	623
71JA3793	F. N. Tebbe and G. W. Parshall; *J. Am. Chem. Soc.*, 1971, **93**, 3793.	672
71JA4027	D. J. Peterson; *J. Am. Chem. Soc.*, 1971, **93**, 4027.	528, 533
71JA4472	B. L. Hawkins, W. Bremser, S. Borcic and J. R. Roberts; *J. Am. Chem. Soc.*, 1971, **93**, 4472.	25
71JA5441	F. A. Cotton and J. R. Pipal; *J. Am. Chem. Soc.*, 1971, **93**, 5441.	673
71JA5469	A. Hassner and A. B. Levy; *J. Am. Chem. Soc.*, 1971, **93**, 5469.	314
71JCS2166	D. H. R. Barton, B. Halpern, Q. N. Porter and D. J. Collins; *J. Chem. Soc.*, 1971, 2166.	206
71JCS(A)3495	W. V. Dahlhoff, T. R. Dick, G. H. Ford, W. S. J. Kelly and S. M. Nelson; *J. Chem. Soc. (A)*, 1971, 3495.	546
71JCS(B)1723	D. S. Ashton and J. M. Tedder; *J. Chem. Soc. (B)*, 1971, 1723.	30
71JCS(C)196	T. Sasaki, A. Kojima and M. Ohta; *J. Chem. Soc. (C)*, 1971, 196.	976
71JCS(C)279	G. C. Barrett, D. M. Hall, M. K. Hargreaves and B. Modarai; *J. Chem. Soc. (C)*, 1971, 279.	3, 37
71JCS(C)511	J. Koketsu and Y. Ishii; *J. Chem. Soc. (C)*, 1971, 511.	405
71JCS(C)780	I. R. Gelling and D. G. Wibberley; *J. Chem. Soc. (C)*, 1971, 780.	413
71JCS(C)925	G. Camaggi and F. Gozzo; *J. Chem. Soc. (C)*, 1971, 925.	737
71JCS(C)1213	M. K. M. Dirania and J. Hill; *J. Chem. Soc. (C)*, 1971, 1213.	177
71JCS(C)1772	M. V. Bhatt, K. M. Kamath and M. Ravindranathan; *J. Chem. Soc. (C)*, 1971, 1772.	54
71JCS(C)1836	I. G. M. Campbell and S. M. Raza; *J. Chem. Soc. (C)*, 1971, 1836.	728
71JCS(C)3031	P. Cooper, R. Fields and R. N. Haszeldine; *J. Chem. Soc.(C)*, 1971, 3031.	111
71JCS(C)3344	M. V. Bhatt, K. M. Kamath and M. Ravindranathan; *J. Chem. Soc. (C)*, 1971, 3344.	58
71JFC(1)123	D. G. Naae and D. J. Burton; *J. Fluorine Chem.*, 1971, **1**, 123.	728
71JFC(1)381	R. W. Vanderhaar, D. J. Burton and D. G. Naae; *J. Fluorine Chem.*, 1971, **1**, 381.	754
71JGU481	A. V. Fokin, Yu. N. Studnev, L. D. Kuznetsova and A. F. Kolomiets; *J. Gen. Chem. USSR (Engl. Transl.)*, 1971, **41**, 481.	135
71JGU2183	S. Kh. Nurtdinov, R. S. Khairullin, T. V. Zykova, V. S. Tsivunin and G. Kh. Kamai; *J. Gen. Chem. USSR (Engl. Transl.)*, 1971, **41**, 2183.	120
71JGU2228	N. I. D'yakonova, É. Kh. Mukhametzyanova and I. M. Shermergorn; *J. Gen. Chem. USSR (Engl. Transl.)*, 1971, **41**, 2228.	121, 126
71JHC241	F. K. Lautenschlaeger, M. Myhre, F. Hopton and J. Wilson; *J. Heterocycl. Chem.*, 1971, **8**, 241.	979
71JOC37	G. M. J. Slusarczuk and M. M. Joullié; *J. Org. Chem.*, 1971, **36**, 37.	728
71JOC575	G. A. Boswell, A. L. Johnson and J. P. McDevitt; *J. Org. Chem.*, 1971, **36**, 575.	9
71JOC818	D. R. Strobach and G. A. Boswell, Jr.; *J. Org. Chem.*, 1971, **36**, 818.	8, 9
71JOC1015	L. A. Paquette and R. W. Houser; *J. Org. Chem.*, 1971, **36**, 1015.	80
71JOC1379	D. Seyferth, R. S. Marmor and P. Hilbert; *J. Org. Chem.*, 1971, **36**, 1379.	489
71JOC1613	J. D. Wilson, J. S. Wager and H. Weingarten; *J. Org. Chem.*, 1971, **36**, 1613.	972, 988
71JOC1685	D. J. Woodman, N. Tontapanish and J. V. van Ornum; *J. Org. Chem.*, 1971, **36**, 1685.	894
71JOC2205	J. C. Martin, R. D. Burpitt, P. G. Gott, M. Harris and R. H. Meen; *J. Org. Chem.*, 1971, **36**, 2205.	836, 887
71JOC2211	J. C. Martin, K. C. Brannock, R. D. Burpitt, P. G. Gott and V. A. Hoyle, Jr.; *J. Org. Chem.*, 1971, **36**, 2211.	887
71JOC2351	D. J. Burton and H. C. Krutzsch; *J. Org. Chem.*, 1971, **36**, 2351.	762
71JOC2357	T. H. Fife and E. Anderson; *J. Org. Chem.*, 1971, **36**, 2357.	189
71JOC2731	F. A. Carey and J. R. Neergaard; *J. Org. Chem.*, 1971, **36**, 2731.	252
71JOC3112	M. A. Ratcliff and J. K. Kochi; *J. Org. Chem.*, 1971, **36**, 3112.	300
71JOC3324	W. C. Baird, Jr., H. H. Surridge and M. Buza; *J. Org. Chem.*, 1971, **36**, 3324.	14
71JOC3386	R. L. Soulen, D. B. Clifford, F. F. Crim and J. A. Johnston; *J. Org. Chem.*, 1971, **36**, 3386.	762, 771
71JOC3442	L. De Vries; *J. Org. Chem.*, 1971, **36**, 3442.	979
71JOC3520	J. J. Eisch and M. W. Foxton; *J. Org. Chem.*, 1971, **36**, 3520.	1063
71JOC3566	K. Field and P. Kovacic; *J. Org. Chem.*, 1971, **36**, 3566.	14
71JOC3843	D. A. Nicholson and H. Vaughn; *J. Org. Chem.*, 1971, **36**, 3843.	338
71JOC3846	G. J. McDonald and M. E. Hill; *J. Org. Chem.*, 1971, **36**, 3846.	440, 442, 443
71JOM(26)215	D. J. Peterson; *J. Organometal. Chem.*, 1971, **26**, 215.	961
71JOM(26)357	U. Zucchini, E. Albizzati and U. Giannini; *J. Organomet. Chem.*, 1971, **26**, 357.	725
71JOM(29)409	E. J. Bulten and J. G. Notes; *J. Organomet. Chem.*, 1971, **29**, 409.	626
71JOM(30)C57	I. M. Pribytkova, A. V. Kisin, Y. N. Luzikov, N. P. Makoveyeva, V. N. Torocheshnikov and Y. A. Ustynyih; *J. Organomet. Chem.*, 1971, **30**, C57.	603, 625, 637, 701
71JOM(30)C64	M. Paul and E. Frainnet; *J. Organomet. Chem.*, 1971, **30**, C64.	832
71JOM(30)151	D. Seyferth and S. B. Andrews; *J. Organomet. Chem.*, 1971, **30**, 151.	154, 155, 156
71JOM(30)349	K. Tamao and M. Kumada; *J. Organomet. Chem.*, 1971, **30**, 349.	140
71KGS1645	V. M. Pavlov, I. V. Galakhov, L. I. Ragulin, A. A. Alekseev, G. A. Sokol'skii and I. L. Knunyants; *Khim. Geterotsikl. Soedin.*, 1971, **7**, 1645 (*Chem. Abstr.* 1972, **77**, 113 741).	833
71LA(744)42	E. V. Dehmlow and J. Schonefeld; *Justus Liebigs Ann. Chem.*, 1971, **744**, 42.	16, 25
71LA(748)59	H. Bohme and W. Hover; *Justus Liebigs Ann. Chim.*, 1971, **748**, 59.	979
71M118	A. Meller and W. Maringgele; *Monatsh. Chem.*, 1971, **102**, 118.	145
B-71MI 401-01	W. Kirmse; "Carbene Chemistry," 2nd edn., Academic Press, New York, 1971.	7, 16, 24

71MI 403-01	A. Sammour, M. I. B. Selim and G. H. El-sayed; *U.A.R.J. Chem.*, 1971, **14**, 235.	105
71MI 403-02	J. M. Bellama and C. J. McCormick; *Inorg. Nucl. Chem. Lett.*, 1971, **7**, 533.	141, 142
71MI 403-03	J. Villieras; *Organomet. Chem. Rev. A*, 1971, **7**, 81.	147
B-71MI 407-01	R. C. Bertelson; in "Photochromism, Techniques of Chemistry Vol III," ed. G. H. Brown, Wiley-Interscience, New York, 1971, p. 49.	307
71MI 410-01	K. Issleib, H. H. Bruenner and H. Oehme; *Organometal. Chem. Syn.*, 1971, **1**, 161.	454
71MI 412-01	N. F. Orlov and M. S. Sorokin; *Kremniiorg. Mater.*, 1971, 126 (*Chem. Abstr.*, 1973, **78**, 29 925).	570
B-71MI 413-01	*Organometallic Chemistry*, Specialist Periodical Reports, Royal Society of Chemistry, Volumes 1–23, 1971–1993.	591
71MI 413-01	Y. Matsumura and R. Okawara; *Inorg. Nucl. Chem. Lett.*, 1971, **7**, 113.	595
B-71MI 424-01	W. Kirmse; "Carbene Chemistry," 2nd edn, Academic Press, New York, 1971.	1080
71OS(51)39	D. Seebach and A. K. Beck; *Org. Synth.*, 1971, **51**, 39.	250
71RTC1141	D. P. Roelofsen, E. R. J. Wils and H. van Bekkum; *Recl. Trav. Chim. Pays-Bas*, 1971, **90**, 1141.	177
71S89	G. Tsuchihashi, K. Ogura, S. Iriuchijima and S. Tomisawa; *Synthesis*, 1971, 89.	74
71S150	H. Böhme and P.-H. Meyer; *Synthesis*, 1971, 150.	49, 310
71S312	S. Kabusz and W. Tritschler; *Synthesis*, 1971, 312.	193
71SAP7004922	M. P. Collier, B. M. Kingston, M. F. Lappert and M. M. Truelock; *S. Afr. Pat.* 7 004 922 (1971) (*Chem. Abstr.*, 1971, **75**, 129 943n).	656
71SC103	J. C. Hubert, W. Steege, W. N. Speckamp and H. O. Huisman; *Synth. Commun.*, 1971, **1**, 103.	303
71T945	W. A. Sheppard; *Tetrahedron*, 1971, **27**, 945.	8
71T3965	F. Mathey and J. Bensoam; *Tetrahedron*, 1971, **27**, 3965.	10
71T5523	E. H. Braye, I. Caplier and R. Saussez; *Tetrahedron*, 1971, **27**, 5523.	112
71T6109	G. Cainelli, A. Umani Ronchi, F. Bertini, P. Grasselli and G. Zubiani; *Tetrahedron*, 1971, **27**, 6109.	146
71TL231	A. Robert, J. Pommeret and A. Foucaud; *Tetrahedron Lett.*, 1971, **12**, 231.	183
71TL871	J. A. Marshall and J. L. Belletire; *Tetrahedron Lett.*, 1971, **12**, 871.	846
71TL1035	J. C. Combret, J. Villiéras and G. Lavielle; *Tetrahedron Lett.*, 1971, **12**, 1035.	738, 739, 745
71TL1137	K. Hirai; *Tetrahedron Lett.*, 1971, **12**, 1137.	835
71TL1879	R. Fields, R. N. Haszeldine and P. J. Palmer; *Tetrahedron Lett.*, 1971, **12**, 1879.	821
71TL2241	R. Kalish, A. E. Smith and E. J. Smutny; *Tetrahedron Lett.*, 1971, **12**, 2241.	843
71TL2321	L. Schutte; *Tetrahedron Lett.*, 1971, **12**, 2321.	245
71TL3151	K. Ogura and G. Tsuchihashi; *Tetrahedron Lett.*, 1971, **12**, 3151.	197
71TL3475	K. M. Nicholas and R. Pettit; *Tetrahedron Lett.*, 1971, **12**, 3475.	688
71TL3553	R. H. Rynbrandt; *Tetrahedron Lett.*, 1971, **12**, 3553.	69
71TL3869	P. Weyerstahl, G. Blume and C. Müller; *Tetrahedron Lett.*, 1971, **12**, 3869.	35
71USP3592896	P. C. Aichenegg and C. D. Emerson; *US Pat.* 3 592 896 (1971) (*Chem. Abstr.*, 1971, **75**, 98 164c).	777
71USP3597463	D. J. Peterson; *US Pat.* 3 597 463 (1971) (*Chem. Abstr.*, 1971, **75**, 118 399p).	365, 391
71USP3607944	D. J. Daigle, L. H. Chance and G. L. Drake; *US Pat.* 3 607 944 (1971) (*Chem. Abstr.*, 1972, **76**, 15 739).	555
71USP3622654	K. C. Pande; *US Pat.* 3 622 654 (1971) (*Chem. Abstr.*, 1972, **76**, 46 303).	561
71ZOB482	A. P. Sineokov, L. A. Tsareva and V. S. Étlis; *Zh. Org. Khim.*, 1971, **41**, 482.	891
71ZOB719	B. B. Kochetkov, M. D. Stadnichuk and A. A. Petrov; *Zh. Obshch. Khim.*, 1971, **41**, 719.	941
71ZOB1873	O. A. Mukhacheva, V. G. Nikolaeva and A. I. Razumov; *Zh. Obshch. Khim.*, 1971, **41**, 1873 (*Chem. Abstr.*, 1972, **76**, 3200v).	728
71ZOB1964	A. N. Pudovik, M. G. Zimin and A. M. Kurguzova; *Zh. Obshch. Khim.*, 1971, **41**, 1964.	342
71ZOB2186	E. A. Chernyshev, E. F. Bugerenko and A. S. Petukhova; *Zh. Obshch. Khim.*, 1971, **41**, 2186.	570
71ZOB2372	E. E. Nifant'ev and I. V. Shilov; *Zh. Obshch. Khim.*, 1971, **41**, 2372 (*Chem. Abstr.*, 1972, **76**, 113 314t).	476
71ZOB2575	M. B. Gazizov, D. B. Sultanova, A. I. Razumov, L. P. Ostanina and A. M. Rusalkina; *Zh. Obshch. Khim.*, 1971, **41**, 2575.	340
71ZOR1670	A. D. Nikolaeva, A. I. Sitkin and G. Kh. Kamai; *Zh. Org. Khim.*, 1971, **7**, 1670 (Engl. Transl. 1734).	993, 994
72ACH363	J. Szabo, I. Varga and E. Vinkler; *Acta Chim. Acad. Sci. Hung.*, 1972, **25**, 363.	319
72ACR65	D. Seyferth; *Acc. Chem. Res.*, 1972, **5**, 65.	8, 25
72ACS1258	K. Jensen and S. Hammersum; *Acta Chem. Scand.*, 1972, **26**, 1258.	437
72ACS1659	J. Becher; *Acta Chem. Scand.*, 1972, **26**, 1659.	436
72ACS1735	K. Andersson; *Acta Chem. Scand.*, 1972, **26**, 1735.	28
72AG(E)301	D. Seebach and D. Enders; *Angew. Chem., Int. Ed. Engl.*, 1972, **11**, 301.	530
72AG(E)311	V. Schöllkopf and R. Schröder; *Angew. Chem., Int. Ed. Engl.*, 1972, **11**, 311.	908
72AG(E)473	G. Köbrich; *Angew. Chem. Int. Ed. Engl.*, 1972, **11**, 473.	145, 741, 816, 817, 820
72AG(E)930	A. Salzer and H. Werner; *Angew. Chem., Int. Ed. Engl.*, 1972, **11**, 930.	694
72AJC1737	P. M. Pojer and I. D. Rae; *Aust. J. Chem.*, 1972, **25**, 1737.	303
72BCJ198	Y. Hayasi and H. Nozaki; *Bull. Chem. Soc. Jpn.*, 1972, **45**, 198.	860
72BCJ2023	G.-I. Tsuchihashi and K. Ogura; *Bull. Chem. Soc. Jpn.*, 1972, **45**, 2023.	233
72BCJ2794	T. Numata and S. Oae; *Bull. Chem. Soc. Jpn.*, 1972, **45**, 2794.	233
72BRP1299180	Agripat SA; *Br. Pat.* 1 299 180 (1972) (*Chem. Abstr.*, 1973, **78**, 137 853).	309
72BSF811	G. Coindard, J. Braun and P. Cadiot; *Bull. Soc. Chim. Fr.*, 1972, 811.	145

72BSF1361	P. Mazerolles and H. Cousse; *Bull. Soc. Chim. Fr.*, 1972, 1361.	141
72CA(77)126770j	G. M. Kosolapoff; *Chem. Abstr.*, 1972, **77**, 126 770j.	1023
72CB511	D. Seebach and N. Peleties; *Chem. Ber.*, 1972, **105**, 511.	399, 400
72CB2233	H. Böhme and M. Haake; *Chem. Ber.*, 1972, **105**, 2233.	478, 487
72CB3173	H. Schmidbaur and W. Vornberger; *Chem. Ber.*, 1972, **105**, 3173.	571
72CB3261	E. Lindner and H. Beer; *Chem. Ber.*, 1972, **105**, 3261.	563
72CC443	N. S. Isaacs and D. Kirkpatrick; *J. Chem. Soc., Chem. Comm.*, 1972, 443.	738
72CC613	M. R. Collier, C. Eaborn, B. Jovanovic, M. F. Lappert, L. Manojlovic-Muir, K. W. Muir and M. M. Truelock; *J. Chem. Soc., Chem. Commun.*, 1972, 613.	656
72CC734	M. Cinquini, D. Landini and A. Maia; *J. Chem. Soc., Chem. Commun.*, 1972, 734.	233
72CC814	J. Schwartz; *J. Chem. Soc., Chem. Commun.*, 1972, 814.	683
72CC841	K. I. Thé and L. K. Peterson; *J. Chem. Soc., Chem. Commun.*, 1972, 841.	437
72CC863	A. Belanger and P. Brassard; *J. Chem. Soc., Chem. Commun.*, 1972, 863.	827
72CC993	P. Hong, N. Nishii, K. Sonogashira and N. Hagihara; *J. Chem. Soc., Chem. Commun.*, 1972, 993.	686
72CCC1392	M. Krumpolc and V. Chvalovsky; *Collect. Czech. Chem. Commun.*, 1972, **37**, 1392.	362
72CEN55	B. L. Van Duuren, S. Laskin and N. Nelson; *Chem. Eng. News*, 1972, **50**, 55.	47
72CEN62	B. L. Van Duuren, S. Laskin and N. Nelson; *Chem. Eng. News*, 1972, **50**, 62.	47
72CI(L)380	T. Yamaguchi, Y. Shimizu and T. Suzuki; *Chem. Ind. (London)*, 1972, 380.	321
72CJC1078	H. J. Callot and C. Benezra; *Can. J. Chem.*, 1972, **50**, 1078.	489
72CJC2902	S. Wolfe, J.-B. Ducep, G. Kannengiesser and W. S. Lee; *Can. J. Chem.*, 1972, **50**, 2902.	417
72CPB2123	T. Watanabe, F. Hamaguchi and S. Ohki; *Chem. Pharm. Bull.*, 1972, **20**, 2123.	211
72CR(C)1357	M. Levas and E. Levas; *C. R. Hebd. Seances Acad. Sci., Ser. C*, 1972, 1357.	746
72CRV705	W. P. Jencks; *Chem. Rev.*, 1972, **72**, 705.	294
72CZ691	K. Krueger and L. Bauer; *Chem.-Ztg.*, 1972, **96**, 691.	492
72DOK(202)362	A. N. Nesmeyanov, V. A. Sazonova and N. N. Sedova; *Dokl. Akad. Nauk SSSR.*, 1972, **202**, 362.	723
72GEP(O)2060217	H.-J. Kleiner; *Ger. Pat. Offen.* 2 060 217 (1972) (*Chem. Abstr.*, 1972, **77**, 101 894g).	119, 121
72HCA249	M. Rosenberger, D. Andrews, F. DiMaria, A. J. Duggan and G. Saucy; *Helv. Chim. Acta*, 1972, **55**, 249.	166
72HCA594	V. Rautenstrauch; *Helv. Chim. Acta*, 1972, **55**, 594.	950
72IC408	C. H. Van Dyke, E. W. Kifer and G. A. Gibbon; *Inorg. Chem.*, 1972, **11**, 408.	638, 641
72IC1150	J. Rathke and R. Schaeffer; *Inorg. Chem.*, 1972, **11**, 1150.	142, 143
72IC2274	C. Couret, J. Satgé and F. Couret; *Inorg. Chem.*, 1972, **11**, 2274.	338
72IC2279	P. C. Ellgen; *Inorg. Chem.*, 1972, **11**, 2279.	692
72IC2340	D. A. Wheatland, C. H. Clapp and R. W. Waldron; *Inorg. Chem.*, 1972, **11**, 2340.	557
72IJ293	F. Heydenreich, A. Mollbach, G. Wilke, H. Dreeskamp, E. G. Hoffmann, G. Schroth, K. Seevogel and W. Stempfle; *Isr. J. Chem.*, 1972, **10**, 293.	583
72IZV215	Yu. L. Kopaevich, G. G. Belen'kii, Z. A. Stumbrevichute, L. S. German and I. L. Knunyants; *Izv. Akad. Nauk SSSR, Ser. Khim.*, 1972, 215.	134
72IZV1347	M. V. Urushadze, E. G. Abduganiev, E. M. Rokhlin and I. L. Knunyants; *Izv. Akad. Nauk SSSR, Ser. Khim.*, 1972, **6**, 1347 (*Chem. Abstr.*, 1972, **77**, 125 843y).	894
72IZV1600	A. N. Nesmeyanov, L. G. Makarova and V. N. Vinogradova; *Izv. Akad. Nauk SSSR, Ser. Khim.*, 1972, 1600.	685
72IZV1823	A. N. Nesmeyanov, A. A. Lubovich and S. P. Gubin; *Izv. Akad. Nauk SSSR, Ser. Khim.*, 1972, 1823.	706, 723, 724
72IZV1895	V. A. Dorokhov and B. M. Mikhailov; *Izv. Akad. Nauk SSSR, Ser. Khim.*, 1972, **8**, 1895 (*Chem. Abstr.*, 1972, **77**, 164 794c).	884
72IZV2594	E. G. Perevalova, D. A. Lemenovskii, O. B. Afanasova, V. P. Dyadchenko, K. I. Grandberg and A. N. Nesmeyanov; *Izv. Akad. Nauk SSSR, Ser. Khim.*, 1972, 2594.	723
72IZV2612	V. Yu. Kovtun, V. A. Gilyarov and M. I. Kabachnik; *Izv. Akad. Nauk SSSR, Ser. Khim.*, 1972, 2612.	563
72JA597	P. Beak and J. W. Worley; *J. Am. Chem. Soc.*, 1972, **94**, 597.	255, 394
72JA1012	J. C. Philips and M. Oku; *J. Am. Chem. Soc.*, 1972, **94**, 1012.	779
72JA2020	E. A. Noe and J. D. Roberts, *J. Am. Chem. Soc.*, 1972, **94**, 2020.	9
72JA3058	W. G. Bentrude, W. D. Johnson and W. A. Khan; *J. Am. Chem. Soc.*, 1972, **94**, 3058.	914
72JA4784	A. Wright, D. Ling, P. Boudjouk and R. West; *J. Am. Chem. Soc.*, 1972, **94**, 4784.	368
72JA5115	T. J. Maricich and C. K. Harrington; *J. Am. Chem. Soc.*, 1972, **94**, 5115.	232
72JA5702	W. R. Cullen, L. D. Hall and J. E. H. Ward; *J. Am. Chem. Soc.*, 1972, **94**, 5702.	134
72JA5837	D. N. Roark and G. J. D. Peddle; *J. Am. Chem. Soc.*, 1972, **94**, 5837.	614, 621
72JA5897	R. E. Ireland and R. H. Mueller; *J. Am. Chem. Soc.*, 1972, **94**, 5897.	831
72JA6854	M. W. Rathke and R. Kow; *J. Am. Chem. Soc.*, 1972, **94**, 6854.	664
72JA7827	A. F. Kluge, K. G. Untch and J. H. Fried; *J. Am. Chem. Soc.*, 1972, **94**, 7827.	187, 199
72JA8795	R. R. Fraser, F. J. Schuber and Y. Wigfield; *J. Am. Chem. Soc.*, 1972, **94**, 8795.	391
72JA8929	S. Hanessian, G. Yang-Chung, P. Lavallee and A. G. Pernet; *J. Am. Chem. Soc.*, 1972, **94**, 8929.	198
72JA8932	E. J. Corey and S. W. Walinsky; *J. Am. Chem. Soc.*, 1972, **94**, 8932.	1074
72JCS(D)2649	J. F. Malone and W. S. McDonald; *J. Chem. Soc., Dalton Trans.*, 1972, 2649.	698
72JCS(P1)41	D. B. J. Easton, D. Leaver and T. J. Rawlings; *J. Chem. Soc., Perkin Trans. 1*, 1972, 41.	850
72JCS(P1)1103	D. H. R. Barton, P. D. Magnus and J. I. Okogun; *J. Chem. Soc., Perkin Trans. 1*, 1972, 1103.	1079
72JCS(P1)1225	T. P. C. Mulholland, R. Foster and D. B. Haydock; *J. Chem. Soc., Perkin Trans. 1*, 1972, 1225.	824

72JCS(P1)1883	M. Cinquini and S. Colonna; *J. Chem. Soc., Perkin Trans. 1*, 1972, 1883.	75, 77, 233
72JCS(P1)1886	M. Cinquini, S. Colonna, R. Fornasier and F. Montanari; *J. Chem. Soc., Perkin Trans. 1*, 1972, 1886.	75
72JCS(P2)296	M. Cinquini, S. Colonna and D. Landini; *J. Chem. Soc., Perkin. Trans. 2*, 1972, 296.	74, 75
72JHC175	M. von Strandtman, D. Connor and J. Shavel Jr.; *J. Heterocycl. Chem.*, 1972, **9**, 175.	221
72JHC715	D. J. Daigle, A. B. Pepperman, Jr. and F. L. Normand; *J. Heterocycl. Chem.*, 1972, **9**, 715.	458
72JHC827	G. Pifferi, A. Vigevani, P. Consonni and G. G. Gallo; *J. Heterocycl. Chem.*, 1972, **9**, 827.	441
72JHC1145	V. A. Marquez, T. Hirata, L.-M. Twanmoh, H. B. Wood Jr. and J. S. Driscoll; *J. Heterocycl. Chem.*, 1972, **9**, 1145.	420
72JHC1295	L. M. Trefonas and J. N. Brown; *J. Heterocycl. Chem.*, 1972, **9**, 1295.	458
72JOC215	D. E. Butler and S. M. Alexander; *J. Org. Chem.*, 1972, **37**, 215.	539
72JOC521	W. W. Zajak and K. J. Byrne; *J. Org. Chem.*, 1972, **37**, 521.	199
72JOC683	E. T. McBee, E. P. Wesseler, D. L. Crain, R. Hurnaus and T. Hodgins; *J. Org. Chem.*, 1972, **37**, 683.	743
72JOC922	C. O. Parker and T. E. Stevens; *J. Org. Chem.*, 1972, **37**, 922.	439
72JOC939	F. A. Carey and A. S. Court; *J. Org. Chem.*, 1972, **37**, 939.	574
72JOC1526	W. G. Phillips and K. W. Ratts; *J. Org. Chem.*, 1972, **37**, 1526.	71
72JOC1902	A. R. Friedman and D. R. Graber; *J. Org. Chem.*, 1972, **37**, 1902.	864
72JOC1960	H. A. P. de Jongh, C. R. M. I. de Jonge, H. J. M. Sinnige, W. J. de Klein, W. G. B. Huysmans, W. J. Mijs, W. J. van den Hoek and J. Smidt; *J. Org. Chem.*, 1972, **37**, 1960.	894
72JOC2662	D. R. Dimmel, C. A. Wilkie and F. Ramon; *J. Org. Chem.*, 1972, **37**, 2662.	619
72JOC2752	A. W. Frank and G. L. Drake Jr.; *J. Org. Chem.*, 1972, **37**, 2752.	453, 456, 458, 475
72JOC2885	W. A. Pryor and H. T. Bickley; *J. Org. Chem.*, 1972, **37**, 2885.	220
72JOC3226	E. Klingsberg; *J. Org. Chem.*, 1972, **37**, 3226.	852
72JOC3304	S. Rengaraju and K. D. Berlin; *J. Org. Chem.*, 1972, **37**, 3304.	473
72JOC3332	R. A. De Marco, D. A. Couch and J. M. Shreeve; *J. Org. Chem.*, 1972, **37**, 3332.	52, 59
72JOC4136	R. W. Begland and D. R. Hartter; *J. Org. Chem.*, 1972, **37**, 4136.	438
72JOC4397	J. P. Berry, A. F. Isbell and G. E. Hunt; *J. Org. Chem.*, 1972, **37**, 4397.	476
72JOM(34)C18	J. Satgé, M. Lesbre, P. Rivière and S. Richelme; *J. Organomet. Chem.*, 1972, **34**, C18.	521
72JOM(35)367	K. Yasufuku and H. Yamazaki; *J. Organomet. Chem.*, 1972, **35**, 367.	711
72JOM(36)267	N. Wiberg, W. Baumeister and P. Zahn; *J. Organomet. Chem.*, 1972, **36**, 267.	699
72JOM(36)277	N. Wiberg and W. Baumeister; *J. Organomet. Chem.*, 1972, **36**, 277.	699
72JOM(39)169	E. Sappa, O. Gambino, L. Milone and G. Cetini; *J. Organomet. Chem.*, 1972, **39**, 169.	683
72JOM(39)365	H. Sakurai and J. Hayashi; *J. Organomet. Chem.*, 1972, **39**, 365.	689
72JOM(40)115	R. D. Brasington and R. C. Poller; *J. Organomet. Chem.*, 1972, **40**, 115.	156
72JOM(42)C21	T. J. Barton and E. Kline; *J. Organomet. Chem.*, 1972, **42**, C21.	610
72JOM(42)C35	K. A. Ostoja Starzewski, H. tom Dieck, K. D. Franz and F. Hohmann; *J. Organomet. Chem.*, 1972, **42**, C35.	580
72JOM(43)117	T. Abe and R. Okawara; *J. Organomet. Chem.*, 1972, **43**, 117.	153
72JOM(43)139	M. Bolourtchian, P. Bourgeois, J. Dunogues, N. Duffaut and R. Calas; *J. Organomet. Chem.*, 1972, **43**, 139.	1014
72JOM(43)157	J. Dunogues, M. Bolourtchian, R. Calas, N. Duffaut and J.-P. Picard; *J. Organometal. Chem.*, 1972, **43**, 157.	933
72JOM(43)333	S. I. Sato, Y. Matsumura and R. Okawara; *J. Organomet. Chem.*, 1972, **43**, 333.	595
72JOM(46)51	O. A. Kruglaya, G. S. Kalinina, I. B. Petrov and N. S. Vyazankin; *J. Organomet. Chem.*, 1972, **46**, 51.	660
72JOM(46)73	C. Ainsworth and Y.-N. Kuo; *J. Organomet. Chem.*, 1972, **46**, 73.	832
72JPR66	H. Oehme, K. Issleib and E. Leissring; *J. Prakt. Chem.*, 1972, **314**, 66.	337
72LA(755)67	A. Grote, A. Haag and G. Hesse; *Justus Liebigs Ann. Chem.*, 1972, **755**, 67.	522
72LA(759)84	H. Zondler and W. Pfleiderer; *Justus Liebigs Ann. Chem.*, 1972, **759**, 84.	409
72LA(763)208	U. Schollkopf and P. Hanssle; *Justus Liebigs Ann. Chem.*, 1972, **763**, 208.	950
72LA(765)55	K. Eiter, K.-F. Hebenbrock and H.-J. Kabbe; *Justus Liebigs Ann. Chem.*, 1972, **765**, 55.	312
72LA(766)130	U. Schöllkopt, R. Schröder and E. Blume; *Justus Liebigs Ann. Chem.*, 1972, **766**, 130.	331, 908
72MI 403-01	L. Maier; *Phosphorus*, 1972, **1**, 249.	120, 122
72MI 404-01	V. Zikán, L. Vrba, B. Kakác and M. Semonsky; *Collect. Czech. Chem. Commun.*, 1972, **38**, 1091.	211
B-72MI 407-01	G. M. Kosolapoff and L. Maier; "Organic Phosphorus Compounds," Wiley Interscience, New York, 1972, vols. 1–6.	336
B-72MI 408-01	H. C. Brown; "Boranes in Organic Chemistry," Cornell University Press, Ithaca, NY, 1972.	364
72MI 408-02	D. J. Peterson; *Organomet. Chem. Rev. A*, 1972, **7**, 295.	377, 390, 394, 398
B-72MI 410-01	G. M. Kosolapoff and L. Maier (eds); "Organic Phosphorus Compounds," Wiley, New York, 1972, vols 1–7.	451
B-72MI 412-01	G. M. Kosolapoff and L. Maier; "Organic Phosphorus Compounds," Wiley, New York, 1972–1974, vol. 1–7.	550
72MI 414-01	D. J. Peterson; *Organomet. Chem. Rev. A*, 1972, **8**, 295.	608, 621, 622, 624, 646
72MI 414-02	R. Damrauer; *Organomet. Chem. Rev. A*, 1972, **8**, 67.	610
B-72MI 419-01	H. O. House; "Modern Synthetic Reactions," 2nd edn, Benjamin/Cummings, New York, 1972, p. 520.	824
72OR(19)279	R. A. Sheldon and J. K. Kochi; *Org. React.*, 1972, **19**, 279.	27
72PS(1)237	L. Maier; *Phosphorus Sulfur*, 1977, **1**, 237.	344
72RTC209	T. Olijnsma, J. B. F. N. Engberts and J. Strating; *Recl. Trav. Chim. Pays-Bas*, 1972, **91**, 209.	331
72S30	E. E. Gilbert; *Synthesis*, 1972, 30.	424

72S259	M. Cinquini and S. Colonna; *Synthesis*, 1972, 259.	74, 75
72S419	D. P. Roelofsen and H. van Bekkum; *Synthesis*, 1972, 419.	177
72SC361	T. A. Bryson; *Synth. Commun.*, 1972, **2**, 361.	210
72SRI239	H. Werner and A. Salzer; *Synth. React. Inorg. Metal-Org. Chem.*, 1972, **2**, 239.	694
72SRI317	K. Moedritzer; *Synth. React. Inorg. Metal-Org. Chem.*, 1972, **2**, 317.	463, 470
72T2587	H. Oehme, K. Issleib and E. Leissring; *Tetrahedron*, 1972, **28**, 2587.	337
72T3545	H. Khedija, H. Strzelecka and M. Simalty; *Tetrahedron*, 1972, **28**, 3545.	863
72T4039	R. Leblanc, E. Corre and A. Foucaud; *Tetrahedron*, 1972, **28**, 4039.	800
72T5055	E. Corre, M. F. Chasle and A. Foucaud; *Tetrahedron*, 1972, **28**, 5055.	996
72T5149	M. Svilarich-Soenen and A. Foucaud; *Tetrahedron*, 1972, **28**, 5149.	800
72T6013	M. A. Ruveda and S. A. de Licastro; *Tetrahedron*, 1972, **28**, 6013.	338
72TL445	H. Meier and I. Menzel; *Tetrahedron Lett.*, 1972, **13**, 445.	874
72TL949	S. Hammerum; *Tetrahedron Lett.*, 1972, **13**, 949.	436, 437
72TL1353	A. Hassner, A. S. Miller and M. J. Haddadin; *Tetrahedron Lett.*, 1972, **13**, 1353.	829
72TL1383	K. Ogura and G. Tsuchihashi; *Tetrahedron Lett.*, 1972, **13**, 1383.	856
72TL2367	A. M. van Leusen, G. J. M. Boerma, R. B. Helmholdt, H. Siderius and J. Strating; *Tetrahedron Lett.*, 1972, **13**, 2367.	331, 538
72TL2477	K. Hovius and J. B. F. N. Engberts; *Tetrahedron Lett.*, 1972, **13**, 2477.	235
72TL3595	D. F. Hunt and G. T. Rodeheaver; *Tetrahedron Lett.*, 1972, **13**, 3595.	193
72TL3769	E. J. Corey and P. L. Fuchs; *Tetrahedron Lett.*, 1972, **13**, 3769.	745
72TL4055	E. M. Gaydou; *Tetrahedron Lett.*, 1972, **13**, 4055.	187
72TL4517	Y. Kanaoka and K. Koyama; *Tetrahedron Lett.*, 1972, **13**, 4517.	1077
72TL5133	J. Kagan, J. T. Przybytek, B. E. Firth and S. P. Singh; *Tetrahedron Lett.*, 1972, **13**, 5133.	183
72USP3632634	L. Maier; *US Pat.* 3 632 634 (1972) (*Chem. Abstr.*, 1972, **76**, 113 371).	563
72USP3692798	S. Barcza; *US Pat.* 3 692 798 (1972) (*Chem. Abstr.*, 1973, **78**, 43 692e).	520
72ZAAC(391)219	G. Fritz, M. Berndt and R. Huber; *Z. Anorg. Allg. Chem.*, 1972, **391**, 219.	621
72ZC60	W. Wegener and P. Scholz; *Z. Chem.*, 1972, **17**, 60.	492
72ZC103	W. Wegener and P. Scholz; *Z. Chem.*, 1972, **12**, 103.	554
72ZC178	W. Wegener and P. Scholz; *Z. Chem.*, 1972, **12**, 178.	343
72ZC334	W. Wegener and K. Schlippes; *Z. Chem.*, 1972, **17**, 334.	492
72ZOB112	Z. S. Novikova, S. N. Zdorova and I. F. Lutsenko; *Zh. Obshch. Khim.*, 1972, **42**, 112.	569
72ZOB293	N. I. Liptuga, V. V. Vasil'ev and G. I. Derkach; *Zh. Obshch. Khim.*, 1972, **42**, 293.	574
72ZOB346	I. L. Knunyants, E. G. Bykhovskaya and Y. A. Sizov; *Zh. Obshch. Khim.*, 1972, **42**, 346.	1023
72ZOB479	I. A. Dormidonov, V. F. Martynov and V. E. Timofeev; *Zh. Obshch. Khim.*, 1972, **42**, 479.	340
72ZOB631	V. F. Mironov, E. M. Stepina and V. I. Shiryaev; *Zh. Obshch. Khim.*, 1972, **42**, 631 (*Chem. Abstr.*, 1972, **77**, 88 609d).	660
72ZOB870	B. B. Kochetkov, M. D. Stadnichuk and A. A. Petrov; *Zh. Obshch. Khim.*, 1972, **42**, 870.	941
72ZOB1227	Yu. Yu. Samitov, R. D. Gareev, L. A. Stabrovskaya and A. N. Pudovik; *Zh. Obshch. Khim.*, 1972, **42**, 1227 (*Chem. Abstr.*, 1972, **77**, 100 550n).	489
72ZOB1361	V. F. Mironov, T. K. Gar and A. A. Buyakov; *Zh. Obshch. Khim.*, 1972, **42**, 1361 (*Chem. Abstr.*, 1972, **77**, 114 484f).	609
72ZOB1521	T. K. Gar, A. A. Buyakov and V. F. Mironov; *Zh. Obshch. Khim.*, 1972, **42**, 1521 (*Chem. Abstr.*, 1972, **77**, 140 216q).	628, 639, 640
72ZOB1647	M. B. Gazizov, D. B. Sultanova, A. I. Razumov and L. P. Ostanina; *Zh. Obshch. Khim.*, 1972, **42**, 1647.	340
72ZOB1727	A. N. Pudovik, T. M. Sudakova, O. E. Raevskaya and V. A. Fedechkina; *Zh. Obshch. Khim.*, 1972, **42**, 1727 (Engl. Transl. 1715).	1000, 1001
72ZOB1861	A. N. Pudovik and R. D. Gareev; *Zh. Obshch. Khim.*, 1972, **42**, 1861.	343
72ZOB2418	N. B. Karlstedt, M. V. Proskurnina and I. F. Lutsenko; *Zh. Obshch. Khim.*, 1972, **42**, 2418.	344
72ZOB2450	G. K. I. Magomedov, V. G. Syrkin and A. S. Frenkel; *Zh. Obshch. Khim.*, 1972, **42**, 2450.	724
72ZOR1329	S. I. Radchenko; *Zh. Org. Khim.*, 1972, **8**, 1329 (*Chem. Abstr.* 1972, **77**, 125 852).	841
72ZOR2457	A. L. Fridman, F. A. Gabitov and V. D. Surkov; *Zh. Org. Khim.*, 1972, **8**, 2457 (Engl. Transl. 2505).	995
73ACS779	S. Hammerum; *Acta Chem. Scand.*, 1973, **27**, 779.	436
73ACS2242	S. Gronowitz and T. Frejd; *Acta Chem. Scand.*, 1973, **27**, 2242.	872
73AG(E)415	H. Schmidbaur, J. Adlkofer and W. Buchner; *Angew. Chem., Int. Ed. Engl.*, 1973, **12**, 415.	585
73AG(E)416	H. Schmidbaur and R. Franke; *Angew. Chem., Int. Ed. Engl.*, 1973, **12**, 416.	586
73AG(E)502	K. Burger, J. Fehn and W. Thenn; *Angew. Chem., Int. Ed. Engl.*, 1973, **12**, 502.	311, 493
73AG(E)853	H. H. Karsch and H. Schmidbaur; *Angew. Chem., Int. Ed. Engl.*, 1973, **12**, 853.	583
73AG(E)854	E. Kurras, U. Rosenthal, H. Mennenga and G. Oehme; *Angew. Chem., Int. Ed. Engl.*, 1973, **12**, 854.	580
73AJC2491	D. St. C. Black and K. G. Watson; *Aust. J. Chem.*, 1973, **26**, 2491.	232
73AP(306)161	G. Zinner and W. Kliegel; *Arch. Pharm.* (*Weinheim, Ger.*), 1973, **306**, 161.	435
73AP(306)389	G. Seitz and H. Moennighoff; *Arch. Pharm.* (*Weinheim, Ger.*), 1973, **306**, 389.	865
73BSF581	F. Clesse and H. Quiiou; *Bull. Soc. Chim. Fr.*, 1973, **2**, 581.	852, 969
73BSF2989	P. Hullot and T. Cuvigny; *Bull. Soc. Chim. Fr.*, 1973, 2989.	535
73BSF3499	Y. Le Floc'h, A. Brault and M. Kerfanto; *Bull. Soc. Chim. Fr.*, 1973, 3499.	322
73CA(78)45000b	P. Golborn and J. L. Dever; *Chem. Abstr.*, 1973, **78**, 45 000b (*Ger. Pat.* 2 222 489 to Hooker Chemical Corp.).	912
73CA(78)85856s	P. Golborn; *Chem. Abstr.*, 1973, **78**, 85 856s (*Ger. Pat.* 2 230 132 to Hooker Chemical Corp.).	920, 921
73CAR(29)209	A. Hasegawa and H. G. Fletcher, Jr.; *Carbohydr. Res.*, 1973, **29**, 209.	184

73CB69	E. Fluck and P. Meiser; *Chem. Ber.*, 1973, **106**, 69.	308
73CB1107	K. Schank, H. Hasenfratz and A. Weber; *Chem. Ber.*, 1973, **106**, 1107.	838
73CB1601	G. Köbrich, B. Kalb, A. Mannschreck and R. A. Misra; *Chem. Ber.*, 1973, **106**, 1601.	806
73CB2277	D. Seebach, M. Kolb and B.-T. Grobel; *Chem. Ber.*, 1973, 2277.	849
73CB2733	M. Wieber and B. Eichhorn; *Chem. Ber.*, 1973, **106**, 2733.	114, 128
73CB2742	M. Wieber and B. Eichhorn; *Chem. Ber.*, 1973, **106**, 2742.	135
73CB3421	E. Burger and E. Einhellig; *Chem. Ber.*, 1973, **106**, 3421.	493
73CC4	F. Jung and T. Durst; *J. Chem. Soc., Chem. Commun.*, 1973, 4.	75, 77
73CC49	J. S. Grossert and R. F. Langler; *J. Chem. Soc., Chem. Commun.*, 1973, 49.	63, 78
73CC50	J. S. Grossert, W. R. Hardstaff and R. F. Langler; *J. Chem. Soc., Chem. Commun.*, 1973, 50.	85
73CC224	T.-L. Ho and C. M. Wong; *J. Chem. Soc., Chem. Commun.*, 1973, 224.	220
73CC258	E. W. Abel, I. W. Nowell, A. G. J. Modinos and C. Towers; *J. Chem. Soc., Chem. Commun.*, 1973, 258.	1042
73CC319	C. G. Venier and H. J. Barager, III; *J. Chem. Soc., Chem. Commun.*, 1973, 319.	78
73CC323	I. M. T. Davidson and A. B. Howard; *J. Chem. Soc., Chem. Commun.*, 1973, 323.	617
73CC601	A. G. Anastassiou and R. Elliott; *J. Chem. Soc., Chem. Commun.*, 1973, 601.	303
73CCC1522	J. Pola, Z. Papouskova and V. Chvalovsky; *Collect. Czech. Chem. Commun.*, 1973, **38**, 1522.	352
73CI(L)331	N. Altabev, R. D. Smith and N. S. I. Suratwala; *Chem. Ind. (London)*, 1973, 331.	28, 39, 40
73CJC1136	A. F. Janzen and O. C. Vaidya; *Can. J. Chem.*, 1973, **51**, 1136.	337, 344
73CJC2448	K. I. Thé, L. K. Peterson and E. Kiehlmann; *Can. J. Chem.*, 1973, **51**, 2448.	437
73CL111	T. Sato, K. Yamamoto and K. Fukui; *Chem. Lett.*, 1973, 111.	892
73CPB2257	H. Sakai, K. Ito and M. Sekiya; *Chem. Pharm. Bull.*, 1973, **21**, 2257.	322, 323
73CPB2770	Y. Tominaga, R. Natsuki, Y. Matsuda and G. Kobayashi; *Chem. Pharm. Bull.*, 1973, **21**, 2770.	844
73CR(C)395	G. Sturtz and B. Corbel; *C. R. Hebd. Seances Acad Sci., Ser. C*, 1973, **277**, 395.	923
73CSC347	B. L. Barnett and C. Krüger; *Cryst. Struct. Commun.*, 1973, **2**, 347.	711
73DOK(209)869	A. N. Nesmeyanov, E. G. Perevalova, L. I. Khomik and L. I. Leont'eva; *Dokl. Akad. Nauk SSSR*, 1973, **209**, 869.	719
73DOK(209)1113	A. N. Nesmeyanov, T. P. Tolstaya and V. V. Korol'kov; *Dokl. Akad. Nauk SSSR*, 1973, **209**, 1113.	726
73EGP94992	H. Köhler, I. Gerats, B. Eichler, G. Erfurt and W. Kockmann; *Ger. (East) Pat.* 94 992 (1973) (*Chem. Abstr.*, 1973, **79**, 42 685).	576, 587
73EGP99803	A. Tzschach, D. Drohne and H. Jocachim; (*East*) *Ger. Pat.* 99 803 (1973) (*Chem. Abstr.*, 1973, **80**, 83 244h).	495, 497
73GEP2228928	F. Krüger; *Ger. Pat.* 2 228 928 (1973) (*Chem. Abstr.*, 1974, **80**, 83 237).	561
73HCA944	A. Niederhauser, A. Frey and M. Neuenschwander; *Helv. Chim. Acta*, 1973, **56**, 944.	885, 976, 977
73HCA1117	W. Fink; *Helv. Chim. Acta*, 1973, **56**, 1117.	308, 511
73HCA1318	A. Niederhauser and M. Neuenschwander; *Helv. Chim. Acta*, 1973, **56**, 1318.	885
73HCA1331	A. Niederhauser and M. Neuenschwander; *Helv. Chim. Acta*, 1973, **56**, 1331.	885
73HCA2166	M. Schlosser, B. Schaub, B. Spahic and G. Sleiter; *Helv. Chim. Acta*, 1973, **56**, 2166.	951
73IC294	L. J. Guggenberger; *Inorg. Chem.*, 1973, **12**, 294.	672
73IC2396	S. A. Gardner, P. S. Andrews and M. D. Rausch; *Inorg. Chem.*, 1973, **12**, 2396.	689, 690
73IZV955	S. V. Rogozhin, V. A. Davankov and Yu. P. Belov; *Izv. Akad. Nauk SSSR, Ser. Khim.*, 1973, 955 (*Chem. Abstr.*, 1973, **79**, 42 610k).	461
73IZV1112	B. A. Arbusov, N. A. Polezhaeva and V. S. Vinogradova; *Izv. Akad. Nauk. SSSR, Ser. Khim.*, 1973, 1112 (Engl. Transl. 1067).	1006
73IZV2796	A. N. Nesmeyanov, L. G. Makarova and V. N. Vinogradova; *Izv. Akad. Nauk SSSR, Ser. Khim.*, 1973, 2796.	669, 684, 722
73IZV2815	A. N. Nesmeyanov, K. N. Anisimov, N. E. Kolobova and Yu. V. Makarov; *Izv. Akad. Nauk SSSR, Ser. Khim.*, 1973, 2815.	726
73JA182	R. D. Norris and G. Binsch; *J. Am. Chem. Soc.*, 1973, **95**, 182.	31, 34
73JA1811	R. B. King and M. S. Saran; *J. Am. Chem. Soc.*, 1973, **95**, 1811.	821
73JA2694	K. Oshima, K. Shimoji, H. Takahashi, H. Yamamoto and H. Nozaki; *J. Am. Chem. Soc.*, 1973, **95**, 2694.	956, 957
73JA3078	T. J. Barton, E. A. Kline and P. M. Garvey; *J. Am. Chem. Soc.*, 1973, **95**, 3078.	627
73JA3356	N. Kornblum, M. M. Kestner, S. D. Boyd and L. C. Cattran; *J. Am. Chem. Soc.*, 1973, **95**, 3356.	332
73JA3420	F. Jung, N. K. Sharma and T. Durst; *J. Am. Chem. Soc.*, 1973, **95**, 3420.	393
73JA3635	A. McKillop, J. D. Hunt, F. Kienzle, E. Bigham and E. C. Taylor; *J. Am. Chem. Soc.*, 1973, **95**, 3635.	194
73JA5829	E. J. Corey and D. J. Beames; *J. Am. Chem. Soc.*, 1973, **95**, 5829.	845
73JA6962	S. E. Jacobson and A. Wojcicki; *J. Am. Chem. Soc.*, 1973, **95**, 6962.	267
73JA7156	R. C. Kelly and I. Schletter; *J. Am. Chem. Soc.*, 1973, **95**, 7156.	168
73JA7431	P. Calzavara, M. Cinquini, S. Colonna, R. Fornasier and F. Montanari; *J. Am. Chem. Soc.*, 1973, **95**, 7431.	76
73JA7813	B. M. Trost, R. M. Cory, P. H. Scudder and H. B. Neubold; *J. Am. Chem. Soc.*, 1973, **95**, 7813.	99
73JA8467	D. J. Burton and D. G. Naae; *J. Am. Chem. Soc.*, 1973, **95**, 8467.	7
73JCS(D)445	M. R. Collier, M. F. Lappert and R. Pearce; *J. Chem. Soc., Dalton Trans.*, 1973, 445.	656
73JCS(P1)1066	D. H. Coy, R. N. Haszeldine, M. J. Newlands and A. E. Tipping; *J. Chem. Soc., Perkin Trans. 1*, 1973, 1066.	794
73JCS(P1)2272	P. F. Jones, M. F. Lappert and A. C. Szary; *J. Chem. Soc., Perkin Trans. 1*, 1973, 2272.	849

Ref	Citation	Pages
73JFC(3)63	R. C. England, L. Soloman and C. G. Krespan; *J. Fluorine Chem.*, 1973, **3**, 63.	731
73JFC(3)329	R. D. Bagnall, P. L. Coe and J. C. Tatlow; *J. Fluorine Chem.*, 1973, **3**, 329.	3
73JGU679	M. G. Voronkov, A. M. Skiyanova, E. O. Tsetlina and R. G. Mirskov; *J. Gen. Chem. USSR (Engl. Transl.)*, 1973, **43**, 679.	813
73JHC11	A. Shafiee and I. Lalezari; *J. Heterocycl. Chem.*, 1973, **10**, 11.	852
73JOC17	C. G. Venier, H.-H. Hsieh and H. J. Barager, III; *J. Org. Chem.*, 1973, **38**, 17.	74, 75, 76, 78
73JOC153	G. A. Ungefug and C. W. Roberts; *J. Org. Chem.*, 1973, **38**, 153.	22
73JOC338	I. Lalezari, A. Shafiee and M. Yalpani; *J. Org. Chem.*, 1973, **38**, 338.	874
73JOC437	N. Finch and C. W. Gemenden; *J. Org. Chem.*, 1973, **38**, 437.	413
73JOC479	R. L. Soulen, S. C. Carlson and F. Lang; *J. Org. Chem.*, 1973, **38**, 479.	739
73JOC554	J. L. Isidor and R. M. Carlson; *J. Org. Chem.*, 1973, **38**, 554.	184
73JOC834	E. J. Corey and R. A. Ruden; *J. Org. Chem.*, 1973, **38**, 834.	181
73JOC1949	J. A. Zoltewicz, T. E. Oestreich, J. K. O'Halloran and L. S. Helmick; *J. Org. Chem.*, 1973, **38**, 1949.	413
73JOC2091	J. L. Webb and J. E. Corn; *J. Org. Chem.*, 1973, **38**, 2091.	30
73JOC2160	G. E. Wilson and R. Albert; *J. Org. Chem.*, 1973, **38**, 2160.	65
73JOC2251	D. Ben-Ishai and Z. Inbal; *J. Org. Chem.*, 1973, **38**, 2251.	98
73JOC2747	G. H. Posner and D. J. Brunelle; *J. Org. Chem.*, 1973, **38**, 2747.	863
73JOC3172	Y. Okamoto, K. L. Chellappa and R. Homsany; *J. Org. Chem.*, 1973, **38**, 3172.	240
73JOC3288	A. T. Nielsen, R. L. Atkins, D. W. Moore, R. Scott, D. Mallory and J. M. LaBerge; *J. Org. Chem.*, 1973, **38**, 3288.	295, 407, 412
73JOC3878	I. L. Doerr and R. E. Willette; *J. Org. Chem.*, 1973, **38**, 3878.	54, 57
73JOC3924	W. J. Middleton; *J. Org. Chem.*, 1973, **38**, 3924.	97
73JOC3935	G. L. Larson and A. Hernandez; *J. Org. Chem.*, 1973, **38**, 3935.	188
73JOC3937	S. J. Valenty and P. S. Skell; *J. Org. Chem.*, 1973, **38**, 3937.	385
73JOC4031	D. Seyferth and R. A. Woodruff; *J. Org. Chem.*, 1973, **38**, 4031.	137
73JOM(47)67	C. Couret, J. Satgé and F. Couret; *J. Organomet. Chem.*, 1973, **47**, 67.	338
73JOM(47)145	R. B. King, I. Haiduc and A. Efraty; *J. Organomet. Chem.*, 1973, **47**, 145.	692
73JOM(49)117	D. Seyferth and G. J. Murphy; *J. Organomet. Chem.*, 1973, **49**, 117.	756
73JOM(50)47	M. R. Ort and E. H. Mottus; *J. Organomet. Chem.*, 1973, **50**, 47.	698
73JOM(51)1	M. D. Rausch, G. A. Moser and C. F. Meade; *J. Organomet. Chem.*, 1973, **51**, 1.	706, 706
73JOM(51)77	D. Seyferth and S. P. Hopper; *J. Organomet. Chem.*, 1973, **51**, 77.	35, 137
73JOM(51)167	S. Kato, T. Kato, M. Mizuta, K. Itoh and Y. Ishii; *J. Organomet. Chem.*, 1973, **51**, 167.	588
73JOM(54)35	D. S. Matteson and P. G. Allies; *J. Organomet. Chem.*, 1973, **54**, 35.	697
73JOM(55)419	A. J. Leusink, G. van Koten, J. W. Marsman and J. G. Nolte; *J. Organomet. Chem.*, 1973, **55**, 419.	708
73JOM(56)269	H. Bürger and C. Kluess; *J. Organomet. Chem.*, 1973, **56**, 269.	719
73JOM(57)C5	R. Corriu and J. Masse; *J. Organomet. Chem.*, 1973, **57**, C5.	648, 649
73JOM(57)261	A. W. Jarvie and R. J. Rowley; *J. Organomet. Chem.*, 1973, **57**, 261.	140, 607
73JOM(57)269	P. J. Davidson, M. F. Lappert and R. Pearce; *J. Organomet. Chem.*, 1973, **57**, 269.	656
73JOM(57)423	B. L. Dyatkin, B. I. Martynov, L. G. Martynova, N. G. Kizim, S. R. Sterlin, Z. A. Stumbrevichute and L. A. Fedorov; *J. Organomet. Chem.*, 1973, **57**, 423.	152
73JOM(59)53	E. M. Kaiser, R. D. Beard and C. R. Hauser; *J. Organomet. Chem.*, 1973, **59**, 53.	393
73JOM(59)237	D. Seyferth and R. S. Marmor; *J. Organomet. Chem.*, 1973, **59**, 237.	563, 740, 741, 801
73JOM(60)95	A. Tzschach, D. Drohne and J. Heinicke; *J. Organomet. Chem.*, 1973, **60**, 95.	496, 498, 499
73JOM(60)C57	R. B. King and M. N. Ackermann; *J. Organomet. Chem.*, 1973, **60**, C57.	681
73JOM(61)27	R. C. Larock; *J. Organomet. Chem.*, 1973, **61**, 27.	697
73JOM(61)133	J. Grobe and G. Heyer; *J. Organomet. Chem.*, 1973, **61**, 133.	566
73JOU418	B. S. Drach, I. Yu. Dolgushina and A. V. Kirsanov; *J. Org. Chem. USSR (Engl. Transl.)*, 1973, **9**, 418.	104
73JPR65	V. A. Tzschach and J. Heinicke; *J. Prakt. Chem.*, 1973, **315**, 65.	495, 496, 497
73JPR526	V. H. Oehme and R. Thamm; *J. Prakt. Chem.*, 1973, **315**, 526.	453, 454, 481
73LA247	F. Boberg and W. von Gentzkow; *Justus Liebigs Ann. Chem.*, 1973, 247.	905
73LA443	W. Walter and J. Krohn; *Justus Liebigs Ann. Chem.*, 1973, 443.	898
73LA573	H. Wamhoff and C. Materne; *Justus Liebigs Ann. Chem.*, 1973, 573.	972
73LA708	H. Lehmkuhl, O. Olbrysch and H. Nehl; *Justus Liebigs Ann. Chem.*, 1973, 708.	698
73LA715	H. Lehmkuhl and O. Olbrysch; *Justus Liebigs Ann. Chem.*, 1973, 715.	699
73MI 409-01	D. Mathies and R. Wolff; *Pharm. Acta. Helv.*, 1973, **48**, 44.	417
73MI 410-01	K. Issleib and P. von Malotki; *Phosphorus*, 1973, **3**, 141.	454
B-73MI 415-01	"Gmelins Handbuch der anorganischen Chemie; Supplement: Kobalt-Organische Verbindungen," Verlag Chemie, Weinheim, 1973, vol. 6, pt. 2, pp. 89–101.	687
73MI 417-01	C. Raulet; *Compt. Rend.*, 1973, **276**, 903.	739
73MI 418-01	L. Maier; *Phosphorus*, 1973, **2**, 229.	803
73MI 420-01	I. F. Lutsenko, M. V. Proskurnina and N. B. Karlstedt; *Phosphorus*, 1973, **3**, 55.	966
B-73MI 424-01	J. K. Kochi (ed.); "Free Radicals," Wiley, New York, 1973, vols 1 and 2.	1077
B-73MI 424-02	M. Jones, Jr. and R. A. Moss (ed.); "Carbenes," Wiley, New York, 1973–1975, vols 1 and 2.	1080
73NKK381	S. Miyano, T. Ohtake, H. Tokumasu and H. Hashimoto; *Nippon Kagaku Kaishi*, 1973, 381 (*Chem. Abstr.*, 1973, **78**, 159 784).	696
73OSC(5)145	I. A. Koten and R. J. Sauer; *Org. Synth., Coll. Vol.*, 1973, **5**, 145.	56
73OSC(5)191	W. D. Emmons and A. S. Pagano; *Org. Synth., Coll. Vol.*, 1973, **5**, 191.	312
73OSC(5)218	R. Schmutzler; *Org. Synth., Coll. Vol.*, 1973, **5**, 218.	47
73OSC(5)221	H. Gross and W. Bürger; *Org. Synth., Coll. Vol.*, 1973, **5**, 221.	47, 48
73OSC(5)231	L. A. Paquette and L. S. Wittenbrook; *Org. Synth., Coll. Vol.*, 1973, **5**, 231.	85

73OSC(5)239	D. C. England and L. R. Melby; *Org. Synth., Coll. Vol.*; 1973, **5**, 239.	32
73OSC(5)387	A. J. Speziale and R. C. Freeman; *Org. Synth., Coll. Vol.*, 1973, **5**, 387.	791
73OSC(5)390	S. A. Fuqua, W. G. Duncan and R. M. Silverstein; *Org. Synth., Coll. Vol.*; 1973, **5**, 390.	9
73OSC(5)959	W. A. Sheppard; *Org. Synth., Coll. Vol.*; 1973, **5**, 959.	9
73PS(3)1	R. D. Bertrand, J. Rathke and J. G. Verkade; *Phosphorus Sulfur*, 1973, **3**, 1.	342
73PS(3)55	I. F. Lutsenko, M. V. Proskurnina and N. B. Karlstedt; *Phosphorus Sulfur*, 1973, **3**, 55.	344
73RTC11	J. W. Scheeren, R. J. F. M. Staps and R. J. F. Nivard; *Recl. Trav. Chim. Pays-Bas*, 1973, **92**, 11.	830
73RTC117	K. Balenovic and A. Deljac; *Recl. Trav. Chim. Pays-Bas*, 1973, **92**, 117.	259
73RTC1047	J. W. De Leeuw, E. R. De Waard, T. Beetz and H. O. Huisman; *Recl. Trav. Chim. Pays-Bas*, 1973, **92**, 1047.	179
73RTC1067	J. Meijer, P. Vermeer, H. J. T. Bos and L. Brandsma; *Recl. Trav. Chim. Pays-Bas*, 1973, **92**, 1067.	843
73RZC929	W. Szczepaniak and J. Siepak; *Rocz. Chem.*, 1973, **47**, 929.	470
73S37	G. Zweifel, R. P. Fisher and A. Horng; *Synthesis*, 1973, 37.	631
73S151	J. W. Scheeren and W. J. M. Tax; *Synthesis*, 1973, 151.	203
73S169	J. H. van Boom and J. D. M. Herschied; *Synthesis*, 1973, 169.	186
73S211	A. Senning; *Synthesis*, 1973, 211.	282
73S243	H. Petersen; *Synthesis*, 1973, 243.	423, 427, 492
73S247	H. Petersen; *Synthesis*, 1973, 247.	301
73S302	B. Unterhalt and D. Thamer; *Synthesis*, 1973, 302.	313
73S303	D. Thamer and B. Unterhalt; *Synthesis*, 1973, 303.	327
73S547	P. Golborn; *Synthesis*, 1973, 547.	912, 913
73S669	M. Mikolajczyk and A. Zatorski; *Synthesis*, 1973, 669.	347
73S703	M. Tramontini; *Synthesis*, 1973, 703.	315, 322
73S787	L. N. Markovskii, V. E. Pashinnik and A. V. Kivsanov; *Synthesis*, 1973, 787.	10
73SC189	W. J. M. van Tilborg, H. Steinberg and Th. J. de Boer; *Synth. Commun.*, 1973, **3**, 189.	299
73SC197	D. G. Naae and D. J. Burton; *Synth. Commun.*, 1973, **3**, 197.	733, 734
73SRI75	K. Moedritzer; *Synth. React. Inorg. Metal-Org. Chem.*, 1973, **3**, 75.	470
73T707	F. Mathey; *Tetrahedron*, 1973, **29**, 707.	341
73T4049	M. K. Eberle and G. G. Kahle; *Tetrahedron*, 1973, **29**, 4049.	415
73T4195	J. Burdon and A. Ramirez; *Tetrahedron*, 1973, **29**, 4195.	439
73T4225	G. Bauduin and Y. Pietrasanta; *Tetrahedron*, 1973, **29**, 4225.	191
73TL173	G. Lavielle, M. Carpentier and P. Savignac; *Tetrahedron Lett.*, 1973, **14**, 173.	340
73TL289	C. Y. Meyers and G. J. McCollum; *Tetrahedron Lett.*, 1973, **14**, 289.	75
73TL611	P. Weyerstahl, R. Mathias and G. Blume; *Tetrahedron Lett.*, 1973, **14**, 611.	36
73TL633	U. Schöllkopft and R. Schröder; *Tetrahedron Lett.*, 1973, **14**, 633.	487
73TL1319	P. Crabbe and A. Cervantes; *Tetrahedron Lett.*, 1973, **14**, 1319.	8
73TL1367	L. Skattebøl, G. A. Abskharoun and T. Greibrokk; *Tetrahedron Lett.*, 1973, **14**, 1367.	25
73TL1599	G. R. Newkome, J. D. Sauer and G. L. McClure; *Tetrahedron Lett.*, 1973, **14**, 1599.	183
73TL1889	L. A. Neiman, S. V. Zhukova and V. A. Tyurikov; *Tetrahedron Lett.*, 1973, **14**, 1889.	312
73TL2389	S. Holm and A. Senning; *Tetrahedron Lett.*, 1973, **14**, 2389.	862
73TL3255	F. Mathey; *Tetrahedron Lett.*, 1973, **14**, 3255.	1026
73TL3831	R. C. Ronald; *Tetrahedron Lett.*, 1973, **14**, 3831.	71
73TL4193	H. Sakurai, K. Nishiwaki and M. Kira; *Tetrahedron Lett.*, 1973, **14**, 4193.	620, 653
73TL4237	L. Duhamel, P. Duhamel and J.-M. Poirier; *Tetrahedron Lett.*, 1973, **14**, 4237.	15
73TL4395	R. H. Mitchell; *Tetrahedron Lett.*, 1973, **14**, 4395.	64
73TL4645	R. W. Ratcliffe and B. G. Christensen; *Tetrahedron Lett.*, 1973, **14**, 4645.	463
73TL4707	J. L. Herrmann, J. E. Richman, P. J. Wepplo and R. H. Schlessinger; *Tetrahedron Lett.*, 1973, **14**, 4707.	857
73TL4711	J. L. Herrmann, G. R. Kieczykowski, R. F. Romanet, P. J. Wepplo and R. H. Schlessinger; *Tetrahedron Lett.*, 1973, **14**, 4711.	855
73TL5201	M. Soroka and P. Mastalerz; *Tetrahedron Lett.*, 1973, **14**, 5201.	476
73USP3734954	L. Maier; *US Pat.* 3 734 954 (1973) (*Chem. Abstr.*, 1973, **79**, 78 960).	556
73USP3751492	S. Y. Delavarenne (Union Carbide Corp.); *US Pat.* 3 751 492 (1973) (*Chem. Abstr.*, 1973, **79**, 91 561).	819
73YZ612	G. Kobayashi, Y. Matsuda, R. Natsuki, Y. Tominaga and M. Sone; *Yakugaku Zasshi*, 1973, **93**, 612.	979
73ZC55	H. G. Henning and T. Forner; *Z. Chem.*, 1973, **13**, 55.	556
73ZC139	K. Issleib, H. Oehme and K. Mohr; *Z. Chem.*, 1973, **13**, 139.	453, 454
73ZC254	J. Rachon and C. Wasielewski; *Z. Chem.*, 1973, **13**, 254.	460
73ZN(B)98	W. Althoff and M. Fild; *Z. Naturforsch., Teil B*, 1973, **28**, 98.	282, 563
73ZN(B)224	M. Baudler, J. Vesper and H. Sandmann; *Z. Naturforsch., Teil B*, 1973, **28**, 224.	545
73ZN(B)508	H.-J. Schmitt and M. L. Ziegler; *Z. Naturforsch., Teil B*, 1973, **28**, 508.	681
73ZOB211	A. D. Sinitsa and B. S. Drach; *Zh. Obshch. Khim.*, 1973, **43**, 211 (Engl. Transl. 211).	966
73ZOB534	Z. N. Mironova, E. N. Tsvetkov, A. V. Nikolaev and M. I. Kabachnik; *Zh. Obshch. Khim.*, 1973, **43**, 534.	339
73ZOB625	A. A. Buyakov, T. K. Gar and V. F. Mironov; *Zh. Obshch. Khim.*, 1973, **43**, 625 (*Chem. Abstr.*, 1973, **79**, 53 477y).	639, 640
73ZOB801	A. A. Buyakov, T. K. Gar and V. F. Mironov; *Zh. Obshch. Khim.*, 1973, **43**, 801 (*Chem. Abstr.*, 1973, **79**, 42 628w).	660, 700
73ZOR39	V. P. Kukhar, V. I. Pasternak and G. V. Pesotskaya; *Zh. Org. Khim.*, 1973, **9**, 39 (Engl. Transl. 37).	971

Ref	Citation	Page
73ZOR69	A. M. Aleksandrov, Yu. V. Samusenko, A. G. Bratolyubova and L. M. Yagupol'skii; *Zh. Org. Khim.*, 1973, **9**, 69 (*Chem. Abstr.*, 1973, **78**, 84 215v).	777
74ACR85	B. M. Trost; *Acc. Chem. Res.*, 1974, **7**, 85.	162
74AC(R)305	A. Tajana, D. Nardi and R. Cappelletti; *Ann. Chim. (Rome)*, 1974, **64**, 305 (*Chem. Abstr.* 1975, **83**, 178 945).	852
74ACS539	J. Becher and P. E. Iverson; *Acta Chem. Scand.*, 1974, **28**, 539.	436
74AG(E)83	B-T. Gröbel and D. Seebach; *Angew. Chem., Int. Ed. Engl.*, 1974, **13**, 83.	620
74AG(E)202	D. Seebach and K. H. Geiss; *Angew. Chem., Int. Ed. Engl.*, 1974, **13**, 202.	392
74AG(E)479	K. H. Geiss, B. Seuring, R. Pieter and D. Seebach; *Angew. Chem., Int. Ed. Engl.*, 1974, **13**, 479.	392
74AG(E)540	W. Malisch and H. Schmidbaur; *Angew. Chem., Int. Ed. Engl.*, 1974, **13**, 540.	573
74AG(E)676	G. Höfle; *Angew. Chem., Int. Ed. Engl.*, 1974, **13**, 676.	56
74AG(E)701	M. Schlosser; *Angew. Chem., Int. Ed. Engl.*, 1974, **13**, 701.	949
74AG(E)806	D. Seebach and A. K. Beck; *Angew. Chem., Int. Ed. Engl.*, 1974, **13**, 806.	394
74AJC679	J. M. Coxon, M. P. Hartshorn and B. L. S. Sutherland; *Aust. J. Chem.*, 1974, **27**, 679.	182
74AJC1579	K. Chamberlain and L. A. Summers; *Aust. J. Chem.*, 1974, **27**, 1579.	327
74AOC323	R. S. Dickson and P. J. Fraser; *Adv. Organomet. Chem.*, 1974, **12**, 323.	687
74BCJ785	T. Kitazume and N. Ishikawa; *Bull. Chem. Soc. Jpn.*, 1974, **47**, 785.	415
74BSF331	L. Duhamel, P. Duhamel and N. Mancelle; *Bull. Soc. Chim. Fr.*, 1974, 331.	405, 406
74BSF1533	J. Ficini and A. Duréault; *Bull. Soc. Chim. Fr.*, 1974, 1533.	15
74BSF2263	H. Christol, H.-J. Cristau and J.-P. Joubert; *Bull. Soc. Chim. Fr.*, 1974, 2263.	921
74CB93	H. H. Karsch, H.-F. Klein and H. Schmidbaur; *Chem. Ber.*, 1974, **107**, 93.	583
74CB102	H. Schmidbaur and K.-H. Räthlein; *Chem. Ber.*, 1974, **107**, 102.	587
74CB1488	W. Steglich, K. Burger, M. Dürr and E. Burgis; *Chem. Ber.*, 1974, **107**, 1488.	297
74CB2345	N. H. Nilsson and A. Senning; *Chem. Ber.*, 1974, **107**, 2345.	862
74CB3562	E. Schaumann and W. Walter; *Chem. Ber.*, 1974, **107**, 3562.	956
74CB3674	H. Schmidbaur and H.-J. Füller; *Chem. Ber.*, 1974, **107**, 3674.	588
74CB3684	H. H. Karsch and H. Schmidbaur; *Chem. Ber.*, 1974, **107**, 3684.	583
74CB3692	H. H. Karsch, H.-F. Klein, C. G. Kreiter and H. Schmidbaur; *Chem. Ber.*, 1974, **107**, 3692.	582
74CB3697	H. Schmidbaur, J. Adlkofer and M. Heimann; *Chem. Ber.*, 1974, **107**, 3697.	585
74CB3706	D. J. Brauer, C. Krüger, P. J. Roberts and Y.-H. Tsay; *Chem. Ber.*, 1974, **107**, 3706.	582, 583
74CC29	C. R. Bennett and D. C. Bradley; *J. Chem. Soc., Chem. Commun.*, 1974, 29.	657
74CC405	L. A. Feiler, R. Huisgen and P. Koppitz; *J. Chem. Soc., Chem. Commun.*, 1974, 405.	829
74CC751	M. G. Miles, J. D. Wilson, D. J. Dahm and J. H. Wagenknecht; *J. Chem. Soc., Chem. Commun.*, 1974, 751.	850
74CC863	R. J. Hughes, A. Pelter and K. Smith; *J. Chem. Soc., Chem. Commun.*, 1974, 863.	373
74CCC2616	V. B. Puchnarevič, J. Včelák, M. G. Veronkov and V. Chvalovský; *Collect. Czech. Chem. Commun.*, 1974, **39**, 2616.	32
74CJC761	T. Durst, M. J. LeBelle, R. Van den Elzen and K. C. Tin; *Can. J. Chem.*, 1974, **52**, 761.	391
74CJC951	J. W. Bunting and W. G. Meathrel; *Can. J. Chem.*, 1974, **52**, 951.	43
74CL15	T. Mukaiyama and M. Hayashi; *Chem. Lett.*, 1974, 15.	192
74CL659	K. Ogura, S. Furukawa and G. Tsuchihashi; *Chem. Lett.*, 1974, 659.	856, 857
74CPB2767	E. Yoshii, Y. Kobayashi, T. Koizumi and T. Oribe; *Chem. Pharm. Bull.*, 1974, **22**, 2767.	831
74CR(C)221	D. Papillon-Jegou, B. Barriuo and M. Kerfanto; *C. R. Hebd. Seances Acad. Sci., Ser. C*, 1974, **279**, 221.	309
74CR(C)695	L. Léger and M. Saquet; *C. R. Hebd. Seances Acad. Sci., Ser. C*, 1974, **279**, 695.	255
74FCF73	H. H. Wasserman, G. M. Clark and P. C. Turley; *Fortschr. Chem. Forsch.*, 1974, **47**, 73.	162
74GEP2331675	F. J. Muller and K. Eicken (BASF A.-G.); *Ger. Pat.* 2 331 675 (1974) (*Chem. Abstr.* 1974, **81**, 63 153).	830
74H(2)177	T. Sasaki, S. Eguchi and O. Hiroaki; *Heterocycles*, 1974, **2**, 177.	165
74HCA1042	W. Fink; *Helv. Chim. Acta*, 1974, **57**, 1042.	508
74ICA5	P. L. Stanghellini, R. Rossetti, O. Gambino and G. Cetini; *Inorg. Chim. Acta*, 1974, **10**, 5.	692
74IZV358	E. G. Abduganiev, E. A. Avetisyan, E. M. Rokhlin and I. L. Knunyants; *Izv. Akad. Nauk SSSR, Ser. Khim.*, 1974, 358 (*Chem. Abstr.*, 1974, **81**, 25 007t).	894
74IZV2545	N. S. Mirzabekyants, M. D. Bargamova, Yu. A. Cheburkov and I. L. Knunyants; *Izv. Akad. Nauk SSSR, Ser. Khim.*, 1974, 2545 (*Chem. Abstr.*, 1975, **82**, 72 503g).	966
74IZV2645	A. N. Nesmeyanov, E. G. Perevalova, L. I. Leont'eva, S. A. Eremin and O. V. Grigor'eva; *Izv. Akad. Nauk SSSR, Ser. Khim.*, 1974, 2645.	719
74IZV2861	N. S. Nametkin, V. M. Vdovin, V. A. Poletaev, V. I. Svergun and M. B. Sergeeva; *Izv. Akad. Nauk SSSR, Ser. Khim.*, 1974, 2861 (*Chem. Abstr.*, 1975, **82**, 112 151z).	656, 658
74JA311	H. C. Brown and N. R. DeLue; *J. Am. Chem. Soc.*, 1974, **96**, 311.	143
74JA925	G. A. Olah, M. Nojima and I. Kerekes; *J. Am. Chem. Soc.*, 1974, **96**, 925.	10
74JA945	P. R. Moses and J. Q. Chambers; *J. Am. Chem. Soc.*, 1974, **96**, 945.	846
74JA1230	C. P. Casey and R. L. Anderson; *J. Am. Chem. Soc.*, 1974, **96**, 1230.	955
74JA1620	K. Shimoji, H. Taguchi, K. Oshima, H. Yamamoto and H. Nozaki; *J. Am. Chem. Soc.*, 1974, **96**, 1620.	648
74JA1941	G. D. Stucky, A. M. McPherson, W. E. Rhine, J. J. Eisch and J. L. Considine; *J. Am. Chem. Soc.*, 1974, **96**, 1941.	1068
74JA1960	K. Ogura and G. Tsuchihashi; *J. Am. Chem. Soc.*, 1974, **96**, 1960.	326, 856, 857
74JA2580	N. Kornblum, S. D. Boyd and N. Ono; *J. Am. Chem. Soc.*, 1974, **96**, 2580.	328
74JA3010	H. Taguchi, H. Yamamoto and H. Nozaki; *J. Am. Chem. Soc.*, 1974, **96**, 3010.	19

74JA3214	A. Wright and R. West; *J. Am. Chem. Soc.*, 1974, **96**, 3214.	359, 362, 371
74JA3222	A. Wright and R. West; *J. Am. Chem. Soc.*, 1974, **96**, 3222.	368, 371
74JA3227	A. Wright and R. West; *J. Am. Chem. Soc.*, 1974, **96**, 3227.	358, 371
74JA4280	S. Iriuchijima, K. Maniwa and G. Tsuchihashi; *J. Am. Chem. Soc.*, 1974, **96**, 4280.	221
74JA5623	Y. Nakadaira, S. Kanouchi and H. Sakurai; *J. Am. Chem. Soc.*, 1974, **96**, 5623.	612
74JA7105	R. D. Bush, C. M. Golino and L. H. Sommer; *J. Am. Chem. Soc.*, 1974, **96**, 7105.	611
74JA7108	R. E. Davis, B. L. Barnett, R. G. Amiet, W. Merk, J. S. McKennis and R. Pettit; *J. Am. Chem. Soc.*, 1974, **96**, 7108.	681
74JA7125	J. E. Baldwin, G. A. Höfle and O. W. Lever, Jr.; *J. Am. Chem. Soc.*, 1974, **96**, 7125.	949
74JA7586	E. C. Baker, K. N. Raymond, T. J. Marks and W. A. Wachter; *J. Am. Chem. Soc.*, 1974, **96**, 7586.	703
74JA7588	J. L. Adcock and R. J. Lagow; *J. Am. Chem. Soc.*, 1974, **96**, 7588.	52
74JCS(D)607	M. Basato and A. J. Poë; *J. Chem. Soc., Dalton Trans.*, 1974, 607.	687
74JCS(D)1856	M. Basato, J. P. Fawcett, S. A. Fieldhouse and A. J. Poë; *J. Chem. Soc., Dalton Trans.*, 1974, 1856.	687
74JCS(D)2537	F. Glockling, M. A. Lyle and S. R. Stobart; *J. Chem. Soc., Dalton Trans.*, 1974, 2537.	141
74JCS(P1)1279	K. Condo, Y. Liu and D. Tunemoto; *J. Chem. Soc., Perkin Trans. 1*, 1974, 1279.	347
74JCS(P1)1611	J. Clark, M. Curphey and I. W. Southon; *J. Chem. Soc., Perkin Trans. 1*, 1974, 1611.	979
74JCS(P1)1644	C. B. C. Boyce, S. B. Webb, L. Phillips and I. R. Ager; *J. Chem. Soc., Perkin Trans. 1*, 1974, 1644.	804
74JCS(P1)2357	G. Guanti, C. Dell'Erba, G. Leandri and S. Thea; *J. Chem. Soc. Perkin Trans. 1*, 1974, 2357.	907
74JCS(P1)2470	M. Elliott, N. F. Janes and D. A. Pulman; *J. Chem. Soc. Perkin Trans. 1*, 1974, 2470.	746
74JGU1386	O. I. Kolodyazhnyi; *J. Gen. Chem. USSR (Engl. Trans.)*, 1974, **44**, 1386.	115
74JHC407	D. J. Daigle, A. B. Pepperman Jr. and S. L. Vail; *J. Heterocycl. Chem.*, 1974, **11**, 407.	458, 477, 481
74JHC1085	D. J. Daigle, A. B. Pepperman Jr. and G. Boudreaux; *J. Heterocycl. Chem.*, 1974, **11**, 1085.	482
74JMC1225	D. Seebach and D. Enders; *J. Med. Chem.*, 1974, **17**, 1225.	511
74JOC77	A. DeBoer and R. E. Ellwanger; *J. Org. Chem.*, 1974, **39**, 77.	209, 213
74JOC167	D. L. Coffen, J. P. DeNoble, E. L. Evans, G. F. Field, R. I. Fryer, D. A. Katonak, B. J. Mandel, L. H. Sternbach and W. J. Zally; *J. Org. Chem.*, 1974, **39**, 167.	412
74JOC209	G. H. Birum; *J. Org. Chem.*, 1974, **39**, 209.	466
74JOC501	C. G. Venier, C. G. Gibbs and P. T. Crane; *J. Org. Chem.*, 1974, **39**, 501.	742
74JOC643	B. B. Jarvis and M. M. Evans; *J. Org. Chem.*, 1974, **39**, 643.	75
74JOC1349	A. T. Nielsen, R. L. Atkins, J. DiPol and D. W. Moore; *J. Org. Chem.*, 1974, **39**, 1349.	412
74JOC1420	P. V. Roling and M. D. Rausch; *J. Org. Chem.*, 1974, **39**, 1420.	706, 724
74JOC1449	W. E. Truce, L. A. Murau, P. J. Smith and F. Young; *J. Org. Chem.*, 1974, **39**, 1449.	86
74JOC1710	R. Sarges and J. R. Tretter; *J. Org. Chem.*, 1974, **39**, 1710.	420
74JOC1965	T. George, D. V. Mehta and D. A. Dabholkar; *J. Org. Chem.*, 1974, **39**, 1965.	105
74JOC2010	M. G. Missakian, R. Ketcham and A. R. Martin; *J. Org. Chem.*, 1974, **39**, 2010.	223, 225
74JOC2516	F. G. Bordwell, M. D. Wolfinger and J. B. O'Dwyer; *J. Org. Chem.*, 1974, **39**, 2516.	68, 71, 81, 83, 84
74JOC2521	F. G. Bordwell and M. D. Wolfinger; *J. Org. Chem.*, 1974, **39**, 2521.	80
74JOC2526	F. G. Bordwell and E. Doomes; *J. Org. Chem.*, 1974, **39**, 2526.	83
74JOC2648	B. M. Trost and R. A. Kunz; *J. Org. Chem.*, 1974, **39**, 2648.	64, 71
74JOC2815	V. I. Stenberg and D. A. Kubik; *J. Org. Chem.*, 1974, **39**, 2815.	177
74JOC3145	J. Kagan and B. E. Firth; *J. Org. Chem.*, 1974, **39**, 3145.	183
74JOC3198	M. Komatsu, N. Nishikaze, M. Sakamoto, Y. Ohshiro and T. Agawa; *J. Org. Chem.*, 1974, **39**, 3198.	979
74JOC3215	J. J. Zeilstra and J. B. F. N. Engberts; *J. Org. Chem.*, 1974, **39**, 3215.	333
74JOC3264	T. H. Chan and E. Chang; *J. Org. Chem.*, 1974, **39**, 3264.	646, 651, 653
74JOC3412	T. Minami, Y. Tsumori, K. Yoshida and T. Agawa; *J. Org. Chem.*, 1974, **39**, 3412.	326
74JOC3745	K. W. Ratts and J. P. Chupp; *J. Org. Chem.*, 1974, **39**, 3745.	425
74JOC3906	M. H. Ghandehari, D. Davalian, M. Yalpani and M. H. Partovi; *J. Org. Chem.*, 1974, **39**, 3906.	874
74JOM(64)271	E. W. Randall, E. Rosenberg, L. Milone, R. Rossetti and P. L. Stanghellini; *J. Organomet. Chem.*, 1974, **64**, 271.	692
74JOM(65)131	A. N. Nesmeyanov, E. G. Perevalova, K. I. Grandberg, D. A. Lemenovskii, T. V. Baukova and O. B. Afanassova; *J. Organomet. Chem.*, 1974, **65**, 131.	696
74JOM(67)C61	R. J. Bertino and G. B. Deacon; *J. Organomet. Chem.*, 1974, **67**, C61.	697
74JOM(69)53	D. S. Matteson and P. B. Tripathy; *J. Organomet. Chem.*, 1974, **69**, 53.	814
74JOM(70)C21	S. Numata, H. Kurosawa and R. O. Kawara; *J. Organomet. Chem.*, 1974, **70**, C21.	659
74JOM(71)377	J. Dunogues, E. Jousseaume and R. Calas; *J. Organomet. Chem.*, 1974, **71**, 377.	608
74JOM(72)C4	A. Bongini, D. Savoia and A. Umani-Ronchi; *J. Organomet. Chem.*, 1974, **72**, C4.	698, 708
74JOM(74)C23	G. Nardin, L. Randaccio and E. Zangrando; *J. Organomet. Chem.*, 1974, **74**, C23.	585
74JOM(77)C15	W. Malisch; *J. Organomet. Chem.*, 1974, **77**, C15.	571
74JOM(78)C35	K. H. Pannell and J. R. Rice; *J. Organomet. Chem.*, 1974, **78**, C35.	614, 656
74JOM(78)395	J. L. Wardell and S. Ahmed (in part); *J. Organomet. Chem.*, 1974, **78**, 395.	154
74JOM(78)405	S. Kato; *J. Organomet. Chem.*, 1974, **78**, 405.	709
74JOM(80)C25	D. Bourgeois, J. Dunogues and N. Duffaut; *J. Organometal. Chem.*, 1974, **80**, C25.	933
74JOM(80)C39	J. L. Davidson and D. W. A. Sharp; *J. Organomet. Chem.*, 1974, **80**, C39.	692
74JOM(81)187	A. Tzschach and P. Franke; *J. Organomet. Chem.*, 1974, **81**, 187.	454, 495, 496, 501, 502
74JOM(82)21	Y. Sato, T. Aoyama and H. Shirai; *J. Organomet. Chem.*, 1974, **82**, 21.	509
74JOU244	V. N. Fetyukhin, A. S. Koretskii, V. I. Gorbatenko and L. I. Samarai; *J. Org. Chem. USSR (Engl. Transl.)*, 1977, **10**, 244.	329

74JOU1565	L. I. Samarai, V. I. Gorbatenko, V. N. Fetyukhin, L. F. Lur'e and N. V. Mel'nichenko; *J. Org. Chem. USSR (Engl. Transl.)*, 1974, **10**, 1565.	314, 329
74JPR304	Von B. Hesse and R. Moll; *J. Prakt. Chem.*, 1974, **316**, 304.	215
74JPR851	V. A. Tzschach and K. Kellner; *J. Prakt. Chem.*, 1974, **316**, 851.	483, 499, 500
74LA44	U. Schollkopf, R. Schroder and D. Stafforst; *Justus Liebigs Ann. Chem.*, 1974, 44.	1005
74LA539	K. Clauss, D. Grimm and G. Prossel; *Justus Liebigs Ann. Chem.*, 1974, 539.	424
74LA751	H.-J. Kleiner, *Justus Liebigs Ann. Chem.*, 1974, 751.	121
74LA1851	E. Fluck and H. Schultheiss; *Justus Liebigs Ann. Chem.*, 1974, 1851.	437
74M684	A. Meller and W. Gerger, *Monatsh. Chem.*, 1974, **105**, 684.	1015
B-74MI 401-01	J. S. Pizey; "Synthetic Reagents, Vol. II," 1974, Ellis Horwood, Chichester, p. 1.	21
74MI 410-01	C. Laurenco, D. Bernard and R. Burgada; *Compt. Rend.*, 1974, **278**, c1301.	478
74MI 411-01	S. L. Ioffe, L. M. Leont'eva, O. P. Shitov, B. N. Khasapov and V. A. Tertakovskii; *Tezisy Vses. Soveshch, Khim. Nitrosoedinenii, 5th*, 1974, 49 (*Chem. Abstr.*, 1974, **87**, 23 357g).	524
74MI 420-01	V. L. Isaev, V. M. Izmailov, A. A. Zamanskii, A. A. Listov, R. N. Sterlin and I. L. Knunyants; *Zh. Vses. Khim. Obshchest.*, 1974, **19**, 353 (*Chem. Abstr.*, 1974, **81**, 63 091y).	894
B-74MI 424-01	N. S. Isaacs (ed.); "Reactive Intermediates in Organic Chemistry," Wiley, New York, 1974.	1072
B-74MI 424-02	D. C. Nonhebel and J. C. Walton; "Free-radical Chemistry," Cambridge University Press, Cambridge, 1974.	1077
74OPP77	C. G. Venier and H. J. Barager, III; *Org. Prep. Proceed. Int.*, 1974, **6**, 77.	73
74OR(21)1	G. A. Boswell, Jr., W. C. Ripka, R. M. Scribner and C. W. Tullock, *Org. React.*, 1974, **21**, 1.	8
74PS(4)109	E. Evangelidou-Tsolis, F. Ramirez, J. F. Pilot and C. P. Smith; *Phosphorus Sulfur*, 1974, **4**, 109.	336
74RCR984	K. A. Petrov, V. A. Chauzov and T. S. Erokhina; *Russ. Chem. Rev. (Engl. Transl.)*, 1974, **43**, 984.	451, 469, 472, 473, 475, 493
74RTC11	J. J. Zeilstra and J. B. F. N. Engberts; *Recl. Trav. Chim. Pays-Bas*, 1974, **93**, 11.	331
74RTC184	J. C. Wesdorp, J. Meijer, P. Vermeer, H. J. T. Bos, L. Brandsma and J. F. Arens; *Recl. Trav. Chim. Pays-Bas*, 1974, **93**, 184.	843
74RTC242	J. Meijer and P. Vermeer; *Recl. Tav. Chim. Pays-Bas*, 1974, **93**, 242.	245
74RTC294	W. J. M. van Tilborg, H. Steinberg and T. J. de Boer; *Recl. Trav. Chim. Pays-Bas*, 1974, **93**, 294.	299
74RZC1119	M. Soroka and P. Mastalerz; *Rocz. Chem.*, 1974, **48**, 1119.	460
74S23	M. Anteunis and C. Becu; *Synthesis*, 1974, 23.	185
74S433	N. H. Nilsson and J. Sandström; *Synthesis*, 1974, 433.	901
74S672	H. Ahlbrecht and J. Eichler; *Synthesis*, 1974, 672.	530
74S705	I. Hori, T. Hayashi and H. Midorikawa; *Synthesis*, 1974, 705.	197
74S816	K. Burger, E. Burgis and P. Holl; *Synthesis*, 1974, 816.	493
74S862	J. M. McIntosh and H. B. Goodbrand; *Synthesis*, 1974, 862.	912
74S888	J. Hartmann, M. Stähle and M. Schlosser; *Synthesis*, 1974, 888.	950
74SRI191	K. Issleib, H. Winkelmann and H. P. Abicht; *Synth. React. Inorg. Metal-Org. Chem.*, 1974, **4**, 191.	454
74SRI221	S. O. Grim and J. D. Mitchell; *Synth. React. Inorg. Met-Org. Chem.*, 1974, **4**, 221.	551
74SRI453	H. Oehme, K. Issleib, E. Leißring and A. Zschunke; *Synth. React. Inorg. Metal-Org. Chem.*, 1974, **4**, 453.	455
74SRI515	D. Seyferth and S. C. Vick; *Synth. React. Inorg. Metal-Org. Chem.*, 1974, **4**, 515.	700
74T1283	F. C. V. Larsson and S.-O. Lawesson; *Tetrahedron*, 1974, **30**, 1283.	903
74T2661	H. Hiyama, Y. Ozaki and H. Nozaki; *Tetrahedron*, 1974, **30**, 2661.	17
74T3127	F. Mathey; *Tetrahedron*, 1974, **30**, 3127.	1026
74TL303	G. Märkl, J. Advena and H. Hauptmann; *Tetrahedron Lett.*, 1974, **15**, 303.	134
74TL909	J. T. Groves and K. W. Ma; *Tetrahedron Lett.*, 1974, **15**, 909.	42, 57
74TL1019	J. W. Scheeren and R. W. M. Aben; *Tetrahedron Lett.*, 1974, **15**, 1019.	830
74TL1403	S. L. Hartzell, D. F. Sullivan and M. W. Rathke; *Tetrahedron Lett.*, 1974, **15**, 1403.	648
74TL1549	T. Fukuyama, S. Nakatsuka and Y. Kishi; *Tetrahedron Lett.*, 1974, **15**, 1549.	218
74TL3171	D. Seebach, M. Kolb and B.-T. Gröbel; *Tetrahedron Lett.*, 1974, **15**, 3171.	254
74TL3625	T. Nakai, H. Shiono and M. Okawara; *Tetrahedron Lett.*, 1974, **15**, 3625.	392
74TL4085	R. Curci and F. DiFuria; *Tetrahedron Lett.*, 1974, **15**, 4085.	237
74TL4157	J. C. Philips, M. Aregullin, M. Oku and A. Sierra; *Tetrahedron Lett.*, 1974, **15**, 4157.	779
74TL4405	T. Hayashi, K. Yamamoto and M. Kumada; *Tetrahedron Lett.*, 1974, **15**, 4405.	1032, 1033
74USP3845169	L. Maier; *US Pat.* 3 845 169 (1974) (*Chem. Abstr.*, 1975, **82**, 86 403).	556
74ZAAC(404)204	G. Schott and K. Golz; *Z. Anorg. Allg. Chem.*, 1974, **404**, 204.	569, 570
74ZAAC(406)131	R. Appel and I. Ruppert; *Z. Anorg. Allg. Chem.*, 1974, **406**, 131.	563
74ZC152	J. Rachon, C. Wasielewski and A. Sobezak; *Z. Chem.*, 1974, **14**, 152.	476
74ZC243	K. Issleib, M. Lischewski and A. Zschunke; *Z. Chem.*, 1974, **14**, 243.	455
74ZN(B)485	H. Schmidbaur and M. Heimann; *Z. Naturforsch., Teil B*, 1974, **29**, 485.	569
74ZOB276	Z. S. Novikova, S. N. Mashoshina and I. F. Lutsenko; *Zh. Obshch. Khim.*, 1974, **44**, 276.	552, 562
74ZOB806	V. F. Mironov, V. I. Shiryaev, V. V. Yankov, A. F. Gladchenko and A. D. Naumov; *Zh. Obshch. Khim.*, 1974, **44**, 806 (*Chem. Abstr.*, 1974, **81**, 49 766r).	660
74ZOB1029	E. S. Kozlov and A. I. Sedlov; *Zh. Obshch. Khim.*, 1974, **44**, 1029.	344
74ZOB1712	B. S. Drach, E. P. Sviridov and Y. P. Shaturskii; *Zh. Obshch. Khim.*, 1974, **44**, 1712 (Engl. Transl. 1681).	
74ZOB2067	M. L. Petrov, V. G. Salishchev and A. A. Petrov; *Zh. Obshch. Khim.*, 1974, **44**, 2067.	913
74ZOR36	V. P. Kukhar and N. G. Pavlenko; *Zh. Org. Khim.*, 1974, **10**, 36. (Engl. Transl. 33).	987, 988

Ref	Citation	Page
74ZOR1986	S. I. Radchenko and A. A. Petrov; *Zh. Org. Khim.*, 1974, **10**, 1986 (*Chem. Abstr.*, 1974, **81**, 135 379y).	786
74ZOR2269	S. I. Radchenko, I. G. Savich and A. A. Petrov; *Zh. Org. Khim.*, 1974, **10**, 2269 (*Chem. Abstr.* 1975, **82**, 139 195).	841
74ZOR2456	S. I. Radchenko, I. G. Sulimov and A. A. Petrov; *Zh. Org. Khim.*, 1974, **10**, 2456 (*Chem. Abstr.* 1975, **82**, 86 049).	872
75ACR62	H. Schmidbaur; *Acc. Chem. Res.*, 1975, **8**, 62.	578
75ACS(B)461	S. Gronowitz and A. Maltesson; *Acta Chem. Scand., Ser. B*, 1975, **27**, 461.	814
75AG(E)15	D. Seebach and D. Enders; *Angew. Chem., Int. Ed. Engl.*, 1975, **14**, 15.	530
75AG(E)350	W. Dumont and A. Krief; *Angew. Chem., Int. Ed. Engl.*, 1975, **14**, 350.	400
75AG(E)473	J. M. Conia and M. J. Robson; *Angew. Chem., Int. Ed. Engl.*, 1975, **14**, 473.	162
75AG(E)700	D. Van Ende, W. Dumont and A. Krief; *Angew. Chem., Int. Ed. Engl.*, 1975, **14**, 700.	287
75AG(E)734	N. Kornblum; *Angew. Chem., Int. Ed. Engl.*, 1975, **14**, 734.	428
75AG(E)801	R. Appel; *Angew. Chem., Int. Ed. Engl.*, 1975, **14**, 801.	738
75BCJ357	S. Tanimoto, R. Taniyasu and M. Okano; *Bull. Chem. Soc. Jpn.*, 1975, **48**, 357.	416
75BCJ2103	M. Fukuda, K. Kan, Y. Okamoto and H. Sakurai; *Bull. Chem. Soc. Jpn.*, 1975, **48**, 2103.	1003, 1004
75BEP821282	C. H. Tieman, W. D. Kollmeyer and S. A. Roman, *Belg. Pat.* 821 282 (1975).	983
75BSF196	M. Kerfanto, Λ. Brault, F. Venien, J. M. Morvan and A. Le Rouzic; *Bull. Soc. Chim. Fr.*, 1975, 196.	405, 411
75BSF657	L. Léger and M. Saquet; *Bull. Soc. Chim. Fr.*, 1975, 657.	255
75BSF1439	D. Paquer; *Bull. Soc. Chim. Fr.*, 1975, 1439.	255
75BSF1797	J. Villieras, C. Bacquet and J.-F. Normant; *Bull. Soc. Chim. Fr.*, 1975, 1797.	22
75BSF2089	M. Hedayatullah and G. Motavaze; *Bull. Soc. Chim. Fr.*, 1975, 2089.	889
75BSF2259	J. Berlan, M. L. Capmau and W. Chodkiewicz; *Bull. Soc. Chim. Fr.*, 1975, 2259.	802
75C512	D. Scholz and H. G. Viehe; *Chimia*, 1975, **29**, 512.	796
75CA(83)10238	L. I. Zakharkin, V. V. Gavrilenko, L. L. Ivanova, B. A. Palei, V. S. Kolesov and M. P. Semenova; *Chem. Abstr.*, 1975, **83**, 10 238.	698
75CB174	J. Gante and G. Mohr; *Chem. Ber.*, 1975, **108**, 174.	900
75CB314	D. Seebach and A. K. Beck; *Chem. Ber.*, 1975, **108**, 314.	290, 402
75CB919	R. Appel and I. Ruppert; *Chem. Ber.*, 1975, **108**, 919.	564
75CB1321	H. Schmidbaur and R. Franke; *Chem. Ber.*, 1975, **108**, 1321.	586, 656
75CB2368	M. Braun, R. Dammann and D. Seebach; *Chem. Ber.*, 1975, **108**, 2368.	395
75CB2649	H. Schmidbaur, W. Richter, W. Wolf and F. Kohler; *Chem. Ber.*, 1975, **108**, 2649.	598
75CC107	J. C. Philips, M. Penzo and G. T. S. Lee; *J. Chem. Soc., Chem. Commun.*, 1975, 107.	777
75CC320	L. Bernardi, R. de Castiglione and U. Scarponi; *J. Chem. Soc., Chem. Commun.*, 1975, 320.	327
75CC432	D. H. R. Barton, C. C. Dawes and P. D. Magnus; *J. Chem. Soc., Chem. Commun.*, 1975, 432.	190
75CC451	M. Green, J. A. K. Howard, A. Laguna, M. Murray, J. L. Spencer and F. G. A. Stone; *J. Chem. Soc., Chem. Commun.*, 1975, 451.	691
75CC468	W. Althoff, M. Fild, H. Koop and R. Schmutzler; *J. Chem. Soc., Chem. Commun.*, 1975, 468.	564
75CC519	R. K. Boekman, Jr., K. J. Bruza, J. E. Baldwin and O. W. Lever, Jr.; *J. Chem. Soc., Chem. Commun.*, 1975, 519.	953
75CC536	D. L. Fogel, A. M. Rennert and C. Steel; *J. Chem. Soc., Chem. Commun.*, 1975, 536.	648
75CC615	A. Ducruix, H. Felkin, C. Pascard and G. K. Turner; *J. Chem. Soc., Chem. Commun.*, 1975, 615.	693
75CC621	D. G. H. Ballard and R. Pearce; *J. Chem. Soc., Chem. Commun.*, 1975, 621.	725
75CC634	S. O. Grim and J. D. Mitchell; *J. Chem. Soc., Chem. Commun.*, 1975, 634.	553
75CC640	Y. Sato, T. Toyo'oka, T. Aoyama and H. Shirai; *J. Chem. Soc., Chem. Commun.*, 1975, 640.	509
75CC671	E. M. Engler and V. V. Patel; *J. Chem. Soc., Chem. Commun.*, 1975, 671.	868
75CC867	H. K. Spencer, M. V. Lakshmikantham, M. P. Cava and A. F. Garito; *J. Chem. Soc., Chem. Commun.*, 1975, 867.	869
75CC877	J. Klein and A. Medlik-Balan; *J. Chem. Soc., Chem. Commun.*, 1975, 877.	668
75CC899	K. Oguro, M. Wada and R. Okawara; *J. Chem. Soc., Chem. Commun.*, 1975, 899.	955
75CJC332	Λ. R. Bassindale, A. G. Brook, P. F. Jones and J. M. Lennon; *Can. J. Chem.*, 1975, **53**, 332.	137
75CJC1922	H. J. Reich and J. E. Trend; *Can. J. Chem.*, 1975, **53**, 1922.	89
75CL607	M. Watanabe, T. Kobayashi, S. Kajgaeshi and S. Kanemasa; *Chem. Lett.*, 1975, 607.	421
75CL1259	Y. Oosawa, T. Saito and Y. Sasaki; *Chem. Lett.*, 1975, 1259.	552
75CPB2390	Y. Tominaga, Y. Morita, Y. Matsuda and G. Kobayashi; *Chem. Pharm. Bull.*, 1975, **23**, 2390.	842
75CPB2749	G. Kobayashi, Y. Matsuda, Y. Tominaga and K. Mizuyama; *Chem. Pharm. Bull.*, 1975, **23**, 2749.	844
75CSR189	P. D. Kennewell and J. B. Taylor; *Chem. Soc. Rev.*, 1975, **4**, 189.	330
75G939	R. P. Ferrari and G. A. Vaglio; *Gazz. Chim. Ital.*, 1975, **105**, 939.	682
75GEP2503114	M. Mitzlaff; *Ger. Pat.* 2 503 114 (1975), (*Chem. Abstr.*, 1976, **85**, 168 885).	303
75GEP(O)2412800	B. Lippsmeier, K. Hestmann and H. Neumaier; *Ger. Pat. (Offen)*, 2 412 800 (1975) (*Chem. Abstr.*, 1976, 84, 5147x).	119
75HCA1316	C. Couret, F. Couret, J. Satge and J. Escudie; *Helv. Chim. Acta*, 1975, **58**, 1316.	455, 481, 998
75IC78	M. Tsutsui, N. Ely and A. Gebala; *Inorg. Chem.*, 1975, **14**, 78.	719, 726
75IC656	S. O. Grim, L. C. Satek, C. A. Tolman and J. P. Jesson; *Inorg. Chem.*, 1975, **14**, 656.	551
75IC1018	*Inorg. Chem.*, 1975, **14**, 1018.	1018

75IC1614	J. A. Morrison and J. M. Bellama; *Inorg. Chem.*, 1975, **14**, 1614.	140
75IC1618	J. M. Bellama and L. L. Gerchman; *Inorg. Chem.*, 1975, **14**, 1618.	361
75ICA155	O. Gambino, R. P. Ferrari, M. Chinone and G. A. Vaglio; *Inorg. Chim. Acta*, 1975, **12**, 155.	682
75ICA(13)79	H. Schmidbaur and R. Franke; *Inorg. Chim. Acta*, 1975, **13**, 79.	586
75IZV1274	E. G. Ter-Galrielyan, E. P. Lur'e, Yu V. Zeifman and N. P. Gambaryan; *Izv. Akad. Nauk SSSR, Ser. Khim.*, 1975, 1274.	894
75IZV1576	M. G. Voronkov, V. M. D'yakov and L. I. Guvanova; *Izv. Akad. Nauk SSSR, Ser. Khim.*, 1975, 1576.	140
75IZV1600	E. D. Babich, M. V. Ushakov, V. M. Vdovin and N. S. Nametkin; *Izv. Akad. Nauk SSSR, Ser. Khim.*, 1975, 1600 (*Chem. Abstr.*, 1975, **83**, 179 197h).	658
75IZV2342	A. A. Onishchenko, T. V. Ternikova, O. A. Luk'yanov and V. A. Tartakovskii; *Izv. Akad. Nauk SSSR, Ser. Khim.*, 1975, 2342 (Engl. Transl. 2227).	995
75JA38	H. L. Carrell, H. M. Berman, J. S. Ricci, Jr., W. C. Hamilton, F. Ramirez, J. F. Marecek, L. Kramer and I. Ugi; *J. Am. Chem. Soc.*, 1975, **97**, 38.	493
75JA2293	G. A. Olah and S. H. Yu; *J. Am. Chem. Soc.*, 1975, **97**, 2293.	42
75JA2702	R. B. King and K. C. Hodges; *J. Am. Chem. Soc.*, 1975, **97**, 2702.	1019, 1020
75JA3822	C. G. Chavdarian and C. H. Heathcock; *J. Am. Chem. Soc.*, 1975, **97**, 3822.	953
75JA4264	T. Shono, H. Hamaguchi and Y. Matsumura; *J. Am. Chem. Soc.*, 1975, **97**, 4264.	303
75JA5434	H. J. Reich, J. M. Renga and I. L. Reich; *J. Am. Chem. Soc.*, 1975, **97**, 5434.	285
75JA5957	J. E. Baldwin, A. Au, M. Christie, S. B. Haber and D. Hesson; *J. Am. Chem. Soc.*, 1975, **97**, 5957.	69
75JA6260	G. Stork and M. Isobe; *J. Am. Chem. Soc.*, 1975, **97**, 6260.	199
75JA7172	M. Tachikawa, J. R. Shapley and C. G. Pierpont; *J. Am. Chem. Soc.*, 1975, **97**, 7172.	682
75JA7182	B. M. Trost and C. H. Miller; *J. Am. Chem. Soc.*, 1975, **97**, 7182.	172
75JA7371	C. M. Golino, R. D. Bush and L. H. Sommer; *J. Am. Chem. Soc.*, 1975, **97**, 7371.	610, 611
75JCS(D)1434	A. M. Devine, P. A. Griffin, R. N. Haszeldine, M. J. Newlands and A. E. Tipping; *J. Chem. Soc., Dalton Trans.*, 1975, 1434.	607, 622
75JCS(D)1614	A. J. Deeming, S. Hasso and M. Underhill; *J. Chem. Soc., Dalton Trans.*, 1975, 1614.	682, 683
75JCS(D)1786	J. L. Wardell; *J. Chem. Soc., Dalton Trans.*, 1975, 1786.	154
75JCS(D)1832	A. M. Devine, R. N. Haszeldine and A. E. Tipping; *J. Chem. Soc., Dalton Trans.*, 1975, 1832.	617, 623, 656, 661
75JCS(D)1837	A. M. Devine, R. N. Haszeldine and A. E. Tipping; *J. Chem. Soc., Dalton Trans.*, 1975, 1837.	654, 658
75JCS(D)2177	R. N. Haszeldine, C. R. Pool and A. E. Tipping; *J. Chem. Soc., Dalton Trans.*, 1975, 2177.	140
75JCS(P1)61	R. N. Butler and W. B. King; *J. Chem. Soc., Perkin Trans. 1*, 1975, 61.	313
75JCS(P1)129	A. Pelter, K. Smith, M. G. Hutchings and K. Rowe; *J. Chem. Soc., Perkin Trans. 1*, 1975, 129.	523
75JCS(P1)138	A. Pelter, M. G. Hutchings, K. Rowe and K. Smith; *J. Chem. Soc., Perkin Trans. 1*, 1975, 138.	523
75JCS(P1)180	D. M. Vyas and G. W. Hay; *J. Chem. Soc., Perkin Trans. 1*, 1975, 180.	271
75JCS(P1)251	C. R. Eck, R. W. Mills and T. Money; *J. Chem. Soc., Perkin Trans. 1*, 1975, 251.	20
75JCS(P1)702	P. Cooper, R. Fields and R. N. Haszeldine; *J. Chem. Soc., Perkin Trans. 1*, 1975, 702.	802
75JCS(P1)772	O. O. Orazi and R. A. Corral; *J. Chem. Soc., Perkin Trans. 1*, 1975, 772.	428
75JCS(P1)1181	J. R. L. Smith and J. S. Sadd; *J. Chem. Soc., Perkin Trans. 1*, 1975, 1181.	445
75JCS(P1)1277	T. Takeshima, N. Fukada, E. Okabe, F. Mineshima and M. Muraoka; *J. Chem. Soc., Perkin Trans. 1*, 1975, 1277.	842
75JCS(P1)1670	K. H. Baggaley, S. G. Brooks and R. M. Hindley; *J. Chem. Soc., Perkin Trans. 1*, 1975, 1670.	308
75JCS(P1)2048	M. Ahmed and J. M. Vernon; *J. Chem. Soc., Perkin Trans. 1*, 1975, 2048.	54
75JFC(6)227	I. L. Knunyants, Yu. V. Zeiffman, T. V. Lushnikova, E. M. Rokhlin, Yo. G. Abduganiev and U. Utebaev; *J. Fluorine Chem.*, 1975, **6**, 227.	731
75JGU1668	A. N. Lavrent'ev, I. G. Maslennikov and E. G. Sochilin; *J. Gen. Chem. USSR* (*Engl. Transl. 1*), 1975, **45**, 1668.	114
75JGU1950	T. A. Zyablikova, N. V. Ivasyuk, É. Kh. Mukhametzyanova and I. M. Shermergorn; *J. Gen. Chem. USSR* (*Engl. Transl.*), 1975, **45**, 1950.	124
75JHC579	D. J. Daigle and A. B. Pepperman Jr.; *J. Heterocycl. Chem.*, 1975, **12**, 579.	458
75JHC595	H. C. Wormser and W.-H. Chiu; *J. Heterocycl. Chem.*, 1975, **12**, 595.	427
75JHC749	J. P. Wineburg, C. Abrams and D. Swern; *J. Heterocycl. Chem.*, 1975, **12**, 749.	205
75JHC981	A. Merle and G. Descotes; *J. Heterocycl. Chem.*, 1975, **12**, 981.	174
75JMC177	P. Aeberli, P. Eden, J. H. Gogerty, W. J. Houlihan and C Penberthy; *J. Med. Chem.*, 1975, **18**, 177.	416
75JMC182	P. Aeberli, P. Eden, J. H. Gogerty, W. J. Houlihan and C Penberthy; *J. Med. Chem.*, 1975, **18**, 182.	416
75JOC148	B. M. Trost and T. N. Salzmann; *J. Org. Chem.*, 1975, **40**, 148.	197
75JOC231	D. Seebach and E. J. Corey; *J. Org. Chem.*, 1975, **40**, 231.	245, 250, 1075
75JOC266	A. Jonczyk, K. Banko and M. Makosza; *J. Org. Chem.*, 1975, **40**, 266.	82, 83, 237
75JOC437	H. Alper and M. S. Wolin; *J. Org. Chem.*, 1975, **40**, 437.	519, 520
75JOC813	H. C. Brown, J.-J. Katz and B. A. Carlson; *J. Org. Chem.*, 1975, **40**, 813.	363
75JOC814	E. Negishi, T. Yoshida, A. Silveira, Jr., and B. L. Chiou; *J. Org. Chem.*, 1975, **40**, 814.	372
75JOC1173	C. G. Stuckwisch; *J. Org. Chem.*, 1975, **41**, 1173.	576
75JOC1371	A. G. Schultz and R. D. Lucci; *J. Org. Chem.*, 1975, **40**, 1371.	167
75JOC2056	A. B. Pepperman, G. J. Boudreaux and T. H. Siddall; *J. Org. Chem.*, 1975, **40**, 2056.	342

75JOC2414	H. H. Seltzman and T. M. Chapman; *J. Org. Chem.*, 1975, **40**, 2414.	297, 304, 325
75JOC2552	D. H. Aue and D. Thomas; *J. Org. Chem.*, 1975, **40**, 2552.	829
75JOC2570	H. J. Reich; *J. Org. Chem.*, 1975, **40**, 2570.	375, 399
75JOC2577	M. G. Miles, J. S. Wager, J. D. Wilson and A. R. Siedle; *J. Org. Chem.*, 1975, **40**, 2577.	850
75JOC2720	B. B. Gavrilenko and S. I. Miller; *J. Org. Chem.*, 1975, **40**, 2720.	984
75JOC2796	D. J. Burton and G. E. Greenlimb; *J. Org. Chem.*, 1975, **40**, 2796.	36
75JOC2962	P. A. Zoretic, P. Soja and W. E. Conrad; *J. Org. Chem.*, 1975, **40**, 2962.	210
75JOC3037	T. Higa and A. J. Krubsack; *J. Org. Chem.*, 1975, **40**, 3037.	71
75JOC3052	P. Beak, J. Yamamoto and C. J. Upton; *J. Org. Chem.*, 1975, **40**, 3052.	394
75JOC3540	M. Ohakara, T. Kojitani, S. Yanagida, M. Okahara and S. Komori; *J. Org. Chem.*, 1975, **40**, 3540.	72
75JOC3807	R. K. Singh and S. Danishefsky; *J. Org. Chem.*, 1975, **40**, 3807.	204
75JOM57	S. Sato and Y. Matsumura; *J. Organomet. Chem.*, 1975, **96**, 57.	595
75JOM393	F. Kober; *J. Organomet. Chem.*, 1975, **94**, 393.	593
75JOM(84)C13	E. Kurras, H. Mennenga, G. Oehme, U. Rosenthal and G. Engelhardt; *J. Organomet. Chem.*, 1975, **84**, C13.	580
75JOM(84)117	G. van Koten, A. J. Leusink and J. G. Noltes; *J. Organomet. Chem.*, 1975, **84**, 117.	695
75JOM(84)129	G. van Koten and J. G. Noltes; *J. Organomet. Chem.*, 1975, **84**, 129.	695
75JOM(86)C10	F. R. Kreissl and W. Held; *J. Organomet. Chem.*, 1975, **86**, C10.	580
75JOM(87)151	J. Dunogues, A. Ekouya, R. Calas and N. Duffaut; *J. Organomet. Chem.*, 1975, **87**, 151.	355
75JOM(88)255	D. Seyferth, R. L. Lambert, Jr. and M. Massol; *J. Organomet. Chem.*, 1975, **88**, 255.	152, 155, 156
75JOM(88)287	D. Seyferth and R. L. Lambert, Jr.; *J. Organomet. Chem.*, 1975, **88**, 287.	155
75JOM(91)C57	S. S. Crawford, G. Firestein and H. D. Kaesz; *J. Organomet. Chem.*, 1975, **91**, C57.	720
75JOM(92)7	D. Seferth, G. J. Murphy and R. A. Woodfuff; *J. Organomet. Chem.*, 1975, **92**, 7.	34, 137
75JOM(93)21	D. S. Matteson and L. A. Hagelee; *J. Organomet. Chem.*, 1975, **93**, 21.	1054
75JOM(93)71	J. Masse and D. Samate; *J. Organomet. Chem.*, 1975, **93**, 71.	646, 654
75JOM(94)327	S. O. Grim and R. C. Barth; *J. Organomet. Chem.*, 1975, **94**, 327.	112
75JOM(96)133	Y. Yamamoto and H. Schmidbaur; *J. Organomet. Chem.*, 1975, **96**, 133.	585
75JOM(96)213	G. S. Kalinina, T. A. Basalgina, N. S. Vyazankin, G. A. Razuvaev, V. A. Yablokov and N. V. Yablokova; *J. Organomet. Chem.*, 1975, **96**, 213.	637, 641
75JOM(97)61	W. P. Fehlhammer, K. Bartel and H. Schmidt; *J. Organomet. Chem.*, 1975, **97**, C61.	1017
75JOM(97)159	E. W. Abel and R. J. Rowley; *J. Organomet. Chem.*, 1975, **97**, 159.	533, 534
75JOM(97)245	W. A. Herrmann, B. Reiter and H. Biersack; *J. Organomet. Chem.*, 1975, **97**, 245.	676
75JOM(97)479	Y. Yamamoto and H. Schmidbaur; *J. Organomet. Chem.*, 1975, **97**, 479.	585
75JOM(99)31	L. E. McCandlish and J. W. Macklin; *J. Organomet. Chem.*, 1975, **99**, 31.	697
75JOM(99)281	J. P. Hickey, J. R. Wilkinson and L. J. Todd; *J. Organomet. Chem.*, 1975, **99**, 281.	681
75JOM(99)353	H. Schmidbaur, H.-J. Füller and F. H. Köhler; *J. Organomet. Chem.*, 1975, **99**, 353.	587
75JOM(99)385	H. Alper and M. S. Wolin; *J. Organomet. Chem.*, 1975, **99**, 385.	519, 520
75JOM(102)437	A. L. Rheingold and J. M. Bellama; *J. Organomet. Chem.*, 1975, **102**, 437.	134, 135
75JOM(102)445	A. L. Rheingold and J. M. Bellama; *J. Organomet. Chem.*, 1975, **102**, 445.	135
75JOM(102)511	J. C. Gaunt and B. L. Shaw; *J. Organomet. Chem.*, 1975, **102**, 511.	720
75JOM(102)551	G. van Koten and J. G. Noltes; *J. Organomet. Chem.*, 1975, **102**, 551.	696
75JOU119	B. S. Drach, V. A. Kovalev and A. V. Kersalov; *J. Org. Chem. USSR (Engl. Transl.)*, 1975, **11**, 119.	794
75JPR137	E. Fanghanel; *J. Prakt. Chem.*, 1975, **317**, 137.	850
75JPR337	H. Schafer and K. Gewald; *J. Prakt. Chem.*, 1975, **317**, 337.	833
75JPR840	D. Gloyna, U. Lachmann and H. G. Henning; *J. Prakt. Chem.*, 1975, **317**, 840.	556
75JPR890	H. Gross, H. Seibt and I. Keitel; *J. Prakt. Chem.*, 1975, **317**, 890.	559
75JST(24)373	E. Ericsson, T. Marnung, J. Sandstrom and I. Wennerbeck; *J. Mol. Struct.*, 1975, **24**, 373.	973
75LA19	W. Walter and H.-W. Meyer; *Justus Liebigs Ann. Chem.*, 1975, 19.	903
75LA266	L. Birkofer and W. Kaiser; *Justus Liebigs Ann. Chem.*, 1975, 266.	313
75LA565	J. J. Eisch and S. G. Rhee; *Justus Liebigs Ann. Chem.*, 1975, 565.	1068
75LA1484	G. Ferdinand, K. Schank and A. Weber; *Justus Liebigs Ann. Chem.*, 1975, 1484.	838
75LA1790	H. Böhme and P. Backhaus; *Justus Liebigs Ann. Chem.*, 1975, 1790.	305, 311
75LA2318	J. Curtze and K. Thomas; *Justus Liebigs Ann. Chem.*, 1975, 2318.	416, 423
75MI 420-01	J. Thiem and H. Paulsen; *Phosphorus*, 1975, **6**, 51.	912
75PAC553	G. Stork; *Pure Appl. Chem.*, 1975, **43**, 553.	357
75RTC14	J. Meijer and P. Vermeer; *Rec. Trav. Chim. Pays-Bas*, 1975, **94**, 14.	963
75RTC209	R. J. Broekma; *Recl. Trav. Chim. Pays-Bas*, 1975, **94**, 209.	827
75RZC397	J. Rachon and C. Wasielewski; *Rocz. Chem.*, 1975, **49**, 397.	460
75S147	D. S. Matteson; *Synthesis*, 1975, 147.	632, 633
75S385	K. Ogura, M. Yamashita and G. Tsuchihashi; *Synthesis*, 1975, 385.	856
75S458	J. Villieras, P. Perriot and J. F. Normant; *Synthesis*, 1975, 458.	740, 741
75S461	D. Seebach and R. Buerstinghaus; *Synthesis*, 1975, 461.	849
75S519	G. E. Vennstra and B. Zwaneburg; *Synthesis*, 1975, 519.	235
75S535	Ph. Savignac, J. Petrova, M. Dreux and Ph. Coutrot; *Synthesis*, 1975, 535.	740, 741
75S720	K. Mori, H. Hashimoto, Y. Takenaka and T. Takigawa; *Synthesis*, 1975, 720.	250
75S789	C. Giordano and A. Belli; *Synthesis*, 1975, 789.	297
75SC33	E. C. Freidrich, S. N. Falling and D. E. Lyons; *Synth. Commun.*, 1975, **5**, 33.	29
75SRI199	A. L. Rheingold and J. M. Bellama; *Synth. React. Inorg. Met.-Org. Chem.*, 1975, **5**, 199.	135
75T327	D. Danneels, M. Anteunis, Z. van Acker and D. Tavernier; *Tetrahedron*, 1975, **31**, 327.	231
75T809	L. Fournier, G. Lamaty, A. Natat and J. P. Roque; *Tetrahedron*, 1975, **31**, 809.	215
75T863	U. Zoller and D. Ben-Ishai; *Tetrahedron*, 1975, **31**, 863.	100, 309, 327

Ref	Citation	Pages
75T891	J. Normant, R. Sauvetre and J. Villieras; *Tetrahedron*, 1975, **31**, 891.	763, 771
75T897	R. Sauvetre, J. Normant and J. Villieras; *Tetrahedron*, 1975, **31**, 897.	763, 771
75T1437	J. C. Hubert, J. B. P. A. Wijnberg and W. N. Speckamp; *Tetrahedron*, 1975, **31**, 1437.	303
75T2387	H. A. Selling; *Tetrahedron*, 1975, **31**, 2387.	861
75TL609	Ph. Savignac, M. Dreux and Ph. Coutrot; *Tetrahedron Lett.*, 1975, **16**, 609.	740
75TL773	J. L. Hahnfeld and D. J. Burton; *Tetrahedron Lett.*, 1975, **16**, 773.	755
75TL925	E. J. Corey and A. P. Kozikowski; *Tetrahedron Lett.*, 1975, **16**, 925.	254
75TL997	Y. Urabe, T. Iwasaki, K. Matsumoto and M. Miyoshi; *Tetrahedron Lett.*, 1975, **16**, 997.	13
75TL1373	G. H. Posner, G. L. Loomis and H. S. Sawaya; *Tetrahedron Lett.*, 1975, **16**, 1373.	745, 746
75TL1613	A. Anciaux, A. Eman, W. Dumont, D. Van Ende and A. Krief; *Tetrahedron Lett.*, 1975, **16**, 1613.	238, 239, 283, 375
75TL2017	E. Vedejs and M. Mullins; *Tetrahedron Lett.*, 1975, **16**, 2017.	222, 365, 393
75TL2923	H. Westmijze, J. Meijer and P. Vermeer; *Tetrahedron Lett.*, 1975, **16**, 2923.	963
75TL3171	G. Märkl, D. Matthes, A. Donaubauer and H. Baier; *Tetrahedron Lett.*, 1975, **16**, 3171.	1026, 1030
75TL3267	E. J. Corey and T. Hase; *Tetrahedron Lett.*, 1975, **16**, 3267.	197, 216
75TL3269	E. J. Corey and M. G. Bock; *Tetrahedron Lett.*, 1975, **16**, 3269.	219
75TL3487	A. M. van Leusen, R. J. Bouma and O. Possel; *Tetrahedron Lett.*, 1975, **16**, 3487.	331
75TL3489	A. P. Bashall and J. F. Collins; *Tetrahedron Lett.*, 1975, **16**, 3489.	190
75TL3579	R. K. Olsen and A. J. Kolar; *Tetrahedron Lett.*, 1975, **16**, 3579.	309
75TL3979	L. M. Weinstock, S. Karady, F. E. Roberts, A. M. Hoinowski, G. S. Brenner, T. B. K. Lee, W. C. Lumma and M. Sletzinger; *Tetrahedron Lett.*, 1975, **16**, 3979.	48
75TL4065	B. H. Bakker and W. N. Speckamp; *Tetrahedron Lett.*, 1975, **16**, 4065.	98
75TL4433	A. J. Mura, Jr., D. A. Bennett and T. Cohen; *Tetrahedron Lett.*, 1975, **16**, 4433.	66, 68
75TL4531	J. P. Marino and R. C. Landick; *Tetrahedron Lett.*, 1975, **16**, 4531.	346
75TL4543	R. M. Munavu and H. Szmant; *Tetrahedron Lett.*, 1975, **16**, 4543.	197
75TL4547	N. H. Andersen, Y. Yamamoto and A. D. Denniston; *Tetrahedron Lett.*, 1975, **16**, 4547.	252
75ZAAC202	F. von Kober; *Z. Anorg. Allg. Chem.*, 1975, **412**, 202.	593, 594
75ZN(B)245	G. Uray and E. Ziegler; *Z. Naturforsch., Teil B*, 1975, **30**, 245.	96
75ZOB12	B. S. Drach, E. P. Sviridov and A. V. Kirsanov; *Zh. Obshch. Khim.*, 1975, **45**, 12 (Engl. Transl. 10).	999, 1000
75ZOB556	A. J. Kostyuk, N. I. Savd'eva, Yu. I. Baukov and I. F. Lutsenko; *Zh. Obshch. Khim.*, 1975, **45**, 556.	915
75ZOB1486	Z. S. Novikova, S. N. Mashoshina and I. F. Lutsenko; *Zh. Obshch. Khim.*, 1975, **45**, 1486.	574, 588
75ZOB1905	B. I. Petrov, G. S. Kalinina and A. Y. Sorokin; *Zh. Obshch. Khim.*, 1975, **45**, 1905 (*Chem. Abstr.*, 1975, **83**, 193 496m).	641
75ZOB2010	M. G. Voronkov, V. M. D'yakov, Yu. A. Lukina, G. A. Samsonova and N. M. Kudyakov; *Zh. Obshch. Khim.*, 1975, **45**, 2010.	569
75ZOB2374	O. E. Nasakin, V. V. Komachev and V. A. Kukhtin; *Zh. Obshch. Khim.*, 1975, **45**, 2374.	768
75ZOB2448	V. F. Mironov, V. I. Shiryaev, E. M. Stepina, V. V. Yankov and V. P. Kochergin; *Zh. Obshch. Khim.*, 1975, **45**, 2448 (*Chem. Abstr.*, 1976, **84**, 74 381f).	660
75ZOB2672	A. A. Buyakov, T. K. Gar, E. M. Berliner and V. F. Mironov; *Zh. Obshch. Khim.*, 1975, **45**, 2672 (*Chem. Abstr.*, 1976, **84**, 150 690y).	621
75ZOR415	V. I. Dronov, R. F. Nigmatullina and V. P. Krivonogov; *Zh. Org. Khim.*, 1975, **11**, 415 (*Chem. Abstr.*, 1975, **82**, 170 541).	864
75ZOR452	A. I. Sitkin, O. Z. Safiulina, R. F. Chernyaeva and A. D. Nikolaeva; *Zh. Org. Khim.*, 1975, **11**, 452 (Engl. Transl. 443).	993
75ZOR961	R. A. Bekker, G. V. Astratyan, B. L. Dyatkin and I. L. Knunyants; *Zh. Org. Khim.*, 1975, **11**, 961 (*Chem. Abstr.*, 1975, **83**, 58 043w).	730
76ACS(B)439	S. Gronowitz and T. Frejd; *Acta. Chem. Scand., Ser (B)*, 1976, **30**, 439.	785
76AG417	C. Wiaux-Zamar, J. P. Dejonghe, L. Ghosez, J. F. Normant and J. Villieras; *Angew. Chem.*, 1976, **88**, 417.	1016, 1017
76AG(E)42	H.-F. Klein and R. Hammer; *Angew. Chem., Int. Ed. Engl.*, 1976, **15**, 42.	582
76AG(E)56	H. J. Padberg and G. Bergerhoff; *Angew. Chem., Int. Ed. Engl.*, 1976, **15**, 56.	594
76AG(E)161	W. Dumont and A. Krief; *Angew. Chem., Int. Ed. Engl.*, 1976, **15**, 161.	375, 652
76AG(E)171	G. N. Schrauzer, R. N. Katz, J. H. Grate and T. M. Vickrey; *Angew. Chem., Int. Ed. Engl.*, 1976, **15**, 171.	1016
76AG(E)270	A. I. Meyers and E. D. Mihelich; *Angew. Chem., Int. Ed. Engl.*, 1976, **15**, 270.	306
76AG(E)367	H. Schmidbaur and M. Heimann; *Angew. Chem., Int. Ed. Engl.*, 1976, **15**, 367.	624
76AG(E)437	D. Seebach, K. H. Geiss and M. Pohmakor; *Angew. Chem., Int. Ed. Engl.*, 1976, **15**, 437.	395
76AG(E)438	D. Seebach and N. Meyer; *Angew. Chem., Int. Ed. Engl.*, 1976, **15**, 438.	378
76AG(E)505	D. Seebach and F. Lehr; *Angew. Chem., Int. Ed. Engl.*, 1976, **15**, 505.	540
76AG(E)541	H. Schmidbaur and H.-J. Füller; *Angew. Chem., Int. Ed. Engl.*, 1976, **15**, 541.	587, 588
76AG(E)542	H. Schmidbaur and O. Gasser; *Angew. Chem., Int. Ed. Engl.*, 1976, **15**, 542.	586, 588
76AG(E)609	R. A. Andersen, A. L. Galyer and G. Wilkinson; *Angew. Chem., Int. Ed. Engl.*, 1976, **15**, 609.	966
76AG(E)619	U. Klingebiel and A. Meller; *Angew. Chem., Int. Ed. Engl.*, 1976, **15**, 619.	606
76AG(E)688	H. D. Scharf and E. Wolters; *Angew. Chem., Int. Ed. Engl.*, 1976, **15**, 688.	207
76AJC2189	R. S. Dickson and S. H. Johnson; *Aust. J. Chem.*, 1976, **29**, 2189.	690
76AP503	H. Möhrle and C.-M. Seidel; *Arch. Pharm. (Weinheim, Ger.)*, 1976, **309**, 503.	416
76BAU853	I. L. Knunyants, U. Utebaev, E. M. Rokhlin, E. P. Lur'e and E. I. Mysov; *Bull. Acad. Sci. USSR, Div. Chem. Sci.*, 1976, 853.	803

76BAU873	I. L. Knunyants, U. Utebaev, E. M. Rokhlin, E. P. Lur'e and E. I. Mysov; *Bull. Acad. Sci. USSR, Div. Chem. Sci.*, 1976, 873.	803
76BCJ256	N. Kunieda, J. Nokami and M. Kinoshita; *Bull. Chem. Soc. Jpn.*, 1976, **49**, 256.	264
76BCJ502	N. Ishikawa and A. Nagashima; *Bull. Chem. Soc. Jpn.*, 1976, **49**, 502.	
76BCJ553	K. Arai and M. Oki; *Bull. Chem. Soc. Jpn.*, 1976, **49**, 553.	66
76BCJ1177	N. Kawabata, M. Naka, T. Komazawa and S. Yamashita; *Bull. Chem. Soc. Jpn.*, 1976, **49**, 1177.	669
76BCJ1931	S. Tanimoto, R. Taniyasu, T. Takahashi, T. Miyake and M. Okano; *Bull. Chem. Soc. Jpn.*, 1976, **49**, 1931.	761, 771
76BCJ2491	T. Kitazume, T. Otaka, R. Takei and N. Ishikawa; *Bull. Chem. Soc. Jpn.*, 1976, **49**, 2491.	731
76BCJ2837	S. Kozima, K. Kobayashi and M. Kawanisi; *Bull. Chem. Soc. Jpn.*, 1976, **49**, 2837.	156
76BCJ3128	H. Yoshida, T. Yao, T. Ogata and S. Inokawa; *Bull. Chem. Soc. Jpn.*, 1976, **49**, 3128.	834
76BSF1482	D. Michelot, R. Lorne, C. Huynh and S. Julia; *Bull. Soc. Chim. Fr.*, 1976, 1482.	323
76BSF2053	D. Clerin, B. Meyer and J. Fleury; *Bull. Soc. Chim. Fr.*, 1976, 2053.	979
76C187	M. Hoffmann, C. Wasielewski and J. Rachon; *Chimia*, 1976, **30**, 187.	460
76CA122034u	F. Seng; *Chem. Abstr.*, 1976, **84**, 122 034u (*Ger. Pat.* 2 431 408 to Bayer AG).	921
76CA192821t	V. G. Salishchev, M. L. Petrov and A. A. Petrov; *Chem. Abstr.*, 1976, **85**, 192 821t.	913
76CAR(46)237	T. van Es; *Carbohydr. Res.*, 1976, **46**, 237.	228
76CAR(52)63	R. J. Ferrier and R. H. Furneaux; *Carbohydr. Res.*, 1976, **52**, 63.	224
76CB473	K.-H. A. O. Starzewski, W. Richter and H. Schmidbaur; *Chem. Ber.*, 1976, **109**, 473.	598
76CB906	E. Schaumann; *Chem. Ber.*, 1976, **109**, 906.	852
76CB2039	G. Maas and M. Regitz; *Chem. Ber.*, 1976, **109**, 2039.	557
76CB2175	P. Entmayr and G. Köbrich; *Chem. Ber.*, 1976, **109**, 2175.	741, 747
76CB3047	D. Hoppe and R. Follman; *Chem. Ber.*, 1976, **109**, 3047.	538
76CB3062	D. Hoppe and R. Follman; *Chem. Ber.*, 1976, **109**, 3062.	538
76CC148	E. M. Engler, D. C. Green and J. Q. Chambers; *J. Chem. Soc., Chem. Commun.*, 1976, 148.	874, 875
76CC174	P. Moreau, G. Dalverney and A. Commeyras; *J. Chem. Soc., Chem. Commun.*, 1976, 174.	820
76CC200	R. A. Bell and M. H. Chisholm; *J. Chem. Soc., Chem. Commun.*, 1976, 200.	822
76CC433	M. P. Doyle and B. Siegfried; *J. Chem. Soc., Chem. Commun.*, 1976, 433.	19
76CC592	A. Kumar, H. Ila, H. Junjappa and S. Mhatre; *J. Chem. Soc., Chem. Commun.*, 1976, 592.	899, 900
76CC593	A. Kumar, H. Ila and H. Junjappa; *J. Chem. Soc., Chem. Commun.*, 1976, 593.	900
76CC623	M. J. S. Gynane, A. Hudson, M. J. Lappert, P. P. Power and H. Goldwhite; *J. Chem. Soc., Chem. Commun.*, 1976, 623.	598
76CC697	C. M. Robbins and G. H. Whitham; *J. Chem. Soc., Chem. Commun.*, 1976, 697.	357
76CC990	R. C. Cookson and P. J. Parsons; *J. Chem. Soc., Chem. Commun.*, 1976, 990.	957
76CCC386	J. Včelák, V. Chvalovský, M. G. Voronkov, V. B. Pukhnarevich and V. A. Pestunovich; *Collect. Czech. Chem. Commun.*, 1976, **41**, 386.	137
76CL203	H. Ishihara and Y. Hirabayashi; *Chem. Lett.*, 1976, 203.	288
76CL769	K. Saigo, M. Osaki and T. Mukaiyama; *Chem. Lett.*, 1976, 769.	193
76CPB102	I. Isaka, K. Nakano, T. Kashiwagi, A. Koda, H. Horiguchi, F. Matsui, K. Takahashi and M. Murakami; *Chem. Pharm. Bull.*, 1976, **24**, 102.	205
76CPB235	Y. Maki; *Chem. Pharm. Bull.*, 1976, **24**, 235.	420
76CR(C)473	J. Massé, E. Parayre and C. Warolin; *C. R. Hebd. Seances Acad. Sci., Ser. C*, 1976, **282**, 473.	508
76CS90	S. Gronowitz and M. Herslof; *Chem. Scr.*, 1976, **10**, 90 (*Chem. Abstr.* 1977, **86**, 171 168).	856
76CS133	T. Frejd; *Chem. Scr.*, 1976, **10**, 133.	785
76CZ336	H. Schumann and S. Hohmann; *Chem.-Ztg.*, 1976, **100**, 336.	578
76DOK(226)1092	A. N. Nesmeyanov, V. A. Sazonova, N. N. Sedova, Yu. V. Volgin and O. V. Dudukina; *Dokl. Akad. Nauk SSSR*, 1976, **226**, 1092.	723
76DOK(231)1138	G. A. Razuvaev, E. N. Gladyshev, V. N. Latyaeva, E. V. Krasil'nikova, A. N. Lineva and Y. N. Novotorov; *Dokl. Akad. Nauk SSSR*, 1976, **231**, 1138 (*Chem. Abstr.*, 1977, **86**, 140 193r).	656
76GEP2523145	W. Herwig and H. J. Kleiner; *Ger. Pat.* 2 523 145 (1976) (*Chem. Abstr.*, 1977, **86**, 90 983).	556
76H(4)963	S. Tamagaki and R. Ichihara; *Heterocycles*, 1976, **4**, 963.	862
76H(5)203	K. Horiki; *Heterocycles*, 1976, **5**, 203.	416
76HCA13	R. Muthukrishnan and M. Schlosser; *Helv. Chim. Acta*, 1976, **59**, 13.	957
76IC90	M. A. Bennett, R. N. Johnson and T. W. Turney; *Inorg. Chem.*, 1976, **15**, 90.	688
76IC97	M. A. Bennett, R. N. Johnson, G. B. Robertson, T. W. Turney and P. O. Whimp; *Inorg. Chem.*, 1976, **15**, 97.	688
76IC107	M. A. Bennett, R. N. Johnson and T. W. Turney; *Inorg. Chem.*, 1976, **15**, 107.	689
76IC743	K. E. Peterman and J. M. Shreeve; *Inorg. Chem.*, 1976, **15**, 743.	107
76IC1697	K. G. Sharp and I. Schwager; *Inorg. Chem.*, 1976, **15**, 1697.	112
76IC1988	B. R. Gragg and G. E. Ryschkewitsch; *Inorg. Chem.*, 1976, **15**, 1988.	534
76IC2567	L. E. Manzer; *Inorg. Chem.*, 1976, **15**, 2567.	578, 579
76IC2671	K. H. Pannell, J. B. Cassias, G. M. Crawford and A. Flores; *Inorg. Chem.*, 1976, **15**, 2671.	685, 722
76IS192	J. Del Gaudio and S. O. Grim; *Inorg. Synth.*, 1976, **16**, 192.	575
76IS195	J. D. Mitchell and S. O. Grim; *Inorg. Synth.*, 1976, **16**, 195.	551, 576
76IZV895	I. L. Knunyants, U. Utebaev, E. M. Rokhlin, E. P. Lur'e and E. I. Mysov; *Izv. Akad. Nauk SSSR, Ser. Khim.*, 1976, 895 (*Chem. Abstr.*, 1976, **85**, 20 576a).	736
76IZV1342	G. S. Kalinina, E. A. Shchupak, N. S. Vyazankin and G. A. Razuvaev; *Izv. Akad. Nauk SSSR, Ser. Khim.*, 1976, 1342 (*Chem. Abstr.*, 1976, **85**, 124 063x).	641
76IZV1416	B. A. Arbuzov, N. P. Bogonostseva and V. S. Vinogradova; *Izv. Acad. Nauk SSSR, Ser. Khim.*, 1976, 1416.	339

76IZV1887	O. A. Kruglaya, I. B. Fedot'eva, B. V. Fedot'eva, I. D. Kalikhman, E. I. Brodskaya and N. S. Vyazankin; *Izv. Akad. Nauk. SSSR, Ser. Khim.*, 1976, 1887 (*Chem. Abstr.*, 1977, **86**, 72 780c).	628
76IZV2200	M. G. Voronkov, V. M. D'yakov, Yu. A. Lukina and M. V. Sigalov; *Izv. Akad. Nauk SSSR, Ser. Khim.*, 1976, **10**, 2200.	140
76JA1627	M. P. Doyle, B. Siegfried and J. J. Hammond; *J. Am. Chem. Soc.*, 1976, **98**, 1627.	19, 27
76JA1774	F. A. Cotton, J. D. Jamerson and B. R. Stults; *J. Am. Chem. Soc.*, 1976, **98**, 1774.	681
76JA1798	H. C. Brown and N. Ravindran; *J. Am. Chem. Soc.*, 1976, **98**, 1798.	631, 635
76JA2008	I. Vlattas, L. Della Vecchia and A. O. Lee; *J. Am. Chem. Soc.*, 1976, **98**, 2008.	956, 957
76JA3037	E. C. Taylor, R. L. Robey, K.-T. Liu, B. Favre, H. T. Bozimo, R. A. Conley, C.-S. Chiang, A. McKillop and M. E. Ford; *J. Am. Chem. Soc.*, 1976, **98**, 3037.	194
76JA3718	T. Hayashi, M. Tajika, K. Tamao and M. Kumada; *J. Am. Chem. Soc.*, 1976, **98**, 3718.	1033
76JA3916	R. C. Wheland and J. L. Gibson; *J. Am. Chem. Soc.*, 1976, **98**, 3916.	869
76JA6046	R. A. Bell, M. H. Chisholm and G. G. Christoph; *J. Am. Chem. Soc.*, 1976, **98**, 6046.	822
76JA6715	S. Danishefsky, T. Kitahara, R. McKee and P. F. Schuda; *J. Am. Chem. Soc.*, 1976, **98**, 6715.	305
76JA6750	E. C. Taylor, C.-S. Chiang, A. McKillop and J. F. White; *J. Am. Chem. Soc.*, 1976, **98**, 6750.	194
76JA6752	P. Warner and S.-L. Lu; *J. Am. Chem. Soc.*, 1976, **98**, 6752.	57
76JA7424	H. Sakurai, Y. Kamiyama and Y. Nakadaira; *J. Am. Chem. Soc.*, 1976, **98**, 7424.	612, 614
76JA7440	Y. Ueno, A. Nakayama and M. Okawara; *J. Am. Chem. Soc.*, 1976, **98**, 7440.	850
76JA7746	T. J. Barton and J. A. Kilgour; *J. Am. Chem. Soc.*, 1976, **98**, 7746.	614
76JA8072	G. P. Pez; *J. Am. Chem. Soc.*, 1976, **98**, 8072.	671
76JA8289	V. W. Day, S. S. Abdel-Meguid, S. Dabestani, M. G. Thomas, W. R. Pretzer and E. L. Muetterties; *J. Am. Chem. Soc.*, 1976, **98**, 8289.	692
76JA8412	W. Priester, R. West and T. Ling Chwang; *J. Am. Chem. Soc.*, 1976, **98**, 8412.	605
76JCS(D)377	P. J. Harris, J. A. K. Howard, S. A. R. Knox, R. P. Phillips, F. G. A. Stone and P. Woodward; *J. Chem. Soc., Dalton Trans.*, 1976, 377.	681
76JCS(D)694	C. J. Attridge, M. G. Barlow, W. I. Bevan, D. Cooper, G. W. Cross, R. N. Haszeldine, J. Middleton, M. J. Newlands and A. E. Tipping; *J. Chem. Soc., Dalton Trans.*, 1976, 694.	136
76JCS(D)1113	G. Bulloch and R. Keat; *J. Chem. Soc., Dalton Trans.*, 1976, 1113.	550, 563
76JCS(D)2204	R. A. Andersen, E. Carmona-Guzman, J. F. Gibson and G. Wilkinson; *J. Chem. Soc., Dalton Trans.*, 1976, 2204.	656, 677
76JCS(P1)54	M. S. Baird; *J. Chem. Soc., Perkin Trans. 1*, 1976, 54.	29
76JCS(P1)323	I. Degani and R. Fochi; *J. Chem. Soc., Perkin Trans. 1*, 1976, 323.	1073
76JCS(P1)416	J. A. Miller and M. J. Nunn; *J. Chem. Soc., Perkin Trans. 1*, 1976, 416.	28
76JCS(P1)513	R. N. Haszeldine, C. R. Pool, A. E. Tipping and R. O'B. Watts; *J. Chem. Soc., Perkin Trans. 1*, 1976, 513.	137
76JCS(P1)783	T. Sato, K. Yamamoto, K. Fukui, K. Saito, K. Hayakawa and S. Yoshiie; *J. Chem. Soc., Perkin Trans. 1*, 1976, 783.	892
76JCS(P1)1004	J. Clark, B. Parvizi and R. Colman; *J. Chem. Soc., Perkin Trans. 1*, 1976, 1004.	978
76JCS(P1)1241	G. Jones and J. R. Phipps; *J. Chem. Soc., Perkin Trans. 1*, 1976, 1241.	892
76JCS(P1)1706	T. Takeshima, N. Fukuda, T. Ishii and M. Muraoka; *J. Chem. Soc., Perkin Trans. 1*, 1976, 1706.	852
76JCS(P1)2349	R. N. Haszeldine, I. Mir and A. E. Topping; *J. Chem. Soc., Perkin Trans. 1*, 1976, 2349.	34
76JCS(P2)996	M. Cinquini, S. Colonna, D. Landini and A. M. Maia; *J. Chem. Soc., Perkin Trans 2*, 1976, 996.	233
76JCS(P2)1564	I. J. Ferguson, A. R. Katritzky and R. Patel; *J. Chem. Soc., Perkin Trans. 2*, 1976, 1564.	409
76JFC(7)1	G. N. Bockerman and R. W. Parry; *J. Fluorine Chem.*, 1976, **7**, 1.	341
76JFC(7)95	K. J. Klabunde; *J. Fluorine Chem.*, 1976, **7**, 95.	152
76JFC(7)153	W. K. Glanville, K. W. Morse and J. G. Morse; *J. Fluorine Chem.*, 1976, **7**, 153.	114
76JFC(7)471	K. Burger, S. Tremmel and H. Schickaneder; *J. Fluorine Chem.*, 1976, **7**, 471.	312, 325
76JFC(7)569	G. P. Gambaretto and M. Napoli; *J. Fluorine Chem.*, 1976, **7**, 569.	31
76JFC(8)305	M. Zupan; *J. Fluorine Chem.*, 1976, **8**, 305.	61
76JGU2076	V. M. D'yakov, Yu. A. Lukina and M. G. Voronkov; *J. Gen. Chem. USSR* (*Engl. Transl.*), 1976, **46**, 2076.	140
76JGU2367	M. A. Kazankova, O. G. Smirnova, I. G. Trostyanskaya and I. F. Lutsenko; *J. Gen. Chem. USSR* (*Engl. Transl.*), 1976, **46**, 2367.	820
76JINC55	G. N. Bockerman and R. W. Parry; *J. Inorg. Nucl. Chem.*, 1976, Supplement, 55.	115
76JOC28	D. J. Scharf; *J. Org. Chem.*, 1976, **41**, 28.	493
76JOC556	B. A. Clement and R. L. Soulen; *J. Org. Chem.*, 1976, **41**, 556.	740
76JOC1303	J. A. Zoltewicz, L. S. Helmick and J. K. O'Halloran; *J. Org. Chem.*, 1976, **41**, 1303.	413
76JOC1668	R. K. Singh and S. Danishefsky; *J. Org. Chem.*, 1976, **41**, 1668.	204
76JOC1883	F. G. Bordwell, M. Van der Puy and N. R. Vanier; *J. Org. Chem.*, 1976, **41**, 1883.	391
76JOC2112	D. B. Miller, P. W. Flanagan and H. Shechter; *J. Org. Chem.*, 1976, **41**, 2112.	796
76JOC2312	J. E. Baldwin, O. W. Lever, Jr. and N. R. Tzodikov; *J. Org. Chem.*, 1976, **41**, 2312.	948
76JOC2874	J. E. Baldwin, O. W. Lever, Jr. and N. R. Tzodikov; *J. Org. Chem.*, 1976, **41**, 2874.	948
76JOC3050	R. J. Boatman and H. W. Whitlock; *J. Org. Chem.*, 1976, **41**, 3050.	421
76JOC3063	W. C. Still; *J. Org. Chem.*, 1976, **41**, 3063.	355
76JOC3221	A. T. Nielsen, D. W. Moore, R. L. Atkins, D. Mallory, J. Dipol and J. M. LaBerge; *J. Org. Chem.*, 1976, **41**, 3221.	444
76JOC3765	J. S. Walia and A. S. Walia; *J. Org. Chem.*, 1976, **41**, 3765.	195

76JOC3975	F. A. Carey, O. D. Dailey, Jr., O. Hernandez and J. R. Tucker; *J. Org. Chem.*, 1976, **41**, 3975.	260
76JOC3979	F. A. Carey, O. D. Dailey, Jr. and O. Hernandez; *J. Org. Chem.*, 1976, **41**, 3979.	263
76JOM(104)145	D. Seyferth and D. Dagani; *J. Organomet. Chem.*, 1976, **104**, 145.	820
76JOM(107)73	R. L. Keiter and E. W. Abel; *J. Organomet. Chem.*, 1976, **107**, 73.	588
76JOM(107)229	E. M. Dexheimer and L. Spialter; *J. Organometal. Chem.*, 1976, **107**, 229.	930
76JOM(108)1	E. J. Stampf and J. D. Odom; *J. Organomet. Chem.*, 1976, **108**, 1.	813
76JOM(108)213	G. Dettlaf and E. Weiss; *J. Organomet. Chem.*, 1976, **108**, 213.	681
76JOM(110)25	D. S. Matteson and P. K. Jesthi; *J. Organomet. Chem.*, 1976, **110**, 25.	633
76JOM(111)C33	E. J. Bulten and H. A. Budding; *J. Organomet. Chem.*, 1976, **111**, C33.	701
76JOM(111)361	P. Bravo, G. Fronza and C. Ticozzi; *J. Organomet. Chem.*, 1976, **111**, 361.	584
76JOM(113)C13	M. Lequan, F. Meganem and Y. Besace; *J. Organomet. Chem.*, 1976, **113**, C13.	534
76JOM(113)115	Y. Sato, Y. Ban and H. Shirai; *J. Organomet. Chem.*, 1976, **113**, 115.	512
76JOM(114)1	D. S. Matteson and R. K. Jesthi; *J. Organomet. Chem.*, 1976, **114**, 1.	665
76JOM(114)281	N. Petragnani, R. Rodrigues and J. V. Comasseto; *J. Organomet. Chem.*, 1976, **114**, 281.	88, 90, 348
76JOM(117)159	C. S. Kraihanzel and P. K. Marples; *J. Organomet. Chem.*, 1976, **117**, 159.	545
76JOM(117)329	E. J. Bulten, H. F. M. Gruter and H. F. Martens; *J. Organomet. Chem.*, 1976, **117**, 329.	699
76JOM(118)1	T. Aoyama, Y. Sato and H. Shirai; *J. Organomet. Chem.*, 1976, **118**, 1.	509
76JOM(120)C13	P. A. Elder, D. J. S. Guthrie, J. A. D. Jeffreys, G. R. Knox, J. Kollmeier, P. L. Pauson, D. A. Symon and W. E. Watts; *J. Organomet. Chem.*, 1976, **120**, C13.	688
76JOM(121)211	W. Beck, H. Brix and F. H. Kohler; *J. Organomet. Chem.*, 1976, **121**, 211.	1019
76JOM(122)31	A. G. Brook, J. M. Duff and G. E. Legrow; *J. Organomet. Chem.*, 1976, **122**, 31.	137, 138, 141, 811
76JOM(122)113	Y. Oosawa, H. Urabe, T. Saito and Y. Sasaki; *J. Organomet. Chem.*, 1976, **122**, 113.	552
76JOM(122)145	M. W. Rathke, E. Chao and G. Wu; *J. Organomet. Chem.*, 1976, **122**, 145.	526
76JOU782	N. G. Pavlenko, V. V. Matsnev, L. S. Kuz'menko, P. P. Kornuta, L. N. Markovskii and V. P. Kukhar; *J. Org. Chem. USSR (Engl. Transl.)*, 1976, **12**, 782.	800
76JOU2252	B. S. Drach and V. A. Kovalev; *J. Org. Chem. USSR (Engl. Transl.)*, 1976, **12**, 2252.	794
76JPR116	Von. H. Gross, B. Costisella, Th. Gnauk and L. Brennecke; *J. Prakt. Chem.*, 1976, **318**, 116.	1003
76JPR403	M. Oswiecimska, B. Costisella, I. Keitel and H. Gross; *J. Prakt. Chem.*, 1976, **318**, 403.	559
76LA36	J. Albanbauer, K. Burger, E. Burgis, D. Marquarding, L. Schabl and I. Ugi; *Justus Liebigs Ann. Chem.*, 1976, 36.	493
76LA1762	H. Meyer, F. Bossert and H. Horstmann; *Justus Liebigs Ann. Chem.*, 1976, 1762.	880
B-76MI 403-01	H. Bohme and H. G. Viehe (eds); "Advances in Organic Chemistry, Part I," Wiley, New York, 1976, vol. 9.	966
B-76MI 410-01	D. Redmore; in "Topics in Phosphorus Chemistry," eds M. Grayson and E. J. Griffith, Wiley, New York, 1976, vol. 8, p. 515.	451
76MI 410-02	A. Petrov, V. A. Chauzov and T. S. Erokhina; *Khim. Elementoorg. Soedin.*, 1976, 200 (*Chem. Abstr.*, 1977, **86**, 5543q).	471
76MI 410-03	K. A. Petrov, V. A. Chauzov, I. V. Pastukhova and N. N. Bogdanov; *Khim. Elementoorg. Soedin.*, 1976, 209 (*Chem. Abstr.*, 1977, **86**, 5545s).	493
76MI 412-01	G. N. Bockerman and R. W. Parry; *Inorg. Nucl. Chem.–Herbert H. Hyman Mem Vol.*, 1976, 55 (*Chem. Abstr.*, 1976, **85**, 160 263).	549
76MI 414-01	G. S. Zaitseva, T. A. Manukina, O. M. Khitrova, Y. I. Baukov and I. F. Lutsenko; *Khim. Elementoorg. Soedin.*, 1976, 76 (*Chem. Abstr.*, 1976, **85**, 177 574w).	637
B-76MI 418-01	L. Ghosez and J. Marchand-Brynaert; in "Iminium Salts in Organic Chemistry," ed. H. Böhme and H. G. Viehe, Wiley, New York, 1976, part 1, p. 421.	790, 791
B-76MI 420-01	L. Ghosez and J. Marchand-Brynaert; in "Advances in Organic Chemistry, Methods and Results. Part 1, Imminium Salts in Organic Chemistry," ed. H. Bohme and H. G. Viehe, Wiley, New York, 1976, p. 421.	891
B-76MI 420-02	L. Ghosez and J. Marchand-Brynaert; in "Advances in Organic Chemistry, Methods and Results. Part 1 Imminium Salts in Organic Chemistry," ed. H. Bohme and H. G. Viehe, Wiley, New York, 1976, p. 474.	905
76MI 421-01	L. Ghosez and J. Marchand-Brynaert; *Adv. Org. Chem.*, 1976, **9**, 421.	971, 990
76MI 421-02	J. G. D. Schulz and A. Onopchenko; *Ind. Eng. Chem. Prod. Res. Dev.*, 1976, **15**, 152.	993
76OR(26)1	H. W. Gschwend and H. R. Rodriguez; *Org. React.*, 1976, **26**, 1.	948
76OS(55)20	G. J. Fox, G. Hallas, J. D. Hepworth and K. N. Paskins; *Org. Synth.*, 1976, **55**, 20.	24
76RCR1052	K. V. Altukhov and V. V. Perekalin; *Russ. Chem. Rev. (Engl. Transl.)*, 1976, **45**, 1052.	428
76RTC202	G. E. Veenstra and B. Zwanenburg; *Recl. Trav. Chim. Pays-Bas*, 1976, **95**, 202.	859
76RZC477	J. Rachon and C. Wasielewski; *Rocz. Chem.*, 1976, **50**, 477.	460
76RZC661	M. Soroka and P. Mastalerz; *Rocz. Chem.*, 1976, **50**, 661.	460
76S107	P. Coutrot, C. Laurenco, J. Petrova and P. Savignac; *Synthesis*, 1976, 107.	347, 773, 778
76S197	P. Savignac and P. Coutrot; *Synthesis*, 1976, 197.	745
76S201	M. Kimura and Y. Ban; *Synthesis*, 1976, 201.	414
76S313	R. J. Kricks and A. A. Volpe; *Synthesis*, 1976, 313.	29
76S404	G. Ferdinand and K. Schank; *Synthesis*, 1976, 404.	838
76S406	G. Ferdinand and K. Schank; *Synthesis*, 1976, 406.	838
76S408	G. Ferdinand and K. Schank; *Synthesis*, 1976, 408.	838
76S413	C. R. Johnson and K. Tanaka; *Synthesis*, 1976, 413.	392
76S489	M. Narita and C. U. Pittman; *Synthesis*, 1976, 489.	851
76S552	H. Yoshida, T. Ogata and S. Inokawa; *Synthesis*, 1976, 552.	847
76S697	M. Hojo, R. Masuda, T. Saeki and S. Uyeda; *Synthesis*, 1976, 697.	78
76S748	G. F. Bettinetti, S. Maffei and S. Pietra; *Synthesis*, 1976, 748.	416
76S766	F. Mathey and P. Savignac; *Synthesis*, 1976, 766.	585

76S797	S. Cabiddu, A. Maccioni and M. Secci; *Synthesis*, 1976, 797.	231
76SA(A)1459	J. W. Macklin; *Spectrochim. Acta, Part A*, 1976, **32**, 1459.	697
76T2089	C. Laurenco and R. Burgada; *Tetrahedron*, 1976, 2089.	478
76T2395	F. Mathey; *Tetrahedron*, 1976, **32**, 2395.	1026
76T2507	D. A. Konen, P. E. Pfeffer and L. S. Silbert; *Tetrahedron*, 1976, **32**, 2507.	842
76TL7	E. J. Corey, D. Enders and M. G. Bock; *Tetrahedron Lett.*, 1976, **17**, 7.	649
76TL65	K. Yamada, K. Kato, H. Nagase and Y. Hirata; *Tetrahedron Lett.*, 1976, **17**, 65.	222
76TL95	E. V. Dehmlow and J. Schmidt; *Tetrahedron Lett.*, 1976, **17**, 95.	190
76TL319	T. H. Chan and B. S. Ong; *Tetrahedron Lett.*, 1976, **17**, 319.	216
76TL423	E. M. Engler and V. V. Patel; *Tetrahedron Lett.*, 1976, **17**, 423.	869
76TL609	I. Hoppe, D. Hoppe and U. Schöllkopf; *Tetrahedron Lett.*, 1976, **17**, 609.	538
76TL613	M. Hojo and R. Masuda; *Tetrahedron Lett.*, 1976, **17**, 613.	74, 75
76TL809	E. J. Corey, J.-L. Gras and P. Ulrich; *Tetrahedron Lett.*, 1976, **17**, 809.	199
76TL895	G. A. Wheaton and D. J. Burton; *Tetrahedron Lett.*, 1976, **17**, 895.	735
76TL943	J. Pfab; *Tetrahedron Lett.*, 1976, **17**, 943.	38, 40
76TL1199	Y. Maki, M. Suzuki and K. Ozeki; *Tetrahedron Lett.*, 1976, **17**, 1199.	413
76TL1553	W. T. Brady and R. A. Owens; *Tetrahedron Lett.*, 1976, **17**, 1553.	807
76TL1725	C. G. Kruse, N. L. J. M. Broekhof and A. Van der Gen; *Tetrahedron Lett.*, 1976, **17**, 1725.	210
76TL2201	A. B. Levy and S. J. Schwartz; *Tetrahedron Lett.*, 1976, **17**, 2201.	934, 949
76TL2405	M. T. Wu, D. Taub and A. A. Patchett; *Tetrahedron Lett.*, 1976, **17**, 2405.	979
76TL2599	G. Märkl and D. Matthes; *Tetrahedron Lett.*, 1976, **17**, 2599.	1038
76TL2731	M. Mikolajczyk, S. Grzejszczak and A. Zatorski; *Tetrahedron Lett.*, 1976, **17**, 2731.	847, 848
76TL2783	K. Oka and S. Hara; *Tetrahedron Lett.*, 1976, **17**, 2783.	71
76TL3361	J. H. Clark, H. L. Holland and J. M. Miller; *Tetrahedron Lett.*, 1976, **17**, 3361.	190
76TL3381	R. A. Olofson, K. D. Lotts and G. N. Barber; *Tetrahedron Lett.*, 1976, **17**, 3381.	60
76TL3739	P. Bentley, G. Brooks and I. Zomaya; *Tetrahedron Lett.*, 1976, **17**, 3739.	205
76TL4041	K. Sachdev; *Tetrahedron Lett.*, 1976, **17**, 4041.	649
76TL4155	F. Mathey; *Tetrahedron Lett.*, 1976, **17**, 4155.	1037
76TL4223	K. Sachdev and H. S. Sachdev; *Tetrahedron Lett.*, 1976, **17**, 4223.	375
76TL4577	E. J. Corey, E. J. Trybulski and J. W. Suggs; *Tetrahedron Lett.*, 1976, **17**, 4577.	20
76TL4775	J.-L. Burgot, J. Masson and J. Vialle; *Tetrahedron Lett.*, 1976, **17**, 4775.	255
76UKZ204	M. M. Kul'chitskii, Y. L. Yagupol'skii and L. M. Yagupol'skii; *Ukr. Khim. Zh.* (*Russ. Ed.*), 1976, **42**, 204 (Engl. Transl. 94).	992
76URP539892	Z. S. Novikova, A. A. Prishchenko and I. F. Lutsenko; *USSR Pat.* 539 892 (1976) (*Chem. Abstr.*, 1977, **86**, 171 596).	546
76USP3932526	R. J. Koshar (Minnesota Mining and Mfg. Co.); *US Pat.* 3 932 526 (1976) (*Chem. Abstr.* 1976, **84**, 106 511).	864
76USP3933914	R. F. Coles and I. H. Skoog (Minnesota Mining and Mfg. Co.); *US Pat.* 3 933 914 (1976) (*Chem. Abstr.* 1976, **84**, 123 412).	864
76USP3984357	R. J. Koshar (Minnesota Mining and Mfg. Co.); *US Pat.* 3 984 357 (1976) (*Chem. Abstr.* 1977, **86**, 30 311).	864
76USP3987191	D. J. Peterson and J. F. Ward; *US Pat.* 3 987 191 (1976) (*Chem. Abstr.*, 1976, **86**, 72 885r).	397
76USP4075001	L. K. Gibbons, *US Pat.* 4 075 001 (1973).	977
76ZAAC(419)157	H. Buerger and K. Wiegel; *Z. Anorg. Allg. Chem.*, 1976, **419**, 157.	607, 621
76ZAAC(421)1	E. Fluck, H. Bayha and G. Heckmann; *Z. Anorg. Allg. Chem.*, 1976, **421**, 1.	574
76ZAAC(422)237	H.-P. Abicht and K. Issleib; *Z. Anorg. Allg. Chem.*, 1976, **422**, 237.	566, 575
76ZAAC(423)112	H. Bürger and C. Kluess; *Z. Anorg. Allg. Chem.*, 1976, **423**, 112.	719
76ZAAC(424)97	K. Issleib, H. Winkelmann and H.-P. Abicht; *Z. Anorg. Allg. Chem.*, 1976, **424**, 97.	454, 481
76ZAAC(426)28	G. Fritz and E. Matern; *Z. Anorg. Allg. Chem.*, 1976, **426**, 28.	621
76ZC64	A. Tzschach, J. Heinicke and W. Gerlich; *Z. Chem.*, 1976, **16**, 64.	495, 496
76ZC65	G. Sonnek and H. Reinheckel; *Z. Chem.*, 1976, **16**, 65.	659
76ZC116	B. Machelett; *Z. Chem.*, 1976, **16**, 116.	685
76ZN(B)153	W. Althoff, M. Fild and H.-P. Rieck; *Z. Naturforsch., Teil B*, 1976, **31**, 153.	282
76ZN(B)721	M. S. Hussain and H. Schmidbaur; *Z. Naturforsch., Teil B*, 1976, **31**, 721.	552, 558
76ZN(B)790	M. Höfler, H. Hausmann and W. Saal; *Z. Naturforsch., Teil B*, 1976, **31**, 790.	545
76ZOB266	R. Z. Aliev and V. K. Khairullin; *Zh. Obshch. Khim.*, 1976, **46**, 266.	556
76ZOB565	I. F. Lutsenko, M. V. Proskurnina and A. L. Checkhun; *Zh. Obshch. Khim.*, 1974, **46**, 565.	913
76ZOB568	I. F. Lutsenko, M. V. Proskurnina and A. L. Chekhun; *Zh. Obshch. Khim.*, 1976, **46**, 568.	552
76ZOB571	Z. S. Novikova, S. Ya. Skorobogatova, A. A. Prishchenko and I. F. Lutsenko; *Zh. Obshch. Khim.*, 1976, **46**, 571.	547
76ZOB575	Z. S. Novikova, S. N. Zdorova, V. N. Kirzner and I. F. Lutsenko; *Zh. Obshch. Khim.*, 1976, **46**, 575.	566
76ZOB594	S. A. Lebedev, L. L. Gervits, S. V. Ponomarev and I. F. Lutsenko; *Zh. Obshch. Khim.*, 1976, **46**, 594 (*Chem. Abstr.*, 1976, **85**, 5813g).	640
76ZOB783	V. K. Byistro, L. A. Krichevskii and Z. M. Muldakhmetov; *Zh. Obshch. Khim.*, 1976, **46**, 783.	343
76ZOB837	T. K. Gar, A. A. Buyakov, A. I. Gusev, M. G. Los, A. V. Kisin and V. F. Mironov; *Zh. Obshch. Khim.*, 1976, **46**, 837 (*Chem. Abstr.*, 1976, **85**, 46 825f).	640
76ZOB1012	V. D. Pak, N. S. Kozlov, I. A. Balykova and G. A. Gartman; *Zh. Obshch. Khim.*, 1976, **46**, 1012 (*Chem. Abstr.*, 1977, **85**, 108 711b).	462
76ZOB1043	V. F. Mironov, V. I. Shiryaev, E. M. Stepina and A. I. Nechaeva; *Zh. Obshch. Khim.*, 1976, **46**, 1043 (*Chem. Abstr.*, 1976, **85**, 108 733k).	660
76ZOB1218	S. N. Zdorova, Z. S. Novikova and I. F. Lutsenko; *Zh. Obshch. Khim.*, 1976, **46**, 1218.	566, 568, 576

76ZOB1246	K. A. Petrov, V. A. Chauzov, I. V. Pastukhova and N. N. Bogdanov; *Zh. Obshch. Khim.*, 1976, **46**, 1246 (*Chem. Abstr.*, 1976, **85**, 94 451c).	478, 493
76ZOB1250	K. A. Petrov, V. A. Chauzov and N. N. Bogdanov; *Zh. Obshch. Khim.*, 1976, **46**, 1250 (*Chem. Abstr.*, 1976, **85**, 143 195).	490
76ZOB1495	K. A. Petrov, V. A. Chauzov and N. N. Bogdanov; *Zh. Obshch. Khim.*, 1976, **46**, 1495 (*Chem. Abstr.*, 1976, **85**, 160 256).	490
76ZOB1652	V. M. Ismailov, T. A. Babaeva and Sh. T. Akhmedov; *Zh. Obshch. Khim.*, 1976, **46**, 1652.	913
76ZOB2385	A. N. Pudovick and G. E. Vershinina; *Zh. Obshch. Khim.*, 1976, **6**, 2385.	913
76ZOR231	V. I. Gorbatenko, N. V. Mel'nichenko, M. N. Gertsyuk and L. I. Samarai; *Zh. Org. Khim.*, 1976, **12**, 231.	441
76ZOR2103	V. I. Gorbatenko, N. V. Mel'nichenko, M. N. Gertsyuk and L. I. Samarai; *Zh. Org. Khim.*, 1976, **12**, 2103	441
77AG(E)1	H. Werner; *Angew. Chem., Int. Ed. Engl.*, 1977, **16**, 1.	694
77AG(E)183	H. Hoberg, R. Krause-Göing, C. Krüger and J. C. Sekutowski; *Angew. Chem., Int. Ed. Engl.*, 1977, **16**, 183.	693
77AG(E)339	U. Schöllkopf; *Angew. Chem., Int. Ed. Engl.*, 1977, **16**, 339.	537
77AG(E)349	H. J. Bestmann; *Angew. Chem., Int. Ed. Engl.*, 1977, **16**, 349.	825, 834
77AG(E)540	W. Dumont and A. Krief; *Angew. Chem., Int. Ed. Engl.*, 1977, **16**, 540.	239, 288
77AG(E)541	W. Dumont, M. Sevrin and A. Krief; *Angew. Chem., Int. Ed. Engl.*, 1977, **16**, 541.	88, 91, 288, 400
77AG(E)710	T. Kauffmann, H. Ahlers, H. J. Tilhard and A. Woltermann; *Angew. Chem., Int. Ed. Engl.*, 1977, **16**, 710.	401
77AG(E)718	I. Ruppert and V. Bastian; *Angew. Chem., Int. Ed. Engl.*, 1977, **16**, 718.	564
77AOC(15)235	J. P. Oliver; *Adv. Organomet. Chem.*, 1977, **15**, 235.	698
77AOC(16)33	M. T. Reetz; *Adv. Organomet. Chem.*, 1977, **16**, 33.	614
77AP(310)30	H. Böhme and K. H. Weisel; *Arch. Pharm.* (*Weinheim, Ger.*), 1977, **310**, 30.	796
77BCJ1353	K. Kobayashi, K. Kunô, M. Kawanisi and S. Kozima; *Bull. Chem. Soc. Jpn.*, 1977, **50**, 1353.	156
77BCJ1588	H. Taguchi, H. Yamamoto and H. Nozaku; *Bull. Chem. Soc. Jpn.*, 1977, **50**, 1588.	19
77BCJ2250	H. Ueda, Y. Kai, N. Yasuoka and N. Kasai; *Bull. Chem. Soc. Jpn.*, 1977, **50**, 2250.	686
77BSF693	A. Alexakis, G. Cahiez, J. F. Normant and J. Villieras; *Bull. Chim. Soc. Fr.*, 1977, 693.	963
77CA121332u	W. D. Kollmeyer (Shell Oil Co); *Chem. Abstr.*, 1977, **86**, 121 332u (*US Pat.*, 1976, 3 996 372).	907
77CA135599v	V. M. Ismailov, V. V. Moskova, T. A. Babaeva, Sh. T. Akmedov and A. I. Razumov; *Chem. Abstr.*, 1977, **87**, 135 599v (*Azerb. Khim. Zh.*, 1976, **4**, 56).	884, 913
77CAR(54)C17	T. Ogawa and M. Matsui; *Carbohydr. Res.*, 1977, **54**, C17.	225
77CAR(58)397	R. J. Ferrier, R. H. Furneaux and P. C. Tyler; *Carbohydr. Res.*, 1977, **58**, 397.	235
77CAR(59)351	K. Blumberg, A. Fuccello and T. van Es; *Carbohydr. Res.*, 1977, **59**, 351.	238, 287
77CB651	T. Kauffmann and R. Eidenschink; *Chem. Ber.*, 1977, **110**, 651.	421
77CB677	H. Schmidbaur, J. Eberlin and W. Richter; *Chem. Ber.*, 1977, **110**, 677.	574
77CB799	F. R. Kreissl and W. Held; *Chem. Ber.*, 1977, **110**, 799.	580
77CB852	B.-T. Gröbel and D. Seebach; *Chem. Ber.*, 1977, **110**, 852.	620, 937, 962
77CB867	B.-T. Gröbel and D. Seebach; *Chem. Ber.*, 1977, **110**, 867.	964
77CB948	H. Poisel; *Chem. Ber.*, 1977, **110**, 948.	304
77CB1312	W. Richter, Y. Yamamoto and H. Schmidbaur; *Chem. Ber.*, 1977, **110**, 1312.	598
77CB1484	A. Roedig, G. Zaby and W. Scharf; *Chem. Ber.*, 1977, **110**, 1484.	851
77CB1748	H. Schmidbaur, A. Wohlleben, F. Wagner, O. Orama and G. Huttner; *Chem. Ber.*, 1977, **110**, 1748.	563
77CB1763	H. Werner, A. Kühn, D. J. Tune, C. Krüger, D. J. Brauer, J. C. Sekutowski and Y.-H. Tsay; *Chem. Ber.*, 1977, **110**, 1763.	693
77CB1833	K.-H. Geiss, D. Seebachand and B. Seuring; *Chem. Ber.*, 1977, **110**, 1833.	367, 392
77CB2200	H. H. Karsch, H.-F. Klein and H. Schmidbaur; *Chem. Ber.*, 1977, **110**, 2200.	582
77CB2382	R. Appel and H. Schöler; *Chem. Ber.*, 1977, **110**, 2382.	112
77CB2659	T. Kauffmann, H. Berg, E. Köppelmann and D. Kuhlmann; *Chem. Ber.*, 1977, **110**, 2659.	535
77CB2880	R. Schlecker, U. Henkel and D. Seebach; *Chem. Ber.*, 1977, **110**, 2880.	943
77CB3034	T. Kauffmann, D. Berger, B. Scheerer and A. Wolterman; *Chem. Ber.*, 1977, **110**, 3034.	536
77CB3235	K. Schank and H. G. Schmitt; *Chem. Ber.*, 1977, **110**, 3235.	236, 838
77CB3501	H. Schmidbaur, O. Gasser and M. S. Hussain; *Chem. Ber.*, 1977, **110**, 3501.	553
77CB3517	H. Schmidbaur, O. Gasser, C. Krüger and J. C. Sekutowski; *Chem. Ber.*, 1977, **110**, 3517.	583, 584, 585, 586, 588
77CB3528	H. Schmidbaur and H.-J. Füller; *Chem. Ber.*, 1977, **110**, 3528.	577, 588
77CB3536	H. Schmidbaur, H.-J. Füller, V. Bejenke, A. Franck and G. Huttner; *Chem. Ber.*, 1977, **110**, 3536.	583, 585, 586, 587
77CB3930	T. Kauffmann and R. Joussen; *Chem. Ber.*, 1977, **110**, 3930.	585
77CC308	J. C. Wilburn and R. H. Neilson; *J. Chem. Soc., Chem. Commun.*, 1977, 308.	571
77CC314	C. Earnshaw, C. J. Wallis and S. Warren; *J. Chem. Soc., Chem. Commun.*, 1977, 314.	343
77CC505	P. Shu, A. N. Bloch, T. F. Carruthers and D. O. Cowan; *J. Chem. Soc., Chem. Commun.*, 1977, 505.	869
77CC513	F. Cooke and P. D. Magnus; *J. Chem. Soc., Chem. Commun.*, 1977, 513.	355
77CC522	B. Harirchian and P. Magnus; *J. Chem. Soc., Chem. Commun.*, 1977, 522.	853, 871, 873, 936, 938, 961, 962
77CC772	D. Ayalon-Chass, E. Ehlinger and P. Magnus; *J. Chem. Soc., Chem. Commun.*, 1977, 772.	648
77CC801	R. H. Everhardus, H. G. Eeuwhorst and L. Brandsma; *J. Chem. Soc., Chem. Commun.*, 1977, 801.	957

77CC835	E. M. Engler, V. V. Patel and R. R. Schumaker; *J. Chem. Soc., Chem. Commun.*, 1977, 835.	869, 874
77CCC2537	V. Tolman; *Collect. Czech. Chem. Commun.*, 1977, **42**, 2537.	3, 32
77CI(L)127	R. Louw and P. W. Franken; *Chem. Ind. (London)*, 1977, 127.	47, 48
77CJC421	J. S. Grossert, W. R. Hardstaff and R. F. Langler; *Can. J. Chem.*, 1977, **55**, 421.	74, 75, 78
77CJC2323	J. F. King, R. P. Beatson and J. M. Buchshriber; *Can. J. Chem.*, 1977, **55**, 2323.	78
77CJC2589	J. M. Duff and A. G. Brook; *Can. J. Chem.*, 1977, 2589.	506
77CJC3031	R. K. Marat and A. F. Janzen; *Can. J. Chem.*, 1977, **55**, 3031.	61, 62, 63
77CJC3190	P. Carlier Y. Gelas-Mialhe and R. Vessière; *Can. J. Chem.*, 1977, **55**, 3190.	779
77CL287	J. Nakayama, M. Ishihara and M. Hoshino; *Chem. Lett.*, 1977, 287.	850
77CL485	M. Sekine, I. Yamamoto, A. Hashizume and T. Hata; *Chem. Lett.*, 1977, 485.	339
77CL1379	K. Tanaka, T. Nakai and N. Ishikawa; *Chem. Lett.*, 1977, 1379.	949
77CPB2964	Y. Terao, K. Matsunaga and M. Sekiya; *Chem. Pharm. Bull.*, 1977, **25**, 2964.	317
77CPB3385	K. Ito, R. Komaki and M. Sekiya; *Chem. Pharm. Bull.*, 1977, **25**, 3385.	317
77CR(C)281	D. Riobe, A. Lebonc and J. Delaunay; *C. R. Hebd. Seances Acad. Sci. Ser. C*, 1977, 281.	950
77CRV691	A. Efraty; *Chem. Rev.*, 1977, **77**, 691.	706
77CZ259	M. Fild, J. Heinze and W. Krüger; *Chem.-Ztg.*, 1977, **101**, 259.	550
77DIS(B)1214	R. A. Owens; *Diss. Abstr. Int. B.*, 1977, **38**, 1214.	807
77DOK(236)371	L. L. Troitskaya, V, I. Sokolov and O. A. Reutov; *Dokl. Akad. Nauk SSSR*, 1977, **236**, 371.	721
77FCR119	D. J. Burton and J. L. Hahnfeld; *Fluorine Chem. Rev.*, 1977, **8**, 119.	7, 32, 35
77G555	M. H. Elnagdi, E. M. Zayed, S. M. Fahmy, M. A. E. Khalifa and S. Amer; *Gazz. Chim. Ital.*, 1977, **107**, 555.	984
77HCA1061	M. Neuenschwander and R. Iseli; *Helv. Chim. Acta*, 1977, **60**, 1061.	53
77HCA1781	M. K. Huber, R. Martin, M. Rey and A. S. Dreiding; *Helv. Chim. Acta*, 1977, **60**, 1781.	979
77IC172	R. J. Restivo, G. Ferguson, T. W. Ng and A. J. Carty; *Inorg. Chem.*, 1977, **16**, 172.	692
77IC636	C. G. Pierpont; *Inorg. Chem.*, 1977, **16**, 636.	682
77IC1770	S. O. Grim and J. D. Mitchell; *Inorg. Chem.*, 1977, **16**, 1770.	546, 550, 551
77IC3187	R. A. Grey and L. R. Anderson; *Inorg. Chem.*, 1977, **16**, 3187.	582
77IC3193	S. S. Crawford and H D. Kaesz; *Inorg. Chem.*, 1977, **16**, 3193.	720
77ICA85	D. S. Ginley, C. R. Bock and M. S. Wrighton; *Inorg. Chim. Acta*, 1977, **23**, 85.	674
77IJ265	K. Tamao, J. Yoshida, S. Okazaki and M. Kumada; *Isr. J. Chem.*, 1977, **15**, 265.	654
77IJ149	C. Krüger, J. C. Sekutowski, R. Goddard, H. J. Füller, O. Gasser and H. Schmidbaur; *Israel J. Chem.*, 1977, **15**, 149.	586
77IJC(B)297	S. Rajappa, R. Sreenivasan, B. G. Advani, R. H. Summerville and R. Hoffmann; *Indian J. Chem., Sect. B*, 1977, **15B**, 297.	973
77IZV596	A. Kh. Filippova, G. S. Lyashenko, O. A. Kruglaya, V. V. Keiko, I. D. Kalikhman and N. S. Vyazankin; *Izv. Akad. Nauk SSSR, Ser. Khim.*, 1977, **3**, 596.	940
77IZV2384	O. A. Luk'yanov, N. I. Shlykova, V. P. Gorelik and V. A. Tartakovski; *Izv. Akad. Nauk SSSR, Ser. Khim.*, 1977, **10**, 2384.	429
77IZV2417	A. N. Nesmeyanov, E. G. Perevalova, E. I. Smyslova, V. P. Dyadchenko and K. I. Grandberg; *Izv. Akad. Nauk SSSR, Ser. Khim.*, 1977, 2417.	152
77JA263	H. J. Reich and S. K. Shah; *J. Am. Chem. Soc.*, 1977, **99**, 263.	399
77JA1631	T. M. Harris and J. V. Hay; *J. Am. Chem. Soc.*, 1977, **99**, 1631.	167
77JA2013	P. R. Jones and T. F. O. Lim; *J. Am. Chem. Soc.*, 1977, **99**, 2013.	611
77JA3184	G. Zweifel and S. J. Backlund; *J. Am. Chem. Soc.*, 1977, **99**, 3184.	1056
77JA3196	D. S. Matteson and R. J. Moody; *J. Am. Chem. Soc.*, 1977, **99**, 3196.	634
77JA3273	R. M. Irwin, J. M. Cooke and J. Laane; *J. Am. Chem. Soc.*, 1977, **99**, 3273.	621
77JA3537	F. Mathey, A. Mitschler and R. Weiss; *J. Am. Chem. Soc.*, 1977, **99**, 3537.	1034
77JA3879	H. Sakurai, Y. Kamiyama and Y. Nakadaira; *J. Am. Chem. Soc.*, 1977, **99**, 3879.	1048
77JA4181	I. Kuwajima, M. Arai and T. Sato; *J. Am. Chem. Soc.*, 1977, **99**, 4181.	932
77JA4405	B. M. Trost and G. S. Massiot; *J. Am. Chem. Soc.*, 1977, **99**, 4405.	220
77JA4536	C. S. Burford, F. Cooke, E. Ehlinger and P. D. Magnus; *J. Am. Chem. Soc.*, 1977, **99**, 4536.	355
77JA4707	T. R. Jack, C. J. May and J. Powell; *J. Am. Chem. Soc.*, 1977, **99**, 4707.	692
77JA4835	F. Nakatsubo, A. J. Cocuzza, D. E. Keeley and Y. Kishi; *J. Am. Chem. Soc.*, 1977, **99**, 4835.	216
77JA5009	D. A. Evans, L. K. Truesdale, K. G. Grimm and S. L. Nesbitt; *J. Am. Chem. Soc.*, 1977, **99**, 5009.	216
77JA5192	G. Zweifel, S. J. Backlund and T. Leung; *J. Am. Chem. Soc.*, 1977, **99**, 5192.	814
77JA5213	P. Beak and B. G. McKinnie; *J. Am. Chem. Soc.*, 1977, **99**, 5213.	382
77JA5317	D. Seyferth, G. J. Murphy and B. Mauzé; *J. Am. Chem. Soc.*, 1977, **99**, 5317.	13
77JA5909	E. M. Engler, B. A. Scott, S. Etemad, T. Penney and V. V. Patel; *J. Am. Chem. Soc.*, 1977, **99**, 5909.	869, 874
77JA6097	C. P. Casey and S. W. Polichnowski; *J. Am. Chem. Soc.*, 1977, **99**, 6097.	580
77JA6435	R. N. Lindquist and A. C. Nguyen; *J. Am. Chem. Soc*, 1977, **99**, 6435.	525
77JA8447	P. R. Jones and T. F. O. Lim; *J. Am. Chem. Soc.*, 1977, **99**, 8447.	606, 611
77JAP7973759	H. Sakurai, H. Umino; *Jpn. Pat.* 7 973 759 (1977) (*Chem. Abstr.*, 1979, **91**, 141 000r).	657
77JCED355	F. Freeman and E. M. Karchefski; *J. Chem. Eng. Data*, 1977, **25**, 355.	203
77JCR(S)116	F. Glocking, N. S. Hosmane, V. B. Mahale, J. J. Swindall, L. Magos and T. J. King; *J. Chem. Res. (S)*, 1977, 116.	656
77JCS(D)812	E. R. Sigurdson and G. Wilkinson; *J. Chem. Soc., Dalton Trans.*, 1977, 812.	656
77JCS(D)999	J. A. J. Jarvis, R. Pearce and M. F. Lappert; *J. Chem. Soc., Dalton Trans.*, 1977, 999.	656, 695
77JCS(D)1328	E. G. Bryan, B. F. G. Johnson and J. Lewis; *J. Chem. Soc., Dalton Trans.*, 1977, 1328.	682
77JCS(P1)189	M. J. Pearson; *J. Chem. Soc., Perkin Trans. 1*, 1977, 189.	447
77JCS(P1)282	R. N. Butler and W. B. King; *J. Chem. Soc., Perkin Trans. 1*, 1977, 282.	313
77JCS(P1)372	N. Furukawa, T. Akasaka, T. Aida and S. Oae; *J. Chem. Soc., Perkin Trans. 1*, 1977, 372.	161

Ref	Citation	Pages
77JCS(P1)550	A. H. Davidson, I. Fleming, I. Grayson, A. Pearce, R. L. Snowden and S. Warren; *J. Chem. Soc., Perkin Trans. 1*, 1977, 550.	650
77JCS(P1)684	P. Lindberg, R. Bergman and B. Wickberg; *J. Chem. Soc., Perkin Trans. 1*, 1977, 684.	299
77JCS(P1)1273	M. Muraoka, T. Yamamoto, S. Yamaguchi, F. Tonosaki, T. Takeshima and N. Fukada; *J. Chem. Soc., Perkin Trans. 1*, 1977, 1273.	852
77JCS(P1)1723	D. H. R. Barton, P. E. Hansen and K. Picker; *J. Chem. Soc., Perkin Trans. 1*, 1977, 1723.	841
77JCS(P1)1811	S. A. Okecha and F. Stansfield; *J. Chem. Soc., Perkin Trans. 1*, 1977, 1811.	321
77JCS(P1)1954	P. M. Hardy and D. J. Samworth; *J. Chem. Soc., Perkin Trans. 1*, 1977, 1954.	417, 424
77JCS(P1)2263	J. I. Grayson and S. Warren; *J. Chem. Soc., Perkin Trans. 1*, 1977, 2263.	346, 347
77JCS(P1)2513	P. Rollin and P. Sinay; *J. Chem. Soc., Perkin Trans. 1*, 1977, 2513.	176
77JCS(P2)775	H. Lumbroso, D. M. Bertin, F. Fringuelli and A. Taticchi; *J. Chem. Soc., Perkin Trans. 2*, 1977, 775.	873
77JFC(10)27	R. N. Haszeldine, D. R. Taylor and E. W. White; *J. Fluorine Chem.*, 1977, **10**, 27.	117
77JFC(10)131	M. J. van Hamme and D. J. Burton; *J. Fluorine Chem.*, 1977, **10**, 131.	753
77JGU2501	K. A. Petrov, V. A. Chauzov, T. S. Erokhina and I. V. Pastukhova; *J. Gen. Chem. USSR (Engl. Transl.)*, 1977, **47**, 2501.	481
77JHC337	V. I. Siele; *J. Heterocycl. Chem.*, 1977, **14**, 337.	832
77JOC38	W. Adam and O. Cueto; *J. Org. Chem.*, 1977, **42**, 38.	490
77JOC376	D. I. C. Scopes, A. F. Kluge and J. A. Edwards; *J. Org. Chem.*, 1977, **42**, 376.	161
77JOC754	V. E. Gunn and J. P. Anselme; *J. Org. Chem.*, 1977, **42**, 754.	167
77JOC1045	W. K. Anderson, E. J. LaVoie and G. E. Lee; *J. Org. Chem.*, 1977, **42**, 1045.	733
77JOC1780	D. G. Naae; *J. Org. Chem.*, 1977, **42**, 1780.	203
77JOC1794	J. A. Marshall and P. G. M. Wuts; *J. Org. Chem.*, 1977, **42**, 1794.	235
77JOC2792	H. A. J. Holterman and J. B. F. N. Engberts; *J. Org. Chem.*, 1977, **42**, 2792.	830
77JOC3128	J. W. Scheeren, R. W. M. Aben, P. H. J. Ooms and R. J. F. Nivard; *J. Org. Chem.*, 1977, **42**, 3128.	212
77JOC3458	A. G. Schultz, J. Erhardt and W. K. Hagmann; *J. Org. Chem.*, 1977, **42**, 3458.	4, 12, 13, 21, 30, 33
77JOC3527	B. Modarai and E. Khoshdel; *J. Org. Chem.*, 1977, **22**, 3527.	433
77JOC4095	R. Bonjouklian and R. A. Ruden; *J. Org. Chem.*, 1977, **42**, 4095.	893
77JOC4159	G. L. Anderson and A. D. Broom; *J. Org. Chem.*, 1977, **42**, 4159.	298
77JOC4217	T. Dominh, A. L. Johnson, J. E. Jones and P. P. Senise, Jr.; *J. Org. Chem.*, 1977, **42**, 4217.	674
77JOM(124)29	R. M. Laine and P. C. Ford; *J. Organomet. Chem.*, 1977, **124**, 29.	689, 690
77JOM(127)93	W.-S. Lee and H. H. Brintzinger; *J. Organomet. Chem.*, 1977, **127**, 93.	648
77JOM(127)281	R. J. P. Corriu, G. Lanneau, J. P. Masse and D. Samate; *J. Organomet. Chem.*, 1977, **127**, 281.	724
77JOM(127)391	P. W. J. De Graaf, J. Boersma and G. J. M. Van der Kerk; *J. Organomet. Chem.*, 1977, **127**, 391.	693
77JOM(129)429	H. Felkin and G. K. Turner; *J. Organomet. Chem.*, 1977, **129**, 429.	580
77JOM(131)49	L. Weber; *J. Organomet. Chem.*, 1977, **131**, 49.	1065
77JOM(131)321	B. Mauze; *J. Organomet. Chem.*, 1977, **131**, 321.	142, 519, 568
77JOM(132)77	J. Grobe and J. Hendriock; *J. Organomet. Chem.*, 1977, **132**, 77.	356
77JOM(132)301	M. de Jesus, O. Rosario and G. L. Larson; *J. Organomet. Chem.*, 1977, **132**, 301.	697
77JOM(136)7	C. Glidewell; *J. Organomet. Chem.*, 1977, **136**, 7.	524
77JOM(137)131	N. E. Miller; *J. Organomet. Chem.*, 1977, **137**, 131.	637, 660
77JOM(137)157	C. Cloutour, D. Lafargue, J. A. Richards and J. C. Pommier; *J. Organomet. Chem.*, 1977, **137**, 157.	696
77JOM(137)217	A. N. Nesmeyanov, Yu. T. Struchkov, N. N. Sedova, V. G. Andrianov, Yu. V. Volgin and V. A. Sazonova; *J. Organomet. Chem.*, 1977, **137**, 217.	677
77JOM(139)C34	L. E. Manzer and L. J. Guggenberger; *J. Organomet. Chem.*, 1977, **139**, C34.	1034
77JOM(139)77	F. Mathey; *J. Organomet. Chem.*, 1977, **139**, 77.	686
77JOM(140)73	W. A. Herrmann, C. Krüger, R. Goddard and I. Bernal; *J. Organomet. Chem.*, 1977, **140**, 73.	1065
77JOM(140)237	J. Auger, C. Courtois and L. Miginiac; *J. Organomet. Chem.*, 1977, **140**, 237.	14
77JOM(141)71	D. Seyferth, G. J. Murphy and R. A. Woodruff; *J. Organomet. Chem.*, 1977, **141**, 71.	707, 720
77JOM(141)85	A. Agarwal, M. J. McGlinchey and T.-S. Tan; *J. Organomet. Chem.*, 1977, **141**, 85.	719
77JOM(141)313	G. A. Razuvaev, G. A. Domrachev, V. V. Sharutkin and D. N. Suvorova; *J. Organomet. Chem.*, 1977, **141**, 313.	694
77JOM(141)339	H. Werner, B. Ulrich and A. Salzer; *J. Organomet. Chem.*, 1977, **141**, 339.	1066, 1067
77JOM(142)9	Y. Frangin and M. Gaudemar; *J. Organomet. Chem.*, 1977, **142**, 9.	806
77JOM(142)39	D. Seyferth, J. L. Lefferts and R. L. Lambert, Jr.; *J. Organomet. Chem.*, 1977, **142**, 39.	355
77JOM(142)C35	A. Ekouya, J. Dunogues, N. Duffaut and R. Calas; *J. Organomet. Chem.*, 1977, **142**, C35.	1053
77JOM(142)C45	M. Ishikawa, T. Fuchikami and M. Kumada; *J. Organomet. Chem.*, 1977, **142**, C45.	657
77JOM(142)C49	H. Sakurai and H. Umino; *J. Organomet. Chem.*, 1977, **142**, C49.	793
77JOU1918	E. A. Chaika, G. I. Matyushecheva and L. M. Yagupol'skii; *J. Org. Chem. USSR (Engl. Transl.)*, 1977, **13**, 1918.	332
77JOU2224	V. M. Neplyuev, T. A. Sinenko and P. S. Pel'kis; *J. Org. Chem. USSR (Engl. Transl.)*, 1977, **13**, 2224.	983, 986
77JPR149	H. Schafer, B. Bartho and K. Gewald; *J. Prakt. Chem.*, 1977, **319**, 149.	970
77JPR545	W. D. Rudorf and M. Augustin; *J. Prakt. Chem.*, 1977, **319**, 545.	340
77LA88	E. Schaumann and F. F. Grabley; *Justus Liebigs Ann. Chem.*, 1977, 88.	
77LA624	K. Burger, J. Albanbauer, F. Käfig and S. Penninger; *Justus Liebigs Ann. Chem.*, 1977, 624.	311, 312, 325
77LA811	D. Seebach and M. Kolb; *Justus Liebigs Ann. Chem.*, 1977, 811.	254

77LA830	D. Seebach, R. Buerstinghaus, B. T. Groebel and M. Kolb; *Justus Liebigs Ann. Chem.*, 1977, 830.	254, 618, 853
77LA846	D. Seebach, N. Meyer and A. K. Beck; *Justus Liebigs Ann. Chem.*, 1977, 846.	238, 284
77LA956	W. Kliegel and G.-H. Franckenstein; *Justus Liebigs Ann. Chem.*, 1977, 956.	408
77LA1116	K. Schank, H.-G. Schmitt, F. Schroeder and A. Weber; *Justus Liebigs Ann. Chem.*, 1977, 1116.	236
77LA1633	R. Rienäcker and D. Schwengers; *Justus Liebigs Ann. Chem.*, 1977, 1633.	698
77LA1676	K. Schank and F. Schroeder; *Justus Liebigs Ann. Chem.*, 1977, 1676.	838
77LA1807	H. Stetter and J. Lennartz; *Justus Liebigs Ann. Chem.*, 1977, 1807.	166
77LA1895	H. Meyer, F. Bossert and H. Horstmann; *Justus Liebigs Ann. Chem.*, 1977, 1895.	972
B-77MI 407-01	R. Richter and H. Ulrich; in "The Chemistry of Cyanates and Their Thio Derivatives," ed. S. Patai, Wiley, Chichester, 1977, p. 619.	315
77MI 412-01	H. B. F. Dixon, M. J. Sparkes and D. Webster; *Biochem. Soc. Trans.*, 1977, **5**, 209 (*Chem. Abstr.*, 1977, **87**, 98 252).	565
77MI 412-02	M. S. Hussain, A. G. Marwat and H. Schmidbaur; *J. Sci. Technol.* (*Peshawar, Pak.*), 1977, 57 (*Chem. Abstr.*, 1979, **90**, 23 170).	585
77NJC369	A. Colens, M. Demuylder, B. Téchy and L. Ghosez; *Nouv. J. Chim.*, 1977, **1**, 369.	790, 791
77NJC371	A. Colens and L. Ghosez; *Nouv. J. Chim.*, 1977, **1**, 371.	889
77OR(25)1	L. A. Paquette; *Org. React.*, 1977, **25**, 1.	79
77OS(56)32	S. R. Sandler; *Org. Synth.*, 1977, **56**, 32.	24
77PS(3)47	G. Hägele and H. Dolhaine; *Phosphorus Sulfur*, 1977, **3**, 47.	804
77PS(3)345	S. Musierowicz and W. T. Waszkuc; *Phosphorus Sulfur*, 1977, **3**, 345.	343
77RTC44	Th. M. Wortel, W. H. Esser, G. van Minnen-Pathuis, R. Taal, D. P. Roelofsen and H. van Bekkum; *Recl. Trav. Chim. Pays-Bas*, 1977, **96**, 44.	177
77RTC68	E. A. Ostveen and M. C. van der Plas; *Rec. Trav. Chim. Pays-Bas*, 1977, **96**, 68.	893
77S239	J. Lukszo and R. Tyka; *Synthesis*, 1977, 239.	461, 463
77S247	G. Simchen and W. West; *Synthesis*, 1977, 247.	834
77S336	H. Ahlbrecht and W. Farnung; *Synthesis*, 1977, 336.	1004
77S357	B.-T. Gröbel and D. Seebach; *Synthesis*, 1977, 357.	197, 244, 1075
77S467	E. C. Taylor and C.-S. Chiang; *Synthesis*, 1977, 467.	191
77S571	J. Oleksyszyn, R. Tyka and P. Mastalerz; *Synthesis*, 1977, 571.	487
77S578	A. A. Frimer; *Synthesis*, 1977, 578.	188
77S625	R. Gancarz and J. S. Wieczorek; *Synthesis*, 1977, 625.	464
77S647	T. Takajo, S. Kambe, K. Saito, T. Hayashi and A. Sakurai; *Synthesis*, 1977, 647.	437, 445
77S765	A. Shafiee, I. Lalezari and F. Savabi; *Synthesis*, 1977, 765.	874
77S791	K. M. More and J. Wemple; *Synthesis*, 1977, 791.	61, 63, 73
77S867	H. Emde and G. Simchen; *Synthesis*, 1977, 867.	831
77S869	R. R. Schmidt and J. Talbiersky; *Synthesis*, 1977, 869.	1016
77S883	J. W. Huber and M. Middlebrooks; *Synthesis*, 1977, 883.	467
77SC409	E. Wenkert and T. E. Goodwin; *Synth. Commun.*, 1977, **7**, 409.	179
77T2019	P. D. Magnus; *Tetrahedron*, 1977, **33**, 2019.	390
77T2949	S. Masson, M. Saquet and A. Thuillier; *Tetrahedron*, 1977, **33**, 2949.	256
77T3105	G. Bauduin, Y. Pietrasanta and B. Pucci; *Tetrahedron*, 1977, **33**, 3105.	191
77TL533	R. A. Holton and R. G. Davis; *Tetrahedron Lett.*, 1977, **18**, 533.	219
77TL591	P. F. Hudrlik, J. P. Arcoleo, R. H. Schwartz, R. N. Misra and R. J. Rona; *Tetrahedron Lett.*, 1977, **18**, 591.	357
77TL695	K. Oka and S. Hara; *Tetrahedron Lett.*, 1977, **18**, 695.	71, 72
77TL851	N. Miyoshi, S. Murai and N. Sonoda; *Tetrahedron Lett.*, 1977, **18**, 851.	240
77TL885	C. G. Kruse, N. L. J. M. Broekhof, A. Wijsman and A. van der Gen; *Tetrahedron Lett.*, 1977, **18**, 885.	64
77TL1225	V. Reutrakul and W. Kanghae; *Tetrahedron Lett.*, 1977, **18**, 1225.	77
77TL1239	W. G. Salmond; *Tetrahedron Lett.*, 1977, **18**, 1239.	738, 739
77TL1617	M. Y. H. Wong and G. R. Gray; *Tetrahedron Lett.*, 1977, **18**, 1617.	253
77TL1823	H. Westmijze, J. Meijer and P. Vermeer; *Tetrahedron Lett.*, 1977, **18**, 1823.	809
77TL1827	E. C. Taylor and C.-S. Chiang; *Tetrahedron Lett.*, 1977, **18**, 1827.	194
77TL1961	T. M. Dolak and T. A. Bryson; *Tetrahedron Lett.*, 1977, **18**, 1961.	365, 391, 392
77TL2893	S. Yanagida, Y. Noji and M. Okahara; *Tetrahedron Lett.*, 1977, **18**, 2893.	52
77TL3007	H. C. Brown and N. R. De Lue; *Tetrahedron Lett.*, 1977, **18**, 3007.	143
77TL3349	A. Seveno, G. Morel, A. Foucaud and E. Marchand; *Tetrahedron Lett.*, 1977, **18**, 3349.	894
77TL3583	R. R. Schmidt and B. Schmid; *Tetrahedron Lett.*, 1977, **18**, 3583.	957
77TL3619	H. Hamberger, H. Reinshagen, G. Schulz and G. Sigmund; *Tetrahedron Lett.*, 1977, **18**, 3619.	983
77TL4187	R. K. Boeckman, Jr. and K. J. Bruza; *Tetrahedron Lett.*, 1977, **18**, 4187.	950
77TL4229	O. Possel and A. M. van Leusen; *Tetrahedron Lett.*, 1977, **18**, 4229.	538
77TL4433	M. A. Pericás and F. Serratosa; *Tetrahedron Lett.*, 1977, **18**, 4433.	767
77TL4519	W. R. Bowman and G. D. Richardson; *Tetrahedron Lett.*, 1977, **18**, 4519.	327
77UKZ1000	D. Aleksiev, I. Mladenov and K. Kurtev; *Ukr. Khim. Zh.* (*Russ. Ed.*), 1977, **43**, 1000 (*Chem. Abstr.*, 1978, **88**, 6479).	797
77URP403314	A. V. Kirsanov, V. V. Vasil'ev and N. I. Liptuga; *USSR Pat.* 403 314 (1977) (*Chem. Abstr.*, 1977, **87**, 135 940).	573
77USP4018810	I. H. Skoog (Minnesota Mining and Mfg. Co.); *US Pat.* 4 018 810 (1977) (*Chem. Abstr.* 1977, **87**, 24 792).	864
77ZAAC(431)76	G. Fritz and W. Hölderich; *Z. Anorg. Allg. Chem.*, 1977, **431**, 76.	545

77ZAAC(433)61	G. Fritz and K. Kreilein; *Z. Anorg. Allg. Chem.*, 1977, **433**, 61.	621
77ZAAC(433)157	J. Grobe and G. F. Scheuer; *Z. Anorg. Allg. Chem.*, 1977, **433**, 157.	566
77ZAAC(434)145	H. Schmidbaur and J. Eberlein; *Z. Anorg. Allg. Chem.*, 1977, **434**, 145.	577, 586
77ZC289	M. Augustin, R. Schmidt and W.-D. Rudorf; *Z. Chem.*, 1977, **17**, 289.	900
77ZN(B)762	H. H. Karsch and H. Schmidbaur; *Z. Naturforsch., Teil B*, 1977, **32**, 762.	545, 546, 547, 575, 582
77ZN(B)858	H. Schmidbaur, W. Scharf and H.-J. Fuller; *Z. Naturforsch., Teil B*, 1977, **32**, 858.	579
77ZN(B)1268	L. Knoll; *Z. Naturforsch., Teil B*, 1977, **32**, 1268.	581
77ZN(B)1490	M. Baudler and B. Carlsohn; *Z. Naturforsch., Teil B*, 1977, **32**, 1490.	545
77ZOB75	V. A. Shokol and N. K. Mikhailyuchenko; *Zh. Obshch. Khim.*, 1977, **47**, 75. (Engl. Transl. 67).	1007, 1010
77ZOB579	K. A. Petrov, V. A. Chauzov and N. Yu. Mal'kevich; *Zh. Obshch. Khim.*, 1977, **47**, 579 (*Chem. Abstr.*, 1977, **87**, 23 418c).	464
77ZOB775	Z. S. Novikova, A. A. Prishchenko and I. F. Lutsenko; *Zh. Obshch. Khim.*, 1977, **47**, 775.	546, 550, 562
77ZOB1422	Yu. G. Gololobov and L. I. Nesterova; *Zh. Obshch. Khim.*, 1977, **47**, 1422 (*Chem. Abstr.*, 1977, **87**, 85 101c).	483
77ZOB1664	V. P. Kochergin, V. I. Shiryaev, E. F. Bugerenko and V. N. Eldikov; *Zh. Obshch. Khim.*, 1977, **47**, 1664.	588
77ZOB1715	V. N. Kushnir, M. I. Shevchuk and A. V. Dombrovskii; *Zh. Obshch. Khim.*, 1977, **47**, 1715.	570
77ZOB1751	V. I. Shiryaev, V. P. Kochergin, E. M. Stepina, T. S. Kuptsova, E. M. Protasov, A. V. Kisin and V. F. Mironiv; *Zh. Obshch. Khim.*, 1977, **47**, 1751 (*Chem. Abstr.*, 1977, **87**, 168 148z).	660
77ZOB1994	B. S. Drach and O. P. Lobanov; *Zh. Obshch. Khim.*, 1977, **47**, 1994 (Engl. Transl. 1823).	1002
77ZOB2636	Z. S. Novikova, A. A. Prishchenko and I. F. Lutsenko; *Zh. Obshch. Khim.*, 1977, **47**, 2636.	555
77ZOB2689	A. A. Prishchenko, Z. S. Novikova and I. F. Lutsenko; *Zh. Obshch. Khim.*, 1977, **47**, 2689.	552
77ZOR254	S. I. Radchenko; *Zh. Org. Khim.*, 1977, **13**, 254 (*Chem. Abstr.*, 1977, **86**, 189 111b).	786
77ZOR457	L. N. Cherkasov; *Zh. Org. Khim.*, 1977, **13**, 457 (*Chem. Abstr.*, 1977, **87**, 5310).	1065
77ZOR504	S. I. Radchenko; *Zh. Org. Khim.*, 1977, **13**, 504 (*Chem. Abstr.*, 1977, **87**, 5333s).	905, 910
77ZOR508	V. M. Neplyuev, T. A. Sinenko and P. S. Pel'kis; *Zh. Org. Khim.*, 1977, **13**, 508 (*Chem. Abstr.* 1977, **87**, 5336).	866
77ZOR723	V. I. Gorbatenko, V. N. Fetyukhin and L. I. Samarai; *Zh. Org. Khim.*, 1977, **13**, 723.	434, 447
77ZOR954	N. D. Bodnarchuk and A. A. Yatsishin; *Zh. Org. Khim.*, 1977, **13**, 954 (Engl. Transl. 877).	980, 987
77ZOR1675	M. A. Kazankova, A. R. Sheffer, E. A. Besolova and I. F. Lutsenko; *Zh. Org. Khim.*, 1977, **13**, 1675.	765
77ZOR1930	N. P. Shusherina, V. L. Lapteva and O. V. Khrashcheva; *Zh. Org. Khim.*, 1977, **13**, 1930.	322
77ZOR2449	V. I. Gorbatenko, V. N. Fetyukhin and L. I. Samarai; *Zh. Org. Khim.*, 1977, **13**, 2449.	436, 441
78ACR407	Y. Kanaoka; *Acc. Chem. Res.*, 1978, **11**, 407.	1077
78ACS(B)141	S. Karlsson and J. Sandström; *Acta Chem. Scand. Ser. B*, 1978, **32**, 141.	902, 979
78AG(E)204	R. R. Schmidt and J. Talbiersky; *Angew. Chem., Int. Ed. Engl.*, 1978, **17**, 204.	1016
78AG(E)526	M. Sevrin, J. N. Denis and A. Krief; *Angew. Chem., Int. Ed. Engl.*, 1978, **17**, 526.	401, 964
78AG(E)936	E. Francotte, R. Merényi and H. G. Viehe; *Angew. Chem., Int. Ed. Engl.*, 1978, **17**, 936.	798
78AHC(23)1	Z. Eckstein and T. Urbanski; *Adv. Heterocycl. Chem.*, 1978, **23**, 1.	302
78AJC1031	P. M. Pojer and S. J. Angyal; *Aust. J. Chem.*, 1978, **31**, 1031.	222
78ANY355	M. P. Cava and M. V. Lakshmikantham; *Ann. N. Y. Acad. Sci.*, 1978, **313**, 355 (*Chem. Abstr.* 1979, **91**, 73 725).	850, 869
78BCJ1397	H. Kurosawa, S. Numaka, T. Konishi and R. Okawara; *Bull. Chem. Soc. Jpn.*, 1978, **51**, 1397.	659
78BCJ2401	N. Ono, T. Yamada, T. Saito, K. Tanaka and A. Kaji; *Bull. Chem. Soc. Jpn.*, 1978, **51**, 2401.	220
78BCJ2614	C. Shin, Y. Yonezawa, K. Suzuki and J. Yoshimura; *Bull Chem. Soc. Jpn.*, 1978, **51**, 2614.	433
78BJ(169)239	D. Webster, M. J. Sparkes and H. B. F. Dixon; *Biochem. J.*, 1978, **169**, 239.	565
78BSB391	J. Gorissen and H. G. Viehe; *Bull. Soc. Chim. Belg.*, 1978, **87**, 391.	794
78BSF(2)83	P. Laurent and L. Bearn; *Bull. Soc. Chim. Fr.*, Part 2, 1978, 83.	302
78BSF(2)595	Y. Le Floc'h; *Bull. Soc. Chim. Fr.*, 1978, **2**, 595.	249
78C253	J. Oleksyszyn, M. Soroka and J. Rachon; *Chimia*, 1978, **32**, 253.	460
78CB705	M. Regitz, W. Illger and G. Maas; *Chem. Ber.*, 1978, **111**, 705.	558
78CB797	H. Schmidbaur and H. P. Schern; *Chem. Ber.*, 1978, **111**, 797.	583
78CB1077	W. A. Herrmann; *Chem. Ber.*, 1978, **111**, 1077.	686
78CB1362	E. J. Corey and D. Enders; *Chem. Ber.*, 1978, **111**, 1362.	649
78CB2630	B. Renger, H. Hügel, W. Wykypiel and D. Seebach; *Chem. Ber.*, 1978, **111**, 2630.	529
78CB2696	H. Schmidbaur and M. Heimann; *Chem. Ber.*, 1978, **111**, 2696.	570
78CB2702	H. Schmidbaur and G. Hasslberger; *Chem. Ber.*, 1978, **111**, 2702.	599
78CB2859	M. Adler and K. Schank; *Chem. Ber.*, 1978, **111**, 2859.	235
78CC74	Y. Ueno, A. Nakayama and M. Okawara; *J. Chem. Soc., Chem. Commun.*, 1978, 74.	847
78CC140	J. L. Atwood, W. E. Hunter, R. D. Rogers, J. Holton, J. McMeeking, R. Pearce and M. F. Lappert; *J. Chem. Soc., Chem. Commun.*, 1978, 140.	656
78CC297	P. D. Magnus and G. Roy; *J. Chem. Soc., Chem. Commun.*, 1978, 297.	355
78CC362	M. Miyashita, T. Kumazawa and A. Yoshikoshi; *J. Chem. Soc., Chem. Commun.*, 1978, 362.	906
78CC475	R. D. Chambers, A. Lindley, H. C. Fielding, J. S. Moilliet and G. Whittaker; *J. Chem. Soc., Chem. Commun.*, 1978, 475.	737
78CC478	I. Kuwajima, M. Kato and T. Sato; *J. Chem. Soc., Chem. Commun.*, 1978, 478.	834, 933
78CC821	R. C. Cookson and P. J. Parsons; *J. Chem. Soc., Chem. Commun.*, 1978, 821.	957

78CC1081	J. Jeffery, M. F. Lappert, N. T. Luong-Thi, J. L. Atwood and W. E. Hunter; *J. Chem. Soc., Chem. Commun.*, 1978, 1081.	656
78CJC1183	J. S. Grossert, M. M. Bharadwaj, R. F. Langler, T. S. Cameron and R. E. Cordes; *Can. J. Chem.*, 1978, **56**, 1183.	258
78CJC2286	A. G. Brook, C. Golino and E. Matern; *Can. J. Chem.*, 1978, **56**, 2286.	1014
78CJC2700	P.-E. Sum and L. Weiler; *Can. J. Chem.*, 1978, **56**, 2700.	197
78CL73	Y. Hori, Y. Nagano, H. Uchiyama, Y. Yamada and H. Taniguchi; *Chem. Lett.*, 1978, **1**, 73.	21
78CL517	H. Okamura, Y. Mitsuhira, M. Miura and H. Takei; *Chem. Lett.*, 1978, 517.	959
78CL1103	J. Lukszo, J. Kowalik and P. Mastalerz; *Chem. Lett.*, 1978, 1103.	461
78CL1243	T. Nakai and K. Mikama; *Chem. Lett.*, 1978, 1243.	852
78CL1253	T. Mukaiyama, Y. Sakito and M. Asami; *Chem. Lett.*, 1978, 1253.	408
78CPB685	S. Yoshimura, S. Takahashi, A. Kawamata, K. Kikugawa, H. Suehiro and A. Aoki; *Chem. Pharm. Bull.*, 1978, **26**, 685.	856
78CRV65	M. M. Campbell and G. Johnson; *Chem. Rev.*, 1978, **78**, 65.	426
78CRV275	P. Beak and D. B. Reitz; *Chem. Rev.*, 1978, **78**, 275.	382, 392
78CRV363	B. M. Trost; *Chem. Rev.*, 1978, 363.	244, 257
78GEP2658127	H. Schmidbaur and H. H. Karsch; *Ger. Pat.* 2 658 127 (1978) (*Chem. Abstr.*, 1978, **89**, 163 734).	546, 575
78GEP2705994	M. Baudler; *Ger. Pat.* 2 705 994 (1978) (*Chem. Abstr.*, 1978, **89**, 197 718).	547
78H(9)831	T. Kametani, K. Kigasawa, M. Hiiragi, N. Wagatsuma, T. Kohagizawa and H. Inoue; *Heterocycles*, 1978, **9**, 831.	325
78HCA512	R. Schlecker, D. Seebach and W. Lubosch; *Helv. Chim. Acta*, 1978, **61**, 512.	382
78HCA1784	D. Bellus, H. Fischer, H. Greuter and P. Martin; *Helv. Chim. Acta*, 1978, **61**, 1784.	830
78HCA2047	M. Neuenschwander, P. Bigler, K. Christen, R. Iseli, R. Kyburz and H. Mühle; *Helv. Chim. Acta*, 1978, **61**, 2047.	53, 54, 56
78HCA2286	M. Petrzilka; *Helv. Chim. Acta*, 1978, **61**, 2286.	51
78HCA2351	P. Dubs and R. Stüssi; *Helv. Chim. Acta*, 1978, **61**, 2351.	256
78HCA2482	P. Mueller, R. Etienne, J. Pfyffer, N. Pineda and M. Schipoff; *Helv. Chim. Acta*, 1978, **61**, 2482.	30
78HCA3075	M. Petrzilka; *Helv. Chim. Acta*, 1978, **61**, 3075.	51
78IC618	E. K. S. Liu and R. J. Lagow; *Inorg. Chem.*, 1978, **17**, 618.	153, 154
78IC625	C. W. De Kock, S. R. Ely, T. E. Hopkins and M. A. Brault; *Inorg. Chem.*, 1978, **17**, 625.	703
78IC1995	M. A. Bennett and P. B. Donaldson; *Inorg. Chem.*, 1978, **17**, 1995.	689
78IC2093	F. A. Cotton, S. A. Koch, A. J. Schultz and J. M. Williams; *Inorg. Chem.*, 1978, **17**, 2093.	673
78IC2324	M. D. Curtis; *Inorg. Chem.*, 1978, **17**, 2324.	622
78ICAL294	G. Thiollet, R. Poilblanc, D. Voigt and F. Mathey; *Inorg. Chim. Acta*, 1978, **30**, L294.	1042
78ICAL331	R. A. Wolcott and J. L. Mills; *Inorg. Chim. Acta.*, 1978, **30**, L331.	1042
78IS135	H. Schmidbaur; *Inorg. Synth.*, 1978, **18**, 135.	569
78IS140	H. Schmidbaur and K. H. Räthlein; *Inorg. Synth.*, 1978, **18**, 140.	585, 587
78IZV1951	R. M. Khomutov and T. I. Osipova; *Izv. Akad. Nauk SSSR, Ser. Khim.*, 1978, 1915 (*Chem. Abstr.*, 1978, **89**, 197 655j).	464
78IZV2654	N. P. Petukhova, N. E. Dontsova, O. M. Sazonova and N. E. Prilezhaeva; *Izv. Akad. Nauk SSSR, Ser. Khim.*, 1978, 2654 (*Chem. Abstr.* 1979, **90**, 103 552).	854
78JA802	W. I. Bailey, Jr., M. H. Chisholm, F. A. Cotton, C. A. Murillo and L. A. Rankel; *J. Am. Chem. Soc.*, 1978, **100**, 802.	673
78JA1164	H. C. Clark, C. R. C. Milne and N. C. Payne; *J. Am. Chem. Soc.*, 1978, **100**, 1164.	1018
78JA1325	D. S. Matteson and K. Arne; *J. Am. Chem. Soc.*, 1978, **100**, 1325.	372, 373, 525
78JA1481	W. C. Still; *J. Am. Chem. Soc.*, 1978, **100**, 1481.	199, 378, 379, 383, 386, 387
78JA1927	W. C. Still and A. Mitra; *J. Am. Chem. Soc.*, 1978, **100**, 1927.	378
78JA1938	G. R. Kieczykowski and R. H. Schlessinger; *J. Am. Chem. Soc.*, 1978, **100**, 1938.	224
78JA2090	E. L. Muetterties, W. R. Pretzer, M. G. Thomas, B. F. Beier, D. L. Thorn, V. W. Day and A. B. Anderson; *J. Am. Chem. Soc.*, 1978, **100**, 2090.	692
78JA2916	E. J. Corey, L. O. Weigel, D. Floyd and M. G. Bock; *J. Am. Chem. Soc.*, 1978, **100**, 2916.	649
78JA3467	D. A. Evans, K. M. Hurst and J. M. Takacs; *J. Am. Chem. Soc.*, 1978, **100**, 3467.	339
78JA3548	M. Y. H. Wong and G. R. Gray; *J. Am. Chem. Soc.*, 1978, **100**, 3548.	253
78JA3611	F. N. Tebbe, G. W. Parshall and G. S. Reddy; *J. Am. Chem. Soc.*, 1978, **100**, 3611.	725
78JA3868	D. Forrest and K. U. Ingold; *J. Am. Chem. Soc.*, 1978, **100**, 3868.	850
78JA3933	G. A. Koppell, L. McShane, F. Jose and R. D. G. Cooper; *J. Am. Chem. Soc.*, 1978, **100**, 3933.	69
78JA4624	B. B. Snider, M. Karras and R. S. E. Conn; *J. Am. Chem. Soc.*, 1978, **100**, 4624.	1060
78JA5428	D. B. Reitz, P. Beak, R. F. Farney and L. S. Helmick; *J. Am. Chem. Soc.*, 1978, **100**, 5428.	392
78JA5562	R. E. Cramer, R. B. Maynard and J. W. Gilje; *J. Am. Chem. Soc.*, 1978, **100**, 5562.	589
78JA5748	F. Mathey, A. Mitschler and R. Weiss; *J. Am. Chem. Soc.*, 1978, **100**, 5748.	1037
78JA5764	W. I. Bailey, Jr., M. H. Chisholm, F. A. Cotton and L. A. Rankel; *J. Am. Chem. Soc.*, 1978, **100**, 5764.	674
78JA6236	W. D. Wulff, W. F. Goure and T. J. Barton; *J. Am. Chem. Soc.*, 1978, **100**, 6236.	614
78JA6362	M. D. Curtis and J. Greene; *J. Am. Chem. Soc.*, 1978, **100**, 6362.	622
78JA6382	M. Walczak, K. Walczak, R. Mink, M. D. Rausch and R. Stucky; *J. Am. Chem. Soc.*, 1978, **100**, 6382.	706
78JA7103	B. M. Trost, M. Ochiai and P. G. McDougal; *J. Am. Chem. Soc.*, 1978, **100**, 7103.	220
78JCR(S)68	B. Cazes, C. Huynh, S. Julia, V. Ratovelomanana and O. Ruel; *J. Chem. Res. (S)*, 1978, 68.	853
78JCR(S)252	H. J. Padberg and G. Bergerhoff; *J. Chem. Res. (S)*, 1978, 252.	594
78JCR(S)368	I. J. Colquhoun and W. McFarlane; *J. Chem. Res. (S)*, 1978, 368.	546, 562

78JCS(D)403	P. J. Harris, J. A. K. Howard, S. A. R. Knox, R. J. McKinney, R. P. Phillips and F. G. A. Stone; *J. Chem. Soc., Dalton Trans.*, 1978, 403.	683
78JCS(D)446	R. A. Andersen, R. A. Jones and G. Wilkinson; *J. Chem. Soc., Dalton Trans.*, 1978, 446.	672
78JCS(D)1314	M. B. Hursthouse, K. M. A. Malik and K. D. Sales; *J. Chem. Soc., Dalton Trans.*, 1978, 1314.	672
78JCS(D)1825	P. M. Bailey, A. Keasey and P. M. Maitlis; *J. Chem. Soc., Dalton Trans.*, 1978, 1825.	693
78JCS(D)1830	A. Keasey and P. M. Maitlis; *J. Chem. Soc., Dalton Trans.*, 1978, 1830.	693
78JCS(P1)422	P. W. L. Bosbury, R. Fields and R. N. Haszeldine; *J. Chem. Soc. Perkin Trans. 1*, 1978, 422.	736
78JCS(P1)947	O. Dahl; *J. Chem. Soc., Perkin Trans. 1*, 1978, 947.	556
78JCS(P1)1113	F. De Sarlo and G. Renzi; *J. Chem. Soc., Perkin Trans. 1.*, 1978, 1113.	894
78JCS(P2)1302	T. Masuda, T. Numata, N. Furukawa and S. Oae; *J. Chem. Soc., Perkin Trans. 2*, 1978, 1302.	221
78JFC(11)441	R. N. Haszeldine, D. R. Taylor and E. W. White; *J. Fluorine Chem.*, 1978, **11**, 441.	116
78JFC(11)527	C. L. Bumgardner and J. C. Wozny; *J. Fluorine Chem.*, 1978, **11**, 527.	30
78JGU711	V. P. Kukhar and N. G. Pavlenko; *J. Gen. Chem. USSR (Engl. Transl.)*, 1978, **48**, 711.	800
78JGU769	V. I. Shiryaev, V. P. Kochergin, A. N. Polivanov, T. F. Slyusarenko, A. V. Kisin and V. F. Mironov; *J. Gen. Chem. USSR (Engl. Transl.)*, 1978, **48**, 769.	156
78JGU1168	V. K. Khairullin, A. A. Bredikhin, S. F. Makhmutova and M. A. Vasyanina; *J. Gen. Chem. USSR (Engl. Transl.)*, 1978, **48**, 1168.	129
78JGU1813	V. K. Khairullin, M. A. Vasyanina and S. F. Makhmutova; *J. Gen. Chem. USSR (Engl. Transl.)*, 1978, **48**, 1813.	128
78JHC949	M. S. Manhas, M. Sugiura and H. P. S. Chawla; *J. Heterocycl. Chem.*, 1978, **15**, 949.	425
78JIC1254	B. N. Ghose; *J. Indian Chem. Soc.*, 1978, **55**, 1254.	114, 123
78JOC43	S. Trofimenko, R. W. Johnson and J. K. Doty; *J. Org. Chem.*, 1978, **43**, 43.	764
78JOC79	W. H. Koster, J. E. Dolfini, B. Toeplitz and J. Z. Gougoutas; *J. Org. Chem.*, 1978, **43**, 79.	227
78JOC82	M. V. Lakshmikantham and M. P. Cava; *J. Org. Chem.*, 1978, **43**, 82.	855
78JOC369	N. C. Gonella and M. P. Cava; *J. Org. Chem.*, 1978, **43**, 369.	847
78JOC438	R. P. Hanzlik and M. Leinwetter; *J. Org. Chem.*, 1978, **43**, 438.	177, 181
78JOC950	D. S. Matteson, M. S. Biernbaum, R. A. Bechtold, J. D. Campbell and R. J. Wilcsek; *J. Org. Chem.*, 1978, **43**, 950.	1054
78JOC992	D. Redmore; *J. Org. Chem.*, 1978, **43**, 992.	462, 464
78JOC996	D. Redmore; *J. Org. Chem.*, 1978, **43**, 996.	470
78JOC1248	T. Fujita, S. Watanabe, K. Suga, R. Yanagi and F. Tsukagoshi; *J. Org. Chem.*, 1978, **43**, 1248.	213
78JOC1256	W. A. Kleschick and C. H. Heathcock; *J. Org. Chem.*, 1978, **43**, 1256.	349
78JOC1595	K. S. Y. Lau and M. Schlosser; *J. Org. Chem.*, 1978, **43**, 1595.	818
78JOC1607	N. Ikota and B. Ganem; *J. Org. Chem.*, 1978, **43**, 1607.	90
78JOC1620	W. E. Fristad, T. R. Bailey and L. A. Paquette; *J. Org. Chem.*, 1978, **43**, 1620.	357
78JOC1697	H. J. Reich, S. Wollowitz, J. E. Trend, F. Chow and D. F. Wendelborn; *J. Org. Chem.*, 1978, **43**, 1697.	92
78JOC1980	S. Yoneda, T. Kawase and Z. Yoshida; *J. Org. Chem.*, 1978, **43**, 1980.	345
78JOC2170	T. F. Crimmins and E. M. Rather; *J. Org. Chem.*, 1978, **43**, 2170.	669
78JOC2203	J. C. Sheehan and T. J. Commons; *J. Org. Chem.*, 1978, **43**, 2203.	70
78JOC2460	C. D. Bedford and A. T. Nielsen; *J. Org. Chem.*, 1978, **43**, 2460.	430, 431
78JOC2500	M. S. Raasch; *J. Org. Chem.*, 1978, **43**, 2500.	905
78JOC2643	G. A. Wheaton and D. J. Burton; *J. Org. Chem.*, 1978, **43**, 2643.	737
78JOC2739	G. Zweifel and W. Lewis; *J. Org. Chem.*, 1978, **43**, 2739.	808, 1059
78JOC3065	F. E. Ziegler and C. M. Chan; *J. Org. Chem.*, 1978, **43**, 3065.	842
78JOC3548	C. G. Kruse, E. K. Poels, F. L. Jonkers and A. Van der Gen; *J. Org. Chem.*, 1978, **43**, 3548.	200, 208
78JOC3794	S. Raucher and G. A. Koolpe; *J. Org. Chem.*, 1978, **43**, 3794.	289, 965
78JOC3838	R. D. Reynolds, D. F. Guanci, C. B. Neynaber and R. J. Conboy; *J. Org. Chem.*, 1978, **43**, 3838.	298
78JOC3922	M. Nakane and C. R. Hutchinson; *J. Org. Chem.*, 1978, **43**, 3922.	853
78JOC3964	H. W. Pinnick and N. H. Lajis; *J. Org. Chem.*, 1978, **43**, 3964.	199
78JOC4136	C. R. Johnson and H. G. Corkins; *J. Org. Chem.*, 1978, **43**, 4136.	86
78JOC4140	C. R. Johnson and H. G. Corkins; *J. Org. Chem.*, 1978, **43**, 4140.	86
78JOC4235	M. J. Taschner and G. A. Kraus; *J. Org. Chem.*, 1978, **43**, 4235.	846
78JOC4252	S. Raucher and G. A. Koolpe; *J. Org. Chem.*, 1978, **43**, 4252.	375, 400, 964
78JOC4255	P. Beak, M. Baillargeon and L. G. Carter; *J. Org. Chem.*, 1978, **43**, 4255.	360, 382, 388
78JOC4276	E. D. Thorsett, E. E. Harris and A. Patchett; *J. Org. Chem.*, 1978, **43**, 4276.	409
78JOC4367	M. S. Newman and P. K. Sujeeth; *J. Org. Chem.*, 1978, **43**, 4367.	18
78JOM(144)155	R. J. P. Corriu, G. F. Lanneau, D. Ledercq and D. Samate; *J. Organomet. Chem.*, 1978, **144**, 155.	648
78JOM(146)C8	O. Rosario, A. Oliva and G. L. Larson; *J. Organomet. Chem.*, 1978, **146**, C8.	357
78JOM(148)C25	L. Knoll; *J. Organomet. Chem.*, 1978, **148**, C25.	580
78JOM(148)247	P. Jutzi and J. Baumgärtner; *J. Organomet. Chem.*, 1978, **148**, 247.	660, 663
78JOM(149)37	M. Ishikawa, T. Fuchikami and M. Kumada; *J. Organomet. Chem.*, 1978, **149**, 37.	613, 614
78JOM(149)167	K. Kellner, B. Seidel and A. Tzschach; *J. Organomet. Chem.*, 1978, **149**, 167.	455, 462, 478, 481, 499, 500, 501
78JOM(150)C1	J. S. Bradley; *J. Organomet. Chem.*, 1978, **150**, C1.	676
78JOM(152)1	A. Mendoza and D. S. Matteson; *J. Organomet. Chem.*, 1978, **152**, 1.	750, 1067
78JOM(152)295	J. V. Comasseto and N. Petragnani; *J. Organomet. Chem.*, 1978, **152**, 295.	348, 349

78JOM(153)59	M. D. Rausch and R. E. Gloth; *J. Organomet. Chem.*, 1978, **153**, 59.	707
78JOM(153)115	A. N. Nesmeyanov, N. N. Sedova, Yu. T. Struchkov, V. G. Andrianov, E. N. Stakheeva and V. A. Sazonova; *J. Organomet. Chem.*, 1978, **153**, 115.	696, 723
78JOM(153)193	T. Aoyama, Y. Sato, T. Suzuki and H. Shirai; *J. Organomet. Chem.*, 1978, **153**, 193.	509
78JOM(153)305	E. J. Bulten and H. A. Budding; *J. Organomet. Chem.*, 1978, **153**, 305.	156
78JOM(153)369	R. W. Hillyard, Jr., C. M. Ryan and C. H. Yoder; *J. Organomet. Chem.*, 1978, **153**, 369.	513
78JOM(154)353	H. Sakurai and K. Mochida; *J. Organomet. Chem.*, 1978, **154**, 353.	140, 141
78JOM(156)C12	J. A. Soderquist and A. Hassner; *J. Organomet. Chem.*, 1978, **156**, C12.	353, 643
78JOM(156)C25	F. DuBoudin and O. Laporte; *J. Organomet. Chem.*, 1978, **156**, C25.	514
78JOM(156)123	A. B. Levy, S. J. Schwartz, N. Wilson and B. Christie; *J. Organometal. Chem.*, 1978, **156**, 123.	935, 949
78JOM(156)299	D. Seyferth and R. E. Mammarella; *J. Organomet. Chem.*, 1978, **156**, 299.	648
78JOM(157)1	J. Y. Becker and J. Klein; *J. Organomet. Chem.*, 1978, **157**, 1.	668
78JOM(157)41	D. S. Ginley, C. R. Bock, M. S. Wrighton, B. Fischer, D. L. Tipton and R. Bau; *J. Organomet. Chem.*, 1978, **157**, 41.	966
78JOM(159)137	D. Seyferth, D. P. Duncan, H. Schmidbaur and P. Holl; *J. Organomet. Chem.*, 1978, **159**, 137.	571
78JOM(159)237	J. A. Soderquist and K. L. Thompson; *J. Organomet. Chem.*, 1978, **159**, 237.	697
78JOM(159)417	K. Oguro, M. Wada and R. Okawara; *J. Organometal. Chem.*, 1978, **159**, 417.	955
78JOM(160)35	E. Kurras and U. Rosenthal; *J. Organomet. Chem.*, 1978, **160**, 35.	580
78JOM(160)41	H. Schmidbaur, G. Blaschke, H. J. Füller and H. P. Scherm; *J. Organomet. Chem.*, 1978, **160**, 41.	582
78JOM(161)C17	E. W. Post and T. F. Crimmins; *J. Organomet. Chem.*, 1978, **161**, C17.	706
78JOM(161)23	M. D. Curtis and R. J. Klingler; *J. Organomet. Chem.*, 1978, **161**, 23.	674
78JOM(161)299	M. Ishikawa, H. Sugisawa, L. Fabry and M. Kumada; *J. Organomet. Chem.*, 1978, **161**, 299.	1047
78JOM(162)C43	H. Sakurai, T. Kobayashi and Y. Nakadaira; *J. Organomet. Chem.*, 1978, **162**, C43.	1052
78JOM(162)1	R. F. Cunico and Y.-K. Han; *J. Organomet. Chem.*, 1978, **162**, 1.	758
78JOM(162)271	V. I. Sokolov, V. V. Bashilov and O. A. Reutov; *J. Organomet. Chem.*, 1978, **162**, 271.	1067
78JOU135	Ya. G. Bal'on and R. N. Moskaleva; *J. Org. Chem. USSR (Engl. Transl.)*, 1978, **14**, 135.	794
78JPR585	W. D. Rudorf and M. Augustin; *J. Prakt. Chem.*, 1978, **320**, 585.	983
78KO686	T. Anthony; *Kirk-Othmer Encyc.*, 1978, **5**, 686.	11
78LA1505	L. Horner and M. Jordan; *Justus Liebigs Ann. Chem.*, 1978, 1505.	427
78LA1928	H. Böhme and A. Ingendoh; *Justus Liebigs Ann. Chem.*, 1978, 1928.	438, 445
B-78MI 408-01	E. Block; "Reactions of Organosulfur Compounds," Academic Press, New York, 1978.	390, 391
78NJC79	C. Bernasconi, L. Cottier, G. Descotes, M. F. Grenier and F. Metras; *Nouv. J. Chim.*, 1978, **2**, 79.	1079
78PAC703	M. F. Lappert; *Pure Appl. Chem.*, 1978, **50**, 703.	656
78PJC321	J. Lukszo and R. Tyka; *Pol. J. Chem.*, 1978, **52**, 321.	461
78PJC959	J. Lukszo and R. Tyka; *Pol. J. Chem.*, 1978, **52**, 959.	461
78PJC1315	C. Wasielewski, K. Antczak and J. Rachon; *Pol. J. Chem.*, 1978, **52**, 1315.	460
78PJC1949	J. Oleksyszyn and R. Tyka; *Pol. J. Chem.*, 1978, **52**, 1949.	467
78PJC2271	J. Zygmunt a ˙ P. Mastalerz; *Pol. J. Chem.*, 1978, **52**, 2271.	473
78PS(4)59	H. Oehme, E. Leissring and A. Zschunke; *Phosphorus Sulfur*, 1978, **4**, 59.	482
78RTC69	R. H. Everhardus, R. Gräfing and L. Brandsma; *Rec. Trav. Chim. Pays-Bas*, 1978, **97**, 69.	956, 957
78S44	K. Burger and R. Ottlinger; *Synthesis*, 1978, 44.	493
78S48	K. S. Kim and W. A. Szarek; *Synthesis*, 1978, 48.	190
78S469	Z. H. Kudzin and W. J. Stec; *Synthesis*, 1978, 469.	466
78S526	K. Burger and S. Penninger; *Synthesis*, 1978, 526.	493
78S588	M. E. Jung, M. A. Mazurek and R. M. Lim; *Synthesis*, 1978, 588.	51
78S593	P. Bigler, H. Mühle and M. Neuenschwander; *Synthesis*, 1978, 593.	53
78S676	J. L. Hubbard and H. C. Brown; *Synthesis*, 1978, 676.	364
78S746	S. Kano, T. Ebata, K. Funaki and S. Shibuya; *Synthesis*, 1978, 746.	648
78S883	A. Jończyk and T. Pytlewski; *Synthesis*, 1978, 883.	84
78SC87	V. Ratovelomanana and S. Julia; *Synth. Commun.*, 1978, **8**, 87.	853
78T179	J. B. P. A. Wijnberg, H. J. Schoemaker and W. N. Speckamp; *Tetrahedron*, 1978, **34**, 179.	303
78T1285	A. Schonberg and E. Singer; *Tetrahedron*, 1978, **34**, 1285.	298
78T1585	G. E. Veenstra and B. Zwanenburg; *Tetrahedron*, 1978, **34**, 1585.	272
78T2399	J. B. P. A. Wijnberg and W. N. Speckamp; *Tetrahedron*, 1978, **34**, 2399.	426
78T3081	M. Mikolajczyk, S. Grzejszczak and A. Zatorski; *Tetrahedron*, 1978, **34**, 3081.	835, 847, 848
78TL5	E. J. Corey and D. L. Boger; *Tetrahedron Lett.*, 1978, **19**, 5.	321
78TL375	K. Ogura, N. Katoh, I. Yoshimura and G. Tsuchihashi; *Tetrahedron Lett.*, 1978, **19**, 375.	326, 856, 857
78TL731	L. G. Wade, Jr., J. M. Gerdes and R. P. Wirth; *Tetrahedron Lett.*, 1978, **19**, 731.	220
78TL1609	J. Rachon and C. Wasielewski; *Tetrahedron Lett.*, 1978, **19**, 1609.	460
78TL2345	S. Ncube, A. Pelter, K. Smith, P. Blatcher and S. Warren; *Tetrahedron Lett.*, 1978, **19**, 2345.	250
78TL2383	P. W. K. Lau and T. H. Chan; *Tetrahedron Lett.*, 1978, **19**, 2383.	655
78TL2407	C. Charrier and F. Mathey; *Tetrahedron Lett.*, 1978, **19**, 2407.	112
78TL2693	J. Lucchetti and A. Krief; *Tetrahedron Lett.*, 1978, **19**, 2693.	399
78TL2717	P. Gosselin, S. Masson and A. Thuillier; *Tetrahedron Lett.*, 1978, **19**, 2717.	256
78TL2749	A. Wissner; *Tetrahedron Lett.*, 1978, **19**, 2749.	832
78TL3267	A. T. Hewson; *Tetrahedron Lett.*, 1978, **19**, 3267.	925

78TL3383	M. van der Liej and B. Zwanenburg; *Tetrahedron Lett.*, 1978, **19**, 3383.	936
78TL3607	S. Y. Dike and J. R. Merchant; *Tetrahedron Lett.*, 1978, **19**, 3607.	304
78TL3861	L. Colombo, C. Gennari and E. Narisano; *Tetrahedron Lett.*, 1978, **19**, 3861.	265
78TL3971	S. Halazy, J. Lucchetti and A. Krief; *Tetrahedron Lett.*, 1978, **19**, 3971.	289
78TL3995	G. H. Posner, P.-W. Tang and J. P. Mallamo; *Tetrahedron Lett.*, 1978, **19**, 3995.	958
78TL4065	B. Cazes and S. Julia; *Tetrahedron Lett.*, 1978, **19**, 4065.	255
78TL4115	Y. Nagao, K. Kaneko and E. Fujita; *Tetrahedron Lett.*, 1978, **19**, 4115.	198
78TL4253	J. B. Lambert, K. Kobayashi and P. H. Mueller; *Tetrahedron Lett.*, 1978, **19**, 4253.	748
78TL4391	T. Kauffmann, K-J. Echsler, A. Hamsen, R. Kregesmann, F. Steinseifer and A. Vahrenhorst; *Tetrahedron Lett.*, 1978, **19**, 4391.	599, 600, 663
78TL4809	K. Tanaka, T. Nakai and N. Ishikawa; *Tetrahedron Lett.*, 1978, **19**, 4809.	735
78TL4995	O. H. Houwen and D. F. Tavares; *Tetrahedron Lett.*, 1978, **19**, 4995.	410
78TL5091	D. Liotta, P. B. Paty, J. Johnston and G. Zima; *Tetrahedron Lett.*, 1978, **19**, 5091.	239
78UKZ183	V. M. Neplyuev, T. A. Sinenko and E. V. Khoina; *Ukr. Khim. Zh.* (*Russ. Ed.*) 1978, **44**, 183 (*Chem.Abstr.* 1978, **88**, 169 714)	866
78URP351438	N. I. Liptuga, V. V. Vasil'ev and G. I. Derkach; *USSR Pat.* 351 438 (1978) (*Chem. Abstr.*, 1978, **89**, 129 702).	573
78USP4075001	L. K. Gibbons; *US Pat.* 4 075 001 (1978).	973, 977
78ZAAC(444)249	K. Issleib, Kl.-P. Döpfer and A. Balszuweit; *Z. Anorg. Allg. Chem.*, 1978, **444**, 249.	461
78ZAAC(447)53	H.-P. Abicht and K. Issleib; *Z. Anorg. Allg. Chem.*, 1978, **447**, 53.	570, 575, 576
78ZC345	M. Augustin, W. Dolling and G. Zomisch; *Z. Chem.*, 1978, **18**, 345.	900
78ZN(B)131	G. V. Roeschenthaler; *Z. Naturforsch., Teil B*, 1978, **33**, 131.	337
78ZN(B)691	M. Baudler and E. Tolls; *Z. Naturforsch., Teil B*, 1978, **33**, 691.	546
78ZN(B)911	W. A. Herrmann and I. Schweizer; *Z. Naturforsch., Teil B*, 1978, **33**, 911.	686
78ZN(B)1208	M. Baudler and F. Saykowski; *Z. Naturforsch., Teil B*, 1978, **33**, 1208.	545
78ZN(B)1325	H. Schmidbaur, J. R. Mandl, A. Wohlleben-Hammer and A. Fügner; *Z. Naturforsch., Teil B*, 1978, **33**, 1325.	586
78ZN(B)1422	H. G. Horn, R. Köntges, F. Kolkmann and H. C. Marsmann; *Z. Naturforsch., Teil B*, 1978, **33**, 1422.	802
78ZOB267	M. A. Kazankova, I. G. Trostyanskaya, T. Ya. Satina and I. F. Lutsenko; *Zh. Obshch. Khim.*, 1978, **48**, 267.	915
78ZOB757	Z. S. Novikova, S. Ya. Skorobogatova and I. F. Lutsenko; *Zh. Obshch. Khim.*, 1978, **48**, 757.	551
78ZOB778	V. P. Kukhar and N. G. Pavlenko; *Zh. Obshch. Khim.*, 1978, **48**, 778 (Engl. Transl. 711).	991
78ZOB803	N. D. Bodnarchuk and A. A. Yatsishin; *Zh. Obshch. Khim.*, 1978, **48**, 803 (Engl. Transl. 733).	990
78ZOB1001	A. N. Pudovik, I. V. Konovalova, G. V. Romanov, V. M. Pozhidaev, N. P. Anoshina and A. A. Lapin; *Zh. Obshch. Khim.*, 1978, **48**, 1001.	343, 561
78ZOB1008	A. N. Pudovik, G. V. Romanov and V. M. Posibaev; *Zh. Obshch. Khim.*, 1978, **48**, 1008 (*Chem. Abstr.*, 1978, **89**, 109 752).	455
78ZOB1422	F. V. Bagrov and N. E. Galkina; *Zh. Obshch. Khim.*, 1978, **48**, 1422.	588
78ZOB1994	B. S. Drach and O. P. Lobanov; *Zh. Obshch. Khim.*, 1978, **48**, 1994 (Engl. Transl. 1815).	1000, 1001
78ZOB2147	G. P. Simirskaya, S. V. Ponomarev and I. Y. Belavin; *Zh. Obshch. Khim.*, 1978, **48**, 2147 (*Chem. Abstr.*, 1978, **89**, 215 508).	1063
78ZOB2653	R. K. Valetdinov, A. N. Zuikova, N. Sh. Yakiminskaya, T. A. Zyablikova, E. N. Ofitserov and A. Vll'yasov; *Zh. Obshch. Khim.*, 1978, **48**, 2653.	342
78ZOR478	F. R. Tantasheva, V. S. Savel'ev, E. A. Berdnikov and E. G. Kataev; *Zh. Org. Khim.*, 1978, **14**, 478 (*Chem. Abstr.* 1978, **89**, 42 654).	861
78ZOR739	Yu. G. Gololobov and L. I. Nesterova; *Zh. Org. Khim.*, 1978, **14**, 739 (*Chem. Abstr.*, 1978, **89**, 59 639f).	483
78ZOR1659	L. N. Markovskii, Yu. G. Shermolovich, V. V. Vasil'ev, I. E. Boldeskul, S. S. Trach and N. S. Zefirov; *Zh. Org. Khim.*, 1978, **14**, 1659.	326
78ZOR1723	V. A. Anfinogenov, V. D. Filiminov and E. E. Sirotkina; *Zh. Org. Khim.*, 1978, **14**, 1723.	310
78ZOR1947	V. M. Neplyuev, V. P. Kukhar, T. A. Sinenko, E. V. Khoina and P. S. Pel'kis; *Zh. Org. Khim.*, 1978, **14**, 1947 (*Chem. Abstr.* 1978, **89**, 215 029).	865
78ZOR2624	V. I. Gorbatenko, V. N. Fetyukhin and L. I. Samarai; *Zh. Org. Khim.*, 1978, **14**, 2624 (Engl. Transl. 2415).	992
79AG(E)782	H. Schmidbaur, E. Weiss and B. Zimmer-Gasser; *Angew. Chem., Int. Ed. Engl.*, 1979, **18**, 782.	564
79AG(E)863	U. Schöllkopf, W. Hartwig and U. Groth; *Angew. Chem., Int. Ed. Engl.*, 1979, **18**, 863.	535
79BCJ466	M. Yamashita, T. Miyano, T. Watabe, H. Inokawa, H. Yoshida, T. Ogata and S. Inokawa; *Bull. Chem. Soc. Jpn.*, 1979, **52**, 466.	854
79BCJ826	T. Iwasaki, H. Horikawa, K. Matsumoto and M. Miyoshi; *Bull. Chem. Soc. Jpn.*, 1979, **52**, 826.	311, 327
79BCJ2013	K. Ogura, Y. Ito and G. Tsuchihashi; *Bull. Chem. Soc. Jpn.*, 1979, **52**, 2013.	856
79BCJ2560	Y. Yamamoto and Z. Kanda; *Bull. Chem. Soc. Jpn.*, 1979, **52**, 2560.	586
79BCJ3342	T. Kawase, S. Yoneda and Z. Yoshida; *Bull. Chem. Soc. Jpn.*, 1979, **52**, 3342.	345
79BCJ3640	K. T. Kang, R. Okazaki and N. Inamoto; *Bull. Chem. Soc. Jpn.*, 1979, **52**, 3640.	979
79CAR(70)217	K. Blumberg, A. Fuccello and T. van Es; *Carbohydr. Res.*, 1979, **70**, 217.	228
79CB28	H. J. Bestmann, R. Engler, H. Hartung and K. Roth; *Chem. Ber.*, 1979, **112**, 28.	843

79CB148	L. Bassignani, B. Biancini, A. Brandt, V. Caciagli, G. E. Bianchi, L. Re, A. Rossodivita and P. Zappelli; *Chem. Ber.*, 1979, **112**, 148.	205, 210
79CB355	G. Saleh, T. Minami, Y. Ohshiro and T. Agawa; *Chem. Ber.*, 1979, **112**, 355.	348
79CB484	K. Klemm, H. Schaefer and E. Daltrozzo; *Chem. Ber.*, 1979, **112**, 484.	987
79CB648	R. Appel, K. Geisler and H. F. Schöler; *Chem. Ber.*, 1979, **112**, 648.	547, 564
79CB936	W. P. Neumann and K. Reuter; *Chem. Ber.*, 1979, **112**, 936.	360
79CB1420	M. Pohmakotr, K.-H. Geiss and D. Seebach; *Chem. Ber.*, 1979, **112**, 1420.	392
79CB1956	A. Q. Hussein and J. C. Jochims; *Chem. Ber.*, 1979, **112**, 1956.	314, 329, 434
79CB2369	F. A. Neugebauer, H. Fischer and C. Krieger; *Chem. Ber.*, 1979, **112**, 2369.	992
79CB2709	L. Knoll and H. Wolff; *Chem. Ber.*, 1979, **112**, 2709.	580
79CB3603	G. Schill, G. Doerjer, E. Logemann and H. Fritz; *Chem. Ber.*, 1979, **112**, 3603.	177
79CC739	C. E. McKenna and J. Schmidhauser; *J. Chem. Soc., Chem. Commun.*, 1979, 739.	130
79CC814	H. Werner and H.-J. Kraus; *J. Chem. Soc., Chem. Commun.*, 1979, 814.	693
79CC1121	W. Ando, M. Takata and A. Sekiguchi; *J. Chem. Soc., Chem. Commun.*, 1979, 1121.	398
79CC1180	A. Devos, J. Remion, A. M. Frisque-Hesbain, A. Colens and L. Ghosez; *J. Chem. Soc., Chem. Commun.*, 1979, 1180.	790
79CJC1153	S. Rakhit, M. Georges and J. F. Bagli; *Can. J. Chem.*, 1979, **57**, 1153.	323
79CL209	V. Reutrakul, A. Tiensripojamarn, K. Kusamran and S. Nimgirawath; *Chem. Lett.*, 1979, 209.	77
79CL785	K. Isobe, M. Fuse, H. Kosugi, H. Hagiwara and H. Uda; *Chem. Lett.*, 1979, 785.	956, 957
79CL983	S. Hayashi, T. Nakai, N. Ishikawa, D. J. Burton, D. G. Naae and H. S. Kesling; *Chem. Lett.*, 1979, 983.	734
79CL1277	K. Suzuki, J. Inanaga and M. Yamaguchi; *Chem. Lett.*, 1979, 1277.	219
79COC(2)61	J. M. Z. Gladych and D. Hartley; *Comp. Org. Chem.*, 1979, **2**, 61.	966
79COC(2)305	R. G. Coombes; *Comp. Org. Chem.*, eds D. H. R. Barton and W. D. Ollis, Pergamon, Oxford, 1979, vol. 2, p. 305.	540
79COC(2)1233	D. J. H. Smith; *Comp. Org. Chem.*, 1979, **2**, 1233.	117
79COC(3)33	G. C. Barrett; *Comp. Org. Chem.*, 1979, **3**, 33.	390
79COC(3)121	T. Durst; *Comp. Org. Chem.*, 1979, **3**, 121.	390
79COC(3)171	T. Durst; *Comp. Org. Chem.*, 1979, **3**, 171.	390
79CPB1691	K. Ito and M. Sekiya; *Chem. Pharm. Bull.*, 1979, **27**, 1691.	325
79CPB2787	S. Sugai and K. Tomita; *Chem. Pharm. Bull.*, 1979, **27**, 2787.	979
79CPB2857	H. Saikachi, T. Kitagawa and H. Sasaki; *Chem. Pharm. Bull.*, 1979, **27**, 2857.	538
79CRV491	D. M. Clode; *Chem. Rev.*, 1979, **79**, 491.	180
79CRV529	L. E. Gusel'nikov and N. S. Nametkin; *Chem. Rev.*, 1979, **79**, 529.	610, 611, 614, 627
79DOK(244)610	B. A. Arbuzov, O. A. Erastov, S. Sh. Khetagurova and T. A. Zyablikova; *Dokl. Akad. Nauk SSSR*, 1979, **244**, 610.	341
79DOK(246)1130	B. A. Arbuzov, E. N. Dianova and N. A. Chadaeva; *Dokl. Akad. Nauk SSSR*, 1979, **246**, 1130 (*Chem. Abstr.*, 1979, **91**, 140 928n).	502
79DOK(247)609	M. G. Voronkov, N. M. Kudyakov, V. M. D'iakov, V. I. Glukhikh and R. K. Valetdinov; *Dokl. Akad. Nauk SSSR*, 1979, **247**, 609.	342
79EGP137839	E. Guenther, W. Kochmann, K. Naumann and T. Roethling; *Ger. (East) Pat.* 137 839 (1979) (*Chem. Abstr.*, 1980, **92**, 180 669).	106
79IC60	A. H. Cowley and R. C.-Y. Lee; *Inorg. Chem.*, 1979, **18**, 60.	564
79IC69	*Inorg. Chem.*, 1979, **18**, 69.	1019
79IC347	J. C. Wilburn and R. H. Neilson; *Inorg. Chem.*, 1979, **18**, 347.	571
79IC2696	H. J. Gysling, H. R. Luss and D. L. Smith; *Inorg. Chem.*, 1979, **18**, 2696.	377
79IC2808	M. P. Brown, J. R. Fisher, R. J. Puddephatt and K. R. Seddon; *Inorg. Chem.*, 1979, **18**, 2808.	691
79IC3453	R. J. Hoxmeier, J. R. Blickensderfer and H. D. Kaesz; *Inorg. Chem.*, 1979, **18**, 3453.	721
79ICAL67	G. Thiollet, F. Mathey and R. Poilblanc; *Inorg. Chim. Acta*, 1979, **32**, L67.	1039
79IJC(B)478	N. Viswanathan, K. R. Ravindranath and P. K. Talwalkar; *Indian J. Chem., Sect. B*, 1979, **17**, 478.	979
79IJC(B)500	S. N. Suryawanshi and V. R. Nayak; *Ind. J. Chem., Sect. B*, 1979, **18**, 500.	749, 750
79IS262	R. A. Andersen and G. Wilkinson; *Inorg. Synth.*, 1979, **19**, 262.	
79IZV866	B. A. Arbuzov, O. A. Erastov and Sh. Khetagurova; *Izv. Akad. Nauk SSSR, Ser. Khim.*, 1979, 866.	341
79IZV1110	R. M. Khomutov, T. I. Osipova, Yu. N. Zhukova and I. A. Gandurina; *Izv. Akad. Nauk SSSR, Ser. Khim.*, 1979, 1110 (*Chem. Abstr.*, 1979, **91**, 141 158y).	472
79IZV1528	L. G. Kuz'mina, Yu. T. Struchkov, L. L. Troitskaya, V. I. Sokolov and O. A. Reutov; *Izv. Akad. Nauk SSSR, Ser. Khim.*, 1979, 1528.	721
79IZV1831	I. L. Knunyants, E. M. Rokhlin and A. Yu. Volkonskii; *Izv. Akad. Nauk SSSR, Ser. Khim.*, 1979, 1831.	697
79IZV1863	B. A. Arbuzov, O. A. Erastov, S. N. Ignat'eva, T. A. Zyablikova and R. P. Arshinova; *Izv. Akad. Nauk SSSR, Ser. Khim.*, 1979, 1863.	343
79IZV2073	O. A. Raevskii, N. G. Mumzhieva, M. M. Gilyazov and A. A. Karelov; *Izv. Akad. Nauk SSSR, Ser. Khim.*, 1979, 2073.	126, 127
79IZV2118	R. M. Khomutov, T. I. Osipova, Yu. N. Zhukova and I. A. Gandurina; *Izv. Akad. Nauk SSSR, Ser. Khim.*, 1979, 2118 (*Chem. Abstr.*, 1980, **92**, 58 881d).	472
79IZV2783	B. E. Ivanov, S. S. Krokhina, T. V. Chichkanova, T. A. Zyablikova and A. V. Il'yasov; *Izv. Akad. Nauk SSSR, Ser. Khim.*, 1979, 2783 (*Chem. Abstr.*, 1980, **92**, 111 115k).	476
79JA83	A. G. Brook, J. W. Harris, J. Lennon and M. El Sheikh; *J. Am. Chem. Soc.*, 1979, **101**, 83.	930
79JA96	H. C. Brown, C. G. Scouten and R. Liotta; *J. Am. Chem. Soc.*, 1979, **101**, 96.	631

79JA371	D. A. Evans, J. M. Takacs and K. M. Hurst; *J. Am. Chem. Soc.*, 1979, **101**, 371.	917
79JA503	K. L. Brown, G. R. Clark, C. E. L. Headford, K. Marsden and W. R. Roper; *J. Am. Chem. Soc.*, 1979, **101**, 503.	150
79JA768	M. F. Semmelhack, J. Bisaha and M. Czarny; *J. Am. Chem. Soc.*, 1979, **101**, 768.	707
79JA1057	P. Helquist and M. S. Shekhani; *J. Am. Chem. Soc.*, 1979, **101**, 1057.	264
79JA1476	J. L. Jensen and W. P. Jencks; *J. Am. Chem. Soc.*, 1979, **101**, 1476.	216
79JA2171	B. Glatz, G. Helmchen, H. Muxfeldt, H. Porcher, R. Prewo, J. Senn, J. J. Stezowski, R. J. Stojda and D. R. White; *J. Am. Chem. Soc.*, 1979, **101**, 2171.	179
79JA2501	A. I. Meyers and P. J. Reider; *J. Am. Chem. Soc.*, 1979, **101**, 2501.	210
79JA3689	D. J. Burton, S. Shinya and R. D. Howells; *J. Am. Chem. Soc.*, 1979, **101**, 3689.	802
79JA4128	M. B. Hursthouse, R. A. Jones, K. M. A. Malik and G. Wilkinson; *J. Am. Chem. Soc.*, 1979, **101**, 4128.	680
79JA4732	A. I. Meyers, D. M. Roland, D. L. Comins, R. Henning, M. P. Fleming and K. Shimizu; *J. Am. Chem. Soc.*, 1979, **101**, 4732.	255
79JA5101	G. Zweifel, W. Lewis and H. P. On; *J. Am. Chem. Soc.*, 1979, **101**, 5101.	758, 816
79JA5440	W.-K. Wong, W. Tam and J. A. Gladysz; *J. Am. Chem. Soc.*, 1979, **101**, 5440.	581
79JA6023	B. B. Snider, D. M. Roush and T. A. Killinger; *J. Am. Chem. Soc.*, 1979, **101**, 6023.	827
79JA6433	D. L. Johnson and J. A. Gladysz; *J. Am. Chem. Soc.*, 1979, **101**, 6433.	581
79JA6452	E. Vedejs and G. R. Martinez; *J. Am. Chem. Soc.*, 1979, **101**, 6452.	517
79JA6638	H. J. Reich, F. Chow and S. K. Shah; *J. Am. Chem. Soc.*, 1979, **101**, 6638.	239, 375
79JA6648	H. J. Reich, S. K. Shah and F. Chow; *J. Am. Chem. Soc.*, 1979, **101**, 6648.	399
79JA6933	G. P. Pez, C. F. Putnik, S. L. Suib and G. D. Stucky; *J. Am. Chem. Soc.*, 1979, **101**, 6933.	671
79JA7104	A. I. Meyers, D. L. Comins, D. M. Roland, R. Henning and K. Shimizu; *J. Am. Chem. Soc.*, 1979, **101**, 7104.	255
79JA7282	C. P. Casey, S. W. Polichnowski, A. J. Shusterman and C. R. Jones; *J. Am. Chem. Soc.*, 1979, **101**, 7282.	580
79JA7410	N. E. Schore; *J. Am. Chem. Soc.*, 1979, **101**, 7410.	566, 568
79JA7617	M. W. E. M. van Tilborg, R. van Doorn and N. M. M. Nibbering; *J. Am. Chem. Soc.*, 1979, **101**, 7617.	50
79JCS(D)45	J. Holton, M. F. Lappert, D. G. H. Ballard, R. Pearce, J. L. Atwood and W. E. Hunter; *J. Chem. Soc., Dalton Trans.*, 1979, 45.	725, 726
79JCS(D)54	J. Holton, M. F. Lappert, D. G. H. Ballard, R. Pearce, J. L. Atwood and W. E. Hunter; *J. Chem. Soc., Dalton Trans.*, 1979, 54.	702
79JCS(D)767	F. Glocking, V. B. Mahale and J. J. Sweeney; *J. Chem. Soc., Dalton Trans.*, 1979, 767.	656
79JCS(D)814	E. W. Abel and C. Towers; *J. Chem. Soc., Dalton Trans.*, 1979, 814.	1042
79JCS(D)1552	E. W. Abel, N. Clark and C. Towers; *J. Chem. Soc., Dalton Trans.*, 1979, 1552.	1034, 1037, 1042
79JCS(D)1801	S. A. R. Knox, R. J. McKinney, V. Riera, F. G. A. Stone and A. C. Szary; *J. Chem. Soc., Dalton Trans.*, 1979, 1801.	683
79JCS(P1)36	V. K. Tripathi and P. S. Venkataramani; *J. Chem. Soc., Perkin Trans 1*, 1979, 36.	213
79JCS(P1)62	D. W. Knight and G. Pattenden; *J. Chem. Soc., Perkin Trans. 1*, 1979, 62.	170
79JCS(P1)249	R. Faragher and T. L. Gilchrist; *J. Chem. Soc., Perkin Trans. 1*, 1979, 249.	307, 799
79JCS(P1)3099	C. Earnshaw, C. J. Wallis and S. Warren; *J. Chem. Soc., Perkin Trans 1*, 1979, 3099.	949
79JFC(13)325	R. D. Bagnall, W. Bell and K. Pearson; *J. Fluorine Chem.*, 1979, **13**, 325.	4
79JFC(13)407	M.-J. van Hamme and D. J. Burton; *J. Fluorine Chem.*, 1979, **13**, 407.	753
79JGU625	M. G. Voronkov, R. G. Mirskov and A. L. Kuznetsov; *J. Gen. Chem. USSR (Engl. Transl.)*, 1979, **49**, 625.	812
79JGU1947	V. P. Kukhar, E. I. Sagina and N. G. Pavlenko; *J. Gen. Chem. USSR (Engl. Transl.)*, 1979, **49**, 1947.	800
79JHC415	A. F. Janzen, G. N. Lypka and R. E. Wasylishen; *J. Heterocycl. Chem.*, 1979, **16**, 415.	323
79JHC1405	I. Lalezari, A. Shafiee and S. Sadeghi-Milani; *J. Heterocycl. Chem.*, 1979, **16**, 1405.	874
79JMC1264	D. E. Nichols and L. J. Kostuba; *J. Med. Chem.*, 1979, **22**, 1264.	188
79JOC155	P. F. Hudrlik, R. H. Schwartz and J. C. Hogan; *J. Org. Chem.*, 1979, **44**, 155.	357
79JOC391	Y. Ozaki, T. Iwasaki, H. Horikawa, M. Miyashi and K. Matsumoto; *J. Org. Chem.*, 1979, **44**, 391.	310
79JOC467	H. C. Brown and J. L. Hubbard; *J. Org. Chem.*, 1979, **44**, 467.	364
79JOC633	C. D. Bedford and A. T. Nielson; *J. Org. Chem.*, 1979, **44**, 633.	995
79JOC656	D. R. Morton and J. Hobbs; *J. Org. Chem.*, 1979, **44**, 656.	248
79JOC930	M. Sato, N. C. Gonnella and M. P. Cava; *J. Org. Chem.*, 1979, **44**, 930.	847, 848
79JOC1057	G. Singh and G. S. Reddy; *J. Org. Chem.*, 1979, **44**, 1057.	569
79JOC1177	L. A. Carpino and J. R. Williams; *J. Org. Chem.*, 1979, **44**, 1177.	71, 78
79JOC1352	A. Mendoza and D. S. Matteson; *J. Org. Chem.*, 1979, **44**, 1352.	849
79JOC1394	D. G. Naae; *J. Org. Chem.*, 1979, **44**, 1394.	34
79JOC1476	D. C. Green; *J. Org. Chem.*, 1979, **44**, 1476.	957
79JOC1736	K. Oka; *J. Org. Chem.*, 1979, **44**, 1736.	72
79JOC1883	D. L. J. Clive and S. M. Menchen; *J. Org. Chem.*, 1979, **44**, 1883.	238, 288
79JOC2326	R. Kupper and C. J. Michejda; *J. Org. Chem.*, 1979, **44**, 2326.	335
79JOC2807	P. Gosselin, S. Masson and A. Thuillier; *J. Org. Chem.*, 1979, **44**, 2807.	255
79JOC2813	M. S. Toy and R. S. Stringham; *J. Org. Chem.*, 1979, **44**, 2813.	46
79JOC2967	M. Mikolajczyk, S. Grzejszczak, A. Chefczynska and A. Zatorski; *J. Org. Chem.*, 1979, **44**, 2967.	346
79JOC3117	M. E. Christy, P. S. Anderson, S. F. Britcher, C. D. Colton, B. E. Evans, D. C. Remy and E. L. Engelhardt; *J. Org. Chem.*, 1979, **44**, 3117.	97
79JOC3279	J. J. Eisch and J. E. Galle; *J. Org. Chem.*, 1979, **44**, 3279.	960

79JOC3601	T. Cohen and R. B. Weisenfeld; *J. Org. Chem.*, 1979, **44**, 3601.	956
79JOC3727	B. S. Bal and H. W. Pinnick; *J. Org. Chem.*, 1979, **44**, 3727.	198
79JOC3842	G. A. Olah, J. T. Welch, Y. D. Vankar, M. Nojima, I. Kerekes and J. A. Olah; *J. Org. Chem.*, 1979, **44**, 3872.	32
79JOC4279	D. L. J. Clive and S. M. Menchen; *J. Org. Chem.*, 1979, **44**, 4279.	288
79JOC4543	O. Tsuge, H. Watanabe, K. Masuda and M. M. Yousif; *J. Org. Chem.*, 1979, **44**, 4543.	422
79JOC4623	R. B. Miller and G. McGarvey; *J. Org. Chem.*, 1979, **44**, 4623.	1059
79JOC4722	J. S. Weinberg and A. Miller; *J. Org. Chem.*, 1979, **44**, 4722.	54
79JOC4744	T. Cohen, R. B. Weisenfeld and R. E. Gapinski; *J. Org. Chem.*, 1979, **44**, 4744.	845
79JOC4825	R. A. Swaringen, Jr., D. A. Yeowell, J. C. Wisowaty, H. A. El-Sayad, E. L. Stewart and M. E. Darnofall; *J. Org. Chem.*, 1979, **44**, 4825.	193
79JOC4847	A. F. Kluge and I. S. Cloudsdale; *J. Org. Chem.*, 1979, **44**, 4847.	340
79JOM(165)C17	W. A. Herrmann, I. Schweizer, M. Creswick and I. Bernal; *J. Organomet. Chem.*, 1979, **165**, C17.	686
79JOM(166)339	E. J. Bulten and H. A. Budding; *J. Organomet. Chem.*, 1979, **166**, 339.	154
79JOM(168)C33	N. M. Loim, N. A. Abramova, Z. N. Parnes and D. N. Kursanov; *J. Organomet. Chem.*, 1979, **168**, C33.	706, 1036
79JOM(168)23	T. J. Barton and W. D. Wulff; *J. Organomet. Chem.*, 1979, **168**, 23.	613
79JOM(168)43	P. Rivière, M. Rivière-Baudet, S. Richelme, A. Castel and J. Satgé; *J. Organomet. Chem.*, 1979, **168**, 43.	521
79JOM(169)123	M. J. van Hamme and D. J. Burton; *J. Organomet. Chem.*, 1979, **169**, 123.	753
79JOM(170)95	Z. Dawoodi, C. Eaborn and A. Pidcock; *J. Organomet. Chem.*, 1979, **170**, 95.	722, 725
79JOM(170)259	D. S. Matteson and D. Majumdar; *J. Organomet. Chem.*, 1979, **170**, 259.	525
79JOM(172)11	T. J. Barton and C. R. Tully; *J. Organomet. Chem.*, 1979, **172**, 11.	647
79JOM(172)293	T. N. Mitchell and M. El-Behairy; *J. Organomet. Chem.*, 1979, **172**, 293.	699, 701
79JOM(174)121	A. Camus, N. Marsich, G. Nardin and L. Randaccio; *J. Organomet. Chem.*, 1979, **174**, 121.	696
79JOM(177)145	M. Obayashi, K. Utimoto and H. Nozaki; *J. Organomet. Chem.*, 1979, **177**, 145.	1061, 1062
79JOM(178)157	L. Maier; *J. Organomet. Chem.*, 1979, **178**, 157.	125, 473
79JOM(179)377	M. Ishikawa, H. Sugisawa, K. Yamamoto and M. Kumada; *J. Organomet. Chem.*, 1979, **179**, 377.	1048
79JOM(179)421	A. Kühn and H. Werner; *J. Organomet. Chem.*, 1979, **179**, 421.	693
79JOM(179)439	H. Werner and A. Kühn; *J. Organomet. Chem.*, 1979, **179**, 439.	693
79JOM(181)425	A. G. Osborne and R. H. Whiteley; *J. Organomet. Chem.*, 1979, **181**, 425.	685, 706, 719
79JOM(182)C20	J. R. Moss and J. C. Spiers; *J. Organomet. Chem.*, 1979, **182**, C20.	584
79JOM(182)77	L. Knoll; *J. Organomet. Chem.*, 1979, **182**, 77.	580
79JOM(182)251	G. Blaschke, H. Schmidbaur and W. C. Kaska; *J. Organomet. Chem.*, 1979, **182**, 251.	585
79JOM(182)455	D. Seyferth and K. R. Wursthorn; *J. Organomet. Chem.*, 1979, **182**, 455.	735
79JOM(182)465	H. Weichmann, B. Ochsler, I. Duchek and A. Tzschach; *J. Organomet. Chem.*, 1979, **182**, 465.	551, 570, 588
79JOM(182)537	V. I. Sokolov, L. L. Troitskaya and O. A. Reutov; *J. Organomet. Chem.*, 1979, **182**, 537.	720
79JOU2207	D. I. Makhon'kov and N. S. Zefirov; *J. Org. Chem. USSR (Engl. Transl.)*, 1979, **15**, 2207.	800
79JPR43	G. Desimoni, G. Malaspina, P. P. Righetti and G. Tacconi; *J. Prakt. Chem.*, 1979, **321**, 43.	986
79JPR699	M. Augustin and G. Jahreis; *J. Prakt. Chem.*, 1979, **321**, 699.	899
79LA278	H. Eckert and I. Ugi; *Justus Liebigs Ann. Chem.*, 1979, 278.	20
79LA492	F. Hammerschmidt and E. Zbiral; *Justus Liebigs Ann. Chem.*, 1979, 492.	561
79LA522	W. Kantlehner, H. Gutbrod and P. Grob; *Justus Liebigs Ann. Chem.*, 1979, 522.	184
79LA842	F. Effenberger and D. Häbich; *Justus Liebigs Ann. Chem.*, 1979, 842.	512
79LA1362	W. Kantlehner and H. Gutbrod; *Justus Liebigs Ann. Chem.*, 1979, 1362.	184
79LA1715	E. Schaumann and F. F. Grabley; *Justus Liebigs Ann. Chem.*, 1979, 1715.	842
79LA2096	W. Kantlehner, H. Jaus, L. Kienitz and H. Bredereck; *Justus Liebigs Ann. Chem.*, 1979, 2076.	894, 973, 974
79MI 407-01	E. K. Gafurov, K. M. Uyzbaev and A. V. Kazantsev; *Izv. Akad. Nauk Kaz. SSR, Ser. Khim.*, 1979, 55 (*Chem. Abstr.*, 1980, **92**, 198 466).	343
79MI 419-01	V. A. Danilenko; *Deposited Doc.*, VINITI 3782, 169 (1979) (*Chem. Abstr.* 1981, **94**, 157 020).	874
B-79MI 420-01	W. Kantlehner; in "Advances in Organic Chemistry, Methods and Results, vol. 9(2): Iminium Salts in Organic Chemistry," ed. H. Bohme and H. G. Viehe, Wiley, New York, 1979, p. 181.	881
B-79MI 424-01	D. C. Nonhebel, J. M. Tedder and J. C. Walton; "Radicals," Cambridge University Press, Cambridge, 1979.	1077
79OR(26)1	H. W. Gschwend and H. R. Rodriguez; *Org. React.*, 1979, **26**, 1.	539
79PJC187	Z. Matacz, H. Piotrowska and T. S. Urbanski; *Pol. J. Chem.*, 1979, **53**, 187.	428
79PJC541	J. Zon; *Pol. J. Chem.*, 1979, **52**, 541.	492
79PJC2327	E. Gruszecka, M. Soroka and P. Mastalerz; *Pol. J. Chem.*, 1979, **53**, 2327.	460
79PS(5)337	M. Fild and W. Handke; *Phosphorus Sulfur*, 1979, **5**, 337.	553
79RTC258	A. M. van Leusen, F. J. Schaart and D. van Leusen; *Rec. Trav. Chim Pays-Bas*, 1979, **98**, 258.	908
79RTC371	C. G. Kruse, F. L. Jonkers, V. Dert and A. van der Gen; *Recl. Trav. Chim. Pays-Bas*, 1979, **98**, 371.	199
79S380	L. Garanti, G. Testoni and G. Zecchi; *Synthesis*, 1979, 380.	440
79S425	R. A. Moss and R. C. Munjal; *Synthesis*, 1979, 425.	18
79S455	R. S. Sukhai and L. Brandsma; *Synthesis*, 1979, 455.	842
79S535	R. Annunziata, M. Cinquini and F. Cozzi; *Synthesis*, 1979, 535.	272
79S701	H. C. Brown, J. L. Hubbard and K. Smith; *Synthesis*, 1979, 701.	364

79S724	R. Sterzycki; *Synthesis*, 1979, 724.	178, 179
79S747	I. A. Lazukina and V. P. Kukhar; *Synthesis*, 1979, 747.	110
79S810	S. Gronowitz and Z. Lidert; *Synthesis*, 1979, 810.	99
79S877	A. Burton, L. Hevesi, W. Dumont, A. Cravador and A. Krief; *Synthesis*, 1979, 877.	288
79S880	K. Ogura, S. Mitamura, K. Kishi and G. Tsuchihashi; *Synthesis*, 1979, 880.	857
79S893	J. Barluenga, P. J. Campos, J. C. Garcia-Martin, M. A. Roy and G. Asensio; *Synthesis*, 1979, 893.	152, 153
79S970	J. A. Amato, S. Karady, M. Sletzinger and L. M. Weinstock; *Synthesis*, 1979, 970.	48
79S985	J. Oleksyszyn, L. Subotkowska and P. Mastalerz; *Synthesis*, 1979, 985.	467
79SC53	J.-P. Duboudin and B. Jousseaume; *Synth. Commun.*, 1979, **9**, 53.	1065
79SC57	R. M. Jacobson and J. W. Clader; *Synth. Commun.*, 1979, **9**, 57.	199
79SC261	M. E. Garst; *Synth. Commun.*, 1979, **9**, 261.	553
79SC341	J. M. Lansinger and R. C. Ronald; *Synth. Commun.*, 1979, **9**, 341.	26
79SC575	N. de Kimpe, L. de Buyck, R. Verhé, F. Wychuyse and N. Schamp; *Synth. Commun.*, 1979, **9**, 575.	12
79T81	B. Davisia; *Tetrahedron*, 1979, **35**, 81.	346
79T551	W.-D. Rudorf, A. Schierhorn and M. Augustin; *Tetrahedron*, 1979, **35**, 551.	969
79T2661	P. Capriel and G. Binsch; *Tetrahedron*, 1979, **35**, 2661.	33, 34, 38
79TL87	C. C. Shen and C. Ainsworth; *Tetrahedron Lett.*, 1979, **20**, 87.	1047
79TL391	P. H. Bentley, G. Brooks, E. Hunt, M. L. Gilpin and I. I. Zomaya; *Tetrahedron Lett.*, 1979, **20**, 391.	824
79TL485	W. Dumont, D. Van Ende and A. Krief; *Tetrahedron Lett.*, 1979, **20**, 485.	375
79TL593	W. C. Still, J. H. McDonald, III, D. B. Collum and A. Mitra; *Tetrahedron Lett.*, 1979, **20**, 593.	378
79TL617	V. Reutrakul and P. Thamnusan; *Tetrahedron Lett.*, 1979, **20**, 617.	77
79TL689	J. M. Gerdes and L. G. Wade, Jr.; *Tetrahedron Lett.*, 1979, **20**, 689.	235
79TL1737	H. Böhme and B. Clément; *Tetrahedron Lett.*, 1979, **20**, 1737.	271
79TL1783	A. Itoh, K. Oshima and H. Nozaki; *Tetrahedron Lett.*, 1979, **20**, 1783.	393
79TL1913	C. W. Jefford, M. Acar, A. Delay, J. Mareda and U. Burger, *Tetrahedron Lett.*, 1979, **20**, 1913.	8
79TL2047	T. Hata, M. Nakajima and M. Sekine; *Tetrahedron Lett.*, 1979, **20**, 2047.	918
79TL2179	I. Paterson and I. Fleming; *Tetrahedron Lett.*, 1979, **20**, 2179.	966
79TL2415	M. Yoshifuji, J. Tagawa and N. Inamoto; *Tetrahedron Lett.*, 1979, **20**, 2415.	346
79TL2925	J. W. Scheeren, F. J. M. Dahmen and C. G. Bakker; *Tetrahedron Lett.*, 1979, **20**, 2925.	825
79TL3329	F. Sales and F. Serratosa; *Tetrahedron Lett.*, 1979, **20**, 3329.	884
79TL3375	G. W. Gokel, H. M. Gerdes, D. E. Miles, J. M. Hufnal and G. A. Zerby; *Tetrahedron Lett.*, 1979, **20**, 3375.	236
79TL3465	M. Isobe, M. Kitamura and T. Goto; *Tetrahedron Lett.*, 1979, **20**, 3465.	937
79TL3643	G. H. Hakimelahi and G. Just; *Tetrahedron Lett.*, 1979, **20**, 3643.	12
79TL4029	T. H. Chan, T. Aida, P. W. K. Lau, V. Gorys and D. N. Harpp; *Tetrahedron Lett.*, 1979, **20**, 4029.	831
79TL4273	R. R. Schmidt, J. Talbiersky and P. Russegger; *Tetrahedron Lett.*, 1979, **20**, 4273.	949
79TL4277	R. R. Schmidt, H. Speer and B. Schmid; *Tetrahedron Lett.*, 1979, **20**, 4277.	958
79TL5011	M. Ogata, H. Matsumoto, S. Kida and S. Shimizu; *Tetrahedron Lett.*, 1979, 5011.	407
79ZAAC(457)203	K. H. Worms and H. Blum; *Z. Anorg. Allg. Chem.*, 1979, **457**, 203.	561
79ZAAC(459)7	R. Appel and M. Huppertz; *Z. Anorg. Allg. Chem.*, 1979, **459**, 7.	112
79ZAAC(459)131	K. Issleib and H. Schmidt; *Z. Anorg. Allg. Chem.*, 1979, **459**, 131.	997, 998
79ZC57	H. Oehme and E. Leissring; *Z. Chem.*, 1979, **19**, 57.	337
79ZC192	E. Fanghaenel and H. Poleschner; *Z. Chem.*, 1979, **19**, 192.	874
79ZC253	C. Wasielewski, K. Antezak and J. Rachon; *Z. Chem.*, 1979, **19**, 253.	460
79ZC416	H. Oehme and E. Leissring; *Z. Chem.*, 1979, **19**, 416.	548
79ZC417	B. Walther, R. Schöps, W. Kolbe and D. Scheller; *Z. Chem.*, 1979, **19**, 417.	338
79ZN(B)31	H. H. Karsch; *Z. Naturforsch., Teil B*, 1979, **34**, 31.	562
79ZN(B)1178	H. H. Karsch; *Z. Naturforsch., Teil B*, 1979, **34**, 1178.	551
79ZN(B)1218	T. E. Fraser, H.-J. Füller and H. Schmidbaur; *Z. Naturforsch., Teil B*, 1979, **34**, 1218.	583
79ZOB332	N. N. Magdesieva and V. A. Danilenko; *Zh. Obshch. Khim.*, 1979, **49**, 332.	928
79ZOB470	Z. S. Novikova, A. A. Prishchenko and I. F. Lutsenko; *Zh. Obshch. Khim.*, 1979, **49**, 470.	552, 562
79ZOB493	R. D. Gareev, G. M. Loginova and A. N. Pudovik; *Zh. Obshch. Khim.*, 1979, **49**, 493 (*Chem. Abstr.*, 1979, **91**, 4894w).	488
79ZOB712	Z. S. Novikova, A. A. Prishchenko and I. F. Lutsenko; *Zh. Obshch. Khim.*, 1979, **49**, 712.	550
79ZOB805	V. A. Ryazantsev and M. D. Stadnichuk; *Zh. Obshch. Khim.*, 1979, **49**, 805.	941
79ZOB812	E. V. Komissarova, N. N. Balyaer and M. D. Stadnichuk; *Zh. Obshch. Khim.*, 1979, **49**, 812.	941
79ZOB829	M. D. Stadnichuk, V. A. Ryazantsev and A. A. Petrov; *Zh. Obshch. Khim.*, 1979, **49**, 829.	948
79ZOB1087	E. P. Nesynov and T. F. Aldokhina; *Zh. Obshch. Khim.*, 1979, **49**, 1087 (Engl. Transl. 945).	992
79ZOB1446	M. M. Kabachnik, A. A. Prishchenko, Z. S. Novikova and I. F. Lutsenko; *Zh. Obshch. Khim.*, 1979, **49**, 1446.	552, 562
79ZOB1470	V. P. Kukhar and E. I. Sagina; *Zh. Obshch. Khim.*, 1979, **49**, 1470.	561
79ZOB1772	F. V. Bagrov and N. E. Galkina; *Zh. Obshch. Khim.*, 1979, **49**, 1772.	588
79ZOB1857	N. N. Magdesieva and V. A. Danilenko; *Zh. Obshch. Khim.*, 1979, **49**, 1857.	928
79ZOB1910	M. V. Proskurnina, N. B. Karlstedt and M. V. Livantsov; *Zh. Obshch. Khim.*, 1979, **49**, 1910.	914
79ZOB1956	V. I. Vysotskii, A. S. Skobun and M. N. Tilichenko; *Zh. Obshch. Khim.*, 1979, **49**, 1956.	339

79ZOB2198	N. N. Magdesieva and V. A. Danilenko; *Zh. Obshch. Khim.*, 1979, **49**, 2198.	928
79ZOB2217	V. P. Kukhar, E. I. Sagina and N. G. Pavlenko; *Zh. Obshch. Khim.*, 1979, **49**, 2217 (Engl. Transl. 1947).	555, 1004, 1008
79ZOB2785	G. S. Bogdanova, G. S. Zaitseva and I. F. Lutsenko; *Zh. Obshch. Khim.*, 1979, **49**, 2785 (*Chem. Abstr.*, 1980, **92**, 198 481y).	616
79ZOR1099	N. S. Zefirov and V. A. Palyulin; *Zh. Org. Khim.*, 1979, **15**, 1099 (Engl. Transl. 981).	970
79ZOR1381	A. A. Yatsishin and N. D. Bodnarchuk; *Zh. Org. Khim.*, 1979, **15**, 1381 (Engl. Transl. 1234).	971
79ZOR1477	V. M. Neplyuev, T. A. Sinenko, V. P. Kukhar and P. S. Pel'kis; *Zh. Org. Khim.*, 1979, **15**, 1477 (*Chem. Abstr.* 1979, **91**, 192 941).	865
79ZOR1824	S. E. Tolchinskii, A. V. Dogadina, I. A. Maretina and A. A. Petrov; *Zh. Org. Khim.*, 1979, **15**, 1824 (Engl. Transl. 1642).	976
79ZOR2356	A. I. Dycenko and P. S. Pel'kis; *Zh. Org. Khim.*, 1979, **15**, 2356 (Engl. Transl. 2134).	995
79ZOR2416	I. I. Maletina, A. A. Miranova, T. I. Savina and Y. U. Yagupol'skii; *Zh. Org. Khim.*, 1979, **15**, 2416 (*Chem. Abstr.* 1980, **92**, 180 766).	864
79ZOR2596	M. L. Petrov, V. Z. Laishev and A. A. Petrov; *Zh. Org. Khim.*, 1979, **15**, 2596 (*Chem. Abstr.* 1980, **92**, 163 912).	877
80ACS(A)577	R. J. S. Beer, A. Hordvik, I. Pederson and H. Singh; *Acta Chem. Scand., Ser. A*, 1980, **34**, 577.	863
80AG208	A. Heintz, R. N. Lichtenthaler and K. Schafer; *Angew. Chem.*, 1980, **92**, 208.	988
80AG632	H. Sakurai, H. Tobita, M. Kira and Y. Nakadaira; *Angew. Chem.*, 1980, **92**, 632.	1052
80AG(E)66	H. A. Staab, J. Ippen, C. Tao-Pen, C. Krieger, and B. Starker; *Angew. Chem., Int. Ed. Engl.*, 1980, **19**, 66.	851
80AG(E)303	D. Hoppe, L. Beckmann and R. Follmann; *Angew. Chem., Int. Ed. Engl.*, 1980, **19**, 303.	938, 957
80AG(E)723	F. Steinseifer and T. Kauffmann; *Angew. Chem., Int. Ed. Engl.*, 1980, **19**, 723.	596, 600
80AG(E)724	F. A. Neugebauer and H. Fisher; *Angew. Chem., Int. Ed. Engl.*, 1980, **19**, 724.	437
80AJC1073	J. R. Cannon, V. Lojanapiwatna, C. L. Raston, W. Sinchai and A. H. White; *Aust. J. Chem.*, 1980, **33**, 1073.	856
80AP(314)65	M.-D. Stachel and G. Papenberg; *Arch. Pharm.* (*Weinheim, Ger.*) 1980, **314**, 65.	889
80AX(B)1344	P. L. Chiche, J. Galy, F. Mathey and G. Thiollet; *Acta Crystallogr.*, Part B, 1980, 1344.	1039
80BCJ789	D. Terunuma, K. Murakami, M. Kokubo, K. Senda and H. Nohira; *Bull. Chem. Soc. Jpn.*, 1980, **53**, 789.	506
80BCJ1061	M. Asaoka, N. Yangida, N. Sugimura and H. Takei; *Bull. Chem. Soc. Jpn.*, 1980, **53**, 1061.	206
80BCJ1138	T. Hayashi, T. Mise, M. Fukushima, M. Kagotani, N. Nagashima, Y. Hamada, A. Matsumoto, S. Kawakami, M. Konishi, K. Yamamoto and M. Kumada; *Bull Chem. Soc. Jpn.*, 1980, **53**, 1138.	1033
80BCJ1661	J. Nakayama, M. Imura and M. Hoshino; *Bull. Chem. Soc. Jpn.*, 1980, **53**, 1661.	850
80BCJ3176	Y. Yamamoto and H. Sugimoto; *Bull. Chem. Soc. Jpn.*, 1980, **53**, 3176.	586, 587
80BCJ3619	K. Tanaka, S. Matsui and A. Kaji; *Bull. Chem. Soc. Jpn.*, 1980, **53**, 3619.	236
80BP2029827	M. Yamabe, S. Kumai and S. Munekata (Asaki Glass Co. Ltd.); *Br. Pat.* 2 029 827 (*Chem. Abstr.*, 1980, **93**, 238 834v).	767
80BSB759	V. M. Thuy, H. Petit and P. Maitte; *Bull. Soc. Chim. Belg.*, 1980, **89**, 759.	181
80CA186471c	A. M. Torgomyan, A. S. Pogosyan, M. Zh. Ovakimyan and M. G. Indzhikyan; *Chem. Abstr.*, 1980, **93**, 186 471c (*Arm. Khim. Zh.*, 1980, **33**, 408).	914
80CA6622a	A. M. Torgomyan, M. Zh. Ovakimyan and M. G. Indzhikyan; *Chem. Abstr.*, 1980, **92**, 6622a (*Arm. Khim. Zh.*, 1979, **32**, 288).	914
80CA71025t	I. L. Knunyants and M. D. Bargamova; *Chem. Abstr.*, 1980, **93**, 71 025t (*USSR Pat.* 686 288).	731
80CAR(80)C17	S. Hanessian, C. Basquet and N. Lehong; *Carbohydr. Res.*, 1980, **80**, C17.	224
80CAR(86)C3	S. Hanessian and Y. Guindon; *Carbohydr. Res.*, 1980, **86**, C3.	224
80CB902	H. Schmidbaur, U. Deschler, B. Zimmer-Gasser, D. Neugebauer and U. Schubert; *Chem. Ber.*, 1980, **113**, 902.	576
80CB1072	H. Werner and H.-J. Kraus; *Chem. Ber.*, 1980, 1072.	693
80CB1226	F. A. Neugebauer and H. Fischer; *Chem. Ber.*, 1980, **113**, 1226.	992
80CB1290	N. Meyer and D. Seebach; *Chem. Ber.*, 1980, **113**, 1290.	378
80CB1480	H. Schmidbaur, G. Müller and G. Blaschke; *Chem. Ber.*, 1980, **113**, 1480.	574
80CB1549	E. Niecke and D.-A. Wildbredt; *Chem. Ber.*, 1980, **113**, 1549.	558
80CC39	D. S. Matteson and D. Majumdar; *J. Chem. Soc., Chem. Commun.*, 1980, 39.	642, 645
80CC158	M. G. Barlow, R. N. Haszeldine and C. J. Peck; *J. Chem. Soc., Chem. Commun.*, 1980, 158.	46
80CC477	A. P. Kozikowski, K. L. Sorgi and R. J. Schmiesing; *J. Chem. Soc., Chem. Commun.*, 1980, 477.	187
80CC506	D. J. Buckley, S. Kulkowit and A. McKervey; *J. Chem. Soc., Chem. Commun.*, 1980, 506.	90
80CC730	M. M. Campbell and N. Carruthers; *J. Chem. Soc., Chem. Commun.*, 1980, 730.	487
80CC749	J. L. Davidson, L. Manojlovic-Muir, K. W. Muir and A. N. Keith; *J. Chem. Soc., Chem. Commun.*, 1980, 749.	714
80CC790	W. E. Bottomley and G. V. Boyd; *J. Chem. Soc., Chem. Commun.*, 1980, 790.	881
80CC866	F. Wudl and D. Nalewajek; *J. Chem. Soc., Chem. Commun.*, 1980, 866.	874
80CC867	L. Y. Chiang, T. O. Poehler, A. N. Bloch and D. O. Cowan; *J. Chem. Soc., Chem. Commun.*, 1980, 867.	874
80CC879	H. M. Colquhoun and T. J. King; *J. Chem. Soc., Chem. Commun.*, 1980, 879.	799
80CC940	C. Bernard and L. Ghosez; *J. Chem. Soc. Chem. Commun.*, 1980, 940.	990
80CC1096	P. J. Kocienski; *J. Chem. Soc., Chem. Commun.*, 1980, 1096.	369
80CC1281	N. M. Boag, M. Green and F. G. A. Stone; *J. Chem. Soc., Chem. Commun.*, 1980, 1281.	692

80CC1284	M. F. Lappert and C. L. Raston; *J. Chem. Soc., Chem. Commun.*, 1980, 1284.	647, 657
80CJC130	E. E. Knaus, K. Avasthi, K. Redda and A. Benderly; *Can. J. Chem.*, 1980, **58**, 130.	949
80CJC878	T. P. Ahern, H. O. Fong, R. F. Langler and P. M. Mason; *Can. J. Chem.*, 1980, **58**, 878.	267
80CJC2780	G. Just, P. Potvin and G. H. Hakimelahi; *Can. J. Chem.*, 1980, **58**, 2780.	847, 848
80CL17	M. Asami and T. Mukaiyama; *Chem. Lett.*, 1980, 17.	408
80CL311	Y. Yamamoto; *Chem. Lett.*, 1980, 311.	586
80CL331	M. Isobe, M. Kitamura and T. Goto; *Chem. Lett.*, 1980, 331.	367, 937
80CL591	T. Yogo and A. Suzuki; *Chem. Lett.*, 1980, 591.	935, 949
80CL617	S. Torii, H. Okumoto and H. Tanaka; *Chem. Lett.*, 1980, 617.	197
80CL1209	H. Takei, H. Sugimura, M. Miura and H. Okamura; *Chem. Lett.*, 1980, 1209.	959
80CPB3098	M. Takamatsu and M. Sekiya; *Chem. Pharm. Bull.*, 1980, **28**, 3098.	885
80CR(C)183	P. Gosselin, S. Masson and A. Thuillier; *C. R. Hebd. Seances Acad. Sci., Ser. C*, 1980, **291**, 183.	255
80CSR477	P. D. Kennewell and J. B. Taylor; *Chem. Soc. Rev.*, 1980, **9**, 477.	330
80EGP142887	A. Tzschach, H. Weichmann, M. Klepel and B. Ochsler; *Ger. (East) Pat.* 142 887 (1980) (*Chem. Abstr.*, 1981, **95**, 81 220).	588
80EUP10396	P. N. Judson and C. R. H. White, *Eur. Pat.* 1980, 010 396.	970, 977, 978
80H(14)15	N. Kawahara and M. Katsuyana; *Heterocycles*, 1980, **14**, 15.	827
80H(14)97	M. Ogata, H. Matsumoto, S. Kida and S. Shimizu; *Heterocycles*, 1980, **14**, 97.	407
80H(14)271	M. V. Lakshmikantham and M. P. Cava; *Heterocycles*, 1980, **14**, 271.	847, 848, 870
80H(14)619	N. Kawahara, T. Itoh, M. Ogura and K. A. Watanabe; *Heterocycles*, 1980, **14**, 619.	893
80HCA102	W. Lubosch and D. Seebach; *Helv. Chim. Acta*, 1980, **63**, 102.	530
80HCA402	T. Winkler and W. L. Bencze; *Helv. Chim. Acta*, 1980, **63**, 402.	922
80HCA555	W. Oppolzer, S. C. Burford and F. Marazza; *Helv. Chim. Acta*, 1980, **63**, 555.	649
80HCA1190	U. Burger and F. Dreier; *Helv. Chim. Acta*, 1980, **63**, 1190.	410, 539
80HCA1947	P. Martin, E. Steiner and D. Belluš; *Helv. Chim. Acta*, 1980, **63**, 1947.	14
80HCA1960	E. Hungerbühler, R. Naef, D. Wasmuth, D. Seebach, H.-R. Loosli and A. Wehrli; *Helv. Chim. Acta*, 1980, **63**, 1960.	197
80IC693	B. H. Freeland, J. E. Hux, N. C. Payne and K. G. Tyers; *Inorg. Chem.*, 1980, **19**, 693.	713
80IC1021	O. T. Beachley Jr. and R. G. Simmons; *Inorg. Chem.*, 1980, **19**, 1021.	642, 659
80IC1982	S. O. Grim and E. D. Walton; *Inorg. Chem.*, 1980, **19**, 1982.	552, 563
80IC3195	S. O. Grim, P. H. Smith, I. J. Colquhoun and W. McFarlane; *Inorg. Chem.*, 1980, **19**, 3195.	550, 551, 576
80IC3207	K. I. Gell and J. Schwartz; *Inorg. Chem.*, 1980, **19**, 3207.	579
80IZV491	E. N. Tsvetkov, T. E. Kron, and M. I. Kabachnik; *Izv. Akad. Nauk SSSR, Ser. Khim.*, 1980, 491.	119, 120, 121
80IZV892	A. F. Eleev, G. A. Sokol'skii and I. L. Knunyants; *Izv. Akad. Nauk SSSR, Ser. Khim.*, 1980, 892 (*Chem. Abstr.* 1980, **93**, 132 018).	867
80IZV952	B. A. Arbuzov, O. A. Erastov and G. N. Nikonov; *Izv. Akad. Nauk SSSR, Ser. Khim.*, 1980, 952 (*Chem. Abstr.*, 1980, **93**, 95 346z).	457
80IZV1125	Yu. P. Belov, G. B. Rakhnovich, V. A. Danankov, N. N. Godovikov, G. G. Aleksandrov and Yu. T. Struchkov; *Izv. Akad. Nauk SSSR, Ser. Khim.*, 1980, 1125 (*Chem. Abstr.*, 1980, **93**, 132 563r).	461
80IZV1147	V. M. Vorontsov, V. S. Khotimskii, S. G. Durgar'yan and A. M. Krapivin; *Izv. Akad. Nauk SSSR, Ser. Khim.*, 1980, 1147 (*Chem. Abstr.*, 1980, **93**, 132 546n).	966
80IZV1438	B. A. Arbuzov, O. A. Erastov and G. N. Nikonov; *Izv. Akad. Nauk SSSR, Ser. Khim.*, 1980, 1438 (*Chem. Abstr.*, 1981, **93**, 204 761d).	457
80IZV1451	D. A. Bravo-Zhivotovskii, V. V. Neretin, O. A. Kruglaya and N. S. Vyazankin; *Izv. Akad. Nauk SSSR, Ser. Khim.*, 1980, 1451 (*Chem. Abstr.*, 1980, **93**, 186 487n).	638
80IZV1626	B. A. Arbuzov, O. A. Erastov, S. Sh. Khetagurova, T. A. Zyablicova, R. P. Arshinova and R. A. Kadyrov; *Izv. Acad. Nauk SSSR, Ser. Khim.*, 1980, 1626.	341
80IZV2417	B. A. Arbuzov, O. A. Erastov and G. N. Nikonov; *Izv. Akad. Nauk SSSR, Ser. Khim.*, 1980, 2417 (*Chem. Abstr.*, 1981, **94**, 47 421u).	453, 456
80IZV2429	G. A. Artamkina, A. B. Tuzikov, I. P. Beletskaya and O. A. Reutov; *Izv. Akad. Nauk SSSR, Ser. Khim.*, 1980, 2429 (*Chem. Abstr.*, 1981, **94**, 65 778v).	642
80IZV2668	N. N. Zemlyanskii, I. V. Borisova, Y. N. Luzikov, N. D. Kolosova, Y. A. Ustynyuk and I. P. Beletskaya; *Izv. Akad. Nauk SSSR, Ser. Khim.*, 1980, 2668 (*Chem. Abstr.*, 1981, **94**, 139 907k).	627
80JA994	G. de Lauzon, B. Deschamps, J. Fischer, F. Mathey and A. Mitschler; *J. Am. Chem. Soc.*, 1980, **102**, 994.	1035
80JA1198	S. W. Baldwin and M. T. Crimmins; *J. Am. Chem. Soc.*, 1980, **102**, 1198.	212
80JA1201	W. C. Still and C. Sreekumar; *J. Am. Chem. Soc.*, 1980, **102**, 1201.	378, 379
80JA1577	J. A. Soderquist and A. Hassner; *J. Am. Chem. Soc.*, 1980, **102**, 1577.	929, 930, 950, 951, 952
80JA1584	T. J. Barton and S. K. Hoekman; *J. Am. Chem. Soc.*, 1980, **102**, 1584.	602, 612, 619, 626, 627, 628, 637, 640, 652, 663
80JA1752	C. E. Sumner, Jr., P. E. Riley, R. E. Davis and R. Pettit; *J. Am. Chem. Soc.*, 1980, **102**, 1752.	679
80JA3095	R. B. Gammill, P. M. Gold and S. A. Mizsak; *J. Am. Chem. Soc.*, 1980, **102**, 3095.	253
80JA3248	S. Murata, M. Suzuki, and R. Noyori; *J. Am. Chem. Soc.*, 1980, **102**, 3248.	192
80JA4970	P. R. Jones, T. F. O. Lim and R. A. Pierce; *J. Am. Chem. Soc.*, 1980, **102**, 4970.	611
80JA5101	T. R. Halbert, M. E. Leonowicz and D. J. Maydonovitch; *J. Am. Chem. Soc.*, 1980, **102**, 5101.	687
80JA6615	J. B. Lambert, P. H. Mueller and P. P. Gaspar; *J. Am. Chem. Soc.*, 1980, **102**, 6615.	748

80JA6900	T. Cohen and J. R. Matz; *J. Am. Chem. Soc.*, 1980, **102**, 6900.	216, 218, 226, 380
80JA7125	A. I. Meyers and W. T. Hoeve; *J. Am. Chem. Soc.*, 1980, **102**, 7126.	530
80JA7588	D. S. Matteson and D. Majumdar; *J. Am. Chem. Soc.*, 1980, **102**, 7588.	362
80JA7590	D. S. Matteson and R. Ray; *J. Am. Chem. Soc.*, 1980, **102**, 7590.	362
80JA7929	B. M. Trost, M. Vaultier and M. L. Santiago; *J. Am. Chem. Soc.*, 1980, **102**, 7929.	329, 392
80JA7979	T. J. Barton and S. A. Jacobi; *J. Am. Chem. Soc.*, 1980, **102**, 7979.	617
80JA7993	E. Vedejs and G. R. Martinez; *J. Am. Chem. Soc.*, 1980, **102**, 7993.	517
80JCR(M)235	A. K. Datta, R. Fields and R. N. Haszeldine; *J. Chem. Res. (M)*, 1980, 235.	1048
80JCS(D)29	M. Berry, N. J. Cooper, M. L. H. Green and S. J. Simpson; *J. Chem. Soc., Dalton Trans.*, 1980, 29.	673
80JCS(D)435	J. A. D. Jeffreys; *J. Chem. Soc., Dalton Trans.*, 1980, 435.	688
80JCS(D)871	D. H. Brown, R. J. Cross and R. Keat; *J. Chem. Soc., Dalton Trans.*, 1980, 871.	552
80JCS(D)1771	R. A. Jones, G. Wilkinson, A. M. R. Galas, M. B. Hursthouse and K. M. A. Malik; *J. Chem. Soc., Dalton Trans.*, 1980, 1771.	680
80JCS(D)2182	N. M. Boag, M. Green, J. A. K. Howard, J. L. Spencer, R. F. D. Stansfield, M. D. O. Thomas, F. G. A. Stone and P. Woodward; *J. Chem. Soc., Dalton Trans.*, 1980, 2182.	692
80JCS(D)2315	R. J. Daroda, G. Wilkinson, M. B. Hursthouse, K. M. A. Malik and M. Thornton-Pett; *J. Chem. Soc., Dalton Trans.*, 1980, 2315.	719
80JCS(D)2428	M. J. S. Gynane, A. Hudson, M. J. Lappert, P. P. Power and H. Goldwhite; *J. Chem. Soc., Dalton Trans.*, 1980, 2428.	597, 598
80JCS(P1)487	A. N. Bell, R. Fields, R. N. Haszeldine and D. Moran; *J. Chem. Soc. Perkin Trans 1*, 1980, 487.	737
80JCS(P1)1407	W. R. Bowman and G. D. Richardson; *J. Chem. Soc., Perkin Trans. 1*, 1980, 1407.	327
80JCS(P1)1627	P.-E. Hansen; *J. Chem. Soc., Perkin Trans. 1*, 1980, 1627.	340
80JCS(P1)2535	T. A. Crab, P. J. Dawson and R. O. Williams; *J. Chem. Soc., Perkin Trans. 1*, 1980, 2535.	165
80JCS(P1)2693	T. Nishiwaki, E. Kawamura, N. Abe and M. Iori; *J. Chem. Soc., Perkin Trans. 1*, 1980, 2693.	850
80JCS(P2)883	G. Behrens, E. Bothe, G. Koltzenburg and D. Schulte-Frohlinde; *J. Chem. Soc., Perkin Trans. 2*, 1980, 883.	1078
80JFC(15)29	L. Field and A. A. Gallo; *J. Fluorine Chem.*, 1980, **15**, 29.	326
80JFC(15)543	D. J. Burton, S. Shinya and R. D. Howells; *J. Fluorine Chem.*, 1980, **15**, 543.	802
80JGU802	V. D. Romanenko, V. I. Tovstenko, É. S. Kozlov and L. N. Markovskii, *J. Gen. Chem., USSR (Engl. Trans.)*, 1980, **50**, 802.	121, 122
80JHC117	A. Shafiee, M. Vosooghi and R. Asgharian; *J. Heterocycl. Chem.*, 1980, **17**, 117.	870
80JHC1163	G. M. Coppola and M. J. Shapiro; *J. Heterocycl. Chem.*, 1980, **17**, 1163.	423
80JHC1655	C. Bak and K. Praefcke; *J. Heterocycl. Chem.*, 1980, **17**, 1655.	232
80JOC541	J. A. Soderquist and A. Hassner; *J. Org. Chem.*, 1980, **45**, 541.	930
80JOC752	R. K. Boeckman, Jr., P. C. Naegely and S. D. Arthur; *J. Org. Chem.*, 1980, **45**, 752.	71
80JOC933	T. E. Young and A. R. Oyler; *J. Org. Chem.*, 1980, **45**, 933.	907
80JOC1046	F. Cooke, R. Moerck, J. Schwindeman and P. Magnus; *J. Org. Chem.*, 1980, **45**, 1046.	369, 938, 942
80JOC1394	D. G. Naae; *J. Org. Chem.*, 1980, **45**, 1394.	52
80JOC1486	E. Gipstein, C. G. Willson and H. S. Sachdev; *J. Org. Chem.*, 1980, **45**, 1486.	286
80JOC1880	J. D. M. Herscheid, R. J. F. Nivard, M. W. Tijhuis, H. P. H. Scholten and H. C. J. Ottenheijm; *J. Org. Chem.*, 1980, **45**, 1880.	304
80JOC2548	A. P. Kozikowski and A. Ames; *J. Org. Chem.*, 1980, **45**, 2548.	538
80JOC2597	F. A. Davis and P. A. Mancinelli; *J. Org. Chem.*, 1980, **45**, 2597.	853
80JOC2698	R. M. Davidson and G. L. Kenyon; *J. Org. Chem.*, 1980, **45**, 2698.	563
80JOC2736	G. Zima, C. S. Barnum and D. Liotta; *J. Org. Chem.*, 1980, **45**, 2736.	401
80JOC2874	H. H. Wasserman, M. J. Hearn and R. E. Cochoy; *J. Org. Chem.*, 1980, **45**, 2874.	162
80JOC2919	R. Kupper and J. Michejda; *J. Org. Chem.*, 1980, **45**, 2919.	896
80JOC3028	W. E. Fristad, T. R. Bailey and L. A. Paquette; *J. Org. Chem.*, 1980, **45**, 3028.	357
80JOC3571	J. A. Soderquist and H. C. Brown; *J. Org. Chem.*, 1980, **45**, 3571.	356, 643
80JOC3634	G. W. Gokel, H. M. Gerdes and D. M. Dishong; *J. Org. Chem.*, 1980, **45**, 3634.	235
80JOC4126	J. G. de Vries and R. M. Kellogg; *J. Org. Chem.*, 1980, **45**, 4126.	234
80JOC4283	W. R. Roush and H. R. Gillis; *J. Org. Chem.*, 1980, **45**, 4283.	179
80JOC4302	E. P. Ahern, K. J. Dignam and A. F. Hegarty; *J. Org. Chem.*, 1980, **45**, 4302.	990
80JOC4683	Y. Kobayashi, S. Fujino, H. Hamana, Y. Hanzawa, S. Morita and J. Kumadaki; *J. Org. Chem.*, 1980, **45**, 4683.	549
80JOC4959	F. T. Oakes and J. F. Sebastian; *J. Org. Chem.*, 1980, **45**, 4959.	949, 950
80JOC5227	H.-J. Reich and W. W. Willis, Jr.; *J. Org. Chem.*, 1980, **45**, 5227.	964
80JOC5333	T. M. Wade and T. Khéribet; *J. Org. Chem.*, 1980, **45**, 5333.	10
80JOM(185)C6	A. N. Nesmeyanov, N. N. Sedova, V. A. Sazonova and S. K. Moiseev; *J. Organomet. Chem.*, 1980, **185**; C6.	723
80JOM(186)427	G. van Koten, C. E. Schaap, J. T. B. H. Jastrebski and J. G. Noltes; *J. Organomet. Chem.*, 1980, **186**, 427.	724
80JOM(187)81	S. Ermer, R. Karpelus, S. Miura, E. Rosenberg, A. Tiripicchio and A. M. M. Lanfredi; *J. Organomet. Chem.*, 1980, **187**, 81.	683
80JOM(187)331	C. Santini-Scampucci and J. G. Riess; *J. Organomet. Chem.*, 1980, **187**, 331.	579
80JOM(188)151	N. Auner and J. Grobe; *J. Organomet. Chem.*, 1980, **188**, 151.	608, 610, 621
80JOM(191)7	D. K. Breitinger, K. Geibel, W. Kress and R. Sendelbeck; *J. Organomet. Chem.*, 1980, **191**, 7.	697

80JOM(193)21	F. Brady, K. Henrick, R. W. Matthews and D. G. Gillies; *J. Organomet. Chem.*, 1980, **193**, 21.	659
80JOM(193)47	L. Knoll; *J. Organomet. Chem.*, 1980, **193**, 47.	580, 581
80JOM(193)307	K. Kellner, A. Tzschach, Z. Nagy-Magos and L. Marko; *J. Organometal. Chem.*, 1980, **193**, 307.	455
80JOM(194)9	G. Sonnek, G. Muller and K. G. Baumgarten; *J. Organomet. Chem.*, 1980, **194**, 9.	397
80JOM(194)147	M. Ishikawa, K. Nishimura, H. Sugisawa and M. Kumada; *J. Organomet. Chem.*, 1980, **194**, 147.	1048
80JOM(197)335	A. Marinetti, E. Sappa, A. Tiripicchio and M. Tiripicchio-Camellini; *J. Organomet. Chem.*, 1980, **197**, 335.	713
80JOM(198)15	G. L. Larson, R. Arguelles, O. Rosario and S. Sandoval; *J. Organomet. Chem.*, 1980, **198**, 15.	357
80JOM(200)287	H. Schmidbaur; *J. Organomet. Chem.*, 1980, **200**, 287. 557, 574, 575, 577, 583, 584, 585, 586, 587, 588	
80JOM(201)C31	A. D. Clauss, P. A. Dimas and J. R. Shapley; *J. Organomet. Chem.*, 1980, **201**, C31.	686
80JOM(201)97	G. Sonnek, H. Reinheckel, S. Pasynkiewicz, M. Boleslawski and T. Dluzniewski; *J. Organomet. Chem.*, 1980, **201**, 97.	698
80JOM(201)197	H. Hillgärtner, W. P. Neumann, W. Schulten and A. K. Zarkadis; *J. Organomet. Chem.*, 1980, **201**, 197.	141, 154, 608, 616
80JOM(201)269	A. A. Pasynskii, Yu. V. Skripkin, V. T. Kalinnikov, M. A. Poraikoshits, A. Santsyshkina, G. G. Sadikov and V. N. Ostrikova; *J. Organomet. Chem.*, 1980, **201**, 269.	717
80JOM(201)477	M. Wada and Y. Koyama; *J. Organometal. Chem.*, 1980, **201**, 477.	955
80JOM(202)C49	J. A. Beck, S. A. R. Knox, G. H. Riding, G. E. Taylor and M. J. Winter; *J. Organomet. Chem.*, 1980, **202**, C49.	674
80JOM(202)C58	V. I. Sokolov, L. L. Troitskaya and O. A. Reutov; *J. Organomet. Chem.*, 1980, **202**, C58.	1033
80JOM(202)C95	A. J. Ashe, III and T. R. Diephouse; *J. Organomet. Chem.*, 1980, **202**, C95.	1042
80JOU1377	I. I. Lapkin, N. N. Pavlova, A. N. Nedugov and G. A. Gartman; *J. Org. Chem. USSR (Engl. Trans.)*, 1980, **16**, 1377.	239
80JOU1762	B. S. Drach, T. P. Popovich, V. N. Kalinin, V. A. Kovalev, G. B. Soifer, A. D. Gordeev and A. A. Kisilenko; *J. Org. Chem. USSR (Engl. Transl.)*, 1980, **16**, 1762.	794
80JPR87	H. Schafer and K. Gewald; *J. Prakt. Chem.*, 1980, **322**, 87.	995
80JPR213	R. Gancarz and J. S. Wieczorek; *J. Prakt. Chem.*, 1980, **322**, 213.	463
80LA1	E. V. Dehmlow and M. Lissel; *Justus Liebigs Ann. Chem.*, 1980, 1.	18
80LA28	D. Stafforst and U. Schöllkopf; *Justus Liebigs Ann. Chem.*, 1980, 28.	538
80LA372	W. Kantlehner, I. C. Ivanov, W. W. Mergen and H. Bredereck; *Justus Liebigs Ann. Chem.*, 1980, 372.	972, 979
80LA1573	H. Griengl, G. Prischl and A. Bleikolm; *Justus Liebigs Ann. Chem.*, 1980, 1573.	415
80LA1665	K. Hartke, G. Henssen and T. Kissel; *Justus Liebigs Ann. Chem.*, 1980, 1665.	229
80LA1765	D. Hoppe, R. Follmann and L. Beckmann; *Justus Liebigs Ann. Chem.*, 1980, 1765.	392
80LA1919	A. Schwarz, S. Uray and H. Junek; *Justus Liebigs Ann. Chem.*, 1980, 1919.	55, 57
80LA2061	H. J. Bestmann and H. Frey; *Justus Leibig's Ann. Chem.*, 1980, 2061.	966
80MI 403-01	M. Momtchev, M. Ivanova and B. Blagoev; *Izv. Khim.*, 1980, **13**, 357 (*Chem. Abstr.*, 1981, **95**, 132 422k).	147
80MI 404-01	P. A. Manthorpe and R. Gigg; *Methods Carbohydr. Chem.*, 190, **8**, 305.	176
B-80MI 404-02	R. G. Bergstorm; in "The Chemistry of Ethers, Crown Ethers, Hydroxy Groups and Sulfur Analogues," ed. S. Patai, Wiley, New York, 1980, vol. 2, p. 881.	176
B-80MI 405-01	K. Pihlaja and P. Pasanen; in "Chemistry of Functional Groups," ed. S. Patai, Wiley, New York, 1980, Suppl. E Pt. 2, p. 821.	229
80MI 417-01	A. Uehara and R. Tsuchiya; *Sci. Rep. Kanazawa Univ.*, 1980, **25**, 83 (*Chem. Abstr.*, 1981, **95**, 168 593f).	763
B-80MI 419-01	"The Chemistry of Ketenes, Allenes and Related Compounds," ed. S. Patai, Wiley Interscience, New York, 1980.	824, 842
80MI 420-01	A. M. van Leusen; *Lectures in Heterocyclic Chemistry*, 1980, **5**, SIII.	908
80PAC615	T. J. Barton; *Pure Appl. Chem.*, 1980, **52**, 615.	1055
80PJC233	J. Zygmunt, U. Walkowiak and P. Masterlerz; *Pol. J. Chem.*, 1980, **54**, 233.	343
80PS(8)67	L. Maier and M. J. Smith; *Phosphorus Sulfur*, 1980, **8**, 67.	470
80PS(9)121	R. Bodalski, T. J. Michalski and J. Monkiewicz; *Phosphorus Sulfur*, 1980, **9**, 121.	585
80RTC39	J. A. Loontjes, M. van der Leijs and B. Zwanenburg; *Recl. Trav. Chim. Pays-Bas*, 1980, **99**, 39.	270, 272
80RTC45	M. van der Leij, H. J. M. Strijtveen and B. Zwanenburg; *Recl. Trav. Chim. Pays-Bas*, 1980, **99**, 45.	859
80S72	F. M. Hauser, R. P. Rhee, S. Prasanna, S. M. Weinreb and J. H. Dodd; *Synthesis*, 1980, 72.	346
80S318	S. Ambasht, S. K. Chiu, P. E. Peterson and J. Queen; *Synthesis*, 1980, 318.	352
80S322	Z. Lidert and S. Gronowitz; *Synthesis*, 1980, 322.	327
80S325	D. van Leusen and A. M. van Leusen; *Synthesis*, 1980, 325.	538
80S554	W. G. Taylor; *Synthesis*, 1980, 554.	739
80S565	P. Messinger and K. Kusuma; *Synthesis*, 1980, 565.	333
80S644	J. Villieras and M. Rambaud; *Synthesis*, 1980, 644.	14, 22
80S736	K. Ogura, Y. Ito and G. Tsuchihashi; *Synthesis*, 1980, 736.	856
80S748	A. Kumar, A. Aggarwal, H. Ila and H. Junjappa; *Synthesis*, 1980, 748.	900
80S823	V. D. Romanenko, V. I. Tovstenko and L. N. Markovski; *Synthesis*, 1980, 823.	122
80S907	E. Schaumann, U. Wriede and J. Ehlers; *Synthesis*, 1980, 907.	851
80S952	N. Ono, H. Miyake, T. Saito and A. Kaji; *Synthesis*, 1980, 952.	249

80S1028	Z. H. Kudzin and A. Kotyński; *Synthesis*, 1980, 1028.	472
80SC83	F. Huet, A. Lechevallier and J. M. Conia; *Synth. Commun.*, 1980, **10**, 83.	188
80SC429	G. H. Hakimelahi and G. Just; *Synth. Commun.*, 1980, **10**, 429.	491
80SC911	R. A. Holton and R. V. Nelson; *Synth. Commun.*, 1980, **10**, 911.	219
80T1345	D. Krois, E. Langer and H. Lehner; *Tetrahedron*, 1980, **36**, 1345.	163
80T1753	M. F. Dube and J. W. Timberlake; *Tetrahedron*, 1980, **36**, 1753.	441
80T1763	A. Pierdet, L. Nédélec, V. Delaroff and A. Allais; *Tetrahedron*, 1980, **36**, 1763.	174
80T1791	W. D. Rudorf; *Tetrahedron*, 1980, **36**, 1791.	900
80T2505	J. Bergman, L. Renström and B. Sjöberg; *Tetrahedron*, 1980, **36**, 2505.	426
80T2531	A. Krief; *Tetrahedron*, 1980, **33**, 2531. 377, 390, 394, 395, 398, 400, 527, 646, 948	
80TL11	E. Ehlinger and P. Magnus; *Tetrahedron Lett.*, 1980, **21**, 11.	648, 655
80TL223	M. Henriet, M. Houtekie, P. Techy, R. Touillaux and L. Ghosez; *Tetrahedron Lett.*, 1980, **21**, 223.	790
80TL395	P. J. Giddings, D. I. John and E. J. Thomas; *Tetrahedron Lett.*, 1980, **21**, 395.	90
80TL429	A. Hassner and J. A. Soderquist; *Tetrahedron Lett.*, 1980, **21**, 429.	614
80TL1357	T. Tsunoda, M. Suzuki and R. Noyori; *Tetrahedron Lett.*, 1980, **21**, 1357.	181, 182
80TL1405	G. Märkl and D. Rudnick; *Tetrahedron Lett.*, 1980, **21**, 1405.	627
80TL1677	D. J. Ager and R. C. Cookson; *Tetrahedron Lett.*, 1980, **21**, 1677.	365
80TL1845	V. G. Märkl, G. Y. Jin and Ch. Schoerner; *Tetrahedron Lett.*, 1980, **21**, 1845.	455
80TL2233	K. Ogura, M. Fujita and H. Iida; *Tetrahedron Lett.*, 1980, **21**, 2233.	262
80TL2555	M. Suda; *Tetrahedron Lett.*, 1980, **21**, 2555.	731
80TL2557	J. Nokami, M. Hatate, S. Wakabayashi and R. Okawara; *Tetrahedron Lett.*, 1980, **21**, 2557.	220
80TL2803	H. J. Tilhard, H. Ahlers and T. Kauffmann; *Tetrahedron Lett.*, 1980, **21**, 2803.	565
80TL2931	T. A. Mastryukova, I. M. Aladzheva, I. V. Leont'eva, P. V. Petrovskii, E. I. Fedin and M. I. Kabachnik; *Tetrahedron Lett.*, 1980, **21**, 2931.	551
80TL2949	H. Kapnang and G. Charles; *Tetrahedron Lett.*, 1980, **21**, 2949.	300
80TL3077	H. Sakurai, Y. Nakadaira, M. Kira and H. Tobita; *Tetrahedron Lett.*, 1980, **21**, 3077.	1052
80TL3089	M. Poje, M. Sikirica, I. Vicković and M. Bruvo; *Tetrahedron Lett.*, 1980, **21**, 3089.	268
80TL3209	I. Kuwajima, S. Hoshino, T. Tanaka and M. Shimizu; *Tetrahedron Lett.*, 1980, **21**, 3209.	400
80TL3339	R. D. Little and S. O. Myong; *Tetrahedron Lett.*, 1980, **21**, 3339.	234
80TL3343	B. H. Lipshutz and J. J. Pegram; *Tetrahedron Lett.*, 1980, **21**, 3343.	199
80TL3451	D. E. Seitz and A. Zapata; *Tetrahedron Lett.*, 1980, **21**, 3451.	652, 660
80TL3463	E. J. Corey and G. Goto; *Tetrahedron Lett.*, 1980, **21**, 3463.	171
80TL3467	G. Markl and G. Yu. Jin; *Tetrahedron Lett.*, 1980, **21**, 3467.	456
80TL3579	K. Oka, A. Dobashi and S. Hara; *Tetrahedron Lett.*, 1980, **21**, 3579.	70
80TL3919	T. Oshima, R. Nishioka and T. Nagai; *Tetrahedron Lett.*, 1980, **21**, 3919.	198
80TL3984	O. I. Kolodyazhnyi; *Tetrahedron Lett.*, 1980, **21**, 3984.	574, 587
80TL4657	P. Beslin and P. Metzner; *Tetrahedron Lett.*, 1980, **21**, 4657.	257
80TL4763	D. J. Ager; *Tetrahedron Lett.*, 1980, **21**, 4763.	250
80USP4224241	H. Neumaier; *US Pat.* 4 224 241 (1980) (*Chem. Abstr.*, 1980, **92**, 146 904u).	125
80ZAAC(464)107	G. Fritz and K. Hohenberger; *Z. Anorg. Allg. Chem.*, 1980, **464**, 107.	622
80ZAAC(469)109	K. Issleib, A. Balszuweit, R. Müller and W. Mögelin; *Z. Anorg. Allg. Chem.*, 1980, **469**, 109.	463
80ZAAC(470)144	G. von Becker and G. Gutekunst; *Z. Anorg. Allg. Chem.*, 1980, **470**, 144.	594
80ZAAC(470)157	G. von Becker and G. Gutekunst; *Z. Anorg. Allg. Chem.*, 1980, **470**, 157.	594
80ZC152	K. Kellner, H. Schultz and A. Tzschach; *Z. Chem.*, 1980, **20**, 152.	477, 481, 503
80ZC437	G. Weber, G. Mann, H. Wilde and S. Hauptmann; *Z. Chem.*, 1980, **20**, 437.	432
80ZN(B)584	H. Schmidbaur and G. Blaschke; *Z. Naturforsch., Teil B*, 1980, **35**, 584.	582
80ZN(B)832	S. O. Grim, L. C. Satek and J. D. Mitchell; *Z. Naturforsch., Teil B*, 1980, **35**, 832.	551
80ZN(B)1376	K.-D. Fuhrmann and F. Huber; *Z. Naturforsch. Teil B*, 1980, **35b**, 1376.	153
80ZOB343	E. A. Stukalo, E. M. Yur'eva and L. N. Markovskii; *Zh. Obshch. Khim.*, 1980, **50**, 343 (Engl. Transl. 278).	1011
80ZOB573	N. N. Magdesieva and V. A. Danilenko; *Zh. Obshch. Khim.*, 1980, **50**, 573.	928
80ZOB692	M. A. Kazankova, V. A. Ilyushin, E. V. Ladeishchikova and I. F. Lutsenko; *Zh. Obshch. Khim.*, 1980, **50**, 692 (*Chem. Abstr.*, 1980, **93**, 239 546).	1055, 1058
80ZOB722	V. A. Ryazantsev and M. D. Stadnichuk; *Zh. Obshch. Khim.*, 1980, **50**, 722.	941
80ZOB751	M. G. Zimin, A. R. Burilov and A. N. Pudovik; *Zh. Obshch. Khim.*, 1980, **50**, 751 (*Chem. Abstr.*, 1980, **93**, 71 873t).	485
80ZOB949	A. A. Prishchenko, Z. S. Novikova and I. F. Lutsenko; *Zh. Obshch. Khim.*, 1980, **50**, 949.	550
80ZOB984	S. E. Tolchinskii, A. V. Dogadina, I. A. Maretina and A. A. Petrov; *Zh. Obshch. Khim.*, 1980, **50**, 984.	885
80ZOB989	Z. S. Novikova, A. A. Prishchenko, S. Ya. Skorobogatova, V. I. Martynov and I. F. Lutsenko; *Zh. Obshch. Khim.*, 1980, **50**, 989.	551
80ZOB1053	V. A. Ryazantsev, M. D. Stadnichuk and A. A. Petrov; *Zh. Obshch. Khim.*, 1980, **50**, 1053.	948
80ZOB1225	K. A. Petrov, V. A. Chauzov, S. V. Agafonov and N. V. Pazhitnova; *Zh. Obshch. Khim.*, 1980, **50**, 1225.	916
80ZOB1427	N. V. Komarov, A. A. Andreev, E. A. Kovtun and V. S. Senichev; *Zh. Obshch. Khim.*, 1980, **50**, 1427 (*Chem. Abstr.*, 1980, **93**, 220 874y).	541
80ZOB1764	T. K. Gar, N. Y. Khromova, V. M. Nosova and V. F. Mironov; *Zh. Obshch. Khim.*, 1980, **50**, 1764 (*Chem. Abstr.*, 1981, **94**, 15 823k).	641
80ZOB1881	A. A. Prishchenko, N. E. Nifant'ev, Z. S. Novikova and I. F. Lutsenko; *Zh. Obshch. Khim.*, 1980, **50**, 1881.	550
80ZOB2044	A. S. Medvedeva, L. P. Safronova and N. S. Vyazankin; *Zh. Obshch. Khim.*, 1980, **50**, 2044 (*Chem. Abstr.*, 1980, **94**, 47 430w).	1055

80ZOB2248	O. P. Lobanov, A. P. Martynyuk and B. S. Drach; *Zh. Obshch. Khim.*, 1980, **50**, 2248 (Engl. Transl. 1816).	999
80ZOB2424	E. S. Kozlov; *Zh. Obshch. Khim.*, 1980, **50**, 2424.	342
80ZOR13	N. A. Bunina, M. L. Petrov and A. A. Petrov; *Zh. Org. Khim.*, 1980, **16**, 13 (*Chem. Abstr.*, 1980, **93**, 8058).	834
80ZOR654	N. M. Magdesieva and V. A. Danilenko; *Zh. Org. Khim.*, 1980, **16**, 654 (*Chem. Abstr.* 1980, **93**, 95 336).	874
80ZOR720	S. I. Radchenko and K. S. Mingaleva; *Zh. Org. Khim.*, 1980, **16**, 720 (*Chem. Abstr.* 1980, **93**, 113 710).	841
80ZOR1748	K. B. Rall and A. I. Vil'davskaya; *Zh. Org. Khim.*, 1980, **16**, 1748.	
81AG(E)104	H.-J. Schmidt and H. J. Schafer; *Angew. Chem., Int. Ed. Engl.*, 1981, **20**, 104.	56
81AG(E)126	T. Hirao, A. Yamada, Y. Ohshiro and T. Agawa; *Angew. Chem., Int. Ed. Engl.*, 1981, **20**, 126.	538
81AG(E)127	R. Hanko and D. Hoppe; *Angew. Chem., Int. Ed. Engl.*, 1981, **20**, 127.	382
81AG(E)131	E. Niecke, W. W. Schoeller and D.-A. Wildbredt; *Angew. Chem., Int. Ed. Engl.*, 1981, **20**, 131.	484
81AG(E)581	L. Roesch, G. Altnau and W. H. Otto; *Angew. Chem., Int. Ed. Engl.*, 1981, **20**, 581.	355
81AG(E)585	J. C. Pommelet, C. Nyns, F. Lahousse, R. Merenyi and H. G. Viehe; *Angew. Chem., Int. Ed. Engl.*, 1981, **20**, 585.	71
81AG(E)586	H. Schmidbaur, U. Deschler and B. Milewski-Mahrla; *Angew. Chem., Int. Ed. Engl.*, 1981, **20**, 586.	583
81AG(E)597	G. Maier, G. Mihm and H. P. Reisenauer; *Angew. Chem., Int. Ed. Engl.*, 1981, **20**, 597.	610
81AG(E)887	D. A. Clemente, B. Rees, G. Bandoli, M. C. Biagini, B. Reiter and W. A. Herrmann; *Angew. Chem., Int. Ed. Engl.*, 1981, **20**, 887.	677
81AG(E)1038	H. J. Bestmann, K. Sühs and T. Röder; *Angew. Chem., Int. Ed. Engl.*, 1981, **20**, 1038.	574
81AG(E)1049	R. Korswagen, R. Alt, D. Speth and M. L. Ziegler; *Angew. Chem., Int. Ed. Engl.*, 1981, **20**, 1049.	679
81AJC181	D. J. Hannahm, R. A. J. Smith, I. Teah and R. T. Weavers; *Aust. J. Chem.*, 1981, **34**, 181.	655
81AOC51	M. Ishikawa and M. Kumada; *Adv. Organomet. Chem.* Academic Press, New York, 1981, vol. 19, p. 51.	1052
81AP65	H. D. Stachel and G. Papenberg; *Arch. Pharm. (Weinheim, Ger.)*, 1981, **314**, 65.	983
81AP(314)218	M. Menge, K. J. Münzenberg and E. Reimann; *Arch. Pharm. (Weinheim, Ger.)*, 1981, **314**, 218.	563
81BCJ817	M. Watanabe, S. Nakamori, H. Hasegawa, K. Shirai and T. Kumamoto; *Bull. Chem. Soc. Jpn.*, 1981, **54**, 817.	226
81BCJ1151	S. Yanagida, Y. Noji and M. Okahara; *Bull. Chem. Soc. Jpn.*, 1981, **54**, 1151.	52
81CA96556w	I. I. Kandror, I. O. Bragina and R. Kh. Freidlina; *Chem. Abstr.*, 1981, **95**, 96 556w (*Izv. Akad. Nauk SSSR, Ser. Khim.*, 1981, **30**, 1167).	899, 902
81CAR284	D. D. Ward and F. Shafizadeh; *Carbohydr. Res.*, 1981, **93**, 284.	42
81CAR(95)308	T. Ogawa, S. Nakabayashi and K. Sasajima; *Carbohydr. Res.*, 1981, **95**, 308.	225
81CAR(C)1	J. P. Praly and G. Descotes; *Carbohydr. Res. (C)*, 1981, **95**, 1.	1079
81CB413	E. Lindner, H.-J. Eberle and S. Höhne; *Chem. Ber.*, 1981, **114**, 413.	581
81CB441	H. Schmidbaur, G. Müller, K. C. Dash and B. Milewski-Mahrla; *Chem. Ber.*, 1981, **114**, 441.	586
81CB684	A. Roedig, C. Ibis and G. Zaby; *Chem. Ber.*, 1981, **114**, 684.	776, 854
81CB1082	W. Althoff, M. Fild and R. Schmutzler; *Chem. Ber.*, 1981, **114**, 1082.	563, 564
81CB1938	K. Schank, R. Blattner, V. Schmidt and H. Hasenfratz; *Chem. Ber.*, 1981, **114**, 1938.	206
81CB1951	K. Schank, R. Blattner and G. Bouillon; *Chem. Ber.*, 1981, **114**, 1951.	43, 55
81CB1958	K. Schank and R. Blattner; *Chem. Ber.*, 1981, **114**, 1958.	55
81CB1976	L. Capuano and K. Djokar; *Chem. Ber.*, 1981, **114**, 1976.	980, 981, 982
81CB2001	K. Klemm, W. Prusse, L. Baron and E. Daltrozzo; *Chem. Ber.*, 1981, **114**, 2001.	987
81CB3421	H. Böhme and E. Raude; *Chem. Ber.*, 1981, **114**, 3421.	99
81CC179	B. L. Chenard, C. Slapak, D. K. Anderson and J. S. Swenton; *J. Chem. Soc., Chem. Commun.*, 1981, 179.	646
81CC377	K. Hiroi and L.-M. Chen; *J. Chem. Soc., Chem. Commun.*, 1981, 377.	365
81CC556	S. V. Attwood, A. G. M. Barrett and J. Florent; *J. Chem. Soc., Chem. Commun.*, 1981, 556.	172
81CC565	J. Nakayama, N. Matsumaru and M. Hoshino; *J. Chem. Soc., Chem. Commun.*, 1981, 565.	850
81CC669	Y. Okamoto and P. S. Wojciechowski; *J. Chem. Soc., Chem. Commun.*, 1981, 669	874
81CC698	N. J. Kermode M. F. Lappert, B. W. Skelton, A. H. White and J. Holton; *J. Chem. Soc., Chem. Commun.*, 1981, 698.	584
81CC806	P. John, B. G. Gowenlock and P. Groome; *J. Chem. Soc., Chem. Commun.*, 1981, 806.	610
81CC862	M. Cooke, D. L. Davies, J. E. Guerchais, S. A. R. Knox, K. A. Mead, J. Roué and P. Woodward; *J. Chem. Soc., Chem. Commun.*, 1981, 862.	680
81CC867	J. C. Jeffery, K. A. Mead, H. Razay, F. G. A. Stone, M. J. Went and P. Woodward; *J. Chem. Soc., Chem. Commun.*, 1981, 867.	712
81CC877	M. Larchevêque and A. Debal; *J. Chem. Soc., Chem. Commun.*, 1981, 877.	648
81CC1149	A. Pelter and J. M. Rao; *J. Chem. Soc., Chem. Commun.*, 1981, 1149.	373
81CC1173	M. Kato and M. Yamabe; *J. Chem. Soc., Chem. Commun.*, 1981, 1173.	117
81CC1255	J. C. Jeffery, I. Moore, H. Razay and F. G, A. Stone; *J. Chem. Soc., Chem. Commun.*, 1981, 1255.	712
81CC1269	G. Johnson and B. C. Ross; *J. Chem. Soc., Chem. Commun.*, 1981, 1269.	421
81CCC1389	J. Svoboda, O. Paleta, F. Liska and V. Dedek; *Collect. Czech. Chem. Commun.*, 1981, **46**, 1389.	970

81CJC27	A. W. Frank; *Can. J. Chem.*, 1981, **59**, 27.	456, 475, 477
81CJC679	N. Y. C. Chu, S. A. Goldstein and P. M. Keehn; *Can. J. Chem.*, 1981, **59**, 679.	436
81CJC3055	C. E. Slemon, L. C. Hellwig, J.-P. Ruder, E. W. Hoskins and D. B. MacLean; *Can. J. Chem.*, 1981, **58**, 3055.	54, 56
81CL447	T. Otsubo, F. Ogura, H. Yamaguchi, H. Higuchi, Y. Sakata and S. Misumi; *Chem. Lett.*, 1981, 447.	402
81CL457	M. Isobe, Y. Ichikawa, M. Kitamura and T. Goto; *Chem. Lett.*, 1981, 457.	937
81CL1059	T. Yogo, J. Koshino and A. Suzuki; *Chem. Lett.*, 1981, 1059.	935
81CPB2496	Y. Ishikawa, Y. Kurebayashi, K. Suzki, Y. Terao and M. Sekiya; *Chem. Pharm. Bull.*, 1981, **29**, 2496.	325
81CRV415	A. K. Bhattacharya and G. Thyagarajan; *Chem. Rev.*, 1981, **81**, 415.	340
81DOK(261)474	I. M. Salimgareeva, N. G. Bogatova, O. Zh. Zhebarov, V. P. Yur'ev and S. R. Rafikov; *Dokl. Akad. Nauk SSSR*, 1981, **261**, 474.	141
81EUP30887	J. Martel, J. Tessier and A. Teche (Roussel Uclaf S.A.); *Eur. Pat.* 0 030 887 (*Chem. Abstr.*, 1981, **95**, 203 410).	757
81EUP40153	C. Cagnon, M. Piteau, J.-P. Senet, R. A. Olofson and J. T. Martz; *Eur. Pat.* 40 153 (1981) (*Chem. Abstr.*, 1982, **96**, 142 281).	58
81H(15)1179	L. Ghosez, P. Notté, C. Bernard-Henriet and R. Maurin; *Heterocycles*, 1981, **15**, 1179.	791
81H(16)1587	J. Gelas and D. Horton; *Heterocycles*, 1981, **16**, 1587.	186
81HCA329	M. Geoffroy and A. Llinares; *Helv. Chim. Acta*, 1981, **64**, 329.	598
81HCA357	B. Weidmann, L. Widler, A. G. Olivero, C. D. Maycock and D. Seebach; *Helv. Chim. Acta*, 1981, **64**, 357.	396
81HCA1208	E. Francotte, R. Merényi, B. Vandenbulcke-Coyette and H. G. Viehe; *Helv. Chim. Acta*, 1981, **64**, 1208.	798
81HCA1247	C. Fehr, J. Galindo and G. Ohloff; *Helv. Chim. Acta*, 1981, **64**, 1247.	166
81IC1328	M. J. Byrne and N. E. Miller; *Inorg. Chem.*, 1981, **20**, 1328.	524
81IC1679	R. H. Neilson; *Inorg. Chem.*, 1981, **20**, 1679.	566
81IC2466	R. E. Cramer, R. B. Maynard and J. W. Gilje; *Inorg. Chem.*, 1981, **20**, 2466.	589
81IC2739	A. B. Burg; *Inorg. Chem.*, 1981, **20**, 2739.	111
81IC3200	N. E. Schore, L. S. Benner and B. E. LaBelle; *Inorg. Chem.*, 1981, **20**, 3200.	566, 568
81IC3734	A. B. Burg; *Inorg. Chem.*, 1981, **20**, 3734.	550
81IJC1003	A. A. Siddiqui, N. H. Khan and Basheeruddin; *Indian J. Chem.*, 1981, **20B**, 1003.	441, 444
81IZV1603	V. M. Vdovin, M. B. Sergeeva, N. B. Bespalova and V. G. Zaikin; *Izv. Akad. Nauk SSSR, Ser. Khim.*, 1981, 1603 (*Chem. Abstr.*, 1981, **95**, 150 758d).	222
81IZV654	G. S. Lynshenko, A. Kh. Filippova, I. D. Kulikhman, V. V. Keiko, O. A. Kruglaya and N. S. Vyazankin; *Izv. Akad. Nauk SSSR, Ser. Khim.*, 1981, **30**, 654.	933
81IZV1113	B. A. Arbuzov, E. N. Dianova and S. M. Shapirova; *Izv. Akad. Nauk SSSR, Ser. Khim.*, 1981, 1113.	558
81IZV1920	V. A. Petrov, G. G. Belen'kii and L. S. German; *Izv. Akad. Nauk SSSR, Ser. Khim.*, 1981, 1920 (*Chem. Abstr.*, 1981, **95**, 219 717j).	730
81JA1276	T. Yoshida and E. Negishi; *J. Am. Chem. Soc.*, 1981, **103**, 1276.	750, 1066
81JA1278	G. R. Steinmetz and G. L. Geoffroy; *J. Am. Chem. Soc.*, 1981, **103**, 1278.	679
81JA2483	B. M. Trost and W. H. Pearson; *J. Am. Chem. Soc.*, 1981, **103**, 2483.	329
81JA2489	K. H. Theopold and R. G. Bergman; *J. Am. Chem. Soc.*, 1981, **103**, 2489.	687
81JA2757	K. Oka, A. Dobashi and S. Hara; *J. Am. Chem. Soc.*, 1981, **103**, 2757.	70
81JA2807	L. E. Overman, L. A. Clizbe, R. L. Freerks and C. K. Marlowe; *J. Am. Chem. Soc.*, 1981, **103**, 2807.	1014
81JA2995	L. J. Krause and J. A. Morrison; *J. Am. Chem. Soc.*, 1981, **103**, 2995.	7
81JA3099	D. A. Evans, J. V. Nelson, E. Vogel and T. R. Taber; *J. Am. Chem. Soc.*, 1981, **103**, 3099.	832
81JA3112	H. J. Reich, S. K. Shah, P. M. Gold and R. E. Olson; *J. Am. Chem. Soc.*, 1981, **103**, 3112.	197
81JA3215	R. B. Woodward, E. Logusch, K. P. Nambiar *et al.*; *J. Am. Chem. Soc.*, 1981, **103**, 3215.	225
81JA3764	A. L. Balch, C. T. Hunt, C.-L. Lee, M. M. Olmstead and J. P. Farr; *J. Am. Chem. Soc.*, 1981, **103**, 3764.	691
81JA5241	D. S. Matteson, K. M. Sadhu and G. E. Lienhard; *J. Am. Chem. Soc.*, 1981, **103**, 5241.	525, 526, 527
81JA5259	T. Takahashi, H. Yamada and J. Tsuji; *J. Am. Chem. Soc.*, 1981, **103**, 5259.	950
81JA5568	Y. Yamakado, M. Ishiguro, N. Ikeda and H. Yamamoto; *J. Am. Chem. Soc.*, 1981, **103**, 5568.	1059, 1060
81JA5598	C. G. Krespan and D. C. England; *J. Am. Chem. Soc.*, 1981, **103**, 5598.	44, 86
81JA5618	T. R. Hoye, D. R. Peck and P. K. Trumper; *J. Am. Chem. Soc.*, 1981, **103**, 5618.	204
81JA5922	K. C. Ott and R. H. Grubbs; *J. Am. Chem. Soc.*, 1981, **103**, 5922.	670
81JA6788	T. J. Barton and M. Vuper; *J. Am. Chem. Soc.*, 1981, **103**, 6788.	614, 616
81JAP(K)56002956	Sagami Chemical Research Center, Japan; *Jpn. Kokai Tokkyo Koho* JP 56 002 956 (1981) (*Chem. Abstr.* 1981, **95**, 97 062).	854
81JCR(M)4016	F. Barbot, L. Poncini, B. Randrianoelina and P. Miginiac; *J. Chem. Res. (M)*, 1981, 4016.	192
81JCS(D)1593	J. C. Jeffery, M. F. Lappert, N. T. Luong-Thi, M. Webb, J. L. Atwood and W. E. Hunter; *J. Chem. Soc., Dalton Trans.*, 1981, 1593.	656, 663, 709, 724
81JCS(P1)78	G. Jones, D. R. Sliskovic, B. Foster, J. Rogers, A. K. Smith, M. Y. Wong and A. C. Yarham; *J. Chem. Soc., Perkin Trans. 1*, 1981, 78.	19, 27
81JCS(P1)785	J. K. Michie and J. A. Miller; *J. Chem. Soc., Perkin Trans. 1*, 1981, 785.	53, 54
81JCS(P1)969	I. Fleming and C. D. Floyd; *J. Chem. Soc., Perkin Trans. 1*, 1981, 969.	620, 1048, 1049
81JCS(P1)991	S. Uemura, H. Miyoshi, M. Okano and K. Ichikawa; *J. Chem. Soc., Perkin Trans. 1*, 1981, 991.	1068
81JCS(P1)1015	K. Sato, S. Inoue, T. Tanami and M. Ohasi; *J. Chem. Soc., Perkin Trans. 1*, 1981, 1015.	167

81JCS(P1)1278	L. Colombo, C. Gennari, C. Scolastico, G. Guanti and E. Narisano; *J. Chem. Soc., Perkin Trans. 1*, 1981, 1278.	856
81JCS(P1)1321	G. E. Gerhardt and R. J. Lagow; *J. Chem. Soc., Perkin Trans. 1*, 1981, 1321.	44
81JCS(P1)1363	A. Szpala, J. C. Tebby and D. V. Griffiths; *J. Chem. Soc., Perkin Trans. 1*, 1981, 1363.	922
81JCS(P1)1569	Ö. Kemal and C. B. Reese; *J. Chem. Soc., Perkin Trans. 1*, 1981, 1569.	317
81JCS(P1)1934	A. P. Davis, G. J. Hughes, P. R. Lowndes, C. M. Robbins, E. J. Thomas and G. H. Whitham; *J. Chem. Soc., Perkin Trans. 1*, 1981, 1934.	519
81JCS(P1)2415	M. J. Carter, I. Fleming and A. Percival; *J. Chem. Soc., Perkin Trans. 1*, 1981, 2415.	649
81JCS(P1)2435	W. R. McKay and G. R. Proctor; *J. Chem. Soc., Perkin Trans. 1*, 1981, 2435.	105, 300
81JCS(P1)2443	W. R. McKay and G. R. Proctor; *J. Chem. Soc., Perkin Trans. 1*, 1981, 2443.	304, 305
81JCS(P1)2527	I. Fleming, T. W. Newton and F. Roessler; *J. Chem. Soc., Perkin Trans. 1*, 1981, 2527.	1059
81JCS(P1)2608	R. C. Cambie, R. C. Hayward, J. L. Jurlina, P. S. Rutledge and P. D. Woodgate; *J. Chem. Soc., Perkin Trans. 1*, 1981, 2608.	27
81JCS(P1)2991	A. M. Damas, R. O. Gould, M. M. Harding, R. M. Paton, J. F. Ross and J. Crosby; *J. Chem. Soc., Perkin Trans. 1*, 1981, 2991.	232
81JCS(P2)143	G. Behrens, E. Bothe, G. Koltzenburg and D. Schulte-Frohlinde; *J. Chem. Soc., Perkin Trans. 2*, 1981, 143.	1078
81JCS(P2)1138	M-ul-Hague, W. Horne, S. E. Cremer, P. W. Kremer and P. K. Kafarski; *J. Chem. Soc., Perkin Trans. 2*, 1981, 1138.	114
81JFC(17)441	V. F. Snegirev, K. N. Makarov and I. L. Knunyants; *J. Fluorine Chem.*, 1981, **17**, 441.	737
81JFC(18)293	D. J. Burton, H. S. Kesling and D. G. Naae; *J. Fluorine Chem.*, 1981, **18**, 293.	734
81JHC587	D. T. Connor and R. J. Sorenson; *J. Heterocycl. Chem.*, 1981, **18**, 587.	226
81JINC629	A. F. Janzen, J. R. Dalziel, S. N. Kay and R. Galka; *J. Inorg. Nucl. Chem.*, 1981, **43**, 629.	455
81JMC1181	I. Antonini, F. Claudi, G. Cristalli, P. Franchetti, M. Grifantini and S. Martelli; *J. Med. Chem.*, 1981, **24**, 1181.	206
81JOC192	T. Tsuda, T. Yazawa, K. Watanabe, T. Fujii and T. Saegusa; *J. Org. Chem.*, 1981, **46**, 192.	696
81JOC196	R. Kaya and N. R. Beller; *J. Org. Chem.*, 1981, **46**, 196.	853
81JOC571	G. A. Olah, S. Yu, G. Liang, G. D. Matseescu, M. R. Bruce, D. J. Donovan and M. Arvanaghi; *J. Org. Chem.*, 1981, **46**, 571.	42
81JOC765	P. A. Wade, H. R. Hinney, N. V. Amin, P. D. Vail, S. D. Morrow, S. A. Hardinger and M. S. Saft; *J. Org. Chem.*, 1981, **46**, 765.	333
81JOC1292	G. Zweifel, R. E. Murray and H. P. On; *J. Org. Chem.*, 1981, **46**, 1292.	808, 811
81JOC1513	W. S. Johnson, B. Frei and A. S. Gopalan; *J. Org. Chem.*, 1981, **46**, 1513.	250
81JOC1828	J. P. Marino and M. P. Ferro; *J. Org. Chem.*, 1981, **46**, 1828.	346
81JOC2069	A. M. van Leusen, H. J. Jeuring, J. Wildeman and S. P. J. M. van Nispen; *J. Org. Chem.*, 1981, **46**, 2069.	331
81JOC2260	R. M. Sandifer, A. K. Bhattacharya and T. M. Harris; *J. Org. Chem.*, 1981, **46**, 2260.	167
81JOC2363	P. Beak and L. G. Carter; *J. Org. Chem.*, 1981, **46**, 2363.	382
81JOC2419	B. H. Lipshutz and M. C. Morey; *J. Org. Chem.*, 1981, **46**, 2419.	184
81JOC2557	W. L. Mock and H. Tsou; *J. Org. Chem.*, 1981, **46**, 2557.	193
81JOC2775	H. J. Reich, W. W. Willis, Jr. and P. D. Clark; *J. Org. Chem.*, 1981, **46**, 2775.	289, 871, 947, 965
81JOC2981	B. C. Barot and H. W. Pinnick; *J. Org. Chem.*, 1981, **46**, 2981.	185
81JOC3273	H. C. J. Ottenheijm, R. M. J. Liskamp, S. P. J. M. van Nispen, H. A. Boots and M. W. Tijhuis; *J. Org. Chem.*, 1981, **46**, 3273.	264
81JOC3340	R. B. Gammill; *J. Org. Chem.*, 1981, **46**, 3340.	302
81JOC3555	R. B. Gammill, D. M. Sobieray and P. M. Gold; *J. Org. Chem.*, 1981, **46**, 3555.	253
81JOC3721	I. L. Reich and H. J. Reich; *J. Org. Chem.*, 1981, **46**, 3721.	91, 92
81JOC3917	T. B. Patrick, J. J. Scheibel and J. L. Cantrel; *J. Org. Chem.*, 1981, **46**, 3917.	10
81JOC4290	W. G. Taylor; *J. Org. Chem.*, 1981, **46**, 4290.	740
81JOC4301	D. Liotta, C. S. Barnum and M. Saindane; *J. Org. Chem.*, 1981, **46**, 4301.	401
81JOC4911	I. W. J. Still and G. W. Kutney; *J. Org. Chem.*, 1981, **46**, 4911.	70
81JOC5041	E. Negishi, C. L. Rand and K. P. Jadhav; *J. Org. Chem.*, 1981, **46**, 5041.	392
81JOC5182	K. S. Kyler and D. S. Watt; *J. Org. Chem.*, 1981, **46**, 5182.	365, 938
81JOC5357	M. C. Croudace and N. E. Schore; *J. Org. Chem.*, 1981, **46**, 5357.	357
81JOC5457	S.-K. Chung; *J. Org. Chem.*, 1981, **46**, 5457.	234
81JOM(204)C21	W. A. Herrmann and C. Bauer; *J. Organomet. Chem.*, 1981, **204**, C21.	679
81JOM(204)13	M. G. Voronkov, S. V. Kirpichenko, E. N. Suslova and V. V. Keiko; *J. Organomet. Chem.*, 1981, **204**, 13.	368
81JOM(205)1	B. Wrackmeyer; *J. Organomet. Chem.*, 1981, **205**, 1.	1068
81JOM(205)301	D. Seyferth, K. R. Wursthorn, T. F. O. Lim and D. J. Sepelak; *J. Organomet. Chem.*, 1981, **205**, 301.	735
81JOM(205)311	A. N. Egorochkin, E. I. Sevast'yanova, S. Ya. Khorshev, S. Richelme and J. Satgé; *J. Organomet. Chem.*, 1981, **205**, 311.	141
81JOM(206)119	G. Váradi, I. T. Horváth, G. Pályi, L. Markó, Yu. L. Slovokhotov and Yu. T. Struchkov; *J. Organomet. Chem.*, 1981, **206**, 119.	688
81JOM(209)1	N. M. D. Brown, F. Davidson and J. W. Wilson; *J. Organomet. Chem.*, 1981, **209**, 1.	1057
81JOM(209)233	N. M. Loim, P. V. Kondrat'ev, N. P. Solov'eva, V. A. Antonovich, P. V. Petrovskii, Z. N. Parnes and D. N. Kursanov; *J. Organomet. Chem.*, 1981, **209**, 233.	706
81JOM(210)C17	A. Amamria and T. N. Mitchell; *J. Organomet. Chem.*, 1981, **210**, C17.	1062
81JOM(212)C31	J.-P. Quintard, B. Elissondo and M. Pereyre; *J. Organomet. Chem.*, 1981, **212**, C31.	388
81JOM(214)191	M. Gielen, S. Simon, and M. van de Steen; *J. Organomet. Chem.*, 1981, **214**, 191.	156
81JOM(216)287	J. V. Comasseto, J. T. B. Ferreira and N. Petragnani; *J. Organomet. Chem.*, 1981, **216**, 287.	90
81JOM(217)51	P. W. Clark and M. J. Mulraney; *J. Organomet. Chem.*, 1981, **217**, 51.	545

81JOM(219)C29	H. Werner, R. Feser, W. Paul and L. Hofmann; *J. Organomet. Chem.*, 1981, **219**, C29.	582
81JOM(221)93	E. Sappa, A. M. M. Lanfredi and A. Tiripicchio; *J. Organomet. Chem.*, 1981, **221**, 93.	713
81JOU578	M. L. Petrov, V. Z. Laishev and A. A. Petrov; *J. Org. Chem. USSR (Engl. Transl.)*, 1981, **17**, 578.	874
81LA1044	M. Regitz, F. Bennyarto and H. Heydt; *Liebigs Ann. Chem.*, 1981, 1044.	558
81LA1105	J. Mattay, W. Thuenker and H. Scharf; *Liebigs Ann. Chem.*, 1981, 1105.	181
81LA2247	M. Braun; *Liebigs Ann. Chem.*, 1981, 2247.	172
B-81MI 407-01	S. Patai and C. J. M. Stirling (eds); "The Chemistry of the Sulphonium Group," Wiley, Chichester, 1981, parts 1 and 2.	330
B-81MI 408-01	E. W. Colvin; "Silicon in Organic Synthesis," Butterworths, London, 1981.	354, 357
81MI 420-01	M. Look and L. R. White; *J. Agric. Food Chem.*, 1981, **29**, 673.	912
B-81MI 420-02	M. Maleki, J. A. Miller and O. W. Lever. Jr,; in "Phosphorus Chemistry, ACS Symposium Series 171" ed. L. D. Quin and J. G. Verkade, American Chemical Society, Washington DC, 1981, p. 145.	917
B-81MI 423-01	E. W. Colvin; "Silicon in Organic Synthesis," Butterworth, London, 1981, p. 44.	1044
81OS(60)66	R. K. Singh and S. Danishefsky; *Org. Synth.*, 1981, **60**, 66.	204
81PAC2307	G. H. Posner, J. P. Mallamo, K. Miura and M. Hulce; *Pure Appl. Chem.*, 1981, **53**, 2307.	960, 964
81PJC411	J. Zygmunt and P. Mastalerz; *Pol. J. Chem.*, 1981, **55**, 411.	473
81PJC643	J. Zon; *Pol. J. Chem.*, 1981, **55**, 643.	462
81PJC713	J. Kowalik, J. Zygmunt and P. Mastalerz; *Pol. J. Chem.*, 1981, **55**, 713.	472, 473
81PS(10)127	R. Dittrich and G. Hägele; *Phosphorus Sulfur*, 1981, **10**, 127.	802, 803
81PS(10)147	A. W. Frank; *Phosphorus Sulfur*, 1981, **10**, 147.	456
81PS(10)163	F. A. Carey, O. D. Dailey, Jr. and T. E. Fromuth; *Phosphorus Sulfur*, 1981, **10**, 163.	268
81PS(10)169	F. A. Carey and O. D. Dailey, Jr.; *Phosphorus Sulfur*, 1981, **10**, 169.	266
81PS(10)207	A. W. Frank; *Phosphorus Sulfur*, 1981, **10**, 207.	475
81PS(11)87	M. R. Marre, J. F. Brazier, R. Wolf and A. Klaebe; *Phosphorus Sulfur*, 1981, **11**, 87.	478
81PS(11)139	L. Maier; *Phosphorus Sulfur*, 1981, **11**, 139.	463, 466
81PS(11)149	L. Maier; *Phosphorus Sulfur*, 1981, **11**, 149.	466
81RTC10	R. S. Sukhai, W. Verboom, J. Meijer, M. J. M. Schoufs and L. Brandsma; *Recl. Trav. Chim. Pays-Bas*, 1981, **100**, 10.	93
81RTC194	R. Jorritsma, H. Steinberg and T. J. de Boer; *Recl. Trav. Chim. Pays-Bas*, 1981, **100**, 194.	49
81S53	S. Tanimoto, S. Jo and T. Sugimoto; *Synthesis*, 1981, 53.	847
81S57	J. Kowalik, L. Kupczyk-Subotkowska and P. Mastalerz; *Synthesis*, 1981, 57.	472
81S137	D. Fishman, J. T. Klug and A. Shani; *Synthesis*, 1981, 137.	188
81S151	T. Takajo and S. Kambe; *Synthesis*, 1981, 151.	445
81S185	G. Solladie; *Synthesis*, 1981, 185.	391
81S282	G. A. Olah, S. C. Narang, D. Meidar and G. F. Salem; *Synthesis*, 1981, 282.	191
81S324	J. Zon; *Synthesis*, 1981, 324.	1005
81S501	F. A. J. Meskens; *Synthesis*, 1981, 501.	176, 177
81S534	Y. Tamura, J. I. Uenishi, H. Maeda, H. D. Choi and H. Ishibashi; *Synthesis*, 1981, 534.	67
81S557	D. E. Seitz and A. Zapata; *Synthesis*, 1981, 557.	660
81S643	Z. H. Kudzin; *Synthesis*, 1981, 643.	466
81S824	J. K. Michie and J. A. Miller; *Synthesis*, 1981, 824.	203
81S878	L. Rene and R. Royer; *Synthesis*, 1981, 878.	193
81S995	V. Baliah, S. Prema, C. B. Jawaharsingh, K. N. Chockalingam and R. Jeyaraman; *Synthesis*, 1981, 995.	272, 277
81SA(A)819	J. Ali, R. Aroca and E. A. Robinson; *Spectrochim. Acta, Part A*, 1981, **37**, 819.	281
81SC673	D. E. Seitz and A. Zapata; *Synth. Commun.*, 1981, **11**, 673.	660
81T753	H. Oehme and E. Leissring; *Tetrahedron*, 1981, **37**, 753.	548
81T1215	H. Singh and P. Singh; *Tetrahedron*, 1981, **37**, 1215.	29
81T1227	B. Costisella, I. Keitel and H. Gross; *Tetrahedron*, 1981, **37**, 1227.	1002
81T1453	S. Rajappa; *Tetrahedron*, 1981, **37**, 1453.	796, 901
81T1487	M. Taniguchi and T. Hino; *Tetrahedron*, 1981, **37**, 1487.	418, 419
81T1875	T. Kozluk, L. Cottier and G. Descotes; *Tetrahedron*, 1981, **37**, 1875.	1079
81T2515	L. Cottier, G. Descotes, M. F. Grenier and F. Metras; *Tetrahedron*, 1981, **37**, 2515.	1079
81T2547	T. Kametani, M. Aizawa and H. Nemoto; *Tetrahedron*, 1981, **37**, 2547.	286
81T4337	L. Vasvari-Debreczy, A. H. Beckett and W. Vutthikongsirigool; *Tetrahedron*, 1981, **37**, 4337.	422
81TL7	T. J. Barton, G. T. Burns, E. V. Arnold and J. Clardy; *Tetrahedron Lett.*, 1981, **22**, 7.	603
81TL191	B. Venugopalan, A. B. Hamlet and T. Durst; *Tetrahedron Lett.*, 1981, **22**, 191.	923
81TL229	G. Märkl and G. Yu Jin; *Tetrahedron Lett.*, 1981, **22**, 229.	456, 482
81TL239	M. Isobe, M. Kitamura and T. Goto; *Tetrahedron Lett.*, 1981, **22**, 239.	937
81TL587	D. J. Ager; *Tetrahedron Lett.*, 1981, **22**, 587.	366, 392, 938
81TL759	M. Suda and A. Fukushima; *Tetrahedron Lett.*, 1981, **22**, 759.	738
81TL1105	G. Märkl and G. Yu Jin; *Tetrahedron Lett.*, 1981, **22**, 1105.	482
81TL1287	B. M. Trost and D. P. Curran; *Tetrahedron Lett.*, 1981, **22**, 1287.	274
81TL1421	M. Suda; *Tetrahedron Lett.*, 1981, **22**, 1421.	734, 764
81TL1495	S. L. Bender, N. F. Haley and H. R. Luss; *Tetrahedron Lett.*, 1981, **22**, 1495.	877
81TL1551	W. R. Bowman and G. D. Richardson; *Tetrahedron Lett.*, 1981, **22**, 1551.	327
81TL1809	M. Suzuki, T. Kawagishi and R. Noyori; *Tetrahedron Lett.*, 1981, **22**, 1809.	193
81TL1821	I. Degani, R. Fochi and V. Regondi; *Tetrahedron Lett.*, 1981, **22**, 1821.	249, 250
81TL1973	B. Loubinoux, G. Coudert and G. Guillaumet; *Tetrahedron Lett.*, 1981, **22**, 1973.	199
81TL2397	Y. Torisawa, M. Shibasaki and S. Ikegami; *Tetrahedron Lett.*, 1981, **22**, 2397.	154, 155

81TL2455	T. Cohen and J. R. Matz; *Tetrahedron Lett.*, 1981, **22**, 2455.	68
81TL2923	D. J. Ager; *Tetrahedron Lett.*, 1981, **22**, 2923.	366, 392, 394, 652
81TL2985	R. J. P. Corriu, C. Guerin and J. M. Boula; *Tetrahedron Lett.*, 1981, **22**, 2985.	655
81TL3097	M. Mikolajczyk, S. Grzcjszczak and K. Korbacz; *Tetrahedron Lett.*, 1981, **22**, 3097.	924
81TL3179	J.-P. Dulcere, J. Grimaldi and M. Santelli; *Tetrahedron Lett.*, 1981, **22**, 3179.	1046
81TL3243	K. Hatanaka, S. Tanimoto, T. Sugimoto and M. Okano; *Tetrahedron Lett.*, 1981, **22**, 3243.	250
81TL3537	J. Oleksyszyn and E. Gruszecka; *Tetrahedron Lett.*, 1981, **22**, 3537.	470
81TL3789	M. Maleki, J. A. Miller and O. W. Lever, Jr.; *Tetrahedron Lett.*, 1981, **22**, 3789.	917
81TL4199	M. V. Lakshmikantham, M. P. Cava, M. Albeck, L. Engman, P. Carroll, J. Bergman and F. Wudl; *Tetrahedron Lett.*, 1981, **22**, 4199.	877
81TL4287	M. Isobe, Y. Ichikawa and T. Goto; *Tetrahedron Lett.*, 1981, **22**, 4287.	937
81TL4499	K. Ogura, J.-I. Watanabe and H. Iida; *Tetrahedron Lett.*, 1981, **22**, 4499.	271
81TL4603	B. H. Lipshutz, J. J. Pegram and M. C. Morey; *Tetrahedron Lett.*, 1981, **22**, 4603.	176
81TL4737	S. Halazy, W. Dumont and A. Krief; *Tetrahedron Lett.*, 1981, **22**, 4737.	375, 400
81USP4348529A	A. L. Borror, E. W. Ellis and C. E. Hammond (Polaroid Corp. USA); *US Pat.* 4 348 529 A (1982) (*Chem. Abstr.* 1983, **98**, 89 372).	855
81ZAAC(473)59	G. Fritz and B. Grunert; *Z. Anorg. Allg. Chem.*, 1981, **473**, 59.	609, 621
81ZAAC(475)18	R. Appel, J. Peters and R. Schmitz; *Z. Anorg. Allg. Chem.*, 1981, **475**, 18.	566
81ZAAC(478)94	G. Fritz and K. Gompper; *Z. Anorg. Allg. Chem.*, 1981, **478**, 94.	621
81ZAAC(479)41	G. Becker, W. Uhl and H.-J. Wessely; *Z. Anorg. Allg. Chem.*, 1981, **479**, 41.	549
81ZAAC(482)65	U. Klingebiel, S. Pohlmann and P. Werner; *Z. Anorg. Allg. Chem.*, 1981, **482**, 65.	649
81ZC341	H.-P. Abicht and K. Issleib; *Z. Chem.*, 1981, **21**, 341.	575
81ZC403	K. Kellner, W. Hanke and A. Tzschach; *Z. Chem.*, 1981, **21**, 403.	317, 499, 500
81ZN(B)188	K.-P. Jensen, D. K. Breitinger and W. Kress; *Z. Naturforsch., Teil B*, 1981, **36**, 188.	697
81ZN(B)1375	H. P. Fritz and W. Kornrumpf; *Z. Naturforsch., Teil B*, 1981, **36**, 1375.	7
81ZN(B)1663	O. J. Scherer and H. Jungmann; *Z. Naturforsch., Teil B*, 1981, **36**, 1663.	585
81ZOB484	A. A. Prishchenko, Z. S. Novikova and I. F. Lutsenko; *Zh. Obshch. Khim.*, 1981, **51**, 484.	546
81ZOB757	G. A. Berkova, G. M. Baranov and L. A. Zhidkova; *Zh. Obshch. Khim.*, 1981, **46**, 757 (*Chem. Abstr.*, 1981, **95**, 96 384p).	490
81ZOB1857	V. S. Zavgorodnii, N. D. Grigor'eva and A. A. Petrov; *Zh. Obshch. Khim.*, 1981, **51**, 1857.	955
81ZOB2140	M. G. Zimin, A. R. Burilov and A. N. Pudovik; *Zh. Obshch. Khim.*, 1981, **48**, 2140.	345
81ZOB2382	V. F. Mironov, V. P. Kozyukov and V. P. Kozyukov; *Zh. Obshch. Khim.*, 1981, **48**, 2382.	295, 313
81ZOR1169	S. E. Tolchinskii and I. A. Maretina; *Zh. Org. Khim.*, 1981, **17**, 1169 (Engl. Transl. 1034).	976
81ZOR1550	V. A. Buevich, N. Z. Nakova and V. V. Perekalin; *Zh. Org. Khim.*, 1981, **17**, 1550.	826
82ACS(B)721	I. O. O. Korhonen; *Acta Chem. Scand., Ser. B*, 1982, **36**, 721.	54
82AG559	R. Gompper, E. Kujath and H. Wagner; *Angew. Chem.*, 1982, **94**, 559.	998
82AG(E)73	P. Jandik, U. Schubert and H. Schmidbaur; *Angew. Chem., Int. Ed. Engl.*, 1982, **21**, 73.	695
82AG(E)376	S. Hietkamp, H. Sommer and O. Stelzer; *Angew. Chem., Int. Ed. Engl.*, 1982, **21**, 376.	546
82AG(E)545	H. J. Bestmann and A. Bomhard; *Angew. Chem., Int. Ed. Engl.*, 1982, **21**, 545.	571
82AG(E)776	U. Schmidt, A. Lieberknecht, U. Schanbacher, T. Beuttler and J. Wild; *Angew. Chem., Int. Ed. Engl.*, 1982, 776.	465
82AJC785	W. L. F. Armarego, P. Waring and B. Paal; *Aust. J. Chem.*, 1982, **35**, 785.	855
82AOC(20)115	Y. Z. Huang and Y. C. Shen; *Adv. Organomet. Chem.*, 1982, **20**, 115.	598
82AOC(20)159	W. A. Herrmann; *Adv. Organomet. Chem.*, 1982, **20**, 159.	669, 677, 709
82AP(315)680	R. Barzen and W. Schunack; *Arch. Pharm. (Weinheim, Ger.)*, 1982, **315**, 680.	970
82AP(315)852	B. Unterhalt and E. Seebach; *Arch. Pharm., (Weinheim, Ger.)*, 1982, **315**, 852.	313
82BAU2498	S. A. Postovoi, Yu. V. Zeifman and I. L. Knunyants; *Bull. Acad. Sci. USSR, Div. Chem. Sci.*, 1982, 2498.	820
82BCJ1163	T. Hirao, Y. Yamada, K. Hayashi, Y. Ohshiro and T. Agawa; *Bull. Chem. Soc. Jpn.*, 1982, **55**, 1163.	517, 538, 1015
82BCJ3025	Y. Yamamoto; *Bull. Chem. Soc. Jpn.*, 1982, **55**, 3025.	574, 588, 589
82BSF43	T. Cuvigny, C. H. du Penhoat and M. Julia; *Bull. Soc. Chim. Fr.*, 1982, **2**, 43.	272, 275
82CA127400p	Kowa Co Ltd.; *Chem. Abstr.*, 1982, **97**, 127 400p (*Jpn. Pat.* 8 264 694 (1982)).	777
82CA215888j	Kowa Co Ltd.; *Chem. Abstr.*, 1982, **97**, 215 888j (*Jpn. Pat.* 82 108 091 (1982)).	777
82CAR(102)99	M. A. Nashed, M. S. Chowdhary and L. Anderson; *Carbohydr. Res.*, 1982, **102**, 99.	176
82CB645	T. Kauffmann, H. Fischer and A. Woltermann; *Chem. Ber.*, 1982, **115**, 645.	600
82CB823	H. H. Karsch; *Chem. Ber.*, 1982, **115**, 823.	112, 546
82CB860	R. Reck and J. C. Jochims; *Chem. Ber.*, 1982, **115**, 860.	434
82CB878	W. A. Herrmann, C. Bauer, G. Kriechbaum, H. Kunkely, M. L. Ziegler, D. Speth and E. Guggolz; *Chem. Ber.*, 1982, **115**, 878.	686
82CB1379	N. Holy, U. Deschler and H. Schmidbaur; *Chem. Ber.*, 1982, **115**, 1379.	580, 581, 584, 585
82CB1733	A. Roedig, K. Grohe and H. Sommer; *Chem. Ber.*, 1982, **115**, 1733.	829, 970
82CB1810	T. Kauffmann, R. Kriegesmann, B. Altpeter and F. Steinseifer; *Chem. Ber.*, 1982, **115**, 1810.	401, 599, 708, 724
82CB1947	K. Diemert, W. Kuchen and J. Kulter; *Chem. Ber.*, 1982, **115**, 1947.	550
82CB3384	M. Heuschmann and H. Quast; *Chem. Ber.*, 1982, **115**, 3304.	122
82CC14	M. F. Lappert, C. L. Raston, B. W. Skelton and A. H. White; *J. Chem. Soc., Chem. Commun.*, 1982, 14.	604, 646, 647
82CC280	P. Pallai and M. Goodman; *J. Chem. Soc., Chem. Commun.*, 1982, 280.	422
82CC282	M. W. Anderson, R. C. F. Jones and J. Saunders; *J. Chem. Soc., Chem. Commun.*, 1982, 282.	414
82CC286	S. Al-Jibori and B. L. Shaw; *J. Chem. Soc., Chem. Commun.*, 1982, 286.	1023, 1028, 1030

Ref	Citation	Pages
82CC287	S. Al-Jibori, W. S. McDonald and B. L. Shaw; *J. Chem. Soc., Chem. Commun.*, 1982, 287.	1028, 1030
82CC335	P. Metzner; *J. Chem. Soc., Chem. Commun.*, 1982, 335.	255
82CC336	K. Lerstrup, D. Talham, A. Bloch, T. Poehler and D. O. Cowan; *J. Chem. Soc., Chem. Commun.*, 1982, 336.	876
82CC358	C. R. Johnson and D. S. Dhanoa; *J. Chem. Soc., Chem. Commun.*, 1982, 358.	396
82CC371	F. W. B. Einstein, B. H. Freeland, K. G. Tyers, D. Sutton and J. M. Waterous; *J. Chem. Soc., Chem. Commun.*, 1982, 371.	713
82CC614	K. A. Azam, A. A. Frew, B. R. Lloyd, L. Manojlovic-Muir, K. W. Muir and R. J. Puddephat; *J. Chem. Soc., Chem. Commun.*, 1982, 614.	585, 691
82CC857	C. C. Fortes, H. C. Fortes and D. C. R. G. Goncalves; *J. Chem. Soc., Chem. Commun.*, 1982, 857.	65
82CC970	D. L. Miller, J. O. Lay and M. L. Gross; *J. Chem. Soc., Chem. Commun.*, 1982, 970.	642, 655
82CC1023	R. I. Papasergio and C. L. Raston; *J. Chem. Soc., Chem. Commun.*, 1982, 1023.	603, 647, 657
82CC1115	A. J. Pratt and E. J. Thomas; *J. Chem. Soc., Chem. Commun.*, 1982, 1115.	387
82CC1184	U. Schubert, W. Neugebauer and P. von R. Schleyer; *J. Chem. Soc., Chem. Commun.*, 1982, 1184.	668
82CC1270	G. M. Blackburn and M. J. Parratt; *J. Chem. Soc., Chem. Commun.*, 1982, 1270.	801
82CC1274	M. A. Nashed and L. Anderson; *J. Chem. Soc., Chem. Commun.*, 1982, 1274.	176
82CC1323	M. F. Lappert, L. M. Engelhardt, C. L. Raston and A. H. White; *J. Chem. Soc., Chem. Commun.*, 1982, 1323.	602, 650
82CJC285	P. Métra and J. Hamelin; *Can. J. Chem.*, 1982, **60**, 285.	444
82CL587	Y. Ohshiro, M. Ishida, J. Shibata, T. Minami and T. Agawa; *Chem. Lett.*, 1982, 587.	66
82CL1555	T. Mukaiyama, T. Sugaya, S. Marui and T. Nakatsuka; *Chem. Lett.*, 1982, 1555.	218
82CL1663	H. Matsumoto, Y. Hoshino and Y. Nagai; *Chem. Lett.*, 1982, 1663.	1044
82CL1697	K. Ogura, M. Fujita, K. Takahashi and H. Iida; *Chem. Lett.*, 1982, 1697.	395
82CL1855	H. Sakurai, Y. Nakadaira and H. Tobita; *Chem. Lett.*, 1982, 1855.	640
82CL1933	M. Yokoyama, M. Tohnishi, A. Kurihara and T. Imamoto; *Chem. Lett.*, 1982, 1933.	900
82COMC-I(1)	G. Wilkinson, F. G. A. Stone and E. Abel (eds); *Comp. Organomet. Chem., 1st edn.*, 1982.	591
82COMC-I(1)221	T. J. Barton; *Comp. Organomet. Chem., 1st ed.*, 1982, **1**, 221.	1051, 1052
82COMC-I(2)365	P. Rivière, M. Rivière-Baudet and J. Satgé; in *Comp. Organomet. Chem., 1st edn.*, 1982, **2**, 365.	519
82COMC-I(2)641	J. J. Eisch; *Comp. Organomet. Chem., 1st ed.*, 1982, **2**, 641.	1068
82COMC-I(2)681	J. L. Wardell; *Comp. Organomet. Chem., 1st. edn.*, 1982, **2**, 681.	344
82COMC-I(7)465	A. McKillop and E. C. Taylor; *Comp. Organomet. Chem., 1st edn.*, 1982, **7**, 465.	194
82CPB63	N. Kawahara, T. Itoh, H. Ogura and K. A. Watanabe; *Chem. Pharm. Bull.*, 1982, **30**, 63.	893
82CPB1036	M. Yamoto and Y. Takeuchi; *Chem. Pharm. Bull.*, 1982, **30**, 1036.	417
82CPB3959	H. Takahata, A. Tomiguchi, A. Hagiwara and T. Yamazaki; *Chem. Pharm. Bull.*, 1982, **30**, 3959.	902
82CRV15	L. D. Freedman and G. O. Doak; *Chem. Rev.*, 1982, **82**, 15.	595
82CSR493	D. J. Ager; *Chem. Soc. Rev.*, 1982, **11**, 493.	365, 366, 392
82EGP154700	T. Czekanski, W. Hartmann, B. Costisella, J. Glöde, H. Gross and B. Johannsen; *Ger. (East) Pat.* 154 700 (1982) (*Chem. Abstr.*, 1983, **98**, 72 241).	553
82EUP49618	R. P. Pioch, *Eur. Pat.* 049 618 (1982).	970
82FES387	A. Balbi, A. Ermili, G. Roma and M. Mazzei; *Farmaco Ed. Sci.*, 1982, **37**, 387.	979
82GEP3111152	G. Hägele; *Ger. Pat.* 3 111 152 (1982) (*Chem. Abstr.*, 1983, **98**, 126 378).	553
82GEP3240287	A. Ruland and W. Reuther (BASF AG); *Ger. Pat.* 3 240 287 (1982) (*Chem. Abstr.* 1984, **101**, 130 439).	826
82H(19)1719	Z. J. Witczak and R. L. Whistler; *Heterocycles*, 1982, **19**, 1719.	238
82HCA1953	A. Vasella and R. Voeffray; *Helv. Chim. Acta*, 1982, **65**, 1953.	489
82HOU(E1/E2)	M. Regitz; *Methoden Org. Chem. (Houben-Weyl)*, 1982, **E1** and **E2**.	336
82IC645	M. Creswick, I. Bernal, B. Reiter and W. A. Herrmann; *Inorg. Chem.*, 1982, **21**, 645.	677
82IC3568	R. H. Neilson and P. Wisian-Neilson; *Inorg. Chem.*, 1982, **21**, 3568.	566
82IC3913	M. R. Churchill and H. J. Wasserman; *Inorg. Chem.*, 1982, **21**, 3913.	581
82IJC(B)1	M. D. Nair, S. Rajappa and J. A. Desai; *Indian J. Chem., Sect B*, 1982, **21**, 1.	94
82IZV161	V. I. Erashko, O. M. Sazonova, A. A. Tishaninova and A. A. Fainzil'berg; *Izv. Akad. Nauk SSSR, Ser. Khim.*, 1982, 161.	335
82IZV1196	B. A. Arbuzov, E. N. Dianova and I. Z. Galeeva; *Izv. Akad. Nauk SSSR, Ser. Khim.*, 1982, 1196.	549
82IZV1425	A. V. Kazantsev, M. G. Meiramov, A. I. Kovredov and L. I. Zakharkin; *Izv. Akad. Nauk SSSR, Ser. Khim.*, 1982, 1425.	143
82IZV1611	R. G. Kostyanovskij, Yu. I. El'natanov, Sh. M. Shikhaliev, S. M. Ignatov and I. I. Chervin; *Izv. Akad. Nauk SSSR Ser. Khim.*, 1982, 1611 (*Chem. Abstr.*, 1984, **98**, 89 480a).	481
82IZV2327	N. P. Petukhova, N. E. Dontsova and E. N. Prilezhaeva; *Izv. Akad. Nauk SSSR, Ser. Khim.*, 1982, 2327 (*Chem. Abstr.* 1983, **98**, 71 447).	854
82IZV2718	I. A. Litvinov, Yu. T. Struchkov, B. A. Arbuzov, E. N. Dianova and I. Z. Galeeva; *Izv. Akad. Nauk SSSR, Ser. Khim.*, 1982, 2718.	549
82JA180	T. Hayashi, M. Konishi, M. Fukushima, T. Mise, M. Kagotani, M. Tajika and M. Kumada; *J. Am. Chem. Soc.*, 1982, **104**, 180.	1033
82JA358	S. Danishefsky, J. F. Kerwin, Jr. and S. Kobayashi; *J. Am. Chem. Soc.*, 1982, **104**, 358.	174
82JA531	K. K. Wang, C. G. Scouten and H. C. Brown; *J. Am. Chem. Soc.*, 1982, **104**, 531.	630, 631
82JA877	A. I. Meyers and G. E. Jagdmann, Jr.; *J. Am. Chem. Soc.*, 1982, **104**, 877.	511
82JA1134	C. P. Casey, P. J. Fagan and W. H. Miles; *J. Am. Chem. Soc.*, 1982, **104**, 1134.	679

82JA1154	F. Wudl and E. Aharon-Shalom; *J. Am. Chem. Soc.*, 1982, **104**, 1154.	876
82JA2072	P. G. Edwards, R. W. Gellert, M. W. Marks and R. Bau; *J. Am. Chem. Soc.*, 1982, **104**, 2072.	696
82JA2223	M. Rule, R. F. Salinaro, D. R. Pratt and J. A. Berson; *J. Am. Chem. Soc.*, 1982, **104**, 2223.	746
82JA2494	W. R. Dolbier, Jr. and S. F. Sellers; *J. Am. Chem. Soc.*, 1982, **104**, 2494.	7
82JA2497	S. D. Harper and A. J. Arduengo, III; *J. Am. Chem. Soc.*, 1982, **104**, 2497.	478
82JA2637	J. A. Gurak, J. W. Chinn, Jr. and R. J. Lagow; *J. Am. Chem. Soc.*, 1982, **104**, 2637.	668
82JA3119	E. Block, E. R. Corey, R. E. Penn, T. L. Renken, P. F. Sherwin, H. Bock, T. Hirabayashi, S. Mohmand and B. Solouki; *J. Am. Chem. Soc.*, 1982, **104**, 3119.	258
82JA4288	H. Sakurai, H. Tobita, Y. Nakadaira and C. Kabuto; *J. Am. Chem. Soc.*, 1982, **104**, 4288.	1050
82JA4955	P. O. Nubel and T. L. Brown; *J. Am. Chem. Soc.*, 1982, **104**, 4955.	677
82JA4997	M. Isobe, M. Kitamura and T. Goto; *J. Am. Chem. Soc.*, 1982, **104**, 4997.	936
82JA5753	T. Shono, Y. Matsumura, K. Inoue, H. Ohmizu and S. Kashimura; *J. Am. Chem. Soc.*, 1982, **104**, 5753.	300, 311
82JA6816	E. J. Corey, B. Pan, D. H. Hua and D. R. Deardorff; *J. Am. Chem. Soc.*, 1982, **104**, 6816.	94
82JA7174	F. E. Ziegler, J.-M. Fang and C. C. Tam; *J. Am. Chem. Soc.*, 1982, **104**, 7174.	849
82JA7325	K. M. Motyl, J. R. Norton, C. K. Schauer and O. P. Anderson; *J. Am. Chem. Soc.*, 1982, **104**, 7325.	679
82JA7357	J. W. Bruno and T. J. Marks; *J. Am. Chem. Soc.*, 1982, **104**, 7357.	658
82JCED94	E. F. Witucki and M. B. Frankel; *J. Chem. Eng. Data*, 1982, **27**, 94	444
82JCED475	S. S. Rawalay and C. F. Beam; *J. Chem. Eng. Data*, 1982, **27**, 475.	449
82JCR(S)116	J. O. Jones and R. S. McElhinney; *J. Chem. Res. (S)*, 1982, 116.	226
82JCR(S)132	M. V. Garad and J. W. Wilson; *J. Chem. Res. (S)*, 1982, 132.	630, 632, 634
82JCR(S)212	J. V. Comasseto, J. T. B. Ferreira, C. A. Brandt and N. Petragnani; *J. Chem. Res. (S)*, 1982, 212.	88
82JCS(D)1155	A. J. Arce and A. J. Deeming; *J. Chem. Soc., Dalton Trans.*, 1982, 1155.	682
82JCS(D)1297	A. F. Dyke, S. A. R. Knox, P. J. Naish and G. E. Taylor; *J. Chem. Soc., Dalton Trans.*, 1982, 1297.	679
82JCS(D)1309	M. Green, R. M. Mills, G. N. Pain, F. G. A. Stone and P. Woodward; *J. Chem. Soc., Dalton Trans.*, 1982, 1309.	686
82JCS(D)1783	S. R. Finnimore, S. A. R. Knox and G. E. Taylor; *J. Chem. Soc., Dalton Trans.*, 1982, 1783.	674
82JCS(D)1915	I. J. Colquhoun and W. McFarlane; *J. Chem. Soc., Dalton Trans.*, 1982, 1915.	1022
82JCS(P1)143	A. R. Katritzky, J. Arrowsmith, N. E. Grzeskowiak, H. J. Salgado and Z. B. Bahari; *J. Chem. Soc., Perkin Trans. 1*, 1982, 143.	540
82JCS(P1)1059	M. Yokoyama, K. Arai and T. Imamoto; *J. Chem. Soc. Perkin Trans. 1*, 1982, 1059.	988
82JCS(P1)1193	F. M. Dean and R. S. Varma; *J. Chem. Soc., Perkin Trans. 1*, 1982, 1193.	307
82JCS(P1)2227	M. J. Broadhurst and C. H. Hassall; *J. Chem. Soc., Perkin Trans. 1*, 1982, 2227.	303
82JCS(P1)2757	P. J. Giddings, D. I. John, E. J. Thomas and D. J. Williams; *J. Chem. Soc., Perkin Trans. 1*, 1982, 2757.	90
82JCS(P1)2771	F. M. Dean, M. Varma and R. S. Varma; *J. Chem. Soc., Perkin Trans. 1*, 1982, 2771.	307
82JCS(P2)881	J. Emsley, V. Gold, M. J. B. Jais and L. Z. Zdunek; *J. Chem. Soc., Perkin Trans. 2*, 1982, 881.	42
82JFC(20)699	N. Redwane, P. Moreau and A. Commeyras; *J. Fluorine Chem.*, 1982, **20**, 699.	812, 820
82JGU392	M. P. Osipova, L. V. Kuz'mina and V. A. Kukhtin; *J. Gen. Chem. USSR (Engl. Transl.)*, 1982, **52**, 392.	485
82JHC1493	M. Artico, F. Corelli, S. Massa and G. Stefancich; *J. Heterocycl. Chem.*, 1982, **19**, 1493.	856
82JIC111	D. Seyferth, C. K. Haas and D. Dagani; *J. Indian Chem. Soc.*, 1982, **59**, 111.	153
82JOC615	C. H. Lin and D. L. Alexander; *J. Org. Chem.*, 1982, **47**, 615.	171
82JOC693	J. A. Marshall and R. D. Royce, Jr.; *J. Org. Chem.*, 1982, **47**, 693.	240
82JOC824	W. F. Berkowitz, S. C. Choudhry and J. A. Hrabie; *J. Org. Chem.*, 1982, **47**, 824.	175
82JOC946	J. A. Panetta and H. Rapoport; *J. Org. Chem.*, 1982, **47**, 946.	175
82JOC1145	N. H. Andersen, A. D. Denniston and D. A. McCrae; *J. Org. Chem.*, 1982, **47**, 1145.	245
82JOC1397	Z. Rappoport and B. Avramovitch; *J. Org. Chem.*, 1982, **47**, 1397.	894
82JOC1618	H.-J. Reich, M. C. Clark and W. W. Willis, Jr.; *J. Org. Chem.*, 1982, **47**, 1618.	947
82JOC1706	T. Nishitani, H. Horikawa, T. Iwasaki, K. Matsumoto, I. Inoue and M. Miyoshi; *J. Org. Chem.*, 1982, **47**, 1706.	424
82JOC2080	H. A. A. Rasoul and H. K. Hall; *J. Org. Chem.*, 1982, **47**, 2080.	827
82JOC2216	S. H. Bertz, G. Dabbagh and P. Cotte; *J. Org. Chem.*, 1982, **47**, 2216.	195
82JOC2251	K. Baum, C. D. Bedford and R. J. Hunadi; *J. Org. Chem.*, 1982, **47**, 2251.	196, 810
82JOC2474	D. A. Cichra and H. G. Adolph; *J. Org. Chem.*, 1982, **47**, 2474.	435
82JOC2582	J. Singh, K. P. Agarwal and G. Singh; *J. Org. Chem.*, 1982, **47**, 2582.	56
82JOC2995	S. Cacchi and D. Misiti; *J. Org. Chem.*, 1982, **47**, 2995.	168
82JOC3027	K. T. Potts, M. J. Cipullo, P. Ralli and G. Theodoridis; *J. Org. Chem.*, 1982, **47**, 3027.	853, 855
82JOC3094	F. T. Oakes, F. A. Yang and J. F. Sebastian; *J. Org. Chem.*, 1982, **47**, 3094.	949, 950
82JOC3140	R. Jacobson, R. J. Taylor, H. J. Williams and L. R. Smith; *J. Org. Chem.*, 1982, **47**, 3140.	171
82JOC3517	R. R. Gallucci and R. C. Going; *J. Org. Chem.*, 1982, **47**, 3517.	208
82JOC3707	R. D. Bach, R. A. Woodward, T. J. Anderson and M. D. Glick; *J. Org. Chem.*, 1982, **47**, 3707.	206
82JOC3808	H. C. Brown, D. Basavaiah and S. U. Kulkarni; *J. Org. Chem.*, 1982, **47**, 3808.	815
82JOC3946	L. Engman and M. P. Cava; *J. Org. Chem.*, 1982, **47**, 3946.	872
82JOC5051	J. J. Eisch, J. E. Galle, A. Piotrowski and M. R. Tsai; *J. Org. Chem.*, 1982, **47**, 5051.	387
82JOC5404	K. Ogura, J. Watanabe, K. Takahashi and H. Iida; *J. Org. Chem.*, 1982, **47**, 5404.	854, 857
82JOM(224)C53	N. N. Sedova, S. K. Moiseev and V. A. Sazonova; *J. Organomet. Chem.*, 1982, **224**, C53.	723

Ref	Citation	Pages
82JOM(224)153	W. P. Fehlhammer, P. Hirschmann and A. Mayr; *J. Organomet. Chem.*, 1982, **224**, 153.	1019
82JOM(224)377	R. S. Dickson, A. P. Oppenheim and G. N. Pain; *J. Organomet. Chem.*, 1982, **224**, 377.	688
82JOM(225)71	E. C. Ashby and R. S. Smith; *J. Organomet. Chem.*, 1982, **225**, 71.	698
82JOM(226)C59	W. A. Herrmann, J. M. Huggins, C. Bauer, M. Smischek, H. Pfisterer and M. L. Ziegler; *J. Organomet. Chem.*, 1982, **226**, C59.	686
82JOM(228)C71	N. J. Kermode, M. F. Lappert, B. W. Skelton, A. H. White and J. Holton; *J. Organomet. Chem.*, 1982, **228**, C71.	584
82JOM(231)191	U. Sicker, A. Meller and W. Maringgele; *J. Organomet. Chem.*, 1982, **231**, 191.	1015
82JOM(232)C78	C. Engelter, J. R. Moss, M. L. Niven, L. R. Nassimbeni and G. Reid; *J. Organomet. Chem.*, 1982, **232**, C78.	585
82JOM(232)33	P. R. Jones and M. E. Lee; *J. Organomet. Chem.*, 1982, **232**, 33.	611
82JOM(232)137	A. Castel, P. Riviere and J. Satgé; *J. Organomet. Chem.*, 1982, **232**, 137.	361
82JOM(233)C28	M. F. Lappert, W. P. Leung, C. L. Raston, A. J. Thorne, B. W. Skelton and A. H. White; *J. Organomet. Chem.*, 1982, **233**, C28.	660
82JOM(235)287	H. Schumann and F. W. Reier; *J. Organomet. Chem.*, 1982, **235**, 287.	578
82JOM(236)C18	W. A. Herrmann, C. Bauer and K. K. Mayer; *J. Organomet. Chem.*, 1982, **236**, C18.	686
82JOM(236)C65	H. Werner, L. Hofmann and W. Paul; *J. Organomet. Chem.*, 1982, **236**, C65.	583
82JOM(236)69	K. Jurkschat and M. Gielen; *J. Organomet. Chem.*, 1982, **236**, 69.	700
82JOM(240)299	M. I. Bruce, J. R. Rodgers, M. R. Snow and F. S. Wong; *J. Organomet. Chem.*, 1982, **240**, 299.	713
82JOU876	I. I. Lapkin, A. N. Nedugov, N. N. Pavlova and G. A. Gartman; *J. Org. Chem. USSR (Engl. Trans.)*, 1982, **18**, 876.	239
82JOU2240	Yu. G. Shermolovich, V. S. Talamov, V. V. Pirozhenko and L. N. Markovskii, *J. Org. Chem. USSR (Engl. Transl.)*, 1982, **18**, 2240.	110
82JPR309	M. Schulz, L. Mogel, W. Riediger, N. Dung and R. Radeglia; *J. Prakt. Chem.*, 1982, **324**, 309.	989
82JPR537	T. Czekanski, H. Gross and B. Costisella; *J. Prakt. Chem.*, 1982, **324**, 537.	553
82JPR832	G. Kempter, W. Ehrlichmann, M. Plesse and H.-U. Lehm; *J. Prakt. Chem.*, 1982, **324**, 832.	410
82JPR915	H. Dehne and P. Krey; *J. Prakt. Chem.*, 1982, **324**, 915.	981
82LA585	H. Böhme, G. Ahrens and W. Krack; *Liebigs Ann. Chem.*, 1982, 585.	903
82LA1022	H. Böhme and G. Ahrens; *Liebigs Ann. Chem.*, 1982, 1022.	860
82LA1946	R. Tacke, H. Lange and M. T. Attar-Bashi; *Liebigs Ann. Chem.*, 1982, 1946.	136
B-82MI 409-01	N. Kornblum; in "The Chemistry of Functional Groups, Suppl. F, Part 1. The Chemistry of Amino, Nitroso and Nitro Compounds and Their Derivatives," ed. S. Patai, Wiley-Interscience, Chichester, 1982, p. 361.	428
B-82MI 409-02	L. Duhamel; in "The Chemistry of Functional Groups, Suppl. F, Part 1. The Chemistry of Amino, Nitroso and Nitro Compounds and Their Derivatives," ed. S. Patai, Wiley-Interscience, Chichester, 1982, p. 849.	404, 405, 406, 408, 410, 411, 412
82MI 420-01	M. H. Akhter; *J. Agric. Food Chem.*, 1982, **30**, 1048.	912
82OM20	D. S. Matteson and R. J. Moody; *Organometallics*, 1982, **1**, 20.	634, 664
82OM219	K. H. Theopold and R. G. Bergman; *Organometallics*, 1982, **1**, 219.	687
82OM225	G. Jaouen, A. Marinetti, J.-Y. Saillard, B. G. Sayer and M. J. McGlinchey; *Organometallics*, 1982, **1**, 225.	713
82OM404	J. R. Hyde, T. J. Karol, J. P. Hutchinson, H. G. Kuivila and J. Zubieta; *Organometallics*, 1982, **1**, 404.	700
82OM553	P. Magnus and G. Roy; *Organometallics*, 1982, **1**, 553.	650
82OM623	D. W. Morton and R. H. Neilson; *Organometallics*, 1982, **1**, 623.	571
82OM830	J. A. Soderquist and G. J.-H. Hsu; *Organometallics*, 1982, **1**, 830.	929, 930, 949, 951, 952
82OM911	S. C. Kao, P. P. Y. Lu and R. Pettit; *Organometallics*, 1982, **1**, 911.	679
82OM1243	P. Magnus and D. A. Quagliato; *Organometallics*, 1982, **1**, 1243.	936
82OM1307	M. Barber, D. Liptak and J. P. Oliver; *Organometallics*, 1982, **1**, 1307.	698
82OM1350	C. E. Sumner, Jr., J. A. Collier and R. Pettit; *Organometallics*, 1982, **1**, 1350.	679
82OM1473	M. Ishikawa, H. Sugisawa, M. Kumada, T. Higuchi, K. Matsui and K. Hirotsu; *Organometallics*, 1982, **1**, 1473.	1053
82OM1597	J. C. Jeffery, C. Sambale, M. F. Schmidt and F. G. A. Stone; *Organometallics*, 1982, **1**, 1597.	712, 716
82OM1607	S. M. Gadol and R. E. Davis; *Organometallics*, 1982, **1**, 1607.	710
82OR(27)1	J.-F. Biellmann and J. B. Ducep; *Org React*, 1982, **27**, 1.	381
82OR(28)1	H. Wynberg and E. W. Meijer; *Org. React.*, 1982, **28**, 1.	17
82POL31	K. W. Chiu, W-K. Wong, G. Wilkinson, A. M. R. Galas and M. B. Hursthouse; *Polyhedron*, 1982, **1**, 31.	656
82PS(13)85	R. Burgada and A. Mohri; *Phosphorus Sulfur*, 1982, **13**, 85.	478
82PS(13)319	M. M. Sidky, M. F. Zayed, K. Praefcke, W. Wong-ng and S. C. Niburg; *Phosphorus Sulfur*, 1982, **13**, 319.	487
82RTC202	A. M. van Leusen and J. Wildeman; *Rec. Trav. Chim. Pays-Bas*, 1982, **101**, 202.	908
82S127	R. P. Fisher, H. P. On, J. T. Snow and G. Zweifel; *Synthesis*, 1982, 127.	756
82S162	A. Cornelis and P. Laszlo; *Synthesis*, 1982, 162.	190
82S173	F. Kober; *Synthesis*, 1982, 173.	593
82S188	C. C. Tam, K. L. Mattocks and M. Tishler; *Synthesis*, 1982, 188.	466
82S193	S. U. Kulkarni, H. D. Lee and H. C. Brown; *Synthesis*, 1982, 193.	815
82S195	H. C. Brown, H. D. Lee and S. U. Kulkarni; *Synthesis*, 1982, 195.	815
82S199	W. Schroth, R. Spitzner and S. Hugo; *Synthesis*, 1982, 199.	793, 970, 971
82S206	R. Spitzner, M. Menzel and W. Schroth; *Synthesis*, 1982, 206.	837, 860
82S261	T. Sakakibara, S. Ikuta and R. Sudoh; *Synthesis*, 1982, 261.	335

82S312	M. Hojo, R. Masuda, K. Yoshinaga and S. Munehira; *Synthesis*, 1982, 312.	226
82S394	S. F. Martin and P. J. Garrison; *Synthesis*, 1982, 394.	924
82S421	A. Liptak, V. A. Olah and J. Kerekgyarto; *Synthesis*, 1982, 421.	190
82S504	G. D. Hartman and R. D. Hartman; *Synthesis*, 1982, 504.	81
82S579	W. Chamchaang, V. Prankprakma B. Tarnchompoo, C. Thebtaranonth and Y. Thebtaranonth; *Synthesis*, 1982, 579.	849
82S653	P. G. Baraldi, M. Guarneri, F. Moroder, G. P. Pollini and D. Simoni; *Synthesis*, 1982, 653.	475
82S725	P. Savignac and F. Mathey; *Synthesis*, 1982, 725.	570, 585, 588
82S748	R. R. Schmidt and R. Betz; *Synthesis*, 1982, 748.	1016
82S767	R. Annunziata and M. Cinquini; *Synthesis*, 1982, 767.	272
82S962	G. A. Olah and A. K. Mehrotra; *Synthesis*, 1982, 962.	203
82S1045	D. Hoppe and A. Broenneke; *Synthesis*, 1982, 1045.	382
82S1078	M. Machida, H. Takechi and Y. Shishido; *Synthesis*, 1982, 1078.	1077
82S1080	T. Takajo and S. Kambe; *Synthesis*, 1982, 1080.	437, 445
82S1089	Y. Kita, H. Yasuda, J. Haruta, J. Segawa and Y. Tamura; *Synthesis*, 1982, 1089.	188
82SC415	C. R. Degenhardt, *Synth. Commun.*, 1982, **12**, 415.	1002
82SC595	J. T. B. Ferreira, J. V. Comasseto and A. L. Braga; *Synth. Commun.*, 1982, **12**, 595.	284
82SC1027	R. Yamaguchi, H. Kawasaki and M. Kawanisi; *Synth. Commun.*, 1982, **12**, 1027.	810
82SUL63	K. Schank and S. Buegler; *Sulfur Letters*, 1982, **1**, 63.	266
82T139	B. Costisella and H. Gross; *Tetrahedron*, 1982, **38**, 139.	1004
82T1673	S. Rajappa, M. D. Nair, R. Sreenivasan and B. G. Advani; *Tetrahedron*, 1982, **38**, 1673.	94
82T1975	P. W. Hickmott; *Tetrahedron*, 1982, **38**, 1975.	294
82T2513	M. M. Campbell, N. I. Carruthers and S. J. Mickel; *Tetrahedron*, 1982, **38**, 2513.	487
82TCC55	H. Siegel; *Top. Curr. Chem.*, 1982, **106**, 55.	145
82TCC125	A. J. Ashe, III; *Top. Curr. Chem.*, 1982, **105**, 125.	1042
82TL353	A. Pelter, R. Al-Bayati and W. Lewis; *Tetrahedron Lett.*, 1982, **23**, 353.	206
82TL603	H. K. Hall, Jr., H. A. A. Rasoul, M. Gillard, M. Abdelkader, P. Nogues and R. C. Sentman; *Tetrahedron Lett.*, 1982, **23**, 603.	827
82TL889	Y. Oikawa, T. Yoshioka and O. Yonemitsu; *Tetrahedron Lett.*, 1982, **23**, 889.	201
82TL1047	B. M. Trost and E. Murayama; *Tetrahedron Lett.*, 1982, **23**, 1047.	250
82TL1075	T. Tanaka, T. Hashimoto, K. Iino, Y. Sugimura and T. Miyadera; *Tetrahedron Lett.*, 1982, **23**, 1075.	842
82TL1279	T. Hiyama, A. Kanakura, Y. Morizawa and H. Nozaki; *Tetrahedron Lett.*, 1982, **23**, 1279.	652
82TL1531	S. L. Bender, M. R. Detty and N. F. Haley; *Tetrahedron Lett.*, 1982, **23**, 1531.	93, 290
82TL1537	R. M. Moriarty and H. Hu; *Tetrahedron Lett.*, 1982, **23**, 1537.	198
82TL1565	J.-M. Alcaraz, A. Breque and F. Mathey; *Tetrahedron Lett.*, 1982, 1565.	341
82TL1945	D. J. Ager; *Tetrahedron Lett.*, 1982, **23**, 1945.	773
82TL2103	T. Hiyama and M. Kai; *Tetrahedron Lett.*, 1982, **23**, 2103.	441
82TL2323	M. Obayashi, E. Ito, K. Matsui and K. Kondo; *Tetrahedron Lett.*, 1982, **23**, 2323.	734
82TL2327	M. Obayashi and K. Kondo; *Tetrahedron Lett.*, 1982, **23**, 2327.	734
82TL2399	H. A. Khan and I. Paterson; *Tetrahedron Lett.*, 1982, **23**, 2399.	66
82TL2539	R. R. Regis and A. M. Doweyko; *Tetrahedron Lett.*, 1982, **23**, 2539.	81
82TL2665	T. Okuyama; *Tetrahedron Lett.*, 1982, **23**, 2665.	1074
82TL3181	M. Sawada, M. Ichihara, Y. Furukawa, Y. Takai, T. Ando and T. Hanafusa; *Tetrahedron Lett.*, 1982, **23**, 3181.	426
82TL3407	J. N. Denis and A. Krief; *Tetrahedron Lett.*, 1982, **23**, 3407.	844
82TL3411	J. N. Dennis and A. Krief; *Tetrahedron Lett.*, 1982, **23**, 3411.	964
82TL3535	F. DiNinno, D. A. Muthard, R. W. Ratcliffe and B. G. Christensen; *Tetrahedron Lett.*, 1982, **23**, 3535.	847
82TL3595	M. Suzuki, A. Yanagisawa and R. Noyori; *Tetrahedron Lett.*, 1982, **23**, 3595.	192
82TL3795	M. S. Baird and P. D. Slowey; *Tetrahedron Lett.*, 1982, **23**, 3795.	742
82TL3835	E. W. Colvin, G. W. Kirby and A. C. Wilson; *Tetrahedron Lett.*, 1982, **23**, 3835.	490
82TL4083	J. Pornet, D. Mesnard and L. Miginiac; *Tetrahedron Lett.*, 1982, **23**, 4083.	1047
82TL4167	S. Warren and A. T. Zaslona; *Tetrahedron Lett.*, 1982, **23**, 4167.	923
82TL4203	E. Block and M. Aslam; *Tetrahedron Lett.*, 1982, **23**, 4203.	368
82TL4371	G. Galambos and V. Simonidesz; *Tetrahedron Lett.*, 1982, **23**, 4371.	240
82TL5003	P. Dureja, J. E. Casida and L. O. Ruzo; *Tetrahedron Lett.*, 1982, **23**, 5003.	743
82TL5083	H. A. Khan and I. Paterson; *Tetrahedron Lett.*, 1982, **23**, 5083.	67
82ZAAC(486)39	M. Baudler and F. Saykowski; *Z. Anorg. Allg. Chem.*, 1982, **486**, 39.	545
82ZAAC(488)75	K. Issleib, H. Schmidt and C. Wirkner; *Z. Anorg. Allg. Chem.*, 1982, **488**, 75.	568
82ZAAC(494)55	H.-P. Abicht, U. Baumeister, H. Hartung, K. Issleib, R. A. Jacobson, J. Richardson, S. M. Socol and J. G. Verkade; *Z. Anorg. Allg. Chem.*, 1982, **494**, 55.	575
82ZC307	G. Paul and E. Herrmann; *Z. Chem.*, 1982, **22**, 307.	555
82ZN(B)284	H. H. Karsch; *Z. Naturforsch., Teil B*, 1982, **37**, 284.	546, 554, 555, 566
82ZN(B)512	A. Malik, N. Afza and S. Siddiqui; *Z. Naturforsch., Teil B*, 1982, **37**, 512.	413
82ZN(B)950	H. Schmidbaur, U. Deschler and D. Seyferth; *Z. Naturforsch., Teil B*, 1982, **37**, 950.	576, 577
82ZN(B)957	F. Preuss and L. Ogger; *Z. Naturforsch., Teil B*, 1982, **37**, 957.	656
82ZN(B)1044	W. P. Fehlhammer, K. Bartel, A. Volkl and D. Achatz; *Z. Naturforsch., Teil B*, 1982, **37**, 1044.	1017
82ZN(B)1146	E. Lindner, F. Bouachir and W. Hiller; *Z. Naturforsch., Teil B*, 1982, **37**, 1146.	583
82ZOB930	N. K. Maidanovich, S. V. Iksanova and Yu. G. Gololobov; *Zh. Obshch. Khim.*, 1982, **52**, 930 (*Chem. Abstr.*, 1982, **97**, 72 446a).	486
82ZOB1122	O. P. Lobanov and B. S. Drach; *Zh. Obshch. Khim.*, 1982, **52**, 1122.	1006, 1010, 1012

82ZOB1386	V. P. Kozyukov, V. P. Kozyukov and V. F. Miranov; *Zh. Obshch. Khim.*, 1982, **52**, 1386.	313
82ZOB2375	S. G. Seredkina, V. E. Kolbina, V. G. Rozinov, A. N. Mirskova, V. I. Donskikh and M. G. Voronkov; *Zh. Obshch. Khim.*, 1982, **52**, 2375.	926
82ZOB2444	R. D. Gareev, A. V. Chernova, E. A. Ishmaeva, E. A. Berdnikov, R. R. Shagidullin, E. N. Strelkova, G. M. Dorozhkina, I. I. Patsanovskii and A. N. Pudovik; *Zh. Obshch. Khim.*, 1982, **52**, 2444 (*Chem. Abstr.*, 1983, **98**, 143 528).	489
82ZOR841	I. V. Koval', V. V. Andrushchenko and M. M. Kremlev; *Zh. Org. Khim.*, 1982, **18**, 841.	326
82ZOR1835	Y. A. Ol'dekop, R. V. Kaberdin and E. E. Buslovskaya; *Zh. Org. Khim.*, 1982, **18**, 1835 (Engl. Transl. 1606).	970
83AG(E)330	C. Elschenbroich, J. Heck, W. Massa and R. Schmidt; *Angew. Chem., Int. Ed. Engl.*, 1983, **22**, 330.	673
83AG(E)500	G. Koltzenburg, G. Behrens and D. Schulte-Frohlinde; *Angew. Chem., Int. Ed. Engl.*, 1983, **22**, 500.	1078
83AG(E)553	H. Schubert and M. Regitz; *Angew. Chem., Int. Ed. Engl.*, 1983, **22**, 553.	889
83AG(E)733	A. Maercker, M. Theis, A. J. Kos and P. von R. Schleyer; *Angew. Chem., Int. Ed. Engl.*, 1983, **22**, 733.	667, 668
83AG(E)907	H. Schmidbaur; *Angew. Chem., Int. Ed. Engl.*, 1983, **22**, 907.	578
83AJC2095	D. J. Collins, P. F. Drygala and J. M. Swan; *Aust. J. Chem.*, 1983, **36**, 2095.	454, 471
83BCJ257	T. Numata, O. Itoh, T. Yoshimura and S. Oae; *Bull. Chem. Soc. Jpn.*, 1983, **56**, 257.	223
83BCJ266	O. Itoh, T. Numata, T. Yoshimura and S. Oae; *Bull. Chem. Soc. Jpn.*, 1983, **56**, 266.	223
83BCJ270	S. Oae, O. Itoh, T. Numata and T. Yoshimura; *Bull. Chem. Soc. Jpn.*, 1983, **56**, 270.	227
83BCJ1772	Y. Yamamoto; *Bull. Chem. Soc. Jpn.*, 1983, **56**, 1772.	577, 587
83BCJ1881	T. Hirao, S. Kohno, Y. Ohshiro and T. Agawa; *Bull. Chem. Soc. Jpn.*, 1983, **56**, 1881.	13
83BCJ2480	T. Migita, K. Nagai and M. Kosugi; *Bull. Chem. Soc. Jpn.*, 1983, **56**, 2480.	14
83CA(98)4614z	G. G. Minasyan, G. Ts. Gasparyan, A. M. Torgomyan, M. Zh. Ovakimyan and M. G. Indzhikyan; *Chem. Abstr.*, 1983, **98**, 4614z (*Arm. Khim. Zh.*, 1982, **35**, 583).	914
83CA(98)16507e	Kowa Co Ltd.; *Chem. Abstr.*, 1983, **98**, 16 507e (*Jpn. Pat.* 5 795 988 (1982)).	777
83CA(99)22296x	J. Amat. Badimas (Barisintex SA); *Chem. Abstr.*, 1983, **99**, 22 296x (*Spanish Pat.* ES508,693).	907
83CB114	R. Appel, M. Huppertz and A. Westerhaus; *Chem. Ber.*, 1983, **116**, 114.	112
83CB348	A. Meyer, A. Hartl and M. Malisch; *Chem. Ber.*, 1983, **116**, 348.	135, 597
83CB473	T. Kauffmann, R. Joussen, N. Klas and A. Vahrenhorst; *Chem. Ber.*, 1983, **116**, 473.	565, 596, 600
83CB690	W. Malisch, H. Blau and U. Schubert; *Chem. Ber.*, 1983, **116**, 690.	581
83CB1001	T. Kauffmann and H. Ahlers; *Chem. Ber.*, 1983, **116**, 1001.	947, 965
83CB1070	E. Lindner, K. A. Starz, N. Pauls and W. Winter; *Chem. Ber.*, 1983, **116**, 1070.	581
83CB1209	E. Lindner, K. A. Starz, H.-J. Eberle and W. Hiller; *Chem. Ber.*, 1983, **116**, 1209.	581
83CB1240	B. Giese and U. Erfort; *Chem. Ber.*, 1983, **116**, 1240.	385
83CB1479	H. Noeth and D. Sedlak; *Chem. Ber.*, 1983, **116**, 1479.	373
83CB2275	R. Appel, U. Baumeister and F. Knoch; *Chem. Ber.*, 1983, **116**, 2275.	559
83CB3141	E. Lindner and E. Tamoutsidis; *Chem. Ber.*, 1983, **116**, 3141.	921
83CB3182	B. Wrackmeyer, C. Bihlmayer and M. Schilling; *Chem. Ber.* 1983, **116**, 3182.	665
83CB3631	N. De Kimpe, L. De Buyck, R. Verhé and N. Schamp; *Chem. Ber.*, 1983, **116**, 3631.	183
83CB3926	A. Mertens, M. Arvanaghi and G. A. Olah; *Chem. Ber.*, 1983, **116**, 3926.	1072
83CC210	T. Livinghouse and R. Smith; *J. Chem. Soc., Chem. Commun.*, 1983, 210.	518
83CC276	P. L. Watson; *J. Chem. Soc., Chem. Commun.*, 1983, 276.	578
83CC721	R. J. Lafrance, Y. Tang, K. Vaughan and D. L. Hooper; *J. Chem. Soc., Chem. Commun.*, 1983, 721.	444
83CC886	G. M. Blackburn and M. J. Parratt; *J. Chem. Soc., Chem. Commun.*, 1983, 886.	131
83CC939	D. H. R. Barton, D. Crich and W. B. Motherwell; *J. Chem. Soc., Chem. Commun.*, 1983, 939.	1077
83CC1040	R. M. Adlington, J. E. Baldwin, J. C. Bottaro and M. W. D. Perry; *J. Chem. Soc., Chem. Commun.*, 1983, 1040.	536
83CC1064	C. W. Jefford, J. Rossier and G. D. Richardson; *J. Chem. Soc., Chem. Commun.*, 1983, 1064.	214
83CC1087	S. Gambarotta, C. Floriani, A. Chiesi-Villa and C. Guastini; *J. Chem. Soc., Chem. Commun.*, 1983, 1087.	696
83CC1156	S. Gambarotta, C. Floriani, A. Chiesi-Villa and C. Guastini; *J. Chem. Soc., Chem. Commun.*, 1983, 1156.	696
83CC1171	E. Niecke, M. Leuer, D.-A. Wildbredt and W. W. Schöller; *J. Chem. Soc., Chem. Commun.*, 1983, 1171.	567
83CC1209	I. Cutting and P. J. Parsons; *J. Chem. Soc., Chem. Commun.*, 1983, 1209.	860
83CC1309	C. Bianchini and A. Meli; *J. Chem. Soc., Chem. Commun.*, 1983, 1309.	851
83CC1322	K. Nishiyama and N. Tanaka; *J. Chem. Soc., Chem. Commun.*, 1983, 1322.	516
83CC1349	C. J. Easton; *J. Chem. Soc., Chem. Commun.*, 1983, 1349.	65
83CC1419	R. I. Papasergio, C. L. Raston and A. H. White; *J. Chem. Soc. Chem. Commun.*, 1983, 1419.	604
83CCC1710	M. Hájek and P. Šilhavý; *Collect. Czech. Chem. Commun.*, 1983, **48**, 1710.	17
83CHE(42-1)133	J. A. Deyrup; *Chem. Heterocycl. Compd.* (*Engl. Transl.*), 1983, **42-1**, 133.	311
83CJC2103	P. Calinaud and J. Gelas; *Can. J. Chem.*, 1983, **61**, 2103.	226
83CJC2354	I. R. Butler and W. R. Cullen; *Can. J. Chem.*, 1983, **61**, 2354.	1034
83CJC2425	P. Kafarski, B. Lejczak and J. Szewczyk; *Can. J. Chem.*, 1983, **61**, 2425.	462
83CL767	K. Ogura, N. Yahata, K. Hashizume, K. Tsuyama, K. Takahashi and H. Iida; *Chem. Lett.*, 1983, 767.	854, 857
83CL933	J. Koshino, T. Sugawa, T. Yogo and A. Suzuki; *Chem. Lett.*, 1983, 933.	935

83CL1131	O. Tsuge, S. Kanemasa and K. Matsuda; *Chem. Lett.*, 1983, 1131.	516
83CL1303	M. Shibasaki, H. Suzuki, Y. Torisawa and S. Ikegami; *Chem. Lett.*, 1983, 1303.	155
83CL1349	S. Torii and T. Inokuchi; *Chem. Lett.*, 1983, 1349.	181
83CPB723	H. Saikachi, H. Sasaki and T. Kitagawa; *Chem. Pharm. Bull.*, 1983, **31**, 723.	442
83CSR53	T. L. Gilchrist; *Chem. Soc. Rev.*, 1983, **12**, 53.	798
83EGP201449	H. A. Lehman, H. Schadow and L. Pfützner; *Ger. (East) Pat.* 201 449 (1983) (*Chem. Abstr.*, 1984, **100**, 139 339).	550
83GEP3241568	R. A. Olofson and J. T. Martz; *Ger. Pat.* 3 241 568 (1983) (*Chem. Abstr.*, 1983, **99**, 53 164).	58
83H(20)1615	H. A. Daboun and S. E. Abdou; *Heterocycles*, 1983, **20**, 1615.	420
83HCA308	D. Seebach, R. Hassig and J. Gabriel; *Helv. Chim. Acta*, 1983, **66**, 308.	816
83HCA809	J.-C. Muller, H. Ramuz and H.-P. Wagner; *Helv. Chim. Acta*, 1983, **66**, 809.	883
83IC575	B.-L. Li, J. S. Engenito, R. H. Neilson and P. Wisian-Neilson; *Inorg. Chem.*, 1983, **22**, 575.	566
83ICA63	R. Usón, A. Laguna, M. Laguna and A. Usón; *Inorg. Chim. Acta*, 1983, **73**, 63.	586
83IZV432	A. M. Kibardin, T. K. Gazizov, K. M. Enikeev and A. N. Pudovik; *Izv. Akad. Nauk SSSR, Ser. Khim.*, 1983, 432 (*Chem. Abstr.*, 1983, **98**, 160 841b).	486
83IZV1374	B. A. Arbuzov, O. A. Erastov, G. N. Nikonov, I. P. Romanova, R. P. Arshinova and R. A. Kadyrov; *Izv. Akad. Nauk SSSR, Ser. Khim.*, 1983, 1374.	342
83IZV1379	O. A. Erastov, G. N. Nikonov and B. A. Arbuzov; *Izv. Akad. Nauk SSSR, Ser. Khim.*, 1983, 1379 (*Chem. Abstr.*, 1983, **99**, 212 595s).	478
83IZV2153	G. S. Lyashenko, A. Kh. Filippova, A. I. Borisova, I. D. Kalikhman and N. S. Vyazankin; *Izv. Akad. Nauk SSSR, Ser. Khim.*, 1983, **9**, 2153.	766
83IZV2568	V. I. Saloutin, I. A. Piterskikh, K. I. Pashkevich and M. I. Kodess; *Izv. Akad. Nauk SSSR, Ser. Khim.*, 1983, 2568.	311
83IZV2802	I. S. Alfer'ev, I. L. Kotlyarevskii, N. V. Mikhalin and V. M. Novikova; *Izv. Akad. Nauk SSSR, Ser. Khim.*, 1983, 2802.	560
83JA619	K. S. Kyler and D. S. Watt; *J. Am. Chem. Soc.*, 1983, **105**, 619.	938, 939
83JA625	W. C. Still, S. Murata, G. Revial and K. Yoshihara; *J. Am. Chem. Soc.*, 1983, **105**, 625.	386
83JA650	D. J. Burton and D. G. Cox; *J. Am. Chem. Soc.*, 1983, **105**, 650.	801
83JA667	P. G. Gassman and S. M. Bonser; *J. Am. Chem. Soc.*, 1983, **105**, 667.	67
83JA1679	Y. C. Lin, J. C. Calabrese and S. S. Wreford; *J. Am. Chem. Soc.*, 1983, **105**, 1679.	678
83JA2077	D. S. Matteson and K. M. Sadhu; *J. Am. Chem. Soc.*, 1983, **105**, 2077.	362
83JA2430	K. C. Nicolaou, S. P. Seitz and D. P. Papahatjis; *J. Am. Chem. Soc.*, 1983, **105**, 2430.	224, 225
83JA2582	J. Edwin, M. Bochmann, M. C. Böhm, D. E. Brennan, W. E. Geiger, C. Krüger, J. Pebler, H. Pritzkow, W. Siebert, W. Swiridoff, H. Wadepohl, J. Weiss and U. Zenneck; *J. Am. Chem. Soc.*, 1983, **105**, 2582.	632
83JA2599	R. S. Herrick, S. J. Nieter-Burgmayer and J. L. Templeton; *J. Am. Chem. Soc.*, 1983, **105**, 2599.	675
83JA2771	T. Harada, E. Akiba and A. Oku; *J. Am. Chem. Soc.*, 1983, **105**, 2771.	17
83JA2905	Ch. Elschenbroich, J. Heck, W. Massa, E. Nun and R. Schmidt; *J. Am. Chem. Soc.*, 1983, **105**, 2905.	672
83JA3411	M. Barfield, E. D. Canada, Jr., C. R. McDaniel, Jr., J. L. Marshall and S. R. Walter; *J. Am. Chem. Soc.*, 1983, **105**, 3411.	24
83JA3661	N. Cohen, B. L. Banner, R. L. Lopresti, F. Wong, M. Rosenberger, Y. Liu, E. Thom and A. A. Liebman; *J. Am. Chem. Soc.*, 1983, **105**, 3661.	168, 170
83JA3720	G. Stork and R. Mook, Jr.; *J. Am. Chem. Soc.*, 1983, **105**, 3720.	175
83JA4846	J. A. King, Jr. and K. P. C. Vollhardt; *J. Am. Chem. Soc.*, 1983, **105**, 4846.	690
83JA5390	D. Seebach, M. Boes, R. Naef and W. B. Schweizer; *J. Am. Chem. Soc.*, 1983, **105**, 5390.	302
83JA5479	A. W. Duff, K. Jonas, R. Goddard, H.-J. Kraus and C. Krüger; *J. Am. Chem. Soc.*, 1983, **105**, 5479.	694
83JA5939	A. F. Hill, W. R. Roper, J. M. Waters and A. H. Wright; *J. Am. Chem. Soc.*, 1983, **105**, 5939.	150, 722
83JA6285	H. C. Brown and T. Imai; *J. Am. Chem. Soc.*, 1983, **105**, 6285.	364
83JA6355	S. J. McLain; *J. Am. Chem. Soc.*, 1983, **105**, 6355.	453
83JA6491	P. L. Watson; *J. Am. Chem. Soc.*, 1983, **105**, 6491.	670, 702
83JA6725	R. R. Jones and M. E. Lee; *J. Am. Chem. Soc.*, 1983, **105**, 6725.	610, 611
83JA6761	E. Negishi and J. A. Miller; *J. Am. Chem. Soc.*, 1983, **105**, 6761.	809
83JCR(S)108	S. C. Cheng, M. L. De S. Fernandez, J. Iley and M. E. N. Rosa; *J. Chem. Res. (S)*, 1983, 108.	444
83JCS(D)1417	A. F. Dyke, S. A. R. Knox, M. J. Morris and P. J. Naish; *J. Chem. Soc., Dalton Trans.*, 1983, 1417.	679
83JCS(D)1441	K. Isobe, A. Vázquez de Miguel, P. M. Bailey, S. Okeya and P. M. Maitlis; *J. Chem. Soc., Dalton Trans.*, 1983, 1441.	687
83JCS(D)1495	S. Pelling, C. Botha and J. R. Moss; *J. Chem. Soc., Dalton Trans.*, 1983, 1495.	582
83JCS(D)2099	R. E. Colborn, A. F. Dyke, S. A. R. Knox, K. A. Mead and P. Woodward; *J. Chem. Soc., Dalton Trans.*, 1983, 2099.	680
83JCS(D)2661	R. E. Colborn, D. L. Davies, A. F. Dyke, A. Endesfelder, S. A. R. Knox, A. G. Orpen and D. Plaas; *J. Chem. Soc., Dalton Trans.*, 1983, 2661.	679
83JCS(P1)861	R. D. Holmes-Smith, R. D. Osei and S. R. Stobart; *J. Chem. Soc., Perkin Trans. 1*, 1983, 861.	566
83JCS(P1)1087	J. Barluenga, F. Aznur, R. Liz and R. Rodes; *J. Chem. Soc., Perkin Trans. 1*, 1983, 1087.	1067
83JCS(P1)1131	D. J. Ager; *J. Chem. Soc., Perkin Trans 1*, 1983, 1131.	392, 938, 957
83JCS(P1)1579	D. M. Hollinshead, S. Christopher Howell, S. V. Ley, M. Mahon, N. M. Ratcliffe and P. A. Worthington; *J. Chem. Soc., Perkin Trans. 1*, 1983, 1579.	165

83JCS(P1)1741	J. A. Elvidge, P. N. Judson, A. Percival and R. Shah; *J. Chem. Soc., Perkin Trans. 1*, 1983, 1741.	970, 977, 978
83JCS(P1)2979	A. G. Cameron and A. T. Hewson; *J. Chem. Soc.*, 1983, 2979.	925
83JCS(P2)1071	K. H. Lee and S. Brumby; *J. Chem. Soc., Perkin Trans. 2*, 1983, 1071.	1078
83JCS(P2)1687	R. W. Hoffmann, W. Barth, L. Carlsen and H. Egsgaard; *J. Chem. Soc., Perkin Trans. 2*, 1983, 1687.	1081
83JFC(22)557	A. F. Janzen, P. M. C. Wang and A. E. Lemire; *J. Fluorine Chem.*, 1983, **22**, 557.	61
83JFC(23)339	D. J. Burton; *J. Fluorine Chem.*, 1983, **23**, 339.	733
83JGU109	V. D. Sheludyakov, V. I. Zhun', S. D. Vlasenko, V. N. Bochkarev, T. F. Slyusarenko, A. V. Kisin and V. M. Nosova; *J. Gen. Chem. USSR (Engl. Transl.)*, 1983, **53**, 109.	140
83JMC129	M. Chorev, E. Rubini, C. Gilon, U. Wormser and Z. Selinger; *J. Med. Chem.*, 1983, **26**, 129.	446
83JOC34	P. E. Bauer, K. S. Kyler and D. S. Watt; *J. Org. Chem.*, 1983, **48**, 34.	938
83JOC162	G. Devincenzis, P. Mencarelli and F. Stegel; *J. Org. Chem.*, 1983, **48**, 162.	415
83JOC298	W. L. Mendelson, J.-H. Liu, L. B. Killmer, Jr. and S. H. Levinson; *J. Org. Chem.*, 1983, **48**, 298.	825
83JOC332	N. Kornblum, H. K. Singh and W. J. Kelly; *J. Org. Chem.*, 1983, **48**, 332.	332, 428, 429
83JOC383	K. S. Kyler, M. A. Netzel, S. Arseniyadis and D. S. Watt; *J. Org. Chem.*, 1983, **48**, 383.	938
83JOC481	J. Charrier, A. Foucaud, M. Person and E. Loukakou; *J. Org. Chem.*, 1983, **48**, 481.	891
83JOC611	M. B. Frankel and D. O. Woolery; *J. Org. Chem.*, 1983, **48**, 611.	444
83JOC623	K. T. Potts, A. J. Ruffini and G. R. Titus; *J. Org. Chem.*, 1983, **48**, 623.	980
83JOC635	K. B. Sloan and S. A. M. Koch; *J. Org. Chem.*, 1983, **48**, 635.	54, 58
83JOC917	G. A. Wheaton and D. J. Burton; *J. Org. Chem.*, 1983, **48**, 917.	735
83JOC1096	J. A. Kaydos and D. L. Smith; *J. Org. Chem.*, 1983, **48**, 1096.	67
83JOC1559	J.-P. Quintard, B. Elissondo and M. Pereyre; *J. Org. Chem.*, 1983, **48**, 1559.	388
83JOC1560	E. Negishi and F.-T. Luo; *J. Org. Chem.*, 1983, **48**, 1560.	954, 1062
83JOC1765	K. S. Kochhar, B. S. Bal, R. P. Deshpande, S. N. Rajadhyaksha and H. W. Pinnick; *J. Org. Chem.*, 1983, **48**, 1765.	203
83JOC2084	P. J. Kropp and N. J. Pienta; *J. Org. Chem.*, 1983, **48**, 2084.	25
83JOC2095	R. H. Smithers; *J. Org. Chem.*, 1983, **48**, 2095.	818
83JOC2098	R. C. Gadwood; *J. Org. Chem.*, 1983, **48**, 2098.	399
83JOC2388	J. E. Saavedra; *J. Org. Chem.*, 1983, **48**, 2388.	529
83JOC3077	T. G. Back, S. Collins and R. G. Kerr; *J. Org. Chem.*, 1983, **48**, 3077.	947
83JOC3382	A. Hassner and J. L. Dillon, Jr.; *J. Org. Chem.*, 1983, **48**, 3382.	742
83JOC4427	S. Hanessian, A. Ugolini and M. Therien; *J. Org. Chem.*, 1983, **48**, 4427.	171
83JOC4432	J. R. Hwu; *J. Org. Chem.*, 1983, **48**, 4432.	234
83JOC4661	W. A. Vinson, K. S. Prickett, B. Spahic and P. R. Ortiz de Montellano; *J. Org. Chem.*, 1983, **48**, 4661.	734
83JOC4713	L.-Y. Chiang, P. Shu, D. Holt and D. Cowan; *J. Org. Chem.*, 1983, **48**, 4713.	850
83JOC4773	E. Vedejs and F. G. West; *J. Org. Chem.*, 1983, **48**, 4773.	517
83JOC4775	L. A. Vargas and A. F. Rosenthal; *J. Org. Chem.*, 1983, **48**, 4775.	555
83JOC5280	N. Bodor, K. B. Sloan, J. J. Kaminski, C. Shih and S. Pogany; *J. Org. Chem.*, 1983, **48**, 5280.	58, 211
83JOC5315	G. Grethe, J. Sereno, T. H. Williams, M. R. Uskokovic; *J. Org. Chem.*, 1983, **48**, 5315.	170
83JOM(243)245	D. K. Breitinger, W. Kress, R. Sendelbeck and K. Ishiwada; *J. Organomet. Chem.*, 1983, **243**, 245.	697
83JOM(244)C21	M. N. Nefedov, V. N. Setkina and D. N. Kursanov; *J. Organomet. Chem.*, 1983, **244**, C21.	707
83JOM(246)141	G. E. Herberich, J. Hengesbach, G. Huttner, A. Frank and U. Schubert; *J. Organomet. Chem.*, 1983, **246**, 141.	695
83JOM(249)335	G. Märkl, H. Hauptmann and A. Merz; *J. Organomet. Chem.*, 1983, **249**, 335.	134
83JOM(250)109	M. Ishikawa, T. Tabohashi, H. Sugisawa, K. Nishimura and M. Kumada; *J. Organomet. Chem.*, 1983, **250**, 109.	615
83JOM(252)47	T. N. Mitchell and A. Amamria; *J. Organomet. Chem.*, 1983, **252**, 47.	1065, 1066, 1068, 1069
83JOM(254)13	K. Tamao, M. Akita and M. Kumada; *J. Organomet. Chem.*, 1983, **254**, 13.	1052
83JOM(255)305	H. Schumann, F. W. Reier and M. Dettlaff; *J. Organomet. Chem.*, 1983, **255**, 305.	578
83JOM(256)37	T. N. Mitchell and A. Amamria; *J. Organomet. Chem.*, 1983, **256**, 37.	813, 1066
83JOM(256)147	W. A. Herrmann, C. Bauer and A. Schäfer; *J. Organomet. Chem.*, 1983, **256**, 147.	689
83JOM(258)81	W. A. Herrmann, C. Bauer, J. M. Huggins, H. Pfisterer and M. L. Ziegler; *J. Organomet. Chem.*, 1983, **258**, 81.	686
83JOM(259)283	R. J. P. Corriu, V. Huynh and J. J. E. Moreau; *J. Organomet. Chem.*, 1983, **259**, 283.	511
83LA290	W. Kantlehner, W. W. Mergen and E. Haug; *Liebigs Ann. Chem.*, 1983, 290.	973, 974
83LA816	H. Emde and G. Simchen; *Liebigs Ann. Chem.*, 1983, 816.	831
83LA1133	U. Schöllkopf, U. Groth, M.-R. Gull and J. Nozulak; *Liebigs Ann. Chem.*, 1983, 1133.	536
83MI 404-01	F. Meskens; *Janssen Chim. Acta*, 1983, **1(2)** 10.	186
83MI 406-01	C. A. Brandt, J. V. Comasseto, W. Nakamura and N. Petragnani; *Quimica Nova (Brazil)*, 1983, **6**, 80 (*Chem. Abstr.*, 1984, **100**, 51 181).	291
B-83MI 408-01	W. P. Weber; "Silicon Reagents for Organic Synthesis," Springer-Verlag, Berlin, 1983.	357
83MI 408-02	H. J. Reich; "Proceedings of the Fourth International Conference of the Organic Chemistry of Selenium and Tellurium," eds. F. J. Berry and W. R. McWinnie, 1983, p. 258.	399
83MI 409-01	M. Chorev and M. Goodman; *Int. J. Peptide Protein Res.*, 1983, **21**, 258.	446
83MI 409-02	P. M. Hardy and I. N. Lingham; *Int. J. Peptide Protein Res.*, 1983, **21**, 406.	446
83MI 410-01	R. Kober and W. Steglich; *Annalen*, 1983, 599.	465
83MI 420-01	M. A. Kira, A. A. Zayed and N. M. Fathy; *Egypt J. Chem.*, 1983, **26**, 253.	882

83OM21	H. Yasuda, T. Nishi, K. Lee and A. Nakamura; *Organometallics*, 1982, **2**, 21.	603, 649
83OM106	T. J. Karol, J. P. Hutchinson, J. R. Hyde, H. G. Kuivila and J. A. Zubieta; *Organometallics*, 1983, **2**, 106.	699, 700
83OM154	G. W. Rice, G. B. Ansell, M. A. Modrick and S. Zentz; *Organometallics*, 1983, **2**, 154.	579
83OM230	D. S. Matteson and D. Majumdar; *Organometallics*, 1983, **2**, 230.	642, 645
83OM236	D. J. S. Tsai and D. S. Matteson; *Organometallics*, 1983, **2**, 236.	642
83OM257	G. Müller, D. Neugebauer, W. Geike, F. H. Köhler, J. Pebler and H. Schmidbaur; *Organometallics*, 1983, **2**, 257.	581, 582, 583
83OM351	M. Ishikawa, T. Tabohashi, H. Ohashi, M. Kumada and J. Iyoda; *Organometallics*, 1983, **2**, 351.	615
83OM726	H. Yamazaki, K. Yasufuku and Y. Wakatsuki; *Organometallics*, 1983, **2**, 726.	714
83OM909	H. G. Kuivila, T. J. Karol and K. Swami; *Organometallics*, 1983, **2**, 909.	701
83OM921	Z.-M. Xie and R. H. Nielson; *Organometallics*, 1983, **2**, 921.	566, 572
83OM1099	J. C. Cochran, A. J. Leusink and J. G. Noltes; *Organometallics*, 1983, **2**, 1099.	701
83OM1167	F. A. Cotton, W. Schwotzer and E. S. Shamshoum; *Organometallics*, 1983, **2**, 1167.	675
83OM1172	R. F. Gerlach, D. N. Duffy and M. D. Curtis; *Organometallics*, 1983, **2**, 1172.	674
83OM1529	D. S. Matteson and D. Majumdar; *Organometallics*, 1983, **2**, 1529.	362, 373, 526
83OM1536	D. S. Matteson, R. Ray, R. R. Rocks and D. J. Tsai; *Organometallics*, 1983, **2**, 1536.	526
83OM1689	M. Rotem and Y. Shvo; *Organometallics*, 1983, **2**, 1689.	206
83OM1846	H. Nakazawa, D. L. Johnson and J. A. Gladysz; *Organometallics*, 1983, **2**, 1846.	581
83OMR(21)64	H.-O. Kalinowski, E. Röcker and G. Maier; *Org. Magn. Reson.*, 1983, **21**, 64.	19, 27, 29
83PAC1799	U. Schöllkopf; *Pure Appl. Chem.*, 1983, **55**, 1799.	535
83PHA449	G. Barnikow, A. Hagen, V. Hagen, E. Göres, D. Richter and K. Fichtner; *Pharmazi*, 1983, **38**, 449.	442
83POL77	S. Aime, D. Osella, L. Milone and A. Tiripicchio; *Polyhedron*, 1983, **2**, 77.	718
83PS(14)171	K. Issleib, K.-P. Döpfer and A. Balszuweit; *Phosphorus Sulfur*, 1983, **14**, 171.	463
83PS(14)295	L. Maier; *Phosphorus Sulfur*, 1983, **14**, 295.	463
83PS(15)93	A. Hinke and W. Kuchen; *Phosphorus Sulfur*, 1983, **15**, 93.	115
83PS(16)345	G. A. Kutyrev, A. A. Kapura, F. Kh. Korataeva, R. A. Cherkasov, Yu. Yu. Samitov and A. N. Pudovik; *Phosphorus Sulfur*, 1983, **16**, 345.	923
83PS(17)73	K. Issleib, U. Kühne and F. Krech; *Phosphorus Sulfur*, 1983, **17**, 73.	458
83PS(18)43	R. H. Nielson; *Phosphorus Sulfur*, 1983, **18**, 43.	567
83PS(18)279	O. Stelzer, S. Hietkamp and H. Sommer; *Phosphorus Sulfur*, 1983, **18**, 279.	550
83S1	P. Brownbridge; *Synthesis*, 1983, 1.	831
83S61	H. Ahlbrecht and H. Simon; *Synthesis*, 1983, 61.	530
83S332	M. Mikolajczyk, P. Kielbasiński and S. Grzejszczak; *Synthesis*, 1983, 332.	923
83S500	S. Kirchmeyer, A. Mertens and G. A. Olah; *Synthesis*, 1983, 500.	315
83S568	J. L. Colin and B. Loubinoux; *Synthesis*, 1983, 568.	199, 314
83S654	S. Kinastowski and S. Wnuk; *Synthesis*, 1983, 654.	306
83S773	E. C. Chukovskaya, R. Kh. Freidlina and N. A. Kuz'mina; *Synthesis*, 1983, 773.	13
83S840	A. M. Richter, C. K. J. Tschoetsch and E. Fanghaenel; *Synthesis*, 1983, 840.	870
83S891	K. S. Petrakis and J. Fried; *Synthesis*, 1983, 891.	161, 212
83S920	T. Sakakibara, M. Manandhar and Y. Ishido; *Synthesis*, 1983, 920.	910
83S926	W. D. Rudorf; *Synthesis*, 1983, 926.	901, 997
83S951	M. Bellassoued, R. Chtara, F. Dardoize and M. Gaudemar; *Synthesis*, 1983, 951.	442
83SC129	D. E. Seitz, J. J. Carroll, P. Claudia, M. Cartaya, S-H. Lee and A. Zapata; *Synth. Commun.*, 1983, **13**, 129.	154
83SC629	M. Tordeux, R. Dorme and C. Wakselman; *Synth. Commun.*, 1983, **13**, 629.	187, 208
83SC1149	J. Koshino, T. Sugawara, T. Yogo and A. Suzuki; *Synth. Commun.*, 1983, **13**, 1149.	935
83T1189	M. Mikolajczyk, B. Costisella and S. Grzejszczak; *Tetrahedron*, 1983, **39**, 1189.	804
83T2023	A. R. Katritzky, C. Jayaram and S. N. Vassilatos; *Tetrahedron*, 1983, **39**, 2023.	539
83T2085	U. Schöllkopf; *Tetrahedron*, 1983, **39**, 2085.	535
83T2493	L. Ghosez, J. Marchand-Brynaert, J. Vekemans and S. Bogdan; *Tetrahedron*, 1983, **39**, 2493.	833
83T3073	K. J. H. Kruithof, R. F. Schmitz and G. W. Klumpp; *Tetrahedron*, 1983, **39**, 3073.	1047, 1058, 1062
83T3189	T. A. van der Knaap and F. Bickelhaupt; *Tetrahedron*, 1983, **39**, 3189.	340
83TL45	G. Altnau, L. Rosch and G. Jas; *Tetrahedron Lett.*, 1983, **24**, 45.	355
83TL87	Yu. L. Yagupolskii, T. I. Savina, D. S. Yufit and Yu. T. Struchkov; *Tetrahedron Lett.*, 1983, **24**, 87.	282
83TL95	D. J. Ager; *Tetrahedron Lett.*, 1983, **24**, 95.	366, 936
83TL117	M. A. McKervey and P. Ratananukul; *Tetrahedron Lett.*, 1983, **24**, 117.	69
83TL139	D. T. W. Chu, J. E. Hengeveld and D. Lester; *Tetrahedron Lett.*, 1983, **24**, 139.	67
83TL237	S. L. Bender, M. R. Detty, M. W. Fichtner and N. F. Haley; *Tetrahedron Lett.*, 1983, **24**, 237.	290
83TL327	I. Fleming and J. Iqbal; *Tetrahedron Lett.*, 1983, **24**, 327.	67
83TL507	M. Yamana, T. Ishihara and T. Ando; *Tetrahedron Lett.*, 1983, **24**, 507.	735
83TL527	V. Reutrakul and K. Herunsalee; *Tetrahedron Lett.*, 1983, **24**, 527.	742
83TL531	V. Reutrakal and P. Poochaivatanon; *Tetrahedron Lett.*, 1983, **24**, 531.	773
83TL623	A. Pelter, B. Singaram, L. Williams and J. W. Wilson; *Tetrahedron Lett.*, 1983, **24**, 623.	664
83TL631	A. Pelter, B. Singaram and J. W. Wilson; *Tetrahedron Lett.*, 1983, **24**, 631.	664
83TL637	M. V. Garad, A. Pelter, B. Singaram and J. W. Wilson; *Tetrahedron Lett.*, 1983, **24**, 637.	642, 645, 1077
83TL725	V. Reutrakul and V. Rukachaisirikul; *Tetrahedron Lett.*, 1983, **24**, 725.	74, 778

83TL767	T. Minami, T. Yamanouchi, S. Takenaka and I. Hirao; *Tetrahedron Lett.*, 1983, **24**, 767.	925
83TL1239	J.-M. Vatele; *Tetrahedron Lett.*, 1983, **24**, 1239.	766
83TL1303	T. A. M. van Schaik, A. van der Gen and A. V. Henzen; *Tetrahedron Lett.*, 1983, **24**, 1303.	829
83TL1975	A. Meriem, J.-P. Majoral, M. Revel and J. Navech; *Tetrahedron Lett.*, 1983, **24**, 1975.	459
83TL2043	J. J. Eisch and A. Piotrowski; *Tetrahedron Lett.*, 1983, **24**, 2043.	696
83TL2361	J. A. Soderquist and W. W.-H. Leong; *Tetrahedron Lett.*, 1983, **24**, 2361.	951
83TL2563	M. Cossement, J. Marchand-Brynaert, S. Bogdan and L. Ghosez; *Tetrahedron Lett.*, 1983, **24**, 2563.	842
83TL3163	E. J. Corey and T. M. Eckrich; *Tetrahedron Lett.*, 1983, **24**, 3163.	380, 389
83TL3165	E. J. Corey and T. M. Eckrich; *Tetrahedron Lett.*, 1983, **24**, 3165.	378, 384, 388
83TL3209	R. W. Hoffmann and B. Landmann; *Tetrahedron Lett.*, 1983, **24**, 3209.	373
83TL3373	M. D. Cooke, K. W. Moore, B. C. Ross and S. E. Turner; *Tetrahedron Lett.*, 1983, **24**, 3373.	834
83TL3591	R. Appel and R. Zimmerman; *Tetrahedron Lett.*, 1983, **24**, 3591.	566
83TL3905	P. J. Kocienski and C. Yeates; *Tetrahedron Lett.*, 1983, **24**, 3905.	953
83TL4113	T. Hiyama, M. Obayashi and M. Sawahata; *Tetrahedron Lett.*, 1983, **24**, 4113.	736
83TL4257	H. Nemoto, X. M. Wu, H. Kurobe, M. Ihara, K. Fukumoto and T. Kametani; *Tetrahedron Lett.*, 1983, **24**, 4257.	387
83TL4303	A. Padwa, G. Haffmanns and M. Tomas; *Tetrahedron Lett.*, 1983, **24**, 4303.	517
83TL4387	D. J. Hart and Y.-M. Tsai; *Tetrahedron Lett.*, 1983, **24**, 4387.	936
83TL4727	A. Hosomi, M. Inaba and H. Sakurai; *Tetrahedron Lett.*, 1983, **24**, 4727.	742
83TL4825	C. B. B. Ekogha, O. Ruel and S. A. Julia; *Tetrahedron Lett.*, 1983, **24**, 4825.	938
83TL4833	J. Lancelin, P. H. A. Zollo and P. Sinaÿ; *Tetrahedron Lett.*, 1983, **24**, 4833.	172
83TL4923	R. K. Boeckman, Jr. and C. J. Flann; *Tetrahedron Lett.*, 1983, **24**, 4923.	188
83TL4979	D. H. R. Barton, D. Crich and W. B. Motherwell; *Tetrahedron Lett.*, 1983, **24**, 4979.	1077
83TL4993	T. Mandai, K. Hara, T. Nakajima, M. Kawada and J. Otera; *Tetrahedron Lett.*, 1983, **24**, 4993.	197
83TL5563	P. N. Confalone, E. M. Huie and N. G. Patel; *Tetrahedron Lett.*, 1983, **24**, 5563.	268
83TL5657	T. Ishihara, M. Yamana and T. Ando; *Tetrahedron Lett.*, 1983, **24**, 5657.	735
83ZAAC(497)134	G. Fritz and G. Brauch; *Z. Anorg. Allg. Chem.*, 1983, **497**, 134.	609
83ZAAC(500)132	N. Auner and J. Grobe; *Z. Anorg. Allg. Chem.*, 1983, **500**, 132.	567, 622, 623
83ZN(B)711	W. Malisch, H. Blau, P. Weickert and K.-H. Griessmann; *Z. Naturforsch., Teil B*, 1983, **38**, 711.	580, 582
83ZN(B)726	E. Lindner and E. Tamoutsidis; *Z. Naturforsch., Teil B*, 1983, **38**, 726.	921
83ZN(B)953	M. Hesse, U. Klingebiel, J. Heinze and J. Mortensen; *Z. Naturforsch., Teil B*, 1983, **38**, 953.	649
83ZN(B)1027	H. H. Karsch; *Z. Naturforsch., Teil B*, 1983, **38**, 1027.	550
83ZN(B)1388	G. E. Herberich and H. Ohst; *Z. Naturforsch., Teil B*, 1983, **38**, 1388.	643
83ZN(B)1399	H. H. Karsch and A. Appelt; *Z. Naturforsch., Teil B*, 1983, **38**, 1399.	566, 588
83ZOB46	M. G. Zimin, A. R. Burilov, R. G. Islamov and A. N. Pudovik; *Zh. Obshch. Khim.*, 1983, **53**, 46.	345
83ZOB119	V. P. Kozyukov, V. P. Kozukov and V. F. Mironov; *Zh. Obshch. Khim.*, 1983, **53**, 119.	313
83ZOB416	N. A. Batyrbaev, V. V. Zorin, A. P. Moravskii, V. F. Shuvalov, S. S. Zlotskii and D. L. Rakhmankulov; *Zh. Obshch. Khim.*, 1983, **53**, 416 (*Chem. Abstr.*, 1983, **98**, 160 120).	1080
83ZOB541	K. A. Petrov, V. A. Chauzov and V. P. Pokatun; *Zh. Obshch. Khim.*, 1983, **53**, 541.	561
83ZOB548	N. K. Mikhailyuchenko, B. N. Kozhushko and V. A. Shokol; *Zh. Obshch. Khim.*, 1983, **53**, 548 (Engl. Transl. 475).	1007, 1010, 1013
83ZOB594	V. I. Boev and A. V. Dombrovskii; *Zh. Obshch. Khim.*, 1983, **53**, 594.	555
83ZOB699	Z. S. Novikova, I. L. Odinets and I. F. Lutsenko; *Zh. Obshch. Khim.*, 1983, **53**, 699.	547
83ZOB1292	L. P. Safronova, A. S. Medvedeva, I. D. Kalikhman and N. S. Vyazankin; *Zh. Obshch. Khim.*, 1983, **53**, 1292.	942
83ZOB1673	Z. S. Novikova, I. L. Odinets and I. F. Lutsenko; *Zh. Obshch. Khim.*, 1983, **53**, 1673.	562
83ZOB1831	V. S. Brovko, N. K. Skvortsov, A. Yu. Ivanov and V. O. Reikhsfel'd; *Zh. Obshch. Khim.*, 1983, **53**, 1831.	569
83ZOB2015	V. S. Brovarets, O. P. Lobanov and B. S. Drach; *Zh. Obshch. Khim.*, 1983, **53**, 2015 (Engl. Transl. 1819).	999
83ZOR1344	B. A. Shainyan and A. N. Mirskova; *Zh. Org. Khim.*, 1983, **19**, 1344.	783
83ZOR1625	R. R. Kostikov, G. S. Varakin and K. A. Oglobin; *Zh. Org. Khim.*, 1983, **19**, 1625.	35
83ZOR2417	L. A. Lazukina, I. L. Mushkalo, V. M. Neplyuev and V. P. Kukhar; *Zh. Org. Khim.*, 1983, **19**, 2417 (*Chem. Abstr.*, 1984, **100**, 53 185).	866
84ACS113	F. S. Jørgensen and T. Thomsen; *Acta Chem. Scand.*, 1984, **B38**, 113.	433
84AG(E)53	H. J. Bestmann and G. Wölfel; *Angew. Chem., Int. Ed. Engl.*, 1984, **23**, 53.	313
84AG(E)224	A. Maercker and R. Dujardin; *Angew. Chem., Int. Ed. Engl.*, 1984, **23**, 224.	1064
84AG(E)239	D. Hoppe and F. Lichtenberg; *Angew. Chem., Int. Ed. Engl.*, 1984, **23**, 239.	384
84AG(E)247	H. Schmidbaur, R. Herr and J. Riede; *Angew. Chem., Int. Ed. Engl.*, 1984, **23**, 247.	1023
84AG(E)313	C. Pues and A. Berndt; *Angew. Chem., Int. Ed. Engl.*, 1984, **23**, 313.	1054, 1063
84AG(E)366	R. Knorr and T. von Roman; *Angew. Chem., Int. Ed. Engl.*, 1984, **23**, 366.	949, 950
84AG(E)381	H. J. Bestmann and T. Arenz; *Angew. Chem., Int. Ed. Engl.*, 1984, **23**, 381.	574
84AG(E)454	B. Pachaly and R. West; *Angew. Chem., Int. Ed. Engl.*, 1984, **23**, 454.	1058
84AG(E)522	H. Schumann, I. Albrecht, F. W. Reier and E. Hahn; *Angew. Chem., Int. Ed. Engl.*, 1984, **23**, 522.	578
84AG(E)723	M. Braun and W. Hill; *Angew. Chem., Int. Ed. Engl.*, 1984, **23**, 723.	956, 957
84AG(E)826	R. Wehrmann, H. Klusik and A. Berndt; *Angew. Chem., Int. Ed. Engl.*, 1984, **23**, 826.	1082

84AG(E)890	G. Seitz, R. Mohr, W. Overheu, R. Allmann and M. Nagel; *Angew. Chem., Int. Ed. Engl.*, 1984, **23**, 890.	229
84AG(E)932	D. Hoppe; *Angew. Chem., Int. Ed. Engl.*, 1984, **23**, 932.	381
84AJC205	D. G. Hewitt and M. W. Tesse; *Aust. J. Chem.*, 1984, **37**, 205.	473
84AP(317)15	W. Löwe, G. Eggersmann and A. Kennemann; *Arch. Pharm. (Weinheim, Ger.)*, 1984, **317**, 15.	234
84BCJ43	Y. Yamamoto; *Bull. Chem. Soc. Jpn.*, 1984, **57**, 43.	574, 583, 585, 586, 587, 588, 589
84BCJ796	K. Mochida; *Bull. Chem. Soc. Jpn.*, 1984, **57**, 796.	637, 641
84BCJ827	M. Kato and I. Kuwajima; *Bull. Chem. Soc. Jpn.*, 1984, **57**, 827.	354
84BCJ1863	T. Takeda, H. Furukawa, M. Fujimori, K. Suzuki and T. Fujiwara; *Bull. Chem. Soc. Jpn.*, 1984, **57**, 1863.	938, 958
84BCJ1876	O. Takazawa, K. Kogami and K. Hayashi; *Bull Chem. Soc. Jpn.*, 1984, **57**, 1876.	192, 193
84BCJ2127	T. Minami, T. Yamanouchi, S. Tokumasu and I. Hirao; *Bull. Chem. Soc. Jpn.*, 1984, **57**, 2127.	925
84BCJ2768	K. Furuta, M. Ishiguro, R. Haruta, N. Ikeda and H. Yamamoto; *Bull. Chem. Soc. Jpn.*, 1984, **57**, 2768.	1061
84C233	G. Solladie; *Chimia*, 1984, **38**, 233.	959
84CB1	H. Ahlbrecht and W. Farnung; *Chem. Ber.*, 1984, **117**, 1.	1003, 1004
84CB383	H. Lehmkuhl, L. Hauschild and M. Bellenbaum; *Chem. Ber.*, 1984, **117**, 383.	654
84CB413	M. Braun and W. Hild; *Chem. Ber.*, 1984, **117**, 413.	397
84CB694	H. tom Dieck and J. Dietrich; *Chem. Ber.*, 1984, **117**, 694.	298
84CB1103	L. Weber and D. Wewers; *Chem. Ber.*, 1984, **117**, 1103.	580
84CB1424	W. Flitsch and W. Lubisch; *Chem. Ber.*, 1984, **117**, 1424.	925
84CB1542	M. Baudler, F. Saykowski, M. Hintze, K. F. Tebbe, T. Heinlein, A. Vissers and M. Feher; *Chem. Ber.*, 1984, **117**, 1542.	1025
84CB1695	A. Schmidpeter, K-H Zirzow, G. Burget, G. Huttner and I. Jibril; *Chem. Ber.*, 1984, **117**, 1695.	112
84CB1877	R. Hassig, D. Seebach and H. Siegel; *Chem. Ber.*, 1984, **117**, 1877.	746
84CB1885	P. Jutzi, C. Otto and T. Wippermann; *Chem. Ber.*, 1984, **117**, 1885.	603
84CB2622	H. Ahlbrecht, W. Farnung and H. Simon; *Chem. Ber.*, 1984, **117**, 2622.	913, 918, 925
84CB2693	R. Appel, C. Casser and F. Knoch; *Chem. Ber.*, 1984, **117**, 2693.	566
84CB3340	H. Fischer and L. Weber; *Chem. Ber.*, 1984, **117**, 3340.	580
84CB3374	H. Schmidbaur, R. Herr and C. E. Zybill; *Chem. Ber.*, 1984, **117**, 3374.	554
84CB3400	S. Hietkamp, H. Sommer and O. Stelzer; *Chem. Ber.*, 1984, **117**, 3400.	550
84CC670	B. J. Banks, A. G. M. Barrett and M. A. Russell; *J. Chem. Soc., Chem. Commun.*, 1984, 670.	906
84CC794	T. Morimoto, T. Takahashi and M. Sekiya; *J. Chem. Soc., Chem. Commun.*, 1984, 794.	313
84CC1028	A. Mottoh and C. B. Reese; *J. Chem. Soc., Chem. Commun.*, 1984, 1028.	841
84CC1130	F. Sato, M. Kusakabe and Y. Kobayashi; *J. Chem. Soc., Chem. Commun.*, 1984, 1130.	1060
84CC1289	C. M. Pant, R. J. Stoodley, A. Whiting and D. J. Williams; *J. Chem. Soc., Chem. Commun.*, 1984, 1289.	448
84CC1634	Y. Y. C. Yeung Lam Ko, R. Carrie, A. Muench and G. Becker; *J. Chem. Soc., Chem. Commun.*, 1984, 1634.	479, 480
84CC1664	H. Kawa, J. W. Chinn Jr. and R. J. Lagow; *J. Chem. Soc., Chem. Commun.*, 1984, 1664.	618
84CC1708	D. Colgan, R. I. Papasergio, C. L. Raston and A. H. White; *J. Chem. Soc., Chem. Commun.*, 1984, 1708.	647
84CCA689	B. Korpar-Colig, Z. Popovic and M. Sikirica; *Croat. Chem. Acta*, 1984, **57**, 689 (*Chem. Abstr.*, 1985, **102**, 220 968).	697
84CHE1231	D. A. Tikhomirov, Yu. V. Schubina and A. V. Eremeev; *Chem. Heterocycl. Compd. (Engl. Transl.)*, 1984, 1231.	793
84CHEC(3)531	H. Neunhoeffer; *Comp. Heterocycl. Chem.*, 1984, **3**, 531.	436
84CHEC(3)995	M. Sainsbury; *Comp. Heterocycl. Chem.*, 1984, **3**, 995.	302
84CHEC(6)177	G. V. Boyd; *Comp. Heterocycl. Chem.*, 1984, **6**, 177.	
84CHEC(6)312	J. V. Metzger; *Comp. Heterocycl. Chem.*, 1984, **6**, 312.	328
84CHEC(6)321	J. V. Metzger; *Comp. Heterocycl. Chem.*, 1984, **6**, 321.	318
84CHEC(6)346	I. Lalezari; *Comp. Heterocycl. Chem.*, 1984, **6**, 346.	333
84CHEC(6)798	D. M. McKinnon; *Comp. Heterocycl. Chem.*, 1984, **6**, 798.	850
84CHEC(6)848	H. Gotthardt; *Comp. Heterocycl. Chem.*, 1984, **6**, 848.	850
84CHEC(6)890	G. W. Fischer and T. Zimmermann; *Comp. Heterocycl. Chem.*, 1984, **6**, 890.	852
84CJC586	I. W. J. Still, W. L. Brown, R. J. Colville and G. W. Kutney; *Can. J. Chem.*, 1984, **62**, 586.	330
84CJC749	H. W. Manning, C. M. Hemens, R. J. Lafrance, Y. Tang and K. Vaughan; *Can. J. Chem.*, 1984, **62**, 749.	444
84CL265	T. Mukaiyama, M. Ohshima and M. Murakami; *Chem. Lett.*, 1984, 265.	187
84CL279	O. Tsuge, S. Kanemasa, S. Kuraoka and S. Takenaka; *Chem. Lett.*, 1984, 279.	519
84CL1599	R. Sato, T. Senzaki, T. Goto and M. Saito; *Chem. Lett.*, 1984, 9, 1599.	419, 420
84CL1747	M. Hayashi, S. Hashimoto and R. Noyori; *Chem Lett.*, 1984, 1747.	45
84CL1751	W. A. Szarek, G. Grynkiewicz, B. Doboszewski and G. W. Hay; *Chem. Lett.*, 1984, 1751.	45
84CL1765	T. Hiyama, K. Nishide and M. Obayashi; *Chem. Lett.*, 1984, 1765.	806, 817
84CPB1829	M. Ochiai, T. Ukita, E. Fujita and S.-I. Tada; *Chem. Pharm. Bull.*, 1984, **32**, 1829.	392
84CRV471	P. Beak, W. J. Zajdel and D. B. Reitz; *Chem. Rev.*, 1984, **84**, 471.	529, 535, 537, 1014, 1015, 1016
84DOK(277)371	I. S. Alfer'ev, L. M. Vainer, A. G. Krainev and N. M. Slynko; *Dokl. Akad. Nauk SSSR*, 1984, **277**, 371.	560
84EGP206557	T. Czekanski, W. Hartmann, B. Costisella, J. Glöde, H. Gross and B. Johannsen; *Ger. (East) Pat.* 206 557 (1984) (*Chem. Abstr.*, 1984, **101**, 192 196).	553

84EGP211356A1	E. Fanghaenel, C. Tschoetsch and A. M. Richter (German Democratic Republic), *E. Ger. Pat.* 211 356 A1 (1982) (*Chem. Abstr.*, 1985, **102**, 80 276).	870
84EGP214610	H. Schadow, H. A. Lehmann, H. Schallschmidt, H. Richter, W. Huth, M. Oertel and R. Kurze; *Ger. (East) Pat.* 214 610 (1984) (*Chem. Abstr.*, 1985, **103**, 196 242).	561
84G405	M. Prato, U. Quintily and G. Scorrano; *Gazz. Chim. Ital.*, 1984, **114**, 405.	304
84GEP3408757A1	R. F. Collins, P. Knowles, L. C. Saunders, F. J. Tierney and P. J. Warne (May and Baker Ltd); *Ger. Pat.* 3 408 757 A1 (1984) (*Chem. Abstr.*, 1985, **102**, 95 541).	866
84H(22)585	T. R. Govindachari, P. Chinnasamy, S. Rajeswari, S. Chandrasekaran, M. S. Premila, S. Natarajan, K. Nagarajan and B. R. Pai; *Heterocycles*, 1984, **22**, 585.	94
84H(22)701	Y. Miki, H. Hachiken, S. Takemura and M. Ikeda; *Heterocycles*, 1984, **22**, 701.	519
84H(22)987	E. Lukevics, V. Gevorgyan, Y. Goldberg, J. Popelis, M. Gavars, A. Gaukhman and M. Shimanska; *Heterocycles*, 1984, **22**, 987.	929
84H(22)1417	K. Matsumoto, S. Hashimoto, Y. Ikemi and S. Otani; *Heterocycles*, 1984, **22**, 1417.	410
84H(22)1955	O. Tsuge, S. Kanemasa, H. Suga and K. Matsuda; *Heterocycles*, 1984, **22**, 1955.	515
84H(22)2351	J. H. Boyer and G. Kumar; *Heterocycles*, 1984, **22**, 2351.	434
84HCA1070	J. Gabriel and D. Seebach; *Helv. Chim. Acta*, 1984, **67**, 1070.	289
84HCA2175	H. Schmidbaur and T. Pollok; *Helv. Chim. Acta*, 1984, **67**, 2175.	548
84IC2063	R. R. Ford, M. A. Goodman, R. H. Neilson, A. K. Roy, U. G. Wettermark and P. Wisian-Neilson; *Inorg. Chem.*, 1984, **23**, 2063.	572
84IC2550	H. Hope, M. Viggiano, B. Moezzi and P. P. Power; *Inorg. Chem.*, 1984, **23**, 2550.	453
84IC3717	J. A. Gurak, J. W. Chinn, Jr., R. J. Lagow, H. Steinfink and C. S. Yannoni; *Inorg. Chem.*, 1984, **23**, 3717.	668
84IJ125	M. Clarembeau, J. L. Bertrand and A. Krief; *Isr. J. Chem.*, 1984, **24**, 125.	396, 401
84IJC(B)1286	D. S. Bhakuni, P. K. Gupta and B. L. Chowdhury; *Indian J. Chem., Sect B*, 1984, **23**, 1286.	306
84IZV194	N. P. Petukhova, N. E. Dontsova and E. N. Prilezhaeva; *Izv. Akad. Nauk SSSR, Ser. Khim.*, 1984, 194 (*Chem. Abstr.*, 1984, **100**, 191 318).	861
84IZV611	N. P. Petukhova, N. E. Dontsova, E. N. Prilezhaeva and V. S. Bogdanov; *Izv. Akad. Nauk SSSR, Ser. Khim.*, 1984, 611 (*Chem. Abstr.*, 1984, **101**, 229 956).	866
84IZV1122	I. S. Alfer'ev, N. V. Mikhalin and I. L. Kotlyarevskii; *Izv. Akad. Nauk SSSR, Ser. Khim.*, 1984, 1122.	560
84IZV2023	I. A. Litvinov, Yu. T. Struchkov, B. A. Arbuzov, E. N. Dianova and I. Z. Galeeva; *Izv. Akad. Nauk SSSR, Ser. Khim.*, 1984, 2023.	549
84JA1010	P. Beak and W. J. Zajdel; *J. Am. Chem. Soc.*, 1984, **106**, 1010.	530
84JA1148	W. C. Still and V. J. Novack; *J. Am. Chem. Soc.*, 1984, **106**, 1148.	165
84JA1809	A. Andrus, J. V. Heck, B. G. Christensen and B. Partridge; *J. Am. Chem. Soc.*, 1984, **106**, 1809.	59
84JA2337	H. R. Allcock, K. D. Lavin, G. H. Riding, P. R. Suszko and R. R. Whittle; *J. Am. Chem. Soc.*, 1984, **106**, 2337.	706
84JA2440	S. Murai, I. Ryu, J. Iriguchi and N. Sonada; *J. Am. Chem. Soc.*, 1984, **106**, 2440.	931
84JA2943	J. E. Lynch and E. L. Eliel; *J. Am. Chem. Soc.*, 1984, **106**, 2943.	231
84JA2954	K. Narasaka, T. Sakakura, T. Uchimaru and D. Guedin-Vuong; *J. Am. Chem. Soc.*, 1984, **106**, 2954.	235
84JA3245	T. Cohen, J. P. Sherbine, J. R. Matz, R. R. Hutchins, B. M. McHenry and P. R. Willey; *J. Am. Chem. Soc.*, 1984, **106**, 3245.	366, 394, 945, 1060
84JA3252	M. Kitamura, M. Isobe, Y. Ichikawa and T. Goto; *J. Am. Chem. Soc.*, 1984, **106**, 3252.	937
84JA3376	J. S. Sawyer, T. L. McDonald and G. J. McGarvey; *J. Am. Chem. Soc.*, 1984, **106**, 3376.	379
84JA3584	J. B. Lambert, R. J. Bosch, P. H. Mueller and K. Kobayashi; *J. Am. Chem. Soc.*, 1984, **106**, 3584.	748
84JA3785	L. Hevesi, S. Desauvage, B. Georges, G. Evrard, P. Blanpain, A. Michel, S. Harkema and C. J. van Hummel; *J. Am. Chem. Soc.*, 1984, **106**, 3785.	1074, 1075
84JA4189	K. C. Nicolaou, R. E. Dolle, D. P. Papahatjis and J. L. Randall; *J. Am. Chem. Soc.*, 1984, **106**, 4189.	45
84JA5544	C. G. Krespan, F. A. Van-Catledge and B. E. Smart; *J. Am. Chem. Soc.*, 1984, **106**, 5544.	1076
84JA5920	R. E. Cramer, A. L. Mori, R. B. Maynard, J. W. Gilje, K. Tatsumi and A. Nakamura; *J. Am. Chem. Soc.*, 1984, **106**, 5920.	589
84JA6015	K. Krageloh, G. H. Anderson and P. J. Stang; *J. Am. Chem. Soc.*, 1984, **106**, 6015.	985
84JA6097	G. Solladie and G. Moine; *J. Am. Chem. Soc.*, 1984, **106**, 6097.	959
84JA6105	L. D. Boardman, V. Bagheri, H. Sawada and E. Negishi; *J. Am. Chem. Soc.*, 1984, **106**, 6105.	1048, 1062
84JA6108	B. Haldon, C. J. Randall and P. J. Stang; *J. Am. Chem. Soc.*, 1984, **106**, 6108.	604, 647
84JA6467	P. von R. Schleyer, T. Clark, A. J. Kos, G. W. Spitznagel, C. Rohde, D. Arad, K. N. Houk and N. G. Rondan; *J. Am. Chem. Soc.*, 1984, **106**, 6467.	378, 391, 816
84JA6794	M. H. Chisholm, K. Folting, D. M. Hoffman and J. C. Huffman; *J. Am. Chem. Soc.*, 1984, **106**, 6794.	675
84JA6806	M. H. Chisholm, D. M. Hoffman and J. C. Huffman; *J. Am. Chem. Soc.*, 1984, **106**, 6806.	676
84JA8174	Y. Kabe, T. Takata, K. Ueno and W. Ando; *J. Am. Chem. Soc.*, 1984, **106**, 8174.	25
84JA8209	D. J. Hart and Y.-M. Tsai; *J. Am. Chem. Soc.*, 1984, **106**, 8209.	937
84JA8303	K. Lerstrup, D. O. Cowan and T. J. Kistenmacher; *J. Am. Chem. Soc.*, 1984, **106**, 8303.	876
84JBC(259)15106	C. A. Kettner and A. B. Shenvi; *J. Biol. Chem.*, 1984, **259**, 15 106.	528
84JCR(M)144	B. Alcaide, G. Escobar, R. Perez-Ossorio, J. Plumet and D. Sanz; *J. Chem. Res. (M)*, 1984, 144.	304
84JCR(S)108	S. C. Cheng, M. L. de Sousa Fernandes, J. Iley and M. E. N. Rosa; *J. Chem. Res. (S)*, 1984, 108.	312

84JCR(S)152	M. Engel and M. Devaud; *J. Chem. Res. (S)*, 1984, 152.	700
84JCR(S)396	E. V. Dehmlow and J. Wilkenloh; *J. Chem. Res. (S)*, 1984, 396.	1080
84JCS(D)311	L. M. Englehardt, R. I. Papasergio, C. L. Raston and A. H. White; *J. Chem. Soc., Dalton Trans.*, 1984, 311.	603, 647
84JCS(D)321	L. M. Engelhardt, W.-P. Leung, C. L. Raston, P. Twiss and A. H. White; *J. Chem. Soc., Dalton Trans.*, 1984, 321.	604, 646, 647
84JCS(D)331	L. M. Engelhardt, W.-P. Leung, C. L. Raston, P. Twiss and A. H. White; *J. Chem. Soc., Dalton Trans.*, 1984, 331.	660
84JCS(D)1563	J. C. Jeffery, J. C. V. Laurie, I. Moore, H. Razay and F. G. A. Stone; *J. Chem. Soc., Dalton Trans.*, 1984, 1563.	712
84JCS(D)1581	J. C. Jeffery, I. Moore, H. Razay and F. G. A. Stone; *J. Chem. Soc., Dalton Trans.*, 1984, 1581.	712
84JCS(D)1801	W.-P. Leung, C. L. Raston, B. W. Skelton and A. H. White; *J. Chem. Soc., Dalton Trans.*, 1984, 1801.	604, 646, 647
84JCS(D)2293	D. L. Davies, S. A. R. Knox, K. A. Mead, M. J. Morris and P. Woodward; *J. Chem. Soc., Dalton Trans.*, 1984, 2293.	679
84JCS(D)2455	M. Green, N. C. Norman, A. G. Orpen and C. J. Schaverien; *J. Chem. Soc., Dalton Trans.*, 1984, 2455.	676
84JCS(D)2553	M. Green, J. A. K. Howard, S. J. Porter, F. G. A. Stone and D. C. Tyler; *J. Chem. Soc., Dalton Trans.*, 1984, 2553.	712
84JCS(D)2859	G. A. Bownaker and H. Schmidbaur; *J. Chem. Soc., Dalton Trans.*, 1984, 2859.	586
84JCS(P1)21	C. D. Foulds, A. A. Jaxa-Chamiec, A. C. O'Sullivan and P. G. Sammes; *J. Chem. Soc., Perkin Trans. 1*, 1984, 21.	285
84JCS(P1)435	I. Yamamoto, K. Okuda, S. Nagai, J. Motoyoshiya, H. Gotoh and K. Matsuzaki; *J. Chem. Soc., Perkin Trans. 1*, 1984, 435.	365
84JCS(P1)681	G. Green, W. P. Griffith, D. M. Hollinshead, S. V. Ley and M. Schröder; *J. Chem. Soc., Perkin Trans. 1*, 1984, 681.	165
84JCS(P1)859	K. R. Lawson, A. Singleton and G. H. Whitham; *J. Chem. Soc., Perkin Trans. 1*, 1984, 859.	843
84JCS(P1)865	K. R. Lawson, A. Singleton and G. H. Whitham; *J. Chem. Soc., Perkin Trans. 1*, 1984, 865.	319
84JCS(P1)1531	R. Ramage, G. J. Griffiths, F. E. Shutt and J. N. A. Sweeney; *J. Chem. Soc., Perkin Trans. 1*, 1984, 1531.	213
84JCS(P1)1561	C.-D. Poon, P.-W. Yuen, T.-O. Man, C.-S. Li and T.-L. Chan; *J. Chem. Soc., Perkin Trans. 1*, 1984, 1561.	1083
84JCS(P1)1643	N. T. Byrom, R. Grigg, B. Kongkathip, G. Reimer and A. R. Wade; *J. Chem. Soc., Perkin Trans. 1*, 1984, 1643.	193
84JCS(P1)2327	W. R. Bowman, D. Rakshit and M. D. Valmas; *J. Chem. Soc., Perkin Trans. 1*, 1984, 2327.	327
84JCS(P1)2641	T. Katada, S. Eguchi and T. Sasaki; *J. Chem. Soc., Perkin Trans. 1*, 1984, 2641.	232
84JCS(P1)2845	E. K. Baylis, C. D. Campbell and J. D. Dingwall; *J. Chem. Soc., Perkin Trans. 1*, 1984, 2845.	464
84JCS(P2)429	D. J. Gulliver, E. G. Hope, W. Levason, S. G. Murray, D. M. Potter and G. L. Marshall; *J. Chem. Soc., Perkin Trans. 2*, 1984, 429.	288
84JFC(24)387	M. Makosza and H. Plenkiewicz; *J. Fluorine Chem.*, 1984, **24**, 387.	736
84JFC(24)503	H. Koroniak; *J. Fluorine Chem.*, 1984, **24**, 503.	1083
84JFC(26)1	H. Lange and D. Naumann; *J. Fluorine Chem.*, 1984, **26**, 1.	152
84JFC(26)29	R. E. Noftle, C. Ellis, G. Johnson and S. F. Bash; *J. Fluorine Chem.*, 1984, **26**, 29.	32
84JFC(26)467	I. Hemer, A. Pošta and V. Dědek; *J. Fluorine Chem.*, 1984, **26**, 467.	33
84JGU421	P. A. Gurevich, V. V. Kiselev and V. V. Moskva; *J. Gen. Chem. USSR (Engl. Transl.)*, 1984, **54**, 421.	465
84JGU635	N. A. Orlova, I. Yu Belavin, V. N. Sergeev, A. G. Shipov and Y. I. Baukov; *J. Gen. Chem. USSR (Engl. Transl.)*, 1984, **54**, 635.	418
84JGU1354	A. A. Prishchenko, A. V. Gromov, Yu. N. Luzikov, A. A. Borisenki, E. I. Lazhko, K. Klaus and I. F. Lutsenko; *J. Gen. Chem. USSR (Engl. Transl.)*, 1984, **54**, 1354.	115
84JGU2250	A. A. Prishchenko, A. V. Gromov, M. I. Kadyko and I. P. Lutsenko; *J. Gen. Chem. USSR (Engl. Transl.)*, 1984, **54**, 2250.	112, 114, 115
84JHC133	C. Kashima, T. Tajima and Y. Omote; *J. Het. Chem.*, 1984, **21**, 133.	899, 979
84JHC1157	S. Mataka, M. Kurisu, K. Takahashi and M. Tashiro; *J. Heterocycl. Chem.*, 1984, **21**, 1157.	21
84JOC113	E. C. Taylor and H. M. L. Davies; *J. Org. Chem.*, 1984, **49**, 113.	422
84JOC168	D. J. Ager; *J. Org. Chem.*, 1984, **49**, 168.	618, 652, 945, 1048, 1049
84JOC263	P. A. T. W. Porskamp, B. H. M. Lammerink and B. Zwanenburg; *J. Org. Chem.*, 1984, **49**, 263.	570, 926
84JOC515	M. J. Cohen and E. McNelis; *J. Org. Chem.*, 1984, **49**, 515.	750
84JOC725	R. B. Miller and M. I. Hassan; *J. Org. Chem.*, 1984, **49**, 725.	809
84JOC806	O. Lerman, Y. Tor, D. Hebel and S. Rozen; *J. Org. Chem.*, 1984, **49**, 806.	44
84JOC821	M. Chorev, S. A. MacDonald and M. Goodman; *J. Org. Chem.*, 1984, **49**, 821.	427
84JOC1084	K. S. Kyler, A. Bahir-Hashemi and D. S. Watt; *J. Org. Chem.*, 1984, **49**, 1084.	94
84JOC1096	Y. Yamamoto, H. Yatagai, Y. Saito and K. Maruyama; *J. Org. Chem.*, 1984, **49**, 1096.	642, 648, 659
84JOC1125	J. J. Sepiol, J. A. Sepiol and R. L. Soulen; *J. Org. Chem.*, 1984, **49**, 1125.	783
84JOC1574	E. D. Sternberg and K. P. C. Vollhardt; *J. Org. Chem.*, 1984, **49**, 1574.	1061
84JOC1653	P. J. Stang, K. A. Roberts and L. E. Lynch; *J. Org. Chem.*, 1984, **49**, 1653.	875, 877
84JOC1691	L. A. G. M. Van den Broek, P. A. T. W. Porskamp, R. C. Haltiwanger and B. Zwanenburg; *J. Org. Chem.*, 1984, **49**, 1691.	365
84JOC1700	M. E. Thompson; *J. Org. Chem.*, 1984, **49**, 1700.	394
84JOC2031	L. Poncini; *J. Org. Chem.*, 1984, **49**, 2031.	192

84JOC2081	R. A. Olofson, J. T. Martz, J.-P. Senet, M. Piteau and T. Malfroot; *J. Org. Chem.*, 1984, **49**, 2081.	58
84JOC2565	J. A. Soderquist, F.-Y. Shiau and R. A. Lemesh; *J. Org. Chem.*, 1984, **49**, 2565.	643, 645, 646
84JOC2688	O. Tsuge, S. Kanemasa and K. Matsuda; *J. Org. Chem.*, 1984, **49**, 2688.	517
84JOC2808	H. Sakurai, K. Sasaki, J. Hayashi and A. Hosomi; *J. Org. Chem.*, 1984, **49**, 2808.	181
84JOC3235	T. Hayama, S. Tomoda, Y. Takeuchi and Y. Nomura; *J. Org. Chem.*, 1984, **49**, 3235.	376
84JOC3314	A. Padwa, G. Haffmanns and M. Tomas; *J. Org. Chem.*, 1984, **49**, 3314.	513
84JOC3517	M. Kitamura, M. Isobe, Y. Ichikawa and T. Goto; *J. Org. Chem.*, 1984, **49**, 3517.	937
84JOC3994	T. Rosen, M. J. Taschner and C. H. Heathcock; *J. Org. Chem.*, 1984, **49**, 3994.	171
84JOC4091	G. J. McGarvey and J. S. Bajwa; *J. Org. Chem.*, 1984, **49**, 4091.	952
84JOC4237	A. Hassner, R. Fibiger and D. Andisik; *J. Org. Chem.*, 1984, **49**, 4237.	314
84JOC4272	G. M. Loudon, A. S. Radhakrishna, M. R. Almond, J. K. Blodgett and R. H. Boutin; *J. Org. Chem.*, 1984, **49**, 4272.	422
84JOC4518	G. W. Gribble, M. G. Saulnier, M. P. Sibi and J. A. Obaza-Nutaitis; *J. Org. Chem.*, 1984, **49**, 4518.	165
84JOC4565	G. Zweifel and G. Hahn; *J. Org. Chem.*, 1984, **49**, 4565.	1061
84JOC4958	W. F. Bailey and A. D. Rivera; *J. Org. Chem.*, 1984, **49**, 4958.	212
84JOC5147	R. L. Willer and R. L. Atkins; *J. Org. Chem.*, 1984, **49**, 5147.	434, 435, 440
84JOM(263)C21	M. Bellassoued and M. Gaudemar; *J. Organomet. Chem.*, 1984, **263**, C21	832
84JOM(264)9	E. E. Aboujaoude, N. Collignon and P. Savignac; *J. Organomet. Chem.*, 1984, **264**, 9.	557, 576
84JOM(264)87	L. Birkofer and A. Kopp; *J. Organomet. Chem.*, 1984, **264**, 87.	513
84JOM(264)155	S. Martin, R. Sauvêtre and J. F. Normant; *J. Organomet. Chem.*, 1984, **264**, 155.	806, 819
84JOM(266)285	C. C. Santini and F. Mathey; *J. Organomet. Chem.*, 1984, **266**, 285.	341
84JOM(267)199	M. Angoletta, P. L. Bellon and F. Demartin; *J. Organomet. Chem.*, 1984, **267**, 199.	688
84JOM(268)C29	E. Lukevics, V. V. Dirnens, Y. S. Goldberg, E. Liepinsh, I. Kalvins and M. V. Shimanska; *J. Organomet. Chem*, 1984, **268**, C29.	514
84JOM(268)31	C. Hu, J.-G. He, D. H. O'Brien and K. J. Irgolic; *J. Organomet. Chem.*, 1984, **268**, 31.	507
84JOM(269)C15	R. Bertani, C. B. Castellani and B. Crociani; *J. Organomet. Chem.*, 1984, **269**, C15.	1018
84JOM(269)C40	A. Maercker and R. Stoetzel; *J. Organomet. Chem.*, 1984, **269**, C40.	647
84JOM(269)21	H. Schumann and F. W. Reier; *J. Organomet. Chem.*, 1984, **269**, 21.	578
84JOM(269)123	N. E. Miller; *J. Organomet. Chem.*, 1984, **269**, 123.	524
84JOM(269)219	B. Fabisch and T. N. Mitchell; *J. Organomet. Chem.*, 1984, **269**, 219.	152
84JOM(269)249	T. N. Mitchell and B. Fabisch; *J. Organomet. Chem.*, 1984, **269**, 249.	656, 663, 700, 724
84JOM(271)107	G. Fritz and J. Thomas; *J. Organomet. Chem.*, 1984, **271**, 107.	1050, 1052
84JOM(272)123	D. Seyferth, D. C. Annarelli and S. C. Vick; *J. Organomet. Chem.*, 1984, **272**, 123.	1051, 1052
84JOM(273)C17	G. R. Clark, W. R. Roper and A. H. Wright; *J. Organomet. Chem.*, 1984, **273**, C17.	583
84JOM(273)C57	A. Maercker and R. Stoetzel; *J. Organomet. Chem.*, 1984, **273**, C57.	650
84JOM(273)195	H. H. Karsch, G. Müller and C. Krüger; *J. Organomet. Chem.*, 1984, **273**, 195.	579
84JOM(273)303	M. Gielen and K. Jurkschat; *J. Organomet. Chem.*, 1984, **273**, 303.	700
84JOM(273)333	D. B. Pourreau, R. R. Whittle and G. L. Geoffroy; *J. Organomet. Chem.*, 1984, **273**, 333.	678
84JOM(276)317	H. Werstmijze, H. Kleijn and P. Vermeer; *J. Organometal. Chem.*, 1984, **276**, 317.	940
84JOM(277)261	J. V. Comasseto, J. T. B. Ferreira and J. A. Fontanillas Val; *J. Organomet. Chem.*, 1984, **277**, 261.	93, 291
84JOM(277)335	R. Willem, M. Gielen, J. Meunier-Piret, M. van Meerssche, K. Jurkschat and A. Tzschach; *J. Organomet. Chem.*, 1984, **277**, 335.	700
84JPR279	S. Schramm, E. Schmitz and E. Gründemann; *J. Prakt. Chem.*, 1984, **326**, 279.	420
84LA450	Y. Koksal, P. Raddatz and E. Winterfeldt; *Liebigs Ann. Chem.*, 1984, 450.	849
84LA1468	E. Müller, R. Kettler and M. Wiessler; *Liebigs Ann. Chem.*, 1984, 1468.	312
84LA1494	W. Plesch and M. Wiessler; *Liebigs Ann. Chem.*, 1984, 1494.	311
84M587	E. Akgün and U. Pindur; *Monatsh. Chem.*, 1984, **115**, 587.	193
84MI 404-01	F. Meskens; *Janssen Chim. Acta*, 1984, **2(1)**, 16.	181
B-84MI 407-01	K. C. Nicolaou and N. A. Petasis; "Selenium in Natural Products Synthesis," CIS Inc., Philadelphia, 1984.	333
B-84MI 407-02	C. Paulmier; "Selenium Reagents and Intermediates in Organic Synthesis," Pergamon, Oxford 1986.	333
84MI 407-03	R. K. Valetdinov, A. Ya. Dorfman, A. A. Kutuev and I. T. Shaikhutdinov; *Khim. Tekhnol. Elementorg. Soedin.*, 1984, 22.	336
84MI 419-01	S. Scheiblich; *Diplomarbeit*, University of Hamburg, 1984.	837
84MI 421-01	H. Brauniger, K. Peseke, M. Bauch, M. Rohde and B. Tech; *Wiss. Z. Wilhelm Pieck Univ. Rostock Naturwiss Reihe*, 1984, **33**, 53 1291.	970
84MI 424-01	T. Durst; *Stud. Org. Chem. (Amsterdam)*, 1984, **5B** (*Compr. Carbanion Chem., Part B.*) (*Chem. Abstr.*, 1984, **101**, 71 800).	1075
B-84MI 424-02	E. V. Dehmlow and S. S. Dehmlow; "Phase Transfer for Catalysis," Monographs in Modern Chemistry, Verlag Chemie, Weinheim, 1984, vol. 11, p. 386.	1080
84MI 424-03	Q. Chen and S. Zhu; *Youji Huaxue*, 1984, 434 (*Chem. Abstr.*, 1985, **102**, 131 301).	1083
84OM1	B. Wrackmeyer; *Organometallics*, 1984, **3**, 1.	665
84OM354	K. R. Pope and P. R. Jones; *Organometallics*, 1984, **3**, 354.	602
84OM476	S. J. Holmes, R. R. Schrock, M. R. Cherchill and H. J. Wasserman; *Organometallics*, 1984, **3**, 476.	581
84OM614	D. S. Matteson and K. M. Sadhu; *Organometallics*, 1984, **3**, 614.	526, 527
84OM825	W. R. Tikkanen, J. Z. Liu and J. W. Egan, Jr.; *Organometallics*, 1984, **3**, 825.	657
84OM1284	D. S. Matteson, P. K. Jesthi and K. M. Sadhu; *Organometallics*, 1984, **3**, 1284.	526, 527
84OM1392	H. C. Brown and I. Imai; *Organometallics*, 1984, **3**, 1392.	815

84OM1438	E. Carmona, E. Gutlérrez-Puebla, A. Monge, J. M. Marín, M. Paneque and M. L. Poveda; *Organometallics*, 1984, **3**, 1438.	583
84OM1660	J. P. Picard, A. A. Elyusufi, R. Calas, J. Dunogues and N. Duffaut; *Organometallics*, 1984, **3**, 1660.	1014
84OM1687	K. Swami, J. P. Hutchinson, H. G. Kuivila and J. A. Zubieta; *Organometallics*, 1984, **3**, 1687.	701
84OM1897	D. Seyferth, S. C. Vick and M. L. Shannon; *Organometallics*, 1984, **3**, 1897.	1051, 1052
84OR(32)375	G. Zweifel and J. A. Miller; *Org. React.*, 1984, **32**, 375.	808, 1062, 1068
84PS(19)61	S. Cabiddu, C. Floris, S. Melis and F. Sotgiv; *Phosphorus Sulfur*, 1984, **19**, 61.	347
84PS(21)59	G. Borisov, V. Vasileva, E. N. Tsvetkov, T. E. Kron and M. I. Kabachnik; *Phosphorus Sulfur*, 1984, **21**, 59.	339
84RRC333	M. Iovu and M. Iqbal; *Rev. Roum. Chim.*, 1984, **29**, 333.	411
84RTC41	D. van Leusen and A. M. van Leusen; *Recl. Trav. Chim. Pays-Bas*, 1984, **103**, 41.	331
84S34	M. Yonovich-Weiss and Y. Sasson; *Synthesis*, 1984, 34.	22, 37
84S227	E. Akgun and U. Pindur; *Synthesis*, 1984, 227.	193
84S245	T. Sakamoto, Y. Kondo, M. Shiraiwa, H. Yamanaka; *Synthesis*, 1984, 245.	195
84S250	A. Rahman, H. Ila and H. Junjappa; *Synthesis*, 1984, 250.	903
84S256	T. Takajo, S. Kambe and W. Ando; *Synthesis*, 1984, 256.	437, 438
84S259	T. Takajo, S. Kambe and W. Ando; *Synthesis*, 1984, 259.	437
84S315	J. G. Schantl, P. Hebeisen and L. Minach; *Synthesis*, 1984, 315.	440
84S369	J. Tsuji; *Synthesis*, 1984, 369.	193
84S384	D. J. Ager; *Synthesis*, 1984, 384.	646
84S439	L. Syper and J. Mlochowski; *Synthesis*, 1984, 439.	288
84S495	J.-P. Quintard, B. Elissondo and B. Jousseaume; *Synthesis*, 1984, 495.	300, 533
84S577	B. Lejczak, P. Kafarski, M. Soroka and P. Mastalerz; *Synthesis*, 1984, 577.	467
84S661	J. Zón; *Synthesis*, 1984, 661.	491
84S683	R. M. Moriarty and K.-C. Hou; *Synthesis*, 1984, 683.	315
84S727	M. Jones; *Synthesis*, 1984, 727.	49
84S757	Y. K. Rao and M. Nagarajan; *Synthesis*, 1984, 757.	277
84SC147	T. Momose, T. Itooka and O. Muraoka; *Synth. Commun.*, 1984, **14**, 147.	165
84SC639	R. Smith and T. Livinghouse; *Synth. Commun.*, 1984, **14**, 639.	516, 537
84SC1073	G. Maas and B. Feith; *Synth. Commun.*, 1984, **14**, 1073.	975
84SC1119	M. Akiba and M. P. Cava; *Synth. Commun.*, 1984, **14**, 1119.	872
84SC1285	R. D. Miller and R. Hassig; *Synth. Commun.*, 1984, **14**, 1285.	773
84SRI21	J. Barrau, G. Rima and J. Satgé; *Synth. React. Inorg. Met.-Org. Chem.*, 1984, **14**, 21.	141, 372
84SUL1	K. Almdal and O. Hammerich; *Sulfur Lett.*, 1984, **2**, 1.	221
84T381	W.-D. Rudorf, E. Gunther and M. Augustin; *Tetrahedron*, 1984, **40**, 381.	837
84T733	B. Costisella, H. Gross and H. Schick; *Tetrahedron*, 1984, **40**, 733.	1004
84T991	T. A. Van Der Knaap, T. C. Klebach, P. Visser, R. Lourens and F. Bickelhaupt; *Tetrahedron*, 1984, **40**, 991.	483
84T1313	D. Seebach, R. Naef and G. Calderari; *Tetrahedron*, 1984, **40**, 1313.	213
84T1333	K.-Y. Ko, W. J. Frazee and E. L. Eliel; *Tetrahedron*, 1984, **40**, 1333.	231
84T1361	A. I. Meyers, L. M. Fuentes and Y. Kubota; *Tetrahedron*, 1984, **40**, 1361.	532
84T2951	A. M. Lamazouere and J. Sotiropoulos; *Tetrahedron*, 1984, **40**, 2951.	854, 855
84T3677	K. Krohn, M. Klimars, H. J. Koehle and E. Ebeling; *Tetrahedron*, 1984, **40**, 3677.	856
84T4963	A. Haas and K. W. Kempf; *Tetrahedron*, 1984, **40**, 4963.	872
84TL3	E. J. Corey, D. H. Hua and S. P. Seitz; *Tetrahedron Lett.*, 1984, **25**, 3.	217
84TL303	T. Fuchikami and I. Ojima; *Tetrahedron Lett.*, 1984, **25**, 303.	810
84TL511	M. T. Reetz, H. Heimbach and K. Schwellnus; *Tetrahedron Lett.*, 1984, **25**, 511.	202
84TL699	C. S. Wilcox and R. E. Babston; *Tetrahedron Lett.*, 1984, **25**, 699.	831
84TL745	R. Lehmann and M. Schlosser; *Tetrahedron Lett.*, 1984, **25**, 745.	378
84TL757	N. Ueda, H. Shimizu, T. Kataoka and M. Hori; *Tetrahedron Lett.*, 1984, **25**, 757.	232
84TL835	R. Tarhouni, B. Kirschleger, M. Rambaud and J. Villieras; *Tetrahedron Lett.*, 1984, **25**, 835.	146
84TL995	C. M. Lindsay, K. Smith, C. A. Brown and K. Betterton-Cruz; *Tetrahedron Lett.*, 1984, **25**, 995.	846
84TL1019	T. B. Patrick and Y.-F. Poon; *Tetrahedron Lett.*, 1984, **25**, 1019.	52
84TL1023	R. A. Moss, D. P. Cox and H. Tomioka; *Tetrahedron Lett.*, 1984, **25**, 1023.	415
84TL1291	C. J. Shishoo, M. B. Devani, V. S. Bhadti, S. Ananthan and G. V. Ullas; *Tetrahedron Lett.*, 1984, **25**, 1291.	988
84TL1353	H. Albrecht and H. Dollinger; *Tetrahedron Lett.*, 1984, **25**, 1353.	528
84TL1769	A. K. Forrest and R. R. Schmidt; *Tetrahedron Lett.*, 1984, **25**, 1769	426
84TL1905	K. Tamao, R. Kanatani and M. Kumada; *Tetrahedron Lett.*, 1984, **25**, 1905.	654
84TL1913	K. Tamao, R. Kanatani and M. Kumada; *Tetrahedron Lett.*, 1984, **25**, 1913.	651, 655
84TL1999	S. Fujikura, M. Inoue, K. Utimoto and H. Nozaki; *Tetrahedron Lett.*, 1984, **25**, 1999.	1060
84TL2021	M. Isobe, Y. Funabashi, Y. Ichikawa, S. Mio and T. Goto; *Tetrahedron Lett.*, 1984, **25**, 2021.	950
84TL2581	D. van Leusen and A. M. van Leusen; *Tetrahedron Lett.*, 1984, **25**, 2581.	908
84TL2705	A. R. Bassindale, R. J. Ellis and P. G. Taylor; *Tetrahedron Lett.*, 1984, **25**, 2705.	653
84TL2883	W. Dumont, C. Vermeyen and A. Krief; *Tetrahedron Lett.*, 1984, **25**, 2883.	917
84TL3047	T. J. Nitz and L. A. Paquette; *Tetrahedron Lett.*, 1984, **25**, 3047.	179
84TL3075	A. Ghribi, A. Alexakis and J. F. Normant; *Tetrahedron Lett.*, 1984, **25**, 3075.	192
84TL3127	R. T. Brown and M. F. Jones; *Tetrahedron Lett.*, 1984, **25**, 3127.	168
84TL3221	H. Oda, Y. Morizawa, K. Oshima and H. Nozaki; *Tetrahedron Lett.*, 1984, **25**, 3221.	141, 811, 812

84TL3539	S. Hackett and T. Livinghouse; *Tetrahedron Lett.*, 1984, **25**, 3539.	834
84TL3629	M. Clarembeau and A. Krief; *Tetrahedron Lett.*, 1984, **25**, 3629.	399
84TL3687	H. H. Karsch, F. H. Kohler and H. U. Reisacher; *Tetrahedron Lett.*, 1984, **25**, 3687.	1026
84TL3805	G. Gil; *Tetrahedron Lett.*, 1984, **25**, 3805.	188
84TL3987	T. Yokozawa, T. Nakai and N. Ishikawa; *Tetrahedron Lett.*, 1984, **25**, 3987.	831
84TL3991	T. Yokozawa, T. Nakai and N. Ishikawa; *Tetrahedron Lett.*, 1984, **25**, 3991.	831
84TL4011	W. Ando, T. Furuhata, Y. Hanyu and T. Takata; *Tetrahedron Lett.*, 1984, **25**, 4011.	836
84TL4195	E. W. Logusch; *Tetrahedron Lett.*, 1984, **25**, 4195.	210
84TL4233	T. Imamoto, Y. Sugiura and N. Takiyama; *Tetrahedron Lett.*, 1984, **25**, 4233.	954
84TL4245	K. Tamao and N. Ishida; *Tetrahedron Lett.*, 1984, **25**, 4245.	654
84TL4249	K. Tamao and N. Ishida; *Tetrahedron Lett.*, 1984, **25**, 4249.	654, 655
84TL4259	B. H. Bakker, D. S. Tjin A-Lim and A. van der Gen; *Tetrahedron Lett.*, 1984, **25**, 4259.	468
84TL4329	J. Fried, S. Kittisopikul and E. A. Hallinan; *Tetrahedron Lett.*, 1984, **25**, 4329.	734
84TL4337	S. Karady, J. S. Amato and L. M. Weinstock; *Tetrahedron Lett.*, 1984, **25**, 4337.	302
84TL4383	P. Knochel and J. F. Normant; *Tetrahedron Lett.*, 1984, **25**, 4383.	357
84TL4425	Y. Shen, Z. Gu, W. Ding and Y. Huang; *Tetrahedron Lett.*, 1984, **25**, 4425.	598
84TL4455	R. A. Holton and R. M. Kennedy; *Tetrahedron Lett.*, 1984, **25**, 4455.	169
84TL4645	T. A. Engler and H. Shechter; *Tetrahedron Lett.*, 1984, **24**, 4645.	647
84TL4797	M. Chmielewski, Z. Kaluza, C. Belzecki, P. Salanski and J. Jurczak; *Tetrahedron Lett.*, 1984, **25**, 4797.	174
84TL5009	P. G. Williard and L. A. Grab; *Tetrahedron Lett.*, 1984, **25**, 5009.	173
84TL5155	T. Fujisawa, K. Kohama, K. Tajima and T. Sato; *Tetrahedron Lett.*, 1984, **25**, 5155.	187
84TL5191	B. J. J. Van de Heisteeg, G. Shat, O. S. Akkerman and F. Bickelhaupt; *Tetrahedron Lett.*, 1984, **24**, 5191.	657, 663, 701
84TL5351	R. D. Miller and R. Hässig; *Tetrahedron Lett.*, 1984, **25**, 5351.	945
84TL5409	Y. Hamada, A. Kawai and T. Shioiri; *Tetrahedron Lett.*, 1984, **25**, 5409.	210
84TL5711	P. E. Peterson and B. L. Jensen; *Tetrahedron Lett.*, 1984, **25**, 5711.	1059
84TL5729	J. L. Belletire, D. R. Walley and S. L. Fremont; *Tetrahedron Lett.*, 1984, **25**, 5729.	846
84TL5797	Y. Tamaru, M. Okada, O. Kitao and Z. Yoshida; *Tetrahedron Lett.*, 1984, **25**, 5797.	230
84URP1067006	K. A. Petrov, V. A. Chauzov and V. P. Pokatun; *USSR Pat.* 1 067 006 (1984) (*Chem. Abstr.*, 1984, **100**, 210 155).	561
84USP4465845A	Y. Okamoto and P. S. Wojciechowski (Koppers Co., Inc., USA); *US Pat.* 4 465 845 A (1984) (*Chem. Abstr.*, 1984, **101**, 211 154).	869
84ZAAC(512)103	G. Fritz and K. P. Woerns; *Z. Anorg. Allg. Chem.*, 1984, **512**, 103.	617
84ZC261	K. Issleib, F. Krech and U. Kühne; *Z. Chem.*, 1984, **24**, 261.	550
84ZC365	K. Kellner and A. Tzschach; *Z. Chem.*, 1984, **24**, 365.	451, 452, 455, 481
84ZN(B)356	H. Kischkel and G.-V. Roeschenthaler; *Z. Naturforsch., Teil B*, 1984, **39**, 356.	455
84ZN(B)798	H. Schumann and R. Mohtacheni; *Z. Naturforsch., Teil B*, 1984, **39**, 798.	708, 726
84ZN(B)1456	H. Schmidbaur, P. Nussstein and G. Müller; *Z. Naturforsch., Teil B*, 1984, **39**, 1456.	565
84ZN(B)1518	H. H. Karsch, L. Weber, D. Wewers, R. Boese and G. Müller; *Z. Naturforsch., Teil B*, 1984, **39**, 1518.	551
84ZN(B)19042	H. Klusik, C. Pues and A. Berndt; *Z. Naturforsch., Teil B*, 1984, **39**, 19 042.	1048, 1054, 1057
84ZOB36	T. A. Mastryukova, I. M. Aladzheva, O. V. Bykhovskaya, I. V. Leont'eva, P. V. Petrovskii and M. I. Kabachnik; *Zh. Obshch. Khim.*, 1984, **54**, 36.	557
84ZOB59	N. I. Liptuga and V. V. Yaremenko; *Zh. Obshch. Khim.*, 1984, **54**, 59.	573
84ZOB116	I. V. Suvorova and M. D. Stadnichuk; *Zh. Obshch. Khim.*, 1984, **54**, 116.	942
84ZOB288	V. S. Brovarets, O. P. Lobanov, T. K. Vinogradova and B. S. Drach; *Zh. Obshch. Khim.*, 1984, **54**, 288 (Engl. Transl. 255).	999
84ZOB309	I. V. Suvorova and M. D. Stadnichuk; *Zh. Obshch. Khim.*, 1984, **54**, 309.	941
84ZOB416	I. L. Rodionov, Yu. N. Luzikov, M. A. Kazankova and I. F. Lutsenko; *Zh. Obshch. Khim.*, 1984, **54**, 416.	914
84ZOB457	S. A. Shilov, M. V. Sendyurev, A. V. Dogadina, B. I. Ionin and A. A. Petrov; *Zh. Obshch. Khim.*, 1984, **54**, 457.	914
84ZOB1324	V. A. Al'fonsov, G. U. Zamaletdinova, I. S. Nizamov, E. S. Batyeva and A. N. Pudovik; *Zh. Obshch. Khim.*, 1984, **54**, 1324.	921
84ZOB1511	A. V. Il'yasov, A. N. Chernov, A. A. Nafikova, I. Z. Galeeva, E. N. Dianova and B. A. Arbuzov; *Zh. Obshch. Khim.*, 1984, **54**, 1511.	549
84ZOB1758	N. G. Khusainova, L. V. Naumova, E. A. Berdnikov and A. N. Pudovik; *Zh. Obshch. Khim.*, 1984, **54**, 1758.	926
84ZOB1817	S. V. Ponomarev, S. N. Nikolaeva, G. N. Molchanova, A. S. Kostyuk and Y. K. Grishin; *Zh. Obshch. Khim.*, 1984, **54**, 1817 (*Chem. Abstr.*, 1984, **102**, 62 308).	1057
84ZOB1842	A. G. Shipov and Y. I. Baukov; *Zh. Obshch. Khim.*, 1984, **54**, 1842 (*Chem. Abstr.*, 1984, **101**, 230 689q).	605
84ZOB1925	L. I. Livantsova, O. P. Perelygina, G. S. Zaitseva and Y. I. Bauker; *Zh. Obshch. Khim.*, 1984, **54**, 1925 (*Chem. Abstr.*, 1985, **102**, 46 045).	827
84ZOB2025	L. I. Rybin, O. A. Vyazankina, N. S. Vyazankin, T. V. Leshina, M. B. Taraban, D. V. Gendin, V. I. Mar'yasova and M. F. Larin; *Zh. Obshch. Khim.*, 1984, **54**, 2025 (*Chem. Abstr.*, 1985, **102**, 46 056a).	656
84ZOB2186	A. P. Martynyuk, V. S. Brovarets, O. P. Lobanov and B. S. Drach; *Zh. Obshch. Khim.*, 1984, **54**, 2186 (Engl. Transl. 1954).	999, 1001, 1002, 1007, 1011
84ZOR972	B. A. Shainyan and A. N. Mirskova; *Zh. Org. Khim.*, 1984, **20**, 972.	783
84ZOR1502	A. N. Mirskova, I. T. Gogoberidze, G. G. Levkovskaya, I. D. Kalikhman and M. G. Voronkov; *Zh. Org. Khim.*, 1984, **20**, 1502.	312

Ref	Citation	Pages
85AG(E)571	A. Maercker and R. Dujardin; *Angew. Chem., Int. Ed. Engl.*, 1985, **24**, 571.	1064
85AG(E)585	L. W. Jenneskens, W. K. De Wolf and F. Bickelhaupt; *Angew. Chem., Int. Ed. Engl.*, 1985, 585.	38
85AG(E)696	J. Vollhardt, H.-J. Gais and K. L. Lukas; *Angew. Chem., Int. Ed. Engl.*, 1985, **24**, 696.	936, 938
85AG(E)788	R. Wehrmann, H. Meyer and A. Berndt; *Angew. Chem., Int. Ed. Engl.*, 1985, **24**, 788.	634
85AG(E)943	W. Siebert; *Angew. Chem., Int. Ed. Engl.*, 1985, **24**, 943.	632
85AG(E)979	G. Zinner and W. P. Fehlhammer; *Angew. Chem., Int. Ed. Engl.*, 1985, **24**, 979.	492
85AG(E)1067	I. Hoppe, U. Schöllkopf, M. Nieger and E. Egert; *Angew. Chem., Int. Ed. Engl.*, 1985, **24**, 1067.	462
85AJC1505	R. D. Grant, J. T. Pinhey, E. Rizzardo and G. C. Smith; *Aust. J. Chem.*, 1985, **38**, 1505.	107
85AP(318)473	D. Matthies, B. Bartsch and U. Blanck; *Arch. Pharm. (Weinheim. Ger.)*, 1985, **318**, 473.	298
85BBA(818)96	A. G. Krainev, L. M. Weiner, I. S. Alferyev and N. M. Slynko; *Biochim. Biophys. Acta*, 1985, **818**, 96.	560
85BCJ1983	T. Satoh, Y. Kaneko, T. Izawa, K. Sakata and K. Yamakawa; *Bull. Chem. Soc. Jpn.*, 1985, **58**, 1983.	74
85BCJ1991	O. Tsuge, J. Tanaka and S. Kanemasa; *Bull. Chem. Soc. Jpn.*, 1985, **58**, 1991.	506, 513
85BSB299	K. Jurkschat and M. Gielen; *Bull. Soc. Chim. Belg.*, 1985, **94**, 299.	700
85BSB475	R. W. Saalfrank, E. Ackermann, M. Fischer, B. Weiss, R. Carrie, D. Danion, K. Peters and H. G. Schnering; *Bull. Soc. Chim. Belg.*, 1985, **94**, 475.	991, 993, 996
85BSF787	A. Duchene, D. Mouko-Mpegna and J.-P. Quintard; *Bull. Soc. Chim. Fr.*, 1985, 787.	388, 389
85BSF825	J. Villieras, R. Tarhouni, B. Kirschleger and M. Rambaud; *Bull. Chim. Soc. Fr.*, 1985, 825.	138, 146, 667
85BSF881	S. Berrada, P. Metzner and R. Rakotonirina; *Bull. Soc. Chim. Fr.*, 1985, 881.	255
85BSF1250	O. Ruel, C. B. B. Ekogha, R. Lorne and S. A. Julia; *Bull. Soc. Chim. Fr.*, 1985, 1250.	938
85CA(102)78589t	E. I. Aoyagi (Chevron Research Co.,); *Chem. Abstr.*, 1985, **102**, 78 589t (*Ger. Pat.* 3 411 465 (1984)).	776
85CA(103)6148w	Kowa Co Ltd.; *Chem. Abstr.*, 1985, **103**, 6148w (*Jpn. Pat.* 59 212 493 (1984)).	778
85CAR(136)177	J.-M. Vatele; *Carbohydr. Res.*, 1985, **136**, 177.	764
85CAR(139)115	H. Lönn; *Carbohydr. Res.*, 1985, **139**, 115.	224
85CAR(144)342	A. Banaszek, X. B. Cornet and A. Zamojski; *Carbohydr. Res.*, 1985, **144**, 342.	176
85CB261	H. Werner, W. Paul, R. Feser, R. Zolk and P. Thometzek; *Chem. Ber.*, 1985, **118**, 261.	582
85CB298	K.-R. Pörschke, G. Wilke and R. Mynott; *Chem. Ber.*, 1985, **118**, 298.	583
85CB370	T. Kauffmann, R. Kriegesmann, A. Rensing, R. König and F. Steinseifer; *Chem. Ber.*, 1985, **118**, 370.	156, 708
85CB391	T. Kauffmann, G. Ilchmann, R. König and M. Wensing; *Chem. Ber.*, 1985, **118**, 391.	139, 141, 146, 156
85CB541	L. Weber and D. Wewers; *Chem. Ber.*, 1985, **118**, 541.	580
85CB814	R. Appel, F. Knoch and R. Zimmerman; *Chem. Ber.*, 1985, **118**, 814.	566
85CB1008	R. Helwig and M. Hanack; *Chem. Ber.*, 1985, **118**, 1008.	799
85CB1039	T. Kauffmann, F. Steinseifer and N. Klas; *Chem. Ber.*, 1985, **118**, 1039.	596, 600
85CB1798	I. Stahl; *Chem. Ber.*, 1985, **118**, 1798.	1073
85CB2137	C. Rücker; *Chem. Ber.*, 1985, **118**, 2137.	1080
85CB2353	T. Kauffmann, B. Altepeter, N. Klas and R. Kriegesmann; *Chem. Ber.*, 1985, **118**, 2353.	135, 592, 596, 599, 600
85CB2365	G. Märkl, W. Weber and W. Weiss; *Chem Ber.*, 1985, **118**, 2365.	1027, 1030
85CB2493	R. W. Hoffmann, A. Riemann and B. Mayer; *Chem. Ber.*, 1985, **118**, 2493.	1049
85CB2822	D. Hoppe, R. Hanko, A. Broenneke, F. Lichtenberg and E. Van Huelsen; *Chem. Ber.*, 1985, **118**, 2822.	382
85CB2852	U. Rheude, R. Schork and W. Sundermeyer; *Chem. Ber.*, 1985, **118**, 2852.	854
85CB3032	W. Paul and H. Werner; *Chem. Ber.*, 1985, **118**, 3032.	582
85CB3105	H. Schmidbaur, R. Herr, T. Pollok, A. Schier, G. Müller and J. Riede; *Chem. Ber.*, 1985, **118**, 3105.	551, 560, 1023
85CB3227	E. Niecke, J. Böske, B. Krebs and M. Dartmann; *Chem. Ber.*, 1985, **118**, 3227.	488
85CB3419	R. Appel and C. Casser; *Chem. Ber.*, 1985, **118**, 3419.	566
85CB3560	L. Weber and D. Wewers; *Chem. Ber.*, 1985, **118**, 3560.	1029
85CB3570	L. Weber, D. Wewers and R. Böse; *Chem. Ber.*, 1985, **118**, 3570.	580
85CB4068	R. Appel, J. Menzel and F. Knoch; *Chem. Ber.*, 1985, **118**, 4068.	567
85CB4248	H. Lehmkuhel, K. Mehler, A. Shakoor, C. Krüger, Y-H. Tsay, R. Benn, A. Rufinska and G. Schroth; *Chem. Ber.* 1985, **118**, 4248.	659
85CC170	N. M. Doherty, C. Elschenbroich, H.-J. Kneuper and S. A. R. Knox; *J. Chem. Soc., Chem. Commun.*, 1985, 170.	673
85CC279	M. Yokoyama, H. Hatanaka and K. Sakamoto; *J. Chem. Soc., Chem. Commun.*, 1985, 279.	900
85CC296	P. Gaviña, S. V. Luis, P. Ferrer, A. M. Costero and J. A. Marco; *J. Chem. Soc., Chem. Commun.*, 1985, 296.	749
85CC571	A. Krief, W. Dumont and J.-N. Denis; *J. Chem. Soc., Chem. Commun.*, 1985, 571.	399
85CC614	G. R. Cooper, F. Hassan, B. L. Shaw and M. Thornton-Pett; *J. Chem. Soc., Chem. Commun.*, 1985, 614.	1023, 1028
85CC618	A. T. Hewson, L. A. March, I. W. Nowell and S. K. Richardson; *J. Chem. Soc., Chem. Commun.*, 1985, 618.	856
85CC678	M. Inbasekaran, N. P. Peet, J. R. McCarthy and M. E. LeTourneau; *J. Chem. Soc., Chem. Commun.*, 1985, 678.	79, 780
85CC686	J. W. Davies, J. R. Malpass and M. P. Walker; *J. Chem. Soc., Chem. Commun.*, 1985, 686.	98
85CC696	D. J. Greig, M. McPherson, R. M. Paton and J. Crosby; *J. Chem. Soc., Chem. Commun.*, 1985, 696.	850

85CC742	P. C. B. Page, M. B. van Niel and P. H. Williams; *J. Chem. Soc., Chem. Commun.*, 1985, 742.	1076
85CC845	R. A. Abramovitch, B. Mavunkel, J. R. Stowers, M. Wegrzyn and C. Riche; *J. Chem. Soc., Chem. Commun*, 1985, 845.	905, 911
85CC870	D. Villemin; *J. Chem. Soc., Chem. Commun.*, 1985, 870.	842
85CC1000	K. M. Nsunda and L. Hevesi; *J. Chem. Soc., Chem. Commun.*, 1985, 1000.	844, 875
85CC1021	P. Patel and J. A. Joule; *J. Chem. Soc., Chem. Commun.*, 1985, 1021.	540
85CC1154	H. Gotthardt and M. Oppermann; *J. Chem. Soc., Chem. Commun.*, 1985, 1154.	835
85CC1183	R. Grigg, H. Q. N. Gunaratne and V. Sridharan; *J. Chem. Soc., Chem. Commun.*, 1985, 1183.	311
85CC1199	S. Kerschl and B. Wrackmeyer; *J. Chem. Soc., Chem. Commun.*, 1985, 1199.	665
85CC1266	N. Furukawa, A. Kawada, T. Kawai and H. Fujihara; *J. Chem. Soc., Chem. Commun.*, 1985, 1266.	94
85CC1278	H. H. Murray, III, J. P. Fackler, Jr. and D. A. Tocher; *J. Chem. Soc., Chem. Commun.*, 1985, 1278.	152
85CC1635	X. L. R. Fontaine, F. S. M. Hassan, S. J. Higgins, G. B. Jacobsen, B. L. Shaw and M. Thornton-Pett; *J. Chem. Soc., Chem. Commun.*, 1985, 1635.	1023, 1028
85CC1684	M. Nakano and Y. Sato; *J. Chem. Soc., Chem. Commun.*, 1985, 1684.	508
85CC1803	K. Tsujimoto, Y. Okeda and M. Ohashi; *J. Chem. Soc., Chem. Commun.*, 1985, 1803.	850
85CCC1507	I. Rosenberg and A. Holý; *Collect. Czech. Chem. Commun.*, 1985, **50**, 1507.	130
85CJC576	A. Jarczewski, G. Schroeder, W. Galezowski, K. T. Leffek and U. Maciejewska; *Can. J. Chem.*, 1985, **63**, 576.	731
85CJC2313	T. G. Back, S. Collins and K. W. Law; *Can. J. Chem.*, 1985, **63**, 2313.	874
85CL939	T. Takata, M. Kuo, Y. Tamura, Y. Kabe and W. Ando; *Chem. Lett.*, 1985, 939.	318
85CL1405	T. Kusumoto and T. Hiyama; *Chem. Lett.*, 1985, 1405.	1045
85CL1933	Y. Masaki, Y. Serizawa and K. Kaji; *Chem. Lett.*, 1985, 1933.	216
85CPB1745	Y. Tsuda and S. Hosoi; *Chem. Pharm. Bull.*, 1985, **33**, 1745.	90
85CPB2256	K. Ogawa, S. Yamada, T. Terada, T. Yamazaki and T. Honna; *Chem. Pharm. Bull.*, 1985, **33**, 2256.	837
85CPB4596	T. Morimoto, Y. Nezu and K. Achiwa; *Chem. Pharm. Bull.*, 1985, **33**, 4596.	509
85CPB5137	T. Taguchi, T. Morikawa, O. Kitagawa, T. Mishima and Y. Kobayashi; *Chem. Pharm. Bull.*, 1985, **33**, 5137.	734
85EUP146838	Y. F. Shealy and C. A. Krauth, *Eur. Pat.* 146 838 (1985) (*Chem. Abstr.*, 1985, **103**, 214 868).	282
85EUP230127	B. G. Jackson and B. A. Slomski (Eli Lilly and Co.); *Eur. Pat.* 230 127 (1985) (*Chem. Abstr.*, 1987, **107**, 19 8308a).	881
85H(23)1127	C. Gonzalez and R. Greenhouse; *Heterocycles*, 1985, **23**, 1127.	410
85H(23)2207	R. Huisgen, G. Mloston and C. Fulka; *Heterocycles*, 1985, **23**, 2207.	328
85H(23)2489	O. Tsuge, S. Kanemasa, T. Yamada and K. Matsuda; *Heterocycles*, 1985, **23**, 2489.	517, 518
85H(23)2509	H. Chikashita, N. Ishimoto, S. Komazawa and K. Itoh; *Heterocycles*, 1985, **23**, 2509.	321
85HCA44	M. E. Scheller and B. Frei; *Helv. Chim. Acta*, 1985, **68**, 44.	359
85HCA135	R. Naef and D. Seebach; *Helv. Chim. Acta*, 1985, **68**, 135.	419, 424
85HCA1243	D. Seebach and A. Fadel; *Helv. Chim. Acta*, 1985, **68**, 1243.	302
85HCA1730	R. Huber, A. Knierzinger, J.-P. Obrecht and A. Vasella; *Helv. Chim. Acta*, 1985, **67**, 1730.	485, 489
85HOU(E111)255	E. Schaumann; *Methoden Org. Chem. (Houben-Weyl)*, 1985, **E111**, 255.	833
85HOU(E111)260	E. Schaumann; *Methoden Org. Chem. (Houben-Weyl)*, 1985, **E111**, 260.	842
85IC642	M. D. Fryzuk, A. Carter and A. Westerhaus; *Inorg. Chem.*, 1985, **24**, 642.	566
85IC1993	R. R. Ford, B.-L. Li, R. H. Nielson and R. J. Thoma; *Inorg. Chem.*, 1985, **24**, 1993.	567
85IC3285	M. Hawkins and L. Andrews; *Inorg. Chem.*, 1985, **24**, 3285.	42
85IJC(A)398	R. Sanehi, R. K. Bansal and R. C. Mehrotra; *Ind. J. Chem., Sect. A*, 1985, **24**, 398.	584, 587
85IZV809	L. I. Zakharkin, A. I. Kovredov and V. A. Ol'shevskaya; *Izv. Akad. Nauk SSSR, Ser. Khim.*, 1985, 809.	143
85IZV1102	S. N. Ignat'eva, G. N. Nikonov, O. A. Erastov and B. A. Arbuzov; *Izv. Akad. Nauk SSSR, Ser. Khim.*, 1985, 1102.	342
85IZV1917	V. M. Vdovin, N. B. Bespalova and M. B. Sergeeva; *Izv. Akad. Nauk SSSR, Ser. Khim.*, 1985, 1917 (*Chem. Abstr.*, 1986, **105**, 60 673t).	610
85JA503	W. D. Wulff and S. R. Gilbertson; *J. Am. Chem. Soc.*, 1985, **107**, 503.	955
85JA675	C. W. Dirk, D. Nalewajek, G. B. Blanchet, H. Schaffer, F. Moraes, R. M. Boysel and F. Wudl; *J. Am. Chem. Soc.*, 1985, **107**, 675.	291
85JA735	J. R. McCarthy, N. P. Peet, M. E. Le Tourneau and M. Inbasekaran; *J. Am. Chem. Soc.*, 1985, **107**, 735.	61, 62, 73, 79, 780
85JA2568	E. Negishi, S. J. Holmes, J. M. Tour and J. A. Miller; *J. Am. Chem. Soc.*, 1985, **107**, 2568.	1061
85JA2811	D. G. Cox, N. Gurusamy and D. J. Burton; *J. Am. Chem. Soc.*, 1985, **107**, 2811.	801
85JA3935	L. R. Robinson, G. T. Burns and T. J. Barton; *J. Am. Chem. Soc.*, 1985, **107**, 3935.	1048
85JA4085	M. Fujita and T. Hiyama; *J. Am. Chem. Soc.*, 1985, **107**, 4085.	819
85JA4260	P. F. Hudrlik, A. M. Hudrlik and A. K. Kulkarni; *J. Am. Chem. Soc.*, 1985, **107**, 4260.	357
85JA4565	W. B. Farnham, B. E. Smart, W. J. Middleton, J. C. Calabrese and D. A. Dixon; *J. Am. Chem. Soc.*, 1985, **107**, 4565.	42
85JA5272	A. L. Balch, L. A. Fossett, M. M. Olmstead, D. E. Oram and P. E. Reedy; *J. Am. Chem. Soc.*, 1985, **107**, 5272.	565
85JA5301	T. Imamoto, T. Kusumoto, N. Suzuki and K. Sato; *J. Am. Chem. Soc.*, 1985, **107**, 5301.	575
85JA6298	M. R. Detty, N. F. Haley, R. S. Eachus, J. W. Hassett, H. R. Luss, M. G. Mason, J. M. McKelvey and A. A. Wernberg; *J. Am. Chem. Soc.*, 1985, **107**, 6298.	93
85JA6639	E. Negishi, D. E. Van Horn and T. Yoshida; *J. Am. Chem. Soc.*, 1985, **107**, 6639.	1066

85JA6729	E. Block and M. Aslam; *J. Am. Chem. Soc.*, 1985, **107**, 6729.	365, 392
85JA7178	L. Hatjiarapoglou, S. Spyroudis and A. Varvoglis; *J. Am. Chem. Soc.*, 1985, **107**, 7178.	867
85JA7219	J. J. Eisch, A. M. Piotrowski, S. K. Brownstein, E. J. Gabe and F. L. Lee; *J. Am. Chem. Soc.*, 1985, **107**, 7219.	1061
85JA7423	F. R. Askham, G. G. Stanley and E. C. Marques; *J. Am. Chem. Soc.*, 1985, **107**, 7423.	550
85JCP1517	B. J. van der Veken, P. Coppens, R. D. Johnson and J. R. Durig; *J. Chem. Phys.*, 1985, **83**, 1517.	130
85JCR(S)38	M. Wada, S. Higashizaki and A. Tsuboi; *J. Chem. Res. (S)*, 1985, 38.	118
85JCR(S)92	G. M. Blackburn, D. Brown and S. J. Martin; *J. Chem. Res. (S)*, 1985, 92.	129
85JCR(S)182	M. S. Baird and H. H. Hussain; *J. Chem. Res. (S)*, 1985, 182.	742
85JCS(D)337	L. M. Engelhardt, W-P. Leung, C. L. Raston and A. H. White; *J. Chem. Soc., Dalton Trans.*, 1985, 337.	646, 647
85JCS(D)905	G. A. Carriedo, J. A. K. Howard, D. B. Lewis, G. E. Lewis and F. G. A. Stone; *J. Chem. Soc., Dalton Trans.*, 1985, 905.	675
85JCS(D)2009	M. R. Awang, R. D. Barr, M. Green, J. A. K. Howard, T. B. Marder and F. G. A. Stone; *J. Chem. Soc., Dalton Trans.*, 1985, 2009.	716
85JCS(D)2483	M. Green, A. G. Orpen, C. J. Schaverien and I. D. Williams; *J. Chem. Soc., Dalton Trans.*, 1985, 2483.	689, 710
85JCS(P1)93	J. E. McCormick and R. S. McElhinney; *J. Chem. Soc., Perkin Trans. 1*, 1985, 93.	103
85JCS(P1)107	F. S. Guziec, Jr., C. J. Murphy and E. R. Cullen; *J. Chem. Soc., Perkin Trans. 1*, 1985, 107.	333
85JCS(P1)587	A. Perter, R. S. Ward, P. Collins, R. Venkateswarlu and I. T. Kay; *J. Chem. Soc., Perkin Trans. 1*, 1985, 587.	230
85JCS(P1)1185	H. Nemoto, X-M. Wu, H. Kurobe, K. Minemura, M. Ihara, K. Fukumoto and T. Kametani; *J. Chem. Soc., Perkin Trans. 1*, 1985, 1185.	154
85JCS(P1)1201	R. B. Boar and A. C. Patel; *J. Chem. Soc., Perkin Trans. 1*, 1985, 1201.	950
85JCS(P1)1567	A. C. Campbell, M. S. Maidment, J. H. Pick and D. F. M. Stevenson; *J. Chem. Soc., Perkin Trans. 1*, 1985, 1567.	56
85JCS(P1)2193	D. J. Buckley and M. A. McKervey; *J. Chem. Soc., Perkin Trans. 1*, 1985, 2193.	90
85JCS(P1)2625	A. T. Hewson and D. T. MacPherson; *J. Chem. Soc., Perkin Trans. 1*, 1985, 2625.	925
85JFC(27)85	D. J. Burton and D. M. Wiemers; *J. Fluorine Chem.*, 1985, **27**, 85.	117
85JFC(27)309	H. Lange and D. Naumann; *J. Fluorine Chem.*, 1985, **27**, 309.	153
85JFC(28)99	K. V. Dvornikova, V. E. Platonov and G. G. Yakobson; *J. Fluorine Chem.*, 1985, **28**, 99.	737
85JFC(29)445	Y. Nakayama, T. Kitazume and N. Ishikawa; *J. Fluorine Chem.*, 1985, **29**, 445.	731
85JFC(29)471	S. Misaki; *J. Fluorine Chem.*, 1985, **29**, 471.	731, 767
85JFC(30)269	K. J. L. Paciorek, J. H. Nakahara and R. H. Kratzer; *J. Fluorine Chem.*, 1985, **30**, 269.	481
85JFC(30)357	Y. Yamazaki, K. Jakita and N. Ishikawa; *J. Fluorine Chem.*, 1985, **30**, 357.	937
85JGU404	Z. S. Novikova, M. M. Kabachnik, N. V. Mashchenko and I. F. Lutsenko; *J. Gen. Chem. USSR (Engl. Transl.)*, 1985, **55**, 404.	462
85JGU1065	A. A. Prishchenko, A. V. Gromov, Yu. N. Luzikov, É. I. Lazhko and I. F. Lutsenko; *J. Gen. Chem. USSR (Engl. Transl.)*, 1985, **55**, 1065.	116
85JGU1079	N. A. Viktorov, T. K. Gar and V. F. Mironov; *J. Gen. Chem. USSR (Engl. Transl.)*, 1985, **55**, 1079.	142
85JGU2497	M. M. Gilyazov and Ya. A. Levin; *J. Gen. Chem. USSR (Engl. Transl.)*, 1985, **55**, 2497.	125
85JHC513	M. Amin, D. G. Holah, A. N. Hughes and T. Rukachaisirikul; *J. Heterocycl. Chem.*, 1985, **22**, 513.	1031
85JHC937	Y. Tominaga and Y. Matsuda, *J. Heterocycl. Chem.*, 1985, **22**, 937.	983, 986
85JHC957	D. E. Gaitanopoulos and J. Weinstock; *J. Heterocycl. Chem.*, 1985, **22**, 957.	102
85JHC1445	A. Monge, J. A. Palop, T. Goñi, A. Martínez and E. Fernández-Alvarez; *J. Heterocycl. Chem.*, 1985, **22**, 1445.	897
85JMC769	S. Salvadori, M. Marastani, G. Balboni, G. P. Santo and R. Tomatis; *J. Med. Chem.*, 1985, **28**, 769.	422
85JMC1917	D. H. Kinder and J. A. Katzenellenbogen; *J. Med. Chem.*, 1985, **28**, 1917.	527
85JOC1019	A. I. Meyers, P. D. Edwards, T. R. Bailey and G. E. Jagdmann, Jr.; *J. Org. Chem.*, 1985, **50**, 1019.	530
85JOC1117	M. D. Broadhurst; *J. Org. Chem.*, 1985, **50**, 1117.	824
85JOC1278	T. Minami, T. Hanamoto and I. Hirao; *J. Org. Chem.*, 1985, **50**, 1278.	488, 489
85JOC1351	A. R. Katritzky, F. Saczewski and C. M. Marson; *J. Org. Chem.*, 1985, **50**, 1351.	323
85JOC1547	T. Tokumitsu and T. Hayashi; *J. Org. Chem.*, 1985, **50**, 1547.	796
85JOC1621	P. Magnus and D. Quagliato; *J. Org. Chem.*, 1985, **50**, 1621.	942
85JOC1699	L. C. Garver, V. Grakauskas and K. Baum; *J. Org. Chem.*, 1985, **50**, 1699.	428
85JOC2195	D. Caine and V. C. Ukachukwu; *J. Org. Chem.*, 1985, **50**, 2195.	818
85JOC2603	A. G. M. Barrett, G. G. Graboski and M. A. Russell; *J. Org. Chem.*, 1985, **50**, 2603.	906
85JOC2736	K. Baum and D. Tzeng; *J. Org. Chem.*, 1985, **50**, 2736.	994
85JOC3255	R. C. Gadwood, M. R. Rubino, S. C. Nagarajan and S. T. Michel; *J. Org. Chem.*, 1985, **50**, 3255.	50, 379, 381, 382, 951, 952
85JOC3402	S. V. Frye and E. L. Eliel; *J. Org. Chem.*, 1985, **50**, 3402.	230
85JOC3627	G. L. Larson and R. Klesse; *J. Org. Chem.*, 1985, **50**, 3627.	188
85JOC3878	G. A. Russell and D. Dedolph; *J. Org. Chem.*, 1985, **50**, 3878.	898, 907, 908, 911
85JOC3946	J. R. Hwu and J. M. Wetzel; *J. Org. Chem.*, 1985, **50**, 3946.	182
85JOC3953	G. Barcelo, J.-P. Senet and G. Sennyey; *J. Org. Chem.*, 1985, **50**, 3953.	59
85JOC4006	A. Padwa, Y.-Y. Chen, W. Dent and H. Nimmesgern; *J. Org. Chem.*, 1985, **50**, 4006.	510
85JOC4227	H. Cho, K. Shima, M. Hayashimatsu, Y. Ohnaka, A. Mizuno and Y. Takeuchi; *J. Org. Chem.*, 1985, **50**, 4227.	420

85JOC4231	H. W. Moore, G. Hughes, K. Srinivasachar, M. Fernandez, N. V. Nguyen, D. Schoon and A. Tranne; *J. Org. Chem.*, 1985, **50**, 4231.	304
85JOC5022	M. Brawner Floyd, M. T. Du, P. F. Fabio, L. A. Jacob and B. D. Johnson; *J. Org. Chem.*, 1985, **50**, 5022.	161
85JOC5167	B. B. Snider and R. A. H. F. Hui; *J. Org. Chem.*, 1985, **50**, 5167.	212
85JOC5379	H. E. Morton and Y. Guindon; *J. Org. Chem.*, 1985, **50**, 5379.	199, 217, 225
85JOC5393	R. L. Danheiser, D. M. Fink, K. Okano, Y. M. Tsai and S. W. Szczepanski; *J. Org. Chem.*, 1985, **50**, 5393.	358
85JOC5410	R. F. Cunico and C.-P. Kuan; *J. Org. Chem.*, 1985, **50**, 5410.	930
85JOC5444	A. Mori and H. Yamamoto; *J. Org. Chem.*, 1985, **50**, 5444.	174
85JOM(280)C31	B. Weinberger, G. Tanguy and H. des Abbayes; *J. Organomet. Chem.*, 1985, **280**, C31.	582
85JOM(280)31	G. Nagendrappa and T. J. Vidyapati; *J. Organomet. Chem.*, 1985, **280**, 31.	357
85JOM(280)313	M. Grignon-Dubois and M. Ahra; *J. Organomet. Chem.*, 1985, **280**, 313.	605
85JOM(281)163	T. N. Mitchell and W. Reimann; *J. Organomet. Chem.*, 1985, **281**, 163.	1060, 1062
85JOM(281)317	H. Werner, L. Hofmann, R. Feser and W. Paul; *J. Organomet. Chem.*, 1985, **281**, 317.	582
85JOM(282)315	J. Barrau, N. Ben Hamida and J. Satgé; *J. Organomet. Chem.*, 1985, **282**, 315.	629
85JOM(285)1	D. Seebach, J. Hansen, P. Seiler and J. M. Gromek; *J. Organomet. Chem.*, 1985, **285**, 1.	531
85JOM(285)121	J. J. Eisch, M. Behrooz and S. K. Dua; *J. Organomet. Chem.*, 1985, **285**, 121.	393
85JOM(286)171	T. G. Back and R. G. Kerr; *J. Organomet. Chem.*, 1985, **286**, 171.	290, 335
85JOM(288)13	J. W. Bruin, G. Schat, O. S. Akkerman and F. Bickelhaupt; *J. Organomet. Chem.*, 1985, **288**, 13.	608, 626, 669, 696, 700
85JOM(288)119	L. V. Rybin, E. A. Petrovskaya, M. I. Rubinskaya, L. G. Kuz'mina, Yu. T. Struchkov, V. V. Kaverin and N. Yu. Koneva; *J. Organomet. Chem.*, 1985, **288**, 119.	583
85JOM(289)141	L. Hofmann and H. Werner; *J. Organomet. Chem.*, 1985, **289**, 141.	148
85JOM(290)125	R. Usón, J. Forniés, R. Navarro, P. Espinet and C. Mendívil; *J. Organomet. Chem.*, 1985, **290**, 125.	584
85JOM(291)277	U. Klingebiel, S. Pohlmann and L. Skoda; *J. Organomet. Chem.*, 1985, **291**, 277.	1050
85JOM(296)411	D. J. Brauer, S. Hietkamp, H. Sommer, O. Stelzer, G. Müller, M. J. Romão and C. Krüger; *J. Organomet. Chem.*, 1985, **296**, 411.	550
85JOM(297)C25	O. A. Reutov, I. V. Barinov, V. A. Chertkov and V. I. Sokolov; *J. Organomet. Chem.*, 1985, **297**, C25.	674
85JOM(297)C30	H. Schumann, F. W. Reier and E. Palamidis; *J. Organomet. Chem.*, 1985, **297**, C30.	577
85JOU806	L. I. Vereshchagin, L. P. Kirillova, G. M. Luzgina and G. A. Gareev; *J. Org. Chem. USSR (Engl. Transl.)*, 1985, **21**, 806.	106
85JPR297	H. Wilde, S. Hauptmann, A. Kanitz, M. Franzheld and G. Mann; *J. Prakt. Chem.*, 1985, **327**, 297.	880
85LA555	U. Schollkopf, I. Hoppe and A. Thiele; *Liebigs Ann. Chem.*, 1985, 555.	1004
85LA2472	E. Akgün and U. Pindur; *Liebigs Ann. Chem.*, 1985, 2472.	193
85M413	P. K. Claus and A. Setzer; *Monatsh. Chem.*, 1985, **116**, 413.	860
85M1051	J. G. Schantl and H. Gstach; *Monatsh. Chem.*, 1985, **116**, 1051.	448
85M1329	J. G. Schantl and H. Gstach; *Monatsh. Chem.*, 1985, **116**, 1329.	329, 444
B-85MI 401-01	R. A. Moss and M. Jones, Jr.; in "Reactive Intermediates," eds. M. Jones, Jr. and R. A. Moss, Wiley, New York, 1985, vol. 3, p. 45.	7, 16, 24
B-85MI 408-01	S. Wolfe; in "Organic Sulfur Chemistry," eds. F. Bernardi, I. G. Csizmadia and A. Mangini, Elsevier, Amsterdam, 1985, p. 133.	390
85MI 408-02	C. R. Johnson; *Aldrichim. Acta*, 1985, **18**, 3.	394
85MI 412-01	E. C. Alyea, R. P. Shakya and A. E. Vougioukas; *Transition Met. Chem. (Weinheim. Ger.)*, 1985, **10**, 435.	566, 573
85MI 424-01	S. Wolfe; *Stud. Org. Chem. (Amsterdam)*, 1985, **19** (*Org. Sulfur Chem.*), 133.	1075
85NJC249	F. Maassarani, M. Pfeffer and J. Fischer; *Nouv. J. Chim.*, 1985, **9**, 249.	647, 656
85OM231	H. H. Karsch, A. Appelt, F. H. Köhler and G. Müller; *Organometallics*, 1985, **4**, 231.	587
85OM355	R. S. Dickson, G. D. Fallon, R. J. Nesbit and G. N. Pain; *Organometallics*, 1985, **4**, 355.	689
85OM581	A. H-B. Cheng, P. R. Jones, M. E. Lee and P. Roussi; *Organometallics*, 1985, **4**, 581.	610, 611, 621
85OM821	D. W. Hawker and P. R. Wells; *Organometallics*, 1985, **4**, 821.	660, 663
85OM1044	T. N. Mitchell, W. Reimann and C. Nettelbeck; *Organometallics*, 1985, **4**, 1044.	661, 701, 1062
85OM1123	M. Mlekuz, P. Bougeard, B. G. Sayer, S. Peng, M. J. McGlinchey, A. Marinetti, J.-Y. Saillard, J. B. Naceur, B. Mentzen and G. Jaouen; *Organometallics*, 1985, **4**, 1123.	713
85OM1208	H. Schmidbauer, R. Herr, G. Muller and J. Riede; *Organometallics*, 1985, **4**, 1208.	94
85OM1233	H. Schmidbaur, E. Weiss and W. Graf; *Organometallics*, 1985, **4**, 1233.	577, 587
85OM1308	E. A. Mintz, A. S. Ward and D. S. Tice; *Organometallics*, 1985, **4**, 1308.	658
85OM1312	K. J. Ahmed, M. H. Chisholm and J. C. Huffman; *Organometallics*, 1985, **4**, 1312.	675
85OM1478	A. J. Ashe, III and F. J. Drone; *Organometallics*, 1985, **4**, 1478.	1039
85OM1690	D. S. Matteson; *Organometallics*, 1985, **4**, 1690.	664, 665
85OM2040	M. Ishikawa, S. Matsuzawa, T. Higuchi, S. Kamitori and K. Hirotsu; *Organometallics*, 1985, **4**, 2040.	1051, 1053
85OM2196	I. R. Butler, W. R. Cullen, J. Ni and S. J. Rettig; *Organometallics*, 1985, **4**, 2196.	706
85OM2206	J. M. Rozell, Jr. and P. R. Jones; *Organometallics*, 1985, **4**, 2206.	603, 637
85OR(34)319	C.-L. J. Wang, *Org. React.*, 1985, **34**, 319.	8
85OS(63)147	J. Salaün and J. Marguerite; *Org. Synth.*, 1985, **63**, 147.	171
85PS(25)39	J. A. Mikroyannidis; *Phosphorus Sulfur*, 1985, **25**, 39.	338, 339
85PS(25)201	R. Hussong, H. Heydt and M. Regitz; *Phosphorus Sulfur*, 1985, **25**, 201.	489
85RTC177	A. M. van Leusen, J. Wildeman, J. Moskal and A. W. van Hemert; *Recl. Trav. Chim. Pays-Bas*, 1985, **104**, 177.	328

85S62	Y. Vo Quang, D. Carniato, L. Quang and F. LeGoffic; *Synthesis*, 1985, 62.	465
85S100	T. Takajo and S. Kambe; *Synthesis*, 1985, 100.	444
85S276	Y. Ohshiro, N. Ando, M. Komatsu and T. Agawa; *Synthesis*, 1985, 276.	328
85S323	N. G. Kundu and S. G. Khatri; *Synthesis*, 1985, 323.	103
85S490	J. Stadlwieser; *Synthesis*, 1985, 490.	49
85S496	F. Huet; *Synthesis*, 1985, 496.	196
85S497	H. Suzuki, M. Yoshinaga, K. Takaoka and Y. Hiroi; *Synthesis*, 1985, 497.	872
85S556	R. H. Smithers; *Synthesis*, 1985, 556.	767
85S676	I. Yamamoto, T. Sakai, S. Yamamoto, K. Ohta and K. Matsuzaki; *Synthesis*, 1985, 676.	773, 781, 783
85S717	R. Anderson; *Synthesis*, 1985, 717.	506, 646
85S751	D. G. Norman, C. B. Reese and H. T. Serafinowska; *Synthesis*, 1985, 751.	190
85S909	A. Cornelis and P. Laszlo; *Synthesis*, 1985, 909.	431
85S958	Y.-A. Heus-Kloos, R. L. P. de Jong, H. D. VerKruijsse, L. Brandsma and S. Julia; *Synthesis*, 1985, 958.	843
85S973	B. Unterhalt and R. Mohr; *Synthesis*, 1985, 973.	447
85S982	T. Imamoto and H. Koto; *Synthesis*, 1985, 982.	122
85S1062	R. W. Saalfrank, F. Schütz and V. Moenius; *Synthesis*, 1985, 1062.	881
85S1148	F. Fülöp and G. Bernáth; *Synthesis*, 1985, 1148.	416, 417
85SC225	R. J. Bass and D. W. Gordon; *Synth. Commun.*, 1985, **15**, 225.	211
85SC829	K. K. Singal and B. Singh; *Synth. Commun.*, 1985, **15**, 829.	422
85SRI321	C. H. Yoder, S. L. Tesno, S. M. Heaney and C. Bohan; *Synth. React. Inorg. Met.-Org. Chem.*, 1985, **15**, 321.	140
85T837	M. Reuman and A. I. Meyers; *Tetrahedron*, 1985, **41**, 837.	306
85T2133	T. Takata, Y. Tamura and W. Ando; *Tetrahedron*, 1985, **41**, 2133.	226
85T2527	S. Pietrre, Z. Janousek, R. Merenyi and H. G. Viehe; *Tetrahedron*, 1985, **41**, 2527.	91, 287, 858, 861
85T3529	A. Padwa, Y.-Y. Chen, U. Chiacchio and W. Dent; *Tetrahedron*, 1985, **41**, 3529.	509
85T3559	R. Smith and T. Livinghouse; *Tetrahedron*, 1985, **41**, 3559.	518
85T4057	P. Martin, E. Steiner, J. Streith, T. Winkler and D. Belluš; *Tetrahedron*, 1985, **41**, 4057.	12
85T4183	C. de Cock, S. Piettre, F. Lahousse, Z. Janousek, R. Merenyi and H. G. Viehe; *Tetrahedron*, 1985, **41**, 4183.	92
85T4793	M. Clarembeau, A. Cravador, W. Dumont, L. Hevesi, A. Krief, J. Lucchetti and D. Van Ende; *Tetrahedron*, 1985, **41**, 4793.	287, 288
85T4861	D. Seebach, G. Calderari and P. Knochel; *Tetrahedron*, 1985, **41**, 4861.	335
85T4979	J. Zygmunt; *Tetrahedron*, 1985, 4979.	130
85T5507	D. H. R. Barton, H. Togo and S. Z. Zard; *Tetrahedron*, 1985, **41**, 5507.	327
85T5525	J. G. Schantl, H. Gstach, P. Hebeisen and N. Lanznaster; *Tetrahedron*, 1985, **41**, 5525.	441
85TL47	S.-F. Chen and P. S. Mariano; *Tetrahedron Lett.*, 1985, **26**, 47.	518
85TL139	P. F. Hudrlik, A. M. Hudrlik and A. K. Kulkarni; *Tetrahedron Lett.*, 1985, **26**, 139.	508
85TL169	P. T. Lansbury, G. E. Bebernitz, S. C. Maynard and C. J. Spagnuolo; *Tetrahedron Lett.*, 1985, **26**, 169.	961
85TL301	E. Sternberg and P. Binger; *Tetrahedron Lett.*, 1985, **26**, 301.	649
85TL377	M. J. Pearson; *Tetrahedron Lett.*, 1985, **26**, 377.	426
85TL411	E. Van Hulsen and D. Hoppe; *Tetrahedron Lett.*, 1985, **26**, 411.	384
85TL695	M. J. O'Donnell, W. D. Bennett and R. L. Polt; *Tetrahedron Lett.*, 1985, **26**, 695.	445
85TL1115	R. L. Kreeger, P. R. Menard, E. A. Sans and H. Shechter; *Tetrahedron Lett.*, 1985, **26**, 1115.	352
85TL1425	E. Block and A. Wall; *Tetrahedron Lett.*, 1985, **26**, 1425.	368
85TL1581	R. Dasgupta and U. R. Ghatak; *Tetrahedron Lett.*, 1985, **26**, 1581.	193
85TL1927	E. Juaristi, B. Gordillo and D. M. Aparicio; *Tetrahedron Lett.*, 1985, **26**, 1927.	230
85TL1931	R. A. Moss and D. P. Cox; *Tetrahedron Lett.*, 1985, **26**, 1931.	1081
85TL2065	R. R. Schmidt and M. Maier; *Tetrahedron Lett.*, 1985, **26**, 2065.	176
85TL2101	K. Utimoto, K. Imi, H. Shiragami, S. Fujikura and H. Nozaki; *Tetrahedron Lett.*, 1985, **26**, 2101.	654
85TL2131	B. F. Bonini, E. Foresti, G. Maccagnani, G. Mazzanti, P. Sabatino and P. Zani; *Tetrahedron Lett.*, 1985, **26**, 2131.	232
85TL2279	P. Dowd, C. Kaufman, P. Kaufman and Y. H. Paik; *Tetrahedron Lett.*, 1985, **26**, 2279.	162
85TL2283	P. Dowd, C. Kaufman, Y. H. Paik; *Tetrahedron Lett.*, 1985, **26**, 2283.	162
85TL2391	R. H. Schlessinger, M. A. Poss, S. Richardson and P. Lin; *Tetrahedron Lett.*, 1985, **26**, 2391.	649
85TL2445	T. Tsushima and K. Kawada, *Tetrahedron Lett.*, 1985, **26**, 2445.	8
85TL2769	H. J. Bestmann, R. Dötzer and J. Manero-Alvarez; *Tetrahedron Lett.*, 1985, **26**, 2769.	571
85TL2861	B. W. Metcalf, E. T. Jarvi and J. P. Burkhart; *Tetrahedron Lett.*, 1985, **26**, 2861.	731, 949
85TL2965	T. Cohen, J. P. Sherbine, S. A. Mendelson and M. Myers; *Tetrahedron Lett.*, 1985, **26**, 2965.	365, 366
85TL3011	Y. Terao, M. Tanaka, N. Imai and K. Achiwa; *Tetrahedron Lett.*, 1985, **26**, 3011.	369
85TL4435	E. E. Aboujaoude, S. Liétge, N. Collignon, M.-P. Teulade and P. Savignac; *Tetrahedron Lett.*, 1985, **26**, 4435.	557
85TL5001	J. D. Buynak, M. N. Rao, R. Y. Chandrasekaran, E. Haley, P. de Meester and S. C. Chu; *Tetrahedron Lett.*, 1985, **26**, 5001.	514
85TL5411	A. S. Kende, and D. J. Wustrow; *Tetrahedron Lett.*, 1985, **26**, 5411.	208
85TL5547	Y. Sato, S.-I. Ninomiya, F.-Z. Liu, N. Shirai and Y. Kawazoe; *Tetrahedron Lett.*, 1985, **26**, 5547.	521
85TL5827	S. Katsumura, S. Fujiwara and S. Isoe; *Tetrahedron Lett.*, 1985, **26**, 5827.	387
85TL6205	Y. Arai, S.-I. Kuwayama, Y. Takeuchi and T. Koizumi; *Tetrahedron Lett.*, 1985, **26**, 6205.	285

85TL6301	S. W. McCombie, B. B. Shankar and A. K. Ganguly; *Tetrahedron Lett.*, 1985, **26**, 6301.	961
85TL6329	H. Altenbach, W. Holzapfel, G. Smerat and S. H. Finkler; *Tetrahedron Lett.*, 1985, 6329.	173
85TL6513	L. Hevesi and K. M. Nsunda; *Tetrahedron Lett.*, 1985, **26**, 6513.	289
85USP4499082	A. B. Shenvi and C. A. Kettner; *US Pat.* 4 499 082 (1985) (*Chem. Abstr.*, 1985, **193**, 71 709u).	528
85USP4501695	V. H. Van Rheenen, *US Pat.* 4 501 695 (1985).	1003, 1004, 1011
85ZAAC(523)7	K. Issleib, U. Kuhne and F. Krech; *Z. Anorg. Allg. Chem.*, 1985, **523**, 7.	458, 477, 481
85ZC172	K. Issleib, E. Leissring and M. Riemer; *Z. Chem.*, 1985, **25**, 172.	453
85ZC369	J. Heinicke, A. Petrasch and A. Tzschach; *Z. Chem.*, 1985, **25**, 369.	497, 498, 499, 500
85ZN(B)182	H. Martin and H. Bretinger; *Z. Naturforsch., Teil B*, 1985, **40**, 182.	698
85ZN(B)352	H. Schmidbaur and R. Pichl; *Z. Naturforsch., Teil B*, 1985, **40**, 352.	579
85ZN(B)968	L. Weber, D. Wewers and E. Lücke; *Z. Naturforsch., Teil B*, 1985, **40**, 968.	580
85ZN(B)1077	F. Zurmühlen, W. Rösch and M. Regitz; *Z. Naturforsch., Teil B*, 1985, **40**, 1077.	565
85ZN(B)1152	U. Fischer and G. Hägele; *Z. Naturforsch., Teil B*, 1985, **40**, 1152.	553
85ZOB331	Z. S. Novikova, M. M. Kabachnik, E. A. Monin, A. A. Borisenko and I. F. Lutsenko; *Zh. Obshch. Khim.*, 1985, **55**, 331.	550
85ZOB838	M. N. Danchenko, I. Yu. Budilova and A. D. Sinitsa; *Zh. Obshch. Khim.*, 1985, **55**, 838.	927
85ZOB1234	I. M. Aladzheva, P. V. Petrovskii, Z. S. Klemenkova, B. V. Lokshin, T. A. Mastryukova and M. I. Kabachnik; *Zh. Obshch. Khim.*, 1985, **55**, 1234.	557
85ZOB1471	B. A. Arbuzov, E. N. Dianova and E. Ya. Zabotina; *Zh. Obshch. Khim.*, 1985, **55**, 1471.	549
85ZOB1629	V. M. Vdovin, N. B. Bespalova, V. S. Nitikin, M. B. Sergeeva, Y. A. Strelenko and D. A. Ivashchenko; *Zh. Obshch. Khim.*, 1985, **55**, 1629 (*Chem. Abstr.*, 1986, **104**, 149 046t).	658
85ZOB1865	R. K. Ismagilov, A. B. Zadornaya and V. V. Moskva; *Zh. Obshch. Khim.*, 1985, **55**, 1865.	559
85ZOB2396	D. A. Bravo-Zhivotovskii, S. D. Pigarev, O. A. Vyazankina and N. S. Vyazankin; *Zh. Obshch. Khim.*, 1985, **55**, 2396 (*Chem. Abstr.*, 1985, **105**, 153 160p).	629, 638
85ZOB2801	S. V. Ponomarev, S. N. Nikolaeva, A. S. Kostyuk and K. Y. Grishin; *Zh. Obshch. Khim.*, 1985, **55**, 2801 (*Chem. Abstr.*, 1985, **106**, 5123).	1057
85ZOR244	S. I. Radchenko, V. Ya. Komarov and B. I. Ionin; *Zh. Org. Khim*, 1985, **21**, 244 (*Chem. Abstr.*, 1985, **102**, 184 510v).	786
85ZOR965	B. A. Shainyan, A. N. Mirskova and V. Yu. Vitkovski; *Zh. Org. Khim.*, 1985, **21**, 965.	783
85ZOR1111	V. V. Perekalin; *Zh. Org. Khim.*, 1985, 21, 1111 (Engl. Transl. 1011).	994
85ZOR2467	A. V. Martynov, A. N. Mirskova and M. G. Voronkov; *Zh. Org. Khim.*, 1985, **21**, 2467 (*Chem. Abstr.*, 1986, **105**, 190 588c).	786
86ACH145	G. Matolcsy, K. Bauer, A. Pal, I. Ujvary, I. Belai, A. Gerlei, M. Kardos, P. Sohar and I. Pelczer; *Acta Chim. Acad. Sci. Hung.*, 1986, **123**, 145.	120
86ACR244	R. M. Moriarty and O. Prakash; *Acc. Chem. Res.*, 1986, **19**, 244.	184
86ACR250	R. R. Schmidt; *Acc. Chem. Res.*, 1986, **19**, 250.	202
86AG1020	G. Markl, H. Sejpka, S. Dietl, B. Nuber and M. L. Ziegler; *Angew. Chem.*, 1986, **98**, 1020.	1047, 1059
86AG(E)160	D. Hoppe and T. Kramer; *Angew. Chem., Int. Ed. Engl.*, 1986, **25**, 160.	382
86AG(E)165	G. E. Herberich, B. Hessner, J. A. K. Howard, D. P. J. Köffer and R. Saive; *Angew. Chem., Int. Ed. Engl.*, 1986, **25**, 165.	717
86AG(E)178	D. Seebach, R. Imwinkelried and G. Stucky; *Angew. Chem., Int. Ed. Eng.*, 1986, **25**, 178.	213
86AG(E)212	R. R. Schmidt; *Angew. Chem., Int. Ed. Eng.*, 1986, **25**, 212.	198
86AG(E)253	A. Schmidpeter and S. Lochschmidt; *Angew. Chem., Int. Ed. Engl.*, 1986, **25**, 253.	550
86AG(E)345	R. Fitzi and D. Seebach; *Angew. Chem., Int. Ed. Engl.*, 1986, **25**, 345.	416
86AG(E)348	H. Schmidbaur and T. Pollok; *Angew. Chem., Int. Ed. Engl.* 1986, **25**, 348.	1023
86AG(E)559	H. J. Bestmann and T. Arenz; *Angew. Chem., Int. Ed. Engl.*, 1986, **25**, 559.	574
86AG(E)571	C. Elschenbroich, J. Kroker, W. Massa, M. Wunsch and A. J. Ashe, III; *Angew. Chem., Int. Ed. Engl.*, 1986, **25**, 571.	1041
86AG(E)734	W. Beck, H.-J. Müller and U. Nagel; *Angew. Chem., Int. Ed. Engl.* 1986, **25**, 734.	710
86AG(E)909	T. Kauffmann, R. Abeln, S. Welke and D. Wingbermühle; *Angew. Chem., Int. Ed. Engl.*, 1986, **25**, 909.	673
86AG(E)925	K. Jonas, W. Rüsseler, C. Krüger and E. Raabe; *Angew. Chem., Int. Ed. Engl.*, 1986, **25**, 925.	671
86AG(E)999	R. Gleiter, G. Krennrich and M. B. Rubin; *Angew. Chem., Int. Ed. Eng.*, 1986, **25**, 999.	162
86AJC1363	M. A. Bennett and G. T. Crisp; *Aust. J. Chem.*, 1986, **39**, 1363.	150
86AP(319)161	H. Mertens, R. Troschutz and H. J. Roth; *Arch. Pharm (Weinheim, Ger.)*, 1986, **319**, 161.	900
86AP(319)954	G. Luputiu and F. Moll; *Arch. Pharm. (Weinheim, Ger.)*, 1986, **319**, 954.	98
86AP(319)1098	K. Hartke and M. Fallert; *Arch. Pharm. (Weinheim, Ger.)*, 1986, **319**, 1098.	984
86BCJ453	T. Okuyama, W. Fujiwara and T. Fueno; *Bull. Chem. Soc. Jpn.*, 1986, **59**, 453.	1074
86BCJ659	M. Hoshi, Y. Masuda and A. Arase; *Bull. Chem. Soc. Jpn.*, 1986, **59**, 659.	1056, 1061
86BCJ1509	Y. Nakadaira, T. Kobayashi and H. Sakurai; *Bull. Chem. Soc. Jpn.*, 1986, **59**, 1509.	616
86BCJ2463	T. Satoh, Y. Kaneko and K. Yamakawa; *Bull. Chem. Soc. Jpn.*, 1986, **59**, 2463.	233
86BCJ2537	O. Tsuge, S. Kanemasa, A. Hatada and K. Matsuda; *Bull. Chem. Soc. Jpn.*, 1986, **59**, 2537.	512, 518
86BCJ2909	M. Yokoyama, K. Kumata, H. Hatanaka and T. Shiraishi; *Bull. Chem. Soc. Jpn.*, 1986, **59**, 2909.	872
86BCJ3293	T. Oshikawa and M. Yamashita; *Bull. Chem. Soc. Jpn.*, 1986, **59**, 3293.	924
86BCJ3625	T. Umemoto and G. Tomizawa; *Bull. Chem. Soc. Jpn.*, 1986, **59**, 3625.	62, 63
86BSF413	M. Bordeau, S. M. Djamei and J. Dunogues; *Bull. Soc. Chim. Fr.*, 1986, 413.	140
86BSF470	J. Villieras, B. Kirschleger, R. Tarhouni and M. Rambaud; *Bull. Soc. Chim. Fr.*, 1986, 470.	139
86BSF937	Y. Yamazaki and N. Ishikawa; *Bull. Soc. Chim. Fr.*, 1986, 937.	937

86CA(104)19636z	G. G. Minasyan, A. M. Torgomyan, M. Zh. Ovakimyan and M. G. Indzhikyan; *Chem. Abstr.*, 1986, **104**, 19 636z (*Arm. Khim. Zh.*, 1985, **38**, 204).	914
86CAR(148)25	N. A. L. Al-Masoudi and N. A. Hughes; *Carbohydr. Res.*, 1986, **148**, 25.	227
86CB162	J. Goerdeler, M. Yunis, H. Puff and A. Roloff; *Chem. Ber.*, 1986, **119**, 162.	979
86CB472	F. M. Atta, R. Betz, B. Schmid and R. R. Schmidt; *Chem. Ber.*, 1986, **119**, 472.	927
86CB535	R. Appel, E. Gaitzsch, K.-H. Dunker and F. Knoch; *Chem. Ber.*, 1986, **119**, 535.	568
86CB1039	R. W. Hoffmann and B. Landmann; *Chem. Ber.*, 1986, **119**, 1039.	363, 373
86CB1725	G. Picotin and P. Miginiac; *Chem. Ber.*, 1986, **119**, 1725.	192
86CB1745	W. Ried and M. A. Jacobi; *Chem. Ber.*, 1986, **119**, 1745.	901
86CB1977	R. Appel, T. Gaitzsch, F. Knoch and G. Lenz; *Chem. Ber.*, 1986, **119**, 1977.	146
86CB2150	T. Kauffmann and P. Schwartze; *Chem. Ber.*, 1986, **119**, 2150.	577, 578, 579, 581
86CB2208	Z. Huang and L. Tzai; *Chem. Ber.*, 1986, **119**, 2208.	973
86CB2387	G. Kaupp and D. Matthies; *Chem. Ber.*, 1986, **119**, 2387.	101
86CB2466	R. Appel, V. Winkhaus and F. Knoch; *Chem. Ber.*, 1986, **119**, 2466.	1026
86CB2832	H. Schmidbaur and S. Schnatterer; *Chem. Ber.*, 1986, **119**, 2832.	550
86CB3150	H. W. Roesky, H. Plenio, K. Keller, M. Noltemeyer and G. M. Sheldrake; *Chem. Ber.*, 1986, **119**, 3150.	800
86CB3253	B. Glaser and H. Nöth; *Chem. Ber.* 1986, **119**, 3253.	642
86CC156	S. Takano, S. Sato, E. Goto and K. Ogasawara; *J. Chem. Soc., Chem. Commun.*, 1986, 156.	168
86CC285	D. G. Harrison and S. R. Stobart; *J. Chem. Soc., Chem. Commun.*, 1986, 285.	685
86CC542	G. R. Doel, N. D. Feasey, S. A. R. Knox, A. G. Orpen and J. Webster; *J. Chem. Soc., Chem. Commun.*, 1986, 542.	675
86CC871	D. V. Griffiths and J. C. Tebby; *J. Chem. Soc., Chem. Commun.*, 1986, 871.	558
86CC882	A. M. Herring, S. J. Higgins, G. B. Jacobsen and B. L. Shaw; *J. Chem. Soc., Chem. Commun.*, 1986, 882.	548, 1023, 1028
86CC1041	J. L. Bookham, W. McFarlane and I. J. Colquhoun; *J. Chem. Soc., Chem. Commun.*, 1986, 1041.	548
86CC1154	G. L. Crocco and J. A. Gladysz; *J. Chem. Soc., Chem. Commun.*, 1986, 1154.	581
86CC1191	A. Berlin, S. Bradamante, R. Ferraccioli and G. A. Pagani; *J. Chem. Soc., Chem. Commun.*, 1986, 1191.	1076
86CC1208	R. Laï, S. LeBot, A. Baldy, M. Pierrot and H. Arzoumanian; *J. Chem. Soc., Chem. Commun.*, 1986, 1208.	580
86CC1794	M. Iyoda, H. Otani, M. Oda, Y. Kai, Y. Baba and N. Kasai; *J. Chem. Soc., Chem. Commun.*, 1986, 1794.	820
86CI(L)490	J. R. Anacona, P. B. Davies and A. H. Ferguson; *Chem. Ind. (London)*, 1986, 490.	30
86CJC1949	P. Wan and S. Muralidharan; *Can. J. Chem.*, 1986, **64**, 1949.	1073
86CL181	T. Nakajima, H. Miyaji, M. Segi and S. Suga; *Chem. Lett.*, 1986, **9**, 181.	931
86CL311	N. Okada, G. Saito and T. Mori; *Chem. Lett.*, 1986, 311.	877
86CL1193	S. Tomoda, Y. Matsumoto, Y. Takeuchi and Y. Nomura; *Chem. Lett.*, 1986, 1193.	516
86CL1623	Y. Hashimoto and T. Mukaiyama; *Chem. Lett.*, 1986, 1623.	850
86CL1655	K. Fuji, Y. Usami, K. Sumi, M. Ueda and K. Kajiwara; *Chem. Lett.*, 1986, 1655.	235, 236
86CL1861	N. Okada, H. Yamochi, F. Shinozaki, K. Oshima and G. Saito; *Chem. Lett.*, 1986, 1861.	873
86CL1895	T. Ishihara, M. Kurasoshi and Y. Okada; *Chem. Lett.*, 1986, 1895.	810
86CL2089	Y. Terao, M. Aono, I. Takahashi and K. Achiwa; *Chem. Lett.*, 1986, 2089.	937
86CPB2362	T. Sakamoto, Y. Kondo and H. Yamanka; *Chem. Pharm. Bull.*, 1986, **34**, 2362.	196
86CPB3273	S.-I. Ninomiya, F.-Z. Liu, H. Nakagawa, K. Kohda, Y. Kawazoe and Y. Sato; *Chem. Pharm. Bull.*, 1986, **34**, 3273.	521
86DOK(286)19	N. A. Kardanov, A. M. Timofeev, N. N. Godovikov, P. V. Petrovskii and M. I. Kabachnik; *Dokl. Akad. Nauk SSSR*, 1986, **286**, 19.	123
86EGP238612	G. Lang, E. Herrmann and I. Schubert; *Ger. (East) Pat.* 238 612 (1986) (*Chem. Abstr.*, 1987, **106**, 156 674).	556
86EUP174832A2	A. H. Al-Shaar, B. J. Broughton, R. K. Chambers and D. W. Gilmour (May and Baker Ltd); *Eur. Pat.* 174 832 A2 (1986) (*Chem. Abstr.*, 1986, **105**, 115 043).	866
86G229	L. Forlani, M. Sintoni and P. E. Todesco; *Gazz. Chim. Ital.*, 1986, **116**, 229.	405
86G471	L. Pandolfo, M. Bressan and G. Paiaro; *Gazz. Chim. Ital.*, 1986, **116**, 471.	701
86GEP3444678	G. Hägele, U. Fischer and H. Blum; *Ger. Pat.* 3 444 678 (1986) (*Chem. Abstr.*, 1986, **105**, 172 736).	553
86H(24)25	Y. Nitta, T. Yamaguchi and T. Tanaka; *Heterocycles*, 1986, **24**, 25.	426
86H(24)1571	Y. Terao, M. Aono and K. Achiwa; *Heterocycles*, 1986, **24**, 1571.	369
86HCA881	R. W. Lang; *Helv. Chim. Acta*, 1986, **69**, 881.	14
86IC376	W. A. Kamil, F. Hapsel-Hentrich and J. M. Shreeve; *Inorg. Chem.*, 1986, **25**, 376.	140
86IC611	R. K. Chadha, J. M. Chehayber and J. E. Drake; *Inorg. Chem.*, 1986, **25**, 611.	377
86IC712	A.-M. Caminade, E. Ocando, J.-P. Majoral, M. Cristante and G. Bertrand; *Inorg. Chem.*, 1986, **25**, 712.	552
86IC1290	C. King, D. M. Roundhill and F. R. Fronczek; *Inorg. Chem.*, 1986, **25**, 1290.	562
86IC4309	G. T. King and N. E. Miller; *Inorg. Chem.*, 1986, **25**, 4309.	143
86ICA51	J. M. Chehayber and J. E. Drake; *Inorg. Chim. Acta*, 1986, **111**, 51.	377, 401
86ICA(115)L29	D. R. Martin, C. M. Merkel and J. P. Ruiz; *Inorg. Chim. Acta*, 1986, **115**, L29.	552
86ICA(119)177	J. R. Moss, M. L. Niven and P. M. Stretch; *Inorg. Chim. Acta*, 1986, **119**, 177.	148
86IS95	C. Tessier-Youngs and O. T. Beachley, Jr.; *Inorg. Synth.*, 1986, **24**, 95.	652
86IS110	R. Appel; *Inorg. Synth.*, 1986, **24**, 110.	566, 568
86IZV91	V. F. Snegirev and K. N. Makarov; *Izv. Akad. Nauk SSSR, Ser. Khim.*, 1986, 91.	731

86IZV640	B. A. Arbuzov, O. A. Erastov, A. Slonkin and S. N. Ignat'eva; *Izv. Akad. Nauk SSSR, Ser. Khim.*, 1986, 640.	343
86IZV810	A. V. Fokin, A. I. Rapkin, V. G. Chilikin, O. V. Verenikin and Yu. N. Studnev; *Izv. Akad. Nauk SSSR, Ser. Khim.*, 1986, 810.	104
86IZV2344	N. P. Gambaryan, D. P. Del'tsova, V. A. Livshits and E. G. Ter-Gabrielyan; *Izv. Akad. Nauk. SSSR, Ser. Khim.*, 1986, 2344 (Engl. Transl. 2147).	979
86IZV2502	B. A. Arbuzov, G. N. Nikonov, A. S. Balueva and O. A. Erastov; *Izv. Akad. Nauk SSSR, Ser. Khim.*, 1986, 2502.	342
86JA40	J. W. Bruno, G. M. Smith, T. J. Marks, C. K. Fair, A. J. Schultz and J. M. Williams; *J. Am. Chem. Soc.*, 1986, **108**, 40.	656, 658
86JA425	C. M. Fendrick and T. J. Marks; *J. Am. Chem. Soc.*, 1986, **108**, 425.	589, 656, 658
86JA1455	J. C. Selover, G. D. Vaughn, C. E. Strouse and J. A. Gladysz; *J. Am. Chem. Soc.*, 1986, **108**, 1455.	150
86JA2096	W. M. Lamanna; *J. Am. Chem. Soc.*, 1986, **108**, 2096.	694
86JA3122	P. R. Jones and T. F. Bates; *J. Am. Chem. Soc.*, 1986, **108**, 3122.	611
86JA3402	E. Negishi and T. Takahashi; *J. Am. Chem. Soc.*, 1986, **108**, 3402.	808, 1059
86JA4229	D. J. Burton and S. W. Hansen; *J. Am. Chem. Soc.*, 1986, **108**, 4229.	821
86JA4568	E. Block, M. Aslam, V. Eswarakrishnan, K. Gebreyes, J. Hutchinson, R. Iyer, J.-A. Lafitte and A. Wall; *J. Am. Chem. Soc.*, 1986, **108**, 4568.	82, 83, 85
86JA5359	R. R. Burch and J. C. Calabrese; *J. Am. Chem. Soc.*, 1986, **108**, 5359.	152
86JA5543	K. Llamas, M. Owens, R. L. Blakeley and B. Zerner; *J. Am. Chem. Soc.* 1986, **108**, 5543.	892, 979
86JA5949	B. Halton, C. J. Randall, G. J. Gainsford and P. J. Stang; *J. Am. Chem. Soc.*, 1986, **108**, 5949.	604, 647
86JA6805	G. M. Smith, J. D. Carpenter and T. J. Marks; *J. Am. Chem. Soc.*, 1986, **108**, 6805.	658
86JA7739	P. J. Wagner, M. J. Thomas and A. E. Puchalski; *J. Am. Chem. Soc.*, 1986, **108**, 7739.	3, 6
86JA7964	P. J. Desrosiers, R. S. Shinomoto and T. C. Flood; *J. Am. Chem. Soc.*, 1986, **108**, 7964.	656
86JCR(M)2843	F. Gavina, S. V. Luis, P. Ferrer, A. M. Costero and J. A. Marco; *J. Chem. Res. (M)*, 1986, 2843.	749
86JCS(D)603	S. I. Bailey, D. Colgan, L. M. Engelhardt, W-P. Leung, R. I. Papasergio, R. L. Raston and A. H. White; *J. Chem. Soc., Dalton Trans.*, 1986, 603.	647, 656
86JCS(D)657	L. Brammer, M. Green, A. G. Orpen, K. E. Paddick and D. R. Saunders; *J. Chem. Soc., Dalton Trans.*, 1986, 657.	676
86JCS(P1)501	R. S. Gairns, C. J. Moody and C. W. Rees; *J. Chem. Soc., Perkin Trans. 1*, 1986, 501.	856
86JCS(P1)683	M. J. Prior and G. H. Whitham; *J. Chem. Soc., Perkin Trans. 1*, 1986, 683.	357
86JCS(P1)733	R. Wolf and E. Steckhan; *J. Chem. Soc. Perkin Trans. 1*, 1986, 733.	743
86JCS(P1)913	G. M. Blackburn and D. E. Kent; *J. Chem. Soc., Perkin Trans. 1*, 1986, 913.	119, 129
86JCS(P1)1077	C. L. Branch and M. J. Pearson; *J. Chem. Soc., Perkin Trans. 1*, 1986, 1077.	447
86JCS(P1)1171	B. R. Fishwick, D. K. Rowles and C. J. M. Stirling; *J. Chem. Soc., Perkin Trans. 1*, 1986, 1171.	798
86JCS(P1)1351	Y. Ueno, O. Moriya, K. Chino, M. Watanabe and M. Okawara; *J. Chem. Soc., Perkin Trans. 1*, 1986, 1351.	187
86JCS(P1)1417	G. M. Blackburn and M. J. Parratt; *J. Chem. Soc., Perkin Trans. 1*, 1986, 1417.	801
86JCS(P1)1425	G. M. Blackburn and M. J. Parratt; *J. Chem. Soc., Perkin Trans. 1*, 1986, 1425.	802
86JCS(P1)1495	M. K. Shepherd; *J. Chem. Soc., Perkin Trans. 1*, 1986, 1495.	21
86JCS(P1)1681	M. J. P. Harger and A. Williams; *J. Chem. Soc., Perkin Trans. 1*, 1986, 1681.	94
86JCS(P1)2207	B. Hanlon and D. I. John; *J. Chem. Soc., Perkin Trans. 1*, 1986, 2207.	90
86JFC(31)405	K. C. Eapen, K. J. Eisentraut, M. T. Ryan and C. Tamborski; *J. Fluorine Chem.*, 1986, **31**, 405.	768
86JFC(32)461	R. E. Banks, R. A. Du Boisson and E. Tsiliopoulos; *J. Fluorine Chem.*, 1986, **32**, 461.	108
86JGC425	R. Yu. Yurchenko, E. E. Lavrova and A. G. Yurchenko; *J. Gen. Chem. USSR*, 1986, **56**, 425.	337
86JGU501	I. I. Patsanovskii, É. A. Ishmaeva, E. N. Sundukova, A. N. Yarkevich and E. N. Tsvetkov; *J. Gen. Chem. USSR (Engl. Transl.)*, 1986, **56**, 501.	119
86JGU632	S. A. Terent'eva, M. A. Pudovik and A. N. Pudovik; *J. Gen. Chem. USSR (Engl. Transl.)*, 1986, **56**, 632.	478
86JGU1258	M. V. Livantsov, A. A. Prishchenko and I. F. Lutsenko; *J. Gen. Chem. USSR (Engl. Transl.)*, 1986, **56**, 1258.	126
86JGU1430	A. N. Yarkevich, S. E. Tkachenko and E. N. Tsvetkov; *J. Gen. Chem. USSR (Engl. Transl.)*, 1986, **56**, 1430.	119
86JGU2144	A. V. Golovanov, I. G. Maslennikov, A. E. Medvedev and A. N. Lavrent'ev; *J. Gen. Chem. USSR (Engl. Transl.)*, 1986, **56**, 2144.	126
86JGU2242	A. V. Golovanov, I. G. Maslennikov, I. V. Gudina, V. V. Lebedev and A. N. Lavrent'ev; *J. Gen. Chem. USSR (Engl. Transl.)*, 1986, **56**, 2242.	126
86JGU2246	A. V. Golovanov, I. G. Maslennikov, V. B. Lebedev, I. V. Gudina and A. N. Lavrent'ev; *J. Gen. Chem. USSR (Engl. Transl.)*, 1986, **56**, 2246.	125
86JGU2394	V. A. Chauzov, Yu. N. Studnev, S. V. Agafonov and A. V. Fokin; *J. Gen. Chem. USSR (Engl. Transl.)*, 1986, **56**, 2394.	119
86JHC1163	H. Ishibashi, M. Okada, A. Akiyama, K. Nomura and M. Ikeda; *J. Heterocycl. Chem.*, 1986, **23**, 1163.	67
86JMC2241	W. K. Anderson and A. S. Milowsky; *J. Med. Chem.*, 1986, **29**, 2241.	517
86JOC117	E. Vedejs and R. G. Wilde; *J. Org. Chem.*, 1986, **51**, 117.	232
86JOC130	R. V. Hoffman, B. C. Jankowski and C. S. Carr; *J. Org. Chem.*, 1986, **51**, 130.	183
86JOC326	C. G. Krespan; *J. Org. Chem.*, 1986, **51**, 326.	767

86JOC508	K. Ogura, N. Yahata, M. Minoguchi, K. Ohtsuki, K. Takahashi and H. Iida; *J. Org. Chem.*, 1986, **51**, 508.	857
86JOC546	T. Shono, O. Ishige, H. Uyama and S. Kashimura; *J. Org. Chem.*, 1986, **51**, 546.	210
86JOC700	K. Ogura, T. Iihama, S. Kiuchi, T. Kajiki, O. Koshikawa, K. Takahashi and H. Iida; *J. Org. Chem.*, 1986, **51**, 700.	857
86JOC872	A. I. Meyers and T. R. Bailey; *J. Org. Chem.*, 1986, **51**, 872.	532
86JOC875	H. J. Cristau, B. Chabaud, R. Labaudiniere and H. Christol; *J. Org. Chem.*, 1986, **51**, 875.	875
86JOC879	S. Hackett and T. Livinghouse; *J. Org. Chem.*, 1986, **51**, 879.	216, 218, 834, 836
86JOC886	Y. Yamamoto, K. Maruyama, T. Komatsu and W. Ito; *J. Org. Chem.*, 1986, **51**, 886.	1057
86JOC891	S. Colonna, S. Banfi and R. Annunziata; *J. Org. Chem.*, 1986, **51**, 891.	262
86JOC902	S. N. Suryawanshi and P. L. Fuchs; *J. Org. Chem.*, 1986, **51**, 902.	174
86JOC951	M. P. Cooke, Jr.; *J. Org. Chem.*, 1986, **51**, 951.	202
86JOC1012	A. G. M. Barrett, G. S. Graboski and M. A. Russell; *J. Org. Chem.*, 1986, **51**, 1012.	906
86JOC1330	J. A. Soderquist and M. R. Najafi; *J. Org. Chem.*, 1986, **51**, 1330.	1056
86JOC1413	M. Ballester, J. Castaner, J. Riera and I. Tabernero; *J. Org. Chem.*, 1986, **51**, 1413.	764
86JOC1427	Y. Kamitori, M. Hojo, R. Masuda, T. Kimura and T. Yoshida; *J. Org. Chem.*, 1986, **51**, 1427.	247
86JOC1537	E. C. Taylor and H. M. L. Davies; *J. Org. Chem.*, 1986, **51**, 1537.	349
86JOC1610	D. P. Phillion, R. Neubauer and S. S. Andrew; *J. Org. Chem.*, 1986, **51**, 1610.	525
86JOC1704	T. V. RajanBabu, B. L. Chenard and M. A. Petti; *J. Org. Chem.*, 1986, **51**, 1704.	170
86JOC1997	O. Tsuge, S. Kanemasa and K. Matsuda; *J. Org. Chem.*, 1986, **51**, 1997.	979
86JOC2228	H. A. Bates, N. Condulis and N. L. Stein; *J. Org. Chem.*, 1986, **51**, 2228.	414
86JOC2276	D. Barillier and M. Vazeux; *J. Org. Chem.*, 1986, **51**, 2276.	233
86JOC2981	H. J. Reich, C. P. Jasperse and J. M. Renga; *J. Org. Chem.*, 1986, **51**, 2981.	71, 88, 288
86JOC3108	A. I. Meyers, T. Sohda and M. F. Loewe; *J. Org. Chem.*, 1986, **51**, 3108.	532
86JOC3123	T. Sheradsky and R. Moshenberg; *J. Org. Chem.*, 1986, **51**, 3123.	105
86JOC3325	C.-W. Chen and P. Beak; *J. Org. Chem.*, 1986, **51**, 3325.	138
86JOC3369	D. L. Fields, Jr. and H. Shechter; *J. Org. Chem.*, 1986, **51**, 3369.	138
86JOC3428	E. Block, J.-A. Laffitte and V. Eswarakrishnan; *J. Org. Chem.*, 1986, **51**, 3428.	223
86JOC3453	S. Spyroudis; *J. Org. Chem.*, 1986, **51**, 3453.	1081
86JOC3508	S. Sondej and J. A. Katzenellenbogen, *J. Org. Chem.*, 1986, **51**, 3508.	9
86JOC3718	M. G. Bock, R. M. Di Pardo and R. M. Friedinger; *J. Org. Chem.*, 1986, **51**, 3718.	417
86JOC3772	T. Hayashi, M. Konishi, Y. Okamoto, K. Kabeta and M. Kumada; *J. Org. Chem.*, 1986, **51**, 3772.	1033
86JOC3811	G. Balina, P. Kesler, J. Petre, D. Pham and A. Vollmar; *J. Org. Chem.*, 1986, **51**, 3811.	203
86JOC3897	S. Shimizu and M. Ogata; *J. Org. Chem.*, 1986, **51**, 3897.	520
86JOC4711	D. A. Goff, R. N. Harris, J. C. Bottaro and C. D. Bedford; *J. Org. Chem.*, 1986, **51**, 4711.	199
86JOC4964	W. G. Dauben, J. M. Gerdes and G. C. Look; *J. Org. Chem.*, 1986, **51**, 4964.	181
86JOC5498	G. A. Russell, W. Jiang, S. S. Hu and R. K. Khanna; *J. Org. Chem.*, 1986, **51**, 5498.	587
86JOM(299)C35	H. Suzuki, T. Tsukui and Y. Moro-Oka; *J. Organomet. Chem.*, 1986, **299**, C35.	657
86JOM(301)1	J. Mink, D. K. Breitinger and W. Kress; *J. Organomet. Chem.*, 1986, **301**, 1.	697
86JOM(301)137	D. C. Snyder; *J. Organomet. Chem.*, 1986, **301**, 137.	510
86JOM(302)1	H. Kleijn and P. Vermeer; *J. Organometal. Chem.*, 1986, **302**, 1.	960, 963
86JOM(303)73	A. Sebald, P. Seiberlich and B. Wrackmeyer; *J. Organomet. Chem.*, 1986, **303**, 73.	1062
86JOM(303)351	R. Sanehi, R. K. Bansal and R. C. Mehrotra; *J. Organomet. Chem.*, 1986, **303**, 351.	588
86JOM(304)283	M.-P. Teulade, P. Savignac, E. E. Aboujaoude, S. Liétge and N. Collignon; *J. Organomet. Chem.*, 1986, **304**, 283.	557, 576, 801
86JOM(306)39	W. Rösch, U. Vogelbacher, T. Allspach and M. Regitz, *J. Organomet. Chem.*, 1986, **306**, 39.	479, 480
86JOM(306)203	H. Fischer and J. Schmid; *J. Organomet. Chem.*, 1986, **306**, 203.	581
86JOM(307)157	A. Sebald and B. Wrackmeyer; *J. Organomet. Chem.*, 1986, **307**, 157.	1054, 1063
86JOM(307)231	C. S. Frampton, K. G. Ofori-Okai, R. M. G. Roberts and J. Silver; *J. Organomet. Chem.*, 1986, **307**, 231.	724
86JOM(308)1	B. J. J. van de Heisteeg, G. Schat, O. S. Akkerman and F. Bickelhaupt; *J. Organomet. Chem.*, 1986, **308**, 1.	669, 670
86JOM(308)153	G. E. Herberich, W. Boveleth, B. Hessner, D. P. J. Köffer, M. Negele and R. Saive; *J. Organomet. Chem.*, 1986, **308**, 153.	695
86JOM(309)C56	R. H. Dawson and A. K. Smith; *J. Organomet. Chem.*, 1986, **309**, C56.	1030
86JOM(309)225	I. J. B. Lin, L. T. C. Kao, F. J. Wu, G. H. Lee and Y. Wang; *J. Organomet. Chem.*, 1986, **309**, 225.	583, 585
86JOM(310)151	B. Wrackmeyer; *J. Organomet. Chem.*, 1986, **310**, 151.	661
86JOM(312)13	G. E. Herberich, U. Büschges, B. Hessner and H. Lüthe; *J. Organomet. Chem.*, 1986, **312**, 13.	695
86JOM(312)283	M.-P. Teulade, P. Savignac, E. E. Aboujaoude and N. Collignon; *J. Organomet. Chem.*, 1986, **312**, 283.	557
86JOM(314)13	H. G. M. Edwards; *J. Organomet. Chem.*, 1986, **314**, 13.	152
86JOM(315)C22	I. Moldes, J. Ros, R. Yañez, X. Solans, M. Font-Altaba and R. Mathieu; *J. Organomet. Chem.*, 1986, **315**, C22.	711
86JOM(315)59	M. Franck-Neumann, D. Martina and M.-P. Heitz; *J. Organomet. Chem.*, 1986, **315**, 59 (*Chem. Abstr.*, 1987, **107**, 198 587).	1076
86JOM(315)135	I. J. B. Lin, L. Hwan, H. C. Shy, M. C. Chen and Y. Wang; *J. Organomet. Chem.*, 1986, **315**, 135.	396
86JOM(315)255	C. Engelter, J. R. Moss, L. R. Nassimbeni, M. L. Niven, G. Reid and J. C. Spiers; *J. Organomet. Chem.*, 1986, **315**, 255.	584, 585

86JOM(316)C4	R. Hacker, P. von R. Schleyer, G. Reber, G. Mueller and L. Brandsma; *J. Organomet. Chem.*, 1986, **316**, C4.	647
86JOM(316)13	P. Coutrot, M. Youssefi-Tabrizi and C. Grison; *J. Organomet. Chem.*, 1986, **316**, 13.	577
86JOM(316)249	E. Lukevics, V. V. Dirnens, Y. S. Goldberg and E. Liepins; *J. Organomet. Chem.*, 1986, **316**, 249.	515
86JPS(A)1839	E. Isobe, T. Masuda, K. Higashimura and A. Yamamoto; *J. Polm. Sci., Polym. Chem.*, Part A, 1986, **24**, 1839.	607, 646
86LA533	M. Mittelbach and H. Junek; *Liebigs Ann. Chem.*, 1986, 533.	883
86M201	M. R. H. Elmoghayar and A. H. H. Elghandour; *Monatsh. Chem.*, 1986, **117**, 201.	984
B-86MI 404-01	R. C. Larock; "Solvomercuration/Demercuration Reactions in Organic Synthesis," Springer-Verlag, Berlin, 1986, p. 162.	171, 188, 194
B-86MI 407-01	S. Patai (ed.); "The Chemistry of Organic Selenium and Tellurium Compounds," Wiley, Chichester, 1986.	333
B-86MI 408-01	C. Paulmier; "Selenium Reagents and Intermediates in Organic Synthesis," Pergamon Press, Oxford, 1986, p. 256.	398
86MI 410-01	H. Kischkel, R. Franke and G.-V. Roeschenthaler; *Rev. Chim. Miner.*, 1986, **23**, 690.	455
86MI 417-01	M. I. Al-Hassan; *J. Chem. Soc. Pak.*, 1986, **8**, 379.	748, 757, 759
86MI 419-01	A. Kunungi, N. Takahashi and T. Hirai; *Mem. Fac. Eng., Osaka City Univ.*, 1986, **27**, 118 (*Chem. Abstr.*, 1988, **109**, 6168).	856
86MI 419-02	J. Jeczalik; *Chem. Inz. Chem.*, 1986, **16**, 171 (*Chem. Abstr.*, 1987, **107**, 23 316).	864
86MI 421-01	M. Ozaki and S. Kuwatsuka; *Nippon Noyaku Gakkaishi*, 1986, **11**, 427.	979
86MI 424-01	A. V. Germash, I. S. Yakupov, S. S. Zlotskii and D. L. Rakhmankulov; *Khim. Fiz.*, 1986, **5**, 1534 (*Chem. Abstr.*, 1987, **106**, 137 884).	1078
86MM2089	P. Wisian-Neilson, R. P. Ford, R. H. Neilson and A. K. Roy; *Macromolecules*, 1986, **19**, 2089.	570
86OM525	M. A. Khan, C. Peppe and D. G. Tuck; *Organometallics*, 1986, **5**, 525.	153, 698
86OM602	M. H. Chisholm, K. Folting, J. C. Huffman and N. S. Marchant; *Organometallics*, 1986, **5**, 602.	676
86OM630	W. L. Olson, D. A. Nagaki and L. F. Dahl; *Organometallics*, 1986, **5**, 630.	148
86OM805	C. H. W. Jones and R. D. Sharma; *Organometallics*, 1986, **5**, 805.	291
86OM877	F. Nief and J. Fischer; *Organometallics*, 1986, **5**, 877.	1036
86OM1197	S. Inoue and Y. Sato; *Organometallics*, 1986, **5**, 1197.	625, 637
86OM1320	I. R. Butler, W. R. Cullen and S. J. Rettig; *Organometallics*, 1986, **5**, 1320.	1034
86OM1551	J. T. B. H. Jastrzebski, G. Van Koten, C. T. Knaap, A. M. M. Schreurs, J. Kroon and A. L. Spek; *Organometallics*, 1986, **5**, 1551.	660
86OM1664	H. H. Karsch, A. Appelt and G. Müller; *Organometallics*, 1986, **5**, 1664.	588
86OM1906	D. J. Ager, G. E. Cooke, M. B. East, S. J. Mole, A. Rampersaud and V. J. Webb; *Organometallics*, 1986, **5**, 1906.	662, 962
86OM1991	T. N. Mitchell and W. Reimann; *Organometallics*, 1986, **5**, 1991.	746, 813, 1069
86OM2030	E. Lindner, P. Neese, W. Hiller and R. Fawzi; *Organometallics*, 1986, 2030.	112
86OM2171	K. J. Ahmed, M. H. Chisholm, K. Folting and J. C. Huffman; *Organometallics*, 1986, **5**, 2171.	675
86OM2220	C.-L. Lee, Y.-P. Yang, S. J. Rettig, B. R. James, D. A. Nelson and M. A. Lilga; *Organometallics*, 1986, **5**, 2220.	545
86OM2274	G. L. Larson, E. Torres, C. B. Morales and G. J. McGarvey; *Organometallics*, 1986, **5**, 2274.	1048, 1058, 1059, 1060, 1062
86OM2384	M. H. Chisholm, B. K. Conroy and J. C. Huffman; *Organometallics*, 1986, **5**, 2384.	676
86OM2413	J. H. Bieri, T. Egolf, W. von Philipsborn, U. Piantini, R. Prewo, U. Ruppli and A. Salzer; *Organometallics*, 1986, **5**, 2413.	683
86OM2457	M. H. Chisholm, B. K. Conroy, K. Folting, D. M. Hoffman and J. C. Huffman; *Organometallics*, 1986, **5**, 2457.	675
86OM2555	R. E. Cramer, J. H. Jeong and J. W. Gilje; *Organometallics*, 1986, **5**, 2555.	581
86OS(64)138	R. M. Moriarty, K.-C. Hou, I. Prakash and S. K. Arora; *Org. Synth.*, 1986, **64**, 138.	184
86OS(64)189	K. T. Potts, P. Ralli and G. Theodoridis; *Org. Synth.*, 1986, **64**, 189.	843
86RTC456	L. Ghosez, F. Sainte, M. Rivera, C. Bernard-Henriet and V. Gouverneur; *Recl. Trav. Chim. Pays-Bas*, 1986, **105**, 456.	897
86S31	T. Allspach, M. Regitz, G. Becker and W. Becker; *Synthesis*, 1986, 31.	484
86S122	E. Napolitano, R. Fiaschi and E. Mastrorilli; *Synthesis*, 1986, 122.	26
86S132	R. J. Schmitt and C. D. Bedford; *Synthesis*, 1986, 132.	807
86S139	W. Ando, T. Takata, L. Huang and Y. Tamura; *Synthesis*, 1986, 139.	318
86S198	E. N. Tsvetkov, N. A. Bondarenko, I. G. Malakhova and M. I. Kabachnik; *Synthesis*, 1986, 198.	546, 553
86S375	H. Chikashita, S. Ikegami, T. Okumura and K. Itoh; *Synthesis*, 1986, 375.	321
86S513	G. A. Olah, P. S. Iyer and G. K. Surya Prakash; *Synthesis*, 1986, 513.	161, 212
86S627	G. Barcelo, J.-P. Senet, G. Sennyey, J. Bensoam and A. Loffet; *Synthesis*, 1986, 627.	58, 59
86S638	F. Babudri and S. Florio; *Synthesis*, 1986, 638.	321
86S643	F. Ladhar, R. El Gharbi, M. Delmas and A. Gaset; *Synthesis*, 1986, 643.	425
86S649	Z. J. Kaminski and M. T. Leplawy; *Synthesis*, 1986, 649.	213
86S737	U. Schöllkopf, D. Pettig, U. Busse, E. Egert and M. Dyrbusch; *Synthesis*, 1986, 737.	536
86S804	A. R. Katritzky, M. Szajda and S. Bayyuk; *Synthesis*, 1986, 804.	317, 416
86S817	A. Vass and G. Szalontai; *Synthesis*, 1986, 817.	446
86S821	C. Yuan and Y. Qi; *Synthesis*, 1986, 821.	467
86S826	A. Amrollah-Madjdabadi, R. Beugelmans and A. Lechevallier; *Synthesis*, 1986, 826.	108

Ref	Citation	Pages
86S828	A. Amrollah-Madjdabadi, R. Beugelmans and A. Lechevallier; *Synthesis*, 1986, 828.	108
86S831	J. Drabowicz; *Synthesis*, 1986, 831.	74, 76
86S934	E. E. Aboujaoude, S. Liétjé, N. Collignon, M. P. Teulade and P. Savignac; *Synthesis*, 1986, 934.	570
86S967	X. Huang and B.-C. Chen; *Synthesis*, 1986, 967.	843
86S1013	M. Hojo, R. Masuda and E. Okada; *Synthesis*, 1988, 1013.	889
86SC233	Y. Arai, S. Kuwayama, Y. Takeuchi and T. Koizumi; *Synth. Commun.*, 1986, **16**, 233.	270, 854, 859, 860
86SC833	J. Kula; *Synth. Commun.*, 1986, **16**, 833.	203
86SC865	A. C. Brouwer and A. M. van Leusen; *Synth. Commun.*, 1986, **16**, 865.	517
86T1389	J. P. Foulon, M. Bourgain-Commerçon and J. F. Normant; *Tetrahedron*, 1986, **42**, 1389.	655
86T1845	T. Butkowskyj-Walkiw and G. Szeimies; *Tetrahedron*, 1986, **42**, 1845.	650
86T2863	M. Isobe, Y. Ichikawa, Y. Funabashi, S. Mio and T. Goto; *Tetrahedron*, 1986, **42**, 2863.	937
86T2979	R. S. Garigipati, R. Cordova, M. Parvez and S. M. Weinreb; *Tetrahedron*, 1986, **42**, 2979.	370
86T3323	B. M. Trost and H. Hiemstra; *Tetrahedron*, 1986, **42**, 3323.	220
86T3731	B. M. Dilworth and M. A. McKervey; *Tetrahedron*, 1986, **42**, 3731.	62, 66, 68
86T4223	J. E. Baldwin, R. M. Adlington, J. C. Bottaro, J. N. Kolhe, M. W. D. Perry and A. U. Jain; *Tetrahedron*, 1986, **42**, 4223.	536
86T4333	S. V. Ley, B. Lygo, F. Sternfeld and A. Wonnacott; *Tetrahedron*, 1986, **42**, 4333.	236
86T4443	J.-M. Vatele; *Tetrahedron*, 1986, **42**, 4443.	766
86TL69	S. Suzuki, Y. Fujita, Y. Kobayashi and F. Sato; *Tetrahedron Lett.*, 1986, **27**, 69.	183
86TL753	K. Jansson, T. Frejd, J. Kihlberg and G. Magnusson; *Tetrahedron Lett.*, 1986, **27**, 753.	176
86TL757	J. P. Cronin, B. M. Dilworth and M. A. McKervey; *Tetrahedron Lett.*, 1986, **27**, 757.	68
86TL795	K. M. Sadhu and D. S. Matteson; *Tetrahedron Lett.*, 1986, **27**, 795.	146
86TL947	K. Hayakawa, S. Ohsuki and K. Kanematsu; *Tetrahedron Lett.*, 1986, **27**, 947.	171
86TL1039	P. Knochel and J. F. Normant; *Tetrahedron Lett.*, 1986, **27**, 1039, 4427, 5727.	709, 724
86TL1043	P. Knochel and J. F. Normant; *Tetrahedron Lett.*, 1986, **27**, 1043.	398
86TL1145	J. Bellan, G. Etemad-Moghadam, M. Payard and M. Koenig; *Tetrahedron Lett.*, 1986, **27**, 1145.	547
86TL1557	S. Raucher and L. M. Gustavson; *Tetrahedron Lett.*, 1986, **27**, 1557.	202
86TL1653	A. N. Tischler and T. J. Lanza; *Tetrahedron Lett.*, 1986, **27**, 1653.	196
86TL1723	M. Clarembeau and A. Krief; *Tetrahedron Lett.*, 1986, **27**, 1723.	288
86TL1757	G. A. Flynn and E. L. Giroux; *Tetrahedron Lett.*, 1986, **27**, 1757.	1004
86TL1903	H. Keller, G. Maas and M. Regitz; *Tetrahedron Lett.*, 1986, **27**, 1903.	562
86TL2191	R. A. Holton and H.-B. Kim; *Tetrahedron Lett.*, 1986, **27**, 2191.	959
86TL2829	E. Negishi, F. E. Cederbaum and T. Takahashi; *Tetrahedron Lett.*, 1986, **27**, 2829.	1061
86TL2957	G. Märkl and H. Seitz; *Tetrahedron Lett.*, 1986, **27**, 2957.	502
86TL3173	T. Fuchikami, Y. Shibata and Y. Suzuki; *Tetrahedron Lett.*, 1986, **27**, 3173.	736
86TL3199	J. Brennan, F. H. S. Hussain and P. Virgili; *Tetrahedron Lett.*, 1986, **27**, 3199.	100
86TL3515	F. Texier-Boullet and M. Lequitte; *Tetrahedron Lett.*, 1986, 3515.	338
86TL3811	A. G. M. Barrett and M. A. Sturgess; *Tetrahedron Lett.*, 1986, **27**, 3811.	1019
86TL3831	D. S. Matteson and A. A. Kandil; *Tetrahedron Lett.*, 1986, **27**, 3831.	362
86TL3869	T. Fuchigami, Y. Nakagawa and T. Nonaka; *Tetrahedron Lett.*, 1986, **27**, 3869.	221
86TL4015	G. Märkl and R. Wagner; *Tetrahedron Lett.*, 1986, **27**, 4015.	628, 639
86TL4281	K. Kirschke, A. Moller, E. Schmitz, R. J. Kuban and B. Schulz; *Tetrahedron Lett.*, 1986, **27**, 4281.	989
86TL4339	T. Sato, H. Matsuoka, T. Igarashi and E. Murayama; *Tetrahedron Lett.*, 1986, **27**, 4339.	708
86TL4355	J. F. Cassidy and J. M. Williams; *Tetrahedron Lett.*, 1986, **27**, 4355.	235, 236
86TL4427	P. Knochel and J. F. Normant; *Tetrahedron Lett.*, 1986, **27**, 4427.	701
86TL4625	S. Cabiddu, C. Floris and S. Melis; *Tetrahedron Lett.*, 1986, **27**, 4625.	391
86TL4635	D. K. Devchand, A. W. Murray and E. Smeaton; *Tetrahedron Lett.*, 1986, **27**, 4635.	918, 1075
86TL5033	A. Pelter, G. Bugden, R. Pardasani and J. Wilson; *Tetrahedron Lett.*, 1986, **27**, 5033.	1077
86TL5611	P. Pellon and J. Hamelin; *Tetrahedron Lett.*, 1986, **27**, 5611.	567
86TL5829	J. Binder and E. Zbiral; *Tetrahedron Lett.*, 1986, **27**, 5829.	916
86TL5907	J. F. Hartwig, M. Jones, Jr., R. A. Moss and W. Lawrynowicz; *Tetrahedron Lett.*, 1986, **27**, 5907.	1083
86TL6201	P. Lesimple, J.-M. Beau, G. Jaurand and P. Sinay; *Tetrahedron Lett.*, 1986, **27**, 6201.	950, 952
86TL6257	O. Bortolini, F. Di Furia, G. Licini, G. Modena and M. Rossi; *Tetrahedron Lett.*, 1986, **27**, 6257.	261
86TL6305	J. A. Soderquist and E. I. Miranda; *Tetrahedron Lett.*, 1986, **27**, 6305.	248
86TL6319	J.-P. Senet, G. Vergne and G. P. Wooden; *Tetrahedron Lett.*, 1986, **27**, 6319.	60
86TL6389	A. Hassner and S. Naidorf; *Tetrahedron Lett.*, 1986, **27**, 6389.	742
86USP4617409	B. Elissando, M. Pereyre and J.-P. Quintard; *US Pat.* 4 617 409 (1986) (*Chem. Abstr.*, 1986, **106**, 5233v).	533
86USP4621078	D. Roush (FMC Corp.); *US Pat.* 4 621 078 (1986) (*Chem. Abstr.*, 1987, **106**, 33 320q).	784
86ZAAC(535)47	H. Schmidt, C. Wirkner and K. Issleib; *Z. Anorg. Allg. Chem.*, 1986, **535**, 47.	566
86ZAAC(536)187	G. Lang and E. Herrmann; *Z. Anorg. Allg. Chem.*, 1986, **536**, 187.	133, 556
86ZN(B)444	S. Lochschmidt, G. Müller, B. Huber and A. Schmidtpeter; *Z. Naturforsch., Teil B*, 1986, **41**, 444.	550
86ZN(B)974	J. Grobe and J. Szameitat; *Z. Naturforsch., Teil B*, 1986, **41**, 974.	112, 116, 117
86ZN(B)1431	M. Herberhold, H. Kniesel, L. Haumeier, A. Gieren and C. Ruiz-Pérez; *Z. Naturforsch., Teil B*, 1986, **41**, 1431.	719
86ZN(B)1527	H. Schmidbaur and J. Ebenhöch; *Z. Naturforsch., Teil B*, 1986, **41**, 1527.	136, 140

86ZOB216	I. S. Nizamov, V. A. Al'Fonsov, A. G. Trusenev, E. S. Batyeva and A. N. Pudovik; *Zh. Obshch. Khim.*, 1986, **56**, 216.	345
86ZOB321	V. S. Brovarets and B. S. Drach; *Zh. Obshch. Khim.*, 1986, **56**, 321 (Engl. Transl. 279).	999, 1000
86ZOB711	A. A. Sobanov, I. V. Bakhtiyarova, M. G. Zimin and A. N. Pudovik; *Zh. Obshch. Khim.*, 1986, **56**, 711.	339
86ZOB805	A. D. Sinitsa, V. S. Krishtal, V. V. Momot, N. K. Maidanovich and V. I. Kal'chenko; *Zh. Obshch. Khim.*, 1986, **56**, 805 (Engl. Transl. 708).	1004
86ZOB1220	I. M. Aladzheva, I. V. Leont'eva, P. V. Petrovskii, T. A. Mastryukova and M. I. Kabachnik; *Zh. Obshch. Khim.*, 1986, **56**, 1220.	557
86ZOB1535	N. A. Viktorov, T. K. Gar, I. S. Nikishina, V. M. Nosova, D. A. Ivashchenko and V. F. Mironov; *Zh. Obshch. Khim.*, 1986, **56**, 1535 (*Chem. Abstr.*, 1986, **106**, 214061).	1058
86ZOB1905	V. A. Al'Fonsov, D. A. Pudovik, E. S. Batyeva and A. N. Pudovik; *Zh. Obshch. Khim.*, 1986, **56**, 1905.	345
86ZOB2000	M. G. Pavlichenko, B. E. Ivanov, B. I. Pantukh, V. N. Eliseenkov and F. B. Gershanov; *Zh. Obshch. Khim.*, 1986, **56**, 2000.	560
86ZOB2681	A. D. Sinitsa, P. P. Onys'ko, T. V. Kim, E. I. Kiseleva and V. V. Pirozhenko; *Zh. Obshch. Khim.*, 1986, **56**, 2681 (Engl. Transl. 2372).	1008, 1009
86ZOB2743	V. D. Sheludyakov, A. I. Korshunov, V. G. Lakhtin, V. S. Timofeev, T. F. Slyusarenko, V. M. Nosova and E. V. Gradova; *Zh. Obshch. Khim.*, 1986, **56**, 2743 (*Chem. Abstr.*, **107**, 217704).	1045
86ZOR582	N. P. Smirnova, I. Ya. Kvitko and A. V. El'tsov; *Zh. Org. Khim.*, 1986, **22**, 582.	1017
86ZOR1923	B. A. Shainyan, A. N. Mirskova and V. K. Bel'skii; *Zh. Org. Khim.*, 1986, **22**, 1923.	94
86ZOR2256	Y. N. Shmatov; *Zh. Org. Khim.*, 1986, **22**, 2256 (Engl. Transl. 2025).	976
86ZOR2327	L. I. Komarova, E. M. Pokrovskaya-Dukhnenko and I. A. Zaiko; *Zh. Org. Khim.*, 1986, **22**, 2327 (*Chem. Abstr.*, 1987, **107**, 58778).	57
86ZOR2457	I. V. Barinov, V. I. Sokolov and O. A. Reutov; *Zh. Org. Khim.*, 1986, **22**, 2457.	674
87ACR207	K. M. Nicholas; *Acc. Chem. Res.*, 1987, **20**, 207.	688
87AG(E)127	H. Wadepohl and H. Pritzkow; *Angew. Chem., Int. Ed. Engl.*, 1987, **26**, 127.	691
87AG(E)229	A. J. Ashe, III, S. Mahmoud, C. Elschebroich and M. Wunsch; *Angew. Chem., Int. Ed. Engl.* 1987, **26**, 229.	1039
87AG(E)236	G. Märkl, H. J. Beckh, K. K. Mayer, M. L. Ziegler and T. Zahn; *Angew. Chem., Int. Ed. Engl.*, 1987, **26**, 236.	549
87AG(E)480	U. Schöllkopf, W. Kühnle, E. Egbert and M. Dyrbusch; *Angew. Chem., Int. Ed. Engl.*, 1987, **26**, 480.	536
87AG(E)546	H. Meyer, G. Baum, W. Massa, S. Berger and A. Berndt; *Angew. Chem., Int. Ed. Engl.*, 1987, **26**, 546.	1082
87AG(E)1011	E. P. O. Fuchs, W. Rösch and M. Regitz; *Angew. Chem., Int. Ed. Engl.*, 1987, **26**, 1011.	549
87AG(E)1187	T. Hiiro, N. Kambe, A. Ogawa, N. Miyoshi, S. Murai and N. Sonoda; *Angew. Chem., Int. Ed. Engl.*, 1987, **26**, 1187.	652
87AG(E)1257	O. Wagner, G. Mass and M. Regitz; *Angew. Chem., Int. Ed. Engl.*, 1987, **26**, 1257.	480
87AR(C)197	J. R. Jones; *Annu. Rep. Prog. Chem., Sect. C.*, 1987, **83**, 197 (*Chem. Abstr.*, 1988, **108**, 149642).	1075
87BCJ1189	Y. Yamamoto; *Bull. Chem. Soc. Jpn.*, 1987, **60**, 1189.	395, 397
87BCJ1523	M. Yamamoto, K. Suzuki, S. Tanaka and K. Yamada; *Bull. Chem. Soc. Jpn.*, 1987, **60**, 1523.	960
87BCJ1831	T. Kawashima, T. Ishii and N. Inamoto; *Bull. Chem. Soc. Jpn.*, 1987, **60**, 1831.	570, 573
87BCJ2263	D. Terunuma, M. Nakamura, E. Miyazawa and H. Nohira; *Bull. Chem. Soc. Jpn.*, 1987, **60**, 2263.	610, 617, 622
87BCJ3291	N. Tokitoh and R. Okazaki; *Bull. Chem. Soc. Jpn.*, 1987, **60**, 3291.	311
87BCJ4079	O. Tsuge, S. Kanemasa, M. Ohe and S. Takenaka; *Bull. Chem. Soc. Jpn.*, 1987, **60**, 4079.	302
87BSF325	H. Benhaqua, J. Piet, R. Danion-Bougot, L. Toupet and R. Carrie; *Bull. Soc. Chim. Fr.*, 1987, 325.	993
87BSF823	J. M. Fabre, L. Giral, A. Gouasmia, H.-J. Cristau and Y. Ribeill; *Bull. Soc. Chim. Fr.*, 1987, 823.	847
87BSJ777	N. Shimizu, F. Shibata and Y. Tsuno; *Bull. Chem. Soc. Jpn.*, 1987, **60**, 777.	806, 818
87CAR(162)145	A. Nudelman, J. Herzig, H. G. Gotlieb, E. Keinen and J. Sterling; *Carbohydr. Res.*, 1987, **162**, 145.	176
87CB213	N. C. Barua, K. Evertz, G. Huttner and R. R. Schmidt; *Chem. Ber.*, 1987, **120**, 213.	936, 950, 951
87CB471	C. Krüger, K. Laakmann, G. Schroth, H. Schwager and G. Wilke; *Chem. Ber.*, 1987, **120**, 471.	604, 647
87CB583	H. Fischer, J. Schmid and S. Zeuner; *Chem. Ber.*, 1987, **120**, 583.	581
87CB789	H. Schmidbaur, R. Pichl and G. Müller; *Chem. Ber.*, 1987, **120**, 789.	570
87CB1281	H. Schmidbaur and P. Nussstein; *Chem. Ber.*, 1987, **120**, 1281.	565, 593
87CB1645	W. Rösch, U. Hees and M. Regitz; *Chem. Ber.*, 1987, **120**, 1645.	479, 480
87CB1695	A. Maercker and R. Stötzel; *Chem. Ber.*, 1987, **120**, 1695.	620, 646, 653
87CB1707	K. P. Langhans and O. Stelzer; *Chem. Ber.*, 1987, **120**, 1707.	546
87CB1897	G. Kaupp and D. Matthies; *Chem. Ber.*, 1987, **120**, 1897.	101, 424
87CB2081	F. H. Kohler, N. Hertkorn and J. Blumel; *Chem. Ber.*, 1987, **120**, 2081.	528, 533, 1020
87CC29	A. Duchene and J.-P. Quintard; *J. Chem. Soc., Chem. Commun.*, 1987, 29.	388
87CC31	I. G. Iovel, Yu. Sh. Goldberg, M. V. Shymansk and E. Lukevics; *J. Chem. Soc., Chem. Commun.*, 1987, 31.	1080

87CC122	M. Kimura, K. Koie, S. Matsubara, Y. Sawaki and H. Iwamura; *J. Chem. Soc., Chem. Commun.*, 1987, 122.	221
87CC469	J. R. McCarthy, C. L. Barney, M. J. O'Donnell and J. C. Huffman; *J. Chem. Soc., Chem. Commun.*, 1987, 469.	1080
87CC826	R. H. Dawson and A. K. Smith; *J. Chem. Soc., Chem. Commun.*, 1987, 826.	1030
87CC1442	A. Hosomi, S. Hayashi, K. Hoashi, S. Kohra and Y. Tominaga; *J. Chem. Soc., Chem. Commun.*, 1987, 1442.	231
87CC1543	T. B. Marder, W. C. Fultz, J. C. Calabrese, R. L. Harlow and D. Milstein; *J. Chem. Soc., Chem. Commun.*, 1987, 1543.	582
87CJC166	W. Ogilvie and W. Rank; *Can. J. Chem.*, 1987, **65**, 166.	432, 433
87CJC282	M. Paveni and J. T. Edward; *Can. J. Chem.*, 1987, **65**, 282.	418
87CJC687	E. Galvez, I. Iriepa, A. Lorente, J. M. Mohedano, F. Florencio and S. Garcia-Blanco; *Can. J. Chem.*, 1987, **65**, 687.	405
87CL1095	T. Koizumi, T. Fuchigami and T. Nonaka; *Chem. Lett.*, 1987, 1095.	221
87CL1569	S. Takano, S. Tomita, M. Takahashi and K. Ogasawara; *Chem. Lett.*, 1987, 1569.	346
87CL1575	R. Okazaki and Y. Itoh; *Chem. Lett.*, 1987, 1575.	311
87CL1963	T. Takeda, Y. Kaneko, H. Nakagawa and T. Fujiwara; *Chem. Lett.*, 1987, 1963.	369
87CL2293	R. Okazaki, M. Unno and N. Inamoto; *Chem. Lett.*, 1987, 2293.	608
87CPB1413	K. Hiroi, H. Sato, L. M. Chen and K. Kotsuji; *Chem. Pharm. Bull.*, 1987, **35**, 1413.	392
87CPB1663	A. Hosomi, S. Kohra, Y. Tominaga, M. Shoji and R. Sakurai; *Chem. Pharm. Bull.*, 1987, **35**, 1663.	650
87CPB1734	Y. Terao, M. Aono, N. Imai and K. Achiwa; *Chem. Pharm. Bull.*, 1987, **35**, 1734.	369
87CPB3959	T. Koizumi, T. Hagi, Y. Horie and Y. Takeuchi; *Chem. Pharm. Bull.*, 1987, **35**, 3959.	79, 781
87CZ149	W. Schwarze, K. Drauz and J. Martens; *Chem.-Ztg.*, 1987, **111**, 149.	100
87CZ247	J. Koeppen, D. Matthies and S. Siewers; *Chem.-Ztg.*, 1987, **111**, 247.	98
87DOK(292)360	V. A. Al'Fonsov, I. S. Nizamov, E. S. Batyeva and A. N. Pudovik; *Dokl. Akad. Nauk SSSR*, 1087, **292**, 360.	345
87DOK(293)1137	A. R. Kudinov and M. I. Rybinskaya; *Dokl. Akad. Nauk SSSR*, 1987, **293**, 1137.	717
87DOK(296)403	V. N. Khabashesku, A. K. Mal'tsev and O. M. Nefedov; *Dokl. Akad. Nauk SSSR*, 1987, **296**, 403 (*Chem. Abstr.*, 1988, **108**, 149 793).	1083
87DOK(297)1132	I. V. Martynov, A. N. Chekhlov, A. Yu. Aksinenko, V. B. Sokolov and A. N. Pushin; *Dokl. Akad. Nauk SSSR*, 1987, **297**, 1132 (*Chem. Abstr.*, 1989, **110**, 95 357y).	494
87EGP251134	A. Tzschach, W. Uhlig, H. Wagner, G. Uhlig, F. Uhlig and U. Thust; *Ger. (East) Pat.* 251 134 (1987) (*Chem. Abstr.*, 1989, **110**, 154 567).	570
87EGP251136	A. Tzschach, W. Uhlig, H. Weichmann, H. Wagner and U. Thust; *Ger. (East) Pat.* 251 136 (1987) (*Chem. Abstr.*, 1989, **110**, 154 568).	570
87G227	M. Barbero, S. Cadamuro, M. Ceruti, I. Degani, R. Fochi and V. Regondi; *Gazz. Chim. Ital.*, 1987, **117**, 227.	846, 854
87H(25)221	N. V. Shah and L. D. Cama; *Heterocycles*, 1987, **25**, 221.	98
87H(26)2381	J. Szabo, G. Bernáth and P. Sohár; *Heterocycles*, 1987, **26**, 2381.	319
87HCA448	D. Seebach, R. Imwinkelried and G. Stucky; *Helv. Chim. Acta*, 1987, **70**, 448.	213
87IZV418	B. A. Arbuzov, O. A. Erastov, A. S. Ionkin and S. N. Ignat'eva; *Izv. Akad. Nauk SSSR, Ser. Khim.*, 1987, 418.	343
87IZV860	I. S. Alfer'ev, N. V. Mikhalin, I. L. Kotlyarevskii and L. M. Vainer; *Izv. Akad. Nauk SSSR, Ser. Khim.*, 1987, 860.	560
87IZV1087	A. A. Kamyshova, V. I. Dostrovalova and E. Ts. Chukhovskaya; *Izv. Akad. Nauk SSSR, Ser. Khim.*, 1987, 1087.	137
87IZV1424	M. G. Voronkov, S. N. Adamovich, S. Y. Khramtsova, B. Z. Shterenberg, V. I. Rakhlin and R. G. Mirskov; *Izv. Akad. Nauk SSSR Ser. Khim.*, 1987, 1424 (*Chem. Abstr.*, 1988, **109**, 23 058h).	617
87IZV1680	I. V. Martynov, A. Yu. Aksinenko, A. N. Chekhlov, A. N. Pushin and V. B. Sokolov; *Izv. Akad. Nauk SSSR Ser. Khim.*, 1987, 1680 (*Chem. Abstr.*, 1988, **109**, 54 850q).	494
87IZV2399	R. K. Alekperov and R. G. Kostyanovskii; *Izv. Akad. Nauk SSSR, Ser. Khim.*, 1987, 2399.	323
87JA279	R. Okazaki, A. Ishii and N. Inamoto; *J. Am. Chem. Soc.*, 1987, **109**, 279.	1049
87JA913	P. R. Jones and T. F. Bates; *J. Am. Chem. Soc.*, 1987, **109**, 913.	1048
87JA931	M. P. Cooke, Jr. and R. K. Widener; *J. Am. Chem. Soc.*, 1987, **109**, 931.	645, 1057
87JA1269	A. B. Smith, III and M. Fukui; *J. Am. Chem. Soc.*, 1987, **109**, 1269.	950
87JA1597	R. A. Holton, R. M. Kennedy, H. Kim and M. E. Krafft; *J. Am. Chem. Soc.*, 1987, **109**, 1597.	191
87JA3318	J. Arnold and T. D. Tilley; *J. Am. Chem. Soc.*, 1987, **109**, 3318.	355
87JA3811	R. A. Moss, M. Wlostowski, J. Terpinski, G. Kmiecik-Lawrynowicz and K. Krogh-Jespersen; *J. Am. Chem. Soc.*, 1987, **109**, 3811.	830
87JA4115	R. D. McCullough, G. B. Kok, K. A. Lerstrup and D. O. Cowan; *J. Am. Chem. Soc.*, 1987, **109**, 4115.	876, 877
87JA4411	C. Couret, J. Escudie, J. Satge and M. Lazraq; *J. Am. Chem. Soc.*, 1987, **109**, 4411.	662
87JA4421	U.-C. Yoon, J.-U. Kim, E. Hasegawa and P. S. Mariano; *J. Am. Chem. Soc.*, 1987, **109**, 4421.	510
87JA4930	D. K. Hutchinson and P. L. Fuchs; *J. Am. Chem. Soc.*, 1987, **109**, 4930.	384
87JA5029	M. A. Kesselmayer and R. S. Sheridan; *J. Am. Chem. Soc.*, 1987, **109**, 5029.	763
87JA5403	P. Beak, J. E. Hunter, Y. M. Jun and A. P. Wallin; *J. Am. Chem. Soc.*, 1987, **109**, 5403.	392
87JA5544	R. W. McClard, T. S. Fujita, K. E. Stremler and C. D. Poulter; *J. Am. Chem. Soc.*, 1987, **109**, 5544.	553, 562
87JA6687	B. Avramovitch, P. Weyerstahl and Z. Rappoport; *J. Am. Chem. Soc.*, 1987, **109**, 6687.	759, 764

Ref	Citation	Pages
87JA7553	R. K. Boekman, Jr., A. B. Charette, T. Asberom and B. H. Johnston; *J. Am. Chem. Soc*, 1987, **109**, 7553.	952
87JCS(D)757	E. W. Abel, S. K. Bhargava, T. E. MacKenzie, P. K. Mittal, K. G. Orrell and V. Sik; *J. Chem. Soc., Dalton Trans.*, 1987, 757.	283
87JCS(D)1647	L. M. Engelhardt, R. I. Papasergio, C. L. Raston, G. Salem, C. R. Whitaker and A. H. White; *J. Chem. Soc., Dalton Trans.*, 1987, 1647.	626, 662
87JCS(D)3085	R. I. Papasergio, C. L. Raston and A. H. White; *J. Chem. Soc., Dalton Trans.*, 1987, 3085.	656
87JCS(P1)181	G. M. Blackburn, D. Brown, S. J. Martin and M. J. Parratt; *J. Chem. Soc. Perkin Trans. 1*, 1987, 181.	131
87JCS(P1)195	P. G. Sammes and R. J. Whitby; *J. Chem. Soc., Perkin Trans. 1*, 1987, 195.	212
87JCS(P1)589	H. Ishibashi, H. Nakatani, Y. Umei, W. Yamamoto and M. Ikeda; *J. Chem. Soc., Perkin Trans. 1*, 1987, 589.	943
87JCS(P1)775	A. R. Katritzky, W. H. Ramer and J. N. Lam; *J. Chem. Soc., Perkin Trans. 1*, 1987, 775.	323
87JCS(P1)781	A. R. Katritzky, S. Rachwal, K. C. Caster, F. Mahni, K. W. Law and O. Rubio; *J. Chem. Soc., Perkin Trans. 1*, 1987, 781.	322
87JCS(P1)791	A. R. Katritzky, S. Rachwal and B. Rachwal; *J. Chem. Soc., Perkin Trans. 1*, 1987, 791.	296, 300, 310
87JCS(P1)799	A. R. Katritzky, S. Rachwal and B. Rachwal; *J. Chem. Soc., Perkin Trans. 1*, 1987, 799.	309, 445
87JCS(P1)811	A. R. Katritzky, W. Kuzmierkiewicz, B. Rachwal, S. Rachwal and J. Thomson; *J. Chem. Soc., Perkin Trans. 1*, 1987, 811.	308, 445
87JCS(P1)967	J. I. Grayson, S. Warren and A. T. Zaslona; *J. Chem. Soc., Perkin Trans. 1*, 1987, 967.	923, 924
87JCS(P1)1017	J. Barluenga, J. J. Martinez-Gallo, C. Nájera and M. Yus; *J. Chem. Soc., Perkin Trans. 1*, 1987, 1017.	808
87JCS(P1)1095	H. Ishibashi, T. Sato, M. Irie, M. Ito and M. Ikeda; *J. Chem. Soc., Perkin Trans. 1*, 1987, 1095.	924
87JCS(P1)1275	R. M. Acheson and P. J. Ansell; *J. Chem. Soc., Perkin Trans. 1*, 1987, 1275.	804, 840
87JCS(P1)1501	H. Takahata, A. Anazawa, K. Moriyama and T. Yamazaki; *J. Chem. Soc., Perkin Trans. 1*, 1987, 1501.	902
87JCS(P1)2017	M. A. Rizzacasa and M. V. Sargent; *J. Chem. Soc., Perkin Trans. 1*, 1987, 2017.	856
87JCS(P1)2189	P. J. Kocienski, S. D. A. Street, C. Yeates and S. F. Cambell; *J. Chem. Soc., Perkin Trans. 1*, 1987, 2189.	953
87JCS(P1)2203	Y. Takeuchi, M. Asahina, K. Nagata and T. Koizumi; *J. Chem. Soc., Perkin Trans. 1*, 1987, 2203.	101
87JCS(P1)2269	I. Fleming, K. Takaki and A. P. Thomas; *J. Chem. Soc., Perkin Trans. 1*, 1987, 2269.	1046
87JCS(P1)2301	P. Ferraboschi, R. Canevotti, P. Grisenti and E. Santaniello; *J. Chem. Soc., Perkin Trans. 1*, 1987, 2301.	168
87JCS(P2)1253	S. Hughes, G. Griffiths and C. J. M. Stirling; *J. Chem. Soc., Perkin Trans. 2*, 1987, 1253.	861
87JFC(35)677	K. Makino and H. Yoshioka; *J. Fluorine Chem.*, 1987, **35**, 677.	101, 105
87JFC(37)223	A. Probst, K. Raab, K. Ulm and K. von Werner; *J. Fluorine Chem.*, 1987, **37**, 223.	730
87JFC(37)313	A. Gregorcic and M. Zupan; *J. Fluorine Chem.*, 1987, **34**, 313.	751
87JGU334	T. K. Gar, O. N. Chernysheva, A. V. Kisin and V. F. Mironov; *J. Gen. Chem. USSR (Engl. Transl.)*, 1987, **57**, 334.	98, 99
87JGU622	A. V. Golovanov, I. G. Maslennikov and A. N. Lavrent'ev; *J. Gen. Chem. USSR (Engl. Transl.)*, 1987, **57**, 622.	128
87JGU844	V. V. Belakhov, E. V. Komarov and B. I. Ionin; *J. Gen. Chem. USSR (Engl. Transl.)*, 1987, **57**, 844.	802
87JHC519	Y. Tominaga, S. Hidaki and Y. Matsuda; *J. Heterocycl. Chem.*, 1987, **24**, 519.	901
87JHC945	J. J. D'Amico, F. G. Bollinger, C. C. Tung and W. E. Dahl; *J. Heterocycl. Chem.*, 1987, **24**, 945.	102
87JMC24	J. C. Aloup, J. Bouchaudon, D. Farge, C. James, J. Deragnaucourt and M. Hardy-Houis; *J. Med. Chem.*, 1987, **30**, 24.	229
87JMC1934	E. Teodori, F. Gualtieri, P. Angeli, L. Brasili and M. Giannella; *J. Med. Chem.*, 1987, **30**, 1934.	232
87JOC101	T. Sheradsky and R. Moshenberg; *J. Org. Chem.*, 1987, **52**, 101.	105
87JOC188	J. R. Hwu, L. Leu, J. A. Robl, D. A. Anderson and J. M. Wetzel; *J. Org. Chem.*, 1987, **52**, 188.	182
87JOC235	A. Padwa and W. Dent; *J. Org. Chem.*, 1987, **52**, 235.	319
87JOC290	J. Leroy, H. Molines and C. Wakselman; *J. Org. Chem.*, 1987, **52**, 290.	731
87JOC365	R. W. M. Aben, R. Smit and J. W. Scheeren; *J. Org. Chem.*, 1987, **52**, 365.	414
87JOC622	D. Askin, C. Angst and S. Danishefsky; *J. Org. Chem.*, 1987, **52**, 622.	174
87JOC782	J. P. Hagen, J. J. Harris and D. Lakin; *J. Org. Chem.*, 1987, **52**, 782.	199, 218
87JOC809	E. Block and A. Wall; *J. Org. Chem.*, 1987, **52**, 809.	138, 393
87JOC844	A. R. Katritzky, W. Kuzmierkiewicz and J. M. Aurrecoechea; *J. Org. Chem.*, 1987, **52**, 844.	392
87JOC855	J.-M. Fang, B.-C. Hong and L.-F. Liao; *J. Org. Chem.*, 1987, **52**, 855.	849
87JOC1256	K. Takaki, M. Yasumura, T. Tamura and K. Negoro; *J. Org. Chem.*, 1987, **52**, 1256.	943
87JOC1351	T. R. Hoye, B. H. Peterson and J. D. Miller; *J. Org. Chem.*, 1987, **52**, 1351.	213
87JOC1353	W. H. Pearson and M. Cheng; *J. Org. Chem.*, 1987, **52**, 1353.	213
87JOC1414	K. Nakamura, M. Higaki, S. Adachi, S. Oka and A. Ohno; *J. Org. Chem.*, 1987, **52**, 1414.	391
87JOC1872	W. R. Dolbier, Jr. and D. M. Al-Fekri, *J. Org. Chem.*, 1987, **52**, 1872.	5
87JOC1885	C. R. Johnson and D. S. Dhanoa; *J. Org. Chem.*, 1987, **52**, 1885.	396
87JOC1972	P. F. Schuda and W. A. Price; *J. Org. Chem.*, 1987, **52**, 1972.	856
87JOC2114	S. Kim, S. S. Kim, S. T. Lim and S. C. Shim; *J. Org. Chem.*, 1987, **52**, 2114.	257

87JOC2427	A. Padwa, P. Eisenbarth, M. K. Venkatramanan and G. S. K. Wong; *J. Org. Chem.*, 1987, **52**, 2427.	512
87JOC2442	J. D. Friedrich; *J. Org. Chem.*, 1987, **52**, 2442.	257
87JOC3232	M. G. Bock, R. M. Di Pardo, B. E. Evans, K. E. Rittle, D. F. Veber, R. M. Friedinger, J. Hirshfield and J. P. Springer; *J. Org. Chem.*, 1987, **52**, 3232.	417
87JOC3243	J. R. Sanderson, E. L. Yeakey, J. J. Lin, R. Duranleau and E. T. Marquis; *J. Org. Chem.*, 1987, **52**, 3243.	1078
87JOC3263	Y.-C. Xu and W. D. Wulff; *J. Org. Chem.*, 1987, **52**, 3263.	955
87JOC3442	R. Fruttero, B. Ferrarotti and A. Gasco; *J. Org. Chem.*, 1987, **52**, 3442.	429
87JOC3466	E. R. Koft; *J. Org. Chem.*, 1987, **52**, 3466.	331
87JOC3713	B. Kahr, S. E. Biali, W. Schaefer, A. B. Buda and K. Mislow; *J. Org. Chem.*, 1987, **52**, 3713.	11
87JOC4258	T. G. Back, S. Collins, M. V. Krishma and K.-W. Law; *J. Org. Chem.*, 1987, **52**, 4258.	953
87JOC4693	A. G. M. Barrett, G. G. Graboski, M. Sabat and S. J. Taylor; *J. Org. Chem.*, 1987, **52**, 4693.	906
87JOC4781	A. P. Marchand, G. S. Annapurna, V. Vidyasagar, J. L. Flippen-Anderson, R. Gilardi, C. George and H. L. Ammon; *J. Org. Chem.*, 1987, **52**, 4781.	428
87JOC4943	D. L. J. Clive, T. L. B. Boivin and A. G. Angoh; *J. Org. Chem.*, 1987, **52**, 4943.	286, 779
87JOC5061	Y. Takeuchi, K. Nagata and T. Koizumi; *J. Org. Chem.*, 1987, **52**, 5061.	108
87JOC5116	D. S. Matteson and M. L. Peterson; *J. Org. Chem.*, 1987, **52**, 5116.	362
87JOC5121	D. S. Matteson and A. A. Kandil; *J. Org. Chem.*, 1987, **52**, 5121.	363
87JOM(319)C51	B. F. G. Johnson, R. Khattar, F. J. Lahoz, J. Lewis and P. R. Raithby; *J. Organomet. Chem.*, 1987, **319**, C51.	682
87JOM(319)9	G. E. Herberich, B. Hessner and R. Saive; *J. Organomet. Chem.*, 1987, **319**, 9.	695
87JOM(320)163	D. C. Snyder; *J. Organomet. Chem.*, 1987, **320**, 163.	510
87JOM(320)171	E. Guibé-Jampel and F. Huet; *J. Organomet. Chem.*, 1987, **320**, 171.	541
87JOM(321)C1	C. Stader and B. Wrackmeyer; *J. Organomet. Chem.*, 1987, **321**, C1.	661
87JOM(321)257	R. Hörlein, W. A. Herrmann, C. E. Barnes, C. Weber, C. Krüger, M. L. Ziegler and T. Zahn; *J. Organomet. Chem.*, 1987, **321**, 257.	709
87JOM(322)141	T. N. Mitchell and W. Reimann; *J. Organomet. Chem.*, 1987, **322**, 141.	1065
87JOM(322)151	T. N. Mitchell and W. Reimann; *J. Organomet. Chem.*, 1987, **322**, 151.	1063, 1065, 1069, 1070
87JOM(326)C9	K. Kellner and W. Hanke; *J. Organomet. Chem.*, 1987, **326**, C9.	455
87JOM(326)159	M. G. Voronkov, S. V. Kirpichenko, A. T. Abrosimova, A. I. Albanov, V. V. Keiko and V. I. Lavrent'yev; *J. Organomet. Chem.*, 1987, **326**, 159.	507
87JOM(328)349	F. Nief, F. Mercier and F. Mathey; *J. Organomet. Chem.*, 1987, **328**, 349.	545
87JOM(330)357	A. Solladié-Cavallo, G. Lapitajs, P. Buchert, A. Klein, S. C. Hona and A. Manfredi; *J. Organomet. Chem.*, 1987, **330**, 357.	540
87JOM(331)181	J. G. Duboudin, M. Ratier and B. Trouve; *J. Organomet. Chem.*, 1987, **331**, 181.	386
87JOM(331)409	J. Vincente, M.-T. Chicote, J. Fernancez-Baeza, J. Martin, I. Saura-Llamas, J. Turpin and P. G. Jones; *J. Organomet. Chem.*, 1987, **331**, 409.	586
87JOM(332)C1	F. Zurmuhlen and M. Regitz; *J. Organomet. Chem.*, 1987, **332**, C1.	341
87JOM(332)99	H. D. Verkruijsse, L. Brandsma and P. von R. Schleyer; *J. Organometal. Chem.*, 1987, **332**, 99.	836, 949
87JOM(333)129	J. Vincente, M.-T. Chicote, I. Saura-Llamas, J. Turpin and J. Fernancez-Baeza; *J. Organomet. Chem.*, 1987, **333**, 129.	586
87JOM(333)155	R. Benn, E. Janssen, H. Lehmkuhl and A. Rufinska; *J. Organomet. Chem.*, 1987, **333**, 155.	698
87JOM(333)269	W. R. Cullen and N. F. Han; *J. Organomet. Chem.*, 1987, **333**, 269.	1033
87JOM(334)117	P. D. Macklin, C. A. Mirkin, N. Viswanathan, G. D. Williams, G. L. Geoffroy and A. L. Rheingold; *J. Organomet. Chem.*, 1987, **334**, 117.	718
87JOM(334)347	M. Herberhold and H. Kniesel; *J. Organomet. Chem.*, 1987, **334**, 347.	719
87JOM(335)283	J. Laduranty, F. Barbot and L. Miginiac; *J. Organomet. Chem.*, 1987, **335**, 283.	321
87JOM(336)C41	H. Andringa, Y. A. Heus-Kloos and L. Brandsma; *J. Organomet. Chem.*, 1987, **336**, C41.	138
87JOM(336)59	N. Auner; *J. Organomet. Chem.*, 1987, **336**, 59.	651
87JOM(336)153	M. Herberhold and P. Leitner; *J. Organomet. Chem.*, 1987, **336**, 153.	290
87JOM(336)187	A. R. Kudinov, M. I. Rybinskaya, Yu. T. Struchkov, A. I. Yanovskii and P. V. Petrovskii; *J. Organomet. Chem.*, 1987, **336**, 187.	694, 717
87JOM(336)413	J. Wolf and H. Werner; *J. Organomet. Chem.*, 1987, **336**, 413.	149
87JOU246	L. M. Yagupol'skii, P. G. Cherednichenko and M. M. Kremlev; *J. Org. Chem. USSR (Engl. Transl.)*, 1987, **23**, 246.	817
87JPR871	H. Teichmann and M. Schnell; *J. Prakt. Chem.*, 1987, **329**, 871.	106
87JPS(A)1967	C. D. Juengst and W. P. Weber; *J. Polym. Sci., Polym. Chem.: Part A*, 1987, **25**, 1967.	570
87JSP(121)304	M. C. Durrant, H. W. Kroto and D. R. M. Walton; *J. Mol. Spectrosc.*, 1987, **121**, 304.	568
87KGS1477	A. K. Shanazarov, N. P. Solov'eva, V. V. Chistyakov, Y. N. Sheinker and V. G. Granik; *Khim. Geterotsikl. Soedin.*, 1987, 1477 (Engl. Transl. 1178).	987
87LA51	R. Tacke, J. Pikies, H. Linoh, R. Rohr-Aehle and S. Gönne; *Liebigs Ann. Chem.*, 1987, 51.	136
87LA311	L. F. Tietze and G. Brill; *Liebigs Ann. Chem.*, 1987, 311.	950
87LA451	G. Singer, G. Heusinger, A. Mosandl and C. Burschka; *Liebigs Ann. Chem.*, 1987, 451.	232
87LA583	B. Rugewitz-Blackholm and M. Weissler; *Liebigs Ann. Chem.*, 1987, 583.	326
87LA607	F. Lieb, U. Niewöhner and D. Wendisch; *Liebigs Ann. Chem.*, 1987, 607.	170
87LA977	R. W. Hoffmann, K. Ditrich and S. Froch; *Liebigs Ann. Chem.*, 1987, 977.	1056
87LA1101	P. B. Sulay and I. C. Ivanov; *Liebigs Ann. Chem.*, 1987, 1101.	981, 982
87M851	J. G. Schantl and H. Gstach; *Monatsh. Chem.*, 1987, **118**, 851.	448
87MI 403-01	R. B. Silverman and M. K. Vadnere; *Bioorganic Chemistry*, 1987, **15**, 328.	141, 142
87MI 405-01	P. Fügedi, P. J. Garegg, H. Lönn and T. Norberg; *Glycoconjugate J.*, 1987, **4**, 97.	223
B-87MI 407-01	D. Liotta ed.; "Organoselenium Chemistry," Wiley-Interscience, New York, 1987.	333

B-87MI 407-02	F. S. Guziec, Jr; in "Organoselenium Chemistry," (ed.) D. Liotta, Wiley-Interscience, New York 1987, p. 277.	333
B-87MI 407-03	R. S. Edmundson (ed.); "Dictionary of Organophosphorus Compounds," Chapman and Hall, London, 1987.	336
B-87MI 408-01	D. S. Matteson; in "The Chemistry of the Metal–Carbon Bond," ed. F. R. Hartley, Wiley, New York, 1987, vol. 4, p. 307.	362, 372, 373
B-87MI 408-02	J. L. Wardell; in "The Chemistry of the Metal–Carbon Bond," ed. F. R. Hartley, Wiley, New York, 1987, vol. 4, p. 1.	377, 390, 398
B-87MI 408-03	H. J. Reich; in "Organoselenium Chemistry," ed. D. Liotta, Wiley, New York, 1987, p. 243.	398
B-87MI 408-04	A. Krief; in "The Chemistry of Organic Selenium and Tellurium Compounds," ed. S. Patai, Wiley, New York, 1987, p. 675.	398
87MI 411-01	J. L. Wardell; in "The Chemistry of the Metal–Carbon Bond," ed. F. R. Hartley, Wiley-Interscience, 1987, vol. 4, p. 1.	527
87MI 414-01	Y. Yamada, T. Yano and N. Itaya; *Nippon Noyaku Gakkaishi*, 1987, **12**, 683 (*Chem. Abstr.*, 1988, **108**, 145415x).	654
87MI 420-01	D. Y. Oh, T. H. Kim and D. H. Kang; *Bull. Korean Chem. Soc.*, 1987, **8**, 219.	924
87MI 421-01	R. L. P. De Jong, H. D. Verkruijsse and L. Brandsma; *Stud. Org. Chem. (Amsterdam)*, 1987, **28** (*Perspect. Org. Chem. Sulfur*), 105.	1017
87MI 424-01	A. Bagno, G. Scottano and R. A. More O'Ferrall; *Rev. Chem. Intermed.*, 1987, **7**, 313.	1075
87OM659	J. Barrau, N. Ben Hamida, A. Agrebi and J. Satgé; *Organometallics*, 1987, **6**, 659.	628, 629
87OM834	M. Austin, K. Gebreyes, H. G. Kuivila, K. Swami and J. A. Zubieta; *Organometallics*, 1987, **6**, 834.	700
87OM861	V. Galamb, G. Pályi, R. Boese and G. Schmid; *Organometallics*, 1987, **6**, 861.	148
87OM959	U. G. Wettermark, P. Wisian-Neilson, G. M. Scheide and R. H. Neilson; *Organometallics*, 1987, **6**, 959.	570
87OM1583	W. M. Lamanna, W. B. Gleason and D. Britton; *Organometallics*, 1987, **6**, 1583.	694
87OM1857	A. Sekiguchi and W. Ando; *Organometallics*, 1987, **6**, 1857.	612, 1049
87OM1861	C.-H. Lin, C.-Y. Lee and C.-S. Liu; *Organometallics*, 1987, **6**, 1861.	616
87OM1992	H. H. Murray, D. A. Briggs, G. Garzon, R. G. Raptis, L. C. Porter and J. P. Fackler; *Organometallics*, 1987, **6**, 1992.	586
87OM2007	J. L. Peterson and J. W. Egan Jr.; *Organometallics*, 1987, **6**, 2007.	657, 659
87OM2110	T. Alonso, S. Harvey, P. C. Junk, C. L. Raston, B. Skelton and A. H. White; *Organometallics*, 1987, **6**, 2110.	654
87OM2164	R. J. Batchelor, F. W. B. Einstein, C. H. W. Jones and R. D. Sharma; *Organometallics*, 1987, **6**, 2164.	291
87OM2489	C. Zybill and G. Müller; *Organometallics*, 1987, **6**, 2489.	118
87OM2568	S. Inoue and Y. Sato; *Organometallics*, 1987, **6**, 2568.	625, 662
87OS(65)119	S. K. Davidsen, G. W. Phillips and S. F. Martin; *Org. Synth.*, 1987, **65**, 119.	99, 465
87OS(65)215	E. L. Eliel, J. E. Lynch, F. Kume and S. V. Frye; *Org. Synth.*, 1987, **65**, 215.	230
87PS(29)287	W. F. Gilmore and J. S. Park; *Phosphorus Sulfur*, 1987, **29**, 287.	127
87PS(30)225	M. Mikolajczyk, P. Balczewski and P. Graczyk; *Phosphorus Sulfur*, 1987, **30**, 225.	346
87PS(30)349	G. Becker, W. Becker, R. Knebl, H. Schmidt, U. Hildenbrand and M. Westerhausen; *Phosphorus Sulfur*, 1987, **30**, 349.	568
87PS(30)523	Y. Y. C. Yeung Lam Ko, P. Cosquer, P. Pellon, J. Hamelin and R. Carrié; *Phosphorus Sulfur*, 1987, **30**, 523.	566
87PS(33)41	L. Maier, W. Kunz and G. Rist; *Phosphorus Sulfur*, 1987, **33**, 41.	105
87PS(33)147	A. K. Roy, U. G. Wettermark, G. M. Scheide, P. Wisian-Neilson and R. H. Neilson; *Phosphorus Sulfur*; 1987, **33**, 147.	129
87RCR859	V. P. Kukhar and V. A. Solodenko; *Russ. Chem. Rev. (Engl. Transl.)*, 1987, **56**, 859.	451, 469, 493
87RTC514	N. J. R. van Eikema Hommes, F. Bickelhaupt and G. W. Klumpp; *Recl. Trav. Chim. Pays-Bas*, 1987, **106**, 514.	668, 699
87RTC545	H. C. P. F. Roelen, G. J. Ligtvoet, G. A. van der Marel and J. H. van Boom; *Recl. Trav. Chim. Pays-Bas*, 1987, **106**, 545.	184
87S49	A. G. Martinez, R. M. Alvarez, A. G. Fraile, L. R. Subramanian and M. Hanack; *Synthesis*, 1987, 49.	214
87S76	R. van der Heiden and L. Brandsma; *Synthesis*, 1987, 76.	792
87S85	A. J. Fatiadi; *Synthesis*, 1987, 85.	168
87S164	M.-A. Kakimoto, T. Seri and Y. Imai; *Synthesis*, 1987, 164.	247
87S169	P. Coutrot, C. Grison and M. Youssefi-Tabrizi; *Synthesis*, 1987, 169.	90, 784
87S223	P. Münster and W. Steglich; *Synthesis*, 1987, 223.	102
87S250	A. M. Maione and A. Romeo; *Synthesis*, 1987, 250.	201
87S357	Z. Huang and Z. Liu; *Synthesis*, 1987, 357.	972
87S477	H. Wójtowicz and P. Mastalerz; *Synthesis*, 1987, 477.	124
87S497	F. Cottet, L. Cottier and G. Descotes; *Synthesis*, 1987, 497.	170
87S589	D. J. R. Massy; *Synthesis*, 1987, 589.	315, 325
87S645	M. Zaidlewicz and C. S. Panda; *Synthesis*, 1987, 645.	653
87S745	A. Hamed, E. Müller, M. Al-Talib and J. C. Jochims; *Synthesis*, 1987, 745.	434
87S837	K. S. Varma, A. Bury, N. J. Harris and A. E. Underhill; *Synthesis*, 1987, 837.	850
87S1020	D. Dauzonne and R. Royer; *Synthesis*, 1987, 1020.	798
87S1027	G. Barcelo, J.-P. Senet and G. Sennyey; *Synthesis*, 1987, 1027.	60
87S1034	V. Fiandanese, G. Marchese, F. Naso and L. Ronzini; *Synthesis*, 1987, 1034.	773
87S1043	T. Mukaiyama and M. Murakami; *Synthesis*, 1987, 1043.	176

87S1099	T. Shono, Y. Matsumura, O. Onomura and Y. Yamada; *Synthesis*, 1987, 1099.	201
87SC1	N. Slougui and G. Rousseau; *Synth. Commun.*, 1987, **17**, 1.	831
87SC1273	N. Kanemoto, S. Inoue and Y. Sato; *Synth. Commun.*, 1987, **17**, 1273.	626, 662
87SC1467	A. Riondel, P. Caubére, J. P. Senet and S. Lecolier; *Synth. Commun.*, 1987, **17**, 1467.	211
87SM647	K. Lerstrup, A. Bailey, R. McCullough, M. Mays, D. Cowan and T. Kistenmacher; *Synth. Met.*, 1987, **19**, 647.	876
87T693	K. Nishiyama, M. Oba and A. Watanabe; *Tetrahedron*, 1987, **43**, 693.	432, 433
87T1793	G. Etemad-Moghadam, J. Bellan, C. Tachon and M. Koenig; *Tetrahedron*, 1987, **43**, 1793.	547
87T4309	S. Piettre, Ch. De Cock, R. Merenyi and H. G. Viehe; *Tetrahedron*, 1987, **43**, 4309.	88, 773, 785
87T4433	A. K. Chakraborti, B. Saha, C. Ray and U. R. Ghatak; *Tetrahedron*, 1987, **43**, 4433.	170
87T4759	Y. Ichikawa, M. Isobe, H. Masaki, T. Kawai, T. Goto and C. Katayama; *Tetrahedron* 1987, **20**, 4759.	937
87T4875	B. Solaja, J. Huguet, M. Karpf and A. S. Dreiding; *Tetrahedron*, 1987, **43**, 4875.	238, 241
87TL405	M.-P. Teulade and P. Savignac; *Tetrahedron Lett.*, 1987, **28**, 405.	557, 576
87TL917	E. Negishi, D. R. Swanson, F. E. Cederbaum and T. Takahasi; *Tetrahedron Lett.*, 1987, **28**, 917.	1061, 1066
87TL989	N. S. Simpkins; *Tetrahedron Lett.*, 1987, **28**, 989.	938, 939, 961
87TL1093	G. Märkl and G. Dorfmeister; *Tetrahedron Lett.*, 1987, **28**, 1093.	549
87TL1299	R. Matsuoka, Y. Horiguchi and I. Kuwajima; *Tetrahedron Lett.*, 1987, **28**, 1299.	1064
87TL1807	T. Kusumoto and T. Hiyama; *Tetrahedron Lett.*, 1987, **28**, 1807.	1045
87TL1811	T. Kusumoto and T. Hiyama; *Tetrahedron Lett.*, 1987, **28**, 1811.	619
87TL1847	A. R. Katritzky and S. Sengupta; *Tetrahedron Lett.*, 1987, **28**, 1847.	383
87TL1851	J. E. MacDonald and G. S. Poindexter; *Tetrahedron Lett.*, 1987, **28**, 1851.	651
87TL1981	W. Kinzy and R. R. Schmidt; *Tetrahedron Lett.*, 1987, **28**, 1981.	174
87TL2225	Y. Guindon, M. A. Bernstein and P. C. Anderson; *Tetrahedron Lett.*, 1987, **28**, 2225.	199
87TL2681	U. Gruseck and M. Heuschmann; *Tetrahedron Lett.*, 1987, **28**, 2681.	414
87TL2693	G. Märkl, W. Hölzl and I. Trötsch-Schaller; *Tetrahedron Lett.*, 1987, **28**, 2693.	567
87TL2723	Y. Ito and T. Ogawa; *Tetrahedron Lett.*, 1987, **28**, 2723.	187
87TL2889	J. van der Louw, J. L. van der Baan, F. Bickelhaupt and G. W. Klumpp; *Tetrahedron Lett.*, 1987, **28**, 2889.	1061
87TL3285	H. H. Mooiweer, H. Hiemstra, H. P. Fortgens and W. N. Speckamp; *Tetrahedron Lett.*, 1987, 3285.	298
87TL3569	N. Kunesch, C. Miet and J. Poisson; *Tetrahedron Lett.*, 1987, **28**, 3569.	176
87TL3719	M. Licchelli and A. Greco; *Tetrahedron Lett.*, 1987, **28**, 3719.	617
87TL3791	D. Seebach and B. Herradon; *Tetrahedron Lett.*, 1987, **28**, 3791.	213
87TL3805	A. Moradpour and S. Bittner; *Tetrahedron Lett.*, 1987, **28**, 3805.	843
87TL3911	R. J. Linderman, A. Godfrey and K. Horne; *Tetrahedron Lett.*, 1987, **28**, 3911.	384
87TL4093	A. Ricci, A. Degl'Innocenti, G. Borselli and G. Reginato; *Tetrahedron Lett.*, 1987, **28**, 4093.	930
87TL4123	S. W. McCombie, B. B. Shanker and A. K. Ganguly, *Tetrahedron Lett.*, 1987, **28**, 4123.	961
87TL4499	D. S. Matteson and E. C. Beedle; *Tetrahedron Lett.*, 1987, **28**, 4499.	526
87TL4793	M. Bhupathy and T. Cohen; *Tetrahedron Lett.*, 1987, **28**, 4793.	222
87TL4925	J. P. Marino and M. W. Kim; *Tetrahedron Lett.*, 1987, **28**, 4925.	240
87TL5121	P. T. Meinke and G. A. Krafft; *Tetrahedron Lett.*, 1987, **28**, 5121.	365, 376, 570
87TL5145	H. Mahler and M. Braun; *Tetrahedron Lett.*, 1987, **28**, 5145.	758
87TL5149	T. Kramer and D. Hoppe; *Tetrahedron Lett.*, 1987, **28**, 5149.	382, 384
87TL5419	A. R. Katritzky and S. Sengupta; *Tetrahedron Lett.*, 1987, **28**, 5419.	520, 540
87TL5489	E. V. Dehmlow and J. Wilkenloh; *Tetrahedron Lett.*, 1987, **28**, 5489.	1080
87TL5595	G. R. Perdomo and J. J. Krepinsky; *Tetrahedron Lett.*, 1987, **28**, 5595.	175
87TL6121	B. A. Boyd, R. J. Thoma and R. H. Neilson; *Tetrahedron Lett.*, 1987, **28**, 6121.	573
87TL6261	B. A. Narayanan and W. H. Bunnelle; *Tetrahedron Lett.*, 1987, **28**, 6261.	655, 656
87TL6317	X.-Y. Li and J.-S. Hu; *Tetrahedron Lett.*, 1987, **28**, 6317.	117, 118
87TL6493	B. J. J. van de Heisteeg, G. Schat, O. S. Akkerman and F. Bickelhaupt; *Tetrahedron Lett.*, 1987, **28**, 6493.	670
87TL6601	J.-L. Gras, R. Nouguier and M. Mchich; *Tetrahedron Lett.*, 1987, **28**, 6601.	185
87TL6709	C. Najera and M. Yus; *Tetrahedron Lett.*, 1987, **28**, 6709.	938, 961
87ZAAC(544)87	M. Baudler, D. Düster and D. Ouzounis; *Z. Anorg. Allg. Chem.*, 1987, **544**, 87.	545
87ZAAC(555)109	M. Fild and W. Handke; *Z. Anorg. Allg. Chem.*, 1987, **555**, 109.	553, 562, 563, 564
87ZN(B)118	M. Herberhold and K. Guldner; *Z. Naturforsch., Teil B*, 1987, **42**, 118.	594
87ZN(B)142	H. Schmidbaur, J. Ebenhoch and G. Muller; *Z. Naturforsch, Teil B.*, 1987, **42**, 142.	1052
87ZN(B)796	E. V. Dehmlow and R. Neuhaus; *Z. Naturforsch, Teil B*, 1987, **42**, 796 (*Chem. Abstr.*, 1988, **108**, 204 236q).	768
87ZOB54	V. A. Chauzov, Yu. N. Studnev, M. G. Iznoskova and A. V. Fokin; *Zh. Obshch. Khim.*, 1987, **57**, 54.	556
87ZOB525	Z. S. Novikova, I. L. Odinets and I. F. Lutsenko; *Zh. Obshch. Khim.*, 1987, **57**, 525.	563
87ZOB831	O. I. Kolodyazhnyi; *Zh. Obshch. Khim.*, 1987, **57**, 831.	572, 575
87ZOB949	Z. S. Novikova, E. A. Monin, M. M. Kabachnik and I. F. Lutsenko; *Zh. Obshch. Khim.*, 1987, **57**, 949.	552
87ZOB1406	A. A. Pritchenko, M. V. Livantsov and I. F. Lutsenko; *Zh. Obshch. Khim.*, 1987, **57**, 1406.	914
87ZOB1741	S. V. Ponomarev, M. V. Erikova, S. N. Nikolaeva, R. Zel and A. S. Kostyuk; *Zh. Obshch. Khim.*, 1987, **57**, 1741 (*Chem. Abstr.*, 1987, **109**, 93 192).	1056, 1059, 1063
87ZOB2138	B. A. Khaskin, T. G. Rymareva and S. N. Golosov; *Zh. Obshch. Khim.*, 1987, **57**, 2138.	561
87ZOB2294	B. M. Gladshtein, V. M. Zimin, I. M. Shishkov and I. A. Revel'skii; *Zh. Obshch. Khim.*, 1987, **57**, 2294.	561

87ZOB2300	E. N. Ryazantsev, D. A. Ponomarev and V. M. Al'bitskaya; *Zh. Obshch. Khim.*, 1987, **57**, 2300.	343
87ZOB2606	Se Phan Min and S. K. Moiseev; *Zh. Obshch. Khim.*, 1987, **57**, 2606.	723
87ZOB2637	Z. S. Novikova, E. A. Monin, A. A. Borisenko, K. S. Zavadskii and I. F. Lutsenko; *Zh. Obshch. Khim.*, 1987, **57**, 2637.	548
87ZOB2640	O. I. Kolodyazhnyi and D. B. Glokhnov; *Zh. Obshch. Khim.*, 1987, **57**, 2640.	572
87ZOR60	A. V. Martynov, A. N. Mirskova and M. G. Voronkov; *Zh. Org. Khim.*, 1987, **23**, 60 (*Chem. Abstr.*, 1987, **107**, 217 182y).	786
88ACS629	L. Teuber and C. Christophersen; *Acta Chem. Scand.*, 1988, **42**, 629.	842
88ACS(B)515	O. Antonsen, T. Benneche and K. Undheim; *Acta Chem. Scand., Ser. B*, 1988, **42**, 515.	232
88AG(E)304	U. J. Vogelbacher, M. Ledermann, T. Schach, G. Michels, U. Hees and M. Regitz; *Angew. Chem., Int. Ed. Engl.*, 1988, **100**, 304.	100
88AG(E)389	G. Märkl, S. Dietl, M. L. Ziegler and B. Nuber; *Angew. Chem., Int. Ed. Engl.*, 1988, **27**, 389.	549
88AG(E)433	U. Schöllkopf, B. Hupfeld, S. Küper, E. Egert and M. Dyrbusch; *Angew. Chem., Int. Ed. Engl.*, 1988, **27**, 433.	536
88AG(E)709	G. Märkl, S. Dietl, M. L. Ziegler and B. Nuber; *Angew. Chem., Int. Ed. Engl.*, 1988, **27**, 709.	565
88AG(E)837	A. R. Barron, A. H. Cowley, S. W. Hall and C. N. Nunn; *Angew. Chem., Int. Ed. Engl.*, 1988, **27**, 837.	549
88AG(E)939	K. Sünkel and D. Motz; *Angew. Chem., Int. Ed. Engl.*, 1988, **27**, 939.	707
88AG(E)1194	U. Schöllkopf, D. Pettig, E. Schulze, M. Klinge, E. Egert, B. Benecke and M. Noltemeyer; *Angew. Chem., Int. Ed. Engl.*, 1988, **27**, 1194.	536
88AG(E)1370	M. Pilz, J. Allwohn, R. Hunold, W. Messa and A. Berndt; *Angew. Chem., Int. Ed. Engl.*, 1988, **27**, 1370.	634
88AG(E)1534	R. Gerhardt, T. Grelbig, J. Buschmann, P. Luger and K. Seppelt; *Angew. Chem., Int. Ed. Engl.*, 1988, **100**, 1534.	868
88AHC(43)2	D. Hewitt; *Adv. Heterocycl. Chem.*, 1988, **43**, 2.	451
88AP(321)863	K. Hartke and H.-G. Muller; *Arch. Pharm. (Weinheim, Ger.)*, 1988, **321**, 863.	895
88AP(321)873	H.-G. Muller, K. Hartke, T. Kampchen, W. Massa and F. Hahn; *Arch. Pharm. (Weinheim, Ger.)*, 1988, **321**, 873.	896
88AP(321)879	H.-G. Muller and K. Hartke; *Arch. Pharm. (Weinheim, Ger.)*, 1988, **321**, 879.	896
88AP(321)903	K. Hartke and A. Afrashteh; *Arch Pharm (Weinheim, Ger.)*, 1988, **321**, 903.	775
88BAU1686	A. A. Kadyrov, E. M. Rokhlin and M. V. Galakov; *Bull. Acad. Sci. USSR, Div. Chem. Sci.*, 1988, 1686.	802, 803
88BCJ609	K. Nishiyama, M. Saito and M. Oba; *Bull. Chem. Soc. Jpn.*, 1988, **61**, 609.	438
88BCJ2693	Y. Ichinose, K. Oshima and K. Utimoto; *Bull. Chem. Soc. Jpn.*, 1988, **61**, 2693.	639
88BCJ3957	S. Kanemasa, H. Kobayashi, J. Tanaka and O. Tsuge; *Bull. Chem. Soc. Jpn.*, 1988, **61**, 3957.	367, 936, 938
88BSF699	L. Germanaud, S. Brunel, V. Chevalier and P. Le Perchec; *Bull. Soc. Chim. Fr.*, 1988, 699.	127
88BSF989	J. Cossy and J. P. Pete; *Bull. Soc. Chim. Fr.*, 1988, 989.	274
88BSF1009	P. Aube, I. Christot, J.-C. Combret and J.-L. Klein; *Bull. Soc. Chim. Fr.*, 1988, 1009.	406
88BSJ3764	M. Hoshi, Y. Masuda and A. Arase; *Bull. Chem. Soc. Jpn.*, 1988, **61**, 3764.	815
88CA(109)211147t	T. Kh. Gazizov, Yu. V. Chugunov, A. G. Abul'khanov and A. N. Pudovik; *Chem. Abstr.*, 1988, **109**, 211 147t (*Dokl. Akad. Nauk SSSR*, 1988, **298**, 112).	915
88CAR(181)246	J. M. Lacombe, N. Rakotomanomana and A. P. Pavia; *Carbohydr. Res.*, 1988, **181**, 246.	176
88CB799	K. Sünkel and D. Motz; *Chem. Ber.*, 1988, **121**, 799.	707
88CB805	W. Ried and M. A. Jacobi; *Chem. Ber.*, 1988, **121**, 805.	901
88CB1165	E. Schaumann, S. Scheiblich, U. Wriede and G. Adiwidjaja; *Chem. Ber.*, 1988, **121**, 1165.	843, 846, 854
88CB1241	H. Schmidbaur, C. Paschalidis, G. Reber and G. Muller; *Chem. Ber.*, 1988, **121**, 1241.	548, 1023
88CB1393	P. Jutzi and M. Meyer; *Chem. Ber.* 1988, **121**, 1393.	652
88CB1413	D. Lentz and H. Michael; *Chem. Ber.*, 1988, **121**, 1413.	683
88CB1509	H. J. Bestmann, T. Röder and K. Sühs; *Chem. Ber.*, 1988, **121**, 1509.	574
88CB1813	J. Barluenga, A. M. Bayón, P. J. Campos, G. Canal, G. Asensio, E. González-Nuñez and Y. Molina; *Chem. Ber.*, 1988, **121**, 1813.	412
88CB1817	R. Neidlein and D. Kikelj; *Chem. Ber.*, 1988, **121**, 1817.	883, 905
88CB2121	R. Neidlein, W. G. Dauben, A. S. Funhoff and R. J. Ollmann; *Chem. Ber.*, 1988, **121**, 2121.	550
88CC189	S. Takano, T. Ohkawa, S. Tamori, S. Satoh and K. Ogasawara; *J. Chem. Soc., Chem. Commun.*, 1988, 189.	827
88CC333	C. Einhorn, C. Allavena and J-L. Luche; *J. Chem. Soc., Chem. Commun.*, 1988, 333.	146
88CC410	J. Barluenga, B. Olano, S. Fustero, M. de la Concepcion Foces-Foces and F. H. Cano; *J. Chem. Soc., Chem. Commun.*, 1988, 410.	408
88CC503	J. B. Verlhac, J.-P. Quintard and M. Pereyre; *J. Chem. Soc., Chem. Commun.*, 1988, 503.	358
88CC560	M. Yamamoto, H. Izukawa, M. Saiki and K. Yamada; *J. Chem. Soc., Chem. Commun.*, 1988, 560.	206
88CC652	S. Harvey and C. L. Raston; *J. Chem. Soc., Chem. Commun.*, 1988, 652.	654
88CC673	N. J. Grist, G. Hogarth, S. A. R. Knox, B. R. Lloyd, D. V. A. Morton and A. G. Orpen; *J. Chem. Soc., Chem. Commun.*, 1988, 673.	1023, 1032
88CC807	G. Bartoli, M. Bosco, R. Dalpozzo and P. E. Todesco; *J. Chem. Soc., Chem. Commun.*, 1988, 807.	646, 653
88CC822	Y. Shiomi, Y. Aso, T. Otsubo and F. Ogura; *J. Chem. Soc., Chem. Commun.*, 1988, 822.	850
88CC867	A. H. Cowley, S. W. Hall, R. A. Jones and C. M. Nunn; *J. Chem. Soc., Chem. Commun.*, 1988, 867.	480

88CC1076	M. Ochiai, M. Kunishima, K. Fuji, M. Shiro and Y. Nagao; *J. Chem. Soc., Chem. Commun.*, 1988, 1076.	808
88CJC187	H. H. Baer and L. Siemsen; *Can. J. Chem.*, 1988, **66**, 187.	176
88CJC2975	J. A. Lopez Sastre, J. Molina Molina, D. Portal Olea and C. Romero-Avila; *Can. J. Chem.*, 1988, **66**, 2975.	858
88CL211	J. Motoyoshiya and K. Hirata; *Chem. Lett.*, 1988, 211.	802
88CL965	H. Sakurai, K. Ebata, K. Sakamoto, Y. Nakadaira and C. Kabuto; *Chem. Lett.*, 1988, 965.	1051, 1052
88CL1101	T. Mukaiyama, K. Wariishi, Y. Saito, M. Hayashi and S. Kobayashi; *Chem. Lett.*, 1988, 1101.	188
88CL1145	K. Okuma, Y. Komiya and H. Ohta; *Chem. Lett.*, 1988, 1145.	415
88CL1317	J. Nakayama and Y. Sugihara; *Chem. Lett.*, 1988, 1317.	88, 89
88CL1441	H. Sakurai, M. Kudo, K. Sakamoto, Y. Nakadaira, M. Kira and A. Sekiguchi; *Chem. Lett.*, 1988, 1441.	613
88CL1887	T. Abe and E. Hayashi; *Chem. Lett.*, 1988, 1887.	794
88CLY402	V. Pouzar; *Chem. Listy*, 1988, 402.	198
88CPL439	I. Sofer and A. Yogev; *Chem. Phys. Lett.*, 1988, **149**, 439 (*Chem. Abstr.*, 1989, **110**, 114 110).	1083
88CRV429	F. Mathey, *Chem Rev.*, 1988, **88**, 429.	1026, 1031, 1035
88CZ69	U. von Allwörden and G. V. Roschenthaler; *Chem.-Ztg.*, 1988, **112**, 69.	94
88CZ107	M. Fild, W. Handke and H. P. Rieck; *Chem.-Ztg.*, 1988, **112**, 107.	866
88CZ146	R. Francke, J. Heine and G. V. Roeschenthaler; *Chem.-Ztg.*, 1988, **112**, 146.	338
88EUP302389	I. Minamida, K. Iwanaga and T. Okauchi, *Eur. Pat.* 0 302 389 (1988).	973, 980, 983, 984, 986
88EUP264183	S. F. Britcher, T. A. Lyle, W. J. Thompson and S. L. Varga (Merck and Co.); *Eur. Pat.* 264 183 (1988) (*Chem. Abstr.*, 1988, **109**, 92 825t).	97
88EUP293881	C. A. Kettner and S. A. Bhikkappa; *Eur. Pat.* 293 881 (1988) (*Chem. Abstr.*, 1988, **112**, 91 790c).	528
88FES103	M. T. Cocco, C. Congiu, A. Maccioni and A. Plumitallo; *Farmaco Ed. Sci.*, 1988, **43**, 103.	972
88G577	L. Pandolfo and G. Paiaro; *Gazz. Chim. Ital.*, 1988, **118**, 577.	628, 701
88GEP3639877A1	H. Wolf, J. Abbink, B. Becker, B. Homeyer, W. Stendel and K. Moriya; *Ger. Pat.* 3 639 877 A1 (1988).	94
88H(27)1929	R. Stevenson and J. V. Weber; *Heterocycles*, 1988, **27**, 1929.	827
88HCA224	M. Gander-Coquoz and D. Seebach; *Helv. Chim. Acta*, 1988, **71**, 224.	302
88HCA1242	R. Neidlein and F. Lucchesini; *Helv. Chim. Acta*, 1988, **71**, 1242.	875
88IC836	H. H. Murray, G. Garzon, R. G. Raptis, A. A. Mazany, L. C. Porter and J. P. Fackler; *Inorg. Chem.*, 1988, **27**, 4038.	586
88IC1787	R. A. Sachleben, J. H. Burns and G. M. Brown; *Inorg. Chem.*, 1988, **27**, 1787.	123
88IC3248	J. B. Sheridan, K. Garrett, G. L. Geoffroy and A. L. Rheingold; *Inorg. Chem.*, 1988, **27**, 3248.	676
88IC3796	C. Pomp and K. Wieghardt; *Inorg. Chem.*, 1988, **27**, 3796.	150
88IC4038	I. G. Phillips, R. G. Ball and R. G. Cavell; *Inorg. Chem.*, 1988, **27**, 4038.	547
88IC4329	R. Winter and G. L. Gard; *Inorg. Chem.*, 1988, **27**, 4329.	868
88IJC(B)912	S. Kumar and H. K. Pujari; *Indian J. Chem., Sect. B*, 1988, **27**, 912.	437
88IZV159	G. N. Nikonov, I. A. Litvinov, S. N. Ignat'eva, O. A. Erastov and B. A. Arbuzov; *Izv. Akad. Nauk SSSR, Ser. Khim.*, 1988, 159.	342
88IZV514	A. N. Mirskova, S. G. Seredkina, V. A. Shagun, I. D. Kalikhman, V. B. Modonov, N. V. Lutskaya and M. G. Voronkov; *Izv. Akad. Nauk Sci. SSSR, Ser. Khim.*, 1988, **3**, 514.	765
88IZV936	A. N. Chekhlov, I. V. Martynov, A. Yu. Aksinenko and V. B. Sokolov; *Izv. Akad. Nauk SSSR, Ser. Khim.*, 1988, 936 (*Chem. Abstr.*, 1989, **110**, 154 401h).	494
88IZV1448	A. V. Kuchin, S. A. Markova and G. A. Tolstikov; *Izv. Akad. Nauk SSSR, Ser. Khim.*, 1988, 1448.	698
88IZV1946	G. Ya. Zueva, T. I. Khaustova and N. V. Serezhkina; *Izv. Akad. Nauk SSSR, Ser. Khim.*, 1988, 1946.	142
88JA313	F. H. Elsner, H. G. Woo and T. D. Tilley; *J. Am. Chem. Soc.*, 1988, **110**, 313.	353
88JA646	E-i Negishi and K. Akiyoshi; *J. Am. Chem. Soc.*, 1988, **110**, 646.	655, 658, 659
88JA842	J. S. Sawyer, A. Kucerovy, T. L. Macdonald and G. J. McGarvey; *J. Am. Chem. Soc.*, 1988, **110**, 842.	379
88JA1547	R. M. Williams, P. J. Sinclair, D. Zhai and D. Chen; *J. Am. Chem. Soc.*, 1988, **110**, 1547.	102
88JA2248	L. E. Overman and A. S. Thompson; *J. Am. Chem. Soc.*, 1988, **110**, 2248.	199
88JA2310	P. J. Fagan and W. A. Nugent; *J. Am. Chem. Soc.*, 1988, **110**, 2310.	1038
88JA2611	S. Bain, S. Ijadi-Maghsoodi and T. J. Baton; *J. Am. Chem. Soc.*, 1988, **110**, 2611.	607, 621
88JA2662	D. R. Mootoo, V. Date and B. Fraser-Reid; *J. Am. Chem. Soc.*, 1988, **110**, 2662.	176
88JA3231	G. A. Olah, O. Farooq, S. M. F. Farnia, M. R. Bruce, F. L. Clouet, P. R. Morton, G. K. S. Prakash, R. C. Stevens, R. Bau, K. Lammertsma, S. Suzer and L. Andrews; *J. Am. Chem. Soc.*, 1988, **110**, 3231.	697, 698
88JA4336	P. Wan and S. Muralidharan; *J. Am. Chem. Soc.*, 1988, **110**, 4336.	1073
88JA4826	M. Nakada, Y. Urano, S. Kobayashi and M. Ohno; *J. Am. Chem. Soc.*, 1988, **110**, 4826.	354
88JA5383	E. Negishi, L. D. Boardman, H. Sawada, V. Bagheri, A. T. Stoll, J. M. Tour and C. L. Rand; *J. Am. Chem. Soc.*, 1988, **110**, 5383.	809
88JA5533	T. K. Hayes, R. Villani and S. M. Weinreb; *J. Am. Chem. Soc.*, 1988, **110**, 5533.	12
88JA5567	J. Barluenga, M. A. Rodriguez, P. J. Campos and G. Asensio; *J. Am. Chem. Soc.*, 1988, **110**, 5567.	749, 819, 1064
88JA5585	S. A. Laneman, F. R. Fronczek and G. G. Stanley; *J. Am. Chem. Soc.*, 1988, **110**, 5585.	550

88JA6070	C. P. Casey, M. Crocker, G. P. Niccolai, P. J. Fagan and M. S. Konings; *J. Am. Chem. Soc.*, 1988, **110**, 6070.	680
88JA6249	R. J. Linderman and A. Godfrey; *J. Am. Chem. Soc.*, 1988, **110**, 6249.	384
88JA6463	A. Igau, H. Grutzmacher, A. Baceiredo and G. Bertrand; *J. Am. Chem. Soc.*, 1988, **110**, 6463.	567
88JA7868	R. M. Bullock, R. T. Hembre and J. R. Norton; *J. Am. Chem. Soc.*, 1988, **110**, 7868.	679
88JA8153	T. Hayashi, K. Hayashizaki, T. Kiyoi and I. Ito; *J. Am. Chem. Soc.*, 1988, **110**, 8153.	1033
88JA8671	P. T. Meinke and G. A. Krafft; *J. Am. Chem. Soc.*, 1988, **110**, 8671.	287, 376, 393
88JA8679	P. T. Meinke and G. A. Krafft; *J. Am. Chem. Soc.*, 1988, **110**, 8679.	287
88JAP267291	M. Fujikawa and H. Haneda; *Jpn. Pat.* 0 267 291 (1988) (*Chem. Abstr.*, 1990, **113**, 59 549r).	660
88JAP63185993	Y. Isomura, S. Sakamoto, M. Yoshida and T. Abe; *Jpn. Pat.* 63 185 993 (1988) (*Chem. Abstr.*, 1989, **110**, 24 090).	563
88JCR(S)34	B. Dhawan and D. Redmore; *J. Chem. Res. (S)*, 1988, 34.	470
88JCR(S)292	M. S. Baird and H. H. Hussain; *J. Chem. Res. (S)*, 1988, 292.	742, 747
88JCS(D)341	R. Usón, A. Laguna, A. Usón, P. G. Jones and K. Meyer-Bäse; *J. Chem. Soc., Dalton Trans.*, 1988, 341.	585
88JCS(D)503	J. L. Bookham, W. McFarlane and I. J. Colquhoun; *J. Chem. Soc., Dalton Trans.*, 1988, 503.	704
88JCS(D)1421	B. F. G. Johnson, R. Khattar, J. Lewis, P. R. Raithby and D. N. Smit; *J. Chem. Soc., Dalton Trans.*, 1988, 1421.	682
88JCS(D)1773	R. McCrindle, G. J. Arsenault, R. Farwaha, M. J. Hampden-Smith, R. E. Rice and A. McAlees; *J. Chem. Soc., Dalton Trans.*, 1988, 1773.	151
88JCS(D)1843	M. Green, N. K. Jetha, R. J. Mercer, N. C. Norman and A. G. Orpen; *J. Chem. Soc., Dalton Trans.*, 1988, 1843.	676
88JCS(D)2403	L. M. Engelhardt, W-P. Leung, C. L. Raston, G. Salem, P. Twiss and A. H. White; *J. Chem. Soc., Dalton Trans.*, 1988, 2403.	604, 647
88JCS(D)2431	M. E. Garcia, J. C. Jeffery, P. Sherwood and F. G. A. Stone; *J. Chem. Soc., Dalton Trans.*, 1988, 2431, 2443.	712
88JCS(D)2781	H. M. Colquhoun, A. E. Crease, S. A. Taylor and D. J. Williams; *J. Chem. Soc., Dalton Trans.*, 1988, 2781.	799
88JCS(P1)269	P. C. Bulman Page, M. B. van Niel and D. Westwood; *J. Chem. Soc., Perkin Trans. 1*, 1988, 269.	849
88JCS(P1)961	M. K. Shepherd; *J. Chem. Soc., Perkin Trans. 1*, 1988, 961.	20
88JCS(P1)975	S. R. Landor, P. D. Landor, A. Johnson, Z. T. Fomum, J. T. Mbafor and A. E. Nkengfack; *J. Chem. Soc., Perkin Trans. 1*, 1988, 975.	979
88JCS(P1)1039	E. F. Birse, A. McKenzie and A. W. Murray; *J. Chem. Soc., Perkin Trans. 1*, 1988, 1039.	918
88JCS(P1)1149	Y. Takeuchi, M. Asahina, K. Hori and T. Koizumi; *J. Chem. Soc., Perkin Trans. 1*, 1988, 1149.	52, 101, 109
88JCS(P1)1631	J. Barluenga, A. M. Bayón, P. Campos, G. Asensio, E. Gonzalez-Nuñez and Y. Molina; *J. Chem. Soc. Perkin Trans. 1*, 1988, 1631.	300
88JCS(P1)1913	G. W. Kirby and A. N. Trethewey; *J. Chem. Soc., Perkin Trans. 1*, 1988, 1913.	290
88JCS(P1)2137	J. H. Boyer, T. Manimaran and L. T. Wolford; *J. Chem. Soc., Perkin Trans. 1*, 1988, 2137.	447
88JCS(P1)2339	A. R. Katritzky and M. Drewniak; *J. Chem. Soc., Perkin Trans. 1*, 1988, 2339.	449
88JCS(P1)2585	I. M. Dawson, J. A. Gregory, R. B. Herbert and P. G. Sammes; *J. Chem. Soc., Perkin Trans. 1*, 1988, 2585.	139
88JCS(P1)2663	O. Meth-Cohn, C. Moore and H. C. Taljaard; *J. Chem. Soc., Perkin Trans. 1*, 1988, 2663.	237
88JCS(P1)2839	L. Hadjiaropoglou, A. Varvoglis, N. W. Alcock and G. A. Pike; *J. Chem. Soc., Perkin Trans. 1*, 1988, 2839.	1081
88JCS(P2)1107	D. R. Clark, J. Emsley and F. Hibbert; *J. Chem. Soc., Perkin Trans. 2*, 1988, 1107.	42
88JFC(40)41	J. G. Morse and J. J. Mielcarek; *J. Fluorine Chem.*, 1988, **40**, 41.	547
88JFC(41)425	M. Shimizu, G.-H. Cheng and H. Yoshioka; *J. Fluorine Chem.*, 1988, **41**, 425.	751
88JFC(41)435	Y. Okada, M. Kuroboshi and T. Ishihara; *J. Fluorine Chem.*, 1988, **41**, 435.	737
88JGU81	V. D. Sheludyakov, A. Kh. Shukyurov, N. I. Kirilina, N. B. Sokova and A. D. Kirilin; *J. Gen. Chem. USSR (Engl. Transl.)*, 1988, **58**, 81.	140
88JGU465	S. E. Tkachenko, A. N. Yarkevich, S. V. Timofeev and E. N. Tsvetkov; *J. Gen. Chem. USSR (Engl. Transl.)*, 1988, **58**, 465.	121, 122
88JGU876	I. V. Konovalova, É. G. Yarkova, L. A. Burnaeva, G. S. Khafizova, É. K. Khusnutdinova, L. F. Il'ina and A. N. Pudovik; *J. Gen. Chem. USSR (Engl. Transl.)*, 1988, **58**, 876.	493
88JGU1148	I. V. Konovalova, Yu. G. Trishin, L. A. Burnaeva, É. K. Khusnutdinova, V. N. Chistokletov and A. N. Pudovik; *J. Gen. Chem. USSR (Engl. Transl.)*, 1988, **58**, 1148.	493
88JGU2192	O. B. Smolii, V. S. Brovarets, V. V. Pirozhenko and B. S. Drach; *J. Gen. Chem. USSR (Engl. Transl.)*, 1988, **58**, 2192.	99
88JHC119	N. Haider, G. Heinisch and D. Lassnigg; *J. Heterocycl. Chem.*, 1988, **25**, 119.	416
88JHC959	S. Kohra, Y. Tominaga and A. Hosomi; *J. Heterocycl. Chem.*, 1988, **25**, 959.	900
88JOC22	A. Hassner, R. Fibiger and A. S. Amarasekara; *J. Org. Chem.*, 1988, **53**, 22.	313, 314
88JOC78	H.-N. Huang, D. F. Persico, R. J. Lagow and L. C. Clark, Jr.; *J. Org. Chem.*, 1988, **53**, 78.	44
88JOC366	D. G. Cox and D. J. Burton; *J. Org. Chem.*, 1988, **53**, 366.	117, 118
88JOC455	H. J. Reich and J. W. Ringer; *J. Org. Chem.*, 1988, **53**, 455.	701
88JOC633	G. L. Larson, V. C. de Maldonado, L. M. Fuentes and L. E. Torres; *J. Org. Chem.*, 1988, **53**, 633.	831
88JOC845	L. V. Dunkerton, N. K. Adair, J. M. Euske, K. T. Brady and P. D. Robinson; *J. Org. Chem.*, 1988, **53**, 845.	224
88JOC1331	R. C. Hahn; *J. Org. Chem.*, 1988, **53**, 1331.	39

88JOC1569	R. J. Linderman and Y. Suhr; *J. Org. Chem.*, 1988, **53**, 1569.	354
88JOC1806	E. Vilsmaier, G. Kristen and C. Tetzlaff; *J. Org. Chem.*, 1988, **53**, 1806.	1080
88JOC2274	D. Spitzner and T. Zaubitzer; *J. Org. Chem.*, 1988, **53**, 2274.	648
88JOC2829	A. M. Piotrowski, D. B. Malpass, M. P. Boleslawski and J. J. Eisch; *J. Org. Chem.*, 1988, **53**, 2829.	698
88JOC2878	R. J. Linderman and A. Ghannam; *J. Org. Chem.*, 1988, **53**, 2878.	387
88JOC2920	U. von der Brüggen, R. Lammers and H. Mayr; *J. Org. Chem.*, 1988, **53**, 2920.	188
88JOC2953	L. A. Paquette and J. A. Oplinger; *J. Org. Chem.*, 1988, **53**, 2953.	224
88JOC2979	P. Garner and J. M. Park; *J. Org. Chem.*, 1988, **53**, 2979.	168
88JOC3089	A. Oku, T. Harada, K. Hatori, Y. Nozaki and Y. Yamaura; *J. Org. Chem.*, 1988, **53**, 3089.	816
88JOC3134	S. Harvey, P. C. Junk, C. L. Raston and G. Salem; *J. Org. Chem.*, 1988, **53**, 3134.	654
88JOC3632	P. T. Meinke, G. A. Krafft and A. Guram; *J. Org. Chem.*, 1988, **53**, 3632.	947
88JOC3647	T. F. Braish, J. C. Saddler and P. L. Fuchs; *J. Org. Chem.*, 1988, **53**, 3647.	948
88JOC4015	C. Gennari and P. G. Cozzi; *J. Org. Chem.*, 1988, **53**, 4015.	834
88JOC4131	C. R. Johnson and J. R. Medich; *J. Org. Chem.*, 1988, **53**, 4131.	387
88JOC4645	T. G. Archibald and K. Baum; *J. Org. Chem.*, 1988, **53**, 4645.	109
88JOC4708	C. Nájera and M. Yus; *J. Org. Chem.*, 1988, **53**, 4708.	938, 961
88JOC5088	M. Cushman, H. Patel and A. McKenzie; *J. Org. Chem.*, 1988, **53**, 5088.	161
88JOC5179	P. P. Castro, S. Tihomirov and C. G. Gutierrez; *J. Org. Chem.*, 1988, **53**, 5179.	230
88JOC5554	S. B. Mahato, N. B. Mandal, A. K. Pal, S. K. Maitra, C. Lehmann and P. Luger; *J. Org. Chem.*, 1988, **53**, 5554.	745
88JOC5574	P. G. Gassman and S. J. Burns; *J. Org. Chem.*, 1988, **53**, 5574.	186
88JOC5584	P. C. M. Chan and J. M. Chong; *J. Org. Chem.*, 1988, **53**, 5584.	386
88JOC5750	W. F. Jarvis, M. D. Hoey, A. L. Finocchio and D. C. Dittmer; *J. Org. Chem.*, 1988, **53**, 5750.	234
88JOC5783	R. C. Hahn; *J. Org. Chem.*, 1988, **53**, 5783.	214
88JOC5974	M. Fujiwara, M. Imada, A. Baba and H. Matsuda; *J. Org. Chem.*, 1988, **53**, 5974.	829
88JOM(338)C31	M. F. M. Al-Dulaymmi, P. B. Hitchcock and R. L. Richards; *J. Organomet. Chem.*, 1988, **338**, C31.	547
88JOM(338)159	O. S. Akkerman and F. Bickelhaupt; *J. Organomet. Chem.*, 1988, **338**, 159.	605, 626, 636, 637
88JOM(338)195	S. Kerschl and B. Wrackmeyer; *J. Organomet. Chem.*, 1988, **338**, 195.	1068
88JOM(338)295	M.-P. Teulade and P. Savignac; *J. Organomet. Chem.*, 1988, **338**, 295.	132, 802
88JOM(339)259	K. Shitara, Y. Sato and R. Nakagawa; *J. Organomet. Chem.*, 1988, **339**, 259.	521
88JOM(339)267	B. Elissondo, J.-B. Verlhac, J.-P. Quintard and M. Pereyre; *J. Organomet. Chem.*, 1988, **339**, 267.	533
88JOM(341)109	G. Fritz, H. Volk, M. Straub, H. G. Von Schering, K. Peters and E. M. Peters; *J. Organomet. Chem.*, 1988, **341**, 109.	609
88JOM(341)125	F. Duboudin, M. Birot, O. Babot, J. Dunoguës and R. Calas; *J. Organomet. Chem.*, 1988, **341**, 125.	624
88JOM(341)225	M. G. Voronkov, N. F. Chernov and E. M. Perlova; *J. Organomet. Chem.*, 1988, **341**, 225.	152
88JOM(341)293	J. J. Eisch and J. E. Galle; *J. Organomet. Chem.*, 1988, **341**, 293.	602
88JOM(342)45	J. Heck and G. Rist; *J. Organomet. Chem.*, 1988, **342**, 45.	673
88JOM(344)61	J. Grobe, J. Szameitat and M. Möller; *J. Organomet. Chem.*, 1988, **344**, 61.	112
88JOM(345)39	J. Barrau, G. Rima, M. El-Amine and J. Satge; *J. Organomet. Chem.*, 1988, **345**, 39.	627
88JOM(346)C1	R. Tarhouni, B. Kirkschleger and J. Villieras; *J. Organomet. Chem.*, 1988, **346**, C1.	146
88JOM(346)1	K. Shitara and Y. Sato; *J. Organomet. Chem.*, 1988, **346**, 1.	521
88JOM(346)297	E. Lukevics, O. Pudova, R. Sturkovich and A. Gaukhman; *J. Organomet. Chem.*, 1988, **346**, 297.	1044
88JOM(346)341	D. W. Hutchinson and D. M. Thornton; *J. Organomet. Chem.*, 1988, **346**, 341.	560
88JOM(348)25	N. Kanemoto, Y. Sato and S. Inoue; *J. Organomet. Chem.*, 1988, **348**, 25.	638, 663
88JOM(349)11	N. E. Miller and D. L. Reznicek; *J. Organomet. Chem.*, 1988, **349**, 11.	524
88JOM(353)C30	R. W. Hoffmann and M. Julius; *J. Organomet. Chem.*, 1988, **353**, C30.	146
88JOM(353)103	M. R. Churchill, J. W. Ziller, J. R. Shapley and W.-Y. Yeh; *J. Organomet. Chem.*, 1988, **353**, 103.	683
88JOM(354)139	R. Tacke and R. Rohr-Aehle; *J. Organomet. Chem.*, 1988, **354**, 139.	352
88JOM(354)147	R. Tacke and B. Becker; *J. Organomet. Chem.*, 1988, **354**, 147.	142, 362
88JOM(354)155	S. Urayama, S. Inoue and Y. Sato; *J. Organomet. Chem.*, 1988, **354**, 155.	604, 625, 638
88JOM(355)71	K. Issleib, H. Schmidt and E. Leissring; *J. Organomet. Chem.*, 1988, **355**, 71.	453
88JOM(356)285	J. Kowalski and J. Chojnowski; *J. Organomet. Chem.*, 1988, **356**, 285.	566, 573
88JOU2003	G. A. Gareev, L. P. Kirillova, V. M. Shul'gina, S. R. Buzilova, L. P. Vologdina and L. I. Vereshchagin; *J. Org. Chem. USSR (Engl. Transl.)*, 1988, **24**, 2003.	106
88KGS852	O. G. Safiev, D. V. Nazarov, V. V. Zorin and D. L. Rukhmankulov; *Khim. Geterotsikl. Soedin.*, 1988, 852 (*Chem. Abstr.*, 1989, **110**, 212 656).	1081
88KGS1144	B. B. Rivkin, I. D. Sadekov and V. I. Minkin; *Khim. Geterotsikl. Soedin.*, 1988, 1144 (*Chem. Abstr.*, 1989, **110**, 192 729).	290, 291
88LA559	L. F. Tietze, A. Goerlach and M. Beller; *Liebigs Ann. Chem.*, 1988, 559.	182
88LA595	J. Pielichowski and D. Bogdal; *Liebigs Ann. Chem.*, 1988, 595.	793
88LA933	K. Hartke and E. Pfleging; *Liebigs Ann. Chem.*, 1988, 933.	834
88LA1169	R. Zschiesche, T. Hafner and H.-U. Reissig; *Liebigs Ann. Chem.*, 1988, 1169.	108
B-88MI 407-01	L. W. Haynes and A. G. Cook; in "Enamines: Synthesis, Structure and Reactivity," (ed.) A. G. Cook, Marcel Dekker, New York, 1988, p. 103.	294

B-88MI 408-01	S. Oae and Y. Uchida; in "The Chemistry of Sulphones and Sulphoxides," ed. S. Patai, Z. Rappoport and C. Stirling, Wiley, New York, 1988, p. 583.	390
B-88MI 410-01	R. S. Edmundson (ed.); "Dictionary of Organophosphorus Compounds," Chapman and Hall, London, 1988.	451
88MI 411-01	K. B. G. Torssell; "Nitrile Oxides, Nitrones and Nitronates in Organic Synthesis," VCH, New York, 1988, p. 95.	540
88MI 412-01	G. H. Robinson, M. F. Self, S. A. Sangokoya and W. T. Pennington; *J. Crystallogr. Spectrosc. Res.*, 1988, **18**, 285.	563
B-88MI 414-01	E. Colvin; "Silicon in Organic Synthesis," 2nd edn., Butterworths, London, 1988.	646
88MI 416-01	A. S. Smirnov, T. G. Kasatkina and A. N. Artemov; *Metalloorg. Khim.*, 1988, **1**, 1169.	715
88MI 417-01	V. Knoppova, P. Vetesnik, D. Vegh and E. Gazurova; *Chem. Papers*, 1988, **42**, 263.	764
B-88MI 419-01	E. W. Colvin; "Silicon Reagents in Organic Synthesis," Academic Press, London, 1988.	831
88MI 419-02	G. Valle, G. Licini and O. de Lucchi; *Z. Krystallogr.*, 1988, **183**, 253.	861
88MI 421-01	K. Peseke, R. M. Bartroli and J. Q. Suarez; *Wiss. Z. Wilhelm-Pieck-Univ-Rostock Naturwiss. Reihe*, 1988, **37**, 46.	967
88MI 424-01	G. O. Pritchard, V. H. Kennedy, G. M. Heldroon, M. L. Piasecki, K. A. Johnson and D. R. Golan; *Int. J. Chem. Kinet.*, 1987, **19**, 963.	1080
88MIP4121	W. Urbaniak and B. Marciniec; *Pol. Pat.* 145 739 (1988) (*Chem. Abstr.*, 1990, **112**, 158 638).	568
88NJC559	G. Conole, K. Henrick, M. McPartlin, A. D. Horton and M. J. Mays; *New J. Chem.*, 1988, **12**, 559.	675
88OM210	P. Ganis, G. Paiaro, L. Pandolfo and G. Valle; *Organometallics*, 1988, **7**, 210.	628, 701
88OM289	J. B. Sheridan, D. B. Pourreau, G. L. Geoffroy and A. L. Rheingold; *Organometallics*, 1988, **7**, 289.	676
88OM572	B. A. Boyd, R. J. Thoma, W. H. Watson and R. H. Nielson; *Organometallics*, 1988, **7**, 572.	566
88OM739	S. Inoue, Y. Sato and T. Suzuki; *Organometallics*, 1988, **7**, 739.	625, 637, 1055, 1058
88OM934	C. P. Casey and P. C. Vosejpka; *Organometallics*, 1988, **7**, 934.	680
88OM997	J. Vincente, M.-T. Chicote, I. Saura-Llamas, P. G. Jones, K. Meyer-Bäse and C. F. Erdbrügger; *Organometallics*, 1988, **7**, 997.	586
88OM1106	H. Werner, L. Hofmann, W. Paul and U. Schubert; *Organometallics*, 1988, **7**, 1106.	149, 583
88OM1208	D. F. Dempsey and S. Gregory; *Organometallics*, 1988, **7**, 1208.	656
88OM1465	R. E. Cramer, M. A. Bruck and J. W. Gilje; *Organometallics*, 1988, **7**, 1465.	589
88OM1516	O. T. Beachley, Jr. and J. C. Pazik; *Organometallics*, 1988, **7**, 1516.	659
88OM1715	J. A. Heppert, M. A. Morgenstern, D. M. Scherubel, F. Takusagawa and M. R. Shaker; *Organometallics*, 1988, **7**, 1715.	722
88OM1828	C. M. Fendrick, L. D. Schertz, V. W. Day and T. J. Marks; *Organometallics*, 1988, **7**, 1828.	656
88OM1882	W. Ando and T. Tsumuraya; *Organometallics*, 1988, **7**, 1882.	629
88OM2082	J. F. Hoover and J. M. Stryker; *Organometallics*, 1988, **7**, 2082.	585
88OM2279	R. Uson, J. Fornies, M. Tomas, J. M. Casas, F. A. Cotton, L. R. Falvello and R. Llusar; *Organometallics*, 1988, **7**, 2279.	692
88OM2415	S. Wang and J. P. Fackler; *Organometallics*, 1988, **7**, 2415.	587
88OM2450	M. J. Chetcuti and K. A. Green; *Organometallics*, 1988, **7**, 2450.	713
88OM2553	A. G. M. Barrett, J. Mortier, M. Sabat and M. A. Sturgess; *Organometallics*, 1988, **7**, 2553.	1019
88OM2566	T. J. Lynch, R. Dominguez and M. C. Helvenston; *Organometallics*, 1988, **7**, 2566.	706
88OR(35)513	M. Hudlický;; *Org. React.*, 1988, **35**, 513.	10
88OR(36)176	R. Engel; *Org. React.*, 1988, **36**, 176.	451
88OS(66)14	R. L. Danheiser, D. M. Fink, K. Okano, Y.-M. Tsai and S. W. Szczepanski; *Org. Synth.*, 1988, **66**, 14.	359
88OSC(4)474	M. Gaudry, Y. Jasor and T. B. Khac; *Org. Synth., Coll. Vol.*, 1988, **4**, 474.	405
88OSC(6)5	J. D. Milkowski, D. F. Verber and R. Hirschmann; *Org. Synth., Coll. Vol.*, 1988, **6**, 5.	296
88OSC(6)64	R. S. Brinkmeyer, E. W. Collington and A. I. Meyers; *Org. Synth., Coll. Vol.*, 1988, **6**, 64.	306
88OSC(6)282	B. Haveaux, A. Dekoker, M. Rens, A. R. Sidani, J. Toye and L. Ghosez; *Org. Synth., Coll. Vol.*, 1988, **6**, 282.	790
88OSC(6)403	L. A. Carpino and L. V. McAdams, III; *Org. Synth., Coll. Vol.*, 1988, **6**, 403.	83
88OSC(6)664	W. J. Middleton and H. D. Carlson; *Org. Synth., Coll. Vol.*, 1988, **6**, 664.	297
88OSC(6)905	I. R. Politzer and A. I. Meyers; *Org. Synth., Coll. Vol.*, 1988, **6**, 905.	306
88OSC(6)981	A. M. van Leusen and J. Strating; *Org. Synth., Coll. Vol.*, 1988, **6**, 981.	331
88OSC(6)987	B. E. Hoogenboom, O. H. Oldenziel and A. M. van Leusen; *Org. Synth., Coll. Vol.*, 1988, **6**, 987.	331
88OSC(6)1033	R. K. Boeckman, Jr., D. M. Blum, B. Ganem and N. Halvey; *Org. Synth., Coll. Vol.*, 1988, **6**, 1033.	807, 1060
88PAC49	A. Alexakis, P. Mangeney, A. Ghribi, I. Marek, R. Sedrani, C. Guir and J. Normant; *Pure Appl. Chem.*, 1988, **60**, 49.	176
88PJC165	B. Krzyzanowska and S. Pilichowska; *Pol. J. Chem.*, 1988, **62**, 165.	1008
88PJC483	J. Pielichowski and D. Bogdal; *Pol. J. Chem.*, 1988, **62**, 483.	761, 764
88POL129	J. L. Bookham and W. McFarlane; *Polyhedron*, 1988, **7**, 129.	548
88POL239	J. L. Bookham and W. McFarlane; *Polyhedron*, 1988, **7**, 239.	548
88POL759	D. H. Berry and J. E. Bercaw; *Polyhedron*, 1988, **7**, 759.	679
88POL919	M. H. Chisholm, J. C. Huffman and N. S. Marchant; *Polyhedron*, 1988, **7**, 919.	676
88POL1953	S. K. Thomson and G. B. Yound; *Polyhedron*, 1988, **7**, 1953.	658
88POL2023	H. Kawa, B. C. Manley and R. J. Lagow; *Polyhedron*, 1988, **7**, 2023.	618, 668
88PS(37)241	H. H. Karsch and H. U. Reisacher; *Phosphorus Sulfur*, 1988, **37**, 241.	559
88PS(39)27	R. W. McClard and S. A. Jackson; *Phosphorus Sulfur*, 1988, **39**, 27.	127

88PS(40)183	D. G. Cameron, H. R. Hudson, I. A. O. Ojo and M. Pianka; *Phosphorus Sulfur*, 1988, **40**, 183.	129
88RTC393	N. J. R. van Eikema Hommes, F. Bickelhaupt and G. W. Klumpp; *Recl. Trav. Chim. Pays-Bas*, 1988, **107**, 393.	669
88S36	M. A. Tius, X. Gu, J. W. Truesdell, S. Savariar and P. P. Crooker; *Synthesis*, 1988, 36.	187
88S95	J. Otera; *Synthesis*, 1988, 95.	197
88S106	K. Nishiyama and T. Yamaguchi; *Synthesis*, 1988, 106.	432, 433
88S111	S. Kinastowski, S. Wnuk and E. Kaczmarek; *Synthesis*, 1988, 111.	306
88S233	A. Thurkauf, A. E. Jacobson and K. C. Rice; *Synthesis*, 1988, 233.	179
88S236	H. Suzuki, M. Aihara, H. Yamamoto, Y. Takamoto and T. Ogawa; *Synthesis*, 1988, 236.	750
88S274	L. F. Tietze, H. Meier and E. Vob; *Synthesis*, 1988, 274.	195
88S407	J.-P. Senet, G. Sennyey and G. P. Wooden; *Synthesis*, 1988, 407.	59
88S482	Y. Tanabe and Y. Sanemitsu; *Synthesis*, 1988, 482.	103
88S547	F. S. Guziec, Jr. and L. J. SanFilippo; *Synthesis*, 1988, 547.	27
88S614	T. Kitazume and T. Ohnogi; *Synthesis*, 1988, 614.	736
88S707	C. Almansa, A. Moyano, M. A. Pericàs and F. Serratosa; *Synthesis*, 1988, 707.	196
88S743	S. C. Suri and R. D. Chapman; *Synthesis*, 1988, 743.	435, 447
88S775	H. Ahlbrecht and D. Kornetzky; *Synthesis*, 1988, 775.	530
88S854	S. Wershofen and H. D. Scharf; *Synthesis*, 1988, 854.	181, 182
88S981	R. Neidlein and D. Kikelj; *Synthesis*, 1988, 981.	824
88S983	B. Cazes, D. Djahanbini, J. Goré, J.-P. Genêt and J.-M. Gaudin; *Synthesis*, 1988, 983.	535
88S988	H. Sugiyama, Y. Sato and N. Shirai; *Synthesis*, 1988, 988.	508
88SC85	J. Augé; *Synth. Commun.*, 1988, **18**, 85.	603, 609
88SC425	R. Tyka, B. Hägele and J. Peters; *Synth. Commun.*, 1988, **18**, 425.	465
88SC1975	M. Letellier, D. J. McPhee and D. Griller; *Synth. Commun*, 1988, **18**, 1975.	512, 518
88SC2289	X.-R. Bu, Q.-J. Meng and X.-Z. You; *Synth. Commun.*, 1988, **18**, 2289.	438
88SC2337	D. Klemm and G. Geschwend; *Synth. Commun.*, 1988, **18**, 2337.	209
88SMB425	A. B. Bailey, R. D. McCullough, M. D. Mays, D. O. Cowan and K. A. Lerstrup; *Synth. Met.*, 1988, **27**, B425.	876
88SRI69	R. Usón, A. Laguna, M. Laguna, A. Usón and M. C. Gimeno; *Synth. React. Inorg. Metal-Org. Chem.*, 1988, **18**, 69.	586
88SRI163	E. C. Alyea, K. J. Fisher, R. P. Shakya and A. E. Vougioukas; *Synth. React. Inorg. Metal-Org. Chem.*, 1988, **18**, 163.	566
88SRI317	J. Barrau, G. Rima, M. El. Amine and J. Satgé; *Synth. React. Inorg. Metal-Org. Chem.*, 1988, **18**, 317.	141
88SRI695	W. Urbaniak and B. Marciniec; *Synth. React. Inorg. Metal-Org. Chem.*, 1988, **18**, 695.	566
88SRI727	E. Fluck and R. Braun; *Synth. React. Inorg. Metal-Org. Chem.*, 1988, **18**, 727.	564
88SRI317	J. Barrau, G. Rima, M. El Amine and J. Satgé; *Synth. React. Inorg. Metal-Org. Chem.*, 1988, **18**, 317.	362
88T281	E. Block and M. Aslam; *Tetrahedron*, 1988, **44**, 281.	364
88T1667	A. Thomas, J. N. Vishwakarma, S. Apparao, H. Ila and H. Junjappa; *Tetrahedron*, 1988, **44**, 1667.	704
88T1827	E. Schaumann; *Tetrahedron*, 1988, **44**, 1827.	328
88T3139	L. A. Paquette, D. T. DeRussy and R. D. Rogers; *Tetrahedron*, 1988, **44**, 3139.	950, 954
88T3781	G. L. Larson, J. A. Prieto and E. Ortiz; *Tetrahedron*, 1988, **44**, 3781.	652
88T4033	J. A. Soderquist and S. J. H. Lee; *Tetrahedron*, 1988, **44**, 4033.	356, 643, 645
88T4087	E. J. Grayson and G. H. Whitham; *Tetrahedron*, 1988, **44**, 4087.	139
88T4113	J. S. Nowick and R. L. Danheiser; *Tetrahedron*, 1988, **44**, 4113.	930, 931
88T4135	M. Fujita, M. Obayashi and T. Hiyama; *Tetrahedron*, 1988, **44**, 4135.	736
88T4277	K. Fugami, J. Hibino, S. Nakatsukasa, S. Matsubara, K. Oshima, K. Utimoto and H. Nozaki; *Tetrahedron*, 1988, **44**, 4277.	1045
88T5403	T. Bretschneider, W. Miltz, P. Münster and W. Steglich; *Tetrahedron*, 1988, **44**, 5403.	99
88T6855	E. P. Kündig and A. F. Cunningham, Jr.; *Tetrahedron*, 1988, **44**, 6855.	274, 275
88T7007	C. Somoza and O. A. Mascaretti; *Tetrahedron*, 1988, **44**, 7007.	100
88T7013	S.-F. Chen, E. Ho and P. S. Mariano; *Tetrahedron*, 1988, **44**, 7013.	518
88TL25	T. Cohen, S.-H. Jung, M. L. Romberger and D. W. McCullough; *Tetrahedron Lett.*, 1988, **29**, 25.	652
88TL313	T. Satoh, T. Oohara, Y. Ueda and K. Yamakawa; *Tetrahedron Lett.*, 1988, **29**, 313.	76
88TL663	Y. Yamada and K. Mukai; *Tetrahedron Lett.*, 1988, **29**, 663.	485
88TL761	W. H. Pearson, D. P. Szura and W. G. Harter; *Tetrahedron Lett.*, 1988, **29**, 761.	537
88TL1355	E. Torres, E. L. Larson and G. J. McGarvey; *Tetrahedron Lett.*, 1988, **29**, 1355.	1060
88TL1603	G. I. Nikishin, M. N. Elinson and I. V. Makhova; *Tetrahedron Lett.*, 1988, **29**, 1603.	197
88TL1657	J. A. Marshall and W. Y. Gung; *Tetrahedron Lett.*, 1988, **29**, 1657.	386
88TL1691	P. Merlin, J. C. Braekman and D. Daloze; *Tetrahedron Lett.*, 1988, **29**, 1691.	414
88TL1899	J. A. Soderquist and A. Hassner; *Tetrahedron Lett.*, 1988, **29**, 1899.	929
88TL2179	A. P. Brunetiere and J. Y. Lallemand; *Tetrahedron Lett.*, 1988, **29**, 2179.	238, 241
88TL2327	H. Fukuda and T. Endo; *Tetrahedron Lett.*, 1988, **29**, 2327.	825
88TL3003	G. Burton, J. S. Elder, S. C. M. Fell and A. V. Stachulski; *Tetrahedron Lett.*, 1988, **29**, 3003.	738
88TL3125	M. Hirama, H. Hioki and S. Ito; *Tetrahedron Lett.*, 1988, **29**, 3125.	858
88TL3269	A. Krief, W. Dumont and A. F. De Mahieu; *Tetrahedron Lett.*, 1988, **29**, 3269.	399
88TL3365	M. L. Boys, E. W. Collington, H. Finch, S. Swanson and J. F. Whitehead; *Tetrahedron Lett.*, 1988, **29**, 3365.	61, 73, 86
88TL3383	S. A. Batcheller and S. Masamune; *Tetrahedron Lett.*, 1988, **29**, 3383.	628

88TL3773	J. C. Medina, M. Salomon and K. S. Kyler; *Tetrahedron Lett.*, 1988, **29**, 3773.	220
88TL3971	T. Sato, E. Yoshida, T. Kobayashi, J. Otera and H. Nozaki; *Tetrahedron Lett.*, 1988, **29**, 3971.	248
88TL4281	E. Schaumann and A. Kirschning; *Tetrahedron Lett.*, 1988, **29**, 4281.	648, 655
88TL4293	J. M. Lacombe, N. Rakotomanomana and A. A. Pavia; *Tetrahedron Lett.*, 1988, **29**, 4293.	225
88TL4583	V. Bolitt and C. Mioskowski; *Tetrahedron Lett.*, 1988, **29**, 4583.	186
88TL4773	M. Isobe, J. Obeyama, Y. Funabashi and T. Goto; *Tetrahedron Lett.*, 1988, **29**, 4773.	231, 940
88TL4889	C. T. Hewkin, R. F. W. Jackson and W. Clegg; *Tetrahedron Lett.*, 1988, **29**, 4889.	858
88TL5233	R. Brückner and B. Peiseler; *Tetrahedron Lett.*, 1988, **29**, 5233.	235
88TL5237	N. J. R. van Eikema Hommes, F. Bickelhaupt and G. W. Klumpp; *Tetrahedron Lett.*, 1988, **29**, 5237.	618
88TL5729	M. J. Robins and S. F. Wnuk; *Tetrahedron Lett.*, 1988, **29**, 5729.	61
88TL5811	F. Camps, J. Coll, A. Llebaria and J. M. Moretó; *Tetrahedron Lett.*, 1988, **29**, 5811.	810
88TL5863	M. P. Doyle, J. Taunton, S.-M. Oon, M. T. H. Liu, N. Soundararajan, M. S. Platz and J. E. Jackson; *Tetrahedron Lett.*, 1988, **29**, 5863.	138, 154
88TL5893	C. Paulmier, F. Outurquin and J.-C. Plaquevent; *Tetrahedron Lett.*, 1988, **29**, 5893.	89
88TL6177	J. Y. Becker, J. Bernstein, S. Bittner, J. A. R. P. Sarma and L. Shahal; *Tetrahedron Lett.*, 1988, 6177.	873
88TL6395	P. F. Hudrlik, P. E. Holmes and A. M. Hudrlik; *Tetrahedron Lett.*, 1988, **29**, 6395.	357
88TL6475	W. O. Moss, R. H. Bradbury, N. J. Hales and T. Gallagher; *Tetrahedron Lett.*, 1988, **29**, 6475.	860
88TL6549	Z. Wu, D. R. Mootoo and B. Fraser-Reid; *Tetrahedron Lett.*, 1988, **29**, 6549.	199
88TL6697	P. Knochel, C. Xiao and M. C. P. Yeh; *Tetrahedron Lett.*, 1988, **29**, 6697.	651, 724
88TL6729	J. Y. Gauthier, T. Henien, L. Lo, M. Thérien and R. N. Young; *Tetrahedron Lett.*, 1988, **29**, 6729.	247
88TL6733	M. Thérien, J. Y. Gauthier and R. N. Young; *Tetrahedron Lett.*, 1988, **29**, 6733.	248
88TL6749	W. R. Dolbier and C. B. Burkholder; *Tetrahedron Lett.*, 1988, **29**, 6749.	1082
88TL6787	N. S. Simpkins; *Tetrahedron Lett.*, 1988, **29**, 6787.	286
88TL6939	M. Kira, T. Hino, Y. Kubota, N. Matsuyama and H. Sakurai; *Tetrahedron Lett.* 1988, **29**, 6939.	618
88TL6975	F. P. J. T. Rutjes, H. Hiemstra, H. H. Mooiweer and W. N. Speckamp; *Tetrahedron Lett.*, 1988, **29**, 6975.	106
88ZAAC(556)23	G. Fritz and J. Honold; *Z. Anorg. Allg. Chem.*, 1988, **556**, 23.	621
88ZAAC(558)55	N. Auner; *Z. Anorg. Allg. Chem.*, 1988, **558**, 55.	611
88ZAAC(558)87	N. Auner; *Z. Anorg. Allg. Chem.*, 1988, **558**, 87.	651
88ZAAC(561)157	M. Fild, R. Fischer and W. Handke; *Z. Anorg. Allg. Chem.*, 1988, **561**, 157.	550, 553, 563
88ZAAC(566)90	M. Fild, D. Bunke and D. Schomburg; *Z. Anorg. Allg. Chem.*, 1988, **566**, 90.	551, 553, 564
88ZAAC(567)23	E. Fluck, B. Neümuller, R. Braun and G. Heckmann; *Z. Anorg. Allg. Chem.*, 1988, **567**, 23.	575
88ZC269	H. Viola, H. Hartenhauer and R. Mayer; *Z. Chem.*, 1988, 269.	256
88ZN(B)31	F. Gol, G. Hasselkuss, P. C. Knüppel and O. Stelzer; *Z. Naturforsch., Teil B*, 1988, **43**, 31.	550
88ZN(B)727	H. Schmidbaur, J. Rott, G. Reber and G. Mueller; *Z. Naturforsch., Teil B*, 1988, **43**, 727.	629
88ZN(B)739	M. Wieber and K. Rudolph; *Z. Naturforsch., Teil B*, 1988, **43**, 739.	599, 600
88ZN(B)1293	K. Endrich, P. Albuquerque, R. P. Korswagen and M. L. Ziegler; *Z. Naturforsch., Teil B*, 1988, **43**, 1293.	580
88ZN(B)1461	M. Wedler, H. W. Roesky, F. T. Edelmann and U. Behrens; *Z. Naturforsch., Teil B*, 1988, **43**, 1461.	719
88ZOB1567	A. V. Kuchin, S. A. Markova, S. I. Lomakina and G. A. Tolstikov; *Zh. Obshch. Khim.*, 1988, **58**, 1567 (*Chem. Abstr.*, 1988, **110**, 154 361).	1068
88ZOB1665	A. P. Marchenko, G. N. Koldan, V. A. Oleinik, I. S. Zal'tsman and A. M. Pinchuk; *Zh. Obshch. Khim.*, 1988, **58**, 1665.	559
88ZOB1987	I. V. Leont'eva, I. M. Aladzheva, T. A. Mastryukova, P. V. Petrovskii, V. V. Negrebetskii, M. Yu. Antipin, Yu. T. Struchkov and M. I. Kabachnik; *Zh. Obshch. Khim.*, 1988, **58**, 1987.	557
88ZOB1998	O. V. Bykhovskaya, I. V. Leont'eva, I. M. Aladzheva, P. V. Petrovskii, M. Yu. Antipin, Yu. T. Struchkov, T. A. Mastryukova and M. I. Kabachnik; *Zh. Obshch. Khim.*, 1988, **58**, 1998.	557
88ZOB2012	I. M. Aladzheva, I. V. Leont'eva, P. V. Petrovskii, T. A. Mastryukova and M. I. Kabachnik; *Zh. Obshch. Khim.*, 1988, **58**, 2012.	557
88ZOB2456	S. Kh. Nurtdinov, R. A. Fakhrutdinova, R. R. Zamaletdinova, F. F. Sadretdinova, R. B. Sultanova, R. Z. Musin and Yu. Yu. Efremov; *Zh. Obshch Khim.*, 1988, **58**, 2456 (*Chem. Abstr.*, 1991, **112**, 56 045s).	463
88ZOR443	A. M. A-Shura, A. V. Anisimov and E. A. Viktorova; *Zh. Org. Khim*, 1988, **24**, 443.	775
88ZOR509	A. V. Martynov, A. N. Mirskova, I. D. Kalikhman and M. G. Voronkov; *Zh. Org. Khim.*, 1988, **24**, 509 (*Chem. Abstr.*, 1989, **110**, 153 824m).	786
88ZOR1945	T. G. Mannafov and E. V. Islyamova; *Zh. Org. Khim.*, 1988, **24**, 1945 (*Chem. Abstr.*, 1989, **110**, 172 796).	56, 57, 91
88ZOR2374	V. V. Kravchenko, A. A. Kotenko, A. F. Popov, L. I. Kostenko, D. Vegh and J. Kovac; *Zh. Org. Khim.*, 1988, **24**, 2374 (Engl. Transl. 2140).	970
89ACS74	T. Benneche, P. Strande and U. Wiggen; *Acta Chem. Scand.*, 1989, **43**, 74.	54, 57
89ACS706	L.-L. Gundersen, T. Benneche and K. Undheim; *Acta Chem. Scand.*, 1989, **43**, 706.	202
89AG(E)225	O. Wagner, M. Ehle and M. Regitz; *Angew. Chem., Int. Ed. Engl.*, 1989, **28**, 225.	110
89AG(E)277	G. Boche; *Angew. Chem., Int. Ed. Engl.*, 1989, **28**, 277.	390, 391, 540

89AG(E)319	G. E. Herberich, I. Hausmann and N. Klaff; *Angew. Chem., Int. Ed. Engl.*, 1989, **28**, 319.	695, 717
89AG(E)621	A. Igau, A. Baceiredo, G. Trinquier and G. Bertrand; *Angew. Chem., Int. Ed. Engl.*, 1989, **28**, 621.	567, 1082
89AG(E)737	G. E. Herberich, B. A. Dunne and B. Hessner; *Angew. Chem., Int. Ed. Engl.*, 1989, **28**, 737.	717
89AG(E)784	M. Pilz, M. Stadler, R. Hunold, J. Allwohn, W. Massa and A. Berdt; *Angew. Chem., Int. Ed. Engl.*, 1989, **28**, 784.	633
89AG(E)1296	M. A. Bennett and H. P. Schwemlein; *Angew. Chem., Int. Ed. Engl.*, 1989, **28**, 1296.	682
89AJC301	H. G. McFadden, R. L. N. Harris and C. L. D. Jenkins; *Aust. J. Chem.*, 1989, **42**, 301.	704
89AJC1307	B. R. Dent and G. J. Gainsford; *Aust. J. Chem.*, 1989, **42**, 1307.	742, 748
89AP(322)263	W. Meichle and H. W. Otto; *Arch. Pharm. (Weinheim, Ger.)*, 1989, **322**, 263 (*Chem. Abstr.* 1989, **111**, 134 025).	838
89AP(322)593	W. Hanefeld and B. Borho; *Arch. Pharm. (Weinheim, Ger.)*, 1989, **322**, 593.	901, 904
89AQ22	I. Tapia, V. Alcazar, J. R. Moran and M. Grande; *An. Quim., Ser. C (Spain)*, 1989, **85**, 22.	266
89AX(C)1751	R. T. Patterson, J. H. Boyer and E. D. Stevens; *Acta Crystallogr. Sect. C*, 1989, **45**, 1751.	995
89B3541	K. Duncan, W. S. Faraci, D. S. Matteson and C. T. Walsh; *Biochemistry*, 1989, **28**, 3541.	528
89B8270	R. V. Talanian, N. C. Brown, C. E. McKenna, T.-G. Ye, J. N. Levy and G. E. Wright; *Biochemistry*, 1989, **28**, 8270.	129
89BAP117	K. Kabzinska and R. Kawecki; *Bull. Acad. Pol. Sci., Ser. Sci. Chim.*, 1989, **37**, 117.	266
89BAP123	J. Pielichowski and D. Bogdal; *Bull. Acad. Pol. Sci., Ser. Sci. Chim.*, 1989, **37**, 123.	793
89BAU635	G. V. Oreshko, G. V. Lagodzinskaya and L. T. Eremenko; *Bull. Acad. Sci. USSR, Div. Chim. Sci.*, 1989, 635.	795
89BCJ1215	H. Chikashita, S. Komazawa, N. Ishimoto, K. Inoue and K. Itoh; *Bull. Chem. Soc. Jpn.*, 1989, **62**, 1215.	319
89BCJ1358	S. Padmanabhan, T. Ogawa and H. Suzuki; *Bull. Chem. Soc. Jpn.*, 1989, **62**, 1358.	286
89BCJ2050	K. Fugami, K. Oshima and K. Utimoto; *Bull. Chem. Soc. Jpn.*, 1989, **62**, 2050.	188
89CAR(188)81	A. Fernandez-Mayoralas, A. Marra, M. Trumtel, A. Veyrières and P. Sinaÿ; *Carbohydr. Res.*, 1989, **188**, 81.	235
89CB331	J. Vermehren and M. Hanack; *Chem. Ber.*, 1989, **122**, 331.	799
89CB453	E. Niecke, M. Leuer and M. Nieger; *Chem. Ber.*, 1989, **122**, 453.	568
89CB463	R. Gerhardt and K. Seppelt; *Chem. Ber.*, 1989, **122**, 463.	868
89CB519	R. W. Saalfrank and U. Wirth; *Chem. Ber.*, 1989, **122**, 519.	991
89CB903	R. W. Hoffmann and S. Dresely; *Chem. Ber.*, 1989, **122**, 903.	363
89CB1757	H. Fritz and W. Sundermeyer; *Chem. Ber.*, 1989, **122**, 1757.	265
89CB1857	H. Schmidbaur, C. Paschalidis, O. Steigelmann, D. L. Wilkinson and G. Muller; *Chem. Ber.*, 1989, **122**, 1857.	1023
89CB1901	H.-J. Müller, K. Polborn, M. Steinmann and W. Beck; *Chem. Ber.*, 1989, **122**, 1901.	710
89CB2023	B. Kruse and R. Brueckner; *Chem. Ber.*, 1989, **122**, 2023.	381
89CB2311	G. Maas, E.-U. Wurthwein, B. Singer, T. Mayer and D. Krauss; *Chem. Ber.*, 1989, **122**, 2311.	321
89CC89	J. D. Buynak, J. B. Strickland, T. Hurd and A. Phan; *J. Chem. Soc., Chem. Commun.*, 1989, 89.	353
89CC178	D. J. Ager and M. B. East; *J. Chem. Soc., Chem. Commun.*, 1989, 178.	212
89CC337	A. R. Katritzky, L. Urogdi and A. Mayence; *J. Chem. Soc., Chem. Commun.*, 1989, 337.	417
89CC418	S. G. Bott, D. L. Clark, M. L. H. Green and P. Mountford; *J. Chem. Soc., Chem. Commun.*, 1989, 418.	675
89CC498	R. D. Brost and S. R. Stobart; *J. Chem. Soc., Chem. Commun.*, 1989, 498.	685
89CC688	G. Conole, K. A. Hill, M. McPartlin, M. J. Mays and M. J. Morris; *J. Chem. Soc., Chem. Commun.*, 1989, 688.	675
89CC957	M. A. Brook, M. A. Hadi and A. Neuy; *J. Chem. Soc., Chem. Commun.*, 1989, 957.	138
89CC988	J. C. Guillemin, M. Le Guennec and J. M. Denis; *J. Chem Soc., Chem. Commun.*, 1989, 988.	111
89CC1159	P. C. Toscano and E. Barren; *J. Chem. Soc., Chem. Commun.*, 1989, 1159.	821
89CC1256	D. J. Ager, J. E. Gano and S. I. Parekh; *J. Chem. Soc., Chem. Commun.*, 1989, 1256.	352
89CC1559	N. Kamigata, T. Fukushima and M. Yoshida; *J. Chem. Soc., Chem. Commun.*, 1989, 1559.	56
89CC1648	A. Höhn, R. J. Geue, A. M. Sargeson and A. C. Willis; *J. Chem. Soc., Chem. Commun.*, 1989, 1648.	499
89CC1872	R. V. Williams and X. Lin; *J. Chem. Soc., Chem. Commun.*, 1989, 1872.	936
89CJC1125	R. Tuloup, R. Danion-Bougot, D. Danion, J. P. Pradère and L. Toupet; *Can. J. Chem.*, 1989, **67**, 1125.	109
89CJC1144	A. R. Katritzky and J. N. Lam; *Can. J. Chem.*, 1989, **67**, 1144.	322
89CL221	M. Yamamoto, S. Irie, M. Miyashita, S. Kohmoto and K. Yamada; *Chem. Lett.*, 1989, 221.	208
89CL659	Y. Masaki, I. Iwata, I. Mukai, H. Oda and H. Nagashima; *Chem. Lett.*, 1989, 659.	199
89CL737	S. Igarashi, Y. Haruta, M. Ozawa, Y. Nishide, H. Kinoshita and K. Inomata; *Chem. Lett.*, 1989, 737.	193
89CL905	T. Abe, E. Hayashi and T. Shimizu; *Chem. Lett.*, 1989, 905.	794
89CL1959	M. Ogima, S. Hyuga, S. Hara and A. Suzuki; *Chem. Lett.*, 1989, 1959.	954
89CPB184	T. Satoh, A. Sugimoto and K. Yamakawa; *Chem. Pharm. Bull.*, 1989, **37**, 184.	233
89CPB526	A. Otaka, H. Morimoto, N. Fujii, T. Koide, S. Funakoshi and H. Yajima; *Chem. Pharm. Bull.*, 1989, **37**, 526.	218
89CRV1535	D. Matteson; *Chem. Rev.*, 1989, **89**, 1535.	144, 362, 373, 525
89CRV1617	F. Perron and K. F. Albizati; *Chem. Rev.*, 1989, **89**, 1617.	171, 176
89CS33	A. R. Katritzky, A. Bieniek and B. E. Brycki; *Chem. Scr.*, 1989, **29**, 33.	317
89CZ320	R. Francke and G. V. Roschenthaler; *Chem.-Ztg.*, 1989, **113**, 320.	338

89CZ349	W. Wolfsberger; *Chem.-Ztg.*, 1989, **113**, 349.	566, 573
89EJM635	G. Grosa, O. Caputo, M. Ceruti, G. Biglino, J. S. Franzone and R. Cirillo; *Eur. J. Med. Chem.*, 1989, **24**, 635.	259
89EUP297872	M. E. Logan; *Eur. Pat.* 0297872 (1989).	975
89EUP315574	H.-W. Kleeman, H.-J. Urbech, D. Ruppert and B. Schölkens; *Eur. Pat.* 315574 (1989) (*Chem. Abstr.*, 1989, **111**, 233679d).	528
89FP2610926	M. Bowman, R. Olofson, T. Malfroot and J.-P. Senet (Societé Nationale des Poudres et Explosives); *Fr. Pat.* 2610926 (*Chem. Abstr.*, 1989, **110**, 212144q).	921
89H(28)1179	P. Coutrot, A. Elgadi and C. Grison; *Heterocycles*, 1989, **28**, 1179.	473
89H(29)865	Y. Ogawa, K. Hosaka, M. Chin, Z. X. Chen and H. Mitsuhashi; *Heterocycles*, 1989, **29**, 865.	408
89H(29)1877	R. Carceller, J. L. Garcia-Navio, M. L. Izquierdo, J. Alvarez-Builla, J. Sanza-Parcio and F. Florencio; *Heterocycles*, 1989, **29**, 1877.	843
89HCA93	U. Burger and A. O. Bringhen; *Helv. Chim. Acta*, 1989, **72**, 93.	97
89HCA690	B. Herradon and D. Seebach; *Helv. Chim. Acta*, 1989, **72**, 690.	213
89HCA1506	D. Guggisberg, P. Bigler, M. Neuenschwander and P. Engel; *Helv. Chim. Acta*, 1989, **72**, 1506.	852
89IC413	K. V. Katti and R. G. Cavell; *Inorg. Chem.*, 1989, **28**, 413.	552
89IC899	D. A. DuBois and R. H. Neilson; *Inorg. Chem.*, 1989, **28**, 899.	572
89IC1872	S. A. Laneman, F. R. Fronczek and G. G. Stanley; *Inorg. Chem.*, 1989, **28**, 1872.	550
89IC3219	B. A. Boyd and R. H. Nielson; *Inorg. Chem.*, 1989, **28**, 3219.	566
89IC3766	R. Winter, D. H. Peyton and G. L. Gard; *Inorg. Chem.*, 1989, **28**, 3766.	868
89IS120	S. Hietkamp, H. Sommer and O. Stelzer; *Inorg. Synth.*, 1989, **25**, 120.	546
89IZV356	V. A. Petrosyan, M. E. Niyazymbetov, A. A. Fainzil'berg, S. A. Shevelev and V. V. Semyonov; *Izv. Akad. Nauk SSSR, Ser. Khim.*, 1989, 356 (*Chem. Abstr.*, 1989, **111**, 214096).	1083
89IZV850	A. N. Bovin and E. N. Tsvetkov; *Izv. Akad. Nauk SSSR, Ser. Khim.*, 1989, 850.	126
89IZV946	G. N. Nikonov, A. A. Karasik, O. A. Erastov and B. A. Arbuzov; *Izv. Akad. Nauk SSSR, Ser. Khim.*, 1989, 946.	342
89IZV1107	G. V. Oreshko and L. T. Eremenko; *Izv. Akad. Nauk. SSSR, Ser. Khim.*, 1989, 1107.	108
89IZV1806	V. A. Dorokhov, M. F. Gordeev, Z. K. Dem'yanets, M. N. Bochkareva and V. S. Bogdanov; *Izv. Akad. Nauk SSSR, Ser. Khim.*, 1989, 1806 (Engl. Transl. 1654).	975
89IZV2562	U. M. Dzhemilev, A. G. Ibragimov, A. B. Morozov, R. R. Muslukhov and G. A. Tolstikov; *Izv. Akad. Nauk SSSR, Ser. Khim.*, 1989, 2562.	669
89JA658	E. Block, V. Eswarakrishnan, M. Gernon, G. Ofori-Okai, C. Saha, K. Tang and J. Zubieta; *J. Am. Chem. Soc.*, 1989, **111**, 658.	223
89JA1127	J. R. McCarthy, E. T. Jarvi, D. P. Matthews, M. L. Edwards, N. J. Prakash, T. L. Bowlin, S. Mehdi, P. S. Sunkara and P. Bey; *J. Am. Chem. Soc.*, 1989, **111**, 1127.	73
89JA1892	R. P. Short and S. Masamune; *J. Am. Chem. Soc.*, 1989, **111**, 1892.	643, 645
89JA2870	S. L. Buchwald and R. B. Nielsen; *J. Am. Chem. Soc.*, 1989, **111**, 2870.	1061
89JA2981	C. A. Broka and T. Shen; *J. Am. Chem. Soc.*, 1989, **111**, 2981.	380
89JA3089	E.-I. Negishi, K. Akiyoshi, B. O'Connor, K. Takagi and G. Wu; *J. Am. Chem. Soc.*, 1989, **111**, 3089.	658
89JA3336	E. Negishi, S. J. Holmes, J. M. Tour, J. A. Miller, F. E. Cederbaum, D. R. Swanson and T. Takahashi; *J. Am. Chem. Soc.*, 1989, **111**, 3336.	809
89JA3748	A. Sekiguchi, T. Nakanishi, C. Kabuto and H. Sakurai; *J. Am. Chem. Soc.*, 1989, **111**, 3748.	618
89JA4127	W. Kirmse and F. Söllenböhmer; *J. Am. Chem. Soc.*, 1989, **111**, 4127.	619, 621
89JA4399	D. S. Matteson, P. B. Tripathy, A. Sarkar and K. M. Sadhu; *J. Am. Chem. Soc.*, 1989, **111**, 4399.	389, 664, 665
89JA4873	J. A. Soderquist, J. C. Colberg and L. Del Valle; *J. Am. Chem. Soc.*, 1989, **111**, 4873.	630, 631, 633, 644
89JA4967	S. Steenken and R. A. McClelland; *J. Am. Chem. Soc.*, 1989, **111**, 4967.	1072
89JA5949	R. Okazaki, N. Kumon and N. Inamoto; *J. Am. Chem. Soc.*, 1989, **111**, 5949.	376
89JA6265	D. P. Curran, M. H. Chen and D. Kim; *J. Am. Chem. Soc.*, 1989, **111**, 6265.	811
89JA6661	R. L. Halcomb and S. J. Danishefsky; *J. Am. Chem. Soc.*, 1989, **111**, 6661.	188
89JA6881	D. Kahne, S. Walker, Y. Cheng and D. Van Engen; *J. Am. Chem. Soc.*, 1989, **111**, 6881.	232
89JA7199	V. J. Shiner, Jr., M. W. Ensinger and J. C. Huffmann; *J. Am. Chem. Soc.*, 1989, **111**, 7199.	139
89JA7285	S. Yamago and E. Nakamura; *J. Am. Chem. Soc.*, 1989, **111**, 7285.	830
89JA8491	N. A. Le, M. Jones, Jr., F. Bickelhaupt and W. H. de Wolf; *J. Am. Chem. Soc.*, 1989, **111**, 8491.	1083
89JA8737	T. H. Chan and P. Pellon; *J. Am. Chem. Soc.*, 1989, **111**, 8737.	137
89JCR(S)146	J.-P. Desverbne, N. Bitit, J.-P. Pillot and H. Bouas-Laurent; *J. Chem. Res. (S)*, 1989, 146.	621
89JCR(S)152	G. Grynkiewicz, I. Fokt, W. Szeja and H. Fittak; *J. Chem. Res. (S)*, 1989, 152.	176
89JCS(D)105	L. T. Byrne, L. M. Engelhardt, G. E. Jacobsen, W.-P. Leung, R. I. Papasergio, C. L. Raston, B. W. Skelton, P. Twiss and A. H. White; *J. Chem. Soc., Dalton Trans.*, 1989, 105.	646
89JCS(D)169	R. Usón, J. Forniés, M. Tomás, J. M. Casas and R. Navarro; *J. Chem. Soc., Dalton Trans.*, 1989, 169.	692
89JCS(D)761	R. McCrindle, G. J. Arsenault, R. Farwaha, A. J. McAlees and D. W. Sneddon; *J. Chem. Soc., Dalton Trans.*, 1989, 761.	150
89JCS(P1)115	I. Fleming, S. K. Patel and C. J. Urch; *J. Chem. Soc., Perkin Trans. 1*, 1989, 115.	365
89JCS(P1)225	A. R. Katritzky, K. Yannakopoulou, P. Lue, D. Rasala and L. Urogdi; *J. Chem. Soc. Perkin Trans. 1*, 1989, 225.	309

89JCS(P1)379	L. Hatjiarapoglou and A. Varvoglis; *J. Chem. Soc., Perkin Trans. 1*, 1989, 379.	866
89JCS(P1)563	M. J. P. Harger and A. Williams; *J. Chem. Soc., Perkin Trans. 1*, 1989, 563.	130, 131
89JCS(P1)691	J. Barluenga, J. L. Fernandez-Simon, J. M. Concellon and M. Yus; *J. Chem. Soc. Perkin Trans. 1*, 1989, 691.	741, 747, 756, 758
89JCS(P1)879	T. Sato, Y. Wada, M. Nishimoto, H. Ishibashi and M. Ikeda; *J. Chem. Soc., Perkin Trans. 1*, 1989, 879.	67
89JCS(P1)1793	E. Dziadulewicz, M. Giles, W. O. Moss, T. Gallagher, M. Harman and M. B. Hursthouse; *J. Chem. Soc., Perkin Trans. 1*, 1989, 1793.	842, 845, 851
89JCS(P1)2009	J. Parrick, A. Yahya, A. S. Ijaz and J. Yizun; *J. Chem. Soc., Perkin Trans. 1*, 1989, 2009.	21
89JCS(P1)2083	B. F. Bonini, G. Mazzanti, P. Zani and G. Maccagnani; *J. Chem. Soc., Perkin Trans. 1*, 1989, 2083.	371, 944
89JCS(P1)2441	P. C. Bulman Page, S. S. Klair and D. Westwood; *J. Chem. Soc., Perkin Trans. 1*, 1989, 2441.	263
89JFC(42)69	C.-M. Hu and Z.-Q. Xu; *J. Fluorine Chem.*, 1989, **42**, 69.	763
89JFC(45)435	Z.-Y. Yang and D. J. Burton; *J. Fluorine Chem.*, 1989, **45**, 435.	139
89JGU200	E. A. Zheltonogova, V. P. Kobzareva, A. G. Shipov and Yu. I. Baukov; *J. Gen. Chem. USSR (Engl. Transl.)*, 1989, **59**, 200.	100
89JGU285	O. I. Kolodyazhnyi; *J. Gen. Chem. USSR (Engl. Transl.)*, 1989, **59**, 285.	133
89JGU1625	A. G. Shipov, O. B. Artamkina and Yu. I. Baukov; *J. Gen. Chem. USSR (Engl. Transl.)*, 1989, **59**, 1625.	104
89JGU1778	A. N. Bovin and E. N. Tsvetkov; *J. Gen. Chem. USSR (Engl. Transl.)*, 1989, **59**, 1778.	124, 126, 128
89JGU2040	S. V. Ponomarev, E. M. Gromova, S. N. Nikolaeva and A. S. Zolotarev; *J. Gen. Chem. USSR (Engl. Transl.)*, 1989, **59**, 2040.	814
89JGU2194	O. I. Kolodyazhnyi and D. B. Golokhov; *J. Gen. Chem. USSR (Engl. Transl.)*, 1989, **59**, 2194.	116
89JHC1335	R. Neeildlein, D. Kikelj and W. Kramer; *J. Heterocycl. Chem.*, 1989, **26**, 1335.	896, 897
89JHC1405	A. Kasahara, T. Izumi, S. Murakami, K. Miyamoto and T. Hino; *J. Heterocycl. Chem.*, 1989, **26**, 1405.	193
89JHC1555	C. O. Kappe and T. Kappe; *J. Heterocycl. Chem.*, 1989, **26**, 1555.	432
89JHC1771	J. P. Chupp and J. M. Molyneaux; *J. Heterocycl. Chem.*, 1989, **26**, 1771.	843
89JMC997	J. R. Sufrin, A. J. Spiess, D. L. Kramer, P. R. Libby and C. W. Porter; *J. Med. Chem.*, 1989, **32**, 997.	61
89JOC91	P. G. McDougal, Y. I. Oh and D. VanDerveer; *J. Org. Chem.*, 1989, **54**, 91.	165
89JOC317	R. F. Horvath and T. H. Chan; *J. Org. Chem.*, 1989, **54**, 317.	520
89JOC610	S. Czernecki and G. Ville; *J. Org. Chem.*, 1989, **54**, 610.	173
89JOC644	A. Padwa, G. E. Fryxell, J. R. Gasdaska, M. K. Venkatramanan and G. S. K. Wong; *J. Org. Chem.*, 1989, **54**, 644.	520
89JOC992	A. G. M. Barrett, M.-C. Cheng, C. D. Spilling and S. J. Taylor; *J. Org. Chem.*, 1989, **54**, 992.	898
89JOC998	J. A. Miller; *J. Org. Chem.*, 1989, **54**, 998.	816
89JOC1135	Y. Nakano, M. Hamaguchi and T. Nagai; *J. Org. Chem.*, 1989, **54**, 1135.	109
89JOC1479	I.-H. Ooi and R. H. Smithers; *J. Org. Chem.*, 1989, **54**, 1479.	767
89JOC1757	M. Yamamoto, T. Takemori, S. Iwasa, S. Kohmoto and K. Yamada; *J. Org. Chem.*, 1989, **54**, 1757.	938
89JOC1784	T. F. Bates and R. D. Thomas; *J. Org. Chem.*, 1989, **54**, 1784.	620
89JOC1789	M. A. Avery, C. Jennings-White and W. K. M. Chong; *J. Org. Chem.*, 1989, **54**, 1789.	359
89JOC2751	J. Appa Rao and M. P. Cava; *J. Org. Chem.*, 1989, **54**, 2751.	168
89JOC2798	J. S. Nowick and R. L. Danheiser; *J. Org. Chem.*, 1989, **54**, 2798.	934
89JOC2869	T. G. Archibald, L. C. Garver, K. Baum and M. C. Cohen; *J. Org. Chem.*, 1989, **54**, 2869.	108
89JOC3002	R. E. Gawley, K. Rein and S. R. Chemburkar; *J. Org. Chem.*, 1989, **54**, 3002.	531
89JOC3140	D. P. Curran and C. T. Chang; *J. Org. Chem.*, 1989, **54**, 3140.	811
89JOC3718	J. M. McNamara, J. L. Leazer, M. Bhupathy, J. S. Amato, R. A. Reamer, P. J. Reider and E. J. J. Grabowski; *J. Org. Chem.*, 1989, **54**, 3718.	247
89JOC4051	J. A. Soderquist, I. Rivera and A. Negron; *J. Org. Chem.*, 1989, **54**, 4051.	353
89JOC4372	R. J. Mills, N. J. Taylor and V. Snieckus; *J. Org. Chem.*, 1989, **54**, 4372.	604, 646
89JOC4549	S. Bengtsson and T. Högberg; *J. Org. Chem.*, 1989, **54**, 4549.	189
89JOC4723	A. G. M. Barrett, J. A. Flygare and C. D. Spilling; *J. Org. Chem.*, 1989, **54**, 4723.	906
89JOC4771	D. StC. Black, D. C. Craig, O. Giitsidis, R. W. Read, A. Salek and M. A. Sefton; *J. Org. Chem.*, 1989, **54**, 4771.	409
89JOC4866	A. A. Frimer, P. Gilinsky-Sharon, G. Aljadeff, V. Marks and Z. Rosental; *J. Org. Chem.*, 1989, **54**, 4866.	205
89JOC5171	P. Bravo, E. Piovosi, G. Resnati and G. Fronza; *J. Org. Chem.*, 1989, **54**, 5171.	170
89JOC5202	P. Knochel, T. S. Chou, H. G. Chen, M. C. P. Yeh and M. J. Rozema; *J. Org. Chem.*, 1989, **54**, 5202.	385
89JOC5453	Y. Takeuchi, K. Nagata and T. Koizumi; *J. Org. Chem.*, 1989, **54**, 5453.	108, 780
89JOC5502	R. R. Soelch, E. McNierney, G. A. Tannenbaum and D. M. Lemal; *J. Org. Chem.*, 1989, **54**, 5502.	160, 755
89JOC5520	W. P. Dailey, P. Ralli, D. Wasserman and D. M. Lemal; *J. Org. Chem.*, 1989, **54**, 5520.	46
89JOC5613	P. F. Hudrlik, E. L. O. Agwaramgbo and A. M. Hudrlik; *J. Org. Chem.*, 1989, **54**, 5613.	619, 620
89JOC5630	T. Kitazume, T. Ohnogi, H. Miyauchi, T. Yamazaki and S. Watanabe; *J. Org. Chem.*, 1989, **54**, 5630.	736
89JOC5651	W. H. Pearson and A. C. Lindbeck; *J. Org. Chem.*, 1989, **54**, 5651.	530
89JOC5695	A. Castaneda, D. J. Kucera and L. E. Overman; *J. Org. Chem.*, 1989, **54**, 5695.	187, 199

89JOC5828	G. K. Cook, W. J. Hornback, C. L. Jordan, J. H. McDonald, III and J. E. Munroe; *J. Org. Chem.*, 1989, **54**, 5828.	951
89JOC5998	M. Sekine and T. Nakanishi; *J. Org. Chem.*, 1989, **54**, 5998.	219
89JOC6022	A. R. Katritzky, S. Rachwal and B. Rachwal; *J. Org. Chem.*, 1989, **54**, 6022.	301, 308
89JOM(359)205	H. des Abbayes, J.-C. Clement, P. Laurent, J.-J. Yaouanc, G. Tanguy and B. Weinberger; *J. Organomet. Chem.*, 1989, **359**, 205.	582
89JOM(359)285	T. F. Bates and R. D. Thomas; *J. Organomet. Chem.*, 1989, **359**, 285.	602, 650
89JOM(361)123	R. D. Kelly and G. B. Young; *J. Organomet. Chem.*, 1989, **361**, 123.	654, 656
89JOM(361)147	I. D. Kalikhman, A. I. Albanov, O. B. Bannikova, L. I. Belousova, M. G. Voronkov, V. A. Pestunovich, A. G. Shipov, E. P. Kramarova and Yu. I. Baukov; *J. Organomet. Chem.*, 1989, **361**, 147.	140
89JOM(361)231	S. J. Tyler and J. M. Burlitch; *J. Organomet. Chem.*, 1989, **361**, 231.	689
89JOM(362)C27	A. Da Rold, Y. Mugnier, R. Broussier, B. Gautheron and E. Laviron; *J. Organomet. Chem.*, 1989, **362**, C27.	720
89JOM(362)117	H. G. Alt, H. E. Engelhardt and R. D. Rogers; *J. Organomet. Chem.*, 1989, **362**, 117.	581
89JOM(363)7	N. Auner, J. Grobe, T. Schäfer, B. Krebs and M. Dartmann; *J. Organomet. Chem.*, 1989, **363**, 7.	611
89JOM(364)105	A. Kabi-Satpathy, C. S. Bajgur, K. P. Reddy and J. L. Peterson; *J. Organomet. Chem.*, 1989, **364**, 105.	657
89JOM(364)407	J. Vincente, M. T. Chicote and J. Fernandez-Baeza; *J. Organomet. Chem.*, 1989, **364**, 407.	584
89JOM(365)163	E. Rosenberg, J. Bracker-Novak, R. W. Gellert, S. Aime, R. Gobetto and D. Osella; *J. Organomet. Chem.*, 1989, **365**, 163.	682
89JOM(366)175	G. C. A. Bellinger, H. B. Friedrich and J. R. Moss; *J. Organomet. Chem.*, 1989, **366**, 175.	148
89JOM(367)1	P. Martinet, R. Sauvêtre and J.-F. Normant; *J. Organomet. Chem.*, 1989, **367**, 1.	806
89JOM(368)167	T. N. Mitchell and H.-J. Belt; *J. Organomet. Chem.*, 1989, **368**, 167.	567
89JOM(368)67	K. Sünkel and D. Steiner; *J. Organomet. Chem.*, 1989, **368**, 67.	707
89JOM(370)43	O. Samuel, B. Ronan and H. Kagan; *J. Organomet. Chem.*, 1989, **370**, 43.	261
89JOM(371)205	M. Herberhold and H. Kniesel; *J. Organomet. Chem.*, 1989, **371**, 205.	719
89JOM(371)329	C. Caballero, J. A. Chávez, O. Göknur, I. Löchel, B. Nuber, H. Pfisterer, M. L. Ziegler, P. Albuquerque, L. Eguren and R. P. Korswagen; *J. Organomet. Chem.*, 1989, **371**, 329.	581
89JOM(372)183	M. I. Al-Hassan; *J. Organomet. Chem.*, 1989, **372**, 183.	807
89JOM(372)53	G. E. Herberich, U. Büschges, B. A. Dunne, B. Hessner, N. Klaff, D. P. J. Köffer and K. Peters; *J. Organomet. Chem.*, 1989, **372**, 53.	695, 717
89JOM(375)265	J. A. Albanese, D. L. Staley, A. L. Rheingold and J. L. Burmeister; *J. Organomet. Chem.*, 1989, **375**, 265.	584
89JOM(375)59	P. Gomez-Sal, M. Mena, F. Palacios, P. Royo, R. Serrano and S. Martinez Carreras; *J. Organomet. Chem.*, 1989, **375**, 59.	656
89JOM(378)1	M. G. Steinmetz and B. S. Udayakumar; *J. Organomet. Chem.*, 1989, **378**, 1.	1054
89JOM(379)201	P. Charreau, M. Julia and J.-N. Verpeaux; *J. Organomet. Chem.*, 1989, **379**, 201.	742, 746
89JPR906	W. M. Abdou, N. M. Abd El-Rahman and M. R. H. Mahran; *J. Prakt. Chem.*, 1989, **331**, 906.	464
89LA187	E. V. Dehmlow and J. Stüetten; *Liebigs Ann. Chem.*, 1989, 187 (*Chem. Abstr.*, 1989, **110**, 94 454).	704
B-89MI 403-01	R. Appel and F. Knoll; in "Advances in Inorganic Chemistry," ed. A. G. Sykes, Academic Press, New York, 1989, vol. 33, p. 259.	111
89MI 403-02	J.-L. Cabioch, B. Pellerin and J.-M Denis; *Phosphorus Sulfur Silicon*, 1989, **44**, 27.	111
89MI 403-03	V. V. Bashilov, E. V. Maskaeva, V. I. Sokolov and O. A. Reutov; *Organometallic Chemistry in the USSR* (*Engl. Transl.*), 1989, **2**, 590.	152
B-89MI 404-01	S. R. Sandler and W. Karo; "Organic Functional Group Preparations," 2nd edn., Academic Press, San Diego, 1989, vol. 3, p. 1.	176
B-89MI 407-01	R. M. Williams; "Synthesis of Optically Active α-Amino Acids," Pergamon, Oxford, 1989, p. 62.	302
89MI 410-01	B. Tangor, C. Malavaud and M. T. Barraus; *Phosphorus, Sulfur and Silicon*, 1989, **45**, 189.	478
89MI 414-01	V. I. Shiryaev, A. A. Grachev, S. N. Tandura, S. I. Androsenko and N. N. Silkina; *Metallorg. Khim.*, 1989, **2**, 764 (*Chem. Abstr.*, 1990, **112**, 179 254s).	660
89MI 417-01	A. H. Gorgues, D. Stephan and J. Cousseau; *Janssen Chim. Acta.*, 1989, 3.	768
89MRC760	P. Sohar, I. Kovesdi, J. Szabo, A. Katocs, L. Fodor, E. Szucs, G. Bernath and J. Tamas; *Magn. Reson. Chem.*, 1989, **27**, 760.	273
89OM67	M. H. Chisholm, B. W. Eichhorn and J. C. Huffman; *Organometallics*, 1989, **8**, 67.	676
89OM80	M. H. Chisholm, B. W. Eichhorn and J. C. Huffman; *Organometallics*, 1989, **8**, 80.	675
89OM412	D. Pufahl, W. E. Geiger and N. G. Connelly; *Organometallics*, 1989, **8**, 412.	676
89OM1114	D. H. Gibson, S. K. Mandal, K. Owens, W. E. Sattich and J. O. Franco; *Organometallics*, 1989, **8**, 1114.	148
89OM1118	A. D. Hunter; *Organometallics*, 1989, **8**, 1118.	715
89OM1192	R. E. Cramer, S. Roth, F. Edelmann, M. A. Bruck, K. C. Cohn and J. W. Gilje; *Organometallics*, 1989, **7**, 1192.	589
89OM1199	J. A. Heppert, M. E. Thomas-Miller, D. M. Scherubel, F. Takusagawa, M. A. Morgenstern and M. R. Shaker; *Organometallics*, 1989, **8**, 1199.	673, 722
89OM1237	S. Inoue and Y. Sato; *Organometallics*, 1989, **8**, 1237.	609, 638
89OM1371	D. Seyferth, L. L. Anderson and W. M. Davis; *Organometallics*, 1989, **8**, 1371.	1015
89OM1585	J. Barrau, N. Ben Hamida, A. Agrebi and J. Satgé; *Organometallics*, 1989, **8**, 1585.	607, 621, 622, 628, 629

89OM1604	R. Fandos, M. Gómez and P. Royo; *Organometallics*, 1989, **7**, 1604.	579
89OM1626	M. H. Chisholm, D. Ho, J. C. Huffman and N. S. Marchant; *Organometallics*, 1989, **8**, 1626.	675
89OM2001	K. H. Theopold, J. Silvestre, E. K. Byrne and D. S. Richeson; *Organometallics*, 1989, **8**, 2001.	691
89OM2077	M. J. Chetcuti, S. R. McDonald and N. P. Rath; *Organometallics*, 1989, **8**, 2077.	713
89OM2286	T. Tsumuraya and W. Ando; *Organometallics*, 1989, **8**, 2286.	629
89OM2461	F. J. Berg and J. L. Petersen; *Organometallics*, 1989, **8**, 2461.	659
89OM2973	J. F. Hoover and J. M. Stryker; *Organometallics*, 1989, **8**, 2973.	151
89OS(67)133	A. Padwa and W. Dent; *Org. Synth.*, 1989, **67**, 133.	509
89PAC1257	R. R. Schmidt; *Pure Appl. Chem.*, 1989, **61**, 1257.	200
89POL57	B. C. Ankianiec and G. B. Yound; *Polyhedron*, 1989, **8**, 57.	658
89POL1293	J. D. Walker and R. Poli; *Polyhedron*, 1989, **8**, 1293.	562
89PS(42)97	S. A. Laneman, F. R. Fronczek and G. G. Stanley; *Phosphorus, Sulfur Silicon*, 1989, **42**, 97.	545
89S97	T. Schrader and W. Steglich; *Synthesis*, 1989, 97.	1001
89S101	M. Mikolajczyk and P. Balczewski; *Synthesis*, 1989, 101.	925
89S128	A. Cambanis, E. Bäuml and H. Mayr; *Synthesis*, 1989, 128.	192
89S188	J. H. Rigby and A. R. Bellemin; *Synthesis*, 1989, 188.	25
89S307	X.-R. Bu, Q.-J. Meng, X-Z. You and S.-H. Sun; *Synthesis*, 1989, 307	438
89S547	M. Soroka; *Synthesis*, 1989, 547.	465
89S647	A. Ricci and A. Degl'Innocenti; *Synthesis*, 1989, 647.	929
89S677	S. D. Sharma and V. Kaur; *Synthesis*, 1989, 677.	302
89S687	G. Agnel and M. Malacria; *Synthesis*, 1989, 687.	140
89S698	F. Freeman and D. S. H. L. Kim; *Synthesis*, 1989, 698.	883
89S775	A. H. H. Elghanandour, M. M. M. Ramiz and M. H. Elnagdi; *Synthesis*, 1989, 775.	883
89SA(A)1011	M. I. Al-Hassan, I. M. Al-Najjar and M. M. Ahmad; *Spectrochim. Acta, Part A*, 1989, **45a**, 1011.	786
89SC21	D. S. Middleton and N. S. Simpkins; *Synth. Commun.*, 1989, **19**, 21.	187, 826
89SC31	B. Labiad and D. Villemin; *Synth. Commun.*, 1989, **19**, 31.	246
89SC197	M. Stebnik; *Synth. Commun.*, 1989, **19**, 197.	208
89SC433	B. Ku and D. Y. Oh; *Synth. Commun.*, 1989, **19**, 433.	246
89SC901	J. Iqbal, R. R. Srivastava, K. B. Gupta and M. A. Khan; *Synth. Commun.*, 1989, **19**, 901.	187
89SC1479	V. Alks and J. R. Sufrin; *Synth. Commun.*, 1989, **19**, 1479.	99
89SC1891	K. Chang, B. Ku and D. Y. Oh; *Synth. Commun.*, 1989, **19**, 1891.	572
89SC2209	J. W. Lee and D. Y. Oh; *Synth. Commun.*, 1989, **19**, 2209.	347
89SC2213	D. I. Han and D. Y. Oh; *Synth. Commun.*, 1989, **19**, 2213.	370
89SC2383	R. B. Perni; *Synth. Commun.*, 1989, **19**, 2383.	247
89SL28	G. Delogu, O. De Lucchi and P. Maglioli; *Synlett*, 1989, 28.	250
89SL52	P. Kocienski and N. J. Dixon; *Synlett*, 1989, 52.	952, 953
89SUL79	M. V. Ramana Reddy, S. Vijayalakshmi, K. N. Ramana Reddy and D. Bhaskar Reddy; *Sulfur Letters*, 1989, **10**, 79.	272, 275
89T337	K. M. Pietrusiewicz, W. Wisniewski and M. Zablocka; *Tetrahedron*, 1989, **45**, 337.	120, 804
89T495	R. J. Lindermann, A. Godfrey and K. Horne; *Tetrahedron*, 1989, **45**, 495.	384
89T1043	J. A. Marshall and W. Y. Gung; *Tetrahedron*, 1989, **45**, 1043.	386
89T1209	T. Sato, J. Otera and H. Nozaki; *Tetrahedron*, 1989, **45**, 1209.	216
89T1801	R. B. Katz, M. B. Mitchell, P. G. Sammes and R. J. Ife; *Tetrahedron*, 1989, **45**, 1801.	900
89T1859	D. S. Matteson; *Tetrahedron*, 1989, **45**, 1859.	525
89T2005	A. Krief, W. Dumont, M. Clarembeau, G. Bernard and E. Badaoui; *Tetrahedron*, 1989, **45**, 2005.	400
89T2023	A. Krief, W. Dumont, M. Clarembeau and E. Badaoui; *Tetrahedron*, 1989, **45**, 2023.	284
89T2819	Y. Gaoni; *Tetrahedron*, 1989, **45**, 2819.	287
89T2957	A. De Meijere, S. Teichman, D. Yu, J. Kopf, M. Oly and N. von Thienen; *Tetrahedron*, 1989, **45**, 2957.	109
89T6019	L. N. Markovski and V. D. Romanenko; *Tetrahedron*, 1989, **45**, 6019.	111
89T6113	H. M. R. Hoffmann, B. Schmidt and S. Wolff; *Tetrahedron*, 1989, **45**, 6113.	99
89T6667	K. Schank, A.-M. A. A. Wahab, P. Eigen and J. Jager; *Tetrahedron*, 1989, **45**, 6667.	838, 1081
89T7161	S. V. Ley, N. J. Anthony, A. Armstrong, M. G. Brasca, T. Clarke, D. Culshaw, C. Greck, P. Grice, A. B. Jones, B. Lygo et al.; *Tetrahedron*, 1989, **45**, 7161.	286
89T7643	P. C. Bulman Page, M. B. van Niel and J. C. Prodger; *Tetrahedron*, 1989, **45**, 7643.	244, 250
89TL257	L. Troisi, L. Cassidei, L. Lopez, R. Mello and R. Curci; *Tetrahedron Lett.*, 1989, **30**, 257.	188
89TL287	G. Stork and K. Zhao; *Tetrahedron Lett.*, 1989, **30**, 287.	197
89TL349	S. Himdi-Kabbab, P. Pellon and J. Hamelin; *Tetrahedron Lett.*, 1989, **30**, 349.	597, 598
89TL403	G. D. Prestwich and C. Wawrzenczyk; *Tetrahedron Lett.*, 1989, **30**, 403.	654
89TL411	M. H. B. Stowell, J. F. Witte and R. W. McClard; *Tetrahedron Lett.*, 1989, **30**, 411.	563
89TL911	J. A. Monn and K. C. Rice; *Tetrahedron Lett.*, 1989, **30**, 911.	107
89TL967	M. Shimizu and H. Yoshioka; *Tetrahedron Lett.*, 1989, **30**, 967.	137
89TL1197	P. Beak and W.-K. Lee; *Tetrahedron Lett.*, 1991, **30**, 1197.	511, 530
89TL1249	N. Tokitoh, T. Suzuki, A. Itami, M. Goto and W. Ando; *Tetrahedron Lett.*, 1989, **30**, 1249.	837
89TL1641	J. Ichikawa, T. Sonoda and H. Kobayashi; *Tetrahedron Lett.*, 1989, **30**, 1641.	736
89TL1833	J.-M. Pons and P. Kocienski; *Tetrahedron Lett.*, 1989, **30**, 1833.	765
89TL2033	M. J. Coghlan and B. A. Caley; *Tetrahedron Lett.*, 1989, **30**, 2033.	59
89TL2091	K. Takahashi, T. Nihira, K. Takase and K. Shibata; *Tetrahedron Lett.*, 1989, **30**, 2091.	847
89TL2179	S. Nambiar, J. F. Daeuble, R. J. Doyle and K. G. Taylor; *Tetrahedron Lett.*, 1989, **30**, 2179.	224

89TL2183	J. A. Marshall and W. Y. Gung; *Tetrahedron Lett.*, 1989, **30**, 2183.	386
89TL2349	A. G. M. Barrett, M.-C. Cheng, S. Sakdarat, C. D. Spilling and S. J. Taylor; *Tetrahedron Lett.*, 1989, **30**, 2349.	897
89TL2445	O. I. Kolodiazhnyi, D. B. Golokhov and I. E. Boldeskul; *Tetrahedron Lett.*, 1989, **30**, 2445.	116, 133
89TL2575	F. Di Furia, G. Licini, G. Modena and O. De Lucchi; *Tetrahedron Lett.*, 1989, **30**, 2575.	261
89TL2665	L. Engman, J. Persson and U. Tilstam; *Tetrahedron Lett.*, 1989, **30**, 2665.	88, 89, 93
89TL2699	A. Carpita, R. Rossi and B. Scamuzzi; *Tetrahedron Lett.*, 1989, **30**, 2699.	946
89TL2703	R. S. Atkinson and B. J. Kelly; *Tetrahedron Lett.*, 1989, **30**, 2703.	515
89TL2915	M. Reggelin, P. Tebben and D. Hoppe; *Tetrahedron Lett.*, 1989, **30**, 2915.	841
89TL3011	S. D. Rychnovsky and D. E. Mickus; *Tetrahedron Lett.*, 1989, **30**, 3011.	380
89TL3019	R. M. Moriarty, T. E. Hopkins, B. K. Vaid and R. K. Vaid; *Tetrahedron Lett.*, 1989, **30**, 3019.	206
89TL3041	T. H. Chan and D. Wang; *Tetrahedron Lett.*, 1989, **30**, 3041.	506
89TL3155	Y. Ichinose, S. Matsunaga, K. Fugami, K. Oshima and K. Utimoto; *Tetrahedron Lett.*, 1989, **30**, 3155.	808, 811, 813
89TL3267	A. G. Sutherland and R. J. K. Taylor; *Tetrahedron Lett.*, 1989, **30**, 3267.	84
89TL3333	T. K. Sarkar and T. K. Satapathi; *Tetrahedron Lett.*, 1989, **30**, 3333.	401
89TL3415	A. Bulpin, S. Masson and A. Sene; *Tetrahedron Lett.*, 1989, **30**, 3415.	847
89TL3919	J. A. Soderquist and I. Rivera; *Tetrahedron Lett.*, 1989, **30**, 3919.	935
89TL3951	W. Schnurr and M. Regitz; *Tetrahedron Lett.*, 1989, **30**, 3951.	115, 567
89TL3963	N. Kopola, B. Friess, B. Cazes and J. Gore; *Tetrahedron Lett.*, 1989, **30**, 3963.	535
89TL4007	B. Pandey, S. Y. Bal and U. R. Khire; *Tetrahedron Lett.*, 1989, **30**, 4007.	260
89TL4165	N. Machinaga and C. Kibayashi; *Tetrahedron Lett.*, 1989, **30**, 4165.	191
89TL4223	W. Adam, L. Hadjiarapoglan, V. Jäger and B. Seidel; *Tetrahedron Lett.*, 1989, **30**, 4223.	209
89TL4449	I. T. Barnish, C. W. G. Fishwick, D. R. Hill and C. Szantay, Jr.; *Tetrahedron Lett.*, 1989, **30**, 4449.	901
89TL4573	J. Barluenga, M. Tomás, A. Ballesteros and L. A. López; *Tetrahedron Lett.*, 1989, **30**, 4573.	421
89TL5153	Y. Satoh, T. Tayano, S. Hara and A. Suzuki; *Tetrahedron Lett.*, 1989, **30**, 5153.	765
89TL5243	V. Broicher and D. Geffken; *Tetrahedron Lett.*, 1989, **30**, 5243.	160
89TL5309	T. Shono, Y. Matsumur, S. Katoh, K. Ikeda, T. Fujita and T. Kamada; *Tetrahedron Lett.*, 1989, **30**, 5309.	197
89TL5413	B. A. Barner and R. S. Mani; *Tetrahedron Lett.*, 1989, **30**, 5413.	382
89TL5693	J. A. Soderquist and B. Santiago; *Tetrahedron Lett.*, 1989, **30**, 5693.	655
89TL5837	J.-C. Cuevas and V. Snieckus; *Tetrahedron Lett.*, 1989, **30**, 5837.	513
89TL5841	J.-C. Cuevas, P. Patil and V. Snieckus; *Tetrahedron Lett.*, 1989, **30**, 5841.	511
89TL6195	M. Hogenbirk, N. J. R. van Eikema Hommes, G. Schat, O. S. Akkerman, F. Bickelhaupt and G. W. Klumpp; *Tetrahedron Lett.* 1989, **30**, 6195.	618
89TL6311	A. De Mesmaeker, P. Hoffmann, B. Ernst, P. Hug and T. Winkler; *Tetrahedron Lett.*, 1989, **30**, 6311.	239
89TL6379	J. Ichikawa, T. Sonoda and H. Kobayashi; *Tetrahedron Lett.*, 1989, **30**, 6379.	736
89TL6497	W. Adam, L. Hadjiarapoglou and X. Wang; *Tetrahedron Lett.*, 1989, **30**, 6497.	188
89TL6697	S. Kim, J. H. Park and S. Lee; *Tetrahedron Lett.*, 1989, **30**, 6697.	216
89TL6701	M. Ochiai, Y. Takaoka, Y. Masaki, M. Inenaga and Y. Nagao; *Tetrahedron Lett.*, 1989, **30**, 6701.	808
89UK1105	V. A. Petrosyan and M. E. Niyazymbetov; *Usp. Khim.*, 1989, **58**, 1105 (*Chem. Abstr.*, 1990, **112**, 54488).	1083
89UK1122	R. R. Kostikov, A. P. Molchanov and A. F. Khlebnikov; *Usp. Khim.*, 1989, **58**, 1122 (*Chem. Abstr.*, 1990, **112**, 54489).	1082
89UKZ1216	I. M. Bazavova, A. N. Esipenko, V. M. Neplyuev and M. O. Lozinskii; *Ukr. Khim. Zh.* (*Russ. Ed.*), 1989, **55**, 1216.	275, 277
89USP4818774	K. M. Kem; *US Pat.* 4818774 (1989) (*Chem. Abstr.*, 1989, **111**, 115590).	563
89USP4877899	R. L. Carney and T. L. Brown (Sandoz); *US Pat.* 4877899 (1989) (*Chem. Abstr.*, 1990, **113**, 5719c).	773
89ZAAC(576)272	U. Schülke, R. Kayser and P. Neumann; *Z. Anorg. Allg. Chem.*, 1989, **576**, 272.	563
89ZC409	H. Weichmann and F. Richter; *Z. Chem.*, 1989, **29**, 409.	563
89ZN(B)96	A. Kraer, H. Pritzkow and W. Siebert; *Z. Naturforsch.*, Teil B, 1989, **44**, 96.	632
89ZN(B)175	U. Althoff, J. Grobe and D. Le Van; *Z. Naturforsch.*, Teil B, 1989, **44**, 175.	112
89ZN(B)285	H. Schmidbaur and J. Rott; *Z. Naturforsch.*, Teil B, 1989, **44**, 285.	629
89ZOB101	Z. S. Novikova, I. L. Odinets and I. F. Lutsenko; *Zh. Obshch. Khim.*, 1989, **59**, 101.	564
89ZOB571	E. B. Silina, B. N. Kozhushko and V. A. Shokol; *Zh. Obshch. Khim.*, 1989, **59**, 571 (Engl. Transl. 505).	1007, 1009, 1011, 1012, 1013
89ZOB955	N. B. Karlstedt, N. V. Boganova and M. A. Kazankova; *Zh. Obshch. Khim.*, 1989, **59**, 955.	766
89ZOB997	V. G. Rozinov, L. P. Izhboldina, V. I. Donskikh, G. V. Ratovskii, L. M. Sergienko, S. V. Dolgushin and M. Yu. Dmitrichenko; *Zh. Obshch. Khim.*, 1989, **59**, 997.	766
89ZOB1686	R. K. Ismagilov, V. V. Moskva and L. Yu. Kopylova; *Zh. Obshch. Khim.*, 1989, **59**, 1686.	559
89ZOB2145	K. V. Kuvaldin, S. A. Klyuchinskii, V. S. Zavgorodini and A. A. Petrov; *Zh. Obshch. Khim.*, 1989, **59**, 2145.	766
89ZOB2435	N. G. Lukyanenko and V. N. Pastushok; *Zh. Obshch. Khim.*, 1989, **25**, 2435.	418
89ZOB2492	A. D. Sinitsa, N. K. Maidanovich, P. P. Onys'ko and A. K. Shurubura; *Zh. Obshch. Khim.*, 1989, **59**, 2492 (Engl. Transl. 2228).	1009
89ZOR1102	I. M. Bazavova, A. N. Esipenko, V. M. Neplyuev and M. O. Lozinskii; *Zh. Org. Khim.*, 1989, **25**, 1102 (*Chem. Abstr.* 1990, **112**, 77070).	865

89ZOR1470	A. V. Martynov, A. N. Mirskova and M. G. Voronkov; *Zh, Org. Khim.*, 1989, **25**, 1470 (*Chem. Abstr.*, 1990, **112**, 20 731y).	786
89ZOR1773	A. V. Martynov, A. N. Mirskova and M. G. Voronkov; *Zh, Org. Khim.*, 1989, **25**, 1773 (*Chem. Abstr.*, 1990, **112**, 138 683n).	786
90ACR49	T. Hosokawa and S. Murahashi; *Acc. Chem. Res.*, 1990, **23**, 49.	193
90AG(E)315	D. Lentz and D. Preugschat; *Angew. Chem., Int. Ed. Engl.*, 1990, **29**, 315.	821
90AG(E)658	E. Baciocchi and M. Crescenzi; *Angew. Chem., Int. Ed. Engl.*, 1990, **29**, 658.	138
90AG(E)686	H. Wadepohl, W. Galm and H. Pritzkow; *Angew. Chem., Int. Ed. Engl.*, 1990, **29**, 686.	690
90AG(E)933	G. Kaupp and K. Sailer; *Angew. Chem., Int. Ed. Engl.*, 1990, **29**, 933.	411
90AG(E)1424	D. Hoppe, A. Carstens and T. Kramer; *Angew. Chem., Int. Ed. Engl.*, 1990, **29**, 1424.	360, 382
90AG(E)1429	G. R. Gillette, A. Baceiredo and G. Bertrand; *Angew. Chem., Int. Ed. Engl.*, 1990, **29**, 1429.	567, 1082
90AKZ537	N. Yu. Grigorian, R. A. Khachatryan, R. V. Petrovskii and M. G. Indzhikyan; *Arm. Khim. Zh.*, 1990, **43**, 537 (*Chem. Abstr.*, 1991, **115**, 8917).	553
90AP(323)619	W. Hanefeld and B. Borho; *Arch. Pharm. (Weinheim, Ger.)*, 1990, **323**, 619.	901, 904
90BCJ51	J. Tanaka, S. Kanemasa and O. Tsuge; *Bull. Chem. Soc. Jpn.*, 1990, **63**, 51.	367, 946
90BCJ166	T. Hosokawa, Y. Ataka and S. Murahashi; *Bull. Chem. Soc. Jpn.*, 1990, **63**, 166.	193
90BCJ466	J. Tanaka, S. Kanemasa, Y. Ninomiya and O. Tsuge; *Bull. Chem. Soc. Jpn.*, 1990, **63**, 466.	263
90BCJ1160	R. Sato, H. Endoh, A. Abe, S. Yamaichi, T. Goto and M. Saito; *Bull. Chem. Soc. Jpn.*, 1990, **63**, 1160.	419
90BCJ1947	T. Kusumoto, K. Nishide and T. Hiyama; *Bull. Chem. Soc. Jpn.*, 1990, **63**, 1947.	809
90BCJ3036	Y. Nagasaki, K. Kurosawa and T. Tsuruta; *Bull. Chem. Soc. Jpn.*, 1990, **63**, 3036.	660
90BCJ3056	T. Okuyama, N. Haga and T. Fueno; *Bull. Chem. Soc. Jpn.*, 1990, **63**, 3056.	1075
90BSF734	F. Di Furia, G. Licini, G. Modena and G. Valle; *Bull. Soc. Chim. Fr.*, 1990, **127**, 734.	249
90CA230883w	O. G. Orazov, O. G. Safiev, D. Kurbanov, V. V. Zorin, A. M. Syrkin, Yu. K. Khekimov and D. L. Rakhmankulov; *Chem. Abstr.*, 1990, **113**, 230 883w (*Izv. Akad. Nauk Turkm. SSR Ser. Fiz-Tekh Khim. Geol. Nauk.*, 1989, 103).	775
90CAR(202)1	B. M. Trost and C. Nübling; *Carbohydr. Res.*, 1990, **202**, 1.	266
90CB115	R. W. Saalfrank, E. Ackermann, M. Fischer, U. Wirth and H. Zimmermann; *Chem. Ber.*, 1990, **123**, 115.	990
90CB177	B. Schuler and W. Sundermeyer; *Chem. Ber.*, 1990, **123**, 177.	266
90CB423	K. Dziwok, J. Lachmann, D. L. Wilkinson, G. Müller and H. Schmidbaur; *Chem. Ber.*, 1990, **123**, 423.	489
90CB541	Z.-T. Huang and X. Shi; *Chem. Ber.*, 1990, **123**, 541.	899
90CB583	E. V. Dehmlow and J. Wilkenloh; *Chem. Ber.*, 1990, **123**, 583.	16, 25
90CB661	H. Bantel, A. K. Powell and H. Vahrenkamp; *Chem. Ber.*, 1990, **123**, 661.	711
90CB739	L. Weber, T. Matzke and R. Böse; *Chem. Ber.*, 1990, **123**, 739.	580, 581
90CB989	G. U. Spiegel and O. Stelzer; *Chem. Ber.*, 1990, **123**, 989.	546
90CB995	K.-P. Langhans, O. Stelzer and N. Weferling, *Chem. Ber.*, 1990, **123**, 995.	112, 546
90CB1143	M. Schmidt, H. Meier, H.-P. Niedermann and R. Mengel; *Chem. Ber.*, 1990, **123**, 1143.	232
90CB1563	G. Opitz, K. Rieth and T. Ehlis; *Chem. Ber.*, 1990, **123**, 1563.	331
90CC431	J. P. Praly, Z. El Kharraf and G. Descotes; *J. Chem. Soc., Chem. Commun.*, 1990, 431.	432, 433
90CC470	M. R. Bryce, A. J. Moore, D. Lorcy, A. S. Dhindsa and A. Robert; *J. Chem. Soc., Chem. Commun.*, 1990, 470.	870
90CC816	M. R. Bryce, G. Cooke, M. C. Petty, A. S. Dhindsa, D. Lorcy, A. J. Moore, M. C. Petty, M. B. Hursthouse and A. I. Karaulov; *J. Chem. Soc., Chem. Commun.*, 1990, 816.	873
90CC1304	H. Nemoto, H. N. Jimenez and Y. Yamamoto; *J. Chem. Soc., Chem. Commun.*, 1990, 1304.	307
90CC1307	R. Bartsch, P. B. Hitchcock and J. F. Nixon; *J. Chem. Soc., Chem. Commun.*, 1990, 1307.	548
90CC1459	S.-Z. Zhu and Q.-Y. Chen; *J. Chem. Soc., Chem. Commun.*, 1990, 1459.	281, 1081
90CJC41	G. B. Richter-Addo, A. D. Hunter and N. Wichrowska; *Can. J. Chem.*, 1990, **68**, 41.	715
90CJC897	Y. Guindon, Y. Girard, S. Berthiaume, V. Gorys, R. Lemieux and C. Yoakim; *Can. J. Chem.*, 1990, **68**, 897.	217
90CJC1408	I. W. J. Still and J. R. Strautmanis; *Can. J. Chem.*, 1990, **68**, 1408.	704
90CL643	K. Toyota, M. Yoshifuji and K. Hirotsu; *Chem. Lett.*, 1990, 643.	559
90CL2019	S. Lee, T. Takata and T. Endo; *Chem. Lett.*, 1990, 2019.	181
90CL2053	M. Sekiguchi, A. Ogawa, S. Fujiwara, I. Ryu, N. Kambe and N. Sonoda; *Chem. Lett.*, 1990, 2053.	334, 909
90CPB902	M. Date, M. Watanabe and S. Furukawa; *Chem. Pharm. Bull.*, 1990, **38**, 902.	936
90CPB1104	Y. Hanzawa, S. Ishizawa, Y. Kobayashi and T. Taguchi; *Chem. Pharm. Bull.*, 1990, **38**, 1104.	736
90CPB3242	N. Katagiri, S. Ise, N. Watanabe and C. Kaneko; *Chem. Pharm. Bull.*, 1990, **38**, 3242.	265, 853, 854
90CSR147	P. C. B. Page, S. S. Klair and S. Rosenthal; *Chem. Soc. Rev.*, 1990, **19**, 147.	929
90EUP402312	S. J. Mickel and G. Von Sprecher; *Eur. Pat.* 402 312 (1990) (*Chem. Abstr.*, 1980, **114**, 185 750h).	123
90G53	L. Pandolfo and G. Paiaro; *Gazz. Chim. Ital.*, 1990, **120**, 53.	550, 701
90G165	F. Di Furia, G. Licini and G. Modena; *Gazz. Chim. Ital.*, 1990, **120**, 165.	261
90G235	E. N. Prilezhaeva, N. E. Donzova, N. P. Petukhova and V. S. Bogdanov; *Gazz. Chim Ital.*, 1990, **120**, 235.	770, 779
90G569	S. Cossu and O. de Lucchi; *Gazz. Chim. Ital.*, 1990, **120**, 569.	770, 779
90G677	M. F. Ismail, E. I. Enayat, F. A. A. El-Bassiony and H. A. Younes; *Gazz. Chim. Ital.*, 1990, **120**, 677.	303
90H(31)1959	M. A. Azzem and E. Steckhan; *Heterocycles*, 1990, **31**, 1959.	206

90H(31)2029	S. Ohta, A. Maruyama, I. Kawasaki, S. Hatakeyama, M. Ichikawa and T. Guro; *Heterocycles*, 1990, **31**, 2029.	448
90HAC135	D. S. Matteson and E. C. Beedle; *Heteroatom Chem.*, 1990, **1**, 135.	813, 934
90HCA1700	R. Heckendorn; *Helv. Chim. Acta*, 1990, **73**, 1700.	448
90HOU(E12b)1	D. Klamann; *Methoden Org. Chem.* (*Houben-Weyl*), 1990, **E12b**, 1.	333
90IC573	O. D. Gupta, R. L. Kirchmeier and J. M. Shreeve; *Inorg. Chem.*, 1990, **29**, 573.	564
90IC808	K. V. Katti, R. J. Batchelor, F. W. B. Einstein and R. G. Cavell; *Inorg. Chem.*, 1990, **29**, 808.	552
90IC1295	M. J. Chetcuti, L. A. DeLiberato, P. E. Fanwick and B. E. Grant; *Inorg. Chem.*, 1990, **29**, 1295.	713
90IC1674	J. Barreau, N. Ben Hamida, H. Ayrebi and J. Satgé; *Inorg. Chem.*, 1990, **29**, 1674.	581
90IC1817	R. Broussier, A. Da Rold, B. Gautheron, Y. Dromzee and Y. Yeannin; *Inorg. Chem.*, 1990, **29**, 1817.	720
90IC4396	D. Lentz and H. Michael-Schulz; *Inorg. Chem.*, 1990, **29**, 4396.	683
90ICA141	J. Barrau and N. Ben Hamida; *Inorg. Chim. Acta*, 1990, **178**, 141.	629
90ICA159	J. Barrau and N. Ben Hamida; *Inorg. Chim. Acta*, 1990, **175**, 159.	629
90IS177	R. E. Cramer, J. H. Jeong, R. B. Maynard and J. W. Gilje; *Inorg. Synth.*, 1990, **27**, 177.	589
90IZV401	V. A. Dorokhov, M. F. Gordeev, A. V. Komkov and V. S. Bogdanov; *Izv. Akad. Nauk SSSR, Ser. Khim.*, 1990, 401 (Engl. Transl. 340).	979
90IZV1133	G. N. Nikonov and A. A. Karasik; *Izv. Akad. Nauk SSSR, Ser. Khim.*, 1990, 1133.	342
90IZV1693	A. V. Martynov, S. G. Seredkina and A. N. Mirskova; *Izv. Akad. Nauk SSSR, Ser Khim.*, 1990, 1693 (*Chem. Abstr.*, 1991, **114**, 61 632b).	787
90IZV1802	M. E. Niyazymbetov, T. Yu. Rudashevskaya, L. V. Adaevskaya and V. A. Petrosyan; *Izv. Akad. Nauk SSSR, Ser. Khim.*, 1990, 1802.	38
90JA607	J. J. Grabowski and R. C. Lum; *J. Am. Chem. Soc.*, 1990, **112**, 607.	271
90JA891	D. H. R. Barton and M. Ramesh; *J. Am. Chem. Soc.*, 1990, **112**, 891.	240
90JA1382	G. Fraenkel, A. Chow and W. R. Winchester; *J. Am. Chem. Soc.*, 1990, **112**, 1382.	603, 648, 649
90JA1999	K. Honda, S. Inoue and K. Sato; *J. Am. Chem. Soc.*, 1990, **112**, 1999.	507
90JA2392	R. J. Linderman and A. Ghannam; *J. Am. Chem. Soc.*, 1990, **112**, 2392.	358, 359
90JA2425	G. D. Stucky, M. M. Eddy, W. H. Harrison, R. Lagow, H. Kawa and D. E. Cox; *J. Am. Chem. Soc.*, 1990, **112**, 2425.	668
90JA2582	G. Fraenkel, A. Chow and W. R. Winchester; *J. Am. Chem. Soc.*, 1990, **112**, 2582.	648
90JA3152	D. Su, C. Y. Guo, R. D. Willett, B. Scott, R. L. Kirchmeier and J. M. Shreeve; *J. Am. Chem. Soc.*, 1990, **112**, 3152.	803
90JA3191	J. R. Cashman, L. D. Olsen and L. M. Bornheim; *J. Am. Chem. Soc.*, 1990, **112**, 3191.	262
90JA3713	C. P. Casey, P. C. Vosejpka and F. R. Askham; *J. Am. Chem. Soc.*, 1990, **112**, 3713.	581
90JA4284	S. Deycard, J. Lusztyk, K. U. Ingold, F. Zerbetto, M. Z. Zgierski and W. Siebrand; *J. Am. Chem. Soc.*, 1990, **112**, 4284.	1078
90JA4873	J. S. Panek and P. F. Cirillo; *J. Am. Chem. Soc.*, 1990, **112**, 4873.	355
90JA5244	T. Imamoto, T. Oshiki, T. Onozawa, T. Kusumoto and K. Sato; *J. Am. Chem. Soc.*, 1990, **112**, 5244.	570, 575
90JA5609	H. J. Reich, R. C. Holtan and C. Bolm; *J. Am. Chem. Soc.*, 1990, **112**, 5609.	285
90JA6118	J. Yamada, H. Abe and Y. Yamamoto; *J. Am. Chem. Soc.*, 1990, **112**, 6118.	390
90JA6139	A. L. Balch and S. P. Rowley; *J. Am. Chem. Soc.*, 1990, **112**, 6139.	453
90JA6263	K. C. Nicolaou, D. G. McGarry, P. K. Somers, B. H. Kim, W. W. Ogilvie, G. Yiannikouros, C. V. C. Prasad, C. A. Veale and R. R. Hark; *J. Am. Chem. Soc.*, 1990, **112**, 6263.	226
90JA6715	N. E. Takenaka, R. Hamlin and D. M. Lemal; *J. Am. Chem. Soc.*, 1990, **112**, 6715.	46
90JA7113	W. E. Geiger, A. Salzer, J. Edwin, W. von Philipsborn, U. Piantini and A. L. Rheingold; *J. Am. Chem. Soc.*, 1990, **112**, 7113.	683
90JA7431	P. Knochel; *J. Am. Chem. Soc.*, 1990, **112**, 7431.	665
90JA8084	D. Scarpetti and P. L. Fuchs; *J. Am. Chem. Soc.*, 1990, **112**, 8084.	273
90JA8189	K. Utimoto, A. Nakamura and S. Matsubara; *J. Am. Chem. Soc.*, 1990, **112**, 8189.	231
90JA9415	J. R. Baran, Jr. and R. J. Lagow; *J. Am. Chem. Soc.*, 1990, **112**, 9415.	668
90JA9620	G. Erker, R. Zwettler, C. Krüger, R. Noe and S. Werner; *J. Am. Chem. Soc.*, 1990, **112**, 9620.	725
90JBC(265)3738	W. W. Bachovchin, A. G. Plaut, G. R. Flentke, M. Lynch and C. A. Kettner; *J. Biol. Chem.*, 1990, **265**, 3738.	527, 528
90JCS(D)761	R. E. Colborn, A. F. Dyke, B. P. Gracey, S. A. R. Knox, K. A. McPherson, K. A. Mead and A. G. Orpen; *J. Chem. Soc., Dalton Trans.*, 1990, 761.	681
90JCS(D)1161	R. I. Papasergio, B. W. Skelton, P. Twiss, A. H. White and C. L. Raston; *J. Chem. Soc., Dalton Trans.*, 1990, 1161.	604, 647, 648
90JCS(D)1553	N. W. Alcock, P. G. Pringle, P. Bergamini, S. Sostero and O. Traverso; *J. Chem. Soc., Dalton Trans.*, 1990, 1553.	151, 585
90JCS(D)2509	I. J. B. Lin, H. C. Shy, C. W. Liu, L.-K. Liu and S.-K. Yeh; *J. Chem. Soc., Dalton Trans.*, 1990, 2509.	583
90JCS(D)2587	M. J. Fernández, J. Modrego, F. J. Lahoz, J. A. López and L. A. Oro; *J. Chem. Soc., Dalton Trans.*, 1990, 2587.	685
90JCS(P1)83	G. J. Hitchings, M. Helliwell and J. M. Vernon; *J. Chem. Soc., Perkin Trans. 1*, 1990, 83.	303
90JCS(P1)139	P. E. Brown, W. Clegg, Q. Islam and J. E. Steele; *J. Chem. Soc. Perkin Trans. 1*, 1990, 139.	742
90JCS(P1)541	A. R. Katritzky, B. Pilarski and L. Urogdi; *J. Chem. Soc., Perkin Trans. 1*, 1990, 541.	406, 410
90JCS(P1)773	B. Beagley, P. H. Crackett, R. G. Pritchard, R. J. Stoodley and C. W. Greengrass; *J. Chem. Soc., Perkin Trans. 1*, 1990, 773.	274

90JCS(P1)1217	A. R. A. S. Deshmukh, T. I. Reddy, B. M. Bhawal, V. P. Shiralkar and S. Rajappa; *J. Chem. Soc., Perkin Trans* 1., 1990, 1217.	901, 904
90JCS(P1)1717	A. R. Katritzky, S. Rachwal and B. Rachwal; *J. Chem. Soc., Perkin Trans. 1*, 1990, 1717.	306, 310
90JCS(P1)1881	M. S. Baird, B. S. Mahli and L. Sheppard; *J. Chem. Soc. Perkin. Trans. 1*, 1990, 1881.	742
90JCS(P1)2035	H. Kato, S. Toda, Y. Arikawa, M. Masuzawa, M. Hashimoto, K. Ikoma, S.-Z. Wang and A. Miyasaka; *J. Chem. Soc., Perkin Trans. 1*, 1990, 2035.	259
90JCS(P1)2251	A. J. Pearson, S.-H. Lee and F. Gouzoles; *J. Chem. Soc., Perkin Trans. 1*, 1990, 2251.	536
90JCS(P1)2775	A. Pelter, R. S. Ward and G. M. Little; *J. Chem. Soc., Perkin Trans. 1*, 1990, 2775.	840, 961
90JCS(P1)2811	E. F. Birse, M. D. Ironside, L. McQuire and A. W. Murray; *J. Chem. Soc. Perkin Trans. 1*, 1990, 2811.	918
90JCS(P1)2919	U. Verfürth and R. Herrmann; *J. Chem. Soc., Perkin Trans. 1*, 1990, 2919.	273
90JCS(P1)2967	A. T. Hewson, S. K. Richardson and D. A. Sharpe; *J. Chem. Soc., Perkin Trans. 1*, 1990, 2967.	856
90JCS(P1)3217	B. Pandey, S. Y. Bal, U. R. Khire and A. T. Rao; *J. Chem. Soc., Perkin Trans. 1*, 1990, 3217.	260
90JCS(P1)3317	T. Ibrahim, T. J. Grattan and J. S. Whitehurst; *J. Chem. Soc., Perkin Trans. 1*, 1990, 3317.	14
90JCS(P2)1987	S. Bien, S. K. Celebi and M. Kapon; *J. Chem. Soc., Perkin Trans. 2*, 1990, 1987.	267
90JFC(46)265	D. Naumann and J. Kischkewitz; *J. Fluorine Chem.*, 1990, **46**, 265.	704
90JFC(49)75	L. G. Sprague, D. J. Burton, R. D. Gunertne and W. E. Bennett; *J. Fluorine Chem.*, 1990, **49**, 75.	802
90JGU397	S. A. Terent'eva, M. A. Pudovik and A. N. Pudovik; *J. Gen. Chem. USSR (Engl. Transl.)*, 1990, **60**, 397.	478
90JGU967	S. V. Ivanov and M. D. Stadnichuk; *J. Gen. Chem. USSR (Engl. Transl.)*, 1990, **60**, 967.	807
90JGU1351	A. N. Yarkevich, S. E. Tkachenko and E. N. Tsvetkov; *J. Gen. Chem. USSR (Engl. Transl.)*, 1990, **60**, 1351.	119
90JGU1536	O. I. Kolodyazhnyi, D. B. Golokhov and S. N. Ustenko; *J. Gen. Chem. USSR (Engl. Transl.)*, 1990, **60**, 1536.	116, 133
90JGU1541	O. I. Kolodyazhnyi, *J. Gen. Chem. USSR (Engl. Transl.)*, 1990, **60**, 1541.	116
90JGU2208	R. L. Yanilkina, P. I. Gryaznov, Yu. Ya. Efremov, A. N. Pudovik and A. M. Kibardin; *J. Gen. Chem. USSR (Engl. Transl.)*, 1990, **60**, 2208.	106
90JGU2516	A. L. Kovalenko, I. V. Tselinskii and Y. V. Serov; *J. Gen. Chem. USSR (Engl. Transl.)*, 1990, 2516.	418
90JHC139	A. R. Katritzky, Z. Wang and R. J. Offerman; *J. Heterocycl. Chem.*, 1990, **27**, 139.	105
90JHC1419	T. Izumi, Y. Nishimoto, K. Kohei and A. Kasahara; *J. Heterocycl. Chem.*, 1990, **27**, 1419.	193
90JHC1433	J. E. Douglass and M. A. Gebhart; *J. Heterocycl. Chem.*, 1990, **27**, 1433.	281
90JMC1510	K. S. Atwal, G. C. Rovnyak, J. Schwartz, S. Moreland, A. Hedberg, J. Gougoutas, M. F. Malley and D. M. Floyd; *J. Med. Chem.*, 1990, **33**, 1510.	905
90JOC247	L. A. Carpino and Y.-Z. Lin; *J. Org. Chem.*, 1990, **55**, 247.	275
90JOC589	W. R. Dolbier, Jr. and C. R. Burkholder; *J. Org. Chem.*, 1990, **55**, 589.	32, 1082
90JOC1323	F. Gasparrini, M. Giovannoli, D. Misiti, G. Natile and G. Palmieri; *J. Org. Chem.*, 1990, **55**, 1323.	260
90JOC1459	A. T. Nielsen, R. A. Nissan, D. J. Vanderah, C. L. Coon, R. D. Gilardi, C. F. George and J. Flippen-Anderson; *J. Org. Chem.*, 1990, **55**, 1459.	298
90JOC1649	A. Togni and S. D. Pastor; *J. Org. Chem.*, 1990, **55**, 1649.	1034
90JOC1768	M. Ramezanian, A. B. Padias, F. D. Saeva and H. K. Hall, Jr.; *J. Org. Chem.*, 1990, **55**, 1768.	993
90JOC1847	V. A. Dang, R. A. Olofson, P. R. Wolf, M. D. Piteau and J.-P. G. Senet; *J. Org. Chem.*, 1990, **55**, 1847.	59
90JOC1857	H. X. Zhang, F. Guibé and G. Balavoine; *J. Org. Chem.*, 1990, **55**, 1857.	813
90JOC2240	M. P. Bowman, R. A. Olofson, J.-P. Senet and T. Malfroot; *J. Org. Chem.*, 1990, **55**, 2240.	58
90JOC2274	R. Soundararajan and D. Matteson; *J. Org. Chem.*, 1990, **55**, 2274.	631, 635
90JOC2311	A. Thenappan and D. J. Burton; *J. Org. Chem.*, 1990, **55**, 2311.	118
90JOC2644	S. W. Lee and W. C. Trogler; *J. Org. Chem.*, 1990, **55**, 2644.	336
90JOC2920	T. G. Archibald, R. Gilardi, K. Baum and C. George; *J. Org. Chem.*, 1990, **55**, 2920.	428
90JOC2972	A. Revis and T. K. Hilty; *J. Org. Chem.*, 1990, **55**, 2972.	831
90JOC2973	D. P. Matthews and J. R. McCarthy; *J. Org. Chem.*, 1990, **55**, 2973.	79
90JOC2999	D. L. Boger, W. L. Corbett and J. M. Wiggins; *J. Org. Chem.*, 1990, **55**, 2999.	305
90JOC3029	W. A. Pieken and J. W. Kozarich; *J. Org. Chem.*, 1990, **55**, 3029.	53
90JOC3244	S. N. Rao and R. A. M. O'Ferrall; *J. Org. Chem.*, 1990, **55**, 3244.	100
90JOC3562	T. G. Archibald and K. Baum; *J. Org. Chem.*, 1990, **55**, 3562.	731
90JOC4063	J. J. G. S. van der Es, K. Jaarsveld and A. van der Gen; *J. Org. Chem.*, 1990, **55**, 4063.	489
90JOC4069	J. J. G. S. van der Es, A. ten Wolde and A. van der Gen; *J. Org. Chem.*, 1990, **55**, 4069.	489
90JOC4448	D. W. Reynolds, P. E. Cassidy, C. G. Johnson and M. L. Cameron; *J. Org. Chem.*, 1990, **55**, 4448.	730
90JOC4634	R. F. Cunico and C. P. Kuan; *J. Org. Chem.*, 1990, **55**, 4634.	354
90JOC4657	R. M. Williams, D. J. Aldous and S. C. Aldous; *J. Org. Chem.*, 1990, **55**, 4657.	102
90JOC4757	S. F. Wnuk and M. J. Robins; *J. Org. Chem.*, 1990, **55**, 4757.	61, 73, 79
90JOC4777	L. Strekowski, R. L. Wydra, M. T. Cegla, A. Czarny, D. B. Harden, S. E. Patterson, M. A. Battiste and J. M. Coxon; *J. Org. Chem.*, 1990, **55**, 4777.	411
90JOC4782	J. T. Welch and R. W. Herbert; *J. Org. Chem.*, 1990, **55**, 4782.	137
90JOC5110	A. G. M. Barrett and S. Sakdarat; *J. Org. Chem.*, 1990, **55**, 5110.	898
90JOC5308	A. Hosomi, Y. Miyashiro, R. Yoshida, Y. Tominaga, Y. Yanagi and M. Hojo; *J. Org. Chem.*, 1990, **55**, 5308.	900

90JOC5406	E. Negishi, D. R. Swanson and C. J. Rousset; *J. Org. Chem.*, 1990, **55**, 5406.	652
90JOC5420	W. R. Dolbier, H. Wojtowicz and C. R. Burkholder; *J. Org. Chem.*, 1990, **55**, 5420.	1083
90JOC5428	R. J. Linderman and B. D. Griedel; *J. Org. Chem.*, 1990, **55**, 5428.	385
90JOC5446	C. E. Tucker, S. A. Rao and P. Knochel; *J. Org. Chem.*, 1990, **55**, 5446.	709
90JOC5515	J.-M. Fang, W.-C. Chou, G.-H. Lee and S.-M. Peng; *J. Org. Chem.*, 1990, **55**, 5515.	263
90JOC5555	J. Kang, D. H. Kim, J. H. Lee, J. G. Rim, Y. B. Yoon and K. J. Kim; *J. Org. Chem.*, 1990, **55**, 5555.	849
90JOC5680	S. Sengupta and V. Snieckus; *J. Org. Chem.*, 1990, **55**, 5680.	841, 950
90JOC5719	W. H. Pearson, S. C. Bergmeier, S. Degan, K.-C. Lin, Y.-F. Poon, J. M. Schkeryantz and J. P. Williams; *J. Org. Chem.*, 1990, **55**, 5719.	22
90JOC5814	M. Kaino, Y. Naruse, K. Ishihara and H. Yamamoto; *J. Org. Chem.*, 1990, **55**, 5814.	174
90JOC5982	M. P. Bowman, J.-P. G. Senet, T. Malfoot and R. A. Olofson; *J. Org. Chem.*, 1990, **55**, 5982.	920, 921
90JOC6071	P. F. Cirillo and J. S. Panek; *J. Org. Chem.*, 1990, **55**, 6071.	354
90JOC6074	T. Fuchigami, M. Shimojo, A. Konno and K. Nakagawa; *J. Org. Chem.*, 1990, **55**, 6074.	62, 63
90JOC6198	M. Balogh, P. Pennetreau, I. Hermecz and A. Gerstmans; *J. Org. Chem.*, 1990, **55**, 6198.	429, 431
90JOM(332)13	J. P. Picard, J. M. Aizpurua, A. Elyusufi and P. Kowalski; *J. Organomet. Chem.*, 1990, **391**, 13.	1013
90JOM(381)69	L. Torres; *J. Organomet. Chem.*, 1990, **381**, 69.	290
90JOM(381)315	V. Broicher and D. Geffken; *J. Organomet. Chem.*, 1990, **381**, 315.	137
90JOM(384)155	N. Deus, G. Hubener and R. Herrmann; *J. Organomet. Chem.*, 1990, **384**, 155.	1033
90JOM(385)351	H. Fischer and D. Reindl; *J. Organomet. Chem.*, 1990, **385**, 351.	580
90JOM(387)C31	C. P. Casey, F. R. Askham and L. M. Petrovich; *J. Organomet. Chem.*, 1990, **387**, C31.	716
90JOM(387)65	J. Barrau, N. Ben Hamida and J. Satgé; *J. Organomet. Chem.*, 1990, **387**, 65.	641
90JOM(388)57	R. Tacke, B. Becker and H. Lange; *J. Organomet. Chem.*, 1990, **388**, 57.	136, 368
90JOM(391)19	T. Suzuki and P. Y. Lo; *J. Organomet. Chem.*, 1990, **391**, 19.	1044
90JOM(391)165	L. Engman and K. W. Tornroos; *J. Organomet. Chem.*, 1990, **391**, 165.	89
90JOM(391)283	J. B. Verlhac and M. Pereyre; *J. Organomet. Chem.*, 1990, **391**, 283.	358
90JOM(393)33	N. Auner and R. Gleixner; *J. Organomet. Chem.*, 1990, **393**, 33.	611
90JOM(394)37	T. Gouyon, R. Sauvetre and J.-F. Normant; *J. Organomet. Chem.*, 1990, **394**, 37.	751
90JOM(395)27	J. Barrau, N. Ben Hamida and J. Satgé; *J. Organomet. Chem.*, 1990, **395**, 27.	629
90JOM(395)305	I. Moldes, J. Ros, R. Yañez, R. Mathieu, X. Solans and M. Font-Bardía; *J. Organomet. Chem.*, 1990, **395**, 305.	713
90JOM(396)115	E. Rotondo, A. Giannetto and S. Lanza; *J. Organomet. Chem.*, 1990, **396**, 115.	151
90JOM(396)299	T. Suzuki and P. Y. Lo; *J. Organomet. Chem.*, 1990, **396**, 299.	1044
90JOM(397)313	S. K. Mandal, D. M. Ho and M. Orchin; *J. Organomet. Chem.*, 1990, **397**, 313.	148, 150, 581
90JOM(400)19	J. F. Normant; *J. Organomet. Chem.*, 1990, **400**, 19.	817, 818, 820
90JOM(400)255	S. A. R. Knox; *J. Organomet. Chem.*, 1990, **400**, 255.	678
90JPR1035	Z. Huang, W. Gan and L. Li; *J. Prakt. Chem.*, 1990, **332**, 1035.	969, 970
90LA745	D. Schulz and G. Simchen; *Liebigs Ann. Chem.*, 1990, 745.	806
90LA965	W. Kantlehner and U. Greiner; *Liebigs Ann. Chem.*, 1990, 965.	975
90MI 403-01	T. Gajda; *Phosphorus Sulfur Silicon*, 1990, **53**, 327.	133
90MI 406-01	J. R. Cashman, L. D. Olsen, C. E. Lambert and M. J. Presas; *Mol. Pharmacol.*, 1990, **37**, 319.	262
B-90MI 406-02	E. Schaumann; in "Perspectives in the Organic Chemistry of Sulfur," ed. B. Zwanenburg and A. J. H. Klunder, Elsevier, Amsterdam, 1990, p. 251.	266
B-90MI 407-01	R. Guglielmetti; in "Photochromism, Molecules and Systems," ed. H. Dürr and H. Bouas-Laurent, Elsevier, Amsterdam, 1990, p. 314.	307
B-90MI 407-02	N. Y. C. Chu; in "Photochromism, Molecules and Systems," ed. H. Dürr and H. Bouas-Laurent, Elsevier, Amsterdam, 1990, p. 493.	307
B-90MI 407-03	F. R. Hartley (ed); "The Chemistry of Organophosphorus Compounds," Wiley, Chichester, 1990, vol. 1.	336
B-90MI 410-01	F. R. Hartley (ed.); "The Chemistry of Organophosphorus Compounds," Wiley, New York, 1990, vols 1–3.	451
B-90MI 412-01	P. Binger; in "Multiple Bonds and Low Coordination in Phosphorus Chemistry," ed. M. Regitz and O. J. Scherer, Thième Verlag, Stuttgart, 1990, p. 90.	547, 549
B-90MI 412-02	M. Regitz; in "Multiple Bonds and Low Coordination in Phosphorus Chemistry," ed. M. Regitz and O. J. Scherer, Thième Verlag, Stuttgart, 1990, p. 58.	549
90MI 416-01	V. N. Postnov, N. N. Meleshonkova and E. I. Klimova; *Metalloorg. Khim.*, 1990, **3**, 710.	719
90MI 417-01	A. M. Al-Shoura, A. A. Bratkov, A. V. Anisimov and E. A. Viktorova; *Vestn. Mosk. Univ. Ser. 2; Khim.*, 1990, **31**, 392 (*Chem. Abstr.*, 1991, **114**, 163 677n).	775
90MI 421-01	M. Bolourtchian, A. Saednya and R. Nouri; *J. Sci. Islamic Repub. Iran*, 1990, **1**, 117.	1014
90OM886	M. L. H. Green and P. Mountford; *Organometallics*, 1990, **9**, 886.	676
90OM1036	V. Guerchais, D. Astruc, C. M. Nunn and A. H. Cowley; *Organometallics*, 1990, **9**, 1036.	581
90OM1222	P. H. M. Budzelaar, J. H. G. Frijns and A. G. Orpen; *Organometallics*, 1990, **9**, 1222.	545
90OM1314	G. Fraenkel and W. R. Winchester; *Organometallics*, 1990, **9**, 1314.	603, 649
90OM1325	S. Inoue and Y. Sato; *Organometallics*, 1990, **9**, 1325.	625
90OM1345	M. J. Chetcuti, B. E. Grant and P. E. Fanwick; *Organometallics*, 1990, **9**, 1345.	709, 712
90OM1355	T. Hiiro, Y. Atarashi, N. Kambe, S. Fujiwara, A. Ogawa, I. Ryu and N. Sonoda; *Organometallics*, 1990, **9**, 1355.	239, 334, 660
90OM1720	M. M. Olmstead and P. P. Power; *Organometallics*, 1990, **9**, 1720.	656
90OM2061	T. Tsumuraya, S. Sato and W. Ando; *Organometallics*, 1990, **9**, 2061.	628, 629
90OM2234	R. Regragni, P. H. Dixneuf, N. J. Taylor and A. J. Carty; *Organometallics*, 1990, **9**, 2234.	711

90OM2677	D. Seyferth, J. L. Robison and J. Mercer; *Organometallics*, 1990, **9**, 2677.	606, 620, 623, 653
90OM2683	J. L. Hubbard and W. K. McVicar; *Organometallics*, 1990, **9**, 2683.	147, 579
90OM2745	R. P. Hughes, S. J. Doig, R. C. Hemond, W. L. Smith, R. E. Davis, S. M. Gadol and K. D. Holland; *Organometallics*, 1990, **9**, 2745.	687
90OM2993	M. Gruselle, C. Cordier, M. Salmain, H. El Amouri, C. Guérin, J. Vaissermann and G. Jaouen; *Organometallics*, 1990, **9**, 2993.	674
90OR(38)655	H. W. Pinnick; *Org. React.*, 1990, **38**, 655.	541
90OR(39)297	T. T. Tidwell; *Org. React.*, 1990, **39**, 297.	222
90OS(68)8	A. G. Barrett, D. Dhanak, G. G. Graboski and S. J. Taylor; *Org. Synth.*, 1990, **68**, 8.	906
90OS(68)25	J. A. Soderquist; *Org. Synth.*, 1990, **68**, 25.	929, 930
90OSC(7)160	A. F. Kluge; *Org. Synth. Coll. Vol.*, 1990, **7**, 160.	338
90PAC473	M. R. Bryce and A. J. Moore; *Pure Appl. Chem.*, 1990, **62**, 473.	870
90PAC1933	P. Kocienski and C. Barber; *Pure Appl. Chem.*, 1990, 1933.	655
90PAC1987	V. Fiandanese; *Pure Appl. Chem.*, 1990, **62**, 1987.	776
90POL277	M. Layh and W. Uhl; *Polyhedron*, 1990, **9**, 277.	698
90POL875	W.-K. Wong, H. Chen and F.-L. Chow; *Polyhedron*, 1990, **9**, 875.	578
90PS(47)327	M. J. Menu, Y. Dartiguenave, M. Dartiguenave, A. Baceiredo and G. Bertrand; *Phosphorus, Sulfur Silicon*, 1990, **47**, 327.	548
90PS(49/50)355	K. Schmidt, K. Issleib and E. Leissring; *Phosphorus, Sulfur Silicon*, 1990, **49/50**, 355.	548
90PS(53)1	A. S. Ionkin, S. N. Ignatéva, V. M. Nekhoroshkov, Yu. Ya. Efremov and B. A. Arbuzov; *Phosphorus, Sulfur Silicon*, 1990, **53**, 1.	569
90RTC305	M. Schakel, M. P. Aarnts and G. W. Klumpp; *Recl. Trav. Chim. Pays-Bas*, 1990, **109**, 305.	528
90RTC399	R. W. M. Aben and J. W. Scheeren; *Recl. Trav. Chim. Pays-Bas*, 1990, **109**, 399.	826
90S36	S. Okazaki and Y. Sato; *Synthesis*, 1990, 36.	512
90S66	D. Dauzonne and P. Demerseman; *Synthesis*, 1990, 66.	798
90S104	M. B. Sassaman, G. K. S. Prakash and G. A. Olah; *Synthesis*, 1990, 104.	216
90S162	Z.-T. Huang and X. Shi; *Synthesis*, 1990, 162.	900
90S195	M. Hojo, R. Masuda, E. Okada, H. Yamamoto, K. Morimoto and K. Okada; *Synthesis*, 1990, 195.	889, 900, 970, 971, 972
90S200	P. B. Tripathy and D. S. Matteson; *Synthesis*, 1990, 200.	144
90S256	C. Yuan and G. Wang; *Synthesis*, 1990, 256.	467
90S271	E. Schaumann, S. Winter-Extra, K. Kummert and S. Scheiblich; *Synthesis*, 1990, 271.	259, 855
90S313	U. Azzena, S. Cossu, T. Denurra, G. Melloni and A. M. Piroddi; *Synthesis*, 1990, 313.	179
90S431	R. M. Moriarty and R. K. Vaid; *Synthesis*, 1990, 431.	1081
90S565	A. R. Katritzky, J. Jiang and L. Urogdi; *Synthesis*, 1990, 565.	329
90S657	R. W. Hoffmann and P. Bovicelli; *Synthesis*, 1990, 657.	25
90S663	A. R. Katritzky and M. S. Chandra Rao; *Synthesis*, 1990, 663.	449
90S717	T. Gajda; *Synthesis*, 1990, 717.	132
90S803	H. Gstach and P. Seil; *Synthesis*, 1990, 803.	440
90S951	M. W. Plath, H.-D. Scharf, G. Raabe and C. Krüger; *Synthesis*, 1990, 951.	425
90S959	R. Neidlein and Z. Sui; *Synthesis*, 1990, 959.	981
90S991	J. A. Goodwin, I. M. Y. Kwok and B. J. Wakefield; *Synthesis*, 1990, 991.	843
90S1048	H. Gstach and P. Seil; *Synthesis*, 1990, 1048.	440
90S1153	T. Schrader and W. Steglich; *Synthesis*, 1990, 1153.	489
90S1159	M. Folkmann and F. J. Lund; *Synthesis*, 1990, 1159.	59, 60
90SC153	R. Miranda, H. Cervantes and P. Joseph-Nathan; *Synth. Commun.*, 1990, **20**, 153.	246
90SC267	D. I. Han and D. Y. Oh; *Syn. Commun.*, 1990, **20**, 267.	943
90SC273	J. W. Lee and D. Y. Oh; *Synth. Commun.*, 1990, **19**, 273.	347, 781, 783, 840
90SC687	S. N. Suryawanshi, A. Mukhopadhyay and D. S. Bhakuni; *Synth. Commun.*, 1990, **20**, 687.	202
90SC925	D. Villemin and A. Ben Alloum; *Synth. Commun.*, 1990, **20**, 925.	281
90SC1671	S. E. Drewes, N. D. Emslie and M. Hemingway; *Synth. Commun.*, 1990, **20**, 1671.	118
90SC1865	O. E. O. Hormi, E. O. Pajunen, A.-K. C. Åvall, P. Pennanen, J. H. Näsman and M. Sundell; *Synth. Commun.*, 1990, **20**, 1865.	553
90SC2527	C. Driss, M. M. Abdelkafi, M. M. Chaabouni and A. Baklouti; *Synth. Commun.*, 1990, **20**, 2527.	190
90SC3245	M. Kabaki, S. Inoue, Y. Nagata and Y. Sato; *Synth. Commun.*, 1990, **20**, 3245.	361
90SC3339	D. Dauzonne, A. Fleurant and P. Demerseman; *Synth. Commun.*, 1990, **20**, 3339.	796
90SL115	S. Kim and Y. C. Kim; *Synlett*, 1990, 115.	918
90SL255	M. J. Ford and S. V. Ley; *Synlett*, 1990, 255.	200
90SL261	J. Barluenga, P. J. Campos, G. Canal and G. Asensio; *Synlett*, 1990, 261.	453, 477
90SL359	C. W. G. Fishwick, A. D. Jones, M. B. Mitchell, D. S. Eggleston and P. W. Baures; *Synlett*, 1990, 359.	277
90SL413	C. J. Moody, C. W. Rees and R. G. Young; *Synlett*, 1990, 413.	539
90SL415	K. Afarinkia, J. I. G. Cadogan and C. W. Rees; *Synlett*, 1990, 415.	461
90SL457	P. C. Bulman Page, E. S. Namwindwa, S. S. Klair and D. Westwood; *Synlett*, 1990, 457.	261
90SL643	H. B. Kagan and F. Rebiere; *Synlett*, 1990, 643.	261
90SL701	A. Herunsalee, M. Isobe, Y. Fukuda and T. Goto; *Synlett*, 1990, 701.	940
90SL769	M. Julia, J.-N. Verpeaux and T. Zahneisen; *Synlett*, 1990, 769.	138
90SRI1373	J. Barrau, N. Ben Hamida and J. Satgé; *Synth. React. Inorg. Metal-Org. Chem.*, 1990, **20**, 1373.	640, 646
90SUL109	B. A. Shainyan; *Sulfur Letters*, 1990, **11**, 109.	771
90SUL157	K. El-Berembally, M. El-Kersh and H. El-Fatatry; *Sulfur Lett.*, 1990, **11**, 157.	237
90T1783	J. R. Moran, I. Tapia and V. Alcazar; *Tetrahedron*, 1990, **46**, 1783.	232

Ref	Citation	Pages
90T1791	M. Tramontini and L. Angiolini; *Tetrahedron*, 1990, **46**, 1791.	96
90T2195	Ch. S. Rao, R. T. Chakrasali, H. Ila and H. Junjappa; *Tetrahedron*, 1990, **46**, 2195.	253
90T2573	P. C. Bulman Page and S. Rosenthal; *Tetrahedron*, 1990, **46**, 2573.	357
90T2587	J. C. Wu and J. Chattopadhyaya; *Tetrahedron*, 1990, **46**, 2587.	276
90T2999	P. Dembech, A. Guerrini, A. Ricci, G. Seconi and M. Taddei; *Tetrahedron*, 1990, **46**, 2999.	358
90T4553	T. Mandai, M. Kohama, H. Sato, M. Kawada and J. Tsuji; *Tetrahedron*, 1990, **46**, 4553.	942
90T5093	W. Ando, H. Sonobe and T. Akasaka; *Tetrahedron*, 1990, **46**, 5093.	258
90T5263	P. Hudhomme and G. Duguay; *Tetrahedron*, 1990, **46**, 5263.	99
90T6645	S. A. Biller and C. Forster; *Tetrahedron*, 1990, **46**, 6645.	557
90T6951	N. S. Simpkins; *Tetrahedron*, 1990, **46**, 6951.	960
90T7197	E. Dominguez and J. C. Carretero; *Tetrahedron*, 1990, **46**, 7197.	271, 272
90T7729	K.-H. König; *Tetrahedron*, 1990, **46**, 7729.	102
90TA143	S.-Y. Po, H.-H. Liu and B.-J. Uang; *Tetrahedron Asymmetry*, 1990, **1**, 143.	232
90TA151	T. Hayashi, Y. Matsumoto, I. Morikawa and Y. Ito; *Tetrahedron Asymmetry*, 1990, **1**, 151.	1033
90TA477	A. Alexakis and P. Mangeney; *Tetrahedron Asymmetry*, 1990, **1**, 477.	176
90TL135	V. K. Aggarwal, I. W. Davies, J. Maddock, M. F. Mahon and K. C. Molloy; *Tetrahedron Lett.*, 1990, **31**, 135.	269
90TL257	H.-H. Liu, E.-N. Chen, B.-J. Uang and S.-L. Wang; *Tetrahedron Lett.*, 1990, **31**, 257.	230
90TL673	H.-C. Cheng and T.-H. Yan; *Tetrahedron Lett.*, 1990, **31**, 673.	946
90TL973	J. R. McCarthy, D. P. Matthews and C. L. Barney; *Tetrahedron Lett.*, 1990, **31**, 973.	52
90TL1043	A. Miller and G. Procter; *Tetrahedron Lett.*, 1990, **31**, 1043.	279
90TL1335	T. K. Chakraborty and G. V. Reddy; *Tetrahedron Lett.*, 1990, **31**, 1335.	516
90TL1405	P. F. de Cusati and R. A. Olofson; *Tetrahedron Lett.*, 1990, **31**, 1405.	59
90TL1567	H.-J. Ha, G.-S. Nam and K. P. Park; *Tetrahedron Lett.*, 1990, **31**, 1567.	467
90TL1759	D. W. Anderson, M. M. Campbell, M. Malik, M. Prashad and R. H. Wightman; *Tetrahedron Lett.*, 1990, **31**, 1759.	472
90TL1819	G. A. Kraus, P. J. Thomas and M. D. Schwinden; *Tetrahedron Lett.*, 1990, **31**, 1819.	1079
90TL1945	R. F. Cunico and C. P. Kuan; *Tetrahedron Lett.*, 1990, **31**, 1945.	354
90TL2127	C. Saluzzo, G. Alvernhe, D. Anker and G. Haufe; *Tetrahedron Lett.*, 1990, **31**, 2127.	787
90TL2169	A. Riera, F. Cabre, A. Moyano, M. A. Pericas and J. Santamaria; *Tetrahedron Lett.*, 1990, **31**, 2169.	851
90TL2197	G. J. P. H. Boons, C. J. J. Elie, G. A. Van der Marel and J. H. Van Boom; *Tetrahedron Lett.*, 1990, **31**, 2197.	176, 199, 352
90TL2287	T. Brigaud and E. Laurent; *Tetrahedron Lett.*, 1990, **31**, 2287.	62
90TL2385	M. Matteucci; *Tetrahedron Lett.*, 1990, **31**, 2385.	219
90TL2541	P. A. Magriotis, T. J. Doyle and K. D. Kim; *Tetrahedron Lett.*, 1990, **31**, 2541.	962
90TL2751	J. Barluenga, M. A. Rodriguez and P. J. Campos; *Tetrahedron Lett.*, 1990, **31**, 2751.	760
90TL2759	M. W. Bredenkamp, C. W. Holzapfel and A. D. Swanepoel; *Tetrahedron Lett.*, 1990, **31**, 2759.	228
90TL2925	K. Tamao, H. Yao, Y. Tsutumi, H. Abe, T. Hayashi and Y. Ito; *Tetrahedron Lett.*, 1990, **31**, 2925.	651
90TL3105	M. Lautens and A. H. Huboux; *Tetrahedron Lett.*, 1990, **31**, 3105.	1062
90TL3121	F. Rebiere, O. Samuel and H. B. Kagan; *Tetrahedron Lett.*, 1990, **31**, 3121.	706
90TL3149	G. Le Guillanton and J. Simonet; *Tetrahedron Lett.*, 1990, **31**, 3149.	277
90TL3931	J. M. Percy; *Tetrahedron Lett.*, 1990, **31**, 3931.	949
90TL4621	K. Ogura, N. Yahata, T. Fujimori and M. Fujita; *Tetrahedron Lett.*, 1990, **31**, 4621.	858
90TL4677	J. A. Soderquist, C. L. Anderson, E. I. Miranda, I. Rivera and G. W. Kabalka; *Tetrahedron Lett.*, 1990, **31**, 4677.	353
90TL4937	V. A. Khripach, V. N. Zhabinskiy and V. K. Olkhovick; *Tetrahedron Lett.*, 1990, 4937.	809
90TL5361	A. N. Bovin, A. N. Chekhlov and E. N. Tsvetkov; *Tetrahedron Lett.*, 1990, **31**, 5361.	134
90TL5381	S. D. Lindell and R. M. Turner; *Tetrahedron Lett.*, 1990, **31**, 5381.	554
90TL5449	J. R. McCarthy, D. P. Matthews, M. L. Edwards, D. M. Stemerick and E. T. Jarvi; *Tetrahedron Lett.*, 1990, **31**, 5449.	61, 79, 780, 782
90TL5571	M. L. Edwards, D. M. Stemerick, E. T. Jarvi, D. P. Mathews and J. R. McCarthy; *Tetrahedron Lett.*, 1990, **39**, 5571.	119, 734
90TL5607	R. F. Cunico; *Tetrahedron Lett.*, 1990, **31**, 5607.	1046
90TL5815	L. Garlaschelli and G. Vidari; *Tetrahedron Lett.*, 1990, **31**, 5815.	246
90TL5829	M. Lautens, A. H. Huboux, B. Chin and J. Downer; *Tetrahedron Lett.*, 1990, **31**, 5829.	661, 702
90TL6137	K. D. Kim and P. A. Magriotis; *Tetrahedron Lett.*, 1990, **31**, 6137.	772, 939, 948
90TL6331	G. Märkl and S. Reithinger; *Tetrahedron Lett.*, 1990, **31**, 6331.	594
90TL6407	T. Itoh and T. Ohta; *Tetrahedron Lett.*, 1990, **31**, 6407.	389
90TL6461	S. Hanessian, Y. L. Bennani and D. Delorme; *Tetrahedron Lett.*, 1990, **31**, 6461.	130
90TL6465	S. Hanessian and Y. L. Bennani; *Tetrahedron Lett.*, 1990, **31**, 6465.	492
90TL6819	W. J. Thompson, T. J. Tucker, J. E. Schwering and J. L. Barnes; *Tetrahedron Lett.*, 1990, **31**, 6819.	279
90TL6999	G. Märkl, C. Dörges, H. Nöth and K. Polborn; *Tetrahedron Lett.*, 1990, **31**, 6999.	484
90TL7375	J. Barluenga, P. J. Campos, F. Lopez, I. Llorente and M. A. Rodriguez; *Tetrahedron Lett.*, 1990, **31**, 7375.	772
90TL7419	R. W. Hoffmann, M. Julius and K. Oltmann; *Tetrahedron Lett.*, 1990, **31**, 7419.	400, 401
90TL7607	S. Himdi-Kabbab and J. Hamelin; *Tetrahedron Lett.*, 1990, **31**, 7607.	502
90UKZ1310	I. M. Bazanova, V. M. Neplyuev and M. O. Lozinskii; *Ukr. Khim. Zh.* (*Russ. Ed.*), 1990, **56**, 1310 (*Chem. Abstr.* 1991, **115**, 49 556).	864

90USP4963655	D. H. Kinder and M. M. Ames; *US Pat.* 4 963 655 (1990) (*Chem. Abstr.*, 1990, **114**, 123 091k).	528
90ZAAC(581)41	H. Lehmkuhl, R. Schwickardi, C. Krüger and G. Raabe; *Z. Anorg. Allg. Chem.*, 1990, **581**, 41.	720
90ZAAC(583)195	E. Popowski, U. Marekowa, T. Reiske, H. Schulz, H. Kelling and H. Jancke; *Z. Anorg. Allg. Chem.*, 1990, **583**, 195.	602, 607, 637, 660
90ZAAC(586)25	H. Köhler, R. Skirl, H. Kretschmann and A. Kolbe; *Z. Anorg. Allg. Chem.*, 1990, **586**, 25.	556, 557, 699
90ZN(B)15	G. Knoerzer and W. Siebert; *Z. Naturforsch., Teil B*, 1990, **45**, 15.	631, 635, 635
90ZN(B)148	J. Grobe, M. Hegemann and D. Le Van; *Z. Naturforsch., Teil B*, 1990, **45b**, 148.	112, 116
90ZN(B)221	B. Bachmann, G. Baum, J. Heck, W. Massa and B. Ziegler; *Z. Naturforsch., Teil B*, 1990, **45**, 221.	672, 673
90ZN(B)961	H. Schmidbaur and J. Rott; *Z. Naturforsch., Teil B*, 1990, **45**, 961.	639
90ZOB227	S. G. Seredkina, A. N. Mirskova and O. B. Bannikova; *Zh. Obshch. Khim.*, 1990, **60**, 227.	772
90ZOB229	P. P. Onys'ko, T. V. Kim, E. I. Kiseleva and A. D. Sinitsa; *Zh. Obshch. Khim.*, 1990, **60**, 229 (Engl. Transl. 200).	1009
90ZOB526	A. S. Nakhmanovich, R. V. Karnaukhova, O. G. Yarosh, V. N. Elokhina, I. D. Kalikman and R. A. Gromkova; *Zh. Obshch. Khim.*, 1990, **60**, 526.	942
90ZOB695	O. I. Kolodyazhnyi and S. N. Yastenko; *Zh. Obshch. Khim.*, 1990, **60**, 695.	570
90ZOB1187	D. M. Malenko, L. I. Nesterova, S. N. Luk'yanenko and A. D. Sinitsa; *Zh. Obshch. Khim.*, 1990, **60**, 1187 (Engl. Transl. 1059).	1009
90ZOB1420	A. A. Prishchenko, M. V. Livantsov and V. S. Petrosyan; *Zh. Obshch. Khim.*, 1990, **60**, 1420.	562
90ZOB1539	N. V. Lukashev, O. I. Artyushin, E. I. Lazhko, E. V. Luzikova and M. A. Kazankova; *Zh. Obshch. Khim.*, 1990, **60**, 1539.	559, 766
90ZOB1718	G. V. Romanov, T. Ya Ryzhiova and A. N. Pudovik; *Zh. Obshch. Khim.*, 1990, **60**, 1718 (*Chem. Abstr.*, 1991, **114**, 42 929n).	455, 481
90ZOB1940	T. Kh. Gazizov and L. K. Sal'keeva; *Zh. Obshch. Khim.*, 1990, **60**, 1940.	921
90ZOB2291	F. R. Garieva, V. I. Gavrilov, R. Z. Musin and Yu. Ya. Efremov; *Zh. Obshch. Khim.*, 1990, **60**, 2291 (*Chem. Abstr.*, 1991, **115**, 136 239).	592
90ZOR740	N. V. Kondratenko, L. A. Khomenko and L. M. Yagupol'skii; *Zh. Org. Khim.*, 1990, **26**, 740.	771
90ZOR978	A. V. Martynov, A. N. Mirskova and M. G. Voronkov; *Zh. Org. Khim.*, 1990, **26**, 978 (*Chem. Abstr.*, 1991, **113**, 171 466n).	787
90ZOR1259	V. V. Shereshovets, N. M. Korotaeva, Yu. I. Puzin, A. A. Elichev, G. V. Leplyanin and G. A. Tolstikov; *Zh. Org. Khim.*, 1990, **26**, 1259 (*Chem. Abstr.*, 1990, **113**, 211 334).	267
90ZOR2056	S. V. Amosova, V. I. Gostevskaya, G. M. Gavrilova, A. V. Afonin and V. A. Potapov; *Zh. Org. Khim.*, 1990, **26**, 2056 (*Chem. Abstr.*, 1991, **115**, 158 464g).	777
91ACR257	T.-Y. Luh; *Acc. Chem. Res.*, 1991, **24**, 257.	244
91AG(E)698	Y. Rubin, C. B. Knobler and F. Diederich; *Angew. Chem., Int. Ed. Engl.*, 1991, **30**, 698.	746
91AG(E)709	H. Grützmacher and H. Pritzkow; *Angew. Chem., Int. Ed. Engl.*, 1991, **30**, 709.	572
91AG(E)998	H. Fischer, J. Hofmann and E. Mauz; *Angew. Chem., Int. Ed. Engl.*, 1991, **30**, 998.	581
91AG(E)1124	J. J. Schneider, R. Goddard, S. Werner and C. Krüger; *Angew. Chem., Int. Ed. Engl.*, 1991, **30**, 1124.	695
91BAU1705	M. D. Bargamova, S. I. Pletnev and I. L. Knunyants; *Bull. Acad. Sci. USSR, Div. Chem. Sci.*, 1991, 1705.	799
91BCJ2751	T. Okuyama, N. Haga, S. Takane, K. Ueno and T. Fueno; *Bull. Chem. Soc. Jpn.*, 1991, **64**, 2751.	1075
91CB175	E. Öhler and E. Zbiral; *Chem. Ber.*, 1991, **124**, 175.	339
91CB199	H. J. Bestmann, T. Röder, M. Bremer and D. Löw; *Chem. Ber.*, 1991, **124**, 199.	574
91CB275	H. Schmidbaur, T. Wimmer, J. Lachmann and G. Müller; *Chem. Ber.*, 1991, **124**, 275.	562
91CB503	B. Wrackmeyer and K. Wagner; *Chem. Ber.*, 1991, **124**, 503.	701
91CB543	W. Zarges, M. Marsch, K. Harms, W. Koch, G. Frenking and G. Boche; *Chem. Ber.*, 1991, **124**, 543.	646
91CB595	R. W. Saalfrank, C. Lurz, J. Hassa, D. Danion and L. Toupet; *Chem. Ber.*, 1991, **124**, 595.	989
91CB1207	O. Wagner, M. Ehle, M. Birkel, J. Hoffmann and M. Regitz; *Chem. Ber.*, 1991, **124**, 1207.	110
91CB1253	A. Schmidt, G. Köbrich and R. W. Hoffmann; *Chem. Ber.*, 1991, **124**, 1253.	146
91CB1259	R. W. Hoffmann and M. Bewersdorf; *Chem. Ber.*, 1991, **124**, 1259.	146
91CB1315	W. W. Dumont, S. Kubiniok, L. Lange, S. Pohl, W. Saak and I. Wagner; *Chem. Ber.*, 1991, **124**, 1315.	88, 288
91CB1379	H. Mahler and M. Braun; *Chem. Ber.*, 1991, **124**, 1379.	758, 818
91CB1511	W. Uhl, M. Lagh and W. Massa; *Chem. Ber.*, 1991, **124**, 1511.	698, 708
91CB1575	J. Okuda, E. Herdtweck and E. M. Zeller; *Chem. Ber.*, 1991, **124**, 1575.	603
91CB1827	K. Müllen, M. Klabunde and V. Enkelman; *Chem. Ber.*, 1991, **124**, 1827.	276
91CB2391	P. Jutzi, R. Krallmann, G. Wolf, B. Neumann and H.-G. Stammler; *Chem. Ber.*, 1991, **124**, 2391.	603
91CB2449	K. Sünkel; *Chem. Ber.*, 1991, **124**, 2449.	822
91CB2489	A. Maercker, R. Schuhmacher, W. Buchmeier and H. D. Lutz; *Chem. Ber.*, 1991, **124**, 2489.	264
91CB2897	A. K. Beck, and D. Seebach; *Chem. Ber.*, 1991, **124**, 2897.	979
91CC181	A. A. Danopoulos, G. Wilkinson and D. J. Williams; *J. Chem. Soc., Chem. Commun.*, 1991, 181.	678
91CC234	K. B. Dillon and T. A. Straw; *J. Chem. Soc., Chem. Commun.*, 1991, 234.	133

91CC297	P. Jankowski, S. Marczak, M. Masnyk and J. Wicha; *J. Chem. Soc., Chem. Commun.*, 1991, 297.	393
91CC302	U. Fleischer, H. Grützmacher and U. Krüger; *J. Chem. Soc., Chem. Commun.*, 1991, 302.	551, 552
91CC462	S. Takano, T. Sato, K. Inomata and K. Ogasawara; *J. Chem. Soc., Chem. Commun.*, 1991, 462.	412
91CC470	J. Hackenberg and M. Hanack; *J. Chem. Soc., Chem. Commun.*, 1991, 470.	1081
91CC524	C. Palomo, J. M. Aizpurua, J. M. Garcia and M. Legido; *J. Chem. Soc., Chem. Commun*, 1991, 524.	514
91CC947	A. J. Bloodworth and A. Shah; *J. Chem. Soc., Chem. Commun.*, 1991, 947.	214
91CC1027	A. Konno, K. Nakagawa and T. Fuchigami; *J. Chem. Soc., Chem. Commun.*, 1991, 1027.	62, 63
91CC1274	P. F. Engel, M. Pfeffer, J. Fischer and A. Dedieu; *J. Chem. Soc., Chem. Commun.*, 1991, 1274.	716
91CC1390	H. Iio, A. Fujii, M. Ishii and T. Tokoroyama; *J. Chem. Soc., Chem. Commun.*, 1991, 1390.	1034
91CC1421	J. Uenishi, M. Motoyama, Y. Nishiyama and S. Wakabayashi; *J. Chem. Soc., Chem. Commun.*, 1991, 1421.	227
91CC1704	J. Barluenga, R. P. Carlón, F. J. González, F. L. Ortiz and S. Fustero; *J. Chem. Soc., Chem. Commun.*, 1991, 1704.	421
91CJC1153	A. R. Katritzky, J. L. Jiang and P. A. Harris; *Can. J. Chem.*, 1991, **69**, 1153.	445
91CL97	D. Terunuma, H. Kizaki, T. Sato, K. Masuo and H. Nohira; *Chem. Lett.*, 1991, 97.	142, 521
91CL121	M. Sekine and T. Nakanishi; *Chem. Lett.*, 1991, 121.	219
91CL415	K. Sasaki, T. Mori, Y. Doi, A. Kawachi, Y. Aso, T. Otsubo and F. Ogura; *Chem. Lett.*, 1991, 415.	238
91CL961	J. Ichikawa, T. Moriya, T. Sonoda and H. Kobayashi; *Chem. Lett.*, 1991, 961.	736
91CL1319	J. Sugimoto, K. Miura, K. Oshima and K. Utimoto; *Chem. Lett.*, 1991, 1319.	737, 743
91CL1775	Y. Nishiyama, S. Nakata and S. Hamanaka; *Chem. Lett.*, 1991, 1775.	238
91CL2213	K. Toyota, Y. Ishikawa, M. Yoshifuji, K. Okada, K. Hosomi and K. Hirotsu; *Chem. Lett.*, 1991, 2213.	559
91COS(1)505	K. Ogura; *Comp. Org. Synth.*, 1991, **1**, 505.	390, 391
91COS(1)629	A. Krief; *Comp. Org. Synth.*, 1991, **1**, 629.	375, 398
91COS(2)604	T. H. Chan; *Comp. Org. Synth.*, 1991, **2**, 604.	831
91COS(2)661	M. D. Bednarski and J. P. Lyssikatos; *Comp. Org. Synth.*, 1991, **2**, 661.	202
91COS(2)893	E. F. Kleinman; *Comp. Org. Synth.*, 1991, **2**, 893.	315
91COS(2)953	H. Heaney; *Comp. Org. Synth.*, 1991, **2**, 953.	301, 316
91COS(3)65	R. E. Gawley and K. Rein; *Comp. Org. Synth.*, 1991, **3**, 65.	526
91COS(3)85	A. Krief; *Comp. Org. Synth.*, 1991, **3**, 85.	390, 398
91COS(3)193	D. R. Cheshire; *Comp. Org. Synth.*, 1991, **3**, 193.	377, 381
91COS(6)33	R. R. Schmidt; *Comp. Org. Synth.*, 1991, **6**, 33.	176, 198, 202
91COS(6)485	W. Kantlehuer; *Comp. Org. Synth.*, 1991, **6**, 485.	881
91COS(6)631	H. Kunz and H. Waldmann; *Comp. Org. Synth.*, 1991, **6**, 631.	176, 180
91COS(7)206	M. Kennedy and M. A. McKervey; *Comp. Org. Synth.*, 1991, **7**, 206.	62, 66
91COS(7)762	S. Uemura; *Comp. Org. Synth.*, 1991, **7**, 762.	330
91CPB1365	Y. Tsuda, S. Hosoi, A. Nakai, Y. Sakai, T. Abe, Y. Ishi, F. Kiuchi and T. Sano; *Chem. Pharm. Bull.*, 1991, **39**, 1365.	90
91CPB3120	Y. Takeuchi, K. Takagi, K. Nagata and T. Koizumi; *Chem. Pharm. Bull.*, 1991, **39**, 3120.	108
91CR(313)517	C. Cellerin, J.-P. Pradere, D. Danion and F. Tonnard; *C. R. Hebd. Seances Acad. Sci.*, 1991, **313**, 517.	328
91CSR95	A. G. M. Barrett; *Chem. Soc. Rev.*, 1991, **20**, 95.	795
91CSR211	H. G. Davies and R. H. Green; *Chem. Soc. Rev*, 1991, **20**, 211.	171, 176
91CSR271	H. G. Davies and R. H. Green; *Chem. Soc. Rev*, 1991, **20**, 271.	176
91EUP413343	S. Sasho, S. Ichikawa, H. Kato, H. Obase, K. Shuto and Y. Oiji, *Eur. Pat.* 0 413 343 (1991).	970
91EUP425030	A. Mete and L. C. Chan; *Eur. Pat.*, 425030 (1991).	970
91EUP488681	T. Shinohara, M. Kudo and K. Matsumura; *Eur. Pat.*, 488 681 (1991) (*Chem. Abstr.*, 1992, **117**, 90 499f).	507
91G471	A. R. Katritsky, P. Lee and N. Malhotra; *Gazz. Chim. Ital.*, 1991, **121**, 471.	370
91H(32)1445	C. W. Holzapfel, J. J. Huyser, T. L. van der Merwe and F. R. van Heerden; *Heterocycles*, 1991, **32**, 1445.	190
91HAC307	I. D. Sadekov, B. B. Rivkin, P. I. Gadzhieva and V. I. Minkin; *Heteroatom Chem.*, 1991, **2**, 307.	290
91HCA27	D. Obrecht; *Helv. Chim. Acta*, 1991, **74**, 27.	277
91HCA451	A. Vasella and R. Wyler; *Helv. Chim. Acta*, 1991, **74**, 451.	912
91HCA579	R. Neidlein and Z. Sui; *Helv. Chim. Acta*, 1991, **74**, 579.	978
91HCA983	T. R. Ward, L. M. Venanzi, A. Albinati, F. Lianza, T. Gerfin, V. Gramlich and G. M. Ramos Tombo; *Helv. Chim. Acta*, 1991, **74**, 983.	566
91HCA1305	P. Renaud; *Helv. Chim. Acta*, 1991, **74**, 1305.	285, 286
91HCA1477	S. Bienz and A. Chapeaurouge; *Helv. Chim. Acta*, 1991, **74**, 1477.	140, 352
91HCA1924	A. R. Katritzky, S. Perumal, W. Kuzmierkiewicz, P. Lue and J. V. Greenhill; *Helv. Chim. Acta.*, 1991, **74**, 1924.	318, 322
91HCA1936	A. R. Katritzky, W. Kuzmierkiewicz and S. Perumal; *Helv. Chim. Acta*, 1991, **74**, 1936.	318
91HOU(E14a/1)1	H. Hagemann and D. Klamman; *Methoden Org. Chem. (Houben-Weyl)*, 1991, **E14a/1**, 1.	176
91HOU(E14a/1)136	H. Hagemann and D. Klamman; *Methoden Org. Chem. (Houben-Weyl)*, 1991, **E14a/1**, 136.	178
91HOU(E14a/1)323	H. Hagemann and D. Klamman; *Methoden Org. Chem. (Houben-Weyl)*, 1991, **E14a/1**, 323.	186
91HOU(E14a/1)412	H. Hagemann and D. Klamman; *Methoden Org. Chem. (Houben-Weyl)*, 1991, **E14a/1**, 412.	202
91HOU(E14a/1)431	H. Hagemann and D. Klamman; *Methoden Org. Chem. (Houben-Weyl)*, 1991, **E14a/1**, 431.	202

91HOU(E14a/1)600	H. Hagemann and D. Klamman; *Methoden Org. Chem.* (*Houben-Weyl*), 1991, **E14a/1**, 600.	163
91HOU(E14a/1)683	H. Hagemann and D. Klamman; *Methoden Org. Chem.* (*Houben-Weyl*), 1991, **E14a/1**, 683.	203, 207
91HOU(E14a/1)785	P. Wimmer; *Methoden Org. Chem.* (*Houben-Weyl*), 1991, **E14a/1**, 785.	229
91HOU(E14a/2)1	W. Rasshofer; *Methoden Org. Chem.* (*Houben-Weyl*), 1991, **E14a/2**, 1.	294, 300, 307, 308
91HOU(E14a/2)170	W. Rasshofer; *Methoden Org. Chem.* (*Houben-Weyl*), 1991, **E14a/2**, 170.	309
91HOU(E14a/2)248	W. Rasshofer; *Methoden Org. Chem.* (*Houben-Weyl*), 1991, **E14a/2**, 248.	704
91IC156	M. C. Kerby, B. W. Eichhorn, L. Doviken and K. P. C. Vollhardt; *Inorg. Chem.*, 1991, **30**, 156.	675
91IC2228	N. E. Miller; *Inorg. Chem.*, 1991, **30**, 2228.	145
91IC4710	M. J. Chetcuti, P. E. Fanwick and J. C. Gordon; *Inorg. Chem.*, 1991, **30**, 4710.	714
91IZV1039	S. V. Basenko, I. A. Gebel, D. D. Toryashinova, V. Yu. Vitkovskii, R. G. Mirskov and M. G. Voronkov; *Izv. Akad. Nauk SSSR, Ser. Khim.*, 1991, 1039.	140
91IZV1849	A. L. Laikhter, T. I. Cherkasova, L. G. Mel'nikova, B. I. Ugrak, A. A. Fainzil'berg and V. V. Semenov; *Izv. Akad. Nauk SSSR, Ser. Khim.*, 1991, **8**, 1849 (Engl. Transl. 1637).	995
91IZV2106	A. V. Martynov and A. N. Mirskova; *Izv. Akad. Nauk SSSR, Ser. Khim.*, 1991, 2106 (*Chem. Abstr.*, 1992, **116**, 6208j).	787
91JA1042	D. Ranganathan and S. Saini; *J. Am. Chem. Soc.*, 1991, **113**, 1042.	311
91JA5735	S. A. Rao and P. Knochel; *J. Am. Chem. Soc.*, 1991, **113**, 5735.	772, 963
91JA7439	J. R. McCarthy, D. P. Matthews, D. M. Stemerick, E. W. Huber, P. Bey, B. J. Lippert, R. D. Snyder and P. S. Sunkara; *J. Am. Chem. Soc.*, 1991, **113**, 7439.	812
91JA7767	M. Matteucci, K.-Y. Lin, S. Butcher and C. Moulds; *J. Am. Chem. Soc.*, 1991, **113**, 7767.	216
91JA8168	W. H. Bunnelle, T. A. Isbell, C. L. Barnes and S. Qualls; *J. Am. Chem. Soc.*, 1991, **113**, 8168.	214
91JA8546	W. H. Pearson and A. C. Lindbeck; *J. Am. Chem. Soc.*, 1991, **113**, 8546.	531
91JA8807	P. A. Wade, P. A. Kondracki and P. J. Carroll; *J. Am. Chem. Soc.*, 1991, **113**, 8807.	108
91JA9377	W.-B. Choi, L. J. Wilson, S. Yeola, D. C. Liotta and R. F. Schinazi; *J. Am. Chem. Soc.*, 1991, **113**, 9377.	230
91JA9406	R. D. Adams, G. Chen, L. Chen, W. Wu and J. Yin; *J. Am. Chem. Soc.*, 1991, **113**, 9406.	955
91JA9796	H. Kuniyasu, A. Ogawa, S.-I. Miyazaki, I. Ryu, N. Kambe and N. Sonoda; *J. Am. Chem. Soc.*, 1991, **113**, 9796.	940, 947
91JA9864	S. Kim and P. L. Fuchs; *J. Am. Chem. Soc.*, 1991, **113**, 9864.	284
91JCS(D)351	M. P. Brown, D. Burns, R. Das, P. A. Dolby, M. M. Harding, R. W. Jones, E. J. Robinson and A. K. Smith; *J. Chem. Soc., Dalton Trans.*, 1991, 351.	1029
91JCS(D)609	A. K. Burrell, G. R. Clark, C. E. F. Rickard, W. R. Roper and A. H. Wright; *J. Chem. Soc., Dalton Trans.*, 1991, 609.	582
91JCS(D)1117	H. Adams, N. A. Bailey, J. T. Gauntlett, I. M. Harkin, M. J. Winter and S. Woodward; *J. Chem. Soc., Dalton Trans.*, 1991, 1117.	580
91JCS(D)2579	J. Vincente, M.-T. Chicote, M.-C. Lagunas and P. G. Jones; *J. Chem. Soc., Dalton Trans.*, 1991, 2579.	586
91JCS(P1)49	Y. Takeuchi, M. Nabetani, K. Takagi, T. Hagi and T. Koizumi; *J. Chem. Soc., Perkin Trans. 1*, 1991, 49.	102
91JCS(P1)323	C. J. Moody, C. W. Rees and R. G. Young; *J. Chem. Soc., Perkin Trans. 1*, 1991, 323.	539
91JCS(P1)662	V. K. Aggarwal, I. W. Davies, R. J. Franklin, J. Maddock, M. F. Mahon and K. C. Molloy; *J. Chem. Soc., Perkin Trans. 1*, 1991, 662.	268
91JCS(P1)897	M. Ashwell, W. Clegg and R. F. W. Jackson; *J. Chem. Soc., Perkin Trans. 1*, 1991, 897.	173, 237
91JCS(P1)1091	I. F. Cottrell, D. Hands, D. J. Kennedy, K. J. Paul, S. H. B. Wright and K. Hoogsteen; *J. Chem. Soc., Perkin Trans. 1*, 1991, 1091.	509
91JCS(P1)1555	S. Yamazaki, W. Mizuno and S. Yamabe; *J. Chem. Soc., Perkin Trans. 1*, 1991, 1555.	947
91JCS(P1)1693	G. T. Bourne, D. Crich, J. W. Davies and D. C. Horwell; *J. Chem. Soc., Perkin Trans. 1*, 1991, 1693.	404, 418, 419, 424, 427
91JCS(P1)2276	C. Evans, R. McCague, S. M. Roberts, A. G. Sutherland and R. Wilson; *J. Chem. Soc., Perkin Trans. 1*, 1991, 2276.	650
91JCS(P1)2575	M. S. Baird, A. G. W. Baxter, A. Hoorfar and I. Jefferies; *J. Chem. Soc. Perkin Trans. 1*, 1991, 2575.	748
91JCS(P1)2789	P. Zanirato; *J. Chem. Soc., Perkin Trans. 1*, 1991, 2789.	515
91JCS(P1)3091	C. T. Hewkin, R. F. W. Jackson and W. Clegg; *J. Chem. Soc., Perkin Trans. 1*, 1991, 3091.	858
91JCS(P1)3103	C. T. Hewkin and R. F. W. Jackson; *J. Chem. Soc., Perkin Trans 1*, 1991, 3103.	237, 783
91JFC(52)1	O. D. Gupta, R. L. Kirchmeier and J. M. Shreeve; *J. Fluorine Chem.*, 1991, **52**, 1.	564
91JGU627	O. I. Kolodyazhnyi and D. B. Golokhov; *J. Gen. Chem. USSR* (*Engl. Transl.*), 1991, **61**, 627.	116, 133
91JGU983	S. G. Seredkina, A. N. Mirskova, O. B. Bannikova and G. V. Dolgushin; *J. Gen. Chem. USSR* (*Engl. Transl.*), 1991, **61**, 983.	804
91JGU1072	A. L. Kovalenko, Y. V. Serov and I. V. Tselinskii; *J. Gen. Chem. USSR* (*Engl. Transl.*), 1991, 1072.	418
91JGU1295	E. T. Bogoradovskii, V. S. Zavgorodnii, E. E. Liepin'sh, I. S. Birgele and A. A. Petrov; *J. Gen. Chem. USSR* (*Engl. Transl.*), 1991, **61**, 1295.	807, 811, 813
91JGU1306	E. T. Bogoradovskii, V. S. Zavgorodnii, E. E. Liepin'sh and A. A. Petrov; *J. Gen. Chem. USSR* (*Engl. Transl.*), 1991, **61**, 1306.	810
91JGU1459	R. R. Shagidullin, V. A. Pavlov, B. I. Buzykin, N. V. Aristova, L. F. Chertnova, I. I. Vandyukova, A. Kh. Plyamovatyi, K. M. Enikeev, M. P. Sokolov and V. V. Moskva; *J. Gen. Chem. USSR* (*Engl. Transl.*)., 1991, **61**, 1459.	132
91JGU1875	N. A. Orlova, A. G. Shipov, I. A. Savost'yanova and Yu. I. Baukov; *J. Gen. Chem. USSR* (*Engl. Transl.*), 1991, **61**, 1875.	98, 99

91JHC485	T. Aotsuka, Y. Okamoto, K. Takagi and M. Hubert-Habart; *J. Heterocycl. Chem.*, 1991, **28**, 485.	979
91JHC1245	Y. Tominaga, H. Okuda, S. Kohra and H. Mazume; *J. Heterocycl. Chem.*, 1991, **28**, 1245.	844
91JHC1863	R. W. Saalfrank, C. Lurz, U. Wirth, H. G. Schnering and K. Peters; *J. Heterocycl. Chem.*, 1991, **28**, 1863.	991
91JIC368	Shafiullah, I. H. Siddiqui and S. A. Ansari; *J. Indian Chem. Soc.*, 1991, **68**, 368.	268
91JMC2525	D. J. Carini, J. V. Duncia, P. E. Aldrich, A. T. Chiu, A. L. Johnson, M. E. Pierce, W. A. Price, J. B. Santella, III, G. J. Wells, R. R. Wexler, P. C. Wong, S.-E. Yoo and P. B. M. W. M. Timmermans; *J. Med. Chem.*, 1991, **34**, 2525.	780
91JOC316	T. R. Walters, W. W. Zajac, Jr. and J. M. Woods; *J. Org. Chem.*, 1991, **56**, 316.	108, 109
91JOC347	S. Inoue and Y. Sato; *J. Org. Chem.*, 1991, **56**, 347.	638, 648, 662
91JOC537	K. Baum, T. G. Archibald, D. Tzeng, R. Gilardi, J. L. Flippen-Anderson and C. George; *J. Org. Chem.*, 1991, **56**, 537.	795
91JOC638	A. G. M. Barrett and J. A. Flygare; *J. Org. Chem.*, 1991, **56**, 638.	139, 652, 655
91JOC650	R. E. Ireland, P. Wipf and J. D. Armstrong, III; *J. Org. Chem.*, 1991, **56**, 650.	831
91JOC657	F. Freeman and D. S. H. L. Kim; *J. Org. Chem.*, 1991, **56**, 657.	883
91JOC717	E. Schaumann, A. Kirschning and F. Narjes; *J. Org. Chem.*, 1991, **56**, 717.	938
91JOC865	Y. H. R. Jois and H. W. Gibson; *J. Org. Chem.*, 1991, **56**, 865.	426
91JOC970	A. B. Cheikh, J. Chuche, N. Manisse, J. C. Pommelet, K.-P. Netsch, P. Lorencak and C. Wentrup; *J. Org. Chem.*, 1911, **56**, 970.	900
91JOC1663	A. Kumar, F. E. Friedli, L. Hsu, P. J. Card, N. Mathur and H. Skechter; *J. Org. Chem.*, 1991, **56**, 1663.	37
91JOC1788	J. Atkinson, P. Morand, J. T. Arnason, H. M. Niemeyer and H. R. Bravo; *J. Org. Chem.*, 1991, **56**, 1788.	175
91JOC1944	R. W. Friessen, C. F. Sturino, A. K. Daljeet and A. Kolaczewska; *J. Org. Chem.*, 1991, **56**, 1944.	651
91JOC1948	Y. Ito, M. Suginome and M. Murakami; *J. Org. Chem.*, 1991, **56**, 1948.	1054
91JOC2193	K. Gawronska, J. Gawronski and H. M. Walborsky; *J. Org. Chem.*, 1991, **56**, 2193.	33, 35, 37
91JOC2276	S. Kiyooka, Y. Kaneko, M. Komura, H. Matsuo and M. Nakano; *J. Org. Chem.*, 1991, **56**, 2276.	202
91JOC3283	A. J. Fry, A. K. Rho, L. R. Sherman and C. S. Sherwin; *J. Org. Chem.*, 1991, **56**, 3283.	203
91JOC3530	F. Benedetti, F. Berti, S. Fabrissin, T. Gianferrara and A. Risaliti; *J. Org. Chem.*, 1991, **56**, 3530.	273, 274
91JOC3556	A. Padwa, W. H. Bullock, A. D. Dyszlewski, S. M. McCombie, B. B. Shanker and A. K. Gangaly; *J. Org. Chem.*, 1991, **56**, 3556.	938
91JOC4322	J. Gonzalez, C. J. Foli and S. Elsheimer; *J. Org. Chem.*, 1991, **56**, 4322.	5
91JOC4439	A. R. Katritzky, J. Pernak, W.-Q. Fan and F. Saczewski; *J. Org. Chem.*, 1991, **56**, 4439.	309
91JOC4529	M. Tiecco, L. Testaferri, M. Tingoli, D. Chianelli and D. Bartoli; *J. Org. Chem.*, 1991, **56**, 4529.	92
91JOC5213	R. J. Linderman, A. Ghannam and I. Badejo; *J. Org. Chem.*, 1991, **56**, 5213.	353
91JOC5445	J. Castaňer and J. Rìera; *J. Org. Chem.*, 1991, **56**, 5445.	17
91JOC5491	R. J. Linderman and B. D. Griedel; *J. Org. Chem.*, 1991, **56**, 5491.	385
91JOC5691	B. Singaram, C. T. Goralski and G. B. Fisher; *J. Org. Chem.*, 1991, **56**, 5691.	524
91JOC5721	A. Ogawa, H. Yokoyama, K. Yokoyama, T. Masawaki, N. Kambe and N. Sonoda; *J. Org. Chem.*, 1991, **56**, 5721.	947
91JOC6038	K. Krohn and G. Borner; *J. Org. Chem.*, 1991, **56**, 6038.	1075
91JOC6386	A. Padwa and P. E. Yeske; *J. Org. Chem.*, 1991, **56**, 6386.	275
91JOC6878	M. J. Robins, S. F. Wnuk, K. B. Mullah and N. K. Dalley; *J. Org. Chem.*, 1991, **56**, 6878.	61, 73, 79
91JOC6987	G. H. Posner, J.-C. Carry, R. D. Crouch and N. Johnson; *J. Org. Chem.*, 1991, **56**, 6987.	764
91JOM(403)29	R. Tacke, S. Brakmann, F. Wuttke, J. Fooladi, C. Syldatk and D. Schomburg; *J. Organomet. Chem.*, 1991, **403**, 29.	353
91JOM(405)59	B. Korpar-Čolig, Z. Popović, M. Sikirica and D. Grdenić; *J. Organomet. Chem.*, 1991, **405**, 59.	697
91JOM(406)123	J. Barrau, N. Ben Hamida and J. Satgé; *J. Organomet. Chem.*, 1991, **406**, 123.	629
91JOM(407)157	J. Ohshita and M. Ishikawa; *J. Organomet. Chem.*, 1991, **407**, 157.	616
91JOM(407)391	M. I. Bruce, G. A. Koutsantonis and E. R. Tiekink; *J. Organomet. Chem.*, 1991, **407**, 391.	711
91JOM(408)297	R. Skoda-Földes, L. Kollár and B. Heil; *J. Organomet. Chem.*, 1991, **408**, 297.	510
91JOM(411)69	B. Mauzé and L. Miginiac; *J. Organomet. Chem.*, 1991, **411**, 69.	704
91JOM(413)55	H. Brunner and S. Limmer; *J. Organomet. Chem.*, 1991, **413**, 55.	548
91JOM(414)295	A. Fürstner, G. Kollegger and H. Weidmann; *J. Organomet. Chem.*, 1991, **414**, 295.	655
91JOM(415)39	H. Hengelsberg, R. Tacke, K. Fritsche, C. Syldatk and F. Wagner; *J. Organomet. Chem.*, 1991, **415**, 39.	506, 513
91JOM(415)181	W. Uhl and M. Layh; *J. Organomet. Chem.*, 1991, **415**, 181.	659, 698
91JOM(415)327	H. Ranaivonjatovo, J. Escudiè, C. Couret and J. Satgè; *J. Organomet. Chem.*, 1991, **415**, 327.	485
91JOM(417)149	H. Werner, O. Schippel, J. Wolf and M. Schulz; *J. Organomet. Chem.*, 1991, **417**, 149.	149, 583
91JOM(418)C21	V. Gevorgyan, L. Borisova and E. Lukevics; *J. Organomet. Chem.*, 1991, **418**, C21.	704
91JOM(418)C24	I. V. Barinov, O. A. Reutov, A. V. Polyakov, A. I. Yanovsky, Yu. T. Struchkov and V. I. Sokolov; *J. Organomet. Chem.*, 1991, **418**, C24.	674
91JOM(421)175	M. Westerhausen, B. Rademacher and W. Poll; *J. Organomet. Chem.*, 1991, **421**, 175.	607, 656
91JOU48	V. I. Potkin, R. V. Kaberdin and Yu. A. Ol'dekop; *J. Org. Chem. USSR (Engl. Transl.)*, 1991, **27**, 48.	797

91JPR333	A. G. A. Elagamey and F. M. A. El-Taweel; *J. Prakt. Chem.*, 1991, **333**, 333.	986
91LA229	E. Öhler and E. Zbiral; *Liebigs Ann. Chem.*, 1991, 229.	339
91LA243	K. Hartke, K.-H. Lee, W. Massa and B. Schwarz; *Liebigs Ann. Chem.*, 1991, 243.	281
91LA655	C. P. Schickli and D. Seebach; *Liebigs Ann. Chem.*, 1991, 655.	102
91LA811	R. W. Hoffmann and M. Julius; *Liebigs Ann. Chem.*, 1991, 811.	146
91MI 403-01	E. V. Maskaeva, V. V. Bashilov and V. I. Sokolov; *Organometallic Chemistry in the USSR (Engl. Transl.)*, 1991, **4**, 295.	152
91MI 403-02	S. N. Gurkova, A. I. Gusev, N. A. Viktorov and V. F. Mironov; *Organometallic Chemistry in the USSR (Engl. Transl.)*, 1991, **4**, 301.	142
B-91MI 404-01	T. W. Greene and P. G. M. Wuts; "Protective Groups in Organic Synthesis," 2nd edn., Wiley, New York, 1991, p. 10.	176, 180, 184, 185, 186, 198
B-91MI 404-02	T. W. Greene and P. G. M. Wuts; "Protective Groups in Organic Synthesis," 2nd edn., Wiley, New York, 1991, p. 175.	176
B-91MI 404-03	T. W. Greene and P. G. M. Wuts; "Protective Groups in Organic Synthesis," 2nd edn., Wiley, New York, 1991, p. 184.	203
B-91MI 404-04	T. W. Greene and P. G. M. Wuts; "Protective Groups in Organic Synthesis," 2nd edn., Wiley, New York, 1991, p. 235.	209
91MI 406-01	K. Manabe, T. Tanaka, S. Kurozumi and Y. Kato; *J. Label. Cmpds. Radiopharm.*, 1991, 1107.	279
91MI 408-01	M. Mladenova and F. Gaudemar-Bardone; *Phosphorus, Sulfur and Silicon*, 1991, **62**, 257.	394
91MI 412-01	V. Jouko, N. Heikki and P. Esko; *J. Labelled Comp. Radiopharm.*, 1991, **29**, 1191 (*Chem. Abstr.*, 1992, **116**, 41 591).	555, 557
91MI 412-02	O. I. Kolodyazhnyi and S. N. Ustenko; *Dokl. Akad. Nauk Ukr. SSR*, 1991, 118 (*Chem. Abstr.*, 1992, **116**, 6623).	570
91MI 413-01	R. L. Wells, C.-Y. Kwag, A. P. Purdy, A. T. McPhail and C. G. Pitt; *Report* 1989, DU/DC/TC-13 (*Chem. Abstr.*, 1991, **114**, 6699).	597
91MI 414-01	V. I. Shiryaev, A. A. Grachev, S. N. Tandura, E. M. Stepina, S. I. Androsenko, N. N. Silkina, V. N. Bochkarev and V. D. Sheludyakov; *Metallorg. Khim.*, 1991, **4**, 453 (*Chem. Abstr.*, 1991, **115**, 50 340z).	662
91MI 419-01	A. N. Mirskova, S. G. Seredkina, A. V. Martynov and O. L. Sizykh; *Otkrytiya, Izobret.*, 1991, 86 (*Chem. Abstr.* 1992, **116**, 106 512).	872
91MI 420-01	F. Freeman, D. S. H. L. Kim and J. W. Ziller; *Acta Cryst.*, 1991, **C47**, 2124.	883
B-91MI 424-01	A. Gilbert and J. Baggott; "Essentials of Molecular Photochemistry," CRC Press, Boca Raton, 1991.	1079
91MI 424-02	P. Wagner and B.-S. Park; *Org. Photochem.*, 1991, **11**, 227.	1079
91NJC301	J. T. B. H. Jastrzebski, P. Van der Schaaf, J. Boersma and G. Van Koten; *Nouv. J. Chim.*, 1991, **15**, 301.	704
91NJC657	J. Barrau and N. Ben Hamida; *Nouv. J. Chim.*, 1991, **15**, 657.	629
91OM369	D. M. Hester and G. K. Yang; *Organometallics*, 1991, **10**, 369.	582
91OM389	H. Fischer, A. Schlageter, W. Bidell and A. Fruh; *Organometallics*, 1991, **10**, 389.	1018
91OM535	M. H. Chisholm, K. Folting, J. C. Huffman and E. A. Lucas; *Organometallics*, 1991, **10**, 535.	675
91OM551	D. Seyferth and H. Lang; *Organometallics*, 1991, **10**, 551.	620
91OM936	T. N. Mitchell and B. S. Bronk; *Organometallics*, 1991, **10**, 936.	700
91OM938	H. Grützmacher and H. Pritzkow; *Organometallics*, 1991, **10**, 938.	588
91OM1278	R. D. Adams, G. Chen and J. Yin; *Organometallics*, 1991, **10**, 1278.	677
91OM1336	C. K. Whitmarsh and L. V. Interrante; *Organometallics*, 1991, **10**, 1336.	606, 654
91OM1907	A. J. Carty, N. J. Taylor, E. Sappa, A. Tiripicchio and M. Tiripicchio-Camellini; *Organometallics*, 1991, **10**, 1907.	713
91OM1960	T. Kobayashi and K. H. Pannell; *Organometallics*, 1991, **10**, 1960.	136, 138, 141
91OM2068	A. J. Ashe, III, T. P. Diephouse J. W. Kampf, and S. M. Al-Taweel; *Organometallics*, 1991, **10**, 2068.	1041
91OM2121	D. W. Macomber, P. Madhukar and R. D. Rogers; *Organometallics*, 1991, **10**, 2121.	934
91OM2159	T. A. Annan, D. G. Tuck, M. A. Khan and C. Peppe; *Organometallics*, 1991, **10**, 2159.	153, 588
91OM2191	D. M. Amorose, R. A. Lee and J. L. Peterson; *Organometallics* 1991, **10**, 2191.	656
91OM2541	R. D. Adams, G. Chen, L. Chen, M. P. Pompeo and J. Yin; *Organometallics*, 1991, **10**, 2541.	678
91OM2695	J. Ohshita, H. Ohsaki and M. Ishikawa; *Organometallics*, 1991, **10**, 2695.	613, 617
91OM2726	A. A. Aradi, F.-E. Hong and T. P. Fehlner; *Organometallics*, 1991, **10**, 2726.	726
91OM2857	N. H. Dryden, P. Leyzdins, J. Trotter and V. C. Yee; *Organometallics*, 1991, **10**, 2857.	656
91OM3003	M. J. Chetcuti, P. E. Fanwick and B. E. Grant; *Organometallics*, 1991, **10**, 3003.	712
91OM3167	L. D. Field, M. G. Gardiner, C. H. Kennard, B. A. Messerle and C. L. Raston; *Organometallics*, 1991, **10**, 3167.	603, 649
91OM3292	A. G. Brook, P. Chiu, J. McClenaghnan and A. J. Lough; *Organometallics*, 1991, **10**, 3292.	613
91OM3392	J. P. Fackler and R. A. Kresinski; *Organometallics*, 1991, **10**, 3392.	587
91OM3791	G. Erker, M. Albrecht, C. Krüger and S. Werner; *Organometallics*, 1991, **10**, 3791.	725
91OM4005	V. Dufaud, J. Thivolle-Cazat, J.-M. Basset, R. Mathieu, J. Jaud and J. Waissermann; *Organometallics*, 1991, **10**, 4005.	722
91OR(40)157	O. De Lucchi, U. Miotti and G. Modena; *Org. React.*, 1991, **40**, 157.	221, 222, 226, 240, 323
91PAC339	D. S. Matteson; *Pure Appl. Chem.*, 1991, **63**, 339.	525
91PHA741	R. Bohm and R. Pech; *Pharmazie*, 1991, **46**, 741.	979
91POL319	R. L. Wells, C.-Y. Kwag, A. P. Purdy, A. T. McPhail and C. G. Pitt; *Polyhedron*, 1990, **9**, 319.	597

91POL1713	F. L. Joslin, J. T. Mague and D. M. Roundhill; *Polyhedron*, 1991, **10**, 1713.	550
91PS(55)41	A. M. M. M. Phillips and T. A. Modro; *Phosphorus, Sulfur*, 1991, **55**, 41.	570
91PS(55)97	M. Topolski and J. Rachon; *Phosphorus Sulfur*, 1991, **55**, 97.	1008
91PS(57)1	R. L. Wells, A. P. Purdy and C. G. Pitt; *Phosphorus, Sulfur*, 1991, **57**, 1 (*Chem. Abstr.*, 1991, **115**, 92 436).	597
91PS(58)207	K. Schank; *Phosphorus Sulfur*, 1991, **58**, 207.	265
91PS(60)131	N. W. Alcock, G. A. Pike, M. Papadopoulos, E. Pitsinos, L. Hatjiarapoglou and A Varvoglis; *Phosphorus, Sulfur*, 1991, **60**, 131.	867
91S26	A. J. Moore and M. R. Bryce; *Synthesis*, 1991, 26.	847, 848
91S31	N. Mimouni, E. About-Jaudet and N. Collignon; *Synthesis*, 1991, 31.	588
91S117	Y. Machida, N. Shirai and Y. Sato; *Synthesis*, 1991, 117.	508
91S125	N. A. Bondarenko, M. V. Rudomino and E. N. Tsvetkov; *Synthesis*, 1991, 125.	552
91S169	K. Marumo, S. Inoue and Y. Sato; *Synthesis*, 1991, 169.	522
91S225	F. Babudri, V. Fiandanese, R. Musio, F. Naso, O. Sciavovelli and A. Scilimati; *Synthesis*, 1991, 225.	476
91S279	A. R. Katritzky, S. I. Bayyuk and S. Rachwal; *Synthesis*, 1991, 279.	301
91S368	S.-B. Lee, S.-D. Lee, T. Takata and T. Endo; *Synthesis*, 1991, 368.	704
91S481	A. Oliva, I. Castro, C. Castillo and G. Leon; *Synthesis*, 1991, 481.	843
91S490	C. Yuan, S. Chen and G. Wang; *Synthesis*, 1991, 490.	470, 488
91S637	H. Ahlbrecht, C. Schmitt and D. Kornetzky; *Synthesis*, 1991, 637.	530
91S661	G. Sturz and J. Guervenou; *Synthesis*, 1991, 661.	560
91S793	N. Petragnani and J. V. Comasseto; *Synthesis*, 1991, 793.	333
91S854	C. Yuan, C. Li and Y. Ding; *Synthesis*, 1991, 854.	801
91S897	N. Petragnani and J. V. Comasseto; *Synthesis*, 1993, 897.	333
91S996	T. Kitano, N. Shirai and Y. Sato; *Synthesis*, 1991, 996.	507
91S1099	E. Fuchs, B. Breit, U. Bergsträsser, J. Hoffmann, H. Heydt and M. Regitz; *Synthesis*, 1991, 1099.	483
91S1147	A. R. Katritzky, I. Takahashi, W.-Q. Fan and J. Pernak; *Synthesis*, 1991, 1147.	323
91S1221	E. Anders, A. Opitz and W. Bauer; *Synthesis*, 1991, 1221.	604, 647
91SC279	K. Lee and D. Y. Oh; *Synth. Commun.*, 1991, **21**, 279.	927, 929
91SC793	X. Morise, P. Savignac, J. C. Guillemin and J. M. Denis; *Synth. Commun.*, 1991, **21**, 793.	131
91SC1039	S. K. Chakraborty and R. Engel; *Synth. Commun.*, 1991, **21**, 1039.	132
91SC1055	R. J. Heffner and M. M. Joullié; *Synth. Commun.*, 1991, **21**, 1055.	160
91SC1213	X. Huang, B.-C. Chen, G.-Y. Wu and H.-B. Chen; *Synth. Commun.*, 1991, **21**, 1213.	900
91SC1369	L. G. Zepeda, H. Cervantes, M. S. Morales-Rios and P. Joseph-Nathan; *Synth. Commun.*, 1991, **21**, 1369.	245
91SC1675	M. Lamothe, M. B. Anderson and P. L. Fuchs; *Synth. Commun.*, 1991, **21**, 1675.	365, 393, 394
91SC1847	A.-M. Chollet-Gravey, L. Vo-Quang, Y. Vo-Quang and F. Le Goffic; *Synth. Commun.*, 1991, **21**, 1847.	487
91SC1951	L. K. Lukanov and A. P. Venkov; *Synth. Commun.*, 1991, **21**, 1951.	488
91SC2231	R. J. Heffner and M. M. Joullié; *Synth. Commun.*, 1991, **21**, 2231.	160
91SL80	P. C. Bulman Page and E. S. Namwindwa; *Synlett*, 1991, 80.	261
91SL84	P. C. Bulman Page and J. C. Prodger; *Synlett*, 1991, 84.	263
91SL87	A. Mertin, T. Thiemann, I. Hanss and A. de Meijere; *Synlett*, 1991, 87.	309
91SL111	R. Gosmini, P. Mangeney, A. Alexakis, M. Commercon and J. F. Normant; *Synlett*, 1991, 111.	405, 408
91SL395	E. Differding, R. O. Duthaler, A. Krieger, G. M. Rüegg and C. Schmit; *Synlett*, 1991, 395.	131
91SL485	K. Smith, A. Small and M. G. Hutchings; *Synlett*, 1991, 485.	296
91SL501	F. Chemla, M. Julia, D. Uguen and D. Zhang; *Synlett*, 1991, 501.	200
91SL503	M. Julia, D. Uguen and D. Zhang; *Synlett*, 1991, 503.	201
91SL557	A. Hosomi, K. Ogata, M. Ohkuma and M. Hojo; *Synlett.*, 1991, 557.	368
91SL565	O. De Lucchi, D. Fabbri and V. Lucchini; *Synlett*, 1991, 565.	272, 854
91SL631	T. J. Michnick and D. S. Matteson; *Synlett*, 1991, 631.	144, 146
91SL911	S. Akai, Y. Tsuzuki, S. Matsuda, S. Kitagaki and Y. Kita; *Synlett*, 1991, 911.	661
91SM2535	G. C. Papavassiliou; *Synth. Met.*, 1991, **42**, 2535.	869
91T323	N. S. Simpkins; *Tetrahedron*, 1991, **47**, 323.	286
91T343	E. Negishi, T. Yoshida, A. Abramovitch, G. Lew and R. M. Williams; *Tetrahedron*, 1991, **47**, 343.	815
91T615	J. Yoshida, S. Matsunaga, T. Murata and S. Isoe; *Tetrahedron*, 1991, **47**, 615.	197
91T1329	D. S. Brown, S. V. Ley, S. Vile and M. Thompson; *Tetrahedron*, 1991, **47**, 1329.	200
91T1525	A. Arcadi, E. Bernocchi, S. Cacchi and F. Marinelli; *Tetrahedron*, 1991, **47**, 1525.	169
91T1547	G. H. Veeneman, G. A. Van Der Marel, H. Van Den Elst and J. H. Van Boom; *Tetrahedron*, 1991, **47**, 1547.	220
91T1925	M. Haber and U. Pindur; *Tetrahedron*, 1991, **47**, 1925.	277
91T2683	A. R. Katritzky, S. Rachwal and G. J. Hitchings; *Tetrahedron*, 1991, **47**, 2683.	323, 406, 410, 445
91T3281	T. Sato, T. Kikuchi, H. Tsujita, A. Kaetsu, N. Sootome, K.-i. Nishida, K. Tachibana and E. Murayama; *Tetrahedron*, 1991, **47**, 3281.	699
91T3431	W. Tong, Z. Xi, C. Gioeli and J. Chattopadhyaya; *Tetrahedron*, 1991, **47**, 3431.	276
91T3727	A. Herunsalee, M. Isobe and T. Goto; *Tetrahedron*, 1991, **47**, 3727.	940
91T6171	D. P. Curran and D. Kim; *Tetrahedron*, 1991, **47**, 6171.	781
91T6539	S. P. Modi, M. A. Michael, S. Archer and J. J. Carey; *Tetrahedron*, 1991, **47**, 6539.	950
91T7091	D. H. R. Barton, J. Boivin, E. Crépon, J. Sarma, H. Togo and S. Z. Zard; *Tetrahedron*, 1991, **47**, 7091.	266

91T8091	W. S. Lee, K. Lee, K. D. Nam and Y. J. Kim; *Tetrahedron*, 1991, **47**, 8091.	260
91TA93	R. Lopez and J. C. Carretero; *Tetrahedron Asymmetry*, 1991, **2**, 93.	864
91TA157	A. Boussoufi, P. Hudhomme, P. Hitchcock and G. Duguay; *Tetrahedron Asymmetry*, 1991, **2**, 157.	101
91TA277	H. C. Brown, R. R. Iyer, V. H. Mahindroo and N. E. Bhat; *Tetrahedron Asymmetry*, 1991, **2**, 277.	815
91TL105	T. Kataoka, M. Yoshimatsu, H. Shimizu and M. Hori; *Tetrahedron Lett.*, 1991, **32**, 105.	288
91TL375	K. Uneyama and K. Kitagawa; *Tetrahedron Lett.*, 1991, **32**, 375.	284
91TL405	I. T. Barnisch, C. W. G. Fishwick and D. R. Hill; *Tetrahedron Lett.*, 1991, **32**, 405.	901
91TL467	A. Kusche, R. Hoffmann, I. Münster, P. Keiner and R. Brückner; *Tetrahedron Lett.*, 1991, **32**, 467.	217
91TL679	T. K. Chakraborty and G. V. Reddy; *Tetrahedron Lett.*, 1991, **32**, 679.	516
91TL855	S. D. Burke, A. D. Piscopio and M. E. Kort; *Tetrahedron Lett.*, 1991, **32**, 855.	355
91TL1035	M. Newcomb and J. L. Esker; *Tetrahedron Lett.*, 1991, **32**, 1035.	1077
91TL1385	C. Alcaraz, J. C. Carretero and E. Dominguez; *Tetrahedron Lett.*, 1991, **32**, 1385.	858, 938, 961
91TL1463	M. Fujita, M. Suzuki, K. Ogata and K. Ogura; *Tetrahedron Lett.*, 1991, **32**, 1463.	73
91TL1539	W. J. Bailey and L.-L. Zhou; *Tetrahedron Lett.*, 1991, **32**, 1539.	826
91TL1899	A. Ricci, A. Degl'Innocenti, A. Capperucci, G. Reginato and A. Mordini; *Tetrahedron Lett.*, 1991, **32**, 1899.	951
91TL1975	T. Tsunoda, K. Fujiwara, Y. Yamamoto and S. Itô; *Tetrahedron Lett.*, 1991, **32**, 1975.	528
91TL2039	H. Tani, K. Masumoto, T. Inamasu and H. Suzuki; *Tetrahedron Lett.*, 1991, **32**, 2039.	247
91TL2049	N. Tokitoh, T. Matsumoto, H. Suzuki and R. Okazaki; *Tetrahedron Lett.*, 1991, **32**, 2049.	653
91TL2125	A. Mercier, Y. Berchadsky, S. P. Badrudin and P. Tordo; *Tetrahedron Lett.*, 1991, **32**, 2125.	485
91TL2177	L. Pasquato, O. de Lucchi and L. Krotz; *Tetrahedron Lett.*, 1991, **32**, 2177.	840, 861
91TL2193	B. M. Trost and J. R. Granja; *Tetrahedron Lett.*, 1991, **32**, 2193.	280
91TL2375	Y. Kita, N. Shibata, N. Yoshida and T. Tohjo; *Tetrahedron Lett.*, 1991, **32**, 2375.	418
91TL2413	C. W. Bird and A. K. Dotse; *Tetrahedron Lett.*, 1991, **32**, 2413.	212
91TL2565	D. Crich and L. B. L. Lim; *Tetrahedron Lett.*, 1991, **32**, 2565.	174
91TL2683	P. D. Berry, A. C. Brown, J. C. Hanson, A. C. Kaura, P. H. Milner, C. J. Moores, J. K. Quick, R. N. Saunders, R. Southgate and N. Whittall; *Tetrahedron Lett.*, 1991, **32**, 2683.	419
91TL2783	M. Taniguchi, K. Oshima and K. Utimoto; *Tetrahedron Lett.*, 1991, **32**, 2783.	944
91TL2971	B. F. Bonini, S. Masiero, G. Mazzanti and P. Zani; *Tetrahedron Lett.*, 1991, **32**, 2971.	232
91TL3163	G. Apitz and W. Steglich; *Tetrahedron Lett.*, 1991, 3163.	100, 311
91TL3285	A. G. M. Barrett and J. M. Hill; *Tetrahedron Lett.*, 1991, **32**, 3285.	660
91TL3313	W. Priebe, G. Grynkiewicz, N. Neamati and R. Perez-Soler; *Tetrahedron Lett.*, 1991, **32**, 3313.	174
91TL3631	M. B. Goli and S. O. Grim; *Tetrahedron Lett.*, 1991, **32**, 3631.	1022, 1076
91TL3687	G. Etemad-Moghadam, C. Tachon, M. Gougou and M. Koenig; *Tetrahedron Lett.*, 1991, **32**, 3687.	1025
91TL3695	G. Solladié, F. Colobert, P. Ruiz, C. Hamdouchi, M. C. Carreno and J. L. Garcia Ruano; *Tetrahedron Lett.*, 1991, **32**, 3695.	270, 863
91TL4279	H. J. Bestmann and H. Lehnen; *Tetrahedron Lett.*, 1991, **32**, 4279.	559
91TL4435	S. Mehta and B. M. Pinto; *Tetrahedron Lett.*, 1991, **32**, 4435.	238
91TL5047	P. A. Magriotis, J. T. Brown and M. E. Scott; *Tetrahedron Lett.*, 1991, **32**, 5047.	962, 963
91TL5159	E. Domínguez, J. C. Carretero, A. Fernández-Mayoralas and S. Conde; *Tetrahedron Lett.*, 1991, **32**, 5159.	938
91TL5683	J. M. Chong and E. K. Mar; *Tetrahedron Lett.*, 1991, **32**, 5683.	389
91TL6085	P. A. Magriotis, M. E. Scott and K. D. Kim; *Tetrahedron Lett.*, 1991, **32**, 6085.	962
91TL6215	H. Ihmels, M. Maggini, M. Prato and G. Scorrano; *Tetrahedron Lett.*, 1991, **32**, 6215.	162
91TL6325	A. H. Djerourou and L. Blanco; *Tetrahedron Lett.*, 1991, **32**, 6325.	352
91TL6563	T. Takeda, F. Kanamori, H. Matsusita and T. Fujiwara; *Tetrahedron Lett.*, 1991, **32**, 6563.	774
91TL6961	S. Birkinshaw and P. Kocienski; *Tetrahedron Lett.*, 1991, **32**, 6961.	934, 935
91TL6973	D. Craig, K. Daniels and A. R. Mackenzie; *Tetrahedron Lett.*, 1991, **32**, 6973.	207
91TL7369	C. Lamberth and M. D. Bednarski; *Tetrahedron Lett.*, 1991, **32**, 7369.	176
91TL7593	S. Zavgorodny, M. Polianski, E. Besidsky, V. Kriukov, A. Sanin, M. Pokrovskaya, G. Gurskaya, H. Lönnberg and A. Azhayev; *Tetrahedron Lett.*, 1991, **32**, 7593.	222
91TL7743	V. K. Aggarwal, R. J. Franklin and M. J. Rice; *Tetrahedron Lett.*, 1991, **32**, 7743.	269
91URP1618747	A. A. Prishchenko, M. V. Livanstov, P. V. Zhutskii and V. S. Petrosyan, *USSR Pat.* 1 618 747 (1991) (*Chem. Abstr.*, 1991, **115**, 71 900x).	459
91USP5023275	D. R. Amick (Rohm & Haas Co.); *US Pat.* 5 023 275 (1991) (*Chem. Abstr.*, 1991, **116**, 6253v).	775
91USP5043489	V. Nocito, L. J. Bedell and M. I. Levinson; *US Pat.* 5 043 489 (1991) (*Chem. Abstr.*, 1992, **116**, 20 682r).	108
91ZAAC(594)66	F. Krech, K. Issleib, A. Zschunke, C. Mügge and S. Skvorcov; *Z. Anorg. Allg. Chem.*, 1991, **594**, 66.	550
91ZAAC(596)139	G. Heckmann, W. Plass and E. Fluck; *Z. Anorg. Allg. Chem.*, 1991, **596**, 139.	552
91ZAAC(601)65	E. Fluck, M. Spahn and G. Heckmann; *Z. Anorg. Allg. Chem.*, 1991, **601**, 65.	562
91ZAAC(605)151	P. Aslanidis and F. Kober; *Z. Anorg. Allg. Chem.*, 1991, **605**, 151.	593
91ZN(B)275	R. Tacke and F. Wiesenberger; *Z. Naturforsch., Teil. B.*, 1991, **46**, 275.	142
91ZN(B)978	J. Grobe, D. Le Van and T. Großpietsch; *Z. Naturforsch., Teil B*, 1991, **46**, 978.	116
91ZN(B)1169	W. Rohde and G. Fendesak; *Z. Naturforsch., Teil B*, 1991, **46**, 1169.	718

91ZN(B)1207	B. Wrackmeyer and K. H. Von Locquenghien; *Z. Naturforsch., Teil B*: 1991, **46**, 1207 (*Chem. Abstr.*, 1992, **116**, 6663d).	665
91ZOB56	V. I. Potkin, R. V. Kaberdin and Yu. A. Ol'dekop; *Zh. Obshch. Khim.*, 1991, **27**, 56.	762
91ZOB874	V. V. Kurg, V. S. Brovarets and B. S. Drach; *Zh. Obshch. Khim.*, 1991, **61**, 874.	771
91ZOB909	M. F. Rostovskaya, V. V. Isakov, M. G. Slabko and V. I. Vysotskii; *Zh. Obshch. Khim.*, 1991, **61**, 909.	339
91ZOB921	S. N. Nikoleva, A. S. Zolotareva, S. V. Ponomarev and V. S. Petrosyan; *Zh. Obshch. Khim.*, 1991, **61**, 921 (*Chem. Abstr.* 1991, **115**, 280 159b).	626
91ZOB1014	N. V. Lukashev, A. A. Fil'chikov, P. E. Zhichkin, M. A. Kazankova and I. P. Beletskaya; *Zh. Obshch. Khim.*, 1991, **61**, 1014.	766
91ZOB1016	A. A. Prishchenko, M. V. Livanstov, D. A. Pisarnitskii and V. S. Petrosyan; *Zh. Obshch. Khim.*, 1991, **61**, 1016 (*Chem. Abstr.*, 1992, **116**, 6649d).	459, 484
91ZOB1084	S. G. Seredkina, A. N. Mirskova, O. B. Bannikova and G. V. Dolgushin; *Zh. Obshch. Khim*, 1991, **61**, 1084.	772
91ZOB1192	S. E. Tolschinskii, M. V. Kormer and I. A. Maretina; *Zh. Obshch. Khim.*, 1991, **27**, 1192.	885
91ZOB1263	Yu. V. Prikhod'ko, A. S. Skobun and M. F. Vysotskii; *Zh. Obshch. Khim.*, 1991, **61**, 1263.	339
91ZOB1315	V. I. Vysotskii and S. V. Levan'kov; *Zh. Obshch. Khim.*, 1991, **61**, 1315.	339
91ZOB1478	A. A. Prishchenko, M. V. Livanstov, D. A. Pisarnitskii and V. S. Petrosyan; *Zh. Obshch. Khim.*, 1991, **61**, 1478 (*Chem. Abstr.*, 1992, **116**, 6655c).	458
91ZOB1600	N. V. Lukashev, A. A. Fil'chikov, A. I. Kozlov, Yu. N. Luzikov and M. A. Kazankora; *Zh. Obshch. Khim.*, 1991, **61**, 1600.	884
91ZOB1893	T. E. Glotova, A. S. Nakhmanovich, O. G. Yarosh, L. S. Romanenko and T. N. Komarova; *Zh. Obshch. Khim.*, 1991, **61**, 1893.	940
91ZOB2698	I. P. Romanova, A. A. Muslinkin and R. Z. Musin; *Zh. Obshch. Khim.*, 1991, **61**, 2698.	561
91ZOR216	I. M. Bazavova, A. N. Esipenko, V. M. Neplyvev and M. O. Lozinskii; *Zh. Org. Khim.*, 1991, **27**, 216.	272
91ZOR382	S. G. Sibrikov, V. N. Kazin, V. V. Kopeiki, G. S. Mironov and T. N. Orlova; *Zh. Org. Khim.*, 1991, **27**, 382 (Engl. Transl. 325).	979
91ZOR1796	V. V. Shhepin, T. Yu. Chuprikova and I. Yu. Petukhova; *Zh. Org. Khim.*, 1991, **27**, 1796.	775
91ZOR2362	B. A. Shainyan and V. K. Bel'skii; *Zh. Org. Khim.*, 1991, **27**, 2362.	704
92ACR182	Y. Huang; *Acc. Chem. Res.*, 1992, **25**, 182.	198
92AG(E)66	S. D. Dietz, N. W. Eilerts and J. A. Heppert; *Angew. Chem., Int. Ed. Engl.*, 1992, **31**, 66.	676
92AG(E)232	M. Lautens, C. Zhang and C. M. Crudden; *Angew. Chem., Int. Ed. Engl.*, 1992, **31**, 232.	661
92AG(E)1055	B. Breit, U. Bergsträsser, G. Maas and M. Regitz; *Angew. Chem., Int. Ed. Engl.*, 1992, **31**, 1055.	549
92AG(E)1335	B. M. Trost, W. Brieden and K. H. Baringhaus; *Angew. Chem., Int. Ed. Engl.*, 1992, **31**, 1335.	206
92AP(325)411	C. Herdeis and W. Engel; *Arch. Pharm.* (*Weinheim, Ger.*), 1992, **325**, 411.	105
92AP(325)551	K. Eger and M. Frey; *Arch. Pharm.* (*Weinheim, Ger.*), 1992, **325**, 551.	979
92BCJ210	H. Uno, K. Sakamoto, F. Semba and H. Suzuki; *Bull. Chem. Soc, Jpn.*, 1992, **65**, 210.	774, 777
92BCJ304	K. Tanemura, T. Horaguchi and T. Suzuki; *Bull. Chem. Soc. Jpn.*, 1992, **65**, 304.	186
92BCJ1513	K. Miura, J. Sugimoto, K. Oshima and K. Utimoto; *Bull. Chem. Soc. Jpn.*, 1992, **65**, 1513.	737, 743
92BCJ2303	S. Hyuga, S. Hara and A. Suzuki; *Bull. Soc. Chem. Jpn.*, 1992, **65**, 2303.	954
92BCJ2366	M. Yamamoto, H. Munakata, K. Kishikawa, S. Kohmoto and K. Yamada; *Bull. Chem. Soc. Jpn.*, 1992, **65**, 2366.	206
92CB143	T. Kauffmann, P. Fiegenbaum, M. Papenberg, R. Wieschollek and J. Sander; *Chem. Ber.*, 1992, **125**, 143.	673
92CB557	R. Gerhardt, R. Kuschel and K. Seppelt; *Chem. Ber.*, 1992, **125**, 557.	868
92CB571	A. Haas and H. W. Praas; *Chem. Ber.*, 1992, **125**, 571.	867
92CB591	R. Tacke, F. Weisenberger, B. Becker, R. Rohr-Aehle, P. B. Schneider, U. Ulbrich, S. M. Sarge, H. K. Cammenga, T. Koslowski and W. von Niessen; *Chem. Ber.*, 1992, **125**, 591.	352, 353
92CB607	R. Tacke, H. Hengelsberg, E. Klingner and H. Henke; *Chem. Ber.*, 1992, **125**, 607.	602, 621, 929
92CB771	P. Becker, H. Brombach, G. David, M. Leuer, H.-J. Metternich and E. Niecke; *Chem. Ber.*, 1992, **125**, 771.	568, 573
92CB801	M. Well, W. Elbers, A. Fischer, P. G. Jones and R. Schmutzler; *Chem. Ber.*, 1992, **125**, 801.	134
92CB1325	I. V. Shevchenko, A. Fischer, P. G. Jones and R. Schmutzler; *Chem. Ber.*, 1992, **125**, 1325.	564
92CB1341	P. Frankhauser, M. Driess, H. Pritzkow and W. Siebert; *Chem. Ber.*, 1992, **125**, 1341.	665
92CB1953	G. Erker, M. Albrecht, S. Werner, M. Nolte and C. Krüger; *Chem. Ber.*, 1992, **125**, 1953.	725
92CB2051	M. Duetsch, F. Stein, R. Lackmann, E. Pohl, R. Herbst-Irmer and A. de Meijere; *Chem. Ber.*, 1992, **125**, 2051.	940, 1014
92CB2081	H. J. Bestmann, R. Dostalek and R. Zimmermann; *Chem. Ber.*, 1992, **125**, 2081.	571
92CB2213	B. Ederer, H. Ederle and H. Nöth; *Chem. Ber.*, 1992, **125**, 2213.	635
92CB2731	R. Hoffmann and R. Brueckner; *Chem. Ber.*, 1992, **125**, 2731.	359
92CC190	T. Minami, M. Nakaya, K. Fujimoto and S. Matsuo; *J. Chem. Soc., Chem. Commun.*, 1992, 190.	349
92CC310	L. A. Brady, A. F. Dyke, S. E. Garner, V. Guerchais, S. A. R. Knox, J. P. Maher, S. M. Nicholls and A. G. Orpen; *J. Chem. Soc., Chem. Commun.*, 1992, 310.	681
92CC487	C. O. Kappe, G. Kollenz, R. Leung-Toung and C. Wentrup; *J. Chem. Soc., Chem. Commun.*, 1992, 487.	894
92CC1031	M. Akita, T. Oku and Y. Moro-oka; *J. Chem. Soc., Chem. Commun.*, 1992, 1031.	679

92CC1048	J. Y. Becker, J. Bernstein, M. Dayan and L. Shahal; *J. Chem. Soc., Chem. Commun.*, 1992, 1048.	873
92CJC1022	Z. Gross and H. Shmaryahu; *Can. J. Chem.*, 1992, **70**, 1022.	993
92CL407	M. Yamashita, A. Iida, K. Ikai, T. Oshikawa, T. Hanaya and H. Yamamoto; *Chem. Lett.*, 1992, 407.	122
92CL819	T. Takeda, S. Sugi, A. Nakayama, Y. Suzuki and T. Fujiwara; *Chem. Lett.*, 1992, 819.	398
92CL1229	K. Narasaka, T. Okauchi and N. Arai; *Chem. Lett.*, 1992, 1229.	252
92CL2047	T. Kusumoto, A. Nakayama, K. Sato, T. Hiyama, S. Takehara, M. Osawa and K. Nakamura; *Chem. Lett.*, 1992, 2047.	654, 655
92CL2173	S. Matsubara, H. Takahashi and K. Utimoto; *Chem. Lett.*, 1992, 2173.	231
92CPB3261	S. Saito, K. Ichinose, Y. Sasaki and S. Sumita; *Chem. Pharm. Bull.*, 1992, **40**, 3261.	766
92CRV175	S. Hermanek, R. E. Williams, B. Stibr, R. N. Grimes, J. Plesek, L. A. Leites and S. Hermanek; *Chem. Rev.*, 1992, **92**, 175.	629
92CSC1116	D. Matković-Čalogovic, Z. Popović and B. Korpar-Čolig; *Cryst. Struct. Commun.*, 1992, **48**, 1116.	697
92CSR271	P. D. Lickiss; *Chem. Soc. Rev.*, 1992, 271.	656
92EUP471651	R. Metternich; *Eur. Pat.* 471 651 (1992) (*Chem. Abstr.*, 1992, **117**, 49 266q).	528
92EUP496413	V. A. Petrov, D. D. DesMarteau and W. Navarrini; *Eur. Pat.* 496 413 (1992) (*Chem. Abstr.*, 1992, **117**, 191 831p).	104
92GEP4134772	P. Hofmann and H. Heiss; *Ger. Pat.* 4 134 772 (1992) (*Chem. Abstr.*, 1992, **117**, 171 685).	546, 577
92H(34)1005	B. Abarca, R. Ballesteros, G. Jones, D. J. Ando and M. B. Hursthouse; *Heterocycles*, 1992, **34**, 1005.	993
92HAC127	N. J. Tyryshkin, A. V. Fuzshenokova, A. N. Vedernikov and B. N. Solomonov; *Heteroatom. Chem.*, 1992, **3**, 127.	922
92HAC223	J. L. Hubbard; *Heteroatom Chem.*, 1992, **3**, 223.	364
92HCA124	R. Neidlein and T. Eichinger; *Helv. Chim. Acta*, 1992, **75**, 124.	890
92HOU(E14a/3)483	S. Pawlenko and S. Lang-Fugmann; *Methoden Org. Chem.* (*Houben-Weyl*), 1992, **E14a/3**, 483.	315, 319
92HOU(E14a/3)545	S. Pawlenko and S. Lang-Fugmann; *Methoden Org. Chem.* (*Houben-Weyl*), 1992, **E14a/3**, 545.	294
92ICA377	M. J. Hostetler, M. D. Butts and R. G. Bergman; *Inorg. Chim. Acta*, 1992, **198–200**, 377.	710
92ICA557	C. P. Casey, Y. Wang, L. M. Petrovich, T. L. Underiner, P. N. Hazin and J. M. Desper; *Inorg. Chim. Acta*, 1992, **198–200**, 557.	716
92ICA689	C. J. Elsevier, W. P. Mul and K. Vrieze; *Inorg. Chim. Acta*, 1992, **198–200**, 689.	716
92ICA741	G. Wilke, H. Benn, R. Goddard, C. Krüger and B. Pfeil; *Inorg. Chim. Acta*, 1992, **198–200**, 741.	676
92IZV196	A. S. Balueva, S. R. Prohorova and G. N. Nikonov; *Izv. Akad. Nauk SSSR, Ser. Khim.*, 1992, 196.	342
92JA360	W. R. Moore, G. L. Schatzman, E. T. Jarvi, R. S. Gross and J. R. McCarthy; *J. Am. Chem. Soc.*, 1992, **114**, 360.	755, 782
92JA372	A. J. Ashe, III, J. W. Kampf and S. M. Al-Taweel; *J. Am. Chem. Soc.*, 1992, **114**, 372.	1038
92JA10078	J. A. Soderquist and E. I. Miranda; *J. Am. Chem. Soc.*, 1992, **114**, 10 078.	354
92JA1329	W. H. Pearson, D. P. Szura and M. J. Postich; *J. Am. Chem. Soc.*, 1992, **114**, 1329.	537
92JA1428	F. A. Davis, R. T. Reddy, W. Han and P. J. Carroll; *J. Am. Chem. Soc.*, 1992, **114**, 1428.	262
92JA2592	S. E. Denmark, J. P. Edwards and S. R. Wilson; *J. Am. Chem. Soc.*, 1992, **114**, 2592.	152
92JA3910	L. A. Paquette, A. M. Doherty and C. M. Rayner; *J. Am. Chem. Soc.*, 1992, **114**, 3910.	284
92JA4436	D. P. Curran and G. Thoma; *J. Am. Chem. Soc.*, 1992, **114**, 4436.	284
92JA5900	W. Oppolzer, O. Tamura, G. Sundarababu and M. Signer; *J. Am. Chem. Soc.*, 1992, **114**, 5900.	107
92JA5902	H. Kuniyasu, A. Ogawa, K.-I. Sato, I. Ryu, N. Kambe and N. Sonoda; *J. Am. Chem. Soc.*, 1992, **114**, 5902.	940
92JA5946	R. J. Parry, Y. Li and E. E. Gomez; *J. Am. Chem. Soc.*, 1992, **114**, 5946.	264
92JA7591	L.-B. Han, K.-I. Ishihara, N. Kambe, A. Ogawa, I. Ryu and N. Sonoda; *J. Am. Chem. Soc.*, 1992, **114**, 7591.	948
92JA7643	S. Rozen, E. Mishani and M. Kol; *J. Am. Chem. Soc.*, 1992, **114**, 7643.	52
92JA8531	G. Erker, M. Albrecht, C. Krüger and S. Werner; *J. Am. Chem. Soc.*, 1992, **114**, 8531.	725
92JA8772	J. R. Cashman, L. D. Olsen, D. R. Boyd, R. A. S. McMordie, R. Dunlop and H. Dalton; *J. Am. Chem. Soc.*, 1992, **114**, 8772.	262
92JA9691	D. Dou, E. N. Duesler, R. T. Paine and H. Noth; *J. Am. Chem. Soc.*, 1992, **114**, 9691.	484
92JA9710	N. Chatani, S. Ikeda, K. Ohe and S. Murai; *J. Am. Chem. Soc.*, 1992, **114**, 9710.	931
92JA9715	E. Nakamura, K. Tanaka and S. Aoki; *J. Am. Chem. Soc.*, 1992, **114**, 9715.	388
92JCS(D)775	M. F. Lappert, W.-P. Leung, C. L. Raston, B. W. Skelton and A. H. White; *J. Chem. Soc., Dalton Trans.*, 1992, 775.	647, 660
92JCS(D)2353	J. L. Bookham, W. McFarlane and M. Thornton-Pett; *J. Chem. Soc., Dalton Trans.*, 1992, 2353.	548
92JCS(D)2423	A. J. Arce, P. Arrojo, A. J. Deeming and Y. De Sanctis; *J. Chem. Soc., Dalton Trans.*, 1992, 2423.	683
92JCS(P1)291	J.-G. Rodriguez, A. Perales, C. A. Ibarra, M. I. López-Sánchez and M. L. Quiroga; *J. Chem. Soc., Perkin Trans. 1*, 1992, 291.	325
92JCS(P1)313	Z.-Q. Xu and D. D. DesMarteau; *J. Chem. Soc., Perkin Trans. 1.*, 1992, 313.	704
92JCS(P1)479	D. V. Griffiths, P. A. Griffiths, B. J. Whitehead and J. C. Tebby; *J. Chem. Soc., Perkin Trans. 1*, 1992, 479.	558

92JCS(P1)1105	D. R. Boyd, N. D. Sharma, J. H. Dorman, R. Dunlop, J. F. Malone, R. A. S. McMordie and A. F. Drake; *J. Chem. Soc., Perkin Trans. 1*, 1992, 1105.	262
92JCS(P1)1179	A. Tsuge, H. Nago, S. Mataka and M. Tashiro; *J. Chem. Soc., Perkin Trans. 1*, 1992, 1179.	273
92JCS(P1)2303	I. Coldham and S. Warren; *J. Chem. Soc., Perkin Trans. 1*, 1992, 2303.	218
92JCS(P1)2371	B. Beagley, M. R. James, R. G. Pritchard, C. M. Raynor, C. Smith and R. J. Stoodley; *J. Chem. Soc., Perkin Trans. 1*, 1992, 2371.	281
92JCS(P1)2813	S. Akai, Y. Tsuzuki, S. Matsuda, S. Kitagaki and Y. Kita; *J. Chem. Soc., Perkin Trans. 1*, 1992, 2813.	661
92JFC(56)341	H. K. Nair and D. J. Burton; *J. Fluorine Chemistry*, 1992, **56**, 341.	151, 152
92JGU39	A. N. Bovin, A. N. Cheklov and E. N. Tsvetkov; *J. Gen. Chem. USSR (Engl. Transl.)*, 1992, **62**, 39.	124, 128
92JGU1222	B. I. Buzykin, M. P. Sokolov, T. A. Zyablikova and L. F. Chertanova; *J. Gen. Chem. USSR (Engl. Transl.)*, 1992, **62**, 1222.	804
92JHC1625	T. Izumi, M. Soutome and T. Miura; *J. Heterocycl. Chem.*, 1992, **29**, 1625.	193
92JHC1631	M. T. Cocco, C. Congiu, A. Maccioni and V. Onnis; *J. Heterocycl. Chem.*, 1992, **29**, 1631.	986
92JMC533	J. A. Secrist, III, R. M. Riggs, K. N. Tiwari and J. A. Montgomery; *J. Med. Chem.*, 1992, **35**, 533.	227
92JOC178	H. B. Wood, H.-P. Buser and B. Ganem; *J. Org. Chem.*, 1992, **57**, 178.	913
92JOC235	K. Baum, S. S. Bigelow, N. V. Nguyen, T. G. Archibald, R. Gilardi, J. L. Flippen-Anderson and C. George; *J. Org. Chem.*, 1992, **57**, 235.	970, 993
92JOC298	A. Padwa, D. N. Kline, S. S. Murphree and P. E. Yeske; *J. Org. Chem.*, 1992, **57**, 298.	275
92JOC386	A. G. M. Barrett, J. M. Hill and E. M. Walace; *J. Org. Chem.*, 1992, **57**, 386.	138, 355
92JOC547	A. R. Katritzky, W.-Q. Fan, M. Black and J. Pernak; *J. Org. Chem.*, 1992, **57**, 547.	309
92JOC742	M. Watanabe, N. Hashimoto, S. Araki and Y. Butsugan; *J. Org. Chem.*, 1992, **57**, 742.	166
92JOC798	M. Lautens and P. H. M. Delanghe; *J. Org. Chem.*, 1992, **57**, 798.	661
92JOC1082	R. Amoroso, G. Cardillo, C. Tomasini and P. Tortoreto; *J. Org. Chem.*, 1992, **57**, 1082.	99
92JOC1202	R. F. Cunico and C. P. Kuan; *J. Org. Chem.*, 1992, **57**, 1202.	359
92JOC1321	J. Yoshida, T. Maekawa, Y. Morita and S. Isoe; *J. Org. Chem.*, 1992, **57**, 1321.	197
92JOC2084	G. Grewal, N. Kaila and R. W. Franck; *J. Org. Chem.*, 1992, **57**, 2084.	187
92JOC2217	J. W. Beach, L. S. Jeong, A. J. Alves, D. Pohl, H. O. Kim, C.-N. Chang, S.-L. Doong, R. F. Schinazi, Y.-C. Cheng and C. K. Chu; *J. Org. Chem.*, 1992, **57**, 2217.	231
92JOC2220	J. M. Chong and S. B. Park; *J. Org. Chem.*, 1992, **57**, 2220.	531, 533
92JOC2981	T. Furuta and Y. Yamamoto; *J. Org. Chem.*, 1992, **57**, 2981.	390
92JOC3331	R. F. Cunico and C. P. Kuan; *J. Org. Chem.*, 1992, **57**, 3331.	979
92JOC3496	A. Castro and T. A. Spencer; *J. Org. Chem.*, 1992, **57**, 3496.	274, 277
92JOC3755	T. Fuchigami, S. Narizuka and A. Konno; *J. Org. Chem.*, 1992, **57**, 3755.	62
92JOC5078	R. J. Linderman and T. V. Anklekar; *J. Org. Chem.*, 1992, **57**, 5078.	361, 379
92JOC5419	T. Usami, N. Shirai and Y. Sato; *J. Org. Chem.*, 1992, **57**, 5419.	513
92JOC5530	V. Martin, H. Molines and C. Wakselman; *J. Org. Chem.*, 1992, **57**, 5530.	736
92JOC5610	S. Yamazaki, H. Fujitsuka, S. Yamabe and H. Tamura; *J. Org. Chem.*, 1992, **57**, 5610.	947
92JOC5805	T. Harada, T. Katsuhira and A. Oku; *J. Org. Chem.*, 1992, **57**, 5805.	758, 819, 821
92JOC5844	B. Santiago and J. A. Soderquist; *J. Org. Chem.*, 1992, **57**, 5844.	954
92JOC6037	Y. S. Jung, W. H. Swartz, W. Xu, P. S. Mariano, N. J. Green and A. G. Schultz; *J. Org. Chem.*, 1992, **57**, 6037.	513
92JOC6107	T. H. Chan and K. T. Nwe; *J. Org. Chem.*, 1992, **57**, 6107.	648
92JOC6354	W. H. Pearson and M. J. Postich; *J. Org. Chem.*, 1992, **57**, 6354.	533, 537
92JOC6390	V. K. Aggarwal, G. Evans, E. Moya and J. Dowden; *J. Org. Chem.*, 1992, **57**, 6390.	269
92JOC6508	G. A. Russell and C. F. Yao; *J. Org. Chem.*, 1992, **57**, 6508.	1004
92JOC6552	P. F. Hudrlik, Y. M. Abdallah, A. K. Kulkarni and A. M. Hudrlik; *J. Org. Chem.*, 1992, **57**, 6552.	355
92JOC6711	Y. Sato, N. Shirai, Y. Machida, E. Ito, T. Yasui, Y. Kurono and K. Hatano; *J. Org. Chem.*, 1992, **57**, 6711.	506
92JOC6903	L. Labaudinière, J. Hanaizi and J.-F. Normant; *J. Org. Chem.*, 1992, **57**, 6903.	809
92JOC6999	R. F. Cunico and C. P. Kuan; *J. Org. Chem.*, 1992, **57**, 6999.	359
92JOC7010	Y.-M. Tsai, H.-C. Nieh and C.-D. Cherng; *J. Org. Chem.*, 1992, **57**, 7010.	930
92JOC7349	B. C. Ranu, S. Bhar and R. Chakraborti; *J. Org. Chem.*, 1992, **57**, 7349.	258
92JOM(423)339	U. Kolb, M. Dräger, E. Fischer and K. Jurkschat; *J. Organomet. Chem.*, 1992, **423**, 339.	570, 585, 660
92JOM(424)273	R. Tacke, F. Wuttke and H. Henke; *J. Organomet. Chem.*, 1992, **424**, 273.	353
92JOM(426)1	N. Auner, W. Ziche and E. Herdtweck; *J. Organomet. Chem.*, 1992, **426**, 1.	651
92JOM(427)C21	M. Albrecht, G. Erker, M. Nolte and C. Krüger; *J. Organomet. Chem.*, 1992, **427**, C21.	725
92JOM(427)231	R. Broussier, A. Da Rold and B. Gautheron; *J. Organomet. Chem.*, 1992, **427**, 231.	720
92JOM(427)293	H. Bürger and P. Moritz; *J. Organomet. Chem.*, 1992, **427**, 293.	136
92JOM(427)335	C. A. Dickson and N. J. Coville; *J. Organomet. Chem.*, 1992, **427**, 335.	712
92JOM(428)1	M. Tanaka, Y. Uchimaru and H.-H. Lautenschlager; *J. Organomet. Chem.*, 1992, **428**, 1.	618
92JOM(429)C14	U. Siemeling; *J. Organomet. Chem.*, 1992, **429**, C14.	603
92JOM(429)311	D. Farah, K. Swami and H. G. Kuivila; *J. Organomet. Chem.*, 1992, **429**, 311.	700
92JOM(429)369	J. L. Hubbard and W. K. McVicar; *J. Organomet. Chem.*, 1992, **429**, 369.	148, 149
92JOM(431)255	M. Biedrzycki, W. H. Scouten and Z. Biedrzycka; *J. Organomet. Chem.*, 1992, **431**, 255.	144
92JOM(433)49	A. Castel, P. Riviere, J. Satgé and D. Desor; *J. Organomet. Chem.*, 1992, **433**, 49.	360
92JOM(436)333	M. Herberhold, W. Feger and U. Kölle; *J. Organomet. Chem.*, 1992, **436**, 333.	719
92JOM(437)C19	J. L. Parrain, J. C. Cintrat and J.-P. Quintard; *J. Organomet. Chem.*, 1992, **437**, C19.	388

92JOM(437)111	R. A. Howie, G. M. Spencer, J. L. Wardell and J. N. Low; *J. Organomet. Chem.*, 1992, **437**, 111.	138, 369, 372
92JOM(438)11	M. J. Rozema, D. Rajagopal, C. E. Tucker and P. Knochel; *J. Organomet. Chem.*, 1992, **438**, 11.	153
92JOM(438)45	R. Tacke, B. Becker, D. Berg, W. Brandes, S. Dutzmann and K. Schaller; *J. Organomet. Chem.*, 1992, **438**, 45.	141
92JOM(440)233	C. Couret, J. Escudié, G. Delpon-Lacaze and J. Satgé; *J. Organomet. Chem.*, 1992, **440**, 233.	647
92JOM(441)81	B. Heim, J. C. Daran, Y. Jeannin, B. Eber, G. Huttner and W. Imhof; *J. Organomet. Chem.*, 1992, **441**, 81.	681
92JOM(441)241	V. P. Maryin; *J. Organomet. Chem.*, 1992, **441**, 241.	723
92JPR190	V. J. Ram, N. Haque and A. Shoeb; *J. Prakt. Chem.*, 1992, **334**, 190.	983
92LA643	R. W. Hoffmann and M. Bewersdorf; *Liebigs Ann. Chem.*, 1992, 643.	288
92LA725	R. W. Hoffmann, T. Rühl and J. Harbach; *Liebigs Ann. Chem.*, 1992, 725.	660
92LA1039	H.-D. Stachel, M. Schorp and T. Zoukas; *Liebigs Ann. Chem.*, 1992, 1039.	259
B-92MI 402-01	C. G. Krespan and D. A. Dixon; in "Synthetic Fluorine Chemistry," ed. G. A. Olah, R. D. Chambers and G. K. S. Prakash, Wiley, New York, 1992, p. 314.	45
92MI 403-01	K.-T. Kang, C. Y. Park and J. S. Kim; *Bull. Korean Chem. Soc.*, 1992, **13**, 48.	139
92MI 403-02	K. V. Pavlov, N. A. Viktorov and V. F. Mironov; *Organometallic Chemistry in the USSR (Engl. Transl.)*, 1992, **5**, 552.	142
B-92MI 404-01	J. March; "Advanced Organic Chemistry," Wiley, New York, 1992, p. 882.	160
B-92MI 404-02	J. March; "Advanced Organic Chemistry," Wiley, New York, 1992, p. 1177.	214
92MI 405-01	K. Naraguchi, S. Saito, H. Tanaka and T. Miyasaka; *Nucleosides Nucleotides*, 1992, **11**, 483.	240
92MI 406-01	C. J. Palmer and J. E. Casida; *J. Agr. Food Chem.*, 1992, **40**, 492.	265
92MI 406-02	V. J. Wacher, R. F. Toia and J. E. Casida; *J. Agr. Food Chem.*, 1992, **40**, 497.	271
B-92MI 407-01	F. R. Hartley (ed); "The Chemistry of Organophosphorus Compounds," Wiley, Chichester, 1992, vol. 2.	336
B-92MI 407-02	P. Mastalerz; in "Handbook of Organophosphorus Chemistry," ed. R. Engel, Marcel Dekker, New York, 1992.	345
92MI 408-01	L. A. Clibze and E. J. Reist; *J. Labelled Comp. Radiopharm.*, 1992, **31**, 951.	389
92MI 411-01	D. S. Matteson and T. J. Michnik; *J. Labelled Comp. Radiopharm.*, 1992, **31**, 567.	527
92MI 411-02	J. Wityak, R. J. Chorvat, R. A. Earl, C. A. Kettner and M. E. Pierce; Abstracts of the 203rd ACS Natl. Meeting, San Francisco, 1992, MEDI160.	528
92MI 414-01	E. A. Monin, A. N. Zharov, A. K. Shestakova and E. A. Chernysher; *Izobreteniya*, 1992, 106 (*Chem. Abstr.*, 1992, **119**, 117 536w).	654
92MI 417-01	I. H. Jeong; *Bull. Korean Chem. Soc.*, 1992, **13**, 468.	775
92MI 419-01	A. Uchida, K. Asahina and S. Sasaoka; *Niihama Kogyo Koto Senmon Gakko Kiyo, Rikogaku-hen*, 1992, **28**, 1 (*Chem. Abstr.* 1992, **116**, 214 389).	863
92MI 419-02	S. Zhu; *Chin. Chem. Lett.*, 1992, **3**, 601 (*Chem. Abstr.* 1993, **118**, 22 192).	864
B-92MI 424-01	J. March; "Advanced Organic Chemistry," 4th edn, Wiley, New York, 1992.	1072, 1075, 1077, 1080
B-92MI 424-02	W. M. Horspool and D. Armesto; "Organic Photochemistry: A Comprehensive Treatment," Ellis Horwood, Chichester, 1992.	1079
92MIP173178	E. A. Monin, A. N. Zharov, A. K. Shestakova and E. A. Chernyshev; *USSR Pat.* 1 731 780 (1992) (*Chem. Abstr.*, 1993, **119**, 117 536w).	6540
92OM154	G. H. Young, R. R. Willis, A. Wojcicki, M. Calligaris and P. Faleschini; *Organometallics*, 1992, **11**, 154.	683
92OM370	F. J. García Alonso, V. Riera, M. A. Ruiz, A. Tiripicchio and M. Tiripicchio Camellini; *Organometallics*, 1992, **11**, 370.	678
92OM613	A. Togni, G. Rihs and R. E. Blumer; *Organometallics*, 1992, **11**, 613.	1034
92OM721	N. Le Berre-Cosquer, R. Kergoat and P. L'Haridon; *Organometallics*, 1992, **11**, 721.	674
92OM859	G. L. Larson, F. K. Cartledge, R. Nunez, R. J. Unwalla, R. Klesse and L. Del Valle; *Organometallics*, 1992, **11**, 859.	138, 139
92OM864	A. D. Hunter, D. Ristic-Petrovic and J. L. McLernon; *Organometallics*, 1992, **11**, 864.	715
92OM989	H. Omori, H. Suzuki, T. Kakigano and Y. Moro-oka; *Organometallics*, 1992, **11**, 989.	681
92OM1137	N. Auner, C. Seidenschwarz and N. Sewald; *Organometallics*, 1992, **11**, 1137.	606
92OM1392	J. K. Gong, T. B. Peters, P. E. Fanwick and C. P. Kubiak; *Organometallics*, 1992, **11**, 1392.	583
92OM1411	B. Deschamps and F. Mathey; *Organometallics*, 1992, **11**, 1411.	1026
92OM1473	R. D. Adams, G. Chen and Y. Chi; *Organometallics*, 1992, **11**, 1473.	677
92OM1480	R. D. Adams, G. Chen, Y. Chi, W. Wu and J. Yin; *Organometallics*, 1992, **11**, 1480.	677
92OM1491	A. J. Ashe, III, J. W. Kampf and S. M. Al-Taweel; *Organometallics*, 1992, **11**, 1491.	1038, 1041
92OM2128	M. J. Chetcuti, K. J. Deck, J. C. Gordon, B. E. Grant and P. E. Fanwick; *Organometallics*, 1992, **11**, 2128.	709
92OM2224	M. Rashidi, G. Schoettel, J. J. Vittal and R. J. Puddephatt; *Organometallics*, 1992, **11**, 2224.	692
92OM2698	W. D. Jones and R. M. Chin; *Organometallics*, 1992, **11**, 2698.	690
92OM2701	K. A. Ostoja Starzewski and L. Born; *Organometallics*, 1992, **11**, 2701.	583
92OM2743	A. J. Ashe, III, J. W. Kampf, D. B. Puranik and S. M. Al-Taweel; *Organometallics*, 1992, **11**, 2743.	1041
92OM2916	D. Lentz, H. Michael-Schulz and M. Reuter; *Organometallics*, 1992, **11**, 2916.	683
92OM3169	C. H. Winter, Y.-H. Han and M. J. Heeg; *Organometallics*, 1992, **11**, 3169.	724
92OM3176	C. Couret, J. Escudié, G. Delpon-Lacaze and J. Satgé; *Organometallics*, 1992, **11**, 3176.	663
92OM3307	M. Namavari and R. T. Conlin; *Organometallics*, 1992, **11**, 3307.	627
92OM3464	D. Seyferth and J. L. Robison; *Organometallics*, 1992, **11**, 3464.	623

92OM3517	G. Erker, M. Albrecht, C. Krüger, S. Werner, P. Binger and F. Langhauser; *Organometallics*, 1992, **11**, 3517.	725
92OM3617	M. S. Kralik, L. Stahl, A. M. Arif, C. E. Strouse and R. D. Ernst; *Organometallics*, 1992, **11**, 3617.	678
92OM3736	D. Seyferth, L. L. Anderson, W. B. Davis and M. Cowie; *Organometallics*, 1992, **11**, 3736.	683
92OM3879	P. Bergamini, E. Costa, S. Sostero, A. G. Orpen and P. G. Pringle; *Organometallics*, 1992, **11**, 3879.	151
92PAC439	A. Ricci, G. Reginato, A. Degl'Innocenti and G. Seconi; *Pure Appl. Chem.*, 1992, **64**, 439.	1018
92PHA754	D. Gravier, G. Hou, F. Casadebaig, J. P. Dupin, H. Bernard and M. Boisseau; *Pharmazi*, 1992, **47**, 754.	420
92PS(66)87	T. S. Hafez; *Phosphorus Sulfur*, 1992, **66**, 87.	320
92PS(72)171	M. Well and R. Schmutzler; *Phosphorus Sulfur*, 1992, **72**, 171.	338
92PS(72)225	C. Ibis and C. Gurun; *Phosphorus Sulfur*, 1992, **72**, 225.	770, 776
92PS(73)67	R. Göbel, F. Richter and H. Weichmann; *Phosphorus, Sulfur*, 1992, **73**, 67.	555
92RCR1220	G. M. Baranov and V. V. Perekalin; *Russ. Chem. Rev. (Engl. Transl.)*, 1992, **61**, 1220.	451, 490, 493
92RRC393	F. Badea, C. Condeiu, M. Gheorghiu, A. Iancu, F. Iordache and C. Simion; *Rev. Roum. Chim.*, 1992, **37**, 393.	152
92RTC163	L. A. G. M. van den Broek, A. J. J. Antonisse, H. C. J. Ottenheijm, A. San Felix, E. Lazaro, J. P. G. Ballesta and P. Lelieveld; *Recl. Trav. Chim. Pays-Bas*, 1992, **111**, 163.	264
92S90	S. Laschat and H. Knuz; *Synthesis*, 1992, 90.	462
92S258	C. Yuan, S. Cui, G. Wang, H. Feng, D. Chen, C. Li, Y. Ding and L. Maier; *Synthesis*, 1992, 258.	343
92S263	L. K. Lukanov and A. P. Venkov; *Synthesis*, 1992, 263.	488
92S367	T. Gajda and M. Matusiak; *Synthesis*, 1992, 367.	491
92S432	H. Westmijze, H. Kleijn, J. Meijer and P. Vermeer; *Synthesis*, 1992, 432.	704
92S495	A. Loffler and G. Himbert; *Synthesis*, 1992, 495.	763
92S710	Q. Wang, J. C. Jochims, S. Köhlbrandt, L. Dahlenburg, M. Al-Talib, A. Hamed and A. E.-H. Ismail; *Synthesis*, 1992, 710.	109
92S787	H. J. Bestmann, A. Bomhard, R. Dostalek, R. Pichl, R. Riemer and R. Zimmermann; *Synthesis*, 1992, 787.	571
92S965	M. Muzard and C. Portella; *Synthesis*, 1992, 965.	252
92S995	B. Dondy, P. Doussot and C. Portella; *Synthesis*, 1992, 995.	353
92S1075	C. S. Rao, O. M. Singh, H. Ila and H. Junjappa; *Synthesis*, 1992, 1075.	253
92S1124	C. Yuan and S. Chen; *Synthesis*, 1992, 1124.	470, 488
92S1295	A. R. Katritzky, S. Jurczyk, B. Rachwal, S. Rachwal, I. Shcherbakova and K. Yannakopoulou; *Synthesis*, 1992, 1295.	323
92SC107	M. Ferrari, G. Jommi, G. Miglierini, R. Pagliarin and M. Sisti; *Synth. Commun.*, 1992, **22**, 107.	462
92SC1359	D. Villemin, A. Ben Alloum and F. Thibault-Starzyk; *Synth. Commun.*, 1992, **22**, 1359.	257
92SC2381	A. Couture, E. Deniau and P. Grandclaudon; *Synth. Commun.*, 1992, **22**, 2381.	487
92SL97	H. Kotsuki; *Synlett*, 1992, 97.	176
92SL133	C. De Lima, M. Julia and J.-N. Verpeaux; *Synlett*, 1992, 133.	147
92SL429	A. Datta and R. R. Schmidt; *Synlett*, 1992, 429.	950
92SL455	T. Satoh and K. Yamakawa; *Synlett*, 1992, 455.	73, 74, 75, 77
92SL483	A. J. Bennett, J. M. Percy and M. H. Rock; *Synlett*, 1992, 483.	949
92SL493	E. Kerbage, M. Malacria and H. Fillion; *Synlett*, 1992, 493.	837
92SL499	A. Degl'Innocenti, P. Ulivi, A. Capperucci, A. Mordini, G. Reginato and A. Ricci; *Synlett*, 1992, 499.	944
92SL565	K. Kato, K. Furuta and H. Yamamoto; *Synlett*, 1992, 565.	208
92SL615	D. L. Comins; *Synlett*, 1992, 615.	299
92SL638	A. Krief, M. Trabelsi and W. Dumont; *Synlett*, 1992, 638.	289
92SL730	V. K. Aggarwal, M. Lightowler and S. D. Lindell; *Synlett*, 1992, 730.	863
92SL739	J. Ichikawa, C. Ikeura and T. Minami; *Synlett*, 1992, 739.	736
92SL764	P. Sommerfeld and D. Hoppe; *Synlett*, 1992, 764.	360, 382
92SL843	J.-I. Yoshida, Y. Morita, M. Itoh, Y. Ishichi and S. Isoe; *Synlett*, 1992, 843.	361
92SL883	A. Degl'Innocenti, P. Ulivi, A. Capperucci, G. Reginato, A. Mordini and A. Ricci; *Synlett*, 1992, 883.	944
92SL886	A. Pirrin, P. Kocienski and S. D. A. Street; *Synlett*, 1992, 886.	962, 963, 964
92SL891	T. Sato, A. Kawase and T. Hirose; *Synlett*, 1992, 891.	155, 700
92SL893	E. Diez-Barra, A. de la Hoz, A. Diaz-Ortiz and P. Prieto; *Synlett*, 1992, 893.	826
92SL977	T. Tsukamoto and T. Kitazume; *Synlett*, 1992, 977.	736
92SUL251	C. Ibis and C. Gurun; *Sulfur Letters*, 1992, **14**, 251.	770
92T1449	J. Otera, N. Dan-oh and H. Nozaki; *Tetrahedron*, 1992, **48**, 1449.	704
92T1485	O. De Lucchi, D. Fabbri and V. Lucchini; *Tetrahedron*, 1992, **48**, 1485.	277, 854
92T2025	S. A. Rao, T. S. Chou, I. Schipor and P. Knochel; *Tetrahedron*, 1992, **48**, 2025.	396
92T5163	D. P. G. Hamon, R. A. Massy-Westropp and P. Razzino; *Tetrahedron*, 1992, **48**, 5163.	102
92T5933	P. C. Bulman Page, S. S. Klair, M. P. Brown, C. S. Smith, S. J. Maginn and S. Mulley; *Tetrahedron*, 1992, **48**, 5933.	249
92T7265	P. C. Bulman Page, A. E. Graham and B. K. Park; *Tetrahedron*, 1992, **48**, 7265.	250
92T7527	A. J. Pearson and H. W. Shin; *Tetrahedron*, 1992, **48**, 7527.	856
92T7551	W. O. Moss, E. Wakefield, M. F. Mahon, K. C. Molloy, R. H. Bradbury, N. J. Hales and T. Gallagher; *Tetrahedron*, 1992, **48**, 7551.	860
92T8377	O. Zschage, J. R. Schwark, T. Kramer and D. Hoppe; *Tetrahedron*, 1992, **48**, 8377.	382

92TA377	T. Yokomatsu and S. Shibuya; *Tetrahedron Asymmetry*, 1992, **3**, 377.	475
92TA1003	S. Pinheiro, A. Guingant, D. Desmaële and J. d'Angelo; *Tetrahedron Asymmetry*, 1992, **3**, 1003.	279
92TA1515	M. Mikolajczyk and W. H. Midura; *Tetrahedron Assymetry*, 1992, **3**, 1515.	346, 924
92TL77	J. P. Genêt, J. Uziel, M. Port, A. M. Touzin, S. Roland, S. Thorimbert and S. Tanier; *Tetrahedron Lett.*, 1992, **33**, 77.	465, 478
92TL85	Y. Tominaga, H. Ueda, K. Ogata, S. Kohra, M. Hojo, M. Ohkuma, K. Tomita and A. Hosomi; *Tetrahedron Lett.*, 1992, **33**, 85.	368
92TL239	T. Sato, Y. Fujita, J. Otero and H. Nozaki; *Tetrahedron Lett.*, 1992, **33**, 239.	225, 238
92TL269	G. Proess, D. Pankert and L. Hevesi; *Tetrahedron Lett.*, 1992, **33**, 269.	288
92TL337	J. Ichikawa, S. Hamada, T. Sonoda and H. Kobayashi; *Tetrahedron Lett.*, 1992, **33**, 337.	736
92TL543	H. Imaniech, P. Quayle, M. Voaden, J. Conway and S. D. A. Street; *Tetrahedron Lett.*, 1992, **33**, 543.	651
92TL717	B. M. Trost, B. A. Vos, C. M. Brzezowski and D. P. Martina; *Tetrahedron Lett.*, 1992, **33**, 717.	280
92TL745	S. S. Magar and P. L. Fuchs; *Tetrahedron Lett.*, 1992, **33**, 745.	273, 275
92TL1221	F. Q. Jin, B. A. Jiang and Y. Y. Xu; *Tetrahedron Lett.*, 1992, **33**, 1221.	735
92TL1321	M. Hamaguchi and T. Nagai; *Tetrahedron Lett.*, 1992, **33**, 1321.	277
92TL1483	T. Satoh, N. Itoh, K. Onda, Y. Kitoh and K. Yamakawa; *Tetrahedron Lett.*, 1992, **33**, 1483.	74, 77
92TL1573	W. R. Baker, S. L. Condon and S. Spanton; *Tetrahedron Lett.*, 1992, **33**, 1573.	279
92TL1831	B. M. Trost and L. Zhi; *Tetrahedron Lett.*, 1992, **33**, 1831.	280
92TL2285	G. J. Angara, P. Bovonsombat and E. McNelis; *Tetrahedron Lett.*, 1992, **33**, 2285.	760
92TL2949	N. V. Nguyen and K. Braum; *Tetrahedron Lett.*, 1992, **33**, 2949.	994
92TL3043	M. Corich, F. Di Furia, G. Licini and G. Modena; *Tetrahedron Lett.*, 1992, **33**, 3043.	261
92TL3161	T. Fuchigami, T. Hayashi and A. Konno; *Tetrahedron Lett.*, 1992, **33**, 3161.	87, 92
92TL3717	J. R. Waas, A. R. Sidduri and P. Knochel; *Tetrahedron Lett.*, 1992, **33**, 3717.	815
92TL3779	J. Ichikawa, T. Minami, T. Sonoda and H. Kobayashi; *Tetrahedron Lett.*, 1992, **33**, 3779.	736
92TL3927	G. M. Blackburn, S. G. Rosenberg and G. Y. Yakovleva; *Tetrahedron Lett.*, 1992, **33**, 3927.	563
92TL4065	P. Bonete and C. Najera; *Tetrahedron Lett.*, 1992, **33**, 4065.	393
92TL4209	S. Elgendy, J. Deadman, G. Patel, D. Green, N. Chino, C. A. Goodwin, M. F. Scully, V. V. Kakkar and G. Claeson; *Tetrahedron Lett.*, 1992, **29**, 4209.	526, 527, 528
92TL4537	J. Weber, L. X. Xu and U. H. Brinker; *Tetrahedron Lett.*, 1992, **33**, 4537.	737
92TL4625	D. C. Humber, M. F. Jones, J. J. Payne, M. V. J. Ramsay, B. Zacharie, H. L. Jin, A. Siddiqui, C. A. Evans, H. L. A. Tse and T. S. Mansour; *Tetrahedron Lett.*, 1992, **33**, 4625.	229
92TL4913	J. Boivin, C. Chauvet and S. Z. Zard; *Tetrahedron Lett.*, 1992, **33**, 4913.	269
92TL5085	M. Kamata, M. Sato and E. Hasegawa; *Tetrahedron Lett.*, 1992, **33**, 5085.	260
92TL5121	C. Cardellicchio, V. Fiandanese, F. Naso and A. Scilimati; *Tetrahedron Lett.*, 1992, **33**, 5121.	776
92TL5399	T. Itoh, H. Hasegawa, K. Nagata, M. Okada and A. Ohsawa; *Tetrahedron Lett.*, 1992, **33**, 5399.	60
92TL5701	Y. Kanda, H. Saito and T. Fukuyama; *Tetrahedron Lett.*, 1992, **33**, 5701.	948
92TL5795	K. Tomooka, T. Igarashi, M. Watanabe and T. Nakai; *Tetrahedron Lett.*, 1992, **33**, 5795.	379
92TL6131	M. Lemarie, Y. Vallee and M. Worrell; *Tetrahedron Lett.*, 1992, 6131.	870
92TL6383	G. Mazzanti, R. Ruinaard, L. A. van Vliet, P. Zani, B. F. Bonini and B. Zwanenburg; *Tetrahedron Lett.*, 1992, **33**, 6383.	859
92TL7031	K. Matsumoto, K. Miura, K. Oshima and K. Utimoto; *Tetrahedron Lett.*, 1992, **33**, 7031.	811
92TL7705	P. Bovonsombat and E. McNelis; *Tetrahedron Lett.*, 1992, **33**, 7705.	760
92TL7895	Y.-M. Tsai, B.-W. Ke, C.-T. Yang and C.-H. Lin; *Tetrahedron Lett.*, 1992, **33**, 7895.	661
92TL8035	A. Datta, D. Datta and R. R. Schmidt; *Tetrahedron Lett.*, 1992, **33**, 8035.	950
92USP5103036	D. R. Magnin and R. B. Sulsky; *US Pat.* 5 103 036 (1992) (*Chem. Abstr.*, 1992, **117**, 151 152).	563
92ZAAC(607)157	S. Kraft and M. Wieber; *Z. Anorg. Allg. Chem.*, 1992, **607**, 157.	595
92ZAAC(607)164	S. Kraft and M. Wieber; *Z. Anorg. Allg. Chem.*, 1992, **607**, 164.	595
92ZAAC(611)68	L. Jager, M. Kretschmann and H. Kohler; *Z. Anorg. Allg. Chem.*, 1992, **611**, 68.	977, 978
92ZN(B)725	G. Hägele, P. Reinemer, M. Batz and D. Mootz; *Z. Naturforsch., Teil B.*, 1992, **47**, 725.	553
92ZN(B)805	N. Auner and E. Penzenstadler; *Z. Naturforsch., Teil B*, 1992, **47**, 805.	611
92ZN(B)1213	I. S. Alferiev and S. Yu. Bobkov; *Z. Naturforsch.*, 1992, **47**, 1213.	561
92ZN(B)1377	E. Penzenstadler and E. Herdtweck; *Z. Naturforsch., Teil B*, 1992, **47**, 1377.	611
92ZOB263	T. A. Mastryukova, I. V. Leont'eva, I. M. Aladzheva, P. V. Petrovskii and M. I. Kabachnik; *Zh. Obshch. Khim.*, 1992, **62**, 263.	551
92ZOB707	V. S. Brovarets, L. V. Budnik and B. S. Drach; *Zh. Obshch. Khim.*, 1992, **62**, 707 (Engl. Transl. 586).	1005
92ZOB946	A. A. Prishchenko, D. A. Pisarnitskii, M. V. Livantsov and V. S. Petrosyan; *Zh. Obshch. Khim.*, 1992, **62**, 946 (*Chem. Abstr.*, **118**, 147 649g).	458
92ZOB1084	V. S. Brovarets, V. V. Kurg, T. K. Vinogradova, O. B. Smolii and B. S. Drach; *Zh. Obshch. Khim.*, 1992, **62**, 1084 (Engl. Transl. 886).	998
92ZOB1268	B. A. Kashimirov, V. N. Osipov, N. F. Savenkov, B. Ya. Chvertkin and P. S. Khokhlov; *Zh. Obshch. Khim.*, 1992, **62**, 1268.	346
92ZOB1423	V. S. Brovarets, R. N. Vydzhak and B. S. Drach; *Zh. Obshch. Khim.*, 1992, **62**, 1423 (Engl. Transl. 1170).	998
92ZOB1472	V. A. Solodenko, T. N. Kasheva, D. A. Mironenko, E. V. Kozlova, V. K. Shvyadas and V. P. Kukhar; *Zh. Obshch. Khim.*, 1992, **61**, 1472 (*Chem. Abstr.*, 1993, **118**, 147 632w).	462

92ZOB2708	V. F. Krutikov, E. V. Sukhanovskaya and I. A. Tsarkova; *Zh. Obshch. Khim.*, 1992, **62**, 2708 (*Chem. Abstr.*, 1993, **119**, 95 658h).	469
92ZOB2730	A. V. Martynov, A. N. Mirskova and S. G. Seredkina; *Zh. Obsch. Khim.*, 1992, **62**, 2730 (*Chem. Abstr.*, 1993, **119**, 95 700r; see also *USSR Pat.* 1 643 534 (*Chem. Abstr.*, 1992, **116**, 106 512x)).	787
92ZOR522	V. V. Shchepin, M. N. Novoselova, G. E. Gladkova, T. Yu. Chuprikova and D. I. Efremov; *Zh. Org. Khim.*, 1992, **28**, 522.	768
92ZOR1711	B. A. Shainyn, V. Y. Vitkovskii and A. G. Azarov; *Zh. Org. Khim.*, 1992, **28**, 1711 (*Chem. Abstr.* 1993, **119**, 8 433).	861
92ZOR1780	V. V. Nosyreva, S. V. Amosova and M. V. Sigalov; *Zh. Org. Khim.*, 1992, **28**, 1780.	775
92ZOR2119	M. V. Vovk; *Zh. Org. Khim.*, 1992, **28**, 2119.	446
93AG(E)105	T. Sato and S. Ariura; *Angew. Chem., Int. Ed. Engl.*, 1993, **32**, 105.	708
93AG(E)295	R. Littger, H. Nöth, M. Thomann and M. Wagner; *Angew. Chem., Int. Ed. Engl.*, 1993, **32**, 295.	632
93AG(E)710	T. A. Waldbach, P. H. van Rooyen and S. Lotz; *Angew. Chem., Int. Ed. Engl.*, 1993, **32**, 710.	720
93AG(E)756	C. C. Cummins, R. R. Schrock and W. M. Davis; *Angew. Chem., Int. Ed. Engl.*, 1993, **32**, 756.	549
93AG(E)893	J. Barluenga, J. M. Gonzalez, I. Llorente and P. J. Campos; *Angew. Chem., Int. Ed. Engl.*, 1993, **32**, 893.	749
93AG(E)923	W. Beck, B. Niemer and M. Wieser; *Angew. Chem., Int. Ed. Engl.*, 1993, **32**, 923.	710
93AG(E)1023	A. Maercker; *Angew. Chem., Int. Ed. Engl.*, 1993, **32**, 1023.	816, 819
93AG(E)1032	G. Boche, M. Marsch, A. Müller and K. Harms; *Angew. Chem., Int. Ed. Engl.*, 1993, **32**, 1032.	816
93AG(E)1048	D. Heineke and H. Vahrenkamp; *Angew. Chem., Int. Ed. Engl.*, 1993, **32**, 1048.	685
93AG(E)1167	M. Soleilhavoup, A. Baceiredo and G. Bertrand; *Angew. Chem., Int. Ed. Engl.*, 1993, **32**, 1167.	548
93AG(E)1359	U. Heim, H. Pritzkow, U. Fleischer and H. Grützmacher; *Angew. Chem., Int. Ed. Engl.*, 1993, **32**, 1359.	572
93AG(E)1623	G. Erker and D. Röttger; *Angew. Chem., Int. Ed. Engl.*, 1993, **32**, 1623.	671
93AG(E)1735	M. Zablocka, F. Boutonnet, A. Igau, F. Dahan, J. P. Majoral and K. M. Pietrusiewicz; *Angew. Chem., Int. Ed. Engl.*, 1993, **32**, 1735.	547, 579
93BCJ1849	M. Onishi, K. Ikemoto, K. Hiraki and R. Koga; *Bull. Chem. Soc. Jpn.*, 1993, **66**, 1849.	656
93BCJ1866	T. Satoh, Y. Hayashi and K. Yamakawa; *Bull. Chem. Soc. Jpn.*, 1993, **66**, 1866.	777
93BCJ2339	T. Satoh, K. Ogura, J. Shishikura, N. Kanetaka, R. Okada and K. Yamakawa; *Bull. Chem. Soc. Jpn.*, 1993, **66**, 2339.	80
93CAR(248)377	S. Houdier and P. J. A. Vottero; *Carbohydr. Res.*, 1993, **248**, 377.	224
93CAR(249)197	J. Thiem and M. Wiesner; *Carbohydr. Res.*, 1993, **249**, 197.	225
93CB79	T. Kauffmann, P. Fiegenbaum, M. Papenberg, R. Wieschollek and D. Wingbermühle; *Chem. Ber.*, 1993, **126**, 79.	673
93CB89	T. Kauffmann, J. Baune, P. Fiegenbaum, U. Hansmersmann, C. Neiteler, M. Papenberg and R. Wieschollek; *Chem. Ber.*, 1993, **126**, 89.	673
93CB373	D. Heineke and H. Vahrenkamp; *Chem. Ber.*, 1993, **126**, 373.	683
93CB575	N. Auner and A. Wolff; *Chem. Ber.*, 1993, **126**, 575.	611
93CB1227	K. Brickmann and R. Brückner; *Chem. Ber.*, 1993, **126**, 1227.	231
93CB1361	B. Wrackmeyer, S. Kundler and R. Boese; *Chem. Ber.*, 1993, **126**, 1361.	665
93CB1397	G. E. Herberich, U. Eigendorf and U. Englert; *Chem. Ber.*, 1993, **126**, 1397.	642
93CB2003	N. Metzler, B. Ederer and H. Nöth; *Chem. Ber.*, 1993, **126**, 2003.	635
93CB2017	B. Hansert and H. Vahrenkamp; *Chem. Ber.*, 1993, **126**, 2017.	683
93CC284	W. Lin, S. R. Wilson and G. S. Girolami; *J. Chem. Soc., Chem. Commun.*, 1993, 284.	680
93CC569	M. Lazraq, J. Escudié, C. Couret, U. Bergsträsser and M. Regitz; *J. Chem. Soc., Chem. Commun.*, 1993, 569.	549
93CC1241	V. Dehmlow and U. Fastabend; *J. Chem. Soc., Chem. Commun.*, 1993, 1241.	1080
93CC1309	J. Kurita, M. Ishii, S. Yasuike and T. Tsuchiya; *J. Chem. Soc., Chem. Commun.*, 1993, 1309.	808
93CC1459	H. Wadepohl, W. Galm, H. Pritzkow and A. Wolf; *J. Chem. Soc., Chem. Commun.*, 1993, 1459.	691
93CC1817	S. Yasuike, H. Ota, S. Shiratori, J. Kurita and T. Tsuchiya; *J. Chem. Soc., Chem. Commun.*, 1993, 1817.	808
93CL17	Y. Masaki, T. Miura and M. Ochiai; *Chem. Lett.*, 1993, 17.	183
93CL267	A. Sekiguchi, M. Ichinine, T. Nakanishi and H. Sakurai; *Chem. Lett.*, 1993, 267.	618
93CL1047	R. Okazaki, M. Minoura and T. Kawashima; *Chem. Lett.*, 1993, 1047.	334
93CL1881	A. Hayashi, M. Yamaguchi, M. Hirama, C. Kabuto and M. Ueno; *Chem. Lett.*, 1993, 1881.	1069
93CPB217	T. Naito, Y. Honda, O. Miyata and I. Ninomiya; *Chem. Pharm. Bull.*, 1993, **41**, 217.	908
93CRV1503	K. Toshima and K. Tatsuta; *Chem. Rev.*, 1993, **93**, 1503.	176, 198, 202, 214
93GEP4122315	D. Hermeling (BASF AG); *Ger. Pat.* 4 122 315 (1993) (*Chem. Abstr.*, 1993, **118**, 233 671).	197
93H(36)1925	R. Neidlein and H. Keller; *Heterocycles*, 1993, **36**, 1925.	491
93HCA211	C. Li and A. Vasella; *Helv. Chim. Acta.*, 1993, **76**, 211.	190
93HCA995	M. Hürzeler, B. Bernet and A. Vasella; *Helv. Chim. Acta*, 1993, **76**, 995.	226
93HCA1779	M. Hürzeler, B. Bernet, T. Mäder and A. Vasella; *Helv. Chim. Acta*, 1993, **76**, 1779.	229
93HOU(E15/2)1599	C. G. Bakker and J. W. Scheeren; *Methoden Org. Chem. (Houben-Weyl)*, 1993, **E15/2**, 1599.	760, 765
93HOU(E15/2)1674	J. W. Scheeren; *Methoden Org. Chem. (Houben-Weyl)*, 1993, **E15/2**, 1674.	824, 829

93HOU(E15/2)1737	A. Pelter; *Methoden Org. Chem.* (*Houben-Weyl*), 1993, **E15/2**, 1737.	832
93HOU(E15/2)1742	S. Pawlenko; *Methoden Org. Chem.* (*Houben-Weyl*), 1993, **E15/2**, 1742.	831
93JA440	G. B. Fisher, J. J. Juarez-Brambila, C. T. Goralski, W. T. Wipke and B. Singaram; *J. Am. Chem. Soc.*, 1993, **115**, 440.	524
93JA10754	C. A. Shook, M. L. Romberger, S.-H. Jung, M. Xiao, J. P. Sherbine, B. Zhang, F.-T. Lin and T. Cohen; *J. Am. Chem. Soc.*, 1993, **115**, 10 754.	652
93JA1551	G. Fraenkel and J. A. Cabral; *J. Am. Chem. Soc.*, 1993, **115**, 1551.	648
93JA1580	S. Raghavan and D. Kahne; *J. Am. Chem. Soc.*, 1993, **115**, 1580.	232
93JA2743	M. J. Hostetler, M. D. Butts and R. G. Bergman; *J. Am. Chem. Soc.*, 1993, **115**, 2743.	710
93JA3322	F. Bernardi, A. Bottoni, M. Olivucci, M. A. Robb and A. Venturini; *J. Am. Chem. Soc.*, 1993, **115**, 3322.	610
93JA3786	M. A. Alvarez, M. E. García, V. Riera, M. A. Ruiz, C. Bois and Y. Jeannin; *J. Am. Chem. Soc.*, 1993, **115**, 3786.	674
93JA4145	R. Usón, J. Forniés, M. Tomás, J. M. Casas, F. A. Cotton, L. R. Falvello and X. Feng; *J. Am. Chem. Soc.*, 1993, **115**, 4145.	692
93JA5430	P. A. Morken, P. C. Bachand, D. C. Swenson and D. J. Burton; *J. Am. Chem. Soc.*, 1993, **115**, 5430.	746, 748, 760, 817, 821
93JA5527	B. K. Campion, R. H. Heyn, T. Don Tilley and A. L. Rheingold; *J. Am. Chem. Soc.*, 1993, **115**, 5527.	656, 657
93JA6025	N. Basso, S. Görs, E. Popowski and H. Mayr; *J. Am. Chem. Soc.*, 1993, **115**, 6025.	637
93JA6065	J. C. Colberg, A. Rane, J. Vaquer and J. A. Soderquist; *J. Am. Chem. Soc.*, 1993, **115**, 6065.	630, 631
93JA6625	H. J. Reich, I. L. Reich, K. E. Yelm, J. E. Holladat and D. Dschneider; *J. Am. Chem. Soc.*, 1993, **115**, 6625.	621
93JA7215	H.-J. Gais, G. Bülow and G. Raabe; *J. Am. Chem. Soc.*, 1993, **115**, 7215.	656
93JA8493	R. Uhrhammer, D. G. Black, T. G. Gardner, J. D. Olsen and R. F. Jordan; *J. Am. Chem. Soc.*, 1993, **115**, 8493.	656
93JA9347	P. Magnus, J. Lacour and W. Weber; *J. Am. Chem. Soc.*, 1993, **115**, 9347.	443
93JCR(S)60	J. A. Lopez Sastre, J. D. Martin-Ramos, A. B. Martinez-Aragon, J. M. Molina, J. Rodriguez Amo and X. Solans; *J. Chem. Res.* (*S*), 1993, 60.	858
93JCR(S)430	O. Prakash, N. Saini and P. K. Sharma; *J. Chem. Res.* (*S*), 1993, 430.	183
93JCS(P1)67	J. Fawcett, S. House, P. R. Jenkins, N. J. Lawrence and D. R. Russell; *J. Chem. Soc., Perkin Trans. 1*, 1993, 67.	960
93JCS(P1)343	P. L. Bailey, W. Clegg, R. F. W. Jackson and O. Meth-Cohn; *J. Chem. Soc., Perkin Trans. 1*, 1993, 343.	938
93JCS(P1)577	P. L. Bailey, C. T. Hewkin, W. Clegg and R. F. W. Jackson; *J. Chem. Soc., Perkin Trans. 1*, 1993, 577.	862
93JCS(P1)795	F. Q. Jin, Y. Y. Xu and W. Y. Huang; *J. Chem. Soc. Perkin Trans. 1*, 1993, 795.	735
93JCS(P1)1493	M. J. Hughes and E. J. Thomas; *J. Chem. Soc., Perkin Trans. 1*, 1993, 1493.	57
93JCS(P1)1945	M. S. Baird, M. F. Shortt, H. H. Hussain and J. R. Al Dulayymi; *J. Chem. Soc., Perkin Trans. 1*, 1993, 1945.	775
93JCS(P1)2303	W. Thomson, D. Nicholls, A. G. Mitchell, J. A. Corner, W. J. Irwin and S. Freeman; *J. Chem. Soc., Perkin Trans. 1*, 1993, 2303.	58
93JCS(P1)2391	R. C. F. Jones and K. J. Howard; *J. Chem. Soc., Perkin Trans. 1*, 1993, 2391.	414
93JFC(60)61	D. I. Rossman and A. J. Muller; *J. Fluorine Chem.*, 1993, **60**, 61.	762
93JFC(62)201	S. Watanabe, K. Sugahara, T. Fujita, M. Sakamoto and T. Kitazume; *J. Fluorine Chem.*, 1993, **62**, 201.	736
93JFC(63)157	C. Gosmini, S. Klein, R. Sauvetre and J.-F. Normant; *J. Fluorine Chem.*, 1993, **63**, 157.	45
93JFC(63)253	S. Munavalli, E. O. Lewis, A. J. Muller, D. I. Rossman, D. K. Rohrbaugh and C. P. Ferguson; *J. Fluorine Chem.*, 1993, **63**, 253.	771
93JFC(63)281	J. Ichikawa, C. Ikeura and T. Minami; *J. Fluorine Chem.*, 1993, **63**, 281.	736
93JFC(65)67	F. J. Weigert; *J. Fluorine Chem.*, 1993, **65**, 67.	730
93JMC305	D. E. Zembower, J. A. Gilbert and M. M. Ames; *J. Med. Chem.*, 1993, **36**, 305.	419, 427
93JOC506	J. P. Hagen; *J. Org. Chem.*, 1993, **58**, 506.	226
93JOC523	J. M. Chong and S. B. Park; *J. Org. Chem.*, 1993, **58**, 523.	155, 534
93JOC546	M. Topolski, M. Duraisamy, J. Rachón, J. Gawronski, K. Gawronska, V. Goedken and H. M. Walborsky; *J. Org. Chem.*, 1993, **58**, 546.	816, 818
93JOC588	P. Knochel, T.-S. Chou, C. Jubert and D. Rajagopal; *J. Org. Chem.*, 1993, **58**, 588.	53, 54, 58
93JOC759	A. P. Marchand, R. Sharma, U. R. Zope, W. H. Watson and R. P. Kashyap; *J. Org. Chem.*, 1993, **58**, 759.	429, 430, 431
93JOC763	G. A. Olah, P. Ramaiah, G. K. S. Prakash and R. Gilardi; *J. Org. Chem.*, 1993, **58**, 763.	428, 429
93JOC1531	A. D. Abell, D. A. Hoult, K. M. Morris, J. M. Taylor and J. O. Trent; *J. Org. Chem.*, 1993, **58**, 1531.	802
93JOC1596	J. C. Carretero and E. Domínguez; *J. Org. Chem.*, 1993, **58**, 1596.	936, 938
93JOC2086	A. R. Katritzky, G.W. Yao, X. F. Lan and X. H. Zhao; *J. Org. Chem.*, 1993, **58**, 2086.	446, 449
93JOC2181	G. Solladié and C. Ziani-Chérif; *J. Org. Chem.*, 1993, **58**, 2181.	245
93JOC2360	H. Ishibashi, N. Uemura, H. Nakatani, M. Okazaki, T. Sato, N. Nakamura and M. Ikeda; *J. Org. Chem.*, 1993, **58**, 2360.	67
93JOC2517	M. Bellassoued and A. Majidi; *J. Org. Chem.*, 1993, **58**, 2517.	604
93JOC3421	B. A. Shainyan and Z. Rappoport; *J. Org. Chem.*, 1993, **58**, 3421.	763, 771
93JOC3800	M. J. Robins and S. F. Wnuk; *J. Org. Chem.*, 1993, **58**, 3800.	62, 63
93JOC4200	S. Narizuka and T. Fuchigami; *J. Org. Chem.*, 1993, **58**, 4200.	62, 63
93JOC4897	T. Harada, T. Katsuhira, D. Hara, Y. Kotani, K. Maejima, R. Kaji and A. Oku; *J. Org. Chem.*, 1993, **58**, 4897.	758, 821

93JOC6949	A. Ashimori, T. Matsuura, L. E. Overman and D. J. Poon; *J. Org. Chem.*, 1993, **58**, 6949.	410
93JOC7274	D. S. Tork, J. J. Figueroa and W. J. Scott; *J. Org. Chem.*, 1993, **58**, 7274.	181
93JOC7382	C. G. Huang, L. A. Rozov, D. F. Halpern and G. G. Vernice; *J. Org. Chem.*, 1993, **58**, 7382.	49
93JOC7474	M. P. Cooke, Jr. and C. M. Pollock; *J. Org. Chem.*, 1993, **58**, 7474.	652, 655
93JOM(444)203	G. Gervasio, E. Sappa and L. Markó; *J. Organomet. Chem.*, 1993, **444**, 203.	688
93JOM(445)133	H. Müller, W. Seidel and H. Görls; *J. Organomet. Chem.*, 1993, **445**, 133.	685
93JOM(447)C1	H. Lang, M. Leise and L. Zsolnai; *J. Organomet. Chem.*, 1993, **447**, C1.	714
93JOM(448)69	A. J. Zapata, C. Fortoul R. and C. Acuña A.; *J. Organomet. Chem.*, 1993, **448**, 69.	660
93JOM(449)203	P. M. Jeffries, S. R. Wilson and G. S. Girolami; *J. Organomet. Chem.*, 1993, **449**, 203.	696
93JOM(451)C1	D. J. Brondani, R. J. P. Corriu, S. El Ayouki, J. J. E. Moreau and M. Wong Chi Man; *J. Organomet. Chem.*, 1993, **451**, C1.	606
93JOM(455)C9	I. V. Barinov, V. A. Chertkov and O. A. Reutov; *J. Organomet. Chem.*, 1993, **455**, C9.	674
93JOM(458)181	K. Sünkel and U. Birk; *J. Organomet. Chem.*, 1993, **458**, 181.	677, 707, 723
93JOM(459)1	G. E. Herberich, U. Englert and D. Pubanz; *J. Organomet. Chem.*, 1993, **459**, 1.	695
93JOM(459)325	J. Müller, T. Akhnoukh, J. Pickardt, M. Siewing and B. Westphal; *J. Organomet. Chem.*, 1993, **459**, 325.	690
93JOM(460)31	M. J. Dabdoub, P. G. Guerrero, Jr. and C. C. Silveira; *J. Organomet. Chem.*, 1993, **460**, 31.	93
93JOM(461)207	W.-Y. Yeh, S.-L. Chen, S.-M. Peng and G.-H. Lee; *J. Organomet. Chem.*, 1993, **461**, 207.	683
93JOM(462)141	W. J. Evans, T. J. Boyle and J. W. Ziller; *J. Organomet. Chem.*, 1993, **462**, 141.	726
93JOM(463)C6	I. R. Butler, W. R. Cullen, F. W. B. Einstein and R. H. Jones; *J. Organomet. Chem.*, 1993, **463**, C6.	715
93JOM(463)163	U. T. Mueller-Westerhoff, Z. Yang and G. Ingram; *J. Organomet. Chem.*, 1993, **463**, 163.	706
93JOM(463)169	K. Hamamura, M. Kita, M. Nonoyama and J. Fujita; *J. Organomet. Chem.*, 1993, **463**, 169.	721
93JPC1576	G. S. Tyndall, T. J. Wallington, M. D. Hurley and W. F. Schneider; *J. Phys. Chem.*, 1993, **97**, 1576.	42
93JPO59	B. A. Shainyan; *J. Phys. Org. Chem.*, 1993, **6**, 59.	861
93LA207	W. Reuther, A. Ruland, S. Gangkofner and U. Baus; *Liebigs Ann. Chem.*, 1993, 207.	890
93LA955	S. Shatzmiller, R. Neidlein and C. Weik; *Justus Liebigs Ann. Chem.*, 1993, 955.	485
93MI 402-01	M. T. Baker, C. K. Chiang and J. H. Tinker; *J. Labelled Compd. Radiopharm.*, 1993, **33**, 801 (*Chem. Abstr.*, 1994, **120**, 133 794).	45
B-93MI 402-02	N. S. Simpkins; "Sulphones in Organic Synthesis," Pergamon, New York, 1993.	79
B-93MI 408-01	N. S. Simpkins; "Sulphones in Organic Synthesis," Pergamon Press, Oxford, 1993.	390, 391, 395
93MI 410-01	C. Ruyu, F. Kesheng, L. Xiaolan, S. Ming and M. Fanming; *Sci. China, Ser. B*, 1993, **36**, 257 (*Chem. Abstr.*, 1993, **119**, 160 403).	488
B-93MI 412-01	W. C. Kaska and K. A. Ostoja Starzewski; in "Ylides and Imines of Phosphorus," ed. A. W. Johnson, Wiley, New York, 1993, p. 485.	578
93MI 417-01	A. Kunugi, K. Yamane, M. Yasuzawa, H. Matsui, H. Uno and K. Sakamoto; *Electrochim. Acta.*, 1993, **38**, 1037.	774, 779
93MI 420-01	I. Minamida, K. Iwanaga, T. Tabuchi, H. Uneme, H. Dantsuji and T. Okauchi; *J. Pesticide Sci*, 1993, **18**, 31.	900
93MM916	J. A. Moore and P. G. Mehta; *Macromolecules*, 1993, **26**, 916.	742, 979
93OM65	M. J. Hostetler, M. D. Butts and R. G. Bergman; *Organometallics*, 1993, **12**, 65.	710
93OM108	M.-H. Cheng, S.-G. Shu, G.-H. Lee, S.-M. Peng and R.-S. Liu; *Organometallics*, 1993, **12**, 108.	711
93OM184	P.-J. Sinnema, A. Meetsma and J. H. Teuben; *Organometallics*, 1993, **12**, 184.	656
93OM289	I. M. T. Davidson and G. H. Morgan; *Organometallics*, 1993, **12**, 289.	614
93OM315	O. C. P. Beers, M. M. Bouman, A. E. Komen, K. Vrieze, C. E. Elsevier, E. Horn and A. L. Spek; *Organometallics*, 1993, **12**, 315.	716
93OM338	E. L. Lyszak, J. P. O'Brien, D. A. Kort, S. K. Hendges, R. N. Redding, T. L. Bush, M. S. Hermen, K. B. Renkema and M. E. Silver; *Organometallics*, 1993, **12**, 338.	650, 656
93OM996	P. J. Stang and D. Cao; *Organometallics*, 1993, **12**, 996.	717
93OM1213	J. Chen, Y. Yu, K. Liu, G. Wu and P. Zheng; *Organometallics*, 1993, **12**, 1213.	716
93OM1514	W. Ando, M. Kako, T. Akasaka and S. Nagase; *Organometallics*, 1993, **12**, 1514.	624
93OM1936	A. K. Hughes, A. Meetsma and J. H. Teuben; *Organometallics*, 1993, **12**, 1936.	656
93OM2227	H. C. L. Abbenhuis, N. Feiken, H. F. Haarman, D. M. Grove, E. Horn, A. L. Spek, M. Pfeffer and G. van Koten; *Organometallics*, 1993, **12**, 2227.	723
93OM2268	B. Hessen, J.-K. F. Buijink, A. Meetsma, J. H. Teuben, G. Helgesson, M. Håkansson, S. Jagner and A. L. Spek; *Organometallics*, 1993, **12**, 2268.	671
93OM2360	I. N. Jung, S. H. Yeon and J. S. Han; *Organometallics*, 1993, **12**, 2360.	609
93OM2423	N. Metzler, H. Nöth and M. Thomann; *Organometallics*, 1993, **12**, 2423.	635
93OM2536	A. Kunai, Y. Matsuo and M. Ishikawa; *Organometallics*, 1993, **12**, 2536.	614
93OM2609	H. J. Heeres, J. Nijhoff, J. H. Teuben and R. D. Rogers; *Organometallics*, 1993, **12**, 2609.	703
93OM2618	W. J. Evans, R. A. Keyer and J. W. Ziller; *Organometallics*, 1993, **12**, 2618.	703
93OM2788	D. Dakternieks, K. Jurkschat, H. Wu and E. R. T. Tiekink; *Organometallics*, 1993, **12**, 2788.	660, 700
93OM2925	M. Akita, S. Sugimoto, A. Takabuchi, M. Tanaka and Y. Moro-oka; *Organometallics*, 1993, **12**, 2925.	674
93OM3187	O. C. P. Beers, C. J. Elsevier, H. Rooijman, W. J. J. Smeets and A. L. Spek; *Organometallics*, 1993, **12**, 3187.	716
93OM3431	R. D. Adams, G. Chen, L. Chen, W. Wu and J. Yin; *Organometallics*, 1993, **12**, 3431.	955
93OM3624	P. A. van der Schaaf, E. Wissing, J. Boersma, W. J. J. Smeets, A. L. Spek and G. van Koten; *Organometallics*, 1993, **12**, 3624.	656

93OM3890	F. J. Berg and J. L. Petersen; *Organometallics*, 1993, **12**, 3890.	659
93OM3955	P. A. van der Schaaf, D. M. Grove, W. J. J. Smeets, A. L. Speck and G. van Koten; *Organometallics*, 1993, **12**, 3955.	656
93OM3979	S. Sharma and K. H. Pannell; *Organometallics*, 1993, **12**, 3979.	639, 656, 663
93OM3998	W. J. Evans, J. Boyle and J. W. Ziller; *Organometallics*, 1993, **12**, 3998.	704
93OM4123	N. Sewald, W. Ziche, A. Wolff and N. Auner; *Organometallics*, 1993, **12**, 4123.	651
93OM4135	N. Auner, C.-R. Heikenwälder and C. Wagner; *Organometallics*, 1993, **12**, 4135.	651
93OM4269	M. D. Butts and R. G. Bergman; *Organometallics*, 1993, **12**, 4269.	710
93OM4545	D. Osella, G. Dutto, G. Jaouen, A. Vessiéres, P. R. Raithby, L. De Benedetto and M. J. McGlinchey; *Organometallics*, 1993, **12**, 4545.	718
93OM4572	N. Brunet, J. D. Debad, P. Legzdins, J. Trotter, J. E. Veltheer and V. C. Yee; *Organometallics*, 1993, **12**, 4572.	656
93OM4708	B. Korpar-Čolig, Z. Popović, D. Matković-Čalogovic and D. Vikić-Topić; *Organometallics*, 1993, **12**, 4708.	697
93OM4930	H. Bürger and P. Moritz; *Organometallics*, 1993, **12**, 4930.	140
93OM4979	G. Erker, M. Albrecht, C. Krüger, M. Nolte and S. Werner; *Organometallics*, 1993, **12**, 4979.	725
93OSC(8)3	S. J. Mickel, S.-N. Hsiao and M. J. Miller; *Org. Synth. Coll. Vol.*, 1993, **8**, 3.	306
93OSC(8)546	L. C. Vishwakarma, O. D. Stringer and F. A. Davis; *Org. Synth. Coll. Vol.*, 1993, **8**, 546.	312
93OSC(8)550	A. G. M. Barrett, D. Dhanak, G. G. Graboski and S. J. Taylor; *Org. Synth. Coll. Vol.*, 1993, **8**, 550.	327
93PAC617	H. J. Bestmann; *Pure Appl. Chem.*, 1993, **65**, 617.	587, 588
93PS(74)373	M. Yoshifuji; *Phosphorus, Sulfur Silicon*, 1993, **74**, 373.	559
93PS(75)23	C. K. McClure, K.-Y. Jung, C. W. Grote and K. Hansen; *Phosphorus, Sulfur Silicon*, 1993, **75**, 23.	561
93PS(75)99	N. Mimouni, E. About-Jaudet, N. Collignon and P. Savignac; *Phosphorus, Sulfur Silicon*, 1993, **75**, 99.	588
93PS(75)143	C. Patois and P. Savignac; *Phosphorus, Sulfur Silicon*, 1993, **75**, 143.	576
93PS(75)233	I. V. Shevchenko and R. Schmutzler; *Phosphorus, Sulfur Silicon*, 1993, **75**, 233.	564
93PS(76)1	M. Regitz, T. Wettling, B. Breit, M. Birkel, B. Geissler, U. Bergsträsser, S. Barth and P. Binger; *Phosphorus, Sulfur Silicon*, 1993, **76**, 1.	549
93PS(76)13	A. Schmidpeter, G. Jochen and M. Thiele; *Phosphorus, Sulfur Silicon*, 1993, **76**, 13.	545, 558
93PS(76)49	M. Soleihavoup, G. Alcaraz, R. Reau, A. Baceiredo and G. Bertrand; *Phosphorus, Sulfur Silicon*, 1993, **76**, 49.	589
93PS(76)57	M. Driess and H. Pritzkow; *Phosphorus, Sulfur Silicon*, 1993, **76**, 57.	567
93PS(77)5	D. Böhn, D. Hu and U. Zenneck; *Phosphorus, Sulfur Silicon*, 1993, **77**, 5.	549
93PS(77)199	S. Jugé, R. Merdès, M. Stéphan and J. P. Genet; *Phosphorus, Sulfur Silicon*, 1993, **77**, 199.	557, 570, 575
93PS(77)254	C. Charrier, M. Sierra, N. Maigrot, L. Ricard and F. Mathey; *Phosphorus, Sulfur Silicon*, 1993, **77**, 254.	545, 565
93PS(77)258	K. Lange, M. Spahn, F. Rosche and E. Fluck; *Phosphorus, Sulfur Silicon*, 1993, **77**, 258.	562
93PS(77)274	R. Bartsch, P. B. Hitchcock and J. F. Nixon; *Phosphorus, Sulfur Silicon*, 1993, **77**, 274.	548
93S45	A. R. Katritzky, L.H. Xie and W.-Q. Fan; *Synthesis*, 1993, 45.	446
93S67	P. Kumar, R. S. Reddy, A. P. Singh and B. Pandey; *Synthesis*, 1993, 67.	246
93S149	H. J. Bestmann, W. Kellermann and B. Pecher; *Synthesis*, 1993, 149.	245
93S202	M. Harmata and B. F. Herron; *Synthesis*, 1993, 202.	655
93S209	Y. H. Kim, H. H. Shin and Y. J. Park; *Synthesis*, 1993, 209.	74, 75
93S229	A. R. Katritzky, W.-Q. Fan and Q.-H. Long; *Synthesis*, 1993, 229.	309, 323
93S285	B. Breit and M. Regitz; *Synthesis*, 1993, 285.	567
93S530	C. E. Tucker and P. Knochel; *Synthesis*, 1993, 530.	709
93S955	C. Yuan, S. Chen, H. Zhou and L. Maier; *Synthesis*, 1993, 955.	472
93S961	H. d'Orchymont; *Synthesis*, 1993, 961.	320
93S1069	P. Kumar, C. U. Dinesh, R. S. Reddy and B. Pandey; *Synthesis*, 1993, 1069.	186
93S1071	B. Bonnet, Y. Le Gallic, G. Plé and L. Duhemel; *Synthesis*, 1993, 1071.	749
93S1218	H. Suzuki, Y. S. Hwang, C. Nakaya and Y. Matano; *Synthesis*, 1993, 1218.	432
93SC1	Y. Shen and B. Yang; *Synth. Commun.*, 1993, **23**, 1.	798
93SC1467	C. Z. Ding and R. B. Silverman; *Synth. Commun.*, 1993, **23**, 1467.	416
93SC1935	A. Diaz-Ortiz, E. Diez-Barra and A. de la Hoz; *Synth. Commun.*, 1993, **23**, 1935.	826
93SC2851	A. Kamabuchi, T. Moriya, N. Miyaura and A. Suzuki; *Synth. Commun.*, 1993, **23**, 2851.	815
93SC2937	T. M. Shoup and G. Zweifel; *Synth. Commun.*, 1993, **23**, 2937.	814
93SL35	S. Hanessian, Y. L. Bennani and Y. Herve; *Synlett*, 1993, 35.	473
93SL111	S. A. Bowles, A. H. Davidson, A. Miller, T. M. Thompson and M. Whittaker; *Synlett*, 1993, 111.	200
93SL189	R. Angell, P. J. Parsons and A. Naylor; *Synlett*, 1993, 189.	807
93SL195	B. V. Yang, D. O'Rourke and J. C. Li; *Synlett*, 1993, 195.	60
93SL273	Y. Kita, H. Maeda, K. Omori, T. Okuno and Y. Tamura; *Synlett*, 1993, 273.	828
93SL349	C. Agami, F. Couty, J. Lin and A. Mikaeloff; *Synlett*, 1993, 349.	202
93SL429	M. Lakshmi Kantam and P. Lakshmi Santhi; *Synlett*, 1993, 429.	185
93SL522	B. Vauzeilles, D. Cravo, J.-M. Mallet and P. Sinaÿ; *Synlett*, 1993, 522.	238
93SR149	I. V. Koval; *Sulfur Reports*, 1993, **14**, 149.	330
93T199	F. Bellesia, M. Boni, F. Ghelfi and U. M. Pagnoni; *Tetrahedron*, 1993, **49**, 199.	249
93T2011	S. Nakatani, J. Yoshida and S. Isoe; *Tetrahedron*, 1993, **49**, 2011.	937
93T2101	J. I. Reddy, B. M. Bhawal and S. Rajappa; *Tetrahedron*, 1993, **49**, 2101.	901, 904

93T2151	A. S. Kiselyov, L. Strekowski and V. V. Semenov; *Tetrahedron*, 1993, **49**, 2151.	245
93T2979	J. W. Wilson, A. Pelter, M. V. Garad and R. Pardasani; *Tetrahedron*, 1993, **49**, 2979.	665
93T2988	A. Pelter, L. Warren and J. W. Wilson; *Tetrahedron*, 1993, **49**, 2988.	664
93T3007	A. Pelter, G. F. Vaughan-Williams and R. M. Rosser; *Tetrahedron*, 1993, **49**, 3007.	704
93T4229	G. Haaima, M. J. Lynch, A. Routledge and R. T. Weavers; *Tetrahedron*, 1993, **49**, 4229.	811
93T6483	M. Pohmakotr and S. Khosavanna; *Tetrahedron*, 1993, **49**, 6483.	962
93T9495	A. Dömling and I. K. Ugi; *Tetrahedron*, 1993, **49**, 9495.	313
93TA1547	G. Solladié and O. Lohse; *Tetrahedron Asymmetry*, 1993, **4**, 1547.	229
93TA2139	P. C. Bulman Page, M. T. Gareh and R. A. Porter; *Tetrahedron Asymmetry*, 1993, **4**, 2139.	262
93TL411	M. Tsukazaki and V. Snieckus; *Tetrahedron Lett.*, 1993, **34**, 411.	768
93TL415	J. Lee, M. Tsukazaki and V. Snieckus; *Tetrahedron Lett.*, 1993, **34**, 415.	841
93TL1085	E. K. Mantus and J. Clardy; *Tetrahedron Lett.*, 1993, **34**, 1085.	245
93TL1141	S. Arseniyadis, R. Rodriguez, J. Camara, E. Guittet and L. Toupet; *Tetrahedron Lett.*, 1993, **34**, 1141.	249
93TL1311	K. Uneyama, Y. Tokunaga and K. Maeda; *Tetrahedron Lett.*, 1993, **34**, 1311.	92
93TL1463	A. P. Marchand and D. Rajapaksa; *Tetrahedron Lett.*, 1993, **34**, 1463.	253
93TL1491	S. Kiyooka, M. Shirouchi and Y. Kaneko; *Tetrahedron Lett.*, 1993, **34**, 1491.	202
93TL2111	D. J. Brondani, R. J. P. Corriu, S. El Ayoubi, J. J. E. Moreau and M. Wong Chi Man; *Tetrahedron Lett.*, 1993, **34**, 2111.	607, 608, 654
93TL2331	T. Satoh, Y. Kitoh, K. Onda and K. Yamakawa; *Tetrahedron Lett.*, 1993, **34**, 2331.	77
93TL3057	D. P. Matthews, S. C. Miller, E. T. Jarvi, J. S. Sabol and J. R. McCarthy; *Tetrahedron Lett.*, 1993, **34**, 3057.	736
93TL3581	J. C. Jung, H. C. Choi and Y. H. Kim; *Tetrahedron Lett.*, 1993, **34**, 3581.	201
93TL4063	Y. Kita, N. Shibata and N. Yoshida; *Tetrahedron Lett.*, 1993, **34**, 4063.	223
93TL4355	S. Stavber and M. Zupan; *Tetrahedron Lett.*, 1993, **34**, 4355.	45
93TL5097	M. Yokoyama, M. Matsushita, S. Hirano and H. Togo; *Tetrahedron Lett.*, 1993, **34**, 5097.	432
93TL5269	S. Ma and L. M. Venanzi; *Tetrahedron Lett.*, 1993, **34**, 5269.	186
93TL5649	G.-J. Boons, D. A. Entwistle, S. V. Ley and M. Woods; *Tetrahedron Lett.*, 1993, **34**, 5649.	236
93TL5769	S. Kim, J. H. Park and J. M. Lee; *Tetrahedron Lett.*, 1993, **34**, 5769.	201, 217
93TL6329	R. C. F. Jones, I. Turner and K. J. Howard; *Tetrahedron Lett.*, 1993, **34**, 6329.	413
93TL6541	J. Belzner and H. Ihmels; *Tetrahedron Lett.*, 1993, **34**, 6541.	1052
93TL6677	T. Axenrod, C. Watnick, H. Yazdekhasti and P. R. Dave; *Tetrahedron Lett.*, 1993, **34**, 6677.	429, 430, 431
93TL7127	H. K. Patney; *Tetrahedron Lett.*, 1993, **34**, 7127.	246
93TL7197	R. S. Gross, S. Mehdi and J. R. McCarthy; *Tetrahedron Lett.*, 1993, **34**, 7197.	755, 803
93TL7335	A. Trehan, A. Vij, M. Walia, G. Kaur, R. D. Verma and S. Trehan; *Tetrahedron Lett.*, 1993, **34**, 7335.	192
93TL7645	D. K. Dubey, R. Nath, R. C. Malhotra and D. N. Tripathi; *Tetrahedron Lett.*, 1993, **34**, 7645.	773
93ZN(B)1558	J. Müller, K. Qiao, R. Schubert and M. Tschampel; *Z. Naturforsch., Teil B*, 1993, **48**, 1558.	690
93ZOB220	M. A. Pudovik, S. A. Terent'eva and A. N. Pudovik; *Zh. Obshch. Khim.*, 1993, **63**, 220 (*Chem. Abstr.*, 1993, **119**, 95 641x).	478
93ZOR56	V. I. Potkin and R. V. Kaberdin; *Zh. Org. Khim.*, 1993, **29**, 56.	771
94AG(E)2202	V. Caliman, P. B. Hitchcock, J. F. Nixon, M. Hofmann and P. von R. Schleyer; *Angew. Chem., Int. Ed. Engl.*, 1994, **33**, 2202.	549
94AHC(61)59	B. A. Arbuzov and G. N. Nikonov; *Adv. Heterocycl. Chem.*, 1994, **61**, 59.	336
94BMC931	M. Jung and R. F. Schinazi; *Bioorg. Med. Chem. Lett.*, 1994, 931.	214
94CC205	F. Diederich, D. Philp and P. Seiler; *J. Chem. Soc., Chem. Commun.*, 1994, 205.	746
94CC749	R. Madsen and B. Fraser-Reid; *J. Chem. Soc., Chem. Commun.*, 1994, 749.	184
94H(37)1951	M. Balasubramanian, J. G. Keay, E. F. V. Scriven and N. Shobana; *Heterocycles*, 1994, **37**, 1951.	310
94JA789	J. K. Stowell and T. S. Widlanski; *J. Am. Chem. Soc.*, 1994, **116**, 789.	769
94JA2153	X. Verdaguer, A. Moyano, M. Pericas, A. Riera, V. Bernardes, A. E. Greene, A. Alvarez-Larena and J. F. Piniella; *J. Am. Chem. Soc.*, 1994, **116**, 2153.	763
94JA4697	A. G. Myers, D. Y. Gin and D. H. Rogers; *J. Am. Chem. Soc.*, 1994, **116**, 4697.	239
94JCS(P1)3	T. Hirao, T. Fujii, T. Tanaka and Y. Ohshiro; *J. Chem. Soc. Perkin Trans. 1*, 1994, 3.	743
94JOC544	M. J. Robins, S. F. Wnuk, K. B. Mullah, N. K. Dalley, C.-S. Yuan, Y. H. Lee and R. T. Borchardt; *J. Org. Chem.*, 1994, **59**, 544.	218
94JOC3721	J. H. Udding, H. Hiemstra and W. N. Speckamp; *J. Org. Chem.*, 1994, **59**, 3721.	327
94JOM(464)191	M. I. Bruce, P. J. Low, B. W. Skelton and A. H. White; *J. Organomet. Chem.*, 1994, **464**, 191.	678
94JOM(468)235	L. Carlton and B. Maaske; *J. Organomet. Chem.*, 1994, **468**, 235.	726
94JOM(471)249	J. Müller, C. Friederich, P. E. Gaede, S. Sodemann, K. Qiao; *J. Organomet. Chem.*, 1994, **471**, 249.	715
94JOM(476)101	M. Wiederhold and U. Behrens; *J. Organomet. Chem.*, 1994, **476**, 101.	679
B-94MI 407-01	N. Petragnani; "Tellurium in Organic Synthesis," Academic Press, London, 1994.	333
B-94MI 407-02	M. Mikolajczyk and P. Balczewski; in "Advances in Sulfur Chemistry," ed. E. Block, Jai Press, London, 1994, p. 41.	347
94MIP9322265	D. C. Woodcock and B. T. Grady; *WO Pat.* 93 22 265 (1994) (*Chem. Abstr.*, 1994, **120**, 76 896).	44
94OM94	C. E. Tucker, B. Greve, W. Klein and P. Knochel; *Organometallics*, 1994, **13**, 94.	723

94OM500	P. R. R. Ratunge-Bandarage, B. H. Robinson and J. Simpson; *Organometallics*, 1994, **13**, 500.	721
94OM511	P. R. R. Ratunge-Bandarage, N. W. Duffy, S. M. Johnston, B. H. Robinson and J. Simpson; *Organometallics*, 1994, **13**, 511.	721
94OM1078	J. Takats, J. Washington and B. D. Santarsiero; *Organometallics*, 1994, **13**, 1078.	710
94OM1899	M. D. Butts and R. G. Bergman; *Organometallics*, 1994, **13**, 1899.	710
94OM1999	C. E. Shuchart, A. Wojcicki, M. Calligaris, P. Faleschini and G. Nardin; *Organometallics*, 1994, **13**, 1999.	711
94OM2244	H. El Amouri, M. Gruselle, Y. Besace, J. Vaissermann and G. Jaouen; *Organometallics*, 1994, **13**, 2244.	674
94OM2309	W. Lin, S. R. Wilson and G. S. Girolami; *Organometallics*, 1994, **13**, 2309.	680
94OM2423	W. Trakarnpruk, A. M. Arif and R. D. Ernst; *Organometallics*, 1994, **13**, 2423.	681
94OM2527	C. P. Casey and G. P. Niccolai; *Organometallics*, 1994, **13**, 2527.	679
94OM2668	M. D. Butts and R. G. Bergman; *Organometallics*, 1994, **13**, 2668.	710
94PC 407-01	B. M. Heron; Personal Communication, 1994.	298
94PS(88)1	J. Guervenou and G. Sturtz; *Phosphorus, Sulfur Silicon*, 1994, **88**, 1.	560
94S23	X.-Y. Jiao, C. Verbruggen, M. Borloo, W. Bollaert, A. De Groot, R. Dommisse and A. Haemers; *Synthesis*, 1994, 23.	461
94S51	C. Hubert, B. Oussaid, G. Etemad-Moghadam, M. Koenig and B. Garrigues; *Synthesis*, 1994, 51.	463
94S107	A. R. Katritzky, J. Jiang and J. V. Greenhill; *Synthesis*, 1994, 107.	329
94S427	N. D. Kimpe, E. Stanoeva and M. Boeykens; *Synthesis*, 1994, 427.	189
94S445	A. R. Katritzky, X. Lan and W.-Q. Fan; *Synthesis*, 1994, 445.	309
94SL152	J. H. Bateson, G. Burton, S. A. Elsmere and R. L. Elliott; *Synlett*, 1994, 152.	208
94T2297	T. G. Gant and A. I. Meyers; *Tetrahedron*, 1994, **50**, 2297.	306
94T2993	D. J. Burton, Z.-Y. Yang and P. A. Morken; *Tetrahedron*, 1994, **50**, 2993.	816, 820
94TA961	K. Iseka, T. Nagai and Y. Kobayashi; *Tetrahedron Asymmetry*, 1994, **5**, 961.	811
94TL57	C. E. McDonald, L. E. Nice and K. E. Kennedy; *Tetrahedron Lett.*, 1994, **35**, 57.	182
94TL89	J. P. Praly, C. Di Stéfano, G. Descotes and R. Faure; *Tetrahedron Lett.*, 1994, **35**, 89.	432
94TL969	B. M. Lillie and M. A. Avery; *Tetrahedron Lett.*, 1994, **35**, 969.	182
94TL1027	D. P. Matthews, R. S. Gross and J. R. McCarthy; *Tetrahedron Lett.*, 1994, **35**, 1027.	812
94TL1825	K. Chauhan, R. K. Bhatt, J. R. Falck and J. H. Capdevila; *Tetrahedron Lett.*, 1994, **35**, 1825.	175
94TL2537	E. Untersteller, Y. C. Xin and P. Sinay; *Tetrahedron Lett.*, 1994, **35**, 2537.	213
95JOM(488)39	P. Leoni, M. Pasquali, T. Beringhelli, G. D'Alfonso and A. P. Minoja; *J. Organomet. Chem.*, 1995, **488**, 39.	693
95JOM(489)C65	I. Moldes, T. Papworth, J. Ros, A. Alvarez-Larena and J. F. Piniella; *J. Organomet. Chem.*, 1995, **489**, C65.	689
B-95MI 424-01	T. Shimizu and W. Ando; "Handbook of Organic Photochemistry and Photobiology," ed. W. M. Horspool, CRC Press, Boca Raton, 1995, p 992.	1082
95T755	A. Dömling, A. Bayler and I. Ugi; *Tetrahedron*, 1995, **51**, 755.	313

Subject Index

Every effort has been made to index as comprehensively as possible, and to standardize the terms used in the index in line with the IUPAC Recommendations. In view of the diverse nature of the terminology employed by the different authors, the reader is advised to search for related entries under the appropriate headings.

The index entries are presented in letter-by-letter alphabetical sequence. Compounds are normally indexed under the parent compound name, with the substituent component separated by a comma of inversion. An entry with a prefix/locant is filed after the same entry without any attachments, and in alphanumerical sequence. For example, 'dienes', '1,3-dienes', and '1-bromo-1,4-dienes' will be filed as:

Dienes
1,3-Dienes
1,4-Dienes, 1-bromo-

The Index is arranged in set-out style, with a maximum of three levels of heading. Location references refer to page number; major coverage of a subject is indicated by bold, elided page numbers: for example,

Halides, alkynyl, synthesis of, **1234–55**
    properties of, 345

See cross-references direct the user to the preferred term; for example,

    Mercaptans *See* Thiols

See also cross-references provide the user with guideposts to terms of related interest, from the broader term to the narrower term, and appear at the end of the main heading to which they refer, e.g.

Thiones
  See also
    Thioketone

Acetal, *N*,*N*-dimethylacetamide diethyl, reactions, with enamines, 987
Acetal, dimethylformamide dimethyl, reactions, with dinitromethane, 995
Acetaldehyde
  lithium enolates of, use in the synthesis of carbon-bridged dicobalt complexes, 687
  reactions, with α-methoxyallylphosphine oxide, 917
Acetaldehyde, trichloro-, reactions, with thiols, 215
Acetaldehyde, trifluoro-, reactions, with hydrogen sulfide, 215
Acetaldimine, α,α-bis(trimethylsilyl)-*t*-butyl, synthesis of, 604
Acetalization
  of aldehydes, 178
  of ketones, 178
S,*N*-Acetal *S*-oxides, synthesis of, 330

Acetals
  deprotection of, 175
  deprotonation of, 961
  from
    acetals, 184
    O,*S*-acetals, 196
    α-acyloxy ethers, 200
    alcohols, 186
    aldehydes, 177
    alkenes, 193
    alkynes, 193
    β-chlorovinyl carbonyl compounds, 196
    cyclic hemiacetals, 201
    1,2-dicarbonyl compounds, 183
    enol acetates, 188
    enol ethers, 186
    ethers, 201

Acetals
    geminal dihalides, 189
    α-halo ethers, 198
    halogenated ketones, 183
    hemiacetals, 200
    ketones, 177
    nitrobenzaldehydes, 183
    ortho esters, 192
    silyl enol ethers, 188
    α-substituted ethers, 198
    α-sulfonyl ethers, 200
    thioacetals, 196
    vinyl epoxides, 183
  irradiation of, 1079
  reactions
    with acetyl chloride, 175
    with acid chloride, 48
    with alcohols, 184
    with amides, 310
    with anhydrides, 211
    with t-butoxy radicals, 1078
    with dimethylcopperlithium, 388
    with β-hydroxycarboxylic acids, 213
    with isocyanates, 426
    with pyrazole, 437
    with sulfinic acids, 236
    with thiols, 249
    with 2,2,2-trichloroethanol, 184
  as starting materials
    for acetals, 184
    for acyl alkyl acetals, 211
    for α-aminoalkylideneamino enamines, 987
    for α-bromo ethers, 50
    for cations, 1072
    for α-chloro ethers, 48
    for dithioacetals, 245
    for hemiacetals, 175
    for monoselenoacetals, 238
    for monotelluroacetals, 238
    for monothioacetals, 215
    for α-phosphoryl enamines, 1004
  stereoselective synthesis of, 182
  synthesis of, 176
    via acetal exchange, 190
    via acetal interchange, 185
    via conjugate addition of alcohols, 195
    via electrochemical methods, 197
    via hetero Diels–Alder reaction, 202
    via palladium(II)-catalyzed oxidation of terminal alkenes, 193
  use in the synthesis of hemiaminals, 309
Acetals, acetaldehyde, synthesis of, 185
Acetals, acyl alkyl
  from
    acetals, 211
    alcohols, 209
    aldehydes, 212
    carboxylic acids, 209
    enol esters, 209
    enol ethers, 208
    ethers, 207
    hemiacetals, 209
    ketones, 212
    α,β-unsaturated ketones, 213
  synthesis of, 207
    via Baeyer–Villiger oxidation, 212
Acetals, aldehyde, transacetalization of, 249
Acetals, alkyl silyl
  from, silylketene acetals, 202
  synthesis of, 202
Acetals, alkynic, synthesis of, 192
Acetals, amide, as starting materials, for ketene aminals, 975
Acetals, arylacetaldehyde, from, styrenes, 197
Acetals, N-(benzenesulfinyl), synthesis of, 325
Acetals, benzophenone, synthesis of, 197, 198
Acetals, 4,6-O-benzylidene, cleavage of, 217
Acetals, bis(methoxyethyl), synthesis of, 829
Acetals, bis(2,2,2-trichloroethyl), synthesis of, 184
Acetals, α-bromo-, as starting materials, for ketene diethyl acetals, 826
Acetals, β-chloroketene, silylation of, 806
Acetals, cyanoketene, synthesis of, 762
Acetals, cyclic, from, methyl acetals, 184
Acetals, diarylketene, synthesis of, 826
Acetals, diaryloxy, synthesis of, 199
Acetals, dicyanoketene, from, tetracyanoethylene, 829
Acetals, dimethyl, synthesis of, 197
Acetals, dipent-4-enyl, for the protection of diols, 184
Acetals, ethylene, synthesis of, 190
Acetals, ethyl glyoxylate, synthesis of, 189
Acetals, formaldehyde, synthesis of, 185, 190, 198
Acetals, α-functionalized, from, enol ethers, 187
Acetals, α-halo
  hydrogen halide elimination from, 826
  as starting materials, for ketene acetals, 826
Acetals, α-hydroxy, synthesis of, 183, 187
Acetals, α-iodo, as starting materials, for ketene acetals, 826
Acetals, ketene
  boryloxy compounds of, synthesis of, 832
  from
    1-alkoxyalkynes, 827
    carboxylic acids, 824
    carboxylic esters, 824
    α,α-difluoro ethers, 826
    gem-dihaloalkenes, 829
    α-haloacetals, 826
    α-halo ortho esters, 827
    α-iodoacetals, 826
    ketenes, 829
    ortho esters, 825
    phosphonium salts, 825
    α-tosyloxyacetals, 826
    triacylmethanes, 824
    trihalomethanes, 826
    α,β-unsaturated acetals, 830
    α,β-unsaturated esters, 827
  reactions
    with amines, 887, 972
    with ammonia, 972, 983
    with dibromo carbene, 162
    with diethylamine, 972
    with hydroxamic vinyl esters, 833
  as starting materials
    for α-hydrazino enamines, 983
    for ketene aminals, 971
    for ketene hemiaminals, 887
  synthesis of, 806, 824
    via [3 + 2] cycloaddition reactions, 830
    via Horner–Wittig reactions, 828
Acetals, ketene diethyl
  from, α-bromoacetals, 826
  reactions, with N-ethylaniline, 887
Acetals, ketene disilyl, from, carboxylic acids, 832
Acetals, ketene silyl
  from
    α-bromo esters, 831
    ester enolates, 831
    ethyl bromoacetate, 831
  reactions, with dibromodifluoromethane, 736
  regioselectivity of, 831
  stereochemistry of, 831
  synthesis of, 831
Acetals, α-keto
  cleavage of, 161, 212
  from, dihalo ketones, 189
Acetals, methoxybenzylidene, synthesis of, 201
Acetals, methyl

as starting materials, for cyclic acetals, 184
synthesis of, 197
Acetals, methylene, from, diols, 190
Acetals, α-phenyl selenenyl, synthesis of, 51
Acetals, silylketene, as starting materials, for alkyl silyl acetals, 202
Acetals, stannyl
  reactions
    with acetyl chloride, 388
    with alkenylaluminums, 388
    with alkynylaluminums, 388
Acetals, symmetrical
  from
    aldehydes, 190
    ketones, 190
  synthesis of, 190
Acetals, tetramethylurea diisopropyl, synthesis of, 974
Acetals, α-tosyloxy, as starting materials, for ketene acetals, 826
Acetals, tributylstannyl, as starting materials, for α-alkoxystannanes, 388
Acetals, α,β-unsaturated
  isomerization of, 830
  reactions, with dimethyl sulfide, 201
  as starting materials, for ketene acetals, 830
Acetals, unsymmetrical
  from, enol ethers, 186, 188
  synthesis of, 198
Acetals, urea, as starting materials, for ketene aminals, 974
N,O-Acetals, ketene
  reactions, with amines, 972
  as starting materials, for α-phosphoryl enamines, 1003
  synthesis of, 972, 974
N,S-Acetals, N-acylated ketene, as starting materials, for α-amido enamines, 981
N,S-Acetals, ketene
  as starting materials
    for α-aminoalkylideneamino enamines, 988
    for α-azido enamines, 990
    for α-hydrazino enamines, 983
    for α-phosphoryl enamines, 1003
N,X-Acetals, ketene, reactions, with dialkyl phosphites, 1003
O,As-Acetals, from, ketones, 344
O,N-Acetals
  from
    aldehydes, 311
    alkenes, 306, 312
    amides, 303
    amino acids, 301
    amino alcohols, 301
    2-aminomethylbenzimidazoles, 296
    hydrazones, 313
    hydroxyamides, 301
    imides, 303
    imines, 304, 312
    iminium salts, 304
    ketones, 311
    nitrones, 312
    oximes, 313
    trichloroacetaldehyde imines, 304
  synthesis of, 293
  See also Hemiaminals
O,N-Acetals, N-acyl, synthesis of, 308
O,N-Acetals, N-nitroso, synthesis of, 312
O,O-Acetals
  reactions, with tris(phenylseleno)borane, 238
  as starting materials, for monothioacetals, 216
O,S-Acetals
  as homologation reagents, 370
  oxidation of, 235
  as starting materials, for acetals, 196
  synthesis of, via S-alkylation, 225
O,S-Acetals, α,β-unsaturated, as starting materials, for ketene monothioacetals, 836

P-Acetals, synthesis of, 340
Se,N-Acetals
  from
    aldehydes, 333
    alkenes, 335
    diazoalkanes, 335
    ketones, 333
    selenocarbonyl compounds, 333
  synthesis of, 333
Se,P-Acetals
  from
    phosphates, 348
    phosphine oxides, 348
    phosphines, 347
  synthesis of, 347
S,N-Acetals
  from
    acid chlorides, 328
    aldehydes, 315, 328
    imines, 319, 328
    iminium salts, 319
    ketones, 315, 328
    sulfones, 332
    sulfonic acid compounds, 331
    thiocarbonyl compounds, 319, 328
  synthesis of, 315
  See also Hemithioaminals
S,N-Acetals, acyclic, synthesis of, 318
S,N-Acetals, N-acyl
  from
    α-chloroalkylamides, 327
    chloromethyl phenyl sulfide, 329
  synthesis of, 325
S,N-Acetals, ketene
  from
    alkynes, 905
    bis(1,1-alkylthio)alkenes, 899
    1-chloro enamines, 904
    nitriles, 905
    thioamides, 899
    α,β-unsaturated thioamides, 901
  synthesis of, 898
S,P-Acetals
  chemistry of, 345
  from
    aldehydes, 345
    alkenes, 346
    diazoalkanes, 346
    ketones, 345
    phosphorus ylides, 346
    thiocarbonyl compounds, 345
Te,N-Acetals
  from, iminium salts, 334
  synthesis of, 333
Te,P-Acetals
  from, phosphorus ylides, 348
  synthesis of, 347
Acetamide, N-acetyltrifluoro-, reactions, with chlorophosphite, 1009
Acetamide, N-alkylidene-, reactions, with phosphorocyanatidites, 493
Acetamide, N-chloro-, reactions, with tristrimethylsilylphosphine, 483
Acetamide, cyano-, deprotonation of, 872
Acetamide, 2,2-dicyano-N-phenyl-, reactions, with secondary α-chloro enamines, 792
Acetamide, N,N-dimethyl-, reactions, with ethyl chloroformate, 882
Acetamide, N,N-dimethylbis(trimethylgermyl)-, synthesis of, 625, 628
Acetamide, N-ethyl-, oxidation of, 303
Acetamide, N-(hydroxymethyl)-, synthesis of, 296
Acetamides, α-bromo-, synthesis of, 99

Acetamides

Acetamides, dicyano-, as starting materials, for ketene aminals, 977
Acetamides, trichloro-, reactions, with tributylphosphine, 791
Acetanilides, α-mercapto-, as starting materials, for α-chloroalkanesulfenyl chlorides, 71
Acetate, as starting material, for hydroxymethylsilanes, 352
Acetate, ammonium, reactions, with salicyl aldehyde, 438
Acetate, t-butyl trimethylgermyl, deprotonation of, 663
Acetate, chromyl, for the oxidation of semicarbazones, 441
Acetate, 6-cyano-2-picolyl-N-oxide, oxidation of, 206
Acetate, dimethylaminomethyl, synthesis of, 305
Acetate, ethyl, from, α-fluoroalkyl esters, 52
Acetate, ethyl aceto-, dichlorination of, 12
Acetate, ethyl bromo-
    reactions, with tris(cyclopentyl)borane, 832
    as starting material, for ketene silyl acetals, 831
Acetate, ethyl cyano-
    reactions
        with carbonyl sulfide, 833
        with ethanethiol, 905
        with isobutylene oxide, 883
Acetate, ethyl diazo-
    reactions
        with benzenetellurenyl bromide, 93
        with dihaloplatinum complexes, 151
        with N-phenylselenophthalimide, 335
Acetate, ethyl dicyano-, reactions, with hydrogen chloride, 792
Acetate, isopropenyl
    reactions
        with alcohols, 188
        with malonic acid compounds, 204
Acetate, isopropyl, reactions, with isopropoxymethanetriamines, 973
Acetate, mercuric
    reactions
        with longifolene, 748
        with 1-propenylmercury acetate, 697
    as reagent, in complete mercuration of cyclopentadiene, 697
Acetate, methoxymethyl, from, dimethoxymethane, 211
Acetate, methyl 2-chloro-2-cyclopropylidene
    reactions
        with diazomethane, 109
        with diphenyldiazomethane, 109
Acetate, methyl chlorodifluoro-, decomposition of, 737
Acetate, methyl cyano-, as starting material, for ketene dithioacetals, 843
Acetate, methyl dichlorofluoro-, reactions, with sodium hydride, 752
Acetate, phenyl-
    reactions
        with phosphorus pentachloride, 766
        with phosphorus trichloride, 742
Acetate, trichlorovinyl dichloro-, from, trichloroacetyl chloride, 768
Acetate, trimethyl orthophenyl-, elimination of methanol from, 825
Acetate, trimethylsilyl (trimethylsilyloxy), reactions, with carbonyl compounds, 213
Acetate, vinyl
    reactions
        with alcohols, 188
        with phosphorus pentachloride, 766
Acetate compounds, arylmethyl, oxidation of, 206
Acetate compounds, propargylic, reactions, with acetic acid, 206
Acetates, azido fluoro-, synthesis of, 109
Acetates, 1-bromocyclopropyl, synthesis of, 57
Acetates, diazo-, photolysis of, 385
Acetates, dicyano-, as starting materials, for ketene aminals, 977

Acetates, enol
    as starting materials, for acetals, 188
    synthesis of, 915
Acetates, fluorinated enol, synthesis of, 768
Acetates, α-fluoroalkyl, synthesis of, 52
Acetates, iododifluoro-, reactions, with trimethyl(vinyl)silanes, 139
Acetates, ketene silyl
    reactions
        with bromotrichloromethane, 743
        with tetrachloromethane, 743
Acetates, phosphonofluoro-, synthesis of, 131
Acetates, vinyl
    reactions
        with arenesulfenyl halides, 56
        with benzeneselenenyl fluoride, 52
        with bromine, 57
    as starting materials
        for α-bromoalkyl esters, 57
        for α-fluoroalkyl esters, 51
O,N-Acetates, ketene, synthesis of, 881
Acetic acid
    reactions
        with bis(methylseleno)alkynes, 841
        with chlorosulfuric acid, 282
        with propargylic acetate compounds, 206
        with 1-selenoalkynes, 841
        with silirenes, 1051
        with 1-telluroalkynes, 841
Acetic acid, 2,2-bis(trifluoromethylthio)-, oxidation and dehydration of, 867
Acetic acid, α-chlorophenyl-, as starting material, for Grignard reagents, 147
Acetic acid, dichloro-
    from
        chloral hydrate, 14
        trichloroacetic acid, 13
Acetic acid, peroxytrifluoro-, reactions, with dienes, 46
Acetic acid, phenyl-, as starting material, for 1,1'-dibromobenzyl sulfone, 83
Acetic acid, trichloro-
    reactions, with ortho esters, 1081
    as starting material, for dichloroacetic acid, 13
    use in the synthesis of acetals, 195
Acetic acid, trifluoro- (TFA)
    reactions
        with alkylidene-1,3-dithianes, 252
        with ortho esters, 1081
    as starting material, for cobaltacyclopentadiene complexes, 689
Acetic acid ethyl ester, benzoylfluoro-, synthesis of, 52
Acetic acids, α-alkyl-α-sulfonyl-, synthesis of, 81
Acetic acids, aryl-, as starting materials, for 2-chloro-2-tosylazostyrene compounds, 799
Acetic acids, α-aryl-α-sulfonyl-, synthesis of, 81
Acetic anhydride
    reactions
        with chlorodiethylphosphine, 911
        with dialkyl phosphites, 921
        with 1-ethoxyisochroman, 212
        with hexafluoroacetone imine, 304
        with 1-methoxy-2-benzopyran-4-one, 212
        with tellurium tetrachloride, 291
        with ylides, 993
    use in the synthesis of 1-aminoalkylphosphonic acids, 470
Acetic anhydride, diphenyl-, reactions, with diphenylketene, 829
Acetimidoyl chlorides, 2,2,2-trihalo-, as starting materials, for *gem*-phosphorylamino phosphorylalkenes, 1008
Acetone
    reactions
        with carbenoids, 146
        with 1-thioglycerol, 229

Acetone, 3,3-dibromo-1,1,1-trifluoro-, reactions, with sodium acetate, 161
Acetone, dichlorotetrafluoro-, reactions, with ethanol, 164
Acetone, 1,3-dichloro-1,1,3,3-tetrafluoro-, reactions, with arenesulfinamides, 311
Acetone, hexafluoro-
  reactions
    with ammonia, 297
    with arsines, 344
    with hydrogen sulfide, 215
    with phosphines, 337
Acetone, nitro-, as starting material, for 2-chloro-2-nitro enamines, 796
Acetone, 1,1,1-trifluoro-
  reactions, with 2-chloroethanol, 164
  as starting material
    for α-bromo alcohols, 42
    for Michael acceptors, 766
Acetonitrile
  as solvent, for the synthesis of ketene aminals, 971
  use in Wittig reactions, 738
Acetonitrile, α-lithiotrimethylgermyl-, stability of, 662
Acetonitrile, monolithio-, reactions, with trimethylsilyl chloride, 604
Acetonitrile, trichloro-, reactions, with propargylic alcohols, 1014
Acetonitrile, trimethylsilyl- (TMS)
  deprotonation of, 648
  oxasilylation of, 358
  synthesis of, 604
Acetonitriles, reactions, with silyloxiranes, 519
Acetonitriles, amino-, lithiation–germylation of, 521
Acetophenone, 2,2-dimorpholino-, reactions, with hydrochloric acid, 309
Acetophenones
  cyclization in, 1079
  reactions, with isothioureas, 980
  as starting materials, for aminals, 406
Acetophenones, α-thio-α-sulfinyl, from, esters, 857
Acetophenones, trifluoro-, synthesis of, 732
Acetoxy groups, reactions, with phosphorus pentachloride, 766
Acetyl chloride
  reactions
    with acetals, 175
    with arsenic(III) oxide, 592
    with imines, 100
    with keto diacids, 204
    with oxaplatinacyclobutane complexes, 151
    with phenylbis(diethyldiamino)methane, 455
    with N-(1-piperidinylmethyl)benzamide, 99
    with stannyl acetals, 388
Acetyl chloride, chloro-
  reactions
    with imines, 100
    with tri(2-chloroethyl) phosphate, 768
Acetyl chloride, diphenyl-, reactions, with triethyl phosphite, 920
Acetyl chloride, fluoro-, reactions, with dimethylphosgeniminium chloride, 794
Acetyl chloride, trichloro-, as starting material, for trichlorovinyl dichloroacetate, 768
Acetyl chlorides, tribromo-, as starting materials, for zinc enolates, 768
Acetyl chlorides, trichloro-, as starting materials, for zinc enolates, 768
Acetylenes *See* Alkynes
Acid anhydrides
  reactions, with bis(dimethylamino)methane, 309
  as starting materials, for bismercurated carboxylic acids, 697
Acid catalysis, for the synthesis of acetals, 186

Acid chlorides
  reactions
    with acetals, 48
    with dimercaptoalkanes, 1073
    with 2-hydroxythiophenol, 1073
  as starting materials
    for S,N-acetals, 328
    for ketene dithioacetals, 846
  thiolation–elimination reactions of, 846
  use in the synthesis of α-chloroalkyl esters, 53
Acid chlorides, α-alkoxy
  decarbonylation of, 49
  as starting materials, for α-chloro ethers, 49
Acid chlorides, α-aryloxy
  decarbonylation of, 49
  as starting materials, for α-chloro ethers, 49
Acid chlorides, β-keto, synthesis of, theoretical, 768
Acid dichlorides, phosphonic, alcoholysis of, 131
Acid diesters, phosphonothioic
  reactions
    with ammonia, 469
    with carbonyls, 469
Acid esters, alkyl chloromethylphosphinic, synthesis of, 126
Acid esters, 2-aryl acetic, oxidation of, 197
Acid esters, aryl chloromethylphosphinic, synthesis of, 126
Acid esters, N,N-bis(diphenylarsinomethyl) amino, synthesis of, via Mannich reactions, 500
Acid esters, α-chlorophosphonic, synthesis of, 131
Acid esters, diboronic, as reagents, for the synthesis of carbon-bridged dimercury complexes, 696
Acid esters, α-halophosphinic, synthesis of, 125
Acid esters, α-halophosphonic, synthesis of, 131
Acid esters, phosphinic, reactions, with nucleophilic halogen, 125
Acid esters, sulfonic, synthesis of, 86
Acid esters, vinyl hydroxamic, synthesis of, 833
Acid halides
  as electrophiles, for dithiocarboxylic enethiolates, 842
  reactions, with N-aminomethylamines, 99
Acid halides, α-halo, as starting materials, for α-phosphorylated enol phosphates, 920
Acridine, reactions, with dimethylketene, 886
Acrolein
  acyclic monothioacetals of, 218
  unsymmetrical acetals of, 199
Acrylate, α-cyano-, reactions, with enol ethers, 827
Acrylate, methyl (E)-β-aziridino-β-bromo-, synthesis of, 793
Acrylate, methyl 3-benzyloxyamino-2-cyano-3-(2-hydroxyethoxy)-, synthesis of, 896
Acrylates, β-alkoxy-, reactions, with alcohols, 195
Acrylates, 3,3-dichloro-, synthesis of, 743
Acrylate salts, (E)-3-bromo-3-lithio-, synthesis of, 818
Acrylic acid, perfluoromethyl-, reactions, with alcohols, 894
Acrylic acids, phosphinico-, from, phosphinic chlorides, 124
Acrylic acids, 2-trifluoromethyl-3-alkoxy-3-fluoro-, as intermediates, in the synthesis of ketene hemiaminals, 894
Acrylonitrile, reactions, with hydroxyalkylphosphorus compounds, 342
Acrylonitrile, 3-benzylamino-3-ethoxy-, synthesis of, 896
Acrylonitrile, β,β-dichloro-, reactions, with alcohols, 762
Acrylonitrile, 1,2,2-trichloro-, reactions, with sodium methoxide, 762
Acrylonitriles, reactions, with sodium toluene-p-sulfinates, 783
Acrylonitriles, dichloro-, reactions, with trifluoromethylthiocopper, 771
Acryloyl chlorides, synthesis of, 125
Acryloyl fluoride, pentafluoromethyl-, synthesis of, 736
Actinium chloride, reactions, with oxaplatina(IV) cyclobutanes, 151
Acylals, 2-(phenylsulfinyl)-, from, phenylvinyl sulfoxides, 207

Acylamines

Acylamines
  reactions, with alkylnitramines, 447
  See also Amides
Acyl anion equivalents, use in the synthesis of acetals, 197
Acyl anions, chiral, as reagents, 269
Acylating agents
  reactions, with acyl phosphonates, 920
  for the synthesis of diacetates, 204
Acylation
  of α-silylimines, 514
  for the synthesis of tricoordinate phosphorus functions, 481
Acyl chlorides, reactions, with 1,3,5-hexatriazines, 99
Acyl chlorides, α-chloro-α-(chlorosulfenyl)-, synthesis of, 71
N-Acyl compounds, tautomerism of, 898
Acyl fluorides, perfluoro-, as starting materials, for α-fluoroalkyl esters, 52
Addition, electrophilic, for the synthesis of α-carbocations of alkyne-bridged dimolybdenum complexes, 674
Addition–elimination, for the synthesis of selenium compounds, 785
A-frame platinum complexes, reactions, with diazomethane, 691
Alane complexes, reactions, with bis(trifluorovinyl)mercury, 815
Alanes, 1,1-bis(boryl)-
  from
    1,1-diborylalkenes, 634
    vinylboranes, 631
  synthesis of, 629
    via changing the groups on the central methylene, 634
    via changing the ligands on boron, 634
    via hydroboration of alkynes, 630
    via hydroboration of vinylboranes, 631
    via quenching a carbanion with a boryl electrophile, 632
Alanes, α-silylmethyl-, synthesis of, 659
Alcohol, propargyl, reactions, with Grignard reagents, 1065
Alcohol complexes, propargyl, protonation of, 688
Alcohols
  from, racemates, 390
  protection of as acetals, 186, 198
  reactions
    with acetals, 184
    with acylimines, 304
    with aldehydes, 177
    with β-alkoxy acrylates, 195
    with alkynes, 194
    with 1,1-bis(diethylamino)ethylene, 888
    with α-bromo-β-selenenyl ethers, 51
    with 3-chloro-1-methoxypropene, 199
    with diazoalkanes, 198
    with α-diazo alkyl phosphonates, 913
    with dibenzoyl peroxide, 220
    with β,β-dichloroacrylonitrile, 762
    with dichlorocarbenes, 767
    with diethylaminosulfur trifluoride, 97
    with dimethylmethyleneammonium chloride, 305
    with 2-diphenylacetoxytetrahydrofuranyls, 200
    with diphenyldiazomethane, 198
    with enol ethers, 186
    with α-haloalkanesulfonyl halides, 86
    with α-haloalkyl carboxylate compounds, 209
    with α-halo ethers, 198
    with α-halo sulfides, 218
    with hydroxy furanone compounds, 211
    with imidoyl chlorides, 312
    with isopropenyl acetate, 188
    with ketones, 177
    with methanesulfonyl chloride, 780
    with nitriles, 883
    with perfluoromethylacrylic acid, 894
    with silirenes, 1051
    with silyl enol ethers, 188
    with thionyl chloride, 97
    with thiophosphinic chlorides, 128
    with vinyl acetate, 188
    with vinyl azides, 894
    with ynamines, 885
  as reagents, for the synthesis of carbon-bridged dialuminum compounds, 699
  as starting materials
    for acetals, 186
    for acyl alkyl acetals, 209
    for α-chloro ethers, 47
  tetrahydropyranylation of, 186
Alcohols, aliphatic, reactions, with 1-alkoxyalkynes, 828
Alcohols, allyl, reactions, with enol ethers, 188
Alcohols, allylic, from, chalcone, 269
Alcohols, amido
  as intermediates, in the synthesis of methylenebisamides, 424
  reactions, with nitriles, 425
Alcohols, amino, as starting materials, for *O,N*-acetals, 301
Alcohols, α-azido, synthesis of, 313
Alcohols, benzyl
  deprotonation of, 387
  reactions, with xenon difluoride, 45
Alcohols, α-bromo
  from
    aldehydes, 42
    levoglucosenone, 42
    1,1,1-trifluoroacetone, 42
    1,1,1-trifluoropentane-2,4-dione, 42
Alcohols, butenolide, as starting materials, for α-chloroalkyl esters, 54
Alcohols, α-chloro, from, 2-hydroxy 1,3-diketones, 43
Alcohols, 2,3-difluoroallylic, as starting materials, for α-fluoro epoxides, 45
Alcohols, erythro, synthesis of, 343
Alcohols, α-fluoro, stability of, 42
Alcohols, α-halo
  stability of, 41
  synthesis of, 41
Alcohols, homoallylic, reactions, with potassium permanganate, 168
Alcohols, α-iodo, from, 4,4-diiodo-1,1-dimethyl-1,4-dihydroquinolinium cation, 43
Alcohols, mercapto
  reactions
    with aldehydes, 229
    with ketones, 229
Alcohols, methoxy, synthesis of, 343
Alcohols, nitro, synthesis of, 906
Alcohols, phthalide, as starting materials, for α-chloroalkyl esters, 54
Alcohols, propargyl, as starting materials, for *gem*-halophosphorus allenes, 802
Alcohols, propargylic
  reactions
    with sodium bis(2-methoxyethoxy)aluminum hydride, 948
    with trichloroacetonitrile, 1014
  as starting materials, for α-silyl enamines, 1014
Alcohols, stannyl, stability of, 386
Aldazines
  reactions
    with ethyl phosphonites, 460
    with iodine compounds, 198
  as starting materials
    for α-aminophosphonic acids, 460
    for α-hydrazinophosphonic acids, 460
Aldazines, aryl, reactions, with dialkyl phosphites, 460
Aldehyde, salicyl
  as electrophile, for dithiocarboxylic enethiolates, 842
  reactions
    with ammonium acetate, 438

with bisiminoalkanes, 438
with 4-dimethylamino benzaldehyde, 438
with enamines, 307
with *N*-styrylmorpholine, 307
Aldehyde 1,1-diols, α-keto, synthesis of, 161
Aldehydes
  acetalization of, 178
  bromoacylation of, 56
  chloroacylation of, 53
  α,α-dichlorination of, 12
  dithioacetalization of, 247
  C-germylation of, 360
  haloalkylation of, 47
  reactions
    with alcohols, 177
    with α-alkoxylead compounds, 390
    with amines, 294, 405
    with (ω-aminoalkyl)phenylphosphines, 453
    with aminosilanes, 509
    with 2-aminothiophenol, 318
    with ammonia, 294
    with ammoniopropanedinitrile toluene-*p*-sulfonate, 883
    with arenesulfonylmethyl isocyanides, 908
    with arylgermyllithiums, 360
    with benzotriazole, 296
    with benzyloxynitromethane, 897
    with bisdiphenylthiophosphinoylmethane, 1022
    with bis-α-hydroxyalkylphosphines, 341
    with bis(methylene)sulfurane, 231
    with bis(trimethylsilyl)bromomethyllithium, 805
    with bis(trimethylsilyl)methyllithium compounds, 936
    with bromonitromethane, 798
    with chloromethyl phenyl sulfoxide, 776
    with chloromethylphosphinates, 126
    with chlorophosphines, 119, 121
    with Clayfen, 431
    with cyclopentadienyltantalum complexes, 355
    with diethyl benzenesulfonylmethyl phosphonate, 925
    with diethyl malonate, 559
    with dimethylphenylsilyllithium, 355
    with diphosphonites, 564
    with 1,3-dipoles, 183
    with disulfonate esters, 864
    with epoxides, 181
    with fluoromethylsulfoximine, 86
    with halonitromethanes, 798
    with halophosphines, 123
    with β-hydroxycarboxylic acids, 213
    with hydroxylamines, 435
    with (3-hydroxypropyl)phenylphosphines, 337
    with α-lithiated amines, 528
    with lithium amides, 299
    with mercaptoalcohols, 229
    with nitro(phenylseleno)methane, 910
    with nitro(phenylthio)methane, 906
    with organoantimony alkoxides, 198
    with ortho esters, 190
    with orthoformates, 190
    with (3-oxoalkyl)phenylphosphines, 454
    with phenyl trimethylsilyl selenide, 239
    with phosgene, 58
    with phosphoramidites, 475
    with phosphorus trichloride, 130
    with silirenes, 1052
    with silyl ethers, 181
    with sodium bisulfite, 237
    with sodium dithionite, 234
    with α-sulfinyl carbanions, 77
    with *N*-sulfinylsulfonamides, 311
    with sulfonamides, 423
    with tetraalkyllead compounds, 390
    with tetrabromomethane, 745
    with tetraiodomethane, 749
    with α-thiol carboxylic acids, 230
    with thiols, 247
    with tributylstannylmagnesium chloride, 386
    with trimethylgermyldimethylphenylsilane, 361
    with (trimethylsilyl)dichloromethyllithium, 742
    with trimethylsilyl sulfides, 216
    with triphenylphosphine, 745, 749
  reduction of, 699
  silyl anion addition to, 354
  C-stannylation of, 386
  as starting materials
    for acetals, 177
    for *O,N*-acetals, 311
    for *Se,N*-acetals, 333
    for *S,N*-acetals, 315, 328
    for *S,P*-acetals, 345
    for acyl alkyl acetals, 212
    for aminals, 405
    for α-aminoalkyl phosphonates, 466
    for α-bromo alcohols, 42
    for bromo dithiacetals, 245
    for α-bromostannanes, 155
    for α-bromo sulfides, 71
    for α-chloro ethers, 47
    for α-chloro sulfides, 69
    for *gem*-diazidoalkanes, 432
    for dibromoalkanes, 25
    for dichloroalkanes, 18
    for difluoroalkanes, 8
    for diselenoacetals, 287
    for dithioacetals, 245
    for *gem*-formamido phosphorylalkenes, 1004
    for geminal diacetates, 203
    for hemiacetals, 164
    for hemiaminals, 294
    for hemidithioacetals, 245
    for hemithioaminals, 324
    for α-iodostannanes, 155
    for ketene monothioacetals, 834
    for monothioacetals, 217
    for phosphorus compounds, 336
    for α-phosphoryl enamines, 1002
    for silylenol ethers, 1049
    for symmetrical acetals, 190
    for thiazolidines, 318
  use in the synthesis of 1,3-azarsa compounds, 495
  use in the synthesis of 1,3-benzazarsolines, 496
Aldehydes, aliphatic, as starting materials, for ketene dithioacetal *S,S*-dioxides, 858
Aldehydes, aryl, reactions, with hexamethylphosphorus triamide, 475
Aldehydes, β-chlorovinyl, reactions, with ethylene glycols, 196
Aldehydes, α-halo-, synthesis of, 132
Aldehydes, *gem*-halo phosphonamido, synthesis of, 804
Aldehydes, α-hydroxy, synthesis of, 231
Aldehydes, keto, reactions, with 3-methoxyphenylmagnesium bromide, 166
Aldehydes, 1,3-keto, synthesis of, 250
Aldehydes, α-keto, reactions, with thiols, 215
Aldehydes, α-oxo, synthesis of, 162
Aldehydes, α,β-unsaturated, acetalization of, 181
4-Aldehydes, pyrrolidone-, reactions, with arenesulfonylmethyl isocyanides, 908
Aldimines
  reactions
    with bis(*p*-methylbenzyl) hydrogen phosphite, 461
    with (chloromethyl)phosphonic diesters, 473
    with dialkyl phosphites, 462
    with dialkyl phosphoriodites, 462
    with diethyl phosphite, 461
    with germanium hydrides, 521
    with (3-oxoalkyl)phenylphosphines, 454

Aldimine salts

with phosphorus trichloride, 488
as starting materials, for α-aminophosphonic acids, 461
Aldimine salts, reactions, with diethyl phosphite, 464
Alkali arsenes
  reactions, with ammonium salts, 499
  use in the synthesis of aminomethylarsines, 501
Alkali metal compounds, trimethylsilyl, synthesis of, 354
Alkali metals, reactions, with bis(trimethylsilyl)mercury, 354
Alkali metal salts, reactions, with dihalomethanes, 147
Alkanamine, N-trityl, reactions, with
    bis(trimethylsilyl)phosphonites, 461
Alkane, diazo-
  reactions
    with cobalt, 686
    with rhodium, 686
  as starting material, for carbon-bridged dicobalt
      complexes, 685
  use in the synthesis of methylene-bridged manganese
      complexes, 676
Alkane compounds, azo-, synthesis of, 440
Alkane compounds, hydrazino-, synthesis of, 440
Alkane compounds, isocyanato-, synthesis of, 441
Alkane compounds, nitramino-, synthesis of, 440
Alkane compounds, nitro-, synthesis of, 439
Alkanes
  bromination of, 19
  chlorination of, 11
  halogenation of, 2
  iodination of, 28
  as starting materials
    for dibromoalkanes, 19
    for dichloroalkanes, 11
    for difluoroalkanes, 2
    for diiodoalkanes, 28
    for *gem*-dinitroalkenes, 993
Alkanes, acylaminoimino-, synthesis of, 449
Alkanes, acylaminomethylazo-, synthesis of, 448
Alkanes, acylaminomethylimino-, synthesis of, 448
Alkanes, α-aluminosilyl-
  from
    organolithium compounds, 659
    organomagnesium compounds, 659
  synthesis of, 659
Alkanes, β-amino dichloro-, as starting materials, for ketene
    aminals, 979
Alkanes, 1-azido 1-methoxy-, synthesis of, 313
Alkanes, 1-(benzotriazol-1-yl)-1-chloro-, displacement
    of, 308
Alkanes, α,ω-bis(bisdialkylaminomethylphosphine)-,
    synthesis of, 452
Alkanes, 1,1-bis(boryl)-
  from, alkynes, 630
  See also Alkanes, *gem*-diboryl-
Alkanes, 1,1-bisgermyl-
  from
    carbon suboxide, 628
    digermenes, 627
    germenes, 627
    ketenes, 628
  synthesis of, 625
    via changing the groups attached to the central
        methylene, 628
    via changing the ligands on germanium, 629
    via oxidative addition, 628
    via quenching a carbanion with a germyl
        electrophile, 625
    via quenching a germyl anion with a carbon
        electrophile, 626
    via rearrangements, 628
Alkanes, bisimino-, reactions, with salicyl aldehyde, 438
Alkanes, 1,1-bis(silyl)-
  from
    bis(silyl)ketenes, 619

di-Grignard reagents, 608
disilanes, 614
disilenes, 609
1,1-disilylated epoxides, 619
1,2-disilylethanes, 616
α-lithiosilylalkanes, 602
α-magnesiosilylalkanes, 606
α-metallosilylalkanes, 602
silenes, 609
1,1,1-tris(trimethylsilyl)alkanes, 619
  as starting materials, for α-lithio compounds, 653
  synthesis of, 602
    via changing the groups attached to the central
        methylene, 620
    via changing the ligands on silicon, 621
    via Diels–Alder reactions, 611
    via hydrosilylation, 617
    via quenching of carbanions with silyl electrophiles, 602
    via quenching silyl anions with carbon electrophiles, 609
    via rearrangements, 614
Alkanes, 1,2-bis(silyl)-, from, vinylsilanes, 617
Alkanes, 1,1-bis(silyl)-1-lithio-, synthesis of, 618
Alkanes, α-borylgermyl-, synthesis of, 646
Alkanes, α-borylsilyl-
  from
    1,3-disilacyclobutanes, 644
    vinylsilanes, 643
  synthesis of, 642
    via changing the groups attached to the central
        methylene, 645
    via changing the groups attached to the metalloids, 645
    via hydroboration of vinylsilanes, 643
    via quenching a carbanion with a boryl electrophile, 642
    via quenching a carbanion with a silyl electrophile, 642
    via rearrangements, 642
Alkanes, bromo-, from, carboxylic acids, 1077
Alkanes, bromochloro-
  from
    bromochlorocarbene, 38
    carboxylic acids, 38
    haloalkanes, 37
    haloalkenes, 38
    ketones, 38
  synthesis of, 37
Alkanes, bromofluoro-
  from
    bromofluorocarbene, 35
    carboxylic acids, 35
    haloalkanes, 33
    haloalkenes, 34
  synthesis of, 33
Alkanes, bromo iodo-, synthesis of, 40
Alkanes, chloro-, from, carboxylic acids, 1077
Alkanes, α-chloroazo-, synthesis of, 109
Alkanes, chlorofluoro-
  from
    alkenes, 30
    carboxylic acids, 33
    chlorofluoroalkenes, 32
    chlorofluorocarbene, 32
    haloalkanes, 30
    imines, 32
  synthesis of, 30
Alkanes, chloroiodo-
  from
    carboxylic acids, 40
    haloalkanes, 39
    haloalkenes, 39
    ketones, 40
  synthesis of, 39
Alkanes, *gem*-chloronitroso-, reactions, with iodine, 40
Alkanes, 1,1-diacetoxy-, synthesis of, 203
Alkanes, dialkoxydialkylamino-, as intermediates, in the
    synthesis of ketene hemiaminals, 881

Alkanes, N^1,N^2-dialkylidene 1,1-diamino-, reactions, with
    dialkyl phosphites, 461
Alkanes, *gem*-diazido-
    from
        aldehydes, 432
        gem-dihaloalkanes, 432
        ketones, 432
        trimethylsilyl azide, 432
    synthesis of, 432
Alkanes, diazo-
    as carbene sources, 669
    reactions
        with alcohols, 198
        with bromochlorocarbene, 759
        with carbenes, 755, 759
        with chlorophosphines, 114
        with haloalkenes, 109
        with halo metal complexes, 147
        with iodine, 29
        with monomeric carbonyls, 686
        with phosphaalkenes, 484
        with selenones, 333
        with thioketones, 328
    as starting materials
        for *Se,N*-acetals, 335
        for *S,P*-acetals, 346
        for carbon-bridged dimetal compounds, 709
        for carbon-bridged dirhodium complexes, 686
        for difluoroalkanes, 10
    for the synthesis of phosphorus-containing
        heterocycles, 488
Alkanes, *gem*-diazo-, synthesis of, 436
Alkanes, diazo nitro-, synthesis of, 439
Alkanes, *gem*-diboryl-
    reactions, with organolithium compounds, 665
    as starting materials, for α-lithioborylalkanes, 664
    See also Alkanes, 1,1-bisboryl-
Alkanes, 1,1-diboryl-1-metallo-, synthesis of, 633
Alkanes, 1,2-diboryl-1-silyl-, synthesis of, via
    borylboration, 644
Alkanes, dibromo-
    from
        aldehydes, 25
        alkanes, 19
        alkenes, 23
        alkynes, 24
        amines, 27
        carboxylic acids, 27
        dibromocarbene, 24
        dihaloalkanes, 22
        hydrazones, 27
        imines, 27
        ketones, 25
        trihaloalkanes, 23
    reactions, with mercuric fluoride, 4
    synthesis of, 19
Alkanes, *gem*-dibromo-, as starting materials, for ketene
    diselenoacetals, 875
Alkanes, dichloro-
    from
        aldehydes, 18
        alkanes, 11
        alkenes, 14
        alkynes, 15
        amines, 19
        dichlorocarbene, 16
        dihaloalkanes, 13
        imines, 19
        ketones, 18
        trihaloalkanes, 13
    as starting materials, for α-chlorophosphonic
        dichlorides, 130
    synthesis of, 11

Alkanes, difluoro-
    from
        aldehydes, 8
        alkanes, 2
        alkenes, 5
        alkynes, 6
        diazoalkanes, 10
        difluorocarbene, 7
        dihaloalkanes, 3
        imines, 10
        ketones, 8
        trihaloalkanes, 5
    synthesis of, 2
Alkanes, *gem*-difluoroamino-
    safety precautions, 439
    synthesis of, 439
Alkanes, dihalo-
    reactions, with trimethyltinlithium, 699
    as starting materials
        for dibromoalkanes, 22
        for dichloroalkanes, 13
        for difluoroalkanes, 3
    synthesis of, **1–40**
Alkanes, 1,1-dihalo-
    Arbuzov reactions of, 555
    reactions
        with 1,2-dimetallodiphosphines, 545
        with metallated cyclopolyphosphines, 545
        with metal phosphides, 545
        with phosphines, 546
        with phosphorus nucleophiles, 553
        with platinumphosphine complexes, 584
    as starting materials, for bis(phosphonium salts), 553
Alkanes, *gem*-dihalo-
    reactions, with lithium, 667
    as starting materials
        for α-chlorophosphonic dihalides, 130
        for *gem*-diazidoalkanes, 432
Alkanes, *gem*-dihydrazino-, synthesis of, 436
Alkanes, *gem*-dihydroxylamino-, synthesis of, 435
Alkanes, *gem*-diimino-
    reactions, with aralkyl ketones, 445
    synthesis of, 437
Alkanes, diiodo-
    from
        alkanes, 28
        alkynes, 29
        diiodocarbene, 29
        haloalkanes, 28
        imines, 29
    as starting materials, for carbon-bridged dicobalt
        complexes, 686
    synthesis of, 28
Alkanes, α,ω-diiodo-, reactions, with
    trimethylsilylalkyne, 810
Alkanes, *gem*-diisocyanato-
    reactions
        with phosphorus ylides, 992
        with triphenylphosphine imides, 441
    as starting materials, for *gem*-alkylideneamino
        isocyanatoalkenes, 992
Alkanes, 1,1-dilithio-, synthesis of, 667
Alkanes, 1,1-dimetallo-
    reactions
        with chlorophosphine oxides, 556
        with chlorophosphines, 546
Alkanes, *gem*-dinitro-
    from
        halo nitroalkanes, 432
        nitroalkenes, 431
        1,1,1-trinitroethane, 431
    synthesis of, 428
Alkanes, *gem*-dinitrosamino-, oxidation of, 440

Alkanes

Alkanes, *gem*-dinitroso-, synthesis of, 428
Alkanes, dithiol-
  reactions
    with acid chlorides, 1073
    with ethanoic anhydrides, 1074
Alkanes, fluoro-, chlorination of, 30
Alkanes, fluorohalo-, synthesis of, 30
Alkanes, fluoroiodo-
  from
    carboxylic acids, 37
    fluoroiodocarbene, 36
    haloalkanes, 36
    haloalkenes, 36
  synthesis of, 36
Alkanes, α-galliosilyl-, synthesis of, 659
Alkanes, geminal diamino-, synthesis of, 404
Alkanes, 1-germanyl 1-silyl-, as starting materials, for 1-germanyl 1-silylalkenes, 1058
Alkanes, 1-germyl-1-silyl-
  from
    disilenes, 639
    silenes, 639
  synthesis of, 636, 639
Alkanes, α-germylsilyl-
  synthesis of, 636
    via changing the groups attached to the metalloids, 640
    via changing the groups attached to the methylene, 640
    via hydrogermylation, 639
    via quenching a carbanion with a germyl electrophile, 636
    via quenching a carbanion with a silyl electrophile, 637
    via quenching a metalloid anion with a carbon electrophile, 638
    via rearrangements, 639
Alkanes, halo-
  as starting materials
    for bromochloroalkanes, 37
    for bromofluoroalkanes, 33
    for chlorofluoroalkanes, 30
    for chloroiodoalkanes, 39
    for diiodoalkanes, 28
    for fluoroiodoalkanes, 36
Alkanes, α-halo azo-
  reactions, with amines, 444
  synthesis of, 109
Alkanes, (halomethyl)silyl-
  reactions
    with tin, 660
    with tin dichloride, 660
Alkanes, halonitro-, as starting materials, for *gem*-dinitroalkanes, 432
Alkanes, α-halonitro-, as starting materials, for α-nitro sulfones, 332
Alkanes, α-hydroxy bromofluoro-, synthesis of, 33
Alkanes, *gem*-isocyanato phosphoryl-, as starting materials, for *gem*-phosphorodiamidoyl ureidoalkenes, 1011
Alkanes, α-keto monofluoro-, bromination of, 33
Alkanes, 1-lithio-1-aryl-1-silyl-, reactions, with tin halides, 660
Alkanes, α-lithioboryl-
  from, *gem*-diborylalkanes, 664
  synthesis of, via deprotonation, 664
Alkanes, α-lithiogermyl-
  synthesis of, 626
    via deprotonation, 662
    via halogen–lithium exchange, 662
Alkanes, 1-lithio-1-silyl-, synthesis of, via deprotonation, 646
Alkanes, α-lithiosilyl-
  as starting materials
    for 1,1-bis(silyl)alkanes, 602
    for 1,3-disilacyclobutanes, 605
  synthesis of, 646
    via addition of an organolithium to a vinylsilane, 651
    via deprotonation, 646
    via halogen–lithium exchange, 652
    via sulfur–lithium exchange, 652
    via transmetallation, 652
  transmetallation of, 660
Alkanes, α-magnesiosilyl-
  as starting materials, for 1,1-bis(silyl)alkanes, 606
  synthesis of, 653, 654
    via addition of a Grignard reagent to a vinylsilane, 654
    via halogen–magnesium exchange, 653
    via rearrangements, 655
    via transmetallation, 655
  transmetallation of, 660
Alkanes, α-metallated nitro-, synthesis of, 540
Alkanes, 1-metallo-1,1-bisgermyl-, synthesis of, 628
Alkanes, 1-metallo-1,1-bis(silyl)-, synthesis of, 620
Alkanes, α-metalloboryl-
  reactions, with silyl electrophiles, 642
  synthesis of, 665
Alkanes, α-metallogermyl-
  as intermediates, for the synthesis of α-germylsilylalkanes, 637
  reactions, with silyl electrophiles, 636
  synthesis of, 662
Alkanes, 1-metallo-1-silyl-, synthesis of, 646
Alkanes, α-metallosilyl-
  reactions
    with boryl electrophiles, 642
    with electrophilic complexes, 656
    with germyl electrophiles, 636
  as starting materials, for 1,1-bis(silyl)alkanes, 602
  transmetallation of, 660
Alkanes, nitro-
  of α-anions, 540
  cycloaddition of diazomethane to, 440
  double deprotonation of, 540
  from, alkyl nitrates, 428
  Henry reaction of, 540
  Nef reaction of, 540
  nitration of, 428
  reactions
    with arenediazonium salts, 439
    with silver nitrate, 428
    with sodium nitrite, 428
    with tetranitromethane, 429
  as starting materials, for *gem*-dinitroalkenes, 994
Alkanes, nitronitroso-, synthesis of, 439
Alkanes, perhalofluoro-, reactions, with ylides, 118
Alkanes, 1-phenylthio-1-silyl-
  reactions, with *N*-chlorosuccinimide, 943
  reductive lithiation of, 652
Alkanes, α-plumbylsilyl-, synthesis of, 660
Alkanes, α-potassiosilyl-, synthesis of, 653
Alkanes, α-selenosilyl-, transmetallation of, 652
Alkanes, α-sodiosilyl-, synthesis of, 653
Alkanes, α-stannylsilyl-
  from, 1-sila-1-stanna-1-alkenes, 661
  synthesis of, 660
    via quenching a stannyl anion with a carbon electrophile, 660
    via transmetallation of α-metallostannylalkanes, 660
    via transmetallation reactions, 660
Alkanes, α-thalliosilyl-, synthesis of, 659
Alkanes, trihalo-
  as starting materials
    for dibromoalkanes, 23
    for dichloroalkanes, 13
    for difluoroalkanes, 5
Alkanes, 1,1,2-trisboryl-, synthesis of, 632
Alkanes, tristannyl-, stability of, 1069
Alkanes, 1,1,1-tris(trimethylsilyl)-, as starting materials, for 1,1-bis(silyl)alkanes, 619

Alkanethiols
  Lewis acid-catalyzed elimination of, 844
  reactions, with arenesulfonylalkynes, 861
Alkanimines, α-chloro, as starting materials, for
    α-phosphoryl enamines, 1004
Alkanols, 2-amino-, reactions, with vinyl ethers, 306
Alkanols, dinitro-, as starting materials, for *gem*-
    dinitroalkenes, 994
Alkenamine compounds, use of term, 898
Alkene, 1,1-bis(bromomagnesio)-, from, alkynes, 1065
Alkene, 1,1-bis(trichlorostannyl)-, synthesis of, 1069
Alkene, *gem*-chloro nitro-, synthesis of, 796
Alkene, 1,1-diplumbyl-, synthesis of, 1069
Alkene, 1-magnesio 1-silyl-, from, alkynes, 1060
Alkene, 1-trimethylsilyl 1-trialkylstannyl-, synthesis of, 1062
Alkene exchange, *gem*-dibromo-, kinetic vs. thermodynamic
    control of, 818
Alkenes
  acetalization of, 193
  alkoxymercuration of, 193
  cyclopropanation of, 152
  double silylation of, 618
  from
    bis(sulfones), 275
    disilanes, 1054
    sulfones, 840
  mercuration of, 1067
  nitration of, 797
  reactions
    with arsenic trichloride, 597
    with azides, 515
    with benzeneselenenyl chloride, 91
    with benzenesulfenyl chloride, 106
    with bromomethanesulfonyl bromide, 83
    with chloromethanesulfonyl bromide, 82
    with chloromethanesulfonyl chloride, 82
    with α-halovinyl selenides, 92
    with [methoxy(phenylthio)methyl]trimethylsilane, 370
    with 2-methyl-2-phenylselenomalonitrile, 284
    with nitrenes, 515
    with nitric acid, 797
    with tetrafluorohydrazine, 104
    with thallium(III) nitrate, 194
  as starting materials
    for acetals, 193
    for *O,N*-acetals, 306, 312
    for *Se,N*-acetals, 335
    for *S,P*-acetals, 346
    for chlorofluoroalkanes, 30
    for dibromoalkanes, 23
    for dichloroalkanes, 14
    for difluoroalkanes, 5
    for *gem*-dinitroalkenes, 993
    for α-fluoro esters, 52
    for 2-haloalkenylphosphorus compounds, 802
    for hemithioaminals, 326
  synthesis of, via pyrolysis of silacyclobutanes, 610
  use in the synthesis of hemiacetals, 167
Alkenes, *gem*-acylhydrazino azido-, from, *gem*-
    diazidoalkenes, 991
Alkenes, *gem*-alkoxycarbonylamino phosphoryl-, synthesis
    of, 1006
Alkenes, 2-alkoxy 1,1-dichloro-, synthesis of, 738
Alkenes, 2-alkoxy 1,1-disulfonyl-, reactions, with
    amines, 866
Alkenes, 1-alkoxy-1-halo-, synthesis of, 912
Alkenes, alkoxy nitro-, synthesis of, 898
Alkenes, 2-alkyl 1,1-distannyl-, synthesis of, 1069
Alkenes, *gem*-alkylideneaminophosphoryl-
  from
    N-acylamides, 1009
    gem-arylideneamino chloroalkenes, 1009
    imidoyl chlorides, 1009
  synthesis of, 1009
Alkenes, 1-alkylthio 1-metallo-, halogenation of, 774
Alkenes, 1-alumino 1-silyl-, synthesis of, 1062
Alkenes, *gem*-amido carboxy-, as starting materials, for *gem*-
    amido phosphonioalkenes, 998
Alkenes, *gem*-amidodiphenylphosphinoyl-, from,
    N-tetrachloroethylamides, 1000
Alkenes, *gem*-amidophosphinoyl-
  reactions, with phosphorus pentachloride, 1001
  as starting materials, for *gem*-chloroarylideneamino
      phosphinoylalkenes, 1001
Alkenes, *gem*-amidophosphonoyl-
  from
    acylaminophosphinic esters, 1001
    N-tetrachloroethylamides, 1002
  synthesis of, 1001
Alkenes, *gem*-amidophosphoramidoyl-, from,
    azalactones, 1010
Alkenes, *gem*-amidophosphoryl-
  from
    amides, 1005
    gem-amido phosphorylalkenes, 1006
    azalactones, 1005
    α-oxo phosphonates, 1005
    N-tetrachloroethylamides, 1006
    N-trichloroethylidene amides, 1006
  as starting materials
    for *gem*-amido phosphorylalkenes, 1006
    for *gem*-arylchloromethyleneamino
        dichlorophosphorylalkenes, 1012
  synthesis of, 1005
Alkenes, *gem*-aminobenzylideneamino
    phosphorodiamidoyl-, from, *gem*-
    chlorobenzylideneamino phosphorylalkenes, 1012
Alkenes, *gem*-aminoborylamino silyl-
  from
    amino haloboranes, 1015
    carbiminosilanes, 1015
Alkenes, *gem*-aminochloromethyleneamino chloro-, as
    starting materials, for α-diaminomethyleneamino
    enamines, 988
Alkenes, *gem*-arsino fluoro-, synthesis of, 805
Alkenes, 2-arylamino-1,1-dinitro-, synthesis of, 994
Alkenes, arylazo-, synthesis of, 799
Alkenes, *gem*-arylazo nitro-, from, nitroformaldehyde
    arylhydrazone, 995
Alkenes, *gem*-arylchloromethyleneamino
    dichlorophosphoryl-, from, *gem*-amido
    phosphorylalkenes, 1012
Alkenes, 2-aryl 1,1-distannyl-, synthesis of, 1069
Alkenes, *gem*-arylideneamino chloro-, as starting materials,
    for *gem*-alkylideneamino phosphorylalkenes, 1009
Alkenes, aryl nitro-, as starting materials, for *gem*-
    dinitroalkenes, 993
Alkenes, 1-arylthio 1-metallo-, halogenation of, 774
Alkenes, *gem*-azido halo-, synthesis of, 799
Alkenes, *gem*-azido phosphazido-, as intermediates, 993
Alkenes, benzyloxynitro-, synthesis of, 897
Alkenes, 2,2-bis(acetoxymercurio)-, synthesis of, 1067
Alkenes, bis(1,1-alkylthio)-
  reactions
    with amines, 899
    with sulfodiimines, 899
    with sulfoximines, 899
  as starting materials, for ketene *S,N*-acetals, 899
Alkenes, 1,1-bis(silyl)-
  from, alkynes, 1044
  reactions, with organolithium compounds, 618
Alkenes, 1,2-bis(silyl)-, from, alkynylsilanes, 1044
Alkenes, 1,1-bis(trimethylstannyl)-, transmetallation
    of, 1065
Alkenes, 1-boro 1-germanyl-, synthesis of, 1058
Alkenes, 1-boronyl 1-silyl-, from, alkynes, 1056

Alkenes

Alkenes, 1-boro 1-stannyl-, from, alkynes, 1063
Alkenes, 1-boro-2-(trimethylstannyl)-, synthesis of, 1068
Alkenes, bromo-
  fluorination of, 34
  reactions
    with bromine, 23
    with hydrogen bromide, 23
Alkenes, bromochloro-, synthesis of, 756, 759
Alkenes, 1-bromo-1-chloro-, synthesis of, 758
Alkenes, *gem*-bromo germyl-, synthesis of, 811
Alkenes, bromo iodo-, synthesis of, 759
Alkenes, *gem*-bromo lithio-
  synthesis of, 818
  See also Lithium compounds, α-bromoalkenyl-
Alkenes, α-bromo nitro-, synthesis of, 795
Alkenes, *gem*-bromo silyl-
  synthesis of, 805
  silylation methods, 806
Alkenes, chloro-
  epoxidation of, 50
  reactions
    with arylamines, 987
    with chlorine, 14
    with chlorosilanes, 806
    with fluorine, 31
  as starting materials, for α-chloro epoxides, 50
Alkenes, *gem*-chloroarylideneamino phosphinoyl-, from, *gem*-amido phosphinoylalkenes, 1001
Alkenes, *gem*-chlorobenzylideneamino phosphoryl-, as starting materials, for *gem*-aminobenzylideneamino phosphorodiamidoylalkenes, 1012
Alkenes, (*Z*)-1-chloro 1-bromo-, synthesis of, 756
Alkenes, 1-chloro-1-dialkylamino-
  reactions
    with phenols, 889
    with silver carboxylates, 889
    with sodium alkoxides, 889
Alkenes, chlorofluoro-, as starting materials, for chlorofluoroalkanes, 32
Alkenes, 1-chloro-1-fluoro-, synthesis of, 753, 755
Alkenes, chloro halo-, synthesis of, 756
Alkenes, *gem*-chloro manganio-, as starting materials, for α-manganio enamines, 1019
Alkenes, *gem*-chloro metallo-, synthesis of, 821
Alkenes, α-chloro nitro-, synthesis of, 795
Alkenes, *gem*-chloro nitroso-, from, α-chloro oximes, 798
Alkenes, 1-chloro 1-phenylsulfinyl-, synthesis of, 777
Alkenes, *gem*-chloro silyl-, synthesis of, silylation methods, 806
Alkenes, (*E*)-*gem*-chloro stannyl-, synthesis of, 812
Alkenes, 1,1-dialumino-, from, alkynes, 1068
Alkenes, *vic*-diamino-, halogenation of, 793
Alkenes, *gem*-diaryl-, as starting materials, for *gem*-dinitroalkenes, 993
Alkenes, 2,2-diaryl-1,1-dinitro-, synthesis of, 993
Alkenes, *gem*-diarylphosphinoylamino phosphoryl-, synthesis of, 1008
Alkenes, *gem*-diazido-
  from, *gem*-dichloroalkenes, 993
  as starting materials
    for *gem*-acylhydrazino azidoalkenes, 991
    for α-azido enamines, 989
    for *gem*-diphosphazidoalkenes, 996
    for *gem*-diphosphiminoalkenes, 996
Alkenes, *gem*-diazo-
  from
    diazonium salts, 992
    3-formyl-1,5-diphenylformazan, 992
    γ-ketocarboxylic acids, 992
Alkenes, 1,1-diboro-
  from, alkynes, 1054
  synthesis of, 1054
Alkenes, *gem*-diboro-, as starting materials, for *gem*-bromo boronalkenes, 814

Alkenes, 1,1-diboryl-
  hydrogenation of, 634
  as starting materials, for 1,1-bisborylalkanes, 634
Alkenes, dibromo-, synthesis of, 745
Alkenes, 1,1-dibromo-
  reactions, with butyllithium, 757
  as starting materials
    for 1,1-diiodoalkenes, 750
    for 1,1-distannylalkenes, 1069
Alkenes, *gem*-dibromo-, lithium–bromine exchange of, 793
Alkenes, (*E*)-1,2-dibromo silyl-, synthesis of, 807
Alkenes, dichloro-, reactions, with tetraethylammonium hydrosulfide, 851
Alkenes, 1,1-dichloro-
  reactions, with 1,2-dipotassium-1,2-di-*t*-butyldiphosphide, 1025
  synthesis of, 737
Alkenes, *gem*-dichloro-, as starting materials, for *gem*-diazidoalkenes, 993
Alkenes, 1,1-di(chloromercurio)-, synthesis of, 1067
Alkenes, *gem*-dichlorophosphoryl isocyanato-
  reactions, with antimony trifluoride, 1012
  as starting materials
    for *gem*-difluorophosphoryl isocyanatoalkenes, 1012
    for *gem*-dihalophosphoryl isocyanatoalkenes, 1013
Alkenes, *gem*-dicyano-, as starting materials, for ketene aminals, 979
Alkenes, 2,2-dicyano bis(1,1-alkylthio)-, reactions, with aziridine, 899
Alkenes, difluoro-
  reactions, with acetyl hypobromite, 52
  synthesis of, 730
    via elimination reactions, 730
    via ylide–carbene reactions, 735
Alkenes, *gem*-difluoro-, as starting materials, for *gem*-diisothioureidoalkenes, 992
Alkenes, *gem*-difluorophosphoryl isocyanato-, from, *gem*-dichlorophosphoryl isocyanatoalkenes, 1012
Alkenes, *gem*-difluorophosphoryl ureido-, from, *gem*-isocyanato phosphorylalkenes, 1012
Alkenes, 1,1-digermanyl-, from, alkynes, 1055
Alkenes, 1,1-digermanyl 2-chloro-, synthesis of, 1055
Alkenes, 1,1-dihalo-, synthesis of, 730
Alkenes, *gem*-dihalo-
  amination of, 793
  reactions, with diphenyltrimethylsilylphosphine, 997
  as starting materials
    for ketene acetals, 829
    for ketene dithioacetals, 851
    for α-phosphino enamines, 997
  substitution by alkoxides, 829
  See also Vinyl halides
Alkenes, *gem*-dihalophosphoryl isocyanato-
  from
    gem-dichlorophosphoryl isocyanatoalkenes, 1013
    dihaloethyl isocyanates, 1012
    haloalkyl isocyanates, 1012
    tetrachloroethyl isocyanates, 1013
  synthesis of, 1012
Alkenes, *gem*-dihydrazino-
  from
    azobenzenes, 985
    bis(dialkylamino)-1-methoxymethane, 985
    tricyanoquinodimethanes, 985
  synthesis of, 985
Alkenes, 1,1-diiodo-
  from, 1,1-dibromoalkenes, 750
  synthesis of, 749
Alkenes, *gem*-diiodo-, synthesis of, 813
Alkenes, *gem*-diisothioureido-
  from
    gem-difluoroalkenes, 992

vinylidene difluorides, 992
Alkenes, 1,1-dilithio-, synthesis of, 1064
Alkenes, 1-dimesitylboronyl 1-trimethylsilyl-, synthesis of, 1057
Alkenes, *gem*-dimethoxazonyl-, from, bismethoxazonylmethane, 996
Alkenes, *gem*-dinitro-
  from
    alkanes, 993
    alkenes, 993
    aryl nitroalkenes, 993
    gem-diarylalkenes, 993
    diaryl carbinols, 993
    1,1-diarylethanes, 993
    diazo compounds, 995
    dinitro alkanols, 994
    dinitromethane, 994
    naphthyl ethanols, 993
    nitroalkanes, 994
    nitroalkenes, 995
  synthesis of, 993
  via nitration reactions, 993
Alkenes, *gem*-diphenylphosphinoyl ureido-, from, *gem*-isocyanato phosphinoylalkenes, 1001
Alkenes, *gem*-diphosphazido-, from, *gem*-diazidoalkenes, 996
Alkenes, *gem*-diphosphimino-, from, *gem*-diazidoalkenes, 996
Alkenes, 1,1-diphosphorus-substituted, addition to, 560
Alkenes, 1,1-diplumbyl-
  from, 1,1-distannylalkenes, 1069
  as starting materials
    for 1-germanyl 1-plumbylalkenes, 1063
    for 1-plumbylvinyllithiums, 1065
  transmetallation of, 1063
Alkenes, 1,1-disilyl-
  from
    alkynesilane, 1048
    haloalkenes, 1048
Alkenes, 1,1-distannyl-
  from
    1,1-dibromoalkenes, 1069
    1-stannylalkynes, 1068
  as starting materials
    for 1,1-diplumbylalkenes, 1069
    for 1-germanyl 1-stannylalkenes, 1063
    for 1-plumbyl 1-stannylalkenes, 1069
  transmetallation of, 1063, 1069
Alkenes, *gem*-distannyl-, halogenation of, 813
Alkenes, 1,1-disulfonyl-
  from, phenyliodinium ylides, 866
  See also Dithioacetal tetroxides, ketene
  See also Dithioacetal *S*,*S*,*S'*,*S'*-tetroxides, ketene
Alkenes, *gem*-disulfoximido-, from, ketene dithioacetals, 997
Alkenes, *gem*-diureido-, from, ketene aminals, 982
Alkenes, fluorinated, reactions, with sodium hypobromite, 45
Alkenes, fluoro-
  bromination of, 34
  reactions
    with alkyllithium, 785
    with chlorine, 31
    with chlorosilanes, 140
    with phosphines, 111
    with tetrafluorodiphosphine, 114
    with trialkylstannane hydrides, 153
  as starting materials, for α-fluoroalkyl esters, 51
Alkenes, fluorobromo-, synthesis of, 751
Alkenes, fluorohalo-
  from, 1-halo 1-metalloalkenes, 754
  isomer ratio of, 752
  synthesis of, 750
Alkenes, *gem*-fluoro nitro-, synthesis of, 795

Alkenes, *gem*-fluoro phenylazo-, synthesis of, 799
Alkenes, 1-fluoro 1-phenylsulfinyl-, synthesis of, 778
Alkenes, *gem*-fluoro potassio-, synthesis of, 819
Alkenes, *gem*-fluoro silyl-
  desilylation of, 819
  synthesis of, 806
Alkenes, *gem*-formamido phosphoryl-
  from
    aldehydes, 1004
    oxazolines, 1005
    tetraethyl formamidomethylenediphosphonate, 1004
  synthesis of, 1004
Alkenes, 1-germanyl 1-plumbyl-, from, 1,1-diplumbylalkenes, 1063
Alkenes, 1-germanyl 1-silyl-, from, 1-germanyl 1-silylalkanes, 1058
Alkenes, 1-germanyl 1-stannyl-
  from
    1,1-distannylalkenes, 1063
    1-germanylalkynes, 1063
Alkenes, halo-
  reactions
    with diazoalkanes, 109
    with germane hydrides, 141
    with hydrogen halides, 2
  as starting materials
    for bromochloroalkanes, 38
    for bromofluoroalkanes, 34
    for chloroiodoalkanes, 39
    for 1,1-disilylalkenes, 1048
    for fluoroiodoalkanes, 36
Alkenes, α-halo acyloxy-, synthesis of, 768
Alkenes, 1-halo-1-dialkylamino-, as starting materials, for ketene hemiaminals, 889
Alkenes, *gem*-halo lithio-, synthesis of, 816
Alkenes, 1-halo 1-metallo-, as starting materials, for fluorohaloalkenes, 754
Alkenes, *gem*-halo nitro-
  synthesis of, 795
    via condensation between aromatic aldehydes and halonitromethanes, 798
    via halogenation, 795
    via nitration, 797
Alkenes, *gem*-halo nitroso-, synthesis of, 798
Alkenes, 1-halo 1-silyl-, lithiation of, 1059
Alkenes, *gem*-halo stannyl-, synthesis of, 812
Alkenes, 1-halo 1-sulfinyl-, synthesis of, 776
Alkenes, 1-halo 1-thioalkyl-, synthesis of, 773
Alkenes, *gem*-iminomethylamino lithio-, from, eneamidines, 1016
Alkenes, *gem*-imino phosphimino-
  from
    α-bromo esters, 996
    1,3-oxazines, 996
Alkenes, 1-iodo 1-alkoxy-, as intermediates, 765
Alkenes, *gem*-iodo lithio-, synthesis of, 819
Alkenes, *gem*-iodo silyl-, synthesis of, 808, 811
Alkenes, 1-iodo 1-stannyl-, as starting materials, for 1-magnesio 1-stannylalkenes, 1066
Alkenes, 1-iodo 1-sulfonyl-, synthesis of, 781
Alkenes, *gem*-iodo trialkylgermyl-, synthesis of, 812
Alkenes, *gem*-isocyanato phosphinoyl-, as starting materials, for *gem*-diphenylphosphinoyl ureidoalkenes, 1001
Alkenes, *gem*-isocyanato phosphonoyl-, as starting materials, for *gem*-phosphonoyl ureidoalkenes, 1002
Alkenes, *gem*-isocyanato phosphoryl-
  from
    1-chlorovinyl isocyanate, 1010
    1,2-dibromoethyl isocyanate, 1009
    tetrachloroethyl isocyanates, 1010
  as starting materials
    for *gem*-difluorophosphoryl ureidoalkenes, 1012
    for *gem*-phosphoramidoyl ureidoalkenes, 1011

Alkenes

for *gem*-phosphorodiamidoyl ureidoalkenes, 1011
for *gem*-phosphoryl ureidoalkenes, 1007
synthesis of, 1009
Alkenes, *gem*-isocyano lithio-, from, vinylidene
isocyanates, 1015
Alkenes, α-isocyano silyl-, from, vinylidene isocyanides, 1014
Alkenes, *gem*-isothiocyanato silyl-, from, isocyanides, 1014
Alkenes, α-lithiated, from, sulfones, 960
Alkenes, lithio-, reactions, with cyanocuprates, 953
Alkenes, 1-lithio-1-diphenylmethylsilyl-, as starting
materials, for 1-trimethylgermanyl-1-
diphenylmethylsilylalkenes, 1058
Alkenes, 1-lithio 1-plumbyl-
as starting materials, for 1-silyl 1-plumbylalkenes, 1063
transmetallation of, 1063
Alkenes, 1-lithio 1-silyl-
from, 1-phenylthio 1-silylalkenes, 1059
stability of, 1059
as starting materials
for 1-magnesio 1-silylalkenes, 1060
for 1-silyl 1-stannylalkenes, 1062
synthesis of, via metal–halogen exchange, 1059
use in the synthesis of 1-cupro-1-silylalkenes, 1061
Alkenes, 1-lithio 1-stannyl-, as starting materials, for
1-magnesio 1-stannylalkenes, 1066
Alkenes, 1-lithio 1-trimethylsilyl-
reactions, with dimesitylfluoroboranes, 1057
stability of, 1059
Alkenes, 1-magnesio 1-silyl-
from, 1-lithio 1-silylalkenes, 1060
as starting materials, for 1-silyl 1-stannylalkenes, 1062
synthesis of, 1060
Alkenes, 1-magnesio 1-stannyl-
from
1-iodo 1-stannylalkenes, 1066
1-lithio 1-stannylalkenes, 1066
Alkenes, 1-magnesio 1-zinco-, synthesis of, 1066
Alkenes, *gem*-metallo silyl-
configurational instability of, 808
stability of, 809
Alkenes, *gem*-methoxycarbonylamino
diphenylphosphinoyl-
from
carbamates, 1000
ethyldiphenylphosphinite, 1000
Alkenes, mixed dihalo-, from, fluorenyldiazomethane, 755
Alkenes, mixed 1,1-dihalo-, from, diazofluorene, 755
Alkenes, monobromo-, metallation of, 818
Alkenes, monosubstituted *trans*-difluoro-, synthesis of, 819
Alkenes, nitro-
dihalogenation of, 795
polyhalogenation of, 797
as starting materials
for *gem*-dinitroalkanes, 431
for *gem*-dinitroalkenes, 995
for α-phosphoryl enamines, 1004
synthesis of, 335, 906
Alkenes, nitroso-, synthesis of, 798
Alkenes, perfluoro-, reactions, with nucleophiles, 1076
Alkenes, 1-phenylthio 1-silyl-, as starting materials, for
1-lithio 1-silylalkenes, 1059
Alkenes, (Z)-1-phenylthio-1-trialkylsilyl-, synthesis of, 942
Alkenes, 1-phenylthio 1-trimethylsilyl-, lithiation of, 1060
Alkenes, *gem*-phosphonio ureido-, from, *N*-tetrachloroethyl
ureas, 999
Alkenes, *gem*-phosphonoyl ureido-, from, *gem*-isocyanato
phosphonoylalkenes, 1002
Alkenes, *gem*-phosphoramidoyl ureido-, from, *gem*-
isocyanato phosphorylalkenes, 1011
Alkenes, *gem*-phosphorodiamidoyl ureido-
from
gem-isocyanato phosphorylalkanes, 1011
gem-isocyanato phosphorylalkenes, 1011

Alkenes, *gem*-phosphorylamino phosphoryl-
from
benzoylhydrazones, 1008
halo tricyanomethanes, 1008
2,2,2-trihalo acetimidoyl chlorides, 1008
synthesis of, 1008
Alkenes, *gem*-phosphoryl ureido-
from
gem-isocyanato phosphorylalkenes, 1007
N-tetrachloroethylureas, 1007
synthesis of, 1007
Alkenes, plumbyl-, synthesis of, 1063
Alkenes, 1-plumbyl 1-stannyl-
from, 1,1-distannylalkenes, 1069
as starting materials, for 1-plumbylvinyllithiums, 1065
Alkenes, polyfluoro-
reactions
with alkoxides, 761
with lithium dialkylamides, 793
with phosphites, 803
Alkenes, polyhalo-, displacement of, 770
Alkenes, 1-seleno 1-nitro-, synthesis of, 910
Alkenes, silyl-, synthesis of, 735, 805
Alkenes, 1-silyl-1-phosphorus substituted, addition to, 572
Alkenes, 1-silyl 1-plumbyl-, from, 1-lithio
1-plumbylalkenes, 1063
Alkenes, 1-silyl 1-stannyl-
from
1-lithio 1-silylalkenes, 1062
1-magnesio 1-silylalkenes, 1062
synthesis of, 1062
transmetallation of, 1060
Alkenes, 1-silyl 1-zinco-, synthesis of, 1061
Alkenes, stannyl-, reactions, with cyanocuprates, 953
Alkenes, sulfonyl-, as starting materials, for
1-sulfonylvinyllithiums, 961
Alkenes, terminal, acetalization of, 193
Alkenes, *gem*-thioalkyl nitromethyl-, as starting materials,
for ketene aminals, 979
Alkenes, (*E*)-tributylstannyl-, synthesis of, 962
Alkenes, 1-trichlorogermanyl 1-silyl-, synthesis of, 1058
Alkenes, 1-trichlorosilyl-1-germanyl-, synthesis of, 1058
Alkenes, 1-trimethylgermanyl-1-diphenylmethylsilyl-,
from, 1-lithio-1-diphenylmethylsilylalkenes, 1058
Alkenes, 1-trimethylsilyl 1-stannyl-, synthesis of, 1062
Alkenes, 1-trimethylsilyl 1-titano-, synthesis of, 1061
Alkenes, trisilyl-, synthesis of, 1051
Alkenes, 1,1,2-trisilyl-, from, 1-silylalkynes, 1045
Alkenes, trisubstituted, synthesis of, 838
Alkenes, trisulfonyl-, synthesis of, 861
1-Alkenes, 1,1-dichloro-, reactions, with difluoroamine, 439
1-Alkenes, 2-metallo 1-silyl-, synthesis of, 1045
1-Alkenes, perfluoro-, reactions, with phenylhydrazine, 799
Alkenyl bromides
reactions
with cadmium, 821
with zinc, 821
Alkenyl compounds, α-chloro-, stability of, 820
Alkenyl compounds, 1-halo-, synthesis of, 814
Alkenyl halides
reactions, with sodium toluene-*p*-sulfinates, 781
as starting materials, for transition metal compounds, 821
Alkoxide ions, use in the synthesis of ketene acetals, 829
Alkoxides
reactions
with alkylthiochloroalkynes, 765
with α-chloroalkyl isocyanates, 314
with chloromethylpentamethyldisilane, 614
with 1,2-dichloroalkyne, 764
with 1,1-dinitro-2,2-diphenylethylene, 898
with geminal dihalides, 189
with α-haloboranes, 362
with α-haloboronates, 362

with α-halo ketones, 183
    with polyfluoroalkenes, 761
    with trihalomethanes, 826
  as starting materials, for difluorocarbene, 1083
Alkoxides, perfluoro-, as starting materials, for α-fluoroalkyl esters, 52
Alkoxy anion, α-fluoro-, tris(dimethylamino)sulfonium salt of, 42
α-Alkoxy carbanions, synthesis of, 377
Alkoxymercuration, of enol ethers, 188
N-Alkoxymethyl compounds, cleavage of, 102
Alkylation
  of carboxylic acids and esters, 824
  for the synthesis of tricoordinate phosphorus functions, 481
Alkylation, intramolecular, for the synthesis of cyclic monothioacetals, 231
S-Alkylation, for the synthesis of O,S-acetals, 225
Alkyl chlorides, α-keto, reactions, with potassium fluoride, 3
Alkyl halides
  displacement of halogen atoms in, 761
  as electrophiles, for dithiocarboxylic enethiolates, 842
  reactions
    with α-alkylthioalkyl phosphonates, 347
    with alkynes, 781
    with 1,1-diselenolates, 875
    with α-lithiated amines, 528
    with nitronate metal salts, 540
    with nucleophiles, 761
    with sulfinate anions, 235
    with tertiary thioamides, 899
    with ylides, 558
  as starting materials
    for α-haloalkenylsilicon compounds, 810
    for ketene dithioacetal S,S-dioxides, 857
Alkyl halides, fluoro-, reactions, with dialkylphosphites, 131
Alkylidene carbenoids
  of lithium, chemistry of, 816
  synthesis of, 816
Alkylidene halides
  displacement of halogen atoms in, 761
  reactions, with nucleophiles, 761
Alkyl iodides, perfluoro-, reactions, with diphenyl diselenides, 284
Alkyl moieties, 1,1-bis(silyl), synthesis of, 624
Alkyl transfer, selectivity reversal, 810
Alkyne, amino-, for the synthesis of manganacyclopentadiene complexes, 677
Alkyne, bis(trimethylsilyl)-, reactions, with nitryl fluoride, 807
Alkyne, chloro-, reactions, with arenesulfenyl chloride, 772
Alkyne, diacetoxy-, reactions, with caesium fluoride, 768
Alkyne, dichloro-
  addition of amines to, 792
  reactions
    with benzeneselenenyl bromides, 787
    with benzeneselenenyl chlorides, 787
    with benzeneselenol, 786
    with nitrogen dioxide, 797
    with nitrogen nucleophiles, 792
    with sodium arenethiolates, 772
  as starting material, for ketene dithioacetals, 851
  synthesis of, 976
Alkyne, 1,2-dichloro-, reactions, with alkoxides, 764
Alkyne, diiodo-
  reactions
    with iodine, 797
    with nitrous acid, 797
Alkyne, di(trimethylsilyl)-, reactions, with boracyclopropane, 1082
Alkyne, ethoxy-
  reactions
    with allylamines, 884
    with α-pyridylaminodialkylboranes, 884
Alkyne, 1-ethoxy-, reactions, with B-iodo-9-borabicyclo[3.3.1]nonane, 765
Alkyne, fluoro-, reactions, with thiols, 772
Alkyne, hexachlorophenyl-, reactions, with sodium methoxide, 764
Alkyne, phenoxy-
  reactions
    with bromine, 765
    with iodine, 765
Alkyne, phenyl-, reactions, with alkyl N,N-dichlorocarbamates, 794
Alkyne, phenyl(trimethylsilyl)-, mercury(II)-catalyzed addition of iodine to, 808
Alkyne, trimethylsilyl-
  reactions
    with allyl bromide, 810
    with α,ω-diiodoalkanes, 810
Alkyne-bridged complexes, from, tungsten–tungsten-bonded complexes, 675
Alkyne chlorides, (alkylthio)-, reactions, with phosphorus pentachloride, 804
Alkynecyclopentadienylnickel–cyclopentadienylmolybdenum complexes, synthesis of, 713
Alkyne exchange, for the synthesis of alkyne-bridged dicobalt complexes, 687
Alkyne insertion
  for the synthesis of carbon-bridged ruthenium complexes, 683
  for the synthesis of carbon-bridged trinuclear iron complexes, 683
  for the synthesis of ferracyclopentadiene complexes, 683
Alkynes
  bishydroboration of, 629
  in cycloaddition reactions, with α,β-unsaturated esters, 827
  diboronation of, 1054
  dihydroboration of, 631
  dihydrosilylation of, 617
  dimercuration of, 1067
  as electrophiles, for dithiocarboxylic enethiolates, 842
  fluorination of, 6
  hydroalumination of, 1068
  hydrostannation of, 701
  photolysis of, 679
  reactions
    with alcohols, 194
    with alkyl halides, 781
    with 2-aminoalkylarsines, 498
    with benzeneselenenyl chloride, 92
    with (benzylideneacetone)tricarbonyliron, 681
    with bromomethanesulfonyl bromide, 84
    with carboxylic acids, 206
    with chloromethanesulfonyl bromide, 82
    with diselenides, 871
    with disulfides, 940
    with dodecacarbonyldicobalt, 687
    with dodecacarbonyltriosmium, 682
    with germanium-bridged cobalt complexes, 687
    with hexacarbonylbis(η^5-cyclopentadienyl)dimolybdenum, 674
    with mercury compounds, 687
    with metal complexes, 712
    with metal imine complexes, 1017
    with metal–metal double-bonded cobalt complexes, 689
    with octacarbonyldicobalt, 687
    with octakis(phosphorus trifluoride)dirhodium, 688
    with organocuprates, 772
    with ortho esters, 192
    with pentacarbonylcyclopentadienylcobaltnickel, 713
    with N-phenylselenophthalimide, 787
    with phosphorus nucleophiles, 561

Alkynes

with phosphorus pentachloride, 772
with rhodium complexes, 689
with seleniumphenyl *p*-tolueneselenosulfonate, 947
with siliranes, 1052
with silirenes, 1051, 1052
with tetramethyldisilene, 1052
with thallium acetate, 1068
with thiols, 939
with tin compounds, 962
with tungsten complexes, 1028
silylalumination of, 1045
silylcupration of, 1045, 1059
silyl manganesation of, 1045
silyltitanation of, 1045
as starting materials
  for acetals, 193
  for allenes, 859
  for 1,1-bisborylalkanes, 630
  for 1,1-bis(bromomagnesio)alkene, 1065
  for 1,1-bis(silyl)alkenes, 1044
  for 1,1-bis(trimethylstannyl)allene, 1068
  for 1-boronyl 1-silylalkenes, 1056
  for 1-boro 1-stannylalkenes, 1063
  for 1,1-dialuminoalkenes, 1068
  for 1,1-diboroalkenes, 1054
  for dibromoalkanes, 24
  for dichloroalkanes, 15
  for difluoroalkanes, 6
  for 1,1-digermanylalkenes, 1055
  for 1,1-digermanylallenes, 1055
  for diiodoalkanes, 29
  for dithianes, 258
  for 1,1-dizincalkene, 1067
  for α-halo enols, 764
  for ketene *S,N*-acetals, 905
  for ketene dithioacetal *S*-oxides, 859
  for 1-magnesio 1-silylalkene, 1060
  for silacyclopropenes, 1048
  for 1-silyl 1-zincoallenes, 1061
  for transition metal compounds, 822
titanation of, 1066
use in the synthesis of carbon-bridged diiron complexes, 681
zirconation of, 1061, 1066
Alkynes, alkoxy-
  reactions
    with dialkylhalophosphines, 914
    with phosphorus compounds, 913
    with phosphorus trichloride, 765
Alkynes, 1-alkoxy-
  reactions
    with aliphatic alcohols, 828
    with hindered phenols, 827
  as starting materials
    for 1-haloenol ethers, 765
    for ketene acetals, 827
Alkynes, alkylthiochloro-
  reactions
    with alkoxides, 765
    with selenols, 787
Alkynes, arenesulfonyl-, reactions, with alkanethiols, 861
Alkynes, 1-(benzenesulfonyl)-, reactions, with 3-phenylglycidol, 840
Alkynes, bis(arenesulfonyl)-, reactions, with methanol, 840
Alkynes, bis(methylseleno)-, reactions, with acetic acid, 841
Alkynes, 1-boro-, hydroboration of, 1054
Alkynes, bromo-, as starting materials, for ketene aminals, 976
Alkynes, 1-bromo-, hydroboration of, 814
Alkynes, ω-bromo 1-silyl-, carbamagnesiation of, 1060
Alkynes, chloro-
  reactions
    with amines, 976

with lithium aluminum hydride, 816
with trialkylaluminum hydrides, 816
as starting materials, for ketene aminals, 976
synthesis of, 976
Alkynes, 1-chloro-
  reactions
    with bis(pyridine)iodine(1) tetrafluoroborate, 760
    with disubstituted boranes, 756
Alkynes, dichloro-, as intermediates, 761
Alkynes, disilyl-, reactions, with dicarbonylcyclopentadienylcobalt, 689
Alkynes, ethoxy-
  reactions, with secondary amines, 884
  as starting materials, for ketene aminals, 979
Alkynes, fluoro-, stability of, 976
Alkynes, 1-germanyl-
  hydrostannation of, 1063
  as starting materials, for 1-germanyl 1-stannylalkenes, 1063
Alkynes, germyl-
  hydroalumination of, 812
  hydroboration of, 812
Alkynes, halo-
  hydroboration of, 814
  reactions
    with bromoborane dimethyl sulfide complexes, 815
    with isopinocampheylboranes, 815
    with selenyl compounds, 786
  as starting materials
    for amidines, 976
    for α-haloalkenyl metal compounds, 822
    for α-halo enols, 764
    for ketene aminals, 976
Alkynes, 1-halo-, aluminum hydride reduction of, 816
Alkynes, iodo-, reactions, with amines, 976
Alkynes, 1-iodo-
  hydroboration of, 814
  reactions
    with bis(pyridine)iodine(1) tetrafluoroborate, 749
    with *N*-bromosuccinimide, 760
    with *N*-iodosuccinimide, 760
Alkynes, metallo-, as starting materials, for α-haloalkenyl metal compounds, 822
Alkynes, nitro-, reactions, with *N*-chloroguanidines, 796
Alkynes, selenyl-substituted, reactions, with halogens, 786
Alkynes, silyl-
  dihydroboration of, 632
  double hydroboration of, 644
  reactions, with diboranes, 1054
Alkynes, 1-silyl-
  hydroboration of, 1056
  hydromagnesiation of, 1060
  hydrosilylation of, 1044
  reactions
    with butyllithium, 1058
    with trialkoxyboranes, 1057
  as starting materials
    for 1-titano 1-silyl allene, 1061
    for 1-trialkylgermanyl 1-silyl allenes, 1058
    for 1,1,2-trisilylalkenes, 1045
Alkynes, silyl(alkylsilyl)-, hydrosilylation of, 1054
Alkynes, stannyl-, reactions, with diboranes, 1054
Alkynes, 1-stannyl-
  hydroboration of, 1063
  hydrostannation of, 1068, 1069
  as starting materials
    for 2-boro borienes, 1054
    for 1,1-distannylalkenes, 1068
Alkynes, substituted, as starting materials, for ketene hemiaminals, 884
Alkynes, 1-sulfonyl-, as starting materials, for ketene monothioacetal *S,S*-dioxides, 840
Alkynes, terminal

bishydroboration of, 697
  as starting materials
    for dichloromethyl ketones, 16
    for ruthenium complexes, 683
  use in the synthesis of carbon-bridged dimanganese
    complexes, 678
Alkynes, 1-trimethylsilyl-, hydroalumination of, 1062
1-Alkynes, reactions, with aluminum hydride, 698
1-Alkynes, 1-chloro-, reactions, with lithium aluminum
    hydride, 758
1-Alkynes, 1-trimethylsilyl-, as starting materials, for
    1,1-bis(trimethylsilyl) allenes, 1047
Alkyn-2-on-1-yl chloride compounds, reactions, with lithium
    trialkylgermanium, 1055
Allenamines, α-lithio, synthesis of, 1015
Allenamines, α-potassio, from, ynamines, 1017
Allene, 1,1-bis(trimethylstannyl)-, from, alkynes, 1068
Allene, 1-titano 1-silyl-, from, 1-silylalkynes, 1061
Allene complexes, use in the synthesis of α-carbocations of
    alkyne-bridged dimolybdenum complexes, 674
Allenes
  as electrophiles, for dithiocarboxylic enethiolates, 842
  from, alkynes, 859
  reactions
    with iodomethane, 1024
    with methanol, 736
  synthesis of, 1045
Allenes, 1,1-(bis)silylvinyl-, synthesis of, 1046
Allenes, 1,1-bis(trimethylsilyl)-, from, 1-trimethylsilyl-1-
    alkynes, 1047
Allenes, 3-bromo-, synthesis of, 794
Allenes, 1-dialkoxyboronyl silyl-, synthesis of, 1057
Allenes, 1,1-digermanyl-, from, alkynes, 1055
Allenes, 1,1-dilithio-, synthesis of, 1064
Allenes, 1,1-dimagnesio-, synthesis of, 1065
Allenes, 1,1-disilyl-
  as starting materials
    for alkyne disilanes, 1046
    for 1-silylbutynes, 1046
Allenes, lithio-, from, 1,3,3-tris(trimethylsilyl)propyne, 1047
Allenes, methoxy-, α-sulfination of, 837
Allenes, silyl-, reactions, with triphenylgermane, 639
Allenes, 1-silyl 1-zinco-, from, alkynes, 1061
Allenes, 1-trialkylgermanyl 1-silyl-, from, 1-
    silylalkynes, 1058
Allenylthiols, reactions, with dodecacarbonyl triirons, 683
Allyl bromide, reactions, with trimethylsilylalkyne, 810
Allyl bromides
  reactions
    with 1,1-dimethylethanethiolate, 373
    with ethanethiolate, 373
Allyl chlorides, deprotonation of, 138
η^1,η^3-Allyl complexes, protonation of, 680
Allyl group, for the protection of anomeric hydroxyl
    groups, 176
Allylic anions, reactions, with carbonyl compounds, 938
Allylic rearrangement, for the synthesis of
    fluorohaloalkenes, 755
Allylmetallation–iodination, for the synthesis of
    α-haloalkenylsilicon compounds, 809
Allylthiols, double deprotonation of, 367
Aluminates, iodination of, 122
Aluminum
  reactions
    with dibromomethane, 698
    with dichloromethane, 698
    with organomercurials, 698
  trialkyls of, synthesis of, 699
Aluminum, (ethoxy)ethyl(iodomethyl)-, synthesis of, 153
Aluminum, (iodomethyl)diethyl-, synthesis of, 153
Aluminum, triethyl-, use in the synthesis of
    tris(cyclooctatetraene)dititanium, 671
Aluminum, triisobutyl-, use in the synthesis of
    hemiacetals, 174

Aluminum, trimethyl-
  photolysis of, 698
  reactions
    with dimethylberyllium, 709
    with dimethylmagnesium, 709
Aluminum alkyls, reactions, with zirconium alkyls, 725
Aluminum amalgam, use in the synthesis of α-aminoalkyl
    phosphonates, 472
Aluminum anions, α-chloroalkenyl-, synthesis of, 816
Aluminum benzeneselenolate, diisobutyl-, use in the
    synthesis of monoselenoacetals, 238
Aluminum chloride
  reactions
    with alkyllithiums, 698
    with chloromethylpentamethyldisilane, 614
    with perfluoro iodo ethers, 768
  use in the synthesis of α-chloro ethers, 49
Aluminum compounds
  reactions
    with 1-metalloalkylphosphines, 587
    with orthoformates, 192
    with phosphorus ylides, 587
    with trimethyl orthoacetate, 192
Aluminum compounds, α-halovinyl-, synthesis of, 815
Aluminum compounds, tris(silylalkyl)-, reactions, with
    sulfur trioxide, 397
Aluminum etherate, tris(trimethylsilyl)-, use in the synthesis
    of α-hydroxysilanes, 355
Aluminum halides, alkyl-, reactions, with alkyllithiums, 698
Aluminum hydride, reactions, with 1-alkynes, 698
Aluminum hydride, diethyl-, use in the synthesis of
    tris(cyclooctatetraene)dititanium, 671
Aluminum hydride, diisobutyl- (DIBAL-H)
  reactions, with 1-trimethylgermanyl-2-phenylethyne, 1063
  for the reduction of ketene dithioacetals, 253
  use in the synthesis of hemiacetals, 169
Aluminum hydrides, trialkyl-, reactions, with
    chloroalkynes, 816
Aluminum hydride–triethylamine, reactions, with
    cobaltacyclopentadiene complexes, 726
Aluminum iodide, diethyl-, reactions, with
    diazomethane, 153
Aluminum–lanthanide compounds, synthesis of, 726
Aluminum methanesulfonate, diisobutyl-, reactions, with
    carbamates, 383
Aluminums, alkenyl-, reactions, with stannyl acetals, 388
Aluminums, α-alkoxy-, synthesis of, 383
Aluminums, alkynyl-, reactions, with stannyl acetals, 388
Aluminums, trialkyl-, dimeric structure of, 698
Aluminums, triaryl-, dimeric structure of, 698
Aluminum thiophenoxide, diethyl-, use in the synthesis of
    monothioacetals, 216
Aluminum tribromide, use in the synthesis of
    1,1-bis(silyl)alkanes, 609
Aluminum trichloride
  reactions
    with α-silylmethyl Grignard reagents, 659
    with ylides, 397
  use in the synthesis of carbon-bridged dialuminum
    compounds, 698
Aluminum trimethylamine complex, tris(trifluorovinyl)-,
    synthesis of, 815
Aluminum–yttrium complexes, synthesis of, 726
Alynes, reactions, with n-butyl t-butyl telluride, 947
Amberlyst A-26 resin, iodide form of, use in the synthesis of
    chloroiodoalkanes, 39
Amidate anions, α,β-unsaturated, reactions, with
    organolithium compounds, 651
Amide, N-chloromethyl-, reactions, with ethyl
    diphenylphosphinite, 487
Amide, diisopropyl-, use in the synthesis of lithium
    compounds, 600
Amide, tertiary, reactions, with phosphorus
    pentachloride, 790

Amide compounds

Amide compounds, α-silyl-, metallation of, 648
Amides
  alkylation of, 881
  from
    bis(chloromethyl)phosphinic chlorides, 127
    β-lactams, 825
  Mannich aminomethylation of, 322
  metallation of, 530
  reactions
    with acetals, 310
    with aminols, 427
    with benzotriazole, 449
    with N-bromosuccinimide, 138
    with chloromethylgermanes, 521
    with dibutyl iodomethylboronates, 525
    with formaldehyde, 416
    with phosgene, 971
    with phosphorus pentasulfide, 424
    with silylmethyl halides, 513
    with tetrakis(dimethylamino)titanium, 972
  as starting materials
    for O,N-acetals, 303
    for gem-amido phosphorylalkenes, 1005
    for dithyreanitrile, 245
    for ketene aminals, 971, 972
    for ketene hemiaminals, 881
    for thioimidates, 517
Amides, N-acyl-, as starting materials, for gem-alkylideneamino phosphorylalkenes, 1009
Amides, N-alkoxymethyl-, as starting materials, for 1-aminoalkylphosphonium salts, 492
Amides, N-alkyl-N-chloromethyl-, synthesis of, 99
Amides, N-alkylthiomethyl-, reactions, with electrophiles, 100
Amides, amino-, condensation of, 417
Amides, β-amino-α,β-unsaturated, as starting materials, for α-lithio enamines, 1016
Amides, α,α-bis(silyl)-, synthesis of, 604
Amides, α-bromo-, from, methyl N-acetyl-α-hydroxyglycinate, 99
Amides, chiral, from, hexatriazines, 99
Amides, α-chloro-
  as starting materials, for N-trimethylsilyloxymethyl lactams, 100
  synthesis of, 98
Amides, N-α-chloroalkyl-, characteristics of, 96
Amides, α-chloroalkyl-, as starting materials, for N-acyl S,N-acetals, 327
Amides, N-chloromethyl-, synthesis of, 99
Amides, α-halo-
  from, α-hydroxy compounds, 98
  synthesis of, 98
    via cleavage of gem-diheteroatomic compounds, 99
    via halogenation, 99
Amides, N-haloalkyl-, synthesis of, 99
Amides, N-(haloalkyl)-, synthesis of, 308
Amides, N-α-haloalkyl-, reactions, with nucleophiles, 96
Amides, hydroxy-
  as starting materials, for O,N-acetals, 301
  use in the synthesis of 1,3-oxazines, 302
Amides, α-hydroxy-
  reactions
    with chlorophosphines, 465
    with tetraphenyl bisphosphines, 465
  as starting materials
    for alkylaminomethyl phosphinic acids, 465
    for α-aminoalkylphosphorus compounds, 465
Amides, N-(hydroxyalkyl)-, synthesis of, 324
Amides, hydroxymethyl-
  reactions, with phosphorus trichloride, 469
  synthesis of, 417
Amides, N-hydroxymethyl-, as starting materials, for 1-aminoalkylphosphonium salts, 492

Amides, β-keto-, use in the synthesis of 1,3-azarsa compounds, 495
Amides, α-magnesio-α-silyl-, synthesis of, 655
Amides, methacrylic, as starting materials, for α-silyl enamines, 1014
Amides, N-methylthiomethyl-, synthesis of, 327
Amides, monoalkyl-, reactions, with dimethylmalonyl chlorides, 881
Amides, N-nitroso-, as starting materials, for N-nitrosoketene hemiaminals, 896
Amides, N-[(S)-1-phenylethyl]-N-chloromethyl-, synthesis of, 99
Amides, phosphinic, synthesis of, 127
Amides, primary
  Hofmann degradation of, 422
  reactions, with 1-(hydroxymethyl)benzotriazole, 449
Amides, N-silyl-, reactions, with silyloxiranes, 513
Amides, silylated, reactions, with ketenes, 886
Amides, tertiary
  acylation of, 881
  as starting materials, for α-phosphoryl enamines, 1003
Amides, N-tetrachloroethyl-
  reactions
    with diethyl phenylphosphinite, 1002
    with ethyl diphenylphosphinite, 1000
  as starting materials
    for gem-amido diphenylphosphinoylalkenes, 1000
    for gem-amido phosphonioalkenes, 999
    for gem-amido phosphonoylalkenes, 1002
    for gem-amido phosphorylalkenes, 1006
Amides, thiol, condensation of, 325
Amides, β-thiol, as starting materials, for thiazines, 325
Amides, γ-thiol, reactions, with paraformaldehyde, 325
Amides, thiophosphinic, synthesis of, 128
Amides, N-trichloroethyl-, as starting materials, for gem-amido phosphonioalkenes, 999
Amides, N-trichloroethylidene, as starting materials, for gem-amido phosphorylalkenes, 1006
Amidines
  from
    haloalkynes, 976
    trimethylsilylmethyl isocyanide, 518
    vinylidene dihalides, 970
    ynamines, 977
  lithiation of, 511
  reactions, with s-butyllithium, 107
  reduction of, for the synthesis of aminals, 413
  silylation of, 511
  as starting materials, for ketene aminals, 979
  as tautomers of ketene aminals, 969
Amidines, acyl, as starting materials, for N-acylaminals, 420
Amidinium salts, reduction of, for the synthesis of aminals, 413
Aminal compounds, ketene, synthesis of, 980
Aminals
  from
    acetophenones, 406
    aldehydes, 405
    amines, 405
    cyclopropanone, 406
    diamines, 407
    dihalo compounds, 411
    dihalomethylbenzenes, 411
    fluorocarbons, 411
    formamidinium salts, 414
    geminal dihalo compounds, 410
    imidazolines, 414
    imines, 411
    iminium salts, 411
    ketones, 405
    selenoaldehydes, 415
    terephthalaldehyde, 406
  solvolysis of, 309

stability of, 404
as starting materials
    for α-amidoalkyl sulfones, 332
    for ketene aminals, 975
    for stannanes, 533
synthesis of, 404
use in the synthesis of S,N-acetals, 322
use in the synthesis of trisaminomethylphosphines, 459
Aminals, acyclic, synthesis of, 405
Aminals, N-acyl
  from
    acylamidines, 420
    α-aminonitriles, 418
  synthesis of, 416
Aminals, β-acylketene, synthesis of, 979
Aminals, N-acyl-N'-sulfonyl alkylidene, synthesis of, 427
Aminals, gem-arylazo nitroketene, synthesis of, 995
Aminals, asymmetrical, synthesis of, 405
Aminals, N-benzoyl, reactions, with 3-methylthiophenol, 323
Aminals, cyanoketene
  from, cyano ynamines, 976
  synthesis of, 972
Aminals, cyclic, synthesis of, 407
Aminals, β,β-diacylketene, reactions, with methanol, 979
Aminals, diaza heterocyclic ketene, synthesis of, 973
Aminals, dicyanoketene
  reactions, with trichloromethyldimethylamine, 987
  synthesis of, 977
Aminals, ketene
  N-acylation of, 975, 980
  deacylation of, 979
  from
    alkoxy tris(dimethylamino)methanes, 973
    amide acetals, 975
    amides, 971, 972
    amidines, 979
    aminals, 975
    β-amino dichloroalkanes, 979
    aminopyridines, 979
    2-amino-2H-pyrido[1,2-a]pyrimidines, 979
    anhydrides, 972
    azines, 979
    benzodiazepines, 979
    1,2-benzodithiole-3-thione, 979
    bis(dialkylamino)malononitriles, 975
    bis(formamidinium) ethers, 975
    bromoalkynes, 976
    carbodiimides, 979
    carboxylic acids, 972
    chloroalkynes, 976
    chloro enamines, 971
    cyanamides, 975
    diaziridines, 979
    dichlorocarbenes, 979
    dicyanoacetamides, 977
    dicyanoacetates, 977
    gem-dicyanoalkenes, 979
    dicyanothioacetates, 977
    1,3-dimethyl-2-ethoxy-4,5-dihydroimidazolium tetrafluoroborate, 975
    1,2-dithiolimines, 979
    enamines, 979
    esters, 972
    ethoxyalkynes, 979
    haloalkynes, 976
    imidazoles, 973
    imidazolines, 979
    imino esters, 972
    5-indolones, 979
    isothioureas, 973
    isoxazoles, 979
    isoxazolium salts, 979
    ketene acetals, 971
    ketene aminals, 979
    ketene dithioacetals, 969
    nitriles, 979
    ortho esters, 975, 979
    oxazoles, 979
    pyrimidines, 978
    tetraamines, 979
    tetrakis(dimethylamino)methane, 975
    tetramethylchloroformamidinium chloride, 974
    N,N,N',N'-tetramethylethoxyformamidinium fluoroborate, 975
    gem-thioalkyl nitromethylalkenes, 979
    thioesters, 979
    thioformamidinium salts, 973
    thioureas, 973
    trichlorocyclopropenes, 979
    tricyanomethanide, 977
    trihalomethanes, 979
    urea acetals, 974
    vinylidene azides, 979
    vinylidene diazoles, 979
    vinylidene dihalides, 970
    ynamines, 976
  reactions, with dibromotriphenylphosphorane, 990
  stability of, 969
  as starting materials
    for α-amido enamines, 980
    for α-aminoalkylideneamino enamines, 987
    for gem-diureidoalkenes, 982
    for ketene aminals, 979
    for ketene hemiaminals, 888
    for α-phosphimino enamines, 990
    for α-phosphoryl enamines, 1003
    for α-ureido enamines, 981
  β substituents, introduction of, 979
  synthesis of, 969
Aminals, methylene
  from, formaldehyde, 410
  synthesis of, 405
Aminals, monoacyl
  from, β-lactams, 419
  synthesis of, 417
Aminals, morpholine, synthesis of, 406
Aminals, nitroketene, synthesis of, 974
Aminals, N-sulfonyl
  from, arenesulfonamides, 423
  synthesis of, 423
Amination, electrophilic, for the synthesis of gem-chloro carbamates, 794
Amine, bis(1-hydroxycyclopropyl)-, synthesis of, 299
Amine, N-chlorodiethyl-, reactions, with lithium phenylalkynide, 976
Amine, cyclohexyl-, reactions, with phosphorus acid, 470
Amine, difluoro-
  addition to aldehydes and ketones, 439
  reactions
    with 1,1-dichloro-1-alkenes, 439
    with methylvinyl ketone, 439
Amine, N-(formylaminomethyl)-, from, N-(formylaminomethyl)benzotriazole, 446
Amine, N-methylene-t-butyl-, addition of amides to, 418
Amine, N-methyl(o-hydroxybenzylidene)-, reactions, with dichlorophenylphosphine, 478
Amine, N-nitroso(methyl)vinyl-, as starting material, for N-nitroso(1-methoxy-2-chloroethyl)methylamine, 896
Amine, tris(chloromethyl)-
  from, hexamethylenetetramine, 307
  reactions, with hydrazines, 437
  silylation of, 511
Amine, tris(phenylthiomethyl)-, synthesis of, 317
Amine functions, α-metallated, synthesis of, 526
Amine hydrobromides, alkyl-, reactions, with silver tricyanomethanide, 977

Amines

Amines
- N-arylthiomethylation of, 316
- cyclopalladation of, 720
- reactions
  - with aldehydes, 294, 405
  - with alkanesulfonyl chlorides, 331
  - with 2-alkoxy 1,1-disulfonylalkenes, 866
  - with N-alkylanilines, 316
  - with alkynic ethers, 884
  - with benzotriazole, 406
  - with bis(1,1-alkylthio)alkenes, 899
  - with bis(chloromethyl) ether, 411
  - with bis-N-(hydroxyalkyl)ureas, 416
  - with α-chloroalkanesulfenyl chlorides, 72
  - with α-chloroalkyl alkyl carbonates, 60
  - with chloroalkylbenzotriazoles, 445
  - with chloroalkynes, 976
  - with 2-(chloromethyl)benzimidazole, 411
  - with chloromethylphosphonic acid, 473
  - with chlorophosphines, 116
  - with 2-cyanobenzaldehydes, 419
  - with cyanogen bromide, 413
  - with 4,6-dichloro-5-substituted pyrimidines, 978
  - with dichlorovinyl ketones, 970
  - with dithionomalonic esters, 895
  - with enamines, 971
  - with 1-fluoro enol ethers, 894
  - with formaldehyde, 295
  - with geminal dihalo compounds, 410
  - with germylmethyl halides, 519
  - with glyoxal, 298
  - with α-haloazoalkanes, 444
  - with halomethylsilanes, 506
  - with imines, 411
  - with iminium salts, 412
  - with iodoalkynes, 976
  - with ketene acetals, 887, 972
  - with ketene N,O-acetals, 972
  - with ketene dithioacetals, 969
  - with ketones, 294, 405
  - with phenylglyoxal, 300
  - with phosphaalkenes, 116
  - with phosphinic anhydrides, 127
  - with phosphinic chlorides, 127
  - with phosphonic dichlorides, 133
  - with potassium salts, 977
  - with selenoalkynes, 910
  - with silyloxiranes, 506
  - with sulfenes, 282
  - with tetrasulfones, 866
  - with α-thiol ketones, 318
  - with thiophosphinic chlorides, 128
  - with vinylidene dihalides, 970
  - with ynamines, 976
  - with yndiamines, 976
- silylmethylation of, 507
- as starting materials
  - for aminals, 405
  - for α-aminoalkylphosphorus compounds, 466
  - for dibromoalkanes, 27
  - for dichloroalkanes, 19
- N-thioalkylation of, 324
Amines, acylvinyl nitro-, reactions, with bromine, 106
Amines, N-(α-alkoxyalkyl)-, synthesis of, 299, 309
Amines, alkyl-, reactions, with ketenimines, 977
Amines, allyl-, reactions, with ethoxyalkyne, 884
Amines, allylic, from, bis(sulfones), 279
Amines, N-amino α-halo-, synthesis of, 105
Amines, aminomethylhydroxyl-, synthesis of, 442
Amines, aromatic, reactions, with chloromethylsilanes, 507
Amines, N-[1-(benzotriazol-1-y-1)alkyl]-, synthesis of, 309
Amines, 1-(benzotriazol-1-yl)alkyl-, synthesis of, 323
Amines, N-benzylidineallyl-, reactions, with trimethylsilyloxyphosphorus(III) compounds, 461

Amines, N-t-butoxy carbonyl allylic, synthesis of, 307
Amines, carbinol-, as starting materials, for α-aminoalkylphosphorus compounds, 465
Amines, N-chloro-
- photolysis of, 414
- as starting materials, for α-chloroamines, 98
Amines, α-chloro-
- from, N-chloroamines, 98
- synthesis of, 97
Amines, α-chloroalkyl-, use in the synthesis of α-aminosilanes, 510
Amines, α-chloro tertiary
- reactions
  - with lithium dialkyl(aryl)phosphines, 455
  - with lithium diphenylphosphide, 455
Amines, cyano-, reduction of, for the synthesis of aminals, 413
Amines, N-cyano-, reduction of, for the synthesis of aminals, 413
Amines, N,N-difluoro α-halo-, synthesis of, 104
Amines, N-ethoxymethyl dialkyl-, reactions, with α-oxophosphonic acid diethyl ester, 458
Amines, fluoro-, synthesis of, 97
Amines, α-fluoro-, from, dicyanodifluoroethenes, 104
Amines, fluoromethyl dialkyl-, characteristics of, 96
Amines, α-halo-
- covalent character of, 96
- in equilibrium, with iminium salts, 96
- from
  - enamines, 100
  - imines, 100
  - methylenediamines, 97
- reactions
  - with electrophiles, 100
  - with trialkyl(aryl)phosphites, 464
- synthesis of, 97
- reviews, 96
Amines, α-halo alkenyl sulfinyl-, synthesis of, 800
Amines, α-hydroxy-, reactions, with phosphorus, 466
Amines, 3-hydroxy alkyl-, synthesis of, 383
Amines, N-hydroxymethyl dialkyl-
- reactions
  - with α,ω-bisphosphinoalkanes, 452
  - with phosphorus, 452
Amines, N-hydroxymethyl-N-nitro-, synthesis of, 311
Amines, α-lithiated
- reactions
  - with aldehydes, 528
  - with alkyl halides, 528
- synthesis of, 527
Amines, methylene-, as starting materials, for N-(phosphinomethyl)arylamines, 453
Amines, monocyclic, as starting materials, for gem-dinitramines, 435
Amines, N-nitro-, reactions, with formaldehyde, 311
Amines, N-nitro α-halo-, synthesis of, 106
Amines, N-nitroso-
- lithiation of, 511
- silylation of, 511
Amines, N-oxy α-halo-, synthesis of, 104
Amines, perfluoro-, bromination of, 794
Amines, N-phosphoryl α-halo-, synthesis of, 106
Amines, primary
- reactions
  - with iodomethaneboronates, 525
  - with tetrakis(hydroxymethyl)phosphonium salts, 456
- as starting materials, for bisphosphinic acids, 470
Amines, primary α-lithio, synthesis of, 529
Amines, secondary
- reactions
  - with 2-(benzeneseleninyl)vinyl, 874
  - with ethoxyalkynes, 884
  - with N-methyl-3,4-dinitropyrrole, 415

with 2-nitrothiophene, 907
with secondary arsines, 500
with tris(dimethylaminomethyl)phosphines, 456
with tris(hydroxymethyl)phosphines, 456
as starting materials, for cyclic aminomethylphosphines, 453
Amines, α-selenenyloxyalkyl-, synthesis of, 311
Amines, silyl-, reactions, with ketenes, 886
Amines, N-silyl-, as nucleophiles, 417
Amines, α-silyl-, synthesis of, 511
Amines, silylated, for the synthesis of aminals, 406
Amines, α-silyloxyalkyl-, synthesis of, 311
Amines, N-substituted α-halo-, synthesis of, 98
Amines, sulfinyl-, synthesis of, 110
Amines, tertiary
    metallation of, 527
    oxidation of, 493
    silylmethylation of, 507
Amines, tertiary α-halo-, synthesis of, 97
Amines, N-thio-α-halo-, synthesis of, 104
Amines, trichlorovinyl-, synthesis of, 791
Amines, N-trifluorovinyl secondary, synthesis of, via elimination of carbon dioxide, 794
Amino acid azides, Curtius rearrangement of, 426, 446
Amino acids
    oxidation of, 311
    as starting materials, for O,N-acetals, 301
    for the synthesis of (α-aminoalkylphosphines, 455
Amino acids, α-fluoro, synthesis of, 101
α-Amino acids
    from, 2-aminoketene dithioacetal monosulfoxides, 857
    phosphonic acid compounds of, synthesis of, 466
    synthesis of, 102
    trifluoroacetylation of, 320
1-Amino-1-alkanols, synthesis of, 294
α-Amino alkoxides, synthesis of, 299
α-Aminoalkyl phosphonites, from, ethyl phenyl phosphonite, 460
α-Amino anions, synthesis of, 527
Amino functions, synthesis of, 495
Amino functions, benzyloxymethyl-, cleavage of, 103
Aminols
    reactions
        with amides, 427
        with sulfonamides, 427
Aminosilyl compounds, N-α-chloroalkyl, synthesis of, 107
Ammonia
    reactions
        with aldehydes, 294
        with benzaldehydes, 437
        with bromomethanesulfonyl chloride, 86
        with chloral, 297
        with di(chloromethyl)phosphinic acid, 473
        with diethyl α-bromovinyl phosphonate, 473
        with dithionomalonic esters, 895
        with formaldehyde, 407
        with hexafluoroacetone, 297
        with iodomethanesulfonyl chloride, 86
        with ketene acetals, 972, 983
        with phosphonothioic acid diesters, 469
        with phthalaldehyde, 298
Ammonia, trimethyl-, reactions, with Simmons–Smith reagents, 534
Ammonias, aldehyde, dehydration of, 407
Ammonium bromide, tetraethyl-, as catalyst, for the synthesis of acetals, 181
Ammonium chloride, benzyltriethyl-, as phase-transfer catalyst, 792
Ammonium chloride, dimethylmethylene-, reactions, with alcohols, 305
Ammonium chloride, N,N-dimethylmethylene-, reactions, with lithium butanetellurolate, 334
Ammonium chloride, methyltrioctyl-, as catalyst, for the oxidation of dithioacetals, 266

Ammonium fluoride, tetrabutyl-, use in the synthesis of gem-isothiocyanato silylalkenes, 1015
Ammonium fluoride, tetra-n-butyl-
    as catalyst, for the synthesis of aminals, 406
    use in the synthesis of fluorobromoalkenes, 751
Ammonium nitrate, ceric (CAN), reactions, with organotin compounds, 252
Ammonium nitrate, ceric triethyl- (CTAN), reactions, with tetrahydrofuran, 201
Ammonium salts
    reactions, with alkali arsenes, 499
    synthesis of, 455
Ammonium salts, alkyl-, synthesis of, 977
Ammonium salts, aryl-, synthesis of, 977
Ammonium salts, dialkoxymethyl-, reactions, with malonates, 193
Ammonium salts, quaternary
    reactions, with diphenylphosphine, 483
    synthesis of, 480
Ammonium salts, N-silyl-, deprotonation of, 512
Ammonium salts, tetraalkyl-, pyrolysis of, 674
Ammonium salts, (trimethylsilylethyl)-, as starting materials, for ylides, 512
Anesthetics, isoflurane, 49
Anhydride, propanoic, use in the synthesis of tris(cyclopentadienyl)dinickel salts, 694
Anhydrides
    reactions, with acetals, 211
    as starting materials, for ketene aminals, 972
Anhydrides, ethanoic, reactions, with dithiolalkanes, 1074
Anhydrides, phosphinic, reactions, with amines, 127
Anhydrosugars, synthesis of, 228
Anilide, thionyl-, reactions, with chlorotrimethylsilane, 370
Aniline
    phenylselenomethylation of, 333
    reactions
        with 2-benzoylbenzoic acid, 299
        with perfluoroacrylic ester, 894
        with phenoxyfluoroxonium (fluorosulfonyl) methylide, 282
        with titanium tetrachloride, 796
        with vinylidene dihalides, 970
        with vinylsilanes, 507
Aniline, benzylidine-, reactions, with 4-hydroxycoumarin, 304
Aniline, p-chloro-, reactions, with N-acylketenimines, 980, 981
Aniline, N-ethyl-, reactions, with ketene diethyl acetals, 887
Aniline, N-ethyl-N-methyl-, oxidation of, 311
Aniline, N-methyl-
    N-methoxymethylation of, 300
    use in the synthesis of aminals, 412
Aniline, N-(phenylselenomethyl)-, synthesis of, problems, 333
Aniline compounds, benzylidene-, reactions, with thiols, 319
Anilines, N-alkyl-
    oxidation of, 311
    reactions, with amines, 316
Anilines, N,N-bis(arylthiomethyl)-, synthesis of, 316
Anilines, N,N-dimethyl-
    reactions
        with iodosylbenzene, 443
        with trimethylsilyl azide, 443
    as starting materials, for azidomethyl compounds, 443
Anilines, N-(methoxymethyl)-, as starting materials, for imines, 412
Anion centres
    flanked by a halogen and an oxygen, 1076
    flanked by oxygen and phosphorus, 1075
    flanked by two halogens, 1076
    flanked by two phosphorus atoms, 1076
    flanked by two sulfur atoms, 1075
Anions

Anions

flanked by a boron and sulfur, 1077
flanked by silicon, 1077
from, α-alkoxy alkylphosphorus compounds, 917
reactions, with electrophiles, 917
synthesis of, 1075
Anions, organometallic, reactions, with ferrocenoyl chloride, 722
Anions, sulfur-stabilized, chemistry of, 390
Anisole, reactions, with sulfuryl chloride, 47
Anisole, p-chloro-, synthesis of, 47
Anisole complexes, reactions, with butyllithium, 707
Anthracene compounds, synthesis of, 258
Anthranilamides, reactions, with ethyl propiolate, 422
Antibiotics
  synthesis of, 205
  tunicamycins, synthesis of, 239
Antimonate, trityl tetrachloro-, reactions, with ortho esters, 1074
Antimony
  α-haloalkenyl compounds of, synthesis of, 805
  synthesis of functions, **591–600**
    with no halogen, chalcogen or nitrogen, **1021–1042**
Antimony, tris(trifluorovinyl)-, synthesis of, 805
Antimony compounds, synthesis of, 598, 1042
Antimony compounds, α-aminoalkyl-, synthesis of, 504
Antimony compounds, α-halo-, synthesis of, 136
Antimony dichloride, phenyl-
  reactions, with 1,1-di-n-butyl-2,5-dimethylstannole, 1041
  use in the synthesis of stiboles, 1039
Antimony functions, synthesis of, 595
Antimony pentachloride, use in the synthesis of dichloroalkanes, 18
Antimony pentafluoride, use in the synthesis of difluoroalkanes, 4
Antimony trichloride
  reactions, with trimethylsilylmethylmagnesium chloride, 598
  use in the synthesis of α-fluoro sulfides, 61
Antimony trifluoride
  reactions, with gem-dichlorophosphoryl isocyanatoalkenes, 1012
  use in the synthesis of chlorofluoroalkanes, 30
  use in the synthesis of difluoroalkanes, 4
Arbuzov reaction
  of 1,1-dihaloalkanes, 555
  of 1-haloalkylphosphorus compounds, 555
  of 1-haloalkylsilanes, 569
  on iodotrifluoroethene, 802
  for the synthesis of gem-alkylideneamino phosphorylalkenes, 1009
  for the synthesis of gem-amido phosphorylalkenes, 1006
  for the synthesis of α-aminoalkylphosphorus compounds, 464
  for the synthesis of α-aminophosphorus compounds, 487
  for the synthesis of carbon–phosphorus bonds, 340
  for the synthesis of phosphines, 913
Arenes, reactions, with [methoxy(phenylthio)methyl]trimethylsilane, 370
Arsabenzene, reactions, with chromium, 1041
Arsacymantrenes
  Friedel-Crafts acetylation of, 1042
  synthesis of, 1042
Arsacymantrenes, 2,5-dimethyl-, synthesis of, 1042
Arsaferrocene, 2,5-dimethyl-, synthesis of, 1039
1,1'-Arsaferrocene, 2,2',5,5'-tetradeuterio-, from, 1,1'-diarsaferrocene, 1039
Arsaferrocenes, synthesis of, 1039
Arsenic
  cyclic compounds, synthesis of, 594
  α-haloalkenyl compounds of, synthesis of, 805
  synthesis of functions, **591–600**
    with no halogen, chalcogen or nitrogen, **1021–1042**
Arsenic, tris(trifluorovinyl)-, synthesis of, 805

Arsenic(III) oxide, reactions, with acetyl chloride, 592
Arsenic compounds, synthesis of, 597, 599, 1042
Arsenic compounds, 1,3-aminoalkyl-
  synthesis of
    via cyclocondensation reactions, 495
    via Mannich reactions, 499
Arsenic compounds, α-aminoalkyl-
  synthesis of
    via alkylation reactions, 501
    via cycloaddition reactions, 502
Arsenic compounds, α-halo-, synthesis of, 134
Arsenic-containing heterocycles, synthesis of, 502
Arsenic dichlorides, alkyl-, reactions, with malonic esters, 594
Arsenic dichlorides, aryl-, reactions, with malonic esters, 594
Arsenic functions, synthesis of, 592
Arsenic oxide, phenyl-, reactions, with dibromomethane, 592
Arsenic trichloride
  reactions
    with alkenes, 597
    with trimethylsilylmethylmagnesium chloride, 597
Arsenides, alkylation of, 134
Arsine, bis(diethylamino)chloro-, reactions, with trimethylsilylmethylmagnesium, 597
Arsine, bis(trimethylsilylmethyl)chloro-, synthesis of, 597
Arsine, dichloro-, dehydrohalogenation of, for the synthesis of arsoranes, 598
Arsine, dichloro(trifluorovinyl)-, synthesis of, 805
Arsine, dichloro(trimethylsilylmethyl)-, synthesis of, 597
Arsine, dilithiophenyl-, reactions, with 1,4-dichlorobuta-1,3-diene, 1039
Arsine, dimethyl-, reactions, with 2,3-dichlorotetrafluoropropene, 805
Arsine, diphenylchloro-
  reactions
    with stibinomethyllithium, 596
    with tributylstannylmethyllithium, 600
Arsine, iodomethyldiphenyl-, as starting material, for diphenylarsinomethyllithium, 599
Arsine, phenyl-, reactions, with t-butyldipropynylphosphine, 1030
Arsine, trimethyl-, reactions, with chloromethyltrimethylsilane, 597
Arsine, tris(dimethylamino)-, use in the synthesis of aminals, 405
Arsine, tris(trimethylsilylmethyl)-, synthesis of, 597
Arsines
  from, bis(diphenylarsino)methane, 593
  reactions, with hexafluoroacetone, 344
  synthesis of, 597
Arsines, aminoalkyl-, reactions, with carbonyls, 495
Arsines, 1-aminoalkyl-, synthesis of, 499
Arsines, 2-aminoalkyl-
  reactions, with alkynes, 498
  as starting materials
    for 1,3-azarsolidines, 498
    for 1,3-benzazarsolines, 498
Arsines, α-(N-arenesulfonamido)benzyl-, synthesis of, 503
Arsines, bicyclic, synthesis of, via dipolar cycloaddition reactions, 502
Arsines, bis(halomethyl)-, synthesis of, 135
Arsines, α-bromo-, synthesis of, 135
Arsines, (bromomethyl)-, synthesis of, 135
Arsines, chloro-
  from, bis(trimethylsilyl)methyllithium, 597
  reactions
    with Grignard reagents, 597
    with trimethylsilylmethyllithium, 597
  synthesis of, 592
Arsines, α-chloro-, synthesis of, 134
Arsines, (chloromethyl)-, synthesis of, 134
Arsines, α-dialkylaminoalkyl-, synthesis of, via Mannich reactions, 500

Arsines, dihalo(halomethyl)-, synthesis of, 135
Arsines, α-dimethylaminomethyl-, synthesis of, 501
Arsines, α-fluoro-, synthesis of, 134
Arsines, α-iodo-, synthesis of, 135
Arsines, (iodomethyl)diphenyl-, synthesis of, 135
Arsines, primary, synthesis of, 495
Arsines, secondary
  reactions
    with N-arenesulfonylbenzaldimines, 503
    with bis(dimethylamino)methane, 501
    with secondary amines, 500
  use in the synthesis of aminomethylarsines, 501
Arsines, secondary aminoalkyl-, synthesis of, 495
Arsines, secondary O-aminophenyl-, synthesis of, via cyclocondensation reactions, 496
Arsines, silyl-, use in the synthesis of aminomethylarsines, 501
Arsines, trialkyl-
  reactions
    with alkyl dihalides, 135
    with chloromethyltrimethylsilane, 597
Arsinic acids, α-halo-, synthesis of, 135
Arsole, 2,5-dimethyl-1-phenyl-
  reactions
    with decacarbonyldimanganese, 1042
    with potassium, 1039
Arsole, 1-phenyl-, synthesis of, 1039
Arsole complexes, tetraphenyl-, synthesis of, 1042
Arsoles
  manganese complexes of, 1042
  reactions, with iron(II) chloride, 1039
  rhenium complexes of, 1042
Arsoles, 1-(1-chloroalkyl)-, synthesis of, 134
Arsolyl anions, reactions, with cyclopentadienyldicarbonyliron iodide, 1039
Arsonane, dibromotris(trimethylsilylmethyl)-, disproportionation of, 135
Arsonic acid, (chloromethyl)-, reduction of, 134
Arsonic acid, O-nitrophenyl-, reduction of, 496
Arsonic acids, α-halo-, synthesis of, 135
Arsonium iodides, (iodomethyl)trimethyl-, synthesis of, 135
Arsonium salts
  from, methylenearsorane, 598
  synthesis of, 598
Arsonium salts, α-halo, synthesis of, 135
Arsorane, methylene-, as starting material, for arsonium salts, 598
Arsoranes
  [2+2]-photodimerization of, 594
  synthesis of, 597
Arylamines, reactions, with chloroalkenes, 987
Arylamines, N(alkoxymethyl)-, synthesis of, 300
Arylamines, N-methoxymethyl-, reactions, with trialkyl phosphites, 467
Arylamines, N-(phosphinomethyl)-
  from, methyleneamines, 453
  synthesis of, 453
N-Arylamines, N-(alkoxymethyl)-, reactions, with secondary phosphines, 453
Aurate, tetraethylammonium tetrabromo-, use in the oxidation of dithioacetals, 260
Azaarenes, ethynyl-, reactions, with sodium methoxide, 195
1-Azabuta-1,3-dienes, reactions, with vinyl ethers, 305
2-Azabutadienes, cycloaddition reactions of, 421
2-Azabutadienes, (E,E)-4-amino-1-aryl-3-cyano-4-methoxy-, synthesis of, 883
Azacarbaphosphaboretanes, synthesis of, 484
Azacrown compounds, as starting materials, for α-aminoalkylphosphines, 453
2-Azacyclobutanes, 1,3-disila-, synthesis of, 606
Aza-1,3-dienes, synthesis of, 518
Azadiphosphetanes, synthesis of, 563

Azagermines, synthesis of, 521
Azalactones
  as starting materials
    for gem-amido phosphoramidoylalkenes, 1010
    for gem-amido phosphorylalkenes, 1005
  synthesis of, 1002, 1006
1-Azapenems, synthesis of, 421
1,3-Azaphosphanes
  from, diallylaminomethylphosphines, 458
  synthesis of, 473
1,3-Azaphosphepane, synthesis of, 458
1,2-Azaphosphetanes, synthesis of, 548
1,3-Azaphosphetanes, synthesis of, via heteroatom exchange, 457
1,3λ^3-Azaphosphinine, synthesis of, 484
1,2-Azaphospholanes, synthesis of, 567
1,3-Azaphospholanes, synthesis of, 453
1,3-Azaphospholan-5-one, 1-amino-, from, carboxyalkylphosphines, 482
Azaphosphoridines, synthesis of, via cycloaddition reactions, 493
1,3-Azaphosphorines, tetrahydro-, synthesis of, 454
1,3-Azarsa compounds, synthesis of, 495
1,3-Azarsinane, use in the synthesis of α-aminoalkylarsenic compounds, 501
1,3-Azarsinanes, synthesis of, via cyclocondensation reactions, 495
Azarsolanes, synthesis of, via alkylation, 496
1,3-Azarsolidines
  from, 2-aminoalkylarsines, 498
  synthesis of, via cyclocondensation reactions, 495
Azasilines, synthesis of, 509
Azasilolines
  from, m-chlorophenylsilanes, 508
  synthesis of, 507
2-Aza-1-thiacephems, desulfurization of, 421
Azetidine, 1,3,3-trinitro-, synthesis of, 429
Azetidine-2-one, 4-acetoxy-, synthesis of, 306
Azetidines, ring-opening of, 891
Azetidines, amino-, synthesis of, 414
Azetidinones, 2-(hydroxyamino)-, synthesis of, 442
Azetidinones, 4-(N-imidazolo)-, synthesis of, 417
Azetidinones, 4-silyl-, synthesis of, 514
Azetidin-2-ones, 4-acetoxy-
  as starting materials
    for 2-oxoazetidin-4-yl phosphinates, 487
    for 2-oxoazetidin-4-yl phosphonates, 487
Azetidin-4-yl phosphinates, 2-oxo-, from, 4-acetoxyazetidin-2-ones, 487
Azetines, reactions, with diphenylketene, 829
Azide, ammonium, for the nucleophilic displacement of halides, 432
Azide, phenyl
  reactions, with mesityl(diphenylmethylene)phosphines, 483
  as starting material, for α-aminoalkylphosphorus compounds, 490
Azide, tetramethyl guanidinium, for the nucleophilic displacement of chloride, 447
Azide, tosyl
  reactions
    with benzene, 483
    with dithioacetals, 860
    with potassium nitronates, 331
Azide, trimethylsilyl
  reactions
    with N,N-dimethylanilines, 443
    with sugar lactones, 432
  as starting material, for gem-diazidoalkanes, 432
Azide, trimethylsilylmethyl- (TMSMA)
  stability of, 516
  synthesis of, 516
Azide anions, reactions, with hexafluoropropene, 799

Azides

Azides
  reactions
    with alkenes, 515
    with aminomethyl halide, 444
  synthesis of, 897
Azides, acyl
  from, α-thioalkanoyl chlorides, 328
  as starting materials, for N-acylated α-ureido enamines, 982
Azides, acylaminomethyl, synthesis of, 447
Azides, 1-alkoxyalkyl, synthesis of, 313
Azides, aminomethyl, synthesis of, 443
Azides, 2,2-dicyanovinyl, as starting materials, for secondary α-chloro enamines, 792
Azides, α-halo, synthesis of, 109
Azides, β-hydroxy-α-silyl, from, silyl oxiranes, 516
Azides, keto, as starting materials, for cations, 1072
Azides, nitroacyl, Curtius rearrangement of, 440
Azides, silyl, synthesis of, 516
Azides, silylmethyl, reduction of, 512
Azides, 1-trimethylsilyloxyalkyl, synthesis of, 313
Azides, vinyl, reactions, with alcohols, 894
Azides, vinylidene, as starting materials, for ketene aminals, 979
Azidomethyl compounds, from, N,N-dimethylanilines, 443
Azines
  as starting materials
    for α-aminoalkylorganophosphorus compounds, 460
    for ketene aminals, 979
Aziridine
  reactions
    with 2,2-dicyano bis(1,1-alkylthio)alkenes, 899
    with methyl bromopropiolate, 793
    with thiols, 899
Aziridine, α-stannyl-, synthesis of, 534
Aziridines, from, acetoxy quinazolones, 515
Aziridines, 2-alkoxy-, from, 1-azirines, 305
Aziridines, carbethoxy-, synthesis of, 514
Aziridines, silyl-, synthesis of, 514
Aziridines, C-silyl-, synthesis of, 514
Aziridines, N-silylmethyl, synthesis of, 515
Azirines, reactions, with diphenylketene, 829
1-Azirines, as starting materials, for 2-alkoxyaziridines, 305
Azobenzenes
  reactions, with 2-methyl-1-propenyl triflate, 985
  as starting materials, for gem-dihydrazinoalkenes, 985
Azo compounds
  from, arylhydrazones of chloral, 799
  metallation of, 536
Azo compounds, aminomethyl, synthesis of, 444
Azoles, N-(chloroalkyl)-, synthesis of, 322
Azoles, N-(α-chloroalkyl)-, use in the synthesis of O,N-acetals, 308
Azoles, N-(hydroxyalkyl)-, acylation of, 310
Azoles, N-(phenylthioalkyl)-, oxidation of, 323
Azomethines, N-t-alkyl-, reactions, with ketene, 421

Baeyer–Villiger oxidation
  of α-alkoxy ketones, 212
  for the synthesis of acyl alkyl acetals, 212
Barbier reaction, for the synthesis of α-fluorovinyltin, 812
Bentonitic earth, as catalyst, for the synthesis of dithioacetals, 246
Benzaldehyde
  reactions
    with S-acetyldiphenylthiophosphinous acid, 345
    with (o-aminobenzyl)phenylphosphine oxide, 471
    with bislithio compounds, 938
    with crotonaldehyde dithioacetals, 849
    with dibromophosphoranes, 745
    with (diethoxymethyl)phosphine oxide, 828
    with dimethyl methylene bis(sulfonylacetate), 277
    with isothiocyanates, 1014
    with 2-lithio-1,1,3,3,5,5-hexamethyl-1,3,5-trisilacyclohexane, 623
    with α-methoxyallylphosphine oxide, 917
    with (methoxymethylene)bisphosphonic esters, 916
    with N-phenylphosphoramidites, 476
    with thiophenol, 216
  as starting material, for 1,1-dichlorostyrene, 738
Benzaldehyde, 4-dimethylamino-, reactions, with salicyl aldehyde, 438
Benzaldehyde, p-(dimethylamino)-, reactions, with disulfonyl fluoride, 864
Benzaldehyde, 4-fluoro-
  reactions
    with morpholine, 298
    with piperidine, 298
Benzaldehyde, 3-methoxy-, reactions, with S-diamines, 408
Benzaldehydes
  reactions
    with alkyldichlorophosphines, 123
    with ammonia, 437
    with aryldichlorophosphines, 123
    with morpholine, 406
    with phosphoroamidothioates, 467
  use in the synthesis of 1,3-benzazarsolines, 496
  use in the synthesis of ketene dithioacetal S-oxides, 856
Benzaldehydes, 2-cyano-, reactions, with amines, 419
Benzaldehydes, nitro-, as starting materials, for acetals, 183
Benzaldimine, N-t-butyl-, oxidation of, 312
Benzaldimines, N-arenesulfonyl-, reactions, with secondary arsines, 503
Benzamide, N-(diethylaminomethyl)-, reactions, with benzylthiol, 322
Benzamide, N,N-dimethyl-, as starting material, for bis(silyl)methylamines, 511
Benzamide, hydro-
  reactions, with dialkyl phosphites, 460
  as starting material, for α-aminoalkylorganophosphorus compounds, 460
Benzamide, N-(piperidinomethyl)-, reactions, with benzylthiol, 322
Benzamide, N-(1-piperidinylmethyl)-, reactions, with acetyl chloride, 99
Benzamides, N-[α-(benzylideneamino)benzyl]-, as starting materials, for N,N'-alkylidene bisamides, 424
Benzamides, N-(α-dialkylaminobenzyl)-, displacement of, 322
Benzamidine
  condensation of, 988
  reactions, with iodomethanesulfonyl chloride, 86
Benzamidine, N-(iodomethanesulfonyl)-, synthesis of, 86
1,3-Benzazarsolines
  from, 2-aminoalkylarsines, 498
  synthesis of
    via alkylation reactions, 497
    via cyclocondensation reactions, 496
Benzazepines, N-tosyl-, halogenation of, 105
Benzenamine, N-methylene-2,6-diethyl-, reactions, with phosgene, 102
Benzene
  reactions
    with hexanitroethane, 994
    with tosyl azide, 483
Benzene, 1,2-bis(dimethylsilyl)-, reactions, with styrene, 618
Benzene, o-(bistrimethylsilyl)-, reactions, with hexachlorides, 290
Benzene, decafluoroethyl-, as starting material, for octafluorostyrene, 730
Benzene, 1,2-diamino-, reactions, with 1,3-dithietanes, 415
Benzene, (dichloroiodo)-
  for the α-chlorination of ethers, 47
  reactions, with ethyl phenyl sulfoxide, 75
Benzene, dichloromethyl-, synthesis of, 11
Benzene, (diphenylthio)methyl-, synthesis of, 255

Benzene, hexafluoro-, as starting material, for α-fluoro
    epoxides, 45
Benzene, iodosyl-
  oxidative azidation of anilines, 443
  reactions
    with alkynylsilanes, 808
    with N,N-dimethylanilines, 443
Benzene, 2-lithio(dimethylaminomethyl)-, reactions, with
    silver bromide, 708
Benzene, 1,3,5-tri[bis(trimethylsilyl)methyl]-, from,
    1,3,5-tri(dichloromethyl)benzene, 608
Benzene, 1,3,5-tri(dichloromethyl)-, as starting material, for
    1,3,5-tri[bis(trimethylsilyl)methyl]benzene, 608
Benzene complexes, chloro-, reactions, with
    butyllithium, 707
Benzene complexes, ruthena-, synthesis of, 680
Benzene compounds, 2-lithio-1-(dimethylaminomethyl)-,
    reactions, with copper(I) bromide, 695
Benzenediazonium hydroxide, reactions with
    γ-ketocarboxylic acids, 992
Benzene dichloride, iodo-, for the oxidation of
    dithiolanes, 260
Benzene oxide, hexafluoro-, synthesis of, 46
Benzenes, dihalomethyl-, as starting materials, for
    aminals, 411
Benzenes, pentahaloethyl-, loss of chlorine from, 730
Benzeneselenenyl bromide
  reactions
    with alkenyllithiums, 785
    with α-diazocycloalkanones, 90
    with (triphenylstannyl)methyllithium, 401
  as starting material, for α-chloro selenides, 90
Benzeneselenenyl bromides, reactions, with
    dichloroalkyne, 787
Benzeneselenenyl chloride
  reactions
    with alkenes, 91
    with alkynes, 92
    with diazo compounds, 90
    with α-diazocycloalkanones, 90
    with ketones, 90
    with silver fluoride, 52
    with vinyl chloride, 91
Benzeneselenenyl chlorides, reactions, with
    dichloroalkyne, 787
α-Benzeneselenenyl enones, synthesis of, 90
Benzeneselenenyl fluoride, reactions, with vinyl acetates, 52
Benzeneselenic acid, use in the synthesis of
    α-selenosilanes, 375
Benzeneselenol
  reactions
    with dichloroalkyne, 786
    with cis-1,2-dichloroethene, 786
    with dimethoxymethane, 238
    with tetrachloroethene, 786
    with trichloroethylene, 786
Benzeneselenyl bromide
  reactions
    with (1-bromovinyl)trimethylsilane, 947
    with phosphonate anions, 929
Benzenesulfenyl chloride
  reactions
    with alkenes, 106
    with cyclopropenes, 775
    with α-diazo ketones, 69
    with 1,3-diketone dimedones, 70
Benzenesulfenyl chlorides, reactions, with phosphorus-
    substituted ethynes, 926
Benzenesulfinamides, N-(hexafluoroisopropylidene)-
  reactions
    with ethanethiol, 325
    with thiophenol, 325
Benzene sulfinate, methyl, reactions, with
    chlorotrimethylsilane, 370

Benzenesulfinic acid
  reactions
    with dihydropyran, 236
    with 2-methoxytetrahydropyran, 236
Benzenesulfonamide, reactions, with formaldehyde, 423
Benzenesulfonyl chloride, reactions, with nitro
    enamines, 907
Benzenetellurenyl bromide, reactions, with ethyl
    diazoacetate, 93
Benzenethiol, 2-amino-, reactions, with silylethynes, 939
Benzene thiosulfinate, phenyl, synthesis of, 232
Benzimidazole
  reactions
    with 1-chloroethyl chloroformate, 60
    with trimethylsilyl cyanide, 426
Benzimidazole, 2-(chloromethyl)-, reactions, with
    amines, 411
Benzimidazole, 2-cyano-1,3-dibenzoyl-2,3-dihydro-,
    synthesis of, 426
Benzimidazole compounds, synthesis of, 323
Benzimidazoles, 2-aminomethyl-, as starting materials, for
    O,N-acetals, 296
Benzimidazoline, synthesis of, 415
Benzimidoyl chloride, N-trichlorovinyl, synthesis of, 794
Benzoate, methyl, reactions, with potassium
    tetrafluorocobaltate, 3
Benzoate, vinyl, reactions, with trifluoromethanesulfonyl
    chloride, 56
Benzoate, zinc, reactions, with diazomethane, 385
Benzoates, as starting materials, for α-chloro sulfides, 69
Benzoates, alkyl, reactions, with s-butyllithium, 382
Benzo-1,3-azaphosphanes, synthesis of, 454
Benzo-1,3-azaphosphepanes, synthesis of, 454
Benzo-1,3-azaphospholanes, synthesis of, 454
Benzobarrelene, reactions, with difluorocarbene, 8
Benzobicyclo[3.3.1]nonanes, synthesis of, 342
2-Benzoboroles, reactions, with lithium 2,2,6,6-
    tetramethylpiperidide, 642
Benzodiazepines, as starting materials, for ketene
    aminals, 979
Benzo-1,3,2-dioxaphospholes, reactions, with Grignard
    reagents, 126
Benzodioxaphosphorinanes
  ethanolysis of, 126
  synthesis of, 128
2H,4H-1,3-Benzodioxin, 4-oxo-, synthesis of, 213
Benzo-1,3-diselenoles, 2-phosphoryl, Horner–Wittig
    reaction of, 875
Benzo-1,3-ditellurole, synthesis of, 290
Benzoditelluroles, synthesis of, 290
Benzodithiepines, as starting materials, for dithioacetals, 250
Benzo-1,3-dithiolane S,S'-dioxides, Mannich reaction
    of, 863
1,3-Benzodithiole, 2-methyl-, oxidation of, 262
Benzodithioles, as starting materials, for dithioacetals, 250
1,3-Benzodithiole tetraoxide, use in the synthesis of
    bis(sulfones), 274
1,2-Benzodithiole-3-thione, as starting material, for ketene
    aminals, 979
Benzo-1,3-dithiole-2-thiones, as starting materials, for
    thiosulfonate esters, 862
Benzo-1,3-dithiole-5-thiones, as starting materials, for
    thiosulfonate esters, 862
Benzoferracyclopentadiene complexes, from,
    tricarbonylbenzocyclobutadieneiron, 681
Benzofurans, synthesis of, 167
Benzoic acid, 2-benzoyl-, reactions, with aniline, 299
Benzoic acid, m-chloro- (MCPBA), for the oxidation of
    dithioacetals, 258, 267, 271
Benzo-1,3,2-oxazaphospholes, synthesis of, 134
1,4,2-Benzooxazaphosphorinanes, hydrolysis of, 124
Benzophenone, reactions, with lithium 1,1-bis(trimethylsilyl)
    alkoxides, 932

Benzophenones, synthesis of, 189
Benzophosphoranes, 3-amino-, from, phosphorus amides, 478
2-Benzopyran-4-one, 1-methoxy-, reactions, with acetic anhydride, 212
Benzopyranones, reactions, with N-bromosuccinimide, 174
Benzopyrano[1,3]oxazines, synthesis of, 304
Benzopyrans, 2-amino-, synthesis of, 307
1,2-Benzoquinone, reactions, with 1-morpholinocyclohexene, 307
1,4-Benzoquinone, 2,3-dichloro-5,6-dicyano- (DDQ), reactions, with methoxybenzyl ethers, 201
p-Benzoquinone, as electrophile, for dithiocarboxylic enethiolates, 842
Benzo-substituted compounds, synthesis of, 1074
Benzotellurole, bromination of, 291
Benzothiazolines, N-substituted, synthesis of, 318
Benzothiazolium salts, 3-substituted, reduction of, 321
2(3H)-Benzothiazolones, 3-chloromethyl-, synthesis of, via Friedel–Crafts reactions, 103
Benzothiete, thermolysis of, 232
Benzotriazole
  aminoalkylation of, 309, 323
  Mannich-type reaction of, 445
  reactions
    with aldehydes, 296
    with amides, 449
    with amines, 406
    with 2,5-dimethoxytetrahydrofuran, 309
    with 9-vinylcarbazole, 323
Benzotriazole, 1-(chloromethyl)-, reactions, with thiols, 322
Benzotriazole, N-(formylaminomethyl)-, as starting material, for N-(formylaminomethyl)amine, 446
Benzotriazole, 1-(hydroxymethyl)-
  as intermediate, for the synthesis of hydroxamic acid compounds, 449
  reactions, with primary amides, 449
Benzotriazole, 1-(phenylthiomethyl)-, lithiation of, 322
Benzotriazole compounds, synthesis of, 449
Benzotriazoles, 1-(1-alkoxyalkyl)-, synthesis of, 300
Benzotriazoles, N-[1-(alkylthio)alkyl]-, synthesis of, 318
Benzotriazoles, aminomethyl-, synthesis of, 445
Benzotriazoles, N-[1-(arylthio)alkyl]-, synthesis of, 318
Benzotriazoles, chloroalkyl-, reactions, with amines, 445
Benzotriazoles, 1-hydroxyalkyl-, synthesis of, 296
Benzotriazoles, 1-(thioalkyl)-, synthesis of, 322
Benzotrifluoride, pentafluoro-, copyrolysis of, with chlorodifluoromethane, 737
Benzoxaphospholes, reactions, with tetraoxaspirononanes, 478
4H-3,1-Benzoxathiines, synthesis of, 232
1,3-Benzoxathiole, synthesis of, 231
Benzoxazaphosphorinanes, synthesis of, 128
1,3-Benzoxazine compounds, synthesis of, 301
3,1-Benzoxazines, synthesis of, 306
1,3-Benzoxazin-4-ones, synthesis of, 302
2H-Benzoxazin-3-ones, 2-hydroxy-, synthesis of, 175
1,2-Benzoxazoles, oxidation of, 206
Benzoyl chloride, reactions, with phospholes, 341
Benzoyl chlorides, methyl-, chlorination of, 11
N-Benzoyl compounds, from, methyl 2-methoxyhippurate, 327
Se-Benzoyl compounds, synthesis of, 239
α-Benzoyloxy compounds, from, 1,4-thiazin-3-ones, 226
Benzthiazoloneimine, 3-methyl-, reactions, with vinylidene difluorides, 992
Benzylamine, reactions, with cyanoacetic thionoesters, 896
Benzylamine, N-benzylidene-, acylation of, 325
Benzylamine, cyclopalladated N,N-dimethyl-, reactions, with carbyne complexes, 716
Benzylamines
  oxidation of, 449
  reactions, with glyoxal, 298

Benzyl bromides, chlorination of, 37
Benzylimine, N-methyl-, reactions, with spirophosphoranes, 478
Benzyl radicals, synthesis of, 1072
Benzyl sulfone, 1,1'-dibromo-, from, phenylacetic acid, 83
Benzylthiol
  reactions
    with N-(diethylaminomethyl)benzamide, 322
    with formaldehyde, 215
    with N-(piperidinomethyl)benzamide, 322
Benzylthiols, double deprotonation of, 367
Benzylthiols, germyl-, from, benzyl germyl thioethers, 371
Benzylthiols, α-silyl-, from, benzyl silyl thioethers, 368
Benzynes, reactions, with α-aminosilanes, 508
Beryllium, dimethyl-, reactions, with trimethylaluminum, 709
Beryllium chloride
  reactions
    with phosphorus ylides, 577
    with ylides, 395
Betaines
  as intermediates, 732
  synthesis of, 342
Betaines, oxaborole, X-ray structure of, 936
Betaine salts, phospho, as intermediates, 752
Bicyclo[2.2.1]hept-5-en-2-one, reactions, with tri-n-butylstannyllithium, 952
Bicyclo[2.2.0]hexenones, synthesis of, 755
Bicyclo[2.2.0]hex-5-en-2-ones, as starting materials, for hydrates, 160
Biphenyl, 2,2'-dilithio-
  synthesis of, 668
  X-ray structure of, 668
Bipyrazoles, ring-opening of, 989
Biradicals, use in the synthesis of radicals, 1077
1,4-Biradicals, synthesis of, via abstraction of a γ-hydrogen, 1079
Bis(acyloxy) compounds, from, carboxylic acids, 205
Bis(η^3-allyl) complexes, reactions, with cyclohexadiene, 694
Bisamides, synthesis of, 418
Bisamides, alkylidene, synthesis of, 424
Bisamides, N,N'-alkylidene
  from, N-[α-(benzylideneamino)benzyl]benzamides, 424
  synthesis of, via condensation of amides with carbonyl compounds, 424
Bisamides, N^1,N^2-alkylidene, synthesis of, 423
Bisanilides, synthesis of, 282
Bisarsenic acid, synthesis of, 592
Bisarsenic oxide, dimeric, as starting material, for bisarsinomethanes, 592
Bisarsenioalkenes, stability of, 1038
Bisarsinomethanes
  from
    bis(dichloroarsino)methane, 593
    dimeric bisarsenic oxide, 592
  synthesis of, 592
1,2-Bis(bromomethyl) compounds, synthesis of, 143
Bis(calcium iodide), methylene-, synthesis of, 669
Biscarbamates, N^1,N^2-alkylidene, synthesis of, 427
Bis(chlorophosphonium salts), hydrolysis of, 554
Bis(dichloroarsino)
  reactions, with 1,2-bis(N-methylamino)ethane, 593
  use in the synthesis of tetraalkyl compounds, 593
Bis(dichlorophosphine), methylene-, reactions of, 550, 562
Bis(dimethylarsines), synthesis of, 134
Bis(diphenylphosphine), methylene-, reactions of, 550, 563
Bisdiphenylphosphinoethene–chromium complexes, structure of, 1030
Bisdithioacetals, from, dicarbonyl compounds, 245
Bishemithioaminals, from, terephthalaldehyde, 317
Bishydrazine, methylene-, synthesis of, 437
Bishydroboration, for the synthesis of carbon-bridged dimercury complexes, 697

Bishydroxylamines, reactions, with ketones, 436
Bisimines, from, diphenylmethanimine, 438
Bisimino complexes, dinuclear, reactions, with ethyne, 683
Bisisonitriles, alkylidene-, chemistry of, 538
Bislithio compounds, reactions, with benzaldehyde, 938
Bismaferrocenes, synthesis of, 1038, 1039
Bis(magnesium bromide), methylene-
   reactions
      with metallocene dichlorides, 670
      with titanocene dichloride, 670
   synthesis of, 669
Bismercuration, of carboxylic acids, 697
Bis(mercuric iodide), methylene-, reactions, with boron tribromide, 632
Bis(mercury halides)
   as starting materials, for methylenedilithium, 668
   transmetallation of, 668
2,2-Bis-(4-methoxybenzoyloxy) compounds, from, 1,3-dicarbonyl compounds, 206
Bismoles
   reactions, with iron(II) chloride, 1039
   synthesis of, 1038
Bismolybdenumtetracarbonyl complexes, synthesis of, 1023
Bismolyl anions
   reactions, with iron(II) chloride, 1041
   synthesis of, 1039
Bismuth
   α-haloalkenyl compounds of, synthesis of, 805
   synthesis of functions, **591–600**
      with no halogen, chalcogen or nitrogen, **1021–1042**
Bismuth, tris(trifluorovinyl)-, synthesis of, 805
Bismuth compounds, synthesis of, 599, 600
Bismuth compounds, α-halo-, synthesis of, 136
Bismuth diiodide, phenyl-, use in the synthesis of bismoles, 1039
Bismuth functions, synthesis of, 595
Bismuthine, synthesis of, 599
Bismuthine, dimethyl(trimethylsilylmethyl)-, synthesis of, 599
Bismuth trichloride, reactions, with trimethylsilylmethylmagnesium chloride, 599
Bis(phenylarsenic acid), methylene-
   from, bis(diphenylarsino)methane, 592
   synthesis of, 592
Bisphosphiminoalkenes, as starting materials, for α-phosphimino enamines, 991
Bisphosphines, synthesis of, 111, 546
Bisphosphines, tetraphenyl, reactions, with α-hydroxyamides, 465
Bisphosphinic acids, from, primary amines, 470
α,ω-Bisphosphinoalkanes, reactions, with N-hydroxymethyl dialkylamines, 452
Bisphosphinoalkenes
   reactions, with organometallic compounds, 1029
   synthesis of, 1028
   transition metal complexes of, synthesis of, 1028
Bisphosphinoalkenes, acyclic
   nomenclature, 1022
   synthesis of, 1022
Bisphosphinoalkenes, cyclic, synthesis of, 1022
1,1-Bisphosphinoalkenes, reactions, with transition metal compounds, 1028
Bis(phosphonates), synthesis of, 553
Bis(phosphonates), methylene, synthesis of, 555
Bis(phosphonium salts)
   from, 1,1-dihaloalkanes, 553
   synthesis of, 545
Bis(phosphonium salts), diphosphirane, synthesis of, 550
Bisselenoesters, synthesis of, 288
Bis(silyl) anions, synthesis of, 618
Bissulfonamides
   from, N-alkyl arylsulfonamides, 428
   synthesis of, 282, 427

Bissulfonamides, N^1,N^2-alkylidene, synthesis of, 427
gem-Bissulfonates, synthesis of, 214
Bis(sulfones)
   electrochemical oxidation of, 277
   from
      alkylidene disulfones, 274
      bis(phenylsulfonyl)methane, 274
      dithioacetals, 272
      ketene dithioacetal tetraoxides, 277
      methylene disulfones, 274
      sulfonyl fluorides, 280
   reactions, with dihalides, 275
   as starting materials
      for alkenes, 275
      for allylic amines, 279
   synthesis of, 272
Bis(sulfones), hydroxycyclopropyl, synthesis of, 277
Bis(sulfones), oxaziridine, synthesis of, 273
Bis(sulfones), quinadyl, synthesis of, 281
Bissulfonic acids, synthesis of, 281
Bis(sulfonylacetate), dimethyl methylene, reactions, with benzaldehyde, 277
Bis(sulfonylacetic acid), methylene-, decarboxylation of, 272
Bissulfoxides
   from
      dithioacetals, 267
      sulfines, 270
Bissulfoxides, methylene, as starting materials, for bissulfoxides, 269
Bisthioamides, N,N'-alkylidene, synthesis of, 424
Bisthiols, reactions, with ketones, 347
Bis(thiophosphoryl chloride), methylene-, reduction of, 553
Bis(tributyltin) compounds, synthesis of, 702
Bis(trifluoroacetate), phenyl iodosyl [PIFA], for the oxidation of amides, 422
1,1-Bis(trimethylsilyl) compounds, synthesis of, 605
Bis(trimethylstannane), methylene-, synthesis of, 699
Bisulfite addition compounds, of aldehydes and ketones, 237
σ-Bonded compounds, synthesis of, 1032
9-Borabicyclo[3.3.1]nonane, as hydroborating reagent, 631
9-Borabicyclo[3.3.1]nonane, B-iodo-, reactions, with 1-ethoxyalkyne, 765
9-Borabicyclo[3.3.1]nonane, B-isopropyl-, bromination of, 143
9-Borabicyclo[3.3.1]nonyl, B-bromo-, bromination of, 143
9-Borabicyclo[3.3.1]nonyl compounds, alkyl-, deprotonation of, 664
1-Boracyclohepta-2,6-diene, 1-phenyl-, reactions, with metal carbonyls, 695
Boracyclopropane, reactions, with di(trimethylsilyl)alkyne, 1082
Boranate, α-amido-, synthesis of, 526
Borane, aminomethyldimethyl-, synthesis of, 524
Borane, bis[(1-trimethylsilyl)isobutyl]-
   hydrolysis of, 645
   methanolysis of, 645
Borane, (1-bromoethyl)diethyl-, synthesis of, 143
Borane, (α-chlorobenzyl)diphenyl-, synthesis of, 142
Borane, chloromethyldimethyl-, synthesis of, 523
Borane, chlorothexyl-, as reagent, for the synthesis of α-haloalkenylboron compounds, 815
Borane, dichloro-, use in the synthesis of 1,1-bisborylalkanes, 631
Borane, (dichloro)vinyl-, reactions, with tetrachlorodiborane, 632
Borane, dicyclohexyl-, reactions, with 1-hexyne, 630
Borane, diethyl-, reactions, with propargyl chloride, 631
Borane, difluorovinyl-, hydroboration of, 632
Borane, dimethoxy(trimethylsilylmethyl)-, synthesis of, 642
Borane, dithexyl-, as reagent, for the synthesis of α-haloalkenylboron compounds, 815
Borane, (fluoromethyl)difluoro-, synthesis of, 143
Borane, thexyl-, use in the synthesis of α-haloalkenylboron compounds, 815

Borane

Borane, tributyl-, reactions, with phenyl isocyanide, 522
Borane, triethyl-
 as catalyst, for the synthesis of α-haloalkenylsilicon compounds, 811
 reactions, with (E)-α-iodoalkenylsilanes, 811
Borane, triisobutyl-, reactions, with 1-methoxyvinyllithium, 935
Borane, trimethyl-, reactions, with chlorine, 142
Borane, tris(cyclopentyl)-, reactions, with ethyl bromoacetate, 832
Borane, tris(phenylseleno)-, reactions, with O,O-acetals, 238
Borane compounds, handling problems, 523
Boranes
 addition of aminotin and alkoxytin compounds to, 701
 reactions, with t-butyllithium, 524
Boranes, acyloxy-, use in the synthesis of ethers, 208
Boranes, α-alkoxy-, synthesis of, 362
Boranes, alkyl-, deprotonation of, 664
Boranes, alkyl dimesityl-, α-metallation of, 664
Boranes, alkyl(α-haloalkyl)-, synthesis of, 142
Boranes, α-amino-, synthesis of, 522
Boranes, aminohalo-, as starting materials, for gem-aminoborylamino silylalkenes, 1015
Boranes, aryl(α-bromoalkyl)(n-butoxy)-, synthesis of, 144
Boranes, aryl(α-haloalkyl)-, synthesis of, 142
Boranes, α-bromoalkyl-, as intermediates, 142
Boranes, (α-bromoalkyl)imino-, synthesis of, 145
Boranes, (α-chloroalkyl)-, oxidation of, 142
Boranes, (α-chloroalkyl)dioxy-, synthesis of, 144
Boranes, (chloromethyl)dimethyl-, synthesis of, 142
Boranes, α,β-dibromo-, synthesis of, 145
Boranes, dimesitylfluoro-, reactions, with 1-lithio 1-trimethylsilylalkenes, 1057
Boranes, disubstituted, reactions, with 1-chloroalkynes, 756
Boranes, α-halo-, reactions, with alkoxides, 362
Boranes, (α-haloalkyl)bis(alkoxy)-
 synthesis of, 144
 use in asymmetric synthesis, 144
Boranes, (α-haloalkyl)oxy-, synthesis of, 144
Boranes, halo(α-haloalkyl)-, synthesis of, 143
Boranes, imino-
 reactions, with diphenyldiazomethane, 145
 synthesis of, 522
Boranes, isopinocampheyl-, reactions, with haloalkynes, 815
Boranes, α-lithio-, synthesis of, 664
Boranes, α-nitro-, synthesis of, 524
Boranes, 1-phosphinoalkyl-, synthesis of, 569
Boranes, α-pyridylaminodialkyl-, reactions, with ethoxyalkyne, 884
Boranes, α-thioalkyl-
 from
  diphenyl dithioacetals, 372
  methyl phenyl thioethers, 372
 synthesis of, 372
Boranes, trialkoxy-, reactions, with 1-silylalkynes, 1057
Boranes, trialkyl-
 reactions
  with α-bromo ester enolates, 832
  with carbon monoxide, 363
  with 1,2-dimethoxy-1-lithioethene, 935
  with dithioacetals, 372
  with isocyanides, 522
  with 5-lithio-2,3-dihydrofurans, 935
  with 1-methoxyvinyllithium, 934
  with nitrogen ylides, 522
  with α-thiocarbanions, 372
  with 2,2,2-trifluoroethyl tosylate, 735
  with trimethylsilylmethyllithium, 642
  with tris(phenylthio)methane anions, 373
 as starting materials, for α-borio enamines, 1015
Boranes, triaryl-, reactions, with nitrogen ylides, 522
Boranes, triarylsilyl-, reactions, with 1,2-bis(trimethylsilyl)ethyne, 1058

Boranes, trihalo-, reactions, with 1-silyl 1-alkyn-2-ol alkyl ethers, 1057
Boranes, (trimethylsilyl)ethyl-, synthesis of, 356
Boranes, vinyl-
 boroboration of, 632
 hydroboration of, 629
 reactions
  with bromine, 145
  with electrophiles, 144
 as starting materials, for 1,1-bisborylalkanes, 631
 synthesis of, 935
Boratabenzene complexes, use in the synthesis of multidecker borole complexes of ruthenium, 695
Borate, bis(pyridine)iodine(1) tetrafluoro-
 reactions
  with alkynyl sulfides, 772
  with 1-chloroalkynes, 760
  with 1-iodoalkynes, 749
Borate, diethoxycarbenium tetrafluoro-, reactions, with ketones, 193
Borate, 1,3-dimethyl-2-ethoxy-4,5-dihydroimidazolium tetrafluoro-, as starting material, for ketene aminals, 975
Borate, 1,3-dithienium fluoro-, synthesis of, 1074
Borate, (hydroxymethyl)triphenylphosphonium tetrafluoro-, reactions, with diethylaminosulfur trifluoride, 117
Borate, 1-methyl-2,5-diphenyl-6-nitro-1,4-dithiinium tetrafluoro-, ring-opening of, 907
Borate, 4-nitrobenzenediazonium tetrafluoro-, reactions, with methylsulfonyldibenzoylmethane, 332
Borate, N,N,N',N'-tetramethylethoxyformamidinium fluoro-, as starting material, for ketene aminals, 975
Borate, trialkyl, reactions, with α-thiocarbanions, 372
Borate, triisopropyl, reactions, with 1-methoxyvinyllithium, 934
Borate, trimethyl
 reactions
  with 1,1-dichloroethane, 813
  with phenylthiomethyllithium, 372
  with trimethylsilylmethylmagnesium chloride, 642
Borate, trityl fluoro-, reactions, with 1,3-dithiane, 1074
Borates, N-alkyl-5-phenylisoxazolium fluoro-, as starting materials, for α-azido enamines, 990
Borate salts, cyano-
 as starting materials, for oxazaborolines, 523
 synthesis of, 523
Borate salts, quaternary, synthesis of, 642
Borate salts, tetrafluoro-
 reactions, with methoxide, 892
 synthesis of, 1073
Borepane, 3,6-dimethyl-, as dihydroborating reagent, 631
Boric acid, fluoro-
 reactions, with nickelocene, 694
 use in the synthesis of carbon-bridged diruthenium complexes, 680
Boric acid, tetrafluoro-, use in the synthesis of carbon-bridged diruthenium complexes, 680
Borienes, 2-boro, from, 1-stannylalkynes, 1054
Borirenes
 from, ethynes, 1058
 synthesis of, 1063
2-Bornanones, as starting materials, for chloro nitroso compounds, 107
exo-Borneol, 10-mercapto-, as starting material, for 1,3-oxathianes, 231
Boroboration, of vinylboranes, 632
Borohydride anions, oligomeric, synthesis of, 524
Borohydrides, α-amino, synthesis of, 522
Borohydrides, trialkyl, carbonylation of, 364
Borolane, β-methoxy-2-(trimethylsilyl)-, synthesis of, 643
Borole complexes, triple-decker, synthesis of, 695
4-Borolene, 1-(diisopropylamino)-2,3-di(trimethylsilyl)-, synthesis of, 643

Boroles, synthesis of, 1063
Boroles, 2-boro, synthesis of, 1054
Boroles, dihydro-, reactions, with metal carbonyls, 695
Boron
   stabilization by, for the synthesis of carbenes, 1082
   synthesis of functions
      bearing a germanium and a boron, 646
      bearing a silicon and a boron, 642
      bearing two borons, 629
Boron, bis(trimethylstannylethynyl)-, reactions, with triethyl boron, 1063
Boron, triethyl-
   reactions
      with bis(trimethylstannylethynyl) boron, 1063
      with 1-(trimethylstannyl)ethyne, 1068
      with trimethylstannylethynyl boron, 1063
Boron, trimethylstannylethynyl, reactions, with triethyl boron, 1063
Boron alkenes, *gem*-bromo, from, *gem*-diboroalkenes, 814
Boronate, α-bromo-, reactions, with sodium methoxide, 363
Boronate, dibutyl α-styryl-, hydroboration of, 631
Boronate, dibutyl β-styryl-, as starting material, for tetrabutyl 2-phenylethylidenebisboronate, 631
Boronate, diisopropyl (1-methoxyvinyl)-
   reactions, with pinanediol, 934
   synthesis of, 934
Boronate, pinacol(1-chloroethyl)-, reactions, with trimethylstannyllithium, 665
Boronate, pinacol(1-iodoethyl)-, reactions, with *t*-butyllithium, 664
Boronate, pinacol[1-(trimethylstannyl)ethyl]-, synthesis of, 665
Boronate, (*S*)-pinanediol (1-methoxyvinyl)-, synthesis of, 934
Boronate, propanediol benzyl-, deprotonation of, 664
Boronate, tetrabutyl 2-phenylethylidenebis-, from, dibutyl β-styrylboronate, 631
Boronate, tetrabutyl 2-phenylethylidene-1,1-bis-, reactions, with ethylene glycol, 634
Boronates
   reactions, with lithium benzyl oxide, 362
   synthesis of, 524, 1056
Boronates, allyl, displacement of halides from, 363
Boronates, aminomethyl-, as chymotrypsin inhibitors, 526
Boronates, azidomethyl-, synthesis of, 526
Boronates, chiral chloro-, synthesis of, 362
Boronates, α-chloro-
   substitution of, with sodium thiophenolate, 373
   synthesis of, 362, 526
Boronates, dibutyl iodomethyl-, reactions, with amides, 525
Boronates, α-halo-
   reactions, with alkoxides, 362
   substitution of, with sulfur nucleophiles, 373
Boronates, α-iodo-, synthesis of, 525
Boronates, α-methoxy-, synthesis of, 363
Boronates, α-phenoxy-, synthesis of, 363
Boronates, α-phenylthiomethyl-, synthesis of, 372
Boronates, 1-phosphonioalkyl-, synthesis of, 574
Boronates, vinyl-
   halogenation of, 814
   reactions, with diborane, 631
Boron bromide, dimethyl-
   reactions, with *O*-glycosides, 225
   use in the synthesis of monothioacetals, 217
Boron bromide, diphenyl, use in the synthesis of monothioacetals, 217
Boron chloride, bis(dimethylamino)-, reactions, with dilithiomethane, 632
Boron compounds
   reactions, with phosphorus ylides, 574
   synthesis of, 1031
Boron compounds, aminomethyl, chemistry of, 522
Boron compounds, (1*S*)-(1-chlorobutyl)-, synthesis of, 144
Boron compounds, α-halo, synthesis of, 142
Boron compounds, α-haloalkenyl-
   synthesis of, 813
      via halogenation reactions, 813
      via hydroboration of haloalkynes, 814
      via ligand exchange, 813
Boron esters, cyclic, reactions, with 1,3,5-triphenyl-1,3,5-triazine, 457
Boron halides, reactions, with perfluorovinyltin compounds, 813
Boron hydrides, (α-haloalkyl)-, synthesis of, 142
Boronic acid, aminomethyl-, synthesis of, 525
Boronic acid, 1-chloro-2-methylpropyl-, synthesis of, 144
Boronic acids
   from, (iodomethyl)boronic esters, 144
   synthesis of, 525
Boronic acids, α-amino-, synthesis of, 522
Boronic acids, α-halo-, synthesis of, 144
Boron tribromide
   reactions, with methylene bis(mercuric iodide), 632
   use in the synthesis of dibromoalkanes, 26
Boron trichloride
   reactions
      with imines, 107
      with silylgermanes, 639
Boron trifluoride, reactions, with mercuric oxide, 195
Boron trifluoride etherate
   as catalyst
      for the alcoholysis of nitriles, 883
      for the condensation of thiols, 245
      for the synthesis of acetals, 181
   reactions
      with alkynylsilanes, 808
      with hydrogen chloride, 885
   use in the synthesis of ketene *S,N*-acetals, 905
Boron trihalides, for the selective deprotection of methyl ethers, 175
Borylboration, for the synthesis of α-borylsilylalkanes, 644
Boryl compounds, dialkyl-, as intermediates, 364
Boryl electrophiles, reactions, with α-metallosilylalkanes, 642
Boryloxy compounds
   from
      α-diazo esters, 832
      α-sulfuranylidene esters, 832
   of ketene acetals, synthesis of, 832
Bridging carbon atoms, in silver and copper complexes, 695
Bromide, reactions, with diazo compounds, 90
Bromide azide
   reactions, with vinylsilanes, 514
   as starting material, for *gem*-diazides, 433
Bromination
   of alkanes, 19
   of ketones, 20
   of vinylgermanes, 811
Bromination–dehydrobromination, stereochemistry of, 779
Bromine
   reactions
      with acylvinyl nitroamines, 106
      with alkynyl sulfides, 772
      with 1,2-bis(hydroxymethyl)dicarbaboranes, 143
      with bromoalkenes, 23
      with α-bromocarboxylic acids, 27
      with 1-cyano-2-tosyl compounds, 105
      with 1-diethylamino-2-phthalimidostyrenes, 793
      with dimethyl benzenesulfamido malonate, 105
      with α-keto acids, 28
      with methylselenobut-3-en-1-yne, 786
      with methylthiomethyl ethers, 216
      with organoaluminums, 758
      with phenoxyalkyne, 765
      with 1-phenyl-2-phospholene 1-oxide, 122
      with 1-(phenylseleno)pent-1-yne, 786

Bromine
    with pyrazolines, 105
    with silver acrylate, 27
    with silver α-fluoro carboxylate, 35
    with sulfides, 71
    with tetrakis(trimethylsilylmethyl)germane, 641
    with 1,1,3,3-tetramethyl-1,3-disilacyclobutane, 622
    with thallium(I) salts, 27
    with 1,2,3-triazolo[1,5-a]pyridine, 27
    with trifluorosilyl(vinyl)silanes, 138
    with trimethylsilyl(vinyl)silanes, 138
    with triphenyl phosphite, 25
    with tris(trimethylsilylmethyl), 597
    with vinyl acetates, 57
    with vinylboranes, 145
    with vinyl-1,2-dicarbadodecaboranes, 143
    with vinylsilanes, 138, **729–788**
Bromine trifluoride, use in the synthesis of α-fluoro ethers, 45
Bromo compounds, reactions, with sodium azide, 109
(R)-(-)-α-Bromo compounds, from, (S)-(-)-methyl phenylvinylphosphine oxide, 804
Bromodecarboxylation, for the synthesis of dibromoalkanes, 28
Bromodifluoromethyl group, as starting material, for difluoromethyl group, 5
Bromoform
    reactions
        with dibromothiolane 1,1-dioxide, 748
        with lithium diisopropylamide, 746
    reduction of, 23
    use in the synthesis of dibromocarbene, 24
1,2-(Bromosilyl) compounds, synthesis of, 1051
Brönsted acids, as catalysts, for the synthesis of dithioacetals, 246
Brook rearrangement
    of α-aminosilanes, 506
    See also Reverse Brook rearrangement
Butadiene, 2,3-dimethyl-
    magnesium compounds of, for the synthesis of trinuclear iridiacyclopentadiene complexes, 690
    reactions
        with (dichloromethyl)phosphine, 111
        with dimethyl dithionooxalate, 229
Butadiene, 1,4-diphenyl-, reactions, with 1,3-diphosphaindene, 548
Butadiene, (E,E)-disilyltetrachloro-, synthesis of, 806
Butadiene, hexachloro-, reactions, with 1,3-propanedithiol, 770
Butadiene, (Z,Z)-1,2,3,4-tetrachloro-, silylation of, 806
1,3-Butadiene, 1,1,-bis(methylthio)-, synthesis of, 852
1,3-Butadiene, 1,4-bis(trimethylsilyl)-, silylation of, 1045
Buta-1,3-diene, 1,4-dichloro-, reactions, with dilithiophenylarsine, 1039
1,3-Butadiene, 2,3-dilithio-, reactions, with bis(chlorodimethylsilyl)methane, 621
1,3-Butadiene, 1-methoxy-3-trimethylsilyloxy-, reactions, with Eschenmoser's salt, 305
1,3-Butadiene, perchloro-, reactions, with arenethiolates, 851
1,3-Butadiene, 1H-perchloro-, nitration of, 797
Buta-1,3-diene, perfluoro 2,3-dimethyl-
    from
        cyclobutanes, 730
        perfluoro 2,3-dimethylbutanes, 730
1,3-Butadiene, 1,2,3,4-tetrakis(diphenylphosphino)-, detection of, 1023
1,3-Butadiene, 1-(trimethylsilyl)-, reactions, with borane dimethyl sulfide, 643
Butadienes, synthesis of, 892, 1028
Butadienes, 1,1-bis(toluene-*p*-sulfinyl)-, synthesis of, 863
Butadienes, 4,4-bis(trimethylsilyl) 1-phenyl-, synthesis of, 1049
1,2-Butadienes, 1,1,3,4-tetrasilyl-, synthesis of, 1045
1,3-Butadienes, 1,1,4,4-tetraalkoxy, from, 1,1,4,4-tetraalkoxy 2-butynes, 830

Butadiyne, bis(trimethylsilyl)-, zirconium-catalyzed *trans*-carboalumination of, 809
Butanal, reactions, with mercury(II) chloride, 697
Butane, 3,4-dibromo-, synthesis of, 23
Butane, 1,3,3-tris(difluoramino)-, synthesis of, 439
Butanes, perfluoro 2,3-dimethyl-, as starting materials, for perfluoro 2,3-dimethylbuta-1,3-diene, 730
Butanoate compounds, 2,2-dihydroxy-3-oxo-, synthesis of, 162
t-Butanol, reactions, with isocyanates, 1007
Butan-2-one, 3,3-dibromo-, synthesis of, 24
Butatriene, 1,1,4,4-tetrakis(trimethylsilyl)-, reactions, with α,α-disilyl ketones, 619
Butatrienes, synthesis of, 1045
1,2,3-Butatrienes, synthesis of, 821
Butene, perfluoroiso-
    reactions, with trimethyltin methoxide, 762
    synthesis of, 737
Butene, polychloro-, reactions, with sodium phenylmethanethiolate, 770
But-1-ene, 4,4-dichloro-, synthesis of, 14
But-2-ene, *trans*-2,3-dibromo-, reactions, with trifluoroperacetic acid, 24
But-2-ene, 2,3-dimethyl-, reactions, with difluorocarbene, 7
2-Butene, (e)-1,4-di(trimethylsilyl)-, dimetallation of, 603
2-Butene, (z)-1,1,4,4-tetra(trimethylsilyl)-, synthesis of, 603
But-3-enoic acid, 2,2-dichloro-, synthesis of, 13
Butenolide, bromination of, 56
Butenolides, bromo, synthesis of, 57
1-Butenyl complexes, 1-amino-3-oxo-, synthesis of, 1018
1-Buten-3-yne, synthesis of, 1045
t-Butoxy chloride, reactions, with morpholinones, 102
t-Butoxy radicals, reactions, with acetals, 1078
Butylamine, reactions, with 1,3-oxazin-6-ones, 981
Butyllithium, reactions, with 1,8-bis(trimethylsilylmethyl)naphthalene, 603
n-Butyllithium–pentamethyldiethylenetriamine, reactions, with tetramethylsilane, 602
n-Butyllithium–potassium *t*-butoxide
    as metallating agent, 650
    See also Superbase
n-Butyllithium–tetramethylethylenediamine, reactions, with ferrocenes, 1034
Butyllithium–tetramethylethylenediamine-[1,2-bis(dimethylamino)ethane], reactions, with 2,3-dihydro-1,1-dimethyl-1-silaphenalene, 603
t-Butylthiol, reactions, with imines, 325
t-Butylthiyl radicals, reactions, with thioamide, 1077
1-Butyne, reactions, with *n*-butyllithium, 605
1-Butyne, 1-phenyl-, dilithiation of, 668
1-Butyne, 1-trimethylsilyl-, lithiation of, 1061
1-Butyne, 1,3,3-tris(trimethylsilyl)-, synthesis of, 605
2-Butyne, reactions, with *n*-butyllithium, 605
But-2-yne
    as starting material, for dihydrido diiridium carbon-bridged complexes, 689
    use in the synthesis of chromacyclopentadienes, 676
2-Butyne, hexafluoro-
    reactions
        with cumulenes, 613
        with silver fluoride, 696
Butynes, reactions, with potassium *t*-butoxide, 1028
Butynes, 1-silyl-, from, 1,1-disilyl allenes, 1046
1-Butynes, 3-bromo 1-silyl, reactions, with silyl cuprates, 1046
2-Butynes, 1,1,4,4-tetraalkoxy, as starting materials, for 1,1,4,4-tetraalkoxy 1,3-butadienes, 830
2-Butynoic acid, polysilylation of, 1047
Butyrate compounds, ethyl 3-oxo-4-alkoxy, oxidation of, 212
γ-Butyrolactols, β-aryl-, synthesis of, 168
γ-Butyrolactones, γ-methylene-, epoxidation of, 209

Cadmium
  reactions, with alkenyl bromides, 821
  use in Wittig reactions, 734
Cadmium, bis(perfluoroisopropyl)-, synthesis of, 152
Cadmium, bis(trifluoromethyl)-, use in the synthesis of difluorocarbene, 7
Cadmium, perfluoroisopropyl-, metathesis of, 151
Cadmium compounds, reactions, with phosphorus ylides, 586
Cadmium compounds, vinyl-, reactions, with copper(I) bromide, 821
Caesium fluoride, reactions, with diacetoxyalkyne, 768
Caesium salt, 1,12-bis(hydroxymethyl)decahydrododecaborate, reactions, with hydrogen iodide, 143
Calcium, reactions, with diiodomethane, 669
Caprolactam
  α,α-dichlorination of, 12
  N-ethylthiomethylation of, 324
Caprothionolactone, reactions, with organolithium compounds, 226
2-Carbaldehyde, pyrrole-, reactions, with piperidine, 406
Carbalumination, of phenylethyne, 1068
Carbamate, ethyl diphenylmethylene, reactions, with hydroxylamines, 448
Carbamates
  from, lithium compounds, 934
  metallation of, 530
  reactions
    with s-butyllithium, 382
    with diisobutylaluminum methanesulfonate, 383
    with tris(dimethylamino)titanium chloride, 383
  as starting materials
    for α-aminoalkylphosphorus compounds, 466
    for gem-methoxycarbonylamino diphenylphosphinoylalkenes, 1000
Carbamates, N-acyl alkylidene, synthesis of, 427
Carbamates, O-alkyl
  deprotonation of, 382
  lithiation of, 382
  silylation of, 360
Carbamates, alkyl N,N-dichloro-, reactions, with phenylalkyne, 794
Carbamates, aryl N-(benzyloxymethyl), hydrogenolysis of, 310
Carbamates, benzyl, as starting materials, for α-aminoalkylphosphorus compounds, 467
Carbamates, benzylic, alkylation of, 382
Carbamates, α-bromo
  synthesis of, 102
  use in the synthesis of α-amino acids, 102
Carbamates, chiral, deprotonation of, 382
Carbamates, gem-chloro, synthesis of, 794
Carbamates, α-chloro-, synthesis of, 102
Carbamates, α-chloroalkyl, synthesis of, 60
Carbamates, enol
  α-metallation of, 841
  α-selenylation of, 841
Carbamates, ethyl
  reduction of, 303
  as starting materials, for α-aminoalkylphosphorus compounds, 467
Carbamates, α-halo
  synthesis of, 101
  via halogenation, 102
Carbamates, lithiated, synthesis of, 530
Carbamates, O-(1-lithiovinyl), reactions, with 1,2-dibromotetrafluoroethane, 768
Carbamates, methyl
  as starting materials, for α-aminoalkylphosphorus compounds, 467
  use in the synthesis of α-aminoalkylphosphines, 456
Carbamates, nitraminomethyl, synthesis of, 447
Carbamates, phenyl, alkylation of, 382
Carbamates, 1-sulfonylalkyl, dehydrochlorination of, 840
Carbamates, 1-sulfonyl allyl, deprotonation of, 841
Carbamates, 1-sulfonylvinyl, synthesis of, 840
Carbamoyl chlorides, synthesis of, 102
Carbamoyl chlorides, chloromethyl-, synthesis of, 102
Carbamoyl compounds, α-halo, synthesis of, 101, 102
Carbanion reaction, for the synthesis of ketene S,N-acetals, 901
Carbanions
  quenching of, for the synthesis of 1,1-bis(silyl)alkanes, 602
  reactions
    with bis(methylthio)methanimines, 901
    with dicarbonyl compounds, 166
    with diiron salts, 680
    with lactones, 171
  selenylation of, 285
  thiocarbamoylation of, 901
Carbanions, dipole-stabilized
  reviews, 382
  use in the synthesis of α-amino compounds, 529
Carbanions, nitrogen-stabilized, reviews, 526
Carbanions, selenium-stabilized
  reactions, with halosilanes, 374
  silylation of, 374
  synthesis of, 398
Carbanions, sulfinyl-stabilized, as nontransferable cuprate ligands, 396
Carbanions, sulfur-stabilized, silylation of, 365
Carbatitanation, of 1-trimethylsilyl-2-phenylethyne, 1061
Carbazate, t-butyl, reactions, with α-chloro esters, 448
Carbazate, ethyl, Mannich-type reaction of, 442
Carbazates, from, hydrazines, 60
Carbazole, as phase-transfer catalyst, 792
Carbazole, N-acetyl-, reactions, with phosphorus pentachloride, 791
Carbazole, N-(1,2-dichlorovinyl)-, synthesis of, 791
Carbazole, N-(phenylthiomethyl)-, synthesis of, 323
Carbazole, 9-vinyl-, reactions, with benzotriazole, 323
Carbazoles, 9-(1-alkoxyalkyl)-, synthesis of, 310
Carbene, bromochloro-
  reactions, with diazoalkanes, 759
  reactivity of, 1080
  as starting material, for bromochloroalkanes, 38
  synthesis of, 38
Carbene, bromofluoro-
  as starting material, for bromofluoroalkanes, 35
  synthesis of, 35
Carbene, chlorofluoro-
  from
    dichlorofluoromethane, 1083
    trichlorofluoromethane, 1082
  reactions, with triphenylphosphonium fluorenylide ylide, 753
  as starting material, for chlorofluoroalkanes, 32
  synthesis of, 32
Carbene, dibromo-
  from
    cyclopropanes, 1083
    phenyl tribromomethylmercury, 748
  reactions, with ketene acetals, 162
  reactivity of, 1080
  as starting material, for dibromoalkanes, 24
  synthesis of, 24
    photochemical methods, 1083
Carbene, dichloro-
  from
    cyclopropanes, 1083
    ketenes, 1083
    tetrachloromethane, 1083
  identification of, 1082
  reactions
    with allyl benzyl sulfide, 775

Carbene

    with diaryldiazomethanes, 740
    with sulfides, 775
    with thioacetals, 775
  reactivity of, 1080
  as starting material
    for dichloroalkanes, 16
    for dichloromethylenetriphenylphosphine, 737
  synthesis of, 16
Carbene, difluoro-
  from
    alkoxides, 1083
    bromodifluoromethylphosphonium salts, 7
    (bromodifluoromethyl)triphenylphosphonium bromide, 737
    chlorodifluoromethane, 8
    trifluoromethylphenylmercury, 1083
  reactions
    with benzobarrelene, 8
    with cyclopropenes, 737
    with 2,3-dimethylbut-2-ene, 7
  reactivity of, 1080
  sources of, 7
  as starting material, for difluoroalkanes, 7
  synthesis of, 1083
    photochemical methods, 1083
Carbene, diiodo-
  as starting material, for diiodoalkanes, 29
  synthesis of, 29
Carbene, dimethoxy-, synthesis of, 1080
Carbene, dinitro-, from, ylides, 1083
Carbene, fluoroiodo-
  as starting material, for fluoroiodoalkanes, 36
  synthesis of, 36
Carbene, fluoro(trifluoromethyl)-, reactions, with phosphines, 111
Carbene, isopropylidene-
  reactions, with *trans*-diaryldiphosphines, 1025
  synthesis of, 985
Carbene complexes, phosphinoiridium-, rearrangement of, 583
Carbene–rhenium complexes, reactions, with zirconocene chlorohydride, 716
Carbenes
  dimerization of, 830
  electrochemical methods, synthesis of, 1083
  flanked by one sulfur and one oxygen, 1081
  flanked by two hetero atoms, 1080
  flanked by two nitrogens, 1080
  flanked by two oxygens, 1080
  flanked by two sulfurs, 1081
  from, diazo compounds, 1082
  as intermediates, in the synthesis of dithioacetals, 250
  reactions
    with diazoalkanes, 755, 759
    with silanes, 137
    with thiocarbonyl compounds, 862
  sources of, 709
  synthesis of, 1080, **1071–1083**
    irradiative methods, 1083
    via α-elimination, 1081
  use in the synthesis of ketene aminals, 975
Carbenes, chloro-
  reactions
    with phosphaalkenes, 115
    with phosphaalkynes, 110
Carbenes, dichloro-
  reactions, with alcohols, 767
  as starting materials, for ketene aminals, 979
Carbenes, dihalo-, synthesis of, 1082
Carbenes, diphosphorus-substituted, addition of, 557
Carbenes, dithiolium-, dimerization of, 851
Carbenes, metallated
  synthesis of, 1083
  See also Carbenoids
Carbenes, phosphino(silyl)-
  addition of, 567
  intramolecular insertion of, 567
Carbenes, α-stannyl-, insertion into C-S bonds, 398
Carbenes, vinyl-
  reactions
    with diphenyl diselenide, 875
    with diphenyl ditelluride, 877
Carbenoids
  definition of, 741
  reactions
    with acetone, 146
    with electrophiles, 817
  synthesis of, 145, 1083
  transmetallation of, 820
  use of term, 816
  See also Carbenes, metallated
Carbinols, synthesis of, 513
Carbinols, diaryl, as starting materials, for *gem*-dinitroalkenes, 993
Carbinols, silyl, synthesis of, 358
Carbinols, stannyl, synthesis of, 533
Carbinols, tertiary, synthesis of, 935
trans-Carboalumination, zirconium-catalyzed, of bis(trimethylsilyl)butadiyne, 809
α-Carbocations
  stability of, 674
  stabilization of, in alkyne-bridged dicobalt complexes, 688
Carbodiimides
  reactions
    with *N*-acylketenimines, 982
    with bis(fluorosulfonyl)methane, 866
  as starting materials, for ketene aminals, 979
Carbodiimides, isocyanatomethyl, synthesis of, 441
Carbodiimides, *N*-silylated, reactions, with α-chloro alkyl isocyanates, 441
Carbohydrate compounds
  cyclization in, 1079
  dehydration of, 858
Carbohydrates, as hemiacetals, 176
β-Carbolines, lithiation of, 532
Carbometallation
  of alkynylsilanes, 809
  of ethynylsilanes, 809
Carbonate, ethylene, as dehydrating agent, for the synthesis of acetals, 178
Carbonate, methyl, chlorination of, 59
Carbonate, 1,2,2,2-tetrachloroethyl *t*-butyl, synthesis of, 59
Carbonate, trichloromethyl, as *in situ* source of phosgene, 58
Carbonates, *O*-alkyl
  lithiation of, 382
  silylation of, 360
Carbonates, α-chloroalkyl, from, chloromethyl chloroformates, 59
Carbonates, chloroalkyl alkyl, synthesis of, 59
Carbonates, α-chloroalkyl alkyl, reactions, with amines, 60
Carbonates, α-haloalkyl alkyl, synthesis of, 59
Carbonates, α-haloalkyl aryl, synthesis of, 59
Carbonates, α-haloalkyl phenyl, synthesis of, 59
Carbon atoms, pentacoordinate, in dimeric bis(cyclopentadienyl) metal–methyl complexes of lanthanides, 702
Carbon–boron bonds, synthesis of, 629
Carbon-bridged complexes, synthesis of, 669
Carbon(η^5-cyclopentadienyl)phosphinerhodium, reactions, with tetracarbonylethyneosmium, 710
Carbon diselenide
  electrochemical reduction of, 875
  reactions, with ethynecarboxylate esters, 874
Carbon disulfide
  reactions
    with dimethylketene, 836

Carboxylic acids

with α-hydrazino enamines, 986
with phenyliodinium ylides, 866
with phosphoranes, 843
Carbon monoxide
  insertion into silylmethyllithium compounds, 931
  reactions
    with iridium complexes, 931
    with silylmethyllithium compounds, 931
    with trialkylboranes, 363
Carbon monoxide insertion, for the synthesis of lactone complexes, 688
Carbon nucleophiles, reactions, with bromotrimethylgermane, 361
Carbon–phosphorus bonds, synthesis of, via Arbuzov rearrangement, 340
Carbons, pentacoordinate, in carbon-bridged dimanganese complexes, 677
Carbons, tetracoordinate aryl, in dimeric dimesityliron complexes, 685
Carbon selenide sulfide, reactions, with dimethyl ethynedicarboxylate, 869
Carbon suboxide
  addition of aminotin and alkoxytin compounds to, 701
  reactions, with chlorodiphenylphosphine, 550
  as starting material, for 1,1-bisgermylalkanes, 628
Carbon–tin bonds, cleavage of, 700
Carbonyl compounds
  hydration of, 160
  reactions
    with allylic anions, 938
    with α-chloro sulfones, 236
    with dialkyl phosphites, 559
    with diethyl 1-fluoro-1-(phenylsulfonyl)methyl phosphonate, 780
    with diphosphinates, 563
    with iminophosphoranes, 518
    with phosphines, 548
    with trimethylsilyl (trimethylsilyloxy) acetate, 213
    with water, 160
  as starting materials
    α-alkoxystannanes, 386
    for cyclic monothioacetals, 229
    for 2-haloalkenylphosphorus compounds, 801
    for α-haloalkenylsilicon compounds, 805
    for monoselenoacetals, 238
    for monotelluroacetals, 238
    for monothioacetals, 215
    for silylmethyl imines, 517
Carbonyl compounds, α-bromo-, chlorination of, 37
Carbonyl compounds, α-chloro, bromination of, 37
Carbonyl compounds, β-chlorovinyl, as starting materials, for acetals, 196
Carbonyl compounds, α-diazo-, reactions, with phosphaalkynes, 479
Carbonyl compounds, β-halo-β-nitro-α,β-unsaturated, synthesis of, 795
Carbonyl compounds, *gem*-halophosphono α,β-unsaturated, synthesis of, 803
Carbonyl compounds, hydroxy
  cyclization of, 164
  as starting materials, for cyclic hemiacetals, 164
Carbonyl compounds, mercapto-, cyclization of, 227
Carbonyl compounds, α-silyl-, metallation of, 648
Carbonyl compounds, α,β-unsaturated, reactions, with phosphines, 339
Carbonyl fluoride, reactions, with cyclobutanones, 59
Carbonyls
  reactions
    with aminoalkylarsines, 495
    with dialkyl phosphites, 469
    with phosphonothioic acid diesters, 469
Carbonyls, hydroxy, synthesis of, 167
Carbonyls, monomeric, reactions, with diazoalkanes, 686
Carbonyls, thiol, cyclization of, 227
Carbonyls, α,β-unsaturated, acetalization of, 179
Carbophilic addition, to allylic Grignard reagents, 255
Carbosilanes
  from
    dimercuracycles, 624
    silacyclobutanes, 610
    tetramethylsilanes, 617
2-Carboxaldehyde, pyridine, reactions, with sulfur dioxide, 237
4-Carboxylate, ethyl 1-hydroxymethylpyrazole-, use in the synthesis of α-fluoroamines, 105
Carboxylate compounds, α-haloalkyl, reactions, with alcohols, 209
Carboxylates, nitro, reactions, with 1,5-diazabicyclo[5.4.0]undec-5-ene, 108
Carboxylates, phosphonium, as intermediates, 571
4-Carboxylates, methyl 2-alkylthiazolidine-, synthesis of, 318
Carboxylic acids
  alkylation of, 205, 824
  bismercuration of, 697
  decarboxylation of, 1077
  mercuration of, 697
  reactions
    with alkynes, 206
    with dialkylboryl triflates, 832
    with dihydropyran, 208
    with enol ethers, 208
    with ethyl enol ethers, 208
    with α-haloalkyl ethers, 209
    with *N*-hydroxy-2-thiopyridone, 266
    with 1-methoxyethyne, 827
  as starting materials
    for acyl alkyl acetals, 209
    for bis(acyloxy) compounds, 205
    for bis(dialkylborylory)ethenes, 832
    for bromoalkanes, 1077
    for bromochloroalkanes, 38
    for bromofluoroalkanes, 35
    for chloroalkanes, 1077
    for chlorofluoroalkanes, 33
    for chloroiodoalkanes, 40
    for dibromoalkanes, 27
    for dithioacetals, 257
    for fluoroiodoalkanes, 37
    for ketene acetals, 824
    for ketene aminals, 972
    for ketene disilyl acetals, 832
    for ketene dithioacetals, 845
Carboxylic acids, bismercurated, from, acid anhydride, 697
Carboxylic acids, α-bromo-
  iodo decarboxylation of, 40
  reactions, with bromine, 27
Carboxylic acids, α-chloro-, reactions, with lead tetraacetate, 40
Carboxylic acids, α-hydroxy-, reactions, with pivaldehyde, 213
Carboxylic acids, β-hydroxy-
  reactions
    with acetals, 213
    with aldehydes, 213
Carboxylic acids, γ-keto-
  reactions, with benzenediazonium hydroxide, 992
  as starting materials, for *gem*-diazoalkenes, 992
Carboxylic acids, α-nitro-, α-fluorination of, 108
Carboxylic acids, 3-oxo-, as starting materials, for α-fluoroalkyl esters, 52
Carboxylic acids, peroxy-, as reagents, for the synthesis of bis(sulfones), 273
Carboxylic acids, α-silyl-, deprotonation of, 648
Carboxylic acids, α-sulfonyl-
  decarboxylation of, 84

Carboxylic acids
    as starting materials, for α-halo sulfones, 81
Carboxylic acids, α-thiol
    reactions
        with aldehydes, 230
        with ketones, 230
Carbyne complexes
    reactions, with cyclopalladated
        N,N-dimethylbenzylamine, 716
    as starting materials, for alkyne-bridged tungsten
        complexes, 675
Carcinogens, bis(chloromethyl) ether, 47
Catalysts, for the synthesis of dithioacetals, 246
Catecholborane, for the reduction of ketene
    dithioacetals, 253
Catechols, methylenation of, 190
Cathepsin G, inhibition by α-aminoboronic acids, 526
Cationic centres
    flanked by an oxygen and a sulfur, 1073
    flanked by a nitrogen and an oxygen, 1072
    flanked by two oxygens, 1072
    flanked by two selenium atoms, 1075
    flanked by two sulfur atoms, 1073
Cations
    from
        acetals, 1072
        keto azides, 1072
    synthesis of, 1072
        group abstraction methods, 1074
        photochemical methods, 1075
Cephems, 2α-methoxy-, synthesis of, 227
Ceric chloride, reactions, with organolithiums, 954
Cerium compounds, with two carbon-bridges, 703
Chalcogen, synthesis of functions, **41–93**
Chalcogens
    and Group 15 elements, synthesis of functions, **293–349**
    synthesis of functions
        with any group other than a halogen or a
            chalcogen, **879–965**
        containing a silicon, germanium, boron or metal,
            351–402
    synthesis of functions bearing two, **823–877**
    synthesis of functions containing two other than
        oxygen, **243–291**
Chalcone, as starting material, for allylic alcohols, 269
Charge transfer complexes, thermolysis of, 870
Chiral auxiliaries, for the synthesis of α-aminophosphonic
    acids, 462
Chloral
    reactions
        with ammonia, 297
        with phosgene, 58
Chloral, aryl hydrazones, as starting materials, for azo
    compounds, 799
Chloral hydrate, as starting material, for dichloroacetic
    acid, 14
Chloramine-T
    reactions
        with (arylthio)dibenzoylmethane, 326
        with sulfur mustards, 773
Chloreal *See* Isocyanuric acid, trichloro-
Chlorination, regioselectivity of, studies, 67
Chlorine
    for the α-chlorination of ethers, 47
    reactions
        with chloroalkenes, 14
        with cyclobutanone oximes, 108
        with 1-diethylamino-2-phthalimidostyrenes, 793
        with ditelluroacetals, 291
        with fluoroalkenes, 31
        with nitroethenes, 108
        with phosphonium salts, 344
        with silver fluoroacetate, 33
        with 1,2,3-triazolo[1,5-*a*]pyridine, 19
        with trimethylborane, 142
Chloro compounds, from, (*S*)-(-)-methyl
    phenylvinylphosphine oxide, 804
α-Chloro compounds, as starting materials, for α-iodo
    ethers, 51
Chloroform
    reactions
        with allyltributyltin, 14
        with potassium *t*-butoxide, 16, 737
        with styrene, 17
        with thiophene 1,1-dioxide, 742
        with trimethyltinsodium, 699
2-Chromanols, synthesis of, 167
Chromium
    dimeric allyls of, synthesis of, 673
    reactions, with arsabenzene, 1041
Chromium, benzene(pentafluorolithiobenzene)-, use in the
    synthesis of metal complexes, 720
Chromium, benzenetricarbonyl-, lithiation of, 707
Chromium, bisallyl-, decomposition of, 673
Chromium, bis(η^6-arsabenzene)-, synthesis of, 1041
Chromium, hexacarbonyl-, reactions, with
    dicarbonylcyclopentadienyl(substituted phenyl)iron
    complexes, 715
Chromium, lithiated tricarbonylthiophene-, reactions, with
    bromopentacarbonylmanganese, 720
Chromium, (lithiobenzene)tricarbonyl-, synthesis of, 705
Chromium,
    pentacarbonyl[(dimethyloxosulfono)methanide]-,
    reactions, with bisdiphenylphosphino ethene, 1029
Chromium, tricarbonyl(chlorobenzene)-, reactions, with
    palladium(0), 722
Chromium, tricarbonyl(lithiobenzene)-, reactions, with
    transition metal halides, 723
Chromium, tricarbonylthiophene-, metallation of, 707
Chromium, tricarbonyl(trisammonia)-, reactions, with
    tungsten hydride complexes, 715
Chromium(0), (methoxy)methylcarbenepentacarbonyl-,
    anion of, structure of, 955
Chromium(III) chloride, reactions, with methyllithium, 708
Chromium carbene complexes, reactions, with
    phosphines, 580
Chromium chloride
    use in the synthesis of carbon-bridged dichromium
        complexes, 673
    use in the synthesis of dimeric allyls of chromium, 673
Chromium complexes, reactions, with *n*-butyllithium, 1030
Chromium complexes, alkyne-derived bridged, synthesis
    of, 676
Chromium complexes, alkynylcarbene-, as starting
    materials, for α-silyl enamines, 1014
Chromium complexes, bromomethyl-, synthesis of, 147
Chromium complexes, chloromethyl-, synthesis of, 147
Chromium complexes, haloarene-, nucleophilic substitution
    of, 715
Chromium complexes, (halomethyl)-, synthesis of, 147
Chromium compounds
    reactions
        with 1-metalloalkylphosphine oxides, 579
        with phosphorus ylides, 580
Chromium compounds, 1-haloalkyl-, reactions, with
    phosphines, 579
Chromium dichloride, pentamethylcyclopentadienyl-, use in
    the synthesis of chromacyclopentadienes, 676
Chromiumphosphinoalkene complexes, synthesis
    of, 1023, 1028
Chromium vapor, reactions, with mesitylene, 694
Chymotrypsin, inhibition by α-aminoboronic acids, 525
Cinnamaldehyde, reactions, with diethyl trimethylsilyl
    phosphite, 918
Cinnamoyl chloride, *p*-chloro-, reactions, with trimethyl
    phosphite, 922
Claisen rearrangement

Eschenmoser modification, for the synthesis of
   bis(tributyltin) compounds, 702
  for the synthesis of ketene silyl acetals, 831
  use in the synthesis of α-stannylsilylalkanes, 661
Claycop, as nitration agent, 431
Clayfen
  reactions, with aldehydes, 431
  as reagent, for the oxidative nitration of oximes, 429
Cobalt
  reactions
    with diazoalkane, 686
    with perfluorocyclooctatetraene, 687
Cobalt, carbonylcyclopentadienyldiiodo-, reactions, with
   sodium amalgam, 690
Cobalt, cyclopentadienyl(1-phenylborole)-, addition
   reactions of, 717
Cobalt, dicarbonylcyclopentadienyl-
  reactions
    with disilylalkynes, 689
    with photopyrone, 690
    with tolan, 689
Cobalt, dimesityl-
  dimeric structure of, 690
  synthesis of, 690
Cobalt(II) chloride
  as catalyst, for the synthesis of geminal diacetates, 203
  reactions, with mesityllithium, 690
Cobaltacyclopentadiene complexes
  from
    nonacarbonyldiiron, 689
    phenylethyne, 689
    sodium tetracarbonylcobaltate, 689
    trifluoroacetic acid, 689
  reactions
    with aluminum hydride triethylamine, 726
    with nonacarbonyldiiron, 714
Cobalt carbonyls
  reactions
    with alkynic esters, 712
    with vinyliron complexes, 711
Cobalt complexes, (chloroacetyl)-, synthesis of, 148
Cobalt complexes, (chloromethyl)-, synthesis of, 148
Cobalt complexes, cyclopentadienyl-, reactions, with
   triphenylstannane, 726
Cobalt complexes, germanium-bridged, reactions, with
   alkynes, 687
Cobalt complexes, metal–metal double-bonded
  reactions
    with alkynes, 689
    with 3,3-dimethylcyclopropene, 689
Cobalt complexes, mono-bridged, synthesis of, 686
Cobalt complexes, tri-bridged, synthesis of, 686
Cobalt complexes, tungstacyclopentadiene-, synthesis
   of, 714
Cobalt compounds
  reactions
    with 1-metalloalkylphosphines, 582
    with phosphorus ylides, 582
Cobalt multidecker sandwich compounds, from,
   pentamethylcyclopentadiene, 695
Cobalt trifluoride, as fluorinating agent, 3
π-Complexes, metallation of, 705
Condensation, via Peterson reactions, 936
Condensation, intramolecular, for the synthesis of
   alkylidenebisamides, 425
Copper, use in the synthesis of 1,1-bis(silyl)alkanes, 609
Copper(I), 1-methoxyvinyl-, from,
   methoxyvinyllithium, 953
Copper(I) ate complexes, α-bromoalkenyl-, as intermediates,
   in the synthesis of radialenes, 820
Copper(I) bromide
  reactions
    with 2-lithioanisole, 696

   with 2-lithio-1-(dimethylaminomethyl)benzene
     compounds, 695
   with vinylcadmium compounds, 821
   with vinylzinc compounds, 821
  use in the synthesis of 1,3-dioxalane compounds, 207
Copper(I) chloride, reactions, with mesitylmagnesium
   bromide, 696
Copper(I) compounds, uses of, 953
Copper(I) compounds, vinyl-, reactions, with S-methyl
   methanethiosulfonate, 940
Copper(I) cyanide, reactions, with pent-4-enylzinc–
   magnesium compounds, 724
Copper(I) iodide, reactions, with sulfoxides, 396
Copper(II) nitrate, montmorillonite-supported, as nitration
   agent, 431
Copper complexes, octanuclear, synthesis of, 696
Copper compounds
  reactions, with phosphorus ylides, 585
  synthesis of, 963
Copper compounds, carbon-bridged, from, metal(I)
   halides, 695
Copper compounds, carbon-bridged pentameric, synthesis
   of, 696
Copper compounds, α-haloalkenyl-, synthesis of, 821
Copper compounds, perfluoroisopropyl-, synthesis of, 151
Copper halides
  reactions
    with 1-metalloalkylphosphine oxides, 585
    with 1-metalloalkylphosphines, 585
    with 1-metalloalkyl phosphonates, 585
Copper iodide, reactions, with 1-ethoxyvinyllithium, 953
Copperlithium, dimethyl-
  reactions
    with acetals, 388
    with (α-phenylseleno) α,β-unsaturated ketones, 401
    with vinyl sulfones, 964
Coppers, α-alkoxy-, synthesis of, 384
Coumarin, dihydro-, reduction of, 170
Coumarin, 4-hydroxy-, reactions, with
   benzylidineaniline, 304
p-Cresol, reactions, with 3,3,3-trichloro-
   2-methylpropene, 745
Crotonates, Reformatsky reaction of, 736
2-Crotyl-1,3-dithiane-1-oxide, synthesis of, 263
Crown ethers, organometallic, synthesis of, 453
Cumulenes
  reactions
    with hexafluoro-2-butyne, 613
    with phenyl isocyanate, 421
  as starting materials, for 1,3-diarsacyclobutanes, 594
  synthesis of, 598, 851
Cuprate, tributylstannyl-, reactions, with
   phenylthioethynes, 962
Cuprate compounds, vinylidene, reactions, with
   acylaminophosphinic esters, 1001
Cuprates
  reactions
    with enones, 385
    with α-phenylselenoheptyllithium, 401
  synthesis of, 1061
Cuprates, α-alkoxy-
  reactions, with α,β-unsaturated ketones, 384
  synthesis of, 384
Cuprates, cyano-
  reactions
    with lithioalkenes, 953
    with selenocyanogen, 947
    with stannylalkenes, 953
  synthesis of, 947
Cuprates, lithio-, synthesis of, 953
Cuprates, α-selenyl, synthesis of, 401
Cuprates, silyl, reactions, with 3-bromo 1-silyl
   1-butynes, 1046

Cupric chloride, reactions, with organostannanes, 774
Cuprous iodide, reactions, with 1-methoxyvinyllithium, 953
Curtius acyl azide rearrangement, for the synthesis of isocyanates, 446
Curtius rearrangement
 of amino acid azides, 426
 for the synthesis of N-acylated α-ureido enamines, 982
 for the synthesis of α-aminoalkylphosphorus compounds, 476, 492
 for the synthesis of gem-diisocyanates, 433
2-Cyanacrylate, 3,3-diazido-, reactions, with triphenylphosphine, 993
Cyanamide, use in the synthesis of bicyclic α-amino organophosphorus compounds, 458
Cyanamides, as starting materials, for ketene aminals, 975
Cyanamides, isocyanatomethyl, synthesis of, 441
Cyanates, aryl
 reactions, with methylene groups, 889
 as starting materials, for ketene hemiaminals, 889
Cyanide
 reactions
  with acylsilanes, 353
  with chlorotrifluoroethene, 5
Cyanide, cuprous, reactions, with tetrakis(1-methoxyvinyl)tin, 954
Cyanide, tosyl, for cyanation processes, 815
Cyanide, trimethylsilyl
 reactions, with benzimidazole, 426
 synthesis of, 369
Cyanide ions, reactions, with tetrafluoroethene, 1076
Cyanides, aryl, reactions, with diethyl methyl phosphonate, 927
Cyanogen bromide, reactions, with amines, 413
Cycloaddition
 for the synthesis of α-aminoalkylarsenic compounds, 502
 for the synthesis of α-aminoalkylphosphorus compounds, 488, 493
 for the synthesis of cyclic monothioacetals, 231
Cycloalkanones, 2-chloro-2-(benzenesulfonyl)-, synthesis of, 80
Cycloalkanones, α-diazo-
 reactions
  with benzeneselenenyl bromide, 90
  with benzeneselenenyl chloride, 90
Cyclobutadienes, tetramethyl-, ring-opening of, 715
Cyclobutane, perfluoro-, synthesis of, 7
Cyclobutane, 1-sila-3-titana-, synthesis of, 657
Cyclobutane, thora-, reactions, with tetramethylsilane, 658
Cyclobutane complexes, oxaplatina-, reactions, with acetyl chloride, 151
Cyclobutanes, as starting materials, for perfluoro 2,3-dimethylbuta-1,3-diene, 730
Cyclobutanes, chlorofluoro-, synthesis of, 32
Cyclobutanes, 1,3-digermyl-, synthesis of, 626
Cyclobutanes, 1,3-dimetallo-, synthesis of, 656
Cyclobutanes, 1,3-disilyl-, ring-openings of, 622
Cyclobutanes, oxaplatina(IV), reactions, with actinium chloride, 151
Cyclobutanes, 1-sila-3-thora-, synthesis of, 658
Cyclobutanols, α-halo-, from, perfluorocyclobutanone, 43
Cyclobutanone, perfluoro-, as starting material, for α-halocyclobutanols, 43
Cyclobutanones, reactions, with carbonyl fluoride, 59
Cyclobutene, 3,3,4,4-tetrafluoro-, synthesis of, 5
Cyclobutenes, dichloro-, ring-opening of, 742
Cyclobutenes, perfluoro-
 reactions
  with trimethylgermane, 141
  with trimethylstannane hydride, 153
Cyclocondensation, for the synthesis of 1,3-aminoalkylarsenic compounds, 495
Cycloheptatriene, synthesis of, 557
Cycloheptatriene complexes, chroma-, synthesis of, 676

Cyclohexadiene, reactions, with bis(η^3-allyl) complexes, 694
1,3-Cyclohexadiene, perfluoro-, reactions, with octacarbonyldicobalt, 687
Cyclohexadienones, chloromethyl-, synthesis of, 17
Cyclohexadienones, tetrabromo-, synthesis of, 23
Cyclohexane, 1-acetoxy-1-nitroso-, synthesis of, 314
Cyclohexane, 1,1-bis(methylthio)-, as starting material, for 1-azido-1-methylthiocyclohexane, 329
Cyclohexane, bis(trimethylenedioxyboryl)methylene-, reactions, with iodine, 750
Cyclohexane, 1,3,5-(dimethylamino)-1,3,5-triboro-, synthesis of, 632
Cyclohexane, 5,5-dimethyl-1,3-dioxo-, oxidation of, 206
Cyclohexane, perfluoro-
 pyrolysis of, 737
 synthesis of, 3
Cyclohexanes, 1-acyloxy-1-nitroso-, synthesis of, 314
Cyclohexanes, (R,R)-diamino-, as starting materials, for (R)-α-chloroalkylphosphonic acids, 129
Cyclohexanes, diazo-, reactions, with 2,2-dimethylpropylidynylphosphine, 480
Cyclohexanes, dilithiomethylene-, synthesis of, 1064
Cyclohexanesulfonyl chlorides, dehydrochlorination of, 331
Cyclohexanol, trans-2-trimethylsilyl(phenylthio)methyl-, synthesis of, 943
Cyclohexanone
 reactions
  with bromochloromethyllithium, 758
  with ethanedithiol, 246
 as starting material
  for enethiol ethers, 247
  for 2-nitro-2-nitroso compounds, 439
Cyclohexanones, reactions, with thiolethanol, 229
Cyclohexene, 1-chloro-, reactions, with hydrogen fluoride, 30
Cyclohexene, 1-morpholino-, reactions, with 1,2-benzoquinone, 307
Cyclohexene, 1-pyrrolidino-, reactions, with o-quinonemethides, 307
Cyclohexene oxide, reactions, with trimethylsilyl(phenylthio)methyllithium, 943
Cyclohexenes, trimethylsilyl-, reactions, with iodine, 139
Cyclohex-2-enone, 3-ethoxy-, use in the synthesis of monothioacetals, 218
Cyclohexenyne complexes, hexafluoro-, synthesis of, 687
Cyclohexylidene compounds, reactions, with lithium 1-dimethylaminonaphthalenide, 945
Cyclooctatetraene
 as starting material
  for bridged iron complexes, 683
  for pentalene-bridged ruthenium complexes, 683
  for ruthenium complexes, 683
 use in the synthesis of carbon-bridged complexes, 703
 use in the synthesis of tris(cyclooctatetraene)dititanium, 671
Cyclooctatetraene, 1,4-dibromo-
 reactions
  with dodecacarbonyltriruthenium, 681
  with nonacarbonyldiiron, 681
  with tetracarbonyldimethylosmium, 681
Cyclooctatetraene, perfluoro-
 reactions
  with cobalt, 687
  with dicobalt, 687
Cyclooctatetraene-bridged complexes, synthesis of, 672
Cyclooctyne, reactions, with octacarbonyldicobalt, 689
Cyclopalladation, of amines, 720
Cyclopentadiene
 mercuration of, 697
 metallation of, 625
 as starting material, for 2-phosphabicyclo[2.2.1]heptenes, 112
 use in the synthesis of carbon-bridged ditin compounds, 701

Cyclopentadiene, 1,1-bis(trimethylgermyl)-, synthesis of, 625
Cyclopentadiene, 1,1-bis(trimethylsilyl)-, synthesis of, 603
Cyclopentadiene, 1-[dimethyl(methoxy)silyl]-1-trimethylsilyl-, synthesis of, 603
Cyclopentadiene, ferra-
    ruthenium analogue of, synthesis of, 683
    synthesis of, 714
Cyclopentadiene, pentamethyl-, as starting material, for cobalt multidecker sandwich compounds, 695
Cyclopentadiene, pentaphenylalumina-, reactions, with nickel(II) bromide, 693
Cyclopentadiene complexes, binuclear metalla-, synthesis of, 689
Cyclopentadiene complexes, rhena-, synthesis of, 677, 678
Cyclopentadiene complexes, rhodia-, synthesis of, 689
Cyclopentadiene complexes, ruthena-, synthesis of, 681, 683
Cyclopentadiene complexes, trinuclear osma-, as intermediates, 682
Cyclopentadienes, reactions, with tetramethylchloroformamidinium chloride, 974
Cyclopentadienes, benzoferra-
    from
        2-bromostyrene, 681
        dodecacarbonyltriiron, 681
        pentacarbonyliron, 681
    photolysis of, 681
Cyclopentadienes, chroma-, synthesis of, 676
Cyclopentadienes, ferra-
    from, monoalkyne complexes, 682
    synthesis of, 681
        via alkyne insertion, 683
    use in the synthesis of diiron complexes, 681
Cyclopentadienes, tetra(phenylsulfonyl)-, synthesis of, 281
Cyclopentadienide, tetrakis(benzenesulfonyl)-, protonation of, 281
Cyclopentadienyl anions, reactions, with trithiocyclopropenium salts, 852
Cyclopentadienyl–cobaltacyclopentadiene complexes, synthesis of, 690
Cyclopentadienyl complexes, iridia-, synthesis of, 690
Cyclopentanes, synthesis of, 654
Cyclopentanone-3,4-dicarboxylate, *trans*-dimethyl-, reactions, with 1,3-hydroxy thiols, 229
Cyclopentenes, diboro-, synthesis of, 632
Cyclopolyphosphines, metallated, reactions, with 1,1-dihaloalkanes, 545
Cyclopropabenzene, 7-lithio-7-trimethylsilyl-, synthesis of, 647
Cyclopropa[*c*]chromene, ring-opening of, 742
1*H*-Cyclopropa[*b*]naphthalene, 1,1-bis(trimethylsilyl)-, synthesis of, 604
Cyclopropa[*b*]naphthalene, reactions, with *n*-butyllithium, 604
Cyclopropane, 1,1-bis(dibenzylamino)-, hydrolysis of, 309
Cyclopropane, 1,1-bis(diphenylphosphino)-, synthesis of, 548
Cyclopropane, bis(phenylseleno)-, synthesis of, 289
Cyclopropane, 1-bromo-1-ethoxy-, synthesis of, 50
Cyclopropane, 1-ethoxy-1-(trimethylsiloxy)-, reactions, with phosphorous tribromide, 50
Cyclopropane, hexamethoxy-, thermolysis of, 1080
Cyclopropane, perfluoro-, synthesis of, 7
Cyclopropane, tetrabromo-, ring-opening of, 747
Cyclopropanecarboxylates, synthesis of, 146
Cyclopropanes
    from, dibromodifluoromethane, 1083
    ring-opening of, 742, 747
    as starting materials
        for dibromocarbene, 1083
        for dichlorocarbene, 1083
    synthesis of, 1082
    use in the synthesis of carbenes, 1083

Cyclopropanes, bis(silyl), synthesis of, 620
Cyclopropanes, *gem*-dibromo-
    reactions
        with alkyllithiums, 146
        with silver acetate, 57
Cyclopropanes, dichloro-, synthesis of, 16
Cyclopropanes, *gem*-dichloro-, reactions, with trimethylsilyl chloride, 605
Cyclopropanes, trihalo-, ring-opening of, 742
Cyclopropanes, vinyl-, reactions, with methacrylate esters, 830
trans-Cyclopropanes, synthesis of, 109
Cyclopropanols, from, dithioesters, 255
Cyclopropanols, α-bromo-, synthesis of, theoretical, 42
Cyclopropanone
    as starting material
        for aminals, 406
        for hemiaminals, 299
    synthesis of, 162
Cyclopropanone hydrate compounds, synthesis of, 162
Cyclopropene, 3,3-dimethyl-
    reactions
        with metal–metal double-bonded cobalt complexes, 689
        with rhodium complexes, 689
    ring-opening of, 710
Cyclopropenes
    reactions
        with benzenesulfenyl chloride, 775
        with difluorocarbene, 737
Cyclopropenes, trichloro-, as starting materials, for ketene aminals, 979
Cyclopropenium bromides, triaryl-
    as starting materials
        for trinuclear palladium complexes, 693
        for trinuclear platinum complexes, 693
Cyclopropenone, diphenyl-, reactions, with dodecacarbonyltriiron, 681
Cyclopropenylium ions, reactions, with sodium azide, 483
Cyclopropyllithium compounds, 1-silyl-, synthesis of, 652
Cyclotriphosphines, synthesis of, 114
Cymantrene, dimethylaminomethyl- *See* Manganese, (dimethylaminomethyl)cyclopentadienyltricarbonyl-
Cymantrene compounds, synthesis of, 1042
Cymantrenes *See* Manganese, cyclopentadienyl-
Cysteine, protecting groups for the sulfur of cysteine, 218, 223
Cystine, cleavage of, 218

Darzens reaction
    for the synthesis of α,β-epoxy sulfones, 236
    for the synthesis of α,β-epoxy sulfoxides, 233
Deacylation, of ketene aminals, 979
Dean–Stark apparatus, for the synthesis of acetals, 177
1-Decalone, failure to react with triphenylphosphine and tetrabromomethane, 745
2-Decalone, reactions, with triphenylphosphine, 745
Decarbonylation, photochemical, for the synthesis of alkyne-bridged diruthenium complexes, 681
Deplancheine, synthesis of, 532
Deprotection, of acetals, 175
Deprotonation
    for the synthesis of α-lithioborylalkanes, 664
    for the synthesis of α-lithiosilylalkanes, 646
    for the synthesis of 1-metallo-1-silylalkanes, 646
Deprotonation–alkylation, of dithiocarboxylic acids and esters, 842
2-Deuteriopentose compounds, 2-deoxy-, synthesis of, 253
Dewar hexafluorobenzene
    reactions, with bis(fluorooxy)difluoromethane, 46
    as starting material, for α-fluoro epoxides, 46
Dextrorphan, synthesis of, 532
Diacetate, acrolein, use in the synthesis of acetals, 190
Diacetate, pyridine-4-carbaldehyde, synthesis of, 206

Diacetate compounds

Diacetate compounds, synthesis of, 206
Diacetates, geminal
  from
    aldehydes, 203
    ketones, 204
gem-Diacetates, allylic, synthesis of, 206
Diacid diesters, synthesis of, 553
Diacyl compounds, synthesis of, 203
gem Diacyl compounds, synthesis of, via oxidation of aromatic methyl and methylene groups, 206
O,O-Diacyl compounds, synthesis of, 206
gem-Diacyloxy compounds, synthesis of, 203
1,3-Dialdehydes, synthesis of, 250
Dialkynes, macrocyclic, in dinuclear nickel complexes, 692
Dialuminum, hexamethyl-
  reactions, with
    dichloro(pentamethylcyclopentadienyl)rhodium dimers, 687
  as reagent, for the reduction of aldehydes and ketones, 699
Dialuminum alkyls, reactions, with mercury(II) chloride, 697
Dialuminum compounds, carbon-bridged, synthesis of, 698
Dialuminum tetrabromide, methylene-, reactions, with butyllithium, 708
Diamides, α-aminophosphonic, synthesis of, 475
Diamides, phosphonic, from, alkyl(diamino)phosphines, 133
Diamides, phosphonous, oxidation of, 133
Diamine, N^1,N^2-bis(benzylidene)phenylmethylene-, synthesis of, 437
Diamine, N,N'-dimethylethylene-, reactions, with glyoxal, 301
Diamines
  reactions
    with bistrifluoromethylketene, 886
    with ketene dithioacetals, 970
    with ketones, 410
    with vinylidene dihalides, 970
  silylmethylation of, 507
  as starting materials
    for aminals, 407
    for α-stannio enamines, 1020
Diamines, chiral, as starting materials, for imidazolidines, 408
Diamines, methylene-, as starting materials, for α-haloamines, 97
gem-Diamines, reactions, with thiols, 249
gem-Diamines, N^1,N^2-dihydroxy, synthesis of, 435
S-Diamines, reactions, with 3-methoxybenzaldehyde, 408
α,α-Diamino acids, synthesis of, 417
α,ω-Diamino components, use in the synthesis of phosphacycloheptanes, 456
Dianils, benzil, reactions, with 2,4-bis(phenylthio)-1,3,2,4-dithiadiphosphetane-2,4-dithione, 320
1,3-Diarsacyclobutanes
  from, cumulenes, 594
  synthesis of, 594
1,1'-Diarsaferrocene, as starting material, for 2,2',5,5'-tetradeuterio-1,1'-arsaferrocene, 1039
1,1'-Diarsaferrocene, 2-acetyl-, synthesis of, 1039
1,1'-Diarsaferrocenes, synthesis of, 1039
Diarsine, tetramethyl-
  reactions
    with chloroethene, 134
    with fluoroethene, 134
    with trifluoroethene, 134
Diazaadamantane, synthesis of, 408
Diazaallyl anions, synthesis of, 536
2,3-Diazabicyclo[2.2.1]heptane compounds, ozonolysis of, 162
Diazabicyclo[4.3.0]nonanes, synthesis of, 414
1,5-Diazabicyclo[5.4.0]undec-5-ene (DBU), reactions, with nitro carboxylates, 108
1,3-Diazabutadienes, Diels–Alder reaction of, 421

1,5-Diaza-3,7-diphosphabicyclo[3.3.0]octanes, synthesis of, 482
1,5-Diaza-3,7-diphosphacyclooctanes, synthesis of, via heteroatom exchange, 457
Diazadiphosphetidines, synthesis of, 478
1,4-Diaza-2,5-diphosphorinanes, synthesis of, 483
1,5-Diaza-3-phosphacycloheptanes, synthesis of, 456
1,2-Diaza-4-phosphacyclopentanes, synthesis of, 482
1,3,5-Diazaphosphanes, from, tetrakis(aminomethyl)phosphonium salts, 458
3,5-Diazaphosphanes, synthesis of, via heteroatom exchange, 457
2,4-Diazaphospholanes, synthesis of, 493
Diazaphospholes
  reactions, with diazo compounds, 549
  synthesis of, 483
1,2,3-Diazaphospholes, reactions, with diazo compounds, 558
1,2,4-Diazaphospholes, synthesis of, via rearrangement reactions, 479
3H-Diazaphospholes, 4,5-dihydro-, synthesis of, 484
3H-1,2,4-Diazaphospholes, synthesis of, 479
1,3,4-Diazaphospholidin-2-one 4-oxides, synthesis of, 488
1,4,2-Diazaphospholines, synthesis of, 493
1,4-Diaza-2,5-phosphorinanes, synthesis of, 486
1H-1,2,4-Diazarsole, reactions, with sulfur ylides, 502
1,2,3-Diazarsoles, 2,5-diphenyl-, reactions, with diarylnitrones, 502
1,2,4-Diazarsoles, reactions, with diphenylnitrones, 502
1,2-Diaza-3-sila-5-cyclopentenes, lithiation of, 649
gem-Diazides
  from, bromide azide, 433
  synthesis of, 432
Diazines, α-haloalkenyl, synthesis of, 799
1,2-Diazines, synthesis of, 105
1,3-Diazinones, synthesis of, 421
Diaziridines, as starting materials, for ketene aminals, 979
Diaziridines, N^1,N^2-diacyl-, homolytic fission of, 427
Diazirine, 3-chloro-3-phenyl-, thermolysis of, 138
Diazirine, phenylchloro-, thermal decomposition of, 154
Diazirines, phenyl halo-, use in the synthesis of acetals, 190
Diazo compounds
  irradiation of, 1082
  photolysis of, 557
  reactions
    with benzeneselenenyl chloride, 90
    with bromide, 90
    with diazaphospholes, 549
    with 1,2,3-diazaphospholes, 558
    with 3,5-diphenylphosphabenzene, 549
    with iodonitroform, 995
    with sulfinyl chlorides, 76
    with sulfinyl halides, 75
  as starting materials
    for carbenes, 1082
    for gem-dinitroalkenes, 995
  synthesis of, 799
Diazo compounds, phosphinyl-, photolysis of, 558
Diazoles, vinylidene, as starting materials, for ketene aminals, 979
1,2-Diazoles, synthesis of, 105
Diazonabicyclo[2.2.2]octane bis(tetrafluoroborate), 1-chloromethyl-4-fluoro-, reactions, with vinylstannanes, 736
Diazonium compounds, synthesis of, 799
Diazonium ions, α-chloroalkenyl, synthesis of, 799
Diazonium salts
  reactions
    with acylaminomalonates, 448
    with hydrazones, 436
  as starting materials, for gem-diazoalkenes, 992
Diazonium salts, arene-
  reactions

with methylamine, 444
with nitroalkanes, 439
with thiols, 225
Dibismaferrocenes, synthesis of, 1041
Dibismaferrocenes, 2,2',5,5'-tetramethyl-, synthesis of, 1041
Dibismuthines, synthesis of, 596
1,3-Diboracyclopent-4-ene, synthesis of, 632
Diboradioxanes, synthesis of, 363
Diborane
  reactions
    with trans-β-bromostyrene, 142
    with vinylboronates, 631
Diborane, 1,2-di t-butyl-1,2-dichloro-, reactions, with
    1,2-bis(trimethylstannyl)ethyne, 1063
Diborane, tetrachloro-
  reactions
    with (dichloro)vinylborane, 632
    with trimethyl(vinyl)silane, 644
Diboranes
  reactions
    with silylalkynes, 1054
    with stannylalkynes, 1054
Diboranes, chloro-, reactions, with
    1,2-bis(trimethylsilyl)ethyne, 1048
Diborapiperazines, synthesis of, 522
1,3-Diboretanes
  from, dilithium compounds, 633
  synthesis of, 632
Diboroles, synthesis of, 635
Diboronate, tetrabromomethylene-, synthesis of, 632
Diboronate, tetrabutyl ethylidene-, synthesis of, 631
Diboronate, tetramethylmethylene-, synthesis of, 632
Diboronation, of alkynes, 1054
Diboronic acids, synthesis of, 631
1,2-Diboronylethenes, trimethylsilyl, synthesis of, 1057
1,2-Dibromides, telluration of, 876
Dibromo compounds, acyclic, synthesis of, 622
Dicarbaboranes, 1,2-bis(hydroxymethyl)-, reactions, with
    bromine, 143
1,2-Dicarbadodecaboranes, vinyl-, reactions, with
    bromine, 143
gem-Dicarbodiimides, synthesis of, 436
Dicarbonyl compounds
  reactions
    with carbanions, 166
    with dialkylzinc compounds, 166
    with organometallic compounds, 166
  reduction of, 165
  as starting materials, for bisdithioacetals, 245
1,2-Dicarbonyl compounds, as starting materials, for
    acetals, 183
1,3-Dicarbonyl compounds
  monoacetals of, 192
  as starting materials, for 2,2-bis-(4-methoxybenzoyloxy)
    compounds, 206
1,3-Dicarbonyl compounds, 2-acyloxy-, as starting
    materials, for 2-acyloxy-2-chloro-1,3-dicarbonyl
    compounds, 55
1,3-Dicarbonyl compounds, 2-acyloxy-2-chloro-
  from
    2-acyloxy-1,3-dicarbonyl compounds, 55
    2-diazo-1,3-dicarbonyl compounds, 55
1,3-Dicarbonyl compounds, 2-diazo-
  as starting materials
    for 2-acyloxy-2-chloro-1,3-dicarbonyl compounds, 55
    for α-bromoalkyl esters, 57
1,5-Dicarbonyl compounds, monoacetals of, 193
Dications, of alkyne-bridged dimolybdenum complexes,
    synthesis of, 674
gem-Dichlorides, as starting materials, for
    dithioacetals, 245, 249
Dichloro complexes, bis(cyclopentadienyl)-, reactions, with
    1,1'-dilithioferrocene, 720

Dichloro compounds, reactions, with potassium
    cyanate, 441
Di(chloromercurio) compounds, synthesis of, 1067
Dichloromethyl groups, from, trichloromethyl groups, 13
Dichromium, tris(mesitylene)-, synthesis of, 694
Dichromium complexes, carbon-bridged, synthesis of, 672
Dicobalt, reactions, with perfluorocyclooctatetraene, 687
Dicobalt, alkynehexacarbonyl-, reactions, with metal
    carbonyls, 711
Dicobalt, dodecacarbonyl-, reactions, with alkynes, 687
Dicobalt, octacarbonyl-
  reactions
    with alkynes, 687
    with cyclooctyne, 689
    with perfluoro-1,3-cyclohexadiene, 687
  as starting material, for hexafluoroisopropylidenedicobalt
    complexes, 685
Dicobalt complexes, alkyne-bridged, synthesis of, 687
Dicobalt complexes, bridged, synthesis of, 690
Dicobalt complexes, carbon-bridged
  from
    diazoalkane, 685
    diiodoalkanes, 686
  synthesis of, 687
Dicobalt complexes, hexacarbonyl-
  as alkyne protecting group, 688
  of small and large-ring alkynes, 688
Dicobalt complexes, hexafluoroisopropylidene-
  from, octacarbonyldicobalt, 685
  synthesis of, 685
Dideuterio compounds, synthesis of, 618
Diels–Alder cyclization, of 2-methyleneimidazolidines, 414
Diels–Alder reaction
  of 1,3-diazabutadienes, 421
  of diselenoesters, 289
  of ketene dithioacetal dioxides, 270
  of ketene dithioacetals, 265
  of ketene dithioacetal tetraoxides, 277
  of ketene dithioacetal trioxides, 272
  of selenothioester S,S-dioxides, 287
  of silenes, 611
  for the synthesis of α-aminoalkylphosphorus
    compounds, 489
  for the synthesis of bissilylated norbornenes, 619
  for the synthesis of tetrahydrophosphorines, 459
  for the synthesis of α-thiosulfoxides, 265
  use in the synthesis of dithioacetals, 257
Dienes
  from, dithionosuccinates, 834
  reactions
    with perfluoro phosphaalkenes, 112
    with peroxytrifluoroacetic acid, 46
    with trifluoroperacetic acid, 46
Dienes, tricyclic, synthesis of, 275
(E)-1,3-Dienes, synthesis of, 737
(Z)-1,4-Dienes, synthesis of, 809
Diesters, (chloromethyl)phosphonic, reactions, with
    aldimines, 473
Diesters, keto, as starting materials, for dithiolanes, 245
Diesters, 1,3-keto, synthesis of, 279
Diesters, α-ketophosphonic, as starting materials, for
    α-phosphoryl enamines, 1004
Diethylamine
  reactions
    with alkylthioalkynes, 905
    with α-bromo dithiocarboxylates, 843
    with (1-chloro-2,2-dicyanovinyl)manganese
      compounds, 1019
    with diethyl 2,4-dinitrobenzylidenemalonate, 306
    with ketene acetals, 972
    with phenyliodinium ylides, 867
Diethylamine, N-hydroxymethyl-, reactions, with
    β-aminoalkylphosphines, 453

Difluoromethyl group, from, bromodifluoromethyl group, 5
1,3-Digermacyclobutane, from, germenes, 627
1,3-Digermacyclobutane, 1,1,3,3-tetramethyl-, synthesis of, 626
Digermacyclopropanes
 ring-expansion of, 629
 synthesis of, 628
2,4-Digermapentane, 2,4-dimethyl-, reactions, with mercuracycles, 629
Digermathiacyclobutanes, synthesis of, 629
Digermene, tetrakis(2,6-diethylphenyl)-, reactions, with diazomethane, 628
Digermenes, as starting materials, for 1,1-bisgermylalkanes, 627
Dihafnium complexes, methylene-bridged, synthesis of, 670
Dihalides
 reactions
  with bis(sulfones), 275
  with disulfones, 275
Dihalides, alkyl, reactions, with trialkylarsines, 135
gem-Dihalides
 reactions
  with selenols, 288
  with telluride anions, 290
 as starting materials
  for diselenoacetals, 288
  for delluroacetals, 290
Dihalo compounds
 reactions, with sodium azide, 432
 as starting materials, for aminals, 411
Dihalo compounds, geminal
 reactions, with amines, 410
 as starting materials, for aminals, 410
Dihalo functions, synthesis of, 730
gem-Diheteroatomic compounds, cleavage of, 99
Dihydrogen trifluoride, tetrabutylammonium, as phase-transfer catalyst, 768
Diimines, silyl-, reactions, with hydrogen chloride, 107
Diindium, tetrachloro(μ-methylene)-, synthesis of, 698
Diiodides
 as starting materials, for dichlorophosphines, 114
 use in the synthesis of carbon-bridged diiron complexes, 679
Diiodo compounds, from, trans-1,2-dinitrospiropentane, 108
Diiridium complexes, dihydrido carbon-bridged
 from
  but-2-yne, 689
  isopropyllithium, 689
Diiridium complexes, methylene-bridged, synthesis of, 685
Diiron, dicyclopentadienyltetracarbonyl-, reactions, with phospholes, 1034
Diiron, dodecacarbonyl-, use in the synthesis of carbon-bridged diiron complexes, 681
Diiron, enneacarbonyl-, reactions, with propargylruthenium complexes, 711
Diiron, nonacarbonyl-
 reactions
  with cobaltacyclopentadiene complexes, 714
  with 1,4-dibromocyclooctatetraene, 681
  with tolan, 681
 as starting material, for cobaltacyclopentadiene complexes, 689
 use in the synthesis of carbon-bridged diiron complexes, 679, 681, 683
Diiron complexes, synthesis of, 679
Diiron complexes, bis-bridged, synthesis of, 681
Diiron complexes, carbon-bridged, synthesis of, 679, 683
Diiron complexes, cationic, synthesis of, 680
Diiron complexes, dichloro-substituted carbon-bridged, synthesis of, 685
Diiron complexes, methylene-, synthesis of, 679
Diiron complexes, mono-bridged, synthesis of, 681

Diiron dianion, octacarbonyl-, reactions, with diiodomethane, 679
Diiron nonacarbonyl, use in the synthesis of diiron complexes, 681
gem-Diisocyanates
 alcoholysis of, 447
 reactions, with triphenylphosphine imides, 436
 synthesis of, 433
gem-Diisocyanides, synthesis of, 437
Diisopropylamine, reactions, with phenylvinyl selenides, 964
1-(Diisopropylamino)-3-borolenes, dimetallation of, 643
Di-O-isopropylidene compounds, as starting materials, for fructopyranose compounds, 222
gem-Diisothiocyanates, synthesis of, 433, 434
Diketones
 reactions
  with nitromethane, 166
  with phosphines, 339
1,3-Diketones, 2-hydroxy, as starting materials, for α-chloro alcohols, 43
[1,4]-Diketones, synthesis of, 287
1,5-Diketones, masked, synthesis of, 255
α-Diketones, as starting materials, for dibromo ketones, 27
Dilead compounds, carbon-bridged, synthesis of, 699
Dilithioferrocene–tetramethylethylenediamine complexes, synthesis of, 1034
Dilithium, diphenylmethylene-, synthesis of, 668
Dilithium, methylene-
 from
  bis(mercury halides), 668
  dibromomethane, 668
  diiodomethane, 668
  methyllithium, 668
 structure of, 668
 synthesis of, 668
Dilithium, phenylmethylene-, synthesis of, 668
Dilithium compounds, as starting materials, for 1,3-diboretanes, 633
Dimanganese, decacarbonyl-
 reactions, with 2,5-dimethyl-1-phenylarsole, 1042
 for the synthesis of manganacyclopentadiene complexes, 677
Dimanganese complexes, carbon-bridged, synthesis of, 677
Dimanganese complexes, methylene-bridged, synthesis of, 676
Dimedones, 1,3-diketone, reactions, with benzenesulfenyl chloride, 70
Dimercuracycles, as starting materials, for carbosilanes, 624
Dimercuration, of alkynes, 1067
Dimercury complexes, carbon-bridged, synthesis of, 696
Dimeric, dimesityliron complexes, synthesis of, 685
1,1-Dimesityl compounds, synthesis of, 1048
Dimetal complexes, synthesis of, 719
1,1-Dimetal complexes, synthesis of, via transmetallation, 669
Dimetal compounds, reactions, with electrophiles, 809
Dimetal compounds, carbon-bridged, from, diazoalkanes, 709
Dimetallic complexes, carbon-bridged, synthesis of, 686
Dimetallic compounds, transmetallation of, 656
Dimethylamine, reactions, with ketene dithioacetals, 969
Dimethylamine, perfluoro-N-bromo-, reactions, with perfluoro-1,1,3-tris(dimethylamino)propadiene, 794
Dimethylamine, trichloromethyl-, reactions, with dicyanoketene aminals, 987
Dimethyl sulfoxide–iodine, for the oxidation of active methylene compounds, 161
Dimolybdenum, hexacarbonylbis(η^5-cyclopentadienyl)-
 reactions, with alkynes, 674
 use in the synthesis of alkyne-bridged dimolybdenum complexes, 674
Dimolybdenum, tris(cyclooctatetraene)-, synthesis of, 673
Dimolybdenum complexes, alkyne-bridged, α-carbocations of, 674

Dimolybdenum complexes, allene-bridged, synthesis of, 673
Dimolybdenum complexes, carbon-bridged, synthesis of, 672
Dimolybdenum complexes, phenylethyne–ethylene diamine, synthesis of, 675
Dimolybdenum complexes, sulfide-bridged dithiocarbamate, synthesis of, 675
Dinickel, bis(pentadienyl)-, synthesis of, 694
Dinickel complexes, synthesis of, 693
Dinickel complexes, alkyne-bridged
 from
  nickel dinuclear carbonyls, 692
  nickelocene, 692
  palladium analogues of, synthesis of, 692
Dinickel salts, tris(cyclopentadienyl)-, synthesis of, 694
Dinitramine, methyl, as nitrating agent, 429
Dinitramine, methylene-, synthesis of, 435
Dinitramines, synthesis of, 435
gem-Dinitramines
 from, monocyclicamines, 435
 synthesis of, 434
Dinitrogen pentoxide, for the oxidation of gem-dinitrosamines, 435
Dinitrogen tetroxide
 as oxidative nitrating agent, 429
 as starting material, for α-aminoalkylphosphorus compounds, 490
 use in the synthesis of gem-dinitroalkenes, 993
gem-Dinitrosamines
 from, imidazolidines, 434
 oxidation of, 435
 synthesis of, 434
Dinucleotides, synthesis of, 216
Diolithium salts, as starting materials, for ketene dithioacetals, 847
Diols
 bisaminomethylation of, 300
 oxidation of, 164
 protection of, 186
 protection as cyclic acetals, 184
 reactions
  with dichlorosilanes, 342
  with 2-methoxypropene, 186
 as starting materials, for methylene acetals, 190
Diols, amino, synthesis of, 507
1,2-Diols
 protection as acetals, 180
 synthesis of, 231
1,3-Diols, protection as acetals, 180
α-Diols
 from, strained cyclic ketones, 162
 synthesis of, 160
  via hydration of carbonyl compounds, 160
  via oxidation, 161
1,3-Dione, indane-, oxidation of, 161
Diorganolithium compounds
 reactions, with dimethylgermanium dichloride, 636
 synthesis of, 618
Diorganothallium compounds, synthesis of, 1068
Diosmium complexes, carbon-bridged, synthesis of, 679
Dioxaborinanes
 homologation of, 364
 reactions, with methoxy(phenylthio)methyllithium, 364
1,2,3,5-Dioxadisilolane, 3,3,5,5-tetramesityl-, synthesis of, 624
1,3-Dioxalane compounds, synthesis of, 207
1,4-Dioxane, 2,3-dibromo-, stability of, 51
1,4-Dioxane, 2,3-dichloro-, synthesis of, 49
3,6-Dioxanonanoyl fluoride, perfluoro-9-azido-2,5-dimethyl-8-n-propoxy-, pyrolysis of, 766
3,6-Dioxanon-1-ene, perfluoro-9-azido-5-methyl-8-n-propoxy-, synthesis of, 767
1,3-Dioxans, 5-substituted-2,2-dimethyl-4,6-dioxo-, synthesis of, 204

1,3,5,2-Dioxaphosphaborinanes, synthesis of, 342
1,3,5,2-Dioxaphosphaborinane salts, synthesis of, 342
1,3,5-Dioxaphosphorinanes, synthesis of, 337, 341
1,4,2-Dioxaphosphorinanes, synthesis of, 340
1,3,5-Dioxathianes, synthesis of, 229
1,5,3-Dioxazepines, synthesis of, 300
Dioxazoles, from, bis(trifluoromethyl)ketene, 833
1,3-Dioxides, synthesis of, 268
S,S'-Dioxides, optically pure, synthesis of, 269
trans-1,3-Dioxides, synthesis of, 268
Dioxirane, dimethyl-
 for the oxidation of diazo carbonyl compounds, 162
 use in the synthesis of epoxy esters, 209
 use in the synthesis of glycoside bonds, 188
1,3-Dioxolane, 2-(methoxycarbonylcyanomethylene)-
 reactions
  with O-benzylhydroxylamine, 896
  with 1,1-dimethylhydrazine, 897
1,3-Dioxolanes, synthesis of, 181
1,3-Dioxolanes, 4-oxo-
 from
  α-hydroxy acids, 213
  β-hydroxy acids, 213
1,3-Dioxolan-4-ones, 5-alkyl-2-t-butyl-, synthesis of, 213
Dioxygen compounds, synthesis of, 203
Dipalladium complexes, alkyne-bridged, from, hex-3-yne, 692
Diphenylphosphido–cyclopentadienyl compounds, lithiation of, 706
1,3-Diphosphaalkenes
 cycloaddition to, 548
 reactions of, 557
1,3-Diphosphaallenes
 hydrolysis of, 559
 isomerization of, 548
 reactions, with sulfur, 559
Diphosphachiropteradienes, synthesis of, 549
Diphosphaferrocene, diacetyltetraphenyl-, synthesis of, 1036
Diphosphaferrocenes
 chemistry of, 1035
 from, 1,2,3-triphenylphosphole, 1035
Diphosphaferrocenes, tetraphenyl-, synthesis of, 1035
1,3-Diphosphaindene, reactions, with 1,4-diphenylbutadiene, 548
Diphosphenes, carbene addition to, 547
1,3-Diphosphetanes, synthesis of, 547, 549, 558, 1026
1,2-Diphosphetes, reactions, with lithium, 545
1,3-Diphosphetes, hydrolysis of, 562
Diphosphinates, reactions, with carbonyl compounds, 563
Diphosphine, methylene-, reactions of, 550, 562
Diphosphine, tetrafluoro-, reactions, with fluoroalkenes, 114
Diphosphine, tetramethyl-, reactions, with hexafluoropropene, 802
Di(phosphine oxides), alkylidene-
 alkylation on carbon, 563
 synthesis of, 553
Diphosphines, carbene insertion into, 547
Diphosphines, alkylidene-
 oxidation of, 552, 562
 synthesis of, 545
Diphosphines, trans-diaryl-, reactions, with isopropylidenecarbene, 1025
Diphosphines, 1,2-dimetallo-, reactions, with 1,1-dihaloalkanes, 545
Di(phosphine sulfides), alkylidene-, synthesis of, 553
1,1-Diphosphinoalkanes
 oxidation of, 552, 562
 synthesis of, 545, 548
1,1-Diphosphinoalkenes, addition to, 548
Diphosphinocarbenes, intramolecular insertion of, 548
Diphosphinomethane, reactions of, 550, 562
Diphosphiranes, synthesis of, 545, 547
Diphosphiranes, di(chlorophenyl)methylene-, synthesis of, 1025

Diphosphiranes, methylene-, synthesis of, 1025
1H-1,3-Diphosphole, 1-t-butyl-2-ethylidene-2,3-dihydro-4-methyl-3-phenyl-, synthesis of, 1026
Diphospholes, synthesis of, 1026, 1030
1,3-Diphospholes, synthesis of, 563
Diphosphonate, tetraethyl dialkylaminomethylene, as starting material, for α-phosphoryl enamines, 1002
Diphosphonate, tetraethyl dichloromethane, reactions, with n-butyllithium, 801
Diphosphonate, tetraethyl formamidomethylene, as starting material, for gem-formamido phosphorylalkenes, 1004
Diphosphonate compounds, amino-, reactions, with sodium hydride, 1002
Diphosphonates, alkene, synthesis of, 803
Diphosphonates, alkylidene
  alkylation on carbon, 563
  reduction of, 550
  synthesis of, 553
Diphosphonates, alkyne, hydrogen halide hydrolysis of, 803
Diphosphonates, benzylidene, synthesis of, 559
Diphosphonates, tetraphenyl, synthesis of, 555
1,1-Diphosphonatoalkanes, reduction of, 550
Diphosphonic acids, (Z)-haloalkene, synthesis of, 803
Diphosphonites, reactions, with aldehydes, 564
Diphosphonites, alkylidene-, Arbuzov reactions of, 552
Diphosphonium salts, alkylidene-, synthesis of, 553
Diphosphoranes, alkylidene-, synthesis of, 564
1,1-Diphosphoranoalkanes, synthesis of, 564
Diphosphorines, synthesis of, 1027
Di(phosphoryl chloride), methylene-
  reactions of, 563
  reduction of, 550
Diplatinum complexes, carbene-bridged, from, perfluoropropene, 691
Diplatinum complexes, methylene-bridged, synthesis of, 691
Dipolar cycloaddition, for the synthesis of α-aminoalkylphosphorus compounds, 479
1,3-Dipoles
  reactions
    with aldehydes, 183
    with ketones, 183
Dipotassium cyclooctatetraenide
  use in the synthesis of carbon-bridged dichromium complexes, 673
  use in the synthesis of carbon-bridged diruthenium complexes, 683
  use in the synthesis of cyclooctatetraene-bridged complexes, 672
  use in the synthesis of tris(cyclooctatetraene)dimolybdenum complexes, 673
  use in the synthesis of tris(cyclooctatetraene)ditungsten complexes, 673
1,2-Dipotassium-1,2-di-t-butyldiphosphide, reactions, with 1,1-dichloroalkenes, 1025
Dirhenium, decacarbonyl-, synthesis of, 677
Dirhenium complexes, flyover, synthesis of, 678
Dirhodiacyclobutane complexes, cyclopentadienyl, use in the synthesis of alkyne-bridged dirhodium complexes, 688
Dirhodium, octakis(phosphorus trifluoride)-, reactions, with alkynes, 688
Dirhodium complexes, alkyne-bridged, reactions, with diazomethane, 689
Dirhodium complexes, bismethylene-bridged, synthesis of, 687
Dirhodium complexes, carbon-bridged, from, diazoalkanes, 686
Dirhodium complexes, carbon-bridged ketonic, from, hex-3-yne, 690
Dirhodium complexes, methylene-bridged, synthesis of, 689
Diruthenium, tetracarbonyldicyclopentadienyl-, use in the synthesis of diruthenium complexes, 679

Diruthenium complexes
  synthesis of, 679
  via photochemical decarbonylation, 681
Diruthenium complexes, alkyne-bridged, synthesis of, 680
Diruthenium complexes, carbon-bridged, synthesis of, 678, 683
Diruthenium complexes, chloride-bridged, synthesis of, 680
Diruthenium complexes, methylene-bridged, synthesis of, 680
Diruthenium complexes, trismethylene-, synthesis of, 680
Diselenide, dimethyl
  reactions
    with sodium, 283
    with tetrachloroethene, 786
Diselenide, diphenyl
  reactions
    with trimethylsilylethyne, 947
    with vinyl carbenes, 875
Diselenides
  reactions, with alkynes, 871
  as starting materials
    for diselenoacetals, 290
    for α-halo selenides, 88
Diselenides, diphenyl
  reactions
    with perfluoroalkyl iodides, 284
    with sulfone carbanions, 286
Diselenium compounds, synthesis of, 287, 874
Diselenoacetals
  cleavage of, 284
  deprotonation of, 289
  from
    aldehydes, 287
    diselenides, 290
    gem-dihalides, 288
    ketones, 287
    selenocarbonyl compounds, 289
  reactions, with butyllithium, 400
  as starting materials, for thioselenoacetals, 284
  synthesis of, 287
Diselenoacetals, ^{13}C-labeled, synthesis of, 289
Diselenoacetals, diphenyl, transmetallation of, 401
Diselenoacetals, ketene
  from
    gem-dibromoalkanes, 875
    malononitrile, 875
  reactions, with n-butyllithium, 964
  synthesis of, 875
Diselenoesters, Diels–Alder reaction of, 289
1,1-Diselenolates, reactions, with alkyl halides, 875
1,3-Diselenoles, 2-benzylidene, synthesis of, 874
1,3-Diselenoles, 2-vinylidene, synthesis of, 874
Diselenothionocarbonate, bis(trifluoromethyl), photolysis of, 872
1,3-Disilabicyclo[1.1.0]butanes, synthesis of, 1049
1,3-Disilabutane, 3-methyl-1,1,3-trichloro-, synthesis of, 611
1,3-Disilabutanes, synthesis of, 1054
1,2-Disilabut-3-enes, cleavage of, 1051
1,3-Disilacyclobutane, synthesis of, 1053
1,3-Disilacyclobutane, 1,3-dichloro-1,3-dimethyl-, reactions, with organometallics, 621
1,3-Disilacyclobutane, 1,1,3,3-tetrachloro-, reduction of, 621
1,3-Disilacyclobutane, 1,1,3,3-tetramethyl-
  metallation of, 620
  reactions
    with bromine, 622
    with methyllithium, 606
    with organomercury carbenoid, 623
    with sulfur dioxide, 624
    with sulfur trioxide, 624
    with trichlorosilane, 617
    with trimethylsilanol, 623

with zirconium tetrachloride, 610
ring-opening of, 653
synthesis of, 608
1,3,Disilacyclobutane, 1,1,3,3-tetramethyl-2,4-dineopentyl-, synthesis of, 611
Disilacyclobutanes
from, bis(chloromethyl)dimethylsilane, 605
metallation of, 623
1,3-Disilacyclobutanes
from
chloromethyldimethylsilyl halides, 654
disilenes, 614
α-lithiosilylalkanes, 605
silenes, 610
hydrosilylation of, 617
ring-opening of, 622, 658, 661
as starting materials
for α-borylsilylalkanes, 644
for tetrasilylethenes, 1049
synthesis of, 608, 1049
via pyrolysis of silacyclobutanes, 610
1,3-Disilacyclobutanes, 1,1,3,3-tetramethyl-, ring-opening of, 658
3,4-Disilacyclobut-3-ene, 3,4-bis(silyl), synthesis of, 1052
1,2-Disilacyclohexadienes, as starting materials, for silenes, 612
1,4-Disilacyclohexa-2,5-dienes, synthesis of, 1053
1,3-Disilacyclohexane, 1,1-dichloro-3,3-dimethyl-, synthesis of, 608
1,4-Disilacyclopentanes, synthesis of, 621
1,3-Disilacyclopentene, 1,1,3,3-tetramethyl-, from, 1,2-diallyl-1,1,2,2-tetramethyldisilane, 617
Disilacyclopentenes, from, silirenes, 1052
1,3-Disilacyclopentenes
from
molybdenum complexes, 616
silacyclopropenes, 616
Disilane, 1,2-bis(bromomethyl)tetramethyl-, reactions, with magnesium, 654
Disilane, 1,2-bis(1-naphthyl)tetramethyl-, photolytic transformation of, 617
Disilane, chloromethylpentamethyl-
reactions
with alkoxides, 614
with aluminum chloride, 614
with potassium acetate, 614
Disilane, 1,2-diallyl-1,1,2,2-tetramethyl-, as starting material, for 1,1,3,3-tetramethyl-1,3-disilacyclopentene, 617
Disilane, hexachloro-, reactions, with diazomethane, 140
Disilane, hexamethyl-
reactions, with methyllithium, 355
thermolysis of, 616
Disilane, hexaphenyl-, synthesis of, 609
Disilane, 2-phenylethenylpentamethyl-, irradiation of, 614
Disilane, 1-phenyl-2-vinyltetramethyl-, photolysis of, 613
Disilanes
as starting materials
for alkenes, 1054
for 1,1-bis(silyl)alkanes, 614
Disilanes, alkenyl, 1,3-silyl migrations of, 613
Disilanes, alkyl, photolysis of, 1052
Disilanes, alkyne
from, 1,1-disilylallenes, 1046
rearrangement of, 1046
Disilanes, styryl, rearrangement of, for the synthesis of 1,1-bis(silyl)alkanes, 614
2,4-Disilapentane, 1-chloro-4-fluoro-2,2,4-trimethyl-, reactions, with magnesium, 608
2,4-Disilapentane, 1-lithio-2,2,4,4-tetramethyl-, reactions, with trimethylsilyl chloride, 606
2,4-Disilapentane, 2-methoxy-2,4,4-trimethyl-
from, methoxytrimethylsilane, 602

reactions, with t-butyllithium, 620
synthesis of, 623
2,4-Disilapentane, 2,2,4-trimethyl-, synthesis of, 616
1,4-Disilapentanes, synthesis of, 1054
2,4-Disilapentanes, 2-chloro-2,4,4-trimethyl-, synthesis of, 614
1,4-Disilapent-2-enes, cleavage of, 1051
Disilazane, hexamethyl- (HMDS), as ammonia substitute, in the synthesis of gem-imines, 438
Disilene, tetramethyl-, reactions, with alkynes, 1052
Disilenes
as intermediates, in the synthesis of 1,1-bis(silyl)alkanes, 614
as starting materials
for 1,1-bis(silyl)alkanes, 609
for 1,3-disilacyclobutanes, 614
for 1-germyl-1-silylalkanes, 639
synthesis of, 614
Disiloxane, from, α-silyl silenes, 612
Disiloxane, chloromethylpentamethyl-, as starting material, for α-silyl Grignard reagents, 654
Disiloxane, di(chloromethyl)-, cyclization of, 609
Disiloxane, hexamethyl-, deprotonation of, 650
Disiloxanes, synthesis of, 623
Disilver complexes, dilithio-, synthesis of, 708
Displacement reaction, for the synthesis of ketene hemiaminals, 893
Distannacycloheptanes, synthesis of, 700
Distannanes, telluration of, 876
1,1'-Distibaferrocene, 2,2',5,5'-tetramethyl-, synthesis of, 1041
Distibaferrocenes, synthesis of, 1041
Distibines, synthesis of, 595
Disulfide, diphenyl
reactions
with diethyl ethylphosphonate, 346
with sodium hydroxide, 284
with sulfolanes, 266
Disulfides
from, dithioacetals, 260
reactions, with alkynes, 940
as starting materials, for α-chloroalkanesulfenyl chlorides, 71
Disulfides, dialkyl, for the α-sulfenylation of enol ethers, 836
Disulfones
reactions
with dihalides, 275
with 2-ethoxyvinyl sulfones, 865
Disulfones, alkylidene, as starting materials, for bis(sulfones), 274
Disulfones, methylene
as starting materials, for bis(sulfones), 274
synthesis of, 272
Disulfones, spiro, synthesis of, 275
Disulfonium salts, synthesis of, 863
Disulfonyl fluoride, reactions, with p-(dimethylamino)benzaldehyde, 864
1,3-Ditelluretanes, trans-dibenzylidene-, synthesis of, 877
Ditelluride, diferrocenyl, reduction of, 290
Ditelluride, diphenyl
reactions
with dichloromethane, 93
with vinyl carbenes, 877
Ditellurides
reactions, with diazomethane, 290
reduction of, 291
as starting materials, for ditelluroacetals, 290
Ditellurides, diaryl, reactions, with sodium hydroxide, 291
1,2-Ditellurides
condensation of, 877
synthesis of, 876
Ditellurium compounds, synthesis of, 290
Ditelluroacetals

Ditelluroacetals

deprotonation of, 290
from
  gem-dihalides, 290
  ditellurides, 290
reactions, with chlorine, 291
Ditelluroacetals, ketene, synthesis of, 876
1,3-Ditelluroles, deprotonation of, 290
Ditelluromethane, diphenyl-, from, lithium tellurides, 290
Diterpenes, lathrane, synthesis of, 948
1,3,2-Dithiaborinane–dimethyl sulfide, reactions, with tin(II) chloride, 257
Dithiacetals, bromo, from, aldehydes, 245
1,3,2,4-Dithiadiphosphetane-2,4-dithione, 2,4-bis(phenylthio)-
  reactions
    with benzil dianils, 320
    with benzil monoanils, 320
Dithiadiselenafulvalenes, synthesis of, 868
1,3-Dithiane
  reactions
    with sodium periodate, 268
    with trityl fluoroborate, 1074
1,3-Dithiane, 2-carbethoxy-, oxidation of, 269
1,3-Dithiane, 2,2-dimethyl-, synthesis of, 262
1,3-Dithiane, 5,5-dimethyl-, reactions, with sodium periodate, 268
1,3-Dithiane, 2-ethyl-, deprotonation of, 261
1,3-Dithiane, 2-methyl-, oxidation of, 260
1,3-Dithiane, 2-methylene-
  reactions, with alkyllithium compounds, 254
  as starting material, for dithioacetals, 254
1,3-Dithiane, 2-methyl-2-trimethylsilyl-, oxidation of, 260
1,3-Dithiane-S,S'-dioxide, enantiomerically pure, synthesis of, 269
1,3-Dithiane-1-oxides, 2-lithio-, reactions, with electrophiles, 263
Dithianes
  from, alkynes, 258
  oxidation of, with Sharpless reagent, 261
  reactions, with methanesulfonic acid, 252
  as starting materials, for dithioacetals, 250
  uses of, 244
1,3-Dithianes, alkylidene-, reactions, with trifluoroacetic acid, 252
1,3-Dithianes, β-keto-, synthesis of, 258
1,3,5-Dithiaphosphorinanes, synthesis of, 347
Dithiepines, as starting materials, for dithioacetals, 250
Dithietane
  from, 1,3-dithietane-1-oxide, 258
  synthesis of, 852
1,3-Dithietane-1-oxide, as starting material, for dithietane, 258
Dithietanes, oxidation of, 854
1,3-Dithietanes, reactions, with 1,2-diaminobenzene, 415
Dithioacetal, formaldehyde, from, dithiols, 249
Dithioacetal, γ-tosyloxy, Grob fragmentation of, 846
Dithioacetal S,S-dioxides, ketene
  from
    aliphatic aldehydes, 858
    alkyl halides, 857
    methyl methylthiomethyl sulfones, 857
    methylthiomethyl p-tolyl sulfone, 857
    thiocarbonyl compounds, 862
    vinyl sulfones, 861
  isomerization of, 857
  oxidation of, 854
  synthesis of, 853, 856
Dithioacetal S,S'-dioxides, ketene, synthesis of, 853, 862
Dithioacetalization
  of aldehydes, 247
  of ketones, 247
Dithioacetal monosulfoxides, ketene
  synthesis of, 856
  via elimination reactions, 858
Dithioacetal monoxides, 2-acetoxy, as intermediates, in the synthesis of ketene dithioacetal monoxides, 857
Dithioacetal monoxides, ketene, (E)–(Z) isomerization of, 856
Dithioacetal S-oxides, ketene
  from
    alkyl alkylthiomethyl sulfoxides, 857
    α-alkylthiosulfines, 859
    alkynes, 859
    methyl methylthiomethyl sulfoxides, 856
    phosphonate esters, 860
    α-substituted dithioacetals, 855
    vinyl sulfoxides, 858
  oxidation of, 854
Dithioacetals
  asymmetrically-substituted, synthesis of, 255
  cyclization of, 228
  enantiomerically enriched, synthesis of, 248
  enantioselective oxidation of, 263
  from
    acetals, 245
    aldehydes, 245
    carboxylic acids, 257
    α-chloro sulfides, 258
    gem-dichlorides, 245, 249
    dideuteriodichloromethane, 249
    dithioacetals, 249
    ketene dithioacetals, 252
    ketones, 245
    2-methylene-1,3-dithiane, 254
    thiocarbonyl compounds, 255
    thioglycolic acid, 272
    trithioorthoformates, 250
  as intermediates, in the syntheses of gingerols, 245
  methylation of, 216
  oxidation of, 853
    for the synthesis of α-thiosulfoxides, 258
  S-oxide, synthesis of, 261
  reactions
    with n-butyllithium, 1075
    with tosyl azide, 860
    with trialkylboranes, 372
  as reagents, 269
  reviews, 244
  as starting materials
    for α-alkylthiosulfoxides, 258
    for bis(sulfones), 272
    for bissulfoxides, 267
    for disulfides, 260
    for dithioacetals, 249
    for dithioketals, 250
    for epoxy-bis(sulfone), 273
    for α-sulfoxide sulfones, 271
  synthesis of, 244, 245
  for the synthesis of radical centres flanked by two sulfur atoms, 1080
Dithioacetals, α-acetoxy, oxidation of, 855
Dithioacetals, acyclic, as starting materials, for cyclic dithioacetals, 228
Dithioacetals, acyl, synthesis of, 247
Dithioacetals, chloral, oxidation of, 855
Dithioacetals, crotonaldehyde, reactions, with benzaldehyde, 849
Dithioacetals, cyclic, from, acyclic dithioacetals, 228
Dithioacetals, cyclopropyl, from, β-oxo dithio esters, 255
Dithioacetals, deuterium-labeled, synthesis of, 249, 250
Dithioacetals, diphenyl
  reductive lithiation of, 380, 394
  as starting materials, for α-thioalkylboranes, 372
Dithioacetals, fluoro, synthesis of, 252
Dithioacetals, hydroxy, from, dihydropyran, 246
Dithioacetals, ketene

alkylation of, 860
chlorination of, 743
Diels–Alder reaction of, 265
from
  acid chlorides, 846
  carboxylic acids, 845
  carboxylic esters, 845
  dichloroalkyne, 851
  gem-dihaloalkenes, 851
  dithiocarboxylates, 843
  dithiocarboxylic acids, 842
  1,3-dithioles, 850
  dithiolium salts, 847
  methyl cyanoacetate, 843
  α-thiodithioacetals, 846
  thioketenes, 852
  p-tosylhydrazones, 852
  trithioortho esters, 844
  α,β-unsaturated dithioesters, 843
as intermediates, 969
monosulfoxidations of, 853
oxidation of, 853
polysulfoxidation of, 853
reactions
  with amines, 969
  with diamines, 970
  with dimethylamine, 969
  with methyl fluorosulfonate, 863
as starting materials
  for *gem*-disulfoximidoalkenes, 997
  for dithioacetals, 252
  for α-hydrazino enamines, 983
  for ketene aminals, 969
  for ketene dithioacetal S-imides, 860
  for ketene hemiaminals, 888
  for ketene monothioacetals, 836
synthesis of, 842
  alkenation methods, 847
  via dehydrogenation, 847
  via double bond formation by elimination, 844
  via Peterson alkenation reactions, 849
  via rearrangement of α,β-unsaturated dithioacetals, 849
Dithioacetals, keto, from, dithioorthoformates, 250
Dithioacetals, nitroketene, reactions, with hydrazines, 983
Dithioacetals, non-polar, synthesis of, 261
Dithioacetals, β-oxo, from, α-oxoketene dithioacetals, 253
Dithioacetals, oxoketene, reactions, with hydrogen, 253
Dithioacetals, α-oxoketene, as starting materials, for β-oxo dithioacetals, 253
Dithioacetals, S-polyfluoroketene, synthesis of, 868
Dithioacetals, α-substituted, as starting materials, for ketene dithioacetal S-oxides, 855
Dithioacetals, theophylline, synthesis of, 259
Dithioacetals, trifluoromethyl
reactions
  with lithium biphenylide, 775
  with lithium naphthalenide, 775
Dithioacetals, α,β-unsaturated, rearrangement of, 849
Dithioacetals, unsymmetrical, synthesis of, 247
Dithioacetal tetroxides, ketene
cycloaddition reactions of, 277
Diels–Alder reaction of, 277
as electrophiles, 277
from
  disulfonylmethanes, 864
  vinyl ethers, 865
as starting materials, for bis(sulfones), 277
synthesis of, 855, 864
See also Alkenes, 1,1-disulfonyl-
Dithioacetal S,S,S',S'-tetroxides, ketene
synthesis of, 853
See also Alkenes, 1,1-disulfonyl-
Dithioacetal trioxides

deprotonation of, 271
synthesis of, 271
Dithioacetal trioxides, ketene, Diels–Alder reaction of, 272
Dithioacetal S,S,S'-trioxides, ketene, synthesis of, 853, 864
Dithioamides, use in the synthesis of distibines, 595
Dithioate, hexafluoro, reactions, with trifluoromethanethiol, 256
Dithioates, furfuryl, reactions, with ethylmagnesium bromide, 255
Dithiobenzoate, phenyl, reactions, with phenyllithium, 255
Dithiocarbenes, trimethylene-, synthesis of, 250
Dithiocarbonimidates, diethyl, reactions, with 2-acetylthiophene, 980
Dithiocarboxylates
as starting materials, for ketene dithioacetals, 843
synthesis of, *in situ*, 843
Dithiocarboxylates, α-bromo, reactions, with diethylamine, 843
Dithiocarboxylation, of Grignard reagents, 843
Dithiocarboxylic acids
oxidative dimerization of, 852
as starting materials, for ketene dithioacetals, 842
Dithio cations, synthesis of, 1073
1,1-Dithio-1,3-dienes
reactions
  with ethyllithium, 255
  with organolithium compounds, 254
Dithioester S-oxides, reactions, with thallium ethoxide, 859
Dithioesters
reactions
  with Grignard reagents, 255
  with maleic anhydride, 843
  with organolithium compounds, 255
  with organometallic compounds, 255
as starting materials
  for cyclopropanols, 255
  for hemidithioacetals, 245, 256
use in the synthesis of distibines, 595
Dithioesters, β-oxo, as starting materials, for cyclopropyl dithioacetals, 255
Dithioesters, α,β-unsaturated, as starting materials, for ketene dithioacetals, 843
Dithioisobutyrate, methyl, reactions, with diethylphosphinous chloride, 927
Dithioketals, from, dithioacetals, 250
Dithiolane S,S'-dioxide, synthesis of, 268
Dithiolanes
double oxidation of, 266
from
  keto diesters, 245
  ketones, 246
oxidation of, 260
1,3-Dithiolanes, (haloalkyl)-, synthesis of, 249
1,2-Dithioles, reactions, with Grignard reagents, 905
1,3-Dithioles, as starting materials, for ketene dithioacetals, 850
1,2-Dithiolimines, as starting materials, for ketene aminals, 979
Dithiolium salts, as starting materials, for ketene dithioacetals, 850
Dithiols
cyclization of, 1073
as starting materials, for formaldehyde dithioacetal, 249
gem-Dithiols, synthesis of, 244
Dithiomalonate compounds, synthesis of, 842
Dithioniadiboratacyclohexanes, synthesis of, 373
Dithionooxalate, dimethyl, reactions, with 2,3-dimethylbutadiene, 229
Dithionosuccinates, as starting materials, for dienes, 834
Dithioorthoformates
reactions, with silyl enol ethers, 250
as starting materials, for keto dithioacetals, 250
Di(thiophosphoryl chloride), methylene-, reduction of, 550

Dithioselenocarbonate, dimethyl, as starting material, for ketene thioselenoacetals, 872
Dithizone, dehydro-, reactions, with phosphorus ylides, 490
Ditin compounds, carbon-bridged
   synthesis of, 699
      via replacement reactions, 701
(Ditin)compounds, methylene-, halogen replacement of, 700
Dititanium, tris(cyclooctatetraene)-, synthesis of, 670
Dititanium complexes, methylene-bridged
   from, tetramethylenetitanocene, 670
   synthesis of, 670
Ditungsten, tris(cyclooctatetraene)-, synthesis of, 673
Ditungsten complexes, alkyne-bridged, synthesis of, 674
Ditungsten complexes, carbon-bridged, synthesis of, 672
Divanadium complexes, carbon-bridged
   pentacoordinate carbon atoms in, 671
   synthesis of, 671
1,1-Dizincalkene, from, alkynes, 1067
Dizinc compounds, carbon-bridged, synthesis of, 696
Dizirconium complexes, methylene-bridged, synthesis of, 670
Double deprotonation, of nitroalkanes, 540
1,1-Dyotropic rearrangement, use in the synthesis of 1,1-bis(silyl)alkanes, 614

Einhorn reaction
   for the synthesis of N-acyl aminals, 416
   for the synthesis of monoacylated aminals, 417
Elastase, inhibition via α-aminoboronic acids, 526
Electrochemical methods, for the synthesis of acetals, 197
Electron transfer, for the synthesis of cations, 1072
Electrophiles
   reactions
      with N-alkylthiomethylamides, 100
      with anions, 917
      with carbenoids, 817
      with dichlorophosphines, 125
      with dimetal compounds, 809
      with enamines, 98
      with α-haloamines, 100
      with imines, 98
      with 2-lithio-1,3-dithiane-1-oxides, 263
      with phosphaferrocenes, 1035
      with phosphino ylides, 551
      with α-sulfonyl carbanions, 236
      with tris(trimethylsilyl)methyllithium, 1048
      with vinylboranes, 144
      with vinylsilanes, 139
      with ynamines, 791
   use in the synthesis of acetals, 187
Electrophiles, silyl
   reactions
      with α-metalloborylalkanes, 642
      with α-metallogermylalkanes, 636
      with α-metallomethyl transition metals, 657
      with organolithium compounds, 602
Electrophilic complexes, reactions, with α-metallosilylalkanes, 656
Elimination, of hydrogen halides, 730
α-Elimination, for the synthesis of carbenes, 1081
Elimination reaction
   for the synthesis of bromo iodoalkenes, 759
   for the synthesis of chlorohaloalkenes, 756
   for the synthesis of 1,1-dichloroalkenes, 742
   for the synthesis of difluoroalkenes, 730
   for the synthesis of fluorohaloalkenes, 750
   for the synthesis of α-haloenols, 766
   for the synthesis of 1-halo-1-thioalkylalkenes, 773
   for the synthesis of ketene dithioacetals, 844
   for the synthesis of tetracoordinate sulfur compounds, 779
Ellipticines, synthesis of, 164
Enals, synthesis of, 188
Enamides, hydrosilylation of, 510

Enamimes, as starting materials, for α-aminoalkylphosphonites, 458
Enamine compounds, α-hydrazino, synthesis of, 986
Enamines
   cycloaddition of, 414
   hydroboration of, 524
   nucleophilic addition to, 426
   reactions
      with amines, 971
      with N,N-dimethylacetamide diethyl acetal, 987
      with S,S-dioxide ketenethioselenoacetals, 286
      with electrophiles, 98
      with salicyl aldehyde, 307
      with sodium dimethylamide, 1017
      with tetrachloromethane, 742
      with trichlorosilane, 510
   as starting materials
      for α-aminoalkylideneamino enamines, 987
      for dichloromethyl ketones, 15
      for α-haloamines, 100
      for ketene aminals, 979
      for α-sodio enamines, 1017
   tautomers of, 880
   use in the synthesis of O,N-acetals, 307
Enamines, N-acylated α-hydrazino, from, α-hydrazino enamines, 986
Enamines, N-acylated α-ureido
   from
      acyl azides, 982
      N-acylketenimines, 982
   synthesis of, 982
Enamines, 1-acyloxy, synthesis of, 889
Enamines, 1-alkoxy, tautomers of, 880
Enamines, α-alkoxycarbonylamino, from, N-acylketenimines, 981
Enamines, β-alkyl-α-phosphoryl, synthesis of, 1003
Enamines, α-amido
   from
      N-acylated ketene N,S-acetals, 981
      N-acylketenimines, 980
      isothioureas, 980
      ketene aminals, 980
      ynamines, 980
   synthesis of, 980
Enamines, α-aminoalkylideneamino
   from
      acetals, 987
      enamines, 987
      iminium perchlorates, 987
      ketene N,S-acetals, 988
      ketene aminals, 987
   as intermediates, in the synthesis of pyrimidines, 988
   synthesis of, 987
Enamines, α-azido
   from
      N-alkyl-5-phenylisoxazolium fluoroborates, 990
      α-chloro enamines, 990
      gem-diazidoalkenes, 989
      ketene N,S-acetals, 990
   as starting materials, for α-phosphimino enamines, 990
   synthesis of, 989
Enamines, α-azo, from, pyrazoles, 989
Enamines, α-bis(methylthio)methyleneamino, from, N,N-dimethylthiourea, 988
Enamines, α-borio
   from
      trialkylboranes, 1015
      ynamines, 1015
Enamines, bromo, synthesis of, 790
Enamines, chloro
   as intermediates, in the synthesis of ketene aminals, 970
   reactions
      with Grignard reagents, 15

with lithium amides, 971
as starting materials, for ketene aminals, 971
Enamines, 1-chloro
  reactions
    with metal salts of thiols, 904
    with thiophenols, 904
  as starting materials, for ketene S,N-acetals, 904
Enamines, gem-chloro
  reactions, with sodium diphenylphosphide, 997
  as starting materials, for α-isocyano enamines, 990
Enamines, α-chloro
  from, dimethylphosgeniminium chloride, 794
  as starting materials
    for α-azido enamines, 990
    for α-ferrio enamines, 1019
    for α-lithio enamines, 1016
    for α-magnesio enamines, 1017
    for α-rhenio enamines, 1020
    for α-sodio enamines, 1017
  synthesis of, 791
Enamines, 2-chloro-2-nitro, from, nitroacetone, 796
Enamines, α-cuprio, from, ynamines, 1018
Enamines, N,N-dialkyl α-halo, synthesis of, 790
Enamines, α-diaminomethyleneamino, from, gem-aminochloromethyleneamino chloroalkenes, 988
Enamines, α-ferrio
  from
    α-chloro enamines, 1019
    iron vinylidene complexes, 1019
  synthesis of, 1019
Enamines, fluoro, synthesis of, 790
Enamines, α-fluoro, synthesis of, 793
Enamines, halo, as starting materials, for α-trialkylphosphino enamines, 998
Enamines, gem-halo, as starting materials, for α-phosphino enamines, 997
Enamines, α-halo
  NMR spectra of, 792
  as starting materials, for α-phosphoryl enamines, 1004
  synthesis of, 790
    via addition of halide, 790
    via amination, 792
    via electrophilic attack, 793
Enamines, α-hydrazino
  from
    imino esters, 983
    ketene acetals, 983, 983
    ketene N,S-acetals, 983
    ketene dithioacetals, 983
    nitriles, 984
    thioamides, 984
    thiosemicarbazides, 984
    α-trichloromethyl enamines, 984
    tricyanoquinodimethanes, 985
    ynamines, 984
  as intermediates, in the synthesis of heterocycles, 983
  reactions
    with carbon disulfide, 986
    with methyl chloroformate, 986
    with methyl iodide, 986
    with methyl isocyanate, 986
  as starting materials, for N-acylated α-hydrazino enamines, 986
  synthesis of, 983
Enamines, α-hydrazino β-nitro, use in the synthesis of α-hydrazino enamine compounds, 986
Enamines, iodo, synthesis of, 790
Enamines, α-isocyano, from, gem-chloro enamines, 990
Enamines, α-lithio
  from
    β-amino-α,β-unsaturated amides, 1016
    β-amino-α,β-unsaturated esters, 1016
    α-chloro enamines, 1016

synthesis of, 1016
Enamines, α-magnesio, from, α-chloro enamines, 1017
Enamines, α-manganio, from, gem-chloro manganioalkenes, 1019
Enamines, 1-(methylseleno), synthesis of, 909
Enamines, α-molybdenio, as intermediates, in the synthesis of dicyanoketeneimmonium compounds, 1018
Enamines, nitro, reactions, with benzenesulfonyl chloride, 907
Enamines, 2-nitro, reactions, with N-halosuccinimides, 796
Enamines, α-palladio, from, isocyanides, 1017
Enamines, phenylhydrazino, synthesis of, 983
Enamines, α-phosphimino
  from
    α-azido enamines, 990
    bisphosphiminoalkenes, 991
    ketene aminals, 990
  synthesis of, 990
Enamines, α-phosphinoyl
  from
    dialkyl phosphinous acids, 1000
    malonitrile, 1000
Enamines, α-phosphorodiamidoyl
  from
    alkoxyphosphorus diamines, 1011
    imino chlorides, 1011
Enamines, α-phosphoryl
  alkylation of, 1004
  from
    acetals, 1004
    aldehydes, 1002
    1-aminoalkyl phosphonates, 1003
    α-chloro alkanimines, 1004
    1-dimethylamino-1-cyanomethanephosphonic acid, 1004
    α-halo enamines, 1004
    iminium esters, 1004
    ketene N,O-acetals, 1003
    ketene N,S-acetals, 1003
    ketene aminals, 1003
    α-ketophosphonic diesters, 1004
    nitroalkenes, 1004
    N-phosphorylated α-phosphoryl enamines, 1004
    α-phosphoryl enamines, 1004
    tertiary amides, 1003
    tetraethyl dialkylaminomethylene diphosphonate, 1002
    ynamines, 1003
  reactions, with t-butyllithium, 1004
  as starting materials, for α-phosphoryl enamines, 1004
  synthesis of, 1002
Enamines, N-phosphorylated α-phosphoryl, as starting materials, for α-phosphoryl enamines, 1004
Enamines, α-platino, from, isocyanides, 1017
Enamines, polyhalogenated α-chloro, synthesis of, 793
Enamines, primary α-halo, synthesis of, 792
Enamines, α-rhenio, from, α-chloro enamines, 1020
Enamines, secondary α-chloro
  from, 2,2-dicyanovinyl azides, 792
  reactions, with 2,2-dicyano-N-phenylacetamide, 792
Enamines, α-silyl
  from
    alkynylcarbenechromium complexes, 1014
    cyanohydrins, 1013
    epoxynitriles, 1014
    methacrylic amides, 1014
    nitriles, 1014
    propargylic alcohols, 1014
  synthesis of, 1013
Enamines, α-sodio
  from
    α-chloro enamines, 1017
    enamines, 1017
Enamines, α-stannio, from, diamines, 1020

Enamines, substituted, synthesis of, 927
Enamines, α-thiophosphinoyl
 from
  malonitrile, 1001
  phosphinothious acid, 1001
Enamines, α-thiophosphoryl, from, ynamines, 1010
Enamines, α-trichloromethyl, as starting materials, for
  α-hydrazino enamines, 984
Enamines, α-(N,N',N''-triphenylphosphorimidic triamido),
  from, phosphorimidic tribromide, 991
Enamines, α-tungstenio, as intermediates, in the synthesis of
  dicyanoketeneimmonium compounds, 1018
Enamines, α-ureido
 from
  imino esters, 982
  ketene aminals, 981
  1,3-oxazin-6-ones, 981
 synthesis of, 981
Enantioselectivity, in the synthesis of α-thiosulfoxides, 258
Eneamidines, as starting materials, for *gem*-
  iminomethylamino lithioalkenes, 1016
Enethiolates
 from
  O,O'-diethyl trithiopyro carbonate, 834
  dimethyloxosulfonium methylide, 834
 synthesis of, 834
Eniminium ions, use in the synthesis of ketene aminals, 976
Enolate anions, carbonyl, reactions, with germyl
  chlorides, 625
Enolate compounds, reactions, with ortho esters, 192
Enolates, α-bromo ester, reactions, with trialkylboranes, 832
Enolates, ester
  silylation of, 831
  as starting materials, for ketene silyl acetals, 831
Enolates, α-germyl, synthesis of, 662
Enolates, phosphino silyl, alkaline hydrolysis of, 1026
Enolates, α-silyl, synthesis of, 648
Enolates, α-silyl ester, C-stannylation of, 660
Enolates, thiol ester
  O-alkylation of, 833
  O-silylation of, 834
Enols, α-halo
 from
  alkynes, 764
  haloalkynes, 764
 synthesis of, 760
Enols, *gem*-iodo phosphonamido, synthesis of, 804
Enones
  acetalization of, 179
  α-dialkoxyalkylation of, 193
  reactions
   with cuprates, 385
   with dimethyl phosphite, 339
Enones, α-halo, synthesis of, 90
Enyl chlorides, as byproducts, 738
Enynes, reactions, with ethanethiol, 872
1,3-Enynes, use in the synthesis of carbon-bridged ditin
  compounds, 701
3-En-1-ynes, 1-silyl, monosilylation of, 1046
Enzymes, for the oxidation of dithioacetals, 262
Episulfone, 1-bromo-1-methylethene, from,
  α-bromoethanesulfonyl chloride, 83
Episulfones, chloro, synthesis of, 82
Epoxidation, of vinylsilanes, 357
Epoxides
  cleavage of, 343
  reactions
   with aldehydes, 181
   with ketones, 181
  synthesis of, 146
Epoxides, α-chloro, from, chloroalkenes, 50
Epoxides, 1,1-disilylated, as starting materials, for
  1,1-bis(silyl)alkanes, 619
Epoxides, disulfonyl, synthesis of, 853
Epoxides, α-fluoro
 from
  Dewar hexafluorobenzene, 46
  2,3-difluoroallylic alcohols, 45
  hexafluorobenzene, 45
 synthesis of, 45
Epoxides, vinyl
  reactions, with bis(phenylsulfonyl)methane, 280
  as starting materials, for acetals, 183
2,3-Epoxy alcohols, ring-opening of, 516
Epoxybis(sulfone), from, dithioacetals, 273
Epoxycyclohexane, 1-chloro-, rearrangement of, 50
Epoxy esters, synthesis of, 209
Epoxynitriles, as starting materials, for α-silyl
  enamines, 1014
2,3-Epoxynorbornane, *exo*-2-chloro-, synthesis of, 50
Epoxyphosphonates, synthesis of, 341
Epoxysilanes
  alcoholysis of, 357
 from
  chloromethyltrimethylsilane, 355
  α-methyl-α-chlorotrimethylsilane, 355
  vinylsilanes, 357
 hydrolysis of, 357
α,β-Epoxysilanes
  cleavage of, 139
  as starting materials, for α-fluoro-β-hydroxysilanes, 137
Epoxy sulfides, oxidation of, 237
α,β-Epoxy sulfones, synthesis of, 236
α,β-Epoxy sulfoxides, synthesis of, 233
Erbium compounds, reactions, with phosphorus ylides, 578
Eschenmoser's salt
  characteristics of, 96
  for the condensation of alkenes, 913
  reactions, with 1-methoxy-3-trimethylsilyloxy-
   1,3-butadiene, 305
Eschenmoser modification, of Claisen rearrangement, for the
  synthesis of bis(tributyltin) compounds, 702
Eschweiler–Clarke reaction, for the synthesis of
  S,N-acetals, 318
Esermethole, synthesis of, 410
Ester, perfluoroacrylic, reactions, with aniline, 894
Ester, polyphosphoric acid trimethylsilyl (PPSE), as
  condensation agent, for the synthesis of
  dithioacetals, 247
Esters
  chlorination of, 54
  deprotonation of, 831
  from, racemates, 390
  reduction of, 169
  as starting materials
   for hemiacetals, 169
   for ketene aminals, 972
   for α-thio α-sulfinyl acetophenones, 857
Esters, acetoacetic
  oxidation of, 162
  use in the synthesis of 1,3-benzazarsolines, 496
Esters, acylaminophosphinic
  reactions, with vinylidene cuprate compounds, 1001
  as starting materials, for *gem*-amido
   phosphonoylalkenes, 1001
Esters, alkene-1,1-diboronic, reactions, with mercuric
  chloride, 1067
Esters, 2-alkoxyvinyl phosphonic, reactions, with acetyl
  nitrate, 490
Esters, O-alkyl
  lithiation of, 382
  silylation of, 360
Esters, alkyl β-fluoro-β-nitroacrylate, synthesis of, 795
Esters, alkylidene biscarbamic, synthesis of, 426
Esters, alkynic, reactions, with cobalt carbonyls, 712
Esters, (E)-3-amino-3-alkoxythioacrylic-O-alkyl, synthesis
  of, 895

Esters, β-amino-α,β-unsaturated, as starting materials, for α-lithio enamines, 1016
Esters, N-benzylidene-α-(diphenylphosphinoyl)glycine, as starting materials, for azomethine ylides, 489
Esters, bissulfonic, synthesis of, 282
Esters, borinic
    reactions, with phosphines, 342
    synthesis of, 815
Esters, boronate, reactions, with dichloromethyllithium, 526
Esters, boronic ethyl, synthesis of, 815
Esters, α-bromo
    reactions, with phenoxides, 199
    as starting materials
        for *gem*-imino phosphiminoalkenes, 996
        for ketene silyl acetals, 831
Esters, α-bromoalkyl
    from
        2-diazo-1,3-dicarbonyl compounds, 57
        vinyl acetates, 57
    synthesis of, 56
Esters, bromomethanediphosphonate, thallium salt of, 801
Esters, carboxylic
    alkylation of, 824
    as starting materials
        for ketene acetals, 824
        for ketene dithioacetals, 845
Esters, cephalosporin, synthesis of, 48
Esters, α-chloro, reactions, with *t*-butyl carbazate, 448
Esters, α-chloroalkyl
    from
        butenolide alcohols, 54
        α-phenylthioalkyl esters, 54
        phthalide alcohols, 54
    synthesis of, 53
Esters, α-chloroboronic, as starting materials, for α-alkoxystannanes, 389
Esters, 1-chloro-1-(chlorosulfenyl), synthesis of, 72
Esters, α-chloro-α-(chlorosulfenyl)carboxylic, as starting materials, for α-chloro sulfides, 70
Esters, ω-chloro-α-iodoalkenylboronate, synthesis of, 815
Esters, chlorophosphine, reactions, with triethylamine, 483
Esters, 1-chlorovinylboronate, from, 1,1-dichloroethane, 813
Esters, 2-cyano, oxidation of, 894
Esters, α-diazo, as starting materials, for boryloxy compounds, 832
Esters, diethylene boronate, synthesis of, 634
Esters, α,α-digermylated, synthesis of, 625
Esters, diphenyl, synthesis of, 467
Esters, disulfonate, reactions, with aldehydes, 864
Esters, α,α-dithioboronate, use in the synthesis of ketene dithioacetals, 849
Esters, dithionomalonic
    reactions
        with amines, 895
        with ammonia, 895
        with *t*-butoxy-bis(dimethylamino)methane, 896
Esters, α,α-dithiophosphonate, reactions, with ketones, 847
Esters, enol
    epoxidation of, 209
    reactions, with peracids, 209
    as starting materials, for acyl alkyl acetals, 209
Esters, 1-ethoxyethyl, synthesis of, 208
Esters, ethynecarboxylate
    reactions
        with carbon diselenide, 874
        with selenolate salts, 874
Esters, α-fluoro, from, alkenes, 52
Esters, α-fluoroalkyl
    from
        fluoroalkenes, 51
        3-oxocarboxylic acids, 52
        perfluoroacyl fluorides, 52
        perfluoro alkoxides, 52
        vinyl acetates, 51
    as starting materials, for ethyl acetate, 52
    synthesis of, 51
Esters, α-fluorobenzyl phosphonate, synthesis of, 131
Esters, α-fluoro-α-iodo, reactions, with *n*-butyllithium, 137
Esters, α-fluoro-α-trimethylsilyl, synthesis of, 137
Esters, α-haloalkyl, synthesis of, 51
Esters, α-halo boronic
    chemistry of, 362
    reactions, with zinc, 665
Esters, halomethanediphosphonate, use in the synthesis of 2-haloalkenylphosphorus compounds, 801
Esters, hydroxamic vinyl, reactions, with ketene acetals, 833
Esters, α-hydroxyalkyl phosphonic, synthesis of, 338
Esters, α-hydroxybenzylphosphonate, reactions, with diethylaminosulfur trifluoride, 129
Esters, imidic, tautomers of, 880
Esters, iminium
    reactions, with sodium diethyl phosphite, 1003
    as starting materials, for α-phosphoryl enamines, 1004
Esters, imino
    reactions, with benzoylhydrazines, 986
    as starting materials
        for α-hydrazino enamines, 983
        for ketene aminals, 972
        for α-ureido enamines, 982
Esters, α-iodoalkyl, synthesis of, 58
Esters, (iodomethyl)boronic, as starting materials, for boronic acids, 144
Esters, α-keto
    use in the synthesis of 1,3-azarsa compounds, 495
    use in the synthesis of 1,3-benzazarsolines, 496
Esters, β-keto, use in the synthesis of 1,3-azarsa compounds, 495
Esters, δ-keto, use in the synthesis of 1,3-benzazarsolines, 496
Esters, α-keto perfluoroalkane carboxylic, reactions, with hydrazine, 311
Esters, malonic
    reactions
        with alkylarsenic dichlorides, 594
        with arylarsenic dichlorides, 594
Esters, methacrylate, reactions, with vinyl cyclopropanes, 830
Esters, (methoxymethylene)bisphosphonic, reactions, with benzaldehyde, 916
Esters, methoxyvinyl, synthesis of, 827
Esters, *p*-methylbenzyl, cleavage of, 478
Esters, orthocarbamate, reactions, with methylene groups, 894
Esters, phenylthioalkyl, bromination of, 57
Esters, α-phenylthioalkyl, as starting materials, for α-chloroalkyl esters, 54
Esters, phenylthiomethyl, synthesis of, 221
Esters, phosphonate
    as starting materials
        for ketene dithioacetal *S*-oxides, 860
        for thiaselenafulvenes, 870
Esters, silyl, reactions, with trialkylsilanes, 832
Esters, α-sily-α-stannyl, from, silylketenes, 661
Esters, α-sulfenyl, electrolysis of, 197
Esters, sulfonyl, deprotonation of, 391
Esters, α-sulfuranylidene, as starting materials, for boryloxy compounds, 832
Esters, thiobenzimidate, synthesis of, 1001
Esters, thiol, oxidation of, 259
Esters, thiophenyl
    reactions, with lithium diisopropylamide, 932
    as starting materials, for trimethylsilyl enol ethers, 932
Esters, thiosulfonate
    from
        benzo-1,3-dithiole-2-thiones, 862

benzo-1,3-dithiole-5-thiones, 862
Esters, trialkylsilyl, use in the synthesis of ketene disilyl acetals, 832
Esters, trimethylsilyl, methanolysis of, 129
Esters, α,β-unsaturated
  cycloaddition reactions of, 827
  as starting materials
    for α-aminophosphinic acids, 460
    for α-aminophosphonic acids, 460
    for ketene acetals, 827
Esters, α,β-unsaturated β-amino, synthesis of, 496
Esters, valine methyl, synthesis of, 536
Esters, vinylboronate, hydroboration of, 631
Ethanal, benzoyl(phenyl)-, condensation of, 894
Ethane, electrochemical fluorination of, 3
Ethane, 1-aryl-1,2,2-trichloro-2-fluoro-, as starting material, for 1-aryl-2-chloro-2-fluoroethenes, 750
Ethane, 1-benzoyloxy-1-ethoxy-, from, diethyl ether, 207
Ethane, 1,2-bis(dichloroboryl)-1-(trimethylsilyl)-, synthesis of, 644
Ethane, 1,1-bis(difluoroboryl)-, synthesis of, 632
Ethane, 1,2-bis(N-methylamino)-, reactions, with bis(dichloroarsino), 593
Ethane, 1,2-dianilino-, use in the synthesis of aldehydes, 408
Ethane, diazo-, use in the synthesis of diruthenium complexes, 681
Ethane, 1,2-dibromo-1-ethoxy-, reactions, with triphenylphosphine, 912
Ethane, 1,2-dibromo-1-methoxy-, reactions, with trimethyl phosphite, 912
Ethane, 1,2-dibromotetrafluoro-, reactions, with O-(1-lithiovinyl) carbamates, 768
Ethane, 1,1-dichloro-
  reactions
    with lithium diisopropylamide, 813
    with sodium diphenylarsenic, 592
    with trimethylborate, 813
  as starting material, for 1-chlorovinylboronate esters, 813
Ethane, 1,2-dichloro-1,2-diisopropoxy-, reactions, with potassium t-butoxide, 767
Ethane, 1,2-dichloro-1,2-dimethoxy-, synthesis of, 767
Ethane, 1,1-dichlorophenyl-, synthesis of, 13
Ethane, 1,1-dideuterio-, synthesis of, 668
Ethane, 2,2-diiodo-1,1-dinitro-, synthesis of, 993
Ethane, N,N'-dimethyl-1,2-diamino-, reactions, with glyoxal, 409
Ethane, 1,1-dinitro-, alkylation of, 432
Ethane, 1,1-disodio-1-phenyl-, synthesis of, 669
Ethane, hexachloro-, reactions, with alkyldiphenylphosphines, 112
Ethane, hexanitro-, reactions, with benzene, 994
Ethane, 1-nitro-, deprotonation of, 335
Ethane, 1,1,2,2-tetrachloro-, reactions, with sodium thiophenoxide, 771
Ethane, 1,1,2,2-tetramorpholino-
  reactions, with methanol, 309
  synthesis of, 300
Ethane, 1,1,2,2-tetranitro-, fluorination of, 795
Ethane, 1,2,2-tribromo-1,2,2-trinitro-, pyrolysis of, 994
Ethane, 1,1,1-trichloro-, reactions, with allyltributyltin, 14
Ethane, 1,1-trichlorodifluoro-, reactions, with alkenyllithiums, 758
Ethane, 1,1,1-trinitro-, as starting material, gem-dinitroalkanes, 431
Ethane, 1,1,2-tris(dichloroboryl)-, synthesis of, 632
Ethanedithiol, reactions, with cyclohexanone, 246
Ethanes, 1-alkoxy-2,2,2-tribromo-1-chloro-, reactions, with sodium ethoxide, 767
Ethanes, 1,1-bis(silyl)-, from, vinylsilanes, 617
Ethanes, 1,1-diaryl-, as starting materials, for gem-dinitroalkenes, 993
Ethanes, 1,1-diaryl-1,2,2-trichloro-2-fluoro-, as starting materials, for 1,1-diaryl-2-chloro-2-fluoroethenes, 750

Ethanes, α,β-dihalo-, synthesis of, 804
Ethanes, 1,2-disilyl-, as starting materials, for 1,1-bis(silyl)alkanes, 616
Ethanes, phosphaazo-, synthesis of, 110
Ethanes, polychlorinated, reactions, with sodium phenoxide, 761
Ethaneselenol, 2-amino-, reactions, with glyoxylic acid, 333
Ethanesulfinic acid, 1-chloro-
  from, α-chlorosulfonyl chloride, 78
  reactions, with thionyl chloride, 78
Ethanesulfonyl chloride, bromo-, reactions, with sodium sulfite, 85
Ethanesulfonyl chloride, α-bromo-, as starting material, for 1-bromo-1-methylethene episulfone, 83
Ethanethiol
  reactions
    with enynes, 872
    with ethyl cyanoacetate, 905
    with N-(hexafluoroisopropylidene)-benzenesulfinamides, 325
    with penta-O-acetyl-β-D-glucopyranose, 224
    with 4-substituted 2-trifluoromethyl-5(4H)-oxazolones, 320
Ethanethiolate, reactions, with allyl bromides, 373
Ethanol
  reactions
    with 1-[bis-(2-chloroethyl)amino]-1,2,2-tricyanoethene, 893
    with dichlorotetrafluoroacetone, 164
    with ethyl propiolate, 195
    with methane disulfonyl chloride, 282
    with nitriles, 883
    with ynamines, 885
Ethanol, 2-chloro-
  reactions
    with dipyridyl ketones, 183
    with methanesulfonyl chloride, 282
    with 1,1,1-trifluoroacetone, 164
Ethanol, 2-(methylamino)-, reactions, with glyoxal, 302
Ethanol, thiol-, reactions, with cyclohexanones, 229
Ethanol, 2-thiol-, reactions, with 1,1,3,3-tetramethoxypropane, 230
Ethanol, 2,2,2-trichloro-, reactions, with acetals, 184
Ethanol, 2,2,2-tris(trimethylgermanyl)-, as starting material, for 1,1-bis(trimethylgermanyl)ethene, 1055
Ethanolamines
  condensation of, 888
  reactions, with 7,7,8,8-tetracyanoquinodimethane, 893
Ethanols, 2-arylthio-2,2-dichloro-, reactions, with zinc, 775
Ethanols, naphthyl, as starting materials, for gem-dinitroalkenes, 993
Ethanone, 2-methanesulfonyl-1-(2-chlorophenyl)-, reactions, with diethyl chlorothiophosphate, 784
1-Ethanone, 1-(t-butyldimethylsilyl)-2-chloro-, reactions, with trimethyl phosphite, 934
Ethene, 2-anilino-1-bromo-1-nitro-, stability of, 796
Ethene, 1-benzenesulfinyl-1-nitro-2,2-diphenyl-, synthesis of, 907
Ethene, 1,1-bis(benzenesulfonyl)-, reactions, with 3H-indolium ylides, 277
Ethene, 1-[bis-(2-chloroethyl)amino]-1,2,2-tricyano-, reactions, with ethanol, 893
Ethene, bis(dibutoxyphosphino)-, synthesis of, 913
Ethene, 1,1-bis(diphenylphosphino)-2,2-bistrifluoromethyl-, synthesis of, 1023
Ethene, 1,1-bis(sulfonyl)-
  Michael addition to, 277
  reactions, with 2,3-dihydropyran-4-one, 277
Ethene, 1,1-bis-(toluene-p-sulfinyl)-, as chiral dienophile, 270
Ethene, 1,1-bis(trimethylgermanyl)-, from, 2,2,2-tris(trimethylgermanyl)ethanol, 1055
Ethene, 1,1-bis(trimethylsilyl)-, synthesis of, 1048

Ethene, 1,1-bis(trimethylsilyl)-2-phenyl-, from, phenylmethanal, 1049
Ethene, 1,1-bis(trimethylstannyl)-, synthesis of, 1069
Ethene, 1-bromo-1,2-diisopropoxy-, synthesis of, 767
Ethene, 1-(bromo)iodo-2-bromo(iodo)-1-phenoxy-, synthesis of, 765
Ethene, bromotrifluoro-, reactions, with tin halides, 812
Ethene, 1-bromo-1-trimethylsilyl-
  reactions, with magnesium, 1060
  synthesis of, 807
Ethene, chloro-, reactions, with tetramethyldiarsine, 134
Ethene, 1-chloro-1,2-diisopropoxy-, synthesis of, 767
Ethene, 1-chloro 2,2-diphenyl-, deprotonation of, 741
Ethene, chlorotrifluoro-
  reactions, with cyanide, 5
  use in the synthesis of chlorofluoroalkanes, 32
Ethene, 2,2-dibromo-1,1-dinitro-, synthesis of, 994
Ethene, 1,1-dibromo-2-phenoxy-, synthesis of, theoretical, 761
Ethene, 1,2-dibromo-1-phenoxy-, synthesis of, 761
Ethene, cis-1,2-dichloro-, reactions, with benzeneselenol, 786
Ethene, cis-1,2-dichloro-1,2-di-(p-tolylthio)-, synthesis of, 770
Ethene, 1,2-dichloro-1-ethoxy-, synthesis of, 761
Ethene, 1,1-dichloro-2-fluoro-, metallation of, 817
Ethene, 1,2-dichloronitro-, synthesis of, 797
Ethene, 1,1-dicyanobis(trifluoromethyl)-
  reactions
    with molybdenum chlorides, 800
    with tungsten chlorides, 800
Ethene, 1,1-difluoro-
  reactions
    with hydrogen bromide, 5
    with hydrogen iodide, 5
Ethene, (E)-1,2-difluoro-1,2-dinitro-, synthesis of, 795
Ethene, 1,2-diisopropoxy-, synthesis of, 767
Ethene, 1,1-dilithio-, stability of, 1064
Ethene, 1-(dimesitylboryl)-1-trimethylsilyl-1-, nucleophilic addition to, for the synthesis of α-borylsilylalkanes, 644
Ethene, 1,2-dimethoxy-1,2-diphenoxy-, synthesis of, 830
Ethene, 1,2-dimethoxy-1-lithio-, reactions, with trialkylboranes, 935
Ethene, 1,1-dinitro-, synthesis of, in situ, 994
Ethene, 1,1-dinitro-2,2-diphenyl-
  reactions
    with potassium phenyl selenide, 911
    with sulfur nucleophiles, 906
Ethene, 1,1-diphenyl-2-fluoro-, halogenation of, 751
Ethene, 1-ethoxy-1,2-dichloro-, reactions, with triethyl phosphite, 912
Ethene, 1-ferrocenyl-2-nitro-
  chloromercuriation of, 797
  iodination of, 796
Ethene, fluoro-, reactions, with tetramethyldiarsine, 134
Ethene, 1-fluoro-1-(benzenesulfonyl)-, as starting material, for α-fluoro sulfones, 79
Ethene, iodotrifluoro-
  Arbuzov reaction of, 802
  reactions, with nitric oxide, 798
Ethene, nitrosotrifluoro-, synthesis of, 798
Ethene, nitrotriiodo-, synthesis of, 797
Ethene, phenoxytriiodo-, synthesis of, 765
Ethene, tetrachloro-
  reactions
    with benzeneselenol, 786
    with dimethyl diselenide, 786
    with sodium arenethiolates, 770
    with sodium thiolate, 770
Ethene, tetrafluoro-
  dimerization of, 5
  reactions
    with cyanide ions, 1076
    with ethyne, 5
    with methanol, 5
  reactivity of, 802
  for the synthesis of α-fluoro ethers, 44
Ethene, tetraiodo-, nitration of, 993
Ethene, tetrakis(trimethylsilyl)-, reactions, with lithium, 618
Ethene, tetramethoxy-, as byproduct, in the synthesis of carbenes, 1080
Ethene, tetramethyl-, reactions, with dibromodinitromethane, 995
Ethene, tetranitro-, synthesis of, 994
Ethene, 1,1,2,2-tetrasilyl-, from, 1,2-bis(silyl)ethyne, 1045
Ethene, tetra(trimethylsilyl)-, synthesis of, 1050
Ethene, 1,2,2-tribromo-1-fluoro-, reactions, with thiophenol, 771
Ethene, 1-tribromogermanyl-1-trimethylsilyl-, synthesis of, 1058
Ethene, 1,1,2-tri(t-butyldimethylsilyl)-1-(trimethylsilyl)-, x-ray analysis of, 1051
Ethene, trichloro-
  reactions
    with lithium amides, 976
    with sodium arenethiolates, 772
Ethene, 1,1,2-trichloro-, reactions, with sodium arenethiolates, 770
Ethene, 1,2,2-trichloro-1-phenoxy-, synthesis of, 766
Ethene, trifluoro-
  reactions
    with fluoronitroso sulfates, 104
    with tetramethyldiarsine, 134
  use in the synthesis of α-fluoroalkenyllithium compounds, 817
Ethene, 1,2,3-trifluoro-1-methoxy-, synthesis of, 761
Ethene, 1-trimethylsilyl-, synthesis of, 1061
Ethene, 1-trimethylsilyl-1-trimethylstannyl-, hydrostannylation of, 661
Ethene, triodonitro-, reactions, with trifluoromethylthiocopper, 771
Ethene, 1,1,2-tris(trimethylsilyl)-, synthesis of, 1049
Ethenes, 1-alkoxy-2,2-dibromo-1-chloro-, synthesis of, 767
Ethenes, 1-alkylthio-2-seleno-2-chloro-, synthesis of, 787
Ethenes, 1-aryl-2-chloro-2-fluoro-, from, 1-aryl-1,2,2-trichloro-2-fluoroethane, 750
Ethenes, bis(dialkylboryloxy)-, from, carboxylic acids, 832
Ethenes, 1-chloro-2,2-diaryl-, metallation of, 817
Ethenes, 2,2-diamino-1,1-disulfonyl-, synthesis of, 866
Ethenes, 1,1-diaryl-, reactions, with mercuric nitrate, 1067
Ethenes, 1,1-diaryl-2-chloro-2-fluoro-, from, 1,1-diaryl-1,2,2-trichloro-2-fluoroethanes, 750
Ethenes, 1,1-diboro 2,2-bis(trimethylsilyl)-, synthesis of, 1048
Ethenes, dicyanodifluoro-, as starting materials, for α-fluoroamines, 104
Ethenes, 1,1-disilyl 2-phenyl-, synthesis of, 1048
Ethenes, germanyl-substituted, synthesis of, 827
Ethenes, gem-halo(phosphono)-, synthesis of, 804
Ethenes, nitro-, reactions, with chlorine, 108
Ethenes, 1-phenylthio-1-triethylgermyl-, synthesis of, 940
Ethenes, silyl-substituted, synthesis of, 827
Ethenes, tetraalkoxy-, synthesis of, 830
Ethenes, tetrasilyl-
  from
    bis(silyl) dichloromethanes, 1050
    1,3-disilacyclobutanes, 1049
Ethenes, 1-thioalkyl-1-nitro-2,2-diphenyl-, synthesis of, 906
Ethenes, 1,1,2-trisilyl-
  from, silirenes, 1051
  synthesis of, 1044
Ethene-1-sulfonyl fluoride, 2,2-difluoro-, reactions, with sulfur trioxide, 867
Ether, benzyl chloromethyl, reactions, with tributylstannyllithium, 389
Ether, bis(chloromethyl)

Ether
  reactions, with amines, 411
  synthesis of, 47
Ether, bis(fluoromethyl)
  as intermediate, in the synthesis of α-fluoro ethers, 44
  synthesis of, 44
Ether, bis(methoxymethyl), synthesis of, 232
Ether, bis(trimethylsilyl)vinyl, synthesis of, 1049
Ether, *t*-butyl methyl, deprotonation of, 378
Ether, 1-chlorocyclopropyl methyl, synthesis of, 49
Ether, chloromethyl methyl
  reactions
    with lithium bis(trimethylsilyl)amide, 313
    with magnesium, 383
    with tributylstannylmagnesium chloride, 389
    with trichlorogermane, 361
  synthesis of, 48, 49
Ether, chloromethyl phenyl, from, sodium chloromethyl sulfonate, 49
Ether, 18-crown-6, as catalyst, for the oxidation of dithioacetals, 266
Ether, 2,2-dibromovinyl ethyl, reactions, with *n*-butyllithium, 818
Ether, di(2,2-dibromo-1-chlorovinyl), synthesis of, 767
Ether, diethyl, as starting material, for 1-benzoyloxy-1-ethoxyethane, 207
Ether, dimethyl, lithiation of, 378
Ether, iodomethyl methyl, synthesis of, 51
Ether, methylvinyl, reactions, with *t*-butyllithium, 948
Ether, silyl enol, reactions, with α-chlorosilanes, 365
Ether, 1,2,2-trichlorovinylphenyl, synthesis of, 742
Ethers
  α-alkoxylation of, 201
  α-chlorination of, 47
  metallation of, 378
  reactions
    with acyl peroxides, 207
    with chlorotrialkylsilanes, 358
  as starting materials
    for acetals, 201
    for acyl alkyl acetals, 207
Ethers, α-acyloxy, as starting materials, for acetals, 200
Ethers, α-acyl α-sulfonyl, acylation of, 838
Ethers, α-alkoxy bromo, reactions, with *t*-butyllithium, 381
Ethers, α-*s*-alkyl, fluorination of, 44
Ethers, alkyl 1-azidoalkyl, synthesis of, 314
Ethers, alkyl chloromethyl, as starting materials, for dialkylaminomethylalkyl ethers, 309
Ethers, alkynic, reactions, with amines, 884
Ethers, allyl, oxidation of, 208
Ethers, allyl enol, as starting materials, for acetoxy tetrahydrofuran compounds, 208
Ethers, allylic
  lithiation of, 381
  silylation of, 358
Ethers, azacrown, as catalysts, for the synthesis of α-aminoalkylphosphorus compounds, 469
Ethers, azidoalkyl alkyl, synthesis of, 314
Ethers, benzylic
  lithiation of, 381
  silylation of, 358
Ethers, bis(formamidinium), as starting materials, for ketene aminals, 975
Ethers, bislactim, synthesis of, 535
Ethers, bis(trimethylsilyl) silyl enol, attempted isolation of, 1048
Ethers, 1-boro 1-germanyl alkenol, synthesis of, 1059
Ethers, α-bromo
  from
    acetals, 50
    α-chloro ethers, 51
    vinyl ethers, 51
  synthesis of, 50
Ethers, *gem*-bromo lithio enol
  as starting materials, for (*E*)-2-ethoxyvinyl bromide, 818
  synthesis of, 818
Ethers, α-bromo-β-selenenyl, reactions, with alcohols, 51
Ethers, *t*-butoxymethyl, synthesis of, 201
Ethers, *t*-butyl, cleavage of, 176
Ethers, α-chloro
  from
    acetals, 48
    alcohols, 47
    aldehydes, 47
    α-alkoxy acid chlorides, 49
    α-aryloxy acid chlorides, 49
    α-dichloro ethers, 49
    ketones, 47
    vinyl ethers, 49
  reactions, with thiolates, 218
  replacement of, 340
  as starting materials, for α-bromo ethers, 51
  synthesis of, 46
  via α-chlorination of ethers, 47
  use in the synthesis of acetals, 198
Ethers, 2-chloroethyl, synthesis of, 305
Ethers, chloromethyl
  displacement of, 389
  as starting materials
    for α-alkoxystannanes, 389
    for Grignard reagents, 383
Ethers, chloromethyl aryl, synthesis of, 47
Ethers, (1-chloro-1-methyl)ethyl methyl, synthesis of, 49
Ethers, chloromethyl methyl
  metallation of, 381
  synthesis of, 47
Ethers, α-cyano, lithiation of, 381
Ethers, cyclopropyl silyl, thermolysis of, 616
Ethers, dialkylaminomethylalkyl, from, alkyl chloromethyl ethers, 309
Ethers, α-dichloro, as starting materials, for α-chloro ethers, 49
Ethers, dichlorovinyl, structure of, 761
Ethers, α,α-difluoro, as starting materials, for ketene acetals, 826
Ethers, dimethylaminomethyl alkyl, synthesis of, 305
Ethers, enethiol
  from, cyclohexanone, 247
  selenylation of, 871
Ethers, enol
  alkoxymercuration of, 188
  as dienophiles, for the synthesis of acetals, 202
  hydration of, 171, 174
  hydroxy mercuration of, 171
  metallation of, 949
  osmylation of, 174
  oxidation of, 187
  reactions
    with alcohols, 186
    with allyl alcohols, 188
    with carboxylic acids, 208
    with α-cyanoacrylate, 827
    with (triseleno)orthoformates, 289
  as starting materials
    for acetals, 186
    for acyl alkyl acetals, 208
    for 1-alkoxyalkyl isocyanates, 314
    for α-functionalized acetals, 187
    for hemiacetals, 171
    for monoselenoacetals, 238
    for monotelluroacetals, 238
    for unsymmetrical acetals, 186, 188
  α-sulfenylation of, 836
  synthesis of, 749
Ethers, ethoxyethyl (EE), use in the synthesis of acetals, 185
Ethers, ethyl enol, reactions, with carboxylic acids, 208
Ethers, ethylvinyl

deprotonation of, 949
nitration of, 208
Ethers, fluorinated 1-lithiovinyl, synthesis of, 949
Ethers, α-fluoro
  from
    acetyl hypofluorite, 44
    paraformaldehyde, 44
    perfluoroallyl fluoro sulfate, 44
    polypropylene oxide, 44
  stability of, 44
  synthesis of, 44
Ethers, 1-fluoro enol, reactions, with amines, 894
Ethers, fluoromethyl aryl, synthesis of, 45
Ethers, 1-germanylalkyne, reactions, with germanium tetrachloride, 1055
Ethers, α-halo
  from
    methoxyethoxymethyl ethers, 199
    methoxymethyl ethers, 199
  reactions
    with alcohols, 198
    with t-butyllithium, 381
    with thiols, 218
  as starting materials
    for acetals, 198
    for monoselenoacetals, 239
    for monotelluroacetals, 239
    for monothioacetals, 218
  synthesis of, 43, 760
Ethers, α-haloalkyl, reactions, with carboxylic acids, 209
Ethers, 1-halo enol, from, 1-alkoxyalkynes, 765
Ethers, halomethyl, reactions, with metal germanes, 360
Ethers, gem-halo silyl enol, synthesis of, 806
Ethers, α-halo α-sulfonyl, dehydrohalogenation of, 838
Ethers, α-iodo
  from, α-chloro compounds, 51
  synthesis of, 51
Ethers, lithiated enol, synthesis of, 817
Ethers, methoxybenzyl, reactions, with 2,3-dichloro-5,6-dicyano-1,4-benzoquinone, 201
Ethers, methoxyethoxymethyl (MEM), as starting materials, for α-halo ethers, 199
Ethers, methoxymethyl (MOM)
  as starting materials, for α-halo ethers, 199
  synthesis of, 185
  use in the synthesis of acetals, 185
Ethers, methylthiomethyl (MTM)
  protection of alcohols as, 219
  reactions, with bromine, 216
  synthesis of, 219
    via oxidation, 220
    via Pummerer reaction, 222
Ethers, α-monochloro, synthesis of, 49
Ethers, perfluoro, synthesis of, 44
Ethers, perfluoro iodo, reactions, with aluminum chloride, 768
Ethers, phenyl, fluorination of, 44
Ethers, phenylthiomethyl, synthesis of, 219, 221
Ethers, propynol, reactions, with allylzinc bromide, 809
Ethers, silyl
  reactions
    with aldehydes, 181
    with hydrochloric acid, 160
    with ketones, 181
  synthesis of, 620
Ethers, α-silyl, synthesis of, 358
Ethers, 1-silyl 1-alkyn-2-ol alkyl, reactions, with trihaloboranes, 1057
Ethers, silyl enol
  from, aldehydes, 1049
  reactions
    with alcohols, 188
    with bromomethanesulfonyl bromide, 84

with diiron salts, 680
    with dithioorthoformates, 250
    with hydrazoic acid, 314
  as starting materials, for acetals, 188
Ethers, (E)-silyl enol, synthesis of, 918
Ethers, 1-silyl ethyn-2-ol, reactions, with germanium tetrachloride, 1058
Ethers, stannylvinyl, synthesis of, via reverse Diels–Alder reaction, 952
Ethers, α-substituted, as starting materials, for acetals, 198
Ethers, α-sulfonyl
  condensation of, with aldehydes, 838
  reactions, with methyl formate, 838
  C-silylation of, 838
  as starting materials
    for acetals, 200
    for ketene monothioacetal S,S-dioxides, 838
  use in Horner–Wittig reactions, 840
Ethers, α-sulfonyl β-halo, dehydrohalogenation of, 838
Ethers, 2-tetrahydrofuryl, synthesis of, 199
Ethers, thioenol, oxidation of, 233
Ethers, trifluoromethyl alkyl, synthesis of, 1083
Ethers, 1-trimethylgermanyl-1-alkynol, hydroboration of, 1059
Ethers, trimethylsilyl, reactions, with hydrogen chloride, 150
Ethers, trimethylsilyl enol
  from, thiophenyl esters, 932
  reactions, with α-tributylstannyl thioacetals, 398
Ethers, tris(trimethylsilyl)ethyl, synthesis of, 1048
Ethers, vinyl
  deprotonation of, 929
    reagents for, 949
  from, disulfonylmethanes, 865
  geometry of, 949
  metallation of, 929
  reactions
    with 2-aminoalkanols, 306
    with 1-azabuta-1,3-dienes, 305
    with t-butyllithium, 929
    with nucleophiles, 306
  as starting materials
    for α-bromo ethers, 51
    for α-chloro ethers, 49
    for ketene dithioacetal tetroxides, 865
Ethoxyethyl compounds, from, α-hydroxy sulfones, 236
Ethylamine, (S)-1-phenyl-, reactions, with formalin, 99
Ethylamine hydrobromide, 1-ethylthio-2,2,2-trifluoro-, synthesis of, 320
Ethylamine hydrochloride, 1-methoxy-2,2,2-trifluoro-, synthesis of, 305
Ethylamines, ferrocenyl-, synthesis of, 720
Ethylene, 1,1-bis(diethylamino)-, reactions, with alcohols, 888
Ethylene, 1,2-dimethoxy-, chlorination of, 767
Ethylene, 1,1-dinitro-2,2-diphenyl-
  reactions
    with alkoxides, 898
    with phenoxides, 898
Ethylene, 1-ethoxy-1-diethylamino-, synthesis of, 888
Ethylene, 1-ethoxy-1-(N-ethylanilino)-, synthesis of, 887
Ethylene, tetracyano- (TCNE), as starting material, for dicyanoketene acetals, 829
Ethylene, tetrafluoro-
  from, sodium perfluoropropionate, 731
  reactions, with sodium methoxide, 761
Ethylene, tribromo-, reactions, with potassium phenolate, 761
Ethylene, trichloro-
  reactions
    with benzeneselenol, 786
    with phenols, 761
    with sodium ethoxide, 761
    with sodium phenolate, 761

Ethylenediamine, tetramethyl- (TMEDA)
  for the deprotonation of vinyl ethers, 949
  lithiation of, 528
  reactions, with *t*-butyllithium, 623
  use in the synthesis of carbon-bridged diindium compounds, 698
  use in the synthesis of lithioferrocenes, 706
Ethylenes, 1,1-bis(dimethylamino)-2-aryl-2-cyano-, synthesis of, 974
Ethylenes, 1,1-bis(silyl)-, reactions, with lithium, 618
Ethylenes, *trans*-1,2-dichloro-1-phenoxy-, synthesis of, 761
Ethylenes, polyhalogenated, reactions, with sodium phenoxide, 761
Ethylenes, tetramino-, synthesis of, 975
Ethylidenebis(mercury chloride), reactions, with lithium, 668
Ethyne
  reactions
    with dinuclear bisimino complexes, 683
    with tetrafluoroethene, 5
Ethyne, benzoyl-, reactions, with salicyclic acid compounds, 213
Ethyne, 1,2-bis(silyl)-, as starting material, for 1,1,2,2-tetrasilylethene, 1045
Ethyne, bis(trimethylsilyl)-, reactions, with 1,1-dimethyl-2,3-bis(trimethylsilyl)silirane, 1052
Ethyne, 1,2-bis(trimethylsilyl)-
  diboronation of, 1048
  reactions
    with chloro diboranes, 1048
    with hydrosilanes, 1045
    with triarylsilylboranes, 1058
Ethyne, 1,2-bis(trimethylstannyl)-, reactions, with 1,2-di-*t*-butyl-1,2-dichloro diborane, 1063
Ethyne, di-*t*-butyl-, use in the synthesis of diiron complexes, 681
Ethyne, diphenoxy-, reactions, with sodium bis(trimethylsilyl)amine, 884
Ethyne, diphenyl-, as bridge, in binuclear nickel complexes, 692
Ethyne, ethoxy-
  reactions
    with triethylplumbyllithium, 955
    with triethylplumbylsodium, 955
Ethyne, 1-ethoxy-
  reactions
    with alkoxy(trimethylsiloxy)phosphines, 914
    with diethyl phosphite, 913
    with tribromophosphine, 914
    with tributylphosphine, 914
    with triethyl phosphite, 913
Ethyne, 1-methoxy-, reactions, with carboxylic acids, 827
Ethyne, phenyl-
  carbalumination of, 1068
  reactions, with triphenylstannane, 701
  as starting material, for cobaltacyclopentadiene complexes, 689
Ethyne, 1-trimethylgermanyl-2-phenyl-, reactions, with diisobutylaluminum hydride, 1063
Ethyne, trimethylsilyl-
  reactions
    with diphenyl diselenide, 947
    with organocopper compounds, 809
Ethyne, 1-trimethylsilyl-2-phenyl-
  carbatitanation of, 1061
  reactions, with silyl silirenes, 1053
Ethyne, 1-(trimethylstannyl)-, reactions, with triethyl boron, 1068
Ethyne, triphenylsilyl-, reactions, with organocopper compounds, 809
Ethyne complexes, diiridium diphenyl-, synthesis of, 688
Ethynedicarboxylate, dimethyl (DMAD)
  reactions
    with 6-amino-1,3-dimethyl uracil, 893

    with α-aminosilanes, 510
    with carbon selenide sulfide, 869
Ethynes, as starting materials, for borirenes, 1058
Ethynes, bis(alkylthio)-, reactions, with phosphorus pentachloride, 926
Ethynes, 1,2-bis(silyl)-
  hydrosilylation of, 1044
  reactions, with chloro methyl hydrosilanes, 1045
Ethynes, phosphorus-substituted, reactions, with benzenesulfenyl chlorides, 926
Ethynes, silyl-, reactions, with 2-aminobenzenethiol, 939
Ethyn-2-ols, 1-germanyl, reactions, with silicon tetrachloride, 1058

Ferracyclopentadiene–nickel complexes, synthesis of, 715
Ferrocene
  lithiation of, 706
  metallation of, 720
  reactions, with pentylsodium, 706
Ferrocene, bromo-
  reactions, with butyllithium, 706
  as starting material, for ferrocenylmagnesium bromide, 709
Ferrocene, chloro-, metallation of, 706
Ferrocene, 2-chlorolithio-, reactions, with chlorocyclopentadienyl(triphenylphosphine)nickel, 719
Ferrocene, (chloromercuri)-, transmetallation of, 706
Ferrocene, 1,1'-dichloro-, lithiation of, 706
Ferrocene, 1',2-dichlorolithio-, use in the synthesis of metal complexes, 719
Ferrocene, 1,1'-dilithio-
  reactions, with bis(cyclopentadienyl)dichloro complexes, 720
  use in the synthesis of metal complexes, 719
Ferrocene, (dimethylaminomethyl)-, cyclometallation of, 720
Ferrocene, (*R,S*)-1-diphenylphosphino-2-(1-dimethylaminoethyl)-, reactions, with *n*-butyllithium, 1033
Ferrocene, disodio-, reactions, with magnesium bromide, 709
Ferrocene, iodo-
  mercuration of, 724
  reactions, with butyllithium, 706
Ferrocene, lithio-, reactions, with transition metal halides, 723
Ferrocene, 2-lithio(dimethylaminomethyl)-, reactions, with transition metal halides, 723
Ferrocene compounds, ethyl-, synthesis of, 1033
Ferrocenes, reactions, with *n*-butyllithium–tetramethylethylenediamine, 1034
Ferrocenes, acyl-, reactions, with pentacarbonyls, 720
Ferrocenes, lithio-, synthesis of, 705
Ferrocenes, vinyl-, synthesis of, 1033
Ferrocenophanes
  reactions, with phenyllithium, 1034
  synthesis of, 1034
Ferrocenoyl chloride, reactions, with organometallic anions, 722
Ferroles, tetrasubstituted, synthesis of, 681
Finkelstein reaction, use in the synthesis of α-iodo alkyl esters, 58
Fischer carbene complexes, as starting materials, for α-alkoxystannanes, 388
Flame retardants, synthesis of, 456
Flash vacuum pyrolysis, of Meldrum's acid compounds, 894
Flash vacuum thermolysis, for the synthesis of carbenes, 1080
Fluorene, diazo-, as starting material, for mixed 1,1-dihaloalkenes, 755
Fluorene, dichloromethylene-, synthesis of, 742
Fluorenes, diphosphatetrahydro-, synthesis of, 548
Fluorenes, 9-silyl-, deprotonation of, 647

Fluorination
  for the synthesis of α-halo nitro compounds, 108
  of 1,1,3,3-tetranitropropane, 795
Fluorine, reactions, with chloroalkenes, 31
Fluorine–fluorine bond strength, effect in elimination, 730
Fluorocarbons, as starting materials, for aminals, 411
Formaldehyde
  reactions
    with β-allylaminoalkylphosphines, 458
    with amides, 416
    with amines, 295
    with 2-aminopropan-1,3-diol, 302
    with ammonia, 407
    with benzenesulfonamide, 423
    with N-benzylglycine, 466
    with benzylthiol, 215
    with chlorodiphenylphosphine, 119
    with dialkyl phosphinous acids, 119
    with dibenzylphosphine oxide, 471
    with dichlorophosphines, 125
    with 2,4-dimethyl-2-thiolpentan-3-one, 317
    with 1,3-dimethylurea, 301
    with hydrazines, 436
    with hydrofluoric acid, 44
    with β-hydroxy-thioamides, 230
    with isocyanic acid, 313
    with methylamine hydrochloride, 317
    with methylsilylphosphines, 337
    with morpholine, 315
    with nitramines, 442
    with N-nitroamines, 311
    with styrene, 827
    with sulfonamides, 427
    with trialkylsilyl isocyanates, 313
    with trimethylhydrazine, 437
  as starting material, for methylene aminals, 410
Formalin, reactions, with (S)-1-phenylethylamine, 99
Formamide, silylmethylation of, 516
Formamide, N,N-bis(trimethylsilyl)-, reactions, with Schiff bases, 419
Formamide, dimethyl (DMF), as solvent, for the synthesis of ketene aminals, 971
Formamides
  reactions, with bis(benzenesulfonyl)methane, 866
  as starting materials, for isonitriles, 908
Formamides, N-(1-arenesulfonylalkenyl)-, synthesis of, 908
Formamides, N-(arylthiomethyl)-, synthesis of, 328
Formamidine, N,N-dimethyl-N'-thiobenzoyl-, S-alkylation of, 328
Formamidines, lithiation of, 531
Formamidinium chloride, tetramethylchloro-
  reactions, with cyclopentadienes, 974
  as starting material, for ketene aminals, 974
Formamidinium chlorides, chloro-
  reactions, with sodium isopropoxide, 974
  as starting materials, for isopropoxymethanetriamines, 973
  synthesis of, 974
Formamidinium salts, as starting materials, for aminals, 414
Formate, 1-chloroethyl chloro-, reactions, with benzimidazole, 60
Formate, ethyl chloro-, reactions, with N,N-dimethylacetamide, 882
Formate, methyl, reactions, with α-sulfonyl ethers, 838
Formate, methyl chloro-
  chlorination of, 59
  reactions, with α-hydrazino enamines, 986
Formates, chloro-, from, 2-bromo-2-methylpropanoyl phosphonates, 921
Formates, α-chloroalkyl chloro-, synthesis of, 58
Formates, α-chloroethyl chloro-, reactions, with mesityl oxide, 59
Formates, chloromethyl chloro-
  reactions, with thiols, 60
  as starting materials, for α-chloroalkyl carbonates, 59
  synthesis of, 59
Formates, α-fluoroalkyl fluoro-, synthesis of, 59
Formates, α-haloalkyl halo-, synthesis of, 58
Formazan, 3-formyl-1,5-diphenyl-, as starting material, for gem-diazoalkenes, 992
Formimidate, ethyl N-cyclohexyl-, reactions, with chlorocyanoketene, 304
Formimidates, reactions, with dinitromethane, 994
Formyl fluoride, reactions, with perfluoroalkoxides, 59
Fructopyranose compounds, from, di-O-isopropylidene compounds, 222
β-L-Fructosylamine, use in the synthesis of α-aminophosphonic acids, 462
Fulvenes, synthesis of, 835
Fulvenes, 6,6-bis(dimethylamino)-, synthesis of, 974
Fumarate, dimethyl chloro-, reactions, with diphenyldiazomethane, 109
Fumaric dichloride, difluoro-, reactions, with sodium tetracarbonylcobaltate, 821
Furan, 2,3-dihydro-, deprotonation of, 949
Furan, 2,5-dimethoxytetrahydro-, reactions, with benzotriazole, 309
Furan, tetrahydro-, reactions, with ceric triethylammonium nitrate, 201
Furan compounds
  oxidation of, 206
  reactions
    with n-butyllithium, 846
    with lead(IV) acetate, 206
  as starting materials, for acetoxy furanone compounds, 206
Furan compounds, acetoxy tetrahydro-, from, allyl enol ethers, 208
Furan compounds, 2-tri-n-butylstannyl-, oxidation of, 206
Furanone compounds, acetoxy, from, furan compounds, 206
Furanone compounds, hydroxy, reactions, with alcohols, 211
Furanones, synthesis of, 824
Furanones, alkoxy, synthesis of, 211
Furans, synthesis of, 961
Furans, dihydro-
  synthesis of, 961
  use in the synthesis of acetals, 186
Furans, 5-lithio-2,3-dihydro-, reactions, with trialkylboranes, 935
Furans, 2-silyloxy-, oxidation of, 206
Furans, tetrahydro-, synthesis of, 168
Furanyl ethers, tetrahydro-, synthesis of, 188
Furanyl ethers, 2-tetrahydro-, synthesis of, 201
Furanyls, 2-diphenylacetoxytetrahydro-, reactions, with alcohols, 200

Gabriel method, for the synthesis of α-aminoalkylphosphorus compounds, 473
Gadolinium compounds, reactions, with phosphorus ylides, 578
Galactosylamine, use in the synthesis of α-aminophosphonic acids, 462
Gallium, (iodomethyl)diethyl-, synthesis of, 153
Gallium, tris(trimethylsilylmethyl)-, synthesis of, 659
Gallium chloride, bis(trimethylsilylmethyl)-, synthesis of, 659
Gallium compounds, reactions, with phosphorus ylides, 587
Gallium dichloride, trimethylsilylmethyl-, synthesis of, 659
Gallium trichloride
  reactions
    with trimethylsilylmethylmagnesium chloride, 659
    with ylides, 397
Geminal dihalides
  reactions, with alkoxides, 189
  as starting materials, for acetals, 189

Germacycles

Germacycles, synthesis of, 626
Germacyclobutanes, pyrolysis of, 627
Germafulvene, diethyl-, as intermediate, in the synthesis of 1,1-bisgermylalkanes, 627
Germane, (acetoxymethyl)trimethyl-, synthesis of, 362
Germane, benzylthiotrimethyl-, reactions, with *t*-butyllithium, 371
Germane, bis(chloromethyl)dimethyl-, reactions, with sodium acetate, 362
Germane, bis(hydroxymethyl)dimethyl-, chlorination of, 141
Germane, bis(iodomethyl)-, as starting material, for thiagermatanes, 372
Germane, bromotrimethyl-, reactions, with carbon nucleophiles, 361
Germane, bromo(trivinyl)-, synthesis of, 141
Germane, (but-2-en-1-ynyl)trimethyl-, reactions, with areneselenyl chlorides, 948
Germane, chloromethyltrimethyl-, reactions, with dimethylbismuthinosodium, 599
Germane, chloromethyl(trimethyl)-, reactions, with lithium, 626
Germane, (chloromethyl)trimethyl-, reactions, with potassium acetate, 362
Germane, chlorotrimethyl-
  reactions
    with di-Grignard reagents, 626
    with 1-methoxy-1-vinyllithium, 929
  as starting material, for α-alkoxygermanes, 361
Germane, dimethyl(divinyl)-
  hydroboration of, 646
  reactions, with borane dimethyl sulfide, 646
Germane, (hydroxymethyl)trimethyl-, synthesis of, 362
Germane, (iodomethyl)trimethyl-
  reactions, with lithium phenyl selenide, 377
  synthesis of, 141
Germane, tetrachloro-, as starting material, for hydroxymethylgermanes, 362
Germane, tetrakis(trimethylsilylmethyl)-, reactions, with bromine, 641
Germane, tetravinyl-, chlorination of, 141
Germane, tribromo(bromomethyl)-, synthesis of, 142
Germane, trichloro-, reactions, with chloromethyl methyl ether, 361
Germane, trichloro(chloromethyl)-
  reactions, with nucleophiles, 142
  reduction of, 141
  synthesis of, 142
Germane, trichloro(iodomethyl)-, synthesis of, 142
Germane, trichloro(methyl)-, chlorination of, 142
Germane, triethyl-, reactions, with phenylthioethynes, 940
Germane, triiodo(iodomethyl)-, synthesis of, 142
Germane, trimethyl-, reactions, with perfluorocyclobutenes, 141
Germane, triphenyl-, reactions, with silylallenes, 639
Germane, tris(trimethylsilylmethyl)-, reactions, with diethylmercury, 641
Germane, tris(trimethylsilylmethyl)bromo-, synthesis of, 641
Germane, vinyltrimethyl-, reactions, with arenesulfenyl chlorides, 372
Germane hydrides, reactions, with haloalkenes, 141
Germanes, alkoxy-
  reverse Brook rearrangement of, 361
  as starting materials, for α-hydroxygermanes, 361
Germanes, α-alkoxy-
  from, chlorotrimethylgermane, 361
  synthesis of, 360
Germanes, alkyl arylsilyl-, as starting materials, for germyl anions, 361
Germanes, alkyl(α-haloalkyl)-, synthesis of, 141
Germanes, α-amino-, synthesis of, 519
Germanes, aryl(α-haloalkyl)-, synthesis of, 141
Germanes, bis(chloromethyl)-, synthesis of, 141
Germanes, bis(halomethyl)dialkyl-
  reactions
    with hydrogen sulfide, 372
    with sodium sulfide, 372
Germanes, bromo-, reactions, with chloromethyllithium, 141
Germanes, bromo(triphenyl)-, synthesis of, 663
Germanes, chloro-
  alkylation of, 141
  synthesis of, 519
Germanes, (chloromethyl)-
  halogen exchange in, 141
  reactions, with amides, 521
  as starting materials
    for Grignard reagents, 371
    for (iodomethyl)germanes, 141
Germanes, α,α-(dibromomethyl)-, reduction of, 141
Germanes, halo-
  nucleophilic displacement of, 361
  reactions, with phosphorus ylides, 574
Germanes, α-haloalkyl-, synthesis of, 141
Germanes, halo(α-haloalkyl)-
  alkylation of, 141
  synthesis of, 142
Germanes, halomethyl-
  reactions
    with oxygen nucleophiles, 362
    with sulfur nucleophiles, 371
Germanes, (halomethyl)trimethyl-, displacement of, 663
Germanes, α-hydroxy-
  from, alkoxygermanes, 361
  synthesis of, 360
Germanes, hydroxymethyl-, from, tetrachlorogermane, 362
Germanes, (hydroxymethyl)-, reactions, with thionyl chloride, 141
Germanes, (iodomethyl)-, from, (chloromethyl)germanes, 141
Germanes, 1-phosphinoalkyl-, synthesis of, 568
Germanes, 1-phosphonatoalkyl-, synthesis of, 574
Germanes, 1-phosphonioalkyl-, synthesis of, 574
Germanes, 1-phosphonitoalkyl-, synthesis of, 568
Germanes, silyl, reactions, with boron trichloride, 639
Germanes, vinyl-
  bromination of, 141, 811
  electrophilic sulfenylation of, 372
  reactions
    with halogens, 141
    with organolithium compounds, 663
Germanium
  reactions, with trichloro(chloromethyl)silane, 639
  synthesis of functions, 1063
    bearing a germanium and a boron, 646
    bearing a silicon and a germanium, 636
    bearing two germaniums, 625
Germanium anions, as intermediates, for the synthesis of α-germylsilylalkanes, 638
Germanium–carbon bonds, synthesis of, 625
Germanium compounds, synthesis of, 599, 1031
Germanium compounds, α-halo-, synthesis of, 141
Germanium compounds, α-haloalkenyl-, synthesis of, 811
Germanium compounds, trifluorovinyl-, synthesis of, 811
Germanium dichloride, bis(trichlorosilylmethyl)-, synthesis of, 639
Germanium dichloride, dimethyl-, reactions, with diorganolithium compounds, 636
Germanium diiodide, reactions, with diiodomethane, 628
Germanium halides, reactions, with trifluorovinylmagnesium halides, 811
Germanium hydrides, reactions, with aldimines, 521
Germanium–phosphorus heterocycles, synthesis of, 485
Germanium tetrachloride
  reactions
    with 1-germanylalkyne ethers, 1055

with 1-silyl ethyn-2-ol ethers, 1058
Germanium trichloride, (trichlorosilylmethyl)-, synthesis of, 639
1-Germa-3-silacyclobutane, 1,1,3,3-tetramethyl-, synthesis of, 638
1-Germa-3-silacyclobutanes, synthesis of, 636
Germazenes, synthesis of, 1055
Germenes
   dimerization of, 627
   as starting materials
      for 1,1-bisgermylalkanes, 627
      for 1,3-digermacyclobutane, 627
   synthesis of, 627
Germyl anions, from, alkyl arylsilylgermanes, 361
Germyl anions, trialkyl-, synthesis of, 626
C-Germylation, mechanism, 636
Germyl chlorides
   reactions, with carbonyl enolate anions, 625
   for the synthesis of germanium–carbon bonds, 625
Germyl electrophiles, reactions, with α-metallosilylalkanes, 636
Gingerols, synthesis of, 245
D-Glucal, 3,4,6-tri-*O*-*t*-butyldimethylsilyl--, reactions, with *t*-butyllithium, 651
β-D-Glucopyranose, penta-*O*-acetyl-, reactions, with ethanethiol, 224
β-D-Glucopyranosides, aryl 1-thio-, synthesis of, 225
Glucosamine compounds, synthesis of, 224
D-Glucose, 1-seleno-, synthesis of, 239
Glutaraldehyde, reactions, with Grignard reagents, 166
Glycals, azidonitration of, 174
Glycidol, 3-phenyl-, reactions, with 1-(benzenesulfonyl)alkynes, 840
Glycinate, *t*-butyl *N*,*N*-bis(benzyloxycarbonyl)-α-fluoro-, synthesis of, 101
Glycinate, *t*-butyl *N*-*t*-butoxycarbonyl-, halogenation of, 102
Glycinate, methyl *N*-acetyl-α-hydroxy-, as starting material, for α-bromoamides, 99
Glycinate, 8-phenylmenthyl *N*-butoxycarbonyl-, bromination of, 102
Glycinates, 2-alkyl, synthesis of, 102
Glycinates, benzyloxycarbonyl)-α-hydroxy-, chlorination of, 102
Glycinates, α-chloro, synthesis of, 101
Glycinates, ethyl fluoro, synthesis of, 101
Glycinates, (*S*)-2'-thiazinylidene-, reactions, with *N*-chlorosuccinimide, 101
Glycine, *N*-benzyl-
   reactions
      with dichlorophosphines, 466
      with formaldehyde, 466
   as starting material, for α-aminophosphonites, 466
Glycine, *N*-phenyl-, reactions, with paraformaldehyde, 302
Glycine carbanion synthons, as lithiated imines of glycine, 535
Glycine compounds, α-bromo-, reactions, with trimethyl phosphite, 465
Glycines, α-alkoxy-, reactions, with phosphorus trichloride, 465
Glycines, α-amino-, monoacyl compounds, synthesis of, 417
Glycines, aryl-, synthesis of, 466
Glycine synthons, synthesis of, 535
Glycol, ethylene, reactions, with tetrabutyl 2-phenylethylidene-1,1-bisboronate, 634
Glycols, ethylene, reactions, with β-chlorovinyl aldehydes, 196
Glycopyranosides, 1-*O*-acetyl, reactions, with thiols, 224
Glycoside bonds
   cleavage of, 176
   synthesis of, 188
Glycoside coupling
   for the synthesis of acetals, 198, 200
   for the synthesis of unsymmetrical acetals, 202
Glycosides, alkyl, deprotection of, 176
O-Glycosides, reactions, with dimethylboron bromide, 225
Glycosylation, use of sulfoxides for, 232
Glycosyl fluorides
   from, phenyl thioglycosides, 45
   as starting materials, for thioglycosides, 225
Glycosyl halides
   reactions, with sulfur nucleophiles, 223
   as starting materials, for glycosyl sulfones, 236
Glycosyl radicals, from, glycosyl tellurides, 240
Glycosyl tellurides, synthesis of, 240
Glyoxal
   reactions
      with amines, 298
      with benzylamines, 298
      with 1,3-diaminopropane, 409
      with *N*,*N*'dimethyl-1,2-diaminoethane, 409
      with *N*,*N*'-dimethylethylenediamine, 301
      with 2-(methylamino)ethanol, 302
      with morpholine, 300
Glyoxal, phenyl-
   reactions
      with amines, 300
      with toluene *p*-sulfinic acid, 234
Glyoxaldehyde, phenyl-, condensation of, 311
Glyoxal hydrate compounds, aryl-, from, phenacyl bromides, 161
Glyoxylic acid, reactions, with 2-aminoethaneselenol, 333
Gold complexes, (chloromethyl)-, synthesis of, 152
Gold complexes, methylene-bridged, synthesis of, 695
Gold compounds
   reactions
      with 1-metalloalkylphosphine sulfides, 586
      with phosphorus ylides, 586
Gold–copper compounds, synthesis of, 724
Gold salts, aryl-bridged, synthesis of, 696
Gold salts, ferrocenyl-bridged, synthesis of, 696
Gold salts, vinyl-bridged, synthesis of, 696
Gold–zinc compounds, synthesis of, 724
Grignard reagents
   dithiocarboxylation of, 843
   from
      (1-bromovinyl)trimethylsilane, 947
      chloromethyl ethers, 383
      chloromethylgermanes, 371
      chloromethyl methyl sulfide, 395
      α-chlorophenylacetic acid, 147
      chlorotrimethylsilane, 370
      perfluorovinyl bromide, 820
      sulfones, 395
      trichloro(chloromethyl)silane, 606
   reactions
      with benzo-1,3,2-dioxaphospholes, 126
      with chloroarsines, 597
      with chloro enamines, 15
      with chloro(methyl)silanes, 137
      with dithioesters, 255
      with 1,2-dithioles, 905
      with glutaraldehyde, 166
      with hexahydropyrimidines, 410
      with imidazolidines, 410
      with imides, 303
      with ortho esters, 192
      with orthoformates, 192
      with phosphinic chlorides, 120
      with propargyl alcohol, 1065
      with sulfonyl fluorides, 280
      with trimethyl(vinyl)silane, 654
      with trithiocarbonate *S*,*S*-dioxides, 862
      with vinylsilanes, 654
   synthesis of, 1017
   thiophilic addition of, 245

Grignard reagents
    use in the synthesis of 1,1-bisgermylalkanes, 626
    use in the synthesis of 1,1-bis(silyl)alkanes, 606
    use in the synthesis of 1,1-dimetal complexes, 669
Grignard reagents, allylic, carbophilic addition to, 255
Grignard reagents, α-amino-, synthesis of, 531
Grignard reagents, aryl-, reactions, with oxazolinium salts, 306
Grignard reagents, bis-, use in the synthesis of carbon-bridged dimercury complexes, 696
Grignard reagents, di-
    reactions
        with chlorotrimethylgermane, 626
        with dimethyltin dichloride, 700
        with tetrachlorosilane, 608
    as starting materials, for 1,1-bis(silyl)alkanes, 608
    synthesis of, 618, 669
Grignard reagents, α-haloalkyl-, synthesis of, 147
Grignard reagents, α-silyl-
    from, chloromethylpentamethyldisiloxane, 654
    synthesis of, 653
Grignard reagents, α-silylmethyl-
    reactions, with aluminum trichloride, 659
    synthesis of, 655
Grignard reagents, (trialkylsilyl)methyl-, synthesis of, 654
Grob fragmentation, of γ-tosyloxy dithioacetal, 846
Group 1 metals
    functions containing two, synthesis of, 667
    synthesis of functions, 705
Group 2 metals
    functions containing two, synthesis of, 669
    synthesis of functions, 709
Guanidines, *N*-chloro-, reactions, with nitroalkynes, 796
Guanidinium salts, synthesis of, 973

H2 antagonist ranitidine, synthesis of, 907
Halides, nucleophilic displacement of, 346
Haloalkylation
    of aldehydes, 47
    of ketones, 47
(α-Haloalkylsilanes), alkyl-, synthesis of, 136
(α-Haloalkylsilanes), aryl-, synthesis of, 136
Halobenzene–tricarbonylchromium complexes, reactions, with anionic organometallics, 722
Halo compounds, mixed, synthesis of, 750
Halodestannylation, for the synthesis of *gem*-halo stannylalkenes, 812
Halogen
    displacement of, for the synthesis of α-germylsilylalkanes, 640
    synthesis of functions, **41–93**
Halogen, nucleophilic, reactions, with phosphinic acid esters, 125
Halogenation
    of organometallics, 774
    of sulfoxides, 75
    for the synthesis of, α-halo nitro compounds, 108
    for the synthesis of α-haloamides, 98, 99
    for the synthesis of α-halo imides, 98, 99
Halogen compounds, synthesis of, 769
Halogen–lithium exchange
    for the synthesis of α-alkoxycarbanions, 381
    for the synthesis of α-lithiogermylalkanes, 662
    for the synthesis of α-lithiosilylalkanes, 652
    for the synthesis of manganese compounds, 707
    for the synthesis of α-thiocarbanions, 395
Halogen–magnesium exchange, for the synthesis of α-magnesiosilylalkanes, 653
Halogen–metal exchange
    for the synthesis of 1,1-bisgermylalkanes, 626
    for the synthesis of α-selenolithiums, 400
    use in the synthesis of 1,1-bis(silyl)alkanes, 602
Halogen replacement, of methylene(ditin)compounds, 700
Halogens
    reactions
        with selenyl-substituted alkynes, 786
        with vinylgermanes, 141
    stabilization by, for the synthesis of carbenes, 1082
    synthesis of functions, **789–822**
        and another heteroatom group other than a chalcogen, **95–157**
        incorporating two halogens or a halogen and a chalcogen, **729–788**
Heck reaction, for the synthesis of β-aryl-γ-butyrolactols, 168
Hemiacetal, cyclopropanone ethyl, synthesis of, 171
Hemiacetals
    acylation of, 209
    alkylation of, 183
    from
        acetals, 175
        aldehydes, 164
        enol ethers, 171
        esters, 169
        ketones, 164
    silylation of, 202
    as starting materials
        for acetals, 200
        for acyl alkyl acetals, 209
    synthesis of, 163
        via addition of carbanions, 171
        via reduction of esters, 169
Hemiacetals, acyclic
    from, α-halo ketones, 164
    synthesis of, 164
Hemiacetals, 2-azido-, synthesis of, 174
Hemiacetals, cyclic
    from, hydroxy carbonyl compounds, 164
    as starting materials, for acetals, 201
    synthesis of
        via addition of carbanions to dicarbonyl compounds, 166
        via oxidation of diols, 164
        via reduction of dicarbonyl compounds, 165
Hemiaminals
    from
        aldehydes, 294
        cyclopropanone, 299
        ketones, 294
    synthesis of, 293
    See also *O*,*N*-Acetals
Hemiaminals, *N*-aminoketene, synthesis of, 897
Hemiaminals, *N*-hydroxylketene, synthesis of, 896
Hemiaminals, ketene
    from
        amides, 881
        aryl cyanates, 889
        dialkynic ynamines, 885
        1-halo-1-dialkylaminoalkenes, 889
        ketene acetals, 887
        ketene aminals, 888
        ketene compounds, 887
        ketene dithioacetals, 888
        ketene hemiaminals, 888
        ketenes, 886
        ketenimines, 894
        nitriles, 883
        substituted alkynes, 884
        thioesters, 895
    as starting materials, for ketene hemiaminals, 888
    synthesis of, 880
Hemiaminals, *N*-nitrosoketene
    from, *N*-nitroamides, 896
    synthesis of, 896
Hemiaminals, *N*-sulfonyl, synthesis of, 312
Hemidithioacetals
    from

aldehydes, 245
dithioesters, 245, 256
synthesis of, 245
Hemithioacetals
from, dihydropyran, 223
with the oxygen in a ring, 223
with the sulfur in a ring, 226
synthesis of, 216
Hemithioaminals
from
aldehydes, 324
alkenes, 326
alkylthiomethyleneiminium salts, 321
imines, 325
ketones, 324
tricoordinate sulfur compounds, 329
N-functionalized, synthesis of, 323
synthesis of, 315
See also S,N-Acetals
Hemithioaminals, N-acyl, synthesis of, 325
Hemithioaminals, N-formamido, synthesis of, 327
Hemithioaminals, N-nitroso, synthesis of, 325
Hemithioaminals, N-tosyl, synthesis of, 326
Henry reaction
of nitroalkanes, 540
See also Nitro-aldol reaction
Heterafulvalenes, synthesis of, 870
Heteroatom exchange, for the synthesis of 3,5-diazaphosphanes, 457
Heterocycles, reactions, with dimethylketene, 886
Heterocycles, N-(silylmethyl), synthesis of, 519
Hetero Diels–Alder reaction, for the synthesis of acetals, 202
Heterogeneous catalysts, for the synthesis of acetals, 178
Hexaalkylditin, as catalyst, for the synthesis of α-haloalkenylsilicon compounds, 811
Hexachlorides, reactions, with o-(bistrimethylsilyl)benzene, 290
Hexakis(phosphines), synthesis of, 547, 550
Hexane, 1,1-bis(dichloroboryl)-, synthesis of, 631
Hexane, 3,3,4-tris(diethylboryl)-, synthesis of, 632
Hexane-3,4-diol, (S,S)-2,5-dimethyl- (DIPED), reactions, with lithium benzyloxide, 362
1-Hexanol, synthesis of, 630
Hexathioethane, dialkyl disulfide elimination from, 846
Hexatriazines, as starting materials, for chiral amides, 99
1,3,5-Hexatriazines, reactions, with acyl chlorides, 99
Hex-1-ene, 1-bromo 1-trimethylsilyl 2-ethyl-, bromination–dehydrobromination of, 748
3-Hexene, cis-3,4-bis(diethylboryl)-, hydroboration of, 632
3-Hexene, cis-3,4-di(diethylboryl)-, pyrolysis of, 632
1-Hexyne, reactions, with dicyclohexylborane, 630
Hex-3-yne
as starting material
for alkyne-bridged dipalladium complexes, 692
for carbon-bridged ketonic dirhodium complexes, 690
for tetranuclear carbon-bridged rhodiacyclopentadiene complexes, 689
HF pyridine, reactions, with N-bromosuccinimide, 45
Hippurate, methyl 2-methoxy-, as starting material, for N-benzoyl compounds, 327
Hippuric acid, synthesis of, 525
Histamines, H2 antagonist ranitidine, 907
Hofmann degradation, for the synthesis of bromochloroalkanes, 38
Hofmann reaction, for the synthesis of α-aminoalkylorganophosphorus compounds, 472
Hofmann rearrangement, for the synthesis of α-aminoalkylphosphorus compounds, 476
Holmium compounds, reactions, with phosphorus ylides, 578
Horner–Emmons reagents, use in the synthesis of ketene dithioacetals, 847
Horner–Wittig reaction

of 2-phosphoryl benzo-1,3-diselenoles, 875
for the synthesis of ketene monothioacetals, 835
for the synthesis of thiaselenafulvenes, 870
use of α-sulfonyl ethers in, 840
use in the synthesis of ketene acetals, 828
Hunsdiecker reaction
for the synthesis of bromofluoroalkanes, 35
for the synthesis of chlorofluoroalkanes, 33
for the synthesis of dibromoalkanes, 27
for the synthesis of fluoroiodoalkanes, 37
Hydantoin, 5-chloro-, synthesis of, 103
Hydantoins, reduction of, 414
Hydralumination
of alkynes, 1068
of 1-trimethylsilylalkynes, 1062
Hydrates
from, bicyclo[2.2.0]hex-5-en-2-ones, 160
synthesis of, 160
Hydrates, trifluoromethyl, synthesis of, 160
Hydrazine, reactions, with α-keto perfluoroalkane carboxylic esters, 311
Hydrazine, 1,1-dimethyl-
acylation of, 897
reactions, with 2-(methoxycarbonylcyanomethylene)-1,3-dioxolane, 897
Hydrazine, phenyl-, reactions, with perfluoro-1-alkenes, 799
Hydrazine, tetrafluoro-, reactions, with alkenes, 104
Hydrazine, trimethyl-, reactions, with formaldehyde, 437
Hydrazines
acetylation of, 897
N-acylated, 986
reactions
with formaldehyde, 436
with nitroketene dithioacetals, 983
with tetranitromethane, 995
with thioamides, 984
with tris-(chloromethyl)amine, 437
as starting materials, for carbazates, 60
Hydrazines, acylaminomethyl-, synthesis of, 448
Hydrazines, aminomethyl-, synthesis of, 441, 444
Hydrazines, benzoyl-, reactions, with imino esters, 986
Hydrazines, bis-, synthesis of, 437
Hydrazines, N^1,N^2-bisphosphinomethylene-N^1,N^2-dimethyl-, synthesis of, 482
Hydrazines, N^1,N^2-dialkyl-, reactions, with bis(hydroxymethyl)phosphine, 482
Hydrazines, hydroxylamino-, synthesis of, 441
Hydrazines, hydroxylaminomethyl-, reduction of, 444
Hydrazines, methoxyaminocarbonyl, synthesis of, 986
Hydrazines, methoxycarbonyl-, synthesis of, 986
Hydrazines, N-methoxymethyl-, chlorination of, 105
Hydrazines, nitromethylacyl-, diazotization of, 440
Hydrazino compounds, aminomethyl, synthesis of, 444
Hydrazoic acid
reactions
with imines, 444
with silyl enol ethers, 314
Hydrazone, acetone 4-chlorophenyl-, bromination of, 329
Hydrazone, acetophenone phenyl-, oxidation of, 314
Hydrazone, cinnamaldehyde benzoyl-, reactions, with diethyl phosphite, 1008
Hydrazone, 2,4-dinitrophenyl-, oxidation of, 313
Hydrazone, nitroformaldehyde aryl-, as starting material, for gem-arylazo nitroalkenes, 995
Hydrazones
electrophilic halogenation of, 109
reactions
with carboxyalkylphosphines, 482
with iodine, 29
with O-mesitylenesulfonyl hydroxylamine, 444
as starting materials
for O,N-acetals, 313
for dibromoalkanes, 27

Hydrazones

Hydrazones, benzoyl-, as starting materials, for *gem*-phosphorylamino phosphorylalkenes, 1008
Hydrazones, *t*-butyl-, metallation of, 536
Hydrazones, nitro-, synthesis of, 440
Hydrazones, phenyl-
  oxidation of, 314
  reactions, with hydroxylamines, 441
Hydrazones, *p*-tosyl-, as starting materials, for ketene dithioacetals, 852
Hydride, trimethyltin, photochemical addition of, to propargyl acetate, 701
Hydrins, bromo-, synthesis of, 122, 139
Hydrins, chloro-, synthesis of, 146
Hydrins, cyano-
  silylation of, 359
  as starting materials, for α-silyl enamines, 1013
Hydrins, epihalo-, use in the synthesis of acetals, 181
Hydrins, halo-, synthesis of, 146
Hydroalumination, of alkynylsilanes, 808
Hydroboration
  of 1-bromoalkynes, 814
  of 1-iodoalkynes, 814
  of 1-silylalkynes, 1056
  of vinylsilanes, 356, 643
Hydrochloric acid
  reactions
    with 2,2-dimorpholinoacetophenone, 309
    with silyl ethers, 160
  See also Hydrogen chloride
Hydrochloride salts, synthesis of, 478
Hydrochloride salts, imino ester, reactions, with urea, 982
Hydrofluoric acid, reactions, with formaldehyde, 44
Hydrogen
  addition to ketene dithioacetals, 252
  reactions, with oxoketene dithioacetals, 253
γ-Hydrogen
  abstraction of, for the synthesis of 1,4-biradicals, 1079
  for the synthesis of biradicals, 1079
δ-Hydrogen, for the synthesis of biradicals, 1079
Hydrogen abstraction, intermolecular, mechanism, 1078
Hydrogenation, ionic, for the synthesis of α-aminoalkyl phosphonates, 491
Hydrogen bromide
  reactions
    with bromoalkenes, 23
    with 1,1-difluoroethene, 5
    with propyne, 24
    with triacetates, 121
  use in the synthesis of α-bromo ethers, 50
Hydrogen chloride
  reactions
    with 1-alkylselenobut-3-en-1-ynes, 786
    with boron trifluoride etherate, 885
    with *t*-butyl isocyanide, 894
    with ethyl dicyanoacetate, 792
    with ethynylplatinum complexes, 822
    with germyl(silyl)methane, 641
    with (methoxymethyl)iron complexes, 148
    with 1-methylselenohexa-1,3-diyne, 786
    with methyl trimethylsilyl ketone, 944
    with silyl diimines, 107
    with trimethylsilyl ethers, 150
    with vinylrhodium complexes, 149
    with zirconium compounds, 547
  See also Hydrochloric acid
Hydrogen fluoride
  reactions, with 1-chlorocyclohexene, 30
  use in the synthesis of fluoro steroids, 9
Hydrogen halides
  elimination of, 730
  reactions
    with (alkoxymethyl) metal complexes, 147
    with alkylvinyl selenides, 91
    with arylvinyl selenides, 91
    with haloalkenes, 2
Hydrogen iodide
  reactions
    with 1,12-bis(hydroxymethyl)decahydrododecaborate caesium salt, 143
    with 1,1-difluoroethene, 5
    with phthalimides, 101
Hydrogen peroxide
  for the oxidation of dithioacetals, 259, 271, 272
  for the oxidation of oximes, 429
  reactions
    with 1-hydroxy-2-nitro-2-phenylselenononane, 335
    with α,β-unsaturated sulfones, 237
Hydrogen phosphite, bis(*p*-methylbenzyl), reactions, with aldimines, 461
Hydrogen sulfide
  reactions
    with acylsilanes, 371
    with bis(halomethyl)dialkylgermanes, 372
    with hexafluoroacetone, 215
    with ketones, 244
    with trifluoroacetaldehyde, 215
1,6-Hydrogen transfers, mechanism, 1079
Hydrogermylation, for the synthesis of α-germylsilylalkanes, 639
Hydrolysis, enzymic, for the synthesis of α-aminophosphonic acids, 462
Hydrosilylation
  of alkynes, 1044
  of alkynylsilanes, 1044
  catalysts for, 1044
  of enamides, 510
  of enamines, 510
  of silyl(alkylsilyl)alkynes, 1054
  for the synthesis of 1,1-bis(silyl)alkanes, 617
Hydrostannation
  of 1-stannylalkynes, 1068
  of 1-trimethylsilylpentyne, 1062
Hydrosulfide, tetraethylammonium, reactions, with dichloroalkenes, 851
Hydroxamic acid compounds, synthesis of, 449
α-Hydroxy acids
  as starting materials, for 4-oxo-1,3-dioxolanes, 213
  synthesis of, 231
β-Hydroxy acids, as starting materials, for 4-oxo-1,3-dioxolanes, 213
N-Hydroxyalkylation, mechanism, 296
α-Hydroxy compounds
  as starting materials
    for α-aminoalkylphosphorus compounds, 475
    for α-haloamides, 98
    for α-haloimides, 98
Hydroxy groups, oxidation of, 165
Hydroxylamine, *O*-benzyl-, reactions, with 2-(methoxycarbonylcyanomethylene)-1,3-dioxolane, 896
Hydroxylamine, *O*-mesitylenesulfonyl- (MSH), reactions, with hydrazones, 444
Hydroxylamine, phenyl-, condensation of, 894
Hydroxylamines
  reactions
    with aldehydes, 435
    with aryl imines, 442
    with ethyl diphenylmethylene carbamate, 448
    with ketones, 435
    with phenylhydrazones, 441
Hydroxylamines, *N*-acyl-, Mannich-type reactions of, 442
Hydroxylamines, acylaminomethyl-, synthesis of, 448
Hydroxylamines, *N*-aryl-, Mannich-type reactions of, 442
Hydroxylamines, bis-(*N*-thiomethyl)-, synthesis of, 325
Hydroxylamines, *O*-mesitylsulfonyl-, as starting materials, for α-aminoalkylphosphorus compounds, 490

Hydroxyl groups, anomeric
  deprotection of, 176
  protection of, 176
Hydroxymercuration, of enol ethers, 171
2-Hydroxymethyl compounds, 1-bromomethyl-, synthesis of, 143
Hydroxythiols, as starting materials, for oxathiolanes, 229
1,3-Hydroxythiols
  reactions, with trans-dimethylcyclopentanone-3,4-dicarboxylate, 229
  as starting materials
    for 1,3-oxathianes, 229
    for 1,3-oxathiolanes, 229
Hypobromite, acetyl, reactions, with difluoroalkenes, 52
Hypobromite, 3-chlorobenzoyl, reactions, with 3,4-dihydro-2H-pyran, 208
Hypochlorite, t-butyl, for the oxidation of diazo ketones, 162
Hypochlorite, trifluoromethyl, reactions, with vinylsilanes, 140
Hypofluorite, acetyl, as starting material, for α-fluoro ethers, 44
Hypofluorite, methyl, synthesis of, 51
Hypophosphorous acid, reactions, with paraformaldehyde, 125

Imidazoles
  as phase-transfer catalysts, 793
  as starting materials, for ketene aminals, 973
Imidazoles, acyl-, reactions, with chlorotrimethylsilane, 933
Imidazolidines
  from, chiral diamines, 408
  reactions, with Grignard reagents, 410
  as starting materials, for gem-dinitrosamines, 434
  synthesis of, 408
Imidazolidines, N^1,N^2-diacyl-, synthesis of, 425
Imidazolidines, 1,3-dihydroxy-, synthesis of, 435
Imidazolidines, 2-methylene-, Diels–Alder cyclizations of, 414
Imidazolidinethiones, synthesis of, 421
Imidazolidinetrione, reduction of, 103
Imidazolidinones
  from
    dimethylketene-N-phenylimine, 422
    thiohydantoins, 420
  synthesis of, 416, 419, 421
4-Imidazolidinones, N-hydroxy-, synthesis of, 442
Imidazolines
  from, isocyanides, 538
  as starting materials
    for aminals, 414
    for ketene aminals, 979
Imidazolines, N^1,N^2-diacyl-, synthesis of, 426
Imidazolium salts, cycloaddition to, 414
Imidazopyridines, perhydro-, from, 2-(methylamino)methylhexahydropyridine, 408
Imide, N-fluorobis(benzenesulfonyl)-, reactions, with phosphonates, 131
Imide, triphenylphosphine benzotriazole-1-ylmethyl-, synthesis of, 329
Imides
  alkylation of, 881
  reactions, with Grignard reagents, 303
  as starting materials, for O,N-acetals, 303
Imides, α-halo-
  from, α-hydroxy compounds, 98
  synthesis of, 98
    via cleavage of gem-diheteroatomic compounds, 99
    via halogenation, 99
Imides, N-haloalkyl-, synthesis of, 99
Imides, phosphine
  synthesis of, 800
  See also Phosphoranes, imino-
Imides, triphenylphosphine
  reactions
    with gem-diisocyanates, 436
    with gem-diisocyanatoalkanes, 441
S-Imides, ketene dithioacetal
  from, ketene dithioacetals, 860
  synthesis of, 860
Imidophosphinates, (bromomethyl)-, synthesis of, 129
Imidoyl chlorides
  reactions, with alcohols, 312
  as starting materials, for gem-alkylideneamino phosphorylalkenes, 1009
Imine, dimethylketene-N-phenyl-, as starting material, for imidazolidinones, 422
Imine, hexafluoroacetone, reactions, with acetic anhydride, 304
Imine, N-trichlorovinylbenzyl-, synthesis of, 794
Imine functions, α-metallated, synthesis of, 535
Imines
  cycloaddition of, 414
  from, N-(methoxymethyl)anilines, 412
  as heterodienophiles, 304
  nucleophilic addition to, 426
  optically pure, for the synthesis of α-aminophosphonic acids, 462
  reactions
    with acetyl chloride, 100
    with amines, 411
    with bistrifluoromethylketene, 886
    with boron trichloride, 107
    with t-butyl thiol, 325
    with chloroacetyl chloride, 100
    with m-chloroperbenzoic acid, 104
    with chlorophosphonites, 486
    with diaryl phosphites, 463
    with dichlorophosphines, 463
    with electrophiles, 98
    with ethyl phenylphosphonite, 463
    with hydrazoic acid, 444
    with nucleophilic phosphorus compounds, 455
    with phenylphosphine, 455
    with phosphines, 455
    with phosphinic acids, 463
    with phosphonites, 463
    with silylphosphines, 481
    with thioketenes, 905
  reduction of, for the synthesis of N-acylaminals, 420
  silylation of, 518
  silylmethylation of, 517
  as starting materials
    for O,N-acetals, 304, 312
    for S,N-acetals, 319, 328
    for aminals, 411
    for α-aminoalkylphosphinic acids, 463
    for α-aminoalkylphosphonic acids, 463
    for α-aminophosphinic acid compounds, 463
    for α-aminophosphonic acids, 462
    for chlorofluoroalkanes, 32
    for dibromoalkanes, 27
    for dichloroalkanes, 19
    for difluoroalkanes, 10
    for diiodoalkanes, 29
    for α-haloamines, 100
    for hemithioaminals, 325
    for oxaziridines, 312
Imines, acyl-, reactions, with alcohols, 304
Imines, N-acyl-, as starting materials, for α-aminophosphonic acids, 486
Imines, aminomethyl
  from, α-bromo glycine diphenylmethanimine, 445
  synthesis of, 445
Imines, N-arenesulfonyl, reactions, with secondary phosphines, 481
Imines, N-aroyl-, synthesis of, 304

Imines

Imines, aryl-, reactions, with hydroxylamines, 442
Imines, benzophenone, lithiation of, 535
Imines, chloro-
  reactions
    with arenesulfenyl chlorides, 110
    with sulfur dichloride, 110
Imines, N-chloro-, reactions, with Lewis acids, 110
Imines, N-fluorosulfonyl-
  reactions, with diphenylphosphine, 481
  as starting materials, for α-aminophosphonic acids, 486
Imines, polyfluorinated phenyl-, reactions, with bis(trifluoromethyl)thioketene, 905
Imines, propargyl-, metallation of, 518
Imines, N-silyl-, reactions, with Lewis acids, 110
Imines, α-silyl-
  acylation of, 514
  deprotonation of, 648
Imines, silylmethyl-
  from
    α-aminosilanes, 517
    carbonyl compounds, 517
Imines, α-stannyl-, synthesis of, 536
Imines, N-sulfonyl-
  reactions, with water, 312
  synthesis of, 311, 427
Imines, trichloroacetaldehyde, as starting materials, for O,N-acetals, 304
Iminium chloride, dimethylphosgen-
  reactions, with fluoroacetyl chloride, 794
  as starting material, for α-chloro enamines, 794
Iminium chlorides, reactions, with trialkyl phosphite, 1003
Iminium ions
  from, perhydrotriazine, 455
  reactions
    with nucleophilic phosphorus compounds, 455
    with tertiary phosphines, 462, 487
    with trisubstituted phosphines, 478
  synthesis of, 533
  use in the synthesis of ketene hemiaminals, 894
Iminium salts
  as electrophiles, 835
  reactions
    with amines, 412
    with metal dithiocarbamates, 320
    with nucleophilic phosphines, 481
  as starting materials
    for O,N-acetals, 304
    for S,N-acetals, 319
    for Te,N-acetals, 334
    for aminals, 411
  synthesis of, 517
Iminium salts, alkoxymethylene-, synthesis of, 881
Iminium salts, alkylthiomethylene-, as starting materials, for hemithioaminals, 321
Iminium salts, methylene-, reactions, with triphenylphosphine, 487
Iminium salts, oxo-, reactions, with diaryl phosphites, 485
Iminium salts, N-(trimethylsilylmethyl)-, as starting materials, for azomethine ylides, 517
Imino chlorides, as starting materials, for α-phosphorodiamidoyl enamines, 1011
Imino–rhenium complexes, cyclopentadienylidene-bridged, synthesis of, 678
Immonium compounds, dicyanoketene-, synthesis of, 1018
Immonium salts, from, N,N-dimethylthiourea, 988
Indan-1,3-diones, 2-amino-2-hydroxy-, synthesis of, 298
Indazoles, N-methyl, lithiation of, 539
Indium(I) bromide, synthesis of, 153
Indium(I) chloride
  reactions, with dichloromethane, 153
  use in the synthesis of carbon-bridged diindium compounds, 698
Indium(I) iodide, synthesis of, 153

Indium(III) chloride, use in the synthesis of carbon-bridged diindium compounds, 698
Indium chlorides, reactions, with trimethylstannylmethyllithium, 726
Indium complexes, (bromomethyl)dibromo-, synthesis of, 153
Indium complexes, (chloromethyl)dichloro-, synthesis of, 153
Indium compounds, reactions, with phosphorus ylides, 587
Indium compounds, (bromomethyl)dibromo-, synthesis of, 153
Indium compounds, 1-haloalkyl-, reactions, with phosphines, 588
Indium compounds, (iodomethyl)diiodo-, synthesis of, 153
Indium trichloride, reactions, with ylides, 397
Indolenium ions, as starting materials, for hexahydropyrrolo[2,3-b]indole nucleus, 418
Indoles, synthesis of, 653
Indoles, 3-acetyl-N-chloromethyl-, from, N-dimethylaminomethylamines, 97
Indolinones, 3-aminoiso-, synthesis of, 419
5-Indolones, as starting materials, for ketene aminals, 979
Induction, asymmetric, in cyclopalladation, 720
Insecticides
  pluridone, 268
  trichlorofon, 912
Iodides, bridgehead, synthesis of, 107
Iodides, tertiary, reactions, with zinc, 730
Iodine
  reactions
    with bis(trimethylenedioxyboryl)methylenecyclohexane, 750
    with gem-chloronitrosoalkanes, 40
    with diazoalkanes, 29
    with diiodoalkyne, 797
    with hydrazones, 29
    with phenoxyalkyne, 765
    with phenoxyethynylmagnesium bromide, 765
    with trifluoromethylphenylmercury, 1083
    with trimethylsilylcyclohexenes, 139
    with zirconocycles, 1039
Iodine compounds, reactions, with aldazines, 198
Iodine monochloride
  reactions
    with organoaluminums, 758
    with organomercurials, 750
Iodine pentoxide, reactions, with phenylpropiolic acid, 749
Iodoform
  reactions, with sodium sulfite, 281
  reduction of, 29
  as starting material, for methane bissulfonic acid, 281
Iodonium salts, thermolysis of, for the synthesis of carbenes, 1081
Ions, dicoordinated, synthesis of, **1071–1083**
Ions, tricoordinated, synthesis of, **1071–1083**
Iridium, dicarbonyl(μ-pyrazolyl)-, reactions, with diiodomethane, 685
Iridium complexes, reactions, with carbon monoxide, 931
Iridium complexes, (halomethyl)-, synthesis of, 150
Iridium compounds, reactions, with phosphorus ylides, 583
Iron, (benzylideneacetone)tricarbonyl-, reactions, with alkynes, 681
Iron, dicarbonylcyclopentadienylmethyl-, reactions, with bis(cyclopentadienyl)hydridoniobium, 717
Iron, pentacarbonyl-, as starting material, for benzoferracyclopentadiene complexes, 681
Iron, tricarbonylbenzocyclobutadiene-, as starting material, for benzoferracyclopentadiene complexes, 681
Iron, tricarbonylcyclobutadiene-
  lithiation of, 706
  mercuration of, 724
Iron, tricarbonyl(lithiocyclobutadiene)-, reactions, with transition metal halides, 723

Iron(II) chloride
  reactions
    with arsoles, 1039
    with bismoles, 1039
    with bismolyl anions, 1041
    with mesitylmercuric chloride, 685
    with phosphonyl anions, 1026, 1035
    with potassium arsolyls, 1039
    with stiboles, 1039
    with stibolyl anions, 1041
Iron(III) chloride
  as catalyst, for the synthesis of geminal diacetates, 203
  σ-complexes, stability of, 820
Iron(III) nitrate, as reagent, for the oxidative nitration of oximes, 429
Iron carbene complexes, reactions, with phosphines, 581
Iron complexes
  photolytic rearrangement of, 639
  synthesis of, 641
  thermal rearrangement of, 1032
Iron complexes, bridged, from, cyclooctatetraene, 683
Iron complexes, dicarbonylcyclopentadienyl(substituted phenyl)-, reactions, with hexacarbonylchromium, 715
Iron complexes, halo-, reactions, with ferrocenyllithium, 669, 684
Iron complexes, (halomethyl)-, synthesis of, 148
Iron complexes, (methoxymethyl)-, reactions, with hydrogen chloride, 148
Iron complexes, propargyl-, rearrangement of, 683
Iron complexes, vinyl-, reactions, with cobalt carbonyls, 711
Iron compounds, reactions, with phosphorus ylides, 581
Iron compounds, 1-haloalkyl-, reactions, with phosphines, 582
Iron hydride complexes, reactions, with tetracarbonylvanadium, 717
Iron iodide, cyclopentadienyldicarbonyl-, reactions, with arsolyl anions, 1039
Iron iodide, dicarbonylcyclopentadienyl-
  reactions
    with diferrocenylmercury, 685
    with tetraphenylphospholyl anion, 1034
Iron multidecker sandwich compounds, synthesis of, 694
Iron–nickel complexes, synthesis of, 713
Iron nitrosyl complexes, methylene-bridged, synthesis of, 679
Iron pentacarbonyl
  as catalyst, for the synthesis of α-haloalkenylsilicon compounds, 810
  reactions, with dimethylgermyl(dimethylsilyl)methane, 641
  use in the synthesis of carbon-bridged diiron complexes, 681
Iron–phosphinoalkene complexes, synthesis of, 1023
Iron vinylidene complexes, as starting materials, for α-ferrio enamines, 1019
Isatin, N-vinyl-, use in the synthesis of α-haloamines, 101
Isobutylene oxide, reactions, with ethyl cyanoacetate, 883
Isochroman, 1-ethoxy-, reactions, with acetic anhydride, 212
Isocyanate, chlorosulfonyl (CSI), reactions, with vinylsilanes, 514
Isocyanate, 1-chlorovinyl
  reactions, with triethyl phosphite, 1010
  as starting material, for gem-isocyanato phosphorylalkenes, 1010
Isocyanate, 1,2-dibromoethyl
  reactions, with triethyl phosphite, 1009
  as starting material, for gem-isocyanato phosphorylalkenes, 1009
Isocyanate, ethyl, chlorination of, 102
Isocyanate, hydroxymethyl, stability of, 313
Isocyanate, methyl, reactions, with α-hydrazino enamines, 986
Isocyanate, phenyl, reactions, with cumulenes, 421
Isocyanate, phenylthiomethyl, synthesis of, 328
Isocyanate, trimethylsilyl, reactions, with α-chloro isocyanates, 434
Isocyanate, trimethylsilylmethyl, synthesis of, 516
Isocyanates
  nucleophilic addition to, 426
  reactions
    with acetals, 426
    with t-butanol, 1007
    with dimethoxymethane, 310
    with thioacetals, 426
  as starting materials, for 1,3,5-triazines, 426
Isocyanates, acylaminomethyl, synthesis of, 446
Isocyanates, 1-alkoxyalkyl, from, enol ethers, 314
Isocyanates, aryl, reactions, with triethylamine, 446
Isocyanates, azomethyl, addition of Grignard reagents to, 448
Isocyanates, azomethylene, synthesis of, 440, 441
Isocyanates, bis(silyl)methyl, synthesis of, 517
Isocyanates, α-bromoalkyl, reactions, with potassium thiocyanate, 434
Isocyanates, carbamoylmethyl, synthesis of, 447
Isocyanates, α-chloro, reactions, with trimethylsilyl isocyanate, 434
Isocyanates, α-chloroalkyl
  reactions
    with alkoxides, 314
    with phenoxides, 314
    with N-silylated carbodiimides, 441
  tautomers of, 329
Isocyanates, dihaloethyl, as starting materials, for gem-dihalophosphoryl isocyanatoalkenes, 1012
Isocyanates, haloalkyl, as starting materials, for gem-dihalophosphoryl isocyanatoalkenes, 1012
Isocyanates, α-(isothiocyanato), synthesis of, 434
Isocyanates, gem-nitro, synthesis of, 440
Isocyanates, silyloxymethyl, synthesis of, 313
Isocyanates, tetrachloroethyl
  reactions, with triethyl phosphite, 1010
  as starting materials
    for gem-dihalophosphoryl isocyanatoalkenes, 1013
    for gem-isocyanato phosphorylalkenes, 1010
Isocyanates, α-thioalkyl, synthesis of, 329
Isocyanates, trialkylsilyl, reactions, with formaldehyde, 313
Isocyanates, tris(silyl)methyl, synthesis of, 517
Isocyanates, vinylidene, as starting materials, for gem-isocyano lithioalkenes, 1015
Isocyanatoalkenes, gem-alkylideneamino, from, gem-diisocyanatoalkanes, 992
Isocyanato groups, hydrolysis of, 1006
Isocyanic acid, reactions, with formaldehyde, 313
Isocyanide, aminomethyl, synthesis of, 446
Isocyanide, t-butyl
  reactions, with hydrogen chloride, 894
  as trapping agent, 1082
  use in the synthesis of carbon-bridged diiron cation complexes, 680
Isocyanide, methyl, lithiation of, 516
Isocyanide, phenyl, reactions, with tributylborane, 522
Isocyanide, tosylmethyl (TosMIC), synthesis of, 331, 538
Isocyanide, trimethylsilylmethyl
  as starting material
    for amidines, 518
    for 1-aminoalkylphosphonium salts, 492
    for thioimidates, 518
    for trimethylsilylmethyl isothiocyanate, 517
  synthesis of, 516
Isocyanides
  reactions, with trialkylboranes, 522
  as starting materials
    for imidazolines, 538
    for gem-isothiocyanato silylalkenes, 1014
    for oxazolines, 538

Isocyanides
    for α-palladio enamines, 1017
    for α-platino enamines, 1017
    for pyrrole compounds, 538
    for thiazolines, 538
Isocyanides, arenesulfonylmethyl
  reactions
    with aldehydes, 908
    with ketones, 908
    with 17-oxosteroids, 908
    with pyrrolidone-4-aldehydes, 908
  use in the synthesis of substituted alkenes, 908
Isocyanides, lithiated, cyclization of, 538
Isocyanides, α-lithiated, synthesis of, 537
Isocyanides, α-metallated, synthesis of, 537
Isocyanides, sulfoximidoylmethyl, synthesis of, 330
Isocyanides, vinylidene
  reactions, with chlorotrimethylsilane, 1014
  as starting materials, for α-isocyano silylalkenes, 1014
Isocyanuric acid, trichloro-, for the chlorination of sulfides, 68
Isoflurane, synthesis of, 49
Isoindole-1-ones, 3-substituted 3-hydroxy-1,3-dihydro-, synthesis of, 303
Isoindolones, synthesis of, 299
Isomerization
  (E)-to(Z)-, 808
  of ketene dithioacetal S,S-dioxides, 857
  of nitroalkenes, (Z) to (E), 796
Isonitriles, from, formamides, 908
N^1,N^2-Isopropylidene protected dipeptides, synthesis of, 417
Isopropyl iodide, heptafluoro-, reactions, with phosphorus, 114
Isopyrazoles, 4-hydroxy-, synthesis of, 948
Isoquinolines, 3,4-dihydro-, as starting materials, for 1-phosphonic acids, 464
Isoquinolines, tetrahydro-, synthesis of, 531
Isothiazole 1,1-dioxide, 3-methylthio-1,2-benz-, reactions, with 1-diethylaminoprop-1-yne, 905
4-Isothiazolin-3-ones, 5-halo-, reactions, with nucleophiles, 775
Isothiocyanate, trimethylsilylmethyl
  from, trimethylsilylmethyl isocyanide, 517
  synthesis of, 516
Isothiocyanates, reactions, with benzaldehyde, 1014
Isothiocyanates, acylaminomethyl, synthesis of, 446
Isothiocyanates, α-aryloxyalkyl, stability of, 314
Isothiocyanates, azomethylene, synthesis of, 440
Isothiocyanates, α-bromoalkyl, reactions, with potassium thiocyanate, 434
Isothiocyanates, lithiated, synthesis of, 538
Isothiocyanates, α-metallated, synthesis of, 537
Isothiosemicarbazides, synthesis of, 984
Isothioureas
  reactions, with acetophenones, 980
  as starting materials
    for α-amido enamines, 980
    for ketene aminals, 973
Isothiouronium salts, hydrolysis of, 346
Isoxazoles
  photolysis of, 892
  as starting materials, for ketene aminals, 979
Isoxazole salts, ring-opening of, 892
Isoxazolidines, synthesis of, 307
Isoxazolium salts, as starting materials, for ketene aminals, 979

Kabachnik–Fields reaction
  mechanism, 469
  for the synthesis of α-aminoalkylorganophosphorus compounds, 469
  for the synthesis of α-aminoalkylphosphorus compounds, 487

Kaplan–Shechter reaction, for the synthesis of *gem*-dinitroalkanes, 428
Ketals
  reactions
    with methyllithium–lithium bromide, 961
    with pyrazole, 437
  reductive lithiation of, 394
Ketazines, as starting materials, for α-aminoalkylorganophosphorus compounds, 460
Ketene
  reactions
    with N-*t*-alkylazomethines, 421
    with *t*-butyl diethyl phosphite, 915
    with mercuric halide complexes of trialkyl phosphites, 915
Ketene, 2,2-bis(fluorosulfonyl)-, synthesis of, 867
Ketene, 2,2-bis(trifluoromethanesulfonyl)-, synthesis of, 867
Ketene, bis(trifluoromethyl)-
  reactions
    with dialkylphosphines, 1023
    with diamines, 886
    with imines, 886
  as starting material
    for cyclic sulfates, 833
    for dioxazoles, 833
Ketene, bromo-, synthesis of, 807
Ketene, chlorocyano-, reactions, with ethyl N-cyclohexylformimidate, 304
Ketene, dichloro-, use in the synthesis of carbenes, 1083
Ketene, dimethyl-
  reactions
    with acridine, 886
    with carbon disulfide, 836
    with heterocycles, 886
    with quinoline, 886
    with trivalent phosphorus compounds, 914
Ketene, diphenyl-
  reactions
    with azetines, 829
    with azirines, 829
    with diphenylacetic anhydride, 829
    with oxiranes, 829
    with tetramethylallene-1,2-sulfide, 836
    with triethyl phosphite, 914
  use in the synthesis of carbon-bridged dirhodium complexes, 686
Ketene, trimethylsilyl-, bromination of, 807
Ketene compounds
  as starting materials, for ketene hemiaminals, 887
  synthesis of, 831
Ketenes
  reactions
    with silylamines, 886
    with silylated amides, 886
  as starting materials
    for 1,1-bis(germyl)alkanes, 628
    for dichlorocarbene, 1083
    for ketene acetals, 829
    for ketene hemiaminals, 886
Ketenes, bis(silyl)-, as starting materials, for 1,1-bis(silyl)alkanes, 619
Ketenes, bis(thiophosphinyl)-, dimerization of, 559
Ketenes, diacyl-, synthesis of, 829
Ketenes, 1-dihaloboronyl 1-silyl-, synthesis of, 1057
Ketenes, 2,2-disulfonyl-, synthesis of, 867
Ketenes, fluoro-, thermolysis of, 736
Ketenes, metallated, reactions, with dialkylphosphine oxides, 915
Ketenes, silyl-, as starting materials, for α-sily-α-stannyl esters, 661
Ketenimine, triphenyl-, reactions, with pentacarbonylbenzylidene tungsten, 1018
Ketenimines

from, potassium tricyanomethanide, 977
reactions, with alkylamines, 977
as starting materials
for ketene aminals, 979
for ketene hemiaminals, 894
Ketenimines, N-acyl-
reactions
with carbodiimides, 982
with p-chloroaniline, 980, 981
as starting materials
for N-acylated α-ureido enamines, 982
for α-alkoxy carbonylamino enamines, 981
for α-amido enamines, 980
Ketenimines, bis(phosphoryl)-, cycloaddition of, 559
Ketenimines, N-phosphorylated, reactions, with triethyl phosphite, 1008
Keteniminium halides, use in the synthesis of α-halo enamines, 790
Keteniminium ions, as intermediates, in the synthesis of ketene aminals, 971
Keteniminium salts, synthesis of, 790
Ketimines
reactions
with germylphosphines, 998
with silylphosphines, 998
as starting materials
for 1-aminoalkyl phosphinates, 463
for 1-aminoalkyl phosphonates, 463
for α-phosphino enamines, 998
α-Keto acid dithioacetals, synthesis of, 320
α-Keto acids
reactions, with bromine, 28
use in the synthesis of 1,3-azarsa compounds, 495
Keto diacids, reactions, with acetyl chloride, 204
Ketone, isoartemisia, synthesis of, 256
Ketone, methyl trimethylsilyl, reactions, with hydrogen chloride, 944
Ketone, methylvinyl (MVK), reactions, with difluoroamine, 439
Ketone, trimethylsilylvinyl, reactions, with thiocarbonyl compounds, 232
Ketone complexes, 3-carbon-bridged, synthesis of, 676
Ketone hydrates, synthesis of, 162
Ketones
acetalization of, 178
bromination of, 20
bromoacylation of, 56
chloroacylation of, 53
α,α-dibromination of, 21
dithioacetalization of, 247
C-germylation of, 360
haloalkylation of, 47
reactions
with alcohols, 177
with amines, 294, 405
with (ω-aminoalkyl)-phenylphosphines, 453
with arenesulfonylmethyl isocyanides, 908
with benzeneselenenyl chloride, 90
with bishydroxylamines, 436
with bis thiols, 347
with bis(trimethylsilyl)methyllithium compounds, 936
with chloro(diethyl)phosphine, 120
with chlorophosphines, 119, 121, 345
with cyclopentadienyltantalum complexes, 355
with dialkyl phosphites, 338
with diamines, 410
with dibromomethyllithium, 746
with diethoxycarbenium tetrafluoroborate, 193
with dimethylphenylsilyllithium, 355
with 1,3-dipoles, 183
with α,α-dithiophosphonate esters, 847
with epoxides, 181
with fluoromethylsulfoximine, 86
with hydrogen sulfide, 244
with hydroxylamines, 435
with (3-hydroxypropyl)phenylphosphines, 337
with lithiodichloromethane, 741
with mercaptoalcohols, 229
with ortho esters, 190
with orthoformates, 190
with phosphoramidites, 475
with silirenes, 1052
with silyl ethers, 181
with sodium bisulfite, 237
with α-sulfinyl carbanions, 77
with tetrabromomethane, 745
with α-thiol carboxylic acids, 230
with thiols, 247
with tributylstannylmagnesium chloride, 386
with trimethylsilyl sulfides, 216
with triphenylphosphine, 745
with Vilsmeier reagent, 196
reduction of, 699
silyl anion addition to, 354
C-stannylation of, 386
as starting materials
for acetals, 177
for O,As-acetals, 344
for O,N-acetals, 311
for Se,N-acetals, 333
for S,N-acetals, 315, 328
for S,P-acetals, 345
for acyl alkyl acetals, 212
for aminals, 405
for bromochloroalkanes, 38
for α-chloro ethers, 47
for chloroiodoalkanes, 40
for gem-diazidoalkanes, 432
for dibromoalkanes, 25
for dichloroalkanes, 18
for difluoroalkanes, 8
for diselenoacetals, 287
for dithioacetals, 245
for dithiolanes, 246
for geminal diacetates, 204
for hemiacetals, 164
for hemiaminals, 294
for hemithioaminals, 324
for ketene monothioacetals, 834
for phosphorus compounds, 336
for symmetrical acetals, 190
synthesis of, 936
use in the synthesis of 1,3-azarsa compounds, 495
use in the synthesis of 1,3-benzazarsolines, 496
Ketones, α-acetoxy-α-methoxy, synthesis of, 161
Ketones, α-alkoxy, Baeyer–Villiger oxidation of, 212
Ketones, alkynic, reactions, with propanedithiol, 258
Ketones, aralkyl, reactions, with gem-diiminoalkanes, 445
Ketones, aryl, reactions, with methyl methylthiomethyl sulfoxides, 857
Ketones, aryloxy, reactions, with zinc, 167
Ketones, α-bromo, as electrophiles, for dithiocarboxylic enethiolates, 842
Ketones, bromomethanesulfonyl, synthesis of, 84
Ketones, chlorodifluoromethyl, as starting materials, for enol phosphates, 735
Ketones, chlorofluoromethyl, synthesis of, 32
Ketones, α-chloro-α-methoxy, synthesis of, 161
Ketones, α-chloro-α-phenylselenyl, synthesis of, 88
Ketones, diaryl, acetalization of, 179
Ketones, α-diazo, reactions, with benzenesulfenyl chloride, 69
Ketones, dibromo, from, α-diketones, 27
Ketones, β,β-dichloroalkenyl aryl, reactions, with pyrrolidine, 753
Ketones, dichloromethyl

Ketones
  from
    enamines, 15
    terminal alkynes, 16
Ketones, α-dichloromethylene, synthesis of, 742
Ketones, dichlorovinyl, reactions, with amines, 970
Ketones, α-(diethoxymethyl), synthesis of, 193
Ketones, α,α-digermyl, synthesis of, 628
Ketones, α,α-digermylated, synthesis of, 625
Ketones, dihalo, as starting materials, for α-ketoacetals, 189
Ketones, dipyridyl, reactions, with 2-chloroethanol, 183
Ketones, α,α-disilyl, reactions, with 1,1,4,4-tetrakis(trimethylsilyl)butatriene, 619
Ketones, α-halo
  reactions, with alkoxides, 183
  as starting materials, for acyclic hemiacetals, 164
Ketones, α-halo-α-benzeneselenenyl, synthesis of, 90
Ketones, halogenated, as starting materials, for acetals, 183
Ketones, α-hydroxy, synthesis of, 536
Ketones, methyl
  reactions
    with diiminosuccinonitrile, 438
    with trimethyl thiophosphite, 345
Ketones, phenyl
  α,α-dichlorination of, 12
  reactions, with sulfuryl chloride, 12
Ketones, (α-phenylseleno) α,β-unsaturated, reactions, with dimethylcopperlithium, 401
Ketones, α-(phenylthio)
  α-acetoxylation of, 220
  synthesis of, 220
Ketones, β-phenylthio-β-tributylstannyl, synthesis of, 398
Ketones, silylvinyl, reactions, with trimethylphenylthiosilane, 944
Ketones, strained cyclic, as starting materials, for α-diols, 162
Ketones, (E)-γ-substituted, from, titanium ate complexes, 918
Ketones, symmetrical, synthesis of, 523
Ketones, α-thiol, reactions, with amines, 318
Ketones, β-tributylstannyl α,β-unsaturated, synthesis of, 398
Ketones, trichloromethyl, synthesis of, 195
Ketones, trifluoromethyl, reactions, with diphenyl(trimethylsilyl)phosphine, 338
Ketones, α,β-unsaturated
  reactions
    with α-alkoxycuprates, 384
    with (2-hydroxyaryl)mercury chlorides, 168
  as starting materials, for acyl alkyl acetals, 213
β-γ-Ketones, synthesis of, 742
Ketoximes, oxidation of, 314
Knoevenagel condensation
  of bis(sulfones), 275
  for the synthesis of ketene dithioacetal tetroxides, 864
  use of methyl methylthiomethyl sulfoxides in, 856
Koenigs–Knorr reaction, mechanism, 198
Koser's reagent, use in the synthesis of bromo iodoalkenes, 760
Kumada's ligands, synthesis of, 1032

Lactams, alkylation of, 881
Lactams, bicyclic, synthesis of, 496
Lactams, α-bromo-, synthesis of, 99
Lactams, N-chlorosulfonyl-, reductive cleavage of, 306
Lactams, hydroxy-, synthesis of, 303
Lactams, N-trimethylsilyloxymethyl-, from, α-chloroamides, 100
β-Lactams
  analogues, synthesis of, 274
  reactions, with xenon difluoride, 778
  as starting materials
    for amides, 825
    for monoacyl aminals, 419
  synthesis of, 906

β-Lactams, chloro-, as starting materials, for oxazepinones, 824
Lactols
  from, lactones, 171
  synthesis of, 164
δ-Lactols, synthesis of, 166
Lactone complexes, synthesis of, via carbon monoxide insertion, 688
Lactones
  C,O-disilylation of, 831
  reactions
    with carbanions, 171
    with lithium alkynide, 171
  reduction of, 170
  as starting materials, for lactols, 171
Lactones, cephem spiroacetal, synthesis of, 208
Lactones, fluoro-, synthesis of, 52
γ-Lactones, γ-chloro-, from, levulinic acid, 54
γ-Lactones, γ-fluoro-, synthesis of, 52
Lanceol, Julia's synthesis of, 254
Lanthanides, dimeric bis(cyclopentadienyl) metal–methyl complexes of, synthesis of, 702
Lanthanum chloride, as catalyst, for the synthesis of dithioacetals, 246
Lanthanum compounds
  reactions, with phosphorus ylides, 578
  with two carbon-bridges, 703
Lead, tetrakis(trifluorovinyl)-, synthesis of, 813
Lead, tetrakis(trimethylsilylmethyl)-, synthesis of, 660
Lead(IV) acetate
  reactions
    with furan compounds, 206
    with malonic acids, 206
Lead(IV) salts, reactions, with trifluorovinylmagnesium iodide, 813
Lead compounds, reactions, with phosphorus ylides, 589
Lead compounds, α-alkoxy-
  reactions, with aldehydes, 390
  synthesis of, 390
Lead compounds, tetraalkyl-, reactions, with aldehydes, 390
Lead dichloride, reactions, with trimethylsilylmethylmagnesium chloride, 660
Leads, α-haloalkenyl-, synthesis of, 812
Lead tetraacetate
  for the oxidation of semicarbazones, 441
  reactions, with α-chloro carboxylic acids, 40
  use in the synthesis of allylic hemiacetals, 174
Leinamycin, synthesis of, 948
Levoglucosenone, as starting material, for α-bromo alcohols, 42
Levulinic acid, as starting material, for γ-chloro γ-lactones, 54
Lewis acid-catalyzed addition, of acetals to enol ethers, 188
Lewis acids
  as catalysts, for the synthesis of acetals, 178
  reactions
    with N-chloroimines, 110
    with N-silylimines, 110
  use in the synthesis of 1,1-diacetates, 203
Lewisite, uses of, 805
Lipase P, use in the synthesis of α-alkoxystannanes, 389
Lithiation, via deprotonation, 535
Lithiation–germylation, of aminoacetonitriles, 521
2-Lithioanisole, reactions, with copper(I) bromide, 696
Lithio compounds, reactions, with tributylchlorostannane, 961
α-Lithio compounds, from, 1,1-bis(silyl)alkanes, 653
Lithium
  reactions
    with 1,1-bissilylethylenes, 618
    with chloromethyl(trimethyl)germane, 626
    with chloromethyltrimethylsilane, 602
    with dichloromethane, 632

with *gem*-dihaloalkanes, 667
with 1,2-diphosphetes, 545
with ethylidenebis(mercury chloride), 668
with pentaphenylphosphole, 112
with tetrakis(trimethylsilyl)ethene, 618
Lithium, alkyl-, reactions, with fluoroalkenes, 785
Lithium, 9-anthryl-, reactions, with
   bis(chlorodimethylsilyl)methane, 621
Lithium, *exo,exo*-[1,3-bis(trimethylsilyl)allyl]-,
   synthesis of, 603
Lithium, bis(trimethylsilyl)bromomethyl-, reactions, with
   aldehydes, 805
Lithium, bis(trimethylsilyl)methyl-, as starting material, for
   chloroarsines, 597
Lithium, bromochloromethyl-
   reactions, with cyclohexanone, 758
   use in the synthesis of 1-bromo-1-chloroalkenes, 758
Lithium, bromomethyl-, synthesis of, 146
Lithium, butyl-
   as metallating agent, 705
   reactions
      with α-alkoxystannanes, 378
      with anisole complexes, 707
      with bis(phenyltelluro)methanes, 402
      with 1,3-bis(trimethylsilyl)propyne, 1059
      with bromoferrocene, 706
      with bromotrichloromethane, 16
      with 1-*t*-butyl-1-sila-2,4-cyclohexadiene, 603
      with chlorobenzene complexes, 707
      with 1,1-dibromoalkenes, 757
      with 1,2-difluorovinyl chloride, 817
      with 1,1-dimethyl-2,5-diphenyl-1-silacyclohex-3,5-
         diene, 616
      with diselenoacetals, 400
      with iodoferrocene, 706
      with methylenedialuminum tetrabromide, 708
      with (pentachlorocyclopentadienyl)(1,5-
         cyclooctadiene)rhodium, 707
      with selenoketals, 400
      with 1-silylalkynes, 1058
      with trihalostannanes, 380
      with triphenyl(triseleno)orthoformate, 289
Lithium, *n*-butyl-
   reactions
      with α-alkoxysilylstannanes, 358
      with benzyl alcohol, 383
      with bis(phenylseleno)methane, 400
      with α-bromo selenides, 400
      with 1-butyne, 605
      with 2-butyne, 605
      with chlorofluoromethyl phenyl sulfide, 773
      with α-chlorophosphonates, 132
      with chloro-3-(trimethylsilyl)-1-propyne, 648
      with chromium complexes, 1030
      with cyclopropa[*b*]naphthalene, 604
      with 2,2-dibromovinyl ethyl ether, 818
      with (dimethylaminomethyl)cyclopenta-
         dienyltricarbonylmanganese, 1036
      with (*R,S*)-1-diphenylphosphino-2-(1-
         dimethylaminoethyl)ferrocene, 1033
      with dithioacetals, 1075
      with (*E*)-ethoxyvinyl chlorides, 817
      with (*Z*)-2-ethoxyvinyl chlorides, 817
      with α-fluoro-α-iodo esters, 137
      with fluoro sulfonate salts, 369
      with furan compounds, 846
      with (*o*-halobenzyl)phosphines, 575
      with ketene diselenoacetals, 964
      with organostannanes, 964
      with phenylvinyl selenides, 401, 964
      with sulfones, 391
      with tetraethyl dichloromethanediphosphonate, 801
      with tributylstannylmethanol, 378
      with tributyl(trimethylsilylmethyl)tin, 652
      with trifluoromethyltriphenylsilane, 735
      with trimethylsilylmethanol, 383
Lithium, *s*-butyl-
   reactions
      with alkyl benzoates, 382
      with amidines, 107
      with (*E*)-1,3-bis(trimethylsilyl)propene, 603
      with triphenylsilyloxirane, 944
Lithium, *t*-butyl-
   as deprotonating agent, 358
   for the deprotonation of vinyl ethers, 949
   reactions
      with α-alkoxy bromo ethers, 381
      with benzylthiotrimethylgermane, 371
      with benzylthiotrimethylsilane, 368
      with boranes, 524
      with chloro(dimethyl)vinylsilane, 611
      with chloro(methyl)phenyl(vinyl)silane, 611
      with chlorovinylsilanes, 611
      with α-halo ethers, 381
      with 2-methoxy-2,4,4-trimethyl-2,4-disilapentane, 620
      with methylvinyl ether, 948
      with α-phosphoryl enamines, 1004
      with pinacol(1-iodoethyl)boronate, 664
      with tetramethylethylenediamine, 623
      with 3,4,6-tri-*O*-*t*-butyldimethylsilyl-D-glucal, 651
      with vinyl ethers, 929
      with vinyltrimethylsilane, 606
   use in the synthesis of lithioferrocenes, 706
   use in the synthesis of methylenedilithium, 668
Lithium, 1-chloro-2,2-bis(4-chlorophenyl)vinyl-, structure
   of, 816
Lithium, chlorobis(trimethylsilyl)methyl-, synthesis of, 146
Lithium, chlorodiphenylmethyl-, synthesis of, 146
Lithium, chloromethyl-
   reactions
      with bromogermanes, 141
      with chloro(methyl)phenylsilane, 136
      with triphenylsilyl chlorides, 138
   synthesis of, 146
Lithium, dibromomethyl-
   reactions, with ketones, 746
   use in the synthesis of dibromoalkanes, 22
Lithium, 1',2-dichloroferrocenyl-, use in the synthesis of
   dichloro-substituted carbon-bridged diiron
   complexes, 685
Lithium, dichloromethyl-
   reactions, with boronate esters, 526
   use in the synthesis of dichloroalkanes, 19
Lithium, dimesitylborylmethyl-, reactions, with
   trimethylsilyl chloride, 642
Lithium, *o*-*N,N*-dimethyl benzyl-, use in the synthesis of
   carbon-bridged dimanganese complexes, 677
Lithium, dimethylphenylsilyl-
   reactions
      with aldehydes, 355
      with chloromethyl(dimethyl)phenylsilane, 609
      with chlorosilanes, 609
      with ketones, 355
   synthesis of, 355
Lithium, diphenylarsinomethyl-
   from
      bis(diphenylarsino)methane, 599
      iodomethyldiphenylarsine, 599
   reactions, with tributyl chloride, 600
Lithium, diphenylstibinomethyl-, from,
   bis(diphenylstibino)methane, 600
Lithium, 1-ethoxyvinyl-
   reactions
      with copper iodide, 953
      with zinc chloride, 954
Lithium, (*Z*)-2-ethoxyvinyl-, synthesis of, 818

Lithium

Lithium, ethyl-
  reactions
    with 2,5-dichloro-3-iodoselenophene, 785
    with 1,1-dithio-1,3-dienes, 255
    with 2,3,5-tribromoselenophene, 785
Lithium, ferrocenyl-, reactions, with haloiron complexes, 669, 684
Lithium, α-fluorovinyl-, synthesis of, 812
Lithium, heptafluoroisopropyl-, synthesis of, 146
Lithium, iodomethyl-, synthesis of, 146
Lithium, isopropyl-, as starting material, for dihydrido diiridium carbon-bridged complexes, 689
Lithium, mesityl-
  reactions, with cobalt(II) chloride, 690
  use in the synthesis of anions, 1077
Lithium, methoxy(phenylthio)methyl-, reactions, with dioxaborinanes, 364
Lithium, methoxy(trimethylsilyl)methyl-, synthesis of, 650
Lithium, methoxyvinyl-
  as starting material
    for lithium di(1-methoxyvinyl)cuprate, 953
    for 1-methoxyvinylcopper(I), 953
Lithium, 1-methoxyvinyl-
  for the displacement of fluoride, 621
  pure, synthesis of, 935
  reactions
    with chlorometal complexes, 955
    with cuprous iodide, 953
    with trialkylboranes, 934
    with triisobutylborane, 935
    with triisopropyl borate, 934
  synthesis of, 949
Lithium, 1-methoxy-1-vinyl-
  reactions
    with chlorotrimethylgermane, 929
    with chlorotrimethylsilane, 929
Lithium, methyl-
  for the lithiation of metals, 706
  reactions
    with chromium(III) chloride, 708
    with hexamethyldisilane, 355
    with phenylvinyl sulfide, 366
    with sulfonylsulfines, 272
    with 1,1,3,3-tetramethyl-1,3-disilacyclobutane, 606
    with tris(dimethoxyboryl)methane, 633
  as starting material, for methylenedilithium, 668
  use in the synthesis of bismethylene-bridged dirhodium complexes, 687
  use in the synthesis of carbon-bridged diruthenium complexes, 680
  use in the synthesis of trismethylenediruthenium complexes, 680
Lithium, methylthiomethyl-, reactions, with trimethylamine–borane complexes, 373
Lithium, phenyl-
  reactions
    with ferrocenophanes, 1034
    with phenyl dithiobenzoate, 255
    with phospholes, 1031
    with selenoamides, 334, 909
    with thiobenzophenone, 394
  as starting material, for carbon-bridged triosmium complexes, 683
Lithium, (phenyldimethylsilyl)chloromethyl-, reactions, with organoboranes, 357
Lithium, α-phenylselenoheptyl-, reactions, with cuprates, 401
Lithium, phenylthiomethyl-
  reactions, with trimethyl borate, 372
  titanation of, 396
Lithium, iso-propyl-, reactions, with phenylvinyl selenide, 401
Lithium, ruthenocenyl-, as starting material, for diruthenocenylmercury, 723
Lithium, α-silylchloromethyl-, synthesis of, 357
Lithium, stibinomethyl-, reactions, with diphenylchloroarsine, 596
Lithium, p-tolylthiomethyl-, reactions, with menthyl toluene-p-sulfinate, 265
Lithium, tributylstannyl-
  addition to aldehydes and ketones, 386
  reactions
    with aminomethyl phenyl sulfides, 533
    with benzyl chloromethyl ether, 389
    with (chloromethyl)trimethylsilane, 660
Lithium, tri-n-butylstannyl-, reactions, with bicyclo[2,2,1]hept-5-en-2-one, 952
Lithium, tributylstannylmethyl-
  reactions
    with diphenylchloroarsine, 600
    with trimethylsilyl chloride, 660
Lithium, triethylplumbyl-, reactions, with ethoxyethyne, 955
Lithium, trifluorovinyl-, reactions, with phosphorus(III) halides, 802
Lithium, trimethylgermylmethyl-, synthesis of, 663
Lithium, trimethylsilyl-, reactions, with thiobenzaldehydes, 376
Lithium, (trimethylsilyl)bromomethyl-, reactions, with organoboranes, 357
Lithium, trimethylsilylcyclopentadienyl-
  reactions
    with dichlorodimethylsilane, 603
    with trimethylsilyl chloride, 603
Lithium, (trimethylsilyl)dichloromethyl-, reactions, with aldehydes, 742
Lithium, trimethylsilylmethyl-
  from, butyl trimethylsilyl telluride, 652
  reactions
    with chloroarsines, 597
    with stiboranes, 599
    with thallium trichloride, 659
    with trialkylboranes, 642
    with triphenyltin chloride, 660
  synthesis of, 652
Lithium, trimethylsilyl(phenylthio)methyl-, reactions, with cyclohexene oxide, 943
Lithium, trimethylstannyl-, reactions, with pinacol (1-chloroethyl)boronate, 665
Lithium, trimethylstannylmethyl-, reactions, with indium chlorides, 726
Lithium, (trimethylstannylmethyl)-, synthesis of, 708
Lithium, trimethyltin-, reactions, with dihaloalkanes, 699
Lithium, triphenylgermyl-, reactions, with trifluorovinylsilanes, 811
Lithium, triphenylgermylmethyl-, synthesis of, 663
Lithium, triphenyllead-, reactions, with dibromomethane, 699
Lithium, (triphenylstannyl)methyl-, reactions, with benzeneselenenyl bromide, 401
Lithium, tris(trimethylsilyl)methyl-
  reactions
    with electrophiles, 1048
    with styrene oxide, 620
Lithium alanates, synthesis of, 816
Lithium alkoxides, reactions, with sulfones, 235
Lithium alkyls, reactions, with tin tetrachloride, 949
Lithium alkynide, reactions, with lactones, 171
Lithium aluminum deuteride (LAD), for the reduction of ketene dithioacetals, 253
Lithium aluminum hydride (BINAL-H), 2,2'-dihydroxy-1,1'-binaphthyl-modified, for the reduction of acylstannanes, 386
Lithium aluminum hydride (LAH)
  reactions
    with chloroalkynes, 816
    with 1-chloro 1-alkynes, 758
    with 3-chloro-3,3-difluoro-2-trifluoromethylprop-1-ene, 735

with titanium tetrachloride, 1082
with ureas, 414
as reducing agent, for the synthesis of allyl-bridged palladium complexes, 693
use in the reduction of ketene dithioacetals, 253
Lithium amides
reactions
with aldehydes, 299
with chloro enamines, 971
with trichloroethene, 976
Lithium benzyloxide
reactions
with boronates, 362
with (S,S)-2,5-dimethylhexane-3,4-diol, 362
Lithium biphenylide, reactions, with trifluoromethyl dithioacetals, 775
Lithium 1,1-bis(trimethylsilyl) alkoxides, reactions, with benzophenone, 932
Lithium bis(trimethylsilyl)amide, reactions, with chloromethyl methyl ether, 313
Lithium–boron ate complexes, ^{11}B NMR spectrum of, 935
Lithium–bromine exchange, for the synthesis of gem-dibromoalkenes, 818
Lithium butanetellurolate, reactions, with N,N-dimethylmethyleneammonium chloride, 334
Lithium butyl(diisobutyl)aluminum hydride, reactions, with phenyl (1-trimethylsilyl)propadienyl sulfide, 946
Lithium carbenoids
reactions, with metal salts, 820
and transmetallation reactions, 821
Lithium carbenoids, α-haloalkyl-
in asymmetric synthesis, 146
synthesis of, 146
Lithium (E)-1-chloro-1-alkenyl alanates, synthesis of, 758
Lithium chlorodifluoroacetate, use in the synthesis of difluorocarbenes, 732
Lithium compounds
stability of, 956
as starting materials, for carbamates, 934
structure of, 949
synthesis of, 599, 956
use in the synthesis of 1,1-dimetal complexes, 669
Lithium compounds, alkyl-
reactions
with 2-methylene-1,3-dithiane, 254
with thioketenes, 956
x-ray crystallography studies, 647
Lithium compounds, aminoalkyl-, stability of, 528
Lithium compounds, aminomethyl-, silylation of, 511
Lithium compounds, aryl-, reactions, with thioketenes, 956
Lithium compounds, bis(trimethylsilyl)methyl-
reactions
with aldehydes, 936
with ketones, 936
Lithium compounds, α-bromo-, reactions, with mercuric chloride, 152
Lithium compounds, α-bromoalkenyl-
synthesis of, 818
See also Alkenes, gem-bromo lithio-
Lithium compounds, t-butoxymethyl-, transmetallation of, 384
Lithium compounds, chiral α-amino-, synthesis of, 531
Lithium compounds, α-chloroalkenyl-
synthesis of, 817
via metallation of 1-chloro-2,2-diarylethenes, 817
Lithium compounds, α,α-dibromoalkyl-, synthesis of, 22
Lithium compounds, α-fluoroalkenyl-
synthesis of, 817
via metallation of 1,1-dichloro-2-fluoroethene, 817
Lithium compounds, α-haloalkenyl-, synthesis of, 816
Lithium compounds, α-haloalkyl-, synthesis of, 145, 146
Lithium compounds, α-(halomethyl)-, synthesis of, 146
Lithium compounds, α-iodoalkenyl-, synthesis of, 819

Lithium compounds, silylmethyl-, reactions, with carbon monoxide, 931
Lithium compounds, 1-trimethylsilylcyclopropyl-, synthesis of, 652
Lithium compounds, vinyl-, synthesis of, 667
Lithium dialkylamides, reactions, with polyfluoroalkenes, 793
Lithium dicyclohexamide, for the deprotonation of phenylvinyl tellurides, 965
Lithium dicyclohexylamide (LDCA), use in the synthesis of anions, 1077
Lithium diisobutylamide, reactions, with bis(phenylseleno)methane, 401
Lithium diisopropylamide (LDA)
as deprotonation agent, 398
reactions
with bromoform, 746
with 1,1-dichloroethane, 813
with (E)-dimethyl 2-phenylvinyl phosphonate, 927
with di(phenyltelluro)methane, 290
with α-methoxyallylphosphine oxide, 917
with thiophenyl esters, 932
Lithium diisopyramide, for the deprotonation of phenylvinyl selenides, 965
Lithium di(1-methoxyvinyl)cuprate, from, methoxyvinyllithium, 953
Lithium 1-dimethylaminonaphthalenide, reactions, with cyclohexylidene compounds, 945
Lithium 1-(dimethylamino)naphthalenide (LDMAN)
for the reductive lithiation of α-phenyl thioethers, 380
use in the synthesis of α-lithio sulfides, 394
use in the synthesis of α-thiosilanes, 366
Lithium diphenylcuprate, reactions, with α-fluoro sulfoxides, 777
Lithium diphenylphosphide, reactions, with α-chloro tertiary amines, 455
Lithium di(trimethylsilyl)amide, use in the synthesis of anions, 1076
Lithium enolates, reactions, with trialkylsilyl chlorides, 832
Lithium ethoxide, reactions, with 2,3-di(t-butylimino)propiononitrile dihydrochloride, 894
Lithium ethyneselenolates, dimerization of, 874
Lithium–halogen exchange
for the synthesis of α-fluoroalkenyllithium compounds, 817
for the synthesis of gem-halo lithioalkenes, 816
Lithium hexamethyldisilazide (LHMDS), use in the synthesis of azidomethylboronates, 526
Lithium iodide, reactions, with chloromethyl methyl sulfide, 71
Lithium metal vapor, reactions, with dichlorobis(trimethylsilyl)methane, 618
Lithium naphthalenide (LN)
reactions
with bis(phenylthio)bis(trimethylsilyl)methane, 945
with ketene thioacetals, 956
with trifluoromethyl dithioacetals, 775
for the reductive lithiation of α-(phenylthio) ethers, 380
use in the synthesis of α-lithio sulfides, 394
use in the synthesis of α-thiosilanes, 366
Lithium naphthalide, reactions, with sulfides, 1026
Lithium nucleophiles, α-alkoxy-, reactions, with tributyltin chloride, 388
Lithium pentamethylcyclopentadienide, use in the synthesis of carbon-bridged dicobalt complexes, 687
Lithium phenylalkynide, reactions, with N-chlorodiethylamine, 976
Lithium phenyl selenide
reactions
with (iodomethyl)trimethylgermane, 377
with (iodomethyl)trimethylsilane, 377
Lithium phenyl telluride, reactions, with diiodomethane, 290
Lithium phospholides, synthesis of, 112

Lithiums

Lithiums, alkenyl-
  reactions
    with benzeneselenenyl bromide, 785
    with 1,1-trichlorodifluoroethane, 758
Lithiums, α-alkoxy-
  synthesis of, 378
    via transmetallation reactions, 378
Lithiums, α-alkoxyalkyl-, as configurationally stable carbanions, 379
Lithiums, alkyl-
  reactions
    with alkylaluminum halides, 698
    with aluminum chloride, 698
    with *gem*-dibromocyclopropanes, 146
    with sulfones, 936
    with sulfoxides, 391
    with vinylphosphines, 575
    with vinyl selenides, 400
  as reagents, for the synthesis of dialuminum alkyl compounds, 698
Lithiums, 1-alkylthiovinyl-
  from
    ketene thioacetals, 956
    2-methoxyalkyl sulfides, 958
Lithiums, allyl-, silyl-substituted, synthesis of, 648
Lithiums, aminoalkyl-, electrophilic silylation of, 511
Lithiums, arylgermyl-, reactions, with aldehydes, 360
Lithiums, arylthiomethyl-, reactions, with trialkyltin chlorides, 397
Lithiums, germyl-, synthesis of, 641
Lithiums, α-haloalkyl-, configurational stability of, 146
Lithiums, α-methylthio alkyl-, synthesis of, 394
Lithiums, 1-oxyalkenyl-, synthesis of, 948
Lithiums, α-phenylthioalkyl-
  copper promoted allylation of, 396
  synthesis of, 394
Lithiums, 1-plumbylvinyl-
  from
    1,1-diplumbylalkenes, 1065
    1-plumbyl 1-stannylalkenes, 1065
Lithiums, α-selenenyl-, synthesis of, via halogen–metal exchange, 400
Lithiums, α-selenoalkyl-
  dimerization of, 401
  synthesis of, 375
Lithiums, 1-stannylvinyl-, synthesis of, 1065
Lithiums, 1-sulfonylvinyl-, from, sulfonylalkenes, 961
Lithium salts, reactions, with trimethylamine hydrochloride, 373
Lithium tellurides, as starting materials, for diphenylditelluromethane, 290
Lithium tetrakis(trimethylsilyl)aluminate, use in the synthesis of α-hydroxysilanes, 355
Lithium tetramethylaluminate, reactions, with chlorobis(cyclopentadienyl) metal complexes, 726
Lithium 2,2,6,6-tetramethylpiperidide
  for the deprotonation of phenylvinyl tellurides, 965
  reactions, with 2-benzoboroles, 642
Lithium tetramethylpiperidide (LITMP)
  as deprotonation agent, 398
  for the deprotonation of phenylvinyl selenides, 965
  reactions, with (Z)-2-phenyltetrafluoropropenes, 755
Lithium trialkylgermanium, reactions, with alkyn-2-on-1-yl chloride compounds, 1055
Lithium triethyl borohydride
  use in the synthesis of carbon-bridged diruthenium complexes, 679
  use in the synthesis of carbon-bridged trinuclear osmium complexes, 683
Longifolene, reactions, with mercuric acetate, 748
Lossen rearrangement, for the synthesis of α-aminoalkylphosphorus compounds, 492
Lutetium, pentamethylcyclopentadienyl complexes of, synthesis of, 702
Lutetium compounds, reactions, with phosphorus ylides, 578

Magic methyl *See* Sulfonate, methyl fluoro-
Magnesium
  reactions
    with 1,2-bis(bromomethyl)tetramethyldisilane, 654
    with 1-bromo-1-trimethylsilylethene, 1060
    with 1-chloro-4-fluoro-2,2,4-trimethyl-2,4-disilapentane, 608
    with chloromethyl methyl ether, 383
    with dibromodi(trimethylsilyl)methane, 618
    with 1,2-dichloro-1,2,2-trifluoroethyl phenyl selenide, 785
    with polyhalogenated ethyl phenyl selenide, 785
    with trichloro(chloromethyl)silane, 654
    with trimethylsilylmethyl chloride, 653
    with trimethylstannylmethyl halides, 709
  use in the synthesis of carbon-bridged tritin compounds, 700
Magnesium, bis(trimethylsilylmethyl)-
  from, trimethylsilylmethylmagnesium chloride, 655
  use in the synthesis of chloride-bridged diruthenium complexes, 680
  use in the synthesis of methylene-bridged diruthenium complexes, 680
Magnesium, bis(trimethylsilylmethylene)-, use in the synthesis of carbon-bridged dichromium complexes, 672
Magnesium, butadienyl-, use in the synthesis of dihydrido diiridium carbon-bridged complexes, 689
Magnesium, dimethyl-, reactions, with trimethylaluminum, 709
Magnesium, trifluorovinyl-, reactions, with phosphorus(III) halides, 802
Magnesium, trimethylsilylmethyl-, reactions, with bis(diethylamino)chloroarsine, 597
Magnesium amalgam, use in the synthesis of di-Grignard reagents, 669
Magnesium–anthracene, use in the synthesis of Grignard reagents, 654
Magnesium bromide, reactions, with disodioferrocene, 709
Magnesium bromide, allyl-
  use in the synthesis of allyl-bridged palladium complexes, 693
  use in the synthesis of dimeric allyls of chromium, 673
Magnesium bromide, ethyl-
  reactions
    with furfuryl dithioates, 255
    with methyl phenyl sulfone, 395
Magnesium bromide, ferrocenyl-, from, bromoferrocene, 709
Magnesium bromide, mesityl-
  reactions
    with copper(I) chloride, 696
    with silver chloride, 696
Magnesium bromide, 3-methoxyphenyl-, reactions, with keto-aldehydes, 166
Magnesium bromide, phenoxyethynyl-, reactions, with iodine, 765
Magnesium bromide, phenyl-, reactions, with ethoxythietanes, 731
Magnesium bromide, phenylethynyl-, reactions, with thionyl chloride, 863
Magnesium bromide, 1-trimethylsilylvinyl-, reactions, with (-)-menthyl *p*-toluene sulfinate, 946
Magnesium bromide, triphenylsilylallyl-, synthesis of, 654
Magnesium bromide etherate, reactions, with oxiranes, 783
Magnesium chloride, allyl-, use in the synthesis of dimeric allyls of molybdenum, 673
Magnesium chloride, benzylthiomethyl-, from, chloromethyl sulfide, 395
Magnesium chloride, isopropyl-, reactions, with dibromomethane, 147

Magnesium chloride, tributylstannyl-
  reactions
    with aldehydes, 386
    with chloromethyl methyl ether, 389
    with ketones, 386
Magnesium chloride, triethoxysilylmethyl-, stability of, 607
Magnesium chloride, trimethylsilylmethyl-
  availability of, 653
  reactions
    with antimony trichloride, 598
    with arsenic trichloride, 597
    with bismuth trichloride, 599
    with gallium trichloride, 659
    with lead dichloride, 660
    with tin tetrachloride, 660
    with trimethylborate, 642
  as starting material, for
    bis(trimethylsilylmethyl)magnesium, 655
  synthesis of, 653
Magnesium compounds, reactions, with phosphorus ylides, 577
Magnesium compounds, α-amino-, transmetallation of, 531
Magnesium compounds, α-haloalkenyl-, synthesis of, 820
Magnesium compounds, α-haloalkyl-, synthesis of, 147
Magnesium compounds, (halomethyl)-, synthesis of, 147
Magnesium compounds, α-silyl alkyl-, synthesis of, 653
Magnesium halides, α-silylbenzyl-, synthesis of, 654
Magnesium halides, trifluorovinyl-
  reactions, with germanium halides, 811
  as starting materials, for α-fluorovinyltin, 812
Magnesium iodide, methyl-, reactions, with acylphosphine oxides, 343
Magnesium iodide, trifluorovinyl-
  reactions, with lead(IV) salts, 813
  synthesis of, 820
Magnesiums, α-alkoxy-, synthesis of, 383
Makaiyama rearrangement, for the synthesis of ketene silyl acetals, 831
Maleic anhydride, reactions, with dithio esters, 843
Malonate, diethyl, reactions, with aldehydes, 559
Malonate, diethyl chloro-, as starting material, for α-chlorophosphines, 115
Malonate, diethyl 2,4-dinitrobenzylidene-, reactions, with diethylamine, 306
Malonate, dimethyl, dichlorination of, 12
Malonate, dimethyl benzenesulfamido, reactions, with bromine, 105
Malonate, dimethyl diazo-, reactions, with di-*t*-butylthioketene, 837
Malonate, monoethyl acetamido-, oxidation of, 311
Malonate compounds, synthesis of, 824
Malonates, reactions, with dialkoxymethylammonium salts, 193
Malonates, acylamino-, reactions, with diazonium salts, 448
Malonic acids
  diiodination of, 28
  reactions
    with isopropenyl acetate, 204
    with lead(IV) acetate, 206
Malonitrile
  as starting material
    for α-phosphinoyl enamines, 1000
    for α-thiophosphinoyl enamines, 1001
Malononitrile, as starting material, for ketene diselenoacetals, 875
Malononitrile, α-bromo-, as starting material, for *gem*-bromo iminophosphoranes, 800
Malononitrile compounds, synthesis of, 836
Malononitriles, bis(dialkylamino)-, as starting materials, for ketene aminals, 975
Malonyl chlorides, dimethyl-, reactions, with monoalkylamides, 881
Manganacyclopentadiene complexes, synthesis of, 677

Manganese, bromopentacarbonyl-
  reactions
    with dimethylstibolyl anion, 1042
    with lithiated tricarbonylthiophenechromium, 720
Manganese, cyclopentadienyl-, synthesis of, 1036
Manganese, cyclopentadienyldicarbonyl-, use in the synthesis of methylene-bridged dimanganese complexes, 676
Manganese, cyclopentadienyltricarbonyl-, deprotonation of, 1036
Manganese,
  (dimethylaminomethyl)cyclopentadienyltricarbonyl-,
  reactions, with *n*-butyllithium, 1036
Manganese, pentacarbonylmethyl-, in metallation reactions, 721
Manganese, tricarbonyl(cyclopentadienyl)-, lithiation of, 706
Manganese, tricarbonyl(lithiocyclopentadienyl)-
  reactions
    with chlorocyclopentadienyl(triphenylphosphine)nickel, 719
    with chlorotriphenyltin, 725
    with titanocene dichloride, 719
    with transition metal halides, 723
  use in the synthesis of dimetal complexes, 719
Manganese(II) chloride
  use in the synthesis of carbon-bridged dimanganese complexes, 677
  use in the synthesis of manganese η^5-pentadienyl complexes, 678
Manganese(II) iodide, use in the synthesis of carbon-bridged dimanganese complexes, 677
Manganese complexes, synthesis of, 955
Manganese complexes, (halomethyl)-, synthesis of, 148
Manganese complexes, iodomethyl-, synthesis of, 148
Manganese complexes, methylene-bridged, synthesis of, 676
Manganese compounds
  lithiation of, 706
  reactions
    with 1-metalloalkylphosphine oxides, 581
    with phosphorus ylides, 581
  synthesis of, 1037
Manganese compounds, (1-chloro-2,2-dicyanovinyl)-, reactions, with diethylamine, 1019
Manganese compounds, silyl-, use in the synthesis of 1,1,2-trisilylalkenes, 1045
Manganesedicarbonyl, methylcyclopentadienyl-, use in the synthesis of methylene-bridged dimanganese complexes, 676
Manganese dioxide
  reactions, with pyridoxol, 164
  use in the synthesis of cyclic hemiacetals, 164
Manganese pentacarbonyl, reactions, with 1-phenylphosphole, 1037
Manganese η^5-pentadienyl complexes, synthesis of, 678
Manicone, synthesis of, 255
Mannich bases, reactions, with phosphines, 452
Mannich reaction
  of benzo-1,3-dithiolane *S*,*S'*-dioxides, 863
  for the synthesis of *S*,*N*-acetals, 315
  for the synthesis of aminals, 406
  for the synthesis of 1,3-aminoalkylarsenic compounds, 499
  for the synthesis of 1-aminoalkylarsines, 499
  for the synthesis of α-amino organophosphorus compounds, 452
  for the synthesis of α-aminosilanes, 509
  for the synthesis of hemiaminals, 299
  for the synthesis of nitraminoalkane compounds, 440
  for the synthesis of *N*-nitroso *O*,*N*-acetals, 312
  for the synthesis of tricoordinate phosphorus functions, 481
Maylansinoids, synthesis of, 936
Maysine, synthesis of, 255

Maysine, 4,5-deoxy-, synthesis of, 255
Meldrum's acid
    synthesis of, 204
    use in the synthesis of ketene acetals, 827
Meldrum's acid compounds, flash vacuum pyrolysis of, 894
Mercaptans See Thiols
Mercuracycles, reactions, with 2,4-dimethyl-2,4-digermapentane, 629
Mercuration
    of alkenes, 1067
    of carboxylic acids, 697
    of cyclopentadiene, 697
    for the synthesis of transition metal complexes, 724
Mercurials, α-diazo-, synthesis of, 385
Mercuric chloride
    reactions
        with alkene 1,1-diboronic esters, 1067
        with α-bromolithium compounds, 152
    as reagent, in complete mercuration of cyclopentadiene, 697
Mercuric chloride, mesityl-, reactions, with iron(II) chloride, 685
Mercuric fluoride, use in the synthesis of difluoroalkanes, 3
Mercuric nitrate, reactions, with 1,1-diarylethenes, 1067
Mercuri compounds, acetoxy-, synthesis of, 724
Mercuric oxide, reactions, with boron trifluoride, 195
Mercury
    use in the synthesis of carbon-bridged dimercury complexes, 696
    use in Wittig reactions, 734
Mercury, bis(iodomethyl)-, synthesis of, 153
Mercury, bis(trifluorovinyl)-, reactions, with alane complexes, 815
Mercury, bis(trimethylsilyl)-
    reactions
        with alkali metals, 354
        with 1,2,3,3,3-pentafluoro-1-trimethylsilylpropene, 1048
        with trifluorovinyl bromide, 821
Mercury, bis(trimethylsilylmethyl)-, reactions, with potassium sand, 653
Mercury, (chloromethyl)phenyl-, synthesis of, 152
Mercury, diethyl-, reactions, with tris(trimethylsilylmethyl)germane, 641
Mercury, diferrocenyl-
    reactions
        with dicarbonylcyclopentadienyliron iodide, 685
        with tin chloride, 726
    transmetallation of, 706
Mercury, diruthenocenyl-, from, ruthenocenyllithium, 723
Mercury, phenyl(bromodichloromethyl)-, use in the synthesis of dichlorocarbene, 16
Mercury, phenyl(fluorodibromomethyl)-, use in the synthesis of bromofluorocarbene, 35
Mercury, phenyl tribromomethyl-, as starting material, for dibromocarbene, 748
Mercury, phenyl(tribromomethyl)-, use in the synthesis of dibromocarbene, 25
Mercury, phenyltrifluoromethyl-, use in the synthesis of difluorocarbene, 8
Mercury, trifluoromethylphenyl-
    reactions, with iodine, 1083
    as starting material, for difluorocarbene, 1083
Mercury(I) iodide, as catalyst, for the synthesis of carbon-bridged dimercury complexes, 696
Mercury(II) acetate, use in the synthesis of trinuclear carbon-bridged mercury compounds, 697
Mercury(II) chloride
    reactions
        with butanal, 697
        with dialuminum alkyls, 697
        with pentan-2,4-dione, 697
        with propanal, 697

Mercury(II) trifluoroacetate, use in the synthesis of polymeric mercury compounds, 697
Mercury acetate, 1-propenyl-, reactions, with mercuric acetate, 697
Mercury chlorides, (2-hydroxyaryl)-, reactions, with α,β-unsaturated ketones, 168
Mercury complexes, trinuclear carbon-bridged, synthesis of, 697
Mercury compounds
    reactions
        with alkynes, 687
        with 1-metalloalkylphosphine oxides, 587
        with 1-metalloalkylphosphine sulfides, 587
        with phosphorus ylides, 587
Mercury compounds, α-acetoxy-, synthesis of, 385
Mercury compounds, α-alkoxy-, synthesis of, 385
Mercury compounds, (bromomethyl)-, synthesis of, 152
Mercury compounds, carbon-bridged bischloro-, synthesis of, 697
Mercury compounds, (α-chloroalkyl)-, synthesis of, 152
Mercury compounds, (iodomethyl)-, synthesis of, 153
Mercury compounds, perfluoroisopropyl-, synthesis of, 152
Mercury compounds, polymeric, synthesis of, 697
Mercury compounds, silyl-, synthesis of, 821
Mercury difluoride, use in the synthesis of chlorofluoroalkanes, 30
Mercurymethylene complexes, synthesis of, via transmetallation, 669
Mesitylene, reactions, with chromium vapor, 694
Mesityl oxide
    reactions
        with α-chloroethyl chloroformates, 59
        with dibutylphosphine oxides, 339
Metal(I) halides
    as starting materials
        for carbon-bridged copper compounds, 695
        for carbon-bridged silver compounds, 695
Metal carbonyls
    reactions
        with alkynehexacarbonyldicobalt, 711
        with dihydroboroles, 695
        with 1-phenyl-1-boracyclohepta-2,6-diene, 695
Metal complexes, reactions, with alkynes, 712
Metal complexes, acyl, decarbonylation of, 722
Metal complexes, (alkoxymethyl), reactions, with hydrogen halides, 147
Metal complexes, chloro, reactions, with 1-methoxyvinyllithium, 955
Metal complexes, chlorobis(cyclopentadienyl), reactions, with lithium tetramethylaluminate, 726
Metal complexes, halo, reactions, with diazoalkanes, 147
Metal compounds, α-fluoroalkenyl, synthesis of, 821
Metal compounds, α-halo, synthesis of, 145
Metal compounds, α-haloalkenyl
    from
        haloalkynes, 822
        metalloalkynes, 822
Metal compounds, vinyl, reactions, with allylzinc bromide, 709
Metal dithiocarbamates, reactions, with iminium salts, 320
Metal germanes, reactions, with halomethyl ethers, 360
Metal halides, reactions, with transition metal carbene complexes, 722
Metal–halogen exchange
    for the synthesis of alkenes, 1048
    for the synthesis of 1-lithio 1-silylalkenes, 1059
    for the synthesis of α-metallated 1,1-bis(silyl)alkanes, 618
    for the synthesis of transition metal compounds, 821
Metal imine complexes, reactions, with alkynes, 1017
Metallaferrocenophanes, synthesis of, 720
Metalla-heterocycles, synthesis of, 1029
Metallaphosphiranes, synthesis of, 581
Metallation

of π-complexes, 705
of nitrosamines, 529
for the synthesis of *gem*-halo lithioalkenes, 816
Metallic elements, synthesis of functions containing two atoms of, **667–703**
Metallocene dichlorides, reactions, with methylenebis(magnesium bromide), 670
Metallocenes, decamethyl-, use in the synthesis of triple decker sandwich complexes, 717
Metallocycles, synthesis of, 1067
Metalloids
    synthesis of functions, **1043–1070**
    synthesis of functions containing at least one, with another metalloid or metal, **601–665**
Metalloles, synthesis of, 822
Metallonitrene complexes, synthesis of, 800
Metal phosphides
    reactions
        with 1,1-dihaloalkanes, 545
        with 1-haloalkylphosphines, 546
        with 1-haloalkylsilanes, 566
Metals
    synthesis of functions, **1043–1070**
    two dissimilar, synthesis of functions, **705–727**
Metals, α-alkoxy, synthesis of, 381
Metal salts, reactions, with lithium carbenoids, 820
Metal salts, nitronate, reactions, with alkyl halides, 540
Metal salts, thiophilic, reactions, with anomeric sulfides, 200
Metal salts of thiols, reactions, with 1-chloro enamines, 904
Methallylthiols, double deprotonation of, 367
Methanal, phenyl-, as starting material, for 1,1-bis(trimethylsilyl)-2-phenylethene, 1049
Methane, (arylthio)dibenzoyl-, reactions, with chloramine-T, 326
Methane, benzenesulfonylnitro-, monoalkylation of, 333
Methane, benzoylnitro-, reactions, with alkynyltin, 541
Methane, benzyloxynitro-, reactions, with aldehydes, 897
Methane, bis(2-acetylpyrrol-1-yl)-, synthesis of, 410
Methane, bis(benzenesulfinyl)-, dithiocarboxylation of, 863
Methane, bis(benzenesulfonyl)-, reactions, with formamides, 866
Methane, bis(chlorodimethylsilyl)-
    reactions
        with 9-anthryllithium, 621
        with 2,3-dilithio-1,3-butadiene, 621
Methane, bis(dialkylamino)-1-methoxy-, as starting material, for *gem*-dihydrazinoalkenes, 985
Methane, bis(dibromoarsino)-, synthesis of, 593
Methane, bis(dichloroarsino)-, as starting material, for bisarsinomethanes, 593
Methane, 1,1-bis(dichlorophosphino)-, reactions of, 550, 562
Methane, bis(dichlorostibino)-, synthesis of, 595
Methane, bis(dimethylamino)-
    reactions
        with acid anhydrides, 309
        with secondary arsines, 501
    synthesis of, 405
Methane, bis[dimethyl(9-anthryl)silyl]-, synthesis of, 621
Methane, 1,1-bis(dimethylsilyl)-, synthesis of, 614
Methane, bis(diphenylarsino)-
    as starting material
        for arsines, 593
        for diphenylarsinomethyllithium, 599
        for methylene bisphenylarsenic acid, 592
    synthesis of, 592
Methane, bis(diphenylphosphino)-, reactions of, 563
Methane, 1,1-bis(diphenylphosphino)-, reactions of, 550
Methane, bis(diphenylstibino)-, as starting material, for diphenylstibinomethyllithium, 600
Methane, bisdiphenylthiophosphinoyl-, reactions, with aldehydes, 1022
Methane, bis(ethanesulfonyl)-, condensation of, 864

Methane, bis(fluorooxy)difluoro-, reactions, with Dewar hexafluorobenzene, 46
Methane, bis(fluorosulfonyl)-, reactions, with carbodiimides, 866
Methane, bis(formylamino)-, reactions, with phosgene, 439
Methane, bis(methanesulfonyl)-, from, methyl methylsulfinylmethyl sulfide, 274
Methane, bismethoxazonyl-, as starting material, for *gem*-dimethoxazonylalkenes, 996
Methane, bis(methylchloroarsino)-, as starting material, for tetraarsacyclohexanes, 594
Methane, bis(methylsulfonyl)-, synthesis of, 272
Methane, bis(phenylseleno)-
    reactions
        with *n*-butyllithium, 400
        with lithium diisobutylamide, 401
Methane, bis(phenylsulfonyl)-
    metallation of, 274
    reactions
        with quinoline-1-oxide, 281
        with vinyl epoxides, 280
    as starting material, for bis(sulfones), 274
Methane, bis(phenylthio)bis(trimethylsilyl)-, reactions, with lithium naphthalenide, 945
Methane, (+)-(*S*,*S*)-bis(toluene-*p*-sulfinyl)-, condensation reactions of, 863
Methane, (*S*,*S*')-bis-*p*-tolylsulfinyl-, synthesis of, 270
Methane, bis(triiodogermyl)-
    as starting material, for bis(trimethylgermyl)methane, 629
    synthesis of, 628
Methane, bis(trimethylenedioxyboryl)-, deprotonation of, 634
Methane, bis(trimethylgermyl)-
    from, bis(triiodogermyl)methane, 629
    metallation of, 628
    synthesis of, 626
Methane, bis(trimethylgermyl)diazo-, photolysis of, 627
Methane, bis(trimethylsilyl)-
    metallation of, 620
    reactions, with sodium methoxide, 653
    synthesis of, 602
Methane, bis(trimethylsilyl)diazo-, thermolysis of, 612
Methane, bis(trimethylsilyl)phenyl-, synthesis of, 608
Methane, bromonitro-, reactions, with aldehydes, 798
Methane, bromotrichloro-
    reactions
        with butyllithium, 16
        with α-cholorovinyltrimethylsilanes, 758
        with enol sulfonates, 742
        with ethynyltin compounds, 813
        with ketene silyl acetates, 743
    use in the synthesis of dibromoalkanes, 21
    use in the synthesis of 1,1-dichloroalkenes, 738
    use in Wittig reactions, 738
Methane, bromotricyano-, reactions, with triethyl phosphite, 1008
Methane, bromotrifluoro-, reactions, with triphenylphosphine, 754
Methane, *t*-butoxybis(dimethylamino)-, reactions, with dithionomalonic esters, 896
Methane, chlorodifluoro-
    copyrolysis of, with pentafluorobenzotrifluoride, 737
    reactions, with (arylalkylidene)triphenylphosphoranes, 735
    as starting material, for difluorocarbene, 8
Methane, chlorofluoro-, synthesis of, 30
Methane, chloronitro-, synthesis of, 798
Methane, chlorotricyano-, reactions, with triethyl phosphite, 1008
Methane, diazo-
    as alkylating agent, 833
    as carbene sources, 669
    reactions

Methane

    with A-frame platinum complexes, 691
    with alkyne-bridged dirhodium complexes, 689
    with allylic selenides, 277
    with benzeneselenenyl thiocyanate, 335
    with chlorosulfenes, 82
    with 1-cyclobutenyldiphenylphosphine oxide, 488
    with dichloro(phenyl)thallium, 153
    with diethylaluminum iodide, 153
    with ditellurides, 290
    with hexachlorodisilane, 140
    with methyl 2-chloro-2-cyclopropylideneacetate, 109
    with N-phenylselenophthalimide, 335
    with selenenyl bromides, 291
    with sulfinyl halides, 76
    with tetrabromosilane, 140
    with tetrachlorosilane, 140
    with tetrakis(2,6-diethylphenyl)digermene, 628
    with thionyl chloride, 76
    with tin tetrabromide, 155
    with tin tetrachloride, 155
    with trichlorofon, 912
    with zinc benzoate, 385
    for the synthesis of methylene-bridged dimanganese complexes, 676
    use in the synthesis of allene-bridged dimolybdenum complexes, 673
    use in the synthesis of methylene-bridged diruthenium complexes, 681
    use in the synthesis of methylene-bridged iron nitrosyl complexes, 679
Methane, diazo di(trimethylgermanyl)-, flash vacuum pyrolysis of, 1055
Methane, dibismuthino-, synthesis of, 596
Methane, dibromo-
  reactions
    with aluminum, 698
    with (diethoxy)phenylphosphine, 127
    with isopropylmagnesium chloride, 147
    with phenylarsenic oxide, 592
    with sulfinate anions, 83
    with triphenylleadlithium, 699
  as starting material, for methylenedilithium, 668
Methane, dibromochloro-, reactions, with triphenylphosphine, 757
Methane, dibromodifluoro-
  reactions, with ketene silyl acetals, 736
  as starting material, for cyclopropanes, 1083
  use in the synthesis of difluorocarbenes, 7, 733
Methane, dibromodinitro-, reactions, with tetramethylethene, 995
Methane, dibromodi(trimethylsilyl)-, reactions, with magnesium, 618
Methane, dichloro-
  reactions
    with aluminum, 698
    with diphenylbismuthsodium, 596
    with diphenyl ditelluride, 93
    with indium(I) chloride, 153
    with lithium, 632
    with phenylmethylantimonysodium, 595
    with pyrrole, 410
    with sodium iodide, 28
    with trimethyltinsodium, 699
    with trivinyltinsodium, 699
  use in the synthesis of carbon-bridged diindium compounds, 698
  use in the synthesis of carbon-bridged diruthenium complexes, 678
  use in the synthesis of chlorofluoroalkanes, 30
Methane, dichlorobis(trimethylsilyl)-
  hydride reduction of, 619
  reactions, with lithium metal vapor, 618
Methane, dichlorodiphenyl-, use in the synthesis of diphenylmethylenedilithium, 668

Methane, dichlorofluoro-
  reactions, with potassium t-butoxide, 751, 753
  as starting material, for chlorofluorocarbene, 1083
Methane, dichlorophenyl-, use in the synthesis of phenylmethylenedilithium, 668
Methane, dideuteriodichloro-, as starting material, for dithioacetals, 249
Methane, difluorodiphenyl-, synthesis of, 4
Methane, diiodo-
  photolysis of, 668
  reactions
    with calcium, 669
    with dicarbonyl(μ-pyrazolyl)iridium, 685
    with germanium diiodide, 628
    with lithium phenyl telluride, 290
    with octacarbonyldiiron dianion, 679
    with sodium benzene sulfinate, 84
    with sodium phenyl sulfinate, 281
    with thiols, 249
    with tin(II) bromide, 699
    with zinc, 696
  as starting material, for methylenedilithium, 668
  synthesis of, 28
  use in the synthesis of carbon-bridged dimercury complexes, 696
  use in the synthesis of methylenebis(magnesium bromide), 669
Methane, diiododi(trimethylsilyl)-, reactions, with zinc, 618
Methane, 1,1-diisonitrilo-, synthesis of, 439
Methane, dilithio-
  reactions
    with bis(dimethylamino)boron chloride, 632
    with trimethylsilyl chloride, 603
    with trimethyltin chloride, 699
Methane, dilithiodi(trimethylsilyl)-, synthesis of, 618
Methane, (dimesitylboryl)-, metallation of, 634
Methane, dimesitylboryl(trimethylsilyl)-, synthesis of, 642
Methane, dimethoxy-
  reactions
    with benzeneselenol, 238
    with isocyanates, 310
    with trimethylsilyl iodide, 51
  as starting material, for methoxymethyl acetate, 211
Methane, dimethylaminotrimethylsilyloxy-, synthesis of, 295
Methane, (+)-(S)-(1,1-dimethylethanesulfonyl)toluene-p-sulfinyl-, Mannich reaction of, 864
Methane, dimethylgermyl(dimethylsilyl)-, reactions, with iron pentacarbonyl, 641
Methane, di(methylthio)-, as starting material, for dithioacetals, 250
Methane, dinitro-
  reactions
    with amido nitriles, 995
    with dimethylformamide dimethyl acetal, 995
    with formimidates, 994
  as starting material, for *gem*-dinitroalkenes, 994
Methane, diphenyldiazo-
  reactions
    with alcohols, 198
    with dimethyl chlorofumarate, 109
    with iminoboranes, 145
    with methyl 2-chloro-2-cyclopropylideneacetate, 109
  as starting material, for mixed dihaloalkenes, 755
Methane, di(phenyltelluro)-, reactions, with lithium diisopropylamide, 290
Methane, di(phenylthio)-, as starting material, for dithioacetals, 250
Methane, (diphenylthiophosphinoyl)phenyldiazo-, Pummerer rearrangement of, 346
Methane, germyl(silyl)-, reactions, with hydrogen chloride, 641
Methane, iodo-
  reactions

with allenes, 1024
with bisdiphenylphosphino ethene, 1024
with ozone, 42
Methane, lithiodichloro-, reactions, with ketones, 741
Methane, lithiotrichloro-, reactions, with *n*-heptylphenyl sulfone, 741
Methane, methoxy(arylthio)-, synthesis of, 218
Methane, methoxy(phenylthio)-
as starting material, for silanes, 370
synthesis of, 218
Methane, methylsulfonyldibenzoyl-, reactions, with 4-nitrobenzenediazonium tetrafluoroborate, 332
Methane, nitro-
alkylation of, 540
reactions, with diketones, 166
Methane, nitro(phenylseleno)-, reactions, with aldehydes, 910
Methane, nitro(phenylthio)-, reactions, with aldehydes, 906
Methane, phenylbis(diethyldiamino)-, reactions, with acetyl chloride, 455
Methane, 1-silyl-1-chloro-1-lithio-, reactions, with transition metal complexes, 658
Methane, tetrabromo-
electroreduction of, 38
reactions
with aldehydes, 745
with ketones, 745
with triphenylphosphine, 745
Methane, tetrachloro-
for the chlorination of benzyl sulfide, 64
reactions
with alkyldiphenylphosphines, 112
with bis(trimethylsilyl)aminophosphines, 572
with diamino(alkyl)phosphines, 116
with enamines, 742
with ketene silyl acetates, 743
with triphenylphosphine, 737
as starting material, for dichlorocarbene, 1083
Methane, tetra(dimethoxyboryl)-, use in the synthesis of 1,1-bisborylalkanes, 633
Methane, tetraiodo-, reactions, with aldehydes, 749
Methane, tetrakis(dimethylamino)-, as starting material, for ketene aminals, 975
Methane, tetranitro-
as nitrating agent, 429, 431
reactions
with hydrazines, 995
with nitroalkanes, 429
Methane, tosyl diazo-, reactions, with sulfonic acids, 235
Methane, tosylnitro-
alkylation of, 332
synthesis of, 331
Methane, trichlorofluoro-
reactions
with triphenylphosphine, 753
with tris(dimethylamino)phosphine, 753
as starting material, for chlorofluorocarbene, 1082
Methane, trichlorogermyl(trimethylsilyl)-, synthesis of, 636
Methane, tricyano-, potassium salt of, for the synthesis of primary α-halo enamines, 792
Methane, trifluoromethanesulfonyl(phenylthio)-, oxidation of, 273
Methane, tris(dimethoxyboryl)-, reactions, with methyllithium, 633
Methane, tris(ethylenedioxyboryl)-, use in the synthesis of 1,1-bisborylalkanes, 633
Methane, tris(trimethylsilyl)-
reactions, with sodium methoxide, 619
synthesis of, 1049
Methane anions, triboryl-, condensation of, 814
Methane anions, tris(phenylthio)-, reactions, with trialkylboranes, 373
Methane bissulfonic acid

from, iodoform, 281
as starting material, for methane disulfonyl chloride, 281
Methaneboronates, iodo-, reactions, with primary amines, 525
Methanedisulfonyl chloride
from, methane bissulfonic acid, 281
reactions, with ethanol, 282
Methanedisulfonyl fluoride, synthesis of, 282
Methanephosphonic acid, 1-dimethylamino-1-cyano-, as starting material, for α-phosphoryl enamines, 1004
Methanes, alkoxy tris(dimethylamino)-, as starting materials, for ketene aminals, 973
Methanes, arenesulfonyl diazo-, photolysis of, 235
Methanes, (benzoyloxy)iodo-, stability of, 58
Methanes, bis(arylsulfonyl)-, synthesis of, 272
Methanes, bis(dichloroarsino)-, synthesis of, 592
Methanes, bis(dithianyl)-, deprotonation of, 845
Methanes, bis(perfluoroalkanesulfonyl)-, as nucleophiles, 864
Methanes, bis(phenyltelluro)-, reactions, with butyllithium, 402
Methanes, 1,1-bis(silyl)-, synthesis of, 609
Methanes, bis(silyl) dichloro-, as starting materials, for tetrasilyl ethenes, 1050
Methanes, diaryldiazo-, reactions, with dichlorocarbene, 740
Methanes, diazo-, reactions, with phosphaalkynes, 479
Methanes, diboryldilithio-, synthesis of, 634
Methanes, dihalo-
displacement of, 190
reactions
with alkali metal salts, 147
with selenide anions, 88
with zinc, 152
use in the synthesis of methylene-bridged gold complexes, 695
Methanes, disulfonyl-
condensation reactions of, 864
reactions, with ortho esters, 865
as starting materials
for ketene dithioacetal tetroxides, 864
for vinyl ethers, 865
Methanes, germyldiazo-, synthesis of, 521
Methanes, halonitro-, reactions, with aldehydes, 798
Methanes, halo tricyano-, as starting materials, for *gem*-phosphorylamino phosphorylalkenes, 1008
Methanes, triacyl-, as starting materials, for ketene acetals, 824
Methanes, trihalo-
reactions
with alkoxides, 826
with phenoxides, 826
as starting materials
for ketene acetals, 826
for ketene aminals, 979
Methanes, trisilyl-, synthesis of, 1050
Methanesulfenic acid, elimination reactions of, 924
Methane sulfinate, sodium hydroxy-, for the reduction of organic halides, 234
Methane sulfinate, trifluoro-, displacement of, 309
Methanesulfinic acid, chloro-, from, chloromethanesulfinyl chloride, 78
Methanesulfinyl bromide, trichloro-, as starting material, for α-bromo sulfoxides, 77
Methanesulfinyl chloride, chloro-
as starting material, for chloromethanesulfinic acid, 78
synthesis of, 78
Methanesulfonanilides, fluoro-, synthesis of, 86
Methanesulfonate, chloro-, reactions, with phosphorus pentachloride, 85
Methane sulfonate, 2-methylpropyl bromo-, synthesis of, 86
Methanesulfonate, silver trifluoro-, as alkylating agent, 881
Methanesulfonates, alkyl-, reactions, with sodium hexamethyldisilazide, 605

Methanesulfonate salts

Methanesulfonate salts, reactions, with potassium tricyanomethanide, 977
Methanesulfonic acid, reactions, with dithianes, 252
Methanesulfonic acid, methylsulfonyl-, sodium salt of, synthesis of, 282
Methanesulfonyl azide, trifluoro-, as starting material, for α-aminoalkylphosphorus compounds, 491
Methanesulfonyl bromide, bromo-
  reactions
    with alkenes, 83
    with alkynes, 84
    with 2-methylpropanol, 86
    with silyl enol ethers, 84
  as starting material, for alkyl bromomethyl sulfones, 83
Methanesulfonyl bromide, chloro-
  reactions
    with alkenes, 82
    with alkynes, 82
  synthesis of, 85
Methanesulfonyl bromide, iodo-
  as starting material, for α-iodo sulfones, 84
  synthesis of, 85
Methanesulfonyl chloride
  reactions
    with alcohols, 780
    with 2-chloroethanol, 282
Methanesulfonyl chloride, bromo-, reactions, with ammonia, 86
Methanesulfonyl chloride, chloro-
  reactions
    with alkenes, 82
    with sodium sulfite, 85
  synthesis of, 85
Methanesulfonyl chloride, iodo-
  reactions
    with ammonia, 86
    with benzamidine, 86
  synthesis of, 85
Methanesulfonyl chloride, trifluoro-
  reactions, with vinyl benzoate, 56
  use in the synthesis of dichloroalkanes, 12
Methanesulfonyl halides, chloro-, as starting materials, for α-chloro sulfones, 82
Methanethiol, trifluoro-, reactions, with hexafluoro dithioate, 256
Methanethiosulfonate, S-methyl
  reactions
    with vinylcopper(I) compounds, 940
    with vinyl sulfones, 858
Methanide, tricyano-, as starting material, for ketene aminals, 977
Methanide salts, synthesis of, 978
Methanide salts, ammonium tricyano-, as intermediates, in the synthesis of ketene aminals, 977
Methanide salts, tricyano-, as intermediates, in the synthesis of ketene aminals, 971
Methanimine, α-bromo glycine diphenyl-, as starting material, for aminomethyl imines, 445
Methanimine, diphenyl-, as starting material, for bisimines, 438
Methanimine, N-methyl-, reactions, with nitrosyl chloride, 325
Methanimines, bis(methylthio)-, reactions, with carbanions, 901
Methanimines, diaryl-, Mannich-type reaction of, 445
Methaniminium chloride, N,N-dimethylchloro-, reactions, with diethyl tetrathiomalonate, 775
Methanol
  photochemical gas phase chlorination of, 42
  reactions
    with allenes, 736
    with bis(arenesulfonyl)alkynes, 840
    with β,β-diacylketene aminals, 979
    with potassium hydroxide, 748, 906
    with tetrafluoroethene, 5
    with 1,1,2,2-tetramorpholinoethane, 309
Methanol, chloro-, synthesis of, 42
Methanol, fluoro-, synthesis of, 42
Methanol, iodo-, synthesis of, 42
Methanol, phenyldimethylsilyl-, for the protection of anomeric hydroxyl groups, 176
Methanol, tributylstannyl-, reactions, with n-butyllithium, 378
Methanol–hydrochloric acid, reactions, with 4-methyl-2-trifluoromethyl-5(2H)-oxazolone, 305
Methanols, dialkylamino-, reactions, with tristrimethylsilylphosphine, 459
Methide anions, bis(dimethoxyboryl)-, synthesis of, 633
Methionines, phospho-, synthesis of, 466
Methoxide
  reactions
    with halomethylsilanes, 352
    with tetrafluoroborate salts, 892
Methylal
  reactions, with polyols, 185
  use in the synthesis of acetals, 185
Methylamine, reactions, with arenediazonium salts, 444
Methylamine, N-acetoxymethyl-N-nitro-, cleavage of, 106
Methylamine, N-methyl-N-nitrosochloro-, reactions, with thiols, 325
Methylamine, N-nitroso(1-methoxy-2-chloroethyl)-, from, N-nitroso(methyl)vinylamine, 896
Methylamine, trimethylsilyl-, synthesis of, 506, 512
Methylamine hydrochloride, reactions, with formaldehyde, 317
Methylamines, N-amino-, reactions, with acid halides, 99
Methylamines, bis(silyl)-, from, N,N-dimethylbenzamide, 511
Methylamines, N-dimethylamino-, as starting materials, for 3-acetyl-N-chloromethylindoles, 97
Methylamines, germyl-, synthesis of, 519
Methylamines, halo-, reactions, with metallosilanes, 510
Methylamines, N-hydroxy-, use in the synthesis of aminomethylarsines, 501
Methyl anions, α-silyl-, reactions, with transition metals, 655
Methyl carbanions, amino-, from, thioaminals, 528
Methyl chloride, methoxy-, reactions, with potassium, 360
Methyl chloride, trimethylsilyl-, reactions, with magnesium, 653
Methyl complexes, halo-, from, formylmolybdenum complexes, 148
Methyl compounds, cyano-, synthesis of, 509
Methyl compounds, hydroxy-, synthesis of, 1078
Methyl compounds, N-hydroxy-, as intermediates, in the synthesis of α-haloamides, 99
Methyl compounds, N-iodo-, synthesis of, 103
Methyl compounds, N-isocyanido-, synthesis of, 441
Methyl compounds, N-methoxy-, as starting materials, for N-chloromethyltriazacyclohexanes, 97
Methyl dianion, hydroxy-, synthon for, 383
Methylenating reagents, synthesis of, 669
Methylene compounds
  reactions
    with sulfinyl chloride, 266
    with N,N,N',N'-tetramethylmethylthioformamidinium iodide, 973
Methylene compounds, active, reactions, with thionyl chloride, 71
Methylene compounds, 1,1-bis(trimethylsilyl)-, synthesis of, 1049
Methylene compounds, diiodo-, synthesis of, 750
exo-Methylene compounds, synthesis of, 827
Methylene groups
  reactions
    with aryl cyanates, 889
    with orthocarbamate esters, 894

Methylene halides
 as starting materials
  for ethylene-bridged palladium complexes, 691
  for methylene-bridged palladium complexes, 691
Methylene transfer, for the synthesis of carbon-bridged dirhodium complexes, 686
Methyl groups, displacement of, 641
Methyl groups, aromatic, oxidation of, 206
Methyl halide, amino-, reactions, with azides, 444
Methyl halides, germyl-
 reactions
  with amines, 519
  with potassium cyanate, 521
Methyl halides, silyl-, reactions, with amides, 513
Methyl halides, trimethylstannyl-, reactions, with magnesium, 709
N-Methyl heterocycles, metallation of, 539
Methylide, dimethyloxosulfonium, as starting material, for enethiolates, 834
Methylide, lithium tribromo, synthesis of, 746
Methylide, phenoxyfluoroxonium (fluorosulfonyl), reactions, with aniline, 282
Methyl iodide
 reactions
  with α-hydrazino enamines, 986
  with thiosemicarbazides, 984
  with trimethyl phosphite, 485
Methyl iodide, tributylstannyl-
 displacement of iodide from, 387
 as starting material, for α-alkoxystannanes, 387
Methyl ligands, bis(trimethylsilyl)-, lithiation of, 656
Methyl ligands, silyl-, as starting materials, for transition metal complexes, 656
Methyllithium–lithium bromide, reactions, with ketals, 961
1-Methylthioethenamine, N,N-dimethyl-, use of term, 898
Methylthiomethyl (MTM) group, for the protection of carboxylic acids, 219
[[2-(Methylthio)phenyl]thio]methyl (MPTM) protecting group, development of, 219
Mevalonolactone, synthesis of, 167
Michael acceptors, from, 1,1,1-trifluoroacetone, 766
Michael addition
 of phosphites, to perfluoromethacrylate esters, 803
 to bis(sulfonyl)ethene, 277
Michael additions, to nitroalkenes, 431
Michael donors, reactions, with bisdiphenylphosphinoethene, 1028
Michaelis–Arbuzov reaction, for the synthesis of dialkyl alkyl phosphonates, 920
Michaelis–Becker reaction
 for the synthesis of α-aminoalkylphosphorus compounds, 464
 for the synthesis of α-aminophosphorus compounds, 487
Michael reaction, of α-aminosilanes, 510
Microwave irradiation, for the synthesis of dithioacetals, 257
Mitsunobu reaction
 for the synthesis of α-aminoalkylphosphorus compounds, 475, 491
 for the synthesis of thioglycosides, 225
(Molybdenacyclononatriene) complexes, 8-carbon-bridged, synthesis of, 676
Molybdenacyclopentadienes, synthesis of, 676
Molybdenum, dimeric allyls of, synthesis of, 673
Molybdenum(V) chloride, use in the synthesis of dimeric allyls of molybdenum, 673
Molybdenum alkyne complexes, reactions, with pentacarbonylrhenate, 710
Molybdenum carbene complexes, reactions, with phosphines, 580
Molybdenum chlorides, reactions, with 1,1-dicyanobis(trifluoromethyl)ethene, 800
Molybdenum complexes, as starting materials, for 1,3-disilacyclopentenes, 616
Molybdenum complexes, alkyne-bridged, synthesis of, 674
Molybdenum complexes, flyover 6-carbon-bridged, synthesis of, 676
Molybdenum complexes, formyl-, as starting materials, for halomethyl complexes, 148
Molybdenum complexes, (halomethyl)-, synthesis of, 147
Molybdenum complexes, phosphine-bridged, synthesis of, 675
Molybdenum compounds, reactions, with phosphorus ylides, 580
Molybdenum compounds, bromomethyl, synthesis of, 148
Molybdenum compounds, chloromethyl-, synthesis of, 148
Molybdenum halides, use in the synthesis of carbon-bridged dimolybdenum complexes, 672
Molybdenum hexacarbonyl
 reactions
  with phosphaferrocenes, 1026
  with 1,1,4,4-tetrakisdiphenylphosphino-1,3-butadiene, 1023
Molybdenum hexafluoride, use in the synthesis of difluoroalkanes, 10
Molybdenum pentoxide–pyridine–hexamethylphosphoramide complex, use in the synthesis of α-hydroxy sulfones, 234
Molybdenum–phosphinoalkene complexes, synthesis of, 1028
Molybdenum tetrachlorides, use in the synthesis of tris(cyclooctatetraene)dimolybdenum complexes, 673
Monoalkyne complexes, as starting materials, for ferracyclopentadienes, 682
Monoanils, benzil, reactions, with 2,4-bis(phenylthio)-1,3,2,4-dithiadiphosphetane-2,4-dithione, 320
Monoarsaferrocenes, synthesis of, 1039
Monobismaferrocenes, synthesis of, 1041
Mono(chloromercurio) compounds, synthesis of, 1067
Monoperphthalic acid, reactions, with aryl methoxymethyl sulfides, 235
Monophosphinic acids, synthesis of, 470
Monophosphonium salts, synthesis of, 555
Monoselenoacetals
 from
  acetals, 238
  carbonyl compounds, 238
  enol ethers, 238
  α-halo ethers, 239
Monoselenoacetals, ketene, synthesis of, 841
Monoseleno monotelluro compounds, syntheses of, 876
Monostibaferrocenes, synthesis of, 1041
Monosulfones, synthesis of, 854
Monosulfoxides, 2-aminoketene dithioacetal
 from, nitriles, 857
 as starting materials, for α-amino acids, 857
Monotelluroacetals
 from
  acetals, 238
  carbonyl compounds, 238
  enol ethers, 238
  α-halo ethers, 239
Monotelluroacetals, ketene, synthesis of, 841
Monothiatriselenafulvalenes, from, 1,3-diselenole 2-thiones, 869
Monothioacetal S,S-dioxides, ketene
 from
  1-sulfonylalkynes, 840
  α-sulfonyl ethers, 838
 synthesis of, via sulfoxidation, 838
 See also Alkoxyvinyl sulfones
Monothioacetals
 from
  acetals, 215
  O,O-acetals, 216
  aldehydes, 217
  carbonyl compounds, 215

Monothioacetals
α-halo ethers, 218
α-halo sulfides, 218
sulfoxides, 221
oxidation of, 232
as starting materials, for α-alkoxy sulfones, 235
synthesis of, 215
via oxidation of sulfides, 220
Monothioacetals, acyclic, synthesis of, 215
Monothioacetals, cyclic
from, carbonyl compounds, 229
synthesis of, 229
via cycloaddition reactions, 231
via intramolecular alkylation, 231
Monothioacetals, ketene
from
aldehydes, 834
ketene dithioacetals, 836
ketones, 834
monothiocarboxylic acids, 833
α,β-unsaturated O,S-acetals, 836
synthesis of
via Horner–Wittig reactions, 835
via Peterson alkenation, 834
Monothioacetals, ketene O-silyl, synthesis of, 833
Monothiocarboxylic acids, as starting materials, for ketene monothioacetals, 833
Montmorillonite KSF, as catalyst, for the synthesis of dithioacetals, 246
Morpholine
reactions
with benzaldehydes, 406
with 4-fluorobenzaldehyde, 298
with formaldehyde, 315
with glyoxal, 300
Morpholine, N-styryl-, reactions, with salicyl aldehyde, 307
Morpholine-N-oxide, N-methyl-, as reoxidant, 357
Morpholinones, reactions, with t-butoxy chloride, 102
Morpholinones, benzyloxycarbonyl-, bromination of, 102
Muconates, 3-fluoro-, cyclization of, 53
Multiphoton irradiation, for the synthesis of carbenes, 1083
Muscarine analogues, synthesis of, 232

Nafion-H
use in the synthesis of diacetates, 203
use in the synthesis of α-diols, 161
Naphthalene, 1,8-bis(trimethylsilylmethyl)-, reactions, with butyllithium, 603
Naphthoquinone compounds, synthesis of, 901
1-Naphthylamine, N-ethyl-N-methyl-, oxidative azidation of, 443
Nef reaction, of nitroalkanes, 540
Neodymium compounds, reactions, with phosphorus ylides, 578
Nickel, chlorocyclopentadienyl(triphenylphosphine)-
reactions
with 2-chlorolithioferrocene, 719
with tricarbonyl(lithiocyclopentadienyl)manganese, 719
Nickel, dicyclooctadiene-, as starting material, for alkyne-bridged binuclear nickel complexes, 692
Nickel, pentacarbonylcyclopentadienylcobalt-, reactions, with alkynes, 713
Nickel(II) bromide, reactions, with pentaphenylaluminacyclopentadiene, 693
Nickel(II) chloride, reactions, with 1,4-pentadiene, 694
Nickel acetylacetonate, use in the synthesis of ketene aminals, 975
Nickelasilacyclobutenes, as intermediates, 1053
Nickel–cobalt alkyne complexes, synthesis of, 713
Nickel complexes, synthesis of, 955
Nickel complexes, alkyne-bridged binuclear, from, dicyclooctadienenickel, 692
Nickel compounds
reactions
with 1-haloalkylphosphines, 583
with phosphorus ylides, 583
Nickel dinuclear carbonyls, as starting materials, for alkyne-bridged dinickel complexes, 692
Nickelocene
reactions, with fluoroboric acid, 694
as starting material, for alkyne-bridged dinickel complexes, 692
Ninhydrin
acylation of, 55
bromoacetylation of, 57
hemithioacetal of, 221
synthesis of, 160
Niobium, bis(cyclopentadienyl)hydrido-, reactions, with dicarbonylcyclopentadienylmethyliron, 717
Niobium bis(cyclopentadienyl)trihydrides, use in the synthesis of dimeric niobocene, 672
Niobium complexes, dinuclear, synthesis of, 672
Niobocene, dimeric, synthesis of, 672
Nitramine, N-chloromethyl-N-ethyl-, reactions, with sodium alkoxides, 313
Nitramine, cyclotrimethylene-, synthesis of, 434
Nitramines
Mannich-type reaction of, 442
reactions, with formaldehyde, 442
Nitramines, acylaminomethyl-, synthesis of, 447
Nitramines, alkyl-, reactions, with acylamines, 447
Nitramines, N-alkyl-N-chloromethyl-, displacement of, 327
Nitramines, aminomethyl-, synthesis of, 442
Nitramines, aminomethylene-, nitrosation of, 440
Nitrate, acetyl
as nitrating agent, 435
reactions, with 2-alkoxyvinyl phosphonic esters, 490
safety precautions, 208
as starting material, for α-aminoalkylphosphorus compounds, 490
Nitrate, ethyl, as starting material, for α-aminoalkylphosphorus compounds, 490
Nitrates, alkyl, as starting materials, for nitroalkanes, 428
Nitrenes, reactions, with alkenes, 515
Nitric acid
for the oxidation of dithioacetals, 260
reactions, with alkenes, 797
Nitric oxide, reactions, with iodotrifluoroethene, 798
Nitrile, dithyrea-, from, amides, 245
Nitriles
addition of amides to, 418
reactions
with alcohols, 883
with amidoalcohols, 425
with ethanol, 883
with phenylphosphine, 481
with trioxan, 425
as starting materials
for α-aminoalkylorganophosphorus compounds, 464
for 2-aminoketene dithioacetal monosulfoxides, 857
for α-chloroalkanesulfenyl chlorides, 72
for α-hydrazino enamines, 984
for ketene S,N-acetals, 905
for ketene aminals, 979
for ketene hemiaminals, 883
for α-silyl enamines, 1014
synthesis of, 512
Nitriles, amido-, reactions, with dinitromethane, 995
Nitriles, α-amino-, as starting materials, for N-acylaminals, 418
Nitriles, bisgermyl-, synthesis of, 625
Nitriles, α,α-digermylated, synthesis of, 625
Nitriles, α-germyl-, deprotonation of, 662
Nitrilimines, for the synthesis of phosphorus-containing heterocycles, 488
Nitrilium ions, as intermediates, in the synthesis of silyloxazolines, 519

Nitrites, alkyl, use in the synthesis of acetals, 193
Nitro-aldol reaction
  for the synthesis of *gem*-halo nitroalkenes, 798
  See also Henry reaction
α-Nitro anions, as intermediates, for the synthesis of nitroalkanes, 540
Nitro compounds, halogenation of, 108
Nitro compounds, aliphatic, halogenation of, 108
Nitro compounds, bromo
  from
    2-adamantanone oxime, 108
    triazines, 109
Nitro compounds, *gem*-chloro-, synthesis of, 108
Nitro compounds, *gem*-fluoro-, from, 2-nitropropane, 108
Nitro compounds, α-halo
  synthesis of, 108
    via addition of halogens to nitroalkenes, 108
    via halogenation, 108
    via halogenation of nitro compounds, 108
Nitroform, iodo-, reactions, with diazo compounds, 995
Nitrogen
  synthesis of functions, **967–1020**
    with another group 15 element, **451–504**
    one metal, **505–541**
    one metalloid, **505–541**
    with one phosphorus, 451
  two, synthesis of functions, **403–449**
Nitrogen compounds, electrophilic, as starting materials, for α-aminoalkylphosphorus compounds, 490
Nitrogen dioxide, reactions, with dichloroalkyne, 797
Nitrogen heterocycles, silylmethylation of, 519
Nitrogen nucleophiles, reactions, with dichloroalkyne, 792
Nitrogen ylides
  reactions
    with trialkylboranes, 522
    with triarylboranes, 522
Nitronates
  hydrolysis to carbonyl compounds, 541
  reactions, with thioketones, 232
  synthesis of, 540
Nitrone, *N*-benzyl-*C*-phenyl-, oxidation of, 312
Nitrone, *C*,*N*-diphenyl-, reactions, with 1-phenyl-1-pyrrolidinoethene, 307
Nitrones
  reactions, with thioketones, 232
  as starting materials, for *O*,*N*-acetals, 312
Nitrones, chiral, reactions, with dialkyl phosphites, 485
Nitrones, diaryl-, reactions, with 2,5-diphenyl-1,2,3-diazarsoles, 502
Nitrones, diphenyl-, reactions, with 1,2,4-diazarsoles, 502
Nitrones, *N*-glycosyl *C*-dialkoxyphosphoryl, as starting materials, for α-aminophosphonic acids, 489
Nitrosamines, metallation of, 529
Nitrosamines, aminomethyl, synthesis of, 442
Nitrosamines, α-chloroalkyl-, as intermediates, 312
Nitrosamines, methoxyalkyl, lithiation of, 529
Nitroso compounds, synthesis of, 314
Nitroso compounds, chloro, from, 2-bornanones, 107
Nitroso compounds, α-halo, synthesis of, via chlorination of oximes, 107
2-Nitroso compounds, 2-nitro-, from, cyclohexanone, 439
Nitrosyl chloride, reactions, with *N*-methylmethanimine, 325
Nitrous acid, reactions, with diiodoalkyne, 797
Nitroxides, β-phosphorylated, synthesis of, 485
Nitryl fluoride, reactions, with bis(trimethylsilyl)alkyne, 807
Norbornadiene compounds, thermolysis of, 1081
Norbornadiene–molybdenum tetracarbonyl, reactions, with phosphaferrocenes, 1026
Norbornane, tetranitro-, synthesis of, 429
Norbornane epoxides, fragmentation of, 169
Norbornenes, bissilylated, synthesis of, via Diels–Alder reactions, 619

Norcaradiene, synthesis of, 557
Norcarane, 7,7-dichloro-, synthesis of, 16
Norrish type I fission, of acetone compounds, 1080
Norrish type II process, involving 1,5-hydrogen transfer, 1079
Nucleofuges, uses in photochemical processes, 1075
Nucleophiles
  reactions
    with acylsilanes, 930
    with acylstannanes, 951
    with alkyl halides, 761
    with alkylidene halides, 761
    with (bromomethyl)chlorosilanes, 140
    with chloro(chloromethyl)silanes, 140
    with *N*-α-haloalkylamides, 96
    with halo(halomethyl)silanes, 140
    with 5-halo-4-isothiazolin-3-ones, 775
    with ortho esters, 192, 193
    with perfluoroalkenes, 1076
    with sulfines, 859
    with trichloro(chloromethyl)germane, 142
    with vinyl ethers, 306
    with ynamines, 885
Nucleophilic displacement reaction, for the synthesis of aminals, 417
Nucleosides, fluoro, synthesis of, 79

Octacarbonyl complexes, bis(alkyne)-, synthesis of, 682
Octalins, metallation of, 532
Octane, 1,1-difluoro-, synthesis of, 9
Octyne, 1-chloro-, reactions, with tributyltin hydride, 812
1-Octyne, 1-ethoxy-, reactions, with iodotrimethylsilane, 765
1-Octyne, 1-trimethylsilyl-, reactions, with allylzinc bromide, 809
Organic reaction prediction software (OGOR2), for the study of enamine hydroboration, 524
Organoalanes, synthesis of, 659
Organoaluminums
  reactions
    with bromine, 758
    with iodine monochloride, 758
Organoantimony alkoxides, reactions, with aldehydes, 198
Organoboranes
  carbonylation of, 363
  reactions
    with alkynyltin compounds, 665
    with (phenyldimethylsilyl)chloromethyllithium, 357
    with (trimethylsilyl)bromomethyllithium, 357
Organoboranes, α-lithio-, synthesis of, 664
Organocerium compounds, synthesis of, 954
Organocopper compounds
  reactions
    with 1-ethynylsilanes, 940
    with trimethyl silylethyne, 809
    with triphenyl silylethyne, 809
Organocuprates
  reactions
    with alkynes, 772
    with alkynic sulfides, 963
    with ortho esters, 192
Organocuprates, chiral α-alkoxy-, synthesis of, 385
Organolithium compounds
  from, tributylstannanes, 652
  for halogen–lithium exchange reactions, 652
  reactions
    with 1,1-bis(silyl)alkenes, 618
    with caprothionolactone, 226
    with chloro(vinyl)silanes, 651
    with *gem*-diborylalkanes, 665
    with 1,1-dithio-1,3-dienes, 254
    with dithioesters, 255
    with silyl electrophiles, 602
    with thionolactones, 226

Organolithium compounds

    with trimethyl(vinyl)silane, 651
    with triphenyl(vinyl)silane, 651
    with α,β-unsaturated amidate anions, 651
    with vinylgermanes, 663
    with vinylsilanes, 651
    as starting materials, for α-aluminosilylalkanes, 659
    for the synthesis of 1,1-bis(silyl)alkanes, 602
    use in the synthesis of carbon-bridged copper and silver compounds, 695

Organolithium compounds, α-alkoxy-, synthesis of, 378
Organolithium nucleophiles, reactions, with vinyl sulfides, 366
Organolithiums
  reactions
    with ceric chloride, 954
    with α-(phenylthio)vinylsilanes, 366
    with thioketones, 392
    with vinyl selenides, 375, 400
    with vinyl selenones, 400
    with vinyl selenoxides, 400
    with vinylsilanes, 366
    with vinyl sulfides, 392

Organomagnesium compounds
  as starting materials, for α-aluminosilylalkanes, 659
  use in the synthesis of carbon-bridged copper and silver compounds, 695
  use in the synthesis of trifluorovinylsilanes, 806

Organomagnesium compounds, α-germyl, as starting materials, for α-thiogermanes, 371

Organomercurials
  reactions
    with aluminum, 698
    with iodine monochloride, 750

Organomercury carbenoid, reactions, with 1,1,3,3-tetramethyl-1,3-disilacyclobutane, 623
Organomercury compounds, α-alkoxy-, synthesis of, 385
Organomercury halides, reactions, with diethylvinyl phosphonate, 587

Organometallic compounds
  quenching of, 757, 760
  reactions
    with bisphosphinoalkenes, 1029
    with dicarbonyl compounds, 166
    with dithioesters, 255
    with ortho esters, 192
    with β-silylvinyl sulfones, 393
    with sulfines, 272

Organometallics
  halogenation of, 774
  reactions
    with 1,3-dichloro-1,3-dimethyl-1,3-disilacyclobutane, 621
    with phenyl (1-trimethylsilyl)propadienyl sulfide, 946

Organometallics, anionic, reactions, with halobenzene–tricarbonylchromium complexes, 722
Organometal–mercury compounds, synthesis of, 724
Organophosphorus compounds, reviews, 451
Organophosphorus compounds, α-amino-, synthesis of, via Mannich type reactions, 452
Organophosphorus compounds, α-aminoalkyl-
  from
    azines, 460
    hydrobenzamide, 460
    ketazines, 460
    nitriles, 464
    oximes, 464
    α-oxophosphorus compounds, 471
    Schiff's bases, 461
  synthesis of
    via conjugate addition, 460
    via Hofmann reactions, 472
    via Kabachnik–Fields reactions, 469

Organophosphorus compounds, α-nitroalkyl-, synthesis of, 493

Organosilicon compounds, from, α-halosilanes, 136
Organosodium compounds, use in the synthesis of 1,1-bis(silyl)alkanes, 609
Organostannanes
  reactions
    with n-butyllithium, 964
    with cupric chloride, 774
  transmetallation of, 390, 652

Organotin compounds
  reactions, with ceric ammonium nitrate, 252
  synthesis of, 755
Organotin compounds, cyclic, synthesis of, 665

Organozinc compounds
  reactions, with vinyl iodides, 954
  synthesis of, 665

Orthoacetate, trimethyl, reactions, with aluminum compounds, 192
Orthoacetates, trialkyl, condensation of, 912
Orthocarbonate, diethylene, use in the synthesis of ethylene acetals, 190

Ortho esters
  as dehydrating agents, for the synthesis of acetals, 177
  elimination of alcohols from, 825
  irradiation of, for oxygen–carbon bond fission, 1075
  reactions
    with aldehydes, 190
    with alkynes, 192
    with allylsilanes, 192
    with disulfonylmethanes, 865
    with enolate compounds, 192
    with Grignard reagents, 192
    with ketones, 190
    with nucleophiles, 192, 193
    with organocuprates, 192
    with organometallic compounds, 192
    with trichloro-acetic acid, 1081
    with trifluoro-acetic acid, 1081
  as starting materials
    for acetals, 192
    for ketene acetals, 825
    for ketene aminals, 975, 979

Ortho esters, α-bromo, reactions, with sodium sand, 827
Ortho esters, cyclic, ring-opening of, 827
Ortho esters, α-halo, as starting materials, for ketene acetals, 827

Orthoformate, triethyl
  reactions
    with dichlorophosphines, 126
    with hydroxyalkylphosphorus compounds, 342

Orthoformates
  dimerization of, 830
  reactions
    with aldehydes, 190
    with aluminum compounds, 192
    with Grignard reagents, 192
    with ketones, 190
    with phosphorus(III) halides, 340
    with Reformatsky reagents, 193

Orthoformates, (triseleno)-, reactions, with enol ethers, 289
Osmacyclopentadiene complexes, synthesis of, 681
Osmium, tetracarbonyldimethyl-, reactions, with 1,4-dibromocyclooctatetraene, 681
Osmium, tetracarbonylethyne-, reactions, with carbon(η^5-cyclopentadienyl)phosphinerhodium, 710
Osmium complexes, carbon-bridged trinuclear, synthesis of, 679
Osmium complexes, (α-chloromethyl)-, synthesis of, 150
Osmium complexes, trinuclear allyl-bridged, synthesis of, 683
Osmium complexes, trinuclear carbon-bridged, synthesis of, 682
Osmium complexes, trinuclear carbon-bridged benzyne, synthesis of, 682

Osmium compounds, methylene-bridged,
 synthesis of, 683
Osmium tetroxide
 for the cleavage of alkanes, 168
 reactions, with vinylsilanes, 357
Osmylation
 of enol ethers, 174
 for the synthesis of hemiacetals, 168
Oxadiazines, synthesis of, 886
Oxadiazines, dihydro, synthesis of, 982
1,3,5-Oxadiazines, tetrahydro-, synthesis of, 428
1,3,5-Oxadithianes, synthesis of, 229
Oxaloyl chloride, use in the synthesis of dichloroalkanes, 18
Oxaphospholanes, synthesis of, 339
1,3-Oxaphospholanes, synthesis of, 337
1,3-Oxaphospholene oxides, synthesis of, 344
1,2-Oxaphospholenes, 2-oxo-, synthesis of, 129
1,3-Oxaphosphorinanes, synthesis of, 337
1,3-Oxathiafulvenes, synthesis of, 835
1,3-Oxathiane dioxides, synthesis of, 235
1,3-Oxathianes
 as chiral auxiliaries, 230
 from
  1,3-hydroxy thiols, 229
  10-mercapto-exo-borneol, 231
 synthesis of, 229
1,3-Oxathianes, 2-acyl-, hydride reduction of, 231
1,3-Oxathian-4-ones, synthesis of, 230
4H-1,5,2-Oxathiazines, synthesis of, 232
1,4,2-Oxathiazolidines, synthesis of, 232
1,3,4-Oxathiazolines, synthesis of, 232
1,4,2-Oxathiazolines, synthesis of, 232
4H-1,3-Oxathiins, synthesis of, 232
Oxathiolane dioxides
 metallation of, 236
 synthesis of, 235
Oxathiolane nucleosides, synthesis of, 231
Oxathiolanes
 from, hydroxythiols, 229
 synthesis of, 229
1,3-Oxathiolanes
 from, 1,3-hydroxy thiols, 229
 synthesis of, 229
1,3-Oxathiolan-5-ones
 reactions, with m-chloroperbenzoic acid, 837
 as starting materials, for sulfoxides, 837
 synthesis of, 230
Oxathiolanyl nucleosides, synthesis of, 229
Oxazaborolines, from, cyanoborate salts, 523
1,3,5-Oxazaphosphanes, synthesis of, 455
Oxazaphospholines, synthesis of, via cycloaddition
  reactions, 493
Oxazepinones, from, chloro β-lactams, 824
Oxazine, tetrahydro-, synthesis of, 306
1,3-Oxazines
 as starting materials, for gem-imino
  phosphiminoalkenes, 996
 synthesis of, 302, 304
1,3-Oxazines, 2-lithio(silyl)methyl-, synthesis of, 649
1,3-Oxazines, N-substituted tetrahydro-, reactions, with
  chlorodiphenylphosphine, 468
1,3-Oxazines, tetrahydro-, synthesis of, 302, 306
2H-1,3-Oxazines, 5,6-dihydro-, synthesis of, 313
4H-1,3-Oxazines, 5,6-dihydro-, reduction of, 306
1,3-Oxazin-6-ones
 reactions, with butylamine, 981
 as starting materials, for α-ureido enamines, 981
Oxaziridine, $trans$-2-phenylsulfonyl-3-phenyl-,
  synthesis of, 312
Oxaziridines
 from, imines, 312
 for the oxidation of dithianes, 262
Oxaziridines, perfluoro-, synthesis of, 104

Oxazoles
 reduction of, 306
 as starting materials, for ketene aminals, 979
Oxazoles, 2,5-dihydro-, synthesis of, 313
Oxazoles, 4,5-dihydro-, reduction of, 306
Oxazolidines
 lithiation of, 531
 synthesis of, 301
Oxazolidinones, lithiation of, 531
Oxazolines
 from, isocyanides, 538
 lithiation of, 531
 as starting materials, for gem-formamido
  phosphorylalkenes, 1005
Oxazolines, silyl-, synthesis of, 519
Oxazolinethiones, synthesis of, 538
Oxazolinium salts, reactions, with aryl Grignard
  reagents, 306
5(2H)-Oxazolone, 4-methyl-2-trifluoromethyl-, reactions,
  with methanol–hydrochloric acid, 305
5(4H)-Oxazolones, 4-substituted 2-trifluoromethyl-,
  reactions, with ethanethiol, 320
Oxazolo[3,4-c]oxazoles, synthesis of, 302
Oxidation, for the synthesis of α-diols, 161
Oxidation, electrochemical, of bis(sulfones), 277
Oxime, 2-adamantanone, as starting material, for bromo
  nitro compounds, 108
Oximes
 chlorination of, 107
 oxidation of, 108
 oxidative nitration of, 429
 reactions, with phosphinic acid, 464
 reduction of, for the synthesis of α-aminoalkyl
  phosphonates, 472
 as starting materials
  for O,N-acetals, 313
  for N-alkoxy-α-amino phosphonates, 485
  for α-aminoalkylorganophosphorus compounds, 464
  for pseudonitroles, 439
Oximes, α-chloro, as starting materials, for gem-chloro
  nitrosoalkenes, 798
Oximes, cyclobutanone, reactions, with chlorine, 108
Oximes, furan-2-carboxaldehyde, reactions, with triethyl
  phosphite, 485
Oximes, hydroxyiminoethyl phosphonate, reactions, with
  chlorodiphenylphosphine, 1008
Oxindoles, arylimino-, reactions, with trialkyl
  phosphites, 464
Oxirane, triphenylsilyl-, reactions, with s-butyllithium, 944
Oxiranes
 cleavage of, 208
 reactions
  with diphenylketene, 829
  with magnesium bromide etherate, 783
 synthesis of, 146
Oxiranes, silyl-
 reactions
  with acetonitriles, 519
  with amines, 506
  with N-silylamides, 513
 as starting materials, for β-hydroxy-α-silyl azides, 516
 use in the synthesis of α-aminosilanes, 508
cis-Oxiranes, as starting materials, for α-aminosilanes, 508
Oxo acid chlorides, as starting materials, for
  (chloromethyl)thiophosphinic chlorides, 127
Oxone, for the oxidation of dithioacetals, 272
Oxonium ions, as intermediates, in the synthesis of
  acetals, 177
2-Oxopenams, alkylation of, 833
α-Oxophosphinates, as starting materials, for
  α-aminoalkylphosphinates, 471
Oxovanadium(V), for the ring-opening of
  dichlorocyclobutenes, 743

Oxygen

Oxygen
   and sulfur, synthesis of functions, 833
   synthesis of functions
      with another chalcogen, **215–241**
         containing selenium or tellurium, 841
      synthesis of functions bearing two oxygens, **159–214**, 824

Oxygen nucleophiles
   reactions
      with halomethylgermanes, 362
      with halomethylsilanes, 352

Ozone
   reactions
      with iodomethane, 42
      with selenides, 241

Ozonides
   hydrogenolysis of, 162
   synthesis of, 214

Palladium(0), reactions, with tricarbonyl(chlorobenzene)chromium, 722
Palladium(0) catalysis, for the synthesis of bis(sulfones), 279
Palladium(II) catalysts, for the synthesis of α-haloalkenylsilicon compounds, 810
Palladium(II) salts, tetrachloro-, reactions, with sulfur ylides, 396
Palladium complexes, synthesis of, 955
Palladium complexes, allyl-bridged
   from, cyclopentadienylpalladium compounds, 693
   platinum analogues, synthesis of, 693
   synthesis of, 693
Palladium complexes, bisallyl-bridged, synthesis of, 693
Palladium complexes, bis(cyclopentadienyl)-bridged, synthesis of, 693
Palladium complexes, cationic trinuclear, synthesis of, 692
Palladium complexes, ethylene-bridged, from, methylene halides, 691
Palladium complexes, (halomethyl)-, synthesis of, 150
Palladium complexes, methylene-bridged, from, methylene halides, 691
Palladium complexes, trinuclear, from, triarylcyclopropenium bromides, 693
Palladium complexes, trinuclear carbon-bridged, from, thallium acetylacetonate, 693
Palladium compounds
   reactions
      with 1-haloalkyl phosphonates, 583
      with phosphorus ylides, 583
Palladium compounds, cyclopentadienyl-, as starting materials, for allyl-bridged palladium complexes, 693
Palladium compounds, α-sulfinyl-, synthesis of, 396
Palladium–phosphinoalkene complexes, synthesis of, 1023, 1028

Paraformaldehyde
   reactions
      with diethyl phosphite, 338
      with hypophosphorous acid, 125
      with perfluoroalkyldichlorophosphines, 125
      with N-phenylglycine, 302
      with phosphinic acid chlorides, 128
      with γ-thiol amides, 325
   as starting material
      for α-fluoro ethers, 44
      for primary chloro sulfides, 69

Paraldehyde, as starting material, for secondary chloro sulfides, 69
Penicillacylases, use in the synthesis of α-aminophosphonic acids, 462
Penicillanic acid, 6-amino-, synthesis of, 898
Penicillanic acid S,S-dioxide, synthesis of, 898
Penicillin, synthesis of, 897
Penicillin compounds, 6,6-diacylamino-, synthesis of, 426
Penicillin G acylase, use in the synthesis of N-acyl-α-aminosilanes, 513

Pentaarsacyclopentane, pentakis(chloromethyl)-, synthesis of, 135
Pentacarbonyls, reactions, with acylferrocenes, 720
1,4-Pentadiene, reactions, with nickel(II) chloride, 694
Pentadienoic acid nitriles, perchloro-, as starting materials, for triphenoxy compounds, 829
Pentafluoroethyl radical, synthesis of, 1080
Pentan-2,4-dione, reactions, with mercury(II) chloride, 697
Pentane-2,4-dione, 1,1,1-trifluoro-, as starting material, for α-bromo alcohols, 42
Pentan-3-one, decafluoro-, irradiation of, 1080
Pentan-3-one, 2,4-dimethyl-2-thiol-, reactions, with formaldehyde, 317
Pent-1-ene, 4,4-dichloro-, synthesis of, 14
Pentenes, 1,1-bis(silyl), synthesis of, 1050
Pent-4-enylzinc–magnesium compounds
   reactions
      with chlorotrimethylstannane, 724
      with copper(I) cyanide, 724
Pentyne, 1-trimethylsilyl-, hydrostannation of, 1062
Pent-1-yne, 1-(phenylseleno)-, reactions, with bromine, 786
Peracetate, t-butyl
   reactions
      with sulfides, 220
      with tetrahydrothiophene, 226
Peracetic acid, trifluoro-
   reactions
      with $trans$-2,3-dibromobut-2-ene, 24
      with dienes, 46
Peracids
   as oxidation reagents, 854
   reactions, with enol esters, 209
Perbenzoic acid, m-chloro- (mcpba)
   for the oxidation of α-chloro sulfides, 74
   for the oxidation of α-fluoro sulfides, 73
   reactions
      with 2,2-bis(methylthio)-1,3-diphenylpropane, 268
      with imines, 104
      with 1,3-oxathiolan-5-ones, 837
   use in monoxidation reactions, 853
   use in the synthesis of α-alkoxy sulfoxides, 232
   use in the synthesis of 1-benzenesulfinyl-1-nitro-2,2-diphenylethene, 907
   use in the synthesis of α-chloro sulfones, 80
Perchlorate, triethylsilyl, as silylating agent, 831
Perchlorates, safety precautions, 250
Perchlorates, acylium, reduction of, 425
Perchlorates, iminium
   hydrolysis of, 987
   as starting materials, for α-aminoalkylideneamino enamines, 987
Perchlorate salts, synthesis of, 987
Perchloryl fluoride
   as fluorinating agent, 3, 754
   for the fluorination of carbon–hydrogen bonds, 780
Perfluoro alkoxides, reactions, with formyl fluoride, 59
Perfluoroalkyl transfer, selectivity reversal, 810
Perfluorophenyl groups, in anionic platinum and palladium complexes, 691
Perillane, synthesis of, 255
Perkow reaction, for the synthesis of enol phosphates, 920
Peroxide, benzoyl, as catalyst, for the synthesis of α-haloalkenylsilicon compounds, 811
Peroxide, bis(trimethylsilyl)
   as oxasilylating agent, 358
   use in the synthesis of α-hydroxy sulfones, 234
Peroxide, dibenzoyl
   reactions
      with alcohols, 220
      with ethyl phenyl selenide, 240
Peroxide, di-t-butyl
   as catalyst, for the synthesis of α-haloalkenylsilicon compounds, 810

irradiation of, 1078
Peroxides, acyl, reactions, with ethers, 207
Peterson alkenation
  of α-silyl α-sulfonyl ethers, 838
  for the synthesis of ketene monothioacetals, 834
  use in the synthesis of ketene dithioacetals, 849
  use in the synthesis of α-thiosulfoxides, 263
Peterson elimination, for the synthesis of gem-halo phosphonoalkenes, 801
Peterson reaction
  for condensation, 936
  for the synthesis of phosphorylsulfines, 926
  for the synthesis of silylalkenes, 805
  for the synthesis of sulfur functions, 925
  for the synthesis of tin compounds, 962
Phase-transfer catalysts, for the elimination of hydrogen halide from α-haloacetals, 826
Phase-transfer environments, carbenes in, 1080
Phenacyl bromides, as starting materials, for arylglyoxal hydrate compounds, 161
Phenol, o-amino-, use in the synthesis of α-halophosphinic acid compounds, 128
Phenol, 2,4,6-tribromo-, bromination of, 23
Phenols
  cyclization of, 167
  reactions
    with 1-chloro-1-dialkylaminoalkenes, 889
    with trichloroethylene, 761
    with ynamines, 885
Phenols, hindered, reactions, with 1-alkoxyalkynes, 827
Phenoxides
  reactions
    with α-bromo esters, 199
    with α-chloroalkyl isocyanates, 314
    with 1,1-dinitro-2,2-diphenylethylene, 898
    with trihalomethanes, 826
Phenyl groups, displacement of, 640
Pheromones, synthesis of, 1079
Phosgene
  as electrophile, for dithiocarboxylic enethiolates, 842
  reactions
    with aldehydes, 58
    with amides, 971
    with bis(formylamino)methane, 439
    with chloral, 58
    with N-methylene-2,6-diethyl benzenamine, 102
    with triethylamine, 790
Phosgeniminium salts, reactivity of, 794
Phosphaalkenes
  cycloaddition reactions of, 112
  as dienophiles, 112
  dimerization of, 549, 558
  from, α-halophosphines, 111
  as intermediates
    in the synthesis of amino(α-haloalkyl)phosphines, 116
    in the synthesis of chlorotetrahydrophosphorins, 111
  reactions
    with amines, 116
    with chlorocarbenes, 115
    with diazoalkanes, 484
    with o-quinones, 340
  trimerization of, 558
Phosphaalkenes, cyclopropenyl, photochemical rearrangement of, 341
Phosphaalkenes, perfluoro, reactions, with dienes, 112
Phosphaalkenes, 2-silyl, addition to, 566, 572
Phosphaalkynes
  reactions of, 549
  reactions
    with chlorocarbenes, 110
    with α-diazo carbonyl compounds, 479
    with diazomethanes, 479
Phosphaallenes, cyclodimerization of, 1026

Phosphabenzene, 3,5-diphenyl-, reactions, with diazo compounds, 549
Phosphabenzene, 2,4,6-triphenyl-, use in the synthesis of phosphacyclohexadienyl ligands, 1036
2-Phosphabicyclo[2.2.1]heptenes, from, cyclopentadiene, 112
Phosphabicyclooctanes, synthesis of, 339
Phosphacyclohexadienyl ligands, synthesis of, 1036
Phosphacymantrenes, synthesis of, 1037
Phosphaferrocenes
  reactions
    with electrophiles, 1035
    with molybdenum hexacarbonyl, 1026
    with norbornadiene–molybdenum tetracarbonyl, 1026
  synthesis of, 1026, 1032, 1034
Phosphaferrocenes, acetyl-, synthesis of, 1035
Phosphaferrocenes, tetraphenyl-, synthesis of, 1034
Phosphametallocenes, synthesis of, 1034
1,3-Phosphanes, synthesis of, 453
1H-1,3-Phospharsoles, 1-substituted 2-ethylidene-2,3-dihydro-4-methyl-3-phenyl-, synthesis of, 1030
1-Phospha-4-stannabenzene, 1,4-dihydro-, synthesis of, 1038
1H-1,3-Phosphastannole, 1-t-butyl-2-ethylidene-2,3-dihydro-4-methyl-3,3-di-n-butyl-, synthesis of, 1038
Phosphate, diethylvinyl, bromination of, 768
Phosphate, tri(2-chloroethyl), reactions, with chloroacetyl chloride, 768
Phosphates, as starting materials, for Se,P-acetals, 348
Phosphates, alkenyl, reactions, with ylides, 737
Phosphates, bromo enol, synthesis of, 768
Phosphates, 1-chlorothiol, synthesis of, 72
Phosphates, enol
  from, chlorodifluoromethyl ketones, 735
  synthesis of, 836, 920
Phosphates, α-phosphorylated enol, from, α-halo acid halides, 920
Phosphates, 1-sulfinylvinyl, synthesis of, 837
Phosphates, 1-sulfonylvinyl, synthesis of, 838
Phosphates, 1-thiovinyl, oxidation of, 837, 838
1,3-Phosphepanes, synthesis of, 453
Phosphetanes, synthesis of, 114
Phosphetan-1-oxide, reduction of, 114
Phosphinate, ethyl(chloromethyl)methyl, synthesis of, 126
Phosphinate, methyl chloromethyl, synthesis of, 125
Phosphinate, methyl(chloromethyl)methyl, synthesis of, 126
Phosphinate, phenyl bis(chloromethyl), reactions, with potassium iodide, 127
Phosphinate anions, α-fluoroalkyl, alkylation of, 131
Phosphinates
  from
    phosphinic chlorides, 126
    vinylphosphorus compounds, 561
Phosphinates, 2-alkoxyalkenyl, as starting materials, for α-nitrophosphinic acid, 490
Phosphinates, 1-aminoalkyl, from, ketimines, 463
Phosphinates, α-aminoalkyl
  from
    α-oxophosphinates, 471
    phosphorus esters, 464
Phosphinates, α-aminobenzyl, synthesis of, 461
Phosphinates, bis(chloromethyl), synthesis of, 127
Phosphinates, bromomethyl, synthesis of, 127
Phosphinates, chloromethyl, reactions, with aldehydes, 126
Phosphinates, ethyl, as starting materials, for phosphinic chlorides, 125
Phosphinates, ethyl(diethoxy)methyl, synthesis of, 126
Phosphinates, methyl chloromethyl(1-hydroxyalkyl), synthesis of, 126
Phosphine, 1-azidocycloprop-2-ynylmethyl)-, synthesis of, 483
Phosphine, bis(hydroxymethyl)-, reactions, with N^1,N^2-dialkylhydrazines, 482

Phosphine

Phosphine, *t*-butyldiethynyl-, reactions, with
    phenylphosphine, 1027
Phosphine, *t*-butyldipropynyl-
  reactions
    with di-*n*-butyltin hydride, 1038
    with phenylarsine, 1030
Phosphine, chlorodiethyl-, reactions, with acetic
    anhydride, 911
Phosphine, chloro(diethyl)-, reactions, with ketones, 120
Phosphine, chlorodiphenyl-
  reactions
    with carbon suboxide, 550
    with formaldehyde, 119
    with hydroxyiminoethyl phosphonate oximes, 1008
    with 3-lithio-1-diphenylphosphinopropyne, 1023
    with *N*-substituted tetrahydro-1,3-oxazines, 468
Phosphine, dibromo(bromomethyl)-, synthesis of, 115
Phosphine, (dichloromethyl)-, reactions, with
    2,3-dimethylbutadiene, 111
Phosphine, dichloromethylenetriphenyl-, from,
    dichlorocarbene, 737
Phosphine, dichlorophenyl-, reactions, with *N*-methyl(*o*-
    hydroxybenzylidene)amine, 478
Phosphine, dichloro(phenyl)-, reactions, with phosphine
    sulfide, 114
Phosphine, (diethoxy)phenyl-, reactions, with
    dibromomethane, 127
Phosphine, difluoro(iodomethyl)-, synthesis of, 115
Phosphine, dimethyl-, reactions, with
    hexafluoropropene, 802
Phosphine, 2,2-dimethylpropylidynyl-, reactions, with
    diazocyclohexanes, 480
Phosphine, diphenyl-
  reactions
    with 1,2-bisdiphenylphosphino ethyne, 1023
    with *N*-fluorosulfonyl imines, 481
    with quaternary ammonium salts, 483
Phosphine, diphenyltrimethylsilyl-, reactions, with *gem*-
    dihaloalkenes, 997
Phosphine, diphenyl(trimethylsilyl)-, reactions, with
    trifluoromethyl ketones, 338
Phosphine, phenyl-
  reactions
    with *t*-butyldiethynylphosphine, 1027
    with *t*-butyldipropynylphosphorine, 1026
    with imines, 455
    with nitriles, 481
Phosphine, phenyldichloro-, reactions, with
    N-phenylurea, 488
Phosphine, tribromo-, reactions, with 1-ethoxyethyne, 914
Phosphine, tributyl-
  reactions
    with 1-ethoxyethyne, 914
    with trichloroacetamides, 791
  use in Wittig reactions, 732
Phosphine, trichloro-, use in the synthesis of
    α-(benzyloxycarbonylamino)benzylphosphonic
    acids, 467
Phosphine, trimethyl-
  use in the synthesis of carbon-bridged dichromium
    complexes, 672
  use in the synthesis of trismethylenediruthenium
    complexes, 680
Phosphine, triphenyl-
  reactions
    with aldehydes, 745, 749
    with bromotrifluoromethane, 754
    with 2-decalone, 745
    with 3,3-diazido-2-cyanacrylate, 993
    with dibromochloromethane, 757
    with 1,2-dibromo-1-ethoxyethane, 912
    with 1,3-dithiole-2-thiones, 850
    with α-haloalkyl aryl selenides, 347
    with ketones, 745
    with methyleneiminium salts, 487
    with tetrabromomethane, 745
    with tetrachloromethane, 737
    with trichlorofluoromethane, 753
  use in the synthesis of 1,1-dichloroalkenes, 737
  use in the synthesis of 1,1-difluoroalkenes, 732
  use in the synthesis of fluoro haloalkenes, 751
Phosphine, tris(chloromethyl)-, from,
    tetrakis(hydroxymethyl)phosphonium chloride, 113
Phosphine, tris(dimethylamino)-
  reactions, with trichlorofluoromethane, 753
  use in Wittig reactions, 733, 738, 745
Phosphine, tris(2,4,6-trimethoxyphenyl)-, as starting
    material, for (chloromethyl)phosphonium salts, 118
Phosphine, tris(trimethylsilyl)-
  reactions
    with *N*-(chloroacetamide), 483
    with dialkylaminomethanols, 459
Phosphine–borane adducts
  deprotonation of, 557
  reactions of, 564, 570, 575
Phosphinegold chloride, triphenyl-, reactions, with
    ylides, 397
Phosphine imines, 1-lithioalkyl-, synthesis of, 576
Phosphine oxide, (*o*-aminobenzyl)phenyl-, reactions, with
    benzaldehyde, 471
Phosphine oxide, benzyldiphenyl-, deprotonation of, 349
Phosphine oxide, bisbenzyl(phenyl)-, bromination of, 122
Phosphine oxide, bis(chloromethyl)methyl-,
    synthesis of, 120
Phosphine oxide, bis(chloromethyl)phenyl-, as starting
    material, for (trimethylsilyloxy phosphonium
    iodides, 122
Phosphine oxide, bis(chloromethyl)*n*-propyl-,
    synthesis of, 121
Phosphine oxide, bis(iodomethyl)methyl-, synthesis of, 122
Phosphine oxide, bis(iodomethyl)phenyl-, synthesis of, 122
Phosphine oxide, (bromomethyl)dimethyl-, synthesis of, 121
Phosphine oxide, (chloromethyl)dimethyl-, from,
    tetrakis(hydroxymethyl)phosphonium chloride, 119
Phosphine oxide, (chloromethyl)diphenyl-, synthesis of, 119
Phosphine oxide, 1-cyclobutenyldiphenyl-, reactions, with
    diazomethane, 488
Phosphine oxide, dibenzyl-, reactions, with
    formaldehyde, 471
Phosphine oxide, (diethoxymethyl)-, reactions, with
    benzaldehyde, 828
Phosphine oxide, difluoromethyl-, use in Wittig–Horner
    reactions, 734
Phosphine oxide, (α-fluorobenzyl)diphenyl-,
    synthesis of, 118
Phosphine oxide, α-methoxyallyl-
  reactions
    with acetaldehyde, 917
    with benzaldehyde, 917
    with lithium diisopropylamide, 917
Phosphine oxide, (*S*)-(-)-methyl phenylvinyl-
  as starting material
    for (*R*)-(-)-α-bromo compounds, 804
    for chloro compounds, 804
Phosphine oxide, α-nitro-, from, 2-alkoxyalkenylphosphine
    oxides, 490
Phosphine oxide, tris(bromomethyl)-, synthesis of, 122
Phosphine oxide, tris(chloromethyl)-, synthesis of, 121
Phosphine oxide, tris(iodomethyl)-, synthesis of, 122
Phosphine oxides
  reactions
    with haloalkyl phosphonates, 554
    with potassium phthalimide, 473
  reactivity of, 120
  as starting materials, for *Se*,*P*-acetals, 348
Phosphine oxides, (acetoxymethyl)-

chlorination of, 119
  as starting materials, for (chloromethyl)phosphine
      oxides, 119
Phosphine oxides, acyl-, reactions, with methylmagnesium
    iodide, 343
Phosphine oxides, 2-alkoxyalkenyl-, as starting materials,
    for α-nitrophosphine oxide, 490
Phosphine oxides, alkyldiphenyl-, metallation of, 346
Phosphine oxides, α-amino-, from, tertiary phosphines, 477
Phosphine oxides, α-aminoalkyl-, from, phosphorus
    esters, 464
Phosphine oxides, aminomethyl-, synthesis of, 473
Phosphine oxides, anilinomethyl-, synthesis of, 471
Phosphine oxides, bis(chloromethyl) (1-hydroxyalkyl)-,
    synthesis of, 121
Phosphine oxides, α-bromo-, synthesis of, 121
Phosphine oxides, 1-cerioalkyl-, synthesis of, 577
Phosphine oxides, chloro-
  reactions
    with 1,1-dimetalloalkanes, 556
    with 1-metalloalkylphosphorus compounds, 557
    with 1-metalloalkylsilanes, 569
    with phosphorus ylides, 557
Phosphine oxides, α-chloro-, synthesis of, 119
Phosphine oxides, (α-chloroalkyl)-
  phosphinylmethylating agents, 119
  synthesis of, 120
Phosphine oxides, (chloromethyl)-, from,
    (acetoxymethyl)phosphine oxides, 119
Phosphine oxides, chloromethyl(divinyl)-, synthesis of, 120
Phosphine oxides, chlorovinyl-, reactions, with
    thiophenol, 120
Phosphine oxides, dialkyl-, reactions, with metallated
    ketenes, 915
Phosphine oxides, dialkyl(chloromethyl)-, synthesis of, 119
Phosphine oxides, α-diazoalkyl-, for the synthesis of
    phosphorus-containing heterocycles, 489
Phosphine oxides, dibutyl-, reactions, with mesityl oxide, 339
Phosphine oxides, dichloro-, synthesis of, 478
Phosphine oxides, α-fluoro-, synthesis of, 118
Phosphine oxides, α-halo-, synthesis of, 118
Phosphine oxides, α-hydroxy-, chlorination of, 119
Phosphine oxides, (hydroxymethyl)-
  bromination of, 121
  chlorination of, 119, 120, 121
Phosphine oxides, α-iodo-, synthesis of, 122
Phosphine oxides, (iodomethyl)-, synthesis of, 122
Phosphine oxides, 1-lithioalkyl-, synthesis of, 575
Phosphine oxides, 1-metalloalkyl-
  reactions
    with chlorophosphines, 551
    with chlorosilanes, 570
    with chlorostannanes, 588
    with chromium compounds, 579
    with copper halides, 585
    with manganese compounds, 581
    with mercury compounds, 587
    with tantalum compounds, 579
Phosphine oxides, 1-phosphinoalkyl-, synthesis of, 551
Phosphine oxides, 1-potassioalkyl-, synthesis of, 577
Phosphine oxides, 1-silylalkyl-, synthesis of, 569
Phosphine oxides, 1-sodioalkyl-, synthesis of, 576
Phosphine oxides, stibinomethyl-, synthesis of, 565
Phosphine oxides, α,β-unsaturated, addition to, 561
Phosphines
  chloromethylation of, 112
  reactions
    with 1-aminoalkylphosphines, 546
    with borinic esters, 342
    with carbonyl compounds, 548
    with chromium carbene complexes, 580
    with 1,1-dihaloalkanes, 546
    with diketones, 339
    with fluoroalkenes, 111
    with fluoro(trifluoromethyl)carbene, 111
    with 1-haloalkylchromium compounds, 579
    with 1-haloalkylindium compounds, 588
    with 1-haloalkyliron compounds, 582
    with 1-haloalkylphosphines, 546
    with 1-haloalkylphosphorus compounds, 554
    with 1-haloalkylplatinum compounds, 585
    with 1-haloalkylrhodium compounds, 582
    with 1-haloalkylsilanes, 569
    with 1-haloalkylstannanes, 588
    with 1-haloalkylzinc compounds, 586
    with hexafluoroacetone, 337
    with imines, 455
    with iron carbene complexes, 581
    with Mannich bases, 452
    with molybdenum carbene complexes, 580
    with rhenium carbene complexes, 581
    with tungsten carbene complexes, 580
    with α,β-unsaturated carbonyl compounds, 339
  as starting materials, for Se,P-acetals, 347
Phosphines, alkoxy(α-haloalkyl)-, synthesis of, 117
Phosphines, alkoxy(trimethylsiloxy)-, reactions, with
    1-ethoxyethyne, 914
Phosphines, alkylbis(chloromethyl)-, from,
    chlorophosphines, 112
Phosphines, alkyl(diamino)-, as starting materials, for
    phosphonic diamides, 133
Phosphines, alkyldichloro-
  reactions, with benzaldehydes, 123
  as starting materials, for alkylaminomethylphosphinic
      acids, 465
Phosphines, alkyldiphenyl-
  reactions
    with hexachloroethane, 112
    with tetrachloromethane, 112
Phosphines, β-allylaminoalkyl-, reactions, with
    formaldehyde, 458
Phosphines, 1-aminoalkyl-
  reactions
    with diethyl phosphite, 551
    with phosphines, 546
Phosphines, α-aminoalkyl-
  alkylation of, 478
  from, azacrown compounds, 453
  optically pure, synthesis of, 455
  as starting materials, for α-aminoalkyl phosphonium
      salts, 478
  synthesis of, 452, 456
Phosphines, β-aminoalkyl-, reactions, with
    N-hydroxymethyldiethylamine, 453
Phosphines, (ω-aminoalkyl)phenyl-
  reactions
    with aldehydes, 453
    with ketones, 453
Phosphines, amino(chloromethyl)-, synthesis of, 116
Phosphines, amino(α-haloalkyl)-, synthesis of, 116
Phosphines, α-aminomethyl-, synthesis of, 452
Phosphines, α-aminomethyl tertiary, synthesis of, 455
Phosphines, 1-arsinoalkyl-, synthesis of, 564
Phosphines, aryldichloro-, reactions, with
    benzaldehydes, 123
Phosphines, 1-barioalkyl-, synthesis of, 577
Phosphines, bis(chloromethyl)-, synthesis of, 115
Phosphines, bis-α-hydroxyalkyl-, reactions, with
    aldehydes, 341
Phosphines, bis(α-hydroxybenzyl)-, synthesis of, 337
Phosphines, bis(trimethylsilyl)amino-, reactions, with
    tetrachloromethane, 572
Phosphines, (bromomethyl)-, oxidation of, 122
Phosphines, carboxyalkyl-
  reactions, with hydrazones, 482
  as starting materials, for 1-amino-1,3-azaphospholan-5-
      one, 482

Phosphines

Phosphines, chloro-
  reactions
    with aldehydes, 119, 121
    with amines, 116
    with diazoalkanes, 114
    with 1,1-dimetalloalkanes, 546
    with α-hydroxyamides, 465
    with ketones, 119, 121, 345
    with 1-metalloalkylphosphine oxides, 551
    with 1-metalloalkylphosphines, 546
    with 1-metalloalkylphosphine sulfides, 551
    with 1-metalloalkylphosphonamidates, 551
    with 1-metalloalkyl phosphonates, 551
    with 1-metalloalkylphosphorus compounds, 551
    with 1-metalloalkylsilanes, 565
    with 1-trialkylsilylalkylphosphines, 547
  as starting materials
    for alkylbis(chloromethyl)phosphines, 112
    for (di-*t*-butylamino)phosphines, 116
Phosphines, α-chloro-, from, diethyl chloromalonate, 115
Phosphines, α-chloroalkyl-
  from, dichlorophosphoranes, 112
  synthesis of, 111
Phosphines, (α-chloroalkyl)diphenyl-, synthesis of, 112
Phosphines, (chloromethyl)-, oxidation of, 120
Phosphines, chloromethylbis(octyloxy)-, synthesis of, 117
Phosphines, chloromethyldihalo-, synthesis of, 115
Phosphines, (chloromethyl)dimethyl-, synthesis of, 112
Phosphines, cyclic α-aminoalkyl-, synthesis of, 453
Phosphines, cyclic aminomethyl-
  from
    orthodisubstituted aromatic phosphines, 453
    secondary amines, 453
Phosphines, dialkyl-, reactions, with bistrifluoromethylketene, 1023
Phosphines, dialkylhalo-, reactions, with alkoxyalkynes, 914
Phosphines, dialkyl(α-haloalkyl)-, synthesis of, 111
Phosphines, diallylaminomethyl-, as starting materials, for 1,3-azaphosphanes, 458
Phosphines, diamino(alkyl)-, reactions, with tetrachloromethane, 116
Phosphines, diaryl(α-haloalkyl)-, synthesis of, 111
Phosphines, (di-*t*-butylamino)-, from, chlorophosphines, 116
Phosphines, dichloro-
  acid hydrolysis of, 123
  alkylation of, 124
  from, diiodides, 114
  reactions
    with *N*-benzylglycine, 466
    with electrophiles, 125
    with formaldehyde, 125
    with imines, 463
    with triethyl orthoformate, 126
Phosphines, dichloro(chloromethyl)-
  chlorination of, 133
  synthesis of, 114
Phosphines, diethoxy(heptafluoroisopropyl)-, synthesis of, 117
Phosphines, diethylaminomethyl-, substitution of, 546
Phosphines, diiodo-, synthesis of, 114
Phosphines, (dimethylamino)-, synthesis of, 116
Phosphines, fluoro-, synthesis of, 116
Phosphines, α-fluoro-, synthesis of, 111
Phosphines, α-fluoroalkyl-, synthesis of, 111
Phosphines, α-fluoroalkyl(trifluoromethyl)iodo-, synthesis of, 114
Phosphines, germyl-, reactions, with ketimines, 998
Phosphines, halo-
  reactions
    with aldehydes, 123
    with phosphorus ylides, 551
Phosphines, (halo)-, reactions, with thioformaldehyde, 123

Phosphines, α-halo-
  reactions, with sulfur, 128
  stability of, 111
  as starting materials, for phosphaalkenes, 111
Phosphines, 1-haloalkyl-
  reactions
    with metal phosphides, 546
    with nickel compounds, 583
    with phosphines, 546
Phosphines, α-haloalkyl-, stability of, 110
Phosphines, α-haloalkylbis(diisopropylamino)-, synthesis of, 116
Phosphines, (*o*-halobenzyl)-, reactions, with *n*-butyllithium, 575
Phosphines, halo(α-haloalkyl)-, synthesis of, 114
Phosphines, (α-hydroxyalkyl)phenyl-, disproportionation of, 341
Phosphines, α-hydroxymethyl-, synthesis of, 336
Phosphines, (3-hydroxypropyl)phenyl-
  reactions
    with aldehydes, 337
    with ketones, 337
Phosphines, α-keto-, synthesis of, 1023
Phosphines, α-lithio-, synthesis of, 575
Phosphines, 1-lithioalkyl-, synthesis of, 575
Phosphines, lithium dialkyl(aryl)-, reactions, with α-chloro tertiary amines, 455
Phosphines, 1-lutetioalkyl-, synthesis of, 577
Phosphines, mesityl(diphenylmethylene)-, reactions, with phenyl azide, 483
Phosphines, 1-metalloalkyl-
  reactions
    with aluminum compounds, 587
    with chlorophosphines, 546
    with chlorosilanes, 566
    with chlorostannanes, 588
    with cobalt compounds, 582
    with copper halides, 585
    with tantalum compounds, 579
    with thorium compounds, 589
    with zirconium compounds, 579
Phosphines, methoxy-, synthesis of, 117
Phosphines, methylene-
  photolysis of, 459
  as starting materials
    for α-aminoalkylphosphonites, 459
    for tetrahydrophosphorines, 459
Phosphines, methylsilyl-, reactions, with formaldehyde, 337
Phosphines, nucleophilic, reactions, with iminium salts, 481
Phosphines, ortho-disubstituted aromatic, as starting materials, for cyclic aminomethylphosphines, 453
Phosphines, (3-oxoalkyl)phenyl-
  reactions
    with aldehydes, 454
    with aldimines, 454
Phosphines, perfluoroalkyldichloro-
  reactions
    with paraformaldehyde, 125
    with trioxane, 125
Phosphines, perfluorotrialkyl-, synthesis of, 112
Phosphines, (phosphoranomethyl)-, synthesis of, 552
Phosphines, 1-potassioalkyl-, synthesis of, 577
Phosphines, primary α-halo-, synthesis of, 111
Phosphines, secondary
  reactions
    with *N*-(alkoxymethyl)-*N*-arylamines, 453
    with *N*-arenesulfonyl imines, 481
Phosphines, secondary α-halo-, synthesis of, 111
Phosphines, silyl-
  reactions
    with imines, 481
    with ketimines, 998
Phosphines, 1-sodioalkyl-, synthesis of, 576

Phosphines, stibinomethyl-, synthesis of, 565
Phosphines, N-substituted aminomethyl-, synthesis of, 481
Phosphines, tertiary
   reactions, with iminium ions, 462, 487
   as starting materials
      for α-aminophosphine oxides, 477
      for α-aminophosphine sulfides, 477
Phosphines, tertiary α-halo-, synthesis of, 111
Phosphines, trialkyl-
   deprotonation of, 575
   oxidation of, 123
Phosphines, trialkyl(aryl)-, as starting materials, for 1-aminoalkylphosphonium salts, 492
Phosphines, 1-trialkylsilylalkyl-, reactions, with chlorophosphines, 547
Phosphines, tris(aminomethyl)-, synthesis of, 456
Phosphines, trisaminomethyl-
   synthesis of, 459
   use in the synthesis of α-aminophosphine sulfides, 477
Phosphines, tris(dialkylaminomethyl)-, synthesis of, 456
Phosphines, tris(dimethylaminomethyl)-
   reactions, with secondary amines, 456
   thermal displacement of, 453
Phosphines, tris(hydroxymethyl)-
   reactions, with secondary amines, 456
   synthesis of, 336
Phosphines, trisubstituted, reactions, with iminium ions, 478
Phosphines, vinyl-, reactions, with alkyllithiums, 575
Phosphines, (vinyldimethylsilylmethyl)-, synthesis of, 566
Phosphinic acid, reactions, with oximes, 464
Phosphinic acid, (1-aminomethyl) (chloromethyl)-, synthesis of, 124
Phosphinic acid, bis(chloromethyl)-, reactions, with thiourea, 346
Phosphinic acid, α-chloromethyl-, synthesis of, 123
Phosphinic acid, (chloromethyl)phenyl-, synthesis of, 123
Phosphinic acid, (2-cyano-1-fluoroethenyl)n-butyl-, hydrogenation of, 123
Phosphinic acid, di(chloromethyl)-, reactions, with ammonia, 473
Phosphinic acid, (fluoromethyl)phenyl-, from, phenyltetrafluorophosphorane, 123
Phosphinic acid, α-nitro-, from, 2-alkoxyalkenyl phosphinates, 490
Phosphinic acid anhydrides, reactions, with phosphorus pentachloride, 125
Phosphinic acid chlorides, reactions, with paraformaldehyde, 128
Phosphinic acid compounds, α-amino-, from, imines, 463
Phosphinic acids
   from, phosphorinanes, 124
   reactions, with imines, 463
Phosphinic acids, alkylaminomethyl-
   from
      alkyldichlorophosphines, 465
      α-hydroxyamides, 465
Phosphinic acids, α-amino-
   from
      alkyl phosphonites, 460
      α,β-unsaturated esters, 460
Phosphinic acids, 1-aminoalkyl-, synthesis of, 461
Phosphinic acids, α-aminoalkyl-, from, imines, 463
Phosphinic acids, α-aminobenzyl-, synthesis of, 460, 467
Phosphinic acids, bis(chloromethyl)-, synthesis of, 125
Phosphinic acids, α-chloroalkyl-, synthesis of, 124
Phosphinic acids, α-chlorobenzyl-, synthesis of, 123
Phosphinic acids, (chloromethyl)-, synthesis of, 124
Phosphinic acids, chloromethyl(1-hydroxyalkyl)-, synthesis of, 124
Phosphinic acids, α-halo-, synthesis of, 123
Phosphinic acids, halomethyl-, as starting materials, for α-aminoalkylphosphorus compounds, 473
2-Phosphinic acids, 2,3-dihydropyranyl-, synthesis of, 339

Phosphinic acid salts
   chlorination of, 125
   as starting materials, for (chloromethyl)methylphosphinic chloride, 125
Phosphinic amides, α-halo-, synthesis of, 127
Phosphinic anhydrides, (chloromethyl)perfluoroalkyl-, synthesis of, 128
Phosphinic anhydrides, α-halo-, synthesis of, 128
Phosphinic chloride, (chloromethyl)methyl-, from, phosphinic acid salts, 125
Phosphinic chlorides
   alcoholysis of, 126
   alkylation of, 119
   from, ethyl phosphinates, 125
   hydrolysis of, 124
   reactions
      with amines, 127
      with Grignard reagents, 120
      with thiols, 128
   as starting materials
      for (chloromethyl)methylthiophosphinic chloride, 128
      for phosphinates, 126
      for phosphinico acrylic acids, 124
Phosphinic chlorides, alkyl chloromethyl-, synthesis of, 125
Phosphinic chlorides, aryl chloromethyl-, synthesis of, 125
Phosphinic chlorides, bis(chloromethyl)-
   as starting materials, for amides, 127
   synthesis of, 125
Phosphinic chlorides, 2-cyanoethyl-, synthesis of, 125
Phosphinic halides, α-halo-, synthesis of, 125
Phosphinimines, synthesis of, 552
Phosphinimines, bis(trimethylsilyl)-, synthesis of, 563
Phosphinite, diethyl phenyl-, reactions, with N-tetrachloroethylamides, 1002
Phosphinite, ethyl diphenyl-
   reactions
      with N-chloromethylamide, 487
      with chloromethyl methyl sulfide, 347
      with N-tetrachloroethylamides, 1000
Phosphinite, ethyldiphenyl-, as starting material, for gem-methoxycarbonylamino diphenylphosphinoylalkenes, 1000
Phosphinites, synthesis of, via methanolysis of phosphines, 117
Phosphinites, alkyl
   reactions
      with 1-haloalkylphosphorus compounds, 556
      with 1-haloalkylsilanes, 570
Phosphinoalkenes, deprotonation of, 1031
Phosphinoalkenes, gem-amido-, from, α-phosphino enamines, 998
Phosphinoalkenes, gem-chloro-, synthesis of, 997
1-Phosphinoalkenes, 1-silyl-, hydrostannylation of, 567
Phosphinoallene, tetrakisdiphenyl-, synthesis of, 1023
Phosphino-1,3-butadiene, 1,1,4,4-tetrakisdiphenyl-
   detection of, 1023
   reactions, with molybdenum hexacarbonyl, 1023
Phosphinocymantrenes, diphenyl-, synthesis of, 1036
Phosphinoenamines, α-trialkyl, from, halo enamines, 998
α-Phosphinoenamines
   from
      gem-dihaloalkenes, 997
      gem-halo enamines, 997
      ketimines, 998
   as starting materials, for gem-amido phosphinoalkenes, 998
   synthesis of, 997
Phosphinoethene, bisdiphenyl-
   reactions
      with cyclopentadienylbisdiphenylphosphinoruthenium chloride, 1029
      with iodomethane, 1024
      with Michael donors, 1028

Phosphinoethene

with pentacarbonyl[(dimethyloxosulfono)-
methanide]chromium, 1029
with platinum(II) chloride, 1028
Phosphinoethene, 1,1-bisdiphenyl-
alkylation of, 1023
reactions, with cycloocta-1,5-
dienyldimethylplatinum, 1028
Phosphinoethene, trisdiphenyl-, synthesis of, 1023
Phosphinoethyne, 1,2-bisdiphenyl-, reactions, with
diphenylphosphine, 1023
Phosphinoferrocenes
as asymmetric catalysts, in Grignard cross-couplings, 1033
synthesis of, 1032
Phosphinometallocenes, synthesis of, 1032
Phosphinomethyl ylides, synthesis of, 552
2-Phosphinophospholanes, synthesis of, 547
Phosphinophosphorus ylides, protonation of, 551
Phosphinopropene, bisdiphenyl-, synthesis of, 1022
Phosphinopropyne, 3-lithio-1-diphenyl-, reactions, with
chlorodiphenylphosphine, 1023
Phosphino propyne, tetrakisdiphenyl-,
non-detection of, 1023
Phosphinorutheniumcarbene complexes,
rearrangement of, 582
Phosphinoruthenium chloride, cyclopentadienylbisdiphenyl-
reactions
with bischlorodicarbonylrhodium, 1029
with bisdiphenylphosphino ethene, 1029
Phosphinothious acid, as starting material, for
α-thiophosphinoyl enamines, 1001
Phosphinous acids, dialkyl-
reactions, with formaldehyde, 119
as starting materials, for α-phosphinoyl enamines, 1000
Phosphinous chloride, diethyl-, reactions, with methyl
dithioisobutyrate, 927
Phosphinoyl compounds, chloro-, reactions, with
phosphorus pentachloride, 1013
Phosphino ylides
reactions, with electrophiles, 551
synthesis of, 551
Phosphirane oxides, synthesis of, 924
Phosphiranes, dichloro-, synthesis of, 115
Phosphiranes, disilyl-, synthesis of, 567
Phosphiranes, 2-silyl-, synthesis of, 567
1*H*-Phosphirenes, 1-chloro-, synthesis of, 110
2*H*-Phosphirenes, 2-chloro-, as intermediates, in the
synthesis of 1-chloro-1*H*-phosphirenes, 110
Phosphite, *t*-butyl diethyl, reactions, with ketene, 915
Phosphite, chloro-, reactions, with
N-acetyltrifluoroacetamide, 1009
Phosphite, diethyl
reactions
with aldimines, 461
with aldimine salts, 464
with 1-aminoalkylphosphines, 551
with cinnamaldehyde benzoylhydrazone, 1008
with 1-ethoxyethyne, 913
with paraformaldehyde, 338
as starting material, for α-aminoalkylphosphonic
acids, 464
use in the synthesis of dichloromethyls, 13
Phosphite, diethyl trimethylsilyl, reactions, with
cinnamaldehyde, 918
Phosphite, dimethyl, reactions, with enones, 339
Phosphite, trialkyl
reactions
with benzoyl phosphonates, 557
with iminium chlorides, 1003
Phosphite, triethyl
reactions
with bis(trifluoromethyl)thioketenes, 926
with bromotricyanomethane, 1008
with α-chloro thiol esters, 836

with chlorotricyanomethane, 1008
with 1-chlorovinyl isocyanate, 1010
with 1,2-dibromoethyl isocyanate, 1009
with diphenylacetyl chloride, 920
with diphenylketene, 914
with 1-ethoxy-1,2-dichloroethene, 912
with 1-ethoxyethyne, 913
with furan-2-carboxaldehyde oximes, 485
with methylthioacetone, 775
with *N*-phosphorylated ketenimines, 1008
with tetrachloroethyl isocyanates, 1010
Phosphite, trimethyl
reactions
with α-bromoglycine compounds, 465
with 1-(*t*-butyldimethylsilyl)-2-chloro-1-ethanone, 934
with *p*-chlorocinnamoyl chloride, 922
with 1,2-dibromo-1-methoxyethane, 912
with methyl iodide, 485
as starting material, for serines, 487
Phosphite, triphenyl, reactions, with bromine, 25
Phosphites, reactions, with polyfluoroalkenes, 803
Phosphites, dialkyl
reactions
with acetic anhydride, 921
with aldimines, 462
with aryl aldazines, 460
with carbonyl compounds, 559
with carbonyls, 469
with chiral nitrones, 485
with N^1,N^2-dialkylidene 1,1-diaminoalkanes, 461
with fluoroalkyl halides, 131
with hydrobenzamide, 460
with ketene *N,X*-acetals, 1003
with ketones, 338
as starting materials
for α-aminoalkyl phosphonates, 462
for α-aminophosphonic acids, 460
Phosphites, diaryl
reactions
with imines, 463
with oxoiminium salts, 485
Phosphites, trialkyl
alkylation of, 920
reactions
with aryliminooxindoles, 464
with 1,3-dithiole-2-thiones, 850
with 1-haloalkylphosphorus compounds, 555
with 1-haloalkylsilanes, 569
with 1-haloalkylstannanes, 588
with *N*-methoxymethyl arylamines, 467
with thiocarbonyl compounds, 345
Phosphites, trialkyl(aryl)
reactions, with α-haloamines, 464
as starting materials, for α-aminoalkyl phosphonates, 466
Phosphites, trialkyl, mercuric halide complexes, reactions,
with ketene, 915
Phosphole, 1-butyl-3,4-dimethyl-, synthesis of, 1037
Phosphole, pentaphenyl-, reactions, with lithium, 112
Phosphole, 1-phenyl-, reactions, with manganese
pentacarbonyl, 1037
Phosphole, 1,2,3-triphenyl-, as starting material, for
diphosphaferrocenes, 1035
Phosphole complexes, 2,3,4,5-tetraphenyl-,
synthesis of, 1037
2-Phospholene 1-oxide, 1-phenyl-, reactions, with
bromine, 122
Phospholes
reactions
with benzoyl chloride, 341
with dicyclopentadienyltetracarbonyldiiron, 1034
with phenyllithium, 1031
synthesis of, 1026
Phospholes, thiooxo-, use in Diels–Alder reactions, 489

2-Phospholine, 1-phenyl-, hydrozirconation of, 547, 579
Phospholyl anion, tetraphenyl-, reactions, with
    dicarbonylcyclopentadienyliron iodide, 1034
Phosphonamidates, N-t-butyl α-chloro-, synthesis of, 133
Phosphonamidates, α-halo-, synthesis of, 133
Phosphonamidates, 1-metalloalkyl-, reactions, with
    chlorophosphines, 551
Phosphonamides, as starting materials, for
    α-aminophosphonic acid compounds, 473
Phosphonamides, chloromethyl-, reactions, with sodium
    azide, 492
Phosphonamides, 2-ethoxyvinyl-, halogenation of, 804
Phosphonamides, α-silyloxyalkyl-, deprotonation of, 917
Phosphonamidic chlorides, N,N-dialkyl α-haloalkyl-,
    synthesis of, 131
Phosphonamidic fluorides, N,N-dialkyl α-haloalkyl-,
    synthesis of, 131
Phosphonamidic halides, α-halo-, synthesis of, 131
Phosphonate, diethyl, condensation of, 912
Phosphonate, diethyl benzenesulfonylmethyl, reactions, with
    aldehydes, 925
Phosphonate, diethyl 1-(benzenesulfonyl)methyl,
    fluorination of, 79
Phosphonate, diethyl α-bromomethyl, hydrolysis of, 130
Phosphonate, diethyl α-bromovinyl, reactions, with
    ammonia, 473
Phosphonate, diethyl chloro(phenylseleno)methyl,
    synthesis of, 90
Phosphonate, diethyl dibromomethyl, use in Wittig–Horner
    reaction, 745
Phosphonate, diethyl dichloromethyl, use in Wittig–Horner
    reaction, 740
Phosphonate, diethyl difluoromethyl, use in Wittig–Horner
    reaction, 734
Phosphonate, diethyl diiodomethyl
    from
        diethyl iodomethylphosphonate, 749
        diethyl methylphosphonate, 749
Phosphonate, diethyl ethyl, reactions, with diphenyl
    disulfide, 346
Phosphonate, diethyl 1-fluoro-1-(phenylsulfonyl)methyl,
    reactions, with carbonyl compounds, 780
Phosphonate, diethyl 1-hydroxy-2,2-diphenylethenyl,
    synthesis of, 920
Phosphonate, diethyl iodomethyl
    reactions
        with sodium benzeneselenolate, 348
        with thioacetate, 346
    as starting material
        for diethyl diiodomethylphosphonate, 749
        for diethyl phenylselenomethylphosphonate, 784
Phosphonate, diethyl lithiodichloromethyl, use in Wittig–
    Horner reaction, 740
Phosphonate, diethyl methyl
    deprotonation of, 929
    reactions, with aryl cyanides, 927
    as starting material, for diethyl diiodomethyl
        phosphonate, 749
Phosphonate, diethyl methylthiomethyl, use in Peterson
    reaction, 925
Phosphonate, diethyl phenylselenomethyl
    from, diethyl iodomethyl phosphonate, 784
    use in Wittig reaction, 784
Phosphonate, diethyl trichloromethyl, use in Wittig–Horner
    reaction, 740
Phosphonate, diethyl (2-trimethylsilylethoxy)methyl,
    Peterson reaction of, 916
Phosphonate, diethylvinyl
    addition of cuprates to, 585
    halogenation of, 804
    reactions
        with organomercury halides, 587
        with tetraethyl pyrophosphite, 552

Phosphonate, diisopropyl fluoromethyl, from, sodium
    diisopropylphosphites, 131
Phosphonate, dimethyl 1-methoxyvinyl, synthesis of, 912
Phosphonate, (E)-dimethyl 2-phenylvinyl, reactions, with
    lithium diisopropylamide, 927
Phosphonate,
    diphenyl[(iodomethyl)phenoxyphosphinyl]methyl,
    synthesis of, 127
Phosphonate anions, reactions, with benzeneselenyl
    bromide, 929
Phosphonate anions, alkyl, electrophilic fluorination of, 131
Phosphonates
    lithiation of, 129
    reactions
        with N-fluorobis(benzenesulfonyl)imide, 131
        with sodium selenide, 348
    reduction of, 342
    synthesis of, 802
Phosphonates, 1-acetoxyvinyl, synthesis of, 921
Phosphonates, acetyl, synthesis of, 921
Phosphonates, acyl, reactions, with acylating agents, 920
Phosphonates, (N-acyliminomethyl), use in Diels–Alder
    reactions, 489
Phosphonates, α-alkoxyalkyl, synthesis of, 340
Phosphonates, 1-alkoxyallyl, isomerization of, 918
Phosphonates, N-alkoxy-α-amino, from, oximes, 485
Phosphonates, alkyl, bromination of, 132
Phosphonates, alkyl hydrogen (1-benzamido-2,2-
    dichlorovinyl), synthesis of, 1006
Phosphonates, α-alkylthioalkyl
    reactions, with alkyl halides, 347
    synthesis of, 346
Phosphonates, 2-(alkylthio)vinyl, bromination of, 804
Phosphonates, α-amino-, synthesis of, 476
Phosphonates, aminoalkyl
    acylation of, 493
    quaternization of, 493
Phosphonates, 1-aminoalkyl
    from, ketimines, 463
    as starting materials, for α-phosphoryl enamines, 1003
    synthesis of, 466
Phosphonates, α-aminoalkyl
    from
        aldehydes, 466
        dialkyl phosphites, 462
        diethyl phosphoramidate, 470
        α-oxophosphonates, 471
        phosphorus esters, 464
        thioureas, 466
        trialkyl(aryl) phosphites, 466
    synthesis of, via ionic hydrogenation, 491
Phosphonates, α-aminobenzyl, synthesis of, 461
Phosphonates, N-arylaminomethyl, synthesis of, 467
Phosphonates, 1-azidoalkyl, synthesis of, via Mitsunobu
    reaction, 491
Phosphonates, benzoyl, reactions, with trialkyl
    phosphite, 557
Phosphonates, α-bromo-, synthesis of, 132
Phosphonates, α-bromoalkyl, synthesis of, 132
Phosphonates, 2-bromo-2-methylpropanoyl, as starting
    materials, for chloroformates, 921
Phosphonates, 1-cesioalkyl, synthesis of, 577
Phosphonates, α-chloro-, reactions, with n-butyllithium, 132
Phosphonates, α-chloroalkyl, from, trichloromethyl
    phosphonates, 131
Phosphonates, 1-chloromethyl, azidation of, 492
Phosphonates, 1-chlorovinyl, displacement of, 927
Phosphonates, (1,1-dialkoxyethyl), synthesis of, 912
Phosphonates, dialkyl alkyl
    metallation of, 346
    synthesis of, 920
Phosphonates, diazo-, use in the synthesis of
    α-aminophosphonates, 472

Phosphonates

Phosphonates, α-diazoalkyl
  reactions, with alcohols, 913
  for the synthesis of phosphorus-containing heterocycles, 489
Phosphonates, diethyl 1-alkoxyvinyl, synthesis of, 912
Phosphonates, diethyl α-bromoalkyl, synthesis of, 133
Phosphonates, diethyl α-chloroalkyl, synthesis of, 132
Phosphonates, diethyl iodomethyl
  hydrolysis of, 130
  synthesis of, 133
Phosphonates, diethyl α-lithio-α-(trimethylsilyl)(chloromethyl), alkylation of, 131
Phosphonates, diisopropyl α-bromoalkyl, synthesis of, 132
Phosphonates, diisopropylfluoromethyl, hydrolysis of, 129
Phosphonates, dimeric, synthesis of, 478
Phosphonates, diphenyl (1-benzamido-2,2-dichlorovinyl), synthesis of, 1006
Phosphonates, α-fluoro-, synthesis of, 131
Phosphonates, haloalkyl, reactions, with phosphine oxides, 554
Phosphonates, 1-haloalkyl
  reactions
    with palladium compounds, 583
    with platinum compounds, 585
Phosphonates, α-halomethyl, reactions, with dimethyl sulfide, 346
Phosphonates, α-hydroxy, reactions, with phthalimide, 475
Phosphonates, α-hydroxyalkyl
  reactions, with diethylaminosulfur trifluoride, 131
  synthesis of, 343
Phosphonates, 1-N-hydroxyiminoalkyl, oxidation of, 491
Phosphonates, 1-lithioalkyl, synthesis of, 576
Phosphonates, 1-magnesioalkyl, synthesis of, 577
Phosphonates, 1-metalloalkyl
  reactions
    with chlorophosphines, 551
    with chlorosilanes, 570
    with chlorostannanes, 588
    with copper halides, 585
    with titanium compounds, 578
Phosphonates, N-methoxy, synthesis of, 485
Phosphonates, methylsulfinyl, synthesis of, 347
Phosphonates, 1-nitroalkyl
  from, α-oxophosphonates, 491
  synthesis of, 490
Phosphonates, α-oxo
  as starting materials
    for gem-amido phosphorylalkenes, 1005
    for α-aminoalkyl phosphonates, 471
    for 1-nitroalkyl phosphonates, 491
Phosphonates, 2-oxoazetidin-4-yl, from, 4-acetoxyazetidin-2-ones, 487
Phosphonates, 1-phosphinoalkyl, synthesis of, 551
Phosphonates, 1-rubidioalkyl, synthesis of, 577
Phosphonates, 1-silylalkyl, synthesis of, 569
Phosphonates, silylmethyl, synthesis of, 569
Phosphonates, α-thioalkyl, synthesis of, 346
Phosphonates, α-thiocyanatoalkyl, synthesis of, 346
Phosphonates, trichloromethyl, as starting materials, for α-chloroalkyl phosphonates, 131
Phosphonates, α,β-unsaturated, addition to, 561
Phosphonates, vinyl
  as starting materials, for 1-bromo-2-aminoethylphosphonic acids, 130
  for the synthesis of phosphorus-containing heterocycles, 489
Phosphonic acid, 1-amino-1-phenylethyl-, synthesis of, 473
Phosphonic acid, chloromethyl-, reactions, with amines, 473
Phosphonic acid, methylenearsonic-, synthesis of, 565
Phosphonic acid compounds, α-amino-, from, phosphonamides, 473
Phosphonic acid diethyl ester, α-oxo-, reactions, with N-ethoxymethyl dialkylamines, 458

Phosphonic acids, N-alkoxy-α-amino-, synthesis of, 485
Phosphonic acids, 1-amino-
  from, nitrilotriacetic acid, 492
  synthesis of, via sonochemical activation, 463
Phosphonic acids, α-amino-
  from
    N-acylimines, 486
    aldazines, 460
    aldimines, 461
    chiral phosphono alcohols, 475
    dialkyl phosphites, 460
    N-fluorosulfonylimines, 486
    N-glycosyl C-dialkoxyphosphoryl nitrones, 489
    imines, 462
    α-phosphonocarboxylic acids, 476
    α,β-unsaturated esters, 460
  optically pure, synthesis of, 462
  synthesis of, via hydrolysis, 469, 476
Phosphonic acids, 1-aminoalkyl-, from, phosphorus acid, 470
Phosphonic acids, α-aminoalkyl-
  from
    diethyl phosphite, 464
    imines, 463
  optically active, synthesis of, 492
Phosphonic acids, α-aminobenzyl-, synthesis of, 467
Phosphonic acids, 1-aminomethyl-, synthesis of, via hydrolysis, 469
Phosphonic acids, α-aminomethyl-, synthesis of, via Kabachnik–Fields reactions, 469
Phosphonic acids, aziridinyl-, synthesis of, 473
Phosphonic acids, 1-benzamido-(2,2-dichlorovinyl), synthesis of, 1006
Phosphonic acids, α-(benzyloxycarbonylamino)benzyl-, synthesis of, 467
Phosphonic acids, 1-bromo-2-aminoethyl-, from, vinyl phosphonates, 130
Phosphonic acids, bromomethyl-, synthesis of, 130
Phosphonic acids, (R)-α-chloroalkyl-, from, (R,R)-diaminocyclohexanes, 129
Phosphonic acids, α-chloroalkyl-, synthesis of, 129
Phosphonic acids, fluoromethyl-, synthesis of, 129
Phosphonic acids, N-formylaminoalkyl, synthesis of, 488
Phosphonic acids, guanidinoalkyl, synthesis of, 487
Phosphonic acids, α-halo-, synthesis of, 129
Phosphonic acids, halomethyl-, as starting materials, for α-aminoalkylphosphorus compounds, 473
Phosphonic acids, α-hydrazino-, from, aldazines, 460
Phosphonic acids, iodomethyl-, synthesis of, 130
1-Phosphonic acids, from, 3,4-dihydroisoquinolines, 464
Phosphonic diamides, α-halo-, synthesis of, 133
Phosphonic dichloride, α-chloroethyl-, condensation of, 132
Phosphonic dichloride, chloromethyl-, reactions, with sodium fluoride, 130
Phosphonic dichloride, fluoromethyl-, synthesis of, 130
Phosphonic dichloride, iodomethyl-, synthesis of, 131
Phosphonic dichlorides
  monoesterification of, 131
  reactions, with amines, 133
Phosphonic dichlorides, α-chloro-, from, dichloroalkanes, 130
Phosphonic dihalides, α-chloro-, from, gem-dihaloalkanes, 130
Phosphonic dihalides, α-halo-, synthesis of, 130
Phosphonic dihalides, α-haloalkyl-, synthesis of, 130
Phosphonioalkenes, gem-alkoxycarbonylamido, from, tetrachloroethyl urethanes, 999
Phosphonioalkenes, gem-amido-
  from
    gem-amido carboxyalkenes, 998
    gem-amido phosphonioalkenes, 999
    N-tetrachloroethylamides, 999
    N-trichloroethylamides, 999

N-methylation of, 999
as starting materials, for *gem*-amido
phosphonioalkenes, 999
synthesis of, 998
Phosphonioalkenes, *gem*-chloroarylideneamino, from, *gem*-amido phosphonioalkenes, 1000
Phosphonite, ethyl phenyl-
reactions, with imines, 463
as starting material, for α-aminoalkylphosphonites, 460
Phosphonites, reactions, with imines, 463
Phosphonites, alkyl, as starting materials, for
α-aminophosphinic acids, 460
Phosphonites, α-amino-, from, N-benzylglycine, 466
Phosphonites, α-aminoalkyl, from, enamimes, 458
Phosphonites, α-aminoalkyl-, from, methylenephosphines, 459
Phosphonites, bis(trimethylsilyl)-, reactions, with N-trityl alkanamine, 461
Phosphonites, chloro-, reactions, with imines, 486
Phosphonites, dialkyl
reactions
with 1-haloalkylphosphorus compounds, 556
with 1-haloalkylsilanes, 570
with 1-haloalkylstannanes, 588
Phosphonites, ethyl, reactions, with aldazines, 460
Phosphonium bromide, (bromodifluoromethyl)triphenyl-, as starting material, for difluorocarbene, 737
Phosphonium bromide, (bromomethyl)triphenyl-, synthesis of, 118
Phosphonium bromide, tetrakis(bromomethyl)-, synthesis of, 118
Phosphonium bromides, α-halo-, synthesis of, 118
Phosphonium bromides, (α-iodopropyl)-, synthesis of, 118
Phosphonium chloride, tetrakis(chloromethyl)-, synthesis of, 118
Phosphonium chloride, tetrakis(hydroxymethyl)-
as starting material
for (chloromethyl)dimethylphosphine oxide, 119
for tris(chloromethyl)phosphine, 113
Phosphonium chlorides, (hydroxymethyl)-, chlorination of, 119
Phosphonium compounds, (Z)-perfluorovinyl-, synthesis of, 801
Phosphonium fluorenylide ylide, triphenyl-, reactions, with chlorofluorocarbene, 753
Phosphonium iodide, methyl(triphenoxy)-, reactions, with (α-hydroxyalkyl)trimethylsilanes, 139
Phosphonium iodides, (trimethylsilyloxy, from, bis(chloromethyl)phenylphosphine oxide, 122
Phosphonium salts
condensation of, 925
reactions
with chlorine, 344
with zinc, 753
as starting materials, for ketene acetals, 825
synthesis of, 112, 336, 347, 571, 754, 1023
Phosphonium salts, acyl-, synthesis of, 341
Phosphonium salts, α-alkoxyalkyl-, synthesis of, 340
Phosphonium salts, α-amino-, synthesis of, 487
Phosphonium salts, 1-aminoalkyl-
from
N-alkoxymethylamides, 492
N-hydroxymethylamides, 492
trialkyl(aryl)phosphines, 492
trimethylsilylmethyl isocyanide, 492
Phosphonium salts, α-aminoalkyl-
from, α-aminoalkylphosphines, 478
synthesis of, 462
Phosphonium salts, 1-arsinoalkyl-, synthesis of, 564
Phosphonium salts, benzoylmethyl-, reactions, with selenoxides, 928
Phosphonium salts, bromodifluoromethyl-, as starting materials, for difluorocarbene, 7

Phosphonium salts, (chloromethyl)-, from, tris(2,4,6-trimethoxyphenyl)phosphine, 118
Phosphonium salts, 1-cycloalkenyl, for the synthesis of phosphorus-containing heterocycles, 488
Phosphonium salts, (dichloromethyl)-, synthesis of, 112
Phosphonium salts, (ethoxycarbonylfluoromethyl)-, synthesis of, 118
Phosphonium salts, α-halo-
from, phosphoranes, 117
as starting materials, for Wittig reagents, 117
synthesis of, 117
Phosphonium salts, α-hydroxyalkyl-, synthesis of, 336
Phosphonium salts, 1-phosphinoalkyl-, synthesis of, 551
Phosphonium salts, 1-silylalkyl-, synthesis of, 569
Phosphonium salts, stiboniomethyl-, synthesis of, 565
Phosphonium salts, tetrakis(aminomethyl)-, as starting materials, for 1,3,5-diazaphosphanes, 458
Phosphonium salts, tetrakis(hydroxymethyl)-
reactions
with primary amines, 456
with sodium ethoxide, 458
Phosphonium salts, tetrakishydroxymethyl-, as starting materials, for α-aminoalkylphosphorus compounds, 475
Phosphonium salts, tri-*n*-butyl-, deprotonation of, 118
Phosphonium salts, tri-*n*-butyl(fluoromethyl)-, synthesis of, 117
Phosphonium salts, triphenyl-, as starting materials, for thiaselenafulvenes, 870
Phosphonium salts, vinyl, use in the synthesis of cyclopentanoid natural products, 925
Phosphonium ylides, bis-, condensation of, 801
Phosphonium ylides, triphenyl, synthesis of, 773
Phosphonoacetamides, as starting materials, for α-aminoalkylphosphorus compounds, 476
Phosphono alcohols, chiral, as starting materials, for α-aminophosphonic acids, 475
Phosphonoalkenes, *gem*-chloro-, synthesis of, 801
Phosphonoalkenes, *gem*-fluoro-, synthesis of, 801, 803
Phosphonoalkenes, *gem*-halo-, synthesis of, via Peterson elimination, 801
α-Phosphonocarboxylic acids, as starting materials, for α-aminophosphonic acids, 476
Phosphonochloridates, α-halo-, synthesis of, 131
Phosphonodichloridate, 2-chloro-2-(1-chloroethenyloxy)ethyl-, synthesis of, 766
Phosphonodithioate, S,S-diethyl 1-chlorovinyl, reactions, with sodium methanethiolate, 927
Phosphonodithioate, S,S-diethyl 1-ethylthiovinyl, synthesis of, 927
Phosphonodithioformates, use in the synthesis of ketene dithioacetals, 847
Phosphonoketenes, *gem*-chloro-, synthesis of, 802
Phosphonyl anions
reactions, with iron(II) chloride, 1026, 1035
synthesis of, 1031
Phosphoramidate, diethyl, as starting material, for α-aminoalkyl phosphonates, 470
Phosphoramidates, N-α-chloro-N-β-phenylsulfenyl-, synthesis of, 106
Phosphoramidates, α-hydroxy, chlorination of, 106
Phosphoramide, hexamethyl- (HMPA)
for the deprotonation of phenylvinyl selenides, 965
use in the synthesis of α-silyl enamines, 1013
Phosphoramidites
reactions
with aldehydes, 475
with ketones, 475
as starting materials, for α-aminoalkylphosphorus compounds, 475
Phosphoramidites, N-phenyl-, reactions, with benzaldehyde, 476
Phosphorane, cyclopentadienylidenetriphenyl-, reactions, with dodecacarbonyltriruthenium, 685

Phosphorane

Phosphorane, dibromotriphenyl-, reactions, with ketene aminals, 990
Phosphorane, difluorotris(trifluoromethyl)-, use in the synthesis of difluorocarbene, 7
Phosphorane, phenyltetrafluoro-, as starting material, for (fluoromethyl)phenylphosphinic acid, 123
Phosphorane, trimethylbis(trimethylsilyl)methylene-, use in the synthesis of 1,1-bis(silyl)alkanes, 624
Phosphorane complexes, imino-, synthesis of, 800
Phosphoranes
  reactions, with carbon disulfide, 843
  as starting materials
    for 1,3-aminophosphoranes, 478
    for α-halophosphonium salts, 117
  synthesis of, 825
Phosphoranes, acyl-, O-alkylation of, 824
Phosphoranes, 1-alkoxyallyltriphenyl-, isomerization of, 918
Phosphoranes, alkyl(dichloro)diphenyl-, synthesis of, 112
Phosphoranes, (alkylidene)chloro-, 1,2-chlorotropic rearrangement of, 112
Phosphoranes, alkylidenetriphenyl-, synthesis of, 348
Phosphoranes, N-(alkylthiomethyl)imino-, as intermediates, 329
Phosphoranes, 1,3-amino-
  from
    phosphoranes, 478
    Schiff bases, 478
    vinylphosphoranes, 478
Phosphoranes, (arylalkylidene)triphenyl-, reactions, with chlorodifluoromethane, 735
Phosphoranes, gem-bromoimino-, from, α-bromomalononitrile, 800
Phosphoranes, (α-chlorovinyl)tetrachloro-, synthesis of, 804
Phosphoranes, dibromo-, reactions, with benzaldehyde, 745
Phosphoranes, dichloro-
  as starting materials, for α-chloroalkylphosphines, 112
  synthesis of, 112
Phosphoranes, P-fluoro-, synthesis of, 803
Phosphoranes, α-halo-, synthesis of, 133
Phosphoranes, imino-
  reactions, with carbonyl compounds, 518
  See also Imides, pohosphine
Phosphoranes, α-thioacyl-, S-alkylation of, 834
Phosphoranes, tricyclic, synthesis of, 478
Phosphoranes, vinyl-, as starting materials, for 1,3-aminophosphoranes, 478
Phosphorimidic tribromide, as starting material, for α-(N,N',N''-triphenylphosphorimidic triamido) enamines, 991
Phosphorinanes, as starting materials, for phosphinic acids, 124
Phosphorinanes, 2,6-dihydroxy-, synthesis of, 339
Phosphorine, t-butyldipropynyl-, reactions, with phenylphosphine, 1026
Phosphorines, tetrahydro-, from, methylenephosphines, 459
Phosphorins, chlorotetrahydro-, synthesis of, 111
Phosphoriodites, dialkyl, reactions, with aldimines, 462
Phosphorocyanatidites, reactions, with N-alkylideneacetamide, 493
Phosphorodiimidates, Wittig rearrangement of, 923
Phosphorous tribromide, reactions, with 1-ethoxy-1-(trimethylsiloxy)cyclopropane, 50
Phosphorus
  α-halo oxo acids of, 123
  reactions
    with heptafluoroisopropyl iodide, 114
    with α-hydroxyamines, 466
    with N-hydroxymethyl dialkylamines, 452
  stabilization by, 1082
  synthesis of functions, **543–589**
    with nitrogen, 451
    with no halogen, chalcogen or nitrogen, **1021–1042**

Phosphorus(III) compounds, trimethylsilyloxy-, reactions, with N-benzylidineallylamines, 461
Phosphorus(III) halides
  reactions
    with orthoformates, 340
    with trifluorovinyllithium, 802
    with trifluorovinylmagnesium, 802
Phosphorus(V) acids, synthesis of, 106
Phosphorus(V) compounds, vinyl-, halogenation of, 804
Phosphorus acetals, oxidation of, 343
Phosphorus acid
  reactions, with cyclohexylamine, 470
  as starting material, for 1-aminoalkylphosphonic acids, 470
  use in the synthesis of α-aminophosphonic acids, 462
Phosphorus acid esters, α-aminoalkyl-, synthesis of, 478
Phosphorus acids, α-aminoalkyl-
  deesterification of, 478
  esterification of, 478
  transesterification of, 478
Phosphorus acids, tervalent, use in the synthesis of α-aminoalkylphosphorus compounds, 470
Phosphorus allenes, gem-halo-, from, propargyl alcohols, 802
Phosphorus amides
  as starting materials
    for α-aminoalkylphosphorus compounds, 475
    for 3-aminobenzophosphoranes, 478
Phosphorus anions, reactions, with 1-haloalkylphosphorus compounds, 554
Phosphorus chemistry, reviews, 336
Phosphorus compounds
  from
    aldehydes, 336
    ketones, 336
  reactions, with alkoxyalkynes, 913
  synthesis of, 345
Phosphorus compounds, α-alkoxy alkyl-, as starting materials, for anions, 917
Phosphorus compounds, amino-, oxidation of, 493
Phosphorus compounds, α-amino-
  synthesis of
    via Arbuzov reactions, 487
    via Michaelis–Becker reactions, 487
  for the synthesis of tricoordinate phosphorus functions, 480
Phosphorus compounds, α-aminoalkyl-
  from
    acetyl nitrate, 490
    amines, 466
    α-aminoalkylphosphorus compounds, 493
    benzyl carbamates, 467
    carbamates, 466
    carbinolamines, 465
    dinitrogen tetroxide, 490
    electrophilic nitrogen compounds, 490
    ethyl carbamates, 467
    ethyl nitrate, 490
    halomethylphosphinic acids, 473
    halomethylphosphonic acids, 473
    α-hydroxyalkylphosphorus compounds, 456
    α-hydroxyamides, 465
    α-hydroxy compounds, 475
    O-mesitylsulfonylhydroxylamines, 490
    methyl carbamates, 467
    α-oxophosphorus compounds, 491
    phenyl azide, 490
    phosphonoacetamides, 476
    phosphoramidites, 475
    phosphorus amides, 475
    tetrakishydroxymethylphosphonium salts, 475
    thioureas, 466
    trifluoromethanesulfonyl azide, 491

ureas, 466
as starting materials, for α-aminoalkylphosphorus compounds, 493
synthesis of
via Arbuzov reactions, 464
via Curtius rearrangement, 476, 492
via cycloaddition reactions, 488, 493
via Diels–Alder reactions, 489
via dipolar cycloaddition reactions, 479
via Hofmann rearrangement, 476
via Kabachnik–Fields reactions, 487
via Lossen rearrangement, 492
via Michaelis–Becker reactions, 464
via Mitsunobu reactions, 491
Phosphorus compounds, aminoalkyl hexacoordinate, synthesis of, 494
Phosphorus compounds, α-carbonyl-, enolization of, 920
Phosphorus compounds, α-chloroalkyl-, replacement of, 339
Phosphorus compounds, α-halo-, synthesis of, 110
Phosphorus compounds, 2-haloalkenyl-
from
alkenes, 802
carbonyl compounds, 801
Phosphorus compounds, α-haloalkenyl-
synthesis of, 801
halogenation methods, 804
Phosphorus compounds, 1-haloalkyl-
Arbuzov reactions of, 555
reactions
with alkyl phosphinites, 556
with dialkyl phosphonites, 556
with phosphines, 554
with phosphorus anions, 554
with trialkyl phosphites, 555
Phosphorus compounds, hydroxyalkyl-
reactions
with acrylonitrile, 342
with triethyl orthoformate, 342
Phosphorus compounds, α-hydroxyalkyl-
as starting materials, for α-aminoalkylphosphorus compounds, 456
synthesis of, 339
Phosphorus compounds, 1-metalloalkyl-
reactions
with chlorophosphine oxides, 557
with chlorophosphines, 551
Phosphorus compounds, nucleophilic
reactions
with imines, 455
with iminium ions, 455
Phosphorus compounds, α-oxo-
as starting materials
for α-aminoalkylorganophosphorus compounds, 471
for α-aminoalkylphosphorus compounds, 491
Phosphorus compounds, tetracoordinated, synthesis of, 569
Phosphorus compounds, α-trifluorovinyl-, synthesis of, 802
Phosphorus compounds, trivalent, reactions, with dimethylketene, 914
Phosphorus compounds, β,γ-unsaturated α-oxy-, isomerization of, 917
Phosphorus compounds, vinyl-, as starting materials, for phosphinates, 561
Phosphorus-containing heterocycles, synthesis of, 488
Phosphorus diamines, alkoxy-, as starting materials, for α-phosphorodiamidoyl enamines, 1011
Phosphorus esters
as starting materials
for α-aminoalkyl phosphinates, 464
for α-aminoalkylphosphine oxides, 464
for α-aminoalkyl phosphonates, 464
Phosphorus esters, trivalent, reactions, with N-bromomethylphthalimide, 464
Phosphorus functions, dicoordinated
synthesis of, 110, 479, 565
theoretical, 452
Phosphorus functions, tetracoordinated, synthesis of, 117, 460, 485
Phosphorus functions, tricoordinated
synthesis of, 110, 452, 480, 565
via Mannich reactions, 481
Phosphorus nucleophiles
reactions
with alkynes, 561
with 1,1-dihaloalkanes, 553
with vinyl chlorides, 553
for the synthesis of tetracoordinate phosphorus functions, 460
Phosphorus oxides, α-haloalkenyl-, synthesis of, 804
Phosphorus pentabromide
reactions, with sodium iodomethanesulfonate, 85
use in the synthesis of dibromoalkanes, 25
Phosphorus pentachloride
for the α-chlorination of ethers, 47
reactions
with acetoxy groups, 766
with acetylated sugars, 766
with N-acetylcarbazole, 791
with (alkylthio)alkyne chlorides, 804
with alkynes, 772
with gem-amido phosphinoylalkenes, 1001
with bis(alkylthio)ethynes, 926
with chloromethanesulfonate, 85
with chlorophosphinoyl compounds, 1013
with phenylacetate, 766
with phosphinic acid anhydrides, 125
with sodium iodomethanesulfonate, 85
with tertiary amide, 790
with vinyl acetate, 766
use in the synthesis of dichloroalkanes, 18
Phosphorus pentasulfide, reactions, with amides, 424
Phosphorus triamide, hexamethyl-, reactions, with aryl aldehydes, 475
Phosphorus tribromide, use in the synthesis of dibromoalkanes, 25
Phosphorus trichloride
reactions
with aldehydes, 130
with aldimines, 488
with alkoxyalkynes, 765
with α-alkoxyglycines, 465
with hydroxymethylamides, 469
with phenyl acetate, 742
Phosphorus ylides
formation of borane adducts, 574
lithiation of, 576
reactions
with aluminum compounds, 587
with beryllium chloride, 577
with boron compounds, 574
with cadmium compounds, 586
with chlorophosphine oxides, 557
with chromium compounds, 580
with cobalt compounds, 582
with copper compounds, 585
with dehydrodithizone, 490
with gem-diisocyanatoalkanes, 992
with erbium compounds, 578
with gadolinium compounds, 578
with gallium compounds, 587
with gold compounds, 586
with halogermanes, 574
with halophosphines, 551
with halosilanes, 570
with halostannanes, 588
with holmium compounds, 578
with indium compounds, 587

Phosphorus ylides

    with iridium compounds, 583
    with iron compounds, 581
    with lanthanum compounds, 578
    with lead compounds, 589
    with lutetium compounds, 578
    with magnesium compounds, 577
    with manganese compounds, 581
    with mercury compounds, 587
    with molybdenum compounds, 580
    with neodymium compounds, 578
    with nickel compounds, 583
    with palladium compounds, 583
    with platinum compounds, 585
    with praseodymium compounds, 578
    with rhodium compounds, 582
    with ruthenium compounds, 582
    with samarium compounds, 578
    with scandium compounds, 578
    with silver compounds, 585
    with tantalum compounds, 579
    with thallium compounds, 587
    with thorium compounds, 589
    with titanium compounds, 578
    with transition metal compounds, 578
    with tungsten compounds, 580
    with uranium compounds, 589
    with vanadium compounds, 579
    with zinc compounds, 586
    with zirconium compounds, 579
    as starting materials
        for S,P-acetals, 346
        for Te,P-acetals, 348
Phosphorus ylides, silyl-, protonation of, 571
Phosphoryl chloride, (dichlorophosphinomethyl)-, reactions of, 553
Phosphoryl chloride, (trimethylsilylmethyl)-, reactions of, 574
Phosphoryl compounds, α-halo-, synthesis of, 123
Phosphoryl groups, chloro-, hydrolysis of, 1006
Photolysis, for the synthesis of carbenes, 1081
Photopyrone, reactions, with dicarbonylcyclopentadienylcobalt, 690
Photoreactions, for the synthesis of arsines, 598
Photoresists, iminosilanes as, 507
Phthalaldehyde
    reactions
        with ammonia, 298
        with thiourea, 298
Phthalide, bromination of, 56
Phthalides, 3-chloro-, synthesis of, 54
Phthalides, 3-fluoro-, synthesis of, 52
Phthalimide
    irradiation of, 1077
    reactions, with α-hydroxyphosphonates, 475
    silylmethylation of, 513
    use in the synthesis of radicals, 1077
Phthalimide, N-bromomethyl-
    reactions
        with silver nitrocyanamide, 447
        with trivalent phosphorus esters, 464
    synthesis of, 99
Phthalimide, N-chloromethyl-, displacement of chlorine from, 499
Phthalimide, N-phenylseleno- (NPSP)
    reactions
        with alkynes, 787
        with diazomethane, 335
        with ethyl diazoacetate, 335
Phthalimide, N-vinyl-, use in the synthesis of α-haloamines, 101
Phthalimides, reactions, with hydrogen iodide, 101
Phthalimides, α-bromo-, synthesis of, 99
Phthalimides, N-chloromethyl-, synthesis of, 483

Phthalimides, N-phosphinomethyl-, synthesis of, 483
Phthalimidines, synthesis of, 298, 303
Phthalimidoalkenes, *gem*-halo, synthesis of, 794
Phthalimidomethyl fluorides, synthesis of, 101
4-Picoline, oxidation of, 206
α-Picoline, trichloro-, reduction of, 13
Pinanediol, reactions, with diisopropyl (1-methoxyvinyl)boronate, 934
Piperazine, diketo-, synthesis of, 218
Piperazines, synthesis of, 301
Piperidine
    reactions
        with 4-fluorobenzaldehyde, 298
        with pyrrole-2-carbaldehyde, 406
Piperidine, N-acetyl-, synthesis of, 99
Piperidine, 1-benzhydryl-, synthesis of, 334
Piperidine, N-methyl-, metallation of, 527
Piperidine, N-nitroso-, lithiation of, 335
Piperidines, 2-silyl-, synthesis of, 511
Pivalate, iodomethyl, reactions, with zinc, 385
Pivaldehyde
    condensation of, 302
    reactions, with α-hydroxycarboxylic acids, 213
Pivalic acid, demethoxycarbonylation of, 846
Platinic(IV) acid, hexachloro-
    as catalyst, for the synthesis of 1,1-bis(silyl)alkanes, 617
    as hydrosilylation catalyst, 1044
    See also Speier's catalyst
Platinum, cycloocta-1,5-dienyldimethyl-, reactions, with 1,1-bisdiphenylphosphino ethene, 1028
Platinum, ethylenebis(triphenylphosphine)-, reactions, with rhodium bis(triflate) complexes, 716
Platinum(0), tetrakis(triphenylphosphine)-, as hydrosilylation catalyst, 1044
Platinum(II) chloride, reactions, with bisdiphenylphosphino ethene, 1028
Platinum(II) complexes, reactions, with iodomethane, 1029
Platinum(II) complexes, dimethyl-, synthesis of, 1028
Platinum(IV) compounds, synthesis of, 151
Platinum complexes, synthesis of, 955
Platinum complexes, anionic, with perfluorophenyl-bridged groups, 691
Platinum complexes, cationic trinuclear, synthesis of, 692
Platinum complexes, (chloromethyl)-, synthesis of, 151
Platinum complexes, dichloro-, synthesis of, 1028
Platinum complexes, dihalo-, reactions, with ethyl diazoacetate, 151
Platinum complexes, dinuclear, synthesis of, 692
Platinum complexes, ethynyl-, reactions, with hydrogen chloride, 822
Platinum complexes, ferrocenyl-, synthesis of, 722
Platinum complexes, α-haloalkyl-, synthesis of, 151
Platinum complexes, (halomethyl)-, synthesis of, 151
Platinum complexes, methylene-bridged, palladium analogues of, synthesis of, 691
Platinum complexes, trinuclear
    from, triarylcyclopropenium bromides, 693
    synthesis of, 692
Platinum compounds
    reactions
        with 1-haloalkyl phosphonates, 585
        with phosphorus ylides, 585
Platinum compounds, 1-haloalkyl-, reactions, with phosphines, 585
Platinum hydride complexes, cationic, synthesis of, 691
Platinumphosphine complexes, reactions, with 1,1-dihaloalkanes, 584
Platinum–phosphinoalkene complexes, synthesis of, 1023, 1028
Plumbane, chlorotriphenyl-, as starting material, for (iodomethyl)triphenylplumbane, 156
Plumbane, (iodomethyl)triphenyl-, from, chlorotriphenylplumbane, 156

Plumbanes, α-bromo-, synthesis of, 156
Pluridone, synthesis of, 268
Polonovski reaction, of trimethylamine N-oxide, 311
Polyazapolycycles, synthesis of, 298
Polyols, reactions, with methylal, 185
Polyphosphazenes, silylation of, 570
Polypropylene oxide, as starting material, for α-fluoro ethers, 44
Polysilanes, rearrangement of, 624
Polysulfide, ammonium, use in the synthesis of α-aminophosphine sulfides, 477
Ponzio reaction, for the synthesis of aryldinitromethanes, 429
Potassium
   reactions
      with 2,5-dimethyl-1-phenylarsole, 1039
      with methoxymethyl chloride, 360
Potassium, (2,4-dimethylpentadienyl)-, reactions, with cyclopentadienylsodium, 678
Potassium, trimethylsilylmethyl-, synthesis of, 653
Potassium acetate
   reactions
      with chloromethylpentamethyldisilane, 614
      with (chloromethyl)trimethylgermane, 362
Potassium arsolyls, reactions, with iron(II) chloride, 1039
Potassium t-butoxide
   for the deprotonation of vinyl ethers, 949
   reactions
      with chloroform, 16, 737, 740
      with 1,2-dichloro-1,2-diisopropoxyethane, 767
      with dichlorofluoromethane, 751, 753
Potassium carbonate, use in the synthesis of hemiaminals, 300
Potassium cyanate
   reactions
      with dichloro compounds, 441
      with germylmethyl halides, 521
Potassium cyclooctaldraenide, use in the synthesis of cyclooctatetraene-bridged complexes, 672
Potassium dicarbonylcyclopentadienylium, use in the synthesis of carbon-bridged diiron complexes, 679
Potassium enolates, synthesis of, 838
Potassium fluoride
   reactions
      with 2,2-dihalovinyl sulfones, 783
      with α-keto alkyl chlorides, 3
   use in the synthesis of chlorofluoroalkanes, 30
Potassium hexamethyldisilazide, reactions, with selenoesters, 841
Potassium hydrogen difluoride, reactions, with N,N-diethyl-1-propynamine, 790
Potassium hydroxide, reactions, with methanol, 748, 906
Potassium iodide, reactions, with phenyl bis(chloromethyl)phosphinate, 127
Potassiumiron(III) cyanide, for the nitration of nitroalkanes, 428
Potassium metabisulfite, reactions, with sugars, 237
Potassium naphthalenide
   use in the synthesis of carbon-bridged dititanium complexes, 671
   use in the synthesis of carbon-bridged zirconium complexes, 671
Potassium nitronates
   reactions
      with p-toluenesulfonyl iodide, 331
      with tosyl azide, 331
Potassium perfluoro 2-or 3-aminopropanoates, as starting materials, for N-trifluorovinyl secondary amines, 794
Potassium permanganate
   for the oxidation of dithioacetals, 265, 272
   reactions, with homoallylic alcohols, 168
   use in the synthesis of monosulfones, 854
Potassium phenolate, reactions, with tribromoethylene, 761

Potassium phenyl selenide, reactions, with 1,1-dinitro-2,2-diphenylethene, 911
Potassium phthalimide
   reactions
      with phosphine oxides, 473
      with pyridinium salts, 448
   silylmethylation of, 513
Potassiums, α-fluoroalkenyl-, synthesis of, 819
Potassium salt, dibenzyl iminodicarboxylate-, N-alkylation of, 101
Potassium salts, reactions, with amines, 977
Potassium sand, reactions, with bis(trimethylsilylmethyl)mercury, 653
Potassium selenocarboxylates, reactions of, 288
Potassium t-butoxide, reactions, with butynes, 1028
Potassium tetrafluorocobaltate, reactions, with methyl benzoate, 3
Potassium thiocyanate
   reactions
      with α-bromo alkyl isocyanates, 434
      with α-bromo alkyl isothiocyanates, 434
Potassium tricyanomethanide
   reactions, with methane sulfonate salts, 977
   as starting material, for ketenimines, 977
Potassium trifluoromethylfluoroborate, pyrolysis of, 7
Potassium trinitromethanide, reactions, with pyridinium salts, 995
Praseodymium compounds, reactions, with phosphorus ylides, 578
Pregnenolone, construction of side chains on, 939
Prins addition, of formaldehyde to tetrafluoroethene, 44
Prodrugs, ester, of β-lactam antibiotics, synthesis of, 205
Propadiene, 1-bromo-1-(trimethylgermyl)-, synthesis of, 811
Propadiene, perfluoro-1,1,3-tris(dimethylamino)-, reactions, with perfluoro-N-bromodimethylamine, 794
Propadienes, 1,3-bis(trimethylsilyl) 1-silyl 3-phenyl-, synthesis of, 1047
Propadienes, gem-bromosilyl-, synthesis of, 807
Propadienes, gem-halo(trimethylstannyl)-, synthesis of, 813
Propanal, reactions, with mercury(II) chloride, 697
Propanal, S-2-benzyloxy-, use in the synthesis of chiral dihydropyrimidines, 408
2-Propanamine, N,N-dimethyl-1-ferrocenyl-, deprotonation of, 1032
Propan-1,3-diol, 2-amino-, reactions, with formaldehyde, 302
Propane, 1,1-bis(diethylboryl)-3-chloro-, synthesis of, 631
Propane, 2,2-bis(methylthio)-1,3-diphenyl-, reactions, with m-chloroperbenzoic acid, 268
Propane, 2-bromo-2-nitro-, reactions, with thiolates, 327
Propane, 1,3-diamino-, reactions, with glyoxal, 409
Propane, 2,2-dibromo-, synthesis of, 24
Propane, 2,2-dimethoxy-
   as dehydrating agent, for the synthesis of acetals, 178
   use in the synthesis of acetals, 184
Propane, 2-iodo-2-nitro-, reactions, with sodium benzene sulfinate, 332
Propane, 2-nitro-
   reactions, with thiolates, 327
   as starting material, for gem-fluoronitro compounds, 108
Propane, 1,1,3,3-tetramethoxy-, reactions, with 2-thiolethanol, 230
Propane, 1,1,3,3-tetranitro-, fluorination of, 795
Propanediol, 2,2-dinitro-, synthesis of, 428
Propanedithiol, reactions, with alkynic ketones, 258
1,3-Propanedithiol, reactions, with hexachlorobutadiene, 770
Propane-2-thiol, 2-methyl-, irradiation of, 1077
Propanoic acid, 3-chloro-2,2,3-trifluoro-, synthesis of, 5
Propanoic acid, 3,3-dichloro-2,2-dimethyl-, synthesis of, 11
Propanol, 2-methyl-, reactions, with bromomethanesulfonyl bromide, 86
Propanol, 3,3,3-tris(trimethylsilyl)-, deprotonation of, 620

Propanoyl chloride

Propanoyl chloride, 1-chloro-1-(chlorosulfenyl)-, reactions, with sodium ethoxide, 72
Propanoyl chlorides, synthesis of, 125
Propargyl chloride, reactions, with diethylborane, 631
Propene, metallation of, 668
Propene, 1-azidopentafluoro-, synthesis of, 799
Propene, 3-chloro-1-methoxy-, reactions, with alcohols, 199
Propene, 1-chloro-2-methyl-, hydroboration of, 144
Propene, (Z)-1,3-dibromo-1-trimethylsilyl-, synthesis of, 807
Propene, 2,3-dichlorotetrafluoro-, reactions, with dimethylarsine, 805
Propene, (E)-1,3-bis(trimethylsilyl)-, reactions, with s-butyllithium, 603
Propene, hexafluoro-
  reactions
    with azide anions, 799
    with dimethylphosphine, 802
    with tetramethyldiphosphine, 802
Propene, 2-methoxy-, reactions, with diols, 186
Propene, 1,2,3,3,3-pentafluoro-1-trimethylsilyl-, reactions, with bis(trimethylsilyl)mercury, 1048
Propene, perfluoro-
  reactions, with sulfur trioxide, 86
  as starting material, for carbene-bridged diplatinum complexes, 691
Propene, 3,3,3-trichloro-2-methyl-, reactions, with p-cresol, 745
1-Propene, 2,3-bis(benzenesulfonyl)-, displacement of, 275
Prop-1-ene, 3-chloro-3,3-difluoro-2-trifluoromethyl-, reactions, with lithium aluminum hydride, 735
1-Propene, 1,1-dilithio-2-methyl-, synthesis of, 1064
1-Propene, 3-dimethylphenylsilyl-3-triphenylgermyl-, synthesis of, 639
1-Propene, 1-phenylthio-3-(trimethylsilyl)-, deprotonation of, 938
1-Propene, 1,3,3,3-tetrafluoro-2-trifluoromethyl-1-methoxy-, synthesis of, 762
1-Propene, (E)-1,3,3-tris(trimethylsilyl)-, synthesis of, 603
Propenes, (Z)-2-phenyltetrafluoro-, reactions, with lithium tetramethylpiperidide, 755
2-Propenoate, ethyl 3-amino-3-(ethylthio)-, synthesis of, 905
Propenolic acid, 3,3-dichloro-, from, chloropropiolic acid, 743
1-Propenyllithium, 1-ethoxy-, isomerization of, 949
Propiolate, ethyl
  reactions
    with anthranilamides, 422
    with ethanol, 195
Propiolate, methyl bromo-, reactions, with aziridine, 793
Propiolic acid, chloro-, as starting material, for 3,3-dichloropropenolic acid, 743
Propiolic acid, phenyl-, reactions, with iodine pentoxide, 749
Propionate, 1,2-dichloropropenyl-α-chloro-, synthesis of, 768
Propionate, ethyl 3-chloro-, reactions, with sodium, 171
Propionate, ethyl 3,3-diethoxy-, synthesis of, 195
Propionic acids, 2-halo-, synthesis of, 129
Propionic acids, phosphinico-, synthesis of, 124
Propiononitrile dihydrochloride, 2,3-di(t-butylimino)-, reactions, with lithium ethoxide, 894
Propionyl chloride, α-chloro-, reactions, with triethylamine, 768
Propiophenones, synthesis of, 365
1-Propynamine, N,N-diethyl-, reactions, with potassium hydrogen difluoride, 790
Propyne, reactions, with hydrogen bromide, 24
Propyne, 1,3-bis(trimethylsilyl)-, reactions, with butyllithium, 1059
Propyne, 1-chlorotrifluoro-, reactions, with silver trifluoro methoxide, 764
Propyne, 1-phenyl-, dilithiation of, 668
Propyne, 1-(triethylsilyl)-, dihydroboration of, 632
Propyne, 1-trimethylsilyl-, reactions, with zirconocene dichloride, 1038

Propyne, 1,3,3-tris(trimethylsilyl)-, as starting material, for lithioallenes, 1047
1-Propyne, 1-chloro-3-(trimethylsilyl)-, reactions, with n-butyllithium, 648
Prop-1-yne, 1-diethylamino-
  reactions
    with N-benzenesulfonyl-S-phenylthiobenzimidate, 905
    with N-benzoyl-S-phenylthiobenzimidate, 905
    with 3-methylthio-1,2-benzisothiazole 1,1-dioxide, 905
    with 3-phenylseleno-1,2-benzisothiazole 1,1-dioxide, 911
Prop-1-yne, 1-phenoxy-, reactions, with diethylsilane, 933
1-Propyne, 1-trimethylsilyl-, monohydroboration of, 644
Propyne complexes, use in the synthesis of α-carbocations of alkyne-bridged dimolybdenum complexes, 674
Propynes, 1,3-bis(trimethylstannyl)-, γ-halodestannylation of, 813
Propynes, 1-silyl 3-phenyl-, reactions, with trimethylsilyl chloride, 1047
Propynes, 3-titano 1-trimethylsilyl, synthesis of, 1061
Propynes, 3-trimethylsilyl 1-silyl 3-phenyl-, synthesis of, 1047
1-Propyn-2-ol, 1-trimethylsilyl-, hydromagnesiation of, 1062
Prop-2-yn-1-ol, reactions, with vinylzinc bromide, 1067
Prostaglandin I₁, synthesis of, 279
Protic acids, as catalysts, for the synthesis of acetals, 178
Protonation, for the synthesis of tricoordinate phosphorus functions, 480
Pseudonitroles
  from, oximes, 439
  synthesis of, 439
Pulegone, use in the synthesis of 1,3-hydroxythiols, 229
Pummerer reaction
  mechanism, 221
  for the synthesis of α-halo selenides, 88
  for the synthesis of monothioacetals, 226
  for the synthesis of 2-(phenylsulfinyl)acylals, 207
Pummerer rearrangement
  of (diphenylthiophosphinoyl)phenyldiazomethane, 346
  for the synthesis of sulfur functions, 923
Purines, reduction of, 413
Purines, 6-substituted, reactions, with 3,4-dihydro-2H-pyran, 306
Pyran, 2,3-dihydro-, deprotonation of, 949
Pyran, dihydro- (DHP)
  reactions
    with benzenesulfinic acid, 236
    with carboxylic acids, 208
  as starting material
    for hemithioacetals, 223
    for hydroxy dithioacetals, 246
  synthesis of, 827
  use in the synthesis of acetals, 186
Pyran, 2-methoxytetrahydro-, reactions, with benzenesulfinic acid, 236
2H-Pyran, 3,4-dihydro-
  reactions
    with 3-chlorobenzoylhypobromite, 208
    with 6-substituted purines, 306
Pyran compounds, 2-alkoxy, synthesis of, 202
Pyran-4-one, 2,3-dihydro-, reactions, with 1,1-bis(sulfonyl)ethene, 277
Pyrans, 2-(benzenesulfonyl)tetrahydro-, synthesis of, 236
Pyrans, α-bromotetrahydro-, stability of, 51
Pyrans, 2-chlorotetrahydro-, synthesis of, 49
Pyranyl ethers, tetrahydro-, synthesis of, 186
Pyrazines, diboradihydro-, synthesis of, 522
Pyrazole
  reactions
    with acetals, 437
    with ketals, 437
Pyrazoles, as starting materials, for α-azo enamines, 989

Pyrazolines
  reactions
    with bromine, 105
    with N-bromosuccinimide, 109
  synthesis of, 488
Pyrazolines, 3-bis(phenylsulfonyl)-, synthesis of, 277
Pyrazolines, 3-chloro-, synthesis of, 109
Pyrazolines, 3-nitro-, synthesis of, 440
5-Pyrazolones, reactions, with ynamines, 986
Pyridazines, cis-2,5-diphenyl tetrahydro-, irradiation of, 105
Pyridine
  for the chlorination of sulfides, 65
  reactions, with selenium(IV) dichlorides, 88
  silylmethylation of, 519
  use in the synthesis of bis(cyclopentadienyl) metal–methyl complexes of lanthanides, 702
Pyridine, 2-bis(trimethylsilyl)methyl-, synthesis of, 604
Pyridine, 2-dibromomethyl-, synthesis of, 27
Pyridine, 2-dichloromethyl-, synthesis of, 19
Pyridine, 2-(methylamino)methylhexahydro-, as starting material, for perhydroimidazopyridines, 408
Pyridine, 1,2,3-triazolo[1,5-a]
  reactions
    with bromine, 27
    with chlorine, 19
Pyridines, 2-alkoxytetrahydro-, synthesis of, 305
Pyridines, amino-, as starting materials, for ketene aminals, 979
Pyridines, bis(silylmethyl)-, synthesis of, 604
Pyridines, quaternized, amination of, 413
Pyridines, trimethylsilylmethyl-, metallation of, 647
Pyridines, 2-trimethylsilylmethyl-
  deprotonation of, 648
  synthesis of, 604
Pyridinium salts
  ammonolysis of, 448
  reactions
    with potassium phthalimide, 448
    with potassium trinitromethanide, 995
  synthesis of, 519
Pyridinium salts, tetrahydro-, synthesis of, 305
2-Pyridone, 1-methyl-, lithiation of, 540
2-Pyridone, 1-trimethylsilyl-, synthesis of, 540
Pyridones, silylmethyl, desilylation of, 540
2H-Pyrido[1,2-a]pyrimidines, 2-amino-, as starting materials, for ketene aminals, 979
Pyridoxal, synthesis of, 164
Pyridoxol, reactions, with manganese dioxide, 164
Pyrimidinediones, hexahydro-, synthesis of, 425
Pyrimidines
  ring-opening of, 978
  as starting materials, for ketene aminals, 978
  synthesis of, 421, 988
Pyrimidines, 4,6-dichloro-5-substituted, reactions, with amines, 978
Pyrimidines, dihydro-, synthesis of, 408
Pyrimidines, hexahydro-
  reactions, with Grignard reagents, 410
  synthesis of, 408
Pyrimidines, tetrahydro-, synthesis of, 408
Pyrimidine-4-thione compounds, synthesis of, 421
Pyrimidones, synthesis of, 978
Pyrophosphite, tetraethyl
  reactions, with diethylvinyl phosphonate, 552
  use in the synthesis of diethoxy(heptafluoroisopropyl)phosphines, 117
Pyrrole, reactions, with dichloromethane, 410
Pyrrole, N-methyl-3,4-dinitro-, reactions, with secondary amines, 415
Pyrrole compounds, from, isocyanides, 538
Pyrroles, synthesis of, 534
Pyrrolidine, reactions, with β,β-dichloroalkenyl aryl ketones, 793

Pyrrolidine, N-allyl, lithiation of, 511
Pyrrolidine, 1-methoxy carbonyl, methoxylation of, 303
Pyrrolidines, synthesis of, 536
Pyrrolidines, 2-(diphenylphosphinoyl)-, synthesis of, 468
Pyrrolidinoalkenes, gem-chloro, synthesis of, 793
1-Pyrrolidinoethene, 1-phenyl-, reactions, with C,N-diphenylnitrone, 307
Pyrrolidin-2-one, N-chloromethyl-, phosphonylation of, 465
Pyrrolines, synthesis of, 415
Pyrrolo[1,2-a]imidazolinones, synthesis of, 422
Pyrrolo[2,3-b]indole nucleus, hexahydro-, from, indolenium ions, 418
Pyrylium salts, reactions, with dimethyl sulfoxide, 863

Quaternary salts, synthesis of, 597
Quinaldine, tribromo-, reduction of, 23
Quinazolinones, synthesis of, 422
Quinazolones, acetoxy, as starting materials, for aziridines, 515
Quinodimethane, 7,7,8,8-tetracyano-, reactions, with ethanolamines, 893
2,3-Quinodimethane, N-benzoylindole-, use in the synthesis of bis(sulfones), 277
Quinodimethans, tricyano-
  as starting materials
    for gem-dihydrazinoalkenes, 985
    for α-hydrazino enamines, 985
Quinoline
  reactions
    with dimethylketene, 886
    with sodium methoxide, 195
Quinoline-1-oxide, reactions, with bis-(phenylsulfonyl)methane, 281
Quinolines, quaternized, amination of, 413
Quinolinium cation, 4,4-diiodo-1,1-dimethyl-1,4-dihydro-, as starting material, for α-iodo alcohols, 43
Quinone compounds, synthesis of, 901
o-Quinonemethides, reactions, with 1-pyrrolidinocyclohexene, 307
o-Quinones, reactions, with phosphaalkenes, 340
Quinonoid compounds, synthesis of, 855
Quinuclidine, irradiation of, 98

Racemates
  as starting materials
    for alcohols, 390
    for esters, 390
Radialenes, synthesis of, 820
Radical cation, via electron transfer, 1072
Radical centres
  flanked by nitrogen and oxygen, 1077
  flanked by nitrogen and sulfur, 1077
  flanked by two halogens, 1080
  flanked by two oxygens, 1078
  flanked by two sulfurs, 1080
Radicals, synthesis of, 1077, **1071–1083**
Ramberg–Bäcklund reaction, for the synthesis of α-halo sulfones, 79
Ramberg–Bäcklund rearrangement, for the synthesis of 1,1-dichloroalkenes, 742
Rearrangement
  for the synthesis of 1,1-dichloroalkenes, 743
  for the synthesis of ketene hemiaminals, 894
Reduction, electrochemical, for the synthesis of dichloromethyls, 13
Reduction, enzymatic, for the synthesis of cyclic hemiacetals, 165
Reformatsky reagents, reactions, with orthoformates, 193
Reimer–Tiemann reaction, for the synthesis of dichloroalkanes, 17
Reverse Brook rearrangement
  of alkoxygermanes, 361
  of alkoxysilanes, 358

Reverse Brook rearrangement
  of germyl thioethers, 371
  of silyl thioethers, 368
  See also Brook rearrangement
Reverse Diels–Alder reaction, for the synthesis of stannylvinyl ethers, 952
L-Rhamnose compounds, synthesis of, 235
Rhenate, pentacarbonyl-, reactions, with molybdenum alkyne complexes, 710
Rhenium, tricarbonyl(cyclopentadienyl)-, lithiation of, 706
Rhenium carbene complexes, reactions, with phosphines, 581
Rhenium complexes, (α-chlorobenzyl)-, synthesis of, 150
Rhenium complexes, α-haloalkyl-, synthesis of, 150
Rhodium, reactions, with diazoalkane, 686
Rhodium, bischlorodicarbonyl-, reactions, with cyclopentadienylbisdiphenylphosphinoruthenium chloride, 1029
Rhodium, dicarbonylcyclopentadienyl-
  decarbonylation of, 686
  use in the synthesis of carbon-bridged dirhodium complexes, 686
Rhodium, (pentachlorocyclopentadienyl)(1,5-cyclooctadiene)-, reactions, with butyllithium, 707
Rhodium(I), chlorotris(triphenylphosphine)-, as hydrosilylation catalyst, 1044
Rhodium(III) complexes, (chloromethyl)-, synthesis of, 149
Rhodium(III) complexes, (iodomethyl)-, synthesis of, 149
Rhodium analogues, synthesis of, 690
Rhodium bis(triflate) complexes, reactions, with ethylenebis(triphenylphosphine)platinum, 716
Rhodium catalysts, reactions, with silanes, 831
Rhodium complexes
  reactions
    with alkynes, 689
    with 3,3-dimethylcyclopropene, 689
Rhodium complexes, (α-chloroethyl)-, synthesis of, 149
Rhodium complexes, (halomethyl)-, synthesis of, 149
Rhodium complexes, mono-bridged, synthesis of, 686
Rhodium complexes, tri-bridged, synthesis of, 686
Rhodium complexes, vinyl-, reactions, with hydrogen chloride, 149
Rhodium compounds
  reactions, with phosphorus ylides, 582
  synthesis of, via cyclotrimerization of tolan, 689
Rhodium compounds, 1-haloalkyl-, reactions, with phosphines, 582
Rhodium dimers, dichloro(pentamethylcyclopentadienyl)-, reactions, with hexamethyldialuminum, 687
Ribonucleosides, oligo-, synthesis of, 219
L-(+)-Ribose, synthesis of, 362
(R)-Imidazolidinones, reactions, with N-bromosuccinimide, 102
Ring-opening
  for the synthesis of ketene S,N-acetals, 905
  for the synthesis of ketene hemiaminals, 891
Ritter reaction, of nitriles, 883
Ruthenacyclopentadiene complexes, from, dodecacarbonyl triruthenium, 683
Ruthenium–bisphosphinoalkene complexes, synthesis of, 1029
Ruthenium complexes
  from
    cyclooctatetraene, 683
    terminal alkynes, 683
Ruthenium complexes, (halomethyl)-, synthesis of, 149
Ruthenium complexes, neopentyl-, for the metallation of ferrocene, 720
Ruthenium complexes, pentalene-bridged, from, cyclooctatetraene, 683
Ruthenium complexes, propargyl-, reactions, with enneacarbonyldiiron, 711
Ruthenium complexes, tetranuclear carbon-bridged, synthesis of, 683

Ruthenium complexes, trinuclear, synthesis of, 683, 685
Ruthenium compounds, reactions, with phosphorus ylides, 582
Ruthenium hydride complexes, alkyne-bridged, synthesis of, 681
Ruthenium hydridocarbonyls, trinuclear hydrazino-bridged, use in the synthesis of trinuclear bridged ruthenium complexes, 683
Ruthenium multidecker sandwich compounds, synthesis of, 694
Ruthenocene
  lithiation of, 706
  mercuration of, 724
Ruthenocene, 1,1'-dilithio-, use in the synthesis of metal complexes, 719
Ruthenocene, pentamethyl-, mercuration of, 724

Safety precautions
  for acetyl nitrate, 208
  for perchlorates, 250
Salicyclic acid compounds, reactions, with benzoylethyne, 213
Salts, cyclic, synthesis of, 554
Samarium compounds
  reactions, with phosphorus ylides, 578
  with two carbon-bridges, 703
Scandium compounds, reactions, with phosphorus ylides, 578
Schiff bases
  from, hexahydro-1,3,5-triazines, 463
  reactions, with N,N-bis(trimethylsilyl)formamide, 419
  as starting materials
    for α-aminoalkylorganophosphorus compounds, 461
    for 1,3-aminophosphoranes, 478
1,2,3-Selenadiazoles, fragmentation of, 874
1,3,4-Selenadiazoles, 2,5-dihydro-, synthesis of, 333
Selenazaphospholes, synthesis of, via cycloaddition reactions, 493
Selenazolidines, synthesis of, 333
Selenenyl bromides, reactions, with diazomethane, 291
Selenenyl chlorides, arene-, reactions, with arylvinyl selenides, 91
Selenenyl halides, arene-, reactions, with ylides, 348
Selenide, bromomethyl, synthesis of, 291
Selenide, *cis*-β-chlorovinyl phenyl, synthesis of, 786
Selenide, cyanomethyl phenyl, oxidation of, 240
Selenide, 1,2-dichloro-1,2,2-trifluoroethyl phenyl, reactions, with magnesium, 785
Selenide, (*E*)-1,2-difluoro-2-chlorovinyl phenyl, synthesis of, 785
Selenide, ethyl phenyl
  oxidation of, 240
  reactions, with dibenzoyl peroxide, 240
Selenide, phenyl trimethylsilyl, reactions, with aldehydes, 239
Selenide, phenylvinyl
  reactions, with *iso*-propyllithium, 401
  as starting material, for α-selenosilanes, 375
  synthesis of, 401
Selenide, polyhalogenated ethyl phenyl, reactions, with magnesium, 785
Selenide, 1,2,2-trifluorovinyl phenyl, synthesis of, 785
Selenide anions, reactions, with dihalomethanes, 88
Selenides
  deprotonation of, 398
  fluorination of, 87
  oxidation of, 241
  oxidative deselenenylation of, 92
  reactions, with ozone, 241
Selenides, α-acyloxy, synthesis of, via seleno-Pummerer reactions, 240
Selenides, alkylvinyl, reactions, with hydrogen halides, 91
Selenides, allylic, reactions, with diazomethane, 277

Selenides, allyl phenyl, as starting materials, for
 α-selenosilanes, 374
Selenides, arylvinyl
 deprotonation of, 947
 reactions
  with areneselenenyl chlorides, 91
  with arenesulfenyl chlorides, 91
  with hydrogen halides, 91
 sulfenylation of, 871
Selenides, bissugar, synthesis of, 239
Selenides, bromo, reactions, with sodium benzylselenide, 288
Selenides, α-bromo
 reactions, with n-butyllithium, 400
 synthesis of, 88, 90
Selenides, 1-butenyl-3-methyl, deprotonation of, 965
Selenides, chloro, synthesis of, 287
Selenides, α-chloro
 from, benzeneselenenyl bromide, 90
 synthesis of, 88, 90
Selenides, 1-chlorobutadienyl, synthesis of, 786
Selenides, 1-chlorovinyl phenyl, synthesis of, 784
Selenides, α-fluoro
 alkylation of, 87
 synthesis of, 87
Selenides, β-D-glucopyranosyl phenyl, synthesis of, 239
Selenides, α-halo
 from
  diselenides, 88
  selenols, 88
  selenophenols, 88
 synthesis of, 400
 via substitution reactions, 88
Selenides, α-haloalkyl aryl, reactions, with
 triphenylphosphine, 347
Selenides, halomethyl, synthesis of, 88
Selenides, α-halovinyl, reactions, with alkenes, 92
Selenides, α-lithio, synthesis of, 375, 399
Selenides, phenylvinyl
 deprotonation of, 964
 reactions
  with n-butyllithium, 401, 964
  with diisopropylamine, 964
Selenides, α-(trimethylsilyloxy)-, synthesis of, 239
Selenides, vinyl
 reactions
  with alkyllithiums, 400
  with organolithiums, 375, 400
Selenium, synthesis of functions, containing oxygen, 841
Selenium(IV) dichlorides, reactions, with pyridine, 88
Selenium(IV) dichlorides, α-halo-, synthesis of, 92
Selenium compounds, synthesis of, 785, 868, 964
Selenium compounds, allyl-, use in the synthesis of carbon-
 bridged ditin compounds, 701
Selenium compounds, α-halo-, synthesis of, 87
Selenium dioxide, as dehydrating agent, for the synthesis of
 acetals, 178
Selenium–lithium exchange, for the synthesis of
 α-thiocarbanions, 394
Selenium–metal exchange, for the synthesis of
 α-selenolithiums, 399
Selenium nucleophiles, reactions, with
 halomethylsilanes, 377
Selenium tetrafluoride, use in the synthesis of
 difluoroalkanes, 10
Selenoacetals
 silylation of, 375
 transmetallation of, 375
Selenoacetate, ethyl phenyl, fluorination of, 92
Selenoaldehydes
 from, α-silylalkyl selenocyanates, 376
 reactions, with seleno anions, 289
 as starting materials, for aminals, 415
1-Selenoalkenes, 1-lithio, synthesis of, 964

Selenoalkynes, reactions, with amines, 910
Selenoalkynes, 1-alkyl-, addition of ethanethiol to, 872
1-Selenoalkynes, reactions, with acetic acid, 841
Selenoamides
 reactions, with phenyllithium, 334, 909
 as starting materials, for 1-alkyl selenoenamines, 909
Selenoanions, reactions, with selenoaldehydes, 289
Selenoanisole, cleavage of, 398
Seleno-1,2-benzisothiazole 1,1-dioxide, 3-phenyl-, reactions,
 with 1-diethylaminoprop-1-yne, 911
α-Selenoboranes, synthesis of, 377
Selenobut-3-en-1-yne, methyl-, reactions, with bromine, 786
Selenobut-3-en-1-ynes, 1-alkyl-, reactions, with hydrogen
 chloride, 786
α-Selenocarbanions
 reactions, with sulfur nucleophiles, 284
 synthesis of, reviews, 398
Selenocarbonyl compounds
 as starting materials
  for Se,N-acetals, 333
  for diselenoacetals, 289
α-Selenocopper compounds, synthesis of, 401
Selenocyanates, α-silylalkyl
 as starting materials, for selenoaldehydes, 376
 synthesis of, 376
Selenocyanogen, reactions, with cyanocuprates, 947
Selenoenamines, 1-alkyl, from, selenoamides, 909
Selenoesters, reactions, with potassium
 hexamethyldisilazide, 841
Selenoesters, O-alkyl, Se-alkylation of, 841
Selenoethene, 1-phenylthio-1-phenyl-, oxidation of, 873
α-Selenogermanes, synthesis of, 377
Selenoglycosides, synthesis of, 238, 239
Selenoglycosides, L-rhamnopyranosyl, synthesis of, 238
Selenohexa-1,3-diyne, 1-methyl-, reactions, with hydrogen
 chloride, 786
Selenoketals
 reactions, with butyllithium, 400
 silylation of, 375
 transmetallation of, 375
Selenolate salts, reactions, with ethynecarboxylate esters, 874
Selenolithium, phenyl-, reactions, with
 (iodomethyl)trimethylstannane, 401
α-Selenolithiums
 synthesis of, 401
 via selenium–metal exchange, 399
Selenols
 aminomethylation of, 333
 boron derivatives of, 288
 condensation with ketones, 287
 reactions
  with alkylthiochloroalkynes, 787
  with gem-dihalides, 288
 silicon derivatives of, 288
 as starting materials, for α-halo selenides, 88
Selenomalonitrile, 2-methyl-2-phenyl-, reactions, with
 alkenes, 284
Selenomethanes, aryl-, fluorination of, 92
Selenones
 deprotonation of, 399
 reactions, with diazoalkanes, 333
 synthesis of, 343
Selenones, α-halo, synthesis of, 92
Selenones, α-metallo, synthesis of, 399
Selenones, vinyl, reactions, with organolithiums, 400
2-Selenones, 1,3-diselenole-, as starting materials, for
 tetraselenafulvalenes, 874
2-Selenones, 1,3-thiaselenole-
 dimerization of, 868
 as starting materials, for thiaselenafulvenes, 870
Selenonitrosamines, α-phenyl-, synthesis of, 335
Selenononane, 1-hydroxy-2-nitro-2-phenyl-, reactions, with
 hydrogen peroxide, 335

Selenophenates, as starting materials, for
  thioselenoacetals, 283
Selenophene, 2,5-dichloro-3-iodo-, reactions, with
  ethyllithium, 785
Selenophene, 2,3,5-tribromo-, reactions, with
  ethyllithium, 785
Selenophenes, ring-opening of, 785
Selenophenols
  condensation with aldehydes, 287
  as starting materials, for α-halo selenides, 88
Selenophosphinic acid compounds, α-halo-, synthesis of, 127
Selenophosphinic chlorides, synthesis of, 128
Selenophospholes, use in Diels–Alder reactions, 489
Selenopropane, 2-phenylthio-2-phenyl-, synthesis of, 284
Seleno–Pummerer reaction, for the synthesis of α-acyloxy
  selenides, 240
Selenosilanes, phenyl-, synthesis of, 375
α-Selenosilanes
  deprotonation of, 375
  from
    allyl phenyl selenides, 374
    (halomethyl)trimethylsilanes, 377
    phenylvinyl selenide, 375
    vinylsilanes, 376
  synthesis of, 374
Selenosugars, synthesis of, 238
α-Selenosulfones, deprotonation of, 286
Selenosulfoxides, α-phenyl, synthesis of, 285
α-Selenosulfoxides, from, α-sulfoxide carbanions, 285
Selenotelluroacetals, from, telluride anions, 291
Selenothioesters, S,S-dioxides, Diels–Alder reaction of, 287
Selenothiopyrane S-oxides, 2-phenyl-, synthesis of, 285
α-Selenotin compounds, synthesis of, 401
Selenoxide, 1-chloroethyl phenyl, stability of, 92
Selenoxide, 1,1-difluoromethyl phenyl, synthesis of, 92
Selenoxide, dimethyl, use in the synthesis of α-acyloxy
  selenides, 240
Selenoxide, 1-lithiovinyl phenyl, half life of, 965
Selenoxide, 1,1,1-trifluoromethyl phenyl, synthesis of, 92
Selenoxides
  deprotonation of, 399
  elimination reactions of, 924
  reactions, with benzoylmethylphosphonium salts, 928
  stability of, 240
  synthesis of, 241
Selenoxides, α-halo, synthesis of, 92
Selenoxides, α-metallo, synthesis of, 399
Selenoxides, phenylvinyl, deprotonation of, 965
Selenoxides, vinyl, reactions, with organolithiums, 400
Selenuranes, synthesis of, 240
α-Selenyl carbanions, synthesis of, 398
Selenyl chloride, phenyl, reactions, with vinyl sulfones, 287
Selenyl chlorides, arene-
  reactions
    with (but-2-en-1-ynyl)trimethylgermane, 948
    with (but-2-en-1-ynyl)trimethylsilane, 948
Selenyl compounds, reactions, with haloalkynes, 786
Self-condensation, problems, 557
Serines, from, trimethyl phosphite, 487
Sesquihalides of aluminum, reduction of, 698
Sevoflurane, synthesis of, 45
Sharpless reagent
  Di Furia-modified, 261
  Kagan-modified, 261
  for the oxidation of dithianes, 269
  for the oxidation of dithioacetals, 261
Silaboralanes, synthesis of, 643
Silacyclobutane, 1,1-dimethyl-, pyrolysis of, 610
1-Silacyclobutane, 1,1-dimethyl-, thermolysis of, 610
Silacyclobutanes
  pyrolysis of, 610
  reactions, with silacyclobutanes, 610
  as starting materials, for carbosilanes, 610

1-Sila-2,4-cyclohexadiene, 1-t-butyl-, reactions, with
  butyllithium, 603
1-Silacyclohex-3,5-diene, 1,1-dimethyl-2,5-diphenyl-,
  reactions, with butyllithium, 616
Silacyclopentane, 1,1-dimethyl-, attempted
  metallation of, 650
1-Silacyclopentene, 1,1-dimethyl-2,3-benzo-5-
  trimethylsilyl-, synthesis of, 614
Silacyclopropenes
  from, alkynes, 1048
  as starting materials, for 1,3-disilacyclopentenes, 616
Silaethene, 1,1-dimethyl-, trimerization of, 610
Silaethene, 1-methyl-1-phenyl-2-neopentyl-,
  synthesis of, 610
1-Silaethene, synthesis of, 610
Silane, alkyne-, as starting material, for
  1,1-disilylalkenes, 1048
Silane, allyltrimethyl-
  reactions, with N-bromosuccinimide, 807
  synthesis of, 369
Silane, aminoalkyl-, stability of, 506
Silane, benzylthiotrimethyl-, reactions, with
  t-butyllithium, 368
Silane, benzyl(trimethyl)-, bromination of, 138
Silane, bis(chloromethyl)dimethyl-
  as starting material
    for bis(lithiomethyl)dimethylsilane, 602, 652
    for disilacyclobutanes, 605
Silane, bis(lithiomethyl)dimethyl-, from,
  bis(chloromethyl)dimethylsilane, 602, 652
Silane, (α-bromobenzyl)trimethyl-, synthesis of, 138
Silane, bromomethyl-, as starting material, for thiocarbonyl
  ylides, 369
Silane, bromotrimethyl-, loss of, in the synthesis of
  arsoranes, 597
Silane, (1-bromovinyl)trimethyl-
  displacement of bromine from, 946
  reactions
    with benzeneselenenyl bromide, 947
    with trialkylstannyl phenyl sulfide, 946
  as starting material, for Grignard reagents, 947
Silane, (but-2-en-1-ynyl)trimethyl-, reactions, with
  areneselenyl chlorides, 948
Silane, t-butyldimethoxy(trimethylsilylmethyl)-,
  synthesis of, 602
Silane, t-butyldimethyl(vinyl)-, synthesis of, predicted, 611
Silane, n-butyl(trimethyl)-, metallation of, 650
Silane, chloro(chloromethyl)dimethyl-, reactions, with
  sodium, 609
Silane, chloro(dimethyl)vinyl-, reactions, with
  t-butyllithium, 611
Silane, (α-chloroethyl)triethyl-, synthesis of, 137
Silane, (chloromethyl)cyclohexyl(phenyl)-, synthesis of, 136
Silane, chloromethyl(dimethyl)phenyl-, reactions, with
  dimethylphenylsilyllithium, 609
Silane, chloro(methyl)phenyl-, reactions, with
  chloromethyllithium, 136
Silane, chloro(methyl)phenyl(vinyl)-, reactions, with
  t-butyllithium, 611
Silane, chloromethyltrimethyl-
  reactions
    with dimethylbismuthinosodium, 599
    with lithium, 602
    with trialkylarsines, 597
    with trimethylarsine, 597
  as starting material, for epoxysilanes, 355
Silane, (chloromethyl)trimethyl-, reactions, with
  tributylstannyllithium, 660
Silane, chlorotrimethyl-
  reactions
    with acylimidazoles, 933
    with alkyl phenyl sulfoxides, 945
    with dihydrothiazine, 370

with 1-methoxy-1-vinyllithium, 929
with methyl benzene sulfinate, 370
with thionylanilide, 370
with vinylidene isocyanides, 1014
as starting material, for Grignard reagents, 370
Silane, α-chlorovinyl-, as starting material, for silenes, 611
Silane, dichlorodimethyl-, reactions, with
trimethylsilylcyclopentadienyllithium, 603
Silane, di(chloromethyl)dimethyl-, use in the synthesis of
α-germylsilylalkanes, 639
Silane, diethyl-
reactions, with 1-phenoxyprop-1-yne, 933
for the synthesis of bismethylene diruthenium
complexes, 679
Silane, (1-germylpropargyl)-, bromodesilylation of, 811
Silane, (iodomethyl)trimethyl-
reactions, with lithium phenyl selenide, 377
synthesis of, 139
Silane, iodotrimethyl-, reactions, with 1-ethoxy-1-
octyne, 765
Silane, [methoxy(phenylthio)methyl]trimethyl-
reactions
with alkenes, 370
with arenes, 370
Silane, methoxytrimethyl-, as starting material, for
2-methoxy-2,4,4-trimethyl-2,4-disilapentane, 602
Silane, α-methyl-α-chlorotrimethyl-, as starting material, for
epoxysilanes, 355
Silane, phenylthiotrimethyl-, reactions, with
acylstannanes, 951
Silane, phenylthio trimethyl-, use in the synthesis of
thioglycosides, 224
Silane, 1-phenylthiovinyl-, reactions, with titanium
tetrachloride, 369
Silane, tetrabromo-, reactions, with diazomethane, 140
Silane, tetrachloro-
as catalyst, for the synthesis of dithioacetals, 246
reactions
with diazomethane, 140
with di-Grignard reagents, 608
Silane, tetramethyl-
partial deprotonation of, 649
reactions
with n-butyllithiumpentamethyldiethylenetriamine, 602
with thoracyclobutane, 658
Silane, trichloro-
reactions
with enamines, 510
with 1,1,3,3-tetramethyldisilacyclobutane, 617
Silane, trichloro(chloromethyl)-
reactions
with germanium, 639
with magnesium, 654
as starting material, for Grignard reagents, 606
Silane, trichloro(dichlorofluoromethyl)-, as starting
material, for α-fluoromethylsilanes, 136
Silane, trifluoromethyltriphenyl-, reactions, with
n-butyllithium, 735
Silane, trimethylgermyldimethylphenyl-, reactions, with
aldehydes, 361
Silane, trimethylmethoxy-, synthesis of, 610
Silane, trimethylvinyl-, reactions, with arenesulfonyl
chlorides, 942
Silane, trimethyl(vinyl)-
reactions
with Grignard reagents, 654
with organolithium compounds, 651
with tetrachlorodiborane, 644
Silane, triphenyl(vinyl)-, reactions, with organolithium
compounds, 651
Silane, vinyltrimethyl-
hydroboration of, 356, 643
reactions

with arylsulfenyl chlorides, 369
with t-butyllithium, 606
Silanes
chlorination of, 137
from, methoxy(phenylthio)methane, 370
photolysis of, 1048
reactions
with carbenes, 137
with rhodium catalysts, 831
Silanes, acyl-
asymmetric reduction of, 353
enolization of, 930
organometallic addition to, 353
reactions
with cyanide, 353
with hydrogen sulfide, 371
with nucleophiles, 930
reduction of, 353
as starting materials, for α-hydroxysilanes, 353
stereoselective addition to, 354
synthesis of, 929
Silanes, N-acyl-α-amino-, synthesis of, 513
Silanes, α-acyloxy-, synthesis of, 352
Silanes, alkenyl-, isomerization of, 808
Silanes, alkoxy-, deprotonation of, 358
Silanes, α-alkoxy-
from, halomethylsilanes, 352
synthesis of, 352
Silanes, alkoxy(alkyl)-, deprotonation of, 650
Silanes, syn-α-alkoxy-β-(siloxy)acyl-, diastereoselective
addition to, 354
Silanes, alkyl-, deprotonation of, 646
Silanes, alkynyl-
bromination of, 807
carbometallation of, 809
hydroalumination of, 808
hydrosilylation of, 1044
reactions
with boron trifluoride etherate, 808
with iodosylbenzene, 808
as starting materials, for 1,2-bis(silyl)alkenes, 1044
Silanes, allyl-
deprotonation of, 648
reactions, with ortho esters, 192
Silanes, amino-
acylation of, 513
reactions, with aldehydes, 509
Silanes, α-amino-
chemistry of, 506
from
α-aminosilanes, 508
cis-oxiranes, 508
Michael reactions of, 510
reactions
with benzynes, 508
with dimethyl ethynedicarboxylate, 510
as starting materials
for α-aminosilanes, 508
for silylmethyl imines, 517
synthesis of, 506
Silanes, aminomethyl-, stability of, 506
Silanes, α-arylthiovinyl-, electrophilic sulfenylation of, 368
Silanes, benzyl-
bromination of, 138
deprotonation of, 647
metallation of, 653
Silanes, α-boryl-
oxidation of, 357
as starting materials, for α-hydroxysilanes, 357
Silanes, bromo-, synthesis of, 140
Silanes, α-bromo-, synthesis of, 136, 138
Silanes, α-bromoacyl-, synthesis of, 783
Silanes, (bromomethyl)chloro-, reactions, with
nucleophiles, 140

Silanes

Silanes, (bromomethyl)oxy-, synthesis of, 140
Silanes, carbimino-, as starting materials, for *gem*-aminoborylamino silylalkenes, 1015
Silanes, chloro-
  bromination of, 140
  reactions
    with chloroalkenes, 806
    with dimethylphenylsilyllithium, 609
    with fluoroalkenes, 140
    with 1-metalloalkylphosphine oxides, 570
    with 1-metalloalkylphosphines, 566
    with 1-metalloalkylphosphine sulfides, 570
    with 1-metalloalkyl phosphonates, 570
Silanes, α-chloro-
  from, vinylsilanes, 137
  reactions, with silyl enol ether, 365
  synthesis of, 136, 137
Silanes, chloro(chloromethyl)-, reactions, with nucleophiles, 140
Silanes, (chloro)dimethyl-, synthesis of, 513
Silanes, chloro(α-fluoro)-, synthesis of, 140
Silanes, chloromethoxy-, use in the synthesis of alkyl silyl acetals, 202
Silanes, chloromethyl-, reactions, with aromatic amines, 507
Silanes, chloro(methyl)-, reactions, with Grignard reagents, 137
Silanes, (chloromethyl)-, use in the synthesis of 1,1-bis(silyl)alkanes, 609
Silanes, chloromethyl hydro-, reactions, with 1,2-bis(silyl) ethynes, 1045
Silanes, (chloromethyl)oxy-, synthesis of, 140
Silanes, *m*-chlorophenyl-
  lithiation of, 508
  as starting materials, for azasilolines, 508
Silanes, chlorotrialkyl-, reactions, with ethers, 358
Silanes, chlorovinyl-, reactions, with *t*-butyllithium, 611
Silanes, chloro(vinyl)-, reactions, with organolithium compounds, 651
Silanes, α-chlorovinyltrimethyl-, reactions, with bromotrichloromethane, 758
Silanes, cyclopropyl acyl-, ring-expansion of, 931
Silanes, 3,3-diborylpropyl-, synthesis of, 632
Silanes, α,β-dibromo-, synthesis of, 138
Silanes, dichloro-, reactions, with diols, 342
Silanes, α,α-dihalo-, reduction of, 138
Silanes, ethynyl-, carbometallation of, 809
Silanes, 1-ethynyl-, reactions, with organocopper compounds, 940
Silanes, α-fluoro-
  characterization of, 136
  synthesis of, 137
Silanes, α-fluoro-β-hydroxy-, from, α,β-epoxysilanes, 137
Silanes, α-fluoromethyl-, from, trichloro(dichlorofluoromethyl)silane, 136
Silanes, (α-fluoromethyl)-, synthesis of, 137
Silanes, formamido-, synthesis of, 516
Silanes, formyl-, organometallic addition to, 353
Silanes, halo-
  reactions
    with phosphorus ylides, 570
    with selenium-stabilized carbanions, 374
Silanes, α-halo-
  from, α-hydroxysilanes, 138
  as starting materials, for organosilicon compounds, 136
  synthesis of, 136
Silanes, 1-haloalkyl-
  Arbuzov reactions of, 569
  reactions
    with alkyl phosphinites, 570
    with dialkyl phosphonites, 570
    with metal phosphides, 566
    with phosphines, 569
    with sodium dialkylphosphites, 569
    with trialkyl phosphites, 569
Silanes, (α-haloalkyl)oxy-, synthesis of, 140
Silanes, halo(α-haloalkyl)-, synthesis of, 139
Silanes, halo(halomethyl)-
  condensation of, 608
  reactions, with nucleophiles, 140
Silanes, halomethyl-
  reactions
    with amines, 506
    with methoxide, 352
    with oxygen nucleophiles, 352
    with selenium nucleophiles, 377
    with stannyl anions, 660
    with sulfur nucleophiles, 367
  as starting materials
    for α-alkoxysilanes, 352
    for α-thiosilanes, 367
  use in the synthesis of α-aminosilanes, 506
Silanes, (halomethyl)trimethyl-, as starting materials, for α-selenosilanes, 377
Silanes, hydro-, reactions, with 1,2-bis(trimethylsilyl) ethyne, 1045
Silanes, α-hydroxy-
  from
    acylsilanes, 353
    α-borylsilanes, 357
    vinylsilanes, 356
  as starting materials, for α-halosilanes, 138
Silanes, β-hydroxy-, synthesis of, 357, 619
Silanes, (α-hydroxyalkyl)trimethyl-, reactions, with methyl(triphenoxy)phosphonium iodide, 139
Silanes, hydroxymethyl-, from, acetate, 352
Silanes, α-imino-, synthesis of, 517
Silanes, α-iodo-, synthesis of, 139
Silanes, (*E*)-α-iodoalkenyl-
  isomerization of, 808
  reactions, with triethylborane, 811
Silanes, (α-iodoalkyl)-, synthesis of, 139
Silanes, (iodomethyl)-, synthesis of, 136
Silanes, α-lithio-, as intermediates, in the synthesis of 1,1-bis(silyl)alkanes, 611
Silanes, metallo-, reactions, with halomethylamines, 510
Silanes, 1-metalloalkyl-
  reactions
    with chlorophosphine oxides, 569
    with chlorophosphines, 565
Silanes, methoxy-, reduction of, 136
Silanes, α-methoxy-, oxidation of, 197
Silanes, methoxymethyl amino-, synthesis of, 509
Silanes, methylchloro-, chlorination of, 140
Silanes, monochloro(α-fluoro)-, synthesis of, 140
Silanes, β-oxoacyl-, tautomerism of, 931
Silanes, α-(phenylthio)alkyl-, synthesis of, 366
Silanes, α-(phenylthio)cyclopropyl-, synthesis of, 366
Silanes, α-(phenylthio)vinyl-, reactions, with organolithiums, 366
Silanes, 1-phosphinoalkyl-
  oxidation of, 573
  synthesis of, 565
Silanes, 1-phosphoranoalkyl-, synthesis of, 575
Silanes, α-seleno-β-nitro-, synthesis of, 376
Silanes, α-selenyl-, synthesis of, 374
Silanes, α-telluryl-, synthesis of, 374
Silanes, tetraalkoxy-, as dehydrating agents, for the synthesis of acetals, 178
Silanes, tetramethyl-, as starting materials, for carbosilanes, 617
Silanes, trialkyl-, reactions, with silyl esters, 832
Silanes, trichloro(halomethyl)-, synthesis of, 140
Silanes, trifluoro-, cleavage of, 152
Silanes, trifluorosilyl(vinyl)-, reactions, with bromine, 138
Silanes, trifluorovinyl-, reactions, with triphenylgermyllithium, 811

Silanes, trimethylsilyl(vinyl)-, reactions, with bromine, 138
Silanes, trimethyl(vinyl)-
  reactions
    with iododifluoroacetates, 139
    with tosyl iodide, 139
Silanes, vinyl-
  electrophilic selenylation of, 376
  electrophilic sulfenylation of, 368
  epoxidation of, 357
  hydroboration of, 356, 643
  hydrosilylation of, 617
  nucleophilic addition to, for the synthesis of α-metallo silylalkanes, 606
  reactions
    with aniline, 507
    with arenesulfenyl chlorides, 138
    with bromide azide, 514
    with bromine, 138, 751
    with chlorosulfonyl isocyanate, 514
    with electrophiles, 139
    with Grignard reagents, 654
    with organolithium compounds, 651
    with organolithiums, 366
    with osmium tetroxide, 357
    with trifluoromethyl hypochlorite, 140
  as starting materials
    for 1,2-bis(silyl)alkanes, 617
    for 1,1-bissilylethanes, 617
    for α-borylsilylalkanes, 643
    for α-chlorosilanes, 137
    for epoxysilanes, 357
    for α-hydroxysilanes, 356
    for α-selenosilanes, 376
γ-Silanes, synthesis of, 511
Silanol, trimethyl-, reactions, with 1,1,3,3-tetramethyl-1,3-disilacyclobutane, 623
Silanols, for the cleavage of 1,3-disilacyclobutanes, 623
1-Silaphenalene, 2,3-dihydro-1,1-dimethyl-, reactions, with butyllithium–tetramethylethylenediamine[1,2-bis(dimethylamino)ethane], 603
Silaphosphiranes, synthesis of, 567
Sila-Pummerer reaction, mechanism, 222
1-Sila-1-stanna-1-alkenes, as starting materials, for α-stannylsilylalkanes, 661
Silenes
  Diels–Alder reactions of, 611
  from
    α-chlorovinylsilane, 611
    1,2-disilacyclohexadienes, 612
  as starting materials
    for 1,1-bis(silyl)alkanes, 609
    for 1,3-disilacyclobutanes, 610
    for 1-germyl-1-silylalkanes, 639
  synthesis of, via pyrolysis of silacyclobutanes, 610
Silenes, α-silyl, as starting materials, for disiloxane, 612
Silicon
  stabilization by, 1082
  synthesis of functions
    bearing a silicon and a boron, 642
    bearing a silicon and a germanium, 636
    bearing two silicons, 602
Silicon–carbon bonds, cleavage of, 653
Silicon–carbon–silicon bonds, synthesis of, 602
Silicon compounds, synthesis of, 1031
Silicon compounds, α-halo-, synthesis of, 136
Silicon compounds, α-haloalkenyl-
  from
    alkyl halides, 810
    carbonyl compounds, 805
  synthesis of, 805
    halogenation methods, 807
    silylation methods, 806
    via addition of alkyl halides to alkynylsilanes, 810

    via halogenation of α-silylalkenyl metal compounds, 808
Silicon–halogen bonds, use in the synthesis of 1,1-bis(silyl)alkanes, 621
Silicon hydrides
  halogenation of, 139
  reactions, with tin(IV) chloride, 140
Silicon tetrachloride, reactions, with 1-germanyl ethyn-2-ols, 1058
Silirane, 1,1-dimethyl-2,3-bis(trimethylsilyl)-, reactions, with bis(trimethylsilyl)ethyne, 1052
Siliranes
  reactions, with alkynes, 1052
  as starting materials, for silyl silirenes, 1052
Silirenes
  dimerization of, 1053
  hydrolysis of, 1051
  reactions
    with acetic acid, 1051
    with alcohols, 1051
    with aldehydes, 1052
    with alkynes, 1051, 1052
    with ketones, 1052
  as starting materials
    for disilacyclopentenes, 1052
    for 1,1,2-trisilylethenes, 1051
  synthesis of, 1051
Silirenes, silyl
  from, siliranes, 1052
  hydrolysis of, 1051
  methanolysis of, 1051
  reactions, with 1-trimethylsilyl-2-phenylethyne, 1053
Siloxanes
  metallation of, 650
  synthesis of, 623
Silver, mesityl-, synthesis of, 696
Silver, perfluoroisopropyl-, synthesis of, 152
Silver acetate, reactions, with gem-dibromocyclopropanes, 57
Silver acrylate, reactions, with bromine, 27
Silver azide, for the nucleophilic displacement of halides, 432
Silver bromide, reactions, with 2-lithio(dimethylaminomethyl)benzene, 708
Silver carboxylates, reactions, with 1-chloro-1-dialkylaminoalkenes, 889
Silver chloride, reactions, with mesitylmagnesium bromide, 696
Silver complexes, tetranuclear alkenyl, synthesis of, 696
Silver compounds, reactions, with phosphorus ylides, 585
Silver compounds, carbon-bridged, from, metal(I) halides, 695
Silver fluoride
  reactions
    with benzeneselenenyl chloride, 52
    with hexafluoro-2-butyne, 696
  use in the synthesis of chlorofluoroalkanes, 30
  use in the synthesis of difluoroalkanes, 4
Silver fluoroacetate, reactions, with chlorine, 33
Silver α-fluoro carboxylate, reactions, with bromine, 35
Silver nitrate, reactions, with nitroalkanes, 428
Silver nitrocyanamide, reactions, with N-bromomethylphthalimide, 447
Silver tricyanomethanide, reactions, with alkylamine hydrobromides, 977
Silver triflate, use in the synthesis of multidecker borole complexes of ruthenium, 695
Silver trifluoromethoxide
  reactions, with 1-chlorotrifluoropropyne, 764
  synthesis of, 764
Silylation, electrophilic, of aminoalkyllithiums, 511
C-Silylation
  of aldehydes, 354
  examples of, 365

Silyl chloride

of ketones, 354
O-Silylation, mechanism, 636
Silyl chloride, *t*-butyldimethyl-, for the O-silylation of ester enolates, 831
Silyl chloride, trimethyl-
  as electrophile, for dithiocarboxylic enethiolates, 842
  reactions
    with *gem*-dichlorocyclopropanes, 605
    with dilithiomethane, 603
    with dimesitylborylmethyllithium, 642
    with 1-lithio-2,2,4,4-tetramethyl-2,4-disilapentane, 606
    with monolithioacetonitrile, 604
    with 1-silyl 3-phenylpropynes, 1047
    with tributylstannylmethyllithium, 660
    with trimethylsilylcyclopentadienyllithium, 603
Silyl chlorides, trialkyl-
  reactions, with lithium enolates, 832
  as silylating agents, 831
Silyl chlorides, triphenyl-, reactions, with chloromethyllithium, 138
Silylethoxymethyl (SEM) group, as protecting group for anomeric hydroxyl groups, 176
Silyl groups, trimethyl-
  metallation of, 651
  1,2-shift of, 615
Silyl halides, chloromethyldimethyl-, as starting materials, for 1,3-disilacyclobutanes, 654
α-Silyl heterocycles, as starting materials, for ylides, 519
Silyl iodide, trimethyl-
  as catalyst, for the synthesis of acetals, 181
  reactions
    with dimethoxymethane, 51
    with vinyl sulfones, 84
Silylmethylation
  of amines, 507
  of formamide, 516
  of nitrogen heterocycles, 519
  of pyridine, 519
Simmons–Smith process, for the synthesis of metallated carbenes, 1083
Simmons–Smith reagents
  reactions
    with chlorotrimethylstannane, 154
    with dichlorodimethylstannane, 154
    with trimethylammonia, 534
  synthesis of, 152
Singlet oxygen, for the oxidation of dithioacetals, 260
S_N2' reaction, for the synthesis of 1,1-difluoroalkenes, 735
Sodium
  reactions
    with chloro(chloromethyl)dimethylsilane, 609
    with dimethyl diselenide, 283
    with ethyl 3-chloropropionate, 171
    with toluene, 609
Sodium, cyclopentadienyl-
  reactions, with (2,4-dimethylpentadienyl)potassium, 678
  use in the synthesis of carbon-bridged dichromium complexes, 673
  use in the synthesis of carbon-bridged dirhodium complexes, 686
  use in the synthesis of cyclooctatetraene-bridged complexes, 672
  use in the synthesis of cyclopentadienyl-bridged imino–rhenium complexes, 678
Sodium, dimethylbismuthino-
  reactions
    with chloromethyltrimethylgermane, 599
    with chloromethyltrimethylsilane, 599
    with chloromethyltrimethyltin, 600
Sodium, diphenylbismuth-, reactions, with dichloromethane, 596
Sodium, pentyl-, reactions, with ferrocene, 706
Sodium, phenylmethylantimony-, reactions, with dichloromethane, 595
Sodium, triethylplumbyl-, reactions, with ethoxyethyne, 955
Sodium, trimethylsilylmethyl-, synthesis of, 653
Sodium, trimethyltin-
  reactions
    with chloroform, 699
    with dichloromethane, 699
Sodium, trivinyltin-, reactions, with dichloromethane, 699
Sodium acetate
  reactions
    with bis(chloromethyl)dimethylgermane, 362
    with 3,3-dibromo-1,1,1-trifluoro acetone, 161
Sodium alkoxides
  reactions
    with 1-chloro-1-dialkylaminoalkenes, 889
    with *N*-chloromethyl-*N*-ethylnitramine, 313
Sodium amalgam, reactions, with carbonylcyclopentadienyldiiodocobalt, 690
Sodium amide, use in the synthesis of trifluorovinyl anions, 819
Sodium arenethiolates
  reactions
    with dichloroalkyne, 772
    with tetrachloroethene, 770
    with trichloroethene, 772
    with 1,1,2-trichloroethene, 770
Sodium arsenes, synthesis of, 499
Sodium aryl tellurides, synthesis of, 291
Sodium azide
  reactions
    with bromo compounds, 109
    with chloromethylphosphonamides, 492
    with cyclopropenylium ions, 483
    with dihalo compounds, 432
    with fluoromethyl phenyl sulfoxide, 86
    with vinylidene dichlorides, 993
Sodium benzeneselenolate, reactions, with diethyl iodomethyl phosphonate, 348
Sodium benzene sulfinate
  reactions
    with diiodomethane, 84
    with 2-iodo-2-nitropropane, 332
Sodium benzylselenide, reactions, with bromo selenides, 288
Sodium bis(2-methoxyethoxy)aluminum hydride, reactions, with propargylic alcohols, 948
Sodium bis(trimethylsilyl)amine, reactions, with diphenoxyethyne, 884
Sodium bisulfite
  reactions
    with aldehydes, 237
    with ketones, 237
Sodium borohydride
  as reagent
    for the synthesis of *N*-acylaminals, 420
    for the synthesis of pyrimidine compounds, 420
  use in the reduction of acyclic esters, 170
  use in the synthesis of carbon-bridged diiron complexes, 679
  use in the synthesis of cyclic hemiacetals, 165
  use in the synthesis of diruthenium complexes, 679
Sodium bromoethanesulfonate, chlorination of, 85
Sodium chlorodifluoroacetate, use in the synthesis of 1,1-difluoroalkenes, 732
Sodium chlorodifluoroacetates, pyrolysis of, 8
Sodium chloromethyl sulfonate, as starting material, for chloromethyl phenyl ether, 49
Sodium cyclopentadienyldicarbonylruthenium, use in the synthesis of carbon-bridged diruthenium complexes, 678
Sodium dialkyl phosphites, reactions, with 1-haloalkylsilanes, 569
Sodium dichlorofluoroacetate, decarboxylation of, 752
Sodium diethyl phosphite, reactions, with iminium esters, 1003

Sodium diisopropyl phosphites, as starting materials, for diisopropyl fluoromethylphosphonate, 131
Sodium dimethylamide, reactions, with enamines, 1017
Sodium diphenylarsenic, reactions, with 1,1-dichloroethane, 592
Sodium diphenylphosphide, reactions, with *gem*-chloro enamines, 997
Sodium dithionite, reactions, with aldehydes, 234
Sodium ethoxide
    reactions
        with 1-alkoxy-2,2,2-tribromo-1-chloroethanes, 767
        with 1-chloro-1-(chlorosulfenyl)propanoyl chloride, 72
        with tetrakis(hydroxymethyl)phosphonium salts, 458
        with trichloroethylene, 761
Sodium fluoride, reactions, with chloromethylphosphonic dichloride, 130
Sodium hexamethyldisilazide, reactions, with alkylmethanesulfonates, 605
Sodium hydride
    reactions
        with aminodiphosphonate compounds, 1002
        with dimethyl sulfoxide, 391
        with methyl dichlorofluoroacetate, 752
Sodium hydroxide
    reactions
        with diaryl ditellurides, 291
        with diphenyl disulfide, 284
Sodium hypobromite, reactions, with fluorinated alkenes, 45
Sodium iodide, reactions, with dichloromethane, 28
Sodium iodomethanesulfonate
    reactions
        with phosphorus pentabromide, 85
        with phosphorus pentachloride, 85
Sodium isopropoxide, reactions, with chloroformamidinium chlorides, 974
Sodium metaperiodate, for the oxidation of dithioacetal oxides, 268
Sodium methanethiolate, reactions, with *S,S*-diethyl 1-chlorovinylphosphonodithioate, 927
Sodium methoxide
    reactions
        with bis(trimethylsilyl)methane, 653
        with α-bromoboronate, 363
        with ethynyl azaarenes, 195
        with halogenated styrenes, 762
        with hexachlorophenylalkyne, 764
        with quinoline, 195
        with tetrafluoroethylene, 761
        with 1,2,2-trichloroacrylonitrile, 762
        with tris(trimethylsilyl)methane, 619
Sodium methyl selenide, synthesis of, 283
Sodium nitrite, reactions, with nitroalkanes, 428
Sodium perfluoropropionate
    as starting material, for tetrafluoroethylene, 731
    thermal decarboxylation of, 731
Sodium periodate
    for the cleavage of alkanes, 168
    for the oxidation of dithioacetals, 259, 267
    reactions
        with 5,5-dimethyl-1,3-dithiane, 268
        with 1,3-dithiane, 268
Sodium phenolate, reactions, with trichloroethylene, 761
Sodium phenoxide
    reactions
        with polychlorinated ethanes, 761
        with polyhalogenated ethylenes, 761
Sodium phenylmethanethiolate, reactions, with polychlorobutene, 770
Sodium phenyl selenide, reactions, with α-brominated sulfone, 286
Sodium phenyl sulfinate, reactions, with diiodomethane, 281
Sodium salts, synthesis of, 545
Sodium sand, reactions, with α-bromo ortho esters, 827

Sodium selenide, reactions, with phosphonates, 348
Sodium sulfide, reactions, with bis(halomethyl)dialkylgermanes, 372
Sodium sulfite
    reactions
        with bromoethanesulfonyl chloride, 85
        with chloromethanesulfonyl chloride, 85
        with iodoform, 281
Sodium telluride, alkylation of, 377
Sodium tetracarbonylcobaltate
    reactions, with difluorofumaric dichloride, 821
    as starting material, for cobaltacyclopentadiene complexes, 689
Sodium tetrachloropalladate, use in metallation reactions, 720
Sodium tetrahydroborate, use in the synthesis of α-aminoalkyl phosphonates, 472
Sodium thiolate, reactions, with tetrachloroethene, 770
Sodium thiophenoxide
    reactions, with 1,1,2,2-tetrachloroethane, 771
    synthesis of, 284
Sodium toluene-*p*-sulfinates
    reactions
        with acrylonitriles, 783
        with alkenyl halides, 781
Sodium trichloroacetate
    decomposition of, 740
    reactions, with thiofluorenone *S*-oxide, 742
Sonochemical activation, for the synthesis of 1-aminophosphonic acids, 463
Sparsomicin, synthesis of, 264
(-)-Sparteine, use in the synthesis of α-alkoxylithiums, 382
Speier's catalyst
    for the hydrosilylation of 1,2-bis(silyl) ethynes, 1044
    for the synthesis of 1,3-disilabutanes, 1054
    uses of, 1045
    See also Hexachloroplatinic(IV) acid
Spiroacetals, synthesis of, 171
Spiroaminals, synthesis of, 413
Spirobenzopyrans, synthesis of, 307
Spirolactones, synthesis of, 204
Spirooxazines, synthesis of, 307
Spiropentane, *trans*-1,2-dinitro-, as starting material, for diiodo compounds, 108
Spirophosphoranes
    reactions, with *N*-methylbenzylimine, 478
    synthesis of, 134
Stabilization, for the synthesis of carbenes, 1082
1-Stanna-3-boracyclopentenes, synthesis of, 665
Stannacyclopentadiene compounds, 2-stannyl, synthesis of, 1068
Stannane, benzyloxymethyl-, synthesis of, 389
Stannane, bis(bromomethyl)dimethyl-, synthesis of, 154
Stannane, bis(α-chloroethyl)diethyl-, synthesis of, 154
Stannane, bis(silylmethyl)-, synthesis of, 660
Stannane, (bromomethyl)trimethyl-, synthesis of, 154
Stannane, α-chloroalkoxy-, as starting material, for α-alkoxystannanes, 388
Stannane, α-chlorobenzyltri-*n*-butyl-, synthesis of, 154
Stannane, chlorotrimethyl-
    reactions
        with pent-4-enylzinc–magnesium compounds, 724
        with Simmons–Smith reagents, 154
Stannane, dibutyl bis(phenylthio), as reagent, for the synthesis of monothioacetals, 216
Stannane, α,α-dichloro-, reduction of, 154
Stannane, dichlorodimethyl-, reactions, with Simmons–Smith reagents, 154
Stannane, (iodomethyl)trimethyl-, reactions, with phenylselenolithium, 401
Stannane, tributylchloro-, reactions, with lithio compounds, 961
Stannane, tributyl(trimethylsilylmethyl)-

Stannane
  deprotonation of, 662
  synthesis of, 660
Stannane, triphenyl-
  reactions
    with cyclopentadienylcobalt complexes, 726
    with phenylethyne, 701
Stannane hydride, trimethyl-, reactions, with perfluorocyclobutenes, 153
Stannane hydrides, trialkyl-, reactions, with fluoroalkenes, 153
Stannanes
  characteristics of, 951
  from, aminals, 533
  reactions, with vinyl halides, 8855
  transmetallation of, 531
Stannanes, acyl-
  asymmetric reduction of, 380, 386
  reactions
    with nucleophiles, 951
    with phenylthiotrimethylsilane, 951
  reduction of, 386
Stannanes, acylamino-, synthesis of, 533
Stannanes, α-acyloxy-, hydrolysis of, 389
Stannanes, acyltributyl-, enantioselective reduction of, 386
Stannanes, (E)-alkenyl-, synthesis of, 813
Stannanes, α-alkoxy-
  from
    carbonyl compounds, 386
    α-chloroalkoxystannane, 388
    α-chloroboronic esters, 389
    chloromethyl ethers, 389
    Fischer carbene complexes, 388
    tributylstannyl acetals, 388
    tributylstannylmethyl iodide, 387
    tributyltin chloride, 387
  reactions, with butyllithium, 378
  transmetallation of, 384
Stannanes, (alkoxymethyl)trihalo-, as intermediates, in the synthesis of α-alkoxylithiums, 389
Stannanes, α-alkoxysilyl-, reactions, with n-butyllithium, 358
Stannanes, 1-alkoxyvinyl-, synthesis of, 951
Stannanes, amino-, synthesis of, 531
Stannanes, α-amino-
  condensation of, with aldehydes and ketones, 537
  synthesis of, 528, 533
Stannanes, aminomethyl-, transmetallation of, 530
Stannanes, azidomethyl-
  Staudinger reaction of, 537
  synthesis of, 537
Stannanes, gem-bromo-, synthesis of, 813
Stannanes, α-bromo-
  from, aldehydes, 155
  synthesis of, 154
Stannanes, bromo(bromomethyl)-, synthesis of, 155
Stannanes, chiral α-alkoxy-, synthesis of, 386
Stannanes, chloro-
  reactions
    with 1-metalloalkylphosphine oxides, 588
    with 1-metalloalkylphosphines, 588
    with 1-metalloalkylphosphine sulfides, 588
    with 1-metalloalkyl phosphonates, 588
Stannanes, α-chloro-, synthesis of, 154
Stannanes, α-fluoro-, synthesis of, 153
Stannanes, halo-, reactions, with phosphorus ylides, 588
Stannanes, 1-haloalkyl-
  reactions
    with dialkyl phosphonites, 588
    with phosphines, 588
    with trialkyl phosphites, 588
Stannanes, halo(α-haloalkyl)-, synthesis of, 155
Stannanes, halo(halomethyl)-, synthesis of, 156
Stannanes, hydroxy-, bromination of, 154
Stannanes, α-hydroxy-, kinetic resolution of, 389

Stannanes, iodo-, reactions, with iodo(iodomethyl)zinc, 155
Stannanes, α-iodo-
  from, aldehydes, 155
  synthesis of, 155
Stannanes, α-iodopropyl-, synthesis of, 155
Stannanes, nitro-, characterization of, 541
Stannanes, phenylselenyl, for the synthesis of selenoglycosides, 238
Stannanes, 1-phosphonatoalkyl-, reduction of, 588
Stannanes, phthalimidomethyl-, synthesis of, 533
Stannanes, α-selenenyl-, synthesis of, 401
Stannanes, trialkyl(α-haloalkyl)-, synthesis of, 153
Stannanes, trialkyl(iodomethyl)-, synthesis of, 155
Stannanes, triaryl(α-haloalkyl)-, synthesis of, 153
Stannanes, tributyl-, as starting materials, for organolithium compounds, 652
Stannanes, trihalo-
  reactions, with butyllithium, 380
  transmetallation of, 380
Stannanes, vinyl-
  hydrostannation of, 701
  reactions
    with arenesulfenyl chlorides, 154
    with 1-chloromethyl-4-fluorodiazonabicyclo[2.2.2]octane bis(tetrafluoroborate), 736
  synthesis of, 386, 534
Stannic chloride, as catalyst, for the synthesis of acetals, 181
Stannole, 1,1-di-n-butyl-2,5-dimethyl-, reactions, with phenylantimony dichloride, 1041
Stannyl anions, reactions, with halomethylsilanes, 660
Stannyl compounds, synthesis of, 735
Stannyl compounds, anti-7-trimethyl-, synthesis of, 155
Staudinger reaction, of azidomethylstannanes, 537
Steroids, dehydrobromination of, 748
Steroids, gem-dinitro-, synthesis of, 429
Steroids, 17-ethynyl-17-hydroxy-, cobalt carbonyl complexes of, 718
Steroids, fluoro-, synthesis of, 9
Steroids, 17-oxo-, reactions, with arenesulfonylmethyl isocyanides, 908
Steroid side chains, synthesis of, 948
Stevens rearrangement, for the synthesis of α-aminosilanes, 509
Stibacymantrenes, synthesis of, 1042
Stibaferrocenes, synthesis of, 1038, 1039
Stibine, dichloro(chloromethyl)-, synthesis of, 136
Stibine, tris(chloromethyl)-, synthesis of, 136
Stibine, tris(trimethylsilylmethyl)-, synthesis of, 598
Stibines
  reactions, with stiboranes, 599
  synthesis of, 598
Stibole, 1-phenyl-2,5-dimethyl-, synthesis of, 1041
Stiboles
  reactions, with iron(II) chloride, 1039
  synthesis of, 1038
Stiboles, 2,5-bis(trimethylsilyl)-3,4-dimethyl-, synthesis of, 1038
Stibolyl anion, dimethyl-, reactions, with bromopentacarbonylmanganese, 1042
Stibolyl anions
  reactions, with iron(II) chloride, 1041
  synthesis of, 1039
Stiborane, tetramethyliodo-, as starting material, for tetramethylstiboranes, 599
Stiboranes
  reactions
    with stibines, 599
    with trimethylsilylmethyllithuim, 599
  synthesis of, 599
Stiboranes, tetramethyl-, from, tetramethyliodostiborane, 599
trans-Stilbene, synthesis of, 81

Strecker reaction, for the synthesis of cyanomethyl compounds, 509
Styrene
  reactions
    with 1,2-bis(dimethylsilyl)benzene, 618
    with chloroform, 17
    with formaldehyde, 827
Styrene, 2-bromo-, as starting material, for benzoferracyclopentadiene complexes, 681
Styrene, *trans-β*-bromo-, reactions, with diborane, 142
Styrene, 2-bromo-2-nitro-, NMR spectra of, 795
Styrene, *trans* 1-chloro-, deprotonation of, 754
Styrene, (*E*)-1-chloro-1-fluoro-, synthesis of, 754
Styrene, 2-chloro-2-nitro-, NMR spectra of, 795
Styrene, 1-chloroperfluoro-, synthesis of, 737
Styrene, 1,1-dibromo-, synthesis of, 745
Styrene, 1,1-dibromo-2-fluoro-, synthesis of, 748
Styrene, 1,1-dichloro-, from, benzaldehyde, 738
Styrene, 2,2-difluoro-1-chloro-, synthesis of, 737
Styrene, α-nitroso-, reactions, with thiobenzophenone, 232
Styrene, octafluoro-, from, decafluoroethylbenzene, 730
Styrene compounds, 2-chloro-2-tosylazo-, from, arylacetic acids, 799
Styrene oxide, reactions, with tris(trimethylsilyl)methyllithium, 620
Styrenes, as starting materials, for arylacetaldehyde acetals, 197
Styrenes, bromonitro-, synthesis of, 798
Styrenes, 2-bromo-2-nitro-, synthesis of, 798
Styrenes, 2-chloro-2-nitro-, synthesis of, 798
Styrenes, 1-diethylamino-2-phthalimido-
  reactions
    with bromine, 793
    with chlorine, 793
Styrenes, halogenated, reactions, with sodium methoxide, 762
Substitution, for the synthesis of tetracoordinate sulfur compounds, 781
Substitution, nucleophilic, for the synthesis of α-carbocations of alkyne-bridged dimolybdenum complexes, 674
Succinimide, *N*-bromo- (NBS)
  reactions
    with allyltrimethylsilane, 807
    with amides, 138
    with benzopyranones, 174
    with diethylaminosulfur trifluoride, 45
    with HF pyridine, 45
    with (*R*)-imidazolidinones, 102
    with 1-iodoalkynes, 760
    with pyrazolines, 109
    with sulfides, 71
  use in the synthesis of dibromoalkanes, 21
Succinimide, *N*-chloro- (NCS)
  for the chlorination of sulfides, 66
  for the chlorination of sulfoxides, 74
  reactions
    with 1-phenylthio-1-silylalkanes, 943
    with sulfides, 283
    with (*S*)-2'-thiazinylideneglycinates, 101
  use in the synthesis of α-chloro sulfides, 66
Succinimide, *N*-iodo- (NIS), reactions, with 1-iodoalkynes, 760
Succinimide, (*N*-phenylthio)-, reactions, with ylides, 346
Succinimides, *N*-alkyl-, reduction of, 303
Succinimides, *N*-aryl-, reduction of, 303
Succinimides, *N*-chloromethyl-, synthesis of, 483
Succinimides, *N*-halo-, reactions, with 2-nitro enamines, 796
Succinonitrile, diimino-, reactions, with methyl ketones, 438
Sugar lactones, reactions, with trimethylsilyl azide, 432
Sugars, reactions, with potassium metabisulfite, 237
Sugars, acetylated, reactions, with phosphorus pentachloride, 766

Sugars, bis(benzylseleno), synthesis of, 287
Sulfate, bis(trimethylsilyl), as catalyst, for the synthesis of dithioacetals, 246
Sulfate, perfluoroallyl fluoro, as starting material, for α-fluoro ethers, 44
Sulfate adducts, dialkyl, use in the synthesis of acetals, 184
Sulfate radical anion, as oxidising agent, 1072
Sulfates, cyclic, from, bis(trifluoromethyl)ketene, 833
Sulfates, dialkyl, as electrophiles, for dithiocarboxylic enethiolates, 842
Sulfates, fluoronitroso, reactions, with trifluoroethene, 104
Sulfenamides, synthesis of, 843
Sulfenamides, 1-chloro, synthesis of, 72
Sulfene, methanesulfonyl-, as starting material, for thietane dioxides, 281
Sulfenes, reactions, with amines, 282
Sulfenes, chloro-, reactions, with diazomethane, 82
Sulfenyl chloride, arene-, reactions, with chloroalkyne, 772
Sulfenyl chlorides
  oxidation of, 78
  reactions, with vinyl sulfones, 861
  as starting materials, for α-chloro sulfides, 69
Sulfenyl chlorides, alkane-, reactions, with *N,N*-diethyl ynamines, 791
Sulfenyl chlorides, arene-
  reactions
    with arylvinyl selenides, 91
    with chloroimines, 110
    with *N,N*-diethyl ynamines, 791
    with vinylsilanes, 138
    with vinylstannanes, 154
    with vinyltrimethylgermane, 372
Sulfenyl chlorides, aryl-, reactions, with vinyltrimethylsilane, 369
Sulfenyl chlorides, carbamoyl-, from, triazines, 103
Sulfenyl chlorides, α-chloroalkane-
  from
    disulfides, 71
    α-mercaptoacetanilides, 71
    nitriles, 72
  reactions, with amines, 72
  synthesis of, 71
Sulfenyl halides, reactions of, 940
Sulfenyl halides, arene-, reactions, with vinyl acetates, 56
Sulfide, allyl benzyl, reactions, with dichlorocarbene, 775
Sulfide, benzyl methyl, metallation of, 347
Sulfide, borane dimethyl
  reactions
    with dimethyl(divinyl)germane, 646
    with 1-(trimethylsilyl)-1,3-butadiene, 643
Sulfide, *t*-butyl chloromethyl, synthesis of, 66
Sulfide, carbonyl, reactions, with ethyl cyanoacetate, 833
Sulfide, chlorocyclopropyl, synthesis of, 68
Sulfide, chlorofluoromethyl phenyl, reactions, with *n*-butyllithium, 773
Sulfide, chloromethyl, as starting material, for benzylthiomethylmagnesium chloride, 395
Sulfide, chloromethyl methyl
  reactions
    with ethyl diphenylphosphinite, 347
    with lithium iodide, 71
  as starting material, for Grignard reagents, 395
Sulfide, chloromethyl *p*-nitrobenzyl, chlorination of, 64
Sulfide, chloromethyl phenyl
  oxidation of, 74
  as starting material, for *N*-acyl-*S,N*-acetals, 329
  synthesis of, 68
Sulfide, dibenzyl
  α-acetoxylation of, 220
  reactions, with sulfuryl chloride, 64
Sulfide, di(chloromethyl)
  from
    dimethyl sulfide, 63

Sulfide

1,3,5-trithiane, 63
synthesis of, 65
Sulfide, dimethyl
  lithiation of, 391
  reactions
    with α-halomethylphosphonates, 346
    with α,β-unsaturated acetals, 201
  as starting material, for di(chloromethyl) sulfide, 63
Sulfide, iodomethyl phenyl, synthesis of, 71
Sulfide, methyl, reactions, with thionyl chloride, 65
Sulfide, methyl methylsulfinylmethyl, as starting material, for bis(methanesulfonyl)methane, 274
Sulfide, methyl phenyl, as starting material, for α-acetoxyphenylthiomethane, 220
Sulfide, phenyl crotyl, chlorination of, 68
Sulfide, phenyl(trimethylsilyl), use in the synthesis of monothioacetals, 217
Sulfide, phenyl trimethylsilylmethyl, anodic oxidation of, 221
Sulfide, phenyl (1-trimethylsilyl)propadienyl
  reactions
    with lithium butyl(diisobutyl)aluminum hydride, 946
    with organometallics, 946
Sulfide, phenylvinyl
  cycloaddition of, 226
  reactions, with methyllithium, 366
  as starting material, for α-lithio sulfides, 392
Sulfide, phosphine, reactions, with dichloro(phenyl)phosphine, 114
Sulfide, trialkylstannyl phenyl, reactions, with (1-bromovinyl)trimethylsilane, 946
Sulfide, trichloromethyl chloromethyl, chlorination of, 85
Sulfide, 1,1,1-trichloromethyl methyl, synthesis of, 65
1,2-Sulfide, tetramethylallene-, reactions, with diphenylketene, 836
Sulfide complexes, bromoborane dimethyl, reactions, with haloalkynes, 815
Sulfide compounds, synthesis of, 956
Sulfides
  bromination of, 71
  deprotonation of, 391
  electrochemical fluorination of, 62
  fluorination of, 61
  α-fluorination of, 61
  oxidation of, 220
    electrochemical, 220
  reactions
    with bromine, 71
    with N-bromosuccinimide, 71
    with t-butyl peracetate, 220
    with N-chlorosuccinimide, 283
    with dichlorocarbene, 775
    with lithium naphthalide, 1026
    with sulfuryl chloride, 80
  as starting materials, for α-chloro sulfoxides, 74
Sulfides, α-acetoxy
  synthesis of, 220
    via Pummerer reaction, 221
Sulfides, alkenyl, deprotonation of, 956
Sulfides, alkyl
  chlorination of, 63
  as starting materials, for α-chloro sulfides, 63
Sulfides, alkyl aryl, chlorination of, 63
Sulfides, alkyl dialkylaminomethyl, synthesis of, 315
Sulfides, alkyl phenyl, chlorination of, 68
Sulfides, alkynic, reactions, with organocuprates, 963
Sulfides, alkynyl
  reactions
    with bis(pyridine)iodine(1) tetrafluoroborate, 772
    with bromine, 772
Sulfides, allylic, chlorination of, 66, 68
Sulfides, aminomethyl phenyl, reactions, with tributylstannyllithium, 533

Sulfides, α-aminophosphine, from, tertiary phosphines, 477
Sulfides, anomeric, reactions, with thiophilic metal salts, 200
Sulfides, aryl methoxymethyl, reactions, with monoperphthalic acid, 235
Sulfides, α-azido, oxidation of, 330
Sulfides, α-azo, synthesis of, 329
Sulfides, α-bromo
  from
    aldehydes, 71
    thiols, 71
  as starting materials, for α-bromo sulfones, 83
  synthesis of, 71
Sulfides, α-chloro
  from
    aldehydes, 69
    alkyl aryl sulfoxides, 69
    alkyl sulfides, 63
    alkyl sulfoxides, 69
    benzoates, 69
    α-chloro-α-(chlorosulfenyl)carboxylic esters, 70
    sulfenyl chlorides, 69
    thiols, 69
  oxidation of, 74
  safety precautions, 63
  stability of, 62
  as starting materials
    for α-chloro sulfones, 80
    for dithioacetals, 258
  synthesis of, 62
  use in the synthesis of monothioacetals, 227
Sulfides, α-chloroalkyl phenyl, reactions, with zinc, 396
Sulfides, (α-chloroalkyl)phosphine, synthesis of, 123
Sulfides, (chloromethyl)phosphine, synthesis of, 123
Sulfides, α-chlorovinyl, synthesis of, 773
Sulfides, cyanomethyl, electrochemical oxidation of, 221
Sulfides, cyclic phosphine, phosphinylation of, 1026
Sulfides, dialkyl, use in the synthesis of α-(phenylthio) ketones, 220
Sulfides, α-diazo phosphine, photolysis of, 346
Sulfides, dibenzyl, synthesis of, 345
Sulfides, α-fluoro
  from, sulfoxides, 61
  oxidation of, 73
  as starting materials, for α-fluoro sulfones, 79
  synthesis of, 61, 218
Sulfides, α-halo
  reactions, with alcohols, 218
  as starting materials, for monothioacetals, 218
  synthesis of, 61
Sulfides, α-halophosphine, synthesis of, 123
Sulfides, 1-halovinyl, synthesis of, 773
Sulfides, α-heteroarylthionitro, synthesis of, 327
Sulfides, β-hydroxy, synthesis of, 220, 391
Sulfides, α-iodo, synthesis of, 71
Sulfides, α-lithio
  from
    phenylvinyl sulfide, 392
    vinyl sulfides, 392
  synthesis of, 366
    via thiophilic addition, 394
Sulfides, 1-lithioalkylphosphine, synthesis of, 576
Sulfides, α-lithiophenyl, synthesis of, 394
Sulfides, 1-lithiovinyl phenyl, reactions, with tin chlorides, 961
Sulfides, 1-metalloalkylphosphine
  reactions
    with chlorophosphines, 551
    with chlorosilanes, 570
    with chlorostannanes, 588
    with gold compounds, 586
    with mercury compounds, 587
    with tantalum compounds, 579
Sulfides, α-methoxy, synthesis of, 222

Sulfides, 2-methoxyalkyl, as starting materials, for
    1-alkylthiovinyllithiums, 958
Sulfides, nitrile, cycloadditions of, 232
Sulfides, nitro, as starting materials, for sulfones, 908
Sulfides, α-nitro, synthesis of, 326
Sulfides, phenyl fluoromethyl, synthesis of, 61
Sulfides, 1-phosphinoalkylphosphine
    reduction of, 550
    synthesis of, 551
Sulfides, phosphole, reduction of, 1026
Sulfides, primary chloro, from, paraformaldehyde, 69
Sulfides, secondary chloro, from, paraldehyde, 69
Sulfides, 1-silylalkylphosphine, synthesis of, 569
Sulfides, α-silyloxy, synthesis of, 223
Sulfides, trimethylsilyl
    reactions
        with aldehydes, 216
        with ketones, 216
Sulfides, α-(trimethylsilyloxy), synthesis of, 216
Sulfides, 1-trimethylsilylvinyl, synthesis of, 945
Sulfides, unsymmetrical, chlorination of, 66
Sulfides, vinyl
    reactions
        with organolithium nucleophiles, 366
        with organolithiums, 392
    as starting materials, for α-lithio sulfides, 392
Sulfides, vinylphosphine, synthesis of, 1023
Sulfimide complexes, synthesis of, 800
Sulfimides, α-halo alkenyl dichloro-, synthesis of, 800
Sulfinamides, deprotonation of, 391
Sulfinamides, arene-, reactions, with 1,3-dichloro-1,1,3,3-tetrafluoroacetone, 311
Sulfinate, (-)-menthyl p-toluene, reactions, with
    1-trimethylsilylvinylmagnesium bromide, 946
p-Sulfinate, menthyl toluene-, reactions, with
    p-tolylthiomethyllithium, 265
p-Sulfinate, (-)-menthyl (S)-toluene-, use in the synthesis of
    dithioacetal trioxides, 271
Sulfinate anions
    reactions
        with alkyl halides, 235
        with dibromomethane, 83
Sulfinates, α-hydroxy, synthesis of, 234
Sulfines
    reactions
        with nucleophiles, 859
        with organometallic compounds, 272
    as starting materials, for bissulfoxides, 270
Sulfines, α-(allylthio)-, tautomerization of, 859
Sulfines, dithioester-, [4π+2π] cycloaddition of, 265
Sulfines, phosphoryl-, synthesis of, 926
Sulfines, sulfonyl-, reactions, with methyllithium, 272
p-Sulfinic acid, toluene-, reactions, with phenylglyoxal, 234
Sulfinic acids
    reactions, with acetals, 236
    as starting materials, for α-alkoxy sulfones, 235
    synthesis of, 270
Sulfinimidoyl chlorides, N-α-chloroalkyl, synthesis of, 110
α-Sulfinyl carbanions
    alkylation of, 77
    chirality of, 391
    from
        α-bromo sulfoxides, 77
        phenyl fluoromethyl sulfoxides, 74
    reactions
        with aldehydes, 77
        with ketones, 77
    use in the synthesis of α-chloro sulfoxides, 77
Sulfinyl chloride, reactions, with methylene compounds, 266
Sulfinyl chlorides
    reactions, with diazo compounds, 76
    stability of, 78
    as starting materials, for α-iodo sulfoxides, 78

synthesis of, 78
Sulfinyl chlorides, α-chloroalkane, synthesis of, 78
Sulfinyl chlorides, α-chloroalkyl, from, trithianes, 78
Sulfinyl compounds, synthesis of, 958
Sulfinyl halides
    reactions
        with diazo compounds, 75
        with diazomethane, 76
Sulfites, dialkyl-, as dehydrating agents, for the synthesis of
    acetals, 177
Sulfodiimines, reactions, with bis(1,1-alkylthio)alkenes, 899
Sulfolanes, reactions, with diphenyl disulfide, 266
Sulfonamide, N-sulfinyl-p-toluene-, reactions, with β-keto
    sulfoxides, 326
Sulfonamide, p-toluene-, condensation of, 311
Sulfonamides
    deprotonation of, 391
    reactions
        with aldehydes, 423
        with aminols, 427
        with formaldehyde, 427
        with trioxan, 427
Sulfonamides, N-alkyl aryl-, as starting materials, for
    bissulfonamides, 428
Sulfonamides, arene-, as starting materials, for N-sulfonyl
    aminals, 423
Sulfonamides, α-bromo-, synthesis of, 105
Sulfonamides, bromoalkane-, synthesis of, 86
Sulfonamides, chloroalkane-, synthesis of, 86
Sulfonamides, α-halo-, synthesis of, 104
Sulfonamides, α-haloalkane-, from, α-halosulfonyl
    halides, 86
Sulfonamides, iodoalkane-, synthesis of, 86
Sulfonamides, N-methyl alkyl-, use in the synthesis of N-thio
    α-haloamines, 104
Sulfonamides, N-sulfinyl-, reactions, with aldehydes, 311
Sulfonate, methyl fluoro-, reactions, with ketene
    dithioacetals, 863
Sulfonate compounds, silylalkane, synthesis of, 397
Sulfonates, α,α-bis(silyl), synthesis of, 605
Sulfonates, enol, reactions, with
    bromotrichloromethane, 742
Sulfonates, α-fluoroalkane, synthesis of, 86
Sulfonate salts, fluoro, reactions, with n-butyllithium, 369
Sulfone, benzyl phenyl, use in the synthesis of
    S,N-acetals, 332
Sulfone, α-brominated, reactions, with sodium phenyl
    selenide, 286
Sulfone, 1-bromovinyl methyl, synthesis of, 779
Sulfone, t-butoxymethyl phenyl, alkylation of, 200
Sulfone, cyclopropyl 1-chloro-1-phenylmethyl,
    synthesis of, 81
Sulfone, dimethyl
    deprotonation of, 391
    metallation of, 332
Sulfone, 1-fluoro-2-chloroethyl phenyl,
    dehydrochlorination of, 79
Sulfone, fluoromethyl phenyl, synthesis of, 780
Sulfone, α-fluoromethyl phenyl, as starting material, for
    α-sulfonyl carbanions, 79
Sulfone, n-heptylphenyl, reactions, with
    lithiotrichloromethane, 741
Sulfone, iodomethyl phenyl
    as starting material, for α-sulfonyl carbanions, 84
    synthesis of, 84
Sulfone, methoxymethyl p-tolyl, synthesis of, 235
Sulfone, methyl phenyl
    deprotonation of, 271
    reactions, with ethylmagnesium bromide, 395
Sulfone, methylthiomethyl p-tolyl, as starting material, for
    ketene dithioacetal S,S-dioxides, 857
Sulfone, methylvinyl, bromination–dehydrobromination
    of, 779

Sulfone

Sulfone, phenylselenomethyl phenyl, synthesis of, 286
Sulfone, phenyl (p-tolylsulfonylmethyl), condensation of, 858
Sulfone, phenylvinyl, bromination–dehydrobromination of, 779
Sulfone carbanions, reactions, with diphenyl diselenides, 286
Sulfones
  α-chlorination of, 80
  deprotonation of, 391
  free-radical bromination of, 83
  from, nitro sulfides, 908
  oxidative desulfonylation of, 234
  reactions
    with alkyllithiums, 936
    with n-butyllithium, 391
    with lithium alkoxides, 235
  as starting materials
    for S,N-acetals, 332
    for alkenes, 840
    for Grignard reagents, 395
    for α-lithiated alkenes, 960
  synthesis of, 960
Sulfones, α-acyloxy, synthesis of, 235
Sulfones, α-alkoxy
  from
    α-diazo sulfones, 235
    monothioacetals, 235
    sulfinic acids, 235
  synthesis of, 235
  via carbon–carbon bond formation, 236
Sulfones, alkoxyvinyl
  synthesis of, 840
  See also Ketene monothioacetal S,S-dioxides
Sulfones, alkyl bromomethyl, from, bromomethanesulfonyl bromide, 83
Sulfones, alkyl chloromethyl, synthesis of, 81
Sulfones, alkyl haloalkyl, synthesis of, 81
Sulfones, alkyl halomethyl, synthesis of, 81
Sulfones, alkynic, reactions, with thiols, 861
Sulfones, α-amidoalkyl, from, aminals, 332
Sulfones, aryl, reductive lithiation of, 381
Sulfones, aryl chloromethyl, synthesis of, 81
Sulfones, aryl haloalkyl, synthesis of, 81
Sulfones, aryl halomethyl, synthesis of, 81
Sulfones, aryl methoxymethyl, synthesis of, 235
Sulfones, arylvinyl, synthesis of, 797
Sulfones, α-azido, synthesis of, 331
Sulfones, α-bromo
  alkylation of, 83
  from, α-bromo sulfides, 83
  as starting materials, for α-sulfonyl carbanions, 83
  synthesis of, 83
    via halogenative decarboxylation, 83
Sulfones, α-bromoalkyl, synthesis of, 83
Sulfones, bromomethyl, synthesis of, 83
Sulfones, bromomethyl dienyl, synthesis of, 84
Sulfones, α-bromomethylvinyl, synthesis of, 84
Sulfones, (Z)-α-bromovinyl, synthesis of, 783
Sulfones, α-chloro
  from
    chloromethanesulfonyl halides, 82
    α-chloro sulfides, 80
  reactions, with carbonyl compounds, 236
  as starting materials, for α-sulfonyl carbanions, 82
  synthesis of, 80
Sulfones, 1-chloroalkenyl phenyl, synthesis of, 781
Sulfones, 1-chlorocyclopropyl, synthesis of, 80
Sulfones, chloromethyl, synthesis of, 82
Sulfones, α-diazo
  hydrolysis of, 234
  as starting materials, for α-alkoxy sulfones, 235
Sulfones, 2,2-dihalovinyl, reactions, with potassium fluoride, 783

Sulfones, 2-ethoxyvinyl, reactions, with disulfones, 865
Sulfones, α-fluoro
  from
    1-fluoro-1-(phenylsulfonyl)ethene, 79
    α-fluoro sulfides, 79
  synthesis of, 79
Sulfones, α-fluoroalkenyl phenyl, reactions, with tributyltin hydride, 812
Sulfones, (E)-1-fluoro 2-aryl, synthesis of, 780
Sulfones, α-fluoro β-substituted ethyl phenyl, synthesis of, 79
Sulfones, glycosyl
  from, glycosyl halides, 236
  synthesis of, 235
Sulfones, α-halo
  from, α-sulfonylcarboxylic acids, 81
  synthesis of, 79
Sulfones, α-haloalkyl, synthesis of, 81
Sulfones, α-hydroxy
  as starting materials, for ethoxyethyl compounds, 236
  synthesis of, 234
Sulfones, α-iodo
  from, iodomethanesulfonyl bromide, 84
  synthesis of, 84
Sulfones, α-iodoalkyl, synthesis of, 84
Sulfones, isochromanyl, synthesis of, 235
Sulfones, α-lithio
  from, α-tributylstannyl sulfones, 394
  synthesis of, 391
Sulfones, methoxymethyl p-tolyl, synthesis of, 235
Sulfones, 1-methoxyvinyl, synthesis of, 838
Sulfones, methyl methylthiomethyl
  condensation reactions of, 857
  as starting materials, for ketene dithioacetal S,S-dioxides, 857
Sulfones, α-nitro
  α-fluorination of, 108
  from, α-halonitroalkanes, 332
  reactions, with thiolates, 328
  synthesis of, 331
Sulfones, β-silylvinyl, reactions, with organometallic compounds, 393
Sulfones, (E)-styryl, synthesis of, 275
Sulfones, α-sulfonyloxy, synthesis of, 235
Sulfones, gem-sulfoxide, synthesis of, 272
Sulfones, α-sulfoxide, from, dithioacetals, 271
Sulfones, α-tributylstannyl, as starting materials, for α-lithio sulfones, 394
Sulfones, α,β-unsaturated, reactions, with hydrogen peroxide, 237
Sulfones, β-α-unsaturated, synthesis of, 780
Sulfones, vinyl
  Michael additions to, 366
  reactions
    with dimethylcopperlithium, 964
    with S-methyl methanethiosulfonate, 858
    with sulfenyl chlorides, 861
    with trimethylsilyl iodide, 84
  as starting materials, for ketene dithioacetal S,S-dioxides, 861
  structure of, 960
  synthesis of, 858
p-Sulfonic acid, toluene-, as catalyst, for the condensation of thiols, 245
Sulfonic acid compounds, as starting materials, for S,N-acetals, 331
Sulfonic acids
  reactions, with tosyl diazomethane, 235
  synthesis of, 267
Sulfonic acids, α-haloalkane-, salts of, synthesis of, 86
Sulfonic acids, α-hydroxy-, synthesis of, 237
Sulfonium salts
  synthesis of, 347
  use in the synthesis of α-aminoalkylphosphines, 456

Sulfonium salts, acyloxy-, synthesis of, 323
Sulfonium salts, 1-thiovinyl, synthesis of, 860
α-Sulfonyl carbanions
  alkylation of, 82
  bromination of, 83
  from
    α-bromo sulfones, 83
    α-chloro sulfones, 82
    α-fluoromethyl phenyl sulfone, 79
    iodomethyl phenyl sulfone, 84
  iodination of, 84
  reactions, with electrophiles, 236
  structure of, 391
Sulfonyl chloride, α-chloro-, as starting material, for 1-chloroethanesulfinic acid, 78
Sulfonyl chlorides, alkane-, reactions, with amines, 331
Sulfonyl chlorides, arene-, reactions, with trimethylvinylsilane, 942
Sulfonyl fluorides
  reactions, with Grignard reagents, 280
  as starting materials, for bis(sulfones), 280
Sulfonyl fluorides, (2-nitrosotrifluoroethyl)oxy-, synthesis of, 104
Sulfonyl halides, bromoalkane-, synthesis of, 85
Sulfonyl halides, α-halo-
  as starting materials, for α-haloalkanesulfonamides, 86
  synthesis of, 85
Sulfonyl halides, α-haloalkane-, reactions, with alcohols, 86
Sulfonyl iodide, p-toluene-, reactions, with potassium nitronates, 331
Sulfonyloxyamine, mesitylene-
  reactions
    with α-bromo sulfoxides, 86
    with α-chloro sulfoxides, 86
Sulfoxidation, for the synthesis of ketene monothioacetal S,S-dioxides, 838
Sulfoxide, benzhydryl benzyl, reactions, with sulfuryl chloride, 74
Sulfoxide, (+)-(S)-1-bromovinyl p-tolyl, synthesis of, 776
Sulfoxide, chloromethyl methyl, as starting material, for ethoxymethyl methyl sulfoxide, 233
Sulfoxide, chloromethyl phenyl
  deprotonation of, 77
  metallation of, 233
  reactions, with aldehydes, 776
Sulfoxide, dimethyl (DMSO)
  as phase-transfer catalyst, 793
  reactions
    with pyrylium salts, 863
    with sodium hydride, 391
  thiophilic addition of, 270
Sulfoxide, divinyl, chlorination of, 77
Sulfoxide, ethoxymethyl methyl, from, chloromethyl methyl sulfoxide, 233
Sulfoxide, ethyl phenyl, reactions, with (dichloroiodo)benzene, 75
Sulfoxide, 1-fluoro-2-(4-biphenylyl)vinyl phenyl, synthesis of, 777
Sulfoxide, fluoromethyl phenyl, reactions, with sodium azide, 86
Sulfoxide, methoxymethyl phenyl, rearrangement of, 232
Sulfoxide, methoxymethyl p-tolyl, synthesis of, 233
Sulfoxide, methyl methylthiomethyl, use in the synthesis of S,N-acetals, 326
Sulfoxide, (+)-methyl p-tolyl, halogenation of, 75
Sulfoxide, phenyl methyl, reactions, with (dimethylamino)tributyltin, 397
Sulfoxide, trimethylsilylmethyl phenyl, thermal rearrangement of, 222
α-Sulfoxide carbanions, as starting materials, for α-selenosulfoxides, 285
Sulfoxides
  bromination of, 77
  chlorination of, 74, 75
  deprotonation of, 391
  from, 1,3-oxathiolan-5-ones, 837
  halogenation of, 75
    stereochemistry, 75
  Pummerer elimination of, 924
  reactions
    with alkyllithiums, 391
    with copper(I) iodide, 396
    with diethylaminosulfur trifluoride, 73
  as starting materials
    for α-fluoro sulfides, 61
    for monothioacetals, 221
  synthesis of, 959
  use in glycosylation, 232
Sulfoxides, alkenyl, deprotonation of, 961
Sulfoxides, (E)-2-alkenyl, synthesis of, 958
Sulfoxides, α-alkoxy
  from, α-chloro sulfoxides, 232
  synthesis of, 232
Sulfoxides, β-alkoxyethyl, synthesis of, 233
Sulfoxides, alkyl, as starting materials, for α-chloro sulfides, 69
Sulfoxides, alkyl alkylthiomethyl, as starting materials, for ketene dithioacetal S-oxides, 857
Sulfoxides, alkyl aryl, as starting materials, for α-chloro sulfides, 69
Sulfoxides, alkyl phenyl, reactions, with chlorotrimethylsilane, 945
Sulfoxides, allenyl, synthesis of, 837
Sulfoxides, aryl, reactions, with trialkyltinamines, 397
Sulfoxides, aryl fluoromethyl, synthesis of, 73
Sulfoxides, bis-
  from, methylene bissulfoxides, 269
  metallation of, 269
  synthesis of, 267
Sulfoxides, α-bromo
  from, trichloromethanesulfinyl bromide, 77
  reactions, with mesitylenesulfonyloxyamine, 86
  as starting materials, for α-sulfinyl carbanions, 77
  synthesis of, 77
Sulfoxides, α-chloro
  from, sulfides, 74
  reactions, with mesitylenesulfonyloxyamine, 86
  as starting materials, for α-alkoxy sulfoxides, 232
  synthesis of, 74
Sulfoxides, α-chloromethyl, synthesis of, 76
Sulfoxides, α-chloro-α'-polychloro, chlorination of, 78
Sulfoxides, α-dichloro, reactions, with titanium(IV) chloride, 773
Sulfoxides, dicyclohexyl, deprotonation of, 264
Sulfoxides, 1-fluoro, synthesis of, 73
Sulfoxides, α-fluoro, reactions, with lithium diphenylcuprate, 777
Sulfoxides, α-fluoroalkyl aryl, synthesis of, 73
Sulfoxides, 1-fluoro-2-hydroxyalkyl phenyl, synthesis of, 778
Sulfoxides, α-germyl, synthesis of, 371
Sulfoxides, α-halo
  reactions, with trimethylsilyl triflate, 773
  synthesis of, 73
Sulfoxides, 1-haloalkyl aryl, synthesis of, 73
Sulfoxides, α-iodo
  from, sulfinyl chlorides, 78
  synthesis of, 77
Sulfoxides, iodomethyl aryl, synthesis of, 78
Sulfoxides, β-keto
  deprotonation of, 961
  reactions, with N-sulfinyl-p-toluenesulfonamide, 326
Sulfoxides, α-lithio
  synthesis of, 391
  transmetallation of, 397
Sulfoxides, metallated, synthesis of, 391
Sulfoxides, methyl alkyl, chlorination of, 76

Sulfoxides, methyl methylthiomethyl
  condensation of, 856
  reactions, with aryl ketones, 857
  as starting materials, for ketene dithioacetal S-oxides, 856
  uses of, 856
Sulfoxides, phenyl fluoromethyl
  alkylation of, 74
  as starting materials, for α-sulfinyl carbanions, 74
Sulfoxides, phenylvinyl, as starting materials, for
    2-(phenylsulfinyl)acylals, 207
Sulfoxides, α-silyl, use in sila-Pummerer reactions, 222
Sulfoxides, unsymmetrical, chlorination of, 75
Sulfoxides, vinyl
  Michael additions to, 366
  as starting materials, for ketene dithioacetal S-oxides, 858
Sulfoximide, vinyl-, α-sulfenylation of, 862
Sulfoximides, N-alkyl-, chlorination of, 86
Sulfoximides, α-aminoalkyl-, synthesis of, theoretical, 330
Sulfoximides, N-chloro-, chlorination of, 86
Sulfoximides, α-chloroalkyl N-alkyl-, synthesis of, 86
Sulfoximides, chloromethyl phenyl N-chloro-,
    reduction of, 86
Sulfoximides, fluoromethyl-, synthesis of, 86
Sulfoximides, α-haloalkyl-, synthesis of, 86
Sulfoximides, N-methyl-, fluorination of, 86
Sulfoximine, fluoromethyl-
  reactions
    with aldehydes, 86
    with ketones, 86
Sulfoximines
  deprotonation of, 391
  reactions, with bis(1,1-alkylthio)alkenes, 899
Sulfoximines, (S)-(sulfinylmethyl)-, synthesis of, 272
Sulfur
  and oxygen, synthesis of functions, 833
  reactions
    with 1,3-diphosphaallenes, 559
    with α-halophosphines, 128
  synthesis of functions containing two sulfurs, 244, 842
Sulfurane, bis(methylene)-, reactions, with aldehydes, 231
Sulfuranes, aminofluoro-, use in the synthesis of
    difluoroalkanes, 10
Sulfur compounds, synthesis of, 345, 769
Sulfur compounds, dicoordinated, synthesis of, 283
Sulfur compounds, α-halo-, synthesis of, 61
Sulfur compounds, tetracoordinated,
    synthesis of, 286, 778, 838, 873
Sulfur compounds, tetracoordinate α-halo-, synthesis of, 79
Sulfur compounds, tricoordinated
  as starting materials, for hemithioaminals, 329
  synthesis of, 285, 776, 837, 873, 907
Sulfur compounds, tricoordinate α-halo-, synthesis of, 78
Sulfur dichloride, reactions, with chloroimines, 110
Sulfur dioxide
  reactions
    with pyridine 2-carboxaldehyde, 237
    with 1,1,3,3-tetramethyl-1,3-disilacyclobutane, 624
Sulfur functions, dicoordinated, synthesis of, 833, 868
Sulfur heterocycles, functionalization of, 226
Sulfuric acid, as dehydrating agent, for the synthesis of
    acetals, 178
Sulfuric acid, chloro-, reactions, with acetic acid, 282
Sulfur mustards, reactions, with chloramine-T, 773
Sulfur nucleophiles
  reactions
    with 1,1-dinitro-2,2-diphenylethene, 906
    with glycosyl halides, 223
    with halomethylgermanes, 371
    with halomethylsilanes, 367
    with α-selenocarbanions, 284
Sulfur tetrafluoride, use in the synthesis of difluoroalkanes, 8
Sulfur trifluoride, diethylamino- (DAST)
  for the fluorination of sulfoxides, 61
  reactions
    with alcohols, 97
    with N-bromosuccinimide, 45
    with α-hydroxyalkyl phosphonates, 131
    with α-hydroxybenzyl phosphonate esters, 129
    with (hydroxymethyl)triphenylphosphonium
      tetrafluoroborate, 117
    with sulfoxides, 73
  use in the synthesis of difluoroalkanes, 10
Sulfur trifluoride, phenyl-, use in the synthesis of
    difluoroalkanes, 9
Sulfur trifluorides, N,N-dialkylamino-, use in the synthesis
    of difluoro compounds, 10
Sulfur trioxide
  reactions
    with 2,2-difluoroethene-1-sulfonyl fluoride, 867
    with perfluoropropene, 86
    with 1,1,3,3-tetramethyl-1,3-disilacyclobutane, 624
    with tris(silylalkyl)aluminum compounds, 397
Sulfuryl chloride
  for the chlorination of sulfides, 63
  for the chlorination of sulfoxides, 74
  reactions
    with anisole, 47
    with benzhydryl benzyl sulfoxide, 74
    with dibenzyl sulfide, 64
    with methylthioacetone, 775
    with phenylketones, 12
    with sulfides, 80
    with toluene, 11
Sulfur ylides
  reactions
    with 1H-1,2,4-diazarsole, 502
    with tetrachloropalladium(II) salts, 396
Superbase
  as metallating agent, 650
  use in the synthesis of 1,1-bis(silyl)alkanes, 620
  See also n-Butyllithium–potassium t-butoxide

Tantalocene, dimeric, synthesis of, 672
Tantalum, use in the synthesis of α-hydroxysilanes, 355
Tantalum bis(cyclopentadienyl)trihydrides, use in the
    synthesis of dimeric tantalocene, 672
Tantalum complexes, cyclopentadienyl-
  reactions
    with aldehydes, 355
    with ketones, 355
Tantalum compounds
  reactions
    with 1-metalloalkylphosphine oxides, 579
    with 1-metalloalkylphosphines, 579
    with 1-metalloalkylphosphine sulfides, 579
    with phosphorus ylides, 579
L-Tartaric acid anhydride, dibenzoyl-, use in the synthesis of
    α-aminophosphonic acids, 462
Tautomeric equilibrium
  of iminoesters and 1-alkoxyenamines, 880
  of ketene S,N-acetals, 898
Taxane, synthesis of, 220
Tebbe reagent
  for the synthesis of diborocyclopentenes, 632
  use in the synthesis of transition metal complexes, 716
Telluradiazoles, synthesis of, 334
Telluride, α-bromo, as intermediate, in the synthesis of
    α-bromotellurones, 93
Telluride, n-butyl t-butyl, reactions, with alkynes, 947
Telluride, butyl trimethylsilyl, as starting material, for
    trimethylsilylmethyllithium, 652
Telluride, chloromethyl phenyl, synthesis of, 93
Telluride anions
  alkylation of, 93
  reactions, with gem-dihalides, 290
  as starting materials, for selenotelluroacetals, 291

synthesis of, 291
Tellurides, arylvinyl, deprotonation of, 947
Tellurides, dialkyl, synthesis of, 377
Tellurides, glycosyl, as starting materials, for glycosyl radicals, 240
Tellurides, phenylvinyl, deprotonation of, 965
Tellurium, synthesis of functions, containing oxygen, 841
Tellurium compounds, synthesis of, 872
Tellurium compounds, α-halo-, synthesis of, 93
Tellurium tetrachloride, reactions, with acetic anhydride, 291
1-Telluroalkynes, reactions, with acetic acid, 841
α-Telluroboranes, synthesis of, 377
α-Tellurocarbanions, synthesis of, reviews, 398
α-Tellurogermanes, synthesis of, 377
α-Tellurolithium compounds, synthesis of, 402
Tellurols, thienyl, synthesis of, 872
Tellurones, α-bromo, synthesis of, 93
α-Tellurosilanes, synthesis of, 377
Tellurotetrathiafulvalenes, monoalkyl, synthesis of, 873
α-Telluryl carbanions, synthesis of, 398
Terephthalaldehyde
  as starting material
    for aminals, 406
    for bishemithioaminals, 317
Tetraalkyl compounds, synthesis of, 660
Tetraamines, as starting materials, for ketene aminals, 979
Tetraarsaadamantanes, synthesis of, 593
Tetraarsacyclohexanes, from, bis(methylchloroarsino)methane, 594
Tetraazafluorenes, perhydro-, synthesis of, 409
Tetracyano compounds, synthesis of, 850
Tetraheterafulvalenes, synthesis of, 868
9-(Tetrahydropyran-2-yl) compounds, synthesis of, 306
Tetrakis(phosphines), synthesis of, 546
Tetra(methylthio) compounds, synthesis of, 842
Tetramine, hexamethylene- (HMT)
  reactions, with thiophenols, 317
  as starting material, for tris(chloromethyl)amine, 307
  synthesis of, 294, 407
  for the synthesis of N,N'-methylenebisamides, 424
Tetranuclear carbon-bridged, rhodiacyclopentadiene complexes, from, hex-3-yne, 689
Tetraoxaspirononanes, reactions, with benzoxaphospholes, 478
1,1,3,3-Tetraoxide, 1,3-dithiane-, synthesis of, 277
Tetraselenafulvalenes
  from
    1,3-diselenole-2-selenones, 874
    1,3-diselenole-2-thiones, 874
  synthesis of, 874
Tetraselenafulvalenes, (methoxycarbonyl)-, synthesis of, 874
Tetrastannacyclohexane, synthesis of, 700
Tetrasulfides, synthesis of, 866
Tetrasulfones
  reactions, with amines, 866
  synthesis of, 277
Tetratellurafulvalenes, synthesis of, 876
Tetratelluranaphthalenes, synthesis of, 877
Tetrathiafulvalenes
  synthesis of, 850, 851
  telluration of, 873
Tetrathiafulvalenes, tetra(alkyltelluro)-, synthesis of, 873
Tetrathiafulvalenes, tetra(methylthio)-, synthesis of, 846
Tetrathiofulvalenes, synthesis of, 847
Tetrathiomalonate, diethyl, reactions, with N,N-dimethylchloromethaniminium chloride, 775
Tetrazines, reactions, with ethyl thionoformate, 229
Tetrazines, tetrahydro-, synthesis of, 437
1,2,4,5-Tetrazines, tetrahydro-, synthesis of, 436
Tetrazoles, N-alkyl, lithiation of, 539
Thallium, (chloromethyl)chloro(phenyl)-, synthesis of, 153
Thallium, dichloro(phenyl)-, reactions, with diazomethane, 153

Thallium(I) salts, reactions, with bromine, 27
Thallium(III) nitrate, reactions, with alkenes, 194
Thallium acetate, reactions, with alkynes, 1068
Thallium acetylacetonate, as starting material, for trinuclear carbon-bridged palladium complexes, 693
Thallium bromide, dialkyl-, from, thallium tribromide, 659
Thallium chloride, bis(trimethylsilylmethyl)-, synthesis of, 659
Thallium compounds, reactions, with phosphorus ylides, 587
Thallium compounds, (chloromethyl)-, synthesis of, 153
Thallium compounds, (halomethyl)-, synthesis of, 153
Thallium ethoxide, reactions, with dithioester S-oxides, 859
Thallium tribromide, as starting material, for dialkylthallium bromide, 659
Thallium trichloride, reactions, with trimethylsilylmethyllithium, 659
Thiacepham compounds, oxidation of, 274
1-Thiacyclohept-4-yne, 3,3,6,6-tetramethyl-, use in the synthesis of diiron complexes, 681
Thiacyclopentane, bromination of, 65
Thiacyclopentane, α-bromo-, synthesis of, 65
Thiacyclopentane, 2,3-dichloro-, synthesis of, 65
Thiadiazine, synthesis of, 229
1,2,3-Thiadiazoles, synthesis of, 852
1,3,4-Thiadiazoles, 2-alkylidene-2,5-dihydro-, synthesis of, 328
1,3,4-Thiadiazolidines, synthesis of, 318
Thiadiphosphetanes, synthesis of, 559
Thiagermatanes, from, bis(iodomethyl)germane, 372
1,3-Thiagermatanes, 3,3-dialkyl-, synthesis of, 372
Thiagermetanes, pyrolysis of, 627
2-Thiapropane, 1-chloro-, as starting material, for 2-thia-4-selenapentane, 283
Thiaselenafulvenes
  from
    phosphonate esters, 870
    1,3-thiaselenole-2-selenones, 870
    1,3-thiaselenole-2-thiones, 870
    1,3-thiaselenolium salts, 870
    triphenylphosphonium salts, 870
  synthesis of, 870
2-Thia-4-selenapentane, from, 1-chloro-2-thiapropane, 283
1,3-Thiaselenolium salts, as starting materials, for thiaselenafulvenes, 870
Thiazaphospholes, synthesis of, via cycloaddition reactions, 493
Thiazepines, synthesis of, 325
1,3-Thiazetidines, synthesis of, 905
1,2-Thiazetidines 1-oxides, 3-alkoxy-, synthesis of, 312
Thiazine, dihydro-, reactions, with chlorotrimethylsilane, 370
Thiazines, from, β-thiol amides, 325
$4H$-1,3-Thiazines, 2-substituted 5,6-dihydro-, reduction of, 321
1,4-Thiazin-3-ones, as starting materials, for α-benzoyloxy compounds, 226
Thiazoles, 2,3-dihydro-, synthesis of, 318
Thiazoles, 2,5-dihydro-, synthesis of, 317, 328
Thiazoles, 4,5-dihydro-, reduction of, 321
Thiazolidines
  from, aldehydes, 318
  oxidation of, 226
Thiazolidines, 4-oxo-, synthesis of, 319
Thiazolines, from, isocyanides, 538
Thiazolinethiones, synthesis of, 538
Thieno compounds, bis-, synthesis of, 850
Thietane dioxides, from, methanesulfonylsulfene, 281
Thietanes, ethoxy-, reactions, with phenylmagnesium bromide, 731
$2H$-Thiin, 3,4-dihydro-, synthesis of, 65
Thiiranes, desulfurization of, 862
Thioacetals
  alcoholysis of, 196

Thioacetals

reactions
  with 2-aminothiophenols, 899
  with dichlorocarbene, 775
  with isocyanates, 426
  as starting materials, for acetals, 196
Thioacetals, diphenyl, reductive lithiation of, 366
Thioacetals, ketene
  reactions, with lithium naphthalenide, 956
  as starting materials, for 1-alkylthiovinyllithiums, 956
Thioacetals, silylated, reactions, with thiols, 247
Thioacetals, α-tributylstannyl, reactions, with trimethylsilyl enol ethers, 398
Thioacetate, reactions, with diethyl iodomethyl phosphonate, 346
Thioacetates, dicyano-, as starting materials, for ketene aminals, 977
Thioacetic acid, phenyl-, decarboxylation of, 68
Thioacetone, methyl-
  reactions
    with sulfuryl chloride, 775
    with triethyl phosphite, 775
Thioacrolein dianions, transmetallation of, 395
Thioaldehyde, synthesis of, 1049
1-Thioaldose, per-O-acetyl-, synthesis of, 223
Thioalkanes, 1,1-bis(trimethylsilyl)-1-phenyl-, reductive lithiation of, 618
α-Thioalkanoyl chlorides, as starting materials, for acyl azides, 328
Thioalkenes, 1-alkoxy 1-halo 2-alkyl-, synthesis of, 765
Thioalkenes, aryl-, silylation of α-carbanions, 938
Thioalkenes, 1,2-dialkoxy 1-alkyl-, synthesis of, 765
Thioalkenes, 1-halo 1-alkyl-, oxidation of, 776
Thioalkenes, 1-halo 1-phenyl-, synthesis of, via Wittig–Horner reaction, 773
Thioalkenes, nitrophenyl-, synthesis of, 907
1-Thioalkenes, 1-amino-, synthesis of, 898
1-Thioalkenes, 1-halo, synthesis of, 774
1-Thioalkenes, 1-nitro-, synthesis of, 906
α-Thioalkylation, mechanism, 315
Thioalkynes, hydrosilation of, 940
Thioalkynes, alkyl-, reactions, with diethylamine, 905
Thioalkynes, 1-alkyl-, addition of diselenides to, 871
Thioalkynes, phenyl-, reactions, with tributyltin hydride, 962
5-Thio-D-allose, synthesis of, 227
α-Thioaluminums, synthesis of, 397
Thioamide, reactions, with *t*-butylthiyl radicals, 1077
Thioamides
  alkylation of, 899
  reactions, with hydrazines, 984
  as starting materials
    for α-hydrazino enamines, 984
    for ketene *S,N*-acetals, 899
Thioamides, β-hydroxy-, reactions, with formaldehyde, 230
Thioamides, *N*-(hydroxymethyl)-, synthesis of, 296
Thioamides, tertiary, reactions, with alkyl halides, 899
Thioamides, α,β-unsaturated, as starting materials, for ketene *S,N*-acetals, 901
Thioaminals, as starting materials, for aminomethyl carbanions, 528
α-Thioanions, synthesis of, 365
Thioanisole
  deprotonation of, 365
  lithiation of, 391
Thioates, phosphoroamido-, reactions, with benzaldehydes, 467
2-Thiobarbituric acids, reduction of, 426
Thiobenzaldehydes, reactions, with trimethylsilyllithium, 376
Thiobenzimidate, *N*-benzenesulfonyl-*S*-phenyl-, reactions, with 1-diethylaminoprop-1-yne, 905
Thiobenzimidate, *N*-benzoyl-*S*-phenyl-, reactions, with 1-diethylaminoprop-1-yne, 905
Thiobenzophenone
  reactions
    with azomethine ylides, 319
    with α-nitrosostyrene, 232
    with phenyllithium, 394
Thiobenzophenones, reactions, with phenyliodinium ylides, 866
α-Thioberylliums, synthesis of, 395
Thioborane complexes, trimethylamine–methyl-, synthesis of, 373
Thioboronates, α-alkyl-, synthesis of, 373
Thioboronates, α-phenyl-, alkylation of, 373
Thio-γ-butyrolactones, γ-phenyl-, synthesis of, 226
Thiocarbamates, use in the synthesis of distibines, 595
Thiocarbamoylation, of carbanions, 901
α-Thiocarbanions
  reactions
    with trialkylboranes, 372
    with trialkyl borate, 372
  synthesis of, 390
  via lithium exchange reactions, 394
Thiocarbonates, chloromethyl, synthesis of, 60
Thiocarbonates, iodomethyl, synthesis of, 60
Thiocarbonyl compounds
  [2+2] cycloadditions of, 852
  reactions
    with carbenes, 862
    with trialkyl phosphites, 345
    with trimethylsilylvinyl ketone, 232
  as starting materials
    for *S,N*-acetals, 319, 328
    for *S,P*-acetals, 345
    for dithioacetals, 255
    for ketene dithioacetal *S,S*-dioxides, 862
Thiocarboxylic acids, α-phenyl-, decarboxylation of, 71
Thiochroman-4-one 1-oxide, 2-azido-, from, thiochromone 1-oxide, 330
Thiochromone 1-oxide, as starting material, for 2-azidothiochroman-4-one 1-oxide, 330
Thiocopper, trifluoromethyl-
  reactions
    with dichloroacrylonitriles, 771
    with triodonitroethene, 771
α-Thiocoppers, synthesis of, 396
Thiocyanate, benzeneselenenyl, reactions, with diazomethane, 335
Thiocyclohexane, 1-azido-1-methyl-, from, 1,1-bis(methylthio)cyclohexane, 329
Thiodisulfides, α-alkyl, synthesis of, 256
α-Thiodithioacetals, as starting materials, for ketene dithioacetals, 846
Thiodithioacrylate, (*E,Z*)-3-chloro-3-ethyl-, synthesis of, 775
Thio-1,3-dithiolium salts, 2-alkyl-, electrophilic reactions of, 852
Thioenamines, 2-nitro-2-phenyl, synthesis of, 907
Thioesters
  as starting materials
    for ketene aminals, 979
    for ketene hemiaminals, 895
  synthesis of, 1074
Thioesters, alkylidene biscarbamic, synthesis of, 426
Thioesters, iodomethyl, reactions, with zinc, 396
Thioesters, *S*-phenyl, synthesis of, 906
Thioesters, ring-opened, synthesis of, 1075
Thioether, bis(trimethylsilyl) silyl, synthesis of, via Brook type rearrangement, 1049
Thioethers, benzyl germyl, as starting materials, for germylbenzylthiols, 371
Thioethers, benzyl silyl, as starting materials, for α-silylbenzylthiols, 368
Thioethers, β-bromo-β-phosphonovinyl, synthesis of, 804
Thioethers, germyl, reverse Brook rearrangement of, 371
Thioethers, α-germyl, synthesis of, 371

Thioethers, methyl phenyl-, as starting materials, for
    α-thioalkylboranes, 372
Thioethers, α-phenyl, reductive lithiation of, 380
Thioethers, silyl, reverse Brook rearrangement of, 368
Thioethers, α-stannyl, synthesis of, 397
Thioethylamine hydrobromide, 2,2-difluoro-1-ethyl-,
    synthesis of, 320
Thioethynes, phenyl-
  reactions
    with tributylstannylcuprate, 962
    with triethylgermane, 940
Thiofluorenone S-oxide, reactions, with sodium
    trichloroacetate, 742
Thioformaldehyde, reactions, with (halo)phosphines, 123
Thioformamidinium iodide, N,N,N',N'-tetramethylmethyl-,
    reactions, with methylene compounds, 973
Thioformamidinium salts, as starting materials, for ketene
    aminals, 973
Thioformyl chlorides, synthesis of, 71
1-Thio-D-galactose, 2,3,4,6-tetra-O-acetyl-, synthesis of, 223
α-Thiogalliums, synthesis of, 397
α-Thiogermanes
  from, α-germyl organomagnesium compounds, 371
  synthesis of, 371
1-Thioglucopyranose compounds, S-alkylation of, 225
Thioglucosides, S-ethyl, synthesis of, 224
1-Thioglycerol, reactions, with acetone, 229
Thioglycolic acid, as starting material, for dithioacetals, 272
Thioglycosides
  from, glycosyl fluorides, 225
  oxidation of, 232, 235
  synthesis of, 223
Thioglycosides, phenyl, as starting materials, for glycosyl
    fluorides, 45
Thioglycosides, pyridyl-, synthesis of, 224
1-Thioglycosides, synthesis of, 223
α-Thiogolds, synthesis of, 396
Thio groups, alkyl-, nucleophilic displacement of, 899
Thiohippurate, methyl 2-benzyl-, oxidative cleavage of, 309
Thiohydantoins, as starting materials, for
    imidazolidinones, 420
Thiohydroxamates, O-acyl, synthesis of, 326
5-Thio-L-idose, synthesis of, 228
Thioimidates
  from
    amides, 517
    trimethylsilylmethyl isocyanide, 518
  synthesis of, 518
  as tautomers of ketene S,N-acetals, 898
α-Thioindiums, synthesis of, 397
Thioketals, polyfluorinated, reactions, with thiolates, 775
Thioketene, bis(trifluoromethyl)-, reactions, with
    polyfluorinated phenyl imines, 905
Thioketene, di-t-butyl-, reactions, with dimethyl
    diazomalonate, 837
Thioketenes
  reactions
    with alkyllithium compounds, 956
    with aryllithium compounds, 956
    with imines, 905
  as starting materials, for ketene dithioacetals, 852
Thioketenes, bis(trifluoromethyl)-, reactions, with triethyl
    phosphite, 926
Thioketones
  reactions
    with diazoalkanes, 328
    with nitronates, 232
    with nitrones, 232
    with organolithiums, 392
Thiolane 1,1-dioxide, dibromo-, reactions, with
    bromoform, 748
Thiolate, 1,1-dimethylethane, reactions, with allyl
    bromides, 373
Thiolates
  reactions
    with 2-bromo-2-nitropropane, 327
    with α-chloro ethers, 218
    with 2-nitropropane, 327
    with α-nitro sulfones, 328
    with polyfluorinated thioketals, 775
    with prop-2-yniminium triflates, 321
Thiolates, arene-, reactions, with perchloro-1,3-
    butadiene, 851
Thiolates, dithiocarboxylic ene-, electrophiles for, 842
Thiolates, thionoester ene-, S-acylation of, 834
Thiolesters, α-chloro, reactions, with triethyl phosphite, 836
α-Thiolithiums, synthesis of, 391
Thiols
  boron derivatives, for the synthesis of dithioacetals, 248
  condensation of, 244
  reactions of, 940
  reactions
    with acetals, 249
    with 1-O-acetyl glycopyranosides, 224
    with aldehydes, 247
    with alkynes, 939
    with alkynic sulfones, 861
    with arenediazonium salts, 225
    with aziridine, 899
    with benzylideneaniline compounds, 319
    with 1-(chloromethyl)benzotriazole, 322
    with chloromethyl chloroformates, 60
    with gem-diamines, 249
    with diiodomethane, 249
    with fluoroalkyne, 772
    with α-halo ethers, 218
    with α-ketoaldehydes, 215
    with ketones, 247
    with N-methyl-N-nitrosochloromethylamine, 325
    with phosphinic chlorides, 128
    with prop-2-yniminium triflates, 321
    with silylated thioacetals, 247
    with trichloroacetaldehyde, 215
  silicon derivatives, for the synthesis of dithioacetals, 248
  as starting materials
    for α-bromo sulfides, 71
    for α-chloro sulfides, 69
  tin derivatives, for the synthesis of dithioacetals, 248
α-Thiomagnesiums, synthesis of, 395
α-Thio metal compounds, synthesis of, 397
Thiomethane, α-acetoxyphenyl-, from, methyl phenyl
    sulfide, 220
Thione, trimethylsilyl-t-butyl-, synthesis of, 371
3-Thione, 1,2-dithiole-, synthesis of, 834
Thiones, synthesis of, 343
Thiones, N-hydroxymethylthiadiazole-, reactions, with
    thionyl chloride, 105
Thiones, trimethylsilyl-
  S-alkylation of, 371
  cycloaddition of, 371
2-Thiones, 1,3-diselenole, as starting materials, for
    monothiatriselenafulvalenes, 869
2-Thiones, 1,3-diselenole-, as starting materials, for
    tetraselenafulvalenes, 874
2-Thiones, 1,3-dithiole-
  1,3-dipolar additions of, 850
  reactions
    with trialkyl phosphites, 850
    with triphenylphosphine, 850
2-Thiones, 1,3-thiaselenole-
  dimerization of, 868
  as starting materials, for thiaselenafulvenes, 870
Thionoesters
  S-alkylation of, 833, 834
  cycloadditions of, 229
Thionoesters, cyanoacetic, reactions, with benzylamine, 896

Thionoformate, ethyl, reactions, with tetrazines, 229
Thionolactones
  cycloadditions of, 229
  reactions, with organolithium compounds, 226
Thionopropionate compounds, synthesis of, 834
Thionoselenoesters, alkylation of, 870
Thionyl chloride
  for the chlorination of sulfides, 65
  reactions
    with active methylene compounds, 71
    with alcohols, 97
    with 1-chloroethanesulfinic acid, 78
    with diazomethane, 76
    with (hydroxymethyl)germanes, 141
    with N-hydroxymethylthiadiazolethiones, 105
    with methyl sulfide, 65
    with phenylethynylmagnesium bromide, 863
    with 1,3,5-trithiane, 65
  use in the synthesis of α-chloroalkyl esters, 54
  use in the synthesis of α-chloro ethers, 48, 49
  use in the synthesis of dichloroalkanes, 18
Thiophen 1,1-dioxide, 3-dichloromethylene 2,3-dihydro-, synthesis of, 742
Thiophene
  reactions
    with dodecacarbonyltriiron, 681
    with dodecacarbonyltriruthenium, 683
  use in the synthesis of carbon-bridged dicobalt complexes, 690
Thiophene, 2-acetyl-, reactions, with diethyl dithiocarbonimidates, 980
Thiophene, 2-azidotetrahydro-, oxidation of, 330
Thiophene, 2-methyl-, reactions, with dodecacarbonyltriruthenium, 683
Thiophene, 2-nitro-, reactions, with secondary amines, 907
Thiophene, tetrahydro-, reactions, with t-butyl peracetate, 226
Thiophene 1,1-dioxide, reactions, with chloroform, 742
Thiophenes, azido-, synthesis of, 856
Thiophenol
  reactions
    with benzaldehyde, 216
    with chlorovinylphosphine oxides, 120
    with α,α-di(morpholino)xylene, 322
    with N-(hexafluoroisopropylidene)benzenesulfinamides, 325
    with 1,2,2-tribromo-1-fluoroethene, 771
Thiophenol, 2-amino-, reactions, with aldehydes, 318
Thiophenol, 2-hydroxy-, reactions, with acid chlorides, 1073
Thiophenol, 3-methyl-, reactions, with N-benzoyl aminals, 323
Thiophenols
  reactions
    with 1-chloro enamines, 904
    with hexamethylenetetramine, 317
Thiophenols, 2-amino-, reactions, with thioacetals, 899
Thiophilic addition
  of dimethyl sulfoxide, 270
  of organo metallics, 255
Thiophosphate, diethyl chloro-, reactions, with 2-methanesulfonyl-1-(2-chlorophenyl)ethanone, 784
Thiophosphinates, O-alkyl, synthesis of, 128
Thiophosphinates, S-alkyl, synthesis of, 128
Thiophosphinates, O-aryl, synthesis of, 128
Thiophosphinates, S-aryl, synthesis of, 128
Thiophosphinates, S-n-butyl(chloromethyl)-, synthesis of, 128
Thiophosphinic acid compounds, α-halo-, synthesis of, 127
Thiophosphinic chloride, bis(chloromethyl)-, reduction of, 115
Thiophosphinic chloride, (chloromethyl)methyl-, from, phosphinic chlorides, 128
Thiophosphinic chlorides
  reactions
    with alcohols, 128
    with amines, 128
  synthesis of, 128
Thiophosphinic chlorides, (chloromethyl)-, from, oxo acid chlorides, 127
Thiophosphinous acid, S-acetyldiphenyl-, reactions, with benzaldehyde, 345
Thiophosphite, dialkyl, reactions, with ynamines, 1010
Thiophosphite, trimethyl, reactions, with methyl ketones, 345
Thiophosphonic acid compounds, aminomethyl-, synthesis of, 469
Thiophosphonic acid compounds, α-halo-, synthesis of, 133
Thiophosphoramidates, α-hydroxy, chlorination of, 106
Thiophosphoryl chloride, (dichlorophosphinomethyl)-, reactions of, 553
Thiopyrans, 2-amino-, synthesis of, 319
Thiopyrans, dihydro-, synthesis of, 901
2-Thiopyridone, N-hydroxy-, reactions, with carboxylic acids, 266
Thioselenoacetals
  deprotonation of, 285
  from
    diselenoacetals, 284
    selenophenates, 283
  synthesis of, 283, 284
Thioselenoacetals, β,β-dicyano-, synthesis of, 284
Thioselenoacetals, S,S-dioxide ketene, reactions, with enamines, 286
Thioselenoacetals, ketene
  from, dimethyl dithioselenocarbonate, 872
  synthesis of, 870
Thioselenole, 2-alkyl, fragmentation of, 872
Thiosemicarbazides
  reactions, with methyl iodide, 984
  as starting materials, for α-hydrazino enamines, 984
Thiosilane, phenyl-, α-chlorination of, 365
Thiosilane, trimethylphenyl-, reactions, with silylvinyl ketones, 944
Thiosilanes, α-methyl-, from, silyl sulfonium ylides, 369
α-Thiosilanes
  from, halomethylsilanes, 367
  rearrangement of ylides from, 369
  synthesis of, 364
Thiostannanes, use in the synthesis of thioglycosides, 225
Thiosugars, synthesis of, 227
1-Thiosugars, synthesis of, 223
Thiosulfines, α-alkyl-, as starting materials, for ketene dithioacetal S-oxides, 859
Thiosulfolane, 2-phenyl-, synthesis of, 266
α-Thiosulfones
  deprotonation of, 266
  from, α-thiosulfoxides, 265
  synthesis of, 265
α-Thiosulfoxide carbanions, as starting materials, for α-thiosulfoxides, 263
Thiosulfoxides, α-alkyl-, from, dithioacetals, 258
α-Thiosulfoxides
  from, α-thio sulfoxide carbanions, 263
  optically active, synthesis of, 264, 265
  as starting materials, for α-thiosulfones, 265
  synthesis of, 258
Thiotelluroacetals, synthesis of, theoretical, 285
Thiotellurophanes, alkyl-, synthesis of, 873
α-Thiotin compounds, synthesis of, 397
α-Thiotitanium compounds, synthesis of, 396
Thiourea
  reactions
    with bis(chloromethyl)phosphinic acid, 346
    with phthalaldehyde, 298
  use in the synthesis of α-thiosilanes, 368
Thiourea, N,N-dimethyl-

as starting material
for α-bis(methylthio)methyleneamino enamines, 988
for immonium salts, 988
Thiourea compounds, bis hydroxymethylation of, 301
Thioureas
as starting materials
for α-aminoalkyl phosphonates, 466
for α-aminoalkylphosphorus compounds, 466
for ketene aminals, 973
Thioureas, N^1,N^2-diacyl-, desulfurization reactions of, 425
Thiouronium salts, cleavage of, 223
α-Thiozincs, synthesis of, 396
Thorium, butyltris(cyclopentadienyl)-, decomposition of, 703
Thorium compounds
reactions
with 1-metalloalkylphosphines, 589
with phosphorus ylides, 589
Thrombin, inhibition via α-aminoboronic acids, 526
Tin, reactions, with (halomethyl)silylalkanes, 660
Tin, alkynyl-, reactions, with benzoylnitromethane, 541
Tin, allyltributyl-
reactions
with chloroform, 14
with 1,1,1-trichloroethane, 14
Tin, chloromethyltrimethyl-, reactions, with dimethylbismuthinosodium, 600
Tin, chlorotriphenyl-, reactions, with tricarbonyl(lithiocyclopentadienyl)manganese, 725
Tin, (dimethylamino)tributyl-, reactions, with phenyl methyl sulfoxide, 397
Tin, α-fluorovinyl-, from, trifluorovinylmagnesium halides, 812
Tin, tetrakis(1-methoxyvinyl)-, reactions, with cuprous cyanide, 954
Tin, tributyl(trimethylsilylmethyl)-, reactions, with n-butyllithium, 652
Tin, (trimethylsilylmethyl)triphenyl-, synthesis of, 660
Tin, (trimethylsilyl)tributyl-, synthesis of, 660
Tin, trimethyltrifluoromethyl-, pyrolysis of, 7
Tin(II) bromide, reactions, with diiodomethane, 699
Tin(II) chloride, reactions, with 1,3,2-dithiaborinane–dimethyl sulfide, 257
Tin(IV) chloride
reactions, with silicon hydrides, 140
use in the synthesis of tetramethylene-bridged tin compounds, 700
Tinamines, trialkyl-, reactions, with aryl sulfoxides, 397
Tin chloride, reactions, with diferrocenylmercury, 726
Tin chloride, tributyl-
reactions, with α-alkoxylithium nucleophiles, 388
as starting material, for α-alkoxystannanes, 387
Tin chloride, trimethyl-, reactions, with dilithiomethane, 699
Tin chloride, triphenyl-, reactions, with trimethylsilylmethyllithium, 660
Tin chlorides, reactions, with 1-lithiovinyl phenyl sulfides, 961
Tin chlorides, trialkyl-, reactions, with arylthiomethyllithiums, 397
Tin complexes, cyclic tetranuclear, synthesis of, 700
Tin complexes, methylene-bridged pentacoordinated, synthesis of, 700
Tin complexes, tetramethylene-bridged, synthesis of, 700
Tin compounds
reactions, with alkynes, 962
synthesis of, 961, 1038
Tin compounds, alkoxy-, addition to vinyltin systems, 701
Tin compounds, alkylhalo-, addition of oxygen and nitrogen donors to, 700
Tin compounds, alkynyl-, reactions, with organoboranes, 665
Tin compounds, amino-, addition to vinyltin systems, 701
Tin compounds, aminomethyl-, synthesis of, 533

Tin compounds, ethynyl-, reactions, with bromotrichloromethane, 813
Tin compounds, α-fluoroalkenyl-, synthesis of, 812
Tin compounds, perfluorovinyl-, reactions, with boron halides, 813
Tin compounds, α-selenyl-, synthesis of, 401
Tin compounds, trichloro alkynyl-, carbostannation of, 1069
Tin dichloride, reactions, with (halomethyl)silylalkanes, 660
Tin dichloride, dimethyl-, reactions, with di-Grignard reagents, 700
Tin halides
reactions
with bromotrifluoroethene, 812
with 1-lithio-1-aryl-1-silylalkanes, 660
Tin hydride, di-n-butyl-, reactions, with t-butyldipropynylphosphine, 1038
Tin hydride, tributyl-
deprotonation of, 386
reactions
with 1-chlorooctyne, 812
with α-fluoroalkenyl phenyl sulfones, 812
with phenylthioalkynes, 962
use in the synthesis of dichloromethyls, 13
Tin hydride, trimethyl-, use in the synthesis of carbon-bridged ditin compounds, 701
Tin–lithium exchange
for the synthesis of α-alkoxy carbanions, 378
for the synthesis of α-thiocarbanions, 394
Tin methoxide, trimethyl-, reactions, with perfluoroisobutene, 762
Tins, α-alkoxy-, synthesis of, 385
Tins, α-haloalkenyl-, synthesis of, 812
Tins, (sulfinylmethyl)trialkyl-, synthesis of, 397
Tin tetrabromide, reactions, with diazomethane, 155
Tin tetrachloride
reactions
with diazomethane, 155
with lithium alkyls, 949
with trimethylsilylmethylmagnesium chloride, 660
for the ring-opening of 1,3-disilacyclobutanes, 661
Tin tetrachlorides, as catalysts, for the synthesis of ketene dithioacetals, 844
Titanation, of alkynes, 1066
Titanium, reduced, synthesis of, 1082
Titanium, tetrakis(dimethylamino)-, reactions, with amides, 972
Titanium(IV) chloride, reactions, with α-dichloro sulfoxides, 773
Titaniumaluminum compounds, synthesis of, 1066
Titanium ate complexes, as starting materials, for (E)-γ-substituted ketones, 918
Titanium binuclear compounds, synthesis of, 673
Titanium bis(cyclooctatetraene), use in the synthesis of tris(cyclooctatetraene)dititanium, 671
Titanium chloride, tris(dimethylamino)-, reactions, with carbamates, 383
Titanium complexes, carbon-bridged, synthesis of, 671
Titanium compounds
reactions
with 1-metalloalkyl phosphonates, 578
with phosphorus ylides, 578
Titaniummethylene complexes, synthesis of, via transmetallation, 669
Titaniums, α-alkoxy-, synthesis of, 383
Titanium tetrabutoxide, use in the synthesis of tris(cyclooctatetraene)dititanium, 671
Titanium tetrachloride
reactions
with aniline, 796
with lithium aluminum hydride, 1082
with 1-phenylthiovinylsilane, 369
Titanium tetrachlorides, as catalysts, for the synthesis of ketene dithioacetals, 844

Titanium trichloride, chloro(dimethyl)silylmethyl-, synthesis of, 658
Titanocene, tetramethylene-, as starting material, for methylene-bridged dititanium complexes, 670
Titanocene complexes, displacement of, 663
Titanocene dichloride
　reactions
　　with methylenebis(magnesium bromide), 670
　　with tricarbonyl(lithiocyclopentadienyl)manganese, 719
　reduction of, 671
Titanocenes, germylmethyl-, synthesis of, 663
Tolan
　reactions
　　with dicarbonylcyclopentadienylcobalt, 689
　　with nonacarbonyldiiron, 681
Toluene
　reactions
　　with sodium, 609
　　with sulfuryl chloride, 11
p-Tolueneselenosulfonate, seleniumphenyl, reactions, with alkynes, 947
Toluene-p-sulfonate, ammoniopropanedinitrile, reactions, with aldehydes, 883
p-Toluenesulfonate, pyridinium, as catalyst, for the synthesis of acetals, 179
Toluenesulfonyl compounds, light absorption of, 1075
p-Tolyl sulfoxide, 1-chloromethyl, optically active, synthesis of, 76
Tosylate, 2,2,2-trifluoroethyl
　deprotonation of, 735
　reactions, with trialkylboranes, 735
Tosylates, displacement of, 346
Tosylates, alkyl, as electrophiles, for dithiocarboxylic enethiolates, 842
2-Tosyl compounds, 1-cyano-, reactions, with bromine, 105
Tosyl iodide, reactions, with trimethyl(vinyl)silanes, 139
Transacetalization, of aldehyde acetals, 249
Transition metal alkyl complexes, stability of, 655
Transition metal carbene complexes, reactions, with metal halides, 722
Transition metal complexes
　from, silylmethyl ligands, 656
　reactions, with 1-silyl-1-chloro-1-lithiomethane, 658
　synthesis of, 955
Transition metal compounds
　from
　　alkenyl halides, 821
　　alkynes, 822
　reactions
　　with 1,1-bisphosphinoalkenes, 1028
　　with phosphorus ylides, 578
　synthesis of, 147, 820, 1032
　via transmetallation reactions, 820
Transition metal halides
　reactions
　　with 2-lithio(dimethylaminomethyl)ferrocene, 723
　　with lithioferrocene, 723
　　with tricarbonyl(lithiobenzene)chromium, 723
　　with tricarbonyl(lithiocyclobutadiene)iron, 723
　　with tricarbonyl(lithiocyclopentadienyl)manganese, 723
Transition metals
　functions containing two, synthesis of, 669
　multidecker sandwich compounds of, synthesis of, 694
　reactions, with α-silylmethyl anions, 655
　synthesis of functions, 709
Transition metals, α-haloalkenyl, synthesis of, 820
Transition metals, α-metallomethyl, reactions, with silyl electrophiles, 657
Transition metals, α-silyl methyl
　synthesis of
　　via changing the groups attached to the central methylene or to the metal, 659
　　via reaction of an α-metallomethyl transition metal with a silyl electrophile, 657
　　via reaction of a transition metal complex with an α-silylmethyl anion, 656
Transmetallation
　of arsenic compounds, 599
　of bis(mercury halides), 668
　of stannanes, 531
　for the synthesis of α-amino Grignard reagents, 531
　for the synthesis of carbon-bridged dimercury complexes, 697
　for the synthesis of 1,1-dimetal complexes, 669
　for the synthesis of α-magnesiosilylalkanes, 655
　for the synthesis of transition metal compounds, 820
Triacetates, reactions, with hydrogen bromide, 121
Triacetic acid, nitrilo-, as starting material, for 1-aminophosphonic acids, 492
Triacetyl compounds, oxidative nitration of, 435
Triamine, ethoxymethane-, use in the synthesis of ketene aminals, 974
Triamines, alkoxymethane-, use in the synthesis of ketene aminals, 973
Triamines, isopropoxymethane-
　from, chloroformamidinium chlorides, 973
　reactions, with isopropyl acetate, 973
Triazaadamantanes, synthesis of, 408
Triazacyclohexanes, N-chloromethyl-, from, N-methoxymethyl compounds, 97
Triazaphosphaadamantanes, acylation of, 481
Triazaphospholes, synthesis of, 483
Triazenes, aminomethyl-, synthesis of, 444
Triazenes, 3-(hydroxymethyl)-, synthesis of, 312
Triazepines, synthesis of, 411
Triazine, perhydro-, as starting material, for iminium ions, 455
Triazine, 1,3,5-tris[(S)-phenylethyl]hexahydro-, synthesis of, 99
Triazine, tris(trimethylsilylmethyl)-, synthesis of, 509
1,3,5-Triazine, 1,3,5-triacetylhexahydro-, nitration of, 447
1,3,5-Triazine, 1,3,5-triphenyl-, reactions, with cyclic boron esters, 457
Triazines
　as starting materials
　　for bromo nitro compounds, 109
　　for carbamoylsulfenyl chlorides, 103
Triazines, aminomethyl-, synthesis of, 443
1,3,5-Triazines
　from, isocyanates, 426
　synthesis of, 416
1,3,5-Triazines, hexahydro-
　as starting materials, for Schiff bases, 463
　synthesis of, 294, 407, 418, 427
1,3,5-Triazines, perhydro-, synthesis of, 418, 425
Triazoles, N-chloromethyl-, synthesis of, 105
1,2,4-Triazoles, acylaminomethyl-, synthesis of, 448
1,2,4-Triazolidin-3-ones, oxidation of, 440
1,2,4-Triazolidin-3-thiones, oxidation of, 440
Triazolines, as intermediates, in the synthesis of silylaziridines, 515
1,2,4-Triazolones, dihydro-, synthesis of, 440
1,3,5-Triboracyclohexane, 1,3,5-trichloro-, synthesis of, 635
Tributyl chloride, reactions, with diphenylarsinomethyllithium, 600
Tricarboxylic acids, as starting materials, for triphosphonic acids, 561
Trichlorofon, reactions, with diazomethane, 912
Trichloromethyl compounds, reduction of, 13
Trichloromethyl groups, as starting materials, for dichloromethyl groups, 13
Trichloromethyl radical, as chain carrier, 1078
Triethylamine
　reactions
　　with aryl isocyanates, 446

with chlorophosphine esters, 483
with α-chloropropionyl chloride, 768
with phosgene, 790
Triflate, *N*-fluorotrimethylpyridinium (NFPT), for the fluorination of sulfides, 62
Triflate, methyl, as starting material, for carbon-bridged triosmium complexes, 683
Triflate, 2-methyl-1-propenyl, reactions, with azobenzenes, 985
Triflate, trimethylsilyl
as catalyst, for the synthesis of acetals, 181
reactions, with α-halo sulfoxides, 773
Triflates, acyloxyiminium, synthesis of, 881
Triflates, ammonium, synthesis of, 507
Triflates, dialkylboryl, reactions, with carboxylic acids, 832
Triflates, prop-2-yniminium
reactions
with thiolates, 321
with thiols, 321
Triflates, trialkylsilyl, as silylating agents, 831
Triflate salts, use in the synthesis of α-aminosilanes, 507
Trifluoromethyl–metal complexes, decomposition of, 7
1,3,5-Trigermacyclohexane, 1,1,3,3,5,5-hexachloro-, reduction of, 629
Triiron, dodecacarbonyl-
reactions
with allenyl thiols, 683
with diphenylcyclopropenone, 681
with thiophene, 681
as starting material, for benzoferracyclopentadiene complexes, 681
Trimethylamine–borane complexes, reactions, with methylthiomethyllithium, 373
Trimethylamine hydrochloride, reactions, with lithium salts, 373
Trimethylamine *N*-oxide, Polonovski reaction of, 311
Trimethylamine-*N*-oxide, as reoxidant, 357
1-Trimethylsilyl compounds, 1-phenyl-, synthesis of, 658
1-Trimethylsilyl compounds, 1-phenylseleno-, deprotonation of, 947
8,9,10-Trinorcarane, 7,7-dibromo-, use in the synthesis of α-bromostannanes, 155
Trinorcaranes, bromo-, use in the synthesis of α-bromoplumbanes, 156
Triosmium, dodecacarbonyl-
reactions, with alkynes, 682
reduction of, 679
Triosmium complexes, carbon-bridged
from
methyl triflate, 683
phenyllithium, 683
Trioxan
reactions
with nitriles, 425
with sulfonamides, 427
as reagent, for the synthesis of methylenebisamides, 424
Trioxane, reactions, with perfluoroalkyldichlorophosphines, 125
1,2,4-Trioxane, synthesis of, 214
1,2,4-Trioxones
synthesis of, 214
See also Ozonides
Triphenoxy compounds, from, perchloropentadienoic acid nitriles, 829
Triphosphonates, alkylation of, 563
Triphosphonic acids, from, tricarboxylic acids, 561
Triple decker sandwich complexes, synthesis of, 717
Triruthenium, dodecacarbonyl-
reactions
with cyclopentadienylidenetriphenylphosphorane, 685
with 1,4-dibromocyclooctatetraene, 681
with 2-methylthiophene, 683
with phosphonium ylides, 683

with thiophene, 683
as starting material, for ruthenacyclopentadiene complexes, 683
Triselenamonothiafulvalenes, synthesis of, 874
Triselenoortho esters, elimination of methaneselenol from, 875
Triselenoorthoformate, triphenyl, reactions, with butyllithium, 289
1,3,5-Trisilacyclohexane, 1,1,3,3,5,5-hexamethyl-
metallation of, 620
synthesis of, 610
1,3,5-Trisilacyclohexane, 2-lithio-1,1,3,3,5,5-hexamethyl-, reactions, with benzaldehyde, 623
1,3,5-Trisilacyclohexanes, transition metal substituted, synthesis of, 622
Trisilacyclopentanes, synthesis of, 621
2,4,6-Trisilaheptanes, synthesis of, 602
Trisilahexanes, synthesis of, 606
Trisilane, 2-chloroheptamethyl-, thermolysis of, 614
Tris(phosphines), synthesis of, 546
Tris(sulfones), synthesis of, 275
Tris(trimethylsilylmethyl), reactions, with bromine, 597
1,2,4-Tritelluroles, synthesis of, 877
2,4,6-Trithiaheptane, synthesis of, 258
1,3,5-Trithiane
chlorination of, 85
reactions, with thionyl chloride, 65
as starting material, for di(chloromethyl) sulfide, 63
Trithianes
bromination of, 85
chlorination of, 71
as starting materials
for α-chloroalkyl sulfinyl chlorides, 78
for dithioacetals, 250
Trithioalkenes, synthesis of, 851
Trithiocarbonate *S,S*-dioxides, reactions, with Grignard reagents, 862
Trithiocyclopropenium salts, reactions, with cyclopentadienyl anions, 852
1,2,4-Trithioles, 3,5-bis(exomethylene), synthesis of, 852
Trithioortho esters
elimination of thiols from, 844
as starting materials, for ketene dithioacetals, 844
Trithioorthoformates, as starting materials, for dithioacetals, 250
Trithiopyrocarbonate, *O,O*'-diethyl, as starting material, for enethiolates, 834
Tritin compounds, synthesis of, 700
Tungstacyclopentadienes, synthesis of, 675
Tungsten, doubly-bridged complexes of, synthesis of, 675
Tungsten, pentacarbonylbenzylidene-, reactions, with triphenylketenimine, 1018
Tungsten, pentacarbonyl (diphenylvinylidene)-, synthesis of, 1018
Tungsten carbene complexes, reactions, with phosphines, 580
Tungsten carbyne complexes, cyclopentadienyl-, use in the synthesis of 3-carbon-bridged ketone complexes, 676
Tungsten chlorides, reactions, with 1,1-dicyanobis(trifluoromethyl)ethene, 800
Tungsten complexes
photolytic rearrangement of, 639
reactions, with alkynes, 1028
rearrangement of, 934
Tungsten complexes, alkyne-bridged, from, carbyne complexes, 675
Tungsten complexes, (halomethyl)-, synthesis of, 147, 148
Tungsten compounds, reactions, with phosphorus ylides, 580
Tungsten diazo complexes, α-chloroalkenyl-, x-ray crystal structure of, 799
Tungsten halides, use in the synthesis of carbon-bridged ditungsten complexes, 672

Tungsten hydride complexes, reactions, with
  tricarbonyl(trisammonia)chromium, 715
α-Tungstenio enamines, from, ynamines, 1019
Tungsten–phosphinoalkene complexes,
  synthesis of, 1023, 1028
Tungsten tetrachlorides, use in the synthesis of
  tris(cyclooctatetraene)ditungsten complexes, 673
Tungsten–tungsten-bonded complexes, as starting materials,
  for alkyne-bridged complexes, 675
L-Tyrosinamide, N-acetyl-, reactions, with Woodward's
  reagent K, 892

Ultrasound, use in the elimination of hydrogen halide from
  α-haloacetals, 826
Umpolung
  application to the synthesis of dithioacetals, 254
  concept of, 244
Umpolung reagents, diselenoacetals as, 287
Uracil, cleavage of, 103
Uracil, 6-amino-1,3-dimethyl-, reactions, with dimethyl
  ethynedicarboxylate, 893
Uranium, chlorotricyclopentadienyl-, as starting material,
  for ferrocenyluranium compounds, 726
Uranium compounds, reactions, with phosphorus ylides, 589
Uranium compounds, ferrocenyl-, from,
  chlorotricyclopentadienyluranium, 726
Urea
  reactions, with imino ester hydrochloride salts, 982
  thioalkylation of, 325
Urea, 1,3-dimethyl-, reactions, with formaldehyde, 301
Urea, N-phenyl-, reactions, with
  phenyldichlorophosphine, 488
Urea compounds, bis hydroxymethylation of, 301
Ureas
  lithiation of, 531
  reactions, with lithium aluminum hydride, 414
  reduction of, for the synthesis of aminals, 414
  as starting materials, for α-aminoalkylphosphorus
    compounds, 466
  use in the synthesis of α-aminoalkylphosphines, 456
Ureas, bis-N-(hydroxyalkyl)-, reactions, with amines, 416
Ureas, α-germyl-, synthesis of, 521
Ureas, N-tetrachloroethyl-
  as starting materials
    for gem-phosphonio ureidoalkenes, 999
    for gem-phosphoryl ureidoalkenes, 1007
α-Ureidoalkylation, mechanism, 301
Urethanes, tetrachloroethyl, as starting materials, for gem-
  alkoxycarbonylamido phosphonioalkenes, 999
Uridine nuleoside compounds, synthesis of, 275

L-Valine, N-acetyl-, as chiral auxiliary, 720
Vanadium, (η^5-cyclopentadienyl)(η^6-naphthalene)-, use in
  the synthesis of carbon-bridged divanadium
  complexes, 671
Vanadium, tetracarbonyl-, reactions, with iron hydride
  complexes, 717
Vanadium(II) chloride, use in the synthesis of
  cyclooctatetraene-bridged complexes, 672
Vanadium(V) oxide, for the oxidation of dithioacetals, 259
Vanadium binuclear compounds, synthesis of, 673
Vanadium complexes, dialkylcyclopentadienyl-, use in the
  synthesis of carbon-bridged divanadium
  complexes, 671
Vanadium compounds, reactions, with phosphorus
  ylides, 579
Vanadium pentoxide, use in the synthesis of α-chloro
  sulfoxides, 74
Vilsmeier reagent, reactions, with ketones, 196
Vinyl, 2-(benzeneselenyl)-, reactions, with secondary
  amines, 874
Vinyl, α-fluoro-, anion equivalent of, synthesis of, 812
Vinyl anions, trifluoro-, synthesis of, 819

Vinyl bromide, (E)-2-ethoxy-, as starting material, for gem-
  bromo lithio enol ethers, 818
Vinyl bromide, perfluoro-, as starting material, for Grignard
  reagents, 820
Vinyl bromide, trifluoro-
  reactions, with bis(trimethylsilyl)mercury, 821
  use in the synthesis of α-fluoroalkenyllithium
    compounds, 817
Vinyl carbanions, synthesis of, 956
Vinyl chloride, reactions, with benzeneselenenyl chloride, 91
Vinyl chloride, 1,2-difluoro-, reactions, with
  butyllithium, 817
Vinyl chloride, trifluoro-
  lithium–chlorine exchange of, 817
  lithium–halogen exchange of, 817
Vinyl chlorides, reactions, with phosphorus
  nucleophiles, 553
Vinyl chlorides, (E)-ethoxy-, reactions, with
  n-butyllithium, 817
Vinyl chlorides, (Z)-2-ethoxy-, reactions, with
  n-butyllithium, 817
Vinyl halides
  hydroboration of, 142
  reactions, with stannanes, 8855
  See also Alkenes, gem-dihalo-
Vinylidene bromides, synthesis of, 1013
Vinylidene dibromides, use in the synthesis of ketene
  aminals, 970
Vinylidene dichlorides
  reactions, with sodium azide, 993
  use in the synthesis of ketene aminals, 970
Vinylidene difluorides
  reactions, with 3-methylbenzthiazoloneimine, 992
  as starting materials, for gem-diisothioureidoalkenes, 992
  use in the synthesis of ketene aminals, 970
Vinylidene dihalides
  reactions
    with amines, 970
    with aniline, 970
    with diamines, 970
  reactivity of, 970
  as starting materials
    for amidines, 970
    for ketene aminals, 970
Vinylidene diiodides, use in the synthesis of ketene
  aminals, 970
Vinyl iodide, perfluoro-, synthesis of, 751
Vinyl iodides, reactions, with organozinc compounds, 954
Vinyl sulfones, reactions, with phenyl selenyl chloride, 287
(+)-(R)-Vinyl p-tolyl sulfoxide, bromination–
  dehydrobromination of, 776

Wacker oxidation, mechanism, 193
Wadsworth–Emmons reaction, for the synthesis of
  2-haloalkenylphosphorus compounds, 801
Wanzlick's reagent, use in the synthesis of aldehydes, 408
Water
  azeotropic removal of, for the synthesis of aminals, 405
  reactions
    with carbonyl compounds, 160
    with N-sulfonylimines, 312
  removal of, for the synthesis of acetals, 177
Wittig condensation, for the synthesis of
  difluoroalkenes, 730
Wittig–Horner reaction
  for the synthesis of 1,1-dichloroalkenes, 740
  for the synthesis of 1-halo 1-phenylthioalkenes, 773
Wittig reaction
  for the synthesis of chloro haloalkenes, 757
  for the synthesis of dibromoalkenes, 745
  for the synthesis of 1,1-dichloroalkenes, 737
  for the synthesis of fluoro bromoalkenes, 754
  for the synthesis of fluoro haloalkenes, 751

for the synthesis of (Z)-perfluorovinylphosphonium
 compounds, 801
for the synthesis of tetracoordinate sulfur compounds, 780
for the synthesis of thiaselenafulvenes, 870
Wittig reaction, aza--, for the synthesis of carbodiimides, 441
Wittig reagents
 from, α-halophosphonium salts, 117
 use in the synthesis of carbon-bridged diiron
  complexes, 679
 use in the synthesis of carbon-bridged diruthenium
  complexes, 679
 use in the synthesis of ketene dithioacetals, 847
Wittig rearrangement, of phosphorodiimidates, 923
Woodward's reagent K, reactions, with N-acetyl-L-
 tyrosinamide, 892

Xanthenes, synthesis of, 307
Xanthogenates, use in the synthesis of distibines, 595
Xanthone, 1,2,3,4-tetrahydro-, synthesis of, 307
Xenon difluoride
 for the fluorination of sulfides, 61
 reactions
  with benzyl alcohols, 45
  with β-lactams, 778
 use in the synthesis of α-fluoro sulfides, 218
X-ray structure determination, of trialkylaluminum or
 triarylaluminum compounds, 698
Xylene, α,α-di(morpholino)-, reactions, with thiophenol, 322

Ylide, synthesis of, 598
Ylide–carbene reaction, for the synthesis of
 difluoroalkenes, 735
Ylides
 from
  α-silyl heterocycles, 519
  (trimethylsilylethyl)ammonium salts, 512
 reactions
  with acetic anhydride, 993
  with alkenyl phosphates, 737
  with alkyl halides, 558
  with aluminum trichloride, 397
  with areneselenenyl halides, 348
  with beryllium chloride, 395
  with gallium trichloride, 397
  with indium trichloride, 397
  with perhalofluoroalkanes, 118
  with (N-phenylthio)succinimide, 346
  with triphenylphosphinegold chloride, 397
 as starting materials, for dinitrocarbene, 1083
Ylides, arsonium, synthesis of, 598
Ylides, azomethine
 from
  N-benzylidene-α-(diphenylphosphinoyl)glycine
   esters, 489
  N-(trimethylsilylmethyl)iminium salts, 517
 reactions, with thiobenzophenone, 319
Ylides, difluoromethylene, synthesis of, 732
Ylides, disilyl, synthesis of, 571
Ylides, 3H-indolium, reactions, with
 1,1-bis(benzenesulfonyl)ethene, 277
Ylides, iodinium, development of, 281
Ylides, phenyliodinium
 reactions
  with carbon disulfide, 866
  with diethylamine, 867
  with thiobenzophenones, 866
 as starting materials, for 1,1-disulfonylalkenes, 866
Ylides, phosphonium, reactions, with dodecacarbonyl
 triruthenium, 683
Ylides, silyl sulfonium, as starting materials, for
 α-methylthiosilanes, 369
Ylides, sulfonium, synthesis of, 837
Ylides, thiocarbonyl, from, bromomethylsilane, 369

Ynamines
 in cycloaddition reactions, with α,β-unsaturated
  esters, 827
 as intermediates, in the synthesis of ketene aminals, 976
 reactions
  with alcohols, 885
  with amines, 976
  with dialkyl thiophosphite, 1010
  with electrophiles, 791
  with ethanol, 885
  with nucleophiles, 885
  with phenols, 885
  with 5-pyrazolones, 986
 as starting materials
  for amidines, 977
  for α-amido enamines, 980
  for α-borio enamines, 1015
  for α-cuprio enamines, 1018
  for α-hydrazino enamines, 984
  for ketene aminals, 976
  for α-phosphoryl enamines, 1003
  for α-potassio allenamines, 1017
  for α-thiophosphoryl enamines, 1010
  for α-tungstenio enamines, 1019
 use in the synthesis of carbon-bridged diiron
  complexes, 681
 use in the synthesis of ferracyclopentadiene, 681
Ynamines, chloro-, as intermediates, in the synthesis of
 ketene hemiaminals, 885
Ynamines, cyano-, as starting materials, for cyanoketene
 aminals, 976
Ynamines, dialkynic, as starting materials, for ketene
 hemiaminals, 885
Ynamines, N,N-diethyl-
 reactions
  with alkane sulfenyl chlorides, 791
  with arene sulfenyl chlorides, 791
Yndiamines, reactions, with amines, 976
Yttrium, pentamethylcyclopentadienyl complexes of,
 synthesis of, 702
Yttrium compounds, methyl-bridged, synthesis of, 670
Yttrium–indium compounds, synthesis of, 726

Zeolites, as catalysts, for the synthesis of dithioacetals, 246
Zinc
 reactions
  with alkenyl bromides, 821
  with aryloxy ketones, 167
  with 2-arylthio-2,2-dichloroethanols, 775
  with α-chloroalkyl phenyl sulfides, 396
  with dihalomethanes, 152
  with diiododi(trimethylsilyl)methane, 618
  with diiodomethane, 696
  with α-halo boronic esters, 665
  with iodomethyl pivalate, 385
  with iodomethylthio esters, 396
  with phosphonium salts, 753
  with tertiary iodides, 730
 use in the synthesis of dibromoalkenes, 745
 use in Wittig reactions, 734
Zinc, iodo(iodomethyl)-, reactions, with iodostannanes, 155
Zincates, ester, synthesis of, 385
Zinc bromide, allyl-
 reactions
  with propynol ethers, 809
  with 1-trimethylsilyl-1-octyne, 809
  with vinyl metal compounds, 709
Zinc bromide, bromomethyl-, synthesis of, 152
Zinc bromide, vinyl-, reactions, with prop-2-yn-1-ol, 1067
Zinc–bromine exchange, with lithium tributylzincate, 821
Zinc chloride
 as catalyst, for the synthesis of diselenoacetals, 288
 reactions, with 1-ethoxyvinyllithium, 954

Zinc chloride, bis(trimethylaminomethyl)-, synthesis of, 534
Zinc compounds
  reactions, with phosphorus ylides, 586
    reactivity of, 396
    synthesis of, 385, 396, 964
Zinc compounds, α-amino-, synthesis of, 534
Zinc compounds, dialkyl-, reactions, with dicarbonyl compounds, 166
Zinc compounds, 1-haloalkyl-, reactions, with phosphines, 586
Zinc compounds, (halomethyl)-, synthesis of, 152
Zinc compounds, vinyl-, reactions, with copper(I) bromide, 821
Zinc–copper couples, use in Wittig reactions, 734
(E)-Zinc enolate, synthesis of, 760
Zinc enolates
  from
    tribromoacetyl chlorides, 768
    trichloroacetyl chlorides, 768
Zinc ester enolates, O-silylation of, 831
Zincs, α-alkoxy-, synthesis of, 384
Zirconabicyclic systems, synthesis of, 1066
Zirconacycles, halogenation of, 809
Zirconation, of alkynes, 1061, 1066

2-Zirconiophospholanes, synthesis of, 579
Zirconium alkyls, reactions, with aluminum alkyls, 725
Zirconiumaluminum alkenes, synthesis of, 1066
Zirconium complexes, from, zirconocene–alkyne complexes, 725
Zirconium complexes, carbon-bridged, synthesis of, 671
Zirconium compounds
  reactions
    with hydrogen chloride, 547
    with 1-metalloalkylphosphines, 579
    with phosphorus ylides, 579
Zirconium tetrachloride, reactions, with 1,1,3,3-tetramethyl-1,3-disilacyclobutane, 610
Zirconocene–alkyne complexes, as starting materials, for zirconium complexes, 725
Zirconocene chlorohydride, reactions, with carbene–rhenium complexes, 716
Zirconocene dichloride
  reactions, with 1-trimethylsilylpropyne, 1038
  use in the synthesis of carbon-bridged dizirconium complexes, 671
Zirconocenes, rearrangement of, 658
Zirconocycles, reactions, with iodine, 1039